Organic Chemistry

A Learner-Centered Approach

Richard J. Mullins

Xavier University

Pearson

Content Strategy & Management: Jeanne Zalesky, Elizabeth Ellsworth Bell, Chris Hess, Terry Haugen, Harry Misthos
Content Development: Barbara Price, John Murdzek, Matt Walker, Hilair Chism
Content Production: Heidi Allgair, Heidi Aguiar, Kristen Flathman, Chloe Veylit, Meaghan Fallano, Margaret Trombley
Product Management: Chris Hess and Ian Desrosiers
Product Marketing: Candice Madden
Rights and Permissions: Chenley Calites-Bulagao, SPi Global

Please contact https://support.pearson.com/getsupport/s/ with any queries on this content

Cover Design and Image by Alan Hynes

Library of Congress Cataloging-in-Publication Data

Names: Mullins, Richard, author.
Title: Organic chemistry : a learner-centered approach / Richard Mullins, Xavier University.
Identifiers: LCCN 2020050553 | ISBN 9780137369263 (cloth)
Subjects: LCSH: Chemistry, Organic--Study and teaching (Higher)
Classification: LCC QD251 .M85 2021 | DDC 547--dc23
LC record available at https://lccn.loc.gov/2020050553

3 2021

Pearson

Rental
ISBN-10: 0-137-36926-3
ISBN-13: 978-0-137-36926-3

To Maggie and John, my children,
who inspire me to be better,
and to Mary, my everything,
who shows me how.

Brief Table of Contents

Key Features

Chemistry in Context

1-A Problem-Solving Strategies: Differential diagnosis (1.5)
2-A Organic Chemistry: Reducing metals (2.3)
2-B Organic Chemistry: Isomers (2.4)
2-C Organic Chemistry: Alkene addition (2.4)
2-D Organic Chemistry: Enzyme inhibition (2.5)
2-E Organic Chemistry: Organic dyes (2.7)
2-F General Chemistry in Context: Purple rhinos (2.8)
3-A Food Chemistry: Saturated versus unsaturated fats (3.2)
3-B Petroleum Chemistry: Distillation of crude oil (3.3)
3-C Physical Organic Chemistry: Measuring strain (3.6)
4-A Medicine: Treating acid reflux disease (4.1)
4-B The Laboratory: Separating acidic and neutral compounds (4.2)
4-C Medicinal Chemistry: Protonation alters drug solubility (4.3)
4-D Medicinal Chemistry: Electron-withdrawing groups reduce basicity (4.4)
5-A Biochemistry: Enzymes as catalysts (5.1)
5-C Biochemistry: Enzymes as catalysts (5.2)
5-D Biochemistry: Enzymes catalyze reactions in both directions (5.2)
6-A Biochemistry: Enzymes resolve enantiomers (6.6)
8-A Natural Products: Terpene biosynthesis: the isoprene rule (8.1)
8-B Biochemistry: Enzymatic isomerization; reversibility of H^+ addition (8.3)
8-C Biochemistry: Synthesis of honey bee pheromones (8.4)
9-D Food Science: Partially hydrogenated oils (9.2)
9-E Ecology: Preventing gypsy moth infestation (9.3)
10-A Problem-Solving Strategies: Working a jigsaw puzzle (10.4)
10-B Problem-Solving Strategies: Differential diagnosis (10.5)
10-C Problem-Solving Strategies: Random vs. targeted drug design (10.8)
11-A Medicinal Chemistry: Oral bioavailability (11.3)
11-B Medicinal Chemistry: Lipophilicity (11.3)
11-C Environmental Science: Persistent organic pollutants (11.3)
12-A Reaction Analysis: Deceptive kinetic studies (12.3)
12-B Chemical Weapons: Mustard gases in war (12.3)
12-C Medicinal Chemistry: From chemical weapon to chemotherapy (12.7)
15-A Natural Product Isolation: Curacin A (15.2)
15-B Natural Product Isolation: ^{1}H NMR of curacin A (15.4)
15-C Natural Product Identification: Using coupling constants (15.5)
15-D Natural Product Identification: Assigning the ^{13}C NMR spectrum of curacin A (15.7)
16-A Natural Product Synthesis: Grignard reaction (16.2)
16-B Natural Product Synthesis: Cyclopropanation (16.3)
16-C Natural Product Synthesis: Suzuki coupling (16.4)
17-A Biochemistry: The chemistry of a hangover (part 1) (17.2)
17-C Biochemistry: Producing energy from amino acids (17.8)
17-D Asymmetric Synthesis: Enantioselective hydrogenation (17.9)
17-F Biochemistry: Nature's reducing agent (17.9)
17-G Biochemistry: The chemistry of a hangover (part 2) (17.10)
18-B Food Chemistry: Sodium benzoate preservative (18.6)
18-C Biochemistry: Metabolism of sodium benzoate (18.7)
18-D Food Chemistry: A precursor to benzene? (18.8)
19-A Biochemistry: The peptide bond (19.6)
19-C Biochemistry: Building the bacterial cell wall (19.7)
19-D Biochemistry: Irreversible inhibition of transpeptidase (19.7)
19-E Biochemistry: Bacteria fight back! (19.10)
20-B Biochemistry: Synthesis of sucrose by the aldol reaction (20.3)
20-D Biochemistry: Claisen in biological systems (20.4)
20-F Biochemistry: Claisen/aldol reactions (20.6)
21-A Chemistry at the Beach: Sunscreen (21.3)
21-B Laundry Room Chemistry: A world in color (21.4)
22-A Biochemistry: Putative Diels–Alderase? (22.3)
22-B Biochemistry: Finally, a Diels–Alderase (22.3)
23-B Medicinal Chemistry: *In vitro* to *in vivo* (23.5)
23-C Chemistry of Color: Synthesis of azo dyes (23.7)
23-D Medicinal Chemistry: Synthesis of sulfanilamide (23.8)
24-A Medicinal Chemistry: Lipid protection by vitamin E (24.2)
24-C Medicinal Chemistry: Drug metabolism by cytochrome P450 (24.5)
25-A Medicinal Chemistry: Antidepressants by S_N2 (25.5)
25-B Medicinal Chemistry: Citalopram by nitrile reduction (25.5)
25-D Medicinal Chemistry: Synthesis of amoxapine, a TCA (25.7)
26-A Molecular Biology: Humoral control of blood pressure (26.6)
26-B Medicinal Chemistry: Lead compounds from pit viper venom (26.7)
26-C Medicinal Chemistry: Peptidomimetic enzyme inhibition (26.9)
27-A Glycobiology (Part 1): How the influenza virus gains entry into a cell (27.6)
27-B Glycobiology (Part 2): Exit of the influenza virus (27.7)
27-C Cancer Chemotherapy: DNA alkylation (27.9)
27-D Medicinal Chemistry: Pain relief by prostaglandin synthesis inhibition (27.15)
28-A Easing Purification: Reactions on solid support (28.4)

Spectroscopy in Context

14-A Infrared: A handheld near-infrared spectrometer (14.3)
14-B Raman: Counterfeit antimalarial drugs (14.4)
14-C GC/MS: Traditional Chinese medicines (14.7)

Green Chemistry

Contents

Preface

When I first began writing this book in 2013, my plan was to write a book that was different than other chemistry textbooks. It would teach the fundamental principles that govern the reactivity of the various organic molecules one might encounter, but all books do that. The book I have tried to write doesn't just talk about organic chemistry; instead, it talks about *learning* organic chemistry. After all, that's what the student as learner cares about most. This idea was so central to the vision of the project that it is part of the name: "A Learner-Centered Approach."

A learner-centered approach means the book should recognize that all students come to this course with different goals and aspirations, different backgrounds, different experiences, and different plans for learning. Accordingly, this textbook affirms that organic chemistry [*and science in general*] is for everyone, regardless of background, and that all are welcome and supported here. [*In case students actually read this: you belong, you deserve to be here, you can do this, you are worthy of your dreams.*] I wanted the book to happen the way a conversation about organic chemistry would happen, allowing for it to spin off organically [*Ha!*] in a variety of directions. The book is not intended to be a lecture from a podium to an unengaged audience. Instead, it is a lively discussion, a back and forth between student and text, the same kind of discussion that might occur in office hours, in the university cafeteria, while standing over a reaction in chemistry lab, or during a late-night review session.

In each of the scenarios described above, there is usually a pre-existing relationship between both parties to the conversation. Can a similar relationship be established between a chemistry textbook and the learner? [*I certainly hope so.*] What follows is a list of the features that were included to make that happen.

Chapter 1

To many, organic chemistry is a scientific discipline, an area of research with the power to create new materials and medicines that improve society. To others, including many of the students we teach, organic chemistry is a class, a stepping stone on their path to improve lives in their own unique and important way. [*It was once just that to me, an undergraduate student aspiring to become a physician.*] While Chapters 2 through 28 teach about organic chemistry [*and how to learn it; more on that later*], Chapter 1 discusses Organic Chemistry, the course, and makes the case for how it should be approached. It acknowledges that organic chemistry is different things to different people, makes the case that this is a good thing, and begins the process of meeting students where they are. Not found in any other book, Chapter 1 establishes a rapport between the author and the students, identifies two traits that are necessary for achieving success in organic chemistry [*and any other difficult course*], and briefly touches on how information is best retained. In the same way that an instructor establishes expectations for students on the first day of class, Chapter 1 does the same and more. It clearly lays out why the book was written the way it was, tells students what to expect from it, and how they should use it to their greatest benefit. Finally, it concludes by establishing an aspirational mindset, reminding students that organic chemistry is one part of a longer journey toward a more significant goal.

Coaching Tips

While Chapter 1 establishes a rapport between student and author, coaching tips [*most often found in bracketed italics, like these*] continue this relationship. They are the unsolicited advice a student might hear from a caring professor [*this causes students trouble; work extra assessments of this type*], the questions asked in a lecture setting [*how do we know that step occurs instead of another one?*], the reminder from an upper-level student [*make sure you know your* pK_a *values*], or the quiet whisper from the angel on their shoulder [*that *was a confusing paragraph—go back and read it again*]. These coaching tips, a term that doesn't exactly describe all of what they are, provide necessary mental breaks that refocus the attention of students by making them pause and consider what is being said. These coaching tips make clear that the text is a conversation between student and author in an ongoing process aimed at better organic chemistry understanding.

Cartoon Rick and Friends

Organic chemistry, given its reputation as a weed-out course [*to any students reading this: it isn't a weed-out course*], causes anxiety and stress to our students. And reading organic chemistry and working assessments can be a lonely endeavor. In keeping with the theme of a conversation-like textbook that has a friendly rapport between student and author, Cartoon Rick and friends bring a bit of levity to complicated material. This collection of fellow scientists and students sits in the margins, introducing features, reminding students of what they know, directing the reader to other helpful resources, and reminding learners of the presence [*and the importance*] of allies on this journey.

Thinking Like a Chemist

Organic chemistry presented as a collection of facts leads to memorization and, frankly, is boring. Organic chemistry presented as a few simple concepts linked logically through structure and mechanism is as much beautiful art as it is rigorous science. Thinking Like a Chemist invites students to see organic chemistry as the latter. Armed with a few basic concepts, students, even in an introductory course, can make the discoveries, invent the reactions, and elucidate the concepts that represent the foundation of our science. By allowing students to "discover" these new concepts as they learn them for the first time, an emotional connection is made to the material, enabling long-term retention. In discovering novel concepts, students gain confidence that organic chemistry is not the insurmountable course they once believed it was. And, maybe, they see in themselves a future scientist who can make similar discoveries in chemistry or other scientific disciplines.

Chemistry in Context

The students who take organic chemistry come with myriad career plans, including becoming physicians, veterinarians, pharmacists, medicinal chemists, biologists, and environmental scientists, to name just a few. [*Some even want to be organic chemists* ☺.] To help students get excited about the organic chemistry they are learning, Chemistry in Context boxes show the connections between the chemistry and the careers listed above. So that these boxes are not skipped, they are integrated into the main narrative of the text, and they often include assessments that further connect organic chemistry understanding to these issues of perhaps greater relevance to students.

Green Chemistry

With a primary goal of producing a given molecule with a certain medicinal property, synthetic organic chemists haven't always prioritized the sustainability of the reactions and processes used. [*Why?*] It is the author's opinion that this is a result of the traditional organic chemistry education graduate students and future researchers have received, where "making the molecule" is most important. As a result, Chapter 7, not found in any other organic chemistry textbook, introduces students to the concept of green chemistry, shows them some of the most successful outcomes of a more sustainable approach, and encourages students to keep these concepts in mind as they progress through the rest of the textbook. The importance of green chemistry is constantly reiterated, with at least one discussion of green chemistry in each chapter, most often in a Chemistry in Context box but sometimes in the narrative itself. While some argue that green chemistry is best suited for a laboratory course, the direct integration of it into a lecture text reinforces the importance of students seeing chemical discovery through a lens of sustainability.

Solved Assessments

In order to dispel the notion of organic chemistry as a "practice in memorization," students are taught how to work through challenging material through periodic solved assessments embedded within the narrative. In addition to teaching problem-solving strategies, these solved assessments also have the secondary benefit of introducing students to new concepts. In the same vein of Thinking Like a Chemist, a solved assessment might open a section to show students that they can make connections between old and new material, building confidence in, and strengthening, their ability to predict what's to come.

[*Quick note to students about the term "assessment." This term is used intentionally in place of words like practice, problem, or exercise. Consistent with the learner-centered approach, assessments are an opportunity for you to assess what you know. While you may be required to turn in some of them as part of a homework assignment, do these to assess your level of understanding at the point you do them. And use the feedback from that assessment to direct your next efforts.*]

Key Concept Videos

Key concept videos give students on-demand access to further support for some of the more difficult and more important concepts. Featuring the author of the textbook and the coauthor of the Student Supplement [*more on that later*], these videos give expanded explanations using the language of the textbook. Recorded using the lightboard format that includes the instructor on screen, these videos continue building the rapport established between author and student in the textbook.

Solved Assessment Videos

As this is a textbook that emphasizes problem solving over memorization, it would be great if there could be more solved assessments throughout the chapters. But a 1,400-page book is probably long enough already. In addition to a standard solutions manual accessible under the Instructor Resource tab in Mastering, a significant number of assessments are solved using the lightboard video format. These videos, running around 1 minute in length, are hosted on YouTube. By accessing an "answer key" in this format, students can not only confirm whether their answer is correct, but also see where they went astray if their answer is incorrect. And by watching the problem-solving strategies employed by the author and friends, they can refine their own problem-solving abilities in a way that directly aligns with the language of the textbook.

Student Supplement

What about the student who wants to go deeper in their studies, or the student who struggles during the first third of the semester? How can students avoid the common pitfalls that have trapped previous organic students? How will this material eventually be tested on upcoming standardized exams (MCAT, DAT, for example)? There must be common questions that all students should know to ask . . . enter the Student Supplement. Designed to complement the approach used in the textbook, the Student Supplement, featuring Dr. Adam Azman as coauthor, does all of these things and more. By presenting the material in a number of ways that align with the pedagogical goals first presented in Chapter 1, this resource will offer students at all levels of interest and ability something to help them achieve at a higher level. The supplement can be purchased at https://www.pearson.com/en-us/learner.html.

In conclusion, a ton of effort by a ton of people went into bringing this book and the features described above to life. [*I thank those people elsewhere.*] For now, let me thank you for giving this book a chance. A chance to be different. A chance to make organic chemistry fun. A chance to make organic chemistry accessible and engaging. A chance to form a relationship with your students. A chance to help [*all of*] them achieve their goals in organic chemistry and beyond. I look forward to having you as a leader on this journey with us. Please do not hesitate to reach out if I can ever be of assistance (mullinsr@xavier.edu). [*Now the fun begins . . . to Chapter 1!*]

Letter to Faculty

Dear Colleague,

Thanks for your willingness to adopt **Organic Chemistry: A Learner-Centered Approach** for your class. This book has been in the works for quite some time, and I am thrilled to finally see it ready for student use. [*After working on so many drafts in Microsoft Word^{TM}, it finally has become a book.*] I hope you and your students like it.

From the outset, the goal of this project was simple: to create an organic chemistry textbook and suite of resources that **deliver organic chemistry concepts** and practice to students in as close a manner as possible to **the way students learn in the classroom**. Informed by education research, this book reinforces learning by introducing new content in the context of previously learned concepts and reactions. The book's insight into how students learn organic chemistry is also refined by my own experience in the classroom and guidance from student and faculty advisory committees.

The book seeks to make organic chemistry accessible by linking all material covered to two simple concepts: **The ball rolls downhill** and **Negatives attack positives**. And it does so **without sacrificing the depth and breadth of coverage** found in traditional organic texts. It all starts with the premise that even the most difficult concepts in organic chemistry are easy if a solid fundamental base is established. All of that begins in the first chapter with a discussion of why organic chemistry is difficult and how students should approach it in a way that will ensure success.

The text was created alongside a supplement that aims to reach students at all levels, from the A-level student who has aced the first three exams to the C/D student who realizes after the second exam that a better approach is needed. In many ways, my approach to the text is informed by the latter, since I was once that student.

Students at my institution report that this approach works better for them than traditional texts. [*They also want recommendation letters to medical school, so take their opinion with a grain of salt.*] Since switching to draft chapters of the book, however, our students have performed better on the ACS standardized exam. On the basis of these results, I am excited to finally share the work with other students.

Though the book has been in the works for eight years, we aren't done yet. Your, and most importantly, your students' experience with the book will be incredibly valuable for the project's further development. It is my hope that **we can incorporate your feedback** to make an organic chemistry resource that continues to effectively teach students organic chemistry at all levels. **Thanks** again for giving this book a chance.

Yours in Organic Chemistry, Rick

Dear Student,

Here's the thing . . . **this book is for you**. Every decision I've made when writing it has been made with you in mind. I'll let you in on a secret: I was once an organic chemistry student. I read my textbook multiple times, memorized definitions, and made flash cards—and bombed my first two exams. Luckily, I had a professor who cared enough to speak honestly and set me on the right path. **He showed me that studying the right way is just as important as studying a certain amount**. He forced me to learn everything in the context of something else. He made me work complicated problems and number carbons. He made me struggle and identify nucleophiles and electrophiles. He taught me that bond dissociation energies and pK_a values and reaction coordinate diagrams can be used to understand almost anything. He showed me that the ball rolls downhill and that negatives attack positives. I am indebted to him for caring enough to have one harsh conversation with me, that was about me, and hundreds more afterwards about organic chemistry. **Those conversations changed the course of my life**. This book pays that debt forward.

In addition to it being a requirement for chemistry majors and future health professionals, organic chemistry is a pretty awesome subject, once you get it. If you study nightly and approach the material the right way, it eventually becomes incredibly logical and dare I say, simple. [*Can I even call it fun?!*] But **you can't memorize your way to success**. [*Trust me, I tried.*] Instead, you have to get engaged, find relevance in the material, make connections to previous material, work assessments, and solve problems . . . and do it nightly.

Think of this book as a conversation. It's not a book about just organic chemistry. It's a book about how to learn organic chemistry. That doesn't mean we are sacrificing concepts. And we are still learning it in a way that promotes long-term retention. Trust the process and enjoy the ride. It is my honor to accompany you in this conversation.

I want this book to be the best resource for all students to learn organic chemistry. With your feedback we can make this happen. **Thanks for being willing to give it a shot**; I hope you enjoy using it as much as I've enjoyed writing it. See you on the inside!

Yours in Organic Chemistry,

Rick

Acknowledgments

You are the sum total of everything you've ever seen, heard, eaten, smelled, been told, forgot—it's all there. Everything influences each of us, and because of that I try to make sure that my experiences are positive. —Maya Angelou

Believing in people before they have proved themselves is the key to motivating people to reach their potential. —John C. Maxwell

There's a moment near the end of a marathon where you realize you're about to finish something that is really hard. Your breath catches, tears well up, and an overwhelming sense of accomplishment fills your body, propelling you through the last few steps. At this point, your mind flashes to images of the countless hours spent training, the sacrifices you and others made to accommodate these efforts, and the people who inspired you to do it at all. [*Essentially, "One Shining Moment" (Section 1.7) plays in your head.*] Writing this book has been its own eight-year marathon, the culmination of the work of so many people, many of whom have no idea of the important role they played in crafting it.

As humans, we are but an amalgamation of the people who influence us on our journey. Parents, teachers, coaches, friends, bosses, and colleagues all contribute to who we are in the same way that a collection of resonance structures makes up the real structure of a molecule. [*Not a perfect analogy . . . see why in Section 2.8.*] Were it not for the people below, and many others not mentioned who believed in me before I had proven myself, this book would not exist. I am forever indebted to each of them for the role they played [*and to you for bearing witness to my montage.*]

The entire team at Pearson has, quite obviously, been instrumental in bringing this idea from a passing thought of what could be to a real book. Highlighting those who played an especially influential role, it all started with **Jeanne Zalesky**, who took a chance on the vision of a (self-proclaimed) not very good writer and, despite moving away from the project, continued to advocate for it from different, and more influential, positions. **Chris Hess** and **Terry Haugen** shepherded this project through the "dog days," keeping the spirits up and the train ambling along, even when the light at the end of the tunnel was nowhere to be seen. In the end, though, it was **Lizzy Bell** who took over the supervision of this project for the final stretch, excellently coordinating the efforts of a large team while managing deadlines and an emotional author with equal adeptness.

You don't have to be a good writer to write a book as long as you have a great development editor. I have had the privilege of working with the two best in the business. No matter how bad the first draft manuscript of each chapter was, **Barbara Price** was always its biggest fan, expressing full belief in the vision of the work, telling me it was great and then working her magic to make it so. **John Murdzek** came on as the closer, meticulously poring over every chapter and every piece of art to make it just right, continuously advocating on my behalf to ensure the vision of the project was fully realized.

This book was heavily influenced by my students in a number of direct and indirect ways. First and foremost, 16 years' worth of **Xavier University organic chemistry students** taught me how to teach, and they continue to do so today, holding me to the same high standard to which they hold themselves. It is the culmination of "what worked for them" that is contained within this text. In that way, this book was written both by and for students. Contributing more directly were **Trisha Makley**, **Jake Ciricillo**, **Sarah Haas**, **Rick Breidenstein**, **Stephan Freeman**, **Andy LeBlanc**, and **Michaelia Fisher**, whose work rendering figures, finding spectral data, proofreading, etc., was equaled in importance by their support of the vision of the project. Of special note is **Emily Kerr**, currently a Ph.D. student who will assuredly be an excellent organic chemistry teacher herself. Having her as a sounding board for ideas, teaching strategies, and proofreading has made this book substantially better.

The following reviewers have also played an enormously important role in the development of this book.

Vanessa McCaffrey, Albion College
Adam Brunet, American International College
Smitha Pillai, Arizona State University
Bradley Merner, Auburn University
Michael Wentzel, Augsburg College
Caesar Almaraz, Austin Community College-Northridge
Ryan Jeske, Ball State
Rizalia Klausmeyer, Baylor University
Anthony Anderson, Bowie State University
Jeremy Klosterman, Bowling Green State University
Tae Oh, California State University, Northridge
Joshua Lawrence, Centenary College of Louisiana
Jennifer Muzyka, Centre College
Brian Duffy, Chowan University
Daniel Whitehead, Clemson University
Vania De Paoli, Cleveland State University
Tim Barker, College of Charleston
Pamela Kerrigan, College of Mount Saint Vincent
Jerry Easdon, College of the Ozarks
Paul Bonvallett, College of Wooster
Nicholas Shaw, College of Wooster
David Dillon, Colorado State Univ - Pueblo
Sheri Strickland, Converse College
Jordan Fantini, Denison University
Christopher Roy, Duke University
Gregg Wilmes, Eastern Michigan University
Eric Abbey, Eastern Washington University
Dennis Awasabisah, Fitchburg State
Laura Frost, Florida Gulf Coast University
Nasri Nesnas, Florida Institute of Technology
Julia Keller, Florida State College at Jacksonville
Virginie Magiotti, Gaston College
Julia Parades, Georgia Gwinnett College
Richard Pennington, Georgia Gwinnett College
Mai Yin Tsoi, Georgia Gwinnett College
Karelle Aiken, Georgia Southern University
Shainaz Landge, Georgia Southern University
Arpita Saha, Georgia Southern University
Christine Whitlock, Georgia Southern University
Joan Mutanyatta-Comar, Georgia State University
Keith Pascoe, Georgia State University
Rita Majerele, Hamline University
Martin Yes, Henderson State University
Adam Braunschweig, Hunter College
Joshua Pak, Idaho State University
Sebastien Laulhe, Indiana U - Purdue
Giselle Marks, Jones County Junior College
John Haseltine, Kennesaw State University
Andrew Robak, Keuka College
Anne Szklarski, King's College
Sunny Lei, Lamar
Christine Miller, Lee College
Amar Tung, Lincoln University
Michael Sundermann, Lonestar College
Olga Michels, Luther College
Jeffrey Hugdahl, Mercer University Macon
Brian Provencher, Merrimack College
Anne Reeves, Messiah College
Bernadette Corbett, Metropolitan CC
Shayna Burchett, Missouri Southern State University
Jon Rhoad, Missouri Western St. University
Jeannine Christensen, Moraine Valley Community College
June Kearney, Nash Community College
Rajesh Komati, Nicholls State University
John Emrani, North Carolina A&T State University
Maria Gallardo-Williams, North Carolina State University
Joshua Pierce, North Carolina State University
Christopher Gorman, North Carolina State University
Ronald Marks, North Greenville University
Dragos Albinescu, Northeastern State University
Spence Pilcher, Northeastern State University
Cindy Browder, Northern Arizona University
Michael Cross, Northern Essex Community College
Mike Cross, Northern Essex Community College
PATRICIA Calvo, Nova Southeastern University
Roman Debinski, Oakland University
Dinty Musk, Ohio Dominican University
Daniel Turner, Ohio State University
David Boyajian, Palomar College
Staínder Bains, Paradise Valley Community College
Amanda Jones, Polk State College
Gabriele Backes, Portland Community College
Christine Hermann, Radford University
Michael Coleman, Rochester Institute of Technology
Ross Weatherman, Rose-Hulman Institute of Technology
Christopher Nicholson, Rose-Hulman IT
Carolyn Weinreb, Saint Anselm College
Cynthia Gilley, San Diego Miramar College
Deborah Mead, Seminole State College of Florida
Lucas Tucker, Siena College
Bernard Majdi, South Georgia State College
Eric Trump, South Georgia State College
Michael Gesinski, Southwestern University
Marisa Blauvelt, Springfield College
Melisa Evans, St Petersburg College
Thomas Gardner, St. Cloud State University
steve samuel, State University of New York
Lynda McMaster-Schuyler, State University of New York
Lynda McMaster-Schuyler, State University of New York
Amber Schaefer, Texas A&M University
William Hersh, The City University of New York
Dan Esterline, Thomas More University
Maria Twenter, Trident Technical College
Claire Filloux, UC Davis
Marco Bonizzoni, University of Alabama
Stephen Woski, University of Alabama
Diana Leung, University of Alabama
Carl Hoeger, University of California-San Diego

Yitzhak Tor, University of California-San Diego
Susan King, University of California Irvine
Vy Dong, University of California Irvine
Joseph O'Connor, University of California San Diego
Donovan Dixon, University of Central Florida
Andy Frazer, University of Central Florida
Dallas New, University of Central Oklahoma
Deborah Lieberman, University of Cincinnati
Lisa Julian, University of Colorado, Denver
Stefanie Habenicht, University of Florida
Anthony Amaro, University of Jamestown
Ashley Steelman, University of Kentucky
Kelli Slunt, University of Mary Washington
Victoria Volkis, University of Maryland Eastern Shore
Shiv Kumar Sharma, University of Miami
Yftah Tal-Gan, University of Nevada, Reno
Joseph Simard, University of New England
Markus Etzkorn, University of North Carolina
Erik Alexanian, University of North Carolina at Chapel Hill
Hua Zhao, University of North Colorado
David Forbes, University of South Alabama
Chuanbing Tang, University of South Carolina
Nicholas Marshall, University of South Carolina - Aiken
Jeffrey Raker, University of Southern Florida
Alicia Frantz, University of Virginia
Ivana Peralta, Vincennes University
Suzanne Ruder, Virginia Commonwealth University
Ajit Dixit, Wake Forest University
Angela King, Wake Forest University
Michael Fultz, West Virginia State University
Joshua Osbourne, West Virginia University
Nicholas Greco, Western Connecticut State University
Nick Greco, Western Connecticut State University
Jonathan Collins, Whitman College
Kerry Breno, Whitworth University
Amy Bradley, Wilkes University
James Dittami, Worcester Polytechnic Institute
Adam Azman, Xavier University

The influences and experiences that have culminated in this work began long before a contract was ever signed. It is this group that follows who definitely "believed in me before I proved myself" and it goes back a long way. **John Calton**, a high school math teacher, met with me several mornings per week to help me catch up to my peers so that I could take calculus as a senior. He was the first to show me what it means for a teacher to truly "go the extra mile." **Joe Workman**, my first and most influential organic chemistry professor, spent countless hours in his office showing me how to learn organic chemistry, at the same time giving me a role model I still aspire to be like in my own teaching and the way I interact with students. **Jennifer Muzyka**, my second organic chemistry professor, for no other reason than loyalty, bought into the early vision of the text, graciously volunteering her students to help improve its development. **David Williams**, my Ph.D. advisor, exposed me to the beauty of natural product synthesis while teaching me how to think through difficult mechanistic and synthetic problems. Perhaps more importantly, he believed in me as a future teacher, patiently allowing me to (slowly) develop as a researcher. **Jeff Johnston**, as one of the last professors I took a class from, taught me the importance of connecting reactions through, among other ways, analysis of bond energies and pK_a values.

My **colleagues at Xavier University** gave me my first, and thus far only, opportunity to be a college professor. By allowing me to fail early and showing patience as I learned from those failures, they taught me how to be a professional teacher, researcher, advisor, mentor, and colleague. Notable among them are **Steve Mills** and **Kendra Denlinger**, the first adopters of draft chapters of the text. Their belief in the project and the feedback their students generated have positively influenced this book.

Adam Azman deserves his own special mention, as he and I have been discussing organic chemistry since 2004. At different times, I have had the privilege of calling him a student, a mentee, a friend, a fellow organic chemistry professor, a colleague, and, now, a coauthor. The influence of our discussions and what he has taught me about teaching and learning organic chemistry are throughout the text, and even more heavily represented in the associated videos and Student Supplement we have written together.

[*Nobody's montage should really be this long, but bear with me for a couple more paragraphs. I've been working on this project for a long time.*] Before we have teachers or coaches, we have our parents. I am so blessed to have **Mom** and **Dad**, the first people to make me believe that I could do anything, who taught me how to work for it, and who were willing to do whatever they could do to help me get there. Not everyone is blessed to have a great relationship with their in-laws, but I am pleased to be in that minority. **Gene** and **Brenda**, as former educators themselves, were supportive from the beginning, giving my wife and me confidence that the project was worthy of the time it would take.

Before signing the contract to write the book, I promised myself that I wouldn't be an author who used this space to talk about all the time I didn't get to spend with my children, **Maggie** and **John**. My best friend and wife, **Mary**, made sure that I could keep this promise. While I can point to various parts of the book that the people above made possible, it is due to her influence that these words were written at all. Even when it was just a "dumb idea" on a long drive to the beach, her belief in the project and, most importantly, her belief in me has never wavered. There aren't enough words to fully express my appreciation for the sacrifices she has made for us, our children, and *our* book. I am so thankful to have her on this journey [*and as the most prominent person in my montage.*]

In closing, [*if you are still reading*], the last person to acknowledge is you, the student/reader. This book was written for you, and it is yours now. I look forward to hearing about your experience with the text. If you find mistakes, concepts that need clarifying, or a way to improve anything at all, or if you just want to talk organic chemistry, please reach out.

Richard J. Mullins
Xavier University
mullinsr@xavier.edu

About the Author

Richard J. Mullins, born near Detroit, Michigan, grew up in Middle Tennessee, where his love of sports and science was cultivated. He attended Centre College, where he pitched for the college baseball team. Though he entered college intending to pursue a career in medicine, an inspiring organic chemistry professor changed his path, leading him to graduate in 1998 with a B.S. in Chemistry and plans for a career as an organic chemist. Upon enrolling at Indiana University, he considered multiple careers in chemistry before soon realizing that teaching at an undergraduate institution was his calling. After earning his Ph.D. from Indiana University in 2004, specializing in organic chemistry, Rick was hired as a tenure track professor at Xavier University.

He has taught at Xavier University ever since, being recently promoted in 2017 to Professor. His teaching has been recognized with multiple awards, including the Joan G. McDonald Teaching Award, which recognizes excellent teaching in the sciences, as well as the Alpha Sigma Nu Bishop Fenwick Teacher of the Year Award, the most prestigious teaching award given at Xavier University. His research has traditionally been focused on the development and application of new methods for the synthesis of small natural products with potential medicinal applications. Rick lives in Cincinnati, Ohio, with his wife, Mary, daughter, Maggie, and son, John. When not teaching or writing about organic chemistry, Rick still enjoys participating in sports by coaching his son's select baseball team, running marathons and half-marathons, and supporting his favorite college and professional sports teams. Though records can't confirm it, because of his longevity in the sport, Rick is the [*self-proclaimed*] all-time leading scorer in intramural basketball history at Xavier University.

About the Book

Organic Chemistry: A Learner-Centered Approach teaches students *how* to learn Organic Chemistry.

Author Rick Mullins' friendly and conversational style establishes an authentic rapport through a balance of authority and humor. Using pedagogy driven by modern learning science, he continuously moves students away from memorization of facts and superficial shortcuts—leading them to become better problem solvers with a fuller understanding of the fundamental principles that link essential concepts. The interactive and socratic narrative incorporates Rick's personal classroom style—creating a dialog with the students while coaching them through mechanisms to emphasize the why behind chemical reactivity. This unique approach makes difficult concepts clear and accessible, without sacrificing rigor—enabling students to succeed in organic chemistry by understanding the chemical logic behind the problem solving process.

"The book seems as if it was written for someone learning organic chemistry rather than somebody who already knows it."

—**Kameelah Belgrave,**
Seminole State College Student

"I just really felt engaged in the text and guided in a way that textbooks do not always guide you. From each section I knew where I was going next and my brain was already formulating questions based on the example problems. I also liked the structure of the sections. The general overview introduced and related the topics to past learnings and the detailed section explained what new topics were being introduced."

—**Drew Rickman,**
University of Alabama student

Features that challenge students to think critically

Thinking Like a Chemist

Thinking Like a Chemist reminds students of the questions they should be asking about the material—providing practice in analytical thinking while building confidence so they can merge their own knowledge to "invent" reactions, discover "new" concepts, and elucidate reaction mechanisms. These discussions help students approach topics the way a scientist might, preparing them for future courses and careers.

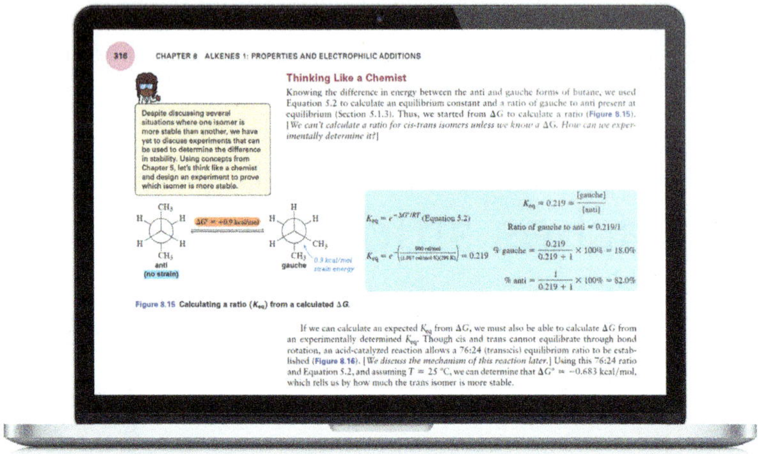

"Really awesome feature that helped to organize a digestible thought process to approach and understand the problems"

—Brittany Carman,
University of Mary Washington Student

Assessment 13.22 The sulfur atom in toluene sulfonyl chloride (TsCl) is strongly electrophilic. Why?

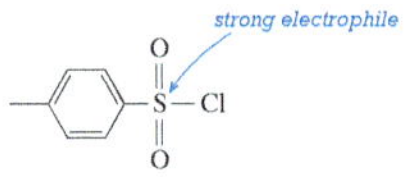

Assessment 13.23 An absent-minded chemist forgot to add triethylamine when attempting to add a tosyl group to the alcohol shown. As a result, the desired product was not isolated. What product might you expect to form? And how could the product's identity be distinguished from the desired product using spectroscopy?

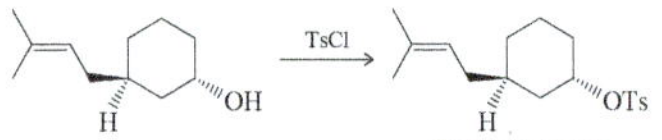

desired product; **not formed**

The importance of the amine base in these reactions is demonstrated by working Assessment 13.23 together.

Solved Assessment 13.23

Regardless of whether there is triethylamine present, the reaction begins the same, with the hydroxyl group displacing the chloride from the sulfonyl group (Figure 13.14). Having brought along its proton, a strong acid is formed. Triethylamine, if present, would deprotonate this intermediate. The only other species that might be able to abstract the proton is the displaced chloride anion. Chloride is a not a very strong base, and HCl, formed when chloride abstracts the proton, is a very strong acid. We have seen HCl used as a reagent in several reactions thus far. [*Do you see a functional group that the HCl could react with?*]

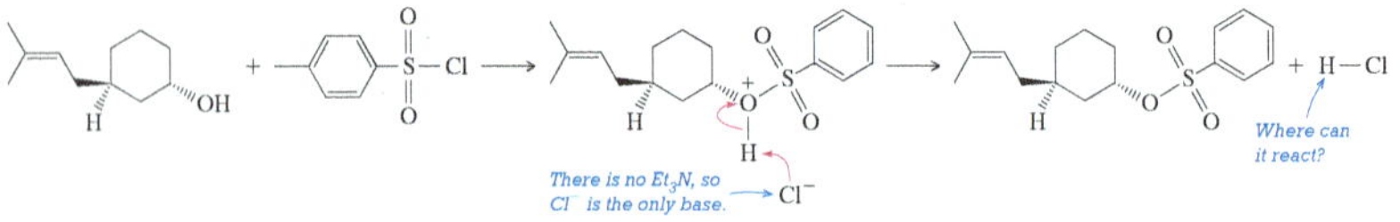

Figure 13.14 With no Et_3N, HCl forms during tosylate formation.

In Chapter 8, we studied the addition of strong electrophiles to nucleophilic alkenes. One of the first reactions (Section 8.3) was hydrohalogenation, a reaction where HBr or HCl adds to alkenes. This molecule, containing a trisubstituted alkene, can undergo this reaction (Figure 13.15), resulting in the formation of a 3° chloroalkane if triethylamine is left out of the reaction. ■

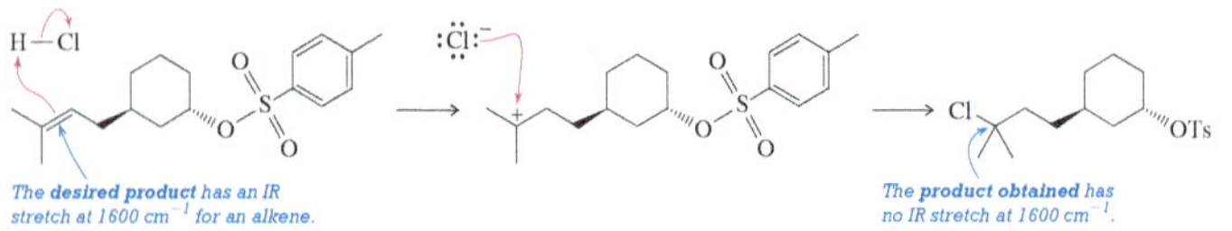

Figure 13.15 HCl can react with other basic functional groups like alkenes.

SPECTROSCOPY INSIGHTS

The desired product contains a C=C bond, exhibiting a stretching band in the IR spectrum around 1600 cm^{-1}. This bond is lost in the reaction with HCl. Thus, the obtained product would be missing this band at 1600 cm^{-1}. [*If you haven't covered IR spectroscopy (Chapter 14) yet, you don't need to worry about this.*]

Coaching Tips

Coaching tips speak directly to students and appear as bracketed comments throughout the text to create an interactive reading experience. These tips are often Socratic in nature, posing questions to students, suggesting ways to process difficult new material, and directing them to resources.

"I found this to be helpful and interactive. It helped slow down my mind and pay more attention to the details. They also were helpful tips that helped me understand the material better."

—Katelyn Winge,
University of Alabama Student

Chemistry in Context

Chemistry in Context boxes help students make direct connections to real-world applications in the areas of biology, pharmacy, medicine and biochemistry.

"It was clear and engaging. I like the personality that is in the text even when only reading small sections, makes it easier to want to read and understand."

—**Karly Gaffrey,**
Clemson University student

13-E CHEMISTRY IN CONTEXT: Green Chemistry

Protecting groups and atom economy

Green Chemistry Principle 2: Reduce waste by maximizing the number of atoms from all reagents that are incorporated into the final product.

Green Chemistry Principle 8: Minimize the use of temporary derivatives such as protecting groups.

While protecting groups are extremely useful in directing reactivity to a particular functional group, they are extremely wasteful. Recall from Section 7.2.2 that atom economy is a way of measuring how much of the mass contained in the reactants and reagents is preserved in the desired (usually final) product. This can be calculated over a single reaction or a series of steps. To see the impact of protecting groups, the atom economy of a protecting group–free reaction is compared to that of an identical reaction of a protected molecule in Figure 13.91.

These calculations imply that 59.2% of the mass that went into making the desired product from only these two steps was wasted. For reactions done on a huge, industrial scale, this amounts to costly waste disposal and potential damage to the environment if the waste is disposed of improperly. This doesn't even take into account the time, energy, and cost of the reagents and solvents that went into putting the protecting group on and taking it off. For these reasons, protecting groups receive special mention in two of the 12 principles of green chemistry.

Though protecting groups continue to be used where absolutely necessary, the design of protecting group–free syntheses has become an important goal of organic and medicinal chemists.

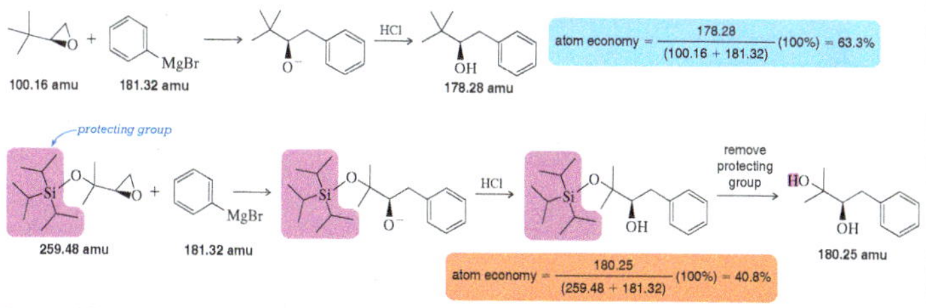

Figure 13.91 Protecting groups reduce the atom economy of a reaction.

Assessment 13.85 Calculate the atom economy of using a reagent like the Dess–Martin periodinane in the following oxidation reaction. Is this a green reaction?

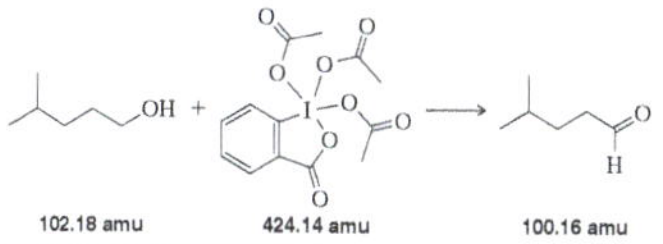

7.2.7 Use renewable feedstocks (GCP 7)

Feedstocks, or the origin of compounds, can be categorized as nonrenewable or renewable. As we study the conversion of organic molecules into other organic molecules, we often disregard the origin of these molecules as being unimportant. That is, we don't think about the feedstock from which our starting materials and reagent molecules are derived. For much of chemical history and continuing today, organic compounds have been derived from oil deposits deep within the ground, isolated as part of petroleum refining and natural gas isolation. [*The next generation of chemists should change this.*] Though these oil deposits were created via the decomposition of living organisms, the fact that they formed millions of years previously makes these functionally nonrenewable resources. Alternatively, some living organisms, like plants and algae, use carbon dioxide in the atmosphere to produce their own organic materials. Compounds derived from these sources are considered to be from a renewable feedstock (Figure 7.13). Green Chemistry Principle 7 encourages the use of renewable feedstocks wherever possible.

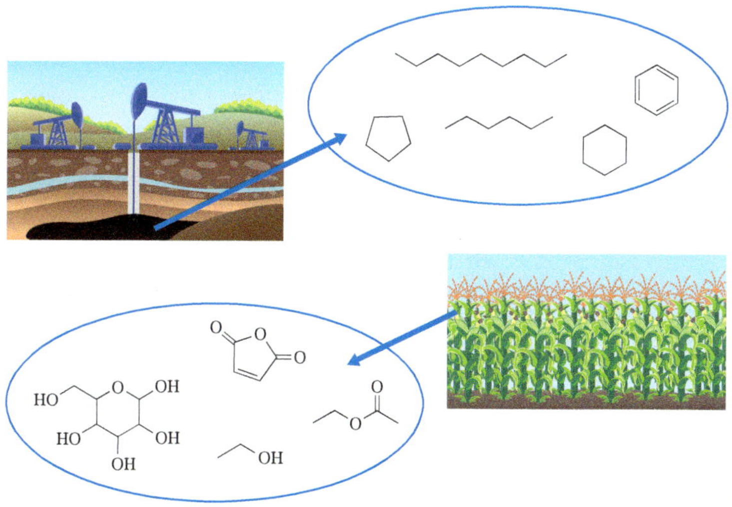

Figure 7.13 Organic compounds can be derived from petroleum (nonrenewable) or bio-based (renewable) feedstocks.

Green Chemistry

Green Chemistry coverage encourages students to consider the implications of the chemical syntheses and processes studied in this course. Students are given case studies and guidelines to think critically about — making them more responsible chemists. In addition to a standalone chapter, Green Chemistry is threaded throughout the entire narrative to show students a more sustainable and environmentally responsible way to do chemistry.

"I would definitely recommend. This book is FOR the reader. Text boxes and pictures make it very easy to understand examples and therefore comprehend the material with ease."

—**Schylar Stiffen,**
Seminole State College student

Mastering Chemistry

Mastering® empowers you to personalize learning and reach every student. This flexible digital platform combines trusted content with customizable features so you can teach your course your way. And with digital tools and assessments, students become active participants in their learning, leading to better results.

Book-Specific Video Series

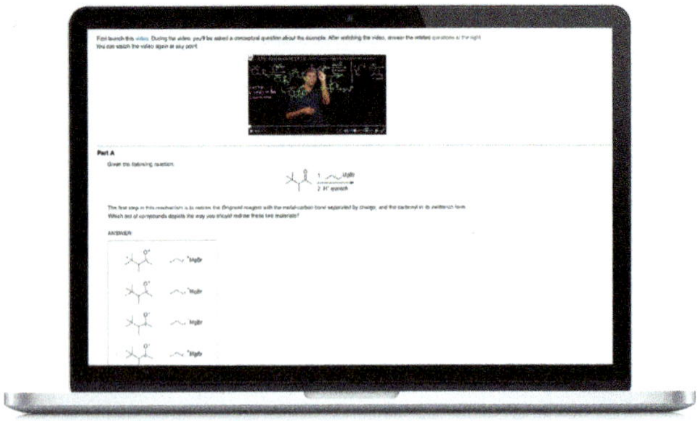

Solved-Assessment Videos help students with problem-solving skills by watching author Rick Mullins and his colleagues work through problems found throughout the book. Access at https://bit.ly/OrganicChemistrybyMullins. [*The link is case sensitive.*]

Key Concept Videos are lightboard-style videos hosted by author Rick Mullins and his colleagues. Embedded throughout the narrative, the videos encourage students to check their understanding of key concepts and reaction mechanisms, prompting them to revisit prior knowledge as a way of helping them achieve long-term retention of information.

End-of-Chapter Problems with Feedback

Enhanced End-of-Chapter Problems provide instructional support when and where students need it, and include links to the eText, math remediation and wrong-answer feedback, and in some cases, fully worked solutions, for homework assignments

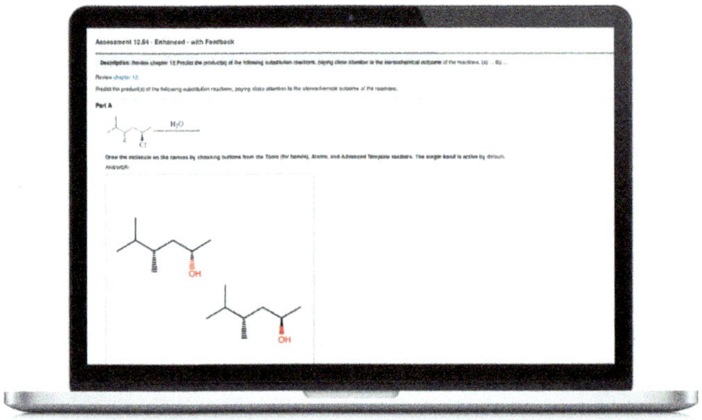

Dynamic Study Modules

Organic Chemistry Dynamic Study Modules focus on general chemistry remediation, acid-base chemistry, functional groups, nomenclature, and key mechanisms. Modules help students study effectively—and at their own pace by keeping them motivated and engaged. The assignable modules rely on the latest research in cognitive science, using methods—such as adaptivity, gamification, and intermittent rewards—to stimulate learning and improve retention. Each module poses a series of questions about a course topic or about a topic specific to the text. These question sets adapt to each student's performance and offer personalized, targeted feedback to help them master key concepts.

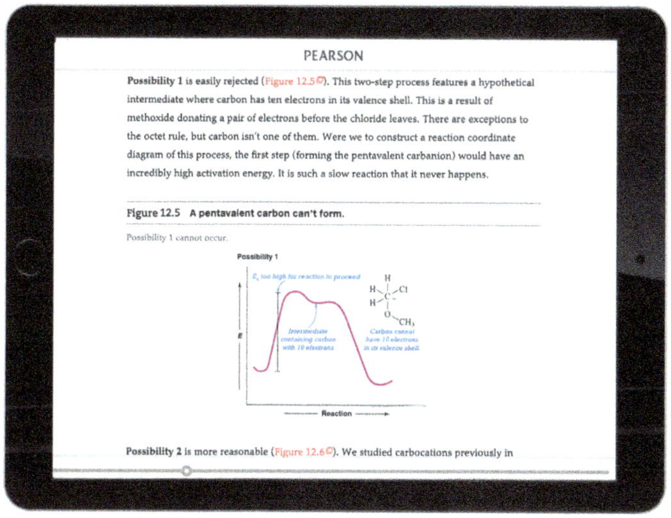

Pearson eText

A mobile-optimized reading experience, Pearson eText is seamlessly integrated with videos, problem sets and other rich media — engaging students and motivating them to keep reading, and keep learning. With the new **Scheduled Reading feature,** easily select and assign specific sections or full chapters through the assignment creation process. Use the Analytics Dashboard to gain insight into how students are working in their eText to plan more effective instruction in and out of class.

Spectroscopy Simulations

NEW! Six NMR/IR Spectroscopy simulations (a partnership with ACD labs) allow professors and students access to limitless spectral analysis with guided activities that can be used in the lab, in the classroom, or after class to study and explore spectra virtually. Activities authored by Mike Huggins, University of West Florida, prompt students to utilize the spectral simulator and walk them through different analyses and possible conclusions.

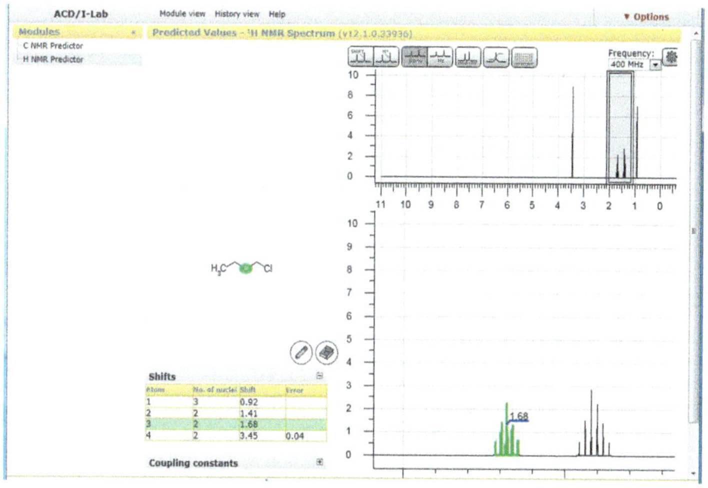

1

Introduction to Organic Chemistry

Introduction

I don't really know how to write the first words of a 6.02×10^{23}-word book, so I'll start simply: Welcome to organic chemistry. When I first imagined writing this book, I wanted it to be different from other books. Most organic chemistry books show the molecules that can be made, the fundamental principles that govern their reactivity, the mechanisms by which they're formed, and cool applications of organic chemistry in other areas of science [*and there are many*]. The book I wrote will do that, too [*at least, it's what I've tried to do*]. But here's the difference . . . I didn't want to write a book that just talks about organic chemistry. I wanted to talk about *learning* organic chemistry. After all, that's what you, as the learner, care about most, and it's the origin of the book's subtitle, "A Learner-Centered Approach."

Additionally, I wanted the book to happen the way a conversation about organic chemistry would happen. This is not me lecturing you from a podium with bright lights shining, where I can't even see the audience as I click through a PowerPoint presentation or draw structures with my back to you. Instead, I intended this book to be a discussion that would occur in office hours, in the university cafeteria, while standing over a reaction in chemistry lab, or during a late-night review session.

A traditional organic chemistry text doesn't have a chapter like this one. With no discussion in Chapter 1 about organic molecules, fundamental concepts, or chemical reactions, what's the point? And yet, given my goals for the book, and likely yours as well, it is perhaps the most important chapter in the whole book. [*Read on to see why, but please, do so with an open mind.*]

Learning Objectives

LO 1.0 Apply the best practices for learning organic chemistry

LO 1.1 Assess the role of grit in successfully completing an organic chemistry course

LO 1.2 Assess the mindset needed to pursue organic chemistry

LO 1.3 Assess strategies for committing information to long-term memory

LO 1.4 Analyze the features of the textbook

LO 1.5 Find sources of assistance for mastering organic chemistry

LO 1.6 Articulate your goal

LO 1.7 Apply the two general rules for organic reactions

"Learn how to handle rejection, work hard, and function outside of your comfort zone. Success will be yours."

—Gonzo Arzuaga

"Experience . . . was merely the name men gave to their mistakes."

—Oscar Wilde

"Whether you think you can, or think you can't, you're right."

—Henry Ford

We can't really have a conversation without knowing each other. And while it is impossible for me to know every one of you, after 18 years of working with students in the classroom, the laboratory, and office hours, coupled with my own 22 years as a student in the classroom, the laboratory, and office hours, I have a pretty good sense of who you are and why you're here. But in the same way that you would argue that you are not typical and that I would be shocked at just how unique you are [*and you'd be right*], I want to give you a bit of detail about me, so you can understand where this book comes from.

While I may love studying, teaching, and doing research in organic chemistry now, that wasn't always my career plan. Nope—I was the pretty typical pre-med student. I took organic chemistry, hoping to avoid being weeded out, to memorize as much as I could, to just get by with my dreams of being a pediatrician still intact. I spent about 6 weeks studying organic chemistry the way I did for all courses: Do whatever homework is required, cram the night before an exam, and voila . . . solid passing grades on my first two exams (C and D, in that order). Luckily, I had a professor who believed I could do more and, most importantly, was willing to help me do that. In sum, he was "learner-centered." Through my interactions with him, and learning organic chemistry the right way, I greatly improved. And, in fact, I avoided being weeded out, finishing two semesters of organic chemistry with my medical school dreams intact.

I continued on the pre-med path, even taking and doing well on the MCAT. The summer before my senior year, however, I chose to do a second summer of organic chemistry research in lieu of volunteering at hospitals and shadowing doctors as required of the pre-med track. During that summer I realized that organic chemistry research and the subject itself was more than a hobby, and that I was more suited to a career in chemistry than one in medicine. So, off to graduate school I went.

Fast forward to now, and I'll sum it up. I wrote this book for you, the student I once was. I have been a biology major taking organic chemistry because I had to. I have been a pre-med worried about getting weeded out of the pre-med track by organic chemistry. I have studied organic chemistry while working in a natural products synthesis lab and wanting a career in chemistry. I have taught organic chemistry while preparing to be a professor. I have taught organic chemistry in the context of making new pharmaceuticals, and while helping prepare students for the MCAT. This book is the culmination of all of those experiences.

From these experiences and from reading the scholarship on teaching and learning, I believe I know what works. I also know that when organic chemistry is truly understood, it is as easy and as logical as any subject you will ever study. It is not an endless string of facts to be memorized, but a logical narrative threaded through every functional group and reaction we will study. And it is difficult. But being a student who struggled initially and eventually made it through, I know that you can, too. So, for this 28-chapter journey you are about to embark on, know that in these words, I'm right here with you.

Test your understanding

Assessment 1.1 True or false: Organic chemistry is a pre-health professions weed-out course.

Assessment 1.2 True or false: The key to success in organic chemistry is to be able to memorize loads of information.

Assessment 1.3 True or false: Some people have a knack for organic chemistry. Others, regardless of how hard they work, will never get it.

Assessment 1.4 Which statement most accurately conveys your feelings about organic chemistry? (a) "I dread taking this course," (b) "I just need to get a good grade in this course," or (c) "I look forward to the challenge of learning organic chemistry."

Assessment 1.5 True or false: The approach I have used for all other classes I've taken should work well here.

Assessment 1.6 True or false: When I study the way I like most, I learn the material better.

1.1 Organic Chemistry: the course

"Old ways won't open new doors."

—Unknown

Organic chemistry is the chemistry of molecules containing carbon. It is the chemistry of pharmaceuticals and plastics. It is the chemistry of DNA, RNA, amino acids, and carbohydrates—it is the chemistry of life. Organic chemistry is a part of virtually everything we do, every product we use, and every medicine we take. These are the things we'll study, but that's for later. For now, let's talk about Organic Chemistry: the course.

I have no doubt you've heard about organic chemistry. I have heard it described as "the hardest course ever," "a weed-out course where the professor wants to keep kids out of medical school," or, more simply, "hell." [*What's its reputation at your school?*] If you watch a contemporary movie or television show and stressed-out college students are being portrayed, they always seem to be studying organic chemistry. While I was writing this book at local coffee shops in Cincinnati, even other patrons crinkled their nose when they found out what I was working on.

To successfully tackle a course like this, we have to understand why organic chemistry [*organic, orgo, or ochem . . . what is it called at your school?*] has the reputation it does. Here are my thoughts on this, in no particular order:

1. Organic chemistry is typically taken by students in their second year of college. If the first year in college is like the fifth year of high school, the second year is when it becomes college. In your first year, applying to medical school and taking the MCAT seems far off. During your second year, your career goals are coming into focus, your GPA is starting to matter more, and the MCAT is right around the corner.

2. Whereas you can get through general chemistry by being a strong math student, and you can get through biology by being able to memorize concepts, there is no math in organic chemistry, and there are so many concepts, reactions, and facts and thousands more ways to test them. In sum, it is virtually impossible to memorize everything.

3. When a course requires consistent discipline and daily effort, it is easy to fall behind. And once you fall behind, the perception can be that it is impossible to catch up. [*This may be true, so do stay caught up. But life happens, and I think there are ways to catch up after falling behind. I'll explain later.*]

4. The course is more about problem-solving, requiring a consistent and logical approach. But not many of the courses you have taken thus far require it in the same way, or to the degree that it's required here. For some students, this makes the course seem harder than it is.

5. Organic chemistry forms the basis of biochemistry, molecular biology, medicine, and pharmaceutical chemistry. Thus, the perceived need to learn it, and understand it well, places a fair amount of pressure on the typical second-year student.

6. The second year in college is tough for science majors. Along with organic chemistry, a chemistry major is probably taking advanced math courses and/or calculus-based physics. With a career in research in mind, a student might be working in a lab or seeking out these experiences. A student interested in a health career is also taking physics, while shadowing doctors and getting involved in extracurricular activities to prepare themselves as a well-rounded student who has a demonstrated passion for medicine and desires a career in medicine for the "right reasons."

7. Organic chemistry is one of a rare set of courses where the second semester is a true extension of the first semester. Most general chemistry classes are structured such that atomic structure, acid–base chemistry, gas laws, stoichiometry, electrochemistry, thermodynamics, and equilibrium are studied separately. So while there is a lot of overlap between topics, it is possible to do poorly on one subject, but then do better on another. In organic chemistry, what is learned on the first day is also used on the 10th, the 50th, and the 175th day.

8. Organic chemistry is a pre-requisite for medical, dental, veterinary, and optometry schools (among many others) and is tested on the MCAT, the DAT, the OAT, and the chemistry GRE. And many schools have organic students take the American Chemical Society standardized exam as a final exam at the end of the second semester. This places pressure on students, because what you learn on day 10 will be tested on exam 1, then reiterated on the next three exams, tested on the Organic I final, then again on the Organic II final. And a year or more later, you'll need it for a standardized exam that may determine whether you get to pursue your desired career. That's a lot of pressure. [*But then, there is a lot of pressure on you as a chemist, a professor, a researcher, a physician, a dentist, a veterinarian, etc.*]

Suppose for a minute that you (or a relative) are suffering from some unknown illness and are about to be treated by a physician. I'm guessing you'd want this physician to be serious, disciplined, confident, and calm under pressure; possess good time management skills; be an excellent and efficient problem-solver; and have a strong collection of medical facts on quick recall. [*And, probably, empathetic with a good bedside manner, but we aren't teaching that here.*]

Would you care if this physician knew organic chemistry? Would it make a difference if she knew that a 3° carbocation was more stable than a 1° carbocation? As she is treating you, will she really need to think about how a strong nucleophile makes an S_N2 reaction more favorable than an S_N1 reaction? The answer to these questions is obviously no. For a student who won't one day become a chemist, this course is a template upon which you will learn valuable problem-solving skills. Yes, you will give yourself a solid foundation that will help you understand the principles of biochemistry that will make molecular biology a bit easier, in college and then again in medical school. But it is these secondary outcomes, beyond knowing more organic chemistry, that are the reason organic chemistry is a prerequisite for all of the pre-health professions. If you can get through organic chemistry, you are set up to get through medical, dental, optometry, or veterinary school.

Quick note: Every semester, I get grief from my students about making every example about medical school when several want to be chemists, biologists, or biochemists, or go into some other (non-medicine) health field (dentistry, pharmacy, optometry, etc.). I understand that there are a variety of interests, and by highlighting medicine, I am not at all suggesting that it is superior to any other field [*I chose to pursue organic chemistry, research, and teaching, after all*]. In this way, "medicine" is simply an easy way of referring to all fields that are not "chemistry."

One more note: To the future chemists—I don't need to tell you why you need this course. You probably already know it is the coolest subject that anyone could ever study. [*Let's work together to convince the pre-health students that it's important too!*]

1.1 SUMMARY

- For a wide variety of future careers, there are many reasons to take organic chemistry. Embracing the reason you are taking it will help you stay motivated.

Assessment 1.7 In your opinion, what is missing from the list of reasons why organic chemistry is such a challenging course?

Assessment 1.8 Why is organic chemistry necessary for your career goals?

1.2 Grit

"You got it on me in nine categories. But if we get on the treadmill together, there's two things: You're getting off first, or I'm going to die. It's really that simple."

—Will Smith

Organic chemistry is not a weed-out course. [*Say it with me.*] Do people take organic chemistry and then decide to no longer pursue the goal of graduate school or medical school? Yes. But every student who takes organic chemistry has a choice in whether or not they are weeded out. Even a student who has fallen hopelessly behind ultimately makes the choice that they aren't willing to do the work necessary to catch back up. Organic chemistry as a course should not be given the power of being a weed-out course. You control your own destiny. If you want to be a dentist, chemist, or physician, organic chemistry will not make that decision for you. If you want it, you'll make it happen. [*How is this true?*]

There are two things every student needs to ensure success in organic chemistry. The first is grit; the second, covered in Section 1.3, is a growth mindset. Note that neither intelligence nor natural talent is listed here. Talent can be helpful, but there is nobody reading this book who is naturally predisposed to succeeding or failing at this course.

Let's talk about grit. Grit is defined as "perseverance and passion for long-term goals." Synonyms of grit include courage, resolve, determination, tenacity, toughness, and endurance. Given the reputation of organic chemistry (Section 1.1) and the fact that so many programs require it for admission, it comes as little surprise that it might be a proving ground for a student's grit. If you have time, you should definitely read *Grit: The Power of Passion*

"I won't just have a job; I'll have a calling. I'll challenge myself every day. When I get knocked down, I'll get back up. I may not be the smartest person in the room, but I'll strive to be the grittiest."

—ANGELA DUCKWORTH

and Perseverance by Angela Duckworth. Much of what is written here is based on that book, a must-read for all college students.

In her book, Duckworth discusses the relationship between talent, effort, and achievement. She acknowledges that some will inevitably have talent that allows them to achieve. But not having talent does not preclude someone from achieving. By her math, skill in a task is accomplished by taking your talent and developing it through practice (repeated and committed effort).

$$\text{Skill} = \text{talent} \times \text{effort} \qquad \text{Equation 1.1}$$

Achievement occurs when your skill is repeatedly, and consistently, applied.

$$\text{Achievement} = \text{skill} \times \text{effort} \qquad \text{Equation 1.2}$$

Put together, achievement in any profession, course, or career goal comes down to the fact that "effort counts twice." [*That has become a bit of a mantra in my family . . . my kids hate that I read Duckworth's book.*]

$$\text{Achievement} = (\text{talent} \times \text{effort}) \times \text{effort} = \text{TE}^2 \qquad \text{Equation 1.3}$$

What does this mean in the context of organic chemistry? Bad news first: If you have zero talent, like absolutely none, then according to Equation 1.3, you cannot possibly achieve. You can't pass organic, you can't get through your chemistry/science curriculum . . . sadly, you will never become a chemist or a biologist or a physician. [*You have talent, you belong here, you can do this!*]

The great, amazing, wonderful news of Equation 1.3 is that as long as there is some non-zero number associated with your natural, or innate talent, then YOU control whether or not you'll achieve. You can't control the amount of talent you have. But you can control the amount of effort you put into the class. You can control the way you study. You decide when to work instead of play. You decide when to hit snooze on the alarm versus getting up to go to class. You decide how to organize your time so that enough of it is dedicated to organic chemistry, your other classes, and [*very importantly*] your emotional and physical wellness [*don't neglect these*]. If success in organic chemistry, or any course, is important to you, then you will make it happen.

I feel compelled to say at this point that **INSERT YOUR CAREER GOAL** is not the only way you can be happy. This idea that you control whether or not you achieve your goal is meant to motivate you. But I sometimes worry that students hear it in a way that puts even more pressure on them. So you can want whatever you want for a career; there are multiple fulfilling careers, and college is about figuring out which one is right for you. Organic chemistry can play a role in helping you figure that out . . . it certainly did for me. But once you know what you want, go after it with grit and determination. [*Remember, effort counts twice.*]

What happens when you aren't successful, when the effort you put in isn't immediately paying off? Here's a scenario that happens often:

> A student works hard but earns a 70 on exam 1. He talks to the instructor about how to change his study habits, showing grit and determination, but earns a 68 on exam 2. He modifies his study habits, continues working daily, and earns only a 75 on exam 3.

What's going on here? It could be a number of things, but here's my guess. Exam 1 is easier than exam 2, which is easier than exam 3. By staying the same [*treading water, if you will*], the student is actually improving. It's hard to recognize this from the student's perspective, but it's true. My advice to that student, and to all students, is to trust the process. Do the work, solicit and take advice, and relentlessly apply it. In other words, you can't get caught up in the results—you simply have to trust the process. Given our culture's emphasis on high grades and achievement, this is really difficult to do, but you can't be results-oriented. Eventually, you will see that spike in grades you are looking for. It may not happen until exam 3 of Organic I, or exam 3 of Organic II, or even exam 2 of Organic I the second time you take the course, but it will happen, if you want it to. [*Effort counts twice.*]

One last thing about grit. Duckworth, using an especially pertinent example, talks about the difference between a student who desperately and passionately *wants* to be a physician versus the student who *likes the idea* of being a doctor. The one who really wants it sees no problem with making the necessary sacrifices because they recognize and even enjoy the hard work associated with the intermediate steps that will allow it to happen. The one who simply likes the idea of being a doctor doesn't recognize and isn't willing to put in the effort to make it happen. [*Which one are you?*]

1.2 SUMMARY

- Effort counts twice!

Assessment 1.9 Talk about a time when you faced a difficult challenge you perhaps thought insurmountable yet through hard work achieved your goal.

Assessment 1.10 What long-term goal brings you to organic chemistry? Do you really want it? What are you willing to do to make it happen? [*You don't need to have figured this out yet . . . .*]

1.3 Mindset

To succeed at organic chemistry . . . be gritty. Seems almost too simple, so why doesn't everyone do it? A lot of this has to do with how our society attributes achievement to talent.

Assessment 1.11 Which of the following statements do you agree with?

(a) By doing well in organic chemistry, I'll prove that I am smart enough to become a **INSERT YOUR CAREER GOAL**.
(b) I'm a naturally gifted test taker.
(c) Math has always come easy to me.
(d) I did well in high school because I was smarter than my classmates.
(e) The person who did best in general chemistry is just naturally smarter than me.
(f) Some people have what it takes to become a doctor; some do not.

This emphasis on being "talented" or "intelligent" or having "natural ability" results in many of us having a **fixed mindset**. [*Another must-read: Mindset by Carol Dweck.*] A fixed mindset is one that thinks of achievement on an exam, for example, as defining us in the same way that other fixed traits do. "I failed my math test, so I'm not good at math." Or "He aced organic chemistry—he just has that knack for organic chemistry." In many ways, the fixed mindset provides a protective barrier, by making your failures out of your control: "I didn't do well on the ACT because I'm bad at standardized exams."

Before we talk about the opposite mindset, let's think about organic chemistry as a weed-out course. After exam 3, fixed-mindset student (FMS) #1, who can't get higher than a C, might say, "Well, I guess I just don't have what it takes to be a doctor. Organic isn't my thing. I'll find another career to pursue." **Figure 1.1** shows how this might have played out from high school to college for FMS #1. Their natural ability just wasn't enough to allow them to accomplish their goal. Alternatively, FMS #2 was blessed with ability higher than what is required at every stage; that's the good news. The bad news: There will always be another level of ability required. Eventually, FMS #2 will not have that ability.

"You do not learn anything by doing everything right."

—Unknown

The opposite of the fixed mindset is the **growth mindset**. The growth mindset sees every challenge as not a judgment of you as a person, but of where you are right now. Every success is attributed to the work you have put in as opposed to who you are. Every failure is a learning opportunity. The key is the word *yet*. To my mind, the fixed mindset is transformed to a growth mindset by the addition of this one word to every statement: "I'm not good at organic chemistry yet." "I don't take standardized exams well yet." "I am not good enough to get into graduate school yet."

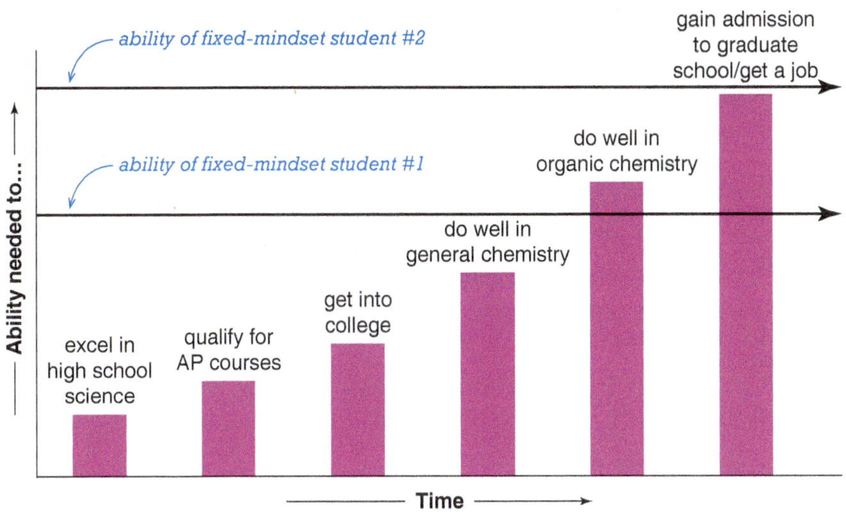

Figure 1.1 A fixed-mindset student has the ability to achieve, or they don't.

Carol Dweck, in *Mindset*, tells of a study where young students were split into two groups. Altogether, they were equivalently successful on a given test. One group was congratulated on their intelligence; the other was congratulated on their hard work. When given a harder test, on which all students scored lower, the "intelligent" group was devastated, whereas the "hard-working" group was inspired to work harder. When the first easy test was re-administered, the intelligent group, their opinions of themselves still shattered, did worse than they did the first time. The hard-working group did better than before. To make matters worse, the intelligent group then lied about the scores they got, probably as a way of protecting their view of themselves. In sum, by attributing our success to a fixed characteristic like natural talent, as opposed to effort, which we control, we make ourselves less intelligent and even dishonest.

Let's solve Assessment 1.12 together.

"For me, survival is the ability to cope with difficulties, with circumstances, and to overcome them."

—Nelson Mandela

Assessment 1.12 In Figure 1.1, which student are you? Do you have what it takes or not?

Solved Assessment 1.12

Assessment 1.12 is an asinine question. You are neither of these students, and you should never define yourself as such. The answer depends on when you answer it. You got into college [*check!*]. You got through general chemistry [*check!*]. Can you pass organic chemistry or get into graduate school? Probably not . . . *yet*. ■

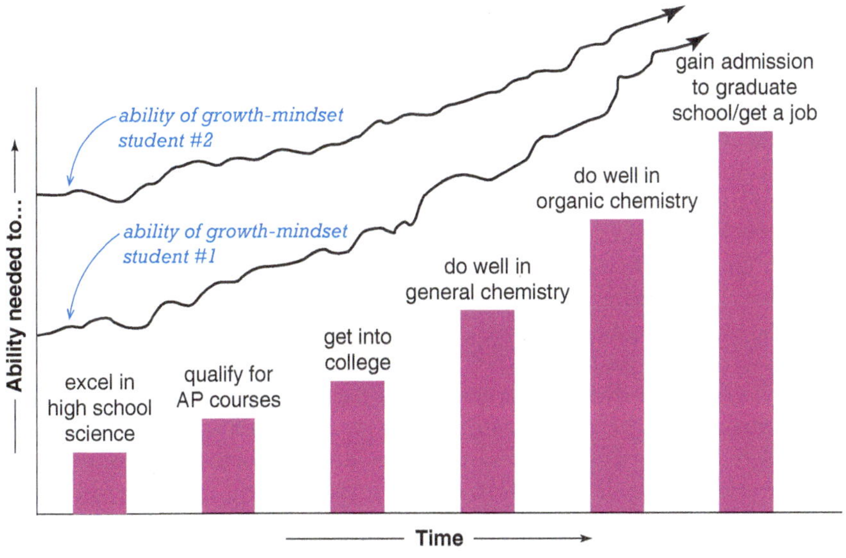

Figure 1.2 A growth-mindset student over time gains the ability to achieve.

How you do on a test, a quiz, or an assessment is a judgment of where you are *right now*. It is not an assessment of where you are going or what you can eventually become. In fact, the assessment is the key to shaping where you will go. By learning from your mistakes, you can figure out what works and what doesn't. You can figure out how to be successful. You can see where you are deficient, and with assistance from friends, parents, advisors, and professors, figure out how to advance to the next level [*how to grow!*]. Figure 1.2 is a re-rendering of how two growth-minded students might advance through their career. Although one seems to start at a higher talent level, both successfully navigate their careers and will likely continue to do so.

With a growth mindset, is there such a thing as a weed-out course? Put simply . . . nope. Conclusion: Have a growth mindset!

1.3 SUMMARY

- A growth mindset allows you to embrace difficulty as an opportunity to grow, improve, and become a better **INSERT YOUR CAREER GOAL**.

Assessment 1.13 Students 1 and 2 both have a growth mindset. Why was student 2 accepted into graduate school when student 1 was not?

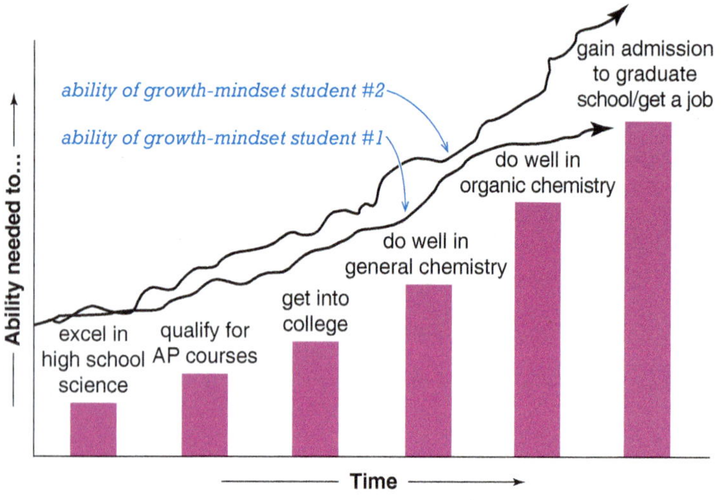

Assessment 1.14 Consider the graph of student X. What happened just after she took general chemistry, indicated by the arrow?

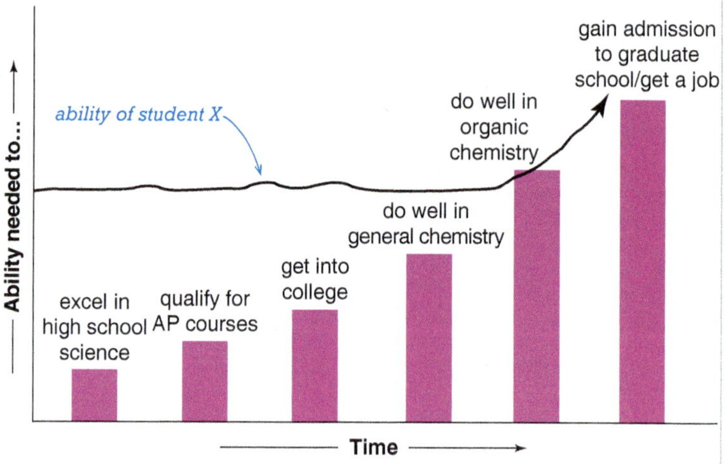

1.4 How the brain works: making it stick

Recent research in neuroscience has given educators and students some insight into how the brain works. Specifically, it has illuminated effective techniques for enhancing understanding and, importantly, long-term recall. While the following section is not a comprehensive review, a brief discussion of how the brain works and how to optimize your efforts in this class is presented in order to maximize your chances of success in organic chemistry. [*It should find broader application in everything you study.*] While there are several books you could read to learn this on your own, most of the information comes from two books: *How the Brain Learns*, 4th ed., by David A. Sousa and *Make It Stick* by Peter C. Brown, Henry L. Roediger III, and Mark A. McDaniel.

While the biology of the brain is discussed elsewhere, we focus on what matters most for this class, which is how the brain processes information. A simple model proposed by Sousa is shown in **Figure 1.3**. The information processing model identifies the manner in which information from the environment is moved through various pathways in and out of the brain. First, information enters the brain through all of five senses. Some of this information is immediately released by the brain, having been deemed unimportant; imagine the background noise at a coffee shop while you study for an exam [*or write a textbook*]. Information that makes it through is immediately stored in short-term, or immediate, memory. Here, information is stored briefly while the brain decides how important it is and whether or not it should be stored. Unimportant information is immediately dumped, preserving neurons and processes for important information. A second type of short-term memory is working memory. Research suggests that information is stored here briefly because the working memory can only hold a few items at a time.

The third type of memory, and the one that matters most to us, is long-term memory. Information that makes it to working memory and has implications for our survival [*wear safety glasses in lab*] is easily stored in long-term memory. Information that causes you to be extremely happy or angry, because it elicits a strong emotional response, is also stored in long-term memory. The concepts covered in organic chemistry, unfortunately, are not

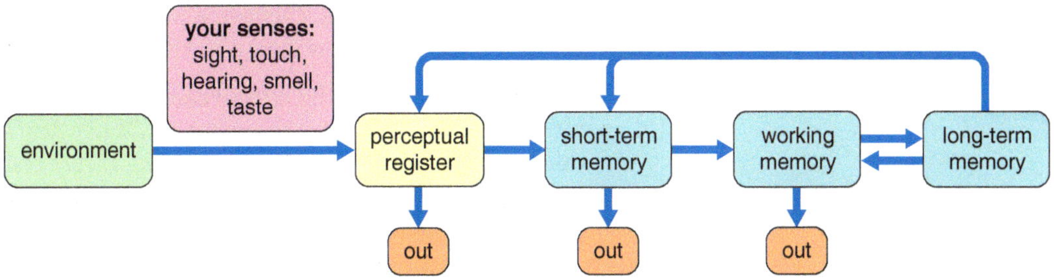

Figure 1.3 The information processing model of how the brain processes information and makes decisions about whether information should be stored in long-term memory or released.

necessary for survival. Nor do they cause extreme emotional responses. [*The course or the instructor might, but the concepts themselves probably do not.*]

The fact that emotional reactions and the need for survival make long-term storage easy helps us understand why learning is so hard. In order for other information to make it to long-term memory, the learner's brain accepts or rejects information on the basis of two questions: "Does this information make sense?" and "Does this information have meaning?" There are some pretty heavy consequences of these two questions in terms of both how you should approach organic chemistry and how this book is organized to assist that. [*Read this paragraph again. Does it make sense that the brain would select information in this way? And importantly, how can you use this knowledge to help you retain information better?*]

Assessment 1.15 Provide a moment from your childhood that you remember as if it happened yesterday.

Assessment 1.16 What two questions must be answered affirmatively for the brain to decide whether information should be held in long-term memory?

Assessment 1.17 When a fire alarm goes off in your dorm, apartment, or house, what are you supposed to do? In spite of the fact that this probably hasn't happened much, why do all students know the answer to this question?

In their book *Make It Stick*, Brown, Roediger, and McDaniel offer research-backed claims about learning that are briefly summarized below, in the context of organic chemistry. [*To me,* Make It Stick *should be required reading of all college students. If you have time over a break, check it out. I wish it had been written when I was a student.*] As you read through these—we'll call them learning claims (LC)—think about that process of moving information from working memory to long-term storage.

LC 1. Learning organic chemistry works better when it is difficult. If learning a particular concept were easy, what would be the motivation for the brain to hang onto it?

LC 2. We don't judge the effectiveness of learning strategies well. When learning feels easy, we think we are gaining information. When learning is hard, we often give up and seek other ways. Unfortunately, easy learning rarely results in long-term retention, whereas hard learning usually does.

LC 3. Rereading and massed practice are not effective strategies for retention of organic chemistry information. When a student sits down to study, one of the first questions that must be answered is what to study. Too often, reading a section and then rereading and then rereading again is a strategy of choice. It turns out that this isn't very effective. When rereading like this, your brain begins to recognize the patterns of the words on the page, and you interpret the pleasure derived from this recognition as a signal that you understand the information. But you probably do not.

On the other end of the spectrum is massed practice, where you actively work the same exact type of assessment over and over again. Suppose you want to learn how to calculate the equilibrium constant using pK_a values. By working an assessment or two, you reveal to your brain the importance of knowing how to do it. By working a few more, you might get the information placed into long-term storage. By then working dozens and dozens more, uninterrupted, your brain quickly tires of the process, feels confident in your ability to do it, and then releases the information. You've simply made the information "easily learned," causing the brain to devalue that information. [*Am I saying you should limit your practice? Of course not . . . read on.*]

LC 4. Retrieval practice is an effective strategy for learning organic chemistry. Practicing the recall of information is important, but is best done in lots of small and varied doses. Back to our equilibrium calculation: Instead of massed practice, you might put pK_a values on flash cards and spend a few minutes going through a list. Then you could work an assessment where you test yourself, allowing yourself to use the values on the cards to help. Then you work through the flash cards again. Then you take a break. When you come back, maybe you work a couple more. And if you can do that, then you probably have the information stored in long-term memory. But you'll want to come back to this retrieval practice a few more times between now and the exam. In this way, you constantly move the information from long-term storage into working memory and then, having reiterated its importance, re-store it. The more often information moves out of long-term storage and into working memory, the more likely the brain will keep this information.

LC 5. You should study organic chemistry each and every day. By bringing information out of long-term storage to use it every day, you reiterate to your brain the importance of long-term storage. This guarantees that the information will be there when you need it on a final exam, the Chemistry GRE, or the DAT.

LC 6. Your study of organic chemistry should occur in brief increments and be separated by your studies in other subjects. This one is the hardest to do and is the most counterintuitive. Interleave your studying of organic chemistry with other subjects. When you pour over organic chemistry for multiple hours at a time, it will become easier. But then, by making it easy on your brain, material learned during those later, "easy hours" will be deemed unimportant, never making it to long-term memory. By switching back and forth every 45 minutes, your brain will struggle with switching gears, thereby recognizing the importance of whatever you keep coming back to.

One mistake students make during a difficult academic year is to prioritize one course over all others. Maybe by doing this, you end up with an A in organic chemistry, but a C in Calculus II and Physics II, which certainly isn't productive. [*I'm speaking from my own experience here.*] Interleaved learning ensures this doesn't happen and strengthens your retention of information in all your classes.

If you plan on taking the MCAT, or working at the interface of two scientific areas, your life will not be organized by subject area. You will, on a daily basis, be jumping from physiology to molecular biology to biochemistry to organic chemistry to general chemistry and then to physics. If we apply our scientific knowledge that way, then why not learn it that way, too? Interleaving helps us with this.

LC 7. Attempting to work an assessment before being taught it helps us retain the new concept better. What better way to establish to the brain that something is important than by trying to figure something out on your own? This is especially true because our brain will notice our disappointment in not doing it correctly the first time. When we finally learn the concepts necessary, the brain will hang on to this information. [*This one requires a growth mindset, which, luckily, we all have.*]

LC 8. Your favorite learning style is probably not as effective as you think it is. If it is easier to learn from one style than from another, then that easier style is probably not effective. Engage all your senses when you study organic chemistry, especially sight, hearing, and touch. [*You probably don't want to taste organic chemistry, but Chapter 8 makes the case that smelling it may be effective.*] Commit to learning in your least favorite styles as often as you do in your favorite styles.

LC 9. By identifying the underlying principles that connect organic chemistry concepts, you will retain them better. The brain, unlike a computer, stores things in sequences. When one aspect of the pattern is recognized later, the whole process is activated, triggering faster recall. Make connections between every concept we study. [*Recall boxes will help with this. See Section 1.5.4.*]

LC 10. Repeated assessment of our knowledge will make sure we know what we know. And as importantly, we will know what we don't know. By taking many "quizzes" in the form of assessments, doing homework, and taking practice exams, we not only show the brain that this information is valued, but we also self-identify concepts that we need more practice with. Yes, it is hard to face our deficiencies, but it is important to do so. [*Even more of a reason to have a growth mindset.*]

LC 11. By making real world connections with organic chemistry information, we retain it better. Remember the question "Does this information have meaning?" By imbuing the concepts with meaning, especially if you can make it personal, the brain will more likely retain the information. [*Chemistry in context asides will help with this. See Section 1.5.9.*]

LC 12. Organization of organic information facilitates quicker recall. There are several ways of organizing organic information. By doing it in multiple ways, you connect different bits of information in lots of different ways. This creates more connections between concepts, giving the brain multiple neural pathways for accessing the information. To use a file storage analogy, information in multiple different files is more likely to be retained.

In a book about organic chemistry, Section 1.4 wouldn't exist. But remember, this book is about *learning* organic chemistry. The 12 research-backed ideas just discussed give us a blueprint for how we are going to learn organic chemistry. And these ideas form the basis for many of the features that are present in the book. By sharing them with you, I hope to convince you of their value so that you will utilize them to their fullest.

1.4 SUMMARY

- Research gives us insight into how and when learning works best. Use these ideas to plan your approach to organic chemistry and all subjects.

Assessment 1.18 Of the 12 ideas presented in Section 1.4, which do you find most believable? Which do you feel is least applicable to you?

Assessment 1.19 Prior to taking organic chemistry, what would you say is your favorite learning style? What is your least favorite? Are you willing to learn using multiple styles?

Assessment 1.20 To become a better free throw shooter, player A lines up at the free throw line and shoots 100 in a row. Player B alternates shooting free throws with dribbling and defensive drills. Which is a more effective way to become a better shooter? Why?

Assessment 1.21 What is the value of answering assessments after reading a section? What is the value of answering assessments prior to reading a section?

1.5 Using the learner-centered approach: how to use this book

The first bit of advice on how to use the book? Read it. [*Obviously you are, so tell your friends to do the same.*] Organic chemistry is always easier to understand the second time. [*I promise this is true.*] If you go to lecture first and then read the book, you'll think the book is better at presenting material. If you read first and then attend lecture, you'll think your instructor is better at presenting material. [*The truth is probably that your instructor is better at presenting it, but there is still value in reading.*] So, read nightly. This doesn't always mean read, highlight, and take notes on every single word the first time you read. But become familiar. Know the important terms; sketch out a reaction, or a mechanism, or two. Doing this will ensure that when you attend class the next day, it won't sound like a foreign language. [*And it may be worthwhile to even work an assessment or two prior to class . . .*]

1.5.1 Assessments

Like every text, this book contains exercises throughout the chapter. Though it is a subtle difference, it is important that these be treated as what they are: assessments. They are not problems or exercises, but assessments. That is, they are an opportunity for ***you to assess what you've learned*** from the reading. If, after reading a section, you are able to answer them successfully, then it is likely that you are ready to move forward. If, on the other hand, you need to look back to answer them, it probably means you didn't understand the material as well as you should have. In this case, you should go back and figure out what you're missing before moving on. Because organic chemistry material builds on itself, if you don't know the answers to assessments at the end of Section 16.5, you aren't likely to understand the material in Section 16.6.

When homework covering a certain number of sections is coming due, you may feel forced to work assessments on material you don't understand well. Consider a problem set covering Assessments 12.1–12.20. Perhaps, for whatever reason, you have to do most of these assessments in a single setting. When this occurs, you may find the first 12 to be pretty straightforward; you can answer 13–16 with a bit of effort, but 17–20 are extremely difficult. This doesn't happen because the last four questions are actually harder; it's just that you didn't understand the previous material as well as you hoped. How do you avoid this situation? Don't wait until the last minute to do homework. Do it nightly, as you read the material. Or do it after you've attended the lecture, which likely comes after you have read the corresponding material. Doing it this way will ensure that you maximize the outcome of the effort you put in. It will also avoid the ineffective strategy of massed practice.

We have already solved Assessment 1.12 together.

1.5.2 Solved assessments

Here's a secret: Getting the correct answer to an assessment is nice, and will help you get credit on a homework assignment, but it is really not the most important outcome. Instead, it's about the process. The reality is you will probably never see that same assessment again.

But what you will encounter, over and over again, is the process that you used to solve it. For this reason, each chapter has solved assessments. These are meant to walk you through the process by which we solve problems. By reading carefully through these, and making sure you understand the process by which problems are solved, you will get better at organic chemistry. [*And better at anything that requires problem solving, to be honest.*]

There's another way that solved assessments are used (**Figure 1.4**). Very often, when a transition is made from one section to the next, you will be able to understand the next section only because you understood the previous. An example of this occurs at the beginning of Section 20.3.4. By understanding the material in Section 20.3.2, a student can basically predict the outcome of Section 20.3.4. Assessment 20.42 illustrates this fact. After you've given it a try [*six months from now*], read through and see how you should have thought about it.

20.3.4 Intramolecular aldol reaction

The aldol reaction can be conducted intramolecularly in an entropically favored process when the reaction is driven to formation of the α,β-unsaturated aldehyde.

Let's learn the mechanism of the intramolecular aldol reaction by working Assessment 20.42, first on your own, and then together.

Assessment 20.42 Suggest a mechanism for the intramolecular aldol reaction. [*Hint: All of the steps are identical to those in the intermolecular version.*]

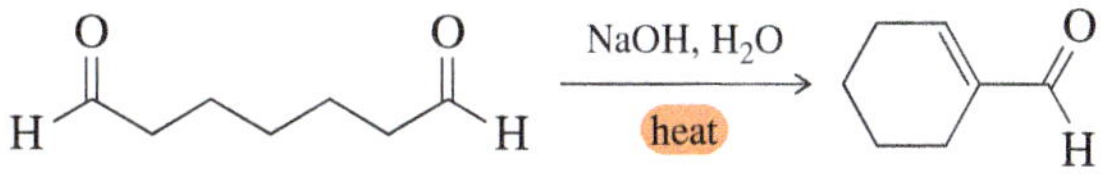

Solved Assessment 20.42

As we work through the mechanism, keep in mind everything we know already about the aldol reaction. Coupling this knowledge to our general organic problem-solving process, a clear picture of the intramolecular aldol reaction should quickly emerge. To start, we number the carbons and identify the bonds formed and broken in **Figure 20.28**.

Figure 1.4 An example of a solved assessment to introduce a section.

1.5.3 Thinking Like a Chemist

There's a way that chemists think, and the sooner you begin thinking this way, the sooner you'll master organic chemistry. Thus, in each chapter are one or two sections entitled "Thinking Like a Chemist." In some cases, these sections are used to show how a reaction might have been invented, how a series of reactions might have been used to solve a problem, or how a concept is best illustrated. Instead of just giving you a fact, or introducing a new reaction, by putting it in the (sometimes historical) context of the way we think about organic chemistry, this section gives the information more relevance. It relates the content to and reviews material you already know that leads to the conclusions on the concept being learned. Learning information in this way may take a bit more time, but in the end, the concept will stick better, allowing you to recall it later.

1.5.4 Purple recall boxes

Everything written in this book has roots in what was learned previously. Even some of the most complicated reactions can be traced back to simple concepts from general chemistry. This will especially be true when we start studying reactions. At some point, you'll realize [*and I'll often remind you*] that a particular reaction is just like one you studied previously. Or that this complicated process is really just a combination of three or four simpler reactions we've seen previously. To make these connections to old material clearer, new reactions are often accompanied by reactions in a purple recall box (**Figure 1.5**). These purple boxed reactions are ones we've introduced previously. By reiterating them in the context of new material, you'll be reminded of what you [*should*] know, reinforcing these concepts. Perhaps as importantly, these recall boxes will demystify reactions that, at first glance, seem complicated. In either case, by connecting the neural networks that contain the new and old reactions, you'll be much more likely to retain material over the long term.

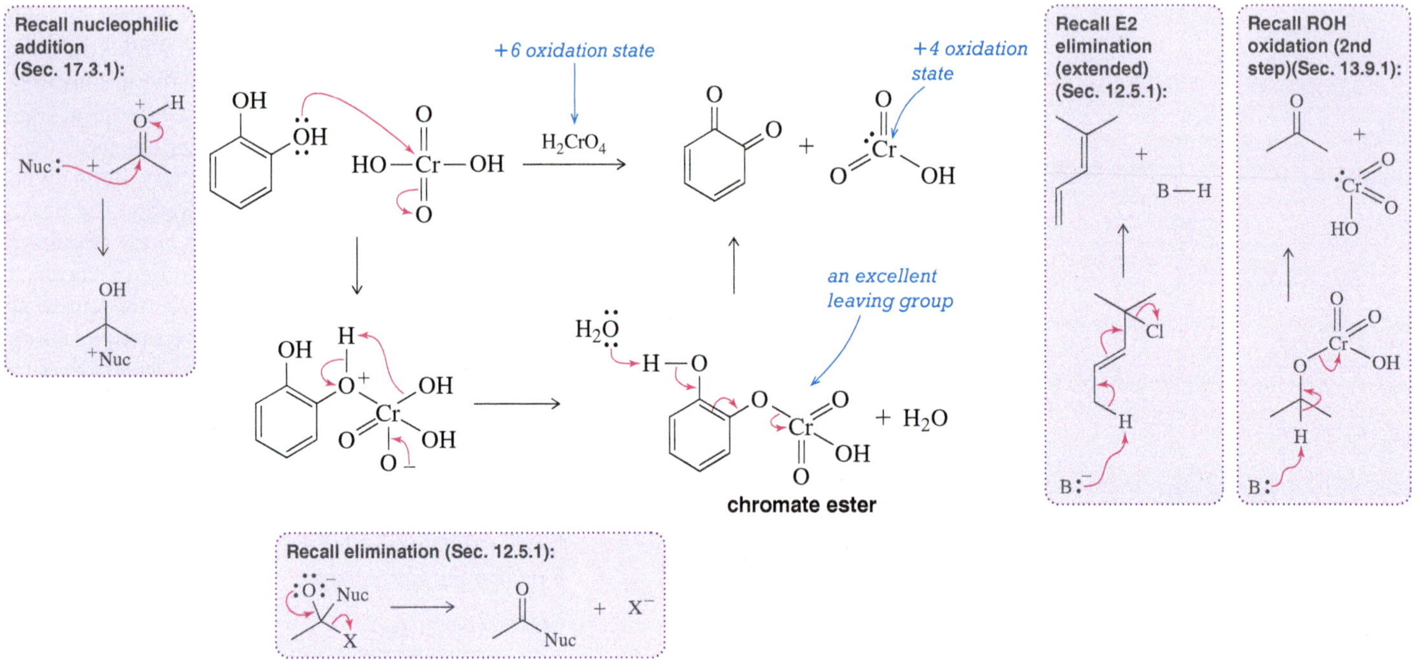

Figure 1.5 **The use of purple boxes in Chapter 24 to remind you of material you have studied previously.**

1.5.5 References to old material

Throughout the book, you'll be reminded (sometimes in parentheses) [*at other times in italics*] of material that you've seen before. We talked in Section 1.4 about how the brain retains information. When you are reminded of the importance of something studied previously, it tells your brain to value the information contained in those neurons. [*See Section 1.4 for a review.*] When you come across these reminders, use them. This doesn't mean you need to invest more than 5 minutes in rereading a given section or reanalyzing a figure, but look back and say, "Yep, I remember studying that." And when you look back, if you think, "I've never seen this before," take some time to study it a bit before continuing. There's no doubt it will lengthen the time you spend studying, but it will be worth it.

1.5.6 Looking back

Some day in the future, you will take an exam that assesses your knowledge on everything you are supposed to know about organic chemistry (ACS standardized exam, Chemistry GRE, DAT, OAT, or MCAT). For organizational purposes, books are laid out in chapters, with sections and subsections that cover related topics. In many ways, this scheme is to the detriment of the student, because it takes a continuous, logical [*and interesting*] narrative of organic chemistry concepts and fragments them. This results in compartmentalization of material, such as "the way we think about this in Chapter 4" as opposed to "the way we should think about this." As a way of blurring the lines between chapters and connecting the neural networks containing related concepts in different chapters, several end-of-chapter (EOC) assessments will be designated as "looking back" or "looking forward." When you work these, be mindful that the tools learned in the chapter will give you a better understanding of something learned previously (looking back). Or, even more empowering, the skills you just gained in the chapter will be applicable to something you will study later (looking forward).

Assessment 1.22 When studying the Diels–Alder reaction (Chapter 22) in lecture, a student asked, "Does this reaction work the same as it does in lab?" What is wrong with this question, and how should the student see the material differently to avoid questions of this type?

1.5.7 [*Comments in italics*]

You may have already seen us, but I wanted to make sure. We are an extension of these italic comments. We will be assisting you on your journey. We will solve assessments together. And when it's time to put on our lab coats and think the way a chemist would to solve a particular problem, you'll find use there as well.

You've probably seen them by now [*and maybe wondered, "What's the point?"*]. The time you spend in the classroom working with your instructor is by far the most valuable time you will spend learning organic chemistry. [*Why?*] Because, during an engaging lecture, you hear other students asking questions. Or, if you're in a larger class, an instructor might say, "Now you might be wondering why that happened." A teaching assistant might say, "You've seen this before," or a supplemental instruction leader might remind you to "Take a few minutes to work this problem." In each of these instances, as a student, you probably were forced to think about something that you hadn't thought about yet. [*That is a huge value of an engaging lecture.*] Or, finally, the discussion leader might make a joke that breaks up the monotony of the material you are listening to. [*The joke probably won't be funny, and you probably won't even know it's a joke, but if the person who tells it could one day write a recommendation letter for you, you'll laugh anyway . . . .*] This entire book is me talking to you. [*When it's in italics, I'm really talking to you.*] This is the purpose of italic comments: not to disrupt the conversation, but to make it more conversation-like.

1.5.8 Ranking of problem difficulty

Don't forget us! We are your fellow students and we will be directing you to helpful hints along the way.

There is never anything wrong with solving an assessment incorrectly. Given our growth mindset, we recognize that each of these miniature failures gives us an opportunity to grow. That said, not all assessments are created equally. When I [*and many organic chemistry professors I speak to*] write exams, I write problems at four different levels. About 25% of the questions are ones that I expect a student deserving of a D to get completely correct. A student who will earn a C would likely get 50% correct. The B student will likely be perfect at 75% of the problems. And finally, an A student should be able to get almost all of them mostly correct. With partial credit available on each question, this results in a grade distribution of 60–90 with an average around 75. The assessments in this book are similar—that is, some are very difficult, and some are relatively straightforward. It is important that students know which problems fall into which categories. If students are going to truly use the assessments to assess themselves, they should know what type of problems they are struggling with. If you are missing the easier questions (•), you should probably spend some time reviewing basic concepts or working more with the Student Supplement to this book. Alternatively, if you are only missing the hardest questions (••••), there is no reason to panic because you are doing well. But instead of reviewing basic concepts, you may need to focus on the subtler aspects of complex problem-solving.

1.5.9 Chemistry in context

You probably don't love organic chemistry . . . yet. Maybe you never will. But education research shows that when material is more relevant to you, it will stick better. As a result, there is a story that begins and ends each chapter. Throughout each chapter there are boxes that continue telling this story of chemistry in context. The context may be medicine, the environment, materials, or biochemistry, or it may have relevance in everyday life. Regardless, spend a few minutes looking over these sections. This doesn't mean you need to learn these things for a test, but if you can see how a concept we are learning relates to something else you are interested in, you'll have a better chance of retaining it. If a Chemistry in Context (CIC) box gets a reaction of "Huh, I never knew that hangovers were a result of an organic chemistry reaction," then its purpose has been served.

1-A CHEMISTRY IN CONTEXT: Problem-Solving Strategies

Differential diagnosis

The last time you visited a doctor, what was that interaction like? Did the physician burst into the room, take one look at you, and immediately start ordering tests and procedures? Probably not. And yet, ordering every possible test to identify your illness would work, but it would be expensive and inefficient.

Aspiring physicians are taught a specific method for diagnosing. Referred to as clinical reasoning, or differential diagnosis, the process is loosely summarized as follows:

1. Acquire data. Examine the patient's vitals; give a physical exam; interview the patient. These steps form the diagnostic reasoning.
2. Develop a problem synthesis statement. This is a concise statement of the problem and includes the context (recurring or constant headache, smoker or nonsmoker) specific to that patient.
3. Identify the cause of the symptoms. This is a list of all possible ailments, prioritized by urgency.
4. Rule out causes. This is done until a single cause is identified by tests and physical examination.
5. Finally, treat the condition. Having identified the most likely cause, the necessary lifestyle changes or drug is prescribed.

Adhering to this process in every physician–patient interaction gives the best odds of identifying the cause of the symptoms and, ultimately, prescribing the correct treatment. A constant and well-defined strategy for diagnosing the problem leaves only the cause of the problem in question, not the method for identifying it.

Assessment 1.23 Go to Section 17.2.2, where you will find CIC 17-A. What cofactor is responsible for oxidizing ethanol to acetaldehyde, the compound that is responsible for the feelings associated with a hangover? [*Luckily, most of you are not old enough to drink, so you don't know what that feels like.*]

1.5.10 The Student Supplement

One of the most difficult conversations I have with students occurs around week 10 of a 15-week semester, when a student hasn't done well on the first two or three exams and is considering dropping the course. Though there are many circumstances where withdrawing from the course is the appropriate move, giving up always seems like the fixed-mindset move. I think much of my trepidation stems from the fact that as a student I once had the "no matter what I do, it isn't working" conversation, but my professor spent a lot of time working with me to get me back on track. That said, there are situations when a student has fallen far behind without the time or the resources to catch up. In these situations, withdrawing from the course is the only choice.

To address the issue of students falling behind, the Student Supplement was created to help students get back on track. It is equal parts recipe and road map to success. Think of it this way: If a driver gets lost on their way to a destination, they don't give up, turn around, and go home or, worse yet, remain lost forever. Instead, they use a map to find their way. [*You've probably not used a map for this purpose, but you've probably at least heard of them. Instead, you'd use the GPS in your phone for directions, but the map analogy works better here.*] It requires work to plan a new route and probably even takes more time, but it at least gives you a chance to make it to your destination. Even if you can't get to your destination on time, you get there eventually. And with a growth mindset, that arrival can be even more fulfilling.

The Student Supplement does things the book doesn't. It provides an extensive review of the basics in the context of every chapter. [*You'll review how to draw Lewis structures for carbonyls at the beginning of Chapter 17, for example.*] It teaches, and reteaches, the specific problem-solving skills you need for certain types of problems. It highlights common mistakes that you might make and discusses how to avoid them. It addresses commonly asked questions—ones you probably have asked, or wish you had. In this way, it provides a pathway to success, regardless of when you take that wrong turn. [*That said, by using a map from the beginning, you are much less likely to get lost, and that is the best course of action here as well.*]

The Student Supplement isn't just a map back to competence; it is also a map to deeper, longer-term understanding while recognizing some of the longer-term goals an organic chemistry student might have. Want to be an organic chemist? You can "Think more deeply" in the Student Supplement. To help you retain information, there is a practice exam and a practice final exam at the end of every chapter. Planning on taking the MCAT? You can find a practice passage at the end of every Student Supplement chapter. With these resources, every student can find something to help them change their trajectory.

1.5.11 Answer keys and videos

Used correctly, an answer key can be an extremely beneficial tool to support your learning. Sadly, answer keys are very often misused, especially by students who are just beginning to learn organic chemistry. I see it happening like this. A student works an assessment, drawing the answer as shown in **Figure 1.6(a)**. Then, upon looking at the answer key [**Figure 1.6(b)**], their thinking goes like this: "Hmm, I was right—or was I? I drew the Br in the wrong spot . . . does that matter? I'd still get a lot of partial credit. If it does matter, then I'll just remember to draw the Br in the other spot next time." [*I've seen this happen* so *many times.*] While the correct answer and the incorrect answer look very similar, there is potentially a huge lack of understanding for the student who writes the incorrect one. Especially when a student has a fixed mindset, they look at answer keys to find out why they were correct (or close to correct) as opposed to figuring out where they had an opportunity to grow.

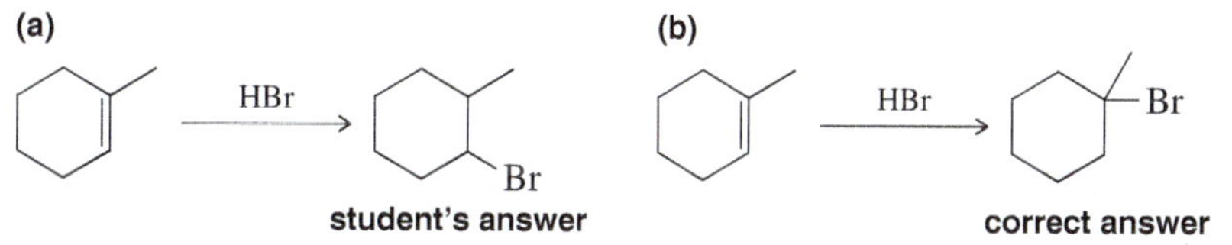

Figure 1.6 When using an answer key goes bad. (a) The answer a student writes. (b) The correct answer. [*No, they are not "close enough."*]

To address this situation, the answers to many assessments are given in videos at www.youtube.com. In these videos, you'll be able to find out whether your answers are right or wrong. If your answer to an assessment is wrong, you can then watch me solve the assessment, allowing you to see where your analysis differed from ours. Not only will this show you where you messed up on this particular assessment, but seeing others solve problems will help you develop your own problem-solving skills.

To help with concepts as you first learn them, Mastering Chemistry includes mini-lecture videos that go along with the text and cover difficult concepts. These are meant only to supplement your reading, not replace your lecture. But if you're studying late at night and need more than the book can provide, these videos are there for you.

Assessment 1.24 Go to www.pearsonmylabandmastering.com and watch Video 13.7, which shows a solution to Assessment 13.22. Write the answer to that problem here.

1.5.12 Index

All books have an index. [*Go to the back of the book and check it out.*] It turns out that the index is one of the more expensive parts of textbook authoring. [*And yet, I once read a statistic that some larger-than-expected percentage of college students didn't know what an index was.*] If you aren't familiar with an index, you can pick any key word in an index and find every page on which it appears in a book. [*It's like Google or Ctrl+f, but on paper.*] This will be really helpful when you are prepping for comprehensive exams, as opposed to when you're doing an assignment that you know to be from a given chapter. [*So take advantage of it.*]

1.5.13 Important tables

The last thing to highlight is not a general feature, but a few specific tables that will help you get through this course. [*If I could put them on a laminated 3 × 5 card and give one to every organic student, I would.*] The first is the pK_a table (Table 4.5, Section 4.3.1), which contains tons of information. It tells how stable an acid is and how stable a base is. We can extrapolate from it to find how nucleophilic a reagent is. Once we get going, you'll want to know most of the values in that table and certainly be able to find them quickly.

The second is Table 5.6 (Section 5.3.1), which lists bond dissociation energies. These values tell us how much energy is required to break many types of bonds. This table will help us understand why one molecule is more stable than another, allowing us to rationalize the outcome of many different reactions.

There is a ton of great stuff in the Appendix. Make sure you check it out as you proceed through the course.

Finally, I've heard of non-organic classes where you are allowed to write as many formulas and equations on a sheet of paper as you can fit and can use that on an exam. Appendix A has my version of that cheat sheet. But instead of facts or concepts or equations, it has a list of the questions you should ask yourself when you get stuck on an assessment. Often, by breaking down a complicated problem into the simpler questions in Appendix A, you can get yourself back on track. You just have to know what questions to ask.

1.5 SUMMARY

- This book was written exclusively for you, the student. It contains a number of tools to help you succeed. Don't wait until it's too late to begin using all it has to offer.

Assessment 1.25 Using the index, find the page where the first description of hyperconjugation appears. In which section is that?

Assessment 1.26 Go to Table 5.2. What is the bond dissociation energy of H—Br? Why is that value positive?

Assessment 1.27 Go to Table 4.5. What molecule has the highest pK_a in the table? What is that pK_a?

Assessment 1.28 If you get stuck on an organic chemistry assessment, give three questions you should ask yourself to get back on track. [*Check out Appendix A for this.*]

1.6 Use your allies

"Important achievements require a clear focus, all-out effort, and a bottomless trunk full of strategies. Plus, allies in learning."

—Carol Dweck

Organic chemistry is a difficult course, and *when* you conquer it, it will be a testament to *your* work ethic, perseverance, grit, and mindset. But you are not alone in this—you have several allies in your learning, and I'm not talking about the book, the Student Supplement, or the videos. I'm talking about your professors. Nobody knows what students at your institution need better than they do. Go to lecture, go to review sessions, go to office hours. [*If it weren't for my organic professor's office hours, I would never have become an organic chemist.*]

I'm also talking about a learning assistance center where you might find individual tutors. I'm talking about classmates and lab partners who you can teach, learn from, and even just share war stories with. If you have recitation sections or supplemental instruction, I'm talking about those, too. Don't wait until you are behind before you utilize these allies. Doing it on your own doesn't make you a "smarter" student. [*This is a fixed mindset.*] In fact, the "smarter" [*and grittier*] student takes advantage of these resources from the beginning. Make sure you can fit these allies into your schedule.

Assessment 1.29 Circle the resources that are available to assist you on your campus.

(a) Office hours (b) Supplemental instruction (c) Tutors
(d) Teaching assistants (e) Learning assistance center

Assessment 1.30 Create a weekly calendar, including your class and work requirements, specifically blocking out time slots during which you can take advantage of your allies in learning.

1.7 "One Shining Moment"

"And all the years, no one knows, just how hard you worked, but now it shows."

—David Barrett

At the end of the NCAA Division I men's basketball championship game, as the network is signing off, the song "One Shining Moment" is played, set to a montage of moments from the previous 3 weeks of basketball. You can watch these montages at www.youtube.com—just search for "One Shining Moment." As the lyrics are sung, you see players diving for the ball (*running for your life*), elation (*you knew . . . you were alive*), and heartbreak (*when it's done, win or lose*), and it concludes with the championship team celebrating their victory (*one shining moment*). This song has always had an emotional pull on me beyond just being a sports fan [*if we ever meet in person, I'll explain*]. In a lot of ways, this song is a metaphor for what you're [*we're*] in the midst of right now. You are on your way to your one shining moment. Organic chemistry, other classes, and your college experience are just one small part of that, but a part nevertheless. At this very moment, you are making your montage. What will you be celebrating when the music plays? Will you have developed a new drug that has just received FDA approval? Will you have treated your first patient in the clinic? Will you have finally completed dental school and be called "Doctor"? Will you have been accepted to a graduate program in environmental chemistry? Will it be your college graduation? [*What will be your "one shining moment"?*]

Here's the reason I ask [*and I strongly encourage you to think carefully about it*]. The scenes in your montage won't always be pretty. But to harness your grit and your mindset, it is helpful to know what you are working toward. You aren't a robot—you need a reason to work every day, to do extra assessments, to attend every class, to block out time for supplemental instruction and tutoring, or to sacrifice a weekend night out to catch up on your studies. It doesn't have to be specific; I only *thought* I knew what I wanted at 19 years old. And over the course of your life, you'll probably have several shining moments. But during the toughest times, it helps to know why you're doing what you're doing.

[*Since you asked, as I was writing this, the "one shining moment" I dreamed about was finishing the book, holding it bound in my hands, and sharing it with those that helped make it happen: my parents, my in-laws, my editors, and, most importantly, my wife and kids. Though it was a long, arduous journey, the effort was worth it. Yours will be, too!*]

Assessment 1.31 What is the "one shining moment" you are working toward now? Where will you be when the music starts playing? What will be shown in your montage? [*Seriously, write this down.*]

Assessment 1.32 Your montage will also have other people present in various clips. Who will those people be?

Concluding thoughts

Are you sufficiently motivated, or terrified? [*Regardless, thanks for reading this far. This was my favorite chapter to write, and not just because it was one of the last I wrote.*] Here's the thing: People think organic chemistry is difficult, and at times it will be. But if you commit to doing things the right way, and you do it with grit and a growth mindset, and you study the material in ways that are shown to "make it stick," and you use the textbook, your instructor, and all other allies to help you on your journey, and you remember your purpose, you can do it. I have no doubt about this.

There are really only two rules you need to know to make sense of virtually everything we will study in this course. And you'll see these rules repeated throughout the text, because they are the secret to success. [*Are you ready for them? Are you sure? OK, but never forget these two rules . . . .*]

Rule 1: The ball rolls downhill

Every process in organic chemistry creates something that is more stable from something that is less stable. In the same way that a ball at the top of a hill has potential energy that makes it roll down a hill, a molecule that is unstable (at high potential energy) reacts to become more stable. This is demonstrated in Figure 1.7.

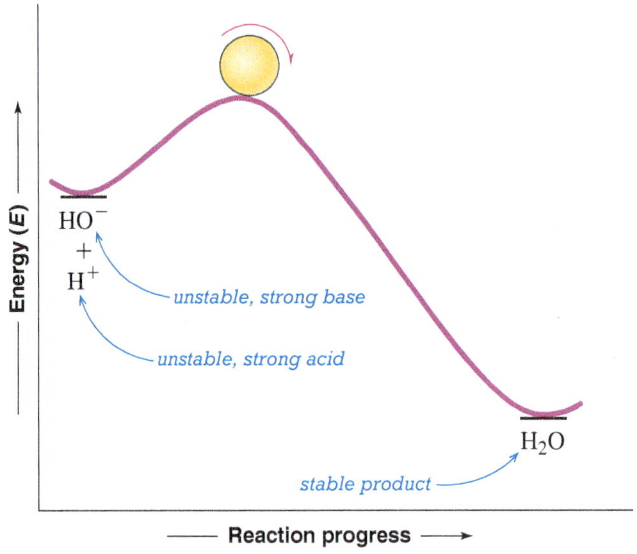

Figure 1.7 **"The ball rolls downhill"** explains why reactions occur.

Rule 2: Negatives attack positives

Opposites attract, and that is definitely true here. But the convention in organic chemistry is that negatively charged molecules attack positively charged molecules. Though chemical reactions are random collisions, we illustrate them with arrows moving from negatively charged (electron-rich) species to positively charged (electron-poor) species, as shown in Figure 1.8.

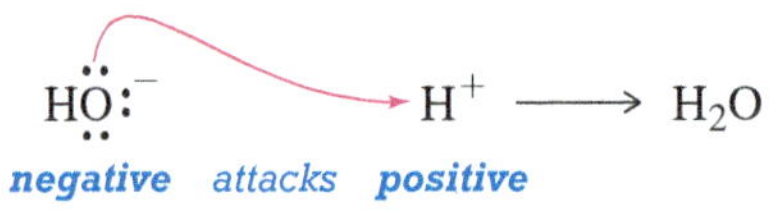

Figure 1.8 **"Negatives attack positives"** explains how reactions occur.

And that's it. It is really that simple. In order to create those multiple neural pathways where we store information to make it easily recalled, we can learn everything in the context of these two rules. After you work a few end-of-chapter assessments, we can get started in Chapter 2 by translating general chemistry to organic chemistry. [*I've been working for seven years to have others on this journey with me, so let's go!*]

End-of-Chapter Assessments

[*Assessments are ranked according to their difficulty; •••• = most difficult.*]

1.33 (•) If you were a thrill-seeking skier, which hill would you want to ski down?

1.34 (•••) LOOKING AHEAD The more stable molecule in the equilibrium reaction is indicated. Place each molecule in the appropriate box on the diagram shown.

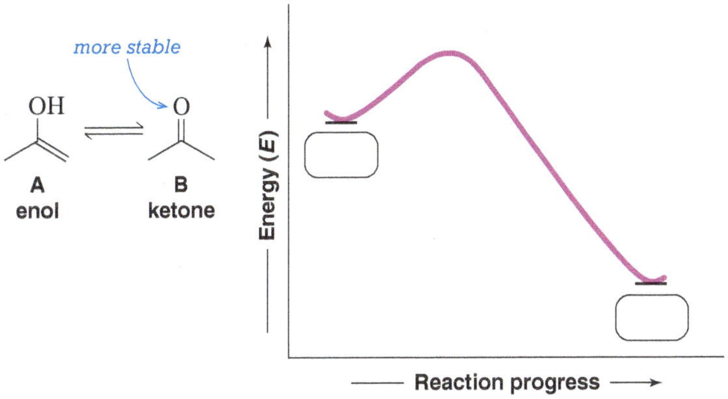

1.35 (••) If you were a runner, which hill do you think you could run over more quickly?

1.36 (••) Several ions you have seen in chemistry classes are shown below. Using rule 2 of organic chemistry, show how these might react with each other.

(a) $H{-}\ddot{\underset{..}{O}}{:}^{-} + H^{+} \longrightarrow$

(b) $:\ddot{\underset{..}{Cl}}{:}^{-} + Na^{+} \longrightarrow$

1.37 (••••) LOOKING BACK As a general chemistry student, you learned that the structure of methane (CH_4) is tetrahedral. Why is it tetrahedral, as opposed to square planar?

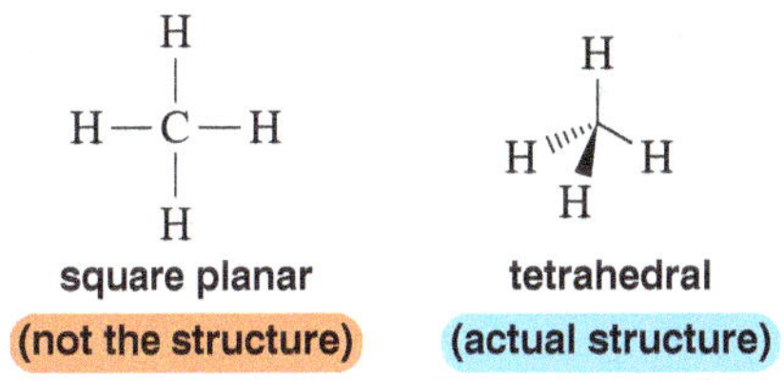

2

General Chemistry Translated: Finding the Electrons

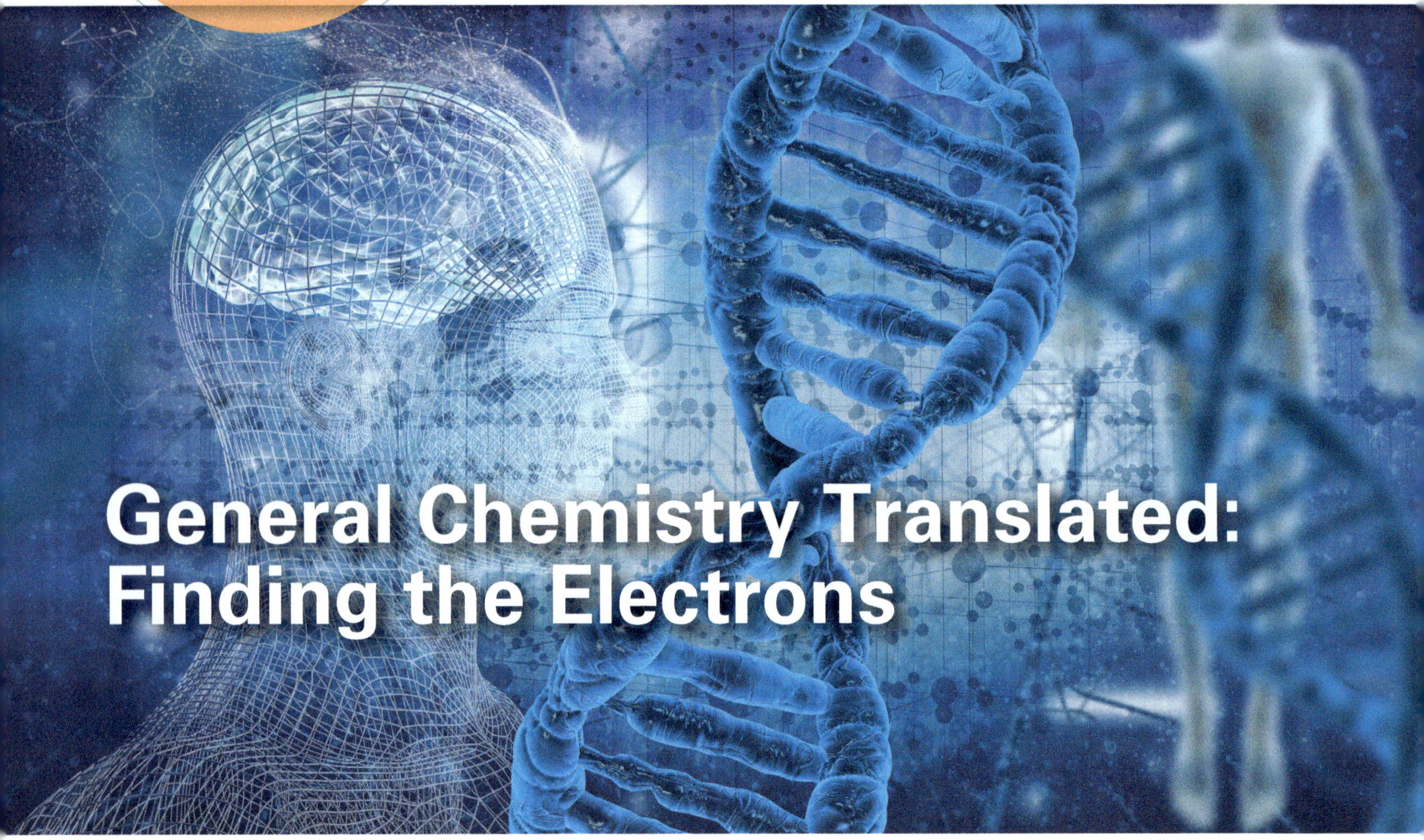

Learning Objectives

LO 2.0 Describe how the electronic structure of the atom relates to the organization of the periodic table

LO 2.1 Explain how differences in the electronegativity of bonded atoms predict whether a bond will be ionic or covalent

LO 2.2 Determine the most stable Lewis structure of a molecule

LO 2.3 Relate the Lewis structure of a molecule to its shape and dipole moment

LO 2.4 Use valence bond theory to describe bonding in molecules

LO 2.5 Use molecular orbital theory to describe bonding in molecules

LO 2.6 Determine the conditions needed for electron pairs to form valid resonance structures

Introduction

Organic chemistry is a far-reaching subject that impacts virtually every aspect of our daily lives. As the chemistry of compounds containing carbon, organic chemistry interfaces with medicine, biochemistry, biology, and materials science, to name just a few. The medicines we take are organic molecules. The plastics that make up the materials we use on a daily basis are organic molecules. The sugars, proteins, and fats we consume are organic molecules. The enzymes and biological molecules that carry out the functions that give us life are organic molecules. Organic chemistry is in our DNA, because DNA is an organic molecule.

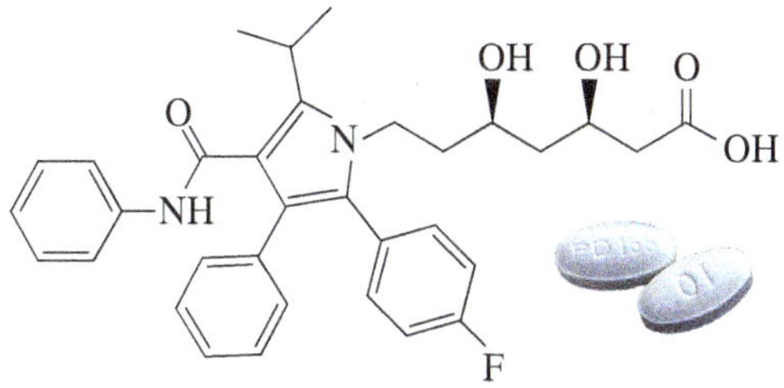

atorvastatin (Lipitor)

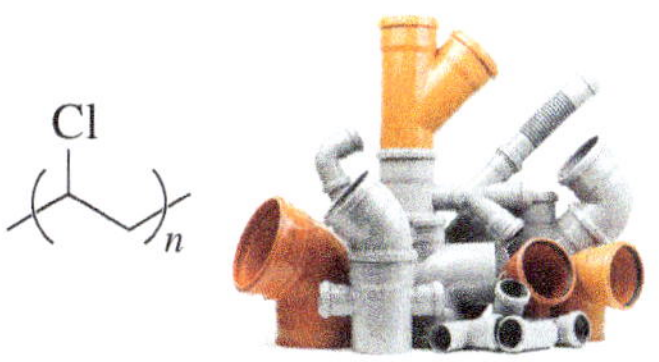

polyvinylchloride (PVC)

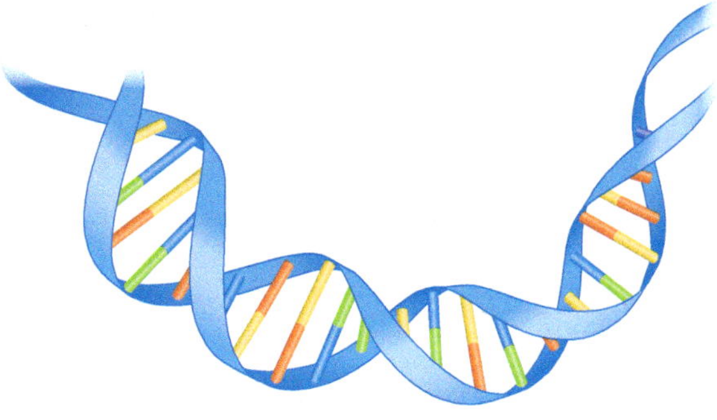

DNA

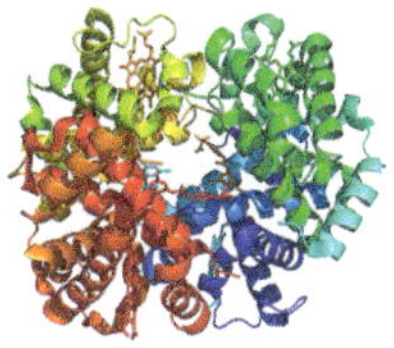

enzyme

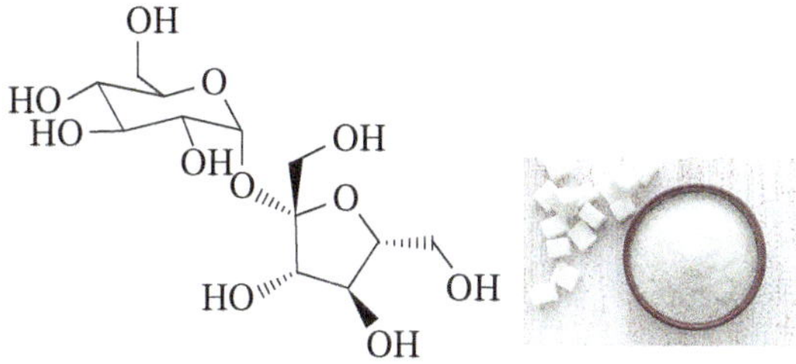

sucrose

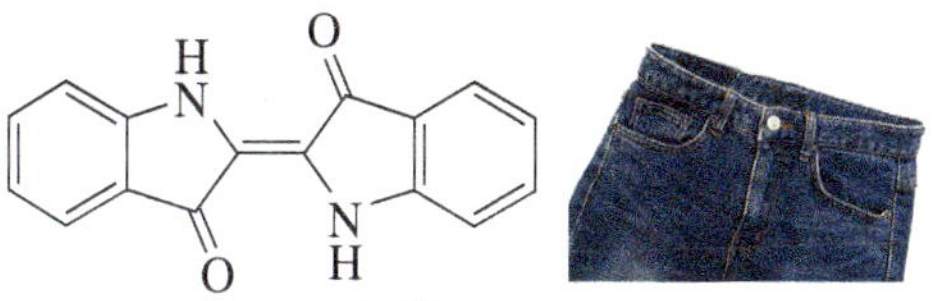

indigo

"What people don't realize is that professionals are sensational because of the fundamentals."

—BARRY LARKIN

Much of what is covered in this chapter you have seen before. Now the fun begins. Though the concepts are the same, the questions we ask in organic chemistry allow us to use our knowledge differently. In general chemistry, for example, you likely learned how to draw Lewis structures. Your homework assignment might have been "Draw the Lewis structure of CH_3Cl." And after drawing the molecule in **Figure 2.1**, and checking your answer, you'd be finished.

Many conversations in organic chemistry begin with a Lewis structure. A Lewis structure can be used to rationalize both the physical and chemical properties of a molecule, and we can begin thinking about chemical reactivity by identifying what is negative and what is positive. [*Negatives attack positives.*] Organic chemistry is actually quite logical once a few foundational principles are established. You may even come to understand general chemistry principles better after you study organic chemistry—in my experience, most students do. [*Why?*] It could be because they are seeing these principles a second time, but more likely it is that organic chemistry gives us a way to visualize concepts that were previously abstract.

Let's start thinking this way right now. Whenever you learn a new concept, ask yourself the following questions:

1. Why does this happen the way that it does?
2. What do I already know that helps me understand this?
3. What insight does this give me for concepts I'll study in the future? [*We preview this with the Chemistry in Context boxes here in Chapter 2.*]

Figure 2.1 **The Lewis structure of CH_3Cl**

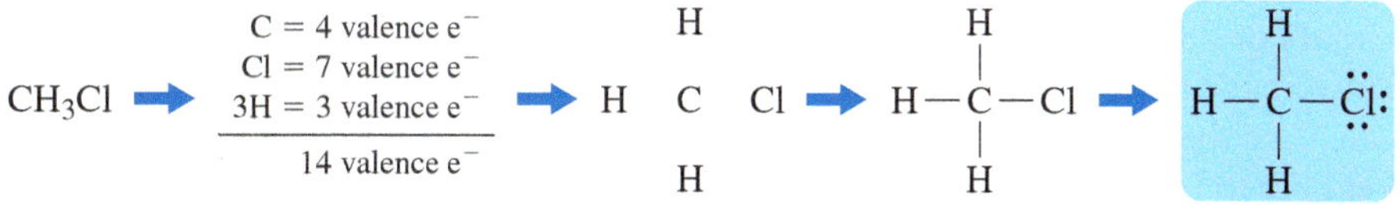

And so, the goals of Chapter 2 are to solidify your understanding of the fundamental principles from your general chemistry course, and then apply and extend that knowledge in this uncharted territory called organic chemistry. We can begin this process by testing our understanding of previous material, as we do in every chapter going forward.

Test your understanding

Assessment 2.1 Give the electron configuration of the following elements.

(a) C (b) N (c) O (d) F (e) H

Assessment 2.2 Why do elements in the same group of the periodic table display similar reactivity?

Assessment 2.3 Rank the following elements from least electronegative to most electronegative.

Na Si F Mg C O N

Assessment 2.4 How many valence shell electrons do each of the following elements contain? How many new bonds can each form?

(a) C (b) N (c) O (d) F (e) H

Assessment 2.5 Why is argon considered to be so stable that it is referred to as a noble gas?

Assessment 2.6 Which way does the ball roll? And what do negatives attack?

2.1 Structure of the atom

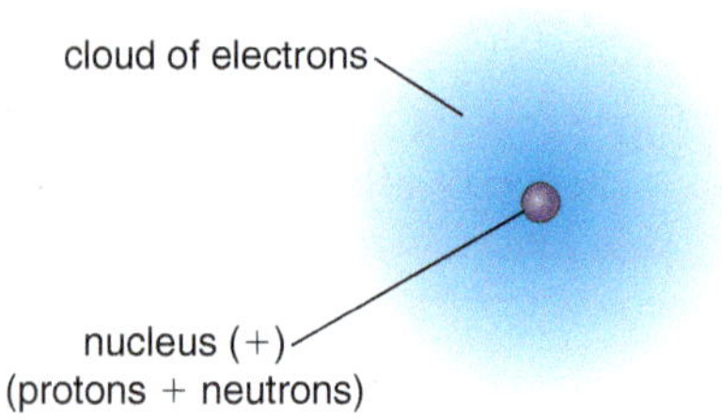

Figure 2.2 The structure of an atom. [*Not to scale.*]

An atom is composed of three fundamental particles: **protons (p^+)**, **neutrons (n^0)**, and **electrons (e^-)**. Protons have a +1 charge and a mass of 1 **atomic mass unit** (amu). Neutrons are neutral particles, also with a mass of 1 amu. Electrons have a −1 charge and a mass of 5.4×10^{-4} amu. Protons and neutrons reside in the nucleus at the center of the atom, while electrons occupy orbitals around the nucleus (Figure 2.2). If the nucleus were as large as that shown in Figure 2.2, the electron cloud would fill an NFL football stadium. Chemical reactivity and bonding are almost exclusively based on the electron, so it receives far greater attention here than either the proton or the neutron.

The different elements are distinguished by their **atomic number** (Z), which is the number of protons in the nucleus. Within neutral elements, the number of electrons is equal to the number of protons, so the negative charge of the electrons neutralizes the positive charge of the protons, making atoms neutral. The **mass number** of an atom is the number of protons and neutrons in its nucleus. Thus, a carbon atom ($Z = 6$) with 6 neutrons has a mass number of 12. Atoms of the same element can contain a different number of neutrons. Two atoms with the same number of protons but a different number of neutrons are referred to as **isotopes**. Thus, ^{12}C (6 protons, 6 neutrons) and ^{13}C (6 protons, 7 neutrons) are isotopes. Because the chemistry of an atom is determined by the electrons, the number of which matches the number of protons, isotopes of a given element tend to show roughly identical reactivity. The atomic mass given in the periodic table is a weighted average of the relative abundance of all isotopes of a given atom. Because only 1% of all carbon atoms are ^{13}C (~99% are ^{12}C), the weighted average atomic mass of carbon, based on relative abundance, is 12.0107 amu (closer to ^{12}C). The weighted average atomic mass can be calculated according to Equation 2.1.

$$\text{Weighted average atomic mass (amu)} = \frac{\left(\text{mass of isotope 1}\right)\left(\%\ \text{abundance}\right) + \left(\text{mass of isotope 2}\right)\left(\%\ \text{abundance}\right)}{100} \qquad \text{Equation 2.1}$$

Assessment 2.7 Bromine-79 (50.7% abundance) has an atomic mass of 78.918 amu, whereas bromine-81 (49.3% abundance) has an atomic mass of 80.916 amu. From these data, calculate the average atomic mass of bromine that you would expect to see in the periodic table.

Assessment 2.8 Chlorine exists as one of two isotopes with atomic masses of 34.969 amu (^{35}Cl) and 36.966 amu (^{37}Cl). Calculate the relative abundance of ^{35}Cl and ^{37}Cl based on the average atomic mass of 35.453 amu.

2.2 Electronic structure

One of the most fundamental principles in organic chemistry is that negatives attack positives. What we are really talking about are the electrons, the negatively charged particles in atoms, molecules, and chemical bonds. Electrons orbit the positively charged nucleus and are found in **orbitals**. Although orbitals can be represented in a simple drawing, they are mathematical descriptions that use the particle and wave nature of electrons to explain the properties of atoms and molecules.

2.2.1 Shells and orbitals

Dmitri Mendeleev first organized the periodic table [*see the inside front cover*] based on reactivity patterns that repeated themselves at well-defined intervals. We know now that this is a consequence of the order in which electron orbitals are filled, starting with the most stable, or lowest energy, orbital. These orbitals, in turn, are contained within "shells" at varying distances from the nucleus (**Figure 2.3**). We describe an atom's electron configuration—where electrons are found in orbitals and shells—using what we call **quantum numbers**, which derive from complex quantum mechanical equations but are easy to understand qualitatively. The **shell** that an electron is in is described by the principal quantum number, n. Elements in row 1 of the periodic table contain all of their electrons in the first shell, identified by the principal quantum number $n = 1$. The first shell can hold only 2 electrons. Elements in row 2 contain electrons in the first ($n = 1$) and second ($n = 2$) shell. Shell 2 can hold a total of 8 electrons. By this count, neon, which has 10 electrons, has 2 electrons in the first shell and 8 in the second. As the shells move farther away from the nucleus, their energy increases and they are able to hold more electrons. Shell 3 ($n = 3$) can hold up to 18 electrons. Fortunately for us, the chemistry in this book focuses largely on the chemistry of the first two rows in the periodic table ($n = 1$ and $n = 2$).

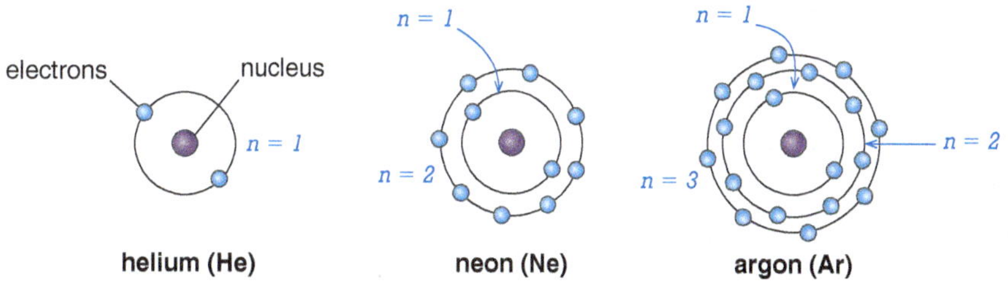

Figure 2.3 Electron shells farther from the nucleus are higher in energy and can hold more electrons. [*This is an oversimplified picture that will be modified in the text that follows.*]

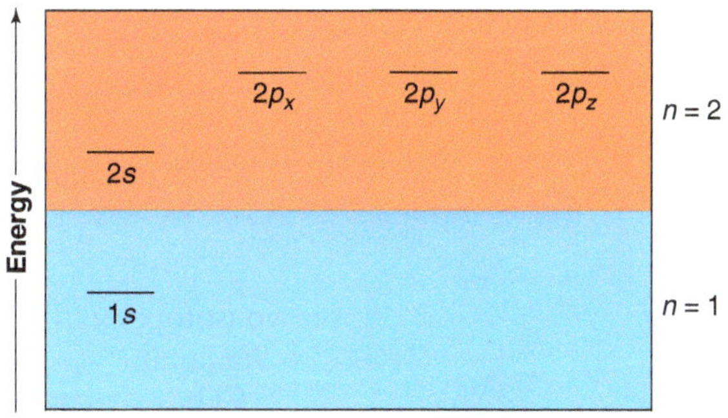

Figure 2.4 Relative energies of orbitals for those in the first two shells ($n = 1$ and $n = 2$).

It is within the individual shells that we find the electrons in orbitals. Though we can't know the precise location of an electron, we can estimate where in space it is 95% likely to appear. This approximation is essentially the orbital. All orbitals can hold a maximum of 2 electrons. **Figure 2.4** shows the relative energies of the orbitals that we encounter most often in organic molecules. [*Section 2.2.2 helps us understand these relative energies.*]

The first electron shell ($n = 1$) contains only the *s* orbital. **Figure 2.5** shows what this orbital looks like in graphical form, as well as in two and three dimensions. The graphical representation shows that electron density reaches its maximum at the nucleus. In three-dimensional form, it takes on the shape of a sphere. That is, the spherical depiction of *s* orbitals is simply the three-dimensional visualization of a wave.

Figure 2.5 Representations of the 1*s* atomic orbital.

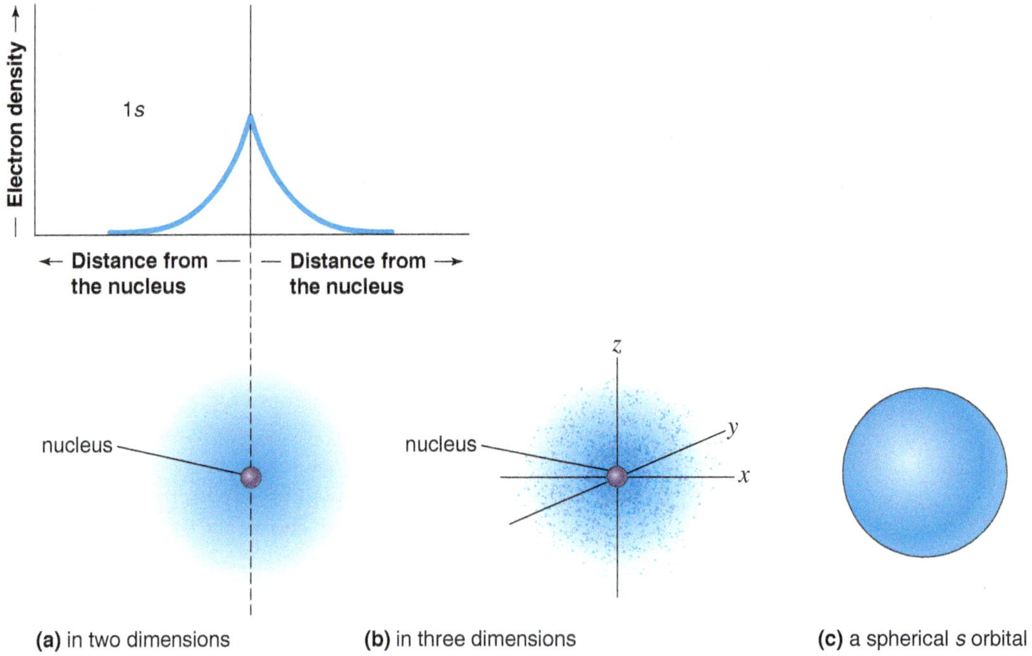

The second shell ($n = 2$) contains both the *s* and *p* atomic orbitals. The 2*s* atomic orbital is similar to the 1*s* atomic orbital in its overall spherical shape (**Figure 2.6**). The only difference is the presence of a **node** separating two regions of high electron density. The node represents a space where there is zero electron density; it arises from the complicated mathematical description of the orbital. [*Confused about nodes? Don't worry about them right now—they'll make sense later on.*] As Figure 2.6 shows, some electron density in the 2*s* atomic orbital is close to the nucleus, but most is farther out, beyond the node.

Figure 2.6 Representations of the 2*s* atomic orbital.

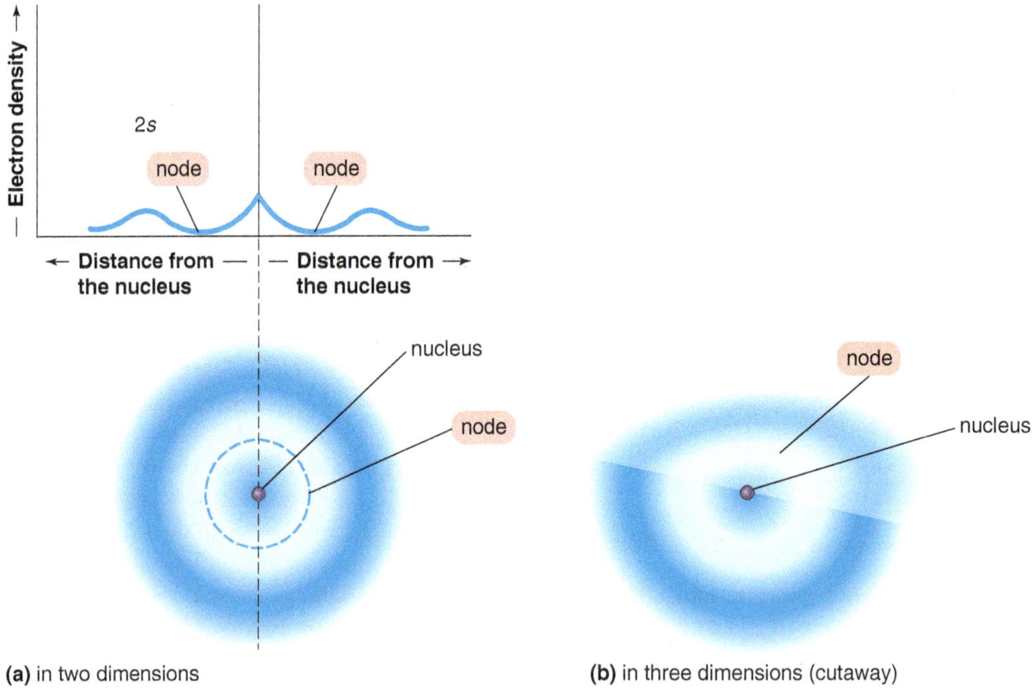

The second shell can hold 8 total electrons and the 2*s* orbital can hold only 2, so three *p* orbitals hold the remaining 6 electrons. Figure 2.7 shows three representations of the 2*p* orbital, with the three-dimensional representation showing the dumbbell shape of the *p* orbital. There is a node at the nucleus. The absence of electron density close to the nucleus means the 2*p* orbital is higher in energy than the 2*s* orbital, which has electron density close to the nucleus. The three *p* orbitals are oriented at 90° angles to each other along the *x*, *y*, and *z* axes. Like the 1*s* and 2*s* orbitals, each of the 2*p* orbitals holds a maximum of two electrons.

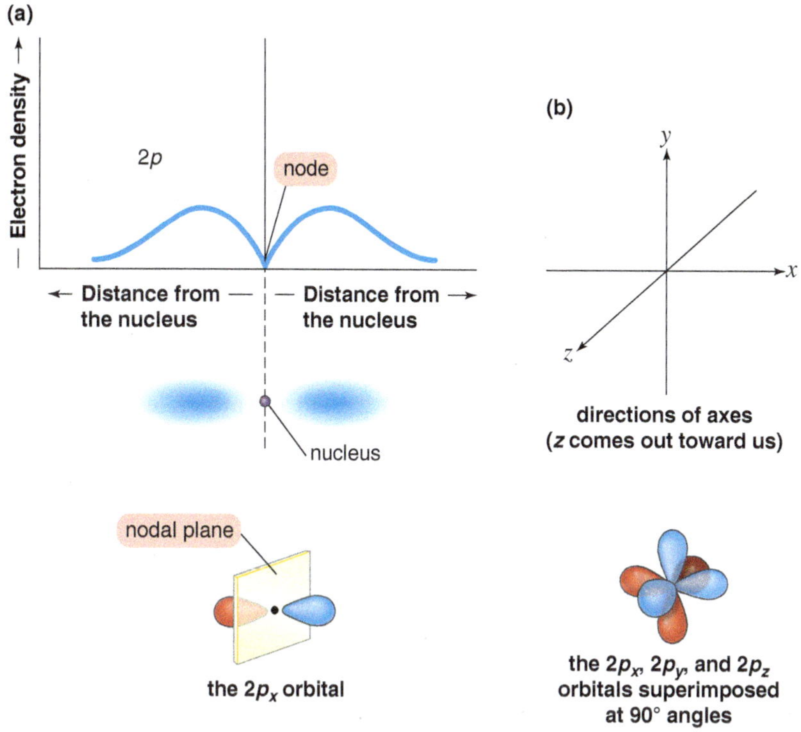

Figure 2.7 Representations of the 2*p* orbitals. (a) Three representations of a 2*p* orbital. (b) The three orbitals are oriented along the *x*, *y*, and *z* axes.

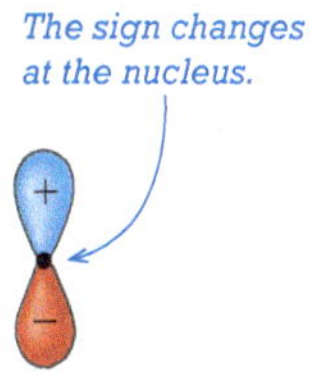

Figure 2.8 The two lobes of a *p* orbital have opposite signs, often indicated by the shading of one lobe.

The shape of the *p* orbital is actually derived from mathematical formulas of quantum mechanics. The dumbbell depiction of *p* orbitals is simply the 3-D visualization of a wave. One lobe of the orbital has a negative sign and the other has a positive sign. A 1*s* orbital has a positive or negative sign, but because 1*s* orbitals have no node, the sign is the same throughout the entire orbital. We generally represent a change in sign by shading one lobe of the *p* orbital red and the other lobe blue, as shown in Figure 2.8. It isn't important what shading corresponds to what sign, as long as it switches at the node. The + and − are *signs*, not *charges*.

2.2.2 Electron configuration

As you move across rows of the periodic table in order of increasing atomic number, each next atom contains an additional proton. To balance the increasing nuclear charge, each atom also picks up an additional electron. Electrons are added to the orbitals in order of increasing energy, with a maximum of two in each (**Pauli exclusion principle**). That is, the lowest, most stable orbitals are filled first [*the ball rolls downhill*]. This "building-up" principle is referred to as the **aufbau principle** (*aufbau* is German for "building up"). Shown in Figure 2.9, the 1*s* orbital is filled first, followed by the 2*s* orbital, and then the 2*p* orbital. Continuing into the third row, 3*s* is filled before 3*p* and is followed by 4*s* and, finally, 3*d*. This continues through each row of the periodic table, but our focus will be only on these first three rows. **Hund's rule** specifies that when two or more orbitals have the same energy, electrons first go into different orbitals before needing to be paired.

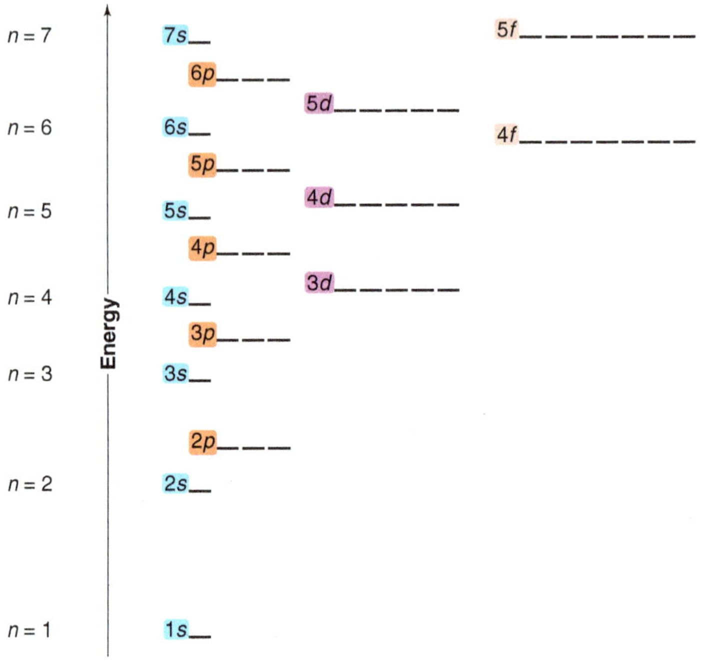

Figure 2.9 Orbitals are filled beginning with the lowest energy first, building up to the higher energy orbitals.

From the aufbau principle, we get the ground state electron configurations for the first three rows of the periodic table (**Figure 2.10**). The reactivity of atoms is determined by the outermost, or **valence, electrons**. The outermost electrons are farthest from the nucleus, so they are held less tightly, and they are at the highest energy. High-energy electrons make atoms reactive because electrons want to be in the lowest possible energy state [*the ball rolls downhill*]. The valence electron configuration is identical for atoms in the same group. These similar valence electron configurations are responsible for the similar reactivity observed by Mendeleev that ultimately led to the creation of the periodic table.

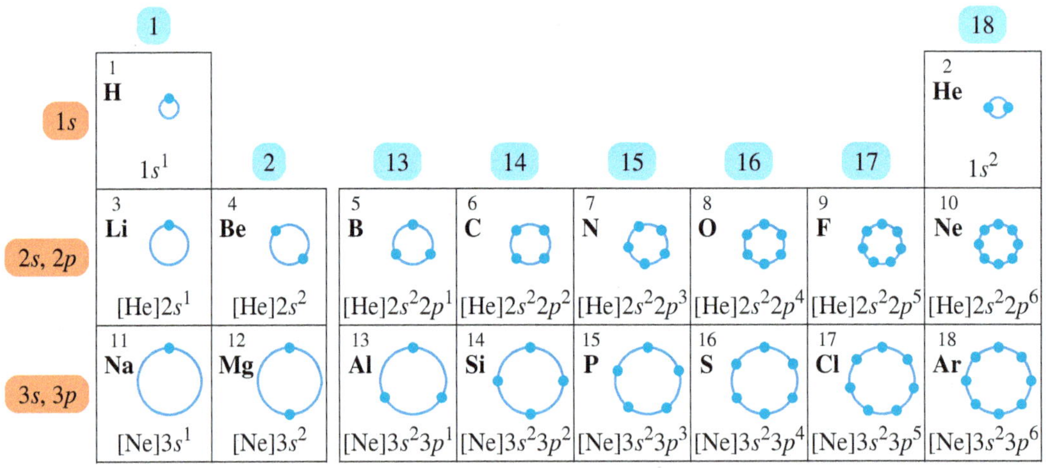

Figure 2.10 Valence shell electron configurations of the first three rows of the periodic table.

Assessment 2.9 To which orbital would you add electrons first, $5p$ or $4f$? Why?

Assessment 2.10 A chemistry student drew the following incorrect electron configuration for carbon. (a) Correct the diagram. (b) Which rule wasn't followed by the student?

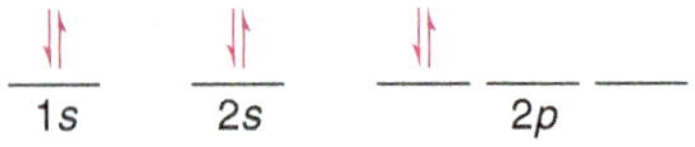

Assessment 2.11 What can you conclude about the relative energy of the $4s$ and $3d$ orbitals given the order in which they are filled?

Assessment 2.12 Would you expect electrons in the $2s$ or $3s$ orbital to be more reactive? Why?

As you move across and down the periodic table, atomic size changes in a predictable manner. Left to right across a row, the addition of each additional proton holds the electrons in a given shell more tightly. Even though the electrons repel each other, the force exerted by the extra protons is stronger. The net result is that the outermost shell is pulled more closely toward the nucleus, decreasing the atomic radius as you move from left to right across a row of the periodic table. Moving down a group, the addition of each additional shell of electrons results in an increased atomic size. This is in spite of the fact that there are more protons holding the electrons more tightly. This periodic trend will be very important in later chapters.

Assessment 2.13 Choose the larger atom in each pair.

(a) Si or C (b) O or F (c) Br or Cl (d) N or S

2.3 Atoms to molecules

The formation of molecules from individual atoms is driven, like much of what we will study, by the ball rolling downhill. With the exception of the noble gases, nonbonded atoms are less stable than bonded atoms. As a result, bond formation is a favorable process. The following sections examine two types of bonding most common in organic chemistry, **ionic bonding** and **covalent bonding**. [*As we study bonding in this section, we step away from the orbital description of electrons, but we return to it in Section 2.6.*]

Because this chapter is largely a review of general chemistry, some students find this material to be easy. Even if this is true for you, continue working hard to build good habits for when the material becomes challenging.

2.3.1 Lewis dot structures and the octet rule

The reactivity of an atom is characterized by its valence electrons. [*In the same way that a book is usually judged by its cover. . . .*] This makes sense given that the valence electrons occupy the outermost shell and are at the highest energy level and are therefore most easily lost or shared in forming new molecules or ions. **Lewis dot structures** are simplified representations of atoms, showing only the valence electrons surrounding the elemental symbol (**Figure 2.11**). Comparison of the Lewis dot structures of atoms in the same group makes clear the similar reactivity that ultimately resulted in Mendeleev's original periodic table.

Group 18 in the periodic table consists of the noble gases. Atoms in this group all have a valence electron configuration of $1s^2$ [*for helium*] or Xs^2Xp^6 (X = 2, 3, 4, 5, or 6) [*for all the others*]. This electron configuration, with 8 total electrons in the valence shell, makes these atoms so stable that they form very few compounds with other elements. [*Stable atoms are at the bottom of the hill—nowhere for them to roll.*] The stability conferred by containing 8 electrons in the valence shell is reflected in Gilbert Lewis's **octet rule**, which states that atoms prefer to bond, ionically or covalently, in ways that give them 8 electrons in their valence shell. The following sections describe how the desire to achieve a noble gas configuration drives both ionic and covalent bonding.

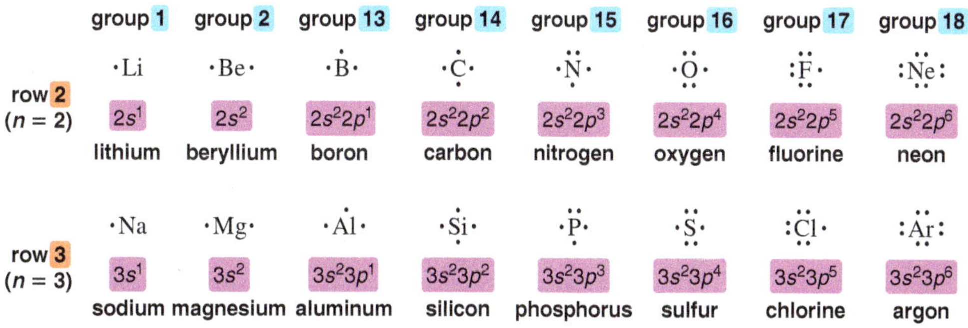

Figure 2.11 Atoms with similar valence configurations show similar reactivity. Group 18 consists of the noble, or inert, gases, the most stable of the elements.

Assessment 2.14 How many electrons does an atom of each of the following elements need to *lose* to achieve a noble gas configuration? By losing that many electrons, which noble gas configuration is achieved?

(a) Beryllium
(b) Aluminum
(c) Magnesium
(d) Potassium

Assessment 2.15 How many electrons does an atom of each of the following elements need to *gain* to achieve a noble gas configuration? By gaining that many electrons, which noble gas configuration is achieved?

(a) Oxygen
(b) Iodine
(c) Phosphorus
(d) Chlorine

2.3.2 Ionic bonding

An **ionic bond** is an electrostatic attraction between two oppositely charged ions resulting from the transfer of electrons between atoms. A standard example is sodium chloride (**Figure 2.12**). Sodium, with a valence configuration of $3s^1$, loses 1 electron to achieve the configuration of the noble gas neon, thus forming a sodium **cation** (+1). Similarly, chlorine, by gaining an extra electron to form a chloride **anion** (−1), achieves the configuration of the noble gas argon with 8 electrons in the valence shell. This process of gaining or losing electrons is called **ionization**. The resulting positively charged sodium cation and negatively charged chloride anion now attract each other, resulting in a strong ionic bond. As shown in Figure 2.12(b), the formation of a stable ionic compound from two unstable atoms releases energy. [*We study these reaction coordinate diagrams in detail later on.*]

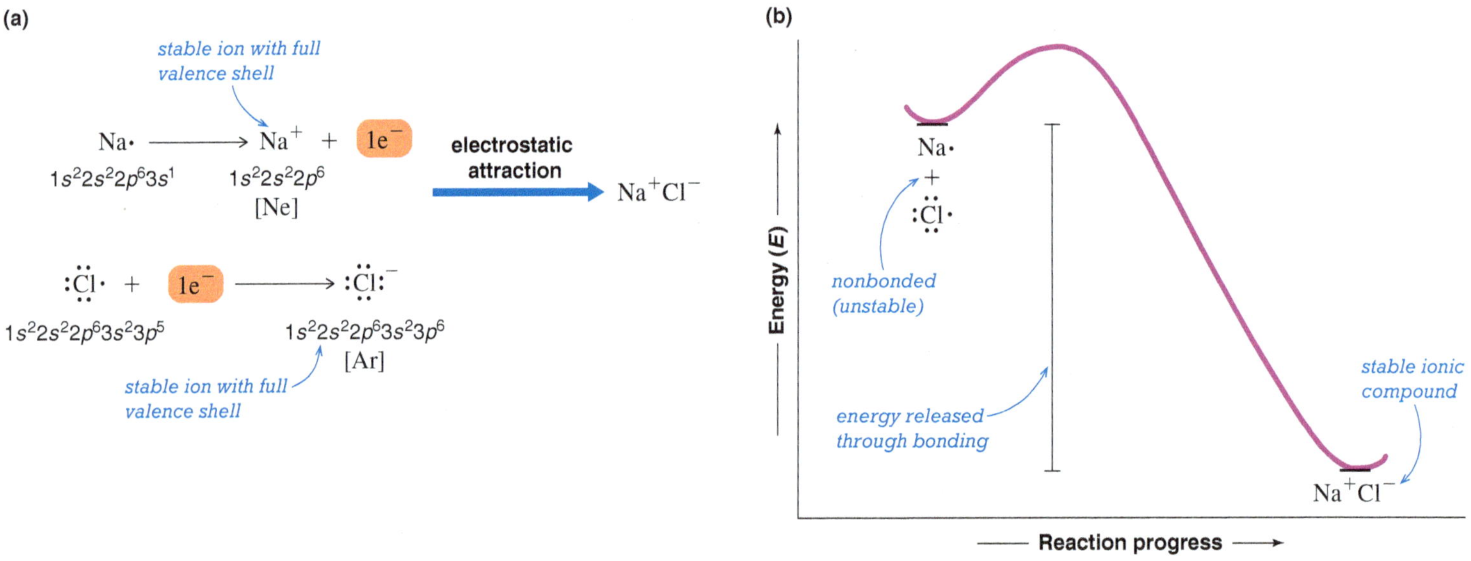

Figure 2.12 **Sodium chloride contains an ionic bond between Na^+ and Cl^-.** Bond formation releases energy and is a favorable process.

A corollary to rule 1 from Chapter 1 (that is, the ball rolls downhill) is that when a ball must go over a hill, it will always go over the smallest one. This implies that, on the way to becoming more stable, an atom (or molecule) will take the easiest pathway. For example, there are technically two ways that a sodium atom could achieve a noble gas configuration. It could lose 1 electron to become like neon, as it did in **Figure 2.13**, or it could gain 7 electrons to become like argon. It is much easier to lose 1 than to gain 7, so this is what happens. Similarly, chlorine could gain 1 electron to become like argon, or it could lose 7 to become like neon. Once more, it is easier to gain 1, so this is what happens.

Figure 2.13 **Atoms take the easiest pathway to becoming more stable ions.**

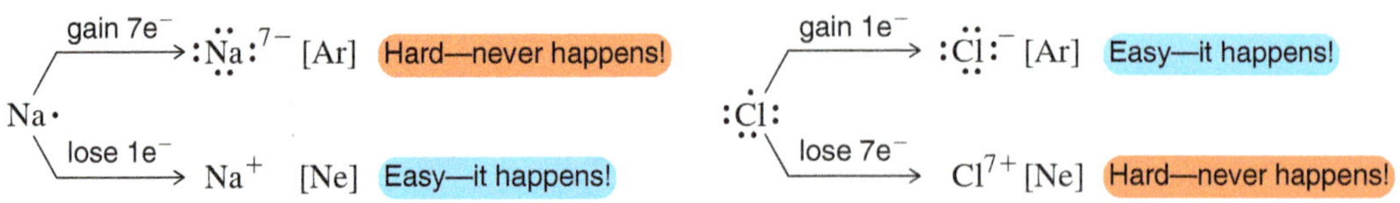

The ideas expressed in Figure 2.13 are responsible for many of the periodic trends we discuss throughout this course. Of principal importance here is that metals, on the left side of the periodic table (groups 1 through 13), tend to lose electrons to achieve a noble gas configuration. As a result, the metals in ionic compounds are typically cations. Nonmetals, on the right side of the periodic table, tend to gain electrons to achieve a noble gas configuration. Therefore, the nonmetals in ionic compounds are typically anions (**Figure 2.14**).

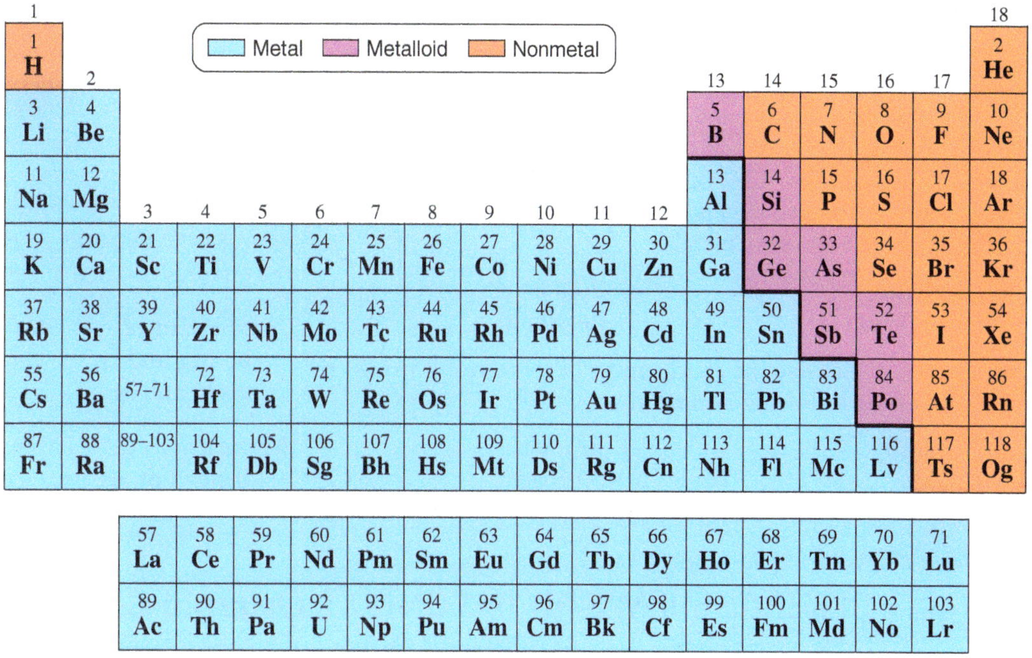

Figure 2.14 **Metals form cations, whereas nonmetals form anions.**

The charge on a cation or anion (or **oxidation number**) depends on how many electrons must be lost or gained in achieving the noble gas configuration (**Figure 2.15**). Elements in group 1 lose 1 electron, gaining a +1 charge. In group 2, 2 electrons are lost, resulting in a +2 charge. Group 6 elements gain 2 electrons, gaining a −2 charge. In group 7, the halogens gain 1 electron, resulting in a −1 charge.

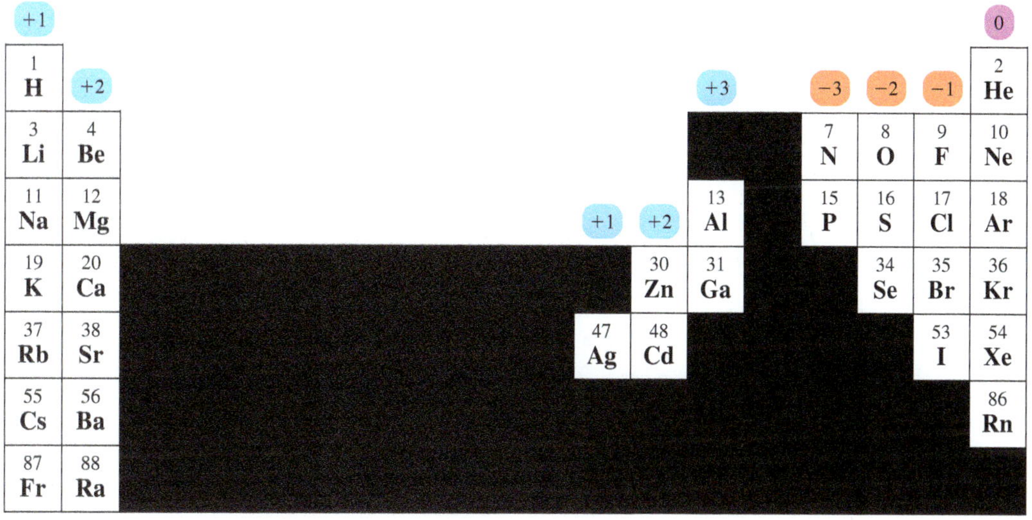

Figure 2.15 **Predicting common oxidation numbers based on the position of an element in the periodic table.**

When metals and nonmetals combine to form salts, they do so in ratios that result in stable, neutral ionic compounds. Thus, as shown in **Figure 2.16**, Al^{3+} combines with 3 F^- to give aluminum fluoride. Similarly, Mg^{2+} combines with one O^{2-} to make magnesium oxide.

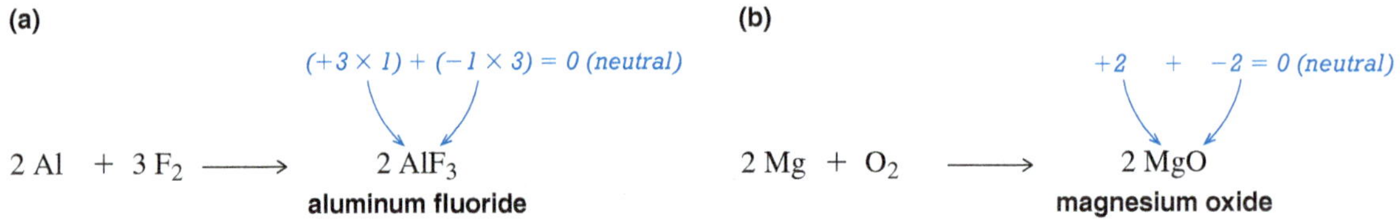

Figure 2.16 Salts consist of ratios of anions and cations that give neutral ionic compounds.

2.3–2.3.2 SUMMARY

- Bonding is driven by forming more stable molecules or ions from less stable atoms. Ionic bonds result from the electrostatic attraction between stable ions formed by the transfer of electrons from metals to nonmetals.

Assessment 2.16 Give the charge most likely to result from ionization of the following metals.

(a) Lithium
(b) Aluminum
(c) Potassium
(d) Beryllium

Assessment 2.17 Give the charge most likely to result from ionization of the following nonmetals.

(a) Chlorine
(b) Sulfur
(c) Bromine
(d) Nitrogen

Assessment 2.18 Show the ionic compound that you would expect to form between the given metal and nonmetal. Label the charges on each species.

(a) Al and F
(b) Mg and Br
(c) Li and N

Assessment 2.19 Fill in the blank. Oxygen ionizes to become −2 instead of −1 because O^{2-} is more ________.

2-A GENERAL CHEMISTRY IN CONTEXT: Organic Chemistry

Reducing metals

[*Note: All of the Chemistry in Context (CIC) boxes in this chapter preview concepts covered in later chapters. Don't worry about the exact details of what is discussed because we focus more on them much later.*]

Much of what we study in this course is the conversion of one molecule into another. Section 2.3.2 showed us that sodium metal transfers an electron to chlorine to make sodium chloride. In this way, we can say that chlorine is reduced by sodium (which is itself oxidized), as shown in **Figure 2.17**.

$$2\,Na^0 + Cl{-}Cl \longrightarrow 2\,NaCl$$

Figure 2.17 Sodium metal reduces chlorine to chloride. In the process, sodium metal is oxidized to sodium cation.

One of the key reactions in Chapter 10 involves this same concept. Instead of transferring an electron to chlorine, sodium, when dissolved in ammonia, transfers an electron to an alkyne (**Figure 2.18**), breaking a carbon–carbon π bond. The resulting radical anion is further reduced, ultimately producing an alkene.

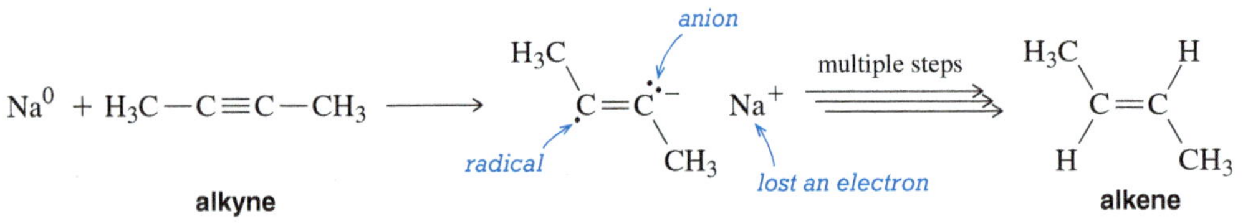

Figure 2.18 Sodium reduces alkynes to alkenes.

2.3.3 Covalent bonding

In contrast to ionic bonding, where two atoms are drawn together by electrostatic attraction resulting from the transfer of electrons, **covalent bonds** form by sharing electrons between two atoms. Consider molecular hydrogen (H_2), which is formed by the reaction of two hydrogen atoms (**Figure 2.19**). On its own, a hydrogen atom, 1 electron short of a full octet [*the octet rule for row 1 is 2 electrons; why might this be?*], is highly unstable. Rather than gain or lose an electron, it instead shares its electron with another hydrogen that does the same. In this way, each hydrogen has a total of 2 electrons surrounding its nucleus. Both electrons are attracted to both nuclei, representing the "glue" that holds the two nuclei together. The resulting H_2 molecule is more stable than the nonbonded hydrogen atoms. [*We talk more about the orbitals involved in this process in Section 2.7.*] When bonded, the 2 electrons are usually represented by a line to represent a two-electron bond. Once more, the reaction coordinate diagram in Figure 2.19(b) shows that energy is released when two unstable atoms form a covalently bonded molecule.

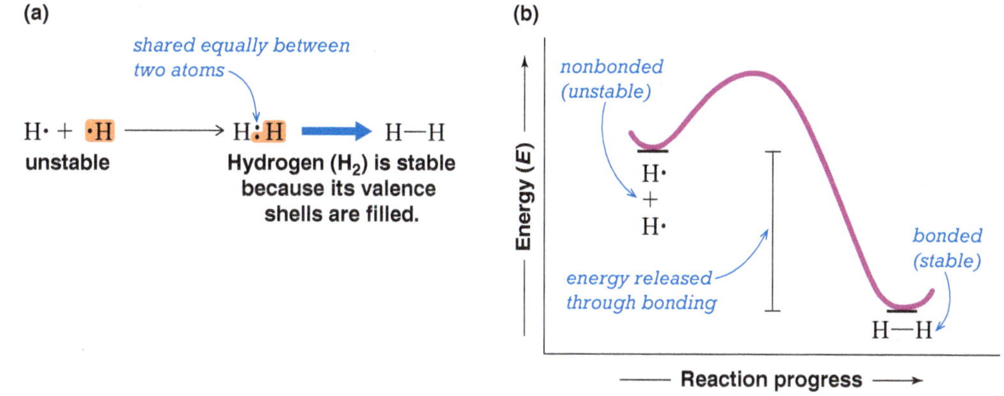

Figure 2.19 Formation of a covalent bond. Electron sharing between two hydrogens is more stable than two nonbonded hydrogen atoms.

2.3.4 Covalent versus ionic bonds

Given these two types of chemical bonds, when is one type used and when is the other? The answer depends on the electronegativity (EN) of the atoms involved. Electronegativity values change in a periodic way. As you move to the right and up in the periodic table, the electronegativity increases. The Pauling electronegativity values shown in **Figure 2.20** are a measure of how much an individual atom attracts electrons. [*See if you can make sense of the electronegativity values in Figure 2.20 before reading further.*]

Figure 2.20 Pauling electronegativity values.

0.1–1.9 | 2.0–2.9 | 3.0–3.9

1	2	3	4	5	6	7	8	9	10	11	12	13	14	15	16	17	18
H 2.1																	H
Li 1.0	Be 1.5											B 2.0	C 2.5	N 3.0	O 3.5	F 4.0	Ne
Na 0.9	Mg 1.2											Al 1.5	Si 1.8	P 2.1	S 2.5	Cl 3.0	Ar
K 0.8	Ca 1.0	Sc 1.3	Ti 1.5	V 1.6	Cr 1.6	Mn 1.5	Fe 1.8	Co 1.8	Ni 1.8	Cu 1.9	Zn 1.6	Ga 1.6	Ge 1.8	As 2.0	Se 2.4	Br 2.8	Kr 3.0
Rb 0.8	Sr 1.0	Y 1.2	Zr 1.4	Nb 1.6	Mo 1.8	Tc 1.9	Ru 2.2	Rh 2.2	Pd 2.2	Ag 1.9	Cd 1.7	In 1.7	Sn 1.8	Sb 1.9	Te 2.1	I 2.5	Xe 2.6
Cs 0.7	Ba 0.9		Hf 1.3	Ta 1.5	W 1.7	Re 1.9	Os 2.2	Ir 2.2	Pt 2.2	Au 2.4	Hg 1.9	Tl 1.8	Pb 1.8	Bi 1.9	Po 2.0	At 2.2	Rn 2.4
Fr 0.7	Ra 0.7		Rf	Db	Sg	Bh	Hs	Mt	Ds	Rg	Cn	Nh	Fl	Mc	Lv	Ts	Og

La 1.1	Ce 1.1	Pr 1.1	Nd 1.1	Pm 1.1	Sm 1.1	Eu 1.1	Gd 1.1	Tb 1.1	Dy 1.1	Ho 1.1	Er 1.1	Tm 1.1	Yb 1.1	Lu 1.2
Ac 1.1	Th 1.3	Pa 1.5	U 1.7	Np 1.3	Pu 1.3	Am 1.3	Cm 1.3	Bk 1.3	Cf 1.3	Es 1.3	Fm 1.3	Md 1.3	No 1.3	Lr

Let's think like a chemist to rationalize the trend in electronegativity values in Figure 2.20.

Thinking Like a Chemist

Good news: You don't need to memorize the electronegativity values in Figure 2.20. However, it is insufficient to just know the periodic trend. As we will get in the habit of doing, we accept nothing as fact unless we understand *why* it is so. One problem-solving method we often employ is to simply identify the difference between the items being compared—in this case, atoms with a high Pauling value and atoms with a low Pauling value. By doing this, we simplify a complicated question into a series of simpler ones.

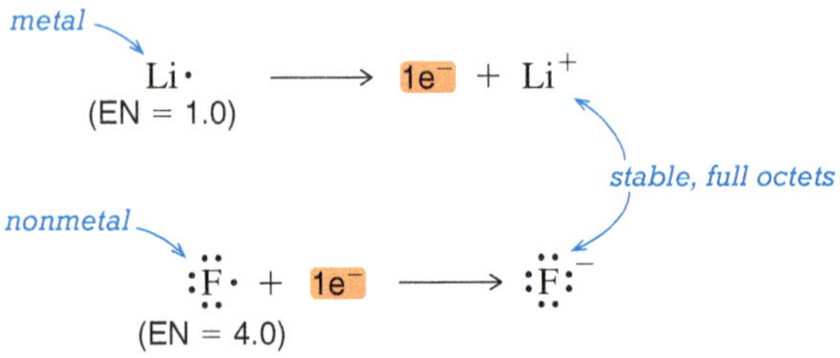

Figure 2.21 Electronegative atoms want electrons. Electropositive atoms do not.

Let's compare fluorine (4.0) and lithium (1.0), which have the highest and lowest Pauling values, respectively, across the same row in (**Figure 2.21**). What's the difference between F and Li? [*Consider Section 2.3.2 in coming up with an answer.*] F is a nonmetal, whereas Li is a metal. [*Other differences? How does F become more stable?*] Fluorine, like all nonmetals, becomes more stable by gaining an electron to become F^-. In this way, fluorine attracts electrons; we say that it is highly electronegative. Lithium (a metal) becomes more stable by losing electrons to become Li^+. In this way, lithium releases electrons; we say that it is electropositive (not electronegative).

And so, by comparing fluorine and lithium, we can generalize that an atom's electronegativity is based on its desire to gain or lose electrons. Because nonmetals, on the right of the periodic table, become stable by gaining electrons, electronegativity increases as you move from left to right in the periodic table. ■

Assessment 2.20 (i) Using the periodic trend, choose the more electronegative atom in each pair. For one pair, you'll need to look at the actual Pauling values. (ii) For which one? (iii) Why?

(a) K vs. Li (b) N vs. O (c) P vs. O (d) B vs. Si

Assessment 2.21 We rationalized why electronegativity increases as you move to the right across a row. Why does it increase as you move up a group? [*Consider the difference between row 2 and row 6 atoms. See Section 2.2.2.*]

Assessment 2.22 Atomic radius decreases as you move left to right in the periodic table. Electronegativity increases as you move left to right in the periodic table. Rationalize these two trends in the context of each other.

Assessment 2.23 Based on your answer to Assessment 2.22, would you expect a larger atom to be more or less electronegative than a smaller atom?

A bond is characterized as being ionic or covalent based on the difference in electronegativity values between the bonded atoms (**Figure 2.22**). When the difference is large (2.0–4.0), the bond is ionic, meaning the electron pair is *not* shared. When the difference is small (0–0.4), the bond is pure covalent, meaning the electrons are shared equally between the two atoms. When the difference is intermediate between the two (0.4–2.0), the bond is polar covalent. The electrons in a polar covalent bond are shared between the two atoms, but unequally. We discuss a consequence of this in Section 2.5.2.

Figure 2.22 How the polarity of a bond correlates with the electronegativity difference of the bonded atoms.

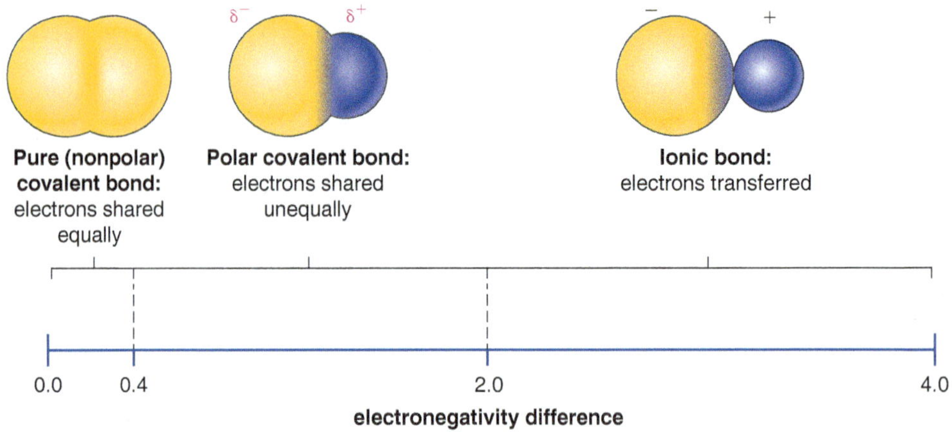

More important than the specific numbers in Figure 2.22, we need to develop the intuition to recognize which class a bond falls into. This will be done by looking at molecules over time, but here we can make a few generalizations. The largest differences in electronegativities occur when molecules on the far right of the periodic table (nonmetals) bond with molecules on the far left (metals). Thus, bonds between a metal and a nonmetal are generally considered ionic. The bond in sodium fluoride (NaF) falls into this class. Bonds between nonmetals of differing electronegativity are polar covalent. The C—O bond falls into this class. Finally, bonds between nonmetals of nearly identical electronegativity are pure covalent. The H—H bond in H_2, the C—C bond [*the organic chemist's favorite*], and the C—H bond fall into this class (**Figure 2.23**).

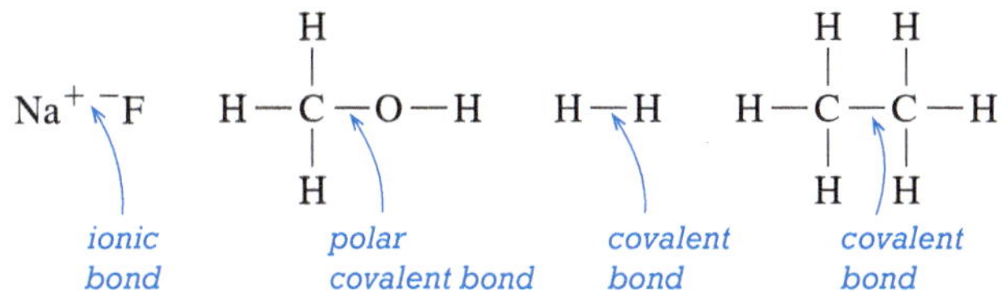

Figure 2.23 Classification of bonds. Ionic bonds form between metals and nonmetals. Polar covalent bonds form between nonmetals of different electronegativity. Covalent bonds form between nonmetals of similar (or identical) electronegativity.

It is best to think of the classification of a bond as ionic, polar covalent, or covalent as being on a spectrum. In this way, we can consider the polarity of two bonds as being relative to one another. [*Assessments 2.24 and 2.25 give you some practice at this.*]

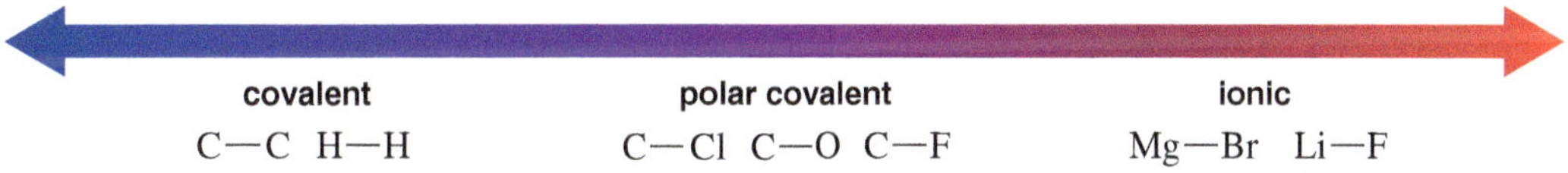

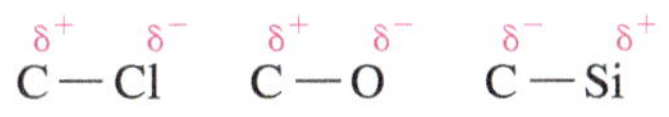

Figure 2.24 Polar covalent bonds possess a bond dipole, where the more electronegative atom has a partial negative charge and the less electronegative atom has a partial positive charge.

Polar covalency can be indicated for a bond by using a lowercase delta (δ) to represent the partial charges caused by the electronegativity difference in bonded atoms, as shown for the C—Cl, C—O and Si—C bonds in **Figure 2.24**. In each, the more electronegative atom, because it holds the electrons more tightly, possesses a partial negative charge. [*We explain the importance of this for overall molecule dipole moments in Section 2.5.1.*]

2.3.3–2.3.4 SUMMARY

- Covalent bonds result from the equal sharing of electrons between atoms, driven by each atom achieving a full octet. Polar covalent bonds result from the unequal sharing of electrons, with the more electronegative atom gaining a partial negative charge (δ^-) and the less electronegative atom gaining a partial positive charge (δ^+).

Assessment 2.24 Without looking at Figure 2.20, use your intuition to estimate whether a bond is ionic, polar covalent, or covalent. [*Really, don't look until you've done all the problems.*] Now, look at Figure 2.20 to check your answers.

(a) Na—Cl (b) C—H (c) O—O (d) C—Cl

Assessment 2.25 Using only your intuition, rank the following covalent bonds in terms of their polarity (1 = most polar; 6 = least polar). [*You can use the periodic trend of electronegativity, but don't use actual numbers.*] Now, look at Figure 2.20 to check your answers.

(a) C—C
(b) C—O
(c) C—H
(d) C—F
(e) C—Cl
(f) C—S

Assessment 2.26 For the polar covalent bonds in Assessment 2.25, use the lowercase delta (δ^- or δ^+) to indicate the partial charge distribution in the bond.

2.4 Drawing Lewis structures

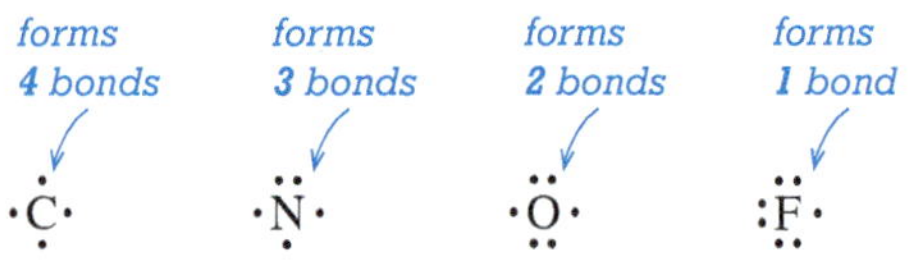

Figure 2.25 The Lewis dot structure of an atom predicts how many bonds an atom will form through electron-sharing.

In studying covalent bonding, the octet rule, coupled with the Lewis dot structure of atoms, gives an indication about how many bonds an atom is likely to form. An atom will gain 1 electron toward a full octet for each bond formed. As a result, carbon, with 4 electrons, most often forms four bonds in stable molecules. Nitrogen, with 5 electrons, forms three bonds. Oxygen, with 6 electrons in its valence shell, forms two bonds, while fluorine, with 7 electrons, forms just one (**Figure 2.25**).

The structure of H_2 shown in **Figure 2.26** is an example of a Lewis structure of a molecule. It is formed by combining the hydrogen atoms in their Lewis dot structure to make the molecule. The sharing of 2 electrons is represented by the H—H bond. Hydrogen's filled valence shell contains 2, as opposed to 8, electrons.

The Lewis structure of methane (CH_4), the simplest organic molecule, can be constructed in a similar way [**Figure 2.27(a)**]. Carbon, with 4 valence electrons, can accommodate 4 more electrons to achieve a full octet. As a result, it can bond with four hydrogen atoms, each of which will share its 1 valence electron. The electrons shared between two atoms can be represented with a dash to indicate the bond. In this stable structure, each hydrogen has 2 electrons in its valence shell. [*For hydrogen, 2 electrons—a duet—represent a full valence shell.*] The carbon now has 8 electrons in the valence shell, so the octet rule is satisfied. These are covalent bonds because the electronegativity difference between carbon (2.5) and hydrogen (2.1) is small. **Figure 2.27(b)** shows that energy is released when unstable atoms form a covalently bonded molecule.

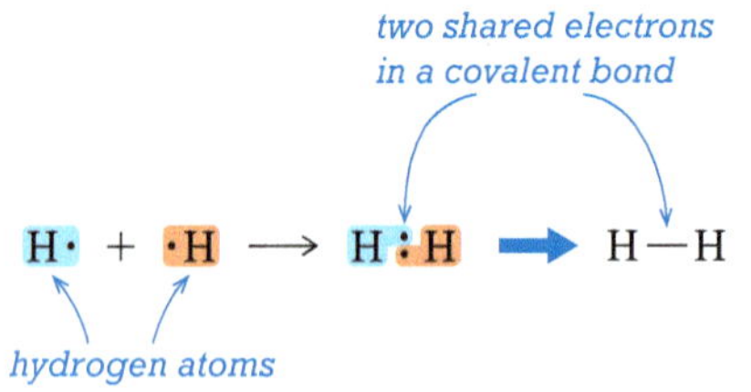

Figure 2.26 Lewis structure of H_2 formed from the Lewis dot structure of two hydrogen atoms.

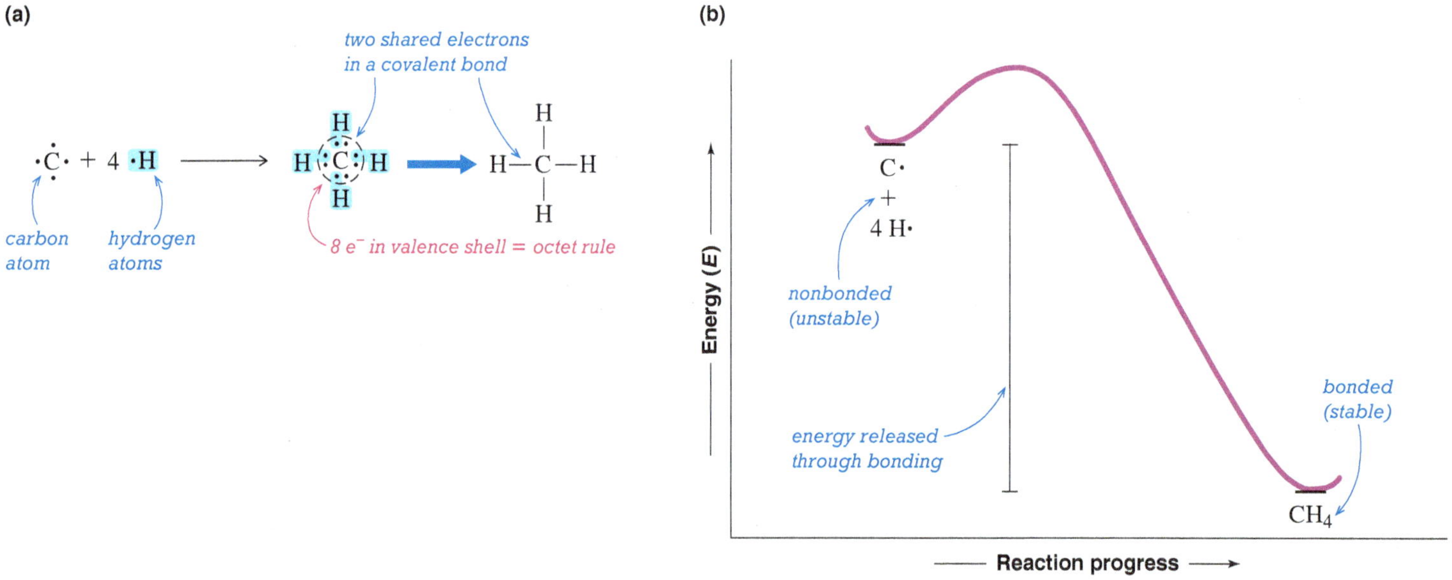

Figure 2.27 Lewis structure of methane formed by overlap of the Lewis dot structures of four hydrogen atoms and one carbon atom.

2.4.1 Drawing Lewis structures for neutral molecules

Given a molecular formula, Lewis structures can be drawn using the rules in **Table 2.1**. The rules are applied to the molecular formulas H_2O and CCl_4. [*By applying these rules to simple molecules, we'll be able to trust them when we look at complicated molecules.*]

Table 2.1 Recipe for drawing Lewis structures.

Steps	Justification	H_2O	CCl_4
1. Count the total number of valence electrons contributed by each atom. This will correspond to the unit of the group number of each atom. [*For example, Group 18 = 8.*]	Only the valence electrons are involved in bonding, so these are the only important ones.	O = 6 valence e^- H = 1 valence e^- H = 1 valence e^- total = 8 valence e^-	C = 4 valence e^- Cl = 7 valence e^- Cl = 7 valence e^- Cl = 7 valence e^- Cl = 7 valence e^- total = 32 valence e^-
1a. For a cation, subtract 1 valence electron for each unit of positive charge. 1b. For an anion, add 1 valence electron for each unit of negative charge.	The electron has a −1 charge, so each unit of charge represents the absence or presence of 1 electron.	Not applicable—H_2O is neutral.	Not applicable—CCl_4 is neutral.
2. Arrange the atoms so that the least electronegative atom (that isn't hydrogen) is in the center.	Hydrogen can only form one bond, so it is not a central atom in molecules. The least electronegative atom is most willing to share electrons, so usually it is in the center. In addition, the least electronegative atom can form more bonds. [*Compare Figure 2.25 to the Pauling electronegativity values in Figure 2.20 to confirm that this is true.*]	*forms 2 bonds* H O H	*forms 4 bonds* Cl Cl C Cl Cl
3. Connect the outer atoms to the central atom using bonds that each represent 2 electrons.	The atoms must be attached to each other to form a molecule.	H—O—H **4 valence e^- used (4 remain)**	Cl Cl—C—Cl Cl **8 valence e^- used (24 remain)**
4. Fill the valence shell of the outer atoms using lone pairs of electrons.	Hydrogen needs no lone pairs because the 2 bonded electrons fill its valence shell. The outer atoms are filled first because they are the most electronegative.	H—O—H **no change**	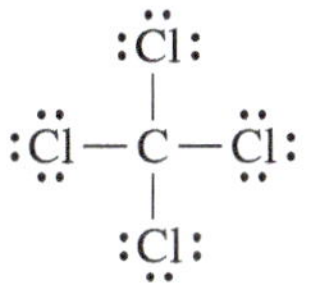 **32 valence e^- used (0 remain)**
5. Fill the valence shell of the central atom using lone pairs of electrons until all valence electrons are used.	The central atom gets electrons last because it is usually the least electronegative.	 **8 valence e^- used (0 remain)**	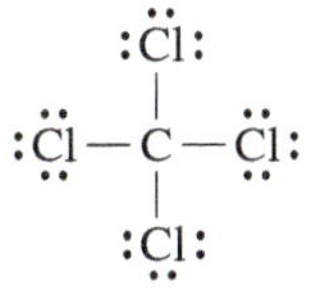 **no change**
6. If all valence electrons are used, and all valence shells are as filled as they can be, you are done.	Be careful not to add extra electrons just to fill valence shells.	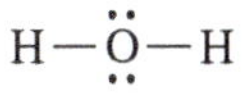 **8 valence e^- used all octets filled**	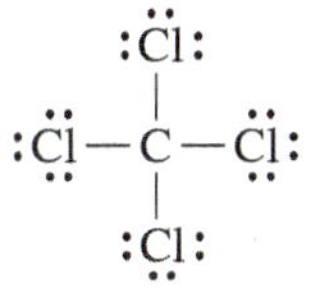 **32 valence e^- used all octets filled**
7. Calculate the formal charge of all atoms to confirm that the best Lewis structure has been drawn. [*See Section 2.4.2.*]	The absence, or presence, of a formal charge provides one measure of a molecule's stability.	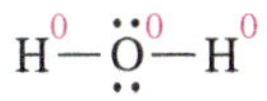 **All formal charges are zero.**	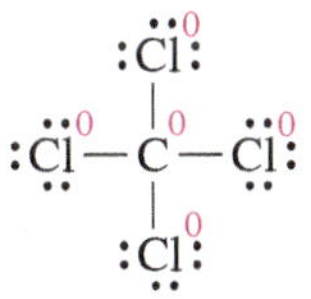 **All formal charges are zero.**

2.4.2 Calculating formal charge

It is important to know where the formal charges exist in a molecule, and one of the goals of drawing Lewis structures is to identify areas of electron deficiency and charge distributions. The **formal charge** is the charge assigned to an atom in a molecule assuming equal sharing of electrons. The formal charge of every atom in a molecule will add up to the total charge in the molecule. The formal charge of a given atom can be calculated using Equation 2.2.

$$\text{Formal charge} = \begin{matrix}\text{Number of}\\ \text{valence electrons}\end{matrix} - \frac{1}{2}\begin{pmatrix}\text{Shared}\\ \text{electrons}\end{pmatrix} - \begin{matrix}\text{Unshared}\\ \text{electrons}\end{matrix} \qquad \text{Equation 2.2}$$

Recall that the unit of the group number is equal to the number of valence electrons in a neutral atom. [*For carbon in group 14, there are 4 valence electrons.*] Nonbonded atomst have no formal charge because the total number of electrons equals the total number of protons in the nucleus (**Figure 2.28**).

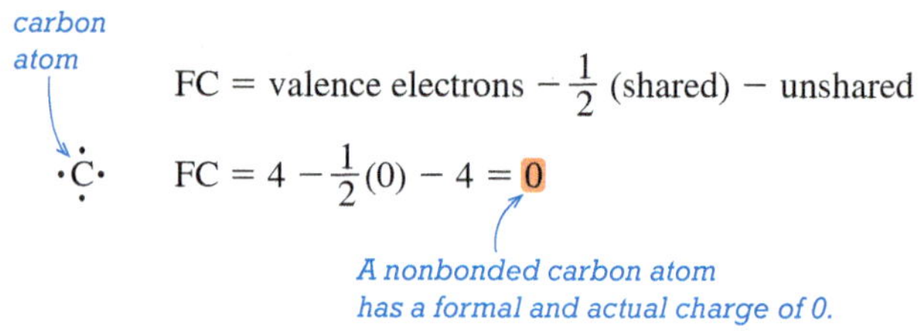

Figure 2.28 **Calculating the formal charge (FC) and actual charge of an atom.**

The formal charge calculation is typically used only for molecules because we can calculate the actual charge for a nonbonded atom, as we did in Figure 2.28. But seeing it this way helps clarify the equation. The formal charge on a particular atom in a molecule represents how many electrons it now "owns" as compared to how many it "owned" as an atom. When an atom shares a bonded pair of electrons, only one of them is technically owned by the atom. For this reason, only half of the bonded electrons are involved in Equation 2.2. Both electrons of a nonbonded pair are owned by the atom and are counted in the Equation 2.2.

The best, meaning the most stable, Lewis structure usually minimizes the formal charges on all atoms. We see this when we calculate the formal charges for each atom in H_2O [**Figure 2.29(a)**] and CCl_4 [**Figure 2.29(b)**]. All of the formal charges are zero and add up to a total charge of zero for each molecule. Because the lack of formal charge is one measure of stability, this implies that the best Lewis structure has been drawn for each. For this reason, the calculation of formal charge is often used as a final step in drawing Lewis structures (see step 7 in Table 2.1).

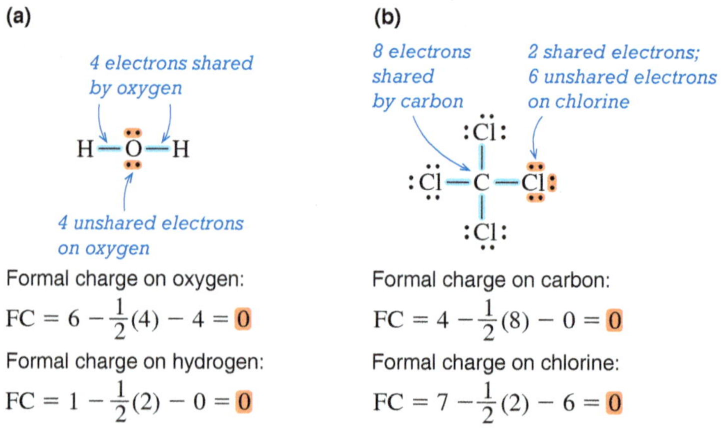

Figure 2.29 **Calculating the formal charge of atoms in a Lewis structure.** (a) Oxygen in H_2O is neutral. (b) Carbon in CCl_4 is neutral, too.

Assessment 2.27 Draw the Lewis structure for the following molecules. Be sure to calculate the formal charge of each atom as a way of confirming your structure is correct. [*To assist you, the central atom has been bolded where you might have trouble figuring it out on your own.*]

(a) CH_3Cl (b) HOBr (c) CF_4 (d) $H_2\mathbf{O}_2$ (e) BH_3

Assessment 2.28 Calculate the formal charge of the indicated atom in the following molecules or ions.

(a) (b) (c) (d) (e) (f)

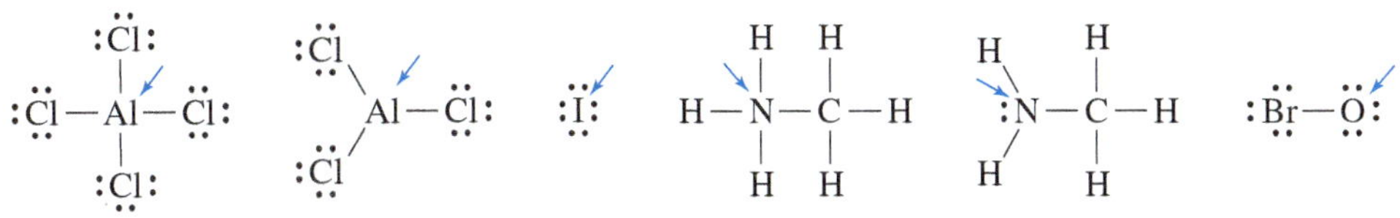

2-B GENERAL CHEMISTRY IN CONTEXT: Organic Chemistry

Isomers

How can a single element in the periodic table (that is, carbon) provide the material for an entire field of chemistry? The answer has to do with the fact that carbon forms four stable bonds, and that it can do so in a variety of different ways. **Figure 2.30** shows seven of many different structures with the molecular formula $C_6H_{12}O$, all of which are known compounds with different properties and chemical reactivity. Molecules that have the same formula but are not identical are referred to as isomers. [*We explain how these are drawn in Chapter 3.*]

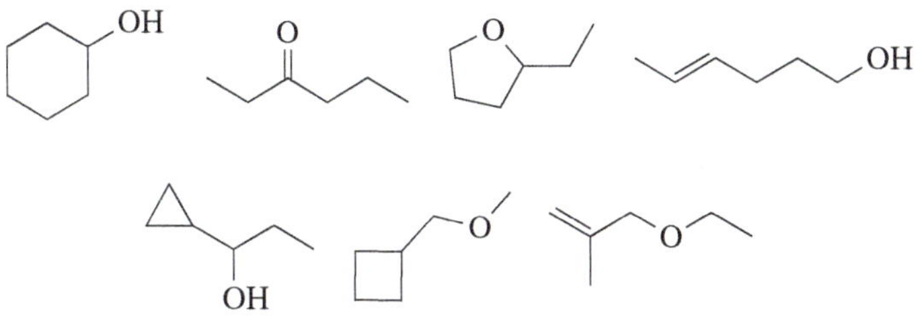

Figure 2.30 Isomers with the molecular formula $C_6H_{12}O$. This is a very incomplete list. [*Can you draw others?*]

2.4.3 Drawing Lewis structures for charged molecules (ions)

Section 2.4.1 discussed drawing Lewis structures for neutral molecules. Many of the most reactive molecules in organic chemistry, though, are reactive because they possess positive or negative charge. [*Does it make sense that neutral molecules are generally more stable?*] Drawing the Lewis structure of charged molecules follows the recipe in Table 2.1, as long as the charge is accounted for when determining the total number of valence electrons in step 1a or 1b. Given that electrons have a −1 charge, a negatively charged ion should have one or more extra electrons. Similarly, a positively charged ion would be expected to be short on electrons.

We can see how this applies to drawing the Lewis structure of the methyl cation (CH_3^+) and the methyl anion (CH_3^-). **Figure 2.31(a)** shows that the methyl cation, after forming the three C—H bonds, leaves the central carbon without a full octet. Calculation of the formal charge in CH_3^+ shows that the central carbon, 2 electrons short of a full octet, has a formal +1 charge. Alternatively, calculation of the formal charge in CH_3^- [**Figure 2.31(b)**] shows that the central carbon, containing 1 more electron than needed, has a formal −1 charge.

(a) CH_3^+

C	$= 4e^-$	
H	$= 1e^- \times 3$ atoms	$= 3e^-$
+1	= subtract $1e^-$	
total	= 6 valence e^-	

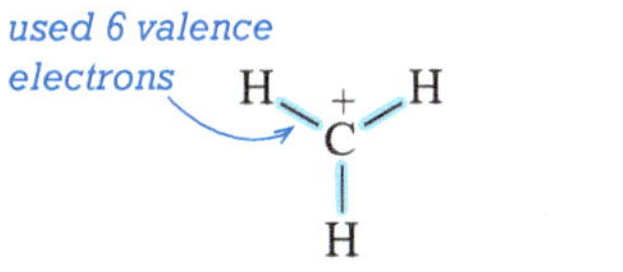

Formal charge on carbon:

$$FC = 4 - \frac{1}{2}(6) - 0 = +1$$

(b) CH_3^-

C	$= 4e^-$	
H	$= 1e^- \times 3$ atoms	$= 3e^-$
−1	= add $1e^-$	
total	= 8 valence e^-	

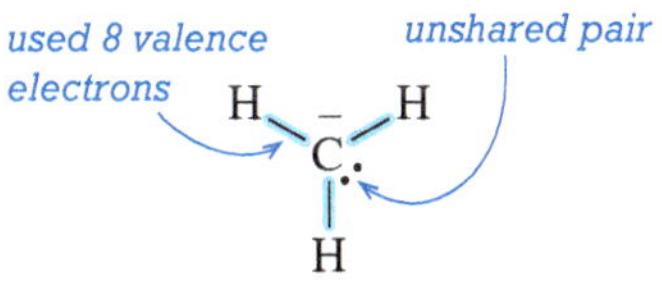

Formal charge on carbon:

$$FC = 4 - \frac{1}{2}(6) - 2 = -1$$

Figure 2.31 Drawing Lewis structures of charged molecules. (a) CH_3^+. (b) CH_3^-.

We mentioned at the beginning of the chapter that Lewis structures were the starting point of our eventual discussion of reactivity. The Lewis structures of CH_3^+ and CH_3^- help illustrate this point. **Figure 2.32** shows how these two ions might react with each other were they to come into contact. The negatively charged carbon can "donate" its pair of electrons to the positively charged carbon [*negatives attack positives*], completing its octet and forming a new molecule, $CH_3—CH_3$, called ethane. This electron pair donation is illustrated by an arrow originating at the electron pair (on the carbon with the negative charge) and pointing to the electron-deficient carbon (the positive charge). This results in both carbons having a formal charge of zero. One carbon now shares 2 electrons instead of possessing both, while the other is sharing 2 electrons it didn't have prior to the reaction. This arrow-pushing formalism is discussed further in Section 4.5.1, but for now it is meant to preview how critical Lewis structures are for understanding chemical reactivity.

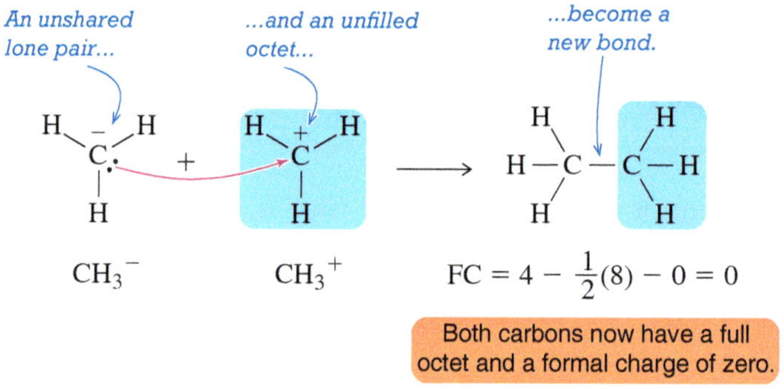

Figure 2.32 Preview of the reaction of two Lewis structures. Negatives attack positives.

Assessment 2.29 Draw the Lewis structure for the following ions. Be sure to calculate the formal charge of each atom to confirm that your structure is correct.

(a) HO^-
(b) H^+
(c) BH_4^-
(d) $AlCl_4^-$
(e) H_3O^+
(f) NH_4^+

Assessment 2.30 (a) Calculate the formal charge of the indicated atoms in the ions shown. (b) Show how the two ions might react to balance the formal charges. (c) Draw the Lewis structure of the resulting product.

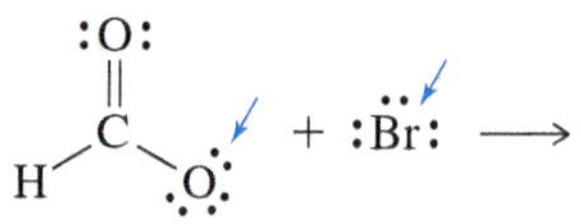

2.4.4 Lewis structures with double and triple bonds

Thus far, the molecules for which we've drawn Lewis structures have only contained single bonds. That won't always be true.

Let's work Assessment 2.31, first on your own, and then together to see how and when a double bond might be necessary.

Assessment 2.31 Draw the Lewis structure for ethene. The molecular formula is C_2H_4.

Solved Assessment 2.31

Following the normal rules for drawing Lewis structures (Table 2.1), you get to a point where you have the two central carbons lacking a full octet, but with only 2 electrons remaining to satisfy them both (**Figure 2.33**). Step 5 says to fill the remaining octets using nonbonded lone pairs, but we have only enough electrons (2) to place a lone pair on just one of the carbons. The Lewis structure shown is *not* the correct answer [*but let's consider it to see what it tells us about how to get to the correct structure*]. The first problem with this structure is that the octet of one of the carbons remains unfilled. Furthermore, calculation of the formal charge shows that one carbon is +1 and one is −1. Because formal charges indicate reactive species, this must *not* be the most stable form of the molecule.

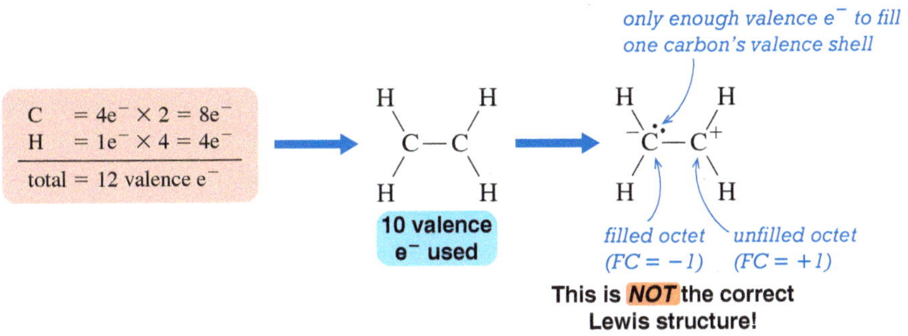

Figure 2.33 **First attempt at ethene's Lewis structure.**

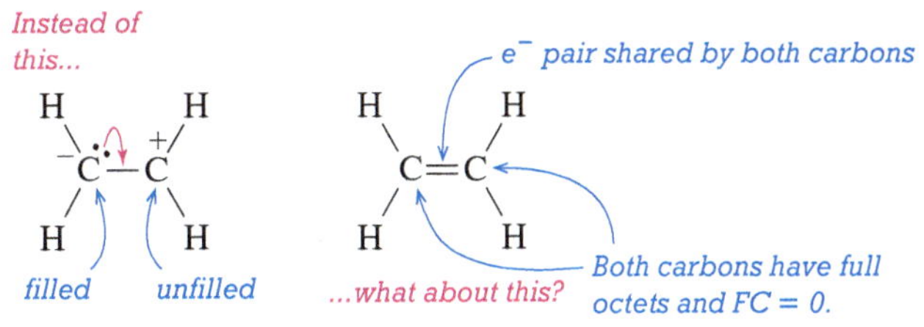

Figure 2.34 **The Lewis structure of ethene** has a double bond between the two carbons.

Instead of localizing the lone pair on one of the carbons, what if we placed it between the two carbons as a second bond, as shown in **Figure 2.34**? Now both carbons have a full octet and [*guess what?*] formal charges of zero. Before we move on, let's analyze what we've done. Using our pencils, technically we erased the lone pair from one carbon and redrew it between the two carbons. What if, instead, we thought about it as sharing the lone pair on one carbon with the other? We could represent this using our arrow-pushing formalism where an arrow starts at the lone pair and finishes as a bond between the two carbons. [*This is the beginning of a concept that will be very important to us as organic chemists.*] Note that sharing a "negative charge" with a "positive charge" results in two neutral atoms. ■

The concept revealed in Solved Assessment 2.31 is important enough to look at in an additional example—namely, drawing the Lewis structure of the nitrate ion (NO_3^-). Following steps 1 through 5 in Table 2.1 will get you the structure shown in **Figure 2.35**. Nitrogen lacks a full octet, and each atom has a nonzero formal charge, so this isn't the best (most stable) Lewis structure. This problem can be solved by sharing a lone pair from one of the oxygens with the nitrogen to form a N═O double bond. Once more, we can use the arrow-pushing formalism to "move" the electron pair to be shared with the positively charged nitrogen. [*Which oxygen shares its lone pair? Good question—we return to this in Section 2.8.*] The resulting structure, containing full octets on all atoms and fewer nonzero formal charges, is the best Lewis structure.

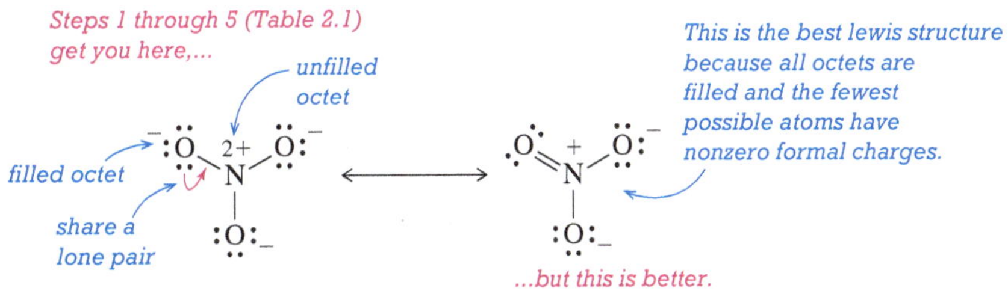

Figure 2.35 **The Lewis structure of the nitrate ion (NO_3^-) has a double bond between the nitrogen and one oxygen.**

Assessment 2.32 Draw the Lewis structures for the following molecules and ion. Be sure to calculate the formal charge of each atom to confirm your structure is correct.

(a) H_2CO
(b) C_2H_2
(c) HCO_2H
(d) CO_3^{2-}
(e) CO_2

Assessment 2.33 Calculate the formal charge on the non-hydrogen atoms in the molecules shown. Use the arrow-pushing formalism to "move" an electron pair such that it is shared between two (formerly) charged atoms. Your arrow should account for the formation of the molecule on the right from the molecule on the left.

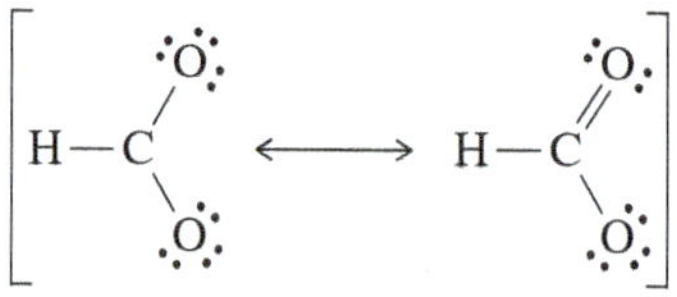

2-C GENERAL CHEMISTRY IN CONTEXT: Organic Chemistry

Alkene addition

Chapters 8 and 9 discuss the reactivity of the carbon–carbon double bond in alkenes. Because it is formed by the overlap of electrons in *p* orbitals to make a π bond (Section 2.4.4), it is an electron-rich bond that is somewhat reactive. Given that negatives attack positives, we will discover that these π electrons attack electron-deficient species (called electrophiles) to form new compounds. Figure 2.36 shows how ethene reacts with hydrobromic acid to make bromoethane. The arrows that illustrate how the reaction works are discussed in detail in Chapter 4.

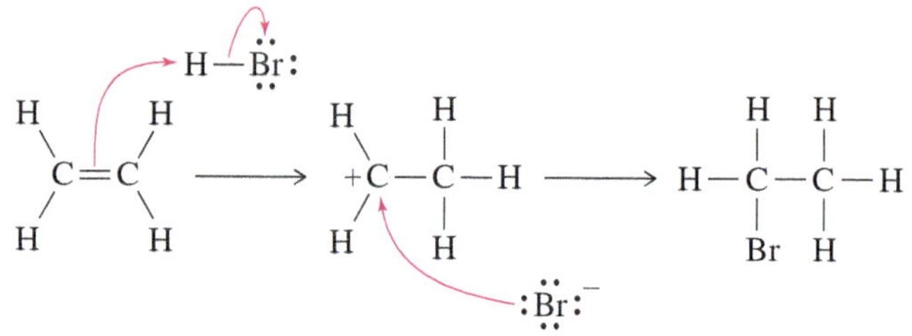

Figure 2.36 Ethene reacts with HBr to make bromoethane.

2.5 Shapes of molecules

Many of the properties of organic molecules arise as a result of their shape. Among these, boiling points, melting points, and even the selectivity of a drug for a particular enzyme or receptor are a result of molecular shape.

2.5.1 VSEPR and molecular shape

Drawing a correct Lewis structure is our first step toward understanding the properties and reactivity of organic molecules. The first property of a molecule that arises from the Lewis structure is its shape. Like most things we will study, a given molecule will adopt whatever shape is most stable. For most molecules, this is driven by orienting the bonded and nonbonded electrons such that they are as far apart as possible. Referred to as **valence shell electron pair repulsion (VSEPR)**, this model recognizes that the like charges of bonded and nonbonded electrons repel each other. Table 2.2 shows a general pattern of predicted shape around a central atom based on the **areas of electron density** emanating from it. These "areas of electron density" include lone pairs, single bonds, double bonds, and triple bonds, each of which is a *single* area of electron density. To be clear, a triple bond is one area of electron density, not three. An example of each, discussed further in Section 2.6, is also shown in Table 2.2.

Table 2.2 The areas of electron density around a central atom predict the molecular shape.

Areas of electron density	Predicted shape around central atom	Average bond angle	Example molecule
4	Tetrahedral	109.5°	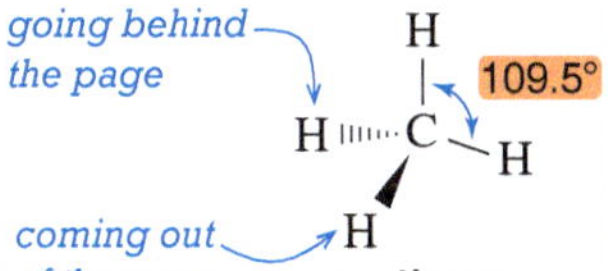 methane
3	Trigonal planar	120°	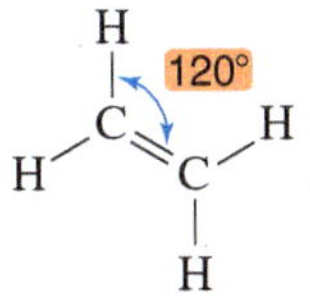 ethene
2	Linear	180°	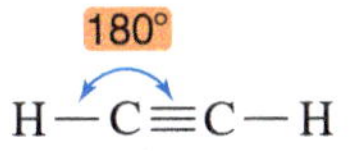 ethyne

In this way, VSEPR only gives us a method for predicting the shape of molecules. These shapes occur because they are the most stable—that is, they minimize energy-increasing repulsion between electrons. [*The ball rolls downhill.*]

Assessment 2.34 You drew the Lewis structures of the following ions in Assessment 2.29. Predict their shapes around the central atom based on the Lewis structure.

(a) BH_4^-
(b) $AlCl_4^-$
(c) H_3O^+
(d) NH_4^+

Assessment 2.35 You drew the Lewis structures of the following compounds and ion in Assessment 2.32. Predict their shapes around the central atom based on the Lewis structure.

(a) H_2CO
(b) C_2H_2
(c) HCO_2H
(d) CO_3^{2-}
(e) CO_2

2-D GENERAL CHEMISTRY IN CONTEXT: Organic Chemistry

Enzyme inhibition

Enzymes catalyze reactions by lowering the activation energy and increasing the rate of chemical reactions. One method for treating disease is to inhibit enzymes that catalyze the formation of harmful molecules. In Chapter 20, we will study cholesterol-lowering drugs that operate by this mechanism. The enzyme HMG-CoA reductase is an enzyme that is used by the body to create cholesterol. By inhibiting it with Lipitor (atorvastatin), cholesterol levels are lowered. Atorvastatin works as an inhibitor because its shape allows it to fit perfectly in the active site of HMG-CoA reductase (Figure 2.37). With atorvastatin occupying the active site, the enzyme is rendered functionally inactive, halting cholesterol synthesis.

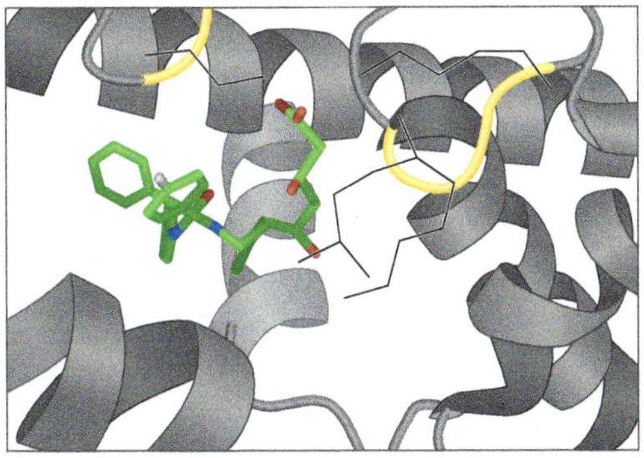

Figure 2.37 The shape of atorvastatin allows it to fit perfectly in the active site of HMG-CoA reductase.

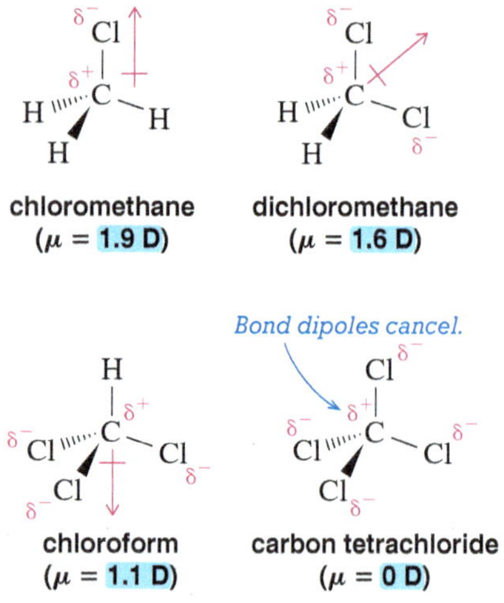

Figure 2.38 The dipole moment of a molecule is the vector sum of its individual bond dipoles.

2.5.2 Polarity of molecules

In Section 2.3.4, we learned that all chemical bonds can be classified as ionic, polar covalent, or covalent based on the electronegativity difference between bonded atoms. Polar covalent bonds within a molecule can combine to give an overall **dipole moment** (μ) to a molecule, measured in units of debye (D). Defined as the net polarity of a molecule, dipole moment is a useful concept when we explore the physical properties, like melting and boiling points, and the chemical reactivity of molecules [*partial negatives attack partial positives*]. The dipole moment is closely correlated to the shape of a molecule because it is the vector sum of all individual bond dipoles in a molecule. [*It is only important that you can recognize when a molecule has a dipole moment and the direction of it.*]

Consider the molecules in Figure 2.38. Chloromethane, which contains a polar C—Cl bond, has a net dipole moment in the direction of the chlorine (μ = 1.9 D). This is indicated by an arrow beginning at the positive end of the dipole and pointing toward the negative end. (This can also be indicated using a lowercase delta, δ, to represent a partial charge.) Dichloromethane, containing two polar C—Cl bonds, is also a polar molecule (μ = 1.6 D), although it is slightly less polar than chloromethane. This is a result of the tetrahedral shape of carbon, which causes the two bond dipoles in dichloromethane to slightly oppose each other, placing the dipole moment on an axis that bisects the two C—Cl bonds. A similar result is observed for chloroform, which contains three polar C—Cl bonds (μ = 1.1 D). In chloroform, the dipole moment points in the opposite direction of the lone C—H bond. Finally, carbon tetrachloride, with four C—Cl bonds extending in four opposite directions, is nonpolar (μ = 0 D) because the individual bond dipoles cancel each other completely.

2.5 SUMMARY

- The molecular shape and the individual bond dipoles combine to give the overall dipole moment (μ) of a molecule.

Assessment 2.36 For the molecules shown, indicate the direction of the dipole moment.

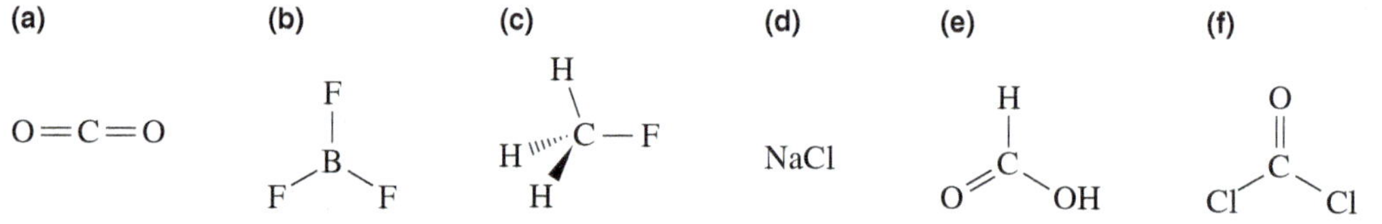

2.6 Bonding in molecules

Bonding in molecules can be described using two related theories: valence bond theory and molecular orbital theory. Though the differences are explained in the following sections, our discussions of bonding throughout the text incorporate features of both, because the combination of the two accurately describes the bonding and reactivity in most organic molecules. [*I promised in Section 2.3 that we'd return to the orbital description of electrons—here we go!*]

2.6.1 Valence bond theory

Imagine molecular hydrogen (H_2) as forming from two hydrogen atoms. Each hydrogen atom, with an electron configuration of $1s^1$, has a single valence electron that it can use for bonding (Figure 2.39). This is a central tenet of valence bond theory and has been discussed in the context of drawing Lewis structures: Atoms combine to form molecules using only valence shell electrons; the inner shell electrons can be ignored. Two unstable hydrogen atoms, each containing a positively charged nucleus, form a bond by sharing their electron with the other positively charged nucleus. With each electron in the bond attracted to both positively charged nuclei, the electrons represent the glue that hold the two positively charged nuclei together.

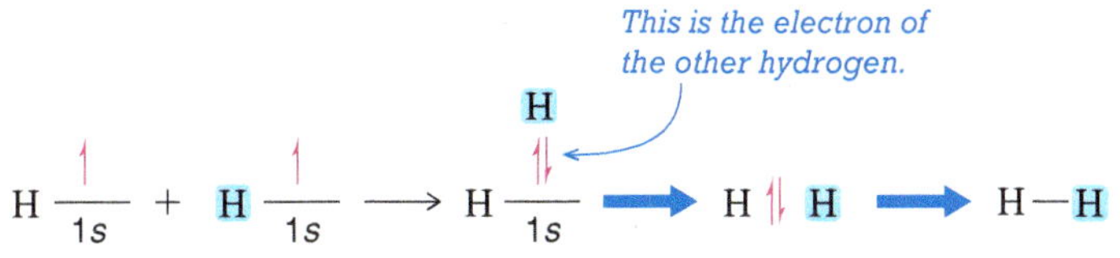

Figure 2.39 Attraction between the shared valence electrons and bound nuclei holds the two nuclei together.

So what actually happens to the atoms when they share electrons? In a nonbonded hydrogen atom, the valence electron is contained in a 1*s* atomic orbital. The general concept of an orbital is used to illustrate the volume in which the electron is 95% likely to appear. The 1*s* orbital, like all *s* orbitals, is spherical, suggesting that the electron occupies a volume immediately surrounding the nucleus (**Figure 2.40**). As the nonbonded hydrogen atoms move closer together, their 1*s* orbitals start to overlap (Figure 2.40). At some point, this overlap is such that we can say the two orbitals are essentially a new orbital belonging to the entire molecule. This new orbital, now contained in a molecule, is referred to as a **molecular orbital**. The electrons can move anywhere in this larger orbital. An important concept in organic chemistry is that, as electrons become more delocalized (that is, as electrons have more space to move around in), they decrease in energy and are therefore more stable. [*Remember this when we get to Section 2.8 on resonance.*] The increased stability of the larger molecular orbital is what causes bonding to be energetically favorable. [*The ball rolls downhill.*] The specific head-to-head overlap of atomic orbitals makes this a sigma (σ) bond. [*We elaborate on this in Section 2.7.*]

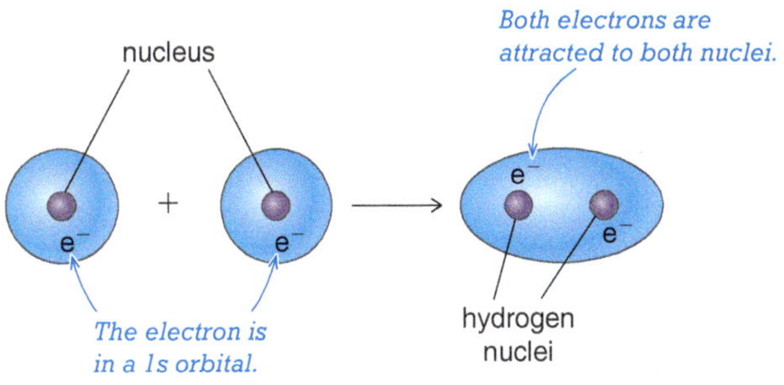

Figure 2.40 Overlap of the 1*s* orbitals of hydrogen results in a sigma (σ) bond.

A similar process occurs when forming F_2 from individual fluorine atoms (**Figure 2.41**). As a result of the electron configuration of fluorine, the F—F sigma (σ) bond is formed by the overlap of two *p* orbitals, each containing a single electron. Although the *p* orbital contains *two* lobes, it is only *one* atomic orbital. That is, each *p* orbital can form only one bond. Because the resulting molecular orbital is formed by head-to-head overlap of *p* orbitals, it is also a σ bond. [*In most organic molecules,* p *orbitals will not form bonds in this way. Section 2.6.2 shows the more common bonding of* p *orbitals.*]

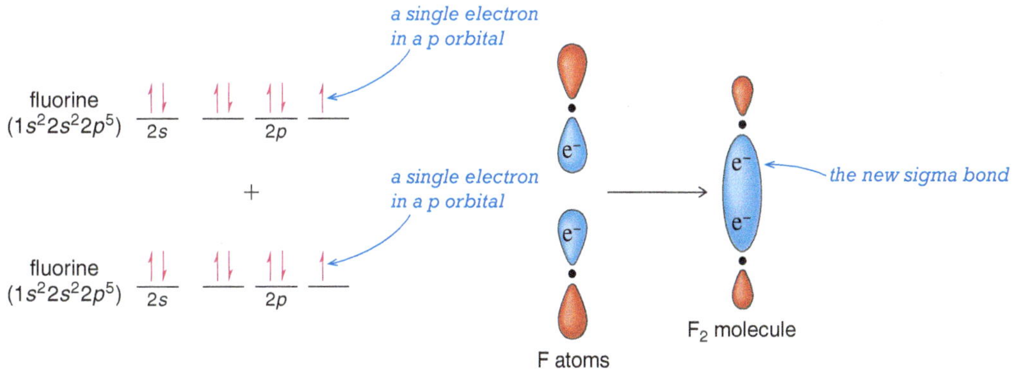

Figure 2.41 Overlap of the 2*p* orbitals of fluorine results in a sigma (σ) bond.

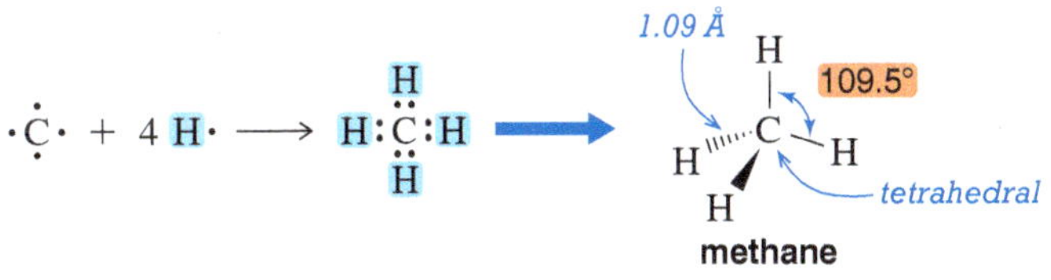

Figure 2.42 One carbon atom bonds with four hydrogen atoms to form methane.

2.6.2 Orbital hybridization (sp^3)

The Lewis structure of methane (CH_4) shows four hydrogen atoms, each sharing 1 electron with a single carbon atom. Each bond, containing 2 electrons, consists of 1 electron from carbon and 1 from hydrogen (**Figure 2.42**). Methane, prevalent in nature (it's a potent greenhouse gas), is more stable than the nonbonded C and H atoms, which are not found in nature in their monatomic forms. [*Stable* *species form from unstable species—the ball rolls downhill.*] We know from experimental data, and expect based on Section 2.5, that methane has a tetrahedral shape around the central carbon.

Whereas H_2 can be thought of as resulting from the interaction of two hydrogen atoms in their ground state electron configuration ($1s^1$), the bonding in methane isn't easily explained in this way. As such, the concept of atomic orbital hybridization is invoked to explain the observed bonding. To understand the need for hybridization, note that the tetrahedral shape indicates that the hydrogens are symmetrically disposed around the central carbon and that the H—C—H bond angle is exactly 109.5° (Figure 2.42). Equally important, the C—H bonds have identical bond lengths (1.09 Å). **Figure 2.43(a)** shows an almost nonsensical picture of forming methane from the ground state configuration of carbon ($2s^2 2p^2$) where one hydrogen shares its electron with the $2p_x$ orbital, one shares with the $2p_y$ orbital, and the other two both share a single electron with the same $2p_z$ orbital. [*This picture recognizes that each orbital can only hold 2 electrons.*] Because a single orbital does not form a bond with two different hydrogen atoms, we need to consider an excited state electron configuration [**Figure 2.43(b)**]. By exciting an electron from $2s$ to $2p_z$, a picture merges of a molecule that can form bonds by sharing 1 electron from each orbital. However, bonding in this way still does not account for the tetrahedral shape. The three p orbitals are often designated as p_x, p_y, and p_z, indicating their relative orientation to one another along the x, y, and z axes. The angles between the three p orbitals are 90°. Thus, each of the three hydrogens that bond with a p orbital would be at 90° angles to one another. In methane, however, the H—C—H bond angles are 109.5° (a perfect tetrahedron). So simply using an excited state cannot fully explain the bonding of carbon.

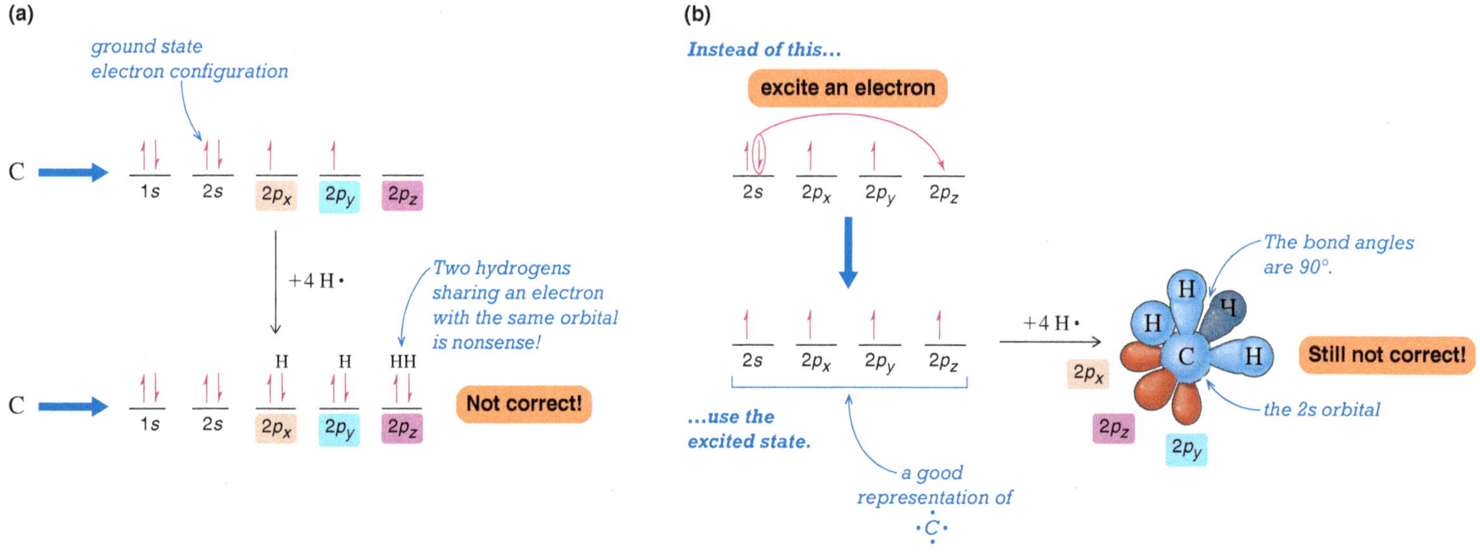

Figure 2.43 Hybridization is needed to explain the bonding of methane. (a) Bonding with the ground state of carbon is nonsensical. (b) Bonding with an excited state would give 90° bond angles instead of 109.5°.

What emerges from this analysis, considering the tetrahedral geometry at carbon, is that the four orbitals used to make the four C—H bonds must be identical. How is it possible that an s orbital and three p orbitals can be identical? The answer is hybridization. **Atomic orbital hybridization** is a human construct that is remarkably useful for understanding the shapes and properties of organic molecules. To make methane, one s orbital is hybridized (mixed/blended) with the three p orbitals. Because four orbitals went into the mixture, four hybridized orbitals must be produced. The hybridization is indicated as sp^3 to account for the number of each orbital type that went into making the hybrid [**Figure 2.44(a)**]. [*The superscript of s, if shown, would be "1," but a superscript of 1 is never shown.*]

You might notice that one lobe of the hybridized orbital is smaller than the other. This is a result of the signs of the two orbitals being hybridized [**Figure 2.44(b)**]. Recall that each lobe of a *p* orbital has an opposite sign. [*See Figure 2.8 in Section 2.2.1 for a review.*] If we assume a + sign for the *s* orbital and a different sign for each lobe of the *p* orbitals, where the + signs overlap, the effect is additive. Just as adding two positive numbers gives you a larger positive number, the lobe increases in size. Where opposite signs overlap, the lobe decreases in size because the + is cancelled by the −. [*This will make more sense when we consider molecular orbital theory.*]

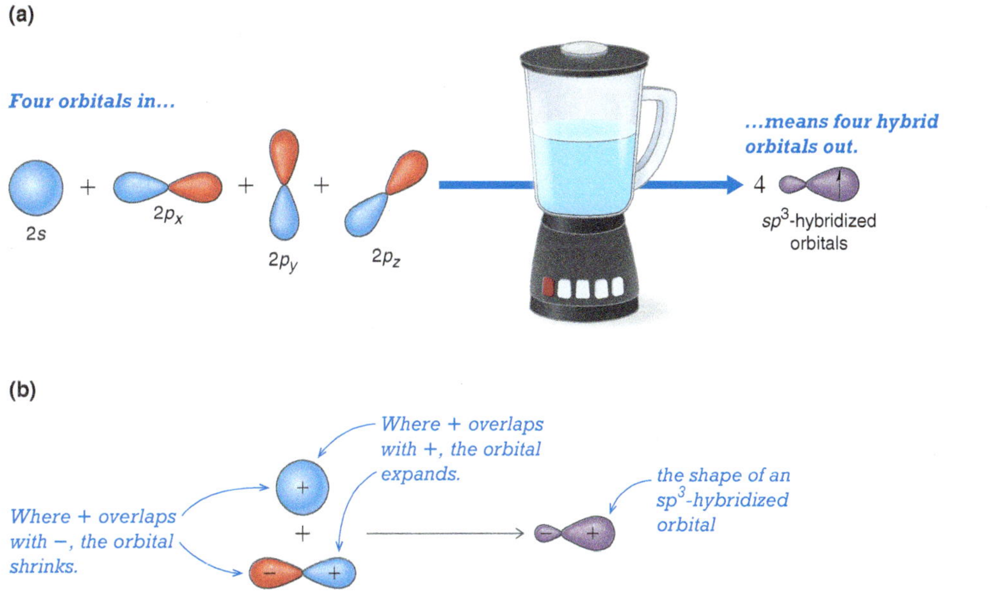

Figure 2.44 Hybridization allows carbon to form bonds to four hydrogen atoms with identical orbitals. (a) Four sp^3-hybridized orbitals are produced. (b) The shape of an sp^3-hybridized orbital.

As shown in **Figure 2.45**, the four sp^3-hybridized orbitals emerge from a central atom in a tetrahedral shape with 109.5° angles between them. The overlap of the four 1*s* orbitals of hydrogen with the four sp^3-hybridized orbitals of carbon forms the new C—H bonds and gives the observed tetrahedral shape of methane.

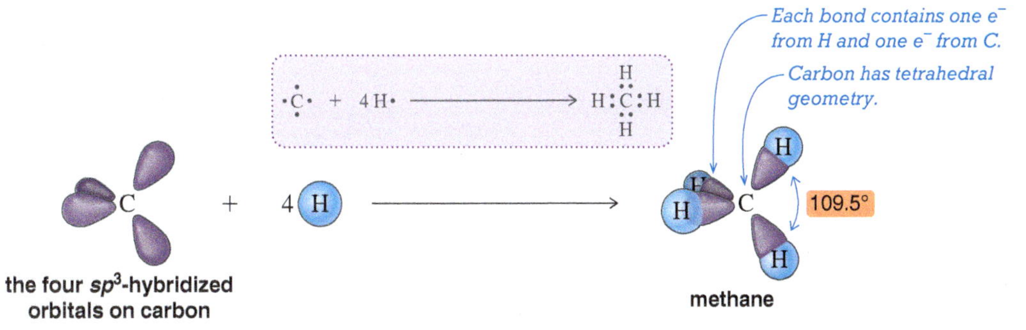

Figure 2.45 Bonding with hybridized orbitals allows us to rationalize the tetrahedral shape of methane. [*The small lobes of the hybridized orbitals have been omitted for clarity.*]

Because hybridization was invented to explain the shapes of molecules, we can now use the shape of a molecule to predict the hybridization. In this way, Table 2.2 from earlier can be modified to show the association between shape and hybridization (**Table 2.3**).

Table 2.3 Correlation of molecular shape and orbital hybridization.

Areas of electron density	Predicted shape around central atom	Average bond angle	Hybridization	Example molecule
4	Tetrahedral	109°	sp^3	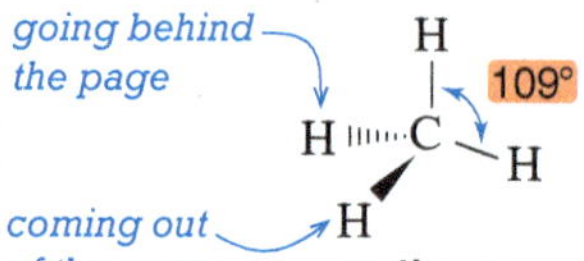 **methane**
3	Trigonal planar	120°	sp^2	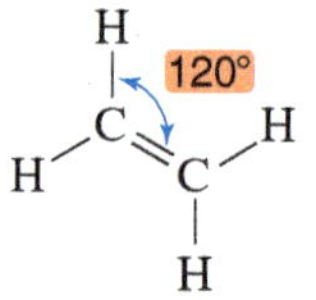 **ethene**
2	Linear	180°	sp	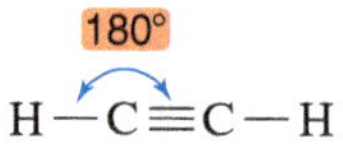 **ethyne**

2.6.3 Orbital hybridization (sp^2)

As shown in Table 2.3, the three hybridizations we focus on are sp^3, sp^2, and sp. Whereas methane uses one s orbital and three p orbitals to make four sp^3-hybridized orbitals, the central carbons in ethene are sp^2-hybridized (**Figure 2.46**). The superscript of 2 implies that only two of the three p orbitals are used to make the hybrid orbitals with the single s orbital [*which has an assumed superscript of 1*]. Once again, the number of orbitals that are used in the hybridization equals the number of hybrid orbitals produced. As such, one s orbital and two p orbitals are hybridized to make a total of three sp^2-hybridized orbitals, leaving one unhybridized p orbital. This is true for both carbons of ethene. Our assumption here, though arbitrary, is that the p_z orbital is the remaining unhybridized p orbital.

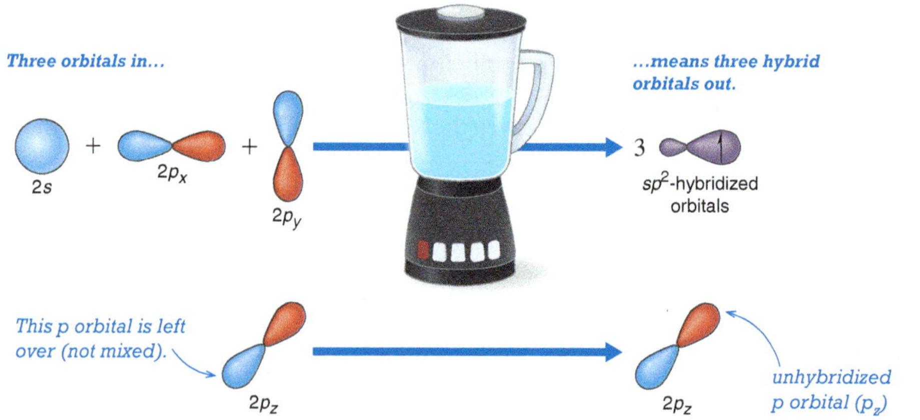

Figure 2.46 Hybridization of one *s* orbital and two *p* orbitals results in three sp^2-hybridized orbitals and one leftover p_z orbital.

Each carbon in ethene has four bonds. Three of these bonds are formed using the three sp^2-hybridized orbitals, with the other bond using the p_z orbital. When using hybridized orbitals to form bonds, the first bonds that form are between orbitals containing s character (σ bonds). Thus, the orbital picture of ethene can be drawn by first connecting the two carbons by overlapping sp^2-hybridized orbitals (**Figure 2.47**). Each of the four C—H σ bonds is then formed by overlap of the sp^2-hybridized orbitals of carbon with the 1*s* orbitals of hydrogen. At this point, there are three bonds to each carbon, and all of the hybridized orbitals have been used. [*Which orbitals haven't?*] The p_z orbitals are used to make the fourth bond to each carbon. Because each p_z orbital runs along the *z* axis going into and coming out of the plane of the page, each must be drawn as such. Moreover, both must be drawn in this way because orbitals must be aligned in order for there to be overlap (or bonding) between them. The overlap between the p_z orbitals occurs above and below (but not along) the axis of the σ bonds. [*This is different from what we saw with* F_2 *(Figure 2.41), but it is the more common bonding pattern observed in organic molecules.*] Bonds of this type, formed in organic chemistry almost exclusively by overlapping *p* orbitals, are called π bonds.

Remember this mnemonic: Orbitals with *s* character make *sigma* bonds, whereas *p* orbitals make *pi* bonds.

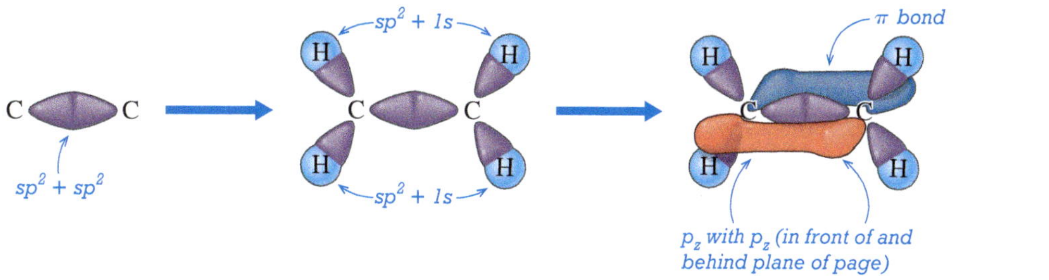

Figure 2.47 Using hybrid orbitals to draw ethene.

Which *p* orbitals you use to form a π bond depends on how the orbital picture is drawn. In **Figure 2.48(a)**, for example, the *p* orbitals are oriented along the *x* axis, so you must use the p_x orbitals for the π bond, while the p_y and p_z oritals are used for the sp^2-hybridized orbitals. Similarly, in **Figure 2.48(b)**, the p_y orbitals are used to make the π bond, in which case the p_x and p_z orbitals are used for the sp^2-hybridized orbitals. Remember: Hybridized orbitals are a human construct, and there is no difference between the molecules in Figures 2.48(a) and 2.48(b)—they can be interconverted by rotating the molecule in space. However, keeping track of which orbitals are used ensures that the incorrect picture in **Figure 2.48(c)** is never drawn. This depiction assumes that the p_y orbitals participate in both the π bond *and* the sp^2-hybridized orbitals, which cannot happen.

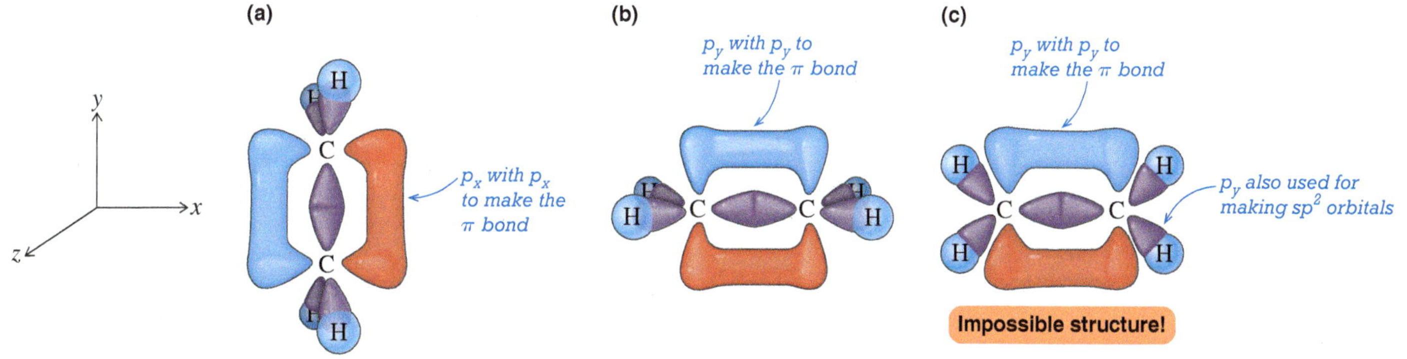

Figure 2.48 Orbital direction as a function of how a molecule is drawn. (a) Using the unhybridized p_x orbitals to make the π bond. (b) Using the p_y orbitals to make the π bond. (c) This structure is incorrect because the p_y orbitals are used for both the π bond *and* the sp^2-hybridized orbitals.

2.6.4 Hybridization (*sp*)

Each carbon atom in ethyne has two areas of electron density. As a result, the shape around these carbons is linear (Table 2.3). The linear shape implies that these carbons are *sp* hybridized. The assumed superscript of 1 implies that only one of the three *p* orbitals is used to make the hybrid orbitals with the single *s* orbital [*also an assumed superscript of 1*]. Once again, the number of orbitals that are used in the hybridization equals the number of hybrid orbitals produced. As such, **Figure 2.49** shows that one *s* orbital and one *p* orbital are hybridized to make a total of two *sp*-hybridized orbitals, leaving two unhybridized *p* orbitals. This is true for both carbons of ethyne. Our assumption here, though arbitrary, is that the p_z and p_y orbitals are the remaining unhybridized *p* orbitals.

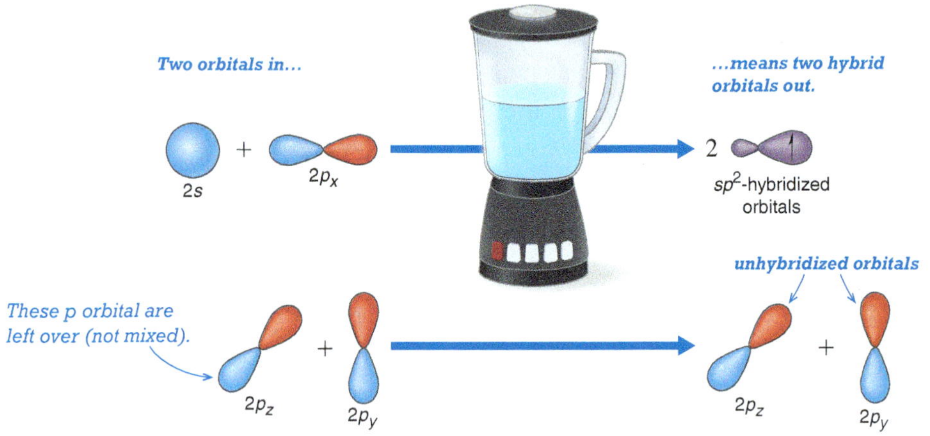

Figure 2.49 **Hybridization of one *s* orbital and one *p* orbital** results in two *sp*-hybridized orbitals and two leftover *p* orbitals (arbitrarily p_y and p_z).

Each carbon in ethyne has four bonds. Two of these bonds are formed using the two *sp*-hybridized orbitals with the other two using the p_z and p_y orbitals. Thus, the orbital picture of ethyne (**Figure 2.50**) can be drawn by first connecting the two carbons by overlapping *sp*-hybridized orbitals. The two C—H σ bonds are formed by overlap of the *sp*-hybridized orbitals of carbon with the 1*s* orbitals of hydrogen. At this point, there are two bonds to each carbon and all of the hybridized orbitals have been used. [*Which orbitals haven't?*] The *p* orbitals are used to make the other two bonds to each carbon. One of the π bonds is formed by overlapping p_y orbitals along the *y* axis [*oriented perpendicular to the σ bond in the plane of the page*]. The second π bond is formed by overlapping p_z orbitals along the *z* axis [*oriented perpendicular to the σ bond going into and coming out of the plane of the page*]. In this way, each of the carbons of ethyne is connected by two π bonds.

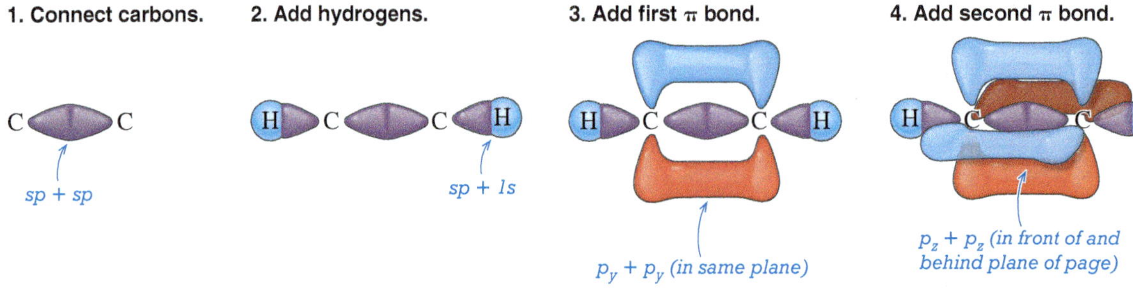

Figure 2.50 **Using hybrid orbitals to draw ethyne.**

2.6–2.6.4 SUMMARY

- Hybridization of atoms is used to explain the shape of molecules. Generally, tetrahedral atoms are sp^3 hybridized, trigonal planar atoms are sp^2 hybridized, and linear atoms are *sp* hybridized.

Assessment 2.37 Hydrogen forms bonds without hybridizing. Why?

Assessment 2.38 Would you expect the nitrogen in ammonia (NH_3) to hybridize? In which orbital would you expect to find the lone pair of electrons?

Assessment 2.39 Based on the way each structure of formaldehyde is drawn, which *p* orbitals on C and O were unhybridized?

(a) (b) (c)

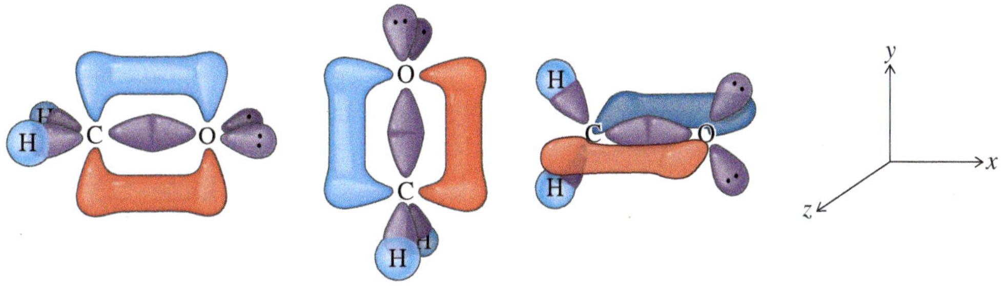

Assessment 2.40 Use hybrid orbitals to draw the following molecules.

(a) (b) (c) (d)

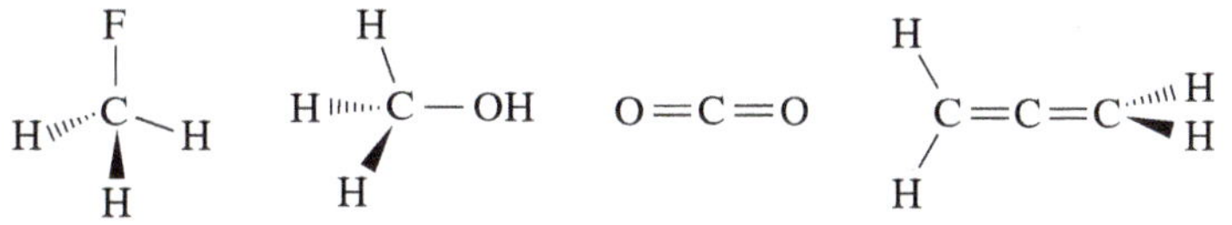

2.6.5 Relative shapes of hybrid orbitals

Sections 2.6.2–2.6.4 showed how to use sp^3-, sp^2-, and *sp*-hybridized orbitals to make the C—H σ bonds in methane, ethene, and ethyne, respectively. **Table 2.4** shows that these bond lengths vary slightly as a function of orbital hybridization. Let's think like a chemist and see how. [*You don't need to know these numbers.*]

Table 2.4 Hybridization and bond length.

Molecule	Hybridization of carbon	C—H bond length (Å)
Methane	sp^3	1.09
Ethene	sp^2	1.08
Ethyne	sp	1.06

We can think like a chemist to rationalize the bond lengths in Table 2.4.

Thinking Like a Chemist

In the same way that the flavor of a strawberry banana smoothie depends on the relative ratio of strawberries and bananas used, the three hybridized orbitals have elements of *s* and *p* orbitals, with the relative amounts of each dictating the overall shape. To understand the data in Table 2.4, our analysis must start with the shapes of the unhybridized *s* and *p* orbitals. The *s* orbital is spherical, while the *p* orbital is dumbbell shaped. The spherical *s* orbital is fatter, with electrons held closer to the nucleus than in the thinner *p* orbital. **Figure 2.51** shows the *s* and *p* orbitals overlaid to illustrate this difference. In sum, electrons in a *p* orbital are farther away from the nucleus than they are in an *s* orbital.

When *s* orbitals (shorter, fatter) and *p* orbitals (longer, thinner) are combined in different ratios, the resulting orbitals display different characteristics. As shown in **Table 2.5**, an *sp*-hybridized orbital has 50% *s* character and 50% *p* character. Moreover, an sp^2-hybridized orbital is 33% *s* and 67% *p*, and an sp^3-hybridized orbital is 25% *s* and 75% *p*. [*Make sure you understand this before continuing.*] As a result, an orbital with maximum *s* character (*sp* hybridized) should be shorter and fatter than an orbital with minimal *s* character (sp^3 hybridized). An sp^2-hybridized orbital is intermediate between the two.

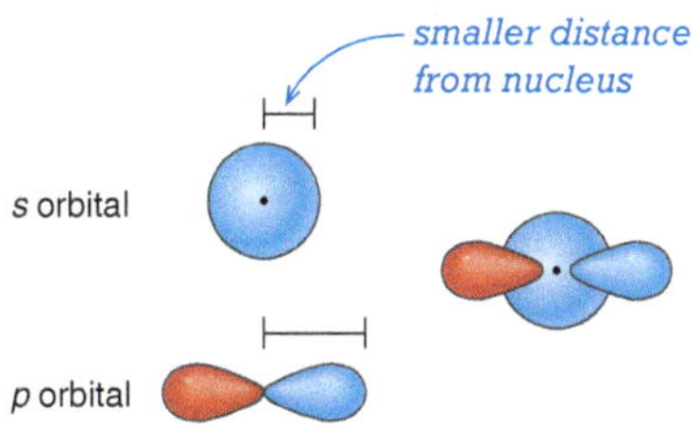

Figure 2.51 Electrons are closer to the nucleus in an *s* orbital than in a *p* orbital.

Table 2.5 Percent *s* character in hybridized orbitals.

s orbitals	*p* orbitals	Hybridization	% *s* character	% *p* character
1	1	sp	50%	50%
1	2	sp^2	33%	67%
1	3	sp^3	25%	75%

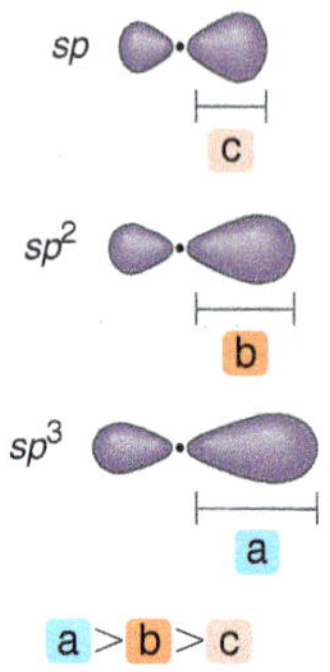

Because the C—H bond is formed by the overlap of the 1s orbital of hydrogen with a hybridized orbital, it makes sense that the bond is shorter when an sp-hybridized orbital is used instead of an sp^3-hybridized orbital. In this way, hybridization accounts for the observed values from Table 2.4 (**Figure 2.52**). ■

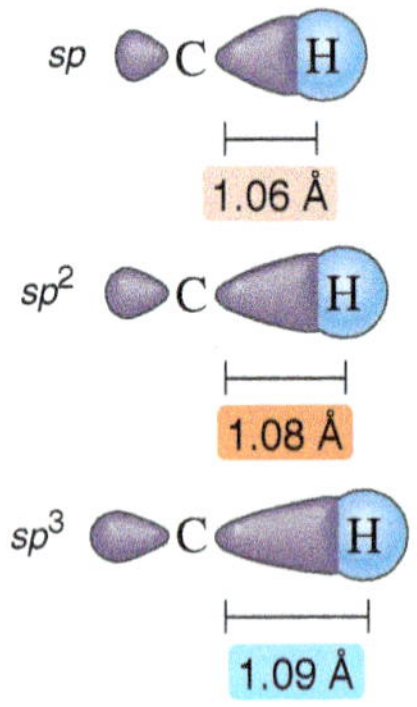

Figure 2.52 Bonds formed from orbitals containing more *s* character are shorter.

Assessment 2.41 Rank the following molecules by the length of the indicated bond from shortest to longest.

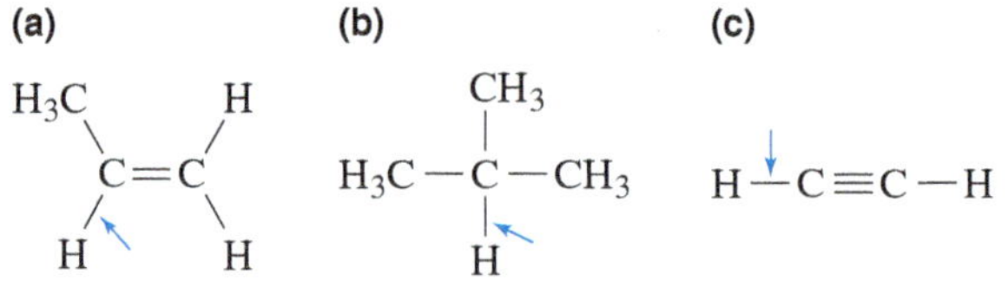

Assessment 2.42 Which would you expect to have longer C—C bonds, a carbocation or a carbanion?

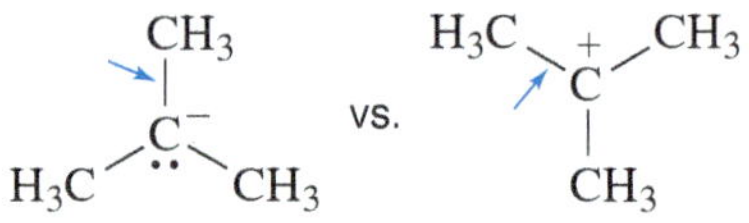

carbanion carbocation

Assessment 2.43 The sulfur and oxygen in methanethiol and methanol are both sp^3 hybridized. Why is the S—H bond longer than the O—H bond?

methanethiol methanol

2.7 Molecular orbital theory

An alternative treatment of bonding is called **molecular orbital (MO) theory**. Described as the **linear combination of atomic orbitals (LCAO)**, the language of molecular orbital theory often borrows some of the language of valence bond theory because similar concepts are involved. Molecular orbital theory is highly mathematical, but here we focus on the qualitative aspects of what the math teaches us.

The primary difference between valence bond theory and molecular orbital theory is that molecular orbital theory views the electron in orbitals that belong to the molecule, rather than a single atom. Recall from our discussion of valance bond theory that we can think of electrons as a wave and the orbital as a wave function that describes its location. The spherical and dumbbell orbital pictures of s and p orbitals, respectively, are simply 3-D visualizations of waves. [*Recall Figure 2.6 for an example of this.*] As shown in **Figure 2.53(a)**, imagine a wave where the location of the nucleus is positioned where the amplitude of the wave is greatest. If this wave is allowed to carve out a path over 360° out of and into the page, a spherical s orbital results. Given that the electron density in this s orbital is close to the nucleus, the s orbital is most stable. Alternatively, imagine a wave in which the location of the nucleus is positioned where the wave crosses zero

[Figure 2.53(b)]. If this wave is allowed to carve out a path over 360° into and out of the page, the dumbbell-shaped *p* orbital results. The node at the nucleus, indicating an area where there is no electron density, explains why the *p* orbital is less stable than an *s* orbital and is therefore filled second. [*We learned these shapes in Section 2.2.1 and in previous chemistry courses, but now we see that they are themselves derived from wave equations.*]

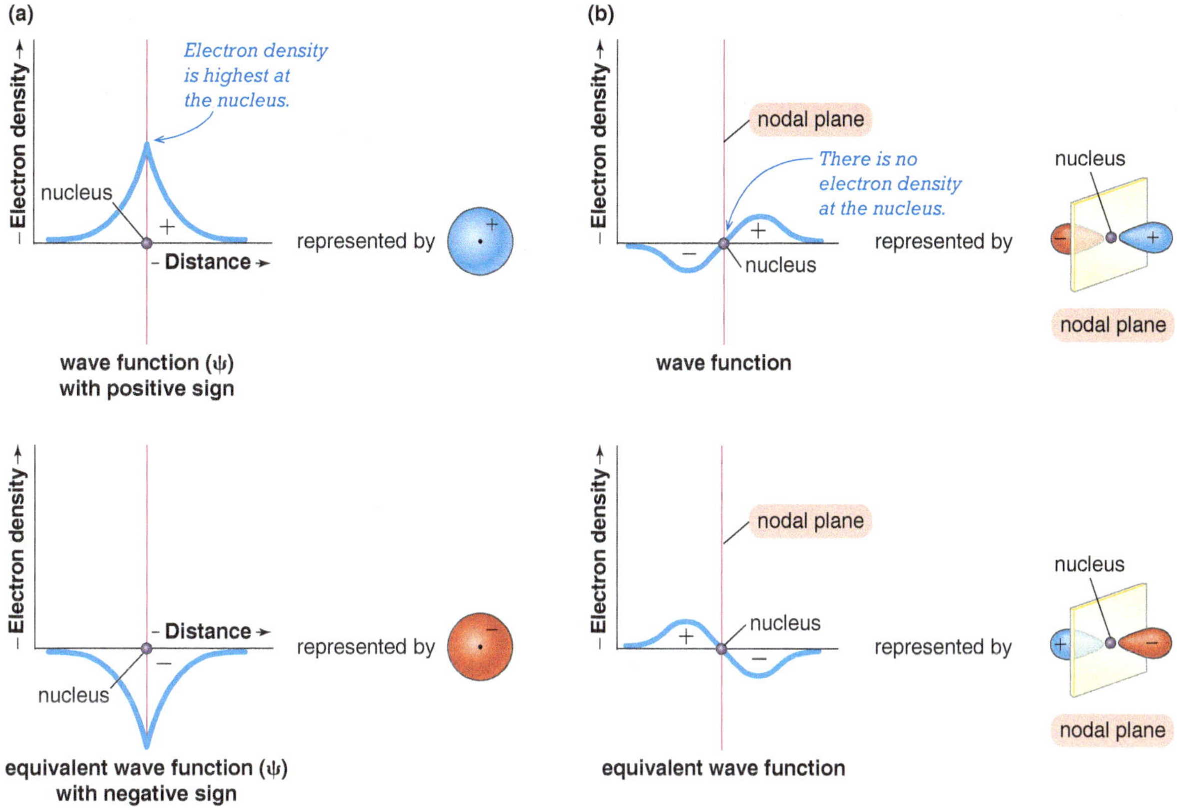

Figure 2.53 The shape of the orbital is a 3-D representation of the wave function. (a) Spherical *s* orbital; (b) dumbbell-shaped *p* orbital.

By recognizing the atomic orbitals as wave functions, it becomes apparent that bonding to form molecular orbitals is simply a result of combining these atomic orbitals. Atomic orbitals always combine in two, and only two, ways: additive (constructive) and subtractive (destructive). Figure 2.54 shows how the 1*s* orbitals of hydrogen combine to make hydrogen gas (H_2). The constructive interaction occurs when the waves are in phase, having the same sign.

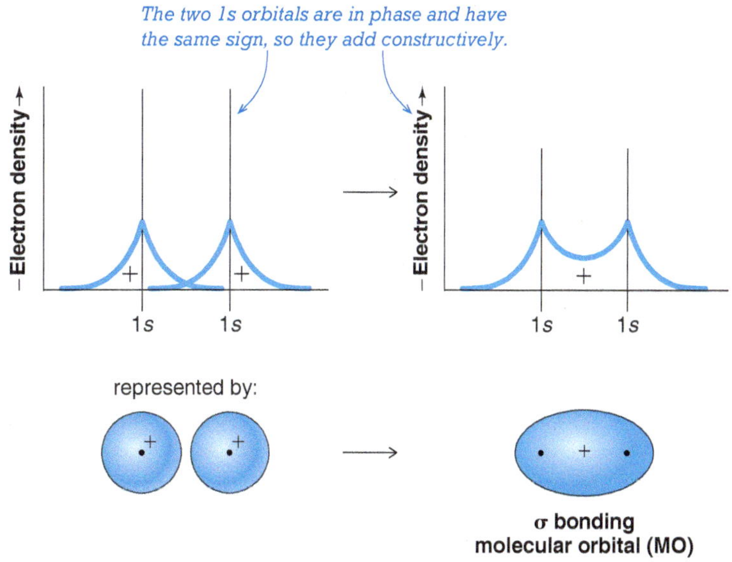

Figure 2.54 Constructive interference between two *s* atomic orbitals leads to a σ bonding molecular orbital.

This results in an expansion of the wave with overlap between the two nuclei. If this resulting molecular orbital wave is allowed to carve out a path over 360° into and out of the page, a molecular orbital results that encompasses both nuclei. Because the electrons can interact closely with both nuclei, this molecular orbital is more stable than the individual atomic orbitals. Recall from Section 2.6.1 that this head-to-head overlap is referred to as the sigma (σ) orbital.

Alternatively, the atomic orbitals can combine via a destructive interaction (**Figure 2.55**). In this case, the waves are out of phase, having opposite signs where they overlap. In the same way that a positive number (for example, +3) cancels the same negative number (for example, −3), the wave function has a value of zero where the two atomic orbitals overlap. In this space, there is no electron density—a situation that we call a node. If this resulting molecular orbital wave is allowed to carve out a path over 360° into and out of the page, a molecular orbital with a node between the two nuclei results. The electrons in this orbital are unable to interact with both nuclei, so this is a less stable molecular orbital. This is what is referred to as the σ^* ("sigma star") orbital.

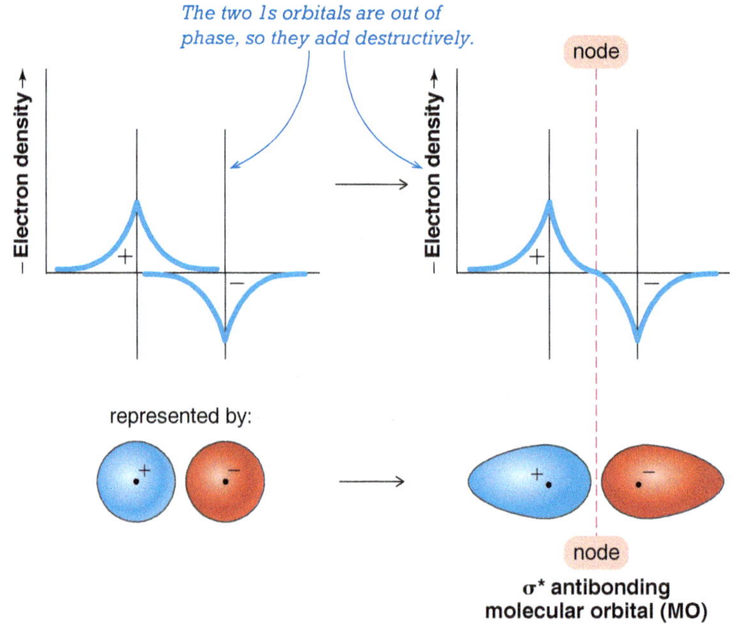

Figure 2.55 Destructive interference between two *s* atomic orbitals gives a σ^* molecular orbital.

The molecular orbital diagram in **Figure 2.56** shows the relative energies of these results, which helps explain why bonding is favorable. [*The ball rolls downhill.*] The combination of the two *s* orbitals happens in two ways to make a σ molecular orbital and a σ^* molecular

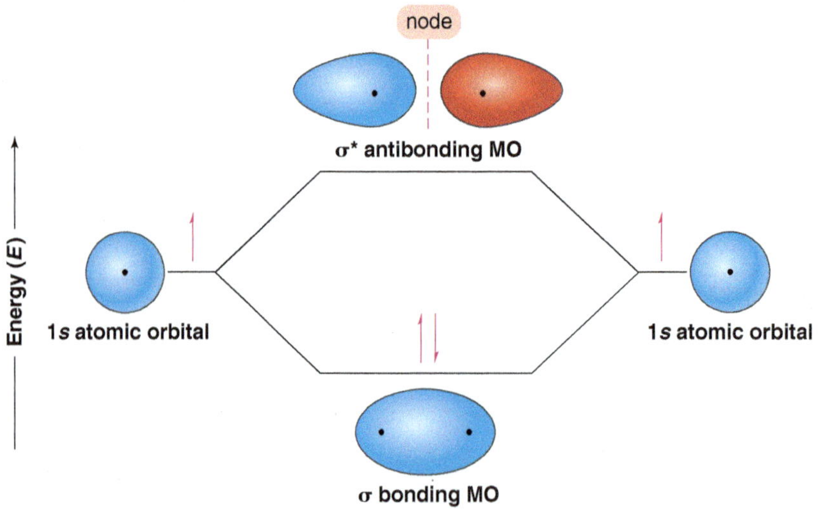

Figure 2.56 The delocalized σ molecular orbital is more stable and is filled first. It is the bonding orbital.

orbital. Each molecular orbital can hold 2 electrons, and each hydrogen atom contributes 1 electron. In the same way that atomic orbitals are filled, starting with the lowest energy orbitals [*remember aufbau?*], electrons are placed in the most stable molecular orbitals first. As a result, we end up with 2 electrons in the σ orbital, making this a filled (occupied) bonding orbital. The σ* antibonding molecular orbital, on the other hand, contains zero electrons (it is unoccupied). In this picture, favorable bonding occurs because the nonbonded hydrogen atoms each have a single electron in a 1*s* atomic orbital that is higher in energy than the σ bonding orbital. Moreover, the σ orbital has maximum overlap, giving the electrons lots of space to move around the two positively charged nuclei. That is to say, the electrons in the σ orbital are delocalized. [*We explain in later sections how enhanced overlap and delocalization result in enhanced stability.*]

Similar to *s* orbitals, *p* orbitals can combine constructively and destructively. [*While it is true that they can combine end to end, in organic chemistry, we are much more interested in their side-to-side overlap to form π bonds.*] The constructive overlap of *p* atomic orbitals is represented by alignment of lobes of like sign, as shown in **Figure 2.57**. Where like signs overlap, the size of the orbital expands, allowing electron density to delocalize in this space. The + lobes become more positive, and the – lobes become more negative. And finally, at the nucleus, there continues to be a value of zero (a node). The constructive interaction represents the π bonding orbital, with electron density located above and below a hypothetical axis connecting the two nuclei. Alternatively, destructive interference results when lobes of opposite sign overlap. In these regions, the values cancel each other out, creating a value of zero wherever overlap occurs. Thus, the second molecular orbital produced via the overlap of *p* orbitals contains a node between the two atoms. This orbital is referred to as π* ("pi star"), an antibonding orbital. The π bonding molecular orbital has greater overlap (better delocalization), so it is more stable (lower in energy) than the π* antibonding molecular orbital.

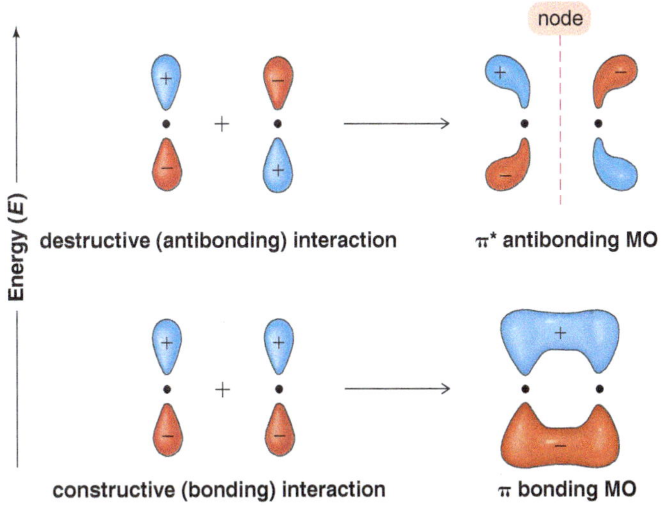

Figure 2.57 The formation of π molecular orbitals by constructive (bonding) interactions and π* molecular orbitals by destructive (antibonding) interactions between *p* atomic orbitals.

Based on the relative stability of the π and π* molecular orbitals, we can now fill the orbitals with the electrons that occupied the original *p* atomic orbitals (**Figure 2.58**). Each *p* orbital contained 1 electron, so we place both electrons into the π molecular orbital, leaving the π* antibonding orbital empty. This picture explains π bonding because the nonbonded *p* atomic orbitals each hold a single electron that is higher in energy than the π bonding molecular orbital. [*Important clarification: Nonbonded* p *orbitals are atomic orbitals and should not be confused with the π* antibonding molecular orbital.*]

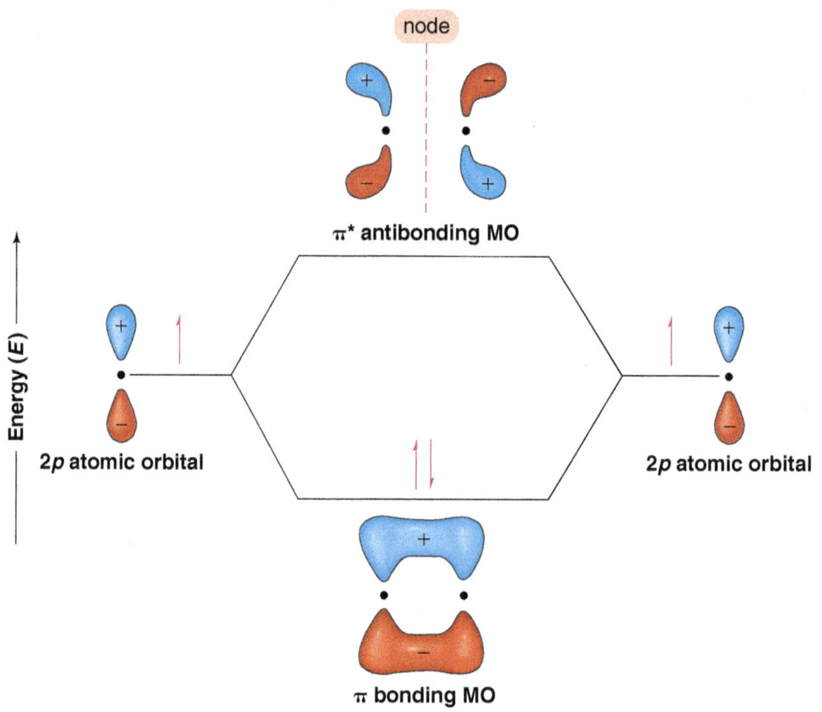

Figure 2.58 The delocalized π molecular orbital has the lowest energy, is the most stable, and is filled first. It is the bonding molecular orbital.

2.7 SUMMARY

- Molecular orbitals form by the constructive (σ and π, bonding) and destructive (σ* and π*, antibonding) combination of atomic orbitals (*s*, sp^3, sp^2, *sp*, and *p*).

2-E GENERAL CHEMISTRY IN CONTEXT: Organic Chemistry

Organic dyes

The dyes that give color to our clothes are organic molecules. Indigo (**Figure 2.59**) is a naturally occurring dye that is used for coloring blue jeans.

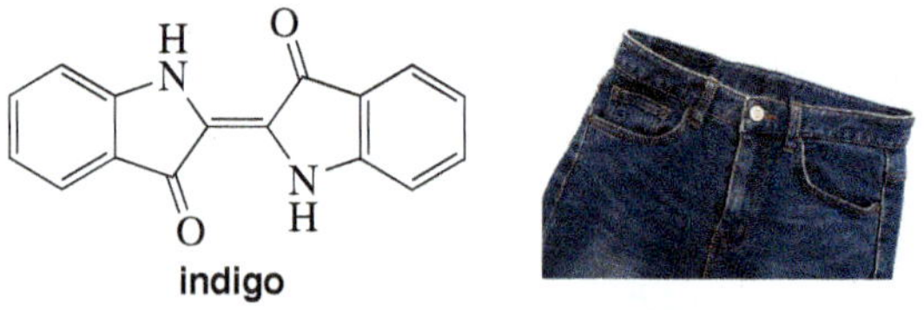

Figure 2.59 Indigo is a naturally occurring dye that gives color to blue jeans.

While most organic molecules are colorless, indigo and other dyes are colored because of their molecular orbitals. Specifically, in molecules that are conjugated [*we explain what this means in Chapter 21*], the π and π* molecular orbitals are close in energy. As a result, absorption of specific wavelengths of visible light causes electrons to jump from π (the ground state) to π* (the excited state), before relaxing back to π again (**Figure 2.60**). In the case of indigo, this absorption of a specific wavelength of visible light results in the conjugated molecule appearing blue.

This same principle explains why carrots are orange, why tomatoes are red, and how sunscreen prevents us from getting skin cancer.

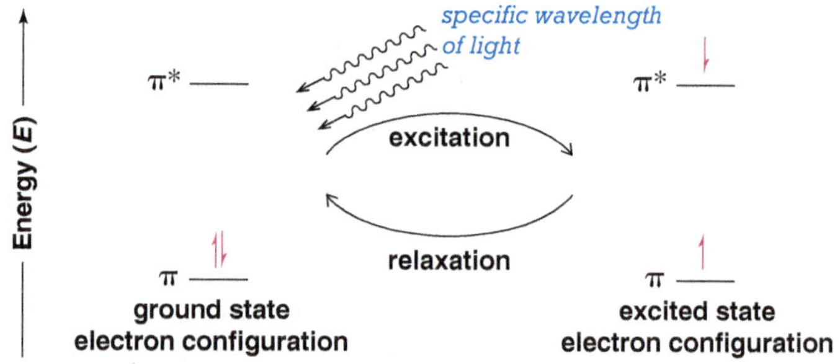

Figure 2.60 Absorption of visible light excites an electron and makes conjugated molecules appear colored.

2.7.1 Hybrid orbitals in molecular orbital theory

In molecular orbital theory, two hybridized atomic orbitals combine in the same way that any other two atomic orbitals do—namely, constructively and destructively. Constructive overlap, as shown in **Figure 2.61** for an sp^3-hybridized orbital, results when lobes of like sign overlap. In this bonding molecular orbital, the electron density is aligned along an axis connecting the two nuclei. Hence, these are σ bonding orbitals. When the sp^3-hybridized orbitals combine via destructive overlap, a node is formed between the two nuclei, making a σ* antibonding molecular orbital. Because the σ orbital has more overlap, and better delocalizes the electron density between the two nuclei, it is more stable than the σ* orbital. Each sp^3-hybridized orbital contained a single electron, so filling the newly formed molecular orbitals, once more, places the 2 electrons in the lowest energy σ orbital first. Regardless of which hybridized orbitals are being used, the picture in Figure 2.61 will look the same.

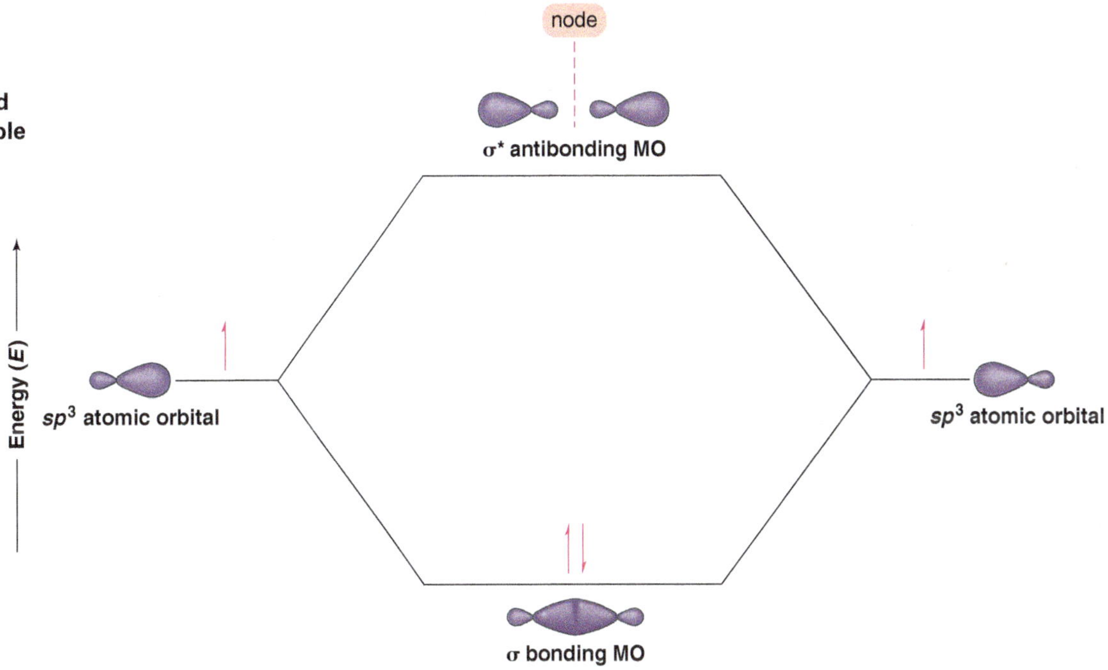

Figure 2.61 Formation of σ and σ* molecular orbitals via the constructive and destructive interaction between sp^3-hybridized orbitals. The σ orbital is more stable and is filled first, making it the bonding molecular orbital.

Using what we have just learned about molecular orbitals, we can draw the molecular orbital picture of any molecule. [*We basically did this in Sections 2.6.2–2.6.4.*] Methane [**Figure 2.62(a)**] is essentially the combination of the sp^3-hybridized orbitals of carbon with the 1*s* orbitals of hydrogen to form the four C—H σ bonds. Similarly, the structure of ethene [**Figure 2.62(b)**] contains four C—H σ bonds, one C—C σ bond, and one C—C π bond. Finally, ethyne [**Figure 2.62(c)**] contains two C—H σ bonds, one C—C σ bond, and two orthogonal (perpendicular) C—C π bonds.

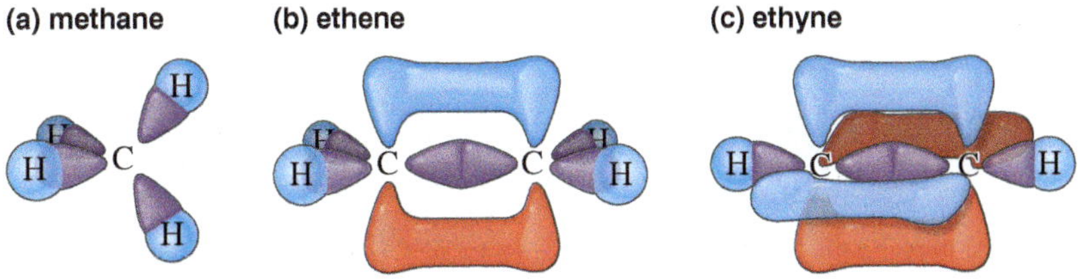

Figure 2.62 Molecular orbital pictures of (a) methane, (b) ethene, and (c) ethyne.

Only the σ and π bonds were shown in Figure 2.62 because these are primarily used to explain the bonding. That is, the antibonding orbitals (σ* and π*) are often left off of the molecular orbital picture. When we study reactivity, however, it will be important to know where these antibonding orbitals are. **Figure 2.63(a)** shows that the σ orbital is between the bonded atoms, whereas σ* is on the outside of the bond, at a 180° angle to the σ bond. This makes sense when you consider that the antibonding orbital forms by destructive interaction. **Figure 2.63(b)** shows, similarly, that the π bond is between the bonded atoms, whereas the four lobes of π* are located on the outside of the bonded atoms in the same plane. Once more, this is consistent with the destructive interaction that forms π*. [*In Chapter 12, we explain how the location of σ* dictates the reactivity of haloalkanes.*]

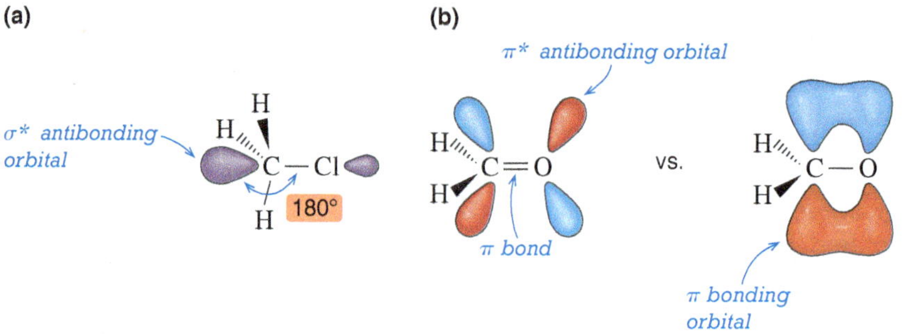

Figure 2.63 Location of the antibonding orbitals. (a) The σ* antibonding orbital of the C—Cl bond in chloromethane. (b) The π* antibonding orbital of formaldehyde versus its π bonding orbital.

Assessment 2.44 A molecular orbital diagram is shown for the C—Cl bond in chloromethane. If two more electrons were added to chloromethane, where would the electrons go?

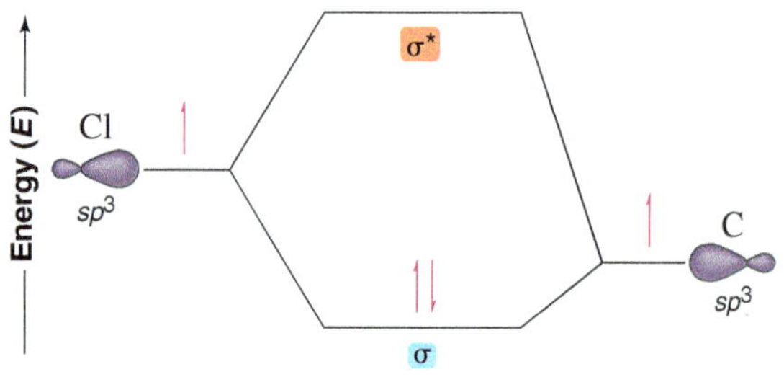

Assessment 2.45 How might electrons be excited from π to π*, based on the molecular orbital diagram shown? [*This will be relevant in Chapter 21.*]

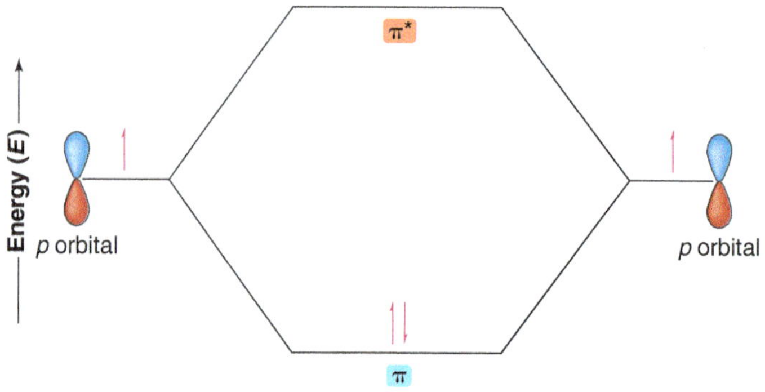

2.7.2 Molecular orbital pictures of charged molecules

Much of the reactivity we will study in this course depends on charged molecules. [*Rule 2: Negatives attack positives.*] Thus, we also need to know how to draw the molecular orbital picture of charged molecules. The Lewis structure and molecular orbital picture of the methyl carbocation (CH_3^+) are shown in **Figure 2.64**. [*What hybridization would you expect for the central carbon given that it has only three bonds and zero lone pairs?*] According to the Lewis structure, the central carbon is sp^2 hybridized and has a formal +1 charge. As a result, the three C—H σ bonds are formed using the three sp^2-hybridized orbitals of carbon. The only orbital not used is the *p* orbital. There are no electrons on the central carbon, so the *p* orbital on the central carbon must be empty. [*Carbocations feature prominently in Chapters 8 and 12, among many others.*]

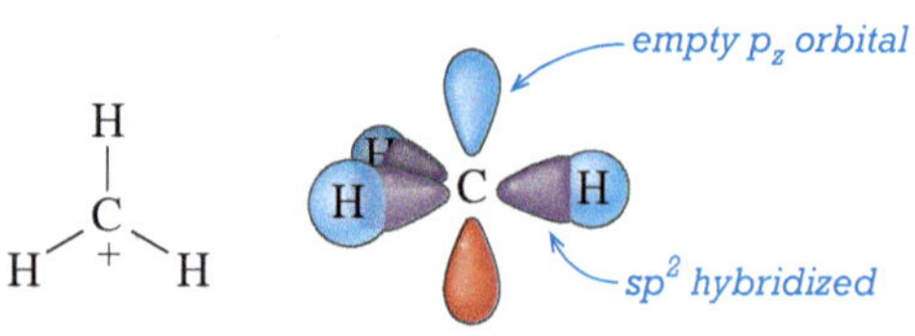

Figure 2.64 Lewis structure and molecular orbital picture of the methyl carbocation.

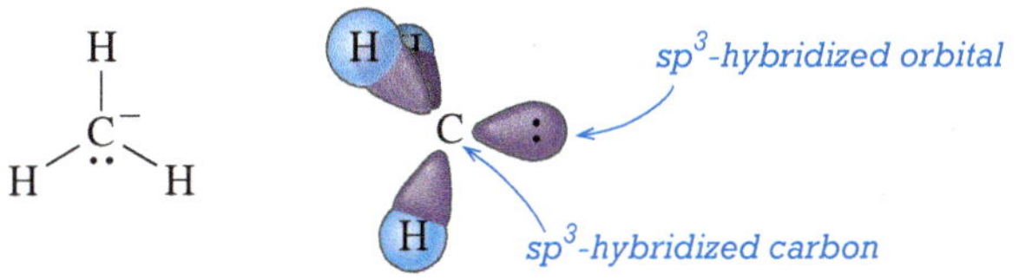

Figure 2.65 Lewis structure and molecular orbital picture of the methyl carbanion.

Figure 2.65 shows the Lewis structure and the molecular orbital picture of the methyl carbanion (CH_3^-). According to the Lewis structure, the carbon atom has four areas of electron density (three σ bonds and one lone pair) with a formal −1 charge, so it is sp^3 hybridized. The three C—H σ bonds are formed using three sp^3-hybridized orbitals of carbon. The remaining sp^3-hybridized orbital houses the lone pair of electrons. By adopting the tetrahedral shape, the four areas of electron density are kept as far apart as possible.

Though it may be getting ahead of ourselves, how do you think a methyl carbocation and a methyl carbanion would react if they came into contact with each other? [*Look back at Figures 2.64 and 2.65.*] The carbocation is electron deficient with a +1 charge, whereas the carbanion is electron rich with a −1 charge. Neutral molecules are more stable than charged molecules, so both of these molecules are unstable. Moreover, negatives attack positives, so the negatively charged carbanion would "attack" the positively charged carbocation. This can be illustrated (see **Figure 2.66**) with a curved arrow that starts at the negatively charged carbon and points toward the positively charged carbon. This **arrow-pushing formalism** is elaborated upon throughout the text [*and you don't need to know it just yet*]. The lone pair in the sp^3-hybridized orbital of the carbanion is donated to the empty p orbital of the carbocation, which rehybridizes to a neutral sp^3-hybridized carbon. The resulting molecule is the neutral (more stable) molecule ethane, in which both carbons are sp^3 hybridized. [*This idea of an unstable negatively charged molecule reacting with an unstable positively charged molecule to make a stable neutral molecule is a basic tenet of organic chemistry. If it makes sense to you, then you'll have no difficulty in this course. If it doesn't right now, no worries—we are just getting started.*]

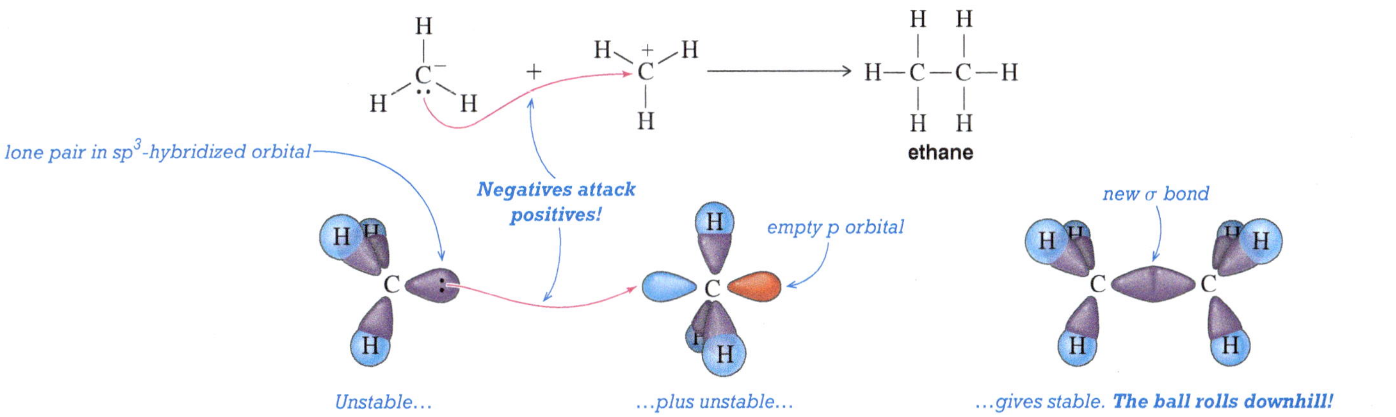

Figure 2.66 Molecular orbital pictures help us understand reactivity. The methyl carbanion (electron rich) attacks the methyl carbocation (electron poor) to make ethane.

2.8 Resonance

Back in Section 2.4.4 (see Figure 2.35), we drew the Lewis structure of another charged molecule, the nitrate ion (NO_3^-). The average bond lengths for the N—O and N=O bonds are 1.4 Å and 1.2 Å, respectively. [*Double bonds are always shorter than single bonds.*] X-ray crystallography, however, has shown that all three bonds in the nitrate ion have identical bond lengths (**Figure 2.67**). Furthermore, mapping the charges in the nitrate ion shows that the central nitrogen has a +1 charge while each oxygen has a −2/3 charge. Neither of these two facts is consistent with the Lewis structure shown in Figure 2.67.

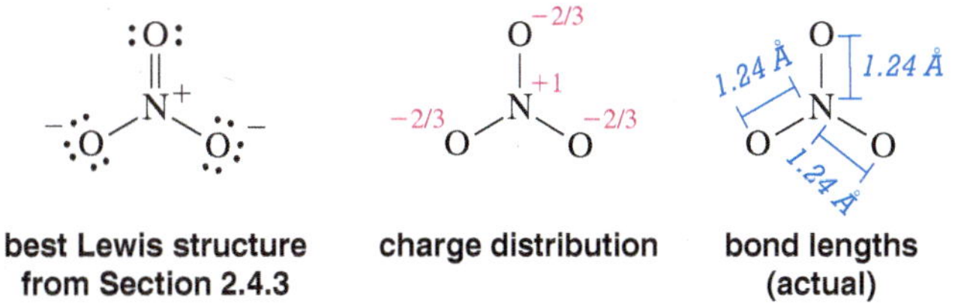

Figure 2.67 The best Lewis structure of the nitrate ion (NO_3^-) doesn't rationalize its properties.

Let's think like a chemist to figure out what's going on here.

Thinking Like a Chemist

Back when the Lewis structure of the nitrate ion was first being drawn, we had to choose which oxygen would share its lone pair with the electron-deficient nitrogen. We arbitrarily chose the one labeled as 1 in **Figure 2.68**. What if we had chosen oxygen 2 or 3? The structures that result from each of those possibilities, represented using our arrow-pushing formalism, are shown in Figure 2.68, too. [*Do any of these structures match the data from Figure 2.67?*]

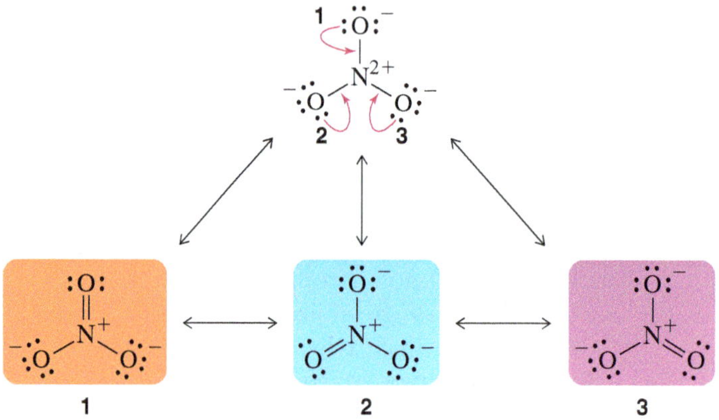

Figure 2.68 Other representations of the nitrate ion use a different oxygen to share the electron pair but are in essence the same.

The Lewis structures in Figure 2.68 follow the widely accepted rules for drawing them, but they don't match the properties of the nitrate ion shown in Figure 2.67. Recall that each oxygen has a −2/3 charge, and each bond distance is 1.24 Å, which is intermediate between the distance for an N—O single bond (1.4 Å) and an N═O double bond (1.2 Å). This is exactly the type of problem contemporaries of Lewis might have encountered using the representations that bear his name. [*How cool is it that we can find ourselves in the same situation as the greatest chemists of all time? This is one of the reasons we think like a chemist.*] It turns out that this is a limitation of Lewis structures, but it is one we can easily address. The three best Lewis structures in **Figure 2.69**, when averaged, perfectly explain the properties. These structures, referred to as **resonance structures**, differ only in the location of the π electrons (as opposed to the σ electrons—this is clarified in Section 2.8.5). Combined, they make up the **resonance hybrid** that rationalizes the properties of the molecule. In two of the three structures, the oxygens labeled as a, b, and c have a −1 charge. On average, then, each oxygen has a −2/3 charge. Similarly, each N—O bond is a single bond in two of the three structures, accounting for the fact that each bond is the same length and intermediate between a single and double bond. ■

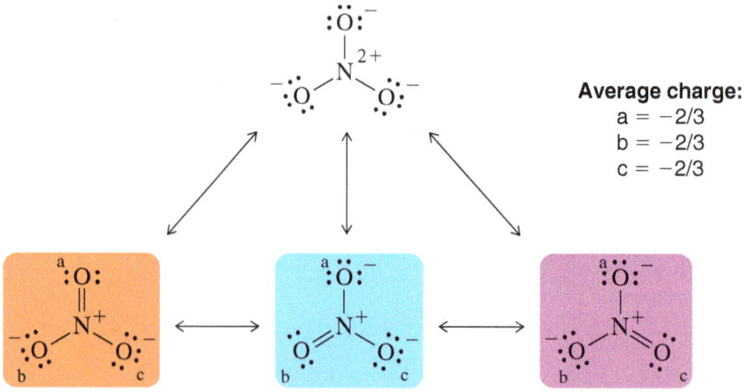

Figure 2.69 The actual structure is the weighted average of the resonance structures.

The concept of resonance is used throughout the course to explain the properties and reactivity of a variety of molecule types. There are three important aspects of resonance structures, which are discussed in detail in Sections 2.8.1–2.8.3. First, we discuss the electron-pushing formalism used to illustrate the flow of electrons in resonance structures [*it is used later to describe chemical reactions*]. Second, rules are given that explain when a molecule can be described by more than one valid resonance structure. And third, for molecules for which resonance structures can be drawn, we discuss how to determine which resonance structure contributes most to the actual structure.

2.8.1 Arrow-pushing formalism

When drawing the Lewis structure for the nitrate ion (NO_3^-), we reached a point where there was a negative charge on all three oxygens and the central nitrogen was 2 electrons short of an octet. Nevertheless, there were no more valence electrons to place. And so, as we learned in general chemistry, we randomly chose an oxygen, erased one of the lone electron pairs, and made one of the N—O single bonds a double bond [**Figure 2.70(a)**]. In doing so, we actually drew a new resonance structure. As organic chemists concerned with the flow of electrons, we instead use the **arrow-pushing formalism** (**electron-pushing formalism**) as a way to generate, and justify the existence of, new resonance structures [**Figure 2.70(b)**]. This formalism assumes that the lone pair of electrons moves to create the new N—O π bond. Placing the tail of the arrow at the lone pair of electrons, we draw the arrow with a slight curve pointing to the bond that will become a new double bond. [*Arrows always, always, always start at a pair of electrons, either as a lone pair or a π bond. Why? Because that's the formalism.*] We describe this process as "The lone pair collapses to the nitrogen to form a double bond." [*This is* not *a chemical reaction, so it isn't really forming a new bond—I'll explain later.*]

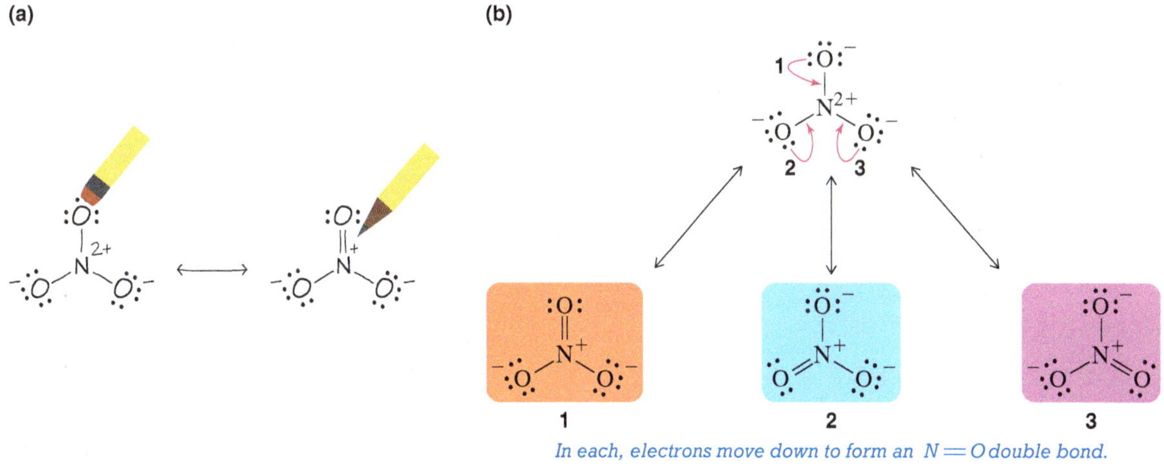

Figure 2.70 Illustrating movement of electrons. (a) In introductory chemistry, we erased a lone pair and drew a double bond. (b) In organic, we'll "push" electrons to show their movement.

The formation of resonance structures can always be reversed. That is, if an arrow can be used to explain the formation of the three best Lewis structures of nitrate from the intermediate structure (Figure 2.70), then there must be a single arrow that forms the intermediate structure from the three others. This is demonstrated in **Figure 2.71(a)** with a curved arrow that begins in the middle of the N=O bond and points toward the oxygen, indicating the movement of the π electrons onto the oxygen. We describe this process as "The π electrons move onto the oxygen." Using multiple arrows, resonance structures can be rationalized as arising from each other [**Figure 2.71(b)**]. By starting at the oxygen lone pair [*always start at a pair of electrons*], the first curved arrow points to the bond between nitrogen and oxygen, indicating formation of the new N=O double bond. If we stopped here, nitrogen would now have 10 electrons, violating the octet rule. Thus, the electrons in the already-present N=O π bond get out of the way, themselves moving onto the oxygen of the former N=O bond. This is indicated by a curved arrow that starts at the middle of the π bond and finishes at the oxygen. We describe this process as "The lone pair collapses to the nitrogen, pushing out the π electrons and forming a new double bond."

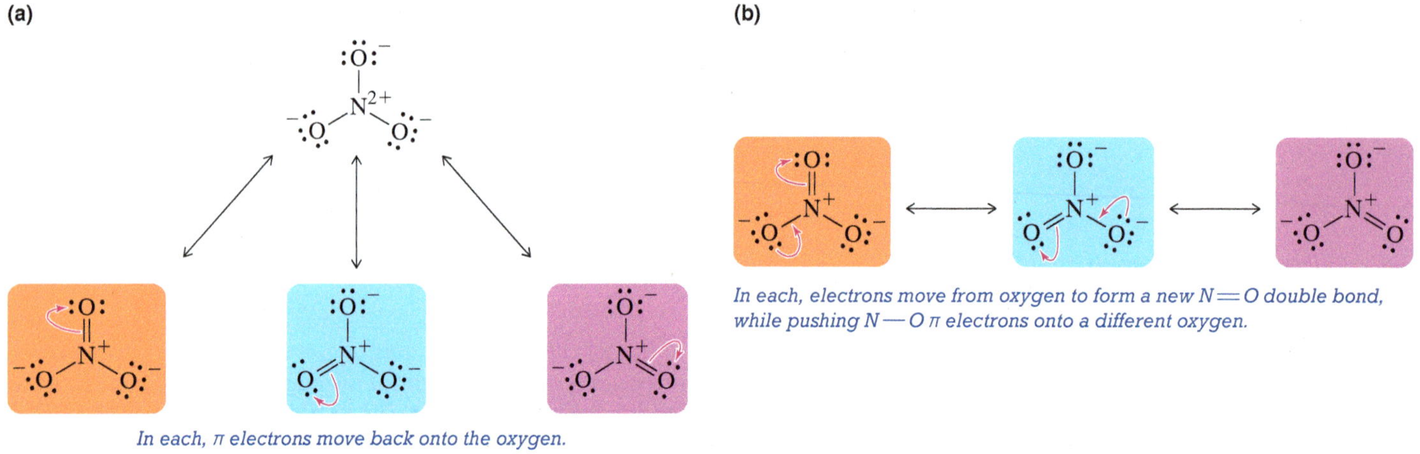

Figure 2.71 More arrow pushing in the nitrate resonance structure. (a) Pushing the N═O π electrons onto the oxygen. [*This is the reverse of Figure 2.72.*] (b) Multiple arrows to interconvert between resonance structures.

Assessment 2.46 Each pair of structures represents two valid resonance structures. Use the arrow-pushing formalism to justify the formation of the one on the right from the one on the left.

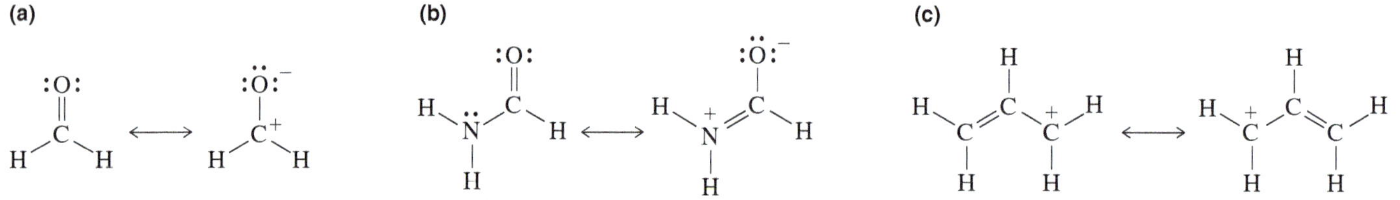

Assessment 2.47 Each pair of structures represents two valid resonance structures. Use the arrow-pushing formalism to justify the formation of the one on the left from the one on the right. [*This is the reverse of what we did in Assessment 2.46.*]

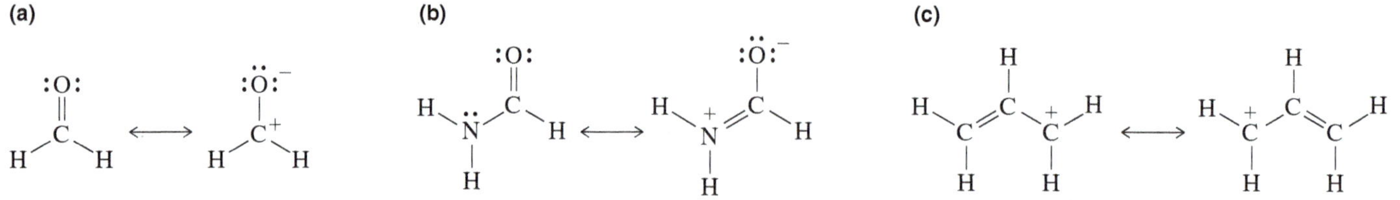

Assessment 2.48 Draw the resonance structure that would result from the indicated movement of electrons.

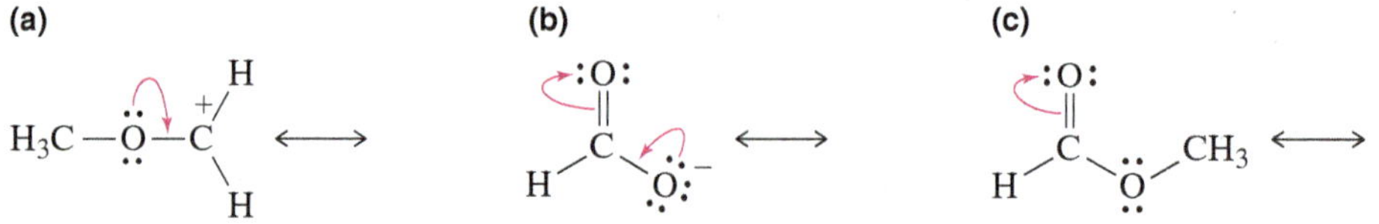

2.8.2 Valid resonance structures

Assessment 2.46 used the term "valid resonance structures." Every molecule has one Lewis structure that can be drawn, so the term "valid," as used here, refers to whether we are allowed to draw a second resonance structure. There are a few rules that need to be followed in order to determine whether or not a second resonance structure can be drawn. If these rules are followed in drawing a new structure, then what has been drawn is considered to be a valid resonance structure. To draw valid resonance structures, you must:

1. Have the same number of paired electrons from one structure to the next,
2. Have the same number of σ bonds (you can't break or form new σ bonds), and
3. Only move electrons in *p* orbitals or π bonds.
4. The two resonance structures will always have the same net charge.

This list of rules might not make much sense at first. Nor may the next set of rules in Table 2.6. But when you apply them to the figures, or to the assessments you work, they should begin to make much more sense.

Another way to determine the validity of resonance structures is to think about the process of going from one to the other. This is also useful in predicting what "new" resonance structures can be drawn. Using the terminology from our electron-pushing formalism, we can look for pushable electron pairs and places to push them. When we have a pushable pair that also has a place to be pushed, then a new, valid resonance structure can be drawn. **Table 2.6** lists what qualifies as each. As we explain in Section 2.8.5, the electrons that can be moved are, in every case, in a *p* orbital or are themselves part of a π bond. In other words, resonance is specifically the movement of electrons in the π system of the molecule. The σ bonded framework never changes.

Table 2.6 Rules for drawing new resonance structures.

To draw a new resonance structure, you must have:	Examples
1. Pushable pairs	Lone pair of electrons, π bonds
2. A place to push them	Empty *p* orbital, electronegative atom, atom with a pushable pair

The use of these rules is illustrated in drawing the resonance structures in **Table 2.7**. Consider the structures on the left and, after identifying the pushable pairs and the places to push them, draw the resonance structures on the right. [*Try them on your own before looking at the answers.*]

Table 2.7 Drawing valid resonance structures using the rules in Table 2.6.

Entry	Original Lewis structure	Analysis (orange = pushable pair; blue = place to push electrons)
1	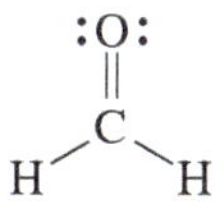	The C—O π electrons are a pushable pair. We have a choice to push them onto carbon or oxygen. Because oxygen is more electronegative than carbon, the oxygen is the best place to push them. A lone pair of electrons on oxygen also represents a pushable pair, but there is no place to push them, so no additional valid resonance structures can be drawn using these.
		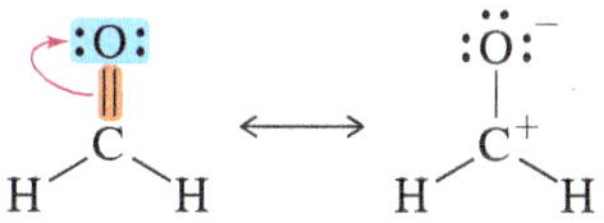
2	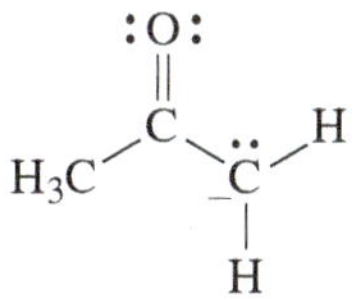	The lone pair on the negatively charged carbon is a pushable pair. The neighboring carbon is a place to push them because the π electrons attached to that neighboring carbon can be moved at the same time.
		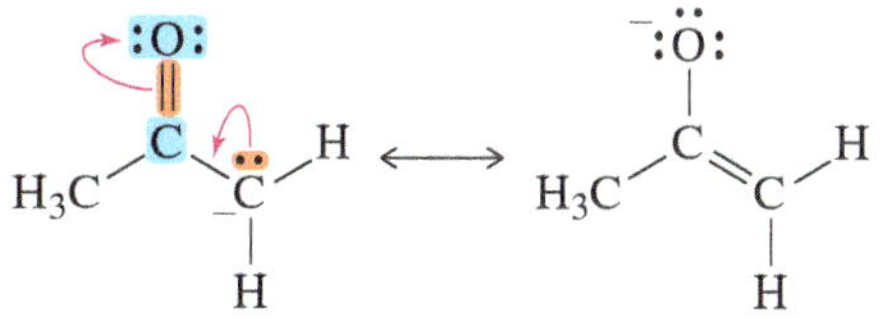

(continued)

Table 2.7 *Continued*

Entry	Original Lewis structure	Analysis (orange = pushable pair; blue = place to push electrons)
3	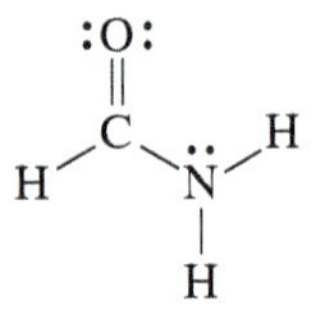	The C—O π bond is a pushable pair that can be moved onto the electronegative oxygen. This creates an empty *p* orbital on carbon (a place to push electrons), so the neighboring nitrogen's lone pair (a pushable pair) can also move to form a new C—N double bond. [*This is similar to what we did using a single step in Entry 2.*]
	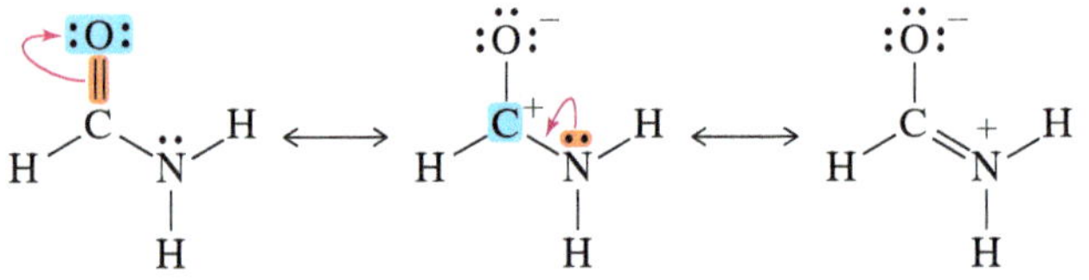	
4	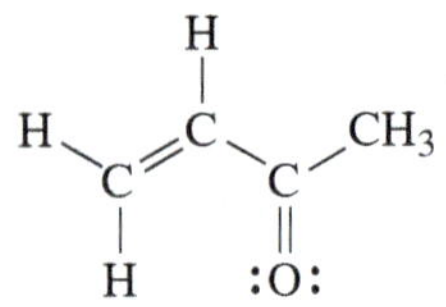	The C—O π electrons are a pushable pair and can be moved onto the electronegative oxygen. This creates an empty *p* orbital on carbon (a place to push electrons), so the neighboring C—C π electrons (a pushable pair) can also move.
	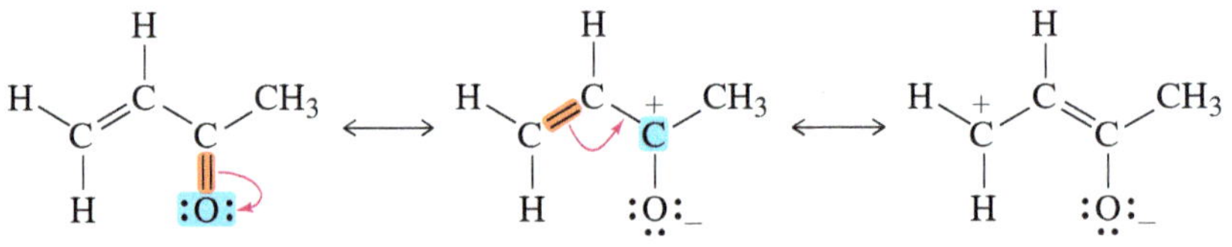	
5	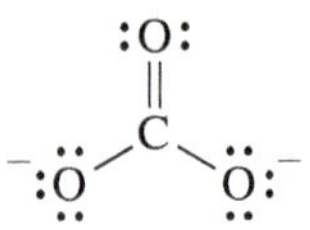	A lone pair on either of the single-bonded oxygen atoms is a pushable pair of electrons. The neighboring carbon is a place to push them because the π electrons of the C═O bond can be moved at the same time onto the electronegative oxygen. This same process can be repeated to generate a third Lewis structure from the pushable pair on the other single-bonded oxygen atom.
	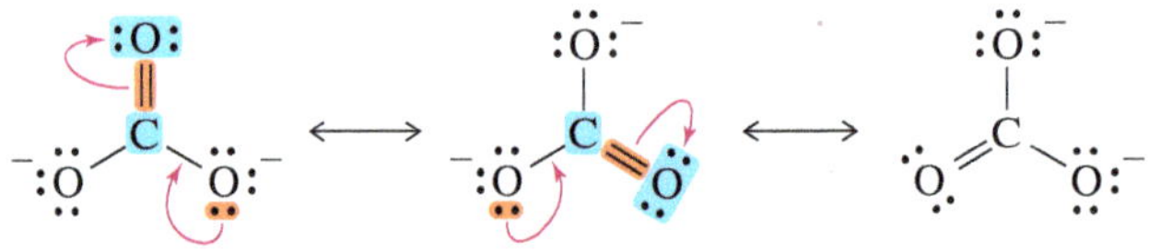	
6	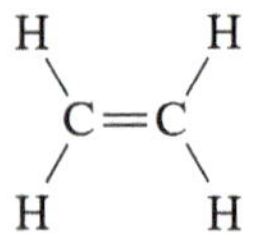	There is a pushable pair of π electrons, and the carbon is a place to push them. However, because this places a negative charge on an atom that is only slightly electronegative, it isn't a great resonance structure. [*We'll learn about the fate of valid, but not great, resonance structures in Section 2.8.3.*]
	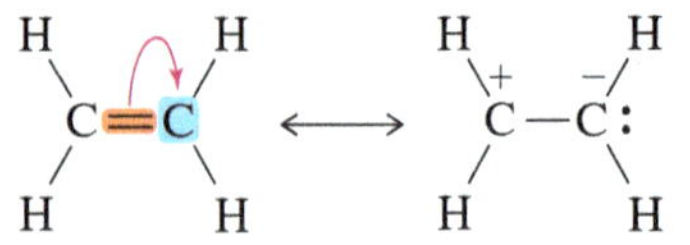	*This is a valid resonance structure, but it doesn't contribute much to the real structure.*

Assessment 2.49 For each of the molecules shown, do the following: (i) Identify all pushable pairs. (ii) Identify all places where electrons can be pushed. (iii) Draw one valid resonance structure.

(a)

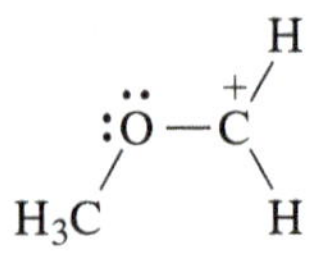

(b)

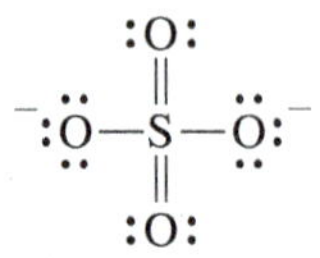

(c)

$H_3C-C\equiv N:$

(d)

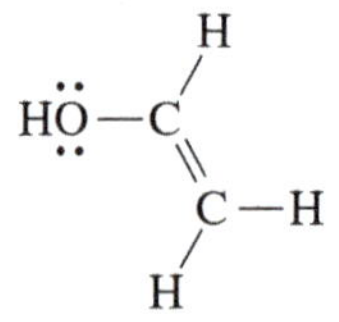

(e)

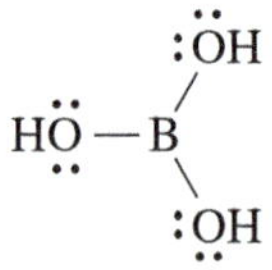

(f)

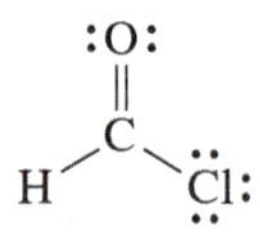

2-F GENERAL CHEMISTRY IN CONTEXT: Organic Chemistry

Purple rhinos

The jump from Lewis structures to resonance structures is not an easy one to make. This is likely a result of the abstract nature of a concept in which a single molecule needs multiple structures to describe its properties. Throughout the text (and the Student Supplement), analogies are used that relate organic chemistry concepts to everyday observations as a way of bringing clarity to the subject. In the case of resonance, it is best understood by looking at two analogies, the first of which is not exactly correct.

If you asked an elementary school student to describe the color purple, they might explain it as a combination of red and blue. And, in fact, combining red and blue paint does make purple (**Figure 2.72**) in the same way that two resonance structures combine to describe the properties of a molecule. This analogy, while close to correct, fails because red and blue themselves are real colors. And thus, we have described a real color (molecule) using two real colors (molecules).

Resonance, however, is invoked when two (or more) imaginary, unreal, nonexistent Lewis structures are used to describe a real molecule. Thus, a better analogy (**Figure 2.73**) would be to describe a rhinoceros (a real animal) as a combination of a unicorn (the horn) and a dragon (the scaly skin). [*Sadly, neither unicorns nor dragons exist.*]

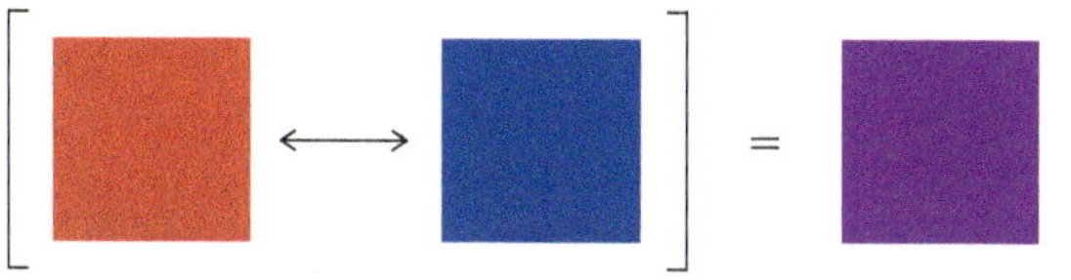

Figure 2.72 The combination of red and blue describes purple.

Figure 2.73 The combination of a unicorn and a dragon describes a rhinoceros.

2.8.3 Contributing resonance structures

The carbonate ion can be represented by the three valid resonance structures shown in **Figure 2.74**. Neither structure individually characterizes the carbonate ion. That is, carbonate is not structure A, structure B, or structure C. Instead, all three of these nonexistent, imaginary structures (A, B, and C) combine to make up the "real" picture of the carbonate ion. This is true for all molecules where multiple valid resonance structures can be drawn.

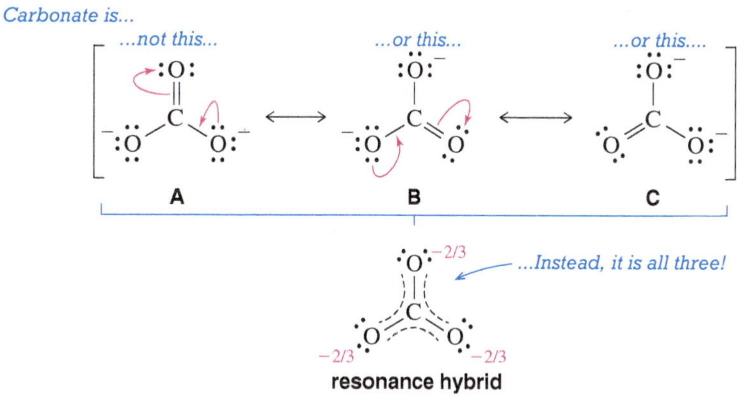

Figure 2.74 The overall resonance hybrid accounts for the "real" structure of a molecule.

You can think of the overall resonance hybrid shown in Figure 2.74 as a weighted average of the resonance structures, based on the degree to which each resonance structure contributes to the resonance hybrid. The contribution of each structure results from the inherent stability of the individual structures, as generally defined by the rules in **Table 2.8**.

Table 2.8 Choosing the more contributing resonance structure.

Entry	The more contributing resonance structure . . .	Example
1	. . . has the maximum number of covalent bonds.	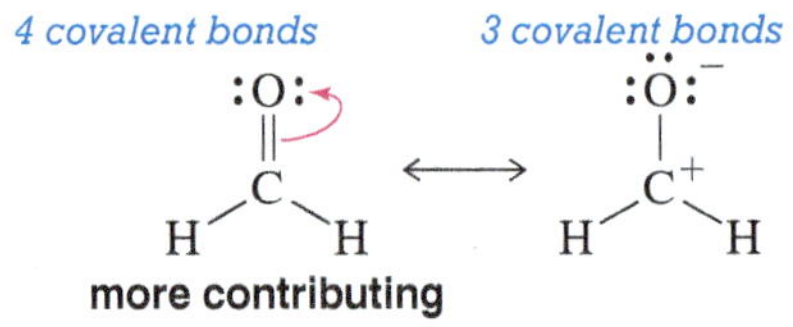
2	. . . has fewer atoms with incomplete valence shells (octets).	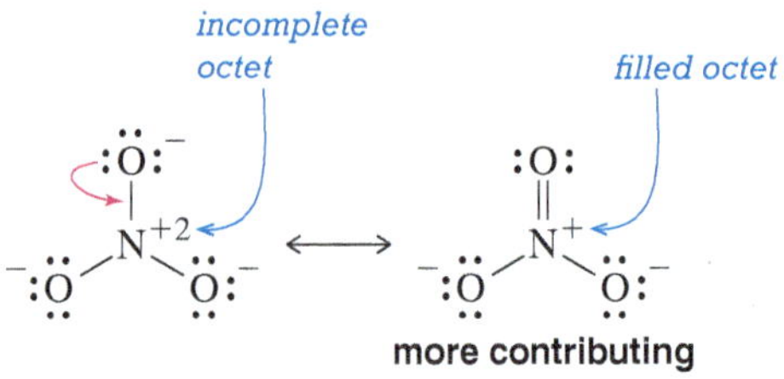
3	. . . has less separation of unlike charges.	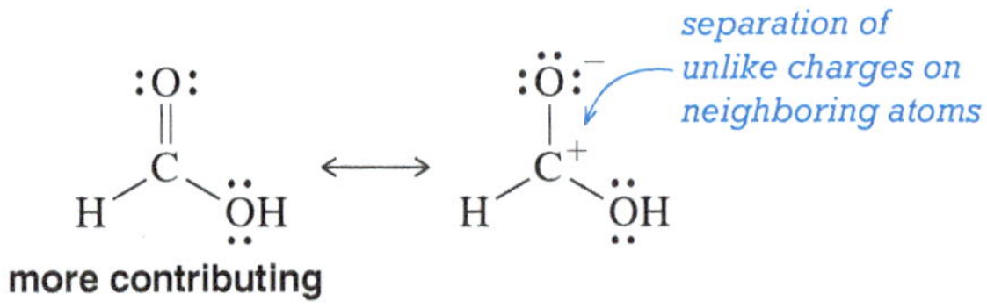
4	. . . has negative charges on the more electronegative atom.	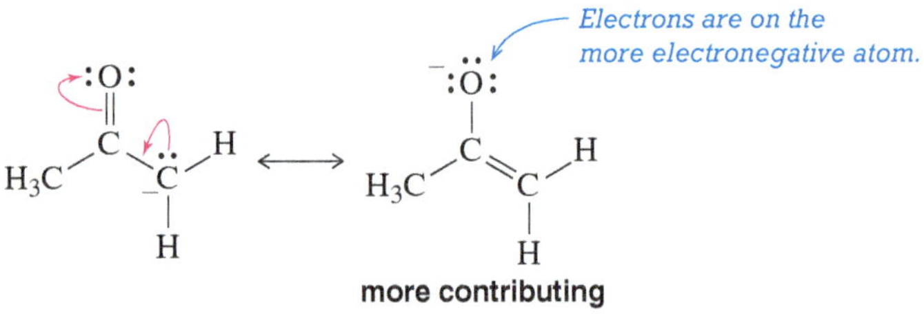

Based on the rules in Table 2.8, there is nothing that distinguishes the carbonate resonance structures from each other. When a molecule is made up of identical resonance structures, each contributes the same to the structure of the molecule. It makes sense, then, that the carbonate ion has an equivalent −2/3 charge on each oxygen (**Figure 2.75**).

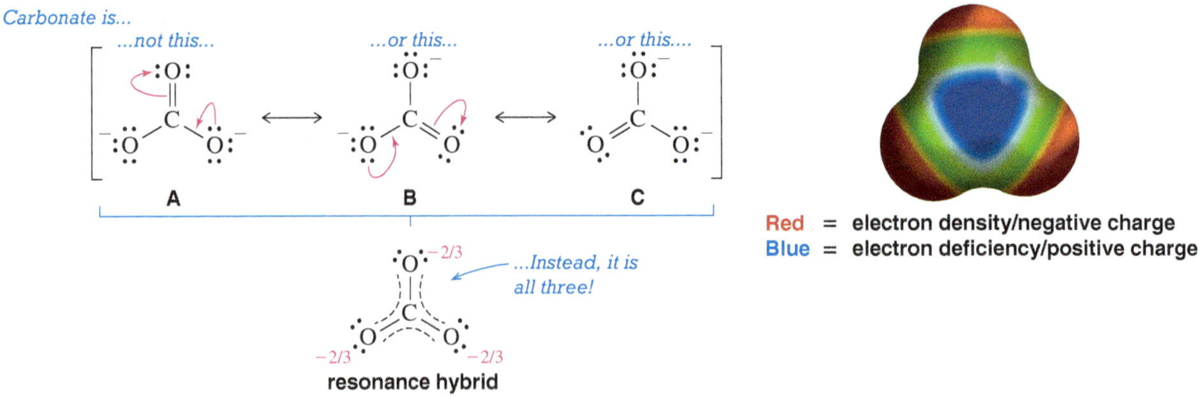

Figure 2.75 Identical structures contribute equally to the resonance hybrid.

Alternatively, in entry 4 of Table 2.8, we show a structure whose resonance hybrid is composed of two different structures. In comparing these two, they each have the same number of covalent bonds. Every atom in both structures has a full valence shell. There is only one negative charge in each. The lone difference between the two is the location of the negative charge—it is on oxygen in one structure and on carbon in the other. Oxygen is more electronegative than carbon, so the structure with the negatively charged oxygen is the more stable and, therefore, the more contributing resonance structure. That the electron density is more highly localized around the oxygen is shown in the resulting electron density map (**Figure 2.76**).

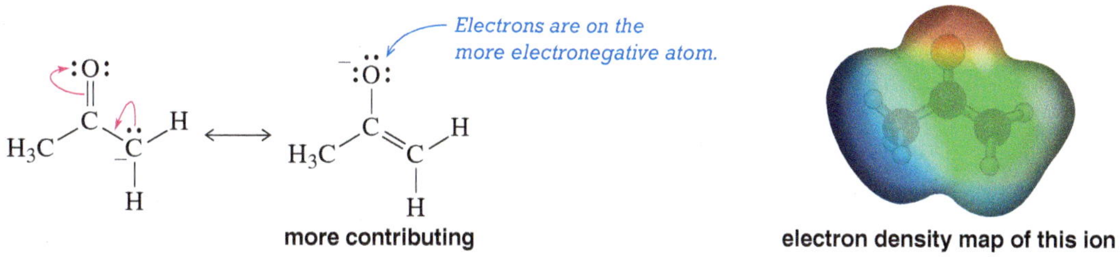

Figure 2.76 The more stable structure contributes more to the resonance hybrid.

Assessment 2.50 (i) Which of the following resonance structures represents the "actual" structure of the molecule shown? (ii) Which contributes more to the resonance hybrid? (iii) Why?

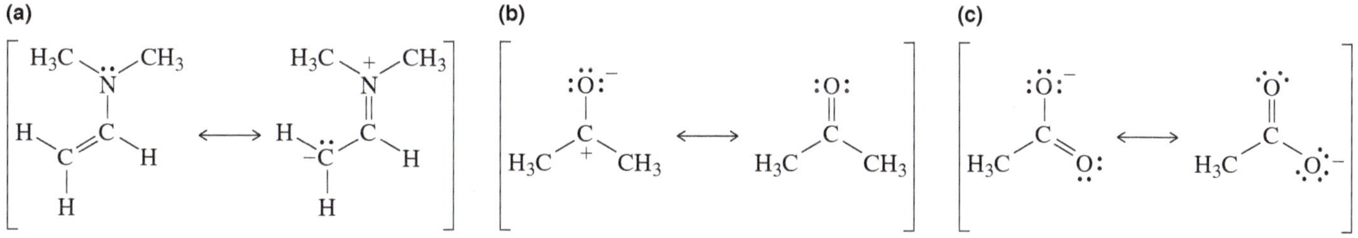

2.8.4 How resonance structures can be used (one example)

To review, valid resonance structures are ones that can be drawn. Each valid resonance structure contributes to some degree to the overall resonance hybrid. The level of contribution is a function of the relative stabilities of the individual structures. The more stable structure contributes more. A common mistake students make is to focus too much attention on the most contributing structure. While this is understandable, given that the most contributing structure is the one we usually draw first, the second-best resonance structure often contains tons of chemical reactivity information.

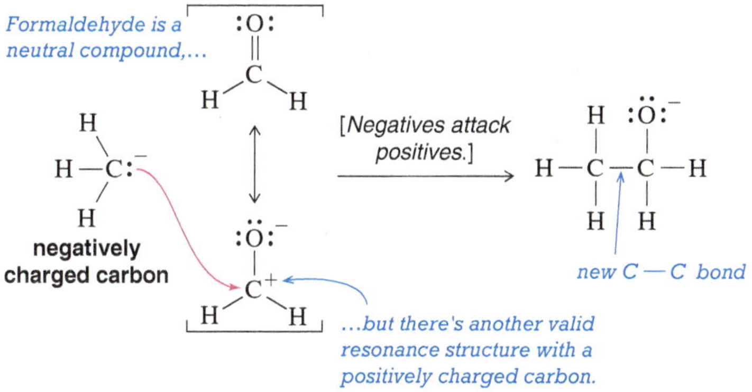

Figure 2.77 The second-best resonance structure helps us understand chemical reactivity.

Consider, for example, the hypothetical reaction shown in **Figure 2.77** between a negatively charged carbon and formaldehyde. [*We study this reaction in Chapter 17. Though it seems early to be discussing it now, you'll see that you can already make sense of it.*] We know that organic chemistry is defined by "negatives attacking positives." And yet, there is nothing positive here—unless we draw all of the resonance structures of formaldehyde. The C—O π electrons can be pushed onto the electronegative oxygen to create a resonance structure containing a positively charged carbon and a negatively charged oxygen. [*Where would we expect the negatively charged carbon* *to attack now?*] The negatively charged carbon would attack the positively charged carbon, forming a new C—C bond in the process. You don't need to worry about understanding this reaction now, but it demonstrates the utility of the "second-best" resonance structure.

Assessment 2.51 To which atom of formaldehyde would you expect H^+ to add?

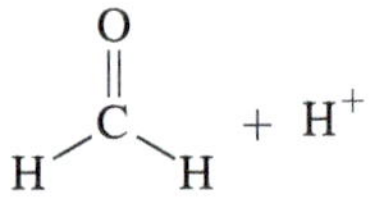

formaldehyde

Assessment 2.52 Which atoms in the enol ether would you expect to react with Br^+?

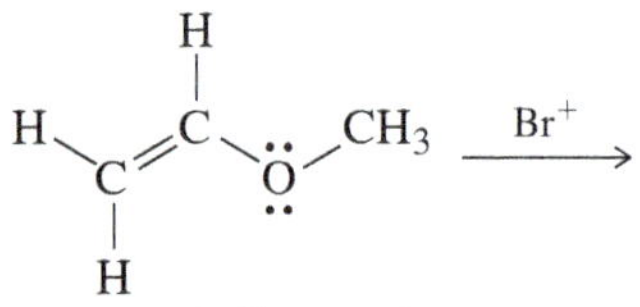

enol ether

2.8.5 VSEPR modified/resonance and molecular orbitals

For electrons to move from one atom to another across resonance structures, they must either be π electrons or sit in *p* orbitals.

Let's work Assessment 2.53 together to see how and why this must be true.

Assessment 2.53 Assign the hybridization of the nitrogen in each resonance structure of acetamide.

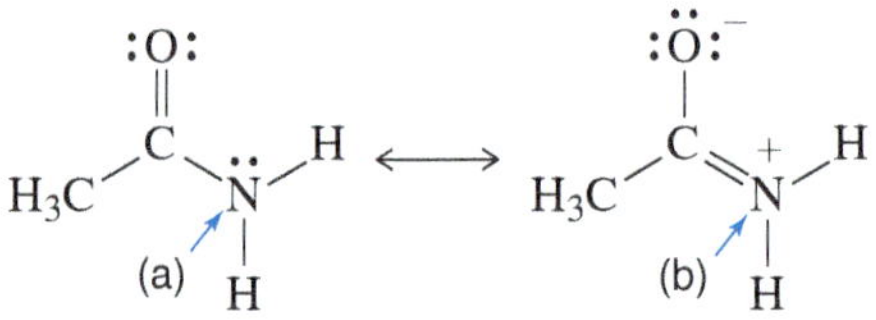

Solved Assessment 2.53

Figure 2.78 shows how we might have determined the shape of the nitrogen atom in each of two resonance structures of acetamide. Using the traditional VSEPR analysis, the nitrogen in the most contributing resonance structure has four areas of electron density. This corresponds to a tetrahedral shape and sp^3 hybridization. Alternatively, in the second resonance structure, this nitrogen now has three areas of electron density, which corresponds to a trigonal planar shape and sp^2 hybridization.

Recall from our understanding of resonance structures that neither resonance form individually represents the real structure of acetamide. Instead, the combination of both of these structures represents the actual structure. The double-headed arrow separating the two structures does not imply that the molecule changes from one to the other. Instead, it implies that the molecule is a combination of both at the same time. By this analysis, how is it possible that the hybridization could change going from one structure to the next? The answer: It can't. The hybridization of an atom is not different from one resonance structure to the next. It turns out that the hybridization of the nitrogen, regardless of which resonance structure is drawn, is always sp^2. [*But isn't VSEPR a result of making a molecule as stable as possible? Read on.*] ■

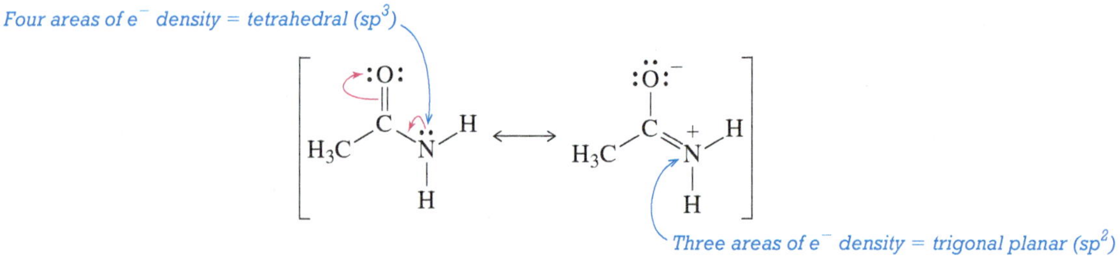

Figure 2.78 **VSEPR analysis predicts different shapes for the same nitrogen.**

Bond	Average length
C—N	1.47 Å
C═N	1.28 Å

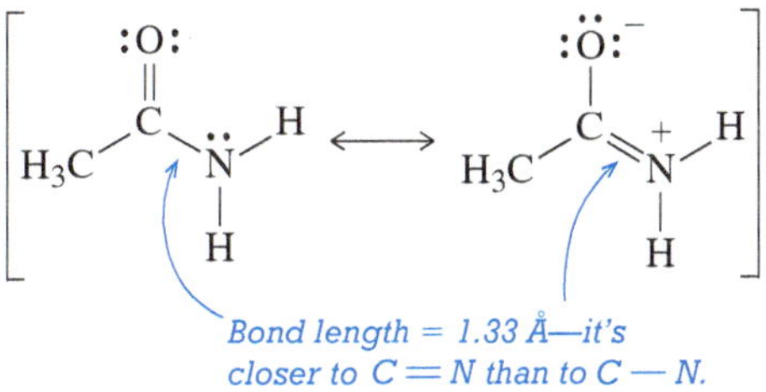

Figure 2.79 The partial C—N double bond character is only explained by the *sp*^{2} hybridization of the acetamide nitrogen.

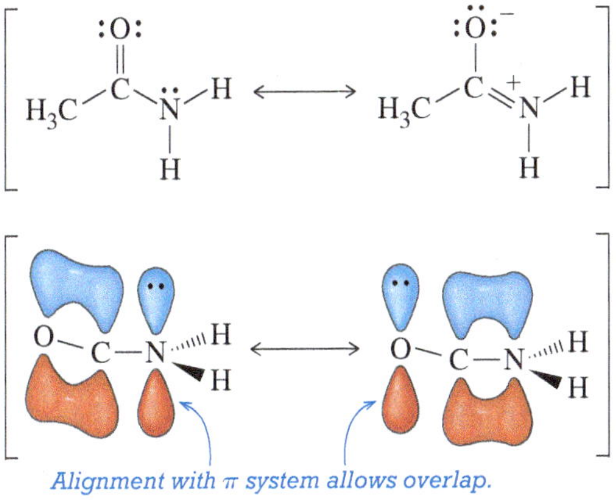

Figure 2.80 Resonance is the delocalization of electron density throughout the π system. [*Extraneous atoms and lone pairs have been removed for clarity.*]

The sp^2 hybridization of the nitrogen in acetamide makes sense when you consider that the actual structure has partial double bond character (Figure 2.79). While the average C—N single bond is 1.47 Å and the average C═N double bond is 1.28 Å, the C—N bond length in acetamide is intermediate between the two (1.33 Å). This indicates that there is partial double bond character in the C—N bond. The nitrogen must be sp^2 hybridized, using one p orbital to form this partial C—N π bond.

To understand why an atom hybridizes differently if resonance is possible, it is helpful to look at the molecular orbital picture. In Figure 2.80, we draw what the molecular orbital picture would look like for each resonance form. (This isn't really correct, though, because the resonance structures combine to represent one structure, so acetamide should be represented by only one set of molecular orbitals, not two.) By placing the lone pair in a p orbital in the resonance structure on the left, it is able to align with the C—O π bond. This alignment is precisely what allows the delocalization of electron density onto the oxygen atom, represented by the molecular orbital picture of the resonance structure on the right. The oxygen in this resonance structure also keeps one lone pair in a p orbital, maintaining this alignment. The combination of these resonance structures can then be understood by recognizing that this electron pair is delocalized throughout the π system.

In conclusion, any atom with an electron pair that participates in resonance must be hybridized such that the electron pair occupies a p orbital. This allows the electron density to be delocalized throughout the π system. Recognizing this also helps us understand why only electron pairs in p orbitals or π bonds participate in forming valid resonance structures (Table 2.6).

2.8.5 SUMMARY

- Electrons that are delocalized by resonance must be contained in p orbitals or π bonds.

Assessment 2.54 Assign the hybridization of the indicated oxygen in the carboxylate and alkoxide shown. Which is more stable, a carboxylate or an alkoxide?

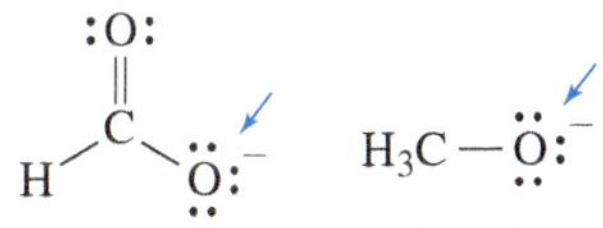

Concluding thoughts

Moving forward, we will begin applying our general chemistry knowledge of bonding to organic molecules. Chapter 2 is of central importance to all we do going forward in organic chemistry. Resonance, for example, will help us understand why certain molecules are strong acids while similar compounds are weak (Section 4.4.3). Understanding the molecular orbital picture of atoms will help us understand which atoms are electron rich and which are electron poor (CIC Box 2-C and Chapter 8, among others). And finally, knowing that carbon can form four bonds, as suggested by the Lewis dot structure, helps us understand why carbon can form a multitude of different compounds with unique functional groups that allow its chemistry to be covered in a 28-chapter book (Chapters 3–28).

Let's start this next phase now, by studying the chemistry of alkanes in Chapter 3.

End-of-Chapter Materials

These EOC materials are intended to direct your studying and provide a quick review of the important information within the chapter. They are not intended to replace reading the chapter or working the assessments. They are presented in multiple formats to help you organize these concepts in your head. It probably does not need to be stated here, but do not simply memorize what is here; make sure you understand what is presented. And if you don't, refer back to the original sections for a more complete review.

Important Concepts

[*These are the major concepts covered in each section, not including reactions themselves, which are summarized in a later section.*]

2.1–2.2: Atoms are composed of protons, neutrons, and electrons. Electrons in the first two shells are held in either *s* or *p* orbitals. The *s* orbital is spherical, while *p* orbitals have a dumbbell shape. The electron configuration of an atom is generated by filling the lowest energy orbitals first.

2.3: Bonding in molecules is driven by forming more stable molecules from less stable atoms. Lewis dot structures are used to represent the valence electrons in atoms. Achieving a full octet in the valence shell explains the stability gained when atoms form ionic and covalent bonds. The type of bond formed (ionic or covalent) depends on the electronegativity difference between bonded atoms.

2.4–2.5: Lewis structures are used to represent the covalent bonding in molecules. The formal charge is the charge assigned to each atom in a molecule based on equal sharing of electrons. Molecules containing fewer atoms with formal charge are more stable. The increased stability gained by maximizing the distance between valence shell electron pairs (VSEPR) predicts the shape of most molecules. The dipole moment (μ) of a molecule is a combination of the individual bond dipoles and is closely correlated to the shape of the molecule.

2.6: The shape of molecules can be explained by the hybridization of atoms. Atoms in the second row can hybridize as sp^3, sp^2, or sp. An atom that is sp^3 hybridized uses four sp^3-hybridized orbitals. An atom that is sp^2 hybridized uses three sp^2-hybridized orbitals and one *p* orbital. An atom that is *sp*-hybridized uses two *sp*-hybridized orbitals and two *p* orbitals. Orbitals with more *s* character yield shorter bonds.

2.7: Molecular orbital theory explains the formation of molecular orbitals from atomic orbitals as the constructive (additive or in-phase) and destructive (subtractive or out-of-phase) combination of wave functions. σ (bonding) and σ^* (antibonding) molecular orbitals are formed via on-axis overlap of atomic orbitals with *s* character (sp^3, sp^2, and sp). π (bonding) and π^* (antibonding) molecular orbitals are formed via overlap of *p* atomic orbitals above and below the axis of the σ bond.

2.8: Resonance is used to explain the bonding of delocalized electrons in a way that can't be explained by Lewis structures alone. Resonance structures are rationalized using the arrow-pushing formalism and are possible when there is a place to push a pushable pair. Valid resonance structures contribute to the overall structure based on their relative stability. Electron pairs that are delocalized through resonance must be contained in *p* orbitals. As a result, atoms involved in resonance will hybridize differently than predicted by VSEPR.

End-of-Chapter Assessments

Lewis structure boot camp

Be sure to take breaks while doing these to maximize retention of the material. See Section 1.4 for why this is important. Because this is a review of general chemistry, move on once you feel comfortable with these assessments.

Assessments 2.55–2.62 steadily increase in difficulty while offering a repetitive experience that ensures you leave Chapter 2 with the ability to draw Lewis structures.

2.55 (•) How many valence electrons does each of the following contribute to a Lewis structure?

(a) C (b) O (c) B (d) F (e) Br (f) N (g) S (h) I (i) Cl (j) Al (k) Si (l) −1 charge

2.56 (•) Draw the Lewis dot structure for each of the following atoms.

(a) C (b) O (c) B (d) F (e) Br (f) N (g) S (h) I (i) Cl (j) Al (k) Si (l) P

2.57 (•) Why are the Lewis dot structures that you drew for N and P in Assessment 2.56 similar?

2.58 (•) Show how the Lewis dot structure for each of the following atoms would overlap to form a single bond.

(a) H and H (b) C and O (c) C and H (d) C and N (e) N and H (f) O and H

2.59 (•) Which atom in each pair would you expect to be the central atom in a Lewis structure?

(a) H vs. O (b) C vs. O (c) H vs. N (d) C vs. N (e) C vs. H (f) C vs. F

2.60 (•) Given the atoms involved and the number of valence electrons remaining, complete the Lewis structure by placing bonds between atoms such that each has a full octet.

(a) (b) (c) (d) (e)

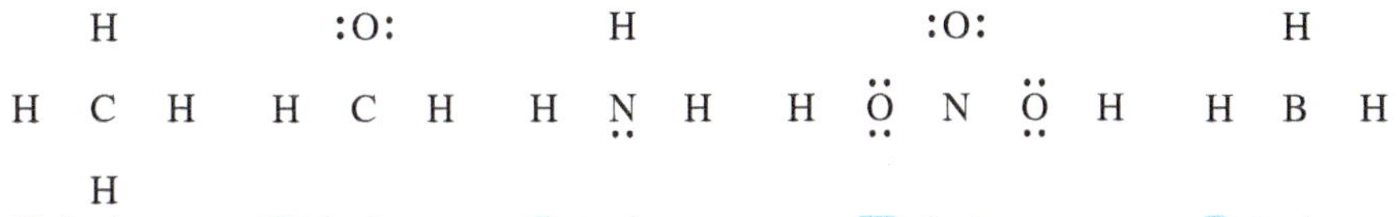

8 electrons | 8 electrons | 6 electrons | 12 electrons | 6 electrons

2.61 (••) By moving an electron pair, draw a better Lewis structure that minimizes formal charges.

(a) (b) (c) (d) (e)

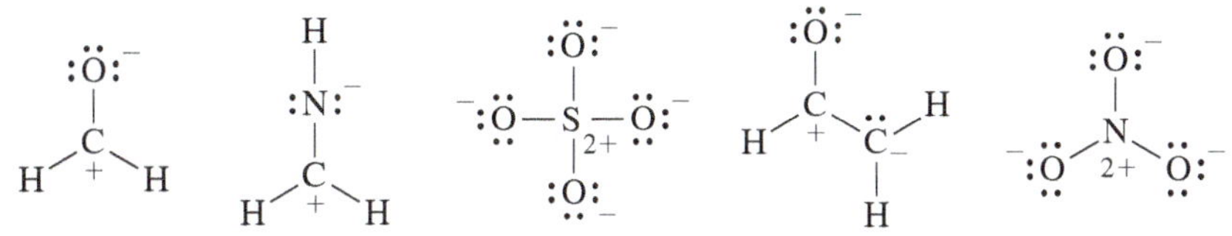

2.62 (••) Draw the Lewis structure for the following molecular formulas.

(a) C_2H_6
(b) C_3H_8
(c) H_3O^+
(d) H_3S^+
(e) CH_4O
(f) CH_2O
(g) CHO_2^-
(h) C_2H_4
(i) C_2H_2
(j) H_2SO_4
(k) H^+
(l) SO_4^{2-}
(m) HNO_3
(n) NO_3^-
(o) NO_2^-
(p) CH_3^+
(q) AlH_3
(r) BH_3
(s) CCl_4
(t) CH_2Br_2
(u) NH_4^+
(v) CH_5N
(w) CO_2
(x) C_2H_2O

Resonance structure boot camp

Assessments 2.63–2.68 steadily increase in difficulty while offering a repetitive experience that ensures you leave Chapter 2 with the ability to draw resonance structures.

2.63 (••) In the following molecules, identify all pushable electron pairs.

(a) (b) (c) (d) (e)

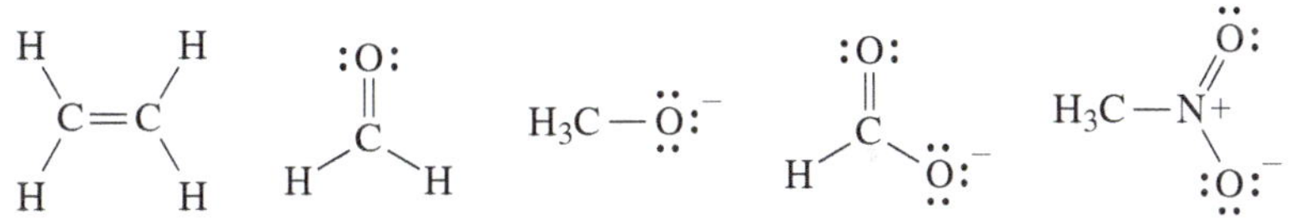

2.64 (••) Two resonance structures are shown for each molecule. Use the arrow-pushing formalism to represent the electron flow from the structure on the left to the one on the right.

[*One arrow is required in each of these.*]

(a) (b) (c)

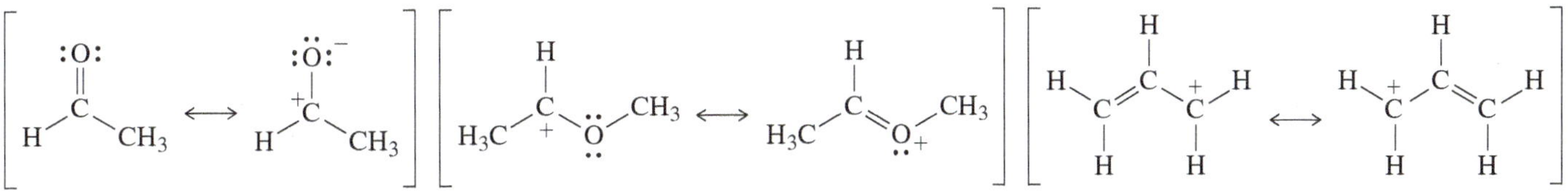

[*Two arrows are required in each of these.*]

(d) (e) (f)

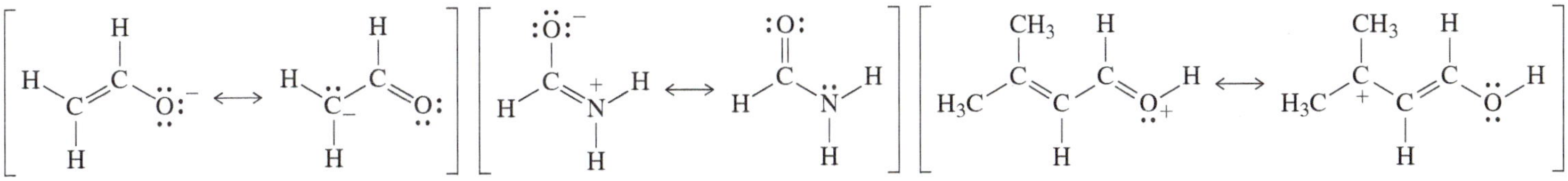

2.65 (••) For the resonance structures in Assessment 2.64, use the arrow-pushing formalism to represent the electron flow from the structure on the right to the one on the left.

2.66 (••) For the resonance structures in Assessment 2.64, (a) identify the actual structure of the molecule and (b) choose the resonance structure that contributes most to the resonance hybrid.

2.67 (••) Identify the resonance structure that will be produced given the molecule shown and the electron flow indicated.

(a)

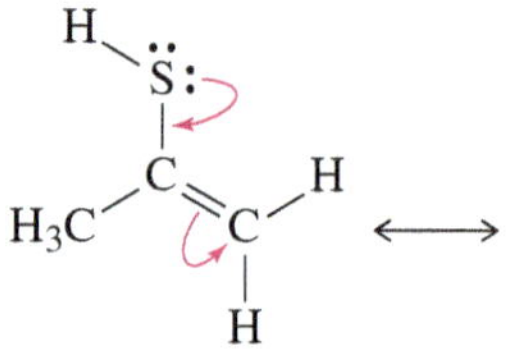

(b)

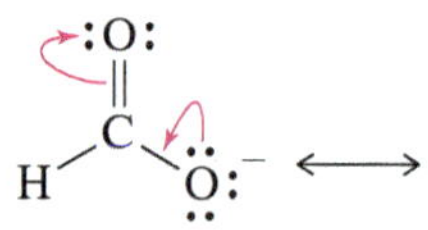

(c)

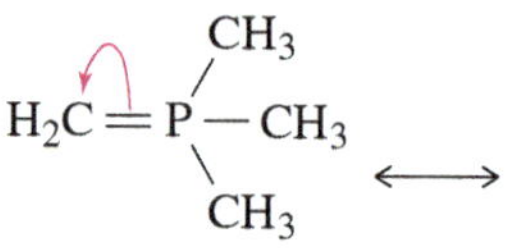

(d)

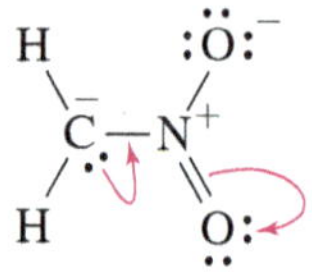

2.68 (••) Draw all resonance structures of the following molecules.

(a)

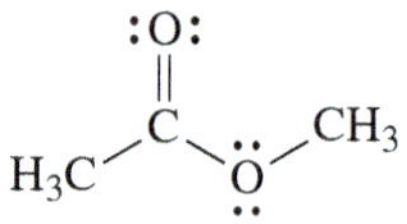

(b)

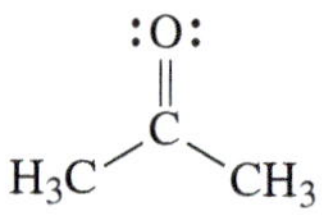

(c)

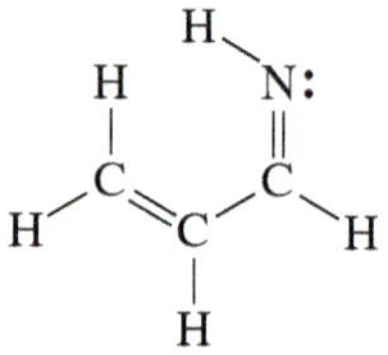

(d)

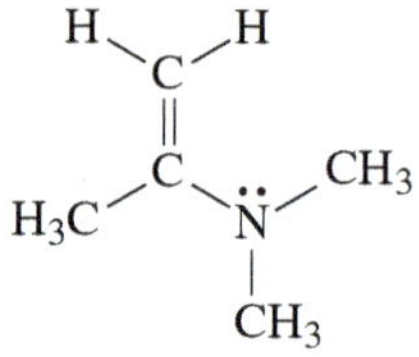

(e)

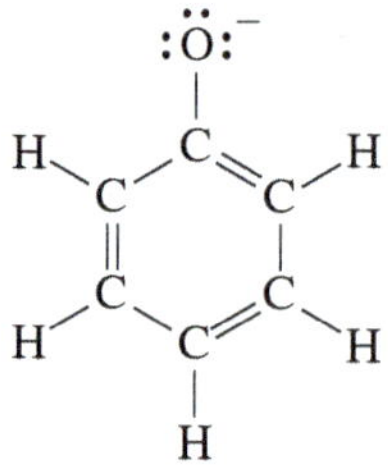

(f)

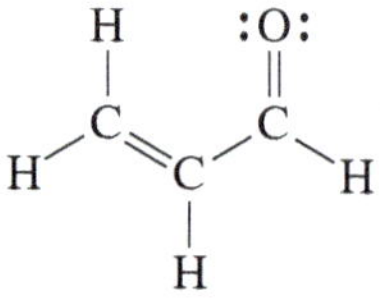

(g)

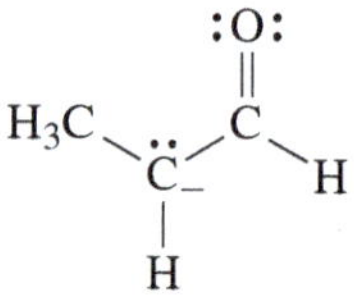

(h)

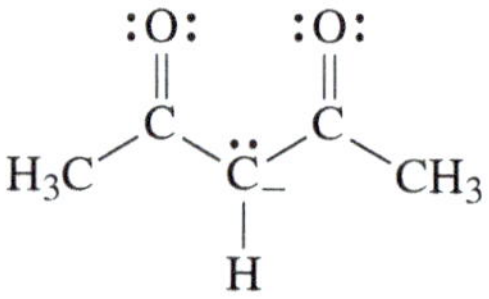

(i)

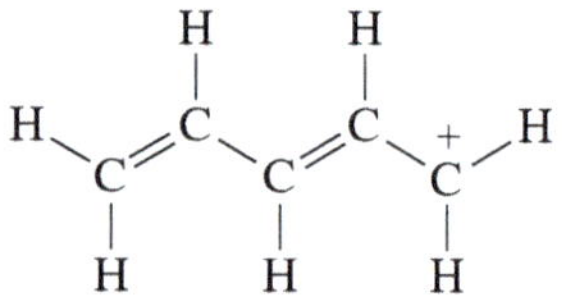

Molecular orbital boot camp

Assessments 2.69–2.73 and 2.76 refer to the following Lewis structures.

(a)

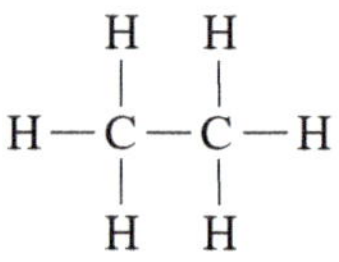

(b)

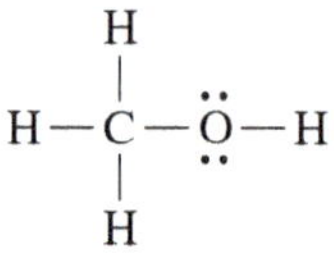

(c)

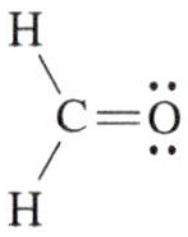

(d)

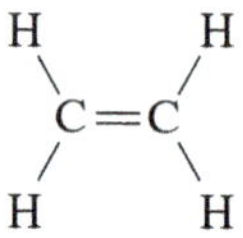

(e)

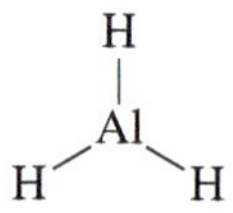

(f)

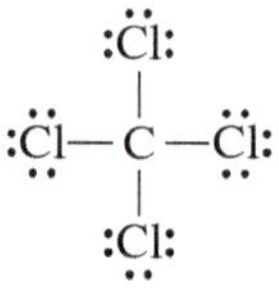

(g)

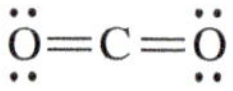

2.69 (••) Predict the shape of all non-hydrogen atoms.

2.70 (••) Predict the hybridization of all non-hydrogen atoms.

2.71 (••) If an atom is sp^3 hybridized, how many sp^3-hybridized orbitals does it use for bonding? How many p orbitals?

2.72 (••) If an atom is sp^2 hybridized, how many sp^2-hybridized orbitals does it use for bonding? How many p orbitals?

2.73 (••) If an atom is sp hybridized, how many sp-hybridized orbitals does it use for bonding? How many p orbitals?

2.74 (••) Connect the atoms using the indicated hybrid orbitals.

(a) C (sp^3) with H (s)

(b) C (sp^2) with C (sp^2)

(c) C (sp^2) with O (sp^3)

(d) C (sp) with N (sp^2)

2.75 (••) For the following partial structures, the σ bond is shown. Add the indicated number of π bonds, being sure to specify the orientation (that is, x, y, or z axis) of the p orbitals used.

(a)

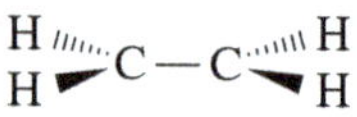

(1 π bond)

(b)

(1 π bond)

(c)

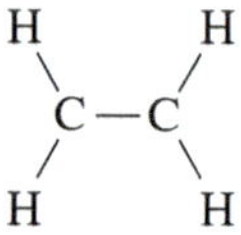

(1 π bond)

(d)

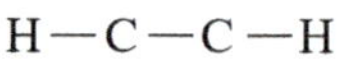

(2 π bonds)

(e)

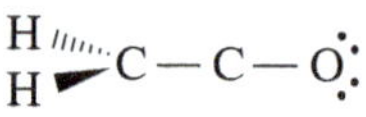

(2 π bonds)

2.76 (••) Draw the molecular orbital picture for the molecules (a)–(g) referred to in Assessments 2.69–2.73.

Applications

[Assessments like these are automatically more challenging than Assessments 2.55–2.76 simply because you have to figure out which concept is necessary to solve a problem. Hints will be given, but try to get to the point where you don't need them.]

2.77 (•••) Given the Lewis structures, indicate the direction of the dipole moment, if there is one. [*See Assessment 2.63.*]

(a) (b) (c) (d) (e)

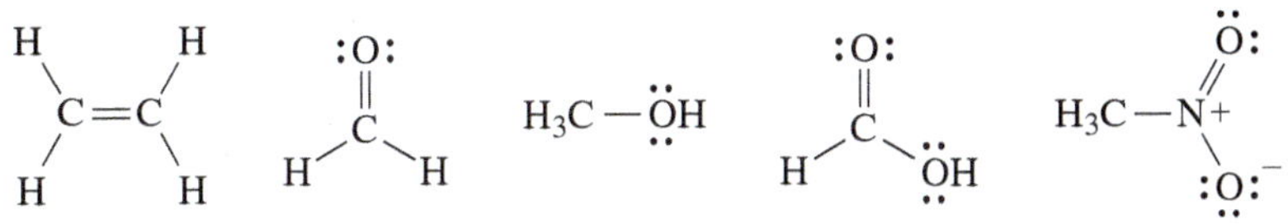

2.78 (••••) Carbon dioxide (CO_2) has no dipole moment (μ = 0 D). The related molecule sulfur dioxide (SO_2) does have a dipole moment (μ = 1.6 D). Explain this observation. [*See Assessments 2.62 and 2.69.*]

2.79 (••••) There is free rotation around the C—C bond in ethane. There is an extremely high barrier to rotation around the C=C bond in in ethene. Explain. [*See Assessments 2.75 and 2.76.*]

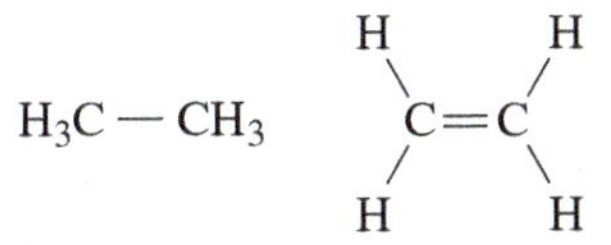

ethane ethene

2.80 (••••) The C—H σ bond length in ethane is 1.09 Å. The C—H σ bond in ethene is 1.07 Å. Explain. [*See Assessment 2.70.*]

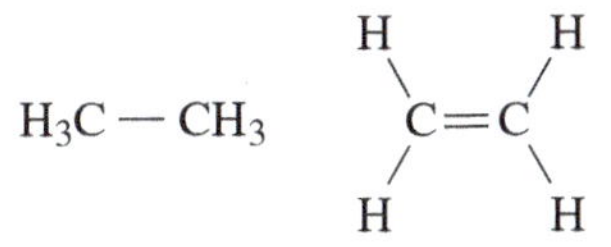

ethane ethene

2.81 (••••) In propene, the indicated C—C bond length is 1.51 Å. In the allyl cation, the indicated C—C bond length is 1.41 Å. Explain. [*See Assessment 2.68.*]

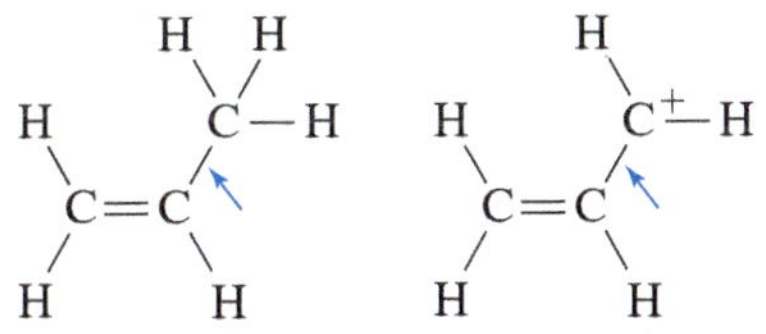

propene allyl cation

2.82 (••••) LOOKING AHEAD In Chapter 4, we explain that molecules like CH_3^+ are Lewis acids or electron pair acceptors. Into which orbital would the new electron pair go? [*See Assessment 2.70.*]

2.83 (••••) LOOKING AHEAD In comparison to CH_3^+ in Assessment 2.82, the related molecule H_3O^+ is not a Lewis acid at oxygen. Why? [*See Assessment 2.76.*]

2.84 (••••) The C—N bond in the following amide is much stronger than the C—N bond in the amine. Explain. [*See Assessment 2.68.*]

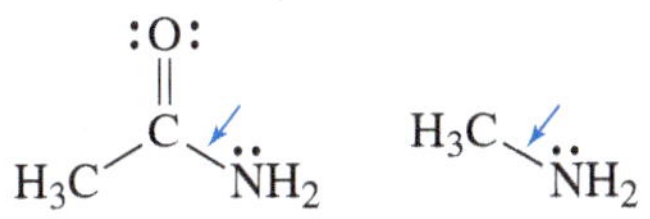

amide amine

2.85 (••••) LOOKING AHEAD In Chapter 3, we describe how the formal charge on an atom can be used to predict the number of lone pairs. Given the charge, or lack of charge, on each atom, fill in the electron pairs.

(a) (b) (c) (d) (e)

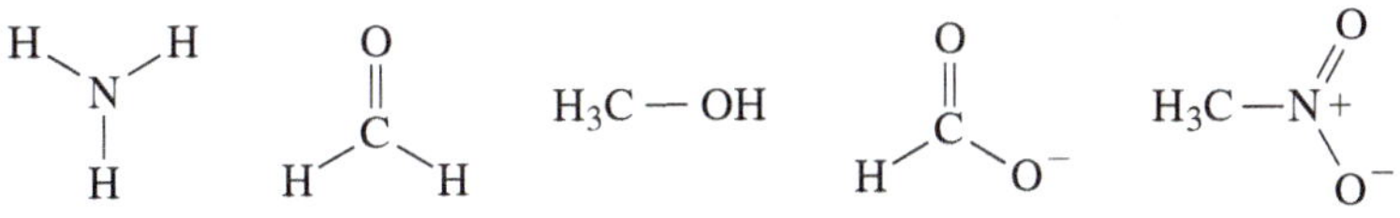

2.86 (••••) LOOKING AHEAD In Chapter 4, we explain that molecules like CH_3^- are Lewis bases or electron pair donors. What makes it a Lewis base? [*See Assessment 2.62.*]

2.87 (••••) LOOKING AHEAD In comparison to CH_3^- in Assessment 2.86, the related molecule BH_4^- is itself not a Lewis base at boron. Why? [*See Assessment 2.62.*]

3

Alkanes and Cycloalkanes: Properties and Conformational Analysis

Learning Objectives

LO 3.0 Analyze the physical properties of alkanes in terms of number of carbons, branching vs. straight chain, and the effects of intermolecular forces

LO 3.1 Compare the structural, hybrid, and condensed structural formulas and line-angle drawings for alkanes

LO 3.2 Name the isomers of alkanes

LO 3.3 Use Newman projections to explain the relative stability of conformational isomers created by rotation around a single bond

LO 3.4 Name the isomers of cycloalkanes

LO 3.5 Determine the effect of ring strain on the physical properties of cycloalkanes

(continued)

Introduction

The presence of hydrocarbons on Earth can be explained by the biogenic theory of oil formation. Hundreds of millions of years ago, dead marine life sank to the bottom of Earth's oceans and became buried by layers of silt and sand. Over millions of years, these organic remains were buried deeper and deeper and eventually were transformed by high levels of heat and pressure into kerogen, a waxy organic material that is a major component of oil shale. Subjected to more heat, kerogen transforms into the lighter hydrocarbons, oil and natural gas. As these lighter compounds move through porous rock toward Earth's surface, their migration upward can be capped by impermeable rock, at which point the oil and gas accumulates. By drilling into these oil-rich rock formations, crude oil is extracted from Earth and refined by distillation into petroleum products, such as fuels and other useful organic compounds.

○ ○ ○ ○ ○ ○ ○

In Chapter 1, we prepared ourselves for Organic Chemistry, the course. Chapter 2 reminded us of everything we once knew from General Chemistry and showed us how it would apply in organic chemistry. Organic chemistry starts now. [*Let's go!*] But first, let's test our understanding of what came before.

"As long as you're comfortable, it feels like freedom."

—Billy Bragg

LO 3.6 Compare the relative stabilities of nonsubstituted, monosubstituted, and disubstituted cycloalkanes using conformational analysis

LO 3.7 Categorize common functional groups in organic chemistry

Test your understanding

Assessment 3.1 Give the Lewis dot structure for the following atoms.

(a) Oxygen (b) Carbon (c) Hydrogen (d) Nitrogen

Assessment 3.2 What is the ground state electron configuration of carbon? How many bonds does carbon usually form?

Assessment 3.3 What is the hybridization of carbon in each of the following molecules?

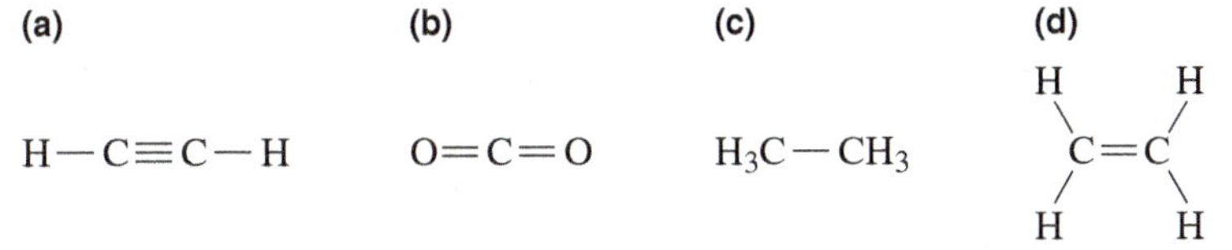

Assessment 3.4 Methane is tetrahedral. What is the H—C—H bond angle?

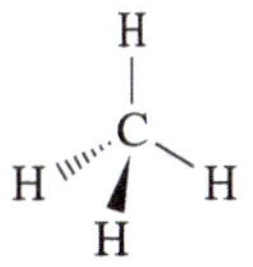

methane

Assessment 3.5 Calculate the formal charge of the indicated atoms. [*All lone pairs are shown.*]

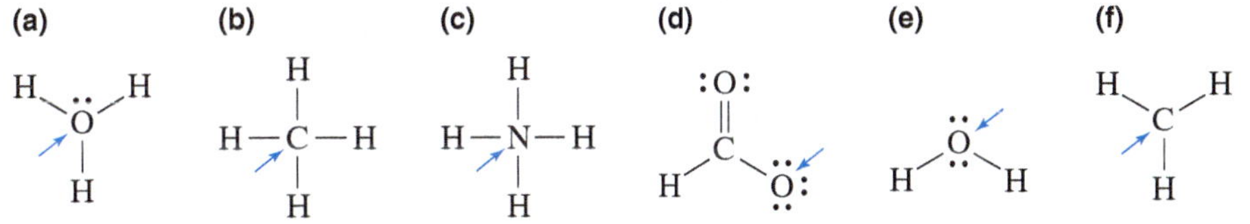

3.1 Alkanes: an overview

In Chapter 2, we saw that carbon, as a result of having 4 valence electrons, is most stable when it has four covalent bonds. Most often, carbon forms these four bonds with some combination of hydrogen and other carbon atoms. If we start with the simplest organic molecules, those containing just carbon and hydrogen, the first molecular class we will study are **alkanes**. Alkanes are **saturated hydrocarbons**, meaning that there is the maximum number of hydrogens per atom of carbon. [*Saturated fats, anyone?*] Methane (Figure 3.1) is the simplest alkane, containing four hydrogens bonded to a single carbon (CH_4). Ethane, with two carbons, has three hydrogens on each carbon for a total of six. And propane, with three carbons, contains two carbons with three hydrogens and a central carbon with two hydrogens, for a total of eight.

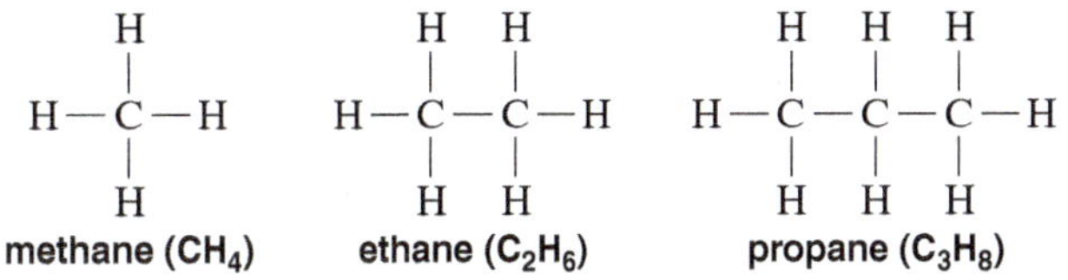

Figure 3.1 Saturated hydrocarbons contain the maximum number of hydrogens per carbon.

The general molecular formula for an alkane is C_nH_{2n+2}, where n = number of carbons. In contrast, alkenes, alkynes, and aromatics, shown in Table 3.1, are **unsaturated hydrocarbons**. Due to their π bond(s), monoalkenes have two fewer and monoalkynes have four fewer than the maximum number of hydrogens. The general formula for an alkene is C_nH_{2n} and that for an alkyne is C_nH_{2n-2}. [*The chemistry of alkenes (Chapters 8 and 9), alkynes (Chapter 10), and aromatics (Chapters 23 and 24) is discussed in the chapters indicated.*]

Table 3.1 Classification of saturated and unsaturated hydrocarbons.

Compound type	Functional group		Example
Alkanes	None (no double or triple bonds)		$CH_3-CH_2-CH_3$ (propane)
Alkenes	$>C=C<$	double bond	$CH_2=CH-CH_3$ (propene)
Alkynes	$-C\equiv C-$	triple bond	$H-C\equiv C-CH_3$ (propyne)
Aromatics	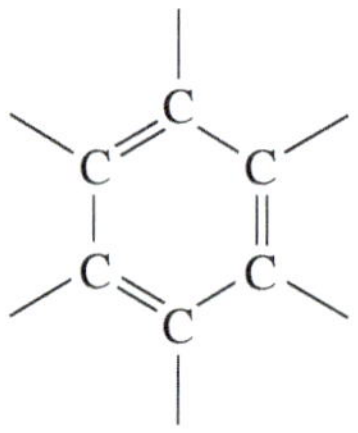	Benzene ring	CH_2CH_3 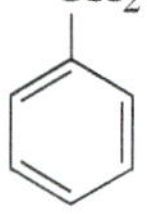 (ethylbenzene)

3.1.1 Straight-chain alkanes

Because carbon forms very stable σ bonds with other carbons, an unlimited number of alkanes are possible. Table 3.2 shows the first 20 straight-chain alkanes, each featuring the addition of an extra carbon. These are often referred to as the *n*-alkanes, where the *n* stands for *normal*, to distinguish them from their branched counterparts. [*We study chain branching in Section 3.1.3.*]

Table 3.2 Structure and physical properties of straight-chain alkanes.

	Alkane	Number of carbons	Structure	Formula	Boiling point (°C)	Melting point (°C)	Density[a]	Uses
gases	Methane	1	CH_4	CH_4	−164	−183	0.55	Natural gas
gases	Ethane	2	CH_3CH_3	C_2H_6	−89	−183	0.51	Production of ethylene
gases	Propane	3	$CH_3CH_2CH_3$	C_3H_8	−42	−189	0.50	Cooking fuel
gases	Butane	4	$CH_3(CH_2)_2CH_3$	C_4H_{10}	0	−138	0.58	Lighters, fuel for camping
liquids	Pentane	5	$CH_3(CH_2)_3CH_3$	C_5H_{12}	36	−130	0.63	Blowing agent for foam production
liquids	Hexane	6	$CH_3(CH_2)_4CH_3$	C_6H_{14}	69	−95	0.66	Nonpolar organic solvent
liquids	Heptane	7	$CH_3(CH_2)_5CH_3$	C_7H_{16}	98	−91	0.68	Nonpolar organic solvent
liquids	Octane	8	$CH_3(CH_2)_6CH_3$	C_8H_{18}	126	−57	0.70	Hydrocarbon in gasoline
liquids	Nonane	9	$CH_3(CH_2)_7CH_3$	C_9H_{20}	151	−51	0.72	Found in kerosene
liquids	Decane	10	$CH_3(CH_2)_8CH_3$	$C_{10}H_{22}$	174	−30	0.73	Component of gasoline
liquids	Undecane	11	$CH_3(CH_2)_9CH_3$	$C_{11}H_{24}$	196	−26	0.74	
liquids	Dodecane	12	$CH_3(CH_2)_{10}CH_3$	$C_{12}H_{26}$	216	−10	0.75	
liquids	Tridecane	13	$CH_3(CH_2)_{11}CH_3$	$C_{13}H_{28}$	235	−5	0.76	
liquids	Tetradecane	14	$CH_3(CH_2)_{12}CH_3$	$C_{14}H_{30}$	254	6	0.76	
liquids	Pentadecane	15	$CH_3(CH_2)_{13}CH_3$	$C_{15}H_{32}$	271	10	0.77	
liquids	Hexadecane	16	$CH_3(CH_2)_{14}CH_3$	$C_{16}H_{34}$	287	18	0.77	
liquids	Heptadecane	17	$CH_3(CH_2)_{15}CH_3$	$C_{17}H_{36}$	303	23	0.76	
waxes	Octadecane	18	$CH_3(CH_2)_{16}CH_3$	$C_{18}H_{38}$	317	28	0.76	Phase change material
waxes	Nonadecane	19	$CH_3(CH_2)_{17}CH_3$	$C_{19}H_{40}$	330	32	0.78	Phase change material
waxes	Eicosane	20	$CH_3(CH_2)_{18}CH_3$	$C_{20}H_{42}$	343	37	0.79	
waxes	Triacontane	30	$CH_3(CH_2)_{20}CH_3$	$C_{30}H_{62}$	>450	66	0.81	

[a]Densities are given in g/mL at 20 °C, except for methane and ethane, whose densities are given at their boiling points.

Alkanes are highly flammable and undergo combustion to produce energy. Thus, many of the alkanes in Table 3.2 are useful as fuels. Methane is the primary component of natural gas and is itself a greenhouse gas with a greater warming potential than carbon dioxide. Propane and butane are used as cooking fuels. The C_4—C_{12} alkanes are primary components of gasoline. The C_{10}—C_{15} alkanes are present in diesel fuel. The higher alkanes ($>C_{20}$) are waxes, malleable solids near room temperature that become liquid with slight heating. Finally, hexane and pentane are standard fare in organic chemistry labs as low boiling, nonpolar solvents.

3.1.2 Molecular orbital picture of alkanes

$H_3C—CH_3$
ethane

The shape and properties of alkanes are a consequence of the molecular orbital picture. [*Try drawing the molecular orbital picture of ethane on your own.*]

Ethane (**Figure 3.2**) is composed of two tetrahedral, sp^3-hybridized carbons. Each C—H σ bond is formed by the overlap of an sp^3-hybridized orbital from carbon and the 1*s* orbital of hydrogen. The σ bond between the two carbons is composed of overlapping sp^3-hybridized orbitals.

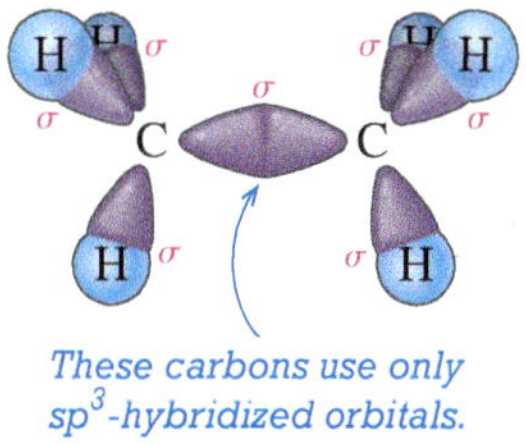

Figure 3.2 **Putting the individual molecular orbitals together to give the full MO picture of ethane.**

Assessment 3.6 How many hydrogens would you expect a 24-carbon compound from each of the following molecular classes to have?

(a) Alkane (b) Alkene (c) Alkyne

Assessment 3.7 Draw the molecular orbital picture for propane. Your picture should clearly show the shape and hybridization of the carbons. Label all σ bonds.

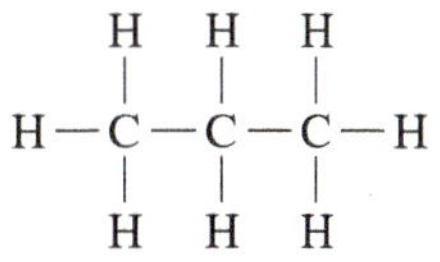

propane

3.1.3 Properties of alkanes

Table 3.2 lists the boiling point, melting point, and density of 21 different alkanes. Each of these physical properties is directly correlated with the strength of their **intermolecular forces**, which depend on the degree to which molecules interact with each other. Molecules with strong attractive forces between them will have high melting points, high boiling points, and high densities.

The four intermolecular forces most commonly studied in an introductory chemistry course are **ionic, hydrogen-bonding, dipole–dipole,** and **van der Waals.** [*Covalent bonds are not considered intermolecular forces because they are bonds within molecules, not forces* *between molecules.*] These four intermolecular forces are listed in order of decreasing strength in Table 3.3. The strong interactions in ionic compounds explain the extremely high melting point for salts like sodium chloride (801 °C) and strong hydrogen bonding explains many of the properties of water (bp = 100 °C, density = 1.00 g/mL).

Table 3.3 Four types of intermolecular forces we'll study in organic chemistry.

Intermolecular force	Description	Example
Ionic	The interaction between oppositely charged ions, as a result of electrostatic attraction	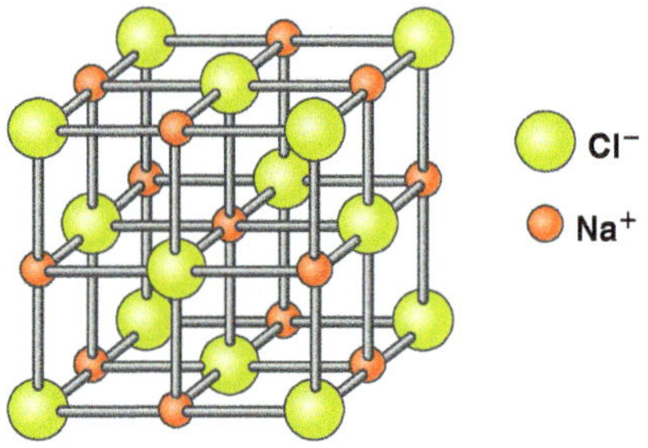
Hydrogen bonding	The formation of partial bonds between electronegative atoms (N, O, and F) and hydrogens attached to electronegative atoms, such as N, O, and F (do not confuse a hydrogen bond with a σ bond to hydrogen)	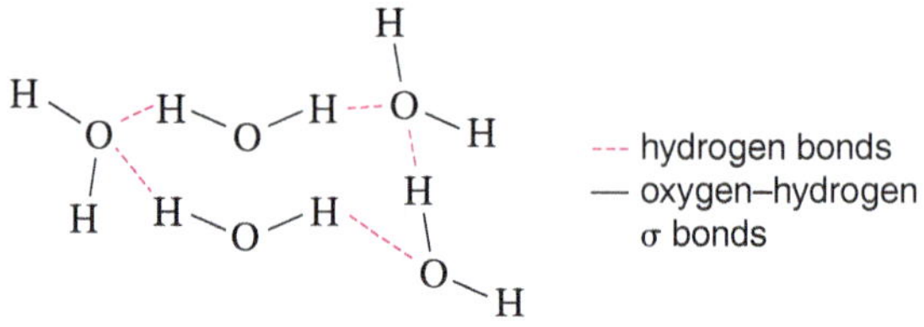
Dipole–dipole	The interaction between opposite partial charges resulting from polar bonds	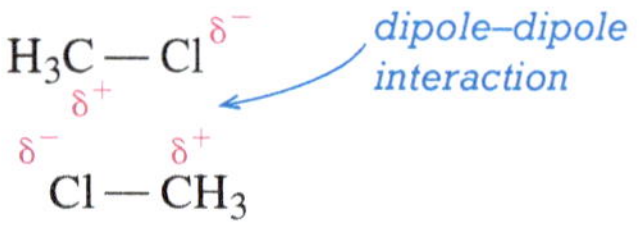
van der Waals/London dispersion forces	The interaction of induced dipoles in otherwise nonpolar molecules ()	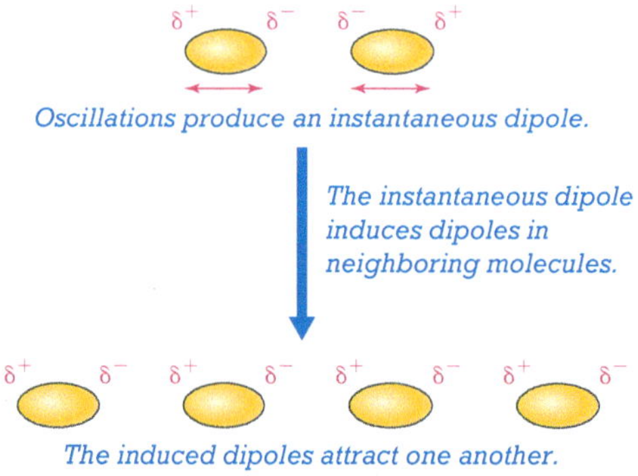

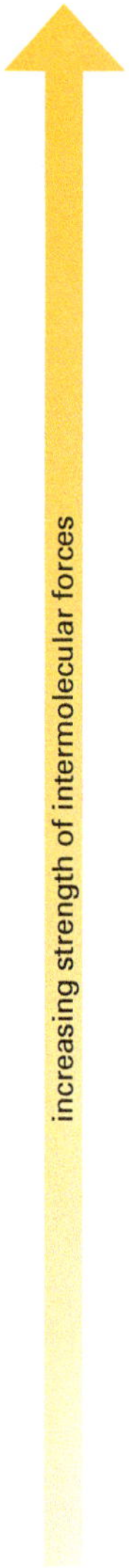

Alkanes are nonpolar molecules because the electronegativity values of carbon (2.55) and hydrogen (2.20) are so similar. As a result, London dispersion forces are the major intermolecular forces between alkanes. The observed boiling point of methane (−164 °C) is a result of the individual dispersion forces being weak [Figure 3.3(a)]. In larger alkanes, however, the cumulative effect of multiple dispersion interactions can result in extremely high boiling points, as in dodecane (216 °C) [Figure 3.3(b)]. The higher boiling points, melting points, and densities of the larger alkanes are a result of having more contact points for these dispersion forces.

In addition to the straight-chain alkanes in Table 3.2, alkanes can also be branched. [*We study these in more detail in Section 3.3.1.*] Chain branching has a dramatic effect on the intermolecular forces, lowering the boiling point and melting point, while increasing water solubility, as shown in the comparison of butane and isobutane in Table 3.4.

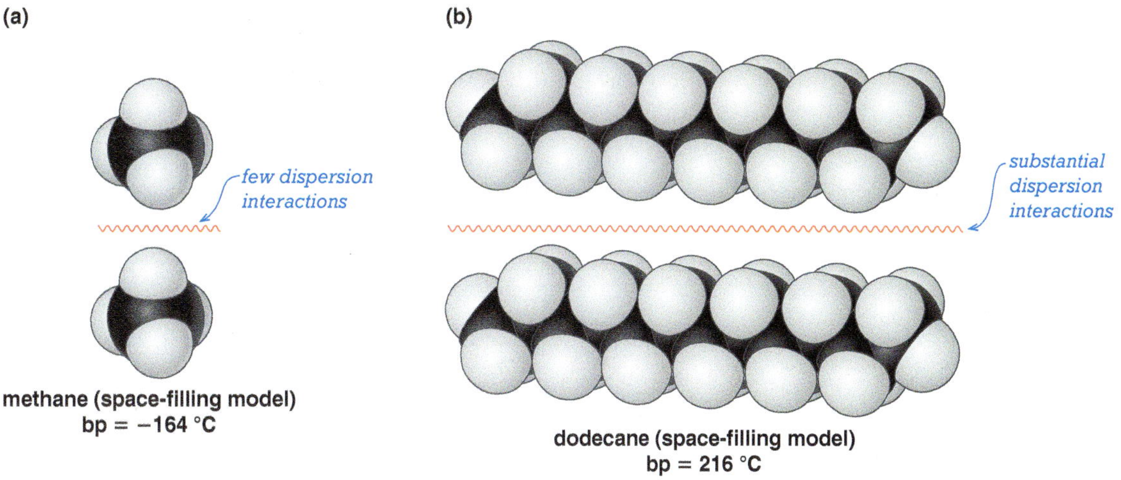

Figure 3.3 Dispersion forces in alkanes. (a) Methane is small and has very few dispersion interactions. (b) Dodecane is large and has many dispersion interactions.

When you come across a table, can you rationalize or identify the trend that is illustrated? This is a key to making the connections necessary for true understanding.

Table 3.4 The effect of chain branching on physical properties.

Molecule	Melting point (°C)	Boiling point (°C)	Water solubility (mg/L H_2O)
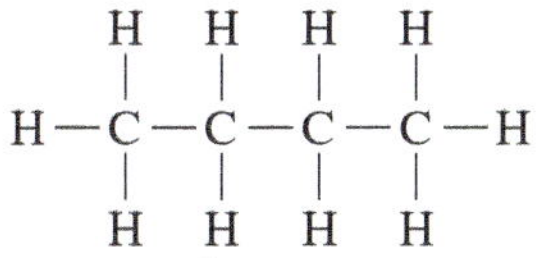 butane	−138	0	16
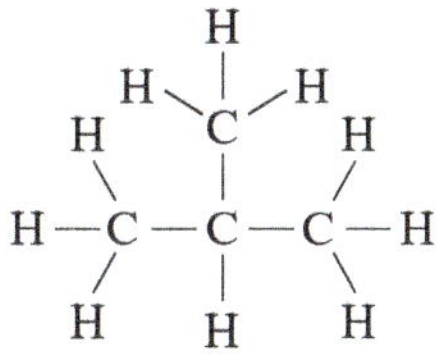 isobutane	−160	−12	49

Branched alkanes are more compact, so they have a much smaller surface area compared to their linear counterparts (Figure 3.4). As a result, there are far fewer contact points for the dispersion forces in branched alkanes. With fewer contact points, branched alkanes are not as attracted to each other and thus boil and melt at lower temperatures than straight-chain alkanes with the same number of carbon atoms.

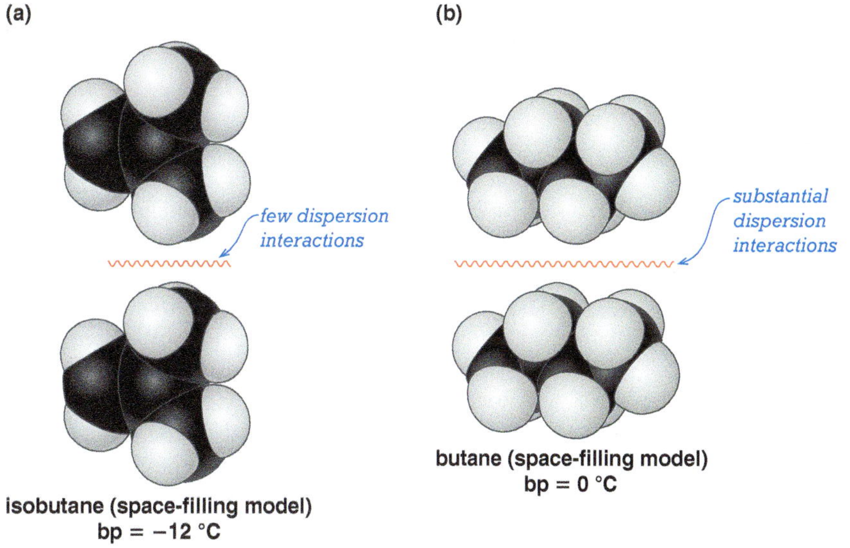

Figure 3.4 Branched alkanes boil at lower temperatures. (a) Isobutane has few dispersion interactions. (b) Butane has more.

Because alkanes are nonpolar, they are essentially insoluble in water (which is polar). Smaller alkanes are more soluble than larger alkanes (**Table 3.5**). Branched alkanes, having less surface area, are more soluble than their straight-chain homolog (Table 3.4).

Table 3.5 Water solubility decreases with increasing alkane size.

Molecule	Structure	Water solubility (mg/L H_2O)
Butane	$CH_3CH_2CH_2CH_3$	61
Pentane	$CH_3CH_2CH_2CH_2CH_3$	40
Hexane	$CH_3CH_2CH_2CH_2CH_2CH_3$	9

3.1.3 SUMMARY

- London dispersion forces are the major intermolecular interaction between alkanes, because they are nonpolar molecules. Alkanes with a larger surface area have stronger intermolecular interactions and therefore increased boiling points, melting points, densities, and decreased water solubilities.

Assessment 3.8 Choose the molecule in each pair you'd expect to have the higher boiling point. Explain your reasoning.

(a) eicosane ($C_{20}H_{42}$) vs. pentadecane

(b)

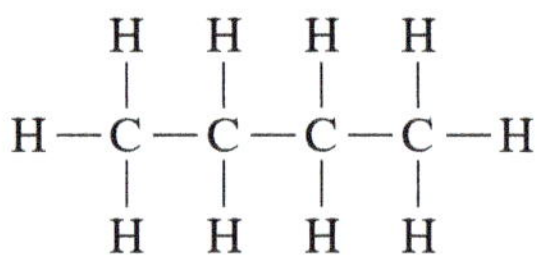

n-butane

vs.

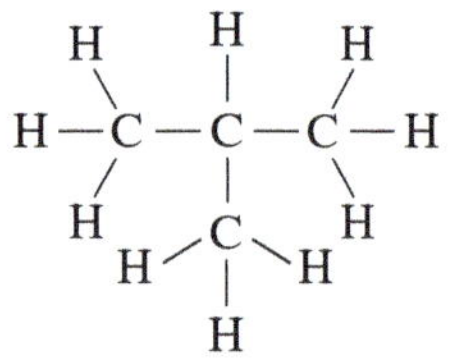

isobutane

Assessment 3.9 Rank the following molecules from least water soluble to most water soluble. Explain your reasoning.

(a)	(b)	(c)	(d)	(e)
$H_3C—OH$	LiCl	$H_3C—CH_3$	$H_3C—F$	$CH_3CH_2CH_2CH_2CH_3$

3.2 Drawing organic molecules

We have drawn several organic molecules in the preceding sections. If you have been taking notes as you read [*always a good idea*] or have been working along with the assessments [*an even better idea*], you have likely noticed that drawing larger organic molecules takes a lot of time. [*There has to be an easier way, right?*]

There are several ways that organic molecules can be represented, each simpler than the next. [*Along with the nomenclature we study in Section 3.3, drawing molecules in an efficient and standardized way makes communicating about organic chemistry easier.*] Shown in **Figure 3.5** are structural representations that are essentially the Lewis structures we learned to draw in Chapter 2. The structural formula shows all bonds (C—C and C—H). At the other end of the spectrum are condensed structural formulas, which show only the atoms in the order they are bonded (but does not show any bonds.) Intermediate between the two is a hybrid structure that shows all C—C bonds but does not show the individual C—H bonds.

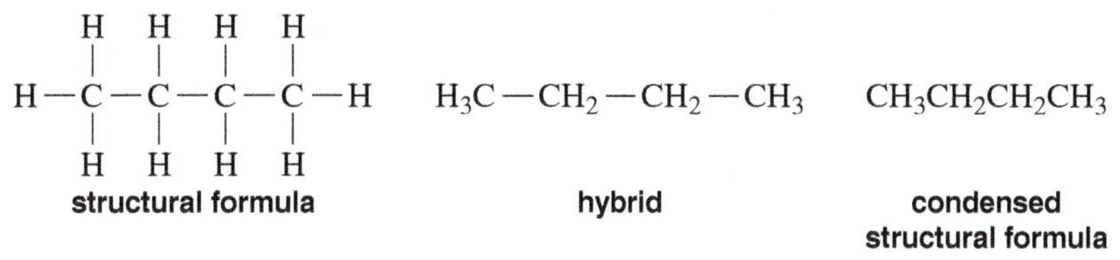

Figure 3.5 Representations of butane. Structural, hybrid, and condensed structural formulas.

Compared to the structural formula, both the hybrid and condensed structural formula save time by leaving the viewer to "fill in" the missing information. While condensed structural formulas are easier to write for small molecules, they are time consuming, and not especially efficient for larger, branched molecules (**Figure 3.6**). They are also not useful for the rings we'll study in Section 3.5. For branched molecules, substituents coming off a ring are indicated by placing it in parentheses. The inefficiencies of the condensed structural formulas are shown by comparison with the hybrid formula for 2,4-dimethylhexane [*We learn how to name in Section 3.3*]. Even for seasoned chemists, the condensed structural formula, especially when other functional groups are included for complex molecules, can be confusing. [*The condensed structural formula arose during the time of typewriters, when molecules needed to be displayed in a single line of text.*]

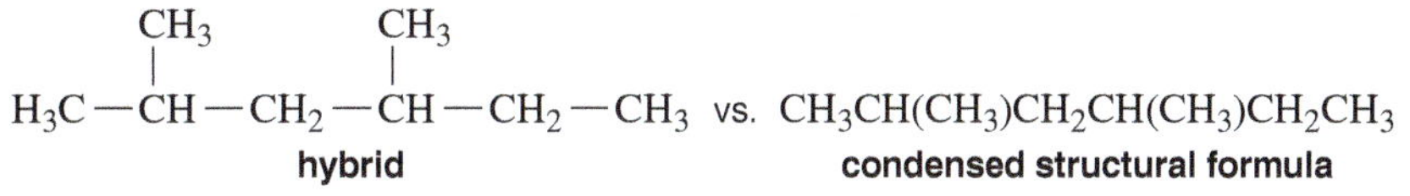

Figure 3.6 Hybrid versus condensed structural formula for 2,4-dimethylhexane.

The problems of the complexity and time required for condensed structural formulas are solved by using **line-angle drawings**. The line-angle drawing for 2,4-dimethylhexane is shown in **Figure 3.7**. Line-angle drawings require the least amount of time to draw, yet they clearly and fully illustrate the structural features of a molecule. That said, they do require the viewer to make some assumptions. First, a carbon is assumed to be present whenever a line ends, begins, or changes direction. Second, because carbon usually has four bonds, and hydrogens are ever present in organic molecules, any bond not shown is assumed to be to a hydrogen. All hydrogens bonded to non-carbon atoms are shown. [*This comes up in Section 3.8.1 when we draw functional groups.*]

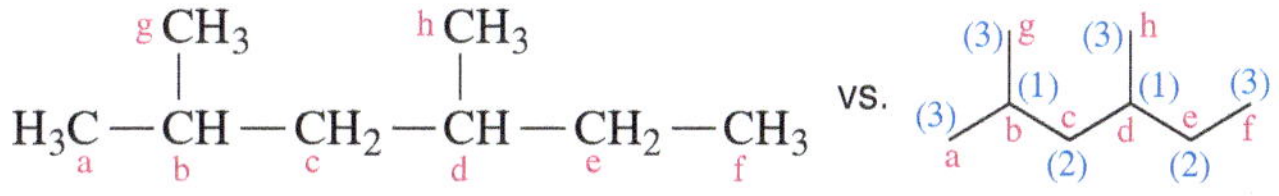

Figure 3.7 A comparison of the hybrid structural formula and the line-angle drawing of 2,4-dimethylhexane shows that both contain the same structural features, but the line-angle drawing is simpler to draw. To help you convert between the two, carbons are identified with letters, and the number of hydrogens is indicated in parentheses.

As shown in **Figure 3.8**, line-angle drawings are great for showing rings, alkenes, alkynes, and other functional groups [*see Section 3.8.1*]. Before reading further, see if you can convert the structural formulas on the left into line-angle drawings on the right. With alkynes, to conserve the 180° bond angle of the *sp*-hybridized carbon, it appears that one line continues in the same direction. The beginning of the two π bonds indicates the presence of the *sp*-hybridized carbons. We use only line-angle drawings in this text going forward, so it is very important that you quickly figure out how to represent molecules in this way. [*Assessments 3.10 and 3.11 can help you with this.*]

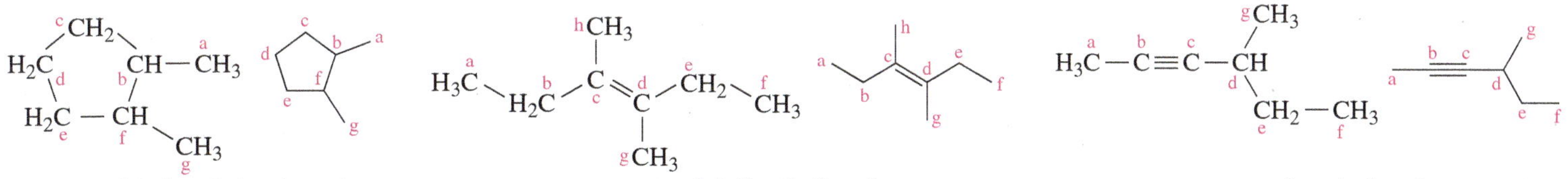

1,2-dimethylcyclopentane 3,4-dimethylhex-3-ene 4-methylhex-2-yne

Figure 3.8 Line-angle drawings for cycloalkanes, alkenes, and alkynes.

3.2 SUMMARY

- Learning how to draw and interpret line-angle drawings is key to succeeding in organic chemistry.

Assessment 3.10 For each of the following line-angle drawings, (i) give the number of carbons, (ii) label the carbons, (iii) tell how many hydrogens are on each carbon, and (iv) draw the hybrid structural formula.

(a) (b) (c) (d)

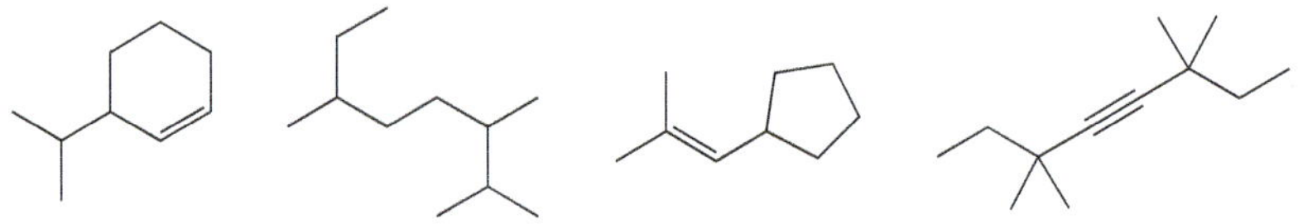

Assessment 3.11 Convert the following hybrid structural formulas into the line-angle drawings.

(a) (b) (c)

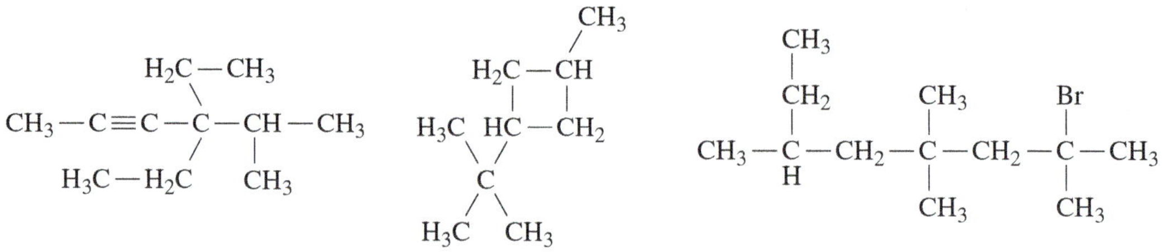

Assessment 3.12 **(LOOKING AHEAD)** In Section 3.3.3, we explain that a 1° carbon is attached to one carbon, a 2° to two, a 3° to three, and a 4° to four. Label each of the carbons in 4-ethyl-1,1-dimethylcycloheptane as 1°, 2°, 3°, or 4°.

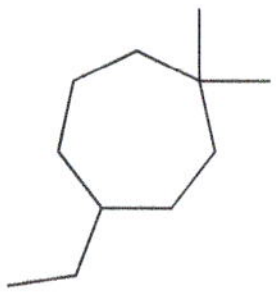

4-ethyl-1,1-dimethylcycloheptane

3-A CHEMISTRY IN CONTEXT: Food Chemistry

Saturated versus unsaturated fats

Section 3.1 introduced the term *saturated* when describing alkanes having the maximum number of hydrogens possible per carbon. Molecules that are *unsaturated* have *fewer* hydrogens than the maximum possible. The terms saturated and unsaturated are used in everyday life in the context of saturated and unsaturated fats. Though it may not be obvious, these terms on nutrition labels mean exactly what they mean in organic chemistry.

Although we study fats, oils, and fatty acids at other points in the text, Figure 3.9 shows that soybean oil, a triglyceride

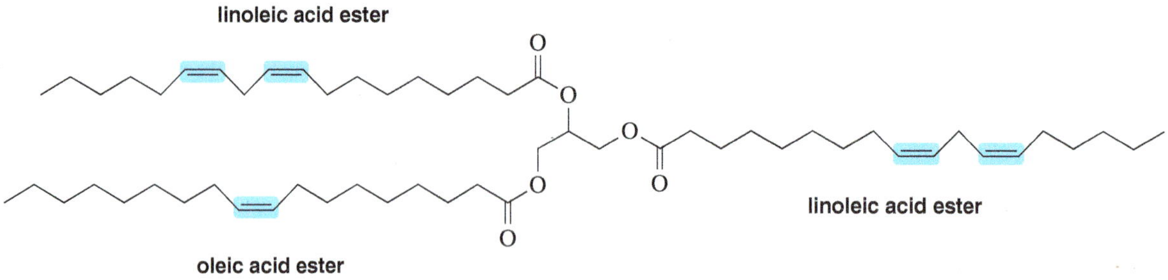

Figure 3.9 Structure of soybean oil, an unsaturated fat. The sites of unsaturation are highlighted in blue.

(continued)

composed of esters of linoleic and oleic acid, contains multiple C=C double bonds. For each double bond, the molecule has two fewer hydrogens than is maximally allowed. Thus, these are unsaturated fats. [*Carboxylic acids and esters are examples of functional groups we preview in Table 3.15.*]

In contrast, triglycerides from animal fats contain zero C—C double bonds (Figure 3.10). As a result, these fats are saturated—they contain the maximum number of hydrogens possible.

For reasons that we explain in later Chemistry in Context boxes, unsaturated fats are better for you than saturated fats.

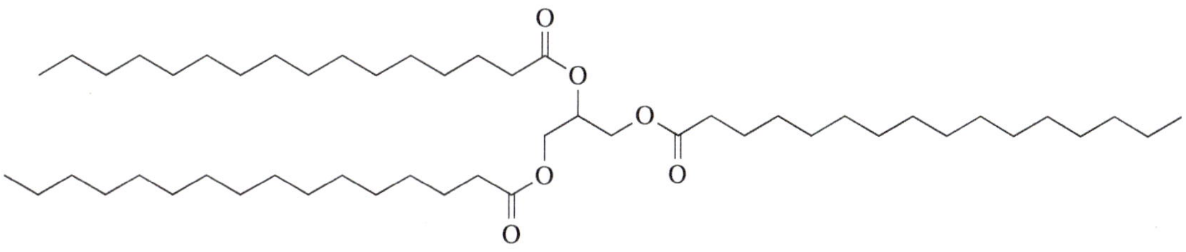

Figure 3.10 Structure of a saturated animal fat. There are no sites of unsaturation.

3.3 Alkane nomenclature

In reading the first few sections of this chapter, it becomes apparent that there are a *huge* number of alkane structures that can be imagined, especially when considering both branched and straight-chain isomers. The following sections will help us understand constitutional isomerism, the need for a systematic approach to nomenclature, and how alkanes [*and all future molecules*] are named.

3.3.1 Constitutional isomerism

This is the first [*of many*] times we'll use the term **isomer** to define relationships between similar molecules. Two molecules are considered to be isomers if they have the same molecular formula but some other feature that makes them different. The prefix "iso-" means equal, while "-mer" implies part. Thus, isomers are "equal in part." Figure 3.11 shows examples of isomeric relationships we'll study in this book. [*You don't need to know these now, but should be able to recognize how the molecules in each pair are similar yet different.*]

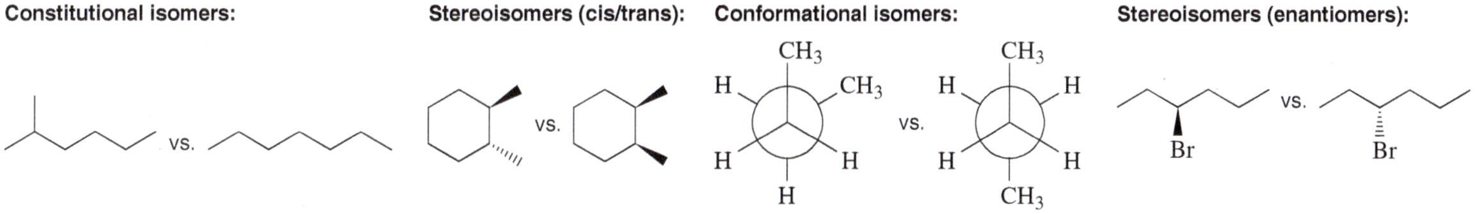

Figure 3.11 Examples of isomerism.

The fact that carbon forms four bonds in a tetrahedral arrangement results in many of the isomeric relationships in Figure 3.11. Important to our studies of alkanes is constitutional isomerism. **Constitutional isomers** are molecules that have the same molecular formula but different connectivity. While methane (CH_4), ethane (C_2H_6), and propane (C_3H_8) don't have constitutional isomers, butane and pentane have two and three, respectively (Figure 3.12). The prefix iso- refers to the constitutional isomer with a methyl group on the second carbon of the longest chain (isobutane and isopentane). Neo- is used when there are two methyl groups on the second carbon of the longest chain (neopentane).

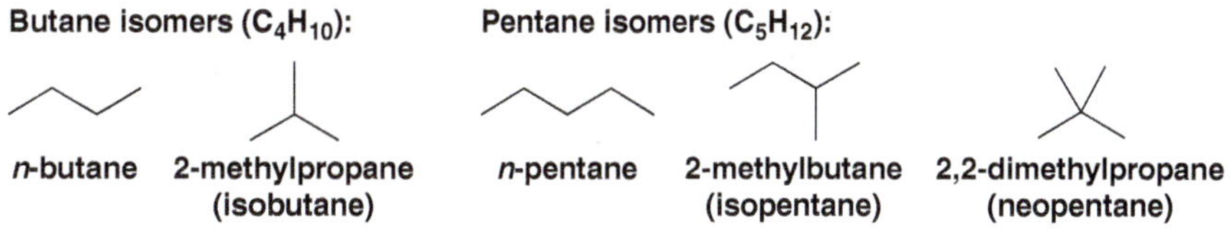

Figure 3.12 Constitutional isomers of butane and pentane.

Table 3.6 Constitutional isomers as a function of carbon number.

C_x alkanes	Constitutional isomers
1	1
5	3
10	75
20	366,319
100	$> 5.92 \times 10^{39}$

As you increase the number of carbons, the number of possible isomers increases exponentially. While a ten-carbon alkane has 75 constitutional isomers, there are 366,319 C_{20} alkane isomers and 5.92×10^{39} C_{100} isomers (Table 3.6). [*Can you imagine the muscle cramps if your instructor asks you to draw all of them?*]

Assessment 3.13 There are five alkane constitutional isomers with the molecular formula C_6H_{14}. Draw them.

Assessment 3.14 There are nine alkane constitutional isomers with the molecular formula C_7H_{16}. Draw them.

3.3.2 IUPAC system

When studying these rules, some people approach them as if they are completely arbitrary and unrelated. Instead, imagine that you were on the committee charged with writing these rules. With each increase in complexity, think about how you would craft the rule such that (1) it would make logical sense and (2) it would differentiate between closely related, but different, compounds.

For smaller alkanes (C_4 and C_5), the use of the prefixes iso- and neo- are sufficient to discriminate between the constitutional isomers. For a C_{10} alkane, however, there are not 75 different prefixes that could be used to differentiate between the constitutional isomers. Even if there were, learning them all would be a fruitless effort in mass memorization. And yet, for the sake of communication between chemists across continents, there is a need for a systematic way to name these different isomers. [*A German chemist, French chemist, and American chemist walked into a bar . . . .*] The International Union of Pure and Applied Chemistry (IUPAC) has developed rules for naming organic compounds. We use this system to name alkanes, cycloalkanes (Section 3.5), and every new functional group we study in later chapters. [*. . . Because they knew IUPAC nomenclature, they could discuss the structures on their napkins.*]

The IUPAC system is best illustrated by applying the rules as we learn them, starting with the simplest cases and moving to the more complex. Two examples of each rule are given in Table 3.7. [*Nothing will help you understand this more than applying the rules on your own in Assessment 3.15.*]

Table 3.7 IUPAC nomenclature for alkanes (part 1).

IUPAC rules	Example 1	Example 2
1. For straight-chain alkanes, use the prefix to indicate the number of carbons. [*See Table 3.2 for the prefixes.*] The infix -an- indicates that it is fully saturated. The suffix -e indicates that the molecule is a hydrocarbon. [*See Assessment 3.15(a, b) for practice.*]	 octane	 decane
2. Branched alkanes are named as if a substituent is coming off of a parent chain. The parent chain is the longest continuous chain and represents the root name of the molecule. The substituent is named using the appropriate prefix to indicate the number of carbons in the substituent and the suffix -yl to indicate that it is a substituent. The location of the substituent is indicated by numbering the carbons such that the substituent comes off the lowest numbered carbon. [*See Assessment 3.15(c, d) for practice.*]	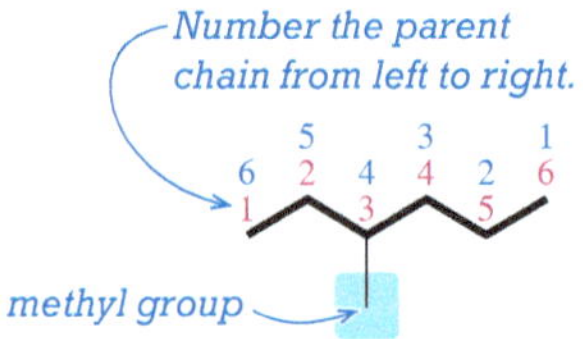 3-methylhexane	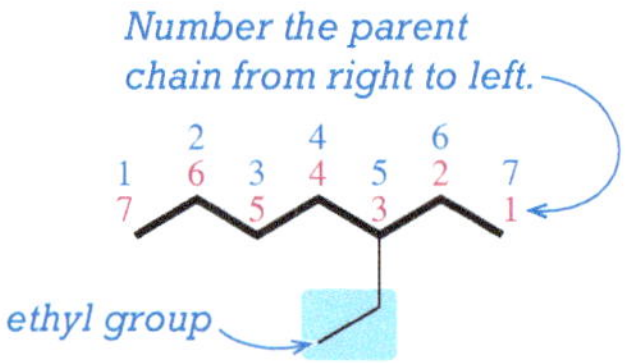 3-ethylheptane

(*continued*)

Table 3.7 ***Continued***

IUPAC rules	Example 1	Example 2
Important note: When writing the names, the IUPAC system is very particular. The location of the substituent is indicated by writing "number-substituent." The name of a molecule is a single word. Thus, there is no space between the final substituent and the root name.		
3. When there is more than one of the same substituent on the parent chain, the prefixes di-, tri-, tetra-, penta-, etc. are used to indicate how many substituents there are. The parent chain is numbered such that the lowest possible number can be assigned to the first substituent. [*See Assessment 3.15(e, f) for practice.*]	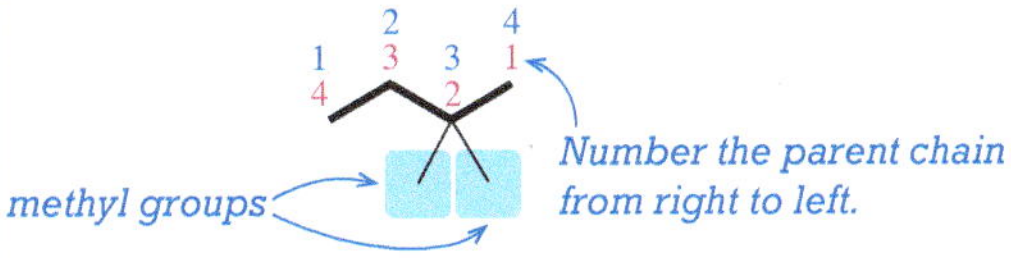 **2,2-dimethylbutane**	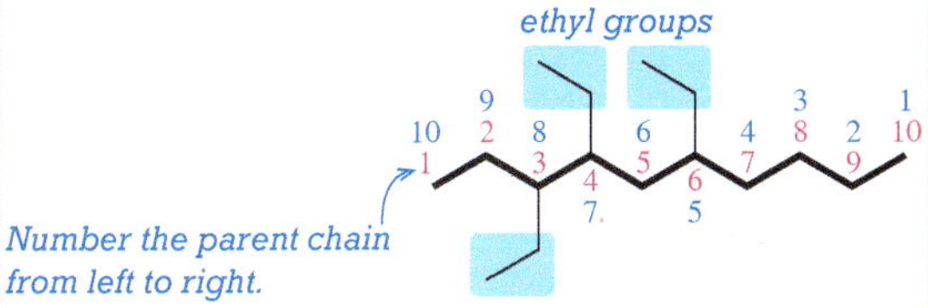 **3,4,6-triethyldecane**
4. When a parent chain has two different substituents attached, the chain is numbered such that the first substituent encountered has the lowest possible number. The names of the substituents are written alphabetically. The prefixes di-, tri-, tetra-, penta-, etc. are *not* alphabetized. [*See Assessment 3.15(g, h) for practice.*]	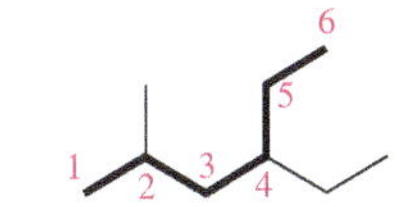 **4-ethyl-2-methylhexane**	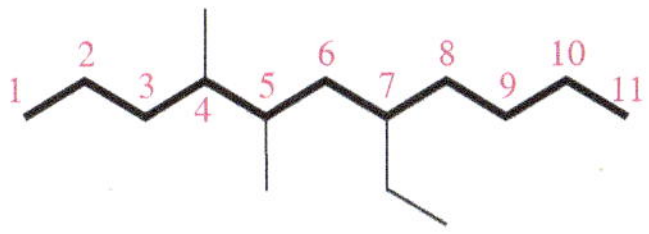 **7-ethyl-4,5-dimethylundecane**
Important note: Numbers that indicate the location of the substituent are separated by commas, as in "number,number-substituent." When a substituent is followed by another substituent, the name is separated from the next number by a dash, as in "number-substituent-number-substituent."		
5. When numbering the parent chain in either direction would give the same number to either of two different substituents, give the lower number to the next substituent encountered to break the tie. If the tie cannot be broken by the "next substituent," give the lower number to the one that comes first alphabetically. [*See Assessment 3.15(i, j) for practice.*]	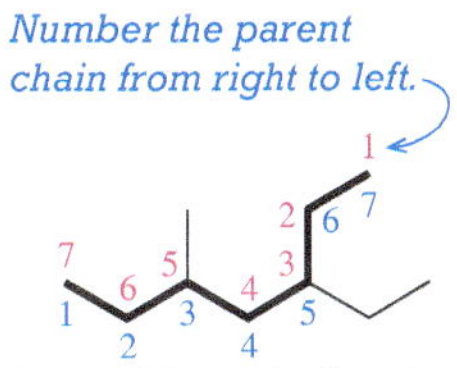 **3-ethyl-5-methylheptane**	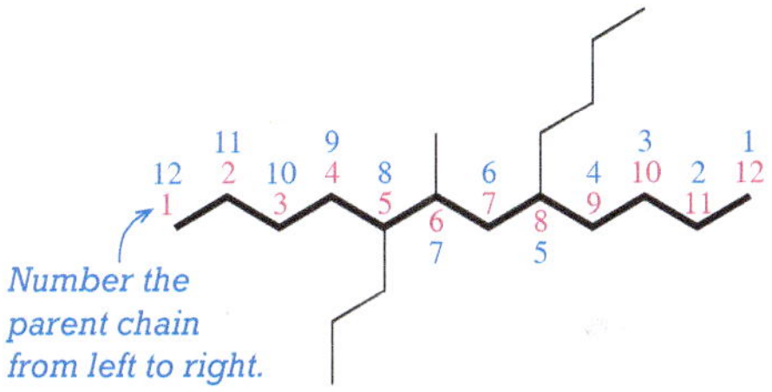 **8-butyl-6-methyl-5-propyldodecane**

The substituents coming off of the parent chain need not be the small, simple substituents like methyl, ethyl, and propyl. The substituents can also be branched themselves. Branched substituents are named in a manner consistent with IUPAC nomenclature. The longest carbon chain of the substituent is named as the parent chain (of the substituent), with the carbon where it attaches to the main chain being assigned the lowest number. All other substituents on the branched substituent are assigned numbers to indicate their position. Several examples of common substituents are named via IUPAC nomenclature in **Figure 3.13** (R = the parent chain). Common names of these substituents [*see Section 3.3.4*] are also given. [*See Assessment 3.19 for practice.*] The IUPAC rules for naming alkanes with branched substituents are listed in **Table 3.8**.

1-methylpropyl (*sec*-butyl) | **2-methylpropyl** (isobutyl) | **1,1-dimethylethyl** (*tert*-butyl) | **1-methylethyl** (isopropyl) | **1,1-dimethylpropyl** (*tert*-pentyl) | **1,2-dimethylpropyl**

Figure 3.13 IUPAC nomenclature of branched substituents. R indicates the parent chain to which the branched substituent is attached.

Table 3.8 IUPAC nomenclature for alkanes with branched substituents (part 2).

IUPAC rules	Example 1	Example 2
6. When branched substituents appear on a parent chain, number them consistent with previous numbering rules and alphabetize them based on the first letter of the substituent. The branched substituent name ends in -yl and is placed in parentheses. [*See Assessment 3.15(k, l) for practice.*]	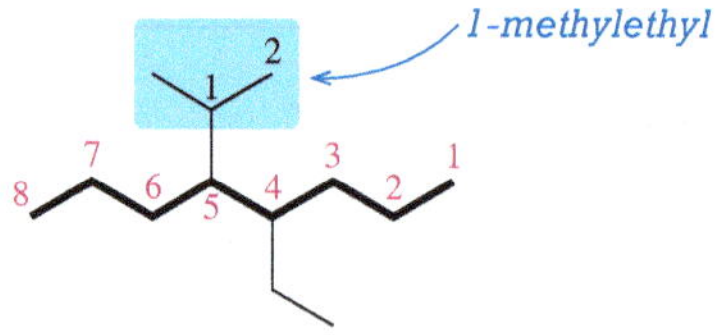 **4-ethyl-5-(1-methylethyl)octane**	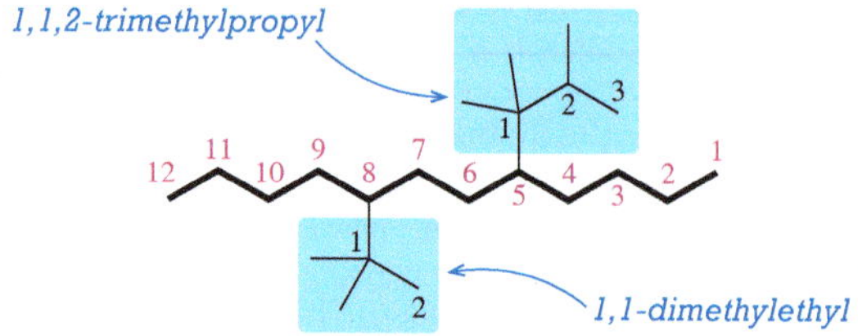 **8-(1,1-dimethylethyl)-5-(1,1,2-trimethylpropyl)undecane**
7. When there are two possible parent chains of equal length, choose the parent chain that has the most substituents in order to avoid using the more complex names for branched substituents. [*See Assessment 3.15 (m, n) for practice.*]	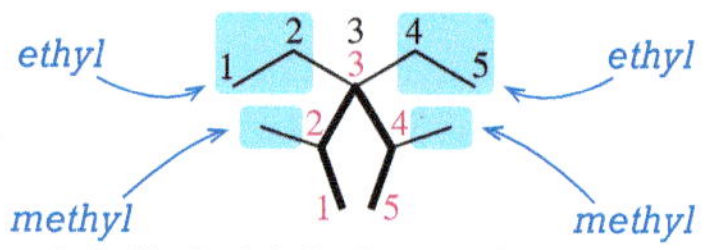 **3,3-diethyl-2,4-dimethylpentane**	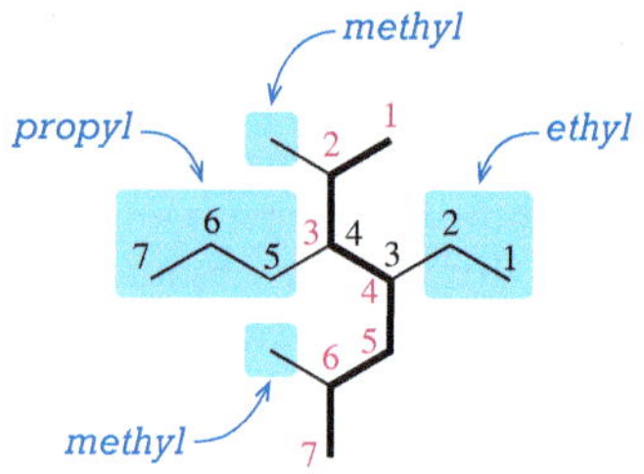 **4-ethyl-2,6-dimethyl-3-propylheptane**

3.3.2 SUMMARY

- The IUPAC rules for nomenclature are logical and will find application for every functional group we encounter. They are best learned through practice.

Assessment 3.15 Name the following alkanes using the IUPAC system of nomenclature. [*Each molecule exemplifies one of the nomenclature rules in Tables 3.7 and 3.8.*]

(a) rule 1

(b) rule 1

(c) rule 2

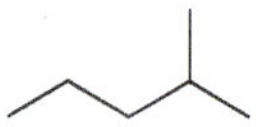

(d) rule 2

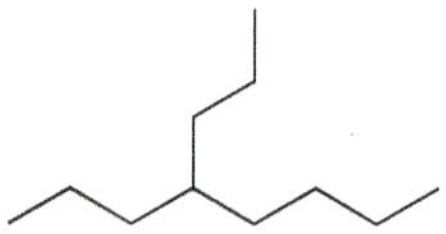

(e) rule 3

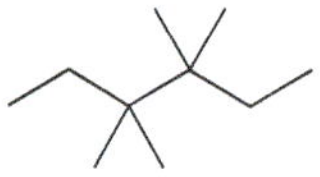

(f) rule 3

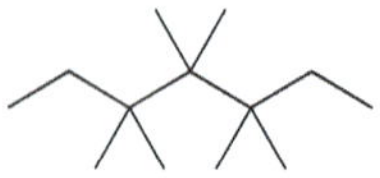

(g) rule 4

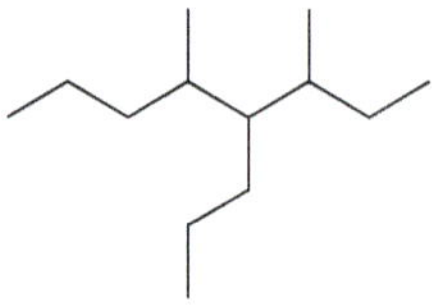

(h) rule 4

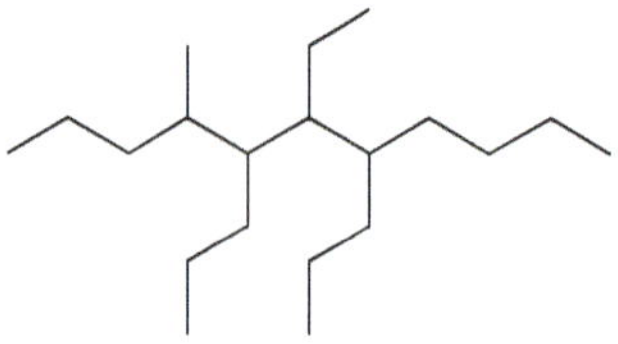

(i) rule 5

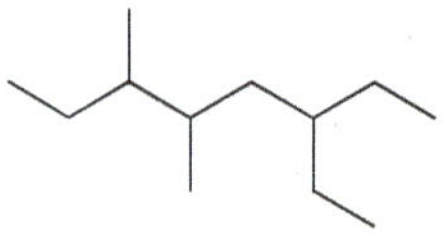

(j) rule 5

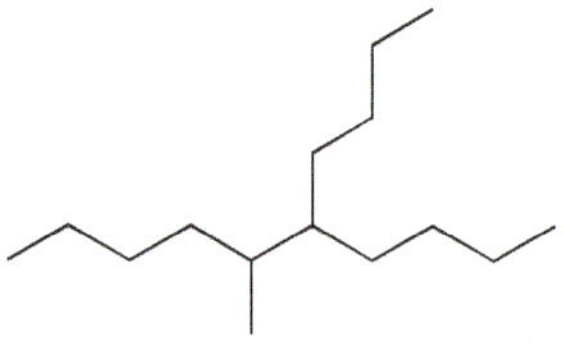

(k) rule 6

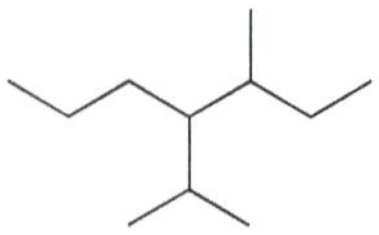

(l) rule 6

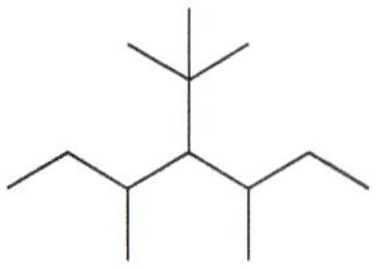

(m) rule 7

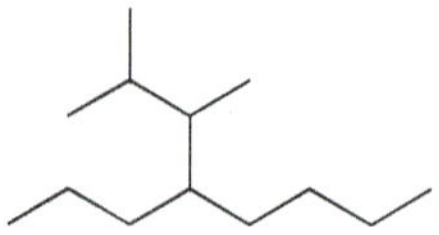

(n) rule 7

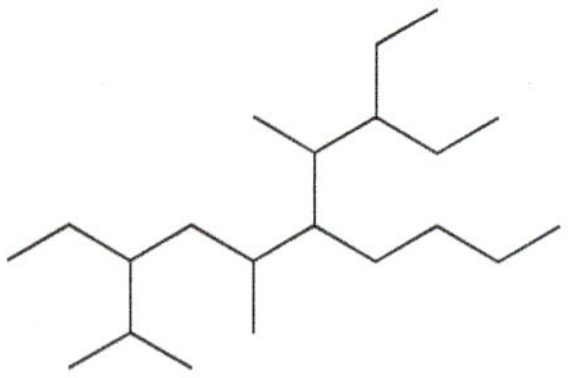

3.3.3 Classifying carbon substitution

There are many times when the reactivity of a molecule will depend on the type of carbon to which a functional group is bound. As a result, it is important to be able to easily distinguish between the different types of carbons, based on the substitution. The substitution of a carbon is indicated using 1° (**primary**), 2° (**secondary**), 3° (**tertiary**), and 4° (**quaternary**) to indicate that one, two, three, or four carbons are attached, respectively. The application of these designations is summarized in **Figure 3.14** and assessed in Assessments 3.16–3.18.

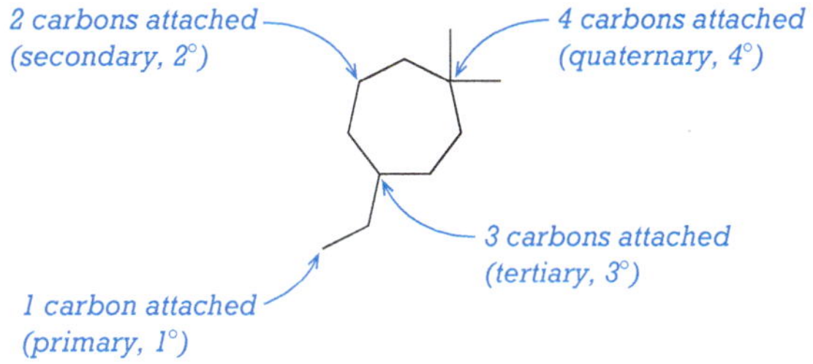

Figure 3.14 Designation of carbon substitution.

How would we use carbon substitution?
Student: To which carbon are you referring?
Me: The 3° (tertiary) carbon!

Assessment 3.16 When we begin studying functional groups, the designation of substitution will be especially important. Label the following bromoalkanes as 1°, 2°, 3°, or 4° based on the carbon to which the bromine is attached.

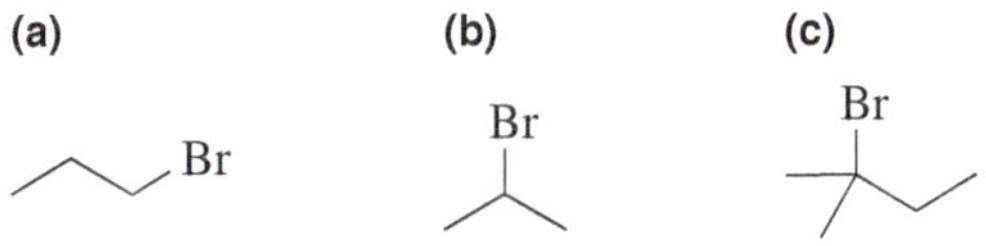

Assessment 3.17 When we begin studying reactive intermediates, the designation of substitution will be especially important. Label the following carbocations as 1°, 2°, 3°, or 4° based on the carbon bearing the positive charge.

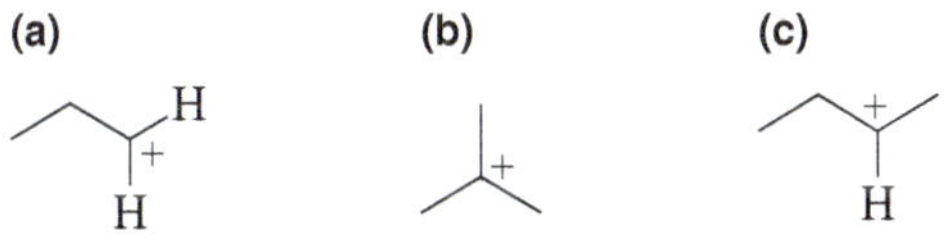

Assessment 3.18 Why is it not possible to have a 4° bromoalkane or carbocation? [*Use the structures in Assessments 3.16 or 3.17 to help you answer this question.*]

3.3.4 Common substituent names

Naming branched substituents in Section 3.3.2 was difficult and complicated the naming of molecules containing them. To mitigate this difficulty, the IUPAC system allows for the use of common substituent names. The most often used common names for C_3, C_4, and C_5 substituents are shown in **Figure 3.15** (R = the parent chain). It is unnecessary to have common names for larger substituents since, except when attached to rings, the parent chain of the substituent often becomes a part of the parent chain of the main molecule. See rule 7 of Table 3.8.

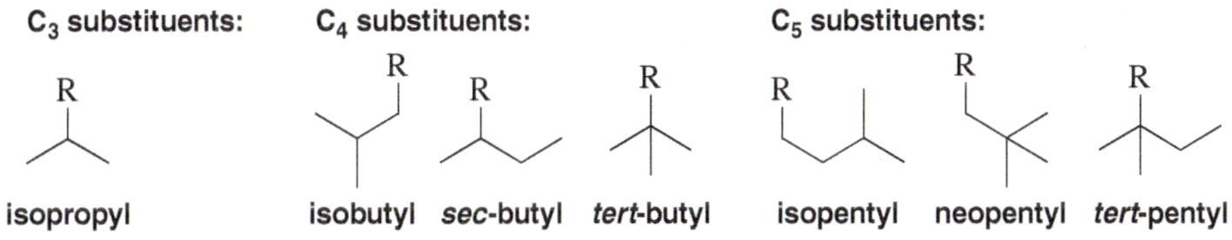

Figure 3.15 Common names of branched substituents. R indicates the parent chain to which the branched substituent is attached.

When using these common names in IUPAC nomenclature, substituents beginning with iso- and neo- should be alphabetized based on "i" and "n," but those beginning with *sec*- and *tert*-, the italicized ones, should be alphabetized based on "butyl" or "propyl." **Figure 3.16** shows two examples of naming using common substituents.

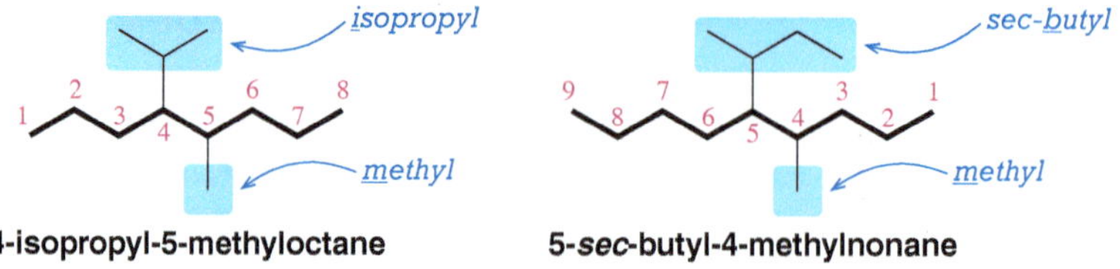

Figure 3.16 Naming with common names for branched substituents.

Assessment 3.19 Name the following alkanes using the common names for branched substituents.

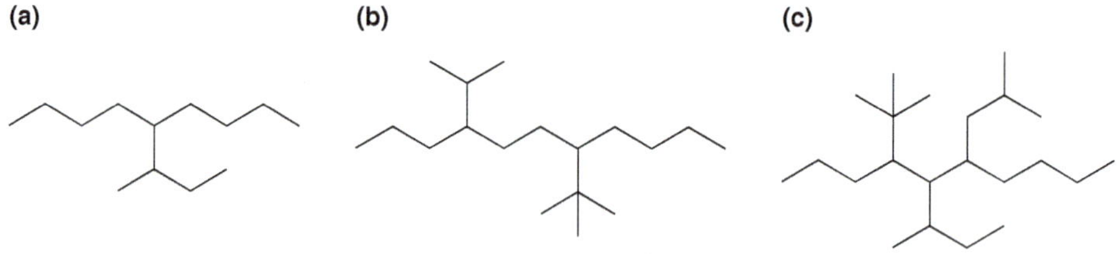

3-B CHEMISTRY IN CONTEXT: Petroleum Chemistry

Distillation of crude oil

When crude oil is removed from the ground, it must be refined into its useful components. This refining process essentially purifies the crude oil through distillation, a technique you will certainly use in the laboratory, though on a much smaller scale. Using a large fractionating column, as shown in Figure 3.17, the crude oil is heated to very high temperatures, causing the compounds to volatilize. As the gases move higher up, the column gets cooler, causing different compounds to condense along the temperature gradient. Larger compounds condense at the higher temperatures at the bottom of the column, while smaller molecules condense at the lower temperatures near the top of the column.

Through this process, refinery gases like methane (natural gas) and propane (for grilling meats and vegetables, among other uses) are obtained highest up the column (lowest boiling point). Gasoline, consisting of a mixture of $C_4—C_{12}$ hydrocarbons, is obtained lower down, at slightly higher temperatures. Approximately 72 liters of gasoline is obtained by refining 159 liters (one barrel) of crude oil. Within this same range, the straight-chain and branched alkanes we study in this chapter are isolated. Other fuels are obtained at still higher boiling points. The tar-like residue that is left over after the refining process is used for filling potholes and sealing roofs.

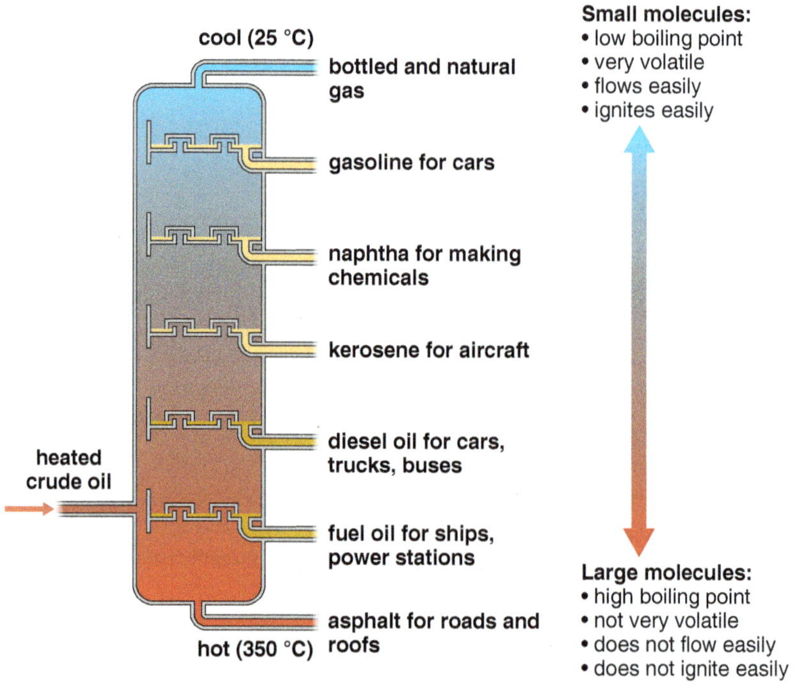

Figure 3.17 Distillation of crude oil into its usable components.

3.4 Conformational analysis

[*When you sleep, do you lie on your back, your stomach, or your side? Do you tuck your arms under your pillow or under your body? How do you make these decisions each night? Think about this question as we study conformational analysis.*]

In Chapter 1, we learned the two rules of organic chemistry as (1) the ball rolls downhill, and (2) negatives attack positives. When we talk about the ball rolling downhill, we are really talking about a potential energy hill (**Figure 3.18**). All chemical processes proceed in a direction of decreasing potential energy. In the same way that a ball at the top of a hill (high potential energy) rolls downhill (to a state of low potential energy), high-energy/unstable species react to form low-energy/stable ones. As we explain in Chapter 4, when we ask *how* a reaction occurs, the answer will be "negatives attack positives." And when we ask *why* a reaction/process occurs, the answer will be "because the ball rolls downhill." The potential energy contained within bonds is discussed in Chapter 5. Here in Chapter 3, we focus exclusively on the potential energy that results from **strain** on a molecule.

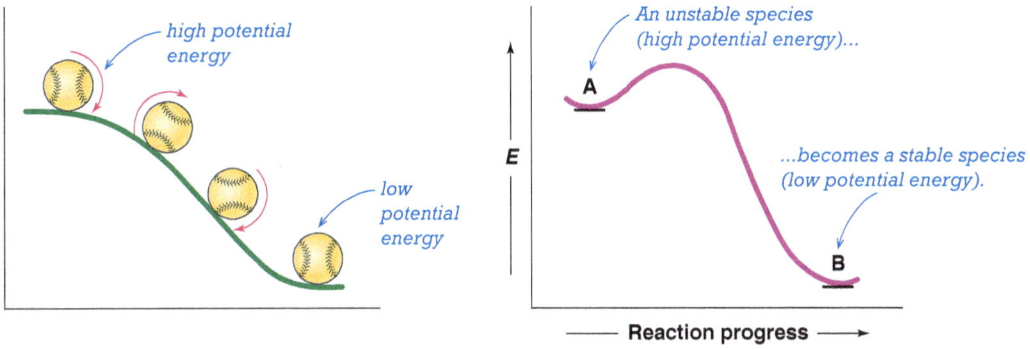

Figure 3.18 The ball rolls downhill. Unstable (high potential energy) species react to become more stable (low potential energy) species.

When we draw a molecule, like ethane, for example, we draw it as if it is completely static when, in fact, it is undergoing 10^{13} bond vibrations per second [**Figure 3.19(a)**] while also rotating around the C—C σ bond 10^{10} times per second [**Figure 3.19(b)**] at room temperature. [*What would 10^{10} rotations per second look like?*] We discuss molecular vibrations in Chapter 14, but we focus on the bond rotation here.

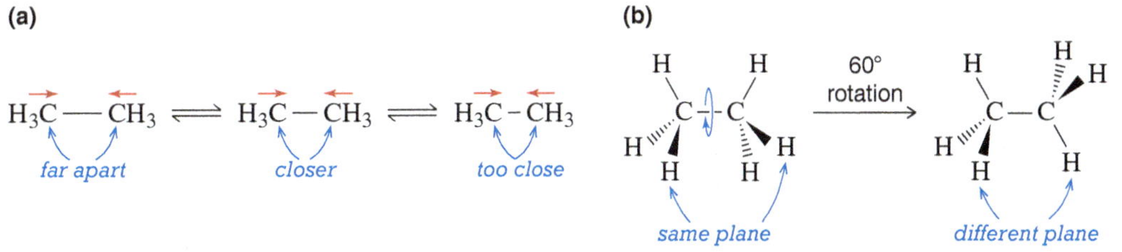

Figure 3.19 A molecule of ethane undergoes constant motion. (a) Bond vibration and (b) bond rotation.

Rotation occurs around the C—C σ bond of alkanes formed by overlapping sp^3-hybridized orbitals. Recall that σ orbitals are aligned along an axis connecting the two carbons. Overlap along this axis is maintained even as the bond rotates. Because rotation around the σ bond does not result in bond breakage, it is allowed and is fast (**Figure 3.20**).

Figure 3.20 Rotation around the single bond in ethane doesn't affect orbital overlap, so it is allowed.

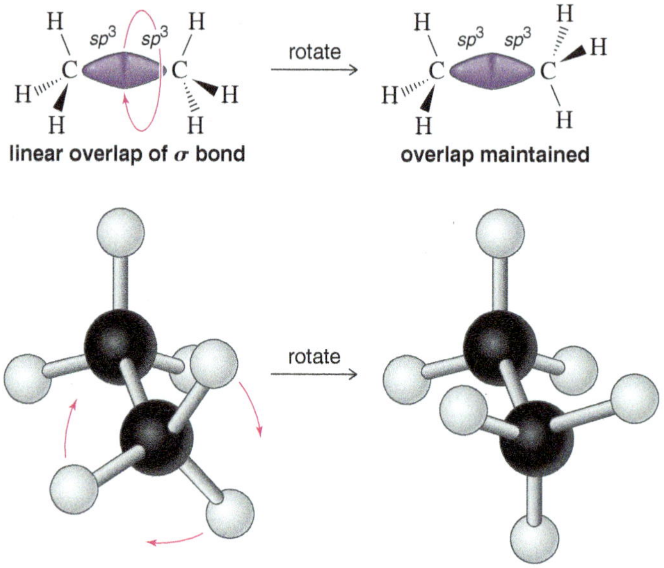

Isomers that differ by bond rotation, as in Figure 3.20, are conformational, not constitutional isomers.

Assessment 3.20 **(LOOKING AHEAD)** In Chapter 8, we study the chemistry of alkenes, like ethene. In contrast to ethane, there is no free rotation around the C=C double bond of ethene. Explain this in the context of the molecular orbital picture of ethene.

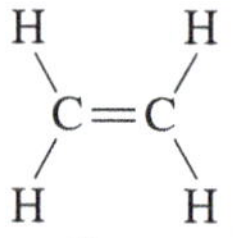

ethene

3.4.1 Newman projections: ethane

The rotation around ethane in Figure 3.20 showed two different conformations. Because it can be difficult to represent 3-D processes on a two-dimensional page, we use Newman projections to help us visualize these conformations. **Figure 3.21** shows the Newman projections for each of the two ethane conformations. Newman projections are drawn by placing your eye into the plane of the page, looking down the single bond of interest, and drawing what is seen. The carbon you see first (C_1) is drawn on top of a circle that hides the back carbon (C_2). Hydrogens in the plane of the page (H_a and H_b) are shown in the vertical plane (aligned with your eyeball). Wedged hydrogens, coming out of the plane of the page, would be on your right and are drawn as such in the Newman projection. Dashed hydrogens, going

Figure 3.21 Using Newman projections to represent conformational isomers.

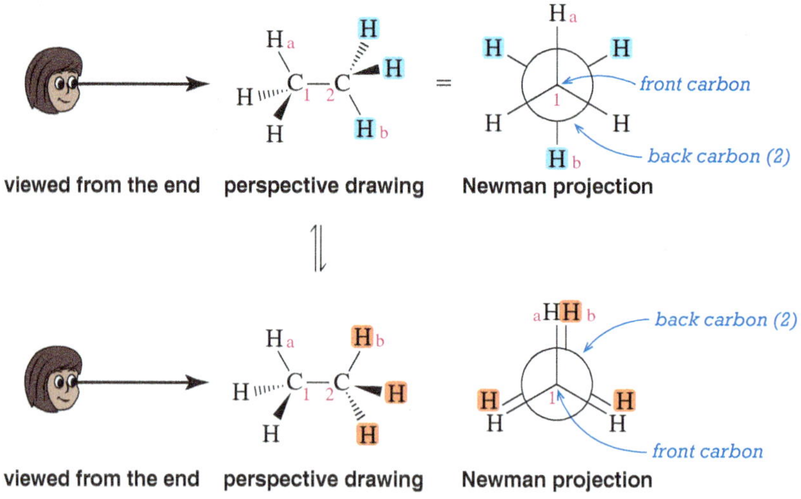

back into the plane of the page, when viewed as shown in Figure 3.21, would be on your left and are drawn as such in the Newman projection. [*Between the two conformations shown in Figure 3.21, one is more stable. Which do you think it is?*]

In order to be able to use Newman projections, you must be able draw them correctly. The ability to do this, especially for those who struggle with visualizing two-dimensional images in three dimensions, is gained only by practicing. For this reason, work Assessments 3.21 and 3.22 before continuing.

Assessment 3.21 For each molecule, draw the Newman projection you would observe if you looked down the $C_a—C_b$ bond in the direction indicated.

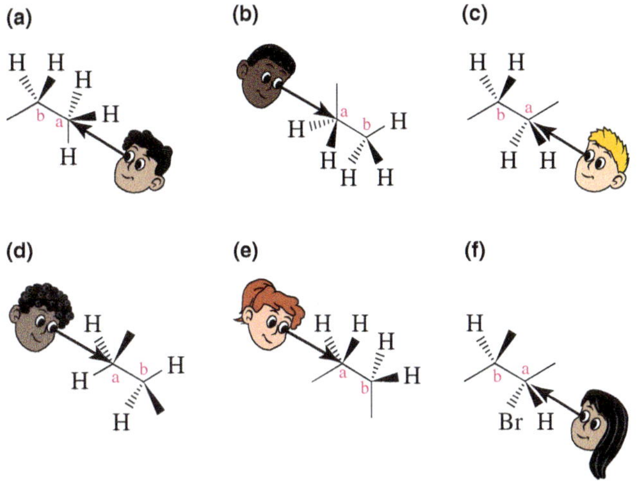

Assessment 3.22 Using the Newman projections shown, draw each molecule in its line-angle drawing with all hydrogens and substituents shown. [*Carbon b is behind carbon a in these structures.*] Wedges and dashes should be used to indicate whether a substituent is coming out of, or going into, the plane of the page. [*Your answers to these will look like the molecules in Assessment 3.21.*]

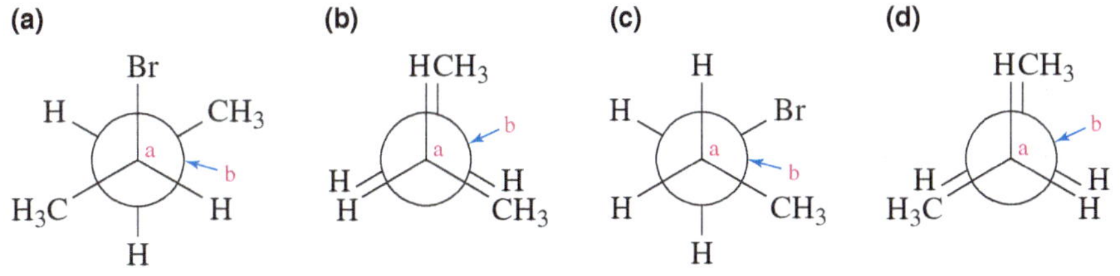

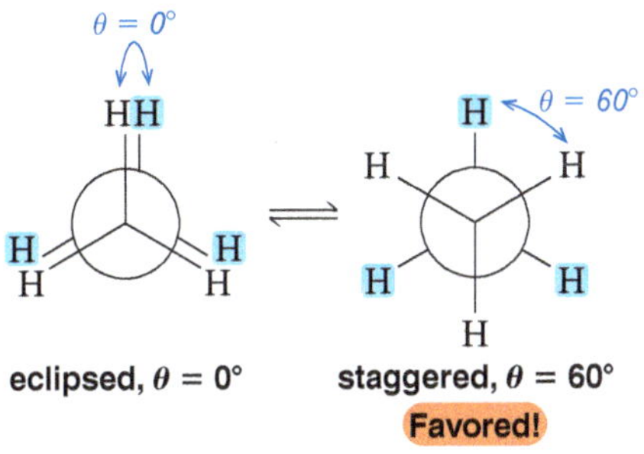

Figure 3.22 Dihedral angles shown in Newman projections of the eclipsed ($\theta = 0°$) and staggered ($\theta = 60°$) conformations of ethane. The staggered conformation has less torsional strain and is therefore more stable.

Once you have mastered the drawing of Newman projections, you can more easily examine the two conformations. Specifically, the Newman projection shows us the **dihedral angle** between atoms on adjacent carbons. Unlike bond angles, which involve three atoms, dihedral angles involve four atoms. **Figure 3.22** shows the measurement of dihedral angles in both the eclipsed ($\theta = 0°$) and staggered ($\theta = 60°$) conformations. The staggered conformation is the more stable of the two because it maximizes the dihedral angle of neighboring C—H bonds ($\theta = 0°$ vs. $\theta = 60°$). We therefore say that the eclipsed conformation has more **torsional strain**. Torsional strain is defined as strain arising from the repulsion of bonded electron pairs. Ethane spends more time in the staggered conformation to minimize torsional strain [*in the same way that you spend more time sleeping in the most comfortable position*].

Ethane doesn't simply oscillate back and forth between the eclipsed and staggered conformations. Instead, it rotates 360° around the C—C σ bond. The change in energy can be mapped as a function of the dihedral angle, as shown in **Figure 3.23**. Each eclipsed conformation is 3.0 kcal/mol (12.6 kJ/mol) less stable than each staggered conformation. Because each eclipsed conformation has a total of three H/H eclipsing interactions, each eclipsing interaction accounts for 1.0 kcal/mol (4.2 kJ/mol) of strain. [*You don't need to know a lot of numbers, but 1.0 kcal/mol (4.2 kJ/mol) for H/H eclipsing interactions is one you might want to remember.*] By this analysis, the staggered conformation is essentially strain free.

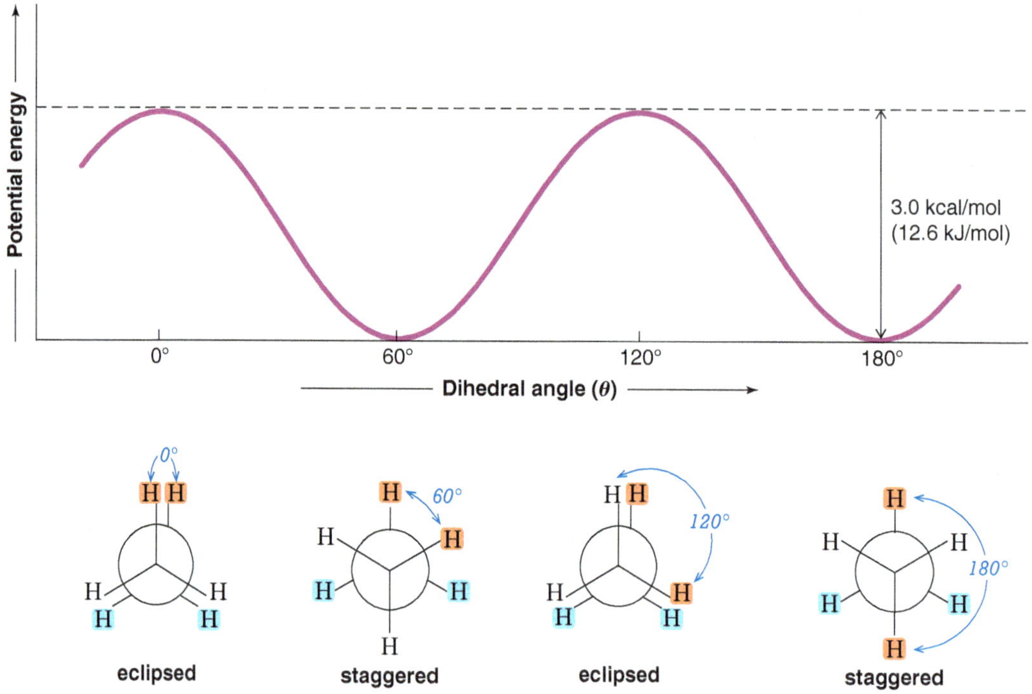

Figure 3.23 The eclipsed conformation is less stable than the staggered conformation by 3.0 kcal/mol (12.6 kJ/mol).

Assessment 3.23 Calculate the dihedral angle (θ) for the conformations shown.

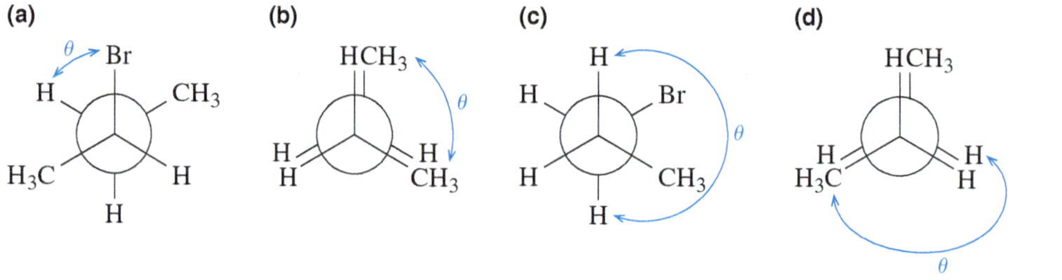

Assessment 3.24 Why is it *incorrect* to say that the dihedral angle shown is 120° or even 109°?

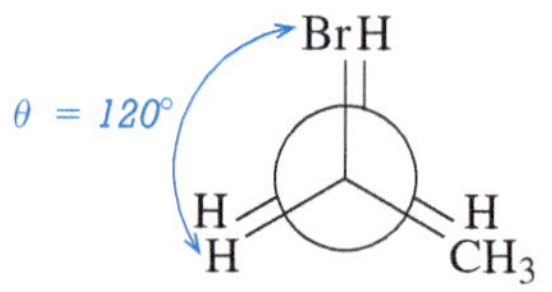

3.4.2 Newman projections: propane

We can do the same analysis of the conformations of propane, once more using Newman projections. **Figure 3.24** shows a plot of the potential energy as a function of dihedral angle. The graph looks very similar to the one drawn for ethane (Figure 3.23), with a minor difference. [*Do you see what it is?*] Whereas the eclipsed conformations in ethane were 3.0 kcal/mol (12.6 kJ/mol) less stable than the staggered conformations, they are 3.3 kcal/mol (13.8 kJ/mol) less stable in propane. This difference of 0.3 kcal/mol (1.2 kJ/mol) is due to the larger size of the methyl group. Owing to the larger size of the methyl group, the CH_3/H eclipsing interactions introduce more strain [1.3 kcal/mol (5.4 kJ/mol)] than does an H/H eclipsing interaction. Known as **steric strain**, this is the first (of many) times we will see that the interaction of larger groups is destabilizing. In contrast to torsional strain, steric strain results from the interaction of nonbonded nuclei.

Figure 3.24 The eclipsed conformation is less stable than the staggered conformation by 3.3 kcal/mol (13.8 kJ/mol).

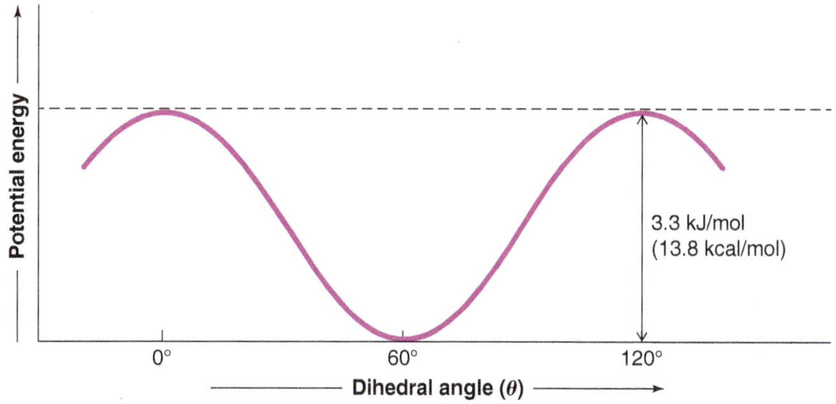

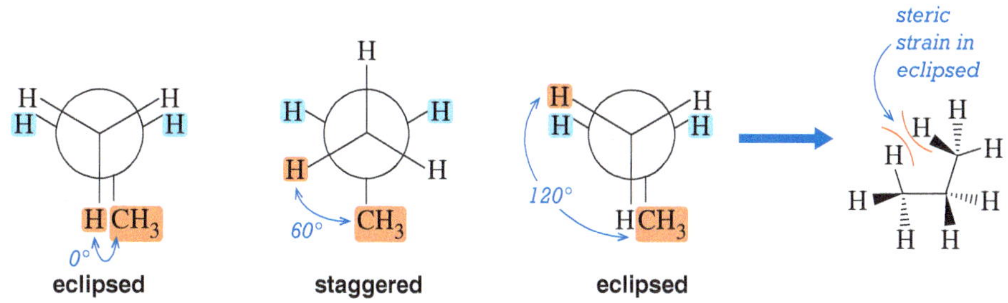

Assessment 3.25 Given that the total strain in the eclipsed conformation of propane is 3.3 kcal/mol, how would you calculate the value of an H/CH_3 eclipsing interaction to be 1.3 kcal/mol?

Assessment 3.26 Calculate the total strain in the eclipsed conformations shown.

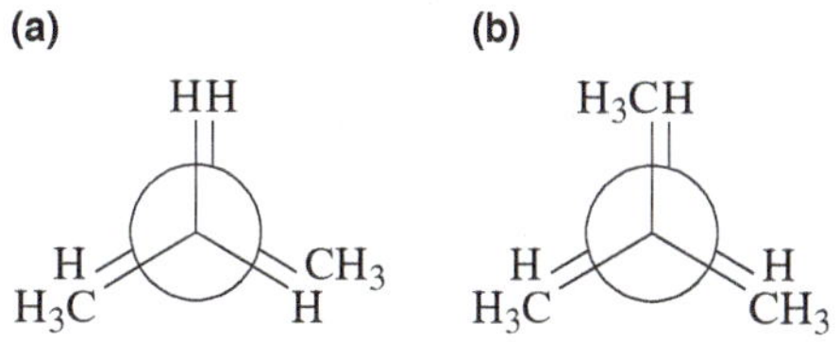

Assessment 3.27 Would you expect a CH_3/CH_3 eclipsing interaction to be larger or smaller than 1.3 kcal/mol?

Once you have worked Assessment 3.28, we'll do it together to see how conformational analysis of butane is different.

3.4.3 Newman projections: butane

The conformational analysis of the C_2—C_3 bond of butane changes a bit, as compared to propane and ethane.

Assessment 3.28 Starting with the given Newman projection, (a) show the other five conformations of butane arising from rotation around the C_2—C_3 bond. (b) Label each conformation as eclipsed or staggered. (c) Of the staggered conformations, which are more stable?

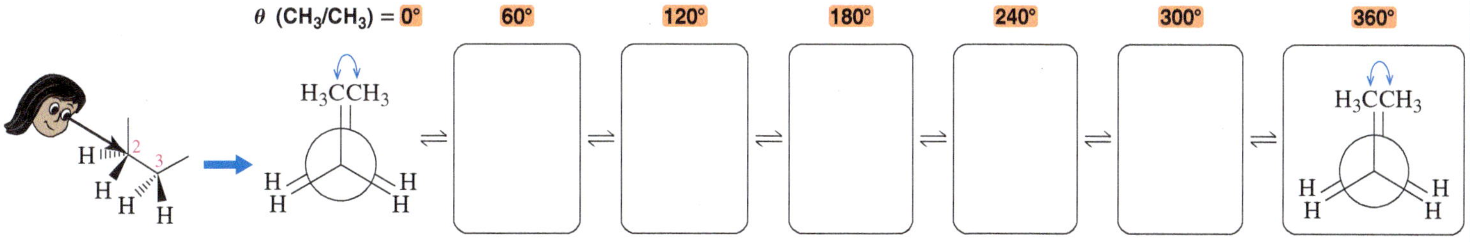

Solved Assessment 3.28

Starting with the given conformation ($\theta = 0°$), the remaining Newman projections can be drawn by holding the back carbon in place and rotating the front carbon 60° at a time (**Figure 3.25**). Because we started with an eclipsed conformation, the next conformation ($\theta = 60°$) should be staggered. Continued 60° rotation alternates between eclipsed and staggered. When a "new" conformation repeats one drawn earlier, you know that you are finished. In other words, a dihedral angle of 360° gives the same conformation as a dihedral angle of 0°.

Figure 3.25 answers parts (a) and (b), leaving part (c), which asks which staggered conformations are more stable. For ethane and propane, *all* of the staggered conformations had essentially zero strain and thus identical potential energy. [*Check out Figures 3.23 (ethane) and 3.24 (propane) to confirm this.*] Do you see a difference between the three staggered conformations of butane in Figure 3.25? There is a 60° [*or* −60°; *same thing*] dihedral angle between the methyl groups in two of the staggered conformations, whereas the other has a dihedral angle of 180°. The conformations with the bulky methyl groups closer to each other are, as you might expect, less stable than the one with them farther apart. The conformations where $\theta = 60°$ and 300° (−60°) between the two methyl groups are referred to as **gauche** conformations (**Figure 3.26**). When $\theta = 180°$, the conformation is referred to as **anti**. Using these new terms, the gauche conformation is less stable than the anti conformation. [*In French, gauche means "left" and is often translated as "clumsy" or "awkward."*] Put another way, a CH_3/CH_3 gauche interaction is a destabilizing interaction. Of note, *all* staggered conformations are *always* more stable than eclipsed conformations. ■

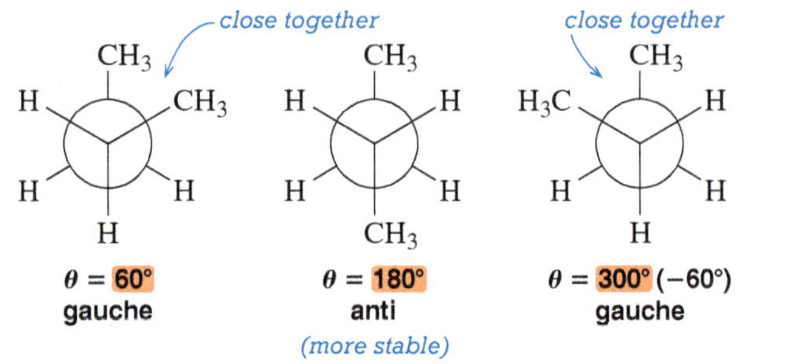

Figure 3.26 Staggered conformations of butane around the C_2—C_3 bond. The anti conformation is more stable than the gauche.

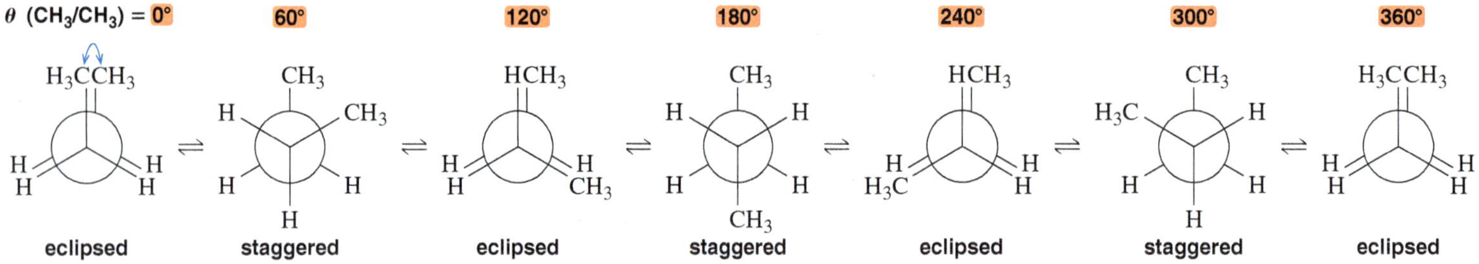

Figure 3.25 All conformations of butane around the C_2—C_3 bond.

Like we did for ethane and propane, we can plot the potential energy for butane as a function of dihedral angle, as shown in **Figure 3.27**. There's lots of information in this graph. First, it confirms that indeed, the eclipsed conformations, always at the top of the hill, are less stable than the staggered conformations. Furthermore, it confirms that the anti conformation is 0.9 kcal/mol (3.8 kJ/mol) more stable than the gauche conformation. The anti conformation is considered to be strain free, while each CH_3/CH_3 gauche interaction results in 0.9 kcal/mol (3.8 kJ/mol) of destabilization. [*0.9 kcal/mol (3.8 kJ/mol) for the CH_3/CH_3 gauche interaction is another number worth remembering.*] Finally, the conformation where the bulky methyl groups are eclipsed ($\theta = 0°/360°$) is least stable. This is due to the increased steric strain of the methyl groups trying to share the same space.

We can now use these numbers and this kind of analysis to evaluate the relative strain for conformations around any single bond. For example, the three staggered conformations of 2-methylbutane are shown in **Figure 3.28**. [*We don't usually need to show the eclipsed conformations because those are so unstable.*] Conformations A and B each have a single CH_3/CH_3 gauche interaction, accounting for a total of 0.9 kcal/mol of strain in each. Conformation C, on the other hand, has two CH_3/CH_3 gauche interactions, resulting in a total of 1.8 kcal/mol of strain. Thus, conformations A and B are 0.9 kcal/mol more stable than conformation C. [*Or, closing with the analogy that began Section 3.4, 2-methylbutane sleeps more comfortably in conformations A and B.*]

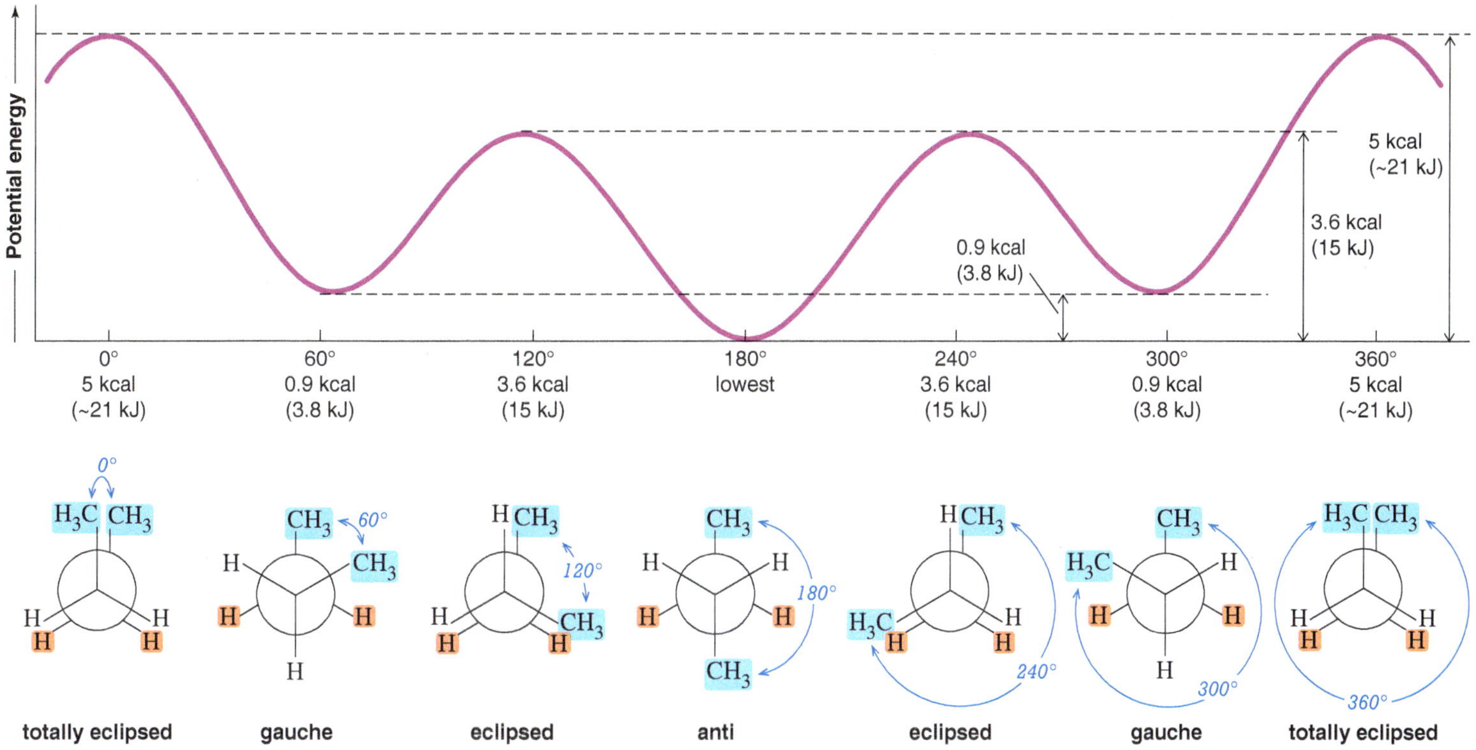

Figure 3.27 Conformational analysis for the C_2—C_3 bond of butane. The staggered, anti conformation is most stable, whereas the conformation with eclipsed methyl groups is least stable.

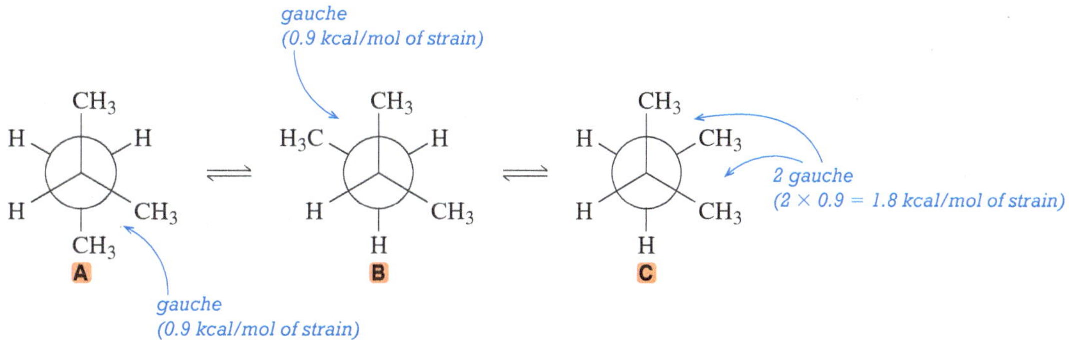

Figure 3.28 Fewer gauche interactions result in more stable staggered conformations.

A quick note about using line-angle drawings to represent organic compounds: It turns out that when we draw molecules in this way, we are in effect drawing them in their most stable conformation, as shown in **Figure 3.29**.

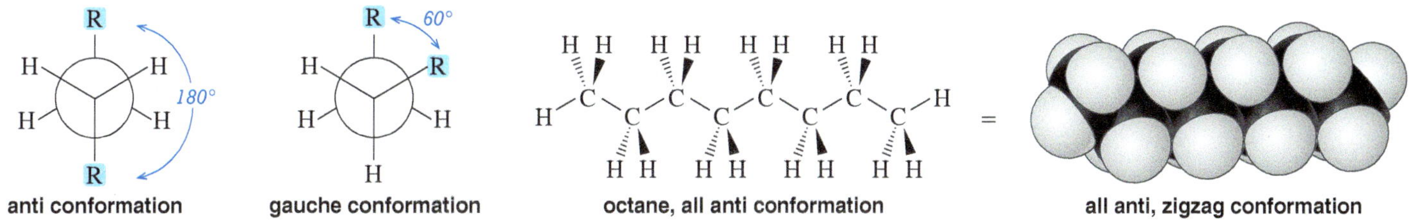

Figure 3.29 Line-angle drawings show the C—C σ bonds of alkanes in their most stable anti conformations.

3.4 SUMMARY

- Molecules will exist in the most stable conformation that keeps the electron pairs in neighboring bonds as far apart as possible. For acyclic alkanes, staggered conformations are more stable than eclipsed. The anti staggered conformation is more stable than the gauche staggered conformation.

Assessment 3.29 Choose the conformation in each pair that is most stable. If both are equally stable, then write "no difference."

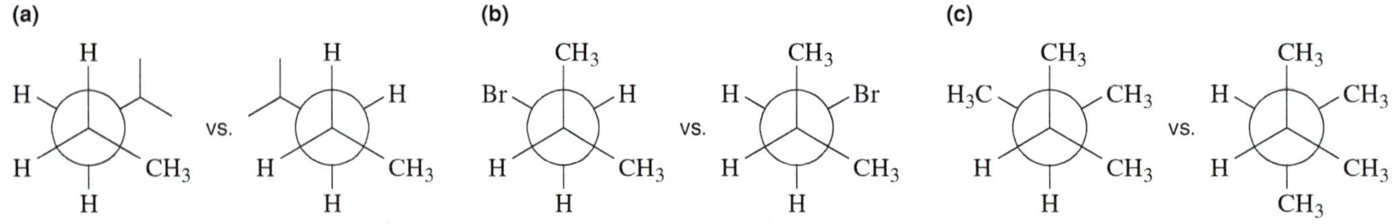

Assessment 3.30 For each of the pairs in Assessment 3.29, which conformation would you expect to be more prominent at equilibrium?

Assessment 3.31 Draw the three most stable conformations of 2,2,4-trimethylpentane looking down the C_a—C_b bond. Of the three, which is most stable?

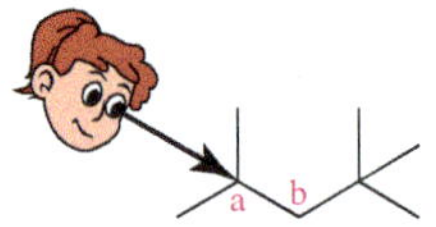

2,2,4-trimethylpentane

3.5 Cycloalkanes: an overview

Cycloalkanes are cyclic alkanes. This means that in spite of having a general molecular formula of C_nH_{2n}, they are saturated hydrocarbons. Some representative cycloalkanes are shown in Figure 3.30.

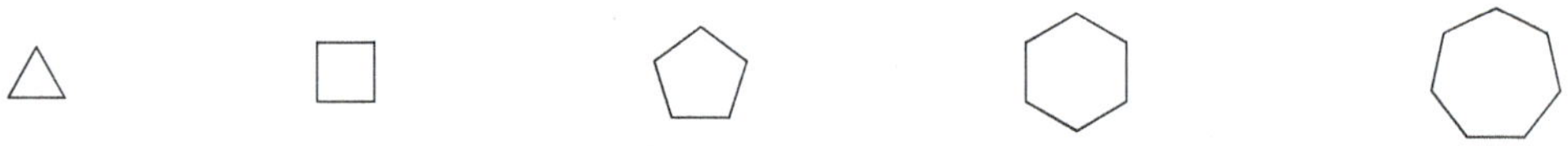

cyclopropane C_3H_6	cyclobutane C_4H_8	cyclopentane C_5H_{10}	cyclohexane C_6H_{12}	cycloheptane C_7H_{14}

Figure 3.30 Representative cycloalkanes

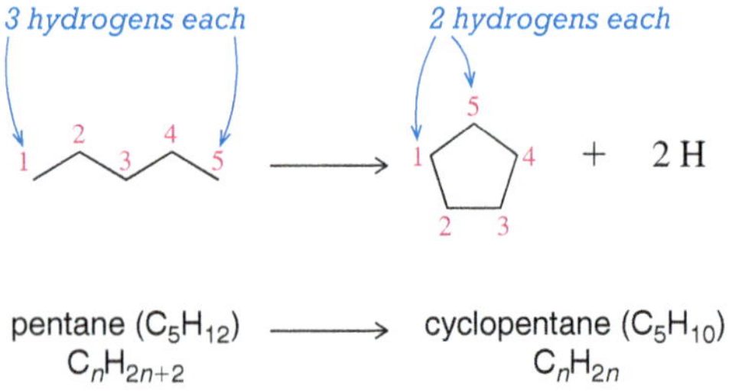

Figure 3.31 A purely hypothetical reaction that shows how the general molecular formula for alkanes and cycloalkanes differs by two hydrogens.

The general molecular formula can be understood in the context of a purely hypothetical reaction of an alkane to form a cycloalkane. [*This is NOT a real reaction.*] In order to form cyclopentane from pentane (see Figure 3.31), C_1 and C_5 would need to be connected. But each carbon already has four bonds, so one bonding site would need to be freed up at each carbon by removing a hydrogen from each. By removing a total of two hydrogens, the general molecular formula is reduced from C_nH_{2n+2} to C_nH_{2n}. [*This concept is used again in Chapter 8 when we discuss alkenes and the index of hydrogen deficiency.*]

Some physical properties of cycloalkanes are listed in Table 3.9. [*Do the properties change as you'd expect?*]

Table 3.9 Physical properties of cycloalkanes.

Cycloalkane	Formula	Boiling Point (°C)	Melting Point (°C)	Density (g/mL)
Cyclopropane	C_3H_6	−33	−128	0.72
Cyclobutane	C_4H_8	−12	−50	0.75
Cyclopentane	C_5H_{10}	49	−94	0.75
Cyclohexane	C_6H_{12}	81	7	0.78
Cycloheptane	C_7H_{14}	118	−12	0.81
Cyclooctane	C_8H_{16}	148	14	0.83

In general, cycloalkanes have higher boiling points and melting points than their alkane homologs. Figure 3.32 compares pentane (bp = 36 °C; mp = −130 °C) to cyclopentane (bp = 49 °C; mp = −94 °C). Recall that alkanes can freely rotate around the σ bonds. This creates a dynamic situation where the packing of molecules required in the liquid and (more so)

Figure 3.32 Cycloalkanes pack more tightly, so they melt and boil at higher temperatures.

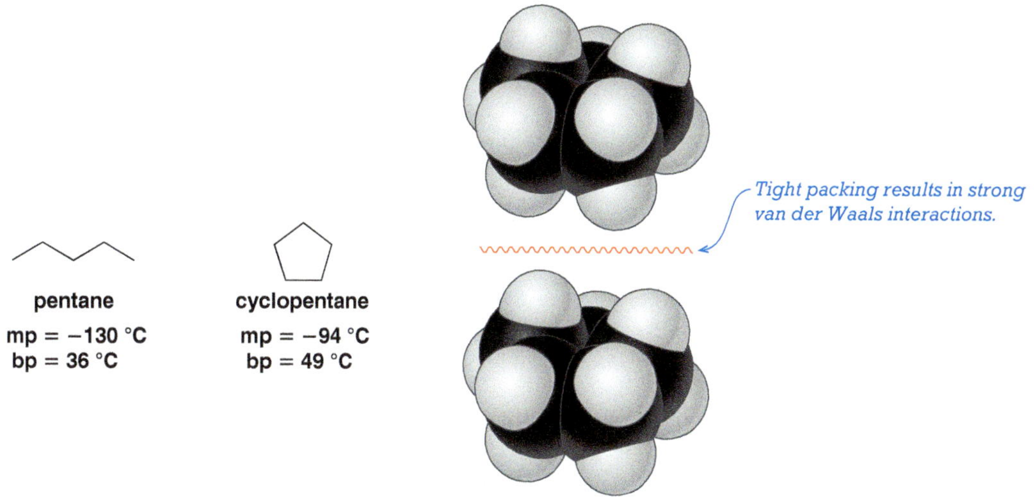

the solid states is disrupted by the freely rotating alkyl chains. Rotation in cycloalkanes, however, is prevented by being constrained in a ring. Thus, cycloalkanes are flatter and pack more tightly, increasing their intermolecular van der Waals interactions.

3.5.1 Nomenclature of cycloalkanes

Cycloalkane nomenclature uses the same IUPAC rules for alkanes, but with slight modifications that make it more straightforward. These are summarized in **Table 3.10**, with two examples of each rule given.

Table 3.10 IUPAC nomenclature for cycloalkanes.

IUPAC rules	Example 1	Example 2
1. For unsubstituted cycloalkanes, use the prefix cyclo- to indicate that it is a ring. The prefixes used to indicate the number of carbons in alkanes are now used as infixes to indicate the number of carbons in the ring. [*See Table 3.2 for the prefixes.*] The infix -an- indicates that this is a fully saturated molecule. The suffix -e indicates that the molecule is a hydrocarbon. [*See Assessment 3.32(a, b) for practice.*]	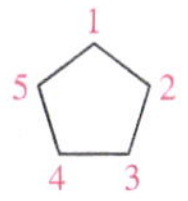 **cyclopentane**	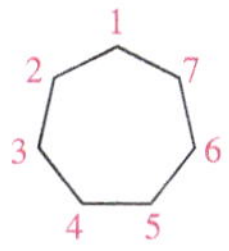 **cycloheptane**
2. For monosubstituted cycloalkanes, substituent names are applied identically as they are for alkanes. The root of the name is the cycloalkane. [*By analogy to alkanes, the ring is the parent chain unless there is a longer alkyl group attached. Then the straight-chain alkane is the parent chain.*] Because the first substituent is always assumed to be on carbon 1, there is no need to place the number in the name. [*See Assessment 3.32(c, d) for practice.*]	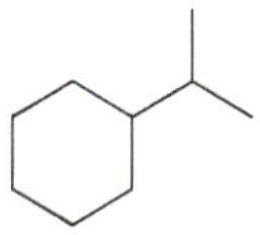 **isopropylcyclohexane**	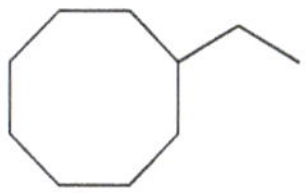 **ethylcyclooctane**
3. For disubstituted cycloalkanes, the two substituents are numbered such that the substituent that comes first alphabetically is on carbon 1, with the other placed on the lowest possible number. The names are then written in alphabetical order. [*See Assessment 3.32(e, f) for practice.*]	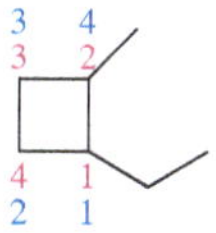 **1-ethyl-2-methylcyclobutane**	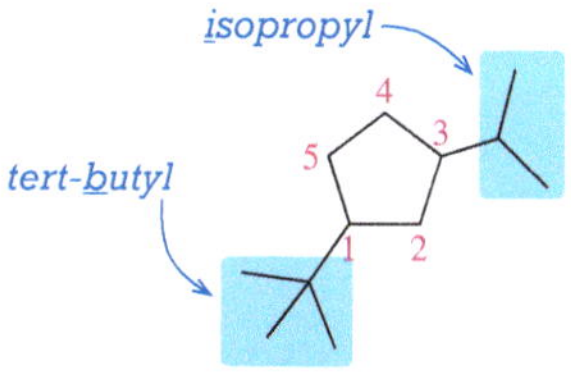 **1-*tert*-butyl-3-isopropylcyclopentane**
4. With three or more substituents on the cycloalkane, the ring is numbered in the way that gives the lowest combination of substituent numbers. The substituent names are still alphabetized in the name, however. [*See Assessment 3.32(g, h) for practice.*]	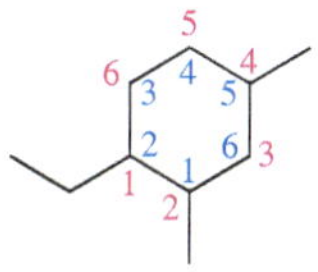 **1-ethyl-2,4-dimethylcyclohexane (*not* 1,5-dimethyl-2-ethylcyclohexane)**	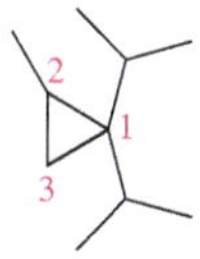 **1,1-diisopropyl-2-methylcyclopropane**

Assessment 3.32 Name the following cycloalkanes using the IUPAC system of nomenclature. [*Hint: Each molecule exemplifies one of the cycloalkane nomenclature rules in Table 3.10.*]

(a) rule 1 **(b)** rule 1 **(c)** rule 2 **(d)** rule 2

(e) rule 3 **(f)** rule 3 **(g)** rule 4 **(h)** rule 4

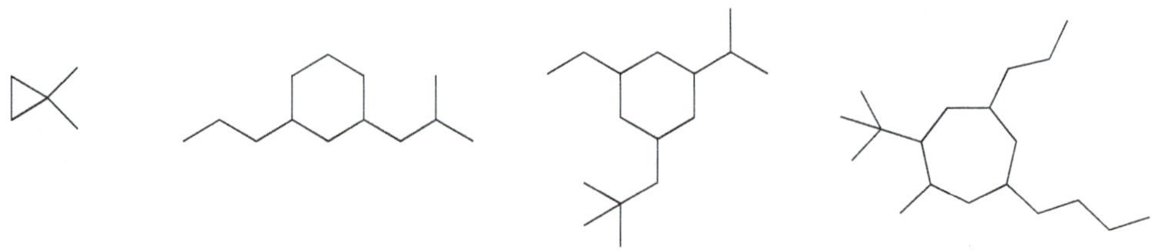

3.5.2 Cis–trans isomerism

A new type of isomerism, referred to as cis–trans isomerism, is revealed by our study of cycloalkane nomenclature.

The solution to Assessment 3.33 illustrates a problem with IUPAC nomenclature. Let's figure it out together.

Assessment 3.33 Name the following cycloalkanes using the IUPAC system of nomenclature.

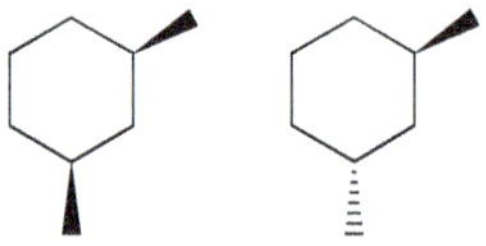

Solved Assessment 3.33

Both structures contain a six-membered ring, so the root name for each will be cyclohexane. They each contain two methyl groups, so that would add dimethyl to the name. And, because numbers are assigned to give the lowest combination of numbers, the two methyl groups would appear at carbons 1 and 3 for both. At this point, the names for these two different compounds would be the same (**Figure 3.33**).

The IUPAC system ensures that every distinct compound has a distinct name. Read on to see how this is addressed. ■

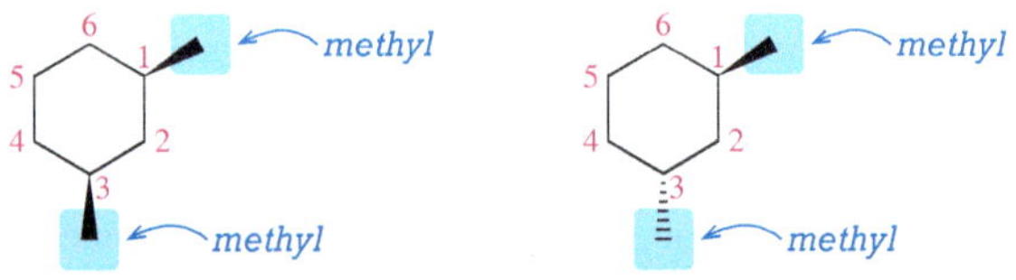

Figure 3.33 The cyclohexanes in Assessment 3.33 have identical names.

But first, let's discuss the relationship between the two compounds. With the same molecular formula, but different structures, they are isomers. They have the same molecular formula and the same connectivity, so they are not constitutional isomers. In order to go from one to the other, because they are constrained by the ring, you can't simply rotate around a bond. So they are not conformational isomers. To convert one to the other, you would have to break the methyl and hydrogen off of C_3 and place the hydrogen on the wedged line/methyl on the dashed line (**Figure 3.34**). Isomers that differ in their spatial connectivity are called **stereoisomers**. We study stereoisomers in more detail in Chapter 6.

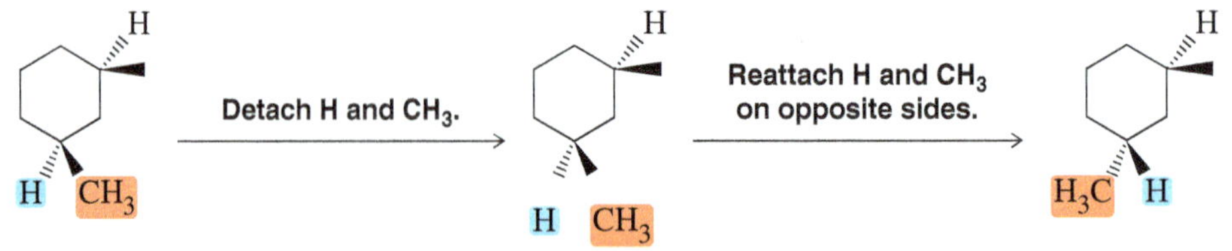

Figure 3.34 The 1,3-dimethylcyclohexanes are stereoisomers since bonds must be broken to convert one to the other.

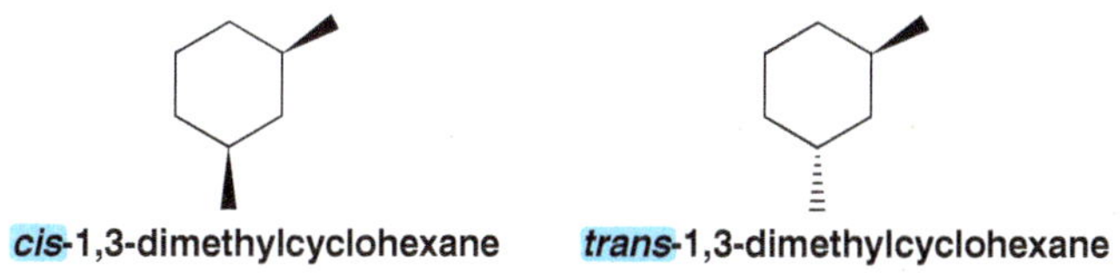

Figure 3.35 ***Cis-* and *trans*-1, 3-dimethylcyclohexane.**

To name this type of stereoisomers, the terms *cis* and *trans* are used in front of the names. When the substituents come off the ring in the same direction (both into or out of the plane of the page), they are referred to as cis. When the substituents are on opposite sides of the ring, they are referred to as trans. Using this nomenclature, the compounds from Assessment 3.33 are named in Figure 3.35.

Assessment 3.34 Name the following cycloalkanes using the IUPAC system of nomenclature, being sure to indicate whether they are cis or trans.

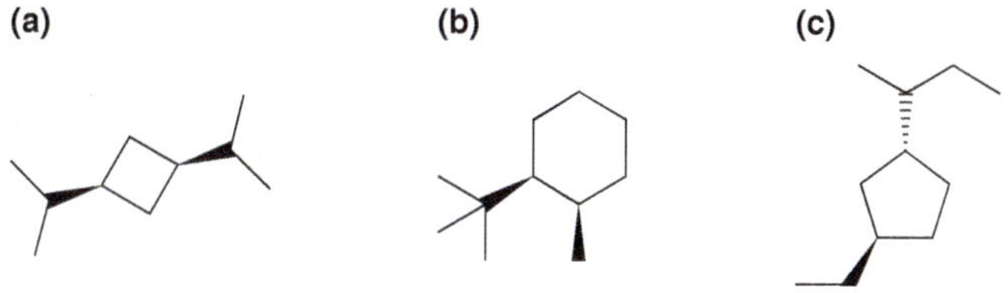

3.6 Strain in rings: conformational analysis

[*Suppose you prefer to sleep on your right side with your right arm tucked under your body. But while walking on campus one day, thinking about an organic chemistry problem, you trip and dislocate your right shoulder. How will you sleep now?*]

Just as we did with alkanes, we can discuss the strain energies of cycloalkanes in their various conformations. When talking about strain in cycloalkanes, we use the more general term ***ring strain*** to represent the total strain in the molecule. Ring strain typically includes both **angle strain** (a new type of strain) and torsional strain [*from Section 3.4*]. Recall that tetrahedral carbons have ideal bond angles of 109.5°. When an sp^3-hybridized carbon is constrained in a ring, preventing it from achieving the desired 109.5° bond angle, angle strain results. The amount of strain is directly correlated to the deviation from ideal bond angle. Cyclopropane, the smallest possible cycloalkane, has lots of angle strain because each carbon is forced to have bond angles of 60° (Figure 3.36).

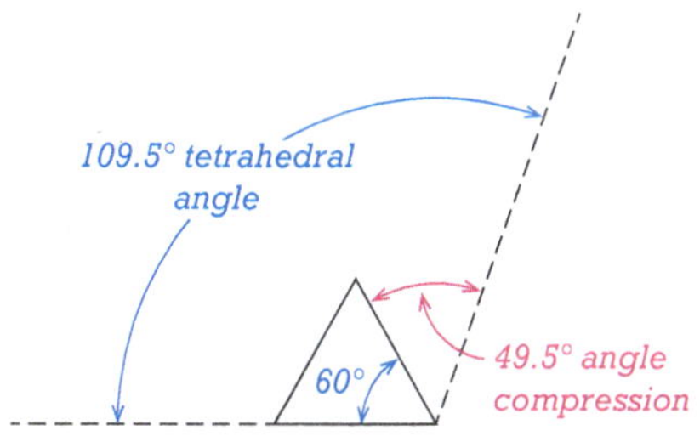

Figure 3.36 Cyclopropane is a highly angle-strained molecule.

As we discuss the strain in rings, note that there is a delicate balance between angle strain and torsional strain. Thus, it might be worth taking on a bit more angle strain if it diminishes torsional strain [*. . . in the same way that sleeping on your left side might reduce the overall discomfort from sleeping on your dislocated right shoulder*].

3.6.1 Cyclopropane

Cyclopropane has the most strain of any of the cycloalkanes. As discussed earlier, much of this strain is a result of angle strain caused by the 60° bond angles required by a three-membered ring. However, that doesn't account for all of it. Because cyclopropane is planar [*geometry: three points define a plane*], there are six H/H eclipsing interactions (Figure 3.37). These eclipsing interactions, which can be visualized by looking down each bond, result in a total of 6 × 1 = 6 kcal/mol (25.2 kJ/mol) worth of torsional strain. Given that the total ring strain in cyclopropane is around 28.0 kcal/mol, the torsional strain accounts for almost 21%. [*The rest is the angle strain.*]

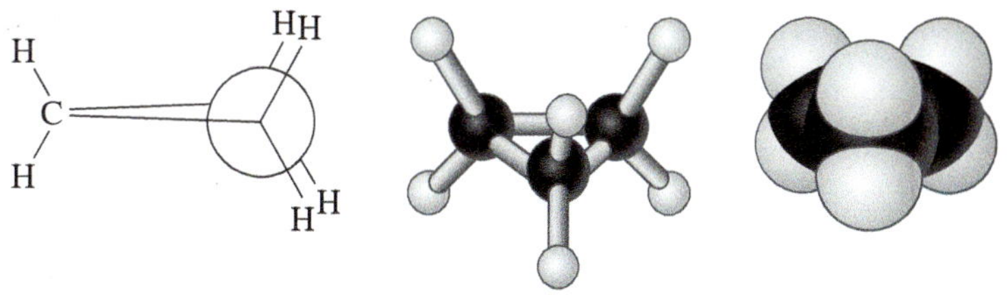

Figure 3.37 Cyclopropane has angle strain and torsional strain.

3.6.2 Cyclobutane

Cyclobutane is a four-membered ring. If it were planar, its bond angles would be 90°. As shown in **Figure 3.38**, planar cyclobutane would have eight H/H eclipsing interactions, resulting in 8.0 kcal/mol (33.6 kJ/mol) of torsional strain. These can be seen by looking down the $C_1—C_4$, $C_2—C_3$, $C_3—C_4$, and $C_1—C_2$ bonds.

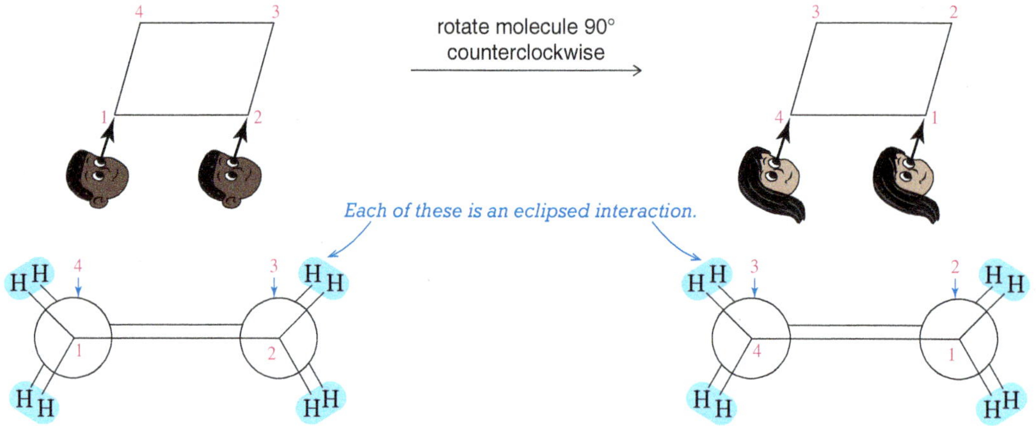

Figure 3.38 Planar cyclobutane has eight H/H eclipsing interactions.

Unlike cyclopropane, the four carbons of cyclobutane do not have to remain in the same plane. Instead, seeking the most stable, strain-minimizing conformation, cyclobutane puckers, as shown in **Figure 3.39**. While this puckering shrinks the bond angles to around 88°, creating more angle strain, it partially removes the H/H eclipsing interactions, lessening the torsional strain. Balancing the increased angle strain with reduced torsional strain, puckered cyclobutane has less overall strain than planar cyclobutane. [*You don't love sleeping on your left side, but it if avoids the pain of sleeping on your injured right shoulder, you'll do it.*]

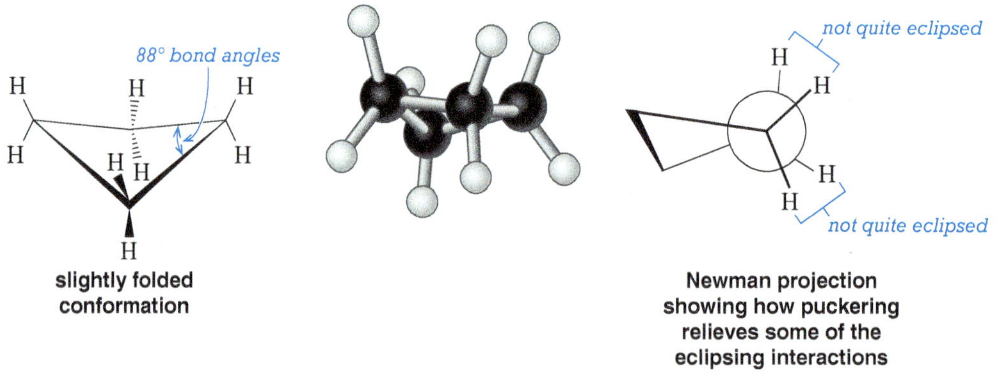

Figure 3.39 Cyclobutane is most stable in a puckered conformation.

3.6.3 Cyclopentane

Continuing our analysis, planar cyclopentane would have virtually no angle strain because a regular pentagon has angles of 108°. [*How much torsional strain would planar cyclopentane have?*] There are ten H/H eclipsing interactions in planar cyclopentane, so it would have 10 kcal/mol (42 kJ/mol) of torsional strain. To minimize this strain, cyclopentane puckers into what is referred to as the envelope conformation (**Figure 3.40**). This removes all but four of the eclipsing interactions but does so by shrinking the bond angles and increasing the angle strain. Once more, we see that a molecule is willing to take on one type of strain if it can diminish more of another type—finding the most stable conformation [*sleeping position*] is a delicate balance.

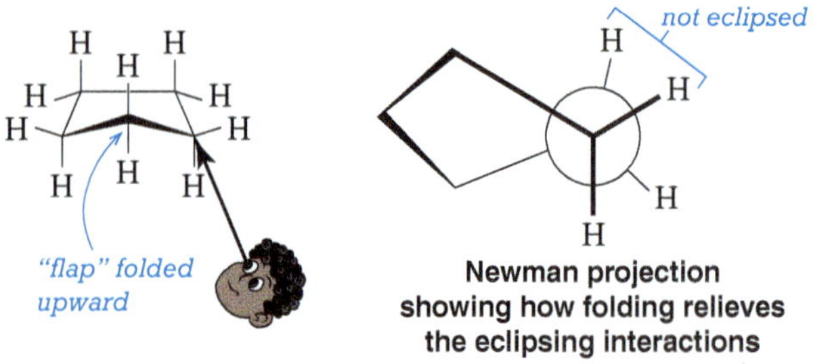

Figure 3.40 Cyclopentane is most stable in its envelope conformation.

3.6.4 Cyclohexane

This brings us to cyclohexane, which is, in some ways, nature's perfect ring. To see how, let's imagine what planar cyclohexane would look like. First, because a regular hexagon has bond angles of 120°, planar cyclohexane would have a considerable amount of angle strain. Remember that angle strain results from *any* deviation from ideal, not just when angles are smaller. And so, like cyclobutane and cyclopentane, cyclohexane puckers (**Figure 3.41**). The resulting chair conformation shrinks the bond angles to 109.5° (absolutely perfect—no angle strain) and completely eliminates all H/H eclipsing interactions (absolutely perfect—no torsional strain). Cyclohexane, then, is virtually strain free, making this six-membered ring by far the most common ring size in nature.

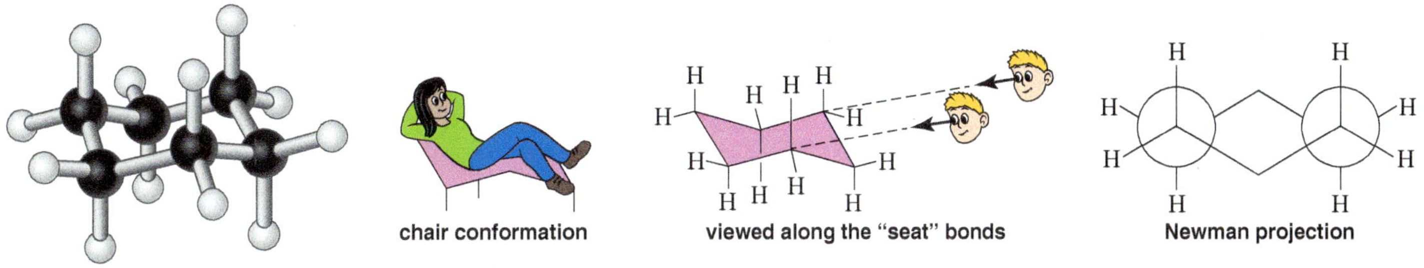

Figure 3.41 Representations of the chair conformation.

3.6 SUMMARY

- The most favorable conformation in cycloalkanes is the one that minimizes the total strain (angle strain plus torsional strain). Cyclohexane, in the chair conformation, is essentially strain free.

3-C CHEMISTRY IN CONTEXT: Physical Organic Chemistry

Measuring strain

The strain present in cycloalkanes, up to this point simply given as approximate values, can be determined experimentally by measuring the heat given off when they are burned (combusted). Combustion analysis consists of burning defined amounts of the compound in a bomb calorimeter, a device shown schematically in **Figure 3.42**. Ignition of the flammable compound causes the temperature of the surrounding water to change. This change in temperature corresponds to the heat given off as a result of combustion. The equation for combustion, using methane as an example, is as follows:

$$CH_4 + 2\,O_2 \rightarrow CO_2 + 2\,H_2O$$

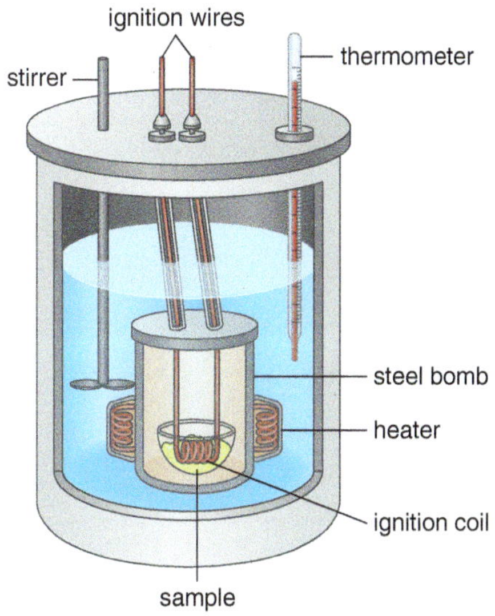

Figure 3.42 Schematic of a calorimeter.

In practice, the relative strain in a series of cycloalkanes can be calculated using combustion analysis. The results of that analysis are listed in **Table 3.11**. The compounds have different molecular formulas, though, so their heats of combustion are compared on a per CH_2 basis. That is, the total measured heat of combustion is divided by the number of CH_2 groups present. Each CH_2 group should have an identical heat of combustion, so any difference between these values will be attributed to ring strain. Cyclopropane has the highest heat of combustion per CH_2, indicating that it has the highest amount of ring strain.

From these values, the total strain energy of cyclopropane is calculated to be 27.6 kcal/mol (3 × 9.2 kcal/mol) and is attributable to both angle strain and torsional strain as described in Section 3.6. Cyclohexane, because it is able to achieve

(*continued*)

a chair conformation with no angle strain and no torsional strain, is considered to be strain free.

Recall that the ball rolls downhill for favored organic processes. We can visualize the impact of strain energy by plotting the values from Table 3.11 on a **reaction coordinate diagram** (**Figure 3.43**). We do more with reaction coordinate diagrams in upcoming chapters, but for now, note that they simply account for the change in energy as a reaction progresses from reactants to products—in this case, from the cycloalkane and O_2 to CO_2 and H_2O. On a per CH_2 basis, these reactions end at the same energy level, so the strained rings must start at a higher energy level. This difference in starting points is the strain energy contained within the rings. Cyclohexane has no ring strain, so it starts at the same energy as an acyclic alkane.

Table 3.11 Heats of combustion of cycloalkanes.

Compound	Heat of combustion [kcal/mol (kJ/mol)]	Heat of combustion per CH_2 [kcal/mol (kJ/mol)]	Strain relative to cyclohexane per CH_2 [kcal/mol (kJ/mol)]	Total strain [kcal/mol (kJ/mol)]
 cyclopropane	499.8 (2089)	166.6 (696)	9.2 (38.5)	27.6 (115)
 cyclobutane	656 (2742)	164 (686)	6.6 (27.6)	19.8 (82.8)
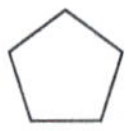 cyclopentane	793.5 (3316)	159 (663)	1.3 (5.4)	3.9 (16)
 cyclohexane	944.4 (3946)	157 (658)	0	0
 hexane	—	157 (658)	0	0

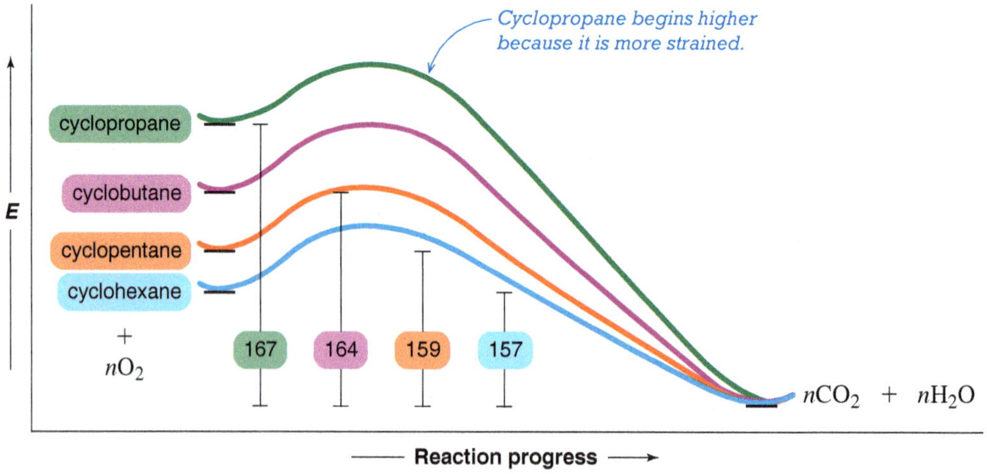

Figure 3.43 Plot of heat of combustions per CH_2 shows that cyclopropane possesses the greatest amount of ring strain (units in kcal/mol).

Assessment 3.35 Propane is a common gas used for outdoor grills. Suppose cyclopropane was available in sufficient quantities to replace propane for this purpose. Would you be able to grill your food more or less efficiently using cyclopropane? Why?

Assessment 3.36 Rank the following compounds in order of their total heat of combustion. These compounds are constitutional isomers, each with a molecular formula of C_6H_{12}.

propylcyclopropane C_6H_{12} | ethylcyclobutane C_6H_{12} | methylcyclopentane C_6H_{12} | cyclohexane C_6H_{12}

3.7 Cyclohexane conformations

Cyclohexane is such an important, and stable, ring that its conformational analysis deserves more study [*and its own section*]. It turns out that there are two chair conformations, two boat conformations, and two twist-boat conformations. [*I'm not making this up.*] Before we look at them in detail, let's see how each is formed. [*A set of molecular models is really helpful here.*] Let's start with the chair conformations. Imagine planar cyclohexane, shown in **Figure 3.44**, with C_1 and C_4 labeled. The chair conformation screened in blue can be produced by pushing C_1 up and C_4 down. Pushing C_1 down and C_4 up forms the chair screened in pink.

Figure 3.44 Forming two chairs from planar cyclohexane.

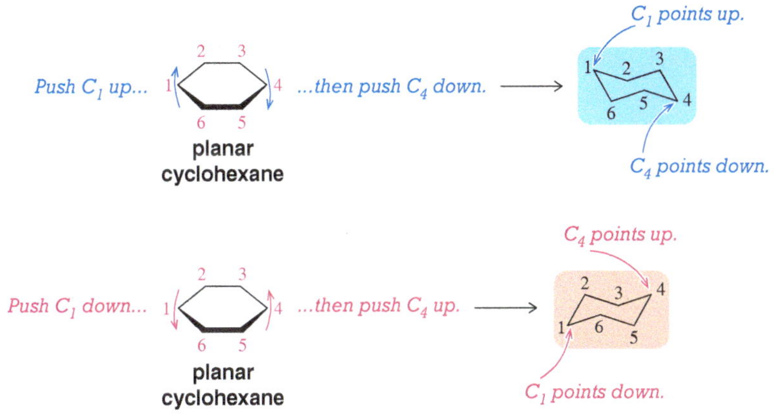

Figure 3.45 shows that the boat conformation can be produced from planar cyclohexane by pushing both flaps (C_1 and C_4) toward each other. To generate the twist-boat, take C_1 and C_4 and twist them in opposite directions. Twisting C_1 forward and C_4 back generates one twist-boat. Twisting C_1 back and C_4 forward generates the other twist-boat conformation.

Figure 3.45 Forming the boat and two twist-boats from planar cyclohexane.

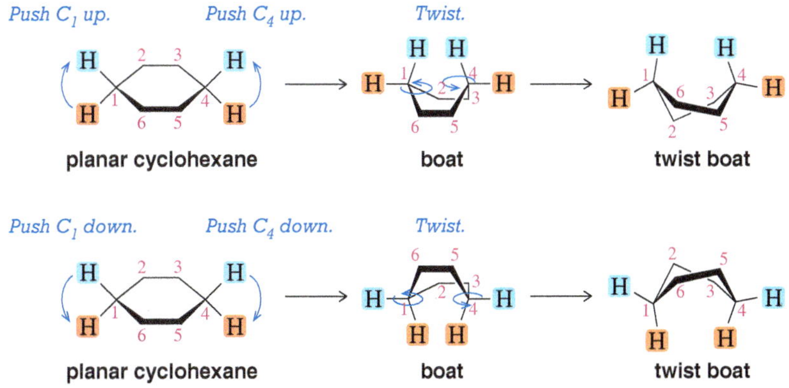

The graph in **Figure 3.46** shows how the energy of cyclohexane changes as it converts from one chair conformation to the other. The chair conformations are more stable than the twist-boat conformations, which are more stable than the boat conformation. In the boat conformation, the hydrogens at C_1 and C_4 point at each other, sharing the same space. The steric clash [*a great name for a rock band*] that results from this **flagpole interaction** destabilizes the boat conformation such that it serves only as a transition from one conformation to the next. Similarly, the half-chair conformations are highly unstable because they each still have the torsional strain of six H/H eclipsing interactions and some angle strain from the still-planar carbons.

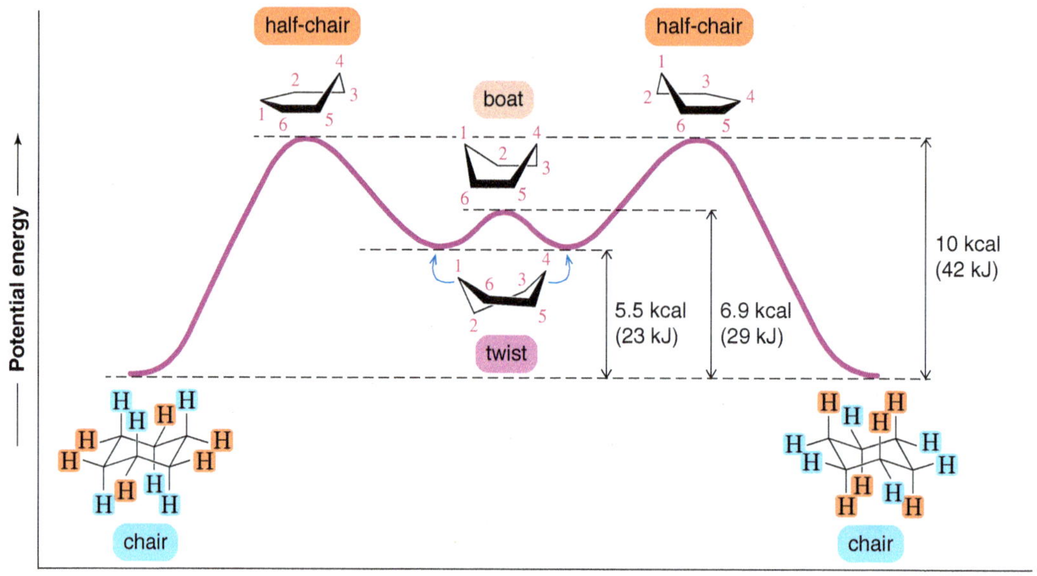

Figure 3.46 Conformational analysis of cyclohexane. The chair conformations are most stable, followed by the twist-boat.

Within the chair conformation, there are two types of hydrogens, **axial** and **equatorial**. Shown in **Figure 3.47**, axial hydrogens are aligned with an axis that cuts through the center of the ring, while equatorial hydrogens point away from the ring around what would be its equator. Upon flipping the chair, all axial hydrogens become equatorial and all equatorial hydrogens become axial.

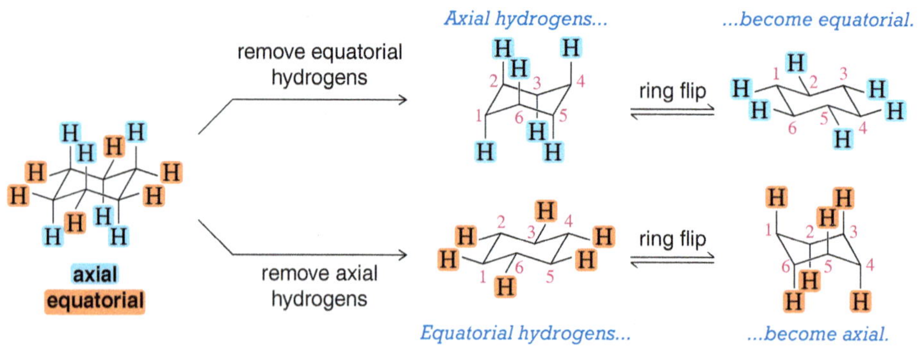

Figure 3.47 Axial and equatorial hydrogens in the chair conformation of cyclohexane.

3.7.1 Drawing chairs

Although we have considered several different conformations of cyclohexane, our attention here is focused most strongly on the chairs. In order to be able to discuss the chemistry of the chair conformation, we have to be able to draw them. For the artistically inclined, this comes pretty naturally, especially if a set of models is present. For the rest of us, it can be difficult. [*Go ahead and try drawing a chair on your own . . . . How did you do?*]

If you ask fifty organic chemists how to draw a chair conformation, you might get fifty different answers. **Table 3.12** lists the steps I use to draw a chair. [*This only generates one of the two chairs—we draw the other chair conformation in Table 3.13.*]

Table 3.12 How to draw a chair cyclohexane structure.

Step	Description	Action
1	Draw two parallel lines, angled slightly downward on the page. The bottom line should start and finish slightly to the right of the top line.	 *Both lines are parallel.*
2	Starting at opposite corners (upper left and upper right), draw two lines parallel to each other. On the left, angle the line downward; on the right, angle it upward.	 *Lines labeled 2 are parallel.*
3	Connect the ends of the lines on each side. This should result in two lines that are parallel to each other. This completes the carbon skeleton of the chair. [*Practice getting to this point before attempting to add the hydrogens.*]	 *Lines labeled 3 are parallel.*
4	Draw the axial hydrogens at each carbon parallel to an axis that runs through the center of the chair. The direction (up or down) of the axial hydrogen can be inferred from the direction of the vertex (the point marking the location of the carbon). If the vertex is pointing up, the axial hydrogen will be up at that point. If the vertex is pointing down, the axial hydrogen will be down at that point. The axial hydrogens alternate up and down around the ring.	*All axial H bonds (4) are parallel to this imaginary axis.* 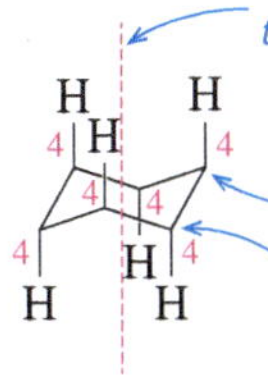 *This C points up, so its axial H is up.* *This C points down, so its axial H is down.*
5	Draw the equatorial hydrogens by simply filling in the empty space, ensuring a 109.5° bond angle. Check your work by making sure that each equatorial bond is parallel to two other bonds in the ring. [*These parallel relationships are indicated at right with the labels a, b, and c.*]	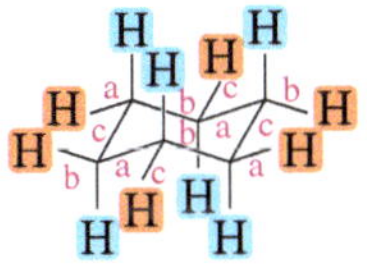 *Each a bond is parallel, as is each b bond and each c bond. Therefore, each equatorial H (orange screens) is parallel to 2 ring bonds and 1 other equatorial C—H bond.*

The other chair can be drawn by following the same process. All that changes, as shown in **Table 3.13**, is step 1, in which the first parallel lines are angled upward.

Table 3.13 How to draw the flipped-chair cyclohexane structure.

Step	Description	Action
1	Draw two parallel lines, angled slightly upward on the page. The bottom line should start and finish slightly to the left of the top line.	 *Both lines are parallel.*
2	Starting at opposite corners (upper left and upper right), draw two lines parallel to each other. On the left, angle the line upward; on the right, angle it downward.	 *Lines labeled 2 are parallel.*
3	Connect the ends of the lines on each side. This should result in two lines that are parallel to each other. This completes the carbon skeleton of the chair. [*Practice getting to this point before attempting to add the hydrogens.*]	 *Lines labeled 3 are parallel.*
4	Draw the axial hydrogens at each carbon parallel to an axis that runs through the center of the chair. The direction (up or down) of the axial hydrogen can be inferred from the direction of the vertex. If the vertex is pointing up, the axial hydrogen will be up at that point. If the vertex is pointing down, the axial hydrogen will be down at that point. The axial hydrogens alternate up and down around the ring.	*All axial H bonds (4) are parallel to this imaginary axis.* *This C points up, so its axial H is up.* *This C points down, so its axial H is down.* 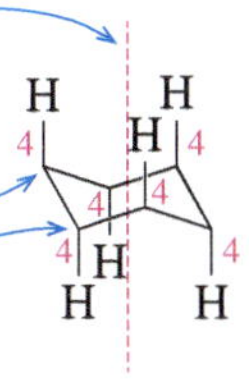
5	Draw the equatorial hydrogens by simply filling in the empty space, ensuring a 109.5° bond angle. Check your work by making sure that each equatorial bond is parallel to two other bonds in the ring. [*These parallel relationships are indicated at right with the labels a, b, and c.*]	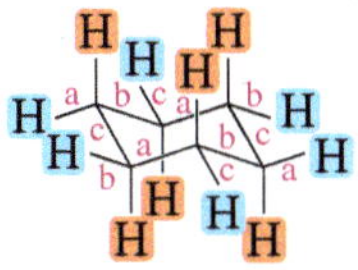 *Each a bond is parallel, as is each b bond and each c bond. Therefore, each equatorial H (blue screens) is parallel to 2 ring bonds and 1 other equatorial C—H bond.*

3.7.1 SUMMARY

- The chair is the most stable conformation of cyclohexane. Being able to draw a chair (and a flipped chair) conformation is an important skill for success in organic chemistry.

Assessment 3.37 Following the instructions in Table 3.12, draw as many chairs as it takes for yours to look like the ones in the book.

Assessment 3.38 Following the instructions in Table 3.13, draw as many flipped chairs as it takes for yours to look like the ones in the book.

3.7.2 Monosubstituted cyclohexane

The importance of the two chair conformations becomes apparent when we consider a substituted cyclohexane. **Figure 3.48** shows methylcyclohexane, drawn first with the methyl in the axial position. When the chair is flipped, going through the boat conformation, all bonds that were axial become equatorial, and all bonds that were equatorial become axial. As a result, the new chair contains the methyl group in the equatorial position. With two different conformational isomers possible, which do you think is more stable?

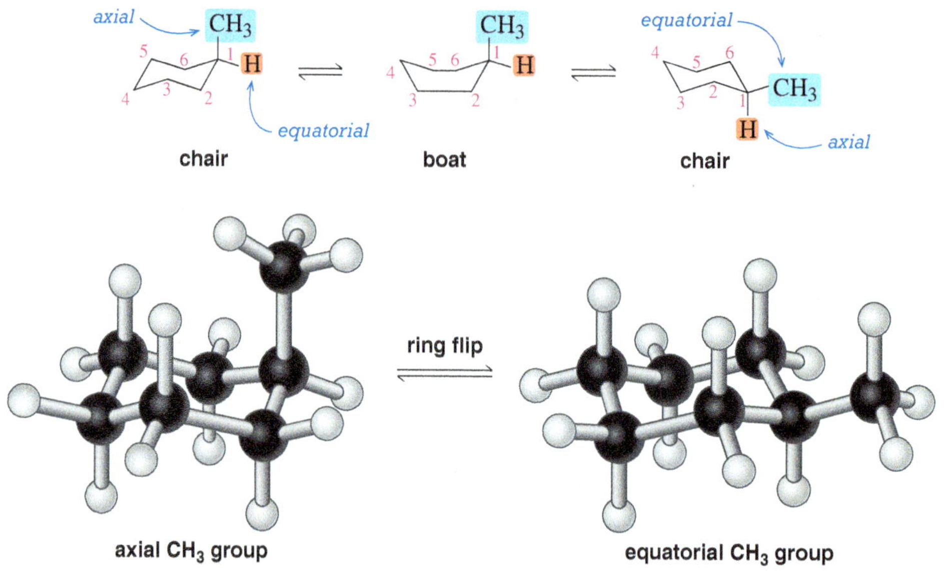

Figure 3.48 Upon ring flipping, a new conformation is produced. Equatorial substituents become axial, and axial substituents become equatorial.

The stability of the two conformations can be analyzed by looking at Newman projections along the impacted bonds. **Figure 3.49(a)** shows the Newman projections when the axial methyl group is viewed along the $C_1—C_2$ and $C_1—C_6$ bonds. Along the $C_1—C_2$ bond, there is a

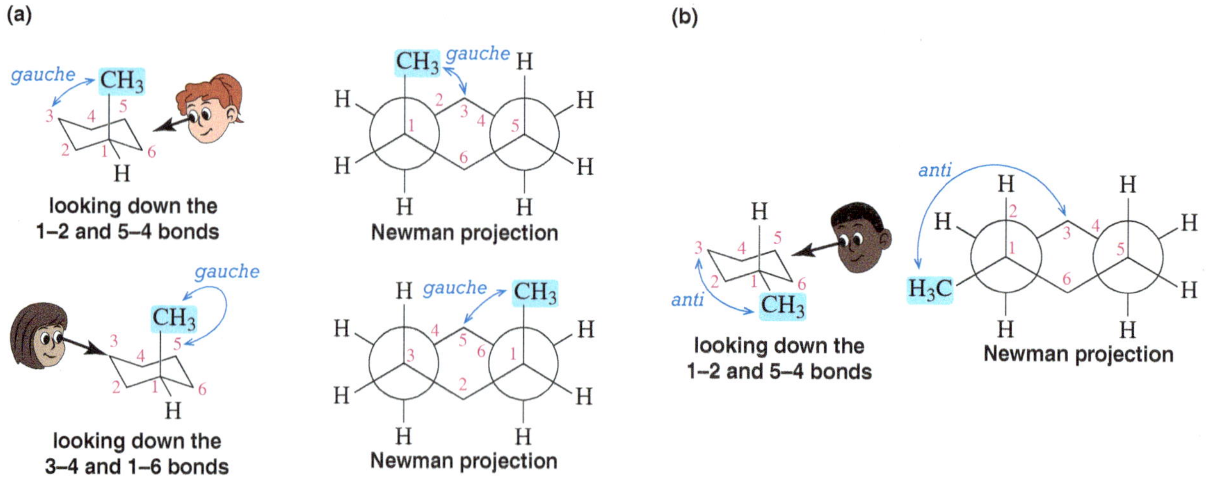

Figure 3.49 Newman projections of the chair conformations. (a) The axial methyl group has two gauche CH_3/C interactions, worth 0.9 kcal/mol of strain each. (b) The equatorial methyl group has zero gauche interactions between the methyl group and the ring.

gauche interaction between the axial methyl group and C_3 of the cyclohexane ring. Similarly, along the $C_1—C_6$ bond, there is a gauche interaction between the axial methyl group and C_5 of the cyclohexane ring. When the methyl group is equatorial, as shown in **Figure 3.49(b)**, there are no gauche interactions between the substituent and the ring. Because each gauche interaction is worth 0.9 kcal/mol of strain, the axial methyl group is 1.8 kcal/mol *less* stable than the equatorial methyl group.

How does the information in Figure 3.49 affect the equilibrium between the axial and equatorial conformations? Such an equilibrium favors placing the larger substituent in the equatorial position in order to avoid the unfavorable **1,3-diaxial interaction** (**Figure 3.50**).

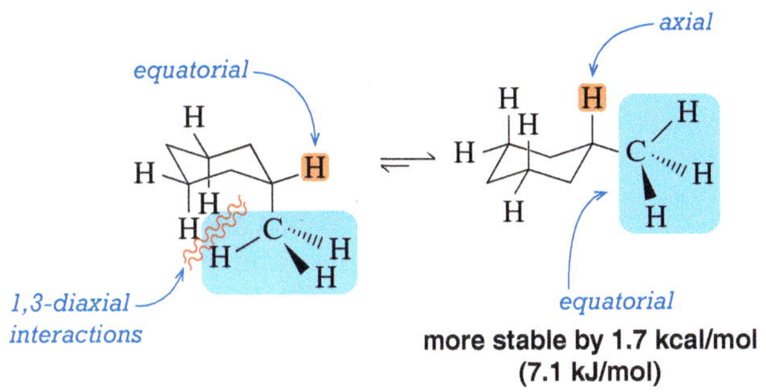

Figure 3.50 The conformation that places the larger group in the equatorial conformation is favored (more stable).

This trend is applicable to more than just the methyl substituent, as shown in **Table 3.14**. We talk more about ΔG (the change in Gibbs free energy) in Chapter 5, but for now, note that ΔG in Table 3.14 is the difference between axial and equatorial. A higher value suggests that there is a greater preference for the equatorial position. As expected, larger substituents have a greater preference for the equatorial position.

Table 3.14 Larger substituents have a greater preference for the equatorial position.

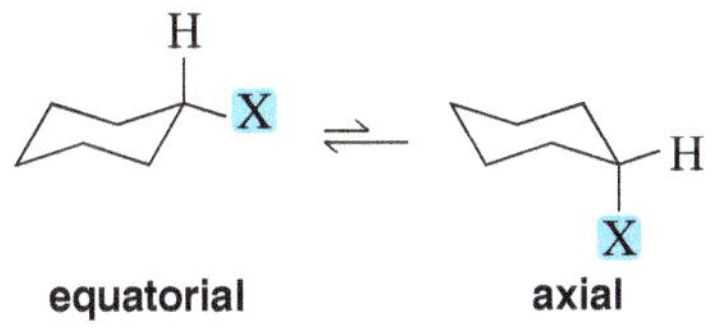

X	ΔG (axial–equatorial)		X	ΔG (axial–equatorial)	
	(kcal/mol)	(kJ/mol)		(kcal/mol)	(kJ/mol)
—F	0.2	0.8	—COOH	1.4	5.9
—CN	0.2	0.8	$—CH_3$	1.8	7.6
—Cl	0.5	2.1	$—CH_2CH_3$	1.9	7.9
—Br	0.6	2.5	$—CH(CH_3)_2$	2.1	8.8
—OH	1.0	4.1	$—C(CH_3)_3$	5.4	23

Assessment 3.39 Draw two chair conformations of the molecules below. Indicate which is most stable.

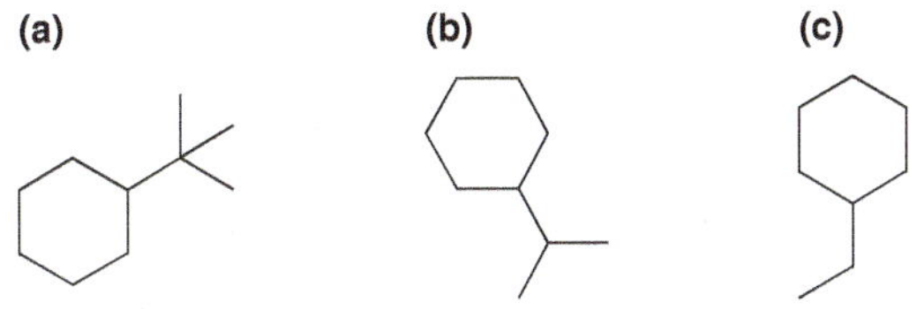

3.7.3 Disubstituted cyclohexane

Conformational analysis is slightly more complicated for disubstituted cyclohexanes, despite the underlying principles being the same. Complications arise because you must consider the relative stereochemistry of the two substituents and then transfer that information to the chair conformation. Consider *cis*-1,3-dimethylcyclohexane, shown in **Figure 3.51**, where both methyl groups are drawn on wedges, indicating that they are coming out of the page, or up. [*The methyl groups could also have been drawn on dashes to indicate they are both going away from the viewer.*] When we convert this planar cyclohexane to a chair conformation, we have two choices. First, do we place the methyl groups axial or equatorial? And second, do we place them up or down? "Up" and "down" do not directly correlate with "axial" or "equatorial." That is, we can draw either of two conformations of *cis*-1,3-dimethylcyclohexane. In one, both substituents are coming toward the viewer (the eyeball), or "up," in the axial position. In the other, they are still coming toward the viewer, or "up," but in the equatorial position. These two representations are ring-flipped conformations of the same molecule, where the one with the two substituents in the equatorial position is favored. To get the "up" or "down" part right, imagine viewing the chair conformation from above, and place substituents appropriately. [*Assessments 3.40 and 3.41 give you practice at this.*]

Figure 3.51 The conformation with two equatorial substituents is favored over the one with two axial substituents.

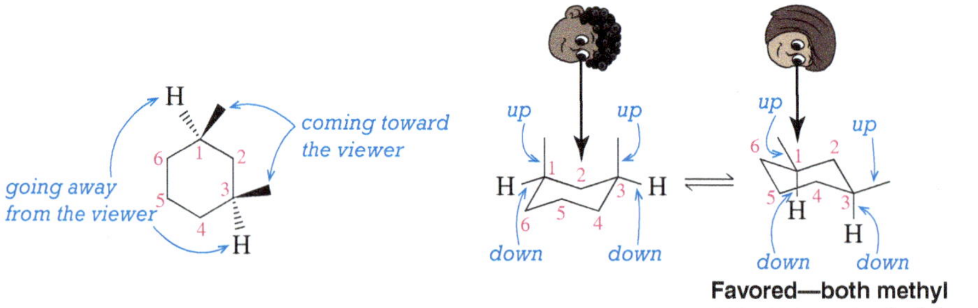

Assessment 3.40 Given the following chair conformations, draw each in its planar form as if you were viewing it from above.

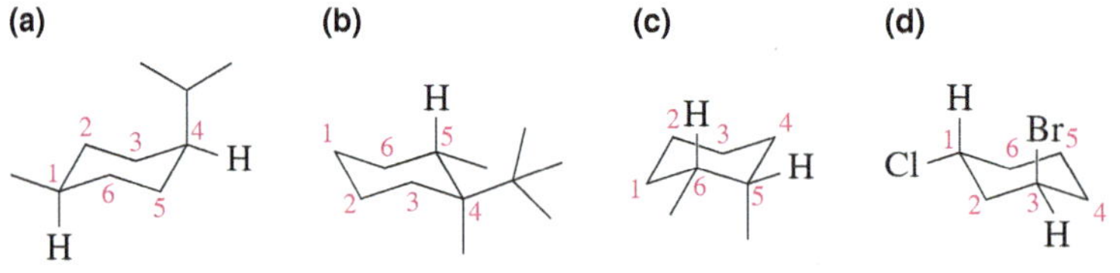

Assessment 3.41 Draw two different chair conformations for each of the following molecules. Make sure that your drawings clearly show the cis–trans stereochemistry.

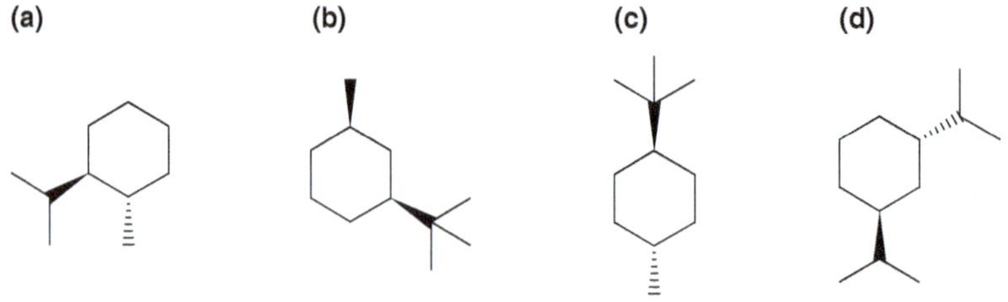

Once the substituents are placed correctly on the ring, the most stable conformation will have all substituents in the equatorial position, if possible. If it is not possible to have all substituents in the equatorial position, the more stable chair conformation minimizes 1,3-diaxial interactions often by placing the larger substituent in the equatorial position. **Figure 3.52** shows some examples.

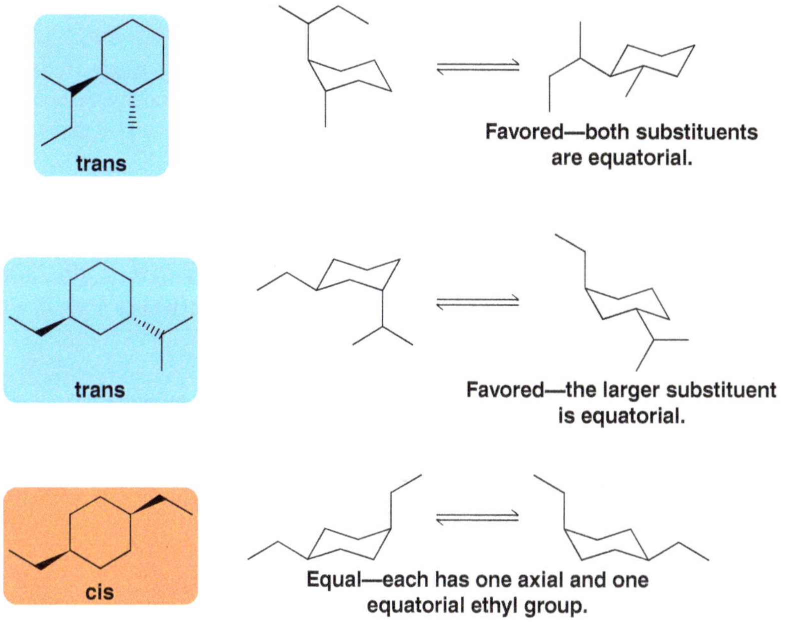

Figure 3.52 The more stable conformation has the larger group in the equatorial position.

3.7.2–3.7.3 SUMMARY

- The more stable chair conformation is the one that minimizes 1,3-diaxial interactions between substituents on the ring.

Assessment 3.42 For each pair of conformations you drew in Assessment 3.41, indicate which is most stable.

Assessment 3.43 Draw the most stable chair conformation for the following trisubstituted cyclohexane.

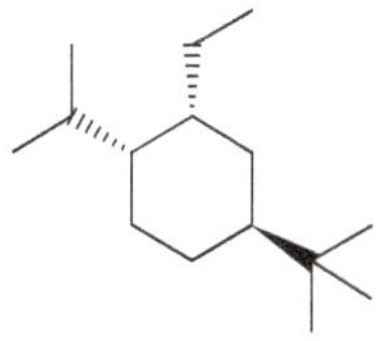

3.8 Tour of functional groups

While this chapter has focused on alkanes and cycloalkanes (the simplest organic molecules), now is a good time to introduce the molecular classes presented in the rest of the book. In order to classify similar molecules, organic chemists focus on their **functional groups.** These groups help us to understand the chemical and physical properties (the function) of the molecules that contain them. We discuss the functional groups listed in Table 3.15 in detail later on, but for now you should familiarize yourself with their structures. In Table 3.15, and throughout the book, "R" is often used to represent a general alkyl group.

3.8.1 More drawing: functional groups

Back in Section 3.2, we learned how to draw organic molecules using line-angle drawings as a way of maximizing the information shown but minimizing the time spent drawing. You may have noticed in Chapter 2, or when drawing the functional groups in Table 3.15, that organic molecules can have lots of lone pairs. Knowing where these lone pairs are will be really helpful when "negatives attack positives," but organic chemists, to save time, don't always show them. Instead, we allow the formal charge, or lack of a formal charge, to tell us how many electron pairs are present. In essence, this is the opposite of how we were first introduced to formal charge. Back in Chapter 2, you were given a Lewis structure and asked to calculate the formal charge on an atom, as shown in Figure 3.53.

$$\text{FC} = \text{valence electrons} - \frac{1}{2}(\text{shared e}^-) - \text{unshared e}^-$$

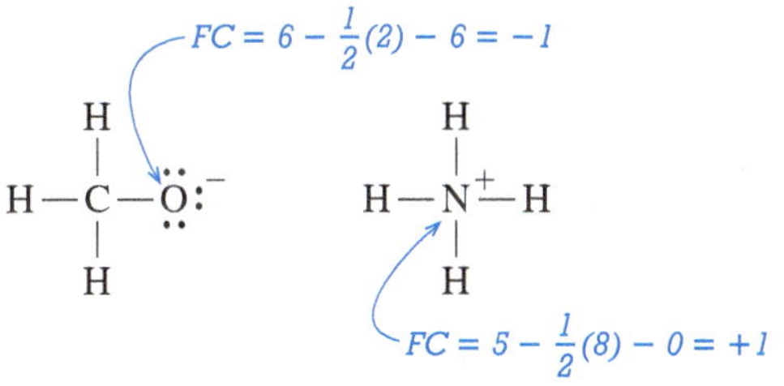

Figure 3.53 Calculating the formal charge (FC) from a Lewis structure.

Table 3.15 Common functional groups.

Molecular class	Functional group	Description	Example
Alkene	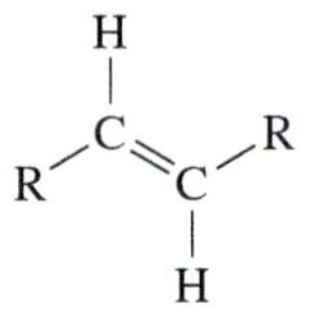	Contains a carbon doubly bonded to another carbon (1 σ and 1 π bond)	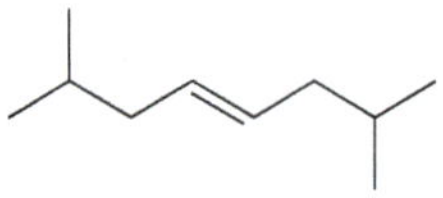
Alkyne	R—C≡C—R	Contains a carbon triply bonded to another carbon (1 σ and 2 π bonds)	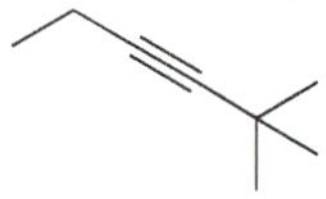
Haloalkane	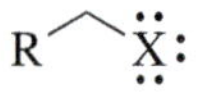 X = F, Cl, Br, or I	Contains a halogen (X) bonded to an sp^3-hybridized carbon	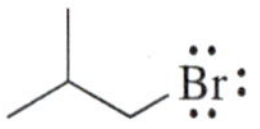
Alcohol	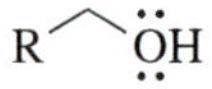	Contains an oxygen singly bonded to both an alkyl group (R) and a hydrogen; the OH group is referred to as a hydroxyl group	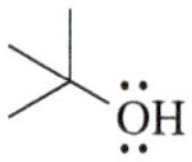
Ether	R—$\ddot{\text{O}}$—R	Contains an oxygen singly bonded to two alkyl groups (R)	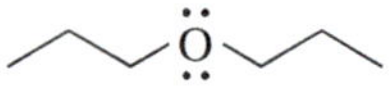
Amine	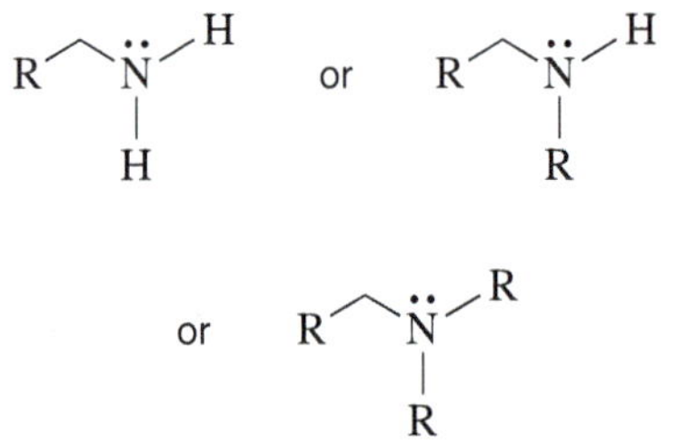	Contains a nitrogen with three bonds to either sp^3-hybridized carbons or hydrogens; the NH_2 group is referred to as an amino group	
Aldehyde	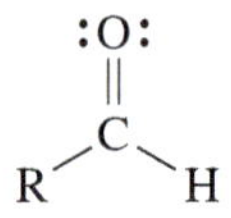	Contains a carbonyl group (C═O) between hydrogen and an alkyl group (R) (or H for formaldehyde)	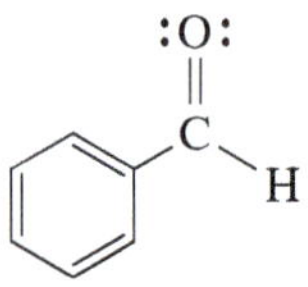
Ketone	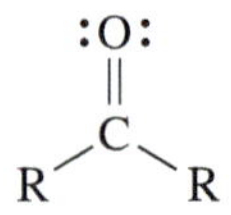	Contains a carbonyl group (C═O) between two alkyl groups (R)	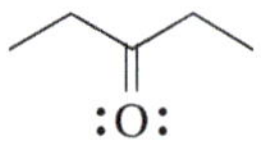
Carboxylic acid	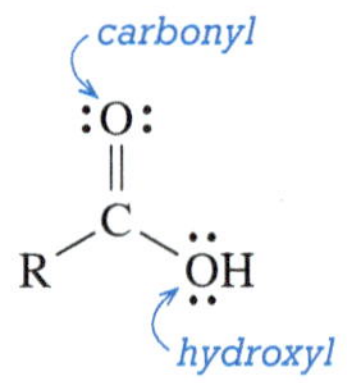	Contains a carbonyl group (C═O) between a hydroxyl group (OH) and an alkyl group (R) (or H for formic acid)	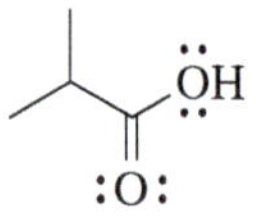
Ester	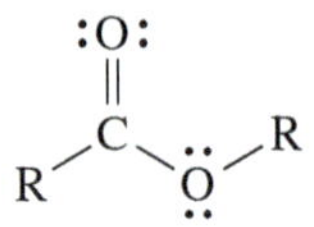	Contains a carbonyl group (C═O) between an alkoxy group (OR) and an alkyl group (R) (or H for formate esters)	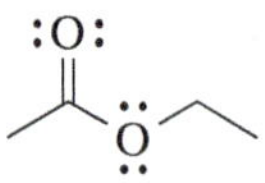
Acid chloride	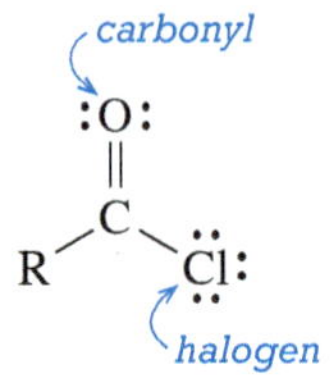	Contains a carbonyl group (C═O) between a chlorine atom and an alkyl group (R) (or H for formyl chloride)	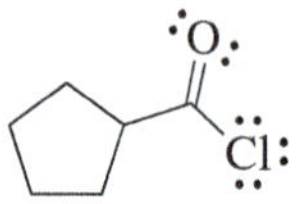

(continued)

Table 3.15 ***Continued***

Molecular class	Functional group	Description	Example
Amide	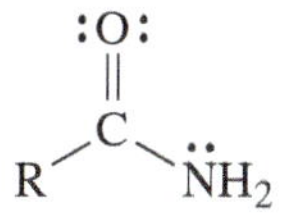	Contains a carbonyl group (C=O) between an amino group (NH_2, NHR, or NR_2) and an alkyl group (R) (or H for formamides)	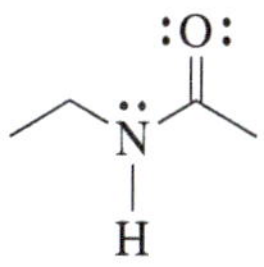
Anhydride	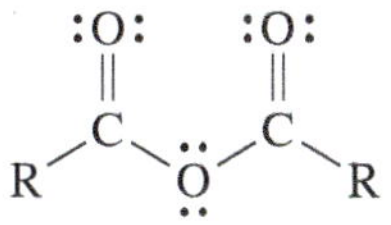	Contains two carbonyl groups (C=O) separated by an oxygen	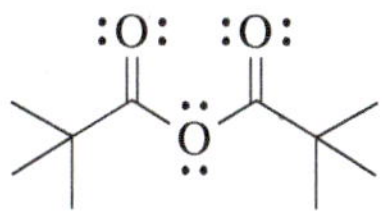
Nitrile	R—C≡N:	Contains a cyano group (C≡N) attached to an alkyl group (R)	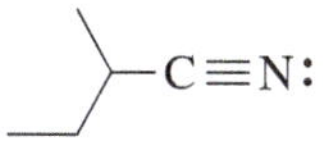
Imine	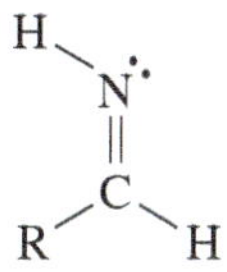	Contains an imino group (C=N) between two alkyl groups (R) or one alkyl group (R) and one hydrogen	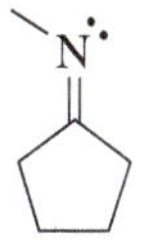
Phosphate	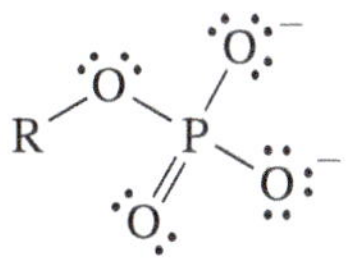	Contains an alkyl group (R) attached to an OPO_3^{2-} group	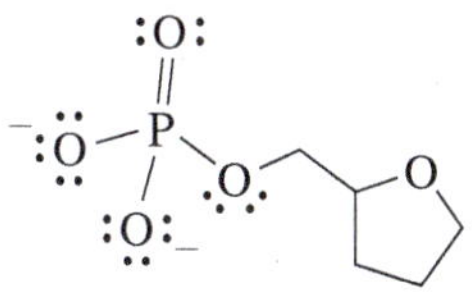

With enough time spent drawing Lewis structures, you may have begun noticing patterns where bonded and nonbonded pairs on a given atom result in the atom possessing a certain formal charge. These patterns are summarized in Table 3.16.

Table 3.16 Correlation between structure and formal charge.

Atom	Formal charge +1	Formal charge 0	Formal charge −1
Carbon	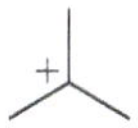 3 bonds 0 lone pairs	 4 bonds 0 lone pairs	 3 bonds 1 lone pair
Nitrogen	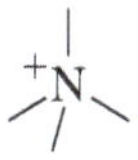 4 bonds 0 lone pairs	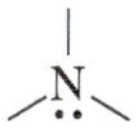 3 bonds 1 lone pair	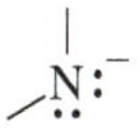 2 bonds 2 lone pairs
Oxygen	 3 bonds 1 lone pair	 2 bonds 2 lone pairs	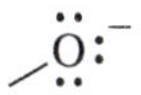 1 bond 3 lone pairs
Halogens (F, Cl, Br, and I)	 2 bonds 2 lone pairs  0 bonds 3 lone pairs	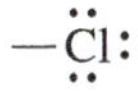 1 bond 3 lone pairs	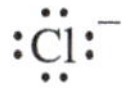 0 bonds 4 lone pairs

Although you will eventually stop drawing lone pairs in every molecule, it might be helpful to continue drawing them for now. Showing them comes in very handy when struggling with a difficult problem.

We can use these patterns to make drawing organic molecules simpler. Instead of drawing a molecule and showing all lone pairs, we show only the nonzero formal charges. Every atom without a formal charge shown is assumed to be neutral, allowing us to predict how many lone pairs are present. When a formal charge is given for an atom, we can predict how many lone pairs (and if carbon, how many hydrogens) are also present. Table 3.17 lists some examples of how we should "see" the molecules we draw. Assessment 3.45 gives you practice drawing and interpreting molecular structure without showing all lone pairs.

Table 3.17 Drawing organic molecules without showing lone pairs.

When this is drawn . . .	. . . you should see this.
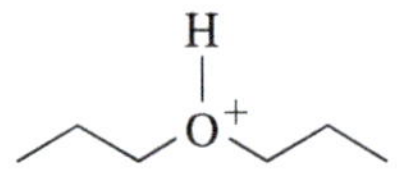	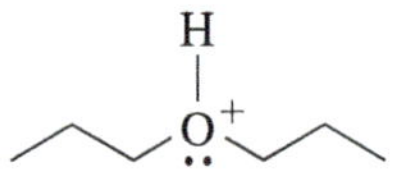
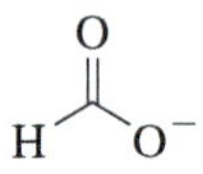	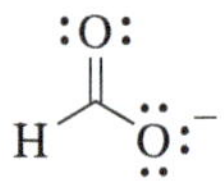
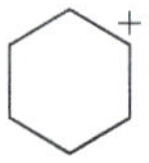	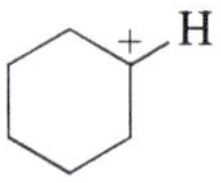
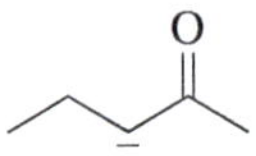	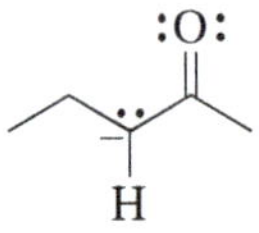
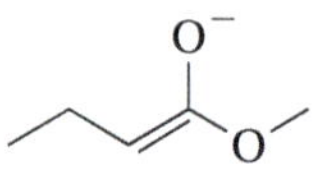	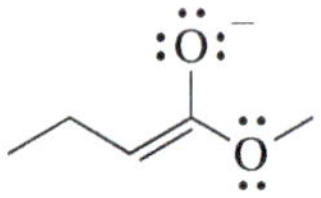

Assessment 3.44 Based on the formal charge, determine how many lone pairs are on each indicated atom.

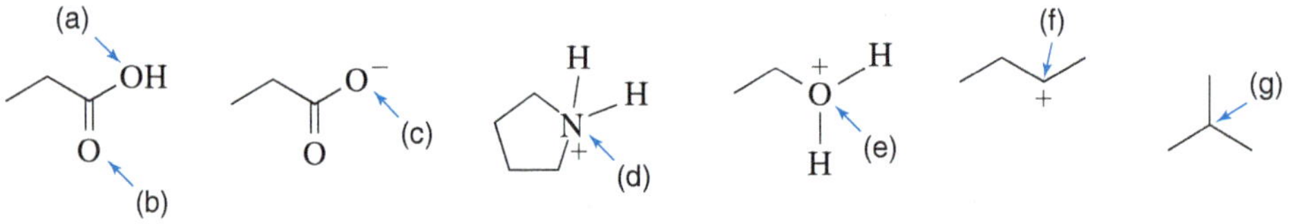

Assessment 3.45 Modify the following line-angle drawings to show all lone pairs.

(a) (b) (c)

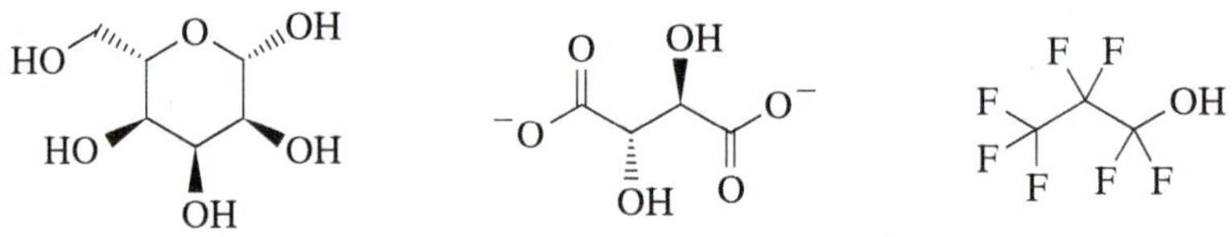

Concluding thoughts

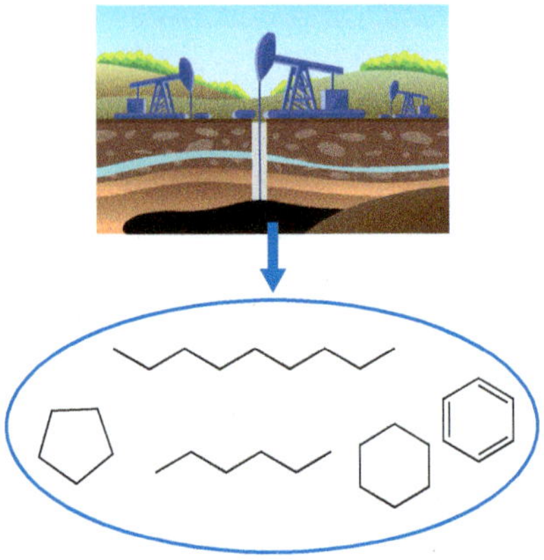

Figure 3.54 Organic compounds can be derived from petroleum (nonrenewable) feedstocks. These resources are finite, so Earth will eventually run out of them.

Organic chemistry deals with the chemistry of carbon, the backbone of all molecules of life. However, in our study of organic chemistry we often dismiss the origin of these molecules as being unimportant. That is, we don't think about the origin of the feedstock—the source of the organic compounds from which our starting materials and reagent molecules are derived. Feedstocks can be categorized as either nonrenewable or renewable. As discussed at the beginning of this chapter, organic compounds have traditionally been isolated from oil deposits deep within the ground, as part of petroleum refining [CIC 3-A] and natural gas isolation (Figure 3.54). Although these oil deposits were created via the decomposition of living

With our first entrée into the structure of organic molecules and functional groups complete, we are ready to study the reactions of organic molecules. One of the principle mechanisms by which organic molecules react is through acid–base reactions. Because the goal of this text is to unify the reactions and principles we present, getting a strong handle on acid–base chemistry early on is vitally important. Let's work toward that beginning in Chapter 4.

organisms (a process that continues to this day), the fact that it took millions of years for them to form makes them functionally nonrenewable resources.

One of the aims of this book is to show, through the framework of green chemistry, how environmentally friendly and sustainable alternatives can be incorporated into the science of organic chemistry. For example, some living organisms, like plants and algae, use carbon dioxide in the atmosphere to produce their own organic materials. Compounds derived from these sources are considered to be from a renewable feedstock (**Figure 3.55**). As we explain in Chapter 7, Green Chemistry Principle (GCP) 7 encourages the use of these renewable feedstocks wherever possible.

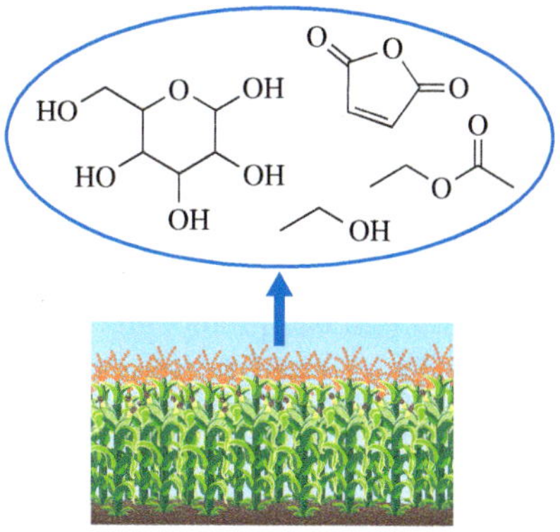

Figure 3.55 Organic compounds can be derived from plant (renewable) feedstocks.

End-of-Chapter Materials

Important Concepts

[*These are the major concepts covered in each section, not including reactions themselves, which are summarized in a later section.*]

3.1: Alkanes are saturated hydrocarbons containing the maximum number of hydrogens per carbon. The general formula for alkanes is C_nH_{2n+2}, where n = number of carbons. Every carbon in an alkane is sp^3 hybridized, forming C—C and C—H bonds by overlapping sp^3-sp^3 and sp^3-1s orbitals, respectively. Larger alkanes have more London dispersion forces and, as a result, higher boiling points, melting points, and diminished water solubility. Branched alkanes have fewer intermolecular interactions relative to straight chain alkanes with the same molecular formula.

3.2–3.3: Organic molecules can be drawn using several representations: structural, hybrid, condensed, and line-angle drawings. Line-angle drawings require the least amount of time to draw yet have all of the necessary information. Constitutional isomerism results when two molecules have the same molecular formula but different connectivity. The large number of alkane constitutional isomers requires the use of the IUPAC system of nomenclature. Carbons can be classified as 1° (primary), 2° (secondary), 3° (tertiary), and 4° (quaternary) to indicate that one, two, three, or four carbons are attached, respectively.

3.4: Molecules undergo constant rotation and vibration, yet spend more time in the more stable conformation that minimizes the strain on the molecule. Torsional strain is the strain caused by the interaction of bonded electron pairs on neighboring carbons. The staggered conformation is more stable than the eclipsed conformation because torsional strain is minimized. Within staggered conformations, the anti conformation (dihedral angle of 180°) is more stable than the gauche conformation (dihedral angle of 60°).

3.5–3.6: Cycloalkanes are cyclic alkanes and have the general molecular formula of C_nH_{2n}, where n = number of carbons. Because there is no free rotation, disubstituted cycloalkanes can exhibit cis-trans isomerism. The favored conformation in cycloalkanes is achieved by minimizing angle and torsional strain. Angle strain arises from a bond angle being forced away from ideal. Cyclopropane, with a high amount of angle and torsional strain, is the least stable cycloalkane. Cyclohexane, with 109.5° bond angles and fully staggered C—H bonds, is essentially strain free in the chair conformation.

3.7: Cyclohexane can exist in either of two stable chair conformations. Axial hydrogens are aligned with an axis that cuts vertically through the center of cyclohexane, while equatorial hydrogens are oriented around the "equator." Monosubstituted cyclohexanes adopt the conformation that places the substituent in the equatorial position. Larger substituents have a greater preference for the equatorial over the axial position in order to minimize steric strain, or the unfavorable interaction of nonbonded nuclei. Disubstituted cycloalkanes will adopt the conformation that places the larger substituent in the equatorial position.

3.8: The chemical and physical properties of organic molecules are determined by the functional groups present.

End-of-Chapter Assessments

Structure drawing boot camp

Be sure to take breaks while doing these to maximize retention of the material. See Section 1.4 for why this is important. Once you feel comfortable with these, move on.

Assessments 3.46–3.49 steadily increase in difficulty while offering a repetitive experience that ensures you leave Chapter 3 with the ability to draw and interpret line-angle drawings.

3.46 (•) Given the line-angle drawings shown, answer the following questions: (i) How many carbons are in each molecule? (ii) How many hydrogens are at the circled carbon? (iii) Is the indicated (→) carbon 1°, 2°, 3°, or 4°?

(a)

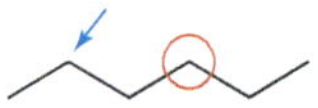

(b)

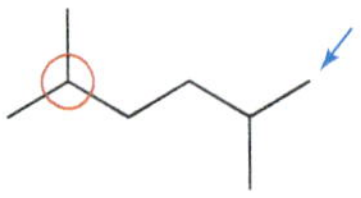

(c)

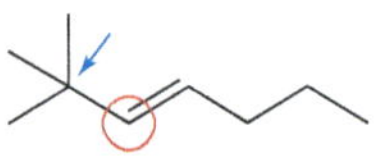

(d)

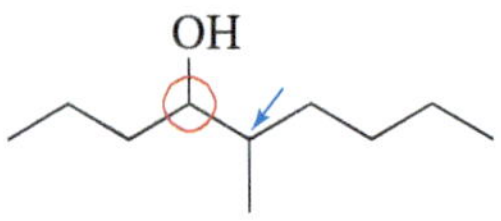

(e)

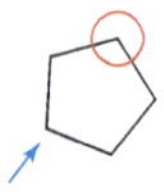

(f)

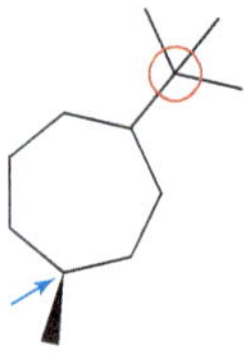

(g)

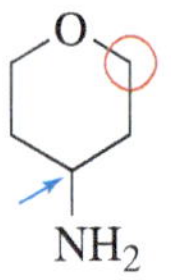

(h)

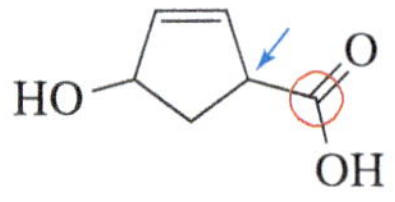

3.47 (•) Represent each of the following condensed structural formulas using a line-angle drawing.

(a) $CH_3CH_2CH_2CH_2CH_2CH_2CH_3$

(b) $(CH_3)_3CCH_2CH_2CH_3$

(c) $CH_3CHOHCH_2CHBrCH_3$

(d) $CH_3CH_2CH_2CCCH_2OH$

(e) $(CH_3)_2CHCH_2CH_2CH_2CH_2CHO$

(f) $CH_3CHCHCHBrCH_3$

3.48 (•) For each molecular formula, represent all constitutional isomers using line-angle drawings.

(a) C_2H_6 (b) C_3H_8 (c) C_4H_{10} (d) C_5H_{12} (e) C_6H_{14}

3.49 (•) Draw in all missing lone pairs for the following molecules.

(a)

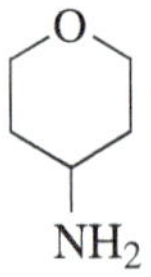

(b)

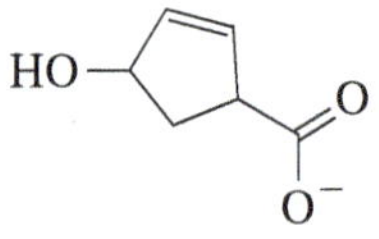

(c)

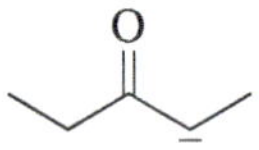

(d)

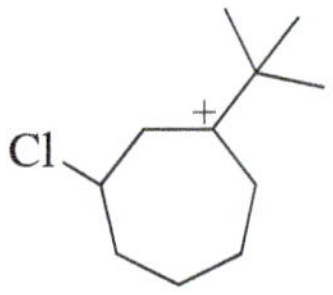

(e)

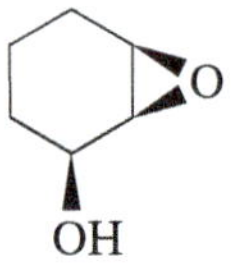

Newman projections boot camp

Assessments 3.50–3.55 steadily increase in difficulty while offering a repetitive experience that ensures you leave Chapter 3 with the ability to draw and interpret Newman projections.

3.50 (•) Given the first Newman projection and the direction and degree of rotation, fill in the resulting Newman projection. [*One substituent has been labeled for you.*]

(a)

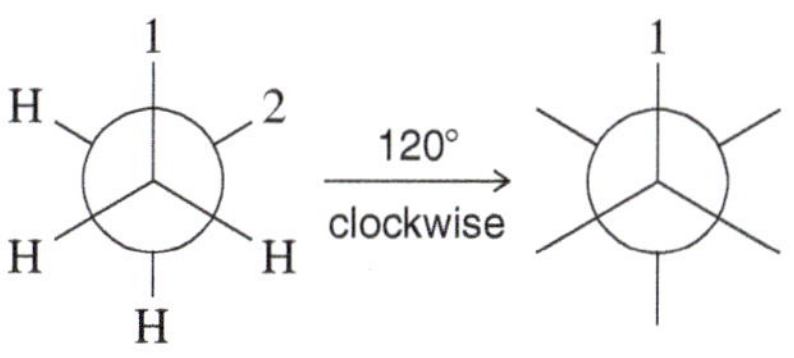

(b)

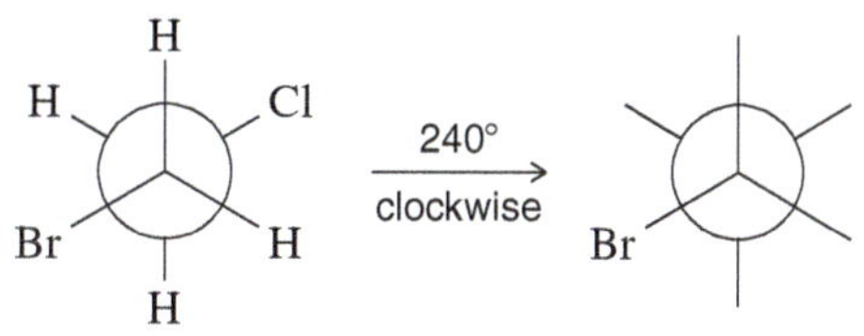

(c)

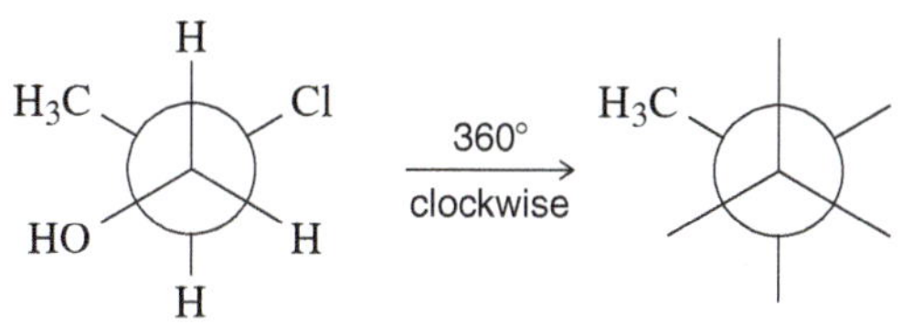

(d)

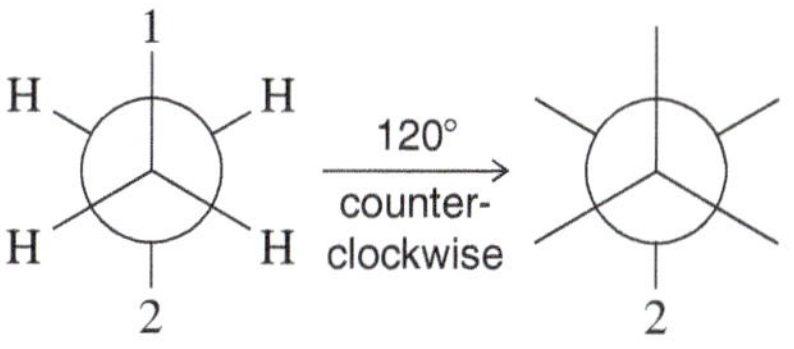

(e)

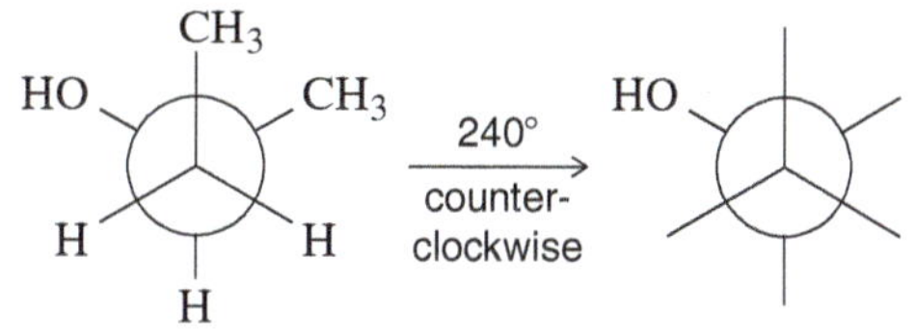

(f)

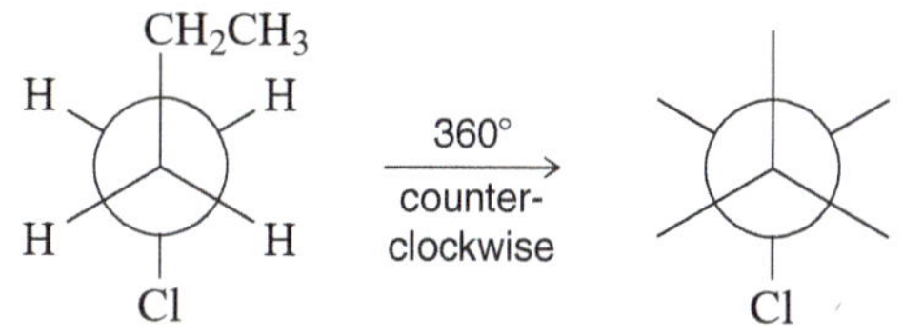

(g)

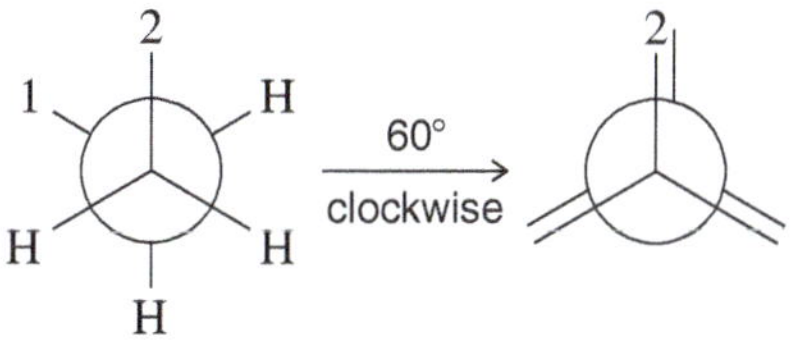

(h)

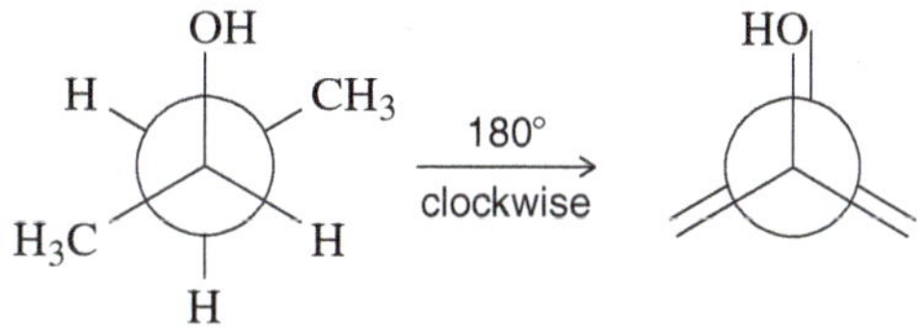

(i)

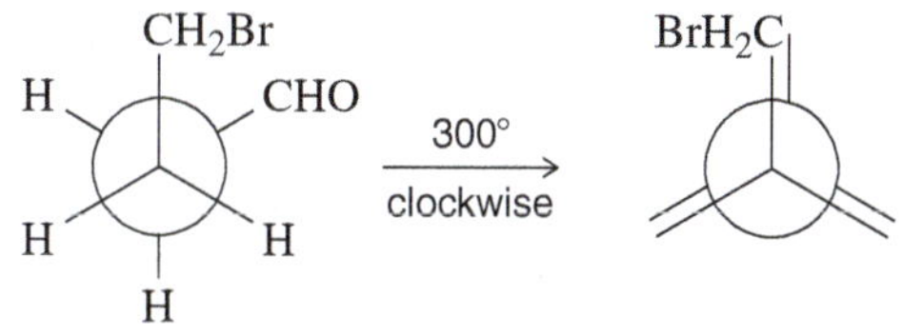

(j)

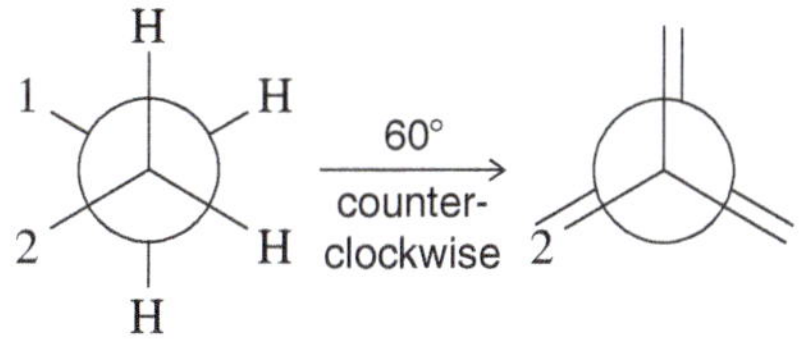

(k)

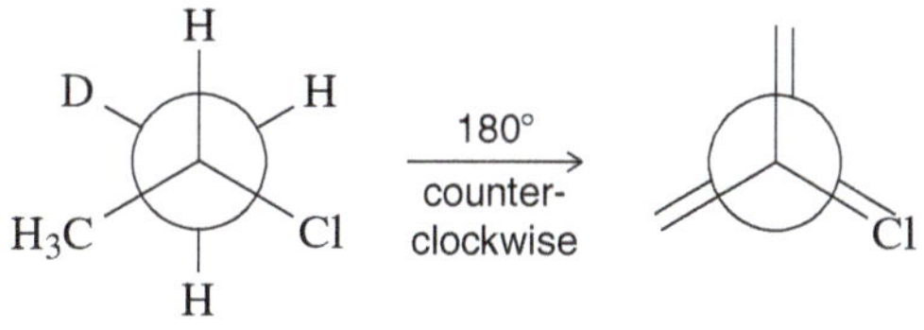

(l)

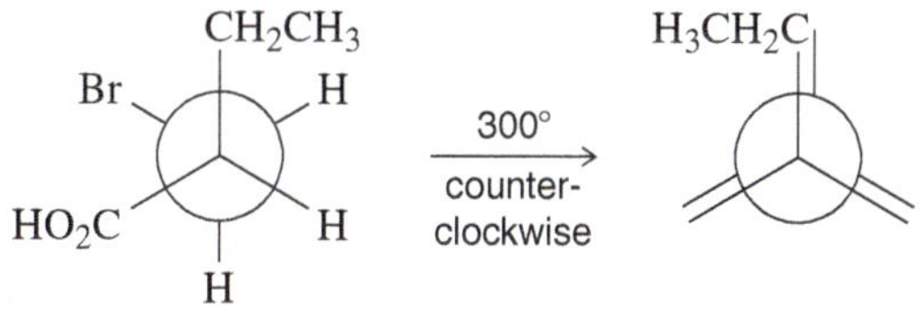

3.51 (•) Drawing all staggered Newman projections will start with one of the following skeletons. What is the difference between A and B? [*How would you reorient A so that you see B?*]

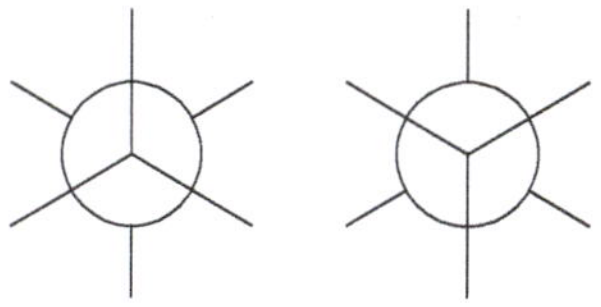

A B

The dots next to the assessment numbers tell you how difficult it is (•••• = most difficult).

3.52 (••) For each of the following structures, which staggered Newman projection skeleton from Assessment 3.51 should you draw first to show what is seen when looking down the indicated bond?

(a)

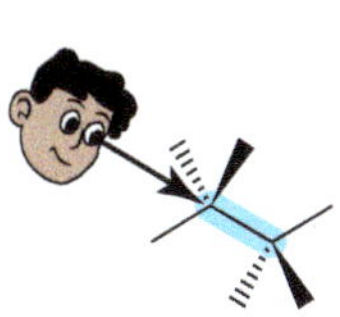

(b)

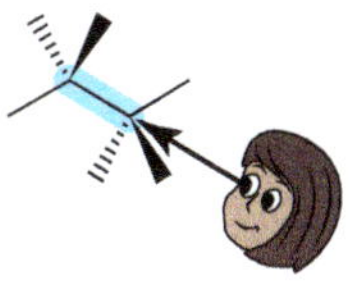

(c)

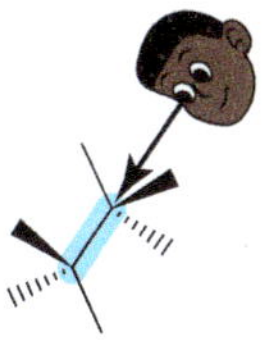

(d)

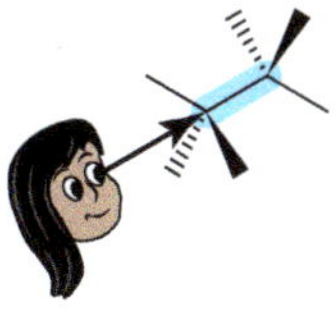

(e)

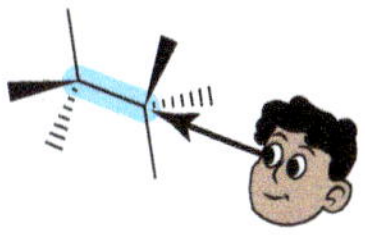

(f)

(g)

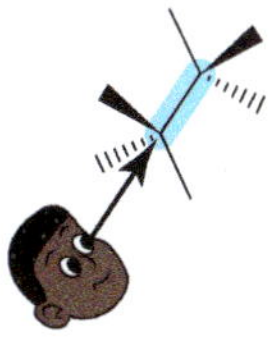

(h)

3.53 (••) Given the following structures, show the Newman projection that would result from looking down the indicated bond in the direction shown. [*Orient yourself as if you were the eyeball looking down the bond. Some of the examples have been partially completed for you to fill in the rest.*]

(a)

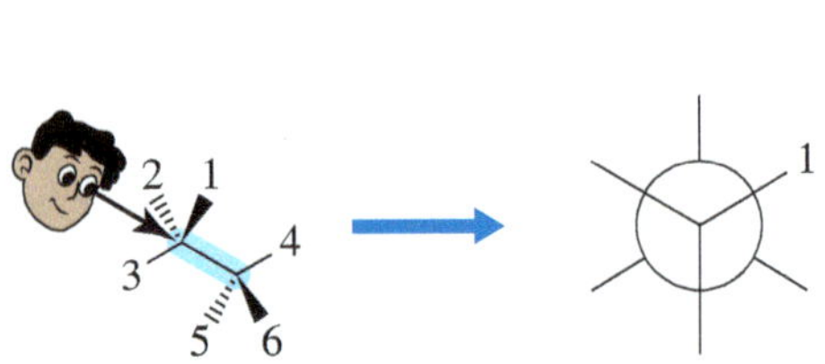

(b)

(c)

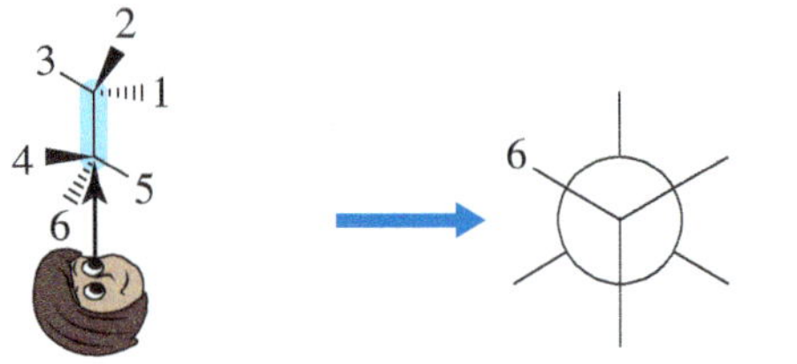

(d)

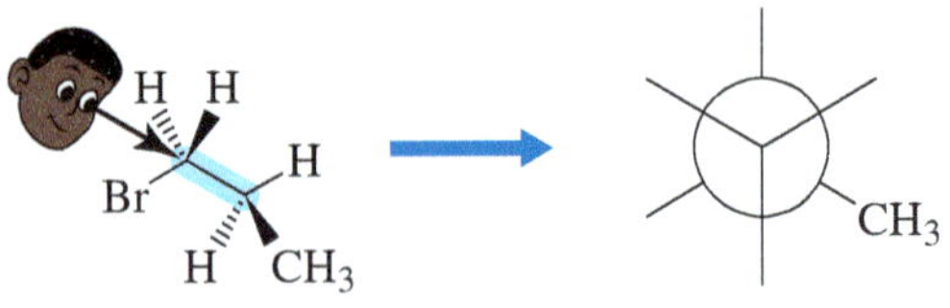

(e)

(f)

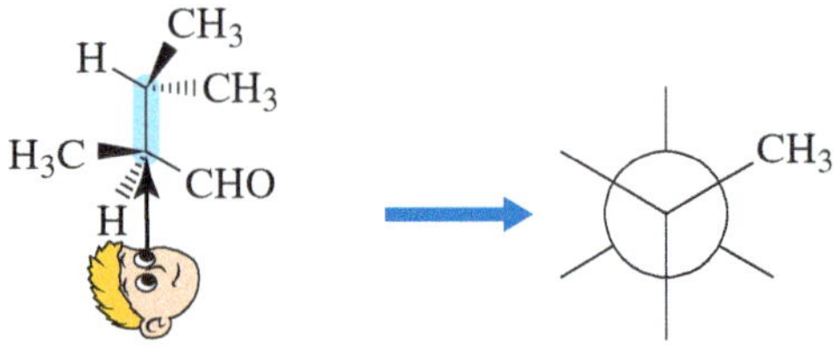

(g)

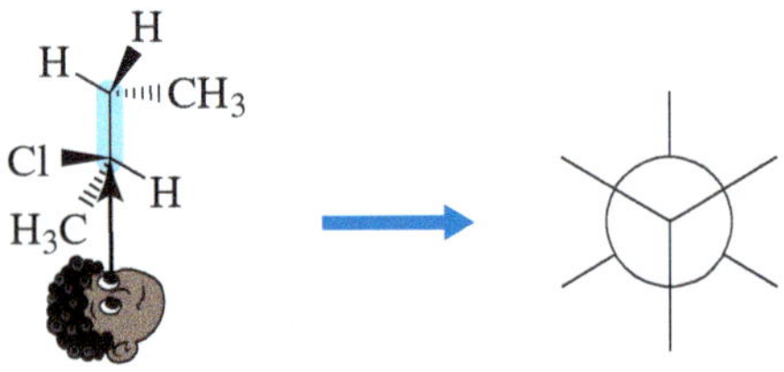

(h)

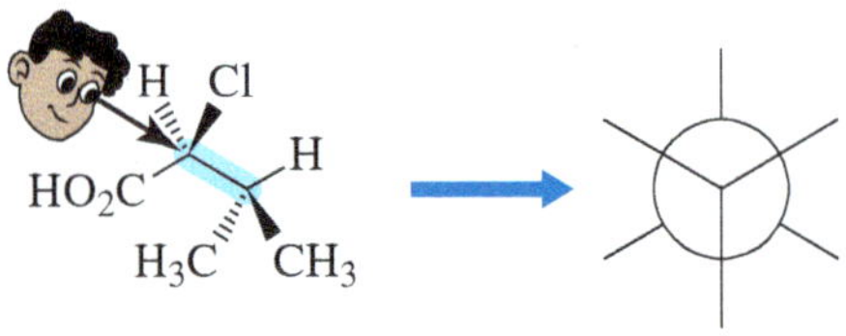

(i)

[*On these last three, you need to use the line angle drawings without hydrogens or carbons shown.*]

(h)

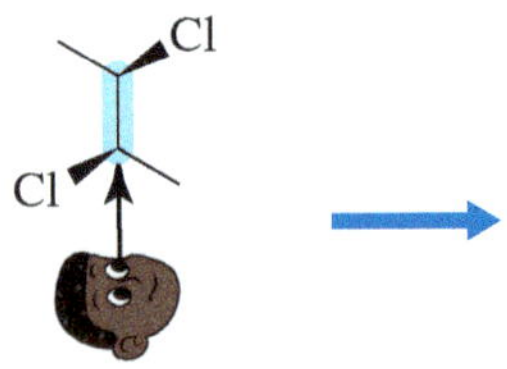

(i)

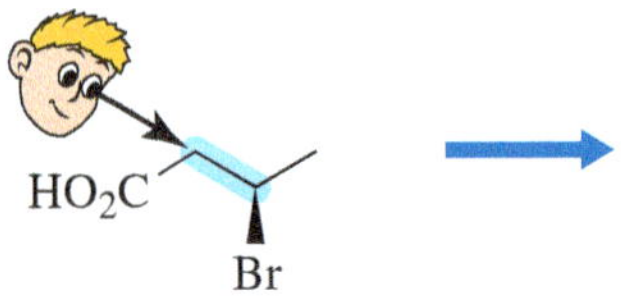

(j)

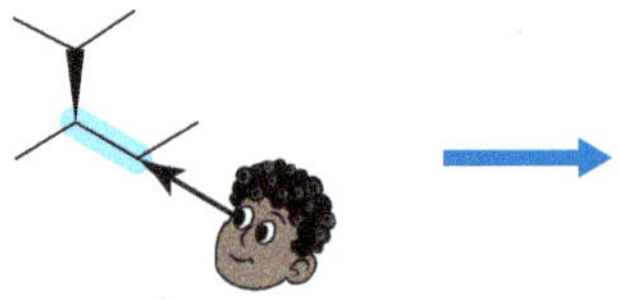

3.54 (••) For each pair of conformations shown, choose which is most stable. [*If both conformations have the same number of gauche interactions, choose the one where the interactions are between smaller groups.*]

(a)

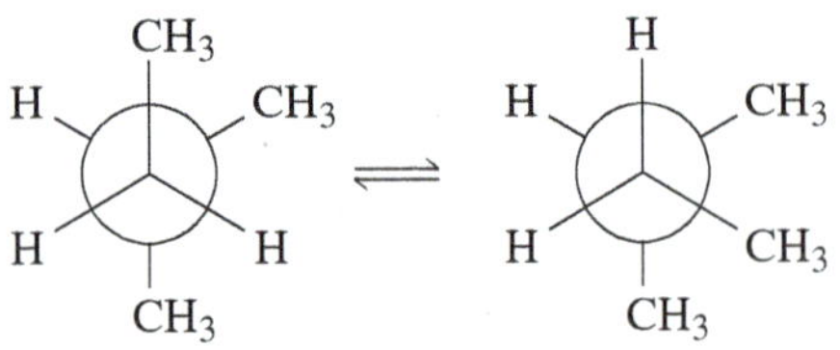

(b)

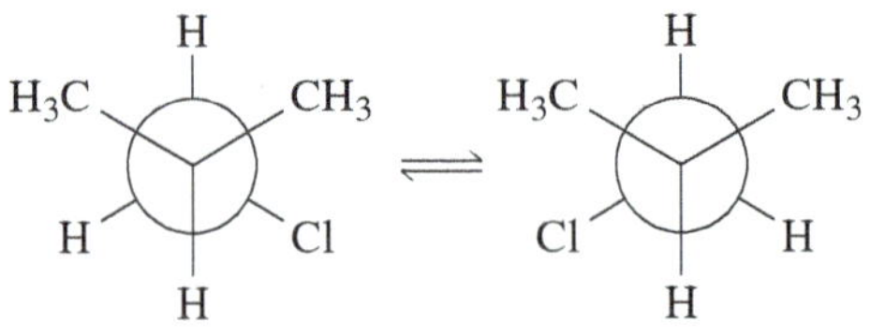

(c)

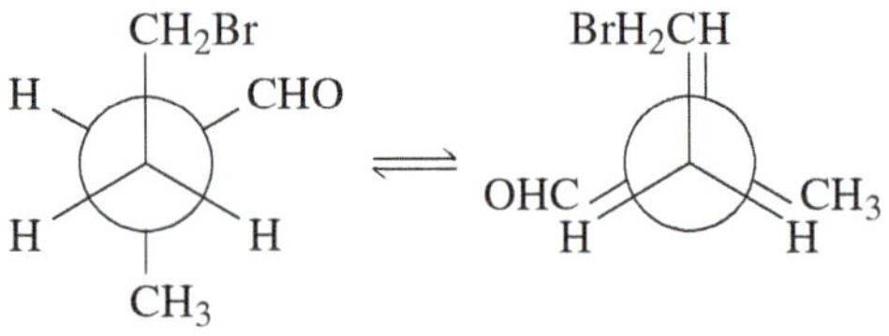

(d)

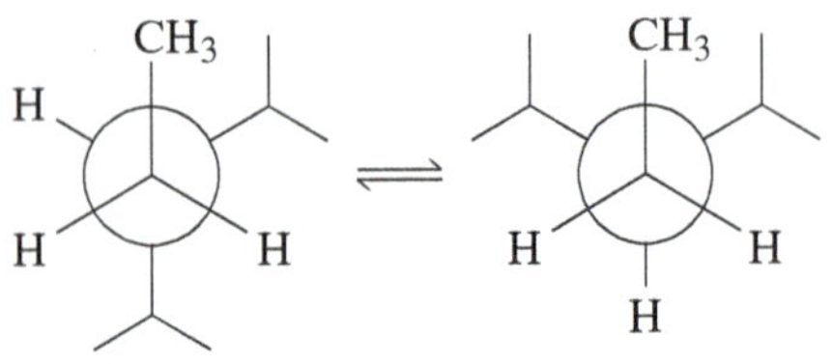

(e)

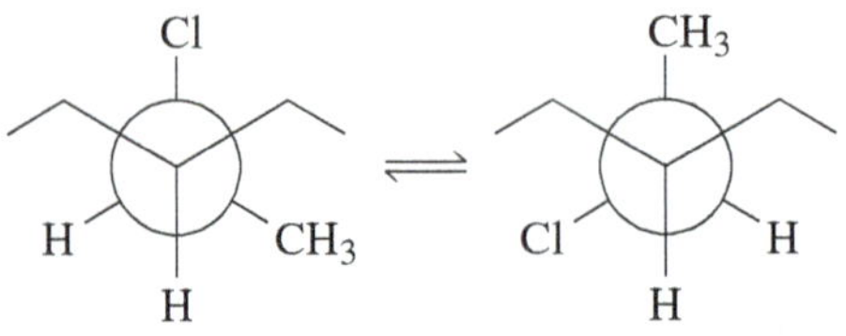

(f)

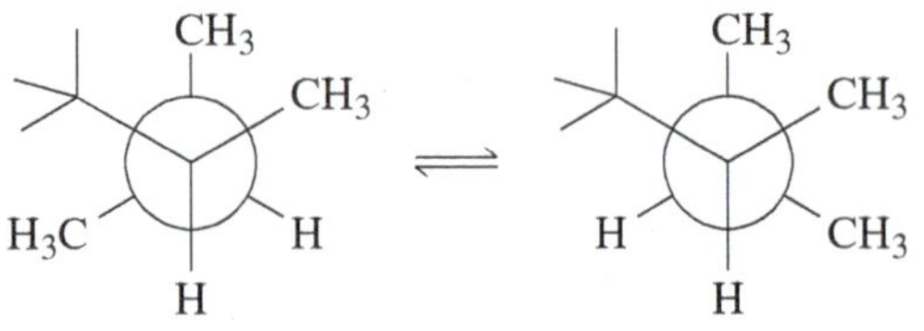

3.55 (•••) Looking down the indicated bond, show the three most stable conformations and choose the one that is most stable. Be sure that the first Newman projection you show is the one you see initially (before rotation). [*Why should none of your three Newman projections show eclipsed conformations?*]

(a)

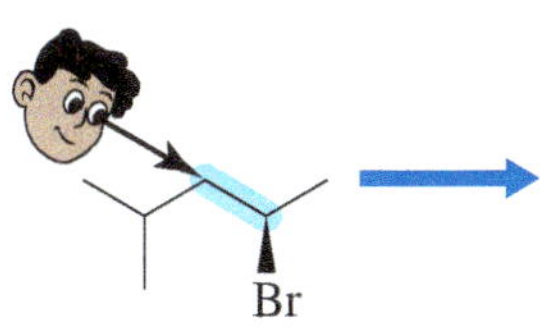

(b)

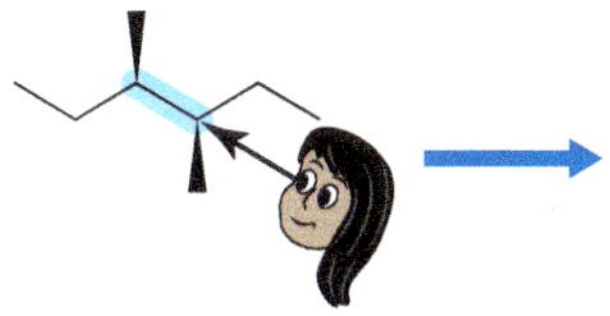

(c)

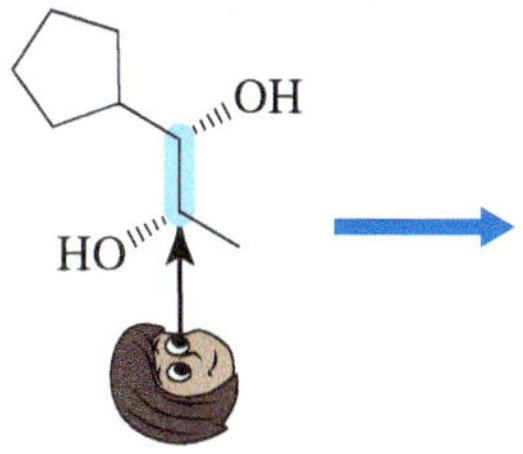

(d)

(e)

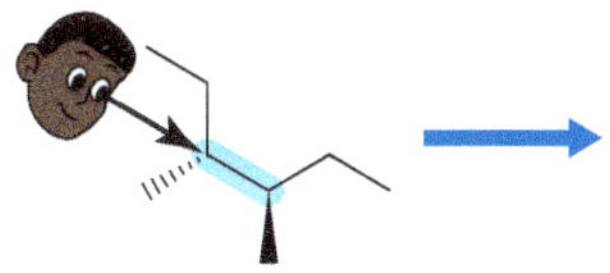

Chair conformation boot camp

Assessments 3.56–3.60 steadily increase in difficulty while offering a repetitive experience that ensures you leave Chapter 3 with the ability to draw and interpret chair conformations.

3.56 (•) Using the numbers shown in the chair conformation on the left, label the carbons of the flipped chair on the right. [*Assume that the angle through which you view the chair conformation doesn't change.*]

3.57 (••) For each chair on the left, place the substituents on the flipped chair. [*Recall that the axial/equatorial designation changes from one chair to the next, but the carbon to which the substituent is attached does not.*]

(a)

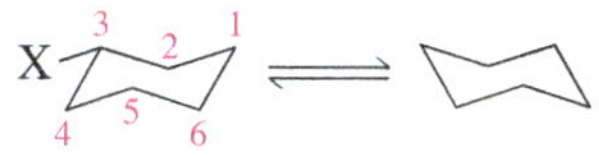

(b)

(c)

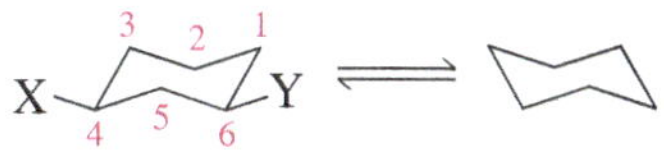

(d)

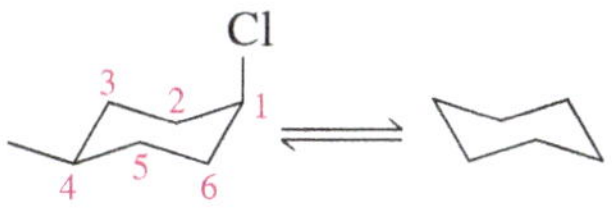

(e)

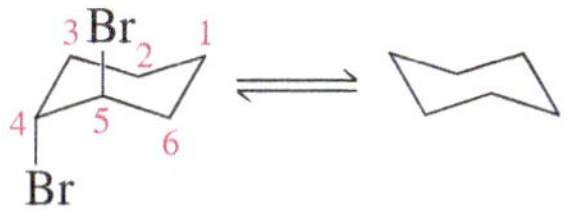

(f)

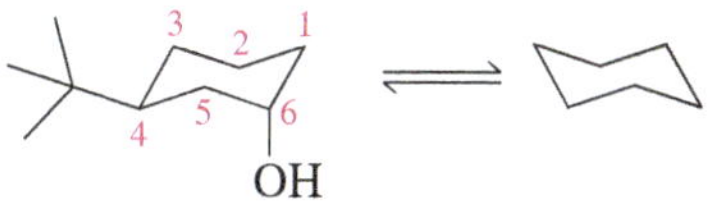

(g)

(h)

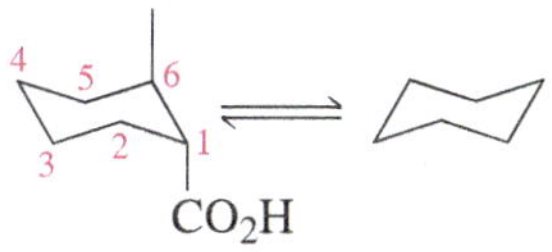

(i)

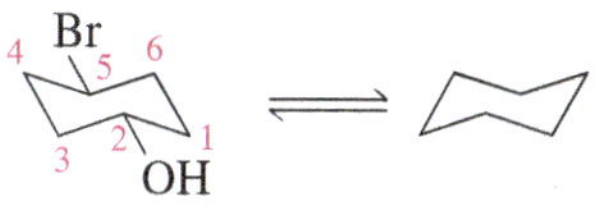

3.58 (••) What is the mistake that was made in drawing each of the flipped chairs on the right from the chair on the left? [*In these, assume that the angle through which you view the chair conformations doesn't change.*]

(a)

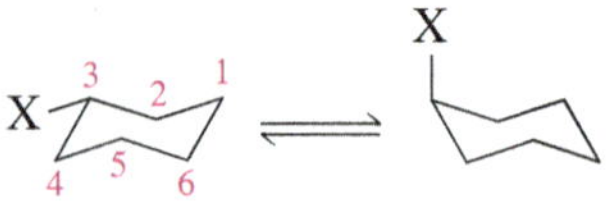

(b)

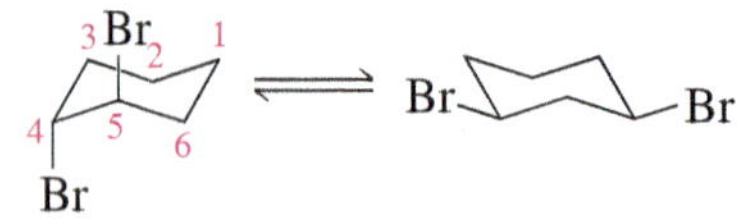

(c)

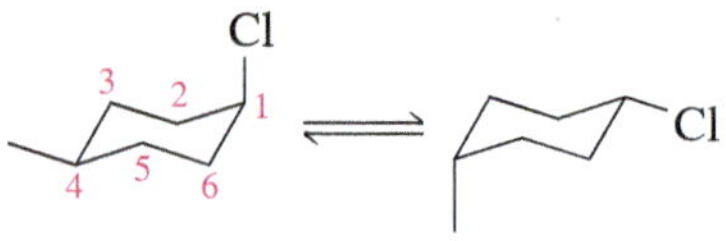

3.59 (••) For each pair of conformations shown, choose which is most stable. If both are equally stable, then write "no difference." [*If both conformations have the same number of axial substituents, choose the one with the smallest axial substituents.*]

(a)

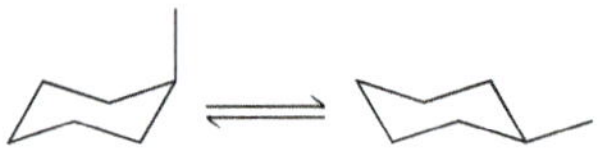

(b)

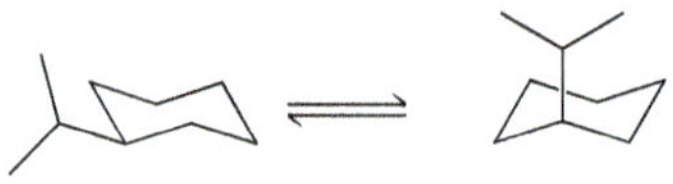

(c)

(d)

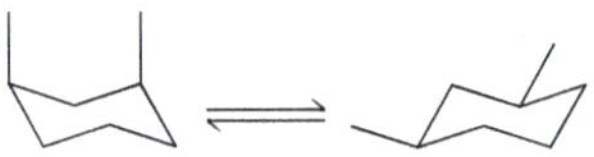

(e)

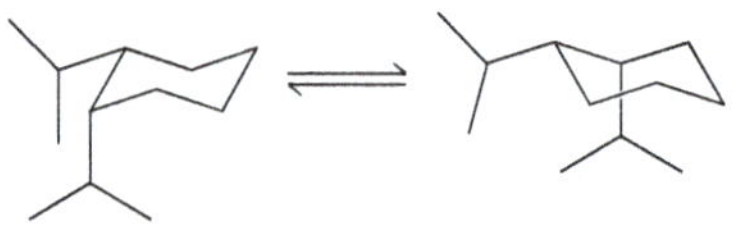

(f)

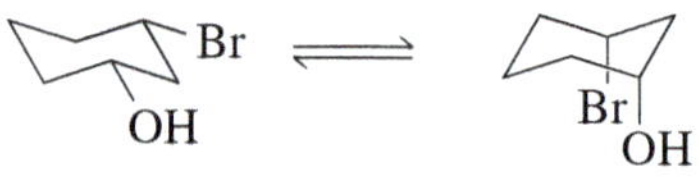

(g)

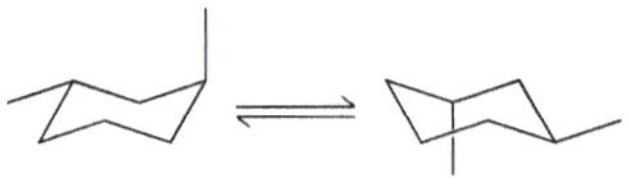

(h)

(i)

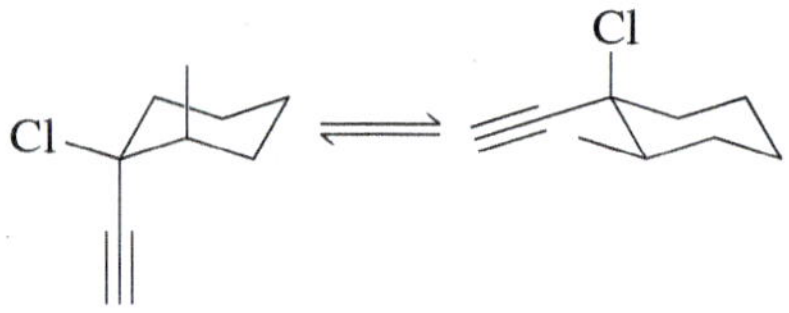

3.60 (•••) For each structure shown, draw the two chair conformations and choose which is most stable. Be sure that your second chair is the flipped version of the first. [*Make sure that wedged substituents are up in the chair, regardless of whether up is equatorial or axial.*]

(a)

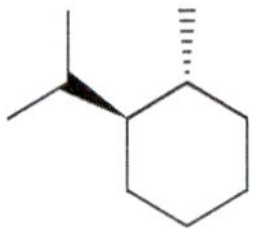

(b)

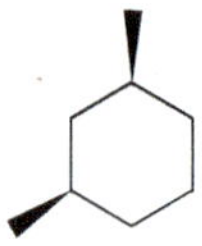

(c)

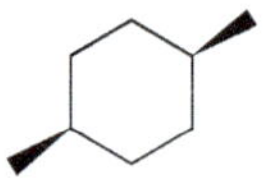

(d)

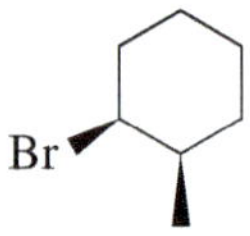

(e)

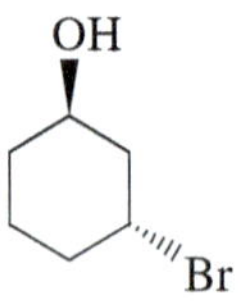

(f)

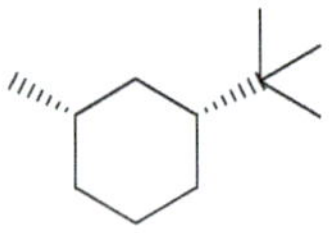

(g)

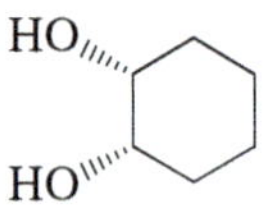

(h)

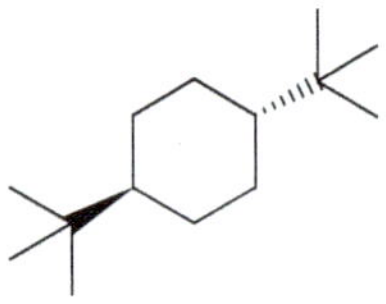

Application assessments

The remaining assessments in this chapter will require the application of the principles learned in this chapter.

3.61 (••••) Calculate the energy difference between each pair of conformations shown by drawing and comparing Newman projections down the indicated bonds in each.

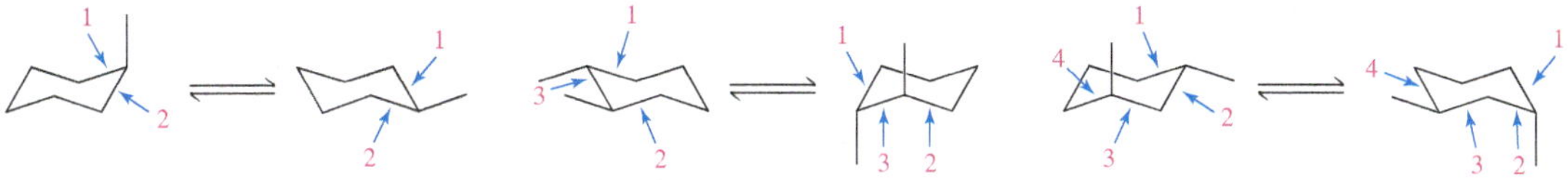

3.62 (•••) In Assessment 3.61, why were those particular bonds chosen and not others?

3.63 (•••) In contrast to ethane and other alkanes studied in this chapter, there is no free rotation around any C—C bonds in cyclopentane (shown below). Why?

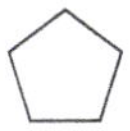

cyclopentane

3.64 (•••) LOOKING AHEAD In Chapter 5, we explain that the equilibrium constant (K_{eq}) for a reaction can be calculated based on the difference in energy between reactants and products, according to the following equation:

$$K_{eq} = e^{-\Delta E/RT}$$

Using this equation, calculate the equilibrium constant for the "reaction" shown. [*For the rest of the book, if not otherwise specified, assume room temperature (298K).*]

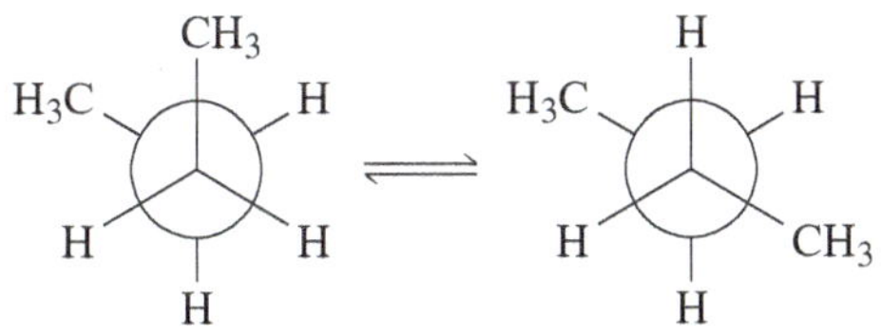

3.65 (••••) LOOKING AHEAD Using the equation in Assessment 3.64, calculate the equilibrium constant for the "reactions" shown in Assessment 3.61.

3.66 (••••) THINKING DEEPLY *Cis*-1,4-di-*tert*-butylcyclohexane exists in the twist-boat conformation. Explain this observation. [*Draw the alternate conformations and see what's wrong with each.*]

3.67 (••••) The normal C(sp^3)—C(sp^3) bond length is 1.54 Å. The normal bond angle for an sp^3-hybridized carbon is 109.5°. The following molecule experiences large deviations from these normal values. Explain these deviations. [*Molecular models would be helpful here.*]

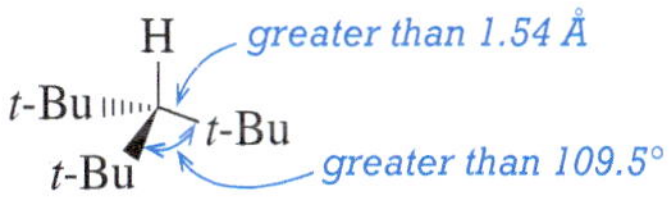

3.68 (•••) When trying to calculate the strain in a given molecule, a misguided chemist drew Newman projections around the indicated bond. Why is this a waste of time?

3.69 (••••) LOOKING AHEAD In Chapter 5, we introduce reaction coordinate diagrams as a plot of potential energy versus the progress of a reaction. Consider the reaction coordinate diagram drawn for the "reaction" of conformation A becoming conformation B. Which structure is present at the top of the hill?

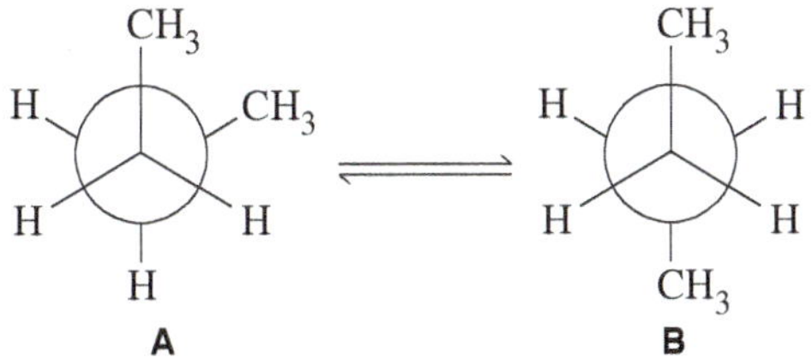

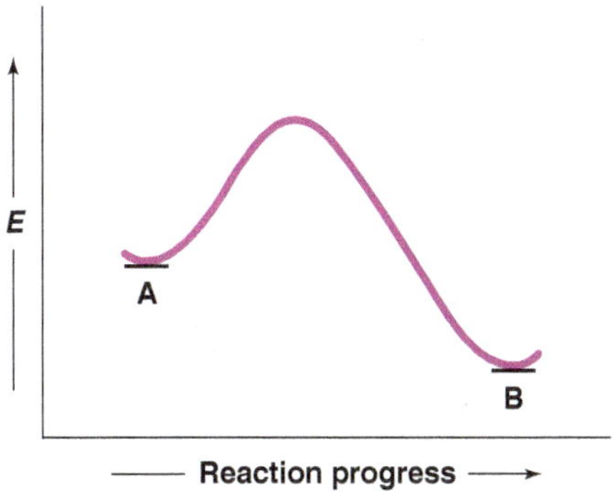

3.70 (••) **CONNECTING CONCEPTS:** In Chapter 2, we learned that the formation of a σ bond also results in the formation of σ*, an antibonding orbital that is located at a 180° angle to the σ bond. (a) On the following Newman projections, draw in σ* for the C—Cl σ bond. (b) In which conformation is σ* closer to the C—H bond? [*This becomes important when studying elimination reactions in Chapter 12.*]

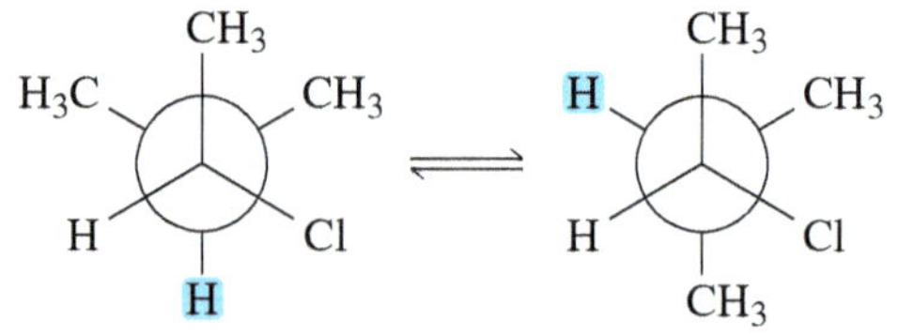

3.71 (•••) Correct the following incorrect names using standard IUPAC nomenclature. [*Draw a compound that corresponds to the incorrect name, and then rename it.*]

(a) 4-methylhexane
(b) 1,5-dimethylcyclohexane
(c) 6-ethyl-3-methyloctane
(d) 4-butyldecane
(e) 2,6-diethyl-1-methylcycloheptane

3.72 (•••) Identify the mistakes contained within the following structures.

(a)

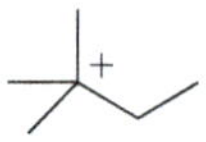

(b)

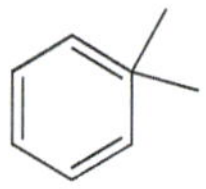

(c)

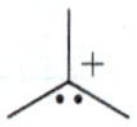

3.73 (•••) Despite methylcyclohexane (98.2 amu) having a higher molecular weight than toluene (92.1 amu), toluene melts at a higher temperature. Why? [*Think about how the molecules can interact with each other based on their shape.*]

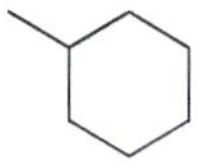

methylcyclohexane

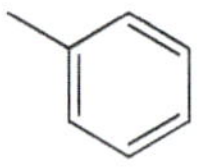

toluene

3.74 (•••) Ionic compounds like sodium acetate have a high melting point. Despite this, they are highly soluble in water. Why?

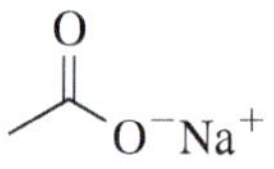

sodium acetate
mp = 324 °C

3.75 (•••) Identify the functional groups in each of the following molecules. [*The number of functional groups has been given to assist you.*]

(a) [*Three*]

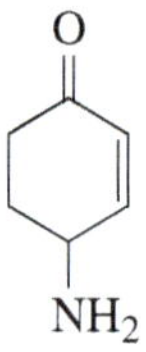

(b) [*Two*]

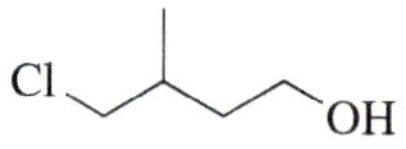

(c) [*Three*]

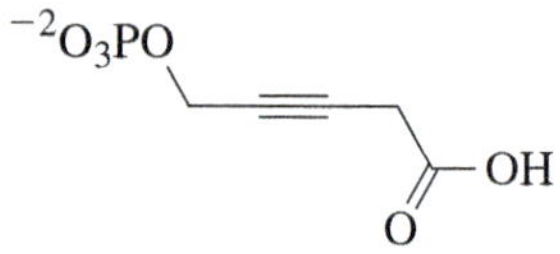

(d) [*Two*]

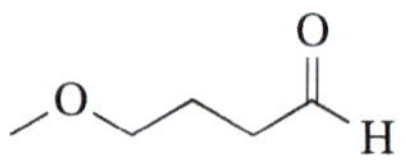

4

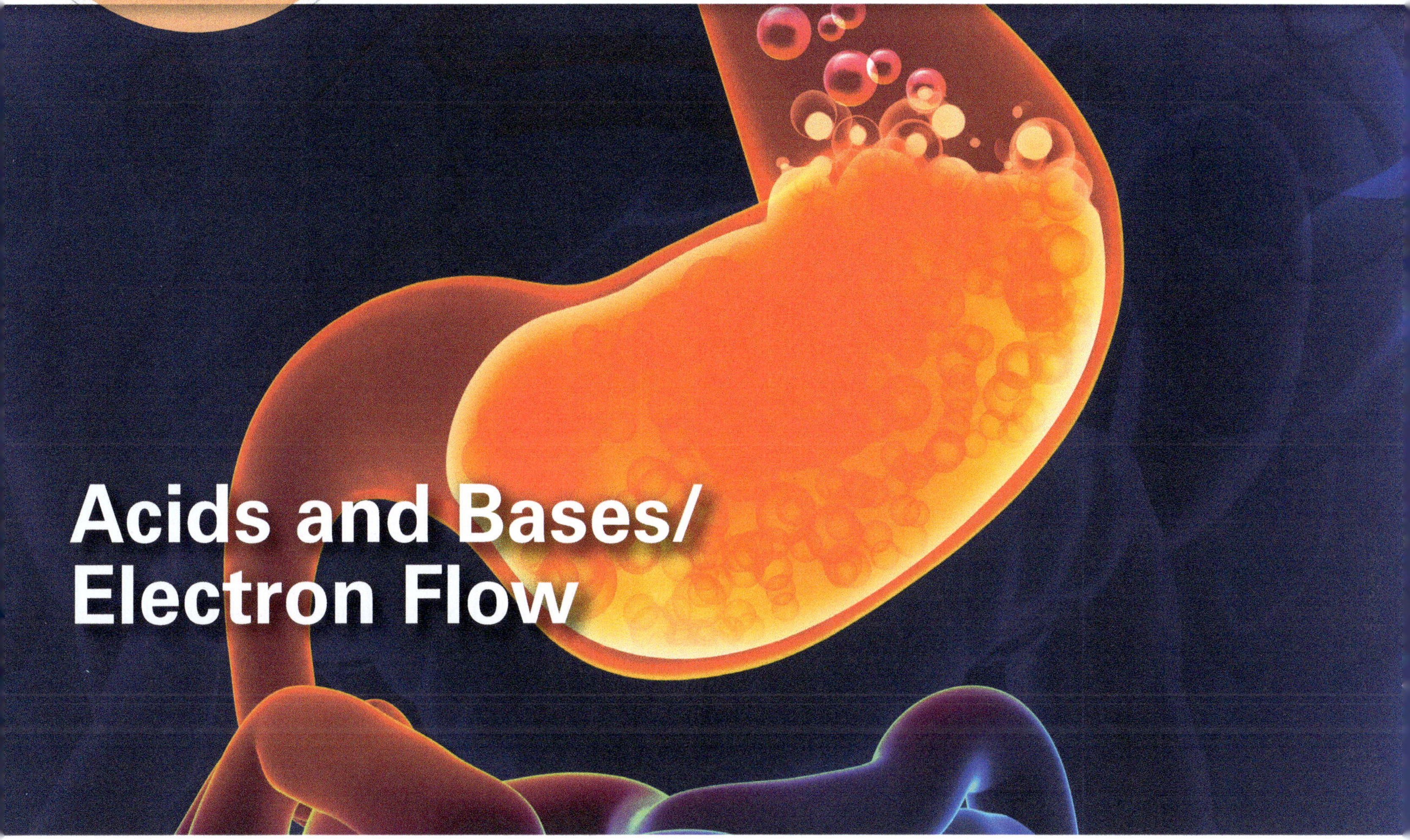

Acids and Bases/ Electron Flow

Introduction

Upon digestion, food passes from the mouth to the esophagus, through the stomach and into the intestines. The pH of the esophagus and the small intestine is neutral (pH = 7), but the stomach is quite acidic (pH $\approx$ 1.5–3.5). For comparison, stomach (or gastric) acid is slightly less acidic than battery acid but more acidic than vinegar.

Gastric acid (composed of 0.1 M HCl, plus smaller amounts of NaCl and KCl) plays several roles in digestion. It helps break down food clumps into smaller particles, facilitating absorption of nutrients. Under acidic conditions, the enzyme pepsin is activated to digest proteins. Acid that enters the small intestine initiates contractions of the intestines to crush and move food through the GI tract. Finally, gastric acid kills potentially harmful bacteria in our foods.

∘ ∘ ∘ ∘ ∘ ∘ ∘ ∘

So far, this text has focused mostly on relatively straightforward aspects of organic chemistry (drawing molecules, representations, etc.). Those concepts are important, but I'll be direct: Chapter 4 is the most important chapter in this entire book. [*The. Most. Important.*] The chemistry of acids and bases represents the foundation of organic chemistry and is incorporated in every remaining chapter.

Learning Objectives

LO 4.0 Distinguish between Arrhenius, Brønsted–Lowry, and Lewis acids and bases

LO 4.1 Evaluate Brønsted–Lowry acid–base reactions

LO 4.2 Draw a reaction coordinate diagram for an acid–base reaction for which K_{eq} is > 1, < 1, and = 1

LO 4.3 Predict the outcome of complex acid–base reactions

LO 4.4 Compare acidity and conjugate base stability in molecules based on structure

LO 4.5 Analyze Lewis acid–base reactions

"The willingness to share does not make one charitable; it makes one free."

—Robert Brault

Let's assess what you should already know in order to succeed in this chapter.

The good news is that acids and bases were an integral part of your introductory chemistry coursework and the material covered here makes sense in light of the two rules of organic chemistry [*namely, negatives attack positives and the ball always rolls downhill*]. Because of its importance to the remainder of organic chemistry [*the next 7 months of your life*], it is important to know this material deeply, to relate new concepts to earlier ones, and to anticipate future applications of this knowledge. [*I'll help you with all of these things!*]

How to do this? Work as many assessments as you can. [*This chapter has a higher number than usual.*] While working these, don't simply get an answer and move on. Instead, ask questions: Does the answer make sense? Can you relate it to material you have studied? How might it be applied going forward? In sum, reflect on how you got your answer and why it is right or wrong.

Test your understanding

Assessment 4.1 If a positively charged particle were placed in close proximity to a negatively charged particle, what would you expect to happen?

Assessment 4.2 Which is more stable, free hydrogen atoms or the diatomic H_2 molecule? Why do you know this to be true?

Assessment 4.3 Draw the Lewis structure of the methoxide ion (CH_3O^-). Draw the Lewis structure of a proton (H^+). To which atom of methoxide would you expect a proton to add?

Assessment 4.4 Draw all contributing resonance structures of the following molecules and ions.

(a)

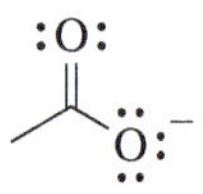

(b)

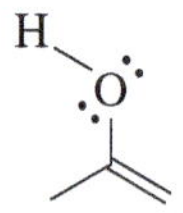

(c)

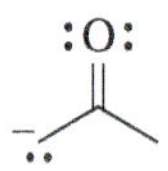

(d)

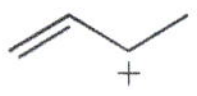

(e)

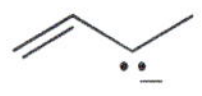

Assessment 4.5 Draw the molecular orbital picture of the following molecules and ions. In each, how many electrons are in the *p* orbital on the central atom?

(a) CH_3^+ (b) BH_3 (c) $AlCl_3$

Assessment 4.6 In each pair of atoms, which has the larger atomic radius? Which is more electronegative?

(a) F vs. Cl
(b) C vs. O
(c) O vs. S
(d) Si vs. P

4.1 Acids and bases: three definitions

There are three important definitions of acids and bases: **Arrhenius, Brønsted–Lowry**, and **Lewis**. The Brønsted–Lowry and Lewis definitions are most important, but it is important to understand all three in the context of each other. Let's consider each definition with respect to hydrochloric acid (Table 4.1) and sodium hydroxide (Table 4.2).

Table 4.1 Hydrochloric acid as an Arrhenius, Brønsted–Lowry, and Lewis acid.

Acid type	Definition	Example
Arrhenius acid	According to Arrhenius, an acid is a species that produces H^+ ions (**protons**) when dissolved in water. Because H^+ is unstable, it associates with H_2O to make H_3O^+ (**hydronium ion**). Thus, an Arrhenius acid is any species that increases the concentration of H_3O^+ when added to water.	$HCl + H_2O \longrightarrow H_3O^+ + Cl^-$ *HCl dissociates to increase the concentration of H_3O^+.*
Brønsted–Lowry acid	A Brønsted–Lowry acid is any molecule capable of donating a proton, without regard to the species that accepts it. HCl is a Brønsted–Lowry acid because it will donate a proton to any molecule able to accept it.	$HCl + H_2O \longrightarrow H_3O^+ + Cl^-$ *HCl donates H^+ to H_2O.* $HCl + CH_3OH \longrightarrow CH_3\overset{+}{O}H_2 + Cl^-$ *HCl donates H^+ to CH_3OH.* 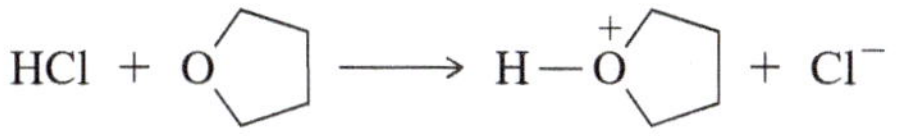 **THF** *HCl donates H^+ to THF.*
Lewis acid	A Lewis acid is an electron-pair acceptor. In order to form H_3O^+, HCl accepts an electron pair from the water molecule. This general acid–base definition, which focuses on the flow of electrons, will be most helpful when we study how organic reactions occur.	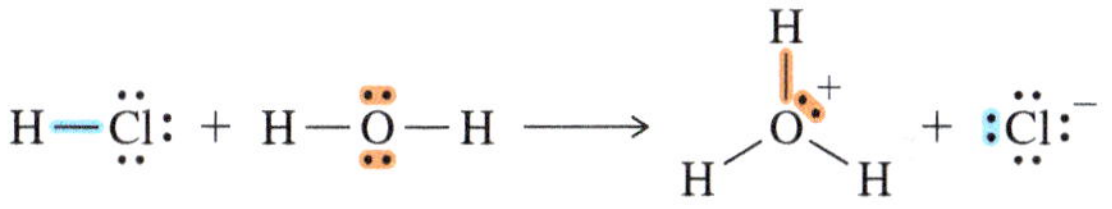 *The proton of HCl accepts the electron pair of H_2O.*

Table 4.2 Sodium hydroxide as an Arrhenius, Brønsted–Lowry, and Lewis base.

Base type	Definition	Example
Arrhenius base	When dissolved in water, an Arrhenius base increases the concentration of **hydroxide ion**. When sodium hydroxide dissolves in water, the concentration of hydroxide increases.	$NaOH + H_2O \longrightarrow Na^+ + {}^-OH + H_2O$ *NaOH dissociates to increase the concentration of ^-OH.*
Brønsted–Lowry base	Brønsted–Lowry bases are proton acceptors. Hydroxide removes (or accepts) a proton from water or any molecule willing to donate it.	$Na^+ + {}^-OH + H{-}OH \longrightarrow Na^+ + HOH + {}^-OH$ *Hydroxide accepts a proton from H_2O.* $Na^+ + {}^-OH + H{-}OCH_3 \longrightarrow Na^+ + HOH + {}^-OCH_3$ *Hydroxide accepts a proton from $HOCH_3$.* 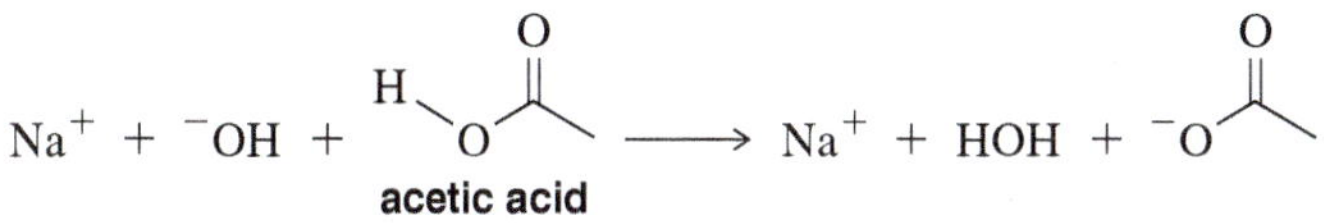 **acetic acid** *Hydroxide accepts a proton from acetic acid.*
Lewis base	Lewis bases are electron-pair donors. A hydroxide ion donates (or shares) a lone pair of electrons to (or with) a proton to make water.	$H\ddot{\underset{..}{O}}\!:^- + H^+ \longrightarrow H\ddot{\underset{..}{O}}{-}H$ *Hydroxide donates an electron pair to H^+.*

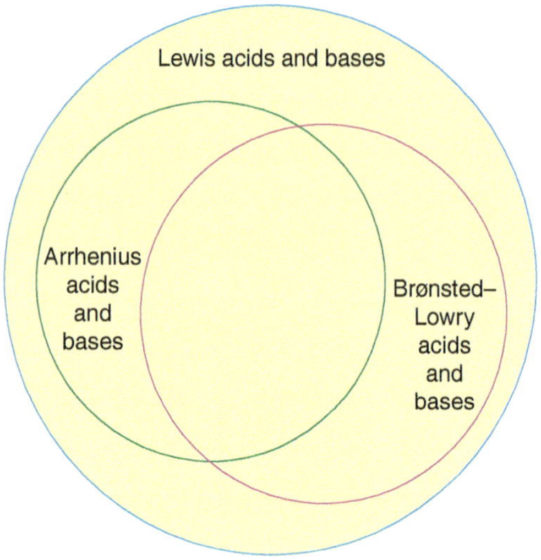

Figure 4.1 The relationship between the three kinds of acids and bases.

In each of the scenarios described in Table 4.1, HCl is an acid. As summarized in Figure 4.1, however, all Arrhenius and Brønsted–Lowry acids are Lewis acids, but only when Brønsted–Lowry acids are dissolved in water are they also Arrhenius acids. [*Think about a base like sodium hydroxide in these three contexts before reading further.*]

Assessment 4.7 Sodium amide ($NaNH_2$) dissociates to give a sodium cation (Na^+) and amide ion (NH_2^-), a very strong base. In the following three equations, identify which definition of base is being exemplified.

(a)

(b)

$$NaNH_2 + H_2\ddot{O} \longrightarrow H\ddot{O}:^- + Na^+ + :NH_3$$

(c)

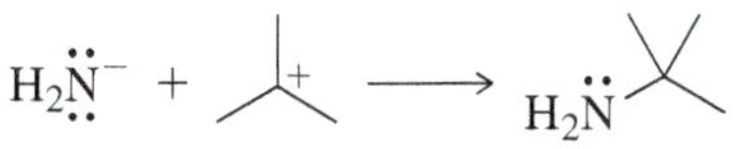

Assessment 4.8 Sulfuric acid is a very strong acid. Show how the following equation can be used to explain that sulfuric acid is at once an Arrhenius, Brønsted–Lowry, and Lewis acid.

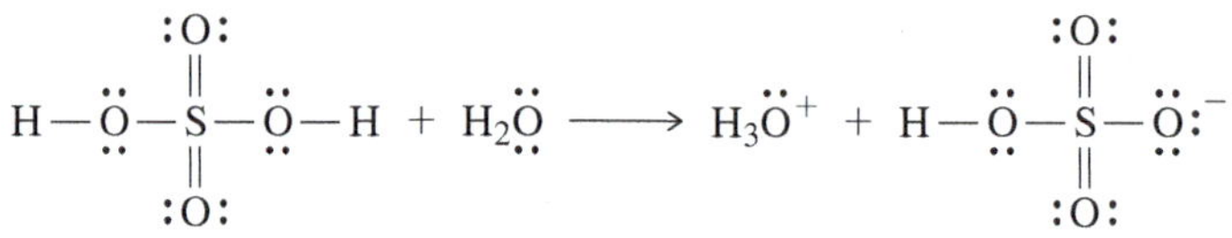

4-A CHEMISTRY IN CONTEXT: Medicine

Treating acid reflux disease

Acid reflux disease occurs when the stomach's lower esophageal sphincter does not close completely. Gastric acid then splashes up into the esophagus, causing a burning sensation in the chest (heartburn). The first line of treatment is to neutralize stomach acid at the onset of heartburn. The most common ingredients in over-the-counter antacids are inorganic bases such as aluminum hydroxide, magnesium hydroxide, and calcium carbonate. Antacids neutralize the acid, increase the pH of the stomach, and reduce the burn of acid leaking into the esophagus. Unfortunately, antacids do not address the cause of the disease.

$$2\,HCl\,(aq) + Mg(OH)_2\,(aq) \longrightarrow 2\,H_2O\,(l) + MgCl_2\,(aq)$$

Overuse of antacids can damage the intestines. Additionally, some patients don't respond well to their use. Alternative strategies that address the cause of acid reflux include histamine-2 (H2) blockers and proton pump inhibitors (PPIs) (Figure 4.2). H2 blockers (competitive antagonists) work by preventing the binding of histamine to parietal cells in the stomach, halting the release of acid. By a different mechanism, PPIs bind irreversibly with the enzyme-mediated proton pump in parietal cells, reducing stomach acidity by as much as 99%. More effective than H2 blockers, PPIs have become the largest selling class of drugs in the world.

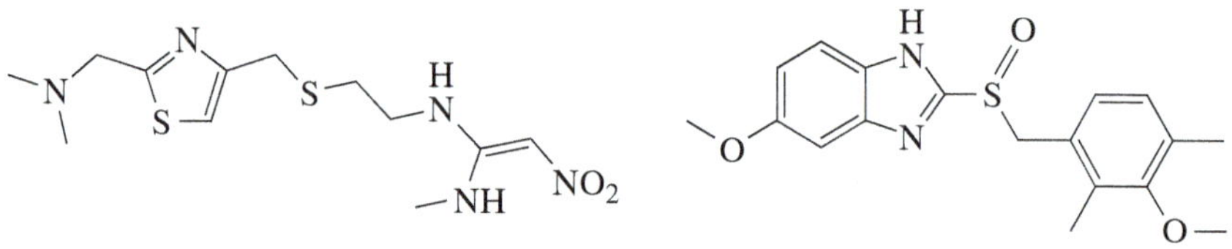

nizatidine
histamine-2 blocker

omeprazole
proton pump inhibitor

Figure 4.2 Example medications for acid reflux.

In Chapter 3, we learned that lone pairs don't need to be drawn for us to know they are there. In Chapter 4, they will be shown in places where it is really important that you don't miss them. In future chapters, you will only see them drawn when they are used in a mechanism. When in doubt, feel free to draw them in yourself. Works on tests, too!

4.2 Brønsted–Lowry acid–base reactions: a detailed analysis

"Acid–base reaction" most often refers to the Brønsted–Lowry definition, whereby a proton (H^+) is transferred from a proton donor (the acid) to a proton acceptor (the base). By definition, for every proton donor (the acid), there must be a proton acceptor (the base). In the reaction of HCl and water, either HCl (acid) donates a proton to water (base) or water (base) removes the proton from HCl (acid). Both statements are correct.

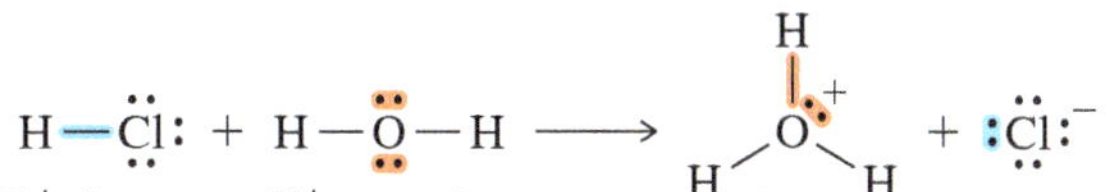

Though the Brønsted–Lowry definition is proton-centered, organic chemists speak of bases as "removing a proton" from an acid, as opposed to having a "proton donated to them." Consistent with the Brønsted–Lowry definition, this terminology will help you draw reaction mechanisms (Section 4.2.1). Going forward, we'll emphasize that Brønsted–Lowry bases "attack/remove/grab protons" from Brønsted–Lowry acids.

Following removal of the proton from HCl, the solution contains the **hydronium ion** and a chloride ion [**Figure 4.3(a)**]. The hydronium ion, containing the donated proton, is the **conjugate acid of the base** (or simply, the **conjugate acid**). The chloride ion, having lost the donated proton, is the **conjugate base of the acid** (or simply, the **conjugate base**.) This terminology is consistent with the fact that most acid–base reactions are reversible [**Figure 4.3(b)**]. If the reaction were reversed, the chloride ion (base) would remove the proton from the hydronium ion (acid), producing HCl and H_2O.

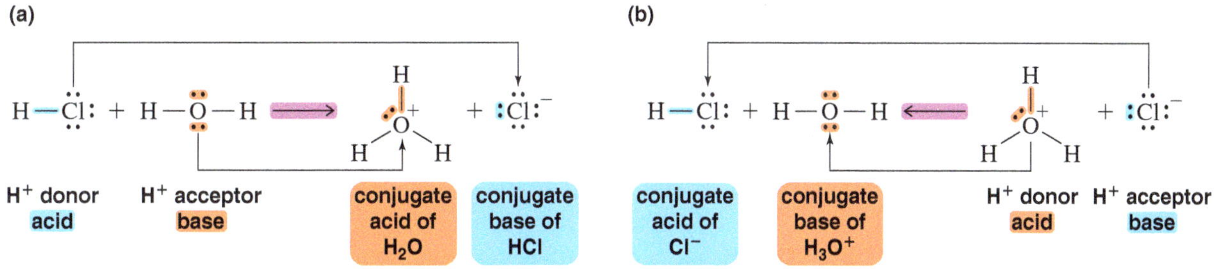

Figure 4.3 Describing an acid–base reaction. (a) The relationship between the acid, base, conjugate acid, and conjugate base. (b) Acid–base reactions are reversible.

Thus, every acid is coupled to its conjugate base (**Figure 4.4**). [*HCl (acid) loses its proton to become Cl^- (conjugate base).*] Every base is coupled to its conjugate acid. [*H_2O (base) gains a proton to become H_3O^+ (conjugate acid).*] More generally, HA represents an acid with conjugate base A^-, whereas B^- represents a base with conjugate acid BH.

$$\underset{\text{acid}}{H{-}A} \longrightarrow H^+ + \underset{\text{conjugate base}}{A^-} \qquad H^+ + \underset{\text{base}}{B^-} \longrightarrow \underset{\text{conjugate acid}}{B{-}H}$$

$$\underset{\text{acid}}{HCl} \longrightarrow H^+ + \underset{\text{conjugate base}}{Cl^-} \qquad H^+ + \underset{\text{base}}{H_2O} \longrightarrow \underset{\text{conjugate acid}}{H_3O^+}$$

Figure 4.4 Relation between conjugate pairs.

Because a proton (H^+) has a +1 charge, the charge of a conjugate base should be one less than that of its acid. If a neutral acid (H_2O) loses a proton, its conjugate base will be negative (HO^-). If a positively charged acid (H_3O^+) loses a proton, its conjugate base will be neutral (H_2O). [*You can always recalculate formal charge. Recognizing and balancing charges will minimize future mistakes.*] **Figure 4.5** shows some common acid–conjugate base pairs.

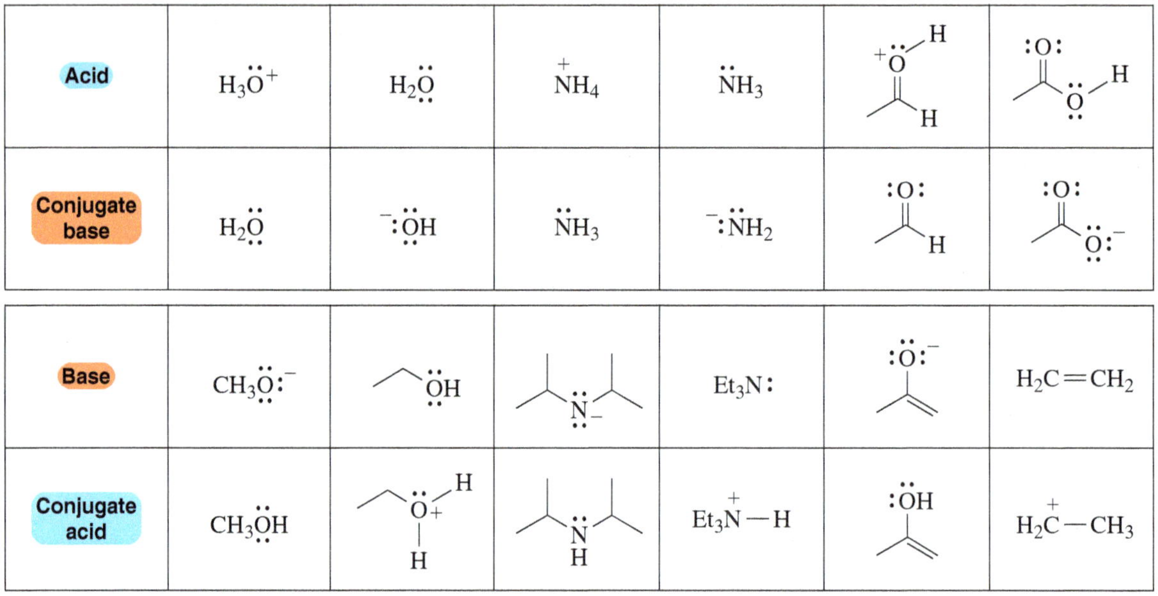

Figure 4.5 Representative acid–base conjugate pairs. [*The charge in each pair differs by 1.*]

Assessment 4.9 What is the conjugate base of each of the following acids? [*The most acidic proton is indicated.*]

(a) (b) (c) (d)

Assessment 4.10 What is the conjugate acid of each of the following bases? [*The most basic atom is indicated.*]

(a) (b) (c) (d)

Assessment 4.11 Identify the acid from which the following conjugate bases were formed. [*The most basic atom is indicated.*]

(a) (b) (c) (d)

Assessment 4.12 Identify the base from which the following conjugate acids were formed. [*The most acidic proton is indicated.*]

(a) (b) (c) (d)

Assessment 4.13 When an atom loses a proton in an acid–base reaction, its formal charge is decreased by one. Thinking about the equation for calculating formal charge, explain this observation in your own words.

To identify the species in an acid–base reaction, look for the molecule that loses a proton (the acid) and the one that gains it (the base). Once the acid and base are identified, the conjugate acid and base differ from them by the loss or gain of a proton. **Figure 4.6** shows a few examples of this.

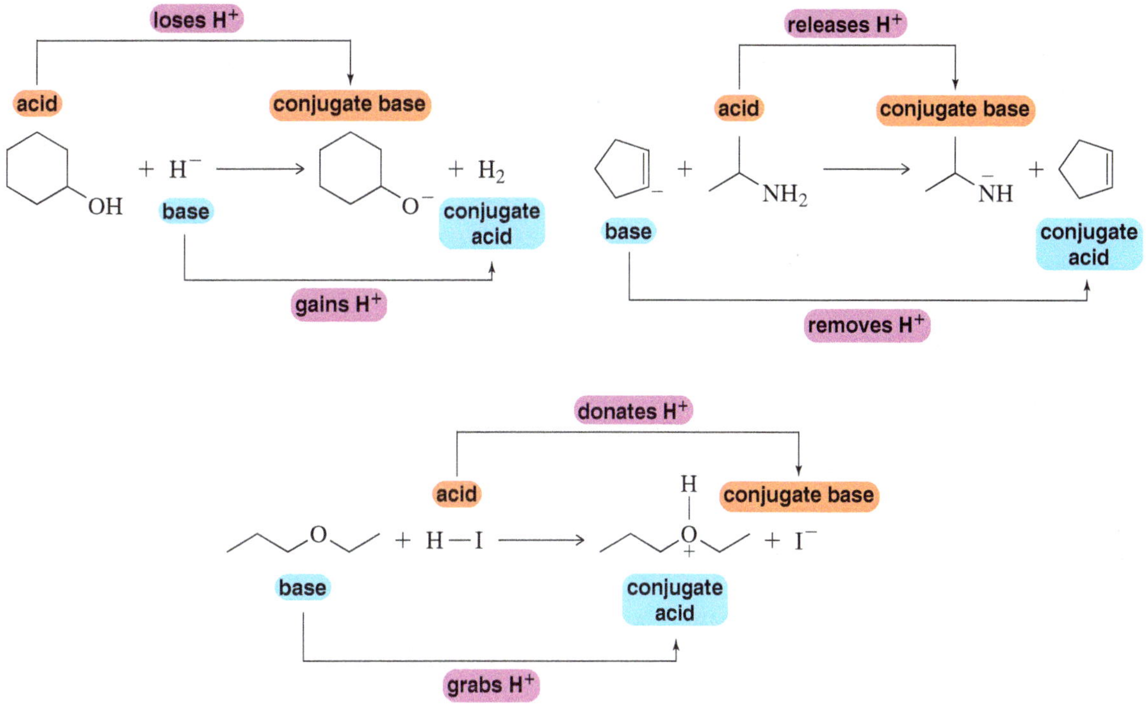

Figure 4.6 Identifying species: acid–base reactions.

The difference between an acid and a conjugate acid, or a base and a conjugate base, is not always obvious. [*Students sometimes struggle with this.*] It's just a matter of perspective. Reactions are typically discussed as they proceed from left to right. The acid and the base are on the left side of the equation, and conjugates are on the right. If a reaction is favored in the reverse direction, however, it would also be correct to have the acid and the base on the right side. The only rule is that the acid and base are on the same side.

Assessment 4.14 Using the convention that the acid and base are on the left side of the chemical equation, label the acid, base, conjugate acid, and conjugate base in the following reactions.

(a)

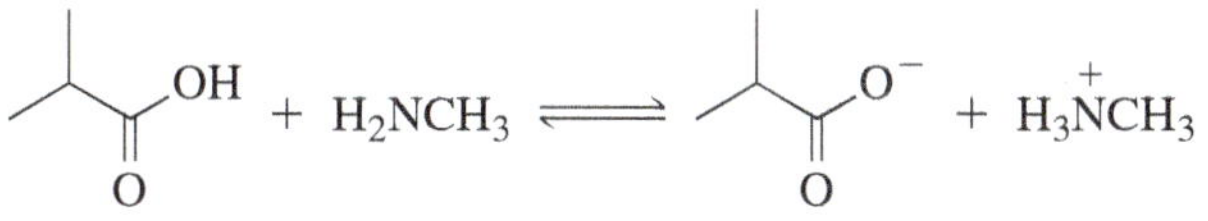

(b)

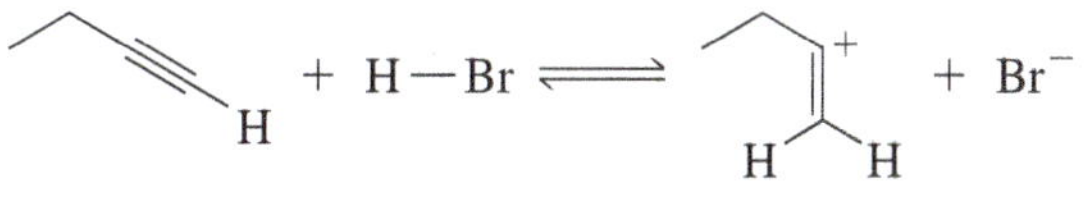

(c)

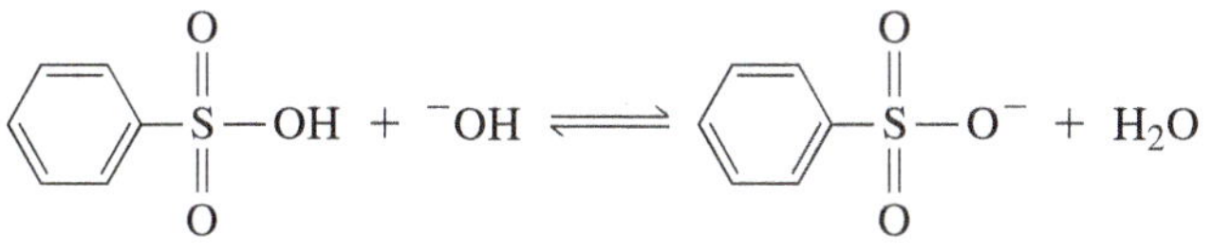

Assessment 4.15 Based on the one species that is identified for you, label the remaining molecules as acid, base, conjugate acid, or conjugate base.

(a)

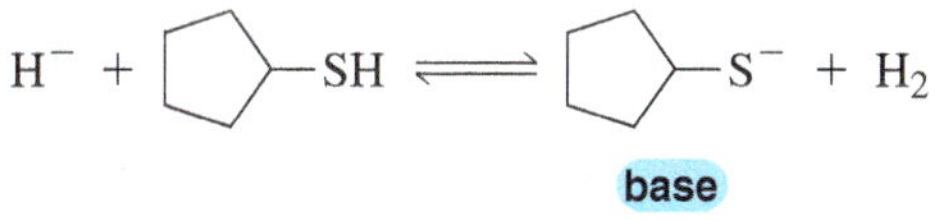

(b)

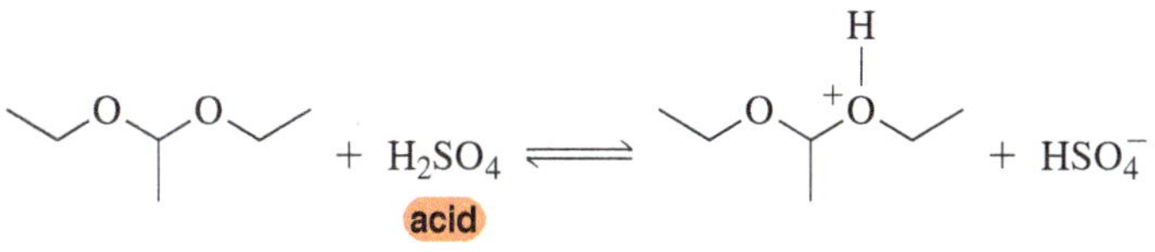

(c)

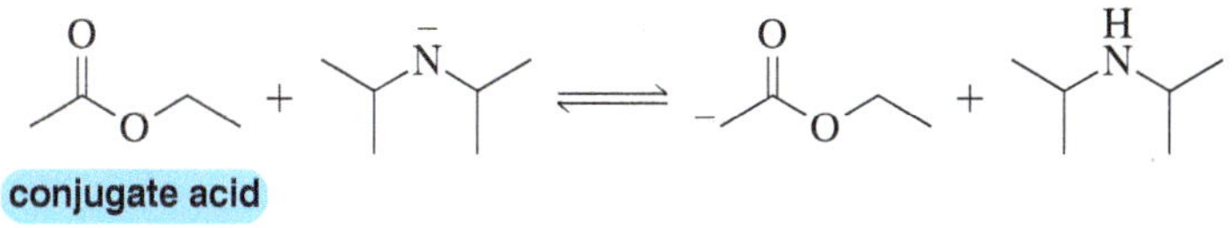

4.2.1 Arrow-pushing formalism in Brønsted–Lowry reactions

We can use the **arrow-pushing formalism** (also referred to as **electron pushing**) to illustrate how acid–base reactions work (the reaction **mechanism**).

Recall from the discussion of resonance structures (Section 2.8) that the movement of electrons can be illustrated by the arrow-pushing formalism. The arrows always start at an electron pair and point toward areas of electron deficiency (**Figure 4.7**). This formalism illustrates rule 2 of organic chemistry: Negatives attack positives. [*Other ways of saying the same thing include electron-rich attacks electron-poor, electron pairs attack empty orbitals, bases attack acids, and electron density attacks electron deficiency.*] The arrow you draw *always* starts at the electrons.

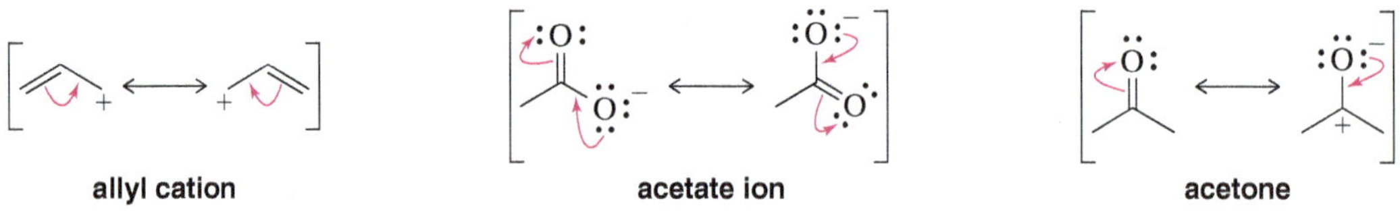

Figure 4.7 Arrow pushing: resonance.

The arrows used for electron pushing in acid–base reactions always start at an electron pair of the base and point toward the acidic proton. Recall that a base attacks/grabs/removes a proton from an acid. The way this process is illustrated informs this language. [*If molecules had a personality, those that are electron-rich are aggressors in chemical reactions, at least by our convention.*]

To simplify, consider the arrow-pushing mechanisms for each half of an acid–base reaction separately. [*This is merely hypothetical; you should never show a bond breaking to form a single H^+ ion.*] Simple electron counting shows an $H{-}Cl$ σ bond and three lone pairs on the Cl (**Figure 4.8**). After losing a proton, H^+ has zero lone pairs, Cl has four, and the σ bond is broken. [*We are like accountants.*] Thus, the electronegative chlorine has taken the two electrons of the σ bond. This is illustrated by drawing an arrow starting at the σ-bonded electrons pointing toward the Cl. For the reaction of water (base), H_2O consists of two $O{-}H$ σ bonds with two lone pairs on the oxygen. After reacting with H^+ (no lone pairs, no bonds), H_3O^+ has three $O{-}H$ σ bonds and only one lone pair. A lone pair on oxygen attacks the electron-deficient H^+ and becomes the new $O{-}H$ σ bond. This is shown by drawing an arrow starting at the lone pair and finishing at H^+.

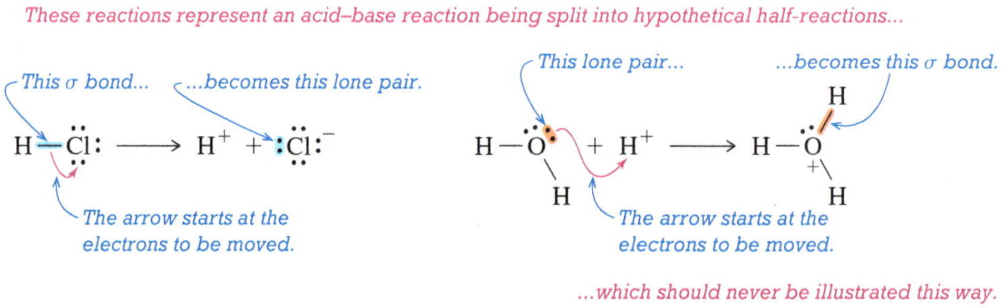

Figure 4.8 Arrow pushing: acid–base reactions.

Assessment 4.16 Provide an arrow-pushing mechanism for the following hypothetical acid half-reactions. [*These are only intended to help you learn about arrow pushing in acid–base reactions.*]

(a)

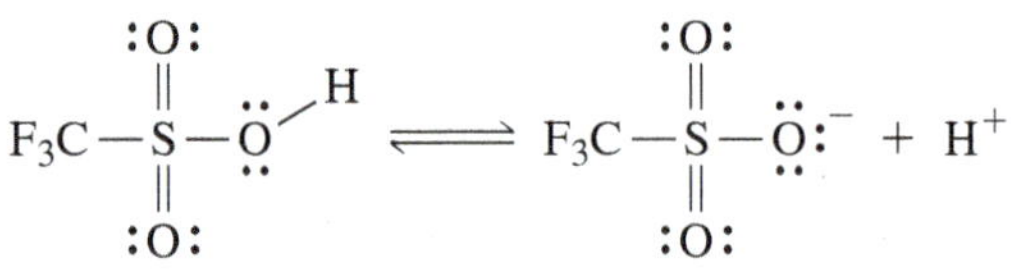

(b)

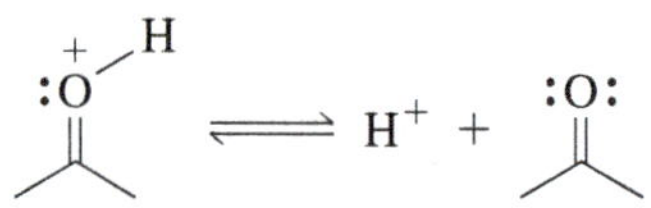

(c)

Assessment 4.17 Provide an arrow-pushing mechanism for the following hypothetical base half-reactions. [*These are only intended to help you learn about arrow pushing in acid–base reactions.*]

(a)

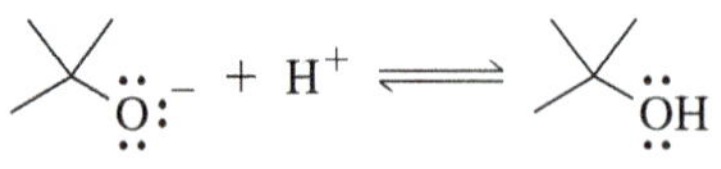

(b)

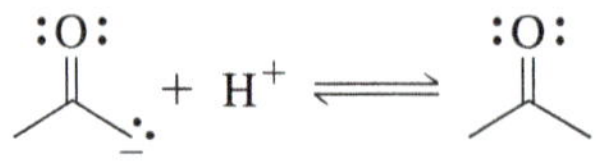

(c)

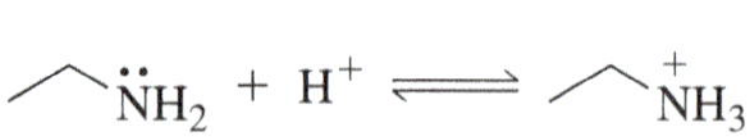

And so, to illustrate the mechanism of an acid–base reaction, identify the proton transferred and how the electron distribution changes. [*The electron pushing for the two half-reactions is combined into a single step.*] Let's consider the reaction of the base ammonia (NH_3) with HBr, a strong acid.

The proton is transferred from HBr to NH_3, producing bromide (Br^-) and ammonium (NH_4^+) ions [**Figure 4.9(a)**]. In HBr, bromine has three lone pairs and a σ bond; in Br^-, it has four lone pairs. NH_3 starts with one lone pair (and three σ bonds), but NH_4^+ has zero (and four σ bonds). Two electrons originally in the H—Br σ bond are now a lone pair on Br^-. The original lone pair on NH_3 is now the new N—H σ bond of NH_4^+. This analysis allows the arrow-pushing mechanism to be drawn as in **Figure 4.9(b)**. The arrow starts at the nitrogen lone pair and points toward the hydrogen. Because the electrons in the H—Br σ bond move onto the bromine, an arrow originates at the σ bond and points toward Br. **Figure 4.10** shows more examples of this process.

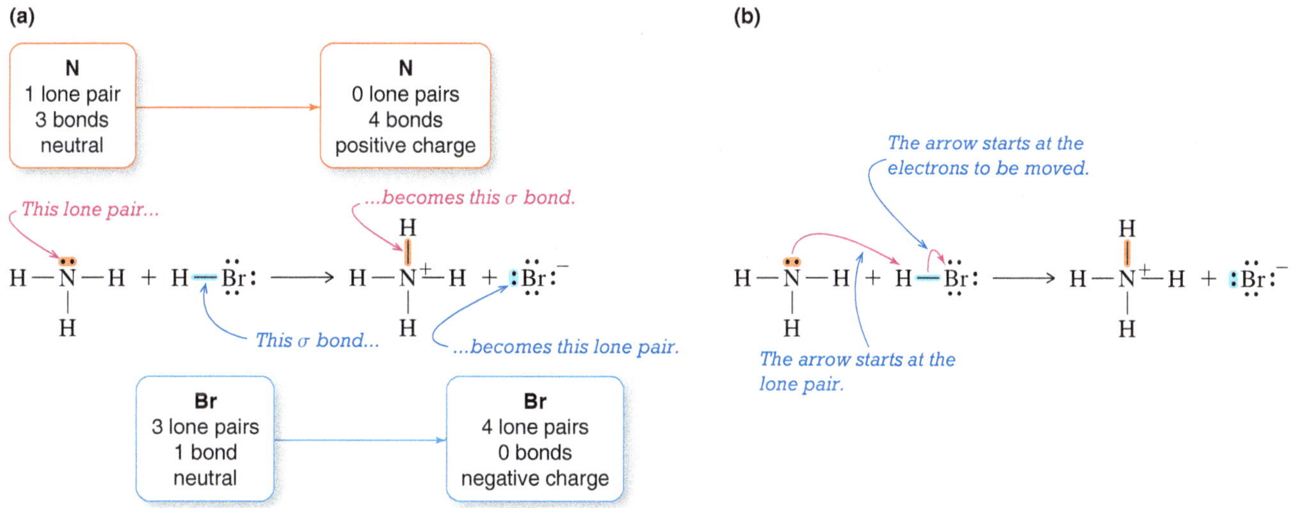

Figure 4.9 The mechanism of an acid–base reaction. (a) Electron accounting tells you where the electrons are. (b) The arrow-pushing mechanism tells you how they got there.

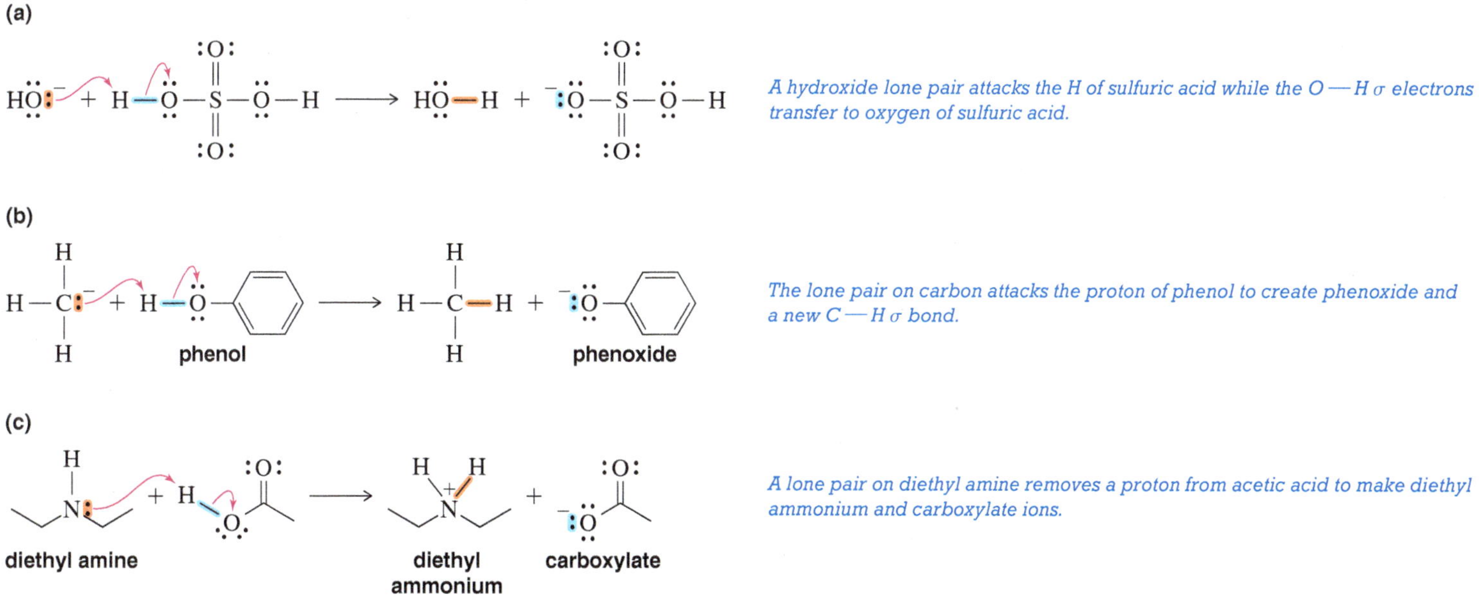

Figure 4.10 More arrow pushing: acid–base reactions. In all cases, the arrow starts at a lone pair of the base and attacks the proton of the acid.

Assessment 4.18 Provide an arrow-pushing mechanism for the following acid–base reactions.

(a)

$$H_3N: + H-\ddot{\underset{..}{Cl}}: \rightleftharpoons H_4N^+ + :\ddot{\underset{..}{Cl}}:^-$$

(b)

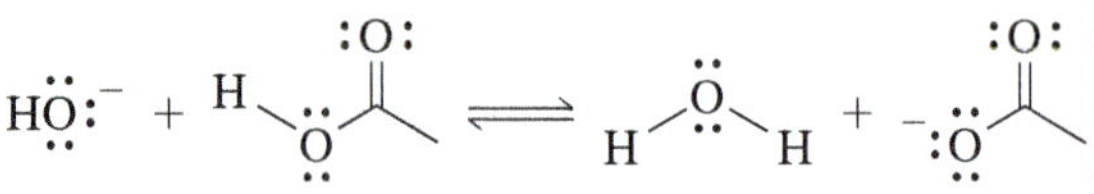

(c)

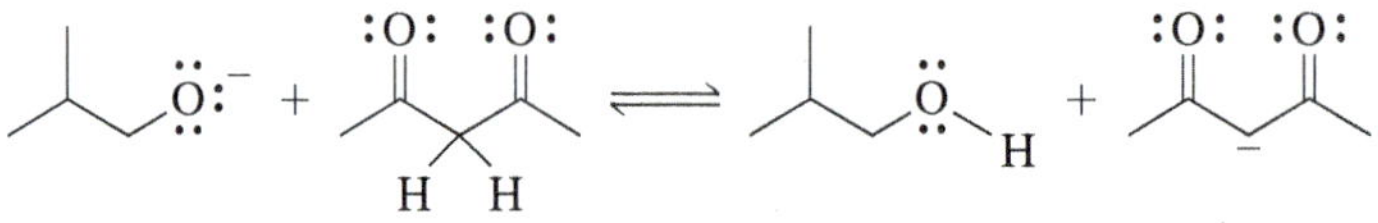

(d)

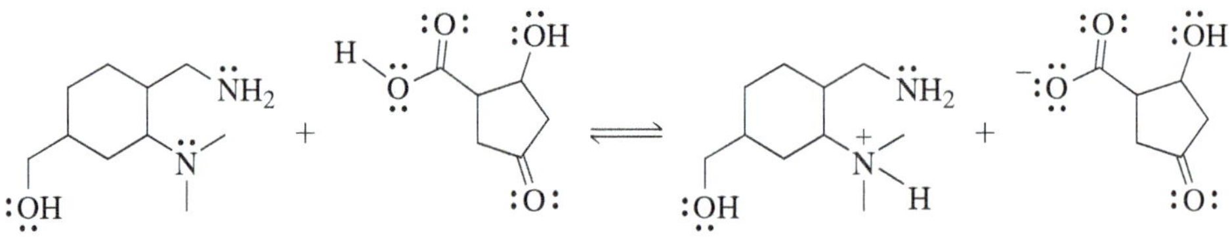

Assessment 4.19 Acid–base reactions are reversible. Show a mechanism for the reverse of the reactions in Assessment 4.18.

4.2.2 Predicting the outcome of Brønsted–Lowry acid–base reactions

To predict the outcome of acid–base reactions, the most basic atom of the base and the most acidic proton of the acid must be identified. Consider the reaction of diisopropyl amide (a strong base) and hydrofluoric acid [Figure 4.11(a)]. The nitrogen, with two lone pairs and a negative charge, is the most electron-rich atom. HF has one hydrogen, so that must be the proton that will be transferred. Mechanistically, a lone pair on nitrogen attacks the proton of H—F while the electrons in the H—F σ bond move onto fluorine. Knowing the flow of electrons leads directly to the products [Figure 4.11(b)]. After reacting, diisopropylamine has a neutral nitrogen, two alkyl groups, a hydrogen, and one lone pair, while fluoride has four lone pairs.

(a) (b)

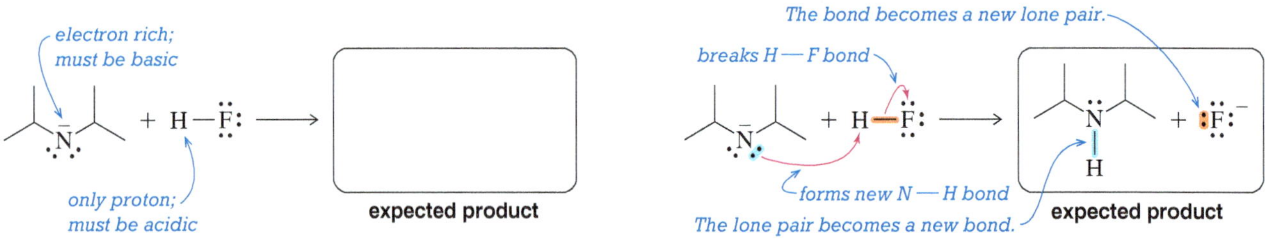

Figure 4.11 Predicting the outcome of an acid–base reaction. (a) Identify the acid and the base. (b) Use the electron flow to predict the products.

Assessment 4.20 In the following reactions, (i) identify the acid and base, (ii) identify the most electron-rich atom in the base, (iii) identify the most acidic hydrogen in the acid, (iv) provide an arrow-pushing mechanism of the proton transfer that will occur, and (v) predict the product of the reactions. [*You'll need to provide the lone pairs here.*]

(a) $H-C\equiv C^- + HBr \rightleftharpoons$

(b)

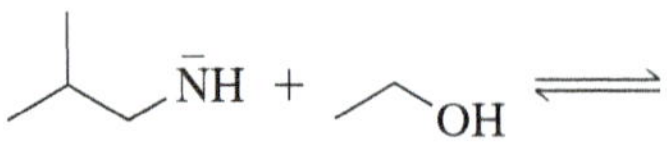

(c)

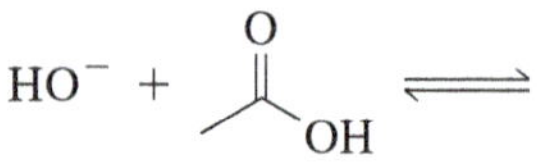

(d) $H_2O + HCl \rightleftharpoons$

4.2.3 Equilibrium constant (K_{eq})

Acid–base reactions are reversible, and arrow-pushing mechanisms can be drawn in both directions (Assessments 4.18 and 4.19). How do we know which direction is favored? Experimentally, the **equilibrium constant (K_{eq})** can be measured in the laboratory. The value of K_{eq} (the point at which the rate of the forward and reverse reactions are equal) tells us which side of a reversible reaction is favored.

All equilibrium expressions for a reaction, including K_{eq}, are the product of the concentrations of the products divided by the product of the concentrations of the reactants, with each concentration raised to a power corresponding to the stoichiometric coefficient for that substance in the balanced chemical equation for the reaction. As a ratio of positive quantities, all equilibrium constants are greater than 0. The inflection point is at 1, which means that products are favored when $K_{eq} > 1$, reactants are favored when $K_{eq} < 1$, and neither is favored when $K_{eq} = 1$. Reactions proceed from least stable to more stable (i.e., the ball always rolls downhill), so K_{eq} also tells us which side of the reaction is more stable (Figure 4.12).

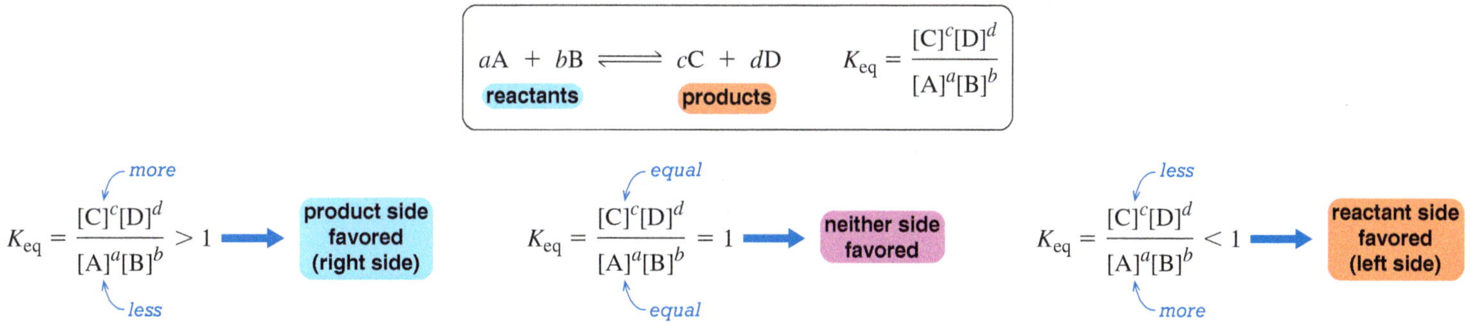

Figure 4.12 The favored side of a reaction determines K_{eq}.

For acid–base reactions, the favored side of the equilibrium is the one containing the least reactive (more stable) acid and base. The disfavored side of the equilibrium is the one containing the more reactive (less stable) acid and base. Imagine a reversible acid–base reaction as a competition between two bases for a proton: The stronger (more reactive) base wins the proton. The weaker (less reactive) base loses the proton [Figure 4.13(a)]. Alternatively, acid–base reactions can be described as two acids fighting to *not* have the proton. The stronger (more reactive) acid will be the one that forces the other (the weaker, less reactive) acid to keep the proton [Figure 4.13(b)].

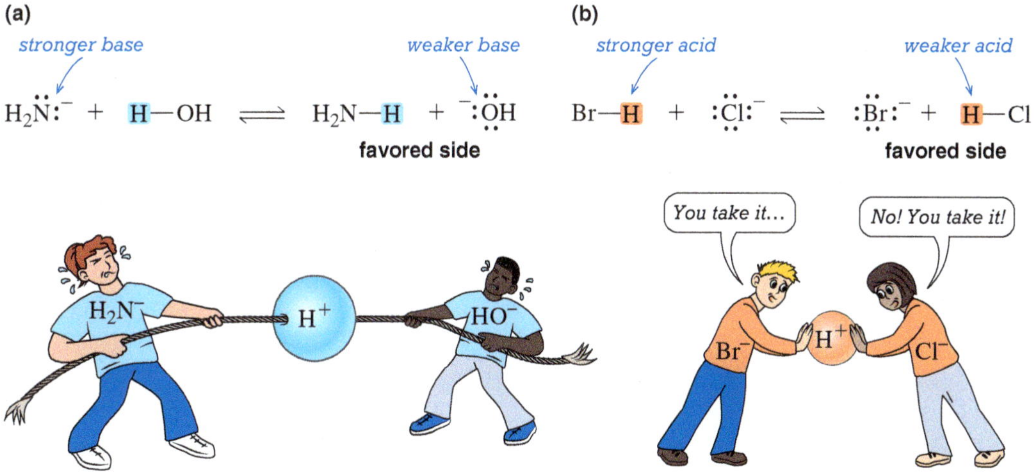

Figure 4.13 (a) The stronger base wins the battle for the proton. (b) The stronger acid gives the proton away.

Thinking of reversible reactions in multiple ways shows the importance of recognizing acid–conjugate base and base–conjugate acid pairs. A strong acid readily loses a proton to form a stable conjugate base, whereas a weak acid reluctantly loses a proton to form an unstable conjugate base (Table 4.3). A strong base readily grabs a proton to form a stable conjugate acid, whereas a weak base reluctantly grabs a proton to form an unstable acid (Table 4.4). [*You may need to read this a time or two to understand it fully.*]

Table 4.3 Relationship between acids and their conjugate bases.

Acid	...dissociates to form...		Conjugate base
Strong	HCl	Cl^-	Stable
Weak	$\diagup\diagdown$OH	$\diagup\diagdown O^-$	Unstable

Table 4.4 Relationship between bases and their conjugate acids.

Base	...protonates to form...		Conjugate acid
Strong	H_2N^-	H_3N	Stable
Weak	H_2O	H_3O^+	Unstable

From this discussion, we see that several terms can be correlated. ***Strong*** bases are ***unstable*** and highly ***reactive***, favorably grabbing a proton. ***Weak*** bases are ***stable*** and ***unreactive***. ***Strong*** acids are ***unstable*** and highly ***reactive***, favorably donating a proton. ***Weak*** acids are ***stable*** and ***unreactive***. [*This is very important to what is coming, so you may want to reread this section.*]

In summary, reversible reactions move away from ***strong***, ***reactive***, and ***unstable*** species toward ***weak***, ***unreactive***, and ***stable*** conjugate species. The equilibrium constant tells us, then, about the relative strengths of the species involved. For the reaction of H_2O and HCl [Figure 4.14(a)], a $K_{eq} > 1$ affirms that HCl is a stronger, more reactive, and less stable acid than H_3O^+ and that Cl^- is a weaker, less reactive, and more stable base than H_2O. For the reaction of water and acetic acid [Figure 4.14(b)], $K_{eq} < 1$, so the reactant side is favored. This implies that H_2O is a weaker base than the acetate ion and that H_3O^+ is a stronger acid than acetic acid.

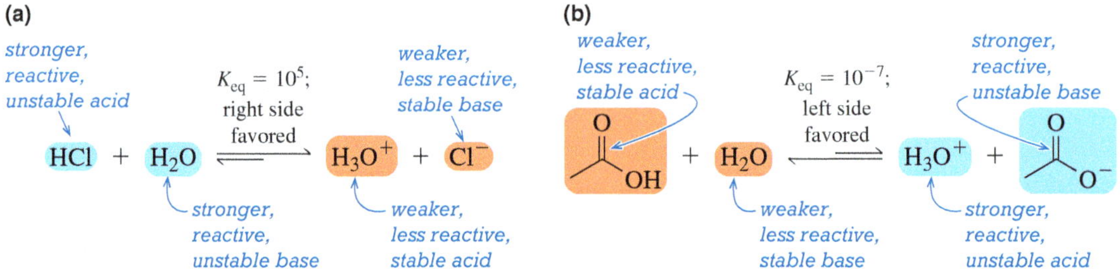

Figure 4.14 Direction of equilibrium. (a) The reaction moves away from the less stable (more reactive) species. (b) The reaction moves toward the more stable (less reactive) species.

4.2.3 SUMMARY

- The terms **strong**, **reactive**, and **unstable** are different ways of saying the same thing. The terms **weak**, **unreactive**, and **stable** are similarly correlated.

Assessment 4.21 Write the K_{eq} expression for the following acid–base reactions. [*You don't need to calculate K_{eq} here.*]

(a)

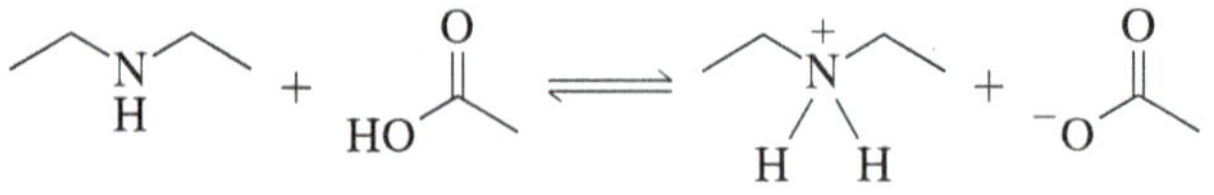

(b)

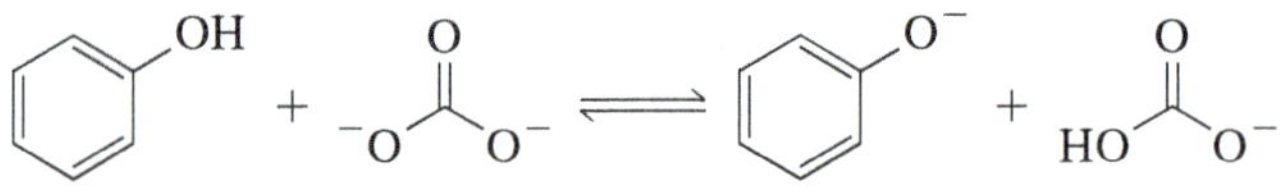

(c)

Assessment 4.22 Given the value of K_{eq} for the following acid–base reactions, identify the weakest acid and the weakest base.

(a)

$$H_3N + H{-}C{\equiv}C^- \underset{K_{eq} = 10^{-13}}{\rightleftharpoons} H_2N^- + H{-}C{\equiv}C{-}H$$

(b)

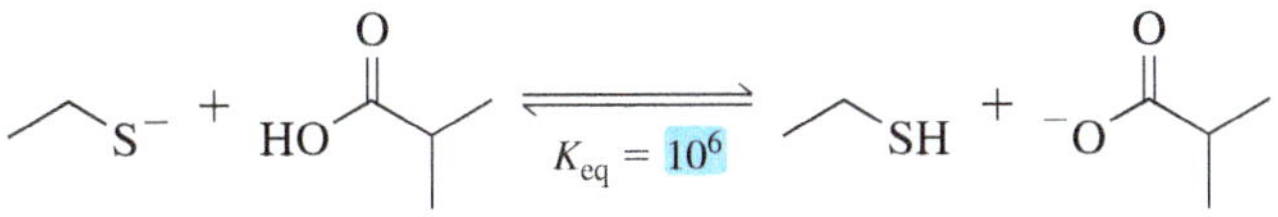

(c)

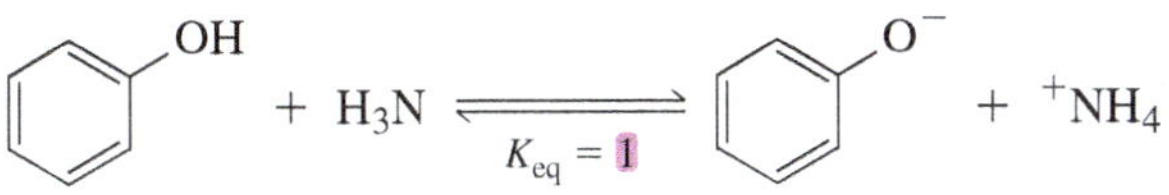

Assessment 4.23 Given the K_{eq} values for the following acid–base reactions, identify the strongest acid and the strongest base.

(a)

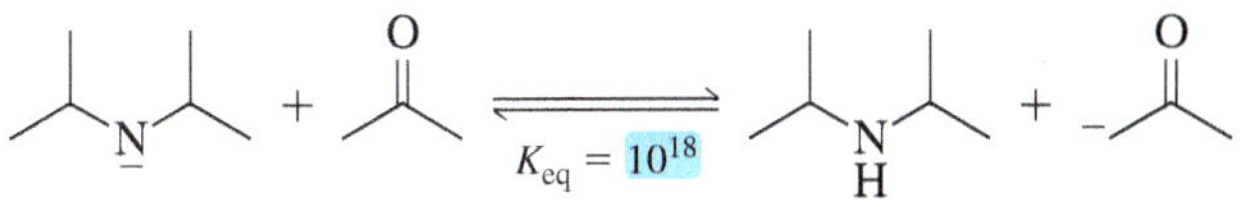

(b)

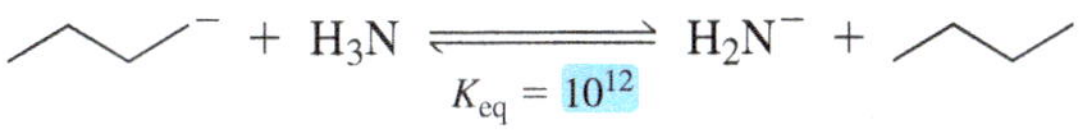

(c)

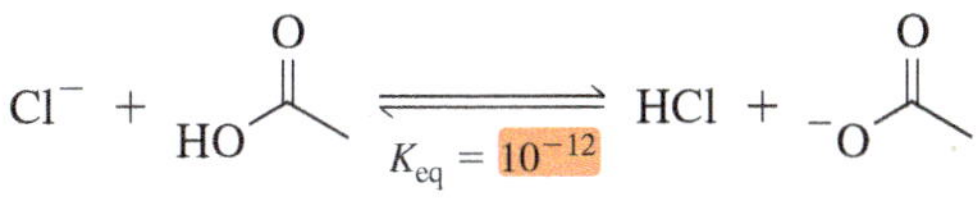

Assessment 4.24 An unknown base (B^-) has been identified as very weak. What does this tell you about the strength of its conjugate acid, **HB**? Is it stable or unstable? Is it reactive or unreactive?

Assessment 4.25 An unknown acid (**HA**) has been identified as very strong. What does this tell you about the stability of the conjugate base, A^-? Is it strong or weak? Is it reactive or unreactive?

4-B CHEMISTRY IN CONTEXT: The Laboratory

Separating acidic and neutral compounds

Nonpolar and high-molecular-weight polar compounds are typically insoluble in water. High-molecular-weight carboxylic acids are examples of insoluble polar compounds. However, after deprotonation with NaOH, a carboxylic acid becomes a sodium carboxylate ion. Now ionic, the carboxylate is very soluble in water. The common laboratory scenario in **Figure 4.15** uses this altered water solubility to separate neutral and acidic compounds.

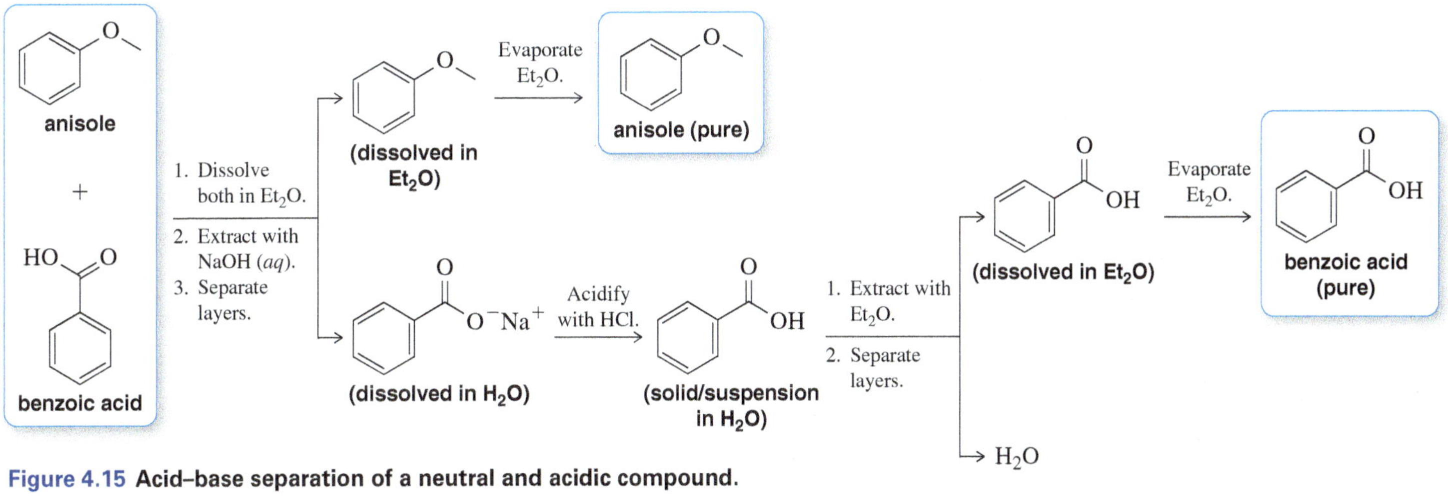

Figure 4.15 **Acid–base separation of a neutral and acidic compound.**

Assessment 4.26 Design an acid–base extraction scheme to separate a mixture of the basic amine *N,N*-dimethylaniline and naphthalene.

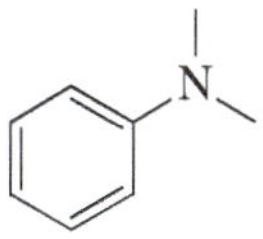

***N,N*-dimethylaniline**

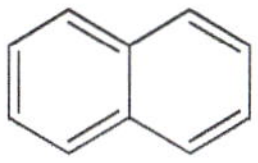

naphthalene

4.2.4 Equilibrium and reaction coordinate diagrams

The rule that the ball always rolls downhill is illustrated by a reversible acid–base reaction because the equilibrium favors the side that is most stable (least reactive). The stability of molecules can be discussed in the context of energy. All molecules have potential energy that is released when they participate in chemical reactions. More reactive (less stable) molecules have more potential energy. Thus, reactive molecules are often referred to as "high [*potential*] energy" species.

An early example dealt with why hydrogen atoms combine to form hydrogen gas (H_2). [*See Section 2.3 for a complete explanation.*] Hydrogen atoms are very unstable (high potential energy) and combine to get to a lower energy (more stable) state. Like a ball at the top of a hill with high potential energy, a reactive species (in this case, atoms) sits at the top of a hill (Figure 4.16). When released, the ball rolls down the hill, its potential energy converted to kinetic energy, until it reaches the bottom. A high-energy, unstable species reacts to achieve a lower energy state (more stable molecule), with its potential energy converted to heat. For hydrogen, this is the heat released upon bond formation and is the reason bonds form between any two atoms.

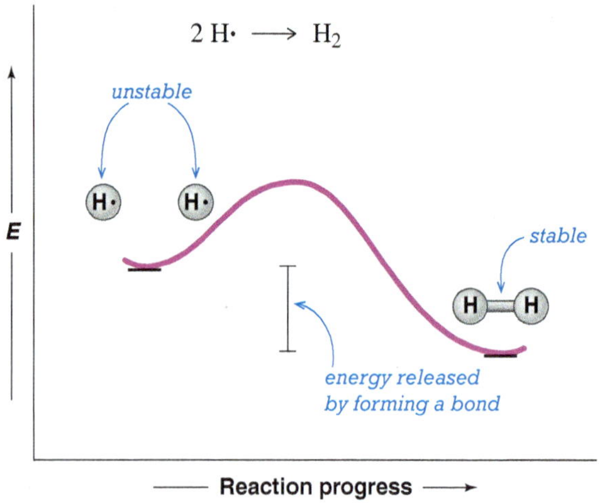

Figure 4.16 Reactions move from unstable to stable.

For an acid–base reaction, each side of the equilibrium is in an energy "valley" on either side of a hill. The lower valley is the favored side—the side that contains the **weakest, most stable, lowest energy,** or **least reactive** acid and base. The higher valley is the side containing the **strongest, least stable, highest energy,** or **most reactive** acid and base. As a graph, with energy on the y axis and reaction progress (reactant → product) on the x axis, this representation is referred to as a **reaction coordinate diagram** (Figure 4.17). It shows how the energy changes as a reaction progresses from reactants to products.

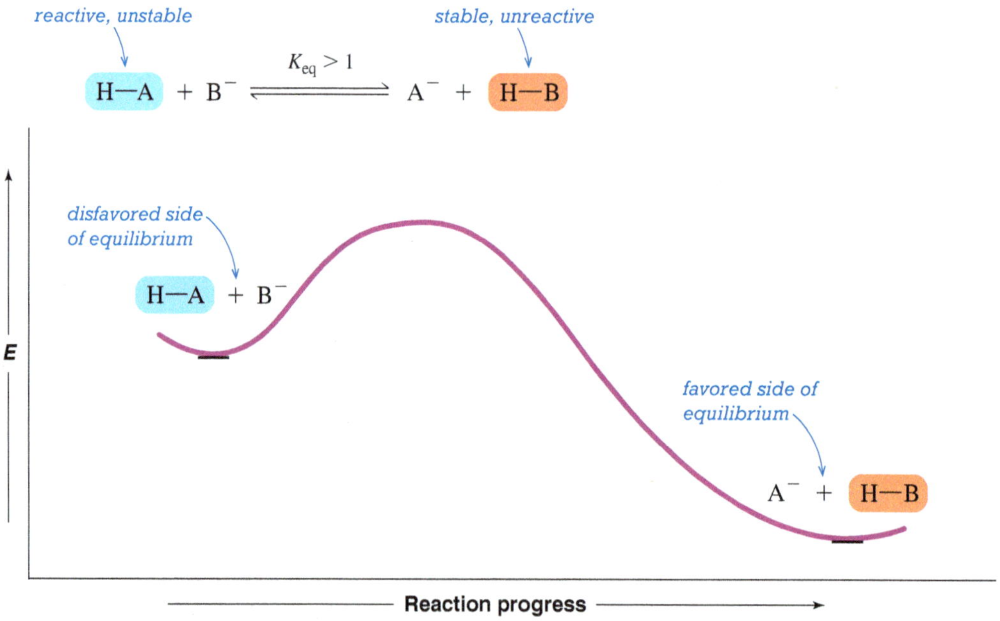

Figure 4.17 Reaction coordinate diagram: acid–base reactions move from unstable to stable.

Given an equilibrium constant, the change in energy for that reaction can be visualized using a reaction coordinate diagram (Figure 4.18).

When $K_{eq} > 1$, the product side is favored/lower in energy.

When $K_{eq} < 1$, the reactant side is favored/lower in energy.

When $K_{eq} = 1$, neither side is favored; the sides have identical energies.

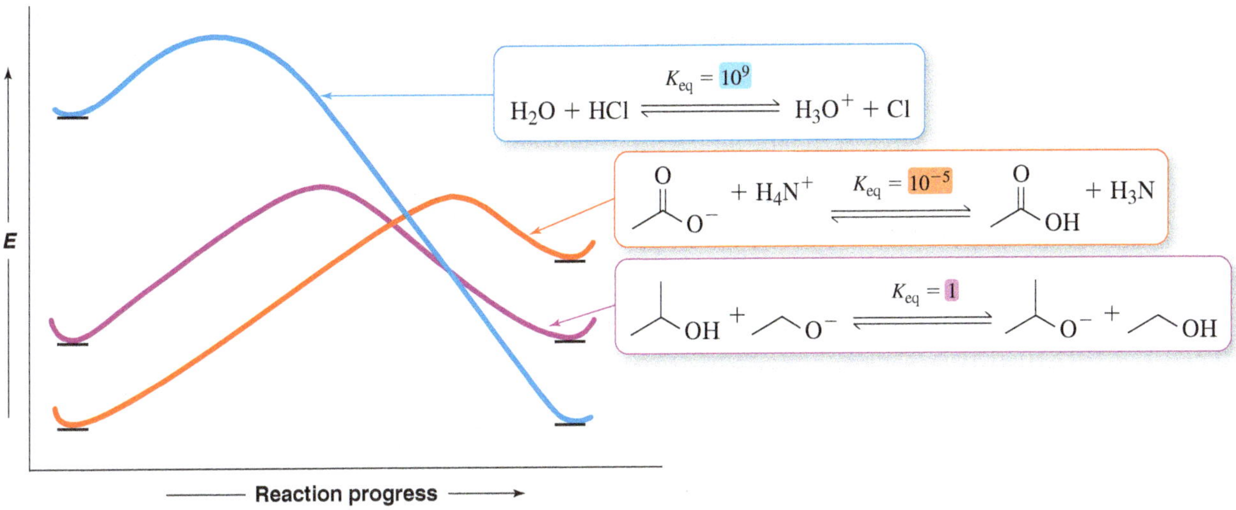

Reaction coordinate diagrams, important to a fundamental understanding of organic reactivity, are discussed in greater detail in Chapter 5.

Figure 4.18 More favorable acid–base reactions are more downhill.

Assessment 4.27 Below are four possible reaction coordinate diagrams. Match them to the acid–base reaction they most likely represent.

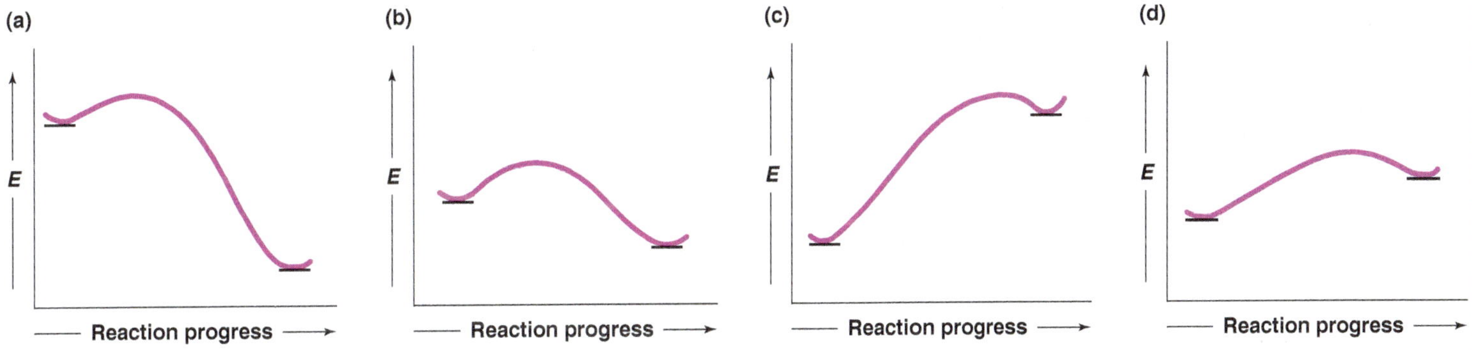

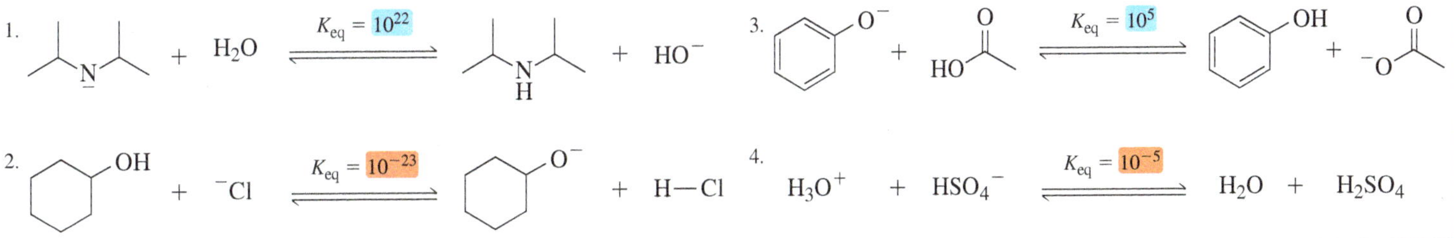

4.3 Measuring acid strength

Predicting the outcome of an acid–base reaction is as simple as moving a proton from the acid to the base. But what about a situation where we have an acid and a base and it isn't clear which is which (**Figure 4.19**)?

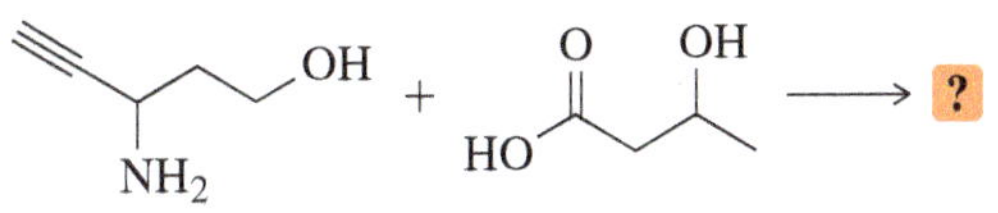

Figure 4.19 Predict the outcome: a complex acid–base reaction.

Though we can experimentally determine the equilibrium constant of a reaction, this only tells us about relative acid and base strength. There are situations where a base is only considered strong in reactions with a strong acid.

Methoxide (CH_3O^-), for example, is a relatively strong base (**Figure 4.20**). When reacting with acetic acid, methoxide is considered a very strong base because the equilibrium lies far to the right ($K_{eq} = 10^{11}$). When reacting with acetylene (HC≡CH), however, methoxide is too weak to deprotonate it, so the equilibrium lies far to the left ($K_{eq} = 10^{-9}$). As a result, base strength depends on the acidity of the proton being removed. It turns out there is an absolute scale by which the strength of acids can be listed—a scale that allows us to predict reaction equilibrium constants.

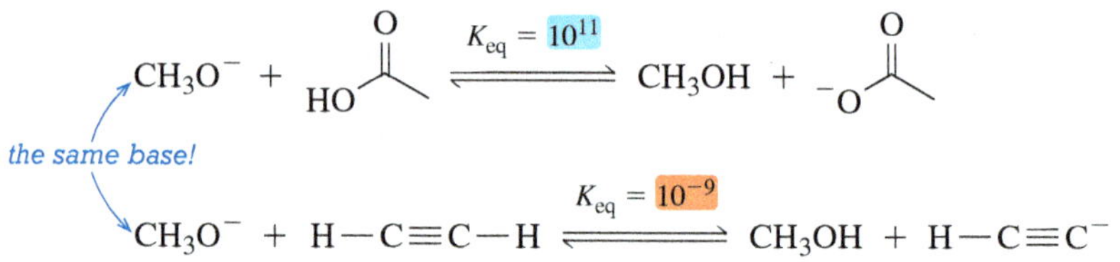

Figure 4.20 Base strength is relative to the strength of the acid being deprotonated.

Before you see the acidity scale, let's think like a chemist and see how it came about. Take a minute to develop your own solution before reading further.

Thinking Like a Chemist

The need: a method by which the relative acidity and basicity of two molecules can be known without conducting an experiment to measure K_{eq}. Because acid strength tells us about base stability, and vice versa, we only need to know acid *or* base strength. Most commonly, organic chemists use relative acid strength. [*We do the same here.*] In general terms, the relative strength of two acids (**A** and **B**) can be determined by measuring the equilibrium constant. Suppose **A** is a stronger acid than **B**. How might the strength of a new acid **C** relate to **A** and **B**? We could measure the equilibrium constant of the acid–base reaction involving **A** + **C** and/or **B** + **C**. [*Wait—and/or? Is it "and" or is it "or"? As you will see, this is the crux of the problem, and the solution to that problem.*]

If we need to compare **A** *and* **C** and then **B** *and* **C**, then the discovery of every new acid requires its comparison (via measuring K_{eq}) with every other known acid. [*This is waaaaaaaaay too complicated and time-consuming!*] But we don't have to do that. Recall the transitive property from mathematics that relates three numbers, X, Y, and Z. If X > Y and Y > Z, then X > Z. [*How does this help us compare relative acid strengths?*]

If **A** is stronger than **B** and **B** is stronger than **C**, then **A** must be stronger than **C**. In this way, for each newly discovered acid, only one experiment is needed to determine relative acidity against all known acids. That is, when all acids are compared to the same acid, a relative scale of acidities results.

We can demonstrate this by comparing the acidity of ethanol, acetic acid, and HCl. Measuring the equilibrium constants for each in an acid–base reaction with water reveals their relative acidity (**Figure 4.21**).

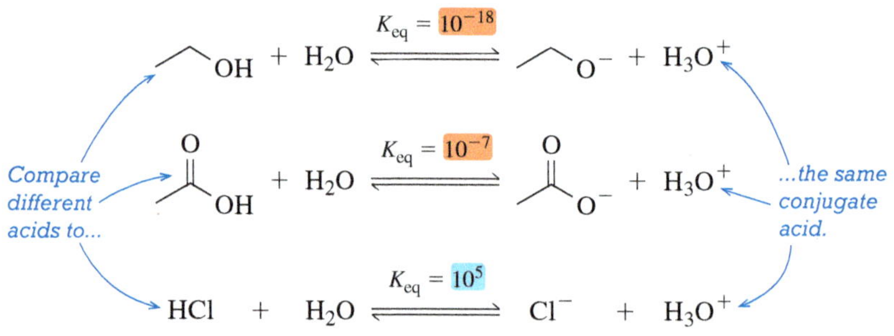

Figure 4.21 Comparing acidity against the same conjugate acid creates an acidity scale.

1. Ethanol and water: $K_{eq} = 10^{-18}$. Ethanol is not strong enough to protonate water; H_3O^+ is a much stronger acid.
2. Acetic acid and water: $K_{eq} = 10^{-7}$. The reaction is still favored to the left; H_3O^+ is a stronger acid than acetic acid.

3. Hydrochloric acid and water: $K_{eq} = 10^5$. The reaction is favored to the right. HCl is a stronger acid than H_3O^+.

From the data (Figure 4.22), HCl is a stronger acid than H_3O^+, while ethanol and acetic acid are weaker. It is also clear that HCl is stronger than acetic acid, which is stronger than ethanol. The overall ranking of acidities is HCl > H_3O^+ > acetic acid > ethanol. The relative acidity can be quantified by dividing the K_{eq}(HCl) by the K_{eq}(other acids). HCl is 10^5 times more acidic than H_3O^+ (based on $K_{eq} = 10^5$), 10^{12} times more acidic than acetic acid, and 10^{23} times more acidic than ethanol. As you shall see, in doing this for every acid discovered, a numerical acidity scale (pK_a values) is created. ■

Figure 4.22 Relative acidity of four acids.

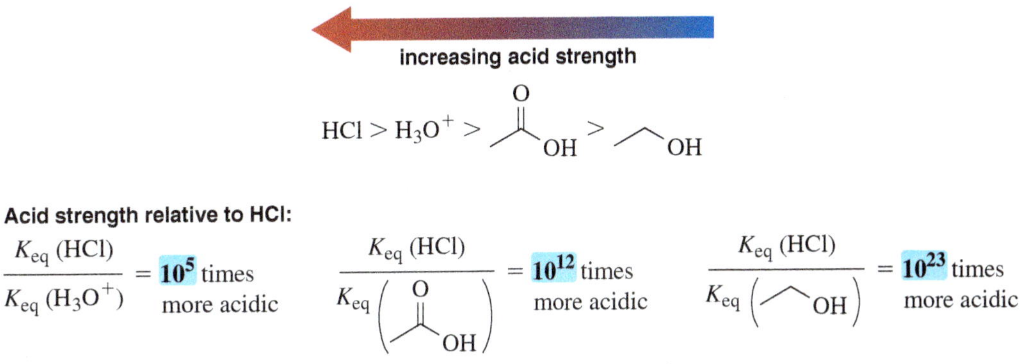

Assessment 4.28 Given the following K_{eq} values, how much stronger of an acid is methanol (CH_3OH) than acetylene (HC≡CH)?

$$CH_3OH + H_2O \overset{K_{eq} = 10^{-18}}{\rightleftharpoons} CH_3O^- + H_3O^+$$

$$H{-}C{\equiv}CH + H_2O \overset{K_{eq} = 10^{-27}}{\rightleftharpoons} H{-}C{\equiv}C^- + H_3O^+$$

Assessment 4.29 Given the following K_{eq} values, how much stronger of a base is acetate ($CH_3CO_2^-$) than chloride (Cl^-)?

$$HCl + H_2O \overset{K_{eq} = 10^{5}}{\rightleftharpoons} Cl^- + H_3O^+$$

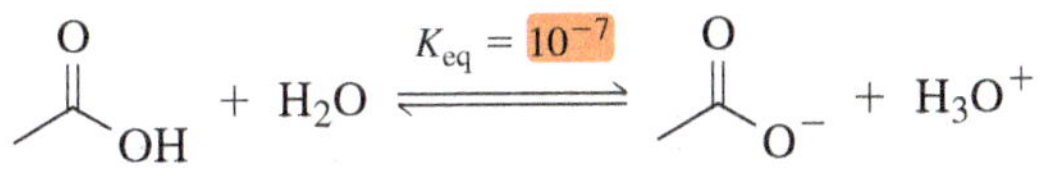

The strength of an acid HA can be described by its propensity to dissociate in water (a weak base). This provides a simple means for comparing the strengths of multiple acids through the dissociation equilibrium expression K_{eq} [Figure 4.23(a)]. Because these experiments use water as the solvent for all acids, the high concentration of water is constant. This means both sides can be multiplied by [H_2O], removing the water from the equilibrium expression [Figure 4.23(b)]. This gives you the **acid dissociation constant** (**K_a**) [Figure 4.23(c)], a measure of an acid's willingness to give up its proton to water. To make these exponential numbers easier to handle, the negative logarithm of the K_a is used [Figure 4.23(d)]. This is the **pK_a value.**

Figure 4.23 Derivation of the pK_a value.

$$HA + H_2O \longrightarrow H_3O^+ + A^-$$

(a) $$K_{eq} = \frac{[A^-][H_3O^+]}{[H_2O][HA]}$$

(b) $$[H_2O] \times K_{eq} = \frac{[A^-][H_3O^+]}{\cancel{[H_2O]}[HA]} \times \cancel{[H_2O]} = K_a$$

(c) $$K_a = \frac{[A^-][H_3O^+]}{[HA]}$$

(d) $$pK_a = -\log K_a$$

4.3.1 pK_a values: a measure of acid strength

Table 4.5 is a comprehensive list of pK_a values. [*I don't say this often, but you should memorize these values. You may wonder why, but there are several reasons. Read on.*]

Table 4.5 Selected pK_a values. As the pK_a value increases, the strength of the acid decreases and the strength of the conjugate base increases.

	Acid dissociation							K_a	pK_a
stronger	HCl hydrochloric acid	+	H_2O	⇌	H_3O^+	+	Cl^- chloride ion	1×10^{7}	−7
strong acids	$HO-S(=O)_2-OH$ sulfuric acid	+	H_2O	⇌	H_3O^+	+	$HO-S(=O)_2-O^-$ hydrogen sulfate (weaker bases)	1×10^{3}	−3
	H_3O^+ hydronium ion	+	H_2O	⇌	H_3O^+	+	H_2O water	55.6	−1.7
	HF hydrofluoric acid	+	H_2O	⇌	H_3O^+	+	F^- fluoride ion	6.8×10^{-4}	3.17
	$H-C(=O)-OH$ formic acid	+	H_2O	⇌	H_3O^+	+	$H-C(=O)-O^-$ formate ion	1.7×10^{-4}	3.76
	$CH_3-C(=O)-OH$ acetic acid	+	H_2O	⇌	H_3O^+	+	$CH_3-C(=O)-O^-$ acetate ion	1.8×10^{-5}	4.74
	$H-C\equiv N$ hydrocyanic acid	+	H_2O	⇌	H_3O^+	+	$^-C\equiv N$ cyanide ion	6.0×10^{-10}	9.22
weak acids	$^+NH_4$ ammonium ion	+	H_2O	⇌	H_3O^+	+	NH_3 ammonia	5.8×10^{-10}	9.24
	H_2O water	+	H_2O	⇌	H_3O^+	+	HO^- hydroxide ion	1.8×10^{-16}	15.7
	CH_3CH_2-OH ethyl alcohol	+	H_2O	⇌	H_3O^+	+	$CH_3CH_2O^-$ ethoxide ion	1.3×10^{-16}	15.9
very weak	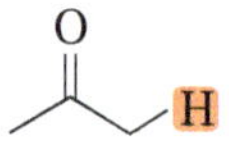 acetone	+	H_2O	⇌	H_3O^+	+	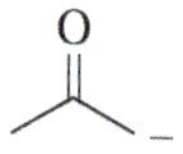 enolate	10^{-20}	20
	$H-C\equiv C-H$ acetylene	+	H_2O	⇌	H_3O^+	+	$H-C\equiv C^-$ acetylide	10^{-25}	25
not acidic	NH_3 ammonia	+	H_2O	⇌	H_3O^+	+	$^-NH_2$ amide ion	10^{-38}	38
weaker	CH_4 methane	+	H_2O	⇌	H_3O^+	+	$^-CH_3$ methyl anion (stronger bases)	10^{-50}	50

To better understand what the pK_a value means, consider CH_4, with a pK_a of 50 [**Figure 4.24(a)**]. A high pK_a indicates a very small K_a. A very small K_a implies CH_4 hardly dissociates (one molecule of CH_3^- for every 10^{50} molecules of CH_4). In essence, CH_4 does not give up a proton; it is an extremely weak acid—one of the weakest you'll see. Thus, *weak* acids have *high* pK_a values. Now consider HCl [**Figure 4.24(b)**], with a pK_a of −7, which is one of the lower values in Table 4.5. A very low pK_a indicates a large K_a. A large K_a means HCl completely dissociates (10^7 molecules of Cl^- for every one molecule of HCl). HCl very easily gives up its proton; it is an extremely strong acid—one of the strongest you'll see. Thus, *strong* acids have *low* pK_a values.

(a)

$$CH_4 + H_2O \xrightleftharpoons{pK_a = 50 \text{ for } CH_4} {}^-CH_3 + H_3O^+$$

$$-\log K_a = 50$$
$$\log K_a = -50$$
$$10^{\log K_a} = 10^{-50}$$

a very small value

$$K_a = 10^{-50} = \frac{1}{10^{50}} = \frac{[CH_3^-][H_3O^+]}{[CH_4]}$$

a very large value

$$[CH_3^-]:[CH_4] = \sqrt{1}:10^{50} = 1:10^{50}$$

Thus, a high pK_a value means the acid is weak.

(b)

$$HCl + H_2O \xrightleftharpoons{pK_a = -7 \text{ for HCl}} {}^-Cl + H_3O^+$$

$$-\log K_a = -7$$
$$\log K_a = 7$$
$$10^{\log K_a} = 10^7$$

a very large value

$$K_a = 10^7 = \frac{10^7}{1} = \frac{[Cl^-][H_3O^+]}{[HCl]}$$

a very small value

$$[Cl^-]:[HCl] = \sqrt{10^7}:1 = 3.2 \times 10^3:1$$

Thus, a low pK_a value means the acid is strong.

Figure 4.24 pK_a values correlate to acid strength. (a) High pK_a values indicate weak acids. (b) Low pK_a values indicate strong acids.

Assessment 4.30 Given that the indicated pK_a values correspond to the acid dissociation reactions shown, calculate the ratio of acid to conjugate base for the reactions shown.

(a)

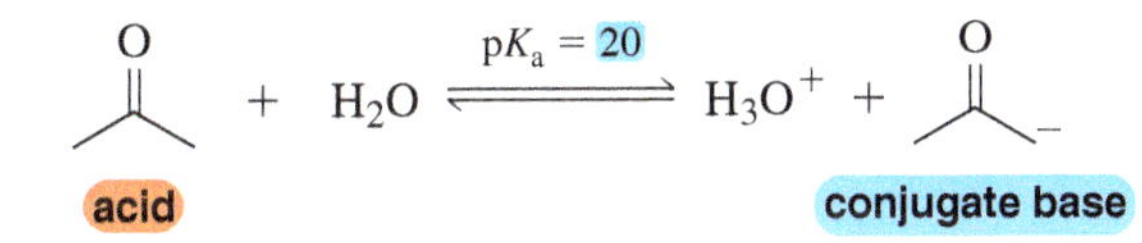

(b)

$$H{-}C{\equiv}N + H_2O \xrightleftharpoons{pK_a = 9} H_3O^+ + {}^-C{\equiv}N$$

(c)

$$HF + H_2O \xrightleftharpoons{pK_a = 3} H_3O^+ + F^-$$

Assessment 4.31 Hydrogen gas (H_2) has a relatively high pK_a value. Is it a stable or unstable acid? Do you expect it to participate in acid–base reactions?

4.3.2 pK_a values: an indirect measure of base strength

Though pK_a values are derived from *acid* dissociation constants, they give as much information about base strengths. A weak acid rarely gives up a proton because its conjugate base is unstable/reactive/strong. Conversely, a strong acid readily gives up a proton because its conjugate base is stable/unreactive/weak. [*Make sure you understand the last two sentences—an important concept!*] Thus, the strength of an acid will tell us whether the base is strong or weak. The instructions for doing this are shown in **Table 4.6** for several examples. [*Look back at Table 4.5; it shows us not only acid strength, but base strength, too!*]

Table 4.6 Using pK_a values to determine the relative strength of a base.

Given a certain base . . .	Identify its conjugate acid . . .	From the table, determine the pK_a value of the conjugate acid . . .	Test: If the pK_a value is low (high), then the conjugate acid is a strong (weak) acid.	Conclusion: The base is weak (strong).
NH_2^-	NH_3	38	NH_3 is a very weak acid.	NH_2^- is a very strong base.
HO^-	H_2O	16	H_2O is a weak acid.	HO^- is a strong base.
NH_3	NH_4^+	10	NH_4^+ is a moderately strong acid.	NH_3 is a moderately strong base.
H_2O	H_3O^+	−1.7	H_3O^+ is a very strong acid.	H_2O is a very weak base.

Assessment 4.32 How do you know that a proton with a low pK_a value is acidic (besides "I just remember")?

Assessment 4.33 If a base has a conjugate acid with a high pK_a value, is it stable or unstable? How do you know this is true (besides "I just remember")?

Assessment 4.34 In each pair, identify the more reactive base.

(a) 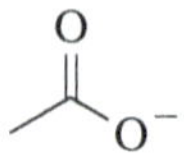 vs. Cl^- (b) NH_3 vs. H_2N^- (c) HO^- vs. $^-C{\equiv}CH$

Assessment 4.35 In each pair, identify the more stable base.

(a) Cl^- vs. H_2O (b) $N{\equiv}C^-$ vs. NH_3 (c) 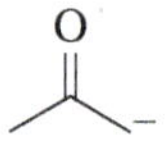 vs. $^-NH_2$

4.3.3 pK_a values: calculating K_{eq} for acid–base reactions

pK_a values allow us to calculate K_{eq} for acid–base reactions not yet studied, as in the reaction of acetylide ion ($HC{\equiv}C^-$) with hydrogen (H_2). There is a formula for this, but any time a formula is used, it is very important first to qualitatively predict the answer. [*This qualitative analysis improves comprehension of the topic and allows you to check the answer the formula produces.*]

$$\underset{\text{hydrogen}}{H{-}H} + \underset{\text{acetylide}}{^-C{\equiv}C{-}H} \rightleftharpoons \underset{\text{hydride}}{H^-} + \underset{\text{acetylene}}{H{-}C{\equiv}C{-}H}$$

Acid-base reactions go toward the more stable species and away from the less stable species. The ball rolls downhill.

Qualitatively, the favored side can be predicted based on either the relative acidity of H_2 (acid) and $HC{\equiv}CH$ (conjugate acid), or the relative basicity of $HC{\equiv}C^-$ (base) and H^- (conjugate base). Both approaches should give the same answer (Figure 4.25). If we split the reaction into two half-reactions, the lower pK_a value for acetylene (pK_a = 25) versus H_2 (pK_a = 36) [Figure 4.25(a)] suggests that acetylene is the stronger acid [Figure 4.25(b)]. Because reactions move away from stronger acids, this reaction is favored to the left. Similarly, the pK_a values of the acids tell us that acetylide is a more stable base than hydride (H^-).

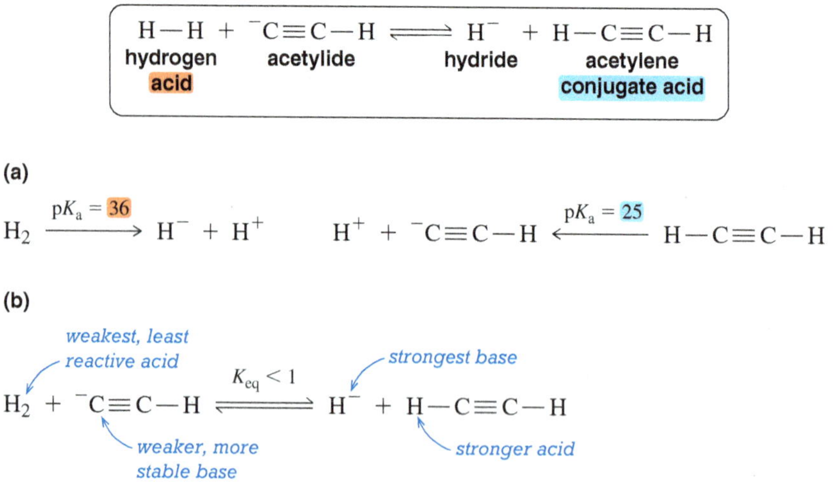

Figure 4.25 Using pK_a values to estimate the direction of equilibrium. (a) Split the reaction into half-reactions and use pK_a values for the acid and conjugate acid to (b) determine the weaker and stronger acids and bases.

Reactions move toward the more stable base. In both cases, we expect the equilibrium to favor the left (reactants) with $K_{eq} < 1$. [*But how much less than 1? How much is the left side favored? Think about it before continuing.*]

Assessment 4.36 Using qualitative reasoning for the acid–base reactions shown, (i) which is stronger, the acid or the conjugate acid? (ii) Which side of the reaction is favored? (iii) Would you expect a K_{eq} greater than, equal to, or less than 1?

(a)

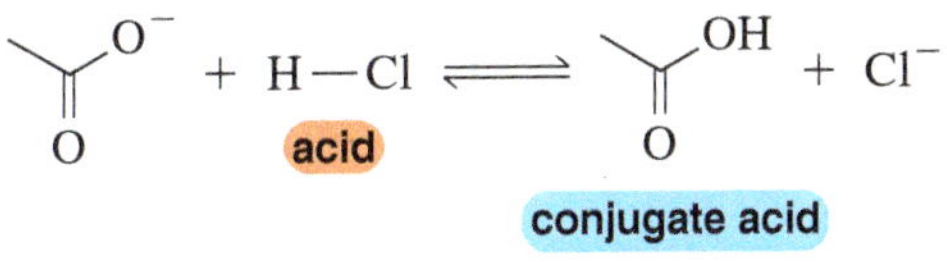

(b)

$CH_4 + {}^{-}NH_2 \rightleftharpoons {}^{-}CH_3 + NH_3$

CH_4: acid; NH_3: conjugate acid

(c)

$H_2SO_4 + H_2O \rightleftharpoons H_3O^+ + HSO_4^-$

H_2SO_4: acid; H_3O^+: conjugate acid

Assessment 4.37 Using qualitative reasoning for the acid–base reactions shown, (i) which is stronger, the base or the conjugate base? (ii) Which side of the reaction is favored? (iii) Would you expect a K_{eq} greater than or less than 1?

(a)

$H{-}C{\equiv}C^- + NH_3 \rightleftharpoons {}^{-}NH_2 + H{-}C{\equiv}C{-}H$

$H{-}C{\equiv}C^-$: base; ${}^{-}NH_2$: conjugate base

(b)

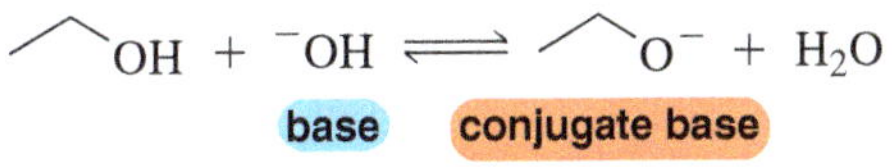

(c)

$Cl^- + H_3O^+ \rightleftharpoons H_2O + HCl$

Cl^-: base; H_2O: conjugate base

Equation 4.1 can be used to calculate an exact value of K_{eq} by comparing the pK_a values for the acids on either side of an acid–base equilibrium. This equation aligns with the qualitative analysis just discussed (Figure 4.25) because it compares the strengths of the two acids involved in the reaction.

$$pK_{eq} = pK_a\,(\text{acid}) - pK_a\,(\text{conjugate acid}) \qquad \text{Equation 4.1(a)}$$

or

$$pK_{eq} = pK_a\,(\text{left-side acid}) - pK_a\,(\text{right-side acid}) \qquad \text{Equation 4.1(b)}$$

Applying Equation 4.1 to the reaction between acetylide ion and hydrogen gives a K_{eq} of 10^{-11}, which verifies that the left side of the equilibrium is favored (by about 100 billion to 1) (**Figure 4.26**). This is consistent with the direction suggested by the qualitative analysis in Figure 4.25. Thus, even for reactions where water is *not* the solvent, pK_a values can be used to calculate K_{eq} using Equation 4.1.

Figure 4.26 Calculation of K_{eq} using pK_a values.

left-side acid (acid) *right-side acid (conjugate acid)*

$$H{-}H + H{-}C{\equiv}C{:}^- \rightleftharpoons H{:}^- + H{-}C{\equiv}C{-}H$$

$pK_a = 36$ (H–H); $pK_a = 25$ (H–C≡C–H)

$$pK_{eq} = pK_a\,(\text{acid}) - pK_a\,(\text{conjugate acid})$$
$$pK_{eq} = 36 - 25 = 11 = -\log(K_{eq})$$
$$\log K_{eq} = -11$$
$$K_{eq} = 10^{-11}$$

The left side is favored by a lot!

Assessment 4.38 Calculate K_{eq} for these acid–base reactions.

(a)

$H{-}C{\equiv}C{-}H + {}^{-}NH_2 \rightleftharpoons NH_3 + H{-}C{\equiv}C^-$

(b)

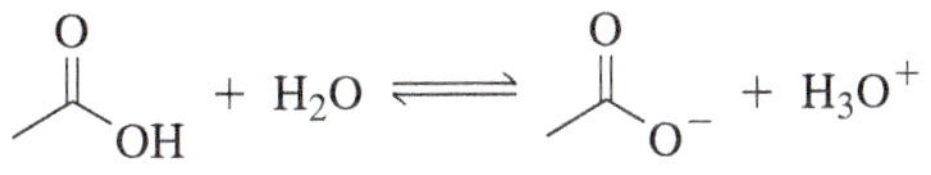

(c)

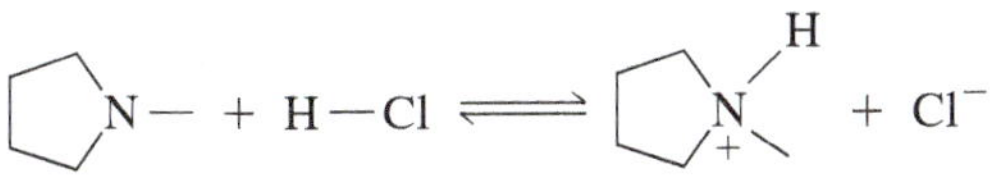

(d)

4.3.4 pK_a values that aren't in the table

Equation 4.1 makes it possible to estimate K_{eq} only when pK_a values are known. Because it's impractical to memorize pK_a values for every acid, ones that are "close enough" can be used instead. Extrapolation of the known pK_a values in Table 4.5 allows the calculation of a "close-enough" K_{eq} in most cases.

Although the values in Table 4.5 are for specific molecules, they also approximate pK_a values for molecules containing similar functional groups. [*This is one reason that functional group classification is so important.*] Consider the hydroxyl proton of ethanol, which has pK_a = 16. Based on this argument, any alcohol [**Figure 4.27(a)**] can be expected to have a pK_a ≈ 16. There can be structural effects that increase or decrease the value a bit, but these subtle differences rarely impact the concept being studied. [*When they do, Section 4.4 shows us how to explain them.*] Similarly, acetic acid, a carboxylic acid, has pK_a = 5. Every carboxylic acid will be close to this value [**Figure 4.27(b)**].

Estimating pK_a values in this way is a little more complicated for ammonium ions. NH_4^+ has pK_a = 10 (Table 4.5). Despite appearing structurally different, the pK_a values of the ammonium ions in **Figure 4.27(c)** are similar. [*Why?*] The nitrogen in NH_4^+ has zero lone pairs, four bonds to hydrogen, and a positive charge. [*How would you describe nitrogen in the other compounds?*] In the other ammonium ions, nitrogen has zero lone pairs, four bonds (at least one to hydrogen), and a positive charge. [*Sound familiar?*] Described this way, the pK_a values should be similar. Similarly, the hydronium ion (pK_a = −1.7), protonated alcohols, and protonated ethers have similar pK_a values. [*Why do you think this is true?*] In each, the oxygen has a single lone pair, three bonds (at least one to hydrogen) and a positive charge [**Figure 4.27(d)**].

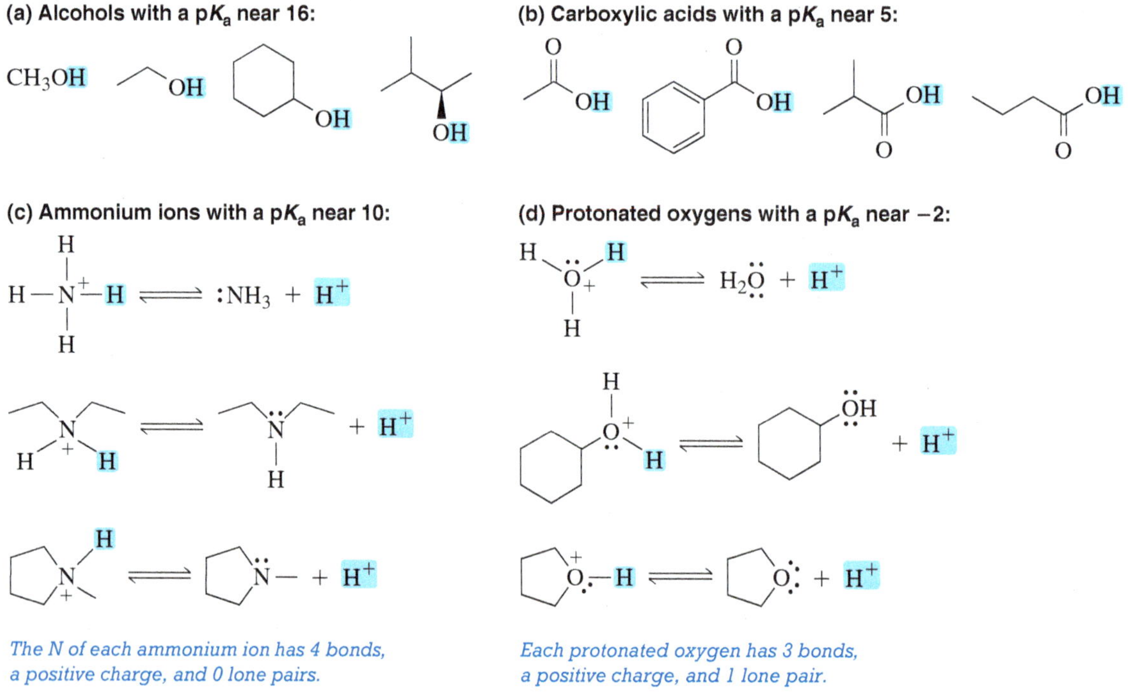

Figure 4.27 Approximation of pK_a values based on structural similarities. (a) Similar alcohols have similar pK_a values. (b) Similar carboxylic acids have similar pK_a values. (c) Similar ammonium ions have similar pK_a values. (d) Similar oxonium ions have similar pK_a values.

And so, when using pK_a values, you simply need to find the pK_a value that is closest based on the structural features that make a compound acidic. [*We haven't talked about those features yet, but we do in Section 4.4. First, practice finding the "best" pK_a value in Assessments 4.39 and 4.40.*]

Assessment 4.39 For each indicated proton, suggest an approximate pK_a value from Table 4.5. Rationalize your choice.

(a) (b) (c) (d) (e) (f)

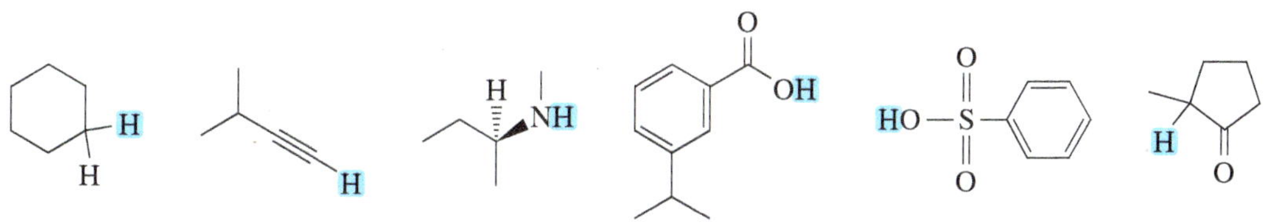

Assessment 4.40 For the bases shown, draw the conjugate acid and identify a pK_a value from Table 4.5 that would help you accurately estimate its stability.

(a) (b) (c) (d) (e) (f)

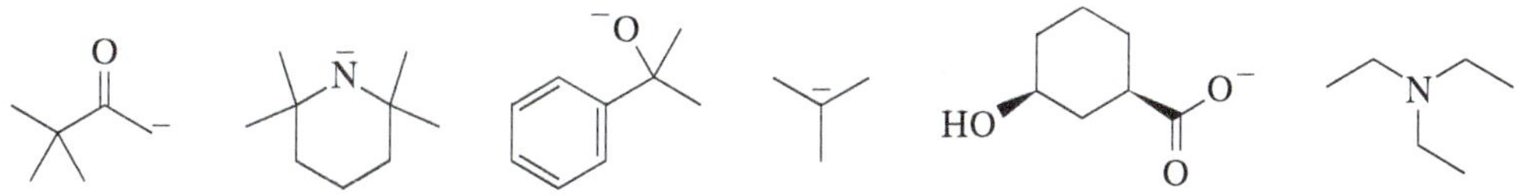

4-C CHEMISTRY IN CONTEXT: Medicinal Chemistry

Protonation alters drug solubility

Pharmacokinetics (see Chemistry in Context 11-A) is an important area of medicinal chemistry (the design and synthesis of medicines). To ensure that a molecule reaches its target in the body, it must easily move between the polar aqueous phase (70% of the body) and nonpolar cell membranes. Highly water-soluble molecules are rapidly excreted, while highly nonpolar molecules get stuck in cell membranes and don't make it to their target.

For drugs containing acidic or basic functional groups, water–lipid solubility depends on the pH of the surrounding medium. Take, for example, the antihypertensive drug metoprolol (Toprol-XL®). **Table 4.7** shows the change in log D for metoprolol as a function of pH. [*Log D is a measurement of water–lipid solubility. When log D < 0, the molecule is highly polar/water soluble; when log D > 0, the molecule is nonpolar/lipid soluble.*] Under acidic conditions (low pH), the molecule is soluble in water. Under basic conditions (high pH), the molecule is more soluble in the nonpolar membrane. At low pH, protonation of the amine creates a positively charged ion with enhanced water solubility. At high pH, the neutral amine is more lipid soluble (**Figure 4.28**).

Table 4.7 Water solubility depends on the protonation state of metoprolol.

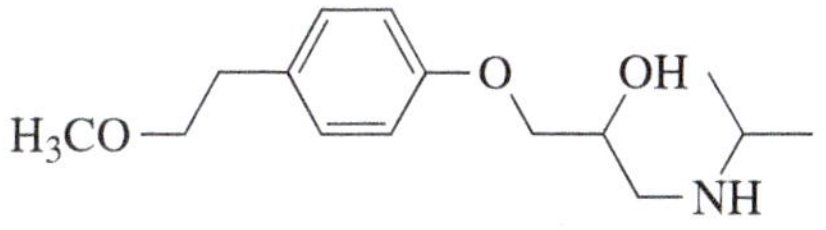

metoprolol

pH	Log *D*
2.0	−1.31
3.0	−1.31
4.0	−1.31
5.0	−1.28
5.5	−1.21
6.0	−1.05
6.5	−0.75
7.0	−0.34
7.5	0.12
8.0	0.59
8.5	1.03
9.0	1.39
10.0	1.73

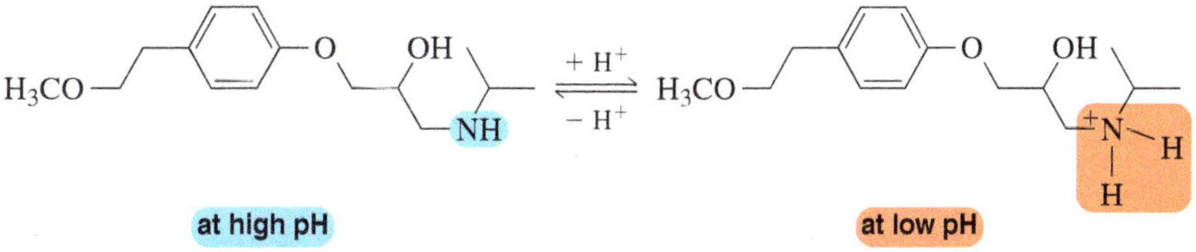

Figure 4.28 The amine is soluble in water at low pH.

4.4 What makes an acid acidic or a base basic? Structural effects on acidity/conjugate base stability

pK_a values are used throughout organic chemistry to explain various reaction outcomes, resulting in you eventually having many of them committed to memory. What makes a compound acidic, and why is a particular base more stable? [*Read on.*]

Sometimes subtle differences in structure are responsible for big differences in acidity. Consider the pK_a values of acetic acid (5) and ethanol (16). The oxygen in each has two lone pairs, has two bonds (at least one to hydrogen), and is neutral. And yet acetic acid is a whopping 100,000,000,000 times more acidic (**Figure 4.29**). [*Begin thinking about why acetic acid is so much more acidic.*]

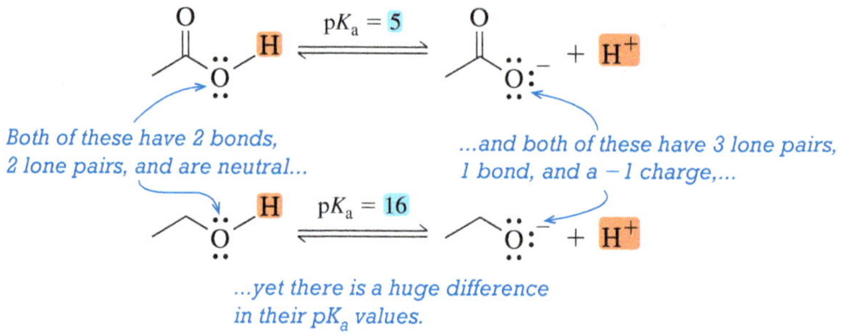

Figure 4.29 Similar structures, but different pK_a values. Why?

Recall that a relatively strong and unstable acid is deprotonated to form a weak and stable conjugate base. Alternatively, a weak and stable acid is deprotonated to form a strong and unstable conjugate base. As a result, when assessing relative acidity, examining the stability of the conjugate base is key. We'll do so by looking at the following structural effects: electronegativity, size, resonance, induction, and hybridization. Note that for every relationship described, "stronger" and "weaker" are relative.

4.4.1 Within a row: electronegativity matters

Figure 4.30 shows a series of acids made up of elements in the second row of the periodic table. Judging from their pK_a values, HF (3) is significantly more acidic than CH_4 (50). Thus, F^- must be significantly more stable than CH_3^-. Because fluorine is a more electronegative atom than carbon, it holds its electrons more tightly and is a stable anion. Carbon, on the other hand, does not want a lone pair; its conjugate base is very unstable. Within a row of the periodic table, the more electronegative atom generally forms the most stable conjugate base, making the acid from which it formed more acidic.

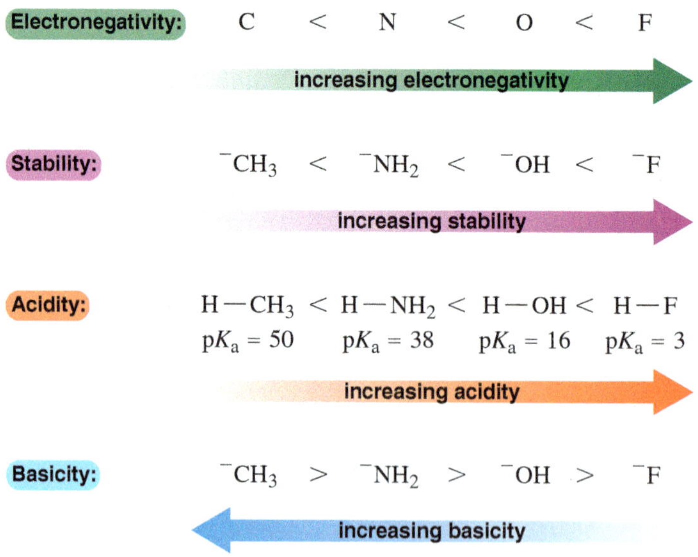

Figure 4.30 As electronegativity increases, a conjugate base is stabilized and acidity increases.

Assessment 4.41 Without using pK_a values, pick out the *more* acidic compound in each pair. Explain your answer.

(a)

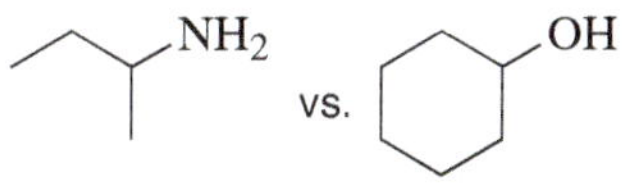

(b)

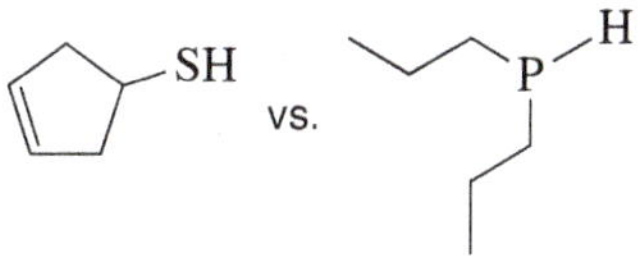

(c)

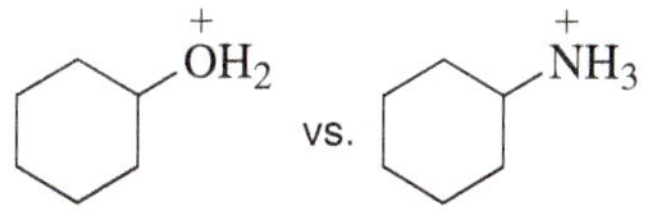

Assessment 4.42 Without using pK_a values, pick out the *more* reactive (least stable) base in each pair. Explain your answer.

(a)

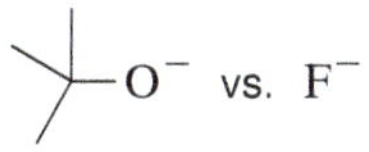

(b)

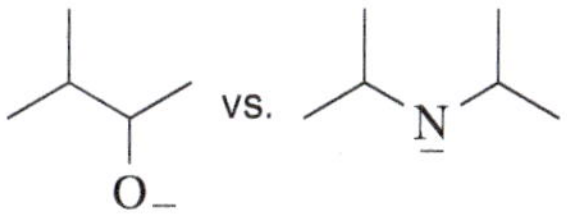

(c)

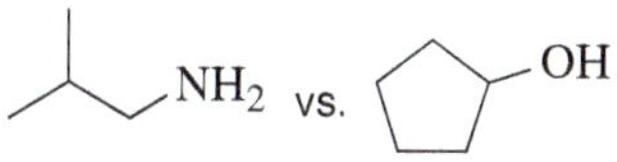

4.4.2 Within a group: size matters

Figure 4.31 shows a series of acids made up of the halogens—the elements in group 17 of the periodic table. From their pK_a values, HI (-9) is more acidic than HF (3). Thus, I^- must be more stable than F^-. Each successive row represents the addition of another shell, each one farther from the nucleus. The valence electrons of I^- and F^- are in the fifth and second shells, respectively. Thus, I^- has a larger atomic radius than F^-. Larger **anions** spread the negative charge over a larger volume and are therefore more stable. Because I^- is more stable than F^-, HI has a lower pK_a value and is 10^{12} times more acidic.

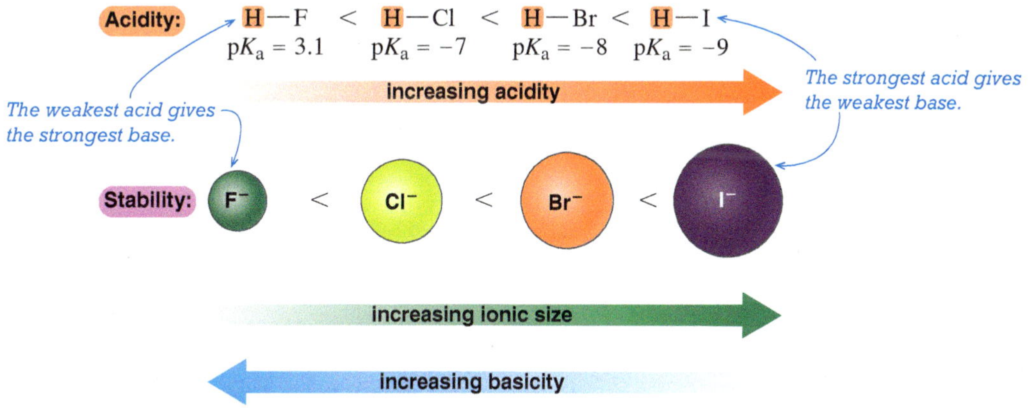

Figure 4.31 As size increases, a conjugate base is stabilized and acidity increases.

Another way to think about this trend [*it will help with what comes next*] is to think about charge per volume. Having a -1 charge spread over a larger volume means there is less charge at any single point. Charge is generally bad [*unstable*], so spreading it out helps to stabilize it. [*Like telling your roommates about a bad day, sharing the negativity makes you feel better (more stable).*]

Assessment 4.43 Without using pK_a values, pick out the *least* acidic compound in each pair. Explain your answer.

(a)

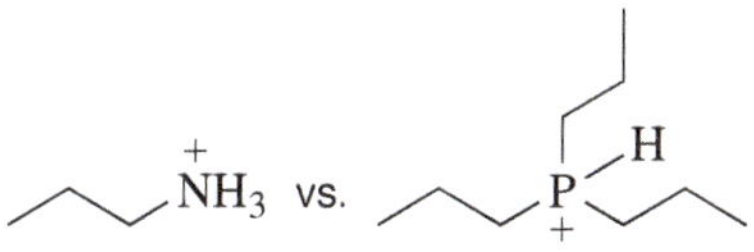

(b)

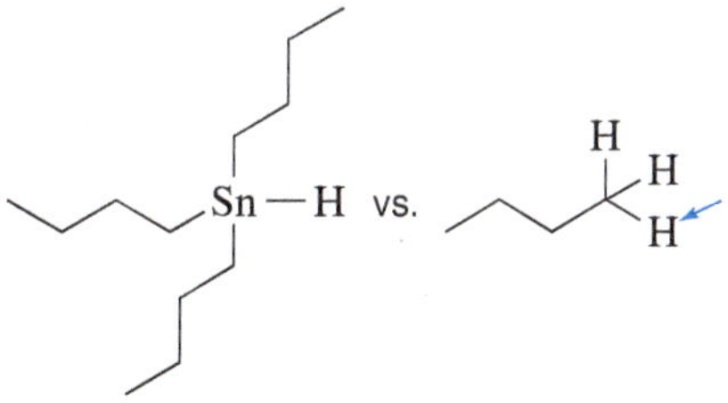

(c)

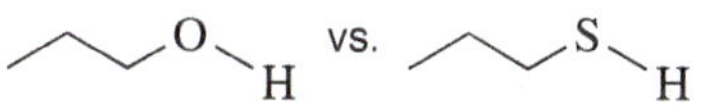

Assessment 4.44 Without using pK_a values, pick out the *least* reactive (most stable) base in each pair. Explain your answer.

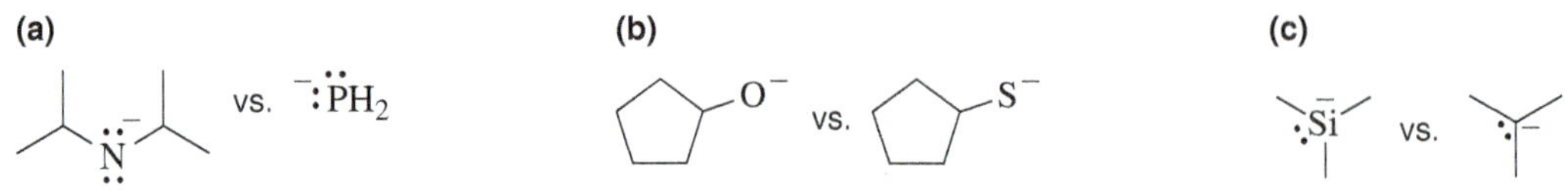

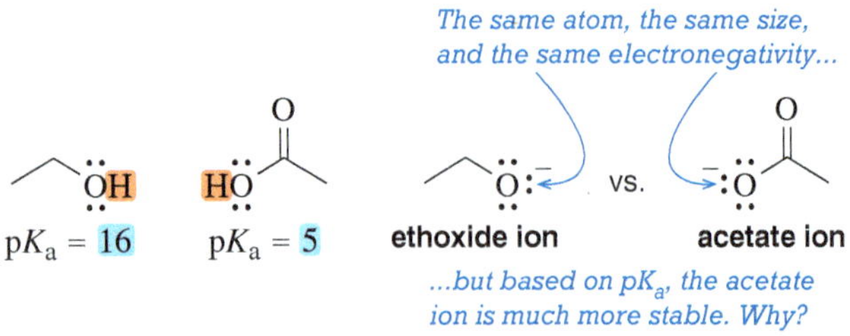

Figure 4.32 Similar structures, but different pK_a values. Why?

4.4.3 Delocalization: resonance rules

We began this section by trying to explain the large difference in acidity between ethanol and acetic acid (see Figure 4.29). Because the negative charge in both conjugate bases is on oxygen, neither electronegativity (Section 4.4.1) nor atomic size (Section 4.4.2) allows us to rationalize the difference in pK_a values. Thus, we need to more closely examine the two conjugate bases, looking for differences (**Figure 4.32**). [*Do you see one?*]

Ethoxide and acetate have three lone pairs on oxygen, yet the acetate anion is more stable (based on pK_a values). Looking closely at the acetate ion, the electrons on oxygen can participate in resonance with the neighboring C—O π bond (**Figure 4.33**). [*Review Section 2.8 to recall how we know this. Is there a pushable pair of electrons? Answer: There is a lone pair of electrons, so YES! Is there a place to be pushed? Answer: There is an atom with a π bond attached, so YES!*] With a second resonance structure, the −1 charge on acetate is spread over two oxygens, each bearing −1/2 charge. Spreading out the charge by resonance stabilizes the conjugate base, as is true for increasing atomic radius (Section 4.4.2). Delocalization of charge by resonance is usually best at stabilizing charged intermediates.

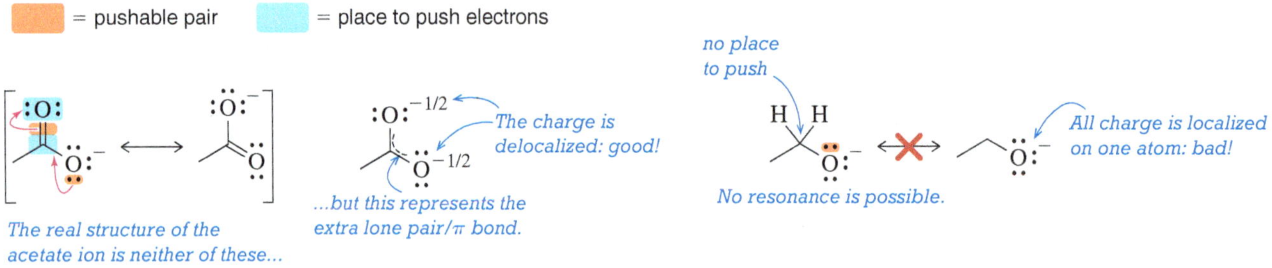

Figure 4.33 Delocalization by resonance stabilizes the conjugate base of acetate and makes acetic acid more acidic than ethanol.

Resonance stabilization of the conjugate base also accounts for the differing acidity for the acid pairs shown in **Figure 4.34**.

Figure 4.34 Resonance stabilization of conjugate bases.

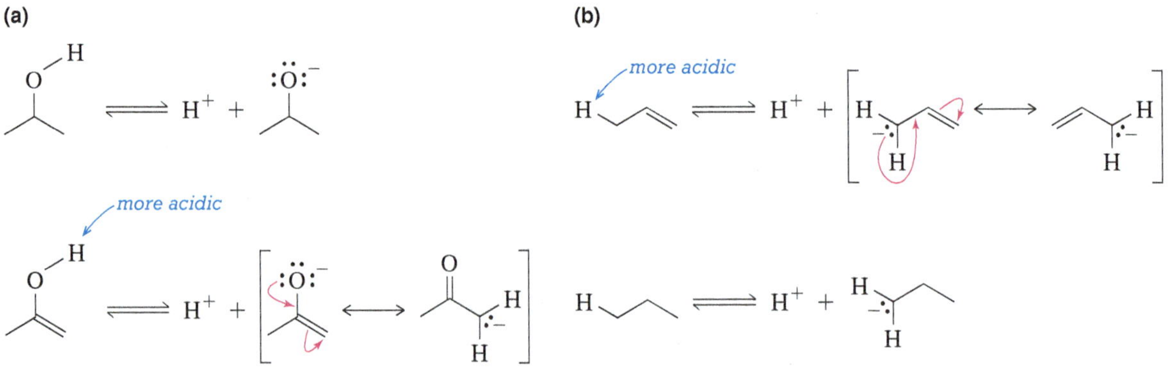

4.3.1–4.3.3 SUMMARY

- Within a row, anions with the negative charge on the more electronegative atom are more stable. Within a group, anions with the negative charge on larger atoms are more stable. Resonance stabilizes conjugate bases, making the acids from which they are formed more acidic.

Assessment 4.45 Which anion in each pair would you expect to react more quickly with H^+?

(a) (b) (c)

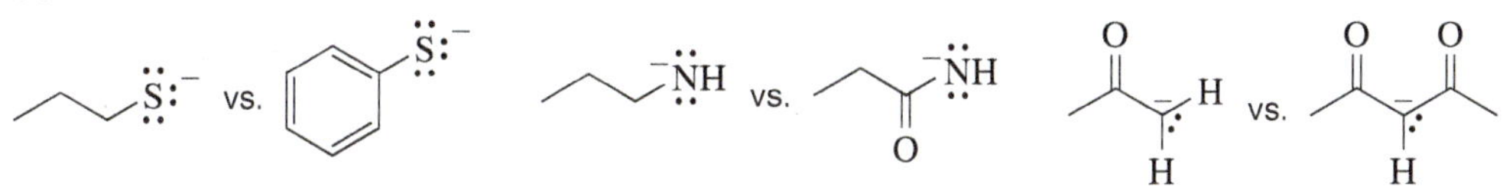

Assessment 4.46 Which acid in each pair would you expect to more readily donate a proton to a basic compound?

(a) (b) (c)

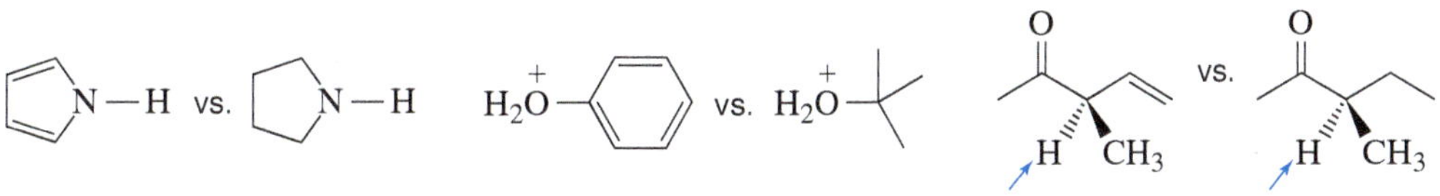

4.4.4 Electronegativity at a distance: induction is important

Induction stabilizes conjugate bases, making the acids from which they form more acidic.

We'll learn about induction by solving Assessment 4.47 together.

Assessment 4.47 Rationalize the rather large difference in pK_a values for the two carboxylic acids shown.

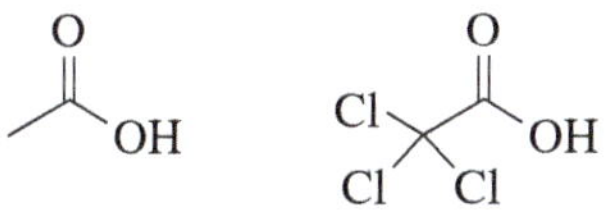

acetic acid pK_a = 4.74 — trichloroacetic acid pK_a = 0.66

Solved Assessment 4.47

Organic molecules with common functional groups share common reactivity patterns. At first glance, the two carboxylic acids, acetic and trichloroacetic acid, should be strong proton donors. [*After all, they have "acid" in their name.*] Based on their pK_a values, though, trichloroacetic acid is a significantly stronger acid (**Figure 4.35**).

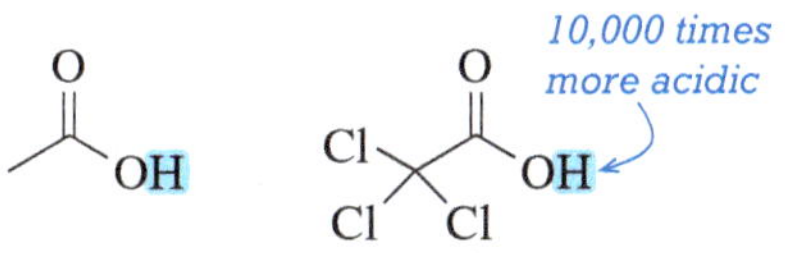

acetic acid pK_a = 4.74 — trichloroacetic acid pK_a = 0.66

Figure 4.35 Both compounds have similar structures, but different pK_a values. Why?

[*What do the pK_a values say about the stability of the conjugate bases?*] From the pK_a values, trichloroacetate (the conjugate base of trichloroacetic acid) is more stable than acetate (the conjugate base of acetic acid). In each case, the negative charge is on the same atom and is resonance stabilized (**Figure 4.36**). [*Are there any differences?*] The only difference is the three chlorines in trichloroacetic acid that are not in acetic acid. [*Looking for any differences in structure is a good problem-solving technique.*]

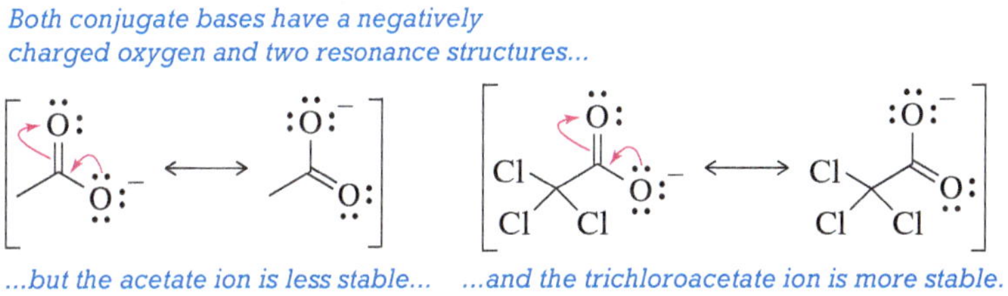

Figure 4.36 Both conjugate bases have two resonance structures, but one base is more stable than the other.

The chlorines stabilize the conjugate base of trichloroacetic acid through **induction.** The highly electronegative chlorine atom polarizes the C—Cl bond, creating a partial negative charge (δ^-) on chlorine and a partial positive charge (δ^+) on carbon. The three δ^+ on carbon are near the carboxylate anion. This interaction stabilizes the conjugate base, making trichloroacetic acid more willing to give up its proton (**Figure 4.37**). ■

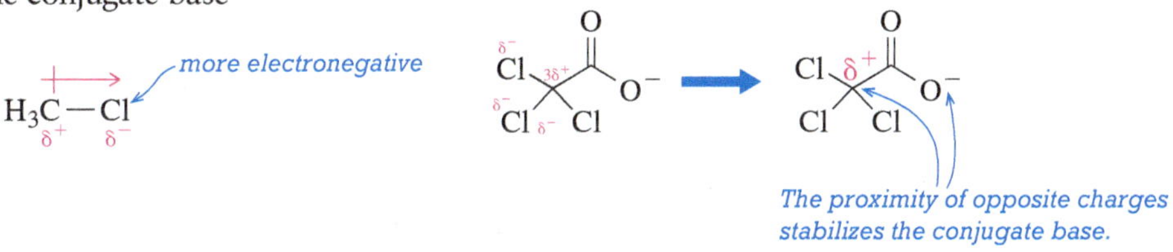

Figure 4.37 Electron-withdrawing chlorines stabilize the conjugate base by induction.

More electronegative atoms have a stronger inductive effect, as shown by chloroacetic and fluoroacetic acid in Figure 4.38. The more electronegative fluorine induces a larger δ^+, resulting in greater stabilization of the carboxylate ion.

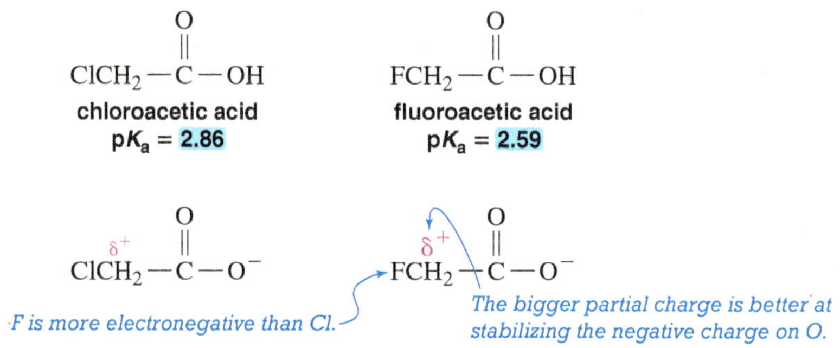

Figure 4.38 Increasing electronegativity leads to a stronger inductive effect.

You can think of the inductive effect as "electronegativity at a distance" because it occurs even when the electronegative atom is located away from the reactive center. In the series of butanoic acids in Figure 4.39, for example, the effect of the chlorine is diminished as it moves farther away from the functional group, but it still influences the stability of the carboxylate ions. To best understand this, consider the role of chlorine in each. For 2-chlorobutanoic acid, the chlorine creates a partial positive charge (δ^+) that stabilizes the conjugate base. For 4-chlorobutanoic acid, the chlorine creates a partial positive charge (δ^+) that pulls electron density from the neighboring carbon (C_4), which in turn creates a smaller partial positive charge (δ^+) that pulls electron density from C_3, ultimately creating an even smaller partial positive charge (δ^+) on C_2. In this way, the carboxylate ion of 4-chlorobutanoic acid is stabilized less, and 4-chlorobutanoic acid is less acidic than 2-chlorobutanoic acid.

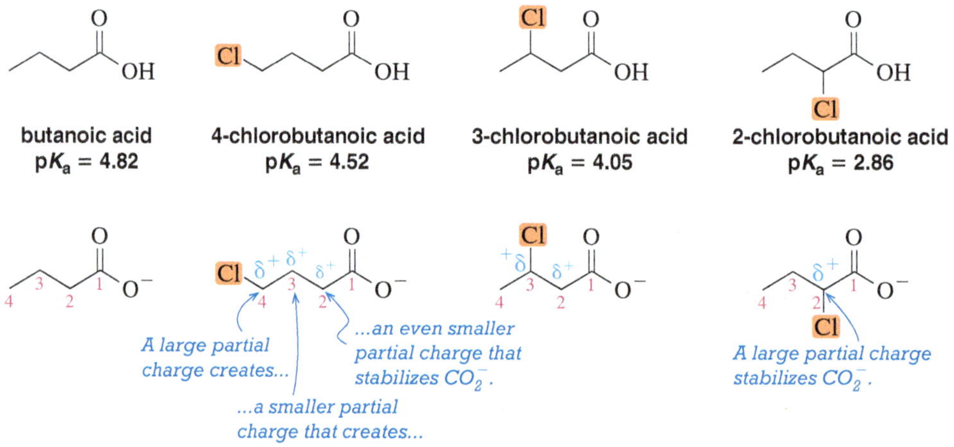

Figure 4.39 Inductive effect: diminishes with distance.

4.4.4 SUMMARY

- Induction, an electron-withdrawing effect of an electronegative atom, creates a partial positive charge that stabilizes an anionic species. The effect diminishes as the electronegative atom is moved farther away from the reactive center. Stabilizing the conjugate base makes the acid more acidic.

Assessment 4.48 Rank the following alcohols in order of descending pK_a value. Explain your ranking.

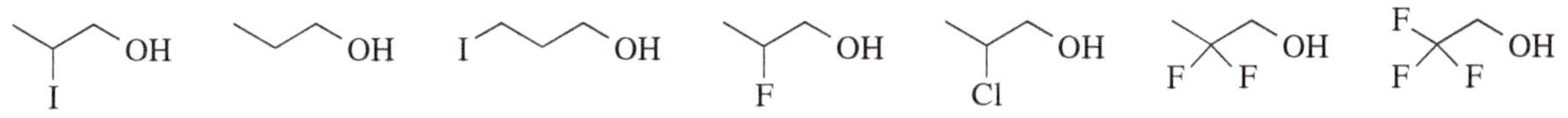

Assessment 4.49 Rank the following amines in order of their basicity (strongest base = 1; weakest base = 6).

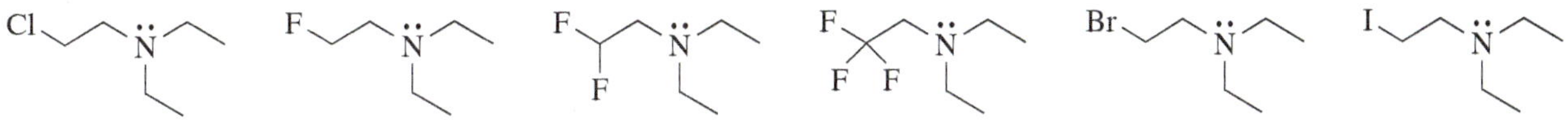

4.4.5 Hybridization

Figure 4.40 compares the pK_a values for ethane, ethene, and ethyne (acetylene). What can account for the conjugate base of ethyne being more stable than the others? The conjugate bases are all **carbanions**, resonance is not possible, and there is no inductive stabilization. [*Do you see any differences between the three conjugate bases?*] The only notable difference between the three is the hybridization of the carbanion.

Figure 4.40 Similar structures, but different pK_a values. Why?

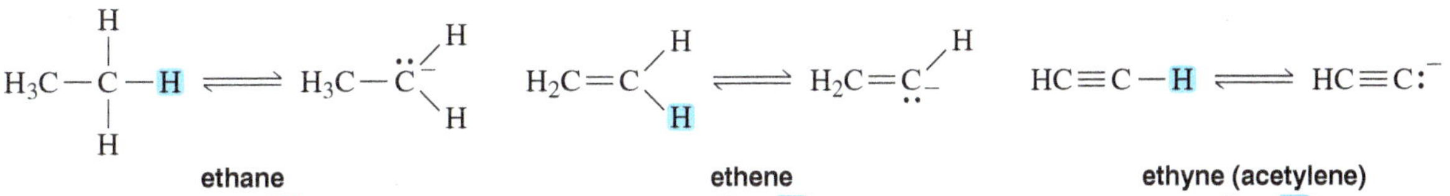

As hybridization changes from *sp* to sp^2 to sp^3, the shape of the orbital containing the lone pair changes. An sp^3-hybridized orbital is one part *s* orbital and three parts *p* orbital, whereas an sp^2-hybridized orbital is one part *s* orbital and two parts *p* orbital, and an *sp*-hybridized orbital is one part *s* orbital and one part *p* orbital. The *sp*-hybridized orbital, then, has the most *s* character. This means it is more spherical around the nucleus, in which case the electrons are held more closely to the nucleus (**Figure 4.41**).

Figure 4.41 Hybridization of one *s* and one *p* orbital makes an *sp*-hybridized orbital shorter than an sp^3-hybridized orbital.

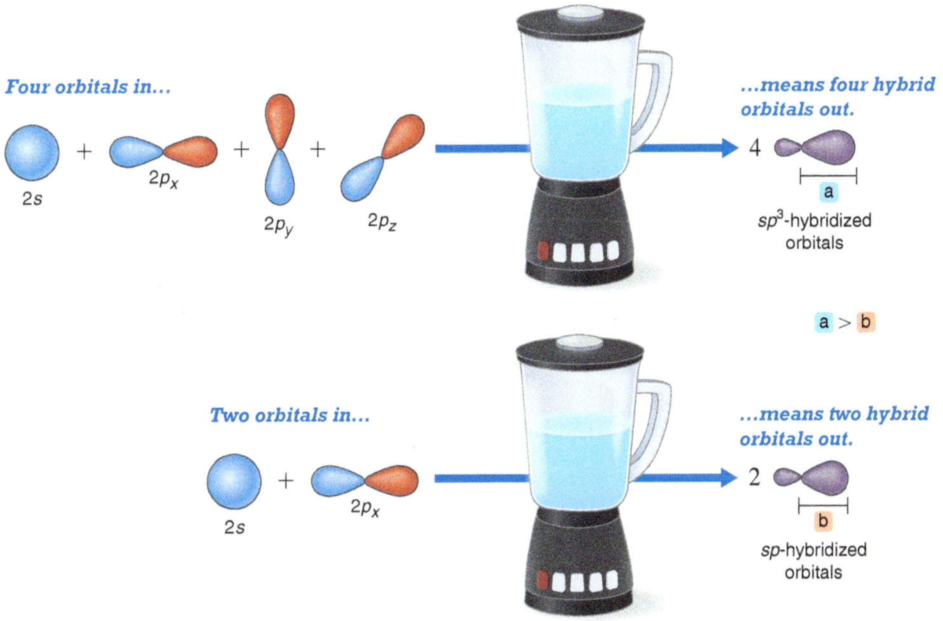

A lone pair in an *sp*-hybridized orbital is held closer to the positively charged nucleus than it is in the conjugate bases formed from ethane (sp^3-hybridized lone pair) and ethene (sp^2-hybridized lone pair) (**Table 4.8**). Having the [*negatively charged*] electrons in closer proximity to the [*positively charged*] nucleus results in enhanced stability.

Table 4.8 Acidity as a function of hybridization. As the percent *s* character of the hybrid orbital holding the lone pair of electrons increases, the lone pair is held closer to the nucleus, making the conjugate base more stable. [*The pK_a of ammonia and alcohols are shown for perspective.*]

Compound	Conjugate base	Hybridization	*s* character	pK_a	
H_3C-CH_2-H	$H_3C-CH_2^-$	sp^3	25%	50	weakest acid
$H_2C=CH-H$	$H_2C=CH^-$	sp^2	33%	44	
$:NH_3$	$^-:\ddot{N}H_2$	(ammonia)		38	
$H-C\equiv C-H$	$H-C\equiv C^-$	*sp*	50%	25	
R—OH	R—$\ddot{O}:^-$	(alcohols)		16–18	stronger acid

4.4.5 SUMMARY

- Lone pairs held in orbitals with more *s* character are less basic because they are closer to the positively charged nucleus.

Assessment 4.50 If reacted with a strong base, which of the labeled protons would you expect to be removed first?

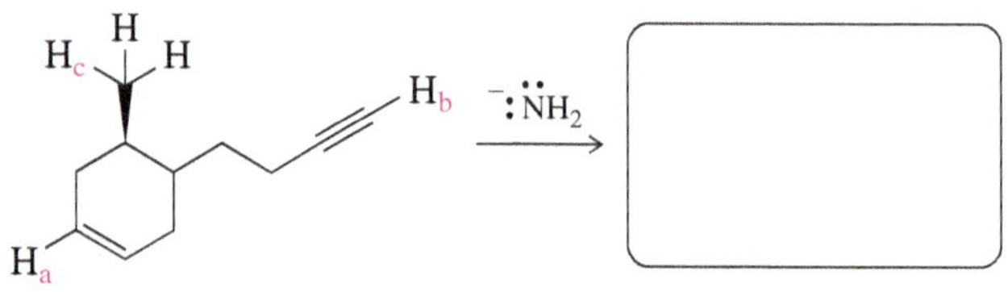

4.4.6 Weighing the effects: building your intuition

Five different structural features have been examined for their impact on acidity and conjugate base stability (Sections 4.4.1–4.4.5). These are summarized in **Table 4.9**.

In Table 4.9, specific examples were chosen to illustrate a particular effect. All variables but the one being discussed were kept constant. But what happens when two acids are different in multiple ways? For one acid, maybe *resonance* is possible on an *electronegative* atom, but for the other, there is *no resonance* on a *large* atom. How is acidity predicted in that situation?

With no definitive rules, the effects discussed can be prioritized as follows to identify stable conjugate bases and more acidic acids:

1. **Atomic size:** Because this involves the atom bearing the negative charge directly, spreading it out results in a more stable conjugate base.
2. **Resonance:** Similar to atomic size, delocalization of charge by resonance means there is less charge per unit volume, especially when the charge is spread among two (or more) electronegative atoms.
3. **Electronegativity:** When the atom bearing the charge is electronegative, it becomes very stable, especially compared to . . .

4. **Induction:** Because induction is "electronegativity through space" and diminishes with distance, it is a weaker effect than electronegativity.
5. **Hybridization:** Though hybridization has a dramatic effect, it is most commonly observed in the ethane/ethene/ethyne analysis discussed previously. Thus, it need not be considered as often.

Table 4.9 Summary of the structural effects on acidity and conjugate base stability.

Structural effect	Representative acids	Conjugate bases	Description
Electronegativity (same row)	H_2S pK_a = 7; HCl pK_a = −7	HS^- Cl^- most stable	Chloride is more electronegative. The lone pair is held more tightly.
Size (same group)	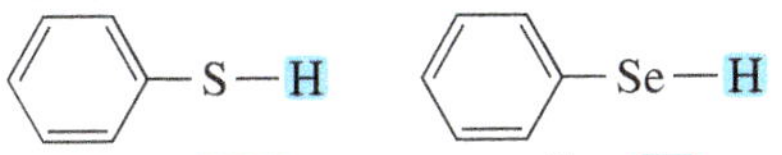 pK_a = 10.3; pK_a = 7.1	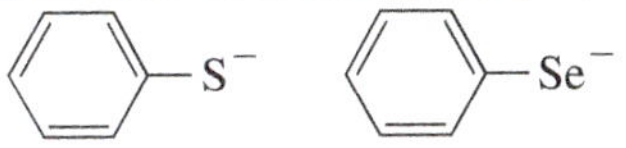 most stable	Selenium is larger. The negative charge is spread over a greater volume.
Resonance	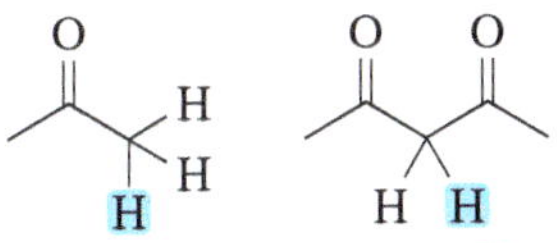 pK_a = 20; pK_a = 9	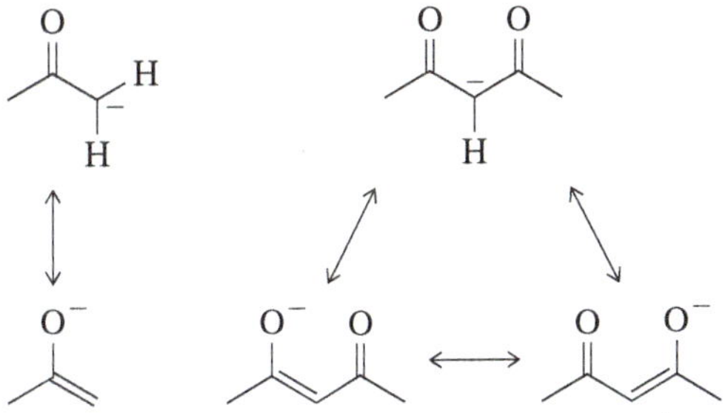 more stable	More resonance structures better delocalize the electrons/negative charge.
Induction	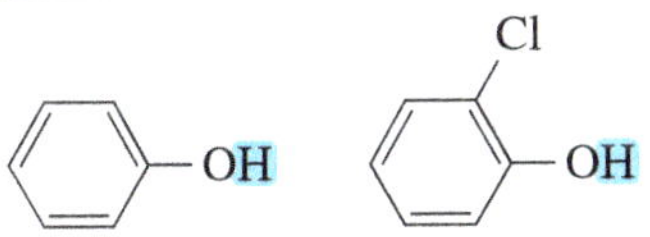 pK_a = 10; pK_a = 8.6	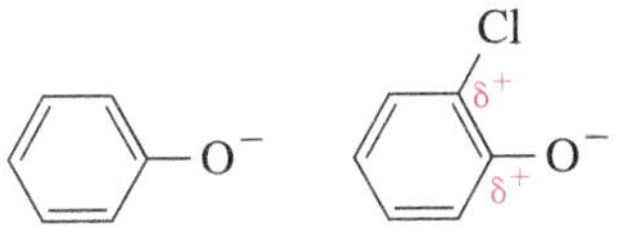 more stable	Electronegative atoms withdraw electron density, stabilizing nearby anions.
Hybridization	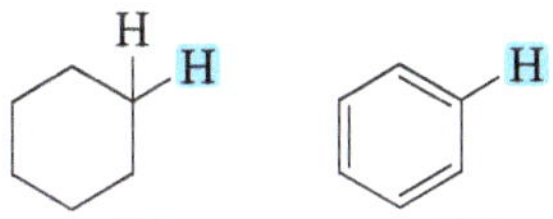 pK_a = 50; pK_a = 43	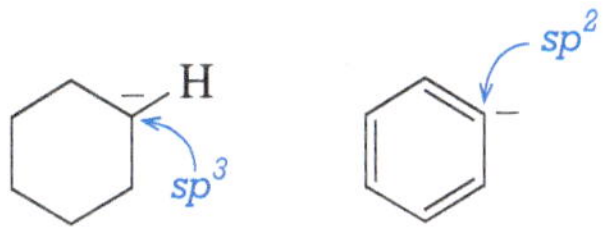 more stable	Orbitals with more *s* character hold electrons closer to the nucleus.

In Assessments 4.51 and 4.52, analyze the situation, make your best guess, check your answer, and, if incorrect, refine your analysis. [*Prepare to be wrong in the beginning. It's no big deal as long as each incorrect answer causes you to think more deeply and refine your analysis. That's how you grow.*]

Assessment 4.51 Identify the most stable conjugate base in each pair. Tell which structural features you analyzed and why you weighted them as you did in picking one answer.

(a)

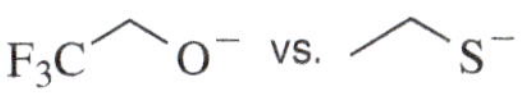

(b)

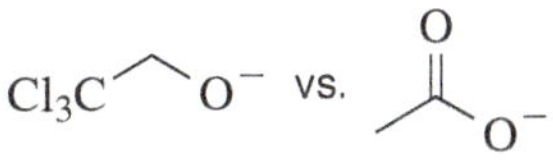

(c)

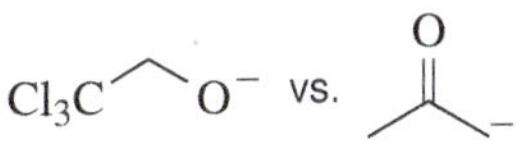

(d)

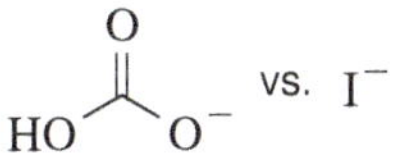

(e)

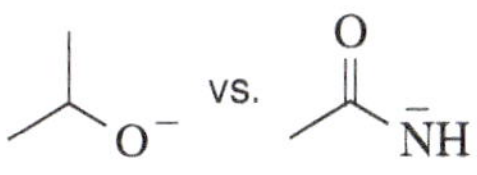

(f)

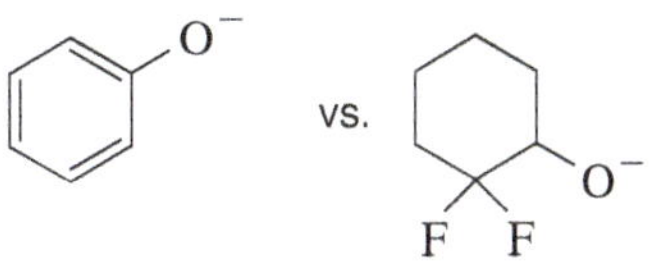

(g)

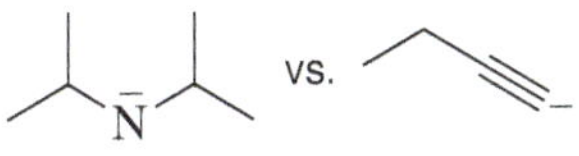

(h)

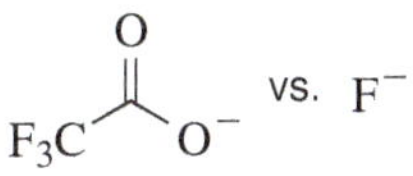

Assessment 4.52 Identify the most acidic proton in each pair. Tell which structural features you analyzed and why you weighted them as you did in picking one answer. [*Always start by drawing the conjugate base.*]

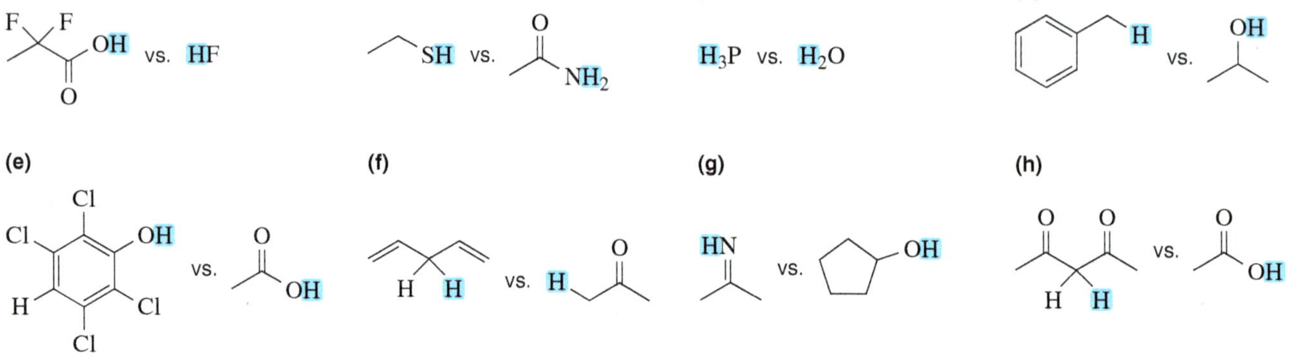

4-D CHEMISTRY IN CONTEXT: Medicinal Chemistry

Electron-withdrawing groups reduce basicity

The acidity and basicity of a molecule can be manipulated to decrease a drug's water solubility, rendering it more likely to reach its target. One example of this effect occurred with neuronal nitric oxide synthase inhibitors—molecules that need to cross the nonpolar blood–brain barrier. The original compound was protonated at two basic amine sites at physiological pH (**Figure 4.42**). Because this dication was so soluble in water, it was excreted by the kidneys before reaching its target.

To address this issue, a modification was sought that would reduce the basicity of one of the amino groups (**Figure 4.43**). By introducing two electron-withdrawing fluorine atoms, a partial positive charge would be created near the amino group, decreasing its ability to share its lone pair. By preventing protonation at physiological pH in this manner, the compound would exist mostly as a monocationic species. The less polar monocation was able to remain in the body longer, giving it a better chance of passing through the nonpolar blood–brain barrier.

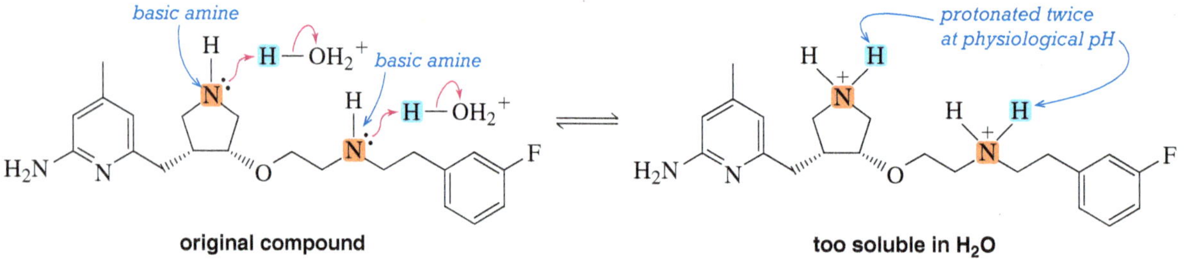

Figure 4.42 **Both basic amines are protonated at physiological pH.**

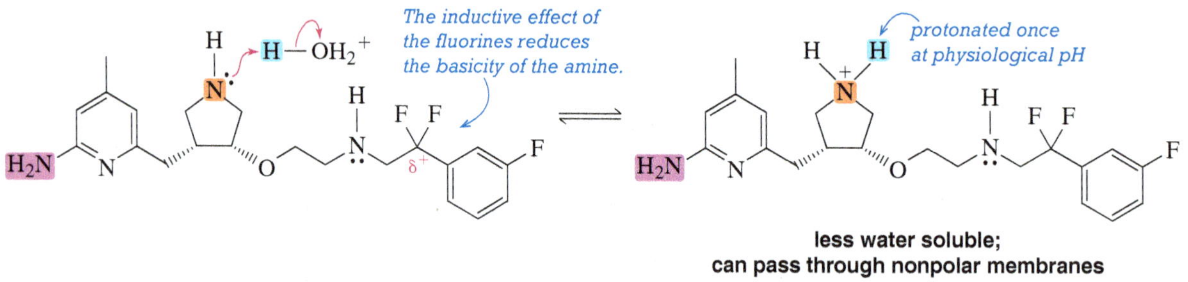

Figure 4.43 **Electron-withdrawing groups** stabilize the basic nitrogen.

Assessment 4.53 The nitrogen screened in purple in Figure 4.43 is not protonated at physiological pH. Why?

4.5 Lewis acid–Lewis base definition

Whereas the Brønsted–Lowry definition is proton centered, the more general Lewis definition focuses on the flow of electrons in an acid–base reaction. [*The Lewis definition explains about 90% of the chemistry that is covered in a standard organic course.*] A Lewis base is an electron-pair donor, whereas a Lewis acid is an electron-pair acceptor.

The flow of electrons in an acid–base reaction is shown in **Figure 4.44** for the reaction of a hydroxide ion with a proton. Hydroxide donates an electron pair (Lewis base), while H^+ accepts the electron pair (Lewis acid).

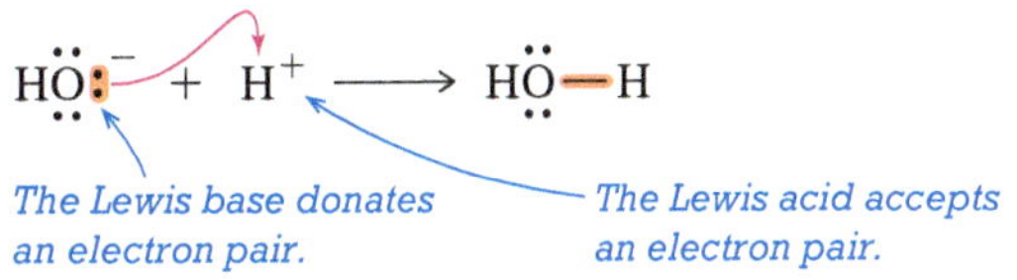

Figure 4.44 An application of the Lewis acid–Lewis base definition.

Thus far, every Brønsted–Lowry acid or base we have studied has also been a Lewis acid or base. Is there no need, then, for the Lewis definition? Organic chemistry is about more than just bases grabbing protons [*just one example of negatives attacking positives*]. Indeed, it is about bases attacking *any* electron-deficient species.

Ammonia (NH_3) reacting with borane (BH_3) is a Lewis acid–base reaction that *is not* also a Brønsted–Lowry acid–base reaction. In this reaction (**Figure 4.45**), the electron-rich species, ammonia, donates a pair of electrons to the empty *p* orbital of the electron-deficient borane. Ammonia is an electron-pair donor (Lewis base), while borane is an electron-pair acceptor (Lewis acid).

To avoid confusion, chemists use a different set of terms to describe the reactions between Lewis acids and Lewis bases. Lewis acids (electron-pair acceptors) "love electrons" and are referred to as **electrophiles**. Lewis bases, being electron-rich species that "love positively charged nuclei," are referred to as **nucleophiles**. **Figure 4.46** shows how this terminology is applied to a Lewis acid–base reaction.

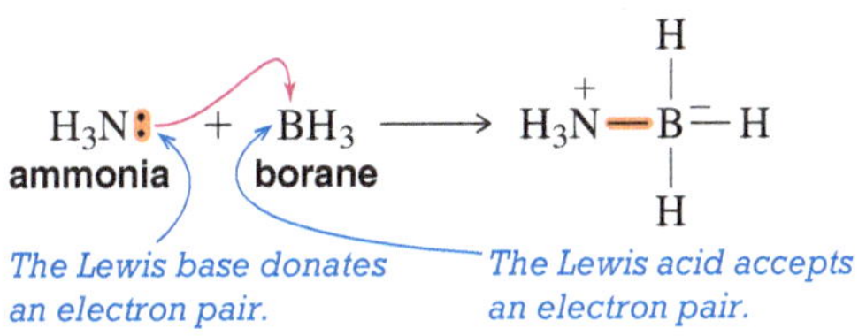

Figure 4.45 A Lewis acid–base reaction that is *not* a Brønsted–Lowry acid–base reaction.

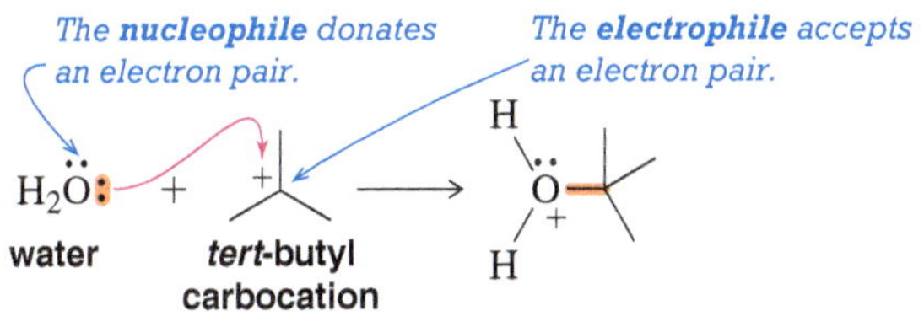

Figure 4.46 Nucleophiles (Lewis bases) react with electrophiles (Lewis acids).

Assessment 4.54 Identify the Lewis acid and the Lewis base in each of the following reactions.

(a)

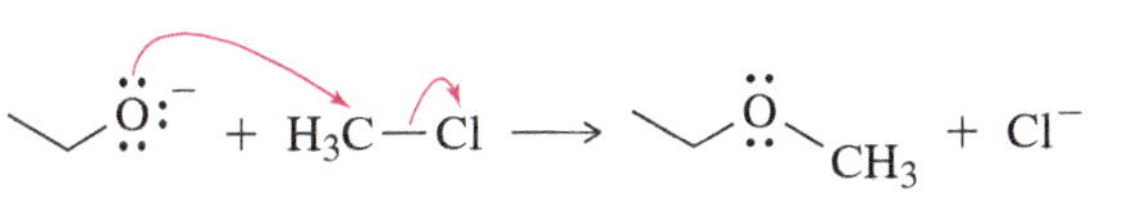

(b)

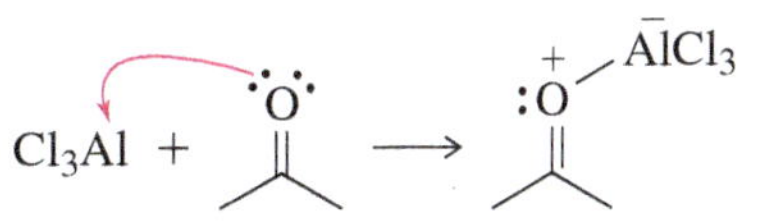

(c)

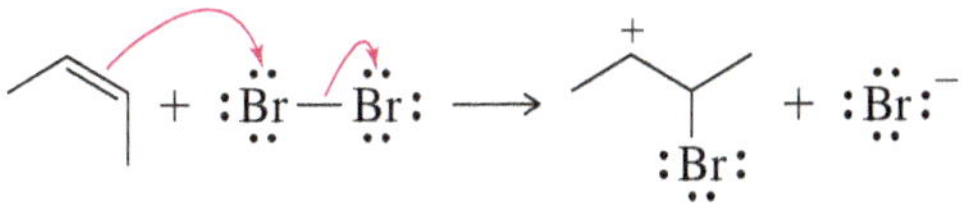

Assessment 4.55 Identify the nucleophile and the electrophile in each of the following reactions.

(a)

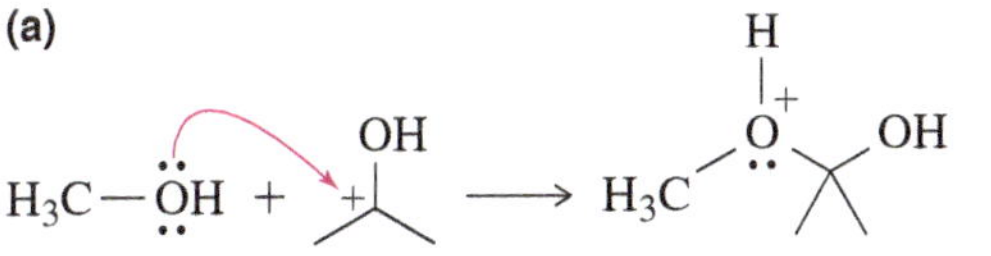

(b)

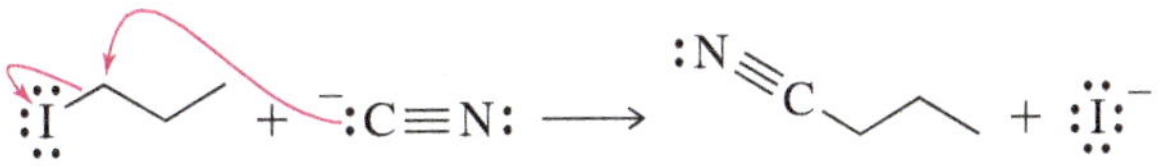

(c)

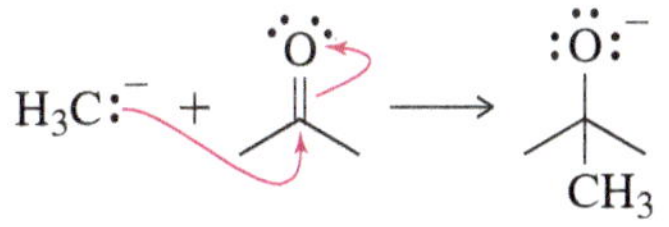

4.5.1 Arrow-pushing formalism in Lewis acid–base reactions

Correct arrow pushing can help you identify Lewis acids and Lewis bases. In each reaction in Section 4.5, the arrow started at the electron-pair donor (the nucleophile) and finished at the electron-pair acceptor (the electrophile). That is, it started at the Lewis base and finished at the Lewis acid. By starting at the electron-rich atom and finishing at the electron-poor atom, a negative attacked a positive [*rule 1 of organic chemistry*].

Now let's see how tracking electrons helps us understand chemical reactions. If a reaction is a story, tracking the formation and breaking of these bonds over the course of a reaction is like a plot summary [*SparkNotes*©]. Arrow pushing tracks the electron flow, showing how the plot [*reaction*] moves from beginning to end. So, to understand reactions, let's start thinking about how electrons flow. [*The next 24 chapters will tell us why they move the way they do.*]

There are three skills you need to master:

1. Given the reactants and products, propose an arrow-pushing mechanism for an individual step (Section 4.5.1.1)
2. Given the reactants, with the arrows pushed for you, predict the products that would have formed (Section 4.5.1.2)
3. Given a number of examples, identify molecules that participate in Lewis acid–base reactions (Section 4.5.1.3)

[*As you analyze the reactions in these sections, take note of what type of molecules are involved as nucleophiles or electrophiles. What is necessary to be electron-rich or electron-poor?*]

4.5.1.1 Arrow pushing: reactants to products

To draw a mechanism for a reaction, start by identifying the bonds that formed and broke. Next, determine how the electrons have been redistributed. The sum of those two processes is the mechanism. [*Try the example at the top of Figure 4.47 on your own before reading further.*]

When drawing molecules in Chapter 3, we saw that it was unnecessary to draw all of the lone pairs, given the formal charge. When drawing arrow-pushing mechanisms, however, *always* draw in *all* lone pairs. [*ALWAYS!*] **Figure 4.47** shows this process. In going from reactants to products, we break the Cl_a—Cl_b σ bond and form the O—Cl_a σ bond. Oxygen starts with three lone pairs, but finishes with two. Cl_a starts and finishes with three lone pairs, but Cl_b starts with three lone pairs and finishes with four.

Figure 4.47 Electron accounting: a Lewis acid–base reaction.

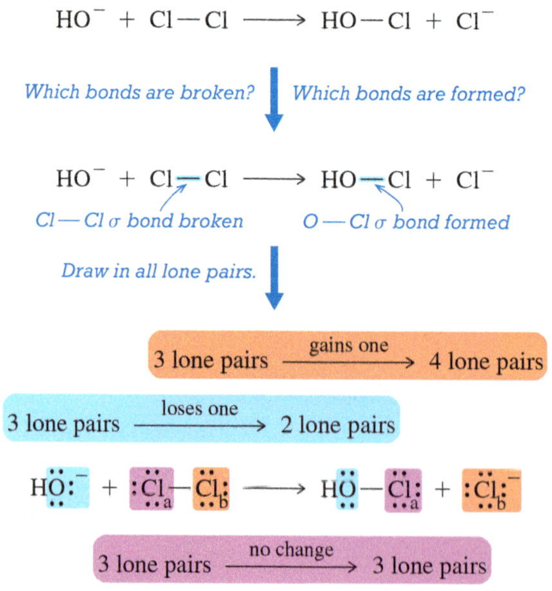

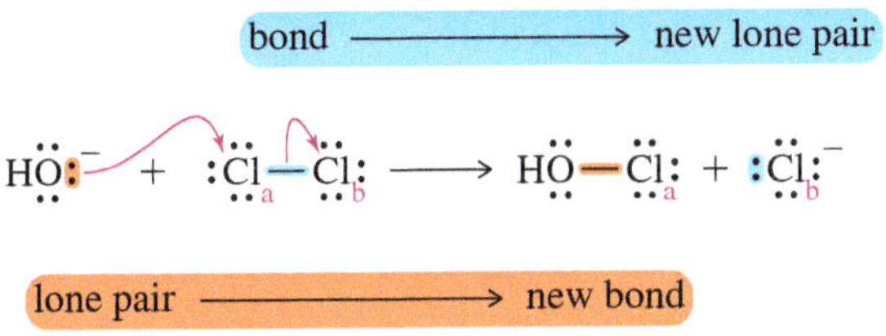

Figure 4.48 **Arrow pushing**: in a Lewis acid–base reaction, based on electron accounting.

The mechanism for this reaction is shown in **Figure 4.48**. Oxygen donates one of its lone pairs to Cl_a, forming the O—Cl_a σ bond. The Cl_a—Cl_b σ bond breaks, giving Cl_b its fourth lone pair and a negative charge. Our arrow pushing illustrates this mechanism. This mechanism shows that what was a lone pair on oxygen is now the C—O σ bond. Similarly, the two electrons of the Cl—Cl σ bond are now a lone pair on Cl_b.

Table 4.10 shows several examples of this analysis, each leading to a logical arrow-pushing mechanism. Look carefully at the reaction on the left; try to figure out the mechanism before looking at the arrow-pushing formalism on the right. [*Remember, this is about the logic needed to draw an arrow-pushing mechanism. You don't need to worry yet about why the reaction happens this way.*]

Table 4.10 **Examples of electron accounting** that lead to arrow-pushing mechanisms of Lewis acid–base reactions.

Reasoning	Result
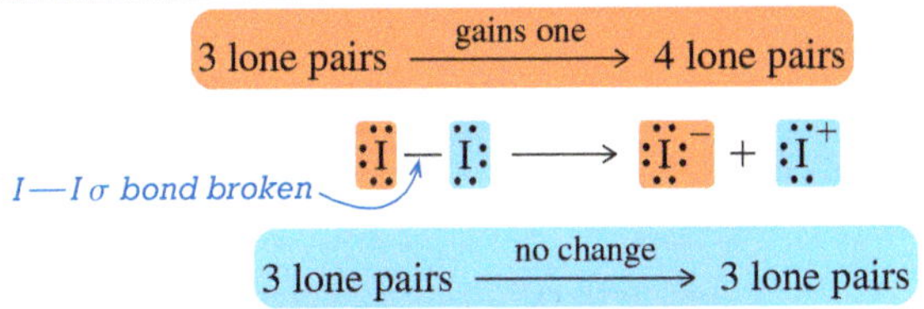	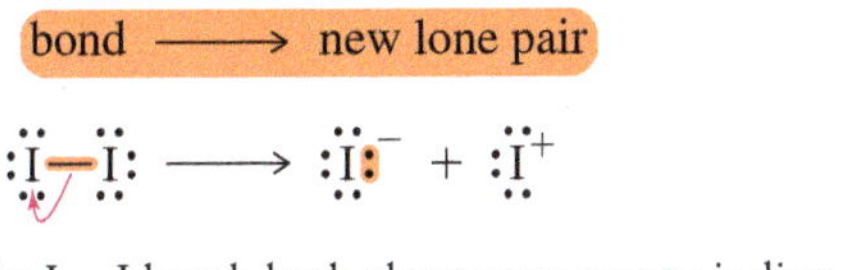 Arrow starts at the I—I bond; both electrons move to iodine.
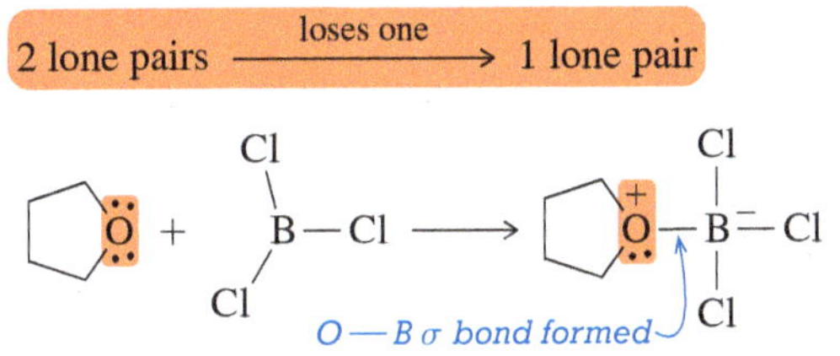	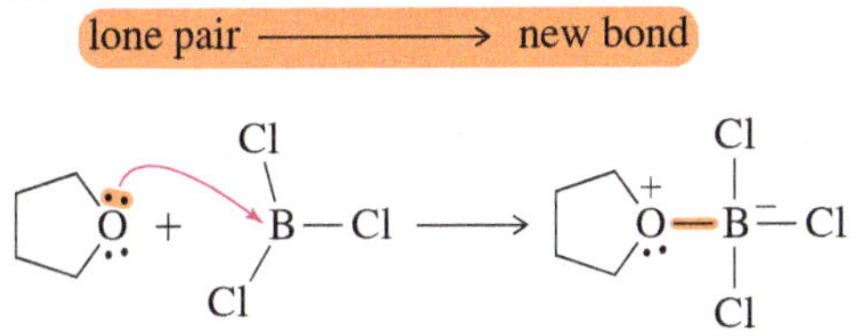 Arrow starts at the oxygen lone electron pair and attacks the empty orbital of boron.
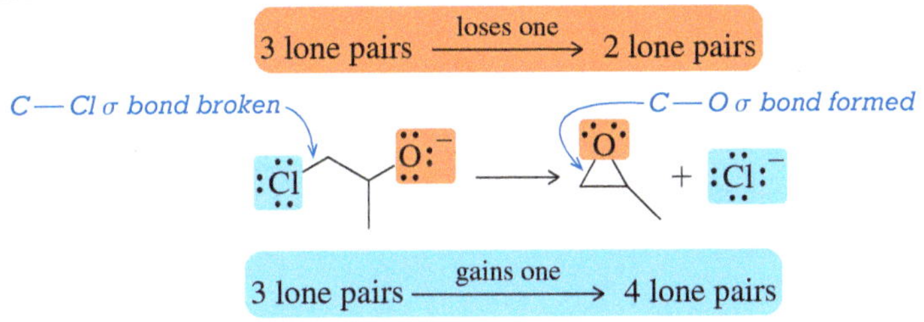	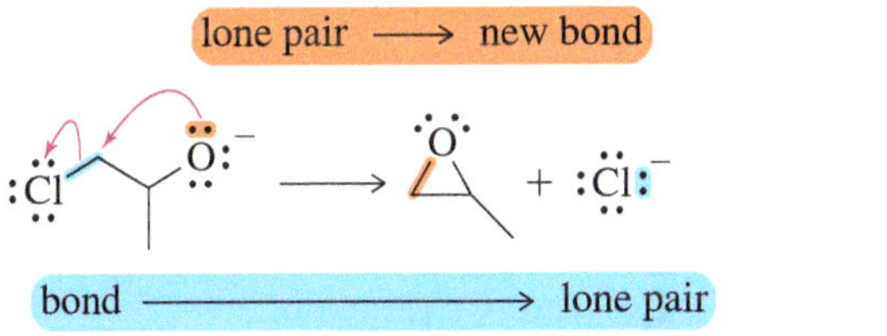 Arrow starts at the oxygen lone electron pair and attacks carbon, kicking out chloride with a lone electron pair.
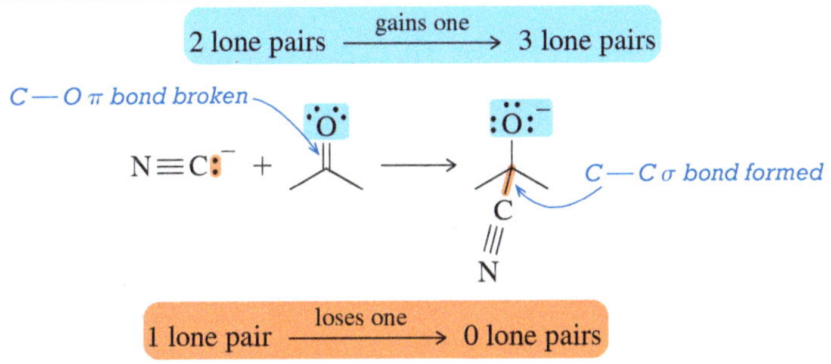	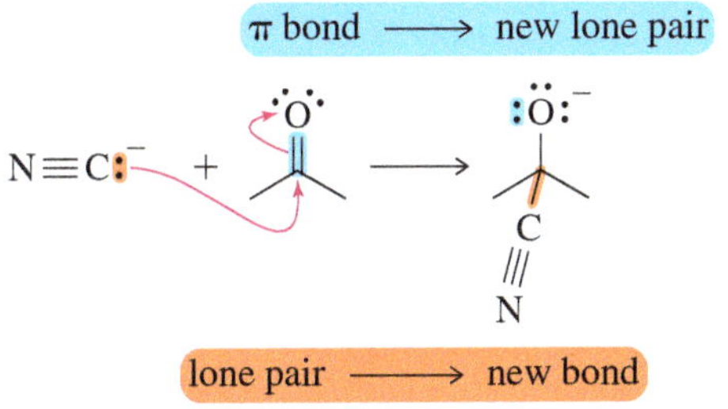 Arrow starts at the carbon lone electron pair and attacks carbon, pushing π electrons onto the oxygen.
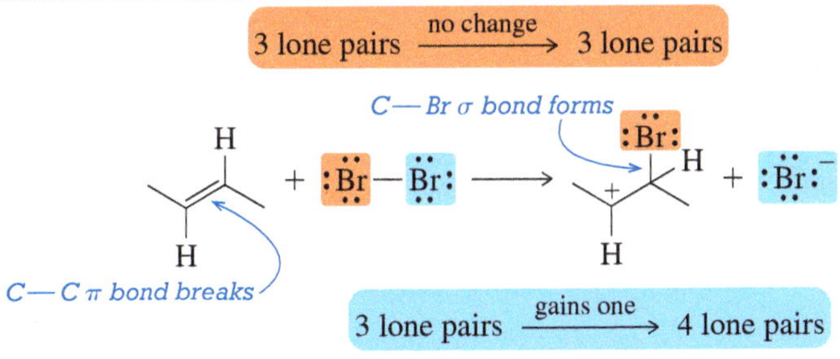	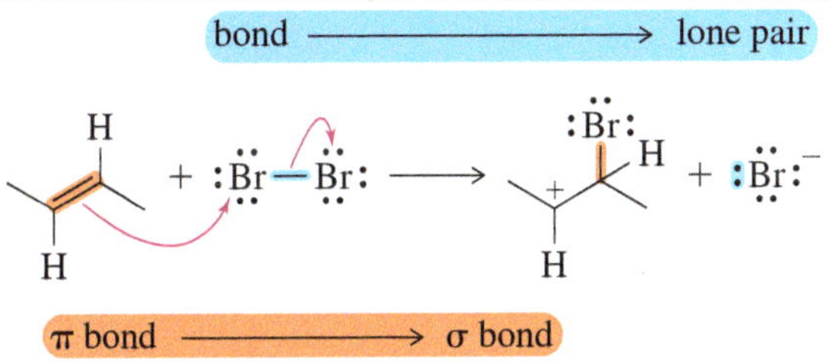 Arrow starts at the electron-rich π bond and attacks the bromine, kicking out bromide with a pair of electrons.

Understanding this process takes practice and is best learned by working assessments. Assessments 4.56–4.58 walk you through a variety of exercises, from simple to complex. [*Use the same analysis shown in Table 4.10 until it becomes second nature.*]

Assessment 4.56 Show an arrow-pushing mechanism that forms the product on the right from the reactant at left. Only one arrow is necessary in each reaction. [*Don't forget to draw in the lone pairs on this and the next two assessments.*]

(a)

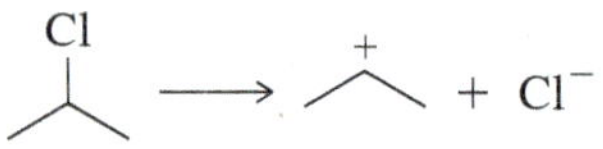

(b)

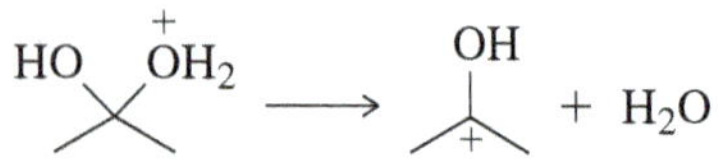

(c)

(d)

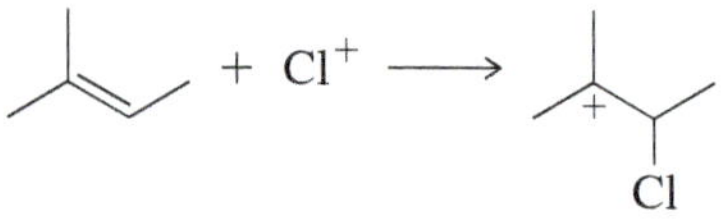

Assessment 4.57 Show an arrow-pushing mechanism that forms the product on the right from the reactant at left. Two arrows are necessary in each reaction.

(a)

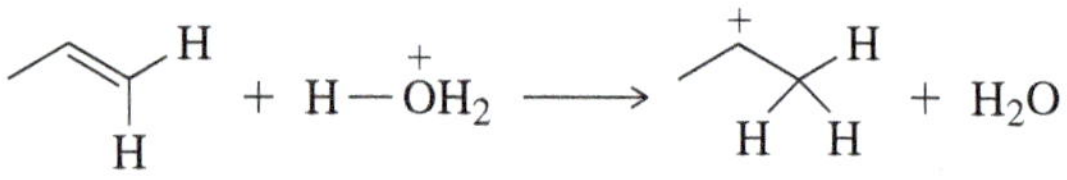

(b)

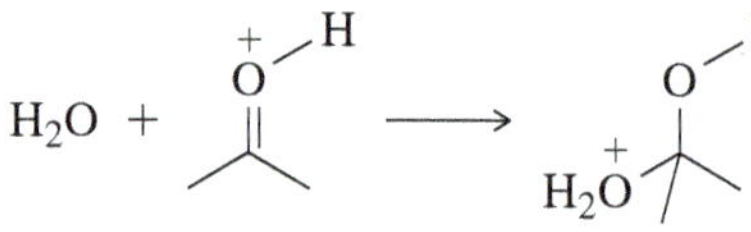

(c)

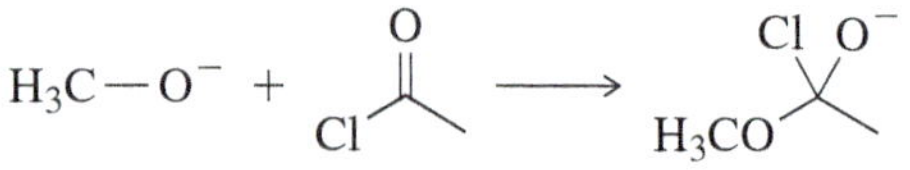

(d)

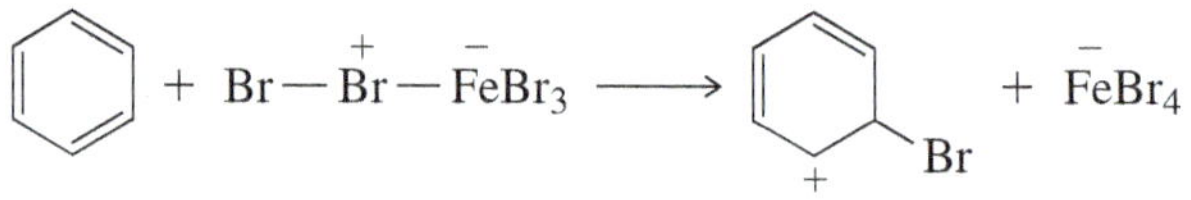

Assessment 4.58 Show an arrow-pushing mechanism that forms the product on the right from the reactant at left. Here, three arrows are necessary in each reaction.

(a)

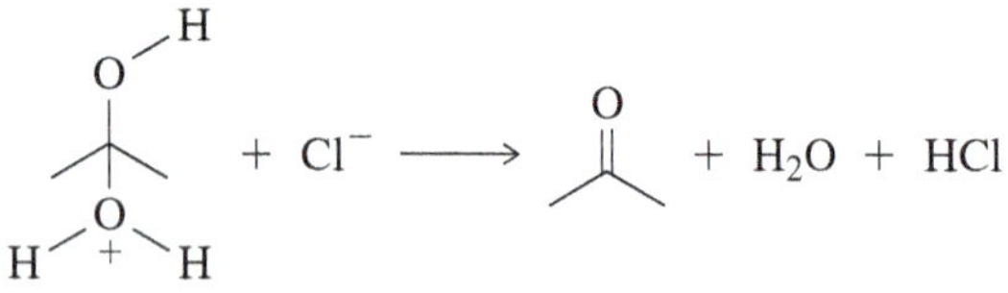

(b)

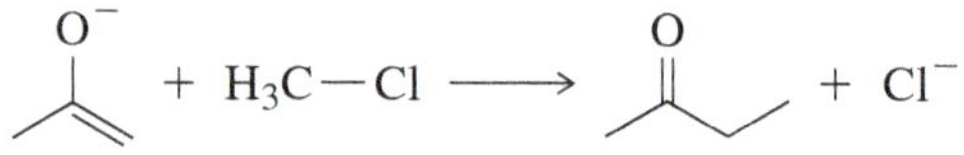

(c)

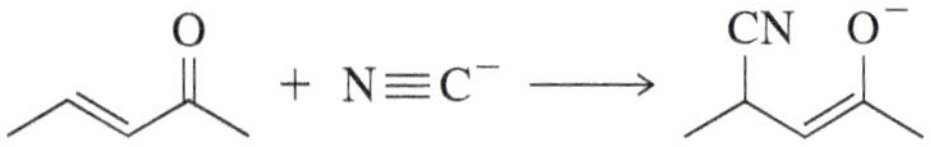

(d)

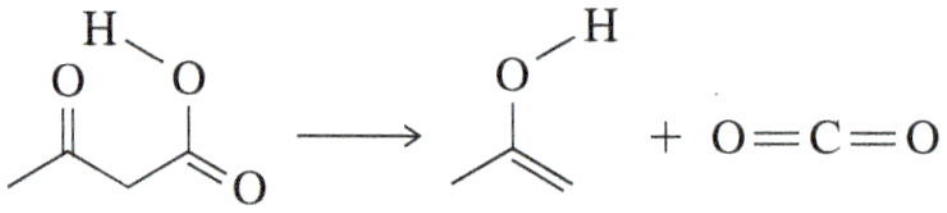

4.5.1.2 Arrow pushing in Lewis acid–Lewis base reactions: predicting products

Organic chemistry is not just about filling in events [*pushing electrons*] based on how a story [*reaction*] ends. We also need the ability to predict how a story ends [*predict the products of a reaction*] based on the way it starts [*reactants*]. This section gives us practice predicting products, given the reactants and pushed electrons. [*If you worked through Assessments 4.56–4.58 with ease, this will be simple for you.*]

Given a set of arrows, the process requires only a few rules. An arrow represents the flow of 2 electrons [*for now*], regardless of whether it starts at a bond or a lone pair. The 2 electrons will remain together, as a new bond or lone pair. There are just a few types of arrows drawn in Lewis acid–Lewis base reactions (**Table 4.11**). In each, the charges are balanced on both sides of a reaction. [*Tracking the balance of charges is a helpful way to avoid mistakes.*]

Table 4.11 Common arrow types seen in organic chemistry. [*Note that type 1 and type 2 arrows are the reverse of each other, as are type 3 and type 5 arrows.*]

Arrow type	Examples
Type 1: Nucleophilic attack by a single atom This arrow is used when a Brønsted–Lowry base, like OH^-, attacks H^+. Similarly, Cl^- (Lewis base) can attack a **carbocation**. Or an oxygen can attack an empty *p* orbital on BH_3. In each case, what was a lone pair of electrons becomes a new σ bond.	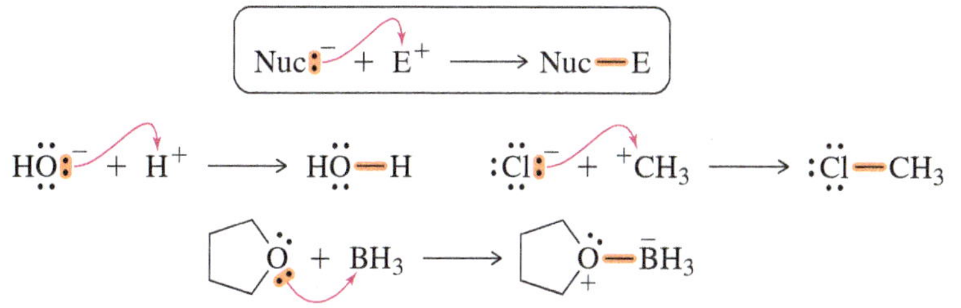
Type 2: Heterolytic σ bond cleavage The opposite of type 1, this arrow is used when an acid dissociates to H^+ and a conjugate base. In the first example given, the electrons in the bond between hydrogen and chlorine move to the chlorine to give H^+ and Cl^-. In the second example, the H—B σ bond becomes a lone pair on hydrogen, leaving behind a neutral boron atom. And finally, a C—O bond can break to give water and a carbocation. In each case, the electrons that were a σ bond become a lone pair.	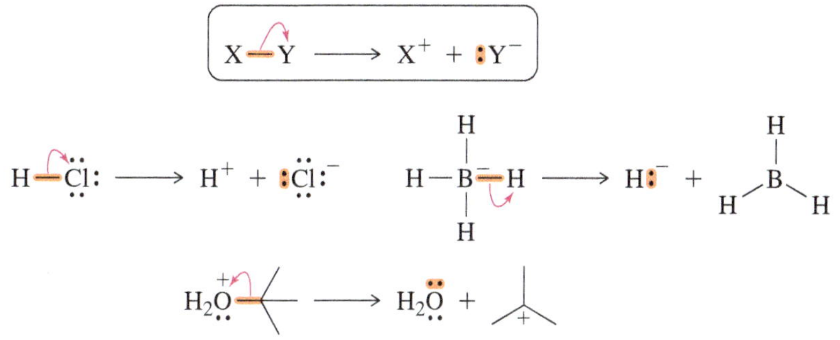
Type 3: Nucleophilic attack by a π bond This is exemplified when an alkene C—C π bond breaks to grab a proton. Similarly (second example), an alkyne π bond attacks a positively charged bromine. Finally, a π bond of benzene attacks a carbocation with this arrow. In each case, the original π bond becomes a new σ bond.	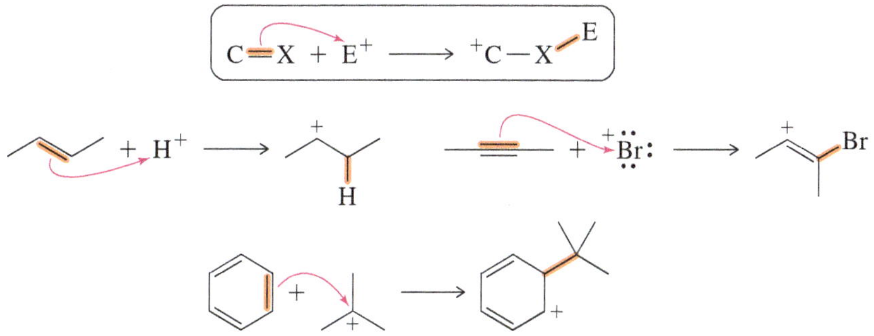
Type 4: π bond dissociation This is an arrow that represents the formation of resonance structures, but is often seen as part of arrow-pushing mechanisms.	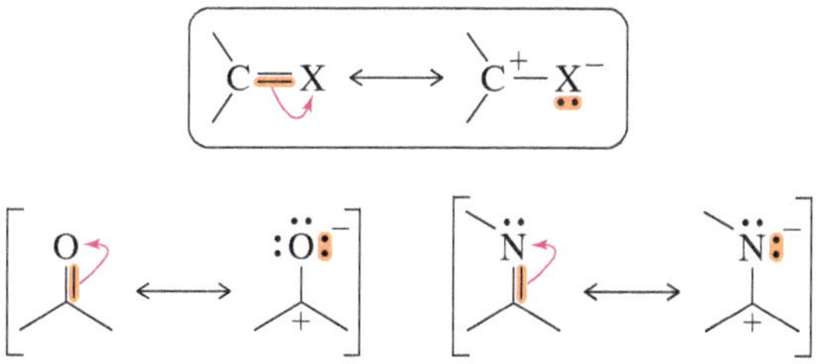
Type 5: π bond formation These arrows, accounting for the formation of π bonds, are essentially the opposite of type 3 arrows. They rarely happen on their own because they are usually accompanied by another reaction [*the gray arrows*].	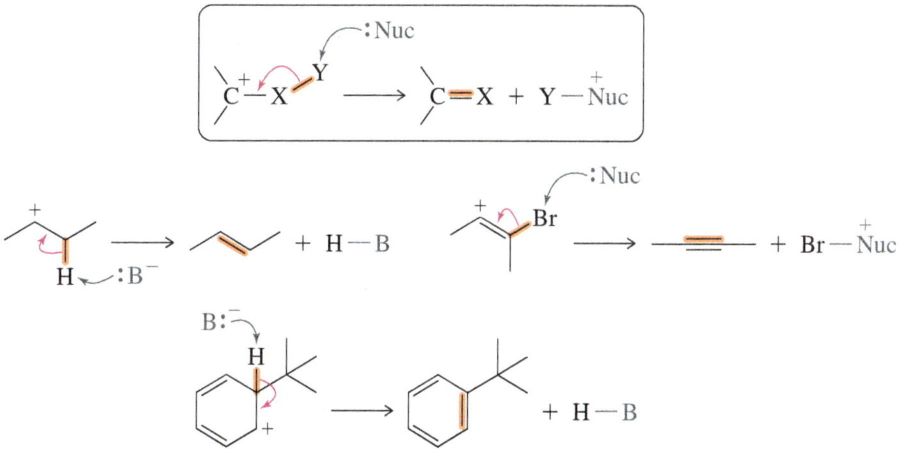

Even if a reaction in Assessments 4.59 and 4.60 shows multiple arrows in a single step, are there any that don't fit one of these types shown in Table 4.11?

Viewed in this way, predicting the products, given a particular arrow, is straightforward because we have an expected outcome based on the arrow type.

Assessment 4.59 Identify the arrow types that are shown in each of these arrow-pushing mechanisms.

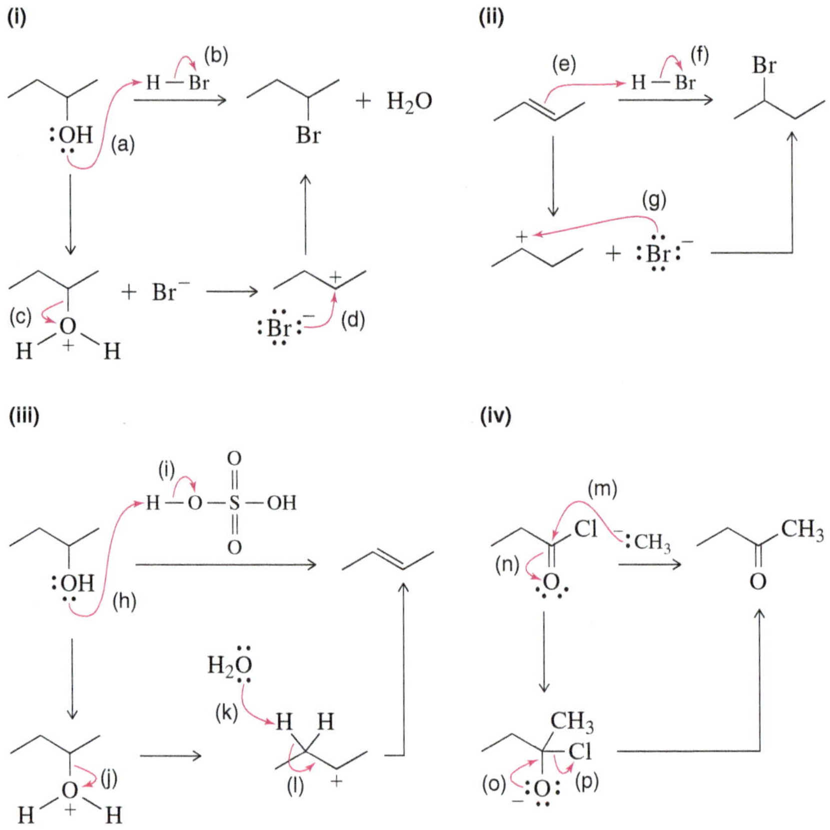

Assessment 4.60 Identify the arrows shown by type and predict the product that should result.

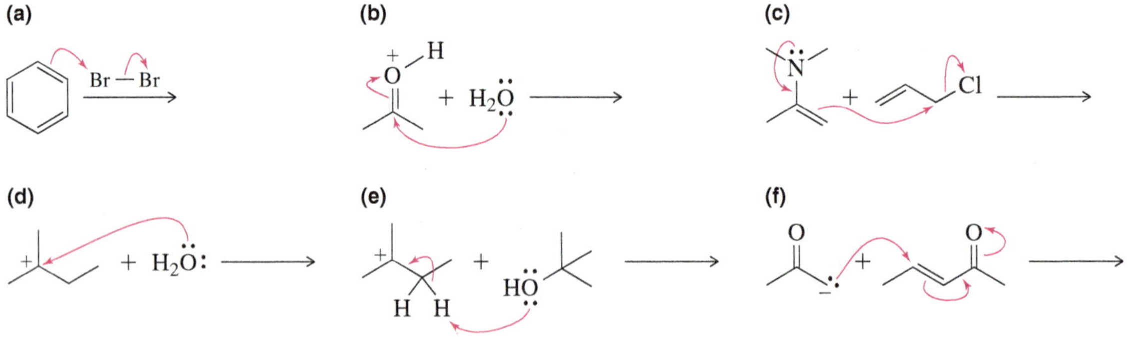

4.5.1.3 Nucleophiles and electrophiles

It is important to be able to look at a molecule and determine whether it is nucleophilic or electrophilic. The key is to look for electron-rich molecules (nucleophiles) and electron-poor molecules (electrophiles). Because nucleophiles and electrophiles come in several forms, a few common ones are listed in **Tables 4.12** and **4.13**.

Table 4.12 Common nucleophile (Lewis base) types seen in organic chemistry.

Nucleophile type	Examples
Negatively charged atoms with a lone pair of electrons: The negative charge makes the molecule want to donate a pair of electrons to an electrophile (E^+).	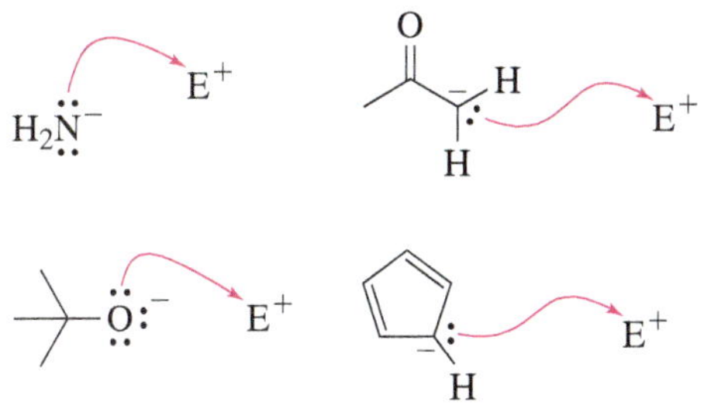
Neutral atoms with a lone pair of electrons: Though not as reactive as negatively charged molecules, these species still donate electrons to reactive electrophiles.	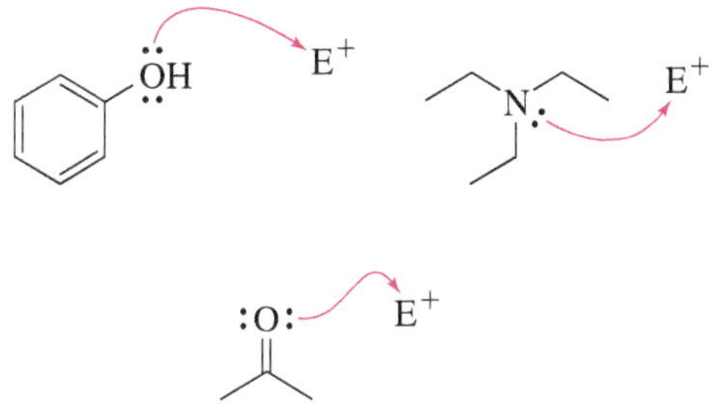
Functional groups with π bonds: Pi bonds are electron-rich and somewhat easily broken. Thus, π electrons are donated to strong electrophiles.	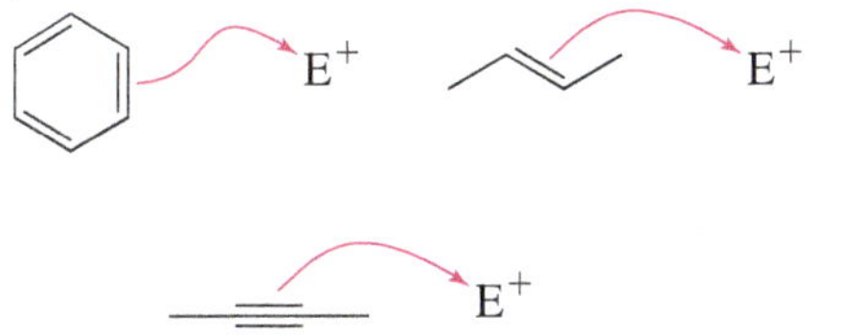

Table 4.13 Common electrophile (Lewis acid) types seen in organic chemistry.

Electrophile type	Examples
Positively charged atoms with empty orbitals: The most reactive of all electrophiles, these need two new electrons from the nucleophile (:Nuc) to complete their octet and become neutral again.	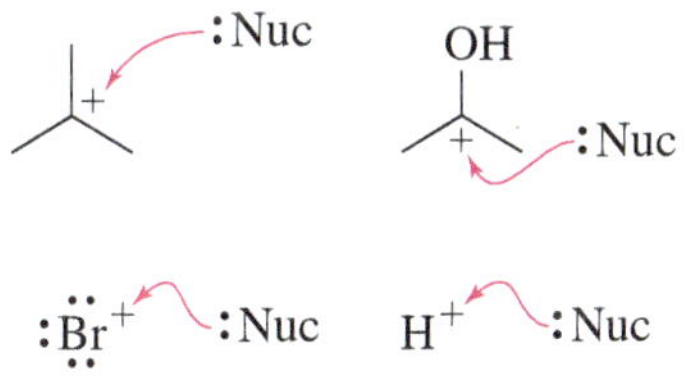
Neutral atoms with empty orbitals: Being neutral makes these atoms a bit less reactive, but new electrons are needed to give them a full octet.	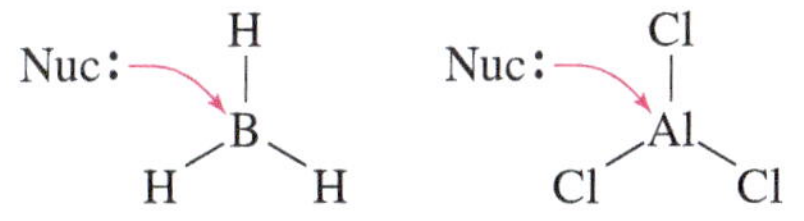
Atoms that can lose σ-bonded electrons to form a stable conjugate base: These neutral electrophiles, usually with a full octet of valence electrons, feature a σ bond whose electrons can be donated to an attached atom (usually large and electronegative) to make a stable conjugate base. In essence, these electrophiles react using type 1 and type 2 arrows at the same time.	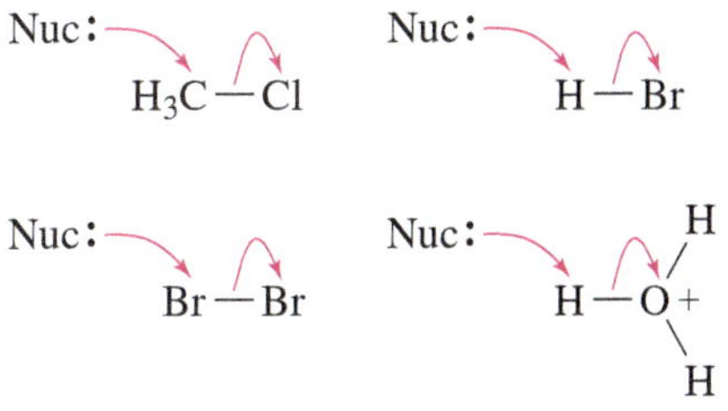
Atoms that can lose π-bonded electrons to an electronegative atom: These are also neutral, with full octets, but upon a nucleophile donating a pair of electrons, they release attached π electrons to become a lone pair on an electronegative atom like oxygen or nitrogen.	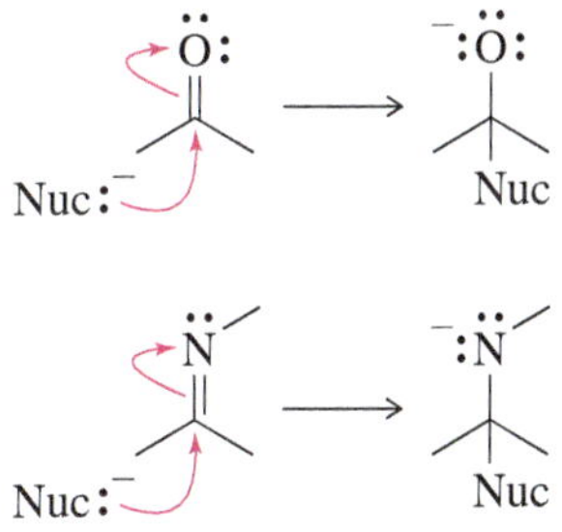

Assessment 4.61 Would you expect the following species to be electrophiles or nucleophiles? Some may be both. Explain your answer.

(a) (b) (c) (d) (e) (f) (g) (h) (i)

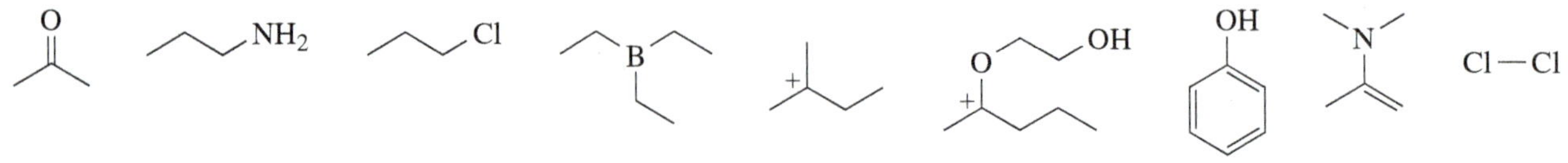

Concluding thoughts

This chapter on the foundational principles of acids and bases emphasizes how small differences in molecular structure can result in large differences in acidity, which in turn can affect drug absorption, water solubility, and the effectiveness of medications used to treat acid reflux disease. The change in the reactivity of drugs as a function of pH is the root cause of many unfavorable drug–drug interactions for patients taking acid reflux medications.

Ideal medicines are ones that can be taken orally. Typically, drugs administered this way must traverse both aqueous environments and nonpolar lipid membranes to reach their targets. Drug design is optimized to accommodate normal stomach acidity; however, patients on acid reflux medications have altered stomach acidity. Absorption of a drug can be different for these patients.

Taking triazolam (for treatment of insomnia) with ranitidine (Zantac®) for heartburn causes a 30% *increase* in absorption of triazolam and a potential 30% increase in potency. The effect goes in the other direction as well, as in the case of erlotinib (Tarceva®), a treatment for non-small cell lung cancer. When erlotinib is taken with omeprazole (Prilosec®), its absorption is *reduced* by 46%. One study suggests that taking erlotinib with cola (acidic) restores absorption to normal levels. Both of these results are attributed to the change in pH altering the protonation state of the drugs, likely at the basic nitrogens in each molecule. Practically speaking, these results demonstrate the importance of physicians knowing what medications you take before prescribing others so that they get the dose exactly as intended by the organic chemists who created them.

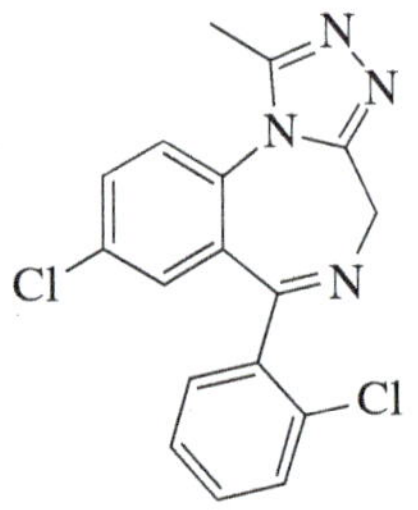

triazolam

increased absorption at lower pH

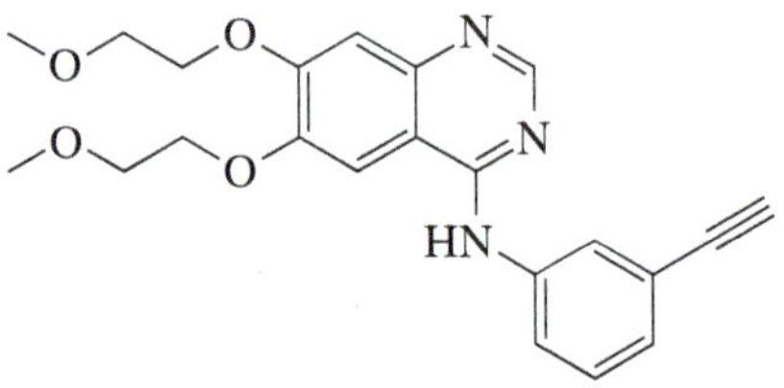

erlotinib

decreased absorption at lower pH

End-of-Chapter Materials

Important Concepts

4.1: The three main definitions of acids and bases are named Arrhenius, Brønsted–Lowry, and Lewis. Brønsted–Lowry acids are proton donors, whereas Brønsted–Lowry bases are proton acceptors. Lewis acids are electron-pair acceptors, whereas Lewis bases are electron-pair donors. All Brønsted–Lowry acids are Lewis acids, but all Lewis acids are not Brønsted–Lowry acids.

4.2: When an acid loses its proton, its conjugate base is formed. When a base grabs a proton, its conjugate acid is formed. Identifying acid–conjugate base and base–conjugate acid pairs is helpful in predicting the outcome of acid–base reactions. Strong, reactive, and unstable acids are deprotonated to form weak, unreactive, and stable conjugate bases. Weak, unreactive, and stable acids are deprotonated to form strong, reactive, and unstable bases.

4.2.1–4.2.2: Following the general rule of negatives attack positives, the mechanism of acid–base reactions can be shown using the arrow-pushing formalism. The arrow starts at the electron-rich base and points toward the acidic proton. Electron accounting makes it possible to predict acid–base reaction outcomes and mechanisms.

4.2.3–4.2.4: The equilibrium constant (K_{eq}) shows which side of the acid–base reaction is favored. A reaction coordinate diagram can be used to illustrate the change in potential energy that occurs as bonds form and break in an acid–base reaction. Because the ball rolls downhill, the side of the acid–base reaction that is lowest in energy or most stable will be favored.

4.3: By standardizing the strength of all acids against the hydronium ion, acid dissociation constants (K_a), and the related pK_a values, provide a relative scale of acidity. pK_a values provide not only a means to assess acidity but can also give information about the stability of conjugate bases. Acids with a high pK_a are weak, and their conjugate bases are strong. Strong acids have a low pK_a, and their conjugate bases are stable. pK_a values can be used to calculate the equilibrium constant for acid–base reactions using Equations 4.1(a) and 4.1(b).

4.4: The relative acidity of two hydrogens can be predicted based on the manner in which the structure stabilizes or destabilizes the conjugate base. Within a row in the periodic table, the more electronegative atom has a more stable conjugate base. Within a group in the periodic table, larger atoms are better able to delocalize charge, stabilizing the conjugate base. A conjugate base made up of multiple resonance structures is more stable than a conjugate base without resonance. The inductive effect of neighboring electron-withdrawing groups stabilizes conjugate bases. Finally, conjugate base stability increases as hybridization changes from sp^3 to sp^2 to sp because the amount of s character increases, so electrons in an sp-hybridized orbital are held more closely to the positively charged nucleus. Each of these structural effects stabilizes the conjugate base, resulting in a more acidic acid.

4.5: A Lewis base is an electron-pair donor, whereas a Lewis acid is an electron-pair acceptor. Lewis bases, because they love nuclei, are commonly referred to as nucleophiles. Lewis acids, because they love electrons, are referred to as electrophiles. Lewis acid–Lewis base reactions account for as much as 90% of the reactions in this book. The mechanism of Lewis acid–Lewis base reactions can be shown using the arrow-pushing formalism with the arrow starting at the Lewis base (the nucleophile) and pointing toward the Lewis acid (the electrophile). In conjunction with electron accounting and identifying common types of nucleophiles and electrophiles, the arrow-pushing formalism also makes it possible to predict the products of Lewis acid–Lewis base reactions.

Important Reactions and Mechanisms

[*In each of these, the charge is balanced on both sides of the equation.*]

1. **Anionic base with cationic acid**
   A lone pair on a negatively charged base grabs a proton from a positively charged acid, resulting in two neutral molecules.

General:

$$B\!:^{-} + H\!-\!A^{+} \rightleftharpoons B\!-\!H + :\!A$$

Specific:

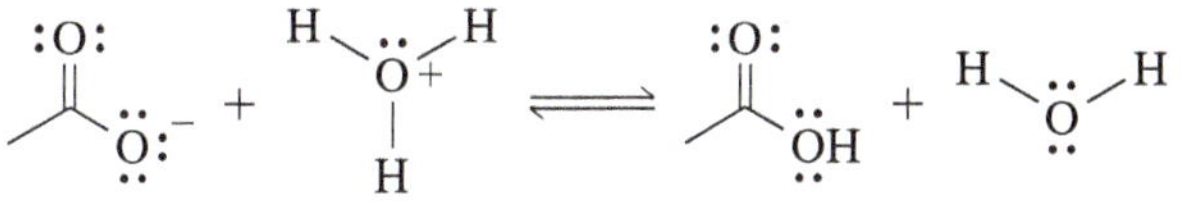

Mechanism (forward):

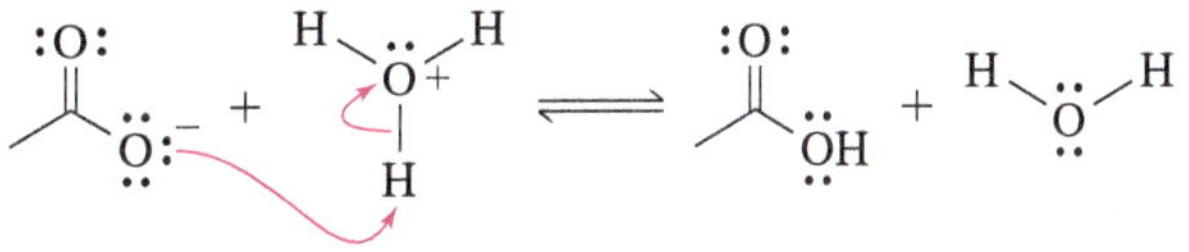

2. Anionic base with neutral acid A lone pair on a negatively charged base grabs a proton from a neutral acid, resulting in a neutral conjugate acid and a negatively charged conjugate base.

General:

$$B{:}^- + H{-}A \rightleftharpoons B{-}H + {:}A^-$$

Specific:

Mechanism (forward):

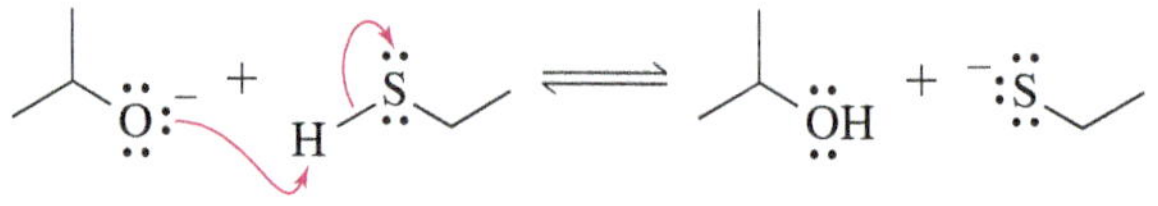

---

3. Neutral base with neutral acid A lone pair on a neutral base grabs a proton from a neutral acid, resulting in a positively charged conjugate acid and a negatively charged conjugate base.

General:

$$B{:} + H{-}A \rightleftharpoons \overset{+}{B}{-}H + {:}A^-$$

Specific:

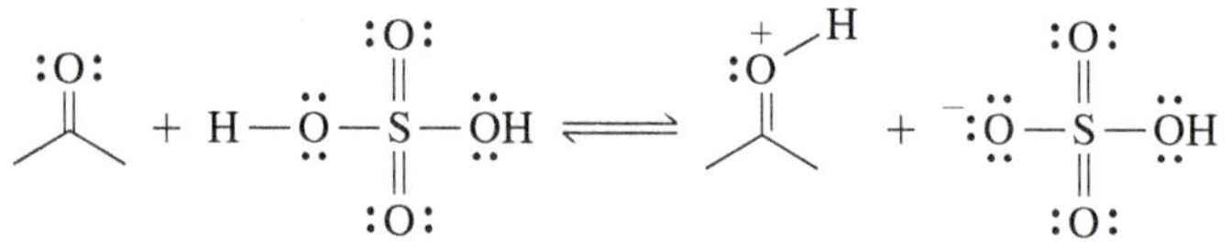

Mechanism (forward):

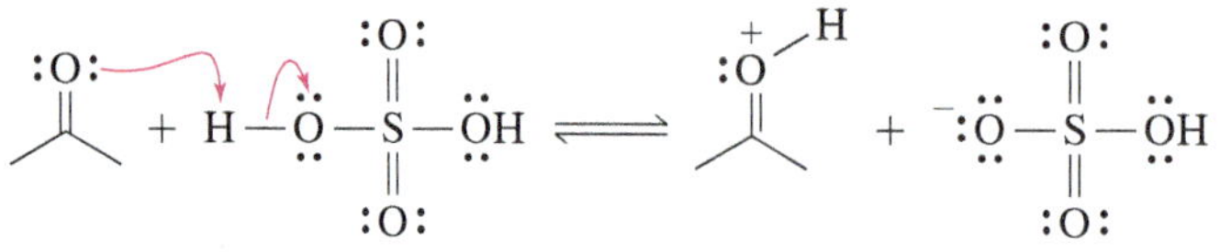

---

4. Neutral base with cationic acid A lone pair on a neutral base grabs a proton from a positively charged acid, resulting in a positively charged conjugate acid and a neutral conjugate base.

General:

$$B{:} + H{-}A^+ \rightleftharpoons \overset{+}{B}{-}H + {:}A$$

Specific:

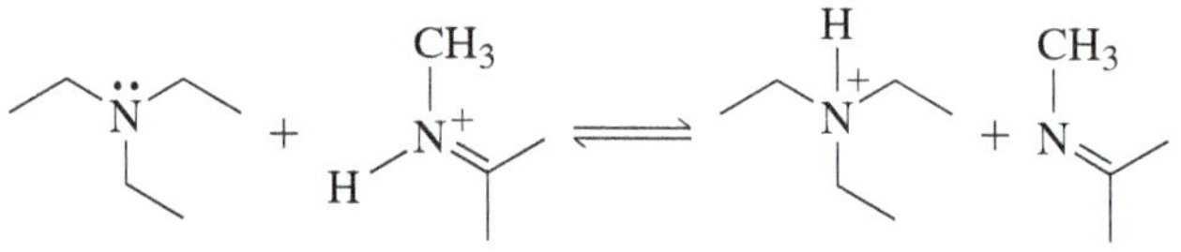

Mechanism (forward):

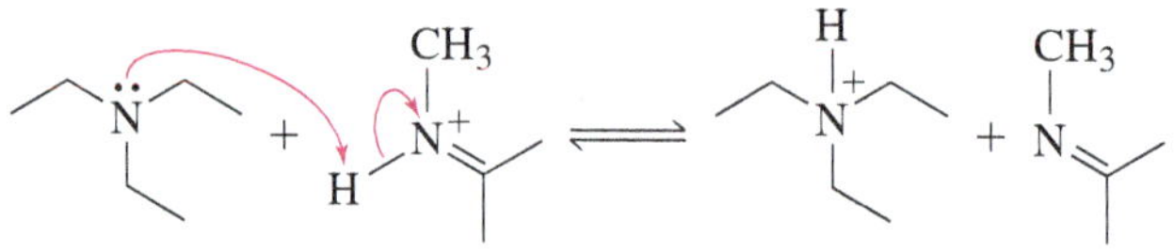

End-of-Chapter Assessments

[*Assessments are ranked according to their difficulty; •••• = most difficult.*]

4.62 (••) For the following acid–base pairs, (i) complete the reaction; (ii) identify the acid (**A**), base (**B**), conjugate acid (**CA**), and conjugate base (**CB**); (iii) predict the favored side of equilibrium; (iv) calculate K_{eq}; (v) show a mechanism for the reaction; and (vi) draw a reaction coordinate diagram.

(a)

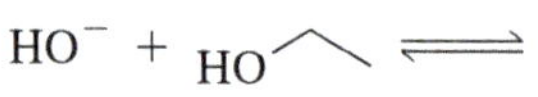

(b)

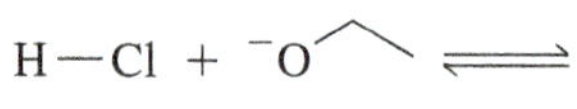

(c)

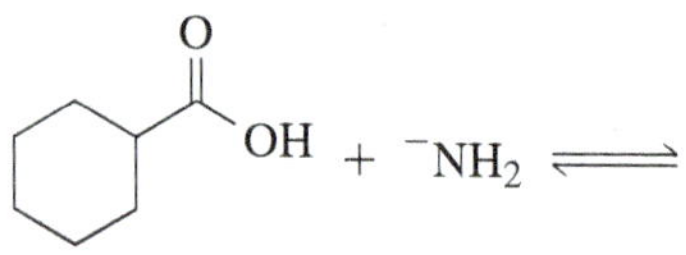

(d)

$H_3N + H{-}C{\equiv}C^- \rightleftharpoons$

(e)

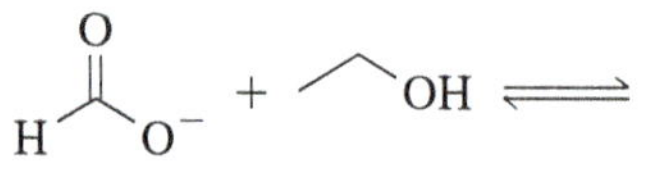

(f)

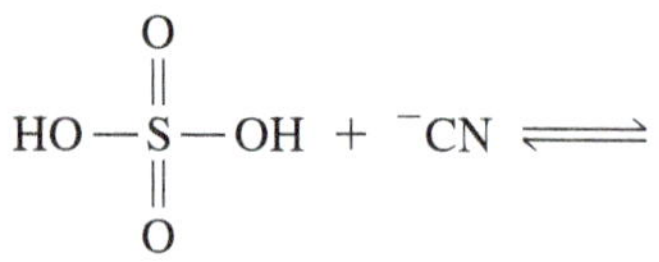

(g)

$H_2O + HCl \rightleftharpoons$

(h)

$NH_3 + HCl \rightleftharpoons$

4.63 (••) Identify the stronger base in each pair. Explain your choice. Citing pK_a values is not an acceptable answer.

(a)

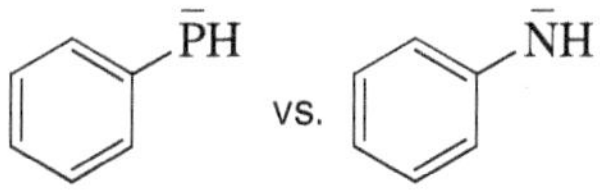

(b)

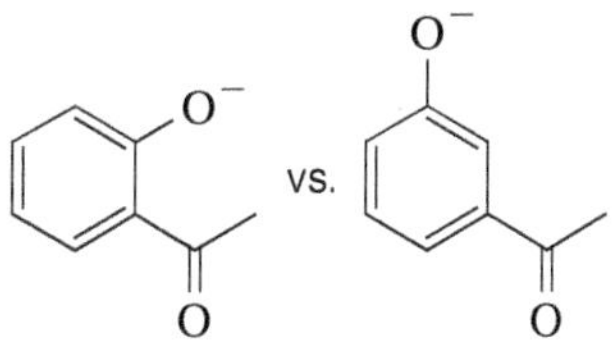

(c)

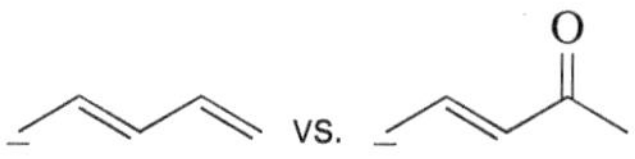

(d)

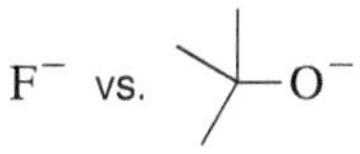

(e)

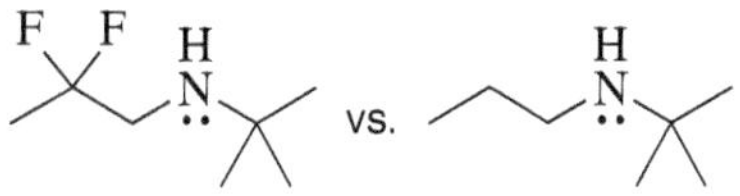

(f)

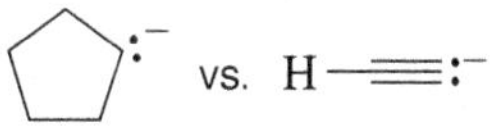

4.64 (••) Identify the stronger acid in each pair. Explain your choice. Citing pK_a values is not an acceptable answer.

(a)

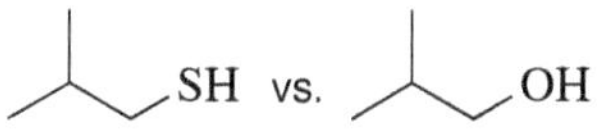

(b)

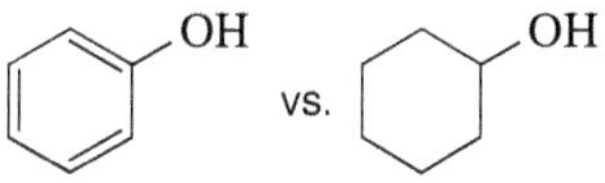

(c)

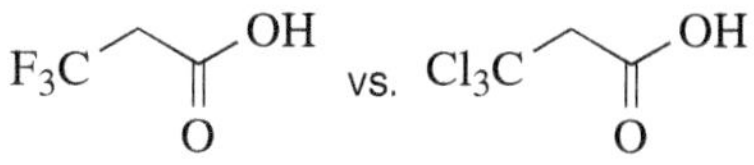

(d)

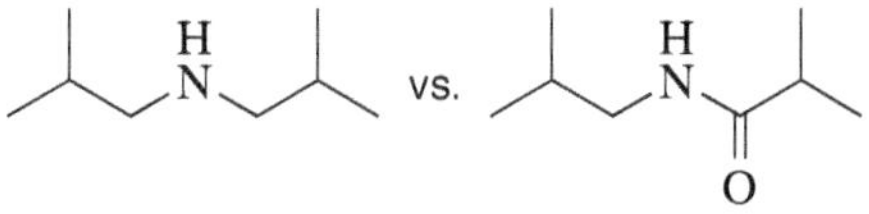

(e)

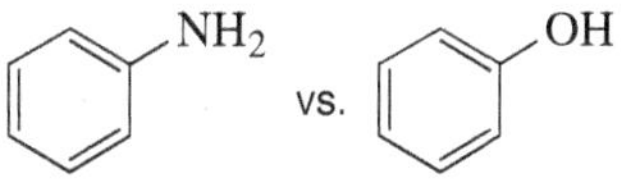

(f)

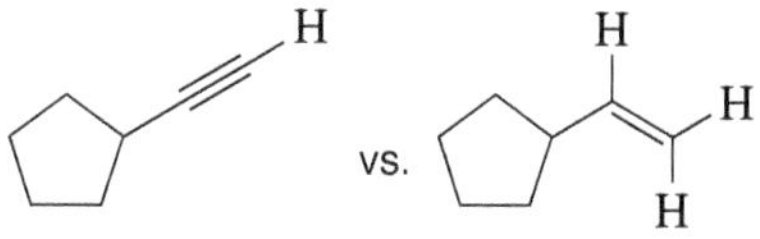

4.65 (•••) Benzoic acid and phenol are insoluble in water. When sodium bicarbonate is added to the water, benzoic acid dissolves, but phenol does not. The addition of sodium hydroxide causes both to dissolve. On the basis of these observations, estimate the pK_a of phenol. [*The* pK_a *of* H_2CO_3 *is 6.4.*]

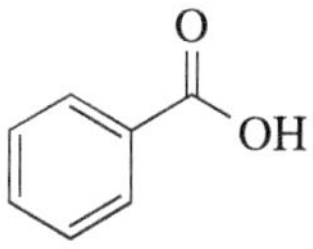

benzoic acid

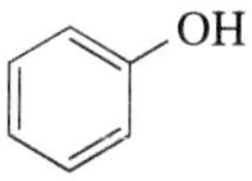

phenol

4.66 (•••) For each of the following molecules, predict the product that would form upon reaction of a single equivalent of a strong base.

(a)

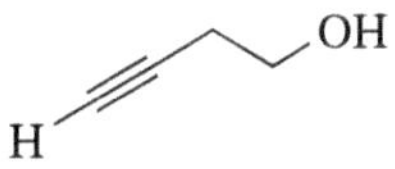

(b)

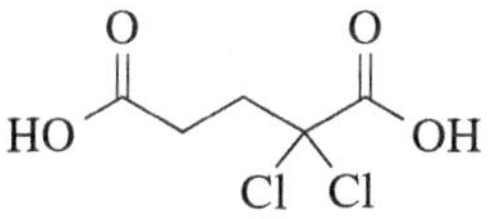

(c)

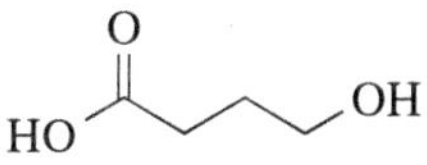

(d)

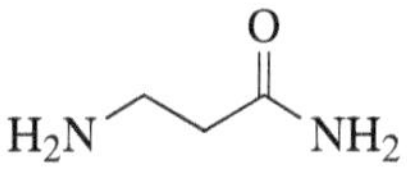

(e)

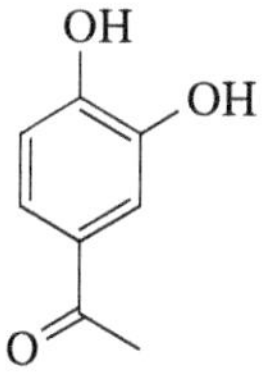

(f)

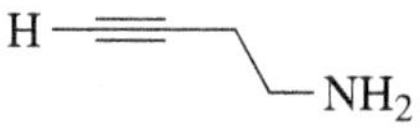

4.67 (•••) Suggest a scheme by which *N*-methylmorpholine, salicylic acid, and naphthalene can be separated. [*Recall Chemistry in Context 4-B.*]

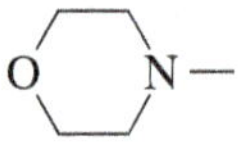
N-methylmorpholine

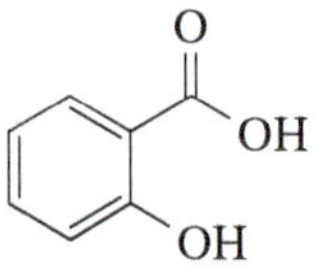
salicylic acid

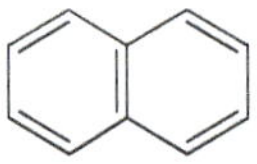
naphthalene

4.68 (•••) **THINKING AHEAD** A medicinal chemist needed to deprotonate acetylene (HC≡CH) for use in a coupling reaction. Among the options given, which base(s) could be used for this process?

NaOH
sodium hydroxide

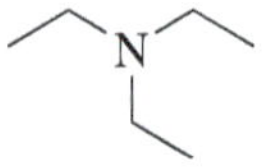
triethylamine

KH
potassium hydride

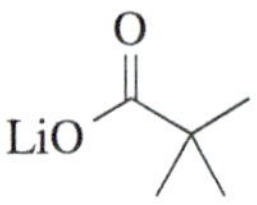
lithium pivalate

$NaNH_2$
sodium amide

4.69 (••••) **THINKING AHEAD** Amino acids exist predominantly in one of the following forms. Which is it? Explain your answer.

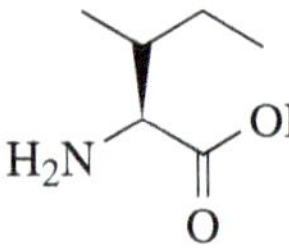

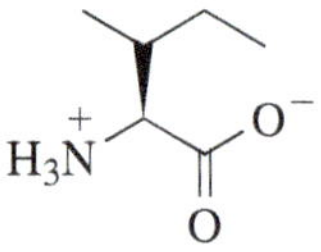

4.70 (•••) **THINKING AHEAD** Using pK_a values for the conjugate acids of the bases on each side of the reaction arrow, identify which side of the equilibrium would be favored in the following hypothetical reactions.

(a)

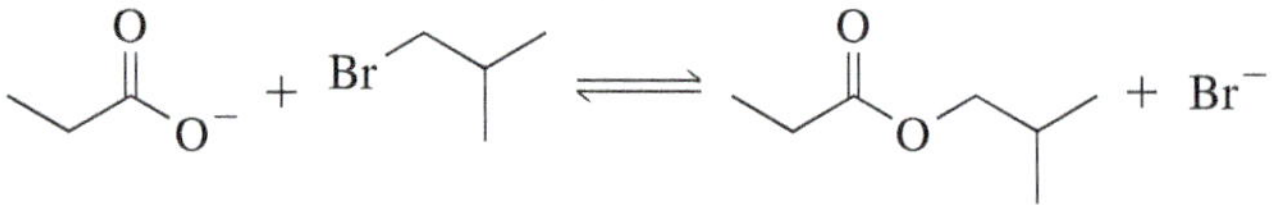

(b)

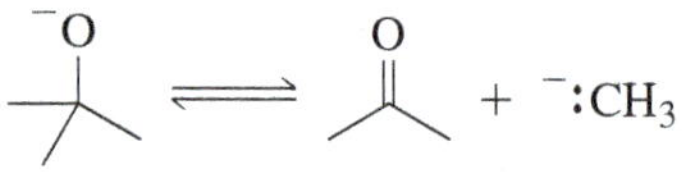

(c)

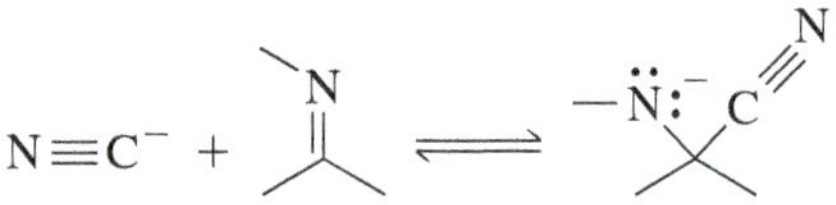

4.71 (••) Predict the value of K_{eq} for each of the reactions in Assessment 4.70.

4.72 (•••) Predict the product of the following Lewis acid–Lewis base reactions.

(a)

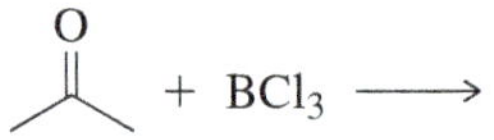

(b)

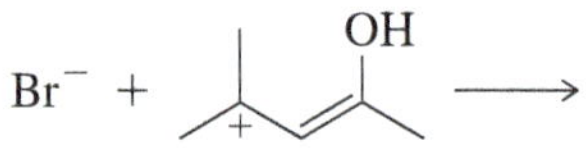

(c)

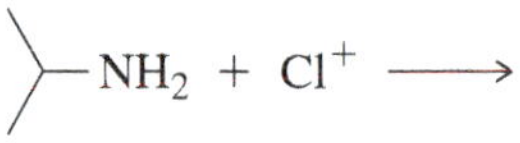

(d) [*an intramolecular reaction*]

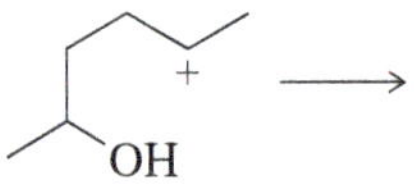

4.73 (••) Suggest an arrow-pushing mechanism for each of the following acid–base reactions.

(a)

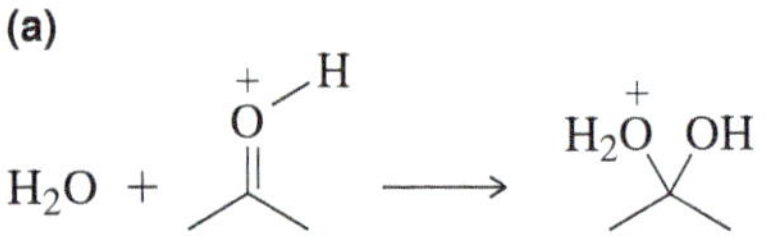

(b)

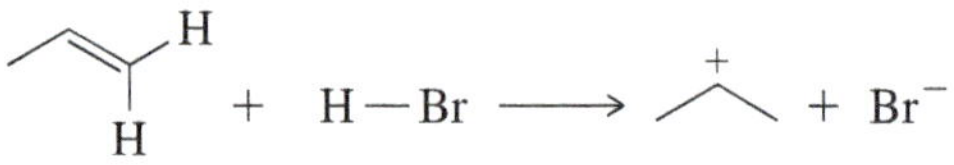

(c)

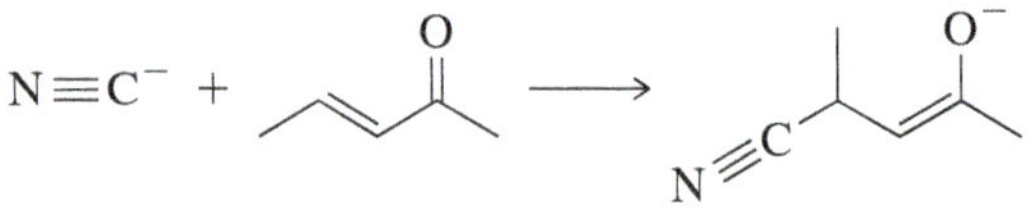

(d) [*an intramolecular reaction*]

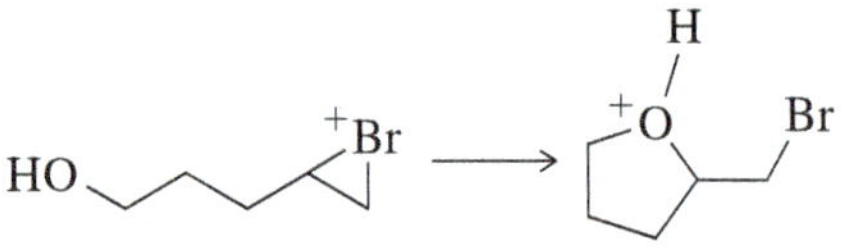

4.74 (••••) **THINKING AHEAD** One of the more reactive species we will study in organic chemistry is the carbene (molecular formula of CH_2). The carbene can react as both a Lewis acid and a Lewis base. Explain these properties.

4.75 (••••) **THINKING AHEAD** BCl_3 has an empty *p* orbital, so it is a strong Lewis acid. Would you expect an amide to react with BCl_3 at nitrogen or oxygen?

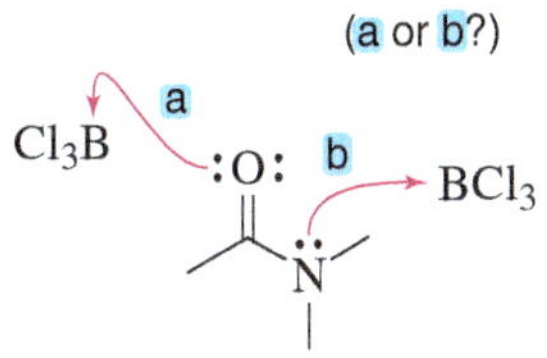

4.76 (•••) Which of the following carbanions would you expect to react most quickly with the Lewis acid BF_3? Explain your answer.

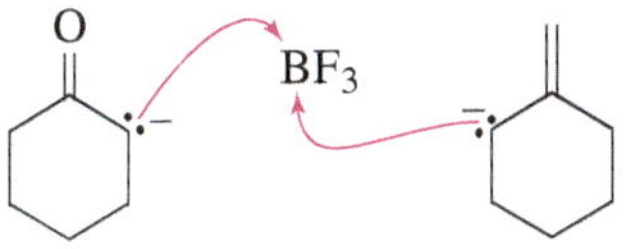

4.77 (•••) Consider the following drugs used to treat the indicated diseases. Would you expect the activity of these drugs to be impacted by a patient taking other medicines for acid reflux disease?

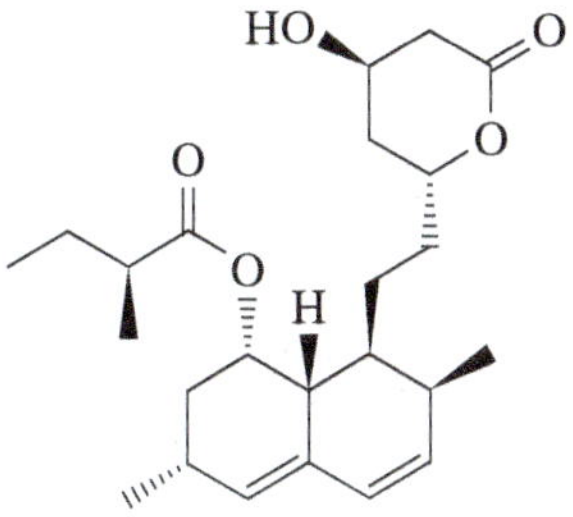

lovastatin (high cholesterol)

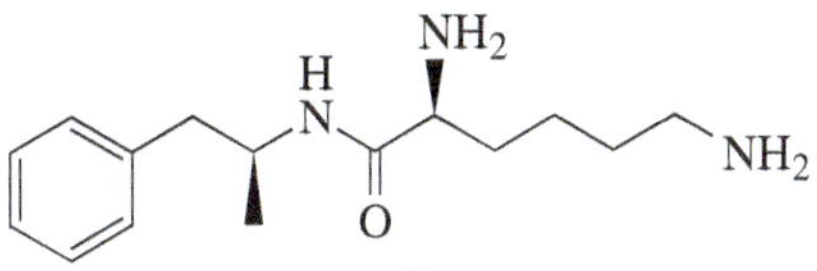

Vyvanse® (ADHD)

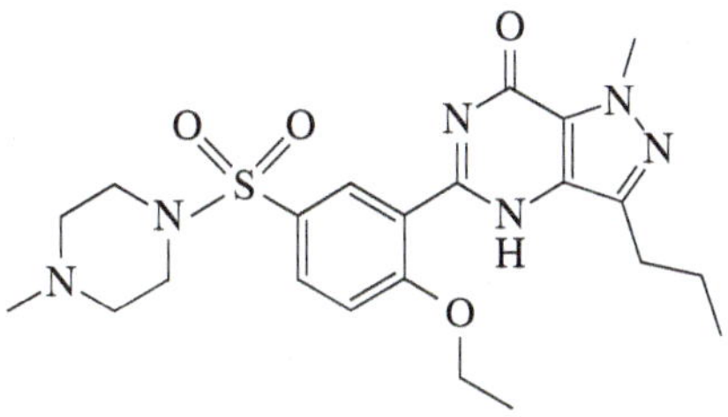

Viagra® (erectile dysfunction)

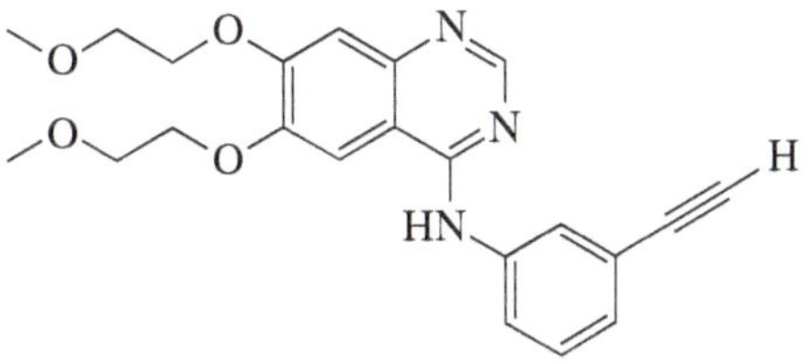

Tarceva (lung cancer)

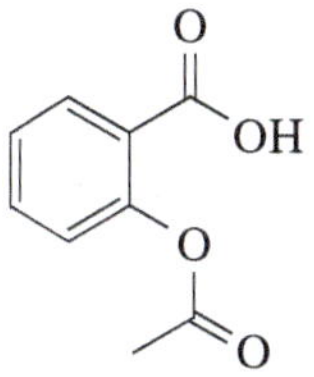

aspirin (pain/inflammation)

4.78 (••••) **THINKING AHEAD** Imidazole has two potentially basic sites (a and b). Of the two, where would you expect a proton to add preferentially?

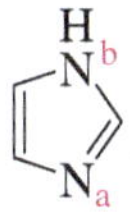

imidazole

5

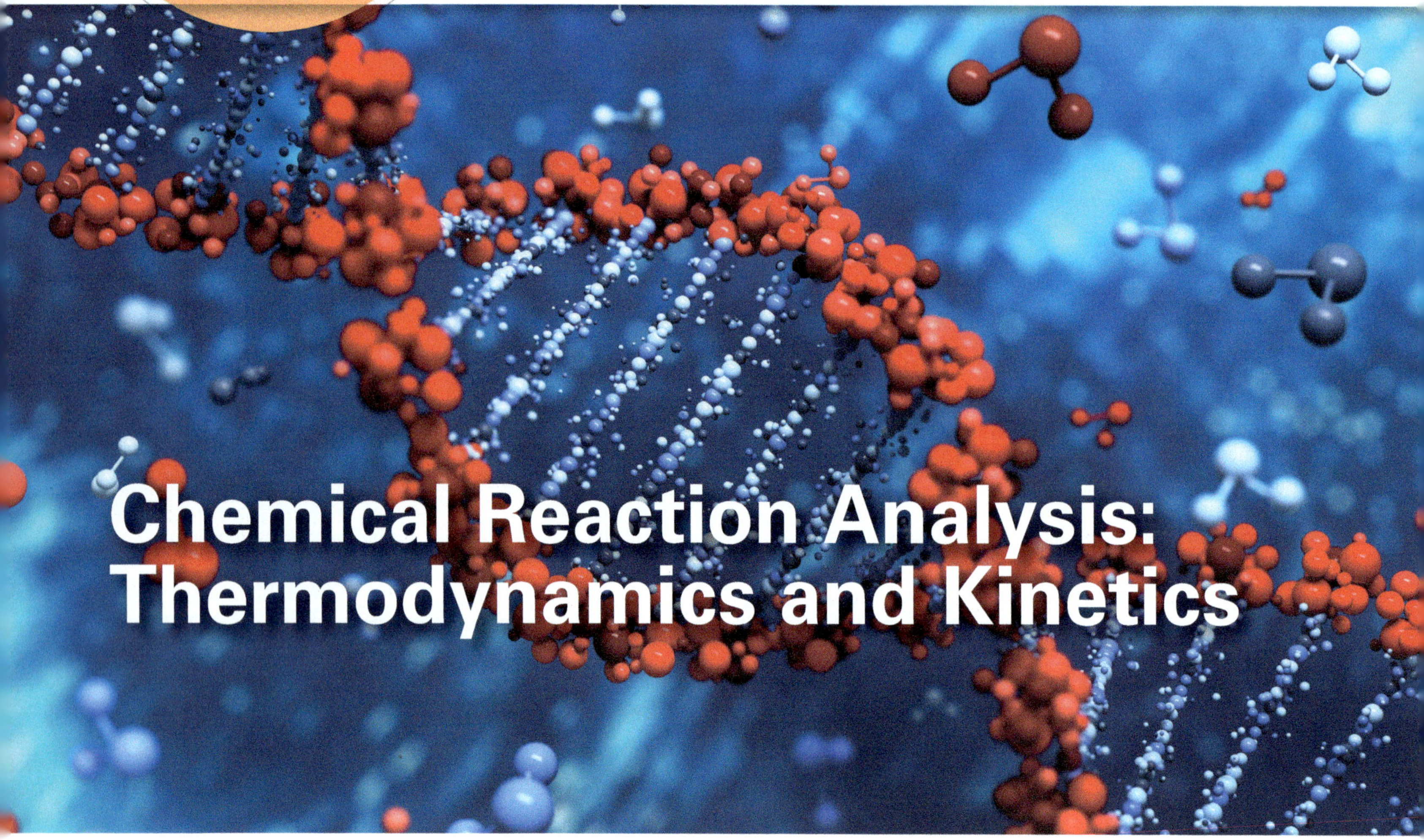

Chemical Reaction Analysis: Thermodynamics and Kinetics

Learning Objectives

LO 5.0 Explain the relationship between $\Delta G°$, $\Delta H°$, $\Delta S°$, and K_{eq} in a chemical reaction

LO 5.1 Explain how E_a, A, and T in the Arrhenius equation affect the rate of a chemical reaction

LO 5.2 Describe the role of the transition state in the path from reactant to product

LO 5.3 Use reaction analysis to determine how and why acid–base reactions occur

LO 5.4 Analyze radical halogenation reactions of alkanes in terms of $\Delta G°$, the rate-determining step, a possible transition state, the rate law, and mechanism

LO 5.5 Analyze conditions under which radical halogenation is nonselective

Introduction

Lysosomal storage diseases are genetically inherited disorders in which the cell's lysosomes lack properly working enzymes. Lysosomes are cytoplasmic organelles containing digestive enzymes that break down (or lyse) materials such as bacteria, viruses, and toxins into smaller molecules, which are then recycled or excreted by the cell. Mutations of lysosomal enzymes render them ineffective at this task, causing the buildup of unwanted molecules. One of the more important examples of lysosomal storage disease is Gaucher disease.

Gaucher disease, present in around one in 20,000 live births, is caused by a deficiency in the enzyme glucocerebrosidase. Glucocerebrosidase is a hydrolase enzyme responsible for the cleavage of the glucosidic linkage of the molecule glucocerebroside (**Figure 5.1**). In Gaucher patients, the buildup of glucocerebroside results in the enlargement of the spleen and the liver, along with symptoms of anemia, fatigue, easy bruising, and a swollen stomach due to the enlarged liver or spleen.

Enzymes act as biological catalysts, increasing the rate of a reaction by lowering the reaction's activation energy without affecting the reaction's $\Delta G°$ or equilibrium constant K_{eq}. Should you continue in biochemistry or medicinal chemistry after your work in this course, how enzymes catalyze reactions will be an important part of your studies. The discussion we begin here represents the foundation upon which these advanced topics are built.

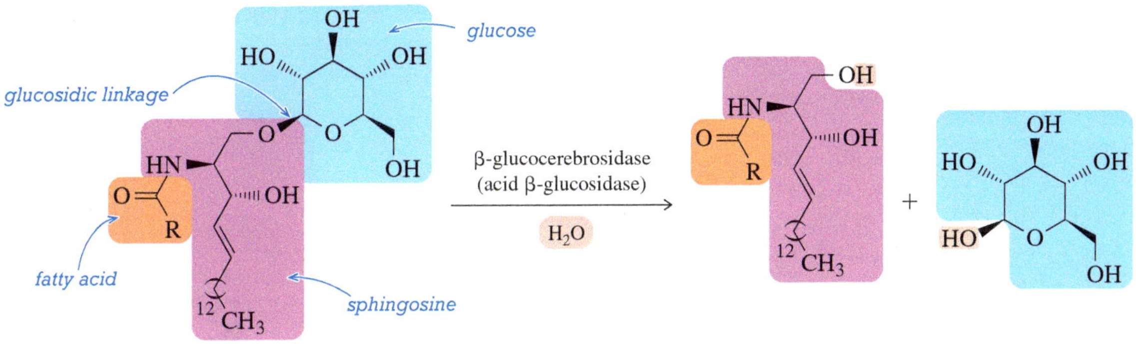

β-glucocerebroside ceramide glucose

Figure 5.1 **Deficiency of the enzyme glucocerebrosidase** is responsible for the symptoms of Gaucher disease.

"To us . . . the only acceptable point of view appears to be the one that recognizes both sides of reality—the quantitative and the qualitative . . . and can embrace them simultaneously."

—Wolfgang Pauli

Organic chemistry is not a heavily quantitative science. However, you are going to see more numbers and equations in this chapter than in any other. The equations and their application are used to strengthen your understanding of previously studied concepts, including conformational analysis (Chapter 3) and acid–base chemistry (Chapter 4). We start by taking simple "reactions" you have seen before and apply to them a more thorough, quantitative analysis. Once we have mastered that material, we apply our analytical approach to reactions of increasing complexity. Chemical reaction analysis is quite logical, regardless of the reaction being studied. With practice, you will gain confidence that you can use the same analytical approach to study reactions introduced later in the course.

Let's get started as we usually do, by assessing ourselves on material studied previously.

Test your understanding

Assessment 5.1 Draw the Newman projection for the most stable conformation around the C_2—C_3 bond for each of the following.

(a)

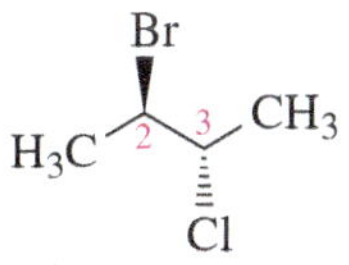

(b)

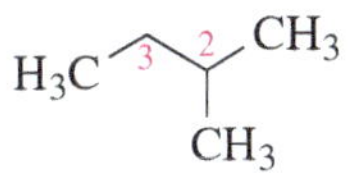

Assessment 5.2 Predict which side of the equilibrium is favored for each of the chair–chair interconversions (ring flips) shown.

(a)

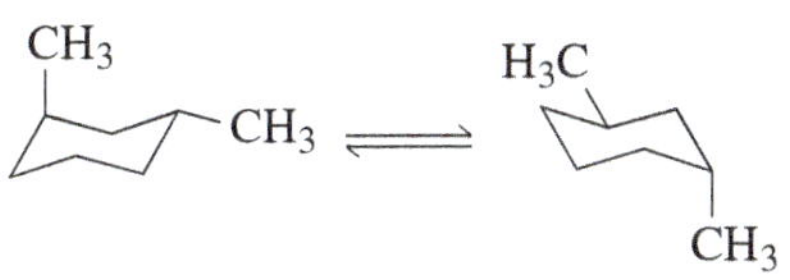

(b)

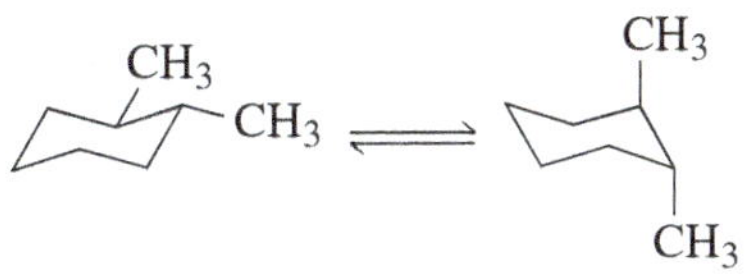

(c)

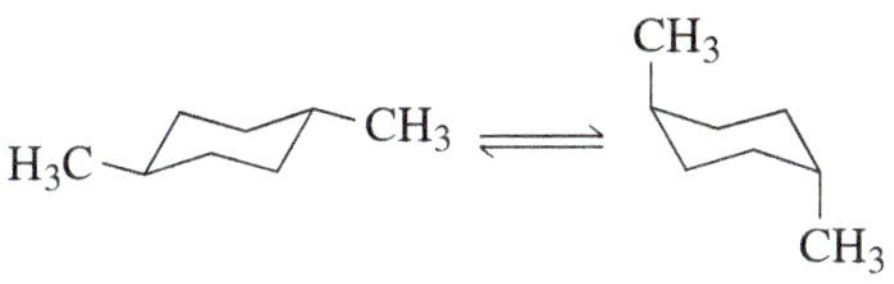

Assessment 5.3 Which is the most stable base in each pair?

(a)

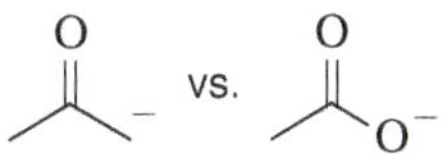

(b) HS^- vs. HO^-

(c) NH_3 vs. H_2O

Assessment 5.4 Which is the most acidic compound in each pair?

(a)

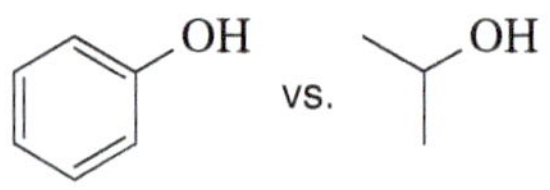

(b)

HF vs. HCl

(c)

H_3C-CH_3 vs. $H-C\equiv C-H$

Assessment 5.5 At which restaurant will our very hungry chemist most likely go to eat?

Assessment 5.6 Provide a reasonable arrow-pushing mechanism for the following Lewis acid–Lewis base reactions.

(a)

$H_3N\!: + BF_3 \longrightarrow H_3\overset{+}{N}-\overset{-}{B}F_3$

(b)

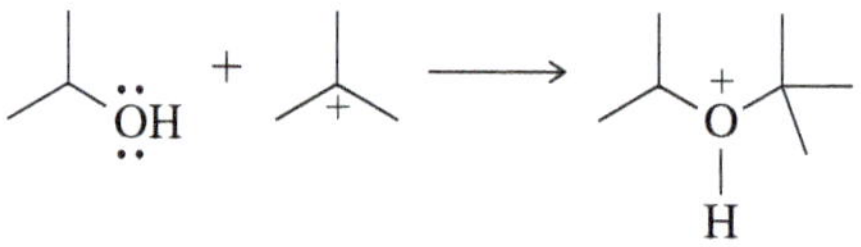

(c)

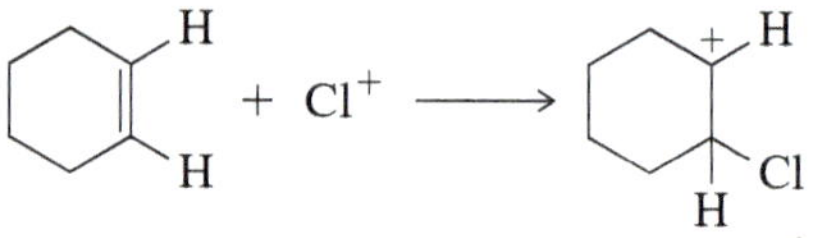

5.1 Chemical reaction analysis: rotation around a single bond

Let's begin by analyzing the rotation around a single (σ) bond, using butane as our model. Recall from Chapter 3 that butane exists primarily in two distinct, staggered conformations, called gauche (less stable) and anti (more stable). Now imagine that the process of going from the anti conformation to the gauche conformation is a chemical reaction, as shown in Figure 5.2. The anti conformation is more stable than the gauche conformation by 0.9 kcal/mol. We can calculate a change in energy for the process shown by subtracting the energy of the reactant from the energy of the product using Equation 5.1.

$$\Delta E = E_{\text{product}} - E_{\text{reactant}} \qquad \text{Equation 5.1}$$

These results imply that going from an effectively strain-free anti conformation to a strained gauche conformation is uphill ($\Delta E = +0.9$ kcal/mol). That is, we go from a low energy state (stable) to a higher energy state (unstable). This difference in energies is more specifically identified as the difference in the **Gibbs free energy** between the two conformations.

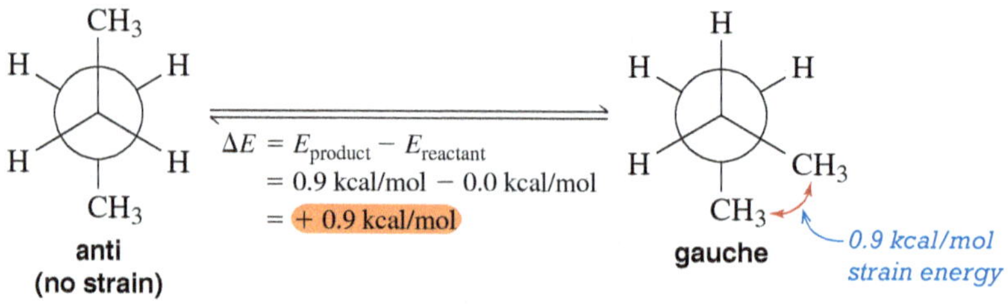

Figure 5.2 **Equilibrium between the anti and gauche conformations of butane around the C_2-C_3 bond.**

5.1.1 Gibbs free energy

The change in Gibbs free energy (ΔG) provides information about the favorability, or spontaneity, of a given process. When reactants and products are in their standard state and the reaction is run under standard conditions, we use the standard change in free energy ($\Delta G°$) to predict in which direction a reaction will proceed. [*We explain later* why *this is a viable approximation, even when reactions are not run under standard conditions.*] But what exactly does favorability/spontaneity mean in the context of rotation around the single bond in butane? Which side of the **equilibrium** would you expect to be favored for the reaction in Figure 5.2, given that ΔE (or $\Delta G°$) = +0.9 kcal/mol? [*Come up with an answer before reading further.*]

5.1.2 Favorability versus equilibrium

Figure 5.2 shows that the gauche conformation is less stable than the anti conformation. At a given time, for a large collection of butane molecules, we expect to find more anti than gauche conformations. [*The ball rolls downhill toward more stable molecules/conformations.*] In our early chemistry studies, we learned that the ratio of products to reactants can be described as the **equilibrium constant** (K_{eq}). With more anti than gauche, we estimate the equilibrium constant to be less than 1. Simply put, a fraction with a larger denominator (more anti) than numerator (more gauche) gives $K_{eq} < 1$ (**Figure 5.3**).

$$K_{eq} = \frac{[\text{product}]}{[\text{reactants}]} = \frac{[\text{gauche}]}{[\text{anti}]} = \frac{\text{small number}}{\text{large number}}$$

$K_{eq} < 1$

Figure 5.3 More anti than gauche in Figure 5.2 gives K_{eq} < 1.

The implication here is that the sign of $\Delta G°$ can give insight into the value of the equilibrium constant and, perhaps most importantly, *the side of the reaction favored by the equilibrium constant.* One way to verify this is to evaluate what impact a change in magnitude and sign of $\Delta G°$ has on the equilibrium constant. Assessment 5.7 does just that.

Assessment 5.7 For the following equilibrium processes and the corresponding $\Delta G°$, indicate whether you expect the equilibrium constant to be greater than, equal to, or less than 1. Justify your expectation in words.

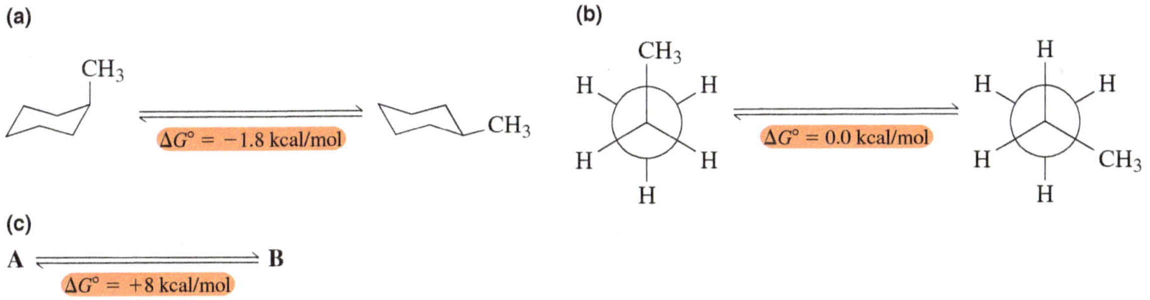

We can work Assessment 5.7 together . . .

Solved Assessment 5.7

There is a correlation between the sign of $\Delta G°$ and K_{eq}. In Assessment 5.7(a), the negative value for $\Delta G°$ suggests that the equatorial methyl group is more stable/less strained than the axial methyl group. [*The ball rolls downhill.*] Thus, we expect to have a higher concentration of molecules in this conformation and $K_{eq} > 1$. When there is no preference, as in Assessment 5.7(b), where $\Delta G° = 0$ kcal/mol, we expect equal amounts of reactants and products, giving $K_{eq} = 1$. Finally, Assessment 5.7(c) shows that it doesn't matter what process/reaction we are referring to, because the fact that B is less stable than A will be reflected in K_{eq} ($K_{eq} < 1$), regardless of the identity of B and A. The generality of this concept is something we continually examine in this chapter and beyond. ■

Given that the sign of ΔG° identifies which side of the equilibrium is favored, the magnitude of ΔG° should correlate to the magnitude of the equilibrium constant. When the absolute value of ΔG° is large, the equilibrium will heavily favor one side of the reaction over the other. A small absolute value, on the other hand, implies that one side of the reaction is only slightly favored. [*Go back and read this section again to ensure that it makes sense up to this point.*]

Assessment 5.8 In Assessment 5.7, you were able to use the sign of ΔG° to predict which side of the reactions was favored. Based on the magnitude of these values, which reaction would you expect to have the highest ratio of favored to unfavored? Explain your answer.

5.1.3 ΔG° and K_{eq}: a quantitative relationship

One challenge of using mathematical equations in a chemistry course is the tendency to memorize them and simply plug in numbers to get the answers. It is critically important, though, that you understand the implications of these equations qualitatively prior to using them quantitatively. Section 5.1.2 established qualitatively a relationship between ΔG° and K_{eq}. [*In other words, it makes sense that they are related quantitatively.*] The quantitative relationship relating the standard Gibbs free energy change (ΔG°) and K_{eq} is shown in Equations 5.2 and 5.3.

If you are asked to solve an equation that requires a temperature, always assume 298 K (25 °C) unless it is otherwise given.

$$\Delta G^\circ = -RT \ln K_{eq} \qquad \text{Equation 5.2}$$

$$\Delta G^\circ = -2.303RT \log_{10} K_{eq} \qquad \text{Equation 5.3}$$

where $R = 1.987 \frac{\text{cal}}{\text{mol} \cdot \text{K}}$ is the gas constant and T is the absolute temperature (in kelvins).

As a reminder, *standard* Gibbs free energy (ΔG°), indicated with the degree symbol (°), refers to reactants and products in their standard states (pure substances at 25 °C and 1 atm pressure). We can now plug in values for ΔG° to calculate K_{eq}, product ratios, and the relative percentages of species present at equilibrium (Table 5.1). [*Always check your answers against your qualitative expectations. If they don't match up, carefully consider whether your use of the equation or your qualitative understanding is incorrect.*]

Table 5.1 Product composition for reaction A ⇌ B as a function of ΔG° at 25 °C. [*The values in the K_{eq} and %B columns were calculated from the values in kcal/mol.*]

Product composition for reaction A ⇌ B as a function of ΔG°

ΔG°			
kJ/mol	kcal/mol	K_{eq}	%B
+4.2	+1.0	0.18	16
+2.1	+0.50	i	ii
0.0	0.00	1.0	50
−2.1	−0.50	2.3	70
−4.2	−1.0	5.4	iii
−8.4	−2.0	29	97
−13.0	−3.0	159	99.4
iv	v	859	99.9
−21.0	−5.0	4600	99.98

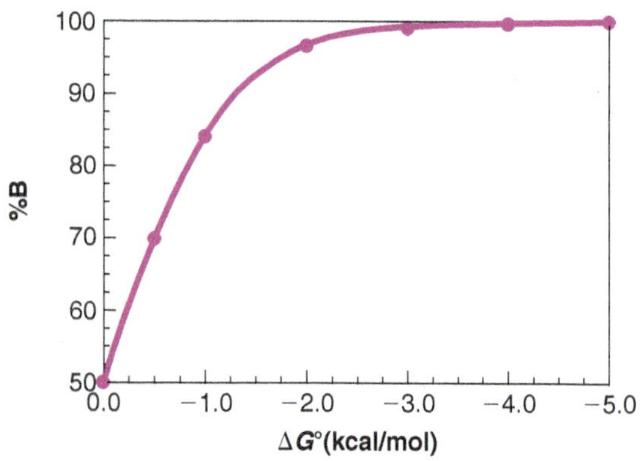

Assessment 5.9 After estimating, use Equation 5.2 to calculate the missing values of ΔG°, K_{eq}, and %B. Recall that 1 cal = 4.184 J and $R = 1.987 \frac{\text{cal}}{\text{mol} \cdot \text{K}}$ $\left(8.314 \frac{\text{J}}{\text{mol} \cdot \text{K}}\right)$.

Let's return to the original reaction we were examining, in which the anti conformation of butane is in equilibrium with its gauche conformation (**Figure 5.4**). Given a value of $\Delta G° = +0.9$ kcal/mol, or +900 cal/mol, we expect $K_{eq} < 1$ because the anti is more stable than the gauche. The calculation using Equation 5.2 shows that $K_{eq} = 0.219$.

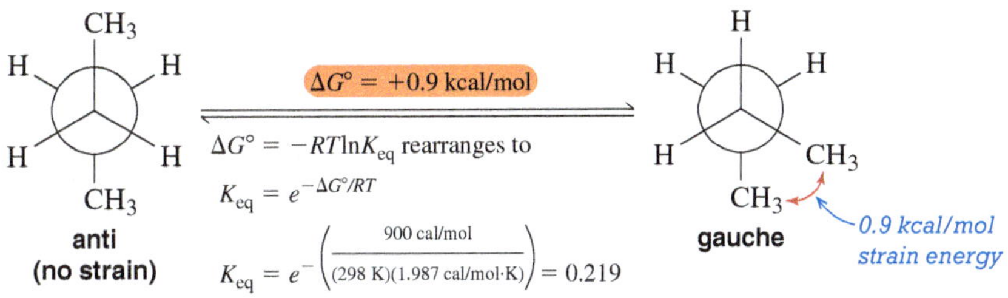

Figure 5.4 The calculation of K_{eq} confirms that the more stable anti conformation is present in higher concentration at equilibrium. [*Note that $\Delta G°$ is expressed in units of calories to match the units of R.*]

This value of K_{eq} represents the ratio of gauche to anti at 298 K, and it makes it possible to calculate the percent of gauche and anti conformations at equilibrium, as shown stepwise in **Figure 5.5**.

(a) $K_{eq} = 0.219 = \dfrac{[\text{gauche}]}{[\text{anti}]}$, so the ratio of gauche to anti = 0.219/1.

Thus, there are 0.219 molecules in the gauche conformation for every 1 in the anti conformation.

(b) General equation: $\%A = \dfrac{A}{A + B} \times 100\%$

(c) A + B = gauche + anti = 0.219 + 1

(d) $\%\ \text{gauche} = \dfrac{0.219}{0.219 + 1} \times 100\% = 18.0\%$

(e) $\%\ \text{anti} = \dfrac{1}{0.219 + 1} \times 100\% = 82.0\%$

Figure 5.5 The conversion of K_{eq} to percent composition at equilibrium. (a) K_{eq} = the ratio of products to reactants, which for this process is the ratio of gauche to anti conformation. (b) General equation for calculating the %A present in a mixture. (c) The total of gauche + anti. (d) Gauche divided by gauche plus anti gives the % gauche. (e) Anti divided by gauche plus anti gives the % anti.

Confirming our analysis that the equilibrium always favors the more stable species, the equations in Figures 5.4 and 5.5 predict more anti (more stable) than gauche (less stable) at equilibrium. Furthermore, the percentages and ratios obtained can be compared to values in Table 5.1, which are based on differences in energy. From Table 5.1, we see that a reaction with a $\Delta G° = +1.0$ kcal/mol and $K_{eq} = 0.18$ has the less stable species making up 16% of the equilibrium mixture. Our calculation, with $\Delta G° = +0.9$ kcal/mol, gave us slightly more of the less stable gauche conformation (18.0%). These similar values should give you confidence that our qualitative approach and quantitative understanding align.

5.1.1–5.1.3 SUMMARY

- $\Delta G°$ is related to K_{eq} through the equation:

$$\Delta G° = -RT \ln K_{eq}$$

- When $\Delta G° < 0$, the reaction to the right is favored, and $K_{eq} > 1$.
- When $\Delta G° > 0$, the reaction to the left is favored, and $K_{eq} < 1$.
- When $\Delta G° = 0$, neither side is favored, and $K_{eq} = 1$.

Assessment 5.10 For the following equilibrium processes and the corresponding $\Delta G°$, calculate (i) K_{eq} and (ii) the % composition of the equilibrium mixture (% reactants, % products) at 298 K.

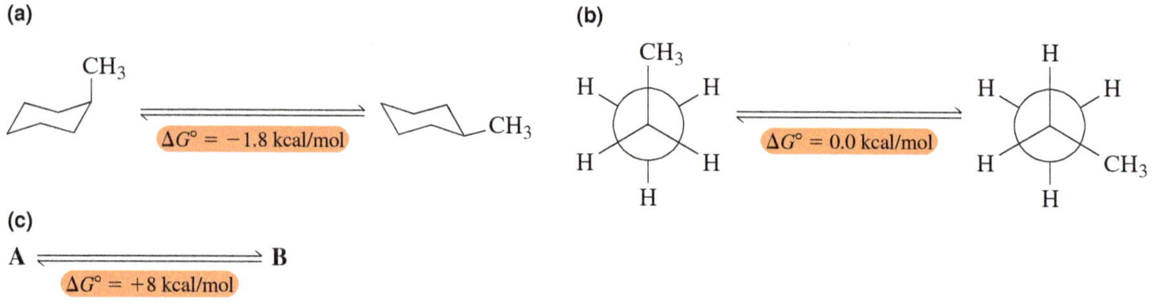

5-A CHEMISTRY IN CONTEXT: Biochemistry

Enzymes as catalysts

Gaucher disease is attributed to the buildup of β-glucocerebroside, which is caused by a defect in the enzyme that breaks it down. When a disease is caused by the accumulation of a substance in the body, one potential therapy is to halt the process associated with creating it. Enzymes, as biological catalysts (see Chemistry in Context 5-B), are involved in the synthesis of most naturally occurring organic molecules. β-Glucocerebroside is synthesized by the reaction of UDP-glucose and ceramide in the active site of glucosylceramide synthase, as shown in Figure 5.6.

In general, enzymes act by lowering the activation energy, resulting in a faster reaction in both directions. Increasing the rate of the forward and reverse reactions leaves the equilibrium constant for an enzyme-mediated reaction unchanged. That is, an enzyme does not change ΔG for a given reaction. It can drive the reaction in either direction, allowing equilibrium to be achieved more quickly, but it does not change K_{eq}. [*This is an important concept, commonly tested on exams like the MCAT®. You will see it repeated throughout the book.*]

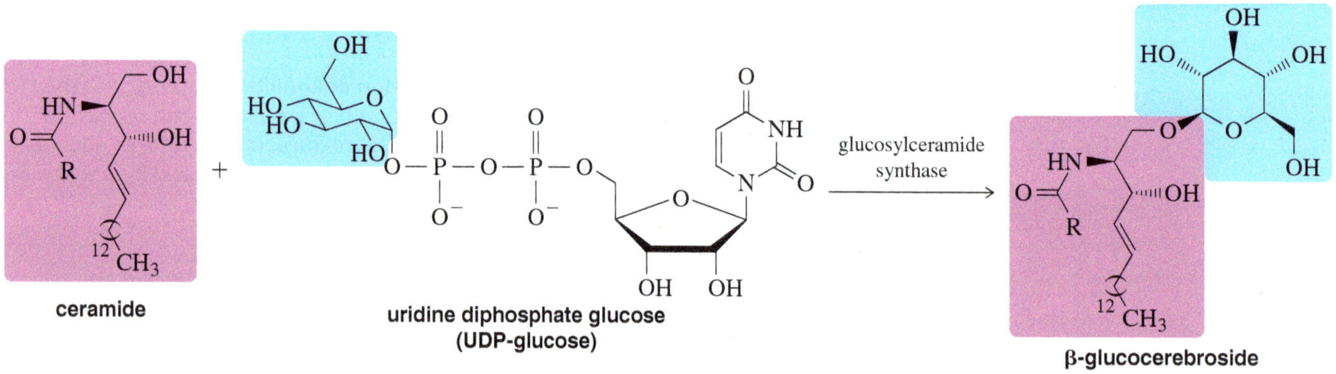

Figure 5.6 The synthesis of β-glucocerebroside is catalyzed by the enzyme glucosylceramide synthase.

5.1.4 Enthalpy and entropy: determining $\Delta G°$

To review, the favorability of a chemical reaction/process is determined by the sign of $\Delta G°$. If $\Delta G° < 0$, the forward reaction is favored (spontaneous). If $\Delta G° > 0$, the forward reaction is disfavored (nonspontaneous). What determines if $\Delta G°$ is positive or negative? We were able to make this determination for rotation around a single bond, but can we predict the value of $\Delta G°$ for more complex reactions?

The Gibbs free energy (G) is made up of two factors, **enthalpy** (H) and **entropy** (S). The change in the Gibbs free energy $\Delta G°$, then, can be related to $\Delta H°$ and $\Delta S°$ by Equation 5.4, where T is the absolute temperature expressed in kelvins:

$$\Delta G° = \Delta H° - T\Delta S° \quad \text{Equation 5.4}$$

Before we talk about what enthalpy and entropy are, let's consider what values of $\Delta H°$ and $\Delta S°$ would result in a favorable process. In the figures that follow, we examine each term individually, later considering how they fit together to determine the favorability of a reaction. In the absence of the other term, then, what sign of $\Delta H°$ or $\Delta S°$ allows $\Delta G° < 0$? We can do this for $\Delta H°$ by imagining a value of zero for $\Delta S°$ (Figure 5.7). In this scenario, where $\Delta G° = \Delta H°$, a favorable process must have a $\Delta H° < 0$. This confirms that negative $\Delta H°$ is favorable and demonstrates that a reaction with no change in entropy ($\Delta S° = 0$) must be driven by a change in enthalpy ($\Delta H° < 0$). [*This concept is very important for future discussions.*]

What the equations say:	What the equations mean:
$\Delta G° = \Delta H° - T\Delta S°$	$\Delta G°$ depends on $\Delta H°$, T, and $\Delta S°$.
If $\Delta S° = 0$,	$\Delta G° < 0$ for a favorable process. If entropy plays no role,
then $\Delta G° = \Delta H°$.	then enthalpy is the only factor that determines Gibbs free energy.
$\Delta H° < 0$ gives $\Delta G° < 0$.	A negative $\Delta H°$ gives a negative $\Delta G°$ and a favorable reaction ($K_{eq} > 1$).

Figure 5.7 A negative change in enthalpy makes a reaction favorable.

We can do the same to determine what values of $\Delta S°$ result in favorable processes (Figure 5.8). By setting $\Delta H° = 0$ (implying that there is no change in enthalpy for a given reaction), we see that $\Delta G° = -T\Delta S°$. Because the temperature is measured in kelvins, by definition, T will always be greater than zero. As a result, for $\Delta G°$ to be less than zero, $\Delta S°$ must be positive.

What the equations say:	What the equations mean:
$\Delta G° = \Delta H° - T\Delta S°$	$\Delta G°$ depends on $\Delta H°$, T, and $\Delta S°$.
If $\Delta H° = 0$,	$\Delta G° < 0$ for a favorable process. If enthalpy plays no role,
then $\Delta G° = -T\Delta S°$.	then temperature and entropy are the only factors that determine Gibbs free energy.
If $-T\Delta S° < 0$, then $\Delta G° < 0$.	The negative product of entropy and temperature gives a negative change in free energy and a favorable reaction.
Because $T > 0$ (Kelvin scale),	Because temperature is always positive on the Kelvin scale,
$\Delta S° > 0$ gives $\Delta G° < 0$.	a positive $\Delta S°$ gives a negative $\Delta G°$ and a favorable reaction ($K_{eq} > 1$).

Figure 5.8 A positive change in entropy makes a reaction favorable.

In conclusion, processes are made favorable when accompanied by negative values of $\Delta H°$ and positive values of $\Delta S°$. The exercises in Figures 5.7 and 5.8 show that you don't need to memorize this. By simply knowing that $\Delta G° < 0$ for a favorable process and that $\Delta G° = \Delta H° - T\Delta S°$, you can figure out which values of $\Delta H°$ and $\Delta S°$ are favorable. Furthermore, if you know the equation relating $\Delta G°$ and K_{eq} ($\Delta G° = -RT \ln K_{eq}$) and recall that a favored process has $K_{eq} > 1$, you can infer that $\Delta G° < 0$ for a favored process. [*Go back and read this paragraph again to make sure you understand it.*]

Table 5.2 Favorability of reactions as a function of $\Delta H°$ and $\Delta S°$.

If...	$\Delta H° > 0$	$\Delta H° < 0$
$T\Delta S° < 0$	$\Delta G° > 0$	$\Delta G° < 0$ when $[\Delta H°] > [T\Delta S°]$
$T\Delta S° > 0$	$\Delta G° < 0$ when $[T\Delta S°] > [\Delta H°]$	$\Delta G° < 0$

There are many situations where a favorable process is not favored by both entropy and enthalpy. Similarly, a disfavored process is not necessarily disfavored by both terms. Because a reaction can be driven forward by changes in entropy or enthalpy alone, it is best to think about it as a competition between the $\Delta H°$ and the $T\Delta S°$ terms. This competition is summarized in Table 5.2. If the negative (favorable) enthalpy term (i.e., the reaction is **exothermic**) is large enough that it overrides the negative (unfavorable) entropy term, then a reaction can still be favored ($\Delta G° < 0$). Similarly, if the positive (favorable) entropy term is large enough to override the positive (unfavorable) enthalpy term (i.e., the reaction is **endothermic**), then the reaction is favored.

Assessment 5.11 For the following values of $\Delta H°$, $\Delta S°$, and T, tell whether the process would be favored.

(a) $\Delta H° = -15$ kcal/mol; $\Delta S° = +37$ cal/mol·K; $T = 273$ K

(b) $\Delta H° = +7.34$ kcal/mol; $\Delta S° = +43$ cal/mol·K; $T = 325$ K

(c) $\Delta H° = -21.3$ kcal/mol; $\Delta S° = -51$ cal/mol·K; $T = 373$ K

(d) $\Delta H° = -8.3$ kcal/mol; $\Delta S° = -12$ cal/mol·K; $T = 298$ K

Assessment 5.12 At what temperature does the entropy change of a process *not* contribute to the favorability of a process?

Assessment 5.13 A certain process has $\Delta H° = 11.7$ kcal/mol and $\Delta S° = +33$ cal/mol·K. That is, this reaction has an unfavorable enthalpy but a favorable entropy term. At what temperature will the process be neither favored nor disfavored?

Assessment 5.14 Considering the process described in Assessment 5.13, will it be favored or disfavored at a temperature higher than the one you calculated? How about at a temperature below what you calculated?

5.1.5 Enthalpy: bond energies and strain

In chemical reactions, a change in enthalpy ($\Delta H°$) results whenever a chemical bond is formed or broken. The $\Delta H°$ of a chemical reaction is the sum of the energy released when bonds are formed minus the sum of the energy absorbed when bonds are broken. [*Review the previous sentence—it represents one of the most important concepts in chemistry, biochemistry, and biology.*] Bond formation is favorable, releases energy, and has $\Delta H° < 0$. This is why elements in the periodic table spontaneously react to form covalent and/or ionic bonds (Chapter 2). The opposite is also true: Breaking bonds is disfavored, requires energy, and has $\Delta H° > 0$.

A lesser, but still important, factor determining the value of $\Delta H°$ is the strain on a molecule. We discussed this in a qualitative sense in Chapter 3 when we examined the torsional strain of an eclipsing H—H interaction (Section 3.4), the angle strain of cyclopropane (Section 3.6.1), and the steric strain of an axial substituent on cyclohexane (Section 3.7.2). An increase in strain results in a decrease in stability and a $\Delta H° > 0$. Conversely, a decrease in strain produces a more stable molecule and a $\Delta H° < 0$.

We can examine the enthalpy change associated with rotation around the central bond of butane. Because this is simply a change in conformation, there is no contribution to enthalpy due to the making or breaking of bonds. However, the gauche conformation has a higher amount of strain than the anti conformation. The specific value of strain energy (0.9 kcal/mol) is really a $\Delta H°$ value, and it is essentially the total free energy change ($\Delta G°$) for the reaction (**Figure 5.9**).

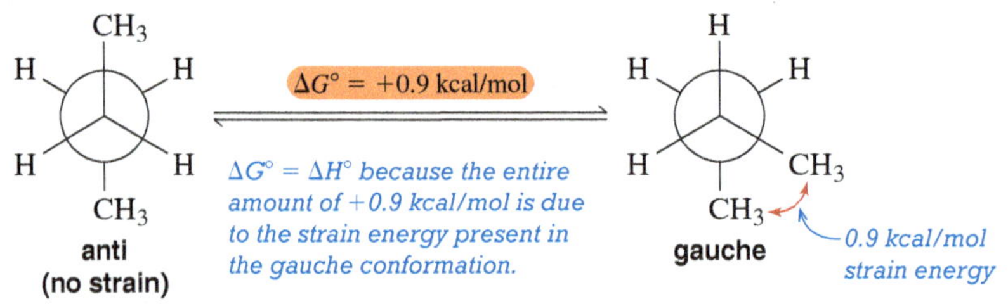

Figure 5.9 An increase in strain results in $\Delta H° \approx 0.9$ kcal/mol for bond rotation in butane.

$\Delta G°$ includes both $\Delta H°$ and $\Delta S°$, so the rotation around the bond must be a nearly entropically neutral process. In Section 5.1.6, we introduce an organic chemist's view of entropy so that we can understand why this is. [*But first, let's assess our understanding of enthalpy as it relates to strain energy.*]

Assessment 5.15 Calculate $\Delta H°$ for the following equilibrium processes.

(a)

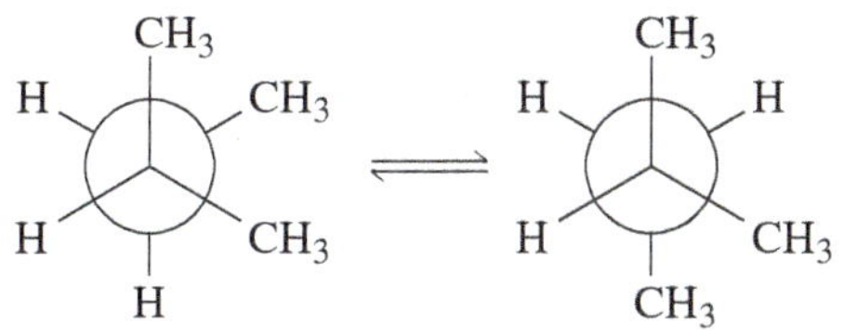

(b)

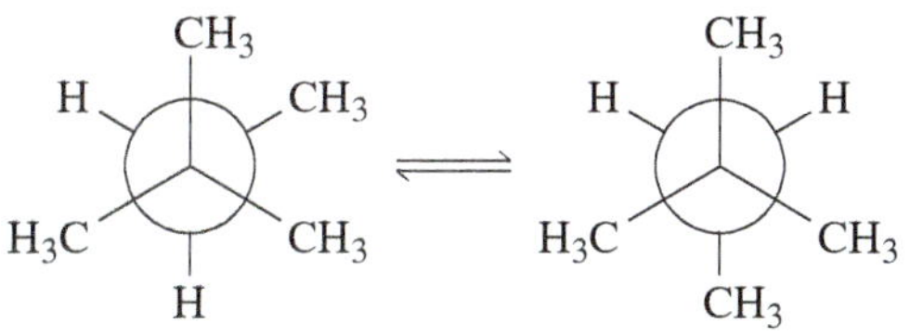

(c)

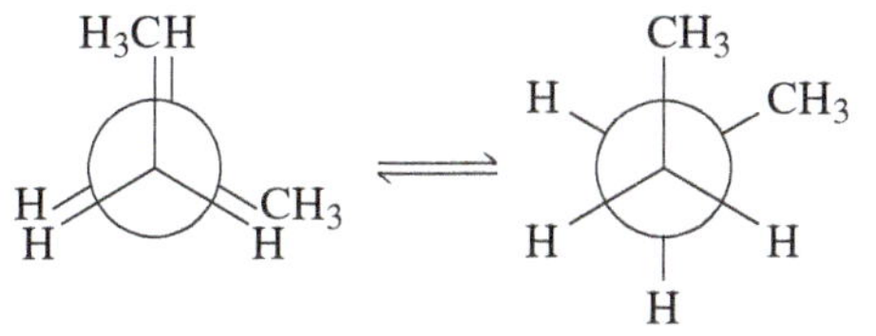

(d)

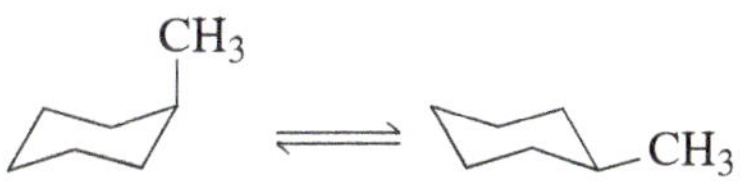

Assessment 5.16 Which conformation in each of the following pairs has the least strain energy?

(a)

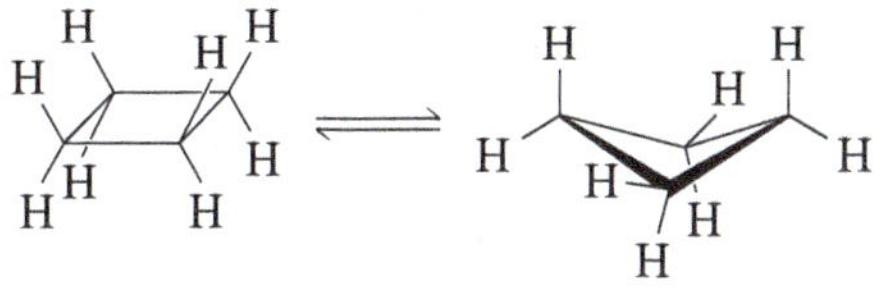

(b)

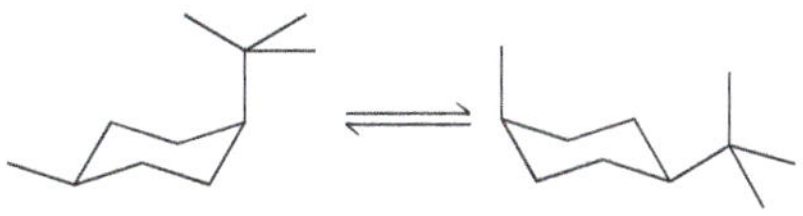

(c)

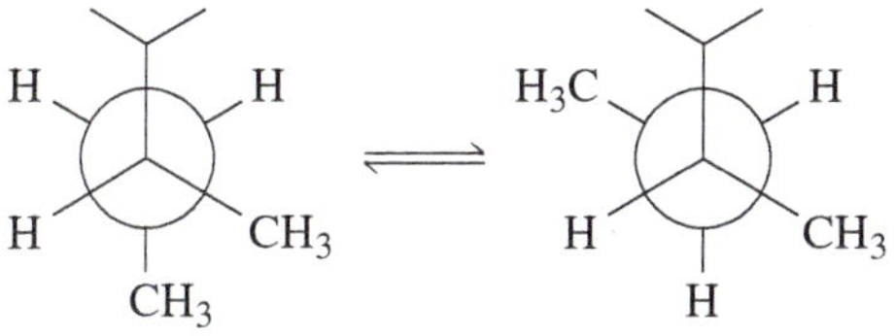

Assessment 5.17 The combustion of alkanes is exothermic ($\Delta H° < 0$). Would you expect the combustion of butane or cyclobutane to be more exothermic?

5.1.6 Entropy: a measure of disorder

Entropy ($\Delta S°$) is best thought of as a measure of *disorder* or *randomness* in a system. Given $\Delta G° = \Delta H° - T\Delta S°$ and the relationships shown in Table 5.2, a positive $\Delta S°$ makes a process more favorable. Thus, a reaction will be favorable in the direction that increases the randomness of a system. Without a reliable way to estimate real values for the change in entropy for a given reaction, we can nevertheless make the generalizations listed in **Table 5.3**.

Table 5.3 Generalized $\Delta S°$ values for selected reactions.

Reaction	Change in entropy result
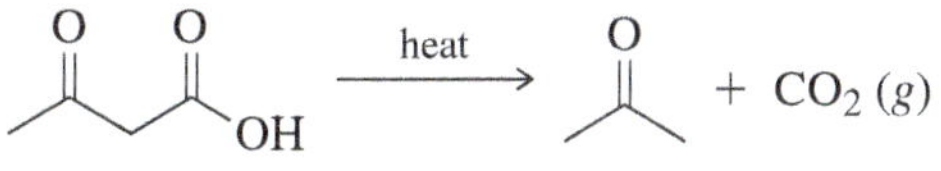 1 molecule → 2 molecules ($\Delta S° > 0$)	The formation of two molecules from one is entropically favored. It is even more favored when one of the products is a gas.
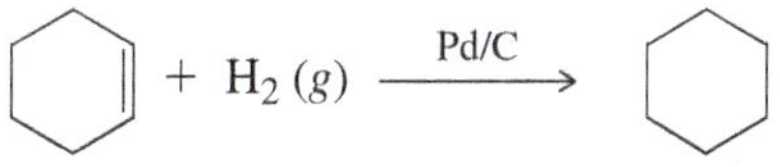 2 molecules → 1 molecule ($\Delta S° < 0$)	The formation of one molecule from two is entropically disfavored, especially if one of the reactants is a gas.
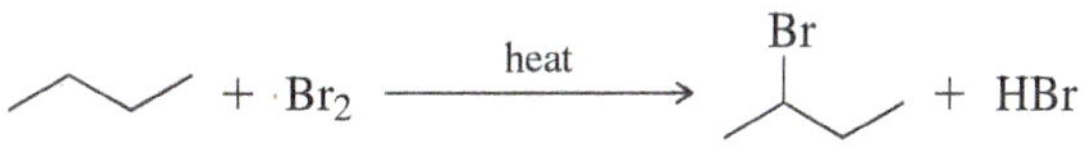 2 molecules → 2 molecules ($\Delta S° \approx 0$, so $\Delta G° \approx \Delta H°$)	A reaction that has the same number of molecules on either side of the equation is entropically neutral, in which case $\Delta S \approx 0$ and $\Delta G° \approx \Delta H°$.

Based on Table 5.3, we use entropy throughout this text as follows: If a reaction results in a greater number of molecules, we assume $\Delta S° > 0$ (favorable). If the opposite is true, we assume $\Delta S° < 0$ (unfavorable). For all other reactions, we assume the change in entropy is negligible.

Assessment 5.18 For each of the following processes, indicate whether you expect $\Delta S°$ to be greater than, less than, or equal to 0. Explain your answer.

(a) Boiling water

(b)

$$2\,C_6H_{14}\,(l) + 19\,O_2\,(g) \longrightarrow 12\,CO_2\,(g) + 14\,H_2O\,(g)$$

(c)

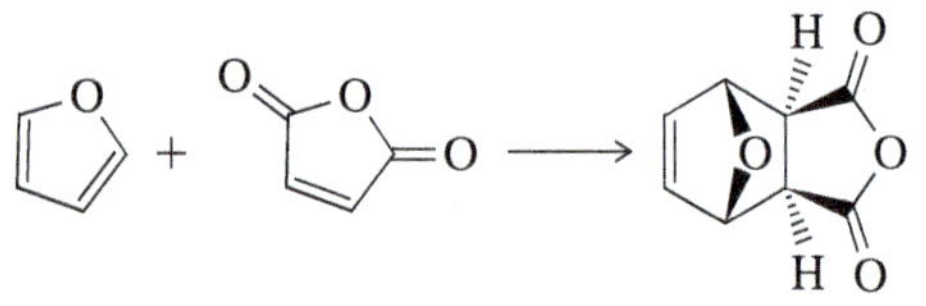

(d)

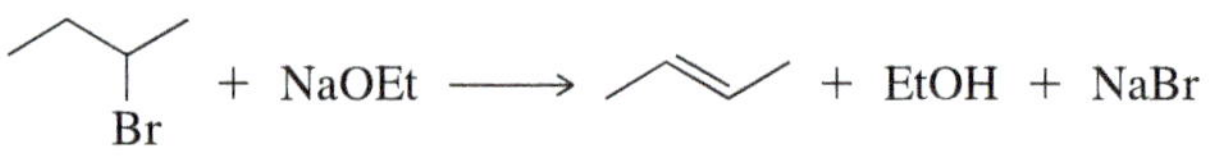

Assessment 5.19 If the following reaction is favorable, what can we say about the sign of $\Delta H°$? Explain your answer.

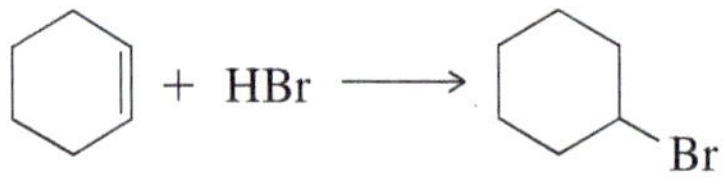

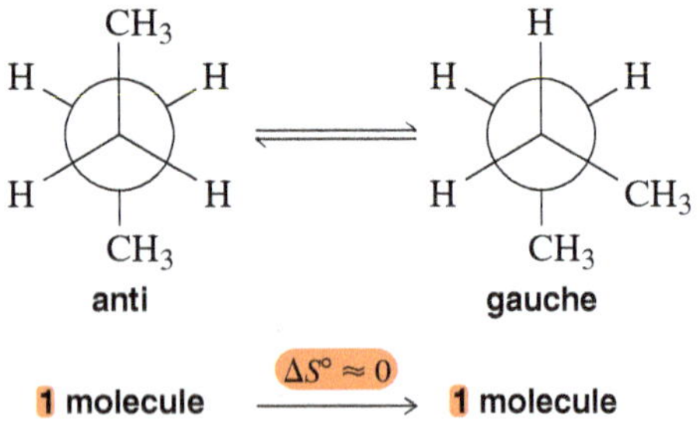

Figure 5.10 **$\Delta S°$ is approximately 0** when one molecule becomes one molecule, an assumption that holds when molecules are in the same phase and of similar complexity.

Rotation around the single bond of butane involves one molecule on each side of the equilibrium arrow. Thus, neither side of the equilibrium appears more random or disordered than the other. This suggests that entropy should not be a factor in favoring one side or the other and that $\Delta S° \approx 0$ (Figure 5.10). This supports our earlier conclusion that the sole contributor to $\Delta G°$ (+0.9 kcal/mol) is $\Delta H°$ (+0.9 kcal/mol). [*In truth, there are two ways to be gauche, so gauche is the entropically favored side. This is a subtle aspect of entropy, though, that won't be important going forward.*]

5.1.4–5.1.6 SUMMARY

- The sign and magnitude of $\Delta G°$ are related to the change in enthalpy ($\Delta H°$) and the change in entropy ($\Delta S°$) through Equation 5.4:

$$\Delta G° = \Delta H° - T\Delta S°$$

- The change in enthalpy is made up of two factors: bonds formed/broken and strain energy. A process with a negative value of $\Delta H°$ ($\Delta H° < 0$) is said to be exothermic (heat releasing) and is favored by the enthalpy term.
- Entropy is a measure of randomness or disorder. Reactions that have an increase in randomness or disorder, which have a positive value of $\Delta S°$ ($\Delta S° > 0$), are favored.

5.1.7 Reaction rates: the Arrhenius equation

We now shift our focus to considering how *quickly* equilibrium is achieved. For example, what is the rate of conversion of the anti conformation of butane to the gauche conformation? Moreover, can we infer anything about the **rate of reaction** based on $\Delta G°$? [*Spoiler: No.*]

Let's compare the experimentally observed relative rates of bond rotation in butane and but-2-ene. $\Delta G°$ for rotation around the C_2—C_3 single bond in butane is +0.9 kcal/mol (Section 5.1.3). $\Delta G°$ for rotation around the C_2—C_3 double bond in but-2-ene is +1.4 kcal/mol. With such similar $\Delta G°$ values, we would expect similar equilibrium constants. Reaction rates, however, couldn't be more different, as shown in **Figure 5.11**. Rotation around the single bond in butane occurs on the order of 10^{10} times per second at 298 K. [*Really fast. Rotation around a single bond is one of the fastest processes we will study in this course.*] On the other hand, the rotation around the C_2—C_3 double bond in but-2-ene is so slow that it virtually never happens at 298 K.

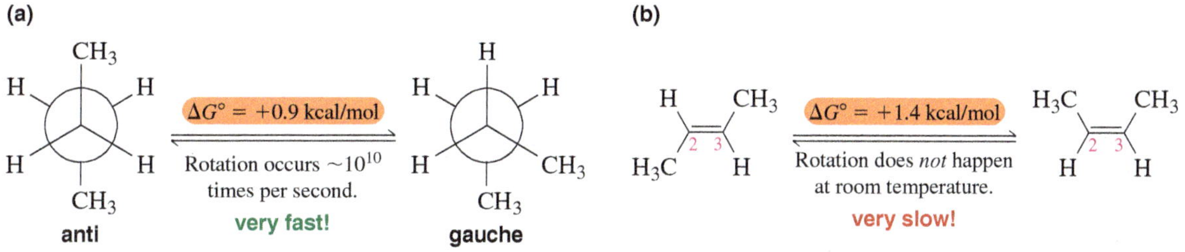

Figure 5.11 **Bond rotation** (a) is very fast around single bonds but (b) doesn't happen around double bonds.

Think about what these equations mean in terms of the concepts we are studying. Don't fall into the trap of just plugging numbers into equations.

Figure 5.11 demonstrates that there is little correlation between $\Delta G°$ and reaction rate. [*We will see some reactions where they are correlated, but it is important to keep equilibrium (based on $\Delta G°$) and reaction rate distinct in your mind.*] The rate of a chemical reaction is influenced by the **Arrhenius equation** (Equation 5.5), which calculates the **rate constant** (k_r) of a reaction as follows:

$$k_r = Ae^{-E_a/RT} \qquad \text{Equation 5.5}$$

where A = a constant called the frequency factor, E_a = activation energy, R = gas constant (1.987 cal/mol·K), and T = absolute temperature (K).

The reaction rate is not the same as the rate constant (k_r). The rate constant is a proportionality coefficient that relates the rate of a reaction at a given temperature to the concentration of reactants. For now, know this: ***A higher rate constant means a faster reaction rate***, all other things being equal. Of the four terms in the Arrhenius equation, there are two important variables, the **frequency factor** (A) and the **activation energy** (E_a). These we discuss in detail in Sections 5.1.8 and 5.1.9.

5.1.8 The frequency factor (A)

In order for a chemical reaction to occur, the reacting molecules must be in physical contact with each other. This happens when molecules in a solution randomly collide. For a chemical reaction, the frequency factor (A) is a measure of the frequency of the collisions that have the orientation required for reaction. Reactions that require two molecules to collide in a very specific way will have a low frequency factor and a low rate constant. A reaction with nonspecific requirements for collisions will have a relatively high frequency factor and a high rate constant.

Consider an analogy of shooting differently sized balls through a basketball hoop, where baskets made represent a successful reaction (**Figure 5.12**). Because ping-pong balls are small relative to the size of the hoop, there are few requirements placed on the trajectory of a ping-pong ball for it to go through the hoop. Few ping-pong balls will bounce off the rim; there will be few "failed reactions." A basketball (larger) requires a more specific trajectory to get through the hoop. There are far more "failed reactions" (missed shots). The ping-pong ball "reaction" has a higher frequency factor (A) and therefore a higher rate constant (k_r).

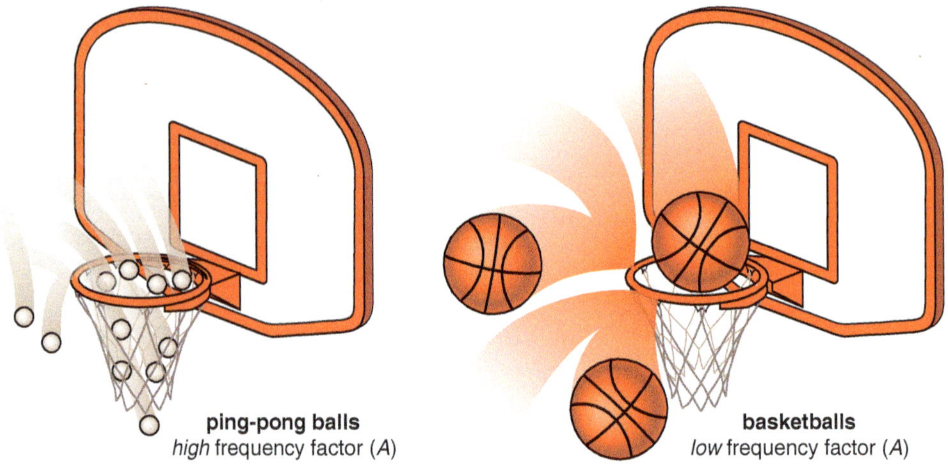

Figure 5.12 The smaller ping-pong balls have a smaller trajectory requirement and therefore go through the hoop with a higher frequency than the larger basketballs.

Is there a specific trajectory required for rotation around the C_2—C_3 single bond in butane and the C_2—C_3 double bond in but-2-ene? [*Think carefully about this before reading further.*]

Rotating around a bond requires very little in the way of a trajectory. Thus, it should happen with a relatively high frequency. As such, the value of A for both of these reactions should be high. And yet, the reaction rates, and therefore the rate constants, as mentioned in Section 5.1.7, are quite different. To understand why, we have to consider the activation energy (E_a).

5.1.9 Activation energy (E_a)

The **activation energy** (E_a) is a measure of the minimum energy needed for molecules to overcome the repulsion between their electron clouds when they collide. As a result, the $e^{-E_a/RT}$ term in the Arrhenius equation (Equation 5.5) represents the fraction of molecules that possess enough energy to allow a reaction to occur. While the value of E_a is typically determined experimentally, we can make good approximations of relative activation energies by taking into account the enthalpy and entropy of the specific collisions that lead to reactions.

The basketball analogy is also helpful here. Consider a scenario where we shoot a basketball through hoops at different heights (10 feet vs. 25 feet; Figure 5.13). The trajectory requirement is the same for both, so they have the same value for A (the frequency factor). It will take more energy, though, to get the ball to a height of 25 feet. The reaction at the higher hoop, with a higher activation energy, will be the slower reaction. That is, there will be fewer molecules (basketballs) with enough energy to react (reach the higher hoop). The lower hoop requires molecules (basketballs) of lower energy; there will be more of them and the reaction will happen more quickly. Given this qualitative analysis, let's correlate activation energy (E_a) and the rate constant (k_r), using the Arrhenius equation, $k_r = Ae^{-E_a/RT}$ (Table 5.4).

Figure 5.13 Scoring points at a lower basket requires less energy/happens more quickly.

Table 5.4 Correlation of E_a with rate constant (T = 298 K; A is constant).

E_a (kcal/mol)	$-E_a/RT$	k_r
0.5	−0.84	0.43 A
1.0	−1.7	0.18 A
2.0	−3.4	0.034 A
5.0	−8.4	2.2×10^{-4} A

Assuming a constant value of A, the rate constant decreases exponentially as the activation energy increases. [*Try to reproduce the values in Table 5.4 to ensure that you can use base* e *in calculations.*]

5.1.10 Temperature dependence

What is the significance of the $e^{-E_a/RT}$ term in the Arrhenius equation? When $E_a = 0$ kcal/mol, then $e^{-E_a/RT} = 1$ and $k_r = A$. [*Confirm this.*] All molecules would have enough energy to react as quickly as allowed by the frequency factor. With an extremely large activation energy, $e^{-E_a/RT}$ will be very small. Only a small fraction of molecules would react as quickly as allowed by A.

If the $e^{-E_a/RT}$ term represents the fraction of molecules that have the requisite activation energy for a successful collision to occur, and the activation energy is constant for a given reaction, then how can we increase the rate constant (k_r) of a reaction? We are raising e to a negative power, so to get a *larger* fraction, we need the E_a/RT term to be as *small* as possible. To make the E_a/RT term smaller, we must raise the temperature if E_a is constant. This should make sense mathematically and intuitively. [*If it doesn't, plug some numbers into a calculator.*] To maximize the number of collisions that occur, the kinetic energy of the molecules in the reaction should be increased. Higher molecular speed guarantees more collisions. This analysis, a Maxwell–Boltzmann type distribution, is demonstrated in **Figure 5.14**. [*These graphs may be familiar to you from previous chemistry classes.*]

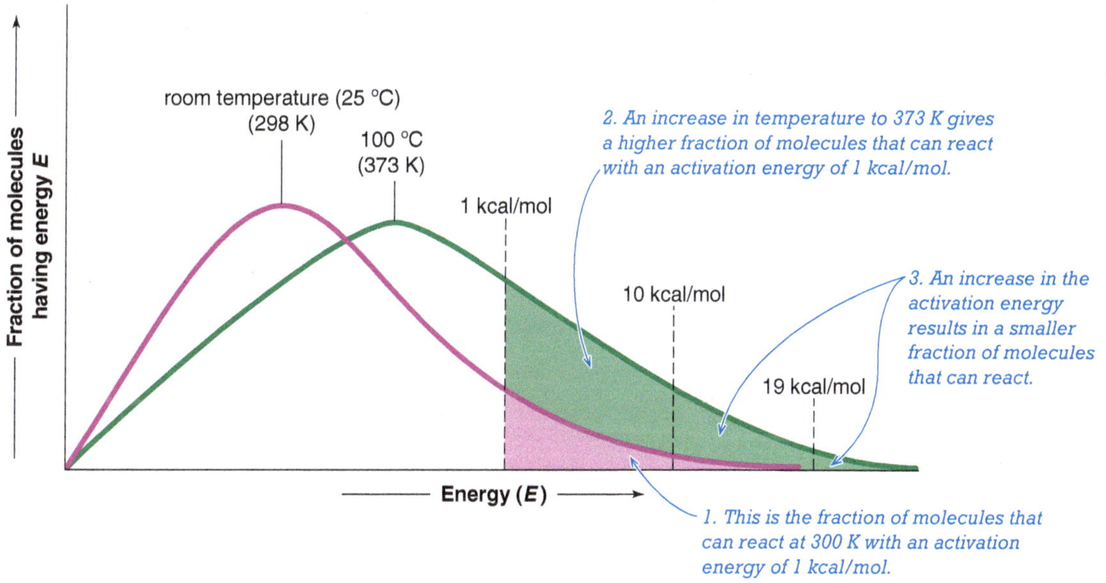

Figure 5.14 Temperature dependence of kinetic energy. The curves show the number of molecules that have a certain kinetic energy at a given temperature. Any molecules that have a kinetic energy higher than the activation energy can react. Thus, the shaded portion represents the fraction of molecules that can react at a given temperature and activation energy.

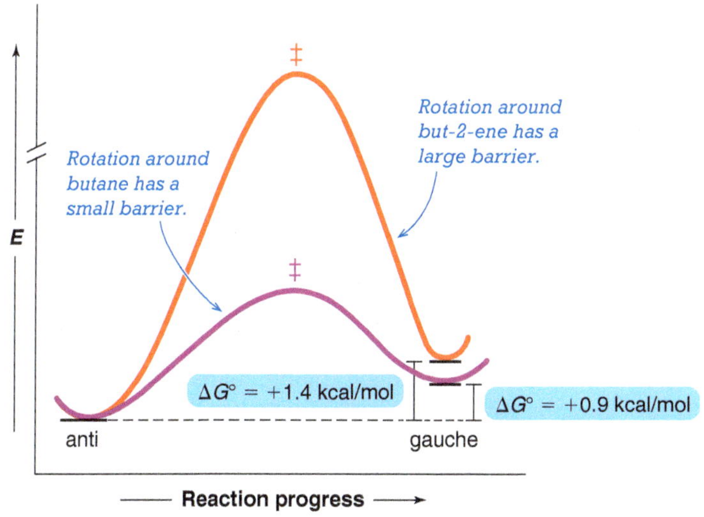

Figure 5.15 A slower reaction has a higher barrier to reaction.

To summarize the information in Figure 5.14, assume that a reaction requires 1 kcal/mol of energy to occur. At room temperature, there will only be a small fraction of molecules with that minimum energy value (shaded in purple). By heating up the reaction, and shifting the curve to the right, you increase the number of molecules that possess that minimum energy (shaded in green + shaded in purple). As a result, the rate constant (k_r) is increased.

Returning to rotation around the $C_2{-}C_3$ bonds of butane and but-2-ene, the activation energy for rotation around but-2-ene must be higher because the reaction rate is so much slower (**Figure 5.15**). Increasing the temperature would increase the rate of either bond rotation due to a larger rate constant. That a certain amount of kinetic energy is required for a molecule to react suggests the presence of a barrier to reaction. To understand this in a more qualitative way, we must ask, "How do molecules change as they *transition* from reactant to product?" [*Enter the transition state.*]

5.1.7–5.1.10 SUMMARY

- The rate of a reaction depends on the rate constant (k_r), a value that is a function of the frequency factor (A), the activation energy (E_a), and the temperature (T) through the Arrhenius equation:

$$k_r = Ae^{-E_a/RT}$$

- The frequency factor (A) is a measure of the frequency of collisions that have the orientation required for a reaction to occur.
- The activation energy (E_a) is the minimum energy needed for molecules to react and can be thought of as the barrier to reactivity.

5.1.11 Transition state

Figure 5.16 In terms of the ball rolling downhill, the transition state represents the top of the hill. The ball either continues to the product side or returns to the reactant side. It cannot sit at the transition state.

The **transition state** of a reaction is the highest energy state in a molecular collision that leads to reaction. Because the activation energy (E_a) is the difference in energy between the reactants and the transition state, the transition state itself represents the barrier that must be overcome for a reaction to proceed from reactants to products. The lifetime of a transition state is on the order of the time it takes for a single bond vibration to occur (Figure 5.16). Unlike products and intermediates, the transition state cannot be observed. A reaction with a high E_a will be slow, whereas a reaction with a low E_a will be fast.

Though it cannot be observed or isolated, we can approximate the appearance of the transition state and make reasonable guesses regarding the activation energy (E_a) and rate constant (k_r) of the reaction. The activation energy is often expressed in terms of the change in free energy of the transition state, or $\Delta G^{\circ\ddagger}$, a value which, as we will see, contains contributions from both $\Delta H^{\circ\ddagger}$ and $\Delta S^{\circ\ddagger}$ according to Equation 5.6. [*The double dagger is used to indicate that we are talking specifically about the transition state.*]

$$\Delta G^{\circ\ddagger} = \Delta H^{\circ\ddagger} - T\Delta S^{\circ\ddagger} \qquad \text{Equation 5.6}$$

Looking first at the faster reaction—namely, rotation around the C_2—C_3 bond in butane—the identity of the transition state is approximated by considering how the molecule moves from the anti to the gauche conformation. As shown in Figure 5.17, the molecule passes through the unstable eclipsed conformation, the highest energy species on the pathway from reactant to product. The energy of the transition state is calculated to be 3.6 kcal/mol by measuring the strain of each eclipsing interaction, using values from Section 3.4. Subtracting the energy of the strain-free anti conformation gives 3.6 kcal/mol as the activation energy of the reaction going from the anti conformation to the gauche conformation [Figure 5.17(a)]. The activation energy is different for the reverse reaction. [*Think about why that might be before continuing.*] The reverse reaction starts with the gauche conformation (0.9 kcal/mol strain) and moves to anti, passing through the same eclipsed transition state (3.6 kcal/mol strain). Thus, the activation energy of the reverse reaction would be 3.6 − 0.9 = 2.7 kcal/mol [Figure 5.17(b)]. This lower activation energy should make sense because going from a stable conformation to a less stable conformation should be slower than the reverse.

The activation energy (E_a or $\Delta G^{\circ\ddagger}$) has contributions from $\Delta H^{\circ\ddagger}$ and $\Delta S^{\circ\ddagger}$. [*Which is more important here?*] As we move from reactant to transition state, we went from one molecule to one molecule, suggesting that there is no change in randomness or disorder. As such, $\Delta S^{\circ\ddagger} \approx 0$, an assumption we are making to simplify this analysis. Thus, the entirety of the activation energy (E_a) arises from the strain, which is accounted for by $\Delta H^{\circ\ddagger}$.

The activation energy for rotation around the C_2—C_3 double bond of but-2-ene will be considerably higher. The C—C π bond is made up of overlapping, aligned *p* orbitals [*Section 2.6.3*]. Rotation around the π bond requires the *p* orbitals to briefly orient orthogonally (perpendicularly) to each other. In this conformation, there is no overlap and therefore no π bond. Thus, rotation around a C—C π bond effectively breaks the π bond (Figure 5.18). The activation energy for this reaction is the energy needed to break the C—C π bond [*shown in Appendix E as approximately 65 kcal/mol.*]

(a) Forward reaction:

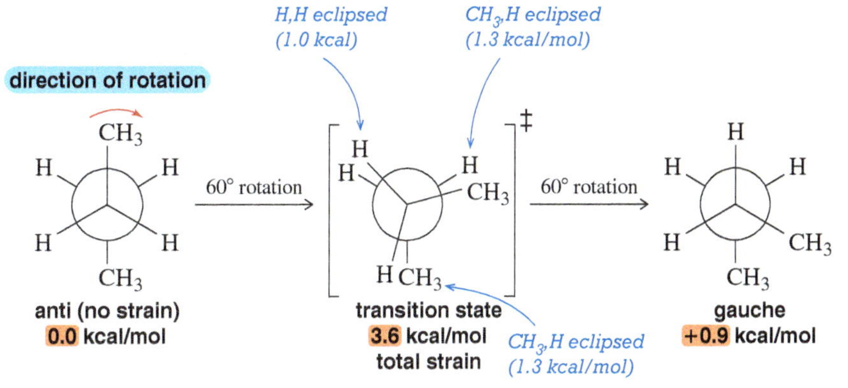

$$\Delta E^{\ddagger} = E^{\ddagger} - E_{\text{reactant}} = 3.6\frac{\text{kcal}}{\text{mol}} - 0.0\frac{\text{kcal}}{\text{mol}} = +3.6\frac{\text{kcal}}{\text{mol}}$$

(b) Reverse reaction:

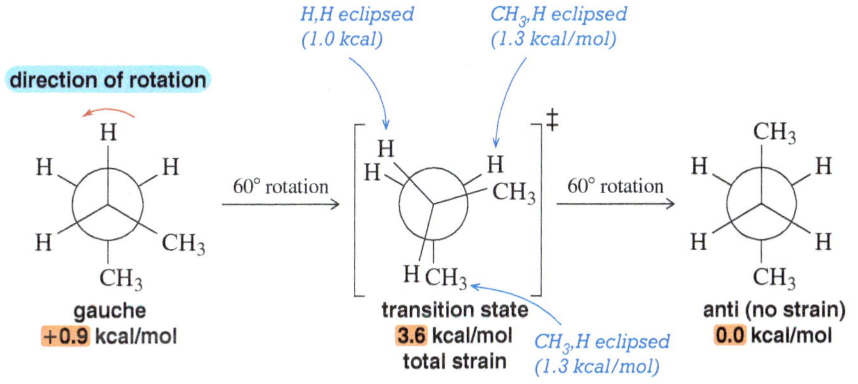

$$\Delta E^{\ddagger} = E^{\ddagger} - E_{\text{reactant}} = 3.6\frac{\text{kcal}}{\text{mol}} - 0.9\frac{\text{kcal}}{\text{mol}} = +2.7\frac{\text{kcal}}{\text{mol}}$$

Figure 5.17 The transition state of bond rotation is the eclipsed conformation. (a) For the forward reaction, the activation energy is 3.6 kcal/mol. (b) For the reverse reaction, the activation energy is 2.7 kcal/mol. [*The transition state is the same in either direction.*]

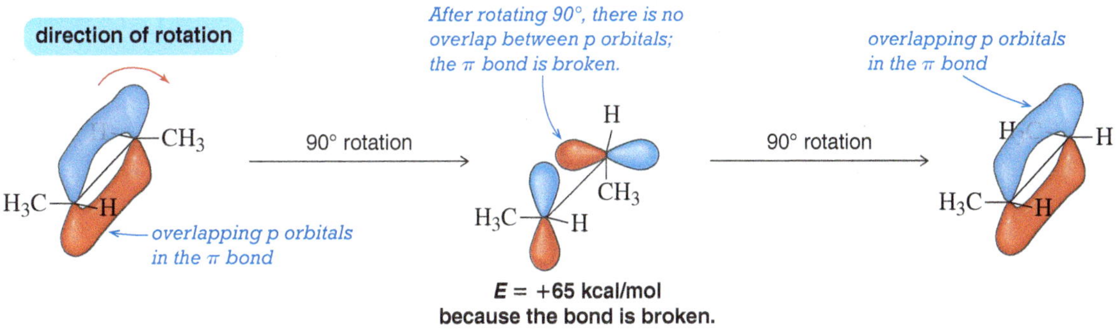

Figure 5.18 Rotation around a π bond requires enough energy to break the bond.

In summary, the rotation of but-2-ene has an extremely high barrier to rotation (~65 kcal/mol) in spite of a very slight difference in energy between the two alkene isomers (~1 kcal/mol). Alternatively, the barrier to rotation in butane is much smaller (3.6 kcal/mol) despite a similar difference in energy between the anti and gauche conformations (0.9 kcal/mol). By plugging these energies into the Arrhenius equation (Equation 5.5), assuming a constant value for A, bond rotation around the C—C single bond is shown to be ~1.1×10^{45} times faster (**Figure 5.19**).

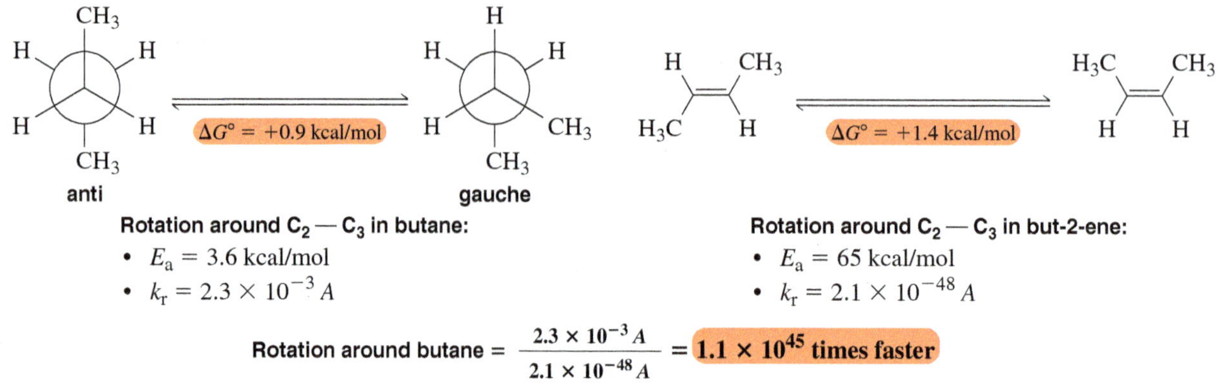

Figure 5.19 **Rotation around a single bond** is significantly faster than rotation around a π bond.

5.1.11 SUMMARY

- The transition state (‡) of a chemical reaction is the short-lived, highest energy chemical species that is the barrier on the pathway from reactant to product.

5.1.12 Bringing it all together: reaction coordinate diagrams

The rule that "the ball rolls downhill" can be graphically illustrated using a **reaction coordinate (energy) diagram,** which is a plot of the change in energy (y axis) against reaction progress (x axis). **Figure 5.20** shows the reaction coordinate diagram for rotation around the C_2—C_3 bond in butane. $\Delta G°$, E_a, the transition state, the reactant (anti), and the product (gauche) have been labeled. [*Study Figure 5.20 before continuing.*]

Figure 5.20 **The reaction coordinate diagram for the rotation around the C_2—C_3 bond in butane** provides a useful summary and graphical interpretation of what drives a chemical reaction.

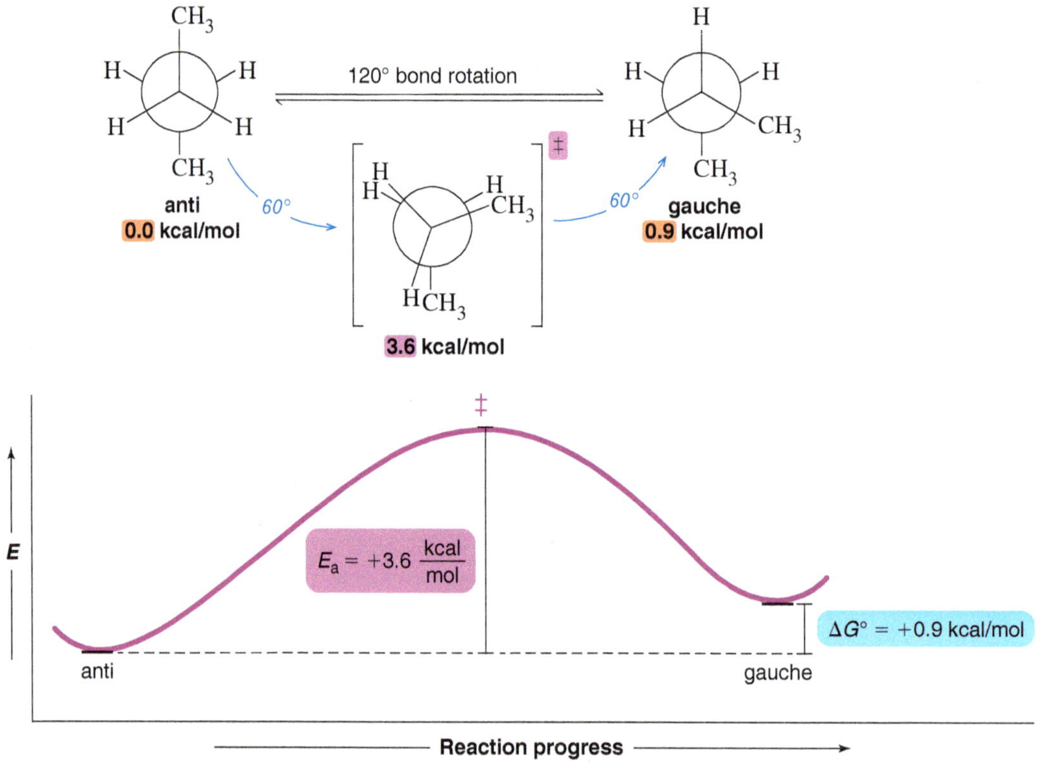

Reaction coordinate diagrams are incredibly useful for studying chemical reactions. Any concept involving the rate of a reaction or the favored side of an equilibrium can be illustrated with a reaction coordinate diagram. To demonstrate, let's ask and answer a few questions about the reaction coordinate diagram in Figure 5.20.

1. *Is the reaction **endergonic** or **exergonic**?* As written from left to right, the reaction is going uphill. Thus, using $\Delta G°(\text{total}) = \Delta G°(\text{products}) - \Delta G°(\text{reactants})$, we see that $\Delta G° > 0$, so the reaction is endergonic.
2. *Is the anti or gauche side of the reaction favored?* Because the gauche side is higher in energy, going from anti to gauche is net uphill, whereas gauche to anti is net downhill. As a result, anti is favored.
3. *Is the forward or the reverse reaction faster?* Because the transition state has a defined energy (at the top of the hill), the higher energy gauche conformation starts at a higher energy. As a result, it has less of a hill to climb. The activation energy is lower, so going from gauche to anti is the faster reaction.
4. *Is rotation around butane or butene faster?* For this, we need to consider what the reaction coordinate diagram would look like with each reaction on the same diagram (**Figure 5.21**). The rotation around butene has a significantly higher activation energy, so rotation around butane is much faster.

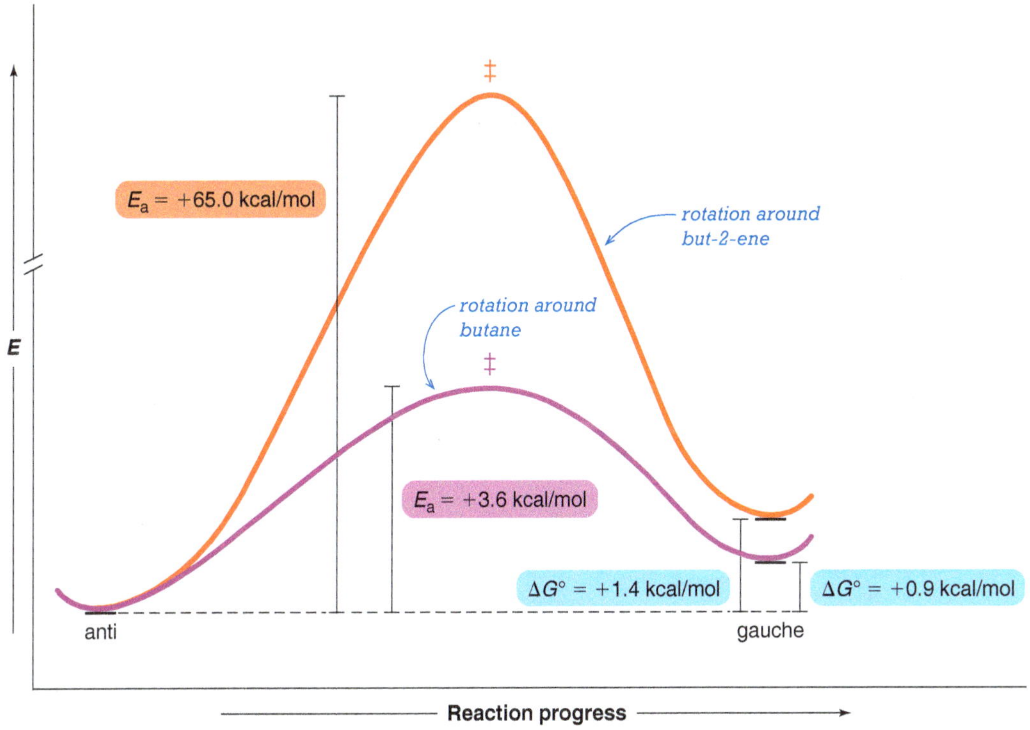

Figure 5.21 **Reaction coordinate diagram comparing the rotation around the C_2—C_3 single bond of butane and the C_2—C_3 double bond of but-2-ene.** [*Not to scale.*]

Reaction coordinate diagrams provide a qualitative picture of quantitative concepts. As we learn new reactions, they will all have different mechanisms. They may feature **reactive intermediates**, such as **carbocations**, **carbanions**, **free radicals**, or **carbenes**. They may be slow or fast, favored to the right or left. They may be acid-catalyzed, base-catalyzed, enzyme-catalyzed, or transition-metal-catalyzed. The vast amount of material to learn might be intimidating. That said, what every reaction has in common is the reaction coordinate diagram and the information it contains ($\Delta G°$, E_a, intermediates, reactants, products, transition states). [*With a little bit of practice, as in Assessments 5.20 and Assessments 5.21, we will be able to construct them on our own for all reactions whose mechanisms are known.*]

Assessment 5.20 Draw a reaction coordinate diagram for a one step reaction that has the following values of E_a and ΔH.

(a) $E_a = 9$ kcal/mol; $\Delta H° = +4$ kcal/mol
(b) $E_a = 2$ kcal/mol; $\Delta H° = -17$ kcal/mol
(c) $E_a = 5$ kcal/mol; $\Delta H° = 0$ kcal/mol

Assessment 5.21 Draw a reaction coordinate diagram for the following conformational equilibrium processes. Be sure to label the transition state. Would you expect (a) or (b) to be faster proceeding from left to right?

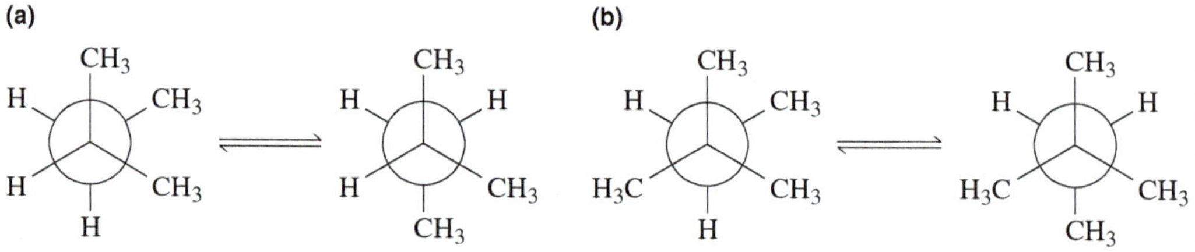

5-B CHEMISTRY IN CONTEXT: Green Chemistry

Catalysis

The rate of a reaction is determined by the activation energy (E_a)—the difference in energy between the reactant and the transition state (Sections 5.1.9–5.1.12). To increase the rate of a reaction, the activation energy needs to be lowered. A **catalyst** lowers the energy of, or stabilizes, the transition state of a reaction, resulting in a faster reaction. A catalyst itself is unchanged in a reaction, so it can be reused/recycled. Consider the enzyme-catalyzed formation of glucocerebroside, shown in **Figure 5.22**. The uncatalyzed reaction has a higher activation energy and is therefore slower.

How does a catalyst work? In many cases, the catalyst lowers the activation energy by providing an alternate route for product formation, often adding multiple steps to the reaction, with each step having a lower activation energy than the original, uncatalyzed reaction. How is it that a reaction can be faster, despite having more steps? As long as each step of the catalyzed reaction has a lower activation energy than the uncatalyzed reaction, the catalyzed reaction will be faster (**Figure 5.23**).

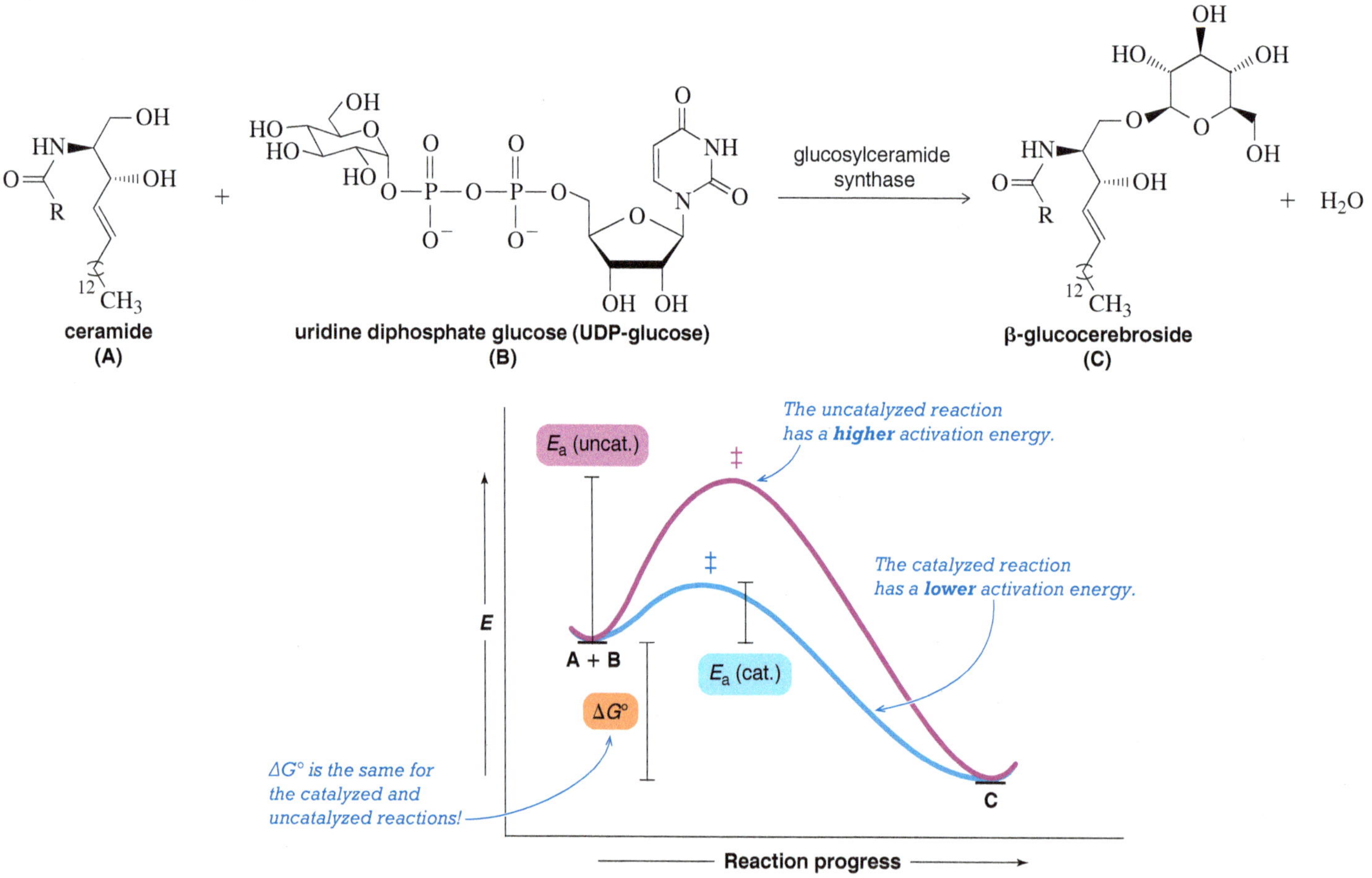

Figure 5.22 Catalysis lowers the activation energy of a reaction.

(continued)

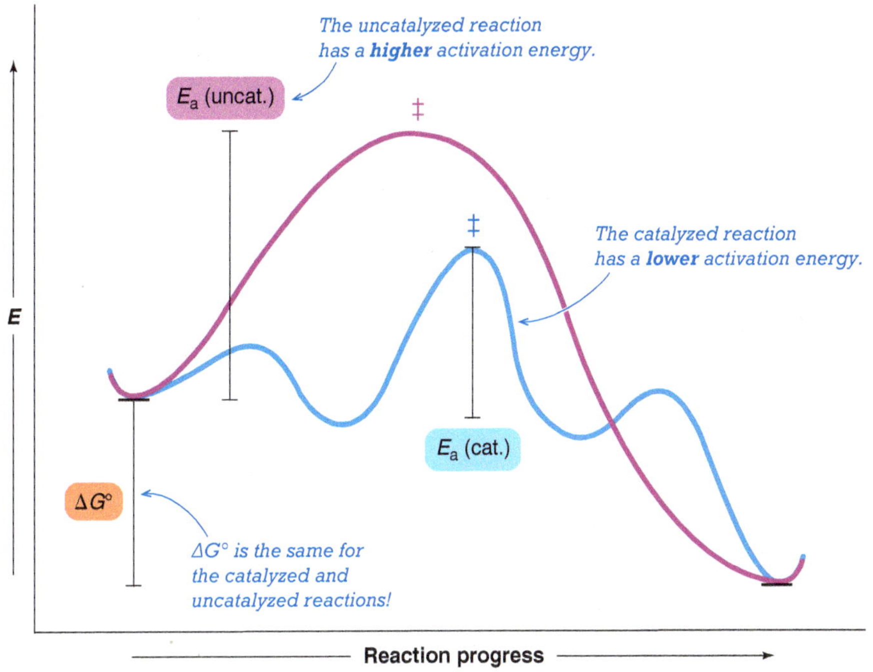

Figure 5.23 Catalysts often add steps, but because the activation energy of the slowest step is lower than the uncatalyzed reaction, the reaction is still faster.

5.2 Chemical reaction analysis: reactions of acids and bases

[*The overarching goal of this chapter is to use reaction analysis to qualitatively and quantitatively understand why and how reactions occur. Earlier, we applied this analysis to the rotation around a single bond; we now use the same approach to consider increasingly complex reactions.*]

Here in Section 5.2, we apply the logic developed in Section 5.1 to acid–base reactions from Chapter 4. [*You may want to return to Chapter 4 for a brief review before we continue.*] We begin with the reaction of ethoxide with water to produce ethanol and hydroxide (**Figure 5.24**). As always, we can draw a mechanism of the reaction, showing the electron-rich ethoxide attacking the acidic proton in water. [*Negatives attack positives; see Chapter 1.*] Our analysis of this reaction is much more quantitative than our analysis of bond rotation was in Section 5.1. [*Don't get intimidated by the numbers. Instead, use the numbers to reinforce your understanding of the chemical principles discussed in Section 5.1.*]

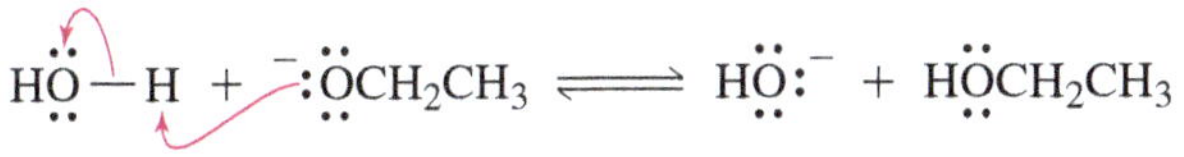

Figure 5.24 The reversible acid–base reaction of ethoxide and water.

5.2.1 $\Delta G°$ and K_{eq}

In Section 5.1, we used our knowledge of unfavorable steric or torsional interactions to calculate $\Delta G°$. Because the reaction of ethoxide and water is a simple acid–base reaction, we can calculate its equilibrium constant using Equation 4.1. From K_{eq} we can then calculate $\Delta G°$.

$$pK_{eq} = pK_a(\text{acid}) - pK_a(\text{conjugate acid}) \qquad \text{Equation 4.1(a)}$$

or

$$pK_{eq} = pK_a(\text{left-side acid}) - pK_a(\text{right-side acid}) \qquad \text{Equation 4.1(b)}$$

In order to use this equation, first identify the acid and base in the reaction. [*If you can't do this with relative ease, go back and review Section 4.2.*] In the forward reaction, water donates the proton, so it is the acid. Ethoxide accepts/grabs the proton, so it is the base.

Hydroxide and ethanol are the conjugate base and conjugate acid, respectively. The pK_a values of 15.7 for water and 16 for ethanol [*Table 4.5*] allow us to calculate the equilibrium constant (**Figure 5.25**). [*Before doing the math, predict qualitatively which side should be favored and by how much.*]

Figure 5.25 Calculation of K_{eq} using pK_a values.

acid | base | conjugate base | conjugate acid

$$\underset{pK_a = 15.7}{HO{-}H} + {}^{-}OCH_2CH_3 \rightleftharpoons HO^- + \underset{pK_a = 16.0}{HOCH_2CH_3}$$

$$pK_{eq} = pK_a\,(\text{acid}) - pK_a\,(\text{conjugate acid})$$
$$= 15.7 - 16.0 = -0.3 = -\log(K_{eq})$$
$$\log K_{eq} = 0.3$$
$$K_{eq} = 10^{0.3} = 2.0$$

Because water has a *lower* pK_a than ethanol, we know that it is a *stronger acid*. Thus, we would expect the equilibrium to favor the right side, where water is dissociated. We can also look at the *lower* pK_a of water to indicate that the *conjugate base* (hydroxide) is *more stable* than the base (ethoxide). The equilibrium favors the side containing the most stable conjugate base. In this reaction, that is hydroxide. Thus, $K_{eq} > 1$ is what we'd have expected for a reaction favored to the right. [*How about its magnitude?*] Given that the two pK_a values are very close, a K_{eq} near 1 is not surprising.

With a K_{eq} of 2.0, we can determine the relative concentration of the species present at equilibrium using the process outlined in **Figure 5.26**.

Figure 5.26 Using K_{eq} to determine the relative concentrations at equilibrium.

$$K_{eq} = \frac{[\text{product}]}{[\text{reactant}]} = \frac{2.0}{1} = \frac{[HO^-]\,[CH_3CH_2OH]}{[H_2O][CH_3CH_2O^-]}$$
$$2.0 = [HO^-]\,[CH_3CH_2OH]$$
$$[HO^-] = [CH_3CH_2OH]$$
$$\text{Thus, } \sqrt{2.0} = 1.41 = [HO^-] = [CH_3CH_2OH]$$
$$1 = [H_2O][CH_3CH_2O^-] \text{ and } [H_2O] = [CH_3CH_2O^-]$$
$$\text{Thus, } \sqrt{1} = 1 = [H_2O] = [CH_3CH_2O^-]$$

As shown in Figure 5.26, the more stable products are present in higher quantity at equilibrium. [*The ball rolls downhill, but how far?*] One way to quantify how far the ball has rolled downhill is by calculating $\Delta G°$ from K_{eq}, using Equation 5.2 (and assuming that $T = 298$ K; $R = 1.987$ cal/mol · K).

$$\Delta G° = -RT \ln K_{eq} \qquad \text{Equation 5.2}$$

Before doing the calculation, let's anticipate the result. [*Do you expect a large $\Delta G°$ based on $K_{eq} = 2.0$? What sign do you expect?*] A reaction favored as written ($K_{eq} > 1$) should have $\Delta G° < 0$. With a K_{eq} close to 1, the slightly favored reaction should be only slightly exergonic. [*Always think about what to expect to ensure your calculation is in the ballpark.*] These expectations are confirmed with the calculations in **Figure 5.27**. The value of $\Delta G° = -0.407$ kcal/mol indicates a slightly favorable acid–base reaction.

Figure 5.27 Conversion of K_{eq} to $\Delta G°$.

$$\Delta G° = -RT\ln K_{eq}$$
$$K_{eq} = 2.0,\ R = 1.987\,\frac{\text{cal}}{\text{mol}\cdot\text{K}},\ \text{and } T = 298\text{ K, so}$$
$$\Delta G° = -\left(1.987\frac{\text{cal}}{\text{mol}\cdot\text{K}}\right)(298\text{ K})\ln 2.0$$
$$= \left(-592\frac{\text{cal}}{\text{mol}}\right)(0.688) = -407\frac{\text{cal}}{\text{mol}} = -0.407\frac{\text{kcal}}{\text{mol}}$$

5.2.1 SUMMARY

- For an acid–base reaction, the **equilibrium constant** (K_{eq}) and $\Delta G°$ can be calculated using the following equation:

$$pK_{eq} = pK_a(\text{acid}) - pK_a(\text{conjugate acid})$$

or

$$pK_{eq} = pK_a(\text{left-side acid}) - pK_a(\text{right-side acid})$$

Assessment 5.22 For the following acid–base reaction, (a) identify the acid and conjugate acid, (b) calculate the equilibrium constant, (c) calculate the ratio of ammonia to propyne at equilibrium, (d) calculate $\Delta G°$ at 298 K, (e) calculate $\Delta G°$ at 273 K, and (f) calculate $\Delta G°$ at 373 K.

propyne ammonia

Assessment 5.23 For the following acid–base reaction, (a) identify the acid and conjugate acid, (b) calculate the equilibrium constant, (c) calculate the ratio of isopropanol to isopropoxide, (d) calculate $\Delta G°$ at 298 K, (e) calculate $\Delta G°$ at 273 K, and (f) calculate $\Delta G°$ at 373 K.

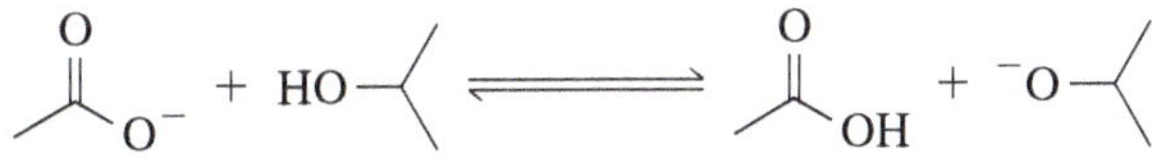

isopropanol isopropoxide

Assessment 5.24 For the following acid–base reaction, (a) identify the acid and conjugate acid, (b) calculate the equilibrium constant, (c) calculate the ratio of butan-2-ol to 2-butoxide, (d) calculate $\Delta G°$ at 298 K, (e) calculate $\Delta G°$ at 273 K, and (f) calculate $\Delta G°$ at 373 K.

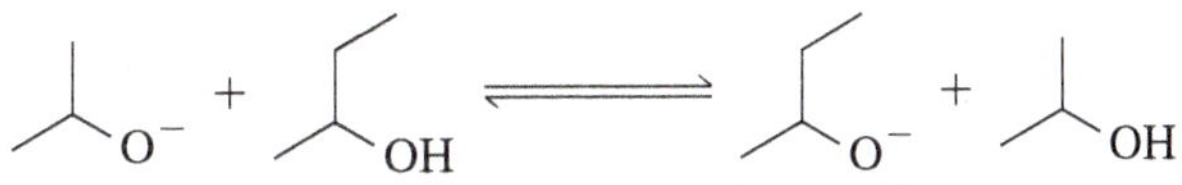

butan-2-ol 2-butoxide

5.2.2 $\Delta H°$ and $\Delta S°$

Knowing $\Delta G°$, we can quickly figure out the contribution of enthalpy and entropy to acid–base reactions using the equation $\Delta G° = \Delta H° - T\Delta S°$. Entropy is the simpler of the two quantities because the number of molecules does not change in every acid–base reaction: Two molecules on the left side of the reaction produce two molecules on the right side. The change in entropy, based on this assumption, is approximately zero (**Figure 5.28**).

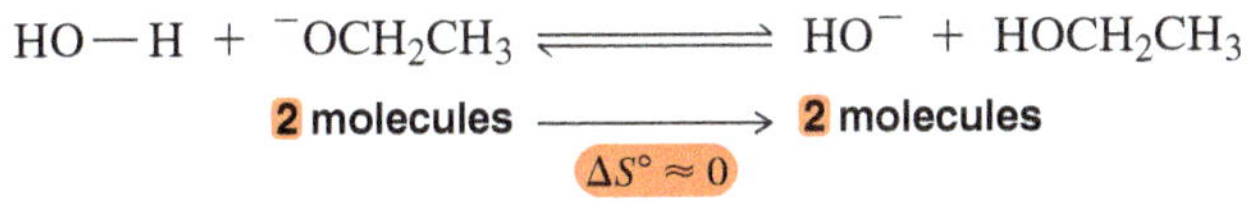

Figure 5.28 $\Delta S°$ **= 0 for Brønsted–Lowry acid–base reactions.**

The situation with $\Delta H°$ is somewhat trickier. With no contribution from entropy, $\Delta G° = \Delta H°$. From the calculations in Section 5.2.1, then, we know that $\Delta H°$ must equal –0.407 kcal/mol at ambient temperature (25 °C/298 K). In the rotation of butane around the $C_2—C_3$ bond, we attributed the change in enthalpy to the strain that was present/absent in one conformation or the other. But in an actual reaction, what is the source of $\Delta H°$?

Recall from our initial discussion about enthalpy that $\Delta H°$ is related to the change in bond energies as well as the change in strain. We know that bond formation releases energy, while bond breakage absorbs energy. In the case of the reaction between ethoxide and water, we are forming an O—H bond (to make ethanol) and breaking an O—H bond (to make hydroxide) (**Figure 5.29**). We are making and breaking very similar bonds, so $\Delta H°$ should be small. The amount of energy absorbed in breaking the first O—H bond should be close to the same amount of energy released when we form the second O—H bond. Why, then, is the value of $\Delta H°$ not exactly zero?

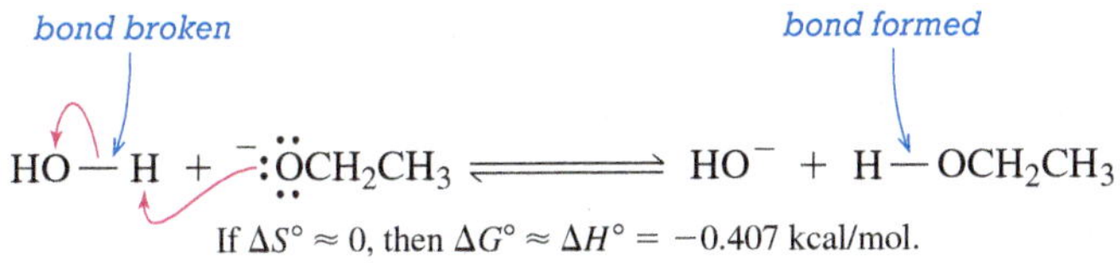

Figure 5.29 Despite a prediction based on bonds formed/broken, $\Delta H° \neq 0$**.**

To understand, we need to recall what we know about acid–base reactions (see **Table 5.5**, which is the same as Table 4.9). Because the strength of an acid is directly related to the stability of its conjugate base, stronger acids result in the formation of more stable conjugate bases. When a weak acid gives up a proton it forms an unstable and therefore more reactive base. The difference in $\Delta H°$ here has to do with the relative stability of the base and the conjugate base (ethoxide and hydroxide, respectively).

Table 5.5 Summary of the structural effects on acidity and conjugate base stability (copied from Table 4.9).

Structural effect	Representative acids	Conjugate bases	Description
Electronegativity (same row)	H_2S pK_a = 7 HCl pK_a = −7	HS^- Cl^- most stable	Chloride is more electronegative. The lone pair is held more tightly.
Size (same group)	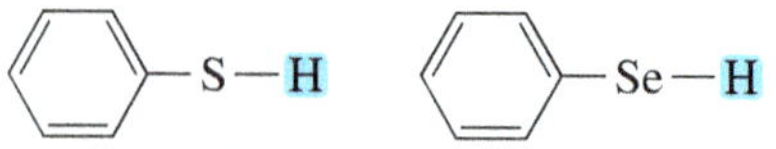 pK_a = 10.3 pK_a = 7.1	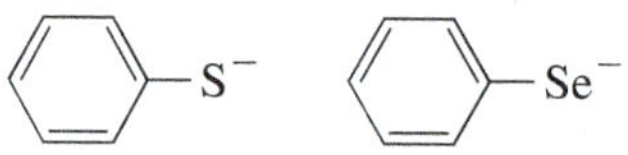 most stable	Selenium is larger. The negative charge is spread over a larger volume.
Resonance	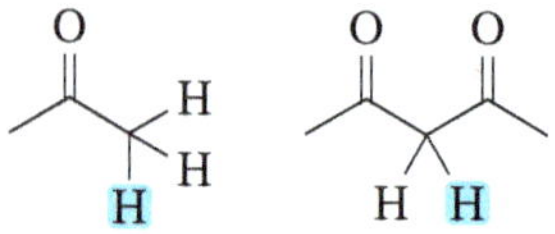 pK_a = 20 pK_a = 9	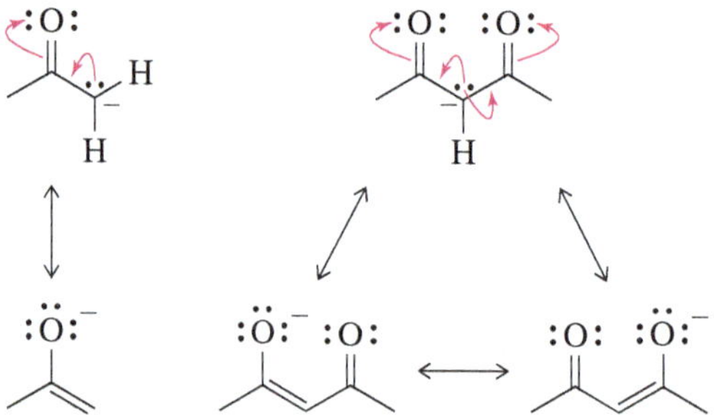 more stable	More resonance structures better delocalize the electrons/negative charge.
Induction	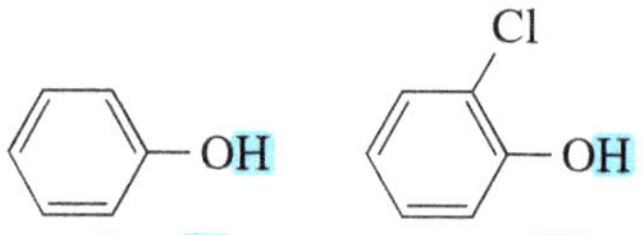 pK_a = 10 pK_a = 8.6	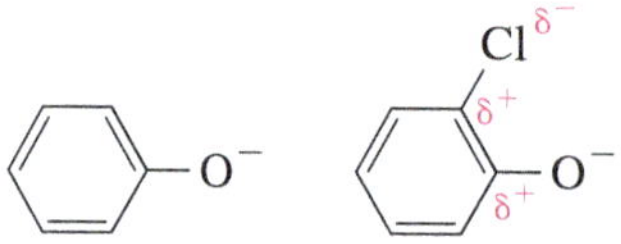 more stable	Electronegative atoms withdraw electron density, stabilizing nearby anions.
Hybridization	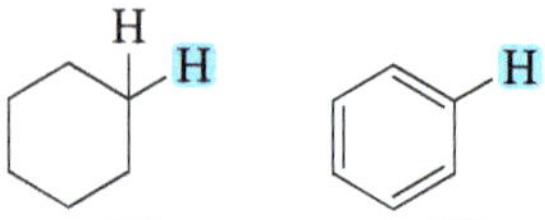 pK_a = 50 pK_a = 43	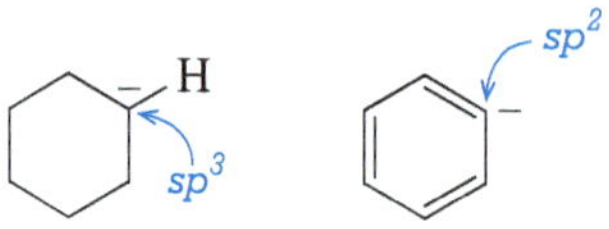 more stable	Orbitals with more *s* character hold electrons closer to the nucleus.

From Table 5.5, the comparison of hydroxide and ethoxide is straightforward, with the only difference being the alkyl group attached. Alkyl groups are electron-releasing by **induction** (and **hyperconjugation**, a concept we discuss in greater detail in Section 5.3.6). By donating electron density to oxygen, the negative charge on oxygen increases, resulting in ethoxide being less stable than hydroxide. The difference in stability between the two bases accounts for the overall negative $\Delta H°$ in this reaction.

This effect becomes even more pronounced when we consider the acid–base reactions in Assessments 5.22 and 5.23. In Assessment 5.22, a large negative value for $\Delta G°$ makes sense because propyne (pK_a = 25) is much more acidic than ammonia (pK_a = 38) and the reaction is heavily favored to the right (**Figure 5.30**). The acetylide ion is the more stable conjugate base because the lone pair of electrons is contained in an *sp*-hybridized orbital. Because the orbital has 50% *s* character, its close proximity to the nucleus results in a stabilizing effect. $\Delta S° \approx 0$, because there are two reactants and two products, so $\Delta G° \approx \Delta H° = -17.7$ kcal/mol, most of which is due to the increased stability of the conjugate base.

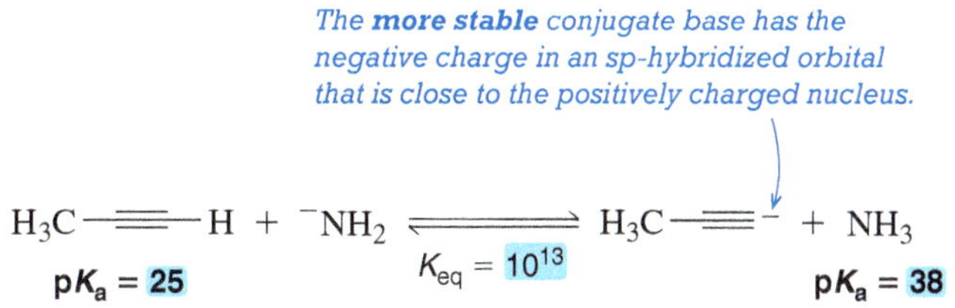

$$\Delta G^\circ = -RT\ \ln K_{eq} = -\left(1.987\frac{\text{cal}}{\text{mol·K}}\right)(298\ \text{K})(\ln 10^{13})$$

$$\Delta G^\circ = \Delta H^\circ = -17{,}700\frac{\text{cal}}{\text{mol}} = -17.7\frac{\text{kcal}}{\text{mol}}$$

Figure 5.30 The negative ΔG° (and ΔH°) results from forming the more stable conjugate base as written.

The acid–base reaction in Assessment 5.23 is unfavorable (**Figure 5.31**). $K_{eq} = 10^{-11}$, so $\Delta G^\circ = +15.0$ kcal/mol. There is no change in entropy for this reaction because we go from two reactants to two products; thus, $\Delta S^\circ \approx 0$ and $\Delta G^\circ \approx \Delta H^\circ = +15.0$ kcal/mol. ΔH° is largely positive due to the relative stability of the conjugate bases. In this case, **resonance** stabilizes the anion in acetate, but not in isopropoxide. As a result, acetate is more stable.

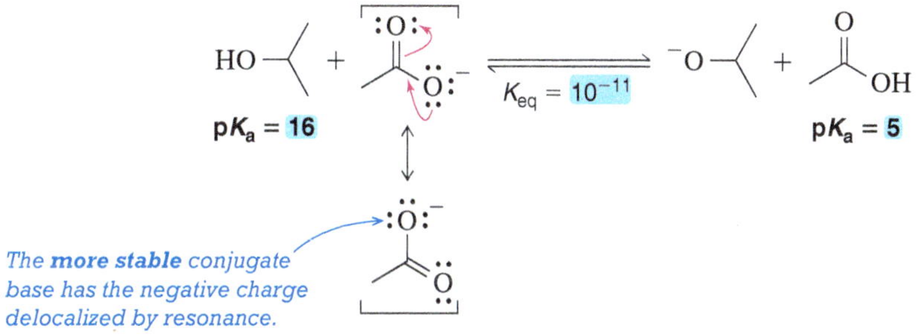

$$\Delta G^\circ = -RT\ \ln K_{eq} = -\left(1.987\frac{\text{cal}}{\text{mol·K}}\right)(298\ \text{K})(\ln 10^{-11})$$

$$\Delta G^\circ = \Delta H^\circ = 15{,}000\frac{\text{cal}}{\text{mol}} = 15.0\frac{\text{kcal}}{\text{mol}}$$

Figure 5.31 The positive ΔG° (and ΔH°) results from forming the less stable conjugate base as written.

5.2.2 SUMMARY

- Because acid–base reactions are essentially entropically neutral ($2 \rightarrow 2$, $\Delta S^\circ \approx 0$), ΔG° ends up being equal to ΔH°. ΔG° and ΔH° calculated from K_{eq} in acid–base reactions are explained by comparing the stability of conjugate bases.

Assessment 5.25 Calculate ΔG°, ΔH°, and ΔS° for the following acid–base reactions. Rationalize the value of ΔH° based on the structure of the conjugate bases. [*Assume T = 298 K.*]

(a)

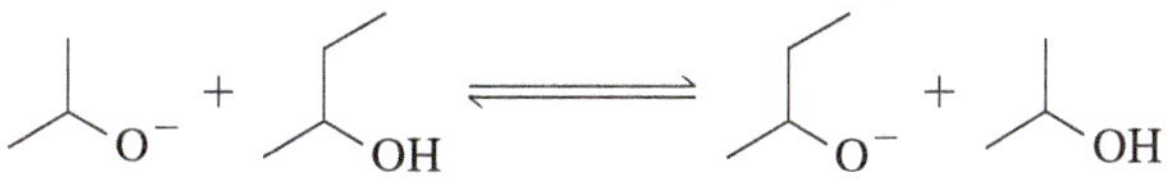

(b)

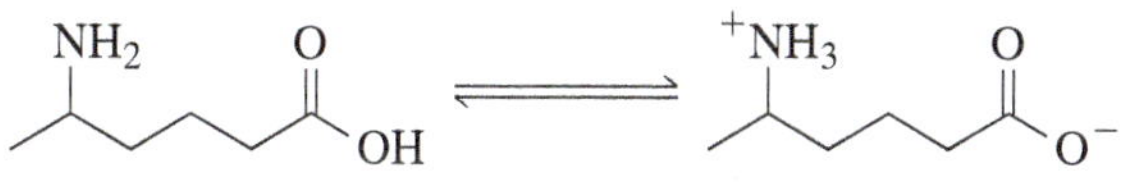

(c)

$$H_3O^+ + Br^- \rightleftharpoons H_2O + HBr$$

5.2.3 Acid–base reactions: rate of reaction, activation energy, frequency factor (*A*), and the transition state

Generally, acid–base reactions are very fast, though not as fast as rotating around a σ bond. Here in Section 5.2.3 we examine why. As a reminder, the Arrhenius equation ($k_r = Ae^{-E_a/RT}$, Equation 5.5) tells us the rate constant of a reaction (k_r) is a function of the activation energy (E_a), the frequency factor (A), and the temperature. The activation energy is measured as the difference in energy between the reactant and the transition state, so the structure of the transition state is important in predicting the rate of a reaction. Because acid–base reactions are the first reactions we've studied for which the frequency factor is important, our analysis begins here.

For bond rotation, the frequency factor is high because no collision is required to progress from reactant to product. A collision *is* required, however, for an acid–base reaction to occur. Think about how an acid–base reaction occurs, using the arrow-pushing formalism. As is the custom, we think of reactions as electron-rich species attacking electron-poor species. This mechanism is illustrated in **Figure 5.32(a)**. [*But what's really happening in solution?*] While the arrow-pushing formalism shows the base seeking out the acid, both species are moving rapidly (and randomly) in solution. Reaction only occurs when a productive collision happens. **Figure 5.32(b)** shows a series of potential collisions. The only one that is productive features the negatively charged oxygen of ethoxide colliding with the proton of water.

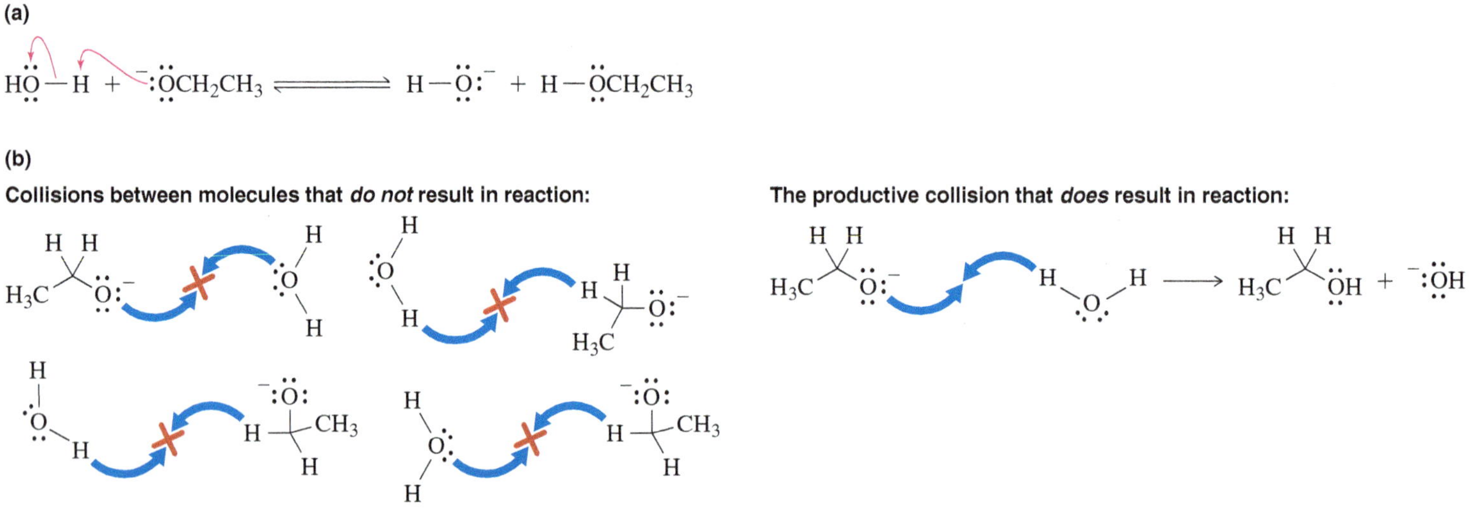

Figure 5.32 The acid–base reaction between ethoxide and water. (a) The arrow-pushing mechanism. (b) Only productive collisions result in reactions. [*The arrows in (b) are not mechanism arrows. They are meant to indicate the movement of the molecules as they collide.*]

There is not much steric hindrance, so a high percentage of collisions result in a productive reaction, and the frequency factor for an acid–base reaction is relatively high. As **Figure 5.33** shows, attack via a number of pathways (using our arrow-pushing formalism) can result in reaction. We will eventually study reactions with specific trajectory and steric requirements. These will have a much lower frequency factor. Assessment 5.26 tests your understanding of this concept.

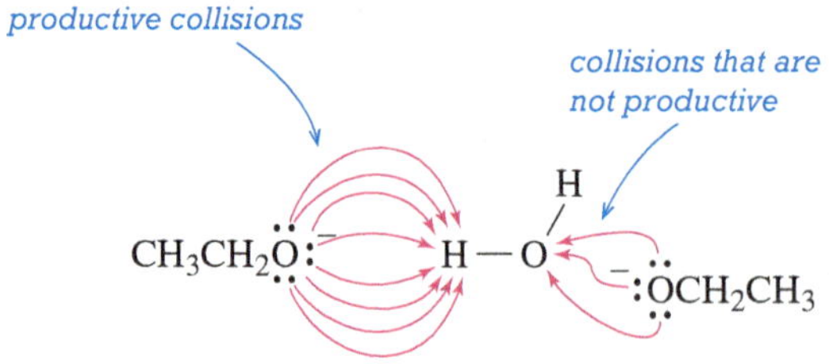

Figure 5.33 Acid–base reactions are fast because there is little steric hindrance. A high percentage of collisions are productive.

Assessment 5.26 Within the following pairs, pick which reaction you would expect to be faster based on having a higher value of the frequency factor (A).

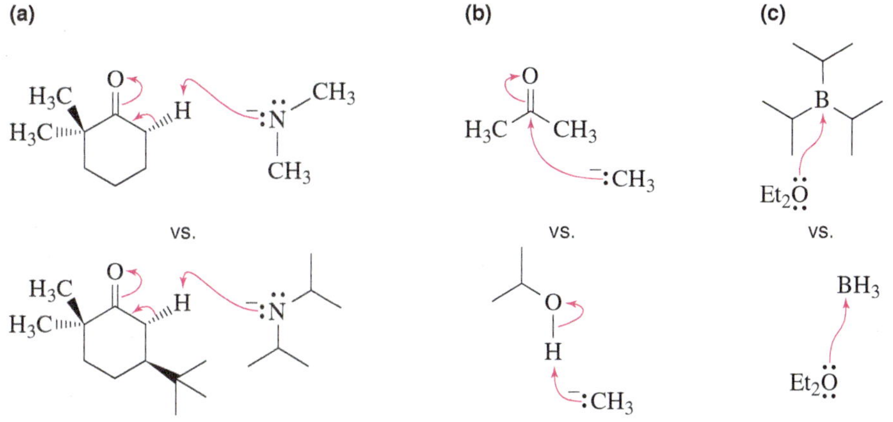

Now let's consider the activation energy, E_a. The first thing we have to look at is the transition state of the reaction. The arrow-pushing formalism is a relatively good predictor of what the transition state might look like. For example, in **Figure 5.34(a)**, the arrow-pushing formalism shows that an O—H bond (ethanol) is formed and an O—H bond (water) is broken. Using dashed lines to show bonds that are partially formed and partially broken, we can depict the transition state as shown in **Figure 5.34(b)**. As the reaction progresses, the −1 charge that was on ethoxide is now a partial charge as the O—H bond forms. Similarly, the oxygen in water is beginning to develop a negative charge as it loses the proton. The complex is shown in brackets with a double dagger (‡) to signify that it is the transition state.

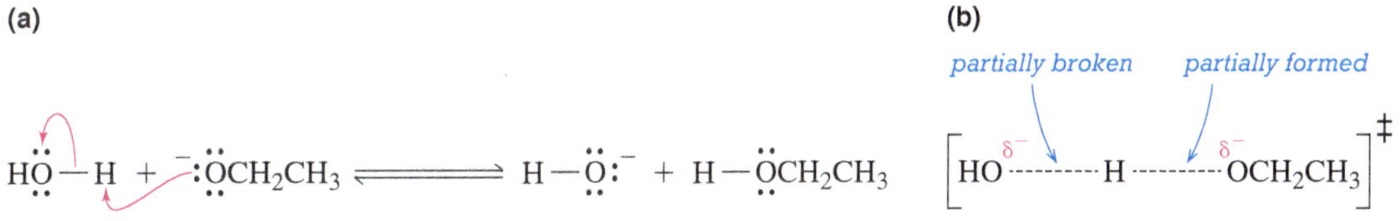

Figure 5.34 The transition state of an acid–base reaction. (a) The arrow-pushing formalism predicts (b) the transition state shown with partially formed and partially broken bonds.

The activation energy, E_a, is the difference in energy between the reactants and the transition state. Put differently, it is the change in energy as we go from reactants to the transition state. With bonds breaking and forming, enthalpy must be involved. The enthalpy of the transition state is denoted as $\Delta H^{\circ\ddagger}$. The entropy ($\Delta S^{\circ\ddagger}$) also changes as two molecules meet to form the transition state complex. The activation energy (E_a) can therefore be expressed as $\Delta G^{\circ\ddagger}$ and calculated using Equation 5.6 ($\Delta G^{\circ\ddagger} = \Delta H^{\circ\ddagger} - T\Delta S^{\circ\ddagger}$).

The transition state features the simultaneous breaking and forming of an O—H bond, suggesting a relatively small change in bond energy compared to the reactants. Furthermore, in this particular case, the partial negative charge is similarly tolerated on both oxygens. In sum, we expect a relatively small value for $\Delta H^{\circ\ddagger}$ for this reaction. Entropically, the transition state is more highly ordered. It can be thought of as a single molecule, despite the molecules being connected only by partial bonds. Thus, two reactant molecules result in one transition state "molecule." This results in a negative (and therefore) disfavored value of $\Delta S^{\circ\ddagger}$.

This reaction is plotted on the reaction coordinate diagram in **Figure 5.35**. It is slightly exergonic ($\Delta G° = -0.407$), and the cumulative effect of a neutral $\Delta H^{°‡}$ and negative $\Delta S^{°‡}$ results in a slight activation barrier for the reaction. [*We can't really predict values for* $\Delta S^{°‡}$, *so we need to approximate its effect.*] The reverse reaction has a slightly greater activation energy because the products are more stable than the reactants (Section 5.1.12). Combined with the fact that there is not a significant steric or directional requirement for productive collisions, this is a fast reaction in either direction.

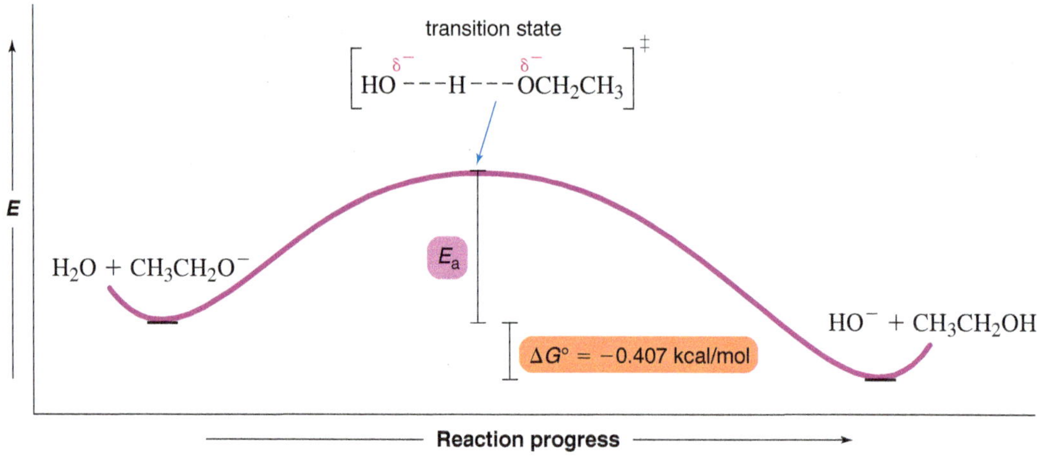

Figure 5.35 A reaction coordinate diagram for a slightly favored acid–base reaction.

5.2.3 SUMMARY

- In terms of the Arrhenius equation (Equation 5.5), acid–base reactions have a high frequency factor and a relatively low activation energy, making them some of the fastest reactions we'll study. The transition state of an acid–base reaction shows the bond forming and bond breaking that occur as the reaction progresses from reactant to product.

Assessment 5.27 For the reaction coordinate diagram shown, is the forward or reverse reaction faster?

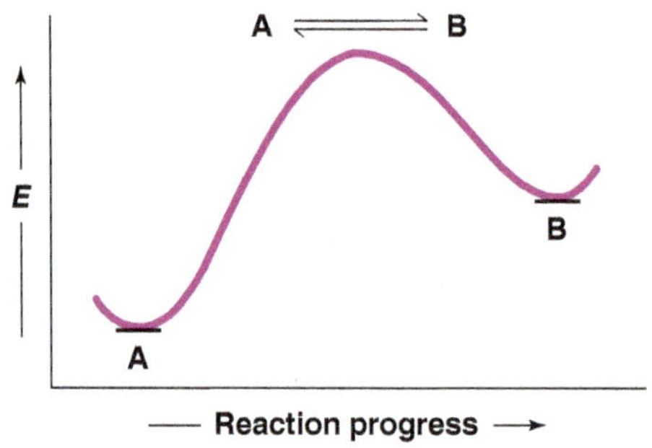

5-C CHEMISTRY IN CONTEXT: Biochemistry

Enzymes as catalysts

The activation energy (i.e., the difference in energy between the reactant and the transition state) determines the rate constant of a chemical reaction. To increase the rate of a chemical reaction, catalysts lower this energy. Enzymes accomplish this in several ways, which are discussed throughout this text (see Chemistry in Context 8-C, 19-B, and 20-B). One way for enzymes to lower the activation energy is to increase the reactivity of the involved functional groups. Consider first the hypothetical, uncatalyzed mechanism in **Figure 5.36**. Ceramide is not strongly Lewis basic, and the leaving group in UDP-glucose is not that good. [*Don't worry about this terminology now; focus on the big picture.*] Thus, this reaction would be slow.

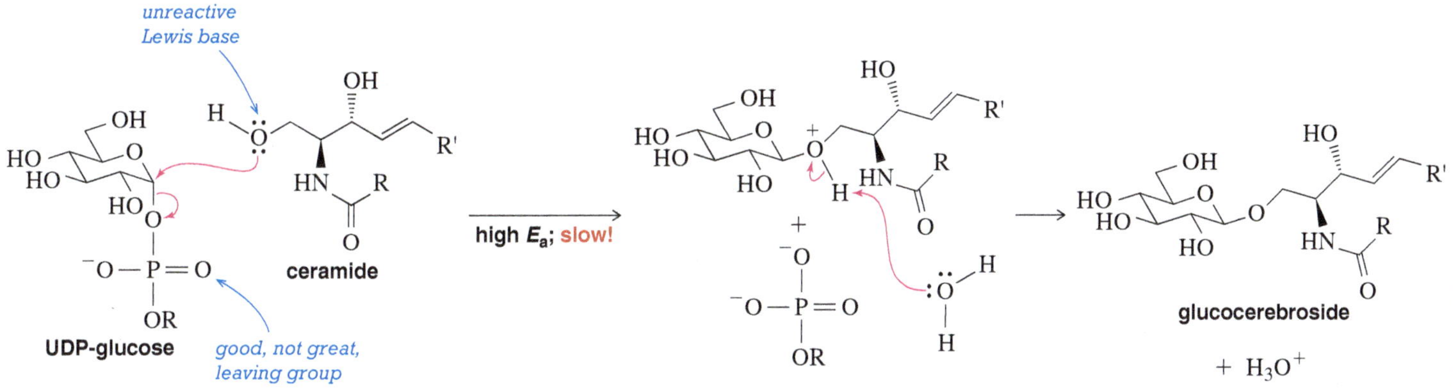

Figure 5.36 Uncatalyzed formation of glucocerebroside by this hypothetical mechanism is slow. [*We introduce more about this reaction in Chapter 12.*]

Figure 5.37 shows a hypothetical mechanism by which glucosylceramide synthase addresses the challenges in the reaction. Based on analogous glycosylation reactions with known mechanisms, an active site base will make the ceramide hydroxyl group more reactive. At the same time, an active site acid will make formation of the UDP leaving group more favorable. In doing these things, the activation energy is decreased and the reaction rate increases.

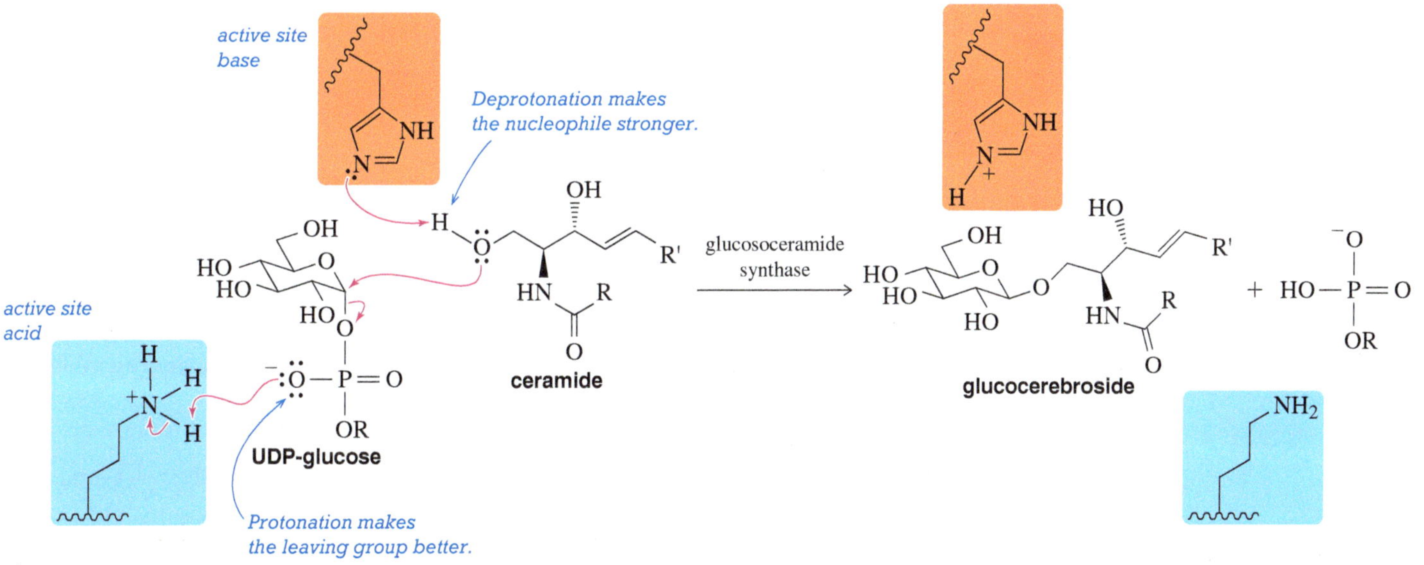

Figure 5.37 Hypothetical mechanism for catalysis by glucosoceramide synthase.

Assessment 5.28 Comparing the catalyzed and uncatalyzed reaction on a reaction coordinate diagram should clearly show that the catalyzed reaction has a lower activation energy and should proceed more rapidly than the uncatalyzed reaction. Compare the reactions in Figures 5.36 and 5.37 on a reaction coordinate diagram.

5.2.4 A closer look at transition states: the Hammond postulate

Figure 5.34 intentionally showed the bonds equally broken and formed in the transition state. [*Confirm that this is true.*] Given that transition states are very short-lived and cannot be visualized with modern instrumentation, how do we know that this picture of the transition state is correct? Can we really know exactly what the transition state looks like?

Figure 5.38 shows three hypothetical transition states for an acid–base reaction. [*See if you can spot the differences.*] The total charge in each possibility adds up to −1, but that charge is unequally distributed in possibilities 1 and 3. In possibility 2, however, the charge is spread equally between the two, with A and B each having a −1/2 charge.

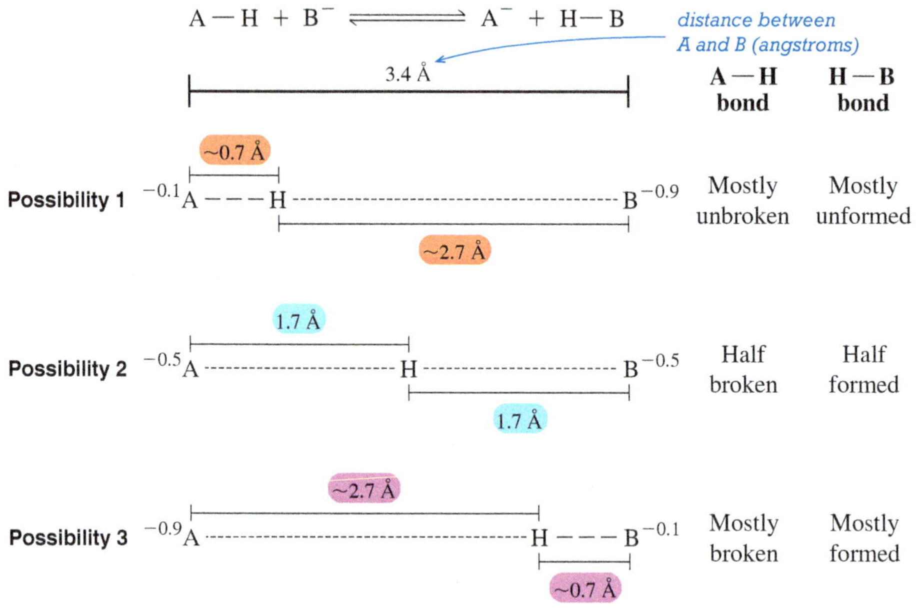

Figure 5.38 Three possible transition states for an acid–base reaction. The total charge in each scenario is −1, but it is distributed differently in each. The length of the bond is meant to illustrate the degree to which it has been broken/formed.

If you accept that these are different transition states [*and they are*], you might wonder whether it really matters which one it is [*it will*] and how we are going to figure out which is correct since we can't visualize them. Enter the **Hammond postulate**, which, as hypothesized by George Hammond in 1953, states that when a transition state leading to an unstable reaction intermediate (product) has nearly the same energy as that intermediate (product), the two are interconverted with only a small reorganization of molecular structure. [*Huh?*] A second, simpler way to say this is that the transition state will bear greater resemblance to the species (reactant, product, or intermediate) that is of higher energy.

A third way of stating the Hammond postulate is as a combination of two rules:

1. For an exothermic reaction (step), the transition state will resemble the reactants, but
2. for an endothermic reaction (step), the transition state will resemble the product (intermediate).

[*Carefully look over these three statements and see that they are really just three ways of saying the exact same thing.*]

What does it mean to resemble, or to be of similar energy to, reactants or products? Return to the three transition states shown in Figure 5.38. For the reaction as written, which transition state resembles the reactants? Which resembles the products? [*Try to answer this yourself before reading on. Hint: Look at the location of the charge.*]

The Hammond postulate is not an easy concept at first. Read through this section again if you don't understand it yet. It will become clearer when applied to a reaction coordinate diagram that shows how close in energy the transition state is to the product or reactant.

Possibility 1, with A—H fully formed and the negative charge almost completely on B, resembles the reactants. This is because, as written, the product of the reaction has H—B fully formed and the negative charge on A. Possibility 1, then, is a transition state for an exothermic reaction. Alternatively, possibility 3, with the H—B bond almost fully formed and the negative charge on A, resembles the products. Possibility 3 is reasonable for an endothermic reaction. Possibility 2 falls right in the middle, bearing little resemblance to either the product or the reactant. You might expect such a transition state when $\Delta G° = 0$, which is very close to the case for the main reaction we have studied in this section (Figure 5.34).

The idea of the transition state being "closer in energy" to products or reactants is best illustrated using reaction coordinate diagrams because these plot energy as a function of reaction progress. Recall Assessments 5.22 and 5.23 where we studied a favorable ($\Delta G° < 0$) and an unfavorable ($\Delta G° > 0$) acid–base reaction. [*What does the Hammond postulate tell us about their transition states?*]

The reaction coordinate diagram for the reaction of propyne and amide (Assessment 5.22) shows a significantly downhill reaction ($\Delta G° < 0$) (**Figure 5.39**). The structure of the transition state is closer in energy to the reactants than it is to the products. Because the energies are so similar, there must be little molecular reorganization in going from the reactants to the transition state. In the transition state the amide has just started to form a bond with the proton of propyne. The amide then has the bulk of the negative charge in this reactant-like transition state.

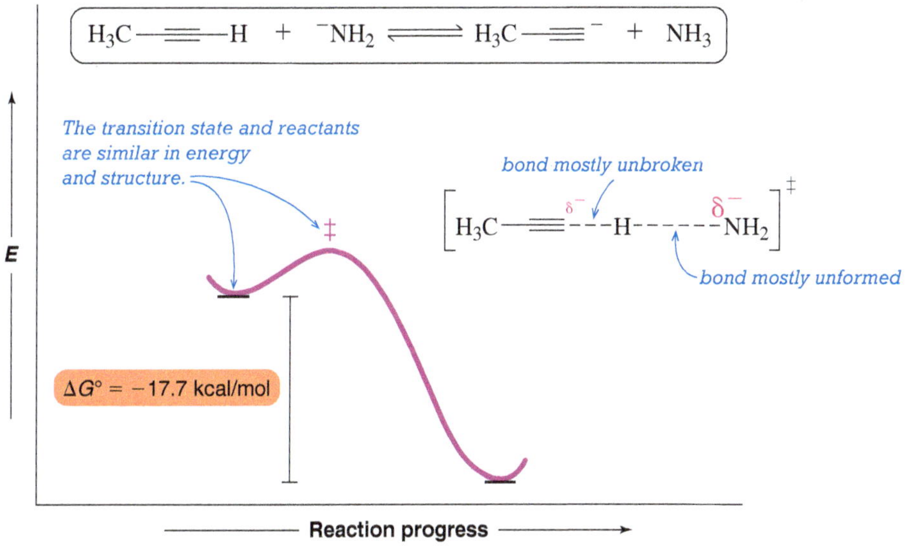

Figure 5.39 For an exothermic reaction, the transition state is closer in energy to the reactants.

Applying similar logic to the disfavored reaction of acetate and isopropanol ($\Delta G° > 0$; Assessment 5.23), the reaction coordinate diagram (**Figure 5.40**) shows a transition state closer in energy to the products. That is, there is little molecular reorganization as the reaction moves to products from the transition state. Within the transition state, the majority of negative charge is on the oxygen of isopropanol.

As long as we know the arrow-pushing mechanism, can identify the bonds that are broken and formed in the reaction step, and can estimate whether the step is endergonic or exergonic, then we can use the Hammond postulate to confidently infer the identity of the transition state.

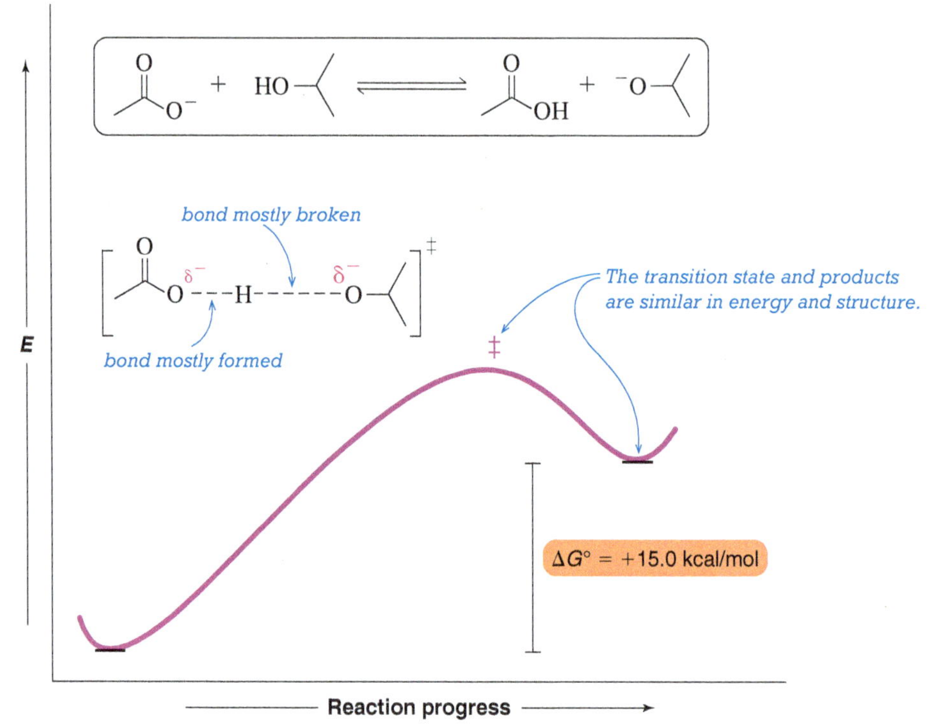

Figure 5.40 For an endothermic reaction, the transition state is closer in energy to the products.

5.2.4 SUMMARY

- An endothermic reaction (step) has a transition state that is of similar structure and energy to the product(s) [intermediate(s)] of that reaction (step). An exothermic reaction (step) has a transition state that is of similar structure and energy to the reactant(s) [intermediate(s)] of that reaction (step).

Assessment 5.29 For the following acid–base reactions studied in Assessment 5.25, draw a likely transition state. Be sure to indicate in your drawing the degree to which bonds are broken or formed.

(a)

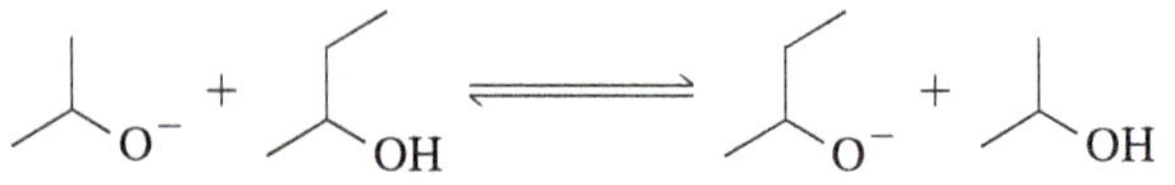

(b)

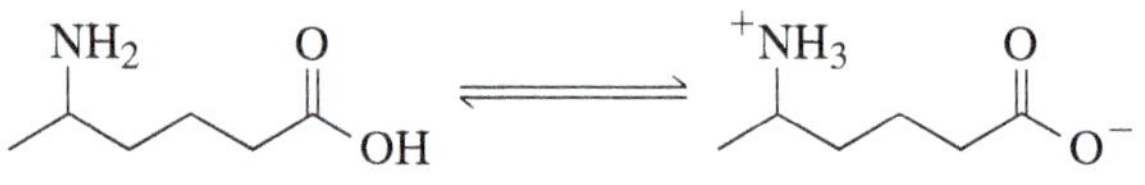

(c)

$H_3O^+ + Br^- \rightleftharpoons H_2O + HBr$

Assessment 5.30 Consider two marathon courses with differing elevation changes. For each, when you reach the highest elevation, do you feel you are (a) closer to having just started or (b) nearing the finish of the race? How does this analogy relate to the Hammond postulate?

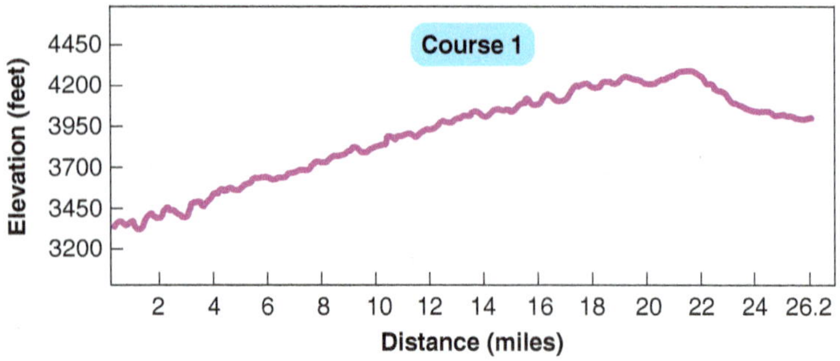

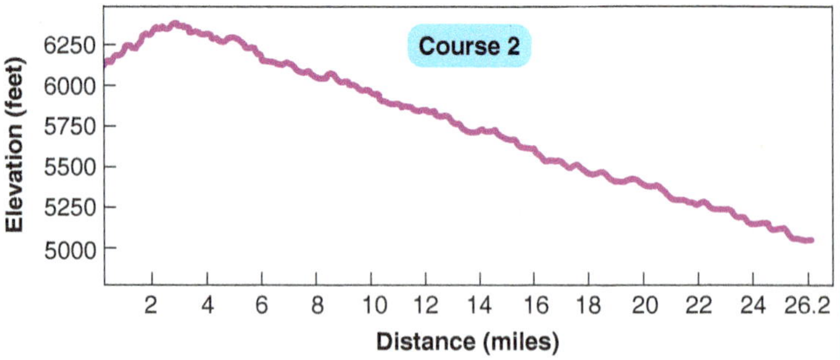

5.2.5 Principle of microscopic reversibility

Suppose you are studying at the library and decide you are hungry for Chinese food. Your study partner suggests a great but out-of-the-way restaurant across town. You ask your partner to give you directions that will get you there as quickly and efficiently as possible. After receiving the complicated instructions, you realize that after eating you will need to get back to the library as quickly as possible to continue studying. You then ask your partner for detailed directions on the fastest way back. Why might your study partner be annoyed at this question? [*How does it relate to the study of chemical reactions?*]

The fastest way back from the restaurant is likely the exact opposite of the fastest way there. Where you went north, go south. Where you turned right, turn left, etc. Assuming no one-way streets, the most efficient way there must be the most efficient way back.

In the context of a chemical reaction, the same principle applies. Although the reactions in this chapter have been shown as reversible by using equilibrium arrows (⇌ not ↔, which is used for resonance), we have not explicitly discussed the reverse reaction. What is the arrow-pushing mechanism in the reverse direction? What about the energy change? The transition state?

The reaction coordinate diagram answers these questions. It is essentially a potential energy map showing how the energy changes as you go from reactants to products and from products to reactants. If the lowest energy pathway is followed from reactants to products, then that same pathway is the lowest energy pathway from products to reactants. We can see how the reaction runs in reverse by simply flipping the reaction coordinate diagram (**Figure 5.41**). What was an uphill process going forward is downhill in

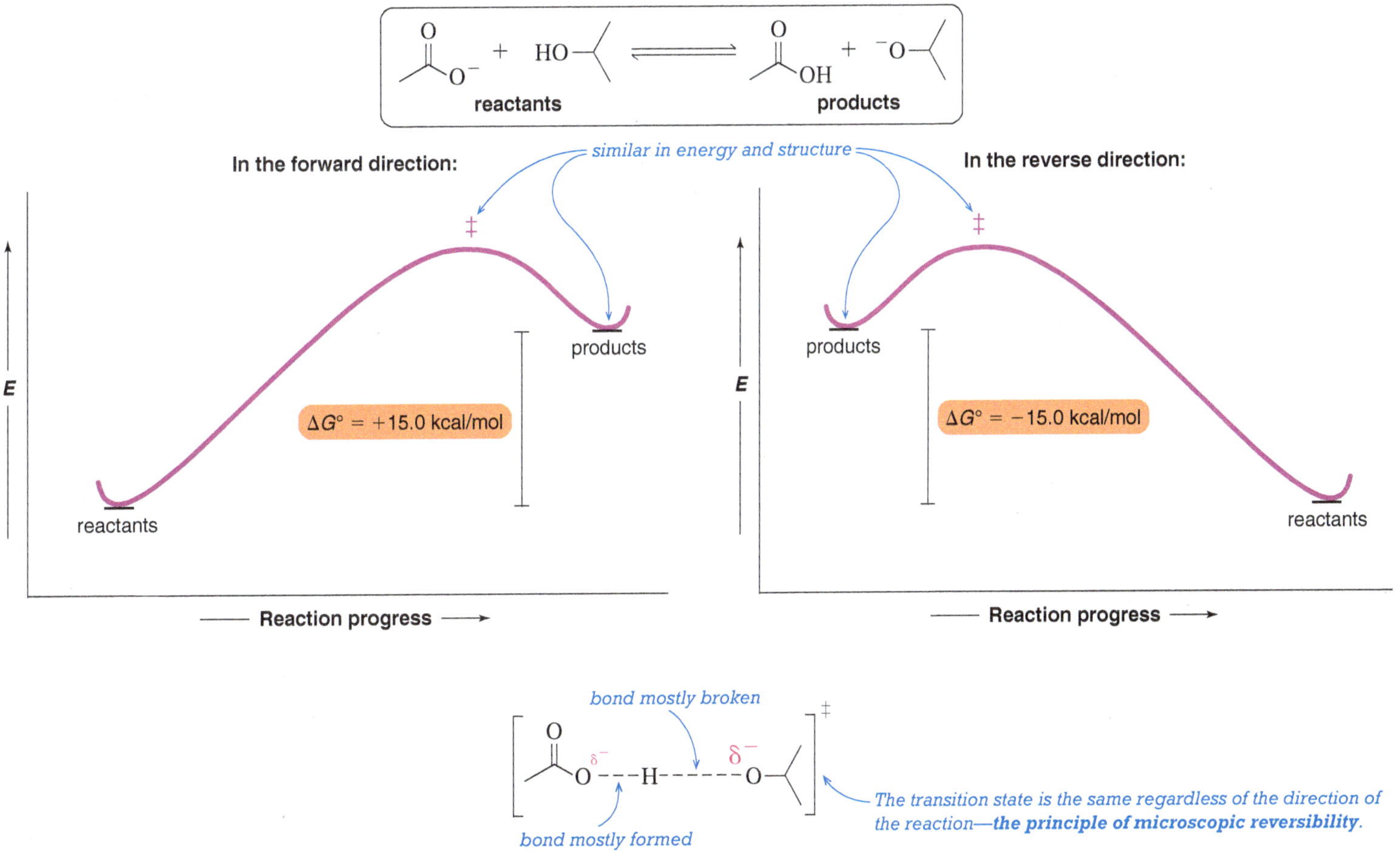

Figure 5.41 The principle of microscopic reversibility. The reaction coordinate diagram of the reverse reaction is the reaction coordinate diagram of the forward reaction flipped. The transition state is the same in both directions.

Lone pairs are shown in Assessment 5.31, but remember, you should know where they are even if they aren't shown.

reverse. The transition state still looks like the products. [*The labels stay the same when we flip the reaction coordinate diagram.*] Regardless of which direction the reaction goes, it passes through the same transition state. This is the **principle of microscopic reversibility**.

The principle of microscopic reversibility will be useful when we draw complex mechanisms of reversible reactions. By knowing the mechanism of the forward reaction, we know the mechanism of the reverse. Assessments 5.31 and Assessments 5.32 give you practice using this concept.

Assessment 5.31 We discuss the following reactions in subsequent chapters. Given the mechanisms shown, draw the mechanism of the reverse reaction.

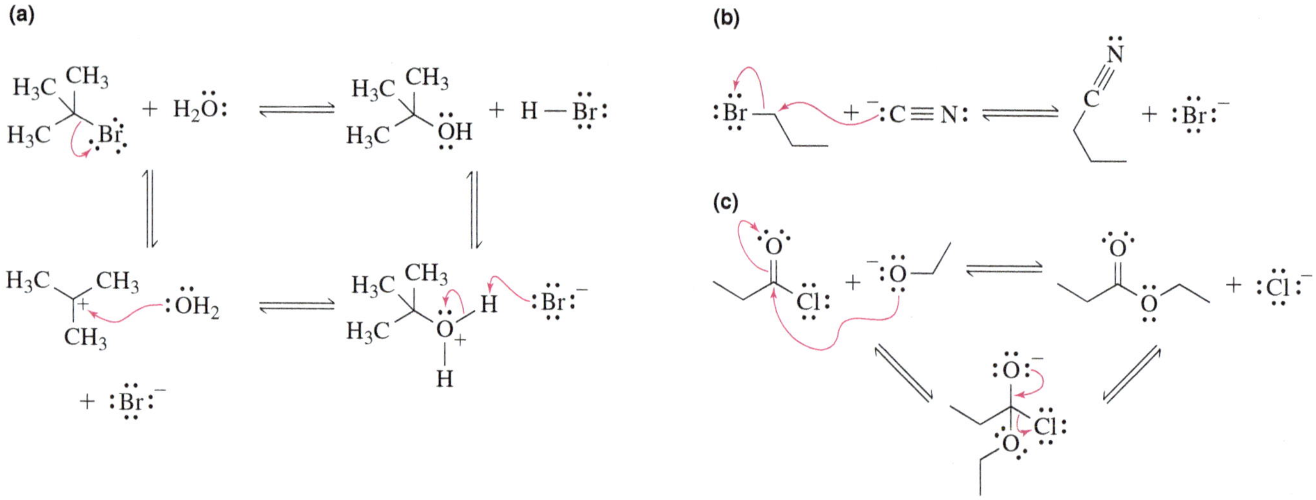

Assessment 5.32 Assuming that $\Delta H° = -15$ kcal/mol for the reaction in Assessment 5.31(b), show the transition state for the forward and reverse reactions.

5-D CHEMISTRY IN CONTEXT: Biochemistry

Enzymes catalyze reactions in both directions

As discussed in Chemistry in Context 5-A, enzymes catalyze reactions in both directions. Consider the catalyzed and uncatalyzed reactions leading to glucosylceramide synthesis shown in **Figure 5.42**. Now compare the reaction coordinate diagram for both reactions. For the catalyzed reaction, the increased nucleophilicity and enhanced leaving group ability are indicated by starting higher on the reaction coordinate diagram.

As shown on the reaction coordinate diagram, the transition state for the forward and reverse reactions are the same. Thus, in either direction, the enzyme-catalyzed reaction is faster than the uncatalyzed reaction. Although it is not explicitly indicated on the reaction coordinate diagrams, $\Delta G°$ for both reactions is also identical. Glucosylceramide synthesis is only useful in the forward direction, but this analysis demonstrates again that catalysis does not change the position of equilibrium; it only increases the rate at which equilibrium is achieved.

(*continued*)

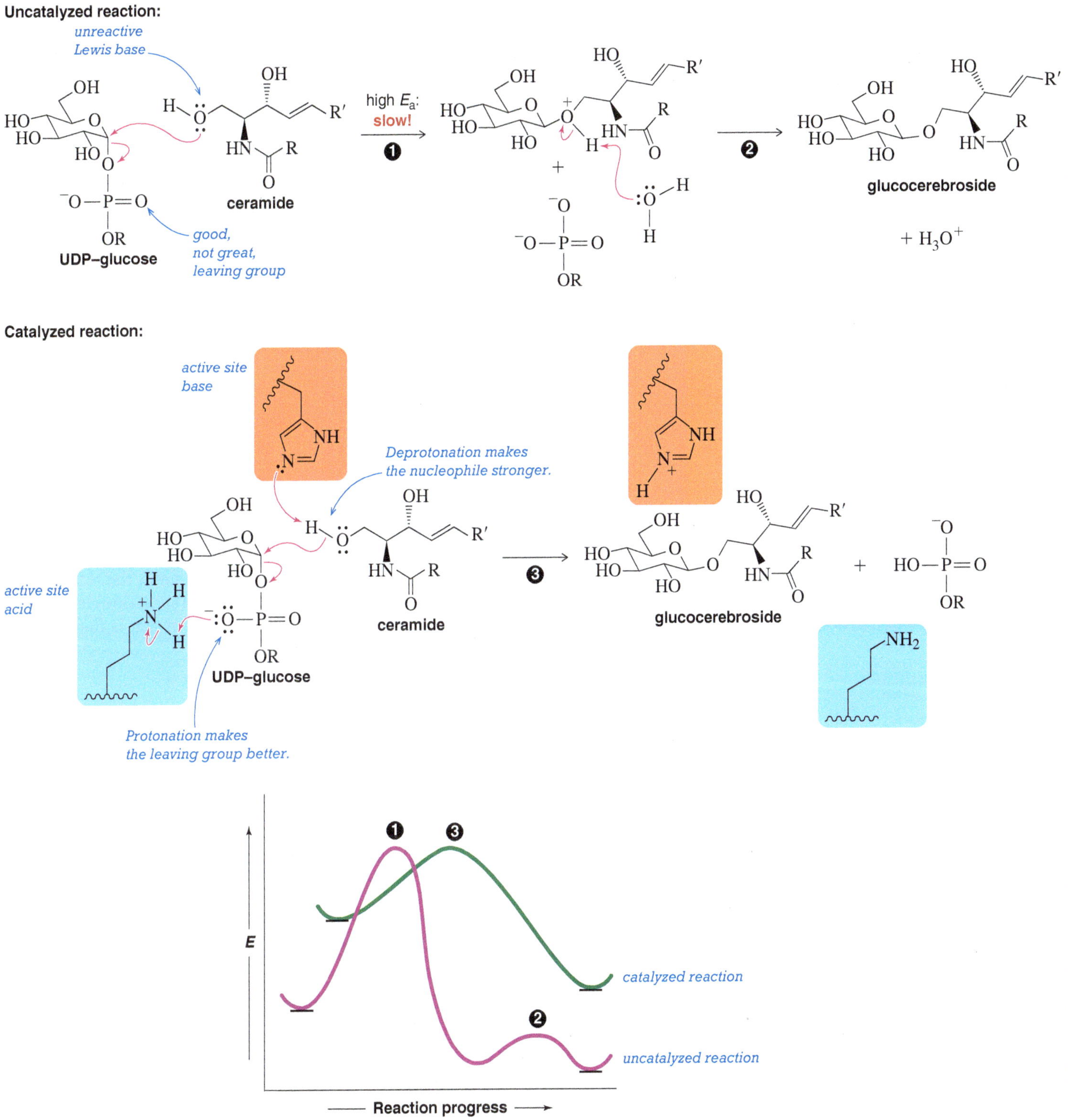

Figure 5.42 Reaction coordinate diagram for the catalyzed and uncatalyzed reactions from Chemistry in Context 5-C. Enzymes catalyze reactions in both directions.

Assessment 5.33 Suggest an arrow-pushing mechanism for the reverse of the catalyzed reaction in Figure 5.42. [*This may seem complicated, but simply reverse the arrows.*]

5.2.6 Rate laws, reaction order, and the transition state

The rate of a reaction depends on the concentration of the reactants. Based on what we have discussed so far, how might increasing concentration make reactions happen more quickly? [*Think about this before you continue reading.*]

Chemical reactions occur via random collisions of molecules on a particular trajectory (A), which have enough energy to overcome the activation barrier (E_a). Increasing the temperature increases both the speed with which molecules move through the solution *and* the likelihood of collision. It follows that an increase in reactant concentration (more molecules in solution) should also make collisions more frequent. This qualitative analysis is confirmed by the **rate law**, or **rate equation**, which says that for a given reaction A + B → C + D:

$$\text{rate} = k_r[\mathrm{A}]^a[\mathrm{B}]^b \qquad \text{Equation 5.7}$$

where a = the order of the reaction with regard to reactant A, b = the order of the reaction with regard to reactant B, and the rate constant = $k_r = Ae^{-E_a/RT}$.

By including the rate constant (k_r), the rate law recognizes the importance of E_a, A, and T, while adjusting this value based on reactant concentration. The concept of **reaction order** is a complicated one, at least initially. Let's connect reaction *order*, as learned in previous courses, to *rate dependence* in organic chemistry.

Reaction order is determined through an experiment in which reaction rate is measured as a function of reactant concentration. Illustrated in **Figure 5.43**, we begin with a certain concentration of A and B and determine a rate of x (the control). For this particular reaction, doubling the concentration of A (trial 1) doubles the rate of the reaction. Thus, the reaction is first order with respect to A. Similarly, doubling the concentration of B (trial 2) doubles the rate of reaction, so the reaction is first order with respect to B, too. A reaction that is first order with respect to A and first order with respect to B is second order overall.

A + B ⟶ C + D

	[A] (M)	[B] (M)	Rate
Control	0.1	0.1	x
Trial 1	0.2	0.1	$2x$
Trial 2	0.1	0.2	$2x$
Trial 3	0.2	0.2	$4x$

$$\text{rate} = k_r[\mathrm{A}]^1[\mathrm{B}]^1$$

1st order with regard to A
1st order with regard to B
2nd order overall

Figure 5.43 The experimental results indicate a second-order reaction.

To achieve a deeper understanding of reaction rates, activation energies, and transition state structures, you need a deeper understanding of the reaction order. We can calculate the energy of the transition state by knowing its structure. The transition state allows us to determine the activation energy, which determines the value of the rate constant, which influences the rate of a reaction. Reaction order is connected to transition state theory as follows: ***The reaction order identifies which molecules are in the transition state of the rate-determining step, which is the slowest step of a multistep reaction.*** [*Go back and read that again; think about what it means.*] For the rate of a reaction to depend on the concentration of a particular reactant, that reactant must be in the transition state of the rate-determining step. This fits perfectly into our analysis, where the transition state structure determines the rate of a reaction (**Figure 5.44**).

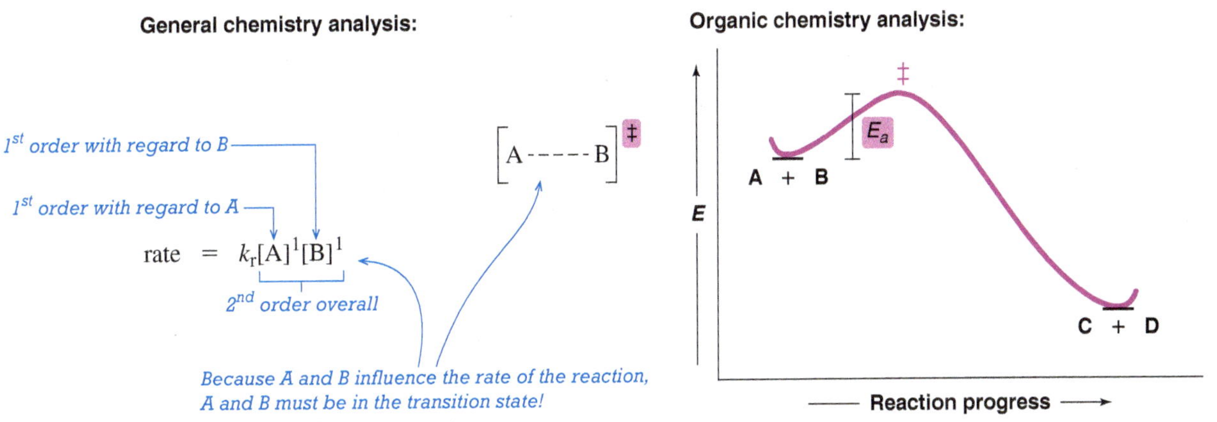

Figure 5.44 A second-order reaction has two molecules in the transition state of the rate-determining step.

Consider the acid–base reaction between water and ethoxide (**Figure 5.45**). From the structure of the transition state, the reaction should be first order with respect to ethoxide, first order with respect to water, and second order overall. We know this because one molecule each of ethoxide and water (two molecules overall) are in the transition state. Thus, the rate of this reaction depends on the concentration of both ethoxide and water.

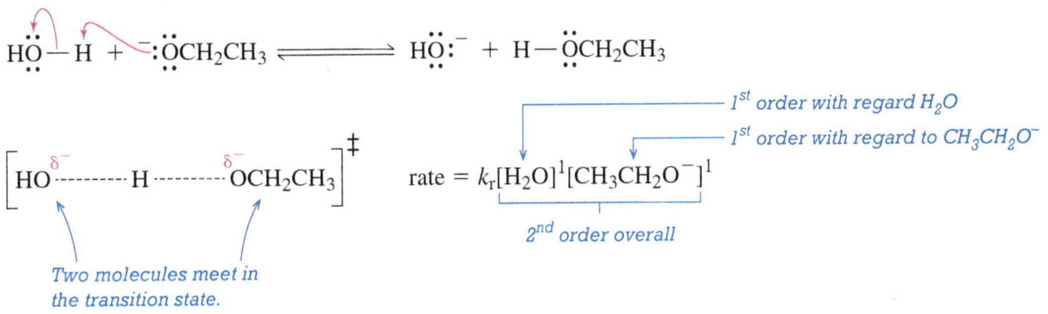

Figure 5.45 The rate of a second-order reaction depends on the concentration of both molecules that meet in the transition state.

Figure 5.45 shows us how the transition state predicts reaction order. The reverse is also true: The experimentally measured reaction order provides insight into how reactions work and the nature of possible transition states. This technique was especially important in the past, when mechanisms were not as easily predicted nor understood.

Let's think like a chemist, using reaction order to analyze a new reaction.

Thinking Like a Chemist

When *t*-butyl bromide is dissolved in water, it is converted to *t*-butanol. **Figure 5.46** shows two possible arrow-pushing mechanisms for the reaction. [*Think for a minute about how you'd use reaction order to determine which possibility is correct.*]

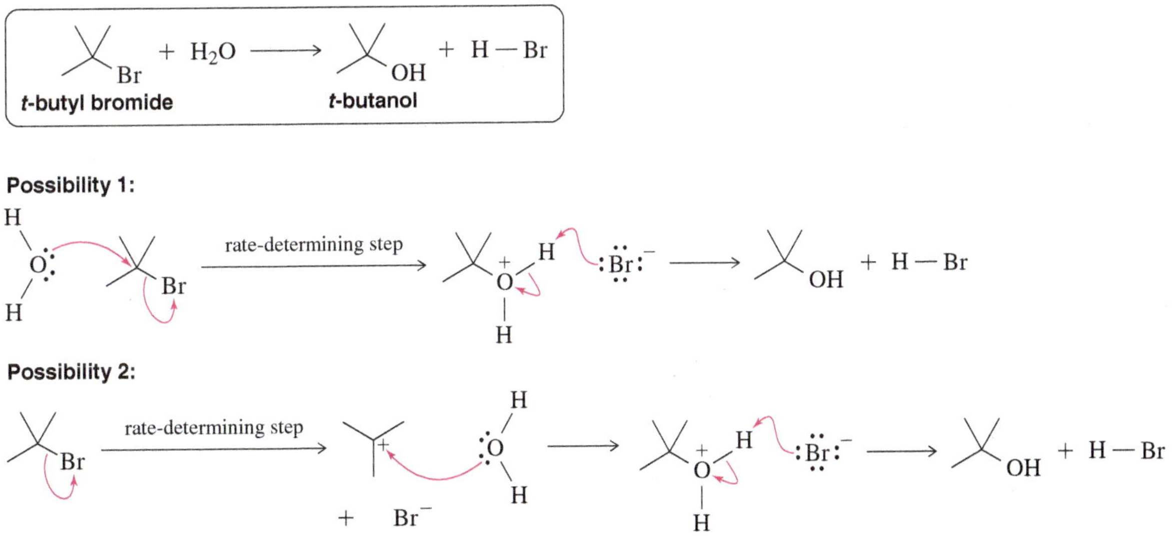

Figure 5.46 Two possible mechanisms for the formation of *t*-butanol from *t*-butyl bromide.

The difference between the two proposed reactions occurs in the first step of each. [*These first steps are characteristic rate-determining steps in the reactions of Chapter 12.*] Possibility 1 has water and *t*-butyl bromide reacting in a single, concerted step, whereas possibility 2 is a stepwise process. We can approach this problem by establishing a baseline rate at a particular concentration—in this case, 0.1 M—for both reactants (**Figure 5.47**). This gives us a rate of *x*. When the concentration of *t*-BuBr is doubled (trial 1), the rate doubles. The reaction, then, is first order with respect to *t*-BuBr, and *t*-BuBr must be in the transition state of the rate-determining step. When we double the concentration of water, there is no change in reaction rate. Thus, the reaction is zeroth order with respect to water, and water is *not* present in the transition state of the rate-determining step. The reaction is first order overall, meaning that it is unimolecular; there is only one molecule present in the rate-determining step. [*Does this match possibility 1 or possibility 2?*]

	[*t*-BuBr] (M)	[H_2O] (M)	Rate
Control	0.1	0.1	*x*
Trial 1	0.2	0.1	2*x*
Trial 2	0.1	0.2	*x*

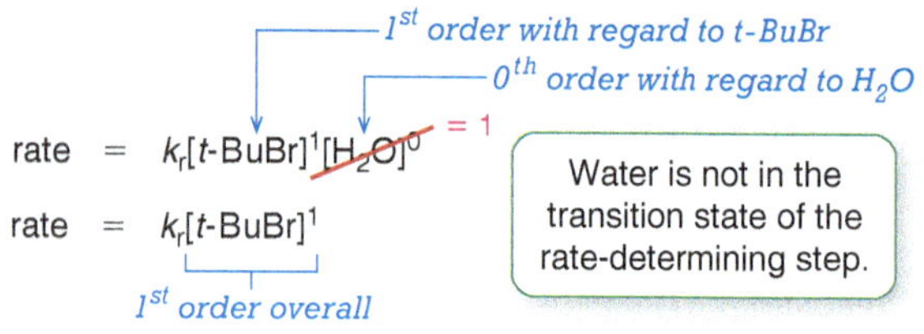

Figure 5.47 The experimental results indicate a first-order reaction. Only *t*-BuBr is present in the transition state of the rate-determining step.

Based on the rate data obtained, the reaction must have only one molecule in the rate-determining step. This eliminates possibility 1, which had water and *t*-butyl bromide reacting together. Possibility 2, featuring only the breaking of the C—Br bond in the first step, must then be the mechanism of the reaction. The transition state of the first step is shown in **Figure 5.48**. [*We learn more about this reaction in Chapter 12. What is important for now is how the reaction order gives us insight into its mechanism.*] ■

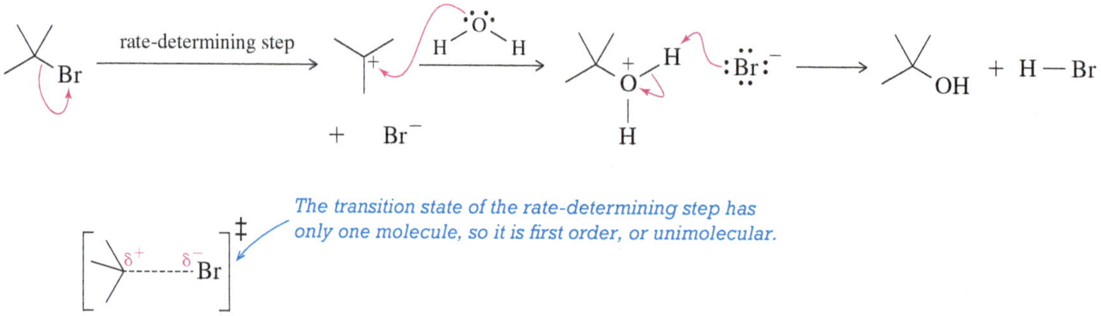

Figure 5.48 The reaction order indicates that possibility 1 is the mechanism.

Assessments 5.34–5.36 test your understanding of how reaction order relates to transition state theory.

5.2.6 SUMMARY

- The order of a reaction identifies how many molecules are present in the transition state of the rate-determining step.

Assessment 5.34 Write the rate law for the following reaction and identify which molecules are present in the rate-determining step. Draw a possible transition state and propose a mechanism.

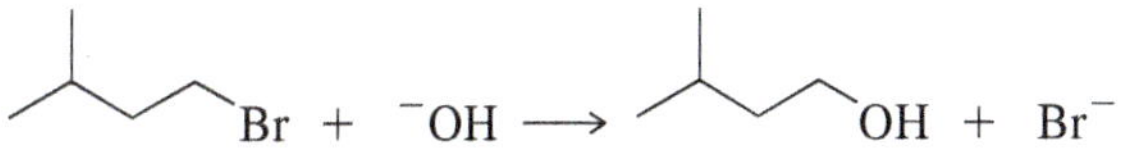

	[RBr] (M)	[HO⁻] (M)	Rate (M/s)
Control	0.1	0.1	1.64×10^{-5}
Trial 1	0.2	0.1	3.28×10^{-5}
Trial 2	0.1	0.2	3.28×10^{-5}
Trial 3	0.2	0.2	6.56×10^{-5}

Assessment 5.35 All things being equal, would you expect a first-order reaction to be faster or slower than a second-order reaction?

Assessment 5.36 Third-order reactions are rare. Why do you think that is?

We have just used two simple processes (rotation around a single bond and an acid–base reaction) to perform relatively complicated analyses. The radical halogenation of alkanes (Section 5.3) is a more complex reaction, but you'll be able to understand it if you use your basic understanding of reaction analysis and adjust your analysis as the new reactions require.

5.3 Chemical reaction analysis: halogenation of alkanes

The synthesis of alkyl halides can be done via **halogenation** of the respective alkane. For example, the reaction of methane with chlorine gas at high temperature results in the formation of chloromethane [Figure 5.49(a)]. Similarly, bromomethane can be synthesized under similar conditions but with bromine instead of chlorine [Figure 5.49(b)]. As we will see in Section 5.4 and Chapter 11, halogenation is an important synthetic technique. Thus, as we learn how to analyze these reactions, know that you are also learning a reaction we will use later. [*Throughout this section, chlorination and bromination are compared by showing the reactions side by side.*]

(a)

$$CH_4 + Cl_2 \xrightarrow{\Delta} CH_3Cl + HCl$$

(b)

$$CH_4 + Br_2 \xrightarrow{\Delta} CH_3Br + HBr$$

Figure 5.49 Alkane halogenation. (a) Chlorination of methane. (b) Bromination of methane.

5.3.1 $\Delta G° = \Delta H°$: bond-dissociation energies

What is the overall $\Delta G°$ for these reactions? Both chlorination and bromination start with two molecules and finish with two molecules, so $\Delta S° \approx 0$ and $\Delta G° \approx \Delta H°$. In contrast to the reactions from Sections 5.1 and 5.2, bond-dissociation energies can be used to calculate $\Delta H°$ for these reactions. The **bond-dissociation energy** is the energy associated with **homolytic bond breaking** (Figure 5.50). Homolysis means that the covalent bond breaks such that each atom takes one of the electrons originally in the bond, forming two **free radicals.** Free radicals are electron-deficient species and are 1 electron short of a full valence shell. Because bond breaking requires/absorbs energy, the bond-dissociation energies (BDEs) in Table 5.6 are all positive numbers. Note the use of **fishhook arrows** to show the flow of single electrons when drawing mechanisms.

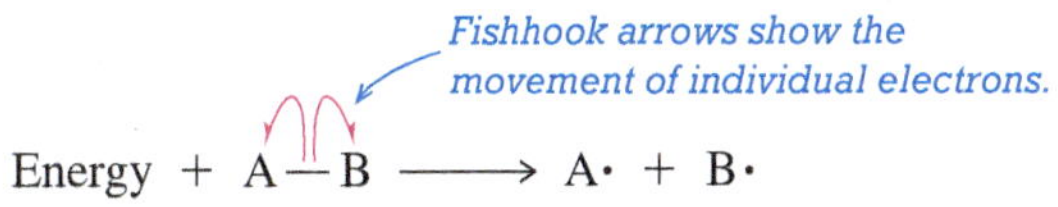

Figure 5.50 Homolytic bond-dissociation requires energy ($\Delta H° > 0$).

Table 5.6 Bond-dissociation energies are all positive values because they involve bond *breaking,* an endothermic process.

$$A{:}B \longrightarrow A\cdot + \cdot B$$

Bond	Bond–dissociation enthalpy kJ/mol	Bond–dissociation enthalpy kcal/mol
H — X bonds and X — X bonds		
$H-H$	435	104
$D-D$	444	106
$F-F$	159	38
$Cl-Cl$	242	58
$Br-Br$	192	46
$I-I$	151	36
$H-F$	569	136
$H-Cl$	431	103
$H-Br$	368	88
$H-I$	297	71
$HO-H$	498	119
$HO-OH$	213	51
$RO-H$	426	102
Methyl bonds		
CH_3-H	435	104
CH_3-F	456	109
CH_3-Cl	351	84
CH_3-Br	293	70
CH_3-I	234	56
CH_3-OH	381	91
Bonds to primary carbons		
CH_3CH_2-H	410	98
CH_3CH_2-F	448	107
CH_3CH_2-Cl	339	81
CH_3CH_2-Br	285	68
CH_3CH_2-I	222	53
CH_3CH_2-OH	381	91

Bond	Bond–dissociation enthalpy kJ/mol	Bond–dissociation enthalpy kcal/mol
Bond to secondary carbons		
$(CH_3)_2CH-H$	397	95
$(CH_3)_2CH-F$	444	106
$(CH_3)_2CH-Cl$	335	80
$(CH_3)_2CH-Br$	285	68
$(CH_3)_2CH-I$	222	53
$(CH_3)_2CH-OH$	381	91
Bonds to tertiary carbons		
$(CH_3)_3C-H$	381	91
$(CH_3)_3C-F$	444	106
$(CH_3)_3C-Cl$	331	79
$(CH_3)_3C-Br$	272	65
$(CH_3)_3C-I$	209	50
$(CH_3)_3C-OH$	381	91
Other C — H bonds		
$PhCH_2-H$ (benzylic)	356	85
$CH_2=CHCH_2-H$ (allylic)	364	87
$CH_2=CH-H$ (vinyl)	464	111
$Ph-H$ (aromatic)	473	113
$HC\equiv C-H$ (acetylenic)	523	125
C — C bonds		
CH_3-CH_3	368	88
$CH_3CH_2-CH_3$	356	85
$CH_3CH_2-CH_2CH_3$	343	82
$(CH_3)_2CH-CH_3$	351	84
$(CH_3)_3CH-CH_3$	339	81
C = Y π bonds		
$C=C$ (estimated)	272	65
$C=O$ (estimated)	356	85

If bond breaking requires energy, then bond formation must release energy. To determine whether a reaction has a favorable enthalpy change ($\Delta H° < 0$), the reaction is analyzed based on the bonds broken and the bonds formed. If the formed bonds are stronger than the broken bonds, the reaction is favored ($\Delta H° < 0$). If the bonds formed are weaker than the bonds broken, the reaction is disfavored ($\Delta H° > 0$).

Which bonds are broken and which are formed in the halogenation of methane? [*Drawing the structural formula of each molecule helps with this.*]

Figure 5.51 shows that the C—H σ bond and the Cl—Cl σ bond are broken, whereas the C—Cl σ bond and the H—Cl σ bond are formed. Table 5.6 lists the BDEs associated with these bonds. Subtracting the BDEs for the bonds formed from the BDEs for the bonds broken gives an overall $\Delta H° = -25$ kcal/mol. Because $\Delta S° = 0$, $\Delta G° = -25$ kcal/mol. The chlorination of methane is heavily favored.

While doing the same calculation for the bromination of methane (**Figure 5.52**), focus on what changes. The C—H BDE (104 kcal/mol) is the same regardless of the reaction conditions. The Br—Br bond, though, is weaker (46 kcal/mol) than the Cl—Cl bond (58 kcal/mol). The BDEs for the formed bonds, C—Br and H—Br, are 70 kcal/mol and 88 kcal/mol, respectively. Subtracting the bonds formed from the bonds broken once more gives $\Delta H° = -8$ kcal/mol. Thus, bromination is also a favored reaction, though not quite as favored as chlorination.

For chlorination:

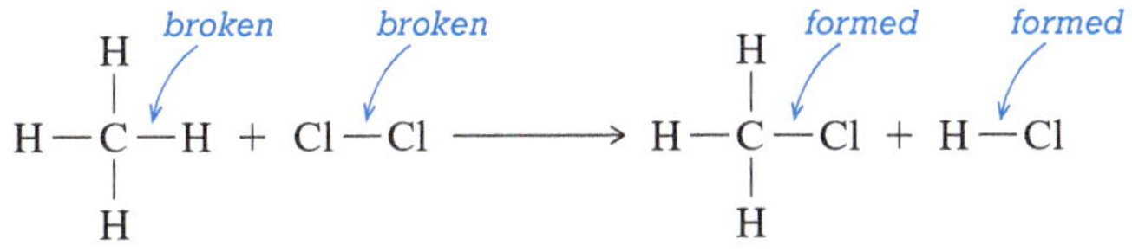

Bonds broken:
C—H σ 104 kcal/mol
Cl—Cl σ 58 kcal/mol
162 kcal/mol

Bonds formed:
C—Cl σ 84 kcal/mol
H—Cl σ 103 kcal/mol
187 kcal/mol

$$\Delta H = \text{BDE}_{\text{bonds broken}} - \text{BDE}_{\text{bonds formed}}$$
$$= 162 \text{ kcal/mol} - 187 \text{ kcal/mol} = -25 \text{ kcal/mol}$$

Figure 5.51 Chlorination of methane is heavily favored.

For bromination:

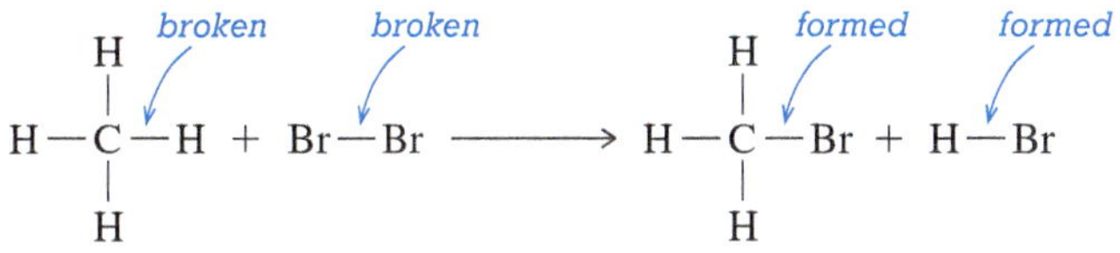

Bonds broken:
C—H σ 104 kcal/mol
Br—Br σ 46 kcal/mol
150 kcal/mol

Bonds formed:
C—Br σ 70 kcal/mol
H—Br σ 88 kcal/mol
158 kcal/mol

$$\Delta H = \text{BDE}_{\text{bonds broken}} - \text{BDE}_{\text{bonds formed}}$$
$$= 150 \text{ kcal/mol} - 158 \text{ kcal/mol} = -8 \text{ kcal/mol}$$

Figure 5.52 Bromination of methane is favored, but not as favored as chlorination.

[Table 5.6 has a number of different C—H and C—Cl bonds listed, but only one number was the "best" number to use. How was this choice made?]

Assessment 5.37 Calculate $\Delta H°$ for the following reactions.

(a)

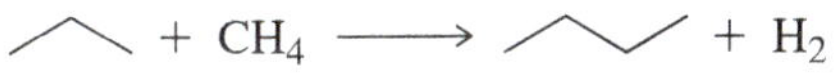

(b) $CH_3Br + HCl \longrightarrow CH_3Cl + HBr$

(c) $CH_3CH_3 + HOOH \longrightarrow CH_3CH_2OH + H_2O$

The choice of which BDE to use from Table 5.6 is made by finding the bond and its resulting free radical that most closely matches the bond being examined.

5.3.2 Bond-dissociation energies: a closer look

As we did with pK_a values, there is value in justifying why a bond is weak (low BDE) or strong (high BDE). [*Why do you think strong bonds have a high BDE?*] What can you infer about the relative stability of the free radicals that result from bond dissociation, when you consider that H—Br has a lower BDE than H—Cl? This analysis can be done by plotting both endothermic bond-breaking processes on a reaction coordinate diagram. Assuming they start at the same point, H—Br (lower BDE) will finish at a lower energy than H—Cl (higher BDE) [**Figure 5.53(a)**]. The 15 kcal/mol difference can only be attributed to Br• being more stable than Cl•. [*H• is not included in this analysis because it is in both reactions.*] Just as we saw in Chapter 4 (when comparing Br^- and Cl^-), spreading the radical character over a larger volume enhances the stability of Br•. The same strategy can be used to compare the BDEs for a 3° C—H (91 kcal/mol) and methyl C—H (104 kcal/mol) bonds [**Figure 5.53(b)**]. Comparing the two reaction coordinate diagrams, we see that the tertiary radical is at a lower energy than the methyl radical. That is, the tertiary radical is more stable. Moreover, the tertiary radical has a lower activation energy and would be likely to form faster.

In conclusion, the bond-dissociation energy directly correlates to the stability of the resulting radical. [*This is very similar to how we used pK_a values to discuss conjugate base stability.*] Assessments 5.38 and 5.39 give you practice using BDEs in this way.

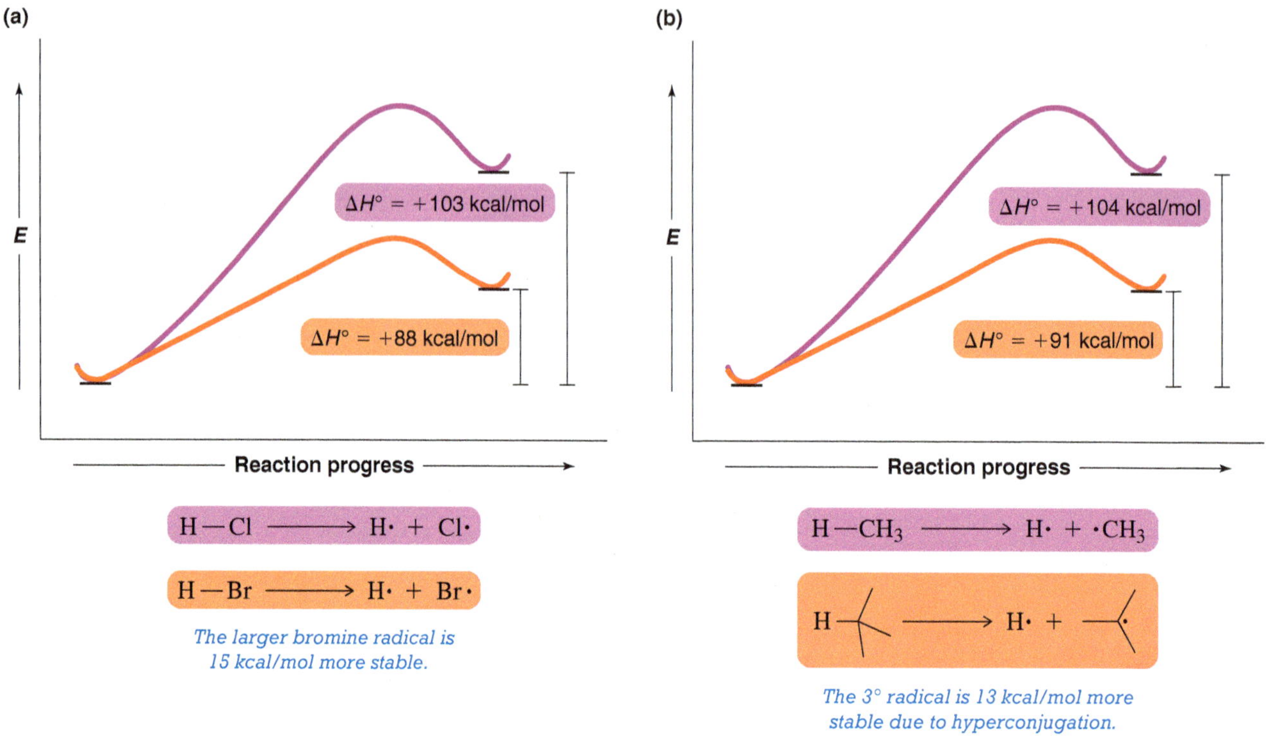

Figure 5.53 **Plotted on a reaction coordinate diagram,** (a) a lower bond-dissociation energy shows that Br• is more stable than Cl• and (b) a 3° radical is more stable than a methyl radical.

Assessment 5.38 Using bond-dissociation energies, identify the most stable radical. Justify the difference in stability based on the structure.

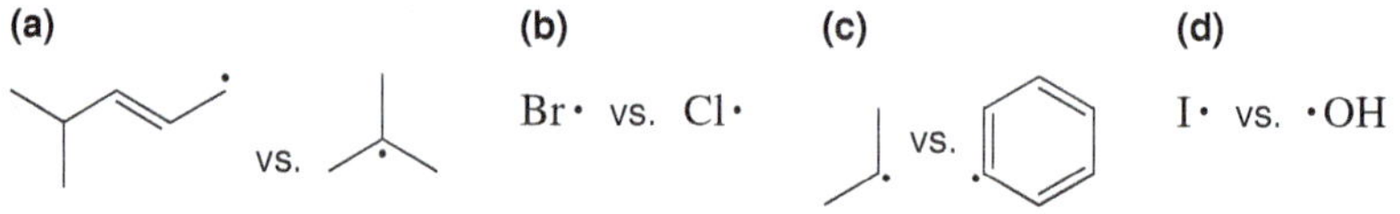

Assessment 5.39 Give the approximate bond-dissociation energy for each indicated bond. [*You will need to make additional inferences to get the answer for (b).*]

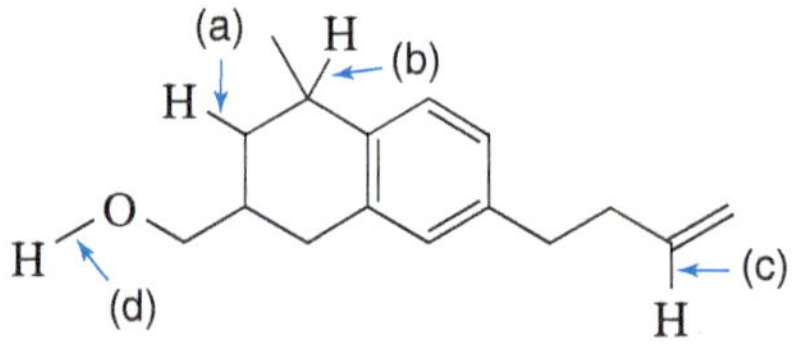

5.3.3 Radical halogenation mechanism

The mechanism of radical halogenation is shown in **Figure 5.54**. Keep in mind a couple of important points as you look at the figure. First, the reaction is initiated by formation of the chlorine radical. Second, each of the mechanistic steps is illustrated using fishhook arrows to show the movement of individual electrons. Finally, this reaction is a **chain reaction** because once initiated, the propagation steps continue until there is no more methane or chlorine. This is why the overall reaction can be illustrated using just the two propagation steps. [*We'll return to this last point.*]

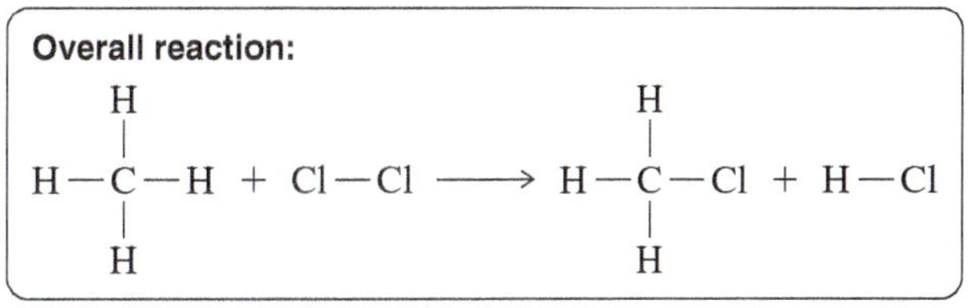

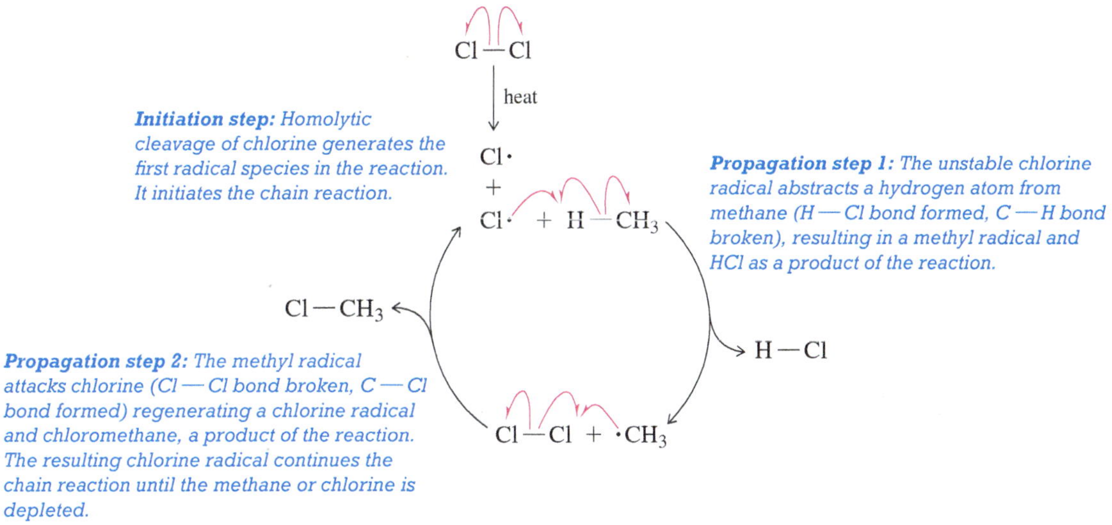

Figure 5.54 Mechanism of radical halogenation, a chain reaction.

While the radical concentration is generally kept low, occasionally two radicals will combine to form a new molecule. Because this event terminates the propagating radical and stops the chain reaction, the reactions in Figure 5.55 are called **termination steps.** Conditions are generally optimized to minimize termination steps so the chain reaction can continue.

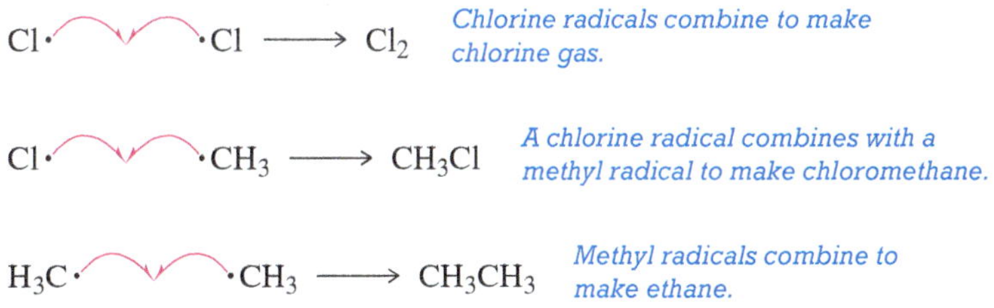

Figure 5.55 Three possible termination steps of radical halogenation.

The mechanism of radical bromination is the same as the mechanism for chlorination.

5.3.4 Energetics of halogenation reactions

The enthalpy change of each step of halogenation can be calculated. The total $\Delta H°$ of the reaction must be equal to the sum of the $\Delta H°$ for each propagation step. Using BDEs from earlier, this is true for chlorination [Figure 5.56(a)] and bromination [Figure 5.56(b)]. The first step is disfavored in both reactions, but significantly more so in the case of bromination. Bromine is a more stable radical, and formation of H—Br does not release as much energy as does formation of H—Cl. [*In both cases, the same C—H bond breaks to form an unstable methyl radical.*] The second propagation step for both reactions, on the other hand, is extremely favorable. Study the reaction coordinate diagrams for a visual comparison of the two reactions. Note that a reaction coordinate diagram should have one hill for each step. The valley between the two hills is considered to be an **intermediate** (in this case, the products of step 1).

(a) For chlorination:

Propagation 1: $Cl\cdot + H{-}CH_3 \longrightarrow H{-}Cl + \cdot CH_3$

ΔH° = 104 kcal/mol − 103 kcal/mol = +1 kcal/mol

Propagation 2: $H_3C\cdot + Cl{-}Cl \longrightarrow H_3C{-}Cl + Cl\cdot$

ΔH° = 58 kcal/mol − 84 kcal/mol = −26 kcal/mol

Total:

$\cancel{Cl\cdot} + \cancel{H_3C\cdot} + H{-}CH_3 + Cl{-}Cl \longrightarrow H_3C{-}Cl + H{-}Cl + \cancel{\cdot CH_3} + \cancel{Cl\cdot}$

$H{-}CH_3 + Cl{-}Cl \longrightarrow H_3C{-}Cl + H{-}Cl$

ΔH°_{total} = +1 kcal/mol + (−26 kcal/mol) = −25 kcal/mol

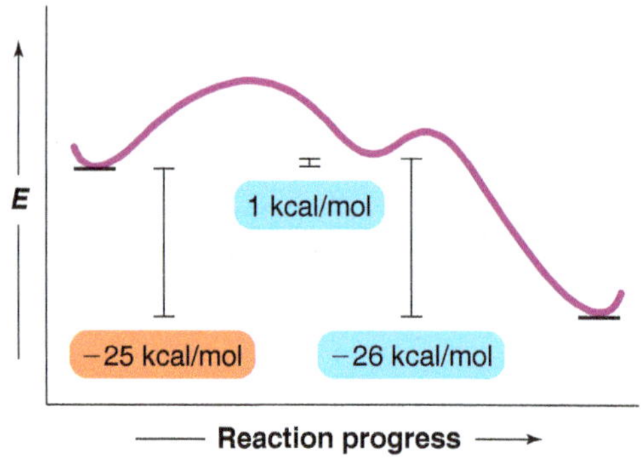

(b) For bromination:

Propagation 1: $Br\cdot + H{-}CH_3 \longrightarrow H{-}Br + \cdot CH_3$

ΔH° = 104 kcal/mol − 88 kcal/mol = +16 kcal/mol

Propagation 2: $H_3C\cdot + Br{-}Br \longrightarrow H_3C{-}Br + Br\cdot$

ΔH° = 46 kcal/mol − 70 kcal/mol = −24 kcal/mol

Total:

$\cancel{Br\cdot} + \cancel{H_3C\cdot} + H{-}CH_3 + Br{-}Br \longrightarrow H_3C{-}Br + H{-}Br + \cancel{\cdot CH_3} + \cancel{Br\cdot}$

$H{-}CH_3 + Br{-}Br \longrightarrow H_3C{-}Br + H{-}Br$

ΔH°_{total} = +16 kcal/mol + (−24 kcal/mol) = −8 kcal/mol

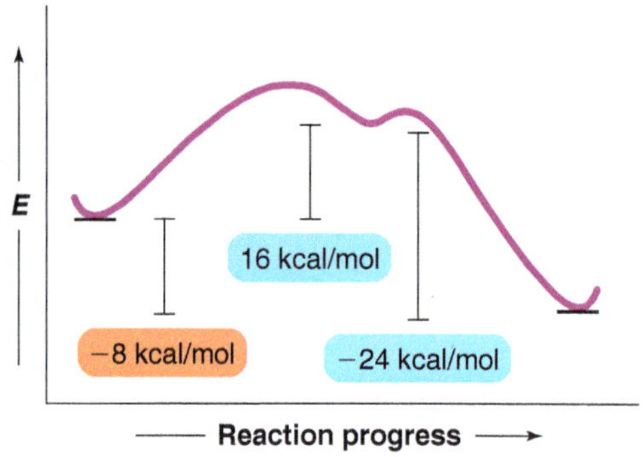

Figure 5.56 The energetics of both steps of methane halogenation. The first step of (a) chlorination is less endothermic than the first step of (b) bromination. [*Note that a two-step reaction has two hills in its reaction coordinate diagram.*]

5.3.5 Rate of halogenation

When a reaction proceeds through multiple steps, the slowest step is **rate-determining**. Based on the reaction coordinate diagrams in Figure 5.56, the endothermic first step is the rate-determining step in radical halogenation. There are a couple ways to think about whether chlorination or bromination is faster overall. Because the bromine radical is larger, it is significantly more stable than the chlorine radical. Thus, we would expect the less stable/more reactive chlorine atom to abstract a hydrogen more quickly. Second, we can look at ΔG° for the rate-determining step of both reactions. Because ΔG° is greater for bromination (16 kcal/mol) than for chlorination (1 kcal/mol), we would anticipate that chlorination should be faster.

Generally, a more uphill reaction step has a higher activation energy and lower reaction rate when reaction mechanisms are similar. Let's think like a chemist to see why this is true.

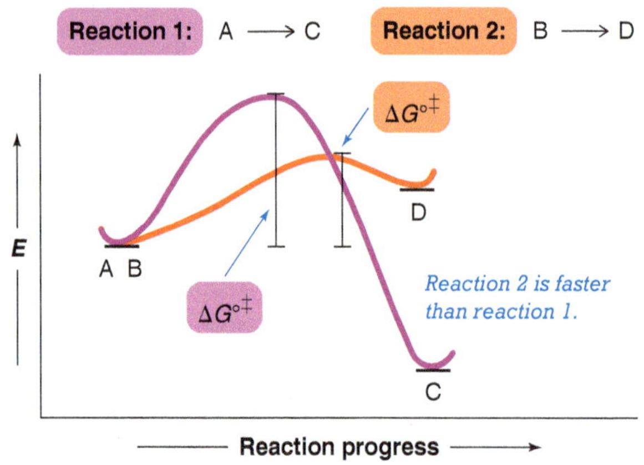

Figure 5.57 E_a **determines the rate of a reaction.**

Thinking Like a Chemist

Our goals here are twofold: (1) to analyze this statement in the context of transition states energies, especially because the Hammond postulate is applicable, and (2) to make sense of the generalization so we can trust it as we go forward.

The reaction coordinate diagram in Figure 5.57 compares two reactions that don't fit the generalization that a net uphill step is slower than a net downhill step. The hypothetical reaction A → C is a slower reaction that produces the more stable C. The hypothetical reaction B → D is a faster reaction that produces the less stable molecule D. [*Why is the downhill step slower?*] Although exaggerated for effect, Figure 5.57 reminds us that the activation energy (E_a or $\Delta G^{\circ\ddagger}$) determines the rate of a reaction step.

Returning to alkane halogenation, to rationalize E_a we need to know about the structure of the transition state. Though we can easily calculate ΔH° for reactions, calculating $\Delta H^{\circ\ddagger}$ is more complicated because in the transition state the bonds are neither fully formed nor broken. We are assuming that $\Delta S^{\circ\ddagger}$ is identical and negative for each transition state. [*Why is* $\Delta S^{\circ\ddagger} < 0$

for the transition state of all bimolecular reactions? Because 2 molecules come together to make one transition state "molecule."] **Figure 5.58** proposes three possible transition states for the formation of the methyl radical in the rate-determining step of both bromination and chlorination. These three transition states differ by the degree to which the C—H bond is broken and the H—Cl/H—Br bond is formed. Specifically, we are considering the transition states when the H—Br/H—Cl bond is 90%, 50%, and 10% formed. [*Make sure you understand what we are looking at before proceeding.*] Considering the BDE of the bonds and the degree of bond breakage, we should be able to estimate $\Delta H^{\circ\ddagger}$ for each possibility.

(a) For chlorination:

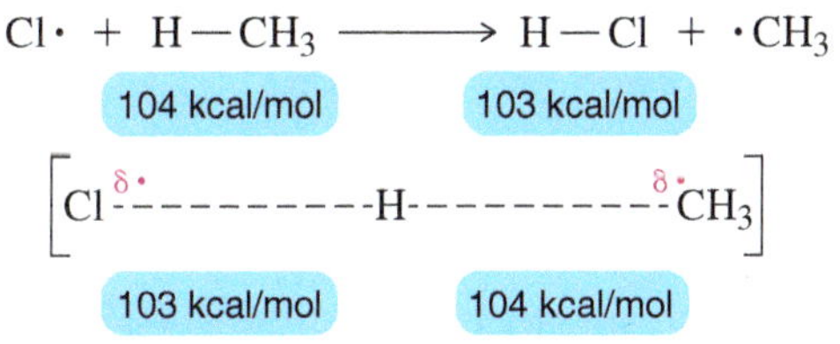

% bond formed	C—H broken (kcal/mol)	H—Cl formed (kcal/mol)	$\Delta H^{\circ\ddagger}$ (kcal/mol)
90	0.90(+104) = +93.6	0.90(−103) = −92.7	+0.9
50	0.50(+104) = +52	0.50(−103) = −51.5	+0.5
10	0.10(+104) = +10.4	0.10(−103) = −10.3	+0.1

(b) For bromination:

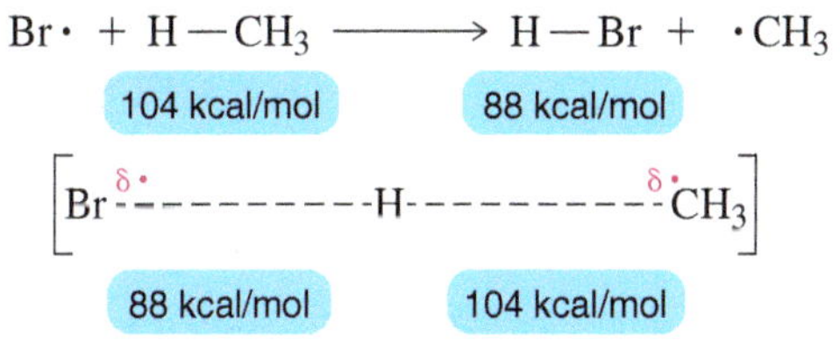

% bond formed	C—H broken (kcal/mol)	H—Br formed (kcal/mol)	$\Delta H^{\circ\ddagger}$ (kcal/mol)
90	0.90(+104) = +93.6	0.90(−88) = −79.2	+14.4
50	0.50(+104) = +52	0.50(−88) = −44	+8
10	0.10(+104) = +10.4	0.10(−88) = −8.8	+1.6

Figure 5.58 Possible transition states for the rate-determining step of (a) chlorination and (b) bromination.

All three cases can't be operating for both reactions, so which one is it? Is the first step 90%, 50%, or 10% complete in the transition state? Enter the Hammond postulate.

For chlorination, the rate-determining step has $\Delta G^\circ = +1$ kcal/mol (nearly 0). According to the Hammond postulate, the transition state will be at a point near 50% completion. From Figure 5.58(a), this would give $\Delta H^{\circ\ddagger} \approx 0.5$ kcal/mol for a transition state like that shown in **Figure 5.59**.

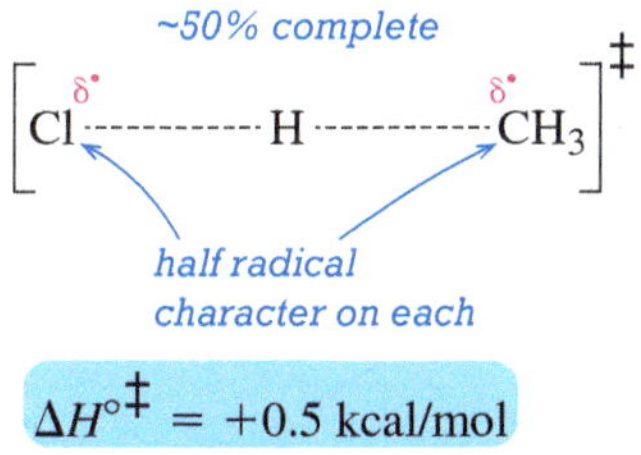

Figure 5.59 The transition state of the endothermic first step of chlorination is halfway between the reactant and product.

For bromination, the rate-determining step has $\Delta G^\circ = +16$ kcal/mol. According to the Hammond postulate, the transition state for an endothermic step will look like the product of that step. We can assume the transition state will be at a point near 90% completion. From Figure 5.58(b), we get $\Delta H^{\circ\ddagger} \approx 14.4$ kcal/mol for a transition state like that shown in **Figure 5.60**.

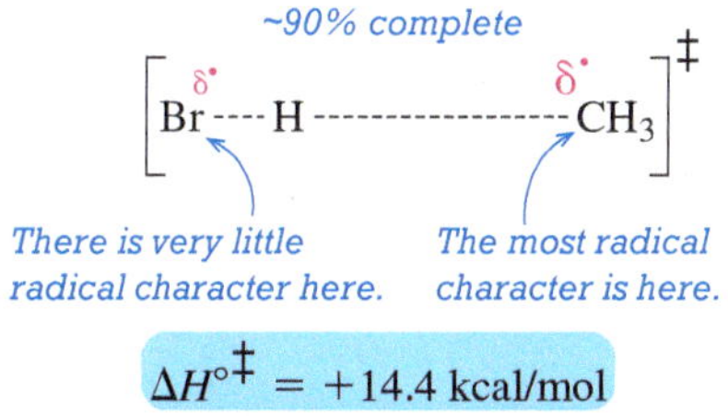

Figure 5.60 The transition state of the endothermic first step of bromination looks like the product.

Assuming the same value for $\Delta S^{\circ\ddagger}$ for both reactions, we can estimate that the activation energy (E_a) of bromination is 13.9 kcal higher in energy than E_a for chlorination. As such, chlorination is significantly faster than bromination. ■

5.3.5 SUMMARY

- The rate of a reaction depends on the structure of the transition state. The Hammond postulate allows us to approximate the identity of the transition state.

Assessment 5.40 For each pair of reactions, predict which will happen more quickly.

[*For (a) and (b), think about the stability of the bases involved.*]

(a)

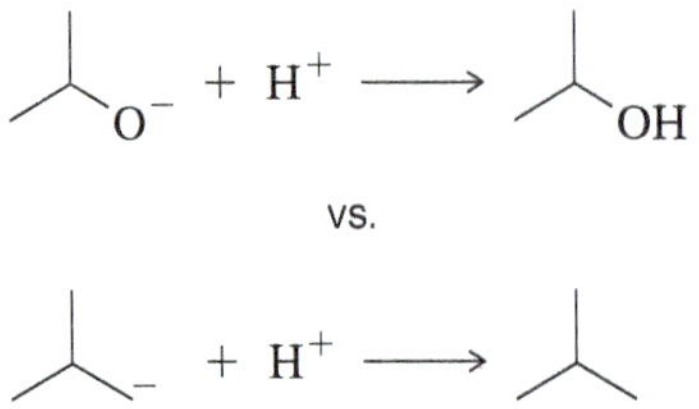

(b)

H—≡—H + $^{-}NH_2$ ⟶ H—≡$^{-}$ + NH_3

vs.

CH_3OH + $^{-}NH_2$ ⟶ CH_3O^{-} + NH_3

[*For (c) and (d), the more substituted carbocation is more stable due to hyperconjugation.*]

(c)

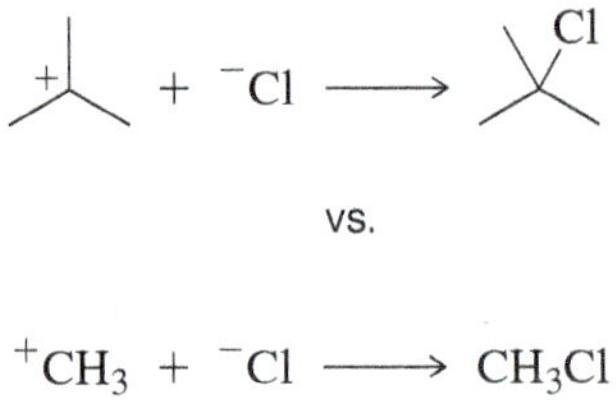

(d)

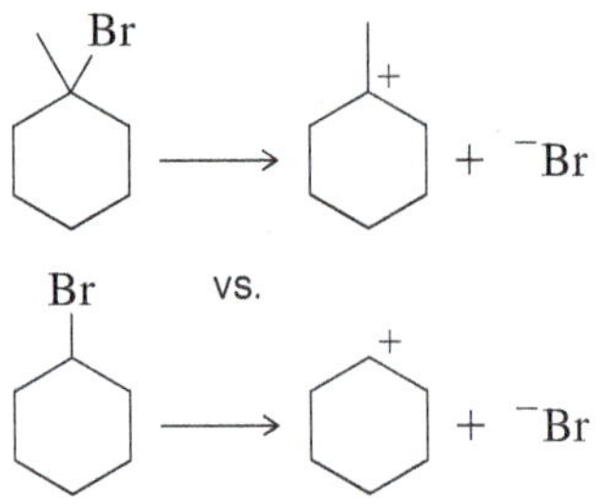

Discovering Selectivity in Halogenation

1. *How selective?*
2. *Origin of selectivity: RDS*
3. *Differing ΔH°*
4. *Hammond postulate*
5. *Bringing it all together*

5.3.6 Selectivity in halogenation

When alkanes larger than methane are halogenated, **regioselectivity** becomes a consideration. As a representative example, monochlorination and monobromination of propane give different product ratios (**Figure 5.61**). Why, though, do both favor formation of the secondary haloalkane? Moreover, why is bromination always more selective? Section 5.3.6, using the guidepost in the margin, discusses this selectivity in detail. [*As you read, pay close attention to the pictures and the calculation of enthalpy changes. As always, reconcile each new "discovery" with information studied previously.*]

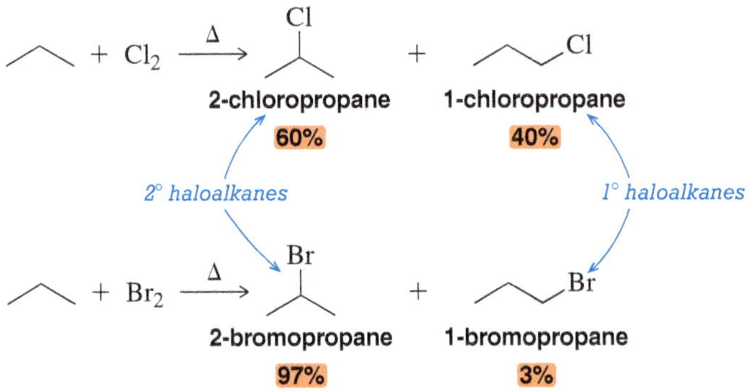

Figure 5.61 The bromination of propane is more selective than chlorination.

In our analysis of halogenation selectivity, we begin by (1) discussing how selective each reaction is, because the percentages don't tell the entire story. Then (2) we discuss the origin of the selectivity in the context of the mechanism and the rate-determining step. Next, (3) we compare the energetics of making the 1- and 2-halopropanes for chlorine and bromine on their own. Finally, (4) in order to understand the difference in selectivity, we further analyze the results in the context of the Hammond postulate. [*What we discover in this section is utilized in several places throughout the text.*] Before proceeding, come up with a hypothesis of your own as to what is going on here.

Discovering Selectivity in Halogenation

1. ***How selective?***
2. *Origin of selectivity: RDS*
3. *Differing* ΔH°
4. *Hammond postulate*
5. *Bringing it all together*

1. How selective?

Selectivity was not an issue for methane halogenation because abstraction of either of the four hydrogens would generate the same product. However, propane has two different types of hydrogens: six of them are primary (1°) and two are secondary (2°). The notation of 1°, 2°, and 3° refers to how many carbons are attached to the carbon to which the hydrogen is bonded (**Figure 5.62**).

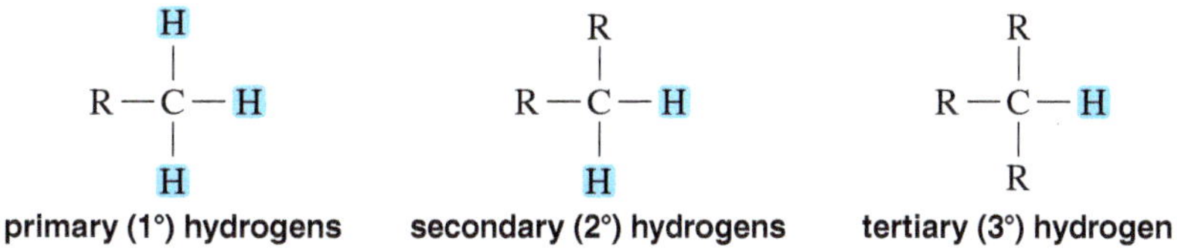

Figure 5.62 The designation of a hydrogen as 1°, 2°, or 3° tells us how many carbons are attached to the carbon it is attached to.

Random collision of a chlorine (or bromine) radical with propane would be expected to give a ratio of 75:25 for 1-chloropropane:2-chloropropane if primary (1°) and secondary (2°) C—H bonds have the same relative reactivity, as shown in **Figure 5.63**. While the calculation in Figure 5.63 may look more complicated than it needs to be, we will eventually encounter the case where 1° and 2° C—H bonds have different reactivities.

Six primary (1°) hydrogens: **Two secondary (2°) hydrogens:**

$$\% \, 1^\circ = \frac{6(x)}{6(x) + 2(y)}(100\%) = \frac{6(1)}{6(1) + 2(1)}(100\%) = 75\%$$

number of 1° C—H (numerator); *number of 1° C—H*, *number of 2° C—H* (denominator)

$$\% \, 2^\circ = \frac{2(y)}{6(x) + 2(y)}(100\%) = \frac{2(1)}{6(1) + 2(1)}(100\%) = 25\%$$

number of 2° C—H (numerator); *number of 1° C—H*, *number of 2° C—H* (denominator)

Figure 5.63 Random collisions would produce a 75:25 ratio of primary:secondary chloropropane. Assume x = relative reactivity of 1° C—H, and y = relative reactivity of 2° C—H. Because this assumes random (nonselective) collisions, the rate of reaction of each are the same. Thus, $x = y = 1$.

Assessment 5.41 The following are all **substitution reactions**, two of which we study in later chapters. With no knowledge of mechanism, what would you expect the ratio of products to be for each reaction, based on a random statistical distribution?

(a) Replacing a hydrogen (H) with deuterium (D):

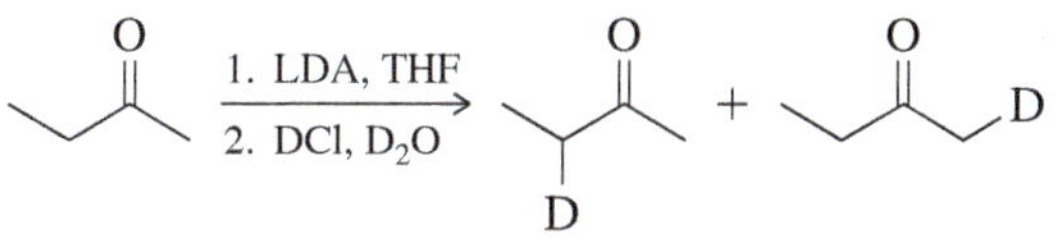

(b) Replacing a hydrogen (H) with bromine (Br):

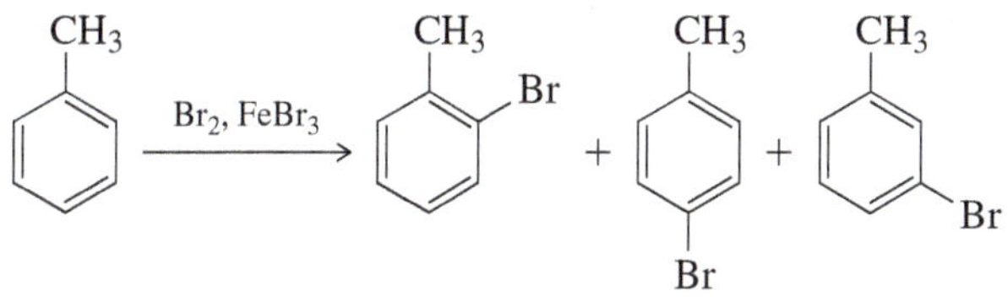

(c) Replacing a hydrogen (H) with chlorine (Cl):

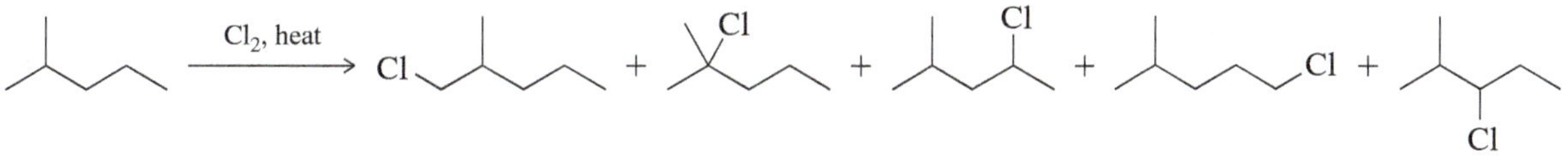

The halogenation reaction, however, does not produce the ratio of products predicted in Figure 5.63. The 2-halopropane is the major product in both chlorination and bromination (Figure 5.61). The expected ratio (75:25) is flipped for chlorination (40:60) and is skewed even more strikingly for bromination (3:97). This result can only be explained if the 2° C—H bond is more reactive than the 1° C—H bond. [*Does this fit the data in Table 5.6?*] Using the equation shown in **Figure 5.64** for chlorination, we can now solve for the relative rates of reaction of the 1° and 2° C—H bonds using the experimental product distribution. The rate for the less reactive 1° C—H bond is set as 1 ($x = 1$) and the relative rate of the 2° C—H bond (y) is calculated. For chlorination, we find that the secondary C—H bond is 4.5 times more reactive than the primary C—H bond.

Assume the reactivity of a 1° C—H bond = 1 and the reactivity of a 2° C—H = *y*. If the % of 2-chloropropane = 60% and the % of 1-chloropropane = 40%, then we can solve for *y* in two different ways.

$$60\% = \frac{2(y)}{2(y) + 6(1)}(100\%)$$

2(y): number of 2°C—H; 6(1): number of 1°C—H, x = 1

Solve for y, the relative reactivity of 2° C—H:

$$1.2y + 3.6 = 2y$$
$$3.6 = 0.8y$$
$$y = 4.5$$

$$40\% = \frac{6(1)}{6(1) + 2(y)}(100\%)$$

6(1): number of 1°C—H, x = 1; 2(y): number of 2°C—H

Solve for y, the relative reactivity of 2° C—H:

$$2.4 + 0.8y = 6$$
$$0.8y = 3.6$$
$$y = 4.5$$

Figure 5.64 **Using the product distribution measured experimentally for the chlorination of propane, we calculate that a 2° C—H bond is 4.5 times more reactive than a 1° C—H bond.**

The preceding calculations were for the *chlorination* of propane. Assessment 5.42 allows you to calculate the relative reactivity of the C—H bonds in the *bromination* of propane.

Before you work the problem, what is your expectation? Does the reactivity of a bond depend on the reaction conditions? Solve it on your own and then we will work it together.

Assessment 5.42 Given the ratio of products obtained in the bromination of propane, calculate the relative reactivity of a 1° C—H bond to a 2° C—H bond under these conditions.

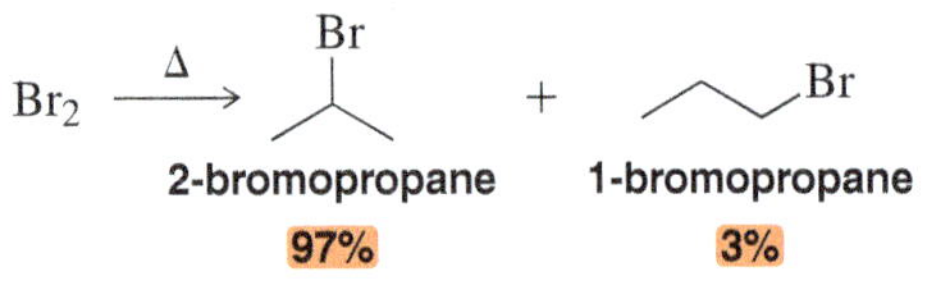

Solved Assessment 5.42

In Assessment 5.42, you have to calculate the relative reactivity of 1° and 2° C—H bonds in the bromination of propane. While your initial hypothesis might be that a 2° C—H bond has the same reactivity, regardless of reaction conditions, the ratio of products in the bromination reaction, as compared to the chlorination reaction, should immediately discredit this hypothesis. Bromination is more selective than chlorination. Thus, the 2° C—H bond is probably more reactive.

We can set up the problem the same way as before but using the ratio of products obtained in the bromination reaction (1°:2° = 3:97). Once again, the 1° C—H bond is less reactive, so we set its value at 1 ($x = 1$), and then solve for the 2° C—H bond (y). Doing this, we find that the 2° C—H bond is 97 times more reactive than the 1° C—H bond (**Figure 5.65**). ■

Assume the reactivity of a 1° C—H bond = 1 and the reactivity of a 2° C—H = *y*. If the % of 2-bromopropane = 97% and the % of 1-bromopropane = 3%, then we can solve for *y* in two different ways.

$$97\% = \frac{2(y)}{2(y) + 6(1)}(100\%)$$

2(y): number of 2°C—H; 6(1): number of 1°C—H, x = 1

Solve for y, the reactivity of 2° C—H:

$$1.94y + 5.82 = 2y$$
$$5.82 = 0.06y$$
$$y = 97$$

$$3\% = \frac{6(1)}{6(1) + 2(y)}(100\%)$$

6(1): number of 1° C—H, x = 1; 2(y): number of 2°C—H

Solve for y, the reactivity of 2° C—H:

$$0.18 + 0.06y = 6$$
$$0.06y = 5.82$$
$$y = 97$$

Figure 5.65 **Using the product distribution measured experimentally for the bromination of propane, we calculate that a 2° C—H bond is 97 times more reactive than a 1° C—H bond.**

Discovering Selectivity in Halogenation

1. *How selective?*
2. ***Origin of selectivity: RDS***
3. *Differing $\Delta H°$*
4. *Hammond postulate*
5. *Bringing it all together*

2. Origin of selectivity: making the most stable radical in the rate-determining step

The **regioselectivity** in alkane halogenation occurs in the formation of the radical in the rate-determining, first propagation step. Once the radical has formed, there is no choice but for the halogen to add to that carbon in the second propagation step. We just learned that the 2° C—H bond is more reactive, but why? [*Think about what Table 5.6 on bond dissociation energies taught us in Section 5.3.1.*]

Radicals are electron-deficient species. Like carbocations, the central carbon of a radical is sp^2-hybridized, containing 1 electron in a partially filled *p* orbital. Overlap of neighboring C—H and C—C σ bonds stabilizes the electron-deficient species through electron donation in a concept called **hyperconjugation** (Figure 5.66). Hyperconjugation has a more stabilizing effect on the 2° radical because there are more neighboring C—H bonds. The order of stability for radicals reflects this with 3° > 2° > 1° > CH_3. [*We talk much more about hyperconjugation in Chapter 8.*]

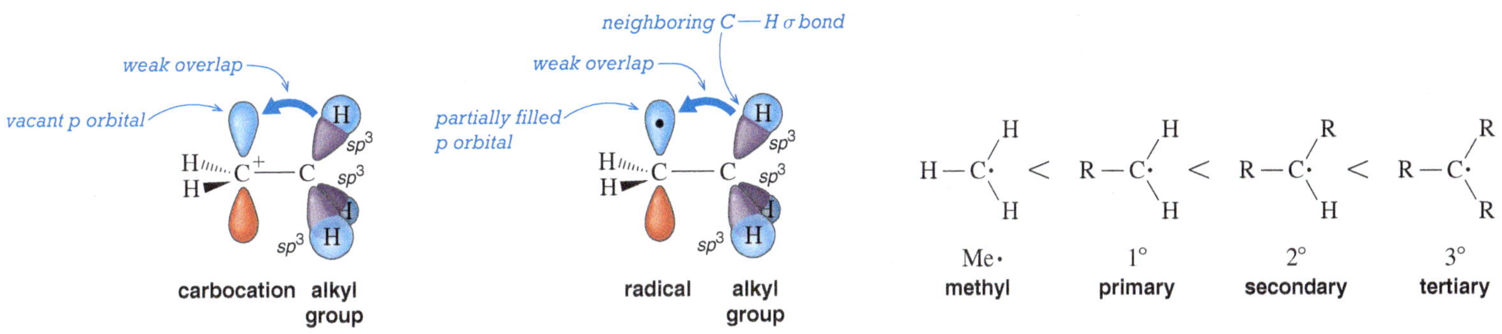

Figure 5.66 Like carbocations, radicals are stabilized by hyperconjugation—the overlap of neighboring σ bonds and the partially filled *p* orbital.

In summary, chlorination preferentially forms 2-chloropropane via the more stable 2° radical. If we went through the same analysis of bromination, we might expect the same result. We don't get the same selectivity, though, so a closer look at the mechanism is needed. Specifically, we next examine the energetics of each possible first step for the halogenation of propane.

Discovering Selectivity in Halogenation

1. *How selective?*
2. *Origin of selectivity: RDS*
3. ***Differing $\Delta H°$***
4. *Hammond postulate*
5. *Bringing it all together*

3. $\Delta H°$ for making primary and secondary haloalkanes

Because the first step determines regioselectivity, this is the step we should examine further. We can use bond-dissociation energies to calculate $\Delta H°$ of the first propagation step for formation of the 1° and 2° radicals in both chlorination and bromination. [*This calculation has been done side by side so you can more easily compare the numbers obtained.*]

Abstraction of a hydrogen by a chlorine atom to form a 1° radical is calculated in Figure 5.67(a). $\Delta H° = -5$ kcal/mol, so the process is slightly *exothermic*.

(a) Primary (1°) hydrogen:

$CH_3{-}CH_2{-}CH_3 + Cl\cdot \longrightarrow CH_3{-}CH_2{-}CH_2\cdot + H{-}Cl$

Energy required to break the $CH_3CH_2CH_2$—H bond	+98 kcal/mol (+410 kJ/mol)
Energy released in forming the H—Cl bond	−103 kcal/mol (−431 kJ/mol)
Total energy for reaction at the primary position:	−5 kcal/mol (−21 kJ/mol)

Abstraction of a hydrogen by a bromine atom to form a 1° radical is calculated in Figure 5.67(b). $\Delta H° = +10$ kcal/mol, so this process is slightly *endothermic*. [*This is an important difference between bromination and chlorination.*]

(b) Primary (1°) hydrogen:

$CH_3{-}CH_2{-}CH_3 + Br\cdot \longrightarrow CH_3{-}CH_2{-}CH_2\cdot + H{-}Br$

Energy required to break the $CH_3CH_2CH_2$—H bond	+98 kcal/mol (+410 kJ/mol)
Energy released in forming the H—Br bond	−88 kcal/mol (−368 kJ/mol)
Total energy for reaction at the primary position:	+10 kcal/mol (+42 kJ/mol)

Figure 5.67 Energetics of radical formation. (a) In chlorination, forming a 1° radical is exothermic. (b) In bromination, forming a 1° radical is endothermic.

(continued)

Abstraction of a hydrogen to make a 2° radical is favored over abstraction to make a 1° radical. This process is still exothermic with $\Delta H° = -8$ kcal/mol [Figure 5.68(a)].

(a) Secondary (2°) hydrogen:

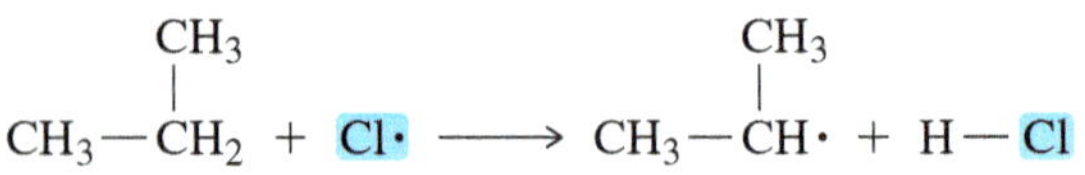

Energy required to break the $CH_3-CH(CH_3)-H$ bond	+95 kcal/mol (+397 kJ/mol)
Energy released in forming the H—Cl bond	−103 kcal/mol (−431 kJ/mol)
Total energy for reaction at the secondary position:	−8 kcal/mol (−34 kJ/mol)

Formation of the 2° radical is favored over formation of the 1° radical, but it is still endothermic [$\Delta H° = +7$ kcal/mol; Figure 5.68(b)]. [*Once more, this is an important difference between bromination and chlorination.*]

(b) Secondary (2°) hydrogen:

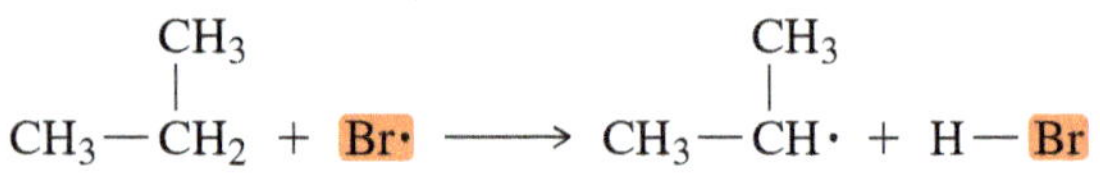

Energy required to break the $CH_3-CH(CH_3)-H$ bond	+95 kcal/mol (+397 kJ/mol)
Energy released in forming the H—Br bond	−88 kcal/mol (−368 kJ/mol)
Total energy for reaction at the secondary position:	+7 kcal/mol (+29 kJ/mol)

Figure 5.68 Energetics of radical formation. (a) In chlorination, forming a 2° radical is more exothermic. (b) In bromination, forming a 2° radical is less endothermic.

Summarizing what we know so far about both chlorination and bromination, the formation of a 2° radical is favored over the 1° radical by 3 kcal/mol. With identical differences of 3 kcal/mol for both, how can bromination be so much more selective? It must have something to do with the fact that the rate-determining step of chlorination is exothermic, whereas that for bromination is endothermic. [*Where have we heard that before?*] Enter, once again, the Hammond postulate.

Discovering Selectivity in Halogenation

1. *How selective?*
2. *Origin of selectivity: RDS*
3. *Differing $\Delta H°$*
4. ***Hammond postulate***
5. *Bringing it all together*

4. Hammond postulate

In trying to rationalize a reaction outcome based on the total $\Delta H°$, we are making a **thermodynamic** argument, one that tells us about the equilibrium of a reaction (K_{eq}), but not about relative reaction rates (E_a) (**kinetics**). If these reactions were at equilibrium, we'd get the same ratio for each using $\Delta G° = -RT \ln K_{eq}$. But this is a kinetic argument, so we have to look at the energy of the transition state to decide which reaction is faster.

To visualize the transition state structures and estimate the relative energy differences, we use the Hammond postulate. [*The analysis of chlorination and bromination is shown side by side so you can directly compare them.*]

The first step of chlorination is slightly exothermic ($\Delta H° = -5$ kcal/mol and -8 kcal/mol, respectively, for the 1° and 2° radical). The Hammond postulate tells us that the transition state for an exothermic step is more like the reactants of that step. The reaction coordinate diagram in Figure 5.69(a) shows that the transition state is closer in structure to propane and Cl•. Thus, there is relatively little radical character on propane in the transition state.

Because there is very little radical character on carbon in the transition state of the exothermic chlorination, there is, by extension, only a small difference in the energy of the two possible reaction outcomes. As a result, Figure 5.70(a) shows that the reaction is only somewhat selective for forming the 2° radical.

The first step of bromination, however, is endothermic ($\Delta H° = +10$ kcal/mol and $+7$ kcal/mol, respectively, for the 1° and 2° radicals). The Hammond postulate tells us that the transition state for an endothermic step is more like the products of that step. The reaction coordinate diagram in Figure 5.69(b) shows that the transition state is closer in structure to a propyl radical and HBr. Thus, there is substantial radical character on propane in the transition state.

Because there is substantial radical character on carbon in the transition state of the endothermic bromination, there is, by extension, a significant difference in the energy of the two possible outcomes. The significant radical character is more stable on a 2° carbon. As a result, the activation energy to make the 2° radical is much lower [Figure 5.70(b)], and formation of the 2° radical is much faster.

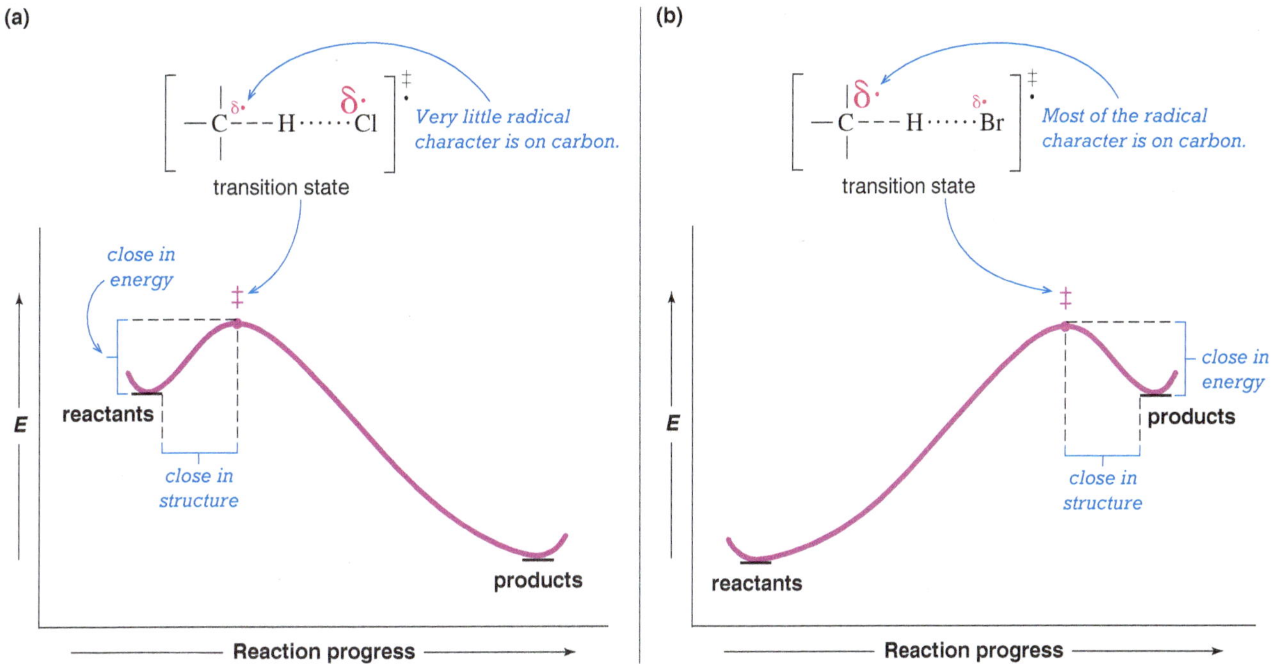

Figure 5.69 Reaction coordinate diagrams. (a) The exothermic first propagation step of chlorination. (b) The endothermic first propagation step of bromination.

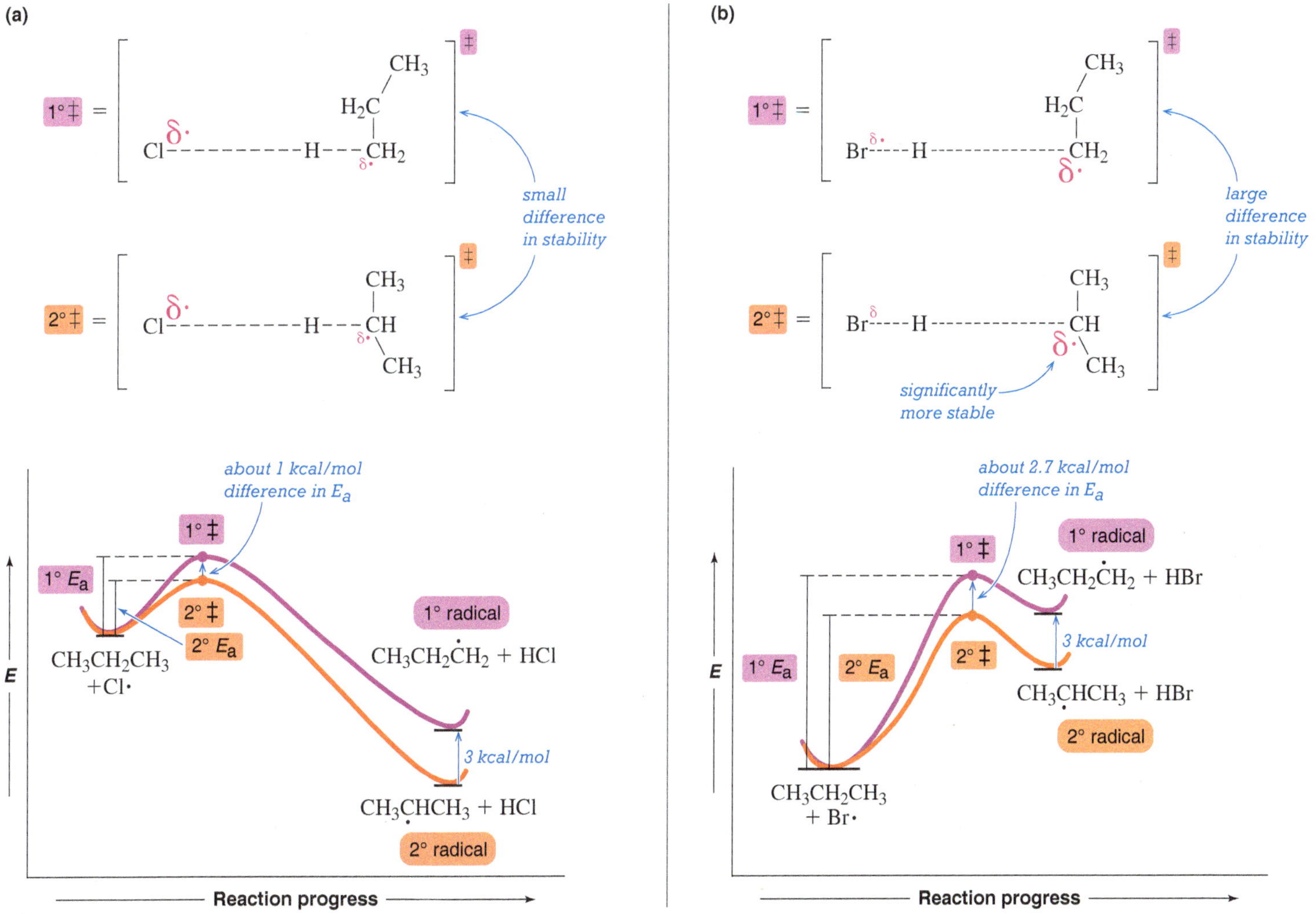

Figure 5.70 The difference in transition state energies correlates with the amount of carbon radical character in the transition state. (a) For chlorination, there is only a small difference in the transition state energies leading to 1° and 2° radicals. (b) For bromination, the more stable 2° radical is formed with a high degree of selectivity.

To summarize, even though the 2° radical product is favored by the same amount for both chlorination and bromination, the two reactions have different degrees of bond forming/breaking in the transition state. Endothermic bromination, with substantial radical character on carbon in the transition state, follows the more favorable transition state, in which the 2° radical is stabilized by hyperconjugation. In contrast, exothermic chlorination has very little radical character on carbon in either transition state. As a consequence, the transition states leading to 1° or 2° radicals are very close in energy and chlorination is not as selective.

Assessment 5.43 Through the course of this chapter, we have discussed only alkane chlorination and bromination, yet there are two other halogens we have not discussed.

(a) Is radical fluorination a favorable reaction? Do you expect it to be selective? Show your calculations.

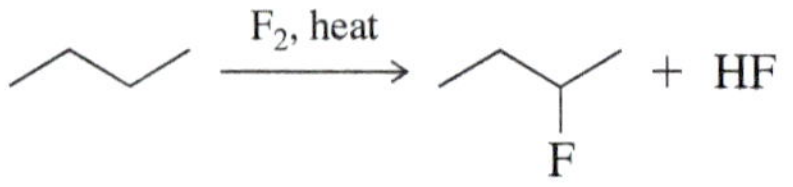

(b) Is radical iodination a favorable reaction? Do you expect it to be selective? Show your calculations.

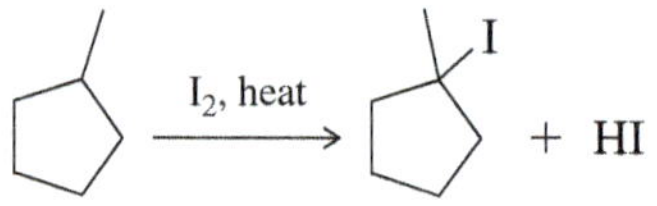

Discovering Selectivity in Halogenation

1. *How selective?*
2. *Origin of selectivity: RDS*
3. *Differing $\Delta H°$*
4. *Hammond postulate*
5. ***Bringing it all together***

5. Bringing it all together

Recall that chlorination is selective for 2° over 1° C—H bonds by a ratio of 4.5:1. This implies that the 2° C—H bond breaks 4.5 times faster than a 1° C—H bond. Recall also from Figure 5.70(a) that the difference between the two transition states leading to 2° and 1° radicals is about 1 kcal/mol. Do these numbers fit with each other? Equation 5.8 relates the activation energy E_a with the selectivity ratio obtained.

$$E_a(1°) - E_a(2°) = RT\ln\left(\frac{k_r(2°)}{k_r(1°)}\right) \qquad \text{Equation 5.8}$$

Using Equation 5.8, we can plug in the ratio of selectivity we got earlier (4.5:1) to calculate a difference in activation energy (at 298 K) of 0.89 kcal/mol (**Table 5.7**), which is very close to the 1 kcal we calculated in Figure 5.70(a).

Table 5.7 Calculation of difference in E_a based on ratio of products.

Justification	Calculation
For chlorination, the selectivity ratio is 4.5:1 for 2° C—H to 1° C—H. This ratio is also the ratio of rate constants.	$\frac{k_r(2°)}{k_r(1°)} = 4.5$
Use Equation 5.8 to determine the difference in activation energies based on this ratio.	$E_a(1°) - E_a(2°) = RT\ln\left(\frac{k_r(2°)}{k_r(1°)}\right)$
Plugging in the value of 4.5 for the ratio, along with values of *R* and *T* . . .	$E_a(1°) - E_a(2°) = \left(0.001987\frac{\text{kcal}}{\text{mol}\cdot\text{K}}\right)(298\text{ K})\ln\left(\frac{4.5}{1}\right)$
. . . gives a difference in activation energy of 0.89 kcal/mol. Thus, the activation energy for breaking the 1° C—H is 0.89 kcal/mol greater than it is for the 2° C—H.	$E_a(1°) - E_a(2°) = 0.89\text{ kcal/mol}$

5.3.6 SUMMARY

- Chlorination, with an exothermic first step, is only slightly selective for forming the most stable radical. Bromination, with an endothermic first step, is very selective for forming the most stable radical.

Assessment 5.44 The following reaction coordinate diagram is for the first step of the bromination of propane. Based on the difference in activation energies for substituting at the 2° versus 1° carbon, calculate the relative reactivity of 2° versus 1° C—H bonds.

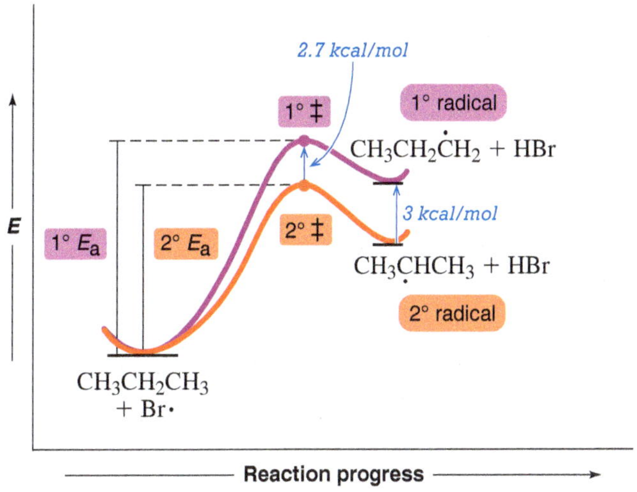

Assessment 5.45 The radical fluorination of 2-methyl propane resulted in a 14:86 ratio of products.

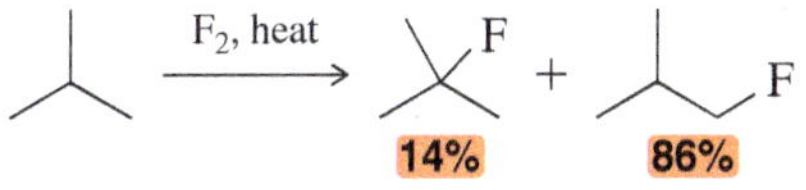

(a) On the basis of this ratio, calculate the relative reactivity of 1° and 3° C—H bonds in the radical fluorination.

(b) From the relative reactivity, calculate the difference in energy between the transition states of the first propagation steps leading to a 1° and 3° radical.

The halogenation reactions we have just studied provide our first look at an important goal of organic chemistry—namely, the synthesis of new molecules. Whether they are medicines, plastics, organoelectronics, or so on, organic chemistry is one of the few scientific fields of study in which things are made. Very often, what organic chemists make are what other scientists study.

5.4 Halogenation as a synthetic technique: a brief preview

Halogenated compounds have important medicinal, herbicidal, and fungicidal activities. It is estimated that 85% of pharmaceutical agents involve chlorine chemistry. Not only does the presence of a chlorine or a bromine enhance the ability of a drug to bind to its receptor [*Chapter 12*], it also enhances the ability of the molecule to find its way to its desired target of action [*Chapter 11*]. A few examples are the antibiotic Lorabid®; Fareston®, a selective estrogen receptor modulator used in the treatment of breast cancer; the herbicide bromoxynil; and vinyl chloride, a monomer used in the synthesis of polyvinylchloride (PVC), a plastic component of numerous building materials (**Figure 5.71**).

Figure 5.71 Important halogenated molecules.

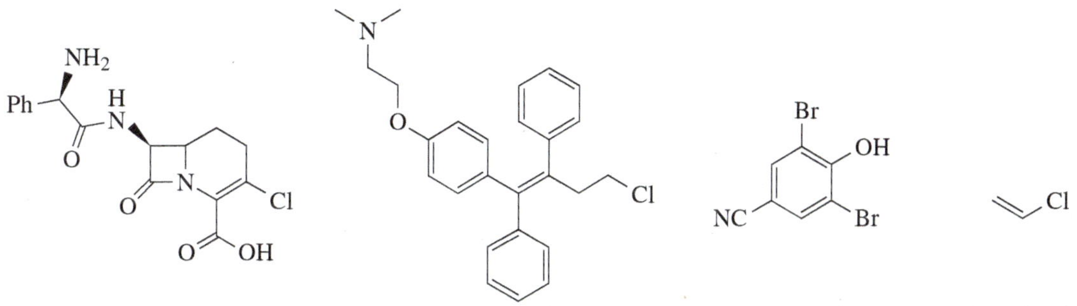

Lorabid (a β-lactam antibiotic)

Fareston (a selective estrogen receptor modulator)

bromoxynil (used for control of broadleaf weeds)

vinyl chloride (used for the synthesis of PVC)

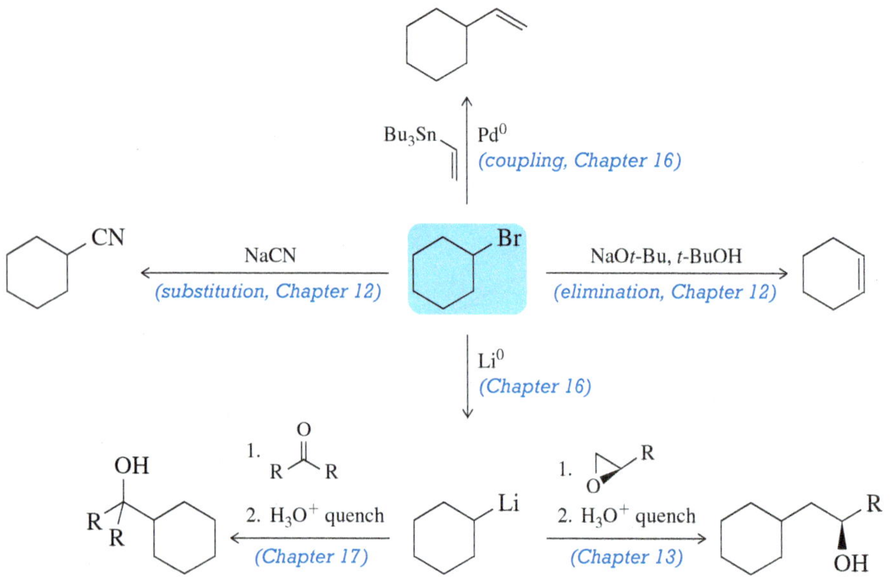

Figure 5.72 Alkyl halides are versatile intermediates used for a variety of important carbon–carbon bond-forming processes.

Halogenated compounds are also versatile intermediates that have utility in the synthesis of other molecules. In Chapter 12 we discuss how alkyl halides can be used as precursors to alkenes and other substitution products, such as alcohols. In Chapter 16, alkyl halides are converted into organolithium compounds. These strongly nucleophilic organometallic reagents can be reacted with electrophilic ketones [*Chapter 17*], aldehydes [*Chapter 17*], or epoxides [*Chapter 13*], among others. Finally, halogenated compounds can be directly attached to other organometallic reagents in palladium coupling reactions to generate new carbon–carbon bonds [*Chapter 16*]. Thus, the synthesis of haloalkanes is important to the study of organic chemistry (**Figure 5.72**).

Unfortunately, the use of radical halogenations as described thus far is complicated by the difficulty in achieving selective synthesis. In order to understand when the reaction is nonselective, it is best to first consider examples of selective halogenation.

The most useful substrates for selective halogenation have only one type of hydrogen. As long as you control the stoichiometry, good yields of a single product can be obtained in bromination or chlorination reactions (**Figure 5.73**).

Figure 5.73 Alkanes with one type of hydrogen can be halogenated with high yield.

Selective halogenation is also possible for substrates with a single carbon that is more substituted than all others. Only bromination is selective enough to produce just a single product with the molecules in **Figure 5.74**.

Figure 5.74 Alkanes with a single more-substituted carbon can be brominated selectively.

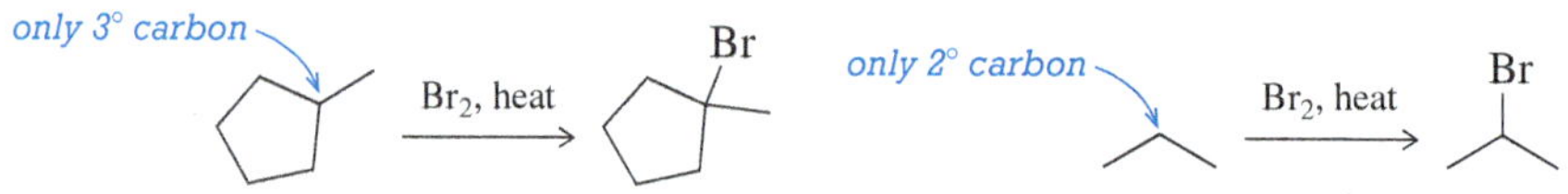

If an alkane does not fall into one of these two categories, then selective halogenation is difficult (**Figure 5.75**). The reaction can still be done, but a complex mixture of products results, requiring complicated purification steps.

Figure 5.75 Alkanes containing multiple similarly substituted carbons give mixtures of halogenated products.

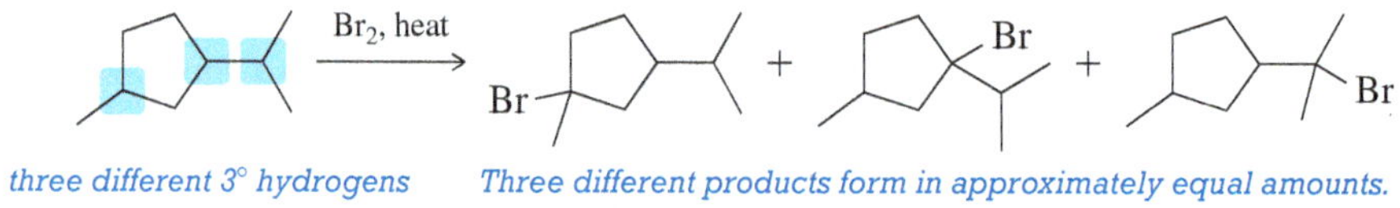

[*We introduce other reliable methods for synthesizing halogenated compounds in Chapters 8, 9, 10, and 11.*]

Assessment 5.46 Predict the major monohalogenation product(s) of the following reactions. Indicate whether you think the reaction will be selective and justify your position.

(a)

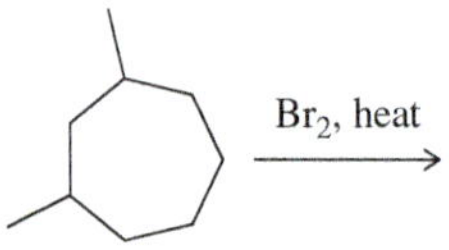

(b)

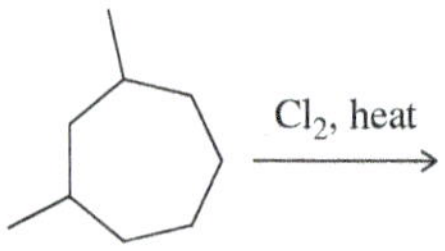

(c)

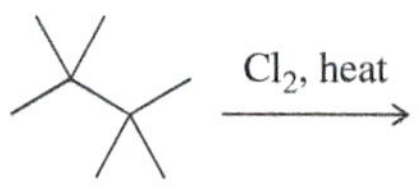

(d)

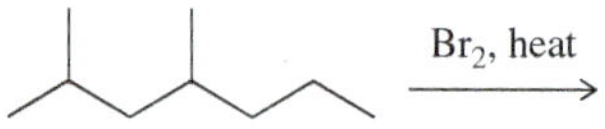

Concluding thoughts

Because Gaucher disease is caused by the buildup of glucocerebroside, there are two potential ways to treat this disease. The first is to limit the amount of glucocerebroside that forms by inhibiting glucosylceramide synthase. Two drugs, miglustat and eliglustat (**Figure 5.76**), have been created for this purpose. When these molecules bind at the active site of glucosylceramide synthase, UDP-glucose and ceramide are denied access. The enzyme is no longer able to catalyze the synthesis of glucosylceramide. Using these medications, the concentration of circulating glucosylceramide can be modulated.

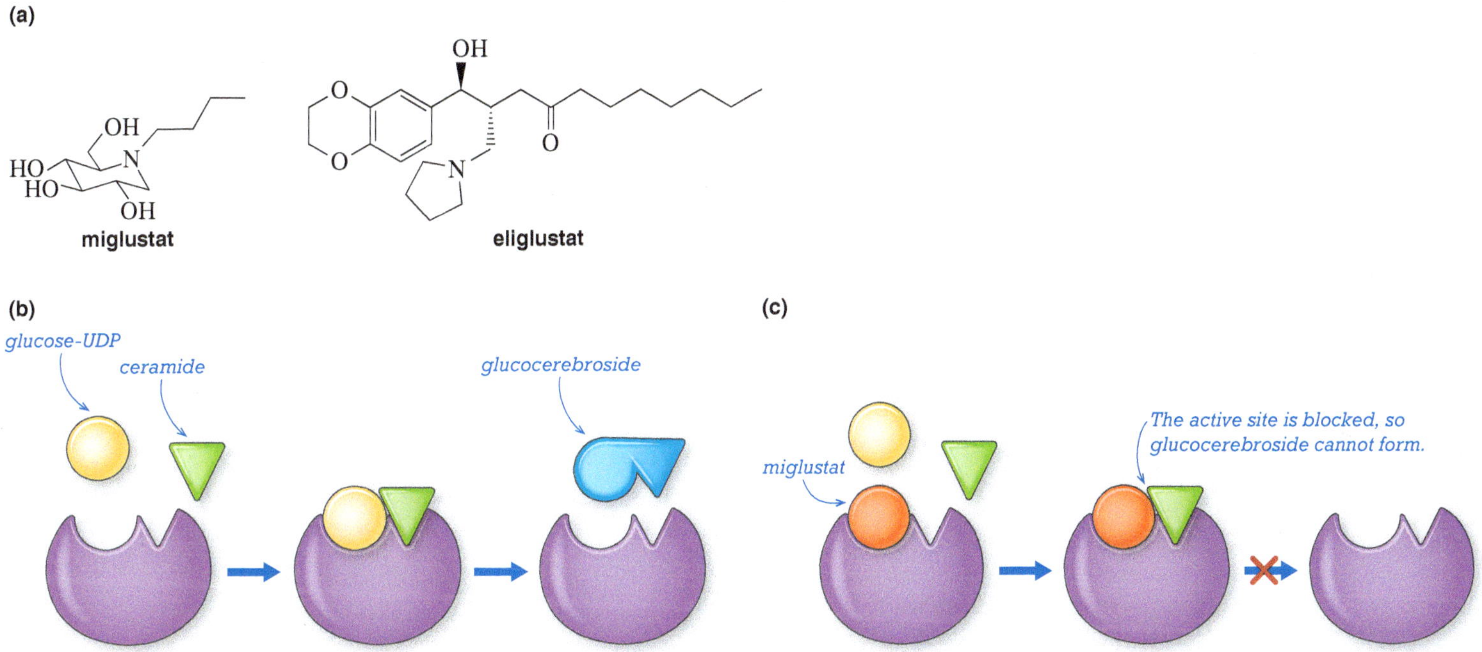

Figure 5.76 Inhibition of glucosylceramide synthase. (a) Miglustat and eliglustat are inhibitors. (b) The enzyme produces glucocerebroside in the absence of inhibitor. (c) Miglustat prevents binding of glucose-UDP, preventing glucocerebroside synthesis.

The second treatment option involves one of the first enzyme replacement therapies. Developed by Genzyme in 1991, imiglucerase (sold as Cerezyme®) is a modified but effective form of glucocerebrosidase (**Figure 5.77**). Because of the complications in producing, isolating, purifying, and administering the drug, Cerezyme is one of the more expensive pharmaceutical agents ever developed, with treatment costing $200,000 per year. In spite of its prohibitive cost, it has been a great commercial success owing to its ability to reverse signs and symptoms of Gaucher disease. In clinical trials patients showed reductions in liver and spleen sizes, increased hemoglobin levels, and decreased bone pain after 2 years of treatment. Patients are injected with the enzyme once every 2 weeks for the duration of their lives to maintain active enzyme in the lysosome, allowing the breakdown and removal of glucocerebroside by normal cellular processes.

Figure 5.77 The use of Cerezyme in place of mutated glucocerebrosidase allows the removal of β-glucocerebroside.

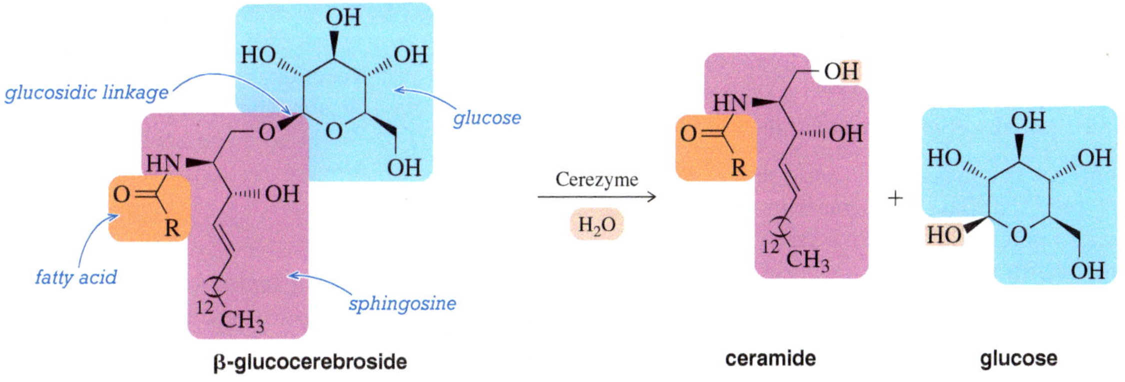

In conclusion, there are relatively few factors that govern the rate and position of equilibrium in a reaction. The phrase "the ball always rolls downhill" is actually quite apt in the context of the position of equilibrium. The caveat that "the ball always goes over the shorter hill" explains which reactions proceed more quickly. Regardless of which reaction we studied here in Chapter 5, the same kinetic and thermodynamic factors were invoked, and recognizing this is key. Going forward, look for the common thread that connects new reactions by relating them to something you've seen before. These similarities will be not only in the arrow-pushing mechanisms, but also in the approach to chemical analysis developed in this chapter. [*And if you aren't sure you will be able to do this, worry not! I'll be here to help you by pointing out these connections whenever possible.*]

End-of-Chapter Materials

Important Concepts

5.1.1–5.1.3: The change in Gibbs free energy ($\Delta G°$) provides a measure of the spontaneity of a chemical process. The sign of $\Delta G°$ tells us about the favorability of one side of an equilibrium process. A negative $\Delta G°$ (exergonic) implies that a reaction is favored as written to the right. A positive $\Delta G°$ (endergonic) implies that a reaction is favored to the left. The magnitude of $\Delta G°$ can be used to calculate the equilibrium constant (K_{eq}):

$$\Delta G° = -RT \ln K_{eq}$$

According to this equation, a negative $\Delta G°$ will result in $K_{eq} > 1$, and a positive $\Delta G°$ will result in $K_{eq} < 1$. When $\Delta G° = 0$, then neither side of a reaction is favored, and $K_{eq} = 1$.

5.1.4–5.1.6: The sign and magnitude of $\Delta G°$ are related to the change in enthalpy ($\Delta H°$) and the change in entropy ($\Delta S°$):

$$\Delta G° = \Delta H° - T\Delta S°$$

The change in enthalpy is made up of two factors: bonds formed/broken and strain energy. It can be determined as the sum of the energy released by bond formation minus the energy absorbed when bonds are broken. An increase in strain results in a positive ΔH, whereas lessening strain results in a negative ΔH. A process with a negative value of $\Delta H°$ ($\Delta H° < 0$) is said to be exothermic (heat releasing) and is favored by the enthalpy term. A process with a positive value of $\Delta H°$ ($\Delta H° > 0$) is said to be endothermic (heat absorbing) and is disfavored by the enthalpy term.

Entropy is a measure of randomness or disorder. Reactions are favored that have an increase in randomness or disorder, accompanied by a positive value of $\Delta S°$ ($\Delta S° > 0$). Processes that result in a decrease in disorder have a negative value of $\Delta S°$ ($\Delta S° < 0$). ΔS in organic reactions becomes important only when there is a difference in the number of molecules in the reactants and products. For all other reactions, the contribution to $\Delta G°$ from $\Delta S°$ is negligible, allowing us to often approximate $\Delta H° \approx \Delta G°$.

5.1.7–5.1.10: Among other factors, the rate of a reaction depends on the rate constant (k_r), whose value is a function of the frequency factor (A), the activation energy (E_a), and the temperature (T) through the Arrhenius equation:

$$k_r = Ae^{-E_a/RT}$$

A high rate constant means the reaction will proceed quickly, all other contributions being equal.

The frequency factor (A) is a measure of the frequency of collisions that have the required orientation to allow a reaction to occur. A reaction with little specificity in the directional requirement of a collision will have a high value of A and therefore give a high rate constant (k_r). A reaction with great specificity in the directional requirement of a collision will have a low value of A and therefore give a lower rate constant (k_r).

The activation energy (E_a) is the minimum energy needed for molecules to react and can be thought of as the barrier to reactivity. A high activation energy (E_a) will decrease the rate constant (k_r), making the reaction slow. A low activation energy (E_a) will result in a higher rate constant (k_r) and therefore a faster reaction.

Reactions proceed more quickly at higher temperature (T) because the molecules now possess a greater amount of kinetic energy to more easily overcome the barrier to reaction. Reactions at low temperature proceed more slowly.

5.1.11: The transition state ($\ddagger$) of a chemical reaction is the short-lived, highest energy chemical species on the pathway from reactant to product. As the barrier that must be overcome to allow a reactant to become product, the difference in energy between the reactant and the transition state is the activation energy (E_a or $\Delta G^{\circ\ddagger}$).

5.1.12: A reaction coordinate diagram is a graphical plot of the change in energy as a reaction proceeds from reactant to product. From a reaction coordinate diagram, you can predict the overall $\Delta G°$, the E_a and therefore the rate constant of a reaction, the number of steps in the reaction, the identity of the intermediates, and the approximate identity of the transition state.

5.2.1: For an acid–base reaction, the equilibrium constant (K_{eq}) can be calculated using the following equation, which relates the equilibrium constant to the pK_a values of the acid and the conjugate acid:

$$pK_{eq} = pK_a(\text{acid}) - pK_a(\text{conjugate acid})$$

$$pK_{eq} = pK_a(\text{left-side acid}) - pK_a(\text{right-side acid})$$

The K_{eq} determined by this equation makes it possible to calculate $\Delta G°$ for acid–base reactions (Equation 4.1).

5.2.2: Acid–base reactions go from two reactant molecules to two product molecules, so the contribution from entropy is very small ($\Delta S° \approx 0$). As a result, $\Delta H° \approx \Delta G°$ for acid–base reactions. In acid–base reactions, the contribution from bond energies is dwarfed by the contribution from going to a more or less stable conjugate base. That is, the magnitude of $\Delta H°$ for a reaction as written is determined by the difference in stability of the base and conjugate base in the reaction (or the difference in acidity of the acid and the conjugate acid). A reaction that produces a conjugate base that is more stable than the base will have a favorable $\Delta H°$ ($\Delta H° < 0$).

5.2.3: Acid–base reactions are some of the fastest reactions in organic chemistry. In terms of the Arrhenius equation (Equation 5.5), acid–base reactions have a high frequency factor, owing to a very small directional requirement and a relatively low activation energy. Both of these factors contribute to the high rate of reaction. The transition state of an acid–base reaction reveals the bond forming and breaking that occurs as the reaction progresses from reactant to product and is approximated by our arrow-pushing mechanism of the reaction. Because the acid and base (two molecules) come together in the transition state to make a single chemical species ("one molecule"), $\Delta S^{\circ\ddagger} < 0$. The activation energy for all reactions is greater than zero.

5.2.4: Because the transition state is so short-lived, we have no way of knowing its exact appearance. The Hammond postulate allows us to predict the form of the transition state on the basis of the exo- or endothermicity of the overall reaction. An endothermic reaction (step) has a transition state that is of similar structure and energy to the product(s) [intermediate(s)] of that reaction (step). An exothermic reaction (step) has a transition state that is of similar structure and energy to the reactant(s) [intermediate(s)] of that reaction (step). A better sense of the identity of the transition state allows us to predict the degree to which bonds are formed or broken in the transition state.

5.2.5: The principle of microscopic reversibility says that for a reversible reaction, the mechanism of the reverse reaction is the exact opposite of the mechanism in the forward direction. This is because the reaction always chooses the lowest energy pathway and will always do so in either direction.

5.2.6: For a reaction A + B → C + D, the rate of a reaction is the product of the concentration of the reactants and the rate constant:

$$\text{rate} = k_r[\text{A}]^a[\text{B}]^b$$

The experimentally determined superscripts a and b refer to the order of the reaction with regard to A and B, which is another way of saying how many molecules of A and B are present in the transition state of the rate-determining step of the reaction and can give insight into the overall mechanism.

5.3.1–5.3.2: The radical halogenation of an alkane is a substitution reaction of the general form A + B → C + D. It goes from two molecules to two molecules, so it is entropically neutral ($\Delta S° \approx 0$) and $\Delta G° \approx \Delta H°$. Bond-dissociation energies can be used to estimate $\Delta H°$ for a number of reactions, according to the general formula:

$$\Delta H° = \text{BDE (bonds broken)} - \text{BDE (bonds formed)}$$

Bond-dissociation energies are a measure of the energy absorbed by the homolytic breaking of a bond, and their magnitude is reflective of the stability of the free radicals produced in the process.

5.3.3: Radical halogenation is a chain reaction. The mechanism features an initiation step, where the radical is first formed, and two propagation steps, which add together to give the net reaction while reforming the propagating radical. Termination steps can occur if the concentration of radicals is high, and these reactions are nonproductive in terms of the overall chain reaction.

5.3.4: Alkane halogenation is a two-step reaction, with the sum of each propagation step adding up to represent the entire reaction. Using bond-dissociation energies, the energy of each step and the entire reaction can be calculated. Bromination features an endothermic ($\Delta H° > 0$) first step and is slightly exothermic ($\Delta H° < 0$) overall. Chlorination has a slightly exothermic ($\Delta H° < 0$) first step and is highly exothermic ($\Delta H° < 0$) overall.

5.3.5–5.3.6: Applying the Hammond postulate to the rate-determining step of each reaction (the first propagation step) reveals that the rate of chlorination is faster than the rate of bromination. The transition state of the highly endothermic first step of bromination is similar in structure and energy to the radical intermediate (the product of the first propagation step). As a result, the transition state of the rate-determining step for bromination has substantial radical character on carbon. The transition state of the mildly exothermic first step of chlorination is closer in structure and energy to the reactants. As a result, the transition state of the rate-determining step for chlorination has less radical character on carbon.

This analysis is also used as the justification for why bromination is significantly more selective than chlorination for forming the most stable radical intermediate and therefore the most substituted haloalkane. The order of C—H reactivity, based on the stability of the resulting radicals, is 3° > 2° > 1°. The order of alkyl radical stability is similar (3° > 2° > 1°), owing to the fact that radicals are stabilized by hyperconjugation.

Knowing the difference in activation energies between two similar reactions (*a* and *b*), we can calculate their relative rates using the following equation:

$$E_a(a) - E_a(b) = RT \ln\left(\frac{k_r(b)}{k_r(a)}\right)$$

5.4: Alkane halogenation is a mildly effective synthetic reaction that is limited by its ability to selectively prepare haloalkanes. It is most efficient when the alkane to be halogenated has only one type of hydrogen because of symmetry (with chlorination or bromination) or has one type of hydrogen that gives a significantly more stable radical (for bromination only). Many reactions in subsequent chapters utilize alkyl halides that can be prepared this way.

Important Reactions and Mechanisms

[Each reaction is shown first as the general form using the R notation to indicate that it can be basically any inert group without changing the reaction outcome. Next, it is illustrated using a real molecule. Finally, a mechanism is shown using the real example.]

1. **Radical halogenation: 5.4** The thermally initiated radical chain reaction, resulting in the substitution of a halogen atom for an alkyl hydrogen.

General:

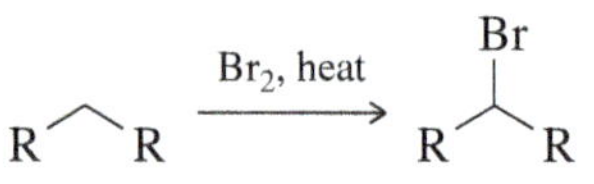

Specific:

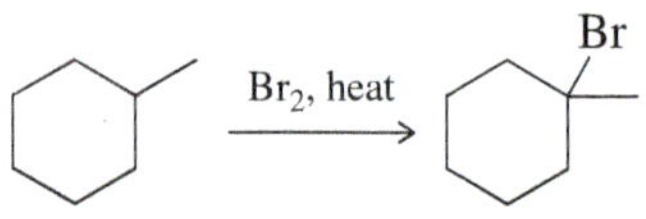

Mechanism:

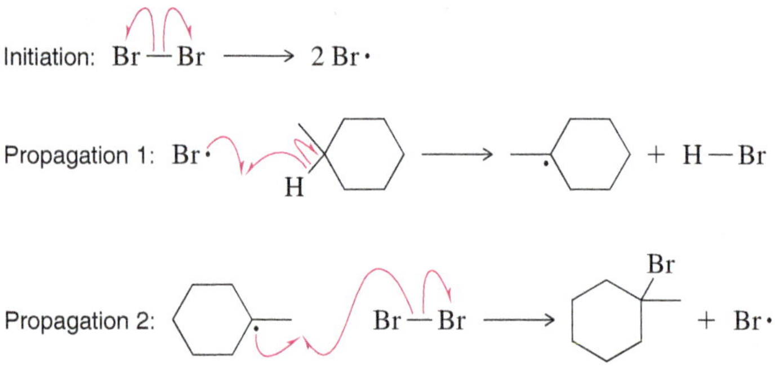

End-of-Chapter Assessments

[Assessments are ranked according to their difficulty; •••• = most difficult.]

Note: If not specified, assume a temperature of 298 K for all problems.

5.47 (••) The following table of strain energies is associated with a variety of 1,2-gauche interactions. Use this table to answer the questions (i)–(iv).

1,2-Gauche	Strain (kcal/mol)	1,2-Gauche	Strain (kcal/mol)
F—CH_3	0.12	Br—CH_3	0.33
CH_3—CH_3	0.90	NH_2—CH_3	0.57
CN—CH_3	0.10	OH—CH_3	0.50

For each of the bond rotations shown, (i) identify which you believe to be more stable, (ii) calculate $\Delta G°$, (iii) calculate the equilibrium constant, and (iv) draw the transition state.

(a)

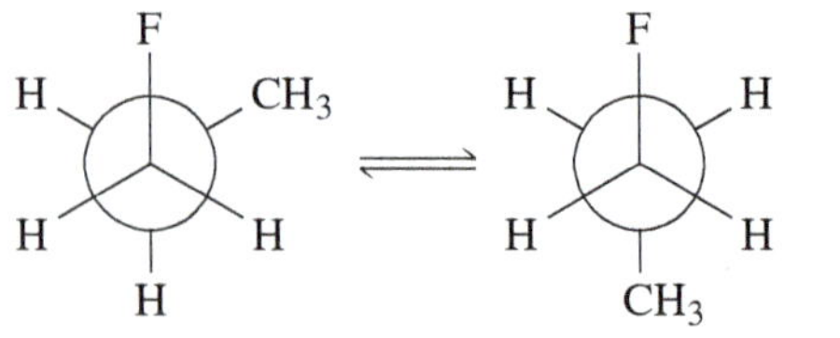

(b)

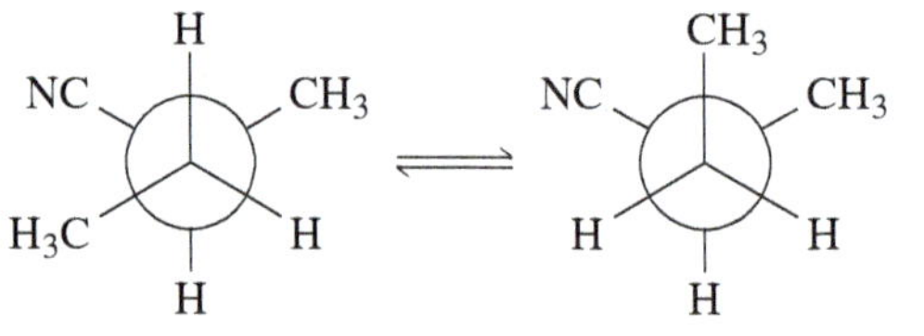

(c)

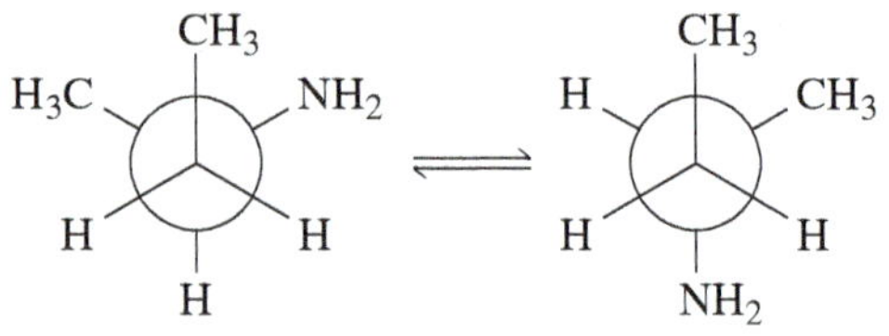

(d)

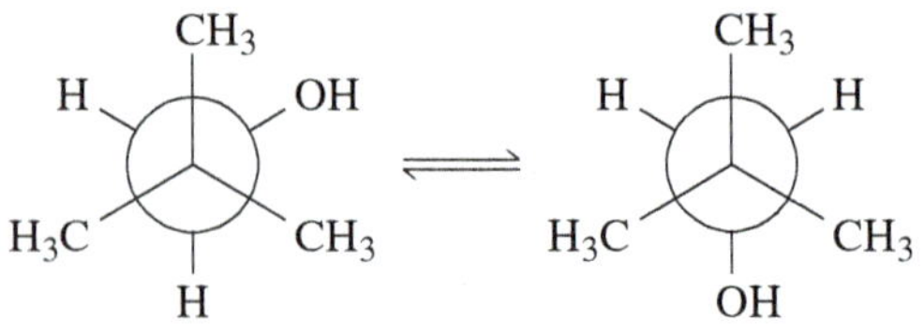

(e)

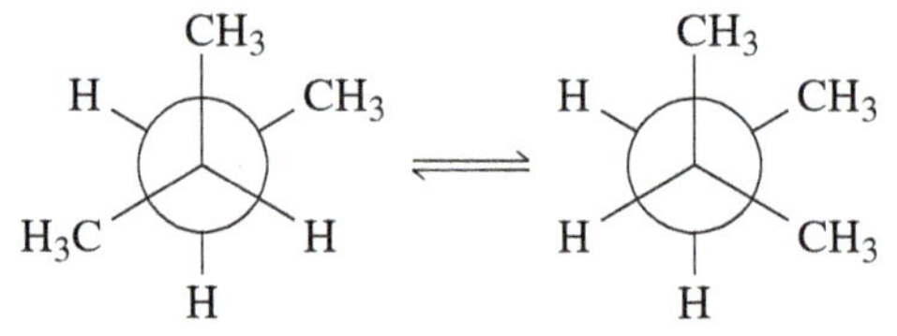

(f)

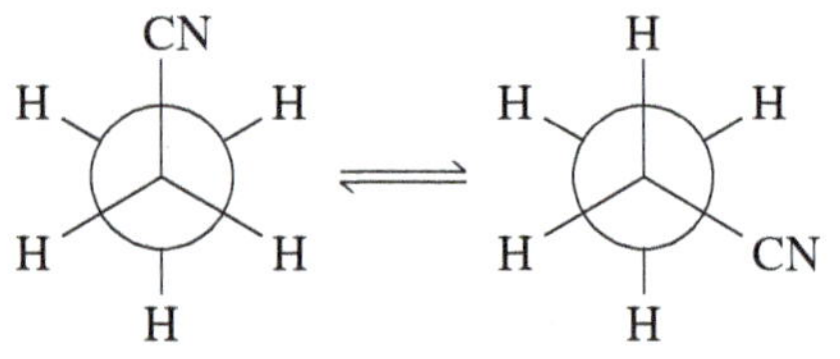

5.48 (•) For each of the following acid–base reactions, (i) predict which side of the reaction you expect to be favored, (ii) calculate K_{eq}, (iii) calculate $\Delta G°$, and (iv) draw the transition state, paying close attention to the degree of bond forming and breaking present in the transition state. If a pK_a is not one of the ten common ones we learned in Chapter 4, it will be given to you.

(a)

(b)

(c)

(d)

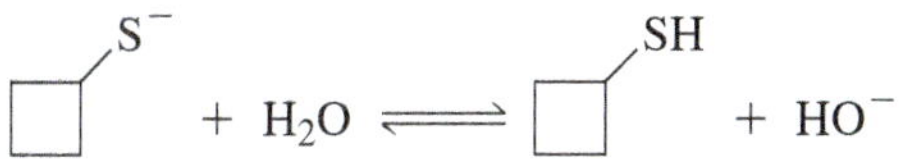

pK_a = 10

5.49 (•) For each of the following reactions, identify the bonds that are broken and formed. Be sure to indicate whether the bond that is broken is a σ bond or a π bond.

(a)

(b)

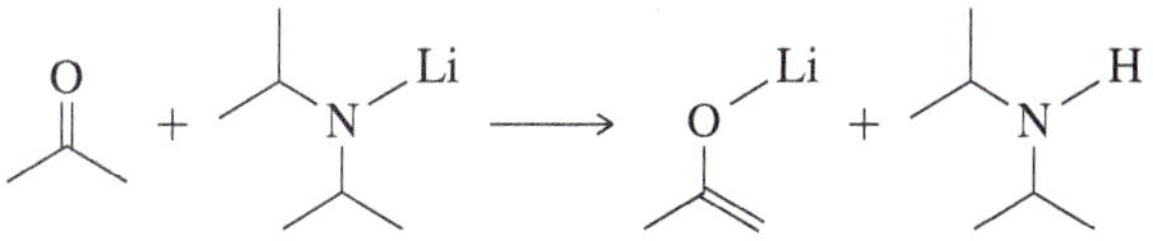

(c)

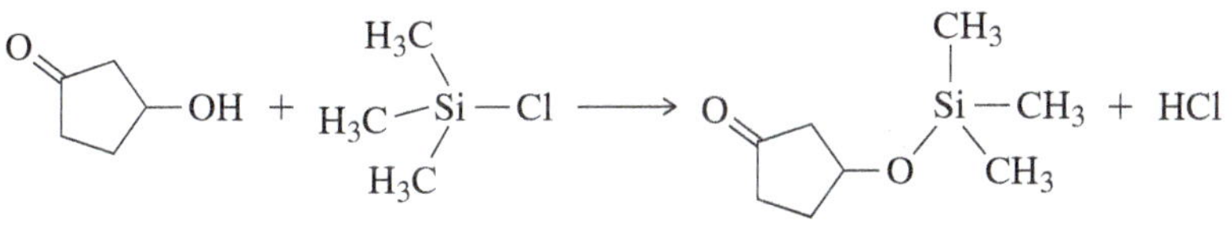

(d)

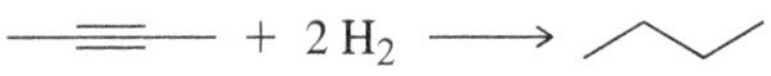

(e)

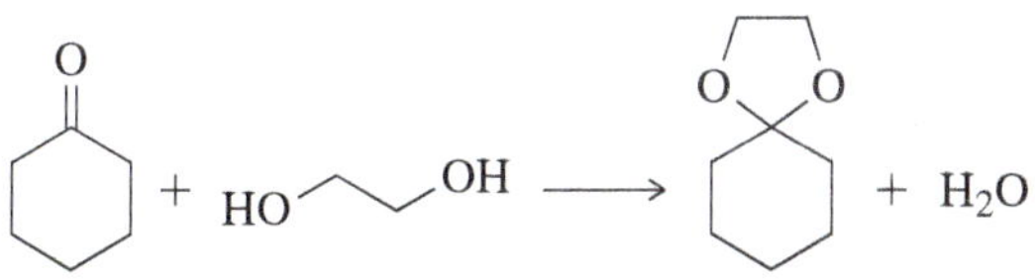

(f)

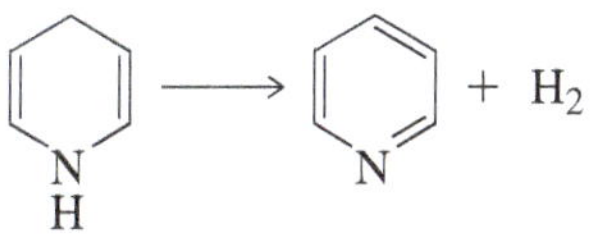

5.50 (••) The A value of a substituent on a cyclohexane ring is essentially the $\Delta G°$ for a substituent going from the equatorial to the axial position in a chair–chair interconversion. Because most substituents prefer to be in the equatorial position, A values are, by definition, positive numbers. Use the table of A values to calculate $\Delta G°$ and K_{eq} for the chair–chair interconversions shown.

(a)

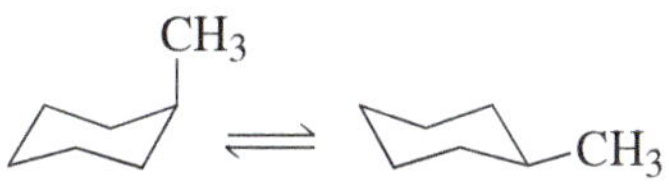

(b)

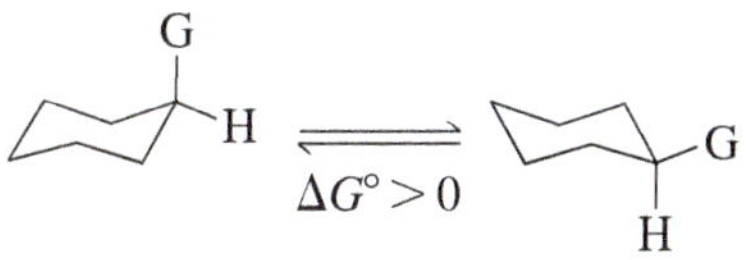

Group (G)	A value (kcal/mol)	Group (G)	A value (kcal/mol)
CH_3–	1.70	Br–	0.38
CH_3O–	0.75	Ph–	3.0
MeO_2C–	1.31	*t*-Bu–	5.1

(c)

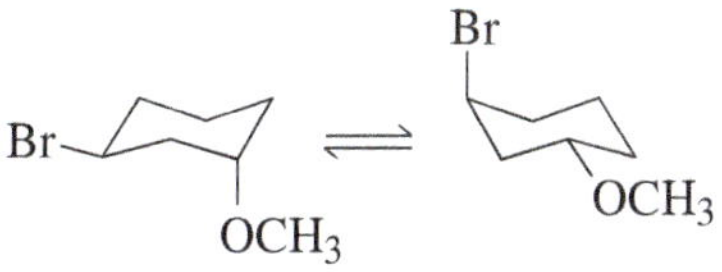

(d)

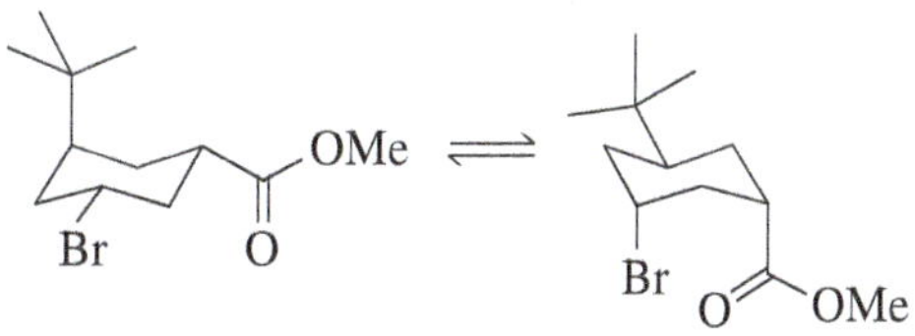

5.51 (••) For each set of reactive intermediates, rank them in order of reactivity (1= most reactive).

(a)

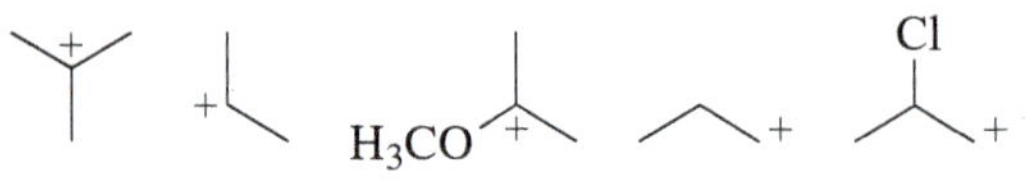

(b)

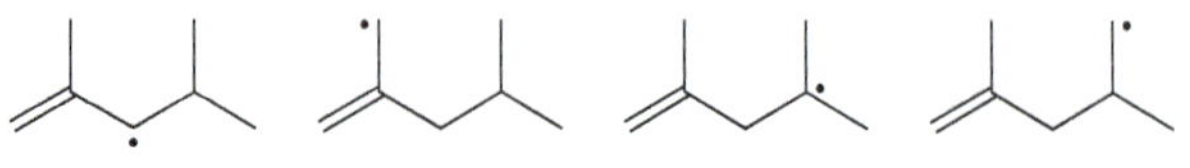

(c)

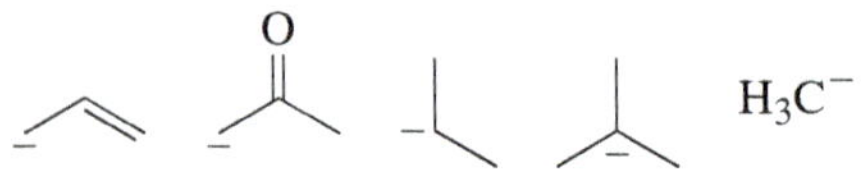

5.52 (••) Predict the product(s) of the following halogenation reactions. Only one equivalent of the halogen is used in each case. If the reaction proceeds selectively, indicate this by only drawing the major product. If the reaction is *not* selective, draw all possible products.

(a)

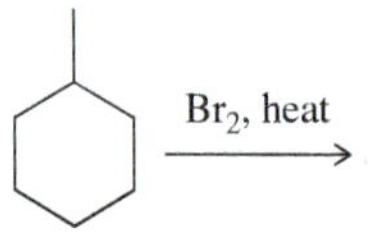

(b)

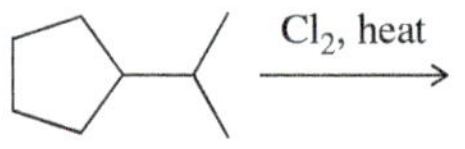

(c)

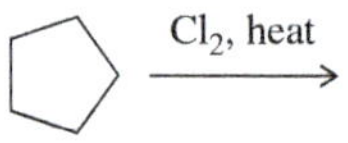

(d)

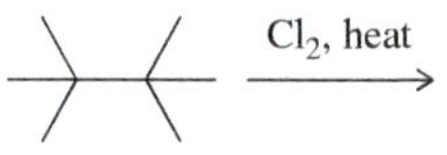

(e)

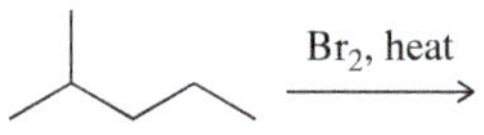

(f)

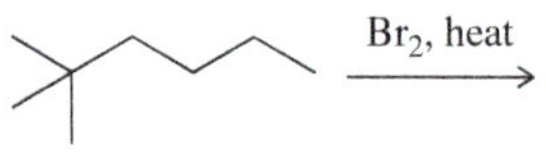

5.53 (••) For each of the reaction coordinate diagrams shown, (i) indicate the number of steps in the reaction, (ii) label the intermediates, (iii) identify the rate-determining step, and (iv) tell whether K_{eq} is greater than, less than, or equal to zero.

(a)

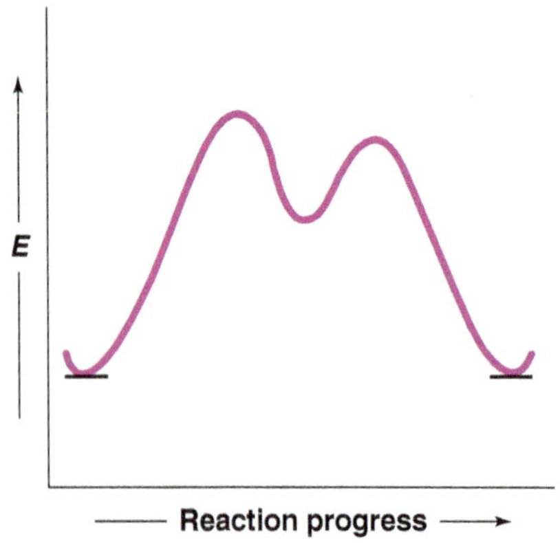

(b)

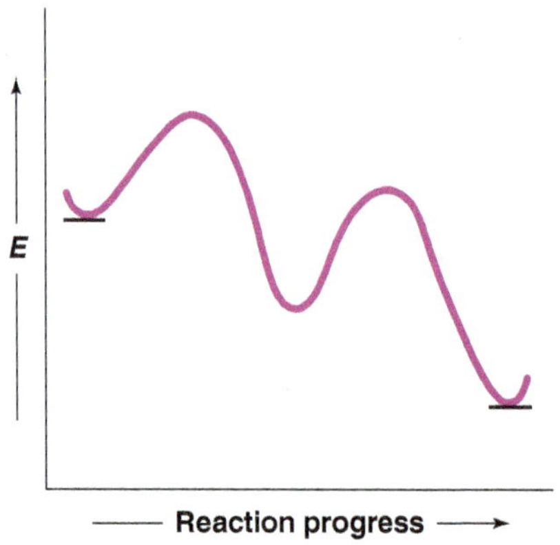

(c)

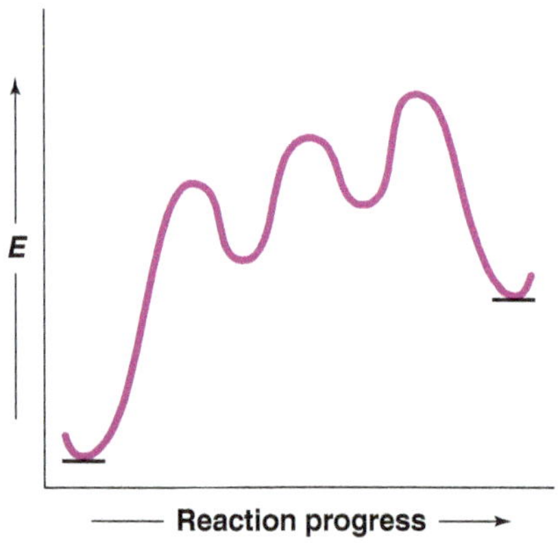

5.54 (••) Draw a reaction coordinate diagram, making sure to label reactants (R), products (P), intermediates (I), transition states (‡), activation energies (E_a), and $\Delta G°$, for each of the following.

(a) an exothermic, one-step reaction

(b) an exothermic, two-step reaction where the second step is rate-determining

(c) a slightly endothermic, three-step reaction where the first step is rate-determining

(d) a slightly exothermic, three-step reaction where the third step is rate-determining.

5.55 (•••) Compound **A** can react in two different ways to produce either compound **B** or compound **C**. Based on the reaction coordinate diagram shown, (a) is the formation of **B** or **C** faster? (b) Assuming a high enough temperature to allow equilibration, will **B** or **C** be the predominant species at equilibrium?

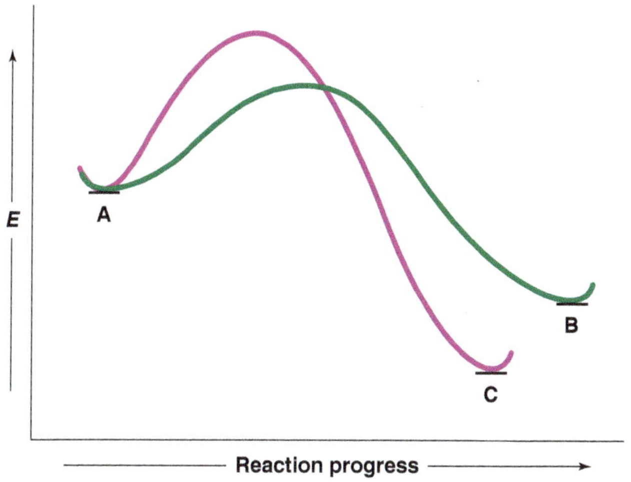

5.56 (••) Calculate $\Delta H°$ for the following alkene addition reaction, one we discuss further in Chapter 7. Predict the sign of $\Delta S°$. (The BDE for C—C π bond is approximately 65 kcal/mol.)

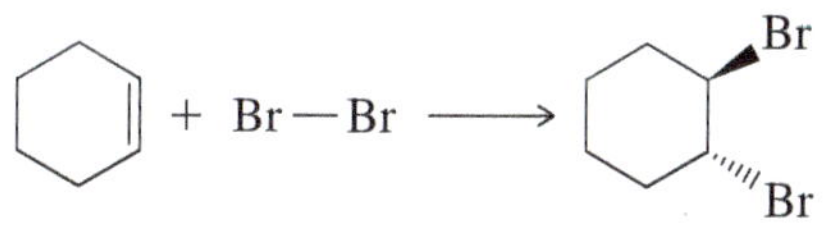

5.57 (••••) We discuss the following reaction in Chapters 12 and 13.

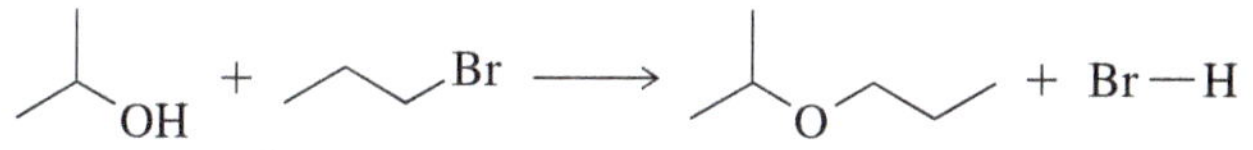

(a) Calculate $\Delta H°$; estimate $\Delta S°$ and $\Delta G°$.

(b) Mechanistically, the reaction occurs as shown below. Why is this reaction favored? Based on the stability of the anions, estimate K_{eq}.

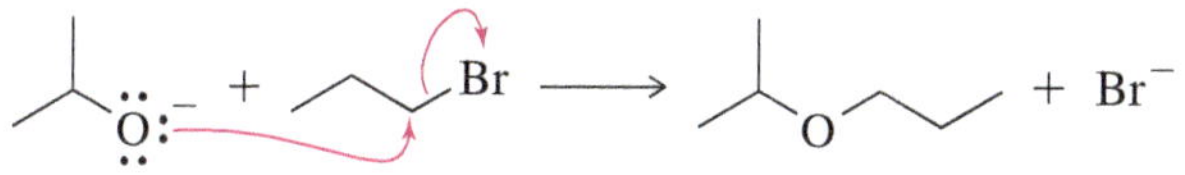

(c) Which would you expect to be faster, the reaction in (b) or the one here in (c)? Which would be more favorable in terms of $\Delta G°$? Explain.

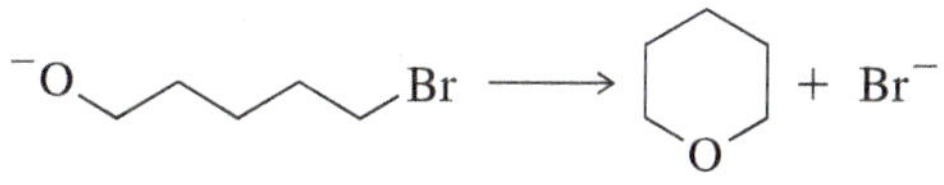

5.58 (•••) A chemist discovered the following reaction and wanted to know more about its mechanism, so she examined the rate of the reaction as a function of the concentration of each reagent.

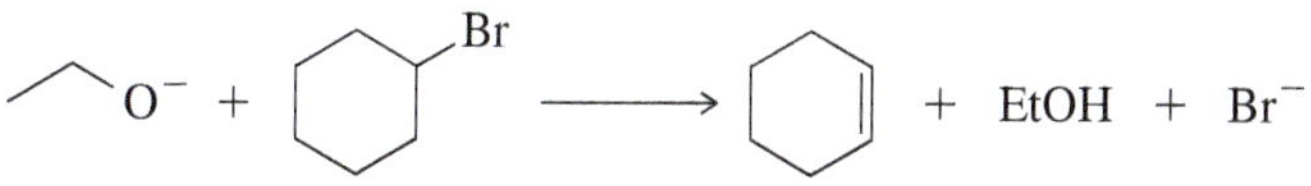

	$[EtO^-]$ (M)	$[C_6H_{11}Br]$ (M)	Rate (M/s)
Trial 1	0.1	0.1	2.33×10^{-2}
Trial 2	0.1	0.2	4.66×10^{-2}
Trial 3	0.2	0.1	4.66×10^{-2}

Based on the data in the table, what can you say about the transition state of the rate-determining step? Suggest a mechanism consistent with these data.

5.59 (••••) Reactions (a) and (b) are disfavored overall ($\Delta G° > 0$), yet they are favored based on $\Delta H°$. Identify the bonds formed and broken for (a) and (b).

(a)

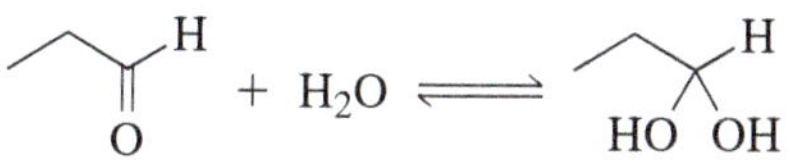

(b)

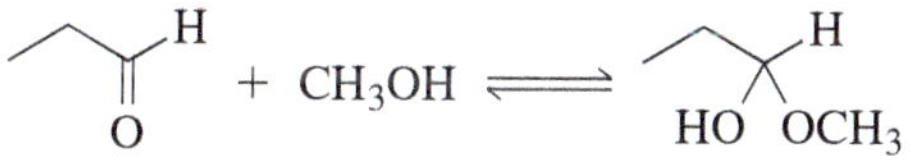

Reaction (c), on the other hand, is favored ($\Delta G° < 0$). Identify the bonds formed and broken and explain this result in light of (a) and (b).

(c)

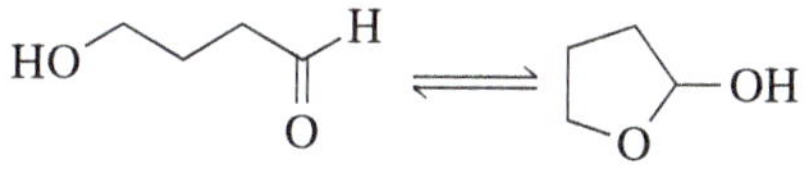

5.60 (•••) Keto–enol tautomerism is a reaction we discuss in detail in Chapter 19. Estimate the equilibrium constant of this reaction (BDE for C—C π bond = 65 kcal/mol; for C—O π bond = 85 kcal/mol).

5.61 (••••) The hydrogenation of alkenes is a reaction we study in Chapter 9.

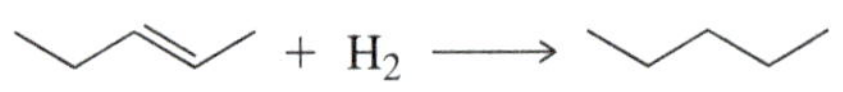

(a) Calculate $\Delta H°$ for this reaction.
(b) Is this reaction favored or disfavored in terms of entropy?
(c) In spite of the large negative value of $\Delta H°$ and overall negative $\Delta G°$, this reaction is very slow unless a catalyst is used. Without worrying about the actual mechanism, sketch the catalyzed and uncatalyzed reactions on a reaction coordinate diagram. [*Assume that the catalyzed and uncatalyzed are a single step.*]

5.62 (•••) The high-temperature thermolysis of cyclobutanol is used in the synthesis of vinyl alcohol.

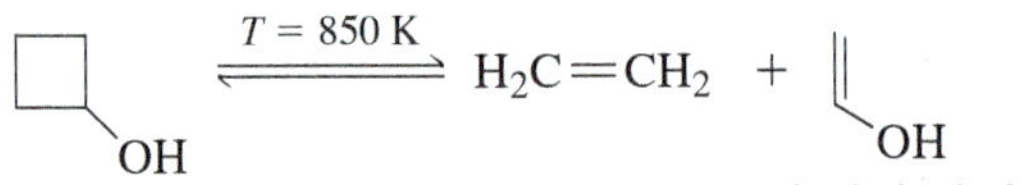

vinyl alcohol

(a) Estimate $\Delta H°$ based on bond-dissociation energies (BDE for C—C π bond = 65 kcal/mol).
(b) Why is the actual value of $\Delta H°$ likely a smaller number than you calculated in part (a)?
(c) Why might this reaction be favored to the right only at high temperatures?

5.63 (••••) Parts (a)–(d) of this assessment assist in the development of what will become a common theme in organic reactions and should be worked in order. [*Think carefully about how each question relates to the others.*]

(a) Calculate $\Delta H°$ for the following reaction (BDE for C—O π bond is approximately 85 kcal/mol).

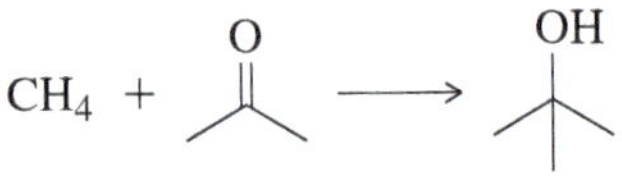

(b) Calculate K_{eq} for the following acid–base reaction.

(c) Without worrying about the mechanism of the reaction, estimate an equilibrium constant for the following carbonyl addition reaction based on the relative stability of the Lewis bases.

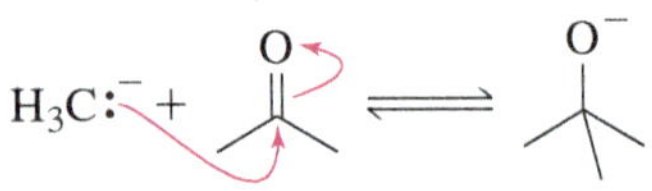

(d) In light of your answers to parts (b) and (c), where both were shown to be quite favorable, imagine a scenario where either reaction is possible. Of the two, which would you expect to be faster? Which would you expect to be more favored? Explain each in the context of the important thermodynamic and/or kinetic parameters.

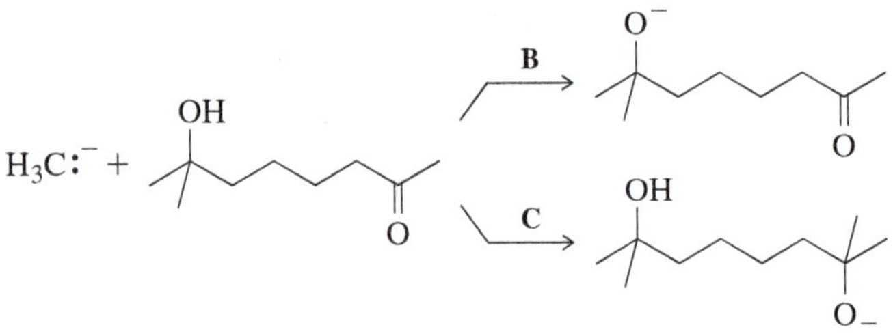

5.64 (•••) Bond-dissociation energies reflect the stability of the radicals produced upon bond breaking. Explain the following discrepancy in BDEs, given that primary radicals are produced in both cases. Your analysis should feature the use of a reaction coordinate diagram to illustrate this concept.

$$H_3CH_2C—CH_2CH_3 \longrightarrow 2\cdot CH_2CH_3 \qquad \Delta H° = +88 \text{ kcal/mol}$$

$\Delta H° = +65$ kcal/mol

5.65 (••••) In Chapter 13, we discuss the ring-opening reactions of epoxides, such as the one shown here.

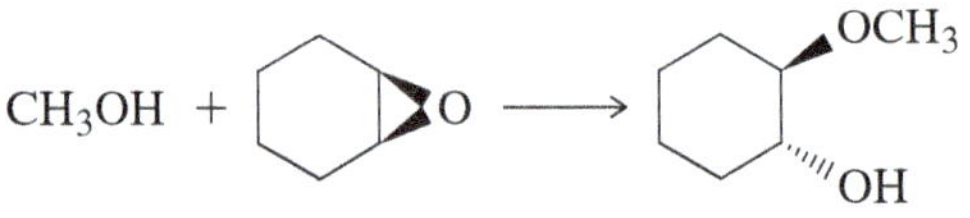

(a) Based on the bonds formed and the bonds broken, calculate $\Delta H°$.
(b) Predict the sign of $\Delta S°$.
(c) Given your answers to parts (a) and (b), why is this reaction favored to the right?

5.66 (•••) "More stable intermediates form first, but react last." Rationalize this statement using a reaction coordinate diagram.

5.67 (••••) When a student attempted a bromination to produce compound **A**, they generated compound **B** instead. Rationalize the formation of **B** using the arrow-pushing formalism.

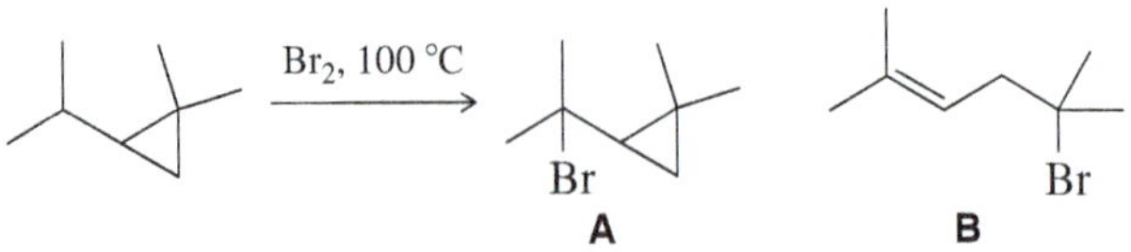

A desired product (not formed)

B (formed)

5.68 (••••) Although bromination is more selective than chlorination, even the chlorination of alkanes is somewhat selective. However, the super-high-temperature (600 °C) monochlorination of propane gives a statistical distribution of products. Based on the selectivity of the chlorine radical for 1° and 2° C—H bonds, (a) what ratio would you have predicted? (b) Why does heating the reaction to 600 °C allow you to get a 3:1 ratio of products?

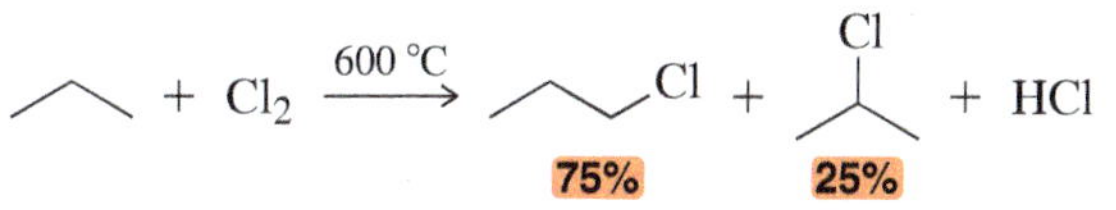

5.69 (••••) Based on the reaction coordinate diagram, (a) calculate the expected ratio of **B**:**C** under thermodynamic control (i.e., at equilibrium) and (b) calculate the ratio of **B**:**C** under kinetic control (i.e., based on relative rates). Assume $T = 298$ K.

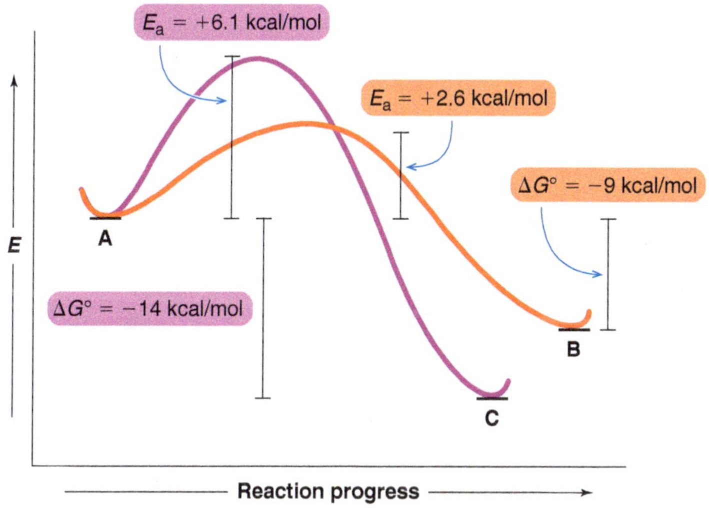

5.70 (••••) An imaginative chemist proposed the following alternate mechanism of methane bromination. [*Note that these two propagation steps add up to the overall reaction.*] How can this mechanism be ruled out in favor of the mechanism we presented in Section 5.3.3?

Propagation 1: $Br\cdot + CH_4 \longrightarrow BrCH_3 + H\cdot$

Propagation 2: $H\cdot + Br_2 \longrightarrow HBr + Br\cdot$

Total: $CH_4 + Br_2 \longrightarrow BrCH_3 + HBr$

5.71 (•••) THINKING AHEAD The halogenation of an alkane when there is an alkene present in the molecule does not proceed with the regioselectivity you might expect. Using principles similar to those developed in this chapter, rationalize the formation of **A** as the only product. We study this reaction further in Chapter 8.

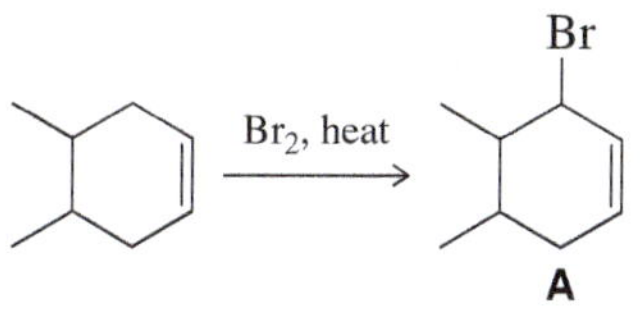

5.72 (•••) Parts (a)–(f) of this assessment refer to the rotation around the single bond of ethane.

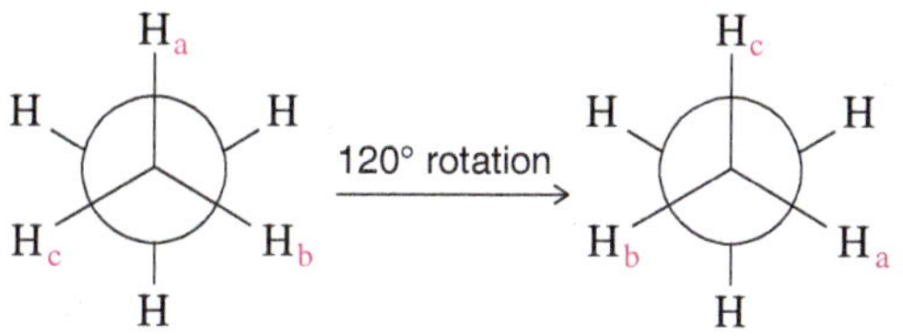

(a) Given that the rate of the reaction is independent of concentration, fill in the missing rates in the following table.

	$[CH_3CH_3]$ (M)	Rate (M/s)
Trial 1	0.1	3.1×10^{10}
Trial 2	0.5	?
Trial 3	1.0	?

(b) What is the order of the reaction with regard to ethane?
(c) Write the rate law for this reaction.
(d) Give the value for the rate constant.
(e) Give the value for the overall rate.
(f) Explain the results you obtained for parts (a)–(e).

5.73 (•••) For rotation around a bond, the rate constant is equal to the reaction rate. Why?

6

Stereoisomerism: Arrangement of Atoms in Space

Learning Objectives

LO 6.0 Assess a pair of molecules to determine if they are enantiomers

LO 6.1 Draw an enantiomer of a molecule

LO 6.2 Name an enantiomer using the Cahn–Ingold–Prelog system

LO 6.3 Determine the enantiomeric composition of a mixture

LO 6.4 Distinguish a diastereomer from other types of stereoisomers

LO 6.5 Describe methods for separating enantiomers

Introduction: the story of thalidomide

In the mid-1940s, the German pharmaceutical company Chemie Grünenthal was trying to develop new antibiotics. When they heated commercially available phthaloylisoglutamine, they obtained the novel compound α-phthalimidoglutarimide (thalidomide) shown in Figure 6.1. [*We will learn this reaction in Chapter 18.*] This compound had virtually no activity against any disease, but a patent was secured nevertheless. Early on, the only notable property of thalidomide was its incredibly high safety profile and lack of side effects in animal models.

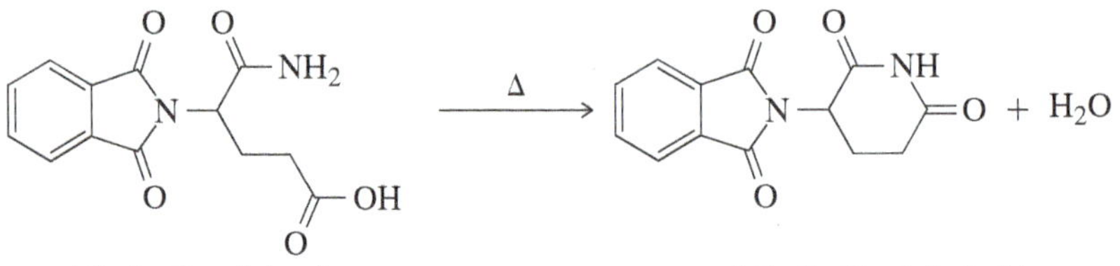

Figure 6.1 **Thalidomide** was produced by an intramolecular condensation reaction (Chapter 18).

"The universe is asymmetric and I am persuaded that life, as it is known to us, is a direct result of the asymmetry of the universe or of its indirect consequences."

—Louis Pasteur

Believing incorrectly that it was structurally similar to barbiturates, a class of tranquilizers, Chemie Grünenthal started human clinical trials with thalidomide. By modern standards, starting clinical trials without successfully demonstrating activity in animal models first is illegal. More disturbingly, the drug was widely distributed to doctors throughout Germany and employees of Grünenthal for self-prescribed use. Thalidomide was found to have a sedative effect in humans, allowing patients to enter long periods of deep sleep. It was immediately released in Germany as an over-the-counter medicine and licensed for sale in the British Isles, Canada, Australia, and New Zealand under the trade name Distaval. Though promoted as safe, thalidomide use had tragic consequences that could have been avoided with an undergraduate-level understanding of stereochemistry. As you will learn, different stereoisomers can exhibit vastly different behaviors in the body.

The story of thalidomide highlights a common theme in chemistry: Slight differences in the structure of a molecule can give it dramatically different properties. **Table 6.1** provides additional examples of this theme.

Table 6.1 Slight changes in structure can give molecules very different properties.

Property of Isomer A	Isomer A	Isomer B	Property of Isomer B
Commercially available anti-inflammatory (Aleve®)	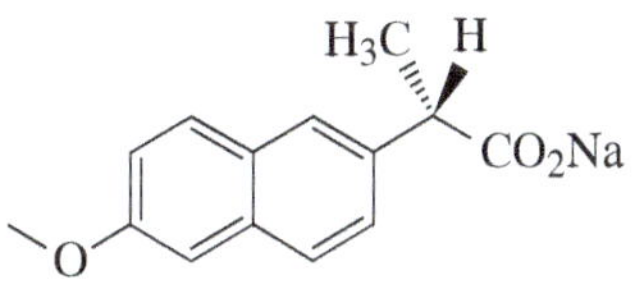 **naproxen sodium**	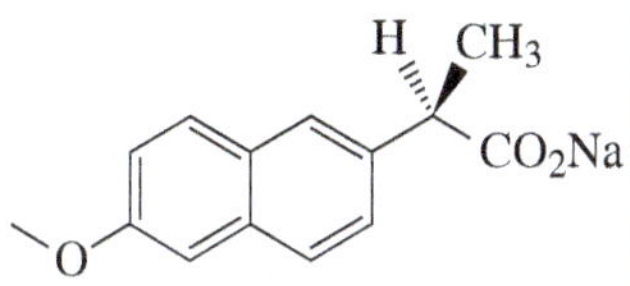	Inactive as anti-inflammatory, severe liver toxicity
160 times more sweet than sugar; commercial artificial sweetener	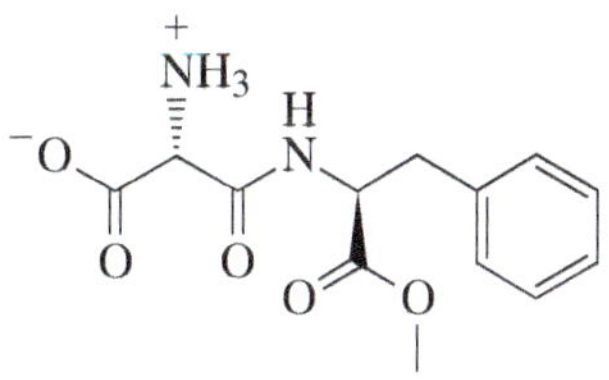 **aspartame**	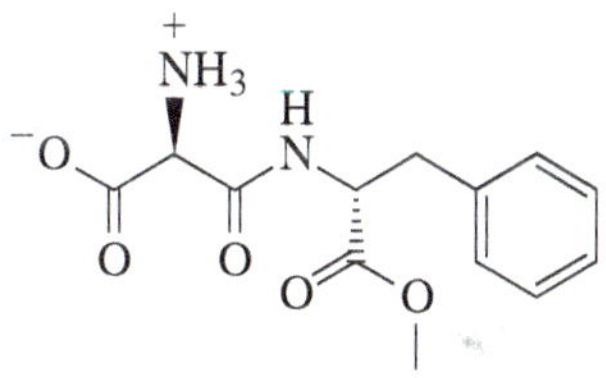	Bitter tasting
Antimycobacterial used in the treatment of tuberculosis	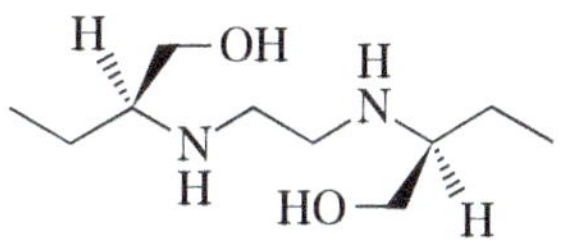 **ethambutol**	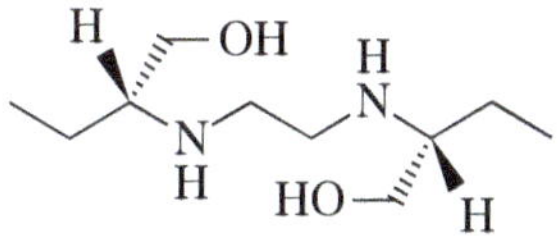	Causes blindness

Table 6.1 suggests two questions that we address here in Chapter 6. First, what is the isomeric relationship between the compounds in each pair? Second, given only slight structural differences between the molecules in each pair, why are certain properties so similar, and others so different? As a preview, although the naproxen isomers have the same melting point and solubility, one is a useful anti-inflammatory drug while the other is a liver toxin.

[*Chapter 6 presents a number of new terms and relationships with similar definitions. Stereoisomerism can be frustrating if approached as a memorization exercise. Rather than memorize these terms, make sure you understand how the definitions differ and how they relate to one another. The story of thalidomide makes this material especially pertinent.*]

Before we get started with something new, let's test our understanding of relevant material studied previously.

Test your understanding

Assessment 6.1 Show all possible constitutional isomers of $C_5H_{12}O$. Label them as functional, positional, or chain isomers.

Assessment 6.2 Convert the following structural formulas to line-angle drawings.

(a) $CH_3CHBrCH_2CH(OH)CH_3$

(b) $(CH_3)_2CHCOCH(CH_2CH_3)CH_3$

(c) $CH_3CH_2NHCH_2CH_2OH$

Assessment 6.3 On the following reaction coordinate diagram, label the reactant (R), product (P), intermediate (I), transition states (‡), activation energies (E_a) of the rate-determining step, and the overall ΔH.

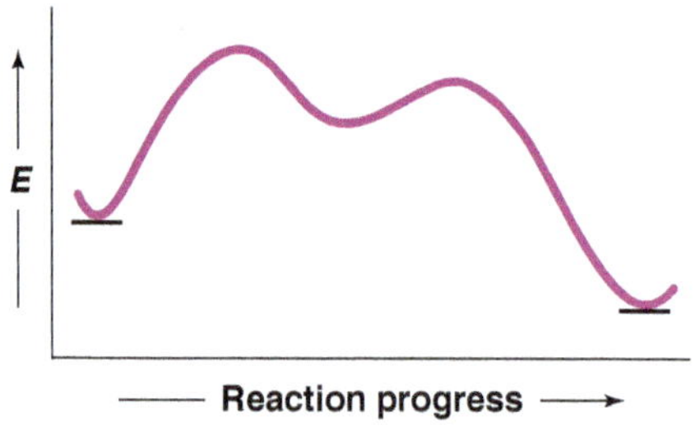

Assessment 6.4 Predict the shape and hybridization of the indicated atoms.

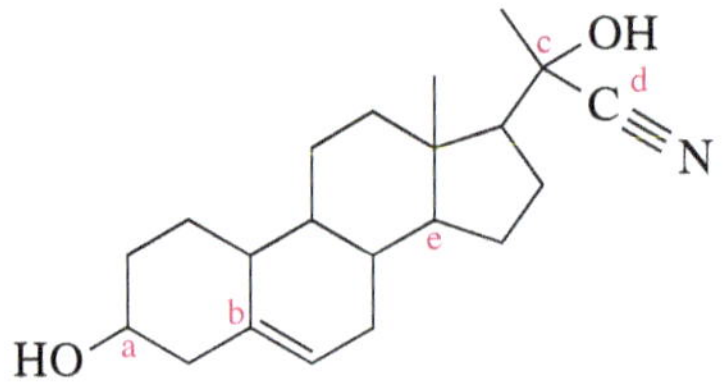

Assessment 6.5 What type of isomerism is indicated by each of the following pairs of molecules? Be as specific as possible.

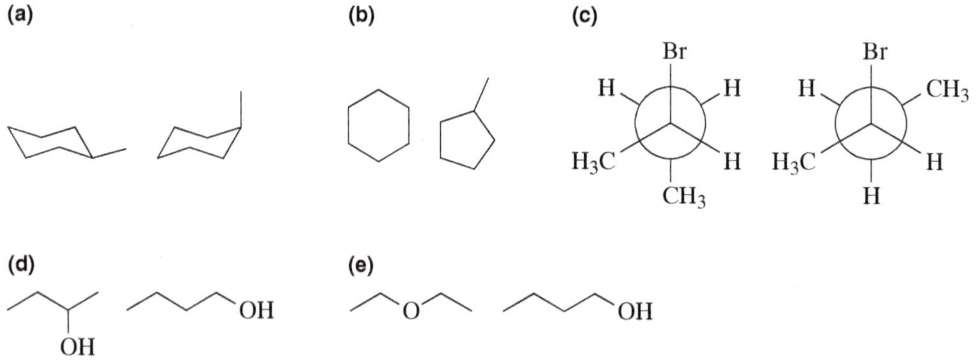

Assessment 6.6 The molecule **A** undergoes a three-step reaction to make **B**, where the first step is rate determining. Conversely, **C** can be produced by **A** in a two-step reaction where the second step is rate determining. Show a reaction coordinate diagram for each reaction, making it clear that the reaction to make **C** is faster than the reaction to make **B**.

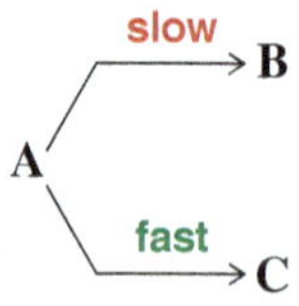

Sales of thalidomide peaked around April of 1961, with two companies, Chemie Grünenthal and Distillers, Ltd., heavily promoting its safety and variety of uses. In the previous 5 years, Grünenthal had informed physicians that thalidomide was the best drug for nausea in pregnant women and nursing mothers, with other distributors following suit. Distillers, in 1961, issued leaflets recommending the use of thalidomide in cases of pediatrics and obstetrics. The safety of thalidomide for pregnant women was being promoted at the same time that the incidence of birth defects in Germany reached historic levels. In 1961, Professor Widukind Lenz, head of the children's clinic at Hamburg University, was alerted to two cases of phocomelia (Greek for "seal limbs") on the same day (Figure 6.2). His investigations into birth records revealed that from 1930 to 1955, there was one case of this type of malformation. Yet in 1960, there were eight cases. The one commonality among these victims was that the mothers all took thalidomide.

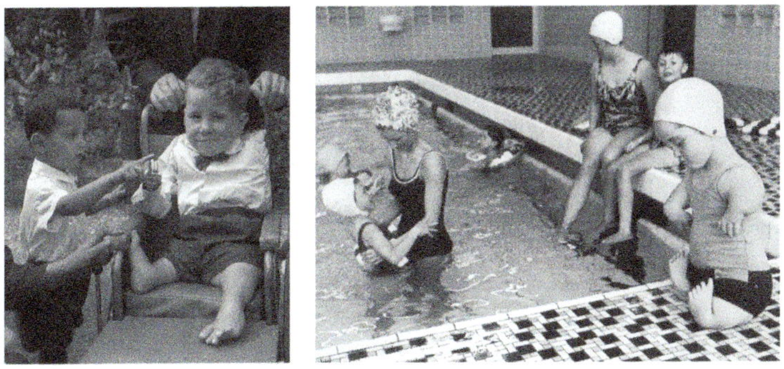

Figure 6.2 Children born with phocomelia.

6.1 Objects and their mirror images

Consider the objects and their mirror images given in Table 6.2. Is each object **superimposable** with its mirror image? That is, can we overlay the image of the object and its mirror image and have them line up perfectly (left-to-right, top-to-bottom, and front-to-back)? In order to determine whether they are superimposable, imagine lifting the mirror image off the page, rotating it parallel to and/or perpendicular to the plane of the page, and sliding it over the original image. If every point lines up perfectly, then the object is superimposable with its mirror image. Of the four objects in Table 6.2, only the spoon and the sled are superimposable with their mirror image. If you perform the same process with the can opener, the crank would be on the opposite side. And for the car, the steering wheel will be on the opposite side. [*Are you superimposable with your mirror image?*]

Table 6.2 Superimposability of objects and their mirror images.

Object	Mirror image

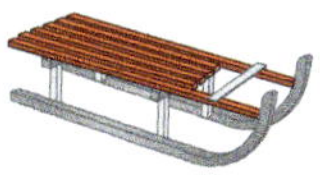	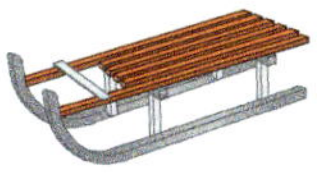
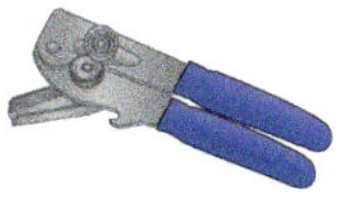	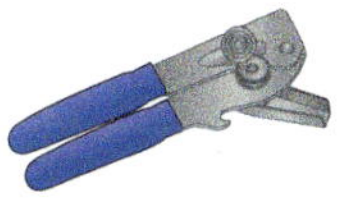

What special property do some objects have that makes them superimposable with their mirror images? [*Look closely at the objects in Table 6.2 and see if you can figure this out.*] The spoon and the sled in Table 6.2 are symmetric. That is, they have a plane of symmetry cutting directly through the middle of the object. A **plane of symmetry** is an imaginary plane that bisects an object into two halves that are mirror images. If you place a mirror (or a darkened smartphone screen) along the bisecting plane of the round-bottomed flask in

Figure 6.3(a), then the image is unchanged, whether the mirror is present or removed. This will be true at any place representing one of these planes of symmetry. The ping-pong ball in Figure 6.3(b) has an infinite number of planes, as shown, since bisecting it at any angle (as long as it's down the middle) results in two halves that are mirror images.

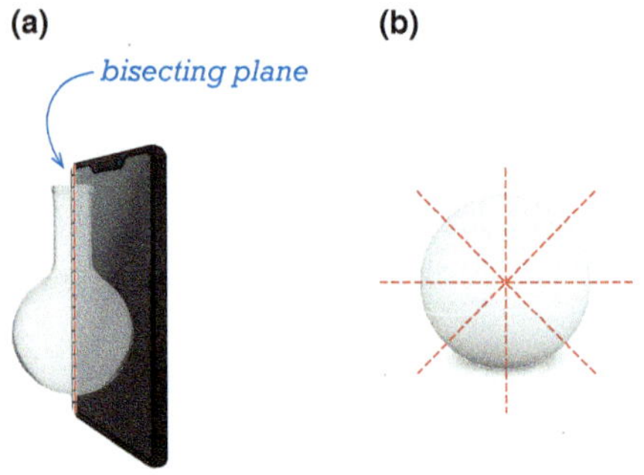

Figure 6.3 **A plane of symmetry makes an object superimposable with its mirror image.** (a) The round-bottomed flask, as shown, has one plane of symmetry. (b) As shown, a ping-pong ball has an infinite number of symmetry planes.

The can opener and the car, on the other hand, are *not* superimposable with their mirror images. Confirm for yourself that neither of these objects possesses a plane of symmetry. Lacking one, these objects could be referred to as asymmetric.

Assessment 6.7 Identify all planes of symmetry in the following objects, if any.

Assessment 6.8 Identify all planes of symmetry in the following molecules/conformations, if any.

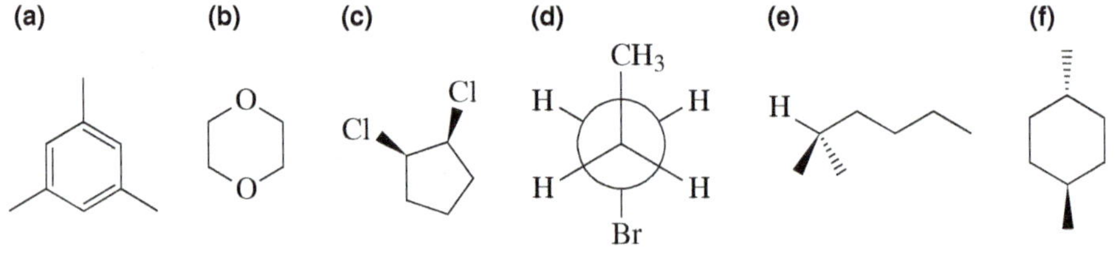

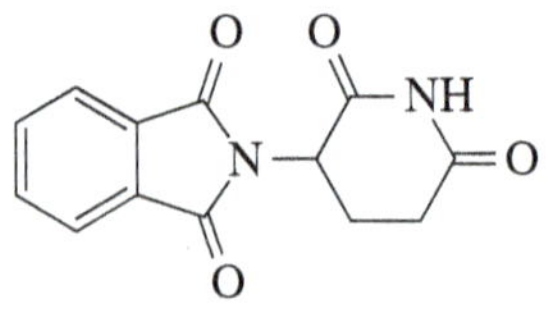

thalidomide

Figure 6.4 **Thalidomide is a mixture of isomers.**

6.2 Drawing tetrahedral atoms in 3-D

Thalidomide, shown at the beginning of the chapter as a single compound, is actually administered as a combination of two different molecules. You can't see that in Figure 6.4, though, because the bonds to the tetrahedral atoms in the structure are all drawn in the same plane.

When we draw tetrahedral atoms in a 3-D representation, two bonds are in the plane of the page, one comes forward, and one goes back. Bonds coming out of the plane of the page are represented by bold, wedge-shaped bonds. Bonds extending behind the plane of the page are represented by hashed wedge-shaped bonds. For methane in **Figure 6.5**, all of the representations present methane equally. [*Models are helpful here. If you have a set, make a model of methane and compare the drawings in Figure 6.5 with different perspectives of your model.*]

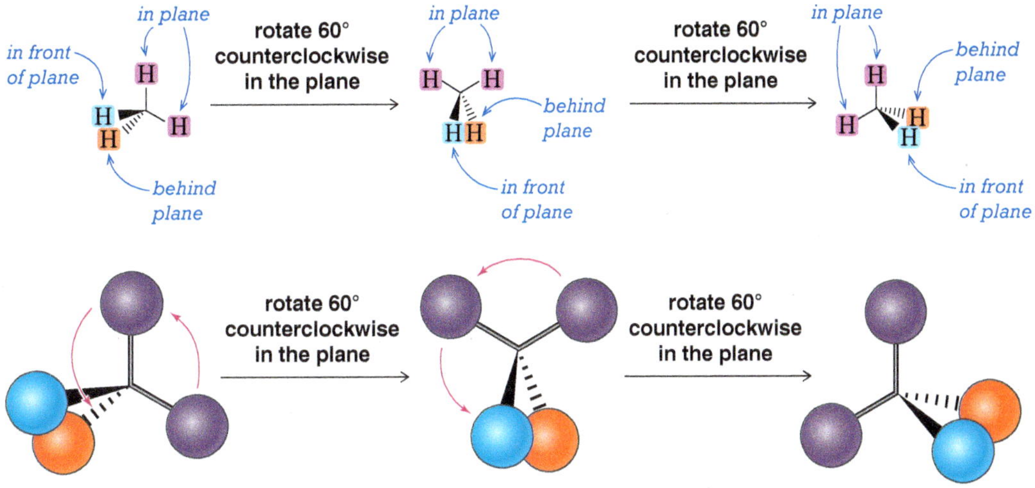

Figure 6.5 Representations of tetrahedral atoms in 3-D.

Let's solve this one together.

Assessment 6.9 For each of the following molecules, draw a 3-D representation.

(a) CH_3OH (b) CH_2Cl_2 (c) $CHBr_3$ (d) CHClBrI

Solved Assessment 6.9(b) and 6.9(d)

Parts (a), (b), and (c) of Assessment 6.9 have similar solutions, so we focus on the one to part (b). Begin by imagining a tetrahedral carbon with four bonds labeled 1–4 (**Figure 6.6**). If we put the chlorines in different places, do we get a different isomer? Compare the three different structures in Figure 6.6 by rotating them so they match each other. [*Making a model helps here.*] You can get from one to the next by simply rotating the molecule on the page [*or in your hand*]. By changing your perspective on the molecule, you'll see each of the three drawn in Figure 6.6.

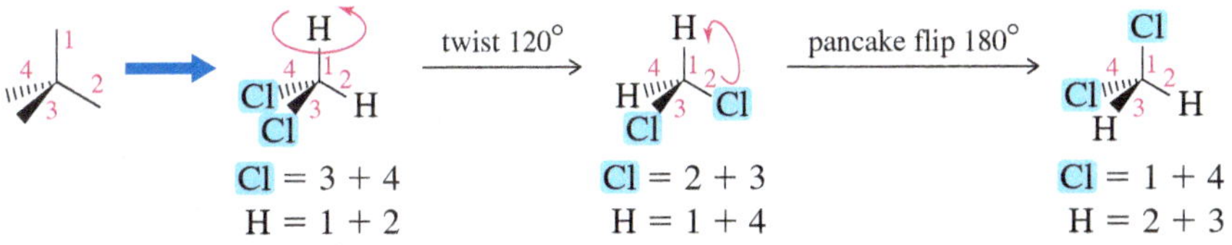

Figure 6.6 Regardless of where the two chlorine atoms are placed, the structures are the same.

If we do the same for part (d), do we get the same result? Only two possibilities are necessary here. Make a model of this compound. Does your model match both compounds A and B in **Figure 6.7**? It doesn't. [*And if you think it does, look more closely. Or better yet, take the model you have just made, and on a second model, keep everything the same (including the perspective) but switch H and I. Now compare the two you have made. Before reading further, make sure you can see that these are not the same.*] And yet, both are acceptable answers if the goal is to simply draw a 3-D representation of CHClBrI. ■

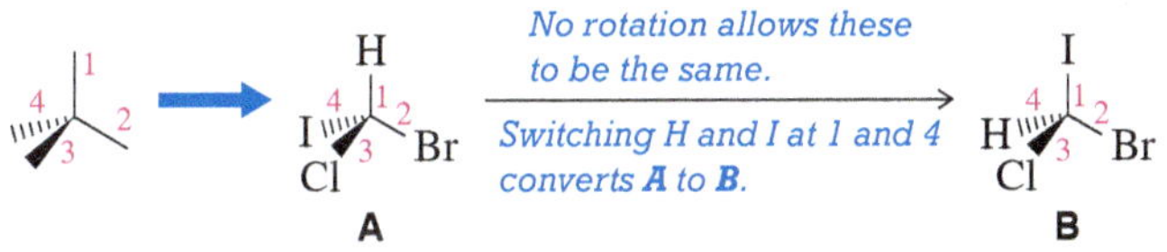

Figure 6.7 With four different groups, switching two of them (H and I) creates a different compound.

The answers to Assessments 6.9(b) and (d) illustrate an important concept in organic chemistry. When a tetrahedral atom has four different atoms/groups attached to it, we can make two possible isomers, depending on where we place the atoms/groups on the tetrahedron. [*We elaborate upon this idea throughout Chapter 6.*] Because of this, it is often important to show the 3-D perspective at atoms where there are four different groups attached. Consider 2-bromocyclohexanol in **Figure 6.8**. For carbons 3 through 6, we do not have four different atoms/groups attached. Switching hydrogens at these carbons does not produce a new isomer. Thus, we do not need to show a 3-D perspective at these carbons. However, at C_1 and C_2, four different molecules are possible when we view the compound using the 3-D perspective.

Figure 6.8 When four different groups are attached at a carbon, it is important to show the 3-D structure.

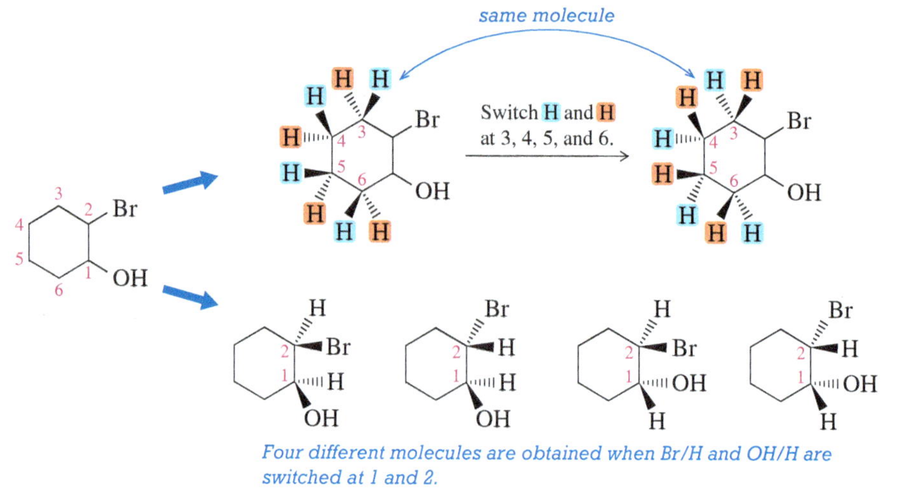

6.2 SUMMARY

- It is only necessary to show the 3-D perspective at atoms with four different groups attached.

Assessment 6.10 For each of the following molecules, draw one 3-D representation at only the necessary atoms (that is, the ones with four different atoms/groups attached).

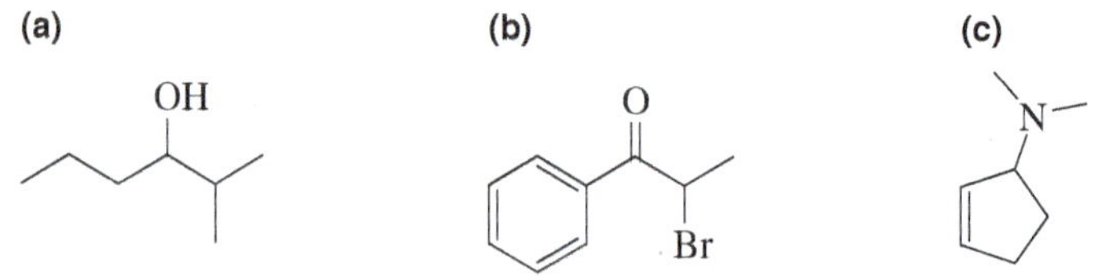

If thalidomide is drawn using the 3-D representation (**Figure 6.9**), then the hydrogen at C_3 can project forward (in front of the plane of the page) or backward (behind the plane of the page). These two representations are actually different isomers. Because scientists were unable to synthesize just one, thalidomide was sold as a mixture containing both. The tragedy of thalidomide is that one isomer is effective as a sedative with few side effects, but the other is highly teratogenic (disturbs the development of an embryo). When taken by pregnant women between the 20th and 36th day after conception, it causes birth defects in the developing fetus. [*Because home pregnancy kits weren't available, most women didn't know they were pregnant this early in the 1950s.*]

Figure 6.9 Thalidomide was sold as a mixture of isomers, each with different biological activity.

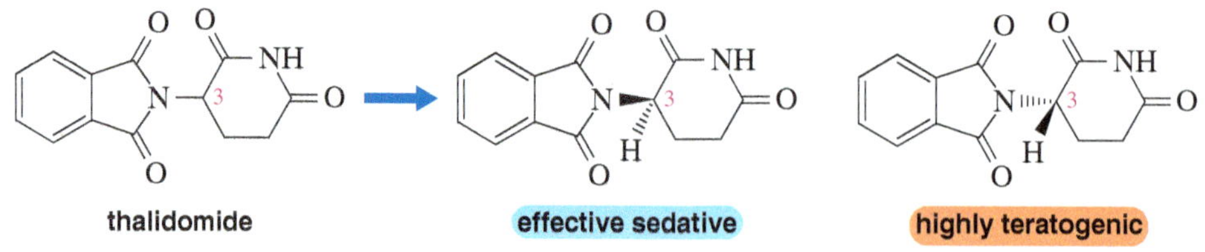

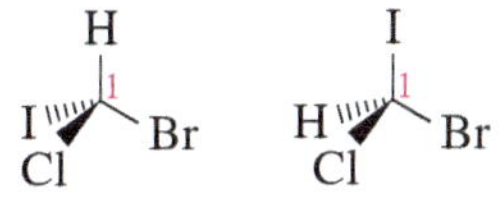

A B

Figure 6.10 Molecules A and B are stereoisomers. C_1 in each is a stereocenter.

6.2.1 Stereoisomerism

Recall from Figure 6.7 that we created two different CHClBrI molecules by switching just two atoms on the tetrahedral carbon. The original (**A**) and the resulting molecule (**B**) have an **isomeric relationship** because they share the same formula but a different arrangement of atoms. Specifically, **A** and **B** are **stereoisomers** (**Figure 6.10**) because they differ only in the spatial arrangement of their atoms. [*Why aren't they constitutional isomers? Because the position of attachment is unchanged.*] In terms used previously, in molecule **A**, H is in the plane of the page and I is going behind the page. In molecule **B**, the H and I are reversed. Thus, the carbon in molecules **A** and **B** is a stereocenter. A **stereocenter** (also called a **stereogenic center**) is an atom containing atoms/groups wherein interchanging any of the two atoms/groups leads to a new stereoisomer.

To see these concepts in action, look at the structure of propane and butan-2-ol in **Table 6.3**. Which of these can exhibit stereoisomerism? If a new molecule is produced when atoms are switched at a given atom, that atom is a stereocenter and the molecule generated is a stereoisomer. [*Try them on your own first.*]

Table 6.3 Does propane or butan-2-ol exhibit stereoisomerism?

Molecule	Switch:	New molecule?	Conclusion
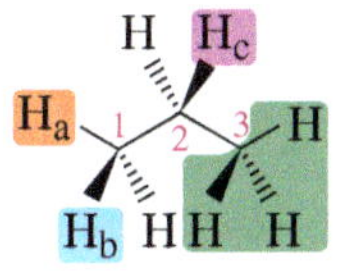 propane	At C_2, switch H_c and C_3: 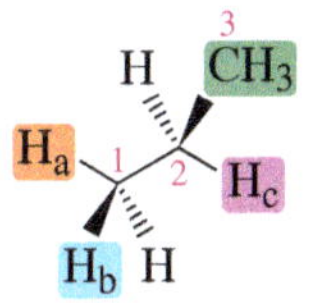	No	C_2 is not a stereocenter.
	At C_1, switch H_a and H_b: 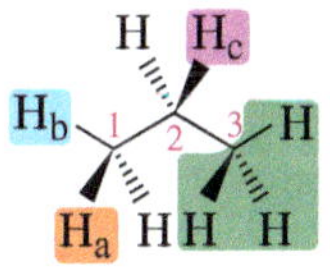	No	C_1 is not a stereocenter.
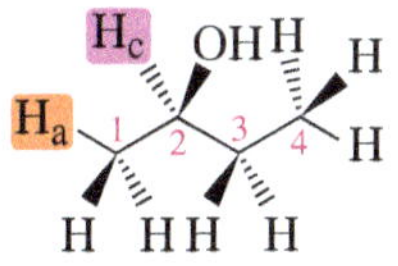 butan-2-ol	At C_1, switch H_a and C_2: 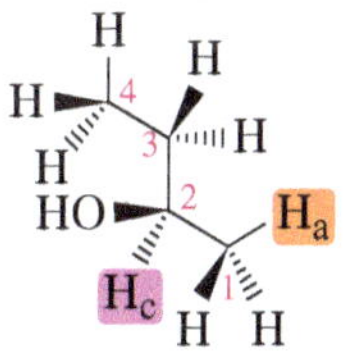	No	C_1 is not a stereocenter.
	At C_2, switch H_c and OH: 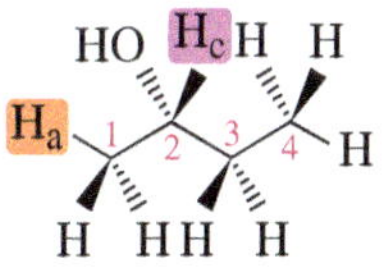	Yes	C_2 is a stereocenter. We also know this because there are four different groups attached.

In order for a molecule to have a stereoisomer, there must be another molecule that has the same formula but a different spatial arrangement of its atoms. The specific site where the two molecules have that different arrangement is the stereocenter. Accordingly, in order to determine if a molecule exhibits stereoisomerism, all we need to do is identify a stereocenter. Neither of the switches in propane creates a new molecule, so propane does *not* have a stereocenter. For butan-2-ol, the switch at C_2 *does* create a new molecule, so C_2 is a stereocenter. Recall from Solved Assessment 6.9 that a carbon is a stereocenter if it has four different groups attached. The groups attached to C_2 of butan-2-ol, are CH_3, H, CH_2CH_3, and OH—four different groups. [*This confirms that we can identify stereocenters and therefore molecules that exhibit stereoisomerism by simply looking for carbons, or other atoms, with four different groups attached.*]

Visualizing 3-D molecules using a 2-D page from a textbook is challenging. If you are having trouble seeing that butan-2-ol exhibits stereoisomerism, make a model. Orient the hydroxyl group forward at C_2 with the hydrogen going back and the carbons of the butyl chain all in the same plane, as shown in **Figure 6.11**. Next make a second four-carbon chain with the hydroxyl group going back and the hydrogen coming forward. No matter how much you rotate and/or twist the two molecules, they are nonsuperimposable, in which case C_2 is a stereocenter.

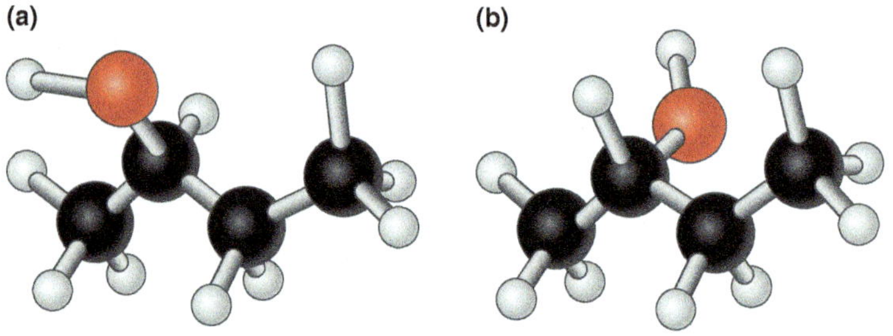

Figure 6.11 Molecular models of the two stereoisomers of butan-2-ol. (a) The original molecule in Table 6.3. (b) After switching OH and H at C_2. These molecules are nonsuperimposable.

We can do the same analysis for propene and but-2-ene in **Table 6.4**. We discuss this type of stereoisomerism more in Chapter 8 when we study alkenes. But for now, notice that but-2-ene exhibits stereoisomerism because switching groups at C_2 or C_3 generates a new molecule. If you could rotate around the C_2—C_3 double bond, these could be the same, but carbon–carbon double bonds don't rotate because doing so would break the π bond formed by the side-by-side overlap of the *p* orbitals (see Chapter 2 and Section 5.1.11).

Table 6.4 Does propene or but-2-ene exhibit stereoisomerism?

Molecule	Switch:	New molecule?	Conclusion
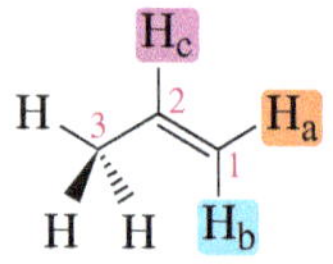 propene	At C_2, switch H_c and C_3: 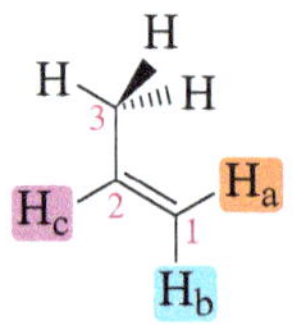	No	C_2 is not a stereocenter.
	At C_1, switch H_a and H_b: 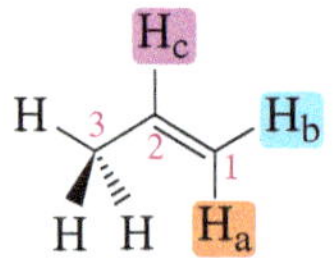	No	C_1 is not a stereocenter.
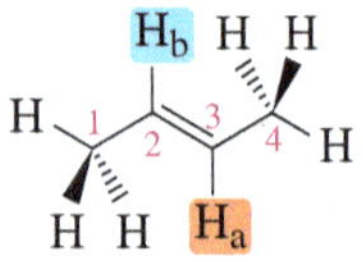 but-2-ene	At C_2, switch H_b and C_1: 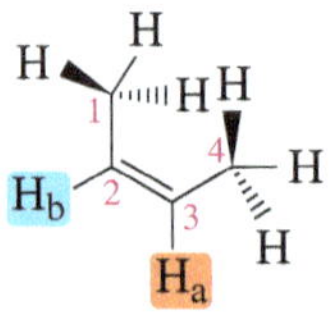	Yes	C_2 is a stereocenter. Technically, the $C_2{=}C_3$ alkene is a stereocenter.
	At C_3, switch H_a and C_4: 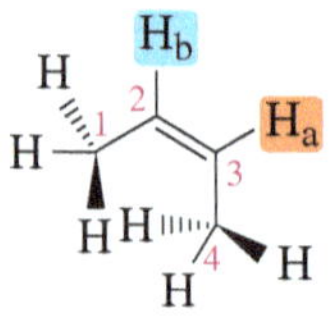	Yes	C_3 is a stereocenter. Technically, the $C_2{=}C_3$ alkene is a stereocenter.

Assessment 6.11 For each of the following molecules, identify all stereocenters and draw all possible stereoisomers.

(a)

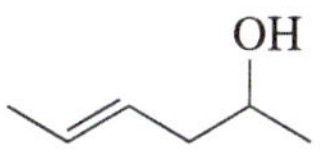

(b)

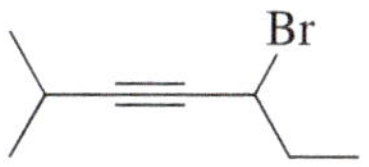

(c)

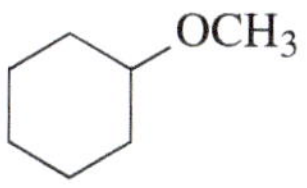

(d)

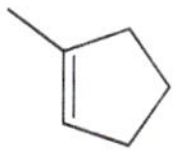

(e)

(f)

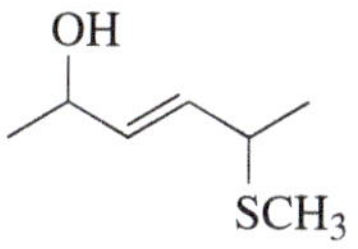

6.2.2 Centers of asymmetry

In Assessment 6.11, you might have drawn a number of molecules, switching substituents in the attempt to generate a new stereoisomer. While any tetrahedral atom with four different groups attached is a stereocenter, it is also referred to as an **asymmetric center**, or a **center of asymmetry**. Recall that symmetric objects are superimposable with their mirror images (see the spoon and sled in Table 6.2). Asymmetric objects are nonsuperimposable with their mirror images. The same principles apply to molecules, as shown in **Table 6.5**. That is, molecules containing an asymmetric center are not symmetric and are therefore nonsuperimposable with their mirror images. [*Try determining whether the molecules in Table 6.5 are symmetric on your own first.*]

Table 6.5 Testing for asymmetry in molecules.

Molecule	Four different groups on any tetrahedral atom?	Mirror image	Superimposable?	Symmetry?
 cyclohexane	No		Yes	Yes, multiple planes of symmetry 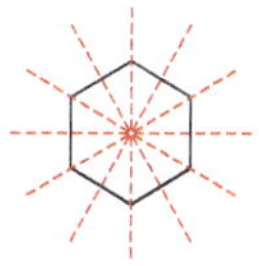
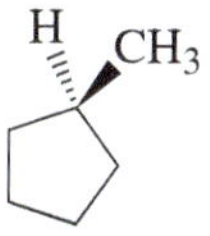 methylcyclopentane	No	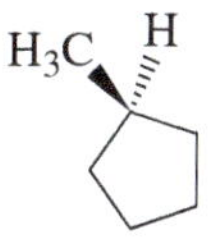	Yes	Yes, one plane of symmetry *H is hidden behind the CH_3.* 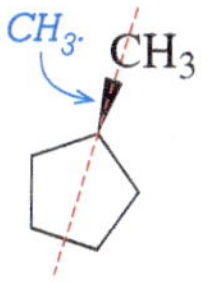
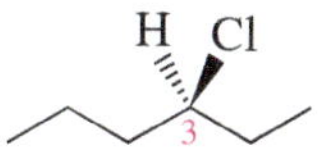 3-chlorohexane	Yes, C_3 is an asymmetric center.	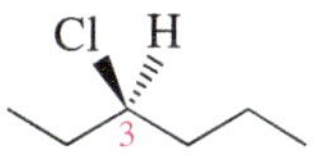	No	No planes of symmetry
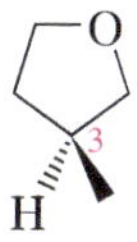 3-methyltetrahydrofuran	Yes, C_3 is an asymmetric center.		No	No planes of symmetry

An **achiral** molecule is one that is superimposable with its mirror image. In Table 6.5, cyclohexane and methylcyclopentane are achiral molecules. A **chiral** molecule is nonsuperimposable with its mirror image. 3-Chlorohexane and 3-methyltetrahydrofuran lack a plane of symmetry. They are nonsuperimposable with their mirror images and are therefore chiral. [*In Table 6.2, the spoon and the sled are achiral, whereas the can opener and the car are chiral.*]

Once again, molecular models can help reveal asymmetry in molecules. If you build a model of 3-chlorohexane and its mirror image (**Figure 6.12**), you will find that you can line up the H and Cl at C_3, or you can line up the ethyl and propyl groups on C_3, but you cannot line up all four. Thus, 3-chlorocyclohexane is nonsuperimposable with its mirror image and is therefore a chiral molecule. Because an asymmetric center makes a molecule chiral, it is commonly referred to as a **chiral center** or **center of chirality**.

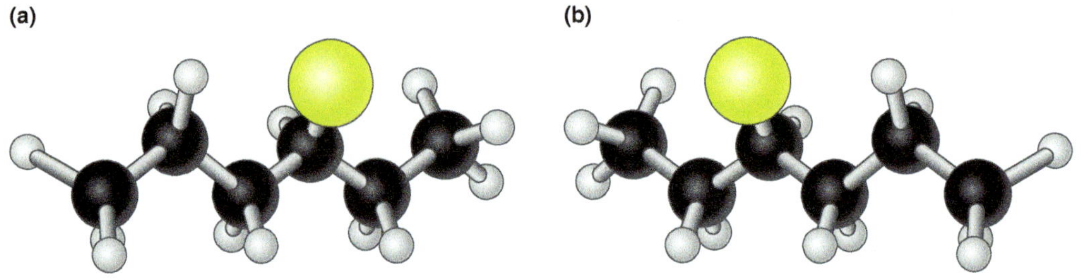

Figure 6.12 Molecular models of the two stereoisomers of 3-chlorohexane. (a) The original molecule in Table 6.5 and (b) its mirror image. These two structures are nonsuperimposable, so 3-chlorohexane has an asymmetric center and is chiral.

6.2.2 SUMMARY

- An atom with four different groups attached is an asymmetric center. Molecules containing an asymmetric center are chiral. Symmetric molecules are achiral.

Assessment 6.12 Identify the following molecules as chiral or achiral. If chiral, draw the nonsuperimposable mirror image and verify its nonsuperimposability.

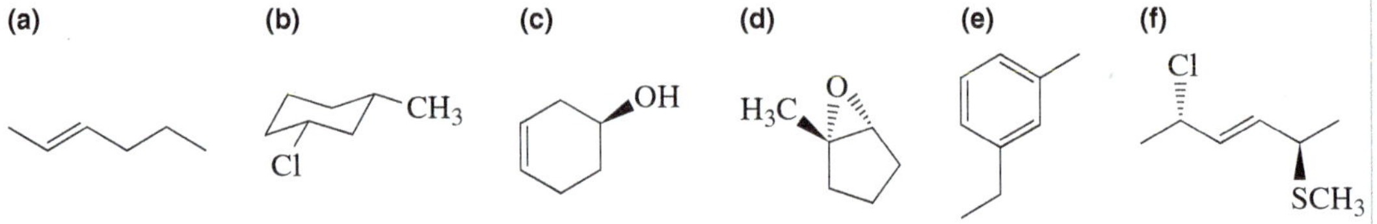

Assessment 6.13 A molecule of the type shown is discussed in greater detail in Section 6.5.1. Although it contains at least one atom with four different groups attached, why is it not a chiral molecule?

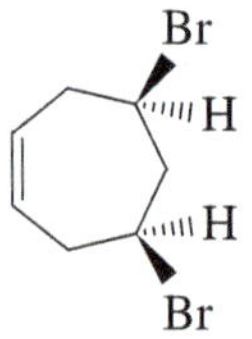

Let's compare a few of the terms studied thus far. A chiral molecule is nonsuperimposable with its mirror image. The mirror image of a chiral molecule and the chiral molecule itself are stereoisomers. Thus, the chirality of a molecule and the existence of a stereoisomer can be confirmed by drawing its mirror image and verifying its nonsuperimposability. Alternatively, Section 6.2.1 showed us that switching two substituents at a carbon bearing four different groups (an asymmetric center) also generates a molecule's stereoisomer. [*Do these techniques match up?*] **Figure 6.13** shows both for 3-chlorohexane to see how the two techniques and the molecules generated relate to each other.

The simplest way of drawing the stereoisomer of 3-chlorohexane is to switch the spatial orientation of the chlorine and hydrogen at C_3. In the structure on the top left (**A**), the chlorine is coming forward, out of the plane of the page (bold wedge), but in the new stereoisomer (**B**), it is going behind the plane of the page (dashed wedge). To draw the mirror image, imagine scanning molecule **A** from right to left, then reproduce what you see, while drawing

The most common mistake students make while drawing mirror images is to also change the spatial orientation at the asymmetric centers. If you do this, you will regenerate the same molecule you started with.

left to right. If you do a 180° pancake flip on B, you get the same stereoisomer you got when you drew the mirror image. [*Confirm this with models if you have them.*]

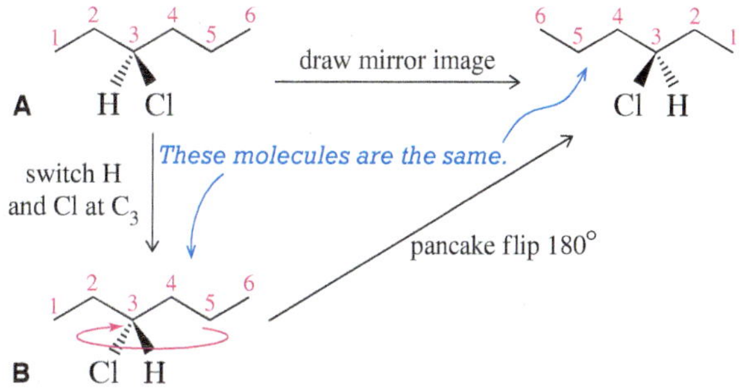

Figure 6.13 Drawing a mirror image and switching two groups at an asymmetric center (chiral center) generates the same stereoisomer.

A chiral molecule and its nonsuperimposable mirror image are called **enantiomers**, and the two enantiomers are said to have an **enantiomeric relationship**. Thus, the two structures of 3-chlorohexane in Figure 6.13 are enantiomers of each other.

Assessment 6.14 For each of the following chiral molecules, obtain the enantiomer (i) by drawing the nonsuperimposable mirror image and (ii) by switching the spatial orientation at each asymmetric center. Confirm (possibly using models) that the structures you drew for (i) and (ii) are the same.

(a) (b) (c) (d) (e) (f)

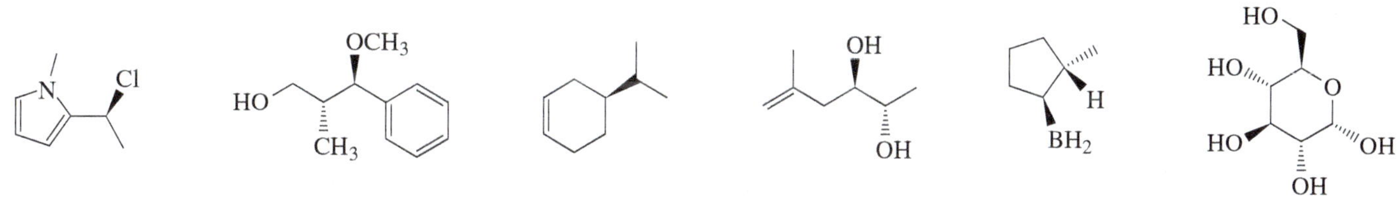

Assessment 6.15 Draw the mirror image of the following molecule. Then, using the mirror image generated, switch the spatial orientation at the asymmetric center. Is the final structure the enantiomer of the original? If not, what is it?

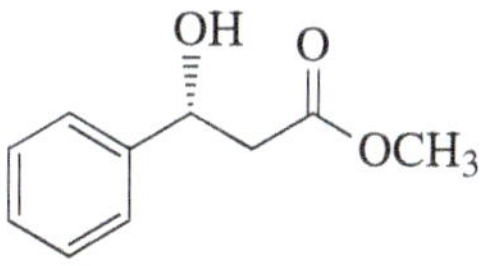

Assessment 6.16 A molecule of the type shown here is discussed in greater detail in Section 6.5.1. [*This is the same molecule from Assessment 6.13.*] Draw the mirror image. Is it superimposable? Switch the spatial orientation at both asymmetric centers. Have you generated a new molecule?

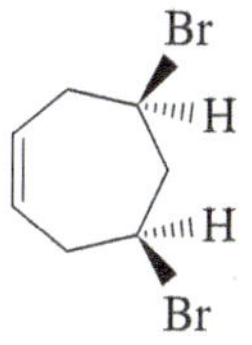

6.2.3 Enantiomers

The three drugs in Table 6.1, which have very similar structures yet dramatically different medicinal activities, have been reproduced in **Table 6.6**. We can now say that isomers A and B of each drug are enantiomers. Table 6.6 also shows the nonsuperimposable mirror image of isomer A. Verify that the mirror image and isomer B are the same for each compound.

Table 6.6 Isomer A and B of the drugs in Table 6.1 are enantiomers.

Isomer A	Mirror image	Isomer B
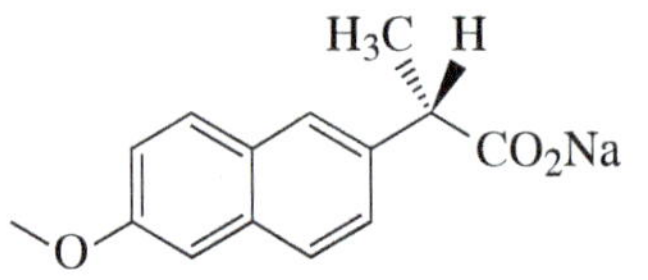	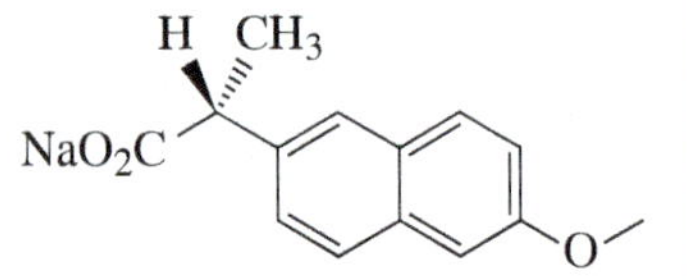	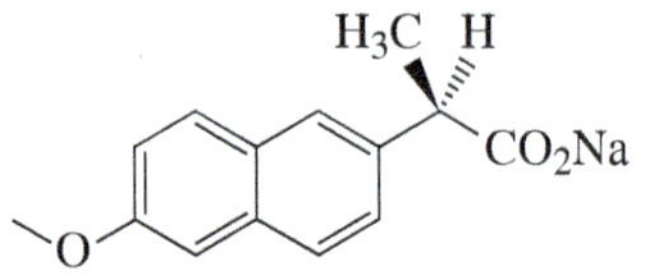
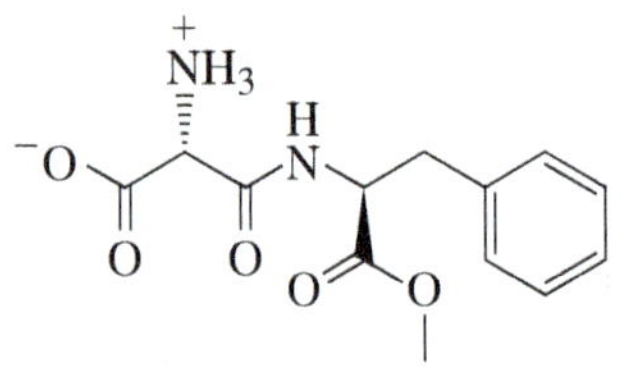	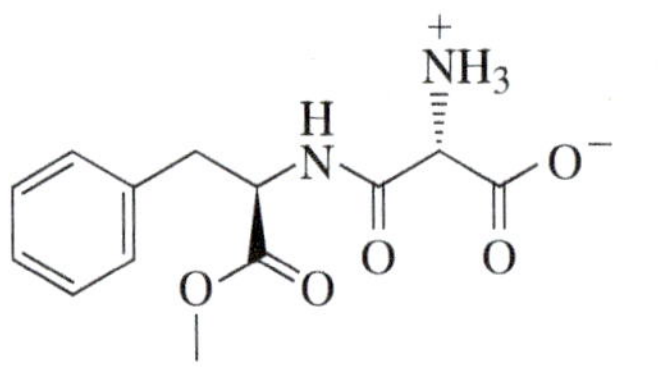	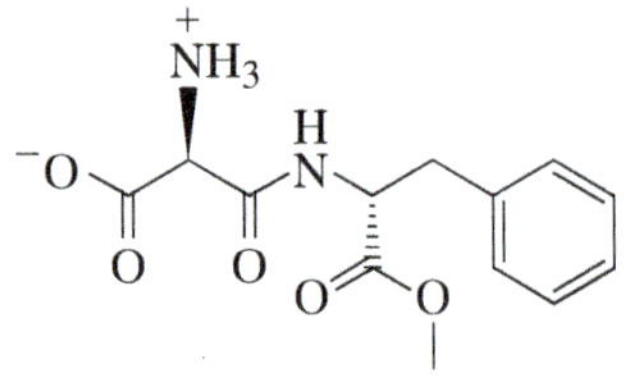
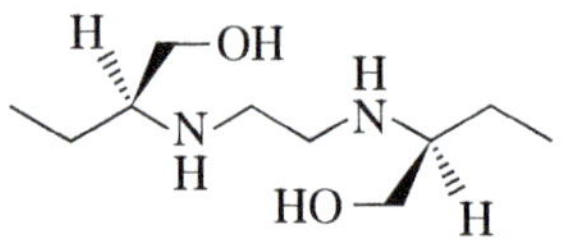	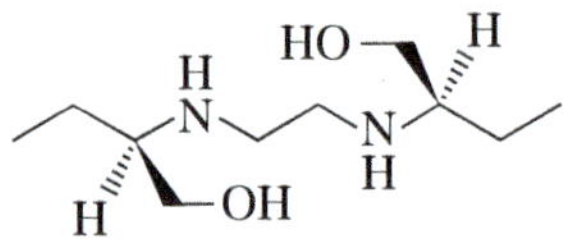	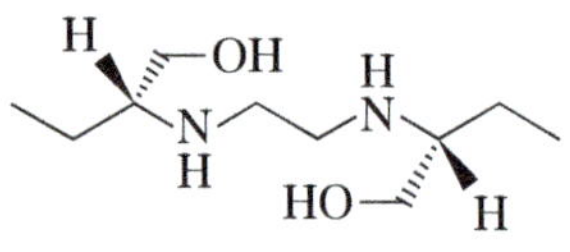

Returning to thalidomide, let's more carefully examine the differences between the safer stereoisomer and the highly teratogenic stereoisomer. **Figure 6.14** shows by switching the orientation at the asymmetric center (C_3) or by drawing the nonsuperimposable mirror image that these stereoisomers are enantiomers. The drug was produced and sold as a mixture of two enantiomers. A mixture containing an equal amount of each of two enantiomers is referred to as a **racemic mixture**.

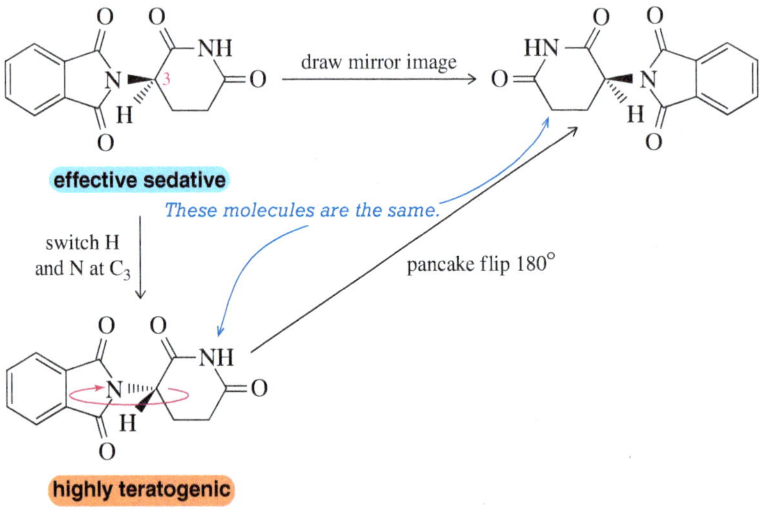

Figure 6.14 The safe stereoisomer and the teratogenic stereoisomer of thalidomide are enantiomers.

6.3 Naming stereoisomers/absolute configuration

Two molecules that are stereoisomers (or enantiomers) of each other must have names that distinguish between them. We discussed this previously in Section 3.5.2 when talking about the relative configuration of cycloalkanes. This is different than **absolute configuration**, which makes clear the exact arrangement of atoms in space. In Section 6.3.1, we study an example where the relative configuration is insufficient for naming.

6.3.1 Relative configuration

The purpose of the IUPAC system is to standardize chemical terms (including names of molecules) and measurements for scientists throughout the world. The name of a molecule must unambiguously describe its structure so that organic chemists can communicate their work with absolute certainty as to what molecule is being described. To illustrate, try drawing the structure of *trans*-1-ethyl-2-methylcyclohexane. [*Really—take out a piece of paper and draw the structure.*]

As shown in **Figure 6.15(a)**, the ethyl and methyl groups are on opposite sides, as indicated by *trans*. If the ethyl is down, then the methyl must be up; if the ethyl is up, then the methyl must be down. Similarly, try drawing the structure of *cis*-1-ethyl-2-methylcyclohexane. As shown in **Figure 6.15(b)**, the two groups that are cis are on the same side of the ring—either both up or both down.

(a) (b)

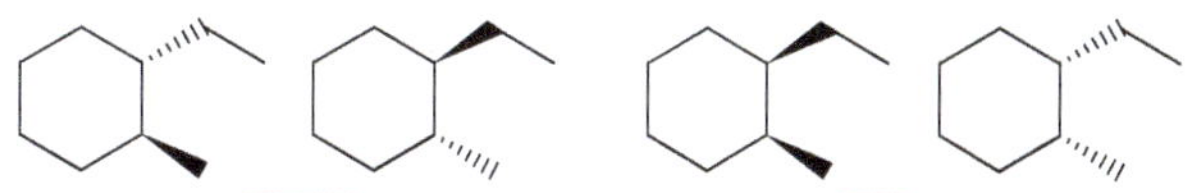

both trans, but different molecules

both cis, but different molecules

Figure 6.15 Cis and trans are insufficient for distinguishing between some molecules. (a) There are two trans molecules and (b) two cis molecules. [*What is the relationship between the two trans molecules? Between the two cis molecules?*]

Assessment 6.17 Identify the following substituted cycloalkanes as cis or trans.

(a) (b) (c) (d)

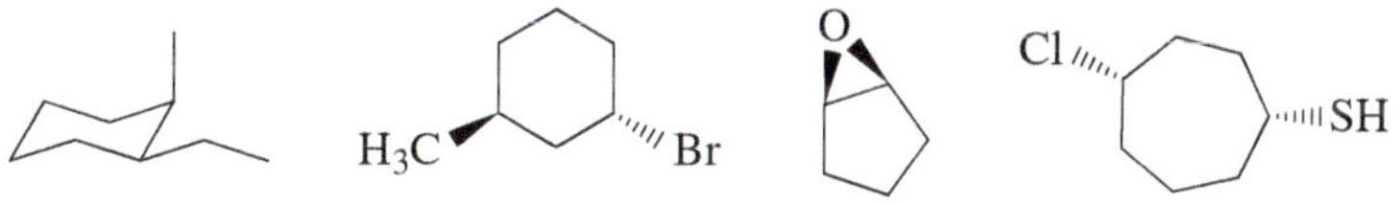

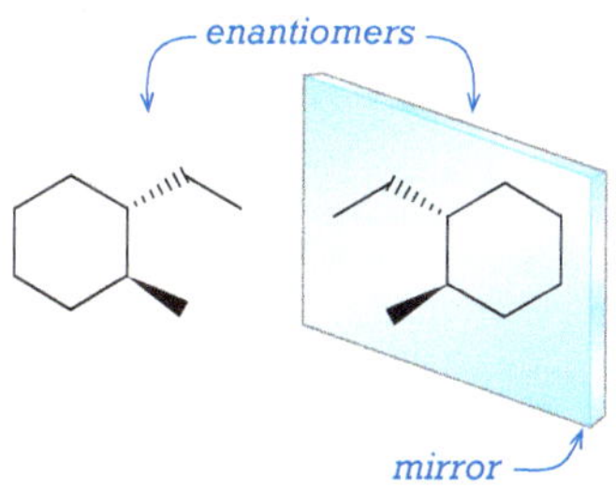

Figure 6.16 The two isomers of *trans*-1-ethyl-2-methylcyclohexane are nonsuperimposable mirror images (enantiomers).

The structures in Figure 6.15 and Assessment 6.17 highlight the problems with a system that only specifies the *relative* configuration. Cis and trans only tell us the spatial orientation of two groups relative to each other. To eliminate this ambiguity, we need a naming system that tells us the *absolute* configuration of the stereocenters. Before we introduce that system, let's identify the relationship between the two trans isomers of 1-ethyl-2-methylcyclohexane. As shown in **Figure 6.16**, they are nonsuperimposable mirror images. That is, they are enantiomers. We explain in Section 6.3.2 that discriminating between two enantiomers by name requires us to determine the absolute configuration.

Assessment 6.18 For the compound 1-ethyl-3-isopropylcyclopentane, (a) draw two different cis and two different trans isomers. (b) What is the stereochemical relationship between the cis isomers? (c) What is the stereochemical relationship between trans isomers? (d) **LOOKING AHEAD** What is the relationship between a cis isomer and a trans isomer?

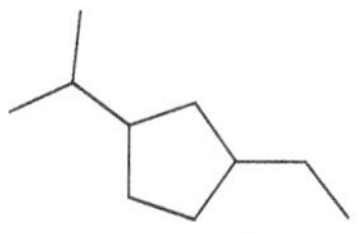

6.3.2 Determining absolute configuration

The Cahn–Ingold–Prelog (CIP) convention is used to determine absolute configuration. This convention unambiguously differentiates between enantiomers. It is logical and simple, yet students may encounter difficulty when using it for two fixable reasons. The first involves not working enough problems. [*Working multiple assessments fixes that.*]

Figure 6.17 Models of enantiomers with atoms labeled 1–4. [*These are used in later sections.*]

The second issue deals with students' difficulty in visualizing molecules represented in 3-D on a flat page. Our model kits will help in this regard. As we begin learning the CIP system, prepare in advance a tetrahedral carbon atom with four different substituents attached. [*Four different substituents = asymmetric carbon and enantiomers.*] Label the atoms 1, 2, 3, and 4. Having made this first model, now make a second model of its enantiomer (**Figure 6.17**). Keep these models handy.

6.3.2.1 Clockwise or counterclockwise

Select one of your models, hold substituent number 4 and face the molecule as shown in **Figure 6.18**. With substituent 4 pointing away from you, trace the path starting at substituent 1, going through substituent 2, and ending at substituent 3. If this path is *clockwise*, then that model is named ***R*** (Latin for *rectus*, meaning "right"). Turning a steering wheel in this direction ($1 \rightarrow 2 \rightarrow 3$) would cause the car to turn to the right. If the path from $1 \rightarrow 2 \rightarrow 3$ goes in a *counterclockwise* direction, however, then this model is named ***S*** (Latin for *sinister*, meaning "left"). In the steering wheel analogy, substituent 4 is the unseen steering column, while substituents 1, 2, and 3 are the steering wheel.

Figure 6.18 With the lowest priority substituent pointing away from you, clockwise is *R* and counterclockwise is *S*.

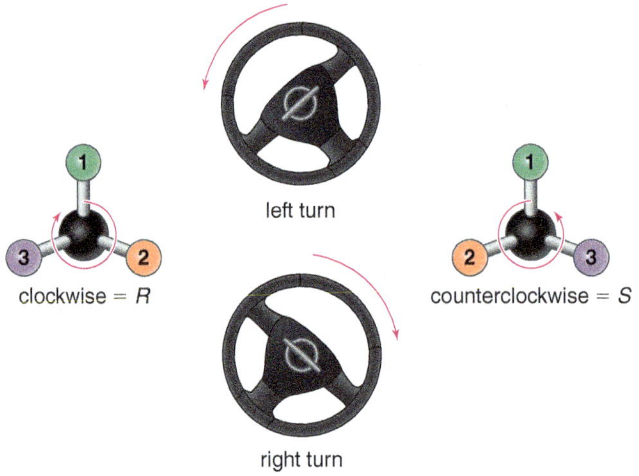

Assessment 6.19 By comparing them to the models you created in Section 6.3.2, label the following chiral centers as *R* or *S*.

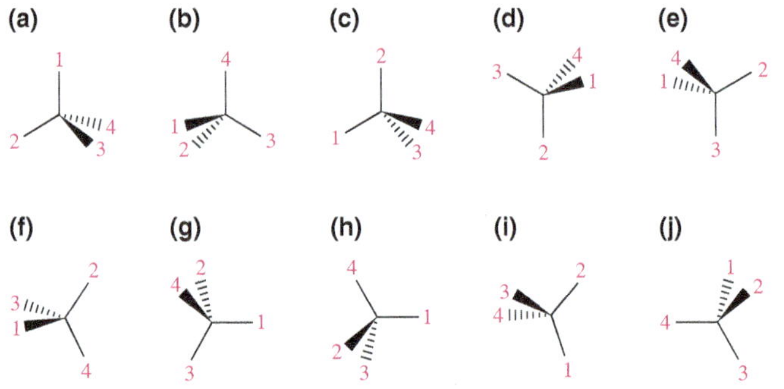

Half the battle of the CIP system is complete. Our models show us that two enantiomers, depending on the directionality of our prioritized substituents (1, 2, 3, and 4) will be either *R* or *S*. [*Convince yourself that there is no other possibility.*] What remains is a system to prioritize the substituents at an asymmetric carbon as 1, 2, 3, and 4.

6.3.2.2 Priority rules

The priority rules for the CIP system are as follows.

1. Prioritize the substituents at an asymmetric carbon based on the atomic number of the four atoms directly attached (**Figure 6.19**). The atom with the highest atomic number has the highest priority (that is, is labeled 1), the second is labeled 2, and so on.

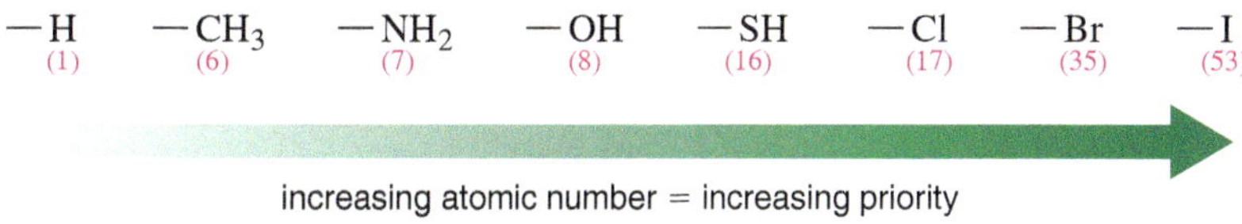

Figure 6.19 Substituents are prioritized by atomic number. Higher atomic number equals higher priority.

This rule is applied to several molecules in **Figure 6.20**, including CHBrClI, the first chiral molecule we encountered. In the first three molecules, all four atoms attached to the chiral center have a different atomic number. As illustrated in the third molecule, isotopes are prioritized by whichever has the higher atomic mass. [*How would priorities be assigned if some of the atoms had the same atomic number, as in the fourth molecule?*]

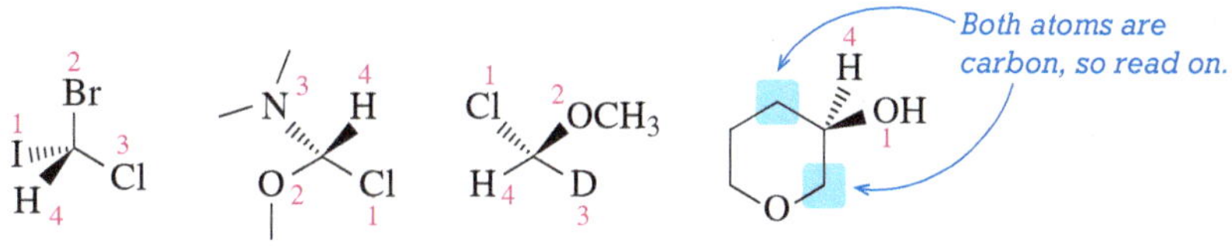

Figure 6.20 Prioritizing substituents at chiral centers based on atomic number.

2. If the priority cannot be assigned based on the atoms directly attached to the asymmetric carbon, move to the second atom in each and assign the priority based on the atomic number (**Figure 6.21**). Do this only for those whose priority could *not* be established using rule 1.

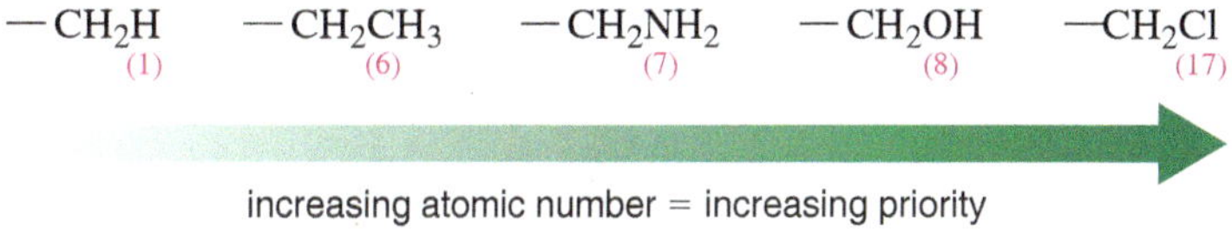

Figure 6.21 If tied, proceed to the next atom and prioritize by atomic number. Higher atomic number equals higher priority.

Once a group's priority has been established, it does not need to be considered further. If the priority cannot be established at the second carbon, continue until the first point of difference, as shown in **Figure 6.22**.

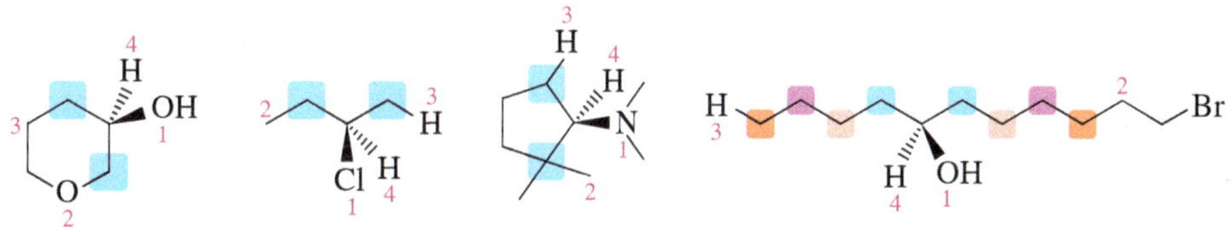

Figure 6.22 Prioritizing substituents at chiral centers based on the atomic number of the first point of difference. The colored screens indicate that similarly colored substituents were tied at that point.

3. A special situation arises when dealing with doubly or triply bonded atoms. For these, we need to invoke the concept of "phantom" atoms, where we assume that for every extra bond, there is an extra atom of that type attached. The vinyl group in **Figure 6.23** illustrates this concept. Instead of viewing C_1 as having just one carbon and one hydrogen attached, we count the π bond to carbon as a second attached carbon. Similarly, C_2 of the vinyl group is treated as having two hydrogens (the ones shown) and an extra carbon (the phantom atom, accounting for the π bond back to C_1).

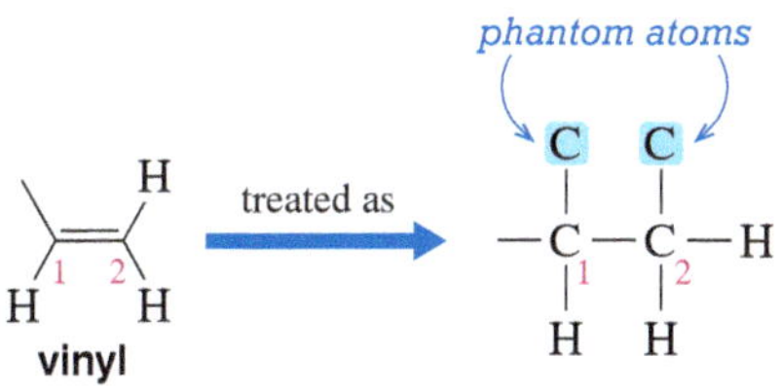

Figure 6.23 Phantom atoms have an assumed extra atom for each π bond.

Because phantom atoms are the most difficult part of these priority rules, **Table 6.7** provides more examples. [*Try them on your own before looking at the answer.*]

Table 6.7 Phantom atom examples. [*Phantom atoms are screened in blue.*]

Substituent	Explanation	Treated as:
$-\underset{1}{C}{\equiv}\underset{2}{C}-H$	C_1 is attached to one real carbon and two phantom carbons (one for each π bond to C_2). C_2 is attached to a hydrogen and two phantom carbons (one for each π bond to C_1).	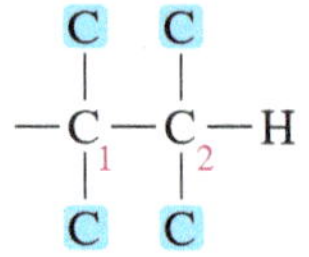
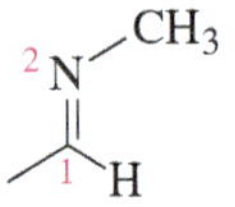	C_1 is attached to a real H and N_2 and one phantom nitrogen (one for the $C_1—N_2$ π bond). N_2 is attached to a real carbon (the methyl group) and to the phantom carbon (for the π bond between N_2 and C_1).	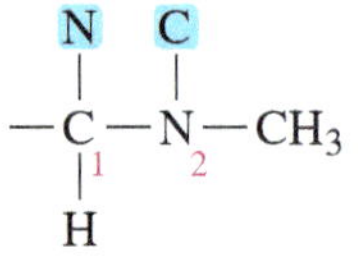
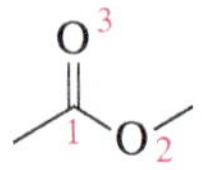	C_1 is attached to the real O_2 and O_3 and one phantom oxygen (one for the $C_1—O_3$ π bond). O_3 is attached to the phantom C (for the π bond between O_3 and C_1).	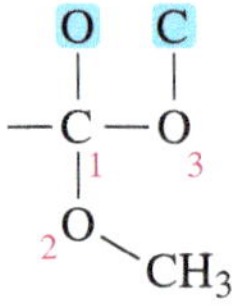

Assessment 6.20 Order the following sets of substituents via their priority using the CIP rules. (*R = position of attachment to the asymmetric center.*)

(a)

(b)

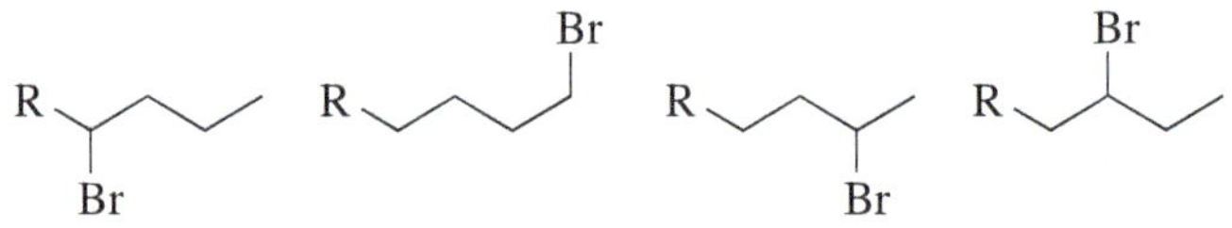

(c)

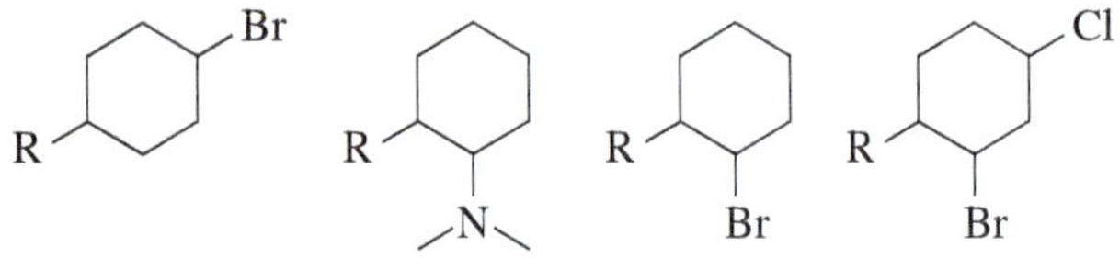

Assessment 6.21 Show how each of the following functional groups would be illustrated using "phantom" atoms. (*R = position of attachment to the asymmetric center.*)

(a)

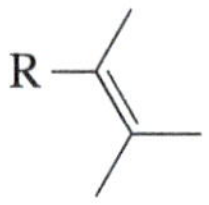

(b) R—C≡N

(c)

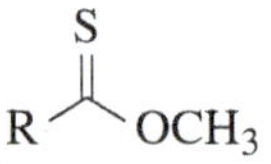

(d)

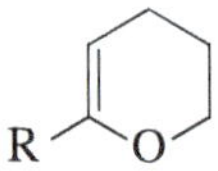

Assessment 6.22 Of the following pairs, identify the higher priority substituent according to the CIP rules. (*R = position of attachment to the asymmetric center.*)

(a)

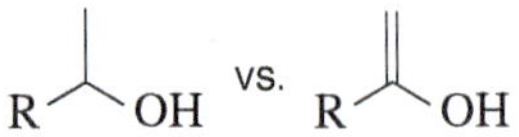

(b)

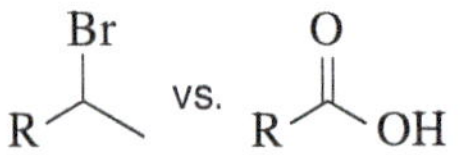

(c)

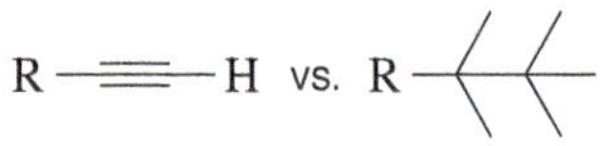

Assessment 6.23 Aldosterone is a steroid hormone that helps regulate blood pressure. Which of the labeled carbons are stereocenters? Of those that are, prioritize the substituents using the CIP rules.

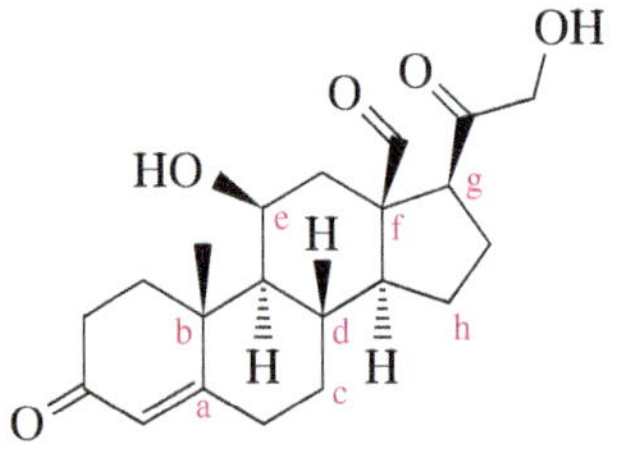

6.3.2.3 Putting it all together—with models

Sections 6.3.2.1 and 6.3.2.2 described how to name a stereocenter as *R* or *S* based on priorities and how to prioritize substituents. To put the two together, identify the asymmetric carbon, prioritize the groups, and then match them up to the picture of the molecular model labeled *R* or *S*. Table 6.8 provides some examples. [*Work Assessment 6.24 on your own for practice.*]

Table 6.8 Using models to assign priority.

Molecule	Priority assigned	Explanation	Model matched
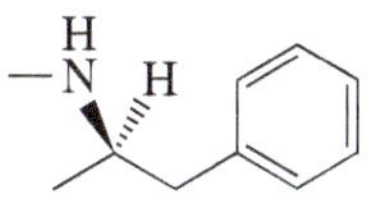	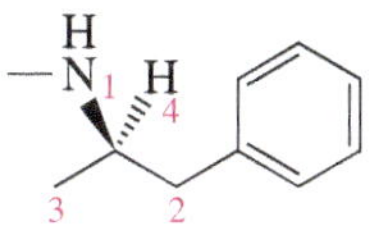	#2 has a second carbon attached, whereas #3 only has hydrogens	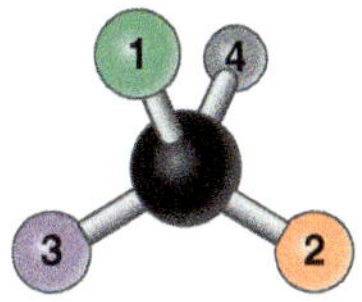 ***R***
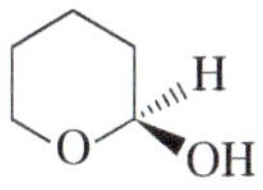	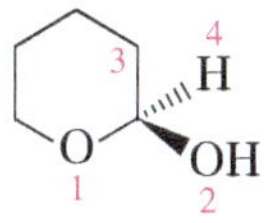	#1 is attached to a carbon, whereas #2 only has a hydrogen	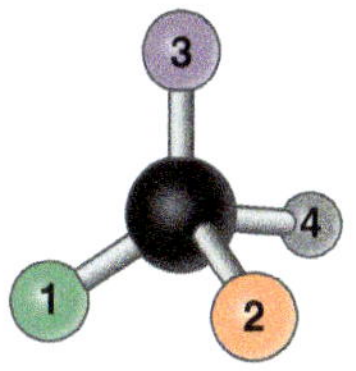 ***S***

Assessment 6.24 Prioritize the substituents at each chiral center and then, by comparing them to the models you created in Section 6.3.2, label the absolute configuration as *R* or *S*.

(a)

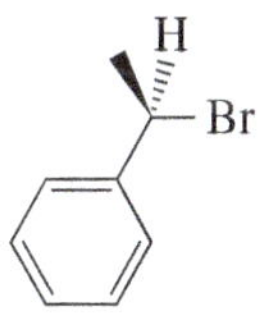

(b)

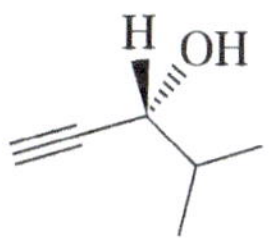

(c)

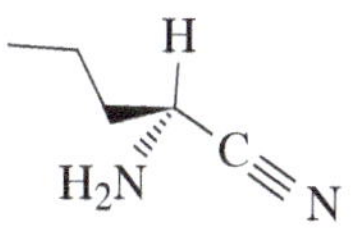

(d)

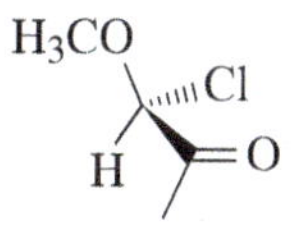

(e)

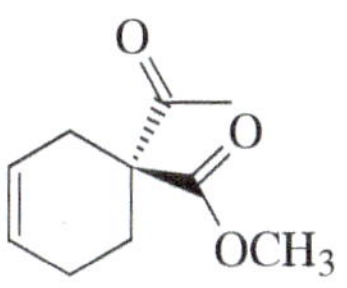

(f)

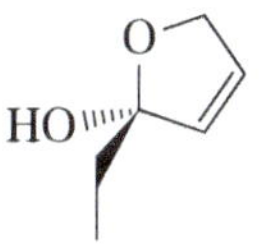

(g)

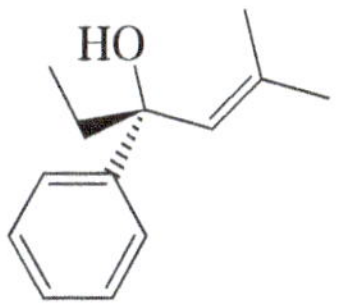

(h)

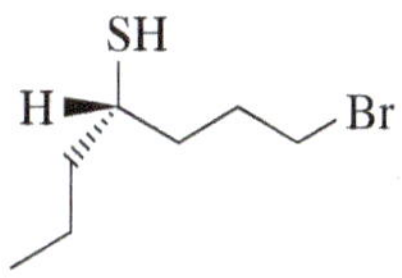

6.3.2.4 Putting it all together—without models

[*Bad news: You won't always have models with you.*] The original statement of the CIP rules assumes that students can visualize 2-D representations as 3-D images. [*The more assessments you work, the easier this will become.*] Here in Section 6.3.2.4, we describe two methods for solving these kinds of problems. Please use both, in combination with models, as a way of learning each method. By seeing that all three give the same answer, you will trust the method you eventually become most comfortable with.

Method 1—Rotate the molecule in your head This method is simple to explain but difficult to execute, especially in the beginning. After prioritizing the groups attached to the chiral carbon, you simply need to orient the molecule such that the priority 4 substituent is going behind the plane of the page (**Figure 6.24**); then trace the path from 1 → 2 → 3. If this path proceeds in a clockwise direction, then the absolute configuration at that carbon is *R*. Recall the steering wheel analogy, where turning a steering wheel clockwise causes the vehicle to turn right. On the other hand, if the path from 1 → 2 → 3 proceeds in a counterclockwise direction, then the absolute configuration is *S*. [*This is how we originally assigned* R *and* S *to our models.*]

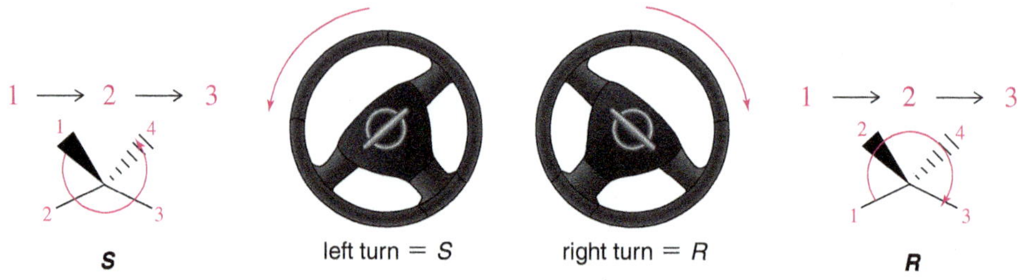

Figure 6.24 Determining *R* and *S* based on orienting the lowest priority substituent behind the plane of the page.

You will encounter three different situations when you use this rotation method. **Table 6.9** shows these situations in order from simple to difficult.

Table 6.9 Situations you might encounter when trying to rotate a molecule in your head.

Molecule with priorities	Situation	Flip necessary?	Result
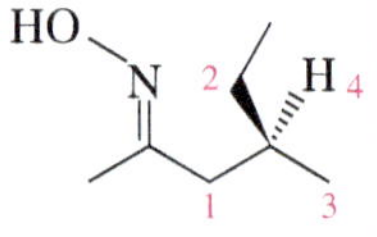	The priority 4 substituent is already going back, so you need only trace the path from 1 → 2 → 3 and assign the stereochemistry accordingly.	No	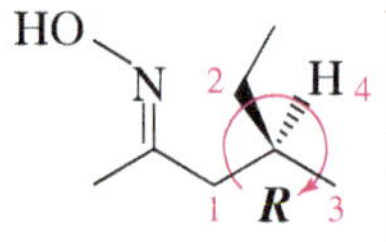
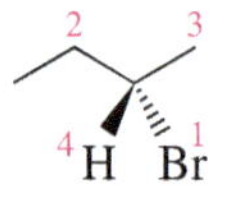	The priority 4 substituent is coming forward, so you must do a pancake flip 180° around an axis in the plane of the page. With this new orientation, you can trace the path from 1 → 2 → 3 and assign the stereochemistry accordingly.	Yes, 180° pancake flip	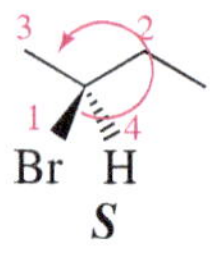
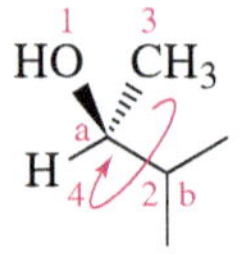	The priority 4 substituent is in the plane of the page, so rotate 120° around the $C_a—C_b$ bond to move substituent 4 back. With this new orientation, you can trace the path from 1 → 2 → 3 and assign the stereochemistry accordingly.	Yes, 120° bond rotation around the $C_a—C_b$ bond	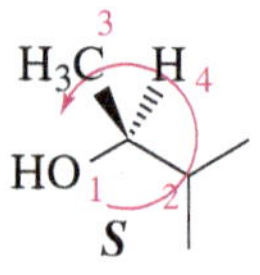

Method 2—The right-hand/left-hand rule The right-hand/left-hand rule method is the one that students have the most success with. Two aspects make this method useful. First, you always have your hands with you (very helpful during exams), and second, you do not need to manipulate the orientation of the molecule. Moreover, this method reinforces important concepts about the behavior of enantiomers in biological systems. [*We explain how in Section 6.8 and Chemistry in Context 6-A.*]

After prioritizing the substituents at the chiral center, use both of your hands, with thumbs extended, to trace the path from 1 → 2 → 3 as in **Figure 6.25**. [*Start with your fingertips at the 1, move them to 2, then to 3.*] To be certain that you are doing this correctly, when doing this with opposite hands, orient your thumbs in opposite directions. Only one of your hands will have your thumb pointed in the direction of the lowest priority substituent. If it is your right hand, the stereochemistry is *R*. If it is your left hand, the stereochemistry is *S*.

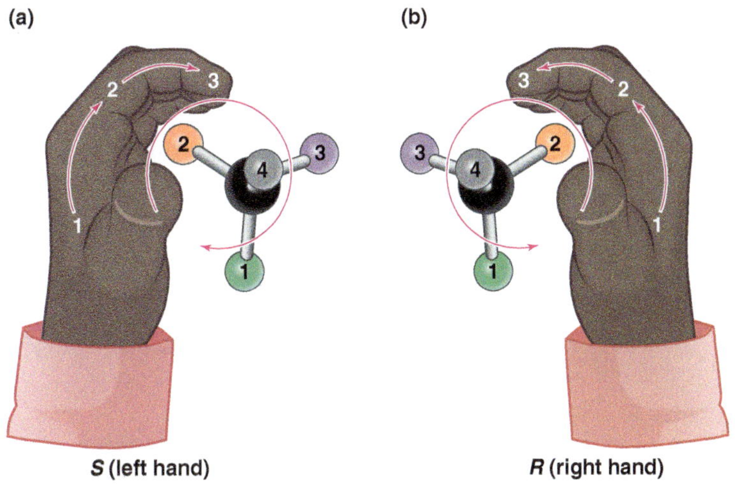

Figure 6.25 Right-hand/left-hand rule. (a) If your fingers trace 1 → 2 → 3 and your left thumb is pointing in the direction of 4, it is *S*. (b) If your fingers trace 1 → 2 → 3 and your right thumb is pointing in the direction of 4, the stereochemistry is *R*. [*The models in the pictures are the enantiomer models we created in Section 6.3.2.*]

This method works equally well when the lowest priority substituent is in the plane of the page. [*It should be easier because it doesn't require any crazy contortions of your hand and wrist. Try it on your own.*] In **Figure 6.26**, substituent 1 is in the plane of the page, 2 is behind the plane of the page, and 3 is in front. Thus, your hands should move as if you are reaching your fingers into the page. Only your left hand has the thumb pointing up—the direction of substituent 4—so this stereocenter is *S*.

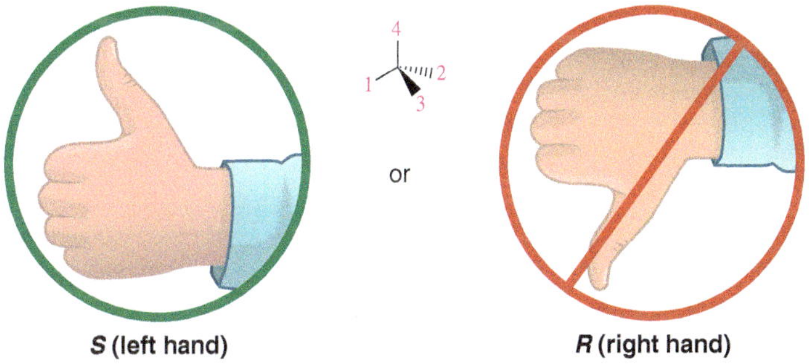

Figure 6.26 The right-hand/left-hand rule determines the (*S*) configuration.

In conclusion, you have been given two ways to figure out *R* and *S* that do not require the use of models. As you practice, try out both ways and compare them to each other and to the use of models. See which one you feel most comfortable with, which one feels more reliable, and which one allows you to get the answer most easily. Try each of them in Assessment 6.25.

Assessment 6.25 Prioritize the substituents at each chiral center and then, by each of the two methods discussed in Section 6.3.2.4, determine the absolute configuration. [*Do not use your models, except to check your answers.*]

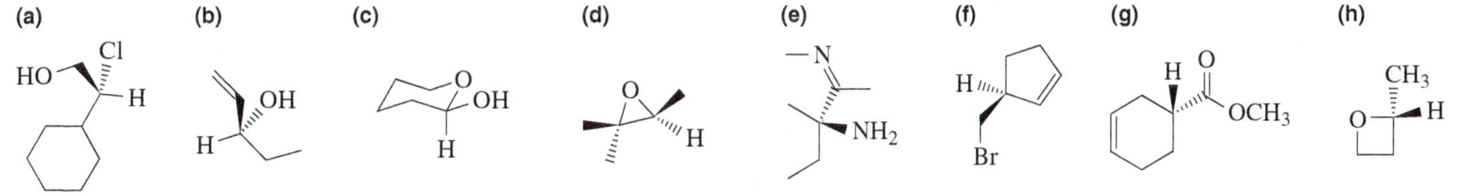

We can now specify the absolute configuration of the thalidomide enantiomers (Figure 6.27). The effective sedative is (*R*)-thalidomide, while the strong teratogen is (*S*)-thalidomide. The birth defects caused by this drug might have been avoided if chemists had synthesized only the *R* enantiomer or, at the minimum, separated *R* from *S*. In the 1950s, administering a drug as a racemic mixture was common because the selective synthesis and separation of enantiomers was quite difficult. We explain why in Section 6.4.

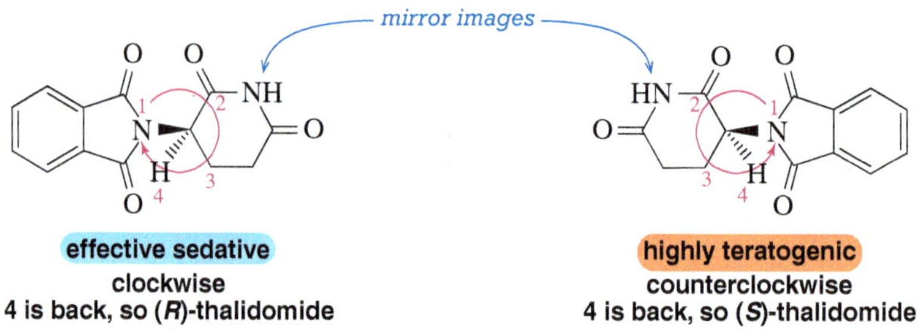

Figure 6.27 Configuration of thalidomide enantiomers.

6.4 Properties of enantiomers

Enantiomers have identical properties, except when in a chiral environment. To illustrate, Table 6.10 lists four common properties for the two enantiomers of 2-bromobutane. Furthermore, their spectroscopic data, as we study in Chapters 14 and 15, are identical, too.

In the context of thalidomide, Table 6.10 shouldn't make sense. (*R*)-thalidomide is a sedative, whereas (*S*)-thalidomide is teratogenic. In their interactions inside the body, these molecules have very different properties. The human body, then, must be a chiral environment. Many important biomolecules are made up of amino acids and nucleotides, all chiral molecules for which nature has selected a single enantiomer (Figure 6.28). As a result, the receptors and enzymes in our bodies are all chiral, resulting in differing interactions between individual enantiomers of molecules like thalidomide. We discuss why we get these different interactions in the following sections.

Table 6.10 Properties of (*R*)- and (*S*)-bromobutane. Enantiomers have identical physical and chemical properties, except in a chiral environment.

	(*R*)-2-bromobutane	(*S*)-2-bromobutane
Boiling point (°C)	91.2	91.2
Melting point (°C)	−112	−112
Refractive index	1.436	1.436
Density (g/mL)	1.253	1.253

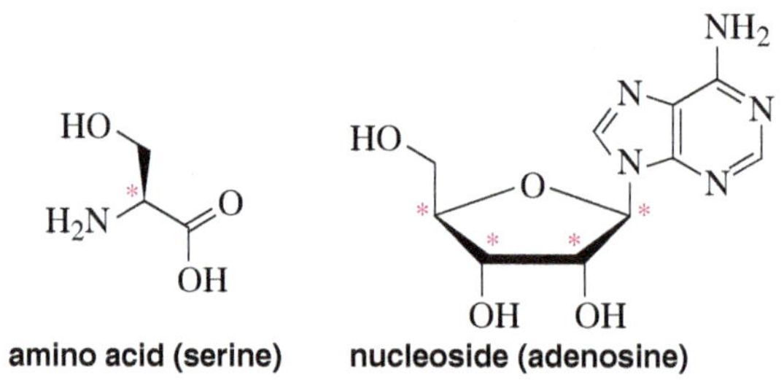

Figure 6.28 Biological molecules are chiral and exist in a single enantiomeric form.

One important way to distinguish between enantiomers takes advantage of the interaction between enantiomers and plane-polarized light. Light is an electromagnetic wave, oscillating randomly in every possible plane from 0 to 360°. When light is passed through a polarizing filter, only light oscillating in one plane emerges. If a second polarizing filter is placed in series along the same axis, light in this plane will pass through [Figure 6.29(a)]. Alternately, if the second polarizing filter is perpendicular to the first, no light will pass through [Figure 6.29(b)]. Polarized sunglasses block glare (on a horizontal plane) by only allowing light on the vertical plane to pass through [Figure 6.29(c)].

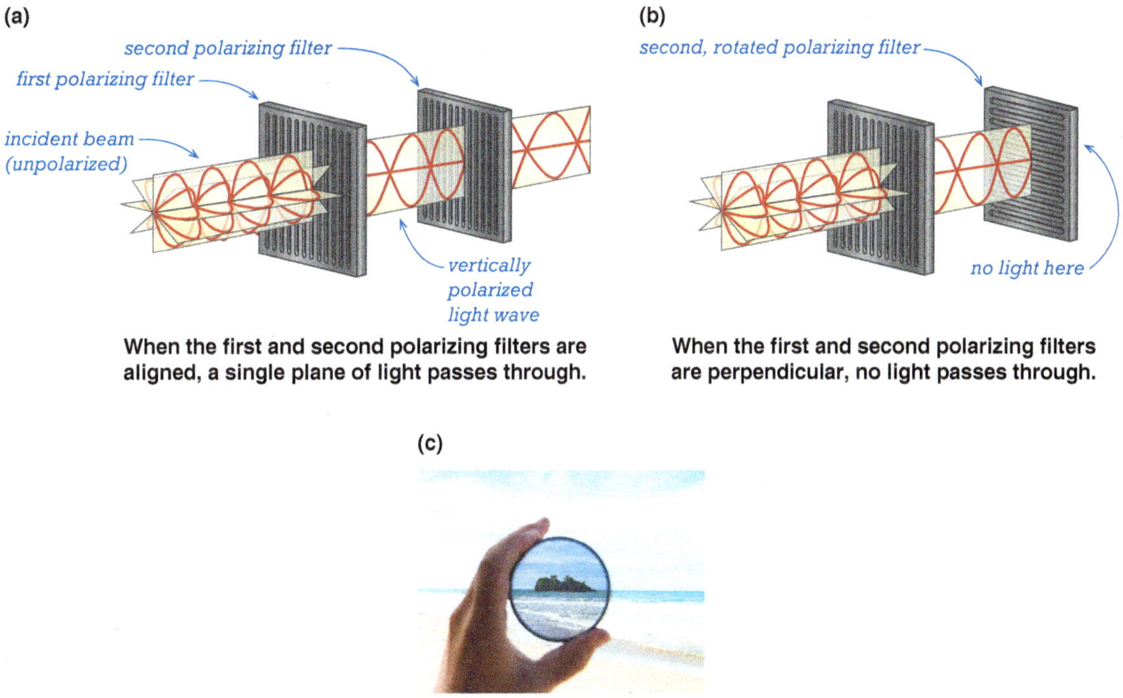

Figure 6.29 A polarizing filter allows only one plane of light to pass through. (a) Alignment of the polarizing filters allows light to pass through. (b) A perpendicular second filter does not allow light to pass through. (c) Blocking the glare gives a clearer image.

If you shine **plane-polarized light** on a single enantiomer, the plane of light is rotated in a certain direction. In the presence of the opposite enantiomer, the plane of light will be rotated in the opposite direction. Consider the two enantiomers, (*R*)- and (*S*)-2-bromobutane in Figure 6.30. (*R*)-2-Bromobutane rotates plane-polarized light to the left (counterclockwise direction), whereas (*S*)-2-bromobutane rotates plane-polarized light to the right (clockwise direction) [Figure 6.30(a)]. When plane-polarized light passes through a solution containing an equal mixture of these molecules (that is, a racemic mixture), the plane is rotated in equal amounts, clockwise and counterclockwise, resulting in no net rotation [Figure 6.30(b)].

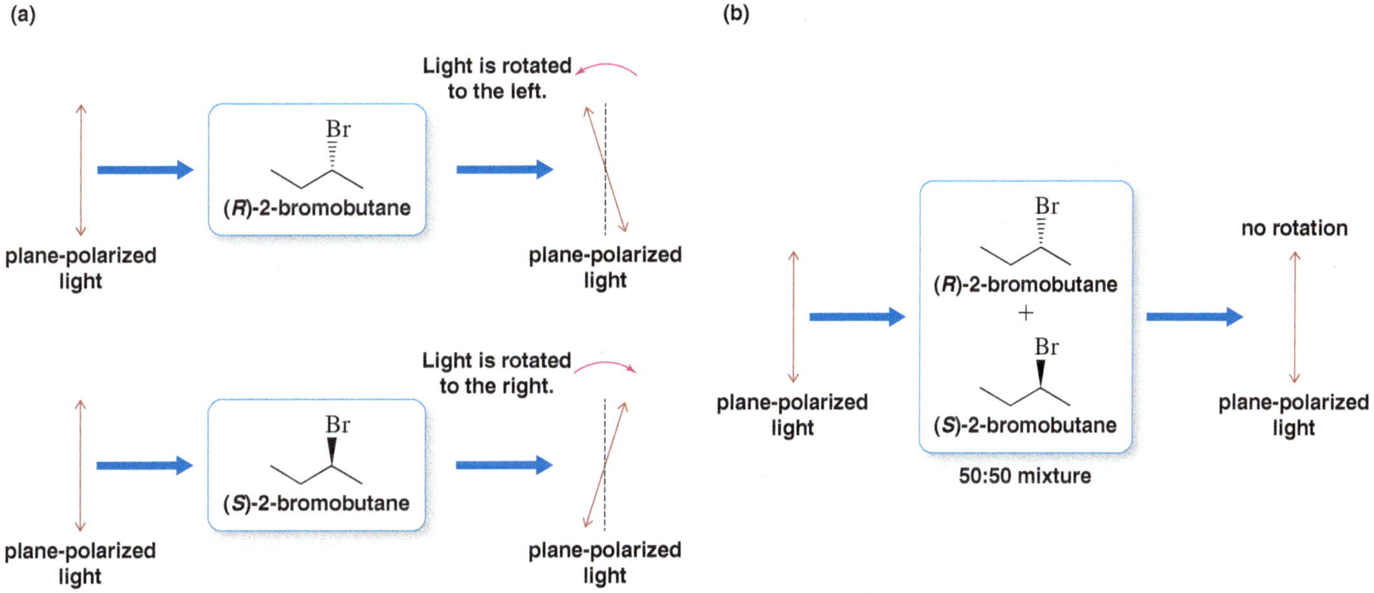

Figure 6.30 Rotation of plane-polarized light. (a) Two pure enantiomers (*R* and *S*) rotate plane-polarized light in equal but opposite directions. (b) A racemic mixture of two enantiomers has no net rotation.

A substance that rotates plane-polarized light is said to be **optically active**, whereas a substance or a mixture of substances that does not is said to be **optically inactive**. For the first time, we see a difference between properties of enantiomers. Given that enantiomers only show different properties in chiral environments, it must be true that plane-polarized light is itself chiral. It turns out that plane polarization can be expressed as a superposition of two opposite circular polarizations, which are themselves enantiomers (Figure 6.31).

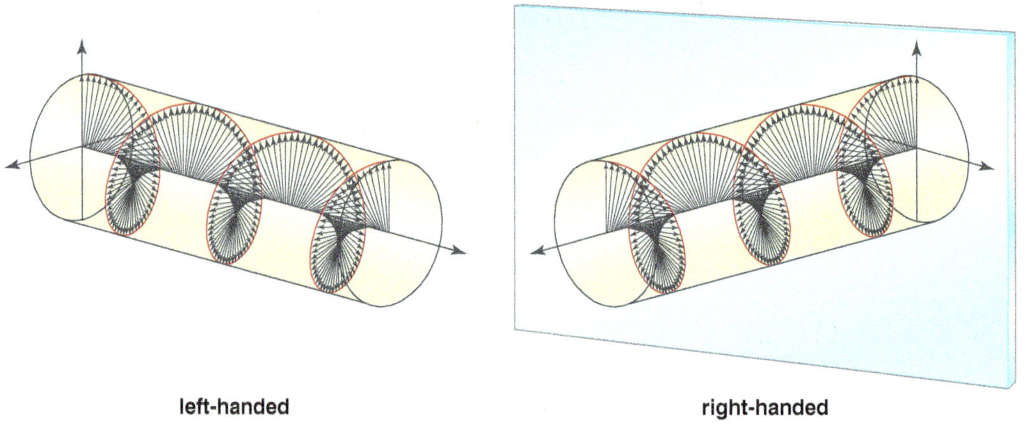

Figure 6.31 Opposite circular polarizations make up plane-polarized light. The two polarizations are enantiomers of each other.

6.4 SUMMARY

- Enantiomers can be distinguished by the direction in which they rotate plane-polarized light. If a pure compound rotates plane-polarized light +15° (clockwise, or to the right), at the same concentration, its enantiomer will rotate plane-polarized light −15° (counterclockwise, or to the left).

6.4.1 Polarimetry

The rotation of plane-polarized light is based on the interaction between light and a chiral compound. Thus, a higher concentration of the compound results in a higher degree of rotation. A **polarimeter** is the instrument used to make these measurements. It is a simple instrument, consisting of the six parts shown in Figure 6.32.

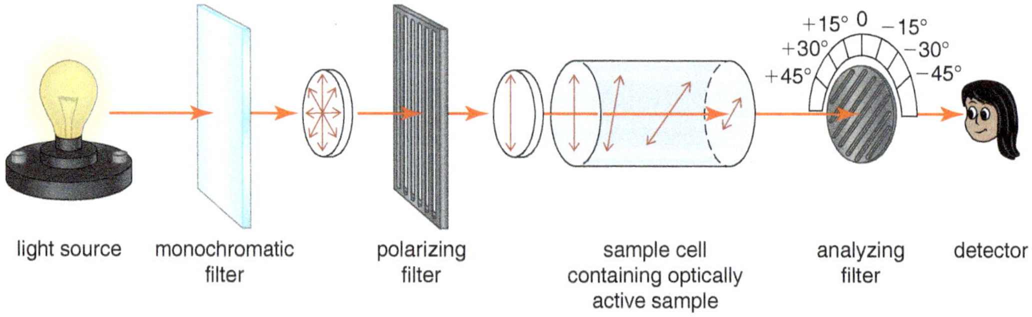

Figure 6.32 Schematic of a polarimeter. The analyzing filter rotates a certain amount to allow the rotated light to reach the detector.

1. Light source: A sodium lamp is most commonly used as the light source because it produces light composed of only a few different wavelengths.
2. Monochromatic filter: Because different wavelengths of light interact differently with a single enantiomer, the sodium light is filtered to give light of a single wavelength. Specifically, a yellow emission line in the sodium spectrum is used, referred to as the sodium D line.
3. Polarizing filter: The first polarizing filter is fixed in place and is used to ensure that only one plane of light passes through the sample.
4. Sample cell: The sample cell holds the chiral compound, typically dissolved in an achiral solvent. The degree to which a compound rotates light depends on the concentration of the compound as well as the relative proportion of enantiomers present. Modern polarimeters use very small sample cells, so not much compound is needed for analysis.

5. Analyzing filter: The second polarizing filter can be rotated to allow the rotated light to pass through. The degree to which the filter is rotated to allow maximum light to pass through is recorded as the observed rotation, α.
6. Detector: The detector is used to measure the amount of light that passes through the analyzing filter.

Table 6.11 lists two ways in which the nomenclature of enantiomers reflects the direction in which they rotate plane-polarized light. These are the *d*/*l* system and the +/− system.

Table 6.11 Terms used to describe the rotation of plane-polarized light.

System	Term	Description
d/*l*	Dextrorotatory (*d*)	Derived from the Greek word *dexios*, meaning "toward the right." A compound is dextrorotatory if it rotates plane-polarized light to the right (clockwise).
	Levorotatory (*l*)	Derived from the Latin word *laevus*, meaning "toward the left." A compound is levorotatory if it rotates plane-polarized light to the left (counterclockwise).
IUPAC (+/−)	(+)	If plane-polarized light is rotated in the clockwise, or positive, direction (see Figure 6.32), the compound is referred to as (+).
	(−)	If light is rotated in the counterclockwise, or negative, direction, it is referred to as (−).

The *S* isomer of 2-bromobutane rotates light in the clockwise, or dextrorotatory, direction. As such, it can be referred to as (+)-2-bromobutane, (*d*)-2-bromobutane, or (+)-(*S*)-2-bromobutane. The *R* enantiomer, which rotates light in the opposite direction, is therefore referred to as (−)-2-bromobutane, (*l*)-2-bromobutane, or (−)-(*R*)-2-bromobutane. A racemic mixture of 2-bromobutane is referred to as (*R*/*S*)-2-bromobutane, (*d*/*l*)-2-bromobutane, or (+/−)-2-bromobutane.

Although the *S* enantiomer of 2-bromobutane is (+) or (*d*), not *all S* enantiomers are (+) or (*d*). *R* or *S* is something we assign using the Cahn–Ingold–Prelog system based on the structure of the molecule. *R* and *S* are therefore unrelated to (+) and (−), or (*d*) and (*l*), both of which are properties of chiral molecules. The only thing we can say for certain is that if the *R* enantiomer is (+) or (*d*), then the *S* enantiomer will always be (−) or (*l*). And, if the *R* enantiomer is (−) or (*l*), then the *S* enantiomer will always be (+) or (*d*). Finally, a racemic mixture will always be (+/−) or (*d*/*l*).

This concept is often confusing and may require rereading to make it clear.

Assessment 6.26 (*R*)-Limonene is a cyclic terpene responsible for the smell of oranges and other citrus fruits. (a) Given that (*R*)-limonene rotates plane-polarized light in the clockwise direction, should it be referred to as (*d*) or (*l*)? (b) Is it (+) or (−)? (c) What direction (*d* or *l*; + or −) would you expect (*S*)-limonene to rotate plane-polarized light?

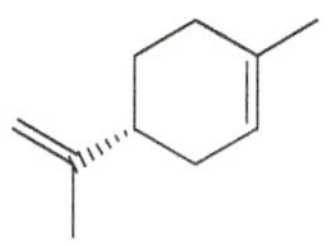

(*R*)-limonene

Assessment 6.27 Naproxen is a commercially available anti-inflammatory sold under the name Aleve. (a) Assign the absolute configuration as *R* or *S*. (b) Given that naproxen rotates plane-polarized light in the clockwise direction, should it be referred to as (*d*) or (*l*)? (c) Is it (+) or (−)? (d) What direction (*d* or *l*; + or −) would you expect the enantiomer of naproxen to rotate plane-polarized light?

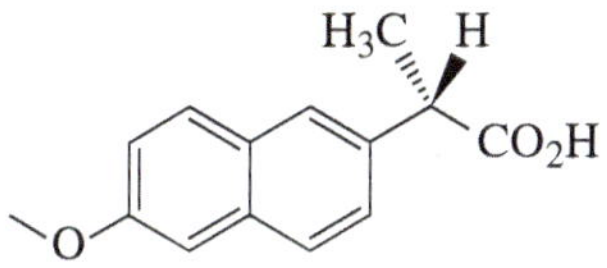

naproxen

Assessment 6.28 Based on the absolute configuration of (*S*)-butan-2-ol, what can you say about the direction it rotates plane-polarized light?

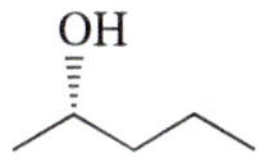

(*S*)-butan-2-ol

Incorporating the direction of rotation of plane-polarized light, the modified names of thalidomide enantiomers are (*R*)-(+)-thalidomide and (*S*)-(−)-thalidomide. In the late 1950s and early 1960s using a polarimeter, you would have seen *no* rotation of plane-polarized light by thalidomide because both enantiomers were present in equal amounts (a racemic mixture). A racemic mixture, therefore, is optically inactive.

Like a boiling point, the degree to which a single enantiomer of a certain molecule will rotate plane-polarized light is a characteristic physical property. The magnitude of rotation is related to concentration, however, because more chiral molecules means there will be more "interactions" between the molecules and plane-polarized light. In lieu of reporting a concentration-dependent **observed rotation** for a given solution, we describe this property in terms of **specific rotation** $[\alpha]$ (reported in degrees), shown in Equation 6.1. The observed rotation is divided by the path length of the cell (l, in units of decimeters) multiplied by the concentration (c). For a pure substance, c = the density of the substance. For a solution, c is in units of g/100 mL. Because the amount of rotation depends on the temperature (T, measured in °C) and the wavelength of the light used (D for sodium D line, 589 nm), it is important to specify this information. By standardizing the measurements in this way, it ensures that, regardless of the instrument being used, the number should come out the same.

$$\text{Specific rotation} = [\alpha]_D^T = \frac{\alpha_{obs}}{l \times c} = \frac{\text{observed rotation (degrees)}}{\text{length (dm)} \times \text{concentration}\left(\frac{g}{100\ mL}\right)} \qquad \text{Equation 6.1}$$

Table 6.12 Specific rotations of a pair of enantiomers.

	(*S*)-2-bromobutane	(*R*)-2-bromobutane
$[\alpha]_D^{25}$	+23.1°	−23.1°

The specific rotations of the enantiomers of 2-bromobutane are shown in Table 6.12. The specific rotations are equal in magnitude but opposite in sign, which is exactly what you should expect based on the relationship between the two enantiomers.

Assessment 6.29 A student wanted to measure the specific rotation of the following propionate derivative (density = 1.12 g/mL).

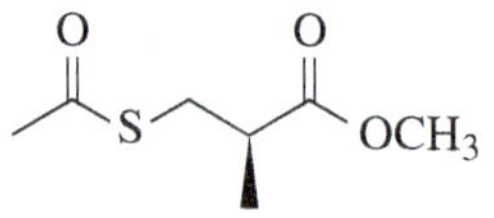

A sample of the pure compound was placed in a 10.0-cm polarimeter tube, and using the sodium D line, the observed rotation at 20 °C was determined to be +46.6°. What is the specific rotation of the propionate derivative?

Assessment 6.30 Cholesterol (50 mg) was dissolved in 10 mL of chloroform and placed in a 1.0-cm polarimeter cell. This solution produced an observed rotation (using the sodium D line at 20 °C) of −1.58°. What is the specific rotation of cholesterol?

Assessment 6.31 Using a 1.0-cm polarimeter cell and a solution prepared by dissolving 2.54 g of glucose in 1 L of H_2O, the specific rotation of D-glucose was calculated to be +52.7° at 20 °C using the sodium D line as the source of light. What was the observed rotation of the solution?

6.4.2 Quantifying the enantiomeric composition of mixtures

Specific rotation can also be used to measure relative amounts of enantiomers in a mixture. For example, the specific rotations of pure (*S*)-2-bromobutane and pure (*R*)-2-bromobutane are +23.1° and −23.1°, respectively. Furthermore, a racemic mixture of the two of them gives a specific rotation of 0° because each enantiomer rotates the light equally, yielding no net rotation. By measuring the specific rotation of a mixture of enantiomers, then, we can determine their relative proportions when the concentration is known.

By plotting the data for 2-bromobutane (Table 6.12) on a range of specific rotations, we can identify the composition of the mixture at each point (Figure 6.33). At point A, there is only (*R*)-2-bromobutane; at point E, there is only (*S*)-2-bromobutane. At point C, we have equal amounts of each. What about at points B and D? Which is there more of at each of these points? [*Think about it before reading on.*]

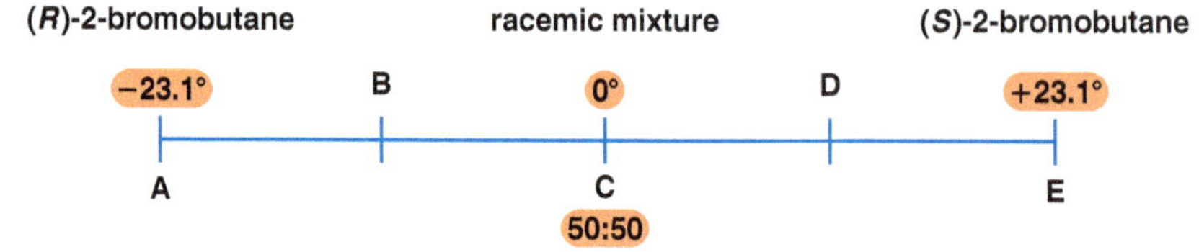

Figure 6.33 A range of specific rotations for pure enantiomers and mixtures of 2-bromobutane.

At point **B**, we have more R than S; at point **D**, we have more S than R. [*Do you see why?*] Can we specify the amounts exactly? The answer is yes, if we know the specific rotation of the mixtures.

Before we see how we can do this, a new term is needed that gives us information about the composition of a mixture of enantiomers. **Enantiomeric excess** (*ee*, expressed as a percentage) tells us how much more of one enantiomer we have than the other. A racemic mixture has an *ee* of 0% because neither enantiomer is in excess. A mixture that has one enantiomer in excess of the other has an *ee* greater than zero. A pure enantiomer, by definition, has an *ee* of 100% because there is only one enantiomer present. Equation 6.2 assumes there is more R than S.

$$\text{Enantiomeric excess}(ee) = |\%R - \%S| \qquad \text{Equation 6.2}$$

We can directly measure the enantiomeric excess using the polarimeter because the *ee* is the same as the optical purity, another term that describes the composition of a mixture of enantiomers. The **optical purity** of a mixture is the ratio of the observed rotation of the mixture to the expected rotation of the pure enantiomer (Equation 6.3).

$$\%\ \text{optical purity} = \frac{[\alpha]_{\text{sample}}}{[\alpha]_{\text{pure enantiomer}}} \times 100 \qquad \text{Equation 6.3}$$

Let's think like a chemist to try and understand optical purity beyond just an equation.

Thinking Like a Chemist

Imagine a mixture of 2-bromobutane enantiomers with a specific rotation of +11.55°. Because $[\alpha] > 0$, there must be more (S)-2-bromobutane (see Figure 6.33)—the sign of $[\alpha]$ matches $[\alpha]$ for pure (S). Because 11.55° is half of 23.1°, we can approximate that half of the mixture must be pure (S). What is the other half? [*Think about this before moving on, because it is the most often misunderstood aspect of the concept of* ee.]

To figure this out, remember that a mixture that is 100% S would have $[\alpha] = +23.1°$, and a mixture that is 100% R/S would have $[\alpha] = 0°$. In the current example, $[\alpha] = 11.55°$, which means that one half of the reaction mixture is pure S. If the other half of the reaction mixture were R [*often a conclusion come to by students studying this for the first time*], the rotation of the S half would be completely cancelled out, resulting in $[\alpha] = 0$. From this, we learn that half of the mixture rotates light in the positive direction, while the other half of the mixture does not rotate light at all. That is to say, the other half is a racemic mixture, equal parts S and R. In other words, a value that is halfway between 0 (racemic) and +23.1 (pure S) is half racemic (S/R) and half S. To complete our qualitative analysis, if the mixture is 50% pure S and the other 50% is racemic (25% R + 25% S), the overall mixture contains 75% S and 25% R.

What would the composition of the mixture be if the specific rotation of the mixture were −11.55°? [*Answer this in the context of the above before reading on, and see if you can come up with an equation on your own that allows you to get the same answer.*]

A specific rotation of −11.55° would arise from a mixture of 75% R and 25% S. Basically, the 25 units of S cancel out the rotation of 25 units of R, leaving 50 units of R responsible for the observed rotation of plane-polarized light. Thus, the enantiomer in excess in a mixture is solely responsible for the direction and magnitude by which the light is rotated. The rest of the mixture is simply racemic. To conclude, then, we can use Equations 6.4 and 6.5 to arrive at the same conclusions:

$$\frac{\text{specific rotation}}{[\alpha]_{\text{pure enantiomer}}} \times 100\% = \%\ ee \qquad \text{Equation 6.4}$$

$$\left(\frac{100 - \%\ ee}{2}\right) + \%\ ee = \%\ \text{of enantiomer in excess} \qquad \text{Equation 6.5}$$

For a specific rotation of +11.55, we know that the mixture is enriched in (*S*)-2-bromobutane, based on the sign. Thus,

$$\frac{+11.55}{+23.1} \times 100\% = 50\%\ ee; \quad \frac{100 - 50}{2} + 50 = 75\%\ (S)$$

$$100 - 75\%\ (S) = 25\%\ (R)$$

Thus, the ratio of *S*:*R* = 75:25. ■

Assessment 6.32 Imagine a sample that is enriched in the *R* enantiomer. If the % *ee* of the sample is 83%, (a) what percent of the mixture is racemic? (b) What is the ratio of *R* to *S*?

Assessment 6.33 (*S*)-Propranolol, a drug used for the treatment of anxiety, has a specific rotation of −25.5°. Attempting to prepare it in enantiopure form, a chemist produced a compound that gave a specific rotation of −18.3°. What is the ratio of (*S*)- to (*R*)-propranolol produced by the chemist?

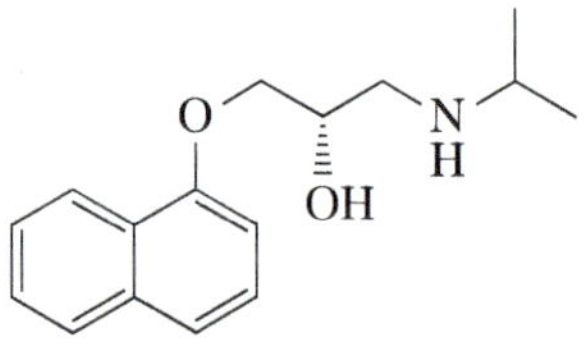

(*S*)-propranolol

Assessment 6.34 (*R*)-Selegiline, a monoamine oxidase (MAO) inactivator, was approved by the FDA in 1989 for the treatment of Parkinson's disease. In pure form, it has a specific rotation, $[\alpha]_D^{20} = -11.0°$. What is the expected specific rotation of a mixture containing 64% *S* and 36% *R*?

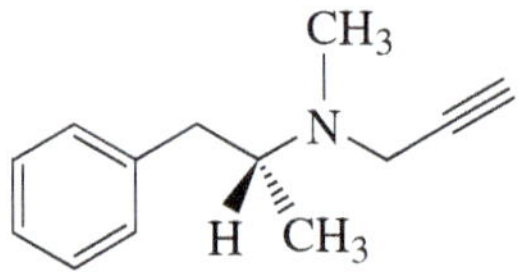

(*R*)-selegiline

The thalidomide pills that were commercially available in the late 1950s had [α] = 0° because they contained equal amounts of *R* and *S* enantiomer. With no enantiomer present in excess, *ee* = 0%. These results are characteristic of a racemic mixture.

6.5 Another type of stereoisomerism—diastereomers

Whereas enantiomers are stereoisomers that are nonsuperimposable mirror images, **diastereomers** are stereoisomers that are *not* mirror images. For example, *cis*- and *trans*-1,2-dimethylcyclohexane (**Figure 6.34**) are diastereomers because they differ in the spatial arrangement of atoms in space, but they are not mirror images.

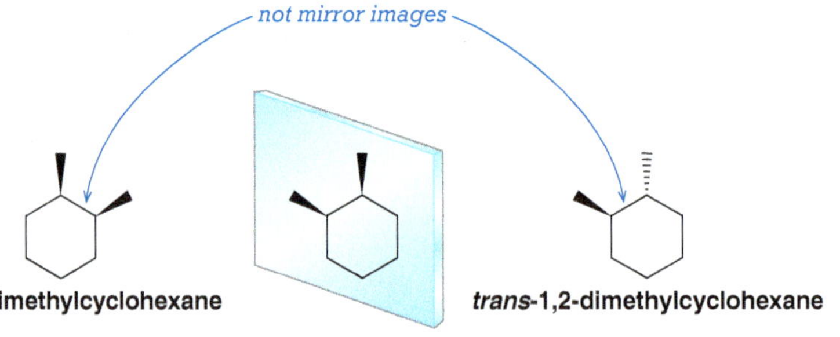

Figure 6.34 Diastereomers are stereoisomers that are not mirror images.

3-Hydroxy-2-methylbutanoic acid has two chiral carbon atoms, which are indicated by red stars in **Figure 6.35**. The four-carbon chain is drawn in the plane of the page, and the different relative configurations of the 2-methyl and 3-hydroxy group are shown. That is, they can both be drawn coming forward or both be drawn going back. Alternatively, the methyl can be forward with the hydroxyl going back, or vice versa. From this, we see that a molecule with two stereocenters can have at most four different stereoisomers. [*We'll explain what is meant by "at most" later.*]

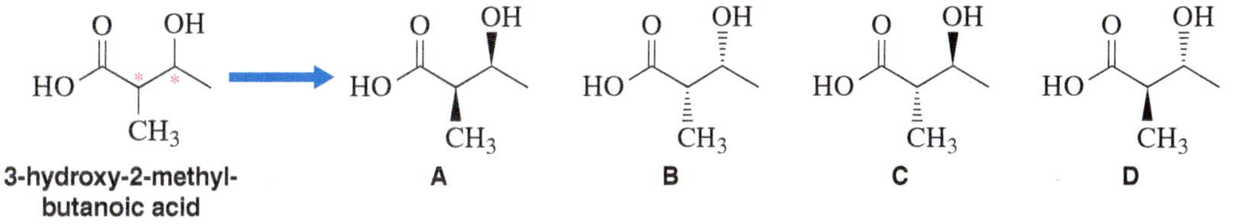

Figure 6.35 **A molecule with two stereocenters can have four stereoisomers.**

What is the relationship between the four stereoisomers (A, B, C, and D) shown in Figure 6.35? Are they enantiomers or diastereomers? Let's start with A and B [Figure 6.36(a)]. Rotating B around a point within the plane of the page shows that A and B are nonsuperimposable mirror images. The same is true for C and D [Figure 6.36(b)].

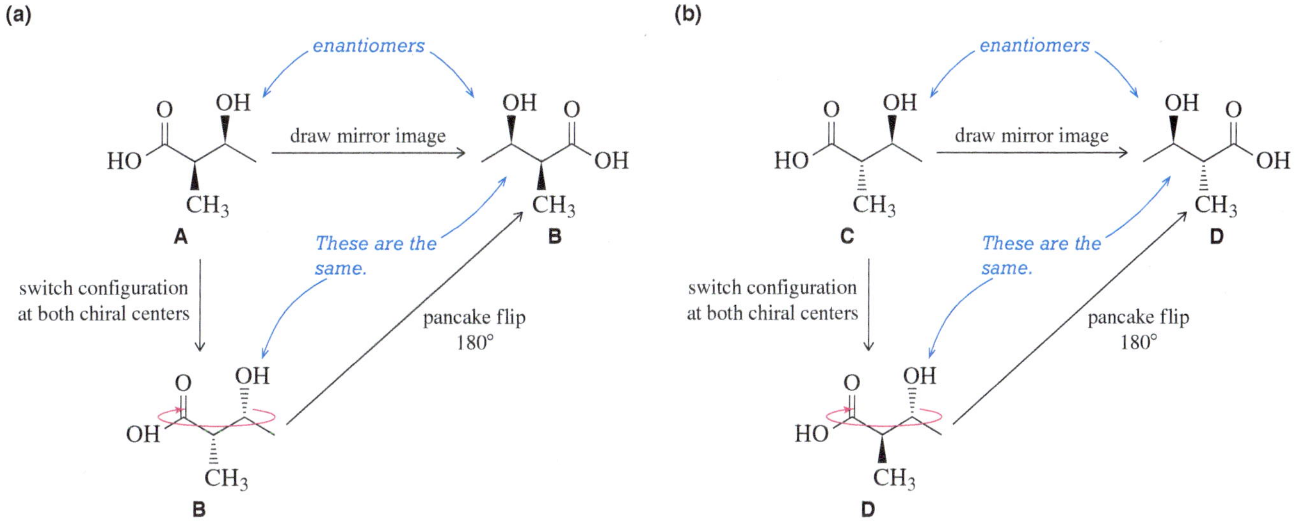

Figure 6.36 **Enantiomeric relationships among stereoisomers.** (a) **A** and **B** are enantiomers. (b) **C** and **D** are enantiomers, too.

But what is the relationship between A and C? Comparing them in the same way we did with A/B and C/D in Figure 6.36 shows that A and C are neither superimposable nor mirror images (Figure 6.37). As such, A and C are diastereomers. Before reading further, confirm for yourself that a similar relationship exists between the pairs A and D, B and C, and B and D.

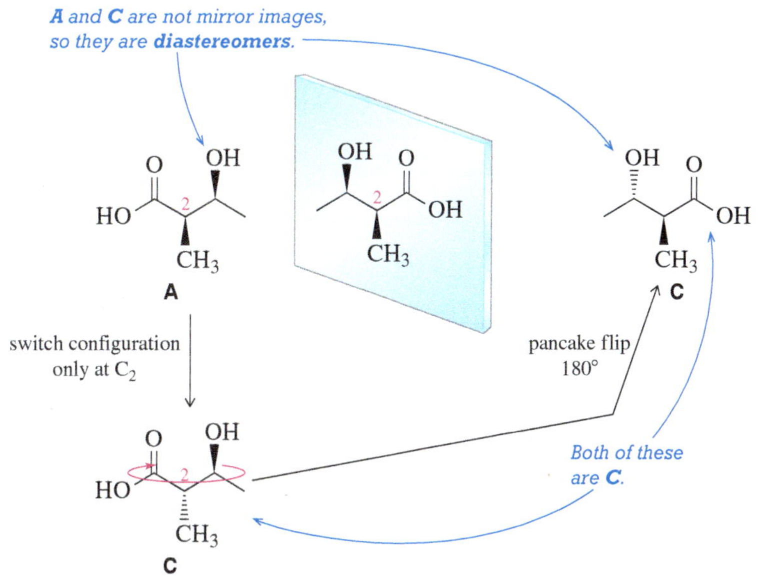

Figure 6.37 **Diastereomeric relationship among stereoisomers.**

These same relationships can be identified using the absolute configuration at each stereocenter. Enantiomers have opposite absolute configurations at each of the chiral centers. Diastereomers, on the other hand, have at least one, but not all absolute configurations different between stereoisomers. **Figure 6.38** confirms the relationships identified previously.

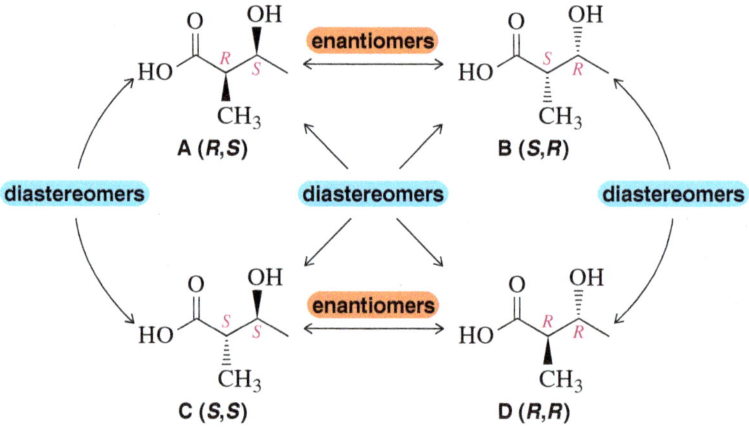

Figure 6.38 Absolute configurations can determine stereoisomeric relationships.

Assessment 6.35 For the molecules shown, (i) count the number of stereocenters present and (ii) draw all possible stereoisomers. (iii) Identify the relationships between stereoisomers as enantiomers or diastereomers.

(a)

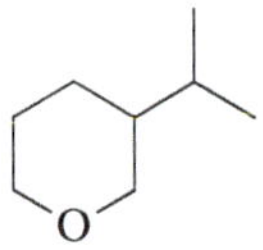

(b)

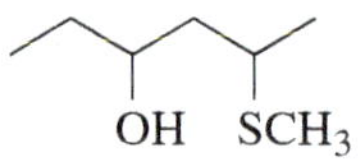

(c)

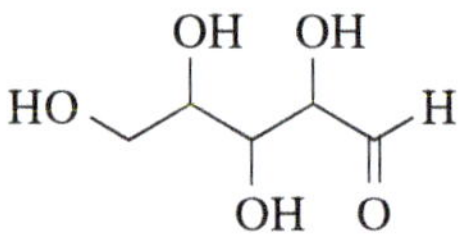

In Assessment 6.35, you may have noticed a relationship between the number of stereocenters in a molecule and the total number of stereoisomers possible. With one stereocenter, there were two stereoisomers. Two stereocenters gave four stereoisomers. And for three stereocenters, eight stereoisomers were possible. This pattern is expressed in Equation 6.6, where there are 2^n stereoisomers for n stereocenters, except in the cases we examine in Section 6.5.1.

$$2^n = \text{number of stereoisomers, where } n = \text{number of stereocenters} \qquad \text{Equation 6.6}$$

Assessment 6.36 How many stereoisomers are possible for each of the following molecules?

(a)

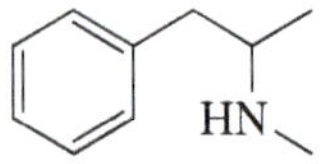

(b)

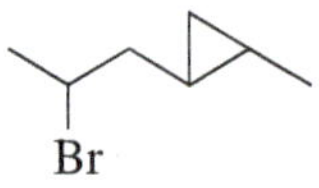

(c)

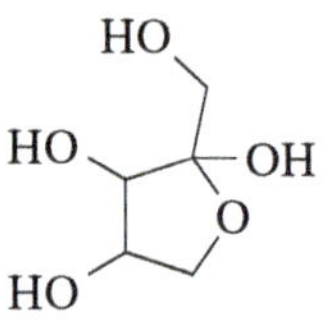

(d)

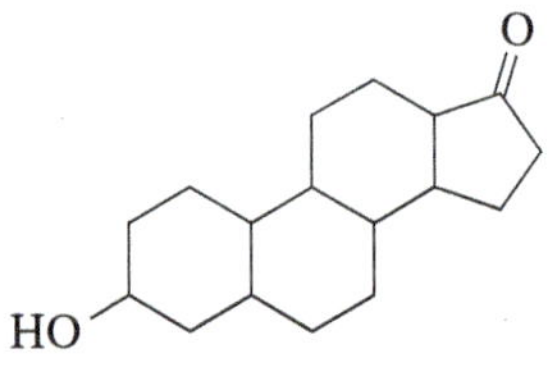

(e)

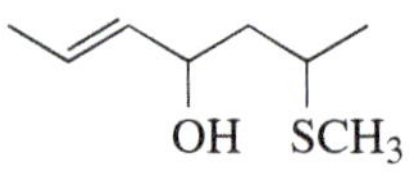

(f)

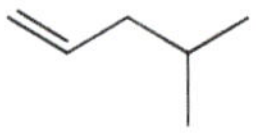

(g)

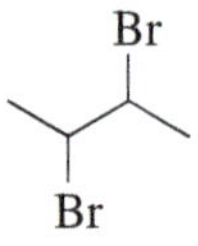

6.5.1 Meso compounds

In Assessment 6.36(g), you likely drew the four stereoisomers shown in **Figure 6.39** with their absolute configurations indicated. What are the stereochemical relationships between these compounds?

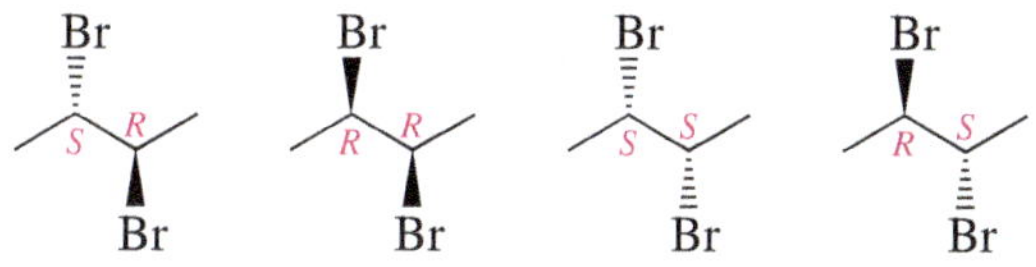

A (2*S*,3*R*) **B** (2*R*,3*R*) **C** (2*S*,3*S*) **D** (2*R*,3*S*)

Figure 6.39 The "four" stereoisomers of 2,3-dibromobutane.

Compounds **B** (2*R*,3*R*) and **C** (2*S*,3*S*) are enantiomers. As shown in **Figure 6.40(a)**, though, a 180° rotation of **A** (2*S*,3*R*) produces **D** (2*R*,3*S*). **A** and **D** are the same molecule. Thus, the number of stereoisomers does not *always* equal 2^n, where n = number of stereocenters; Equation 6.6 does not always apply. [*Think about why this might be true before reading on.*]

Compounds **A** and **D** may have seemed different initially, but they turned out to be superimposable, which indicates that they are actually the same. To see why, look at **Figure 6.40(b)**, where we rotate 180° around the $C_2—C_3$ bond in **A/D** to give a conformation with a plane of symmetry. Recall from Section 6.2.2 that a molecule with a plane of symmetry is achiral. Because it is achiral, it does *not* have an enantiomer. Nor does it rotate plane-polarized light. **A/D** is a **meso compound**, one with asymmetric carbons that is achiral because it has a plane of symmetry.

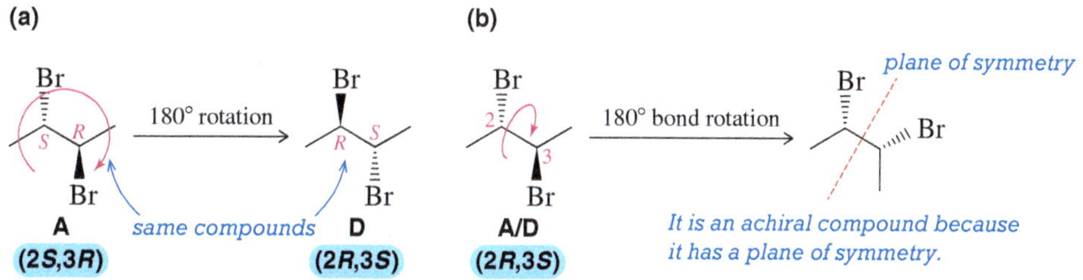

Figure 6.40 Meso compounds are achiral. (a) Molecular rotation shows that **A** and **D** are identical. (b) A symmetry plane makes a molecule containing asymmetric carbons achiral.

Assessment 6.37 Which of the following compounds are meso?

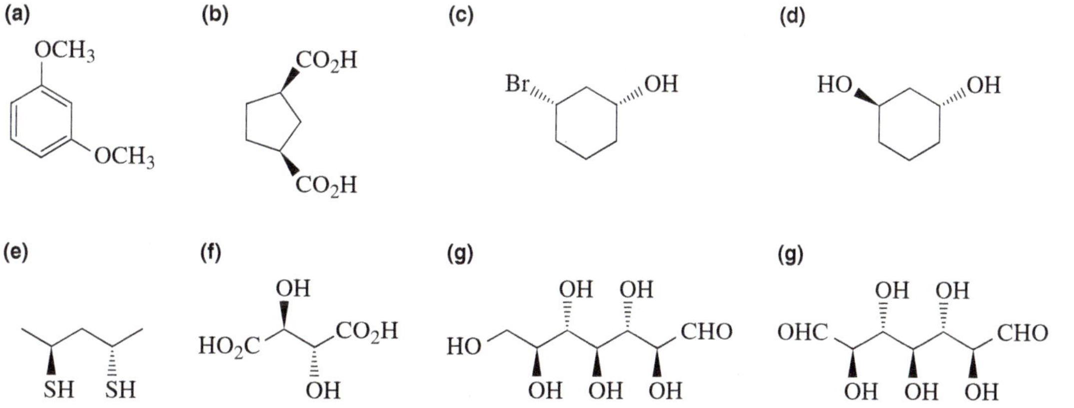

Assessment 6.38 How many stereoisomers are possible for each of the following molecules?

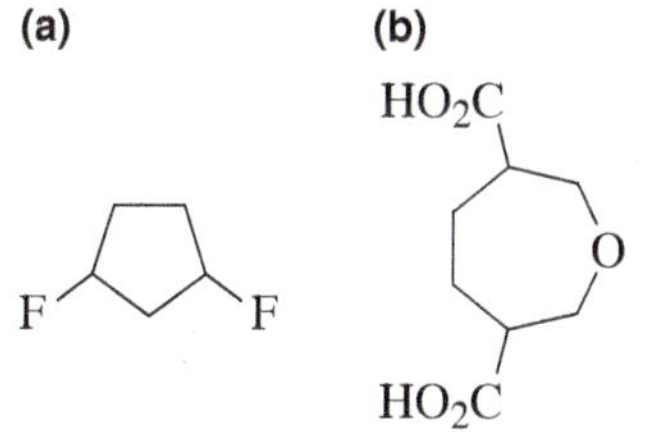

Assessment 6.39 Which of the following diastereomers of hexane-2,5-diol is optically inactive? Why?

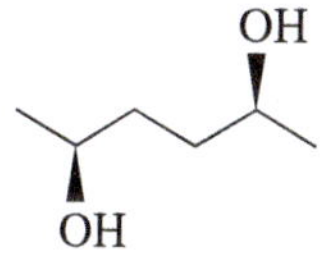

A

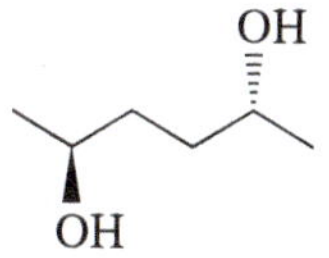

B

6.5.2 Properties of diastereomers

Whereas two enantiomers have essentially the same properties, diastereomers do not. Instead, diastereomers are completely different molecules with completely different properties. Table 6.13 compares four common physical properties of the two diastereomers of tartaric acid. Their properties are significantly different, even though their structures are so similar. In fact, the two diastereomers can be separated by either recrystallization or distillation.

Table 6.13 Properties of diastereomers of tartaric acid. Most physical and chemical properties of diastereomers are different.

	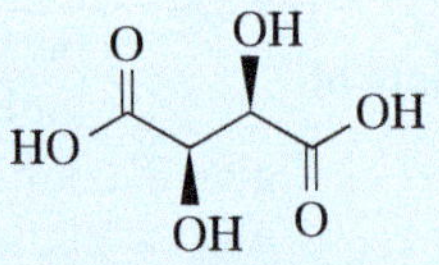 **(*R*,*R*)-tartaric acid**	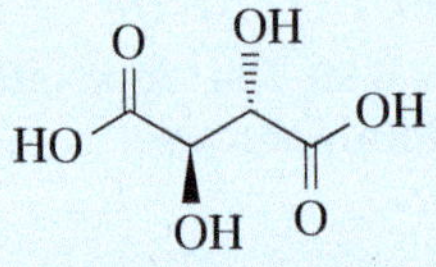 ***meso*-tartaric acid**
Solubility (in H_2O, 20 °C) (g/100 mL)	139	125
Melting point (°C)	171–174	165–166
Refractive index	1.436	1.585
Density (g/mL)	1.7598	1.660

Sometimes the differences in the properties of diastereomers can be quite profound. As an example, the melting point for (+)-2,3-dibromosuccinic acid is nearly 100 °C less than the melting point for the meso-diastereomer (Figure 6.41).

(+)- and (−)-2,3-dibromosuccinic acid
mp for either is 158 °C

***meso*-2,3-dibromosuccinic acid**
mp 256 °C

Figure 6.41 The difference between diastereomer properties can be significant. [*Read on to see what this new way of drawing molecules is.*]

Assessment 6.40 (i) Which of the following pairs of compounds would you expect to have different physical properties? (ii) What is the relationship between each of the pairs? (iii) Assign the absolute configuration of each stereocenter to confirm your answer.

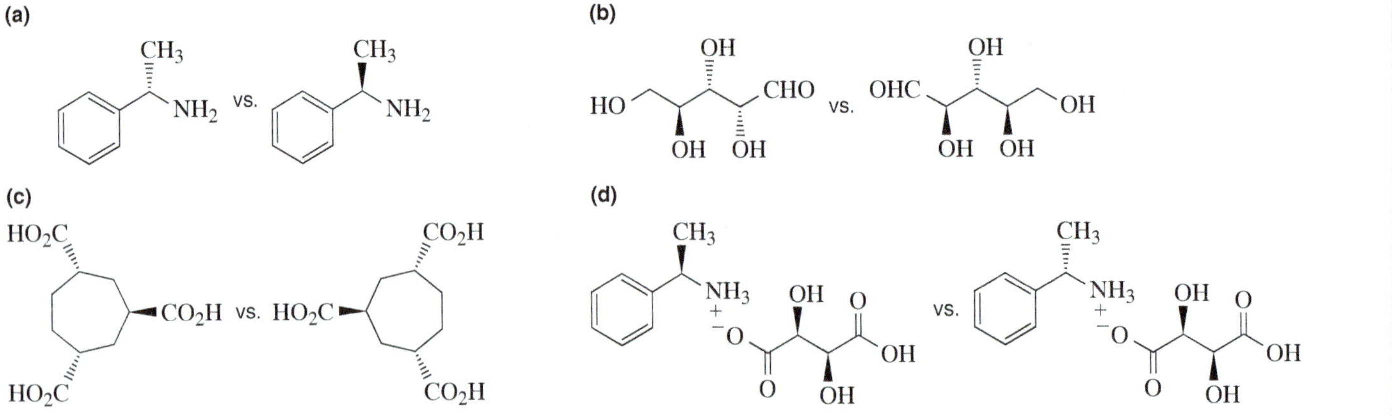

6.5.3 Fischer projections

Figure 6.41 showed the structures of the 2,3-dibromosuccinic acid stereoisomers in a form not yet discussed. These representations, shown again in Figure 6.42, are called **Fischer projections** and are especially useful for chiral carbons. As we show in Chapter 27, these two-dimensional representations of three-dimensional structures are used mostly in carbohydrate chemistry. Within the flat Fischer projection, the horizontal bonds originating at the chiral center are actually coming out of the plane of the page, while the vertical bonds are going back into the page. Figure 6.42 shows how the Fischer projection of both enantiomers of glyceraldehyde would be redrawn in the line-angle drawing format.

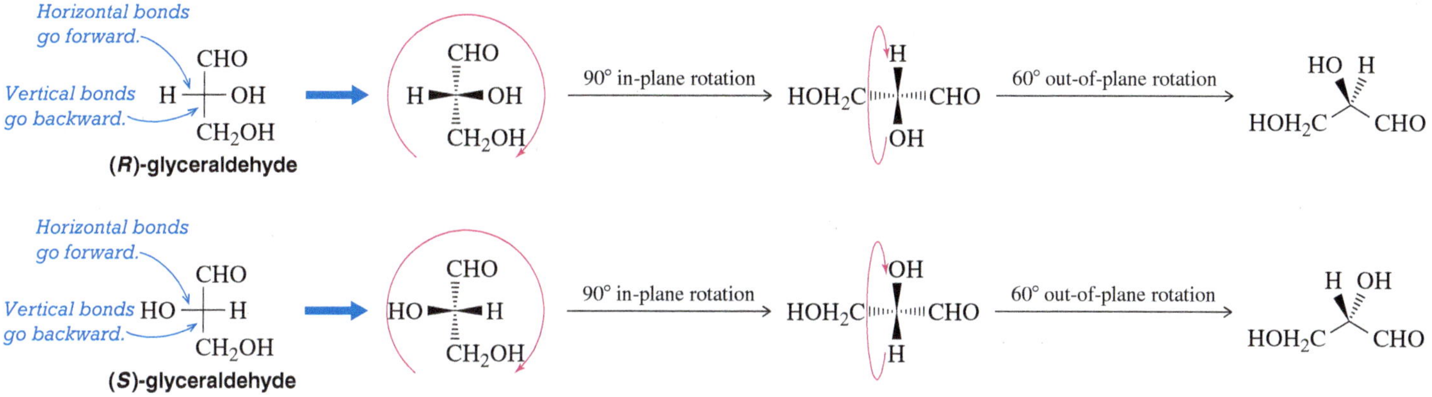

Figure 6.42 Conversion of Fischer projection to line-angle drawings. [*Models will help you visualize this process.*]

Whereas a line-angle drawing shows the carbon chain in its most stable zigzag conformation, a Fischer projection does not (**Figure 6.43**). Instead, it is the equivalent of going back at each new carbon such that the chain essentially circles back on itself. [*This sentence probably doesn't make sense without looking at Figure 6.43.*] Thus, when converting Fischer projections of larger molecules, several bond rotations are necessary to get the molecule into the stable conformation indicated by the line-angle drawing.

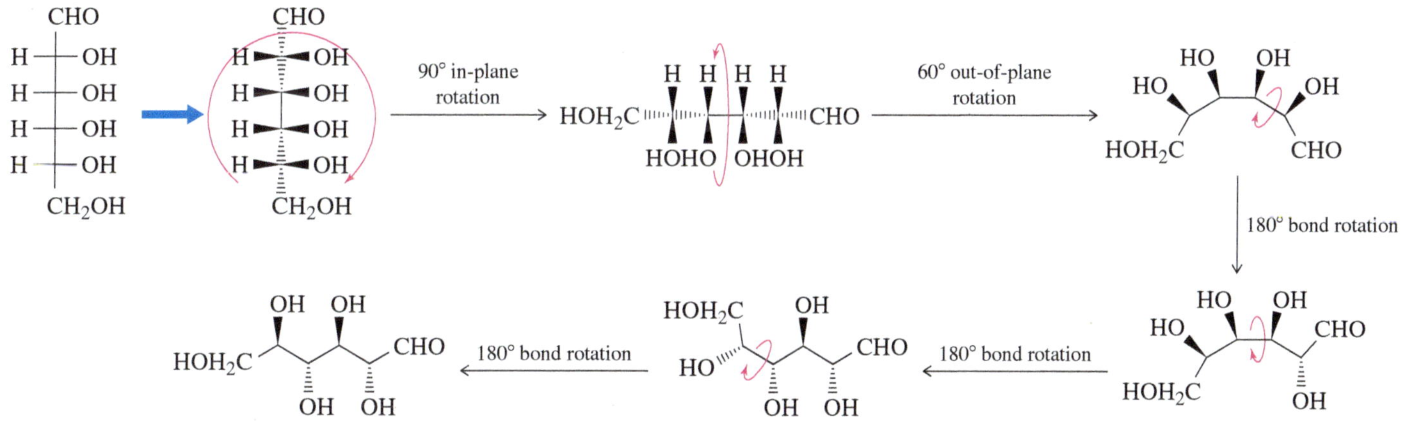

Figure 6.43 Conversion of Fischer projection to line-angle drawing for larger molecules. [*Again, models will help you visualize this process.*]

Assessment 6.41 Convert the Fischer projections into line-angle drawings and assign (*R*) and (*S*) at each chiral center.

(a) (b) (c)

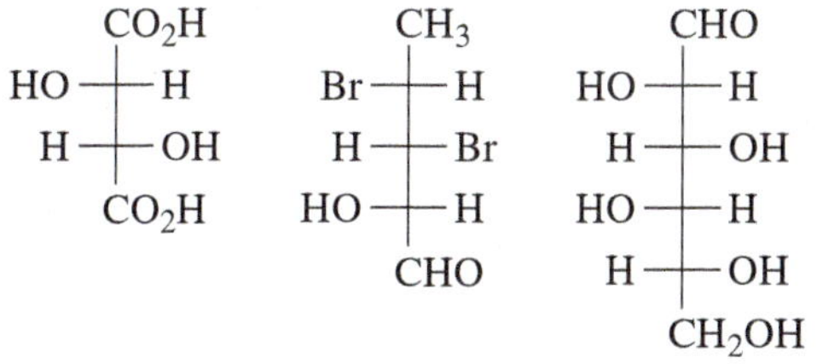

Assessment 6.42 Convert the line-angle drawings into Fischer projections.

(a) (b) (c)

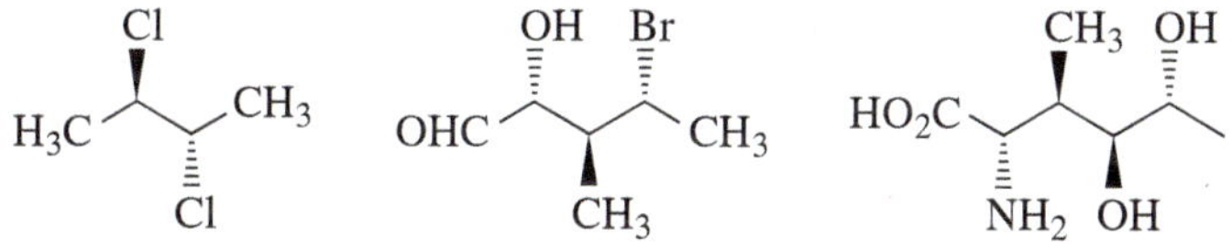

6.5.4 Implications of being diastereomers

Diastereomers have different properties. One important result of this is the different biological activities that enantiomers (for example, thalidomide) have in the body. Enzymes and biological receptors in the body are all chiral, nonracemic molecules. The receptor for the sedative activity of thalidomide, which we'll refer to as the hypothetical (*R*)-thalido-receptor, is not well understood. [*It is a single enantiomer.*] There are two possible complexes this receptor can form with (*R*)- and (*S*)-thalidomide (**Figure 6.44**), one designated (*R*,*R*), the other (*S*,*R*). With one absolute configuration that is the same and one that is different, these two complexes are diastereomers. Thus, there are different affinities for forming each diastereomeric complex and different levels of biological response. It is for this reason that (*R*)-thalidomide is a better sedative than (*S*)-thalidomide.

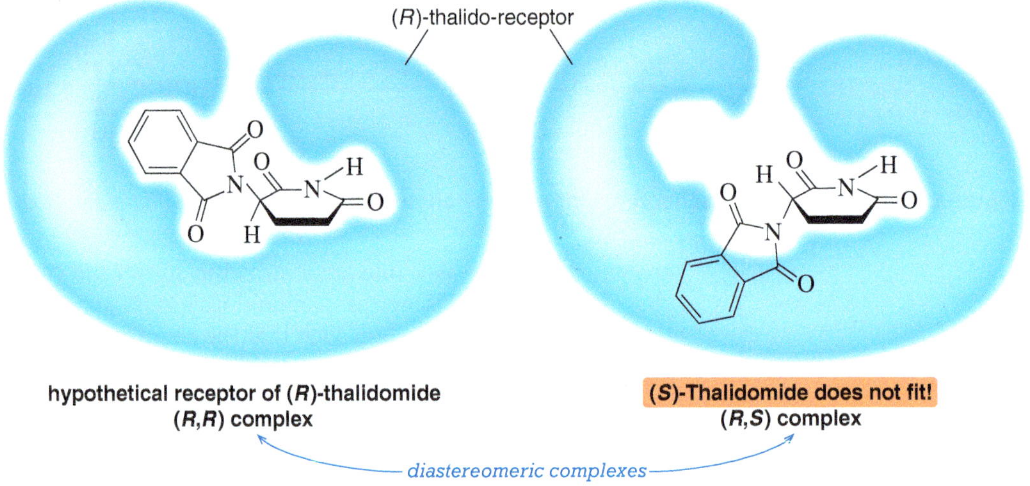

Figure 6.44 **The complex between (*R*)-thalidomide and the (*R*)-thalido-receptor is more stable than the complex between (*S*)-thalidomide and the (*R*)-thalido-receptor.**

Good News! The difference in diastereomer properties described here can be exploited for the synthesis of enantiomerically pure molecules.

Similarly, **Figure 6.45** depicts the hypothetical (*S*)-terato-receptor, which is responsible for the birth defects caused by thalidomide. Once more, the complexes formed between the receptor and (*R*)- and (*S*)-thalidomide are diastereomers. The binding affinities are different and, therefore, the levels of teratogenicity are different. From just this basic picture, we can rationalize why two enantiomers might behave so differently in the body.

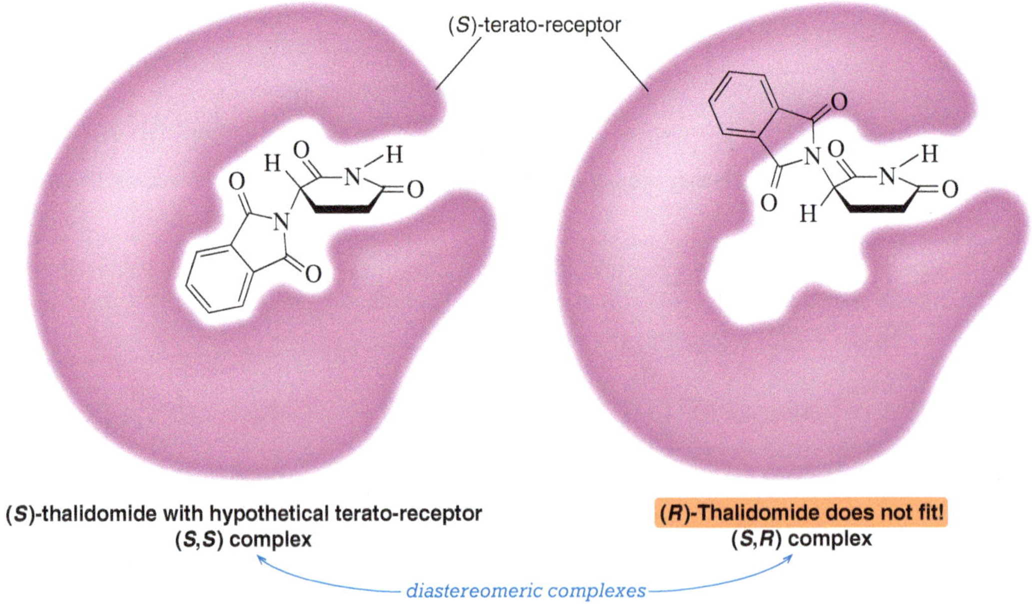

Figure 6.45 **The complex between (*S*)-thalidomide and the (*S*)-terato-receptor is more stable than the complex between (*R*)-thalidomide and the (*R*)-terato-receptor.**

6.6 Enantioselective synthesis—challenges

If a mixture of enantiomers causes a problem, why not just synthesize the desired enantiomer? Had that been possible in the case of thalidomide, the tragedy might have been avoided. To understand the challenge of producing a single enantiomer, we must think about the energetics and rates of competing reactions the way we did in Chapter 5.

The hydrogenation of a ketone using a transition metal catalyst like palladium is a reaction we discuss in greater detail in Chapter 17. Nevertheless, a reasonable arrow-pushing mechanism shows that both hydrogens are added at the same time [**Figure 6.46(a)**]. [*You don't need to know this mechanism now.*] Hydrogenation of ethyl 3-oxobutanoate results in a racemic mixture containing (*R*)-ethyl 3-hydroxybutanoate and (*S*)-ethyl 3-hydroxybutanoate [**Figure 6.46(b)**]. This reaction ends up with a racemic mixture of alcohols, but why?

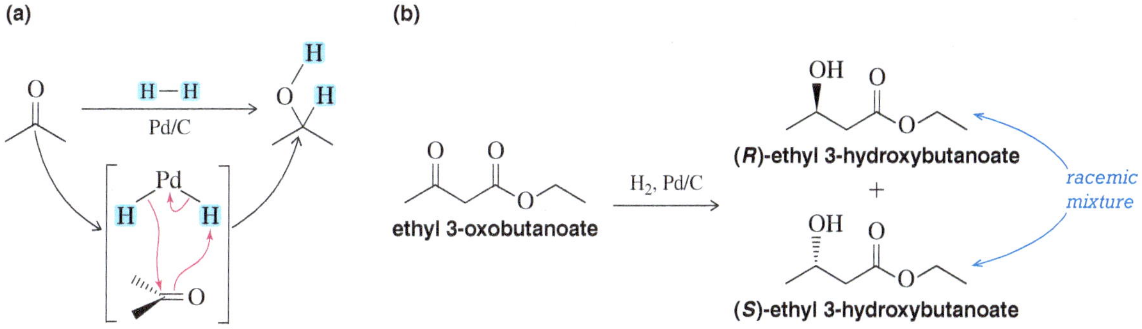

Figure 6.46 Hydrogenation of ketones. (a) Hydrogens add at the same time to give (b) a racemic mixture when a new chiral center is produced.

This reaction is under kinetic control, so the product distribution between (*R*) and (*S*) is going to be based on the relative rate of formation of each. As we explained in Chapter 5, the rate of formation of a product is indirectly proportional to the activation energy of the reactions leading to each. The lower activation energy gives the faster reaction, whereas a higher activation energy results in a slower reaction. This relationship is summarized by Equation 5.5,

$$k_r = Ae^{-E_a/RT} \qquad \text{Equation 5.5}$$

where A = a constant (frequency factor), E_a = activation energy, R = gas constant (1.987 cal/mol·K), and T = absolute temperature. The activation energy is determined by the structure of the transition state.

To predict the product ratio between (*R*)- and (*S*)-ethyl 3-hydroxybutanoate, we must compare the transition states leading to each product. Shown in **Figure 6.47**, the transition state to make (*R*)-ethyl 3-hydroxybutanoate is the nonsuperimposable mirror image (the enantiomer) of the transition state to make (*S*)-ethyl 3-hydroxybutanoate. We can refer to these as *R* and *S* transition states, respectively. Like enantiomers, these two enantiomeric transition states have identical properties in an achiral environment. Importantly, the energies of the two transition states are also the same. As such, the activation energies for making *R* and *S* are the same $[E_a(R) = E_a(S)]$, so the rates $[k_r(R) = k_r(S)]$ for making them are also the same. This is indicated by the reaction coordinate diagram in Figure 6.47.

In some cases, you might find the situation in Figure 6.47 described incorrectly as follows: *Hydrogenation of a ketone produces a racemic mixture because addition to the top and bottom of the ketone is equivalent.* The transition states are not equal but are instead enantiomeric, which results in identical activation energies leading to a racemic mixture of products.

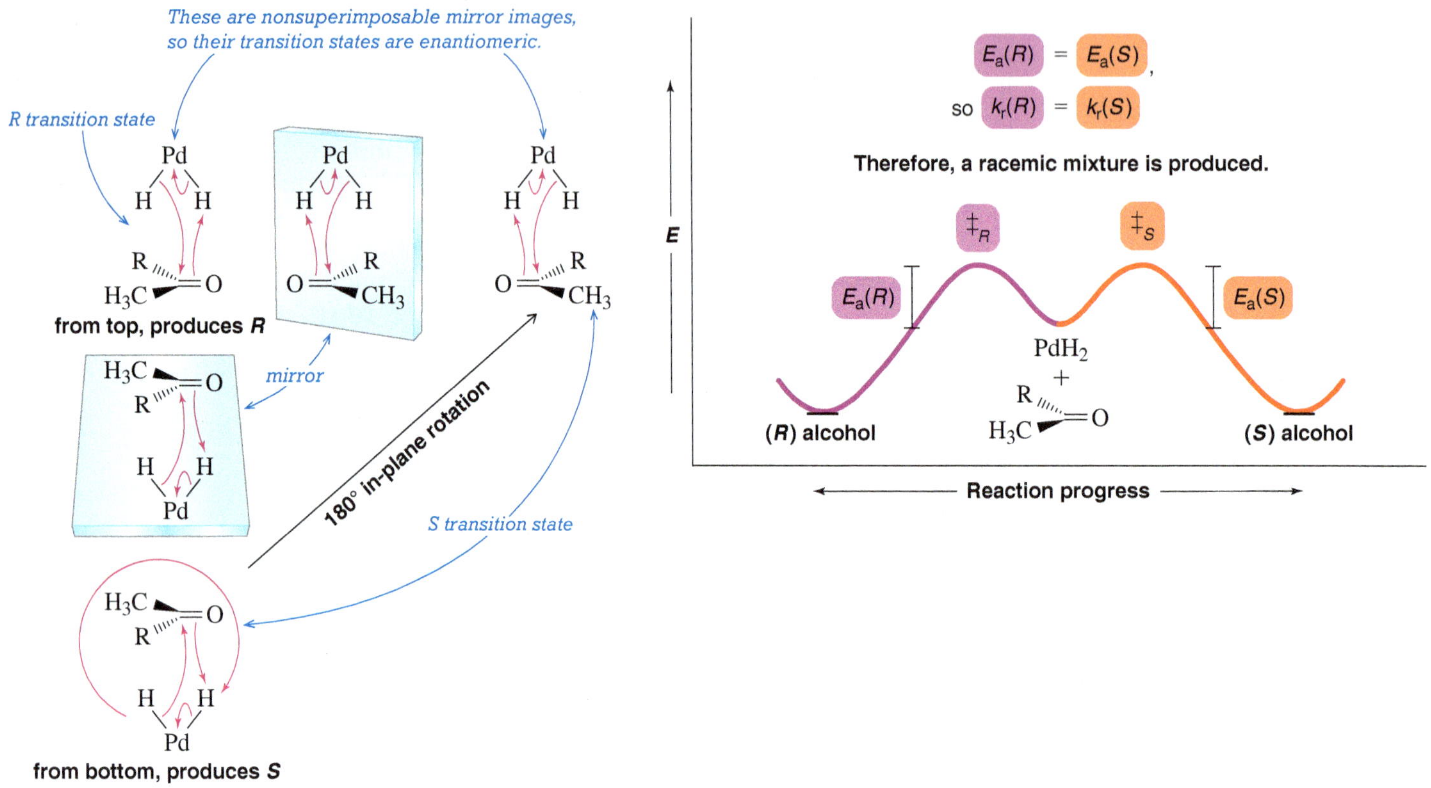

Figure 6.47 **The transition states are enantiomers and therefore have equal activation energies and equal rates.**

6.6 SUMMARY

- Because most reactions are conducted in an achiral environment where approach from the front and the back are enantiomeric transition states with equal activation energies, the majority of reactions we study result in the production of racemic products.

Synthesizing molecules as a single enantiomer can be difficult. This was true for thalidomide, and as a result, commercial thalidomide was sold as a racemic mixture. Modern technology has given us a way to synthesize thalidomide as only the more effective and nonteratogenic isomer. Unfortunately, even if this technology had been available in the 1950s, the tragedy would not have been avoided because (*R*)-thalidomide is converted to (*S*)-thalidomide under physiological conditions by a mechanism we study in Chapter 20.

6.6.1 Special topic: resolution of enantiomers by chiral complexes

If we don't have a way to synthesize compounds in enantiomerically pure form, to avoid tragedies similar to the thalidomide story, we need a way of separating, or **resolving**, the enantiomers.

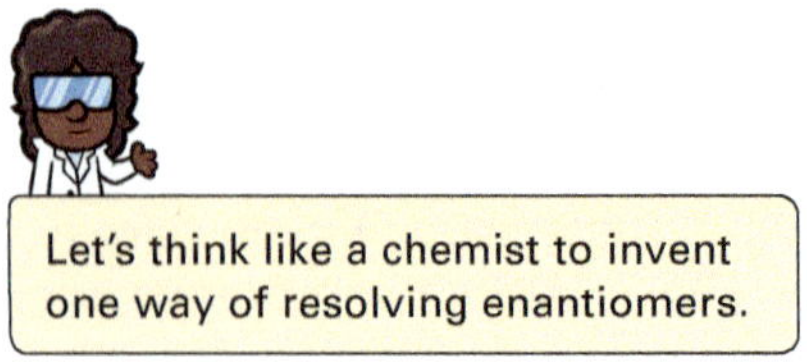

Thinking Like a Chemist

When confronted with a problem like this, we need to take note of what we have learned thus far in crafting a solution to the problem:

- To separate a mixture of two compounds, we need them to have different properties, such as boiling points (distillation), polarity (column or gas chromatography), or tendency to crystallize (recrystallization).
- Enantiomers are identical with regard to all chemical and physical properties, *except when in a chiral environment.*

The key to solving this problem is in the italicized part of the previous sentence. In a chiral environment, two enantiomers should behave differently. In fact, it was mentioned earlier in the context of the thalidomide story that the interaction of the two thalidomide enantiomers with the chiral, nonracemic receptors in the body resulted in diastereomeric complexes. In that context, these diastereomeric complexes resulted in different levels of biological response. Using this logic, what type of molecules would we want to convert enantiomers into in order that they have different properties? [*Think about this before reading further.*]

If we can convert two enantiomers (M_R and M_S) into diastereomeric complexes, then we should be able to separate them. As completely different molecules, diastereomers can be separated by recrystallization, distillation, or column chromatography. If complex formation were reversible, then we could release the enantiomers from the complexes, leaving them in pure single enantiomer form. The general strategy is shown in **Figure 6.48**. ■

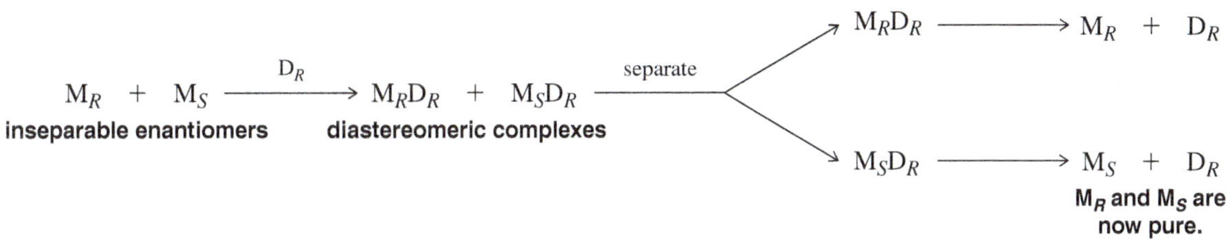

Figure 6.48 Separation of enantiomers by forming diastereomeric complexes. D_R is the stereochemically pure molecule used to make the diastereomeric complexes.

In practice, the diastereomeric complex can be formed via ionic or covalent bonds. A commonly used way to separate enantiomeric amines is to utilize a chiral, nonracemic carboxylic acid as a **resolving agent** in an acid–base reaction (**Figure 6.49**). Following transfer of the proton from the acidic carboxyl group to the basic amine, the oppositely charged species form ionic bonds. These ionically bonded complexes are now diastereomers and are able to be separated by recrystallization. Following their separation, reaction with sodium hydroxide results in release of the amine from the ionic complex. By forming the diastereomeric complexes, we are able to separate, or resolve, the enantiomeric amines.

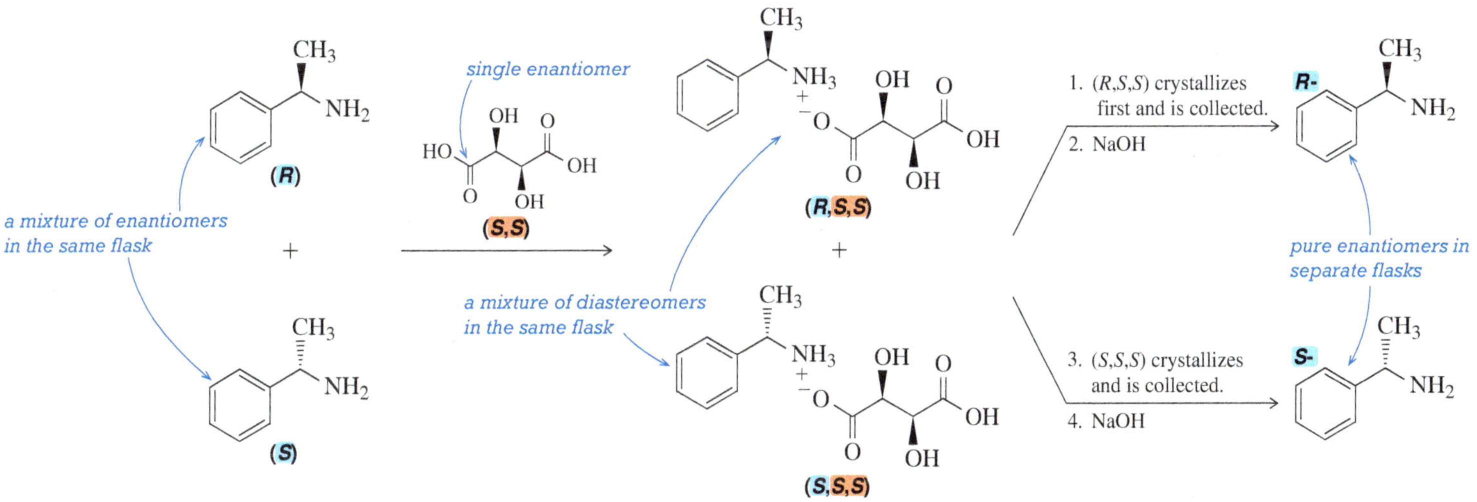

Figure 6.49 Separation of enantiomeric amines. Diastereomeric salts are formed, which can be separately recrystallized. The pure salts can then be converted to the enantiopure amines by treating separately with NaOH.

6-A CHEMISTRY IN CONTEXT: Biochemistry

Enzymes resolve enantiomers

Enzymes are chiral, nonracemic biological macromolecules. As nature's catalysts, they increase the rate of biological reactions by lowering the activation energy. [*See Chapter 5.*] Enzymatic catalysis occurs by the reaction of an enzyme and a substrate [*a biochemistry word for reactant*] to give an enzyme-substrate complex (E-S) (**Figure 6.50**). While complexed to the enzyme, the substrate is converted to the product, resulting in an enzyme-product complex (E-P). The enzyme then releases the product so the process can start again.

As chiral, nonracemic molecules, enzymes can preferentially bind to one of a pair of enantiomers. A chiral binding pocket that can efficiently bind one enantiomer is unable to

(*continued*)

accommodate the other enantiomer. Because the enzyme is a single enantiomer, the two complexes in Figure 6.51 are actually diastereomeric complexes with different binding affinities.

As a result of a chiral binding pocket, enzymes often react more quickly with one of a pair of enantiomers. Organic chemists have exploited this feature, utilizing enzymes to kinetically resolve a pair of enantiomers. The concept of **kinetic resolution**—that is, separating (resolving) two molecules based on their rate of reactivity (kinetic)—requires two things to work. First, the reaction must be slow in the absence of the enzyme. If the reaction happens rapidly in the absence of the enzyme (the chiral environment), both enantiomers should react equally fast. Second, one enantiomer must react more quickly than the other in the active site of the enzyme. For example, the racemic secondary alcohol in Figure 6.52 can be converted to an ester, with the (*R*) enantiomer reacting more quickly in the lipase-catalyzed reaction. When the reaction is complete, the mixture containing the (*S*) alcohol and the (*R*) ester can be easily separated to give two enantiomerically pure products.

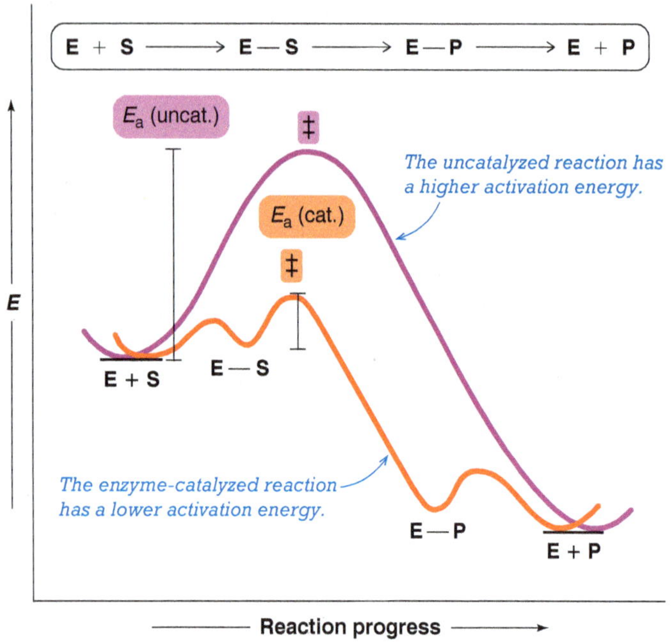

Figure 6.50 Enzymes catalyze reactions by lowering the activation energy.

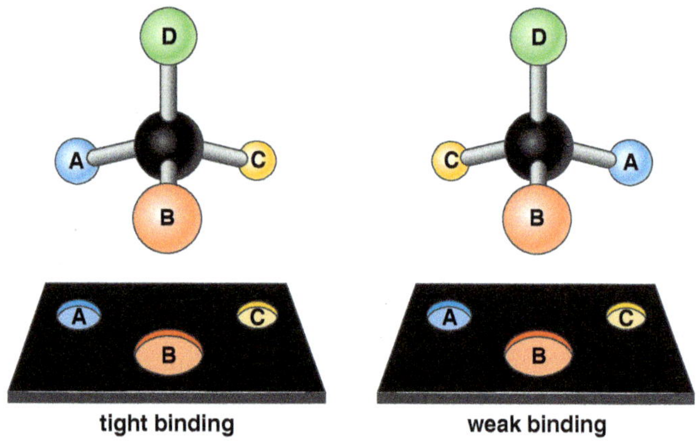

Figure 6.51 Enzymes have a chiral binding pocket that is selective for one enantiomer.

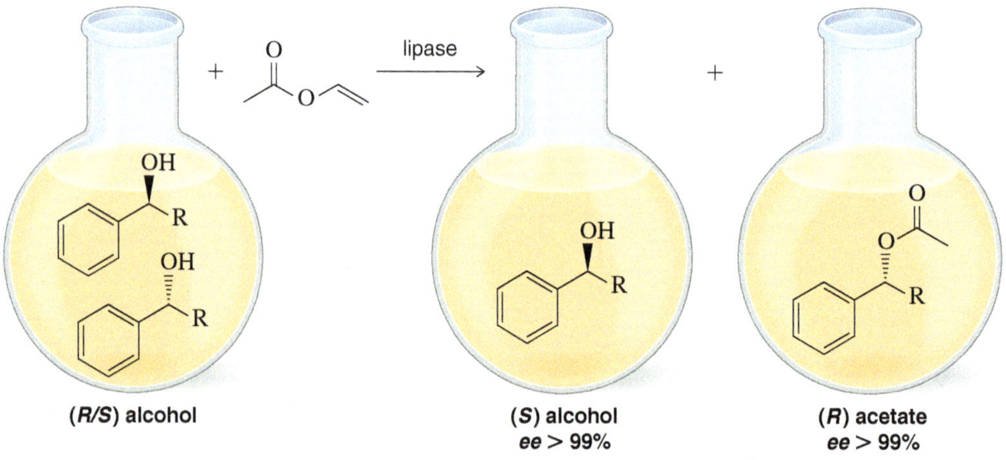

Figure 6.52 Lipase-catalyzed kinetic resolution of secondary alcohols.

Assessment 6.43 How might you separate enantiomers of 2-phenylpropionic acid?

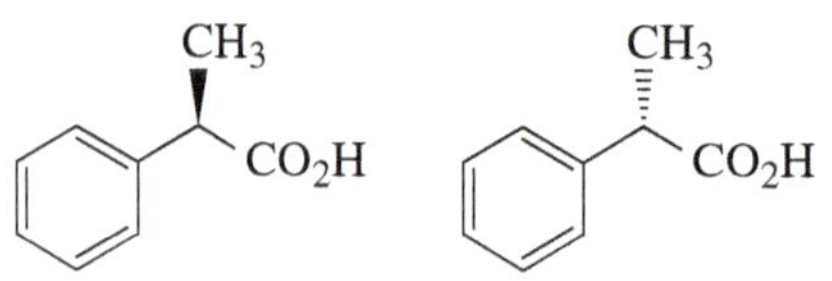

Assessment 6.44 Figure 6.52 shows the lipase-catalyzed kinetic resolution of secondary alcohols. Show a reaction coordinate diagram that rationalizes the results obtained.

6.6.2 Special topic: enantioselective synthesis

In Section 6.6.1, we discussed two ways that two enantiomers can be separated. Inherent in each of these methods is a problem. When syntheses are designed, it is hoped that 100% of the reactants is effectively converted to product. A reaction that gives a higher yield is a "better" reaction, all other things being equal. We often want to synthesize a single enantiomer of a drug, especially if one enantiomer is toxic. If we do a racemic synthesis and then separate the enantiomers, the best yield we can get is 50%. The other 50% of material is waste, which is costly for the company making the molecules and has potentially deleterious environmental impacts. Green chemistry, which we discuss in detail in Chapter 7, seeks to avoid waste in the development of chemical processes. Being able to synthesize single enantiomers in greater than 50% yield is a goal consistent with this movement.

You were challenged to think about how to do an enantioselective synthesis in Section 6.6. [*Take a second to go back and read why most reactions result in racemic mixtures.*] In summary, reactions that result in a racemic mixture do so because *the transition states leading to each enantiomer are themselves enantiomeric.* If enantiomeric transition states give racemic products, then diastereomeric transition states should give products enriched in one enantiomer. This is consistent with creating diastereomeric complexes to separate enantiomers (Section 6.6.1) and a chiral enzyme catalyzing the reaction of one enantiomer to the exclusion of the other (see Chemistry in Context 6-A).

To generate diastereomeric transition states, a reaction is run in the presence of a chiral molecule that is close enough to the reacting centers to influence the path of the reaction. We can look again at the hydrogenation of ethyl 3-oxobutanoate from Figure 6.46 when we first explored enantiomeric transition states. The specific reaction being studied is unimportant; only the general principle matters.

Asymmetric hydrogenation of a ketone makes use of a chiral ruthenium-phosphine complex (Ru-BINAP) that is synthesized by reacting (*R*)-(+)-2,2′-bis(diphenylphosphino)-1,1′-binaphthyl with ruthenium acetate [**Figure 6.53(a)**]. When this complex is stirred with a ketone such as ethyl 3-oxobutanoate under an atmosphere of hydrogen, the asymmetric hydrogenation occurs, giving very high enantioselectivity [**Figure 6.53(b)**].

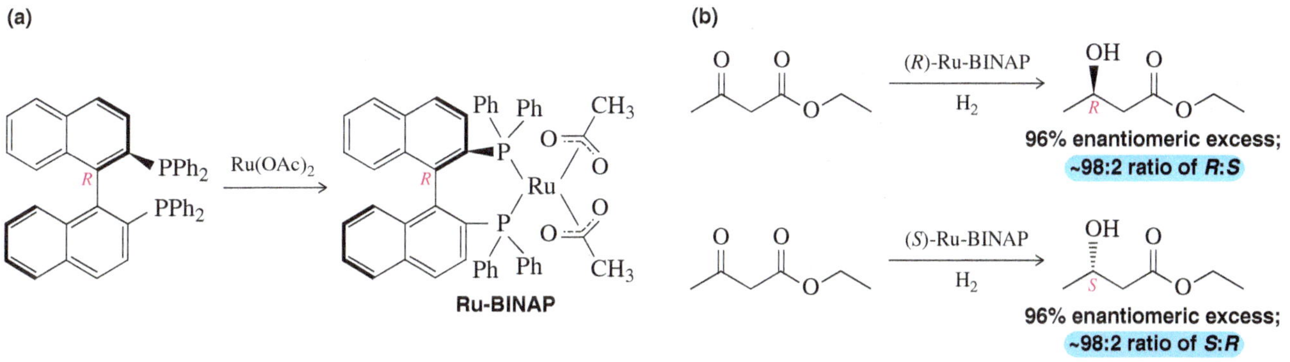

Figure 6.53 Asymmetric hydrogenation (a) with a chiral catalyst (b) gives high enantioselectivity.

Figure 6.54 compares the possible transition states for the reaction in the presence or the absence of the chiral catalyst. In the presence of the chiral catalyst, what were enantiomeric transition states (and therefore equal in energy) are now diastereomeric transition states. Just like diastereomers are different compounds, diastereomeric transition states have different energies. These different activation energies result in one enantiomer being formed more quickly than the other. In this way, enantioselective synthesis is achieved. Once more, we see enantiomers (the transition state without the chiral catalyst) behaving differently in a chiral environment (the transition state with the chiral catalyst).

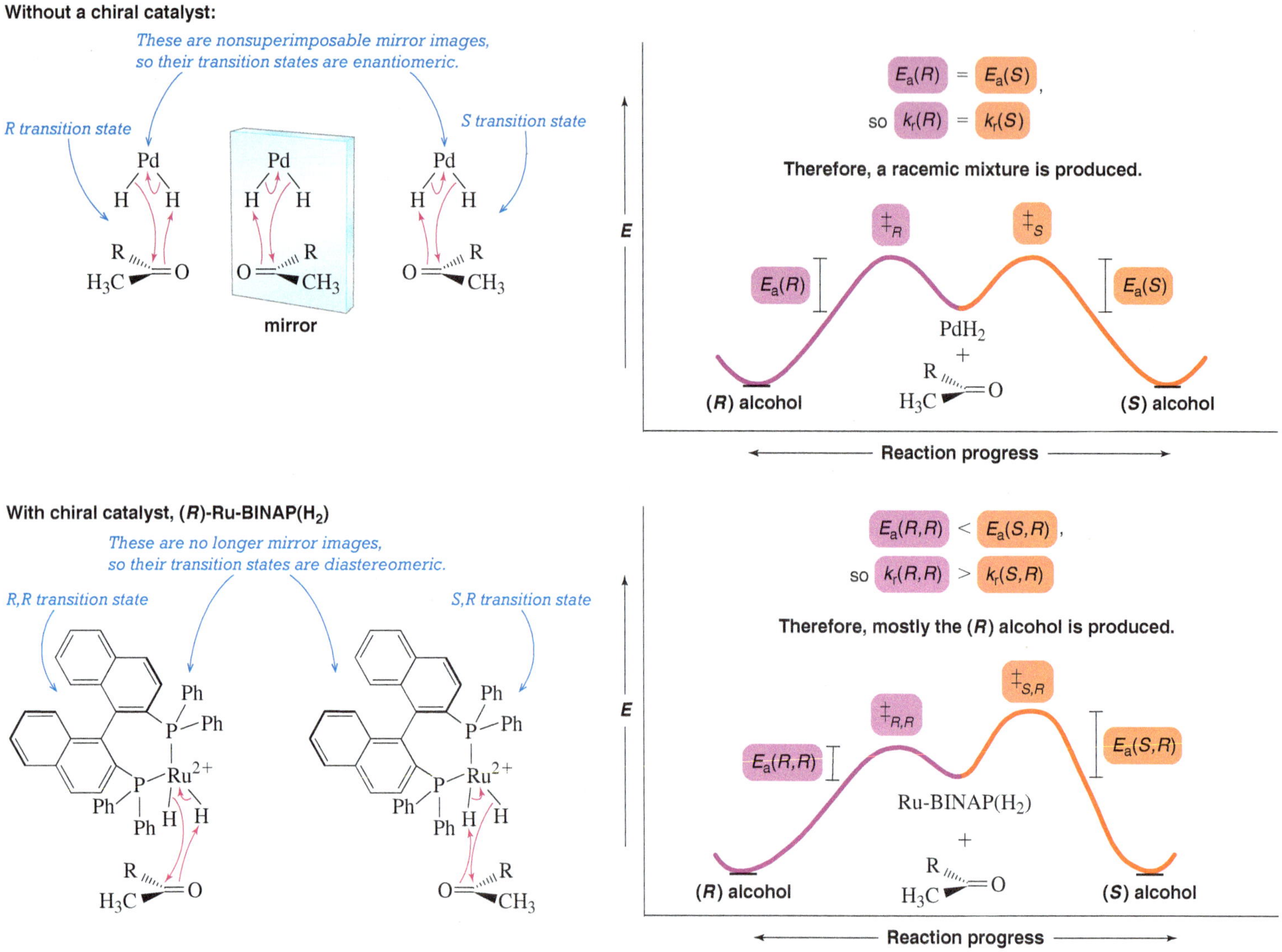

Figure 6.54 Using a chiral catalyst creates diastereomeric transition states that have different activation energies and therefore different reaction rates. The diastereomeric transition states allow for the synthesis of predominantly a single enantiomer.

Assessment 6.45 LOOKING AHEAD We discuss the reaction of Grignard reagents (organomagnesium compounds) to ketones in Chapter 17.

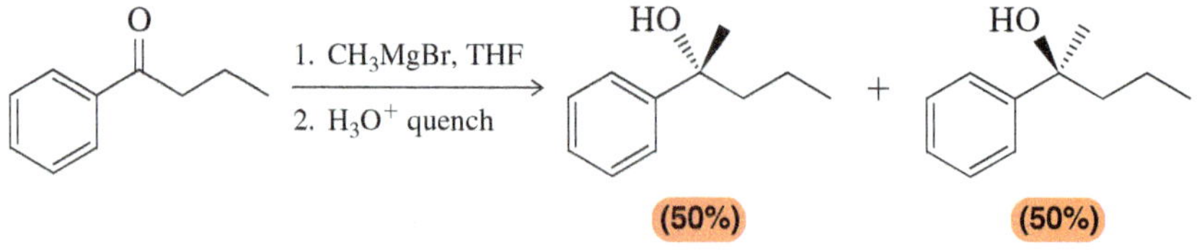

Mechanistically, the reaction proceeds by the nucleophilic addition of a methyl carbanion to the electrophilic carbon of the carbonyl, breaking the C—O π bond, resulting in an alkoxide intermediate that is subsequently protonated to produce the 3° alcohol.

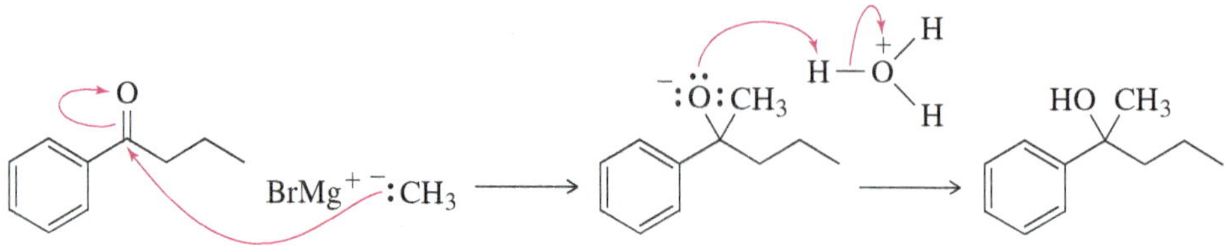

(a) Why does this reaction produce a racemic mixture of 3° alcohols?
(b) Why, when the substrate is modified slightly, does the reaction result in an excess of one stereoisomer?

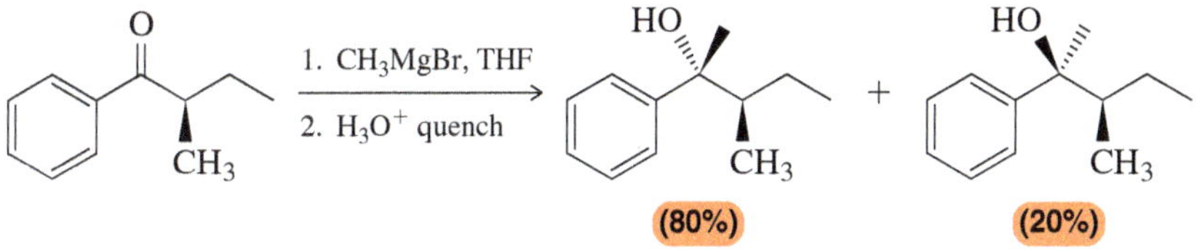

6.7 Chiral compounds without asymmetric atoms

The compound (*R*)-RuBINAP is chiral but without chiral carbons. The chirality in a molecule like BINAP has to do with **atropisomerism**, which results when rotation around a bond is slow enough that individual enantiomeric conformations can be isolated. While there is usually free rotation around single bonds, this is not true in the case of the single bond between the two naphthyl rings in BINAP. If you made models *[Do this!]* and tried to rotate around the single bond, you would encounter a significant amount of steric strain as you get to the eclipsed conformation (**Figure 6.55**). It is this eclipsed conformation and the steric interactions within that prevent rotation. As a result, the (*R*)- and (*S*)-BINAP (atropisomers) are actually just conformational isomers of each other that happen to be enantiomers.

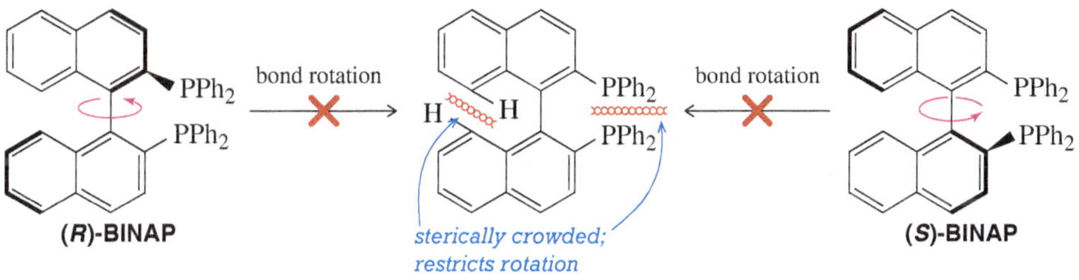

Figure 6.55 Rotation around the central aryl–aryl bond of (*R*)-BINAP, if possible, would produce the enantiomer (*S*)-BINAP.

Figure 6.56 shows a view of the two molecules down the axis of the bond that connects the naphthyl rings. Viewed this way, it is clearer that these conformations are enantiomers.

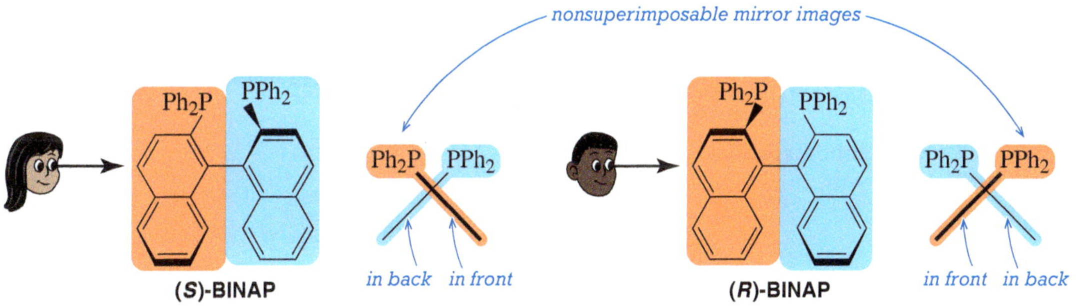

Figure 6.56 Viewed down the aryl–aryl bond, (*R*)- and (*S*)-BINAP are more clearly nonsuperimposable mirror images.

A similar effect happens in allenes, compounds containing two double bonds meeting at an *sp*-hybridized central carbon. When allenes have at least one different substituent at each terminus of the allene, they are chiral. **Figure 6.57(a)** shows the molecular orbital picture of allene. The two π bonds coming off the central carbon are perpendicular to one another. If the π bonds cannot be in the same plane as each other, then the methyl groups at C_2 and C_4 in penta-2,3-diene must also be perpendicular to each other [**Figure 6.57(b)**]. Because there is no free rotation about the double bonds of the allene, we cannot convert between what seem like two different conformations. These apparent conformations, then, are actually two nonsuperimposable mirror images (they are enantiomers).

Figure 6.57 Appropriately substituted allenes can be chiral. (a) Molecular orbital picture of allene. (b) Two allene enantiomers.

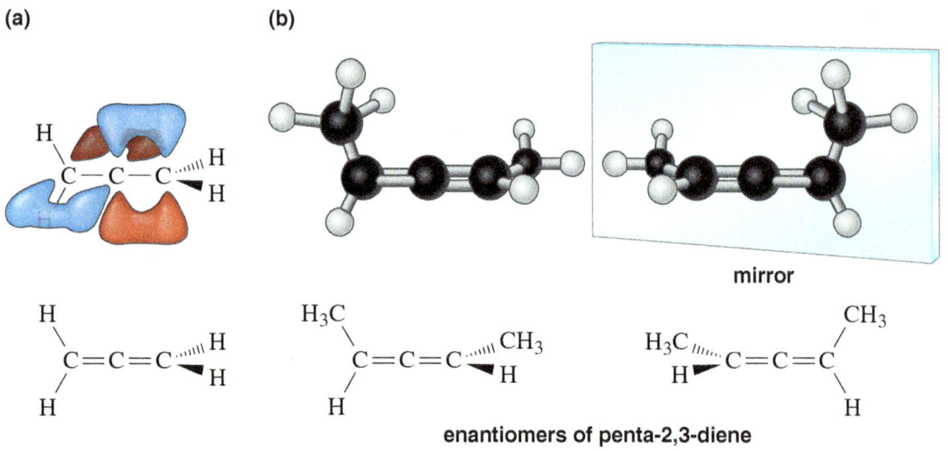

6.8 Right-hand/left-hand rule revisited

We now return to our discussion from Section 6.3.2.4 about how you can use your hands in conjunction with the Cahn–Ingold–Prelog system to assign the absolute configuration of chiral centers. This method is rooted in a conceptual understanding of the properties of enantiomers. To understand why, put your hands out in front of you as if you were pushing against a wall. Notice that your hands are mirror images of each other—nonsuperimposable mirror images, in fact. Your hands are enantiomers of each other.

Enantiomers behave differently in a chiral environment. If your hands are in fact different enantiomers, then each hand should behave differently in a chiral environment. You have probably experienced many instances where this occurs. Have you ever tried to shake somebody's right hand with your left hand? Or tried to put a left-handed baseball mitt on your right hand? Or tried to hit a right-handed golf shot with a left-handed club? All of these are very awkward situations that can be attributed to the fact that your hands are enantiomers of each other and that as enantiomers, they behave differently in a chiral environment.

How, then, does this relate to the right-hand/left-hand rule? This method is based on the fact that your right hand interacted differently with a chiral molecule than your left hand did. Think about what we did as you look at Figure 6.58. We made an effort to use our fingers to trace the direction of priorities $1 \rightarrow 2 \rightarrow 3$ while holding our thumb in the direction of the priority 4 substituent. Only one hand could do this. If it was our right hand, we called the chiral center *R*. If it was our left hand, we called it *S*. In the same way that only one enantiomer fits perfectly in the active site of the enzyme, only one of our enantiomeric hands fits perfectly with the molecule in the manner just described.

Figure 6.58 Our enantiomeric hands interact differently with chiral molecules.

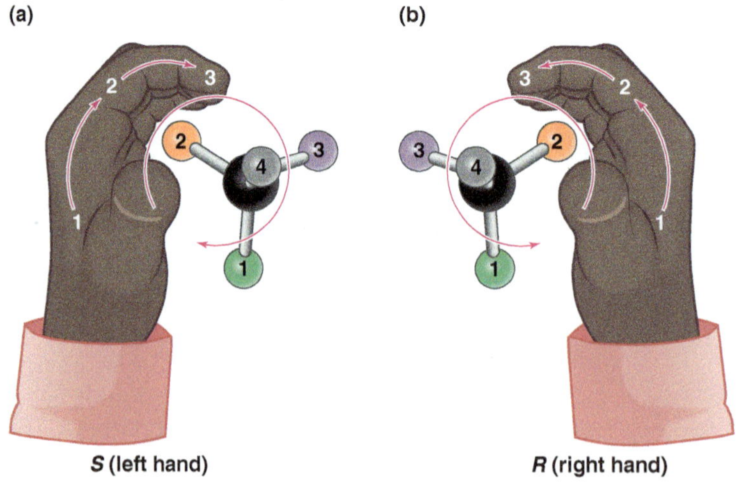

Concluding thoughts

Even if the advances in asymmetric synthesis had been available in the 1950s, it is unlikely that the thalidomide tragedy would have been avoided because thalidomide tends to racemize in the body. Out of this unspeakable tragedy has come some good, though that is probably of little consequence to the victims and their families. In the area of drug regulation, laws were passed in Europe and the United States that place significant restrictions on where, when, and how drugs can be marketed. In recent years, thalidomide [*the previously bad enantiomer, in fact*] has emerged as a frontline treatment for a variety of important illnesses, such as leprosy, multiple myeloma, and other cancers. As a result, pharmaceutical companies have worked to develop more effective and safe analogs (**Figure 6.59**). Worries persist that these drugs will have side effects similar to those of thalidomide as a result of their structural similarity. The teratogenic effects have been mitigated somewhat by placing extremely tight restrictions on distribution, ensuring that pregnant individuals or people living with pregnant individuals do not take these drugs.

Figure 6.59 Medicines derived from the thalidomide structure.

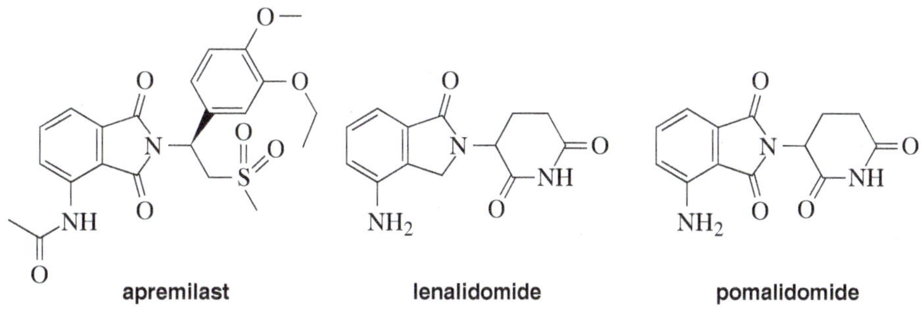

The tragic outcomes of thalidomide exposure were relatively rare in the United States due almost entirely to the diligence of pharmacologist Frances Oldham Kelsey, a medical officer with the U.S. Food and Drug Administration. Newly appointed at the time, Kelsey was charged with reviewing the application from Richardson–Merrell, Inc., in Cincinnati, Ohio, to license and sell thalidomide in the United States. While Merrell hoped to have the application fast-tracked, Kelsey stood firm in her demands that additional clinical trials be conducted to ensure effectiveness as well as safety. Kelsey's insistence helped focus attention on the unfolding tragedy in Germany and England, where the dangers of thalidomide to developing fetuses were just beginning to be understood. With the backing of the FDA, Kelsey almost single-handedly prevented a similar medical disaster in the United States, for which she was awarded the President's Award for Distinguished Federal Civilian Service by President John F. Kennedy in 1962. She was also the first recipient of the Kelsey Award in 2010, given by the FDA to one of its staff members. For a more detailed history of thalidomide, check out *Dark Remedy: The Impact of Thalidomide and Its Revival as a Vital Medicine* by Rock Brynner and Trent Stephens.

Important Concepts

[*These are the major concepts covered in each section, excluding reactions.*]

6.1: An object that possesses a plane or point of symmetry is superimposable with (or the same as) its mirror image. An object without symmetry is said to be chiral and is therefore not superimposable with its mirror image.

6.2: When drawing tetrahedral atoms, two groups are usually shown in the plane of the page, while one group is shown coming forward (bold wedge) and one group is shown going backward (hashed wedge). It is only necessary to show the bold and hashed bonds at carbons with four different groups attached (chiral centers).

6.2.1: Two compounds are stereoisomers if they differ only in the spatial arrangement of their atoms. A stereocenter is an atom containing atoms/groups wherein interchanging any two of the atoms/groups results in a new stereoisomer.

6.2.2: A carbon with four different groups attached is a special type of stereocenter and is referred to as an asymmetric center, a center of asymmetry, or a chiral center. A molecule containing a chiral center is said to be chiral and can exist in either of two enantiomeric forms. Molecules that are enantiomers are nonsuperimposable mirror images.

6.2.3: A racemic mixture consists of two enantiomers in exactly equal amounts. In chiral environments, such as the body, enantiomers can have very different interactions.

6.3–6.3.1: The cis/trans notation can be used to differentiate between two stereoisomers that differ in the relative orientation of substituents at a stereocenter. The R/S system is used to differentiate between stereoisomers that differ in absolute orientation at chiral centers.

6.3.2: The absolute configuration of a chiral center is determined to be *R* or *S* using the Cahn–Ingold–Prelog (CIP) system. This process begins by prioritizing the groups attached to the chiral center based on the atomic number of the directly attached atoms, or at the first point of difference in the chain. Once the priorities have been determined, there are several ways of assigning the configuration as *R* or *S*, including the right-hand/left-hand rule.

6.4: Enantiomers are identical with respect to all physical properties, except for the manner in which they interact with chiral environments, such as the rotation of plane-polarized light.

6.4.1: Polarimetry is an analytical technique that takes advantage of the ability of a single enantiomer to rotate plane-polarized light. By passing a single plane of light through a solution that is enriched in a single enantiomer, the plane of light will be rotated. A racemic mixture does not rotate plane-polarized light and is said to optically inactive.

6.4.2: By measuring the specific rotation of a mixture, the exact enantiomeric composition of a mixture can be quantified.

6.5: Diastereomers are stereoisomers that are not mirror images of each other. A molecule with *n* stereocenters can exist as 2^n stereoisomers.

6.5.1: Meso compounds are stereoisomers (they possess centers of asymmetry), but they are achiral because they have a plane of symmetry that makes them identical to their mirror image.

6.5.2: Diastereomers differ in all physical properties and can be separated on the basis of some of these differences.

6.5.3: Because receptors and enzymes in the body are themselves chiral and present in single enantiomeric form, two enantiomers of a drug can have markedly different biological effects.

6.6: Transition states that are themselves enantiomers have identical activation energies and therefore result in equal reaction rates. As a result, synthesizing a single enantiomer is difficult.

6.6.1: Enantiomers can be separated in a variety of ways, all taking advantage of their conversion, or temporary conversion, to diastereomeric complexes. This can be done using acid–base reactions or by doing a kinetic resolution using enzymes.

6.6.2: In order to do an enantioselective synthesis, reactions need to be run under conditions in which the transition states leading to the two enantiomers are diastereomeric. Diastereomeric transition states have different activation energies and therefore give rise to different reaction rates.

6.7: Molecules without asymmetric atoms can still be chiral as a result of conformational enantiomerism, resulting in two isomers that are nonsuperimposable mirror images.

Important Equations

1. In polarimetry, the specific rotation $[\alpha]$ is defined as the observed rotation found under specific conditions—namely, a 10-cm sample cell and a concentration of 1 g/100 mL using a defined light source (usually the sodium D line) at a defined temperature (usually 25 °C). (6.4.1)

$$\text{Specific rotation} = [\alpha]_D^T = \frac{\alpha_{obs}}{l \times c} = \frac{\text{observed rotation (degrees)}}{\text{length (dm)} \times \text{concentration}\left(\frac{\text{g}}{100\text{ mL}}\right)}$$

2. The enantiomeric excess (*ee*) of a mixture of enantiomers tells how much of one enantiomer is present in excess of the other. A racemic mixture (50% *R* and 50% *S*) has an *ee* of 0. (6.4.2)

$$\text{Enantiomeric excess}(ee) = |\%R - \%S|$$

3. The percent optical purity of a sample is measured using a polarimeter. It can be determined by dividing the measured specific rotation [α] of a mixture of enantiomers by the specific rotation of the pure enantiomer. (6.4.2)

$$\%\ \text{optical purity} = \frac{[\alpha]_{\text{sample}}}{[\alpha]_{\text{pure enatiomer}}} \times 100$$

4. The % *ee* can also be calculated by dividing the specific rotation of a mixture of enantiomers by the expected rotation of a single enantiomer. (6.4.2)

$$\frac{\text{specific rotation}}{\text{expected rotation}} \times 100 = \%\ ee$$

5. The specific amount of the enantiomer in excess can be determined by recognizing that the % *ee* of a mixture represents the portion of the solution that is not racemic (an equal mixture of *R* and *S*). The remainder of the solution is a racemic mixture and is divided by 2 to determine how much (*R*) and (*S*) there is besides the percentage that is in excess. (6.4.2)

$$\left(\frac{100 - \%\ ee}{2}\right) + \%\ ee = \%\ \text{of enantiomer in excess}$$

6. The number of possible stereoisomers can be calculated based on the number of stereocenters in a molecule, as long as one of the stereoisomers is not meso. (6.5)

$$2^n = \text{number of stereoisomers}$$

$$\text{where } n = \text{number of stereocenters}$$

End-of-Chapter Assessments

[*Assessments are ranked according to their difficulty;* •••• = *most difficult.*]

6.46 (•) Classify the following objects and molecules as chiral or achiral.

(a)

(b)

(c)

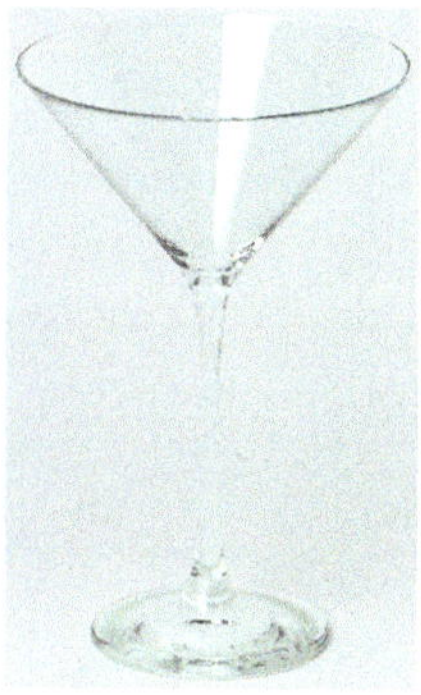

(d)

(e)
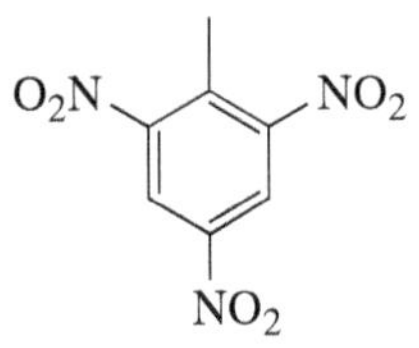

(f)
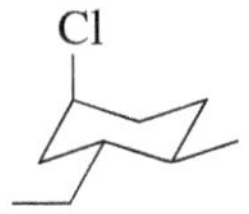

(g)
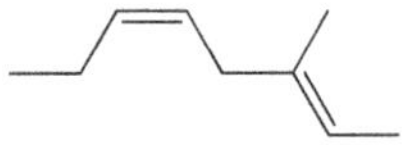

(h)
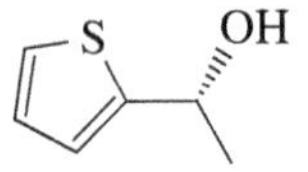

(i)
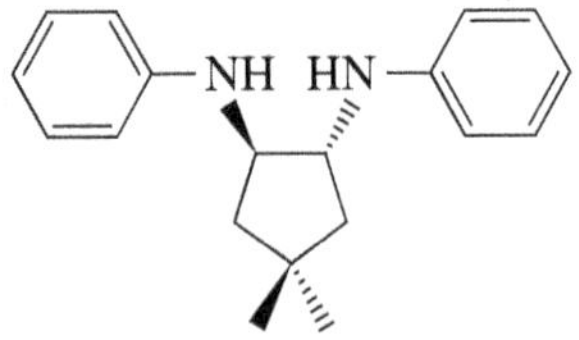

(j)
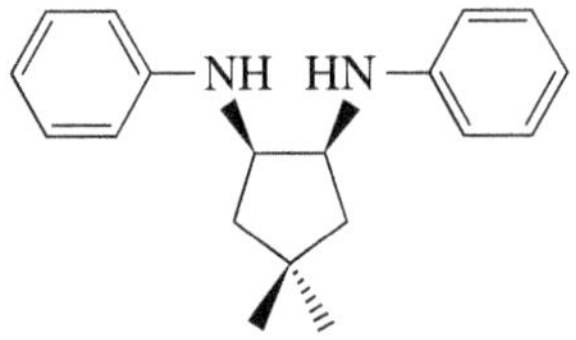

(k)
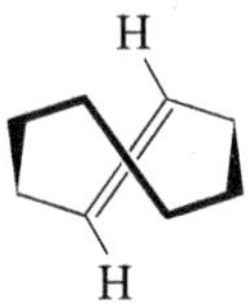

(l)
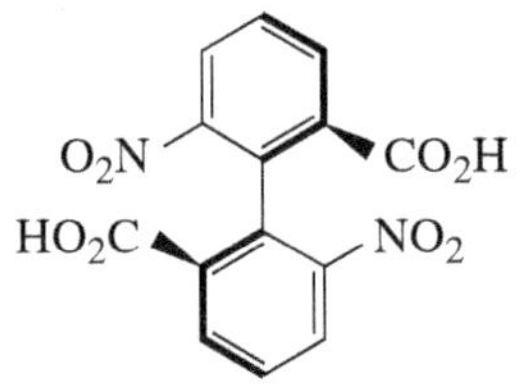

6.47 (•••) Define the relationship between each set of two molecules as chain isomers, positional isomers, functional group isomers, enantiomers, diastereomers, conformational isomers, or identical.

(a)
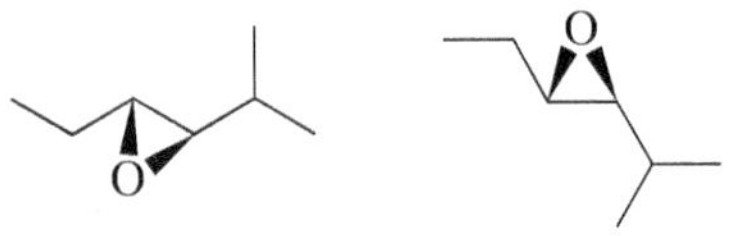

(b)
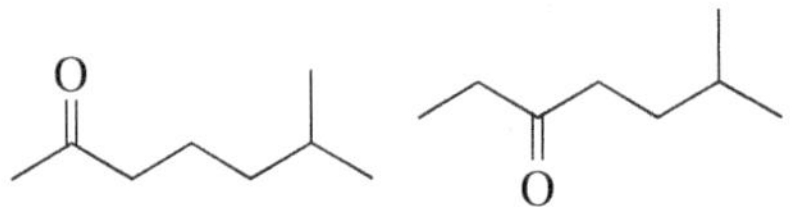

(c)
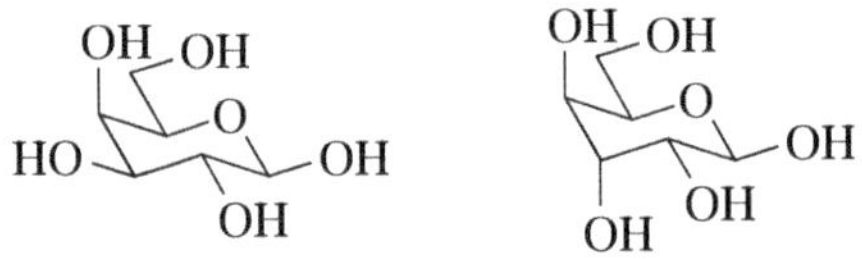

(d)
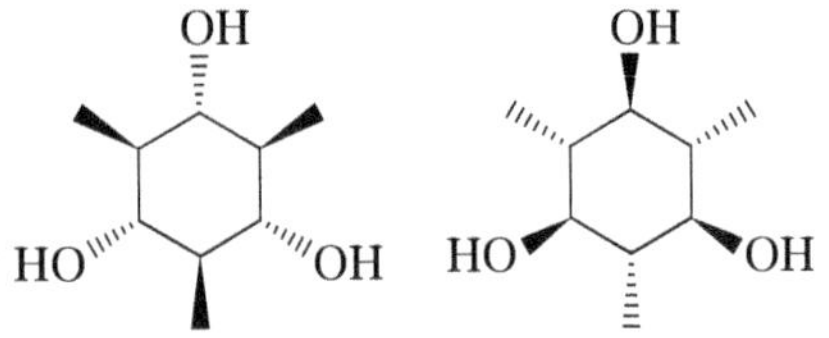

(e)
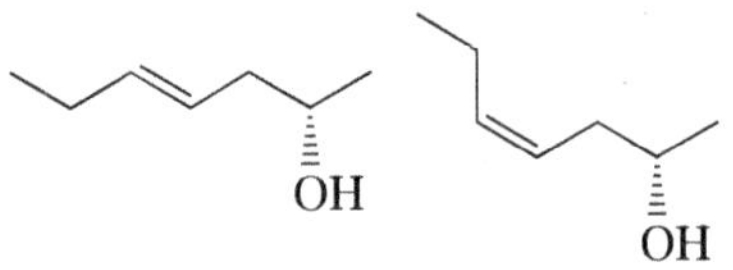

(f)
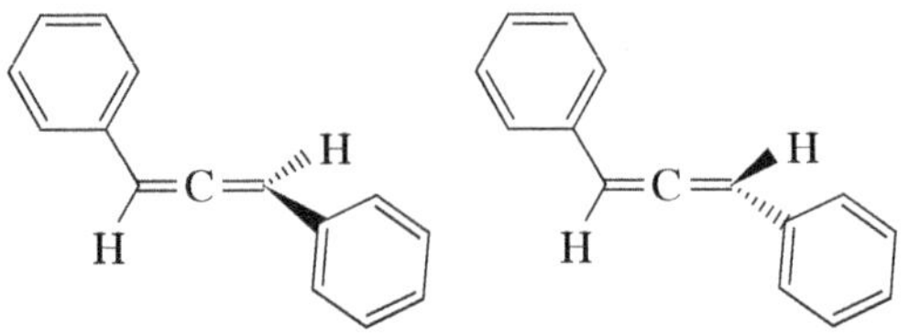

(g)
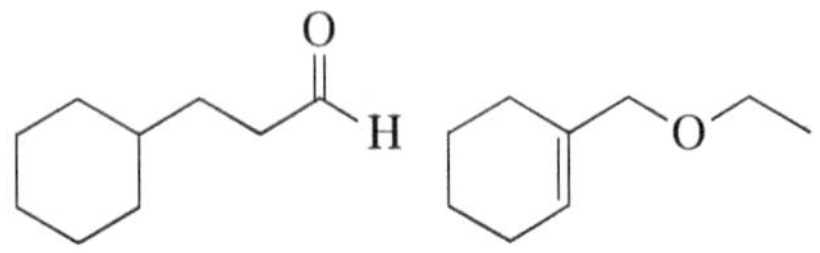

(h)
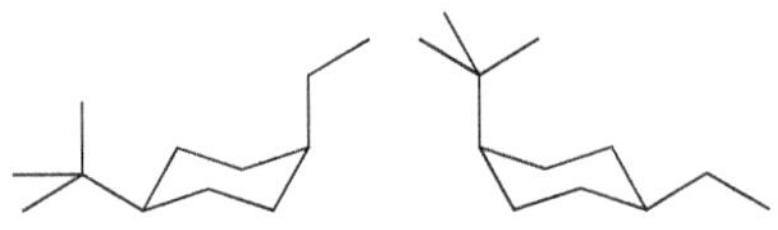

6.48 (•) Identify the following stereocenters as *R* or *S*.

(a)
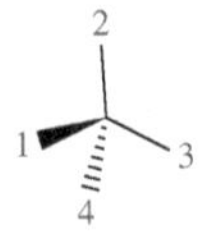

(b)
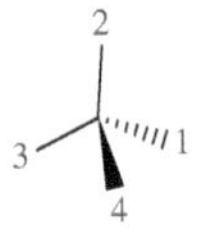

(c)
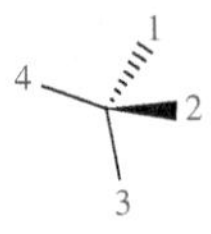

(d)

(e)
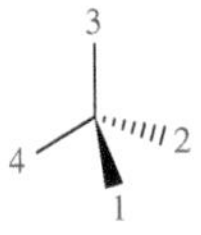

(f)
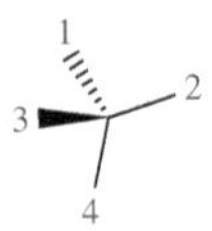

(g)
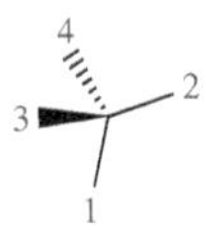

(h)
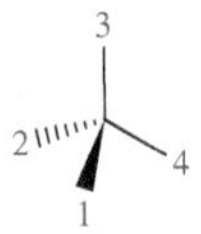

(i)
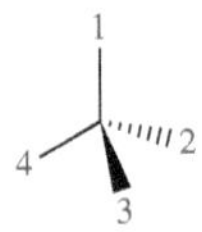

(j)
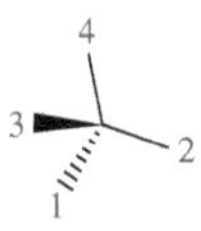

6.49 (••) For each of the following stereocenters, fill in the missing numbers to give the absolute configuration (*R* or *S*) shown.

(a)

R

(b)
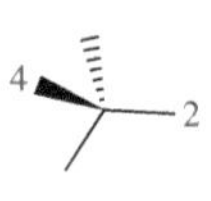
S

(c)
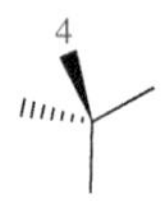
R

(d)

S

(e)

R

(f)

S

(g)
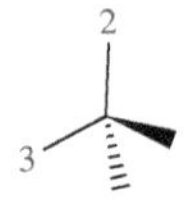
R

(h)
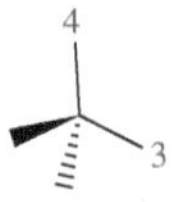
S

(i)
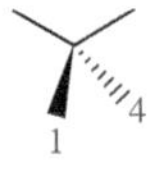
R

(j)

S

6.50 (••) Assign the absolute stereochemistry for each of the following molecules drawn in their Fischer projection.

(a)

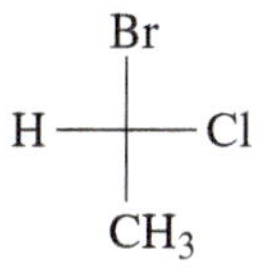

(b)

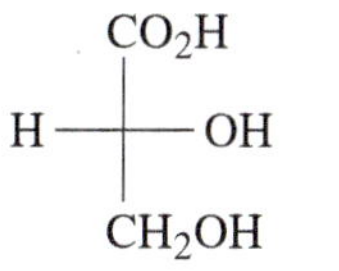

(c)

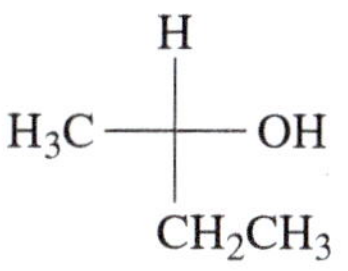

(d)

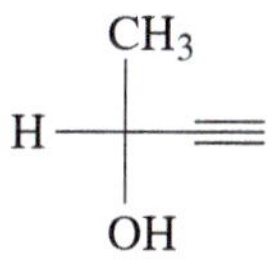

6.51 (••) Assign the absolute stereochemistry for each of the following molecules.

(a)

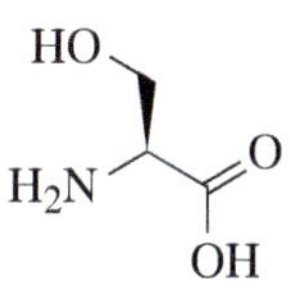

serine (amino acid)

(b)

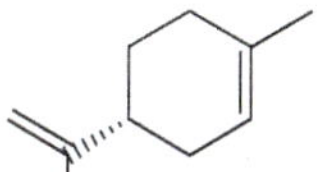

limonene

(c)

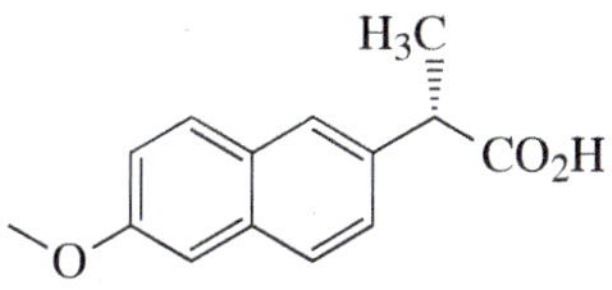

naproxen

(d)

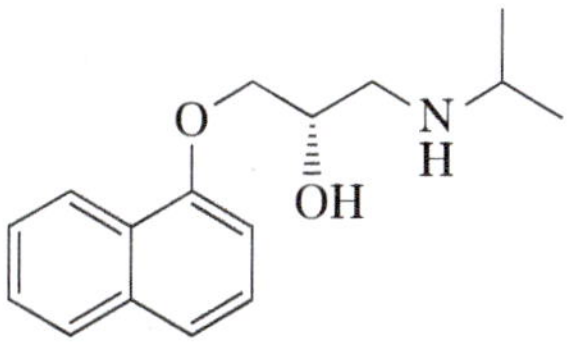

propranolol

(e)

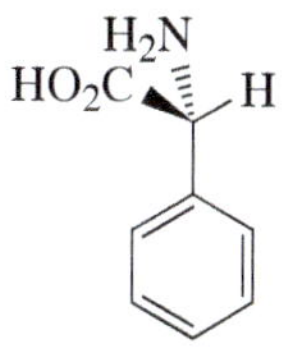

phenylglycine

(f)

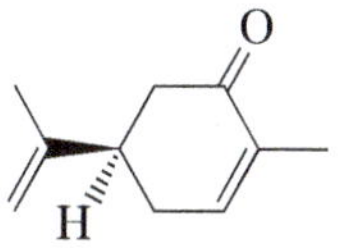

carvone

6.52 (••) Complete the structure of each of these so that it matches the (*R*) or (*S*) configuration associated with the name.

(a)

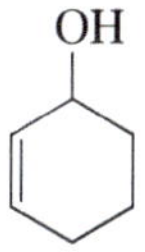

(*R*)-cyclohex-2-en-1-ol

(b)

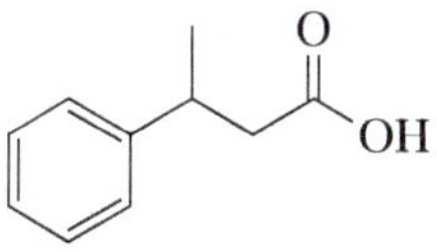

(*R*)-3-phenylbutanoic acid

(c)

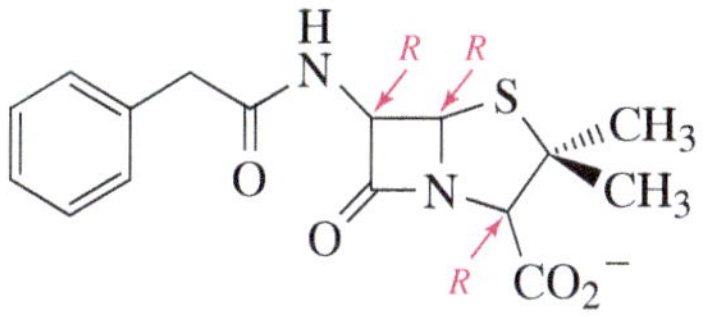

penicillin (antibiotic)

(d)

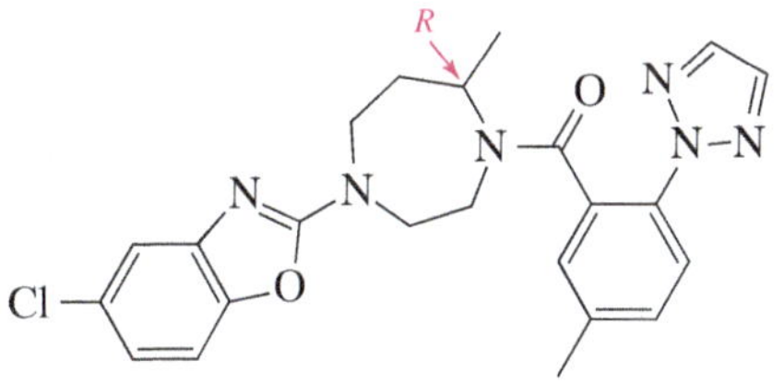

suvorexant (insomnia)

(e)

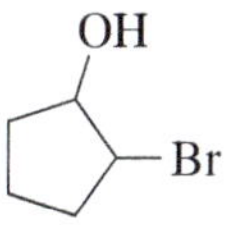

(1*R*,2*S*)-2-bromocyclopentan-1-ol

(f)

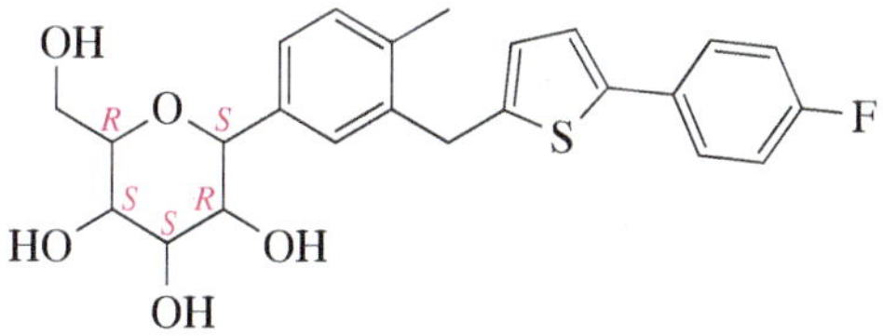

canagliflozin (type 2 diabetes)

6.53 (•••) Draw the enantiomer of each of the molecules you drew in Assessment 6.52.

6.54 (•••) Natural products are organic compounds produced by living organisms, including plants, fungi, and animals. Often referred to as secondary metabolites because they are not required for survival of the organism, natural products have found broad utility as drugs themselves or as lead compounds used in the development of medicines. One such example is paclitaxel, which has been used as a cancer drug. Paclitaxel is isolated from the yew tree, where it is produced as a single stereoisomer (shown). Based on its structure, how many stereoisomers are possible for paclitaxel?

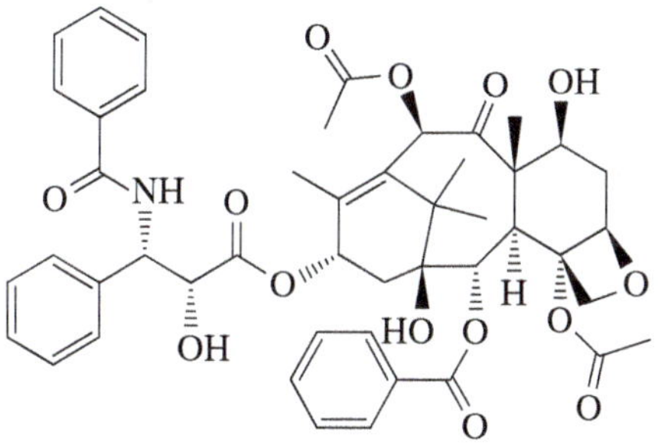

paclitaxel

6.55 (••) As we learned in Chapter 2, we don't need to show hydrogens bonded to carbons when drawing organic molecules using line-angle formulas. At asymmetric centers, however, we often show the hydrogen. Why? When might it be unnecessary to show the hydrogen at a chiral center?

6.56 (•••) LOOKING AHEAD Amino acids, so-called because they possess an amino group and a carboxylic acid separated by a single carbon containing one of 20 unique side chains, are the building blocks of our proteins and enzymes.

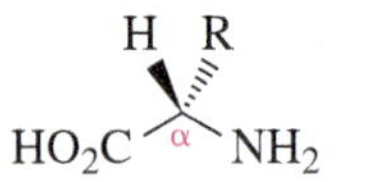

general form of L-amino acids
(R = side chain)

Because the carbon containing the side chain (the α-carbon) has four different groups attached, amino acids are chiral molecules and can therefore exist in either of two enantiomeric forms (D- and L- is an antiquated system that we discuss in Chapter 24), with neither being energetically favored over the other. And yet, virtually all of the amino acids found in nature are of the L configuration. Scientists have yet to conclusively explain why life has evolved to prefer a single enantiomeric form of the amino acids, yet it is postulated that all amino acids are derived from a common precursor produced initially in the presence of a chiral surface. Assuming that this is true and that all amino acids have the general form shown previously, how can you explain the fact that 18 amino acids have the (*S*) configuration, one is neither *R* nor *S*, and still another is *R*? (*See the list of the amino acids in Table 26.1.*)

6.57 (••••) When the following substituted biphenyl was synthesized, it was found to have a specific rotation [α] of −23° at 25 °C. When the specific rotation was measured at 100 °C, the compound had a specific rotation of 0°. Upon cooling back to 25 °C, the specific rotation was measured again, resulting in [α] = 0°. Explain these results.

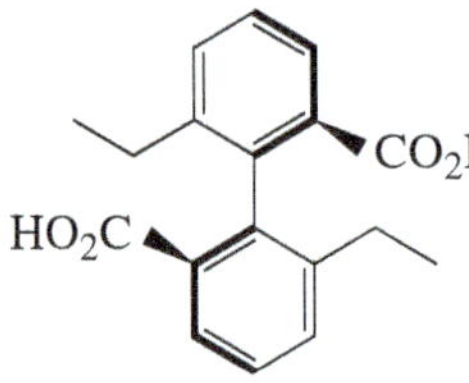

6.58 (•••) A chemist working on the synthesis of (+)-pilocarpine, an alkaloid used in the treatment of dry mouth and glaucoma, produced a mixture of enantiomers that gave a specific rotation $[\alpha]_D = +97°$. Based on the specific rotation of the pure enantiomer, calculate the ratio of (+)- to (−)-pilocarpine produced by the chemist.

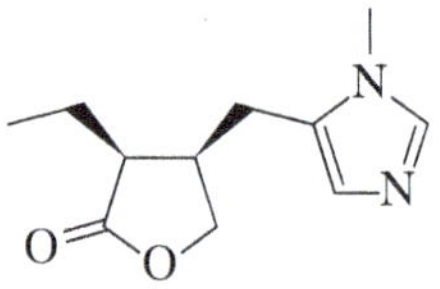

(+)-pilocarpine
$[\alpha]_D = +106°$

6.59 (••••) A compound with two chiral centers that is meso will always have opposite absolute configurations at the two chiral centers. That is, a meso compound will never be (*R*,*R*) or (*S*,*S*); instead, it will be (*R*,*S*). Explain why this must be true.

6.60 (•••) LOOKING AHEAD In Chapter 12, we introduce the S_N2 reaction, a nucleophilic substitution reaction that proceeds with inversion. Confirm that inversion has occurred in each of the following examples by determining the absolute configuration of the chiral center in the reactants and products.

(a)

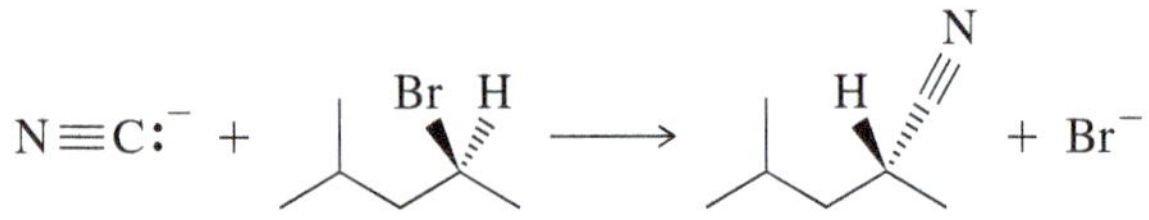

(b)

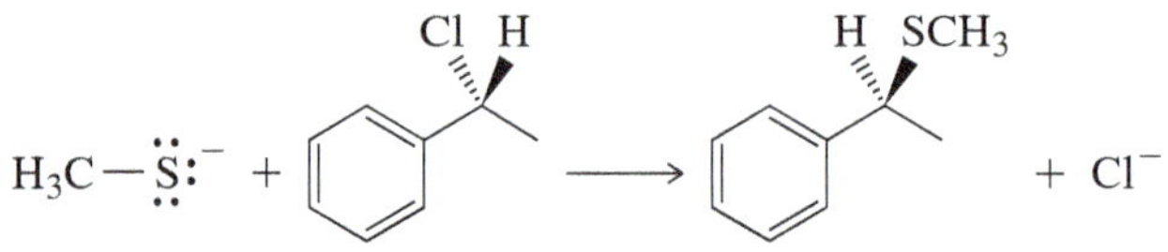

6.61 (••••) The following reaction proceeds by the same S_N2 mechanism as in Assessment 6.60, yet the CIP-assigned absolute configuration does *not* change in going from reactant to product. Verify that the absolute configuration does not change, and then explain why the reaction proceeds with inversion in spite of this observation.

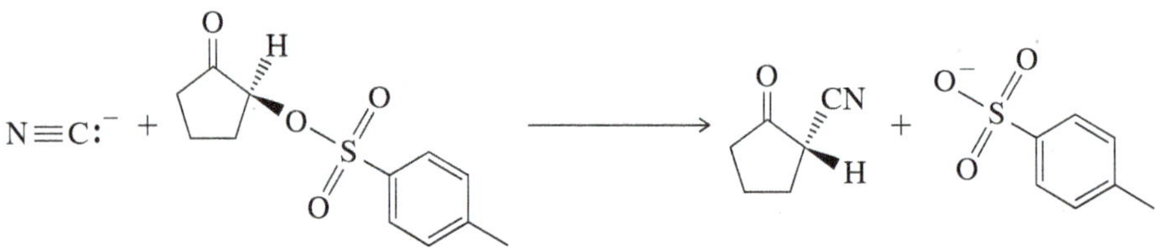

6.62 (••••) LOOKING AHEAD In Chapter 13, we explain how to convert secondary alcohols into ketones using a mild oxidation reaction. When the following enantiomerically pure and optically active secondary alcohol is submitted to these reaction conditions, the product is optically inactive. Explain this observation.

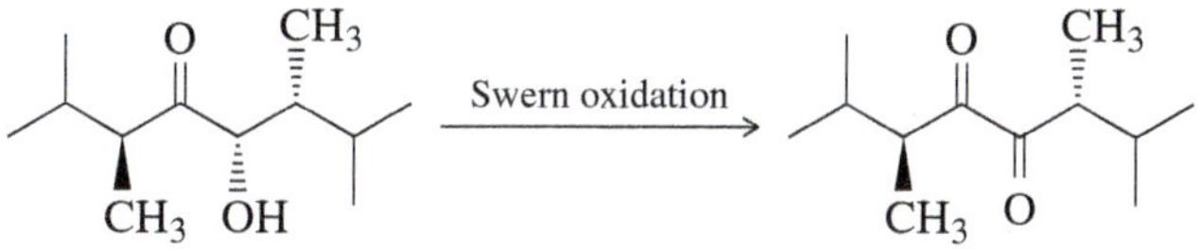

single enantiomer, optically pure [α] = 0°

6.63 (•••) A chemist prepared a racemic mixture of the enantiomeric sulfonic acids shown here. Suggest two ways that these might be separated.

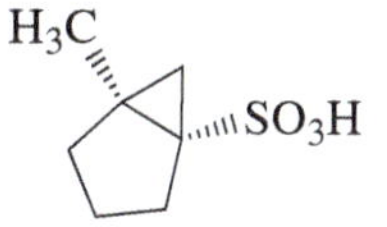

6.64 (•••) In contrast to Assessment 6.63, the following compounds should be easily separable using standard methods. Why?

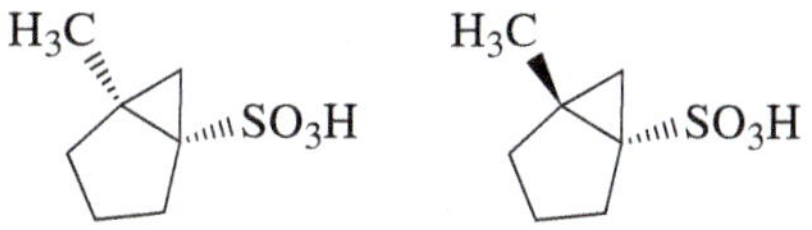

6.65 (••••) Chromatography using a chiral stationary phase is able to separate a pair of enantiomers because one of the enantiomers forms a more tightly bound complex with the stationary phase. Assuming that the (*S*) enantiomer elutes more slowly, rationalize this result on a reaction coordinate diagram.

$E_R + E_S + (-)\text{-}S^* \longrightarrow (-)\text{-}S^*\cdot E_R + (-)\text{-}S^*\cdot E_S$

enantiomers — **chiral stationary phase** — **diastereomeric complexes**

6.66 (••••) LOOKING AHEAD The following reaction, which we introduce in Chapter 20, is referred to as a conjugate addition reaction and results in the formation of a new carbon–carbon bond. When the reaction is performed using methyl 3-phenylprop-2-enoate, a racemic mixture is produced.

(a) Without worrying about the mechanism of the reaction, explain why the reaction produces a racemic mixture.

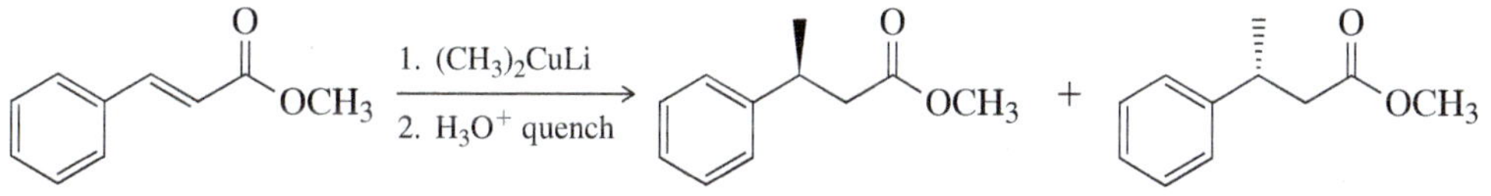

methyl 3-phenylprop-2-enoate **(50%)** **(50%)**

One method of synthesizing the product above in enantiopure form involves the use of a chiral auxiliary, as shown in the following scheme featuring reactions we discuss in later chapters.

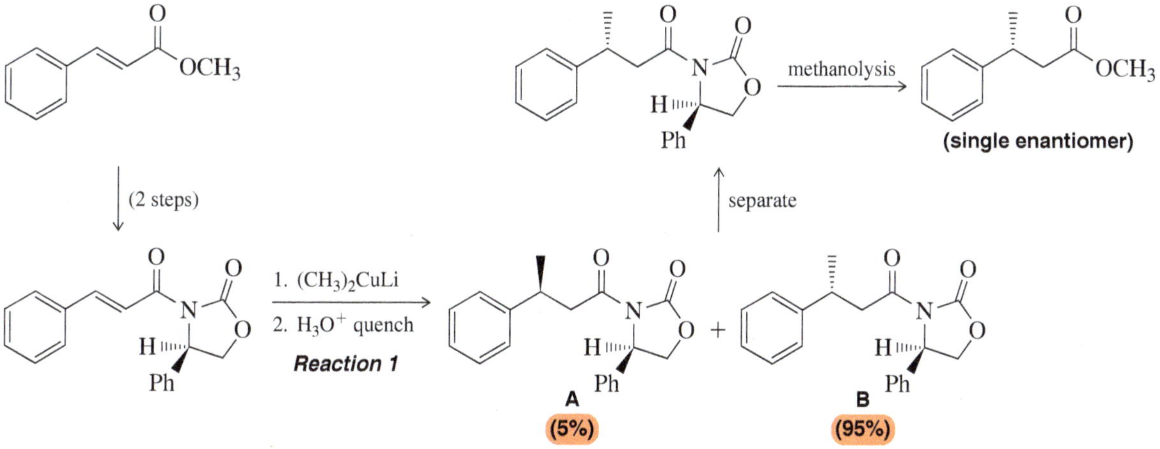

(b) Using a reaction coordinate diagram, suggest why reaction 1 is selective.
(c) **A** and **B** are separable using normal (nonchiral) column chromatography. Why?
(d) Given the 5:95 ratio of **A**:**B**, what is the diastereomeric excess? [*Diastereomeric excess is calculated in the same way as enantiomeric excess.*]

7

Principles of Green (Organic) Chemistry

Learning Objectives

LO 7.0 Apply the 12 principles of green reactions

LO 7.1 Assess a reaction based on the principle of waste reduction

LO 7.2 Assess the atom economy of a reaction

LO 7.3 Compare the use of safe solvents and auxiliaries to less green solvents and auxiliaries

LO 7.4 Examine ways to make a process more energy efficient

LO 7.5 Evaluate the use of renewable and bio-based feedstocks

LO 7.6 Correlate the greenness of a reaction to the use of catalysts

Introduction

In the 1970s, seeking to encourage industrial growth, the Indian government invited Union Carbide Corporation (UCC), an American company, to build a plant in Bhopal, India, where it would produce the commercial pesticide carbaryl (Sevin®). Because the plant was originally intended for the manufacture of small amounts of carbaryl from imported starting materials, it was allowed to be built within the city, in an area zoned for light manufacturing. Despite the initial use of imported methyl isocyanate and 1-naphthol, an evolving business model required synthesis of bulk quantities of starting materials and the final product at the same plant. This process was regarded as being more hazardous, especially since the plant wasn't designed for it.

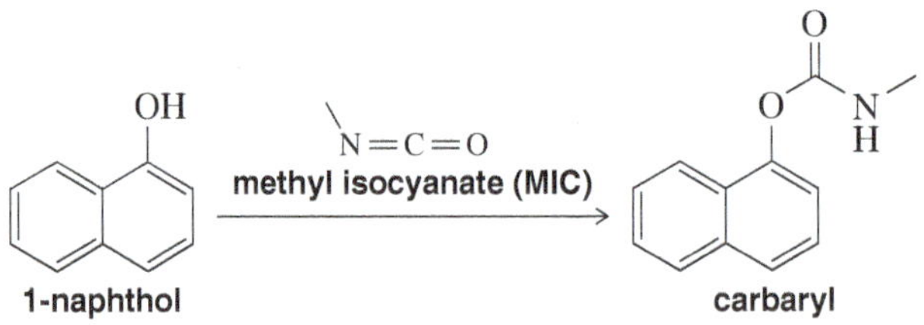

"It is our collective and individual responsibility to protect and nurture the global family, to support its weaker members and to preserve and tend to the environment in which we all live."

— Dalai Lama

With a diminished pesticide market in the early 1980s, UCC began dismantling the Bhopal plant, preparing it for sale, while production of carbaryl continued. Unfortunately, many safety devices either had been turned off or had fallen into disrepair. In the early morning of December 4, 1984, a confluence of circumstances led to the accidental release of over 40 tons of methyl isocyanate into the surrounding community. By morning, more than 3800 Bhopal residents had perished, most of them concentrated in the neighboring poverty-stricken slums. Over the next several years, and even today, adverse effects from this tragedy have continued to accrue with an estimated 10,000 deaths within days of the accident and including up to 20,000 premature deaths being attributed to it. All told, more than 500,000 people were exposed to the gas, with increased morbidity to the affected population in what is regarded as the worst chemical disaster in history.

○ ○ ○ ○ ○ ○ ○

Generations of synthetic chemists have grown up with the goal of making a society-changing compound, be it a polymer, a medicine, or a pesticide. This worthwhile goal has spawned the development of myriad chemical methods that accomplish the task of performing whatever chemical transformation is required, with a presumed lack of regard, or lack of knowledge, of chemical hazards and toxicities. And it trickles down. Society needs a cure for HIV, so someone makes a molecule. A graduate student wants a degree, so they make a molecule. An undergraduate research student wants a recommendation letter, so she makes a molecule. When making the molecule is what is important (and for many legitimate reasons, it is very important), researchers look to the literature and their organic chemistry books for information.

Disasters like Bhopal emphasized that our short-term chemical needs should align with long-term sustainability and environmental responsibility. Thus, green chemistry was born—first as its own discipline, but hopefully as a construct within which all areas of chemistry will one day operate. Presently, a scientist who needs a molecule *right now* has little time to find, or invent, the greenest way of carrying out a reaction. Instead, they resort to the highly optimized syntheses of yore. This chapter [*and stories throughout the book*] aims to answer the question "What would be gained if principles of green chemistry were learned alongside the traditional organic chemistry reactions typically covered in the second-year chemistry curriculum?" Many of the reactions we present in this book are not green, but are shown because they form the fundamental suite of reactions all organic students need to know. But by thinking about green chemistry principles now, or seeing more sustainable reaction alternatives where possible, a generation of future scientists [*that's you*] will be free of the conflict of choosing the efficient reagent over the sustainable reagent, because hopefully they will be one and the same. Before we dive into that, let's test our understanding of material studied previously.

This is why green chemistry deserves its own chapter in an organic chemistry book.

Test your understanding

Assessment 7.1 What is the role of a catalyst in a chemical reaction?

Assessment 7.2 Chapter 5 taught us that chemical reactions are random collisions. Experimentally, how can we make molecules collide more often?

Assessment 7.3 Gasoline fuels our automobile engines via the combustion of hydrocarbons like cyclohexane. (a) Is this reaction an oxidation or a reduction? (b) Balance the combustion equation. (c) Why does this reaction produce energy?

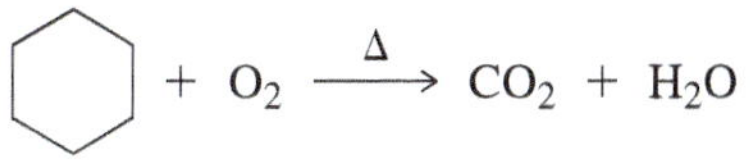

Assessment 7.4 What is the role of a solvent in a chemical reaction? What happens to the solvent at the end of a reaction?

7.1 Twelve Principles of Green Chemistry—an overview

In the years following the Bhopal disaster, governments, nongovernmental organizations (NGOs), industry, and society at large began discussing a more sustainable, and less dangerous, approach to industrial chemical production. Though the United States Environmental Protection Agency (U.S. EPA; created in the 1970s) had largely been concerned with pollution cleanup and toxic chemicals in particular, in 1988, there was a paradigm shift within the organization with the creation of the Office of Pollution Prevention and Toxics. Scientists trained during the environmental movement spawned by the 1962 publication of Rachel Carson's *Silent Spring* began to look for ways to prevent pollution and environmental damage.

Though they followed three decades of new regulations, commissions, and conferences on reducing pollution, the 1990s were the tipping point of the burgeoning green chemistry movement. Between the Pollution Prevention Act of 1990, publication of "Chemistry for a Clean World" by the European Community's Chemistry Council, and the creation of the annual Presidential Green Chemistry Challenge Award, green chemistry was finally a legitimate area of scientific study. The U.S. EPA originally coined the term *green chemistry*, with the mission:

> "to promote innovative chemical technologies that reduce or eliminate the use or generation of hazardous substances in the design, manufacture and use of chemical products."

In 1998, Paul Anastas and John Warner published *Green Chemistry: Theory and Practice*, a must-read for all practicing chemists. In it, the 12 Principles of Green Chemistry were outlined. These 12 principles, listed in **Table 7.1**, form the basis of modern green chemistry research.

Table 7.1 The 12 Principles of Green Chemistry

1. Waste prevention	It is better to prevent waste than to treat or clean up waste after it has been created.
2. Atom economy	Synthetic methods should be designed to maximize the incorporation of all materials used in the process into the final product.
3. Less hazardous chemical synthesis	Wherever practicable, synthetic methods should be designed to use and generate substances that possess little or no toxicity to human health and the environment.
4. Designing safer chemicals	Chemical products should be designed to effect their desired function while minimizing their toxicity.
5. Safer solvents and auxiliaries	The use of auxiliary substances (e.g., solvents, separation agents) should be made unnecessary wherever possible and innocuous when used.
6. Design for energy efficiency	Energy requirements of chemical processes should be recognized for their environmental and economic impacts and should be minimized. If possible, synthetic methods should be conducted at ambient temperature and pressure.
7. Use of renewable feedstocks	A raw material or feedstock should be renewable rather than depleting whenever technically and economically practicable.
8. Reduce derivatives	Unnecessary derivatization (use of blocking groups, protection/deprotection, temporary modification of physical/chemical processes) should be minimized or avoided if possible, because such steps require additional reagents and can generate waste.
9. Catalysis	Catalytic reagents (as selective as possible) are superior to stoichiometric reagents.
10. Design for degradation	Chemical products should be designed so that at the end of their function they break down into innocuous degradation products and do not persist in the environment.
11. Real-time analysis for pollution prevention	Analytical methodologies need to be further developed to allow for real-time, in-process monitoring and control prior to the formation of hazardous substances.
12. Inherently safer chemistry for accident prevention	Substances and the form of a substance used in a chemical process should be chosen to minimize the potential for chemical accidents, including releases, explosions, and fires.

7.2 Application of the 12 principles to organic chemistry

The 12 Principles of Green Chemistry apply generally to all subfields of chemistry. However, given their focus on chemical products, and given that a major focus of organic chemistry research is the synthesis of these products, the 12 principles are especially pertinent to our discipline. Section 7.2 gives an expanded summary of each of the 12 principles, while highlighting an example of a process that has been made more sustainable by adhering to that principle. Very often, a green process will adhere to multiple green chemistry principles. The adherence to these green chemistry principles is noted in parentheses (GCP 1, for example). [*When studying these, don't worry about mechanisms or predicting products. Those are covered in later chapters. For now, focus on the green chemistry principle itself.*]

7.2.1 Prevent waste (GCP 1)

Though the numbers vary, it has been estimated that nearly half of all global waste is generated within industry, of which less than half can be recycled. The problem of waste looms larger for the chemical industry because chemical waste has particular potential to cause environmental problems. As a way of managing the deleterious effects of waste on the environment, under the Resource Conservation and Recovery Act, the U.S. EPA established a "cradle-to-grave" waste management program. This program ensures the safe management of waste from the time it is created through transportation, treatment, and, finally, storage (Figure 7.1).

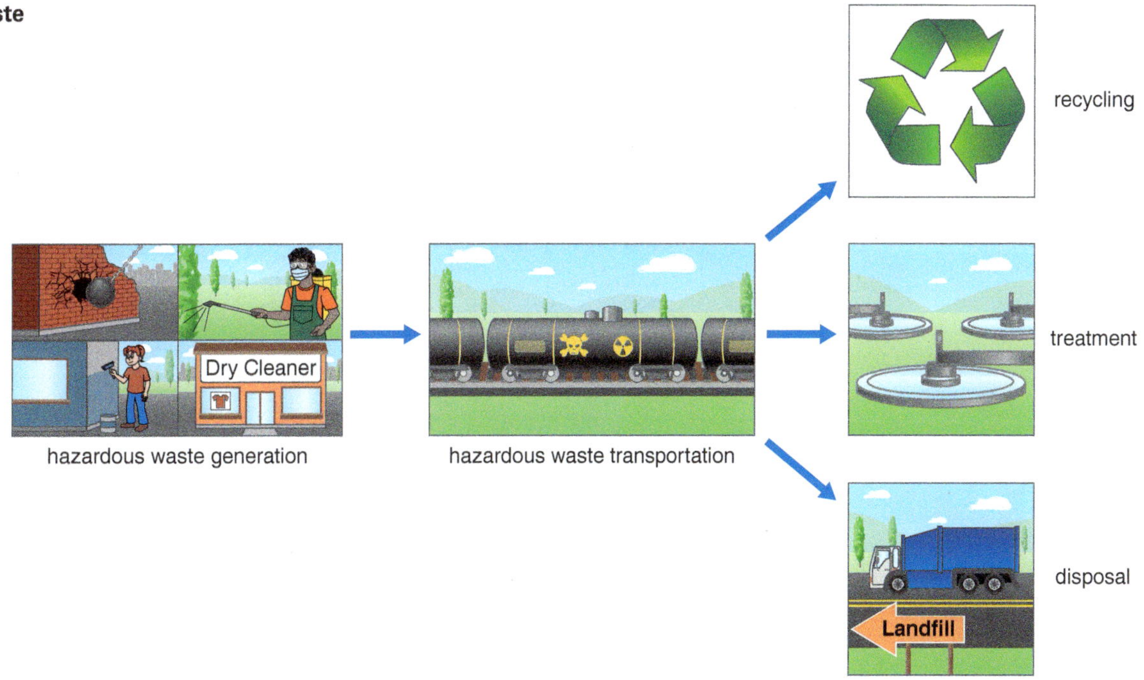

Figure 7.1 Cradle-to-grave waste management.

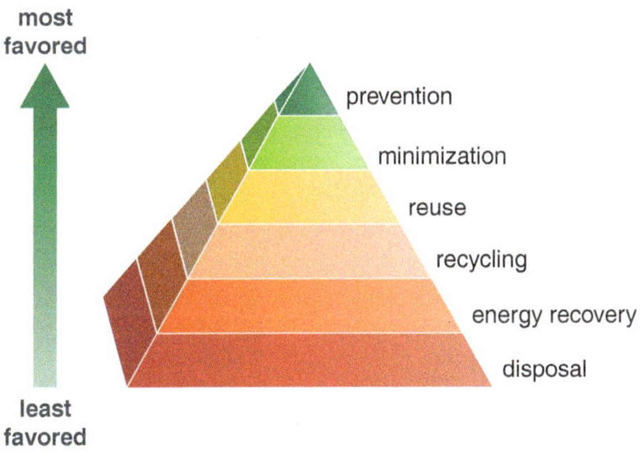

Figure 7.2 Generating less waste is more sustainable than reusing, recycling, or safely storing it.

A hierarchy of green procedures for dealing with waste has emerged. Shown in Figure 7.2, this hierarchy privileges waste reduction over reuse, recycling, and safe storage. If waste is not generated in the first place, it will not need to be dealt with or stored, saving energy, time, and money.

Shown in Table 7.2, one of the initial green chemistry metrics for waste reduction was the **environmental factor (E-factor)**, which is the ratio of waste produced by a chemical process to the amount of product obtained in that process. A smaller ratio represents a greener, more sustainable process. Though it is a simple and reliable metric, one shortcoming of the E-factor is that it doesn't take into account the relative hazards of the specific waste produced. A modification of the E-factor is

Table 7.2 Green chemistry metrics.

E-factor	$E = \dfrac{\text{mass}_{\text{waste}}}{\text{mass}_{\text{final product}}}$
Effective mass yield	$\%\ \text{EMY} = \dfrac{\text{mass}_{\text{product}} \times 100\%}{\text{mass}_{\text{hazardous reagents}}}$
Process mass intensity	$\text{PMI} = \dfrac{\text{mass}_{\text{materials used in synthesis}}}{\text{mass}_{\text{final product}}}$

the **effective mass yield (EMY)**, which is the ratio of product mass to the mass of all hazardous materials used in the process. Finally, the American Chemical Society (ACS) Green Chemistry Pharmaceutical Roundtable proposed the **process mass intensity (PMI)** as the desired, but simple, metric of greenness of a process. It is the ratio of the mass of all materials used in the synthesis of a product to the mass of the product produced. Although it is similar to the E-factor, PMI focuses instead on reducing the amount of materials that are put into making a product as opposed to the E-factor's focus on the waste generated.

A landmark example of waste prevention in the pharmaceutical industry was the redesign of the synthesis of sertraline, the key ingredient in the antidepressant Zoloft®. The original and redesigned (green) routes are compared in **Figure 7.3**. In the original synthesis, highly reactive titanium tetrachloride is used as a dehydrating reagent in an imine formation reaction [*Section 17.8.2*] that generates a substantial amount of solid waste. Each of the three steps in the synthesis requires different solvent combinations, necessitating the isolation of the organic product after each step. The reduction in step 2 [*Sections 9.2 and 25.6.2*] initially resulted in a 6:1 ratio of the desired cis isomer to the trans isomer, which was disposed of as waste. In redesigning the route, all three reactions were carried out using a single (and greener) solvent, in a single reaction vessel. This eliminated the need for product isolation and significantly reduced the amount of solvent wasted. Furthermore, the need for the highly reactive $TiCl_4$ was eliminated, while the palladium-catalyzed reduction

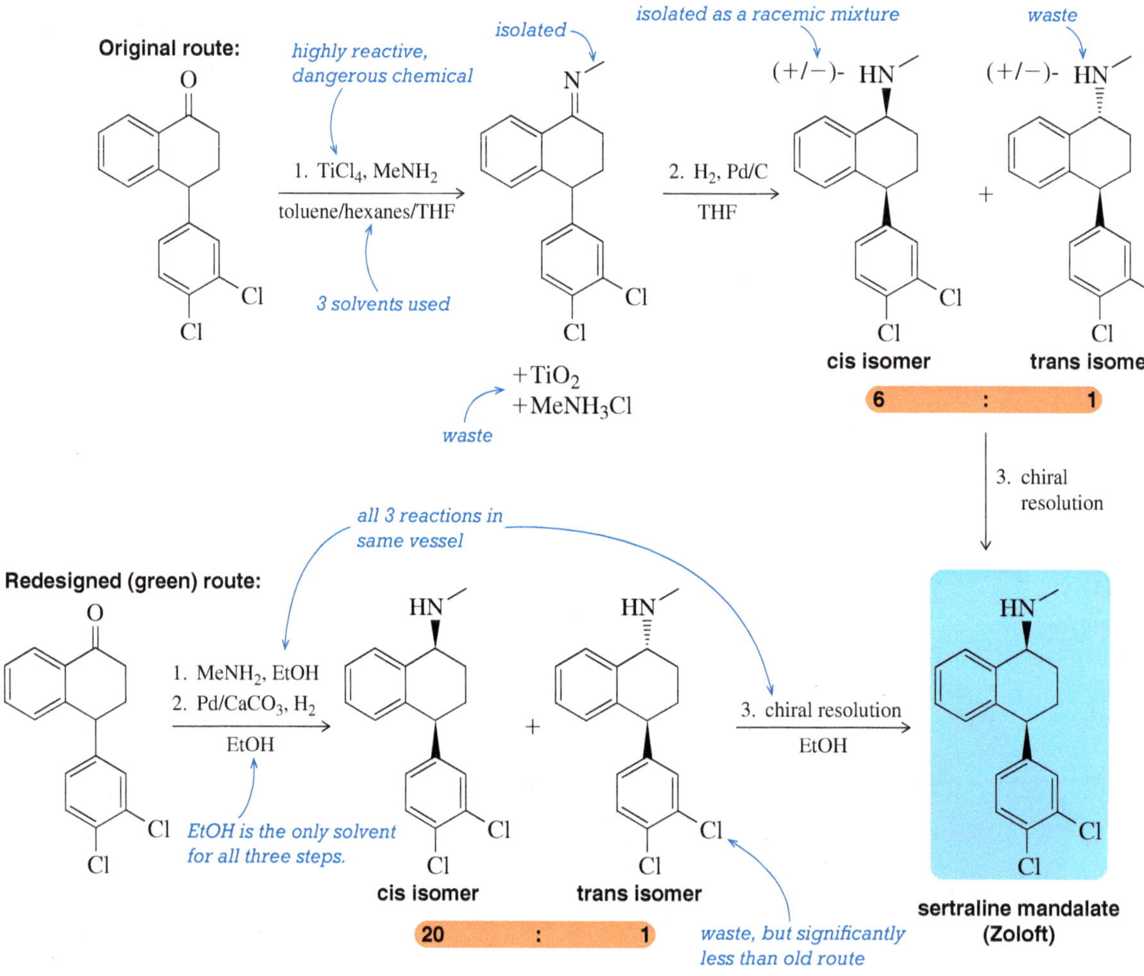

Figure 7.3 **A more sustainable route to sertraline generates less solvent and solid $TiO_2/MeNH_2/HCl$ waste.**

resulted in greater selectivity, thus reducing the amount of trans isomer that would be discarded as waste.

The redesigned route, for which Pfizer received a Presidential Green Chemistry Challenge Award from the EPA in 2002, eliminated the need for 140,000 kg of $TiCl_4$ per year, which enhanced worker safety (GCP 12). By streamlining the three steps in ethanol [a renewable (GCP 7) and safe (GCP 5) solvent], over 74,000 L of solvent waste were eliminated per 1000 kg of sertraline produced. Furthermore, the new route eliminated 680,000 kg of sodium hydroxide, hydrochloric acid, and titanium dioxide waste per year (GCP 1). The improved product ratio of step 3 reduces the amount of raw material needed for sertraline synthesis.

Assessment 7.5 Two different sets of conditions were used for formation of the imine in the sertraline synthesis. Calculate the E-factor and the PMI for both reactions. Which is greener?

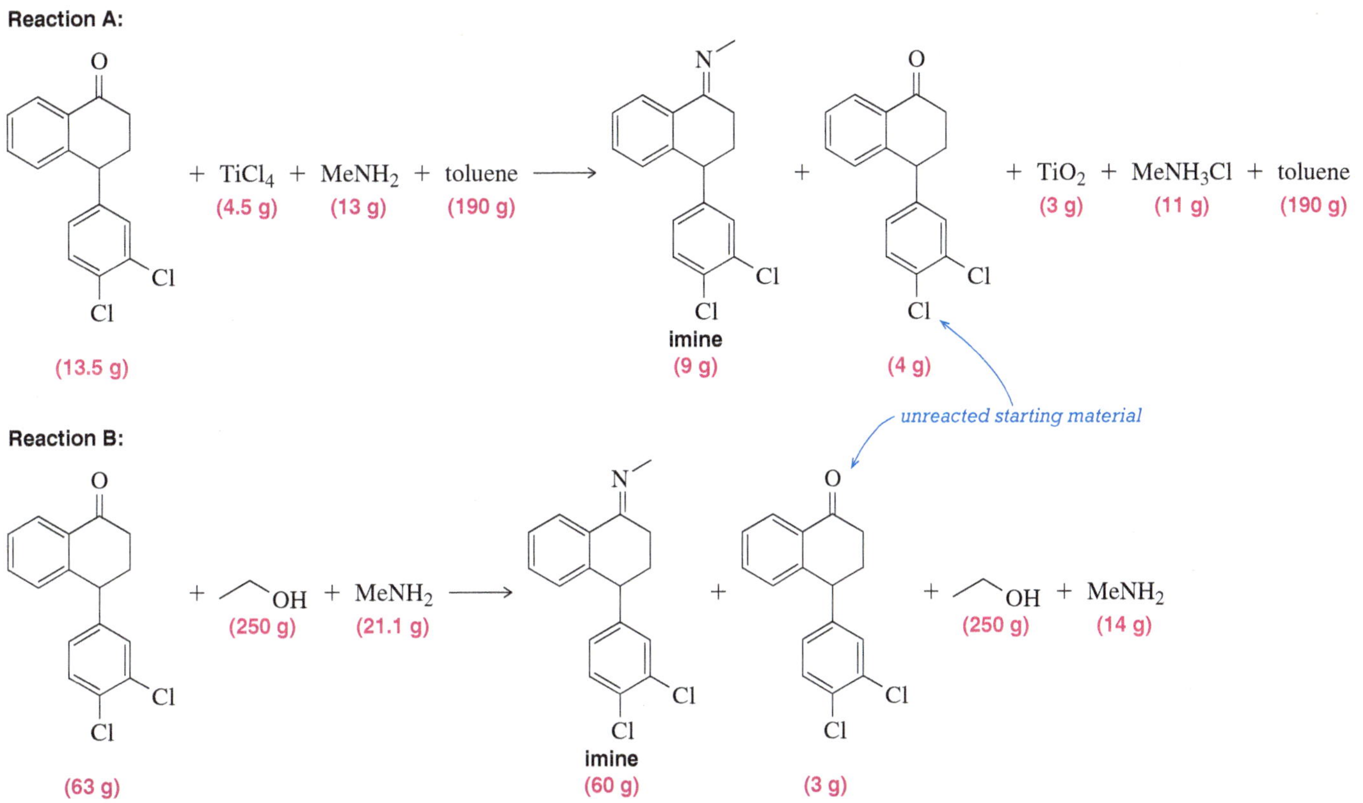

Assessment 7.6 (a) Which route would be better: one that produces a small amount of hazardous waste or one that produces a large amount of nonhazardous waste? (b) Which green chemistry metric accounts for this dilemma?

7.2.2 Maximize atom economy (GCP 2)

Organic chemists have traditionally measured reaction efficiency by the **percent yield**, the actual yield as a percentage of the theoretical yield. A higher percent yield implies that a reaction is more efficient because the amount of product produced is closer to the maximum possible. While this metric is valuable in determining how much of the important organic product is made based on the limiting reagent, it doesn't differentiate between green and non-green reactions. Developed by Barry Trost, the concept of **atom economy** evaluates reactions based on how much of the mass that you put in the reaction flask ends up in the product. Atom economy is calculated using Equation 7.1.

$$\text{Atom economy} = \frac{\text{molecular mass of product}}{\text{molecular mass of reactants}} \times 100\% \qquad \text{Equation 7.1}$$

The Grignard reaction (Sections 16.2 and 17.7.1), shown in **Figure 7.4(a)**, is one of the most important reactions for forming carbon–carbon bonds. Developed in 1900, it involves the reduction of a haloalkane to form an alkyl magnesium species, which then adds to an aldehyde to produce a 2° alcohol following quenching with a strong acid. Even if high percent yields are possible, it isn't a particularly atom-economical reaction. Neither the bromide (of the haloalkane), nor the magnesium, nor the chloride (of HCl) ends up in the product. These are all presumably lost as waste (GCP 1), necessitating extra processing or disposal by industries employing the Grignard reaction. Alternatively, hydrogen transfer-mediated C—C bond formation [**Figure 7.4(b)**] has recently been shown to produce the same 2° alcohol. Although this reaction is beyond the scope of this book, every atom in the starting reactants ends up in the major organic product, giving an atom economy of 100%. This reaction also avoids the use of potentially toxic haloalkanes (GCP 3). Catalysts aren't technically included in the atom economy calculation because they aren't reactants, emerging unchanged from the reaction. Thus, they are often used in significantly smaller quantities than their stoichiometric counterparts. We explain more about the green value of catalysts (GCP 9) in Section 7.2.9. Chemist Michael Krische won a 2007 Presidential Green Chemistry Challenge Award from the EPA for developing this technology.

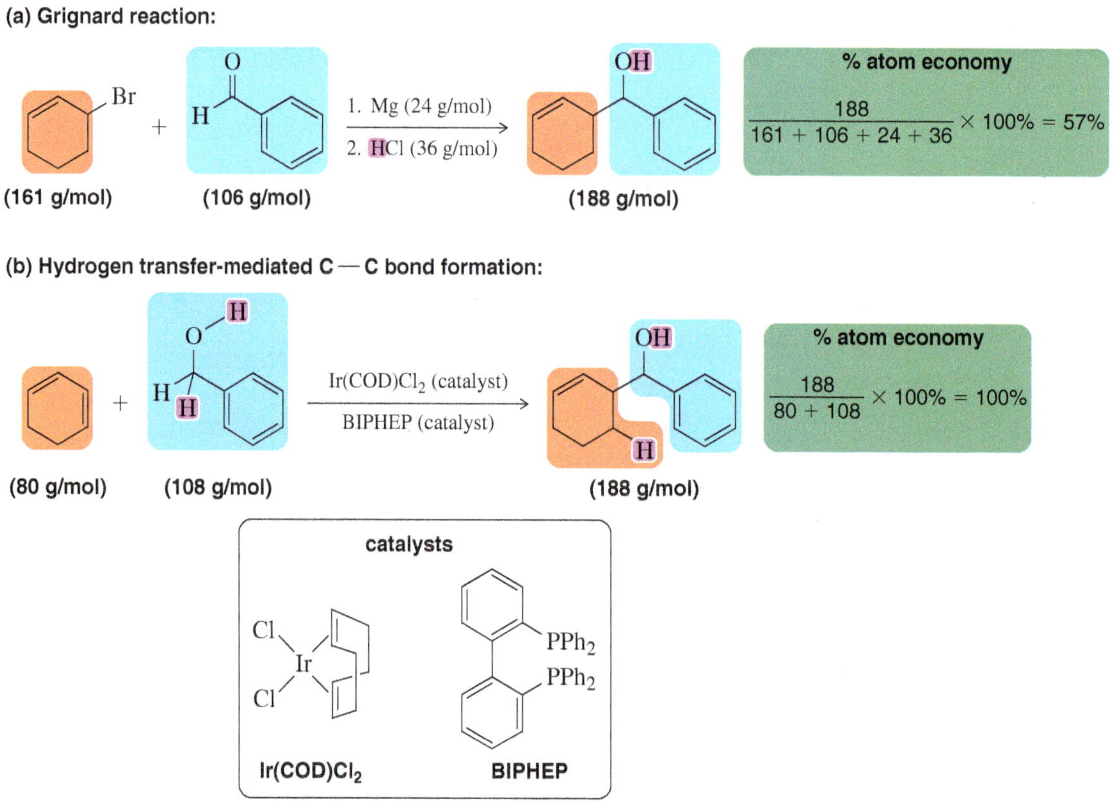

Figure 7.4 A comparison of atom economy. (a) A Grignard reaction of haloalkanes with an aldehyde has a much lower atom economy than (b) hydrogen transfer-mediated C—C bond formation.

Assessment 7.7 Calculate the atom economy for the reactions shown. In each, what happens to the percentage of material that is not incorporated into the major product?

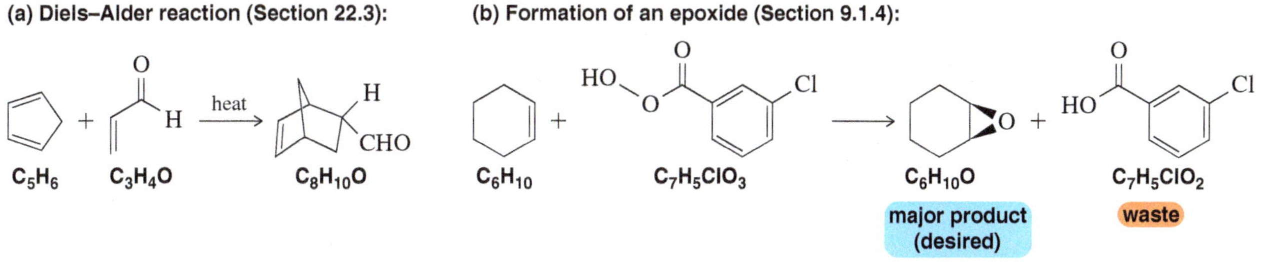

Assessment 7.8 Which is better: a reaction run in a toxic solvent that has a high atom economy or a reaction run in water that has a lower atom economy?

7.2.3 Reduce toxicity in synthesis (GCP 3)

For the sake of efficiency, toxic reagents and synthetic methods are often used to develop chemical products. This might have made sense at one time—as long as the final product wasn't toxic, there was no concern about danger to the public. Yet this view dramatically [*sometimes catastrophically*] ignores the danger of using toxic reagents to the chemical worker, or in the event of an accidental release of toxic reagents, to the public at large. Hence, Green Chemistry Principle 3 directs chemists and the chemical industry to preferentially use synthetic procedures that minimize the use of toxic reagents.

Polyurethanes are a class of polymer with broad industrial applications, including foams, adhesives, and coatings, among others. Since the 1930s, polyurethanes have been synthesized via the use of diisocyanates and aliphatic diols, as shown in **Figure 7.5**. Unfortunately, diisocyanates contain the same functional group as methyl isocyanate, the compound allegedly responsible for the Bhopal disaster discussed at the beginning of the chapter. Even in small, nonlethal quantities, though, isocyanates have been shown to cause asthma; irritation of the eyes, lungs, and nose; as well as cancer upon long-term exposure, so they are clearly a danger to the workers that utilize them.

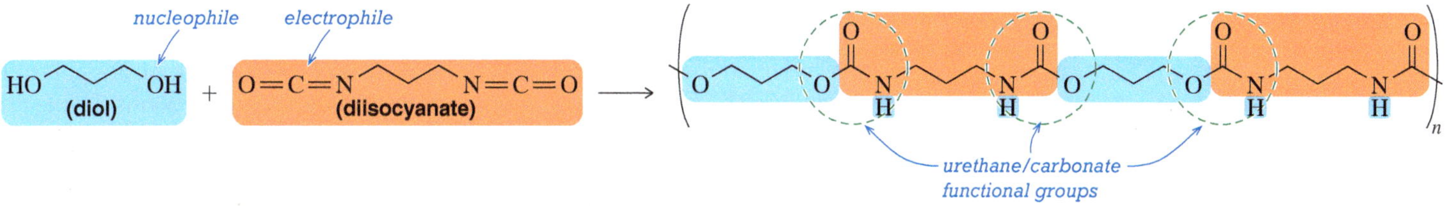

Figure 7.5 Representative synthesis of a polyurethane using diisocyanates and diols.

Recent efforts have been directed to the synthesis of polyurethanes using cyclic carbonates as a replacement for the highly reactive isocyanate functional group (**Figure 7.6**). By replacing the electrophilic isocyanate (with a carbonate) and the diol (with a diamine), the urethane/carbonate functional group can be synthesized using significantly less-toxic reagents (GCP 2). Another advantage to this process is that the resulting polymer contains unreacted hydroxyl groups that can be used to further functionalize the polymer, imbuing it with different properties.

The EPA awarded Hybrid Coating Technologies the 2015 Presidential Green Chemistry Challenge Award for its Green Polyurethane,™ prepared using a process similar to this, noting its high atom economy (GCP 2) and use of renewable resources (GCP 7).

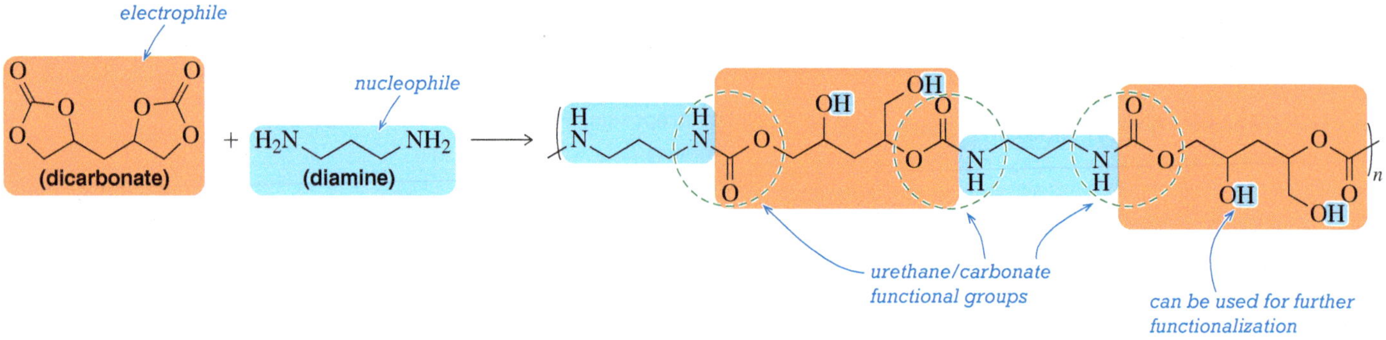

Figure 7.6 **Sustainable synthesis of a polyurethane** using dicarbonates and diamines.

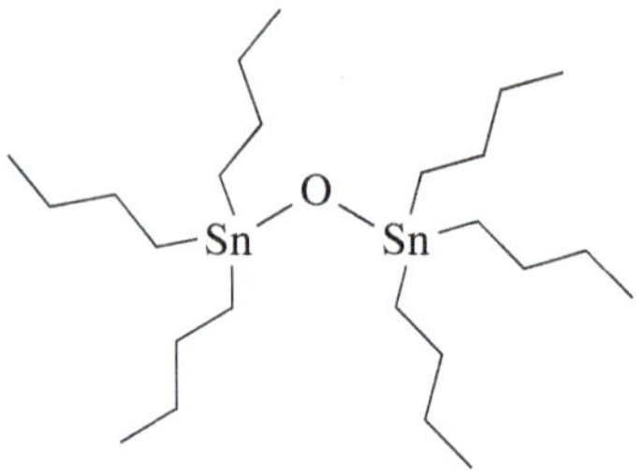

tributyltin oxide

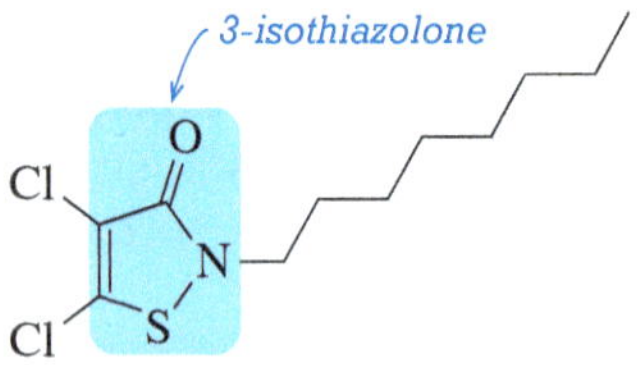

Sea-Nine antifoulant

Recall from Chapter 5 that reactants must collide in order for chemical reactions to occur.

7.2.4 Design safer chemicals (GCP 4)

Often a chemical product may have excellent properties for its primary purpose but possesses secondary, unintended negative environmental consequences. Green Chemistry Principle 4 states that chemists should identify these potential secondary consequences prior to developing a product. The product should then be prepared to serve its primary purpose while avoiding these negative secondary consequences. In essence, chemicals should be designed to carry out their desired effect safely.

The shipping industry is negatively affected by the growth of marine plants and animals on the hulls of ships. Termed fouling, the buildup on the ship's surface results in hydrodynamic drag, reducing transit speeds and increasing fuel consumption. It was estimated in 1996 that fouling costs the shipping industry $36 billion in increased nonrenewable fuel consumption per year that also contributes to climate change. Tributyltin oxide, which decreases the reproductive viability of marine organisms, has been incorporated into paints used on the hulls of ships to prevent fouling. Unfortunately, tributyltin oxide is also a persistent and acute environmental toxin. In France, the use of tributyltin oxide was correlated to an increase in oyster shell malformations in areas of high shipping activity.

In the early 1990s, Rohm and Haas invented Sea-Nine,™ an antifoulant for use in paints applied to ships, whose active ingredient is a substituted 3-isothiazolone. Although it retained the acute toxicity necessary to prevent fouling, it showed no chronic toxicity once leached into the ocean. This diminished toxicity is a result of its quick degradation (GCP 10) in seawater and sediment. Rohm and Haas was awarded the 1996 Presidential Green Chemistry Challenge Award for the development of Sea-Nine.

7.2.5 Use safer solvents/auxiliaries (GCP 5)

Organic reagents can be solids, liquids, or gases and can also be ionic, polar, or nonpolar; in order to bring them into the same phase so that reactions can occur, a variety of solvents are used. Solvents, while necessary, are almost never incorporated into the product being made. [*They are incorporated in the solvolysis reactions of Chapter 12.*] However, solvents have their own environmental issues. For example, halogenated solvents like dichloromethane are suspected carcinogens (GCP 4), while certain ethers are highly flammable/explosive (GCP 12). Solvents like benzene and toluene are derived from nonrenewable feedstocks (GCP 7). While the ideal situation would be to avoid the use of solvents entirely, the next best option is to use water, the safest of all solvents.

Although water is a safe solvent, it does not dissolve nonpolar organic compounds because it is a polar liquid. Furthermore, certain reagents, such as strong bases, are protonated by water, making water incompatible with their reactions. As an example, organozinc halides are quenched by the acid–base reaction in **Figure 7.7**.

Figure 7.7 **Organozinc halides are destroyed by reaction with the acidic proton of water.**

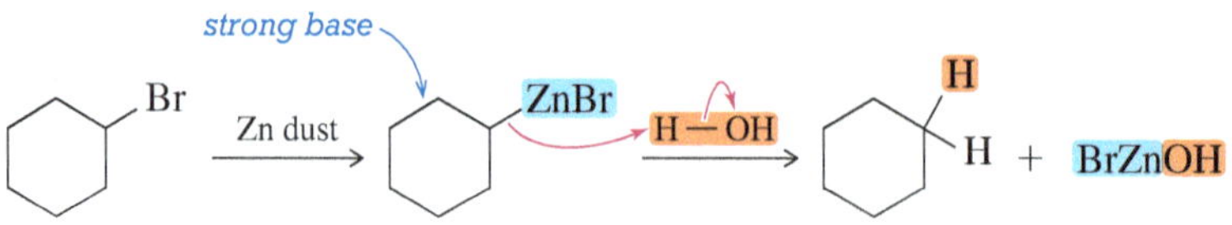

The Negishi coupling, a reaction we introduce in Section 16.4.2, utilizes organozinc halides in a cross-coupling reaction with aryl halides. As shown in **Figure 7.8**, this reaction is almost always conducted in a polar, aprotic solvent like THF. Because THF is a volatile and highly flammable ether, it imposes a potential danger upon the chemist using it (GCP 12).

Figure 7.8 Negishi coupling in THF.

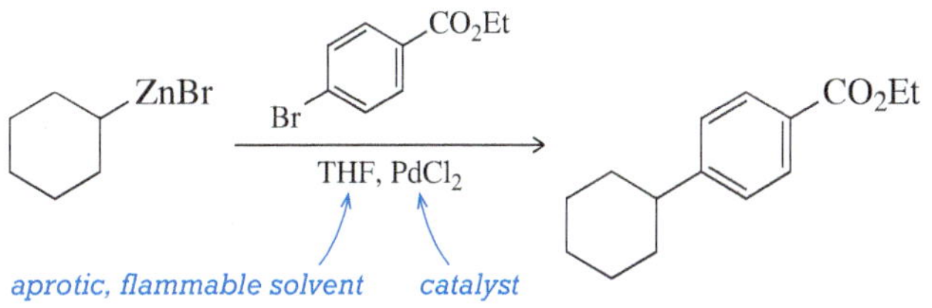

Seeking to make the Negishi (and other) reactions possible in water, Bruce Lipshutz pioneered research on running these reactions in micelles. Micelles are highly organized spherical structures made of ambiphilic molecules, like TPGS-750-M (**Figure 7.9**). Ambiphilic molecules possess a polar (hydrophilic) head group and a nonpolar (hydrophobic) tail. When a small amount of TPGS-750-M is dissolved in water, the molecules orient to form the spherical structure shown in **Figure 7.10**. When the reaction is conducted in this environment, the highly basic organozinc halide forms within the nonpolar (and aprotic) core of the micelle. This allows it to react with the nonpolar bromobenzene prior to being exposed to water. Through this technology, many reactions that previously could not be done in water are now possible in the safest imaginable solvent (GCP 5). Moreover, TPGS-750-M is synthesized from materials that are themselves naturally occurring and environmentally safe (GCP 3 and GCP 4).

In this case, and throughout this chapter, don't get hung up on the chemistry we haven't discussed yet. To summarize all of this micelle talk: Being able to run the reaction in water is excellent!

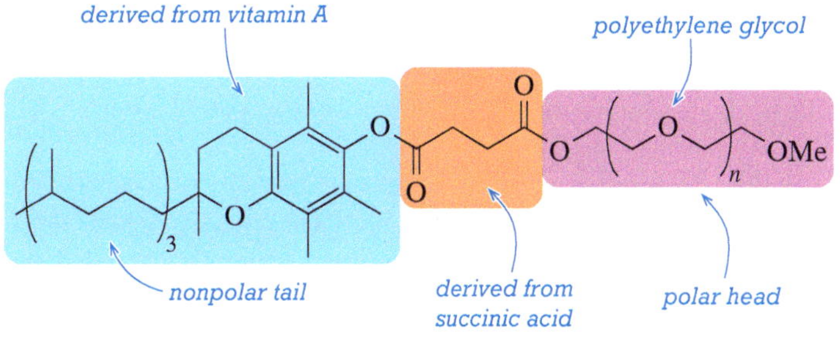

Figure 7.9 TPGS-750-M is composed of environmentally safe compounds.

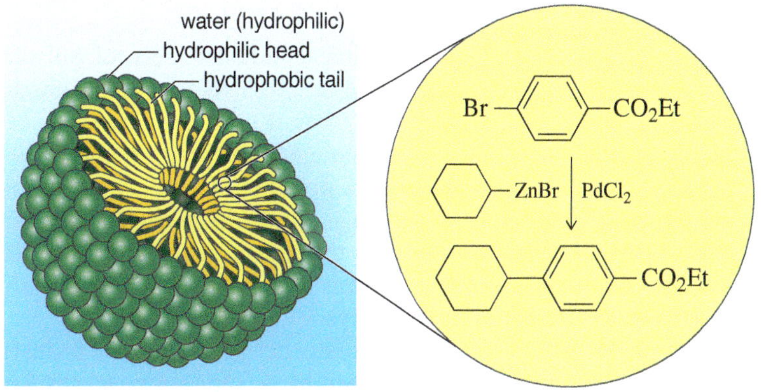

Figure 7.10 Reaction of the highly basic organozinc halide occurs in the aprotic core of the micelle.

A Presidential Green Chemistry Challenge Award was given in 2011 to Bruce Lipshutz for developing this technology and bringing us closer to ending our dependence on organic solvents.

Assessment 7.9 Which of the following reactions would benefit from micellar technology (Figure 7.10), allowing them to be run in water?

(a)

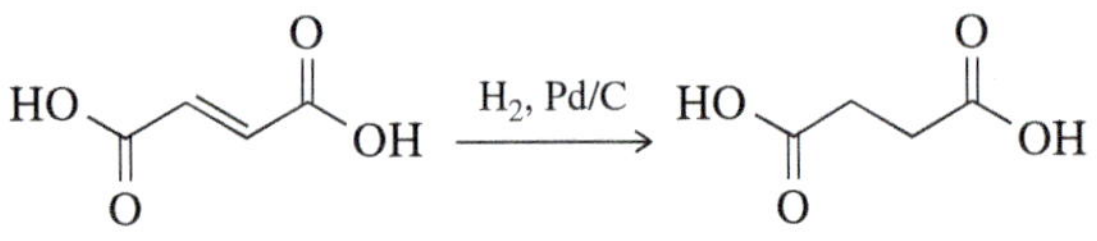

(b)

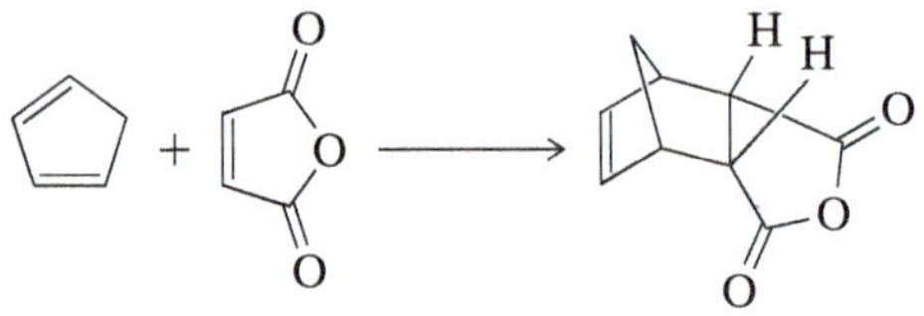

(c)

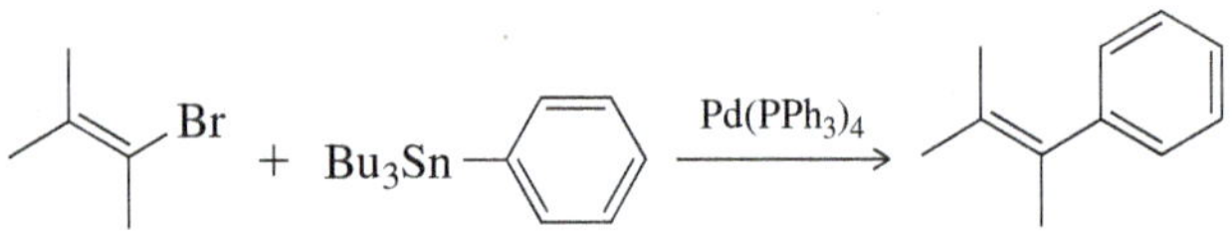

Assessment 7.10 Compare and contrast the "greenness" of the following reactions.

(a)

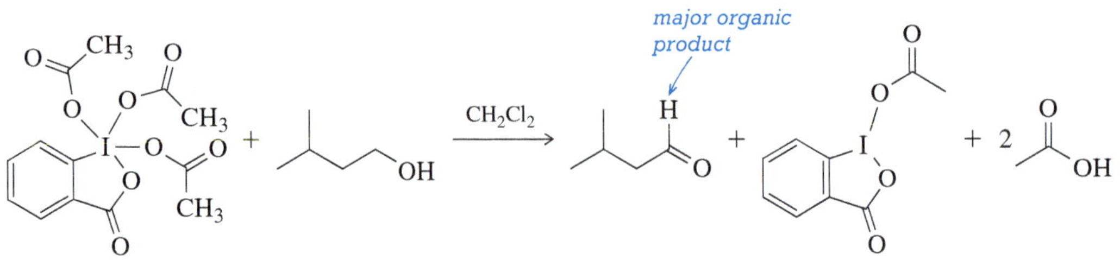

(b)

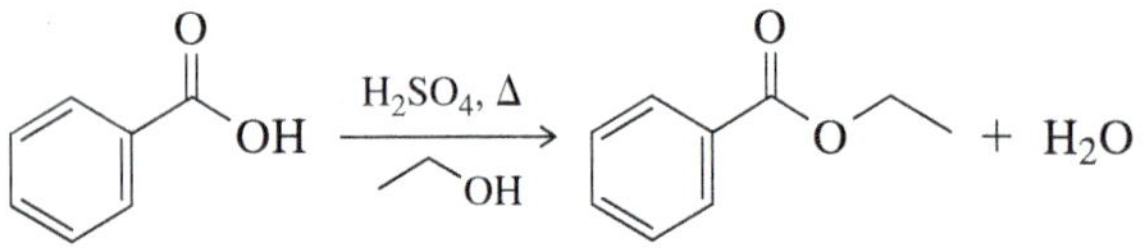

7.2.6 Be energy efficient (GCP 6)

The chemical industry, second only to petroleum refining in terms of energy usage, was estimated in 2013 to have consumed 27% of all energy utilized by the entire industrial sector. Energy is used at every stage of the production process. As a result, Green Chemistry Principle 6 states that chemists should employ chemical processes that require as little energy as possible. These efforts result in conservation of nonrenewable resources (coal, natural gas, and oil), reduced production of CO_2, and monetary savings. As is true for many of the green chemistry principles, then, adherence to Green Chemistry Principle 6 not only is the moral and ethical thing to do, but also makes sense from a business perspective.

The cosmetics and skin care industries use esters derived from fatty acids as ingredients in their products. One such ester is 2-ethylhexyl palmitate, produced from the reaction of palmitic acid and 2-ethylhexan-1-ol. 2-Ethylhexyl palmitate is used as an emollient in skin care products, softening the skin to which it is applied. Traditionally, these esters are produced via a Fischer esterification reaction (**Figure 7.11**), requiring an acid catalyst (GCP 9) to increase the rate of reaction and high temperatures to push the reaction forward. [*We study this reaction in Section 18.7.2.*] On the scale required to support an estimated $1.8 billion per year industry, the energy requirements of reactions like this contribute substantially to global warming and diminishing petroleum reserves. To address this problem, Eastman Chemical Company utilized solvent-free, immobilized enzyme catalysis to produce 2-ethylhexyl palmitate, allowing the reaction temperature to be reduced to 40 °C. By immobilizing the enzyme, the product can be easily purified by filtration and the enzyme can be reused up to 21 times. The energy savings and the reduced environmental footprint of this process resulted in Eastman receiving a 2009 Presidential Green Chemistry Challenge Award.

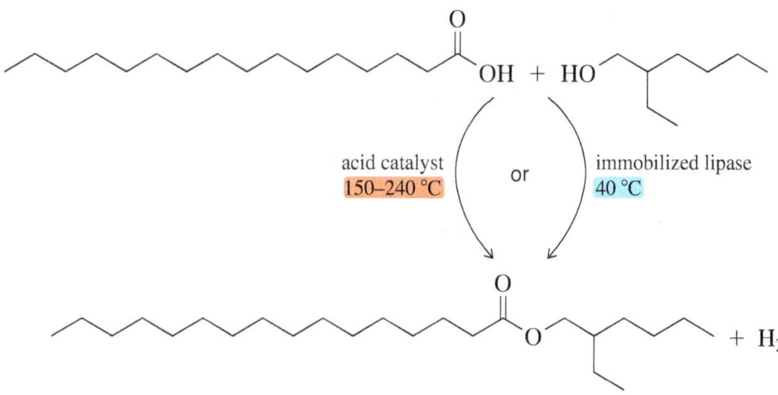

Figure 7.11 The use of enzyme catalysis (immobilized lipase, in this case) dramatically reduces the energy requirements of esterification reactions.

A second example of energy efficiency is the bio-based production of succinic acid, a platform molecule that can be easily derivatized to provide a number of organic starting materials. Succinic acid is traditionally produced from a process that begins with butane isolated from natural gas deposits or as a product of petroleum refining, both energy-intensive processes (**Figure 7.12**). Instead, using an *Escherichia coli* catalyst, succinic acid can be produced via the fermentation of glucose, a renewable resource (GCP 7). Integrated with a water-based purification process (GCP 5), this method requires 60% less energy than production of succinic acid from fossil fuels and actually consumes carbon dioxide rather than producing it, in the fermentation process. In 2011, BioAmber, Inc., received a Presidential Green Chemistry Challenge Award for this technology.

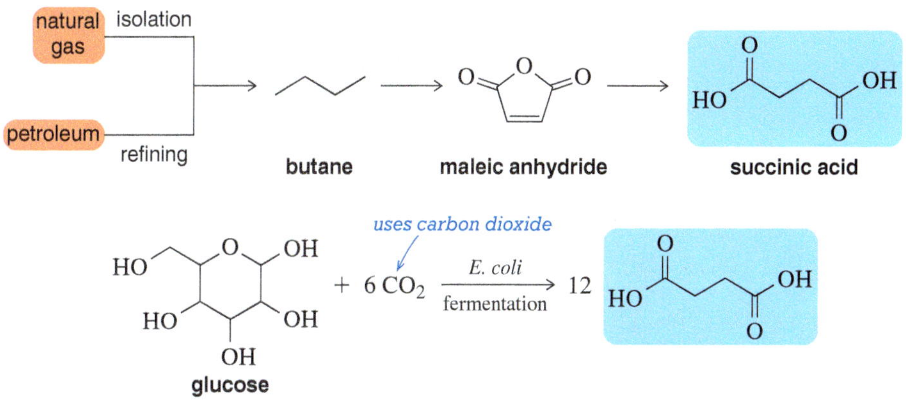

Figure 7.12 Bio-based production of succinic acid uses less energy and removes carbon dioxide from the atmosphere.

Assessment 7.11 In Chapter 5, we learned that a reaction requires energy to overcome an activation barrier in order to proceed. How does a catalyst reduce the energy required by a reaction?

7.2.7 Use renewable feedstocks (GCP 7)

Feedstocks, or the origin of compounds, can be categorized as nonrenewable or renewable. As we study the conversion of organic molecules into other organic molecules, we often disregard the origin of these molecules as being unimportant. That is, we don't think about the feedstock from which our starting materials and reagent molecules are derived. For much of chemical history and continuing today, organic compounds have been derived from oil deposits deep within the ground, isolated as part of petroleum refining and natural gas isolation. [*The next generation of chemists should change this.*] Though these oil deposits were created via the decomposition of living organisms, the fact that they formed millions of years previously makes these functionally nonrenewable resources. Alternatively, some living organisms, like plants and algae, use carbon dioxide in the atmosphere to produce their own organic materials. Compounds derived from these sources are considered to be from a renewable feedstock (Figure 7.13). Green Chemistry Principle 7 encourages the use of renewable feedstocks wherever possible.

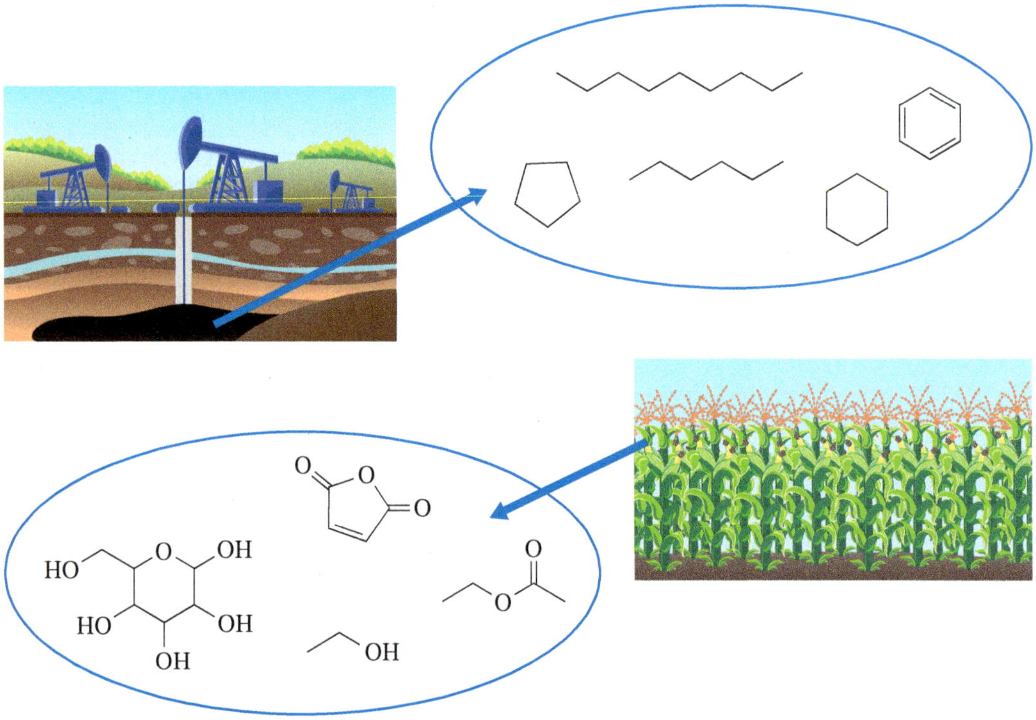

Figure 7.13 Organic compounds can be derived from petroleum (nonrenewable) or bio-based (renewable) feedstocks.

The creation of biodiesel as an alternative fuel source is an excellent example of replacing fuels originating from nonrenewable petroleum feedstocks with those from renewable natural sources. Specifically, biodiesel fuel consists of long-chain esters derived from vegetable oils and animal fats. These esters, containing long saturated hydrocarbon chains, burn in a similar manner to gasoline and other fuels. The synthesis of biodiesel involves the transesterification of triglycerides with ethanol to make the ethyl ester for use as fuel, along with glycerol as a by-product (Figure 7.14). Early on, the glycerol from biodiesel production replaced that from nonrenewable feedstocks in the flavor and perfume industries.

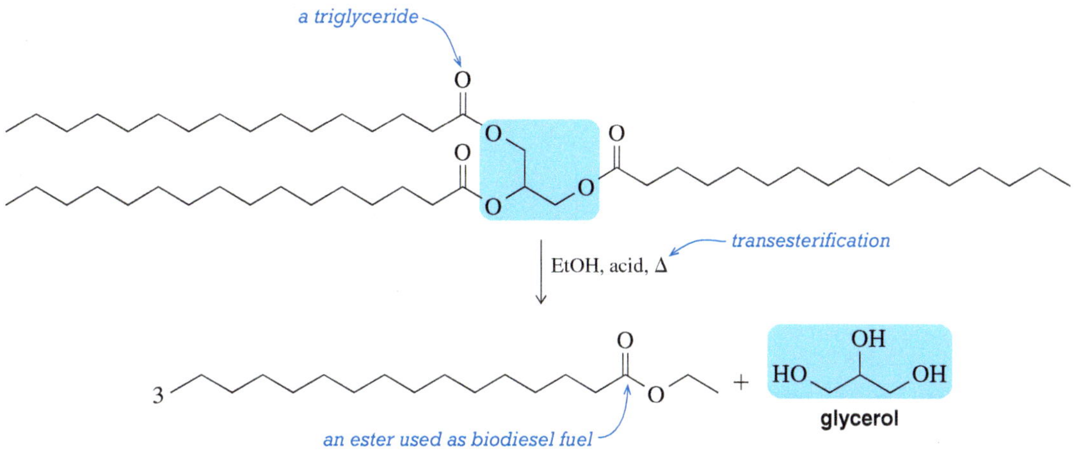

Figure 7.14 Transesterification of animal fats and vegetable oils results in glycerol and long-chain esters, which can be used as biodiesel fuel.

As biodiesel production has increased, more glycerol is being produced than the global market can possibly use. A new market for glycerol, itself derived from a renewable resource, is needed to further bring down the cost of biodiesel production and incentivize its use over other nonrenewable fuel sources. Galen Suppes, in the early 2000s, modified a process for converting glycerol to propylene glycol (**Figure 7.15**). Propylene glycol represents a less toxic, but more expensive, alternative to ethylene glycol, used in antifreeze. Utilizing a new catalyst, the temperature required for this transformation could be reduced slightly (GCP 6). The pressure was similarly reduced, resulting in a safer reaction (GCP 12). The propylene glycol made by this catalytic (GCP 9) process replaces, at lower cost, propylene glycol derived from petroleum feedstocks. Suppes received a 2006 Presidential Green Chemistry Challenge Award for his work in this area. [*200 psi is still a pretty high pressure, and heating to 200 °C requires a significant energy input, so there is potential to further "green" this process.*]

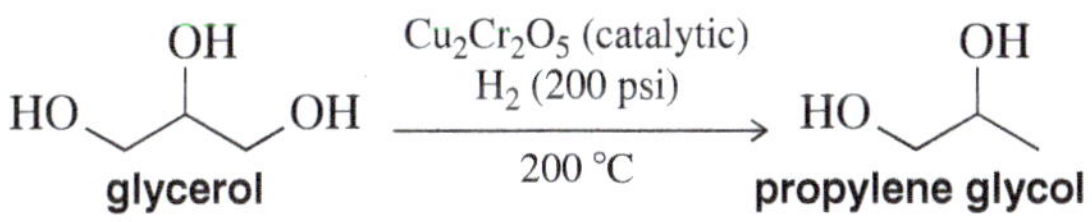

Figure 7.15 Catalytic hydrogenolysis of glycerol gives propylene glycol.

Butane-1,4-diol, a compound used in spandex, running shoes, and other products, has a worldwide market of around $3 billion. It is traditionally prepared from acetylene, a petroleum-based precursor. Genomatica, a bioengineering company, developed a microbe that makes butane-1,4-diol by sugar fermentation (**Figure 7.16**). This fermentation process requires no organic solvents (GCP 5), occurs at ambient temperature and pressure (GCP 6), and like the succinic acid fermentation in Figure 7.12, actually uses CO_2. In this way, CO_2 emissions are reduced by 70% as compared to production of butane-1,4-diol by traditional means from nonrenewable feedstocks.

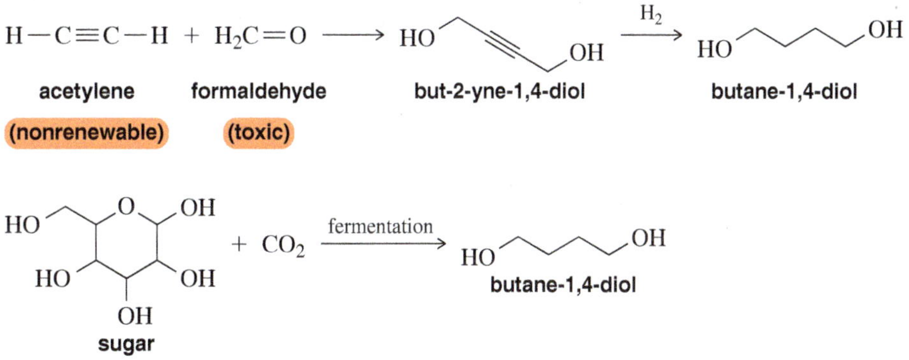

Figure 7.16 Fermentation of sugars generates butane-1,4-diol from a renewable resource.

Assessment 7.12 Based on the fermentation products in Figures 7.14, 7.15, and 7.16, which of the following might someday be similarly accessible from renewable, bio-based feedstocks?

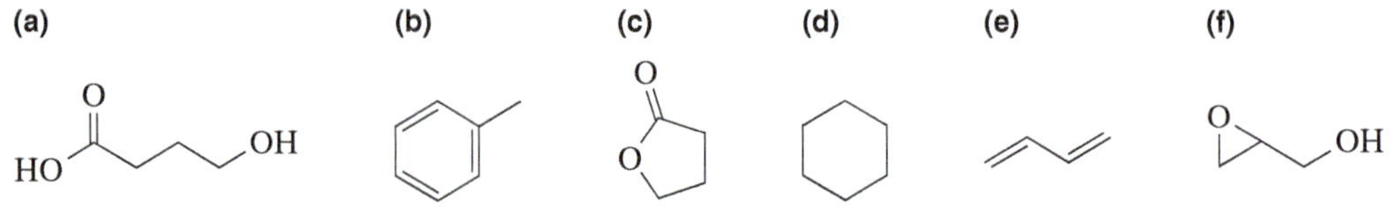

Assessment 7.13 Why should the manufacturing cost of a product from a renewable feedstock matter?

7.2.8 Reduce derivatives (GCP 8)

The synthesis of molecules is challenging, especially when a single molecule contains multiple reactive functional groups. [*You may not appreciate this now, but you will eventually.*] In navigating the synthesis of molecules of ever-increasing complexity, chemists have developed methods by which a functional group can be blocked (preventing reactivity) or enhanced (encouraging reactivity) through covalent modifications. Each of these derivatizations, though, requires additional steps and generates additional waste. If a blocking or a protecting group is installed, it must later be removed. If a group is installed to enhance reactivity, once it has served its purpose, it must be removed, creating an additional purification step. And since each of these activating or blocking groups is not found in the final product, not only has the atom economy been reduced (GCP 2), but these groups must be disposed of or treated as waste (GCP 1). This is why Green Chemistry Principle 8 states that we should avoid the use of derivatives wherever possible. Furthermore, GCP 8 highlights the interconnectedness of the 12 Green Chemistry Principles, in that, by adhering to one, many others are also adhered to.

The Merck synthesis of sitagliptin, the active ingredient in Januvia®, a treatment for type 2 diabetes, is a landmark example of the incorporation of green chemistry principles in pharmaceutical synthesis. In recognition of these efforts, Merck was awarded two Presidential Green Chemistry Challenge Awards (2006 and 2010). [*The analysis here really won't do it justice, but it at least offers a brief overview of its significance.*] **Figure 7.17** highlights three (of many) aspects of the first-generation synthesis of sitagliptin that were problematic. In two different places in the overall eight-step synthesis, a carboxylic acid is derivatized with *N*-(3-dimethylaminopropyl)-*N*′-ethylcarbodiimide (EDC) to promote a nucleophilic acyl substitution reaction that we discuss in Chapter 19. This is an example of using a derivative to enhance reactivity. To control the reactivity of the nitrogen atom installed in the first nucleophilic acyl substitution, it was installed with a protecting group intact. Thus, a final step was required to remove the benzyl ether, producing benzyl alcohol as waste. This is an example of using a derivative to block reactivity. All three of these cases required extra purification steps and thus generated more waste.

Again, don't get distracted by the chemistry that's occurring, just notice the general principles. You'll learn these reactions later on.

Over the next several years, Merck developed a second- and third-generation synthesis of sitagliptin. The third-generation synthesis is shown in **Figure 7.18**. It completely negates the use of the EDC derivatizing reagent and obviates the need for the benzyl ether protecting group (GCP 8). Compared to the previous syntheses, the use of an enzyme-catalyzed (GCP 9) transaminase avoids the use of all metals, avoids a costly chiral purification step (GCP 5), and results in a 56% increase in productivity and a 19% reduction in overall waste generation. With a shortened synthesis (three steps compared to eight originally), a significant amount of raw materials, time, energy, and money are saved. Over the lifetime of sitagliptin's manufacture, it is estimated that the second- and third-generation syntheses will eliminate over 330 million pounds of waste (GCP 1).

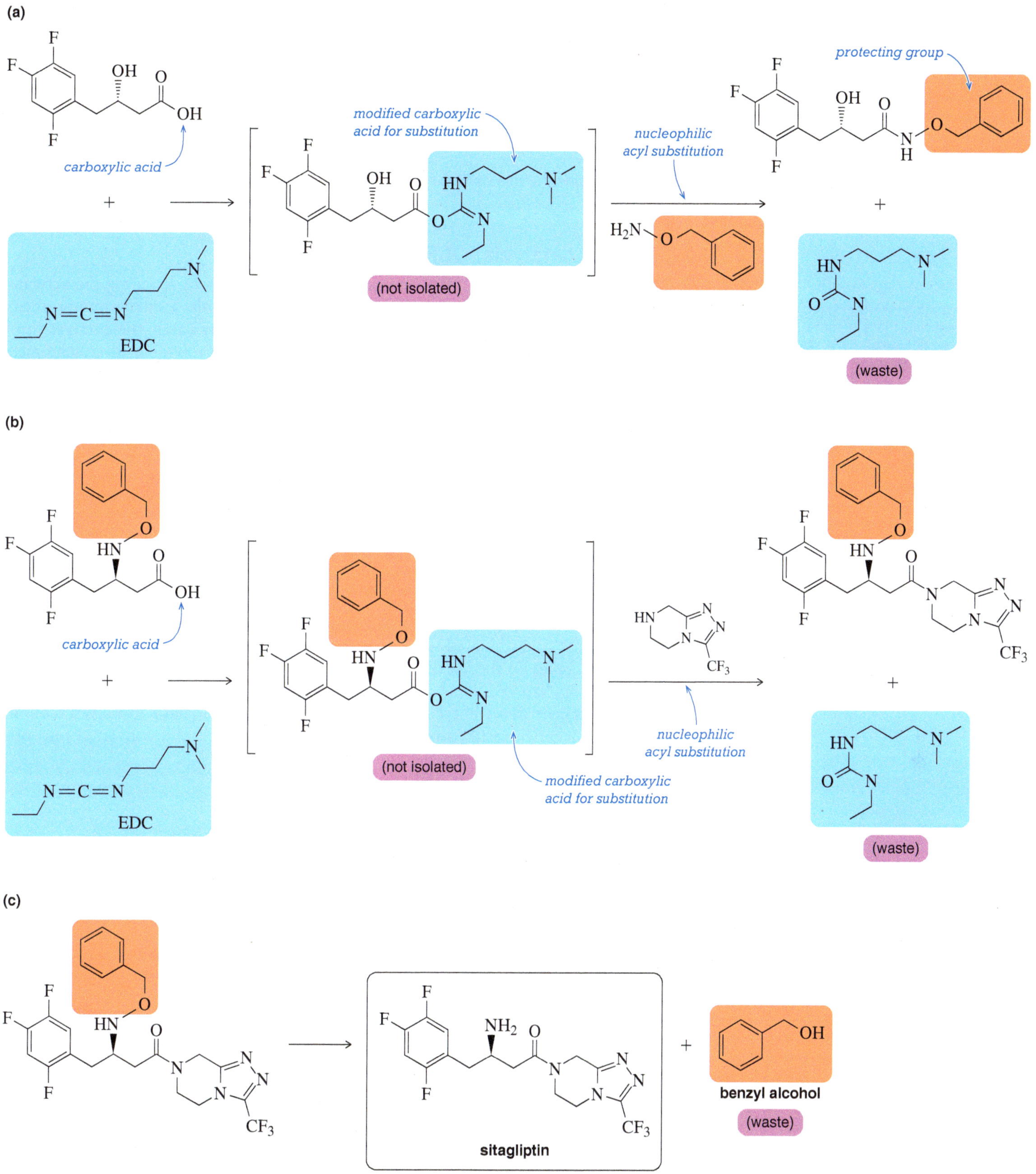

Figure 7.17 First-generation sitagliptin synthesis issues as a result of derivatization. (a) An activating group is used to promote a substitution on a carboxylic acid. (b) The same activating group is used to promote a substitution on a carboxylic acid. (c) A protecting group installed previously must be removed.

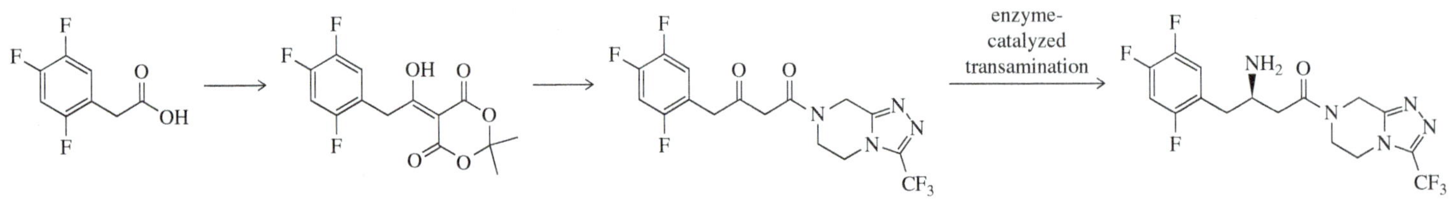

sitagliptin

Figure 7.18 The third-generation synthesis of sitagliptin avoids the use of EDC as an activating agent and uses no protecting groups.

One more time—there is some complicated chemistry here. You will soon know what's going on. For now, focus on how the route is greener.

7.2.9 Employ catalysis (GCP 9)

As discussed in Chapter 5, catalysts increase the rate of reaction by lowering the activation energy (E_a), but they are not consumed in the reaction. As a result, a small amount of catalyst can be used to carry out reactions on a huge scale. In many cases, the catalyst can be recycled and reused more than once. Catalytic reagents are preferable to stoichiometric reagents because they minimize both the resources needed to run a reaction and the amount of waste generated. Often, a catalyst will allow reactions to proceed at lower temperatures and pressures, resulting in safer reaction conditions. Finally, because a small amount is used relative to the amount of product being produced, issues of catalyst toxicity are minimized. For these reasons, wherever possible, Green Chemistry Principle 9 encourages the use of catalytic as opposed to stoichiometric reagents.

The worldwide demand for propylene oxide is estimated at over 14 billion pounds, making it one of the largest-volume chemical intermediates produced in the world. Traditionally, the synthesis of propylene oxide is done in one of two ways, as shown in **Figure 7.19**. Chlorination of an alkene in the presence of water gives rise to a chlorohydrin, by a reaction that we discuss in Section 9.2.3. Treatment with base results in formation of propylene oxide in a two-step process. This process suffers from the use of toxic chlorine and the generation of wastewater streams contaminated with chloride [Figure 7.19(a)]. A second method, shown in Figure 7.19(b), relies on the use of an organic peroxide, like *t*-butyl peroxide, with a catalyst. Although propylene oxide is produced in a single step (an improvement over the chlorohydrin route), it generates *t*-butyl alcohol as a waste product and has a low atom economy. To mitigate this negative result, *t*-butyl alcohol can be collected and converted into other useful chemicals.

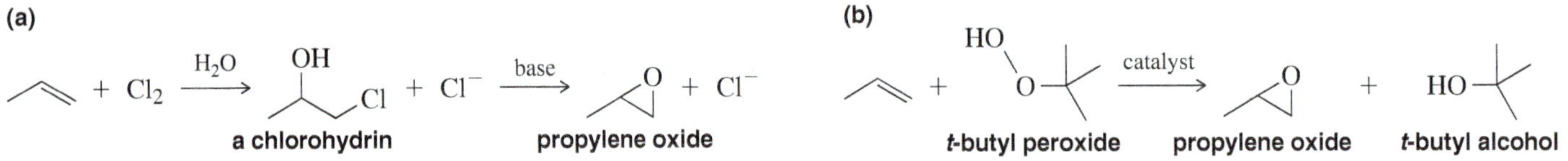

Figure 7.19 Traditional methods for propylene oxide synthesis. (a) Chlorohydrin formation followed by treatment with base. (b) Catalysis with *t*-butyl hydroperoxide.

While developing a process by which waste can be converted into a useful product is consistent with green chemistry principles, even higher on the waste prevention pyramid (Figure 7.2) is avoiding the generation of waste. That is the approach undertaken by Dow and BASF when they developed a catalytic method for converting propylene to propylene oxide using only hydrogen peroxide (**Figure 7.20**). The replacement of *t*-butyl peroxide with hydrogen peroxide results in the formation of only water (GCP 1) as co-product in a highly atom-economical process (GCP 2). Because of the high activity of the catalyst, only moderate temperatures (GCP 6) and pressures (GCP 12) are required, while minimizing how much peroxide is needed for the reaction. DOW and BASF won a 2010 Presidential Green Chemistry Challenge Award for the development of this catalytic process.

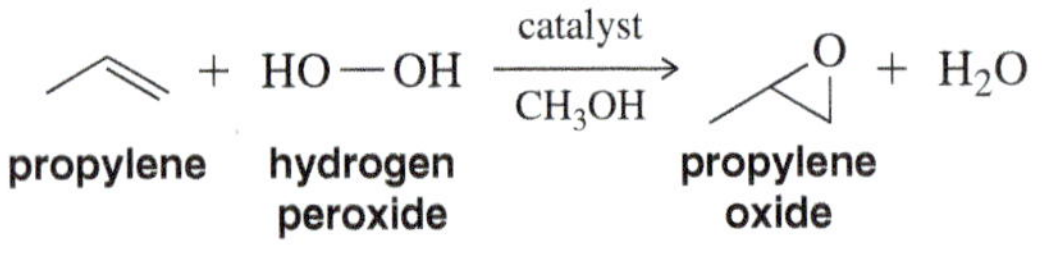

Figure 7.20 Propylene oxide synthesis by catalysis with hydrogen peroxide produces zero waste.

Of the 12 green chemistry principles, catalysis is by far the one used most often in this book. Look for numerous examples in virtually every chapter.

Assessment 7.14 (a) Calculate the atom economy for the reaction shown in Figure 7.20 (epoxidation of propylene). (b) How does it compare to the stoichiometric epoxidation in Assessment 7.8(b)?

Assessment 7.15 (a) On a reaction coordinate diagram, show why a general, one-step reaction is faster when a catalyst is employed. (b) Why does a catalyst often allow a reaction to proceed at lower temperatures?

Assessment 7.16 Osmium tetroxide is a highly toxic compound that is used in the dihydroxylation process shown. Alternatively, it can be used in sub-stoichiometric amounts, when a co-oxidant like NMO is employed. [*After OsO_4 reacts, it is regenerated by NMO to do more dihydroxylations.*] How could this greener alternative process be further improved?

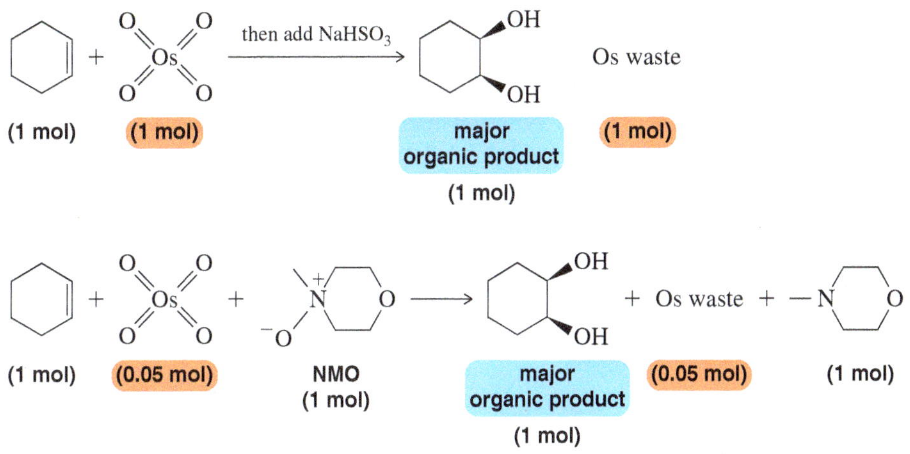

7.2.10 Design for degradation (GCP 10)

When a chemical product is manufactured, the principal concern is how well it will perform its intended function. But once that function has been carried out, what happens next? Will the chemical persist in the environment? Will it accumulate? **Bioaccumulation**, the buildup of chemicals in tissues of living organisms, occurs when an organism's rate of intake of a chemical is greater than its rate of excretion or metabolism. This becomes especially problematic for a chemical compound that is toxic at high concentrations. Nonpolar compounds (which are difficult to excrete due to their insolubility in water) are stored in an organism's nonpolar membranes and fatty tissues, where their concentrations tend to build up. As an example, an individual blade of grass might contain a harmless amount of a toxin, but the cow that eats a field of grass will ultimately store in its tissues the toxin from each and every blade. In turn, when the cow is consumed, those toxins become even more concentrated in the tissues of the human eating the cow. In this way, toxins accumulate up the food chain in a process called bioamplification.

Green Chemistry Principle 4 encourages chemists to design chemicals that, while effectively carrying out their intended function, are nonhazardous in other settings. Green Chemistry Principle 10 takes this one step further. The nontoxic chemical product should be able to degrade into other smaller, nontoxic, or even beneficial chemicals. This not only reduces the chances of a dangerous environmental interaction in the short term, but also ensures that the chemical won't accumulate in higher amounts and cause damage in the long term.

The issue of chemical toxicity is especially pertinent when the intended use of the chemical requires it to be toxic to certain organisms, such as insects and fungi, or even human tumor cells. Such is the case in the insect control industry, where Dow AgroSciences pioneered the

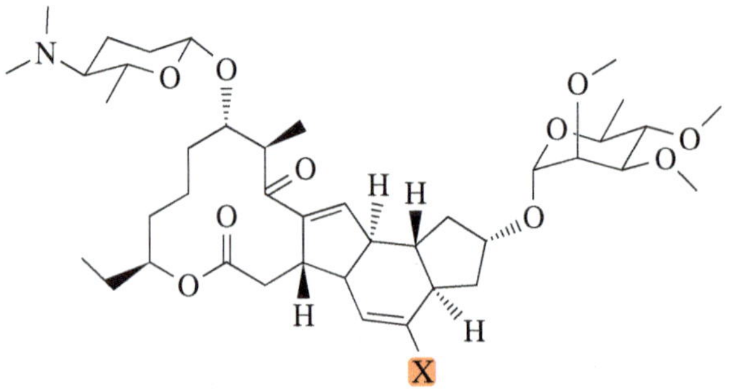

Figure 7.21 Components of spinosad, an environmentally friendly, biodegradable insect control agent.

use of spinosad for controlling chewing insect pests on food crops. Spinosad, which is composed of the natural products spinosyn A and D (**Figure 7.21**), is the product of microbial fermentation from *Saccharopolyspora spinosa*, a microorganism discovered in a Caribbean soil system. Spinosad is considered a green insecticide for several reasons. It exhibits low toxicity to mammals and birds (GCP 4) and shows activity only on chewing insect pests. Most beneficial insects, like wasps, are left unharmed. Because of its high molecular weight and strong adsorption to soils, it neither volatilizes nor leaches into groundwater. Importantly, it degrades by a photochemical mechanism when exposed to light, ensuring that it does not accumulate in the environment (GCP 10). Finally, it is produced via fermentation, so it is derived from renewable feedstocks (GCP 8). The development of Spinosad earned Dow AgroSciences LLC a Presidential Green Chemistry Challenge Award in 1999.

7.2.11 Analyze in real-time (GCP 11)

In a normal laboratory setting, chemists typically only analyze a reaction after it is finished. Nuclear magnetic resonance (NMR) spectroscopy (Chapter 15), infrared (IR) spectroscopy (Chapter 14), and/or mass spectrometry (Chapter 14) are used to confirm the identity of the product formed when it is too late to make changes that will put the reaction on a better path. Luckily, when reactions are run on a small scale, the consequences of a "bad reaction" are minimal.

Now imagine a reaction using multi-kilogram quantities of starting material in 1000-L batch reactors under high pressures and temperatures in an industrial plant. The quantity of waste associated with a reaction like this "going bad" is significant. And what about a situation where a reaction is spontaneously heating up, or pressure is dangerously increasing? Such situations can cause the type of chemical accidents that make the news. Thus, Green Chemistry Principle 11 advocates the use of real-time reaction analysis on timescales that allow problems to be solved as they occur. This can be done according to the scheme shown in **Figure 7.22** by using reaction information to immediately adjust reaction conditions.

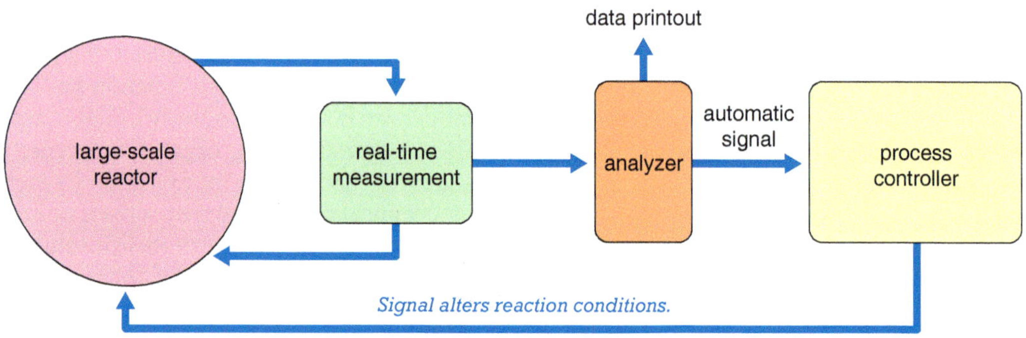

Figure 7.22 General real-time analysis scheme. Image taken from Lancaster, *Green Chemistry*, 3rd edition.

In the biodiesel industry, international standards limit the amounts of glycerol, mono-, di-, and triglycerides, alcohol, moisture, and free fatty acids that can be present in usable fuel (**Figure 7.23**). In a conventional approach, determining the concentrations of these contaminants is done only when the reaction, followed by an extensive purification process, is complete. At that point, huge quantities of time, energy, and resources have been invested in producing this batch of biodiesel. If the analysis is satisfactory, the fuel is sold. If the results are unsatisfactory, the batch must be reprocessed and repurified [*more time, more energy*] or, even worse, discarded as waste. Both of these problems can be avoided with real-time analysis, in which analysis of the components suggests whether the reaction needs increased reaction times, more reagents, or a higher temperature. The Pimentel research group in Brazil has recently developed a process using infrared spectroscopy (Chapter 14) to monitor reactions in real time, thereby reducing costs and time while improving the quality of the final product.

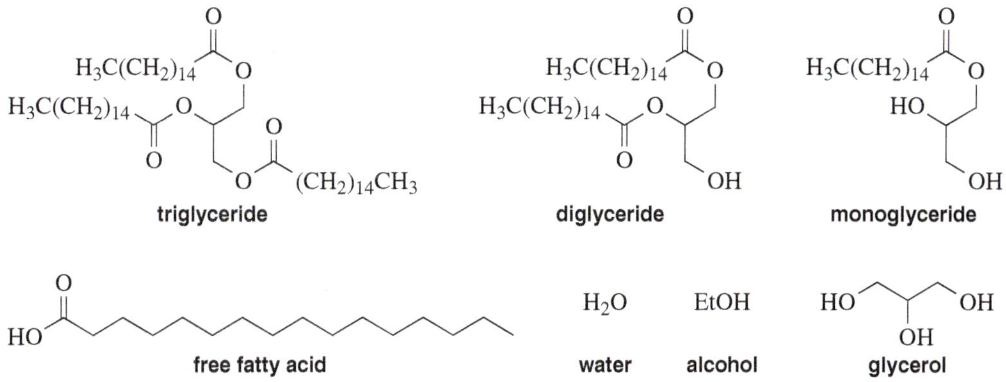

Figure 7.23 **Potential contaminants of biodiesel fuel.**

7.2.12 Prevent accidents (GCP 12)

According to Green Chemistry Principle 12, reactions and manufacturing processes should be chosen to minimize the possibility of accidents. Whereas most green chemistry principles are concerned with the environment and/or longer-term impacts, Green Chemistry Principle 12 first and foremost seeks to protect workers, and in the case of preventing larger accidents, to safeguard the surrounding community. Because accidents by definition are unexpected occurrences, it is difficult to avoid them completely. However, careful analysis prior to using a particular process can prevent a large number of problems. Consider a situation where there are two routes to a particular molecule. One is more expensive but offers fewer hazards than the other; which do you choose? Perhaps there is only one, potentially dangerous, route to a particular chemical product—should it be pursued at all? GCP 12 implores us to consider potentially dangerous reaction conditions and the possibilities for an accident as important factors in determining whether a process should be used. As well, it encourages the development of newer, less dangerous processes to replace older, dangerous processes.

Disodium iminodiacetate is an intermediate in the synthesis of glyphosate, the active ingredient in Monsanto's herbicide Roundup® (**Figure 7.24**). Traditionally, disodium iminodiacetate was synthesized using the Strecker process (Chapter 26), which is employed in the synthesis of amino acids. The Strecker process utilizes highly toxic formaldehyde and hydrogen cyanide and is conducted at high temperatures. Because it is an exothermic reaction, special conditions must be used to ensure that a runaway reaction doesn't occur. For these and other reasons, it was a reaction in need of greening.

Figure 7.24 **The traditional synthesis of disodium iminodiacetate** involves a high degree of risk to workers.

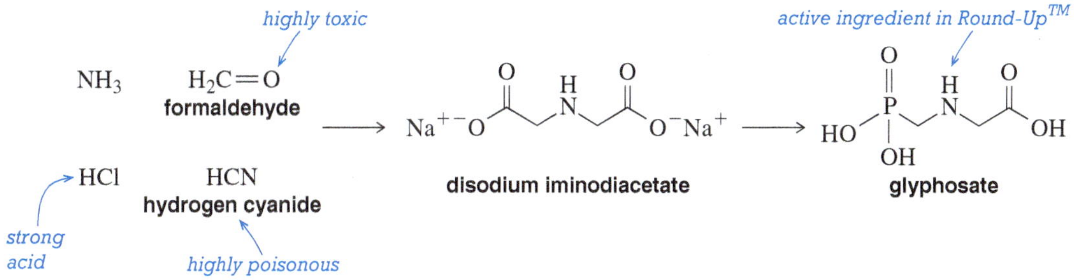

In 1996, Monsanto received a Presidential Green Chemistry Challenge Award for an alternative process that synthesizes disodium iminodiacetate via the copper-catalyzed dehydrogenation of diethanolamine (**Figure 7.25**). The new process requires no purification (GCP 5), produces no waste (GCP 1), and avoids the use of toxic chemicals (GCP 3), and because it is endothermic, no dangerous runaway reaction is possible. This new synthesis not only is better for the environment, but also is safer to run for the worker (GCP 12).

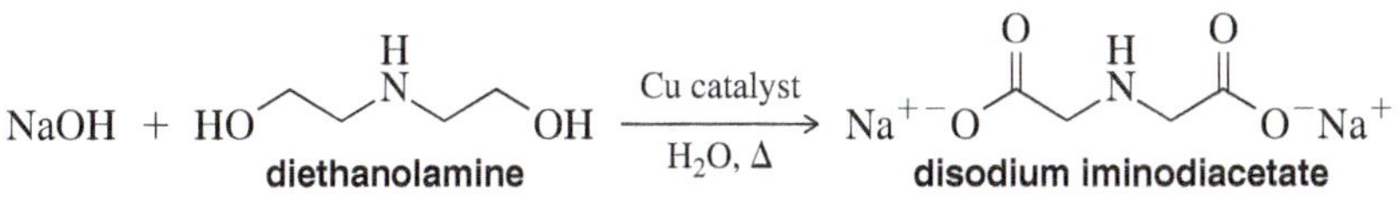

Figure 7.25 **Monsanto's safer process for the synthesis of disodium iminodiacetate.**

Concluding thoughts

It is a moral imperative that we should do all we can to avoid chemical disasters that disproportionately affect the poor and disenfranchised (e.g., the victims in the slums of Bhopal). Despite the chemical industry improving literally every aspect of our lives, its reputation among the population is just above that of the tobacco industry. Moving toward more sustainable processes will allow society to recognize the many benefits provided by the chemical industry.

Chapter 8 begins a run of 16 chapters (out of 18) that focus on reactions and reaction mechanisms that can be used to make new medicines, polymers, and pesticides. Because of the requirements of the standard second-year organic chemistry curriculum, many reactions you will study will be of the nongreen variety. Nevertheless, the idea of green chemistry is threaded throughout the narrative of the book and hopefully will lurk in the subconscious of you, the reader [*and future scientist/health professional/citizen*]. Our study of reactions begins with alkenes—let's go there now!

End-of-Chapter Materials

Important Concepts

7.1: The United States Environmental Protection Agency created the term *green chemistry* "to promote innovative chemical technologies that reduce or eliminate the use or generation of hazardous substances in the design, manufacture and use of chemical products." In 1998, Paul Anastas and John Warner defined the 12 Principles of Green Chemistry.

7.2: The 12 Principles of Green Chemistry are as follows:

Principle	Summary	Section
1. Prevention	It is better to prevent waste than to treat or clean up waste after it has been created.	7.2.1
2. Atom economy	Synthetic methods should be designed to maximize the incorporation of all materials used in the process into the final product.	7.2.2
3. Less hazardous chemical synthesis	Wherever practicable, synthetic methods should be designed to use and generate substances that possess little or no toxicity to human health and the environment.	7.2.3
4. Designing safer chemicals	Chemical products should be designed to affect their desired function while minimizing their toxicity.	7.2.4
5. Safer solvents and auxiliaries	The use of auxiliary substances (e.g., solvents, separation agents) should be made unnecessary wherever possible and innocuous when used.	7.2.5
6. Design for energy efficiency	Energy requirements of chemical processes should be recognized for their environmental and economic impacts and should be minimized. If possible, synthetic methods should be conducted at ambient temperature and pressure.	7.2.6
7. Use of renewable feedstocks	A raw material or feedstock should be renewable rather than depleting whenever technically and economically practicable.	7.2.7
8. Reduce derivatives	Unnecessary derivatization (use of blocking groups, protection/deprotection, temporary modification of physical/chemical processes) should be minimized or avoided if possible, because such steps require additional reagents and can generate waste.	7.2.8
9. Catalysis	Catalytic reagents (as selective as possible) are superior to stoichiometric reagents.	7.2.9
10. Design for degradation	Chemical products should be designed so that at the end of their function they break down into innocuous degradation products and do not persist in the environment.	7.2.10
11. Real-time analysis for pollution prevention	Analytical methodologies need to be further developed to allow for real-time, in-process monitoring and control prior to the formation of hazardous substances.	7.2.11
12. Inherently safer chemistry for accident prevention	Substances and the form of a substance used in a chemical process should be chosen to minimize the potential for chemical accidents, including releases, explosions, and fires.	7.2.12

End-of-Chapter Assessments

[The assessments that follow represent the application of green chemistry principles to each of the remaining chapters. Because you haven't studied many organic reactions yet, you may not be able to say much about the specific reactions. But you can certainly answer the green chemistry portion shown in green type. After you study the chemistry in later chapters, you can return and answer the other aspects of each question. In other words, don't worry about what you aren't supposed to know yet. Assessments are ranked according to their difficulty; •••• = most difficult.]

7.17 (••) LOOKING AHEAD CHAPTERS 8, 17 In Chapters 8 and 17 we learn two reaction for the synthesis of the alcohol shown. (a) Show a mechanism for each of the reactions. (b) If you were designing a synthetic route, which would be considered more sustainable? Consider all factors. [*Assume the starting organic molecules are equally green.*]

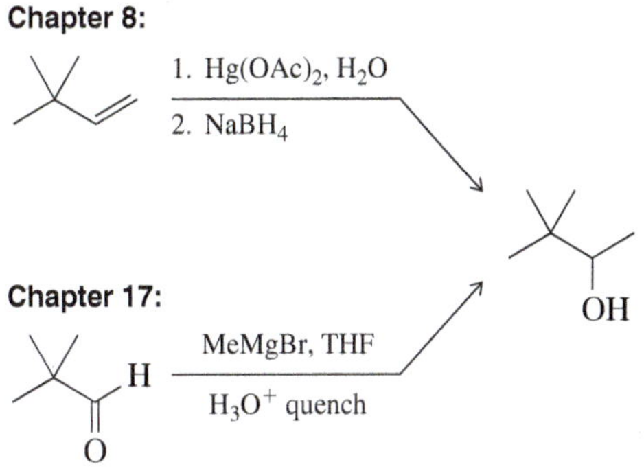

7.18 (•••) LOOKING AHEAD CHAPTERS 9, 13, 27 Dihydroxylated steroids are commonly found in food and may be interesting drug targets. (a) Draw a mechanism for the epoxide opening reaction used to prepare a dihydroxylated steroid. (b) A recent paper reported the first epoxide opening on a steroid scaffold performed in water using Teflon™-coated sand to provide a surface on which the reaction occurs. What might have prevented water from being used in previous syntheses? (c) Identify which green chemistry principles are demonstrated in this reaction.

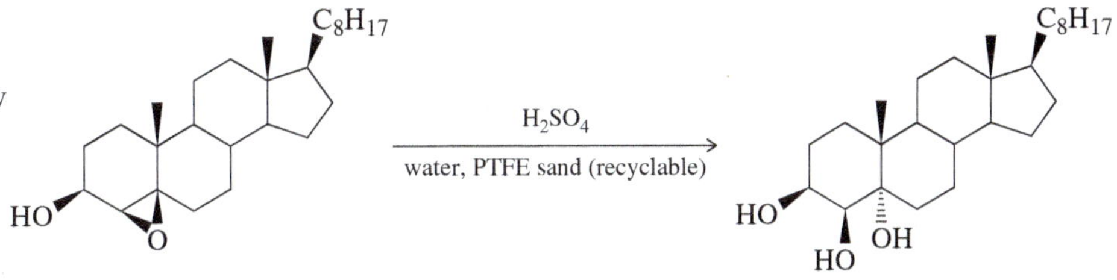

7.19 (•••) LOOKING AHEAD CHAPTER 10 The following reaction was recently reported to have been performed electrochemically.

(a) Identify the reagent and a solvent that could have been used if this reaction was done traditionally. (b) What safety and environmental hazards are improved or worsened when doing this reaction electrochemically?

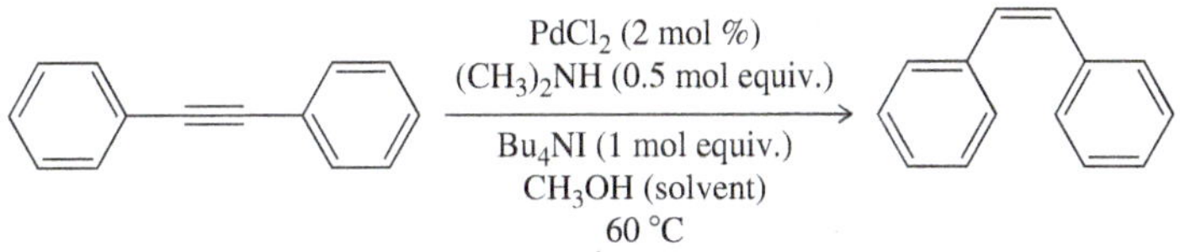

7.20 (••••) LOOKING AHEAD CHAPTER 11 Suppose you intend to synthesize the secondary alkyl halide from each reaction shown. (a) What is the atom economy of each reaction? (b) What is the E-factor, effective mass yield, and process mass intensity? (c) Which reaction would you consider greener? Why?

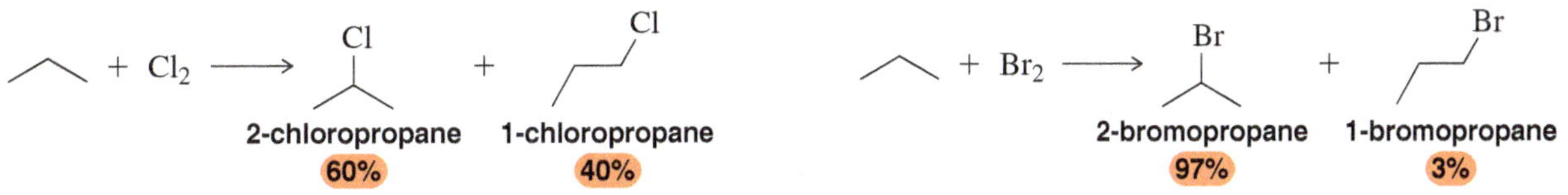

7.21 (•••) **LOOKING AHEAD CHAPTER 12** A 2017 study reported the following enantioselective synthesis. At first glance, this looks like an S_N1 reaction that produces a single enantiomer of product. (a) Based on what you know about S_N1 reactions, why is the enantioselectivity unexpected? In truth, this reaction proceeds by a different mechanism [*you don't need to know it now*], and the enantioselectivity is due to the presence of the chiral ligand. (b) Explain why this catalytic approach is more environmentally friendly than using an S_N1 reaction to make the same product. (c) How does the use of a complex ligand like the one shown reduce the sustainability of a process, despite it being used catalytically?

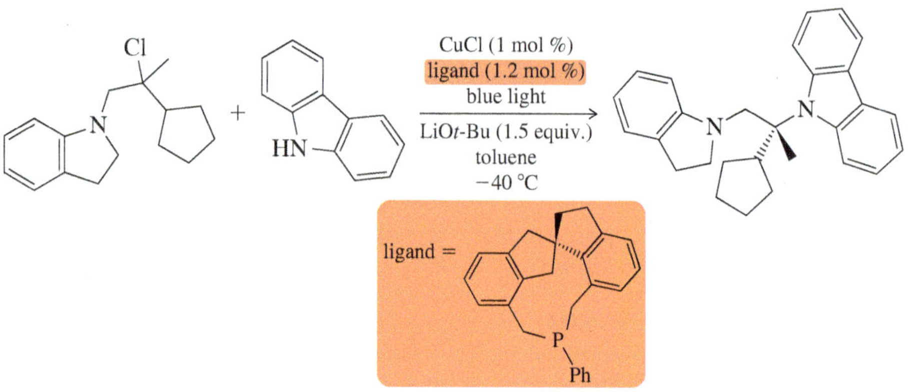

7.22 (••) **LOOKING AHEAD CHAPTER 13** A catalytic system involving a Cu(I) salt and the organic oxidant TEMPO has been utilized for the green oxidation of alcohols. TEMPO performs the oxidation of the alcohol, while the Cu(I) salt reoxidizes TEMPO in the presence of oxygen (O_2). (a) Predict the product of the reaction—is it an aldehyde or a carboxylic acid? (b) Justify your choice. (c) Why is this reaction preferable to the traditional use of PCC for this transformation? (d) How could this reaction be made even more sustainable?

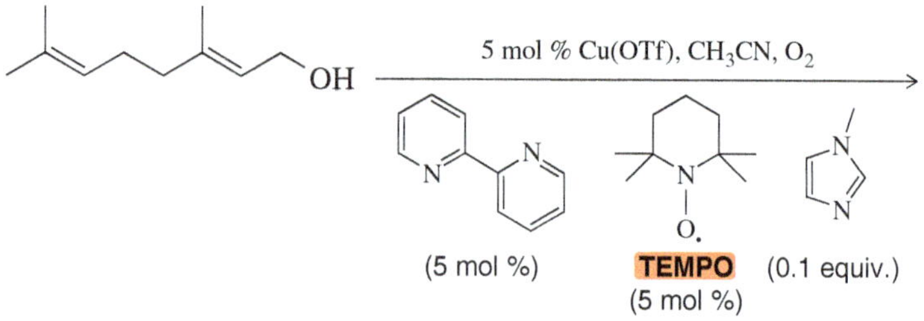

7.23 (••) **LOOKING AHEAD CHAPTER 16** Suzuki couplings provide an easy way to make C—C bonds and so are common in the chemical and pharmaceutical industries. Because of their popularity, looking for environmentally sustainable syntheses using these reactions is an active field of research. A 2015 study found that Suzuki couplings could be carried out in water, rather than an organic solvent, if 0.5% of sulfocalix[8]arene [*image below*] was added to the solution. Explain this result. [*Hint: Revisit the discussion of micelles.*]

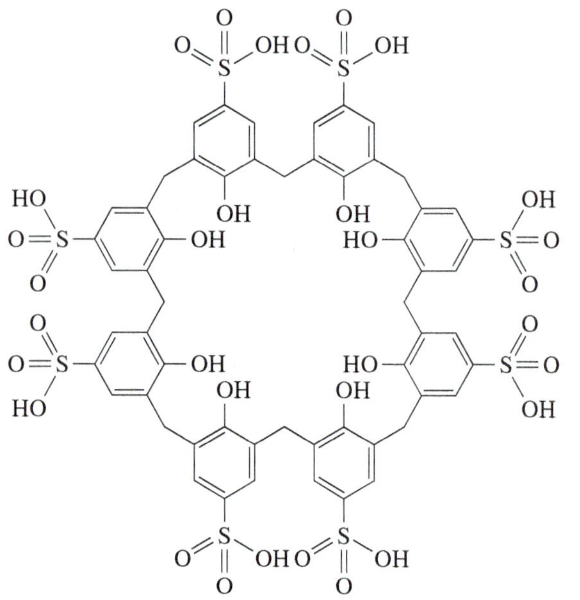

7.24 (••) **LOOKING AHEAD CHAPTER 17** Traditionally run in an organic solvent like THF with a strong base, this Wittig reaction is conducted in water (a safer solvent) using sodium hydroxide as the base. (a) Provide a mechanism for the reaction. (b) Why can sodium hydroxide be used to form the ylide in this reaction? (c) What is the purpose of the hydrochloric acid in the second step? (d) Calculate the atom economy of this reaction.

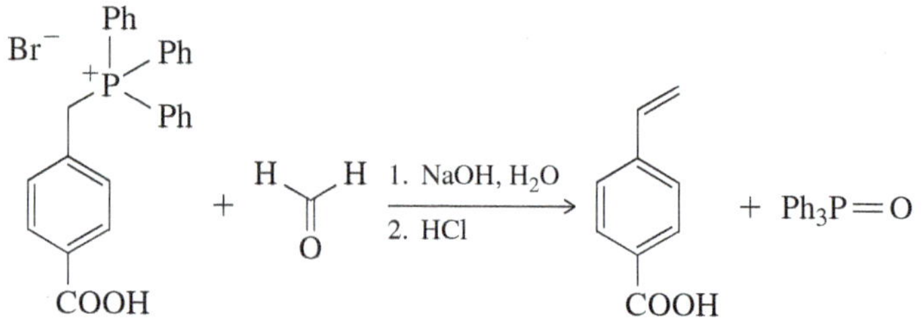

7.25 (•) **LOOKING AHEAD CHAPTERS 14, 18** A reaction can be analyzed in real time using infrared spectroscopy. A specialized IR probe is placed into the reaction solution, and the operator watches the IR spectra change as the starting material is converted to products. If you were the operator for the following reaction, what change would you look for in the IR spectra?

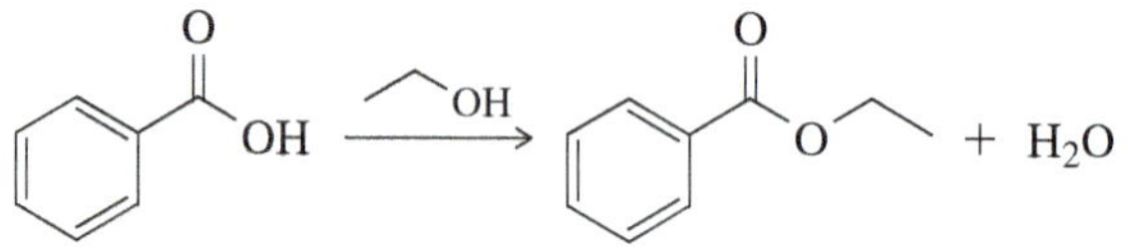

7.26 (••) LOOKING AHEAD CHAPTERS 19, 27 A Presidential Green Chemistry Award was awarded in 2008 for the development of a laser printer ink using a diacid (isolated from soybeans) and a diamine. Following polymerization, the ink easily degrades, ultimately increasing the recyclability of discarded paper. (a) Predict the product and draw a mechanism for the reaction between the diacid and the diamine shown. (b) Propose a method for breaking the polymer bonds to remove the ink when the paper is recycled. (c) What green chemistry principles do you observe in this process?

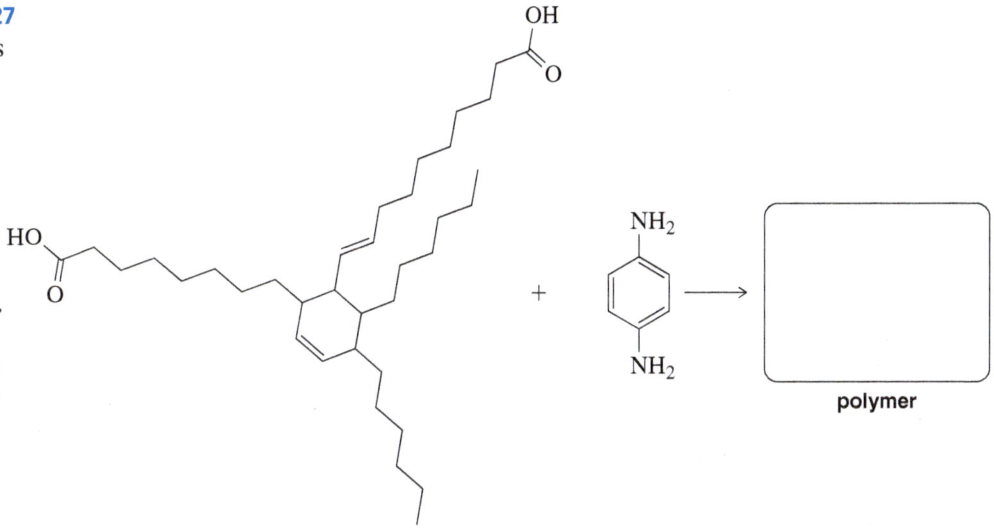

7.27 (••••) LOOKING AHEAD CHAPTER 20 When the chirality of the product is important, substitution at the carbon alpha to a carbonyl can be accomplished by the use of a chiral auxiliary. After these commercially available chiral compounds are attached to the molecule, the desired substitution is carried out at the alpha carbon, and the auxiliary is then removed. (a) Why can't an enantiopure mixture be made directly without the use of the chiral auxiliary? (b) There is significant research into replacing chiral auxiliaries with catalysts. Explain the environmental and economic advantages of using a catalyst in the following reaction.

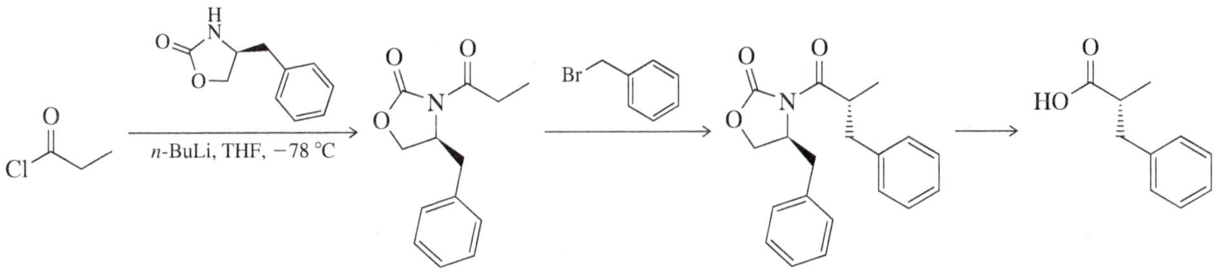

7.28 (•••) LOOKING AHEAD CHAPTERS 17, 21, 25, 26 Conventional organic molecules used in sunscreens have recently been criticized for damaging coral reefs when they leach into the ocean. Molecules produced by the algae living in coral reefs have been examined as alternatives. The final step in the biosynthesis of one of these molecules, shinorine, is shown here. (a) Draw a mechanism for this transformation, assuming the enzyme active site has whatever acids and bases are needed. (b) Which part of this molecule allows it to absorb UV light? (c) What green chemistry principle is illustrated here?

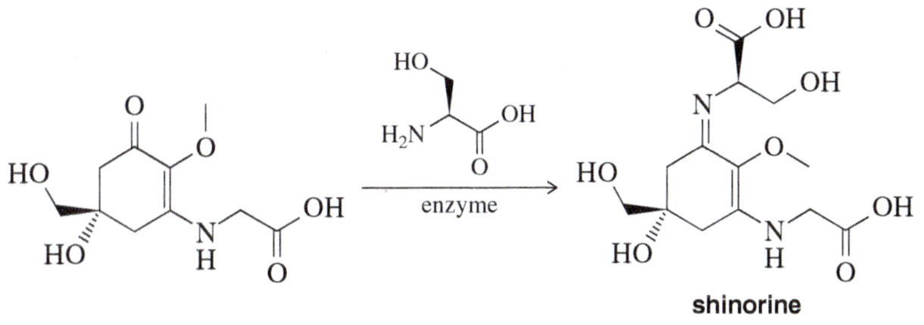

7.29 (•••) LOOKING AHEAD CHAPTER 22 A 2018 study discovered that the terpene farnesene could undergo a pericyclic reaction to give a monomer useful for synthesizing biodegradable polymers. (a) For the synthesis of monomers **A**, **B**, and **C**, identify the type of pericyclic reaction and the starting material that reacted with farnesene. (b) What green chemistry principles do you see in the synthesis and design of these polymers?

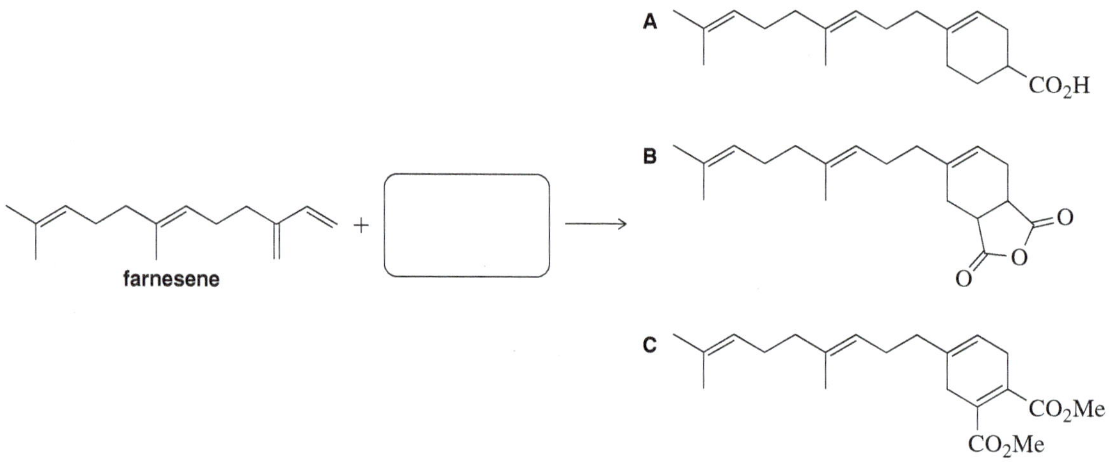

7.30 (••) LOOKING AHEAD CHAPTER 23 Ionic liquids have long held promise as green solvent alternatives. Ionic liquids are organic salts that are liquid at room temperature. As typically used, upon completion of a reaction, extracting the ionic liquid with a solvent like diethyl ether removes the organic product, leaving the ionic liquid unchanged for use in a later reaction. (a) Draw a mechanism for the Friedel–Crafts reaction shown here. [*It is similar to the conjugate addition reaction of Chapter 21.*] (b) Calculate the atom economy of the reaction. (c) What is a limiting factor to the greenness of this reaction?

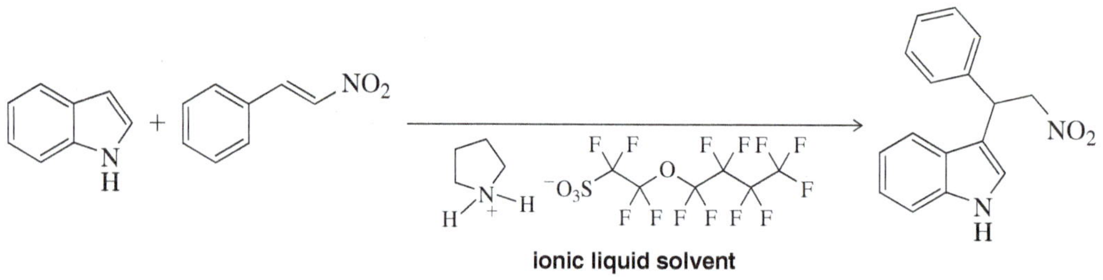

7.31 (••••) LOOKING AHEAD CHAPTER 24 Eastman Chemical Company won a 2009 Greener Synthetic Process Award for the use of enzymes to form esters. The following reaction was used to prepare an active ingredient in medicated skin creams. (a) Propose a synthesis for the aromatic starting material starting from either phenol or toluene. (b) What environmental and economic advantages does enzyme catalysis have over traditional synthesis?

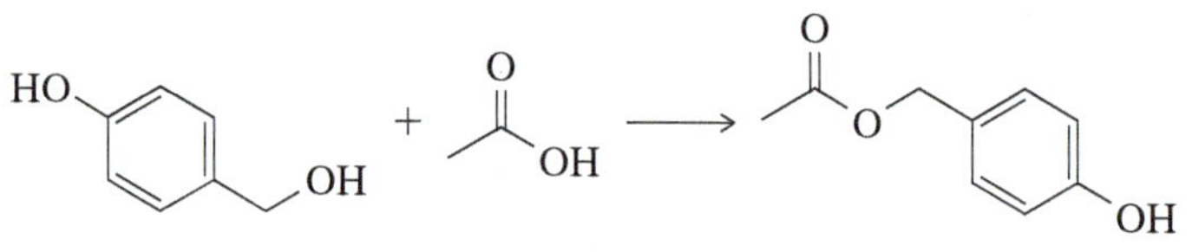

7.32 (••) LOOKING AHEAD CHAPTER 25 The pharmaceutical company AbbVie was interested in developing a fast Buchwald–Hartwig amination. By using the reaction conditions shown here, they were able to perform the reaction very quickly on a variety of substrates. Identify at least two green chemistry principles used in their method.

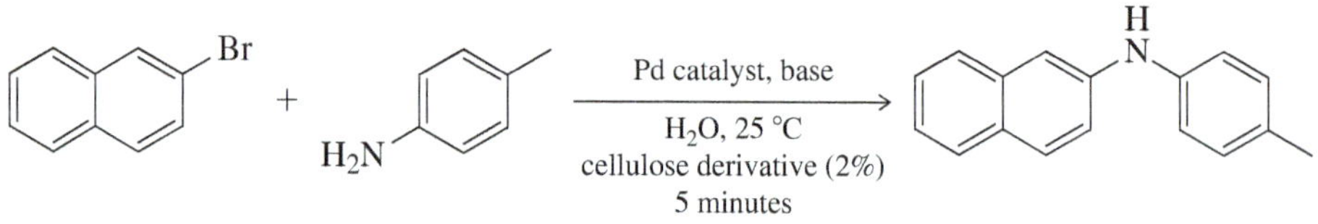

7.33 (••) LOOKING AHEAD CHAPTER 26 Oxytocin (brand name Pitocin®) is a naturally occurring cyclic polypeptide used to induce labor with the following amino acid sequence:

CYIQNCPLG

The cyclic portion of the molecule is closed by a sulfide bond. (a) Draw the molecule. (b) Describe the greenest method to synthesize the peptide chain. [*You may ignore the formation of the ring.*]

7.34 (••••) LOOKING AHEAD CHAPTER 27 A derivative of chitin called chitosan is sometimes incorporated into bandages to aid healing. A dimer of chitosan is shown below. (a) Propose a reaction to convert the chitin dimer into a chitosan dimer. [*Note: The glycoside linkage is unstable in acid.*] (b) What green chemistry principle is demonstrated when using chitosan as a healing agent as opposed to some other synthetic molecule?

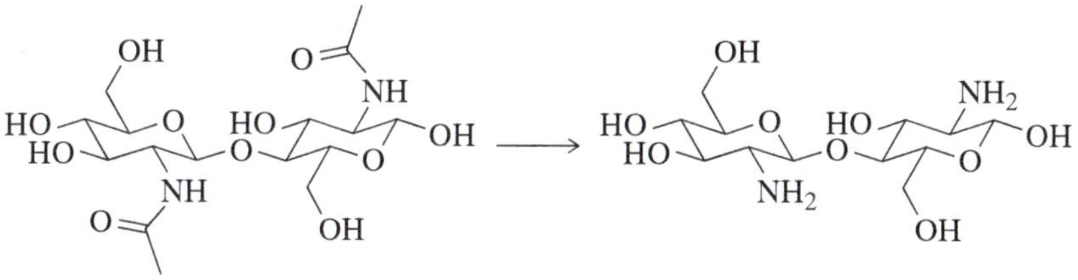

chitin chitosan

7.35 (••) LOOKING AHEAD CHAPTER 28 Ring-opening metathesis polymerization (ROMP) employs the Grubbs catalyst in water for the synthesis of polymers. (a) Draw the polymer you'd expect for the following ROMP reaction. (b) Map the carbons in the starting material and product. (c) Calculate the atom economy for the reaction.

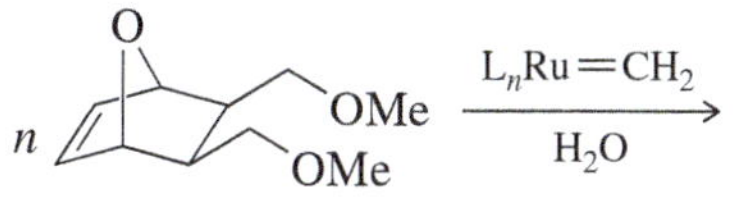

Bouncing Arrows as a Teaching Tool

This is a tool that will help your understanding of mechanisms, but you may not get full credit if you draw mechanisms this way. **Check this with your instructor.**

When studying resonance in Chapter 2, the acid–base reactions of Chapter 4, and the radical reactions of Chapter 5, we began using the electron-pushing formalism. Through the use of arrows that begin at electron-rich atoms/bonds and end at electron-deficient atoms [*negatives attack positives*], this formalism is essentially an accounting system for the movement of electrons in chemical reactions. As demonstrated in **Figure 1** for an acid–base reaction, these arrows allow us to justify product formation, explain how the product is formed, and identify how electrons change places in going from reactant to product. There is relatively little ambiguity in recognizing that the lone pair that used to be localized on nitrogen is sharing its electrons with hydrogen to create a new N—H bond. Simultaneously, the O—H bond is breaking, and the electrons that used to be shared between hydrogen and oxygen are being localized onto oxygen as a new lone pair.

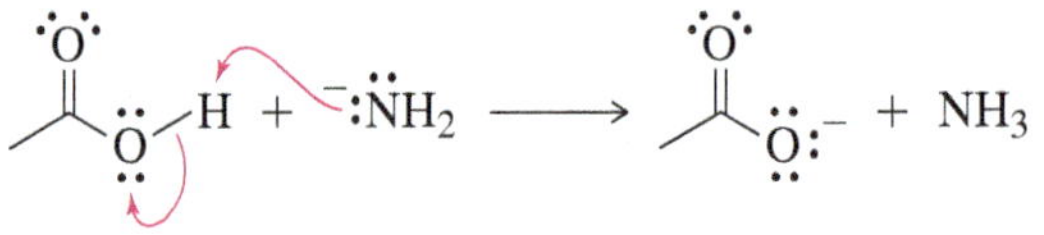

Figure 1 Arrow pushing reveals many details of acid–base reactions unambiguously.

When we begin looking at alkene reactions in Chapter 8, the traditional arrow-pushing formalism will not always be so unambiguous. To see why, consider the arrow drawn in **Figure 2**. Based on the arrow that's shown, which product, **A** or **B**, is forming? [*Note that this is simply an acid–base reaction where the π electrons are grabbing a proton from the hydronium ion.*]

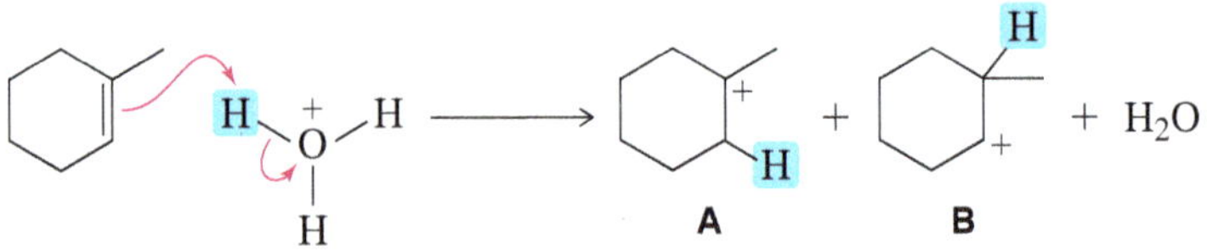

Figure 2 Which product is forming based on the arrow shown?

Unlike in Figure 1, the arrow alone in Figure 2 doesn't make it obvious which product is forming. We show in Chapter 8 that the carbocation in **A** is more stable than the one in **B**, and so the mechanism in **Figure 3** means that **A** forms—except that the arrow isn't what tells you that. The arrow tells you that the π electrons attack. Your knowledge of the relative stability of carbocations tells you that **A** must form. This is how organic chemists have been illustrating alkene addition reactions since the invention of the electron-pushing formalism.

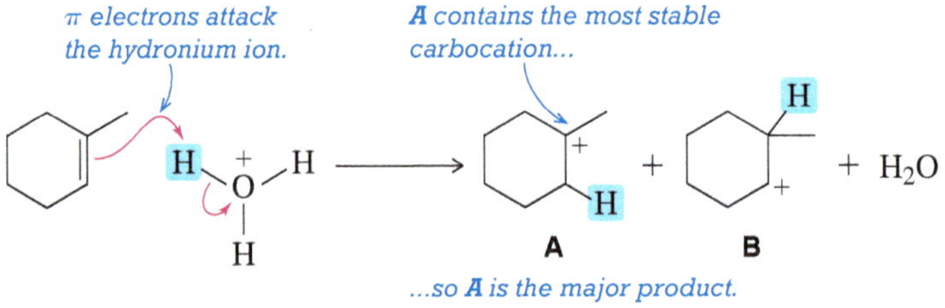

Figure 3 Only because you [*will eventually*] know that carbocation A is more stable does the arrow make clear that A forms using the arrow-pushing formalism.

In the early 2000s, a chemistry professor at Virginia Tech introduced bouncing arrows as a less ambiguous alternative to traditional arrows. The mechanism in Figure 3 is shown again in **Figure 4** but using a bouncing arrow. [*Note that the electrons don't really bounce; this is just an extension of a formalism and should not be understood as what is happening in solution.*] The arrow makes clear that when the π bond attacks the proton, the electrons from the π bond remain attached to C_2. In order for the π electrons to attack from C_2, they must be taken away from C_1, leaving C_1 as a carbocation. [*We haven't studied carbocations yet, but they are carbons with only six electrons in their valence shell.*]

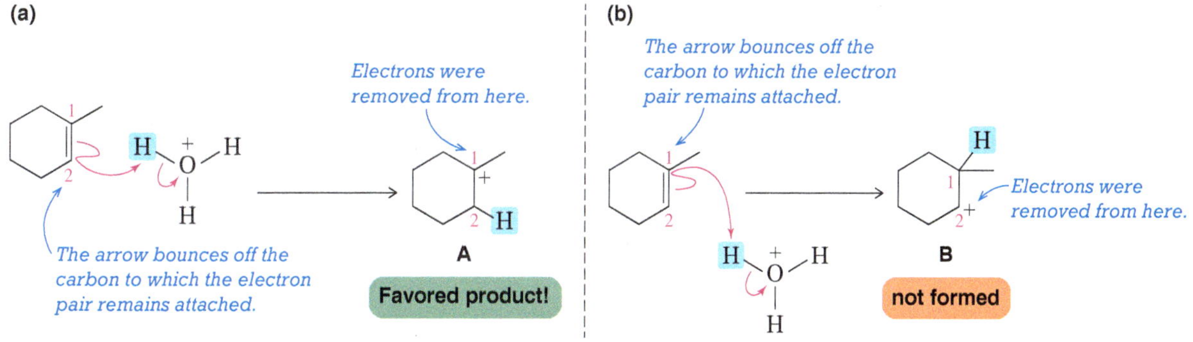

Figure 4 The bouncing arrow gives more information on what product to expect from a given arrow. (a) The new bond forms at C1 in the favored reaction. (b) The new bond forms at C2 in the disfavored reaction.

Directly comparing traditional arrows with bouncing arrows reveals a single difference. Whereas the traditional arrow itself tells you "the π electrons attack," the bouncing arrow adds "from here." Occasionally in this text, when we encounter a situation where bouncing arrows enhance the discussion, the mechanism is modified as shown in **Figure 5**. In these cases, the "extra information" available from the bouncing arrow is highlighted alongside the traditional arrow. [*Though bouncing arrows will enhance your understanding of mechanisms, you'll want to confirm with your instructor whether they will be accepted as viable exam answers.*]

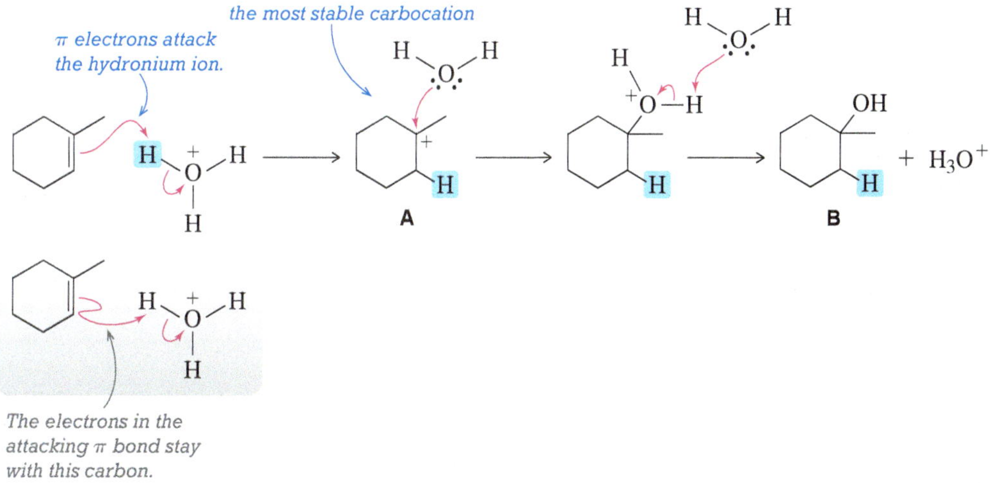

Figure 5 When bouncing arrows are helpful, the traditional mechanism will occasionally be annotated with bouncing arrow steps.

Table 1 gives multiple examples of the information you'll get from the use of bouncing arrows throughout the text. Don't worry about these mechanisms now; you'll encounter all of them again.

<table>
<tr><th>Entry</th><th>Example</th></tr>
<tr><td>1</td><td>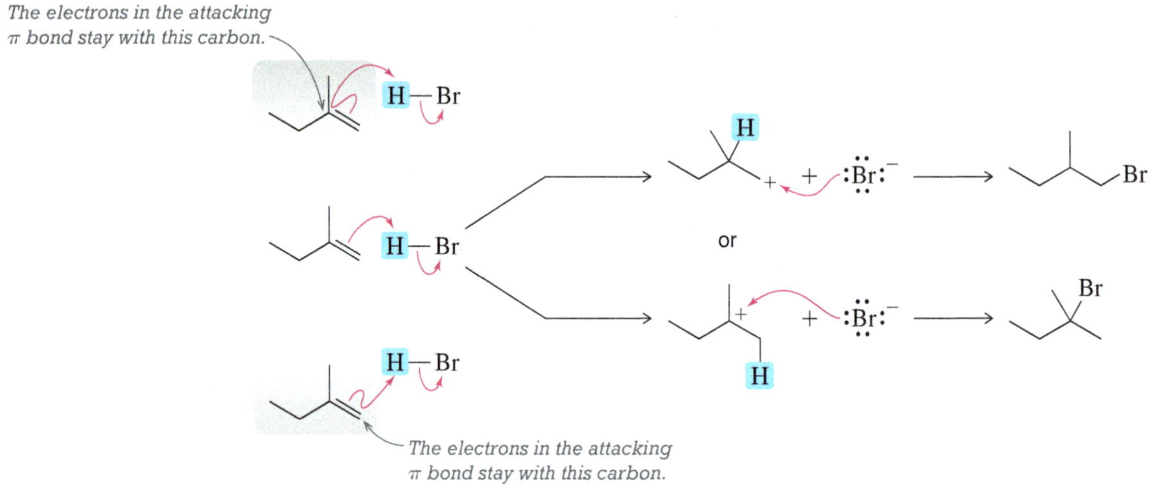
</td></tr>
<tr><td>2</td><td>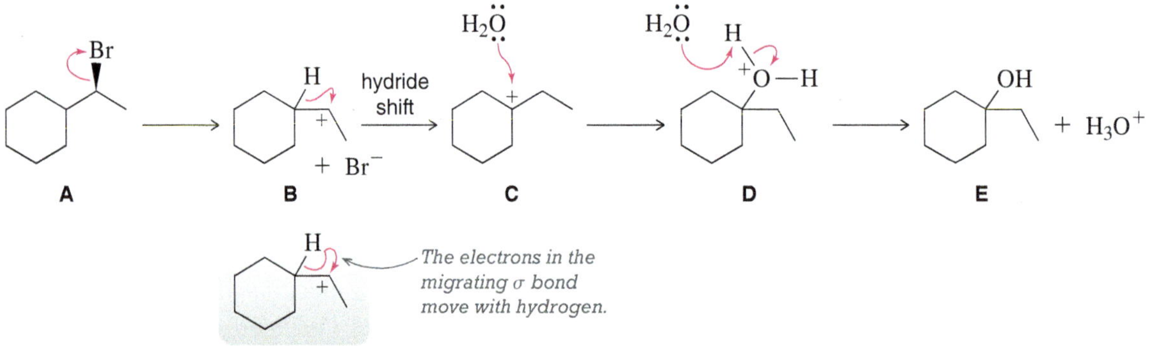
</td></tr>
<tr><td>3</td><td>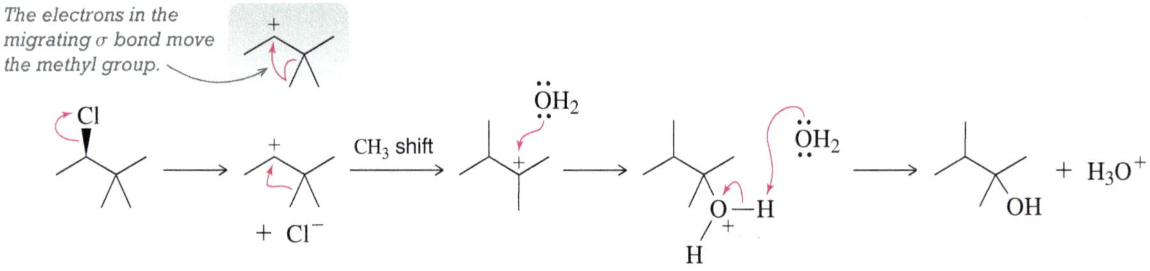
</td></tr>
</table>

(continued)

Don't worry if you don't see bouncing arrows in every mechanism! They only apply in certain reactions.

Table 1 ***Continued***

Entry	Example
4	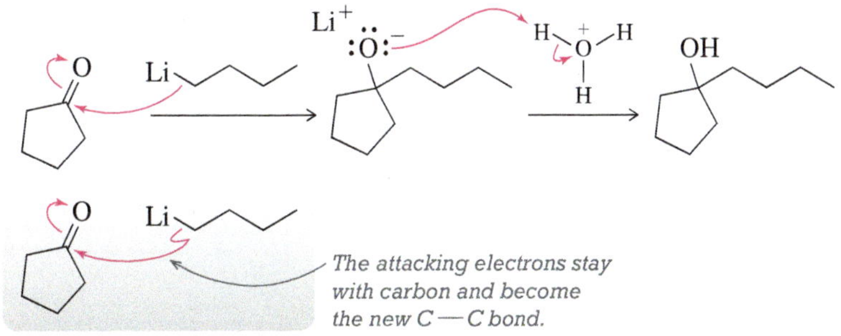
5	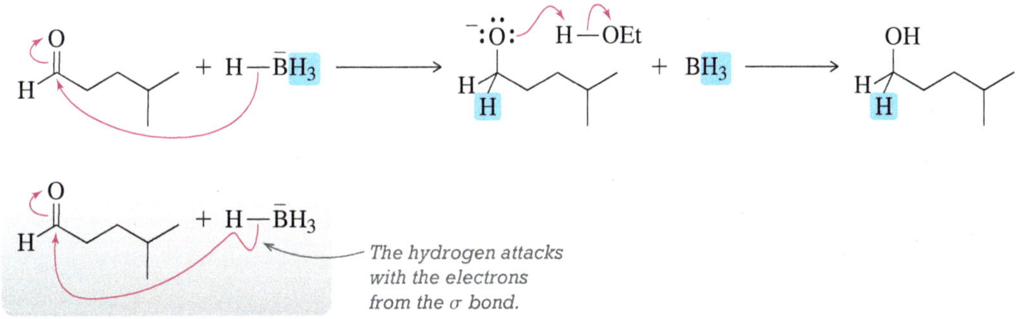

8

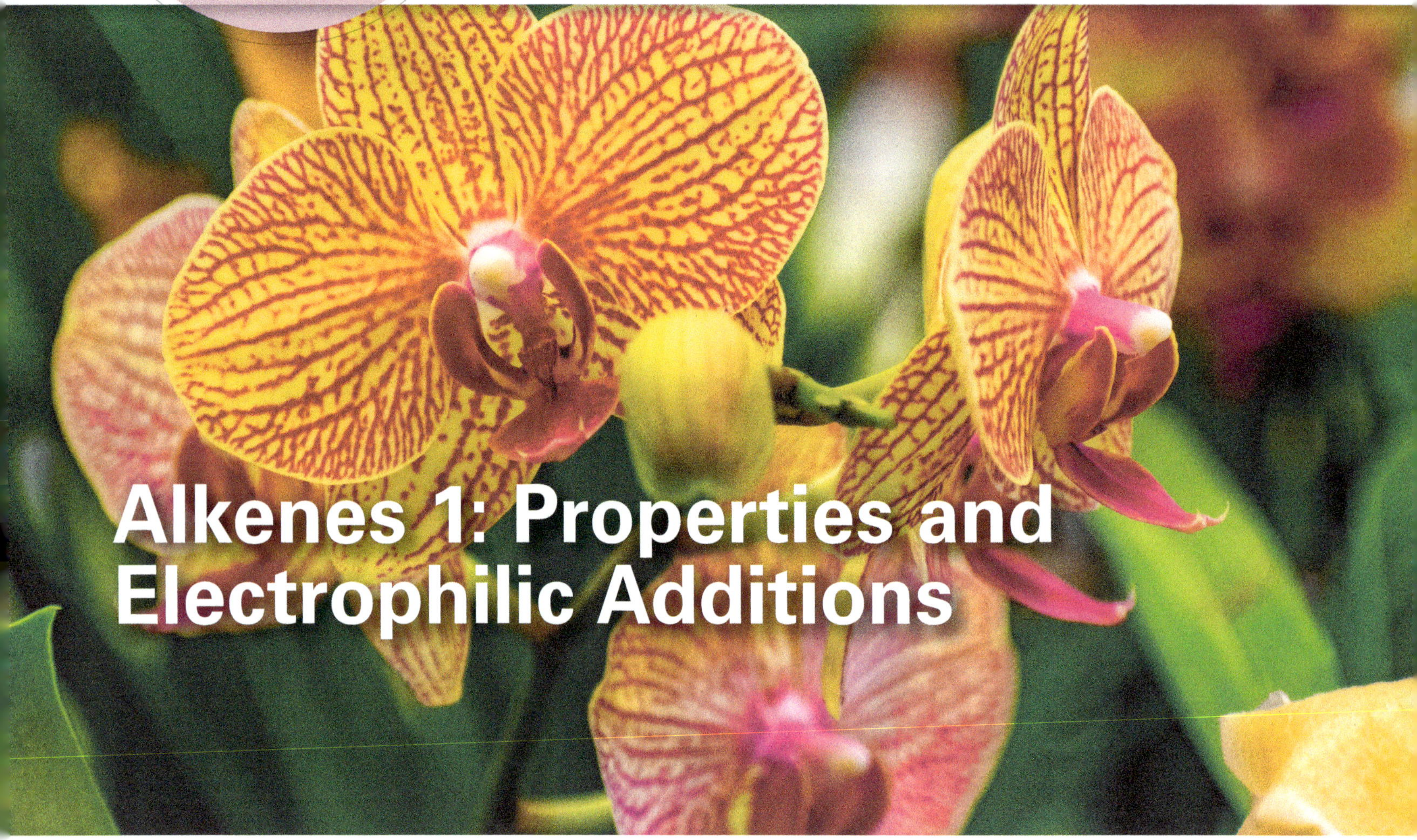

Alkenes 1: Properties and Electrophilic Additions

Learning Objectives

LO 8.0 Analyze the physical properties of alkenes

LO 8.1 Determine the IUPAC names for alkenes and their isomers

LO 8.2 Compare the relative stability of isomers of alkenes

LO 8.3 Apply the isoprene rule

LO 8.4 Analyze addition reactions of alkenes

LO 8.5 Analyze the radical addition reactions of alkenes

LO 8.6 Predict the products of the addition of water to an alkene

Introduction

Have you ever experienced emotional responses to memories triggered by certain scents? They could be good memories, like the scent of Play-Doh reminding you of your childhood or the odor of freshly mowed grass reminding you of outdoor adventures. [*But they are not always positive—the smell of organic molecules on a trip to my graduate school laboratory vividly brought to mind the anxiety, nervousness, and excitement I felt during my doctoral studies.*]

Because only a few synapses separate the olfactory nerve from the amygdala (the part of your brain responsible for emotion) and the hippocampus (the part responsible for your memories), the sense of smell is closely linked to emotional memory. Within the olfactory nerves, nearly 400 different odor receptor proteins become activated by organic molecules [*of course!*] of the appropriate size and shape, transmitting a signal to the brain.

The first chemist on record is Tapputi-Belatekallim, a Babylonian perfumer from around 1220 B.C. Chemicals used in early perfumes were procured from plants and flowers as essential oils. In 800 A.D., Arab philosopher and scientist Al-Kindi experimented with the combination of **terpenes**, or **terpenoids**, isolated from plants, to produce different scents. Still used in the modern perfume industry, terpenes are derived biosynthetically from five-carbon isoprene units (**Figure 8.1**). Due to their widespread occurrence in nature, they have attracted attention as a class of natural products that

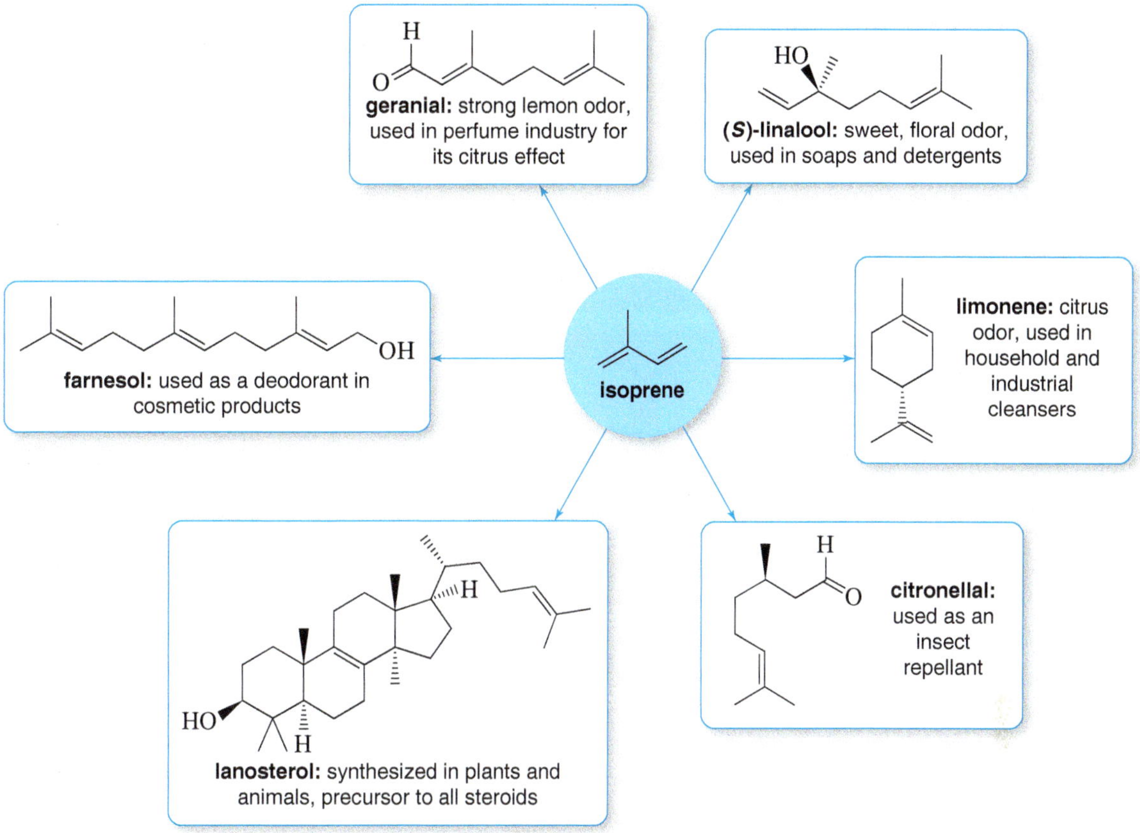

Figure 8.1 **Important terpenes are synthesized by plants and animals from the five-carbon isoprene unit.**

We study elimination to form alkenes in Chapter 12 and substitution of alcohols in Chapter 13.

"A good fragrance is really a powerful cocktail of memories and emotion."

—Jeffrey Stepakoff, The Orchard

can be converted into fragrances, perfumes, flavors, and pharmaceuticals, as well as useful synthetic intermediates and chiral building blocks. Some of these compounds, along with their uses, are shown in Figure 8.1.

Terpenes are unsaturated compounds, often containing the alkene functional group that is introduced here in Chapter 8. Nature produces these molecules using electrophilic addition, one of the more important reactions of alkenes. To understand terpene synthesis, we need to understand elimination (to reform the alkene—see Chapter 12) and substitution (of alcohols—see Chapter 13) reactions. In this way, the story of terpene biosynthesis previews interesting organic chemistry to come.

Using the foundation built in previous chapters [*specifically Chapter 5*], we are now ready to begin our study of organic reactions. Were you to discuss organic chemistry with a former student, they might recall "countless" numbers of reactions they had to remember or memorize. In our approach, it is better to focus, instead, on the number of reaction types we have to understand. [*There are probably ten of these.*] Learning the important foundational concepts that connect each of these reaction types will help you understand the majority of the reactions in the text. To facilitate this, we will look at a new reaction and first consider what we learned previously that will help us understand it. In doing so, organic chemistry becomes a perfectly logical (and hopefully interesting) story. To get started in our study of alkene reactions, let's make sure we understand a few important concepts.

Assessment 8.1 (i) Using bond dissociation energies, calculate $\Delta H°$ for the following reactions. (ii) Which side of the reaction is favored by entropy? (iii) If $\Delta S° = 0$ for these reactions, calculate $\Delta G°$. (Assume $T = 298$ K.) [*BDE for O—H = 110 kcal/mol.*]

(a)

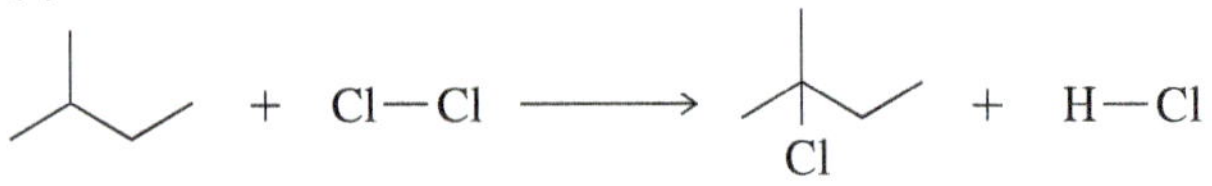

(b)

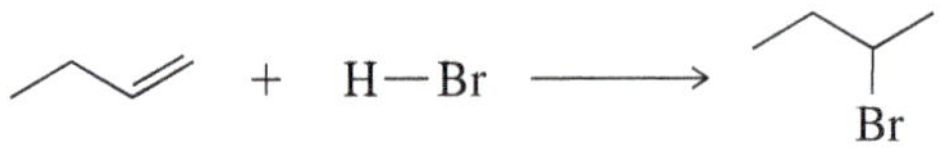

(c)

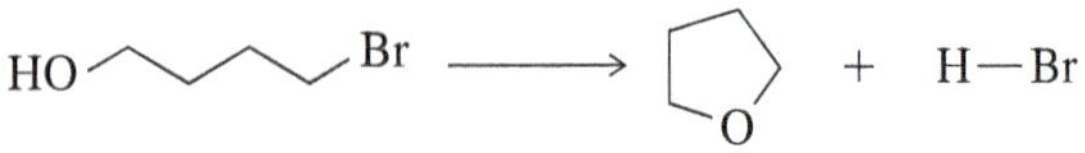

Assessment 8.2 (a) Rank the following carbocations from least stable to most stable. (b) Which would you expect to form first? (c) Which would you expect to react most quickly with a bromide ion (Br^-)? [*Carbocations, like radicals, are electron deficient.*]

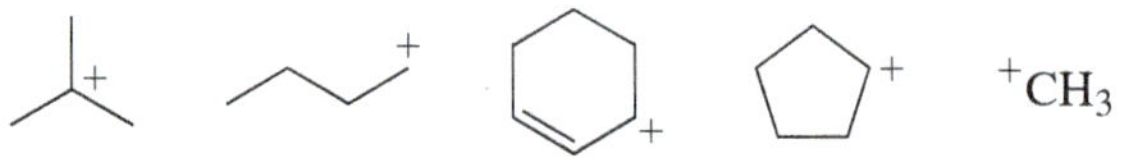

Assessment 8.3 Given that **A** is more stable than **B**, draw a reaction coordinate diagram that rationalizes the fact that **A** forms more quickly than **B** in the following reaction.

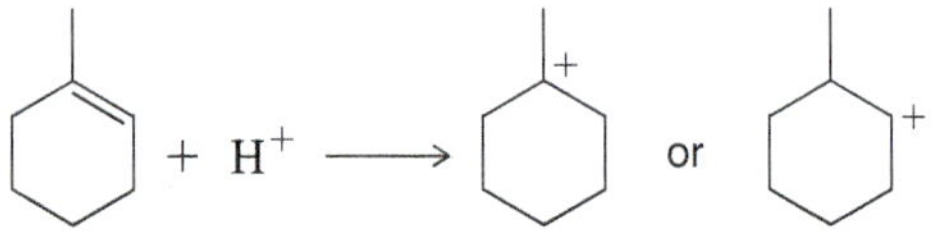

A **B**

Assessment 8.4 Rationalize the fact that reaction **A** results in an unequal mixture of products, but reaction **B** yields an equal mixture of two products.

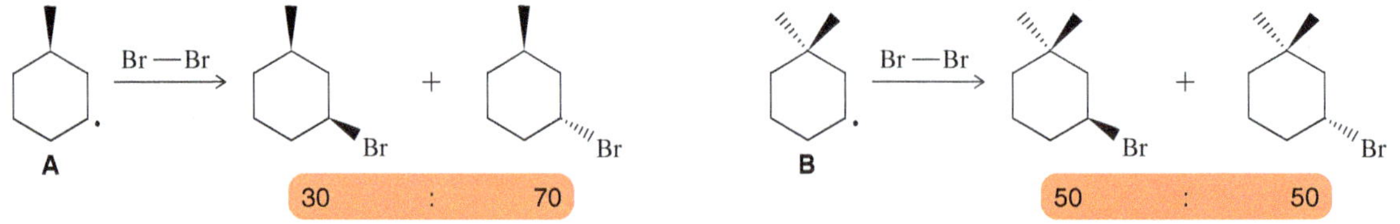

Assessment 8.5 Predict the product that results from the following "pushed electrons."

(a)

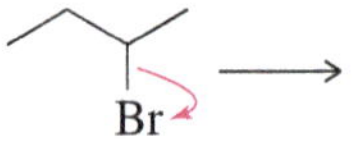

(b)

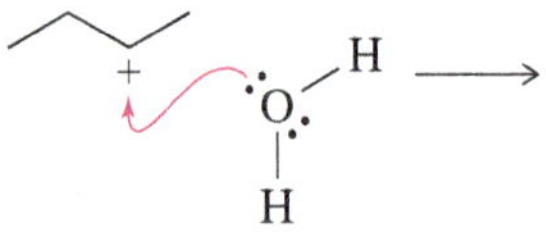

(c)

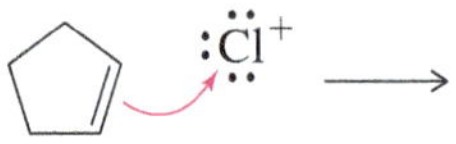

(d)

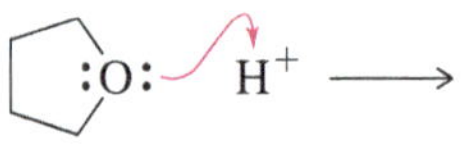

(e)

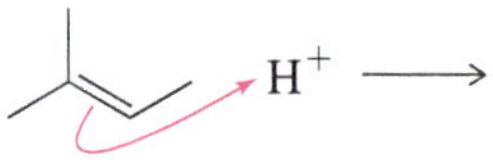

(f)

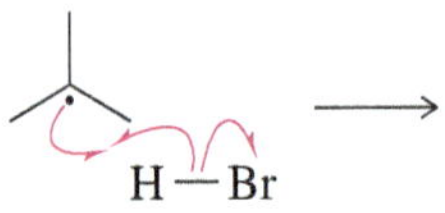

(g)

(h)

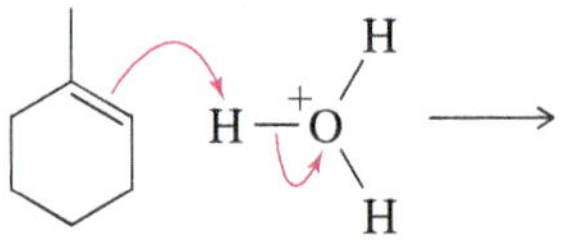

(i)

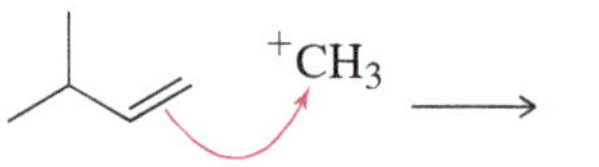

8.1 Properties of Alkenes

Low-molecular-weight **alkenes**, like alkanes, are nonpolar, insoluble in water, and volatile (Table 8.1).

Table 8.1 The properties of alkenes are similar to alkanes.

	Hexane	Hex-1-ene

Boiling point (°C)	69	63
Melting point (°C)	−95	−140
Density (25 °C) (g/mL)	0.659	0.673

Their low boiling points arise from the weak interactive forces between alkene molecules. These weak van der Waals interactions become more substantial as the alkenes become larger (**Figure 8.2**). As a result, alkenes containing four carbons or fewer are gases, while C_5 to C_{19} alkenes are liquids at room temperature. The first straight-chain alkene that is a solid at room temperature (mp 26–30 °C) is eicos-1-ene, a C_{20} alkene.

Figure 8.2 Larger alkenes boil and melt at higher temperatures.

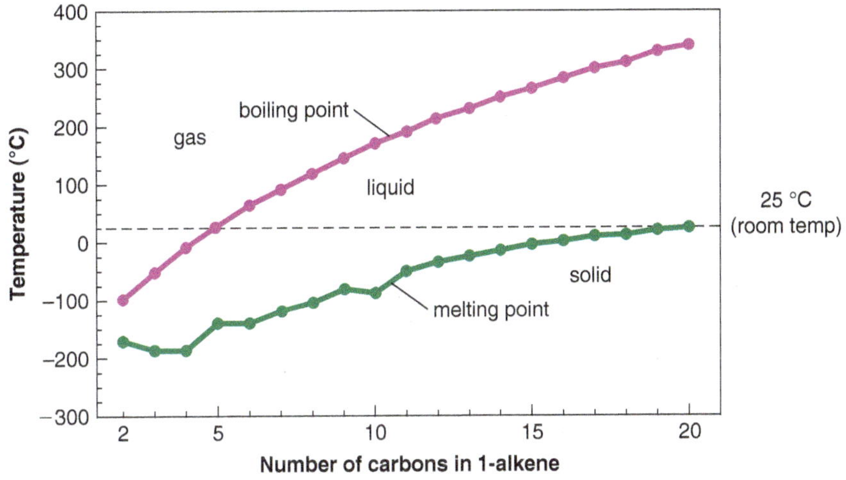

8.1.1 Degrees of unsaturation

Whereas alkanes are completely *saturated* hydrocarbons, alkenes are **unsaturated** (**Figure 8.3**). A saturated molecule has the maximum number of hydrogens it can have, based on the number of carbons. The C_6 alkane hexane, for example, has 14 hydrogens according to the formula C_nH_{2n+2}. The C_6 alkene (hex-1-ene), on the other hand, has only 12 hydrogens. Because hex-1-ene does not have the maximum number of hydrogens possible, it is said to be unsaturated. [*We can add H_2 to alkenes by a reaction we introduce in Chapter 9.*]

Figure 8.3 Alkenes can accommodate more hydrogens, so they are considered to be unsaturated.

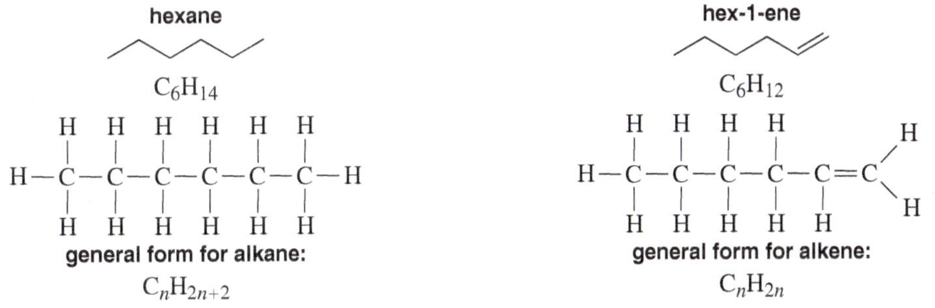

A molecule containing a single carbon–carbon double bond is said to contain one **degree of unsaturation**. One degree of unsaturation, or an **index of hydrogen deficiency (IHD)** of 1, implies that the molecule is two hydrogens (or one H_2) short of being fully saturated. Hex-1-ene has one degree of unsaturation. A molecule that is unsaturated can have more hydrogens added to it.

Alkenes have two fewer hydrogens because each alkene carbon uses one of its four possible bonds to form the π bond, leaving these unavailable to form a bond with hydrogen. Thus, each π bond adds a degree of unsaturation. For mono-alkenes, then, the general formula is C_nH_{2n}. For similar reasons, a ring also represents one degree of unsaturation. In order to convert an alkane into a ring, two hydrogens must be removed to form the new C—C bond (**Figure 8.4**). [*The "reactions" in Figure 8.4 are only hypothetical.*]

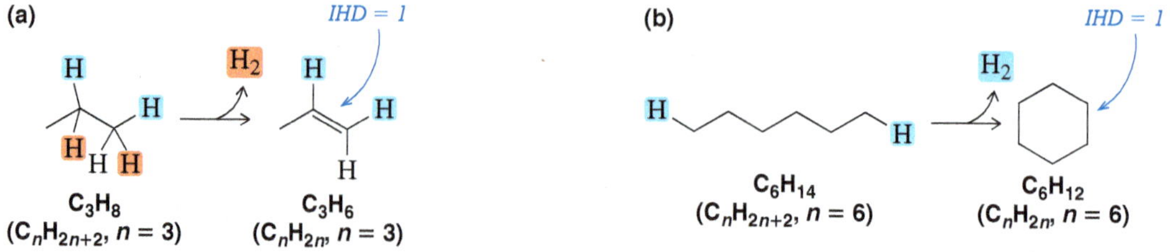

Figure 8.4 **(a) A π bond and (b) a ring replace two hydrogens from a saturated alkane.**

Assessment 8.6 Given the structure, calculate the index of hydrogen deficiency (degrees of unsaturation) of the following molecules.

(a) (b) (c)

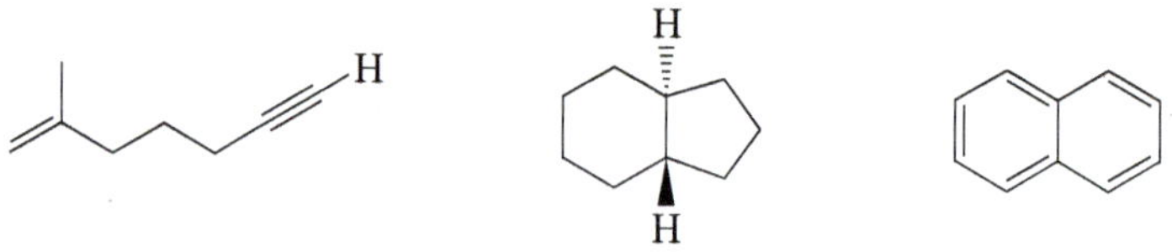

Knowing the index of hydrogen deficiency is particularly helpful when we study spectroscopy in Chapters 14 and 15.

As long as we know the molecular formula of a molecule, we can calculate the index of hydrogen deficiency using Equation 8.1.

$$\text{IHD} = \text{number of carbons} - \frac{\text{number of hydrogens}}{2} - \frac{\text{number of halogens}}{2} + \frac{\text{number of nitrogens}}{2} + 1 \qquad \text{Equation 8.1}$$

Assessment 8.7 What is the index of hydrogen deficiency for each of the following molecular formulas?

(a) $C_6H_{12}O_6$
(b) $C_{12}H_{11}NO_3$
(c) $C_8H_{14}Cl_2$
(d) $C_9H_{12}N_2O_2$
(e) $C_6H_{12}O_2$
(f) $C_7H_{13}NO$

8.1.1 SUMMARY

- Alkenes are unsaturated molecules. Each double bond in the molecule makes it two hydrogen atoms (one H_2) short of being fully saturated.

8.1.2 Molecular orbital picture of alkenes

The chemical reactivity and physical properties of alkenes can be explained as a consequence of the molecular orbital picture. Let's quickly review how we developed the molecular orbital picture of ethylene, the simplest alkene, in Chapter 2. [*See if you can draw it on your own before reading further.*]

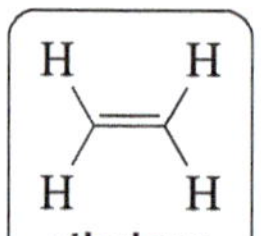

ethylene

We first have to determine which atomic orbitals ethylene uses to make the four C—H σ bonds, the C—C σ bond, and the C—C π bond. The hybridization of an atom can be determined by the shape of the molecule around that particular atom (**Figure 8.5**). Both C_1 and C_2 have three areas of electron density, suggesting that these are both trigonal planar and sp^2 hybridized. Finally, hydrogen uses its 1*s* atomic orbital to form bonds.

Molecular orbitals are formed by the overlap of atomic orbitals. Two atomic orbitals always combine in an additive (constructive) or subtractive (destructive) manner to form two molecular orbitals (**Figure 8.6**). From our analysis in Figure 8.5, the four C—H σ bonds

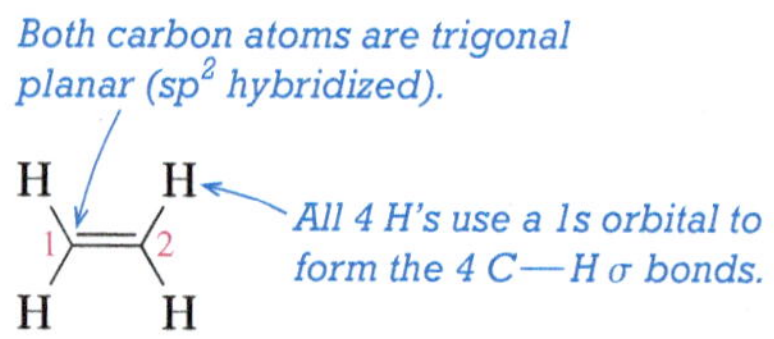

Figure 8.5 **Alkene carbons are *sp*² hybridized.**

are formed by overlap of the $2sp^2$-hybridized orbital of carbon and the 1*s* orbital of H. These two orbitals overlap to give a σ bonding molecular orbital (which represents the σ bond) and a σ* antibonding molecular orbital. The C—C σ bond is formed by overlap of the two $2sp^2$-hybridized orbitals of C_1 and C_2. Once more, these orbitals overlap to give a σ bonding molecular orbital and a σ* antibonding molecular orbital. Finally, the 2*p* orbitals on C_1 and C_2 overlap to give the π and π* molecular orbitals.

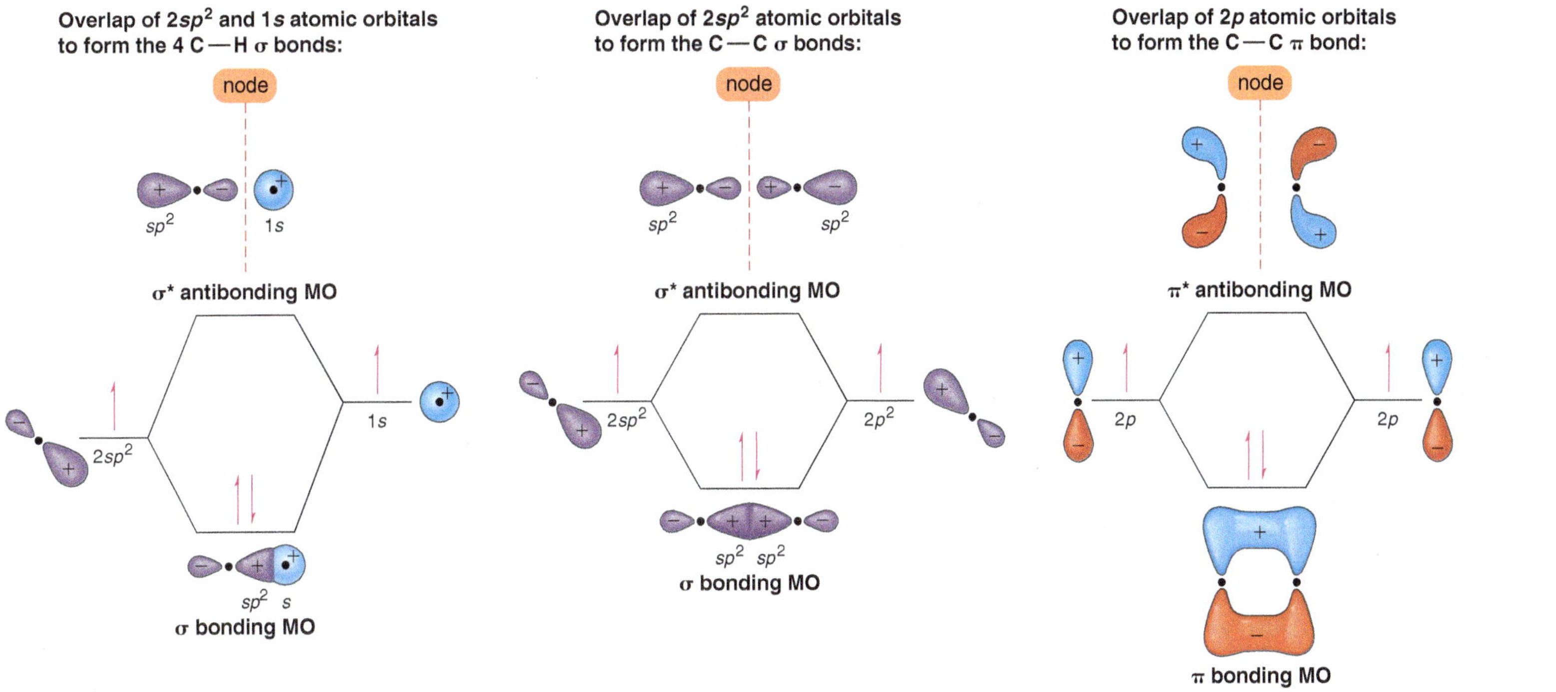

Figure 8.6 Constructing the molecular orbitals of ethylene.

With these molecular orbitals formed, we can now begin building the molecule as shown in **Figure 8.7**. Beginning with the σ bond framework, we can illustrate the σ bonds by showing them as overlapping sp^2-hybridized orbitals. Recall that σ bonds overlap along an axis connecting the two nuclei, whereas the π bond overlaps above and below the plane of the σ bond. Putting it all together, the molecular orbital picture of ethylene includes all of the bonding orbitals (σ and π) in the same structure.

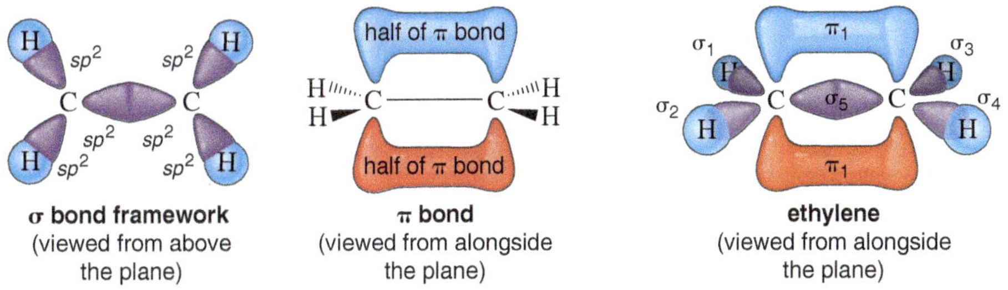

Figure 8.7 Putting the individual molecular orbitals together to give the full molecular orbital picture of ethylene.

[*If you were unable to draw the molecular orbital picture of ethylene on your own, return to Section 2.7 for further review.*] What is pertinent to our discussion here, though, is the manner in which the molecular orbital picture of an alkene impacts its properties and reactivity. For this, we are going to focus specifically on the π bond.

Assessment 8.8 Draw the molecular orbital picture of *trans*-but-2-ene.

trans-but-2-ene

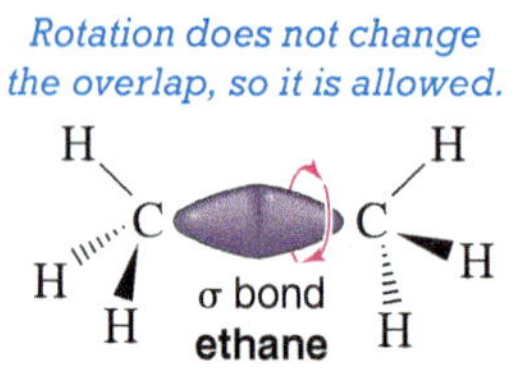

Figure 8.8 There is free rotation around σ bonds.

8.1.3 Implications of the π bond: no free rotation

Recall from Chapter 2 that there is free rotation around the C—C σ bond of ethane because the overlap is between the two carbons along an axis connecting them (**Figure 8.8**). Bond rotation does not change the amount of overlap between the two atomic orbitals, so the bond is maintained through the rotation. Recall from Chapter 5 that bond rotation, especially in ethane, is quite rapid, occurring on the order of 10^{10} times per second.

A π bond forms when two parallel *p* orbitals overlap above and below their σ bond axis. Rotation around the bond would cause the two *p* orbitals to turn perpendicular to each other, essentially breaking the π bond (**Figure 8.9**). Thus, there is *no* free rotation around π bonds, except when an amount of energy equivalent to the strength of (or the energy required to break) the C—C π bond is applied.

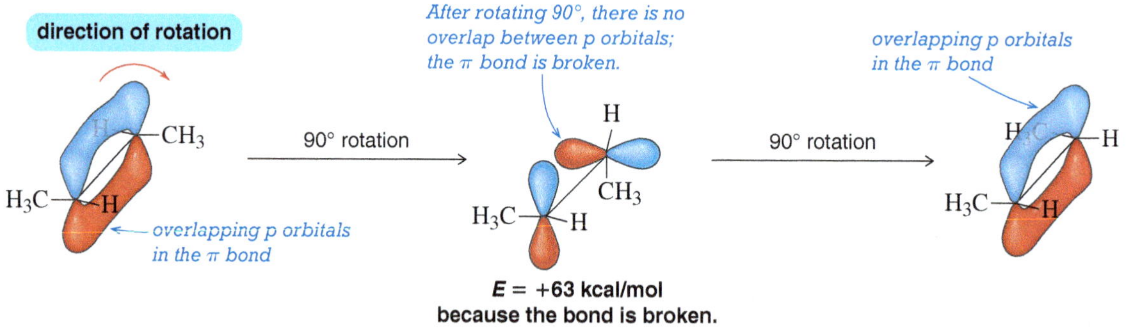

Figure 8.9 Rotation around a π bond is restricted by the aligned *p* orbitals.

8.1.2–8.1.3 SUMMARY

- The C—C π bond is formed by the alignment of overlapping *p* orbitals. There is no free rotation around the π bond because this alignment must be maintained.

8.1.4 Implications of the π bond: stereoisomerism

The inability to rotate about a C—C π bond causes any alkene with two different substituents at each of the alkene carbons to exhibit stereoisomerism. We first explain how to name alkenes, then we discuss the stereoisomerism of but-2-ene.

8.1.4.1 IUPAC nomenclature of alkenes

According to the IUPAC system of nomenclature, the naming of alkenes is similar to the naming of alkanes (Section 3.3). The main difference is in identifying the parent chain. For alkanes, the longest chain is the parent, whereas for alkenes, the parent chain is defined as the longest chain that also contains both alkene carbons. **Table 8.2** applies the rules.

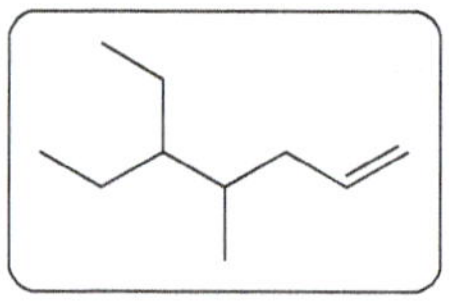

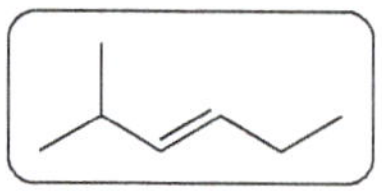

Table 8.2 Rules for naming alkenes according to the IUPAC system of nomenclature.

Rule	Example 1	Example 2
1. Identify the parent chain, which is the longest carbon chain containing both carbons of the alkene. The parent chain will be the root of the name. Replace the infix -an- of the corresponding alkane with -en- for alkenes.	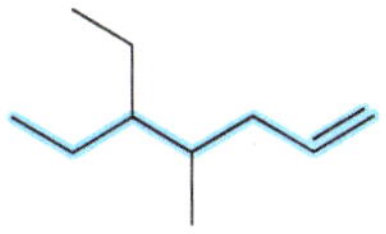 **7-carbon parent chain** **heptene**	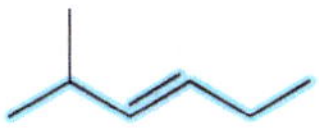 **6-carbon parent chain** **hexene**
2. Number the carbons such that the end of the parent chain closest to the alkene gets the lowest number. The second carbon of the alkene always gets the next number. If the alkene is equidistant from both ends, number so that a substituent gets the lower number.	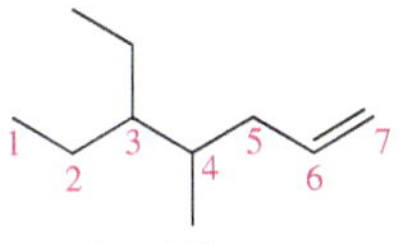 ~~**hept-6-ene**~~ 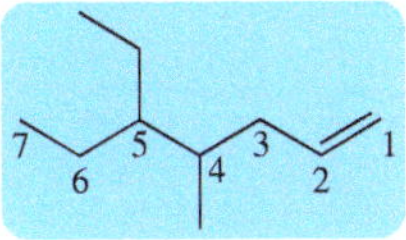 **hept-1-ene**	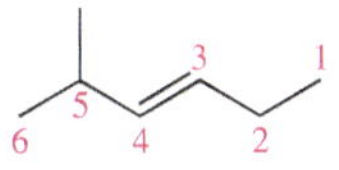 ~~**hex-3-ene**~~ 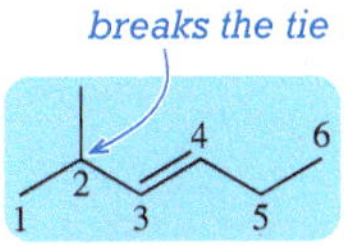 **hex-3-ene**
3. Name the substituents on the alkene parent chain the same way they were named with alkanes.	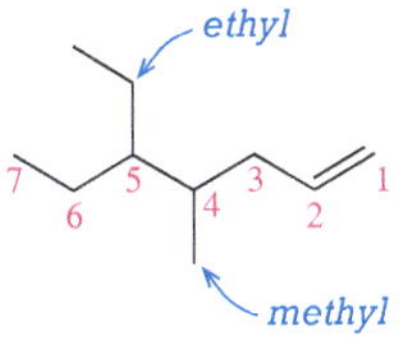	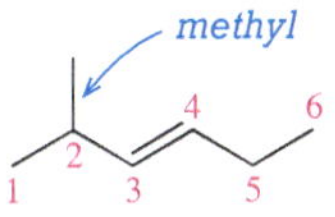
4. With the carbon chain numbered, write the name following the standard IUPAC alphabetization rules, using numbers to indicate the location of the substituents and the first carbon of the double bond.	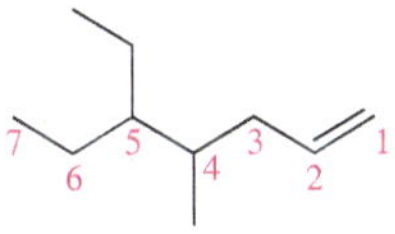 **5-ethyl-4-methylhept-1-ene**	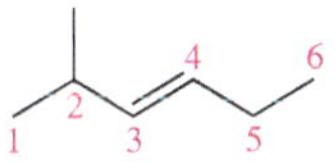 **2-methylhex-3-ene**

The rules from Table 8.2 can be applied to cycloalkenes with one small change (**Figure 8.10**). When the alkene is the highest priority functional group, carbon numbers 1 and 2 are always the alkene carbons. Because the alkene will always be at position 1, it is not necessary to include the alkene number in the name. The direction of numbering should lead to the first encountered substituent getting the lowest number possible.

Figure 8.10 Nomenclature of cycloalkenes.

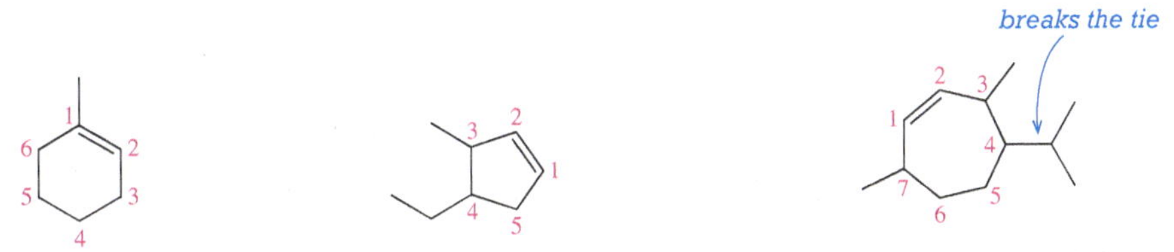

1-methylcyclohexene **4-ethyl-3-methylcyclopentene** **4-isopropyl-3,7-dimethylcycloheptene**

Assessment 8.9 Provide the IUPAC names for the following alkenes.

(a)

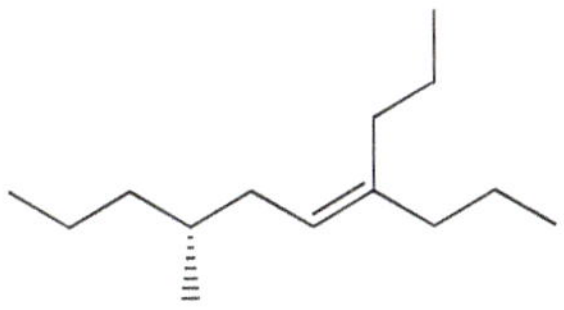

(b)

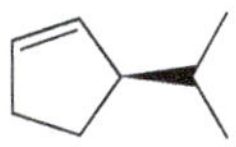

(c)

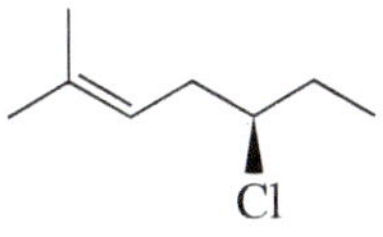

Assessment 8.10 Given the following IUPAC names, draw the corresponding structures.

(a) (*R*)-3-isopropyl-6-methylnon-1-ene
(b) (*R*)-3-chlorocyclobutene
(c) (*S*)-3-fluoropent-1-ene
(d) but-2-ene

8.1.4.2 Cis-trans isomerism

In Assessment 8.10(d), you were asked to draw the structure of but-2-ene. Did you notice that there were two possible molecules that could be named but-2-ene, as shown in **Figure 8.11**? [*What is the relationship between these two structures?*] If there was free rotation around the π bond, these would be conformational isomers. Alignment of the *p* orbitals restricts rotation, though, so these isomers differ in the arrangement of their atoms in space—that is, these two are stereoisomers. More specifically, they are diastereomers, because they are stereoisomers that are *not* mirror images [*recall Section 6.5*]. In one diastereomer, C_1 and C_4 of the butene chain are on the *same* side of a plane defined by the alkene. The IUPAC nomenclature used here is quite simple, referring to this as the cis isomer. In contrast, when C_1 and C_4 of the main chain are on the *opposite* sides of a plane defined by the alkene, we refer to this as the trans isomer.

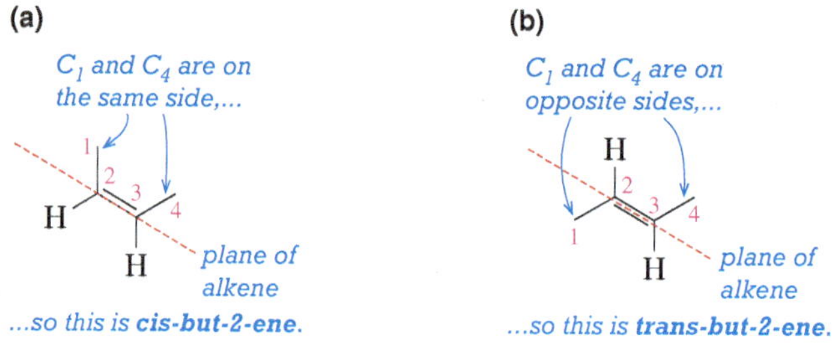

Figure 8.11 **Naming alkenes that exhibit cis-trans isomerism.** (a) *cis*-but-2-ene (b) *trans*-but-2-ene

Assessment 8.11 Name the following alkenes, being sure to specify whether they are cis or trans.

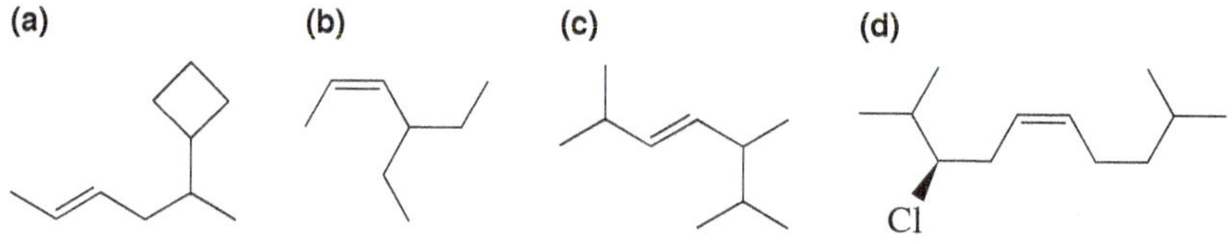

You only need *E*/*Z* and cis/trans when each carbon of the alkene has two different groups coming off them.

8.1.4.3 *E*/*Z* isomerism

The cis-trans system works for disubstituted alkenes like but-2-ene, but it is insufficient for naming trisubstituted alkenes. [*To see why, try assigning cis or trans to 1-bromo-2-chloropent-1-ene.*]

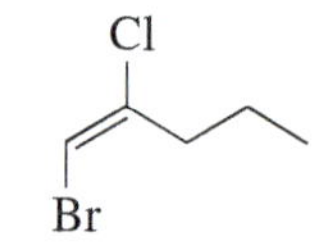

1-bromo-2-chloropent-1-ene

Is it cis because the bromine and the propyl group are on the same side of the alkene, or is it trans because the bromine and the chlorine are on opposite sides? This is similar to the discussion of relative versus absolute configuration in Chapter 6. Cis or trans tells you where the substituents are relative to each other, which is sufficient for disubstituted alkenes. With trisubstituted alkenes, though, we need a way to define the absolute configuration of an alkene. We need the ***E*/*Z* system**, which is closely related to the *R*/*S* system discussed in Chapter 6. [*E/Z is actually simpler than R/S.*]

To start, we use the Cahn–Ingold–Prelog system (see Section 6.3.2.2) to prioritize the two substituents at each carbon of the alkene. As shown in **Figure 8.12(a)**, each carbon of the alkene will have one group that is priority 1 and one that is priority 2. [*We don't number 1 through 4 as we did for chiral centers.*] If the two 1 (higher) priority substituents are on *opposite* sides, then the configuration of the alkene is ***E*** (*entgegen*, meaning "opposite"). If the two 1 (higher) priority substituents are on the *same* side (or *zame zide*), then the configuration of the alkene is ***Z*** (derived from the German word, *zusammen*, meaning "together"). Applied to 1-bromo-2-chloropent-1-ene, the higher priority substituents (Cl and Br) are on opposite sides [**Figure 8.12(b)**]. Thus, the molecule is *E*.

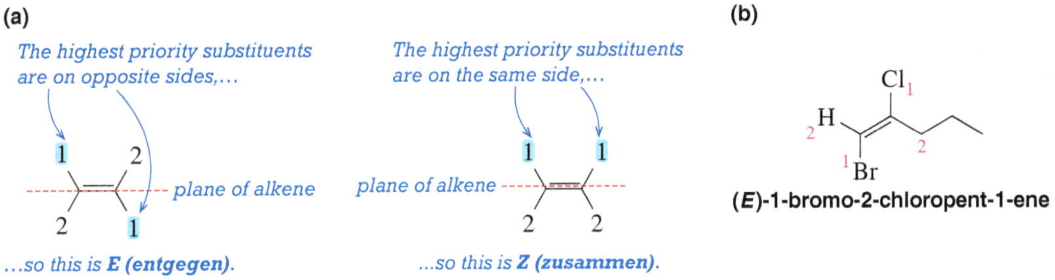

Figure 8.12 Naming using the *E*/*Z* system. (a) Use *Z* when highest priority substituents are on the same side and *E* when they are on opposite sides. (b) Br and Cl are on opposite sides, so this structure is *E*.

8.1.4.2–8.1.4.3 SUMMARY

- Because there is no free rotation around the π bond, alkenes can exhibit stereoisomerism. While cis and trans can be used for disubstituted alkenes, the *E*/*Z* system has broader applicability.

Assessment 8.12 Which of the following alkenes are *E* and which are *Z*?

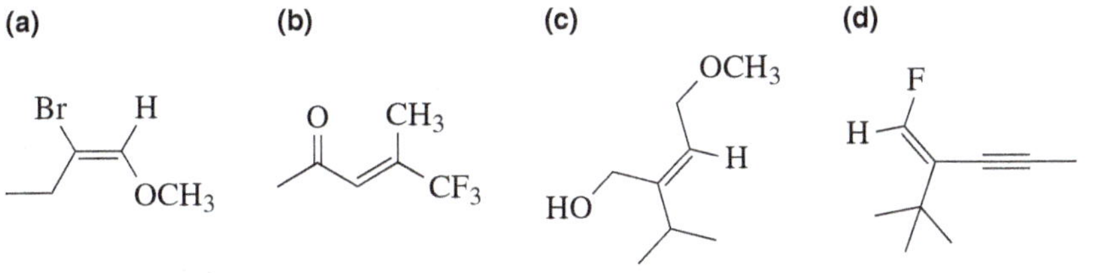

Assessment 8.13 Given the name, draw the structure of the following alkenes.

(a) (*E*)-4-ethyl-5-methyloct-3-ene
(b) (*Z*)-1-cyclohexyl-2-methylhept-2-ene
(c) (*Z*)-3-isopropylhept-3-ene
(d) (*E*)-2-chloro-3-fluoropent-2-ene

Assessment 8.14 Which terms have broader application, cis/trans or *E*/*Z*? Why?

8.1.4.4 Naming polyenes

When naming polyenes, the name of the molecule should indicate the number of alkenes present using *di-*, *tri-*, or *tetra-*, while also specifying their location on the chain. An "a" is inserted before the suffix to prevent two consonants being next to each other [*for example, hexa-1,3-diene instead of hex-1,3-diene*]. Similarly, the configuration of each alkene should be indicated as E or Z. Two examples are shown in **Figure 8.13**. [*Cover up the name and see if you can name them on your own before attempting to do the same in Assessment 8.15.*]

(*Z*)-4-fluoro-2-methylhexa-2,4-diene (4*R*,7*Z*)-4,8-dimethyldeca-1,7-diene

Figure 8.13 Nomenclature of dienes.

Assessment 8.15 Name the following alkenes.

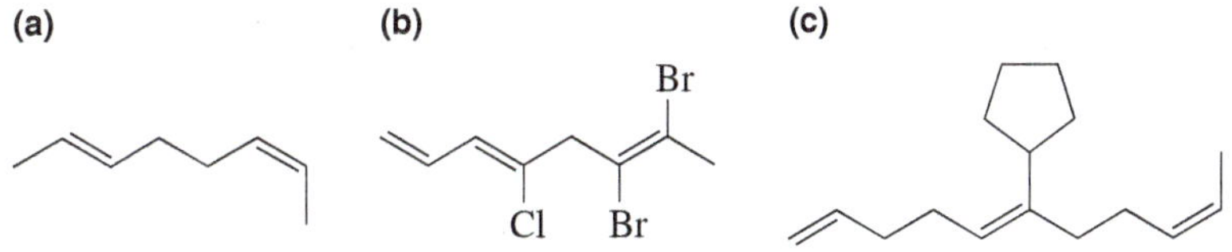

Assessment 8.16 How many stereoisomers are possible (a) for 4-methylnona-2,7-diene? (b) For 6-chloronona-2,4,7-triene?

8.1.4.5 Combustion of alkenes: the relative stability of alkene stereoisomers

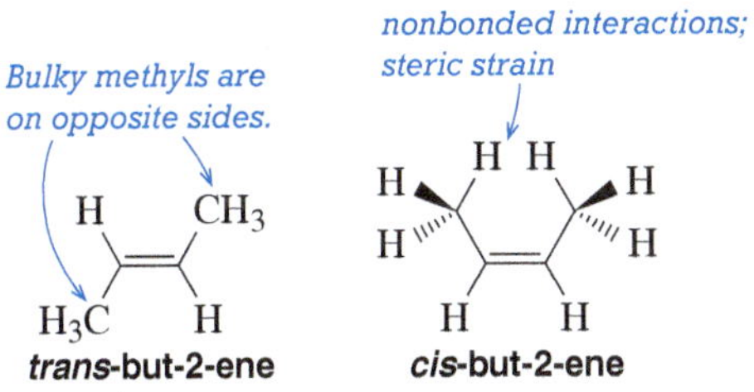

Figure 8.14 *trans*-But-2-ene is more stable than *cis*-but-2-ene because it avoids steric strain.

Like all diastereomers, cis-trans alkene stereoisomers have different properties. Specifically, *cis*-but-2-ene is less stable than *trans*-but-2-ene. [*Any ideas why?*] This difference in stability is attributed to the steric strain present in the cis isomer (**Figure 8.14**), which results from the two methyl groups being held in close proximity by the conformationally rigid alkene.

Despite discussing several situations where one isomer is more stable than another, we have yet to discuss experiments that can be used to determine the difference in stability. Using concepts from Chapter 5, let's think like a chemist and design an experiment to prove which isomer is more stable.

Thinking Like a Chemist

Knowing the difference in energy between the anti and gauche forms of butane, we used Equation 5.2 to calculate an equilibrium constant and a ratio of gauche to anti present at equilibrium (Section 5.1.3). Thus, we started from ΔG to calculate a ratio (**Figure 8.15**). [*We can't calculate a ratio for cis-trans isomers unless we know a* ΔG. *How can we experimentally determine it?*]

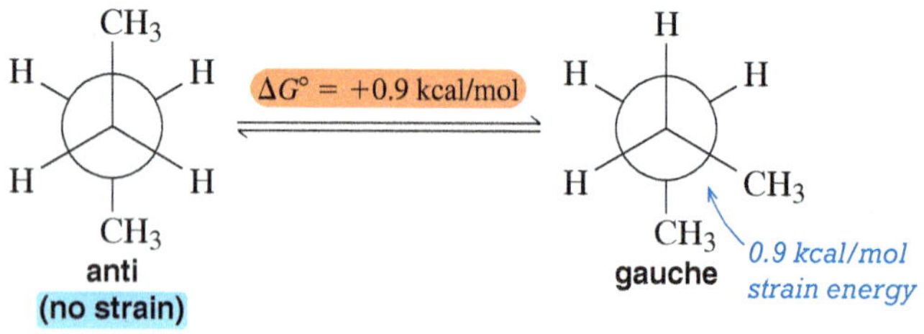

$$K_{eq} = e^{-\Delta G^\circ/RT} \text{ (Equation 5.2)}$$

$$K_{eq} = e^{-\left(\frac{900 \text{ cal/mol}}{(1.987 \text{ cal/mol}\cdot\text{K})(298 \text{ K})}\right)} = 0.219$$

$$K_{eq} = 0.219 = \frac{[\text{gauche}]}{[\text{anti}]}$$

Ratio of gauche to anti = 0.219/1

$$\%\text{ gauche} = \frac{0.219}{0.219 + 1} \times 100\% = 18.0\%$$

$$\%\text{ anti} = \frac{1}{0.219 + 1} \times 100\% = 82.0\%$$

Figure 8.15 Calculating a ratio (K_{eq}) from a calculated ΔG.

If we can calculate an expected K_{eq} from ΔG, we must also be able to calculate ΔG from an experimentally determined K_{eq}. Though cis and trans cannot equilibrate through bond rotation, an acid-catalyzed reaction allows a 76:24 (trans:cis) equilibrium ratio to be established (**Figure 8.16**). [*We discuss the mechanism of this reaction later.*] Using this 76:24 ratio and Equation 5.2, and assuming $T = 25\ °C$, we can determine that $\Delta G^\circ = -0.683$ kcal/mol, which tells us by how much the trans isomer is more stable.

Figure 8.16 Calculating ΔG° from an experimentally determined ratio (K_{eq}).

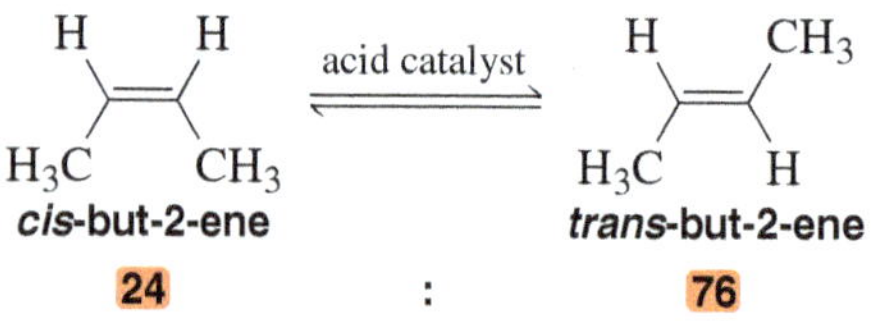

$$K_{eq} = \frac{[\text{products}]}{[\text{reactants}]} = \frac{76}{24} = 3.17$$

$$\Delta G^\circ = -RT\ln K_{eq} = -(1.987\frac{\text{cal}}{\text{mol}\cdot\text{K}})(298 \text{ K})\ln 3.17$$

$$= -683\frac{\text{cal}}{\text{mol}} = -0.683\frac{\text{kcal}}{\text{mol}}$$

The equilibration reaction in Figure 8.16 establishes a ratio of products to reactants that can be measured to get K_{eq} and then ΔG. Another strategy is having *cis*-but-2-ene and *trans*-but-2-ene undergo an identical reaction to make identical products. Because these reactions would finish at the same energy (the same products), any difference in an experimentally determined ΔH would be attributable to a difference in stability. [*Read these last two sentences again to make sure you understand this logic.*] In a **combustion** reaction, both molecules would produce CO_2 and H_2O. If this reaction is run in a calorimeter [**Figure 8.17(a)**], then the heat given off can be measured, giving ΔH for both *cis*- and *trans*-but-2-ene. **Figure 8.17(b)** shows the results of these experiments. In sum, combustion of *cis*-but-2-ene is more exothermic. [*Does this mean it is more or less stable? Think about this before reading on.*]

To see whether a more exothermic reaction results from the more or less stable product, it is helpful to plot the data on a reaction coordinate diagram. This is done for *cis*- and *trans*-but-2-ene in **Figure 8.18**. Because both of these reactions finish with the same products, the reaction coordinate diagram will finish at the same energy. The most exothermic reaction must start at a higher energy. In this case, it is the cis isomer that starts higher, so it is less stable than the trans isomer. ■

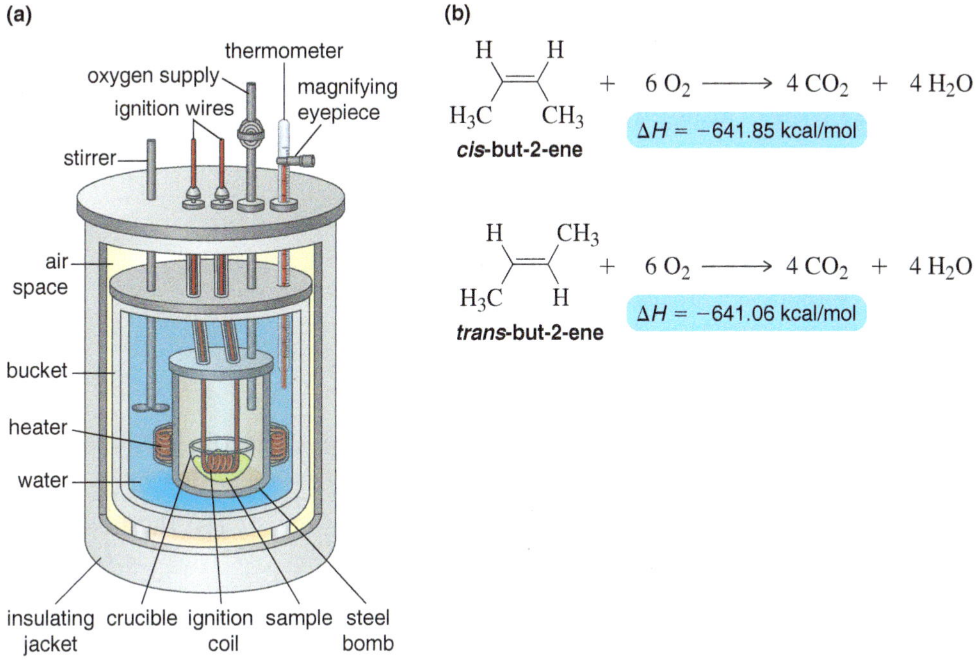

Figure 8.17 Combustion analysis of *cis*- and *trans*-but-2-ene. (a) Schematic of a calorimeter used to measure heat given off in a combustion reaction. (b) The value of ΔH for the combustion can be compared because the products are the same for both reactions.

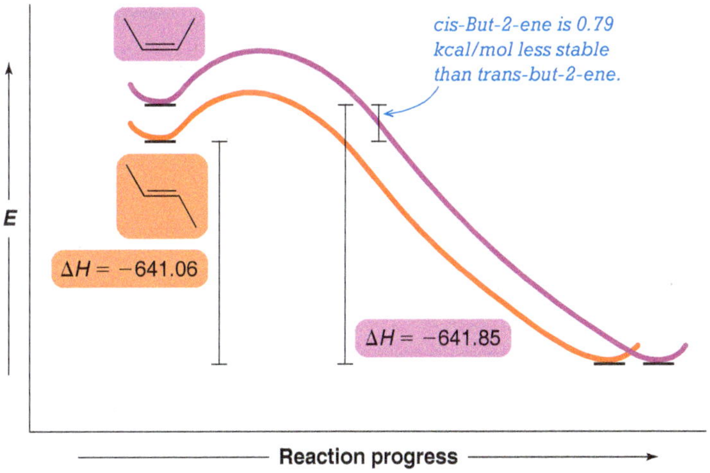

Figure 8.18 *cis*-But-2-ene starts higher on the reaction coordinate diagram, so it gives off more heat when it is combusted. (ΔH values are given in kcal/mol.)

8.1.4.5 SUMMARY

- Combustion analysis can be used to determine the relative stability of compounds having the same molecular formula, but different structures. Trans alkenes are more stable than cis alkenes because the trans arrangement reduces steric strain between groups bonded to the carbons of the double bond.

Assessment 8.17 Which constitutional isomer, A or B, would you expect to have the highest heat of combustion ($\Delta H_{combustion}$)?

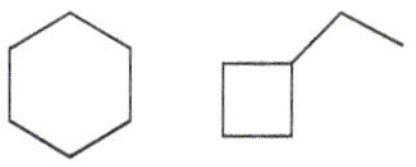

A B

8-A CHEMISTRY IN CONTEXT: Natural Products

Terpene biosynthesis: the isoprene rule

Terpenes are classified according to how many isoprene units are used to make them. Terpenes, containing ten carbons, are made up of two isoprene units. Sesquiterpenes (from sesqui, meaning one and one half) contain 15 carbons and are made from three isoprene units. Diterpenes contain 20 carbons and are derived from four isoprene units. There are three ways for isoprene units to combine (**Figure 8.19**)—namely, head to tail $(1 \rightarrow 4)$, tail to tail $(4 \rightarrow 4)$, and head to head $(1 \rightarrow 1)$. The **isoprene rule** recognizes that in nature terpenes are formed predominantly via the head to tail $(1 \rightarrow 4)$ combination of isoprenes. [*The chemistry of the bond-forming reactions comes in Chemistry in Context 8-C.*]

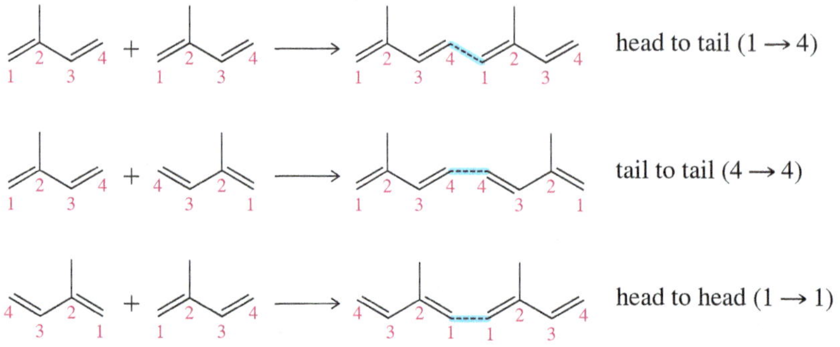

Figure 8.19 Three ways for isoprene units to combine. Head to tail is the most common in nature.

A terpene, regardless of size, can be dissected into its constituent isoprene units (**Figure 8.20**). To do this, ignore the functional groups and focus only on the carbon skeleton. Next, recognize that the branching (typically a methyl group, but sometimes a CH_2) always occurs at C_2 of the four-carbon chain of isoprene. Having identified C_2, C_1—C_4 of the four-carbon chain of the isoprene unit can be labeled. The branched carbon is C_5. To see if you are correct, the four-carbon chain of the next isoprene unit should once again have branching at C_2. [*If this is true, then you have probably successfully identified each of the isoprene units.*]

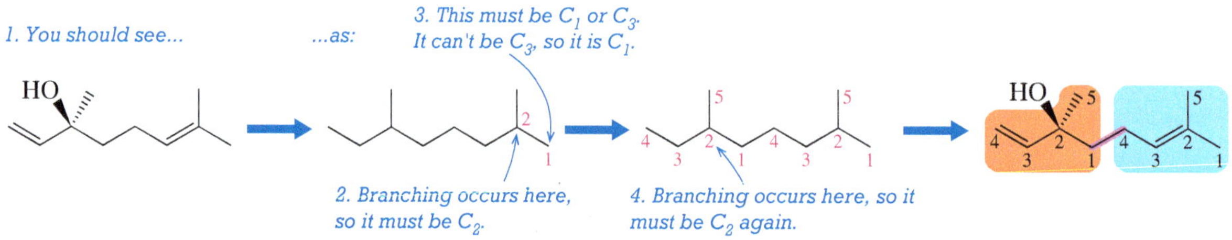

Figure 8.20 Identifying the isoprene units of a terpene.

[*Identifying isoprene units isn't that important as an end skill. But it helps us develop the skill of identifying and labeling carbons as belonging to a particular fragment of a molecule. This skill will come in very handy later. Let's see if you can do it on your own in Assessment 8.18.*]

Assessment 8.18 For the following terpenes, identify the isoprene units. In cross-linked or ring-containing terpenes, linkages can be formed between more than just C_1 and C_4 of isoprene.

(a)

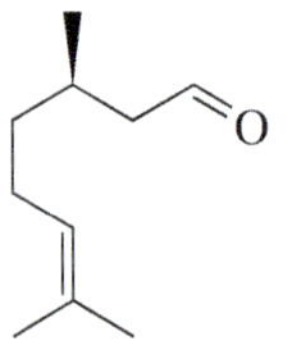

citronellal

(b)

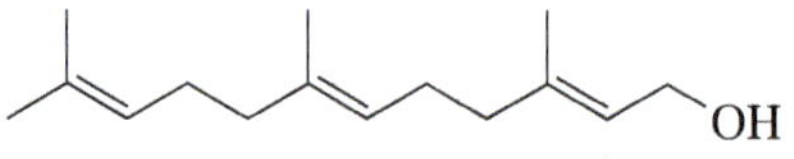

farnesol

(c)

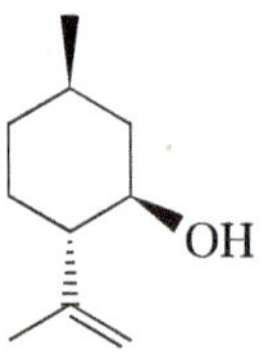

menthol

(d)

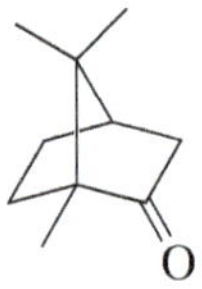

camphor

(e)

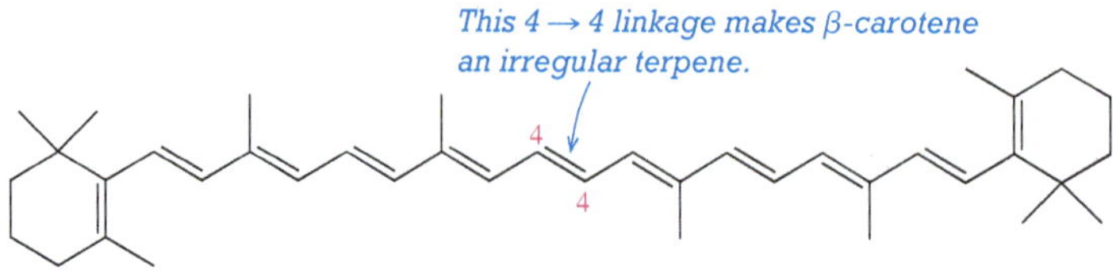

β-carotene

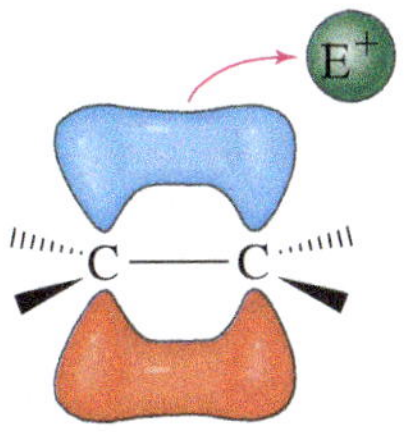

Figure 8.21 **The electron-rich π bond breaks easily** by donating electrons to electrophiles.

8.1.5 Implications of the π bond: the nucleophilicity of alkenes

Although neutral, alkenes react as Lewis bases (nucleophiles)—that is, they are electron-pair donors. This pattern of reactivity is yet another implication of the π bond present in alkenes. Compared to a σ bond, which results from the on-axis overlap of two sp^2-hybridized orbitals, there is not as much overlap between the *p* orbitals that form the π bond. As a result, π bonds are significantly weaker than σ bonds and therefore break more easily. The electron-rich π bond breaks by donating electrons to Lewis acids (electrophiles) (**Figure 8.21**). [*Most of the reactions in Chapter 8 and 9 occur via this mechanism.*]

8.1.5 SUMMARY

- Alkenes are good electron-pair donors and react efficiently with electrophiles in Lewis acid–Lewis base reactions.

8.2 Alkene addition reactions—general overview

This list of reactions can seem intimidating. But if you study this material as if we are learning just one new reaction with slight variations, it will eventually be quite simple.

The most important reactions of alkenes are **addition reactions**, a process generalized by considering the addition of a hypothetical molecule X—Y. As shown in **Table 8.3**, there are a number of reagents that can be X—Y equivalents for the synthesis of haloalkanes (Section 8.3), alcohols (Sections 8.4.1, 8.4.4, and 8.4.5), alkanes (Section 9.2), epoxides (Section 9.1.4), cis or trans diols (Sections 9.1.5 and 9.1.6), and halohydrins (Section 9.1.3). Polymers can also be made using alkene addition reactions (Section 28.4.1). The alkene, then, is a versatile functional group that can be converted into many others, so it plays a key role in organic synthesis.

Table 8.3 **Alkene reactions from Chapters 8 and 9, all of which share mechanistic similarities.**

$$H_2C{=}CH_2 \xrightarrow{X-Y} H_2C(Y)-CH_2(X)$$

Reagent	X	Y	Representative alkene	Product	Process
HBr	H	Br		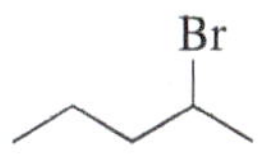	Hydrohalogenation (Section 8.3)
HBr, H_2O_2	Br	H		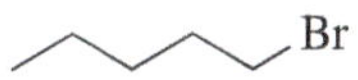	
H_2SO_4, H_2O	H	OH	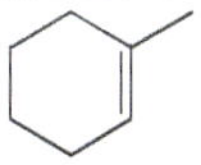	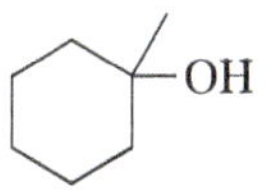	Hydration (Section 8.4.1)
1. $Hg(OAc)_2$, H_2O 2. $NaBH_4$	HgOAc	OH		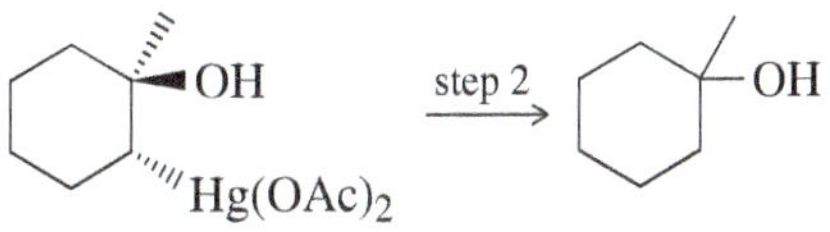	Oxymercuration–reduction (Section 8.4.4)
1. BH_3 2. H_2O_2, NaOH	BH_2	H		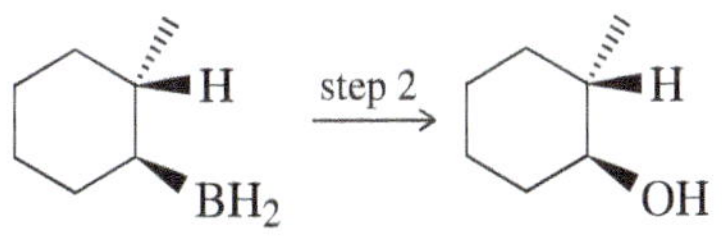	Hydroboration–oxidation (Section 8.4.5)

(continued)

Table 8.3 *Continued*

Reagent	X	Y	Representative alkene	Product	Process
Br_2	Br	Br	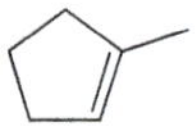	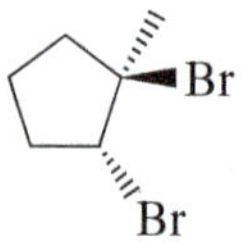	Halogenation (Section 9.1.2)
Br_2, H_2O	Br	HO		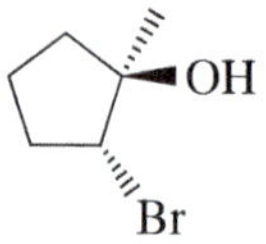	Hydrohalogenation (Section 9.1.3)
Br_2, CH_3OH	Br	CH_3O		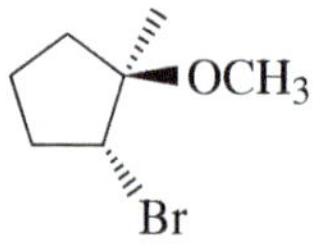	Alkoxyhalogenation (Section 9.1.3)
H_2, Pd/C	H	H	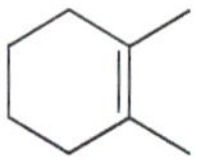	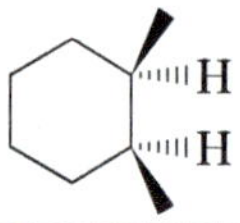	Hydrogenation (Section 9.2)
1. OsO_4 2. $NaHSO_3$	HO	HO		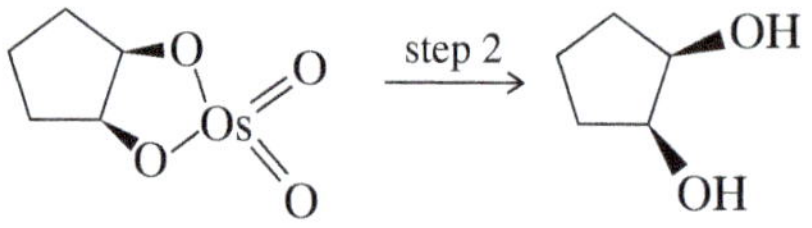	Dihydroxylation (Section 9.1.5/9.1.6)
1. RCO_3H 2. H_2O, H_2SO_4	HO	HO		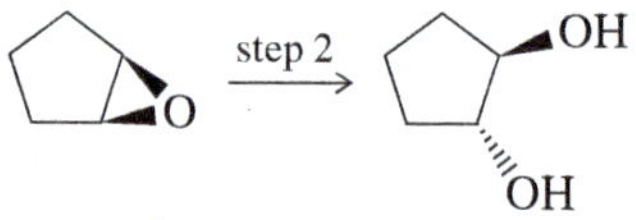	

8.3 Hydrohalogenation—addition of HX

The first electrophilic addition we study is the addition of HBr to alkenes (**Figure 8.22**). For example, addition of HBr to *trans*-but-2-ene gives the alkyl halide (+/−)-2-bromobutane.

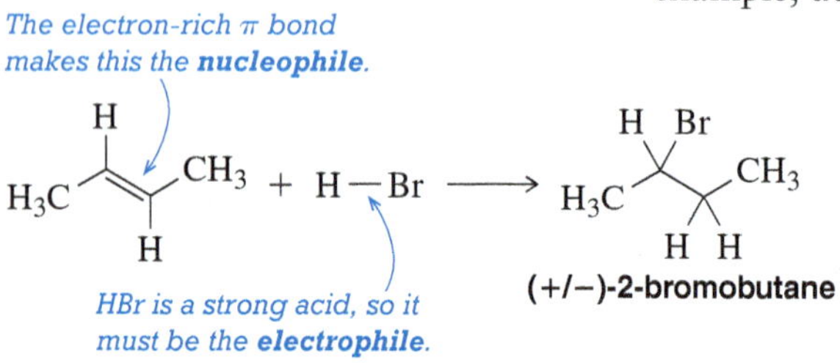

Figure 8.22 Electrophilic addition of HBr to alkenes gives haloalkanes.

Most organic reactions are simply the reactions of nucleophiles (Lewis bases) with electrophiles (Lewis acids), so we start by assigning roles to HBr and *trans*-but-2-ene. Hydrobromic acid (HBr), with a pK_a of −8, is an extremely strong acid. As a proton (H^+) donor, HBr would also be considered a strong electrophile. This implies that the alkene must be the nucleophile, using the electron-rich π bond to attack electrophiles. Section 8.1.5 showed that the diminished overlap of *p* orbitals in π bonds allows them to break easily.

Our arrow-pushing formalism is used in **Figure 8.23** to show how this reaction occurs. [*Pay careful attention to where the electrons go in each step.*] The reaction begins with the π electrons attacking HBr in a Brønsted–Lowry acid–base reaction. [*The arrow always starts at the electron-rich nucleophile and points to the electron-poor electrophile.*] What were the two π electrons are now a new C—H σ bond, while the original H—Br σ electrons have become the fourth lone pair on the highly stable bromide

Figure 8.23 Arrow-pushing mechanism of H—Br addition.

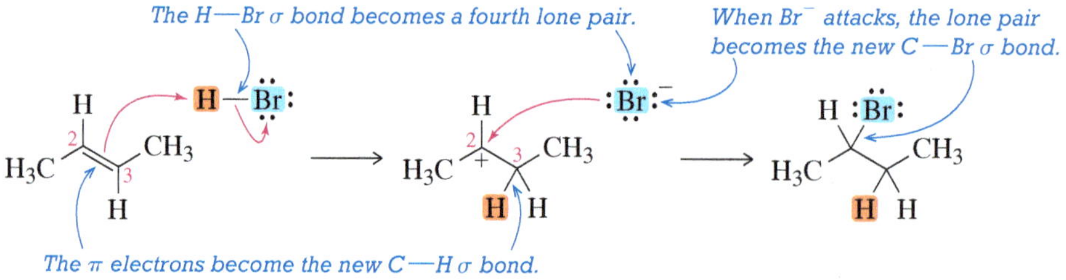

ion (formal charge of −1). Electrons had to be removed from C_2 so that C_3 could bond with the proton. As a result, C_2 now lacks a full octet and possesses a formal charge of +1. This intermediate is referred to as a **carbocation**. In the second step, Br^- donates a pair of electrons to the electron-deficient carbocation to form a new C—Br σ bond.

Looking more deeply at the mechanism in Figure 8.23, notice that the new C—H bond formed at C_3 when the π bond attacked HBr. The C—H bond could have formed at C_2, instead, leaving C_3 as the positively charged carbon, as shown in **Figure 8.24**. But-2-ene is symmetrical, though, so the carbocation, and the ultimate product, is the same regardless of which carbon gets the hydrogen.

Figure 8.24 With a symmetrical alkene, the proton adds equally to either carbon, thus producing the same molecule.

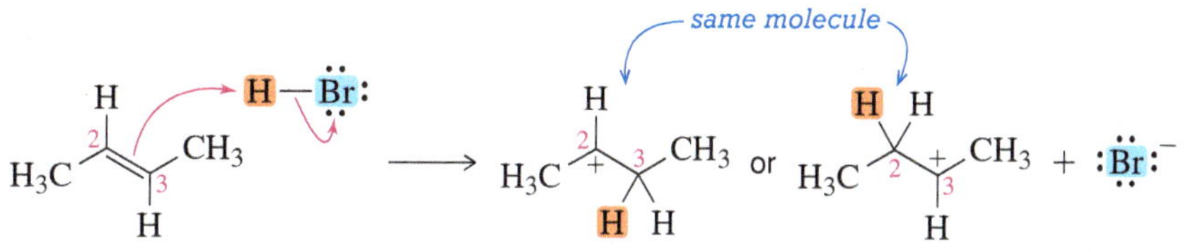

Even though both carbocations form equally in Figure 8.24, we can use the bouncing arrow formalism to show a different arrow for the formation of each. [*The utility and necessity of the bouncing formalism are discussed in detail in the insert between Chapters 7 and 8.*] The mechanism from Figure 8.24 is shown again in **Figure 8.25** but using a bouncing arrow. [*The electrons don't really bounce; this is just an extension of a formalism and should not be understood as what is happening in solution.*] The bouncing arrow in **Figure 8.25(a)** indicates that when the π bond attacks the proton, the electrons from the π bond remain attached to C_2. In **Figure 8.25(b)**, the π electrons remain attached to C_3. In each case, the carbon that lost the π electrons is now positively charged. [*The bouncing arrow is more useful when electrophilic addition can result in two different carbocations (Section 8.3.2).*]

Figure 8.25 The bouncing arrow clarifies which carbon keeps the electrons and which becomes positively charged. In this reaction, both pathways still occur because the resulting carbocations are the same molecule.

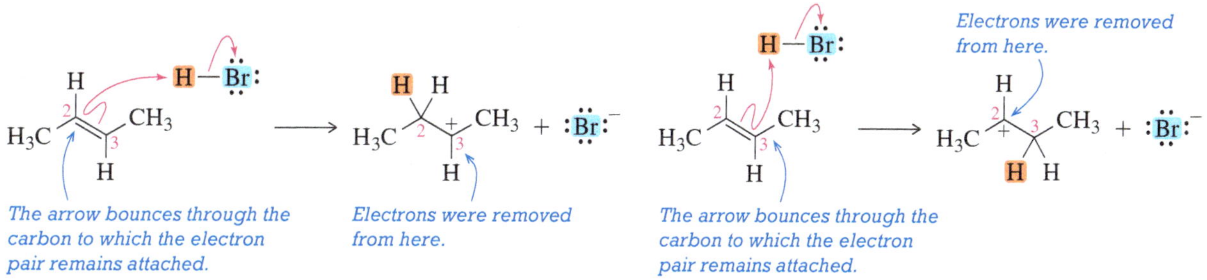

Finally, we examine the second step in a bit more detail. Recall from Figure 8.22 that the product is racemic (+/−)-2-bromobutane. Why is the product racemic? When the bromide attacks the sp^2-hybridized carbocation, it does so by donating electrons to the empty *p* orbital equally from the front or the back. **Figure 8.26** shows attack from the top (a) and (b) bottom on the planar carbocation occurring at equal rates to give a racemic mixture. [*Begin thinking about what this will look like on a reaction coordinate diagram to prepare for the next section.*]

Figure 8.26 Approach from the top and bottom occur at equal rates to give a racemic mixture. [*The carbocation has been turned on its side.*]

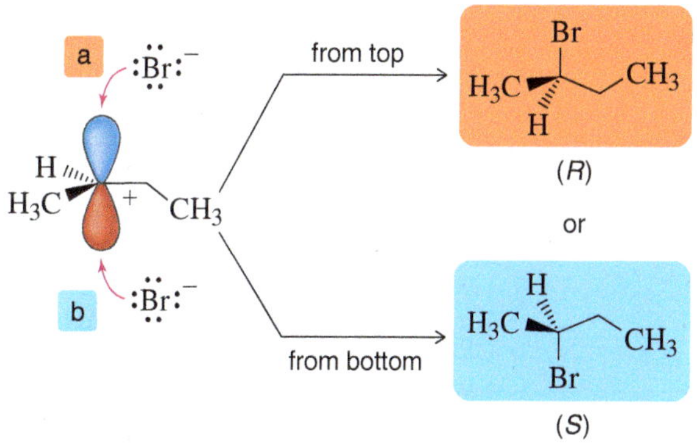

Assessment 8.19 Predict the product(s) of each of the following reactions. If you expect a racemic mixture, draw both enantiomers. [*The rings in (f) are benzene rings and, for reasons we discuss in Chapter 23, do not react under these conditions.*]

(a)

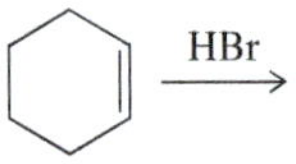

(b)

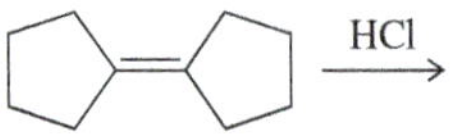

(c)

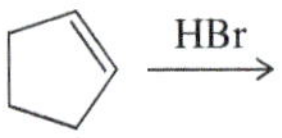

(d)

$H_2C{=}CH_2 \xrightarrow{HCl}$

(e)

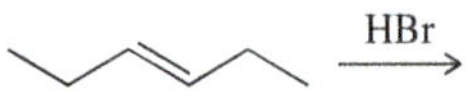

(f)

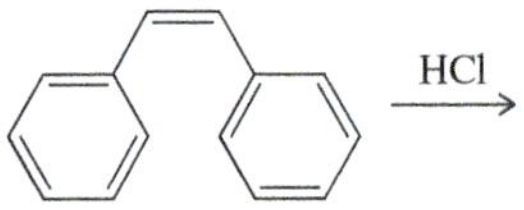

Assessment 8.20 The following reaction steps are shown using conventional electron pushing. (a) Draw the second product whose formation would have been rationalized with this same arrow. (b) Use the bouncing arrow formalism to illustrate the formation of only the product shown.

(a)

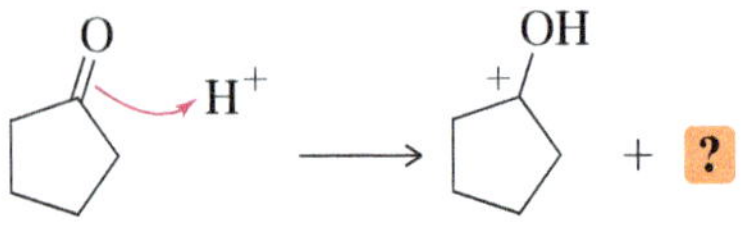

(b)

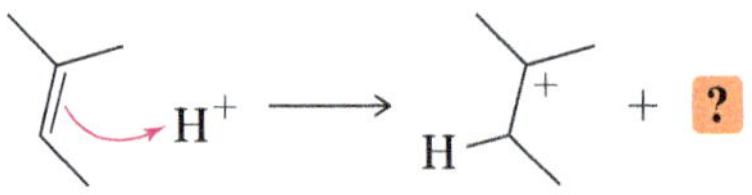

(c)

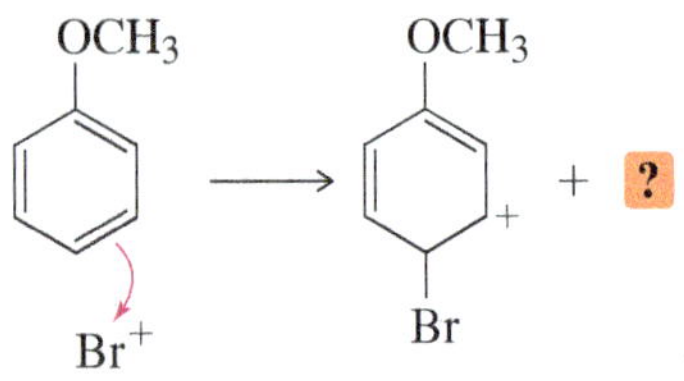

Assessment 8.21 Provide an arrow-pushing mechanism that rationalizes the formation of each alkyl halide in Assessment 8.19.

Assessment 8.22 Suggest an alkene that, upon reaction with the appropriate hydrohalic acid, will produce only the alkyl halide shown. [*Ignore stereochemistry.*]

(a)

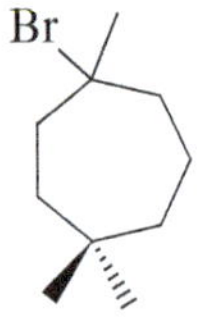

(b)

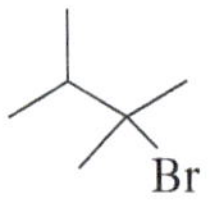

(c)

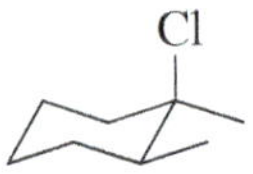

(d)

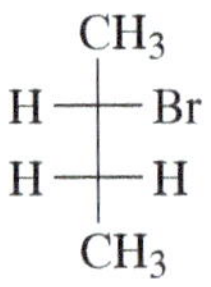

8.3.1 Addition of HX: further analysis

To better understand *why* HBr or HCl reacts with an alkene to give both of the observed products, we can analyze the reaction in the context of ΔG as we did for reactions in Chapter 5. Table 8.4 summarizes this analysis. [*Be sure to carefully consider both* ΔH *and* ΔS *for each step.*]

Table 8.4 Deep analysis of the two steps of HX addition.

The reaction is *favored* because . . .	Reaction step	The reaction is *disfavored* because . . .
a. It's an acid–base reaction, going from 2 molecules to 2 molecules ($\Delta S \approx 0$). b. HBr is an incredibly strong acid. c. Bromide ion is large and able to delocalize the negative charge, making it very stable. [*See Chapter 4.*]	**Step 1** 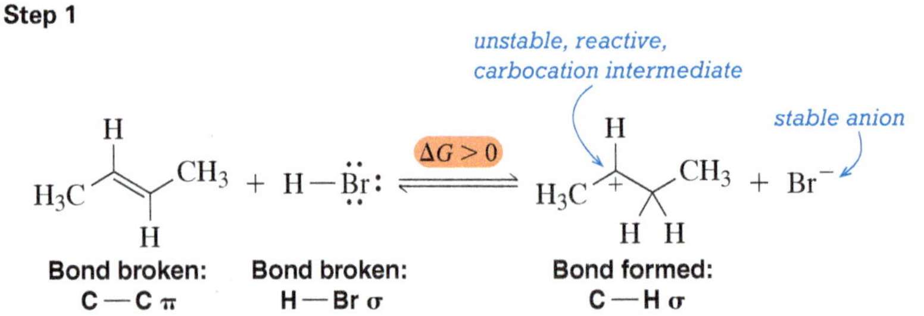 • Two bonds break and one forms, so $\Delta H > 0$. • Acid–base reaction: 2 molecules → 2 molecules, so $\Delta S \approx 0$.	a. A C—C π bond (~65 kcal/mol) and an H—Br σ bond (88 kcal/mol) break, and a C—H σ bond (95 kcal/mol) forms ($\Delta H_{total} = +58$ kcal/mol). b. A reactive, unstable carbocation forms.
Conclusion for step 1: The large, positive ΔH coupled with $\Delta S \approx 0$ makes the first step endergonic, disfavored, and slow ($\Delta G > 0$).		

(*continued*)

Table 8.4 ***Continued***

The reaction is *favored* because . . .	Reaction step	The reaction is *disfavored* because . . .
a. A C—Br σ bond (68 kcal/mol) forms with no bonds breaking ($\Delta H_{total} = -68$ kcal/mol). b. Two charged intermediates react to form a neutral compound.	**Step 2** 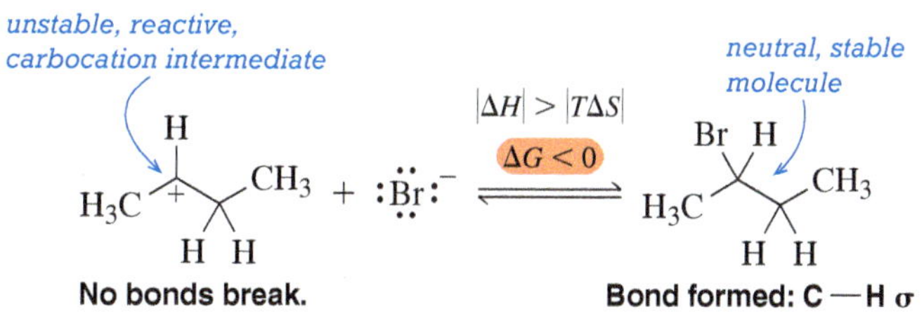 • Zero bonds break and one forms, so $\Delta H < 0$. • Addition reaction: 2 molecules → 1 molecule, so $\Delta S < 0$.	a. Two molecules become 1 ($\Delta S < 0$).
Conclusion for step 2: The large, negative ΔH, even with $\Delta S < 0$, makes the second step exergonic and highly favored ($\Delta G < 0$).		

Based on the information in Table 8.4, the overall reaction is disfavored by entropy but highly favored by enthalpy, resulting in an overall favorable reaction with $\Delta G < 0$. A reaction coordinate diagram (**Figure 8.27**) shows an uphill first step with a steeply downhill second step. We predict, then, that the first step (the most uphill step of the reaction) will have the higher activation energy. Thus, step 1 is slower and rate-determining. [*We analyze the transition state in the next paragraph to confirm this prediction.*]

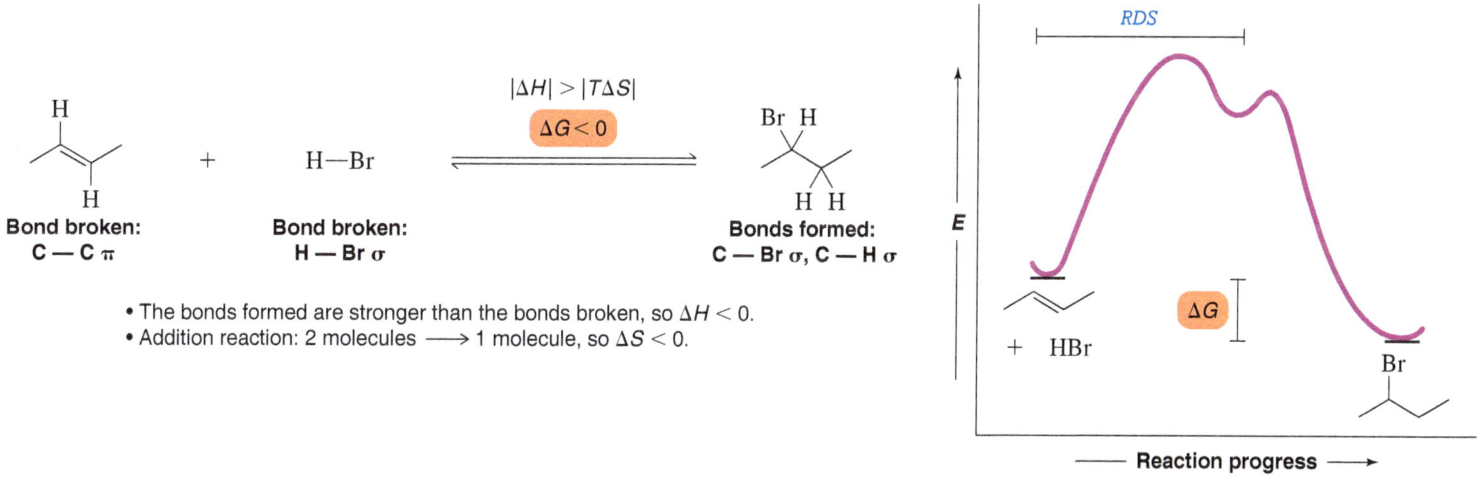

Figure 8.27 Reaction coordinate diagram for the addition of HBr to *trans*-but-2-ene. The first step is slow and rate-determining, but the overall reaction is exergonic ($\Delta G < 0$). RDS = rate-determining step.

The reaction coordinate diagram makes it possible to approximate the identity of the transition state of both steps of HBr addition (**Figure 8.28**). Because step 1 is endothermic, the Hammond postulate (Section 5.2.4) states that the transition state will resemble the products (carbocation and Br^-) [Figure 8.28(a)]. As a result, there is significant carbocation character in the transition state. This will have implications for the regioselectivity of electrophilic additions, which we study in future sections. For step 2, which is exothermic, the Hammond postulate predicts a reactant-like transition state [Figure 8.28(b)]. As such, the new C—Br bond is barely formed, suggesting a high level of charge on carbon (positive) and bromine (negative) in the transition state.

Before we move on, let's revisit the stereoselectivity of the reaction. The stereochemical outcome is determined in the second step, when the bromide attacks the carbocation. Because

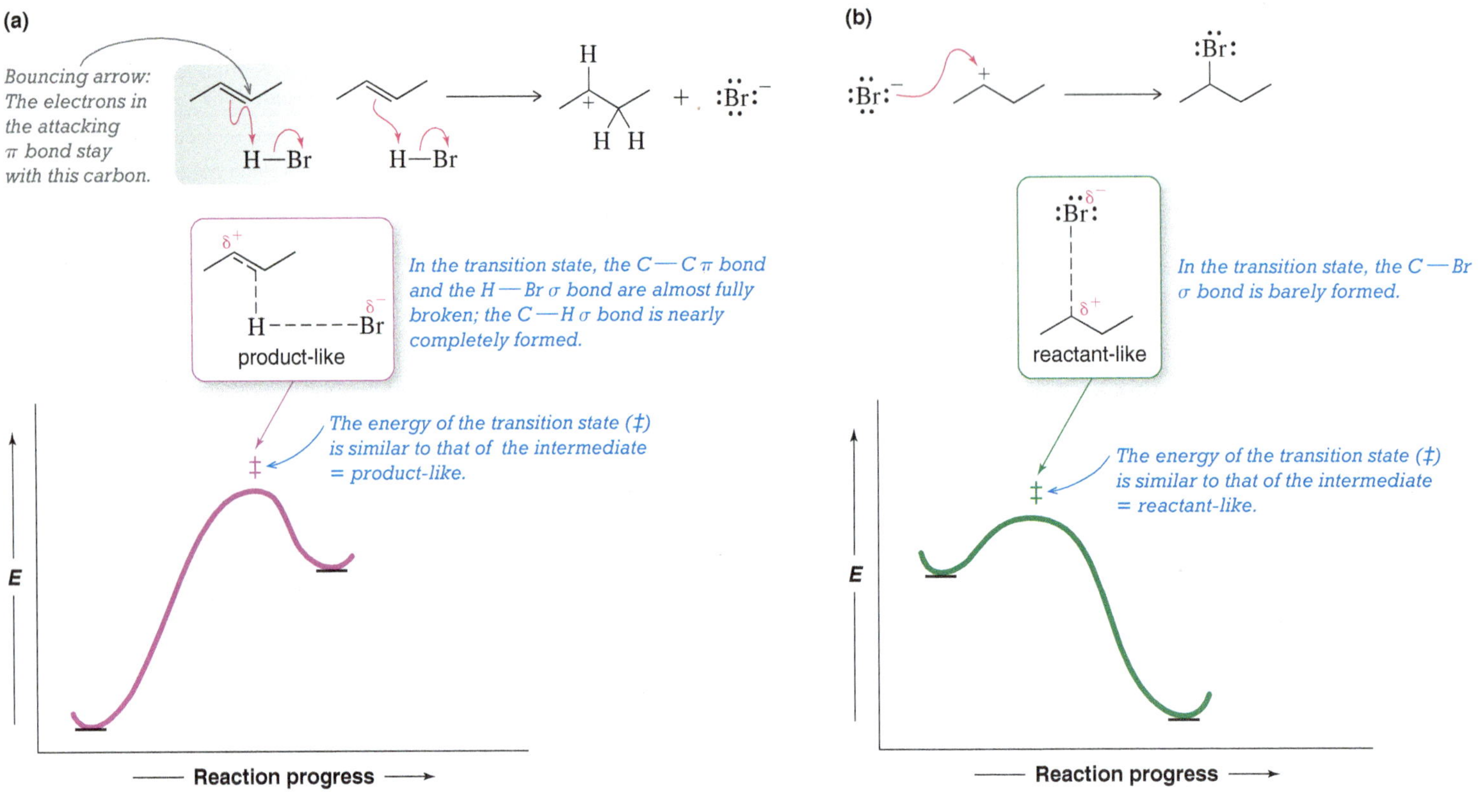

Figure 8.28 The Hammond postulate predicts the identity of the transition state. (a) Step 1 is product-like with substantial cation character on carbon. (b) Step 2 is reactant-like with substantial carbocation character. [*The carbocation and bromide (intermediates of the overall reaction) are the products of step 1 and the reactants of step 2.*]

the transition states are enantiomers, approach from the front or back (top or bottom) occur at equal rates. Because enantiomeric transition states are equal in energy (in an achiral environment), there is no difference in activation energy and therefore no difference in rates (**Figure 8.29**). As a result, the two enantiomers will form in equal amounts, creating a racemic, or optically inactive, mixture. In general, when a planar electrophile is attacked by a nucleophile, both frontside attack and backside attack will usually occur through enantiomeric transition states, resulting in racemic mixtures.

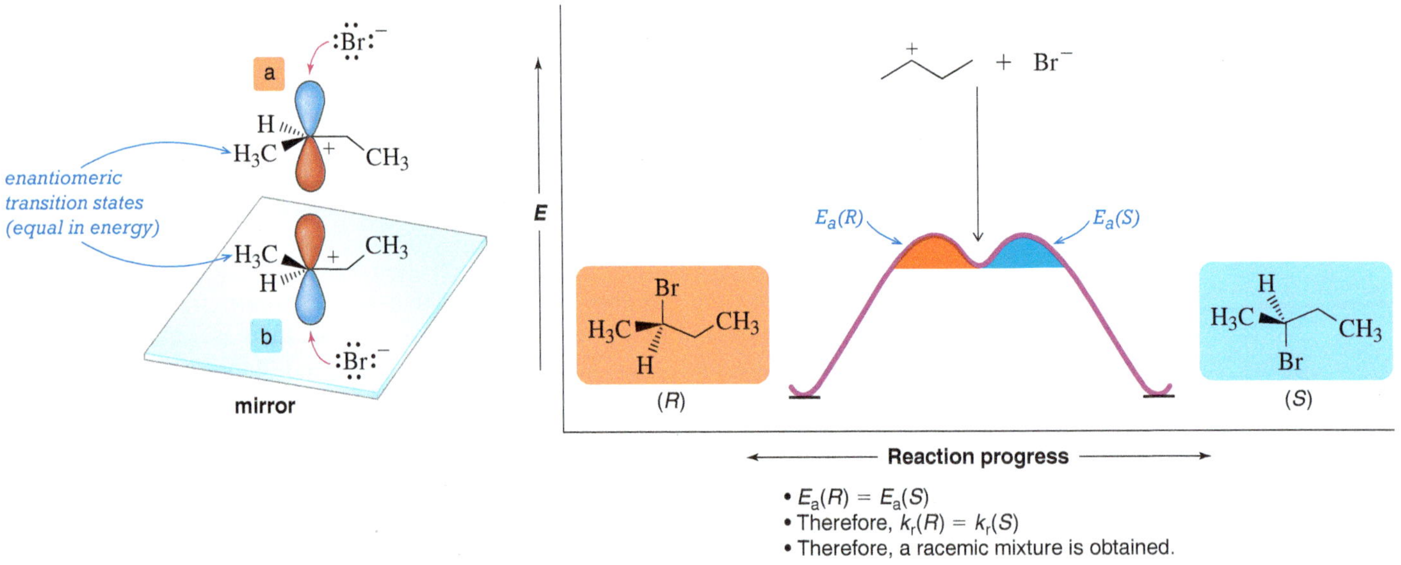

Figure 8.29 Enantiomeric transition states have the same activation energy and therefore react at the same rate.

In Section 8.3.2, we examine the regioselectivity of the addition of HBr to unsymmetrical alkenes. **Regioselectivity** refers to the particular region of a molecule to which a second atom/reagent will add. The Hammond postulate plays prominently in this discussion. [*Before you move on, test your understanding of the Hammond postulate in Assessment 8.25.*]

Assessment 8.23 A wayward chemist proposed the following mechanism for the addition of HBr to an alkene.

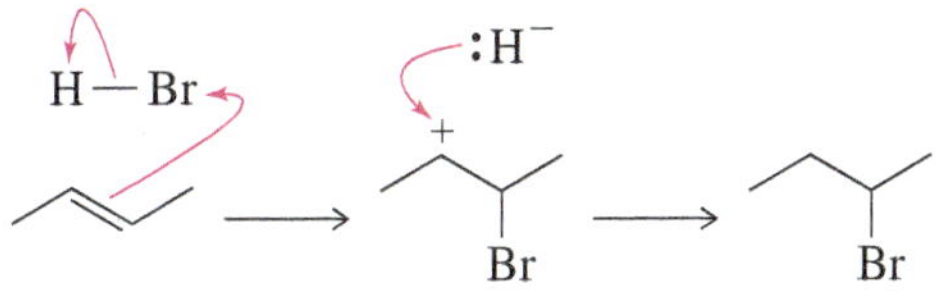

(a) Why is this mechanism unlikely? (b) Compare the reaction coordinate diagrams for the actual mechanism studied in Section 8.3.1 and this alternate mechanism on the same graph.

Assessment 8.24 Would you expect *cis*-but-2-ene or *trans*-but-2-ene to react more quickly in the electrophilic addition of HBr? Show a reaction coordinate diagram that compares the first step and the rate-determining step of the two reactions.

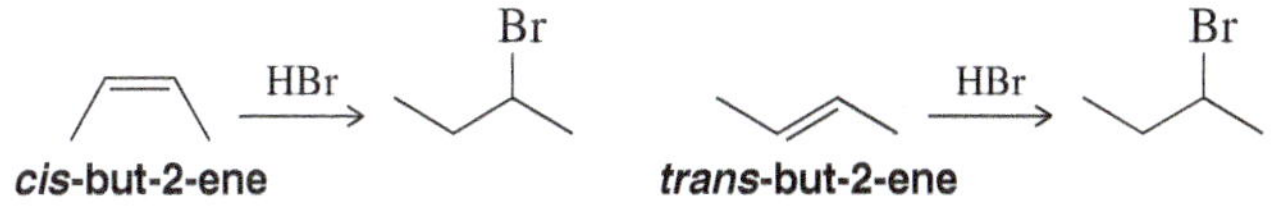

Assessment 8.25 Which is the most likely transition state for the reactions shown? Explain your answer. [*Note the difference in the size of the partial charges or partial unpaired electrons.*]

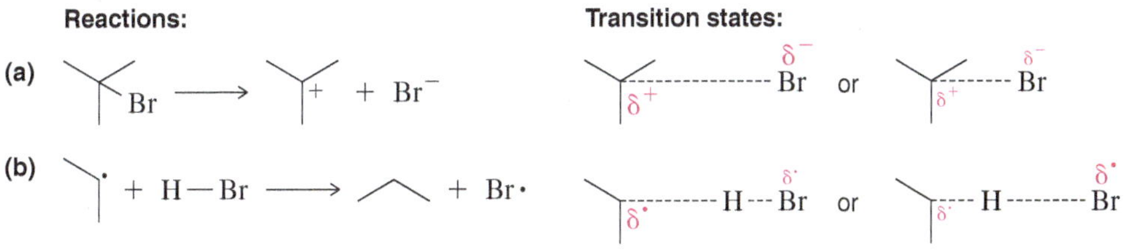

8.3.2 Addition of HX: regioselectivity

In the first example given in Section 8.3.1, a symmetrical alkene was reacted with HBr. In the first step (Figure 8.24), it didn't matter which carbon attacked the proton because attack at either gave the same intermediate species. **Figure 8.30** shows a situation where a different product is produced from addition at either alkene carbon of 2-methylbut-1-ene. When the proton of H—Br is attacked by the π bond at C_1, then the bromide attacks the carbocation at C_2, yielding a tertiary alkyl halide. On the other hand, if the proton is attacked by the π bond at C_2, then the bromide attacks the carbocation at C_1 to give the primary alkyl halide. [*Note how the bouncing arrow illustrates formation of these products more explicitly.*] Which of these might be the more favorable process?

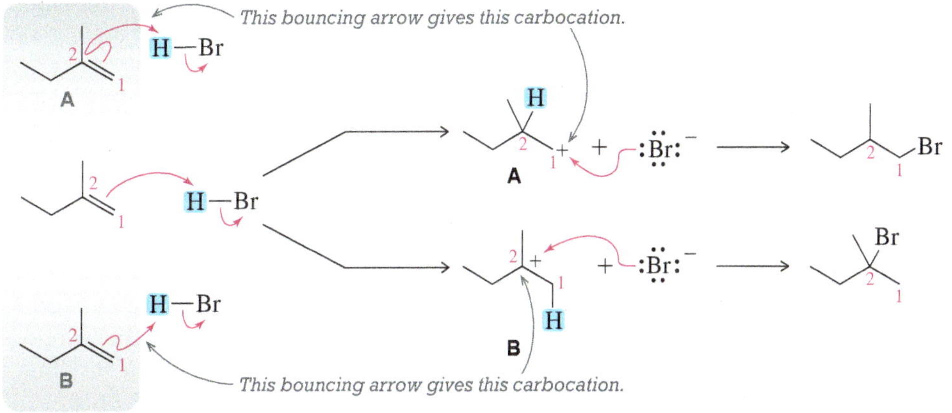

Figure 8.30 Addition to an unsymmetrical alkene can give two possible products.

Formation of the carbocation is the rate-determining step and determines the selectivity of the reaction. Focusing on the first step, there is a choice between forming a 1° carbocation in **A** and a 3° carbocation in **B**. Recall from Chapter 5 that radicals, which are electron-deficient species, are stabilized by **hyperconjugation** [Figure 8.31(a)], the overlap of neighboring σ bonds with empty or partially empty *p* orbitals. Carbocations, which are more electron-deficient than radicals, are similarly stabilized by hyperconjugation. The formation of the more stable 3° carbocation [Figure 8.31(b)] is faster because, as an endothermic step, there is substantial carbocation character in the competing transition states [Figure 8.31(c)]. The transition state leading to the 3° carbocation is more stable, so it has a lower activation energy.

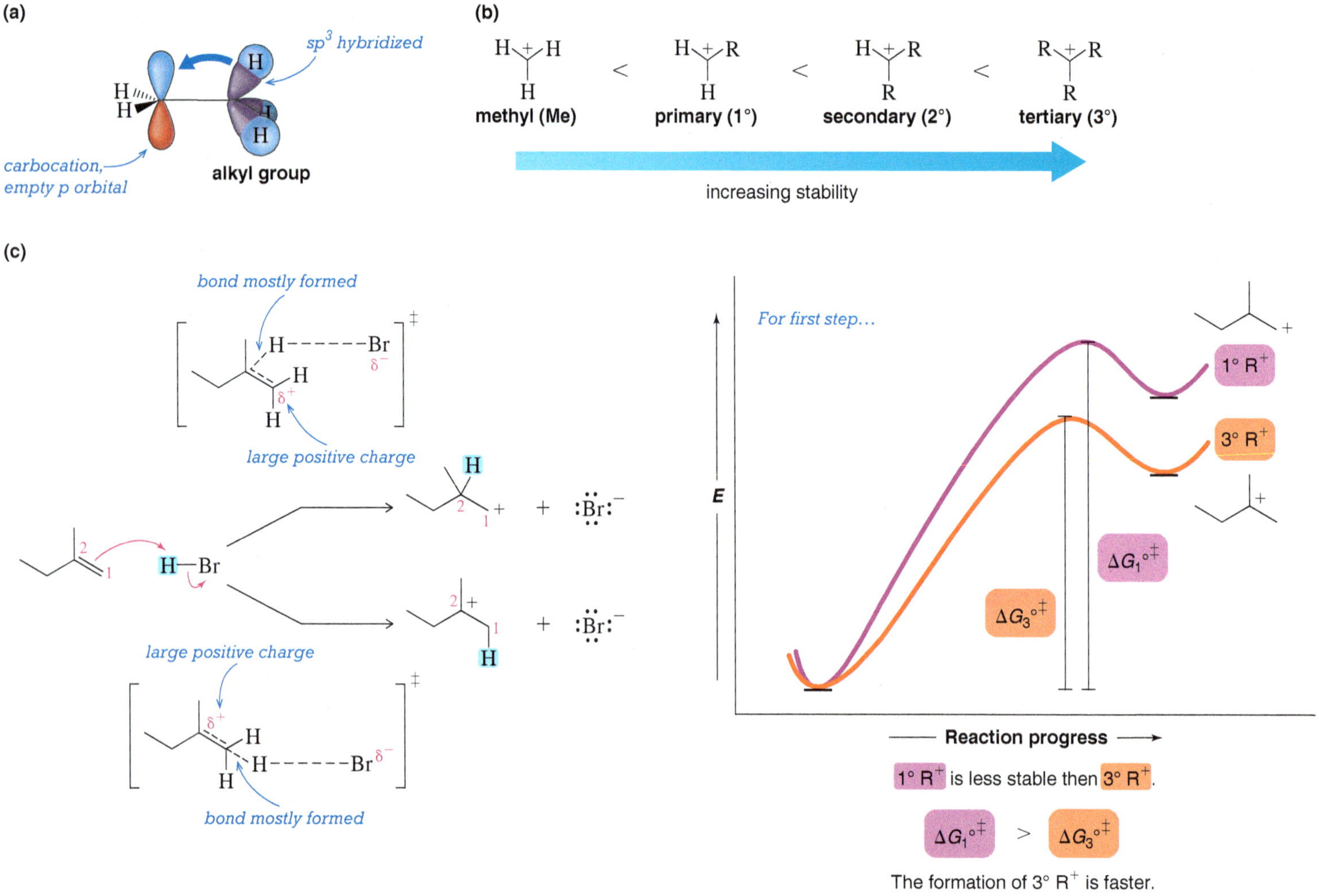

Figure 8.31 Formation of the 3° carbocation is favored. (a) Hyperconjugation stabilizes carbocations as it does radicals. (b) The more-substituted carbocation is more stable. (c) The Hammond postulate suggests that there is substantial carbocation character in the transition state of the rate-determining step. [*Recall* $\Delta G^{\circ\ddagger} = E_a$.]

A trend emerges, as we study organic reactions, that the more stable intermediate forms first, especially in endothermic reactions. It's only a generalization, though, because of what the Hammond postulate has to say about the identity of the transition state of these endothermic reactions. [*Assessments 8.26 and 8.27 further illustrate this concept.*]

Thus, by knowing which intermediate forms first, the product of a variety of alkene addition reactions can be predicted. In the case of 2-methylbut-1-ene, the 3° carbocation forms almost exclusively in the first step. In the second step, the bromide anion only has one carbocation it can attack. And it does so in a very fast step, to form 2-bromo-2-methylbutane. The overall mechanism is shown in Figure 8.32.

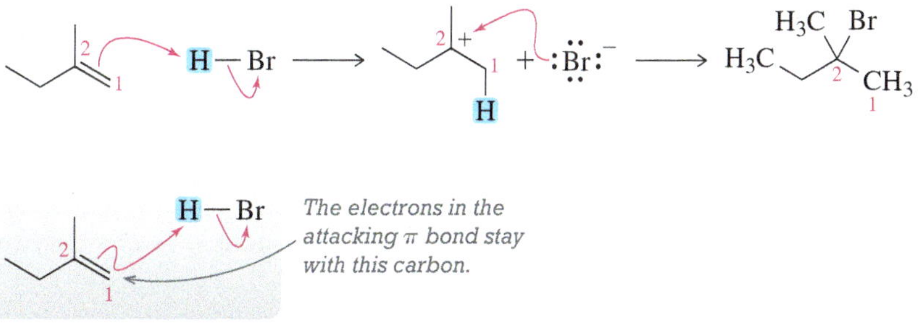

Figure 8.32 Complete mechanism of HBr addition to an alkene. The 3° carbocation is the only one that forms in the first step.

Assessment 8.26 Which would you expect to be more selective for carbocation formation, the electrophilic addition of HF or HBr to 2-methylbut-2-ene? Explain your answer.

Assessment 8.27 Imagine an electrophilic addition where the first step is exothermic. Which carbocation—2°, 3°, or neither—would you expect to form preferentially? Explain.

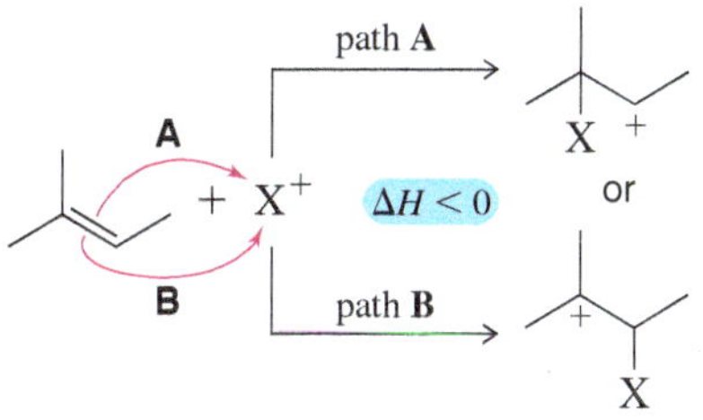

8-B CHEMISTRY IN CONTEXT: Biochemistry

Enzymatic isomerization; reversibility of H^+ addition

Plants synthesize terpenes using dimethylallyl pyrophosphate (DPP), 3-isopentenyl pyrophosphate (IPP), and the enzyme isopentenyl pyrophosphate isomerase (**Figure 8.33**). Because alkenes are more stable when there are more carbon substituents attached to the alkene carbons, DPP is favored at equilibrium. [*We explain why in Chapter 9.*] The conversion of IPP to DPP is a slow reaction, so it requires a catalyst to occur at an appreciable rate. The mechanism in Figure 8.33 imagines using an acid catalyst (H—B) instead of an enzyme. The alkene of IPP is first protonated to give a stable 3° carbocation and the conjugate base (B^-). In the absence of a nucleophilic species to attack the carbocation, the base

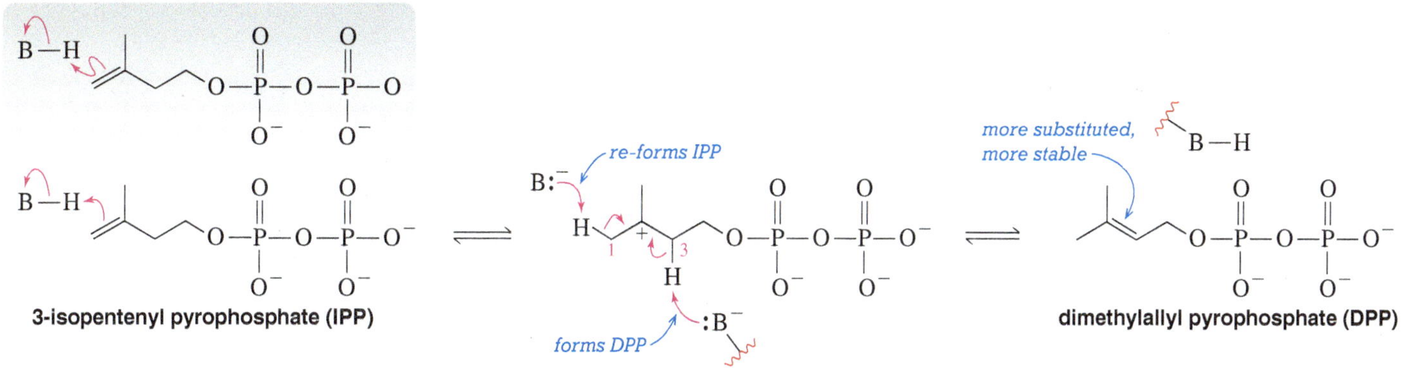

Figure 8.33 The acid-catalyzed conversion of IPP to DPP is reversible.

(continued)

can simply remove the proton from C_1 that it just donated to reform IPP. Or it could remove a proton from C_3 to form the alkene of DPP. With either choice possible, equilibrium is achieved.

The reaction follows a similar pathway in the active site of isopentenyl pyrophosphate isomerase (**Figure 8.34**), where an acid (—B—H) and a base (—B:$^-$) are present. The acid protonates C_1, while the active site base selectively deprotonates C_3. DPP is converted to IPP by this same mechanism in the opposite direction. Although enzymes usually catalyze reactions in both directions, a buildup of IPP drives the equilibrium toward formation of DPP in accordance with Le Châtelier's principle. DPP is then consumed in later reactions, which also shifts the equilibrium in the direction of DPP.

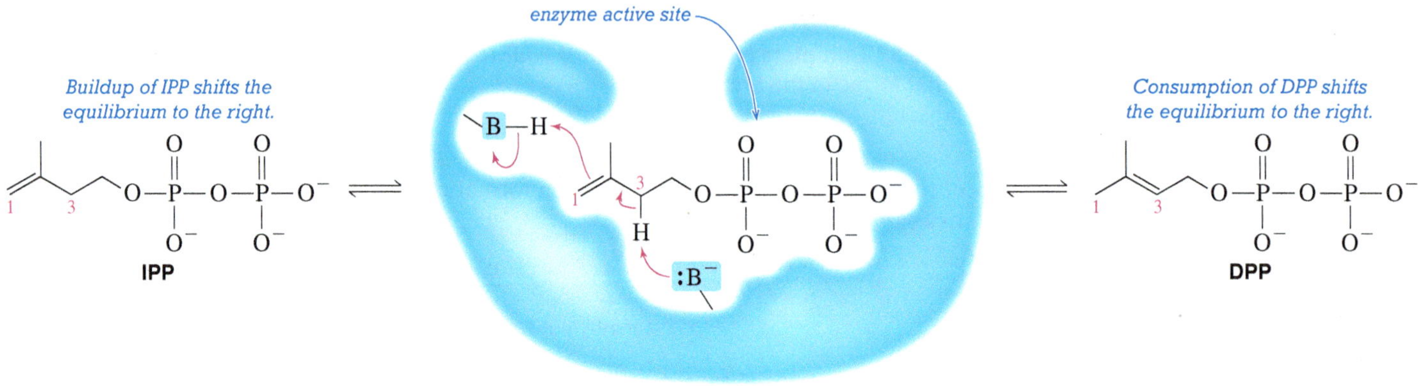

Figure 8.34 The enzyme-catalyzed conversion of IPP to DPP is driven by Le Châtelier's principle.

Assessment 8.28 For the enzyme isopentenyl pyrophosphate isomerase, IPP binds tightly as a result of interactions between the active site amino acid residues and the diphosphate of IPP. Without concerning yourself with the structure of amino acids, what charges might you expect to be present in the active site to hold IPP in place so that the enzymatic reactions can occur?

Assessment 8.29 Draw a mechanism for the acid-catalyzed conversion of DPP to IPP. How do you know that your mechanism is correct?

Assessment 8.30 Using a rough sketch of an enzyme active site like the one depicted in Figure 8.34, show a mechanism for the enzyme-catalyzed conversion of DPP to IPP. How do you know that it proceeds by the mechanism you drew?

8.3.3 On the soapbox—Markovnikov's rule

The reaction of hydrogen halides with alkenes usually forms the alkyl halide that has the halogen on the most substituted carbon. This concept is usually called **Markovnikov's rule** after the Russian chemist Vladimir Markovnikov, who first recognized it in the 1870s. Markovnikov's rule says that in the addition of H—X to an alkene, the hydrogen is added to the least substituted carbon, while the halide is added to the most substituted carbon. This rule allows you to predict the products of the reactions in **Figure 8.35** with 100% success. These products are sometimes referred to as Markovnikov products.

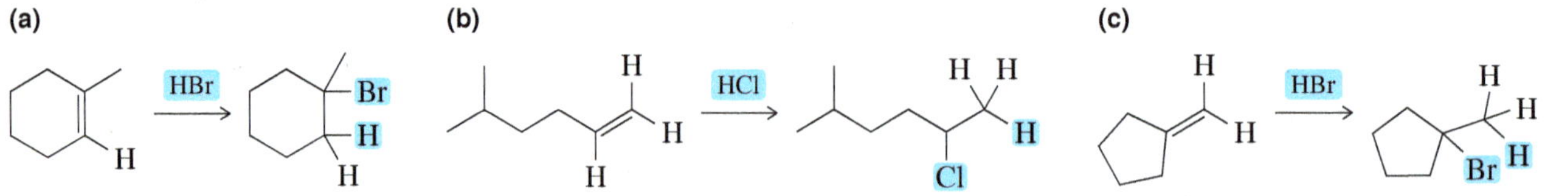

Figure 8.35 The original statement of Markovnikov's rule: The hydrogen adds to the least substituted carbon, and the halogen adds to the more substituted carbon. [*This leads to memorization.*]

It is ill advised to simply memorize Markovnikov's rule, although students will sometimes do that. If learned and applied "correctly," it can be a very valuable rule that finds application in virtually every reaction we study. We prefer to restate Markovnikov's rule as

Markovnikov's rule of making the most stable intermediate will apply when alkenes react with radicals too!

follows: In each step of a reaction, predict a pathway in which the most stable intermediate/product will result. The reactions in Figure 8.35 have been modified in **Figure 8.36** in the context of this restatement. Instead of simply remembering what the product will be, we can use our knowledge of carbocation stability to predict which intermediate will be produced and thereby rationalize the regiochemical outcome (the **regiochemistry**) of the reaction. In each case, the regiochemistry is dictated by forming the more stable intermediate in the rate-determining first step.

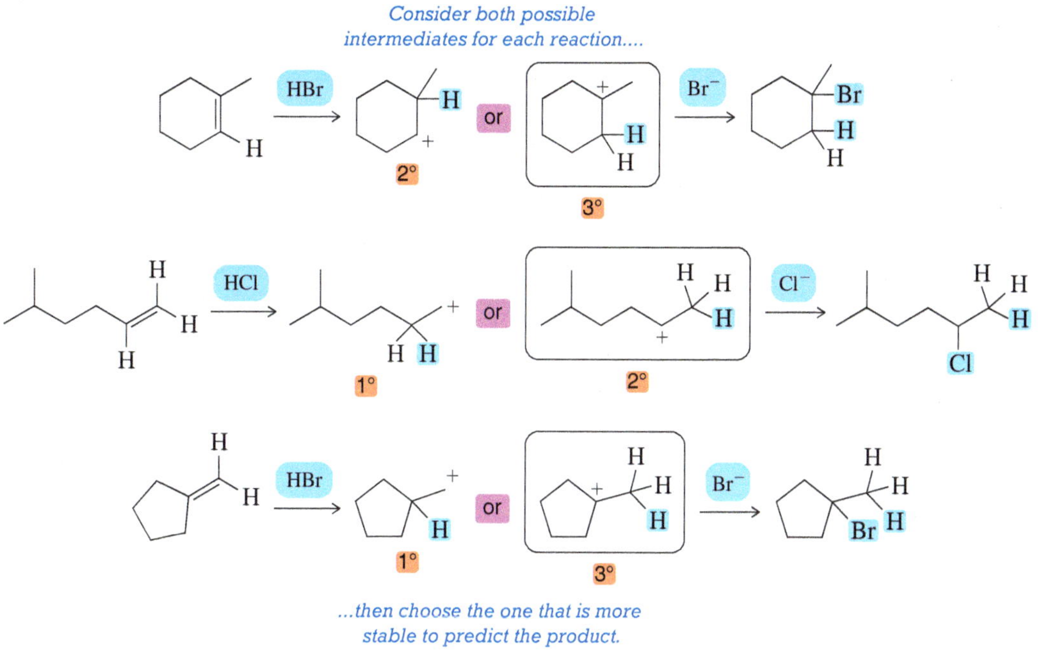

Figure 8.36 The modern statement of Markovnikov's rule: The alkene (π bond) reacts with an electrophile to produce the more stable carbocation intermediate. [*Consider both intermediates and predict the formation of the one that is more stable.*]

8.3–8.3.3 SUMMARY

- Hydrohalogenation is regioselective, proceeding through the more stable carbocation in an endothermic first step. A racemic mixture of haloalkanes is often produced because the halide attacks the planar carbocation from both sides at equal rates.

Assessment 8.31 What is the expected product for the reaction of each of the following alkenes with (i) HBr and (ii) HCl?

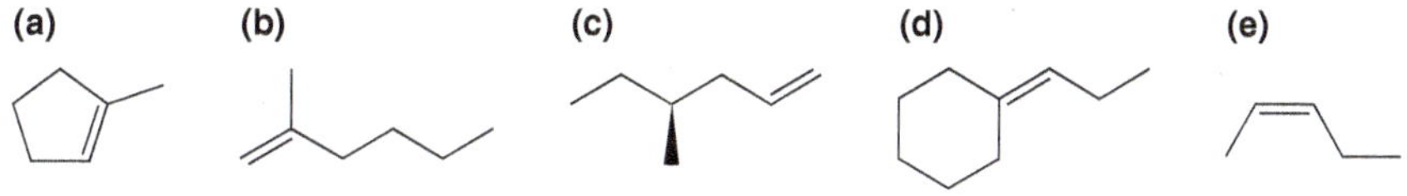

Assessment 8.32 Provide an arrow-pushing mechanism that rationalizes the formation of each of the products you predicted in Assessment 8.31. Make sure your mechanism accounts for all products formed, including stereoisomers and regioisomers, where applicable.

Assessment 8.33 Would you expect but-2-ene or 2-methylbut-2-ene to react more quickly with HBr?

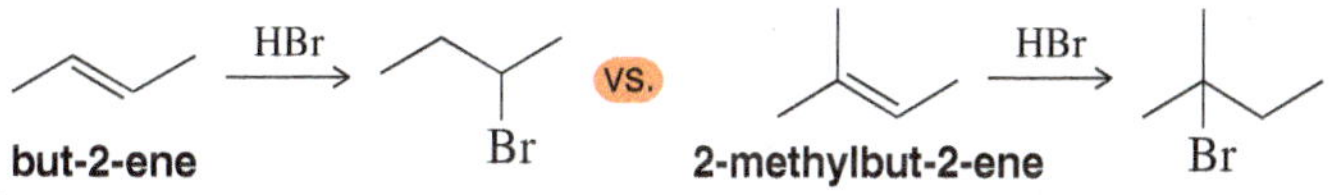

Assessment 8.34 In your own words, explain why it is not possible to make primary alkyl halides, such as 1-bromopentane, using the electrophilic addition of HCl or HBr to an alkene.

8.3.4 Radical addition of HX: "anti-Markovnikov"

Although HBr and HCl can react with alkenes to generate 3° and 2° haloalkanes, Sections 8.3.1–8.3.3 did not provide a way to make 1° haloalkanes. As illustrated in Assessment 8.34, this is because a 1° carbocation will never be the more stable carbocation. Nevertheless, synthetic chemists often need to make 1° haloalkanes. To do so, they treat an alkene with HBr in the presence of a peroxide. Figure 8.37 gives an example of the **peroxide effect,** where hex-1-ene reacts under these conditions to form 1-bromohexane as the major product. The driving force for this reaction, referred to as **anti-Markovnikov addition,** is identical to what drives the Markovnikov reactions from earlier sections. [*This is another reason not to memorize Markovnikov's rule.*]

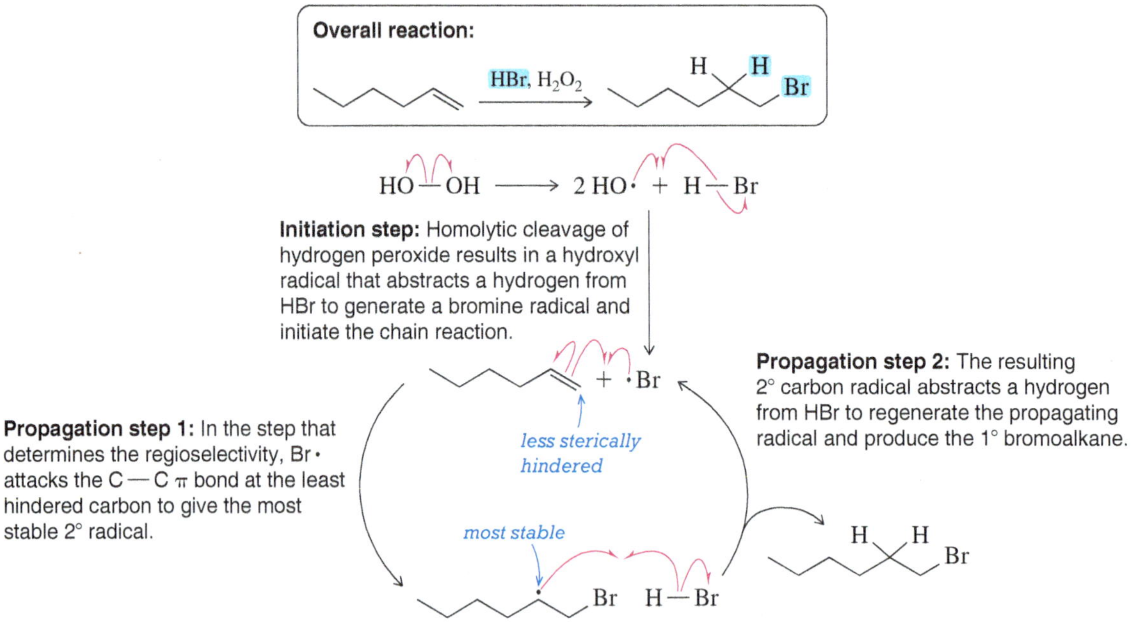

Figure 8.37 The peroxide effect allows a 1° haloalkane to form via HBr addition to an alkene.

Some clarification is probably needed about the mechanism in Figure 8.37. [*Take a second look at it if needed.*] Although it should remind you of radical halogenation (see Chapter 5), the bromine radical attacks an alkene as opposed to abstracting a hydrogen. Moreover, when the bromine attacks, it does so at the *least* substituted carbon of the alkene. Figure 8.38 resolves both of these issues with a detailed analysis of the first propagation step. Recall from Chapter 5 that radicals react with the weakest bond in the molecule. This is true here as well, with the bromine radical attacking the weaker C—C π bond (~65 kcal/mol) rather than breaking a C—H σ bond (~85 kcal/mol) [Figure 8.38(a)]. When the bromine radical attacks the alkene, it can do so to produce a 1° or 2° radical. Once more, the reaction proceeds to make the most stable intermediate—namely, the 2° radical that is better stabilized by hyperconjugation [Figure 8.38(b)]. Finally, bromine is a bulky atom, so attack at the 1° carbon is less hindered and therefore faster than attack at the 2° carbon [Figure 8.38(c)].

There are many similarities between the reactions we discussed in Section 8.3.2 and here in Section 8.3.4. In both cases, the regiochemical outcome is driven by the formation of the most stable intermediate in the rate-determining first step. In other words, the Markovnikov addition has the same reaction principles as the anti-Markovnikov addition. [*Based on the names, this might be surprising. So focus on understanding* how *and* why *reactions occur rather than memorizing names.*]

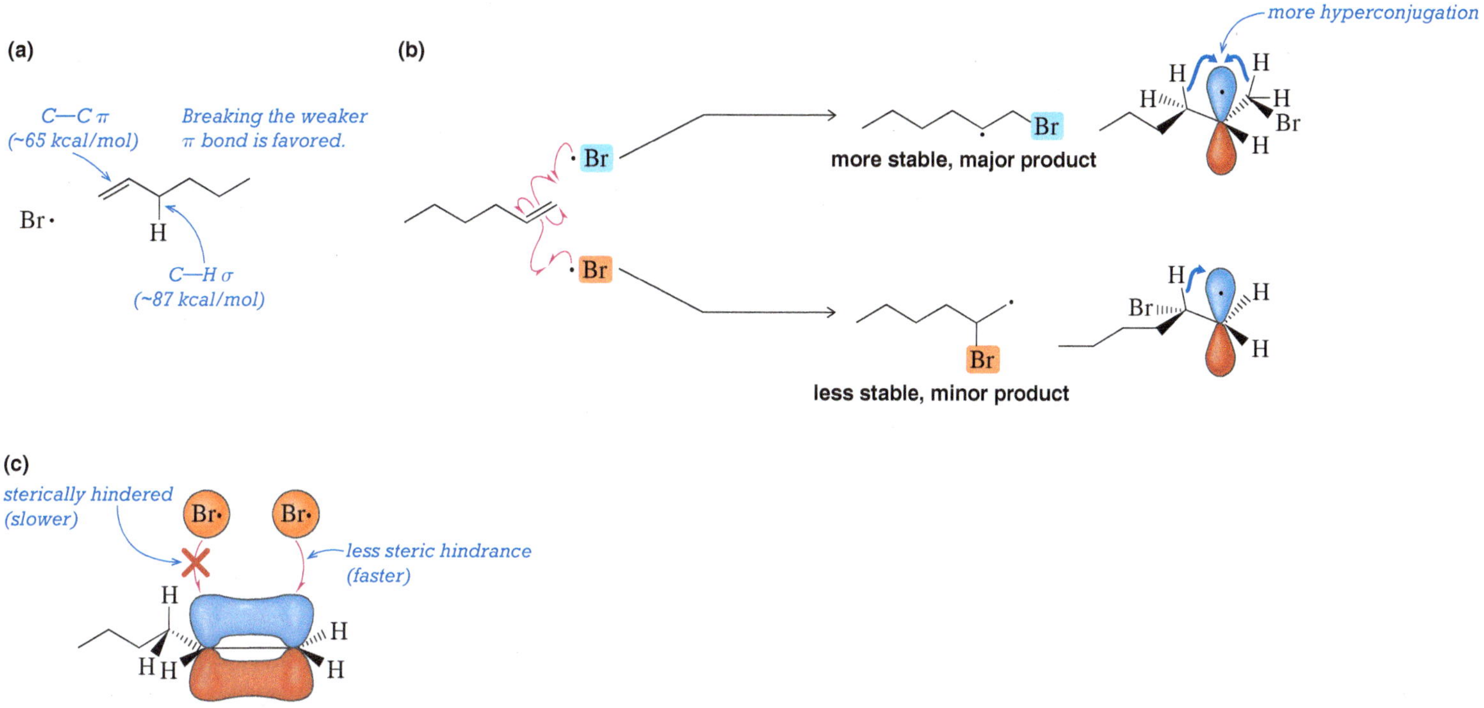

Figure 8.38 The selectivity of addition is determined in the first propagation step. (a) The bromine radical breaks the weaker C—C π bond. (b) Bromine adds to the less substituted carbon to produce the more stable radical and (c) because the 1° carbon is less sterically hindered.

Assessment 8.35 Provide the expected product for the reaction of each of the following alkenes with (i) HBr and (ii) HBr, H_2O_2.

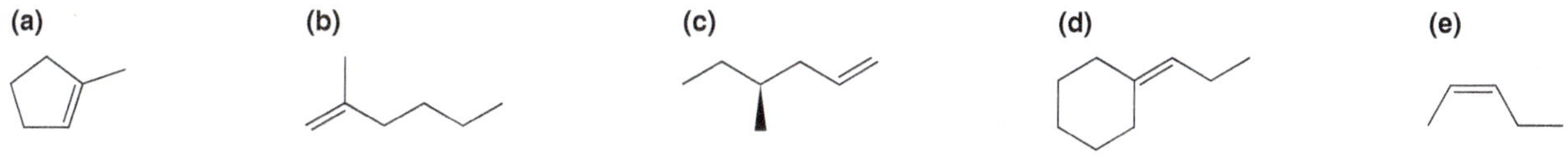

Assessment 8.36 Provide an arrow-pushing mechanism that rationalizes the formation of each of the products you predicted in Assessment 8.35(ii). Make sure your mechanism accounts for all products formed, including stereoisomers and regioisomers, where applicable.

Assessment 8.37 Which reagent system (HBr or HBr, H_2O_2) would you use to carry out the following transformations?

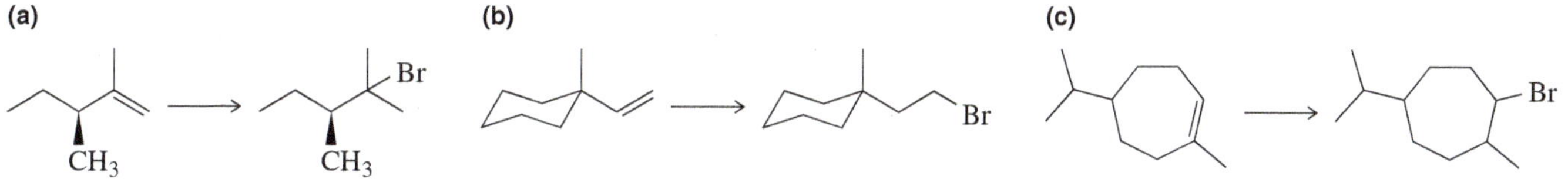

Assessment 8.38 Radical addition to alkenes is not effective for the synthesis of iodo- and chloroalkanes. Using your knowledge of the mechanism of this reaction, along with bond dissociation energies, explain why the radical additions of HI and HCl are not effective. (Assume ΔH = 65 kcal/mol for the C—C π bond.)

Let's solve this one together.

Solved Assessment 8.38

We already know that formation of a bromoalkane is a favorable process ($\Delta H \approx -10$ kcal/mol), with both propagation steps being exothermic. This is important because it ensures that the propagation steps are faster than the termination steps [*when radicals combine; recall Section 5.3.3*], thus keeping the chain reaction going. According to the bond dissociation energies in Table 8.5, the first step for HI is very endothermic ($\Delta H \approx +12$ kcal/mol), making it unfavorable. The second step of HCl addition is endothermic ($\Delta H \approx +8$ kcal/mol) and unfavorable, too. Because these two steps are unfavorable, and because the competing termination steps are faster, radical addition to alkenes is not effective with HI and HCl.

Put another way, the iodine radical is too stable to add in the first propagation step, whereas the chlorine radical is too unstable to form in the second propagation step. Bromine, on the other hand, is unstable enough to attack the alkene, but not too unstable to form in the second step. [*Where do you think the addition of HF would fall on this spectrum of reactivity? HF has a more exothermic first step than HCl and a more endothermic second step than HCl.*] ■

Table 8.5 Energetics of the addition of HI, HBr, and HCl. HI and HCl addition does not work because the first and second propagation steps, respectively, are unfavorable.

Addition of HI	ΔH (kcal/mol)
 65 kcal/mol; 53 kcal/mol	+12
 71 kcal/mol; 95 kcal/mol	−24
Addition of HBr	**ΔH (kcal/mol)**
 65 kcal/mol; 68 kcal/mol	−3
 88 kcal/mol; 95 kcal/mol	−7
Addition of HCl	**ΔH (kcal/mol)**
 65 kcal/mol; 81 kcal/mol	−16
 103 kcal/mol; 95 kcal/mol	+8

8.4 Addition of water

Electrophilic addition to alkenes can be used to make more than just haloalkanes. For example, we can synthesize an alcohol by the **hydration reaction** (addition of water) in Figure 8.39. The reaction is favorable [$\Delta H = -5$ kcal/mol; see Figure 8.39(a)] because two σ bonds form at the expense of one σ bond and one π bond. It is, however, disfavored by entropy ($\Delta S < 0$) because two molecules react to form one. Despite being favorable, mixing water with an alkene does *not* result in reaction. By analogy to the addition of HBr, a likely first step is the acid–base reaction shown in Figure 8.39(b). Analysis of this step shows the

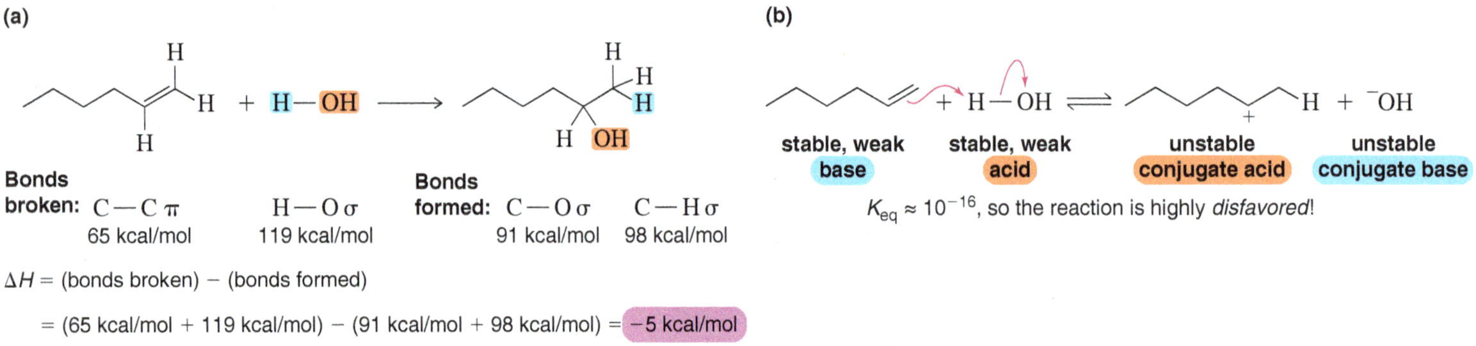

Figure 8.39 Hydration of alkenes. (a) It is favored by ΔH. (b) Water is not strong enough to protonate an alkene.

reaction of two weak bases to make two highly unstable molecules. Because water is not a strong enough acid, this reaction is too uphill to actually occur. [*How could we overcome this limitation?*]

8.4.1 Acid-catalyzed hydration

Taking a lesson from the addition of HBr, if water isn't a strong enough acid to donate a proton to the alkene, can we simply use a stronger acid to protonate the alkene and then have water attack the carbocation? [*Indeed, we can.*] The most common method for hydrating alkenes is to run the reaction under acidic conditions, typically using an acid with a nonnucleophilic conjugate base. Shown in **Figure 8.40**, sulfuric acid is most often used for the **acid-catalyzed hydration.** When a strong acid is dissolved in water, it dissociates to make hydronium (H_3O^+) and HSO_4^-. [*What is* K_{eq} *for the reaction of* H_2SO_4 *and* H_2O*? Recall Section 4.3.3.*] H_3O^+ is a strong enough acid to be attacked by the π electrons of the alkene to make the more substituted 2° carbocation in an endothermic step. The resulting planar carbocation is attacked by water from both sides at an equal rate, resulting in a protonated alcohol. [*See Section 6.6 if you have forgotten why.*] A final acid–base reaction occurs to give the desired alcohol as a mixture of enantiomers, while regenerating H_3O^+.

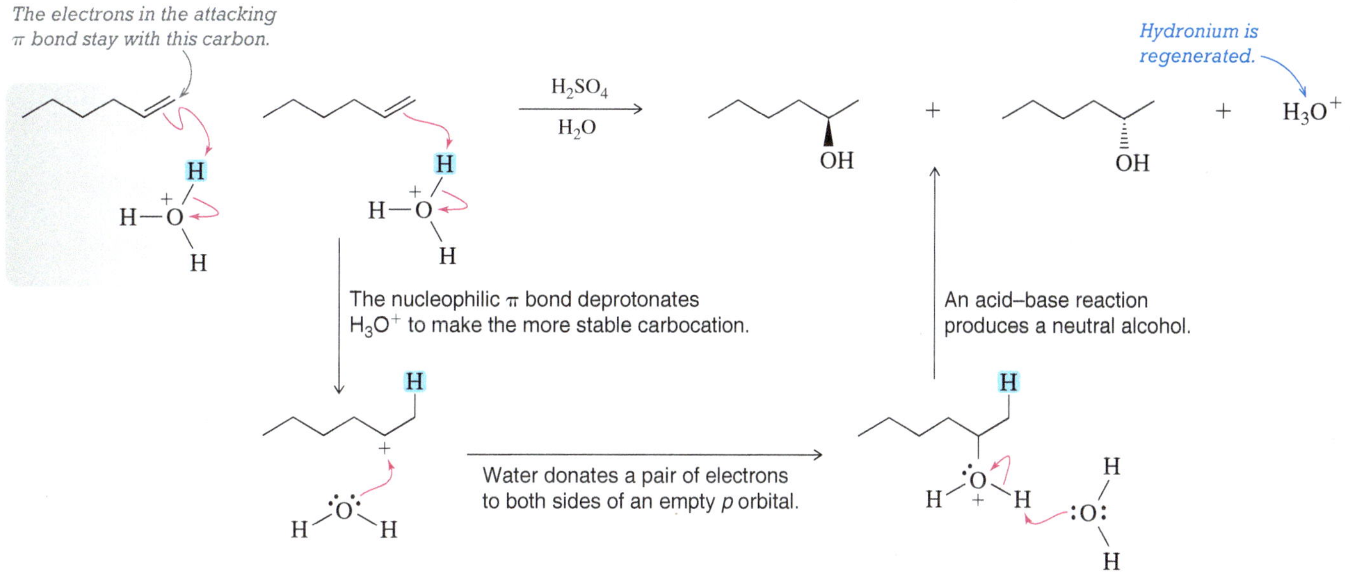

Figure 8.40 The acid-catalyzed hydration of alkenes produces a racemic mixture of alcohols.

To recap, the slow hydration reaction in Figure 8.39(b) was made possible by generating an acid (H_3O^+). H_3O^+ facilitates the reaction, but it is regenerated/not consumed, so this reaction is acid-catalyzed. [*Our ability to draw mechanisms and reaction coordinate diagrams helps us understand how catalysts increase the rate of chemical reactions.*] **Figure 8.41** shows a reaction coordinate diagram containing both the catalyzed and uncatalyzed reactions. The uncatalyzed reaction is not possible because the first acid–base reaction between water and the alkene (the rate-determining step) is too uphill (disfavored) and has a high activation energy. Although the acid-catalyzed reaction consists of three steps instead of two, the rate-determining first step is less disfavored. It is still endothermic, proceeding selectively as predicted by the Hammond postulate, but it is much faster than the uncatalyzed version. Because the activation energy in the rate-determining step is lowered under acidic conditions, the catalyzed reaction is faster.

ΔH for catalyzed and ΔH for uncatalyzed versions of the same reaction are always the same. Catalysis does not make one side more favorable than another—that is, it does not change the equilibrium constant—and it does not change ΔG, ΔH, or ΔS for the reaction. By lowering the activation energy, catalysts only increase the rate of a reaction.

This concept is important enough that I'll say it: Catalysis does not change the equilibrium constant of a reaction—it only increases the rate at which equilibrium is achieved.

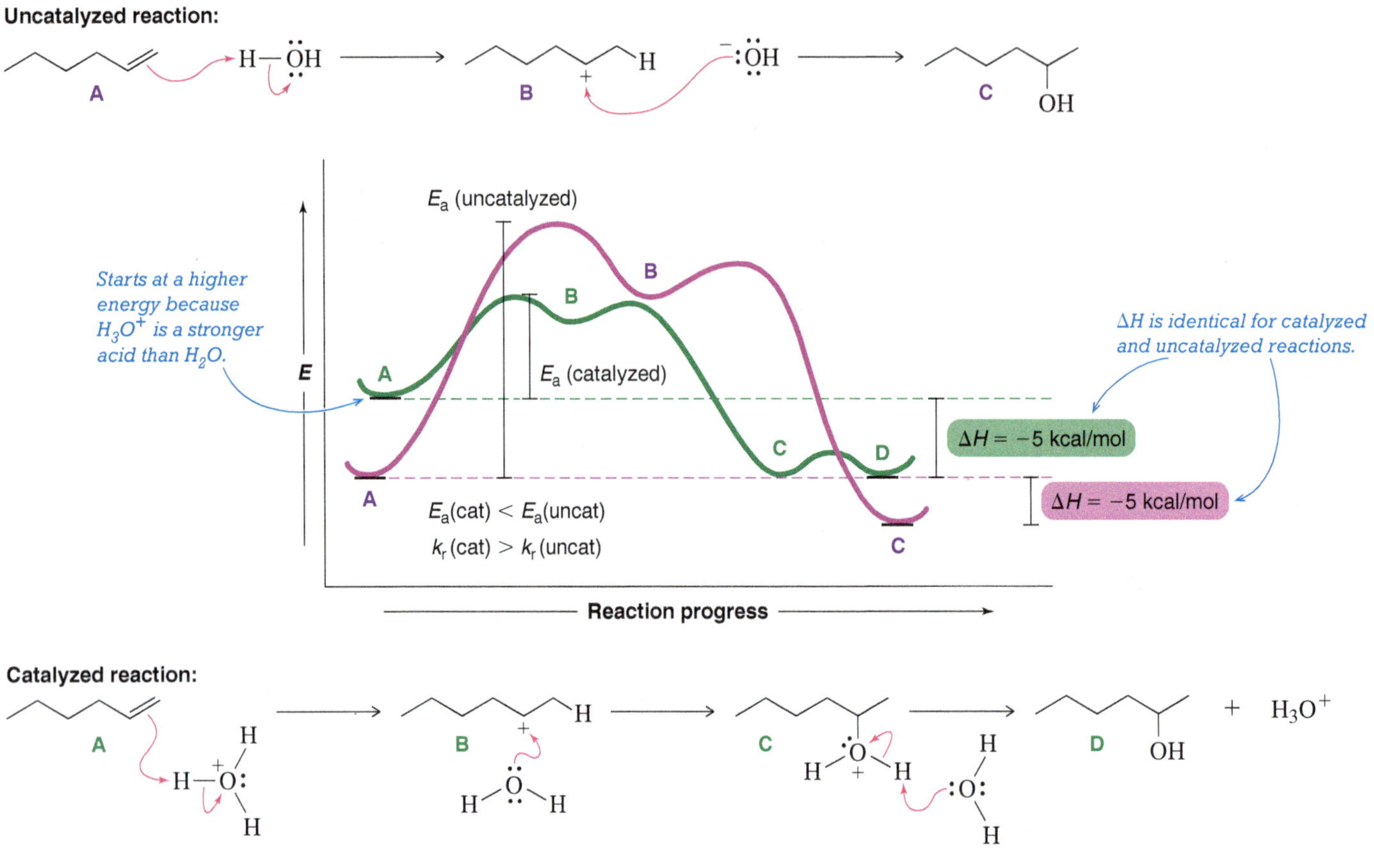

Figure 8.41 **Acid catalysis lowers the activation energy, making the hydration reaction proceed at a reasonable rate.**

Assessment 8.39 Provide the expected product for the reaction of each of the following alkenes with H_2SO_4 and H_2O.

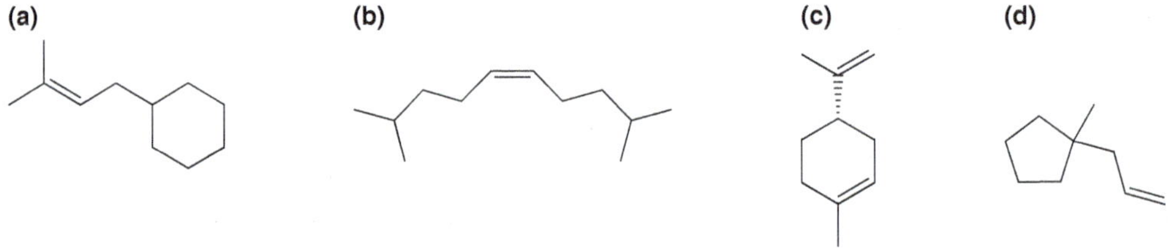

Assessment 8.40 For each of the products you predicted in Assessment 8.39, provide an arrow-pushing mechanism which rationalizes the formation of each product. Make sure your mechanism accounts for all products formed, including stereoisomers and regioisomers, where applicable.

Assessment 8.41 Provide the alkene that would give the following alcohols under acid-catalyzed hydration conditions.

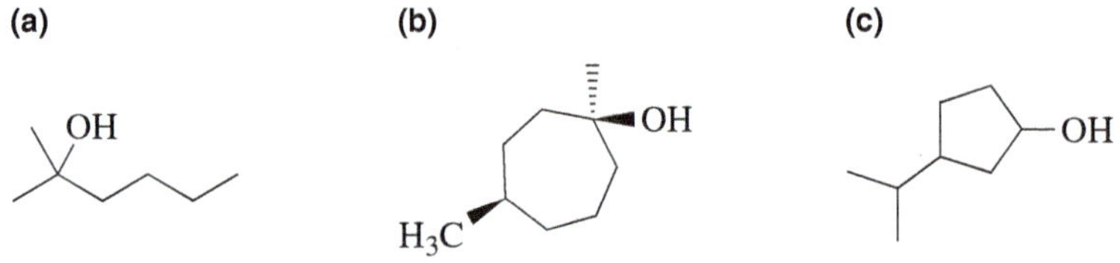

Assessment 8.42 Ethers can be synthesized by substituting ethanol for water in the acid-catalyzed hydration reaction. Suggest an arrow-pushing mechanism which accounts for the formation of the ethyl ether under these conditions.

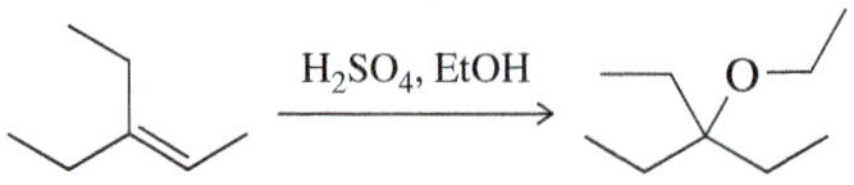

8-C CHEMISTRY IN CONTEXT: Biochemistry

Synthesis of honey bee pheromones

Geraniol and nerol are terpenes extracted from the petals of certain rose species by the perfume industry. They are also vital for communication among worker honey bees. Geraniol and nerol are secreted from the Nasonov gland to attract other colony members to flowers rich in nectar, an unoccupied hive, and significant water sources.

The connection of two isoprene units is a major bond-forming step in the synthesis of geraniol and nerol in the Nasonov gland (**Figure 8.42**). In the active site of the enzyme farnesyl pyrophosphate synthase, the diphosphate group is expelled from dimethylallyl pyrophosphate (DPP) to give a resonance-stabilized carbocation. [*We discuss this reaction in Chapter 12.*] The alkene π electrons of isopentenyl pyrophosphate (IPP) attack the electrophilic carbocation to give a stable 3° carbocation, extending the carbon chain by one isoprene unit. Removal of the hydrogen re-forms the double bond, generating two alkene isomers. [*There's more like this in Chapter 12 also.*] Finally, the pyrophosphate is hydrolyzed to give the diastereomeric terpenes geraniol and nerol. [*This one comes up in Chapter 13—so much more to come!*]

While many species of insects can distinguish between synthetic and natural pheromones based on slight differences in isomeric compositions, produce farmers often use a mixture of geraniol and nerol (and other compounds) to mimic the Nasonov pheromone. Spraying their fields with these compounds attracts more honey bees to the area, increasing pollination and produce yield.

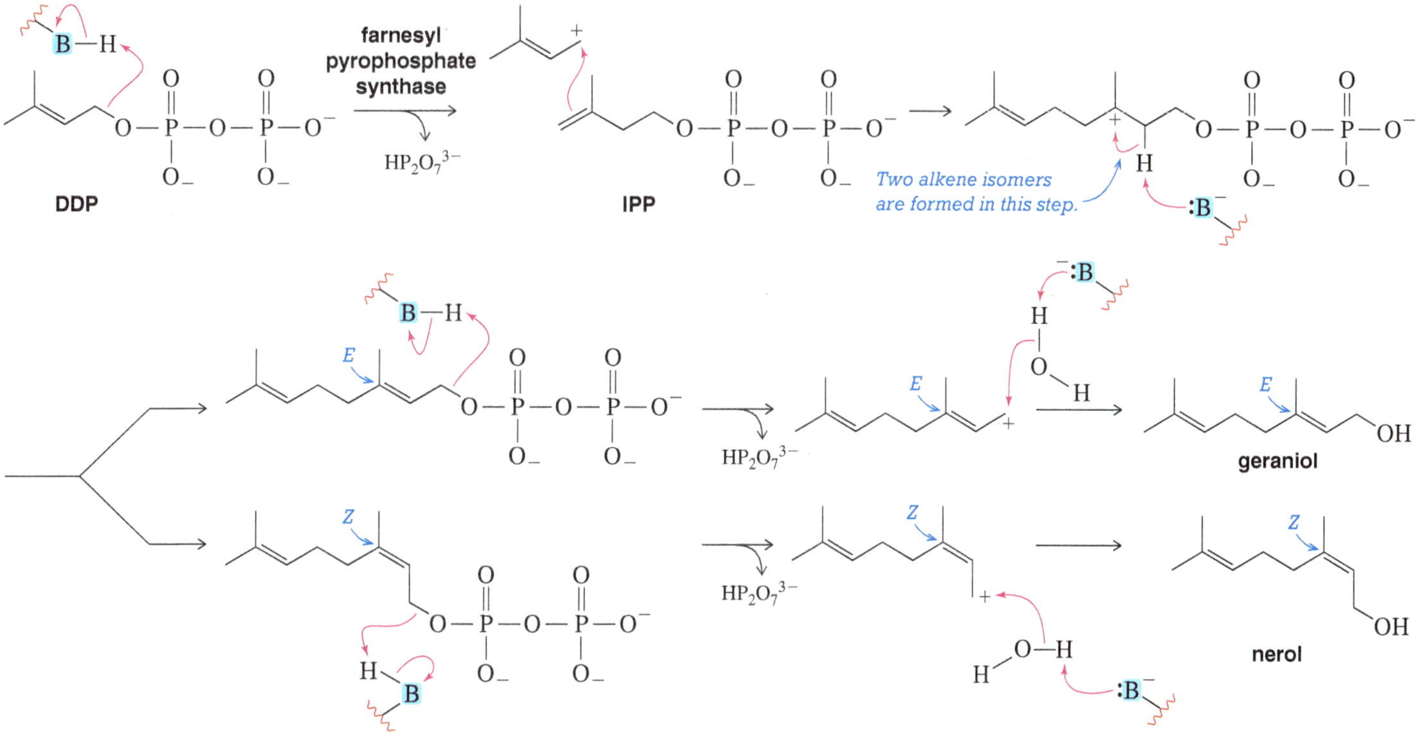

Figure 8.42 The biosynthesis of geraniol and nerol utilizes an alkene addition reaction.

Assessment 8.43 When combining isoprene equivalents, IPP is never the electrophile. Why might this be? What is special about DPP that allows it to function better as an electrophile? [*Draw the carbocation that each becomes.*]

Assessment 8.44 Suggest a mechanism for the synthesis of farnesol beginning with IPP and DPP.

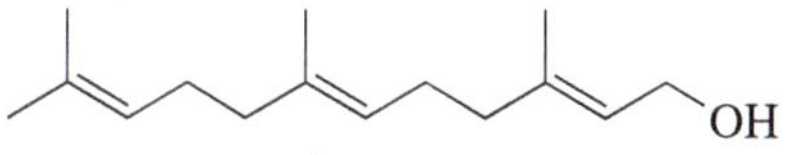

farnesol

8.4.2 Other alcohol syntheses—a hypothetical scenario

We learned one way to make alcohols in Section 8.4.1, and two more are presented here in Section 8.4.2. Why do organic chemists need more reactions to make alcohols? [*It isn't to make students' lives miserable.*] A hypothetical (but realistic) scenario can answer this question.

Imagine you are an industrial medicinal chemist tasked with developing an inhibitor of the enzyme *imaginokinase* [**Figure 8.43(a)**]. [*We are basically thinking like a chemist for the remainder of Section 8.4.*] From a poorly resolved crystal structure of the enzyme, 1-hydroxyethylcyclohexane emerges as a target. It can be made from vinylcyclohexane [**Figure 8.43(b)**], which is helpful because ease of synthesis is an important consideration in the pharmaceutical industry. [*To the lab!*]

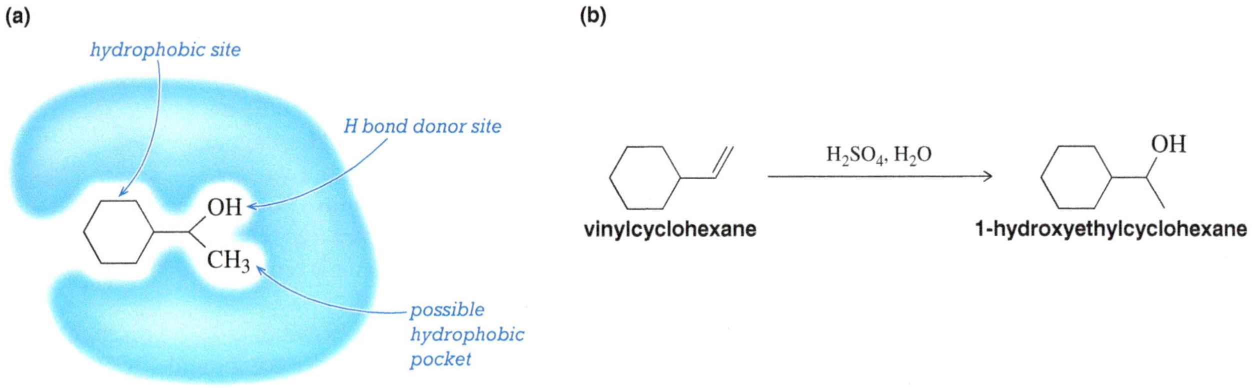

Figure 8.43 Our medicinal chemistry scenario consists of (a) the hypothetical active site of imaginokinase and (b) a proposed synthesis of 1-hydroxyethylcyclohexane (our target) from vinylcyclohexane.

Unfortunately, treating vinylcyclohexane with H_2O in the presence of catalytic H_2SO_4 almost exclusively produces 1-ethylcyclohexanol (**Figure 8.44**), a constitutional isomer of the desired product. Going forward, we will try to answer the following questions: Why and how did this undesired product formed? Can we generalize and predict this reaction outcome in other situations? And, most importantly, can we modify the reaction conditions such that we can produce the desired constitutional isomer?

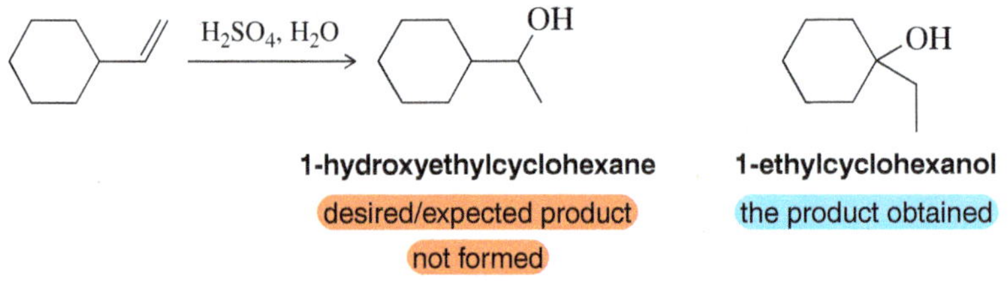

Figure 8.44 Our attempted synthesis results in an unexpected product.

Let's start by thinking like a chemist to discover what happened in this reaction.

Thinking Like a Chemist

By isolating an unexpected product, we have discovered a new [*to us*] pattern of reactivity. The mechanism we'll eventually figure out would be considered tricky for a novice. And yet, using the right process, you are fully capable of figuring out exactly how this reaction happened. Whenever you encounter a difficult problem, it is important to break it down into a series of simpler problems. By solving these, you'll get the answer to the difficult question. [*What matters here is not the answer, but the* process *we use to get it.*]

The steps you can use to figure out a problem like this are listed in **Table 8.6**. [*These are useful for* all *mechanistic problems, not just this one, so bookmark this page.*] As an organic student, I was taught to "Draw, don't think!" You will have to think eventually, but the point of "Draw, don't think!" is to *do the easy steps first*. By doing so, you will find that the next question is just as easy as the previous one. A defined plan also helps you figure out where you become stumped.

Table 8.6 Steps for solving complex mechanism problems.

Step	What the step accomplishes
1. Draw the structures and number the carbons.	By numbering the carbons, we can identify what groups moved where in the reaction. It's like a map of the reaction.
2. Identify which bonds break and which form in the reaction.	Knowing which bonds break and which form in the reaction gives us a big-picture view of how the reaction occurs. It can be helpful to draw in the hydrogens here because they sometimes move around in reactions.
3. Work backward.	Did this reaction finish like a reaction we saw previously? What was the last step? What was the one before that?
4. Work forward.	Given the reagents, how did this reaction probably start? [*Again, you are comparing this to reactions seen previously.*]
5. Meet in the middle.	We know how a reaction starts and how it finishes. How do the "two halves" meet up? Luckily, the answer is "always on the page," meaning, we *know* what has to happen because we *know* the product is formed from the reactant, even if we have to push electrons in a way that we haven't seen before.
6. Justify the outcome.	Does the mechanism we've drawn make sense? Is the ball rolling downhill? Are we going from less stable to more stable? Are negatives attacking positives? Are we pushing electrons toward areas of electron deficiency?

[*Before reading further, apply these six steps to the reaction in Figure 8.44 on your own.*]

Step 1: Numbering in the reactant is completely arbitrary, but should probably be done consecutively as you move around the ring (**Figure 8.45**). When transferring these numbers to the product, you have to make sure that what you call carbon 1 in the reactant is really carbon 1 in the product. Because the basic carbon skeleton stays the same in this reaction, it is safe to assume that the numbering can be done similarly.

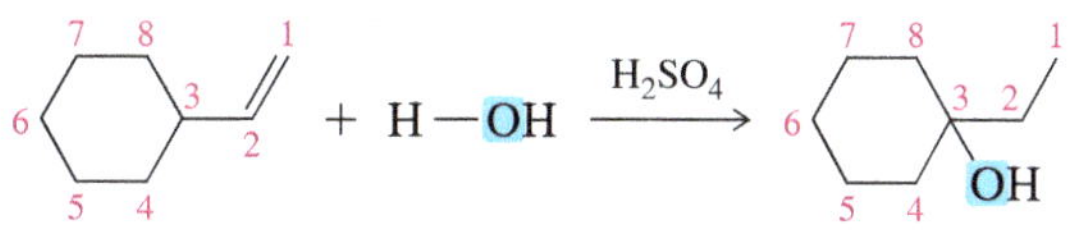

Figure 8.45 Draw the structures and number the carbons.

Check yourself by verifying that, for example, carbons 4–8 are CH_2 groups in both the reactant and the product.

The oxygen in H_2O, because it is the only source of oxygen, is likely the oxygen in the alcohol product.

Step 2: Identifying which bonds are broken and which are formed (**Figure 8.46**) is easier when the carbons are numbered and all C—H bonds close to the reaction center are shown.

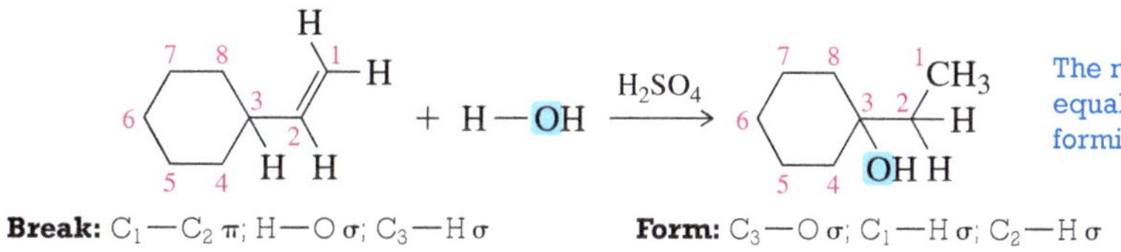

Figure 8.46 Identify which bonds break and which form in the reaction.

The number of bonds breaking equals the number of bonds forming in a given reaction.

Step 3: The reagents are identical, so we can use Section 8.4.1 to predict how this reaction probably finishes (**Figure 8.47**).

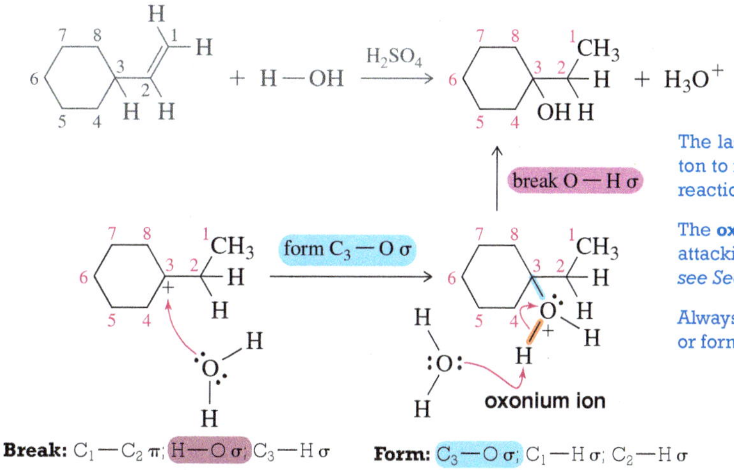

Figure 8.47 Work backward.

The last step is probably removal of a proton to regenerate H_3O^+, the catalyst in this reaction. [*See Section 8.4.1.*]

The **oxonium** ion forms by water attacking a 3° carbocation at C_3. [*Again, see Section 8.4.1.*]

Always confirm that we are breaking and/or forming bonds on the list in step 2.

Step 4: The reagents are identical, so we can use Section 8.4.1 to predict how this reaction probably starts, too (**Figure 8.48**).

Figure 8.48 Work forward.

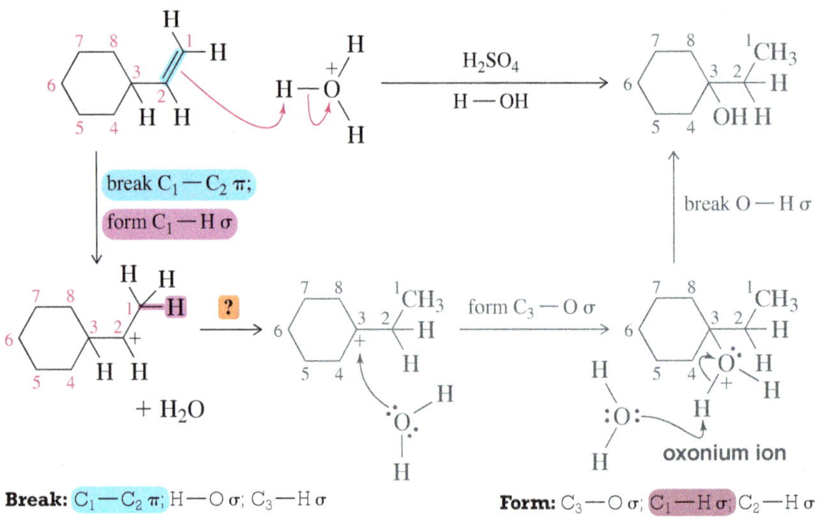

This reaction is probably a normal electrophilic addition to alkenes. [*See Section 8.4.1.*]

H_3O^+ is produced when H_2SO_4 is dissolved in H_2O.

The C—C π bond deprotonates H_3O^+. [*Again, see Section 8.4.1.*]

The more stable 2° carbocation forms; this is why we expected a 2° alcohol as the product (see Figure 8.44).

Step 5: Step 4 says a 2° carbocation should form. Step 3 says the reaction finishes through a 3° carbocation. In this step, we have to connect the forward and backward processes (**Figure 8.49**). If our logic has been sound to this point, whatever we draw here *has* to be what happens. [*We'll worry about why later.*]

Figure 8.49 Meet in the middle.

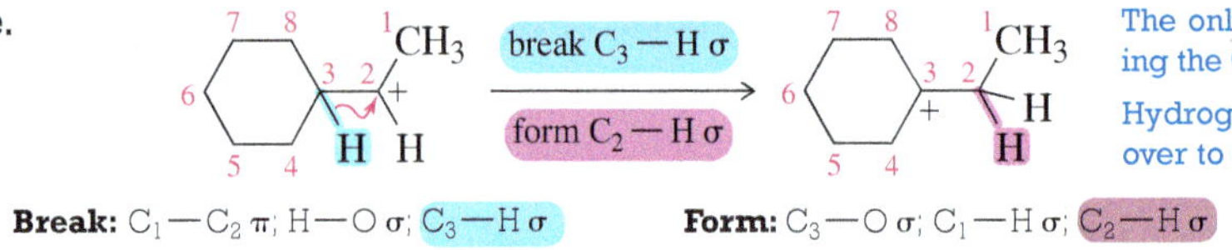

The only two bonds we haven't dealt with are forming the C_2—H bond and breaking the C_3—H bond. Hydrogen at C_3 takes its electron pair and moves over to C_2.

Step 6: Although our mechanism gets us from reactant to product, we aren't finished until we have justified each step. All but the one from step 5 we have seen previously. We can quickly analyze whether step 5 is favorable using concepts studied previously. A C—H σ bond breaks and a C—H σ bond forms, so $\Delta H \approx 0$. [*That's OK.*] The reaction goes from one molecule to one molecule, so $\Delta S \approx 0$. [*That's fine, too.*] A 2° carbocation becomes a more stable 3° carbocation. [*That should be highly favorable.*] Thus, forming the more stable carbocation is the driving force for this mechanistic step. And with that, we can be confident in our suggested mechanism of the reaction. ■

8.4.3 Carbocation rearrangement—a general problem/process

The rearrangement discovered while Thinking like a Chemist is referred to as a **1,2-hydride shift** and is an example of a **carbocation rearrangement**. The overall mechanism is shown in **Figure 8.50(a)**. In essence, a hydrogen removes the σ electrons from one carbon and donates them to a nearby empty *p* orbital in a single step. [*More examples of concerted processes are encountered in Chapter 9 and Chapter 22, among others.*] The "1,2-" signifies that these shifts can only occur from one atom to its direct neighbor. A hydride usually cannot migrate to a carbocation that is two or more carbons away. The reaction coordinate diagram [**Figure 8.50(b)**] isn't much different from the usual acid-catalyzed hydration, except that it has an extra step (the hydride shift, B → C) on the pathway from reactant to product.

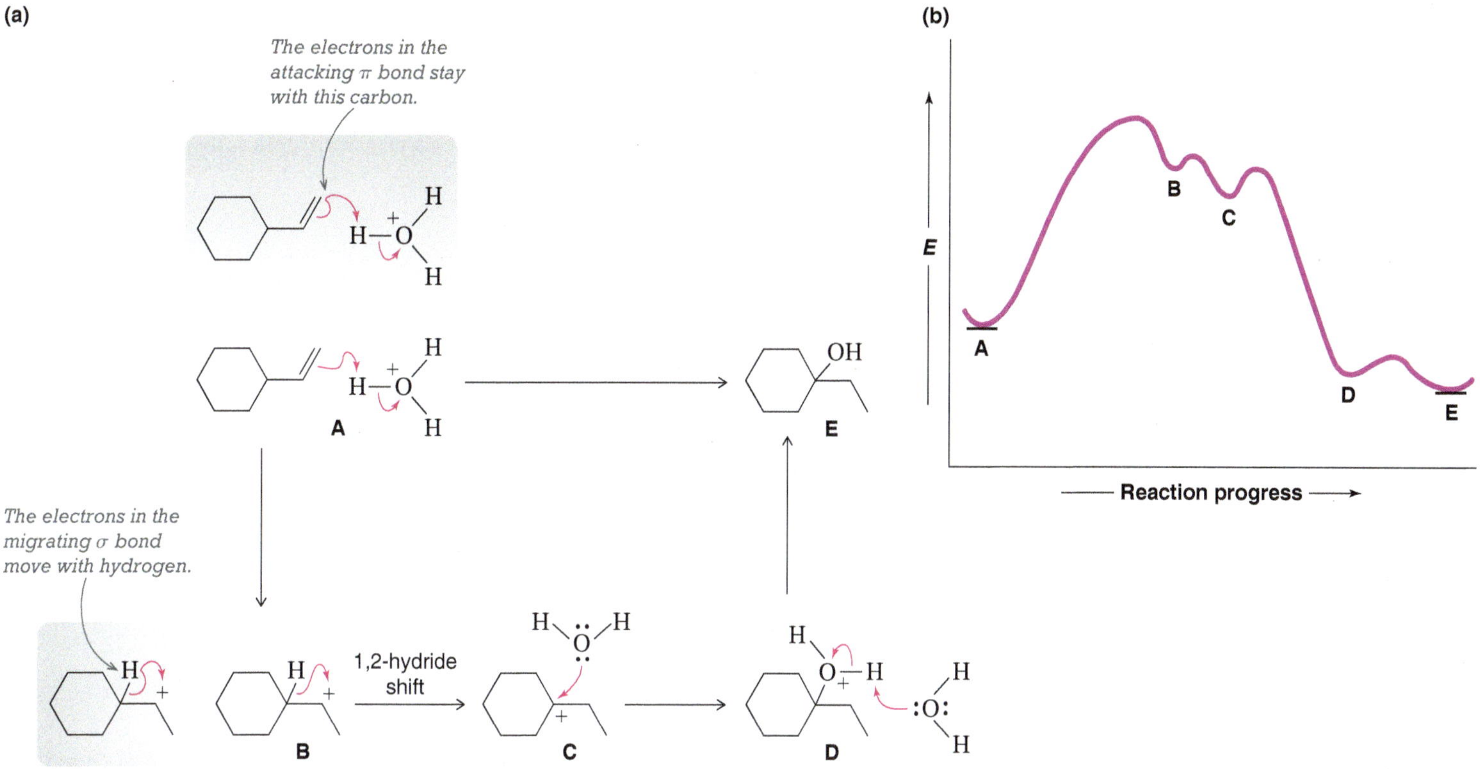

Figure 8.50 An acid-catalyzed hydration that features a carbocation rearrangement. (a) Overall mechanism. (b) The reaction coordinate diagram.

Although none of the secondary alcohol formed in this reaction, at some point the 2° carbocation was present. H_2O never had a chance to attack the 2° carbocation because the hydride shift is a much faster process for two reasons (**Figure 8.51**). First, the hydride is much closer to the carbocation than is the water. Furthermore, the hydride shift has $\Delta S \approx 0$ (1 molecule $\rightarrow$ 1 molecule), while attack by water requires two molecules to collide to make one ($\Delta S < 0$). In general, intramolecular reactions will always be faster than similar intermolecular reactions. [*Remember this!*]

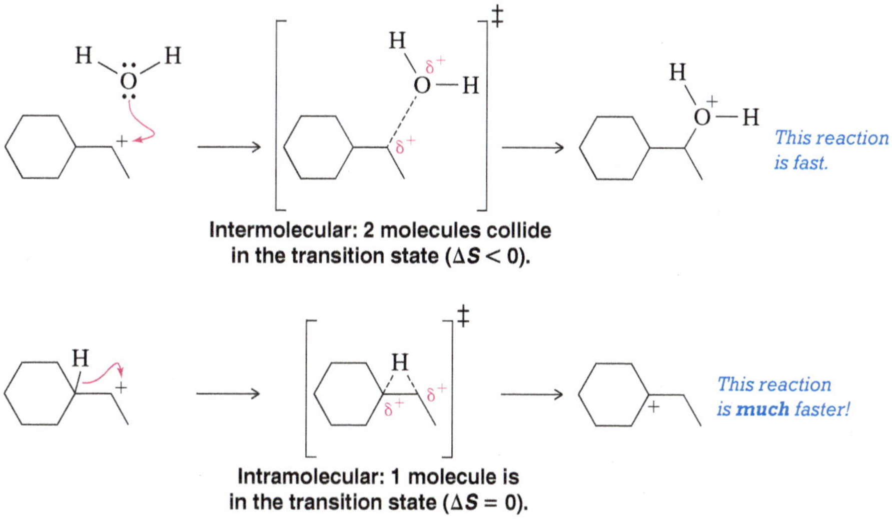

Figure 8.51 Hydride shift is much faster than nucleophilic attack by water.

Rearrangements are not limited to hydride shifts; alkyl groups can also migrate, and ring expansion/contraction can occur. A 1,2-methyl shift occurs during the acid-catalyzed hydration of 3,3-dimethylbut-1-ene (**Figure 8.52**). You will typically only see an alkyl shift when the carbon next to the carbocation is fully substituted with alkyl groups. When there are multiple alkyl groups that can move, the smaller alkyl group generally tends to migrate more quickly.

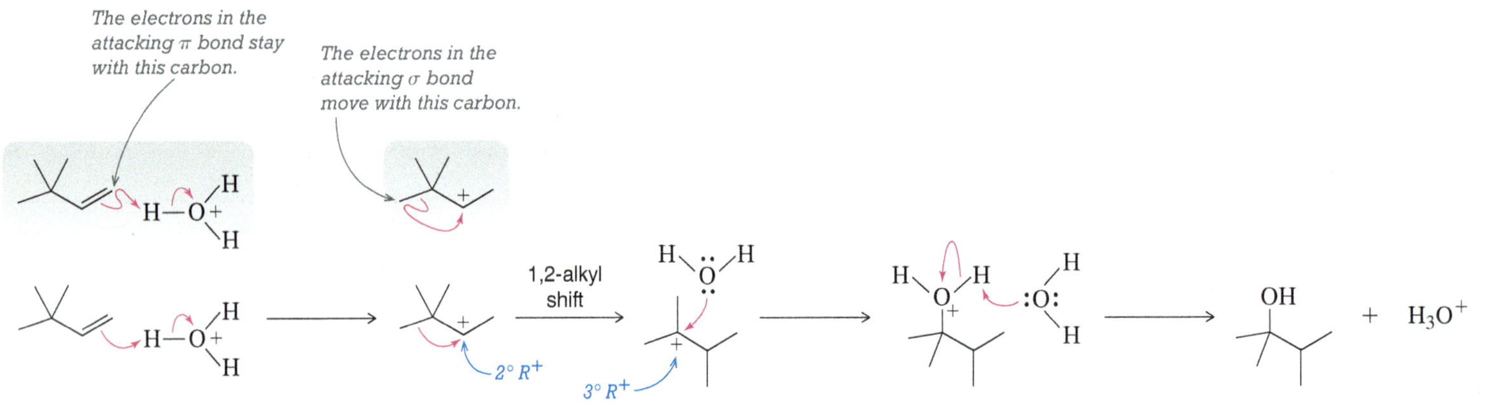

Figure 8.52 1,2-Alkyl shifts can occur if a more stable carbocation is produced.

Rearrangements are quite common and will occur whenever the new carbocation is more stable than the original. A carbocation rearrangement can be predicted when a carbocation is less substituted than either of its neighboring carbons. [*Read that sentence again.*] While there will be circumstances where a carbocation might undergo multiple shifts, avoid predicting more than one rearrangement unless otherwise specified.

8.4–8.4.3 SUMMARY

- The acid-catalyzed hydration reaction is regioselective, proceeding through the most stable carbocation in the rate-determining first step. Rearrangement will occur when the migration of hydride or alkyl anion from a neighboring carbon produces a more stable carbocation. A racemic mixture of products results when a new chiral center forms in the reaction.

Assessment 8.45 Which of the following carbocations would you expect to rearrange? If you expect rearrangement, draw the carbocation you expect to form and the mechanism by which it will form.

Assessment 8.46 Provide a mechanism for the following reactions occurring with rearrangement.

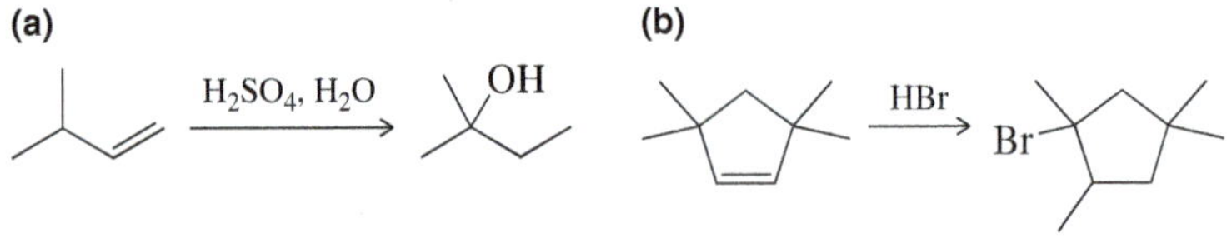

Assessment 8.47 The hydration of three C_5H_{10} alkene isomers can give 2-methylbutan-2-ol. Draw them.

Assessment 8.48 Our wayward chemist from Assessment 8.23 suggested the following stepwise mechanism for a hydride shift. Show this mechanism on a reaction coordinate diagram to illustrate why the concerted mechanism is more likely. Justify the picture you have drawn.

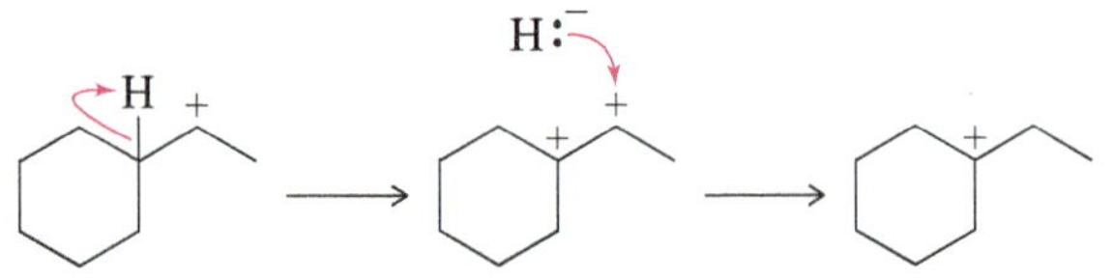

8-D CHEMISTRY IN CONTEXT: Green Chemistry

Limonene: a natural degreaser

Because of its ability to dissolve both polar and nonpolar organic molecules, methylene chloride is an ideal choice as an industrial cleanser. Unfortunately, long-term exposure studies show that it produces lung and liver tumors in mice. As a result of this and other potential hazards, efforts have been made to regulate the use of methylene chloride as well as to discover alternative cleansing agents.

One such alternative is limonene, a terpene by-product from the citrus industry. A component of orange peels, limonene is produced at a rate of about 30,000 tons per year during orange juice production. Limonene is a viable alternative cleansing solvent for two important reasons. First, it does not have the same toxicity profile as methylene chloride. Second, it is produced in nature, so it is a renewable resource, unlike methylene chloride and toluene (another degreaser), which are derived from nonrenewable petroleum feedstocks.

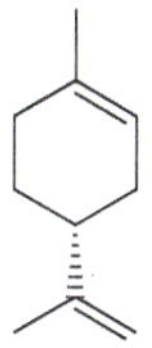

limonene

The synthesis of limonene (**Figure 8.53**) utilizes the same pathway as the synthesis of nerol discussed in CIC 8-C but diverges at the point of hydrolysis of neryl pyrophosphate. [*The carbons are numbered to help you track the bonds that break and the bonds that form.*] Upon forming the carbocation at C_8, the $C_2—C_3$ π bond attacks to form a new $C_3—C_8$ bond and a stable 3° carbocation at C_2. As we saw in CIC 8-C, an active site base ($—B:^-$) removes a hydrogen from C_9 to form the $C_2—C_9$ alkene. This reaction is stereoselective, forming only the (*R*) enantiomer.

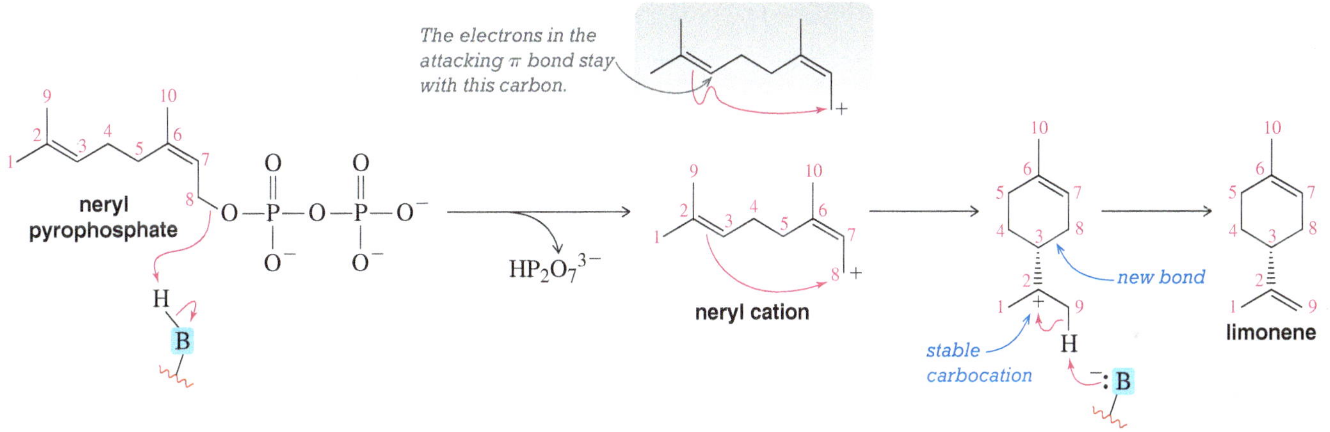

Figure 8.53 **Biosynthesis of limonene.**

Assessment 8.49 Why do you think this reaction is stereoselective?

8.4.4 Oxymercuration–reduction: preventing rearrangement

The rearrangement discussed in Section 8.4.3 prevents us from making 1-hydroxyethylcyclohexane via the acid-catalyzed hydration of vinylcyclohexane. To make the molecule, a reaction is needed that avoids formation of a 2° carbocation. This can be done using a two-step process featuring **oxymercuration** with mercury(II) acetate and **reduction/demercuration** of the resulting oxyorganomercury species with sodium borohydride to produce a racemic mixture of the desired alcohol (**Figure 8.54**). This reaction proceeds in a similar manner to the previous reactions studied in this chapter, but does not result in rearrangement. [*Let's explore why!*]

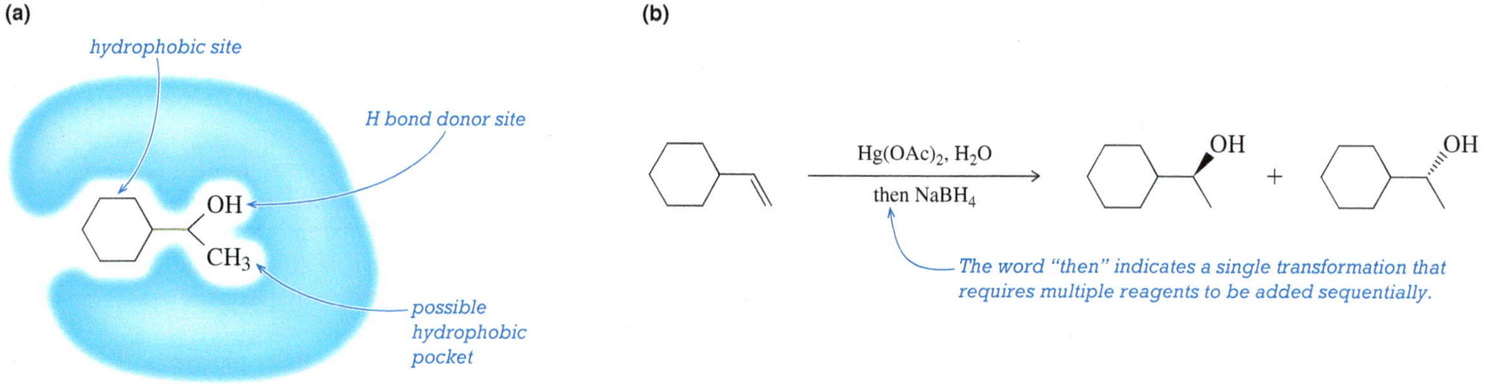

Figure 8.54 Our medicinal chemistry scenario consists of (a) a hypothetical active site of imaginokinase and (b) the oxymercuration–reduction of alkenes to make a 2° alcohol without rearrangement.

While the reaction consists of two parts, our focus is on the oxymercuration step of the reaction, the mechanism of which is shown in **Figure 8.55**.

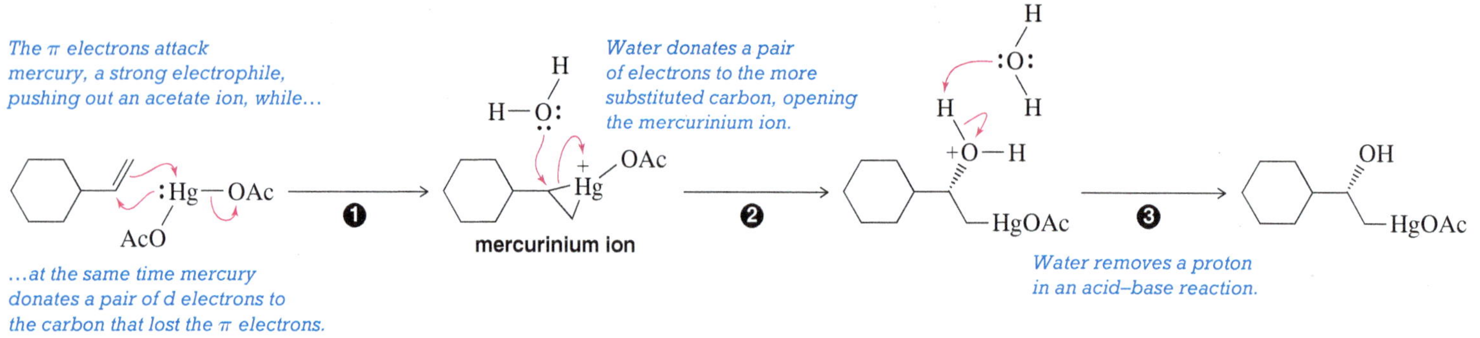

Figure 8.55 The mechanism of oxymercuration. [*Water attacks from the front also to produce a racemic mixture.*]

Though analogous to alkene reactions studied thus far (HBr and H_3O^+), there are a few places where further detail is needed. In particular, we need to answer the following questions: (1) Why is mercury an electrophile? (2) Why does mercury reattack the carbon to form a **mercurinium ion**? (3) Why does water open the mercurinium ion by attacking carbon? (4) Why does it open the mercurinium ion at the more substituted carbon?

1. ***Why is mercury an electrophile?*** Mercury(II) acetate is an ionic salt that dissociates when placed in water to produce a mercury cation and a resonance-stabilized acetate ion. With a +1 charge, the mercury cation is a very good electrophile. In many ways, you can think of a positively charged mercury ion as a big, soft proton [**Figure 8.56(a)**]. While a mechanism can be drawn in which $Hg(OAc)_2$ dissociates prior to attack, organic chemists most often show the acetate ion leaving as the π electrons attack [**Figure 8.56(b)**]. Drawn this way, the mechanistic arrows (1 and 2) resemble those of an acid–base reaction.

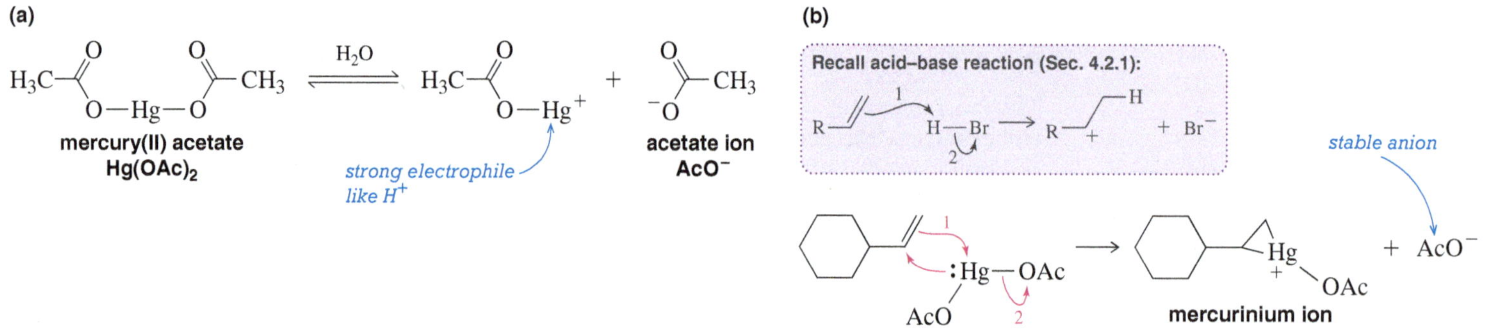

Figure 8.56 Mercury acetate is a strong electrophile. (a) The ionization of $Hg(OAc)_2$ in H_2O yields a mercury cation with a +1 charge. (b) π electrons attack mercury directly.

2. ***Why does mercury reattack the carbon to form a mercurinium ion?*** To understand why the three-membered mercurinium ion forms in a single, concerted step, it is helpful to compare it to a reaction where a three-membered ring doesn't form. **Figure 8.57** directly compares π electrons attacking HBr and $Hg(OAc)_2$. Unlike the proton that added in the HBr reaction, mercury contains *d* electrons. These *d* electrons are donated to the unstable, electron-deficient carbocation to produce an intermediate that is more stable than the carbocation. [*Let me repeat that for emphasis: The mercurinium ion is more stable than a carbocation. This is very important for preventing rearrangement.*]

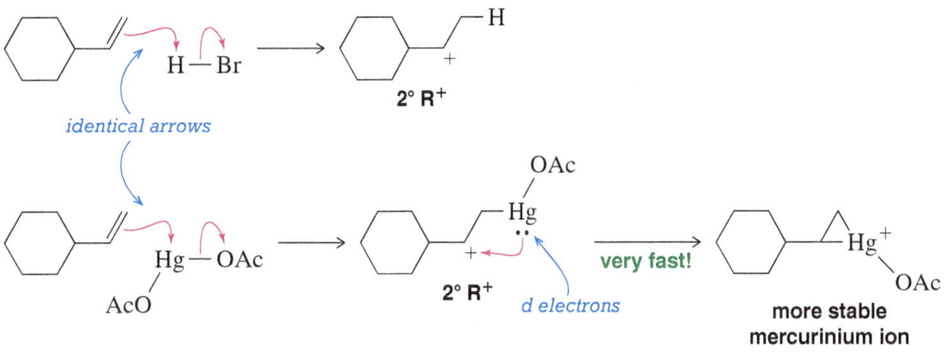

Figure 8.57 Mercury donates a pair of *d* electrons to form the three-membered mercurinium ion.

SIDE CONVERSATION **Concertedness prevents rearrangement** Recall from Figure 8.55 that mercurinium ion formation is **concerted**, meaning the bonds break and form at the same time in a single step. In Figure 8.57, it was drawn as being stepwise so that a more direct comparison could be made to the addition of HBr. To understand what makes it concerted, **Figure 8.58** compares two nucleophiles attacking a carbocation. Bromide can be considered to be far away from the carbocation, whereas mercury is just a few angstroms away. [*Which would attack more quickly?*] In Section 8.3, we decided that the reaction of two charged intermediates (Br^- and a carbocation) to give a neutral molecule is a fast reaction. But the bromide still has to travel a long distance through a solution to reach its target. In comparison, the mercury is right there. [*A general rule is illustrated here—namely, intramolecular (within a molecule) reactions are faster than intermolecular (between molecules) reactions.*] Thus, if attack by bromide is fast, attack by mercury must be really fast. So fast, in fact, that the reaction goes directly to the mercurinium ion, completely bypassing the carbocation.

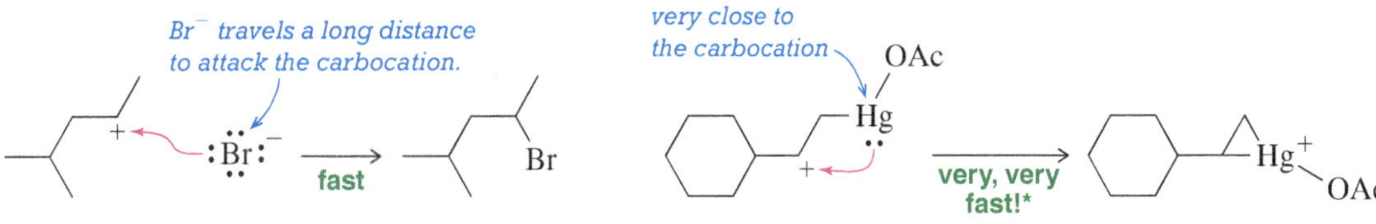

Figure 8.58 Intramolecular reactions are faster than intermolecular reactions. *The intramolecular reaction is so fast that the carbocation never forms in the oxymercuration reaction.

Concertedness has several implications. For one, ring closure prevents rearrangement from occurring. We said earlier that a hydride shift was extremely rapid and would occur any time a carbocation formed on a carbon that is less substituted than its neighbor. The absence of rearrangement here indicates that the carbocation doesn't actually form in oxymercuration reactions (**Figure 8.59**). [*In this way, preventing rearrangement is both an indication of and an implication of concertedness.*] We'll see additional implications of concertedness when we study hydroboration (Section 8.4.5) and bromination (Section 9.1.2)—reactions that proceed by similar mechanisms. ■

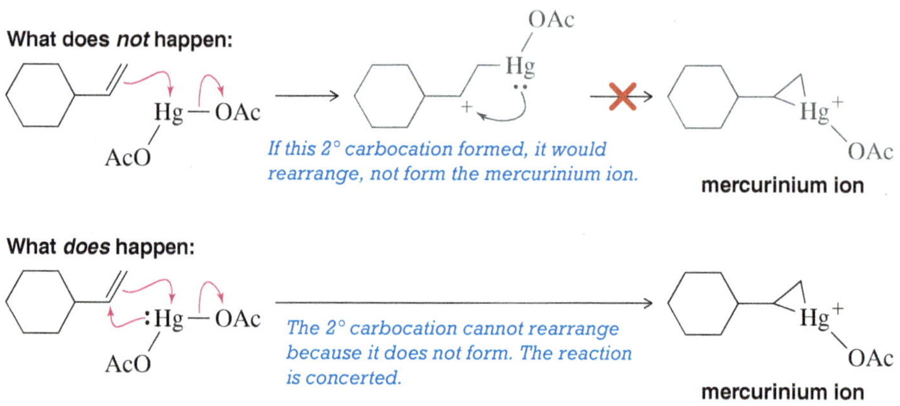

Figure 8.59 The carbocation never forms in a concerted reaction, so no rearrangement is possible.

3. ***Why does water open the mercurinium ion by attacking carbon?*** Once formed, the mercurinium ion (a positively charged, three-membered ring) is still a reactive intermediate (**Figure 8.60**). The mercury can regain the pair of electrons it is sharing if water attacks the more substituted carbon of the mercurinium ion. [*Why the more substituted carbon? The answer is coming.*] This has the beneficial effects of relieving the strain of the three-membered ring and allowing mercury to be neutral again.

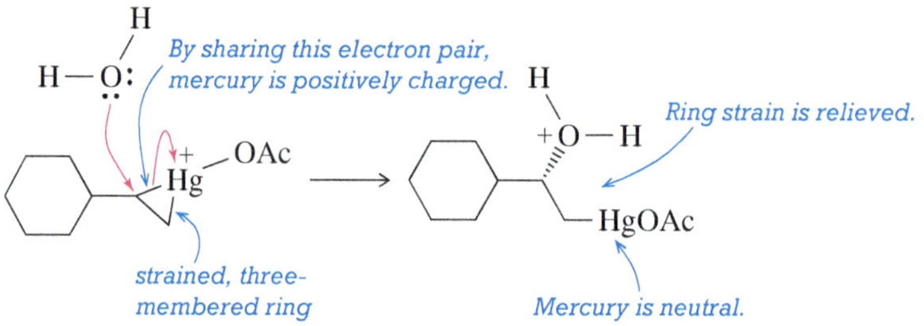

Figure 8.60 Water opens the mercurinium ion to relieve ring strain and make mercury neutral again.

4. ***Why does water attack the mercurinium ion at the more substituted carbon?*** Think of the mercurinium ion as the equilibrium between the three structures in **Figure 8.61**. [*It's not really. If it were, a hydride shift would occur.*] The mercurinium ion is the most stable of the three species in this hypothetical equilibrium; the 1° R^+ is least stable. The mercury bridge shifts *away* from the more stable 2° carbocation and *toward* the less stable 1° carbocation. As a result, the unsymmetrical mercurinium ion has a longer bond between mercury and the 2° carbon and a greater partial positive charge on that carbon. Longer bonds are weaker bonds, so water attacks at the 2° carbon.

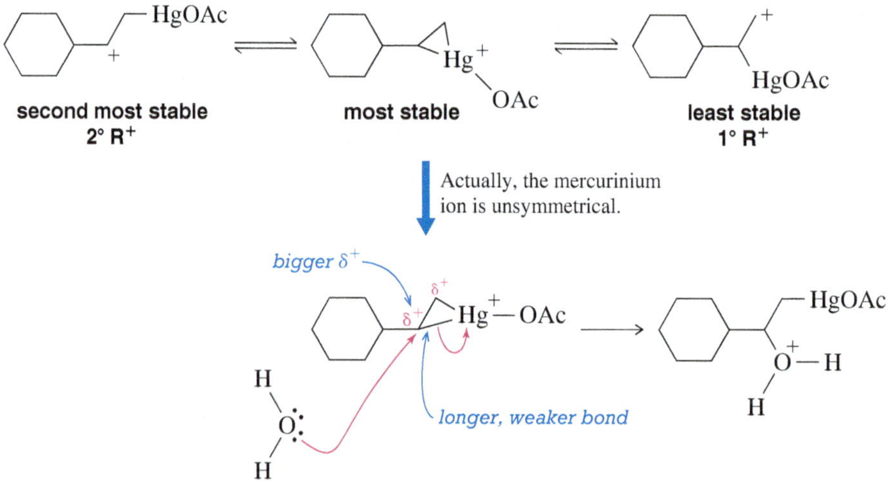

Figure 8.61 Water opens the mercurinium ion at the more substituted carbon.

The reduction step of the oxymercuration process is a complicated, and poorly understood, organometallic mechanism. We study reactions like this in Chapter 16; for now, you can at least follow the electron flow in **Figure 8.62**. Sodium borohydride donates a hydride anion to the mercury atom, displacing an acetate ion. At this point the C—Hg bond breaks while transferring the hydrogen to the carbon in a reaction believed to involve radicals in a non-concerted step. The net result is that a new C—H σ bond forms and the C—Hg bond breaks.

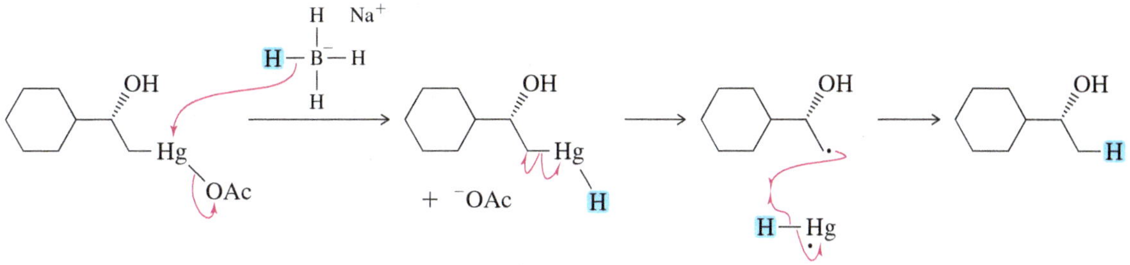

Figure 8.62 Reduction of the C—Hg bond replaces mercury with a hydrogen.

In conclusion, the oxymercuration–reduction is useful for hydrating alkenes because it prevents rearrangement while producing the more substituted alcohol (**Figure 8.63**). Because the mercurinium ion forms on either side of the planar alkene and because demercuration is not concerted, the net result is a racemic mixture of alcohol products.

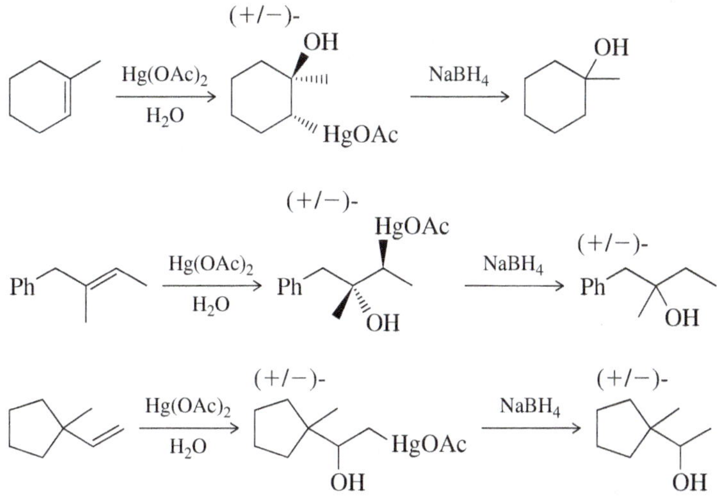

Figure 8.63 Representative examples of alcohol synthesis via oxymercuration–reduction. The (+/−) designation indicates that the product contains a racemic mixture of enantiomers.

8.4.4 SUMMARY

- Oxymercuration–demercuration proceeds through a concerted first step to produce a bridged mercurinium ion that prevents rearrangement. Opening of the mercurinium ion occurs at the more substituted carbon. In a second step, the carbon–mercury bond is cleaved by reduction with sodium borohydride to give the net addition of H_2O.

Assessment 8.50 Predict the products you would get when the following alkenes react under the following conditions: (i) H_2SO_4, H_2O and (ii) 1. $Hg(OAc)_2$, H_2O 2. $NaBH_4$

(a) (b) (c) (d)

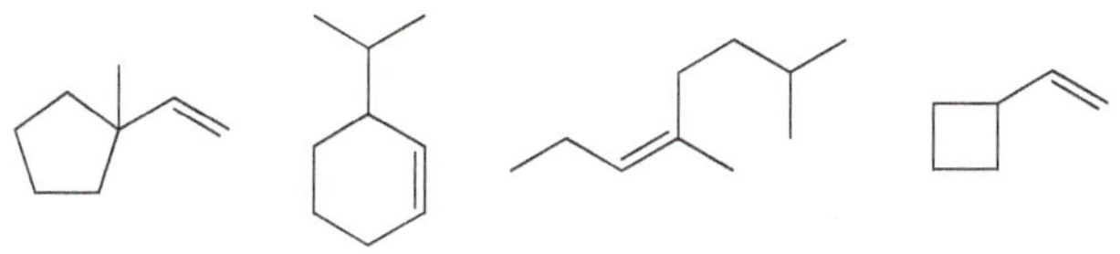

Assessment 8.51 Provide an arrow-pushing mechanism, accounting also for the stereochemical outcome, of the first step (oxymercuration) of the three reactions in Figure 8.63.

Assessment 8.52 Which is more stable, a carbocation or a mercurinium ion? How do you know?

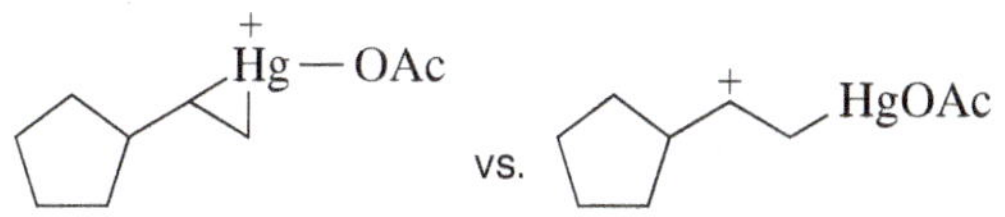

Assessment 8.53 Oxymercuration–reduction, like acid-catalyzed hydration, can be modified to synthesize ethers. Suggest an alkene and the appropriate reaction conditions to synthesize the following ethers.

(a)

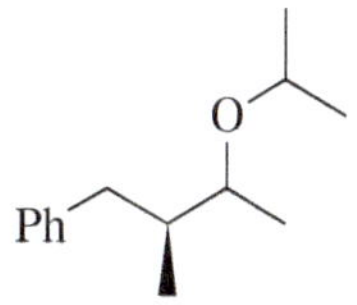

(b)

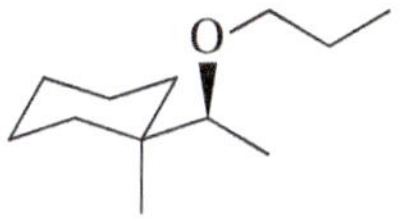

8.4.5 Hydroboration: anti-Markovnikov alcohols

Continuing with our hypothetical, imagine that the alcohol synthesized in Section 8.4.4 was a poor inhibitor, with studies forcing us to revise our picture of the active site to that shown in Figure 8.64(a). From the picture we now have of the active site, we decide the compound 2-hydroxyethylcyclohexane might fit optimally. And since our lab [*for some reason*] only has vinylcyclohexane, we need to be able to perform the reaction shown in Figure 8.64(b). [*Remember, the purpose of this exercise is to justify why we need multiple alcohol syntheses.*] Previous reactions place the alcohol at the more substituted carbon (Markovnikov) but here, the hydroxyl group is located on the least substituted carbon (**anti-Markovnikov**). [*How can we make this using an alkene addition reaction?*]

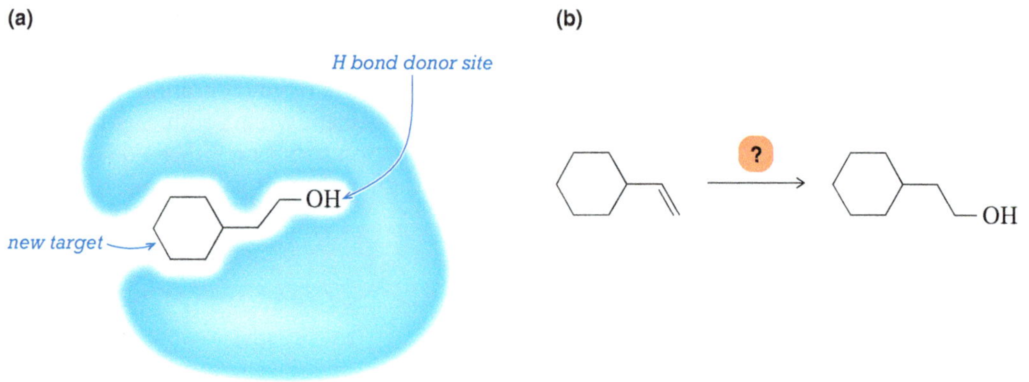

Figure 8.64 Our new medicinal chemistry scenario consists of (a) a revised active site of imaginokinase and a new target alcohol. (b) A reagent system is needed to synthesize an anti-Markovnikov alcohol.

In every reaction studied previously, the first group to be added (that is, the group attacked by the alkene) adds to the least substituted carbon in order to generate the more stable intermediate (radical or carbocation). The second group attacked this intermediate (radical or carbocation) to complete the reaction. Figure 8.65 shows that one way to make a 1° alcohol would require an electrophilic source of hydroxyl (HO^+) and a nucleophilic source of hydrogen (H^-). Although logically sound, this reaction cannot be done. Within water, the electron-rich oxygen almost always acts as a nucleophile, while the hydrogen atom is usually donated as an electron-poor proton. [*Recall Assessments 8.23 and 8.48, where you taught our wayward chemist that forming a hydride ion is an unfavorable process.*]

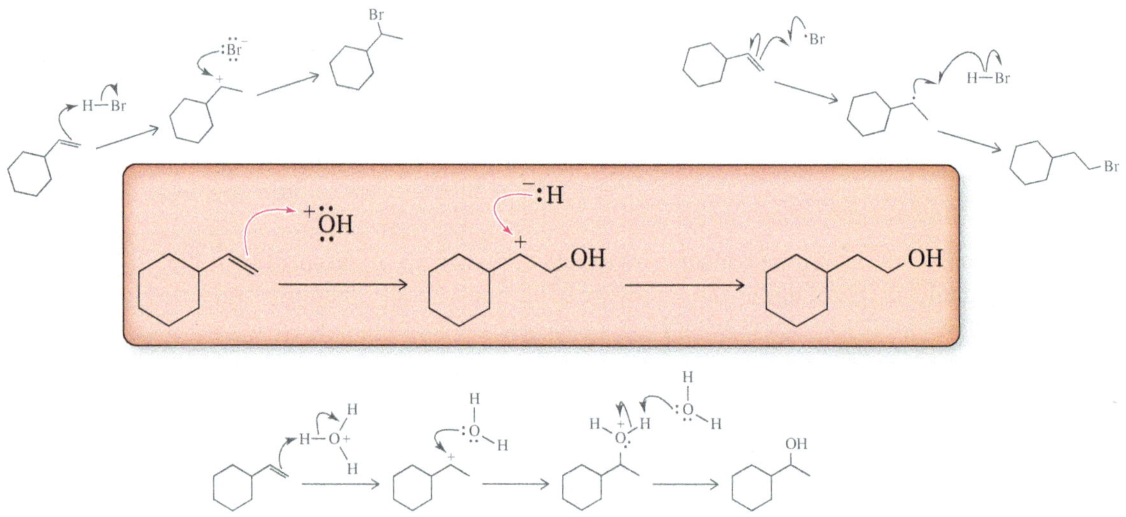

Figure 8.65 The hypothetical synthesis of a 1° alcohol by electrophilic addition. [*This is not a real reaction. And we are ignoring the potential for rearrangement in the "background" reactions.*]

H.C. Brown and coworkers are credited with the discovery of **hydroboration–oxidation** as a useful method for making less-substituted alcohols from alkenes. [*As we learn more details of this reaction sequence, we'll see that our hypothetical idea wasn't so bad.*] Treatment of the alkene with borane makes a trialkyl borane that is oxidized in a basic peroxide solution to generate the primary alcohol (Figure 8.66). [*Our discussion focuses most heavily on the hydroboration part of this reaction.*]

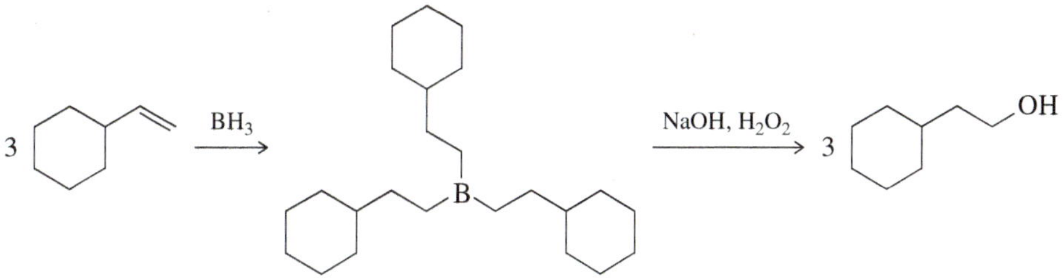

Figure 8.66 Hydroboration–oxidation of alkenes gives the least substituted alcohol.

The alkene, once again, is a nucleophile. In this case, the boron in borane lacks a full octet, is sp^2 hybridized, and contains an empty *p* orbital. As an exception to the octet rule, boron is stable with 6 electrons in its valence shell, but it is still more stable with 8 (**Figure 8.67**). In this way, it is like a carbocation, which is a strong electrophile. Borane, then, is also a strong electrophile/Lewis acid.

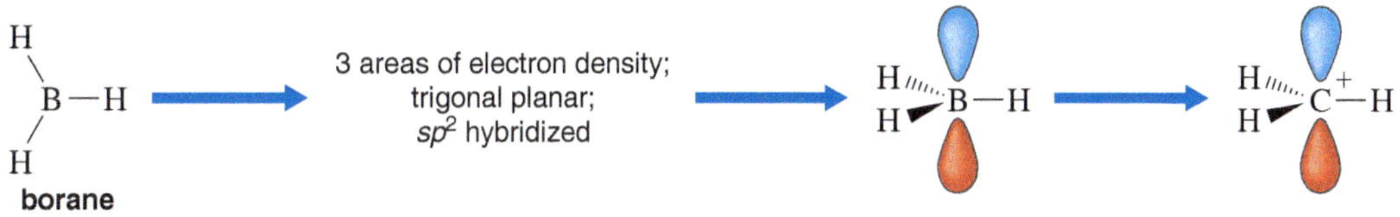

Figure 8.67 Borane, like a carbocation, is a strong Lewis acid.

When the alkene and borane mix, the alkene donates its π electrons to the electrophilic borane (**Figure 8.68**). Rather than form a carbocation, a hydride ion is transferred from boron to carbon in a single concerted step to give an alkyl borane.

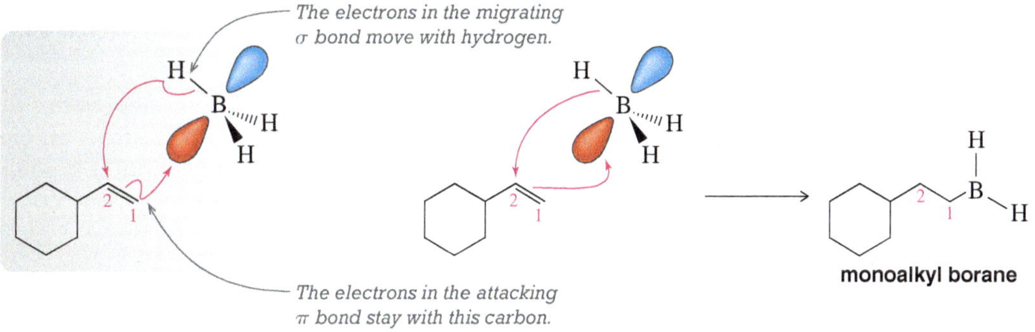

Figure 8.68 Addition of borane to an alkene is concerted.

By looking at hydroboration as if it were a nonconcerted reaction, two things become clear (**Figure 8.69**). First, hydroboration is concerted because a slow, intermolecular, entropically disfavored addition of borane to an alkene to make an intermediate with two charges is followed by an extremely fast, intramolecular, entropically neutral hydride transfer to make a product with no charges. [*That is, a slow reaction followed by a very fast reaction = a concerted reaction!*] Though a carbocation does not form in a concerted reaction, the stepwise mechanism makes clear that addition to the terminal carbon causes a greater buildup of positive charge on the more substituted 2° carbon. [*A carbocation does not form in a concerted step.*]

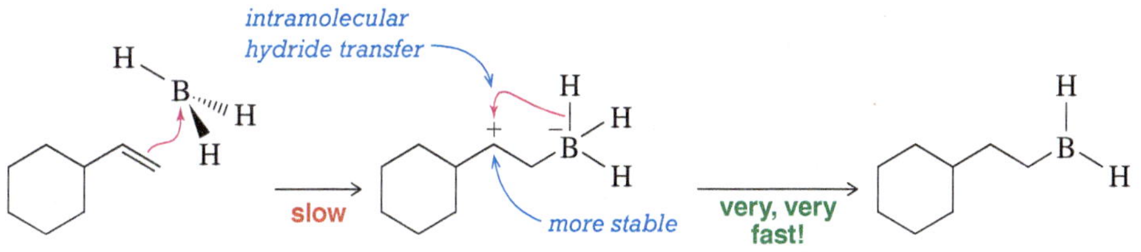

Figure 8.69 A hypothetical stepwise mechanism explains the regioselectivity.

This same analysis can be done on the concerted mechanism. [*The mechanism is concerted. Make sure you always draw it this way.*] Because a partial carbocation is forming in the transition state, we expect the transition state with the more stable partial carbocation to be favored [**Figure 8.70(a)**]. Furthermore, the reaction of borane at the less substituted carbon of the alkene is less sterically hindered, making it the lower energy transition state and, therefore, the faster reaction. Plotting the two possibilities on a reaction coordinate diagram in **Figure 8.70(b)** shows that formation of the 1° alkyl borane is the faster reaction.

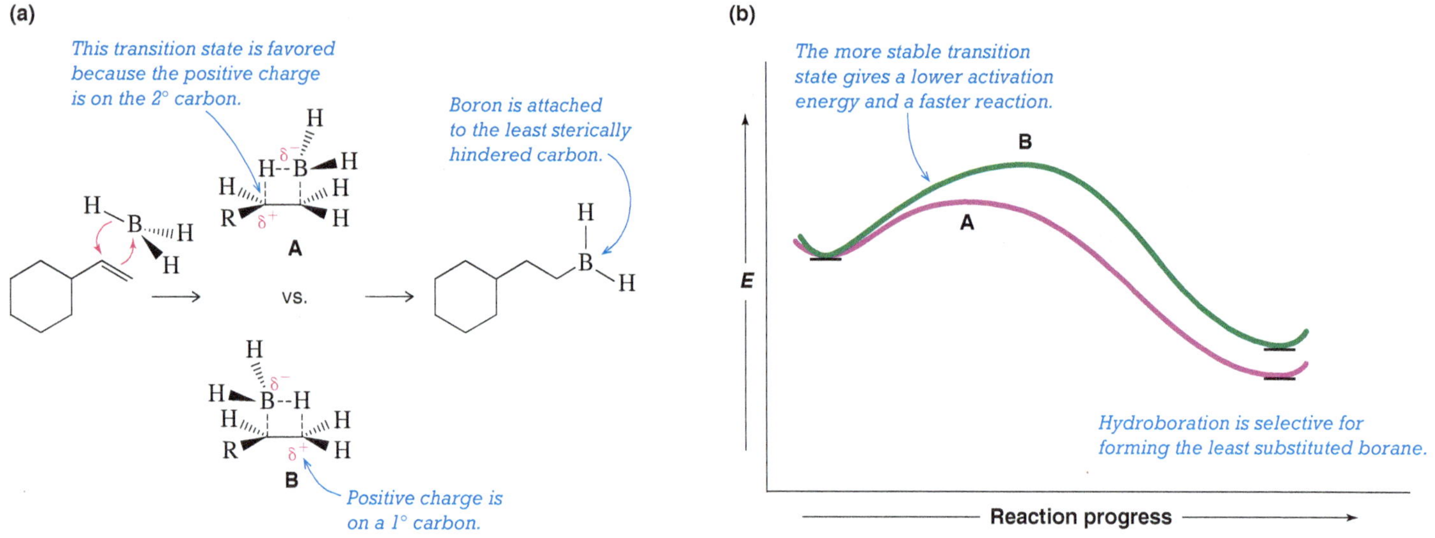

Figure 8.70 The regioselectivity can be explained in the context of concerted hydroboration. (a) The transition state with the partial charge on the more substituted carbon is more stable and places the bulkier boron at the least sterically hindered carbon. (b) The faster reaction proceeds through the more stable transition state.

Before moving on to the oxidation step, notice that the alkyl borane we created can continue to react because it once again contains an empty *p* orbital (**Figure 8.71**). In fact, the hydroboration reaction will continue until it becomes a trialkyl borane, having given up all three of its hydrogens. The mechanism for each of these hydroborations is exactly the same as what we have seen previously.

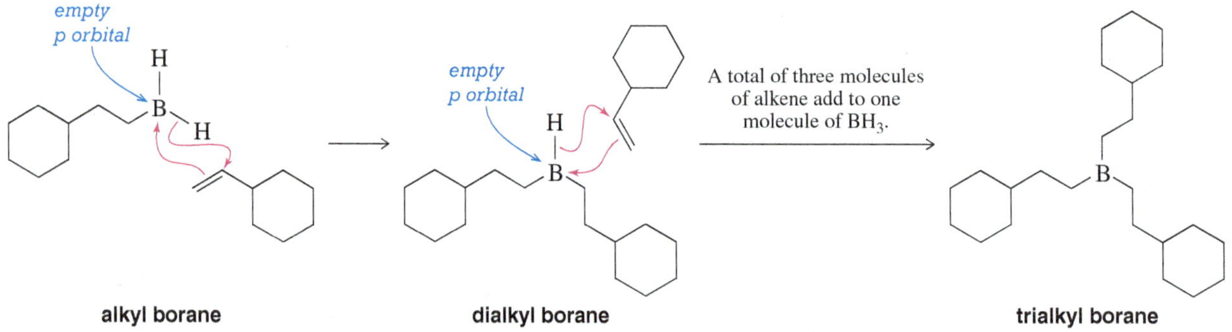

Figure 8.71 Hydroboration continues until a trialkylborane is produced.

The finicky nature of boron helps us understand the mechanism in Figure 8.71. As we said earlier, it is stable with 6 electrons, but would prefer to have 8. Borane (6 electrons) acts as a Lewis acid and is attacked by the alkene, but as the negative charge develops (8 electrons), it transfers a hydride to remain neutral (6 electrons). The oxidation mechanism (the second step of hydroboration–oxidation), shown in **Figure 8.72**, is driven the same way. Boron wants 8 electrons until it has them, and then it finds a way back to having 6. The reaction begins with the deprotonation of hydrogen peroxide by sodium hydroxide. The resulting peroxy anion attacks boron, causing an alkyl rearrangement to produce an alkoxydialkylborane.

Don't tell the honey bees from CIC 8-C, but boric acid has many uses, one of which is as an insecticide.

The process occurs twice more to form a trialkoxy borane. At this point, hydroxide adds to the empty *p* orbital of boron, resulting again in a negatively charged boron. With no carbons or hydrogens to leave, the only way to alleviate the negative charge on boron is to have one of the alkoxy groups leave. After leaving, the resulting alkoxide removes the acidic proton from the hydroxyl group attached to the boron. This happens two more times, ultimately producing the expected alcohol and the conjugate base of boric acid.

a good nucleophile (→ HOO^-)

$$H_2O_2 + HO^- \rightleftharpoons H_2O + HOO^- \qquad K_{eq} = 10^4$$

$pK_a = 12$ (H_2O_2) $pK_a = 16$ (H_2O)

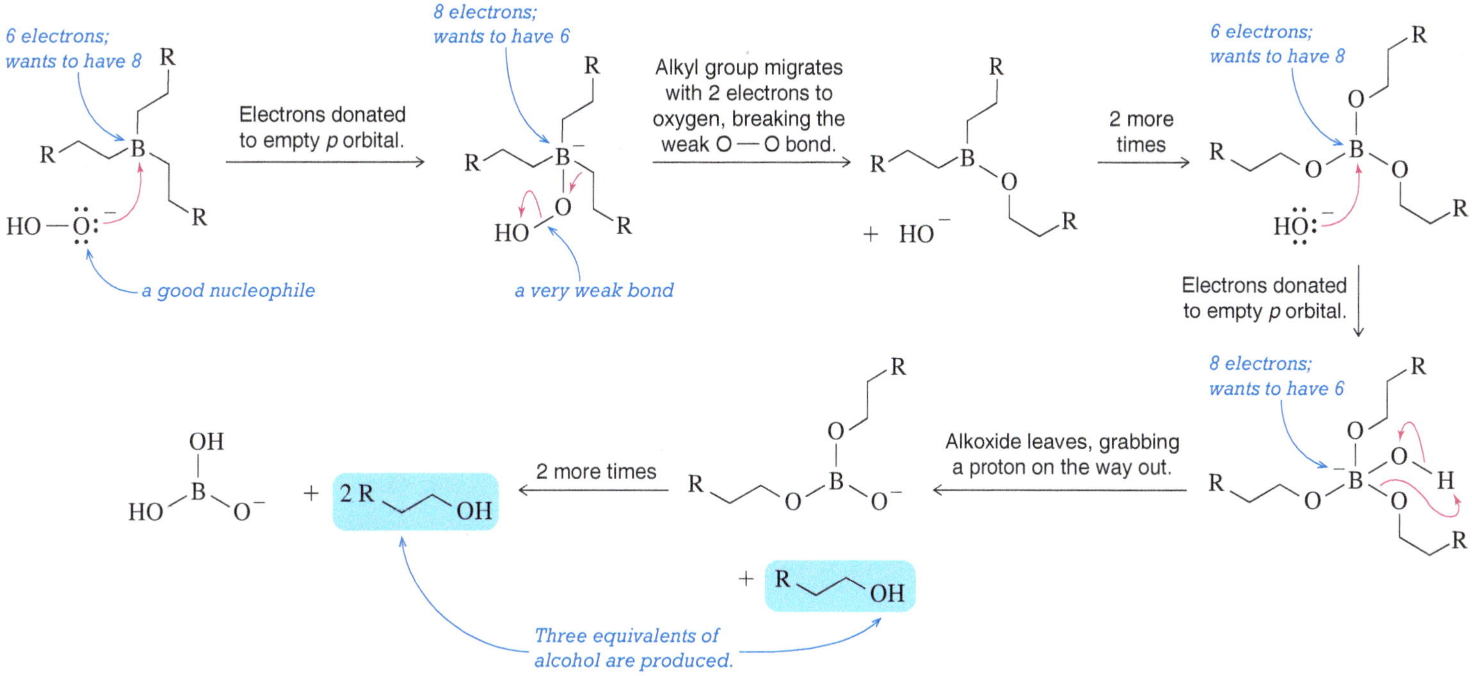

Figure 8.72 The mechanism of the peroxide oxidation of a trialkylborane.

In summary, we can now make primary alcohols, as necessitated by our hypothetical medicinal chemistry exercise. The use of hydroboration–oxidation to make the less substituted alcohols in **Figure 8.73** introduces us to the concept of **stereospecific reactions**.

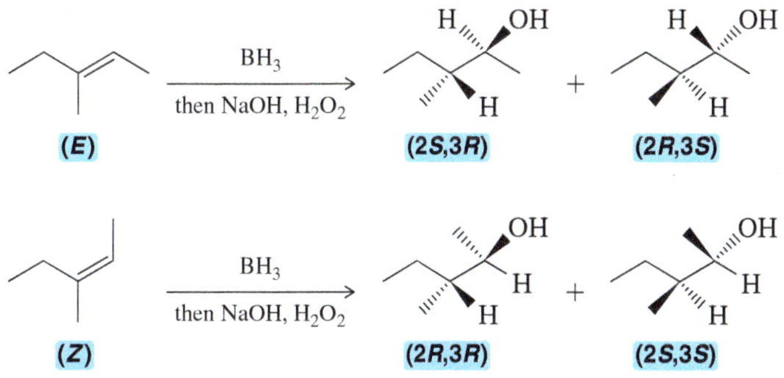

Figure 8.73 Hydroboration is stereospecific. Diastereomeric alkenes give different stereoisomeric products.

SIDE CONVERSATION **Concerted reactions are stereospecific** A reaction is said to be stereospecific if stereoisomeric reactants react under identical conditions to give different stereoisomers of the same product. Consider the hydrobromination of (*E*)- and (*Z*)-3-methylhex-2-ene (**Figure 8.74**). Once the carbocation is formed, the bromide can add to the top or the bottom. Regardless of which alkene isomer we started with, the product is the same racemic mixture of 3-bromo-3-methylhexane.

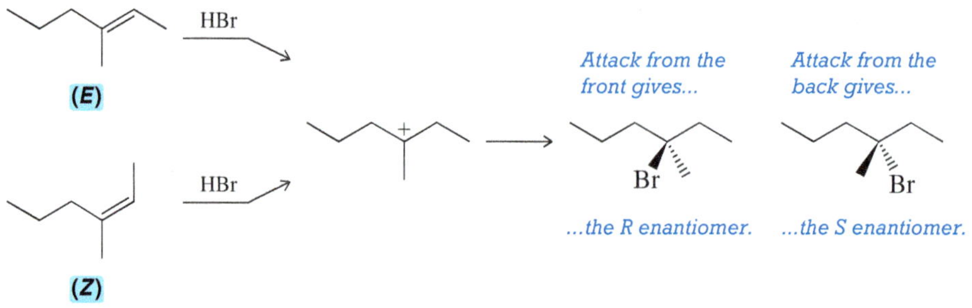

Figure 8.74 Nonconcerted reactions are not stereospecific. Diastereomeric alkenes give the same product mixture.

For hydroboration (a concerted reaction), the boron and the hydrogen add at the same time and on the same side (**Figure 8.75**). This is referred to as **syn addition** and occurs before there can be rotation around the $C_2—C_3$ single bond. That addition of both groups occurs before bond rotation, one of the fastest processes we know, is further evidence of the concertedness of hydroboration. [*Suppose the bond rotation happened* after *addition of boron but* before *hydrogen. What would the product mixture look like?*] Because the replacement of borane with a hydroxyl group proceeds with retention of stereochemistry, the net result is syn addition of water. A racemic mixture is still obtained because borane adds to the top and the bottom of the alkene at equal rates. As a result, the *E* alkene gives one set of racemic enantiomers, whereas the *Z* alkene gives the diastereomeric set of enantiomers. ■

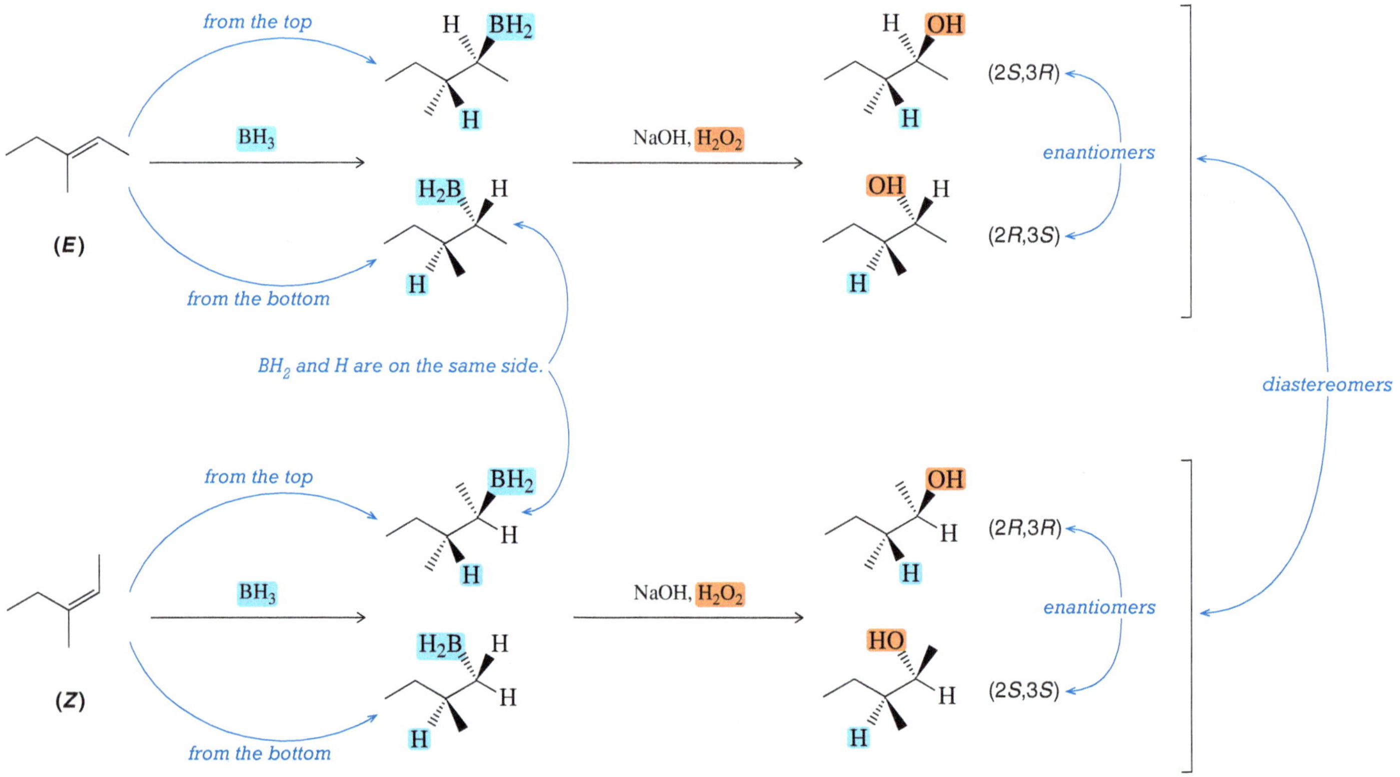

Figure 8.75 The concerted addition of borane is stereospecific.

A similar result is obtained when adding borane to 1-methylcyclohexene, resulting after oxidation in syn addition of water ($H + OH = H_2O$) to the alkene (**Figure 8.76**). Syn addition implies that the two groups were added on the same side. We discuss examples of stereospecific anti addition in Chapter 9.

Figure 8.76 Syn addition of water through hydroboration–oxidation.

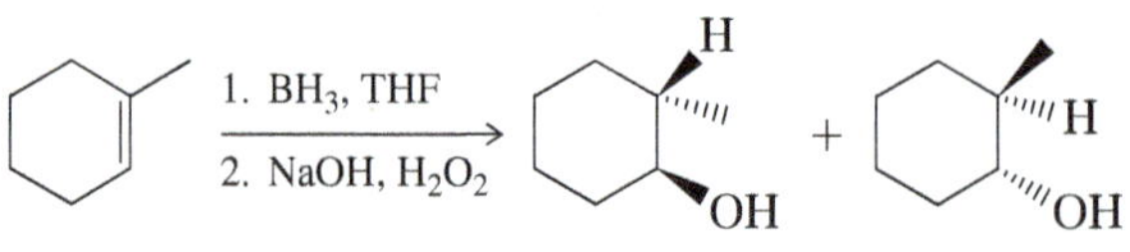

8.4.5 SUMMARY

- Hydroboration is the concerted, stereospecific, syn addition of borane to an alkene. Because boron adds to the least substituted carbon, oxidation of the C—B bond with basic peroxide produces the less substituted alcohol.

Assessment 8.54 Predict the products you would get when the following alkenes undergo (i) hydroboration–oxidation (1. BH_3 2. NaOH, H_2O_2) or (ii) oxymercuration–reduction [1. $Hg(OAc)_2$, H_2O 2. $NaBH_4$].

(a) (b) (c) (d)

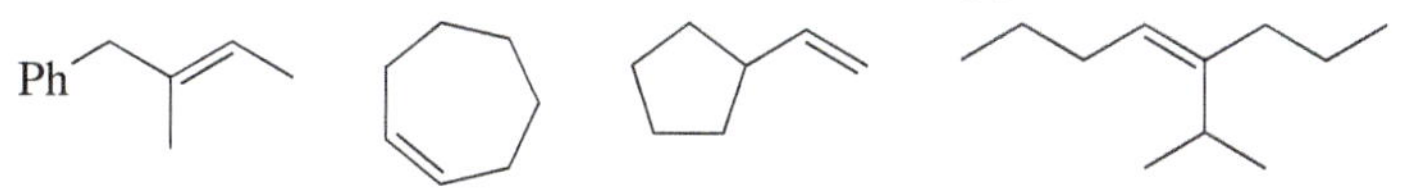

Assessment 8.55 Suggest an alkene to undergo hydroboration–oxidation (1. BH_3 2. NaOH, H_2O_2) to give exclusively the alcohols shown. Pay close attention to the relative (but not absolute) stereochemical outcome.

(a) (b) (c) (d)

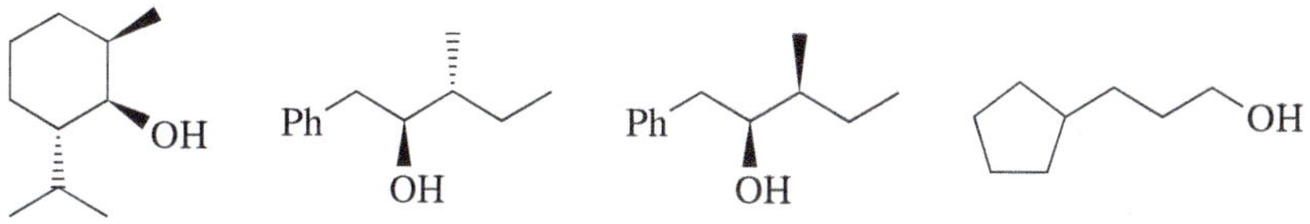

Assessment 8.56 Unlike hydroboration–oxidation, the addition of H_2O catalyzed by H_3O^+ is not stereospecific. Thinking carefully about the mechanism of the reaction, give two reasons why.

Assessment 8.57 The first step of oxymercuration–reduction is stereospecific, and yet this fact wasn't emphasized in that discussion. Show the stereospecificity of the first step for the following alkene and then explain why the stereospecificity becomes unimportant after the second step.

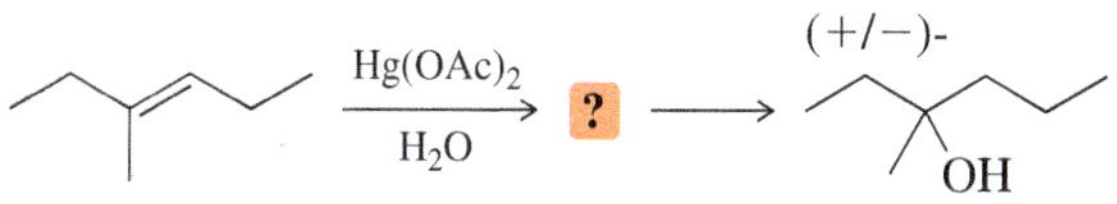

Don't be intimidated by this section. It is only here to demonstrate how nature does organic synthesis. With your level of understanding, you should be able to follow along with each step of this complex process. In doing so, you'll see that many of these reactions are similar to what we have seen previously.

Concluding thoughts

Terpenes larger than three isoprene units are no longer volatile and are therefore odorless. The steroids are examples of higher terpenes and are produced in human cells using pathways similar to those used by plants to synthesize the essential oils referred to throughout this chapter. All steroids in human cells are produced through the common steroid precursor lanosterol (Figure 8.77).

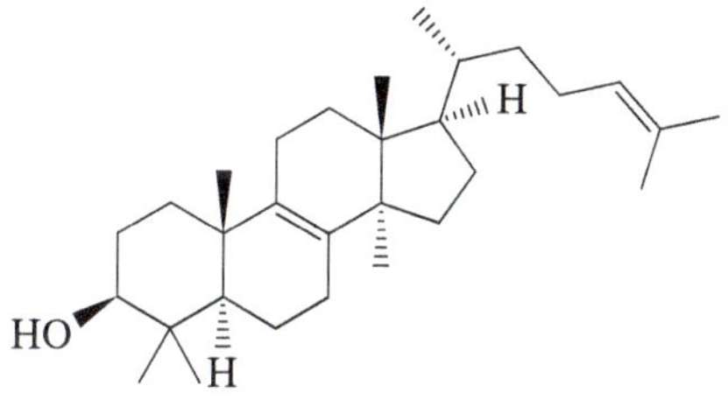

Figure 8.77 Precursor of all human steroids.

The synthesis of lanosterol begins with a farnesyl pyrophosphate coupling to give squalene (Figure 8.78). [*We rationalized the formation of farnesyl pyrophosphate as part of Assessment 8.44.*] Squalene is called an irregular terpene because the coupling of farnesyl pyrophosphates occurred in a head to head or 1→1 linkage. [*For this reason, you don't need to be concerned with the mechanism of this reaction.*]

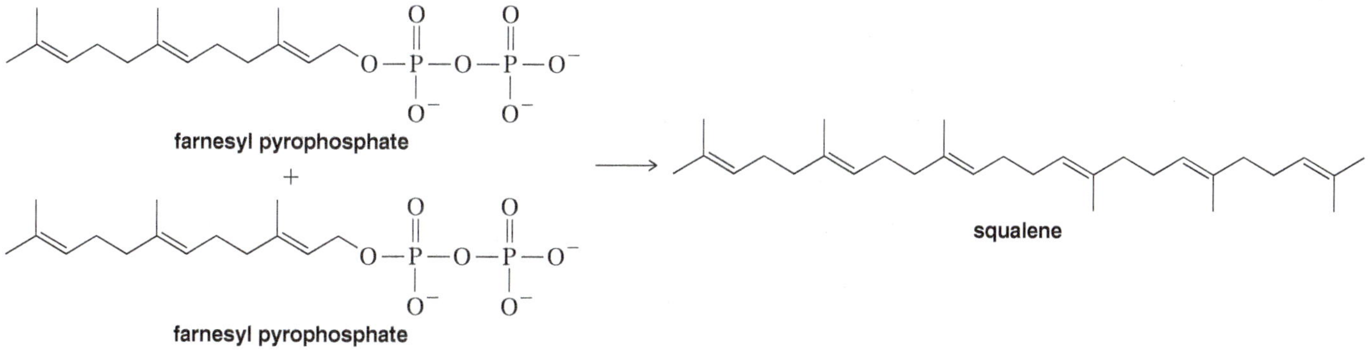

Figure 8.78 The coupling of farnesyl pyrophosphate gives squalene.

The most challenging aspect of lanosterol biosynthesis is the cyclization of squalene to make the four rings. The stage is set for the cyclization reaction only when squalene undergoes an enzymatic epoxidation (**Figure 8.79**). [*This is similar to a reaction we examine in Chapter 9.*]

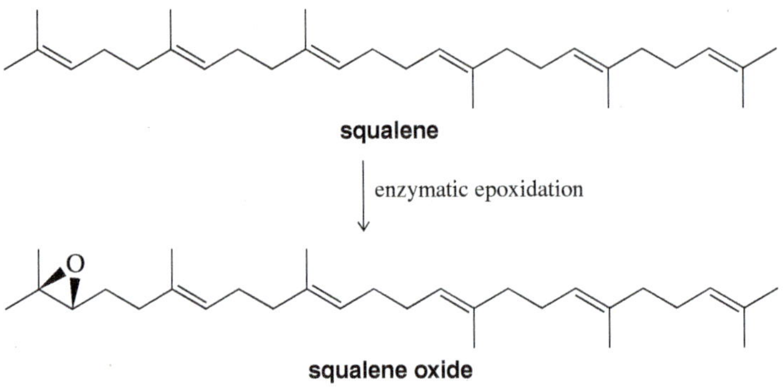

Figure 8.79 The enzymatic epoxidation of squalene.

Protonation of the epoxide with an acid in the active site of the enzyme and ring opening to give the more stable carbocation at C_2 trigger a cyclization cascade that results in the formation of the four rings of lanosterol (**Figure 8.80**).

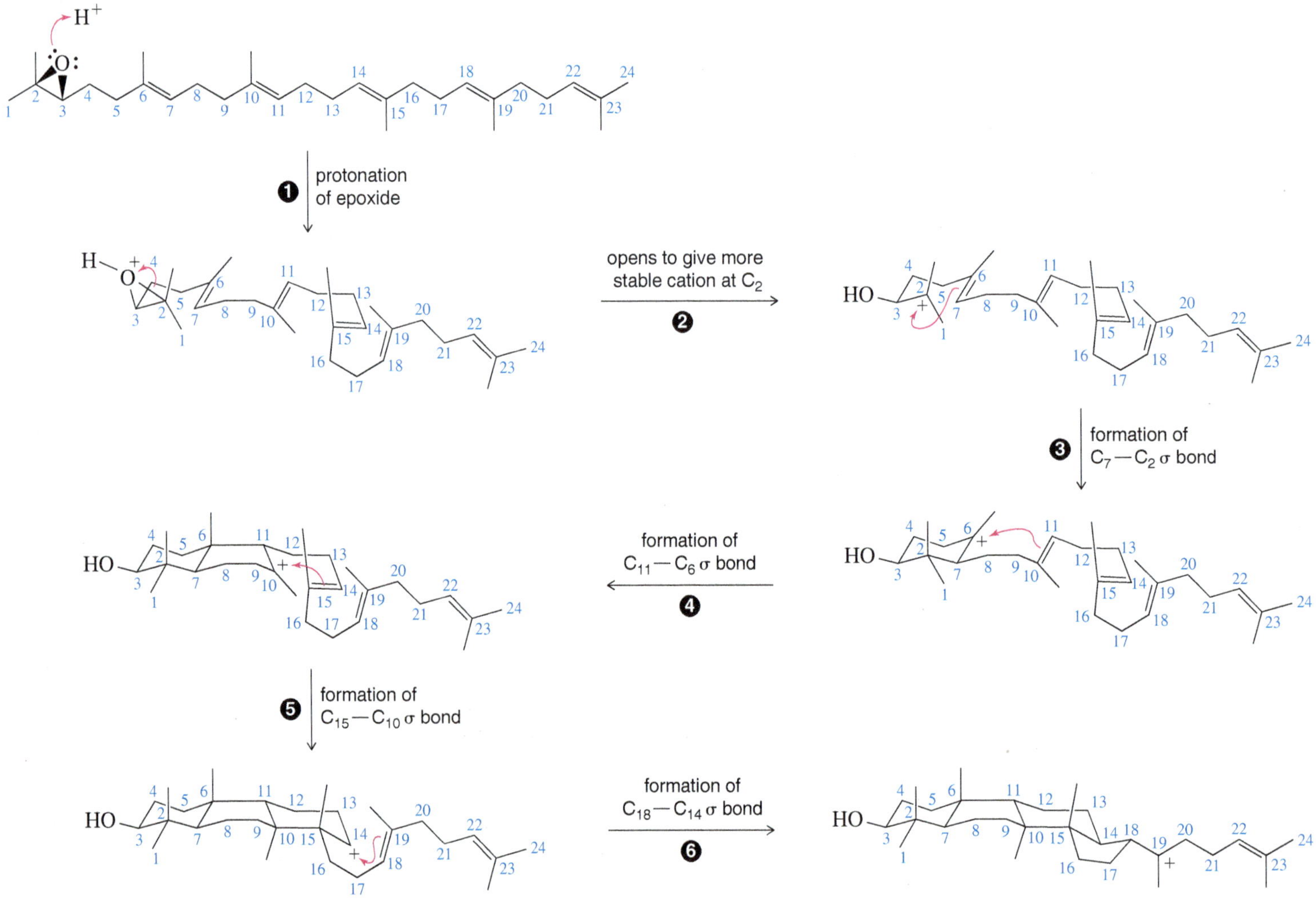

Figure 8.80 Construction of the carbon skeleton of lanosterol.

With the steroid ring skeleton in place, a series of carbocation rearrangements occurs (**Figure 8.81**), ultimately resulting in lanosterol after deprotonation of C_{11} by a base in the enzyme's active site. As mentioned previously, all human steroids are derived from lanosterol.

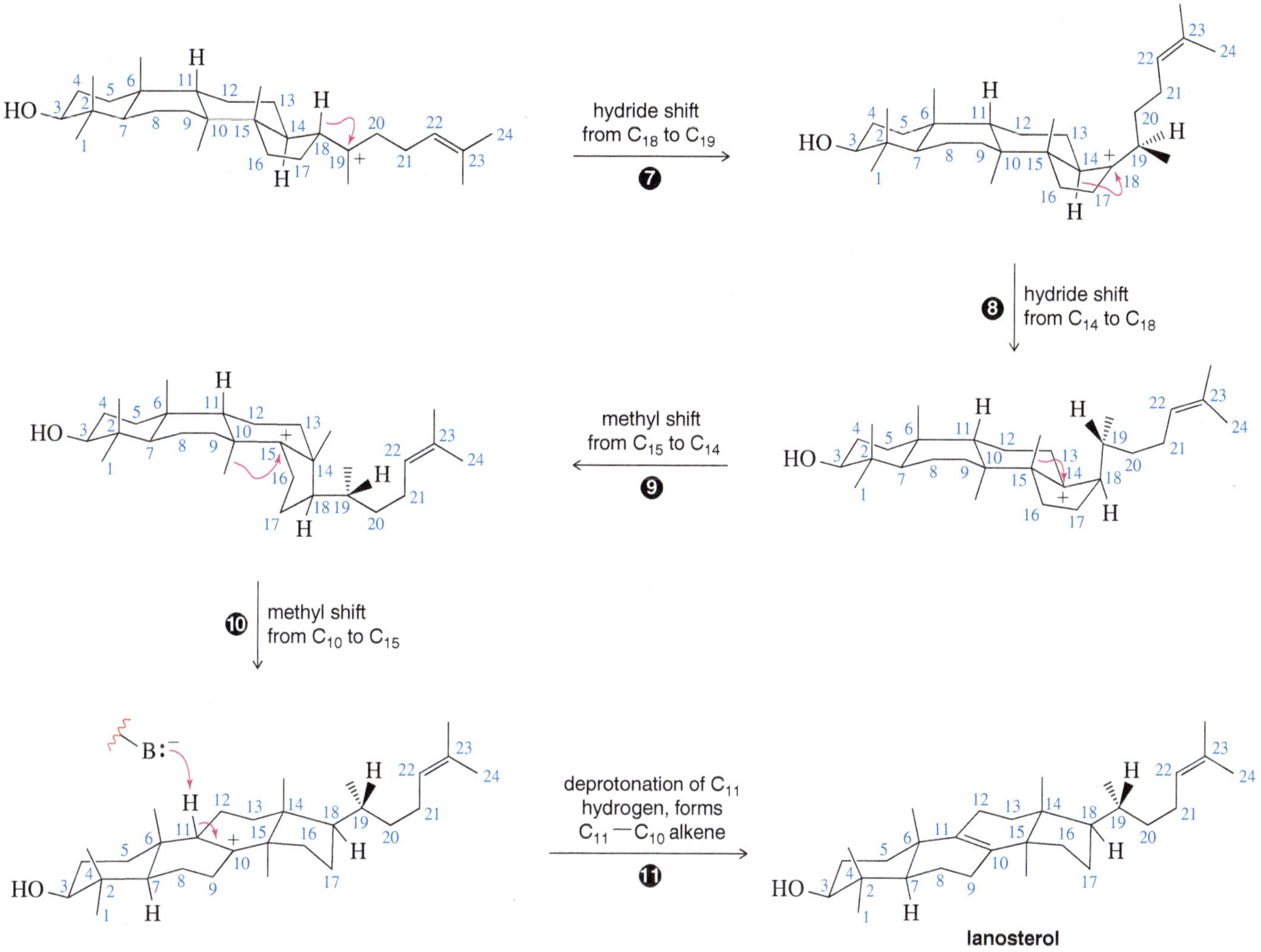

Figure 8.81 Carbocation rearrangements to complete the synthesis of lanosterol.

A slow, careful analysis of the 11-step lanosterol cascade should allow you to see how each reaction in this process is related to reactions we studied in this chapter.

Your ability to successfully analyze this complex reaction should give you confidence as you continue studying organic chemistry. To help you on this path, don't forget to relate every reaction that we study to reactions we have studied previously. We will have the opportunity to do this when we continue the reactions of alkenes in Chapter 9. (*Hint: The reactions we study there are just like the ones in Chapter 8.*)

End-of-Chapter Materials

Important Concepts

[*These are the major concepts covered in each section, excluding reactions.*]

8.1: Alkenes are nonpolar, insoluble in water, and volatile.

8.1.1: The index of hydrogen deficiency can be calculated from the molecular formula to give information about what functional groups might be present in the molecule.

$$\text{IHD} = \text{number of carbons} - \frac{\text{number of hydrogens}}{2} - \frac{\text{number of halogens}}{2} + \frac{\text{number of nitrogens}}{2} + 1$$

An alkene is said to be unsaturated because forming the C—C π bond means there must be two fewer hydrogens in the molecule. The general formula of a monoalkene is C_nH_{2n}.

8.1.2–8.1.4.3: The molecular orbital picture of an alkene consists of a C—C σ bond and a C—C π bond. The σ bond is formed by overlap of two sp^2-hybridized orbitals along an axis that connects the two nuclei. The π bond results from overlap of two $2p$ orbitals above and below the axis of the σ bond. Because overlap requires that the p orbitals be aligned parallel to one another, rotation around the π bond is restricted, essentially requiring the bond to break to allow rotation. The lack of rotation around a π bond means that disubstituted alkenes can be stereocenters, in which case they need names that identify their stereochemistry around the alkene. For disubstituted alkenes, the cis-trans system is used most commonly. The *E*/*Z* system is generally applicable to all alkenes.

8.1.4.5: Combustion analysis can be used to determine the relative stability of compounds having the same molecular formula but different structures. For example, combustion of a cis alkene is more exothermic because it is less stable, so it starts at a higher energy than the corresponding trans alkene.

8.1.5: Based on the orbital picture of the π bond, we know that the π bond is weaker (less overlap) than a σ bond and that the area above and below the plane of the alkene is electron-rich. As a result, alkenes are good electron-pair donors and react efficiently with electrophiles in Lewis acid–Lewis base reactions.

8.2: Alkenes commonly participate in electrophilic addition reactions, which, as the name implies, result in the addition of an electrophile to a nucleophilic π bond. Disfavored by entropy, addition reactions are typically driven by enthalpy because two σ bonds are formed, usually at the expense of one σ bond and one weaker π bond.

8.3–8.3.3: The hydrohalogenation reaction is the addition of a hydrohalic acid (like HBr or HCl) to the alkene. It is regioselective, resulting in the formation of the most stable carbocation in an endothermic first step. Markovnikov's rule says that for any electrophilic addition reaction, you will form the most stable carbocation in the first step. A racemic mixture of alkyl halides is often produced because attack by the halide on the planar carbocation can occur through either of two enantiomeric transition states that are equal in energy.

8.3.4: In the presence of peroxide, HBr adds with opposite regioselectivity. Now proceeding through a radical mechanism, the most stable intermediate in the first step is formed by the attack of the bromine radical at the least substituted carbon, leaving behind the most stable carbon radical. In the second propagation step, the carbon radical abstracts a hydrogen from HBr, allowing the chain reaction to continue.

8.4–8.4.3: The hydration reaction is the addition of water to an alkene and can be accomplished using acid catalysis. With sulfuric acid as the catalyst, this reaction is, like all of the reactions in this chapter, regioselective as a result of forming the most stable carbocation in the first step. The resulting carbocation is then attacked from the top and bottom through enantiomeric transition states to produce a racemic mixture. Because this reaction goes through a carbocation intermediate, rearrangement will occur when, by the migration of hydride or an alkyl anion from a neighboring carbon, a more stable carbocation results.

8.4.4: Hydration of an alkene can also be accomplished using the two-step process of oxymercuration–demercuration. Through a concerted first step, the alkene reacts with mercury acetate to immediately generate a bridged mercurinium ion. This mercurinium ion prevents the rearrangement seen in the acid-catalyzed hydration. In a second step, the carbon–mercury bond is cleaved by reduction with sodium borohydride.

8.4.5: The two-step process of hydroboration–oxidation makes it possible to prepare anti-Markovnikov alcohols. The electrophilic addition of borane to the alkene is a concerted process, resulting in the boron and hydrogen adding on the same face of the alkene (syn addition). In a second step, the reaction of basic hydrogen peroxide cleaves the carbon–boron bond, replacing it stereoselectively with retention of configuration. This, coupled with the stereospecificity of the first step, makes this overall process stereospecific, meaning that two diastereomeric alkenes would be expected to give different diastereomeric alcohols.

Important Reactions and Mechanisms

Most of the reactions in this chapter proceed by the same mechanism, involving a nucleophilic alkene attacking an electrophile or an alkene being attacked by a radical. Make sure you recognize the similarities between each of these reactions, in terms of both the mechanism and their regio- and stereoselectivity.

1. **Hydrohalogenation (Section 8.3):** Electrophilic addition of HBr features an acid–base reaction to produce the more stable carbocation, which is attacked by bromide in the second step. Because attack by bromide in the second step can occur with equal rates to both sides of the carbocation (enantiomeric transition states), a racemic mixture of products results.

General:

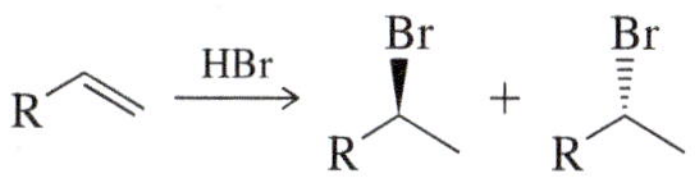

Specific:

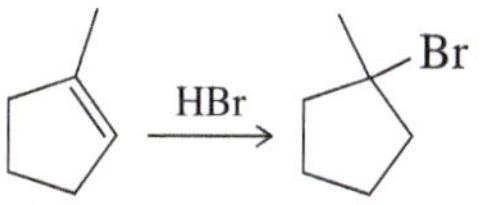

Mechanism:

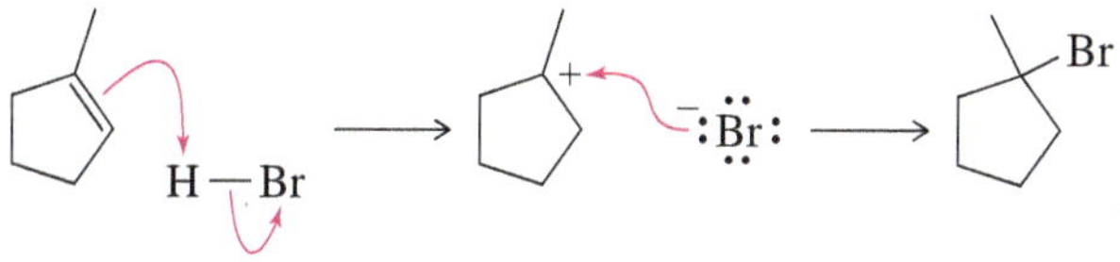

2. **Radical hydrohalogenation (Section 8.3.4):** Addition of HBr in the presence of peroxide results in a radical mechanism. Attack by a bromine radical at the least substituted carbon results in formation of the more stable carbon radical in the first propagation step. The new C—H bond is formed in the second propagation step.

General:

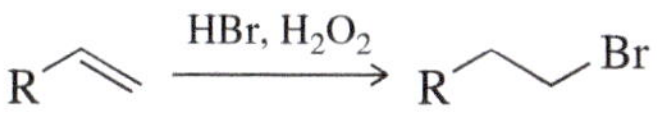

Specific:

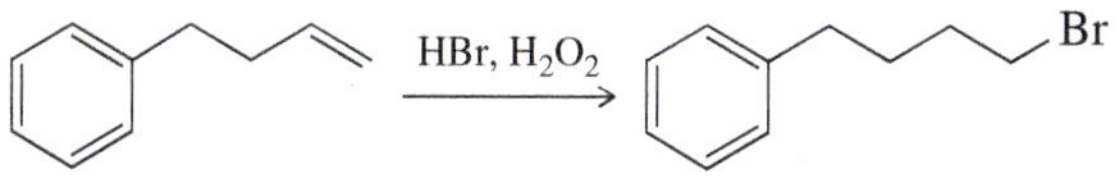

Mechanism:

Initiation:

Propagation 1:

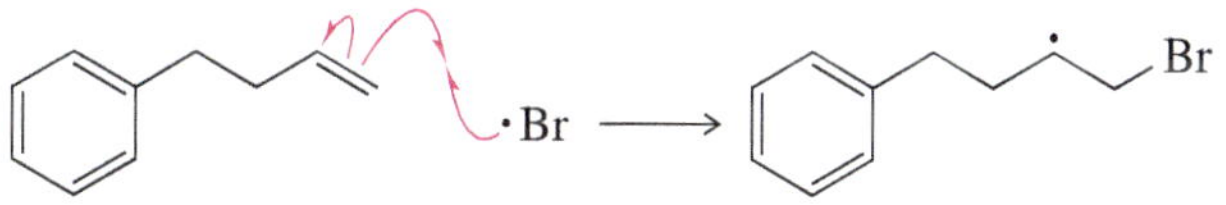

Propagation 2:

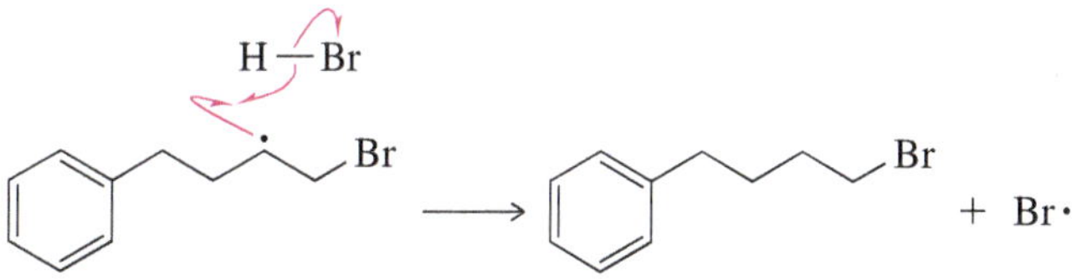

3. **Acid-catalyzed hydration (Section 8.4.1):** Regardless of which sufficiently strong acid is used, protonation of the alkene with H_3O^+ in the first step produces the more stable carbocation, which is attacked by water to give a hydronium ion. Removal of the proton from the hydronium ion regenerates H_3O^+, demonstrating that the reaction is catalytic. Because attack by water in the second step can occur with equal rates on both faces of the carbocation (enantiomeric transition states), a racemic mixture of products results. Under these conditions, rearrangement can occur prior to attack by water if doing so will result in a more stable carbocation.

General:

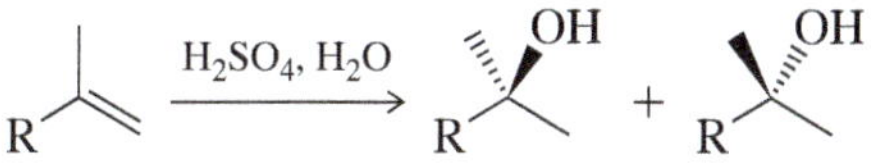

Specific:

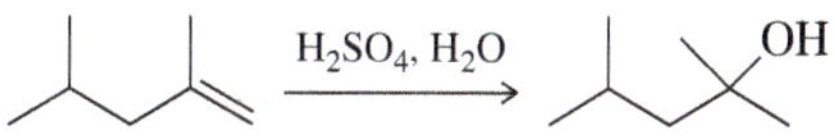

Mechanism:

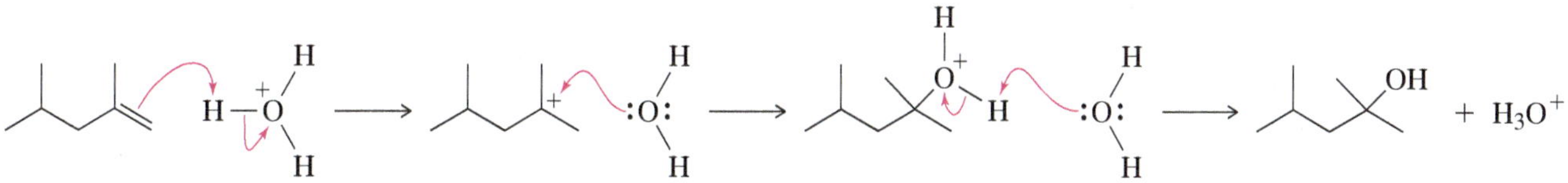

4. **Oxymercuration–reduction (Section 8.4.4):** The alkene attacks an electrophilic mercury cation of mercury acetate. Because of the presence of *d* electrons in mercury, a bridged mercurinium ion forms in a concerted process, preventing any possible rearrangement under these conditions. Water attacks the mercurinium ion at the more substituted carbon on the face opposite the mercury. In the second step, the carbon–mercury bond is reduced by sodium borohydride, replacing the mercury with a hydrogen.

General:

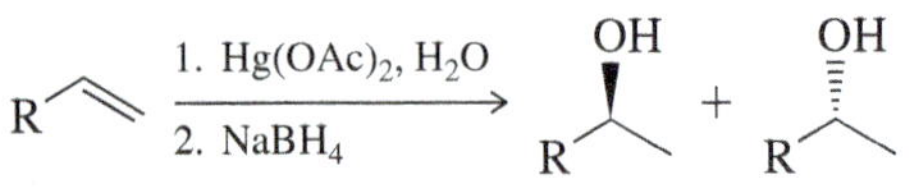

Specific:

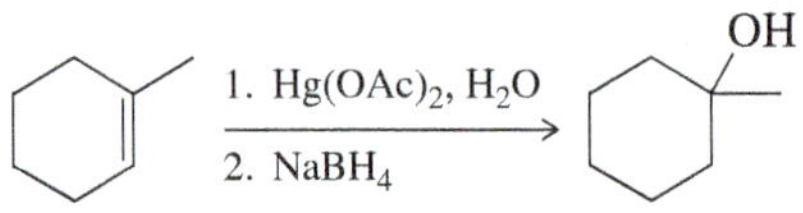

Mechanism:

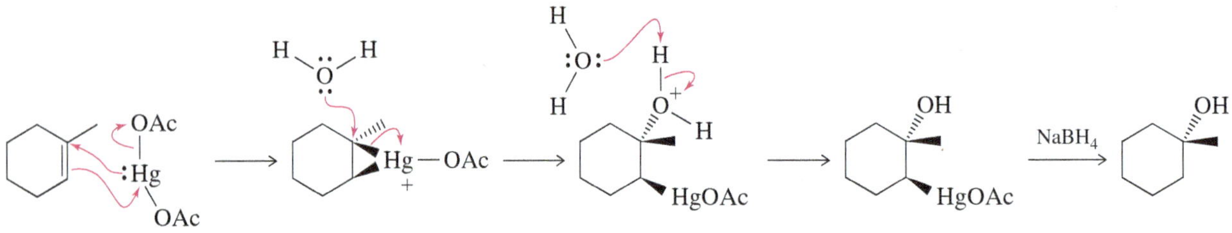

5. **Hydroboration–oxidation (Section 8.4.5):** With an empty *p* orbital, borane is a strong Lewis acid and is attacked by an alkene in a Lewis acid–Lewis base reaction. In order to mitigate the developing negative charge on the borane, hydride is immediately transferred in a single concerted step, resulting in the net addition of BH_3 across the double bond. Because it has three hydrogens, a single molecule of borane can hydroborate three alkene molecules. Oxidative cleavage of the carbon–boron bond with basic hydrogen peroxide proceeds with retention of configuration, resulting in the stereospecific syn addition of water across the alkene.

General:

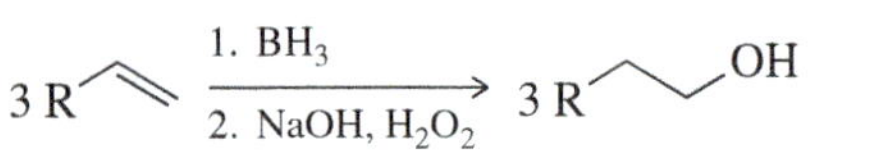

Specific:

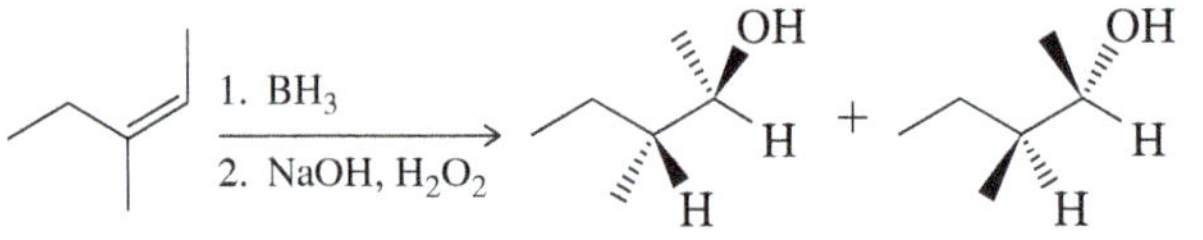

stereospecific, but a racemic mixture

Mechanism:

from front and back, but showing only one

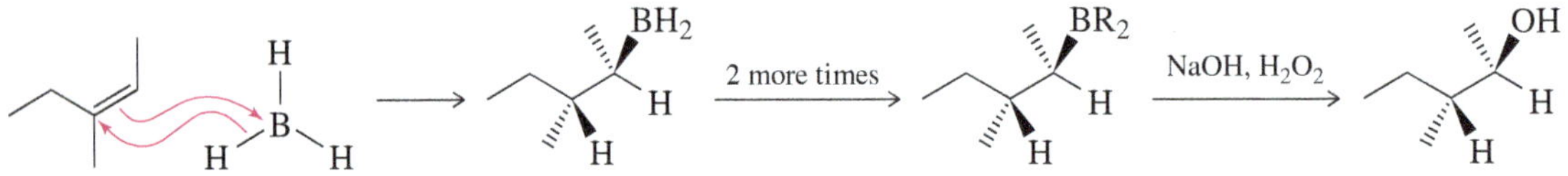

(*Z*)

End-of-Chapter Assessments

[*Assessments are ranked according to their difficulty;* •••• = *most difficult.*]

8.58 (•) Calculate the index of hydrogen deficiency for the following molecular formulas and structures.

(a) 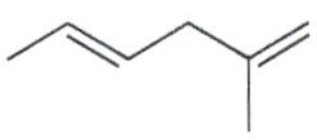

(b) 

(c) 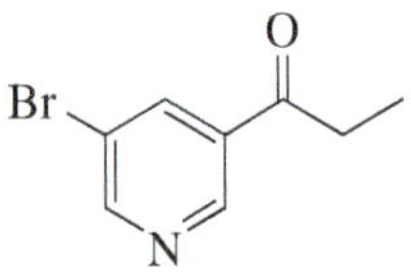

(d) 

(e) $C_9H_{15}N$

(f) $C_6H_8O_2$

(g) $C_8H_{13}ClO_2$

(h) $C_8H_{15}BrO$

8.59 (•) Using IUPAC rules, name the following molecules.

(a) 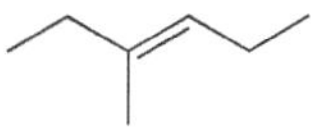 (b) 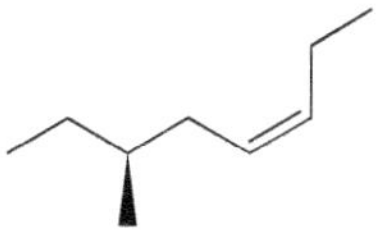 (c) 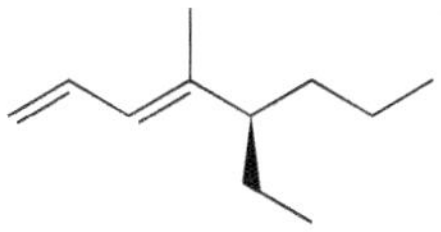 (d) 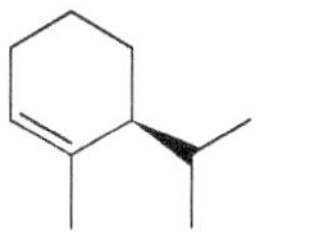

(e) 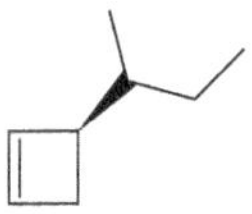 (f)  (g) 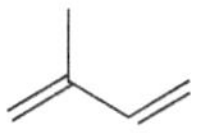 (h) 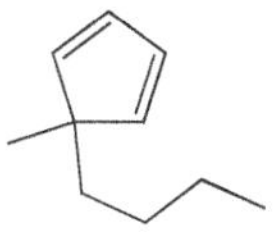

8.60 (••) Given the name, draw the structure of the following compounds.

(a) *cis*-6,7-dimethylcyclohepta-1,4-diene
(b) (4*Z*,8*R*)-8-bromo-5-methylnon-4-ene
(c) (3*S*,6*Z*)-8-ethyl-3-iodododeca-1,5-diene
(d) (*E*)-2-chlorohexa-1,3,5-triene
(e) (3*R*,5*S*)-5-chloro-3-isopropylcycloheptene
(f) (*E*)-2-chlorobut-2-ene

8.61 (•) Predict the product(s) that would result when the alkenes shown here are allowed to react under the following conditions: (i) HBr; (ii) HCl; (iii) HBr, H_2O_2; (iv) H_2SO_4, H_2O; (v) 1. $Hg(OAc)_2$, H_2O 2. $NaBH_4$; (vi) 1. BH_3 2. H_2O_2, NaOH

(a) 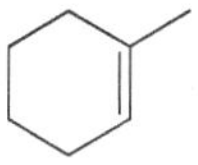 (b) 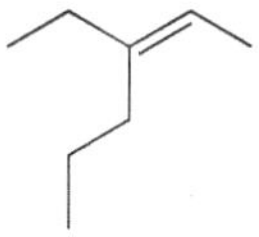 (c) 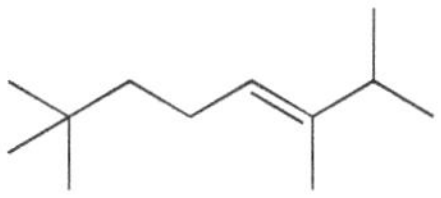 (d) 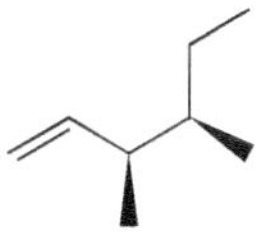 (e) 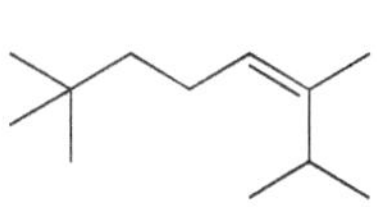 (f) 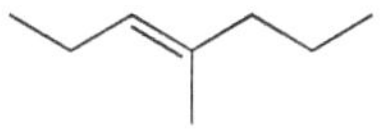

(g) 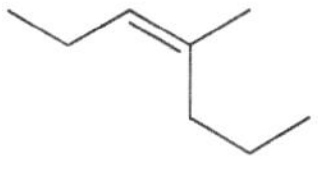 (h) 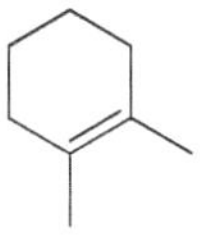 (i)  (j) 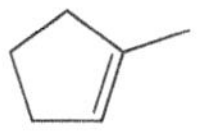 (k) 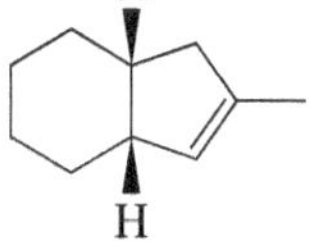

(l) 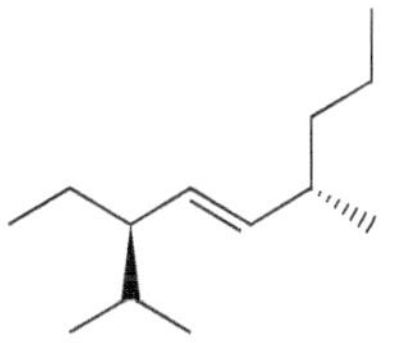

8.62 (••) Which of the following alkenes would you expect to react fastest with HBr in each pair?

(a) 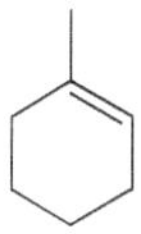 vs. 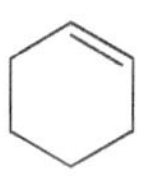

(b) 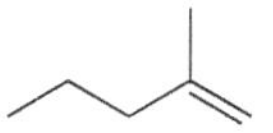 vs. $H_2C{=}CH_2$

(c) 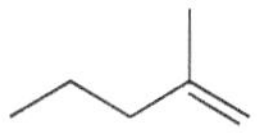 vs. 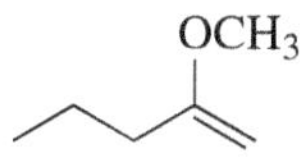

8.63 (•••) Propane is a popular fuel used in engines, portable heaters, and backyard gas grills. Cyclopropane, if used as a fuel, would produce a hotter flame. Why? Justify your answer using a reaction coordinate diagram.

propane

cyclopropane

8.64 (•••) Which of the following cycloalkanes would you expect to produce the least heat upon combustion when measured per CH_2? Explain your answer.

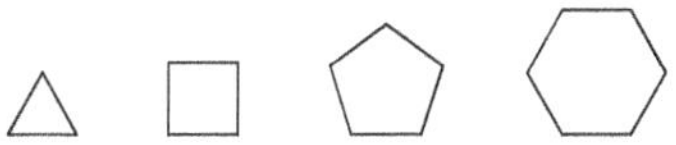

8.65 (•••) Cyclooctene is one of the smaller rings that can form a trans alkene. Would you expect *cis*-cyclooctene or *trans*-cyclooctene to react more quickly in an acid-catalyzed hydration reaction?

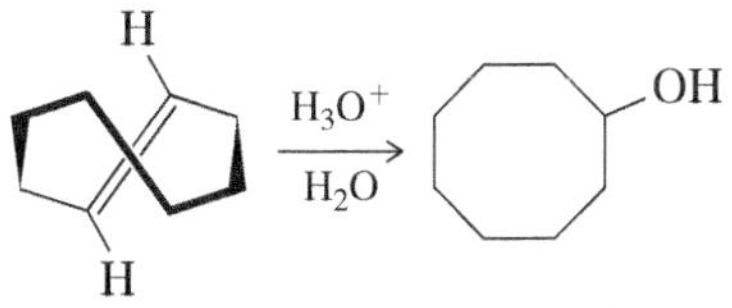

trans-cyclooctene

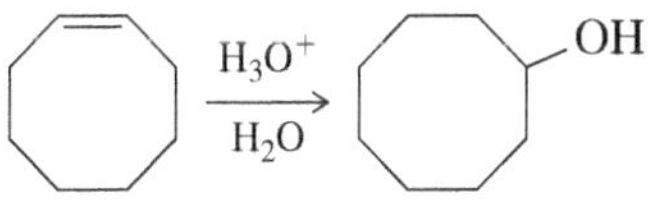

cis-cyclooctene

8.66 (•) How many stereoisomers are possible for the following alkenes?

(a) (b) (c) (d)

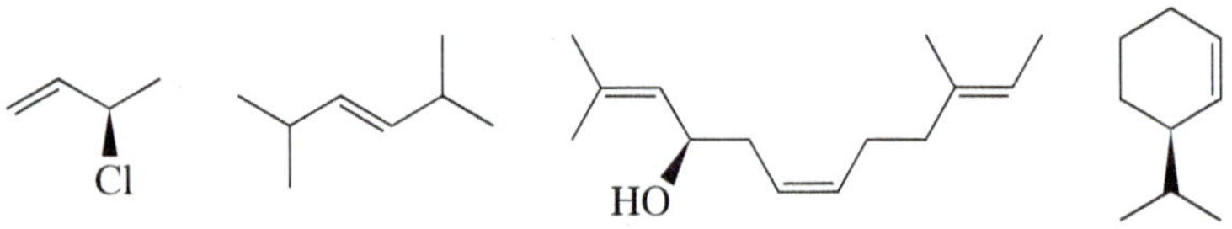

8.67 (••••) Because of the angle strain present in cyclopropanes, they tend to open up in the presence of nearby radicals. Show a mechanism for the following reaction that demonstrates this principle.

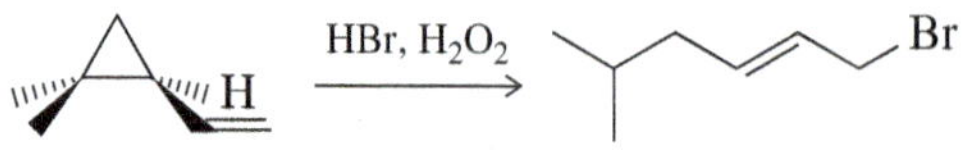

8.68 (•••) When using sulfuric acid, but in the absence of other nucleophiles like water or bromide ion, less stable alkenes can be isomerized to their more stable isomer. Provide a mechanism for these acid-catalyzed isomerization reactions. [*This is one illustration of the principle of microscopic reversibility.*]

(a)

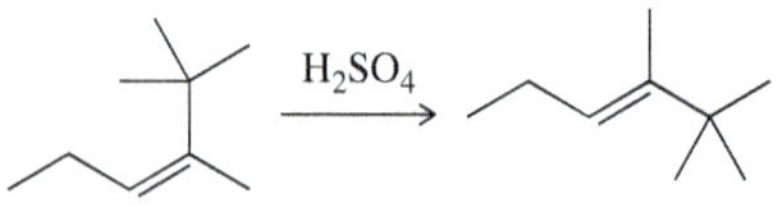

(b)

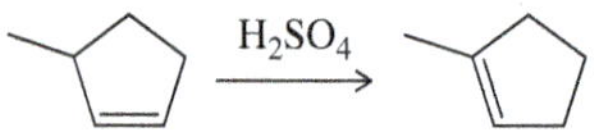

(c)

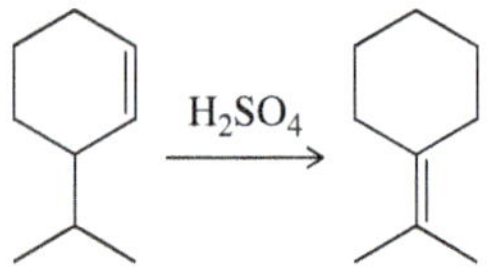

8.69 (••••) THINKING AHEAD In Chapter 19, we study a reaction referred to as a keto–enol tautomerism. Based on your answers to Assessment 8.68, suggest a reasonable mechanism by which the enol is converted to the ketone.

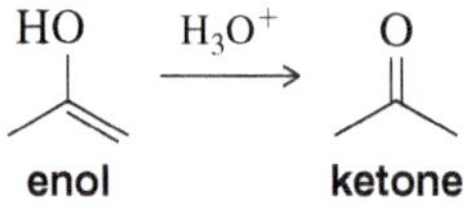

8.70 (•••) If you react carvone with a single equivalent of HBr, only *one* product is produced, even though carvone has *two* carbon–carbon double bonds. Explain this observation.

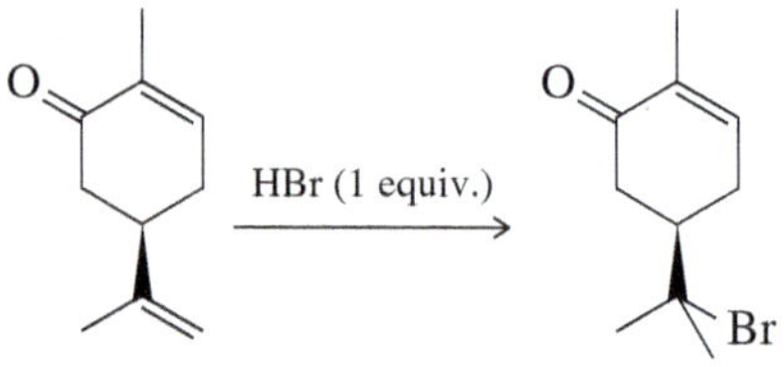

8.71 (•••) FROM THE LITERATURE Predict the product of each of the following hydroboration–oxidation or oxymercuration–reduction reactions used in the modern synthetic organic chemistry literature (modified to use reagents we are used to seeing). [*These molecules, while perhaps more complex than any you have seen previously, are, in the presence of borane or mercury(II) acetate, simply alkenes. Thus, don't get distracted by the other functional groups and abbreviations present and predict the product based on what you have learned here in Chapter 8.*]

(a) Hydroboration–oxidation was used toward the synthesis of *Lycopodium* alkaloids, molecules that contain potent biological activity (*Org. Lett.* **2013**, *15*, 2140–2143).

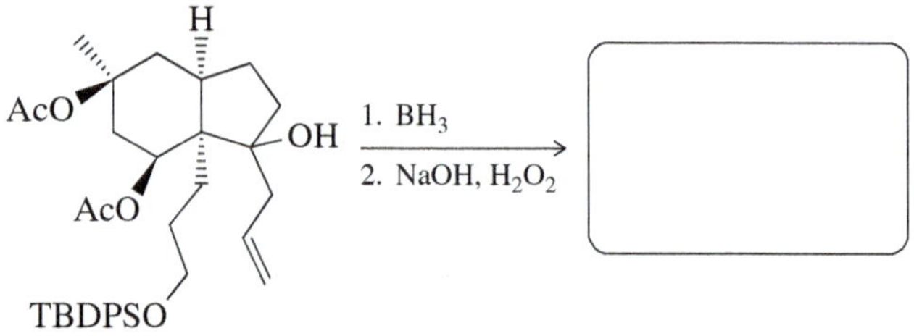

(b) A two-step hydroboration–oxidation was used to prepare a silanediol peptidomimetic as a serine protease inhibitor (*Org. Lett.* **2012**, *14*, 4422–4425).

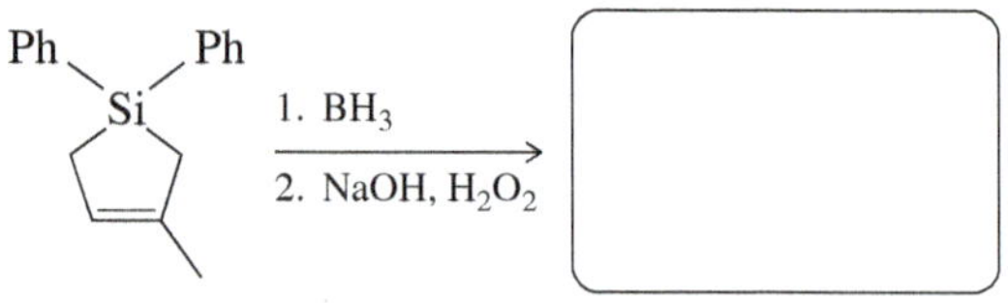

(c) A similar sequence was featured in the synthesis of muricadienin, a proposed precursor in the biosynthesis of solamin (*Org. Lett.* **2014**, *16*, 5886–5889).

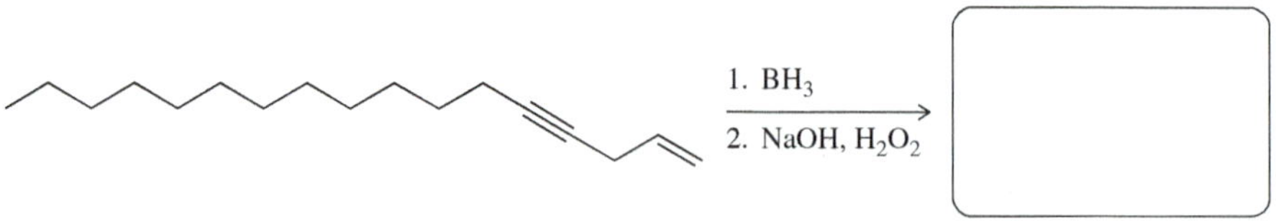

(d) Oxymercuration–reduction was part of a synthetic scheme directed toward the synthesis of anatoxin-*a*, a potent agonist of the nicotinic acetylcholine receptor (*Tetrahedron Lett.* 2004, *45*, 4397–4399).

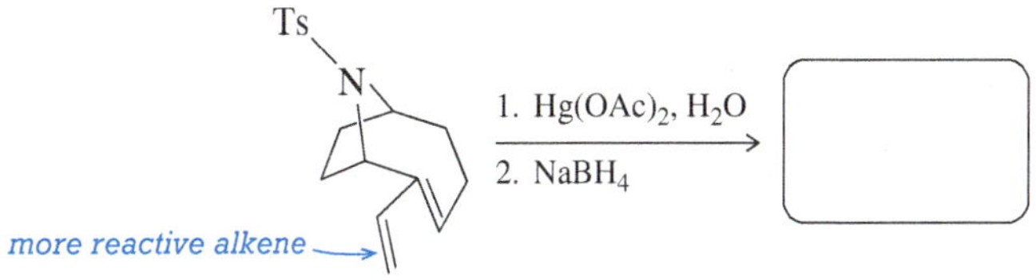

8.72 (••••) FROM THE LITERATURE The following oxymercuration reaction was used in the synthesis of a potential interleukin-1B converting enzyme inhibitor. Unlike most oxymercuration reactions, there is no hydroxyl group in the product. Suggest a mechanism for the formation of the only observed product (*Org. Lett.* 2004, *6*, 4535–4538). [*OTf is similar to OAc.*]

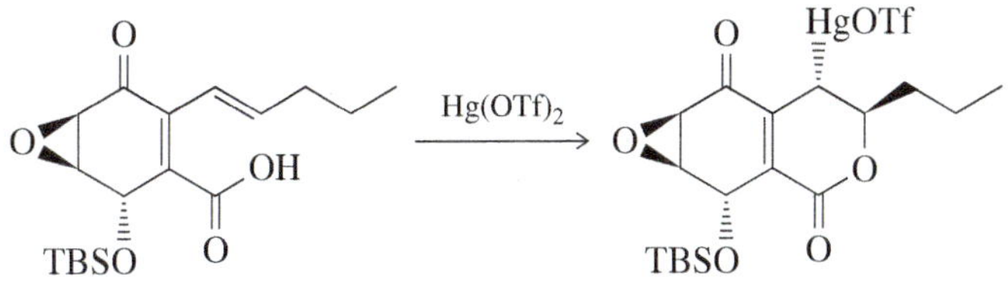

8.73 (••••) A common situation occurs when both carbons of the mercurinium ion are secondary. In a situation like the one shown, to which carbon would you expect water to add? That is, would you expect to produce **A** or **B**? Explain your answer.

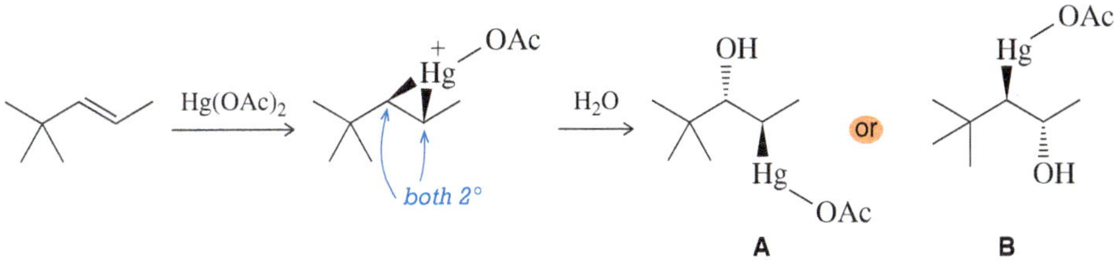

8.74 (•••) FROM THE LITERATURE In light of your answer to Assessment 8.73, predict the product of the following oxymercuration–reduction reactions, each of which results in a single product. [*Don't worry about the absolute stereochemistry, though these reactions are also stereoselective.*]

(a) The following oxymercuration–reduction sequence was performed in efforts toward the synthesis of a high-affinity HIV-1 protease inhibitor (*Tetrahedron Lett.* 2008, *49*, 3409–3412).

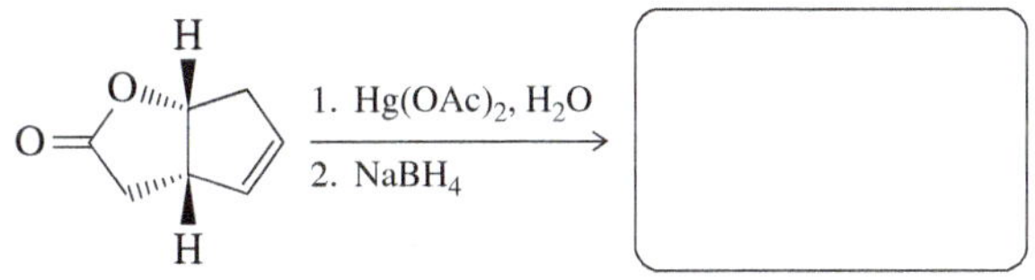

(b) This reaction sequence was used in the synthesis of (+)-lineatin, a monoterpene aggregation pheromone from the female ambrosia beetle *Trypodendron lineatum* (*Org. Lett.* 2004, *6*, 1449–1452).

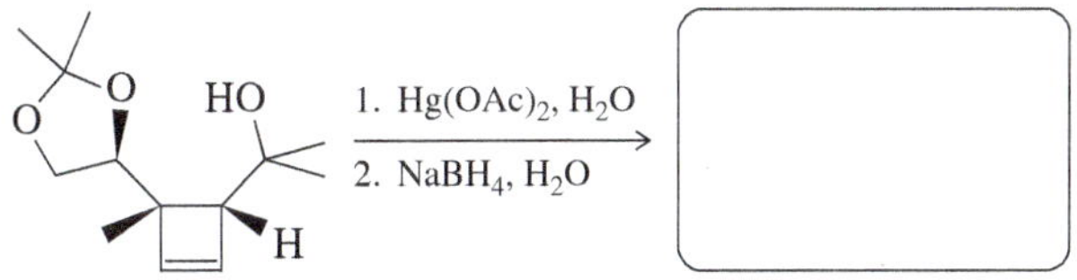

(c) Oxymercuration–reduction was used in the stereoselective synthesis of the macrolactone core of neopeltolide, a marine macrolide isolated from a Caribbean sponge that has potent anticancer activity (*Org. Lett.* 2012, *14*, 2346–2349).

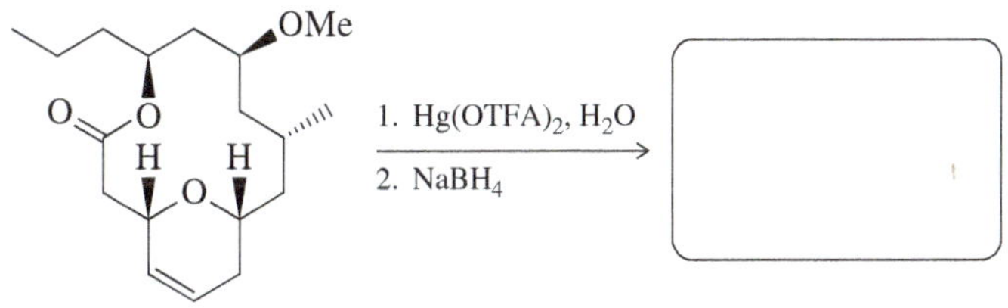

8.75 (••••) FROM THE LITERATURE The formation of five-membered ring ethers is an important goal in synthetic organic chemistry because tetrahydrofurans are contained within a number of antitumor natural products. Toward that end, a one-pot synthesis of a bis-THF containing compound was developed (*Eur. J. Org. Chem.* 2010, 6263–6268). Suggest a mechanism for this transformation.

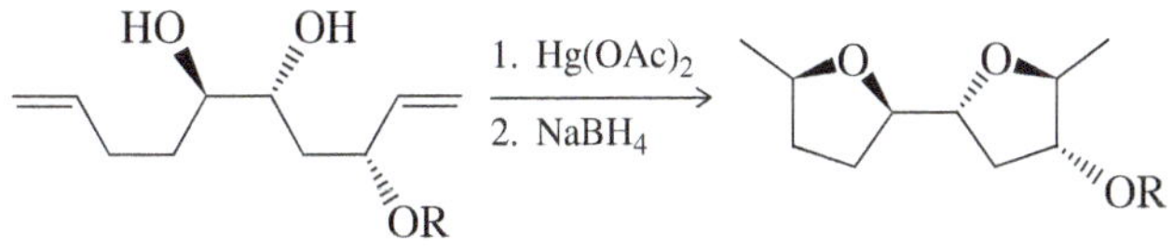

8.76 (••••) North American termite soldiers, when encountering enemy insects, contract their mandibular muscles, expelling a mixture of chemicals that essentially trap their enemies in a glue-like substance. This weapon, built into the face of the termite, is called the fontanellar gun. It releases a mixture of pinene (62%), myrcene (27%), and limonene (11%).

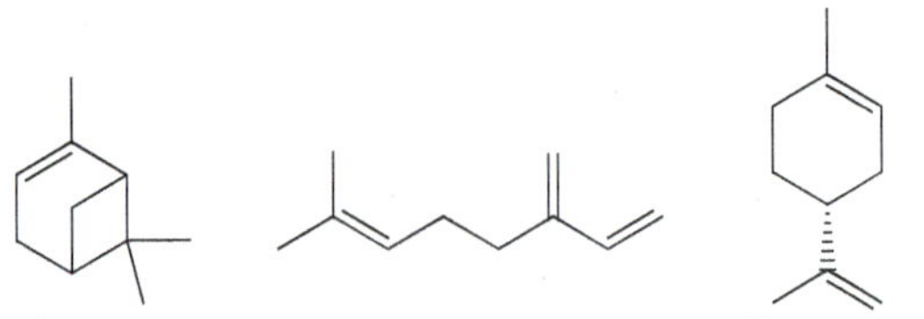

pinene 62% myrcene 27% limonene 11%

(a) Identify the isoprene units in pinene, myrcene, and limonene.
(b) Suggest an acid-catalyzed mechanism by which pinene could be produced from limonene.

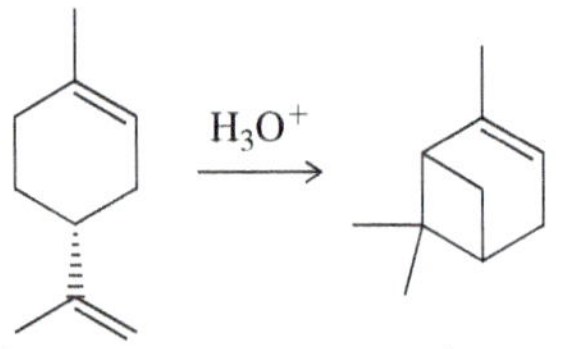

limonene pinene

8.77 (•••) THINKING AHEAD The acid-catalyzed hydration we learned here in Chapter 8 is reversible [*we introduce this reaction in Chapter 13*]:

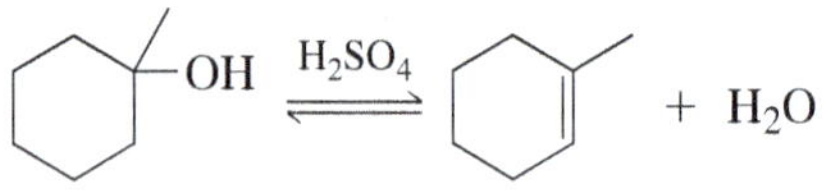

(a) Propose a mechanism for the formation of an alkene from an alcohol.
(b) Without looking ahead, how do you know your mechanism is correct?
(c) Which side of the reaction would be favored by running the reaction at high temperatures?
(d) How might you shift the equilibrium to the right?
(e) How might you shift the equilibrium to the left?

8.78 (•••) The reaction of (*E*)-3-chlorohex-3-ene with HCl results in the selective formation of 3,3-dichlorohexane instead of 3,4-dichlorohexane. Explain this result.

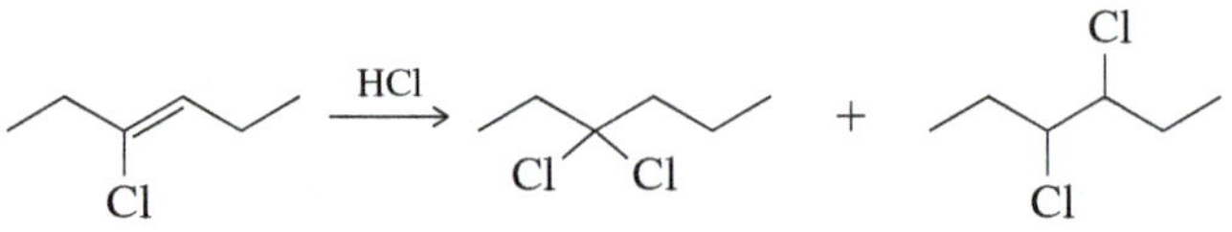

3,3-dichlorohexane formed 3,4-dichlorohexane not formed

8.79 (••••) In addition to radicals, anions, and cations, a fourth class of reactive intermediates is carbenes. A neutral species, the simplest carbene has a molecular formula of CH_2.

(a) Draw the Lewis structure of CH_2.
(b) What are the hybridization and shape of the central carbon of CH_2?
(c) Carbenes are Lewis acids. Based on your answers to (a) and (b), explain this observation.
(d) Carbenes are also Lewis bases. Based on your answers to (a) and (b), explain this observation.
(e) Carbenes have great synthetic utility, especially in the synthesis of cyclopropanes from alkenes. Based on your answers to (a)–(d), show a mechanism for the cyclopropanation of cyclohexene.

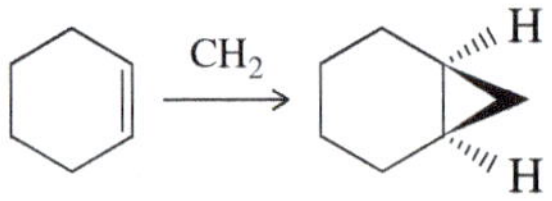

(f) Cyclopropanation of (*E*)- and (*Z*)-3-methylhex-3-ene gives two different products. Rationalize this outcome.

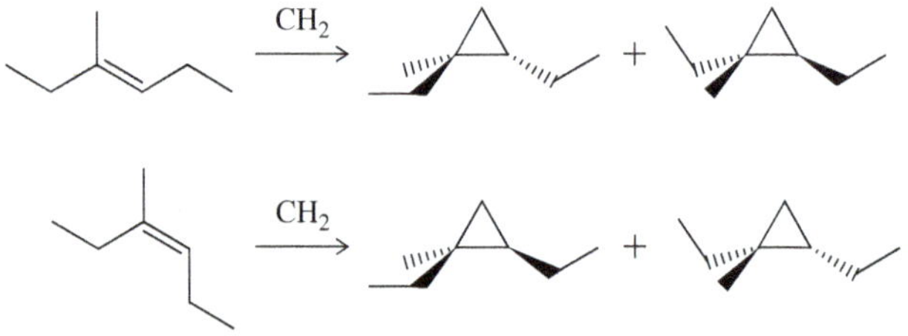

(g) On the basis of your answers to (a)–(f), draw a reaction coordinate diagram for a general cyclopropanation reaction.
(h) Of the reactions studied here in Chapter 8, cyclopropanation is most similar to which?

8.80 (••••) Suggest an arrow-pushing mechanism that accounts for the formation of the following products.

(a)

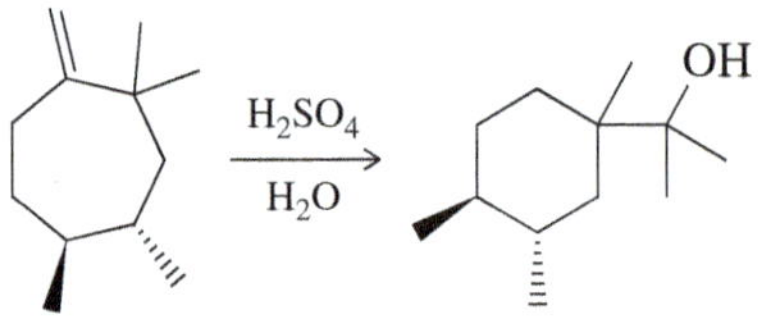

(b)

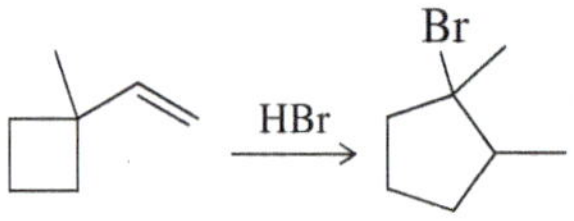

(c)

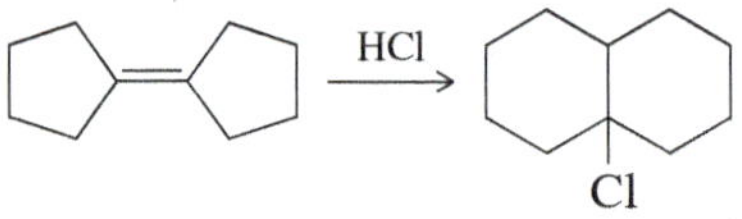

9

Alkenes 2: Oxidation and Reduction

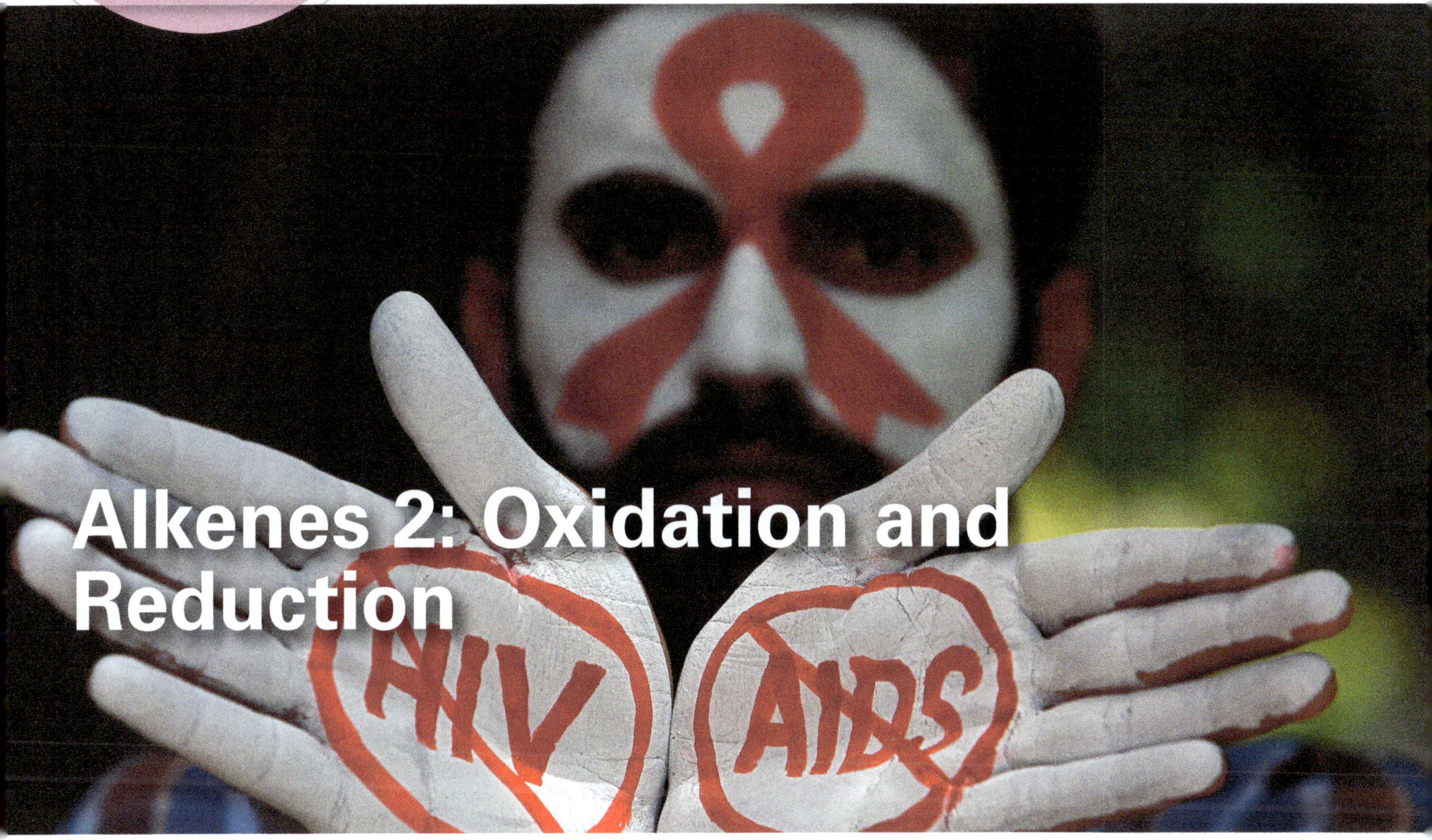

Introduction

Identified in 1981, human immunodeficiency virus (HIV) has become one of the deadliest diseases in modern history. Since its original identification, over 32 million people worldwide have died as a result of this virus. In sub-Saharan Africa, which accounts for 70% of the world's infected patients, 25 million still live with the disease.

HIV is transmitted through direct contact with bodily fluids of an infected patient. Like all viruses, HIV is not able to reproduce on its own and instead uses the cellular machinery (specifically, CD4 cells) of its host to replicate itself, as shown in **Figure 9.1**. There are three places where enzymes are involved in HIV replication. The enzyme *reverse transcriptase* converts viral RNA into viral DNA. *HIV integrase* integrates viral DNA into the DNA of the host cell. And, finally, *HIV protease* cuts long strands of viral proteins into the smaller useful viral proteins as part of the process of assembling new viral particles. Because enzymes often have a rigid, structurally defined active site, medicinal chemists can design molecules that fit perfectly in this active site, preventing the enzyme from catalyzing its native reaction. Using stereoselective reactions of alkenes presented here in Chapter 9, Merck produced Crixivan® (indinavir sulfate), an inhibitor of HIV protease. First marketed in 1996, it has been one of the most effective protease inhibitors on the market and, as part of a cocktail of HIV drugs, has greatly added to the life expectancy of HIV patients, allowing them to "live with" rather than "die from" the disease.

Learning Objectives

LO 9.0 Determine whether a redox reaction has occurred

LO 9.1 Evaluate oxidation reactions of alkenes (including halogenation, halohydrin formation, epoxidation, dihydroxylation, and ozonolysis)

LO 9.2 Evaluate reduction reactions of alkenes

LO 9.3 Assess the stability of alkenes (in terms of the heat of hydration, equilibrium constant, and $\Delta H°$)

LO 9.4 Predict the predominant enantiomer in a redox reaction

LO 9.5 Evaluate the use of asymmetric catalysis in redox synthesis reactions

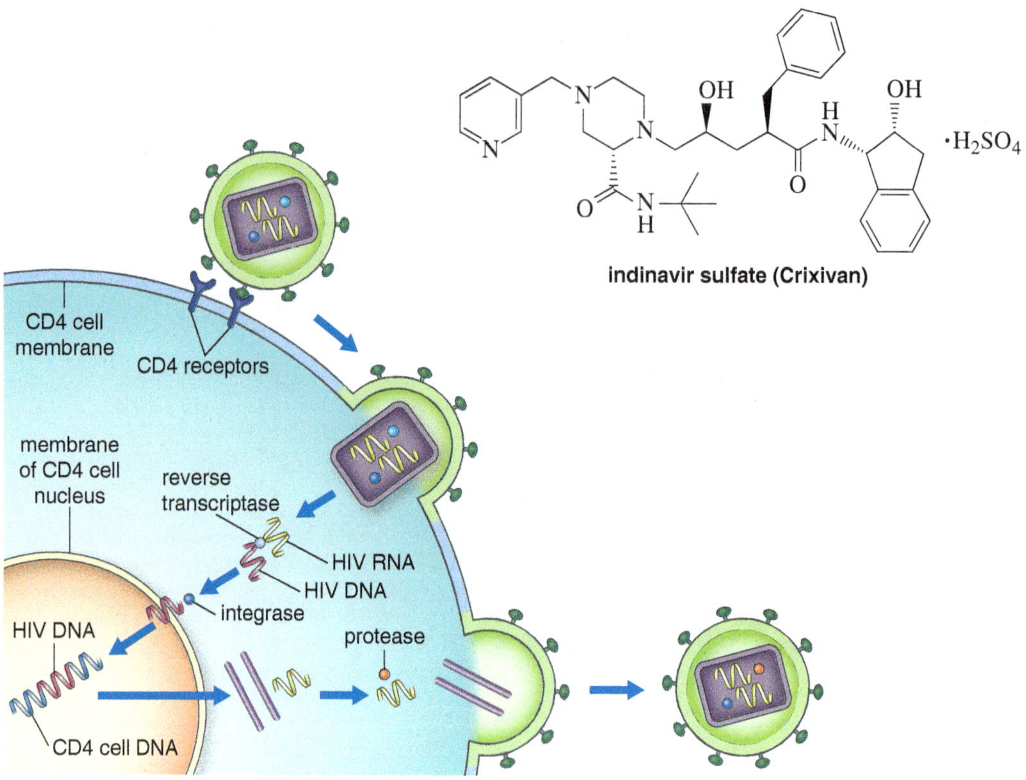

Figure 9.1 HIV uses cellular machinery to reproduce. (a) The HIV life cycle. (b) Indinavir sulfate is an HIV protease inhibitor.

"A planet without AIDS is possible, but to create that planet we must do away with the vestiges of the old planet where testing positive to the HIV virus effectively relegates an individual to the subclass of Human society."

—OCHE OTORKPA, The Unseen Terrorist

Four of the 59 **new molecular entities** (NMEs) approved by the FDA in 2018 are shown in **Figure 9.2**. Each of these NMEs, defined as drugs that have no active constituent previously approved by the FDA, represents the end result of 10–20 years of the drug discovery process. The wide array of functional groups, and the several thousand compounds that did *not* become NMEs, highlights the need for having a variety of organic reactions at our disposal. As students of organic chemistry, we must learn what these reactions produce and how they work. More importantly, learning these reactions provides us a framework for developing problem-solving skills applicable to all careers. The assessments in this chapter, course exams, and standardized exams assess how well we are developing those skills.

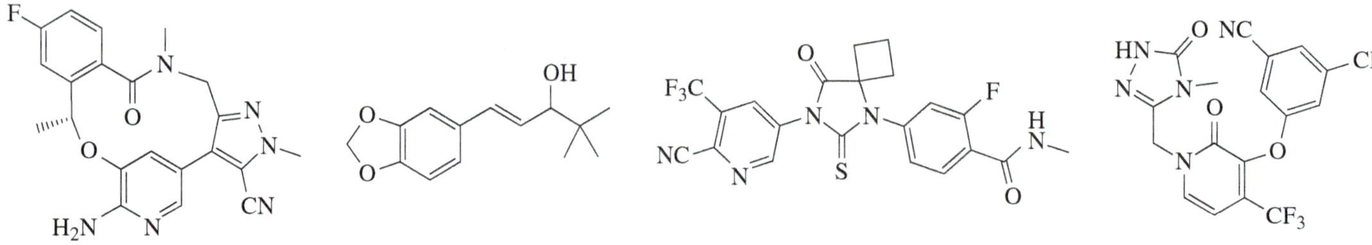

lorlatinib **stiripentol** **apalutamide** **doravirine**

Figure 9.2 New molecular entities (NMEs) approved during 2018.

Table 9.1 summarizes the progress we hope to gain in developing these problem-solving skills. The goal of this text is continued pursuit of level 4 of the hierarchy in the table. For every new reaction we study, then, you need to think about what happens [*predict the outcome*], how it happens [*draw an arrow-pushing mechanism*], why it happens [*rationalize it in the context of a reaction coordinate diagram*], and how your knowledge of old material helped you to understand this material [*make the connections*]. We continue this emphasis in Chapter 9, starting by assessing our understanding of previous concepts.

Table 9.1 Hierarchy of organic chemistry problem solving.

We should be able to . . .	Summary
1. Predict the outcome of a reaction.	Sometimes this results in memorizing "what does what." This is not a way to learn organic chemistry and will ultimately prove unsuccessful. [*Luckily, we learned this several chapters ago.*]
2. Know how a reaction happens.	We must be able to draw a mechanism, showing how electrons flow during the course of a reaction. It is easier to do this with a solid understanding of fundamental organic chemistry reactivity. [*Negatives attack positives.*]
3. Understand why a reaction happens the way that it does.	Chapter 5 helped us analyze the energetics of reactions that explain why they occur and the rate at which they occur. [*The ball rolls downhill.*]
4. Recognize and make connections with other reactions.	This is the pinnacle, where you can relate new reactions to old reactions, predict outcomes of unstudied reactions, "invent" new reagents, and predict how a concept might appear on a test. [*Long term retention is also achieved.*]

Test your understanding

Assessment 9.1 Would you expect the following to produce an equal or unequal mixture of stereoisomers?

(a)

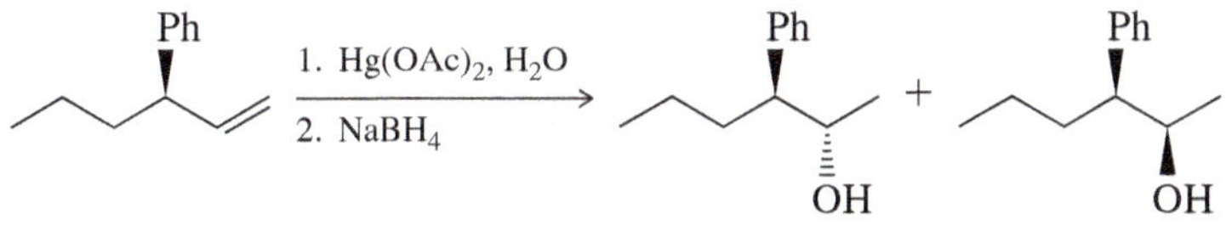

(b)

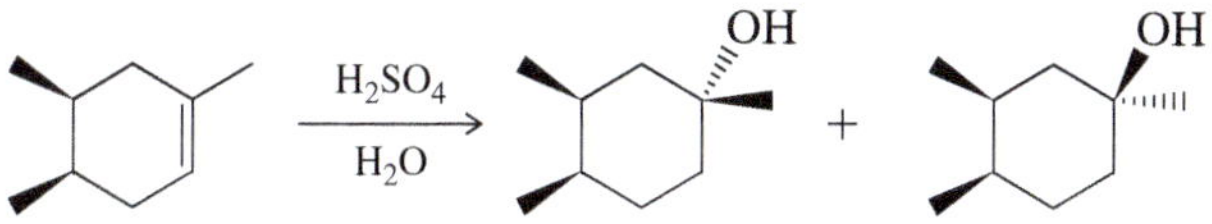

(c)

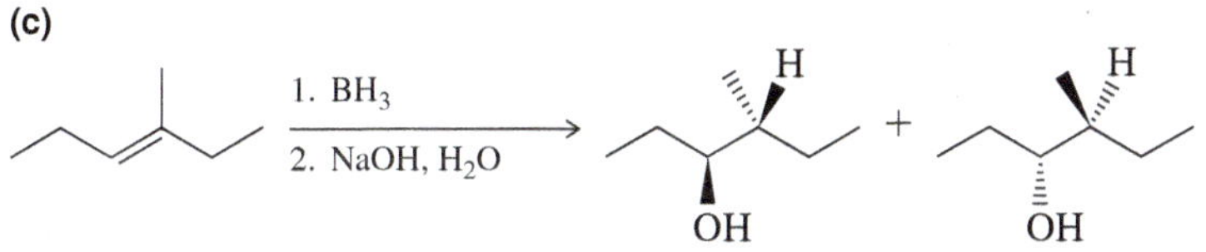

Assessment 9.2 Show a mechanism for the formation of one isomer in each of the reactions in Assessment 9.1. [*We studied these reactions in Chapter 8.*]

Assessment 9.3 Calculate the index of hydrogen deficiency for the following molecules.

(a)

(b)

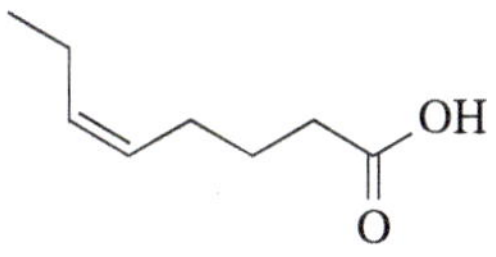

(c)

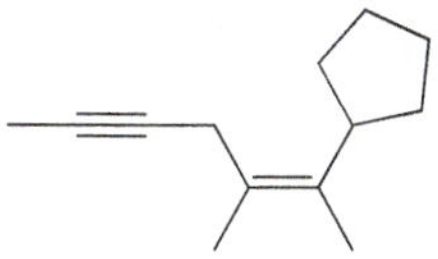

9.1 Oxidation of alkenes

The reactions in Section 9.1 are alkene oxidation reactions. Alkene reduction reactions are then presented in Section 9.2. In order to recognize when an oxidation (or reduction) reaction occurs, we need to know the oxidation numbers of the atoms participating in a given reaction. Section 9.1.1 explains how to do this.

9.1.1 Calculating oxidation numbers: identifying redox reactions

The convention of assigning oxidation numbers begins with the premises that the more electronegative atom "owns" both electrons in a single covalent bond and that all nonbonded atoms have an oxidation number of zero because the number of electrons balances out the number of protons. In order to assign oxidation numbers, then, you need to be able to draw Lewis structures and to understand electronegativity trends. **Table 9.2** walks you through the process for four molecules. In the case of methane, carbon is more electronegative than hydrogen, so carbon owns each of the 2 electrons in the four C—H bonds. Subtracting the "owned" electrons from the valence electrons in the neutral atom gives an oxidation number of −4 for carbon and +1 for each hydrogen. The sum of the oxidation numbers should equal the overall charge on the molecule. When two identical atoms are bonded, they are assumed to equally share the electrons in their mutual bond(s). In ethene, each carbon atom of the double bond owns 1 electron from each of the two bonds, for a total of 2.

Table 9.2 Assigning oxidation numbers.

Lewis structure	Electrons owned	Valence shell electrons	Atom oxidation number	Structure with oxidation numbers
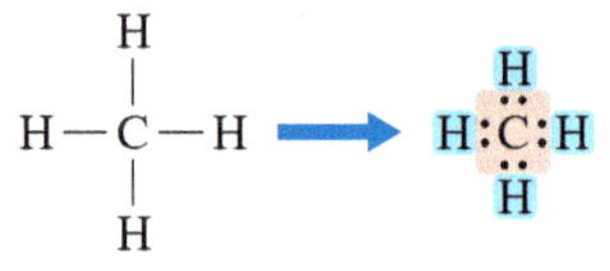 Methane	Carbon: 8 Hydrogen: 0	Carbon: 4 Hydrogen: 1	4 − 8 = −4 1 − 0 = +1	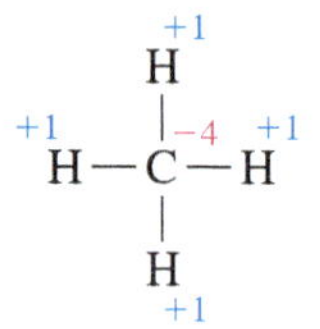
 Carbon dioxide	Carbon: 0 Oxygen: 8	Carbon: 4 Oxygen: 6	4 − 0 = +4 6 − 8 = −2	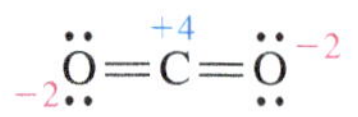
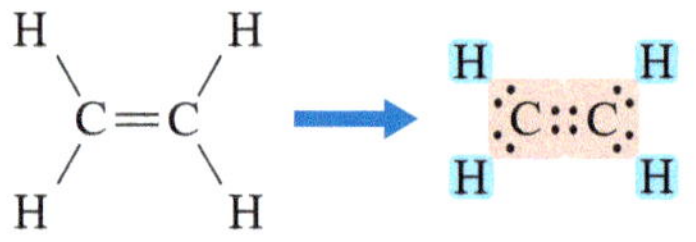 Ethene	Carbon: 6 Hydrogen: 0	Carbon: 4 Hydrogen: 1	4 − 6 = −2 1 − 0 = +1	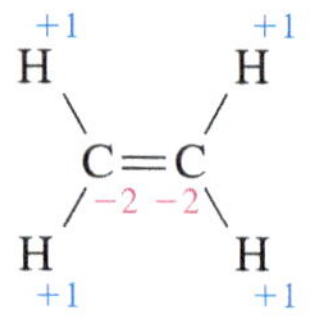
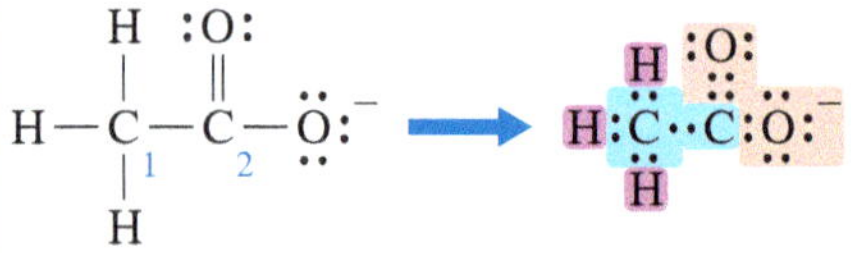 Acetate	Carbon 1: 7 Carbon 2: 1 Oxygen: 8 Hydrogen: 0	Carbon 1: 4 Carbon 2: 4 Oxygen: 6 Hydrogen: 1	4 − 7 = −3 4 − 1 = +3 6 − 8 = −2 1 − 0 = +1	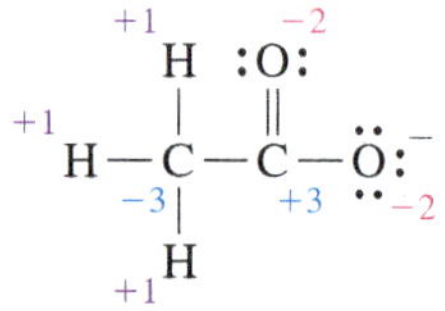

Assessment 9.4 Determine the oxidation number of each carbon indicated by an arrow.

(a)

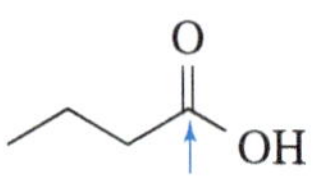

(b)

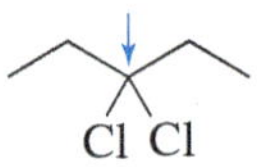

(c)

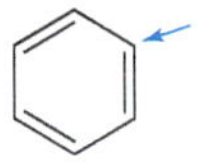

(d)

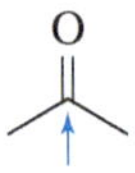

(e)

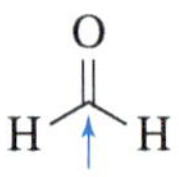

(f)

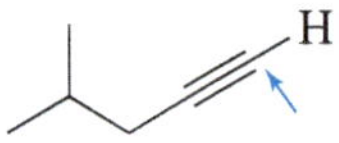

(g)

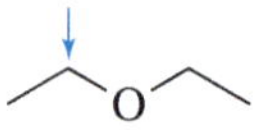

(h)

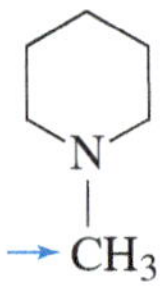

We can now look at the reactions in **Figure 9.3** and determine whether a **redox reaction** has occurred. [*A reduction reaction is always paired with an oxidation.*] If a change in oxidation numbers has occurred in going from reactants to products, then a redox reaction has occurred. In the radical bromination of methane [*Section 5.3*], the oxidation number of carbon changes from −4 to −2, a difference of 2 [**Figure 9.3(a)**]. Thus, carbon "lost" two electrons going from methane to bromomethane. The oxidation number of each bromine atom

Recall that an atom that loses electrons is oxidized. An atom that is reduced gains electrons.

changes from 0 in Br_2 to −1 in bromomethane and HBr. Thus, the two bromines gained a total of 2 electrons, implying that the Br_2 was reduced. This is indeed a redox reaction. Ethene hydration [**Figure 9.3(b)**] is *not* a redox reaction. The only change in oxidation numbers occurs at the carbons, with one changing from −2 to −3, while the other changes from −2 to −1. Since this is simply a reshuffling of electrons within the same molecule, this is *not* a redox reaction.

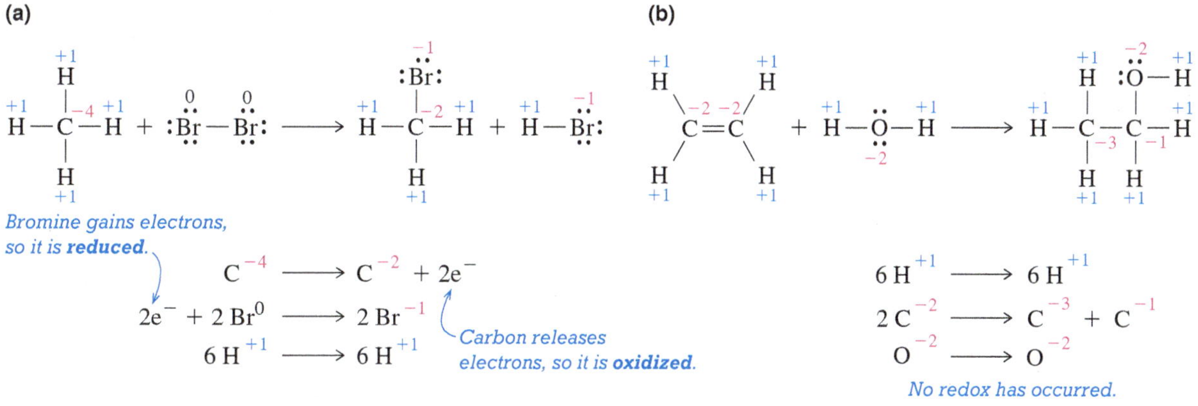

Figure 9.3 Recognizing a redox reaction. (a) Electrons are transferred from one molecule to another, so this is a redox reaction. (b) There is no transfer of electrons, so this is not a redox reaction.

9.1.1 SUMMARY

- Oxidation numbers are hypothetical numbers assigned to atoms in a molecule that make it possible to track the flow of electrons in a redox reaction.

Assessment 9.5 Use oxidation numbers to determine which of the following are redox reactions.

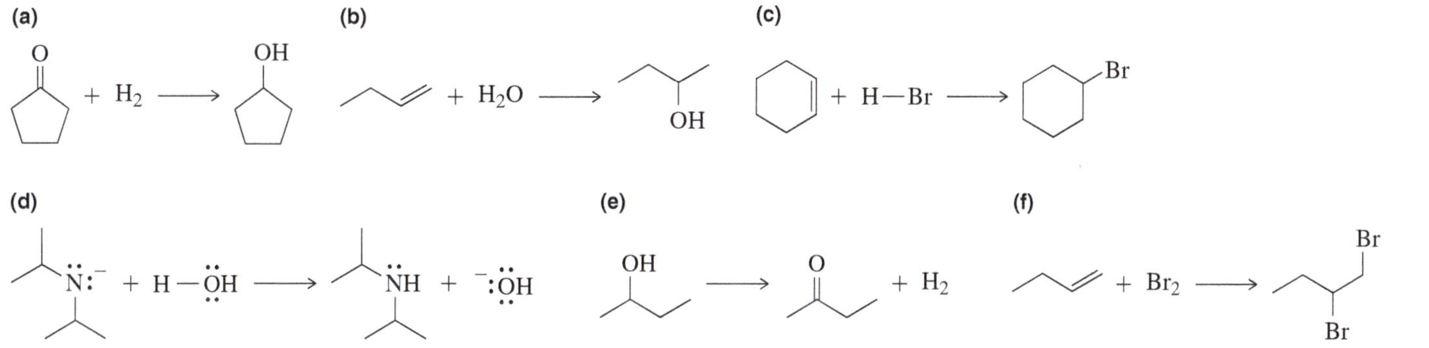

9.1.2 Alkene oxidation—halogenation

Vicinal means the two halides are on neighboring carbons.

Alkenes undergo **halogenation** to produce **vicinal** (1,2) dihalides. Using ethene as the alkene, **Figure 9.4(a)** shows that this is a redox process. Because organic chemists are only concerned with the fate of the organic molecule, it is referred to as an oxidation reaction. Either bromine or chlorine can be used in this stereospecific reaction to produce trans-dihalides, as shown in **Figure 9.4(b)**. Alkene halogenation is favored by enthalpy due to the formation of two new σ bonds (C—Br) at the expense of one σ bond (Br—Br) and one π bond (C—C). It is disfavored by entropy ($\Delta S < 0$), though, because two molecules react to form one.

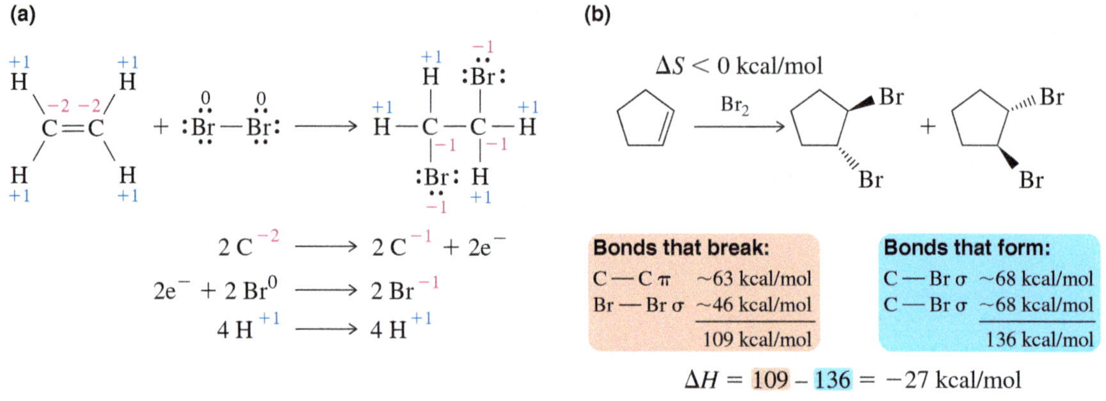

Figure 9.4 Bromination of alkenes. (a) This is a redox reaction because electrons are transferred from ethene to bromine. (b) Bromination is favored by enthalpy.

Br_2 is a very strong electrophile because it contains a long, weak, and polarizable Br—Br bond. [*Bonds between heteroatoms are typically weak.*] Mechanistically (see **Figure 9.5**), the alkene π electrons attack Br_2, pushing out the very stable bromide ion (Br^-) in the process. At the same time, bromine uses a lone pair of electrons to reattack the carbon that lost the π electrons to form a three-membered bromonium ion. [*This reaction should remind you of oxymercuration from Section 8.4.4.*] In the second step, a bromide ion attacks the carbon on the opposite face of the bromonium ion, completing formation of the trans-dibromide. Because the bromonium ion can form on the top or bottom of the alkene, a racemic mixture of products usually results.

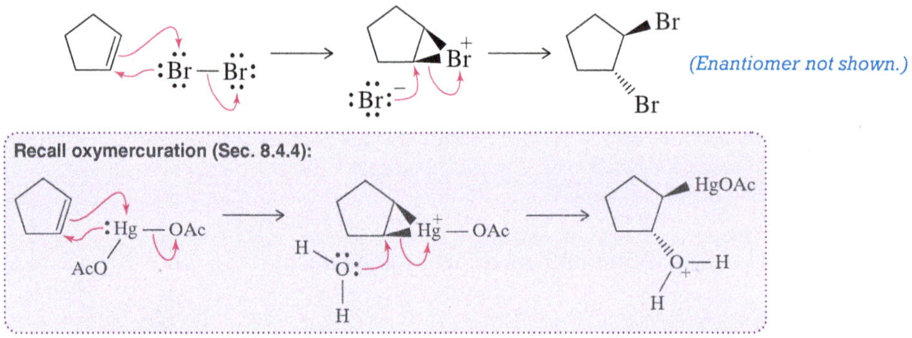

Figure 9.5 The mechanism of alkene halogenation is similar to that of oxymercuration. [*The reaction proceeds identically if Cl_2 is used.*]

Why is formation of the bromonium ion concerted in Figure 9.5? Moreover, why does the reaction preferentially produce the trans-dibromide? [*Think about both of these questions before reading further.*]

To understand why the reaction is concerted, **Figure 9.6** imagines what it would look like if it were not concerted. The alkene attacks the bromine in a slow first step that is followed by a very fast, intramolecular reaction to form the bromonium ion. As with

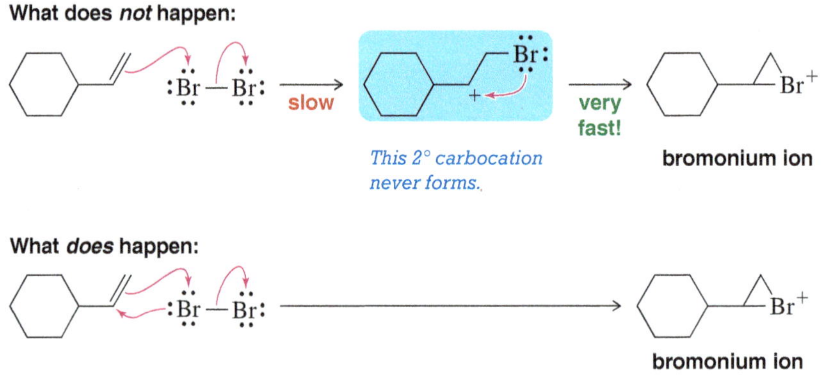

Figure 9.6 Bromonium formation is concerted. The intramolecular ring closing step is so fast that the 2° carbocation does not form.

oxymercuration and hydroboration (see Figure 8.58), this situation results in the reaction being concerted, with both C—Br σ bonds forming and the Br—Br σ bond breaking at the same time. As a result, a carbocation does *not* form and rearrangement is prevented. Like the other concerted reactions we have studied, bromonium ion formation is stereospecific.

SIDE CONVERSATION **Opening the three-membered ring** Figure 9.5 showed that the bromonium ring is opened from the opposite face to give the trans-dibromide. This will be a general phenomenon we see going forward. As such, **Figure 9.7** carefully analyzes this situation. Because bromine is a large atom, it is thought that the bromonium bridge is so large that attack on the same face is sterically impeded [**Figure 9.7(a)**]. A more subtle reason, illustrated in **Figure 9.7(b)**, recognizes the role that σ* plays in reactions of this type. [*Recall from Section 2.7 that formation of a σ bond is always accompanied by formation of a σ* antibonding orbital.*] When a nucleophile attacks carbon, causing another σ bond to break, the electrons are donated to σ*, located at a 180° angle from the σ bond. Thus, to bond with carbon, the bromide has to attack opposite the bromonium ion. The reaction coordinate diagram in **Figure 9.7(c)** shows that attack opposite the bromonium has a lower activation energy for both of these reasons. [*Backside attack plays a more prominent role in the substitution reactions of Chapter 12.*] ■

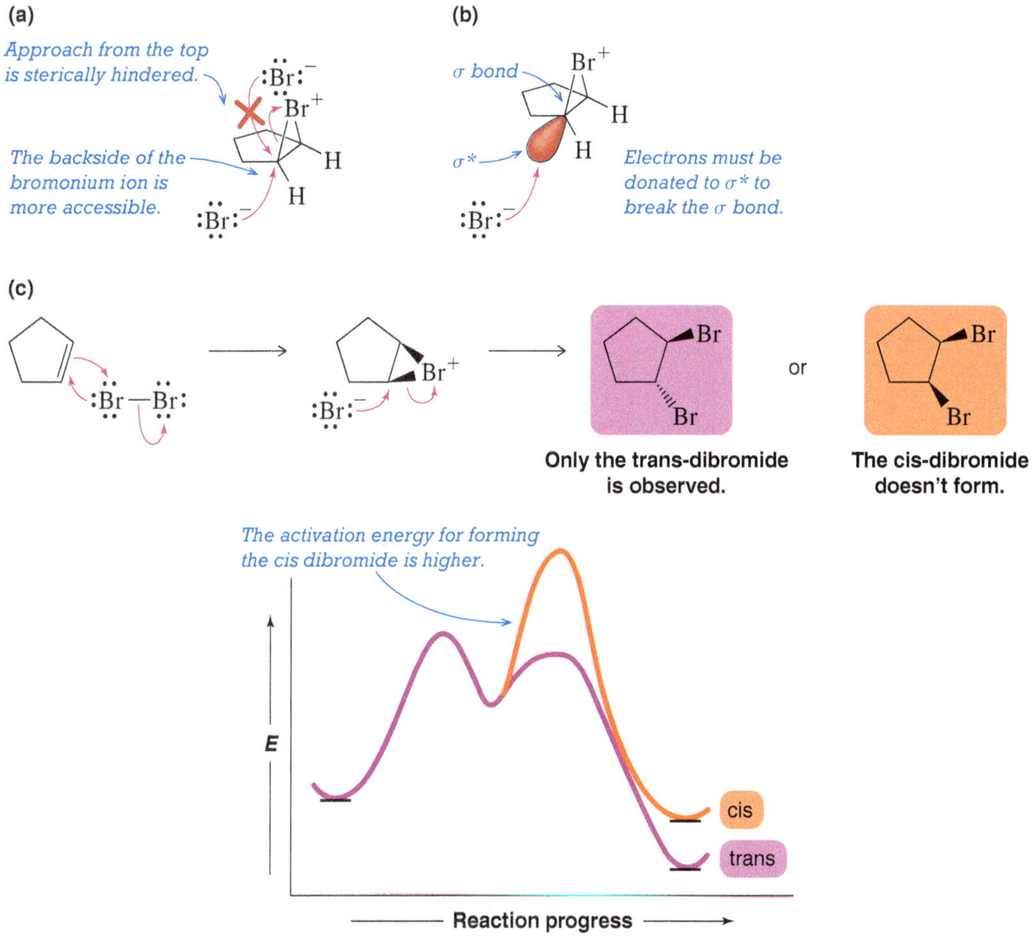

Figure 9.7 The bromonium opens to form the trans-dibromide. (a) Steric hindrance prevents approach from the same side. (b)The σ* orbital is located 180° from the breaking C—Br σ bond. (c) Forming the cis-dibromide, which would occur if Br^- attacked the bromonium ion from the same side, has a higher activation energy.

When the bromonium ion is symmetrical, attack occurs at either carbon equally, resulting in a mixture of enantiomers (**Figure 9.8**). This is true regardless of whether the bromonium ion formed on the top or bottom face. If the bromide attacks at C_1 of the bromonium ion, we get the enantiomer of the dibromide that results from the bromide attacking at C_2. In all cases, the trans-dibromide forms exclusively. That is to say, this reaction is stereospecific (all trans) but produces a racemic mixture of enantiomers.

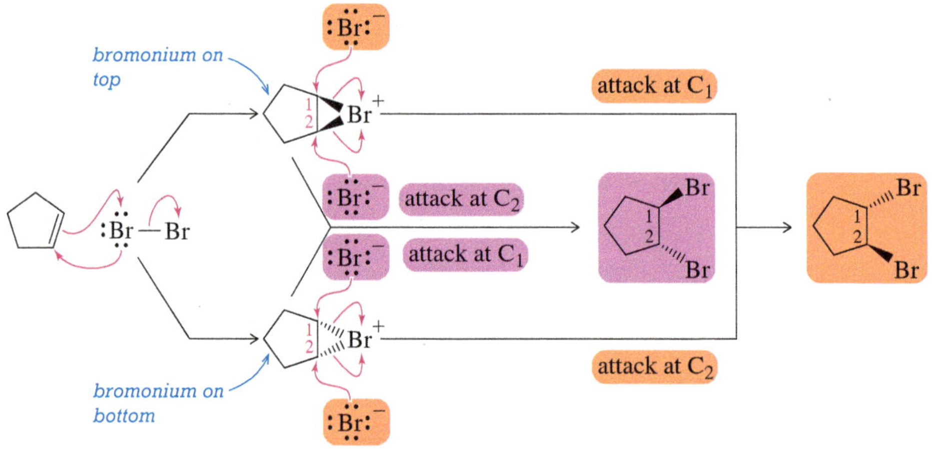

Figure 9.8 Attack of the bromine by the alkene can occur from the front or the back. When it is symmetrical, the bromonium ion can be opened at either carbon. As a result, two different pathways result in a racemic mixture.

In acyclic systems, the concerted formation of the bromonium ion and subsequent backside attack give rise to the stereospecificity in the vicinal dibromide. The *E* and *Z* isomers of hept-3-ene in **Figure 9.9** each give rise to a different diastereomer. For the sake of simplicity, only formation of the bromonium on the top face of the alkene has been shown. In the second step, the bromide attacks at either carbon of the bromonium ion to generate the racemic mixture of diastereomers.

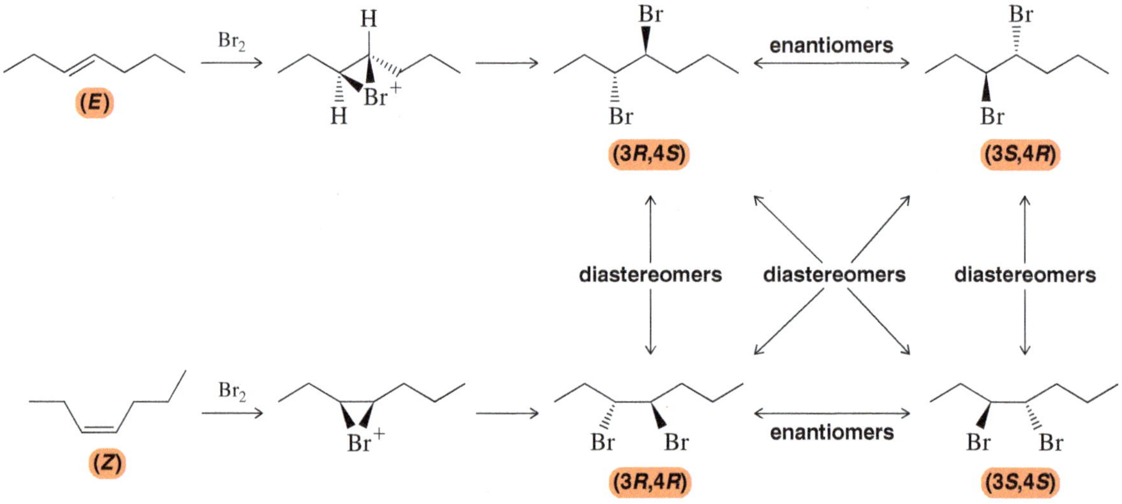

Figure 9.9 Bromination is stereospecific because two alkene stereoisomers give different diastereomeric products. [*Bromonium ion formation on bottom occurs also, but is not shown here.*]

9.1.2 SUMMARY

- The electrophilic addition of Br_2 or Cl_2 results in the stereospecific preparation of vicinal dihalides. Following nucleophilic attack of the alkene on the halogen and formation of the halonium ion, nucleophilic attack of a bromide ion opens the three-membered ring from the opposite side.

Assessment 9.6 Predict the product(s) of each of the following reactions, making sure to indicate the relative stereochemical outcome. Indicate any racemic mixtures by drawing both enantiomers.

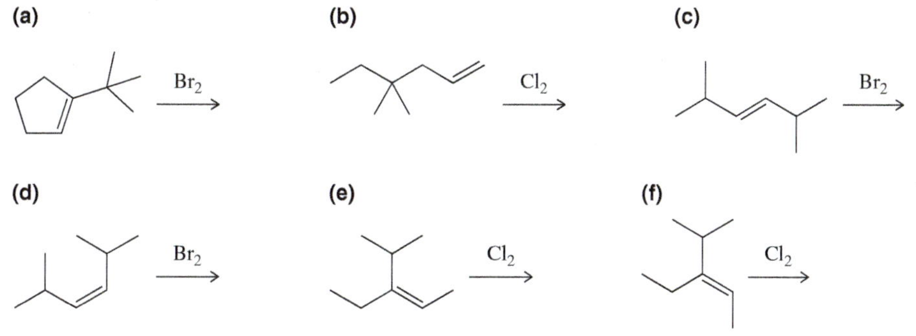

Assessment 9.7 Provide an arrow-pushing mechanism that rationalizes the stereospecific formation of each dihalide in Assessment 9.6.

Assessment 9.8 In which of the two steps in the alkene halogenation mechanism does a redox reaction occur?

Assessment 9.9 In contrast to the addition of Br_2, the addition of HBr [*Section 8.3*] is *not* stereospecific. Why?

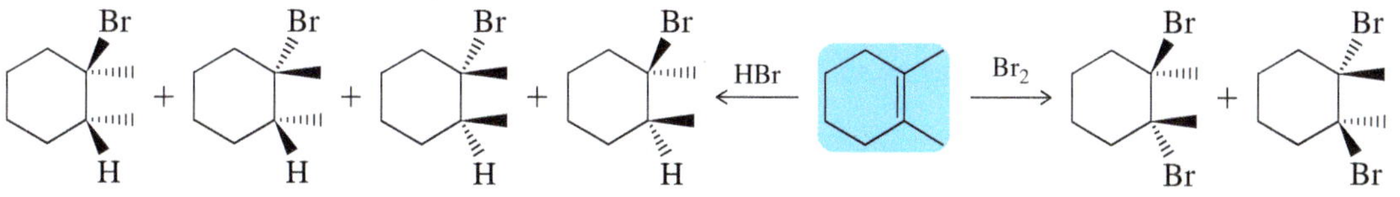

9.1.3 Alkene oxidation—halohydrin formation

If the halogenation reaction from Section 9.1.2 is attempted with water as the solvent, the **halohydrin** in Figure 9.10 is formed. Mechanistically similar to the bromination we studied in Section 9.1.2 [*and oxymercuration from Section 8.4.4*], the only difference is in the nucleophile that opens the bromonium ion. Instead of a bromide ion attacking, water opens the bromonium ion [*in the same way that water opened the mercurinium ion*]. Because water brings along its proton when attacking, a final acid–base reaction occurs to produce the neutral alcohol. [*Br^- is present in solution . . . why doesn't it attack? Think about this before continuing. Hint: We talked about the relative reactivity of Br^- and H_2O in Chapter 4.*]

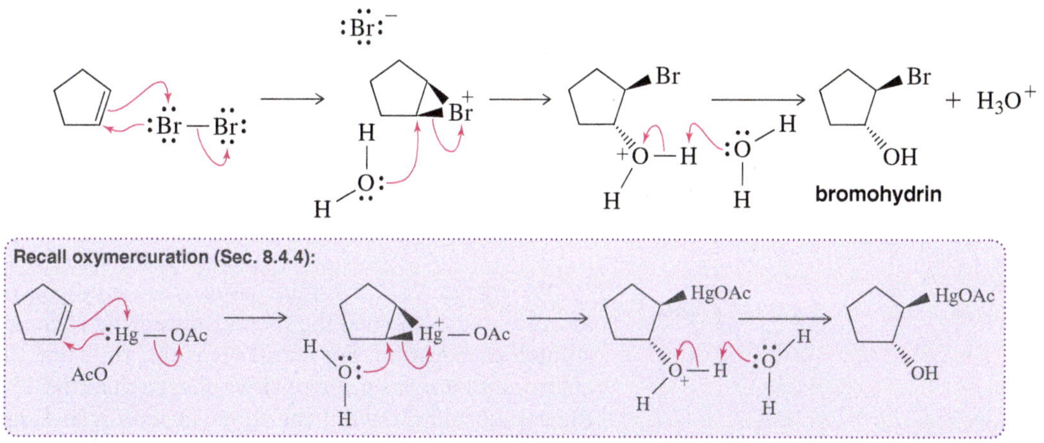

Figure 9.10 Halohydrin formation occurs by the same mechanism as oxymercuration. [*The reaction would proceed identically if Br_2 were replaced with Cl_2.*]

Because water is the solvent, there is far more water present than bromide. Thus, we expect more attack by water on a purely statistical basis. Second, water is a better base than bromide. We know this by comparing the pK_a values of their respective conjugate acids. Hydrobromic acid has a pK_a of −7, while hydronium has a pK_a of −1.7 [**Figure 9.11(a)**]. [*Recall from Section 4.2.3 that the more acidic acid gives rise to the more stable conjugate base.*] Under these conditions, water is a less stable, more reactive base than bromide, so it attacks the bromonium ion more quickly. This concept is demonstrated on the reaction coordinate diagram in **Figure 9.11(b)**, where the reaction of water and the bromonium ion starts at a higher energy. The net effect is that the activation energy for attack by water is lower.

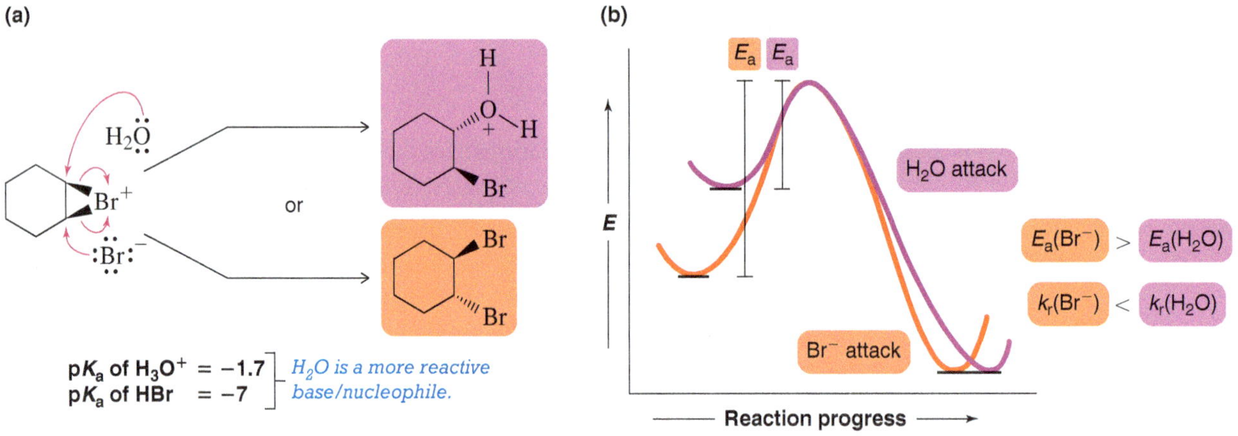

Figure 9.11 Water opens the bromonium ion preferentially. (a) Based on pK_a values, water is the stronger base. (b) Water starts higher on the reaction coordinate diagram, making attack by water faster.

The formation of halohydrins allows us to explore issues of regioselectivity that were not important in dihalide formation. **Figure 9.12** examines the reaction of 2-methylhex-1-ene with Cl_2 and H_2O. After the chloronium ion forms, water can open it at either the 1° or 3° carbon. [*Which carbon do you think water attacks? Can you point to a previously studied reaction that helps us definitively answer this question?*]

Figure 9.12 Water can attack at either of two carbons of the chloronium ion. [*Which do you expect it to attack?*]

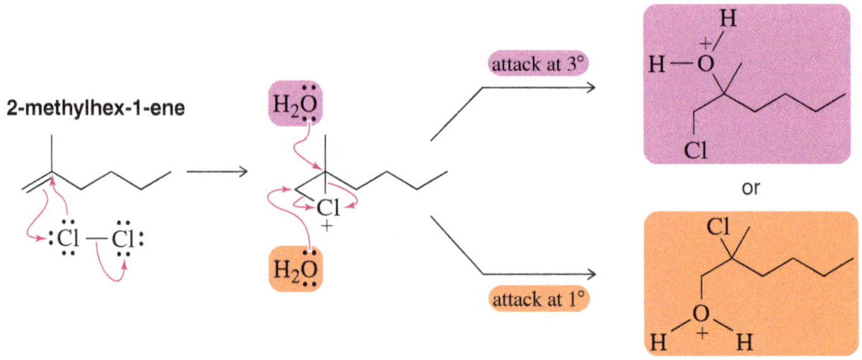

We encountered a similar issue when studying oxymercuration [*Section 8.4.4*]. In that instance, water opened the mercurinium ring at the position corresponding to the most substituted carbocation. For similar reasons, the same thing happens here. The 3° carbon of the chloronium ion has more positive charge than the 1° carbon. Think of the chloronium ion as the equilibrium between the three structures in **Figure 9.13(a)**. [*It's not really.*] The chloronium ion is the most stable of the three species in this hypothetical equilibrium; the 1° R^+ is

By thinking of it as an equilibrium, we learn that there is more positive charge on the more substituted carbon.

the least stable. The chlorine bridge shifts *away* from the more stable 3° carbocation and *toward* the less stable 1° carbocation. As a result, the unsymmetrical chloronium ion has a longer bond between chlorine and the 3° carbon, and the 3° carbon has a greater partial positive charge. Longer bonds are weaker bonds, so in spite of there being greater steric hindrance at that site, water attacks the 3° carbon. The difference in activation energy that makes attack at the 3° carbon faster is shown in **Figure 9.13(b)**.

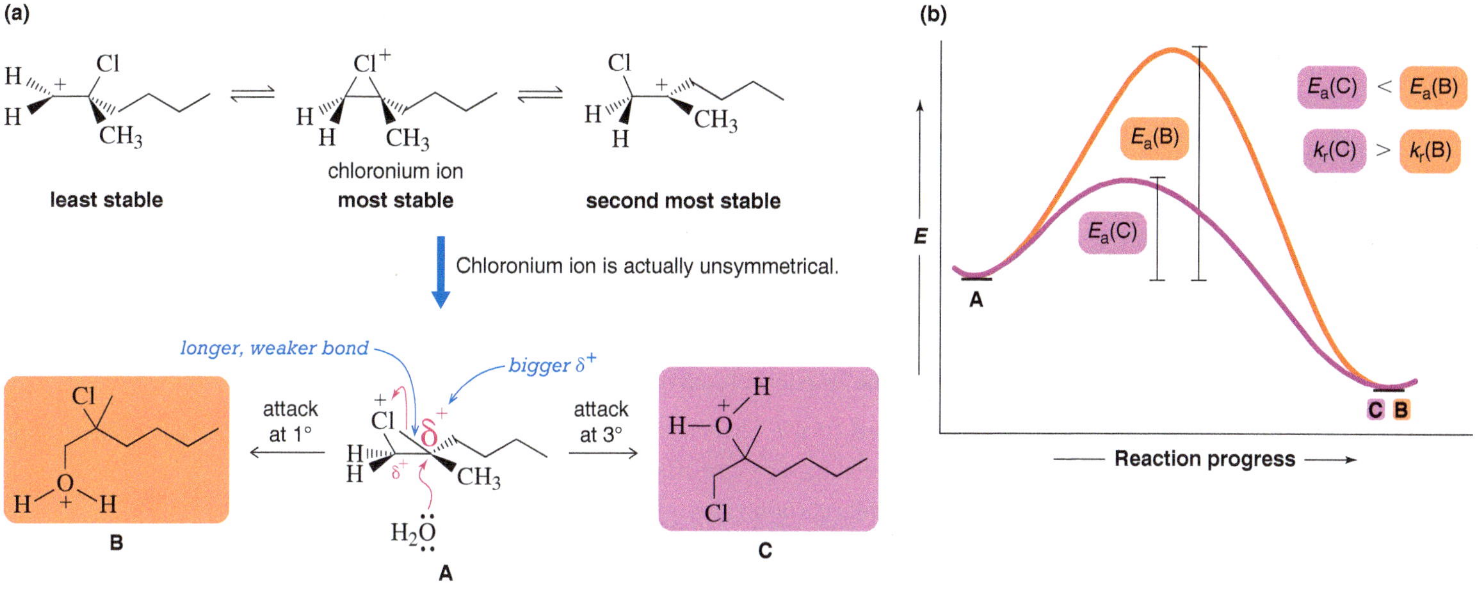

Figure 9.13 Chloronium [*and bromonium*] ions are opened at the more substituted carbon. (a) There is a greater charge on the 3° carbon. (b) Attack at the 3° carbon has a lower activation energy.

Assessment 9.10 Predict the product(s) of each of the following reactions, making sure to indicate the relative stereochemical outcome. Indicate any racemic mixtures by drawing both enantiomers.

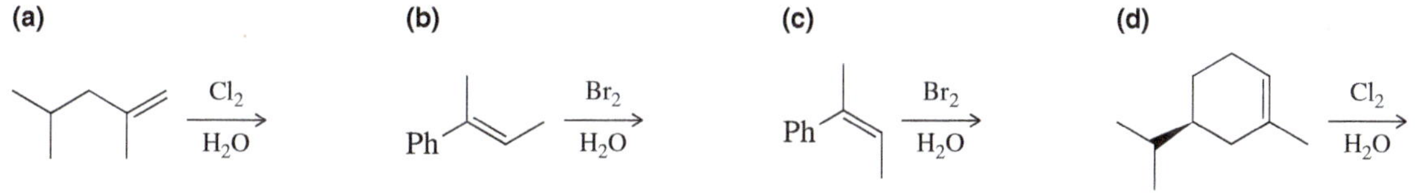

Assessment 9.11 Provide arrow-pushing mechanisms for Assessments 9.10(b) and 9.10(c) that rationalize the regioselective and stereospecific formation of each halohydrin.

Assessment 9.12 Explain why water attacks the carbon of the bromonium ion as opposed to the bromonium ion itself in the second step of halohydrin formation.

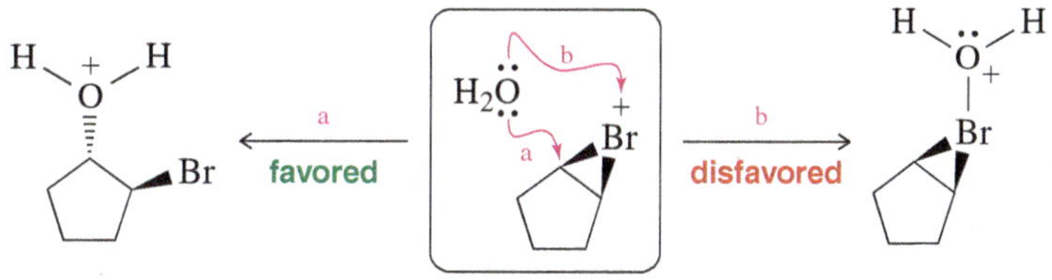

Assessment 9.13 Suggest an alkene that could be used to make each of the following halohydrins.

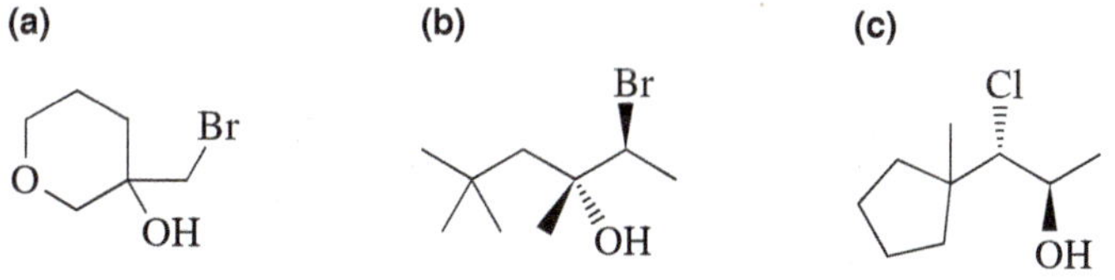

9.1.4 Alkene oxidation—epoxidation

Using what we've learned, let's see if we can "invent" a reaction/reagent allowing us to synthesize an **epoxide** (**oxirane**) starting with an alkene (**Figure 9.14**).

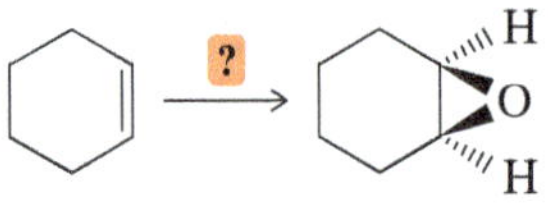

Figure 9.14 Section 9.1.4 shows how we can invent this reaction.

First, you *can* do this, even if you're new to organic chemistry. Second, don't get caught up in trying to identify a particular reagent; we don't yet know what's possible. The goal is to come up with a hypothetical "reagent," which, if it existed, would do the reaction we desire.

Thinking Like a Chemist

Can the reaction happen? A C—C π bond breaks and two C—O σ bonds form, while a strained three-membered ring is made (**Figure 9.15**). Even with the ring strain, we should be able to make the molecule under the right conditions.

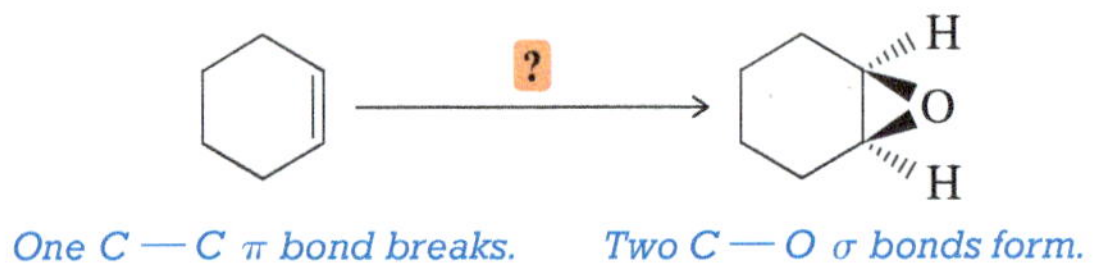

Figure 9.15 The reaction should be favorable.

How do bonds form? There are three ways to form one of the C—O bonds (**Figure 9.16**):

1. a nucleophilic oxygen can attack an electrophilic carbon,
2. a nucleophilic carbon can attack an electrophilic oxygen, or
3. a carbon radical can combine with an oxygen radical.

[*This analysis would be the same for any reaction we were trying to invent.*]

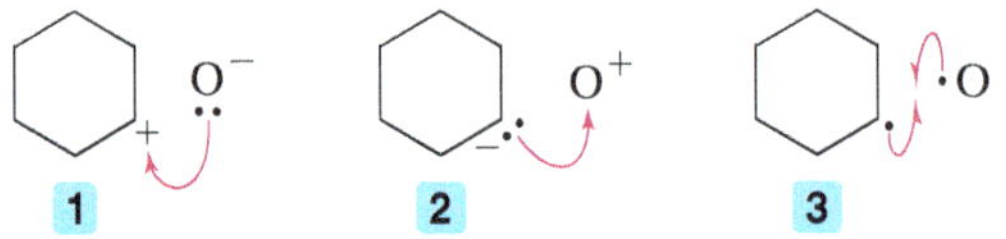

Figure 9.16 Three ways to form the first C—O bond.

Which is most likely here? We already know a lot about alkene reactivity. They are electron-rich, using the π electrons to attack electrophiles. As such, whatever oxygen species we use, it must be electrophilic. A viable first step is shown in **Figure 9.17**. [*Where have we seen a reaction that looks like this before?*]

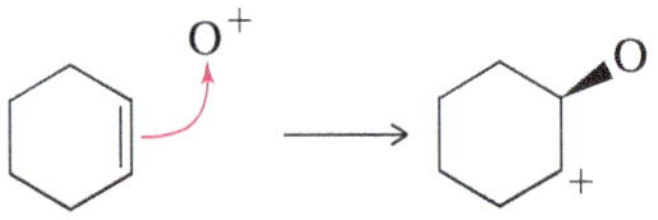

Figure 9.17 Alkenes are nucleophilic. [*This is a hypothetical reaction.*]

What next? If an alkene attacked an electrophilic oxygen to form the first C—O bond, a carbon now has a positive charge. For oxygen to attack and form the second bond, it needs to be nucleophilic. [*Have we seen an atom react intramolecularly to form a three-membered ring? Yes!*] Like bromine (a lone pair) and mercury (*d* electrons), oxygen has lone pairs that can be donated to the empty *p* orbital (**Figure 9.18**).

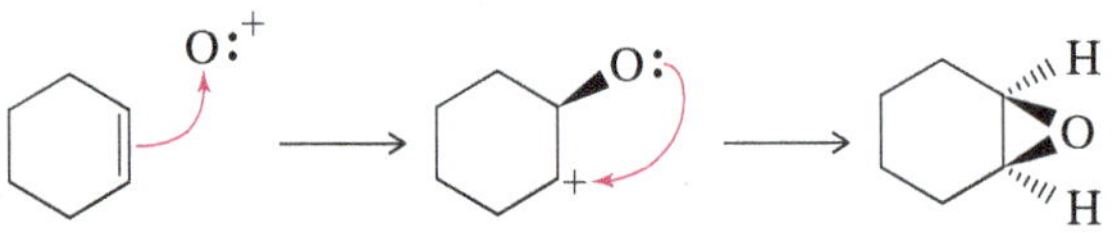

Figure 9.18 Oxygen could reattack if it had a lone pair of electrons. [*This is also hypothetical.*]

There are two lone pairs of electrons on the epoxide oxygen. Thus, in addition to the lone pair needed to reattack the carbocation, our oxygen reagent needs a total of three lone pairs. With this slight change, we aren't using O^+ but, instead, elemental oxygen. Elemental oxygen is so electron-deficient and unstable that it doesn't exist in nature as O; it is diatomic instead. And yet, if we could generate elemental oxygen, it would react with an alkene exactly as shown (**Figure 9.19**). [*We can do better. Think about how the bromonium ion formed.*]

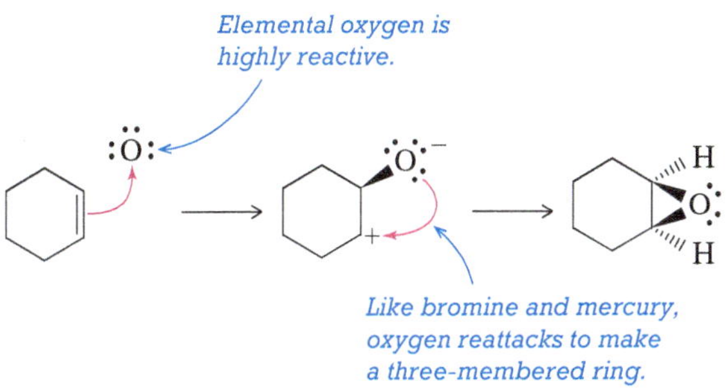

Figure 9.19 Elemental oxygen should react with an alkene to form an epoxide. [*Still hypothetical . . .*]

When the bromonium ion formed in Figure 9.5, did the alkene attack Br^+ or elemental bromine (Br) in the first step? No, it attacked Br_2 instead, thus kicking out a good leaving group (Br^-) in the process. What would our reaction look like if the oxygen was attached to a good leaving group? For one thing, it would be a more reasonable molecule, because the oxygen in question would have a full octet. If one of the lone pairs were instead bonded to hydrogen, we could imagine the mechanism shown in **Figure 9.20**.

To make an epoxide, then, we need an electrophilic oxygen bound to a good leaving group that can also deprotonate the resulting oxonium ion. Let's see how we actually accomplish this. ■

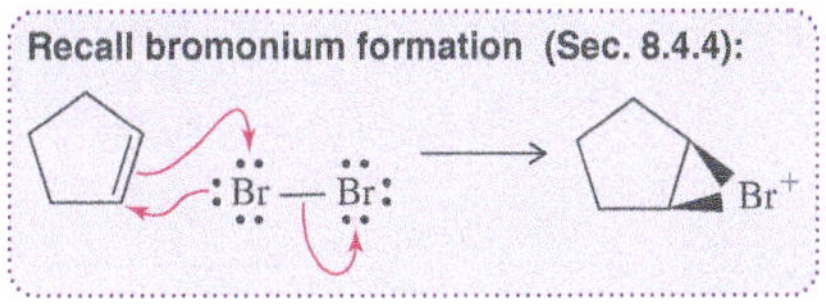

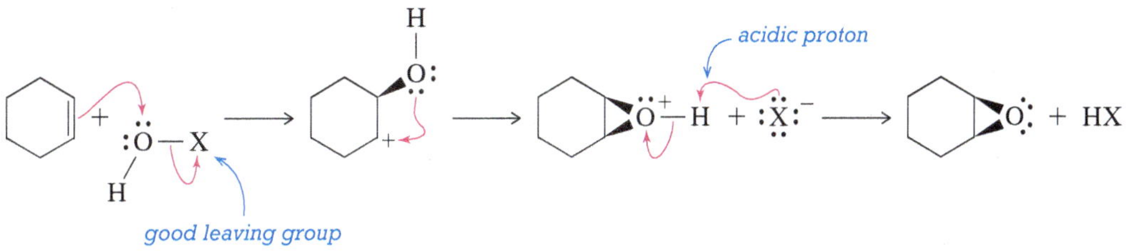

Figure 9.20 Our hypothetical formation of an epoxide could proceed by the bromonium mechanism. [*It's starting to look more real now.*]

Peroxyacids [like *m*-chloroperbenzoic acid (*m*CPBA)] epoxidize alkenes via a mechanism similar to alkene bromination (**Figure 9.21**). The reaction is entropically neutral (2 molecules → 2 molecules; $\Delta S \approx 0$), and it forms two new C—O σ bonds at the expense of a C—C π bond and the very weak O—O σ bond. [*Recall from Section 8.3.4 that the O—O bond in hydrogen peroxide breaks to initiate the radical addition of HBr to alkenes.*] Thus, even while forming a strained three-membered ring, it is a very favorable reaction. That it's an oxidation reaction is confirmed by the change in oxidation state of the alkene carbons from −1 to 0 while two oxygens in *m*CPBA are reduced from −1 to −2. [*When identifying redox, the unchanged parts of the molecule can be ignored.*]

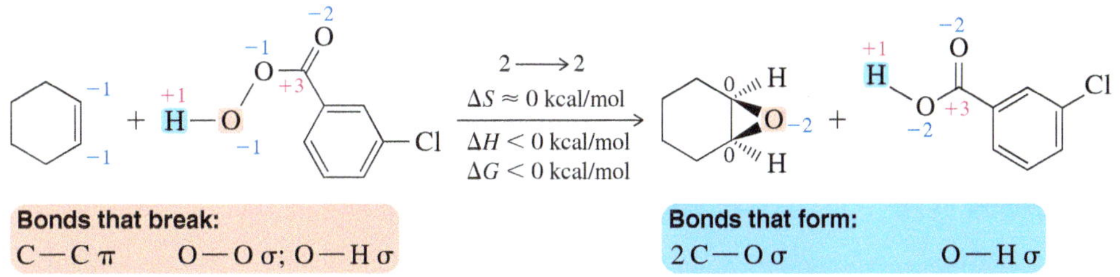

Figure 9.21 Epoxidation with *m*CPBA is an enthalpically favored alkene oxidation reaction.

The epoxide mechanism shown in **Figure 9.22(a)** is similar to that postulated while *Thinking Like a Chemist*, except that it is concerted, requiring five mechanism arrows in one step. The stepwise mechanism in **Figure 9.22(b)** makes this process a bit clearer and helps us understand why the reaction is concerted. The C—C π bond attacks the terminal oxygen, kicking out the resonance-stabilized carboxylate anion in what would be a relatively slow step. A very fast intramolecular ring closure step [*see Section 9.1.2*] is followed by a fast acid–base reaction. [*Acid–base reactions are always fast.*] Consistent with a recurring theme, the formation of the three-membered ring and subsequent deprotonation are so fast that this is actually a concerted mechanism. That is, all three steps occur at the same time.

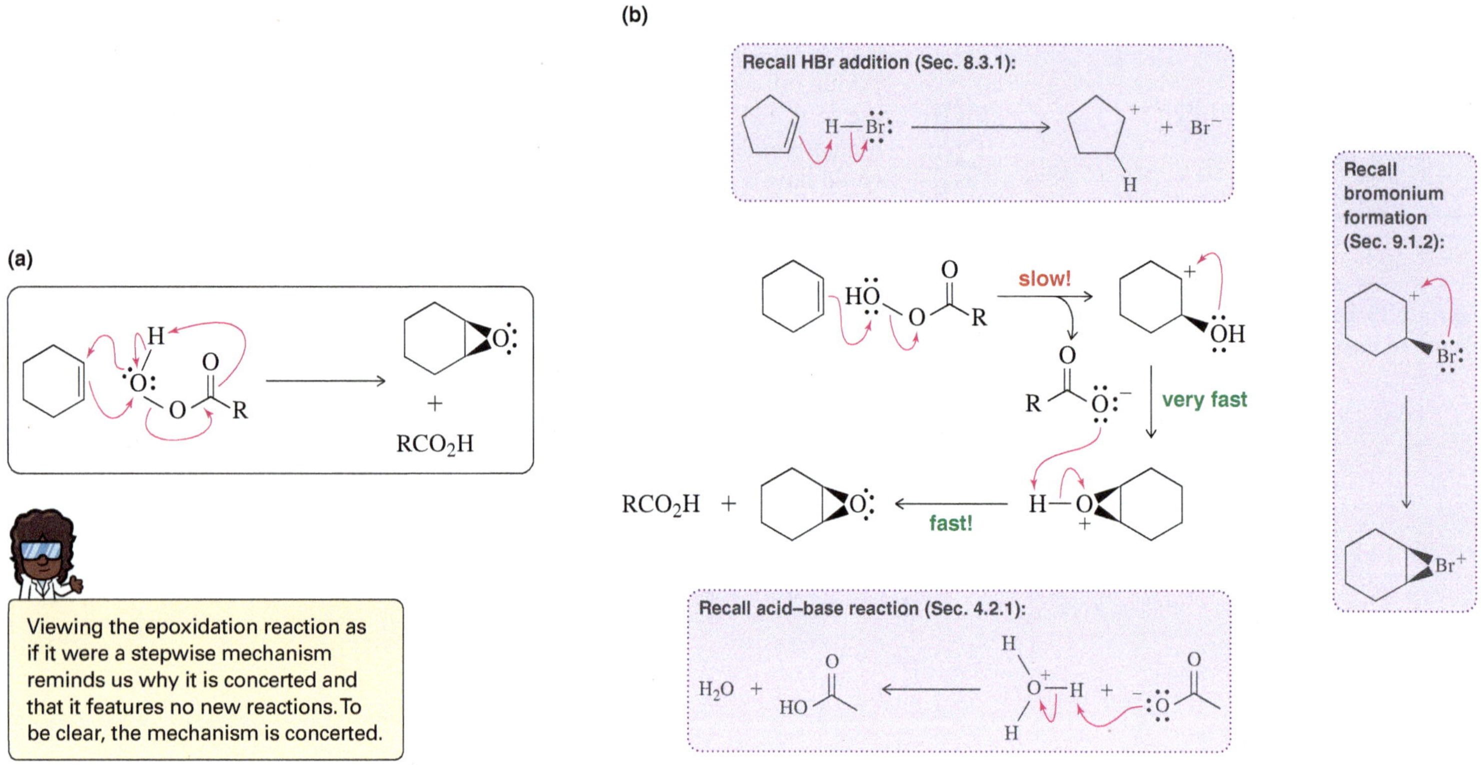

Figure 9.22 Mechanism of epoxide formation. (a) The actual mechanism is concerted. (b) The mechanism is shown stepwise.

As a concerted reaction, epoxide formation is stereospecific, because formation of the epoxide is complete before there is a chance to rotate around the C—C σ bond. As a result, *E* and *Z* isomers of 3-methylhept-2-ene result in different diastereomers (**Figure 9.23**).

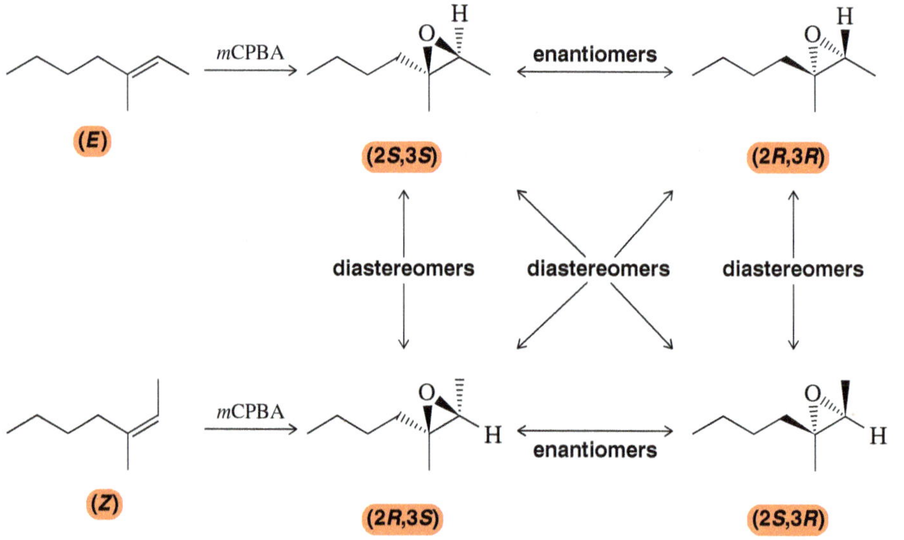

Figure 9.23 Epoxidation is stereospecific because two alkene stereoisomers give different diastereomeric products.

9.1.4 SUMMARY

- In what is essentially the addition of elemental oxygen, alkenes can be oxidized with peroxyacids to give epoxides (oxiranes) in a concerted, stereospecific reaction.

9-A CHEMISTRY IN CONTEXT: Green Chemistry

Activated peroxide for epoxidation

Epoxidation of alkenes with *m*CPBA and other peracids suffers from poor atom economy. Efforts to develop a greener epoxidation have focused on the much smaller and more easily generated molecule hydrogen peroxide (**Figure 9.24**). [*m*CPBA *is made from hydrogen peroxide by a reaction we discuss in Chapter 19.*] As an added benefit, the by-product of hydrogen peroxide as an epoxidizing agent is the environmentally benign molecule water.

Although we can imagine a mechanism where hydrogen peroxide could do the epoxidation on its own, hydroxide is not a good enough leaving group to facilitate attack by the alkene (**Figure 9.25**). Complex formation with iron(III) chloride weakens the O—O bond while making a better leaving group. In this way, hydrogen peroxide is "activated," allowing electrophilic addition to proceed. [*The role of* $FeCl_3$ *is similar to the role played by the carbonyl group in* mCPBA.]

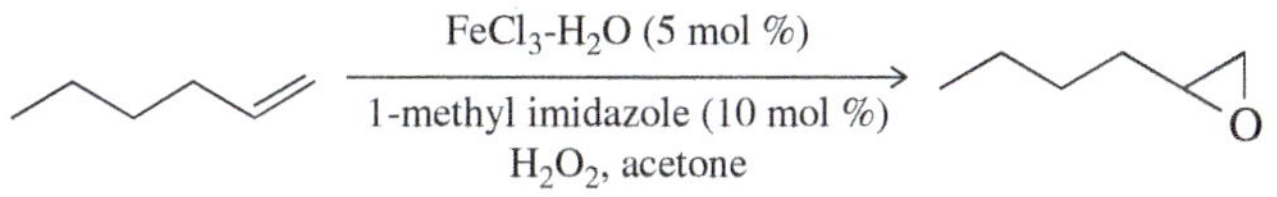

Figure 9.24 **A more sustainable epoxidation using a catalyst.**

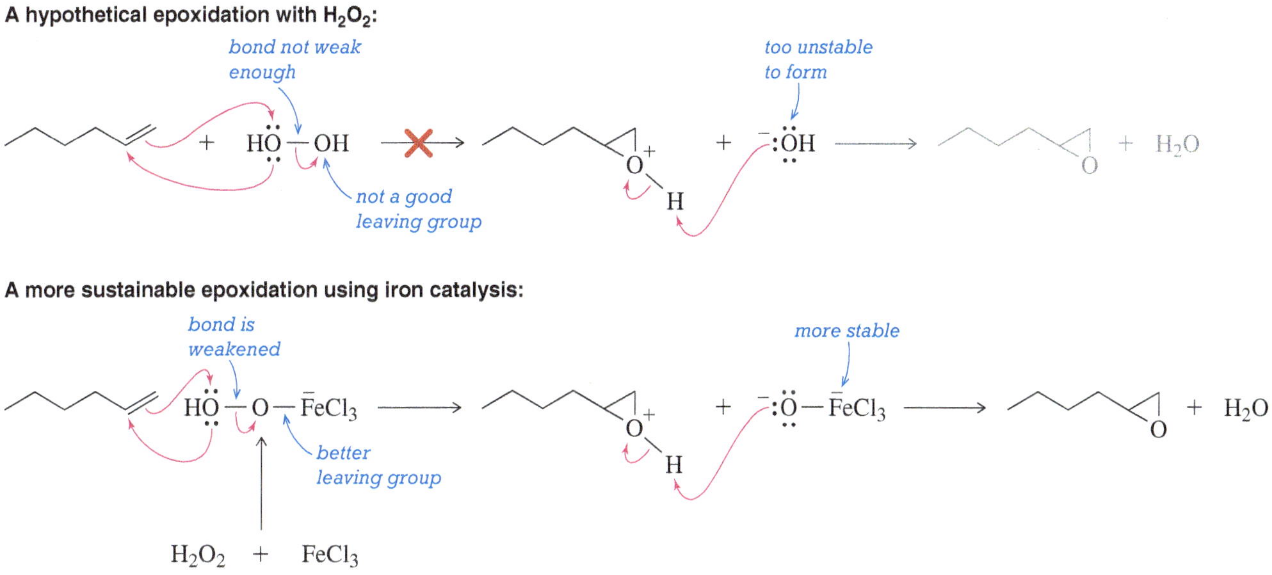

Figure 9.25 **Activated hydrogen peroxide can be used as part of a green epoxidation reaction.**

Assessment 9.14 How does the carbonyl in *m*CPBA weaken the O—O σ bond (i.e., make a better leaving group)?

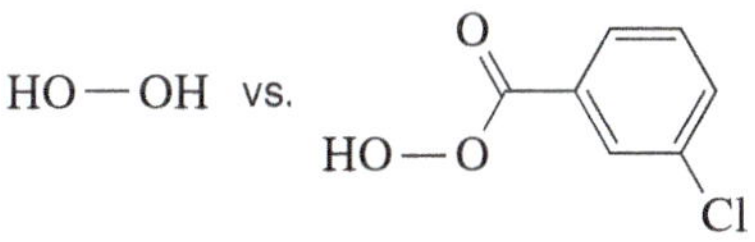

Assessment 9.15 Calculate the atom economy of the reaction in Figure 9.24. [*Catalysts are not included in the atom economy calculation.*]

Assessment 9.16 Predict the product(s) when each of the following are reacted with *m*CPBA, making sure to indicate the relative stereochemical outcome. Indicate any racemic mixtures by drawing both enantiomers.

(a)

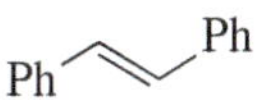

(b)

(c)

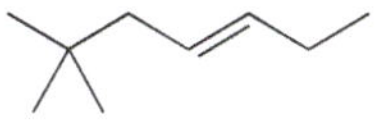

(d)

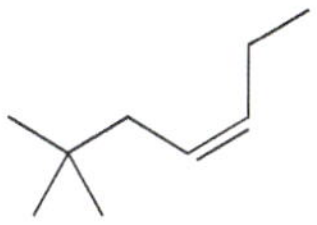

Assessment 9.17 Evidence for the concertedness of epoxide formation comes from the stereospecificity of the reaction. If step 2 of a hypothetical stepwise mechanism were slow enough to make the reaction non-concerted, how would the product distribution change?

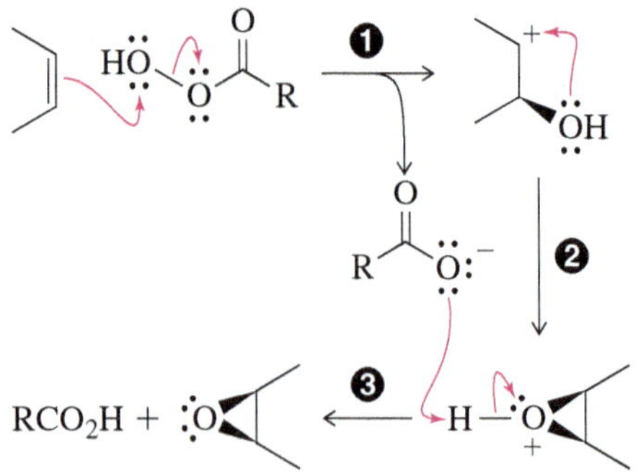

Assessment 9.18 When producing a chiral molecule, epoxide formation still results in a mixture of enantiomers, despite its stereospecificity. (a) Explain why this reaction produces a racemic mixture. (b) How is it that a reaction can be stereospecific while still producing two enantiomers?

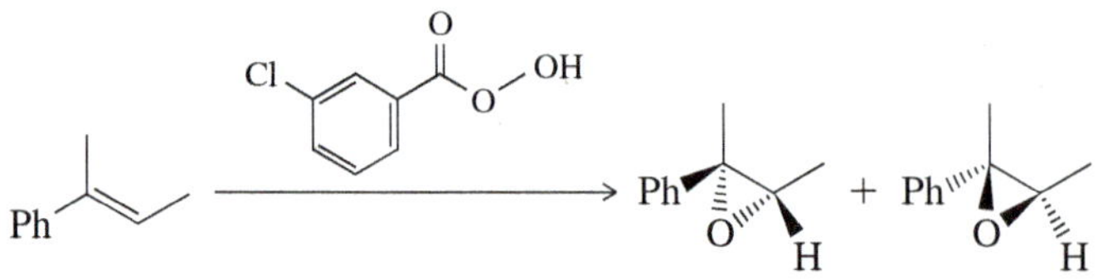

9.1.5 Alkene oxidation—dihydroxylation

Dihydroxylation, shown in Figure 9.26, converts alkenes to cis-diols using similar alkene addition chemistry. Historically, this reaction was done using a stoichiometric amount of osmium tetroxide to make an osmate ester in the first step. Cleavage of the osmate ester using $NaHSO_3$ results in diol formation. While efficient at producing the diol product, osmium tetroxide is a highly poisonous substance, made worse by the fact that it is a volatile solid. [*A greener reagent is needed.*] Though this reaction proceeds by a mechanism we haven't studied yet, there are clues to the mechanism in the product. [*What does the relative stereochemistry of the hydroxyl groups suggest about the mechanism?*]

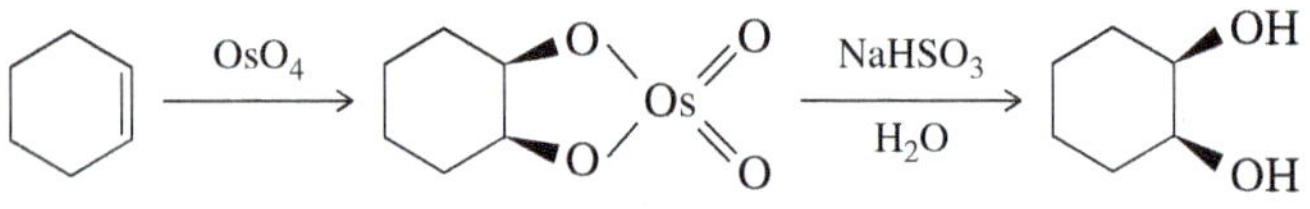

Figure 9.26 Syn dihydroxylation with osmium tetroxide.

The hydroxyl groups added to the same face of the alkene. We have seen "same-face" addition in the following *concerted* reactions: mercurinium ion formation, hydroboration, bromonium formation, and epoxide formation. Thus, the formation of a cis-diol is a strong clue that dihydroxylation is a concerted reaction. Indeed, the mechanism in Figure 9.27 shows a concerted reaction in which three pairs of electrons move at the same time in a cyclic manner. [*This 6-electron, cyclic transition state is quite common and is studied further in Chapter 22.*] In the second step, the osmate ester breaks down via an inorganic chemistry mechanism that is poorly understood and not important to this discussion.

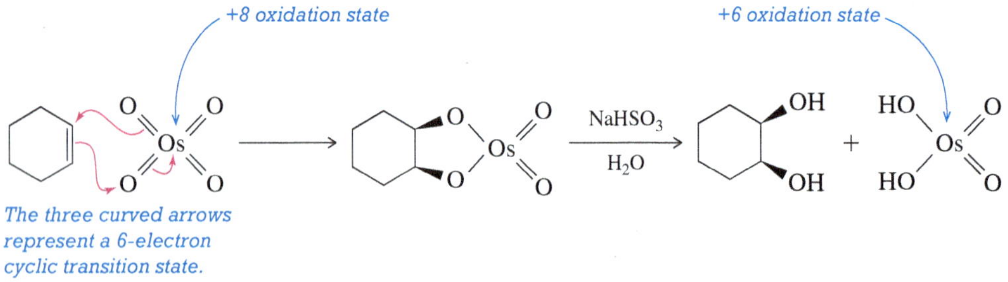

Figure 9.27 Dihydroxylation proceeds through a concerted 6-electron transition state.

The stepwise mechanism shown in Figure 9.28 relates the mechanism to all of the alkene reactions in Chapters 8 and 9. The key to understanding the mechanism is to recognize that osmium is a transition metal. Transition metals can easily change oxidation states. In this reaction, for example, osmium is reduced from +8 to +6.

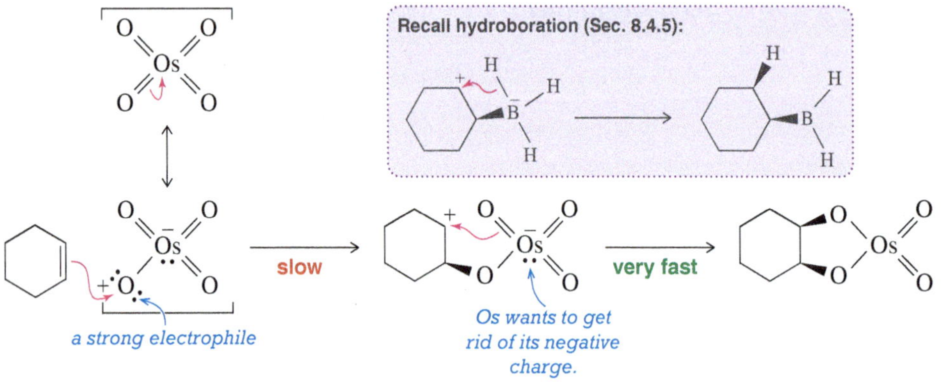

Figure 9.28 Imagining the mechanism as stepwise helps relate it to chemistry we have seen previously. [*The actual mechanism is concerted.*]

An oxidation number of +8 suggests that osmium is highly electron-deficient and makes it somewhat conceivable to draw a resonance structure of the Os—O double bond where one oxygen has a positive charge and osmium is negative. [*Transition metals do weird things when viewed through the lens of organic chemistry and our arrow-pushing formalism.*] Using this resonance structure, the alkene π bond attacks the electrophilic oxygen in what would be a slow *inter*molecular step. At this point, the negatively charged osmium releases the Os—O π electrons to attack the newly formed carbocation in what would be a very fast *intra*molecular step. Once again, a slow first step followed by a very fast second step explains why this reaction is concerted.

Like the concerted reactions studied thus far, dihydroxylation is stereospecific. **Figure 9.29** shows that (*E*)- and (*Z*)-pent-2-ene are oxidized to give diastereomeric diols.

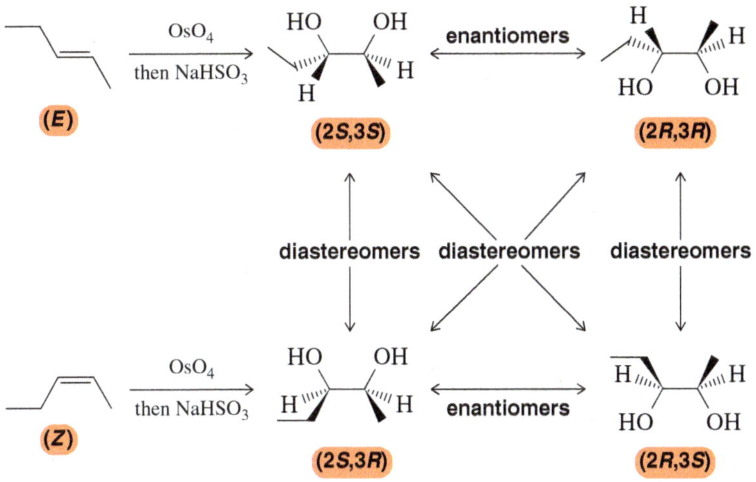

Figure 9.29 Dihydroxylation is stereospecific because two alkene stereoisomers give different diastereomeric products.

9-B CHEMISTRY IN CONTEXT: Green Chemistry

Greener dihydroxylation

The dihydroxylation reaction has been made greener by the development of a system that uses a sub-stoichiometric amount of osmium tetroxide. [*Less compound used = less exposure.*] In order to do this, a **stoichiometric** amount of a co-oxidant like hydrogen peroxide [*here it is again*] is used to regenerate the OsO_4, while producing the diol at the same time (**Figure 9.30**). Another common oxidizing agent used to regenerate OsO_4 is *N*-methylmorpholine-*N*-oxide (NMO).

A catalytic cycle is shown in **Figure 9.31** using NMO to regenerate the OsO_4. The use of a full equivalent of NMO reduces the atom economy of the reaction because the NMM by-product is usually discarded as waste. [*This represents an advantage of using H_2O_2 as co-oxidant because the by-product is simply water.*]

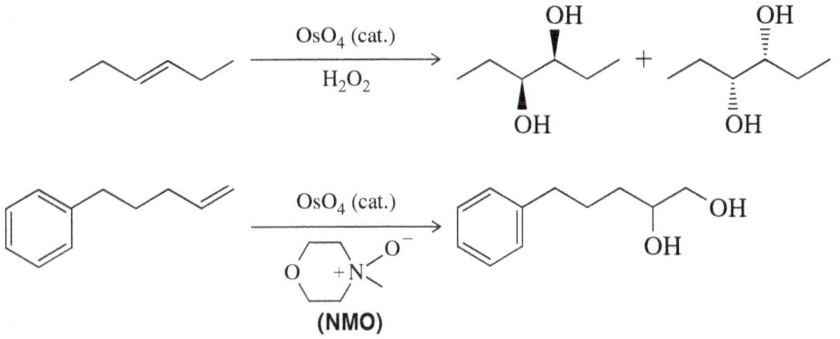

Figure 9.30 Using a co-oxidant allows the use of less OsO_4.

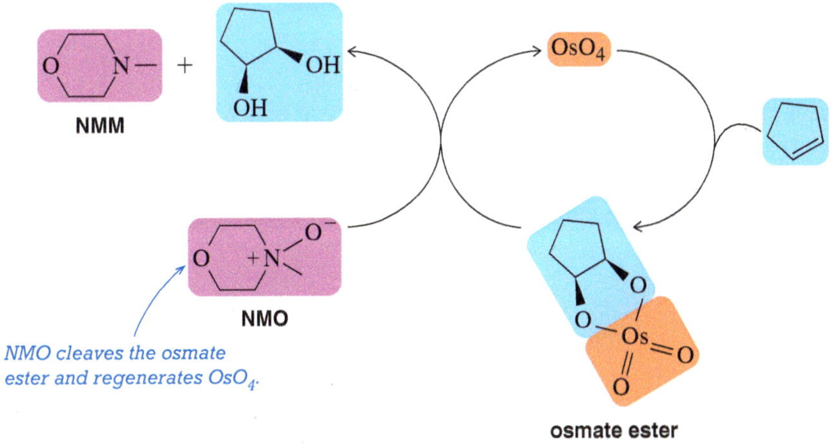

Figure 9.31 A possible catalytic cycle showing how NMO regenerates OsO_4.

Assessment 9.19 How might you make the catalytic cycle in Figure 9.31 more sustainable while still using NMO as the co-oxidant?

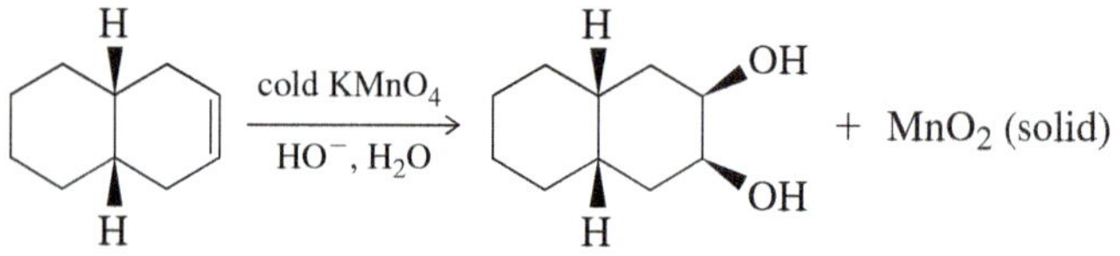

Figure 9.32 Dihydroxylation with $KMnO_4$ avoids the use of toxic OsO_4.

Dihydroxylation can also be accomplished using cold potassium permanganate ($KMnO_4$), which is less toxic than OsO_4 (**Figure 9.32**). Based on the structural similarities between OsO_4 and MnO_4^-, it should not be surprising that these compounds react similarly.

$KMnO_4$ is also used in qualitative tests for alkenes. The deep purple color of $KMnO_4$ persists when an alkane is added, indicating that no reaction has occurred [**Figure 9.33(a)**]. In the presence of an alkene, dihydroxylation occurs, resulting in the formation of a cloudy brown solution [**Figures 9.33(b)** and **9.33(c)**].

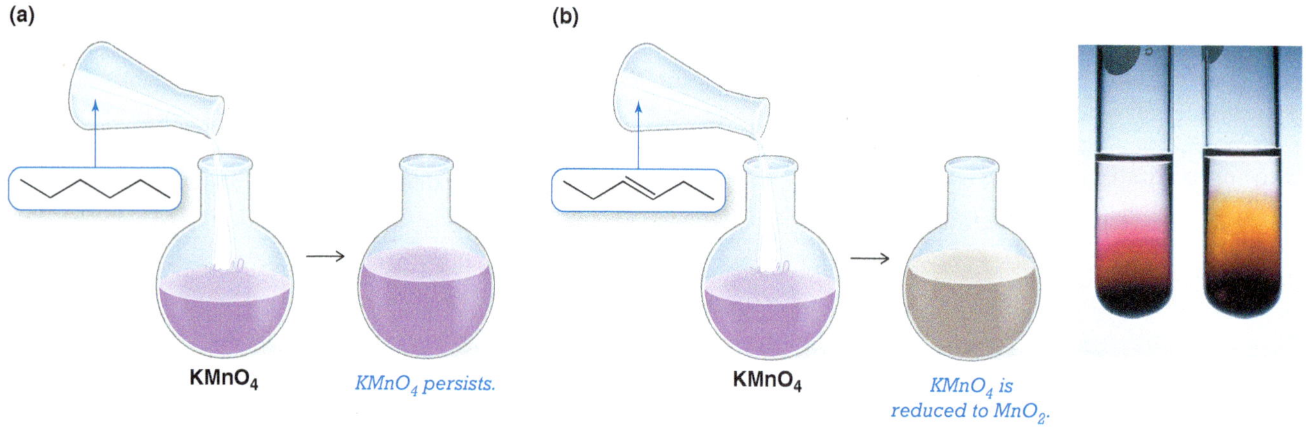

Figure 9.33 $KMnO_4$ is used in the lab to test for the presence of an alkene. (a) The purple color of $KMnO_4$ persists if something other than an alkene is added. (b) The presence of an alkene is indicated by a color change.

9.1.5 SUMMARY

- Alkenes are oxidized to give vicinal diols by reaction with OsO_4 in a concerted reaction that delivers both hydroxyl groups to the same face of the molecule.

Assessment 9.20 Predict the product(s) of each of the following reactions, making sure to indicate the relative stereochemical outcome. Indicate any racemic mixtures by drawing both enantiomers.

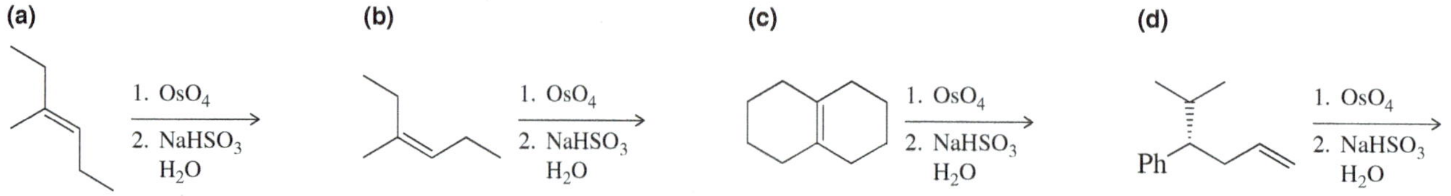

Assessment 9.21 When using a terminal alkene under the conditions shown here, explain why it is unnecessary to show the relative stereochemical outcome in the product.

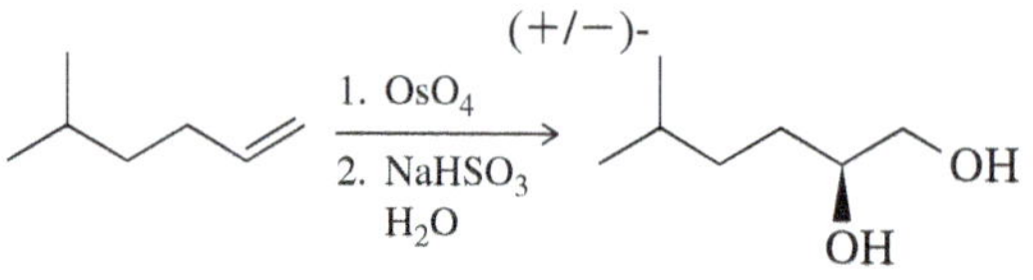

Assessment 9.22 Which of molecules **A–D** would you expect to give a positive permanganate test? That is, which would result in a purple $KMnO_4$ solution turning brown?

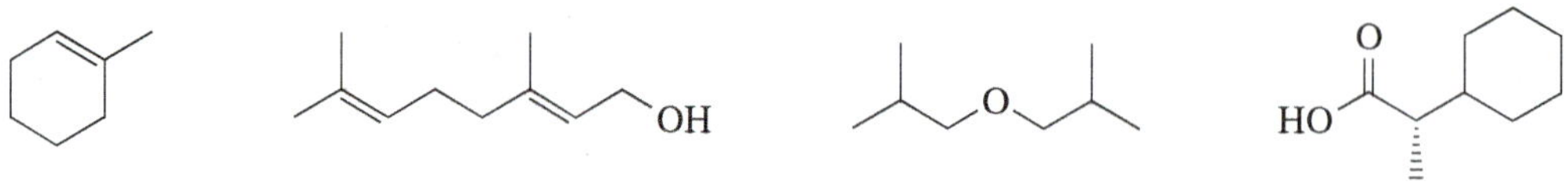

A **B** **C** **D**

Assessment 9.23 The concertedness of the OsO_4 reaction results in both oxygens being added to the same face of the molecule (i.e., syn addition). How might these conditions be modified in order to prepare a trans-diol?

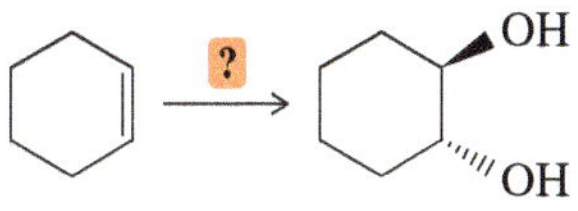

9.1.6 Alkene oxidation—epoxidation and hydrolysis combined

Assessment 9.20 demonstrates that OsO_4 (and $KMnO_4$) can be used to make cis-diols. How, though, could you make trans-diols?

Let's Think Like a Chemist and come up with a valuable addition to the synthetic chemist's toolbox.

Thinking Like a Chemist

As always, our work going forward must be informed by looking back. Figure 9.34 classifies previous reactions based on whether anti or syn addition was the primary reaction outcome. Figure 9.34(a) shows anti addition of bromines in the alkene halogenation reaction (Section 9.1.2), anti addition of bromine and water in halohydrin formation (Section 9.1.3), and anti addition of mercury and water in the oxymercuration reaction (Section 8.4.4). These reactions are all two-step processes, whereby in the second step, attack of the nucleophile on the bromonium or mercurinium occurs on the opposite face. On the other hand, Figure 9.34(b) shows syn addition in hydroboration (Section 8.4.5), dihydroxylation with OsO_4 (Section 9.1.5), and epoxidation (Section 9.1.4). In these reactions, both groups added at the same time on the same face. [*That these reactions may produce a racemic mixture is being ignored in Figure 9.34.*]

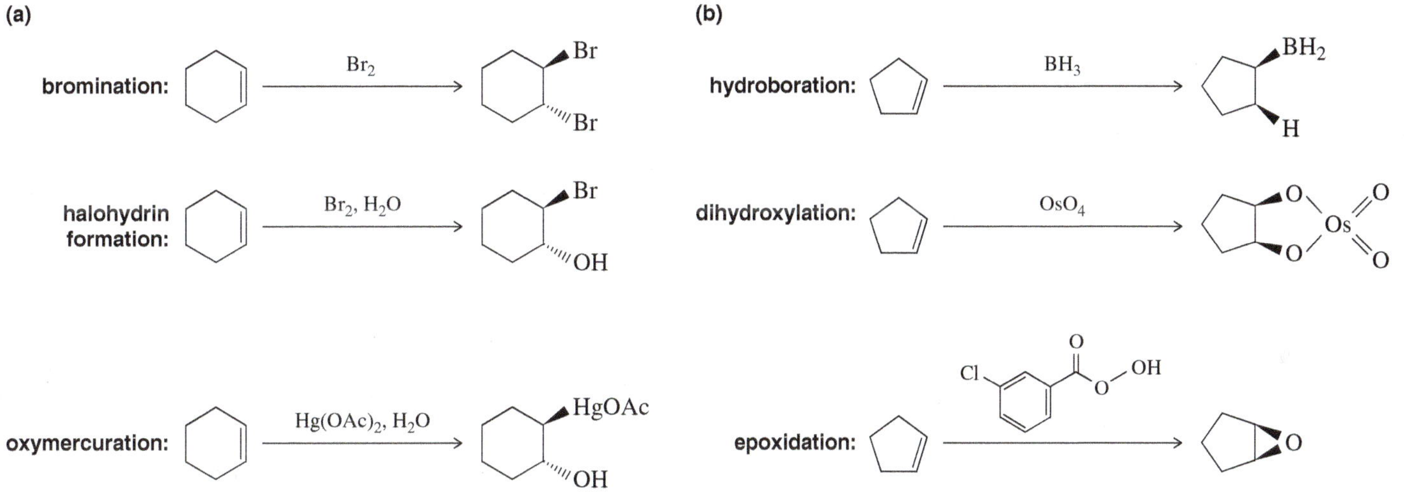

Figure 9.34 Previous reactions categorized by anti or syn addition. (a) Anti addition results from two groups adding in separate steps. (b) Syn addition results from two groups adding at the same time.

Based on the logic depicted in Figure 9.34, you would expect that creating trans-diols requires a protocol in which the two hydroxyl groups add at different times. We can do this by first forming an epoxide and then reacting it with water under acid-catalyzed conditions (Figure 9.35). ■

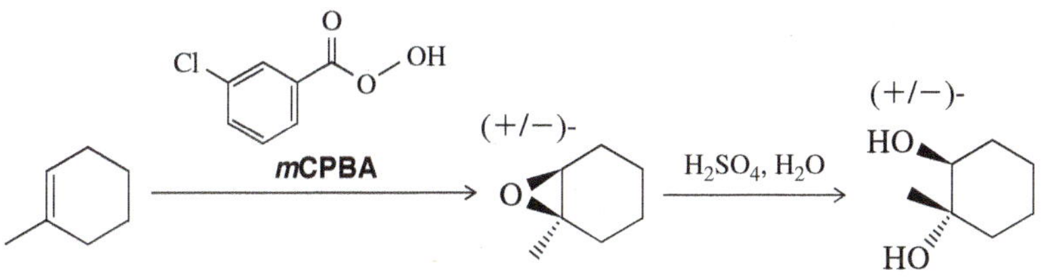

Figure 9.35 Adding the hydroxyl groups at different times results in a trans-diol. [*Reminder: The (+/−)-designation is a shorthand way of saying that the product is a racemic mixture of enantiomers.*]

The reaction begins with protonation of the epoxide to give an intermediate reminiscent of a bromonium or mercurinium ion (Figure 9.36). Similar to those reactions, water opens the oxonium ring from the backside at the carbon that would be the most stable carbocation. In the final step, H^+ is removed from the diol, regenerating H_3O^+. As such, this is an acid-catalyzed process.

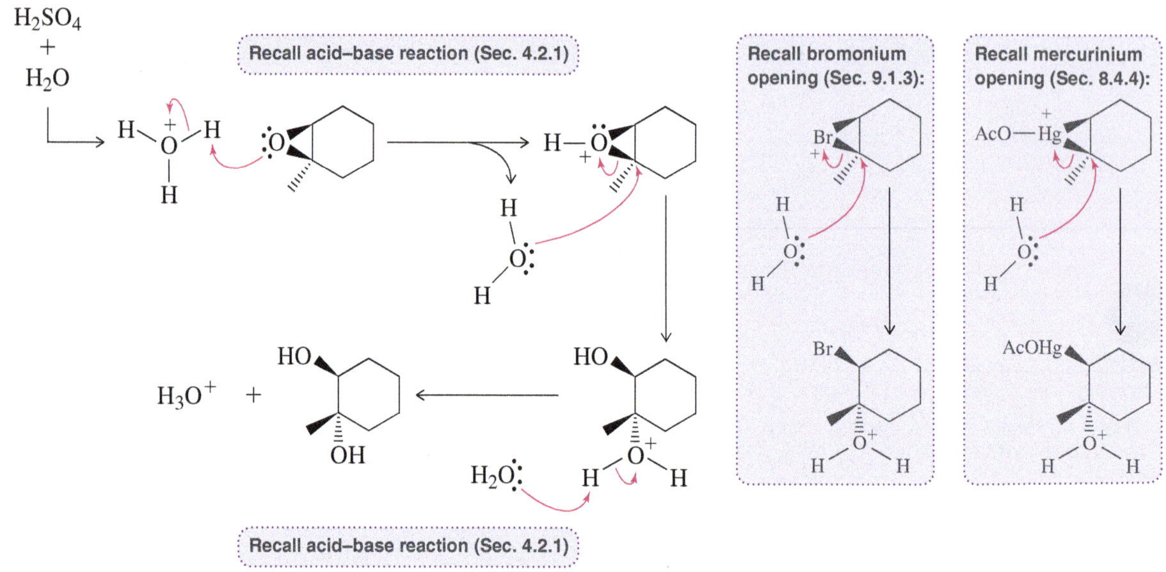

Figure 9.36 Epoxidation followed by acid-catalyzed hydrolysis gives a trans-diol. The ring opening is acid-catalyzed and mechanistically similar to bromonium and mercurinium opening.

9-C CHEMISTRY IN CONTEXT: Green Chemistry

Greener trans dihydroxylation

The combination of epoxidation and hydrolysis has made it possible to develop a more sustainable synthesis of trans-diols in a single step (**Figure 9.37**). This example uses a polymer-bound sulfonic acid catalyst and hydrogen peroxide [*our favorite environmentally friendly oxidizing agent*]. Hydrogen peroxide reacts with the polymer-bound sulfonic acid to generate persulfonic acid. The persulfonic acid epoxidizes the alkene while regenerating sulfonic acid, which catalyzes epoxide opening to give the trans-diol. Reaction of another equivalent of hydrogen peroxide with sulfonic acid restarts the cycle. Notable in this transformation is that it uses a low-molecular-weight oxidizing agent, produces water as a by-product, uses a recyclable catalyst (up to 10 times), and does *not* require the use of any organic solvents.

[In the last three Chemistry in Context boxes, you have seen green modifications of reactions we are learning in this chapter. This is done neither to burden

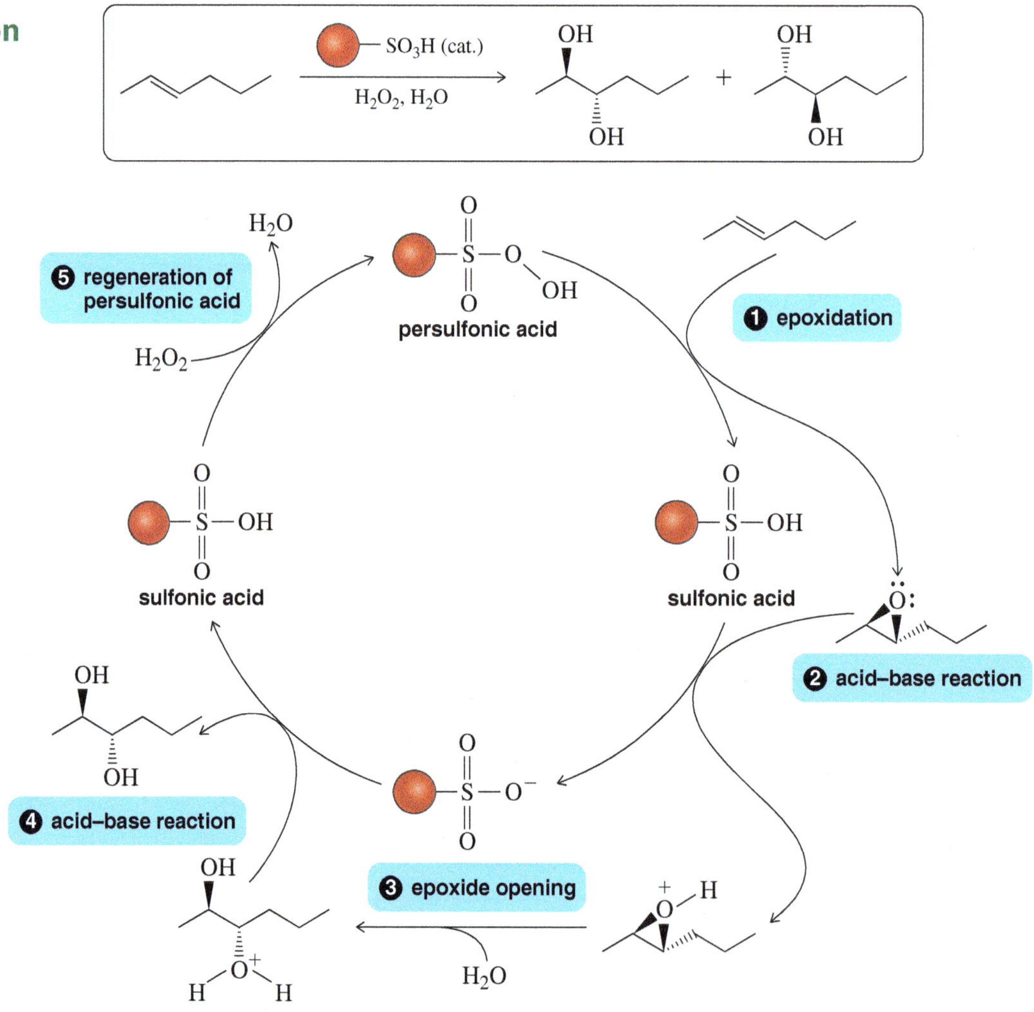

Figure 9.37 A green trans dihydroxylation.

(continued)

you with more reactions nor to suggest that these are the greenest reactions. Instead, the goal is to showcase cleaner reactions that accomplish similar ends to those of their less sustainable counterparts. And, possibly, to inspire you to continue the search for even greener reagents.]

Assessment 9.24 There are two mechanisms by which each of the two enantiomers can form in the reaction shown in Figure 9.37. Show them.

Assessment 9.25 Show an arrow-pushing mechanism for reactions 1–4 in Figure 9.37.

Assessment 9.26 Predict the product(s) of each of the following reactions, making sure to indicate the relative stereochemical outcome. Indicate any racemic mixtures by drawing both enantiomers.

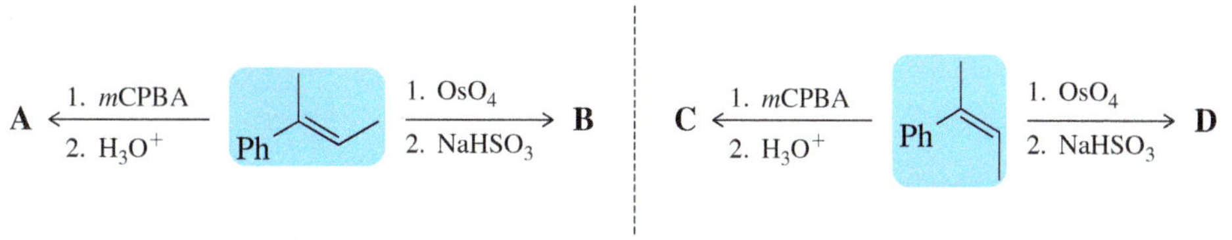

Assessment 9.27 In Assessment 9.26, what is the relationship between the enantiomers of compound A and the enantiomers of compound B?

Assessment 9.28 Despite it being equally favorable, opening of the epoxide does *not* happen in the absence of an acid catalyst. How does acid make the reaction faster? Demonstrate this concept by directly comparing the reaction coordinate diagram for both situations A and B.

Situation A (without acid):

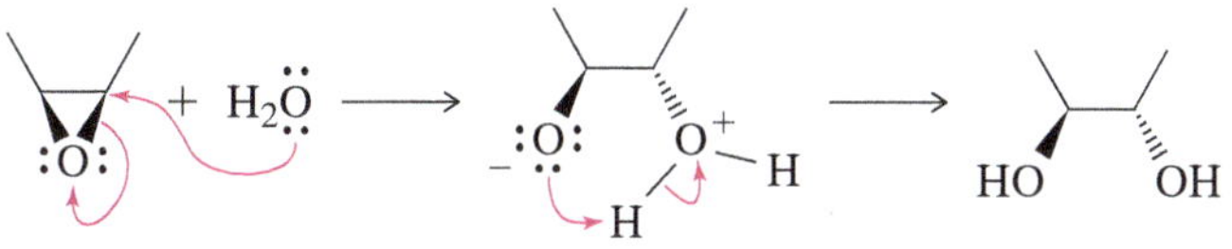

Situation B (with acid):

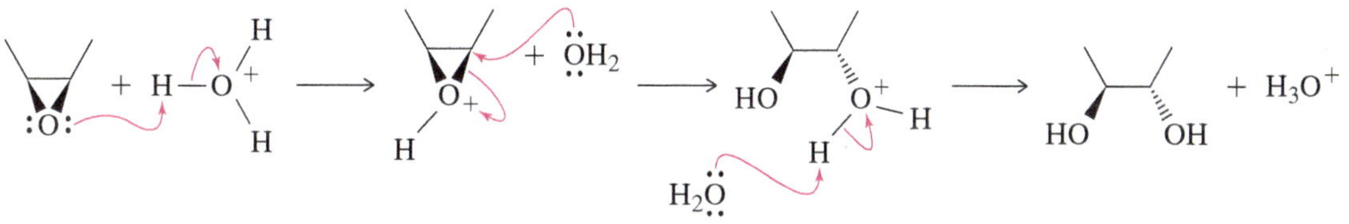

9.1.7 Alkene oxidation—ozonolysis

In Chapters 8 and 9, we have learned how to convert alkenes into 1°, 2°, and 3° alkyl halides; 1°, 2°, and 3° alcohols; ethers; epoxides; syn and anti diols; and halohydrins. We are almost ready to do synthesis, except that the synthesis of larger organic molecules is not possible using only functional group transformations. At some point, we will need a way to make new carbon–carbon bonds. The carbonyl functional group (specifically aldehydes and ketones) is very useful for this purpose. As a result, the **ozonolysis** of alkenes to create aldehydes or ketones is included here (see Figure 9.38), even though its mechanism is somewhat unique among the mechanisms studied thus far. The overall reaction is a two-step process using ozone gas, a powerful oxidizing agent, and a reductive workup to completely cleave an alkene, producing two carbonyls in the process.

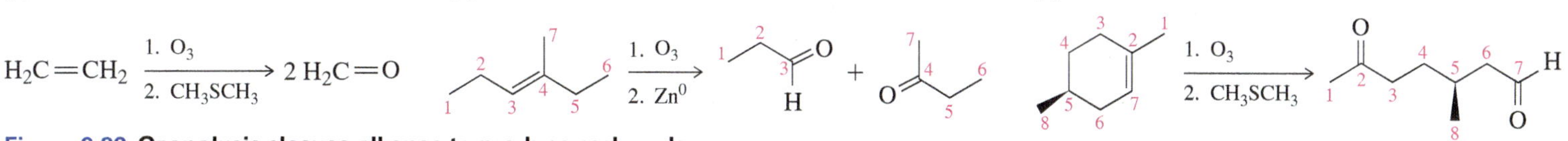

Figure 9.38 Ozonolysis cleaves alkenes to produce carbonyls.

Ozone (O_3), usually found only in the stratosphere (7–20 km above Earth's surface) or in polluted air over cities, can be synthesized in the lab immediately prior to its use (**Figure 9.39**). Oxygen is hit with a high-energy electric spark, which converts a small amount of oxygen into ozone. The ozone is then bubbled into a solution containing the alkene.

Figure 9.39 Ozone gas is generated and passed through the reaction flask to oxidatively cleave alkenes.

Assessment 9.29 Draw the product(s) you'd expect when each of these alkenes is treated first with O_3, then with CH_3SCH_3.

(a)

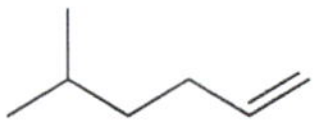

(b)

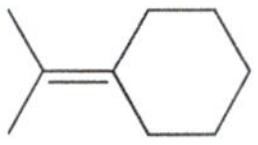

(c)

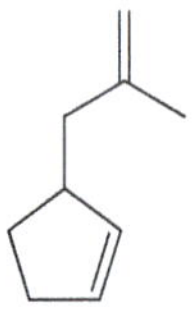

(d)

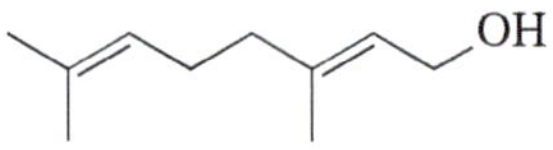

Assessment 9.30 Suggest a synthesis of the following aldehydes or ketones using the ozonolysis reaction of an alkene.

(a)

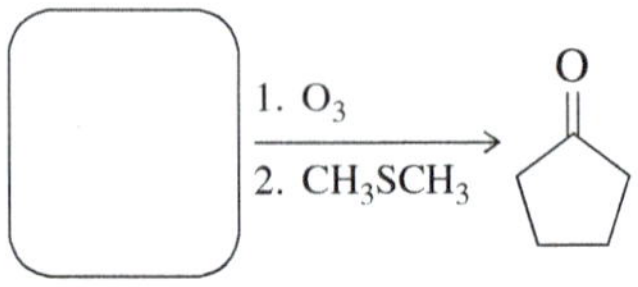

(b)

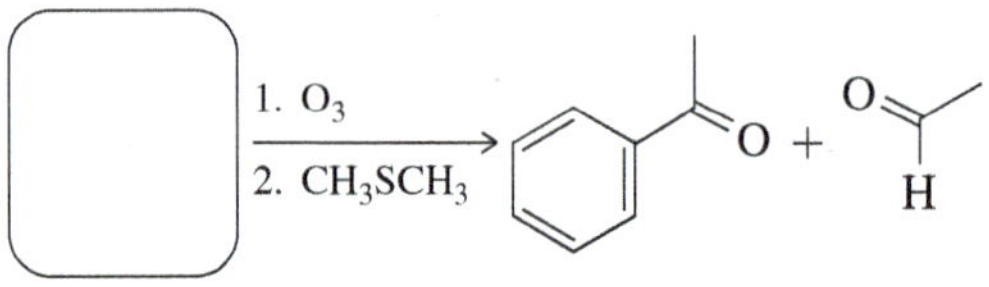

(c)

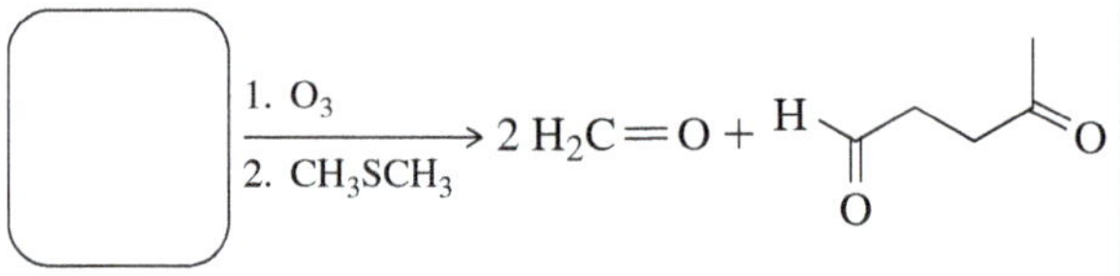

Let's work Assessments 9.29 and 9.30 together.

Solved Assessments 9.29 and 9.30

In Assessment 9.29 we are asked to predict the product of an organic reaction, whereas Assessment 9.30 asks us to work backward and identify what reactant would be used to produce a particular product. Most students have an easier time predicting reaction outcomes. When we begin **retrosynthetic analysis** in Chapter 10, we will need the skill of working backward. [*That these are easy assessments allows us to practice the skill of working backward.*]

Table 9.3 presents a straightforward method for predicting ozonolysis products rooted only somewhat in mechanistic understanding. Ozonolysis involves cleaving both the σ and π bond of the alkene and attaching an oxygen atom from ozone to each of the carbons. [*The other oxygen atom ends up attached to the sulfur in dimethyl sulfide* (CH_3SCH_3).] We are going to use our eraser to cleave the bonds and our pencil to place the oxygens. After working through parts (a) and (b) together, you can do parts (c) and (d) on your own.

Table 9.3 Predicting ozonolysis products for Assessment 9.29(a) and (b).

1. Draw the alkene with an elongated C—C double bond.	2. Cleave the double bond with your eraser.	3. Finally, attach an oxygen to each of the former alkene carbons.	4. And voila! You are done, but you may want to redraw the products in a more aesthetically pleasing way.

Assessment 9.29(a)

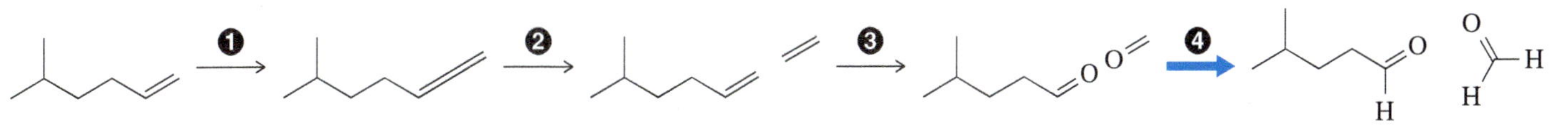

Assessment 9.29(b)

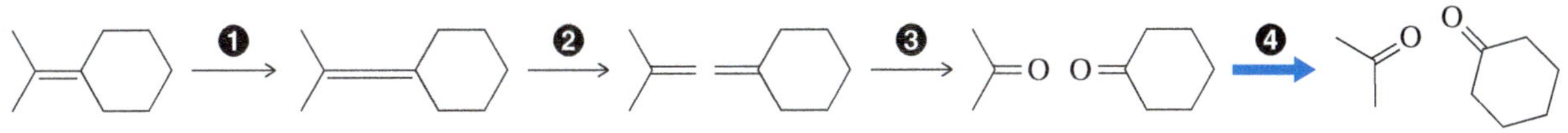

If you know how to solve a problem in the forward direction, then you can work it backward, too. In a nod to the principle of microscopic reversibility [*which we call the principle of microscopic "re-workability"*], simply do the reverse of what you did when working forward. By repeating the four steps from Table 9.3 in reverse order, we will generate the alkenes needed to make the desired carbonyls in Assessment 9.30(a) and (b) (**Table 9.4**). You can attempt part (c) on your own. ■

Table 9.4 Working backward to solve a problem.

4. Draw the desired products such that their oxygens are pointing to each other.	3. Erase the oxygens.	2. Connect the former carbonyl carbons with an elongated alkene.	1. And voila! You are done, but you probably want to redraw the products with appropriately sized alkenes.

Assessment 9.30(a)

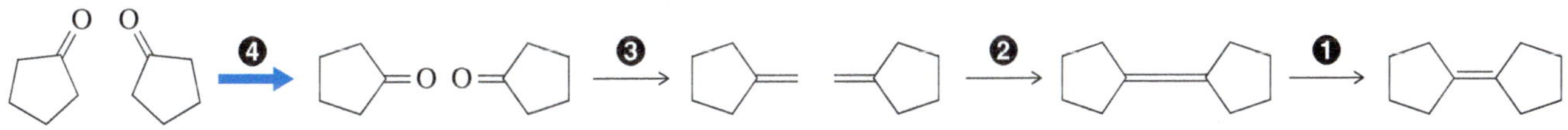

Assessment 9.30(b)

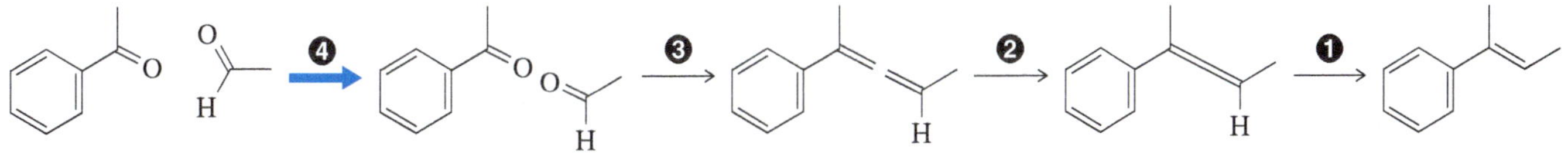

Assessment 9.31 In Solved Assessment 9.30(b), we came up with an alkene that under the conditions of ozonolysis would produce acetophenone and acetaldehyde. There is one other alkene that should produce the same compounds under these conditions. Which is it?

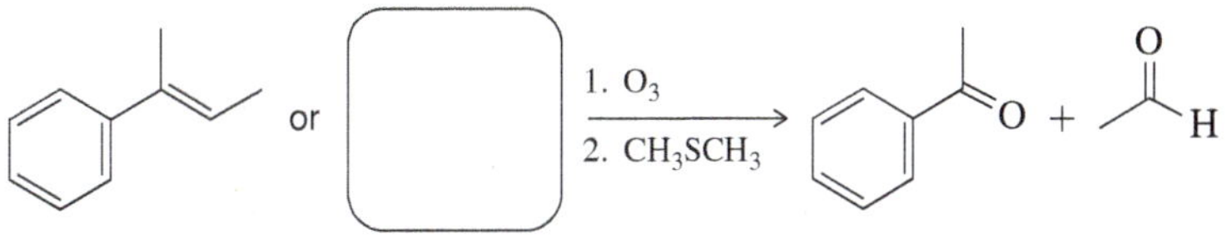

Mechanistically, the first step of ozonolysis proceeds in a manner similar to what we observed for the dihydroxylation reaction with osmium tetroxide (**Figure 9.40**). That is, it proceeds through a concerted cyclic, 6-electron transition state. The molozonide contains two weak O—O σ bonds. As a result, it spontaneously decomposes and recyclizes with different regiochemistry to give an ozonide, in what is referred to as the Criegee rearrangement.

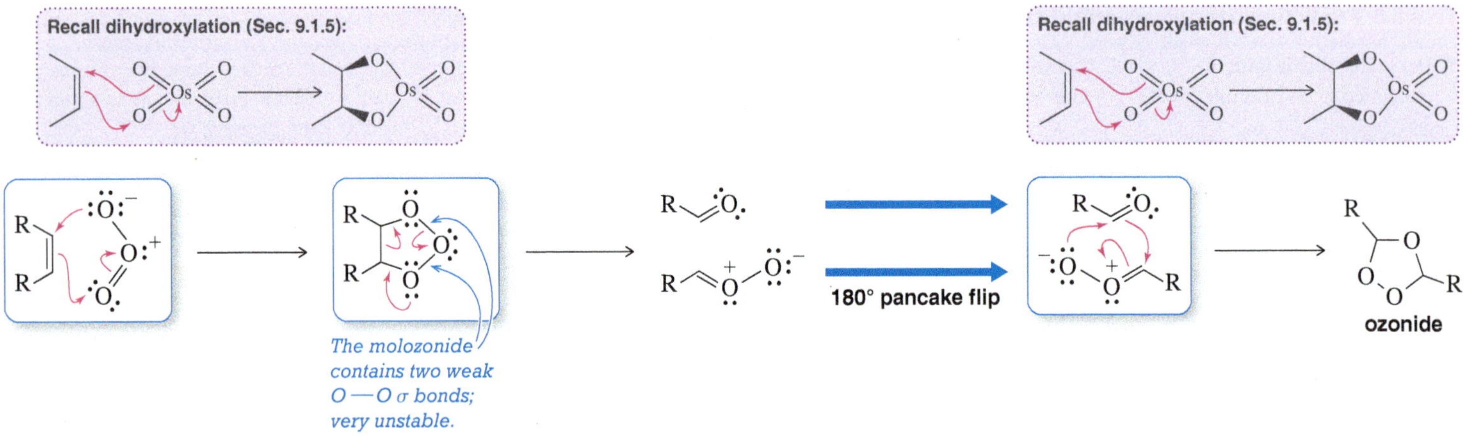

Figure 9.40 The mechanism of ozonide formation proceeds through the three 6-electron cyclic transition states in blue boxes. The reaction is driven by reducing the number of weak O—O bonds.

The intermediate ozonide, still containing an unstable O—O σ bond, is treated in the second step with dimethyl sulfide, a reducing agent (**Figure 9.41**). The nucleophilic sulfur attacks one of the oxygens to break the weak O—O bond. The intermediate alkoxide decomposes to give the two carbonyls and dimethyl sulfoxide. [*The three oxygens of* O_3 *are now attached to sulfur and the two alkene carbons, which are now the carbonyl carbons of the two aldehydes.*]

Figure 9.41 Reduction of the ozonide forms the two carbonyls and dimethyl sulfoxide (DMSO).

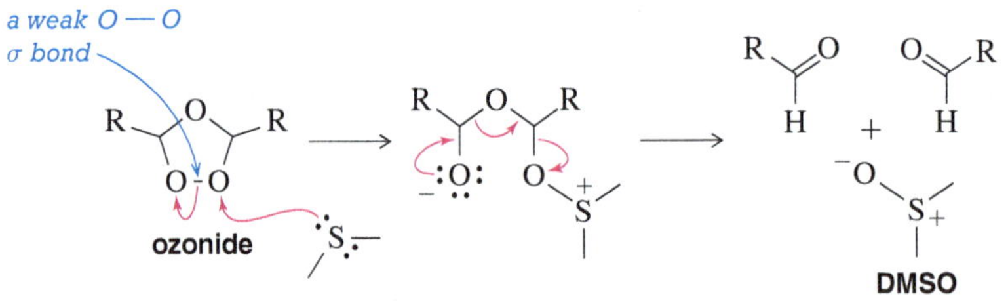

9.1.7 SUMMARY

- The reaction of ozone (O_3) with an alkene, referred to as ozonolysis, results in cleavage of the alkene to give two carbonyls.

Assessment 9.32 Ozonolysis can be broken down into four distinct steps. Which of these steps, (a)–(d), are redox reactions? Justify your answer by calculating the change in oxidation states.

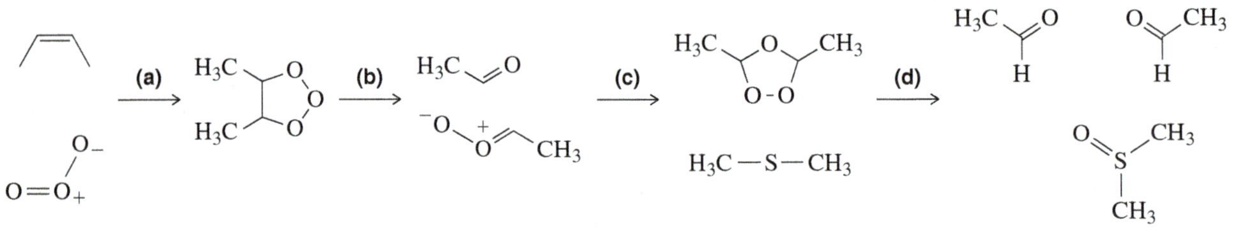

Assessment 9.33 Formation of the molozonide can be expected to proceed stereospecifically. Why is this the case? Show the two different molozonides you would expect to get from ozonolysis of (*E*)- and (*Z*)-3,4-dimethylhept-3-ene.

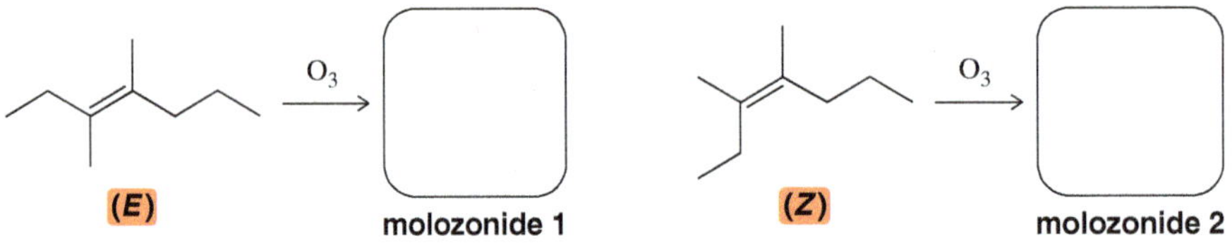

9.2 Reduction of alkenes—catalytic hydrogenation

Oxidation numbers for carbon range from −4 to +4. Ethene (**Figure 9.42**) has each carbon at the −2 oxidation state, which implies that alkenes can also be reduced. This is done by adding more bonds to hydrogen, resulting in an alkane. Were we to invent a reagent to carry out this process by the mechanisms we've studied in Chapters 8 and 9, we'd invent hydrogen gas (H_2). And yet, despite being very favorable ($\Delta H° = -27$ kcal/mol), mixing H_2 and ethene alone results in no reaction [Figure 9.42(a)]. Figure 9.42(b) imagines how this reaction might work, with the alkene deprotonating H_2. Though consistent with previous mechanisms, this reaction fails because the H—H bond is too strong, H_2 is a terribly weak acid, and the hydride (H^-) is an unstable conjugate base.

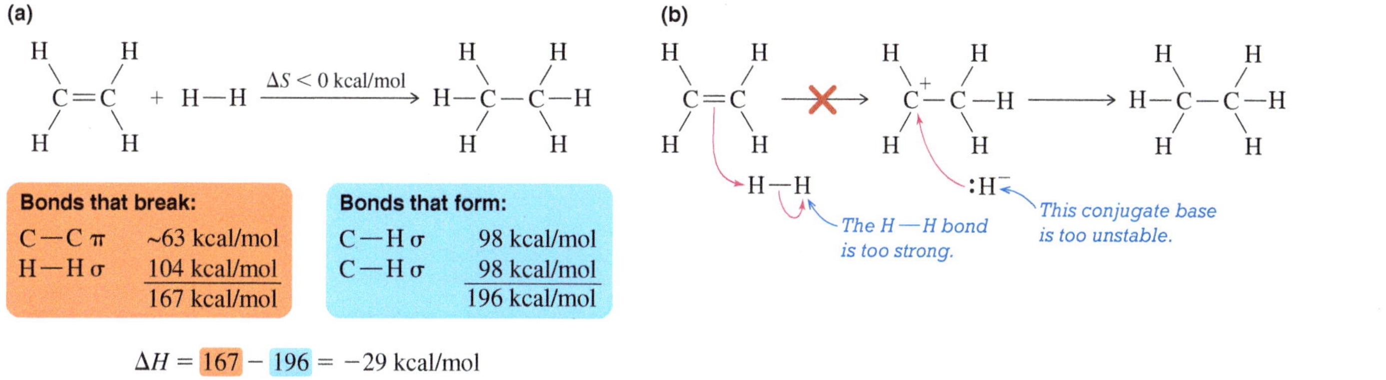

Figure 9.42 **Hypothetical alkene hydrogenation.** (a) Despite being heavily favored by enthalpy, (b) the reaction does not happen. H_2 is not acidic enough to react with an alkene.

Plotting the mechanism in Figure 9.42(b) on a reaction coordinate diagram shows a two-step process where the first step is highly disfavored with a large activation energy (**Figure 9.43**). It is this high barrier to reaction that prevents the reaction from occurring. Luckily, chemists can use catalysis to lower the activation barrier to make **hydrogenation** of alkenes possible. By changing the mechanism, the activation energy is lowered and the reaction proceeds more quickly.

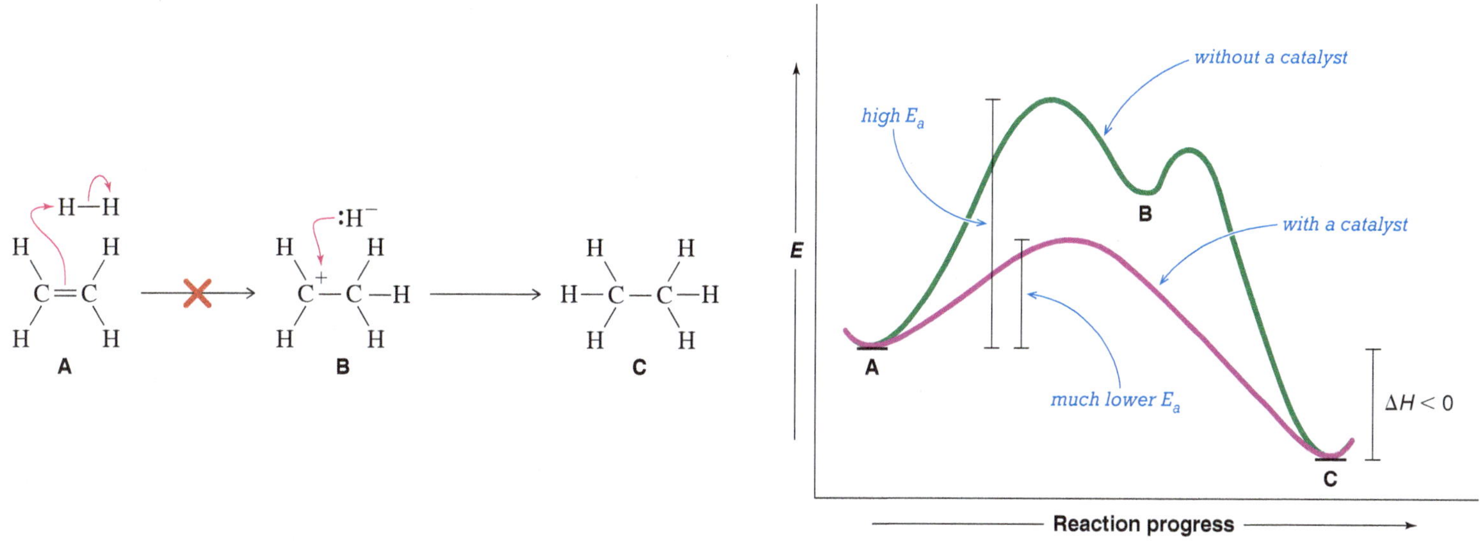

Figure 9.43 **By changing the reaction path, catalysts can lower the activation energy and increase the rate of a reaction.**

Pd/C is spoken as "palladium on carbon."

Hydrogenation of alkenes is achieved using a palladium catalyst adsorbed to a graphite surface (designated as Pd/C). This reaction is classified as **heterogeneous catalysis** because palladium is insoluble under the reaction conditions (i.e., the catalyst is in a different phase than the reactants). **Figure 9.44(a)** shows the experimental setup, where a balloon supplies the hydrogen gas to the alkene/catalyst mixture. The commonly accepted mechanism of this reaction [**Figure 9.44(b)**] involves adsorption of hydrogen gas and the alkene on the palladium surface. In this way, the palladium weakens the H—H σ bond and brings the two reactants into closer proximity, thereby increasing the rate of reaction.

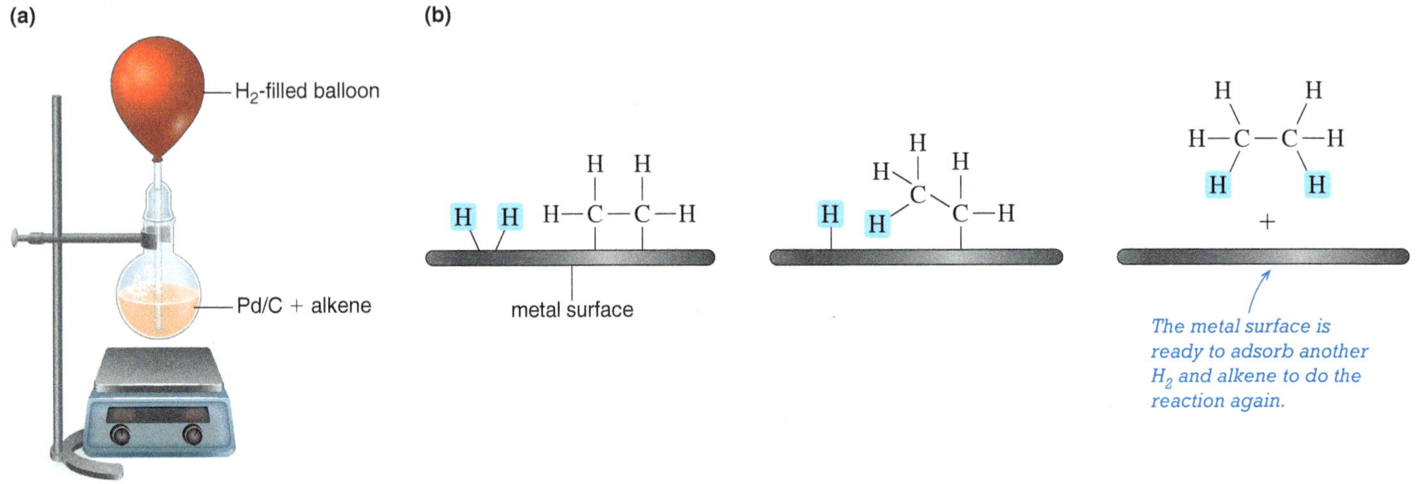

Figure 9.44 In one proposed mechanism of catalytic hydrogenation, transfer of hydrogen to the alkene occurs on the surface of a metal catalyst, resulting in syn addition.

An alternate mechanism is shown in **Figure 9.45**. [*This mechanism will gain relevance when we study organometallic reactions in Chapter 16.*] The adsorption of hydrogen onto palladium is represented by palladium inserting itself into H_2, breaking the H—H σ bond. The alkene removes the two hydrogens, producing an alkane and regenerating the palladium catalyst. The alkene is reduced (each C: −2 → −3) while the hydrogens are oxidized (each H: 0 → +1).

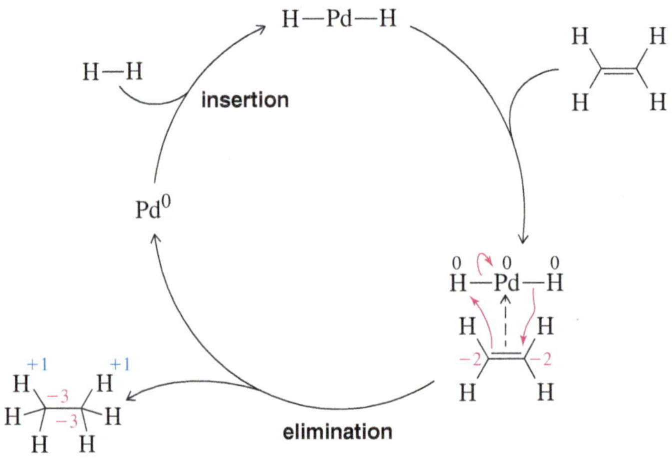

Figure 9.45 An alternate plausible mechanism of catalytic hydrogenation involves insertion and elimination steps. We study this mechanism in more detail in Chapter 16.

A key feature of both mechanisms is that both hydrogens add from the same face of the alkene before bond rotation can occur. As such, the hydrogenation of alkenes is highly stereospecific. The stereospecificity resulting from the syn addition of hydrogen in each of the reactions is shown in **Figure 9.46**.

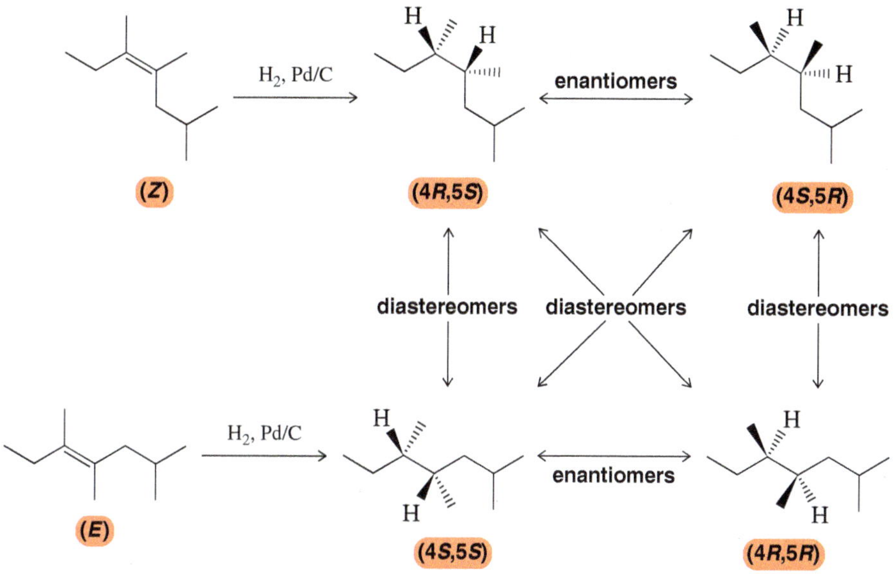

Figure 9.46 Catalytic hydrogenation is stereospecific because two alkene stereoisomers give different diastereomeric products.

Assessment 9.34 What product results when the following molecules are treated with H_2, Pd/C? Be sure to indicate the relative stereochemical outcome. Draw both enantiomers of any racemic mixtures. [*It is difficult to control the stoichiometry of gases so there is enough H_2 to reduce all alkenes.*]

(a)

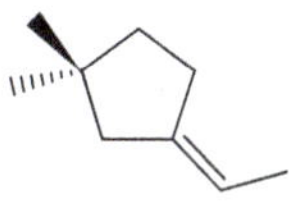

(b)

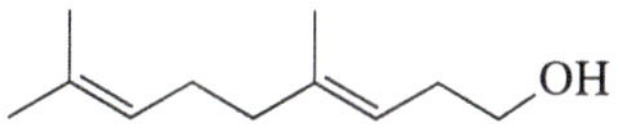

(c)

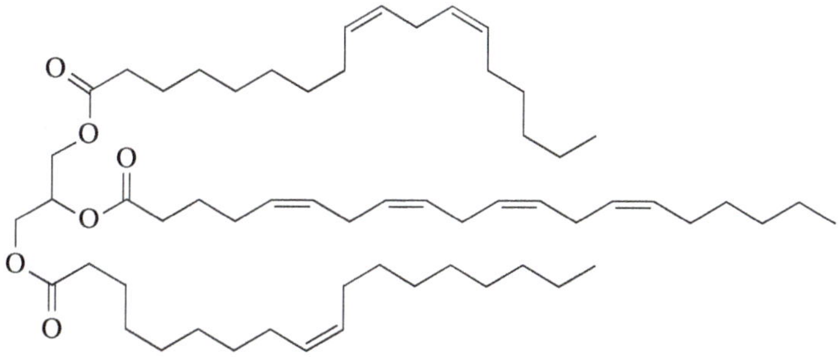

(d)

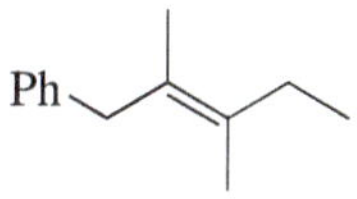

(e)

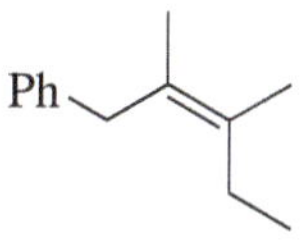

Assessment 9.35 A misguided chemist attempted to synthesize *trans*-1,2-dimethylcyclohexane via the hydrogenation of 1,2-dimethylcyclohexene. Explain why this is not possible.

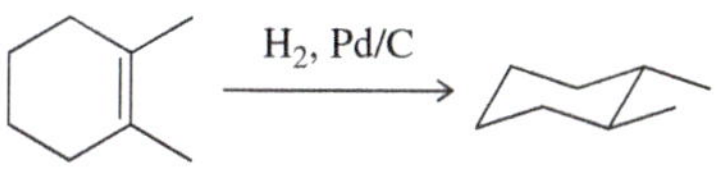

Assessment 9.36 Because deuterium behaves like hydrogen in chemical reactions yet is detected differently (as we explain in Chapter 15), chemists use the incorporation of deuterium to better understand the subtleties of reaction mechanisms. Deuterium is incorporated by replacing H_2 with D_2 in the hydrogenation reaction. Identify the product expected when the alkenes in Assessment 9.34 react with D_2 and Pd/C. [*Don't worry about showing all diastereomers.*]

9-D CHEMISTRY IN CONTEXT: Food Science

Partially hydrogenated oils

Vegetable oils, often extracted from the seeds of plants like corn and soybean, are triglycerides—that is, they are composed of glycerol and three long-chain fatty acids (Figure 9.47). Most naturally occurring vegetable oils are unsaturated, containing only cis-alkenes. These cis double bonds prevent the triglyceride chains from packing tightly together, thus breaking up intermolecular interactions and resulting in most vegetable oils being liquid at room temperature. Soybean oil, for example, is composed principally of triglycerides containing linoleic and oleic acid. These oils have many uses, including cooking and producing biodiesel fuel.

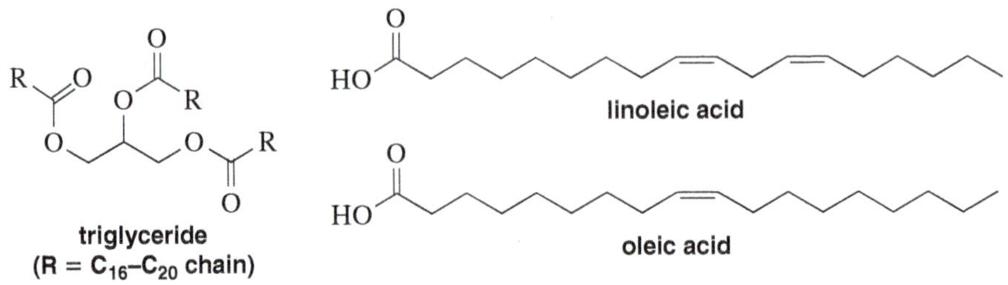

Figure 9.47 **Vegetable oils are unsaturated triglycerides of cis fatty acids.**

While **unsaturated fats** are relatively healthy, the cis-alkene present imparts oxidative instability. [*This is consistent with our understanding of the diminished stability of cis vs. trans double bonds.*] As a result, more stable **saturated fats**, typically procured from animals, found greater utility for use in cooking in the first half of the 20th century. Amidst speculation that saturated fats contributed to the spike in deaths due to heart disease, the food industry began preparing partially **hydrogenated oils**. These are prepared by hydrogenating vegetable oils using the less toxic transition metal nickel as the catalyst. Removing only some of the cis-alkenes increases the stability while maintaining the health benefits of the vegetable oils. Unfortunately, even alkenes that are not hydrogenated come into contact with the nickel (Figure 9.48). This interaction (one we study in Chapter 16) results in the isomerization of cis-alkenes to trans-alkenes, giving rise to the accidental invention of trans fats.

In the body, trans fats increase low-density lipoprotein (LDL) cholesterol. The presence of LDL cholesterol, the "bad" form, results in a buildup of arterial plaques, which contribute to heart disease. Therefore, foods containing partially hydrogenated oils, and by extension, trans fats, should be avoided as much as possible. The problem caused by trans fats has initiated further research into the amounts of saturated, unsaturated, and trans fats that should ideally be a part of a healthy diet.

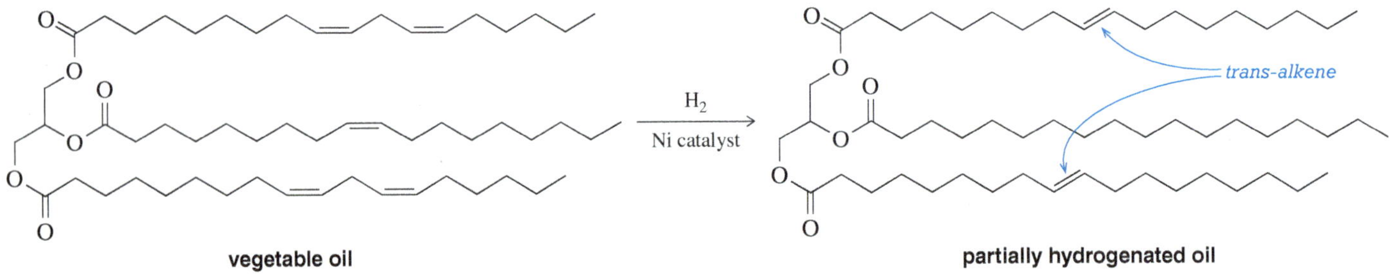

Figure 9.48 **The desired synthesis of partially hydrogenated oils resulted in the invention of trans fats.**

9.2.1 Hydrogenation to assess alkene stability

Analogous to combustion analysis (Section 8.1.4.5), hydrogenation can be used to experimentally determine the relative stability of alkene isomers. Using the bomb calorimeter in Figure 9.49(a) under carefully controlled conditions, the amount of heat released during alkene hydrogenation can be directly correlated to the stability of the π bond. Figure 9.49(b) gives the experimentally determined **heat of hydrogenation** (ΔH_{hyd}) for hydrogenation of but-1-ene and *trans*-but-2-ene. This experiment tells us that but-1-ene is less stable because hydrogenating it releases more heat. [*Does this make sense? Think about why less stable alkenes would give off more heat before reading further.*]

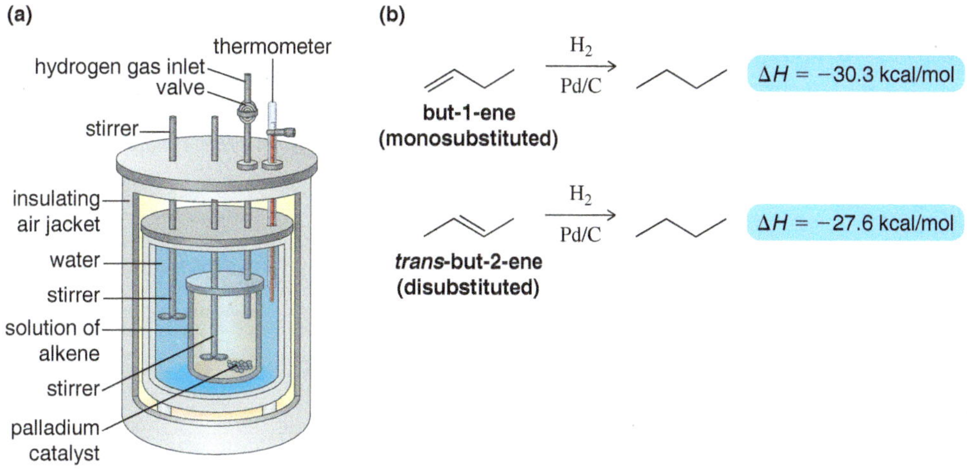

Figure 9.49 Hydrogenation to measure alkene stability. (a) Schematic of a bomb calorimeter. (b) Hydrogenation of monosubstituted alkenes gives off more heat than disubstituted alkenes.

Figure 9.50 compares the two hydrogenation reactions on a reaction coordinate diagram. Because butane is the product in both, the reaction coordinate diagrams finish at the same point. Thus, the difference in ΔH° is due to where the alkenes start. The monosubstituted alkene releases more heat only if it starts at a higher energy. Starting at a higher energy means that a molecule is less stable. From the numbers in Figure 9.49, we can say that a monosubstituted alkene (but-1-ene) is 2.7 kcal/mol less stable than a trans disubstituted alkene (*trans*-but-2-ene).

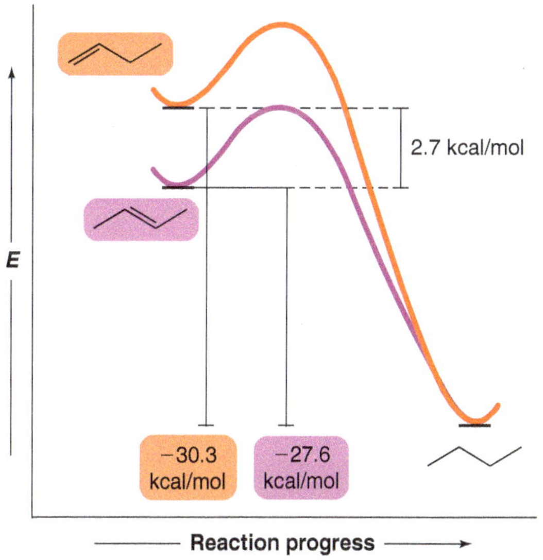

Figure 9.50 Hydrogenation of the less stable alkene releases more energy.

When the same analysis is done for 3-methylbut-1-ene and 2-methylbut-2-ene, the trisubstituted alkene is more stable by 3.4 kcal/mol (**Figure 9.51**).

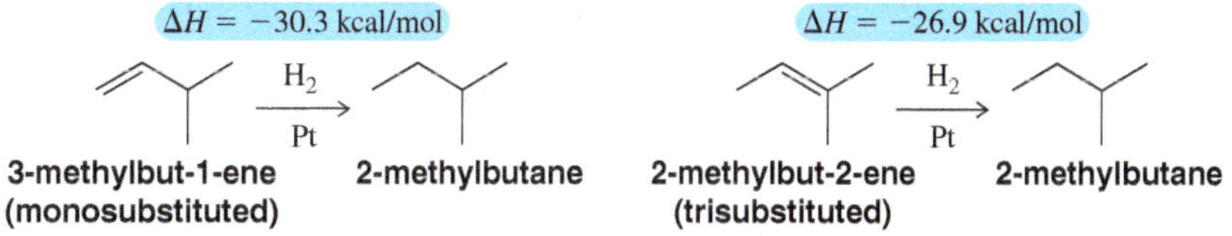

Figure 9.51 Trisubstituted alkenes are more stable than monosubstituted alkenes.

Table 9.5 lists experimentally determined heats of hydrogenation for a variety of alkenes. For similarly substituted alkenes, the values are nearly identical. For example, six different monosubstituted alkenes have ΔH_{hyd} within 0.2 kcal/mol of each other. As a result, we can use the table to approximate the relative stability of any mono-, di-, or trisubstituted alkene. [*Don't worry about knowing the specific values; focus on the trends instead.*]

Table 9.5 Molar heats of hydrogenation of alkenes.

Name	Structure	Molar heat of hydrogenation ($-\Delta H°$) kcal	kJ	General structure
Ethene (ethylene)	$H_2C{=}CH_2$	32.8	137	unsubstituted
Propene (propylene)	$CH_3{-}CH{=}CH_2$	30.1	126	$R{-}CH{=}CH_2$ **monosubstituted**
But-1-ene	$CH_3{-}CH_2{-}CH{=}CH_2$	30.3	127	
Pent-1-ene	$CH_3{-}CH_2{-}CH_2{-}CH{=}CH_2$	30.1	126	
Hex-1-ene	$CH_3{-}(CH_2)_3{-}CH{=}CH_2$	30.1	126	
3-Methylbut-1-ene	$(CH_3)_2CH{-}CH{=}CH_2$	30.3	127	
3,3-Dimethylbut-1-ene	$(CH_3)_3C{-}CH{=}CH_2$	30.3	127	
cis-But-2-ene	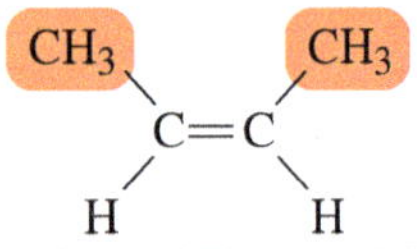	28.6	120	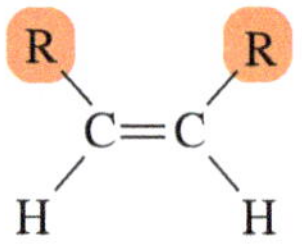 **disubstituted (cis)**
cis-Pent-2-ene	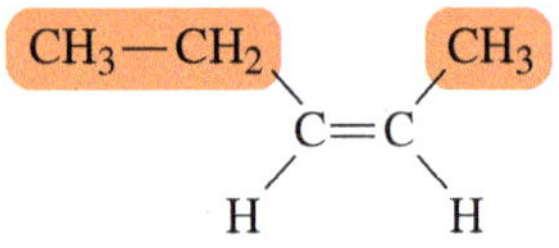	28.6	120	
2-Methylpropene (isobutylene)	$(CH_3)_2C{=}CH_2$	28.0	117	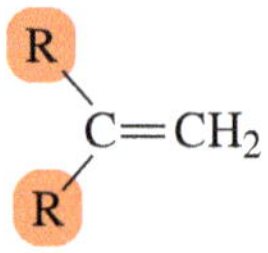 **disubstituted (geminal)**
2-Methylbut-1-ene	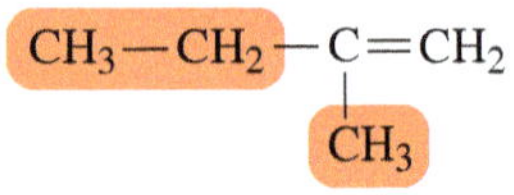	28.5	119	
2,3-Dimethylbut-1-ene	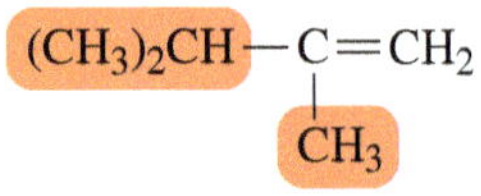	28.0	117	
trans-But-2-ene	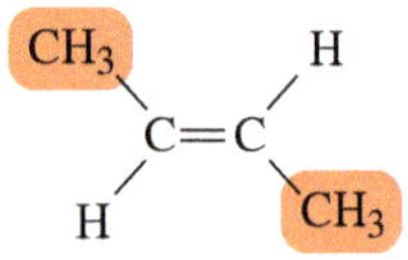	27.6	116	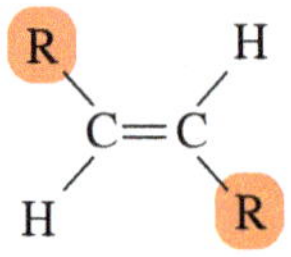 **disubstituted (trans)**
trans-Pent-2-ene	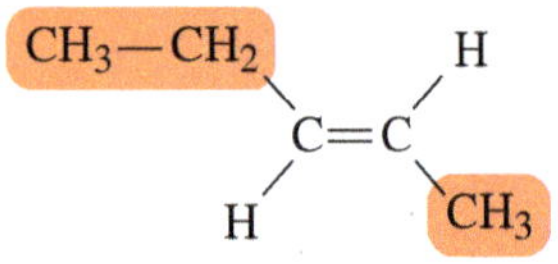	27.6	116	
2-Methylbut-2-ene	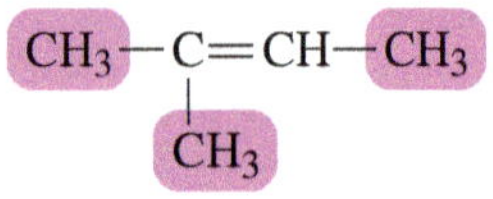	26.9	113	$R_2C{=}CHR$ **trisubstituted**
2,3-Dimethylbut-2-ene	$(CH_3)_2C{=}C(CH_3)_2$	26.6	111	$R_2C{=}CR_2$ **tetrasubstituted**

Note: A lower heat of hydrogenation corresponds to lower energy and greater stability of the alkene.

The general trend that emerges from Table 9.5 is that more substituted alkenes release less energy when hydrogenated and are therefore more stable. There are two reasons for this. **Figure 9.52(a)** shows that more substituted alkenes have more hyperconjugation. Just as neighboring C—H and C—C σ bonds can stabilize radicals and carbocations, alignment with π* of the alkenes results in a slight delocalization of electrons. **Figure 9.52(b)** shows that the more substituted alkene allows the methyl groups at C_3 to be farther apart, with a bond angle of 120°. This minimizes the steric strain compared to the monosubstituted alkene, where they are forced to be closer together (109.5° bond angle).

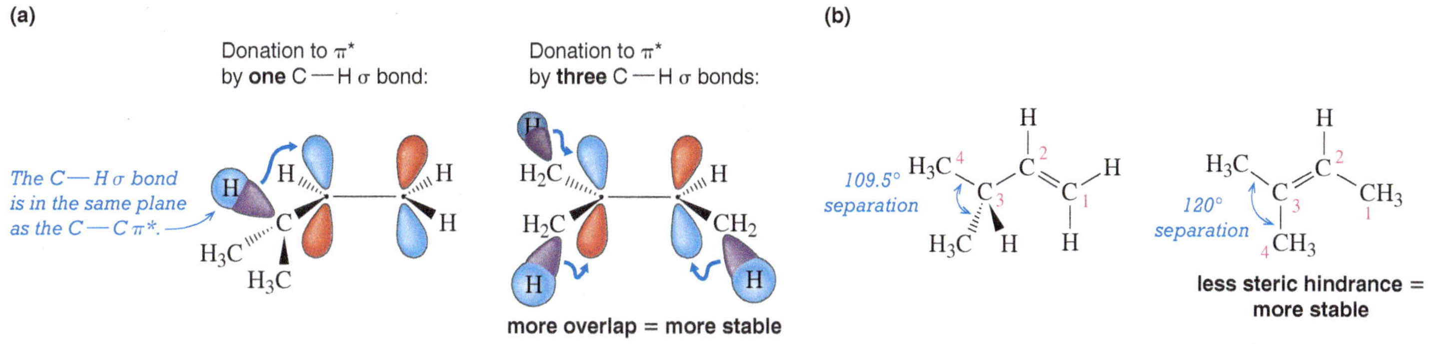

Figure 9.52 More substituted alkenes are more stable. (a) Hyperconjugation is maximized in the more substituted alkene. (b) Steric strain is minimized in the more substituted alkene.

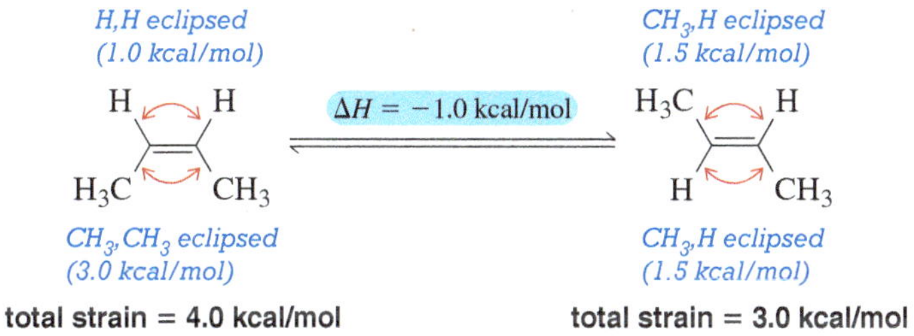

Figure 9.53 Cis-alkenes are less stable than trans-alkenes.

Table 9.5 also verifies our earlier assertion from Section 8.1.4.5 that cis-alkenes ($\Delta H = -28.6$ kcal/mol) are less stable than trans-alkenes ($\Delta H = -27.6$ kcal/mol.) The experimental difference of 1 kcal/mol is consistent with the strain energy we'd have predicted based on the eclipsing conformations present in the planar alkene (**Figure 9.53**).

Figure 9.54 previews how we might use information on alkene stability to predict reaction outcomes. In this case, there are three possible alkene products that might form via an elimination reaction. [*Don't worry about how; we discuss that in Chapter 12.*] In what we will eventually refer to as Zaitsev's rule, elimination reactions typically produce the most stable/most substituted alkene. Thus, the major product of this reaction would be 1,2-dimethylcyclohexene.

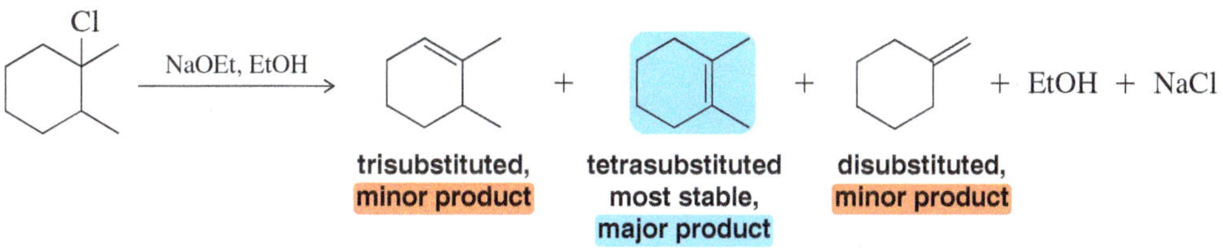

Figure 9.54 The elimination reaction we discuss in Chapter 11 gives the more substituted, more stable alkene.

9.2 SUMMARY

- The metal-catalyzed hydrogenation of alkenes results in the concerted, stereospecific addition of hydrogen to produce a fully saturated alkane. By measuring the heat of hydrogenation ($\Delta H°$), the stability of isomeric alkenes can be calculated.

Assessment 9.37 Without concerning yourself with the mechanism of the reaction, calculate the equilibrium constant for the following equilibrium processes. (Assume $T = 298$ K.)

(a)

(b)

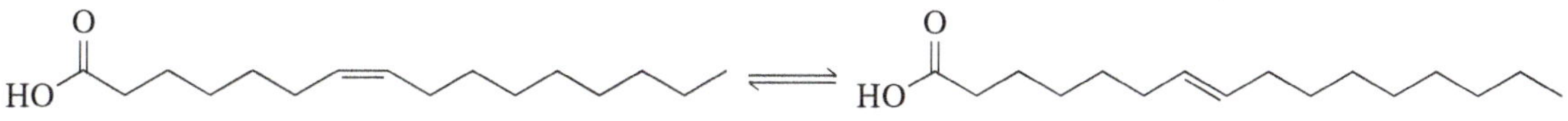

Assessment 9.38 Conjugated dienes, molecules containing two alkenes separated by one single bond, are discussed in detail in Chapter 21. (a) Considering the observed $\Delta H°$ of hydrogenation, is hexa-1,3-diene (conjugated) or hexa-1,4-diene (unconjugated) more stable? (b) How might you account for this difference in stability?

9.3 Special topic: enantioselective oxidation–reduction of alkenes

Recall from Chapter 6 that two enantiomers of a single molecule can have markedly different activities. As a result of the thalidomide tragedy, the FDA requires chiral drugs to be administered as a single enantiomer. And yet a racemic mixture of products resulted anytime we created a new stereocenter in the alkene reactions of Chapters 8 and 9. This is because the approaches from the front and back of a planar alkene represent enantiomeric transition states (Section 6.6) [Figure 9.55(a)]. These transition states are equal in energy, so we end up with identical activation energies and identical reaction rates. For stereoselective synthesis to work, diastereomeric transition states are needed that lead to different enantiomers. As diastereomers, the transition states will have different activation energies, leading to one reaction being faster than the other [Figure 9.55(b)]. The larger the difference between the two transition state energies, the more one stereoisomer will be favored over the other.

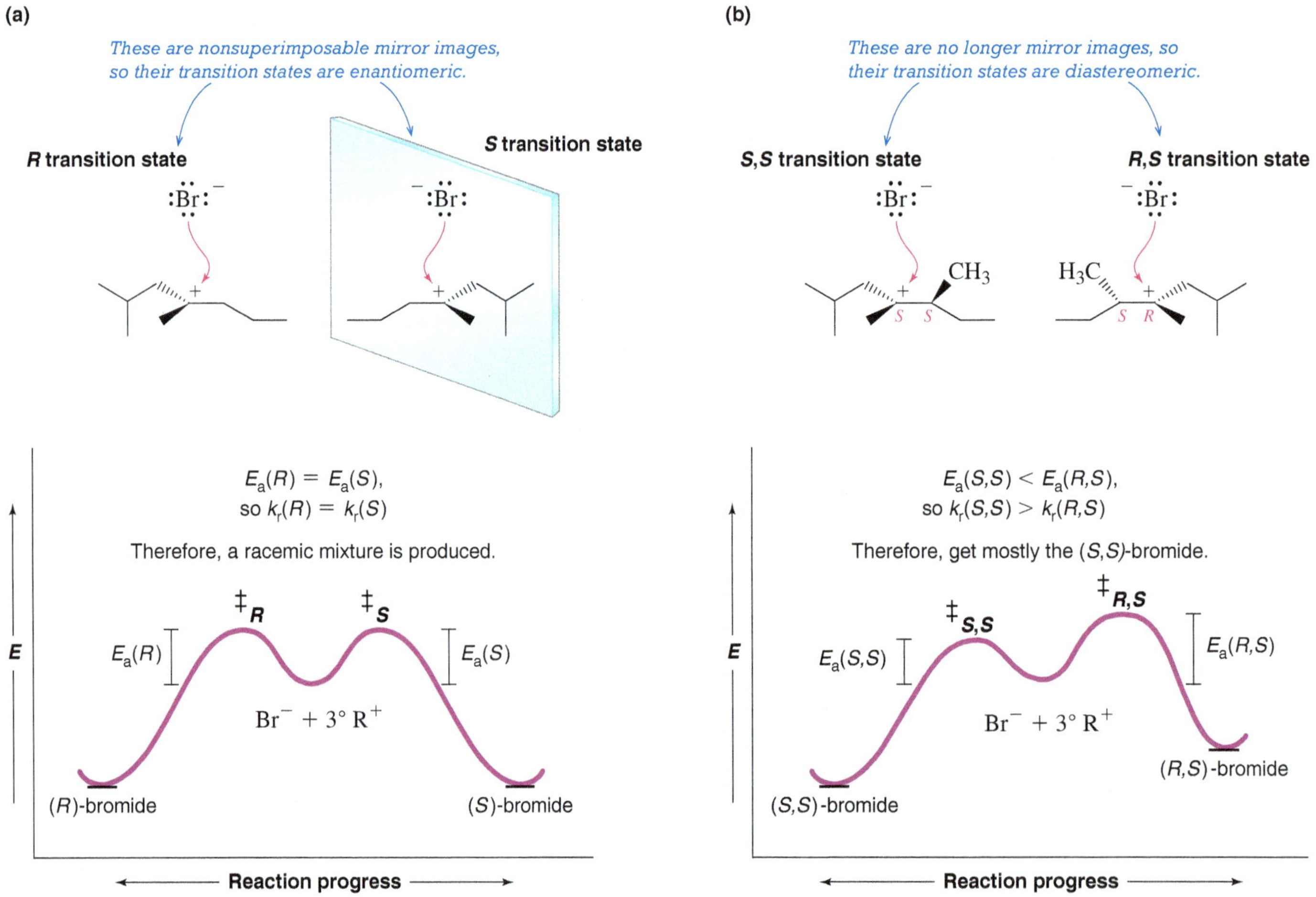

Figure 9.55 Making reactions stereoselective is as simple as turning (a) enantiomeric transition states into (b) diastereomeric transition states.

Here in Section 9.3, we explore the synthesis of the anti-HIV drug indinavir using stereoselective versions of epoxidation and hydrogenation. [*But first, a general commentary on stereoselectivity.*]

9.3.1 Relationship between E_a differences and stereoselectivity

Consider the **asymmetric** epoxidation shown in **Figure 9.56**, which has been simplified to A → B + C. To produce B stereoselectively, formation of B must be faster than formation of C. This will only be true when the activation energy for forming B is lower than the activation energy for forming C. In Section 5.3.6, we encountered an equation that relates the rate constants (k_r) for two reactions to the difference in activation energy between the two reactions. This quantitative relationship, modified to account for our purpose, is shown here as Equation 9.1. [*Don't worry, the math is just to illustrate some concepts.*]

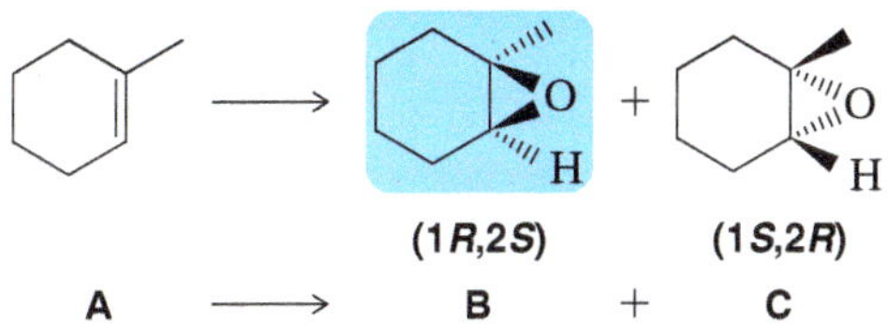

Figure 9.56 The stereoselective synthesis of B is the goal.

$$E_a(\mathrm{C}) - E_a(\mathrm{B}) = RT \ln\left(\frac{k_r(\mathrm{B})}{k_r(\mathrm{C})}\right) \qquad \text{Equation 9.1}$$

where $R = 1.987 \frac{\text{cal}}{\text{mol} \cdot \text{K}}$ and T = temperature.

Using Equation 9.1, assuming a temperature of 298 K, generates the activation energies required to produce several different ratios of products B and C (**Table 9.6**). As expected, more of B is produced when B has the lower activation energy. Also apparent from Table 9.6 is that moderate differences in activation energies produce substantially different product ratios.

Table 9.6 Correlation of activation energy differences with enantioselectivity of a reaction.

A → B + C (B and C are enantiomers)				
E_a(C) − E_a(B) (kcal/mol)	$\frac{k_r(\mathrm{B})}{k_r(\mathrm{C})}$	% B	% C	% *ee*
2.72	99	99	1	98.0 (B)
1.30	9	90	10	80.0 (B)
0.95	5	83.3	16.7	66.6 (B)
0.41	2	66.7	33.3	33.4 (B)
0	1	50	50	0
−0.41	0.5	33.3	66.7	33.4 (C)
−0.95	0.2	16.7	83.3	66.6 (C)
−1.30	0.11	10	90	80.0 (C)
−2.72	0.011	1	99	98.0 (C)

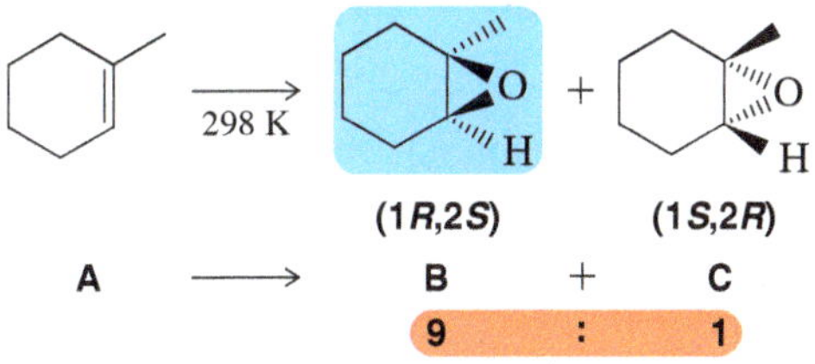

Figure 9.57 A 9:1 ratio implies a difference in activation energies of 1.3 kcal/mol.

Assume that our first attempt at stereoselective epoxidation resulted in a 9:1 ratio of B:C (80% *ee*) (**Figure 9.57**). If B is a potential pharmaceutical agent and there is no way to separate the two enantiomers, then this level of purity is unacceptable. [*Can we improve the selectivity to a usable ratio of 99:1 (98%* ee*) using the same catalyst?*]

To improve the ratio of rate constants, we can't change the E_a difference of 1.30 kcal/mol. The only other variable is the temperature. [*What is the implication of this?*] For a given difference in activation energy, the ratio of products obtained is a function of the temperature at which a reaction is run. The data in **Table 9.7**, generated by plugging different temperatures into Equation 9.2 (which has been rearranged from Equation 9.1), indicate that reactions are more selective at lower temperatures. In other words, lower temperatures increase the impact of the activation energy difference. [*Does this make sense? Jump ahead and scan Figure 9.58 as you think about this.*]

$$\frac{k_r(\mathrm{B})}{k_r(\mathrm{C})} = e^{\frac{E_a(\mathrm{C}) - E_a(\mathrm{B})}{RT}} \qquad \text{Equation 9.2}$$

Table 9.7 At lower temperatures, reactions are more stereoselective.

$A \rightarrow B + C$ (B and C are enantiomers) $E_a(C) - E_a(B) = 1.30$ kcal/mol				
T (K)	$\frac{k_r(B)}{k_r(C)}$	% B	% C	% *ee* (B)
373	5.8	85.2	14.8	70.5
298	9.0	90	10	80
270	11.3	91.9	8.1	83.7
135	16.2	94.2	5.8	88.4
195	28.6	96.6	3.4	93.3
142	100	99	1	98

Figure 9.58 With lots of energy, you'll climb either hill. With less, you will only climb the smaller hill.

By lowering the temperature, there is less energy in the system. Less energy slows down all reactions. But it slows down the reaction with a higher activation energy more, giving a higher ratio of the desired reaction compared to the undesired reaction (Figure 9.58). [*So we can just cool it down more, right? Sadly, no.*] If we continue to lower the temperature, we eventually reach a temperature where neither reaction can overcome their activation energy. Or the reaction might be so slow that it proceeds at an impractical rate. [*Would you wait a week for a reaction to finish? Perhaps if you were going on vacation . . . .*] So the ideal reaction temperature is one where the rate of the desired reaction is reasonable that also minimizes the likelihood of promoting the undesired reaction.

Disappointingly, to produce a ratio of 99:1 (98% *ee*) for a reaction with a difference in activation energies of 1.3 kcal/mol, we need to run the reaction at 142 K (−131 °C), not really possible using standard organic solvents. [*Organic chemists can easily get to −78 °C, though.*] Thus, with a difference in activation energy of only 1.30 kcal/mol, we are unable to achieve a 99:1 ratio with this catalyst. [*Guess we'll need a better catalyst.*]

9.3–9.3.1 SUMMARY

- The enantioselectivity of a reaction depends on the difference in activation energies between the diastereomeric transition states leading to the two different enantiomers. The enantioselectivity also depends on temperature because lower temperatures give higher levels of stereoselectivity.

Assessment 9.39 Questions (a)–(d) all refer to the following reaction, which has been engineered to produce one enantiomer to the exclusion of the other.

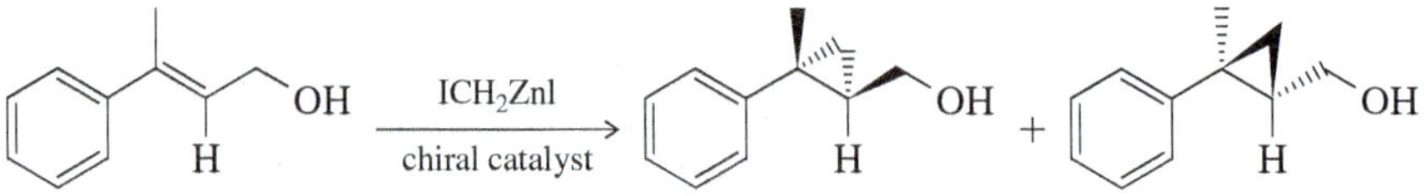

A **B** **C**

(a) If the activation energy for the formation of compound **C** is 2.1 kcal/mol lower than the activation energy for the formation of compound **B**, what ratio (**B**:**C**) would you expect? What is the % *ee* of the reaction ($T = 298$ K)?
(b) What difference in activation energies would give **C** as the predominant enantiomer in 74% enantiomeric excess at 315 K?
(c) Suppose the difference in activation energy is 1.6 kcal/mol. At what temperature would you produce **C** in 99% *ee*?
(d) With an activation energy difference of 1.3 kcal/mol, at what temperature would you get a racemic mixture of **B** and **C**?

9.3.2 Asymmetric epoxidation

Figure 9.59 shows the **retrosynthetic analysis** of indinavir. The necessary fragments are identified by pointing backward to them using a "retrosynthetic arrow." [*We explain more about synthesis when we study alkynes in Chapter 10.*] Our analysis of the synthesis begins with the use of the Sharpless asymmetric epoxidation of allylic alcohols to make the three-carbon epoxide of fragment **A**.

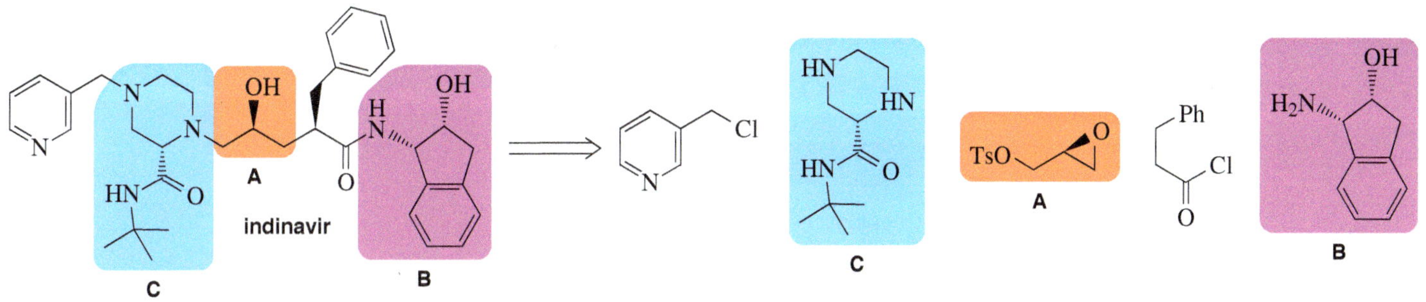

Figure 9.59 Retrosynthetic analysis of indinavir shows the key fragments needed for its synthesis.

9.3.2.1 Sharpless asymmetric epoxidation—synthesis of fragment A

Although not possible using the reactions previously described here in Chapter 9, the enantioselective synthesis of epoxides has been made routine by Nobel Laureate K. Barry Sharpless (**Figure 9.60**). Using a catalyst system involving titanium(IV) isopropoxide and one enantiomer of diethyltartrate, along with the stoichiometric oxidant *t*-butylperoxide, high enantioselectivities are obtained when using **allylic alcohols** as the organic substrate.

Figure 9.60 Sharpless asymmetric epoxidation. Epoxidation in a chiral environment produces a single enantiomer.

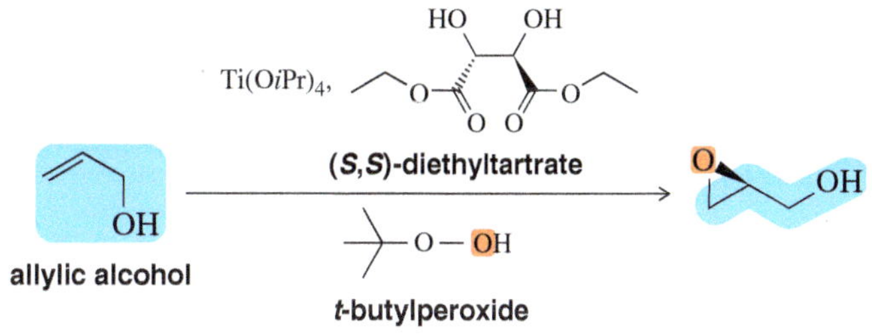

[*Do me a favor. Jump ahead and look at the mechanism in Figure 9.62. Are you back? Looks pretty complicated—some say too complicated. I disagree. Don't get distracted by the flashing lights; it's essentially two reactions. One that we don't know yet is repeated over and over. And the other is identical to one we've seen before. We describe the new one first.*]

Titanium(IV) isopropoxide is a strong Lewis acid. Its most important role in this reaction is bringing the oxidizing agent, the chiral environment [(*S*,*S*)-diethyltartrate], and the allylic alcohol together so that enantioselective oxidation can occur. Six of the reactions in the catalytic cycle are simply substitutions of alkoxy groups on titanium. As shown in **Figure 9.61**, we

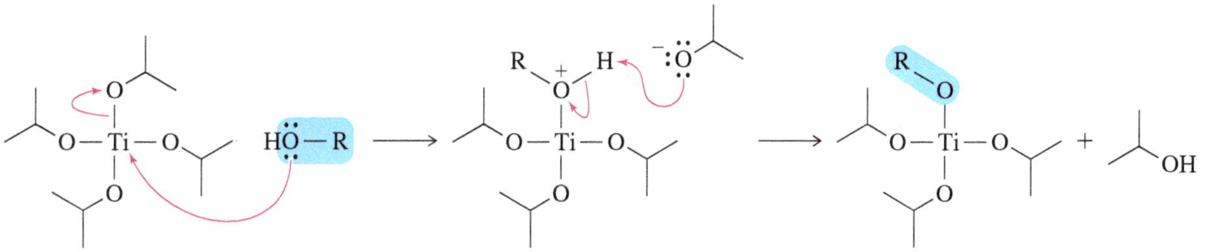

titanium(IV) isopropoxide
Ti(O*i*Pr)$_4$

Figure 9.61 Exchange of one alkoxy group for another. This happens six times in the Sharpless asymmetric epoxidation catalytic cycle. [*In Figure 9.62, these are referred to as alkoxy exchanges.*]

can imagine a substitution reaction proceeding by displacement of isopropoxide and subsequent deprotonation. We discuss more about this reaction in Chapter 12, but having an alkoxy group replace another should seem reasonable. It is through these reactions that the chiral complex is assembled around titanium.

The mechanism of the Sharpless asymmetric epoxidation is shown in **Figure 9.62**. The epoxidation step, highlighted in blue, occurs by an identical mechanism to epoxidation with *m*CPBA. The rest of the steps are simply the alkoxy exchanges first shown in Figure 9.61. Because the reaction happens in the chiral environment assembled around titanium, a single epoxide enantiomer is obtained.

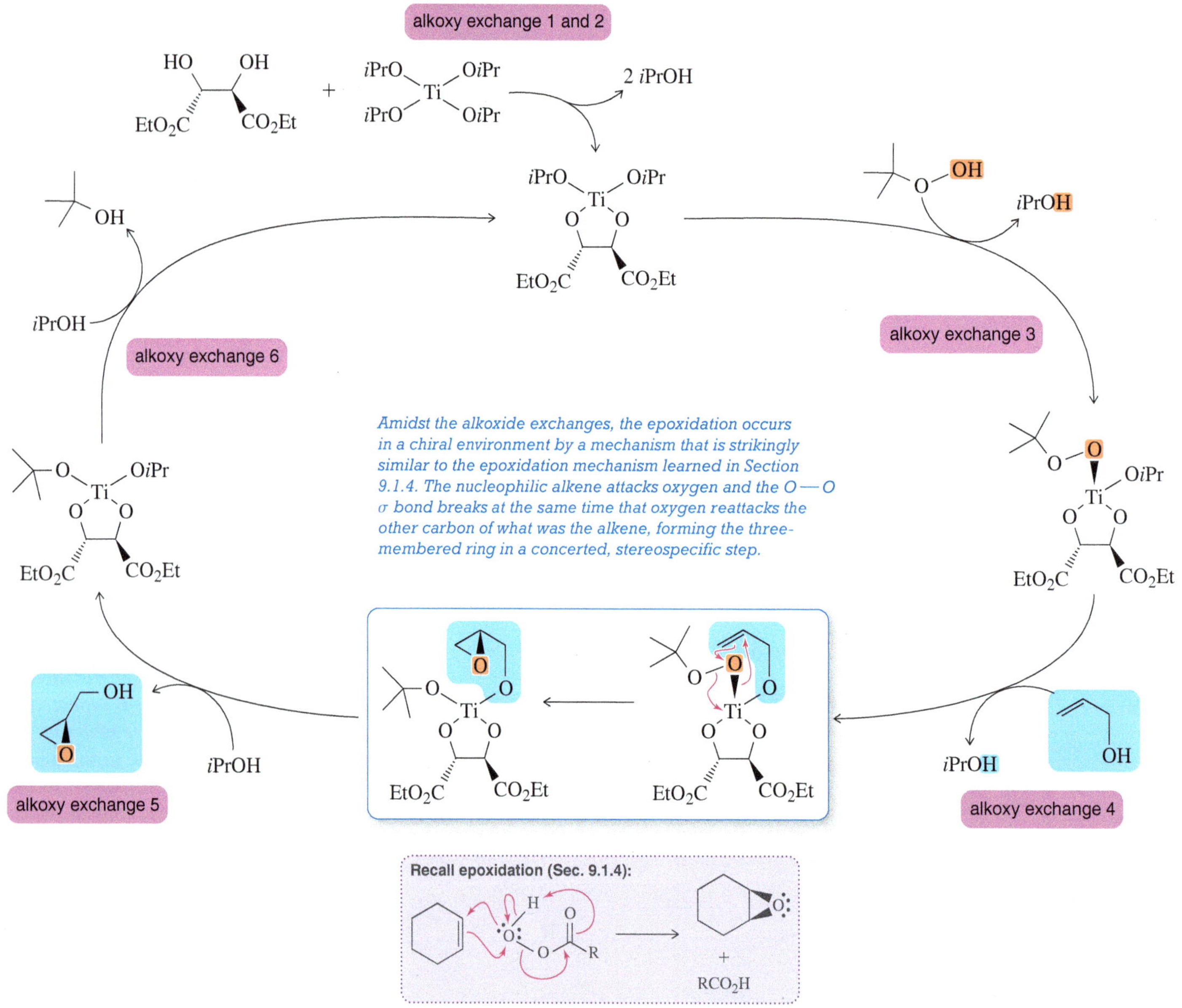

Figure 9.62 **The mechanism for the Sharpless asymmetric epoxidation.**

9-E CHEMISTRY IN CONTEXT: Ecology

Preventing gypsy moth infestation

The gypsy moth (*Lymantria dispar*) was introduced to the United States by a scientist hoping to breed this European insect with native moths to make them more resistant to disease. Unfortunately, some of the imported moths escaped. With no local predator, the gypsy moth has become a major pest to northeastern hardwood trees. The gypsy moth is one of the most destructive insects in the eastern United States, causing a reported $868 million in damages annually. [*A cautionary tale about interfering with nature, to be sure.*]

While pesticides have been used successfully to eradicate large infestations of gypsy moths, a more environmentally friendly method of controlling populations is to disrupt gypsy moth mating patterns. Female gypsy moths use the sex pheromone disparlure (*cis*-7,8-epoxy-2-methyloctadecane) to attract interested males to mate. When signs of gypsy moth invasion are evident, scientists blanket the area with the synthetic pheromone, confusing the male gypsy moths, resulting in diminished mating. Because this pheromone is unique to this species and is basically a natural perfume, it has no deleterious effects on other species, including humans.

The Sharpless asymmetric epoxidation was first used industrially for the synthesis of disparlure (**Figure 9.63**). Beginning with the C_{13} allylic alcohol, the epoxide was produced in greater than 98% enantiomeric excess. [*Less than 1% of the mixture was the undesired enantiomer.*] Notably, this reaction has been run on greater than 10 kg scale. [*Trust me—successful reactions on this scale are impressive.*] Further reactions of the 2,3-epoxy alcohol (not covered in this chapter) resulted in the synthesis of disparlure.

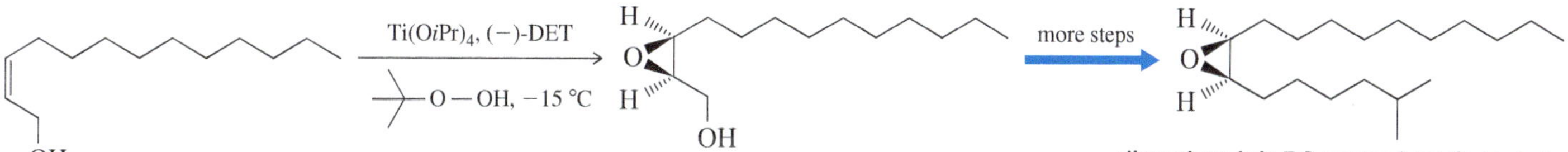

disparlure (*cis*-7,8-epoxy-2-methyloctadecane)
> 98% *ee*

Figure 9.63 **Synthesis of disparlure, a gypsy moth pheromone, using the Sharpless asymmetric epoxidation.**

The synthesis of fragment A of indinavir utilized the Sharpless asymmetric epoxidation of the simplest allylic alcohol, prop-2-en-1-ol, to produce the (*R*)-epoxyalcohol with nearly perfect enantioselectivity (**Figure 9.64**). Fragment A is completed by sulfonylation of the terminal hydroxyl group. [*We introduce this reaction in Chapter 13.*]

Figure 9.64 **Synthesis of fragment A of indinavir using the Sharpless asymmetric epoxidation.**

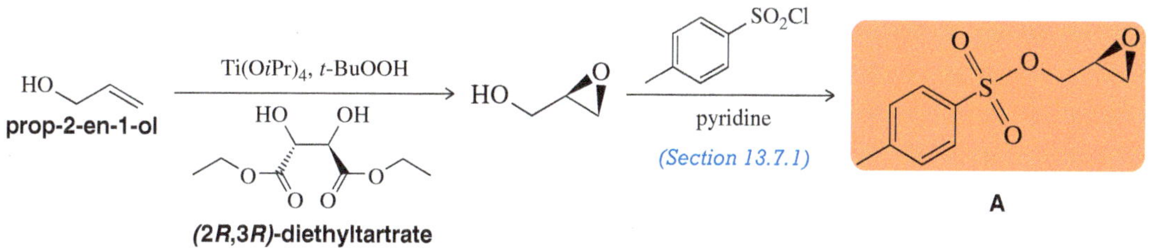

Unfortunately, the Sharpless epoxidation is significantly less stereoselective without a hydroxyl group at the neighboring carbon. [*Does this make sense? What was the role of the neighboring hydroxyl group?*] Complexation between the hydroxyl group and the titanium atom is required to assemble the rigid complex in which the stereoselective epoxidation occurs [*see Figure 9.62*]. As such, the Sharpless reaction only works with allylic alcohols.

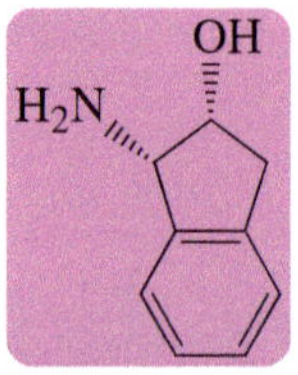

B
(1*S*,2*R*)-(−)-*cis*-1-amino-2-hydroxyindane

9.3.2.2 Jacobsen epoxidation—synthesis of fragment B

The synthesis of indinavir required the use of enantiopure fragment B for two reasons. First, it is important for the protease inhibitor activity of the drug. Second, it provides a chiral environment that controls the stereoselective introduction of the benzyl group at C_2. [*We return to this in Chapter 20.*]

The synthesis of fragment B required the epoxide shown in **Figure 9.65**. With no allylic alcohol to anchor the alkene to the titanium complex, a Sharpless epoxidation would not have given the required selectivity. For asymmetric epoxidations like this, the more versatile catalyst first prepared by Jacobsen and Katsuki is more successful. As an example, the desired epoxide for the synthesis of indinavir was prepared in 88% enantiomeric excess by chemists at Merck. The 88% enantiomeric excess is impressive, given that this is an unfunctionalized alkene. [*I know 88% doesn't seem high, but you'll have to trust me on this one.*]

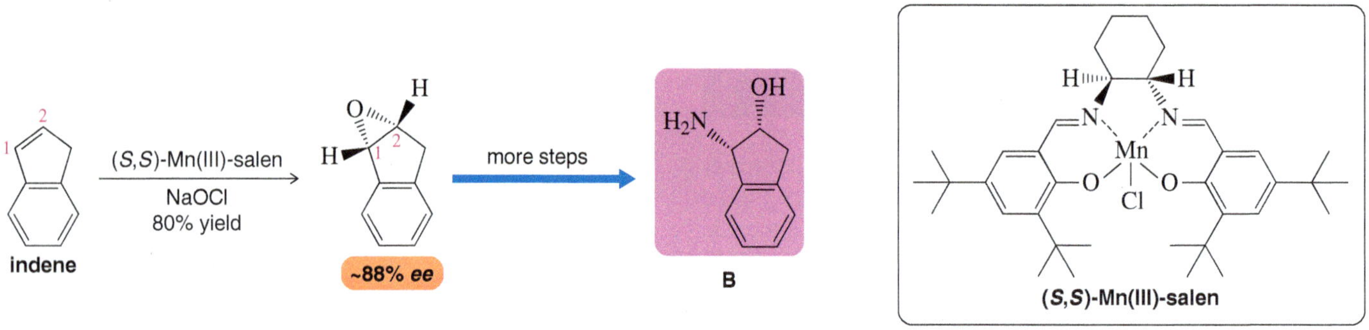

Figure 9.65 The Jacobsen epoxidation is used to make fragment B of indinavir.

Although the mechanism of epoxidation with Jacobsen's catalyst is not well understood, there are still aspects that can be compared to the Sharpless reaction. The catalyst provides the chiral environment [*the Sharpless reaction used diethyl tartrate for this*] that makes the reaction selective and a transition metal [*Ti in Sharpless, Mn in Jacobsen*] is involved. As shown in **Figure 9.66**, the sodium hypochlorite oxidizes the manganese by a familiar mechanism. [*Read it out loud:*] Manganese [*an alkene*] attacks oxygen [*bromine*] and kicks out a

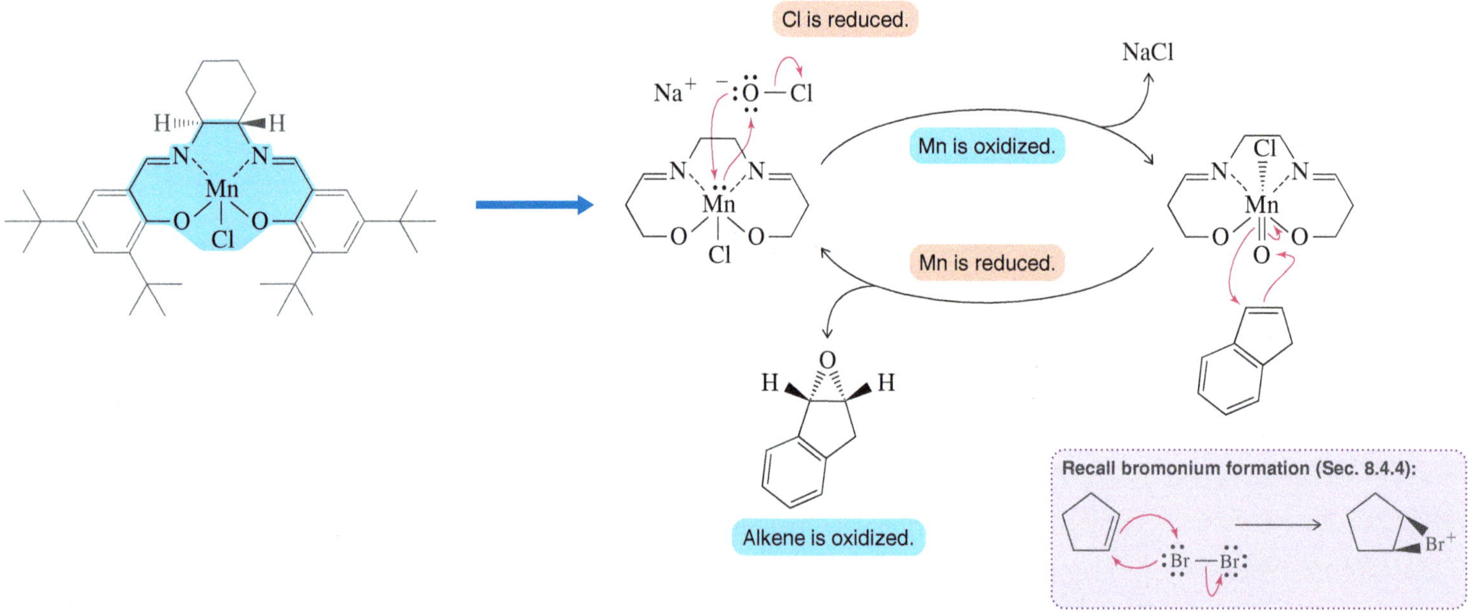

Figure 9.66 The mechanism of the Jacobsen epoxidation. The chiral environment results in a stereoselective reaction.

good leaving group, while at the same time oxygen [*bromine*] reattacks the manganese [*carbon of the alkene*] to produce the active oxidizing agent. [*You may want to read that sentence again.*] When approached by the alkene, the oxygen atom is delivered in a concerted reaction to give the epoxide. [*Once more, a familiar mechanism represented by those same three arrows.*] Because of the chiral environment, this reaction happens on one face of the molecule to give one enantiomer in excess.

Assessment 9.40 Identify the alkene that would react with $Ti(OiPr)_4$, (+)-diethyltartrate, and *t*-butylhydroperoxide to give the following chiral, nonracemic epoxides.

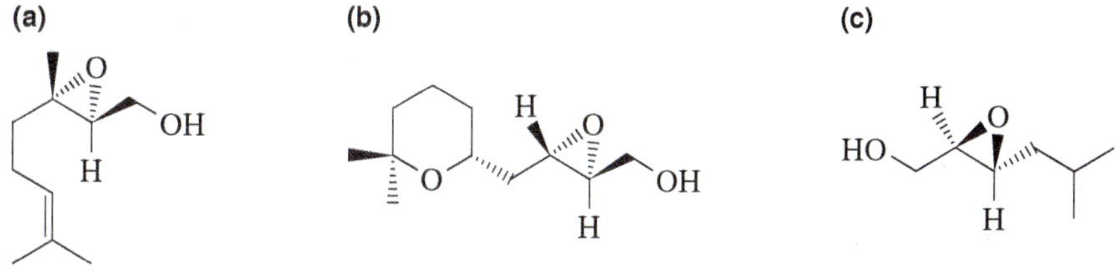

9.3.2.3 Asymmetric hydrogenation—synthesis of fragment C

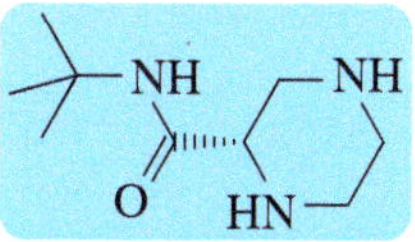

C

Finally, we come to the synthesis of fragment **C**, which can be synthesized using alkene hydrogenation in a chiral environment.

The evolution of enantioselective alkene hydrogenation began in 1965 with the discovery of Wilkinson's catalyst, $RhCl(PPh_3)_3$ (**Figure 9.67**). Inspired by its ability to hydrogenate alkenes (similar to Pd/C in this regard), William Knowles added chiral ligands to rhodium in order to create a chiral environment for the reduction. Using bidentate ligands, a rigid and sterically defined complex allows reduction to occur on one face of the alkene. The development of asymmetric hydrogenation continued when Ryōji Noyori utilized the chiral ligand BINAP and ruthenium(II) acetate for similar transformations.

Don't worry about the specific structures or how they were designed. That is a topic for an organometallics course. Instead, focus on the big picture: These complexes create a chiral environment for the reactions.

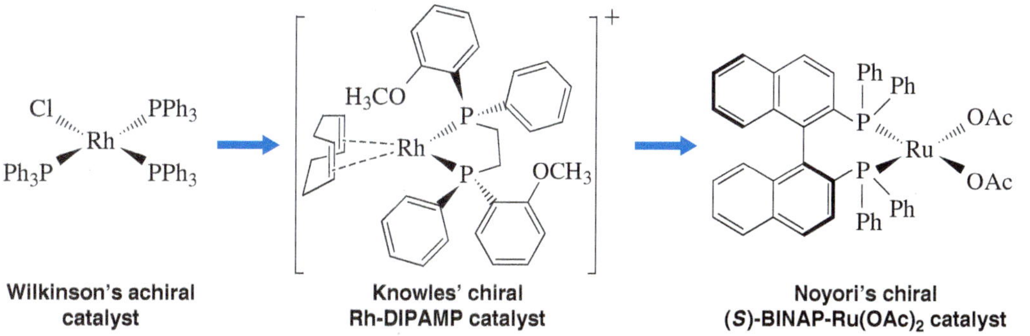

Figure 9.67 Previous work with Wilkinson's catalyst led to the development of Knowles' chiral rhodium hydrogenation catalyst and Noyori's chiral ruthenium catalyst. [*It's cool to see how one idea inspires the next. Perhaps your idea is next?*]

The Rh-DIPAMP catalyst was one of the first asymmetric catalysts to be used industrially. Knowles and coworkers at Monsanto used it to synthesize L-DOPA, a drug that increases the dopamine concentration in the brains of Parkinson's patients [**Figure 9.68(a)**]. Noyori's catalyst has been successfully used for a number of highly enantioselective hydrogenations, including the synthesis of naproxen, the active ingredient in the pain reliever Aleve®, in 97% *ee* [**Figure 9.68(b)**].

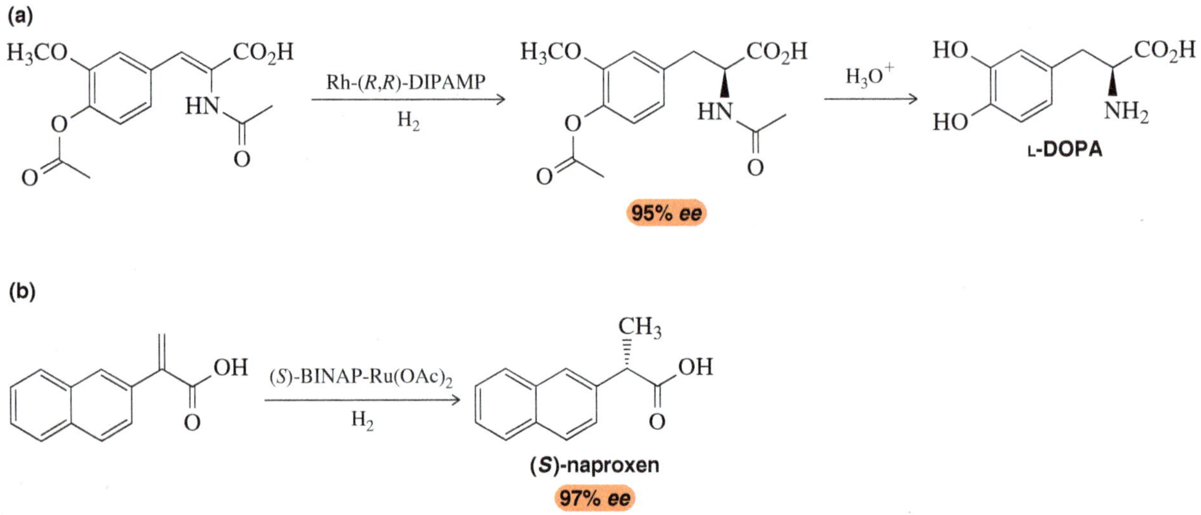

Figure 9.68 Asymmetric hydrogenation in industry. (a) The industrial synthesis of L-DOPA using Knowles' catalyst and (b) the synthesis of naproxen using Noyori's catalyst.

Returning to indinavir, the synthesis of fragment C utilized a chiral rhodium complex (**Figure 9.69**). [*When you encounter complicated molecules with numerous functional groups, ignore the "shiny objects." Treat it as an alkene with a variety of other groups coming along for the ride. Although it will take time, this is an important skill to develop. It has numerous real-life applications.*] Once again, the chiral catalyst provides a chiral environment, allowing the hydrogenation reaction to happen selectively from one side of the alkene.

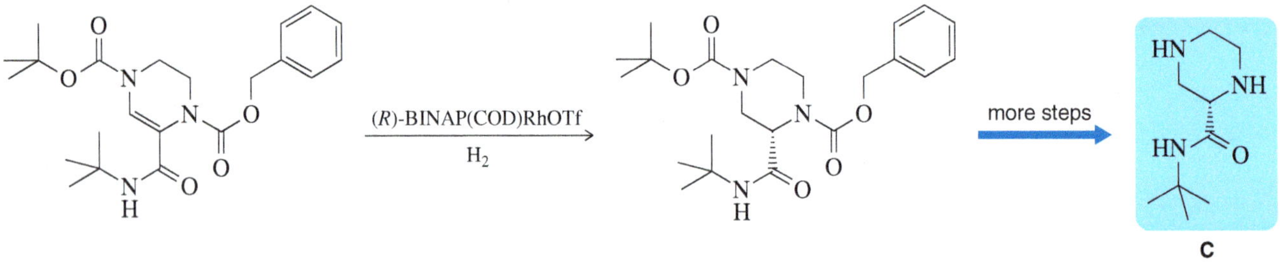

Figure 9.69 The synthesis of fragment C of indinavir using asymmetric hydrogenation.

Assessment 9.41 The chiral catalyst (*R*)-BINAP(COD)RhOTf was used in a hydrogenation reaction as part of the synthesis of fragment C of indinavir. Using the same alkene, predict the product that would be obtained if (*S*)-BINAP(COD)RhOTf were used instead.

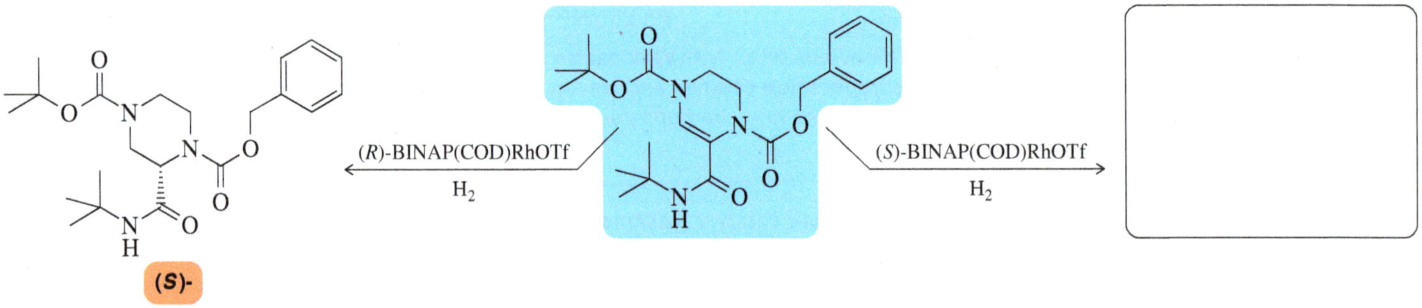

9.3.4 Completion of indinavir synthesis

Figure 9.70 shows the complete synthesis of indinavir. [*Perhaps it inspires, perhaps it intimidates; either way, know that by the end of this book, you will fully understand each of these reactions.*]

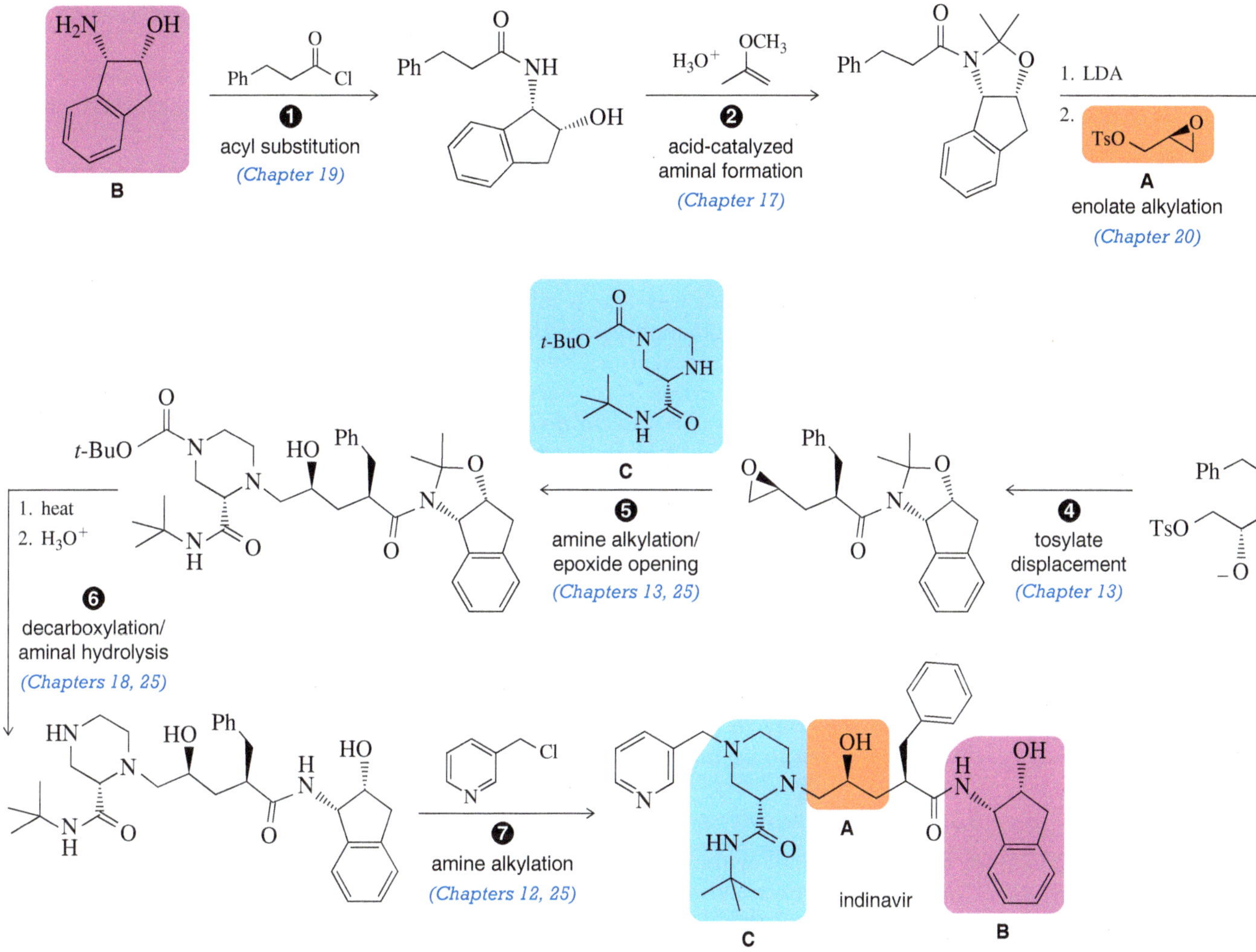

Figure 9.70 **The entire indinavir synthesis.**

9.3.5 Importance of asymmetric synthesis

Even though chemists at Merck successfully synthesized indinavir by the route in Figure 9.70, this is not how it is currently produced. [*Why? I'm glad you asked.*] Within the pharmaceutical industry, several factors guide the decision of which route to use—namely, reproducibility, cost-effectiveness, and the ease of purification (of intermediates and the final product). For whatever reason, it was decided that the route featuring asymmetric epoxidation and hydrogenation was not optimal. Yet by analyzing the route Merck used, we see why asymmetric synthesis is so important. As shown in **Figure 9.71**, a chiral resolution [*recall Section 6.6.1*] was done after three steps to obtain fragment B as a single enantiomer. [*Don't worry about the reactions used, although the first is a greener epoxidation.*]

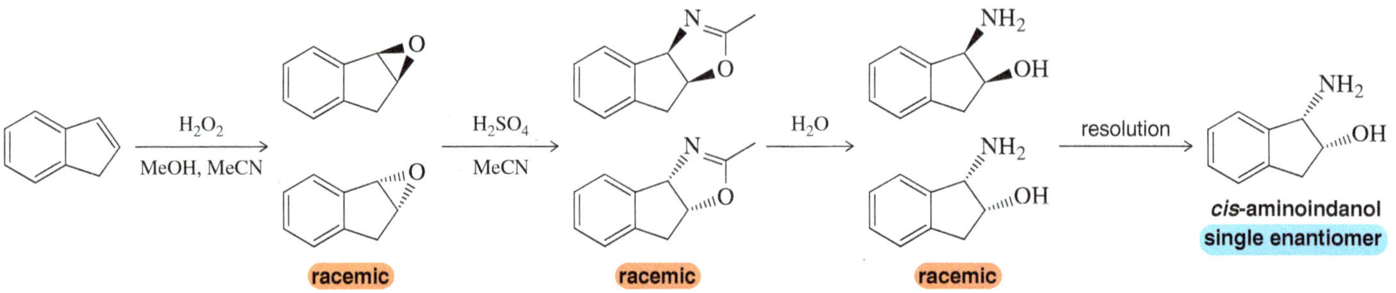

Figure 9.71 An alternate route to fragment B of indinavir.

Only half of the material produced by this route is useful for the synthesis of indinavir. Resolving enantiomers is wasteful; being able to prepare them in high purity and high enantiomeric excess, using cheap, environmentally friendly catalytic methods, is better. Catalytic asymmetric synthesis is the future (and the present, really), offering access to medicines and other useful molecules in a way that is mindful of the environment, minimizes waste, and reduces the buildup of potentially toxic chemicals. [*It would be irresponsible to not point out here the fact that the reagents used in Figure 9.71 are regarded as safe, nontoxic, and inexpensive. With the exception of the last step, it is a generally green route. As discussed in Chapter 7, the evaluation of routes for their greenness is nuanced.*]

Recognizing the importance of asymmetric reactions, Sharpless, Noyori, and Knowles shared the 2001 Nobel Prize in Chemistry for this work. Specifically, for Knowles and Noyori, half of the prize was awarded "for their work on chirally-catalysed hydrogenation reactions." Sharpless was awarded his share of chemistry's highest honor "for his work on chirally-catalysed oxidation reactions." [*It is humorously satisfying that when an oxidation reaction receives the Nobel, it is accompanied by a reduction reaction in the same year.*] Given the many successes of asymmetric reactions, that it wasn't used for the indinavir synthesis is the biggest news of all. It illustrates the ongoing need for more robust, more selective, less expensive, and more environmentally friendly catalysts. [*As a future organic chemist, it's exciting to think about the problems that remain to be solved. Let's hurry through this book so we can begin working on them.*]

9.3.2–9.3.5 SUMMARY

- Reactions that allow the direct preparation of compounds with high levels of enantioselectivity are extremely important. The Sharpless asymmetric epoxidation, the Jacobsen epoxidation, and the Noyori hydrogenation are three examples of asymmetric catalysis that have found broad utility in the chemical industry.

Assessment 9.42 Suggest a method for the separation (resolution) of amino alcohol enantiomers. [*Hint: Look back at Sections 6.6.1 and 6.6.2.*]

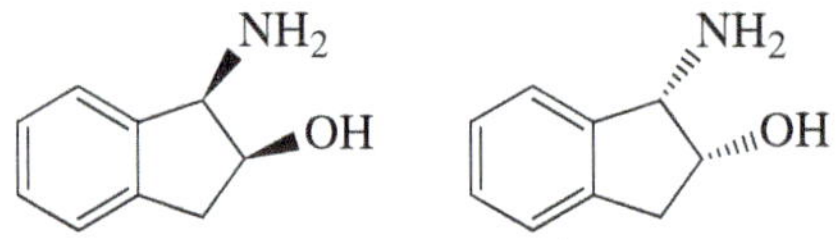

amino alcohol enantiomers

Concluding thoughts: HIV and Crixivan

Upon reaching the market in 1996, Crixivan was the eighth antiretroviral medication approved by the FDA for treatment of HIV. It has been suggested that medications like it changed the course of the HIV/AIDS epidemic by making the disease manageable. Two years after the release of the first protease inhibitors, annual deaths in the United States fell by

Unfortunately, I have no answers to solving these problems. And yet, as organic chemists, future medicinal chemists, biochemists, and healthcare professionals, we do have a role to play. For now, that role is to continue learning the ways that we can use organic chemistry to help address these issues. We return to doing just that in Chapter 10.

more than 64%. Crixivan was one of the first drugs to be used as part of a drug cocktail aimed at multiple enzyme targets concurrently. The success of these and other antiretrovirals might lead you to believe that the HIV crisis has been solved. And the advances made every day in early detection and treatment for the disease give confidence that, at least medically, a solution/cure is imminent. For those in the United States with the means to afford the best medications, it is now possible to live a relatively normal life with HIV. For those *without* access to healthcare, however, the disease is as deadly as ever; this is especially true in developing nations, such as those in sub-Saharan Africa. In many ways, then, HIV/AIDS has moved from a medical problem to a social problem. It is a problem of rich versus poor, affluent versus marginalized, and educated versus uneducated.

End-of-Chapter Materials

Important Concepts

[*These are the major concepts covered in each section, excluding reactions.*]

9.1.1: Oxidation numbers are hypothetical numbers assigned to atoms in a molecule that allow the flow of electrons in a redox reaction to be tracked.

9.1.2: The electrophilic addition of Br_2 or Cl_2 results in the stereospecific preparation of vicinal dihalides. Following nucleophilic attack of the alkene on the halogen and formation of the bromonium ion, nucleophilic attack of a bromide ion opens the three-membered ring from the opposite side.

9.1.3: In the presence of water, the bromonium ion formed by attack of an alkene on Br_2 is attacked by water from the opposite side. Water always attacks at the more substituted carbon to give a high level of regio- and stereoselectivity.

9.1.4: In what is essentially the addition of elemental oxygen, alkenes can be oxidized with peroxyacids to give epoxides (oxiranes). Proceeding through what is referred to as the butterfly transition state, the reaction is highly stereospecific, resulting in the syn addition of oxygen to the alkene. Although the reaction produces a high yield of epoxide, the reaction suffers from a very poor atom economy; as such, research is ongoing to develop greener epoxidations.

9.1.5: Alkenes are oxidized to give vicinal diols by reaction with OsO_4 in a concerted reaction that delivers both hydroxyl groups to the same face of the molecule. Because OsO_4 is a highly toxic reagent, systems have been developed that utilize substoichiometric OsO_4 along with a stoichiometric co-oxidant.

9.1.6: Concerted addition of hydroxyl groups results in syn addition, so it is necessary to add the hydroxyl groups in a stepwise fashion in order to generate vicinal trans-diols. Epoxidation of an alkene, followed by the acid-catalyzed hydrolysis of the epoxide, is used to stereospecifically generate trans 1,2-diols.

9.1.7: The reaction of ozone (O_3) with an alkene, referred to as ozonolysis, results in cleavage of the alkene to give two carbonyls.

9.2: The metal-catalyzed hydrogenation of alkenes results in the stereospecific addition of hydrogen to produce a fully saturated alkane. Proceeding by a concerted mechanism, the reaction is presumed to take place at the surface of an insoluble metal, making this an example of heterogeneous catalysis. By measuring the heat of hydrogenation ($\Delta H°$), the stability of isomeric alkenes can be calculated. The hydrogenation of vegetable oils can be used to make partially hydrogenated oils, which are often contaminated by trans fats.

9.3–9.3.1: The enantioselectivity of a reaction depends on the difference in activation energies between the diastereomeric transition states leading to the two different enantiomers. If the difference is large, the reaction will be highly stereoselective. If the difference is small, there will be very little preference for one stereoisomer. The enantioselectivity also depends on temperature because lower temperatures give higher levels of stereoselectivity, provided that there is enough energy in the system to allow the reaction to proceed at an appreciable rate.

9.3.2–9.3.5: The preparation of stereochemically pure compounds is a necessary and important aim in the development of drugs by the pharmaceutical industry. As a result, reactions that allow the direct preparation of compounds with high levels of enantioselectivity are extremely important. The Sharpless asymmetric epoxidation, the Jacobsen epoxidation, and the Noyori hydrogenation are three examples of asymmetric catalysis that have found broad utility in the chemical industry, as demonstrated by their use in the Merck synthesis of indinavir, an important drug used in the fight against HIV.

Important Reactions and Mechanisms

Major category I: Alkene oxidation reactions

1. **Halogenation (Section 9.1.2):** Electrophilic addition of bromine to give a bromonium ion is followed by bromide opening of the bromonium to give a vicinal trans-dibromide in a stereospecific reaction. Halogenation works in a similar manner with Cl_2. Because the bromonium could form on either side of the alkene with equal rates (enantiomeric transition states), the reaction results in a racemic mixture.

General:

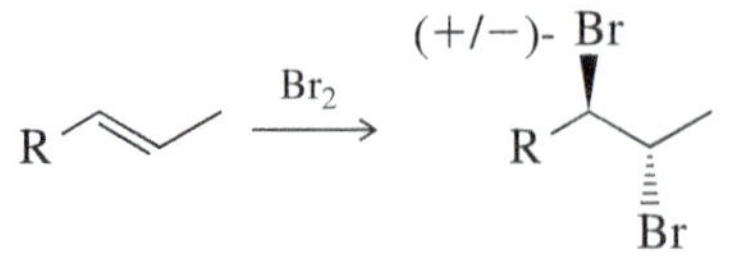

Specific:

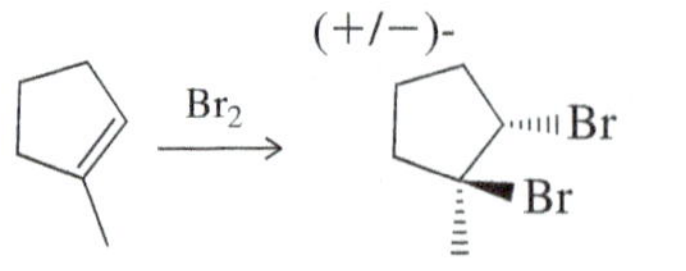

Mechanism:

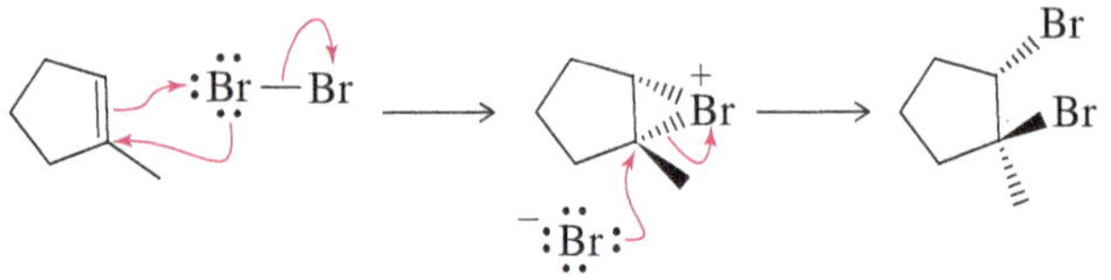

2. **Halohydrin formation (Section 9.1.3):** Electrophilic addition of chlorine to give a chloronium ion is followed by opening of the chloronium ion to give a halohydrin in a regiospecific and stereospecific reaction. Bromohydrins can be formed similarly using Br_2 in water. Because the chloronium ion can form on either side of the alkene with equal rates (enantiomeric transition states), the reaction results in a racemic mixture. Once formed, the chloronium is attacked at the most substituted carbon by the water from the opposite face, ensuring the trans relationship between the vicinal chlorine and the hydroxyl group.

General:

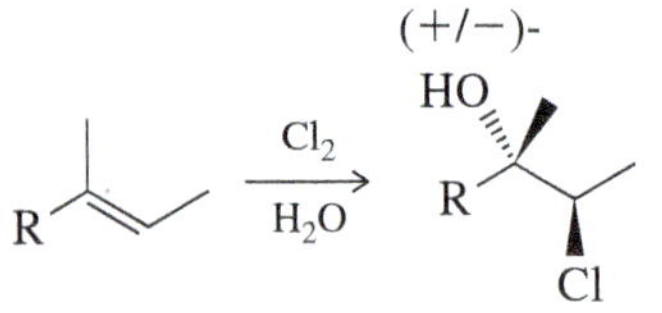

Specific:

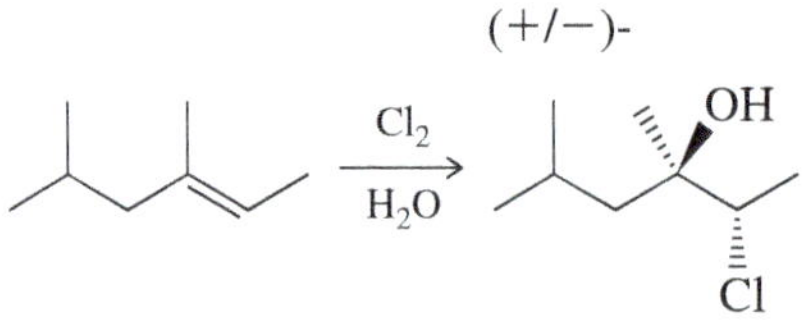

Mechanism:

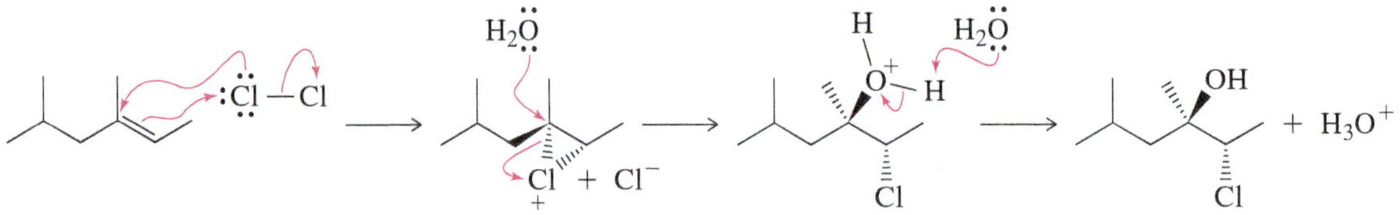

3. **Epoxidation (Section 9.1.4):** The terminal oxygen of a peroxyacid is attacked by an electron-rich π bond, triggering a concerted cascade featuring the cleavage of a weak O—O σ bond, formation of a protonated three-membered ring, and deprotonation of the three-membered ring to yield the epoxide (oxirane) product. Because the reaction is concerted, epoxide formation happens before there is any opportunity to rotate around a σ bond; as a result, the reaction is stereospecific. This concerted mechanism has been referred to as the butterfly transition state. Because the alkene can attack the peroxyacid from either side with equal rates (enantiomeric transition states), the reaction results in a racemic mixture of products. Several asymmetric versions of this reaction have been developed (see Section 9.3.2).

General:

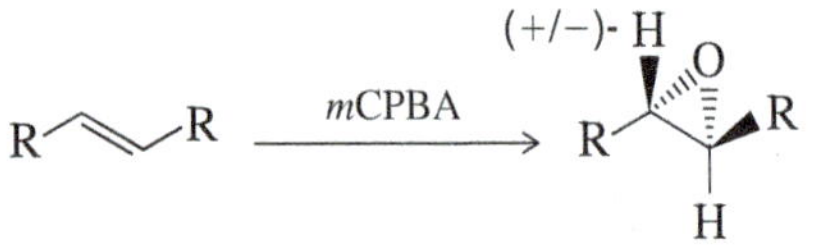

Specific:

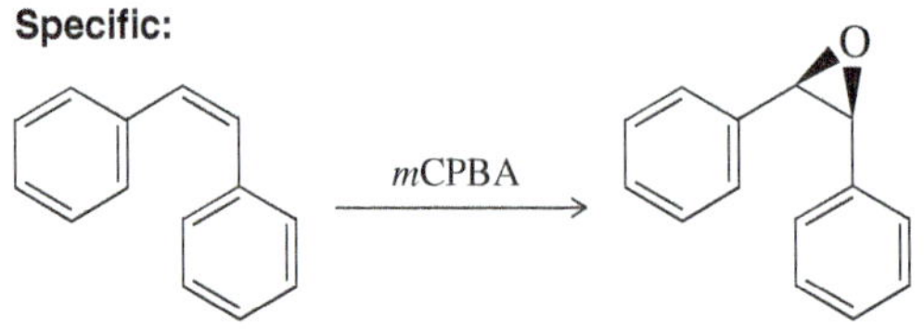

Mechanism:

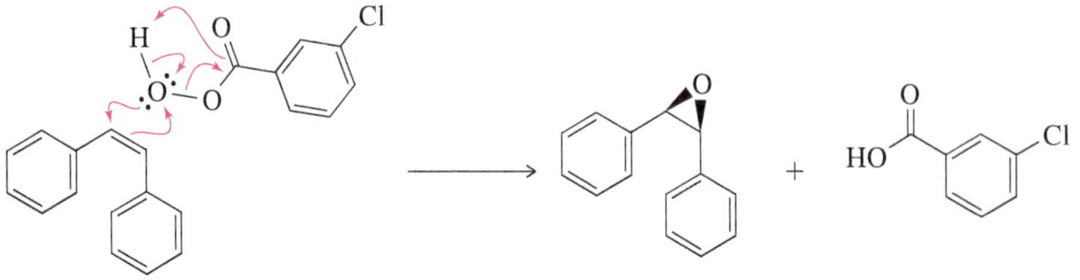

4. **Dihydroxylation (Section 9.1.5):** The strong oxidizing agent OsO_4 delivers one oxygen to each alkene carbon in a concerted reaction featuring a cyclic 6-electron transition state to give the five-membered osmate ester. Because the reaction is concerted, osmate ester formation happens before there is any opportunity to rotate around a σ bond; as a result, the reaction is stereospecific. Because the alkene can attack OsO_4 from either side with equal rates (enantiomeric transition states), the reaction results in a racemic mixture of products. In a second step, $NaHSO_3$ cleaves the osmate ester to generate the vicinal cis-diol via an unimportant and poorly understood mechanism.

General:

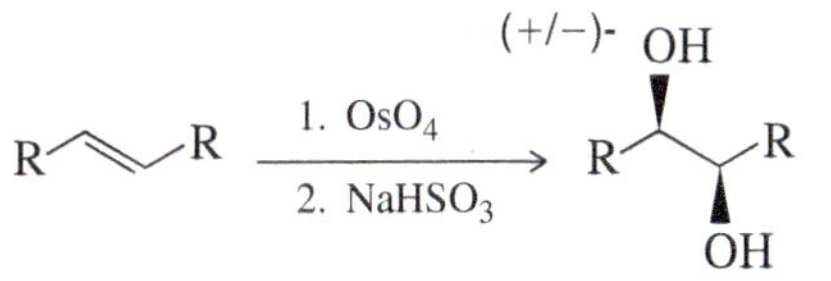

Specific:

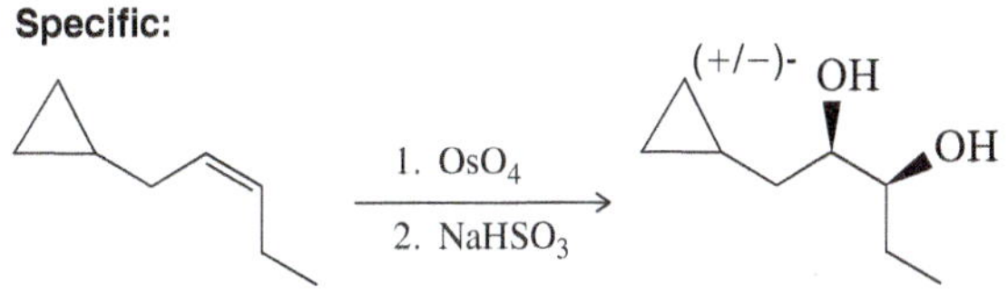

Mechanism:

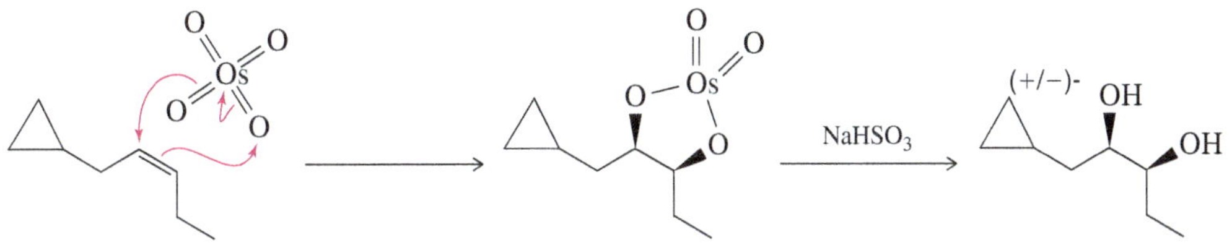

5. **Epoxidation/hydrolysis (Section 9.1.6):** Useful for the synthesis of trans-1,2-diols, epoxidation of an alkene using a peroxyacid (see 3. Epoxidation) is followed by acid-catalyzed ring opening. Protonation of the oxygen of the epoxide ring weakens the C—O bond at the most substituted carbon, initiating nucleophilic attack by water. In an analogous fashion to the chemistry of the bromonium or halonium ion, attack occurs at the more substituted carbon from the opposite face of the oxonium ion, resulting in the preparation of the trans-diol in a stereospecific reaction. Because the epoxide can be formed with equal rates on either side of the alkene (enantiomeric transition states), this reaction produces a racemic mixture of diols.

General:

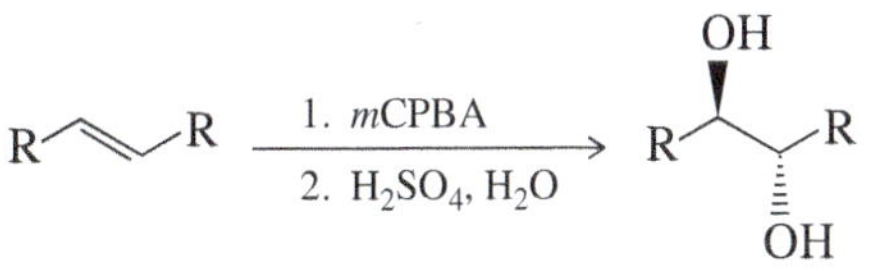

Specific:

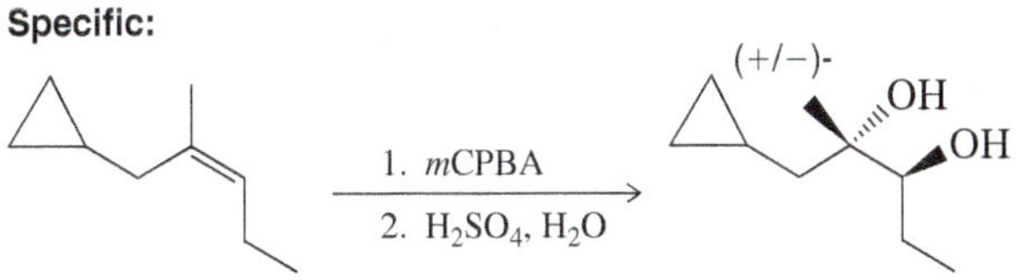

Mechanism:

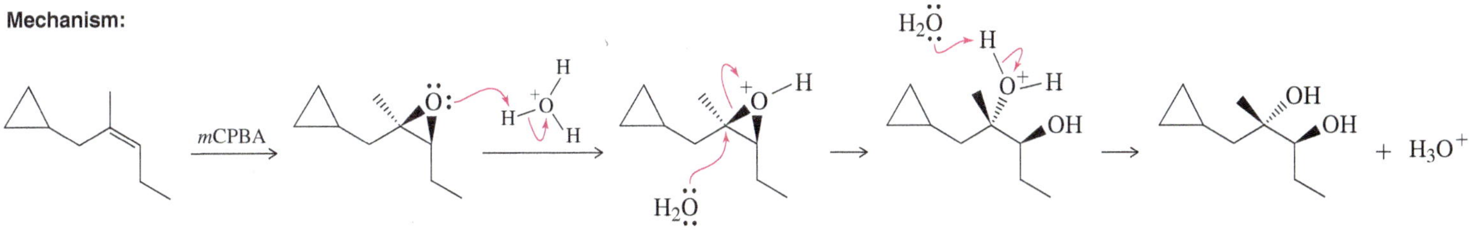

6. **Ozonolysis (Section 9.1.7):** The highly reactive oxidizing gas ozone oxidatively cleaves both the σ and π bonds of an alkene in a three-step process to initially generate a five-membered ozonide. This reactive ozonide is reduced on exposure to dimethyl sulfide, producing carbonyl compounds at each of the original alkene carbons while also producing dimethyl sulfoxide (DMSO).

General:

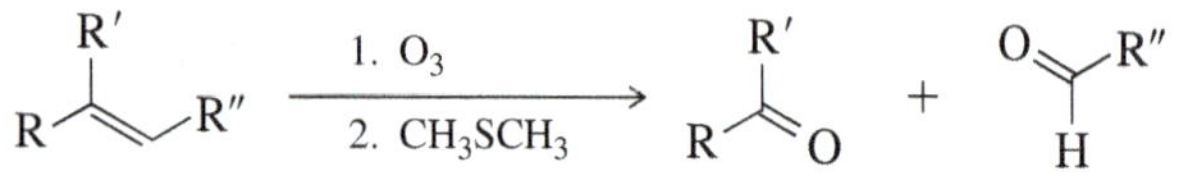

Specific:

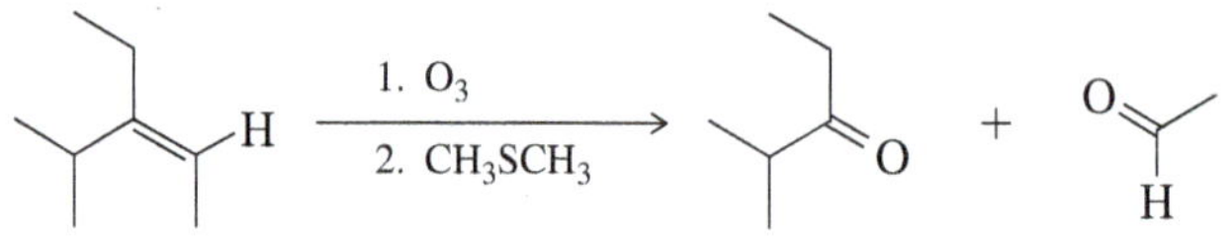

Mechanism (step 1):

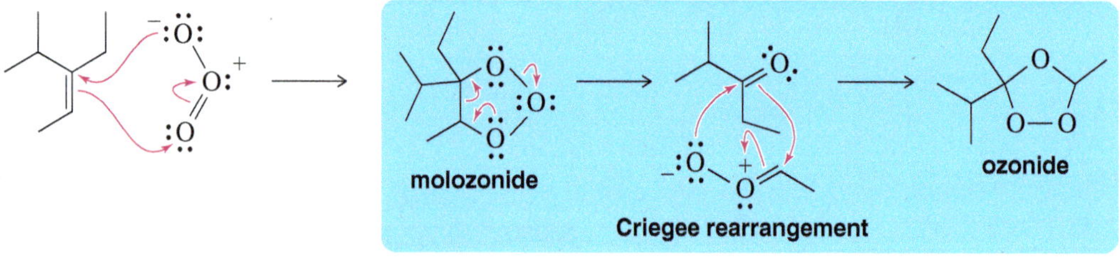

Mechanism (step 2):

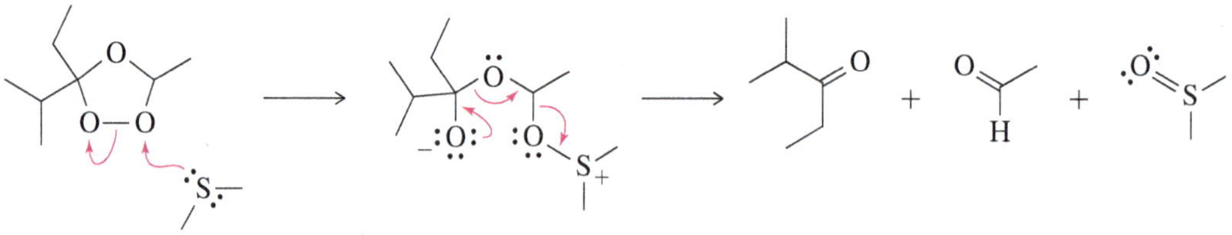

Major category II: Alkene reduction reactions

7. **Catalytic hydrogenation (Section 9.2):** Using a metal catalyst such as Pd, Pt, or Ni, H_2 can be added across a double bond, resulting in a fully saturated compound. Though not well understood, the mechanism of this reaction is assumed to involve the adsorption of hydrogen gas and the alkene to a metal surface, facilitating the rapid addition of a hydrogen atom to each carbon of the original alkene. Each hydrogen is delivered in a concerted manner before there is a chance for bond rotation; as a result, this reaction is stereospecific. Because the alkene can attack the reactive catalyst from either side with equal rates (enantiomeric transition states), the reaction results in a racemic mixture of products. The reaction can be done stereoselectively by the addition of a chiral metal catalyst. The hydrogenation reaction has been utilized in the food industry for the synthesis of partially hydrogenated oils, saturated fats, and, unfortunately, trans fats.

General:

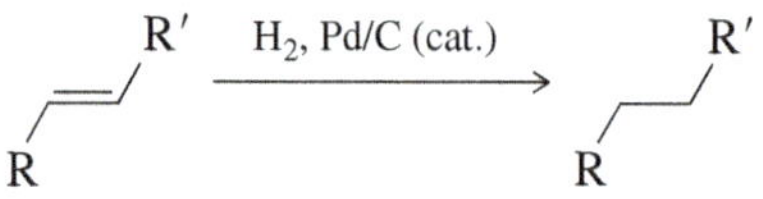

Specific:

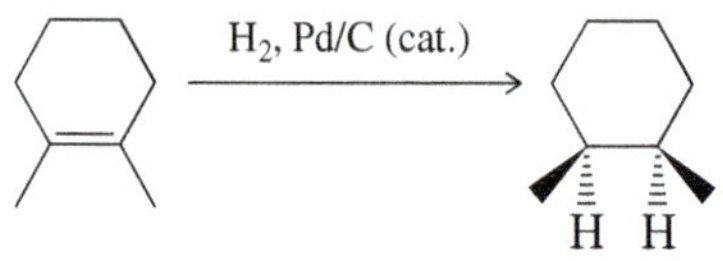

Mechanism:

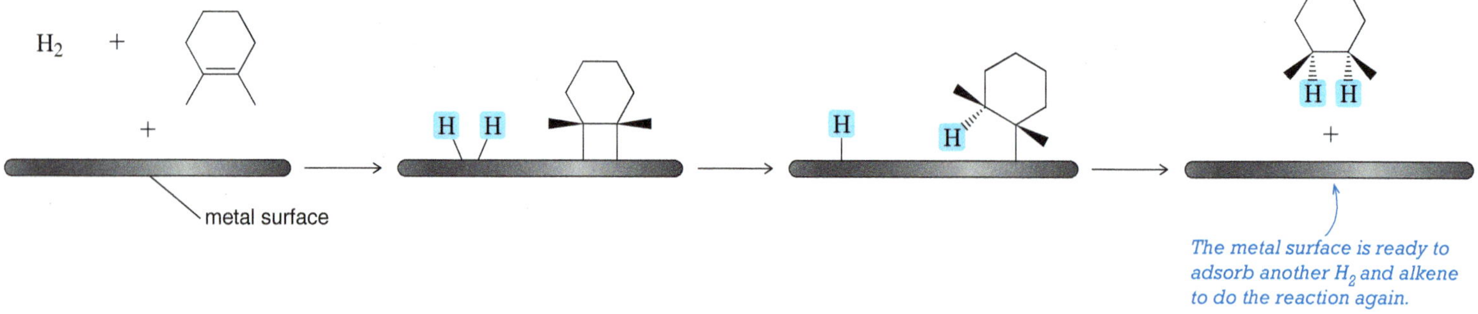

End-of-Chapter Assessments

[Assessments are ranked according to their difficulty; •••• = most difficult.]

9.43 (•) Calculate the oxidation number for the indicated carbons.

(a)

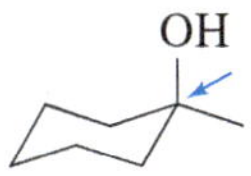

(b)

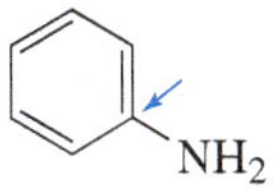

(c)

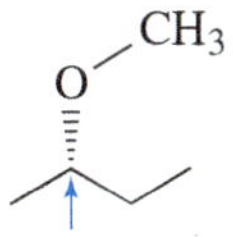

(d)

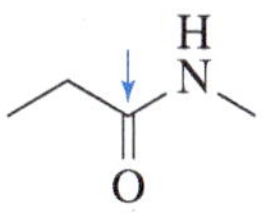

9.44 (••) Determine whether the following reactions are redox reactions. If they are, identify whether the molecule has been oxidized or reduced.

(a)

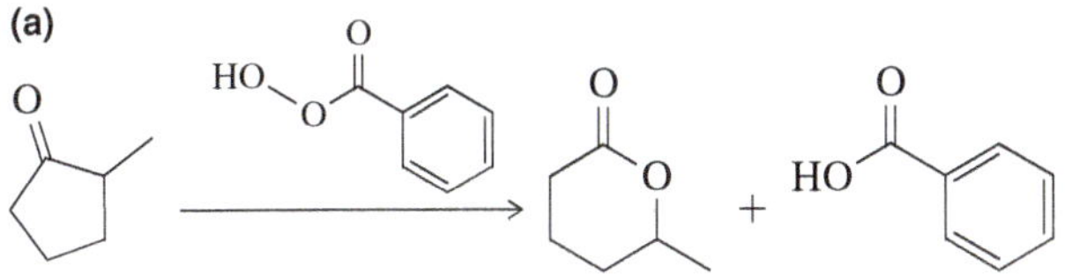

(b)

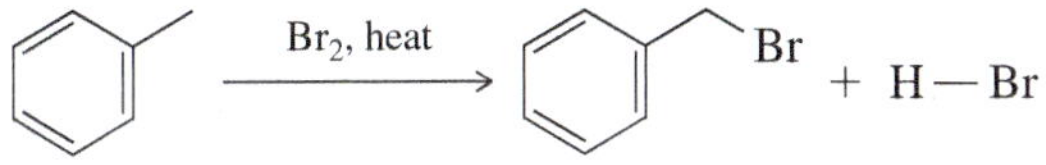

(c)

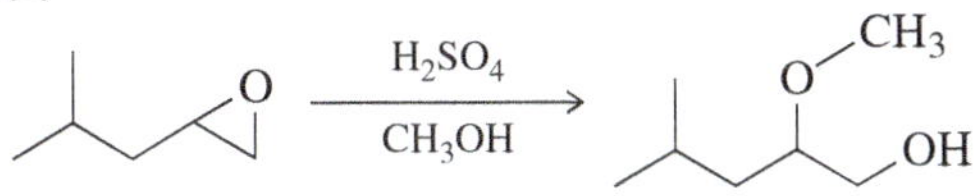

(d)

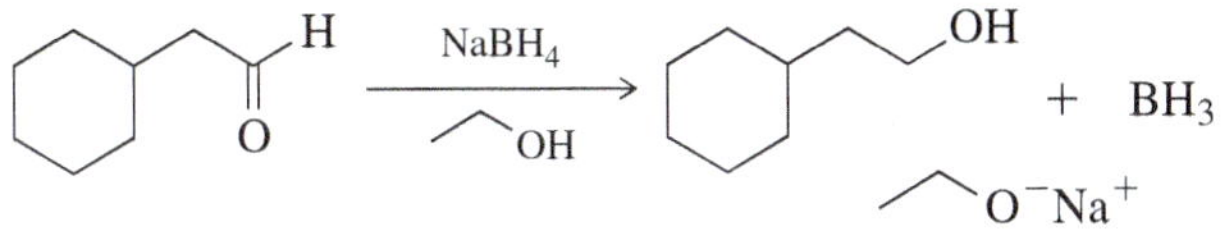

9.45 (••) Predict the product(s) that would result when the alkenes are allowed to react under the following conditions: (i) Br_2; (ii) Cl_2; (iii) Br_2, H_2O; (iv) Cl_2, CH_3OH; (v) *m*CPBA; (vi) 1. OsO_4 2. $NaHSO_3$; (vii) 1. *m*CPBA 2. H_2SO_4, H_2O; (viii) 1. O_3 2. CH_3SCH_3; (ix) H_2, Pd/C; and (x) D_2, Pd/C.

(a)

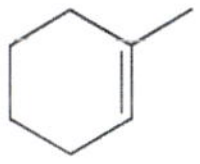

(b)

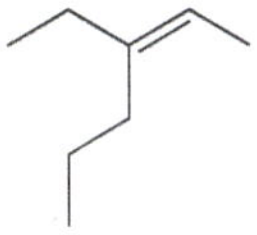

(c)

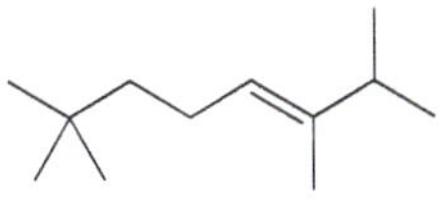

(d)

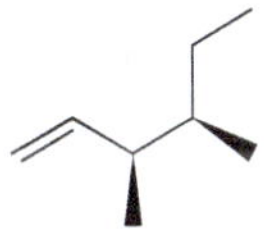

(e)

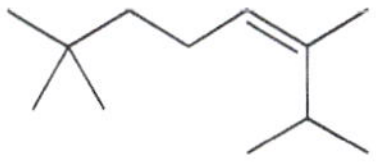

(f)

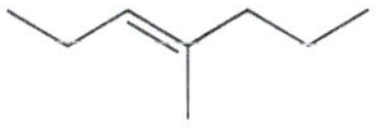

(g)

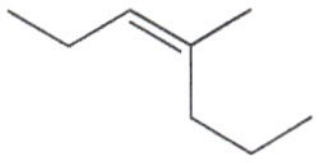

(h)

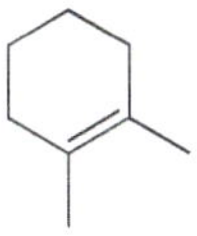

(i)

(j)

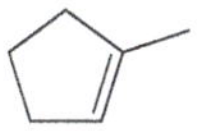

(k)

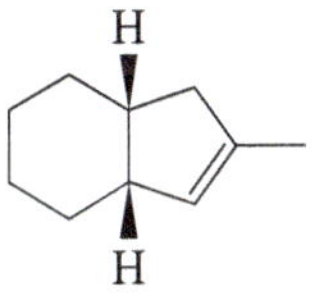

(l)

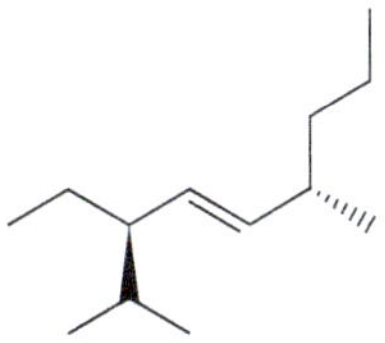

9.46 (•••) Rank the reactivity of the following alkenes with *m*CPBA (1 = most reactive, 5 = least reactive).

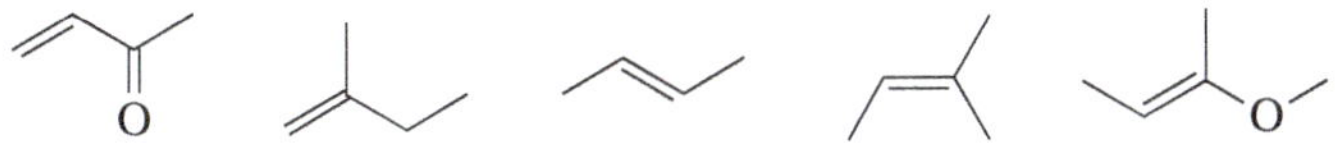

______ ______ ______ ______ ______

9.47 (••) When alkenes react with bromine in water, a halohydrin is produced. When water is replaced with methanol in this reaction, a different product is produced. Suggest a mechanism for the formation of this product.

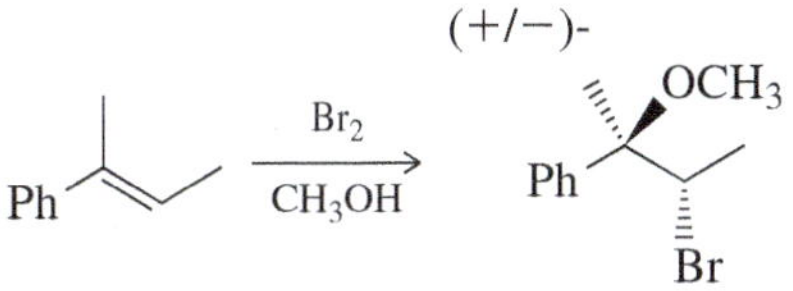

9.48 (•••) In light of your answer to Assessment 9.47, predict the product of the following reactions we have seen previously where an alcohol is substituted for water.

(a)

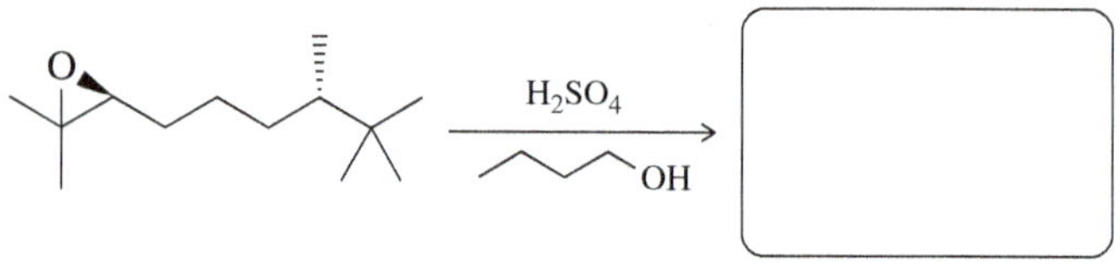

(b)

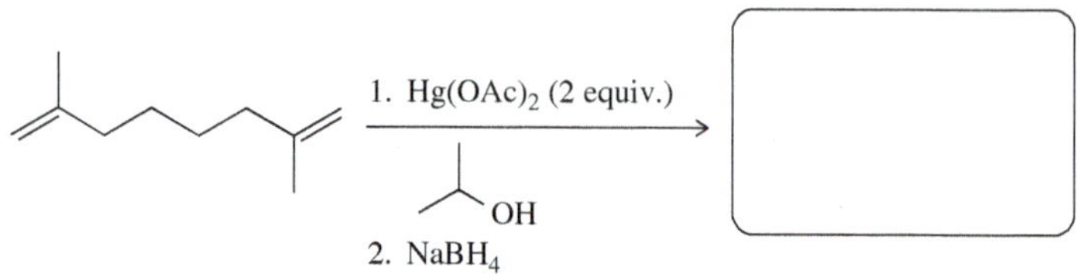

(c)

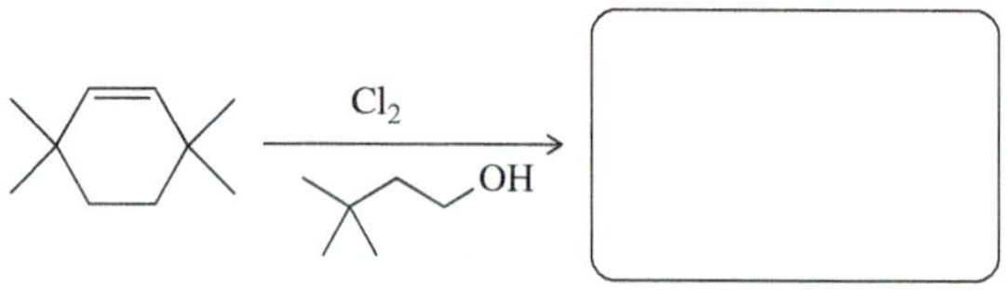

9.49 (••••) In light of your answers to Assessment 9.48, suggest mechanisms for the following ring-forming reactions.

(a)

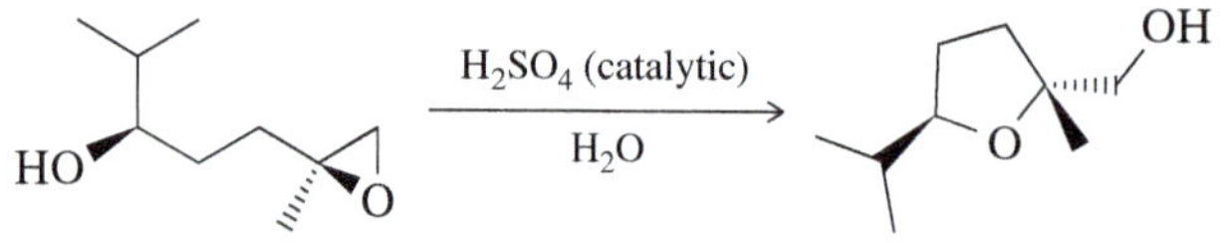

(b)

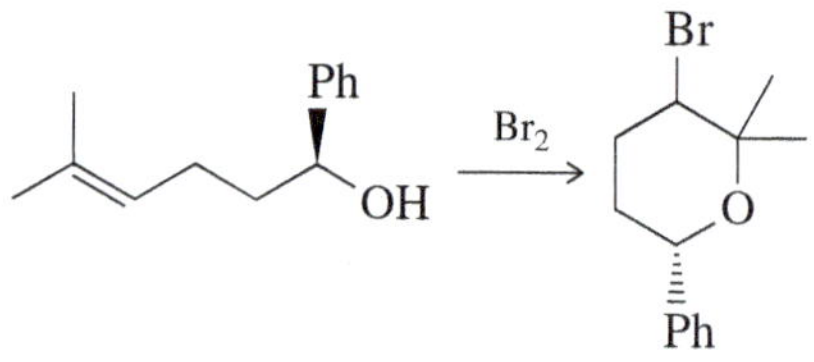

9.50 (••••) One way to think about concerted reactions is to imagine them as being stepwise reactions where, besides the slowest step, all others have infinitesimally small activation energies. Considering the hypothetical stepwise mechanism and actual concerted mechanism of epoxide formation, show what a reaction coordinate diagram might look like for each possibility.

(a) Stepwise, hypothetical mechanism:

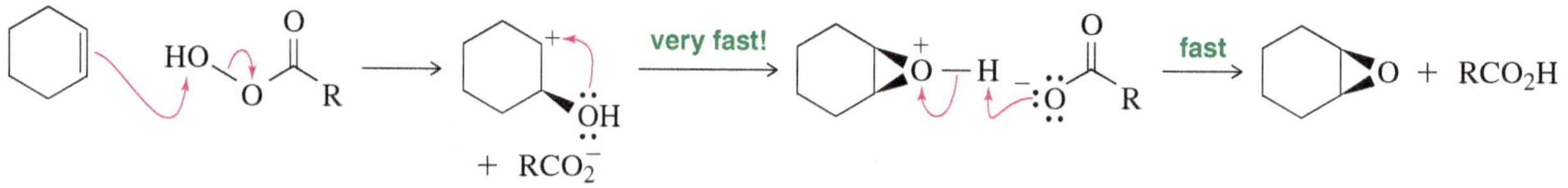

(b) Concerted, actual mechanism (butterfly transition state):

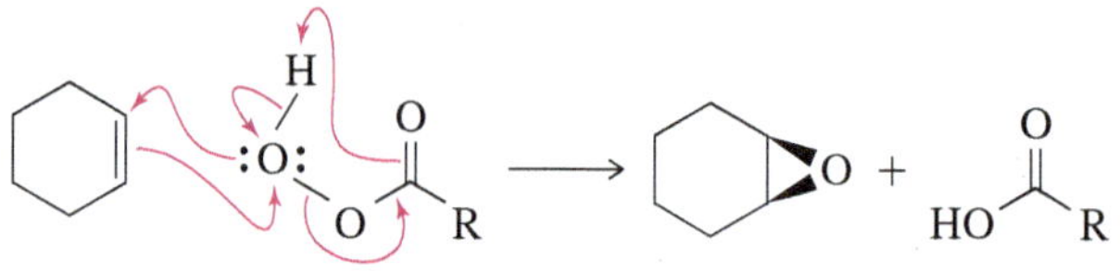

9.51 (•••) LOOKING AHEAD Consider the Cope rearrangement, a reaction we describe in Chapter 20. (a) Using the knowledge we have gained here in Chapter 9, suggest a one-step, concerted mechanism that explains the formation of B from A. (b) Which side of the reaction would you expect to be favored? Justify your answer. (c) Which product, A or B, would you expect to be hydrogenated with the more exothermic heat of hydrogenation?

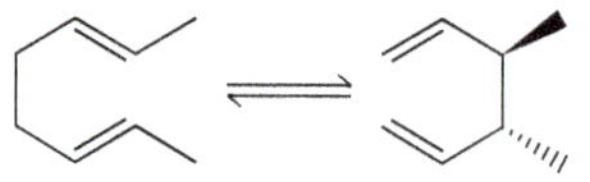

A **B**

9.52 (••••) LOOKING AHEAD Bromination of buta-1,3-diene with a single equivalent of Br_2 can give either of two products. (a) Which of these products (A or B) would you predict to be more stable? Justify your answer. (b) Suggest an experiment you could perform that would allow you to test your hypothesis.

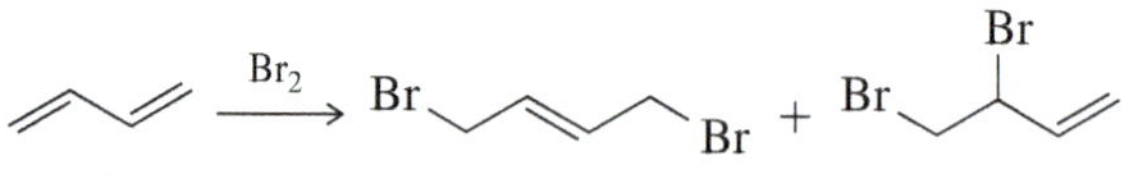

buta-1,3-diene **A** **B**

9.53 (••••) Bromination of a highly electron-rich alkene such as 2-methoxybut-2-ene has been shown to produce approximately equal mixtures of the trans- and cis-dibromide. Suggest an explanation for this observation.

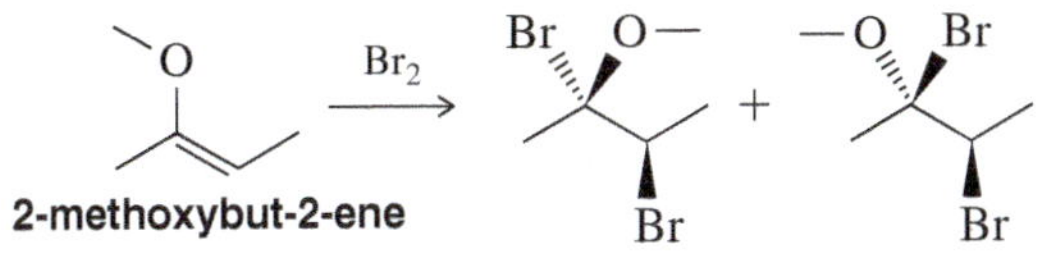

9.54 (•••) LOOKING AHEAD In Chapter 19, we discuss the reaction of enols with bromine. This reaction produces α-bromoketones in good yields. Suggest a mechanism for this reaction and justify its deviation from the dibromide product you might have expected.

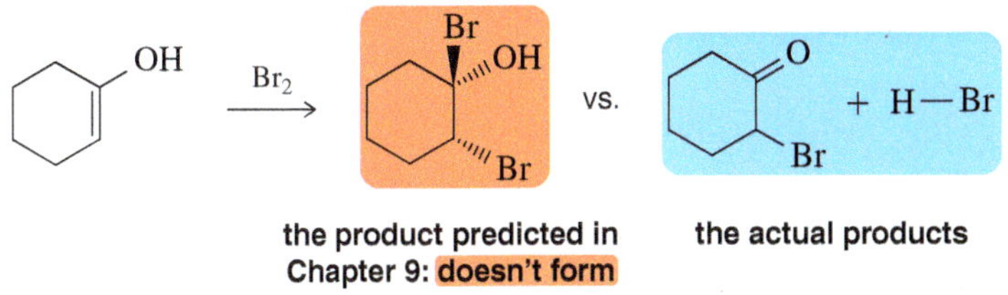

9.55 (••) Predict the product of ozonolysis of the triglyceride shown.

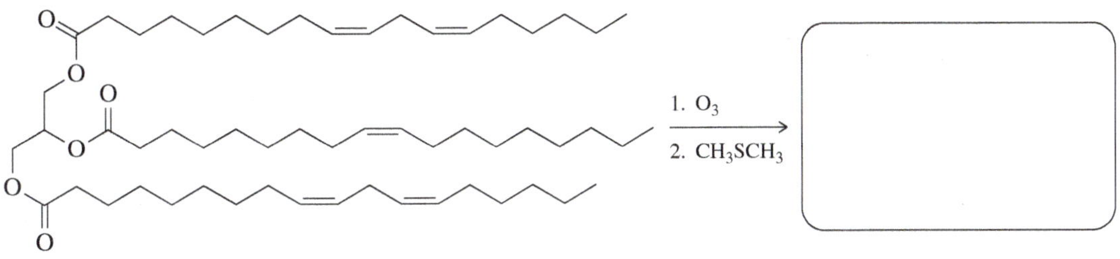

9.56 (•••) Ozonolysis of an unknown alkene **A** gives the products shown. Predict the product that results from hydrogenation of alkene **A**. [*There are multiple answers, but only show the one with the 6-membered ring.*]

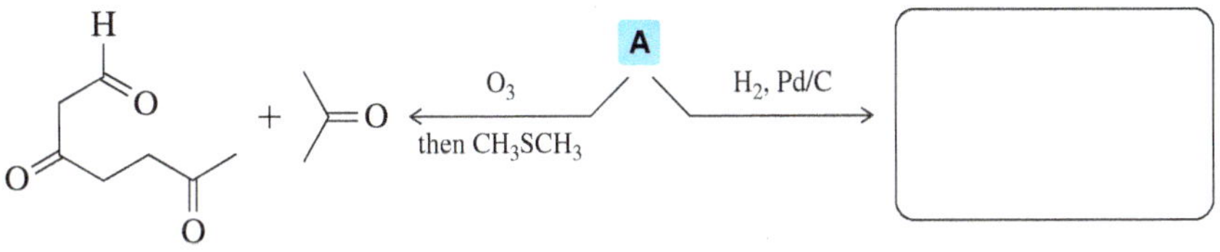

9.57 (••) In spite of being mechanistically similar to some of the reactions we saw in Chapter 8, rearrangement never occurred here in Chapter 9. Why doesn't rearrangement occur in the following bromination reaction despite the proximity of a more substituted carbon?

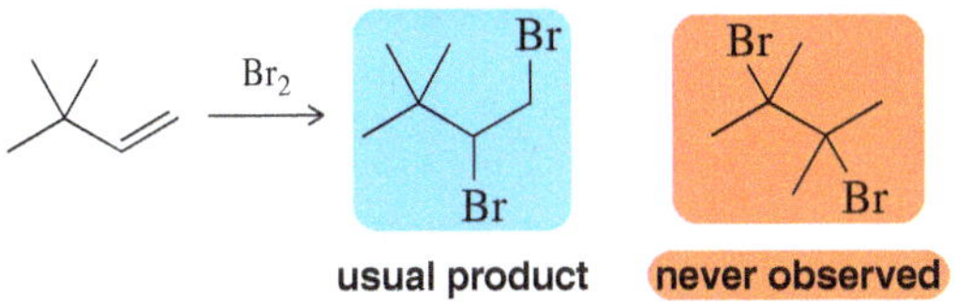

9.58 (••) Enantioselective epoxidation of *trans*-stilbene using the Jacobsen catalyst results in an optically active epoxide. When the same conditions are utilized with *cis*-stilbene as the substrate, however, the product is optically inactive. Explain this observation.

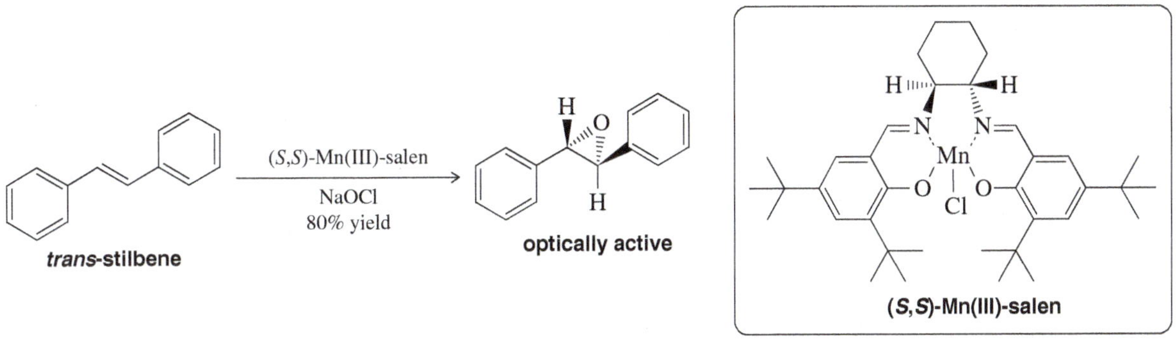

9.59 (••) At the beginning of Chapter 9, we stated that after finishing Chapters 8 and 9, we would have the ability to make a large variety of functional groups using related reactions. Show the reagent(s) necessary to convert 1-isobutylcyclohexene into the following molecules.

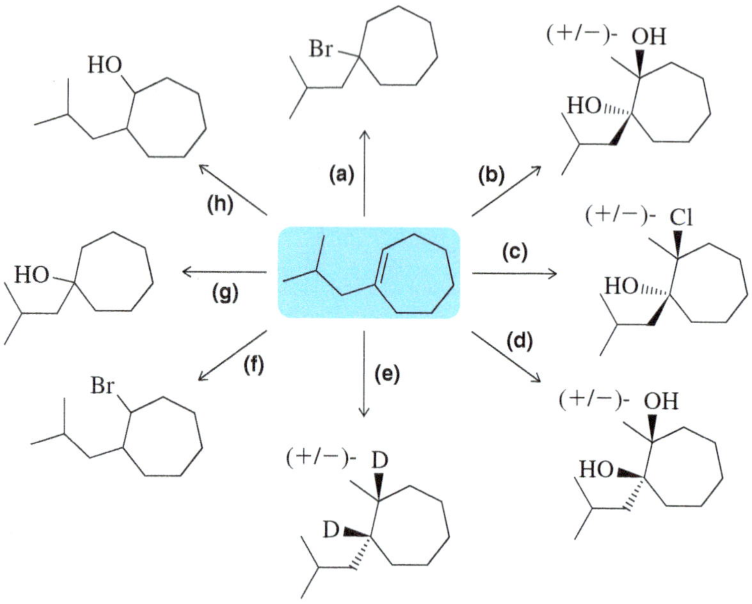

9.60 (•••) Retrosynthetic analysis is the process of working backward to develop the synthesis of a new compound. In Chapter 10, we begin developing multistep syntheses in this manner. For now, try to work backward a single step by suggesting an alkene and a reagent that would give products (a)–(i). [*Your answers should not include alkenes that undergo rearrangement to give the desired products.*]

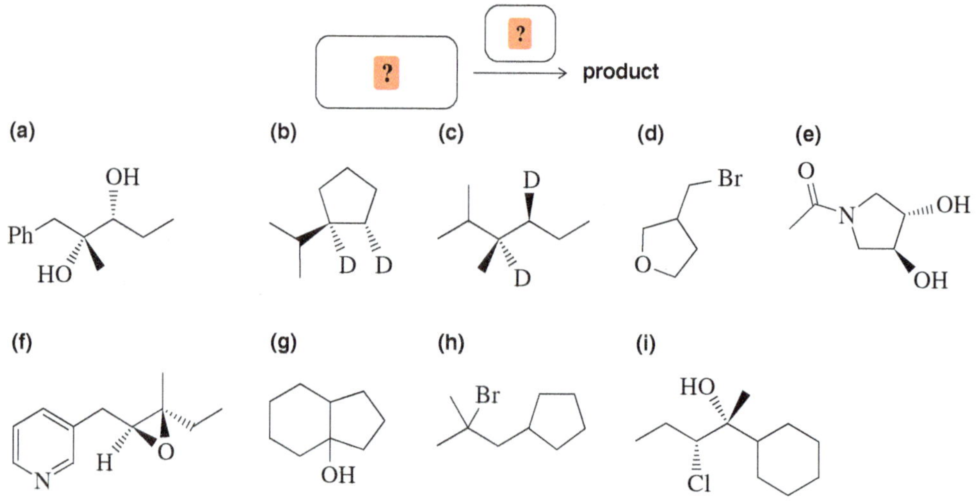

9.61 (••••) Suggest mechanisms for the following reactions, which are similar to the mechanism we saw for lanosterol biosynthesis at the end of Chapter 8.

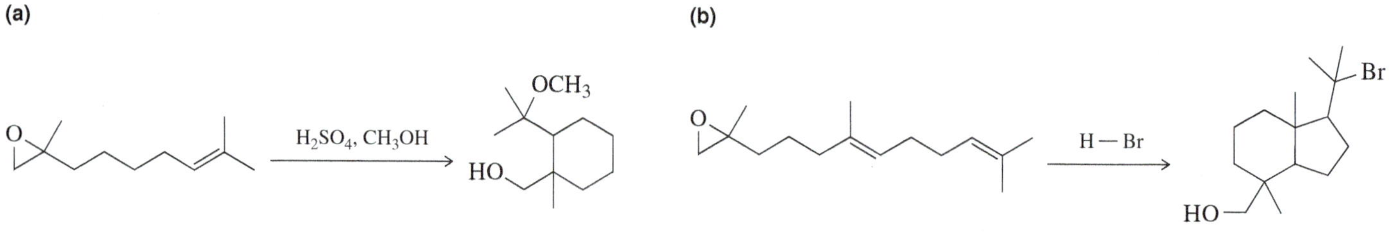

10

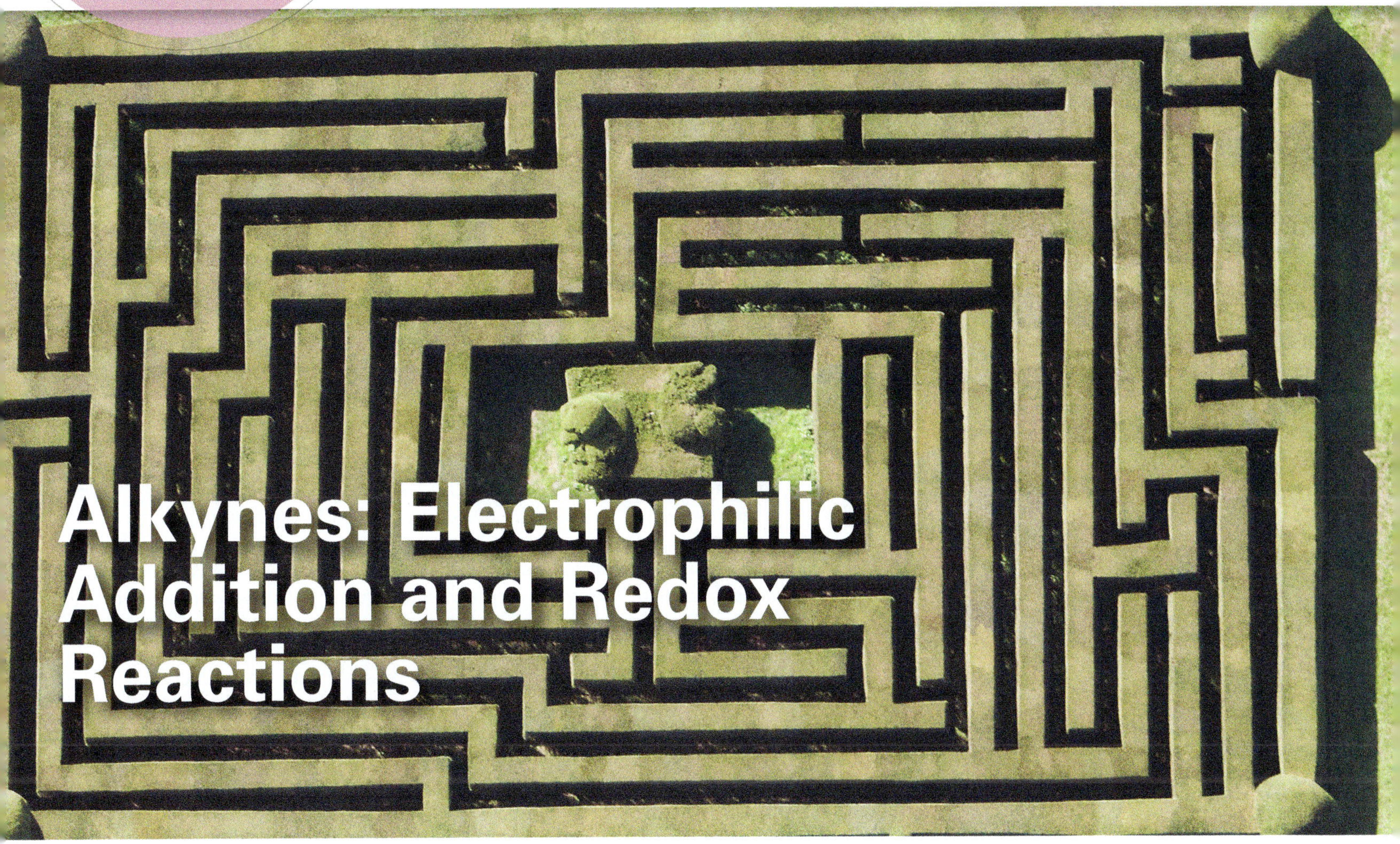

Alkynes: Electrophilic Addition and Redox Reactions

Introduction

One of the distinguishing features of organic chemistry as a science is the ability to make molecules, to build complex structures from simple ones, to prepare molecules that other science disciplines study, and to create compounds that improve, and even save lives. Alkynes, the subject of Chapter 10, are critical for these synthetic endeavors.

In addition to studying reactions of alkynes, a secondary goal is to plan syntheses of complex molecules from simpler ones. Synthesis design is greatly simplified using a logical approach—an approach you likely learned when solving mazes as a child. This chapter illustrates that the strategy for developing a synthesis is as simple as the strategy used to solve one of these mazes. Thus, rather than connecting alkyne chemistry to real-world applications as we usually do, let's work a maze instead. The maze, although not particularly difficult, has been solved using three different strategies. Note that in each method, the pencil never came off the page.

1. *Every possible path* [Figure 10.1(a)]: This strategy uses no real thought. The decision to turn left or right is random, independent of the overall goal, and not focused on the long term. But it works . . . eventually.
2. *Hand on the wall* [Figure 10.1(b)]: This strategy, which uses very little thought, imagines you are walking through the maze while keeping your hand on the right wall as a way of telling you which way to turn. Although still inefficient, this is a more logical approach than that used in method 1.

Learning Objectives

LO 10.0 Analyze the structural properties of alkynes

LO 10.1 Name alkynes and their isomers

LO 10.2 Determine the properties of alkynes using MO theory

LO 10.3 Analyze reactions of alkynes (in terms of K_{eq}, stability, and a possible transition state)

LO 10.4 Predict the outcome of alkyne reactions

LO 10.5 Analyze reduction reactions of alkynes

LO 10.6 Create reaction sheets to organize related organic synthesis reactions

LO 10.7 Analyze electrophilic addition reactions of alkynes

LO 10.8 Deduce the reaction mechanism of new and unknown reactions

Figure 10.1 Three different ways to solve a maze: (a) every possible path, (b) hand on the wall, and (c) working backward.

3. ***Working backward*** [Figure 10.1(c)]: Here, the maze is worked in the reverse direction, starting at the exit. This strategy takes advantage of the forward-looking design of most mazes. That is, most decision-making points are near the entrance. Since this process starts from the end, it is virtually impossible to veer off the correct path. [*Try it for yourself to see that this is true.*]

"If you're walking down the right path and you're willing to keep walking, eventually you'll make progress."

—Barack Obama

As you see, there are many ways to solve a maze, some more efficient than others. Working backward is most efficient because the decisions, especially ones with greater consequences [*a significant distance needs to be traveled to notice the mistake*] are in the forward direction. [*Does this relate to organic chemistry? Or to a profession you might be considering, such as environmental scientist, physician, or contractor overseeing the building of houses?*]

Thus far, the reactions we have studied involve only functional group transformations. Reactions have converted alkanes to alkyl halides and alkenes to alcohols, alkyl halides, or halohydrins, to name a few. Organic synthesis is not possible with only functional group transformations. In order to make large, complex molecules from smaller molecules, the molecules have to be combined. Thinking of reactions as the tools ultimately used to build molecules, we currently lack the most important tool. That necessary tool is a reaction for making carbon–carbon bonds.

Are we talking about building houses or building molecules here?

And yet the tools alone are not enough. A construction worker, well-versed in using the tools of the trade, might lack the strategy needed to build a house. Similarly, having the tools to make carbon–carbon bonds does not make us synthesis experts. A strategy is required to efficiently put together the most stable, protective, and convenient structure possible. With that in mind, let's get started the way we always do—by assessing our understanding of pertinent concepts.

Test your understanding

Assessment 10.1 To which atom would you expect H^+ to add in the following ester?

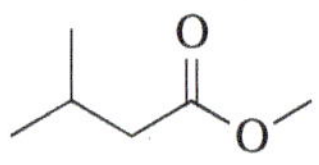

Assessment 10.2 Calculate the oxidation numbers for each indicated atom in norethindrone, a steroid contraceptive.

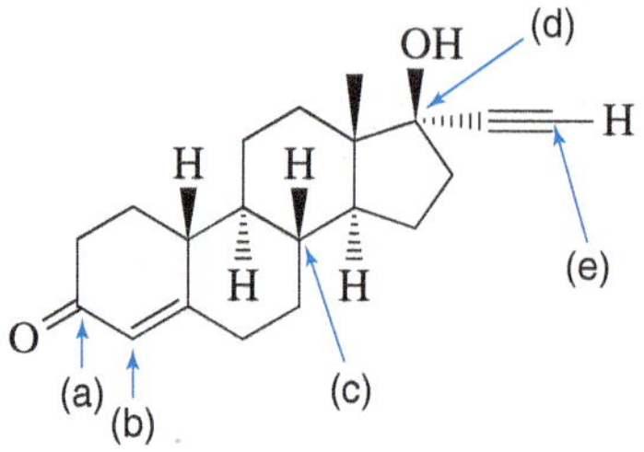

Assessment 10.3 Rank the reactivity of the following anions with a general electrophile from least to most reactive.

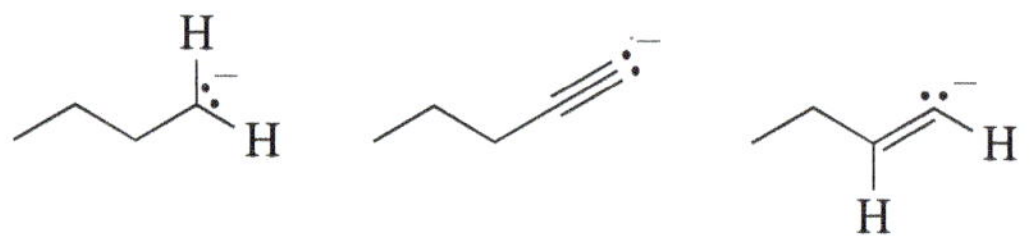

Assessment 10.4 Would amide or acetate most easily deprotonate an alcohol to make the alkoxide anion? Calculate K_{eq} for each possibility.

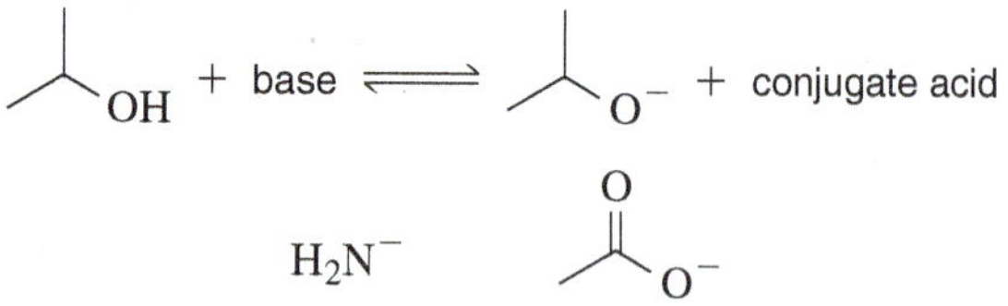

Assessment 10.5 The following reaction gives two different constitutional isomers in nearly equal yields. Why doesn't this reaction produce only one product?

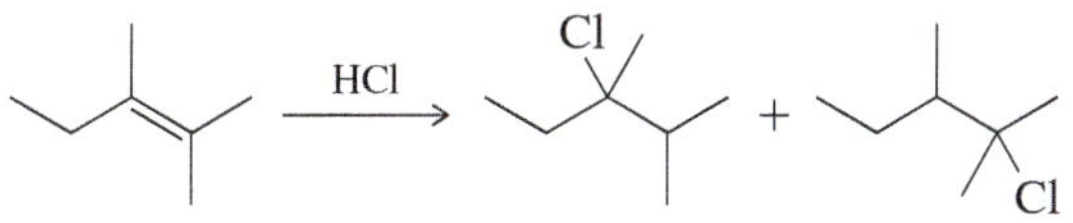

10.1 Properties of alkynes

Alkynes are unsaturated hydrocarbons containing two carbons bound by a triple bond. Like alkenes, they are nonpolar and usually insoluble in water. The linear C—C triple bond results in stronger intermolecular interactions and, consequently, slightly higher boiling points than the corresponding alkene and alkane (Table 10.1).

Table 10.1 **Alkynes have a higher boiling point than the corresponding alkane and alkene.**

C_5 compound	Boiling point (°C)
 pentane	36
 *trans*-pent-2-ene	37
 pent-2-yne	56

10.1.1 Classification of alkynes

There are three classes of alkynes: terminal, internal/symmetrical, and internal/unsymmetrical. A representative example of each is shown in Table 10.2.

Table 10.2 **Classification of alkynes as terminal, internal/symmetrical, and internal/unsymmetrical.**

	General form	Example
Terminal alkynes have the C—C triple bond at the end of the carbon chain.	R—C≡C—H	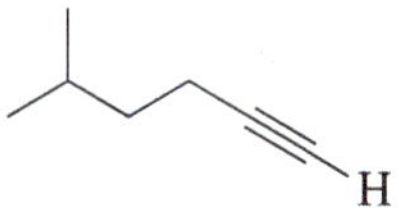
Internal/symmetrical alkynes have the C—C triple bond between equivalent substituents.	R—C≡C—R	
Internal/unsymmetrical alkynes have the C—C triple bond between different substituents.	R—C≡C—R′	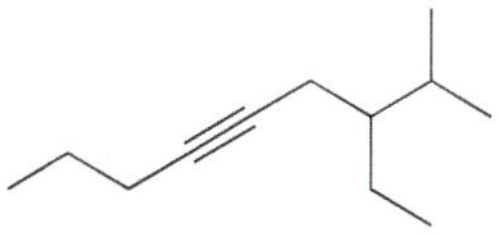

Assessment 10.6 Identify the following alkynes as terminal (T), internal/symmetrical (IS), or internal/unsymmetrical (IU).

(a)

(b)
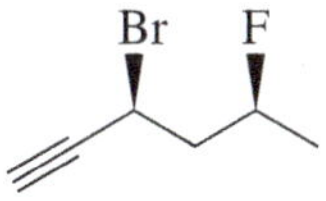

(c)
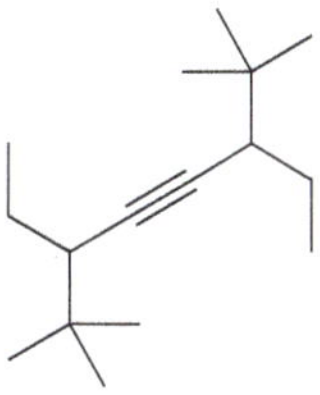

(d)
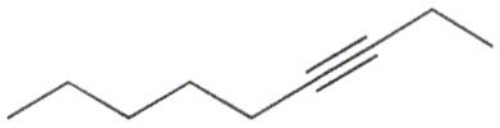

(e)
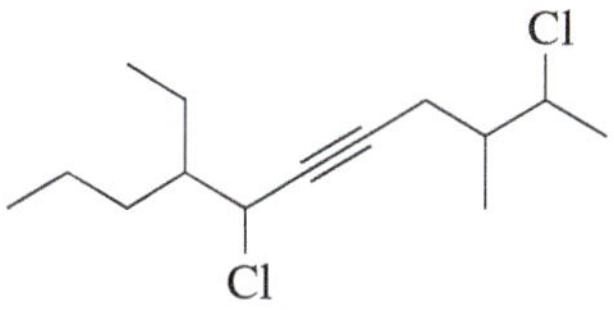

(f)
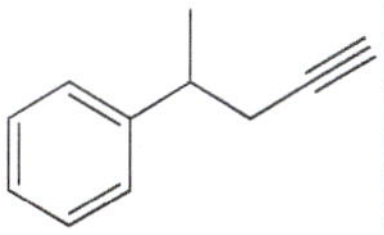

Assessment 10.7 Draw all internal symmetrical alkynes with the molecular formula $C_{10}H_{18}$.

10.1.2 Oxidation states

The carbons in the triple bonds of alkynes are at a higher oxidation state than the carbons in the double bond of alkenes. As a result, alkynes are more reducible than alkenes. In internal alkynes, each alkyne carbon has a zero oxidation state (Figure 10.2). For a terminal alkyne, the internal carbon has an oxidation state of 0, whereas the terminal carbon has an oxidation state of −1. [*See Section 9.1.1 for a review on how to calculate these values.*]

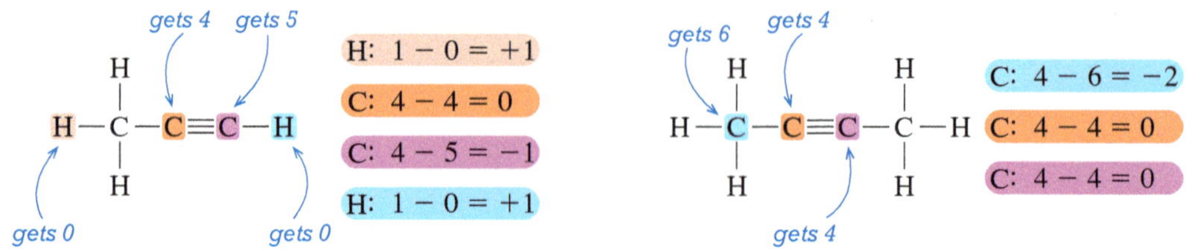

Figure 10.2 To calculate oxidation states, electrons in a covalent bond are assigned to the more electronegative atom.

$H_3C{-}CH_3$	$H_2C{=}CH_2$	$HC{\equiv}CH$
IHD = 0	IHD = 1	IHD = 2

Figure 10.3 Each π bond is worth an index of hydrogen deficiency of 1.

10.1.3 Index of hydrogen deficiency

Alkynes are unsaturated molecules, so they lack the full complement of hydrogens possible. The two π bonds each cause the molecule to be short by two hydrogens (or one molecule of H_2). Thus, the index of hydrogen (H_2) deficiency [*Section 8.1.1*] caused by the alkyne functional group is 2 (**Figure 10.3**).

10.2 Nomenclature of alkynes

The IUPAC system for naming alkynes is shown in **Table 10.3**.

Table 10.3 Rules for naming alkynes according to the IUPAC system of nomenclature.

1. The parent chain is the longest chain containing both carbons of the alkyne and forms the root of the name. The infix -an- of the corresponding alkane is replaced with -yn- for alkynes.	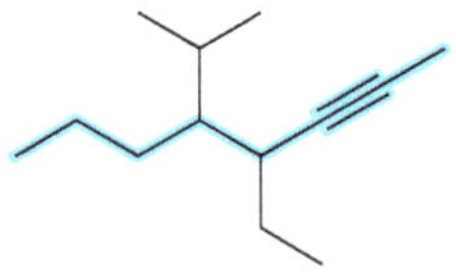 **8-carbon parent chain** **octyne**
2. The parent chain is numbered, giving the first carbon of the alkyne the lowest number.	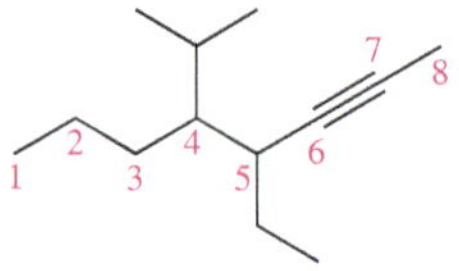 ~~oct-6-yne~~ 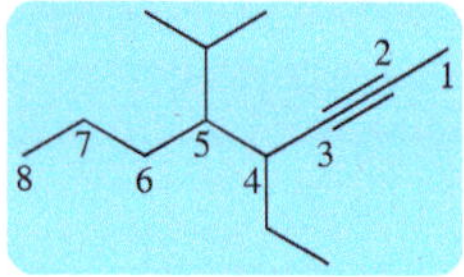 oct-2-yne
3. Substituents on the parent chain are named as has been done previously.	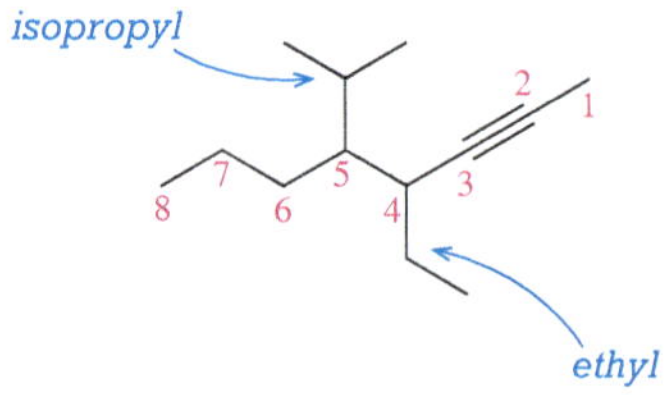
4. The name is written following standard IUPAC alphabetization rules. Numbers are used to indicate the location of the substituents and the triple bond.	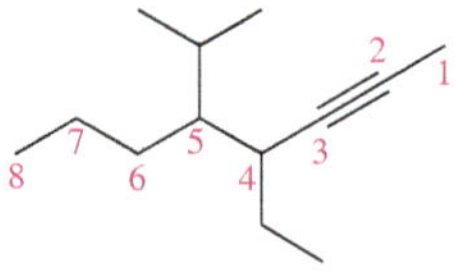 **4-ethyl-5-isopropyloct-2-yne**

Assessment 10.8 Provide the IUPAC names for the following alkynes.

(a)

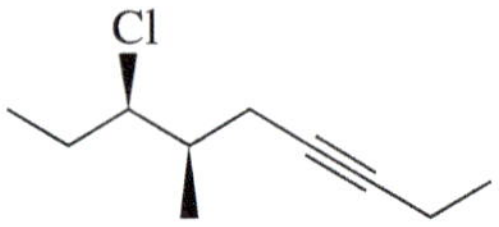

(b)

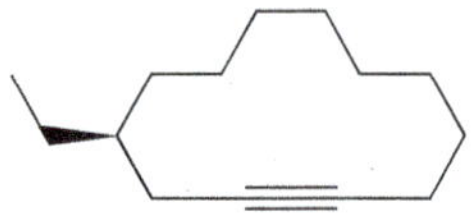

(c)

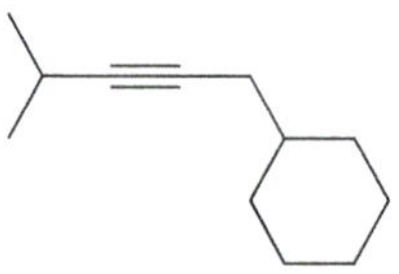

Assessment 10.9 Draw the structures that correspond to the following IUPAC names.

(a) (*R*)-4-isopropyl-6-methylhept-2-yne
(b) 2-methylhept-3-yne
(c) (*S*)-3-fluoropent-1-yne
(d) (*S*)-3-ethyl-2-methyloct-4-yne

10.2.1 Naming with multiple functional groups

Molecules containing multiple functional groups can be named using the IUPAC system of prioritizing functional groups (Table 10.4). [*We are limited at this point to alkenes and alkynes, the only functional groups studied thus far.*] Alkenes and alkynes receive equal priority, so numbering of the chain is based on which functional group comes first. Molecules containing both are always named as —X-en-Y-yne (Figure 10.4).

Table 10.4 Priority of functional groups in nomenclature.

Functional group	Priority	Functional group	Priority
Acids	1 (highest)	Ketones	7
Esters	2	Alcohols	8
Acid chlorides	3	Amines	9
Anhydrides	3	Alkenes, alkynes	10
Amides	4	Alkanes	—
Nitriles	5	Ethers	—
Aldehydes	6	Alkyl halides	(lowest)

(*E*)-3-methyloct-5-en-1-yne (*Z*)-8-bromo-7-isopropyldec-2-en-5-yne

Figure 10.4 Examples of enyne nomenclature.

Assessment 10.10 Provide the IUPAC name for the following molecules.

(a)

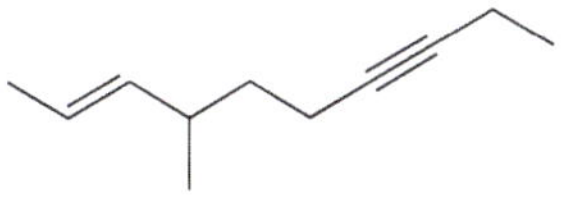

(b)

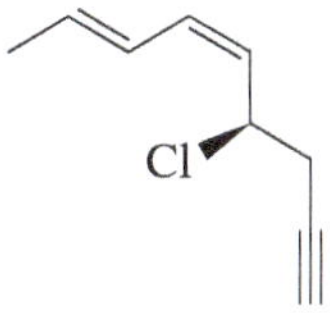

(c)

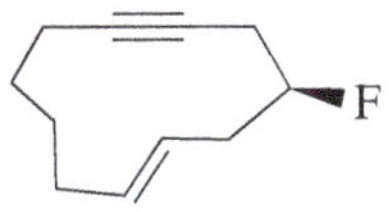

Assessment 10.11 Draw the structures that correspond to the following names.

(a) (*S*)-3-methylhex-5-en-1-yne
(b) (3*Z*,8*S*)-8-ethyl-3-methylundec-3-en-6-yne
(c) (*Z*)-2-chloro-7-methyloct-2-en-4-yne

10.3 Molecular orbital picture of alkyne

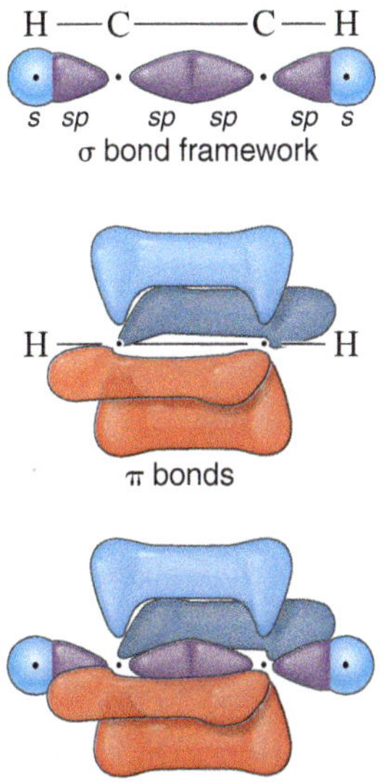

Figure 10.5 The full MO picture of acetylene consists of the σ bond framework and two π bonds that are perpendicular to each other.

acetylene

Many properties of alkynes can be explained by their molecular orbital picture. [*Try drawing the molecular picture of acetylene on your own.*]

Composed of two linear, *sp*-hybridized carbons, each C—H σ bond is formed by overlap of an *sp*-hybridized orbital and the 1*s* orbital of hydrogen. The σ bond between the two carbons is composed of overlapping *sp*-hybridized orbitals. Overlap of two sets of *p* orbitals (above and below; in front and behind) generate the two **orthogonal** π bonds. Putting it all together gives the molecular orbital picture shown in Figure 10.5.

Assessment 10.12 Draw the molecular orbital picture of but-2-yne.

but-2-yne

10.4 Implications of the molecular orbital picture

10.4.1 High potential energy

The instability of alkynes is an important consideration in Section 10.8.

With two electron-rich π bonds in such close proximity, alkynes have a high potential energy. This potential energy can be measured in two different ways (Figure 10.6). Alkynes are highly combustible, releasing a large amount of heat per mole of alkyne burned [Figure 10.6(a)]. For this reason, acetylene is used in welding torches. The instability caused by the proximity of electron density in the two π bonds can also be examined by comparing heats of hydrogenation (for similar experiments, see Section 9.2.1). Hydrogenation of but-2-yne to butane (examined in more detail in Section 10.6.1) releases a total of 65 kcal/mol of energy.

This is a two-step process with *cis*-but-2-ene as an intermediate. Because hydrogenation of *cis*-but-2-ene releases 28.3 kcal/mol of energy, hydrogenation of but-2-yne must release 36.7 kcal/mol. Thus, the alkyne π bond is 36.7 − 28.3 ≈ 8 kcal/mol less stable than the alkene π bond. [Figure 10.6(b)].

(a)

$$CH_3CH_3 + \tfrac{7}{2}O_2 \longrightarrow 2\,CO_2 + 3\,H_2O \quad \Delta H^\circ = -373\text{ kcal}$$

−373 kcal divided by 5 moles of products = −75 kcal/mol of products

$$H_2C{=}CH_2 + 3\,O_2 \longrightarrow 2\,CO_2 + 2\,H_2O \quad \Delta H^\circ = -337\text{ kcal}$$

−337 kcal divided by 4 moles of products = −84 kcal/mol of products

$$HC{\equiv}CH + \tfrac{5}{2}O_2 \longrightarrow 2\,CO_2 + 1\,H_2O \quad \Delta H^\circ = -317\text{ kcal}$$

−317 kcal divided by 3 moles of products = −106 kcal/mol of products

(b)

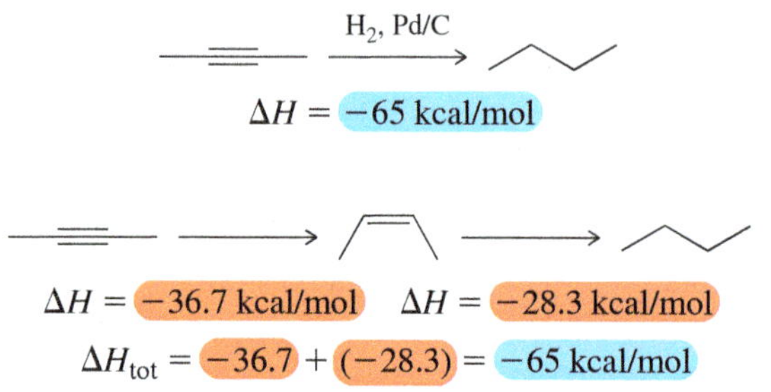

Figure 10.6 Alkynes are less stable than alkenes. (a) They give off more heat than alkenes or alkanes when combusted. (b) Hydrogenation of the first π bond of an alkyne is more favorable than hydrogenation of the alkene π bond.

10.4.2 Bond strengths

The C≡C triple bond of an alkyne is stronger than the C=C double bond of an alkene because there are simply more bonds to break. [*This distinction causes confusion at times.*] The bond-dissociation energy (BDE) of an alkyne represents breaking all three bonds (2 π and 1 σ) (**Figure 10.7**). Conversely, the bond-dissociation energies of ethene (1 π and 1 σ) and ethane (1 σ) result from breaking fewer bonds.

The carbon–carbon bonds in H—C≡C—H are stronger than those in $H_2C{=}CH_2$ and $H_3C{-}CH_3$.

$$H{-}C{\equiv}C{-}H \longrightarrow H{-}\dot{\ddot{C}}\cdot + \cdot\dot{\ddot{C}}{-}H \quad \text{BDE} = 200\text{ kcal/mol}$$

break 1 σ and 2 π bonds

$$H_2C{=}CH_2 \longrightarrow H_2\dot{C}\cdot + \cdot\dot{C}H_2 \quad \text{BDE} = 146\text{ kcal/mol}$$

break 1 σ and 1 π bonds

$$H_3C{-}CH_3 \longrightarrow H_3C\cdot + \cdot CH_3 \quad \text{BDE} = 83\text{ kcal/mol}$$

break 1 σ bond

Figure 10.7 Because there are more bonds to break, the carbon–carbon triple bond of an alkyne is stronger than the corresponding carbon–carbon double and single bonds in alkenes and alkanes, respectively.

The situation is different when comparing the energy needed to break the π bonds one at a time. The relative reactivity of the "first" π bond depends on the conditions. Under the conditions of hydrogenation, the "first" alkyne π bond is weaker/less stable than the alkene π bond (Section 10.4.1). In electrophilic additions, alkynes are *less* reactive than alkenes. For example, *trans*-but-2-ene reacts much more quickly with HBr than does but-2-yne (**Figure 10.8**). The alkyne π bond is more reactive under hydrogenation conditions yet less reactive under electrophilic addition conditions because of the mechanism of the two reactions. [*Can you guess the mechanism of the alkyne reaction with HBr? It is presented in Section 10.8.*]

Favorability in hydrogenation is based on ΔG, whereas relative rates of the reaction are being compared in electrophilic addition. Alkynes react more slowly (higher E_a) in the rate-determining step than do alkenes when reacting with an electrophile. Unless the electrophile is strong, the E_a of the rate-determining step is too high for alkynes to react at all. [*The high* E_a *for alkyne reactions is a recurring theme throughout this chapter.*]

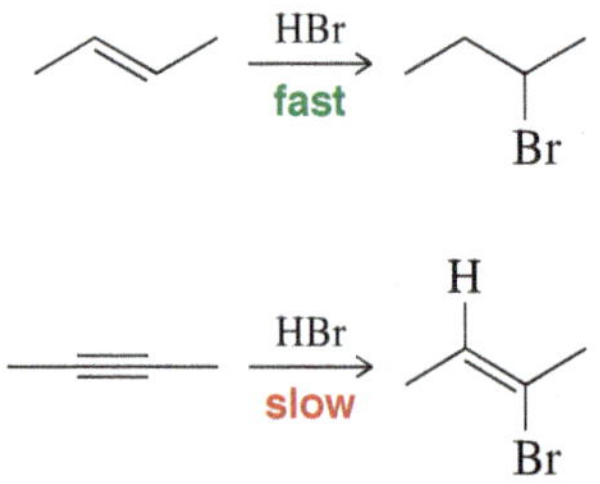

Figure 10.8 Because of the mechanism of the reaction, alkenes react more quickly with electrophiles than do the corresponding alkynes.

10.4.1–10.4.2 SUMMARY

- Even though the alkyne π bond is weaker than the alkene π bond, reaction conditions still dictate their relative reactivity.

The C—H σ bonds of alkynes are stronger than the C—H σ bonds of alkenes, which are stronger than the C—H σ bonds of alkanes. Recall (Section 5.3.1) that bond-dissociation energies refer to the homolytic cleavage of a covalent bond (i.e., cleavage where each atom gets one electron; see **Figure 10.9**). The bond-dissociation energy can be thought of as a measure of relative radical stability (see Chapter 5). Due to diminished hyperconjugation, *sp*-hybridized radicals are significantly less stable than sp^2- and sp^3-hybridized radicals.

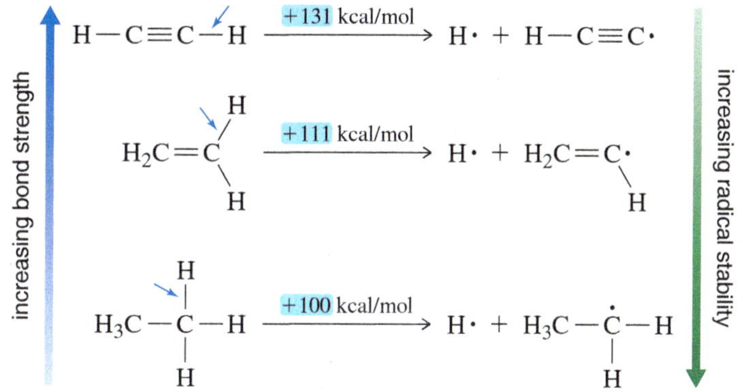

Figure 10.9 The bond-dissociation energy is higher for bonds that break to give less-stable radicals.

The increased strength of the alkyne C—H bond is also a function of increased overlap between the *sp* orbital of carbon and the 1*s* orbital of hydrogen. Hybridization is discussed in Section 10.4.3 in the context of relative C—H bond lengths.

Assessment 10.13 Which of the following would be expected to give a hotter flame during combustion? Explain.

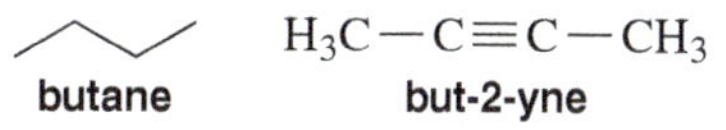

10.4.3 Bond lengths

The C—C triple bond of alkynes is 0.13 Å shorter than the C—C double bond of alkenes and 0.34 Å shorter than the C—C single bond of alkanes (**Figure 10.10**). Because of the two π bonds, the carbon nuclei are held closer together in alkynes than in alkenes and alkanes.

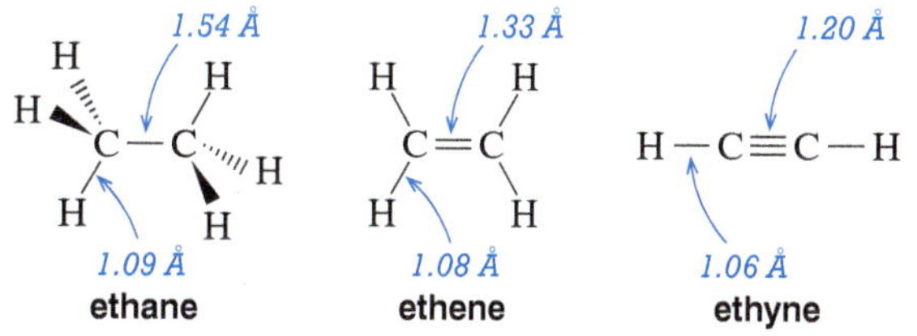

Figure 10.10 The presence of π bonds results in decreased bond length.

There is also a subtle change in bond length for C—H σ bonds as you move from alkanes to alkenes to alkynes (**Figure 10.11**). The C—H bond length is a function of carbon's hybridization. The carbon is sp^3 hybridized in alkanes but is *sp* hybridized in alkynes. Thus, the alkane C—H σ bond uses an atomic orbital (sp^3) that is 25% *s* and 75% *p*, whereas the alkyne C—H σ bond uses an atomic orbital (*sp*) that is 50% *s* and 50% *p*. The *p* orbital (shaped like a dumbbell) extends farther from the nucleus than the *s* orbital (sphere). Thus, an orbital with more *p* character will extend farther from the nucleus. As the amount of *s* character increases, the orbital becomes more spherical and holds electrons closer to the nucleus. Overlap of these hybridized atomic orbitals with the 1*s* orbital of hydrogen results in a shorter and stronger *sp*-hybridized C—H σ bond.

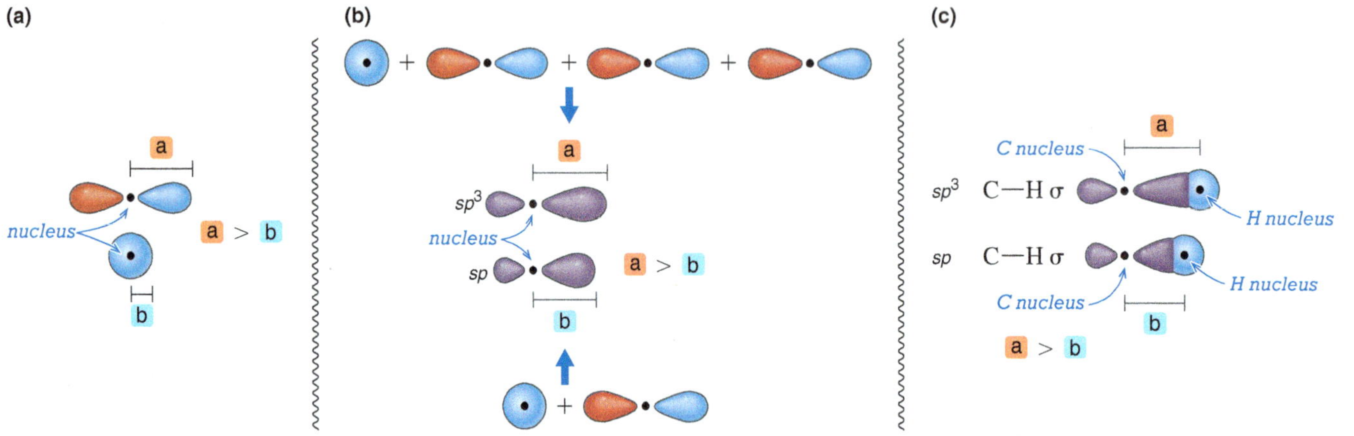

Figure 10.11 Because *sp*-hybridized orbitals have more *s* character, they are held closer to the nucleus. (a) *s* Orbitals are closer to the nucleus than *p* orbitals. (b) Hybridization with more *p* orbitals creates longer hybridized orbitals. (c) This results in shorter σ bonds that have more *s* character.

Assessment 10.14 Rank the indicated C—C bonds from longest to shortest (1 = longest; 3 = shortest). Explain your answer.

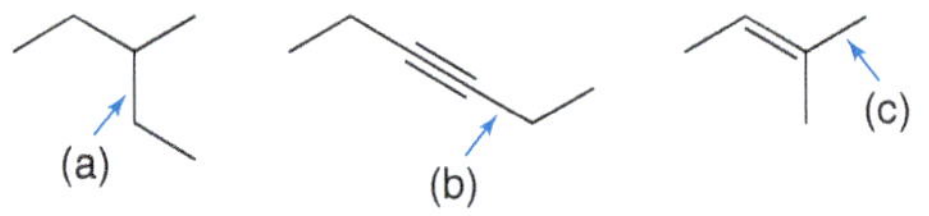

10.4.4 Acidity of the hydrogen of a terminal alkyne

An important consequence of the molecular orbital picture of an alkyne is the enhanced acidity of the alkynyl proton. Unlike alkanes and alkenes, alkynes can be deprotonated by strong bases. The alkyne hydrogen has a pK_a value of 25, which falls right between that of an amine (pK_a = 33) and an alcohol (pK_a = 16). [*See Table 4.5 for these values.*] The explanation for the lower pK_a parallels the description of relative C—H bond lengths (Table 10.5). The lone pair of electrons in the conjugate base of an alkyne are in an *sp*-hybridized orbital. Thus, they are held closer to the positively charged nucleus than in alkenes (sp^2 hybridized) and alkanes (sp^3 hybridized). Having the [*negatively charged*] electrons in closer proximity to the [*positively charged*] nucleus results in enhanced stability.

Table 10.5 More *s* character in the conjugate base means the lone pair is held closer to the nucleus and is therefore more stable. [*The pK$_a$ of ammonia and alcohols are shown for perspective.*]

Compound	Conjugate base	Hybridization	Percent *s* character	pK_a	
H_3C-CH_3	$H_3C-CH_2^-$ (lone pair in orbital)	sp^3	25	50	weakest acid
$H_2C=CH_2$	$H_2C=CH^-$ (lone pair in orbital)	sp^2	33	44	↓
:NH_3	$^-$:$\ddot{N}H_2$	(ammonia)	n/a	38	↓
H—C≡C—H	H—C≡C$^-$ (lone pair in orbital)	*sp*	50	25	↓
R—OH	R—$\ddot{O}$:$^-$	(alcohols)	n/a	16–18	stronger acid

Assessment 10.15 Calculate K_{eq} for the following acid–base reactions. Which is the best base to use to deprotonate acetylene and make the acetylide anion?

(a)

$$HO^- + H{-}C{\equiv}C{-}H \rightleftharpoons {}^-C{\equiv}C{-}H + H_2O$$

(b)

$$H_2N^- + H{-}C{\equiv}C{-}H \rightleftharpoons {}^-C{\equiv}C{-}H + H_3N$$

(c)

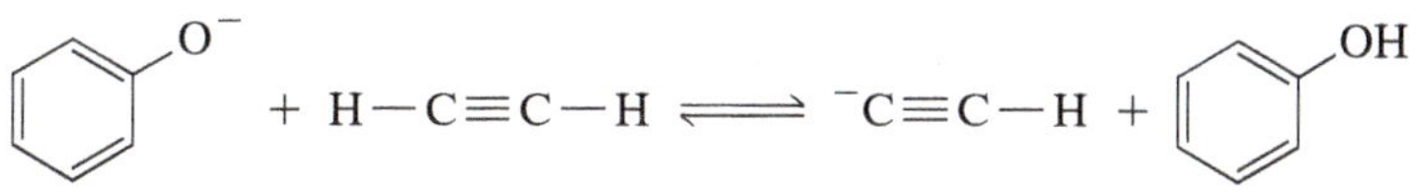

pK_a = 10

10.4.5 Cylinder of electron density

A second reaction type to be studied in this chapter involves the electron-rich C—C triple bond reacting with strong electrophiles. These are similar to the electrophilic addition reactions of alkenes (Section 8.2). The close proximity of the two orthogonal π bonds creates a cylinder of electron density around the σ bond (**Figure 10.12**). This electron density, linear in shape, creates a sterically unhindered environment for reaction with electrophiles.

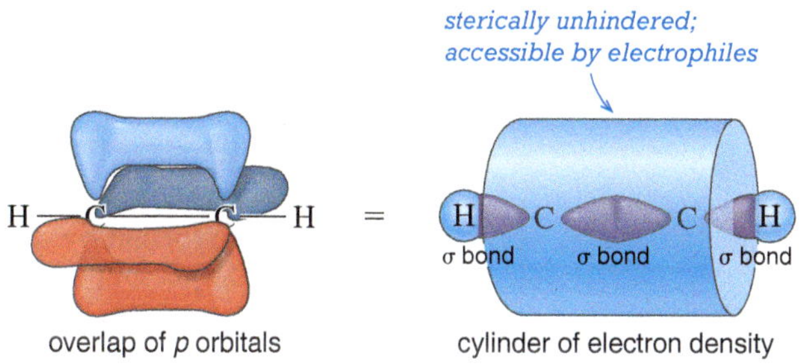

Figure 10.12 The cylinder of electron density makes alkynes reactive with strong electrophiles.

10-A CHEMISTRY IN CONTEXT: Problem-Solving Strategies

Working a jigsaw puzzle

One of the many wonders of parenting was watching my three-year-old work her first puzzle. Realizing that the goal was to connect oddly shaped pieces to make the picture on the box, she set out to achieve her goal by picking up two pieces and trying every which way to get them to fit. When that failed, she picked up another two pieces, attempting to do the same. She repeated this process, failing far more than she succeeded. But, because it was just a 15-piece puzzle, she eventually finished.

Obviously, my daughter's process could have been corrected [*bad parenting, maybe, but she was the first child*]. That her default setting was to start at the beginning and work toward the goal was interesting. There was no thought about strategy, such as ***you*** would employ if confronted with the same problem. The ideas of laying out side and corner pieces, assembling the border, and grouping colors based on where they show up in the picture are in essence a strategy that works backward from the picture on the box. Recognizing the goal allows the goal to lead us. It doesn't get us to a more correct answer; it simply makes the process more efficient and more logical.

10.5 Using the acidic hydrogen: forming C—C bonds

To make complex organic molecules, we need methods to make carbon–carbon bonds. The displacement of a 1° alkyl halide by deprotonated alkynes (acetylide anions) is very useful in this regard.

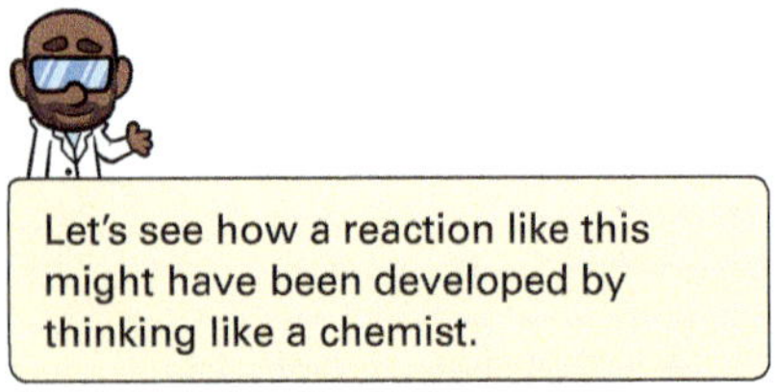

Thinking Like a Chemist

Similar to the discussion in Section 9.1.4, consider the ways a C—C bond might form without regard to specific reagents. In general, what intermediates might come together to make a new covalent bond between two carbons? [*Write down your answer to this before moving on.*]

There are three important reactive intermediates in organic chemistry: cations, anions, and radicals. If R and R′ represent the halves of our eventual C—C bond, then there are only three ways for them to combine (**Figure 10.13**). A negatively charged (electron-rich/nucleophilic/Lewis basic) R can attack a positively charged (electron-poor/electrophilic/Lewis acidic) R′. Conversely, a negatively charged R′ can attack a positively charged R. And, finally, the radicals of R and R′ can combine, each contributing a single electron to the new bond. [*This last reaction is one of the termination steps in radical halogenation from Chapter 5.*]

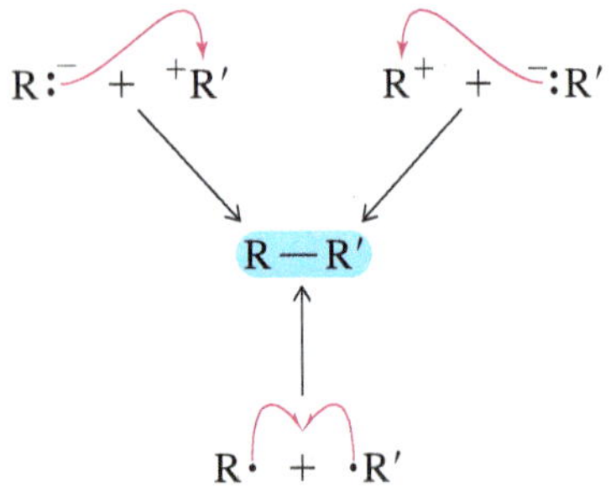

Figure 10.13 Three ways reactive intermediates can come together to make a carbon–carbon bond (R—R′).

The majority of reactions in organic chemistry are Lewis acid–Lewis base reactions. Focusing on the nucleophile/electrophile pathways, a bottle of a **carbanion** or carbocation cannot be obtained from a chemical stockroom. Instead, they must be generated from available, stable reagents. Have we learned a way to generate a carbanion recently? [*Look in Section 10.4.4 if you have forgotten.*]

We can react a terminal alkyne with a strong base like sodium amide (**Figure 10.14**). This generates an electron-rich, nucleophilic species that reacts with moderately strong electrophiles. If the electrophile is a carbon, then a new C—C bond will be made.

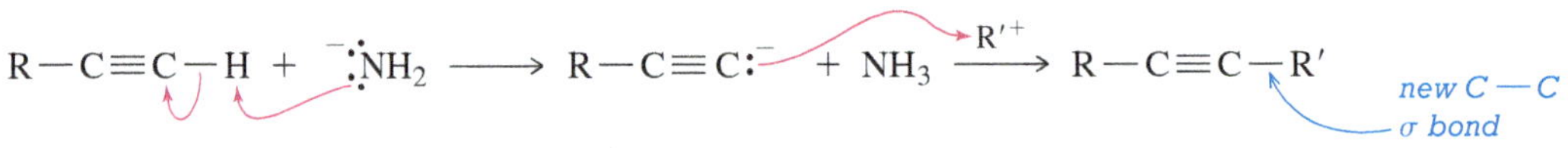

Figure 10.14 Synthesis of a new C—C bond using a deprotonated alkyne.

Because the carbanion is relatively reactive, the electrophile doesn't need a full positive charge to facilitate formation of the new bond. Even a carbon-based electrophile with only a partial positive charge will allow this process to happen. ■

From the analysis in the preceding Thinking Like a Chemist feature, deprotonated alkynes, generally referred to as acetylide anions, react with electrophiles to generate C—C bonds. The ideal electrophile for this process is a 1° alkyl halide, the chemistry of which we explore in Chapters 11 and 12.

To understand why and when this works, consider the overall reaction, breaking down each step of the process (**Figure 10.15**). In the first step, sodium amide deprotonates acetylene in a favorable reaction to generate ammonia and the reactive acetylide nucleophile. Acetylide, with a lone pair/negative charge on carbon, wants to share 2 electrons. Chloromethane, or any 1° alkyl halide, contains a partial positive charge on the carbon. Consistent with rule 2 of organic chemistry [*negatives attack positives*], the acetylide ion attacks the partially positive carbon, pushing out the chloride ion at the same time. As a result, a new C—C bond is formed. Overall, the acetylide ion is substituted for the chloride ion. This reaction type, one we discuss in Chapter 12, is a **substitution reaction**.

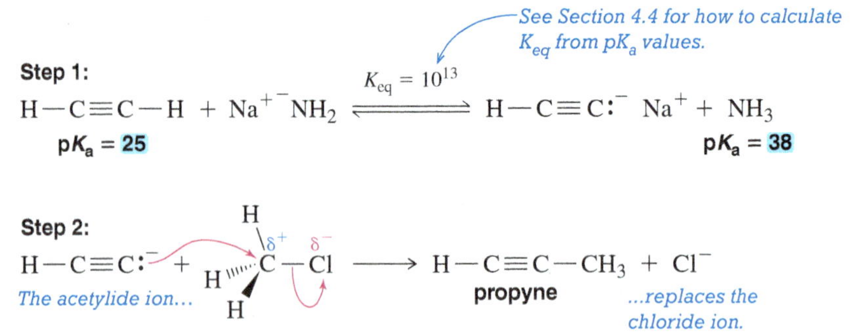

Figure 10.15 Following deprotonation of acetylene, the acetylide ion displaces chloride in a substitution reaction to make a new C—C bond.

Were this a reversible reaction, the favored side of the equilibrium could be approximated by comparison of the bases involved (chloride ion vs. acetylide ion) (**Figure 10.16**). The pK_a values of the conjugate acids (acetylene, pK_a = 25; HCl, pK_a = −7) show us that HCl is a much stronger acid than acetylene (a difference of 32 pK_a units). Thus, the acetylide ion is the stronger/more reactive/less stable base. The equilibrium of a reaction shifts toward the more stable base, which in this case is the chloride ion (Chapters 4 and 5). In sum, this is a highly favorable reaction.

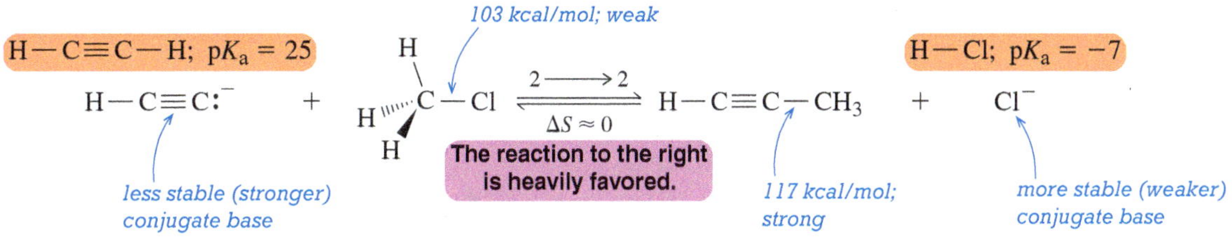

Figure 10.16 **Formation of a stable base makes acetylide alkylation highly favorable.**

Let's work this together.

Assessment 10.16 Using pK_a values, calculate an approximate K_{eq} value for the following substitution reaction.

Solved Assessment 10.16

This substitution reaction is favored on the basis of going from an unstable acetylide ion to a very stable iodide ion. This question is asking us to quantify just how favorable the reaction is. This reaction contains negatively charged species (bases) on each side, so consider the pK_a values of the conjugate acids (**Figure 10.17**).

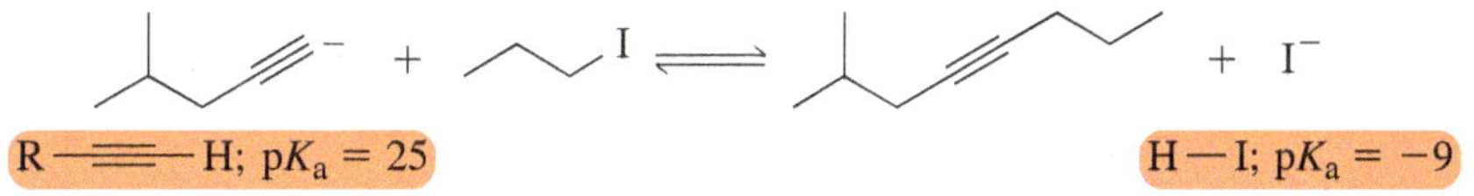

Figure 10.17 **The pK_a values of the conjugate acids suggest that this should be a favorable reaction.**

Because acetylide is a much more reactive/unstable base than iodide, the reaction proceeds to the right. The actual K_{eq} should be approximately the difference in stability of the two anions. The pK_a values suggest a difference in stability of 34 pK_a units; therefore, $K_{eq} \approx 10^{34}$.

Because this is not a true acid–base reaction, it is not appropriate to use Equation 4.1(b) here.

$$pK_{eq} = pK_a(\text{left-side acid}) - pK_a(\text{right-side acid}) \qquad \text{Equation 4.1(b)}$$

To summarize, pK_a values can be used to estimate K_{eq} for a reaction based on the relative stability of the involved bases. ■

A deeper analysis of this new reaction, one we will study in Chapter 12, shows the electron-rich acetylide ion attacking the partially positive carbon bearing the chlorine (**Figure 10.18**). Donation of electrons to this carbon alone without chloride leaving would result in a carbon with 10 electrons in its valence shell. This is a highly unstable intermediate that does not actually form. If it did, however, carbon cannot exceed 8 electrons in its valence shell, so 2 electrons would need to be lost in order to accommodate the new electrons. Although there are three groups that could take 2 electrons, the chloride ion is so stable that it takes a pair of electrons with it as it detaches from carbon. To avoid formation of the unstable intermediate, the chloride leaves as the acetylide ion attacks. Stable halide ions like Cl^- are considered good **leaving groups** in substitution reactions.

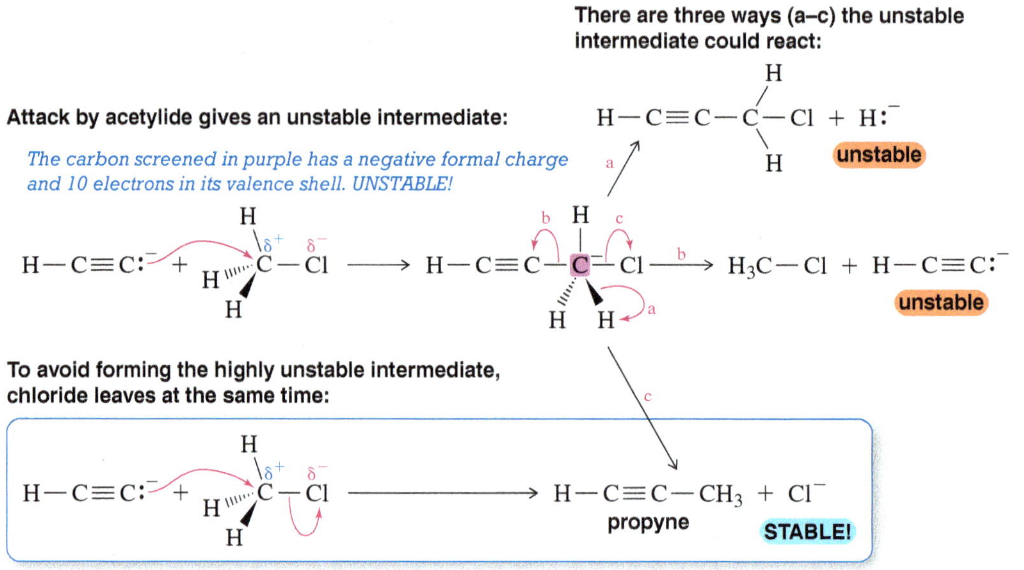

Figure 10.18 Mechanism of a substitution reaction, a reaction we'll study in Chapter 12. [*The actual mechanism is enclosed in a blue box.*]

10.5 SUMMARY

- Deprotonation of the acetylenic hydrogen with a strong base generates a reactive nucleophile.

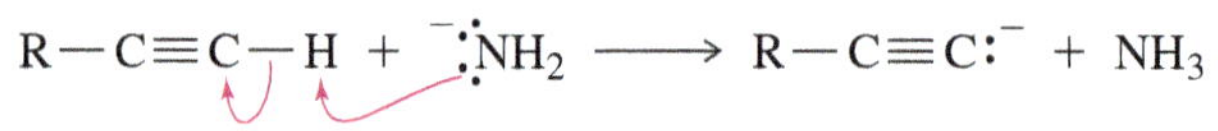

- The acetylide anion attacks the (partially) positively charged carbon, displacing the attached halogen (a good leaving group) to form a new carbon–carbon bond.

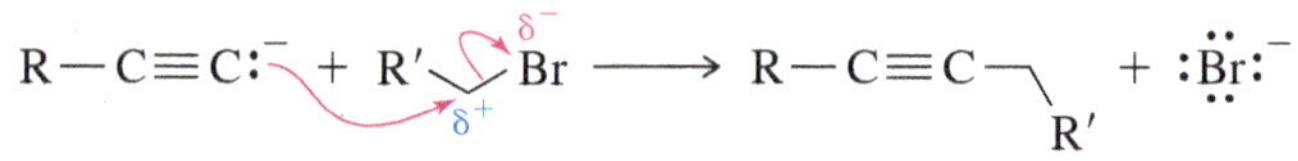

Assessment 10.17 Estimate the K_{eq} for the following reactions based on the stability of the anions involved.

(a)

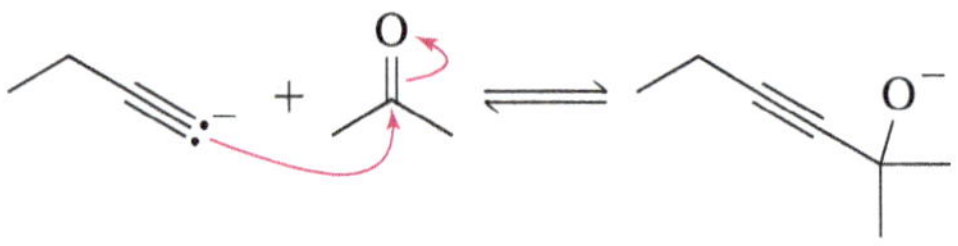

(b)

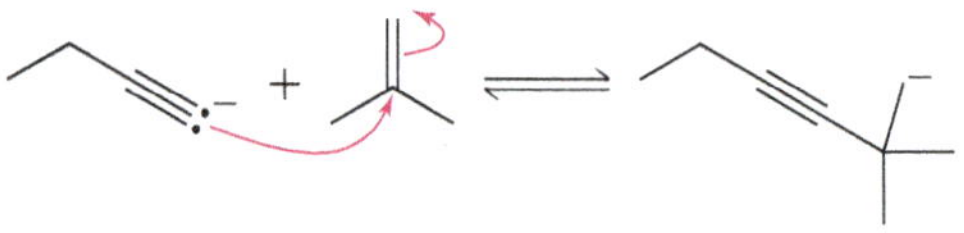

(c)

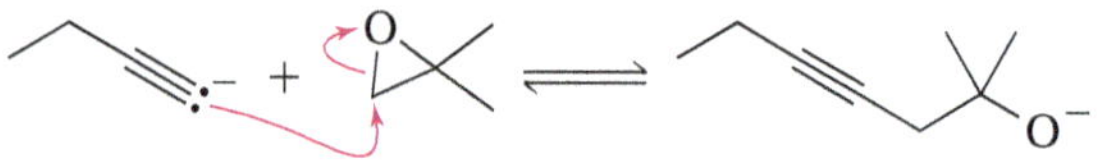

Assessment 10.18 Draw a transition state for the following substitution reaction.

$$H_3C{-}C{\equiv}C^-\,Na^+ + H_3C{-}Cl \longrightarrow H_3C{-}C{\equiv}C{-}CH_3$$

10-B CHEMISTRY IN CONTEXT: Problem-Solving Strategies

Differential diagnosis

The last time you visited a doctor, what was that interaction like? Did the physician burst into the room, take one look at you, and immediately start ordering tests and procedures? Probably not. Ordering every test to identify your illness would work, but it would be expensive and inefficient.

Aspiring physicians are taught a specific method for diagnosing. Referred to as clinical reasoning, or differential diagnosis, the process is loosely summarized as follows:

1. Acquire data. Examine the patient's vitals; give a physical exam; interview the patient. These steps form the diagnostic reasoning.
2. Develop a problem synthesis statement. This is a concise statement of the problem and includes the context (recurring or constant headache, smoker or nonsmoker) specific to that patient.
3. Identify the cause of the symptoms. This is a list of all possible ailments, prioritized by urgency.
4. Rule out causes. This is done until a single cause is identified by tests and physical examination.
5. Finally, treat the condition. Having identified the most likely cause, the necessary lifestyle change or drug is prescribed.

Adhering to this process in every physician–patient interaction gives the best odds of identifying the cause of the symptoms and, ultimately, prescribing the correct treatment. A constant and well-defined strategy for diagnosing the problem leaves only the cause of the problem in question, not the method for identifying it.

10.5.1 Synthetic utility of acetylide alkylation

Acetylide alkylation works with a wide variety of alkyl halide substrates (**Figure 10.19**). Other functional groups are tolerated as long as they are neither acidic nor electrophilic. The halogen needs to be on a 1° or methyl carbon because there is too much steric hindrance for substitution to occur at a 2° or 3° carbon. [*The alternate outcome is discussed in Chapter 12.*]

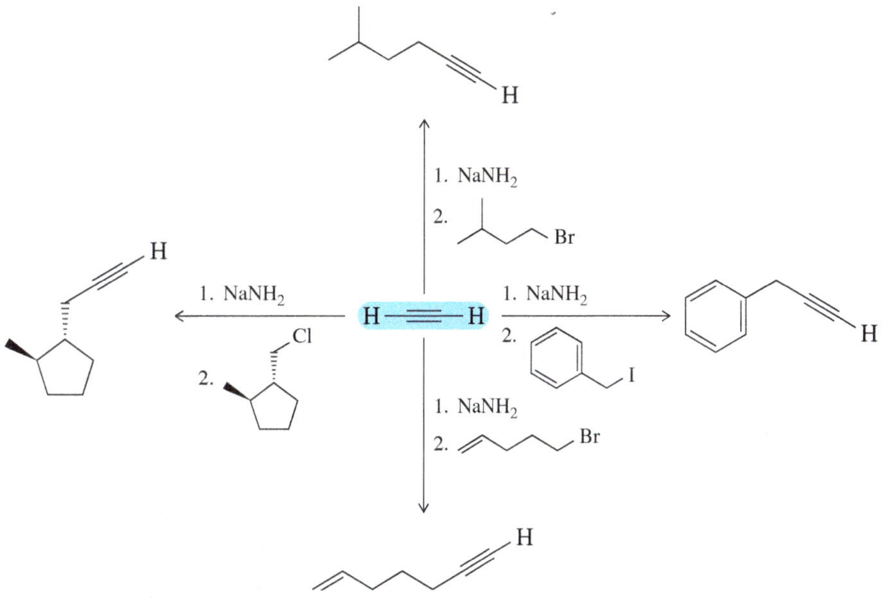

Figure 10.19 A variety of 1° alkyl halides can be displaced by the acetylide ion.

After the first alkylation, an acidic proton remains. Thus, a second alkylation is possible (**Figure 10.20**). In this way, symmetrical and unsymmetrical internal alkynes can be prepared. When alkylating acetylene twice with two equivalents of the same alkyl halide, both alkylations can be performed in the same flask by reacting with two equivalents of base.

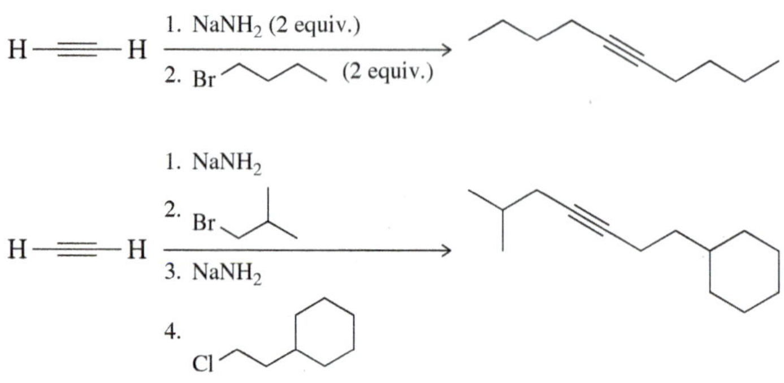

Figure 10.20 Acetylene can be alkylated twice to make internal alkynes.

When predicting products of C—C bond-forming reactions, a common mistake is to miscount the carbons (**Figure 10.21**). [*This is a common occurrence with acetylide alkylation because it is the first C—C bond-forming reaction we have studied.*] Remember that the lone pair of the acetylide ion becomes the new σ bond. And always check your answer by numbering the carbons.

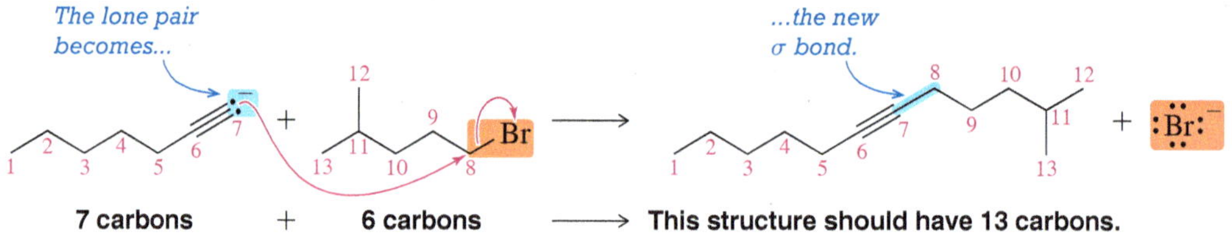

Figure 10.21 Numbering carbons and identifying the lone pair that becomes the new σ bond will help you avoid miscounting carbons when predicting products.

Assessment 10.19 Predict the product of the following acetylide alkylations.

(a)

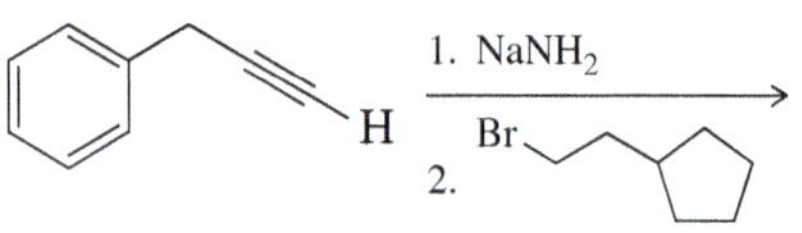

(b)

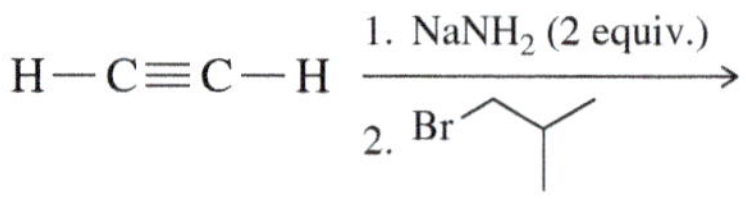

(c)

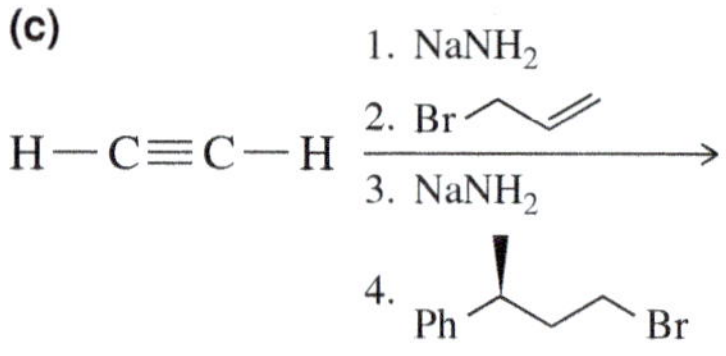

Assessment 10.20 Suggest reagents you might use to generate the product from the given reactant.

(a)

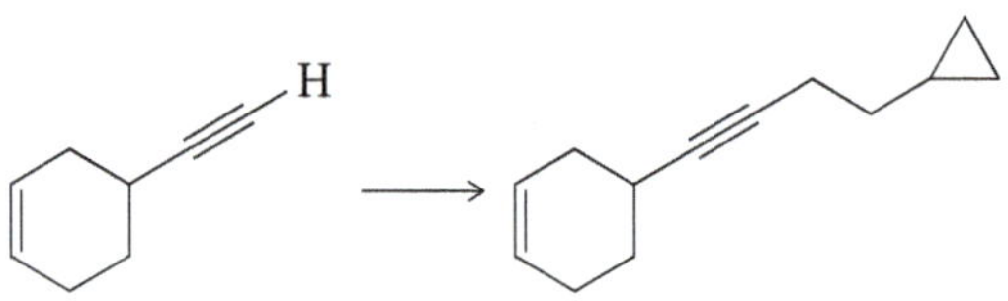

(b)

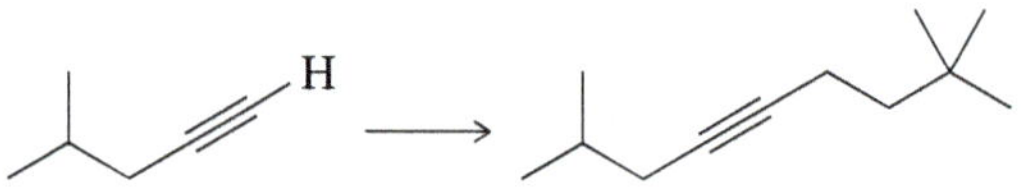

(c)

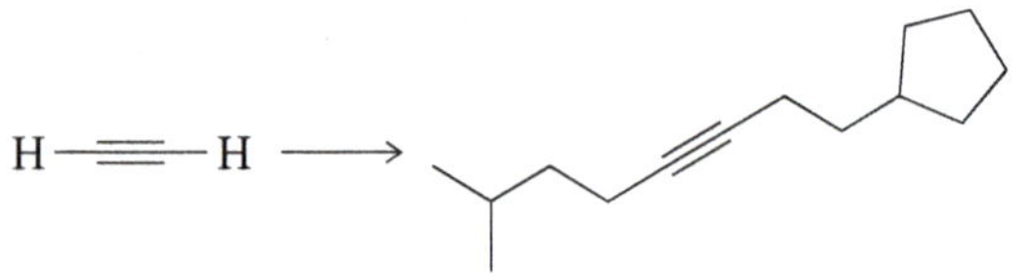

Assessment 10.21 A student provided the product of the following reactions, but made a mistake. Identify the mistake, correct the mistake, and suggest a way for the student to avoid that mistake in the future.

(a)

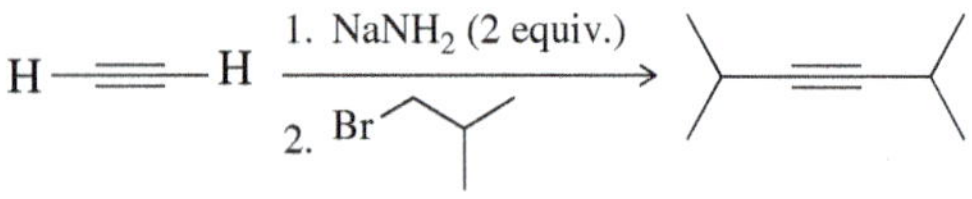

(b)

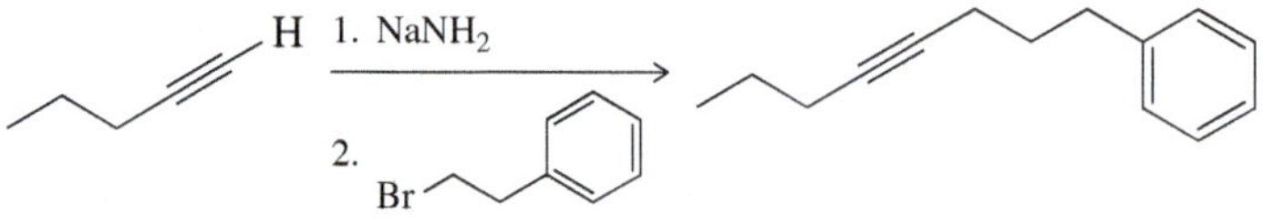

10.6 Reduction of alkynes

How do the slight, but important, differences in reactivity between alkynes and alkenes result in different outcomes? Why do alkynes need slightly different reagents to do similar reactions? These questions and others should be considered while reading further.

With the synthesis of larger alkynes now possible, into which functional groups can alkynes be converted? Their molecular orbital picture is similar to that of alkenes, so the reactions of Chapters 8 and 9 seem like a good place to begin our discussion of alkyne reactivity. The mechanistic rationale of alkyne reactions will be rooted in the same analysis used in those chapters. [*As stated multiple times, organic chemistry is not a series of unrelated reactions. Each reaction you learn helps you understand what comes later.*] As a result, the focus here is on comparing alkyne and alkene reactions.

10.6.1 Hydrogenation of alkynes

Recall from Chapter 9 that alkenes can be reduced by H_2 and a transition metal catalyst, such as palladium, platinum, or nickel. The alkene carbons of *trans*-but-2-ene (−1 oxidation state) were reduced in this way (**Figure 10.22**). Alkynes (0 oxidation state) also contain electron-rich π bonds that can be reduced completely under similar conditions (but-2-yne → butane; see **Figure 10.23**). [*At a higher oxidation state, alkynes can be reduced further* ($0 \rightarrow -2$) *than alkenes* ($-1 \rightarrow -2$).] This reaction consumes two equivalents of hydrogen gas for each alkyne reacted.

Figure 10.22 That alkenes can be reduced suggests that alkyne reduction should also be possible.

This reaction occurs by a poorly understood, but similar mechanism to alkene reduction (compare Figure 10.23 to Figure 9.44). [*Several reactions in this chapter have mechanisms that are not well understood. But, through analogies to reactions studied previously, we can construct reasonable mechanisms.*] The reaction begins with the adsorption of H_2 and the alkyne to a metal surface. Adsorption facilitates the concerted addition of a hydrogen atom to each carbon of the original alkyne, resulting in stereospecific formation of the cis-alkene. Because the palladium catalyst is regenerated and there is usually a huge excess of H_2, the cis-alkene is further reduced to the alkane.

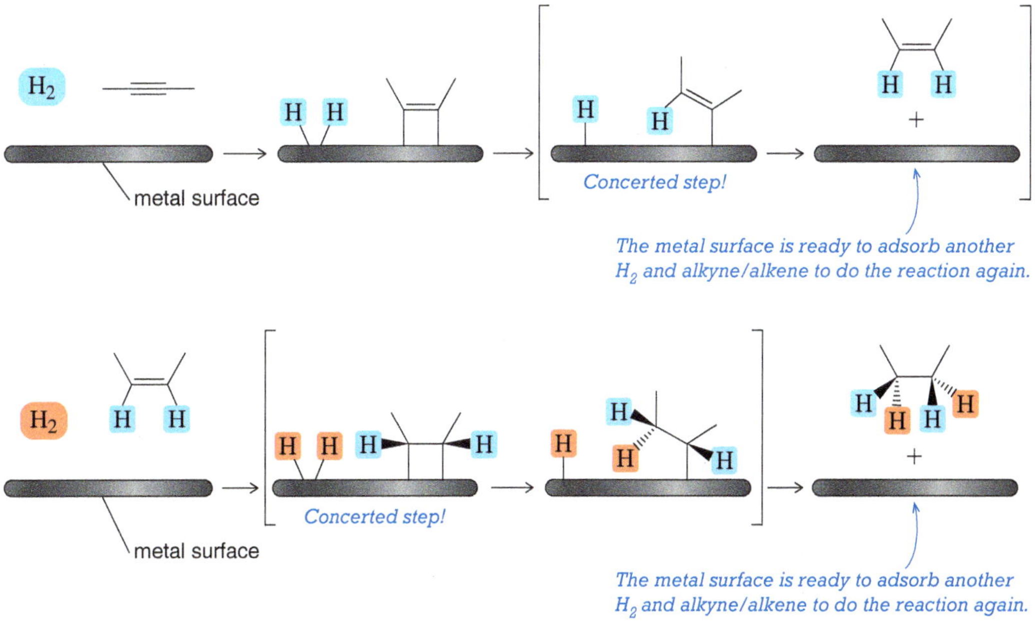

Figure 10.23 Hydrogenation of an alkyne occurs via the concerted delivery of two hydrogens to the same face of the alkyne from a palladium catalyst, just as it does when hydrogenating alkenes to give alkanes.

Assessment 10.22 Predict the product of the following hydrogenation reactions.

(a)

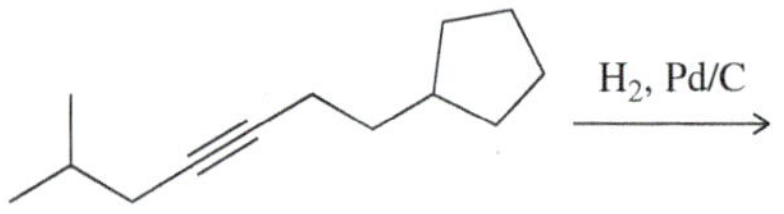

(b)

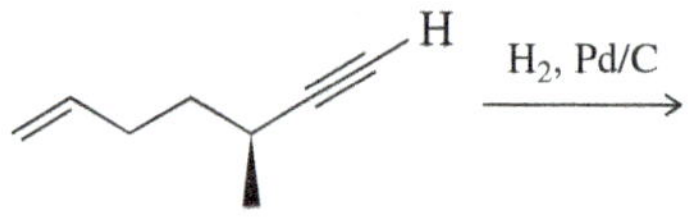

(c)

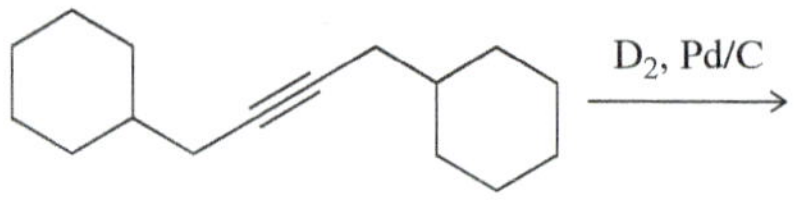

Note: D is an isotope of H and behaves similarly.

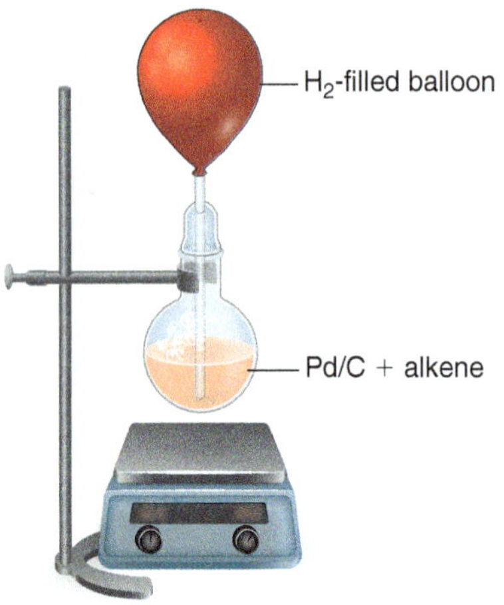

Figure 10.24 When chemists run hydrogenation reactions, the balloon filled with H_2 gas surrounds the reaction with an atmosphere of H_2, resulting in a very large excess of H_2. The reaction is finished when the balloon stops deflating. In this case, it will be when all of the alkyne has been converted to alkane.

10.6.2 Hydrogenation with a poisoned catalyst

[*Three-year-olds, over and over again, ask the simple question, "Why?" Synthetic chemists, on the other hand, develop new reactions by asking, "But can we . . . ?"*]

The reduction of alkynes to alkanes proceeds through an intermediate alkene. Thus, you might ask, "*But can we* stop the reaction at the alkene?" [*Great idea!*] We have learned several alkene reactions, so being able to make them would be synthetically useful. [*Easy game, right?*]

Under the conditions described in Section 10.6.1, the reaction always proceeds to the alkane. The catalyst is constantly regenerated and controlling the stoichiometry of gases in reactions is difficult (Figure 10.24). Thus, there is always another palladium surface with adsorbed hydrogen available to react with the intermediate alkene.

One possible solution relies on the relative stabilities of the alkyne and alkene π bonds. [*Go back and reread Section 10.4.2.*] Under hydrogenation conditions, the alkyne π bond is more reactive than the alkene π bond. [*The alkene is more stable than the alkyne.*] As a result, the reaction can be run such that the alkyne reacts completely, leaving the alkene unreacted. This difference in reactivity was first exploited by Herbert Lindlar in the 1950s. Lindlar observed that heterogeneous catalysts consisting of powdered palladium metal, calcium carbonate, and lead salts were significantly deactivated, or **poisoned**. These poisoned catalysts, referred to as **Lindlar catalysts**, were active enough to reduce alkynes but not reactive enough to react with the resulting alkene (Figure 10.25). As a result, cis-alkenes can be synthesized from alkynes via the concerted, syn addition of H_2 to the same face of the alkyne. This reaction occurs by the mechanism in Figure 10.23 but stops at the alkene.

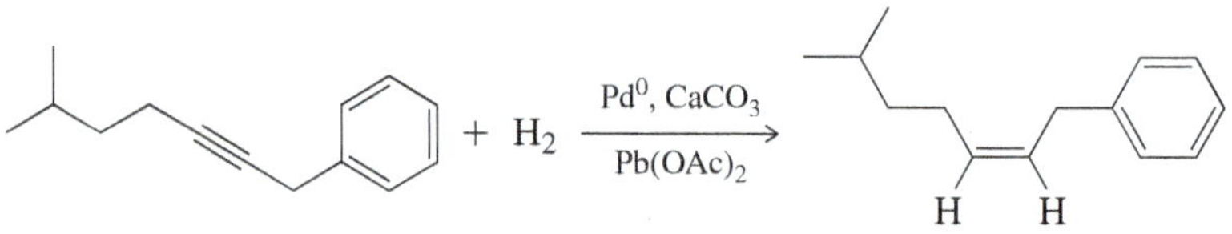

Figure 10.25 Hydrogenation with Lindlar catalysts reduces alkynes to cis-alkenes.

Assessment 10.23 Predict the product of the following hydrogenation reactions run with a poisoned catalyst.

(a)

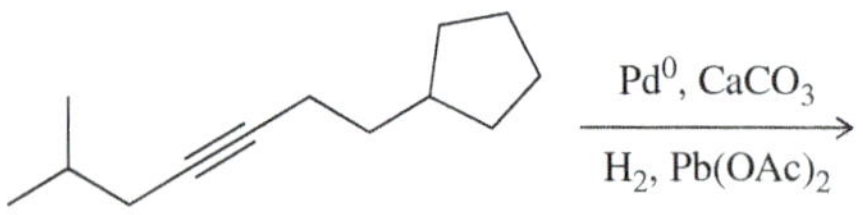

(b)

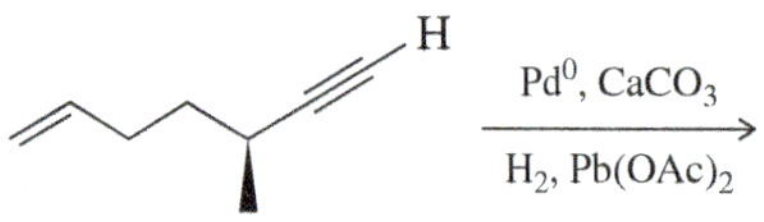

(c)

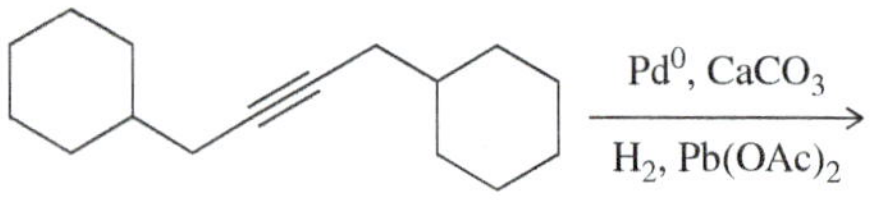

Assessment 10.24 Hydrogenation of which alkynes would produce the following cis-alkenes?

(a)

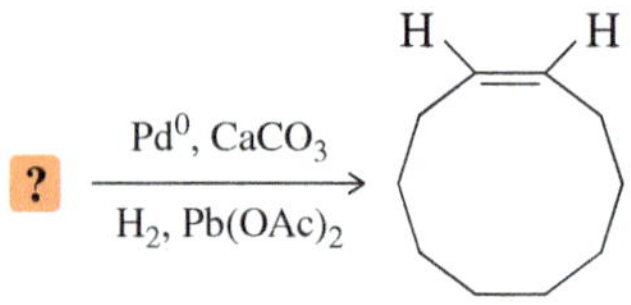

(b)

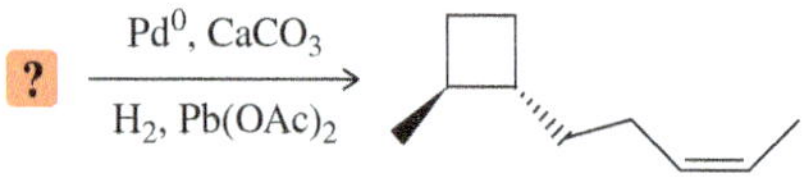

(c)

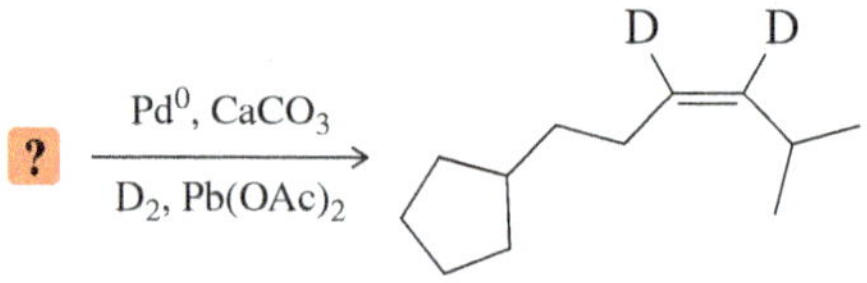

10.6.3 Dissolving-metal reduction

We can prepare cis-alkenes, *but can we* make trans-alkenes in a similar way? [*That would be useful!*] Because of the concerted mechanism of the Lindlar reaction, it could never make a trans-alkene. [*Do you see why?*] Concerted reactions in Chapters 8 and 9 resulted in addition to one side of the molecule (syn addition). Thus, to get trans-alkenes, the hydrogens need to add at different times. This is accomplished by the reaction of an alkyne with sodium dissolved in liquid ammonia (−33 °C) (**Figure 10.26**).

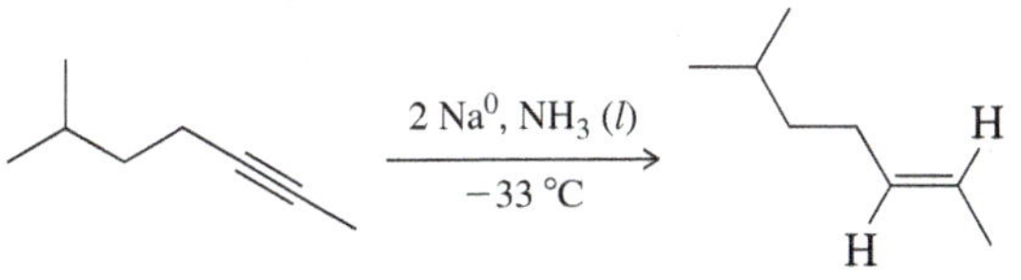

Figure 10.26 The use of sodium and liquid ammonia reduces an alkyne to a trans alkene.

This nonconcerted reduction occurs via an electron-transfer mechanism. Sodium metal (Na^0), when dissolved in a **protic solvent**, ejects an electron, achieving a noble gas configuration (Na^+). The colorless NH_3 becomes dark blue [**Figure 10.27(a)**], a result of electrons dissolved in a protic solvent. These highly reactive electrons react with the alkyne C—C π bond, the weakest bond in solution. As shown in **Figure 10.27(b)**, one π electron meets the incoming electron, forming a carbanion on C_1, while the other π electron goes to C_2, forming an sp^2-hybridized radical. Equilibration of the resulting radical anion gives the more stable trans configuration. Being an unstable, strong base, the radical anion deprotonates NH_3, forming a new C—H σ bond at C_1. The electron-deficient radical is then attacked by another electron in solution to give a second carbanion. Protonation at C_2 completes the reduction, providing the trans-alkene. The reaction stops at the trans-alkene because, once again, the alkene is less reactive than the alkyne.

(a) (b)

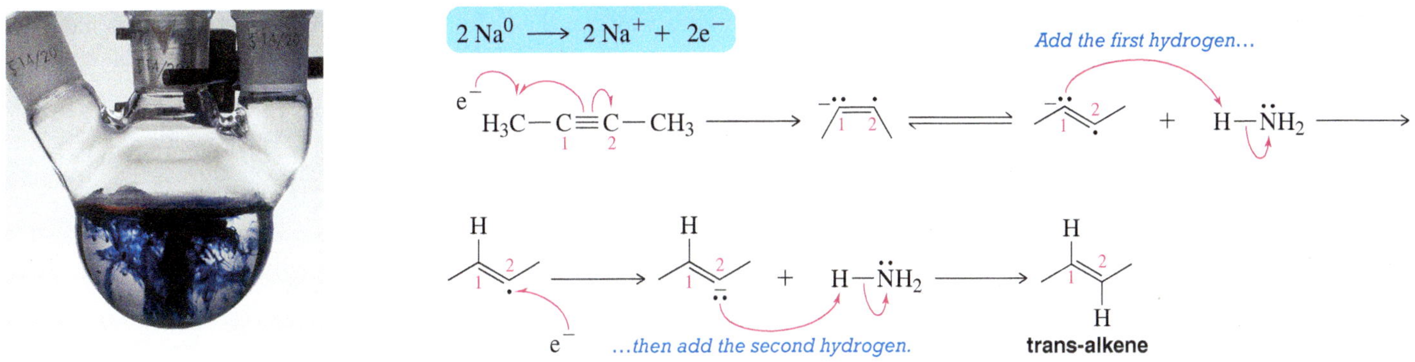

Figure 10.27 Reduction of an alkyne with sodium and ammonia. (a) The blue color is a resulted of solvated electrons. (b) The mechanism of alkyne reduction. Since the C—H bonds are formed at different times, a trans-alkene results.

10.6 SUMMARY

- Alkynes can be completely reduced to give the fully saturated alkane . . .

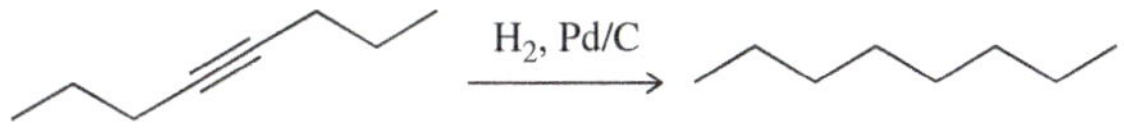

- . . . or, using Lindlar's catalyst, the concerted, syn addition of H_2 stops at the cis-alkene . . .

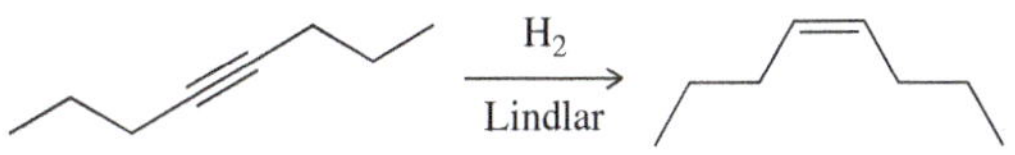

- . . . or, using sodium in liquid ammonia (Na^0/NH_3), the hydrogens add in different steps, resulting in the trans-alkene.

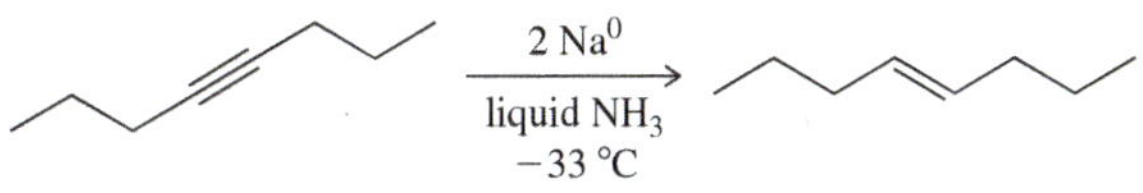

Assessment 10.25 Predict the product of the following alkyne reductions.

(a)

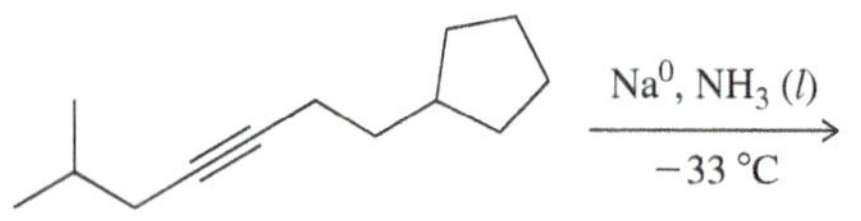

(b)

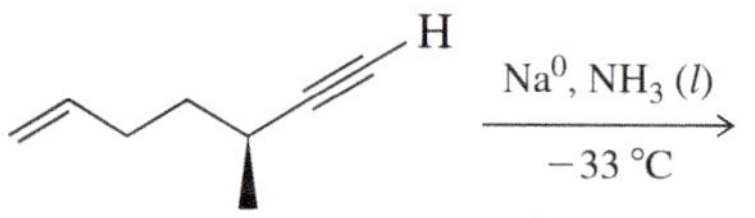

(c)

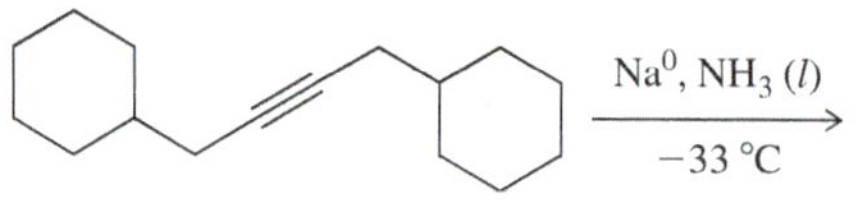

Assessment 10.26 Suggest alkynes that might be used to make the following trans-alkenes.

(a)

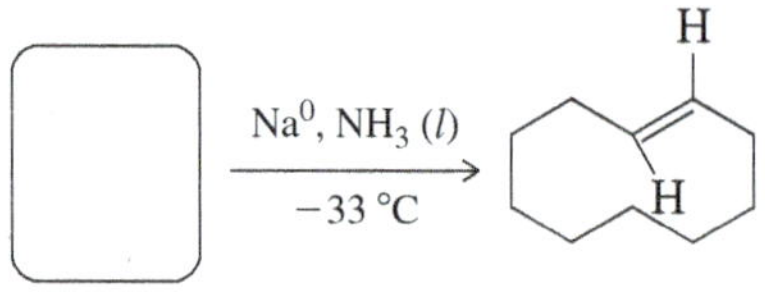

(b)

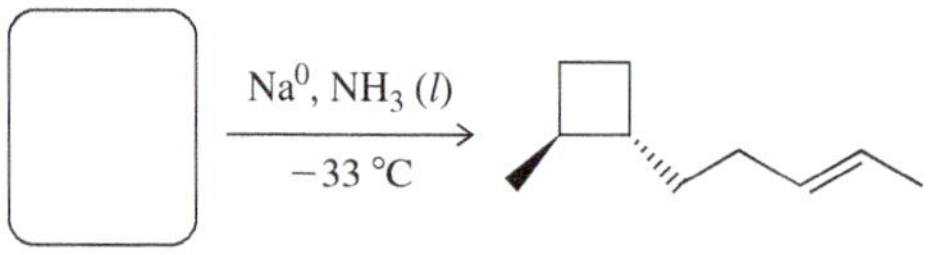

(c)

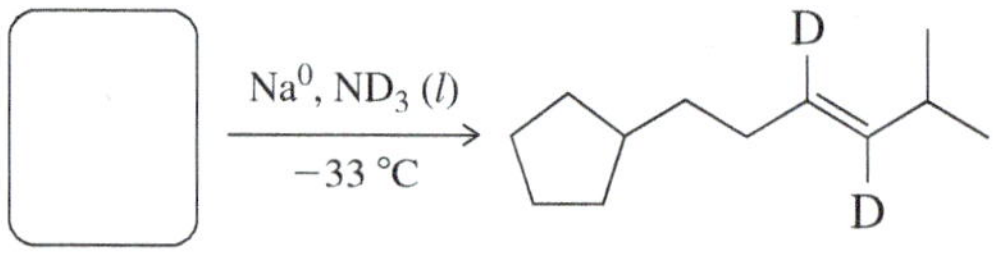

Assessment 10.27 The following deprotonation step occurs during the trans reduction of an alkyne. Calculate K_{eq} for this reaction.

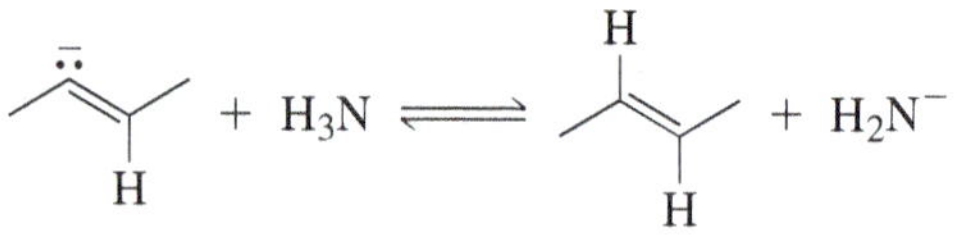

Assessment 10.28 When sodium generates electrons in the presence of ammonia, these electrons persist in solution, giving the blue color. However, electrons do not persist when sodium is added to water. Why?

Assessment 10.29 Sodium amide, the base we use to deprotonate terminal alkynes, is synthesized by reducing ammonia via a mechanism similar to the reduction of alkynes in Figure 10.21. Suggest a mechanism for this reaction.

$$2\ NH_3 \xrightarrow{2\ Na^0} 2\ NaNH_2 + H_2$$

Assessment 10.30 Long, polarized bonds are also reducible in the same way that the C—C π bond of an alkyne is. Show a mechanism by which a C—Br bond might be reduced by sodium metal.

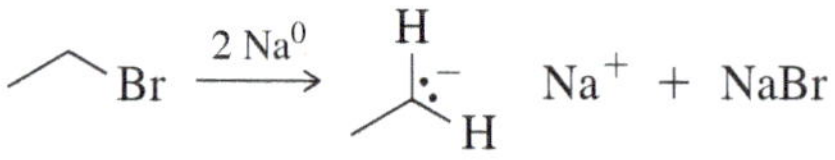

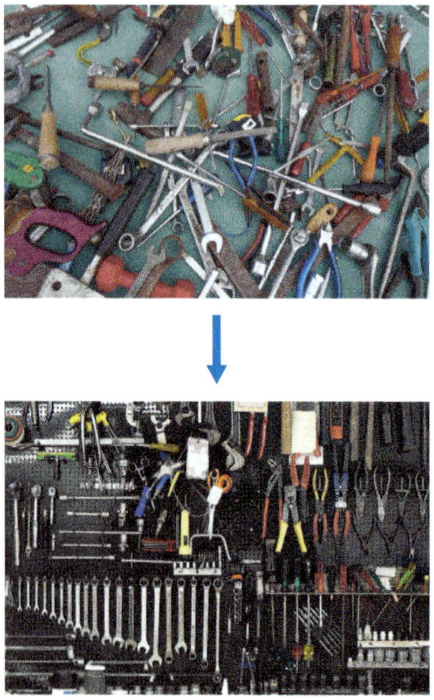

10.7 Organizing our tools: reaction sheets

The construction analogy introduced at the beginning of this chapter identified reactions as tools for building molecules. By my count, 15 reactions have been presented thus far in the text. [*See if you can list them without looking back.*] There are ways to make alcohols (2), alkyl halides (3), C—C bonds (1), aldehydes and ketones (1), halohydrins (1), epoxides (1), diols (2), alkanes (2), alkenes (2) . . . pretty intimidating, yeah? Our synthetic toolbox is filling up, but it must be organized and the reactions easily found when needed. [*What good is a hammer if we don't know where it is?*]

New reactions have continually been connected to each other in terms of how and why they occur. Here, they are organized for efficiently doing organic synthesis on creatively named "reaction sheets." [*An idea I got from my own organic professor—thanks to Joe Workman, Centre College.*] Because synthesis problems require knowledge of how a functional group reacts or how it might be formed, the reaction sheets should be organized such that each functional group gets its own sheet of paper. One side contains the "Reactions of **INSERT FUNCTIONAL GROUP**"; the other should have the "Preparation of **INSERT FUNCTIONAL GROUP**."

In making the reaction sheets, choose real molecules but recognize that each reaction works for a variety of structures. [*It may take more time, but trust me, it will pay off in the end.*] Consider hydroboration (Section 8.4.5) and three ways of representing it (**Figure 10.28**); is one better?

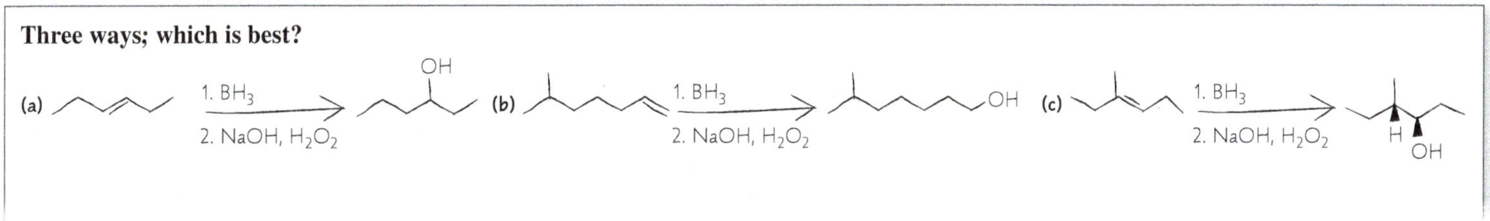

Figure 10.28 Which is the best reaction to include on a "Preparation of alcohols" reaction sheet?

Organizing reactions in this way should not replace trying to understand the how and why of reactions and their mechanisms. Reaction sheets are simply a convenient way to organize reactions for quick access while doing synthesis problems.

Any of the three hydroboration reactions shown will work perfectly. To keep them concise, reaction sheets should only have one reaction for each reagent. Hydroboration is a concerted syn addition resulting in stereospecific synthesis of the least substituted alcohol. Your reaction sheet should show only reaction (c) because all of these details are illustrated. [*This is despite the fact that reaction (b) is how hydroboration is most often used.*]

To help you get started making them, Figure 10.29 shows the hydroboration reaction we have studied on the two reaction sheets it appears on. [*Why on two?*] Because each reaction represents the reaction of one functional group and the preparation of another, hydroboration is on both "Reactions of alkenes" and the "Preparation of alcohols." I recommend writing brief notes in the margins to remind yourself of the important aspects of the reaction. [*In this way, your reaction sheets not only are helpful for synthesis, but also give you a means of quick review when prepping for an exam.*]

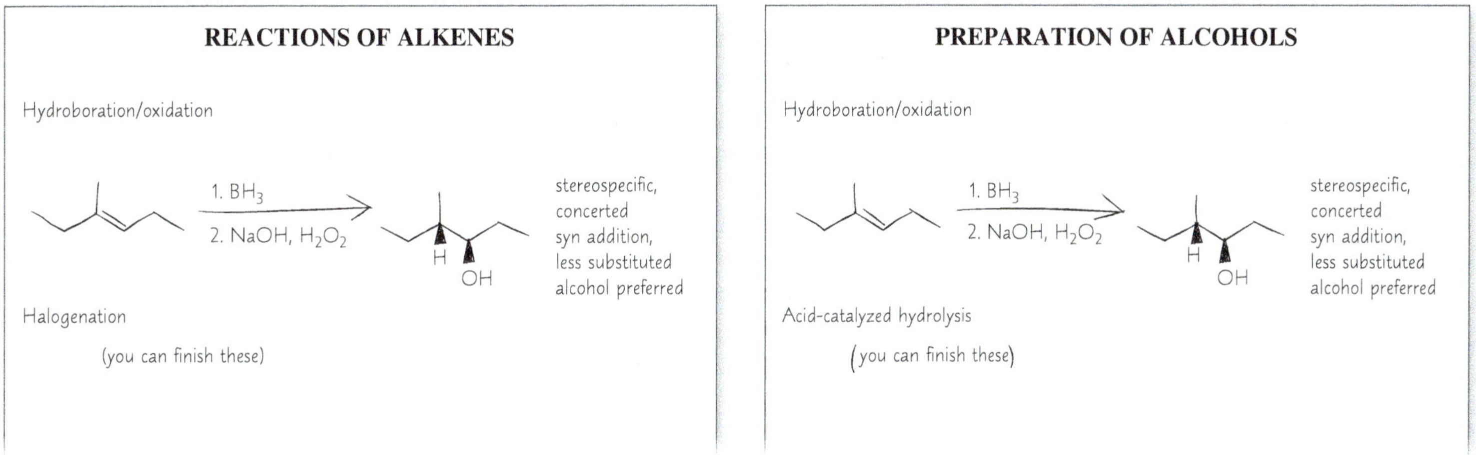

Figure 10.29 Sample reaction sheets. Each reaction should appear on two sheets—one for the reaction it undergoes and one for the functional group it prepares. [*At this point, you should have seven reaction sheets, one for each of the following functional groups: alkanes, alkenes, alkynes, alcohols, haloalkanes/haloalkenes, ethers/epoxides, and aldehyde/ketones.*]

An organized toolbox of reactions makes for a happier chemist!

Assessment 10.31 is meant to help you think about the best way to construct your reaction sheets. [*As you are making your reaction sheets, confirm that you have all of the reactions included by looking at Appendix A-1.*]

Assessment 10.31 We have studied the following reactions in previous chapters. For each, (i) indicate which reaction sheets they should appear on, (ii) show the best structure to use to represent them, and (iii) write the notes you could put in the margin so that the mechanism is implied.

(a) Oxymercuration–reduction of an alkene (b) Radical halogenation of an alkane (c) Ozonolysis of an alkene

10.8 Electrophilic additions to alkynes—general overview

Like alkenes in Chapters 8 and 9, alkynes, with an electron-rich π bond (Section 10.4.5), undergo electrophilic addition. As you read, look for the similarities between the reactions of alkynes and the analogous alkene reaction. [*Connecting new concepts and reactions to old ones will enable deeper understanding.*] Moreover, look for the subtle differences that the alkyne functional group imparts. To begin, these reactions are listed in Table 10.6, which is analogous to Table 8.3 for alkenes, demonstrating that all are simply addition reactions.

Table 10.6 Addition reactions of alkynes, all of which share mechanistic similarities.

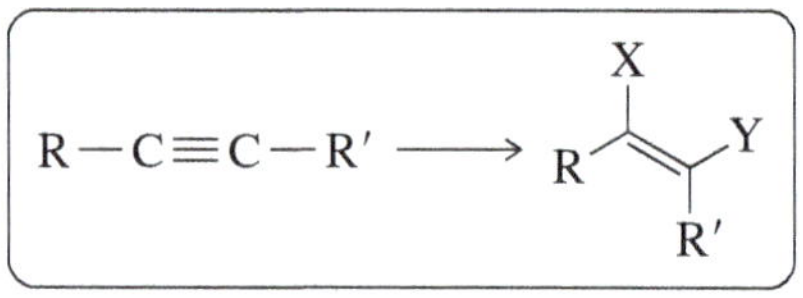

Reagent	X	Y	Representative alkyne	Initial product/ intermediate	Process
HBr	H	Br	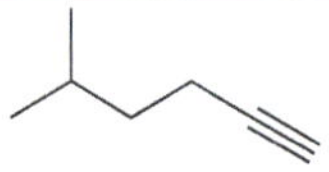	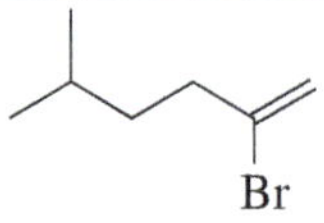	Hydrohalogenation (10.9.2)
$HgSO_4$, H_2O	Hg^+	OH		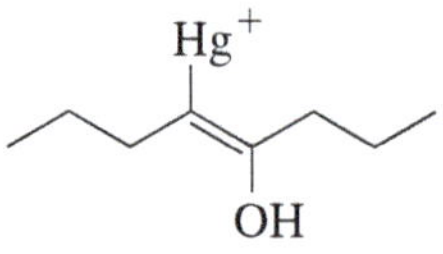	Oxymercuration (10.9.4)
BH_3	H	BH_2	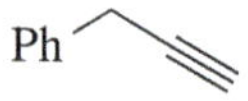	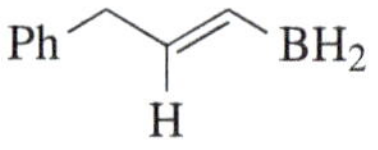	Hydroboration (10.9.5)
Br_2	Br	Br	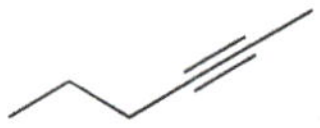	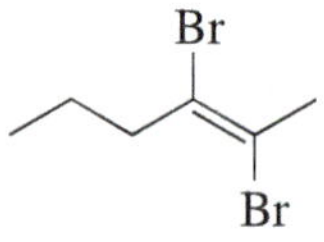	Halogenation (10.9.7)

Table 10.7 Alkenes have a higher rate of reaction than alkynes in electrophilic addition reactions.

Electrophilic reaction	k_{alkene}/k_{alkyne}
Bromination (Br_2)	3.6×10^5
Acid-catalyzed hydration (H_3O^+)	16.6
Chlorination (Cl_2)	7.4×10^3

10.8.1 General considerations

The alkyne π bond is weaker (lower bond-dissociation energy) than the alkene π bond (Section 10.4.2), but the rate data in Table 10.7 show that alkenes react more quickly than alkynes in electrophilic addition reactions. [*How can the less stable alkyne π bond react more slowly?*]

Electrophilic additions result in the formation of carbocations [Figure 10.30(a)]. Thus, the difference in rates can be explained by the diminished stability of alkenyl carbocations (the product of alkyne addition) compared to alkyl carbocations (the product of alkene addition). Though both are secondary, the relative stability of alkyl and alkenyl carbocations results from differences in hyperconjugation caused by their hybridizations. [*Read that sentence again.*] The alkyl carbocation is sp^2 hybridized, whereas the alkenyl carbocation is *sp* hybridized

[**Figure 10.30(b)**]. Fewer bonds overlap via hyperconjugation to stabilize the alkenyl carbocation, resulting in an estimated 15 kcal/mol difference in stability. Because the alkenyl carbocation is less stable, formation of it by electrophilic addition is more endothermic [**Figure 10.30(c)**]. As predicted by the Hammond postulate [*the transition state of the highly endothermic reaction resembles the product*], this difference in stability is reflected in the transition states of the reactions. Alkynes have a higher activation energy, so they react more slowly.

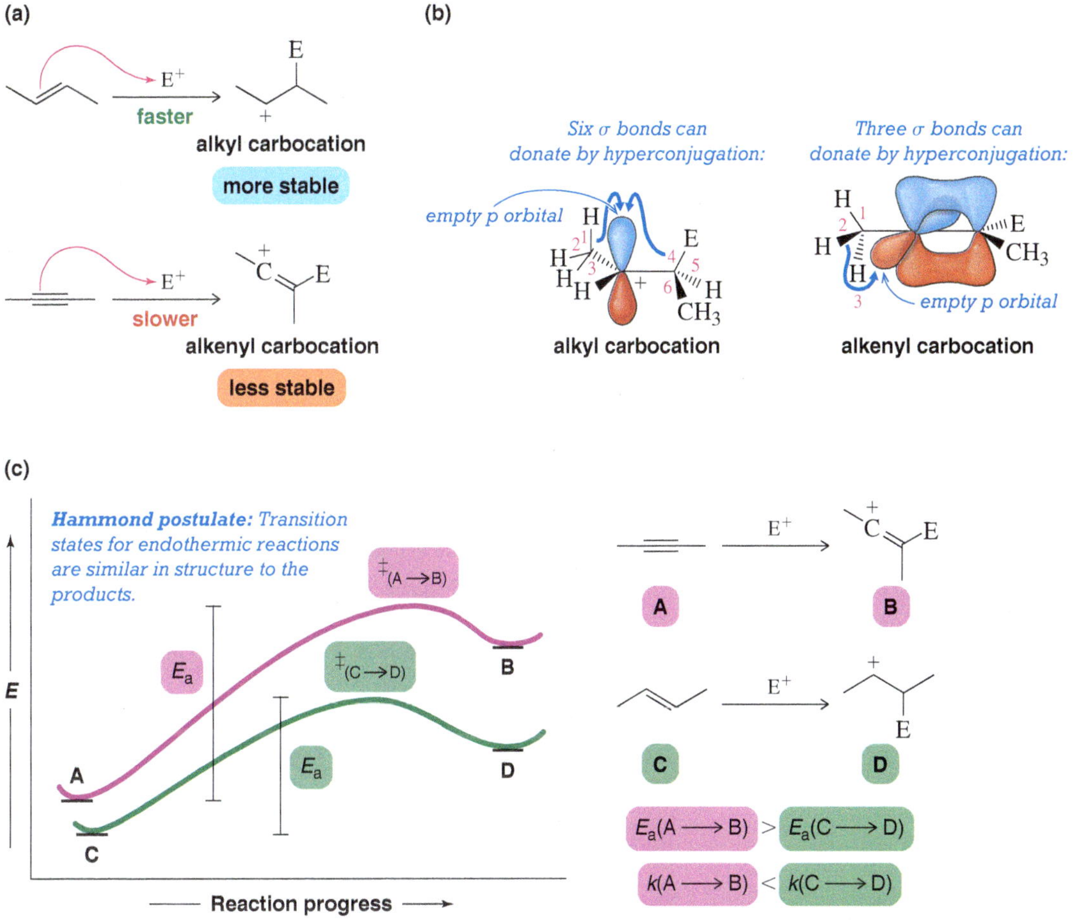

Figure 10.30 Electrophilic addition to alkynes versus alkenes. (a) Alkenyl carbocations are less stable than alkyl carbocations because alkyl carbocations are (b) better stabilized by hyperconjugation. (c) The activation energy to make an alkyl carbocation is lower than the activation energy to make an alkenyl carbocation.

10.8.1 SUMMARY

- Alkynes react more slowly than alkenes in electrophilic addition reactions because the resulting carbocation is less stable.

Assessment 10.32 The alkyl carbocation is estimated to be 15 kcal/mol more stable than the alkenyl carbocation. If this is also the difference in the energies of the transition state leading to each, what is the expected rate difference?

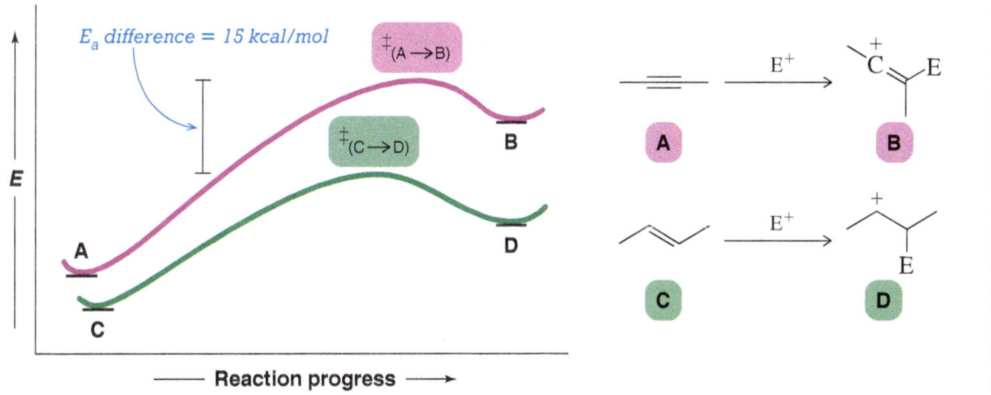

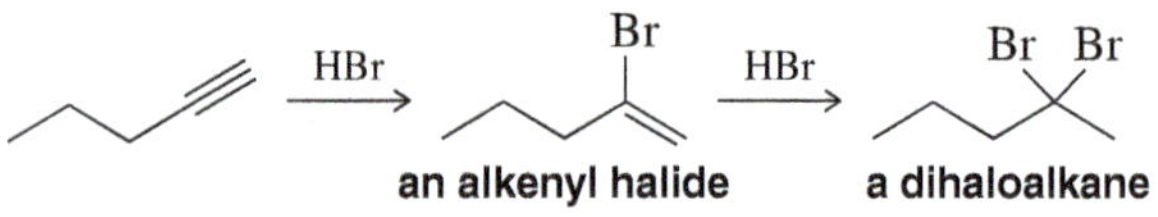

Figure 10.31 Addition of HBr to an alkyne to make a bromoalkene or dibromoalkane.

10.8.2 Addition of HBr and HCl

The addition of one equivalent of HBr (or HCl) to an alkyne results in the preparation of an alkenyl halide (**Figure 10.31**). A second equivalent reacts with the alkenyl halide to produce a dibromoalkane (or dichloroalkane).

Although the mechanism of these and other alkyne addition reactions is not yet settled, they can be analyzed as being analogous to the electrophilic addition reactions of alkenes.

10.8.2.1 HBr addition to a terminal alkyne

As was observed with alkenes, deprotonation of H—Br by the electron-rich π-bond results in the alkenyl carbocation and a bromide ion. Only the more stable 2° alkenyl carbocation is formed (**Figure 10.32**). [*There is uncertainty about whether a carbocation forms here; read on.*] Once formed, the 2° alkenyl carbocation is quickly attacked by the bromide ion to give the bromoalkene. With one equivalent of HBr, this reaction stops at the bromoalkene product. With a second equivalent of HBr, a second halogen adds to the alkene (Section 8.3). The π electrons deprotonate HBr, forming a 2° carbocation that is attacked by bromide to give the **geminal** dibromo compound.

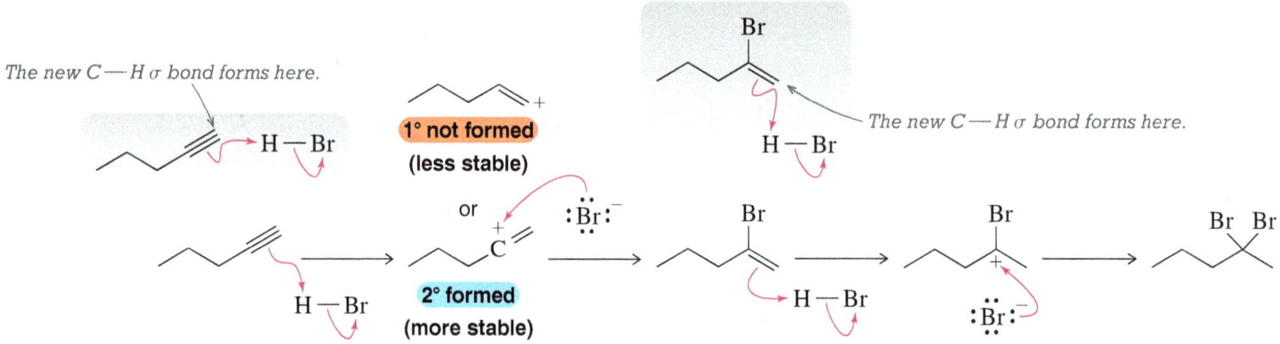

Figure 10.32 HBr addition to alkynes is analogous to alkene addition.

10.8.2.2 HBr addition to an internal unsymmetrical alkyne

When HBr (or HCl) is added to internal alkynes like hex-2-yne, two possible carbocations can form via deprotonation of HBr (**Figure 10.33**). Notably, these are equally stable 2° carbocations in terms of electronic (equivalent hyperconjugation) and steric effects (linear at both carbons, no bulky substituents on either side). Thus, these 2° carbocations form at equal rates. Once formed, both are attacked by the bromide ion to give a nearly equal mixture of 2- and 3-bromohex-2-ene. A reaction that gives only a 50% yield of the desired product in an inseparable mixture is not efficient, useful, or environmentally friendly. [*The inefficiency of electrophilic additions to internal, unsymmetrical alkynes is a theme of this chapter.*]

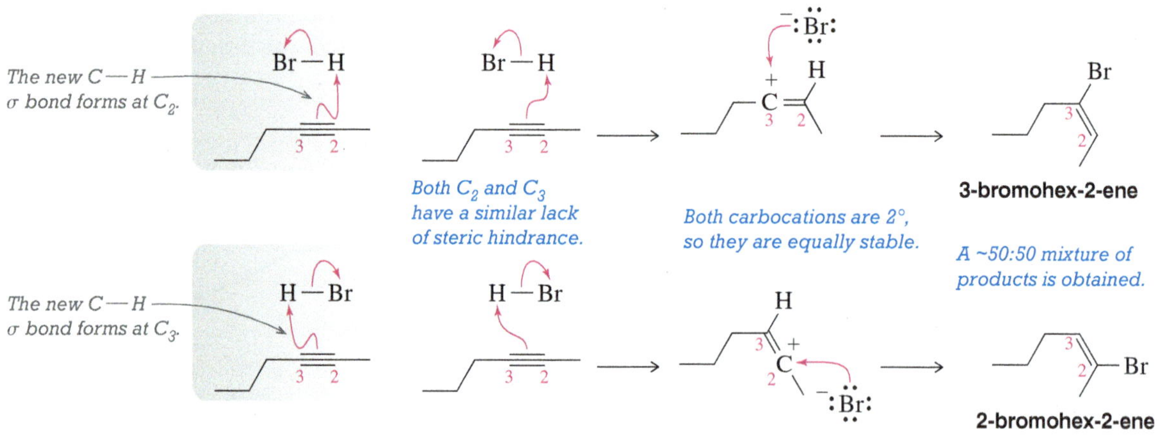

Figure 10.33 Because the carbocations are equal, internal unsymmetrical alkynes always give equal amounts of two different products.

10.8.2.3 HBr addition to an internal symmetrical alkyne

With an internal symmetrical alkyne like hex-3-yne, the situation changes (**Figure 10.34**). Protonation of the alkyne occurs at either C_3 or C_4 to give equivalently stable carbocations. Selectivity is not an issue here because the cations thus formed are identical. Attack of the bromide at either carbocation gives the same product, 3-bromohex-3-ene.

Figure 10.34 Protonation at either carbon of a symmetrical alkyne results in the same product.

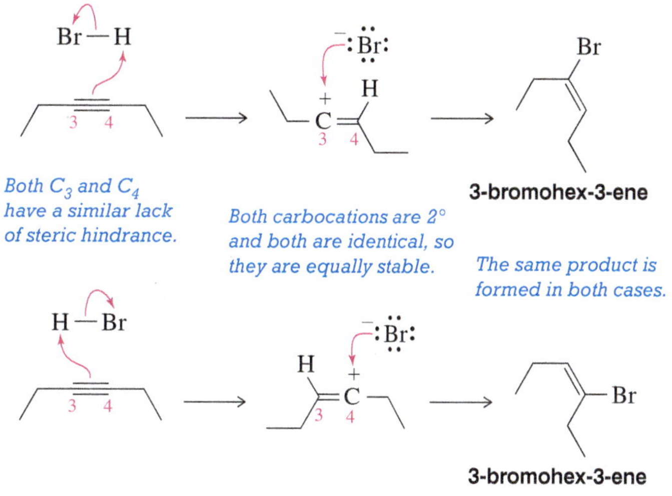

In terms of stereochemical outcome, based on the similarity to alkene electrophilic addition, you might expect that the alkenyl cation is attacked on both sides, in nearly equal quantities (Section 8.3.1). And yet the result is different for alkyne addition (**Figure 10.35**). An intermediate π complex is proposed to form, with the bromide ion attacking the opposite face, generating the (*Z*)-alkene through the net anti addition. (Among chemists, the exact mechanism of this reaction is *not* a settled matter.)

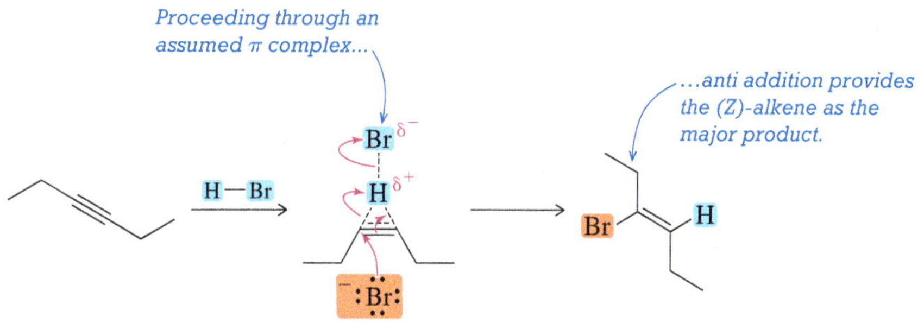

Figure 10.35 Anti addition to alkynes is presumed to occur through an intermediate π complex that avoids formation of an unstable alkenyl carbocation.

10.8.2 SUMMARY

- Addition of HBr or HCl effectively produces high yields of haloalkenes only when the alkyne is terminal or internal and symmetrical. For many of these reactions, anti addition is assumed to occur through an intermediate π complex.

Assessment 10.33 Predict the major products resulting from the addition of one equivalent of HX to the following alkynes.

(a)

(b)

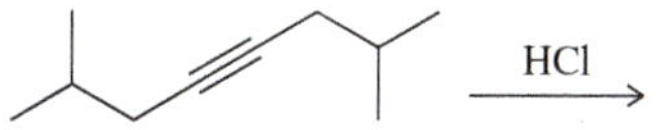

(c)

Assessment 10.34 Predict the alkyne and reactants you might use to make the following haloalkenes. [*Providing the reactant and the reagent is how we start thinking about synthesis.*]

(a)

(b)

10.8.3 Hydration of alkynes: keto–enol tautomerism

Like alkenes, alkynes can be hydrated, albeit under slightly modified conditions. Whereas the product of alkene hydration is an **aliphatic** alcohol (i.e., all of the carbon atoms are sp^3 hybridized and oriented in a chain), the initially formed product of alkyne hydration is an alkenyl, or **vinyl** alcohol (**Figure 10.36**). (*Vinyl* refers to the specific alkenyl group with the form $-CH{=}CH_2$.) This functional group is referred to as an ***enol*** because it contains both an alk***ene*** and an alcoh***ol***. It is presented in detail in Chapter 20, but the **tautomerization** of enols needs to be discussed here in order to understand the ultimate outcome of alkyne hydration.

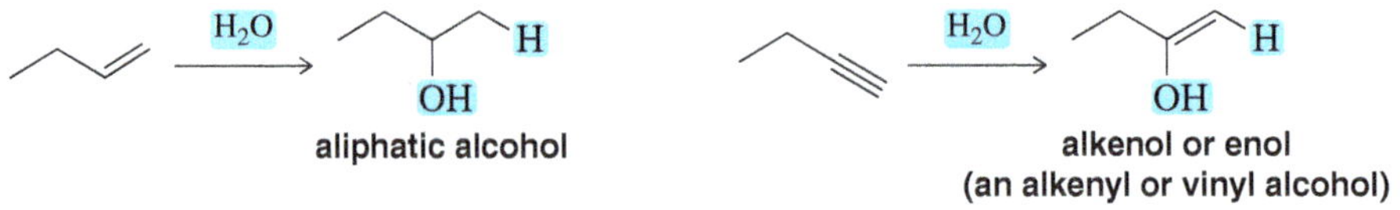

Figure 10.36 Addition of water to an alkyne gives an enol.

In solution, enols are in equilibrium with the corresponding ketone (**Figure 10.37**). The ketone and enol are a form of constitutional isomers differing only in the location of a single atom (**tautomers**). For these tautomers, the hydrogen originally attached to the oxygen in the enol is now attached at C_1 of the ketone. Analyzing the bonds in going from the enol to the ketone, an O—H σ and C—C π bond are broken, whereas a C—H σ and C—O π bond are formed. Thus, the keto form is more stable than the enol form by ~11 kcal/mol. [*C—O π bonds are more stable than C—C π bonds.*] Keto–enol tautomerism is catalyzed by either acid or base. [*Try drawing the acid-catalyzed mechanism on your own before looking at Figure 10.37(b).*] Similar to the chemistry discussed in Chapter 8, alkenes react with strong acids to form carbocations. The resulting 2° resonance-stabilized cation is delocalized over carbon and oxygen. Deprotonation of the protonated carbonyl generates the ketone while regenerating H_3O^+ (acid catalysis).

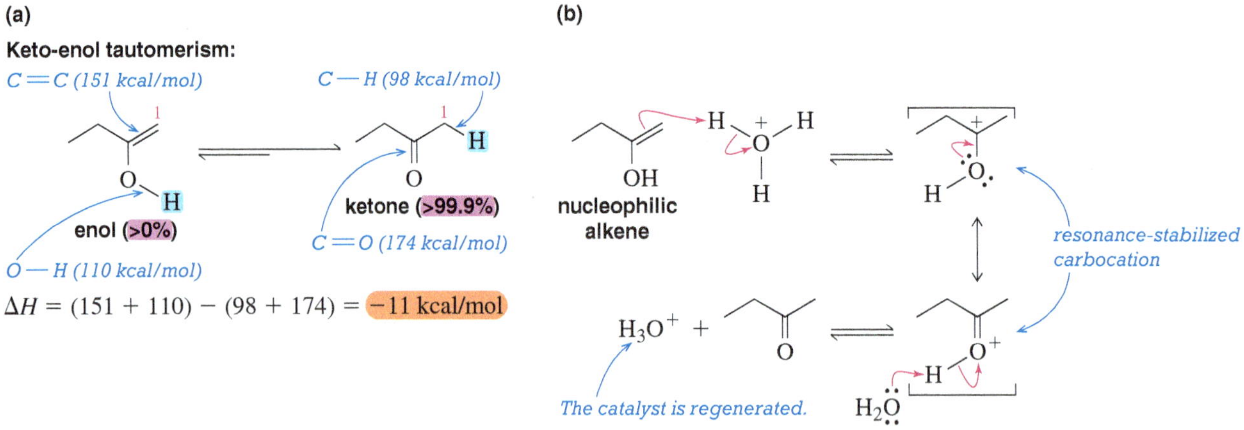

Figure 10.37 Keto–enol tautomerism. (a) The ketone side is favored. (b) Keto–enol tautomerism is acid-catalyzed.

10.8.3 SUMMARY

- In reactions where an enol forms, the enol rapidly equilibrates to the ketone in the presence of an acid catalyst by a mechanism that bears some similarity to reactions studied in Chapter 8.

Assessment 10.35 Classify the following constitutional isomers as being tautomers or not tautomers. If tautomers, identify the single atom that is in a different location in each molecule.

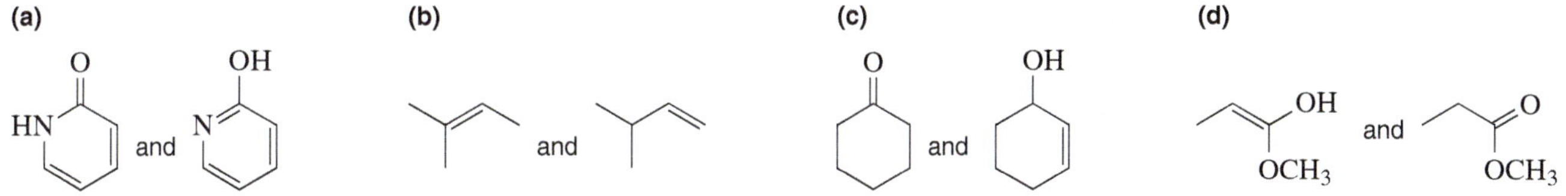

Assessment 10.36 The principle of microscopic reversibility says that a reversible reaction should proceed in both directions by equal but opposite mechanisms. Given this, write a mechanism for acid-catalyzed enol formation.

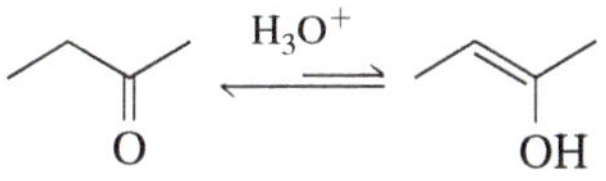

10.8.4 Oxymercuration

In Section 8.4.1, aqueous sulfuric acid was a reasonable choice for hydrating alkenes. Unfortunately, alkyne hydration does not occur under these conditions because the alkenyl carbocation intermediate is too unstable [**Figure 10.38(a)**]. [*Once again, the stability difference between alkyl and alkenyl carbocations changes the outcome of alkyne addition.*] In 1881, the Russian chemist Mikhail Kucherov discovered that the presence of mercury(II) salts in the sulfuric acid/water solution allowed for the electrophilic addition reaction to propyne to occur, with the ultimate product of the reaction being acetone. Like alkene hydration, alkyne hydration results in the net addition of H_2O, even if the location of the added H_2O is different for each [**Figure 10.38(b)**]. For alkenes, one H from H_2O adds to C_1 while the OH adds to C_2. With alkynes, the two hydrogens from H_2O add at C_1 of the alkyne with only oxygen adding to C_2.

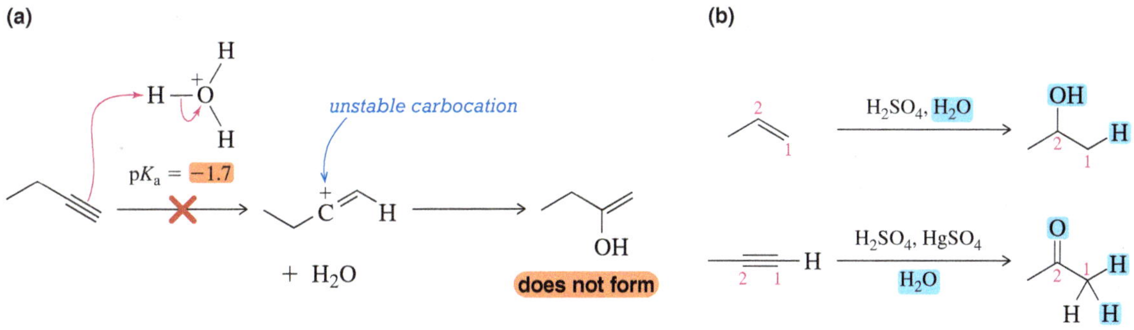

Figure 10.38 Oxymercuration of alkynes. (a) The hydronium ion is not a strong enough acid to allow formation of the unstable alkenyl carbocation. (b) The use of a mercury(II) salt allows hydration of propene to give acetone. In contrast to alkenes, two hydrogens of H_2O add to C_1.

The mechanism in **Figure 10.39** helps us understand why H_2O adds the way it does. [*It's a long mechanism but only has one new step.*] Similar to the oxymercuration of alkenes (Section 8.4.4), Hg^{2+} (a good electrophile) is attacked by the π bond to form a three-membered mercurinium ion. Water opens mercurinium by attacking at the most substituted carbon. The acidic proton of H_2O is removed to give a Hg^+-substituted enol. [*The mechanism becomes more complicated here but is analogous to keto–enol tautomerization (Section 10.8.3).*]

Protonation of the enol alkene forms a resonance-stabilized carbocation. [*Here comes the only new step.*] In the tautomerization, H_2O would remove H^+ from the oxygen to make the α-mercuryketone. But three factors conspire to make removal of the mercury the next step of this reaction. First, the C—Hg bond is long and weak. Second, relative to H^+, Hg^+ is huge; thus, when water randomly collides with this molecule, it collides most often at Hg^+. Finally, Hg^+ is a good electrophile for water to attack. When H_2O attacks the Hg^+ ion, the C—Hg electrons collapse to reform the enol, freeing Hg^{2+} to react with another alkyne molecule [*catalysis*]. The enol tautomerizes to the ketone by the acid-catalyzed mechanism in Section 10.8.3.

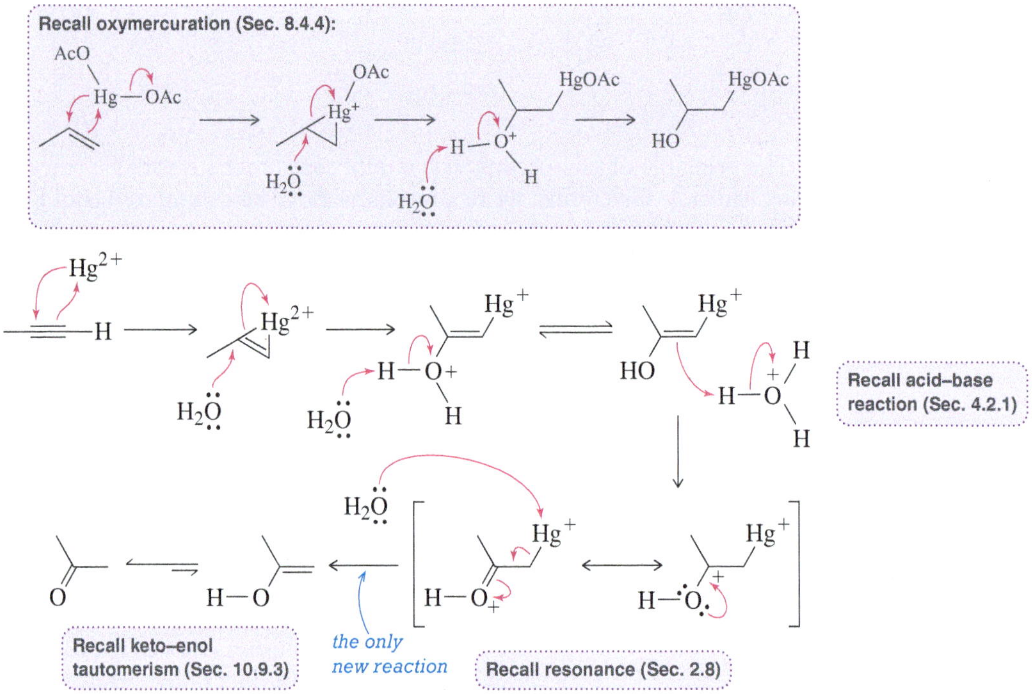

Figure 10.39 The mechanism of propyne oxymercuration to give acetone.

[*Before reading further, think about two questions: (1) How does Hg^{2+} allow this reaction to occur where H_3O^+ could not? (2) Will this reaction work equally well on terminal alkynes, symmetrical alkynes, and internal unsymmetrical alkynes?*]

Recall (Figure 10.38) that H_3O^+ is not a strong enough acid to promote formation of the unstable alkenyl carbocation. The carbocation never actually forms in the presence of Hg^{2+}. The mercurinium ion, then, is stable enough that the activation energy is reduced, allowing addition to occur (**Figure 10.40**). In the same way, the three-membered mercurinium ion prevents carbocation rearrangements in alkene oxymercuration (Section 8.4.4).

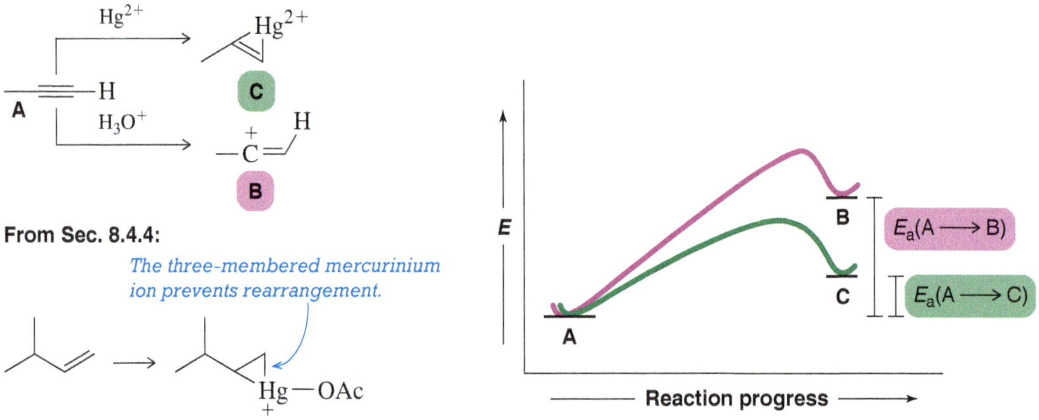

Figure 10.40 The three-membered mercurinium ion is more stable than the alkenyl or alkyl carbocation, so it forms more easily.

In alkyne hydration reactions, different alkyne types give slightly different results (**Figure 10.41**). The three-membered mercurinium ion arising from addition to terminal alkynes is opened selectively at the position of the more stable, more substituted 2° carbocation. As a result, oxymercuration of terminal alkynes gives **methyl ketones.** With symmetrical alkynes, the three-membered ring can be opened at either carbon to produce the same ketone. Internal *un*symmetrical alkynes, on the other hand, form a mercurinium ion with two different 2° carbons, so in the absence of significant steric hindrance, water adds at either carbon at equal rates. Thus, oxymercuration of internal unsymmetrical alkynes gives at best a ~50% yield, which is not useful synthetically.

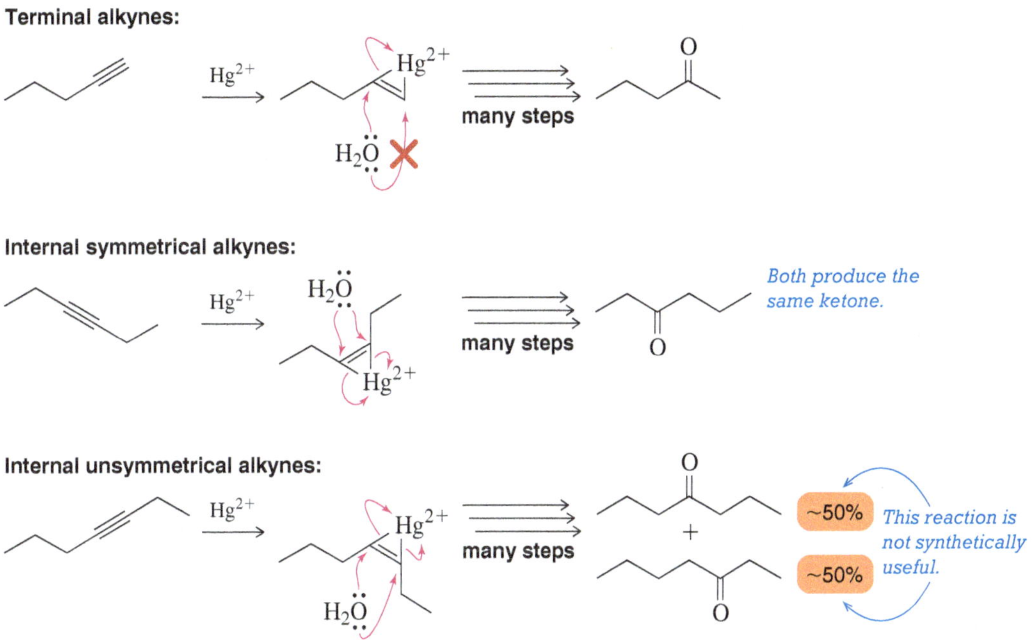

Figure 10.41 Different alkyne types give different outcomes using identical reagents.

Assessment 10.37 Draw the ketone(s) you would expect to form by reacting the following alkynes under the conditions of oxymercuration: (a) 6-methyloct-1-yne, (b) 1,10-dicyclohexyldec-5-yne, and (c) 5-phenylhex-2-yne.

$$\text{alkyne} \xrightarrow[H_2O]{HgSO_4,\ H_2SO_4} \text{ketone}$$

10.8.5 Hydroboration–oxidation

Just like for alkenes (Section 8.4.5), hydroboration of alkynes gives the opposite regioselectivity of oxymercuration (**Figure 10.42**). Mechanistically, boron's empty *p* orbital is attacked by the π bond of the alkyne at the least substituted carbon. At the same time, a hydride is transferred from BH_3 to the more substituted carbon. This syn addition is consistent with a concerted mechanism. As was true for hydroboration of alkenes, boron adds to the least substituted carbon (anti-Markovnikov; Section 8.4.5).

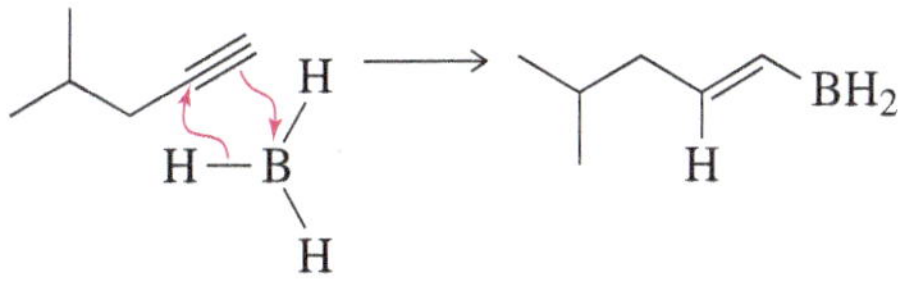

Figure 10.42 Hydroboration of alkynes is analogous to alkene hydroboration.

Hydroboration of terminal alkynes is often difficult to stop after the addition of one equivalent of borane. That is, the product alkenyl borane can undergo a second hydroboration reaction. To avoid this second reaction, chemists have developed a sterically hindered version of borane, containing two bulky isoamyl groups (**Figure 10.43**). Because of the steric hindrance and the fact that there is only one hydrogen to donate, di-*sec*-isoamylborane $[(sia)_2BH]$ adds just once to alkynes, with the resulting alkenylborane too sterically hindered to accommodate a second hydroboration.

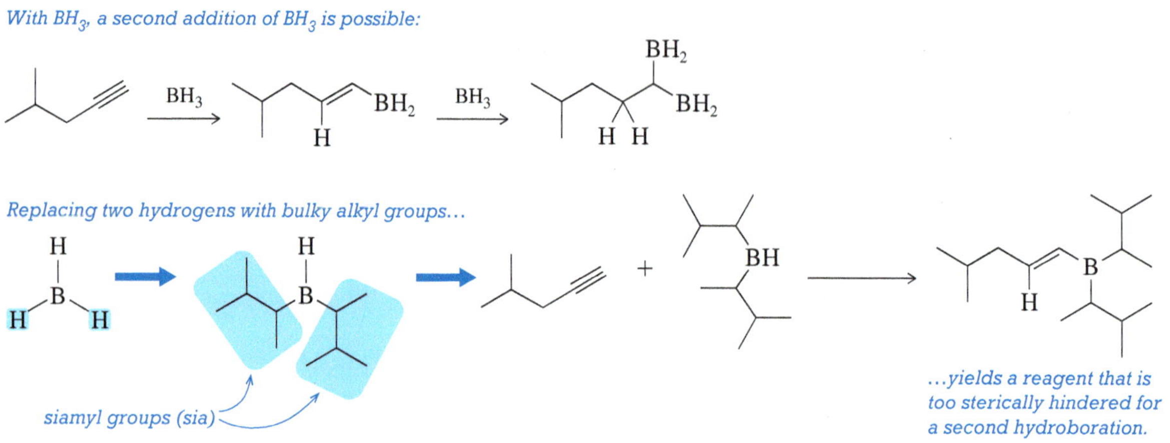

Figure 10.43 The use of a sterically hindered borane reagent prevents the second hydroboration from occurring, stopping the reaction at the alkenyl borane.

The resulting alkenylborane is treated with basic hydrogen peroxide, forming the enol by a mechanism studied in Section 8.4.5. [*Go back and look at this mechanism.*] Just like with alkenes, hydroboration–oxidation of alkynes results in the net syn addition of water. This enol also tautomerizes to the more stable carbonyl—an aldehyde in this case. [*This reaction is called a keto–enol tautomerization regardless of whether the product is a ketone or an aldehyde.*] Compared to the acid-catalyzed mechanism [Figure 10.37(b)], the base-catalyzed keto–enol tautomerization (**Figure 10.44**) simply reverses the order of the protonation/deprotonation steps. Deprotonation of the oxygen first results in an **enolate**, in essence a deprotonated enol. Two resonance forms make up the structure of an enolate, one of which has the negative charge on carbon. Protonation of the enolate at carbon forms the aldehyde, while regenerating hydroxide. Thus, the reaction is catalytic in base. [*Enolates are studied further in Chapter 20.*]

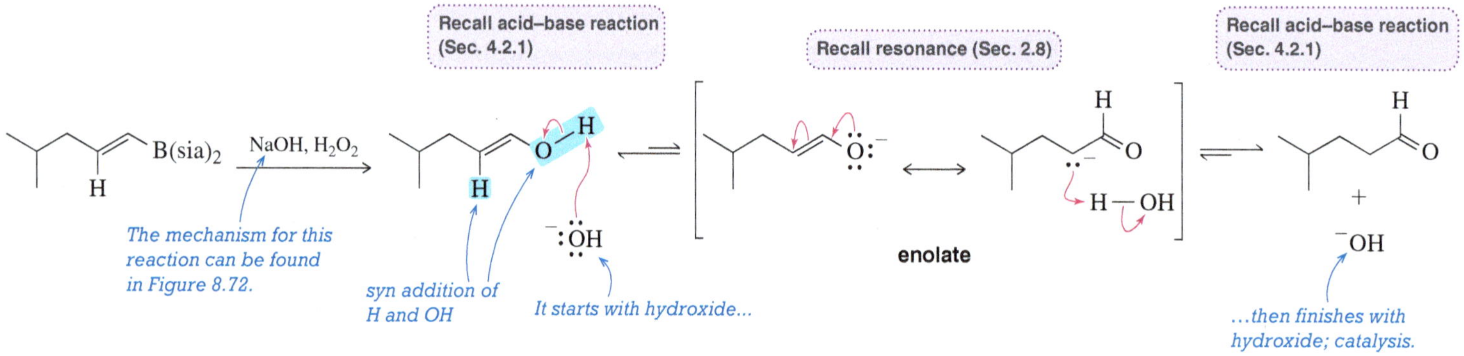

Figure 10.44 Oxidation of the C—B bond gives an enol that undergoes a base-catalyzed tautomerization to the ketone.

The efficiency of hydroboration–oxidation depends on the alkyne undergoing reaction (**Figure 10.45**). For terminal alkynes containing differentially substituted carbons, exclusive regioselectivity is achieved to produce an aldehyde. For internal unsymmetrical alkynes (two secondary carbons), hydroboration–oxidation, like oxymercuration, results in a mixture of products. Symmetrical alkynes give two products that are actually the same. Thus, high yields are possible with internal symmetrical alkynes. With internal alkynes, BH_3 can be used without the possibility of a second addition.

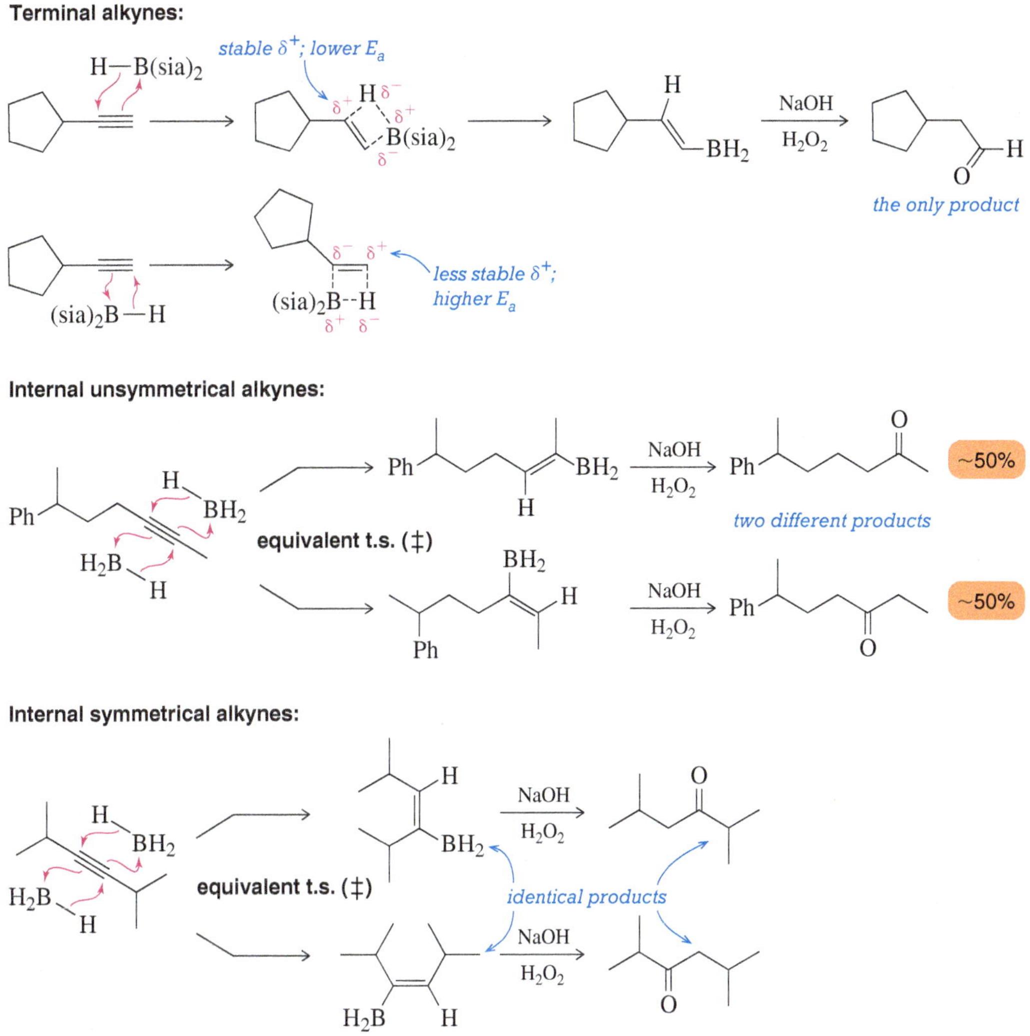

Figure 10.45 **Hydroboration is only an efficient synthetic technique for addition to internal symmetrical and terminal alkynes.**

Assessment 10.38 Draw the ketone(s) you would expect to form by treating the following alkynes under the conditions of hydroboration–oxidation: (a) 6-methyloct-1-yne, (b) 1,10-dicyclohexyldec-5-yne, and (c) 5-phenylhex-2-yne.

$$\text{alkyne} \xrightarrow[\text{2. NaOH, } H_2O_2]{\text{1. } BH_3} \text{ketone}$$

10-C CHEMISTRY IN CONTEXT: Problem-Solving Strategies

Random vs. targeted drug design

In the 1970s, the U.S. Congress and the National Cancer Institute (NCI) declared a "war on cancer" by passing the National Cancer Act of 1971. Within this legislation was the initiative that every compound, natural or synthetic, would be screened against cancer cell lines with the hopes of discovering several compounds that could result in the development of new therapies. Although hundreds of thousands of compounds were randomly screened, there were relatively few hits.

Random screens like those used in the "war on cancer" can be productive when little is known about the molecular cause of disease. On the other end of the spectrum is a more

(continued)

rational approach, one used to develop rotigotine (Neupro®), a treatment for Parkinson's disease. Efforts to develop rotigotine began with the hypothesis that higher dopamine levels would mitigate Parkinson's symptoms. Dopamine is a neurotransmitter produced by the brain that cannot be taken directly as a medicine; thus, it became the lead compound for efforts that resulted in a viable drug. By first identifying the molecular mechanism by which the disease could be treated, researchers could more logically produce the drug. This target-oriented approach (working backward) also resulted in the cytisine-inspired development of varenicline (Chantix®), a smoking-cessation drug.

Natural ligand	Drug	Indication
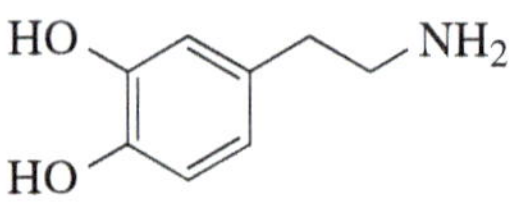 dopamine	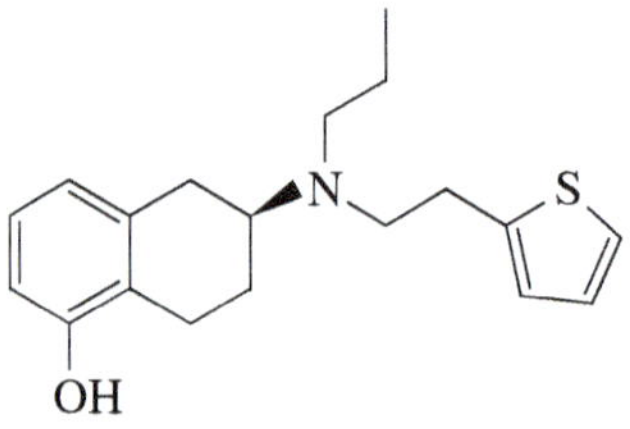 rotigotine	Parkinson's disease and restless leg syndrome
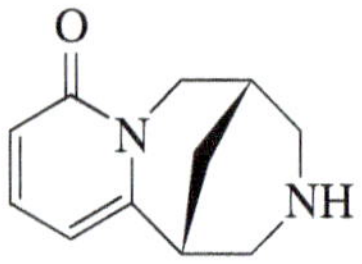 cytisine	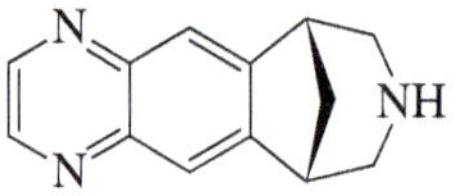 varenicline	Smoking cessation

10.8.6 Hydroboration versus oxymercuration

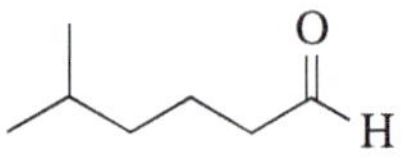

5-methylhexanal

One of the challenges of synthesis (a goal of this chapter) is knowing when to use the correct reagent. How is this choice made between hydroboration–oxidation and oxymercuration, two similar processes? The choice of reagent depends on another important decision—namely, with which alkyne the synthesis begins. [*Thinking backward emerges as a viable skill, as in working mazes (introduction) or developing a drug (Chemistry in Context 10-C).*] The synthesis of 5-methylhexanal, and other examples, illustrates how these choices might be made.

Hydroboration is the only reaction of alkynes that results in an aldehyde (**Figure 10.46**). Importantly, the alkyne must be terminal because the mechanism of the reaction dictates placement of oxygen at the least substituted carbon. In sum, aldehydes can only be prepared by hydroboration–oxidation of terminal alkynes.

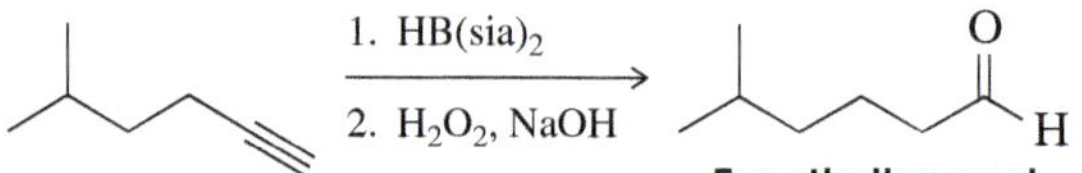

5-methylhexanal

Hydroboration of terminal alkynes gives aldehydes.

Figure 10.46 Hydroboration of terminal alkynes gives aldehydes.

Making a methyl ketone, like 5-methylhexan-2-one, is a bit more complicated. Two different alkynes could give the carbonyl at the appropriate carbon—namely, an internal unsymmetrical alkyne (5-methylhex-2-yne) or a terminal alkyne (5-methylhex-1-yne). The internal unsymmetrical alkyne, regardless of reagent used, always gives two different ketones (5-methylhexan-2-one and 5-methylhexan-3-one) in equal yields [**Figure 10.47(a)**]. On the other hand, oxymercuration of terminal alkynes produces the desired methyl ketone exclusively [**Figure 10.47(b)**].

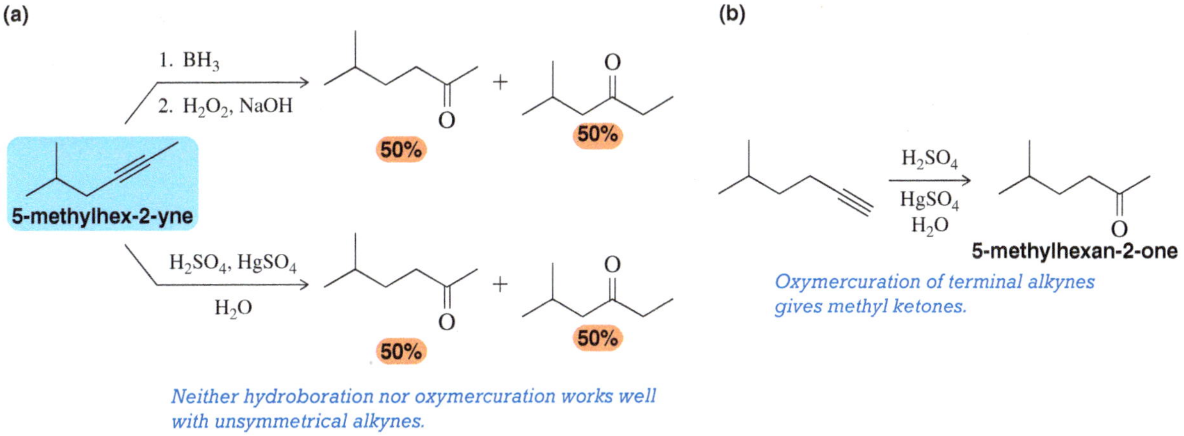

Figure 10.47 Regioselectivity in hydroboration and oxymercuration of alkynes. (a) For internal unsymmetrical alkynes, neither is efficient. (b) Oxymercuration of terminal alkynes gives methyl ketones selectively. [*Compare this to hydroboration of terminal alkynes that gives aldehydes in Figure 10.46.*]

Given the failure of oxymercuration and hydroboration with internal unsymmetrical alkynes, can other ketones be made in these ways? They can, but only when the internal alkyne is symmetrical [Figure 10.48(a)]. In the case of octan-4-one for example, oct-3-yne or oct-4-yne could be used. By working backward to a symmetrical alkyne (oct-4-yne), the correct ketone is produced regardless of the method used. With an unsymmetrical alkyne like oct-3-yne, on the other hand, a reasonable yield of ketone is impossible [Figure 10.48(b)].

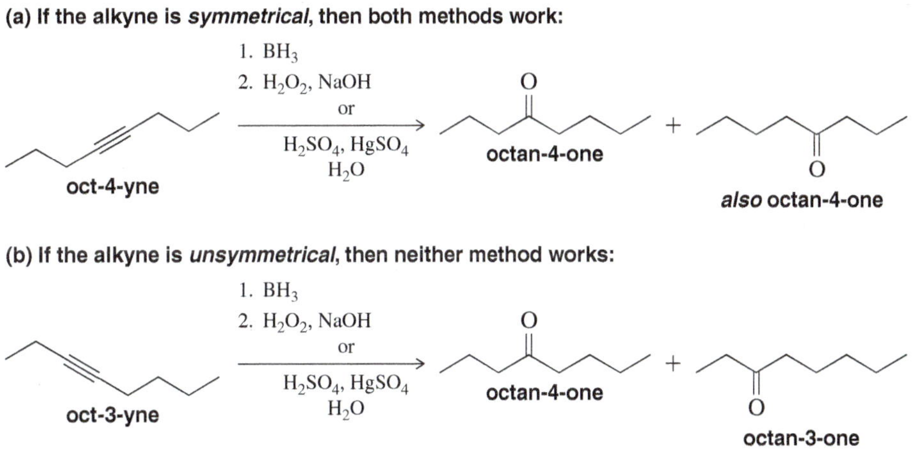

Figure 10.48 Regioselectivity of internal alkyne hydration. (a) Hydroboration and oxymercuration are efficient for symmetrical alkynes. (b) Neither hydroboration nor oxymercuration are efficient for unsymmetrical alkynes.

To conclude, the choice of reagents depends on the alkyne being used to make the desired product. At times a reasonable yield will be possible; at other times it won't. Assessment 10.39 tests you on recognizing these situations.

10.8.4–10.8.6 SUMMARY

- Terminal alkynes selectively give aldehydes using hydroboration and methyl ketones using oxymercuration. Symmetrical alkynes give the same ketone regardless of whether hydroboration or oxymercuration is used. Internal unsymmetrical alkynes, in the absence of some steric bias, give two different ketones in approximately equal yields with both oxymercuration and hydroboration.

Assessment 10.39 For each of the following ketones/aldehydes, indicate whether it is possible to synthesize it from an alkyne as the only compound in good (>50%) yield. If so, how would you do it?

(a)

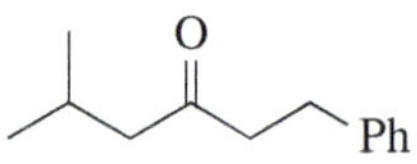

(b)

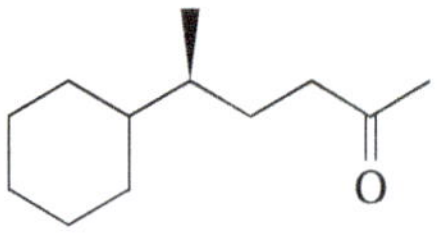

(c)

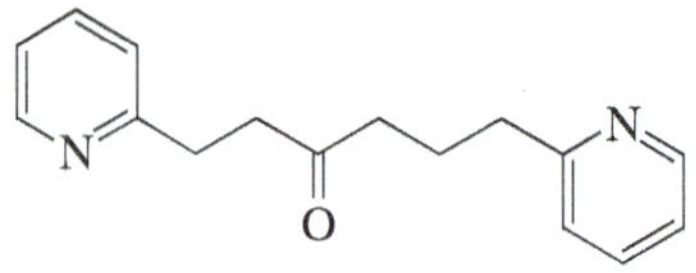

(d)

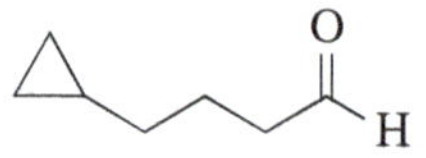

10.8.7 Bromination and chlorination

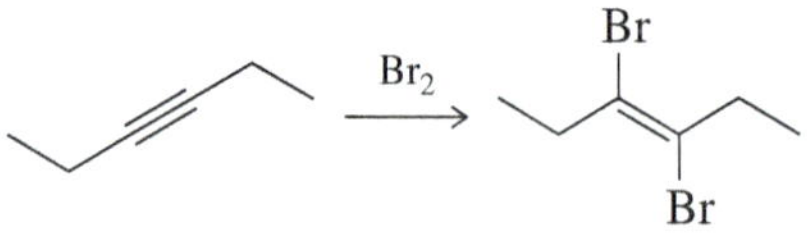

Figure 10.49 Bromination of alkynes gives trans-dibromoalkenes.

Similar to the reaction of alkenes (Section 9.1.2), bromination of an alkyne occurs to give the trans-dibromide (**Figure 10.49**). Because the same atoms are added to both carbons, regioselectivity is not a concern. Terminal alkynes, symmetrical alkynes, and unsymmetrical internal alkynes react to form a single product.

Mechanistically, attack of bromine by the π bond forms a new C—Br σ bond, releasing bromide as a stable conjugate base (**Figure 10.50**). [*This is identical to how Br_2 reacted with alkenes.*] One of the lone pairs on bromine is quickly donated to the developing alkenyl carbocation, resulting in the concerted formation of the bromonium ion. The bromonium ion prevents formation of the unstable alkenyl carbocation. This reaction is faster than other alkyne electrophilic additions. Attack by the bromide anion opposite the bromonium ion gives the trans-dibromide. With excess bromine, a second addition occurs to give a tetrabromoalkane by the usual alkene addition mechanism (Section 9.1.2). Put simply, formation of the bromonium ion is followed by ring opening from the opposite face.

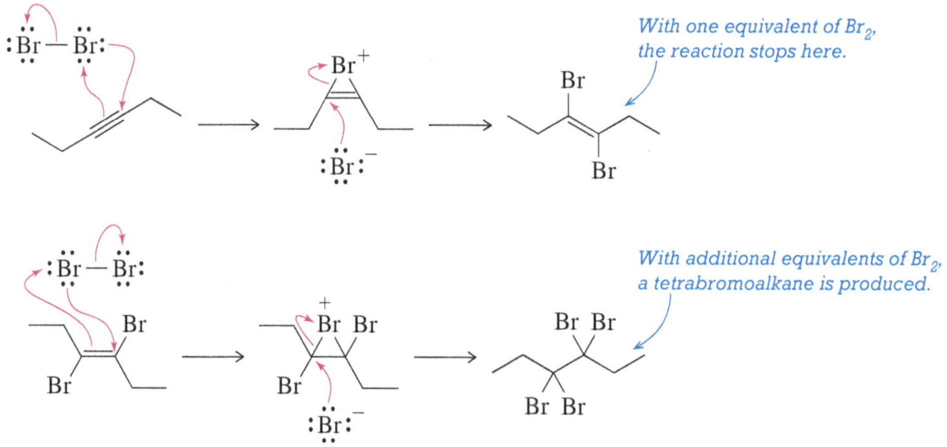

Figure 10.50 The mechanism of alkyne bromination is analogous to alkene bromination.

Before reading Section 10.9, update your reaction sheets. As you practice synthesis, you will use your reaction sheets so often that eventually the reactions will be committed to memory. Having these reactions "committed to memory" as a result of using them over and over allows them to stay in your long-term memory, facilitating their recall.

10.9 Synthesis of organic molecules: putting it all together

With a full, neatly organized toolbox, we can now do what distinguishes organic chemists from other science disciplines: make molecules. To keep it simple at first, think about making pentan-2-ol, starting with propene (**Figure 10.51**).

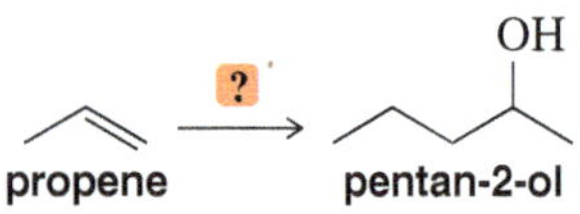

Figure 10.51 How do we convert propene to pentan-2-ol?

Because we're beginning with propene, look at your reaction sheets and identify the reactions of alkenes (**Figure 10.52**). Propene can be converted to propan-2-ol, propan-1-ol, 2-bromopropane, 1-bromopropane, propane-1,2-diol, acetaldehyde/formaldehyde, and 1-bromopropan-2-ol. [*Doing synthesis will clearly help us review reactions.*] Given this extensive list of molecules, the focus turns to what can be made from them.

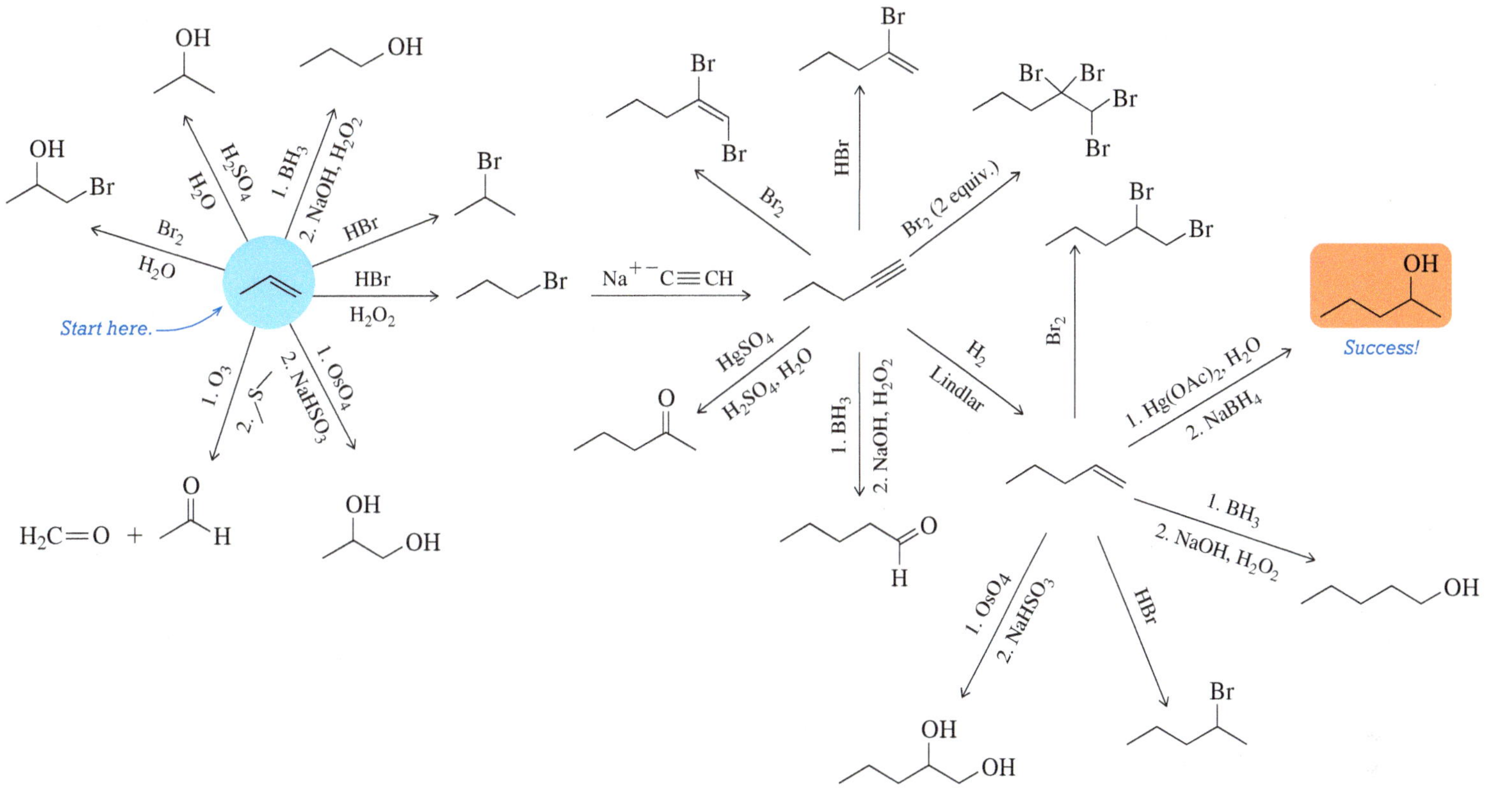

Figure 10.52 One method of doing synthesis: Consider all of the possible reactions you can do until you stumble upon the correct path. A great way to learn lots of reactions, but not very efficient.

We haven't learned much about reactions of alcohols and ketones. Based on our reaction sheets at this point, the only functional group for which we have studied a reaction is the 1° alkyl halide. In this case, 1-bromopropane can alkylate sodium acetylide to form pent-1-yne. [*An especially convenient reaction because it extends the chain to the five carbons needed for pentan-2-ol.*] With pent-1-yne in hand, let's turn our attention to the reactions of alkynes. [*There are many.*] From pent-1-yne, *trans*-1,2-dibromopent-1-ene, 2-bromopent-1-ene, pent-1-ene, pentanal, and pentan-2-one, among others, can be made. With several alkenes as products here, the "synthesis of alkenes" reaction sheet shows us what can be made with these. To expedite the process, pent-1-ene can be used to make pentane-1,2-diol, 1,2-dibromopentane, 2-bromopentane, pentan-1-ol and . . . [*wait for it*] . . . PENTAN-2-OL! [*Success, at long last. Although it was summarized in just a few paragraphs, thinking it through takes a significant amount of time. Is there a better way?*]

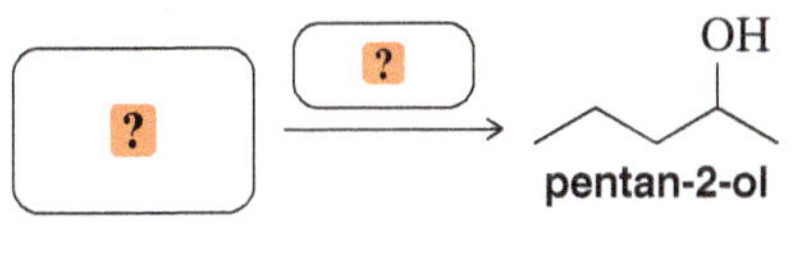

Figure 10.53 Working backward is like a fill-in-the-blanks question with two blanks.

Would working backward make this process more efficient? In our first attempt, we used only one side of the reaction sheets ("Reactions of . . . "). Starting at the desired molecule, what reactions are needed to make pentan-2-ol (**Figure 10.53**)? Problems like this have been done previously; instead of predicting the reactant or the reagent, predict both. As an exam problem, it might have a blank for the reactant and reagents. [*Some find this problem type difficult because they seem vague. However, there is freedom in this vagueness because you can choose whatever organic molecule you want to start with.*]

The reaction sheets guide the decision, especially the one titled "Preparation of alcohols." All of the possible reactions start with alkenes (**Figure 10.54**). To narrow down the choices, a method should be chosen that places the hydroxyl group at the most substituted carbocation. Sulfuric acid and water form a carbocation at the most substituted carbon, which is attacked by water. [*This works!*] Hydroboration places a boron at the least substituted carbon, which is then replaced by the oxygen. [*This does not work.*] Oxymercuration–reduction shows Markovnikov selectivity. [*It works, too!*]

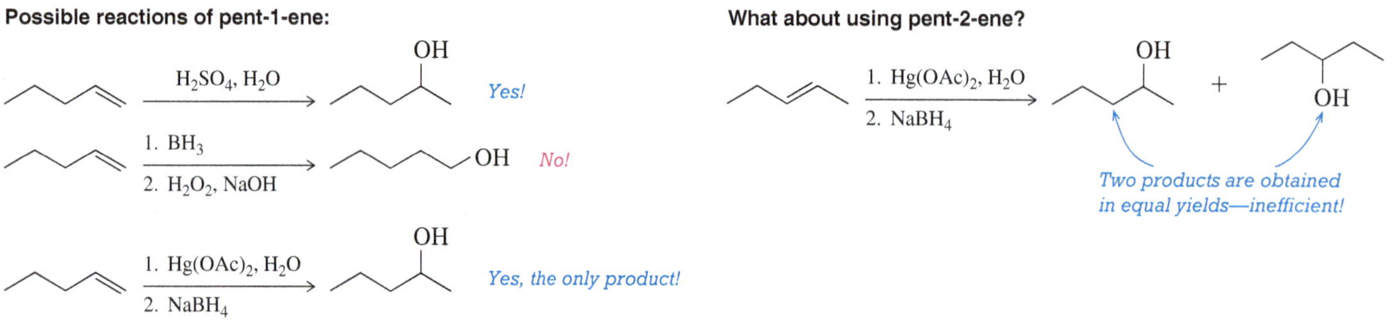

Figure 10.54 Possible syntheses of pentan-2-ol.

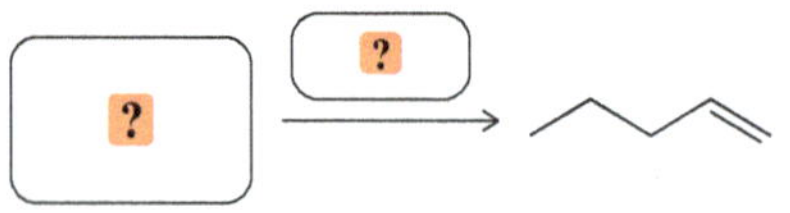

Figure 10.55 Working backward is like a fill-in-the-blanks question with two blanks.

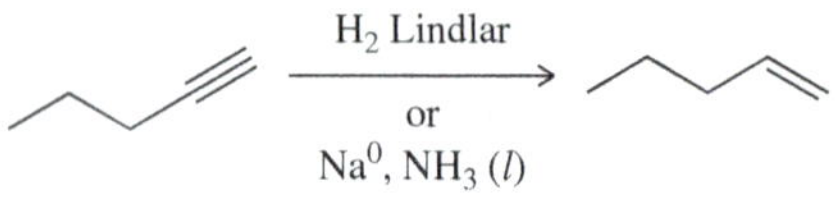

Figure 10.56 Either alkyne reduction produces pent-1-ene.

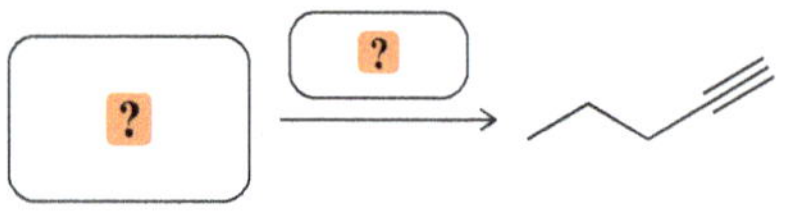

Figure 10.57 Working backward is like a fill-in-the-blanks question with two blanks.

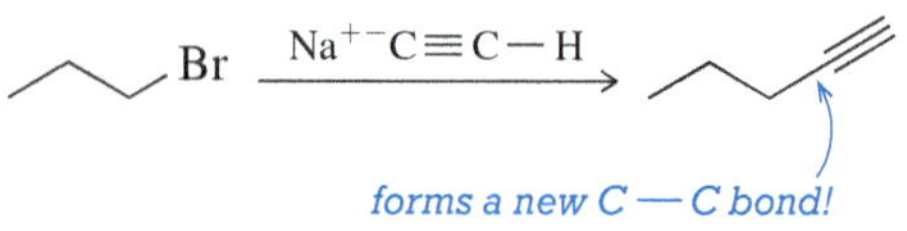

Figure 10.58 Acetylide alkylation forms a new C—C bond.

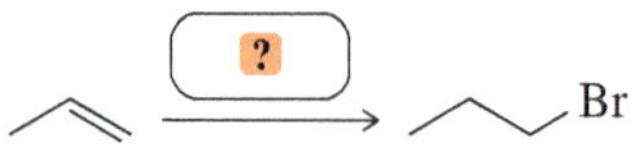

Figure 10.59 What reagent converts an alkene to a 1° bromoalkane?

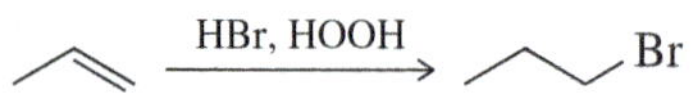

Figure 10.60 Peroxide-mediated addition of HBr produces 1-bromopropane.

For the sake of full analysis, this step could have employed pent-2-ene. Of pent-1-ene and pent-2-ene, is one better? [*This is an important consideration. Syntheses should be designed to give the best possible yield, minimizing waste and expense.*] With pent-2-ene, there is no way to distinguish between the two possible secondary carbocations. Thus, two possible products would form in the oxymercuration–reduction. On the other hand, the alkene carbons of pent-1-ene are 1° and 2°, and water adds exclusively to the more substituted carbon under these conditions. Oxymercuration–reduction, then, is best for the synthesis of pentan-2-ol from pent-1-ene.

At this point, the process gets simpler. How is pent-1-ene made? Once more, this is just a fill-in-the-blanks question with two blanks (**Figure 10.55**).

Looking at the "Preparation of alkenes" reaction sheet, there are only two reactions that make alkenes; both work here. Either Na^0 in liquid NH_3 or the hydrogenation with the Lindlar catalyst accomplishes this transformation (**Figure 10.56**).

This brings us to the key reaction for our synthesis efforts in this chapter. [*Think about why this might be the case.*] An alkyne must be synthesized. Again, consider it a fill-in-the-blanks problem (**Figure 10.57**).

On the "Preparation of alkynes" reaction sheet, there is just one way to make alkynes: alkylation of acetylide with a 1° alkyl halide (**Figure 10.58**). Reacting 1-bromopropane with sodium acetylide should produce only the desired product. This is the key reaction because it is the only way to make C—C bonds studied thus far. Any synthesis that starts with a small molecule and synthesizes a larger one must feature this reaction. [*Always remember: when making an alkyne, do it with the acetylide alkylation reaction because it gets us back to smaller molecules.*]

Each time we get back to a new molecule, be reminded of the overall goal. To which molecule are we working backward? At each stage of synthesis, ask whether the molecule we are interested in can be made from the starting material. [*This hasn't been an issue to this point; 1-bromopropane is the first molecule with the same number of carbons as prop-1-ene.*] So now the fill-in-the-blanks question becomes a predict-the-reagent question (**Figure 10.59**). A reagent is needed to make 1-bromopropane from propene.

Only one reaction from the "Preparation of alkyl halides" reaction sheet produces the less substituted alkyl halide. For this, the radical-initiated addition of HBr completes the synthesis (**Figure 10.60**).

Working backward from propan-2-ol to propene took effort but is a more logical approach than working forward. It allowed us to come up with the single, best path to a molecule without encountering the dead ends encountered when working the maze [*ahem, synthesis*] in the forward direction. To see the improvement, consider **Figure 10.61**, which is like Figure 10.52 but works backward from pentan-2-ol.

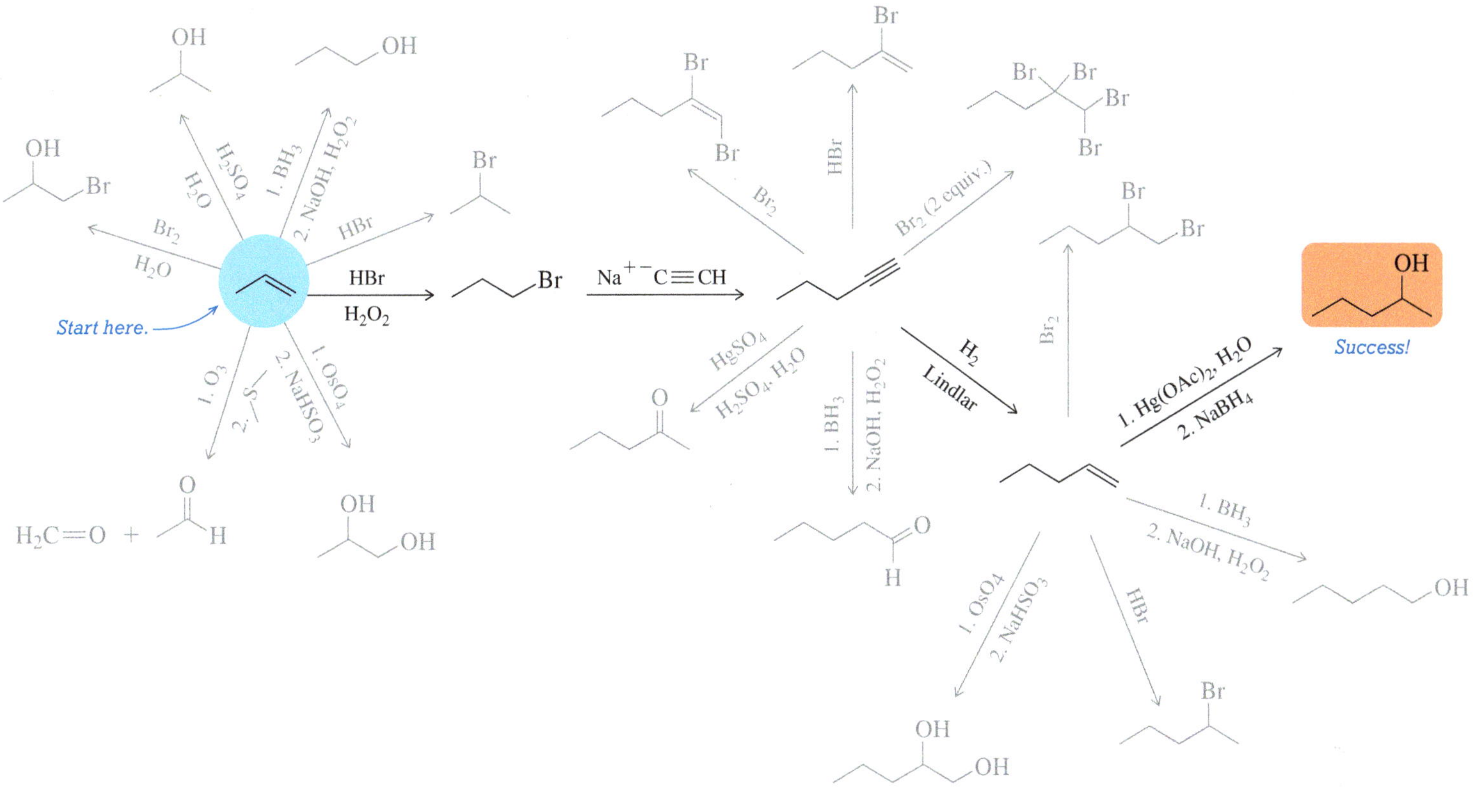

Figure 10.61 **By working backward, our solution emerges in a much more straightforward and logical manner.**

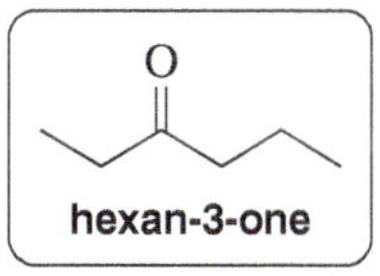

The process of working backward is referred to as **retrosynthesis**, or **retrosynthetic analysis**. It was first elucidated by Professor E. J. Corey of Harvard University, one of the greatest synthetic organic chemists of all time. His retrosynthetic methodology, the syntheses he completed, and the reactions developed to enable these syntheses earned him the 1990 Nobel Prize in Chemistry. What follows is a quick summary of a synthesis of hexan-3-one that you can do in this way. [*Give it a shot before looking at the answer. As you practice, use your reaction sheets; they make up your toolbox.*] Work backwards, starting with reagents and reactants containing three carbons or fewer. [*This is a very common way instructors ask synthesis questions on exams.*]

Quick summary: hexan-3-one synthesis

1. We can make ketones using either oxymercuration or hydroboration of alkynes. [*Whenever possible, always try to get back to an alkyne, because it gets us to the key C—C bond-forming reaction.*] We have to use hex-3-yne here because hex-2-yne is an unsymmetrical internal alkyne and thus would only give us at best a 50% yield of the product.
2. We can now do our key C—C bond-forming reaction (acetylide alkylation) to work backward to but-1-yne.
3. Because but-1-yne has four carbons, we need to work backward to acetylene by repeating the same reaction (acetylide alkylation)—done!

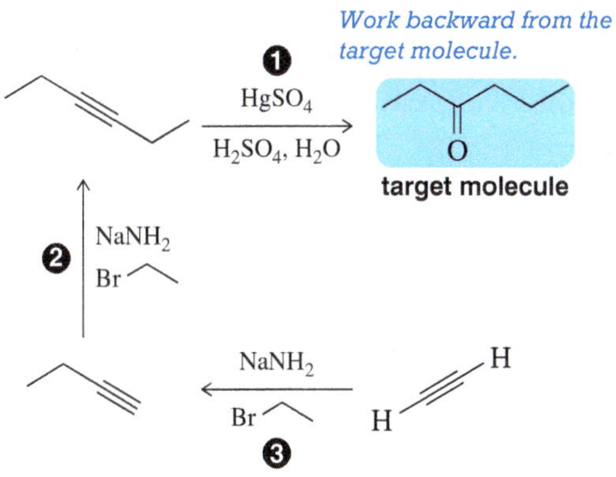

Retrosynthetic analysis gives us a great way of logically working backward to simpler materials with which we can begin our synthesis. [*This will be hard at first, but only because, until this chapter in the book, we have been focused on analyzing reactions in the forward direction. I have encountered numerous students who struggle with this concept.*] The key is to practice. Work every problem you can find, in the chapter, in Mastering, or online. You can even make them up yourself. Simply sketch an organic molecule (keep it as a single, straight chain for now), put on a functional group, and then try and make it. And don't forget to use your reaction sheets. [*Trust me: If you are diligent about this, synthesis will become one of the easier and more fun (YES, FUN!) exercises you will do in organic chemistry. Assessments 10.40 and 10.41 should get you started. Enjoy!*]

Assessment 10.40 When doing synthesis, you will often find yourself repeating the same series of steps. To see this in action, synthesize the following aldehydes beginning with an organic molecule containing three carbons or fewer.

(a)

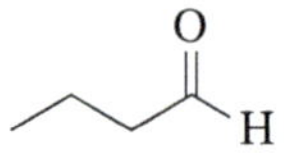

(b)

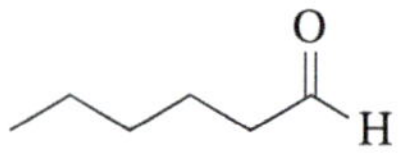

(c)

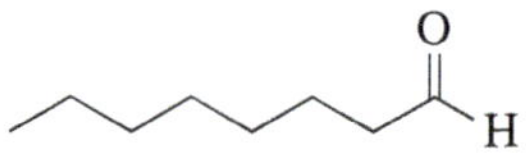

(d)

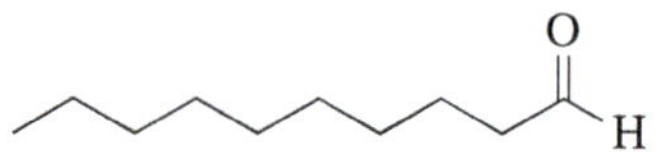

Assessment 10.41 Beginning with the molecules on the left of each chemical equation, synthesize the molecules shown. While there can be multiple ways of doing each synthesis, the minimum number of steps necessary is indicated over each reaction arrow.

(a)

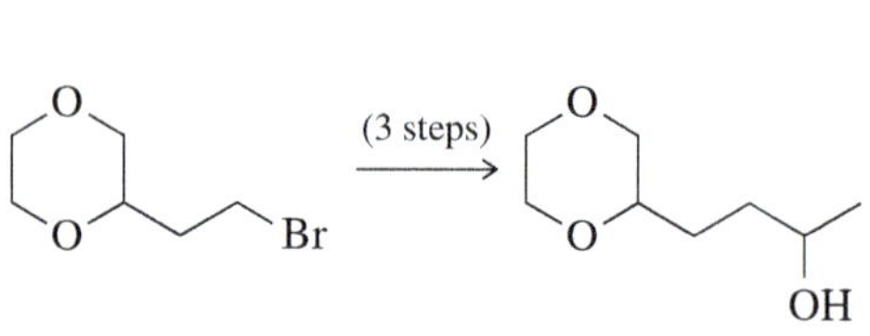

(b)

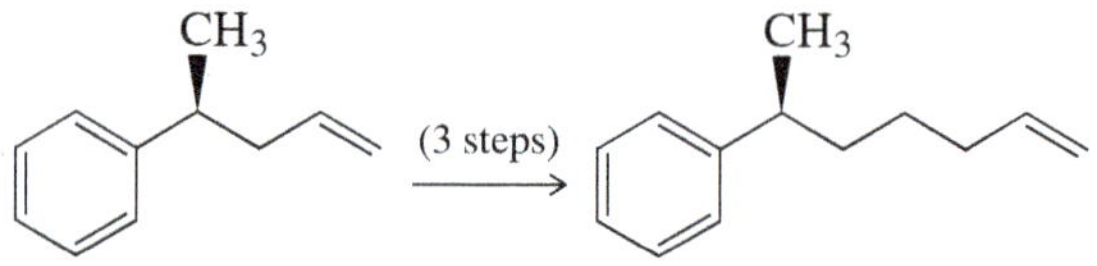

(c)

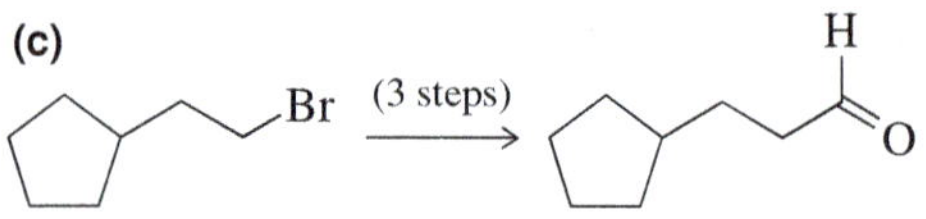

(Hint: $H_2C{=}O$ is also made in this reaction.)

Whatever your career, you will be confronted with problems/questions you need to solve/answer. More important than the answer/solution is the process used to figure it out. Lacking a problem-solving strategy, we are like three-year-olds working a jigsaw puzzle trying to match every single piece with another. With a solid, logical problem-solving strategy, you are likely to arrive quickly at the best answer. Let's continue, then, to develop our problem-solving strategies as we look at the synthesis of and properties of alkyl halides in Chapter 11.

Concluding thoughts

The truth is that many of you reading this book have no desire to become organic chemists. Many students take this course only as a requirement for a biology major, or as a prerequisite for dental, medical, or veterinary school. So what does a course like this have to offer a student with little interest in chemistry or biochemistry, a student who might want to be an environmental scientist and work with the EPA, a student who wants to see patients as a psychiatrist, or a student who wants to work in the marketing department of a pharmaceutical company? The problem-solving skills developed here are used by professionals working in many different jobs: physicians, construction workers, auto mechanics, and recreational puzzle solvers, to name just a few.

Synthesis requires a strategy that starts at the target and works backward to simpler starting materials. This strategy, applied equally to mazes, is very logical. Just as you are unable to efficiently assemble a puzzle without identifying corner and side pieces, you must be aware of where your synthesis needs to go. Working backward ensures mindfulness of the overall goal.

Important Concepts

[*These are the major concepts covered in each section, excluding reactions.*]

10.1: Alkynes, unsaturated compounds containing two carbons connected by a triple bond, are nonpolar, insoluble in water, and volatile. They can be classified as terminal, internal/symmetrical, and internal/unsymmetrical as a way to help us understand their reactivity. Alkynes are at a higher oxidation state than alkenes, so alkynes can be reduced to give both alkenes and alkanes. Because alkynes have two π bonds, they have an index of hydrogen deficiency of 2.

10.2: Alkynes are named similarly to alkenes. Because they are of equal priority according to the IUPAC rules, the direction of numbering in compounds that has both an alkene and an alkyne is based on which of the two will give the lower number. Regardless, the name of the molecule will end in -yne.

10.3–10.4: The molecular orbital picture of an alkyne is composed of a C—C σ bond and two C—C π bonds orthogonal to one another. The σ bond is formed by overlap of two *sp*-hybridized orbitals along an axis that connects the two nuclei. The π bonds result from sideways overlap of two 2*p* orbitals above and below and in front and behind the axis of the σ bond. The proximity of the electron-rich π bonds results in the alkyne possessing a high potential energy, stronger C—C and C—H bonds, and when terminal, a relatively acidic hydrogen. Finally, the cylinder of electron density, arising from the orthogonal π bonds, makes alkynes reactive with strong electrophiles.

10.5: The acidic proton of terminal alkynes can be removed with a strong base like sodium amide ($NaNH_2$). Once removed, the acetylide ion is a strong nucleophile that can be alkylated in a substitution reaction with primary alkyl halides, making it possible to construct new C—C σ bonds.

10.6: Alkynes can be reduced by hydrogenation to give an alkane, or when a poisoned catalyst like the Lindlar catalyst is used, to an alkene. Because the two hydrogens are added at the same time, the alkene produced will be in the cis configuration. Alternatively, the trans-alkene can be generated by reducing the alkyne under dissolving-metal conditions, using sodium dissolved in liquid ammonia to deliver electrons to the reducible π bond.

10.7: As a means of organizing the reactions we have learned (and assisting us with synthesis), reaction sheets can be created for each functional group. The front side should show all reactions that the functional group undergoes, while the reverse side should show all reactions that produce the functional group.

10.8–10.8.1: Alkynes are reactive only with strong electrophiles. Because the alkenyl carbocation is not very stable, electrophilic addition to alkynes are not as fast as the corresponding additions to alkenes.

10.8.2: The addition of one equivalent of HBr and HCl results in the formation of a haloalkene. Because the alkenyl cation is so unstable, the product forms by anti addition to an intermediate π complex. If excess HBr or HCl is used with the alkyne, the result is a geminal dihaloalkane.

10.8.3–10.8.6: Hydration of an alkyne results initially in the formation of an enol. The enol undergoes rapid tautomerization to the more stable ketone. The enhanced stability of the ketone can be attributed to the fact that the C—O π bond is stronger than the corresponding C—C π bond. The hydration reaction can be done using hydroboration to make aldehydes from terminal alkynes or oxymercuration to make methyl ketones from terminal alkynes. When the alkyne is internal and symmetrical, there is no difference between the choice of reagents (hydroboration/oxymercuration) you might use. For internal unsymmetrical alkynes, a mixture of regioisomers is always produced.

10.8.7: Alkynes can be halogenated with one equivalent of bromine or chlorine to produce, via the formation of a bridged halonium ion, an (*E*)-dihaloalkene, Reaction of an alkyne with an excess of bromine or chlorine results in a tetrahalogenated alkane.

10.9: In order to synthesize more complex molecules from simple starting materials, it is important to utilize the process of retrosynthesis, a technique of working backward from the target molecule to the starting point of the synthesis.

Important Reactions and Mechanisms

1. **Acetylide alkylation (Section 10.5):** Deprotonation of the acidic proton of a terminal alkyne by a strong base results in the formation of an acetylide ion. The acetylide ion (a strong nucleophile) can be used to displace a halide [*a good leaving group*] from a primary carbon to form a new carbon–carbon bond.

General:

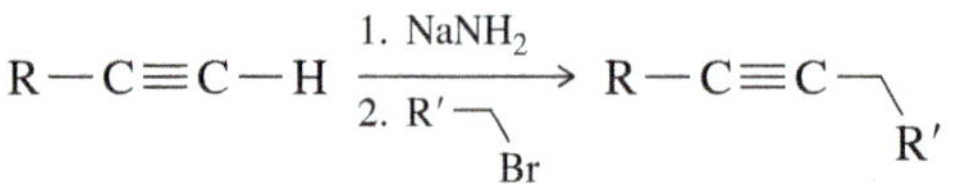

Specific:

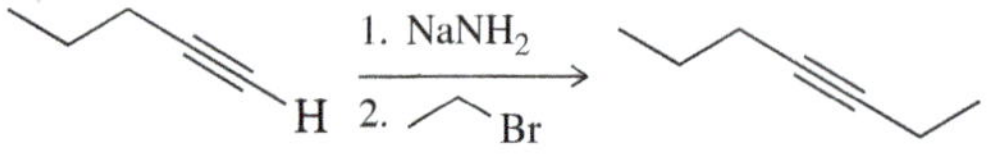

Mechanism:

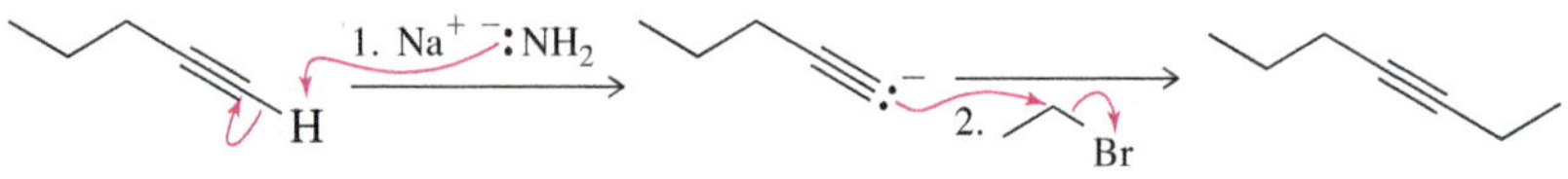

2. **Hydrogenation (Section 10.6.1):** The addition of hydrogen gas (H_2) to an alkyne is catalyzed by a transition metal, like palladium. Similar in mechanism to the hydrogenation of alkenes in Chapter 9, the hydrogens are added from the metal surface in a concerted mechanism to the same side of the alkyne, resulting initially in the formation of an alkene. Because these conditions are identical to alkene hydrogenation, the reduction continues until an alkane is generated.

General:

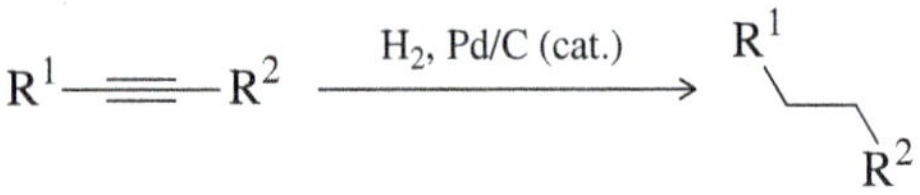

Specific:

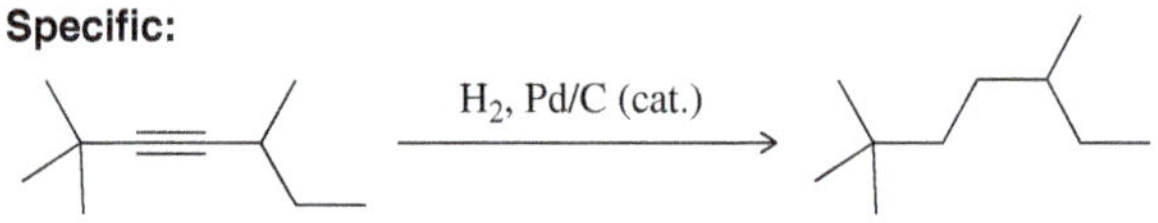

Mechanism:

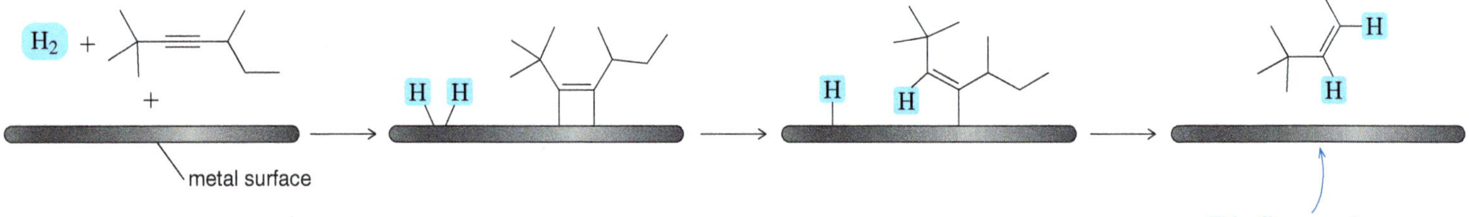

This alkene can be further reduced to an alkane (see Chapter 9).

3. **Hydrogenation/Lindlar (Section 10.6.2):** When the palladium catalyst is poisoned with lead salts, the hydrogenation of alkynes stops at the cis-alkene. This process takes advantage of the fact that under reduction conditions, alkynes are more reactive than the corresponding alkene. The cis product is a result of the concerted, syn addition of hydrogen from the palladium to a single face of the alkyne.

General:

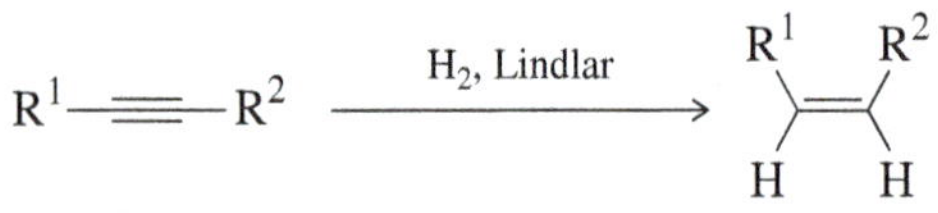

Specific:

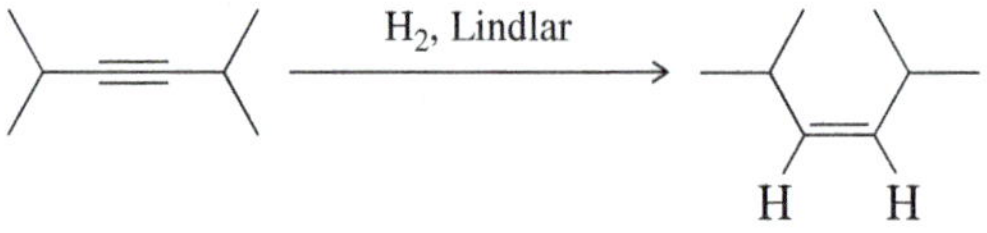

Mechanism:

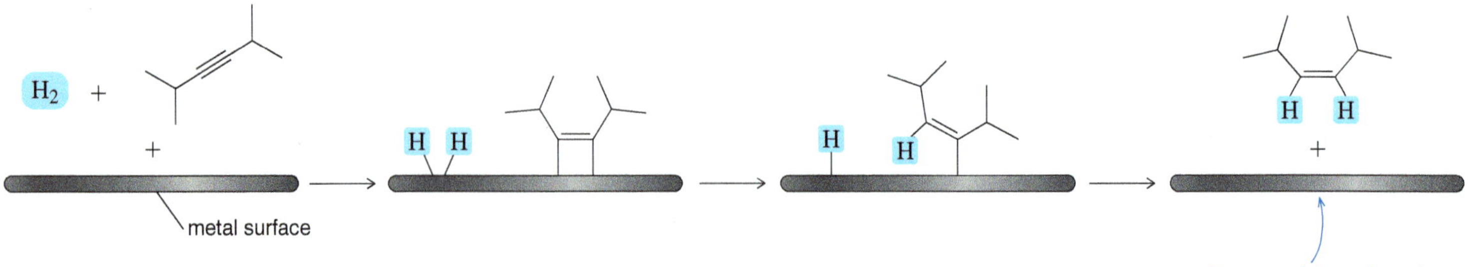

Because of the poisoned catalyst, the reaction stops at the cis-alkene.

4. Dissolving-metal reduction (Section 10.6.3): The alkyne, as a reducible bond, accepts electrons from sodium metal dissolved in ammonia. Proceeding through a radical anion intermediate, the two new C—H bonds are formed on opposite sides, resulting in the formation of a trans-alkene. The reaction stops at the alkene.

General:

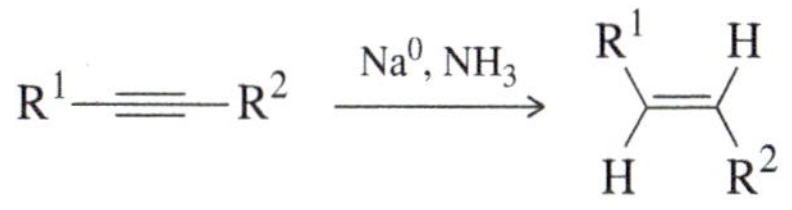

Specific:

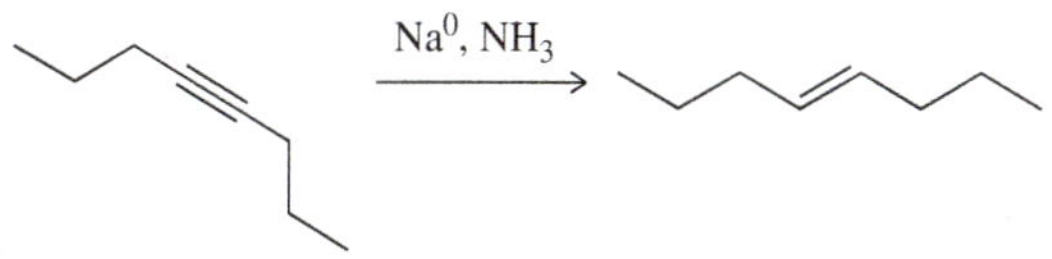

Mechanism:

$Na^0 \longrightarrow Na^+ + e^-$

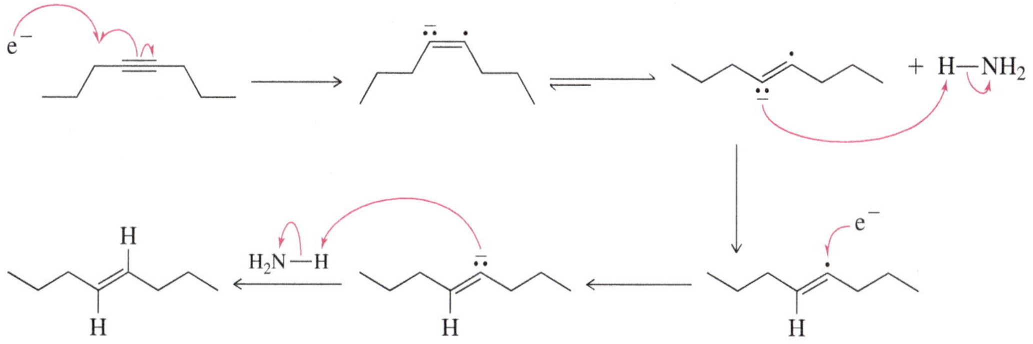

5. Hydrohalogenation (Section 10.8.2): Electrophilic addition of HBr (or HCl) features an acid–base reaction to produce the more stable alkenyl carbocation, which is attacked by bromide in the second step. Proceeding through a proposed intermediate π complex, hydrohalogenation is selective for anti addition. If an excess of HBr (or HCl) is used, a second equivalent can add to the alkene to give a geminal dibromide.

General:

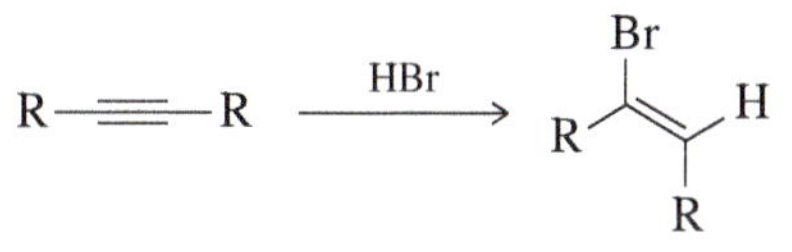

Specific:

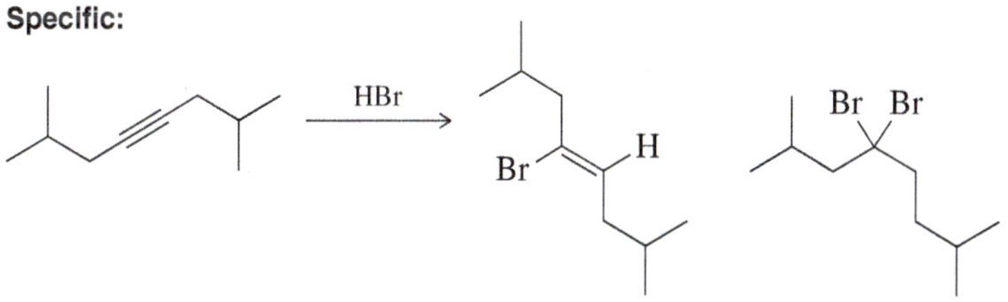

(if 2 equiv. HBr)

Mechanism:

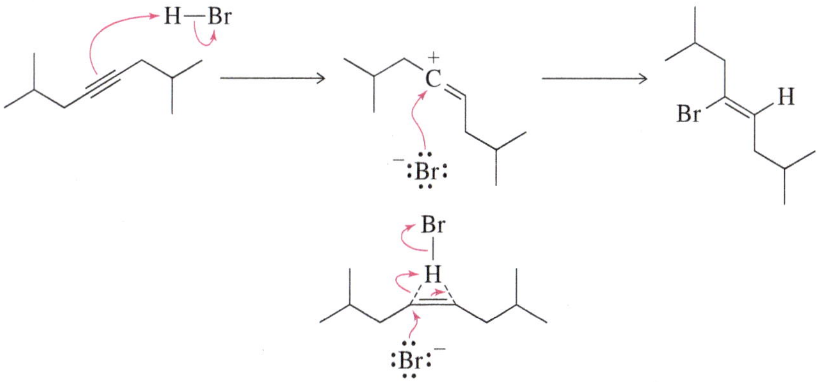

6. **Oxymercuration (Section 10.8.4):** The alkene attacks an electrophilic mercury cation of mercury sulfate. Because of the presence of *d* electrons in mercury, a bridged mercurinium ion forms in a concerted reaction, making this a fast process because the relatively unstable alkenyl carbocation never forms. Water attacks the mercurinium ion at the more substituted carbon on the face opposite the mercury to generate a mercury enol. In consecutive steps, the mercury cation is removed and the enol tautomerizes to the more stable ketone.

General:

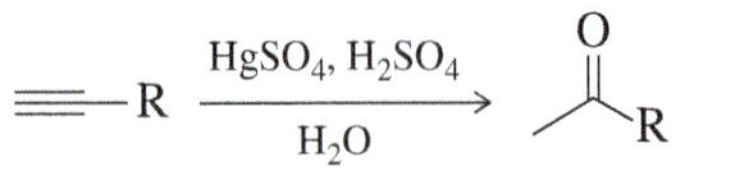

Specific:

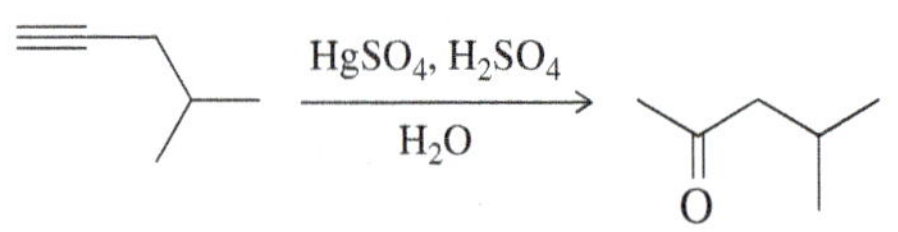

Mechanism:

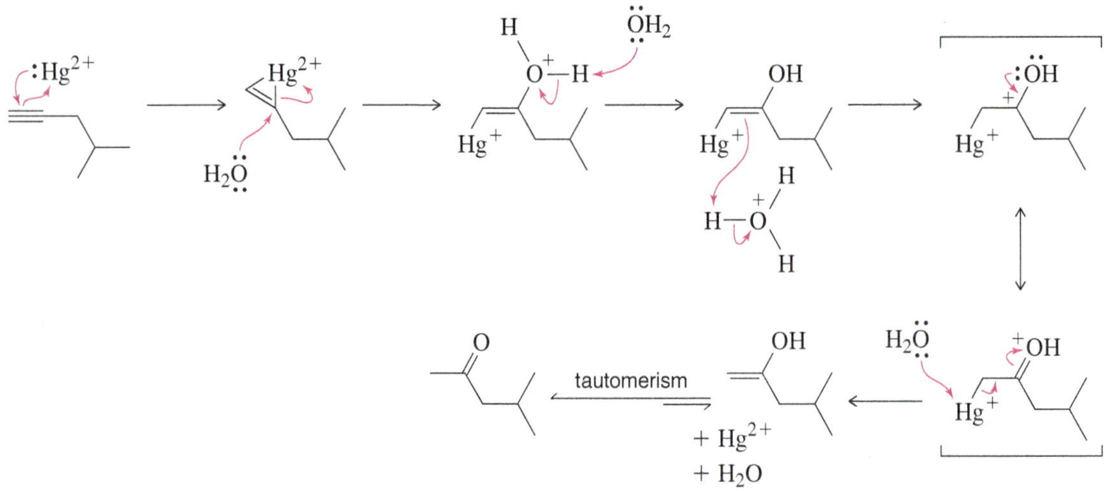

7. **Hydroboration–oxidation (Section 10.8.5):** With an empty *p* orbital, borane is a strong Lewis acid and is attacked by an alkyne in a Lewis acid–Lewis base reaction. In order to mitigate the developing negative charge on the borane, hydride is immediately transferred in a single concerted step, resulting in the net addition of BH_3 across the double bond. Oxidative cleavage of the carbon–boron bond with basic hydrogen peroxide proceeds with retention of configuration, resulting in the stereospecific syn addition of water across the alkyne to give a short-lived enol. Tautomerization of the enol produces the more stable ketone/aldehyde.

General:

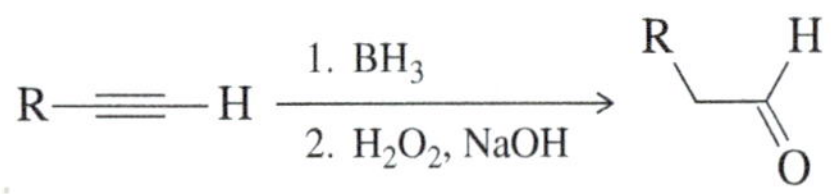

Specific:

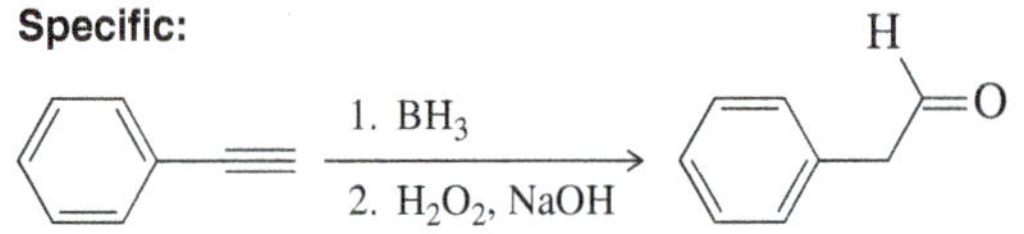

Mechanism:

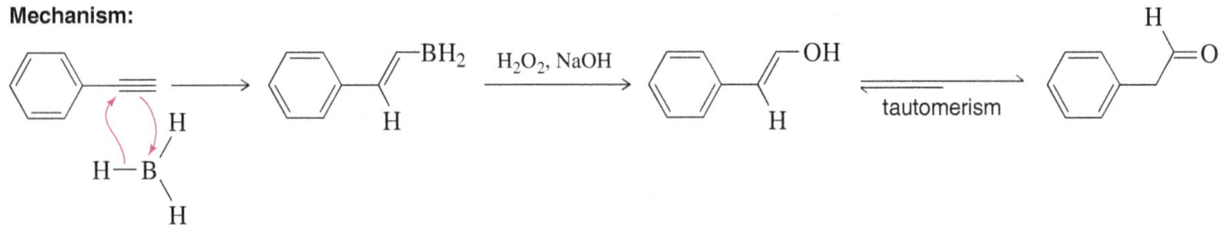

8. Halogenation (Section 10.8.7): Electrophilic addition of bromine to give a bromonium ion is followed by bromide opening of the bromonium to give a trans-dibromoalkene in a stereospecific reaction. Halogenation works in a similar manner with Cl_2. In the presence of additional equivalents of halogen, the reaction can be used to create a tetrahalogenated alkane.

General:

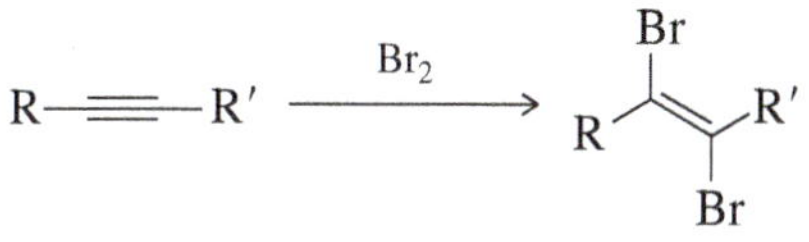

Specific:

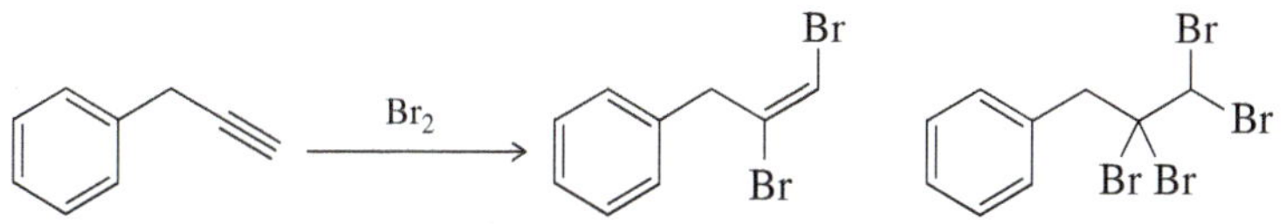

(if 2 equiv. Br_2)

Mechanism:

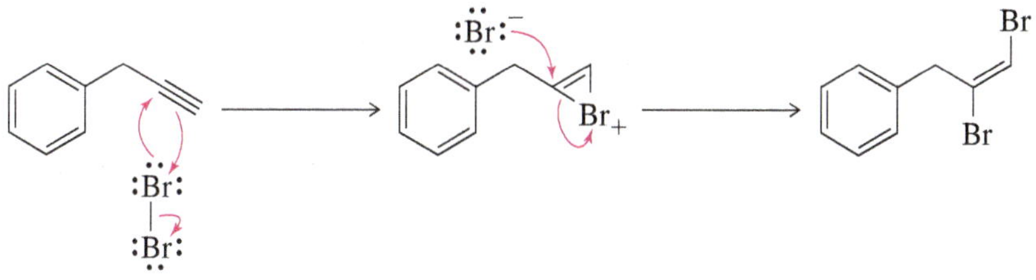

End-of-Chapter Assessments

[*Assessments are ranked according to their difficulty; •••• = most difficult.*]

10.42 (•) Name the following alkynes according to the IUPAC rules of nomenclature.

(a)

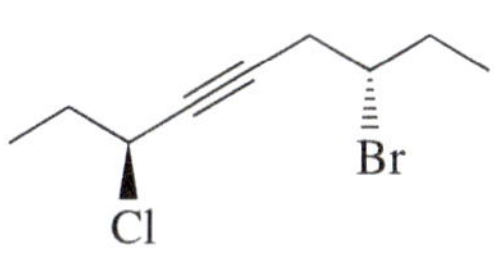

(b)

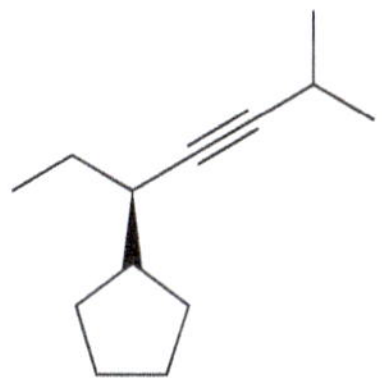

(c)

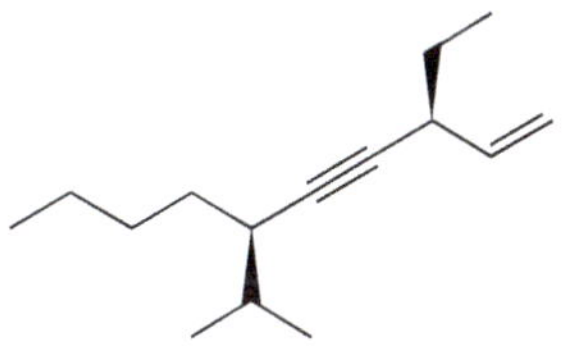

(d)

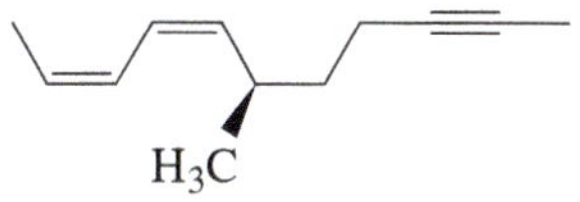

(e)

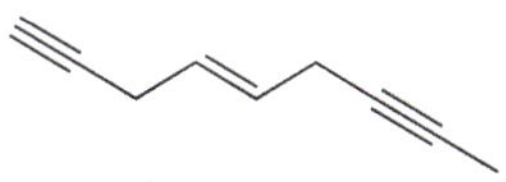

10.43 (•••) Draw the molecular orbital picture of pent-1-en-3-yne. To which carbon would you expect an electrophile to add (1, 3, or 4)? Explain your answer.

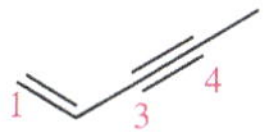

pent-1-en-3-yne

10.44 (••) For the alkynes shows here, show the product(s) expected to form when treated under the following conditions: (i) H_2, Pd/C; (ii) H_2, Pd/C, $Pb(OAc)_2$, $CaCO_3$ (Lindlar's catalyst); (iii) Na^0, NH_3(liquid), −33 °C; (iv) HBr (1 equiv.); (v) HCl (2 equiv.); (vi) H_2SO_4, $HgSO_4$, H_2O; (vii) 1. BH_3, 2. H_2O_2, NaOH; (viii) Cl_2 (1 equiv.); and (ix) Br_2 (2 equiv.). If you expect two products, show both.

(a)

(b)

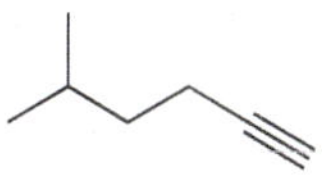

(c)

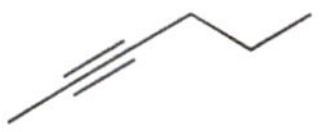

(d)

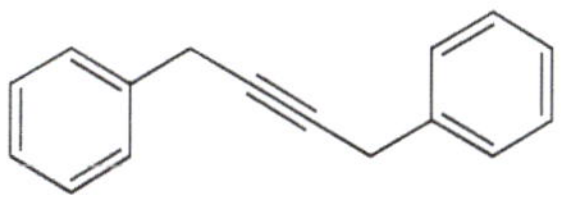

(e)

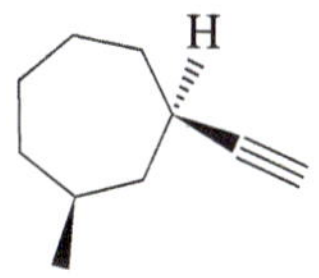

(f)

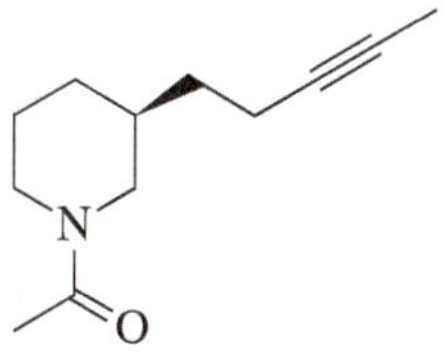

10.45 (•) Show the products of the following acetylide alkylation reactions. [*Make sure your product has the correct number of carbons.*]

(a)

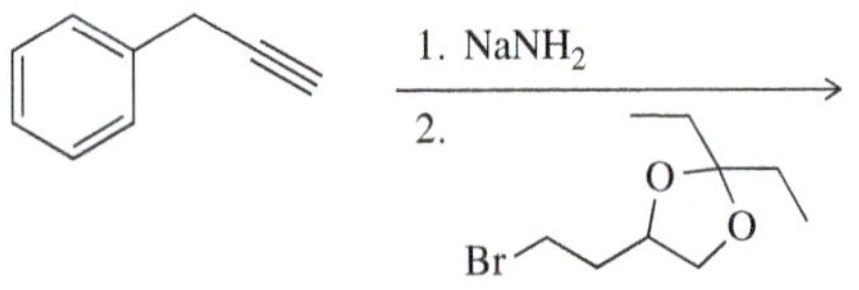

(b)

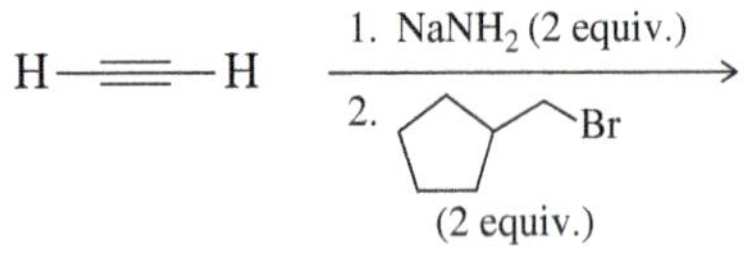

(c)

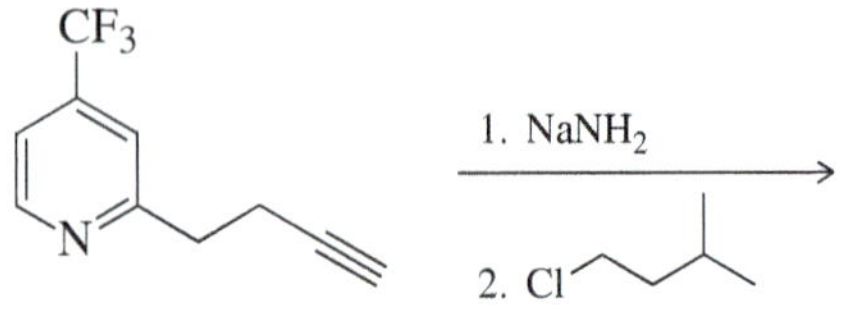

(d)

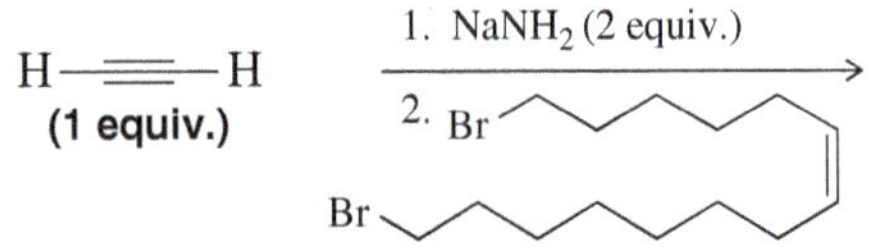

10.46 (••) Complete the following synthesis by providing the necessary reagents.

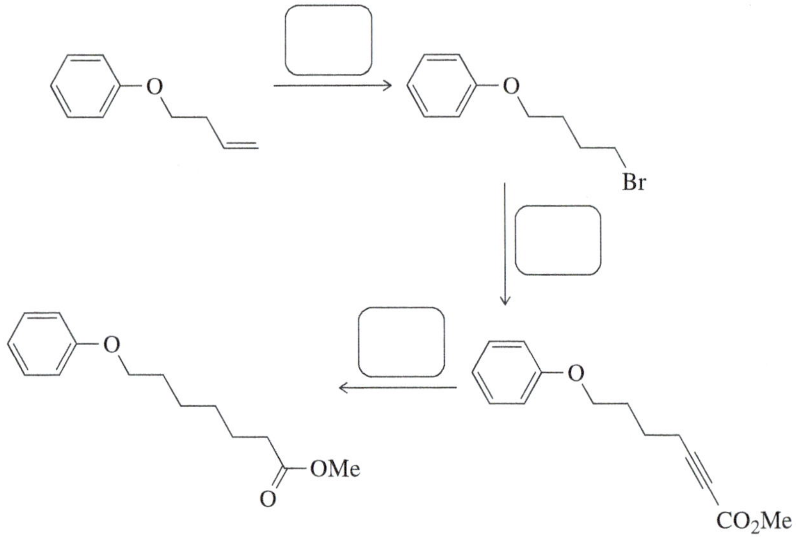

10.47 (••) Suggest a method for synthesizing the following alkynes using an alkyne and an alkyl halide. [*There are two correct answers for each product.*]

(a)

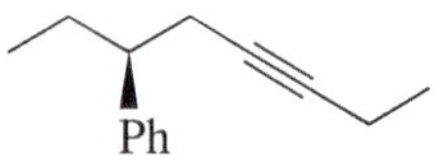

(b)

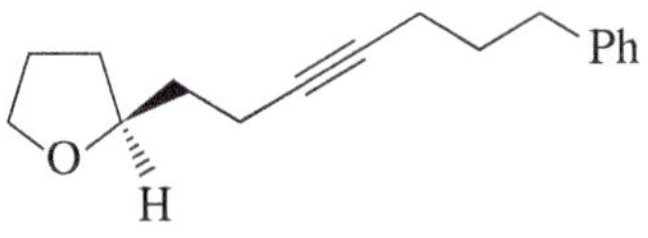

(c)

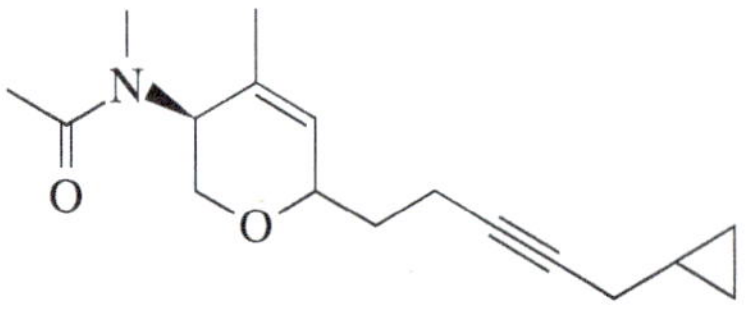

10.48 (•••) Beginning with the molecules on the left, provide a synthesis of the molecule on the right. The ideal number of steps is indicated over the reaction arrow, although there may be alternate routes worth considering.

(a)

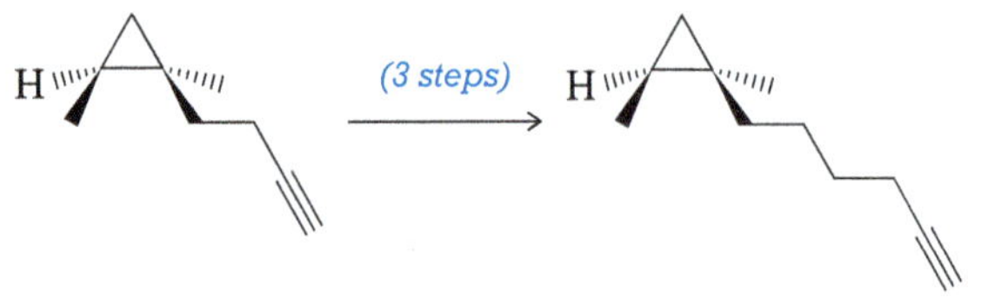

(b)

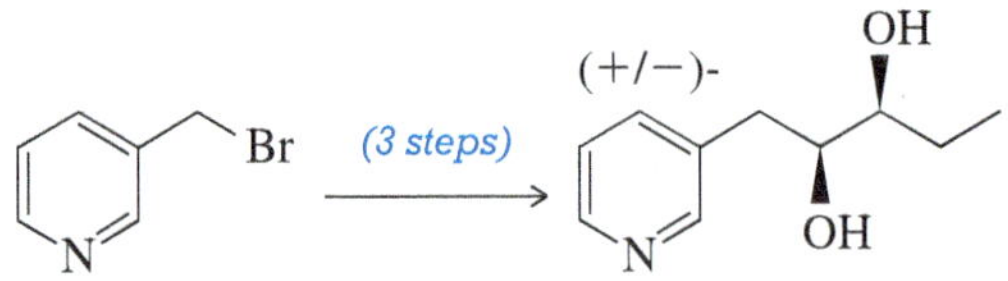

(c)

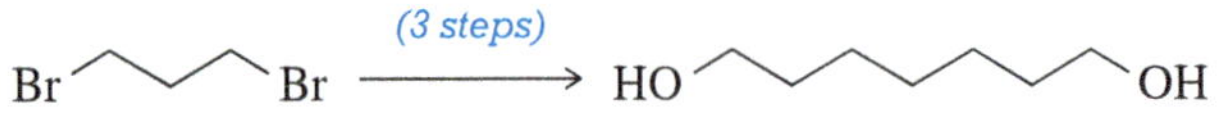

(d)

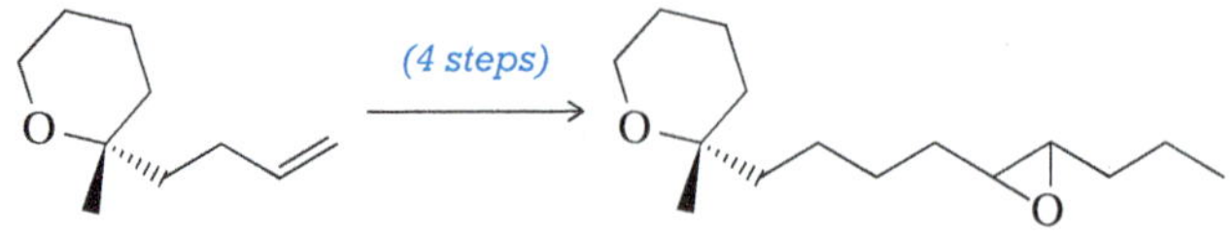

10.49 (•••) Synthesize the following molecules beginning with only organic molecules containing three carbons or fewer.

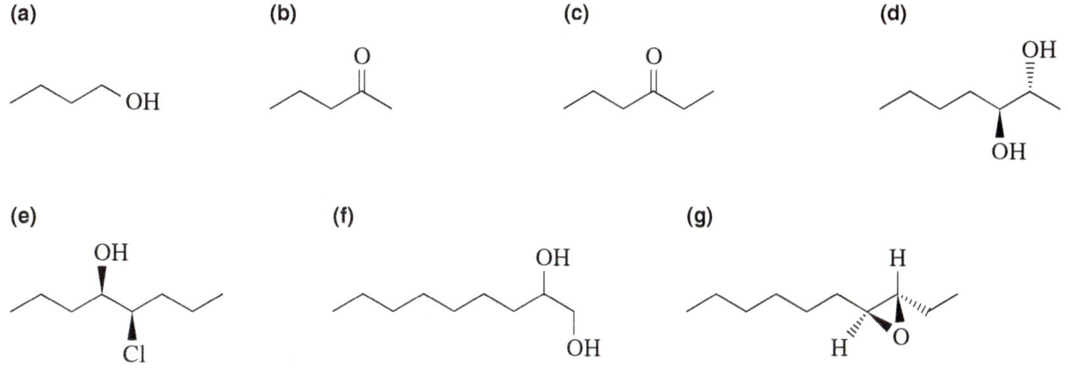

10.50 (••••) In Section 10.8.1, you learned that alkenes react more quickly with electrophiles than do the corresponding alkynes $\left(\frac{k_{alkene}}{k_{alkyne}} > 1\right)$. Explain why there is a greater disparity in the alkene versus alkyne reactivity in the addition of HBr as compared to the addition of Br_2. [*The rate data are not real, but are meant to illustrate a real trend.*]

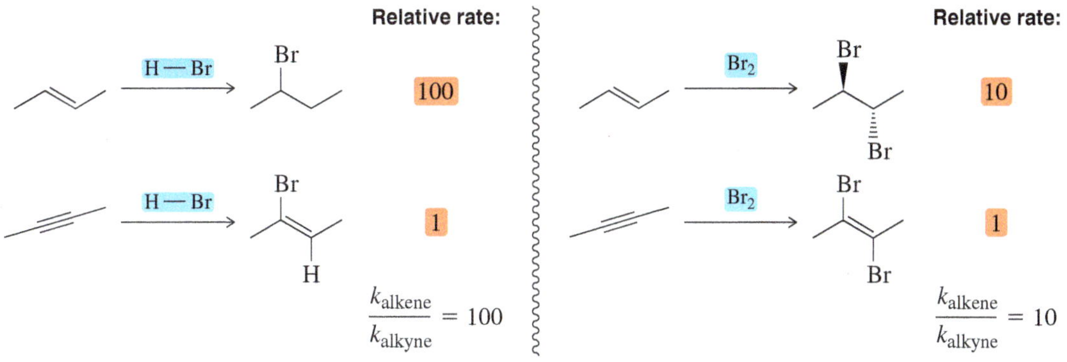

10.51 (••••) Related to Assessment 10.50, explain why the discrepancy in relative rates between alkenes and alkynes reacting with chlorine is larger than the discrepancy in the relative rates of reaction with bromine. [*Again, the data are not real, but the trend is.*]

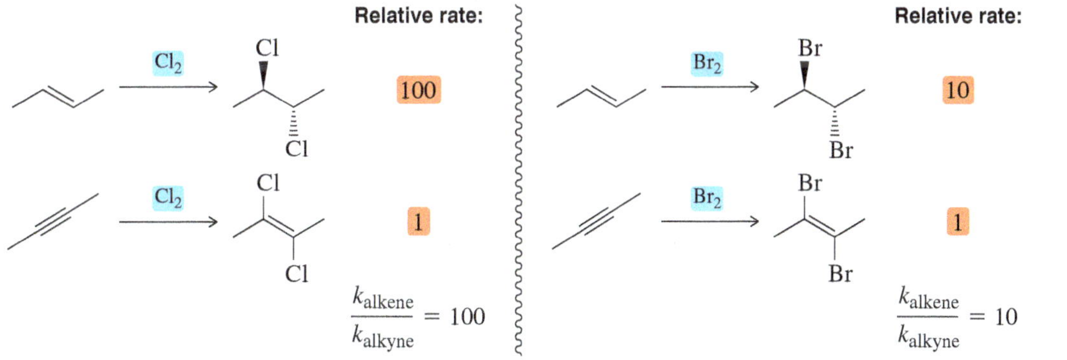

10.52 (••••) Trans addition is heavily favored for the addition of Br_2 and Cl_2 to alkynes. With chlorination, however, more of the syn addition product is formed. Rationalize this fact in light of your answer to Assessments 10.50 and 10.51.

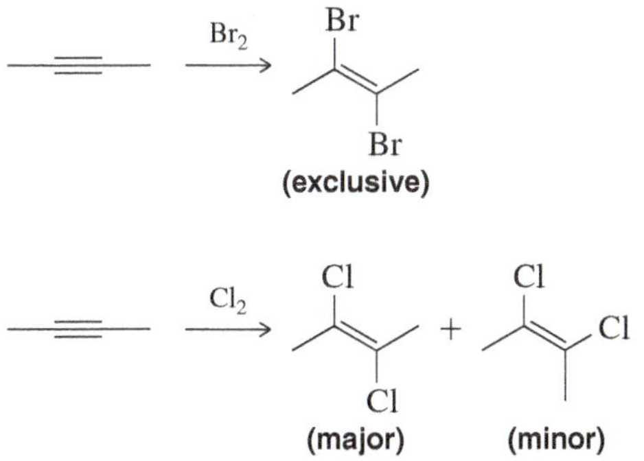

10.53 (••) We learned here in Chapter 10 that enols can equilibrate to the more stable keto form in an acid-catalyzed reaction. This equilibrium also occurs rapidly under basic conditions. Provide a mechanism for the base-catalyzed tautomerization in both directions.

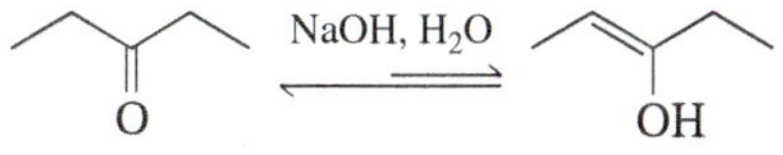

10.54 (••••) The addition of H—X to alkynes has been shown to occur predominately via anti addition:

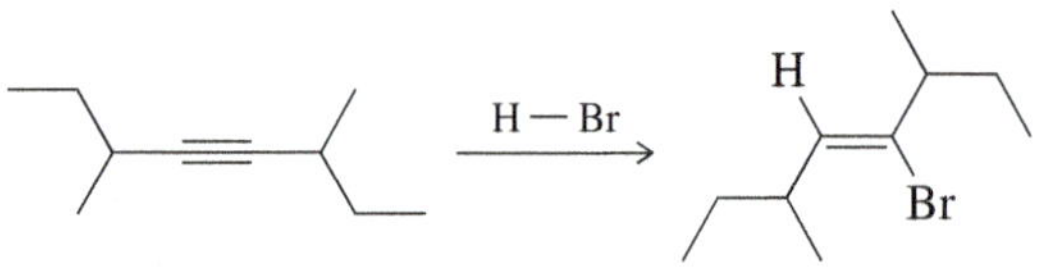

Two chemists disagreed on whether or not anti addition would happen on terminal alkynes as well. Suggest an experiment through which you could resolve this dispute.

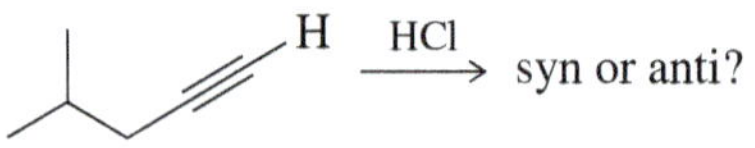

10.55 (•••) **IN THE CHEMICAL LITERATURE** The synthesis of five-membered lactones (cyclic esters) has been accomplished using the electrophilic addition of I—Cl to an alkyne. Suggest a mechanism for this cyclization reaction. (Structure modification of Yao, T.; Larock, R.C. *J. Org. Chem.* **2005**, *70*, 1432–1437.)

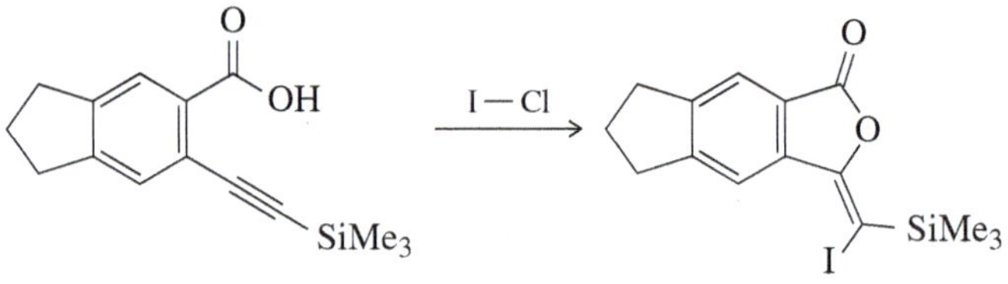

10.56 (•••) When alkynes are treated with water and bromine a bromoketone is produced. Provide a plausible arrow-pushing mechanism that accounts for the formation of this product.

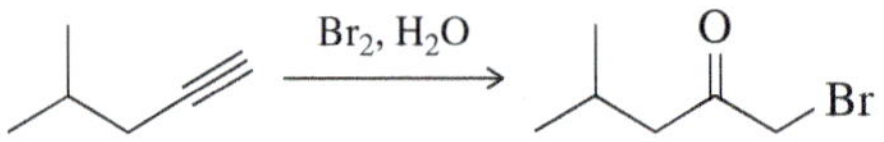

10.57 (••••) In Chapter 8, we learned reactions of alkenes to make the more- and less-substituted alkyl halide. Drawing on that knowledge, suggest reagents for carrying out the following transformation of an alkyne to give a terminal bromoalkene. Also, show a mechanism that rationalizes the regioselectivity.

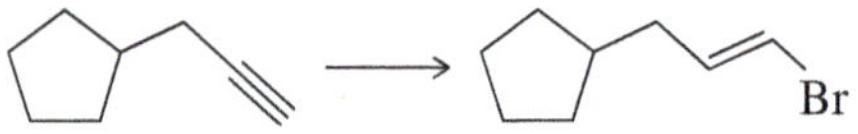

10.58 (••••) A chemist attempted to do the following acetylide alkylation reaction but was unsuccessful in several attempts, producing only the original starting materials in each case. Explain why the reaction didn't work. [*We describe one way to solve this problem in Chapter 13.*]

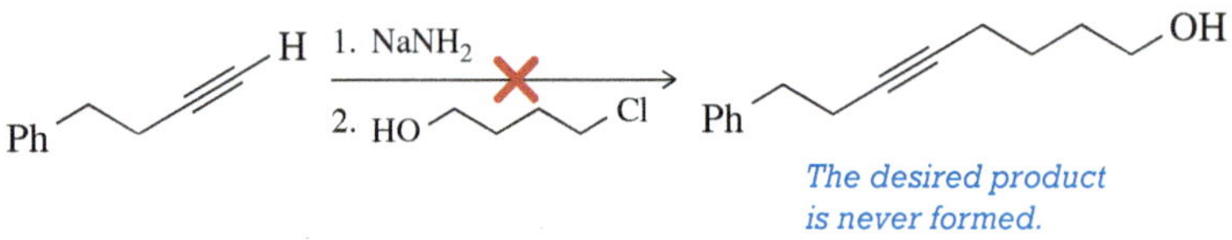

10.59 (••••) **IN THE CHEMICAL LITERATURE** Alkynes can undergo a carboalumination reaction by a mechanism we discussed in this chapter. Suggest a mechanism for this reaction. (Wipf, P.; Lim, S. *Angew. Chem. Int. Ed. Engl.* **1993**, *32*, 1068.) [*The reaction is usually catalyzed by* $(Cp)_2ZrCl_2$*, but your mechanism does not need to use it.*]

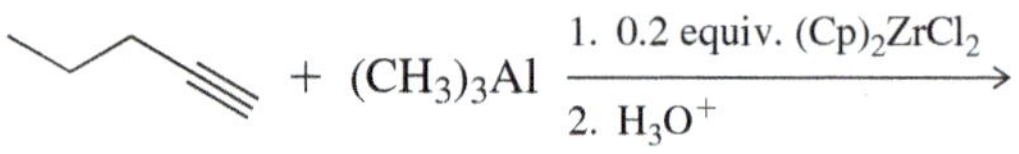

10.60 (••) **IN THE CHEMICAL LITERATURE** In 1973, Caine and Tuller reported a synthesis of racemic oplapanone, a sesquiterpene isolated from *Oplopanax japonicus* (a deciduous shrub) involving a reaction we learned in this chapter. Predict the product of the reaction shown. (Caine, D.; Tuller, F. N. *J. Org. Chem.* **1973**, *38*, 3663.)

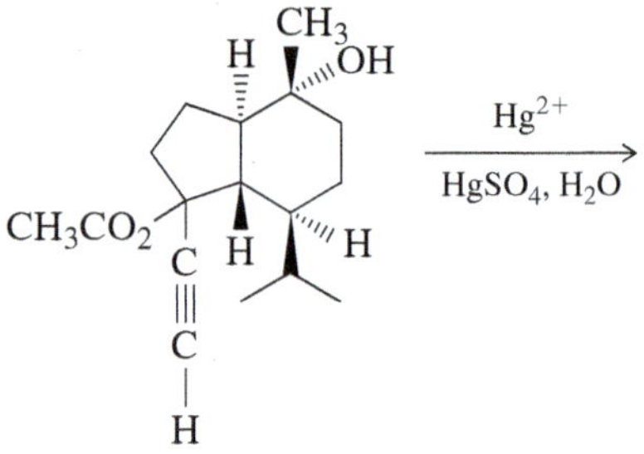

10.61 (••••) **LOOKING AHEAD** An alternate method for the synthesis of alkynes relies on the double elimination of H—Br from a dihaloalkane under basic conditions. Suggest a mechanism for this reaction that we discuss in Chapter 12.

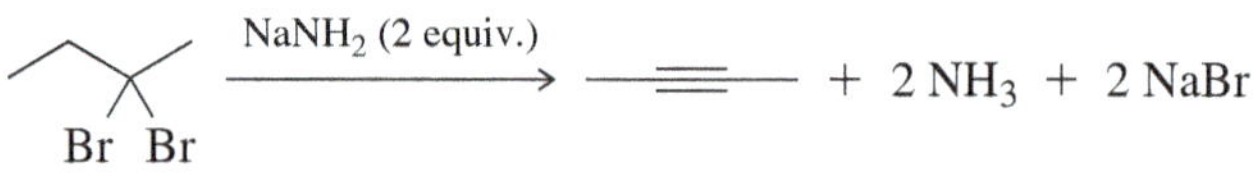

10.62 (••) **LOOKING AHEAD** Ethinylestradiol is a synthetic hormone mimic used as a contraceptive for its ability to prevent ovulation. Suggest a mechanism for the synthesis using sodium acetylide and estrone, followed by quenching with acid.

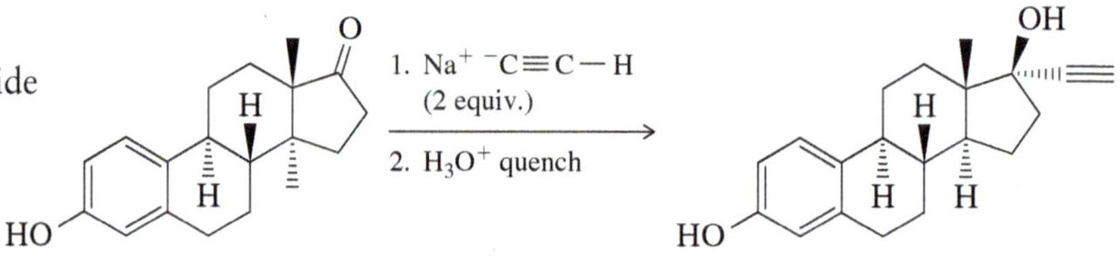

11

Properties and Synthesis of Alkyl Halides: Radical Reactions

Introduction: pharmacodynamics and pharmacokinetics

In drug discovery research, the development of new medicines involves two aspects of pharmacology: pharmacodynamics and pharmacokinetics. Pharmacodynamics deals with the manner in which a molecule interacts with its biological target, which can be an enzyme, DNA, or membrane-bound receptor. The focus of pharmacodynamics, then, is enzyme mechanisms and active site structures, and medicinal chemists attempt to maximize drug–receptor interactions through covalent, ionic, and/or hydrogen bonding. Because these studies are exclusively focused on the drug–receptor interaction, they are often performed in a flask [*in vitro*]. But what if a drug, administered to the patient, orally or by injection, never makes it to the receptor?

This question is addressed by pharmacokinetics, the study of how quickly a drug gets to its target and how long the drug remains in the body at a concentration high enough to have the desired therapeutic effect. Pharmacokinetics can be divided into four domains, defined by the following ADME scheme:

Absorption: the process by which a drug enters the bloodstream

Distribution: the process by which a drug spreads through bodily tissues

Metabolism: the manner in which a drug is metabolized by human enzymes

Excretion: the mechanism by which the body removes a drug as waste

The chemistry of haloalkanes has implications for both pharmacodynamics and pharmacokinetics. In Chapter 12, the role of haloalkanes in the development of anticancer drugs is

Learning Objectives

LO 11.0 Compare the structures of haloalkanes, haloalkenes, and haloarenes

LO 11.1 Explain the physical properties of haloalkanes

LO 11.2 Evaluate the synthetic pathways for radical halogenation of alkanes

LO 11.3 Predict the products formed by halogenation of alkenes

LO 11.4 Compare mechanisms devised by nature and by chemists for preventing radical autoxidation

"A Who's Who of pesticides is therefore of concern to us all. If we are going to live so intimately with these chemicals—eating and drinking them, taking them into the very marrow of our bones—we had better know something about their nature and their power."

—RACHEL CARSON, Silent Spring

discussed. Chapter 11 deals mostly with the properties and preparation of haloalkanes, so it provides a unique opportunity to discuss the role that halogens play in helping a drug reach its target. To be effective, a drug must be broadly distributed throughout the body, and in order for the drug's concentration to remain high, the body must be prevented from destroying it.

Chapters 8–10 focused on the reactivity of the π bond in alkenes and alkynes. Seen as the natural extension of previous acid–base and/or Lewis acid–Lewis base chemistry, we developed a thread connecting these concepts. The introduction of halogenated compounds changes the focus of our attention, but the idea of connecting concepts is the same. Chapters 11 and 12 recall chemistry studied previously, whether it's the use of 1° alkyl halides in acetylide alkylation (Chapter 10), the manner in which the alkenes of Chapters 8 and 9 can be synthesized, or the role that the Hammond postulate plays in predicting the regioselectivity of endothermic reaction steps (Chapter 5). So, although Chapters 11 and 12 emphasize content not studied in detail before, what you learn here is part of the bigger picture, relying on the same fundamentals to justify and predict reaction outcomes.

Test your understanding

Assessment 11.1 Which of the following molecules has the larger dipole moment?

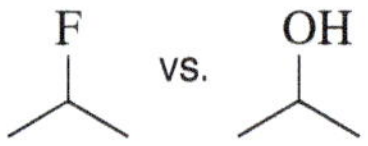

Assessment 11.2 Show the transition state of each step of the following alkene addition reaction. Be sure to indicate whether the transition state in each is reactant-like or product-like.

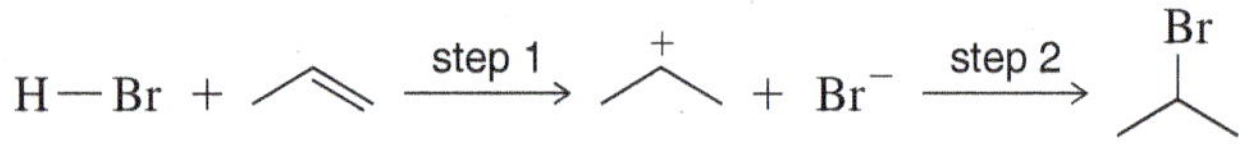

Assessment 11.3 Using the bond-dissociation energies in Table 5.6 (see Section 5.3.1), estimate the equilibrium constant of the following reaction at 298 K.

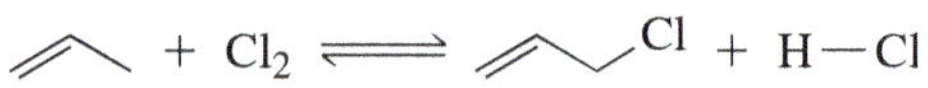

Assessment 11.4 Is the following reaction a redox reaction? Calculate oxidation numbers to justify your answer.

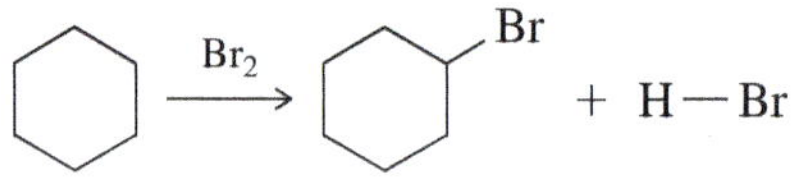

Assessment 11.5 Draw all resonance structures of the following carbanion and carbocation.

(a)

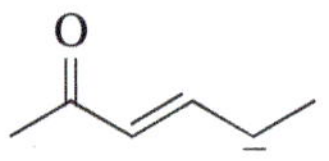

(b)

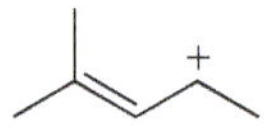

11.1 General structure of haloalkanes

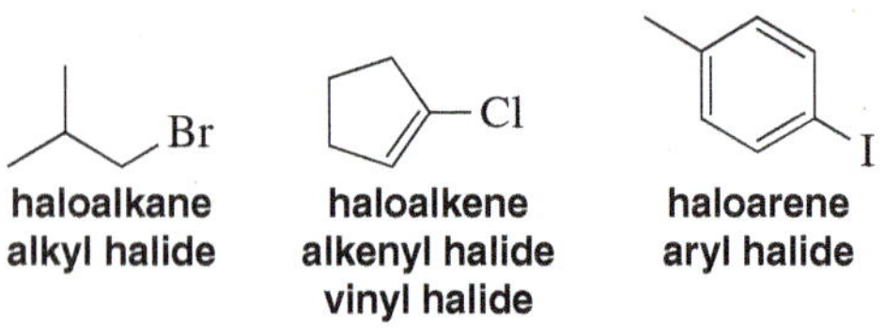

Figure 11.1 Representative halogenated compounds.

Haloalkanes, also called **alkyl halides**, represent a general class of compounds in which at least one hydrogen of a parent alkane has been replaced by a fluorine, chlorine, bromine, or iodine atom. In haloalkanes, abbreviated as R—X, the halogen is attached to a tetrahedral, sp^3-hybridized carbon. **Haloalkenes,** in contrast, have the halogen attached to the sp^2-hybridized carbon of an alkene. Finally, a halogen attached to a benzene ring is referred to as a **haloarene**. Representative halogenated compounds are shown in Figure 11.1. You will see in subsequent chapters that haloalkanes, haloalkenes, and haloarenes show different reactivity patterns.

Assessment 11.6 Identify the following halogen-containing compounds as a haloalkane, haloalkene, or haloarene.

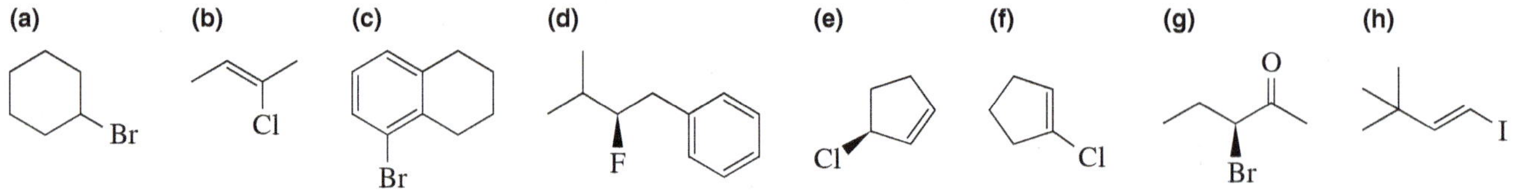

11.2 Nomenclature of haloalkanes

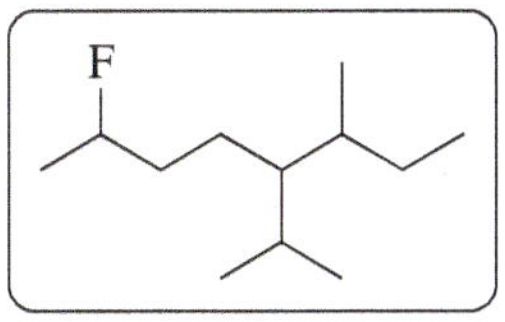

The IUPAC nomenclature of haloalkanes is virtually identical to how we name alkanes (summarized in Table 11.1). In essence, a haloalkane is considered to be an alkane with a halo- substituent attached.

Table 11.1 Rules for naming haloalkanes according to the IUPAC system of nomenclature.

1. Identify the parent chain as the longest unbranched carbon chain. The parent chain is the root of the name. The infix -an- indicates that it is a fully saturated alkane.	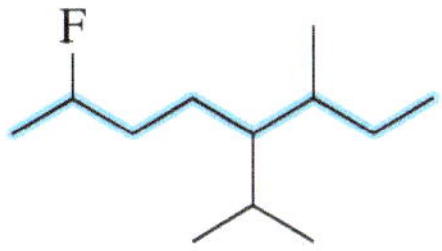 **8-carbon parent chain** **octane**
2. Number the carbons of the parent chain so that the first substituent, as an alkyl group or a halogen, gets the lowest number. If there is a tie, number such that the next substituent receives the lower number. As with alkanes, if each direction gives the same lowest number (or sets of numbers), number such that the first alphabetical substituent gets the lowest number.	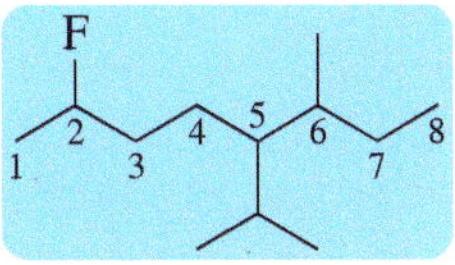 **2-fluorooctane** or 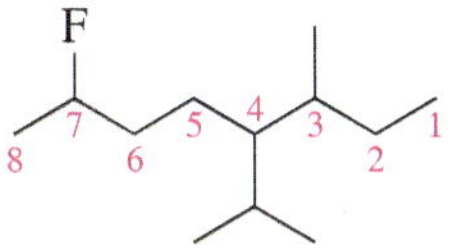 **~~3-methyloctane~~**
3. Substituents on the parent chain are named as before. Halogen substituents are named by dropping -ine from the element name and adding -o. As such, the halo- substituents are named as fluoro-, chloro-, bromo-, and iodo-.	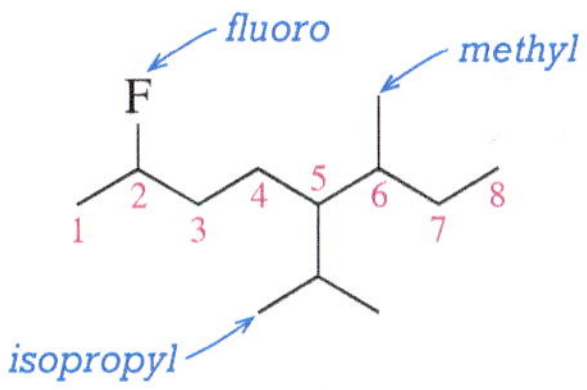
4. The name is then written following standard IUPAC alphabetization rules, with numbers indicating the position of the substituents.	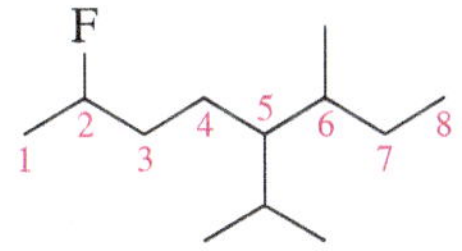 **2-fluoro-5-isopropyl-6-methyloctane**

Halogen-containing molecules with other functional groups are named following rules for that functional group, with the halogen as a substituent (Figure 11.2).

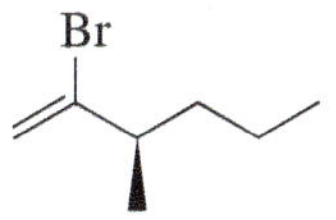
(*R*)-2-bromo-3-methylhex-1-ene

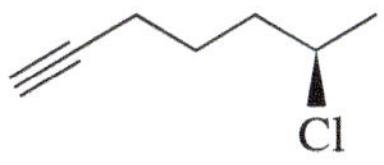
(*R*)-6-chlorohept-1-yne

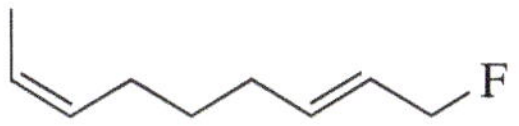
(2*E*,7*Z*)-1-fluoronona-2,7-diene

Figure 11.2 Naming halogenated compounds with multiple functional groups.

Assessment 11.7 Provide the IUPAC names for the following halogen-containing compounds.

(a)

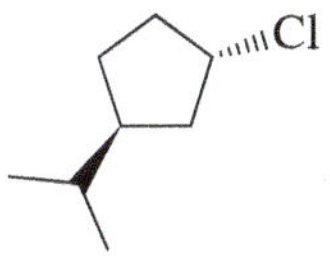

(b)

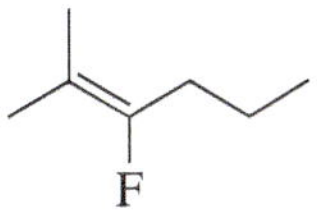

(c)

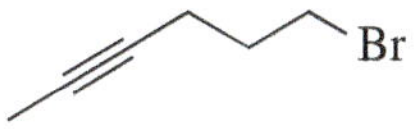

Assessment 11.8 From the IUPAC name, draw the corresponding structure.

(a) (*R*)-6-iodo-3-isopropylnon-1-ene
(b) (1*R*,2*S*)-1-chloro-2-methylcyclobutane
(c) (*S*)-3-ethyl-4-fluorooctane
(d) (*S*)-3-fluorohept-1-yne

11.3 Properties of haloalkanes

There are more tables here than in any previous chapter. Rationalizing information from tables is an important skill for future standardized exams. Thus, when looking at the tables, try to interpret the data on your own, making sure to relate it to concepts studied previously.

The manner in which haloalkanes react, their use as solvents in organic chemistry reactions, and how they impact drug design are largely a result of the properties discussed here in Section 11.3.

11.3.1 Polarity

Because they contain an electronegative halogen, low-molecular-weight haloalkanes are polar molecules. Although electronegativity values might predict differently, fluoroalkanes are less polar than chloroalkanes. This is due to the shorter C—F bond (C—F, 1.38 Å vs. C—Cl, 1.78 Å). [*The dipole moment is a function of both electronegativity difference and bond length.*] For the other halogens in the series, dipole moment (μ) parallels electronegativity differences (Table 11.2).

Table 11.2 Low-molecular-weight haloalkanes are polar molecules.

C—X	μ (debye)	C—X bond length (Å)
CH_3F	1.85	1.39
CH_3Cl	1.87	1.78
CH_3Br	1.81	1.93
CH_3I	1.62	2.14

11.3.2 Boiling and melting points

Even with similar dipole moments, the boiling point of related haloalkanes varies, depending on the halogen (Table 11.3). Fluorine is a small atom, so all of its electrons are localized over a small volume. Iodine, on the other hand, is a large atom, so its electrons are more spread out. This results in the potential for an uneven distribution of electrons in iodine. This enhanced **polarizability** results in significantly stronger London dispersion intermolecular attraction between iodoalkane molecules. [*As discussed in Section 3.1.3, larger molecules have higher boiling points and melting points.*]

Table 11.3 Substitution with larger halogens increases the boiling point.

C—X	bp (°C)
CH_3CH_2F	−38
CH_3CH_2Cl	12
CH_3CH_2Br	38
CH_3CH_2I	72

ethyl fluoride

ethyl chloride

ethyl bromide

ethyl iodide

Similar to trends seen in the alkane series in Table 3.2, larger chloroalkanes boil and melt at higher temperatures as a result of increased London dispersion forces. The greater surface area provides more opportunity to form favorable noncovalent interactions with other molecules, causing a greater resistance to boil and/or melt (Table 11.4).

Table 11.4 Higher molecular weight chloroalkanes boil and melt at higher temperatures.

C—X	bp (°C)	mp (°C)
CH_3CH_2Cl	12.3	−138.7
$CH_3CH_2CH_2Cl$	46.7	−122.8
$CH_3CH_2CH_2CH_2Cl$	78	−123.1
$CH_3CH_2CH_2CH_2CH_2Cl$	108	−99

Finally, with the exception of fluorine, adding more halogens to a molecule increases the boiling point. The addition of each halogen makes the molecule larger and more polarizable, resulting in greater intermolecular attraction (Table 11.5).

Table 11.5 More halogens increase the boiling point.

C—X	bp (°C)
CH_3Cl	−24
CH_2Cl_2	40
$CHCl_3$	61
CCl_4	77

11-A CHEMISTRY IN CONTEXT: Medicinal Chemistry

Oral bioavailability

Many drug candidates that have the desired biological effect (i.e., receptor/enzyme binding) still fail as a result of pharmacokinetic issues. A major problem of this type involves a drug's bioavailability, the degree to which a particular dose reaches the targeted tissue. Estimates suggest that 40% of molecules that fail in clinical trials fail because of poor bioavailability or increased metabolism and excretion. There is a very high cost associated with drug candidates failing animal testing or human trials. Since the most efficient way for a medicine to be taken is by swallowing a pill, it is important for medicinal chemists to assess oral bioavailability early in the discovery process.

Drug action is a two-step process: getting to the target (pharmacokinetics) and acting on the target (pharmacodynamics). Corwin Hansch, the father of computer-assisted molecular design, posited that the first step in a drug's diffusion through the body, the drug making its way from the aqueous environment of the digestive tract to the target site in a cell, is like a random walk. Along the way, the molecule traverses two different environments, the polar, aqueous environment of the stomach/circulatory system and the nonpolar environment of the cell membranes it crosses. As a result, a delicate balance must be maintained with regard to the optimum polarity/water solubility of a drug. If it is too soluble in water, then the drug remains in the aqueous phase and is rapidly excreted before reaching its target. If it is too *in*soluble in water, then the drug is metabolized too quickly, or it gets stuck in a nonpolar membrane, rendering it incapable of reaching the target.

11.3.3 Water solubility and hydrophilicity

Generally speaking, haloalkanes are insoluble in water (**hydrophobic**), in spite of being polar. Methanol and chloromethane are two molecules that have similar dipole moments. Methanol ($\mu = 1.69$ D) is completely **miscible** (i.e., infinitely soluble) in water. Chloromethane ($\mu = 1.87$ D) is more polar but is significantly less soluble (5.3 g/L). [*Why is the less polar molecule more soluble in water?*] Methanol not only is polar, but also has the ability to hydrogen bond with water, as a donor and an acceptor. Halogens are neither hydrogen bond donors nor acceptors, making them unable to break the strong hydrogen bonds between water molecules. As a result, haloalkanes tend to self-aggregate, resulting in their insolubility, especially as the alkyl chain to which they are attached gets larger (Table 11.6).

Table 11.6 Haloalkanes are not very soluble in water, especially as they get larger. [*In contrast, methanol is highly soluble in water.*]

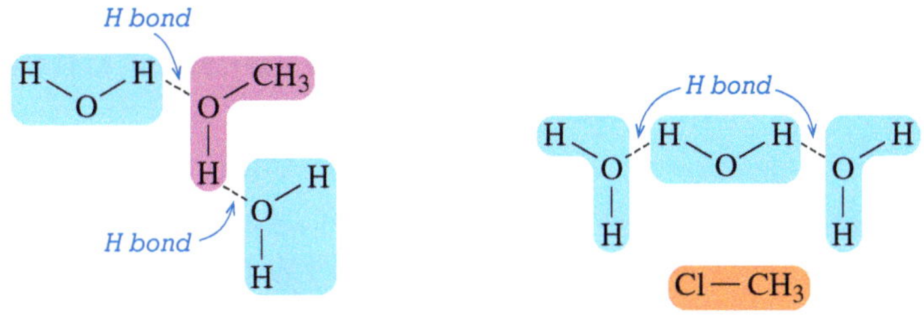

C—X	Water solubility (g/L)
CH_3CH_2Cl	5.74
$CH_3CH_2CH_2Cl$	2.7
$CH_3CH_2CH_2CH_2Cl$	0.5
$CH_3CH_2CH_2CH_2CH_2Cl$	0.197

11.3.4 Lipophilicity

Before we introduce a new concept, let's consider whether the addition of a second chlorine would make a molecule more or less soluble in water. [*Think about it before reading further.*] For one, it's a polar group, so that should make the molecule more soluble. It can't hydrogen bond, however, suggesting that a second chlorine should make a molecule less soluble. This question can be answered using benzene, chlorobenzene, and 1,2-dichlorobenzene, molecules we study in Chapter 23.

Table 11.7 Adding halogens makes a compound less soluble in water.

Molecule	Water solubility (g/L)
Benzene	1.84
Chlorobenzene	0.5
1,2-Dichlorobenzene	0.14

Benzene is nonpolar and only slightly soluble in water. Addition of a first and second chlorine results in molecules that are even less soluble. Thus, the addition of halogens diminishes water solubility, making the molecule more hydrophobic (or less hydrophilic) (Table 11.7).

A discussion of compound solubility is not limited to the use of water as solvent. Many organic reactions are conducted in organic solvents, which are themselves nonpolar or only slightly polar. The favorability of a molecule to be dissolved by a nonpolar solvent is referred to as its **lipophilicity**. Lipophilicity [*lipid-loving*] and hydrophilicity [*water-loving*] are inversely correlated, so making a molecule more lipophilic usually makes it less hydrophilic. As Table 11.7 shows, adding halogens to a molecule decreases their aqueous affinity, making them more lipophilic.

11.3.1–11.3.4 SUMMARY

- In general, the addition of halogens to organic molecules makes them more polar, increases the melting and boiling points, reduces water solubility, and increases lipophilicity.

Assessment 11.9 Classify the following molecules as hydrophilic, hydrophobic, lipophilic, or lipophobic. Each molecule should have two classifications.

(a)

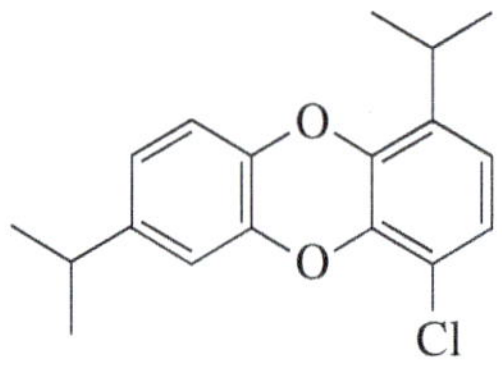

(b)

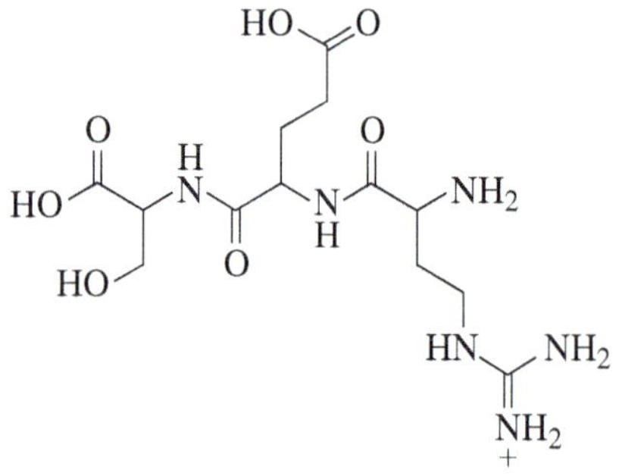

(c)

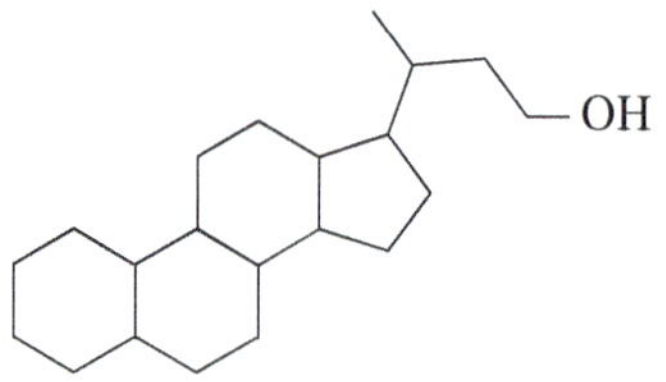

(d)

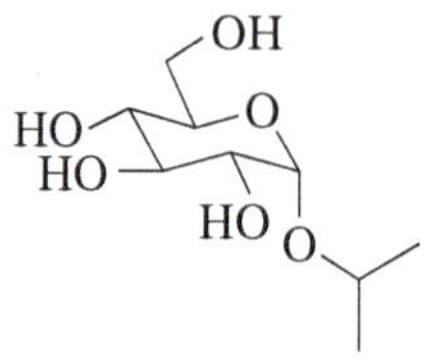

11-B CHEMISTRY IN CONTEXT: Medicinal Chemistry

Lipophilicity

Charles Richet, winner of the 1913 Nobel Prize in Physiology or Medicine for his discovery of anaphylaxis, first documented the role of lipophilicity in drug design in 1893. He discovered that the narcotic action of a series of molecules was inversely proportional to their water solubility, a concept that has proven general across a series of drug classes.

The graph in Figure 11.3 shows that each drug has an optimum lipophilicity for maximum activity. For drug molecules containing a carbon chain, increasing the chain length increases the potency, but only to a point (around C_8). Further lengthening beyond C_8 reduces potency. These results suggest that short-chain analogs are more hydrophilic and less likely to cross the nonpolar lipid bilayer of the cell membrane. At the other extreme, the nonpolar long-chain analogs are too lipophilic, possibly getting stuck in the cell membrane and not making it to the target site inside the cell. In the middle ($\sim C_8$), the molecule is perfectly balanced—nonpolar enough to enter a nonpolar membrane and polar enough to not be trapped there.

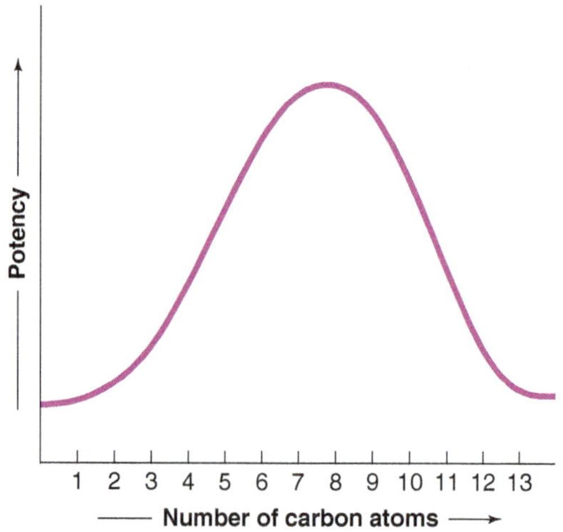

Figure 11.3 The potency of a drug is maximized at a certain balance of hydrophilicity and lipophilicity.

11-C CHEMISTRY IN CONTEXT: Environmental Science

Persistent organic pollutants

The addition of halogens can tune the lipophilicity of drugs in a beneficial manner. However, molecules with multiple halogens can be environmental hazards because they are too lipophilic. This is true for what are referred to as *persistent organic pollutants* (POPs). These compounds each contain multiple halogens, posing many environmental challenges. Halogenated compounds have shown causative links to cancer. [*Strike 1.*] The presence of halogens in molecules hinders their metabolism within organisms. That means that, if ingested, POPs stay in the body for longer periods of time. [*Strike 2!*] Their high lipophilicity causes POPs to bioaccumulate in fat deposits, especially in organisms higher up the food chain. [*Strike 3—you're out!*]

Bioaccumulation (the buildup of chemicals in tissues of living organisms) occurs when the rate of intake is greater than the rate of excretion or breakdown by cellular metabolism. This becomes problematic for a compound that is toxic at high concentrations. Stored [*or stuck*] in nonpolar membranes, lipophilic compounds [*ones with multiple halogens*] tend to bioaccumulate up the food chain. Although an individual blade of grass might contain a harmless amount of a toxin, the cow that eats a field of grass will ultimately store in its body the toxin from each and every blade.

A striking example of the deleterious impacts of a POP on the environment is the insecticide DDT, used in the 1940s and 1950s to prevent the spread of deadly insect-vectored diseases like typhus and malaria. This highly chlorinated molecule had negative consequences for bird populations, especially ones considered apex predators, such as bald eagles. By inhibiting the blood's ability to transport calcium to developing eggshells, DDT is believed to cause thinner, weaker eggshells, lowering the birthrate of infected species. Rachel Carson's book *Silent Spring* discusses the effects of DDT on species and the environment and is credited with launching the modern environmental movement.

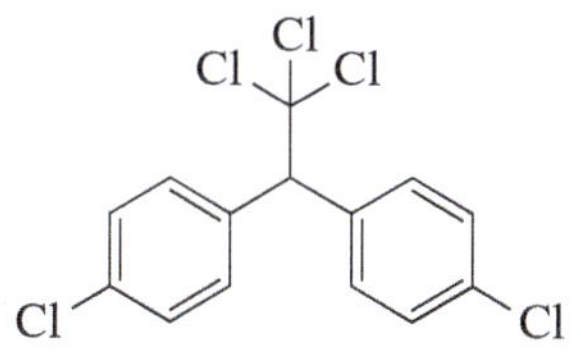

DDT
dichlorodiphenyltrichloroethane

11.3.5 Haloalkanes as solvents

Under many standard reaction conditions, chloroalkanes and chloroarenes are unreactive. That, coupled with their ability to dissolve nonpolar compounds, makes them ideal reaction solvents. [*Most organic molecules and the reagents we use are nonpolar.*] The most commonly used halogenated solvents are dichloromethane, 1,2-dichloroethane, chloroform, and dichlorobenzenes (**Figure 11.4**). In many cases, the choice of solvent is made based on the temperature at which the reaction is to be run, with higher boiling solvents allowing for higher temperature processes.

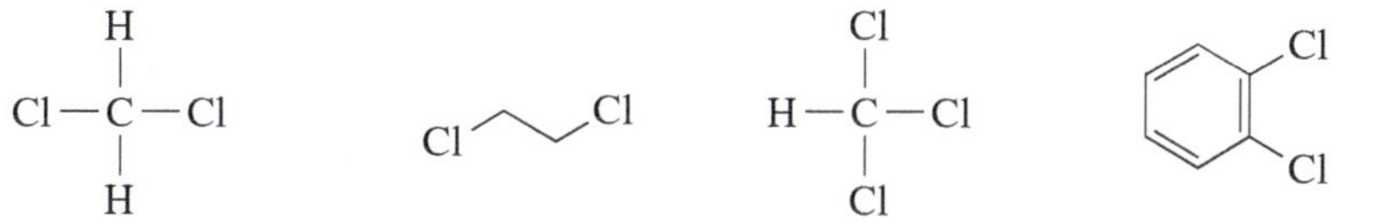

dichloromethane (methylene chloride)	1,2-dichloroethane	chloroform	1,2-dichlorobenzene
bp 39.6 °C	bp 84 °C	bp 61.2 °C	bp 180.5 °C

Figure 11.4 Common organic solvents.

11.3.6 Density

Haloalkanes have a greater density (ρ) than the corresponding alkanes. [*Before reading further, think about why this is true for hexane* ($\rho = 0.65\ g/mL$) *and its halogenated derivative 1-fluorohexane* ($\rho = 0.8\ g/mL$).] A halogen atom, containing a significantly larger atomic mass yet only a slightly larger volume than a hydrogen atom, has a greater mass per unit volume [*density*] (**Table 11.8**). Adding a halogen atom to an alkane thus increases both mass and volume, but the increase in mass is greater than the corresponding increase in volume.

Table 11.8 The small increase in volume but large increase in mass makes halogens more dense compared to hydrogen.

	Hydrogen	Fluorine
	 H	 F
Radius (pm)	25	50
Mass (amu)	1	19

2× increase in radius...

...but 19× more massive.

Proceeding down the halogen family in the periodic table, the density of the halogens increases [*large increase in mass, slight increase in volume*]. This trend is shown in **Table 11.9** for the 1-halohexanes. Iodoalkanes are more dense than the corresponding bromo-, chloro-, and fluoroalkanes.

Table 11.9 Iodoalkanes are the most dense haloalkanes.

Compound	Density (g/mL)
 hexane	0.65
 1-fluorohexane	0.80
 1-chlorohexane	0.87
 1-bromohexane	1.18
 1-iodohexane	1.46

Methylene chloride (ρ = 1.33 g/mL) contains two chlorines and is more dense than water. This makes methylene chloride an ideal solvent for extracting nonpolar organic compounds from water. When using a separatory funnel, methylene chloride, after vigorous shaking, can be removed as the bottom layer, leaving the water layer in the funnel available for a second extraction (**Figure 11.5**).

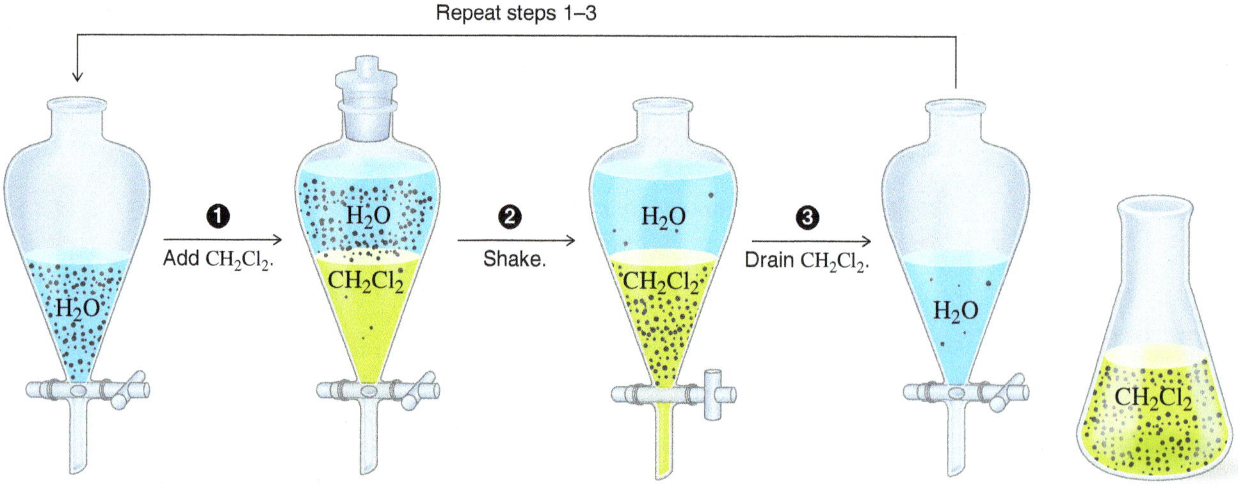

Figure 11.5 The density of methylene chloride makes it an ideal extraction solvent. [· = *nonpolar molecules*]

11-D CHEMISTRY IN CONTEXT: Green Chemistry

Extraction of caffeine

Halogenated solvents, with the ability to dissolve both polar and nonpolar compounds, are very useful for extracting organic compounds from both aqueous solutions and solids. Beginning in the early 1970s, methylene chloride was used for extracting caffeine from coffee beans because of its reduced toxicity relative to other extraction agents (**Figure 11.6**). Additional advantages of methylene chloride were its high volatility [*it is easily removed from the beans*] and that the flavorful ingredients remained.

caffeine

Figure 11.6 Caffeine is extracted from coffee beans using methylene chloride.

When evidence suggested a link between methylene chloride and cancer, it was replaced with environmentally benign supercritical carbon dioxide (sCO_2) (**Figure 11.7**). At temperatures and pressures beyond the critical point, carbon dioxide is a supercritical fluid, in which the properties of the liquid and the gas are indistinguishable from each other. In the extraction process, the sCO_2 is passed through a pressurized container of raw coffee beans. Being gas-like, sCO_2 penetrates deeply into the beans, allowing for efficient extraction. Up to 99% of the caffeine can be removed using this method. The supercritical fluid containing the caffeine is collected. sCO_2 is nontoxic, so upon evaporation, the isolated caffeine can be sold to beverage companies for use in other drinks.

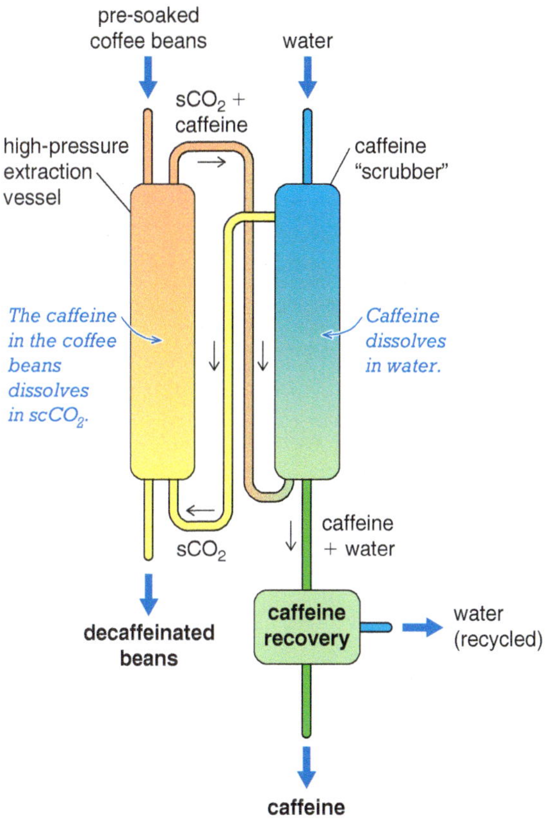

Figure 11.7 Extraction of caffeine with supercritical carbon dioxide.

11.3.7 Bond strengths and bond lengths

Another result of the increasing atomic size as you go down the group is that the carbon–halogen bond length increases (**Table 11.10**). The reverse is true for bond strengths (expressed as bond dissociation energies, or BDEs). The strength of the C—X bond decreases from C—F to C—I, a trend characterized by the generalization that *longer bonds are weaker bonds*.

Table 11.10 Longer C—X bonds are weaker.

Halomethane	C—X bond length (Å)	Bond-dissociation energy (kcal/mol)
$H_3C—F$	1.40	109
$H_3C—Cl$	1.79	84
$H_3C—Br$	1.97	70
$H_3C—I$	2.16	56

There are a couple of reasons the "longer is weaker" generalization holds true. First, the C—X σ bond in each case is formed by overlap of the $2sp^3$-hybridized orbital of carbon and a $2sp^3$-, $3sp^3$-, $4sp^3$-, or $5sp^3$-hybridized orbital of fluorine, chlorine, bromine, or iodine, respectively. As **Figure 11.8(a)** shows, orbitals of similar size and energy levels have better overlap. Thus, the $2sp^3$–$2sp^3$ overlap of the C—F σ bond is greater than the $2sp^3$–$5sp^3$ overlap of the C—I σ bond. The C—F bond is then stronger than the C—I bond. The second reason has to do with the radical formed upon **homolytic** bond dissociation. Recall from Section 5.3.2 that the tendency for a bond to dissociate homolytically (1 electron to each atom) is based on the stability of the resulting radicals. On a reaction coordinate diagram [**Figure 11.8(b)**], a higher hill must be climbed to form the chlorine radical compared to the bromine radical. Because the methyl radical is the same in each case, the difference in bond-dissociation energy can be attributed to the difference in stability between the bromine and chlorine radicals. The bromine radical is more stable because, being larger, the radical character is delocalized over a larger volume.

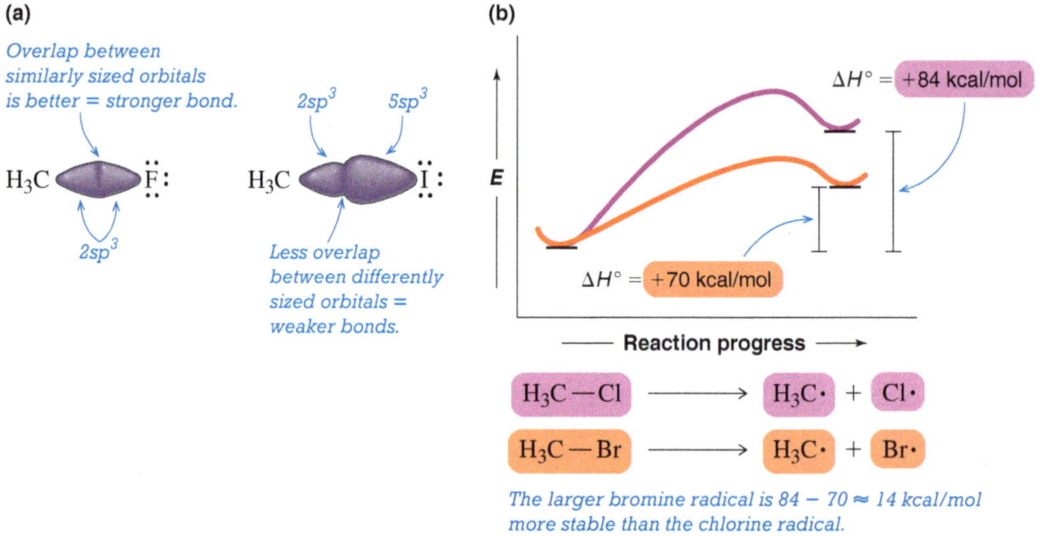

Figure 11.8 Longer C—X bonds are weaker. (a) Overlap between differently sized orbitals makes bonds weaker. (b) The homolysis of weaker bonds forms more stable radicals.

11.3.5–11.3.7 SUMMARY

- Because they readily dissolve organic molecules and are generally unreactive (except under certain conditions), haloalkanes are excellent reaction solvents. The addition of halogens to a molecule increases the density. As the length of a carbon–halogen bond increases, its strength decreases.

11.4 Molecular orbital picture of haloalkanes

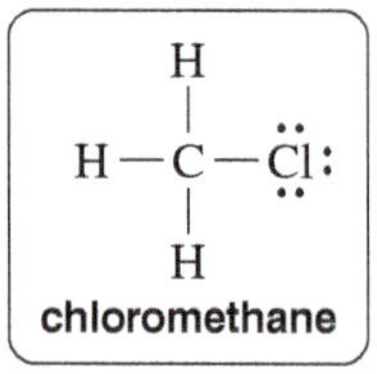

The chemical reactions of haloalkanes, which are presented in Chapter 12, can be explained by examining the molecular orbital picture. [*Try to draw the molecular orbital picture of chloromethane on your own before reading further.*]

Chloromethane has a central sp^3-hybridized carbon connected to three hydrogens and a chlorine via σ bonds. The three C—H bonds form when the sp^3-hybridized orbitals of carbon overlap with the 1*s* orbitals of hydrogen. Chlorine, with three lone pairs and one bond, has four areas of electron density and bonds as if it were sp^3 hybridized. Thus, the C—Cl σ bond is formed by overlap of an sp^3-hybridized orbital of carbon with an sp^3-hybridized orbital of chlorine. Recall that σ bonds overlap along an axis connecting the two nuclei (**Figure 11.9**).

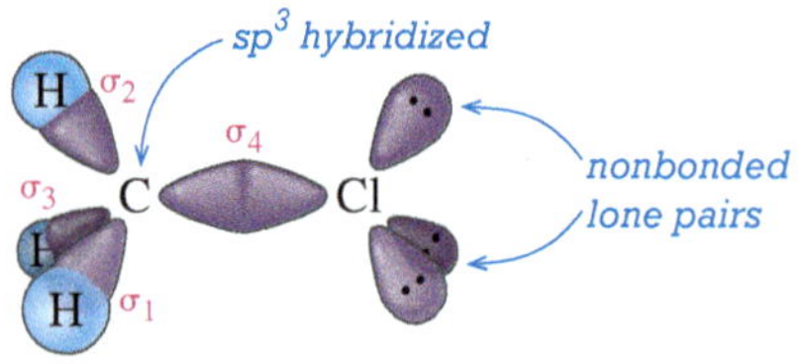

Figure 11.9 Molecular orbital picture of chloromethane.

11.4.1 Reactivity of C—F versus C—Cl: an MO viewpoint

The difference in bond dissociation energies among the series of C—X bonds was attributed in Section 11.3.7 to the degree of overlap between the $2sp^3$-hybridized orbital of C and the $2sp^3$-, $3sp^3$-, $4sp^3$-, or $5sp^3$-hybridized orbitals of F, Cl, Br, or I, respectively. Orbitals of similar energy [*size*] have better overlap, forming stronger bonds. A comparison of the molecular orbital energy diagrams of a C—F and a C—Cl bond shows this (**Figure 11.10**). The $2sp^3$- hybridized orbital of F is closer in energy to the $2sp^3$-hybridized orbital of C than is chlorine's $3sp^3$-hybridized orbital. As a result, the C—F bond is harder to break (stronger) than the C—Cl bond.

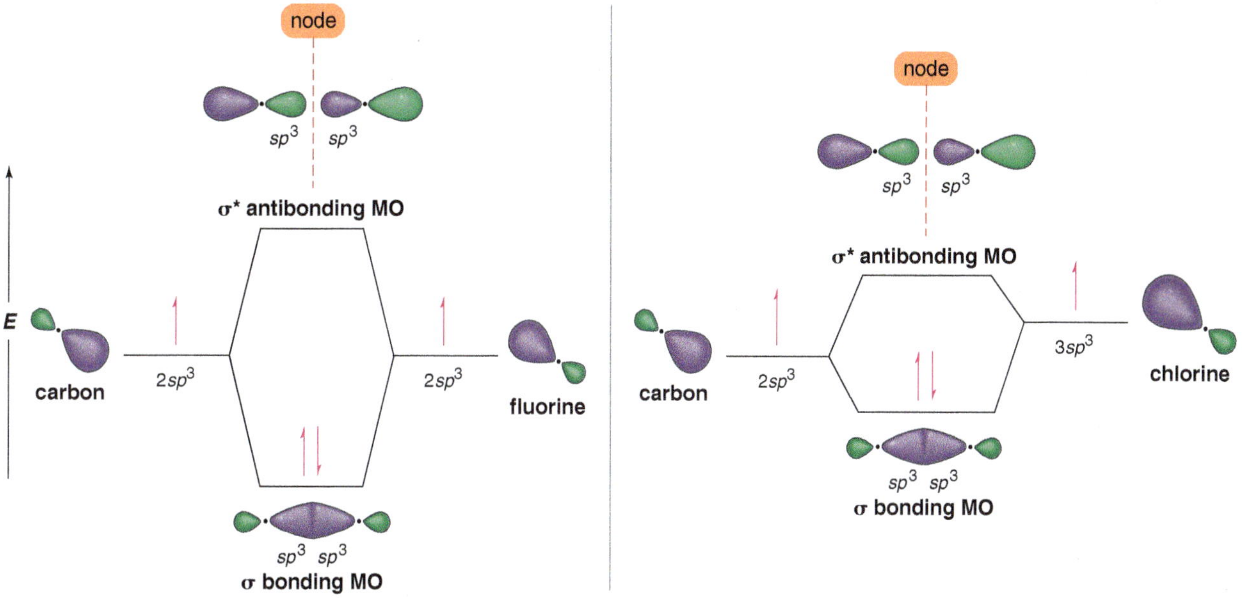

Figure 11.10 Relative orbital energies. The $2sp^3$ orbital of C is closer in energy to the $2sp^3$ orbital of F than it is to the $3sp^3$ orbital of Cl.

11.4.2 Antibonding orbital (σ*)

Most pertinent to our discussion is the manner in which the MO picture of a haloalkane impacts its reactivity. The focus here is on the C—Cl σ* antibonding orbital. Although they possess no electrons in the ground state configuration, the antibonding orbitals are always present. Specifically, the C—Cl σ* is located on the C—Cl bonding axis with a node (no overlap) between the two nuclei (**Figure 11.11**). This arrangement leaves a significant portion of the σ* lobe on opposite sides of carbon and chlorine. The partial positive charge induced by the electronegative halogen as well as the empty σ* orbital results in nucleophiles attacking the carbon of haloalkanes (Chapter 12). The location of σ* (180° angle to the C—Cl bond) forces the nucleophile to add opposite the chlorine atom. This is a significant part of the chemistry covered in Chapter 12. [*Stay tuned!*]

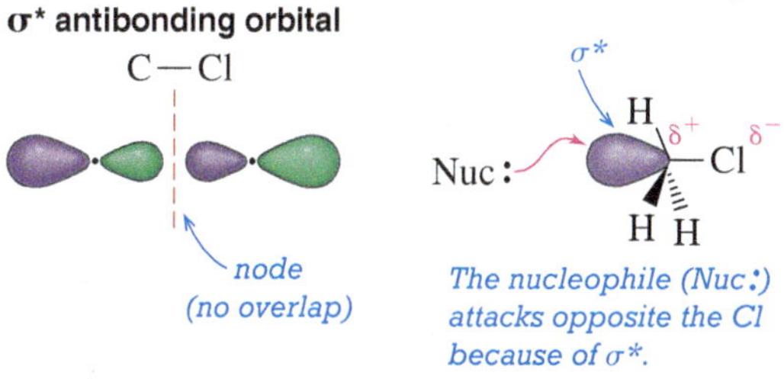

Figure 11.11 Nucleophiles add to σ* when attacking haloalkanes.

11.4 SUMMARY

- Incomplete overlap between orbitals of different sizes explains why C—I bonds are weaker than C—F bonds. The antibonding orbital, σ*, is unfilled, but it plays a crucial role in the substitution and elimination reactions in Chapter 12.

11.5 Preparation of alkyl halides

We have learned multiple ways of synthesizing haloalkanes (**Table 11.11**), two of which are reviewed extensively here in Section 11.5.

Table 11.11 Previous synthesis of haloalkanes.

Radical halogenation (Sections 5.3 and 5.4)	
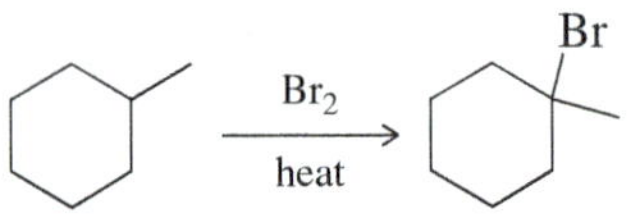	The thermally or photochemically initiated radical chain reaction replaces a hydrogen with a halogen, usually at the more substituted carbon. [*More detail is provided in Section 11.5.1.*]
Hydrohalogenation (Section 8.3)	
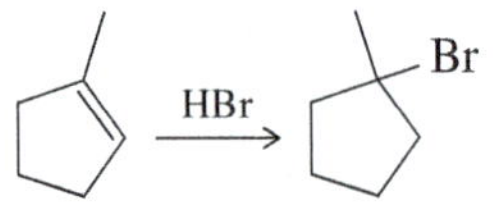	Electrophilic addition of HBr produces the more stable carbocation, which is attacked by bromide in the second step. Because attack by bromide in the second step can occur with equal rates to both sides of the carbocation (enantiomeric transition states), a racemic mixture of products results. The reaction with HCl proceeds in a similar manner to produce substituted chloroalkanes.
Radical hydrohalogenation (Section 8.3.4)	
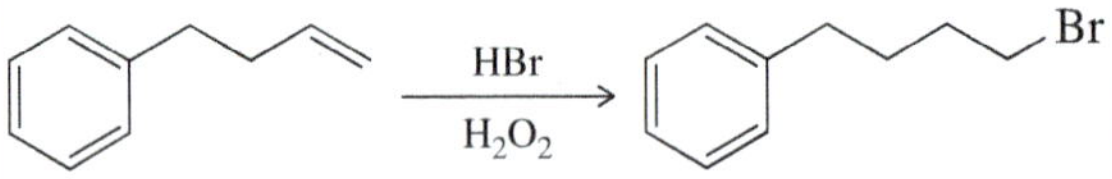	Addition of HBr in the presence of peroxide results in a radical mechanism. Attack by a bromine radical at the least substituted carbon results in formation of the more stable carbon radical in the first propagation step. The new C—H bond is formed in the second propagation step. [*More detail is provided in Section 11.5.2.*]
Alkene halogenation (Section 9.1.2)	
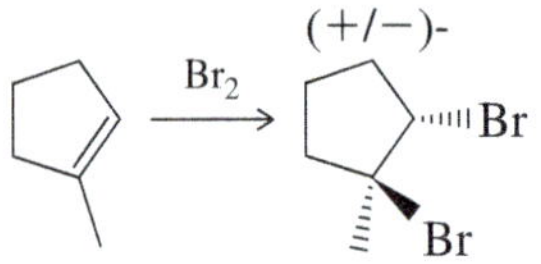	Electrophilic addition of bromine to give a bromonium ion is followed by opening of the bromonium ion, producing a vicinal trans-dibromide in a stereospecific reaction. [*Halogenation works in a similar manner with Cl_2.*] Because the bromonium ion could form on either side of the alkene with equal rates (enantiomeric transition states), the reaction results in a racemic mixture.
Halohydrin formation (Section 9.1.3)	
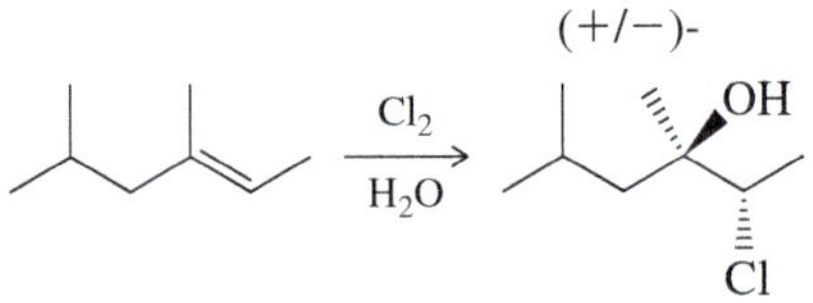	Electrophilic addition of chlorine to give a chloronium ion is followed by opening of the chloronium ion, producing a halohydrin in a regio- and stereospecific reaction. [*Bromohydrins can be formed similarly using Br_2/H_2O.*] Because the chloronium ion can form on either side of the alkene with equal rates (enantiomeric transition states), the reaction results in a racemic mixture. The chloronium ion is attacked at the most substituted carbon by H_2O from the opposite side, ensuring the observed anti addition.

Assessment 11.10 Predict the product of the following haloalkane syntheses.

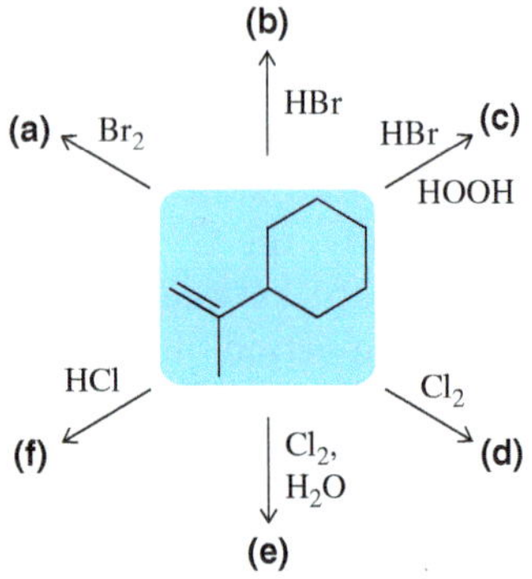

Assessment 11.11 Specify the conditions that would allow the synthesis of the 1° and 3° bromoalkanes from the same starting alkene.

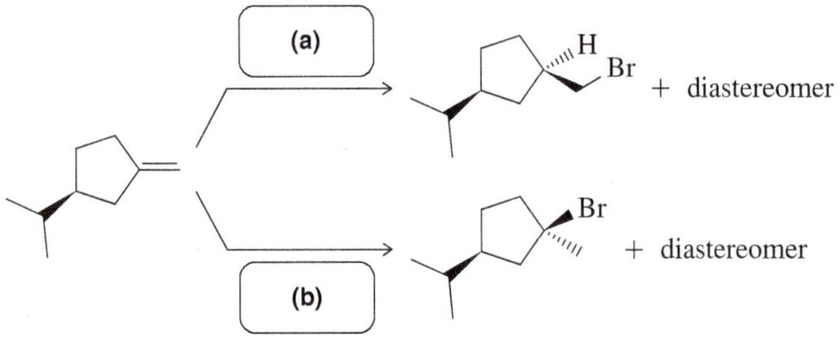

Assessment 11.12 Halohydrin formation is a stereospecific reaction. Identify the products of halohydrin formation of the following diastereomeric alkenes to demonstrate this stereospecificity.

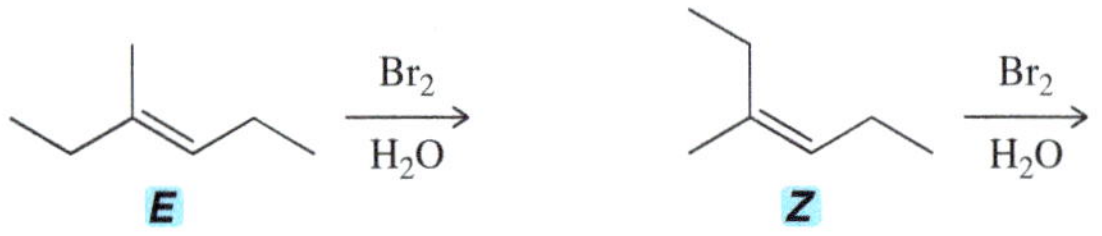

Assessment 11.13 Provide a mechanism for the following hydrohalogenation reaction.

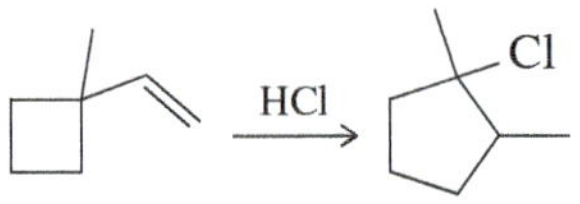

11.5.1 Radical halogenation of alkanes: a detailed review

Radical halogenation was previously covered in Sections 5.3 and 5.4, where the mechanism, mathematics, and energetics of the reaction were emphasized. Although the same topics are discussed here from a more qualitative (less mathematical) perspective, the study of haloalkane synthesis requires a review of these concepts because to determine the synthetic utility of a reaction, you must keep in mind how the reactions proceed. [*For details, including selectivity and the Hammond postulate, you may need to return to Section 5.3.*]

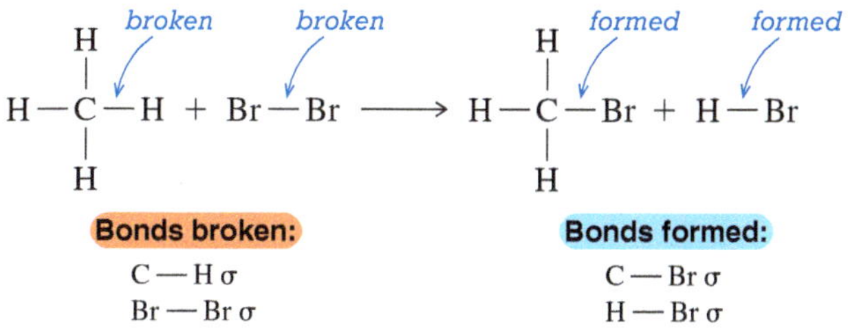

Figure 11.12 The analysis of alkane halogenation begins by identifying the bonds broken and the bonds formed.

11.5.1.1 General mechanism of radical reactions

Starting with the bromination of methane, a general mechanism for the radical halogenation of alkanes emerges. [*This mechanism is applicable regardless of whether chlorine or bromine is being incorporated.*] The mechanistic analysis begins with identifying the bonds formed and broken in the reaction. Based on this, the arrow-pushing mechanism needs to include breaking a C—H and a Br—Br bond while forming a C—Br and a H—Br bond (**Figure 11.12**). Bromination of methane takes place only at elevated temperatures or in the presence of high-energy light. These conditions produce enough energy to cleave Br_2 into two bromine radicals, initiating a **chain reaction.**

These are referred to as chain reactions because once a radical species is generated, it is continually consumed and regenerated as the reaction proceeds. In every radical reaction, there are three processes that can occur: **initiation**, **propagation**, and **termination.** [*Be sure to identify these three processes as we look at three radical reactions used to make haloalkanes, one new and two seen previously.*] **Table 11.12** describes the initiation and propagation steps of radical halogenation.

Table 11.12 Initiation and propagation steps of radical halogenation.

Step	Description	Mechanism
Initiation	Cleavage of Br_2 produces two Br radicals.	$Br{-}Br \xrightarrow{\Delta \text{ or } h\nu} 2\,Br\cdot$
Propagation step 1	The unstable Br• abstracts a hydrogen from methane to give a methyl radical.	$Br\cdot + H{-}CH_3 \longrightarrow H{-}Br + \cdot CH_3$
Propagation step 2	The methyl radical reacts with bromine to form a new C—Br bond and produce another Br•. The newly formed Br• now participates in propagation step 1, allowing the chain reaction to continue until the methane and/or Br_2 is depleted.	$Br{-}Br + \cdot CH_3 \longrightarrow H_3C{-}Br + Br\cdot$

The sum of the two propagation steps represents the entire reaction (**Figure 11.13**). [*This is true because the propagating radicals cancel out.*] Thus, when analyzing the regio- and stereoselectivity of this reaction, only the propagation steps are of consequence.

Propagation 1: $\cancel{Br}\cdot + H{-}CH_3 \longrightarrow H{-}Br + \cdot\cancel{CH_3}$

Propagation 2: $\cancel{H_3C}\cdot + Br{-}Br \longrightarrow H_3C{-}Br + \cancel{Br}\cdot$

Total: $H{-}CH_3 + Br{-}Br \longrightarrow H_3C{-}Br + H{-}Br$

Figure 11.13 The sum of the two propagation steps is equal to the overall reaction.

The radical concentration is usually kept low in these reactions, making collisions between them rare. When two radical species do collide, a reaction occurs, producing a new covalent bond. Following any of the three possible reactions described in **Table 11.13**, no more radicals are available to propagate through the chain reaction. As a result, these reactions end the chain reaction and are therefore referred to as termination steps. [*These are not productive pathways. To ensure a continuous chain reaction, it is important, then, to keep the radical concentration low.*]

Table 11.13 Possible termination steps of radical halogenation.

Step	Description	Mechanism
Possible termination 1	The combination of two Br• generates Br_2.	$Br\cdot \; \cdot Br \longrightarrow Br_2$
Possible termination 2	A Br• combines with a methyl radical to produce bromomethane.	$Br\cdot \; \cdot CH_3 \longrightarrow CH_3Br$
Possible termination 3	Two methyl radicals combine to form a new C—C bond.	$H_3C\cdot \; \cdot CH_3 \longrightarrow CH_3CH_3$

Assessment 11.14 Provide a mechanism for the chlorination of cyclohexane. Be sure to include initiation, propagation, and three possible termination steps.

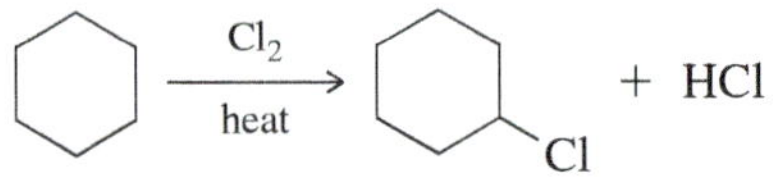

Assessment 11.15 A radical reaction, as it progressed through its propagation steps, involved the following radical species. Suggest which products might form in all possible termination steps (six products are possible).

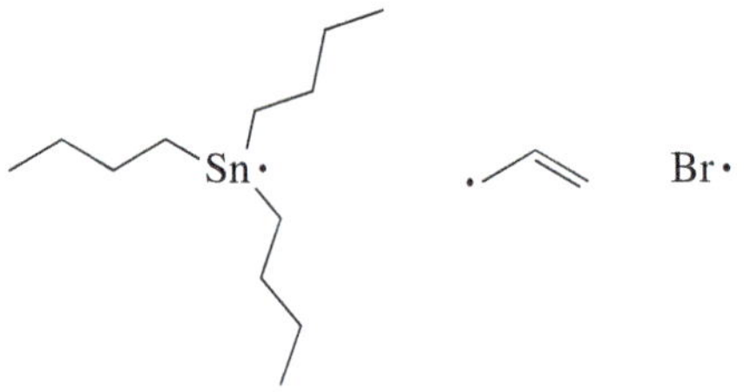

11.5.1.2 Chlorination versus bromination

If a reaction is to be synthetically useful, high yields and selectivity must be possible. Though radical monohalogenation is only selective under stoichiometrically controlled conditions, for reasons discussed in great detail in Section 5.3, bromination is more selective than chlorination. Chlorination of propane results in a 60:40 mixture, whereas bromination gives a synthetically useful 97:3 ratio of products, favoring the more substituted haloalkane (**Figure 11.14**).

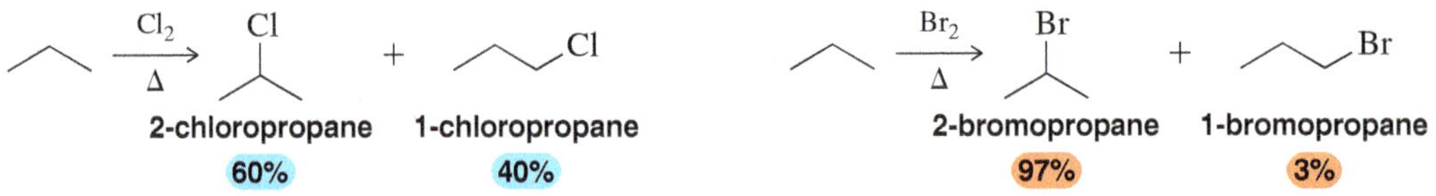

Figure 11.14 Radical bromination has greater synthetic utility than radical chlorination.

The origin of the regioselectivity is in the rate-determining first step. Given the choice, a radical will abstract the hydrogen atom that gives the most stable alkyl radical. As electron-deficient species, radicals are stabilized by hyperconjugation, making the 2° radical more stable than the 1° radical (**Figure 11.15**). This is true regardless of which radical (Br• or Cl•) is doing the hydrogen abstraction.

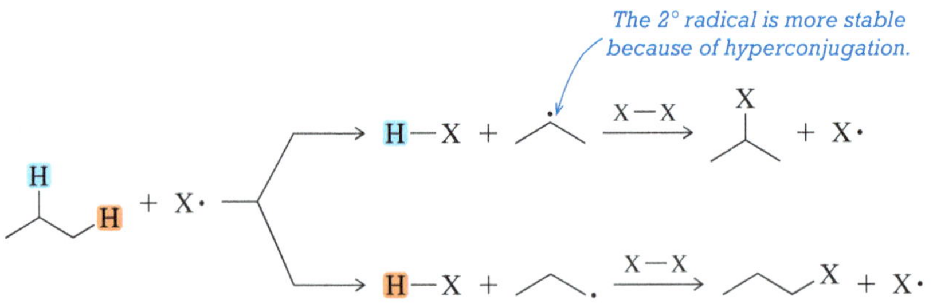

Figure 11.15 Abstraction of a hydrogen by a halogen radical preferentially occurs to give the more stable radical. [*The degree to which it is preferential depends on which halogen is used.*]

Chapter 5 showed that bond-dissociation energies (BDEs) can be used to assess the stability of the resulting radical. Less energy is required to break bonds that form more stable radicals (**Figure 11.16**). This is observed for 1° and 2° C—H bonds. The 2° C—H is more easily broken, requiring only 95 kcal/mol, compared to 98 kcal/mol for the 1° C—H bond. Once again, the energy necessary to break a bond should be independent of which radical (Br• or Cl•) is abstracting the hydrogen. [*And yet bromination is more selective than chlorination. Take a moment to think about why this is.*]

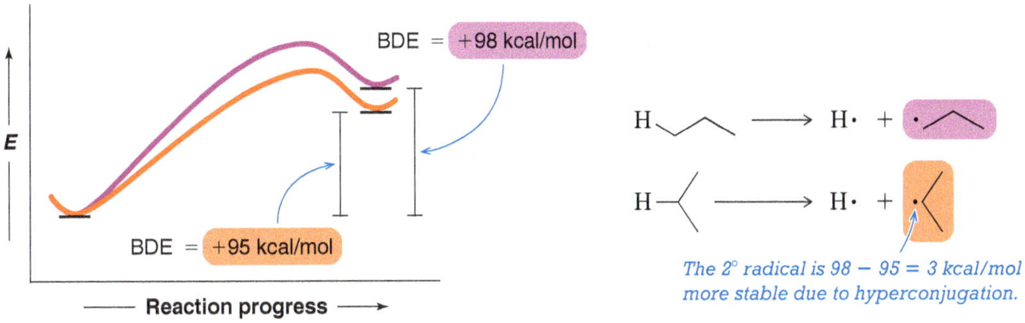

Figure 11.16 **Breaking the weaker bond gives the more stable radical.**

Assessment 11.16 Based on the stability of the radicals produced, predict which bond in each pair would have the higher bond-dissociation energy.

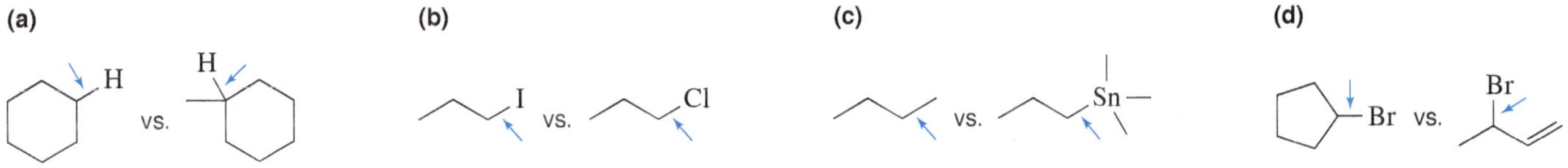

Assessment 11.17 The most stable intermediate forms first. Explain this statement by showing a reaction coordinate diagram for the formation of a 3° carbocation over a 2° carbocation in the following alkene addition reaction.

The Hammond postulate explains the difference in regioselectivity for bromination versus chlorination. It states that an exothermic reaction has a transition state similar in structure to the reactants, whereas an endothermic reaction has a transition state similar in structure to the products. [*We have seen this a few times now; if you didn't understand it the first couple of times, stick with it.*] Having limited our discussion to the selectivity-determining first propagation step, the difference in selectivity is due to bromine, a larger atom, being a more stable radical than chlorine. With better delocalization of its radical character, bromine is less reactive/more selective in choosing which hydrogen to remove. Examining ΔH of the potential propagation steps (to make 1° and 2° carbon radicals) for bromination and chlorination should clarify the differences in transition state structures that result in different levels of selectivity (**Figure 11.17**).

Selectivity is low in chlorination because . . .

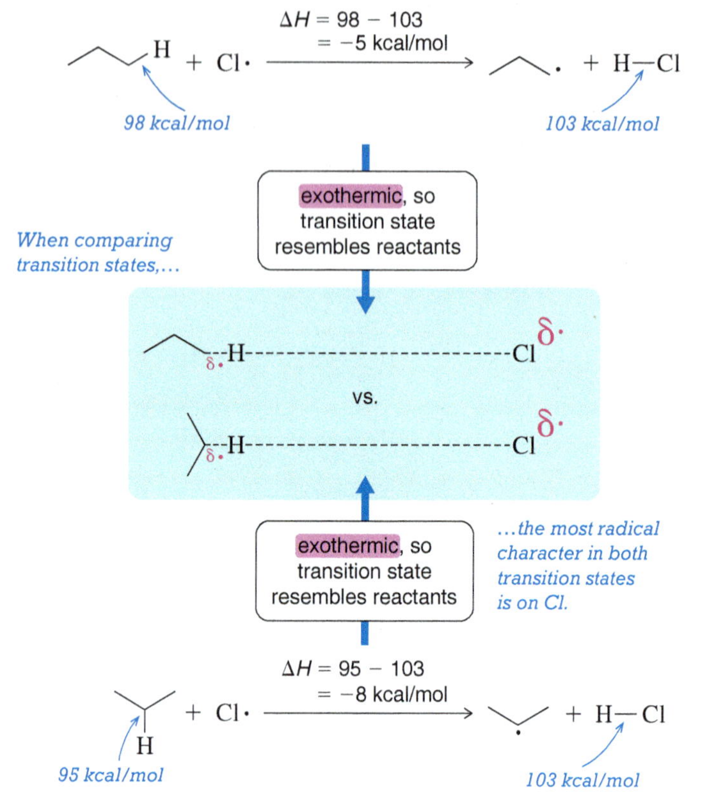

The major difference in the stability of the transition states is a result of the difference in stability of the radicals present. The majority of radical character is on the chlorine atom in both, so these two transition states are nearly identical in energy. As a result, the activation energy for forming each is nearly identical, and the rate of formation of each is nearly identical. As indicated in the reaction coordinate diagram, a less selective reaction is expected.

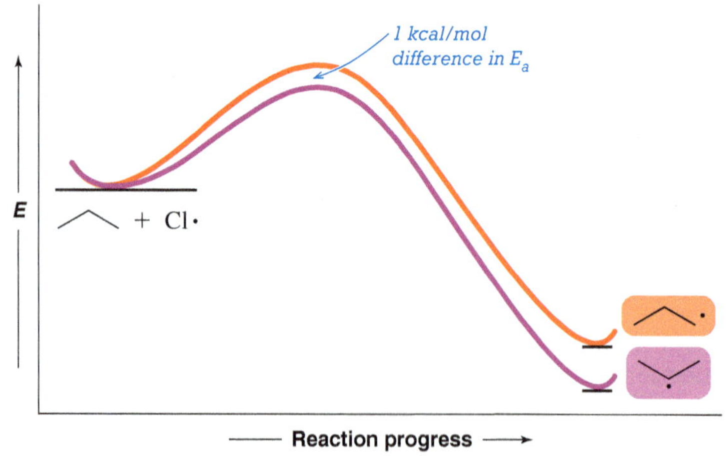

Selectivity is high in bromination because . . .

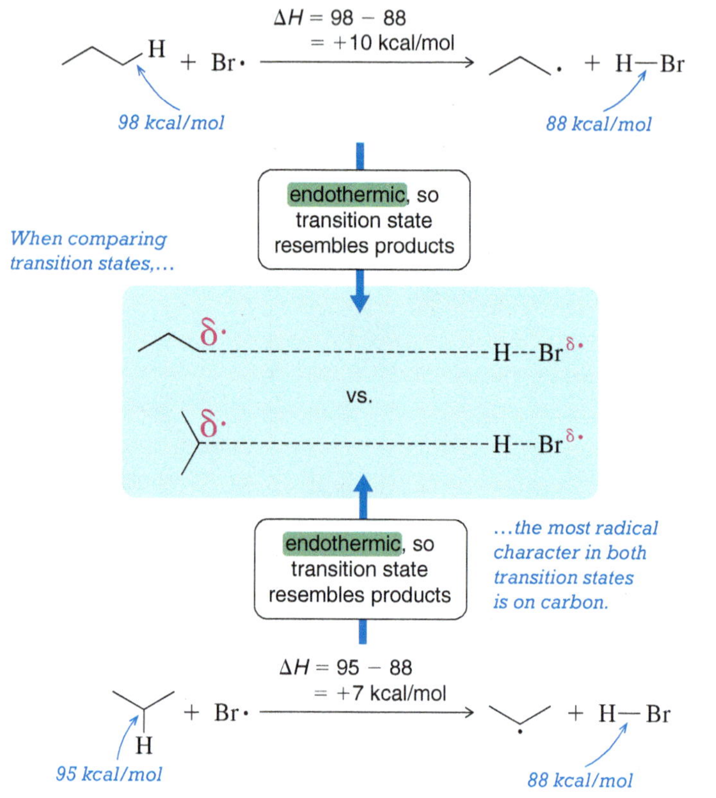

The major difference in the stability of the transition states is a result of the difference in stability of the radicals present. The majority of radical character is on carbon in both, so these two transition states have very different energies, with the 2° radical being more stable than the 1° radical. As a result, the activation energy is lower and the rate of formation faster for forming the more stable 2° radical. As indicated in the reaction coordinate diagram, a more selective reaction is expected.

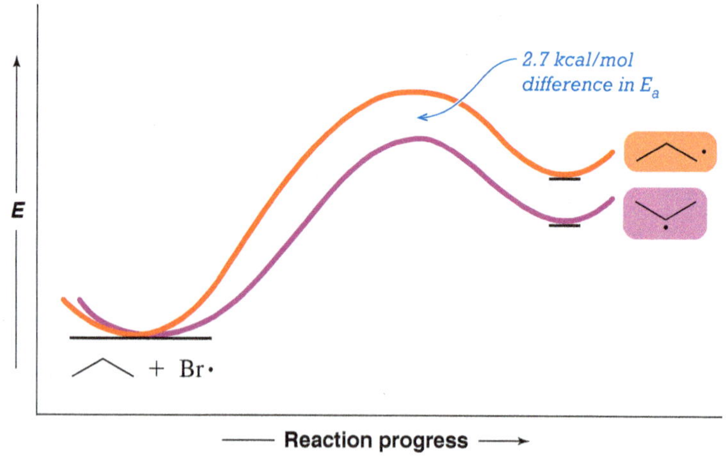

Figure 11.17 Formation of the more stable of two intermediates is only significantly faster when the reaction is endothermic and the transition states are product-like. [*This is not an easy figure. Read it multiple times if necessary.*].

Assessment 11.18 Using the bond-dissociation energies in Table 5.6, (a) predict whether or not a fluorine radical would be selective for forming a single radical from propane. (b) Will the transition state be reactant-like or product-like? Explain your answer.

Assessment 11.19 Using the bond-dissociation energies in Table 5.6, (a) predict whether or not an iodine radical would be selective for forming a single radical propane. (b) Will the transition state be reactant-like or product-like? Explain your answer.

11.5.1.3 Stereochemistry of halogenation

In the bromination of 3-methylhexane, hydrogen atom abstraction occurs at the most substituted tertiary carbon (**Figure 11.18**). In the second propagation step, the alkyl radical (produced in the first propagation step) attacks Br_2 to generate the new C—Br bond along with regenerating another Br• (to be used in the first propagation step). The sp^2-hybridized radical is planar, so it is attacked by bromine to form the new C—Br bond on either side (front or back) of the molecule. If the carbon radical is unsymmetrical, then attack on either side results in a racemic mixture, for reasons that have been discussed in great detail in Sections 6.6, 8.3.1, and 9.3, and as summarized in Figure 11.18. Approach of bromine from the front results in formation of (*R*)-3-bromo-3-methylhexane, whereas approach from the back generates (*S*)-3-bromo-3-methylhexane.

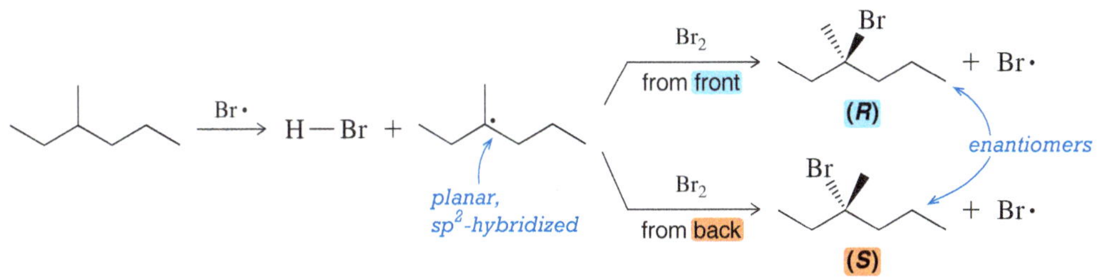

Figure 11.18 Attack from either side in the second propagation step results in a racemic mixture of haloalkanes.

The transition states leading to (*R*)-3-bromo-3-methylhexane and (*S*)-3-bromo-3-methylhexane are nonsuperimposable mirror images (**Figure 11.19**); these are enantiomeric transition states. Enantiomeric transition states are equal in energy, so they form at equal rates. In this way, halogenation of any achiral alkane (like 3-methylhexane) generates a racemic mixture of products when a new stereocenter is produced in the reaction.

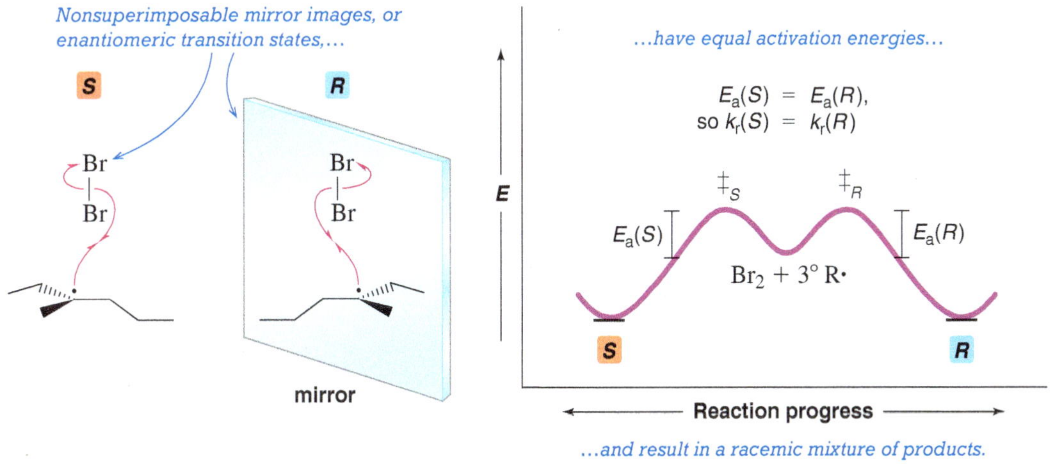

Figure 11.19 Enantiomeric transition states result in the formation of a racemic mixture of bromoalkanes.

Assessment 11.20 When (1*R*,3*S*)-1-*tert*-butyl-1,3-dimethylcyclopentane is halogenated, one stereoisomer is produced in excess. (a) Predict the identity of the major stereoisomer and (b) explain why this reaction did not produce an equal mixture of stereoisomers.

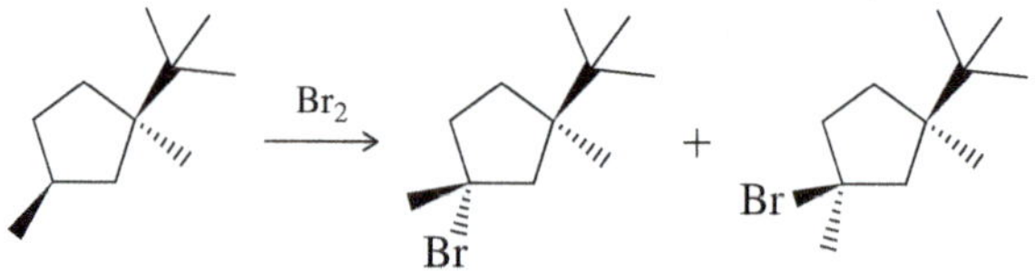

11.5.1.4 Synthetic limitations

In addition to producing racemic mixtures, the utility of alkane halogenation is also limited by the regioselectivity of the reaction. That is, a reaction resulting in the formation of a mixture of products is inefficient and environmentally costly, generates waste, and requires expensive purification procedures. Because the location of the new C—X bond is determined by formation of the most stable radical in the first propagation step, it is important that the molecule to be halogenated has one of two major characteristics that allow regioselectivity in radical formation.

1. ***Substrates with one type of hydrogen:*** Selective halogenation occurs most efficiently for substrates with one type of hydrogen. For ethane and cyclohexane, using chlorination or bromination results in a high yield of a single product, as long as a single equivalent of the halogen (X_2) is used (**Figure 11.20**).

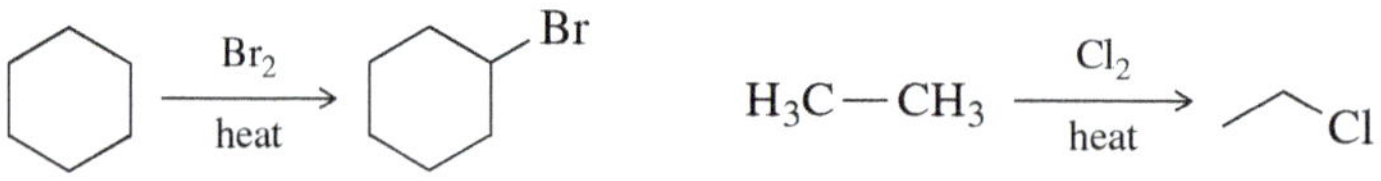

Figure 11.20 Substrates with a single hydrogen type give high yields of a single halogenation product.

2. ***Substrates with only one "most substituted" carbon:*** The second type of alkane that gives selective halogenation are those with a single carbon that is more substituted than all others. Halogenation of this type requires bromination because chlorination is not selective enough to produce a single product (**Figure 11.21**).

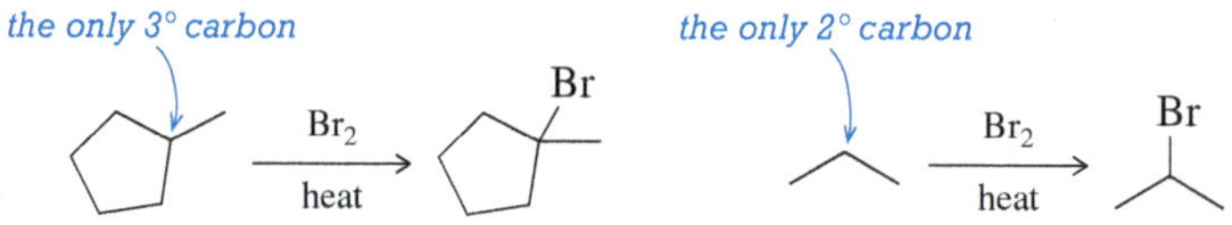

Figure 11.21 Substrates with a single "most substituted" carbon give high yields of a single bromination product.

For alkanes not falling in one of these categories, selective halogenation is difficult. It does not mean that the reaction cannot be done. But there will be a complex mixture of products, requiring complicated purification steps (**Figure 11.22**).

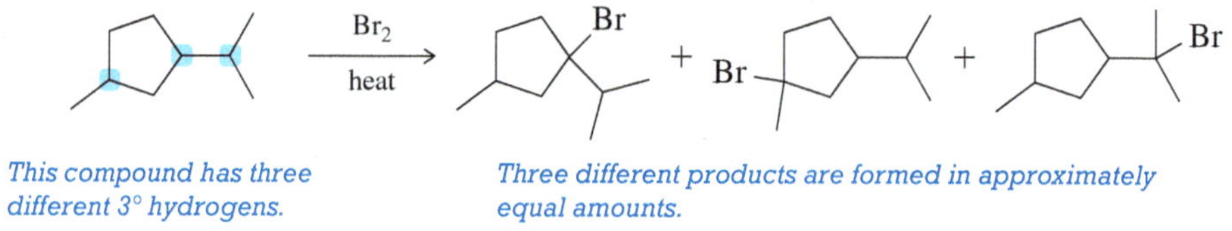

Figure 11.22 Halogenation is not selective when there are multiple "most substituted" carbons.

Assessment 11.21 Suppose a 2-halobutane was needed for a synthetic sequence. Starting your synthesis with butane, would it be best to put a chlorine or bromine at that position? Explain.

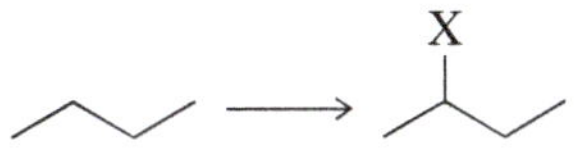

Assessment 11.22 Predict the major products of the following alkane halogenation reactions. [*The number of products shown ignores the formation of racemic mixtures.*]

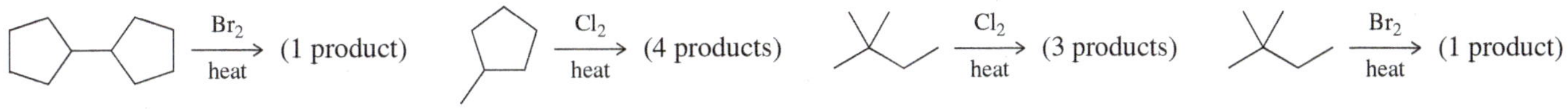

Assessment 11.23 Can you make a 1° bromoalkane like (3-bromopropyl)cyclopentane using alkane halogenation? Why or why not?

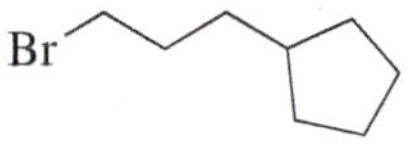

(3-bromopropyl)cyclopentane

By working Assessment 11.23 together, we can begin thinking about how we might form a 1° bromoalkane.

Solved Assessment 11.23

As a means of answering Assessment 11.23, consider each step of the halogenation and see what would be necessary to generate a 1° alkyl halide.

After forming in the initiation step, Br• breaks the weakest C—H bond to form the most stable radical. Using the bond-dissociation energies in Table 5.6, we can specify the BDE for each C—H bond that might be broken [Figure 11.23(a)]. From this, we see that the 3° C—H bond is the weakest and, upon breaking, gives the most stable radical. As a result, if given a choice, Br• will always abstract the 3° hydrogen atom [Figure 11.23(b)].

Is it ever possible to make a 1° bromoalkane using alkane halogenation? Not often, but it can be done. [*Any ideas? Write down your guesses before reading on.*] The only situation where it is possible is when there are no other C—H bonds in the molecule besides the one you are trying to replace with a bromine. An example, shown in Figure 11.24, is 2,2,3,3-tetramethylbutane, which can be halogenated under these conditions to produce a 1° haloalkane. ■

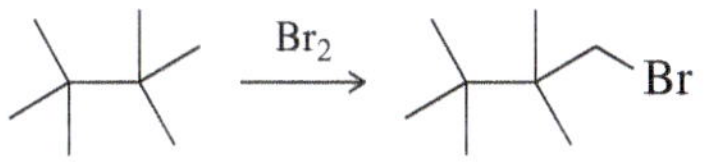

Figure 11.24 Formation of a 1° haloalkane is only possible when there is only one type of C—H bond in the molecule and it is 1°.

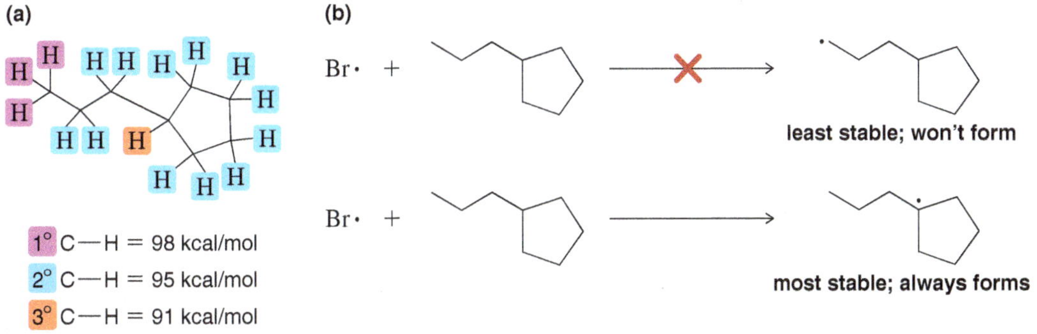

Figure 11.23 A 1° radical doesn't form if there are other options. (a) The weakest bond breaks first to form (b) the most stable radical.

11.5.1 SUMMARY

- Radical halogenation is a mildly effective synthetic reaction. Bromination is more selective than chlorination because the first step is endothermic (Hammond postulate). It is most efficient when the alkane to be halogenated has only one type of hydrogen because of symmetry (with chlorination or bromination) or has one type of hydrogen that gives a significantly more stable radical (for bromination only). When a new stereocenter is formed, a racemic mixture is produced.

11.5.2 Radical addition to alkenes

A major topic of Chapter 10 was synthesis, the building of larger molecules from smaller ones while converting one functional group to another. Your reaction sheets from that chapter provide a way to organize your tools for synthesis. Radical halogenation is one of these tools [*put it on your sheets under "Preparation of haloalkanes/Reactions of alkanes"*]. Solved Assessment 11.23 reveals, however, that only in very rare circumstances can 1° haloalkanes be synthesized in this way. And yet among haloalkanes, 1° haloalkanes are the most useful synthetically because of the many reactions they enable [*as we will see in Chapter 12*]. As such, having a reliable reaction [*tool*] for synthesizing 1° alkyl halides is essential for organic chemists.

Recall from Chapter 8 that 1° bromoalkanes can be made by the addition of HBr to terminal alkenes in the presence of peroxides (radical hydrohalogenation). [*This section is a review of a reaction we studied previously, set in the context of synthesis. For more detailed mechanistic analysis, see Section 8.3.4.*] This reaction, shown in **Figure 11.25**, is subject to the **peroxide effect** and results in the preparation of the less substituted haloalkane. The reaction begins in the typical way, with high-energy light initiating the propagating radical. [*Now you know why the hydrogen peroxide in your medicine cabinet is in a brown bottle.*]

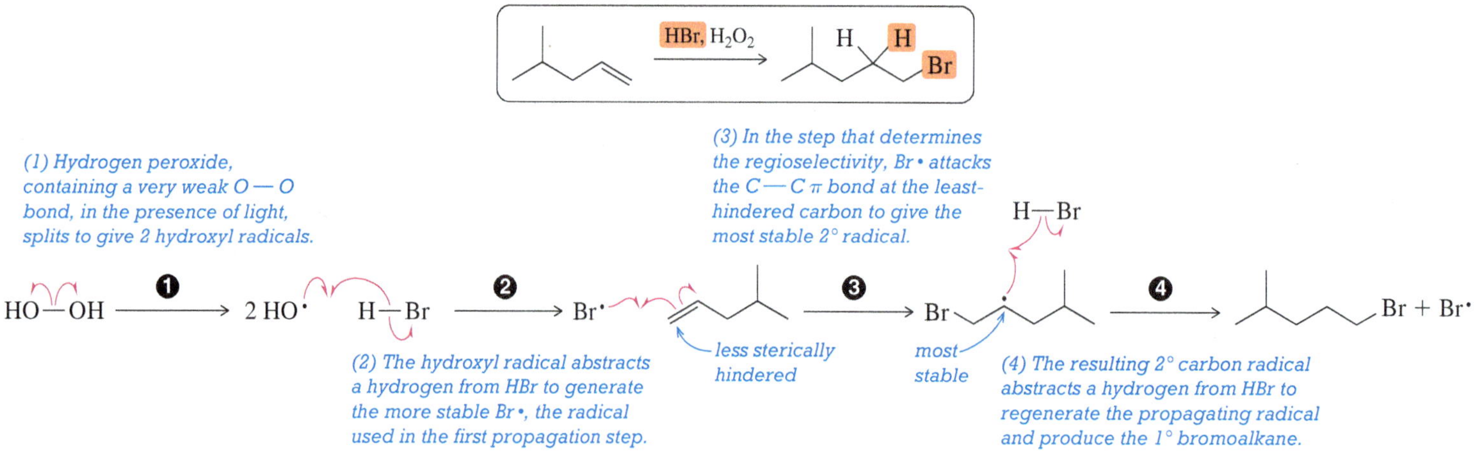

Figure 11.25 **In the presence of peroxides, the addition of HBr to alkenes gives 1° bromoalkanes.**

Assessment 11.24 Why is a 2° carbon radical more stable than a 1° carbon radical?

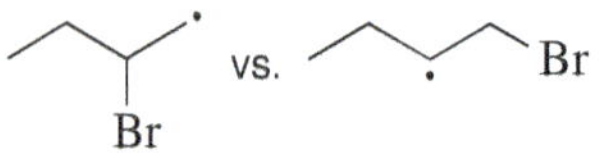

Assessment 11.25 Predict the major products of the following alkene halogenation reactions. [*D is the symbol for deuterium, an isotope of hydrogen.*]

(a)

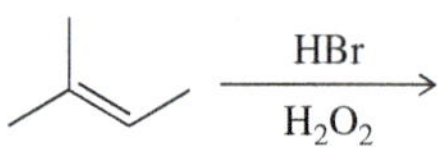

(b)

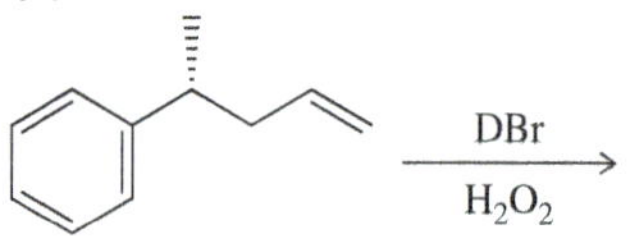

(c)

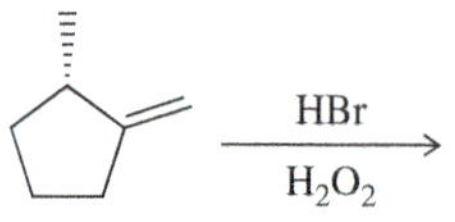

(d)

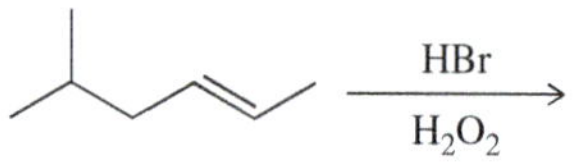

Assessment 11.26 HBr and peroxides are also used to generate 1-bromo-1-alkenes. Suggest a starting reactant that would successfully undergo this reaction.

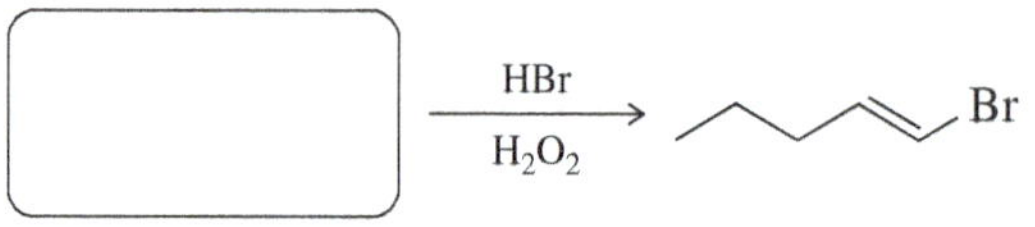

This section discusses a new reagent for a seemingly new reaction. However, it is closely related to reactions just studied. To see how, let's examine a scenario in which a chemist conducts a bromination reaction, expecting one product but getting others.

11.5.3 Allylic halogenation

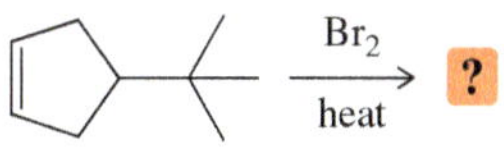

What might happen if radical halogenation were conducted in the presence of other functional groups, such as an alkene? With motivation and access to chemicals, you could head to the lab and find out. So let's do just that. [*To the lab!*] But, before we do, predict the product you would expect to form under these conditions, starting with 4-*tert*-butylcyclopent-1-ene. [*Really! Think about this before proceeding.*]

Based on Section 11.5.1, incorporation of the bromine would be expected at the most substituted carbon (product **A**, **Figure 11.26**) through formation of the 3° radical in the first propagation step. Some **A** is obtained, but significantly more **B** and **C** form. [*There is also a fourth product, to be discussed later. Can you justify the formation of* **B** *and* **C**?]

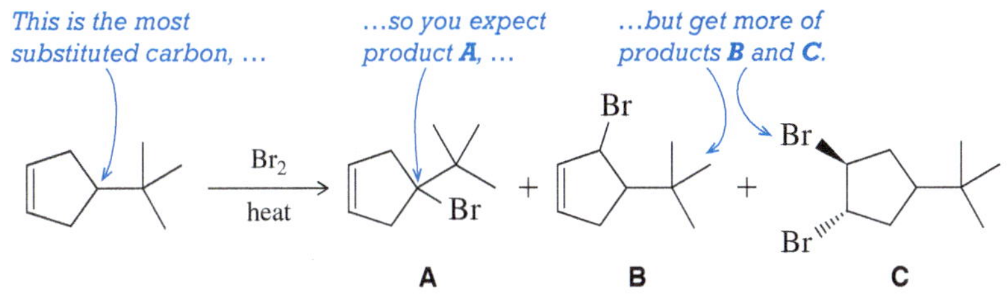

Figure 11.26 Attempted radical bromination produces three products. [*Product C will be a mixture of stereoisomers but that is unimportant to this discussion.*]

The formation of the trans-dibromide (product **C**, **Figure 11.27**) is a good place to start. Looking closely, this is a reaction seen previously. The nucleophilic alkene attacks Br_2, creating a bromonium ion that is opened by a bromide anion. [*For a review, see Section 9.1.2.*] Thus, one complication caused by an alkene is that electrophilic addition of Br_2 becomes a favorable side reaction.

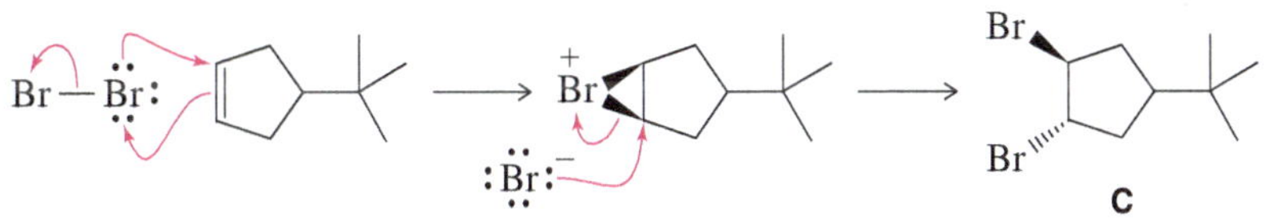

Figure 11.27 Formation of the trans-dibromide (product C) occurs through the halogenation mechanism from Section 9.1.2.

Formation of the 2° bromide (product **B**, **Figure 11.28**) appears to result from substitution of a bromine atom for a hydrogen atom. This is the kind of product you'd expect from a radical reaction. But why this particular radical, given that it is less substituted than the expected 3° radical? Comparing the two, why might the 2° radical form instead of the 3° radical? [*Think about the ways intermediates are stabilized/destabilized, beyond just hyperconjugation. Whenever you consider hyperconjugation, you should always consider how* ***induction*** *or* ***resonance*** *might also play a role.*]

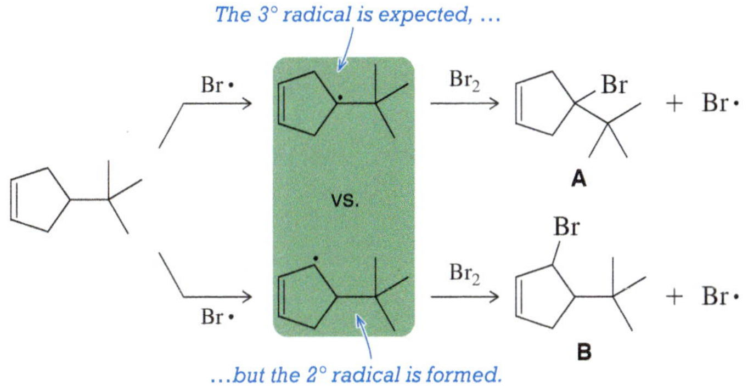

Figure 11.28 Contrary to expectation, the 2° radical forms preferentially over the 3° radical.

The 3° radical (which leads to **A**) is stabilized only by hyperconjugation, whereas the alkene next to the 2° radical (which leads to **B**) allows for an additional resonance structure. The delocalized radical is more stable, forming more quickly in the first propagation step [**Figure 11.29(a)**]. From Section 5.3, the relative bond-dissociation energies (BDEs) of similar bonds are indicative of the relative stability of the resulting radicals. Is this true here? Indeed, it is, with the C—H bond next to the alkene having a BDE of 87 kcal/mol, while the 3° C—H bond has a BDE of 91 kcal/mol [**Figure 11.29(b)**]. Thus, of the C—H bonds in the molecule, the one next to the alkene is the weakest, breaking most quickly in the first propagation step.

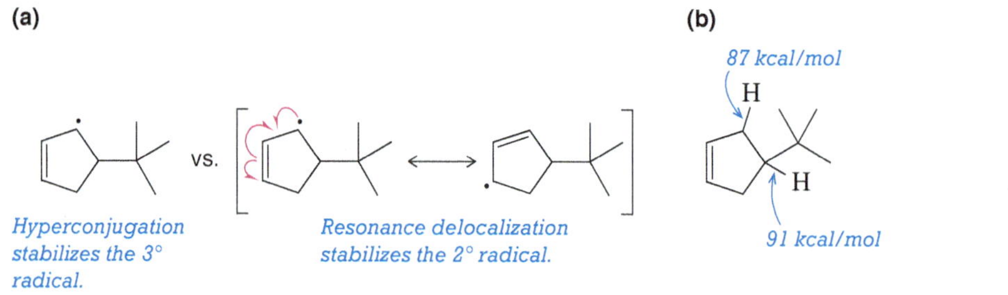

Figure 11.29 The resonance-stabilized radical is (a) more stable than a 3° radical and is formed by (b) breaking a weaker C—H bond.

A couple of important points emerge from this exercise. First, the alkene imparts different properties on the neighboring carbon and neighboring C—H bonds. For this reason, the carbon next to an alkene is referred to as the **allylic** carbon. It is named for allyl, the common name for the substituent of chemical formula $—CH_2CH{=}CH_2$. The neighboring alkene allows resonance structures to be drawn, altering the usual reactivity of allylic substituents, a concept seen in a variety of organic reactions. The substituent attached to the allylic carbon is also referred to as allylic, whether it is hydrogen, oxygen, or bromine, among others (**Figure 11.30**). [*Because of the altered reactivity, it is important that you recognize when a substituent is allylic. Assessments 11.27–11.29 should help with this.*]

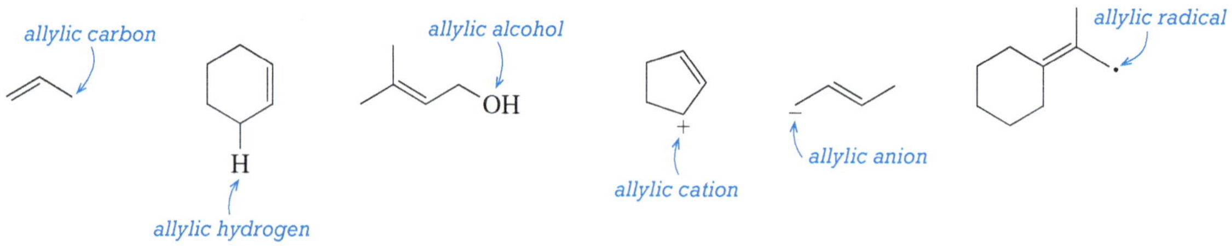

Figure 11.30 Allylic species exhibit different reactivity [*as we explain in more detail in Chapter 21*].

Assessment 11.27 Identify the allylic carbon(s) in the following molecules.

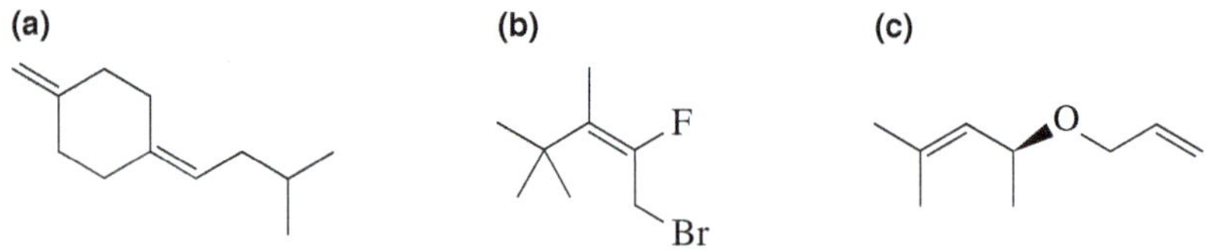

Assessment 11.28 We show in Chapter 12 that C—Br bonds can break to give a carbocation and a bromide anion. For which of the organohalides (**A** or **B**) would you expect this process to be fastest? Explain.

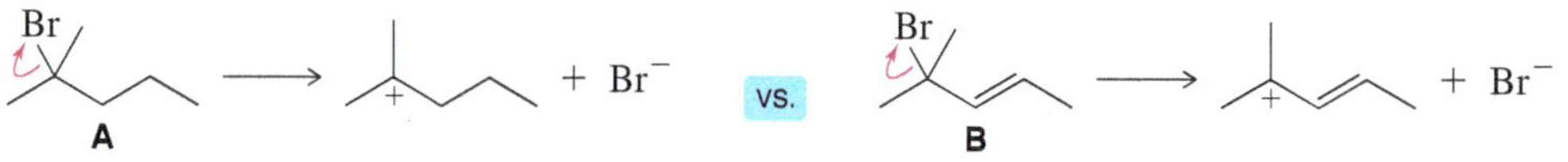

Assessment 11.29 In the following reaction, which C—H bond would be most likely to react with a bromine radical?

The second outcome of this exercise is the "discovery" that the allylic carbon can be functionalized through radical halogenation. In the presence of Br_2, however, electrophilic addition is a major competing reaction. [*How can formation of the trans-dibromide be avoided?*]

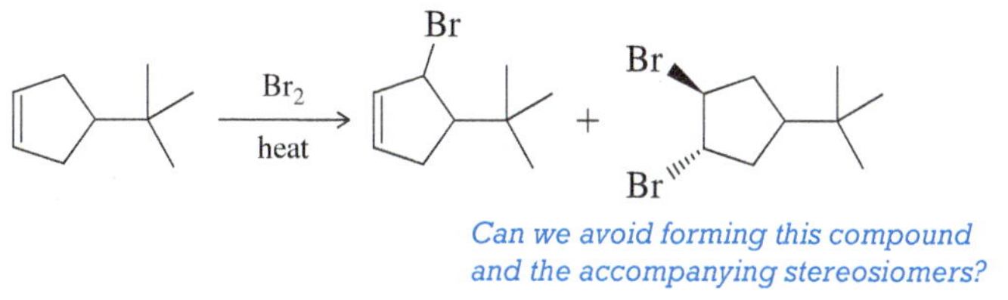

The answer is to keep the concentration of Br_2 low. This seems impossible, given the need for Br_2 in the second propagation step. But the answer lies in the molecule *N*-bromosuccinimide (NBS). The mechanism is shown in **Figure 11.31** using the simpler molecule propene.

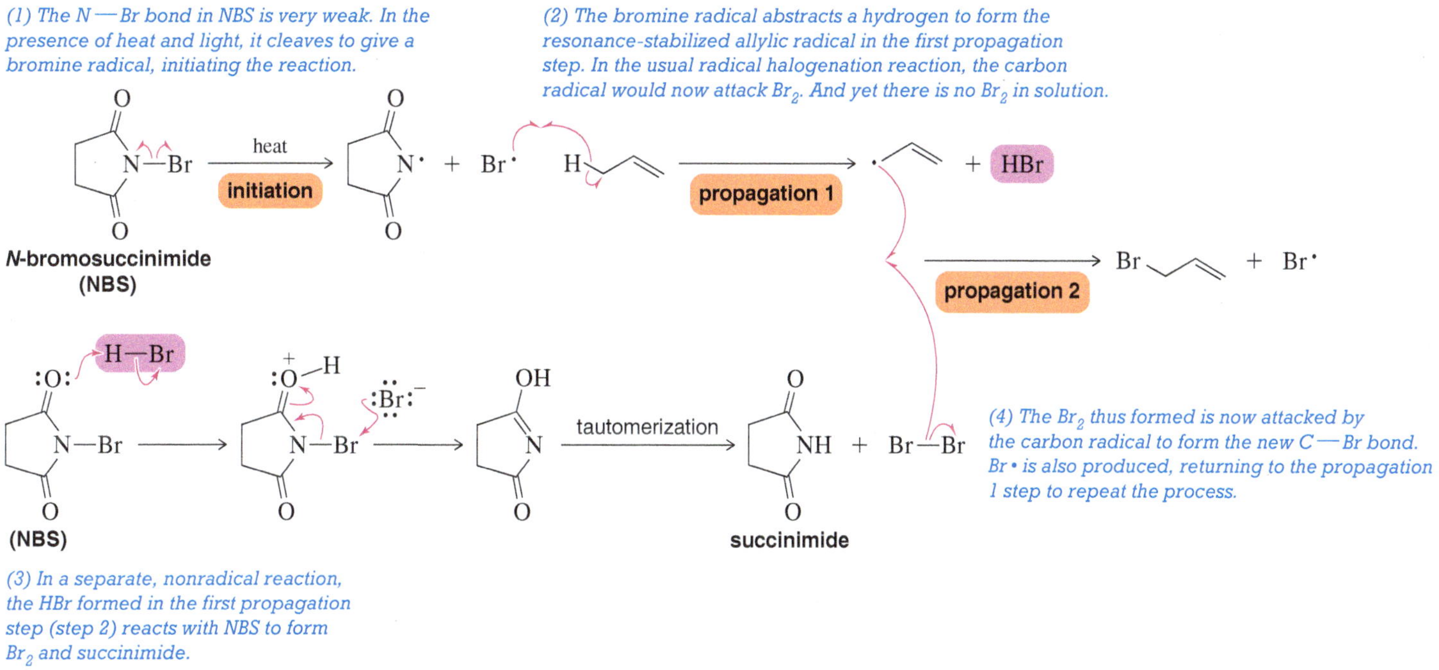

Figure 11.31 ***N*-bromosuccinimide halogenates the allylic position without competing electrophilic addition.**

Notice in Figure 11.31 that Br_2 forms only upon formation of HBr following abstraction of the allylic hydrogen in step 2 (propagation 1). With a low concentration of Br_2, competing electrophilic addition never occurs. The small amount of Br_2 present only reacts with the highly reactive allylic radical.

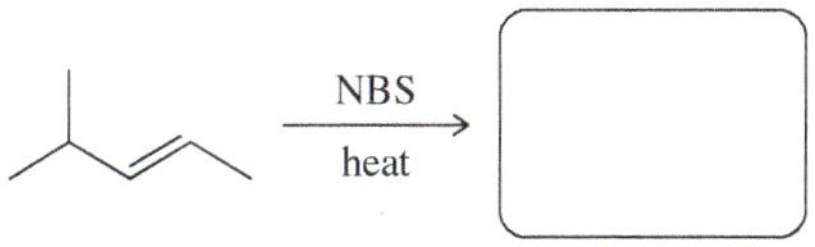

Figure 11.32 Allylic halogenation can result in multiple products. Which is favored?

11.5.3.1 Allylic halogenation: selectivity issues

Two issues arise regarding the selectivity of allylic halogenation. First, if there are multiple allylic hydrogens, which is removed? Second, because the allylic radical is a combination of two resonance structures, at which position does the new C—Br bond form? [*Think through both of these issues with regard to the reaction in* ***Figure 11.32***.]

The first question is the easier of the two and was discussed in detail earlier in this chapter. The most stable radical always forms in the first propagation step. If there are two allylic carbons, the hydrogen is abstracted from the more substituted one (**Figure 11.33**). A 3° resonance-stabilized radical is more stable than a 1° resonance-stabilized radical. [*If this doesn't make sense, reread Sections 11.5.1 and 5.3.6 where these decisions were first made.*]

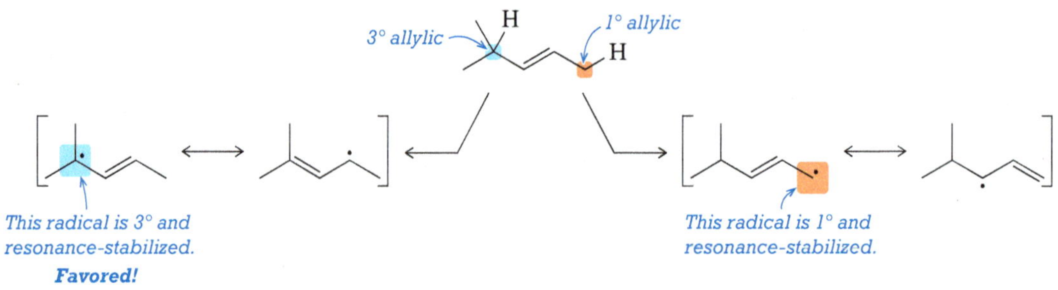

Figure 11.33 The more-substituted resonance-stabilized radical is more stable and forms preferentially.

Formation of the 3° allylic radical clarifies our next dilemma. With the two contributing resonance structures of the allylic radical, from which position will the radical attack Br_2 in the next step? The 3° resonance structure contributes a bit more to the resonance hybrid, but the new C—Br bond will form at the less substituted, 2° carbon. In addition to this being the less sterically hindered position, reaction here gives the more substituted/more stable alkene (**Figure 11.34**).

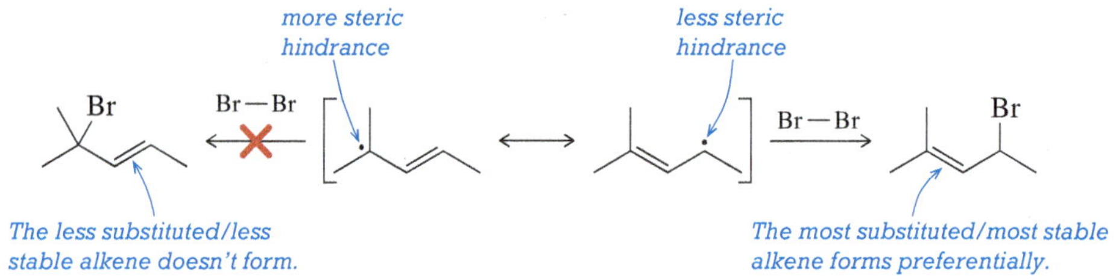

Figure 11.34 Bromine is added to the least sterically hindered carbon of an allylic radical to give the more substituted/more stable alkene.

Generalizing for allylic halogenation, the most stable allylic radical forms in the first propagation step. In the second propagation step, bromine is attacked at the least sterically hindered carbon to form the more substituted alkene as the major product.

11.5.3 SUMMARY

- Allylic halogenation is performed with NBS. The concentration of Br_2 is kept low to avoid competing electrophilic addition. Formation of the more stable radical occurs in the first step, while Br_2 is attacked in the second propagation step at the least sterically hindered carbon, leaving behind the most substituted alkene.

Assessment 11.30 In the following molecules, identify the carbon where the radical is most likely to form in the first propagation step.

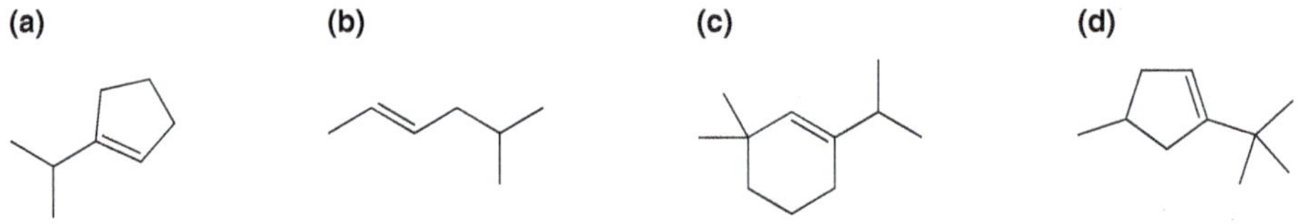

Assessment 11.31 In the following allylic radicals, identify the carbon where the new C—Br bond is most likely to form in the second propagation step.

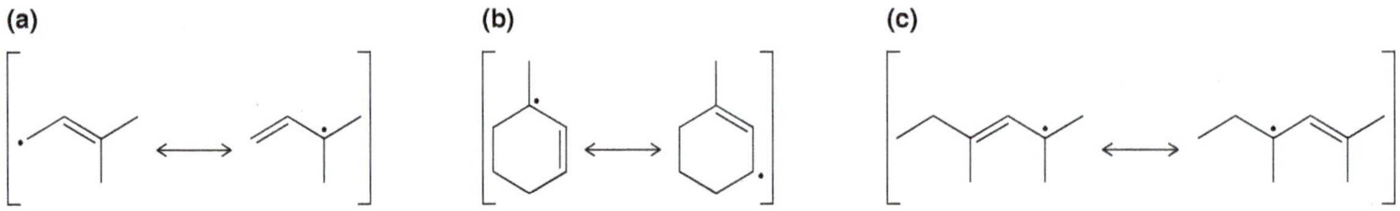

Assessment 11.32 Predict the products of the following allylic halogenation reactions.

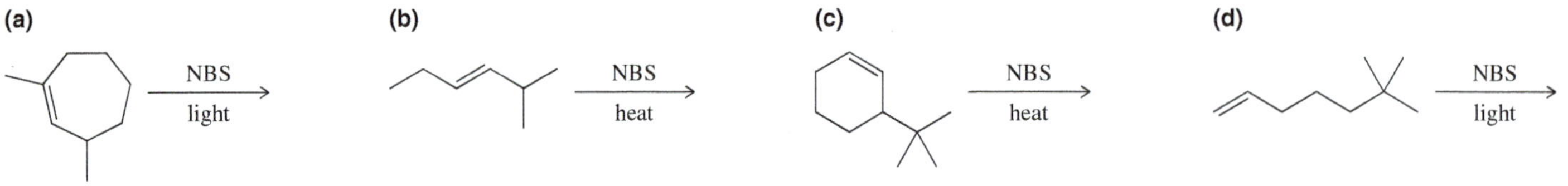

Assessment 11.33 The fact that allylic halogenation results in formation of the most stable alkene suggests that it is under thermodynamic control [*see Chapter 5 for a review.*] Thus, the second propagation step must be reversible. Suggest an arrow-pushing mechanism by which the less stable allylic halide might equilibrate to the more stable allylic halide.

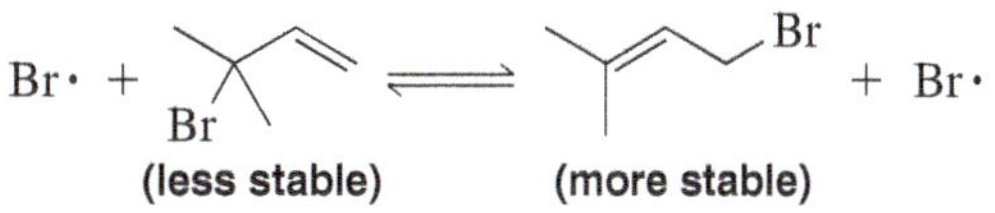

It is important to understand why the similar conditions give different products. You are already equipped with the skills to understand this. Read carefully, think deeply, frame this information in the context of previous concepts and have confidence—you'll get it.

11.5.4 Allylic halogenation versus alkene addition: a mechanistic conundrum

[*You should read this section with a pencil and paper in hand.*] Sections 11.5.2 and 11.5.3 showed two closely related reactions of alkenes proceeding under similar conditions to produce halogenated compounds. And yet, the products obtained were markedly different. In Section 11.5.2, Br• reacted to give the net *addition* of HBr to an alkene. In Section 11.5.3, Br• reacted to give the net *substitution* of H with Br at the allylic position. Taking the "just memorize the products associated with each reagent" view, similar reactions producing different products is disconcerting. One way to mitigate this discomfort is with a deep analysis of the mechanisms at work. To set the stage, let's quickly remind ourselves of the conundrum while using cyclohexene as our example (**Figure 11.35**).

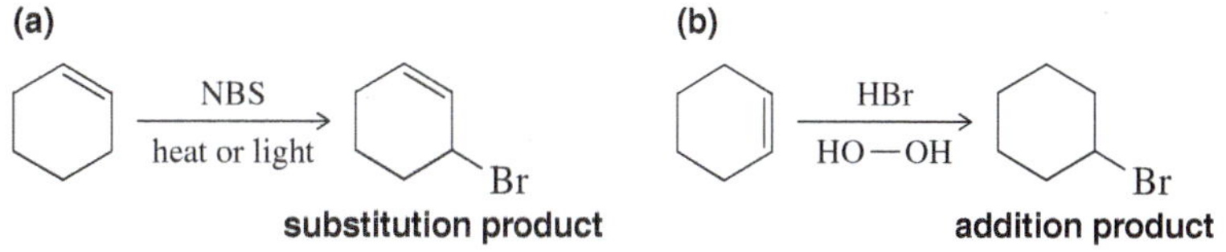

Figure 11.35 Similar conditions give different products with cyclohexene. (a) NBS does allylic halogenation. (b) HBr and peroxide add to the alkene.

Clouding the issue further is the fact that the first propagation step of each features a bromine radical reacting with an alkene in very different ways (**Figure 11.36**).

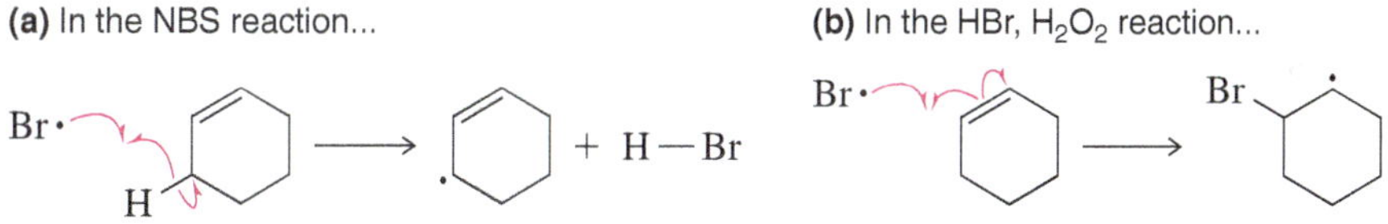

Figure 11.36 Different reactions of the bromine radical: (a) hydrogen abstraction and (b) alkene addition.

[*We sometimes speak as if the Br• thinks about what it wants to react with. Does the Br• really know that it came from NBS and therefore should abstract a hydrogen? Or if the Br• formed from HBr, then it should attack the alkene? Of course not; reactions are the result of random collisions. We have to figure out why a different random collision is favored in each case.*]

Irrespective of how the Br• formed, there are two competing fates—namely, hydrogen abstraction [reaction 1; **Figure 11.37(a)**] and alkene addition [reaction 2; **Figure 11.37(b)**]. Hydrogen abstraction (reaction 1) is slow and irreversible, whereas alkene addition (reaction 2) is fast and reversible. The left side of reaction 2 is favored by entropy, which explains the generalization that β-bromoradicals are unstable, preferentially eliminating via the reverse of addition to give an alkene. Shown on a reaction coordinate diagram [**Figure 11.37(c)**], the transition state energy for hydrogen abstraction is higher than for alkene addition, making the abstraction a slower, but still more favorable, process.

At this point, we have to consider the fate of the radicals formed in reactions 1 and 2 in Figure 11.37, specific to the conditions. That is, when an allylic radical (reaction 1) or a

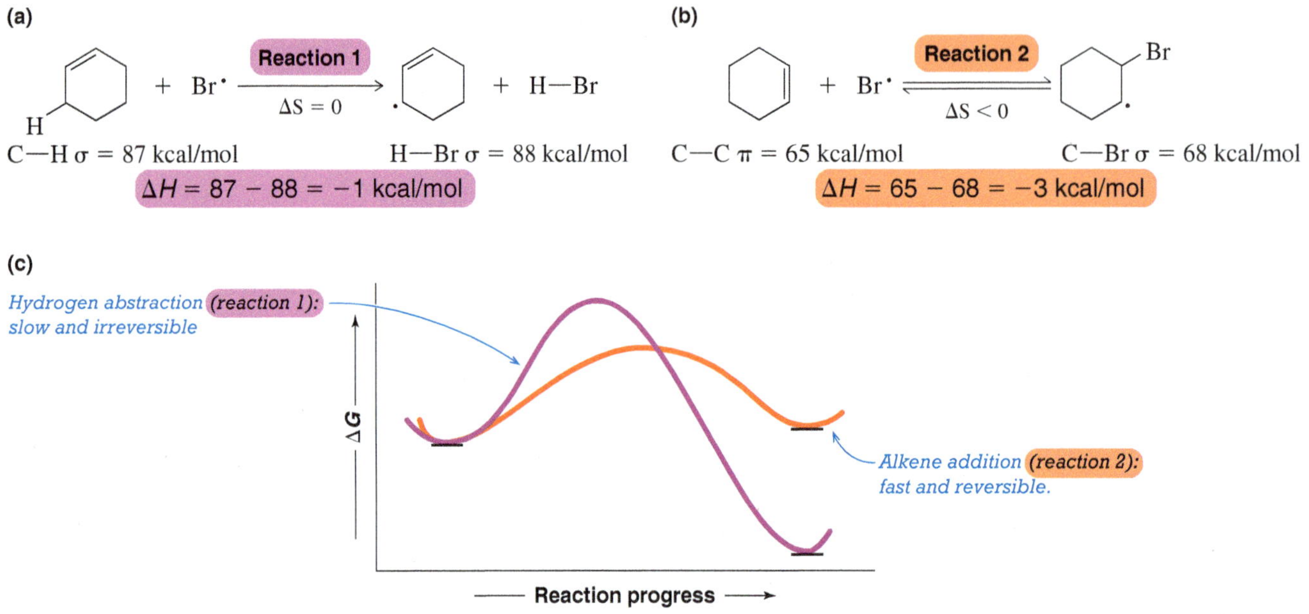

Figure 11.37 Hydrogen abstraction (reaction 1) vs. alkene addition (reaction 2). (a) Hydrogen abstraction is entropically neutral, so it is slow and irreversible. (b) Alkene addition is entropically disfavored, so it is fast and reversible. (c) A reaction coordinate diagram compares the two. [*This diagram also accounts for the approximate entropy of each reaction.*]

β-bromoradical (reaction 2) is formed, under which conditions does it react further (in the presence of NBS or HBr/H_2O_2)? When reacting further, what is/are the possible/expected product(s)? **Figures 11.38** and **11.39** give insight into these questions. [*To understand Figures 11.38 and 11.39, you need to remember what is present under the conditions of NBS. See Section 11.5.3 for a review.*]

(a) Hydrogen abstraction (reaction 1) product in the presence of NBS.

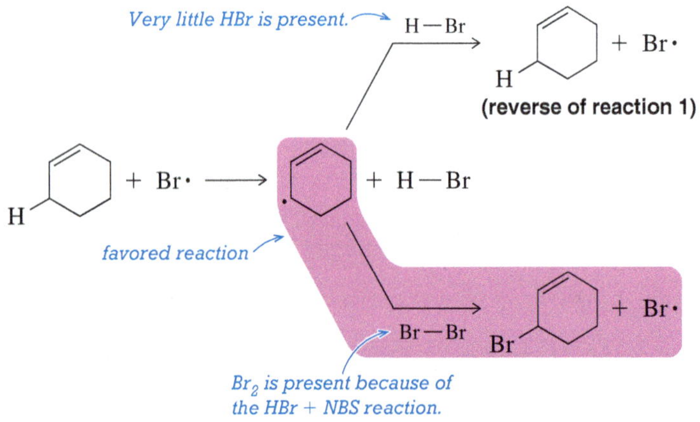

Figure 11.38 Of the two hypothetical outcomes for reaction 1, attack of the radical by Br_2 is the only one that can actually occur.

Once the allylic radical is formed, a hypothetical second reaction is to abstract a hydrogen from H—Br, the reverse of reaction 1. This reaction was described previously as being slow. [*Assessment 11.34 will help you understand why.*] The second, favored possibility is to react with the Br_2 that formed when HBr from the first step (the product of reaction 1) reacted with NBS. In sum, because HBr is consumed in the side reaction to make Br_2, there isn't much of it around, making reaction with the highly reactive Br_2 the favored process for the allylic radical.

(b) Alkene addition (reaction 2) product in the presence of NBS.

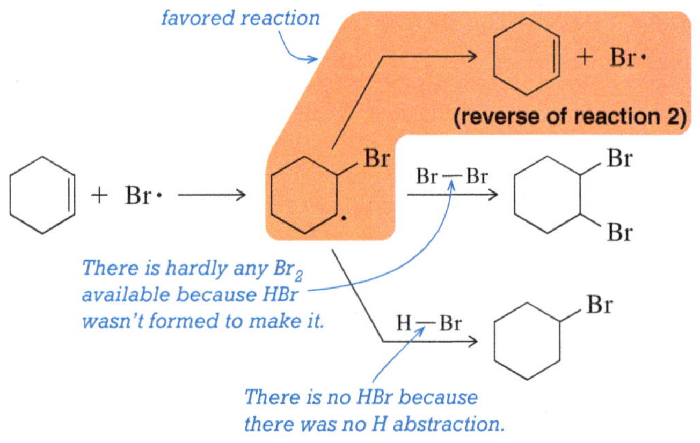

Figure 11.39 Of the three hypothetical outcomes for reaction 2, the reverse of reaction 2 (elimination of Br•) is the most likely to occur.

Occasionally, reaction 2 (addition of Br• to the alkene) will occur. What happens to this β-bromoradical in the presence of NBS? The first and most likely possibility is that it simply reverts back to the alkene, kicking out the very stable Br• in an entropically favored reaction. [*Reaction 2 is fast and reversible.*] The other two possibilities are for the β-bromoradical to react further; we have only seen radicals react with either HBr or Br_2 so far. How much Br_2 and HBr are even present? [*Think about the conditions.*] Br_2 only forms when HBr reacts with NBS. And HBr only forms when a bromine radical abstracts a hydrogen (reaction 1). So when reaction 2 occurs, there is very little Br_2 and even less HBr around to allow either of these possible reactions to occur. Left with no other options, the β-bromoradical simply reverts to the alkene.

In sum (**Figure 11.40**), reactions 1 and 2 can and do occur in the presence of NBS. Because the β-bromoradical has nothing to react with, the only possible outcome is for the allylic radical to react with bromine to produce the allylic bromide.

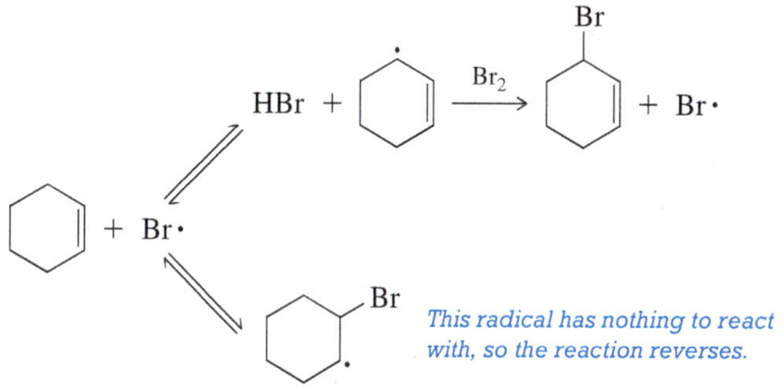

Figure 11.40 The addition of a bromine radical to an alkene reverses, leading ultimately to hydrogen abstraction.

Assessment 11.34 (a) Using bond-dissociation energies (Table 5.6), which of the indicated bonds should break most easily? (b) How does that help you explain the results shown in Figure 11.40?

H—Br vs. Br—Br

The reaction of alkenes with HBr/H_2O_2 can also be analyzed by looking at each possible outcome of reactions 1 and 2 under these conditions (**Figures 11.41** and **11.42**).

(a) Hydrogen abstraction (reaction 1) product in the presence of HBr.

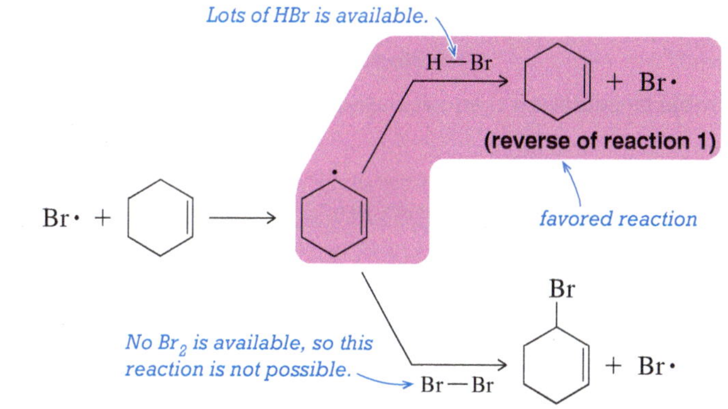

Figure 11.41 Of the two hypothetical outcomes for reaction 1, the reverse reaction is the only one that can happen.

After the allylic radical forms by hydrogen abstraction, we can consider its fate seen for reaction 1 previously. First, attacking the hydrogen of H—Br is a favorable reaction, especially with a high concentration of HBr. As the reverse of reaction 1, this simply regenerates the starting material. The second possibility is for the allylic radical to react with Br_2 to give an allylic bromide. [*Except . . . ?*] Under these conditions, there is no Br_2 around. So on the rare occasion that reaction 1 occurs, it simply reverses.

(b) Alkene addition (reaction 2) product in the presence of HBr.

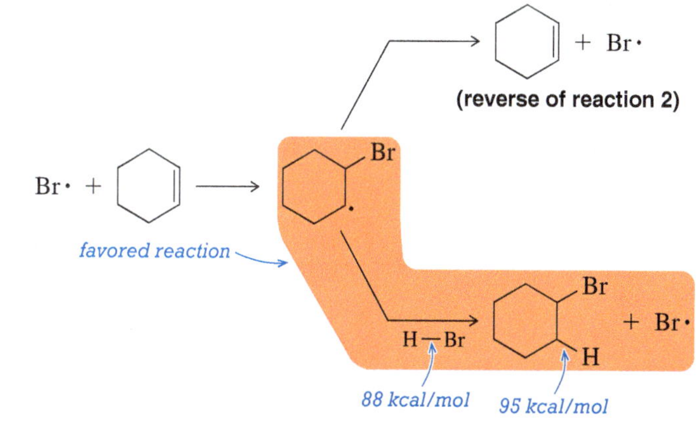

Figure 11.42 Of the two hypothetical outcomes of reaction 2, hydrogen abstraction is the only one that leads to reaction.

Addition of Br• to the alkene generates the β-bromoradical, leaving two possibilities for what comes next. First, it can reverse to the starting alkene, a reaction we know is fast and favorable. Alternatively, it can react with HBr (of which there is plenty) to give the bromoalkane. Bromoalkane formation is the favored process, forming a C—H bond (95 kcal/mol) at the expense of the H—Br bond (88 kcal/mol).

To summarize (**Figure 11.43**), while both reaction 1 and 2 can happen with HBr/H_2O_2, the allylic radical, which forms slowly anyway, has nothing to react with. Instead, the faster addition of the bromine radical allows for a very favorable hydrogen abstraction ($\Delta H = -7$ kcal/mol) from HBr to give the bromoalkane product.

The preceding analysis is complicated. But understanding the subtler details leaves you better equipped to fully understand and appreciate future reactions. If this is still unclear, it might be worthwhile to take a few minutes to read through it again.

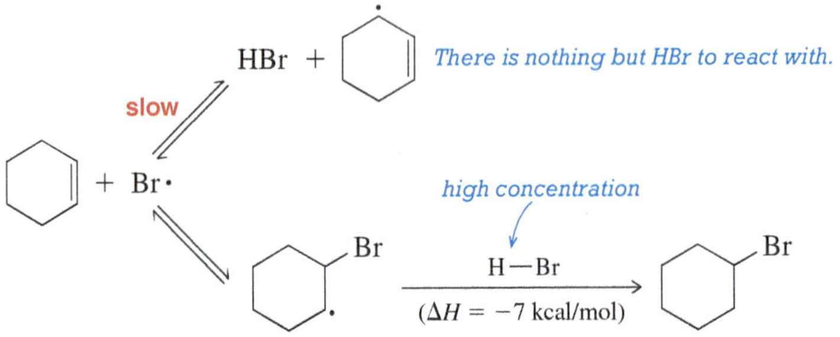

Figure 11.43 Addition of a bromine radical to an alkene is followed by hydrogen abstraction, resulting in net addition of HBr to the alkene.

11.5.4 SUMMARY

- HBr/H_2O_2 with alkenes only gives alkene addition because, without Br_2 present, there is no other productive pathway. NBS only gives allylic halogenation because hydrogen abstraction is the only productive pathway leading to a new product.

11.6 Radical autoxidation

Chemistry in Context 9-D explained that cis-fats, while healthy for human consumption, are not useful as cooking oils because of their oxidative instability. This is especially the case for polyunsaturated fatty acids like linoleic acid (Figure 11.44), which contains two cis-alkenes separated by a single CH_2 group. As a result, saturated oils (such as stearic acid) are the oils of choice for cooking.

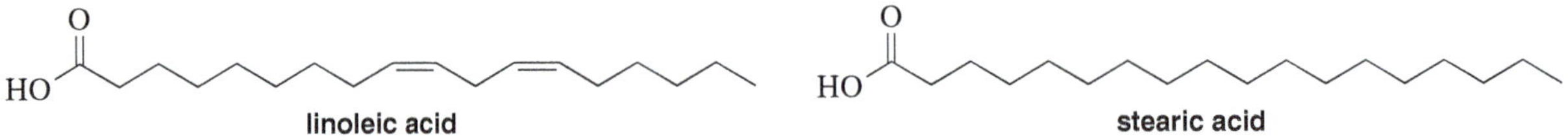

Figure 11.44 Cis unsaturated fats are more susceptible to oxidation than saturated fats.

Equipped only with the knowledge that oxidation in fats occurs by a radical process, can you rationalize the trend in oxidative susceptibility for the fatty acids shown in Table 11.14? [*Think through this before reading further.*]

Table 11.14 Relative rate of oxidation of a series of fatty acids.

Fatty acid	Structure	Relative rate of oxidation
Stearic	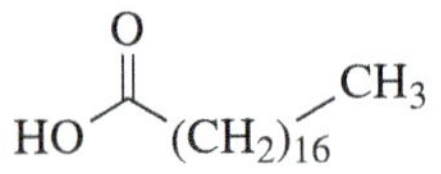	1
Oleic	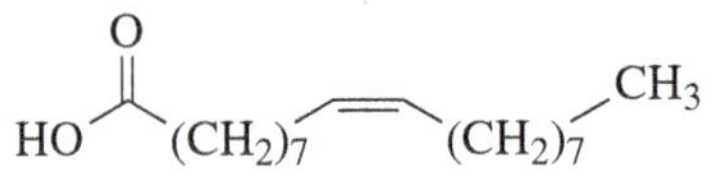	100
Linoleic	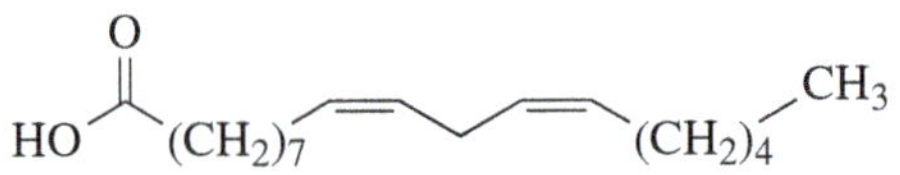	1200
α-Linoleic	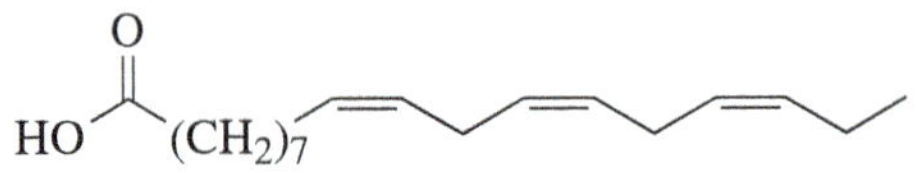	2500

As a radical process, the difference must be attributable to the propensity of each fatty acid to form a radical. Stated another way, polyunsaturated fats have weaker/more reactive C—H bonds. The relative rates in Table 11.14 correlate to the stability of the radicals that form upon hydrogen abstraction. More unsaturation allows for enhanced stabilization by multiple contributing resonance structures (**Figure 11.45**).

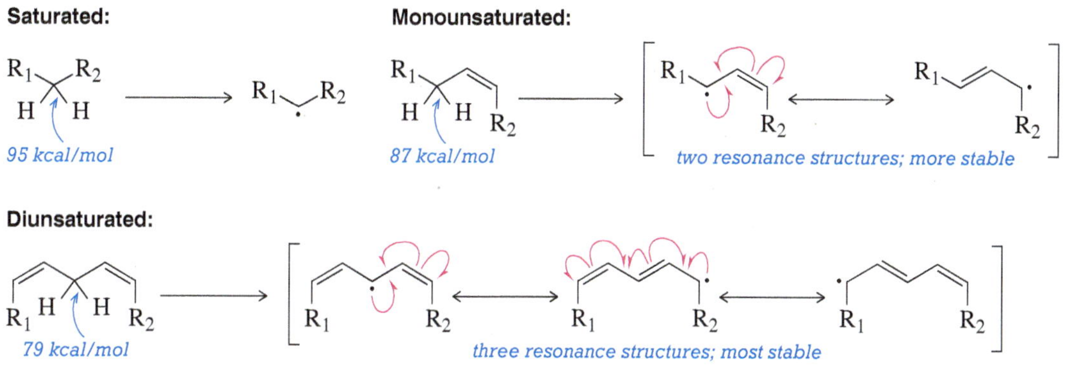

Figure 11.45 Oxidative susceptibility of fatty acids is a function of how stable the resulting radical is. (Alkyl chains of fatty acids have been removed for clarity.)

Proceeding by a radical chain reaction, **autoxidation** of fatty acids occurs, producing a hydroperoxide intermediate (**Figure 11.46**). Autooxidation is defined as the spontaneous oxidation of a molecule by oxygen at room temperature.

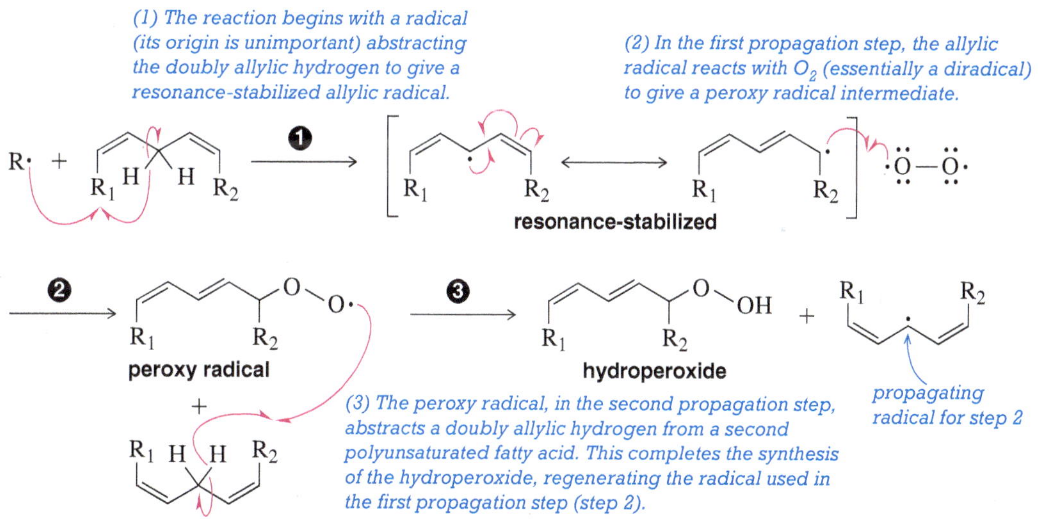

Figure 11.46 Radical mechanism of fatty acid autoxidation.

The hydroperoxides formed here are relatively unstable molecules, reacting further to produce a variety of aldehydes and carboxylic acids, giving rancid oils their distinctive smell. Similar autoxidation processes are known to spoil meat and other food products and also have deleterious effects on the structural integrity of rubber and plastics. As such, preventing autoxidation would be an important problem that a chemist (and/or nature) might have solved.

Assessment 11.35 The following polyunsaturated fat, which does not exist in nature, is oxidized at a rate similar to oleic acid (Table 11.14), despite having two cis-alkenes. Why?

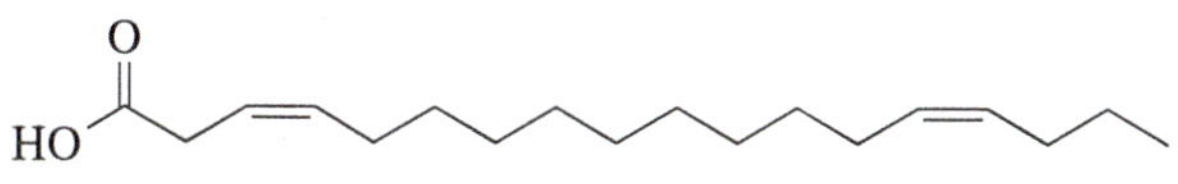

After working Assessment 11.35, let's think like a chemist and see if our understanding of radical reactions allows us to prevent autoxidation.

11.7 Radical inhibition

Thinking Like a Chemist

Autoxidation (Section 11.6) is a destructive process with negative consequences for the food and materials industry, even negatively impacting human health. As such, inhibiting this process is a goal worth pursuing. One viable, but impractical, solution is to exclude oxygen, a necessary component of autoxidation. Given what we know about radical chain reactions, can we design a molecule that can prevent these reactions from occurring?

Designing a molecule to interrupt the chain reaction requires careful thought about why the doubly allylic C—H bond is broken. A reasonable assumption is that it occurs because it produces a more stable radical. Thus, in the initiation step, the radical abstracts the doubly allylic H because that C—H bond is *weakest*. After forming an oxygen radical in the first propagation step, the doubly allylic H is abstracted in the second propagation step (**Figure 11.47**). Once more, this is because that C—H bond is *weakest*. [*Carefully read over this paragraph again—what should our inhibitor possess to disrupt the chain reaction?*]

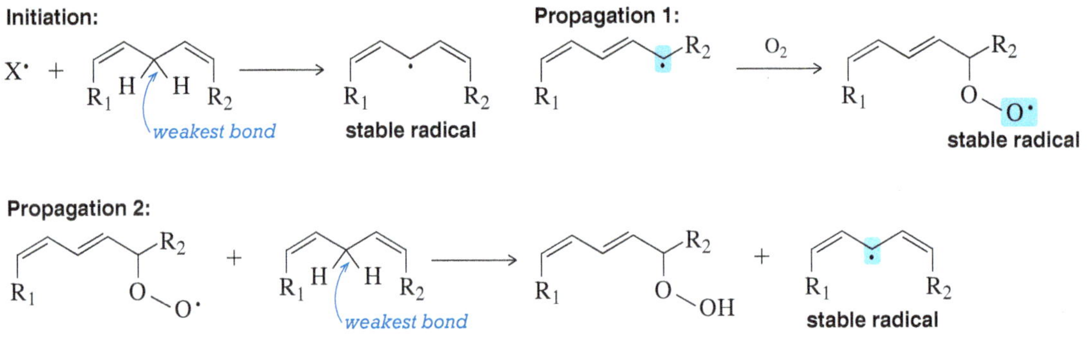

Figure 11.47 **Mechanism of radical autoxidation.**

Our inhibitor should have a bond that is weaker than any bond present in the lipid/polymer/material being protected from autoxidation. Coming up with a molecule on your own might be a stretch, but oxygen radicals are relatively stable, and resonance helps to stabilize radicals. A phenol, such as BHT (butylated hydroxytoluene), would be a good inhibitor of radical processes (**Figure 11.48**).

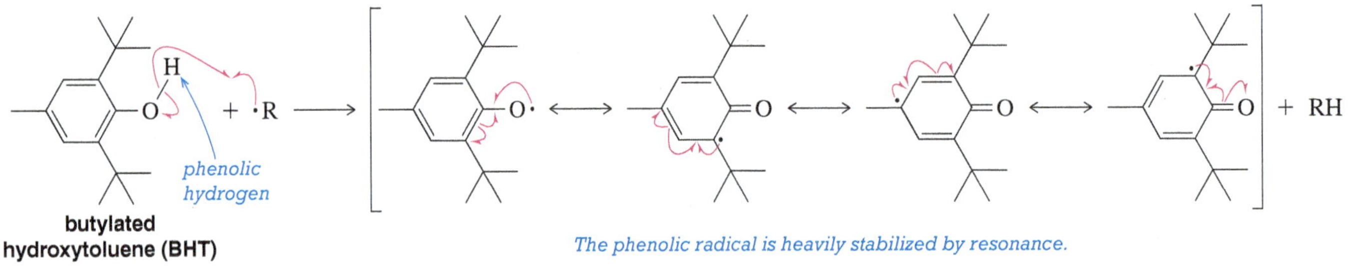

Figure 11.48 **Abstracting the phenolic hydrogen gives a very stable radical.**

The radical formed by abstraction of the phenolic H is highly stabilized by the resonance of the benzene ring. Furthermore, because of the bulky *tert*-butyl groups, there is a significant amount of steric hindrance around the radical, making it resistant to further reaction. As a result, a molecule like BHT interrupts the chain reaction by interfering with the second propagation step. BHT is very useful in the food and plastics industry as an oxidation inhibitor (**Figure 11.49**). ■

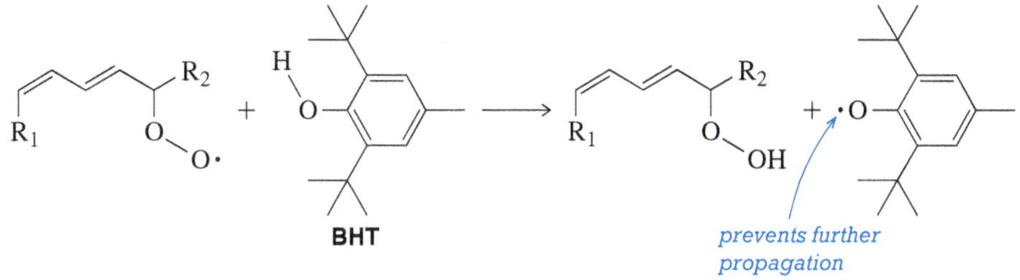

Figure 11.49 Formation of the stable radical prevents further propagation, thereby inhibiting autoxidation.

Many of the cooking oils in use today originate in plants. In their own oxygen-rich environment, plants have evolved a way to protect the fatty acid esters (lipids) they produce from oxidation, using vitamin E (**Figure 11.50**) as a natural version of BHT. With the similarity in structures, it is likely that vitamin E inspired the development of inhibitors like BHT.

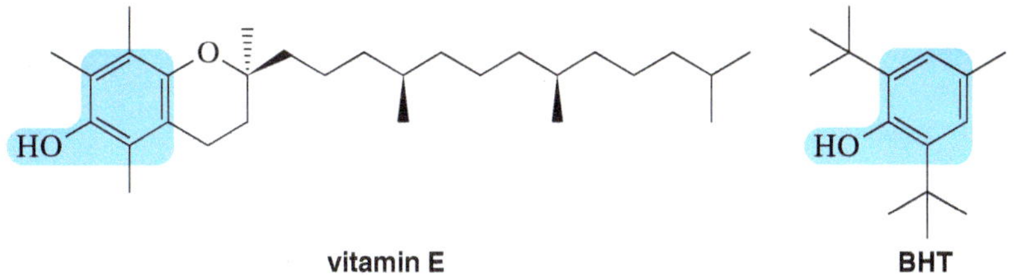

Figure 11.50 Vitamin E and BHT have similar structures.

Mechanistically, vitamin E interrupts the chain reaction in a manner similar to BHT (**Figure 11.51**). By donating a hydrogen atom to the peroxy radical, the resulting phenoxy radical is now available to react (through a contributing resonance structure) with a second peroxy radical. Because this is the combination of two radical species, you should recognize it as a chain termination step. No radicals are produced, so the chain reaction is halted, autoxidation stops, and the plant's lipids are protected. [*Why doesn't naturally occurring vitamin E protect cooking oils? Unfortunately, it is removed during their purification.*]

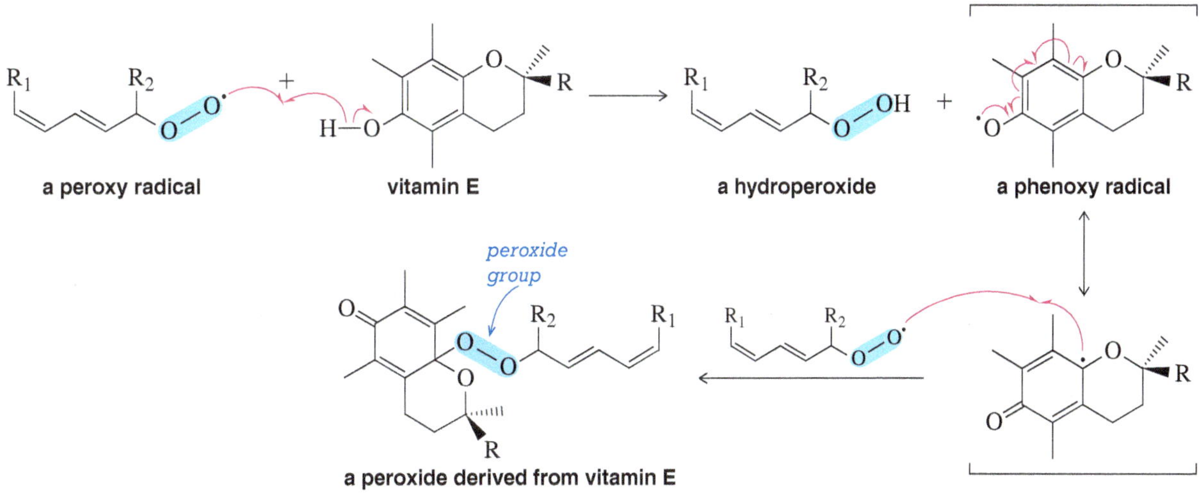

Figure 11.51 Vitamin E acts as an inhibitor by trapping the resulting peroxy radicals.

11.7 SUMMARY

- Autoxidation is inhibited by quenching the peroxy radical (formed in the second step of autoxidation) with synthetic (BHT) or natural phenols (vitamin E).

Assessment 11.36 Vitamin E, especially the B ring and the side chain, is a terpenoid derivative. Identify the isoprene units (*Chemistry in Context 8-A*) nature used to make it.

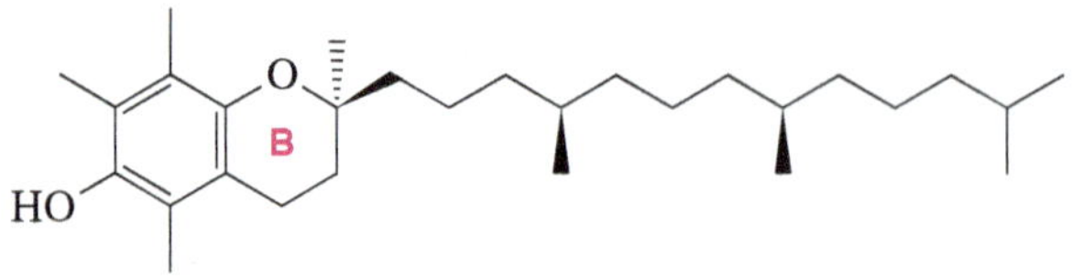

11-E CHEMISTRY IN CONTEXT: Green Chemistry

Natural Preservatives

Among the general public, there is often a fear of synthetic molecules [*sometimes rationally, sometimes not*]. Unnatural molecules are perceived to have negative environmental and health impacts. Recently included in this debate are synthetic preservatives such as BHT. Though it has received the "generally regarded as safe (GRAS)" designation in the United States and research has never shown a conclusive link, it has been suggested that BHT contributes to increased cancer risk, asthma, and other health issues. Scientists are equipped to carefully evaluate claims regarding the safety or danger of a particular compound; it is a moral imperative that we do so.

Whether the claims regarding BHT are valid or invalid [*further research is needed*], a significant portion of society has chosen to eat organic foods, with very little processing and using only natural preservatives. One natural preservative was discovered when extracts of rosemary (used as a flavoring agent) extended the shelf life of meat. Rosemary extracts were found to contain carnosic acid and rosmarinic acid, each of which contains the phenol structure present in BHT and vitamin E (**Figure 11.52**). As a result, these natural preservatives inhibit oxidation of fats and oils by the mechanism discussed in Section 11.8.

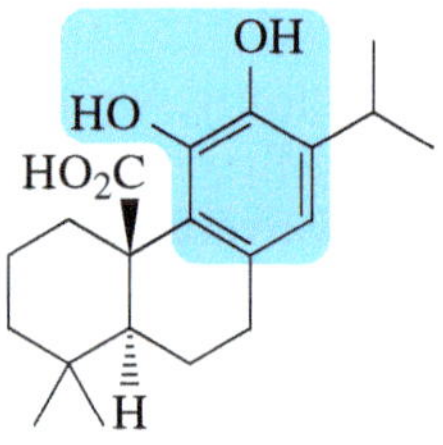

carnosic acid

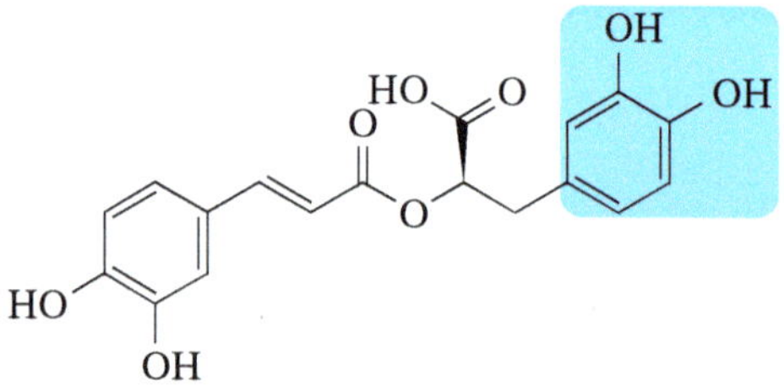

rosmarinic acid

Figure 11.52 Naturally occurring phenol food preservatives.

Seeing a market for natural preservatives, companies have begun extracting and selling carnosic and rosmarinic acid for use in other food products.

Concluding thoughts: halogens in drug design

We study the role that haloalkanes play in syntheses in Chapter 12 and other chapters to come. Here in Chapter 11, the value and impact of halogens on the properties of medicines, insecticides, and environmental contaminants have been discussed. Because adding halogens to molecules increases their lipophilicity, halogenated compounds tend to persist in the environment, accumulating within the tissues of animals that ingest or consume them, ultimately to the detriment of predators higher in the food chain.

That said, the effects of added halogens can sometimes be quite beneficial. For example, fluorine, a small but highly electronegative atom, is widely incorporated in active pharmaceutical agents. It is estimated that 20% of all pharmaceuticals on the market contain one or more fluorine atoms. The incorporation of fluorine can have a variety of positive effects, including enhanced binding, modification of pK_a values of nearby acid residues, and increased bioavailability. One example of this is the macrolide antibiotic Flurizic® (flurithromycin ethyl succinate). By increasing the lipophilicity of the antibiotic erythromycin, incorporation of a fluorine atom results in a higher concentration of the drug reaching the desired target (**Figure 11.53**).

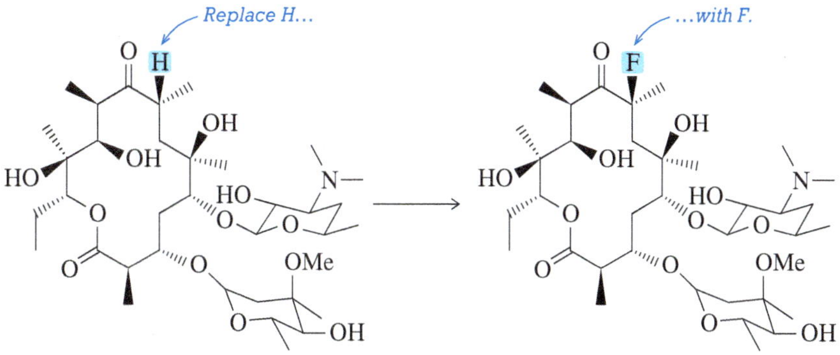

erythromycin **flurithromycin (Flurizic)**

Figure 11.53 Incorporation of fluorine increases the bioavailability of the erythromycin-derived antibiotics.

Similarly, the negative impacts that a halogen has on an organism's ability to metabolize and excrete an environmental toxin can be harnessed in a positive way for the development of more-potent drugs. Consider tolbutamide, one of the very early antidiabetic medications in the sulfonylurea class [**Figure 11.54(a)**]. Tolbutamide is metabolized rapidly by the body, so higher doses are required to maintain the therapeutic effect. Because the metabolism likely occurs via the formation of a resonance-stabilized benzylic radical by the action of cytochrome P450, replacement of the methyl group by a chlorine atom eliminated this mechanism of metabolism. As a result, chlorpropamide lasts about eight times longer in the body, resulting in a longer-lasting therapeutic benefit. A similar modification was made to a D_2 antagonist as part of the development of aripiprazole (Abilify®), an antidepressant [**Figure 11.54(b)**]. Replacement of the methyl groups with chlorines once again reduced metabolism, allowing it to last longer in the body.

In Chapter 12, the story changes slightly, as the reactions of haloalkanes reveal the pharmacodynamic effects this functional group has had when used in chemical weapons for wars between countries and also in the "war" on cancer.

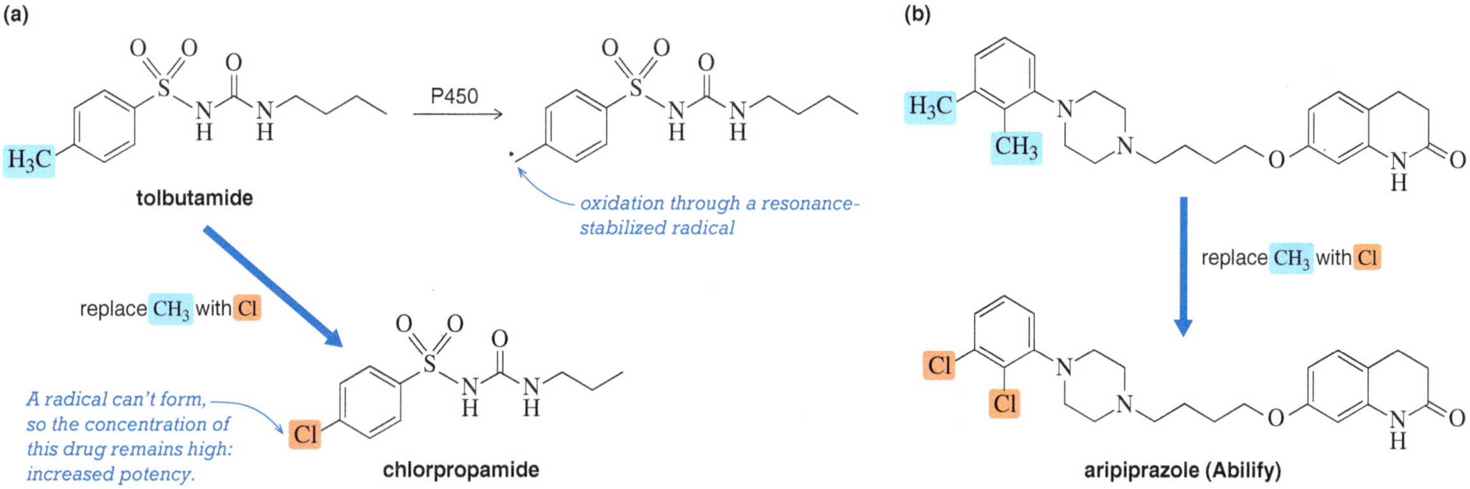

Figure 11.54 Halogens prevent radical oxidation of drugs. Replacement of methyl groups increases the half-lives of (a) chlorpropamide, an antidiabetic, and (b) aripiprazole, an antidepressant.

End-of-Chapter Materials

Important Concepts

[These are the major concepts covered in each section, excluding reactions.]

11.1: Haloalkanes (alkyl halides) are a class of organic compounds where at least one hydrogen of the parent alkane is replaced by a halogen. The general formula is R—X, where X=F, Cl, Br, or I, and R represents an alkyl group, alkene (haloalkene), or an aromatic ring (haloarene).

11.2: To name haloalkanes, simply treat the halogen as a substituent on the parent alkane. Halo substituents are named by dropping -ine from the halogen name and adding -o (fluorine → fluoro).

11.3: Small haloalkanes are polar molecules, due to the electronegativity difference in the C—X bond. Because they are unable to hydrogen bond, halogens have relatively low boiling and melting points, and they are insoluble in water (hydrophobic) but soluble in nonpolar solvents (lipophilic). The ability to dissolve polar and nonpolar molecules makes haloalkanes ideal solvents for chemical reactions and extractions. C—X bond lengths and strengths are inversely correlated, making the C—I bond the longest and weakest.

11.4: The C—X σ bond is formed by overlap between an sp^3-hybridized orbital of carbon and an sp^3-hybridized orbital of the halogen. Located at a 180° angle to the C—X bond, the antibonding orbital (σ*) is important to the chemistry of haloalkanes as electrophiles in reactions that are presented in Chapter 12.

11.5.1: Radical halogenation is a mildly effective synthetic reaction that is limited by its ability to selectively prepare haloalkanes. It is most efficient when the alkane to be halogenated has only one type of hydrogen because of symmetry (with chlorination or bromination) or has one type of hydrogen that gives a significantly more stable radical (for bromination only).

11.5.2: In the presence of peroxide, HBr adds to an alkene with anti-Markovnikov regioselectivity. Proceeding through a radical mechanism, the most stable intermediate in the first step is formed by the attack of the bromine radical at the least substituted carbon, leaving behind the most stable carbon radical. In the second propagation step, the carbon radical abstracts a hydrogen from HBr, allowing the chain reaction to continue.

11.5.3: The allylic carbon (next to an alkene) can be halogenated using *N*-bromosuccinimide (NBS). Abstracting the most substituted allylic hydrogen (first propagation step) gives a resonance-stabilized radical that attacks Br_2, forming the C—Br bond in the second propagation step. NBS helps avoids the side reaction of alkene addition (Section 9.1.2) by keeping the concentration of Br_2 low. When there are two possible alkene products, the more substituted/more stable alkene forms selectively.

11.6–11.7: The autoxidation of cis unsaturated fatty acids proceeds by a radical chain reaction to produce unstable hydroperoxide intermediates. Autoxidation can be inhibited using butylated hydroxytoluene (BHT), which interrupts the second propagation step through formation of a phenolic radical.

Important Reactions and Mechanisms

1. **Radical halogenation (Section 11.5.1):** The thermally initiated radical chain reaction results in the substitution of a halogen atom for an alkyl hydrogen. Br• abstracts the most substituted hydrogen to form a carbon radical in the first propagation step. The carbon radical attacks Br_2 to form the C—Br bond in the second, while regenerating Br•. The reaction can also be done with Cl_2, but it is more selective when using Br_2.

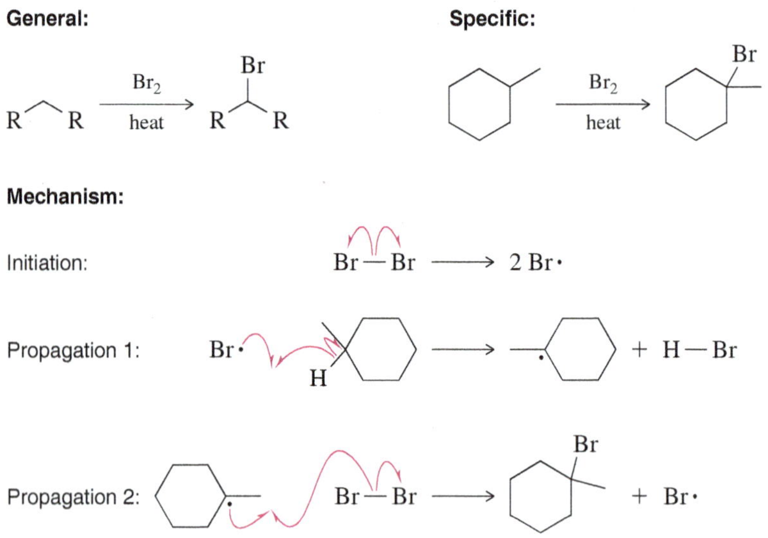

2. Radical hydrohalogenation (Section 11.5.2): Addition of HBr in the presence of peroxide proceeds by a radical mechanism. Attack by a bromine radical at the least substituted carbon results in formation of the more stable carbon radical in the first propagation step. The new C—H bond is formed in the second propagation step.

General:

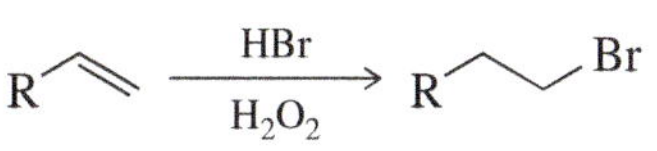

Specific:

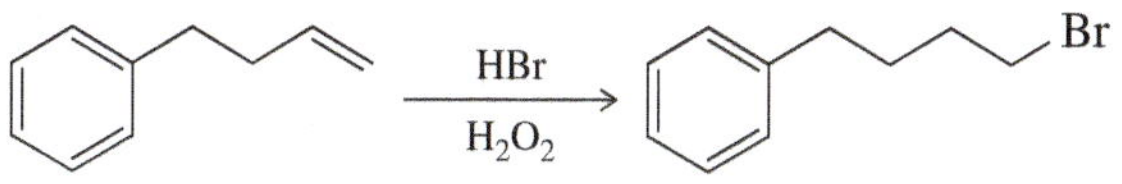

Mechanism:

Initiation:

Propagation 1:

Propagation 2:

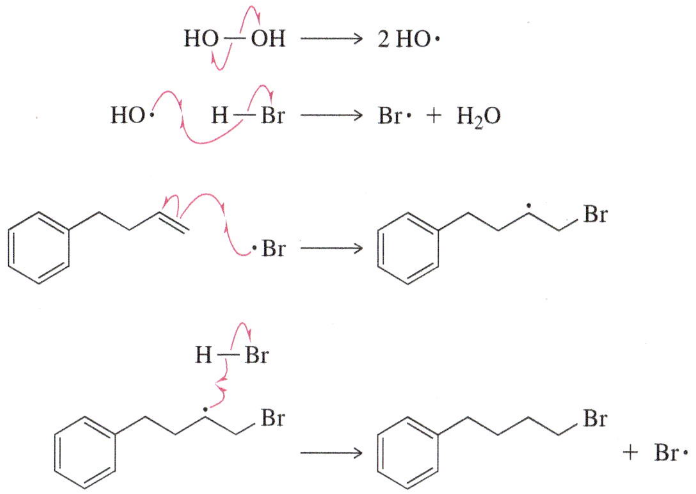

3. Allylic bromination (Section 11.5.3): NBS, in the presence of heat, replaces allylic hydrogens with a bromine atom. Abstracting the most substituted allylic hydrogen (first propagation step) gives a resonance-stabilized radical that attacks Br_2, forming the C—Br bond in the second propagation step. The side reaction of HBr and NBS generates Br_2 in a nonradical reaction.

General:

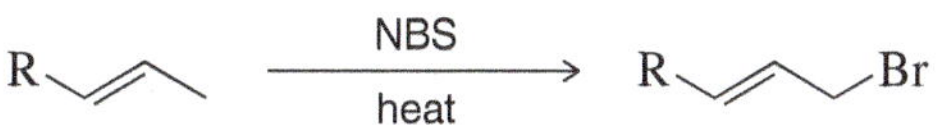

Specific:

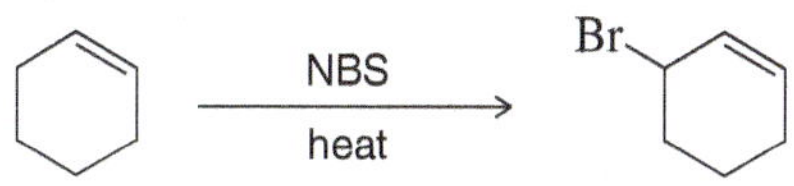

Mechanism:

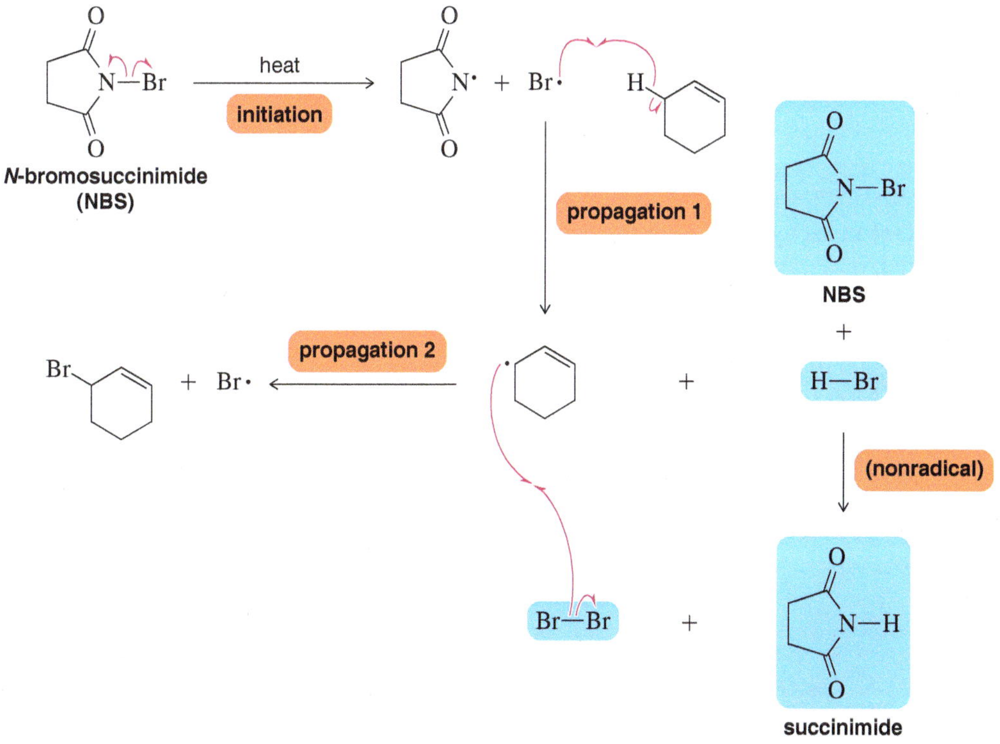

4. **Autoxidation (Section 11.6):** The oxidation of natural fatty acids (and cooking oils), which occurs by a radical reaction, is not synthetically useful. The formation of a resonance-stabilized radical is followed by reaction with O_2 to give a peroxy radical. The peroxy radical abstracts a hydrogen from another lipid, thereby continuing the chain reaction.

General:

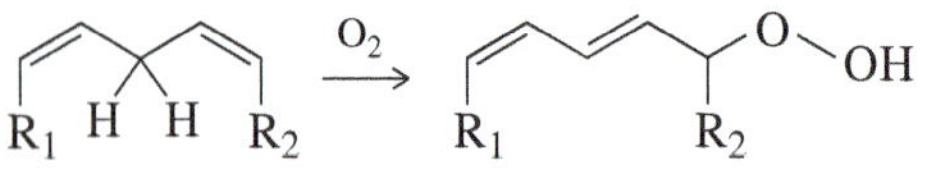

Specific:

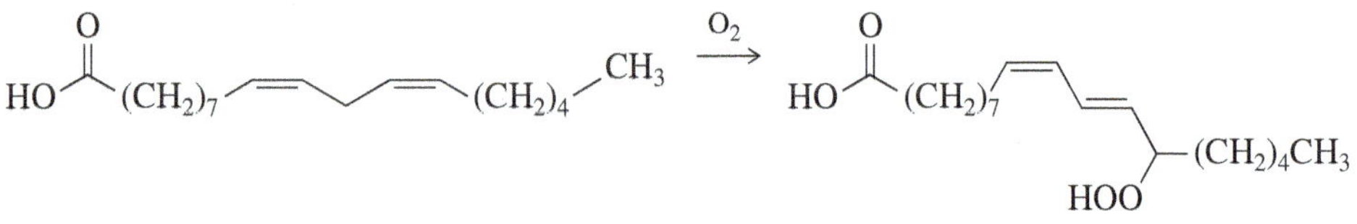

Mechanism:

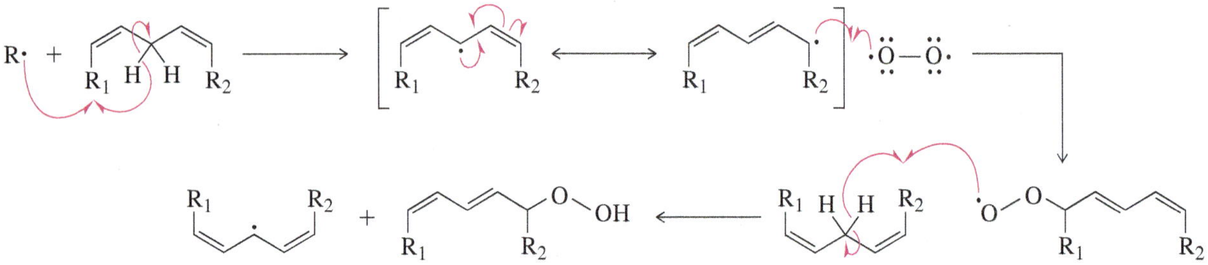

End-of-Chapter Assessments

[*Assessments are ranked according to their difficulty;* •••• *= most difficult.*]

11.37 (•) Name the following halogenated compounds according to the IUPAC rules of nomenclature.

(a)

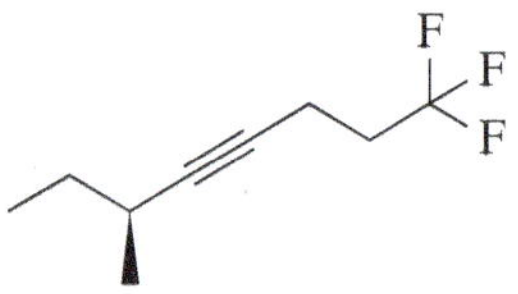

(b)

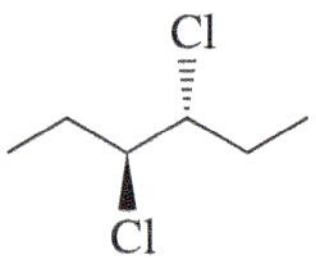

(c)

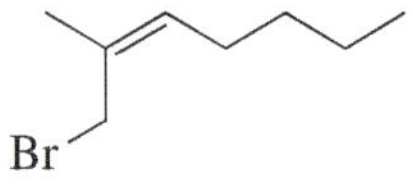

(d)

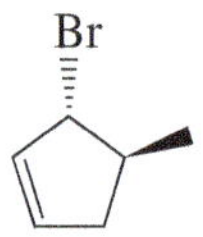

11.38 (•) Draw the structure that corresponds to the compound names shown.

(a) (*S*)-3-bromo-3-ethylcyclohex-1-ene
(b) (5*R*,6*E*)-5-bromooct-6-en-1-yne
(c) (*Z*)-5,5,5-trichloropent-2-ene
(d) 1,1,1,2,2-pentafluoro-6-methylheptane

11.39 (••) In each pair, choose the most acidic compound. Justify your answer. The most acidic proton in each compound has been indicated.

(a)

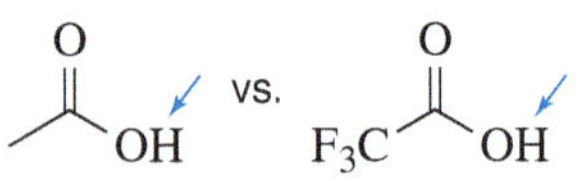

(b)

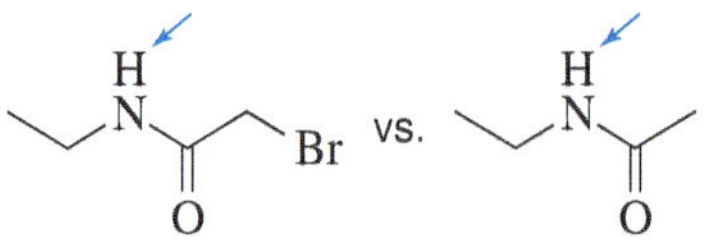

(c)

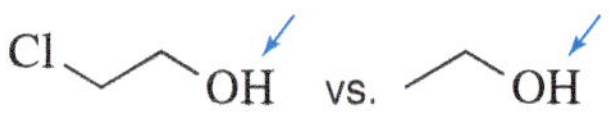

(d)

H—F vs. H—I

11.40 (••) In each pair, choose the most basic compound. Justify your answer.

(a)

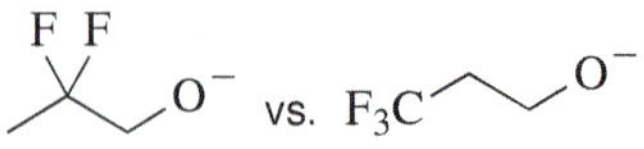

(b)

Cl^- vs. Br^-

(c)

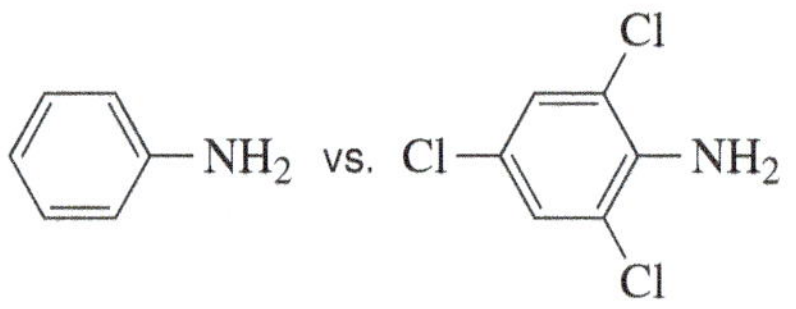

(d)

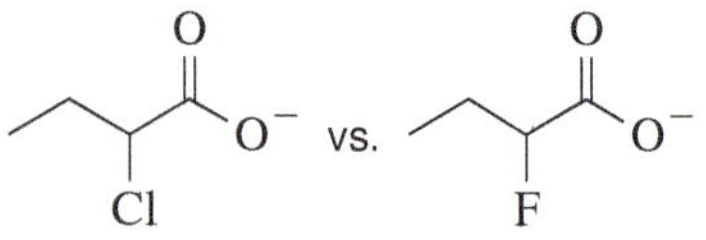

11.41 (••) Which of the reactions studied in this chapter result in oxidation of the organic molecule? Justify your answer.

(a)

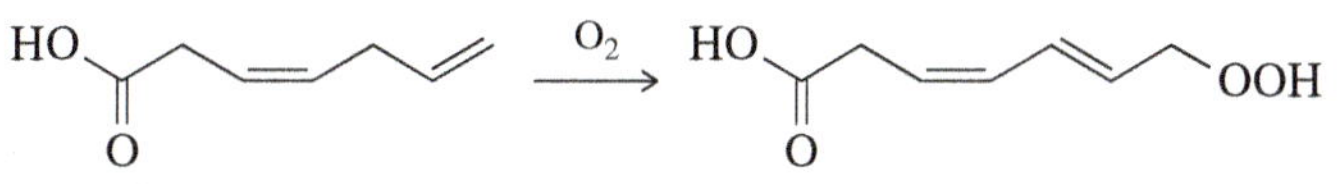

(b)

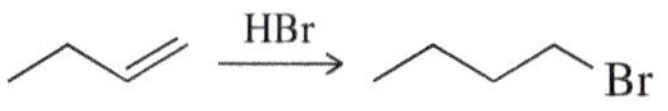

(c)

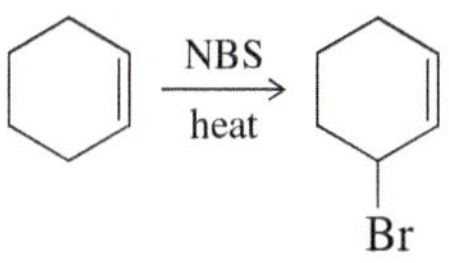

(d)

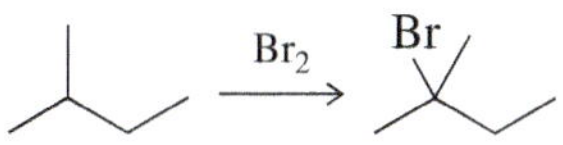

11.42 (•) In each pair, (i) choose the compound that is most soluble in water. (ii) Within each pair, which is most soluble in nonpolar solvents?

(a)

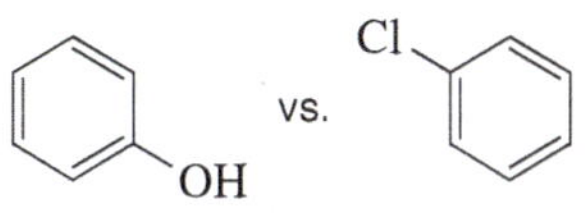

(b)

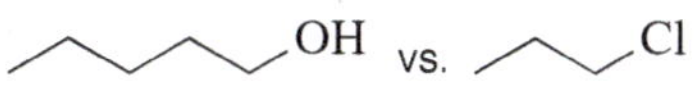

(c)

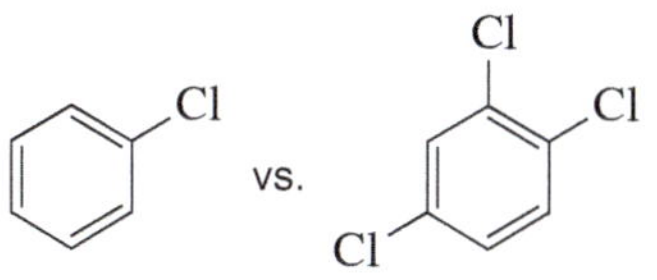

11.43 (•) In each pair, choose the compound you would expect to have the highest boiling/melting point.

(a)

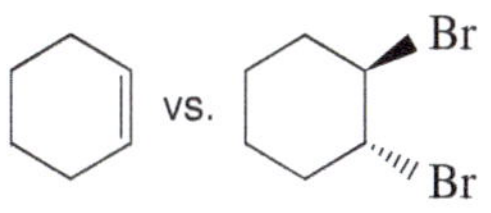

(b)

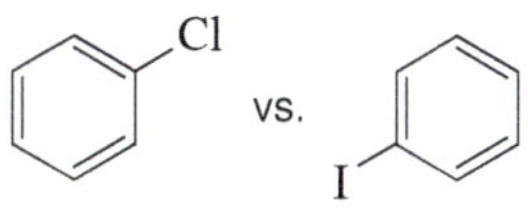

(c)

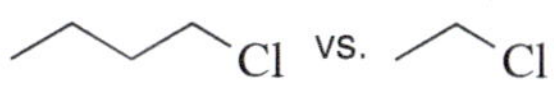

11.44 (••••) Contrary to expectation, the addition of more fluorines to ethane *lowers* the boiling point. Explain this phenomenon.

C—X	bp (°C)
CH_3CH_2F	−32
CH_3CHF_2	−25
CH_3CF_3	−47
CF_3CF_3	−78

11.45 (•) Predict the major product of the following bromination reactions.

(a)

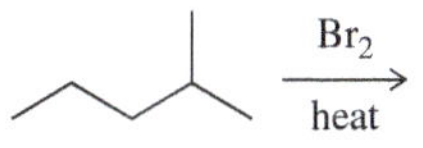

(b)

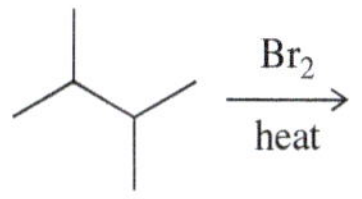

(c)

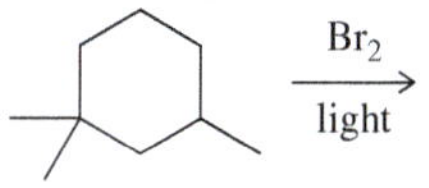

(d)

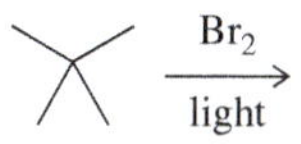

11.46 (•) Predict the product of the following alkene addition reactions.

(a)

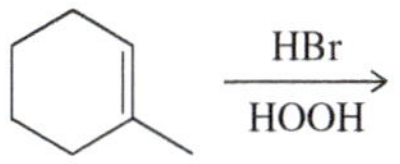

(b)

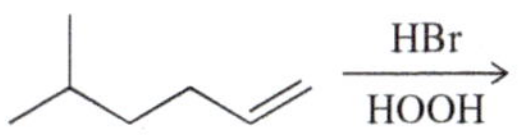

(c)

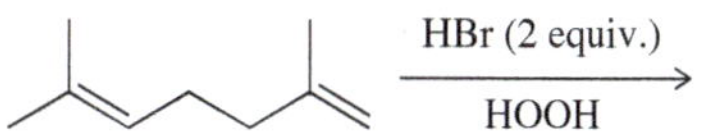

(d)

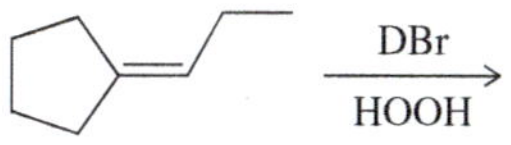

11.47 (••) Predict the major product(s) of the following allylic bromination reactions.

(a)

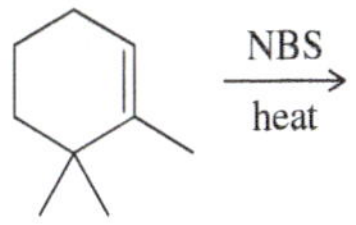

(b)

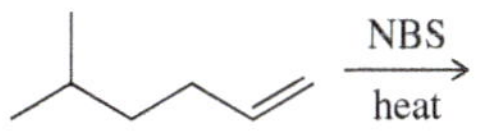

(c)

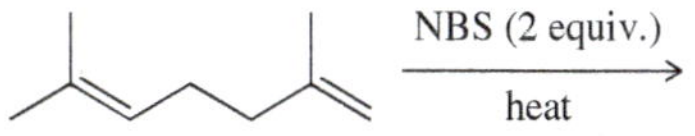

(d)

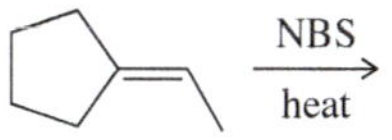

11.48 (•••) Suggest a mechanism for the following reactions.

(a) (b) (c)

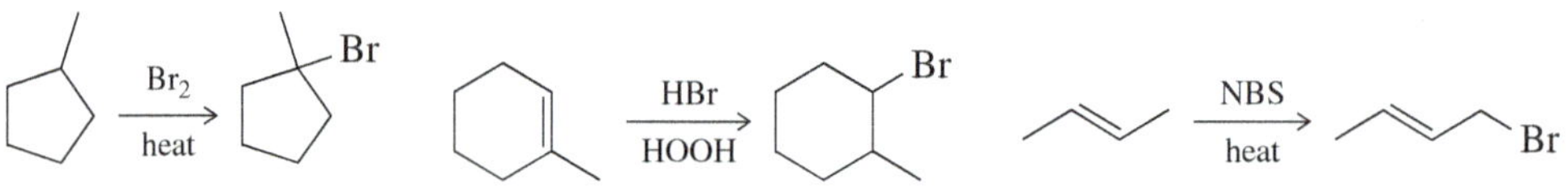

11.49 (•••) The following haloalkanes can all be prepared starting from the same alkene. Give the alkene and the reagents needed for each.

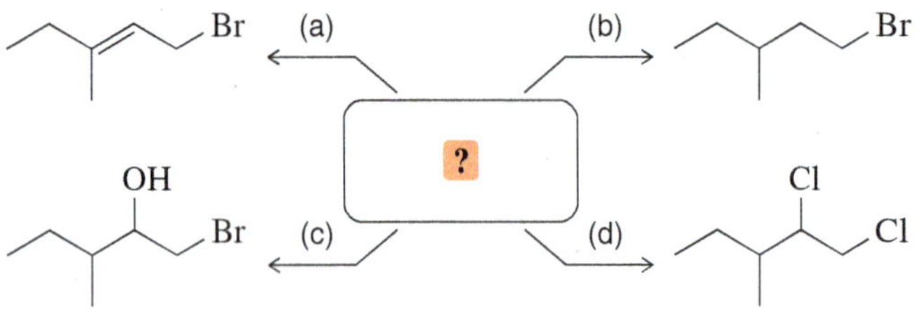

11.50 (••) Identify the two different alkenes that undergo allylic halogenation to prepare the compound shown.

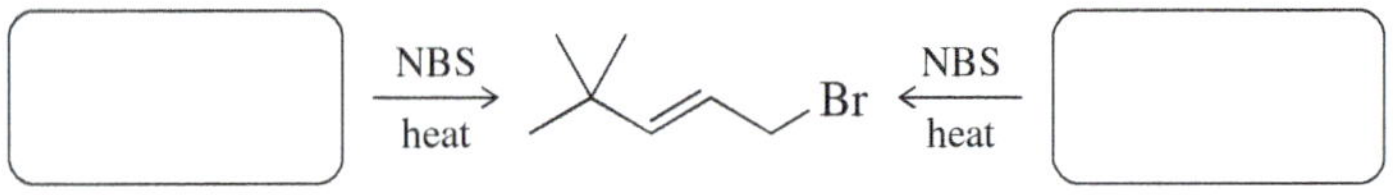

11.51 (•••) For the following reaction, answer questions (a)–(d).

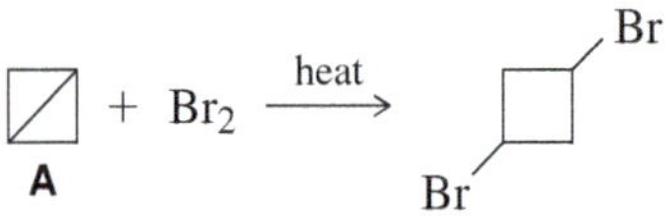

(a) Give an arrow-pushing mechanism, including the initiation step and two propagation steps.
(b) Which is the weakest bond in molecule **A**?
(c) What clue in the reaction helped you figure out the answer to (b)?
(d) Given your choice in (b), why is it the weakest bond?

11.52 (••••) Provide the mechanism of the radical reactions shown.

(a)

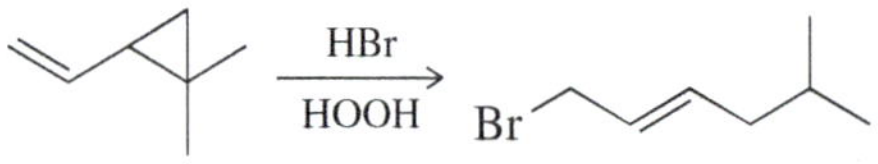

(b)

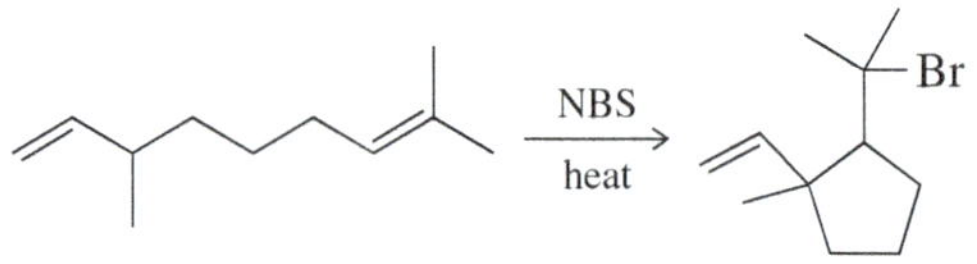

11.53 (••••) One danger associated with storing ether solvents is their tendency to form explosive peroxides when exposed to oxygen. Suggest a mechanism by which the hydroperoxide might form. You can assume the presence of X• to start the reaction.

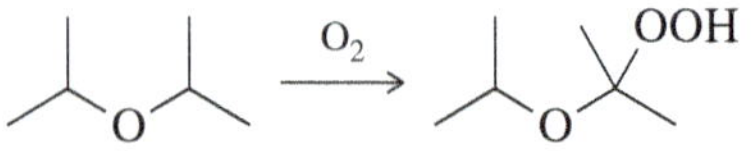

11.54 (•••) The following ethers are ranked according to their ability to form explosive peroxides. Explain this ranking based on your knowledge of the reaction mechanism.

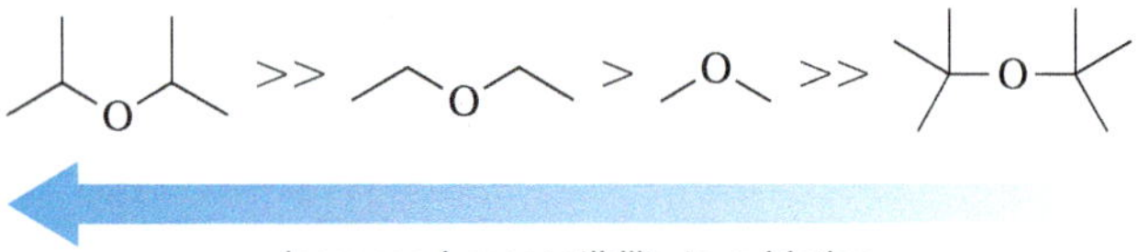

increased susceptibility to oxidation

11.55 (••••) The solvent tetrahydrofuran (THF) is often sold with a small amount of BHT added. Provide a mechanism that explains why this might be so. [*See Figure 11.48 for the structure of BHT.*]

tetrahydrofuran (THF)

11.56 (•••) If a small amount of a moderately nonpolar poisonous compound was added to a pond, why would it be safer to drink the water than it would be to eat the fish that live there?

Learning through assessments:

Assessments 11.57–11.61 should be answered in order. Each question should be used to help you answer the next. At the end of these assessments, you should have learned a new skill.

11.57 (•) Other molecules can be used as initiators in radical reactions. One such molecule is 2,2′-azobisisobutyronitrile (AIBN). Show an arrow-pushing mechanism that rationalizes the formation of the following radical species. What are the driving forces of this reaction?

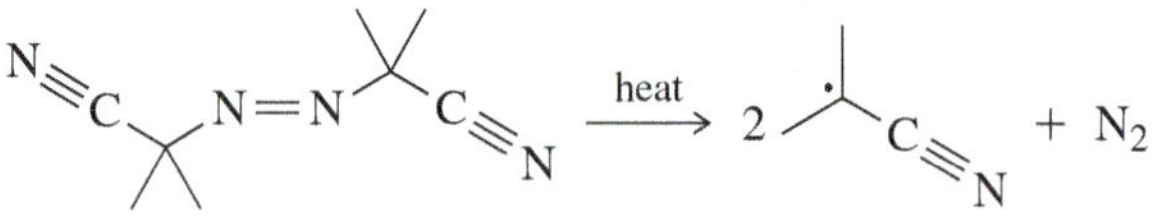

11.58 (••) Tributyltin hydride (Bu_3SnH) is often used as a "radical carrier" in radical reactions. Which bond would you expect to be weaker, Sn—H or C—H? How might this relate to radical stability? Explain your answer.

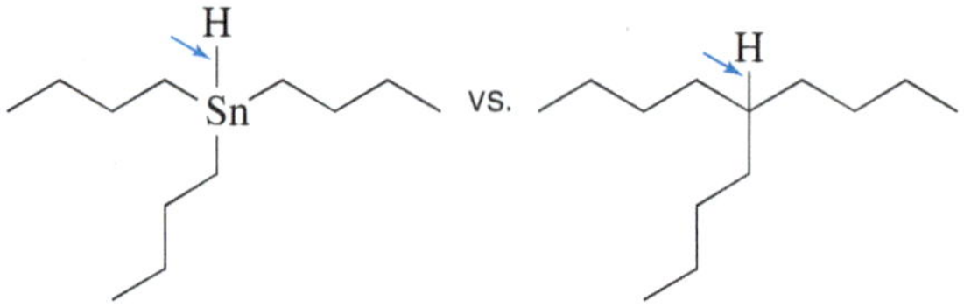

(Bu_3SnH)

11.59 (•••) Given a choice, which hydrogen in Assessment 11.58 would you expect to be abstracted by the radical formed from AIBN? Explain your choice.

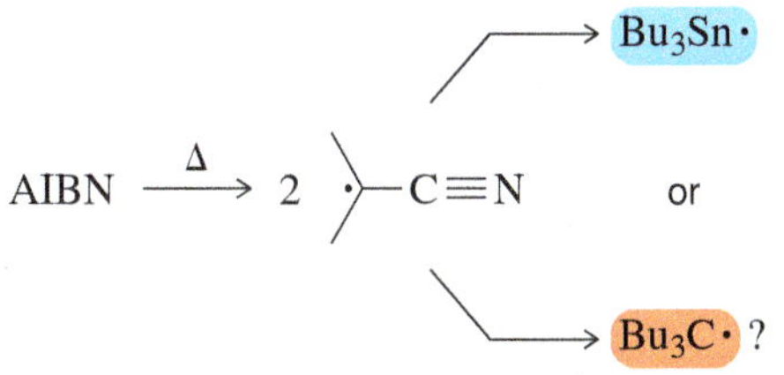

11.60 (••••) Haloalkanes can be reduced to alkanes using radical reactions involving Bu_3SnH and a catalytic amount of AIBN. Suggest a mechanism for this reaction. [*Pay close attention to the bonds formed and broken in developing your mechanism.*]

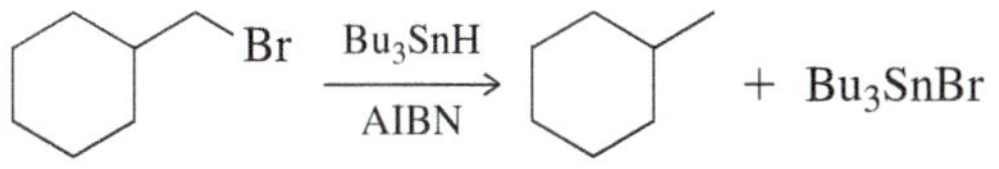

11.61 (••••) Cyclizations can be carried out under radical conditions involving Bu_3SnH and a catalytic amount of AIBN. Suggest a mechanism for this reaction. [*Pay close attention to the bonds formed and broken in developing your mechanism. It may be helpful to number the carbons.*]

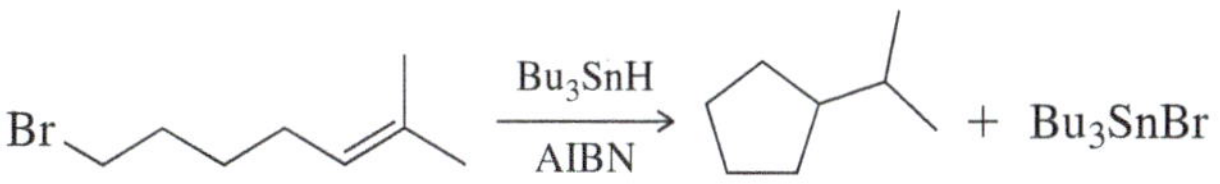

Learning through assessments:

Assessments 11.62–11.65 should be answered in order.

11.62 (•••) A halogenation intended to make compound **A** formed **B** instead.

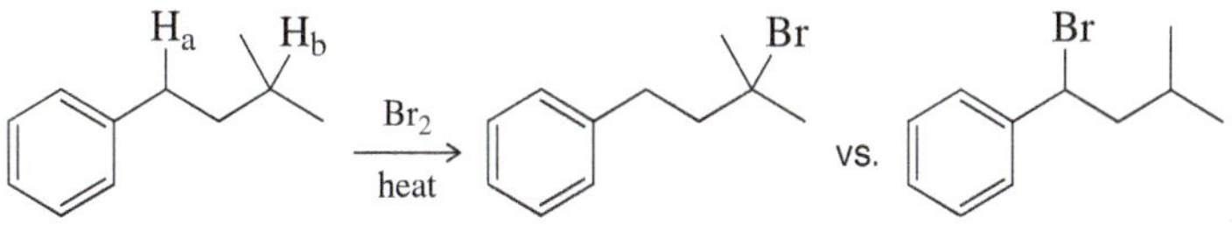

A is *not* formed. **B** is formed.

(a) Suggest a mechanism for the intended formation of **A**.
(b) Suggest a mechanism that rationalizes the formation of **B**.
(c) Given the two mechanisms you drew, why might **B** have formed selectively?
(d) Without looking it up, would you expect $C—H_a$ or $C—H_b$ to have the lower bond-dissociation energy?

11.63 (•••) Predict the product of the following benzylic bromination reactions.

(a) **(b)** **(c)**

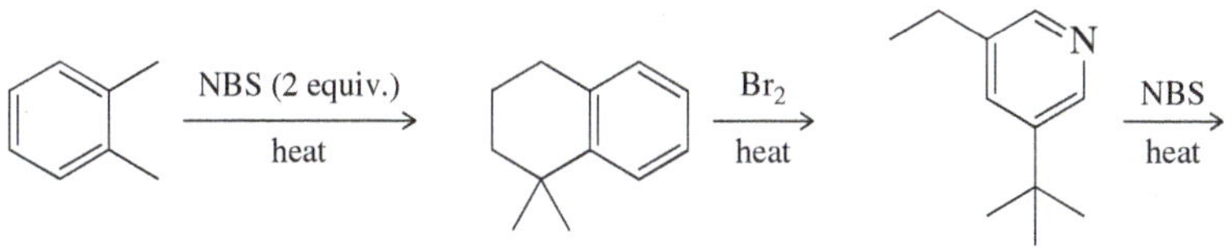

11.64 (••••) The human body can excrete drugs and other exogenous molecules by converting them into polar, water-soluble compounds by a reaction similar to the autoxidation described in Section 11.6. Why are the following drug candidate molecules susceptible to oxidation by this pathway?

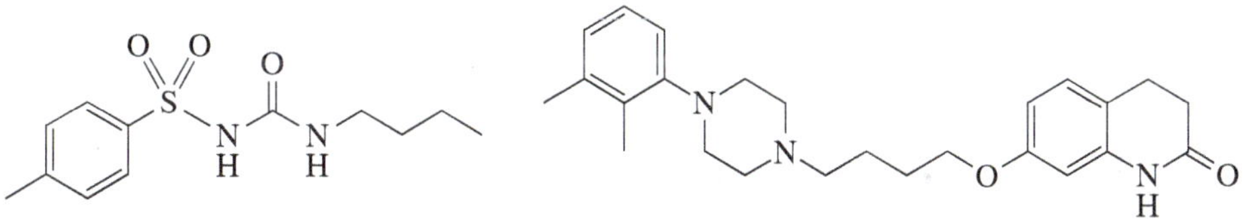

11.65 (••••) Molecule **A** held promise as a treatment for rheumatoid arthritis except that it remained in the body for too long (~120 hours in rat models). Suggest a minor modification you could make to increase the rate at which it might be metabolized and therefore excreted by the body. [*If you can solve this problem, you will have invented the drug Celebrex®.*]

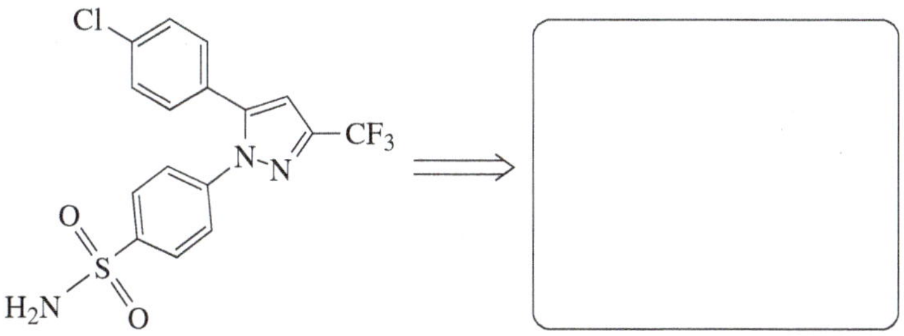

A **Celebrex**

12

Substitution and Elimination: Reactions of Haloalkanes

Learning Objectives

LO 12.0 Explain the difference between S_N2 and S_N1 reactions

LO 12.1 Predict which mechanism a substitution reaction will follow

LO 12.2 Rationalize the different outcomes that result when a substitution reaction proceeds through an S_N2 or S_N1 mechanism

LO 12.3 Explain the differences between E1 and E2 reactions

LO 12.4 Predict which mechanism an elimination reaction will follow

LO 12.5 Rationalize the different outcomes if an elimination reaction proceeds through an E1 or E2 mechanism

LO 12.6 Predict whether a reaction proceeds through an S_N2, S_N1, E2, or E1 mechanism

Introduction: chemical weapons to chemotherapy

Cancer, a broad term describing a number of related diseases, is characterized by uncontrolled cell division, which requires cells to rapidly produce and use DNA. As such, many anticancer drugs interfere with DNA function or synthesis. Because the structure of DNA in human and in cancer cells is identical, it is challenging to achieve selectivity in designing anticancer DNA-interactive agents.

Fortunately, this rapid cell division provides a quantifiable difference between healthy and tumor cells. Rapidly dividing cells have a greater need for DNA precursor molecules than do healthy cells. Accordingly, tumor cells more quickly absorb nutrients, metabolites, and drug molecules from the bloodstream. Taking advantage of the "cellular greed" of tumor cells, medicinal chemists can somewhat selectively target tumor cells. Unfortunately, some normal cells that also divide rapidly, such as healthy cells of the stomach lining and hair follicles, also absorb toxic doses of the chemotherapeutic agent. The unintentional destruction of stomach and hair follicle cells results in two side effects of chemotherapy, nausea, and hair loss, respectively.

In Chapter 11, the pharmacokinetic role of halogens in drug design was discussed, specifically, how halogenation affects water/lipid solubility and metabolic susceptibility. In Chapter 12, we discuss haloalkanes and their chemistry in the context of the pharmacodynamic aspects of drug development. By studying the substitution reactions of haloalkanes, we can trace the deployment of chemical weapons in war to the development of potent anticancer drugs.

"If you can make the diagnosis, the treatment is easy and the damage can be reversed. But making the diagnosis is tricky."

—Ralph Green

The transferability of skills learned in organic chemistry to other professions has been a theme of this text. As a result, you may have wondered how this course will feature in your everyday life as a researcher, dentist, or veterinarian—whatever career goal that required that you take organic chemistry. Chapter 10 discussed the similarities between puzzle solving, organic synthesis, and making a medical diagnosis. In this chapter, we examine these same types of thought processes during our study of substitution and elimination reactions of haloalkanes.

There are only four reactions in this chapter, labeled for now as S_N2, S_N1, E2, and E1. However, for some students [*especially for those intent on memorizing*], this material can be difficult. Thus, it is critically important to take the correct approach to learning this material. In the same way that a physician examines a patient, examining all symptoms before making a diagnosis, we will diagnose reactions in terms of (1) the *solvent*, (2) the *nucleophile*, and (3) the *substitution* of the haloalkane to decide how the reaction must proceed. In learning how these three factors impact a reaction, we continue to build on our organic chemistry fundamentals. [*Let's check what we remember from previous chapters, and then get to it!*]

Test your understanding

Assessment 12.1 Which of the following molecules are capable of being hydrogen bond donors? Which of the following molecules are polar/nonpolar?

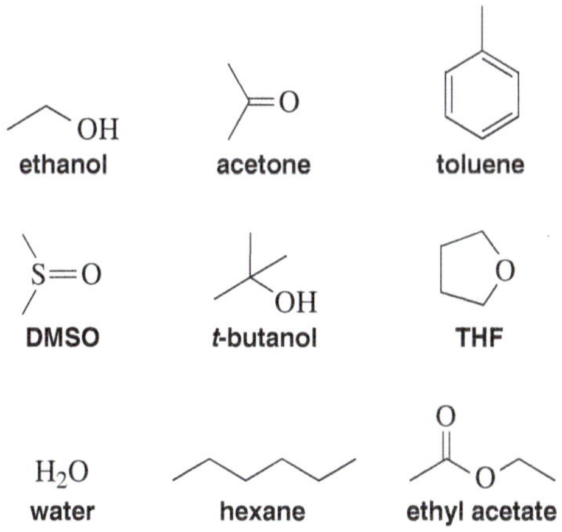

Assessment 12.2 Protonation of which of the following alkenes would be faster?

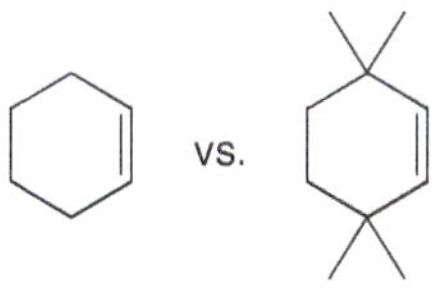

Assessment 12.3 Which of the following indicated atoms would you expect to be most basic?

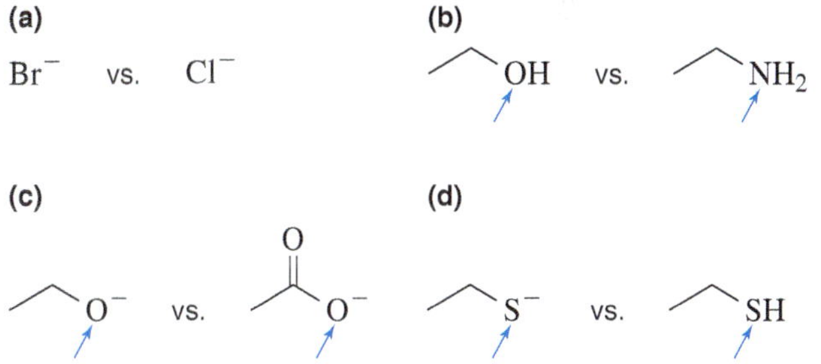

Assessment 12.4 Which of the following carbocations would you expect to rearrange?

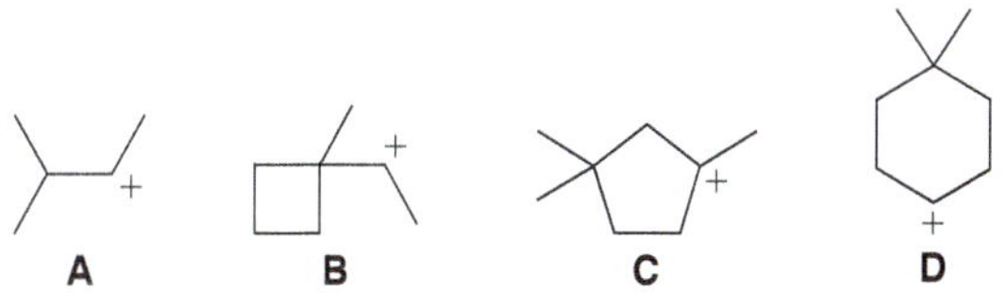

Assessment 12.5 Given that step 1 of the following reaction is rate determining, write the rate law for the overall reaction.

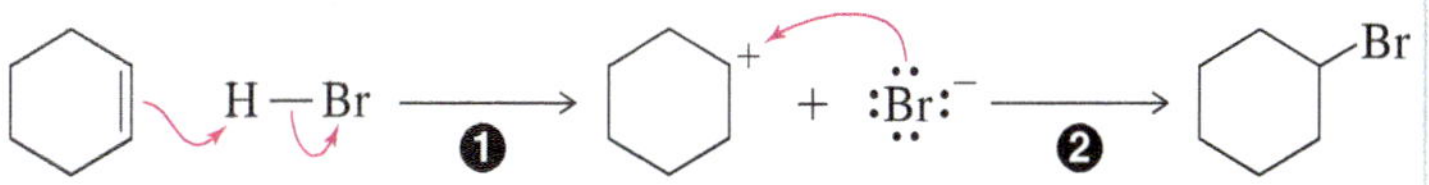

Assessment 12.6 For the following molecule, draw the Newman projection (around the 2,3-bond) with a dihedral angle of 180° between the bromine and hydrogen.

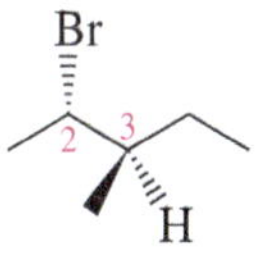

12.1 Substitution and elimination: a general overview

In your introductory chemistry course, you likely learned about four different reaction types. In addition to **single replacement** [*not shown*], there is **synthesis** (addition), **double replacement** (substitution), and **decomposition** (elimination) reactions (**Figure 12.1**). In organic chemistry, we learn why and how these reactions occur. For example, in Chapters 8–10 we studied the reactions of alkenes and alkynes. These can be considered synthesis reactions, because two molecules (A + B) become one (C).

The substitution reactions that we study in Chapter 12 are double replacement reactions (AB + CD → AC + BD). For example, in the reaction between methanol and chloromethane, the hydrogen of methanol and the chlorine of chloromethane exchange partners. Finally, the opposite of synthesis/addition—elimination reactions (Sections 12.5–12.7)—are organic examples of decomposition reactions (C → B + A).

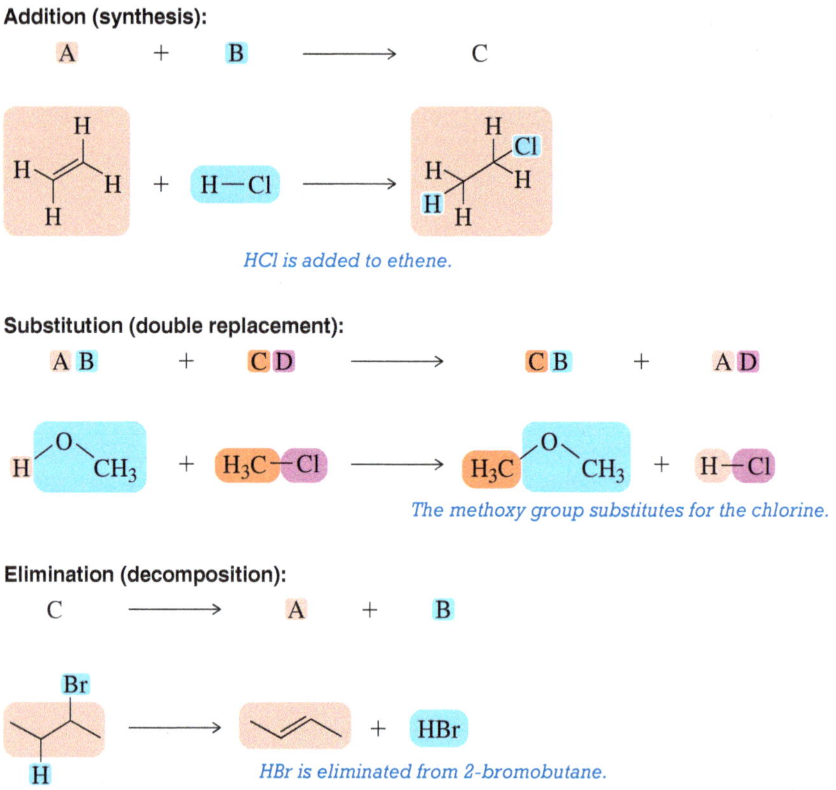

Figure 12.1 **Organic reactions provide concrete examples of reactions previously studied in an introductory chemistry course.**

Take a few minutes now to think about the why and how of the substitution reaction. Why should it be favorable? How does it happen?

12.2 Substitution: a detailed overview

Favorability in substitution reactions can be assessed in the same way it was done previously (Sections 5.3.1 and 8.3.1). Consider the reaction in **Figure 12.2**. If it were an equilibrium reaction, the value of K_{eq} (>1 or <1), would be determined by the sign of ΔG. Because $\Delta G = \Delta H - T\Delta S$ (Equation 5.4), the reaction should be analyzed in the context of

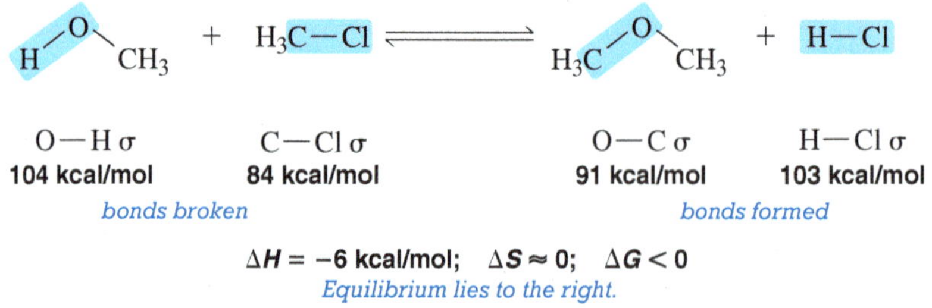

Figure 12.2 **Substitution reactions** are spontaneous based on favorable enthalpy change and near-zero entropy change.

ΔH and ΔS. To approximate ΔH, examine the bonds formed and broken through the course of the entire reaction. On the left side of the reaction, the O—H and C—Cl σ bonds are broken; the right side reveals the formation of C—O and H—Cl σ bonds. [*The formed and broken bonds also tell the story of the reaction. Keep this in mind.*] Bond dissociation energies allow ΔH to be approximated, revealing a reaction that is favored by ~6 kcal/mol. Because there are two reactants and two products, substitution reactions have a $\Delta S \approx 0$. [*In truth,* ΔS *is slightly greater than zero. HCl is a strong acid and dissociates, resulting in a more disordered solution on the product side.*]

Assessment 12.7 In the following reactions, identify the bonds formed and the bonds broken.

(a)

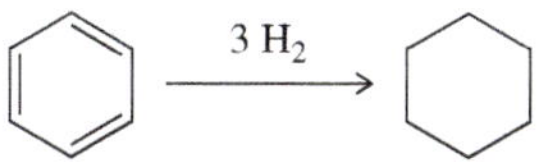

(b)

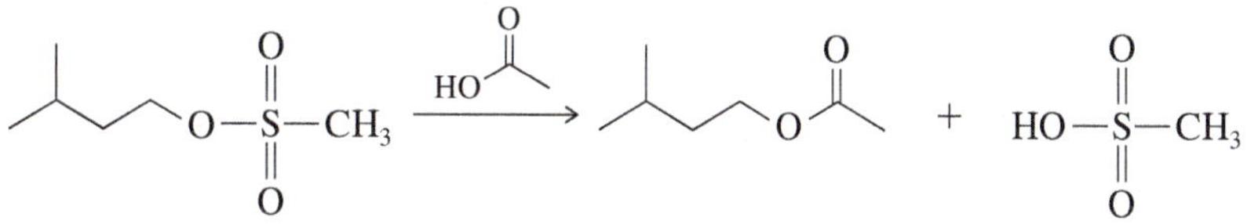

(c)

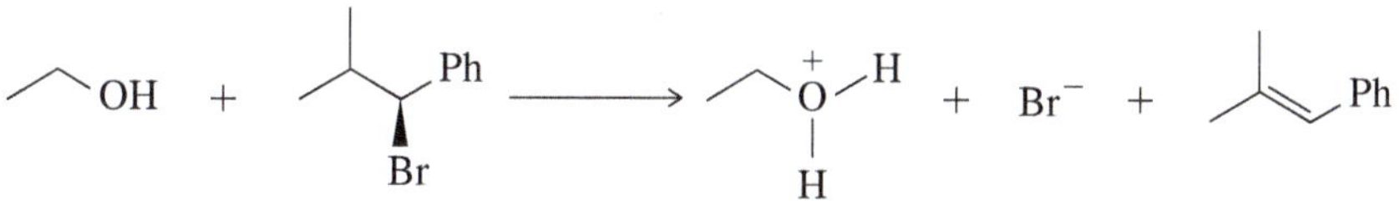

The substitution in Figure 12.2, while favored, is slow but can be made faster under basic conditions. Instead of methanol, deprotonated methanol (methoxide) reacts with chloromethane to give dimethyl ether and chloride as the by-product. [*Why do you think this reaction is faster than the first one?*] The reaction is still favored, with one bond breaking (C—Cl σ) and one bond forming (C—O σ), and being entropically neutral (2 → 2; $\Delta S \approx 0$). Furthermore, it is favored by formation of a more stable conjugate base (Cl^- vs. CH_3O^-) (**Figure 12.3**). The stable chloride anion, willing to "leave" the carbon, is referred to as a good **leaving group**.

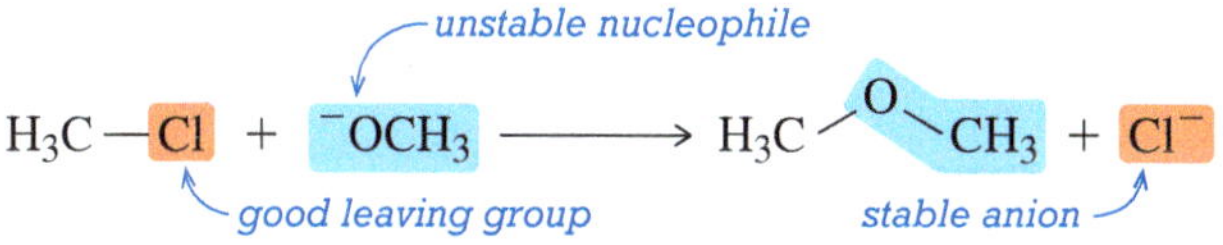

Figure 12.3 The formation of a stable conjugate base makes substitution favorable.

Assessment 12.8 Based on the stability of the bases involved (the nucleophile and the leaving group), estimate K_{eq} for the following reactions.

(a)

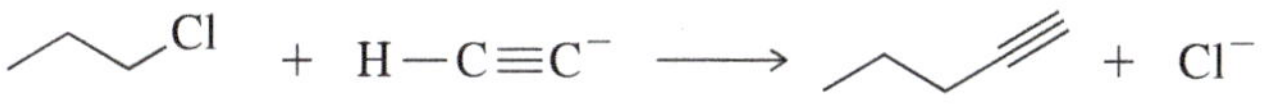

(b)

(c)

Assessment 12.9 Based on your answer to Assessment 12.8, which bases would you expect to be good leaving groups?

12.2.1 Substitution: two mechanisms

The electron distribution in the reactants and products suggests there are only three ways for the reaction in Figure 12.3 to occur. [*Can you identify them?*] Think about what happens in the reaction. Is it "the C—O bond forms, ***and then*** the C—Cl bond breaks" (possibility 1), "the C—Cl bond breaks, ***and then*** the C—O bond forms" (possibility 2), or "the

C—O bond forms *at the same* time that the C—Cl bond breaks" (possibility 3)? **Figure 12.4** shows how electrons can be pushed to demonstrate these three possibilities. [*Note that the statements above with the phrase "and then" indicate that the mechanism has two steps.*] Only one possibility is correct, but which one?

Figure 12.4 Three hypothetical substitution mechanisms.

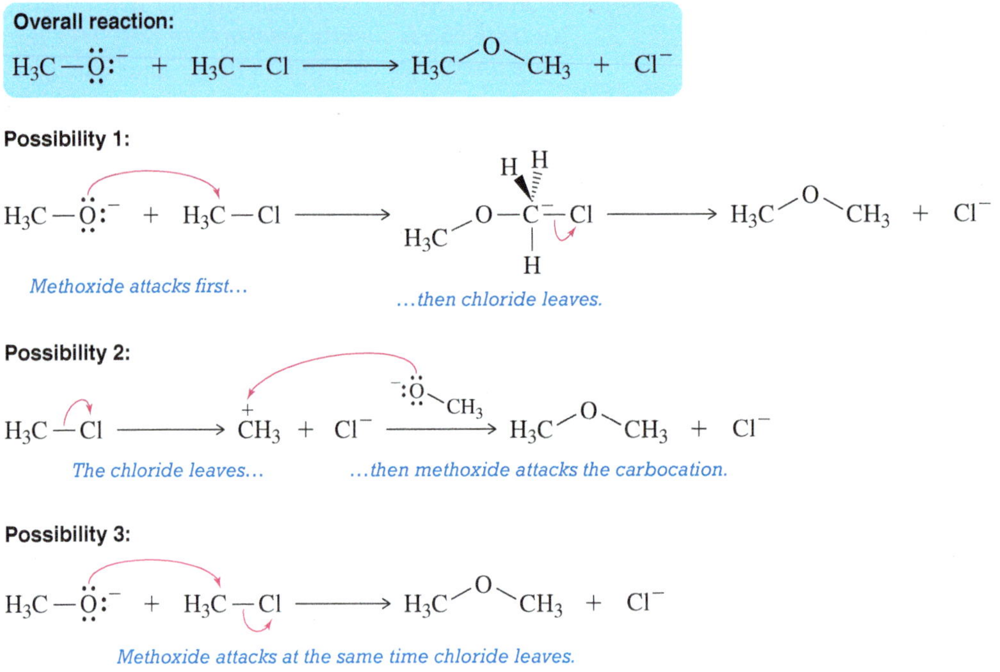

Possibility 1 is easily rejected (**Figure 12.5**). This two-step process features a hypothetical intermediate where carbon has ten electrons in its valence shell. This is a result of methoxide donating a pair of electrons before the chloride leaves. There are exceptions to the octet rule, but carbon isn't one of them. Were we to construct a reaction coordinate diagram of this process, the first step (forming the pentavalent carbanion) would have an incredibly high activation energy. It is such a slow reaction that it never happens.

Possibility 2 is more reasonable (**Figure 12.6**). We studied carbocations previously in alkene addition reactions (Chapters 8 and 9). Carbocations are not particularly stable; were one to form, it would be attacked right away by an electron-rich alkoxide.

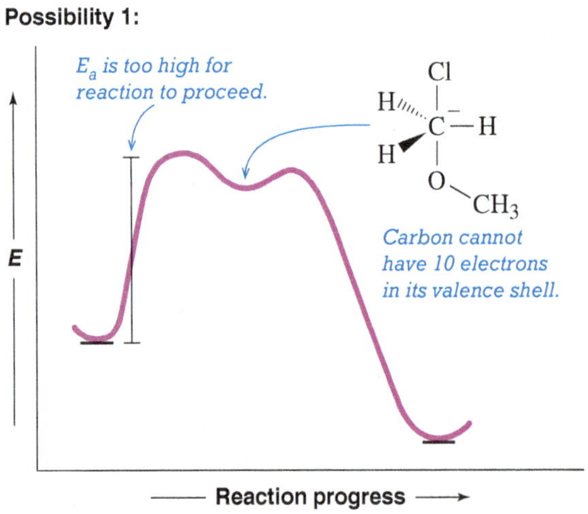

Figure 12.5 A pentavalent carbon can't form. Possibility 1 cannot occur.

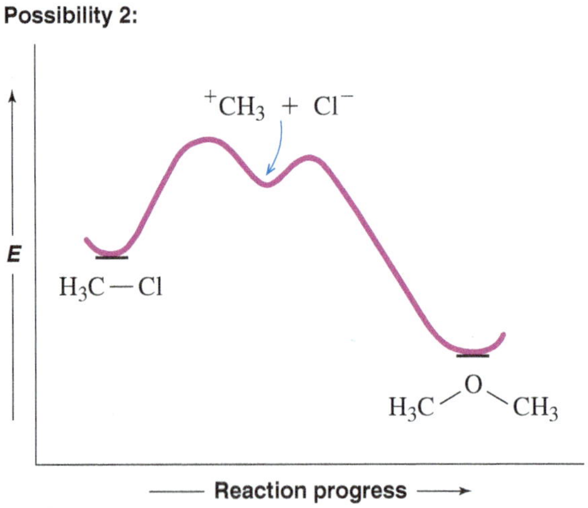

Figure 12.6 Possibility 2 is a reasonable two-step mechanism.

Possibility 3:

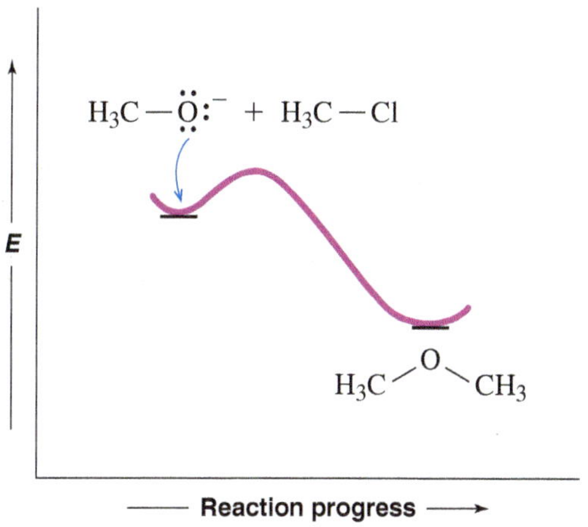

Figure 12.7 Possibility 3 is a reasonable one-step reaction.

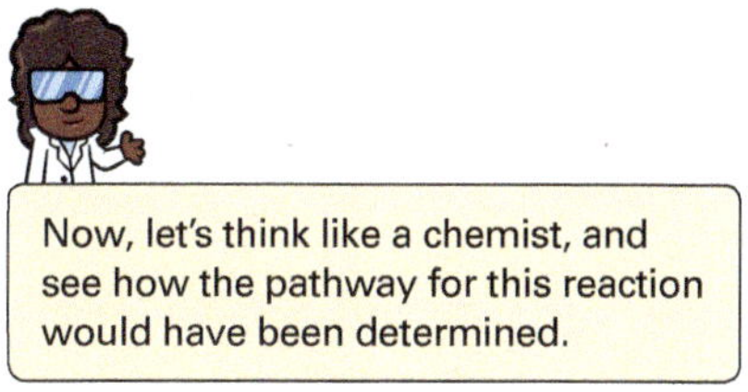

Possibility 3 looks like an acid–base reaction (**Figure 12.7**). Instead of attacking a proton and kicking out a chloride anion, here the base attacks the methyl group. [*Possibility 3 has been discussed previously; do you remember when?*]

Determining which is the more reasonable mechanism for this reaction would have been an exciting challenge for physical organic chemists of the 1930s, and they likely would have had animated conversations about how the reaction proceeds. Luckily, methods for determining mechanisms existed then and now. [*For a fun (and revealing) experience, as you read this section, imagine that you were the chemist who first discovered this reaction.*]

Thinking Like a Chemist

The two plausible mechanisms in **Figure 12.8** have different rate-determining steps. In **possibility 2**, bond breaking to form the unstable carbocation and the chloride ion is likely rate determining. **Possibility 3** has only one step; it must be rate determining. Determination of the rate law (and reaction order) of the reaction can tell us what species are present in the rate-determining step. [*We did this while Thinking Like a Chemist in Chapter 5.*]

Possibility 2:

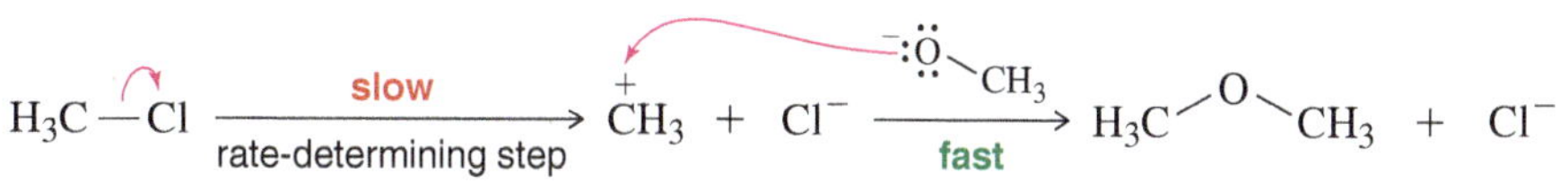

Possibility 3:

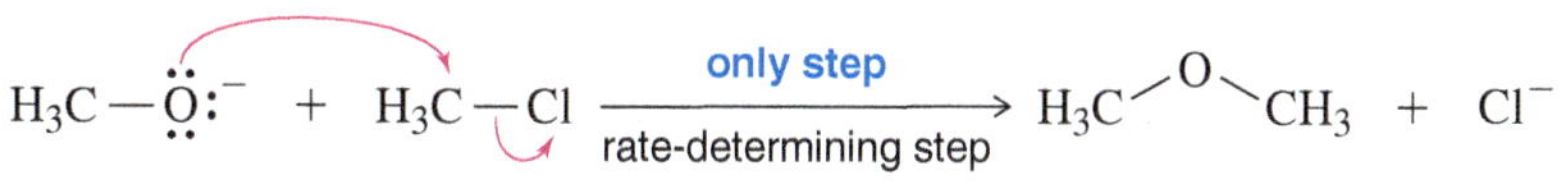

Figure 12.8 Rate-determining steps for our two possible substitutions.

To determine reaction order, we need to see how a change in concentration of the reagents affects the rate of the reaction. A higher concentration results in more collisions, increasing the reaction rate only if the reagents collide during the rate-determining step. The first step is to establish a baseline rate at a particular concentration [0.1 M, for both reactants (CH_3O^- and CH_3Cl)]. This gives us a rate of x (**Table 12.1**). Upon doubling the concentration of CH_3Cl (control → trial 1), the rate doubles to $2x$. Thus, the reaction is first order with regard to CH_3Cl; there must be one molecule of CH_3Cl in the transition state of the rate-determining step. [*For review, see Section 5.3.6.*]

Doubling the concentration of CH_3O^- (control → trial 2) also doubles the rate. Thus, the reaction is also first order with respect to CH_3O^-; there must be one molecule of methoxide in the transition state of the rate-determining step. The reaction is second order overall (**bimolecular**), meaning there are two molecules in the rate-determining step (**Figure 12.9**).

Table 12.1 The rate law is determined by varying the concentration and measuring the rate of the reaction.

$$H_3C-\ddot{O}:^- + H_3C-Cl \longrightarrow H_3C-O-CH_3 + Cl^-$$

	$[CH_3O^-]$ (M)	$[CH_3Cl]$ (M)	Rate
Control	0.1	0.1	x
Trial 1	0.1	0.2	$2x$
Trial 2	0.2	0.1	$2x$

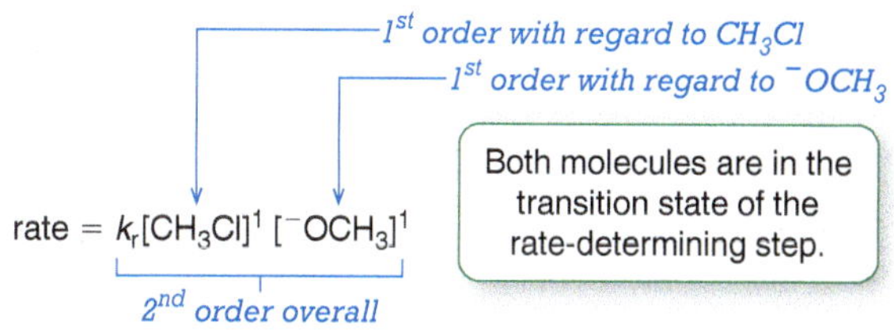

Figure 12.9 The reaction is 1st order with regard to each reactant, making it 2nd order overall.

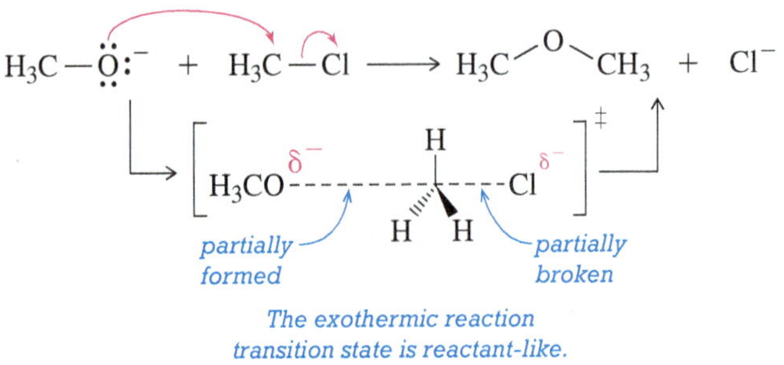

Figure 12.10 This substitution reaction proceeds by possibility 3.

Of the two plausible possibilities (**2** and **3**), only in **possibility 3** are CH_3O^- *and* CH_3Cl present in the transition state of the rate-determining step. Thus, this reaction occurs by **concerted** attack of methoxide, displacing the chloride anion (**Figure 12.10**). There are no intermediates in a concerted reaction; all bond breaking and bond forming occurs in a single step. This mechanism is consistent with a transition state where the new C—O σ bond is partially formed and the C—Cl σ bond is partially broken. ■

The rate-determining step in the mechanism above features the *Substitution* of a *Nucleophile* with a *bimolecular (2 molecules)* rate-determining step. Substitution reactions occurring by this mechanism are termed S_N2 **reactions.** [*For students who memorize, the fact that a one-step reaction is referred to as* S_N2 *may be confusing. Thus, keep in mind what* S_N2 *means, rather than just using it as a label.*]

Assessment 12.10 Given the following rate data, how many molecules are in the rate-determining step of this reaction? [*Does this fit with the mechanism we learned in Chapter 8?*]

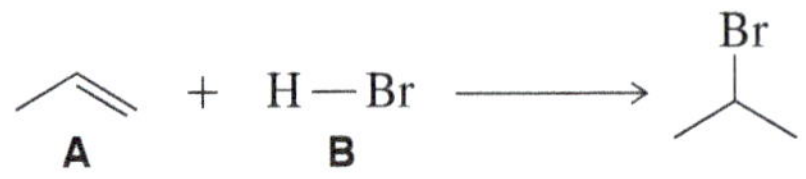

	[A] (M)	[B] (M)	Rate
Control	0.05	0.25	x
Trial 1	0.10	0.25	$2x$
Trial 2	0.05	0.50	$2x$

Though the reaction above (Thinking Like a Chemist) proceeds by mechanistic **possibility 3**, some substitution reaction mechanisms do follow **possibility 2**. The formation of *t*-butyl alcohol by the reaction of water and 2-bromo-2-methylpropane proceeds by this alternate mechanism.

Consider the rate data for the reaction given in **Table 12.2**. Doubling the concentration of *t*-BuBr (control → trial 1) results in doubling of the rate, whereas doubling the concentration of H_2O has no effect. The reaction is first order with regard to *t*-BuBr and zeroth order with regard to H_2O. Though H_2O is involved in the reaction, only *t*-BuBr is present in the rate-determining step (**Figure 12.11**).

Table 12.2 The rate law is determined by varying the concentration and measuring the rate of the reaction.

	[*t*-BuBr] (M)	[H_2O] (M)	Rate
Control	0.1	0.1	x
Trial 1	0.2	0.1	$2x$
Trial 2	0.1	0.2	x

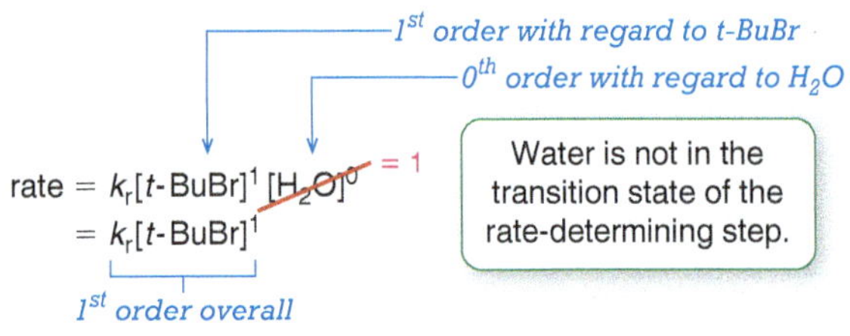

Figure 12.11 The reaction is 1st order. Only one molecule is in the rate-determining step.

These results conclusively show that this reaction proceeds by **possibility 2**. As shown in **Figure 12.12**, Br^- leaves in step 1, the rate determining step, producing a carbocation. [*The transition state of this step involves just one molecule (t-BuBr).*] In step (2), the highly reactive carbocation is attacked by water (a weak nucleophile), forming a hydronium ion. Finally, in step (**3**), deprotonation of the hydronium ion completes the formation of *t*-BuOH. [*When neutral nucleophiles attack carbocations, as discussed in Chapter 8, it will usually be followed by a deprotonation step.*]

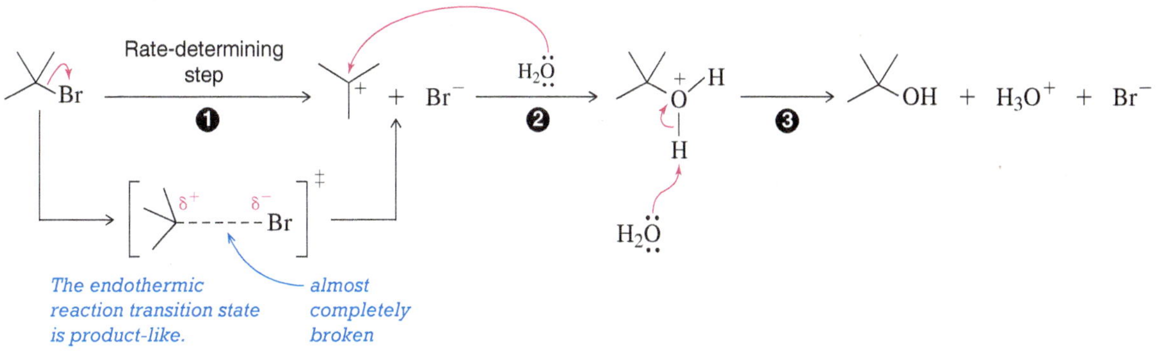

Figure 12.12 **This substitution reaction proceeds by possibility 2.**

This mechanism features the *Substitution* by a *Nucleophile* with a ***unimolecular*** *(1 molecule)* rate-determining step. Substitution reactions occurring by this mechanism are termed $\mathbf{S_N1}$ **reactions.** [*The fact that a two-step reaction is referred to as* S_N1 *can be confusing. Keep in mind what the label is telling you about the mechanism (the number of molecules in the transition state of the rate-determining step) to avoid confusion.*]

12.2.1 SUMMARY

$\mathbf{S_N2}$ **reaction:** The concerted substitution reaction of a haloalkane by a nucleophile, proceeding in a single bimolecular step. The nucleophile attacks at the same time the leaving group leaves.

$\mathbf{S_N1}$ **reaction:** The two-step substitution reaction of a haloalkane by a nucleophile, proceeding with a unimolecular rate-determining step. The leaving group leaves; the resulting carbocation is attacked by the nucleophile in step 2.

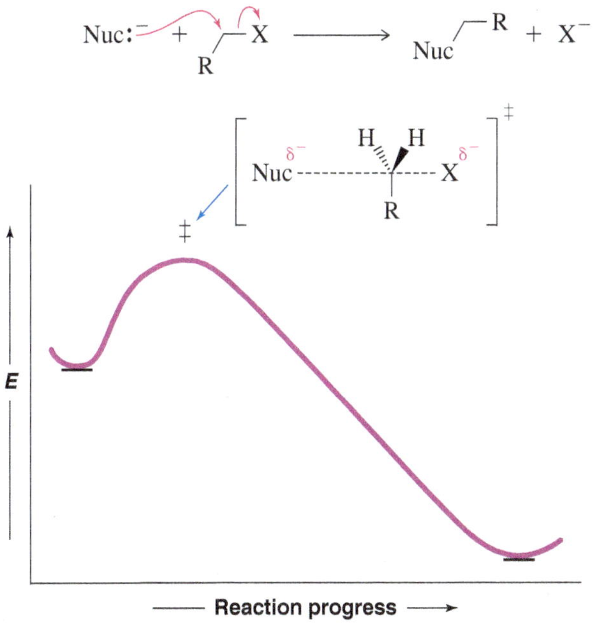

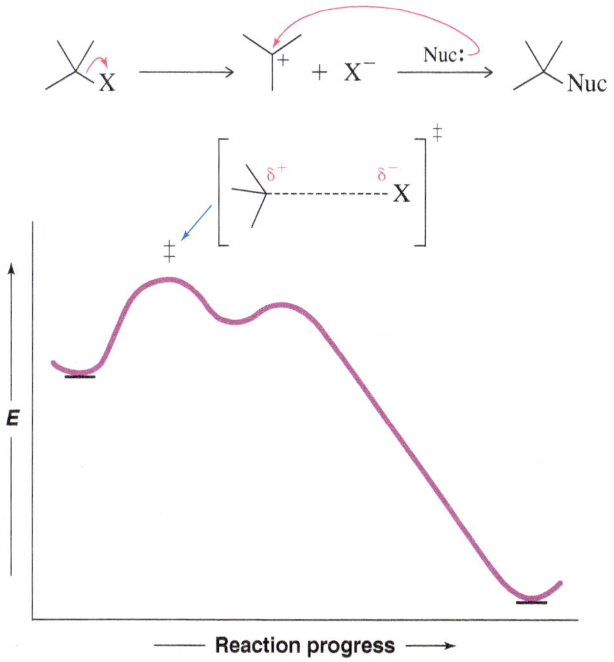

Assessment 12.11 Practice your electron-pushing skills by drawing a mechanism for the following S_N2 reactions.

(a)

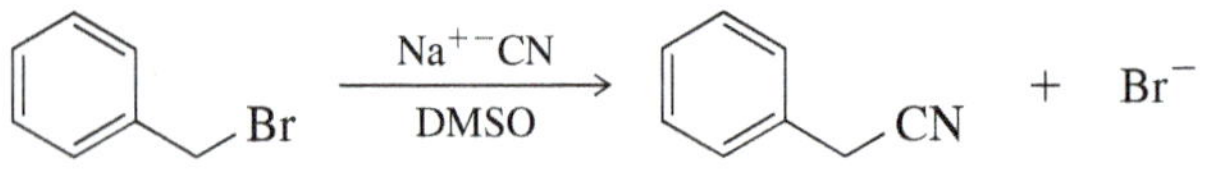

(b)

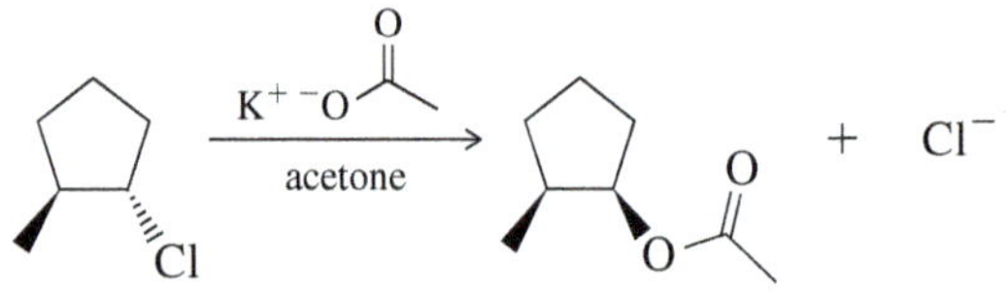

(c)

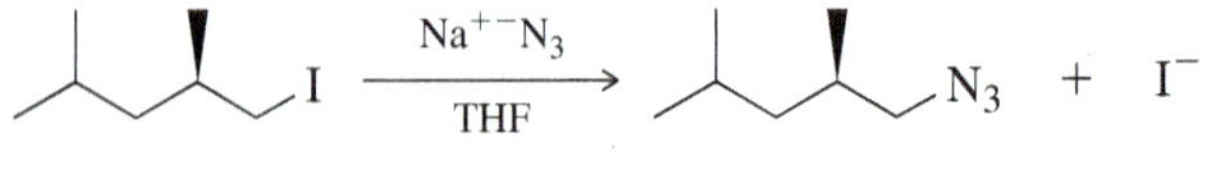

Assessment 12.12 Practice your electron-pushing skills by drawing a mechanism for the following S_N1 reactions.

(a)

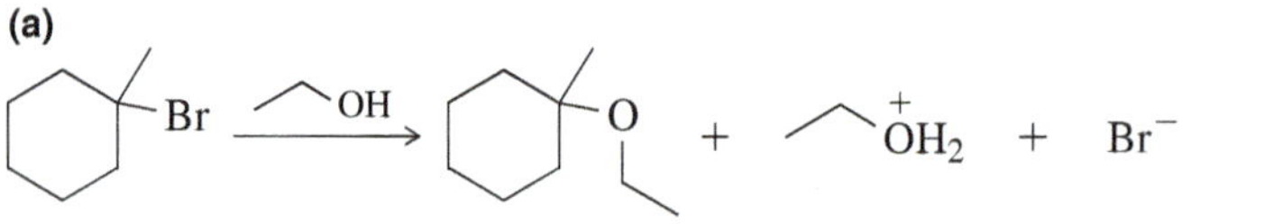

(b)

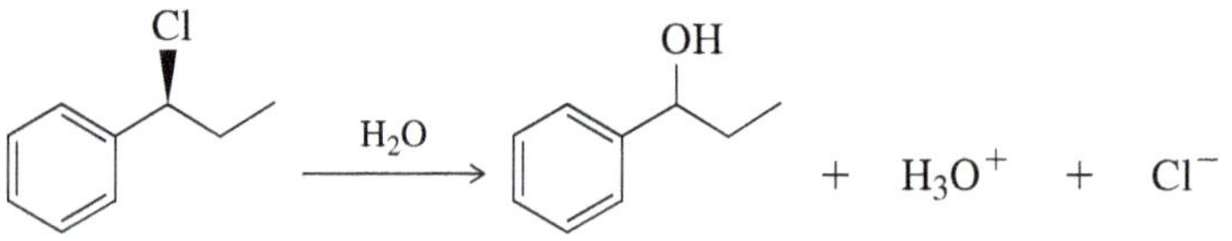

(c)

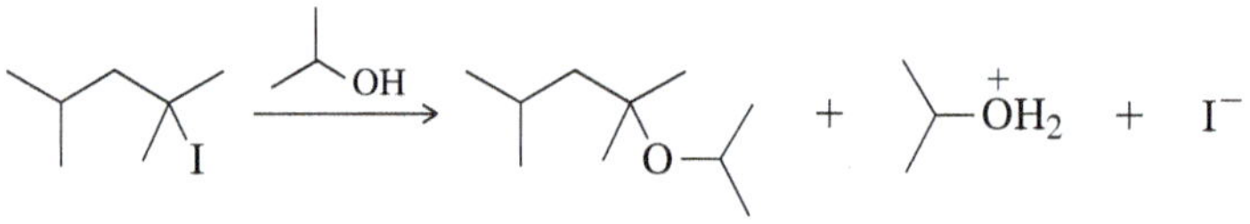

12.2.2 S_N2 and S_N1: a spectrum of reactivity

The rate law (Section 12.2.1) tells us whether a reaction proceeds by the S_N2 or S_N1 mechanism [*a single reaction usually proceeds by just one mechanism*]. However, it is helpful to think of the S_N2 and S_N1 mechanisms being on a *spectrum of reactivity*, in which changing conditions can shift the reaction from one side of the spectrum to the other (**Figure 12.13**). The concept of a spectrum of reactivity helps us understand the subtle situations where reactions have characteristics of both mechanisms. [*These situations will be clear when we get there.*]

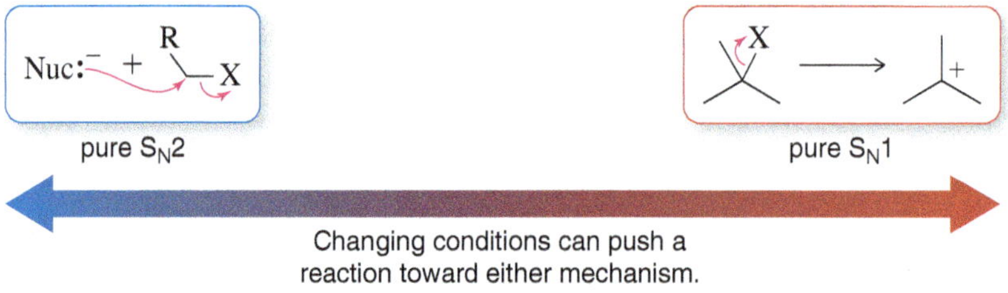

Figure 12.13 All substitution reactions lie on a spectrum between pure S_N2 and pure S_N1, depending on the conditions of the reaction.

12.3 Distinguishing between S_N2 and S_N1

We can predict, without doing the rate experiment, the mechanism by which a substitution reaction will proceed. To start, consider the substitution of the 1° and 3° haloalkanes in the S_N2 and S_N1 reactions shown in **Figure 12.14** and **Figure 12.15**, respectively. We will examine why a particular substitution pattern favors one mechanism and why the other pathway is disfavored by that same substitution pattern. [*Make sense? Read on, it will!*]

S_N2 reaction

Figure 12.14 **A 1° haloalkane** undergoes S_N2 substitution preferentially.

S_N1 reaction

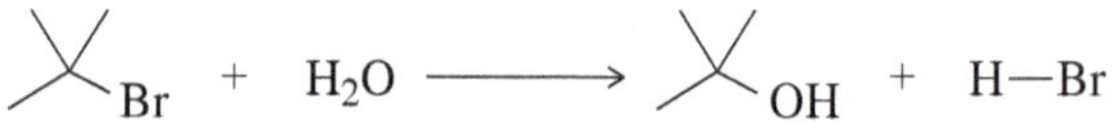

Figure 12.15 **A 3° haloalkane** undergoes S_N1 substitution preferentially.

12.3.1 S_N2 vs. S_N1: where do the electrons go?

We must first acknowledge one of the fundamental differences between the S_N2 and S_N1 mechanisms. In the process of forming the new bond, where does the nucleophile place its electrons?

S_N2

S_N2: In the only step, the nucleophile attacks an sp^3-hybridized carbon containing a full octet of electrons. The empty orbital that accepts the electron pair is the antibonding orbital σ*, opposite the C—Br σ bond (**Figure 12.16**). [*Recall the molecular orbital picture drawn in Section 11.4.*] Thus, to donate electrons to σ*, the nucleophile attacks at a 180° angle to the leaving bromide. This is referred to as **backside displacement.**

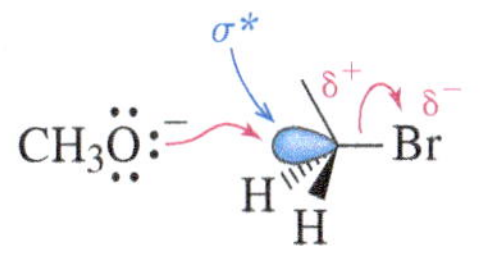

Figure 12.16 The nucleophile donates electrons to σ* in the S_N2 reaction.

S_N1

S_N1: After the bromide leaves in the rate-determining step, the resulting carbocation is sp^2 hybridized and trigonal planar with an empty p orbital. Thus, when the nucleophile attacks, it donates a pair of electrons to the empty p orbital, forming the new σ bond (**Figure 12.17**).

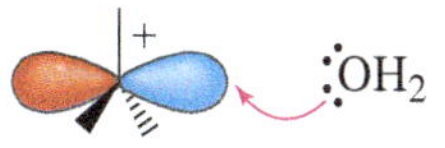

Figure 12.17 The nucleophile donates electrons to an empty *p* orbital in the S_N1 reaction.

There are significant ramifications of attacking either σ* or an empty p orbital. [*That is a topic for later; for now, let's identify which reaction will occur.*]

12.3.2 S_N2 vs. S_N1: substitution of the haloalkane

Knowing where the nucleophile attacks in the S_N2 and in the S_N1 clarifies why substitution reactions of differently substituted haloalkanes proceed by a particular mechanism.

Proceeds by S_N2

Why S_N2? A 1° haloalkane gives the nucleophile sterically unhindered access to σ*, making the concerted attack a low-activation-energy pathway (**Figure 12.18**).

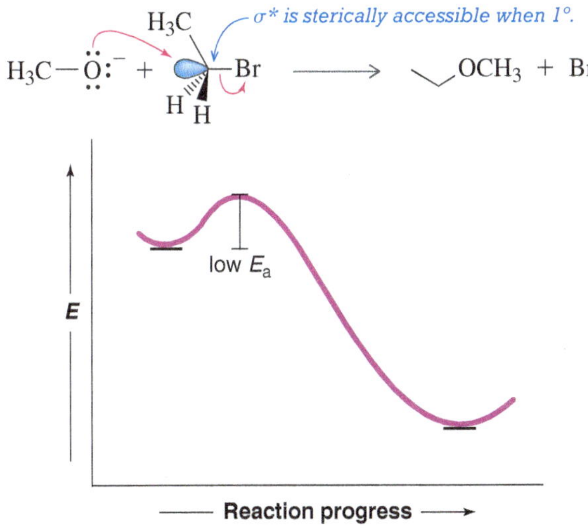

Figure 12.18 σ* is sterically accessible in a 1° haloalkane.

Proceeds by S_N1

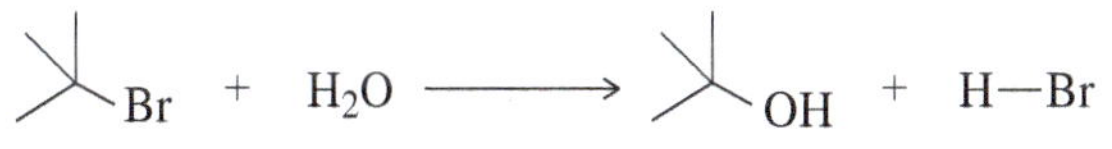

Why not S_N2? A 3° haloalkane is very hindered, with three alkyl groups blocking access to σ*. The steric strain encountered as the nucleophile approaches increases the energy of the transition state. Thus, S_N2 at a 3° carbon has such a high E_a pathway that it never happens (**Figure 12.19**).

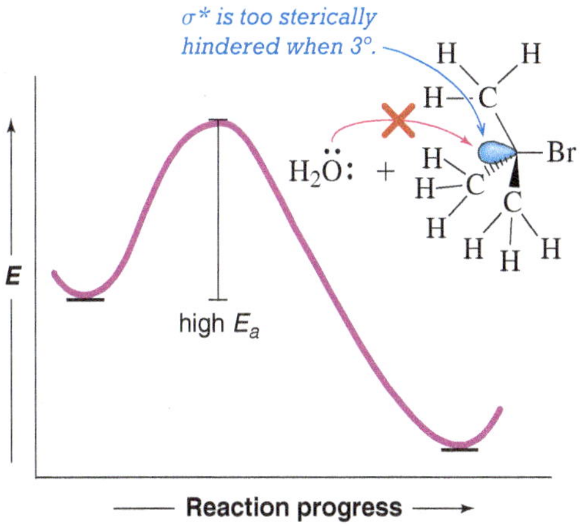

Figure 12.19 σ* is sterically hindered in a 3° haloalkane.

Why not S_N1? With little hyperconjugation to help stabilize them, 1° carbocations are very unstable. Formation of a 1° carbocation is an uphill process (high E_a) and therefore a very slow reaction (**Figure 12.20**).

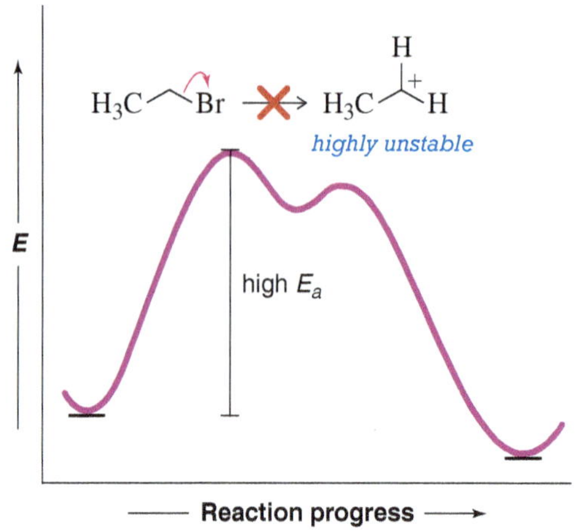

Figure 12.20 **1° carbocations are too unstable to form,** so the S_N1 reaction cannot happen.

Why S_N1? The 3° carbocation has three alkyl groups to stabilize it by hyperconjugation. Thus, formation is relatively easy by a low-activation-energy pathway (**Figure 12.21**).

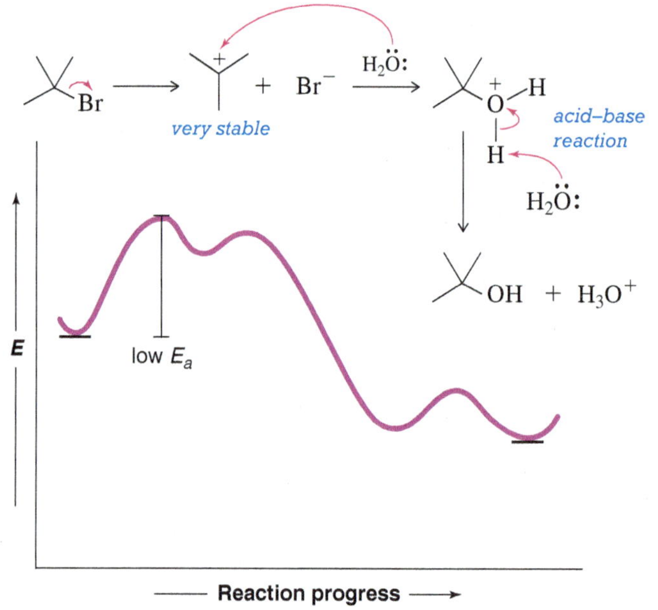

Figure 12.21 **3° carbocations are very stable and form easily,** so the S_N1 reaction is favored.

In general, 1° alkyl halides provide sterically unhindered access to σ*, so they readily participate in S_N2 reactions. On the other hand, 1° halides leave to form very unstable carbocations; thus, S_N1 reactions are impossible with 1° haloalkanes. 1° haloalkanes will *only* undergo substitution by the S_N2 mechanism. [*If a 1° haloalkane is also allylic, meaning the resulting carbocation can be stabilized by resonance, then S_N1 is possible.*]

Alternatively, 3° haloalkanes are too hindered to allow the nucleophile to reach σ*, so S_N2 reactions of 3° haloalkanes are impossible. 3° halides leave to produce highly stable carbocations; thus, S_N1 reactions are very favorable with 3° haloalkanes. 3° haloalkanes will *only* undergo substitution by the S_N1 mechanism.

Intermediate between the two, 2° haloalkanes provide moderate access to σ* while also forming moderately stable carbocations, so S_N2 and S_N1 reactions are both possible. [*If voting for a mechanism, 2° haloalkanes would be undecided, needing further analysis before making a decision.*] The ideas expressed here are summarized in **Table 12.3**.

Table 12.3 The rate of S_N2 reaction as a function of haloalkane substitution.

Haloalkane	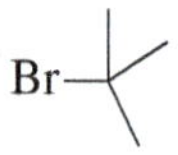	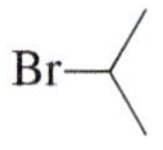	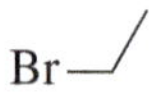	Br—CH_3
Relative rate	<0.002	1	80	4000

Low → High
relative S_N2 reactivity

Carbons that are directly attached to alkenes are termed **allylic.** Leaving groups on allylic carbons undergo especially fast S_N1 reactions because the resulting carbocation is stabilized by resonance. **Figure 12.22** shows a greater than 100× increase in rate as compared to the 3° chloroalkane, which is stabilized only by hyperconjugation. In Chapter 21, we discuss more situations where being allylic changes the reactivity profile.

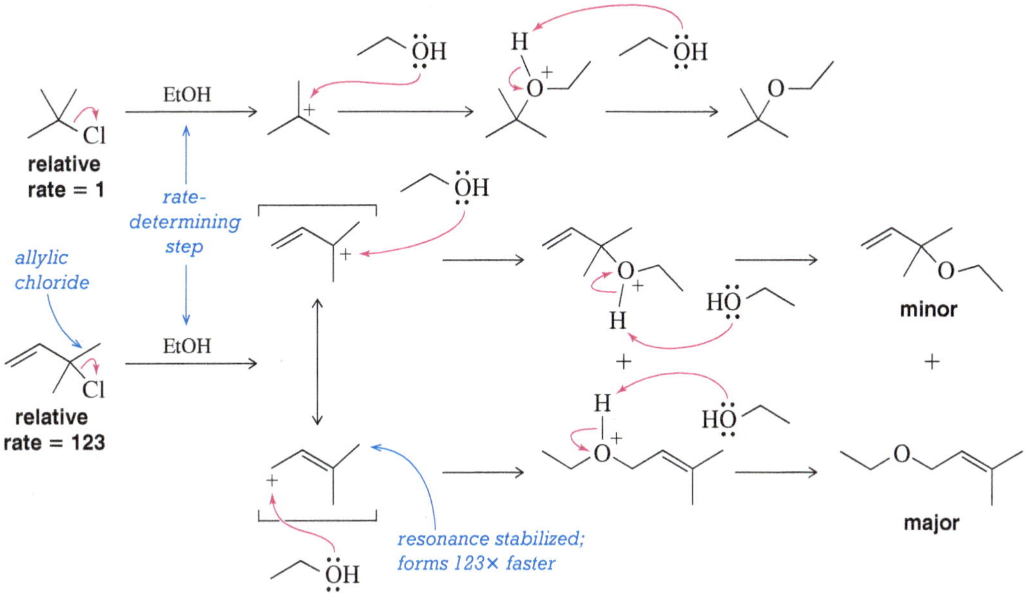

Figure 12.22 Allylic carbocations (conjugated) are stabilized by resonance and thus form more quickly.

Figure 12.23 makes clear that the order of the S_N1 reaction rate parallels the relative stability of carbocations. This is consistent with an endothermic first step where the carbon that is losing the leaving group has significant cationic character in the transition state. [*Recall the Hammond postulate; Section 5.2.4.*]

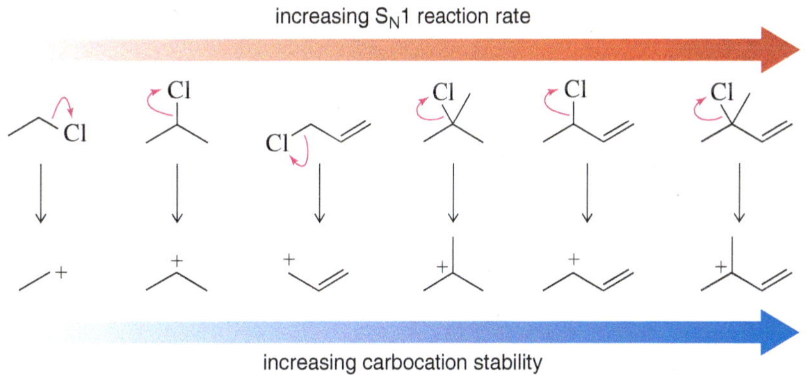

Figure 12.23 S_N1 reactions are faster when the resulting carbocation is stabilized.

Steric hindrance at neighboring carbons slows down reactions proceeding by the S_N2 mechanism. Consider the rates of S_N2 reactions at a 1° carbon as a function of substitution at the neighboring carbon (**Table 12.4**). As substitution increases, the reaction rate decreases. In fact, when the neighboring carbon is fully substituted (quaternary, or 4°), the S_N2 reaction essentially cannot occur.

Table 12.4 The rate of an S_N2 reaction as a function of substitution at the neighboring carbon.

Haloalkane	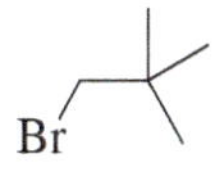	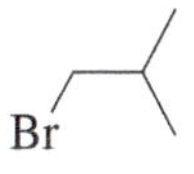	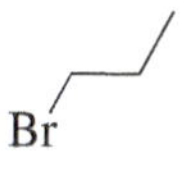	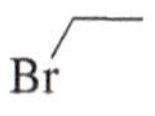
Relative rate	0.0006	1	22	28

These results demonstrate that steric hindrance at a neighboring carbon can block the nucleophile's access to σ*. In a molecular model of neopentyl bromide, the steric hindrance is obvious (**Figure 12.24**). [*Make a model on your own to see this.*] When you encounter a 1° carbon, verify that the neighboring carbon is not fully substituted before predicting an S_N2 reaction.

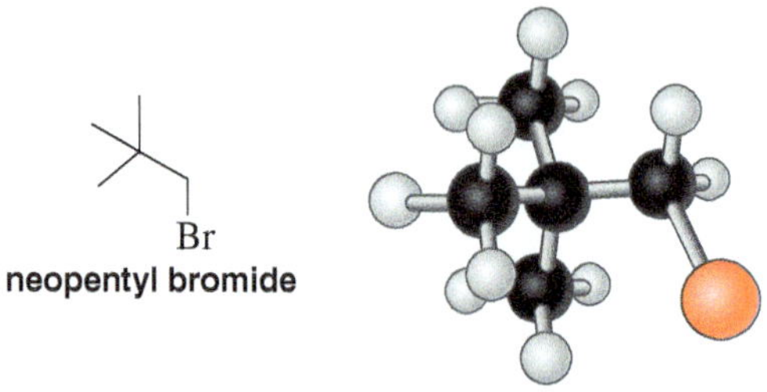

Figure 12.24 Neopentyl substitution blocks σ*, even for 1° haloalkanes.

12.3.2 SUMMARY

Substitution of the haloalkane impacts the mechanism of substitution reactions in the following ways:

- 1° R—X: Unhindered, forms unstable carbocations; always S_N2; only S_N1 if resonance is possible.
- 2° R—X: Moderately unhindered, forms somewhat stable carbocations; can be S_N2 or S_N1, depending on other factors.
- 3° R—X: Extremely hindered, forms very stable carbocations; never S_N2, always S_N1.
- Steric hindrance at a neighboring carbon slows down S_N2 reactions.
- Resonance-stabilized carbocations form quickly in S_N1 reactions and allow for the formation of two products.

Assessment 12.13 Which of the following S_N2 reactions should proceed at a faster rate? Justify your answer on a reaction coordinate diagram.

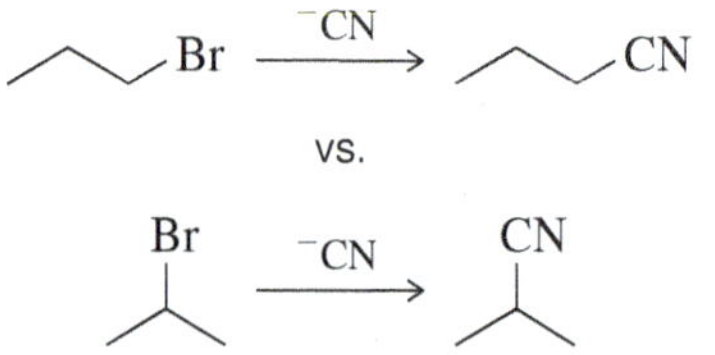

Assessment 12.14 Which of the following S_N1 reactions should proceed at a faster rate? Justify your answer on a reaction coordinate diagram.

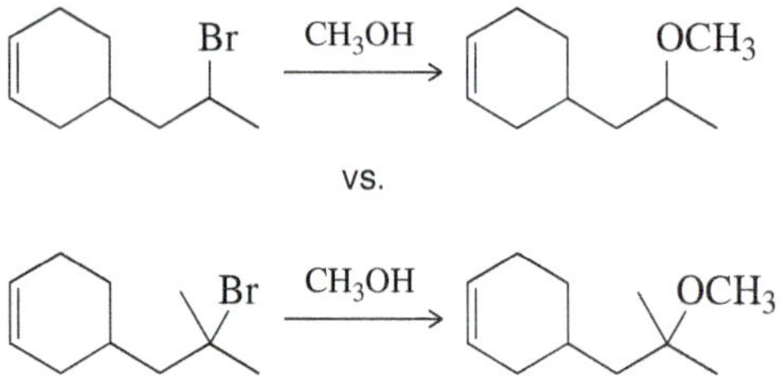

Assessment 12.15 The allylic bromide below gives two S_N1 products. Justify the formation of each.

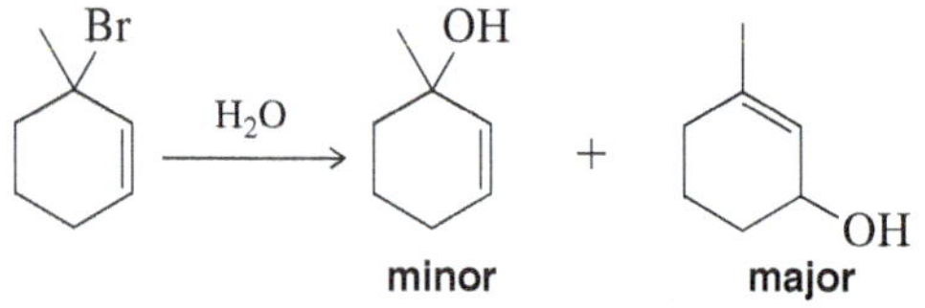

Assessment 12.16 Even with an excess of cyanide, only one equivalent will react with the following dibromoalkane. To which carbon will the cyanide add? Predict the product and explain your choice.

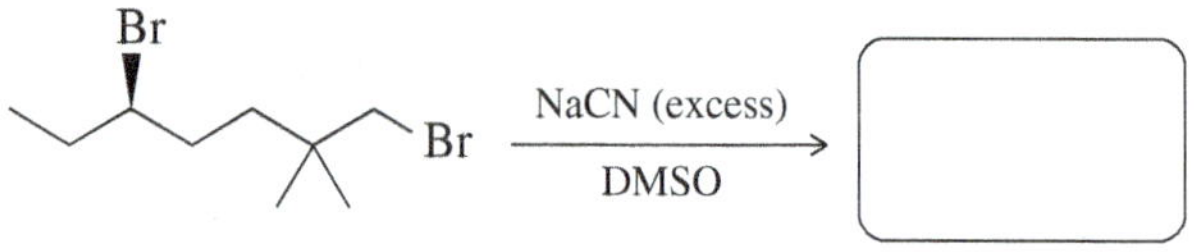

The reactivity of allylic systems can be discovered by working Assessment 12.15 together.

Solved Assessment 12.15

As we do when we work mechanism problems, we can number the carbons (**Figure 12.25**). This shows that in addition to the expected product (minor) where water has added to C_1, we have an additional product (major) where water has added to C_3 with the alkene having moved over. Following cleavage of the C—Br bond in the first step of the S_N1 reaction, the resulting carbocation is stabilized by resonance. Recall from Chapter 2 that this means that the positive charge is delocalized over C_1 and C_3. Water, as a weak nucleophile, can now attack at either carbon, resulting in the formation of two products. The major product is the one containing the more substituted alkene. ■

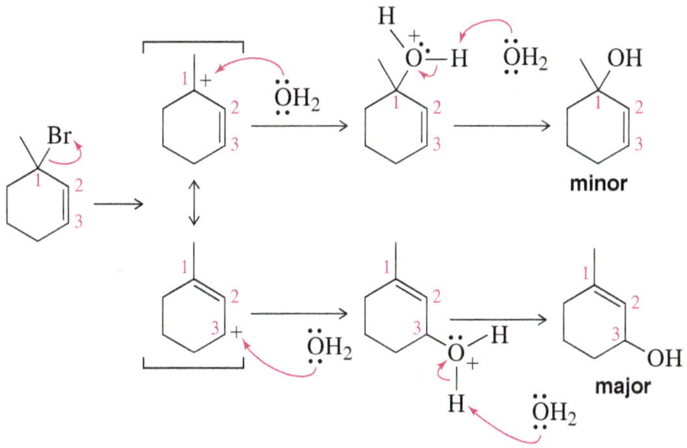

Figure 12.25 **Attack at C_1 and C_3 of the resonance-stabilized carbocation results in two products.**

12-A CHEMISTRY IN CONTEXT: Reaction Analysis

Deceptive kinetic studies

The kinetic studies from Section 12.2.1 don't always give such exact results. Such is true for a class of compounds containing a Lewis basic atom attached at the β-carbon of a haloalkane. As an example, chloroethyldiethylamine is a 1° haloalkane. Kinetic studies of its reaction with water suggest a unimolecular rate-determining step (an S_N1 reaction) (**Figure 12.26**). Yet 1° haloalkanes never form the carbocation intermediate of an S_N1 reaction. The data tell that there is one molecule in the transition state of the rate-determining step, but say nothing about how the electrons move.

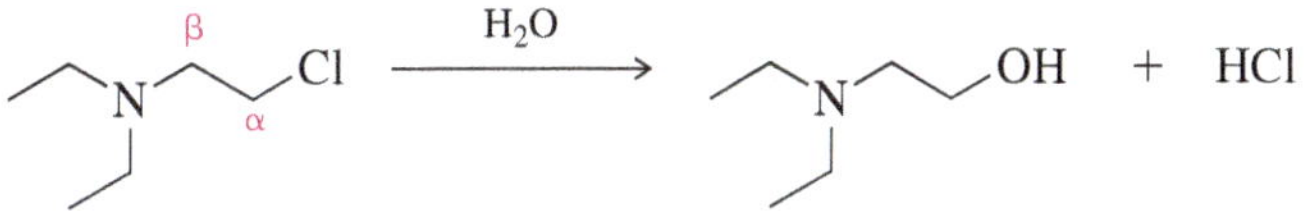

chloroethyldiethylamine (A)

	[A] (M)	[H_2O] (M)	Rate
Control	0.1	0.1	x
Trial 1	0.2	0.1	$2x$
Trial 2	0.1	0.2	x

1st order with regard to A

0th order with regard to H_2O

$$\text{rate} = k_r[\mathbf{A}]^1\,[H_2O]^0 \quad (= 1)$$
$$= k_r[\mathbf{A}]^1$$

1st order overall

Water is not in the transition state of the rate-determining step.

Figure 12.26 **Kinetic studies show a reaction that is first order overall.** Thus, one molecule is present in the transition state of the rate-determining step.

Molecules of this type, while showing an S_N1 profile, participate in substitution reactions by doing two consecutive concerted (S_N2-like) reactions (**Figure 12.27**). (1) The Lewis basic nitrogen donates an electron pair to σ*, kicking out the chloride anion. This is termed **neighboring group participation (anchimeric assistance)**, where the neighboring nitrogen assists with the expulsion of the chloride ion. (2) This three-membered ammonium ring is a good electrophile and is attacked by water at either primary carbon, returning the lone pair of electrons to the nitrogen. (3) Removal of a proton from the hydronium ion completes the mechanism. Water, while doing an S_N2 reaction here, is not involved in the rate-determining first step, yielding kinetic data consistent with an S_N1 mechanism.

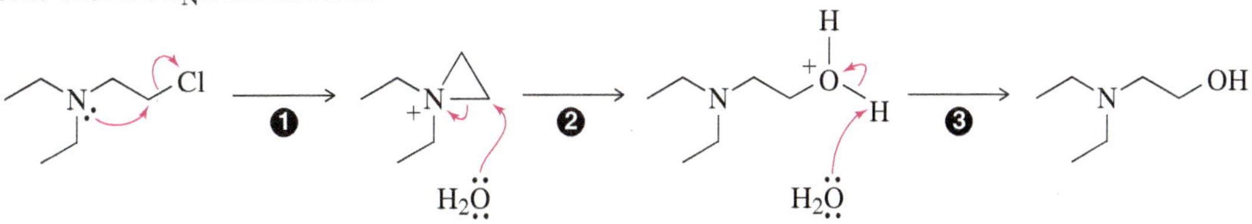

Figure 12.27 **Despite the kinetic data, the reaction proceeds by consecutive S_N2-like reactions.**

12.3.3 S_N2 vs. S_N1: nucleophilic strength and solvent dependence

In both S_N2 and S_N1 reactions, a nucleophile displaces a good leaving group. Similar to the effects of substitution, nucleophilic strength plays a role in determining the mechanistic outcome (S_N2 vs. S_N1). The nucleophile can be any Lewis basic molecule (electron pair donor). Several nucleophiles are shown in Table 12.5.

As Table 12.5 shows, nucleophilic strength (and relative reaction rates) parallels basicity, with negatively charged (anionic) nucleophiles considered strong. Neutral nucleophiles, donating a lone pair from a neutral atom, are considered weak. The exception to this generalization [*there is always at least one*] is the amine, which, while neutral, is considered moderately strong.

Table 12.5 Nucleophiles that can participate in substitution reactions.

$Nuc^{:-} + H_3C{-}X \longrightarrow H_3C{-}Nuc + X^-$		
Name	**Product**	**Relative rate of S_N2 reaction**
H_2O	$H_3C{-}\overset{+}{O}H_2$	1
$H_3C{-}CO_2^-$	$CH_3CO_2CH_3$	500
NH_3	$H_3C{-}\overset{+}{N}H_3$	700
Cl^-	$H_3C{-}Cl$	1000
^-OH	$H_3C{-}OH$	10,000
CH_3O^-	$H_3C{-}OCH_3$	25,000
I^-	$H_3C{-}I$	100,000
^-CN	$H_3C{-}CN$	125,000
^-SH	$H_3C{-}SH$	125,000

Assessment 12.17 Order the following molecules on the basis of their nucleophilic strength using the pK_a values of their conjugate acids.

Assessment 12.18 For each pair, choose the more reactive nucleophile.

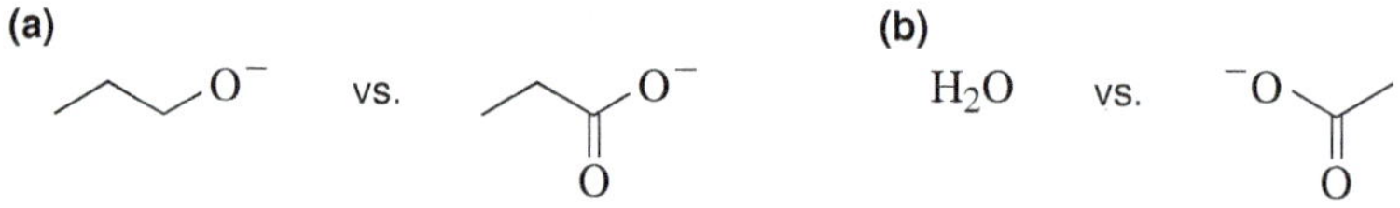

Assessment 12.19 Which S_N2 reaction would you expect to be faster? Explain your answer.

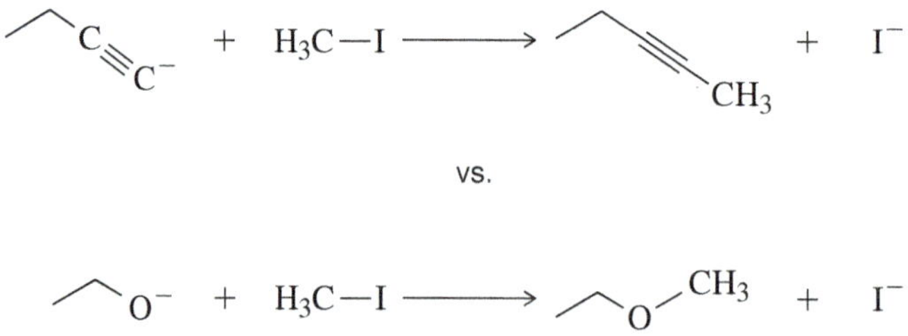

To examine the importance of nucleophilic strength, consider the substitution reaction of 2-bromobutane, a 2° haloalkane that can react via either the S_N2 or S_N1 reaction (Figure 12.28).

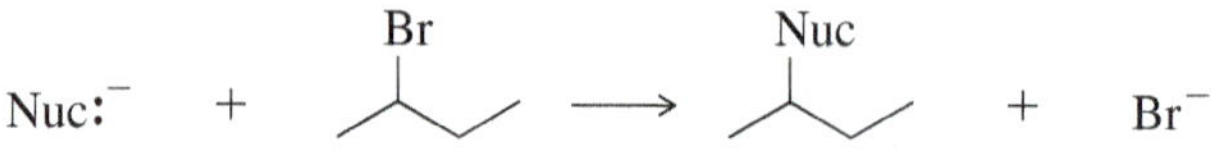

Figure 12.28 Substitution of a 2° haloalkane.

Strong nucleophiles react by the S_N2 mechanism.

Because stronger nucleophiles are less stable, reactions using stronger nucleophiles should start at a higher energy on a reaction coordinate diagram. As the strength of the nucleophile increases from H_2O to ^-OAc to ^-CN, it becomes easier to reach the transition state (**Figure 12.29**). [*This assumes that the transition state for each is the same—a close, but not exact, approximation.*]

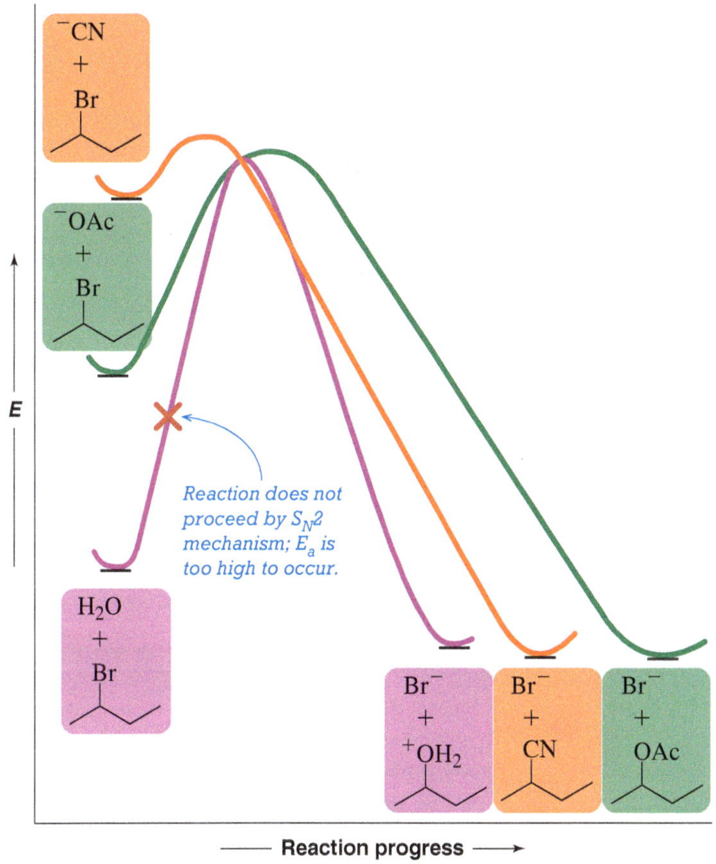

Figure 12.29 Stronger nucleophiles react more quickly in S_N2 reactions.

Weak nucleophiles react by the S_N1 mechanism.

Weak nucleophiles are too stable to react with a neutral electrophile in the S_N2 reaction. For a weak nucleophile to react, the electrophile must be stronger, like a carbocation. On a reaction coordinate diagram showing both an S_N1 and S_N2 reaction, the activation energy for water to react directly (S_N2) is higher than the activation energy for the bromide to leave on its own (S_N1) (**Figure 12.30**).

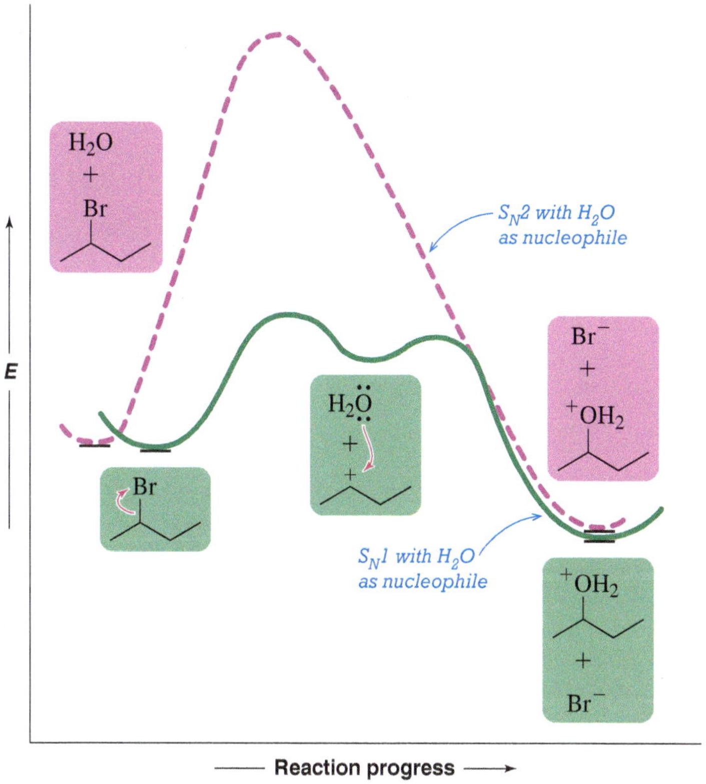

Figure 12.30 Weak nucleophiles wait for a carbocation to form in S_N1 reactions.

12.3.3 SUMMARY A

- In the absence of other considerations, anionic nucleophiles and amines react directly, donating a pair of electrons to σ^* in S_N2 reactions. Neutral nucleophiles are too weak to react in the S_N2 reaction. Instead, a more reactive carbocation forms so an electron pair can be donated to an empty p orbital in an S_N1 reaction.

Assessment 12.20 Would the following nucleophiles be more likely to participate in an S_N1 or S_N2 reaction?

(a)

(b) F^-

(c)
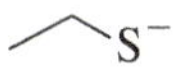

(d) NH_3

(e) H_2O

(f)
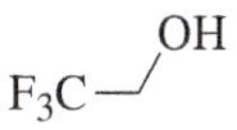

To understand the importance of solvent on reactions, consider sodium cyanide (the source of cyanide, a nucleophile) added to a haloalkane. Larger haloalkanes are generally

Table 12.6 Examples of polar protic and polar aprotic solvents.

Polar protic	Polar aprotic
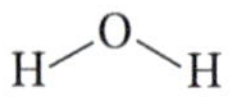 water	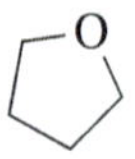 tetrahydrofuran (THF)
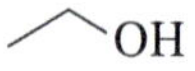 ethanol	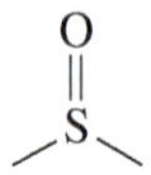 dimethyl sulfoxide (DMSO)
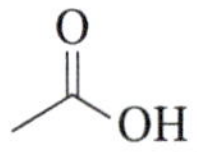 acetic acid	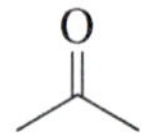 acetone

nonpolar molecules insoluble in water (Section 11.3.3). On the other hand, NaCN is an ionic solid. Thus, NaCN does not dissolve in 2-bromobutane. With the reactants in different phases (solid and liquid), few collisions occur. In the absence of solvent, reactions rarely occur [*though green chemists are working to change this*]. As a result, reactions are run in a solvent capable of dissolving both reactants; within the same phase, collisions are now possible.

With solvents as another variable, how might a solvent alter the reactivity of a nucleophile? There are two main solvent types to consider: **polar protic** and **polar aprotic** (Table 12.6). Polar protic solvents form hydrogen bonds by sharing a hydrogen with an electron-rich atom. As you learned in previous chemistry courses, O, N, and F can serve as hydrogen bond donors. Thus, water, ethanol, and acetic acid, containing a hydrogen attached to an electronegative oxygen, are examples of polar protic solvents.

THF, DMSO, and acetone (polar aprotic solvents), with electronegative atoms, are polar molecules. Because there are no hydrogens attached to the electronegative atoms, they are unable to participate as hydrogen bond donors.

Assessment 12.21 Identify the following solvents as polar protic, polar aprotic, or nonpolar.

(a) (b) (c) (d) (e) (f)

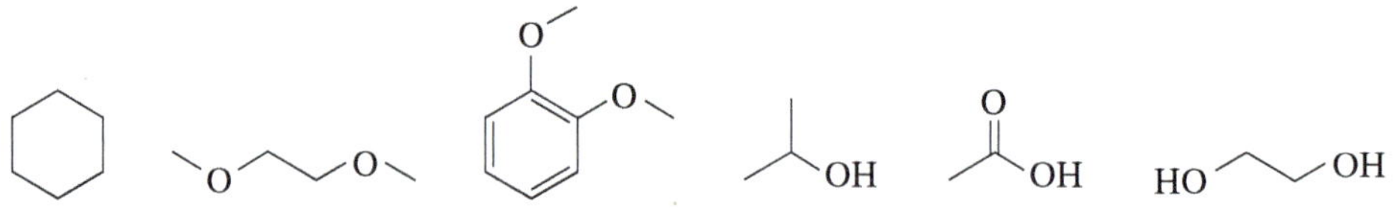

Consider how a negatively charged molecule might interact with a polar protic solvent. [*Draw it, using cyanide* ($^-$CN) *as the nucleophile and water as the polar protic solvent.*] How might a polar aprotic solvent interact with the same nucleophile? [*Again, draw it, using cyanide* ($^-$CN).]

Polar aprotic solvents make nucleophiles stronger.

Solvents that cannot hydrogen bond are unable to stabilize electron-rich species. At minimum, they have no stabilizing effect on anionic species like cyanide. But cyanide is usually part of an ionic compound like sodium cyanide, which changes the situation slightly. As an ionic solid, $^-$CN is close to Na^+. Thus, it is relatively happy (stable). However, when dissolved in a polar aprotic solvent (acetone), Na^+ is surrounded by electron-rich oxygens; $^-$CN is left alone. Stripping the cation away makes $^-$CN more reactive (Figure 12.31).

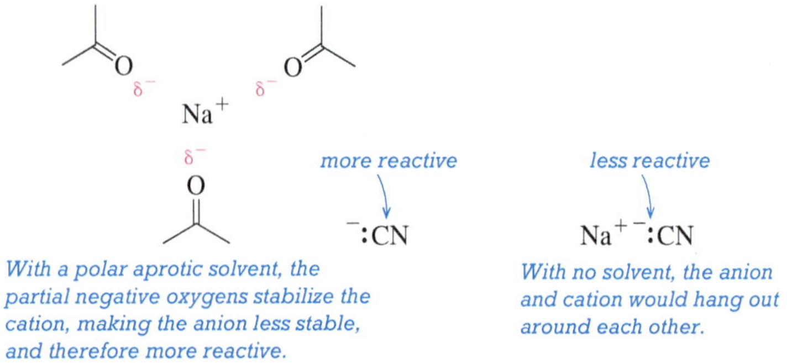

Figure 12.31 Polar aprotic solvents solvate cations but not anions. The nucleophile becomes more reactive.

Polar protic solvents make nucleophiles weaker.

The ability to hydrogen bond is due to the δ+ on hydrogens attached to electronegative atoms in a polar covalent bond. Thus, hydrogen bonding stabilizes electron-rich species. Surrounding $^-$CN with hydrogen bonds makes it less interested in reacting with another electrophile (Figure 12.32).

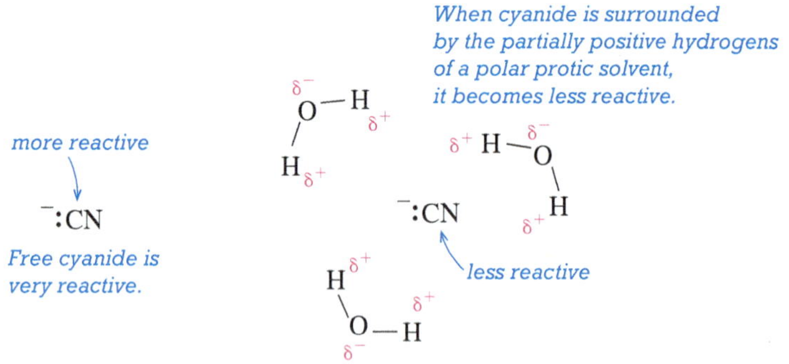

Figure 12.32 Polar protic solvents solvate anions, making them less reactive nucleophiles.

12.3.3 SUMMARY B

- By separating anions from cations, **polar aprotic** solvents make nucleophiles more reactive, favoring S_N2 reactions. By donating hydrogen bonds to anions, **polar protic** solvents make nucleophiles less reactive, favoring S_N1 reactions.

Figure 12.33 shows how nucleophilic strength changes in a polar aprotic solvent, decreasing as you move down the periodic table. Polar aprotic solvents have no effect on the anion itself; all that matters is the way in which the atom itself stabilizes the negative charge. As described in Section 4.4.2, the larger anion (iodide) has the negative charge more spread out and is therefore more stable. [*As far as the anion is concerned, having a polar aprotic solvent is like having no solvent at all.*]

In a polar protic solvent, one that *can* hydrogen bond, the trend reverses. The smaller anion (fluoride) is now less reactive/more stable. And the larger anion (iodide), normally more stable, is now the better nucleophile. This reversal is a result of the enhanced ability of a solvent to hydrogen bond with smaller anions (**solvation**) (**Figure 12.34**). The fluoride anion is more easily surrounded by hydrogen bond donors than the larger iodide anion. As a result, the stabilizing effect is greater for fluoride than for iodide, resulting in a net reversal of nucleophilic strength. [*This is a general trend; smaller anions are more easily solvated.*]

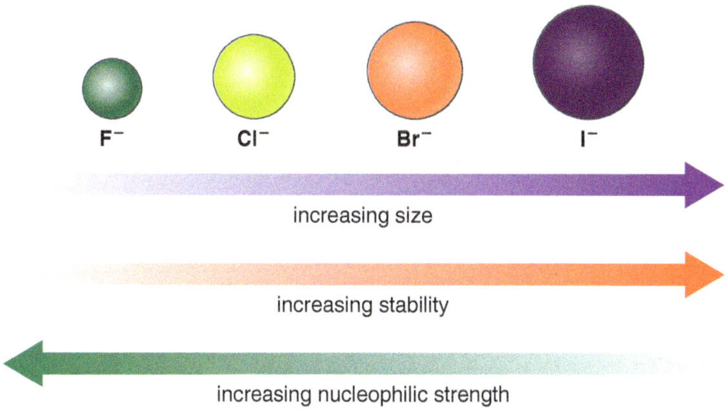

Figure 12.33 Nucleophilic strength in a polar aprotic solvent decreases with increasing size.

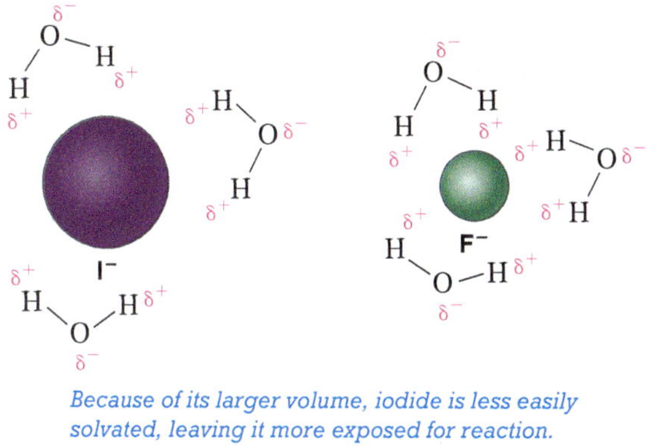

Figure 12.34 Smaller anions are more easily solvated, becoming less reactive in polar protic solvents.

Assessment 12.22 Which nucleophile would be more reactive in the solvent given?

(a)

(b)

Cl^- vs. Br^- in DMSO

(c)

F^- vs. I^- in 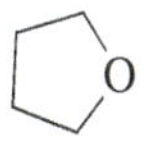

Assessment 12.23 Will the following S_N2 reaction proceed more rapidly in DMSO or H_2O?

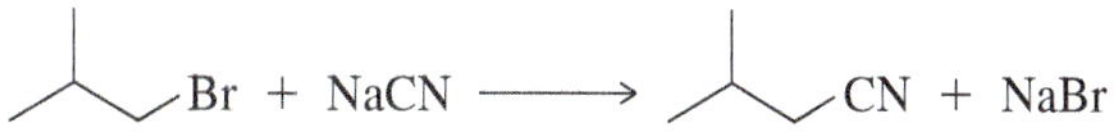

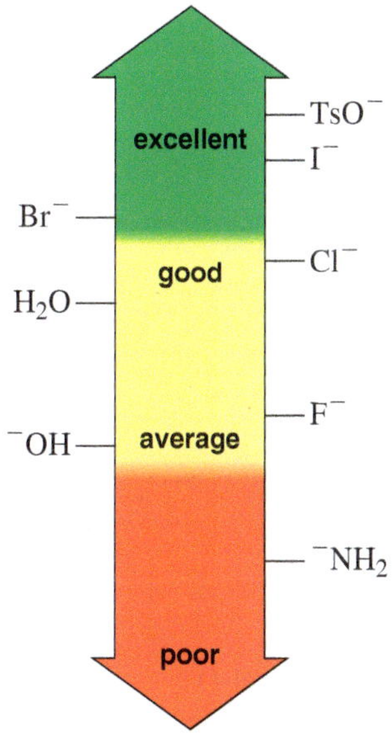

Figure 12.35 Stable bases make good leaving groups.

12.3.4 S_N2 vs. S_N1: leaving group

A third variable to consider in substitution reactions is the leaving group (**Figure 12.35**). The nature of the leaving group has little impact on whether the reaction proceeds by the S_N1 or S_N2 mechanism. However, for either to occur there must be a leaving group. Halogens are very good leaving groups. Thinking about why this is the case will help us understand why other functional groups are, or are not, good leaving groups.

The most important characteristic of a good leaving group is to be stable after taking the electron pair from what was formerly the C—X σ bond. Because the leaving groups are conjugate bases, their relative stability can be inferred by considering the pK_a of their conjugate acids (from Section 4.3.2). In addition, the stability can be rationalized based on structural effects (from Section 4.4). From this, it is clear that the halide anions (except F^-) have conjugate acids with low pK_a values; they are good leaving groups. Iodide, the largest halide, delocalizes the negative charge over a large volume; it is the best leaving group of the halides (**Table 12.7**).

Table 12.7 Leaving group ability is correlated to base stability.

Reaction	Relative rate of reaction	pK_a values of HX
$^-OH + RCH_2I \longrightarrow RCH_2OH + I^-$	30,000	−10
$^-OH + RCH_2Br \longrightarrow RCH_2OH + Br^-$	10,000	−9
$^-OH + RCH_2Cl \longrightarrow RCH_2OH + Cl^-$	200	−7
$^-OH + RCH_2F \longrightarrow RCH_2OH + F^-$	1	3.2

As the pK_a of the conjugate acid of a leaving group increases, the leaving group ability decreases (**Table 12.8**). Sulfonate anions (conjugate acid $pK_a = -5$), which we study in Chapter 13, are very good leaving groups. The chemistry of water as an excellent leaving group (pK_a of $H_3O^+ = -1.7$) is discussed in Chapter 13. Alcohols (pK_a of $ROH_2^+ = -1$) feature prominently as good leaving groups in the chemistry of carbonyls (Chapter 17). By the same logic, hydroxide (pK_a of $H_2O = 16$) and alkoxides (pK_a of ROH = 17) are not good leaving groups. Though condition dependent, if the pK_a of the conjugate acid is greater than 5, the group is generally considered to be a poor leaving group. [*After studying Table 12.8, work Assessments 12.24 and 12.25 to see how well you understand this concept.*]

Table 12.8 More stable bases are better leaving groups.

Reaction (leaving group screened in blue)	Conjugate acid	pK_a
Sulfonate ion 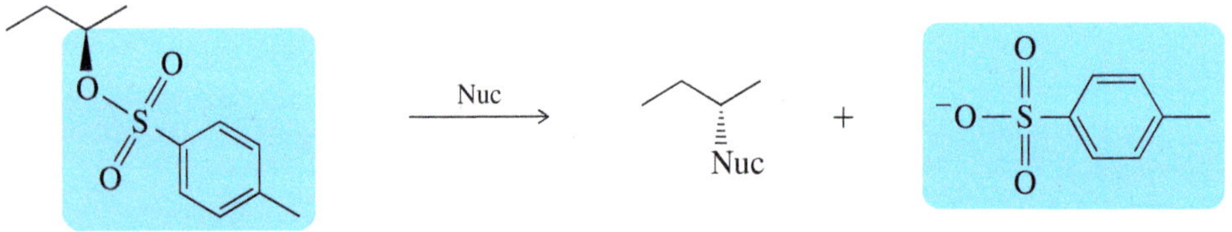	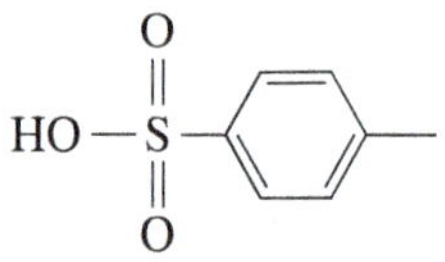	−5
Water 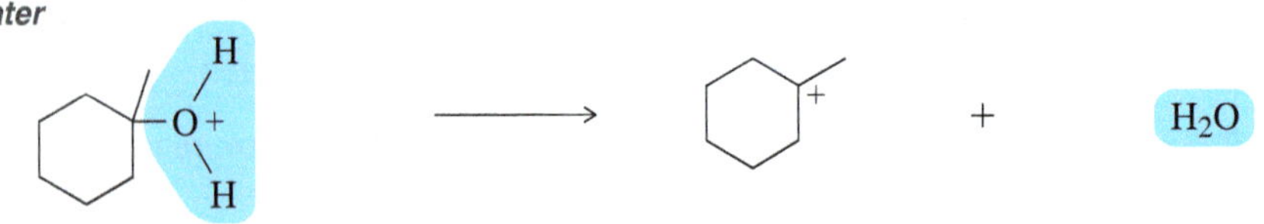	H_3O^+	−1.7
Alcohols 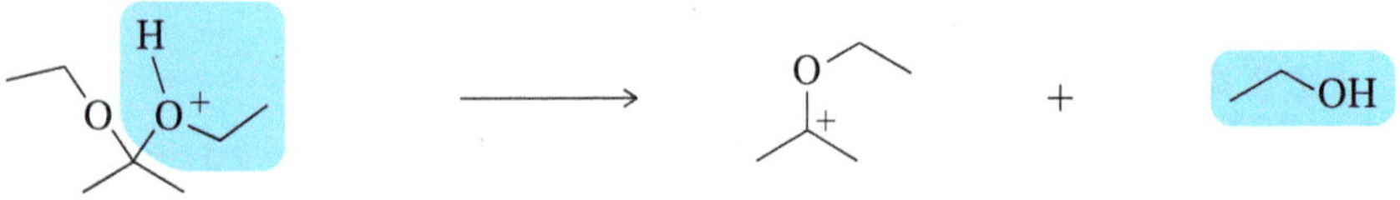	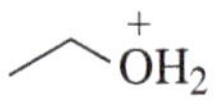	−1
Hydroxide 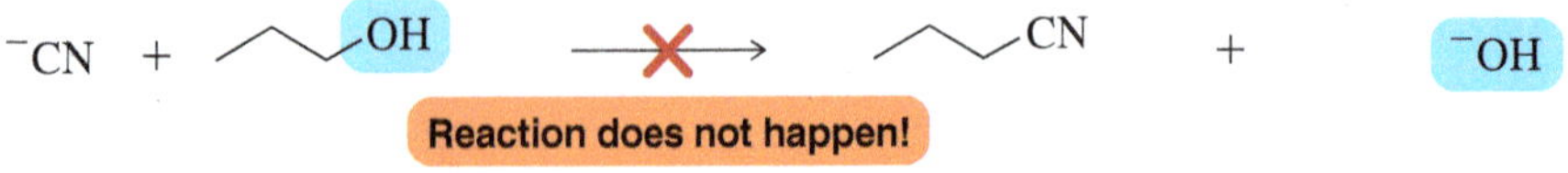	H_2O	16

Assessment 12.24 Formation of the carbocation should be fastest for which leaving group?

(a)

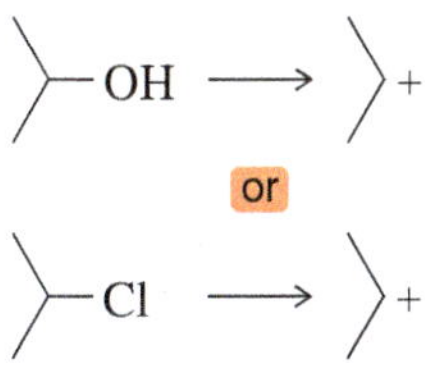

(b)

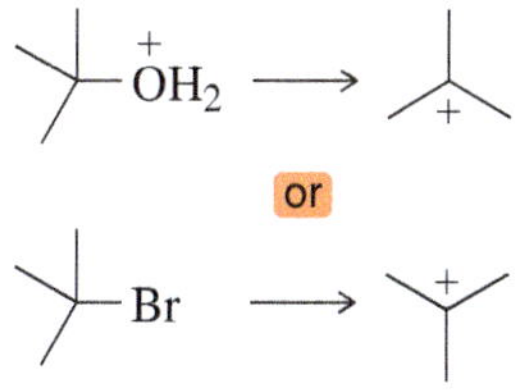

(c)

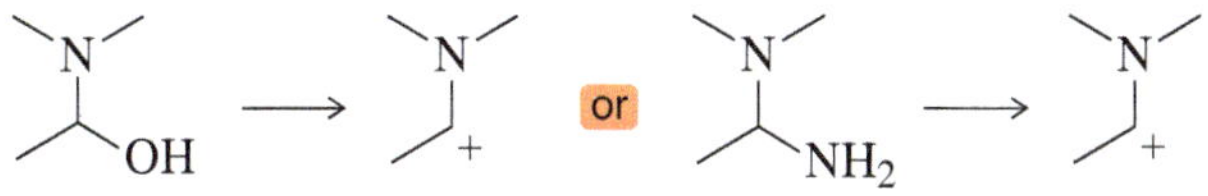

Assessment 12.25 Which of the following is the better leaving group in a polar aprotic solvent?

(a)

HO^- vs. F^-

(b)

H_2O vs. 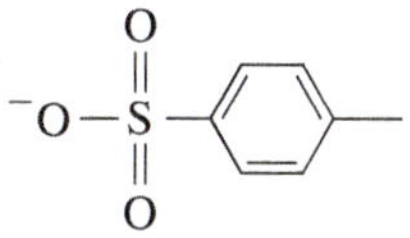

(c)

Br^- vs. I^-

Polar protic solvents stabilize anionic leaving groups by hydrogen bonding [*just as they stabilize nucleophiles*]. Hydrogen bonding, by solvating the nucleophile, reduced its reactivity (Section 12.3.3). Solvating the leaving group, on the other hand, stabilizes it, making it leave more readily (**Figure 12.36**). Polar aprotic solvents, because they cannot hydrogen bond, have no ability to stabilize leaving groups. We will see the impact of this on the substitution mechanism in Section 12.3.5.

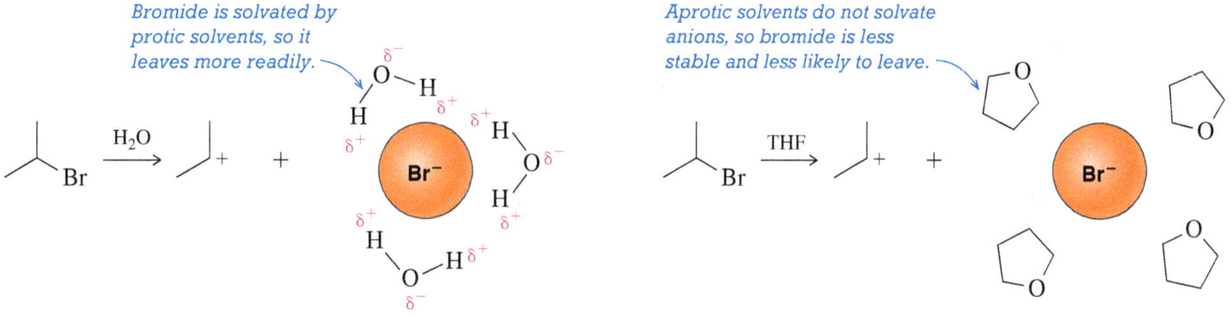

Figure 12.36 Polar protic solvents stabilize leaving groups through solvation.

Assessment 12.26 Which reaction would be faster, the one with DMSO as the solvent or the one with ethanol (EtOH)?

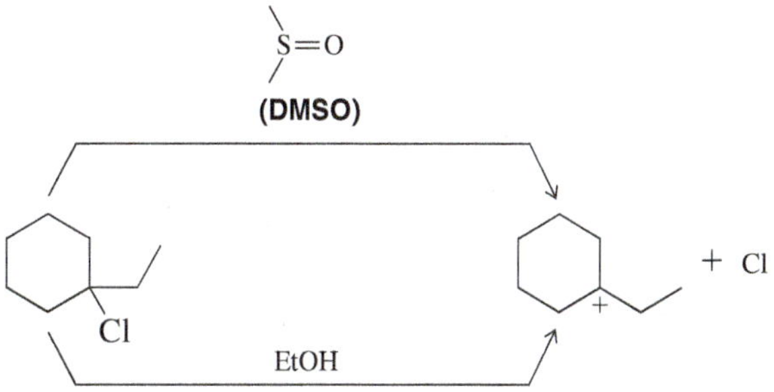

12.3.5 S_N2 vs. S_N1: solvent

The past two sections (12.3.3 and 12.3.4) suggest that the solvent used in a substitution reaction can play a significant role in influencing its mechanism. To see how, again consider the reaction of 2-bromobutane, since it can do either mechanism. [*For the* S_N2 *reaction, the nucleophile is cyanide; for the* S_N1 *reaction, the nucleophile is water.*]

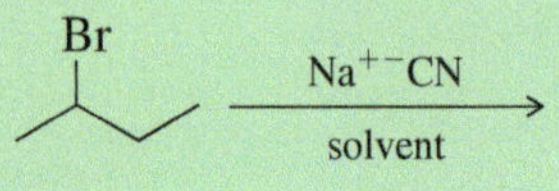

Polar aprotic solvents favor the S_N2 mechanism.

Analysis involving a reaction coordinate diagram will demonstrate how a particular solvent favors or disfavors the S_N2 reaction (**Figure 12.37**).

1. Without solvent, Na^+ and ^-CN remain associated with each other. This situation has a high activation energy, resulting in a slow reaction rate.
2. A protic solvent (like CH_3OH) separates Na^+ and ^-CN, allowing substitution to occur when the ^-CN collides with σ* of the C—Br bond. As a hydrogen bond donor, methanol stabilizes ^-CN. Due to the electron-rich oxygen, methanol stabilizes Na^+. [*It also stabilizes* Br^-. Br^- *is already pretty stable, so the effect is minimal.*]
3. A polar aprotic solvent only solvates Na^+, leaving the ^-CN as a more reactive "naked" anion. [Br^- *is also not solvated but is already stabilized by delocalizing its negative charge over a large volume.*] The net destabilizing of the anion results in a lower activation energy, increasing the rate of the S_N2 reaction.

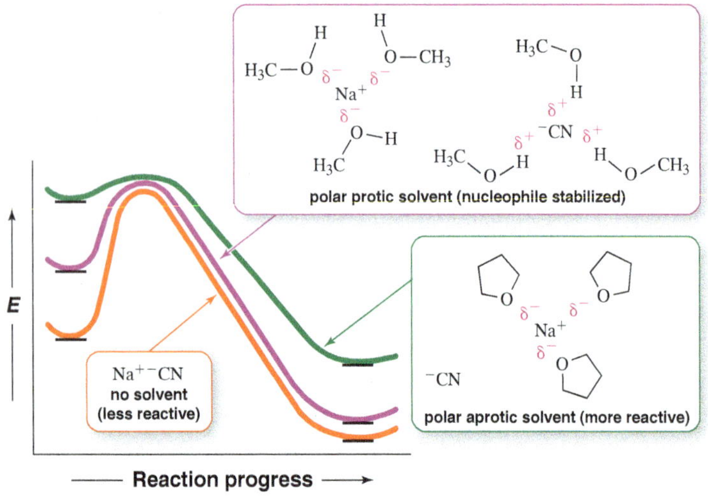

Figure 12.37 **Polar aprotic solvents increase the rate of S_N2 reactions** by destabilizing the nucleophile.

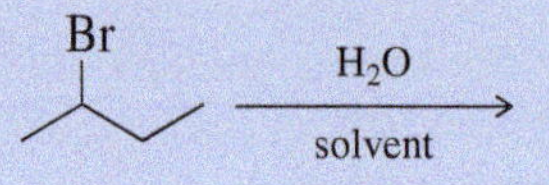

Polar protic solvents favor the S_N1 mechanism.

How does the rate of carbocation formation (the rate-determining step) change in different solvents (**Figure 12.38**)?

1. Without solvent, bromide leaving is a function of the stability of the carbocation and bromide that form in the first step. [*This establishes a baseline to which the other two scenarios can be compared.*]
2. A protic solvent (like H_2O) stabilizes the bromide anion leaving group. This lowers the energy of the intermediate. Because the transition state of this endothermic reaction is like the intermediate, the activation energy is also reduced, increasing the rate.
3. A polar aprotic solvent is unable to solvate Br^-, so the reaction coordinate diagram resembles the solvent-free situation (1). As a result, the S_N1 mechanism is favored under conditions that can stabilize the leaving group.

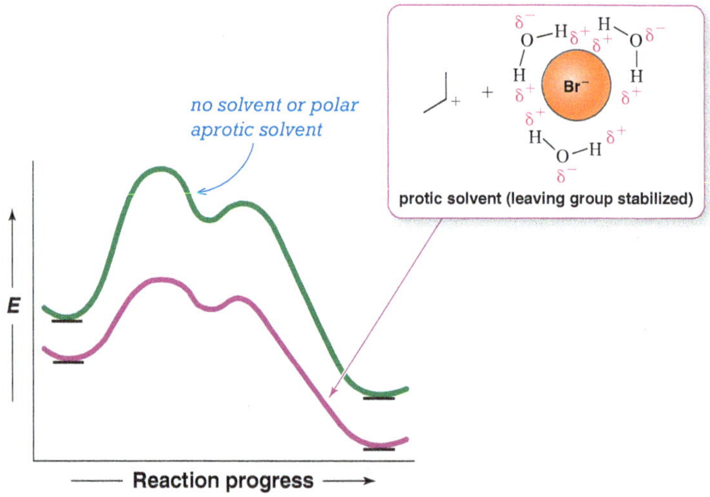

Figure 12.38 **Polar protic solvents increase the rate of the S_N1 reactions** by stabilizing the leaving group.

The situation where a weak nucleophile like water participates in an S_N1 reaction deserves special mention. Referred to as a **solvolysis reaction**, it is the most common S_N1 reaction [*lysis (breaking) by solvent = solvolysis*]. Under these conditions, water is both the polar protic solvent and the nucleophile. Experimentally, these reactions involve simply adding a small amount of the alkyl halide to a large excess of solvent. There is typically enough solvent present to serve both roles: It is the protic, leaving-group-stabilizing solvent as well as the weak, waiting-around-for-the-reactive-carbocation-to-form nucleophile. When the nucleophile is the polar protic solvent, an S_N1 mechanism usually occurs. [*When might a reaction in a polar protic solvent still proceed by the S_N2 mechanism? Assessment 12.27 asks that very question.*]

12.3.5 SUMMARY

- By destabilizing the nucleophile, polar aprotic solvents favor S_N2 reactions. By stabilizing the leaving group and making nucleophiles less reactive, polar protic solvents favor S_N1 eactions.

Assessment 12.27 The following reaction, though run under standard solvolysis conditions, occurs via an S_N2 reaction. Why?

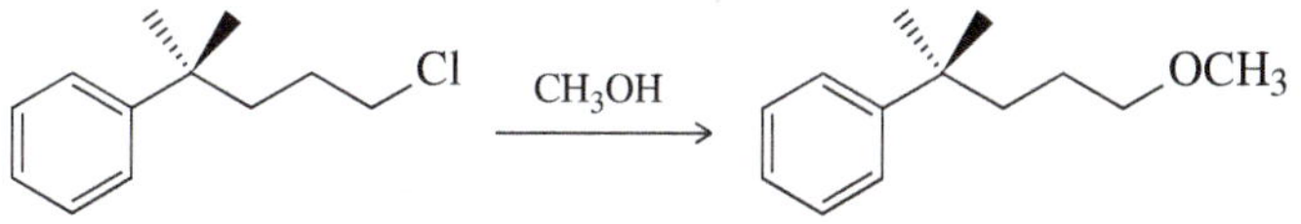

Assessment 12.28 For each solvent, indicate the most likely substitution reaction to take place.

(a)
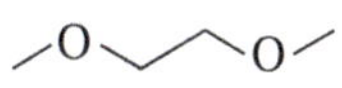

(b)
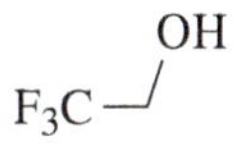

(c)
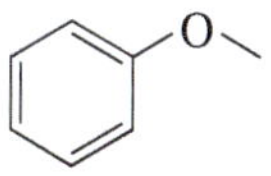

(d)

(e)
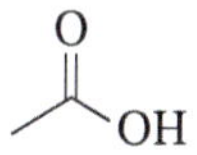

(f)
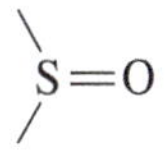

(g)
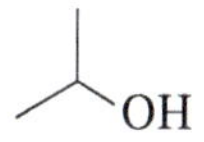

(h)
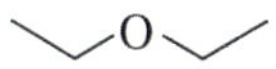

12.3.6 Putting it all together: making a diagnosis

The previous sections (12.3.1–12.3.5) examined the effect a number of factors have on the mechanism by which a substitution occurs: the substitution of the leaving group, the nucleophile, the leaving group, and the solvent. [*Though this is a large amount of information and can be overwhelming, you do not have to simply memorize. Careful analysis with reaction coordinate diagrams shows that the impact of each variable makes sense in terms of the rules of organic chemistry. The ball rolls downhill; negatives attack positives.*] **Table 12.9** summarizes our results to this point.

Table 12.9 The particular substitution of the leaving group, nucleophile, and solvent influence the mechanism by which substitution reactions occur.

Substitution	**1° or methyl:** very little steric hindrance; can't form carbocations	S_N2 only
	2°: not too much steric hindrance; can form relatively stable carbocations	S_N1 or S_N2
	3°: σ* is sterically blocked; can form very stable carbocations	S_N1 only
Nucleophile	**Strong:** anionic nucleophiles and amines	S_N2 favored
	Weak: neutral nucleophiles	S_N1 favored
Solvent	**Polar aprotic:** solvates inorganic ions, but not anions/nucleophiles or leaving groups	S_N2 favored
	Polar protic: solvates inorganic ions, nucleophiles, and, importantly, leaving groups	S_N1 favored

When confronted with a potential substitution reaction, it is important to be able to diagnose how the reaction occurs, even when the product is unknown. As Table 12.9 shows, three factors can be identified for diagnosis (substitution, nucleophile, and solvent), with each making a case for either the S_N2 or S_N1 mechanism. Thus, by examining these categories, an overall diagnosis of reaction mechanism can be made. In some cases, all three categories will point to a single mechanism; in others, it will be a split decision, requiring more subtle analysis. A few worked examples are provided in **Table 12.10**. [*Try to figure out the ones in Table 12.10 before looking at the solution.*]

Table 12.10 Several examples of substitution reactions and the mechanism by which they proceed.

Reaction	Reaching a diagnosis
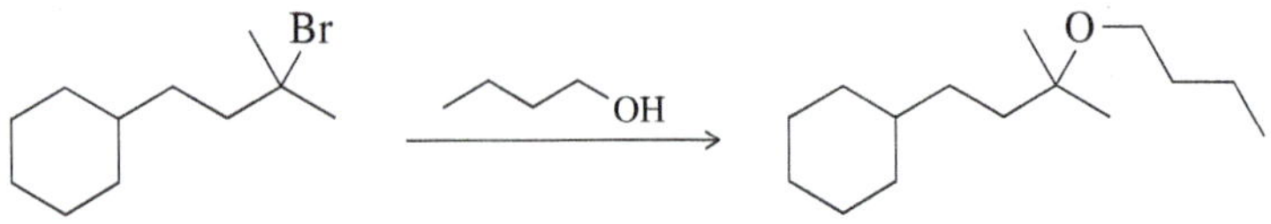	1. ***Substitution***: 3° R—X; can only be S_N1 2. ***Nucleophile***: neutral alcohol . . . weak nucleophile → favors S_N1 3. ***Solvent***: polar protic; favors S_N1 ***Diagnosis***: All categories suggest S_N1. Because the solvent is the nucleophile, this is a solvolysis reaction, proceeding by the S_N1 mechanism.
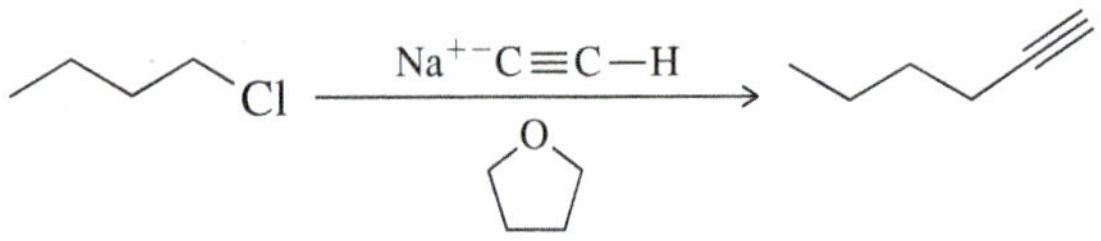	1. ***Substitution***: 1° R—X; can only be S_N2 2. ***Nucleophile***: anionic . . . strong nucleophile → favors S_N2 3. ***Solvent***: polar aprotic; favors S_N2 ***Diagnosis***: Based on analysis of all categories, this substitution proceeds by the S_N2 mechanism.
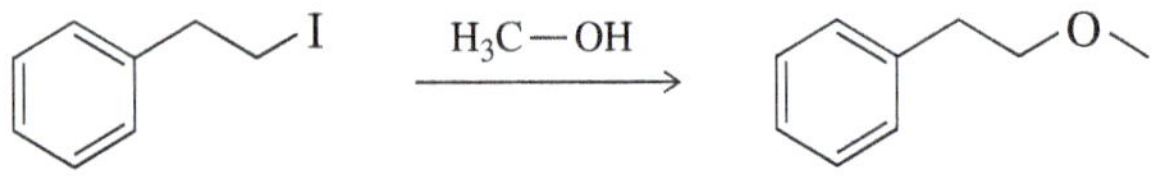	1. ***Substitution***: 1° R—X; can only be S_N2 2. ***Nucleophile***: neutral alcohol . . . weak nucleophile → favors S_N1 3. ***Solvent***: polar protic; favors S_N1 ***Diagnosis***: In this case, there are two "votes" for S_N1, but a 1° R—X never undergoes S_N1 reactions. Thus, the substitution of the haloalkane dictates that this reaction proceeds by the S_N2 mechanism [*albeit slowly, given that methanol is a weak nucleophile*].
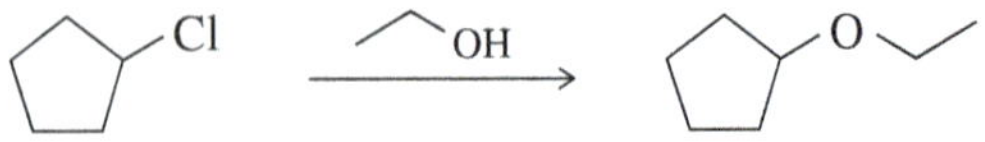	1. ***Substitution***: 2° R—X; can be S_N1 or S_N2 2. ***Nucleophile***: neutral alcohol . . . weak nucleophile → favors S_N1 3. ***Solvent***: polar protic; favors S_N1 ***Diagnosis***: Here, there are two "votes" for S_N1. The haloalkane substitution essentially abstains. In this situation, the weak nucleophile is most important. Because it is a weak nucleophile, it waits for a carbocation to form, reacting by the S_N1 mechanism.

In the examples given in Table 12.10, a few definitive statements emerge that can be used to diagnose the trickier situations:

1. A 3° R—X never undergoes S_N2; a 1° R—X never undergoes S_N1.
2. Anionic nucleophiles do not wait around for carbocations to form.
3. When the solvent is the nucleophile, it is likely S_N1 [*as long as it's not a 1° R—X*].

[*Over time, your intuition will help you diagnose trickier situations. The only way to develop this intuition is to practice as much as possible, using the strategies in Table 12.9. By giving Assessments 12.29 and 12.30 a legitimate effort, only consulting the answer key to verify your answers, you will see this intuition begin to develop.*]

Assessment 12.29 Would you expect the following conditions to favor S_N1 or S_N2?

(a)

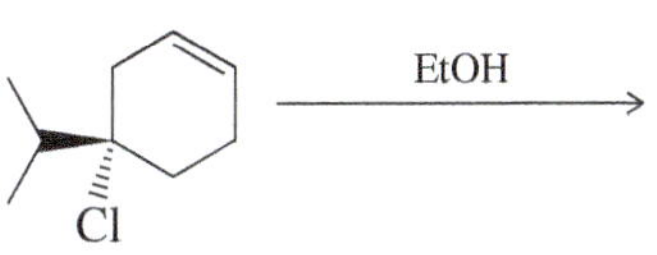

(b)

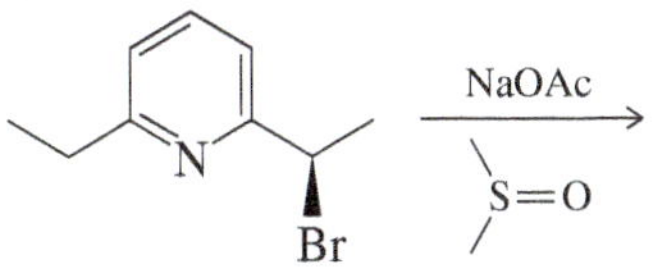

(c)

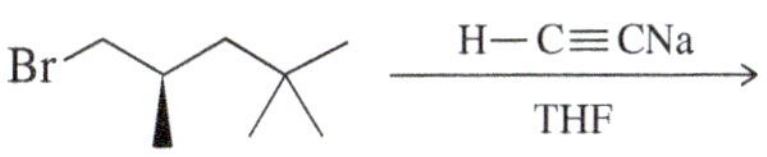

(d)

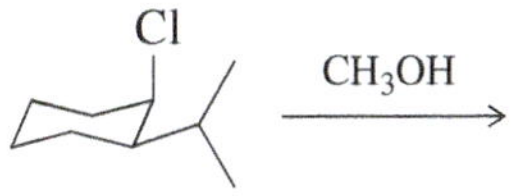

Assessment 12.30 Suggest a mechanism for the following substitution reactions.

(a)

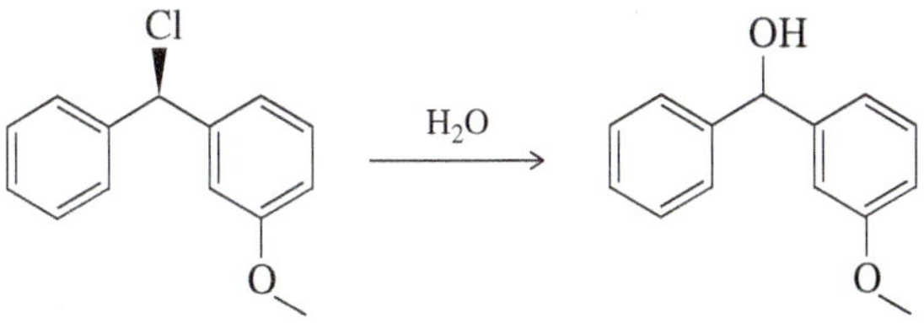

(b)

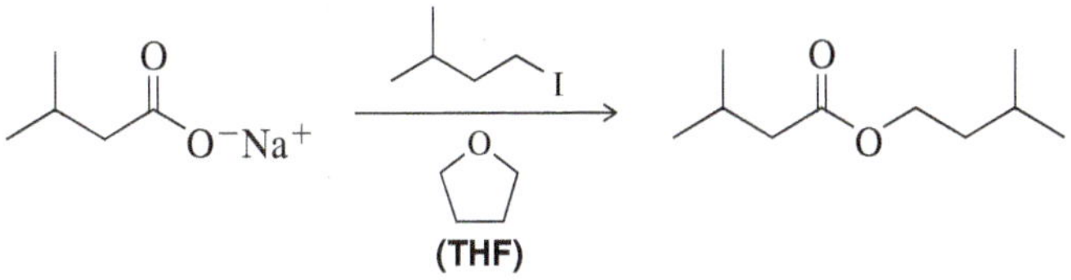

(c)

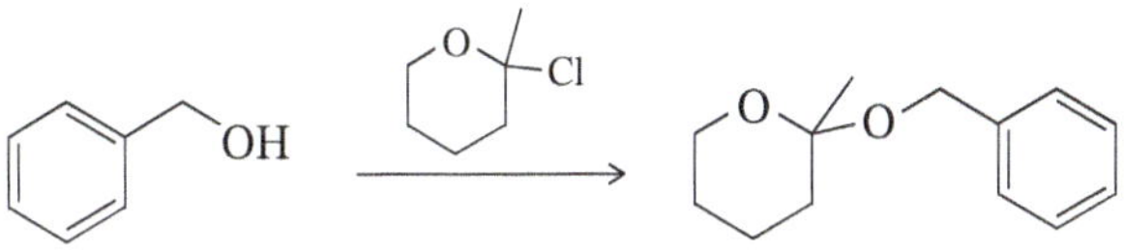

12-B CHEMISTRY IN CONTEXT: Chemical Weapons

Mustard gases in war

Molecules of the type discussed in Chemistry in Context 12-A have a checkered past, owing to their use as chemical weapons during World War I. Because of their smell, these molecules are referred to as mustard gases—they contain a sulfur atom at the β-carbon of a haloalkane (**Figure 12.39**). This sulfur serves as the Lewis basic site and causes their lethal reactivity. Released as aerosols by military weapons, these compounds incapacitated enemy troops. Inhalation of the vapor or absorption through the skin causes severe irritation and painful blistering and can be deadly in large concentrations.

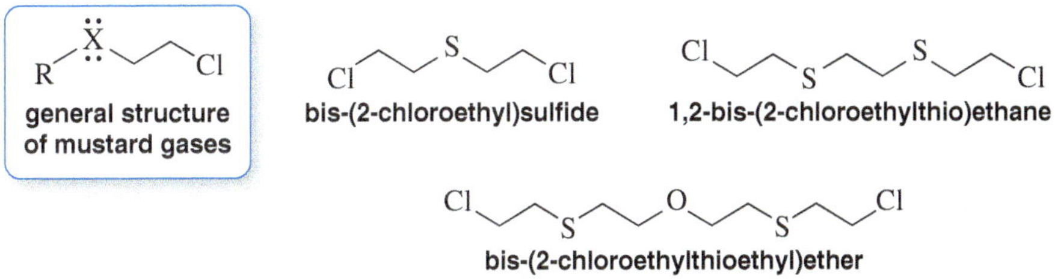

Figure 12.39 **Examples of sulfur mustards used as chemical weapons.**

The toxicity of these compounds is attributed to their participation in substitution reactions. By the same mechanism discussed earlier, these compounds are alkylating agents for a variety of nucleophiles found in the body (**Figure 12.40**). The Lewis basic sulfur displaces the chloride anion by an intramolecular backside displacement reaction, giving a highly reactive sulfonium ion. This sulfonium ion is attacked at the terminal carbon by, among others, the thiol group on the side chain of proteins or amino groups found in the nitrogenous bases of deoxyribonucleic acids (DNA). Cells in the lungs and skin are most affected.

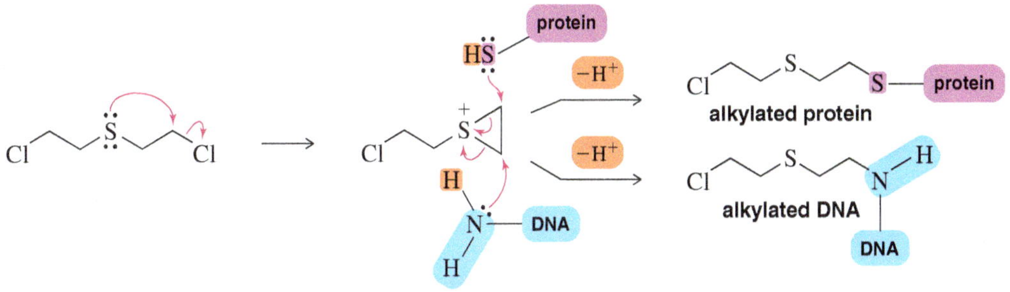

Figure 12.40 **Sulfur mustards react with biological molecules.**

12.4 S_N2 vs. S_N1 outcomes

At this point, you might wonder why it matters how a substitution occurs. In addition to being significant for reactions in biological systems and in drug design, in certain cases the two mechanisms give very different products. Sections 12.4.1 and 12.4.2 show how.

12.4.1 S_N2 vs. S_N1 outcomes: stereochemistry

For this analysis, we examine reactions where, following both S_N1 and S_N2 mechanisms, the stereochemical outcome matters.

S_N2 reaction

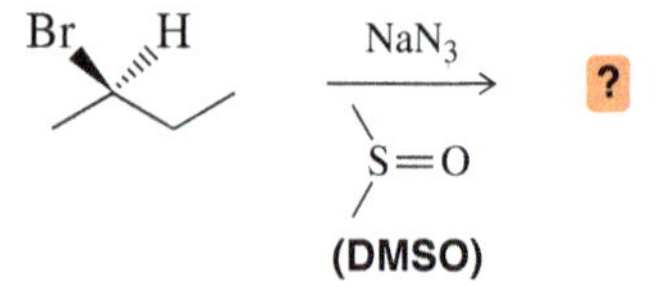

In the S_N2 reaction, the nucleophile donates a pair of electrons to σ*, breaking the C—Br σ bond. Because σ* is always at a 180° angle from the leaving bromide, the nucleophile only adds to the back (**Figure 12.41**).

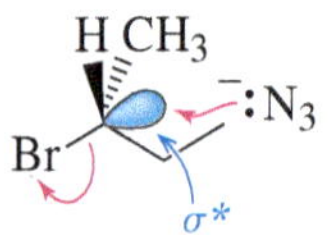

Figure 12.41 Nucleophiles displace leaving groups via backside attack.

As the nucleophile approaches σ*, the substituents have to move out of the way to make room for it. Similarly, when the leaving group departs, space opens for the substituents opposite the nucleophile. In this way, the stereocenter flexes while proceeding from reactants to transition state to products (**Figure 12.43**). [*This movement is analogous to the way an umbrella is turned inside out by a strong wind.*]

Figure 12.43 S_N2 reactions proceed with inversion of stereochemistry.

When an S_N2 reaction occurs at a chiral center, the stereocenter is **inverted**. Originally discovered by Paul Walden, inversion often implies that a molecule that was *R* will now be *S*. More specifically, it means that wherever the leaving group was in the reactant, the nucleophile will be on the opposite side in the product. [*Making models can help you visualize this inversion of stereochemistry.*]

S_N1 reaction

The stereochemistry is determined in the second step of the S_N1 reaction, when the oxygen of ethanol attacks the carbocation. The planar carbocation can be attacked from either side (top or bottom). These two possible modes of attack are nonsuperimposable mirror images of each other, meaning that the two transition states are enantiomers (**Figure 12.42**).

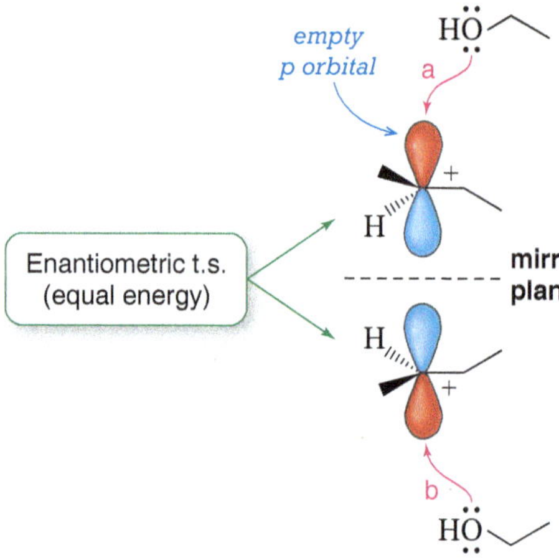

Figure 12.42 Enantiomeric transition states are of equal energy.

Enantiomeric transition states, in the absence of a chiral environment, have equal activation energies, resulting in identical rates of formation of the enantiomeric ethers (**Figure 12.44**).

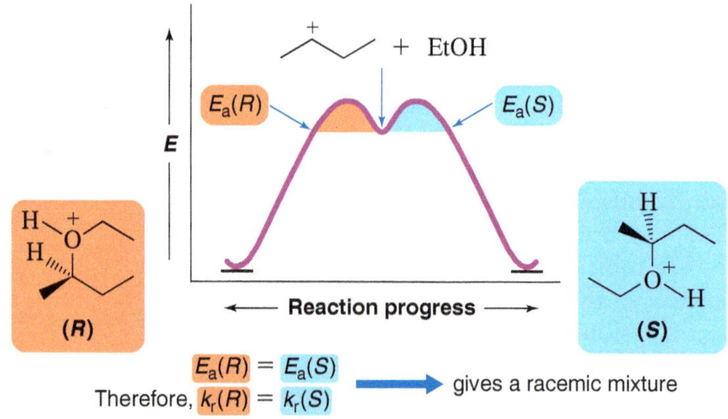

Figure 12.44 Enantiomeric transition states result in a racemic mixture.

Thus, because of proceeding through carbocations (like alkene addition reactions), S_N1 reactions typically result in a racemic mixture of products, even if the reaction begins with a single stereoisomer.

12.4.1 SUMMARY

S_N2 reaction	S_N1 reaction
Attack of the nucleophile at σ* results in inversion of stereochemistry.	Attack of the nucleophile at an empty *p* orbital results in a racemic mixture.
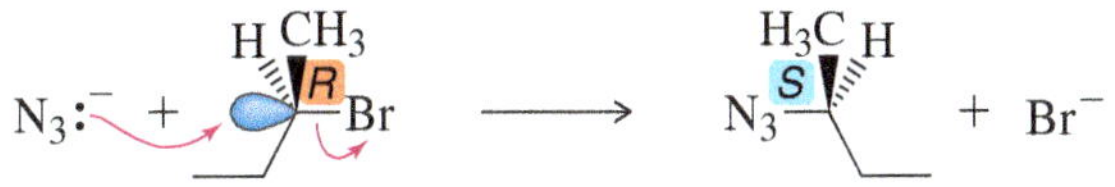	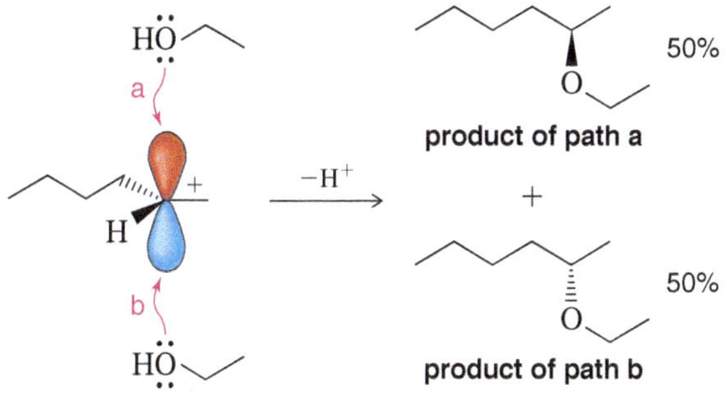

Assessment 12.31 Predict the product of the substitution reactions, paying attention to the stereochemical outcome.

(a)

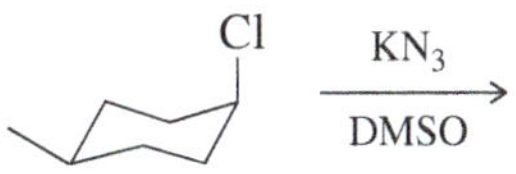

(b)

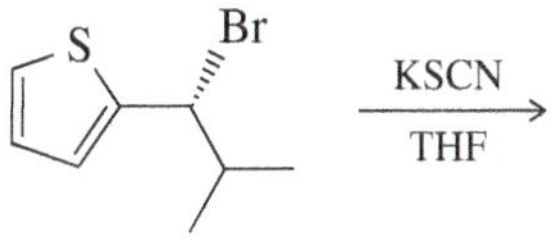

(c)

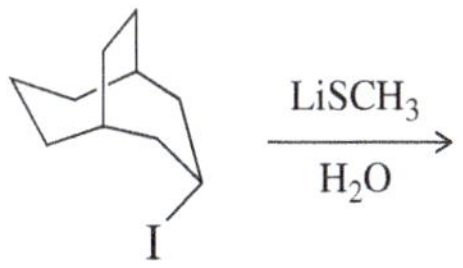

(d)

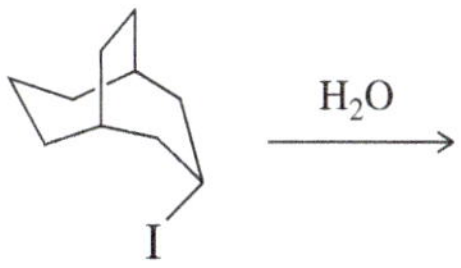

12.4.2 S_N2 vs. S_N1: rearrangement

Like all reactions involving carbocations, there is the possibility for rearrangement [*only*] in the S_N1 reaction. As we saw in Section 8.4.3, a carbocation positioned next to a more substituted carbon rearranges to give a more stable carbocation. A comparison of the same carbocation formed by alkene addition and S_N1 reaction (**Figure 12.45**) makes clear that

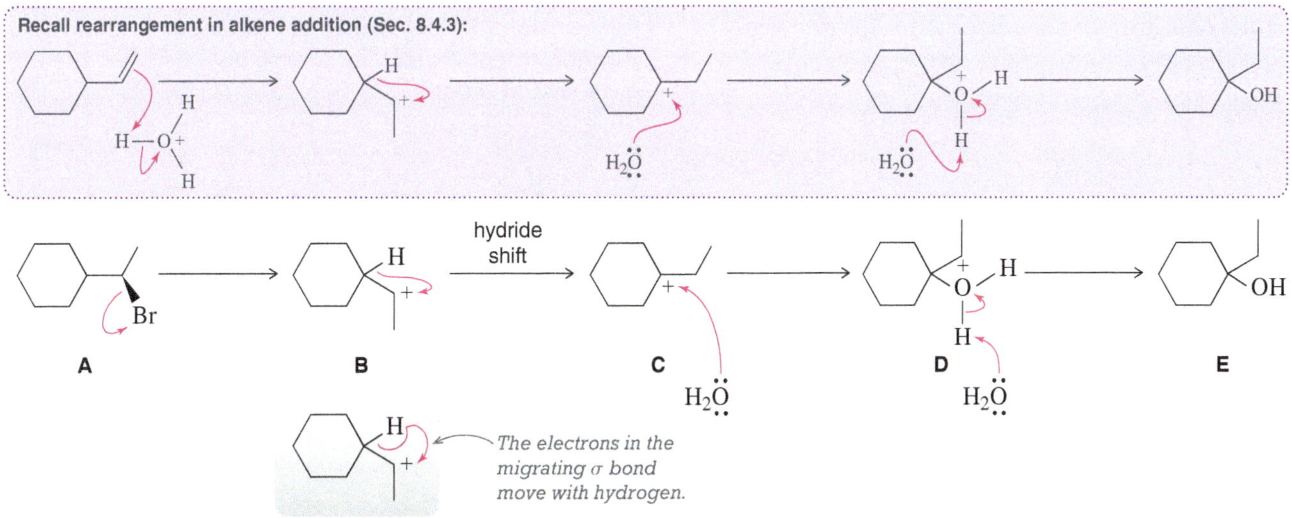

Figure 12.45 **Carbocations, regardless of how they form, will rearrange if a more stable carbocation results.**

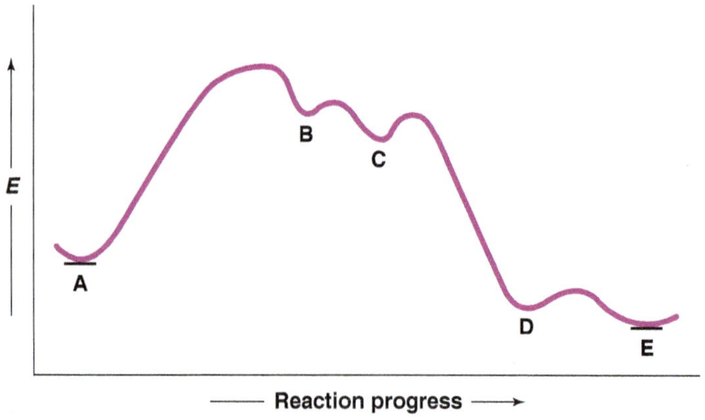

Figure 12.46 The hydride shift (B → C) is fast, resulting in a more stable intermediate.

rearrangement will happen here as well. [*After all, a carbocation knows not how it forms.*] In each case, the identical 2° carbocation is attached to a 3° carbon. Following the rapid, intramolecular 1,2-hydride shift, a more stable 3° carbocation is produced, which is then attacked by the nucleophile in solution.

On a reaction coordinate diagram (**Figure 12.46**), this extra step is represented by a low-activation-energy downhill step (B → C), consistent with formation of a more stable carbocation. The hydride shift, an entropically neutral and intramolecular reaction, is faster than intermolecular attack by water. Thus, the hydride shift always occurs before the nucleophile has a chance to attack.

If the neighboring carbon is 4°, as in **Figure 12.47**, then a 1,2-alkyl shift occurs to generate the more stable carbocation. [*If a hydrogen and an alkyl group are present, only a hydride shift occurs. Why?*]

Figure 12.47 Alkyl shifts occur in S_N1 reactions.

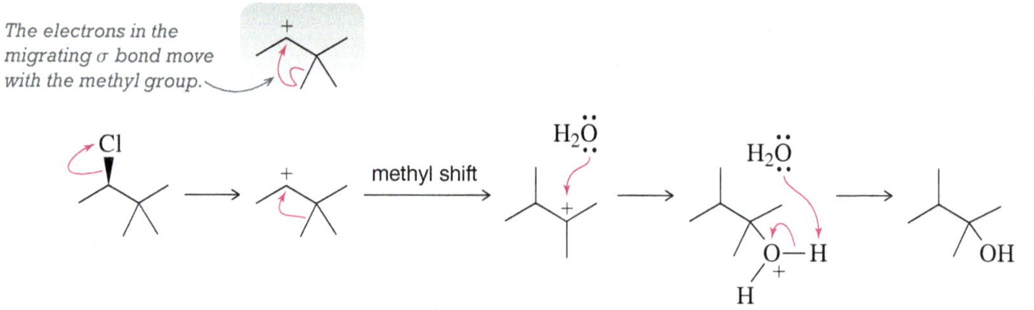

S_N2 reactions, because they do not proceed through a carbocation, do not undergo rearrangements.

12.4.2 SUMMARY

- Because they occur through the formation of a carbocation, S_N1 reactions can rearrange by a hydride or a methyl shift when the initially formed carbocation is less substituted than the neighboring carbon.

Assessment 12.32 Provide a mechanism for the following S_N1 reactions that feature a rearrangement.

(a)

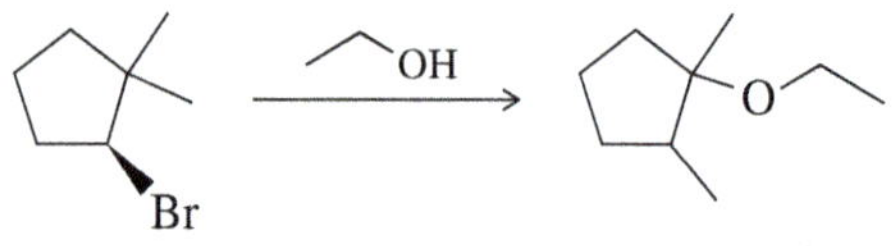

(b)

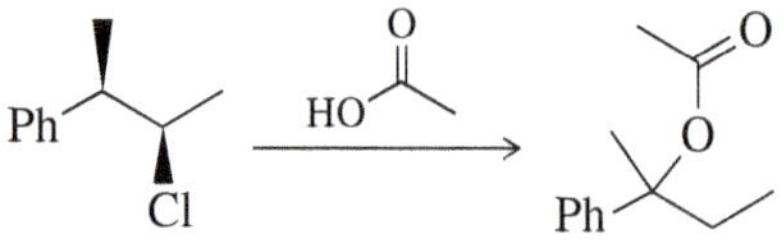

Assessment 12.33 Predict the product of the following substitution reactions, making sure to note whether a rearrangement should occur.

(a)

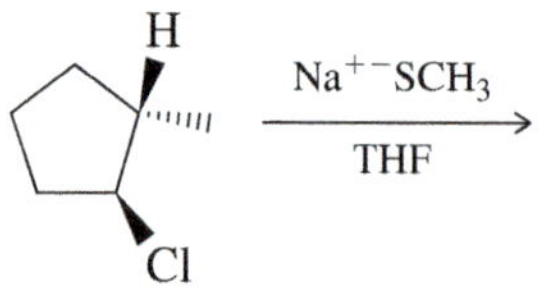

(b)

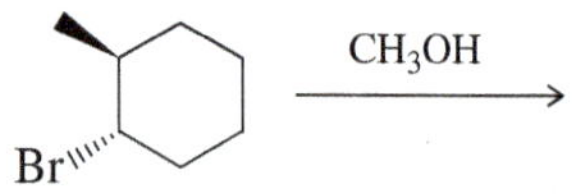

(c)

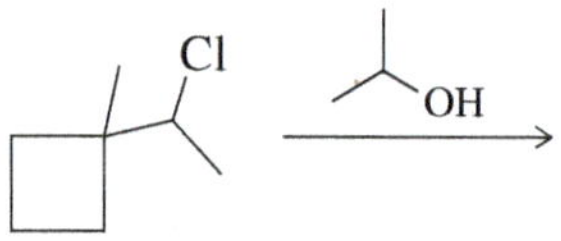

(d)

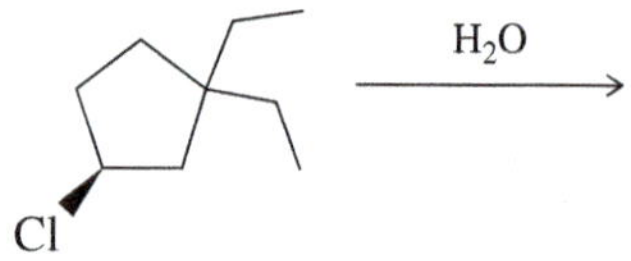

12.5 Elimination: a detailed overview

The second reaction type we discuss in this chapter is elimination, so named because a hydrogen and a good leaving group are eliminated from a molecule. In introductory chemistry classes, these are classified as decomposition reactions (**Figure 12.48**). [*At first glance, this is the opposite of alkene addition.*] The reactions in this chapter are considered **β-elimination** reactions because the hydrogen is removed from the carbon (**β-carbon**) next to the carbon bearing the leaving group (**α-carbon**).

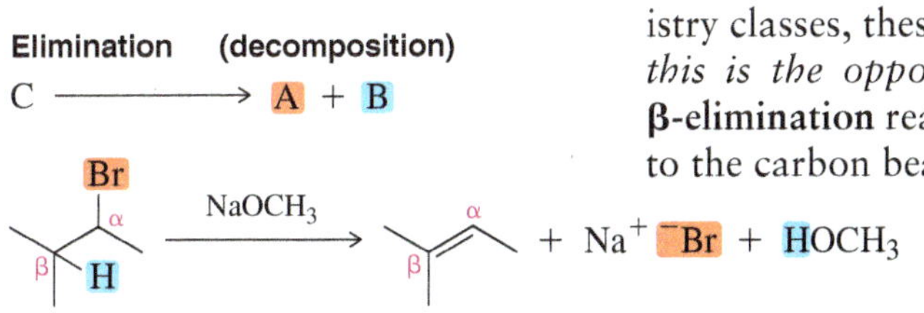

Figure 12.48 Elimination reactions were called decomposition reactions in introductory chemistry classes.

Our study of elimination reactions will parallel that of substitution reactions in analyzing the substitution of the leaving group, the nucleophile, and the solvent. [*Using the same concepts again reinforces our understanding of this material and will prove helpful in future chapters.*]

As shown in **Figure 12.49**, the elimination reaction between methoxide and 2-bromopropane is favorable. This reaction features the breaking of a C—Br σ bond and a C—H σ bond, while forming an O—H σ bond and a C—C π bond. Although only slightly favorable based on bonds formed and broken, other factors make elimination favorable. The reaction begins with a strong base [pK_a of CH_3OH (conjugate acid) is 16] and forms a very stable bromide anion [pK_a of HBr (conjugate acid) is −9]. Based on this factor alone, the reaction should be extremely favorable. Finally, unlike substitution ($\Delta S \approx 0$), entropy is favorable ($\Delta S > 0$), because elimination produces three molecules from two.

Figure 12.49 Elimination reactions are favored by ΔH and ΔS.

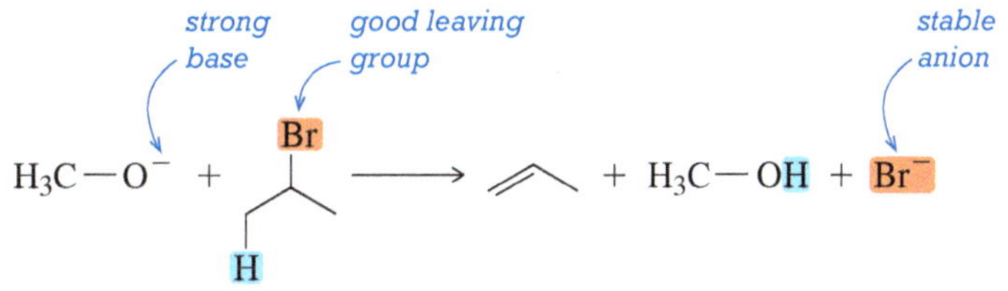

Bonds broken:
C—Br σ 68 kcal/mol
C—H σ 98 kcal/mol

Bonds formed:
O—H σ 104 kcal/mol
C—C π 65 kcal/mol

$\Delta H = (68 + 98) - (104 + 65) = -3$ kcal/mol

ΔS: 2 ⟶ 3
$\Delta S > 0$

$\Delta G = \Delta H - T\Delta S$
$\Delta G < 0$

Assessment 12.34 Identify the bonds that break and form in the following elimination reactions.

(a)

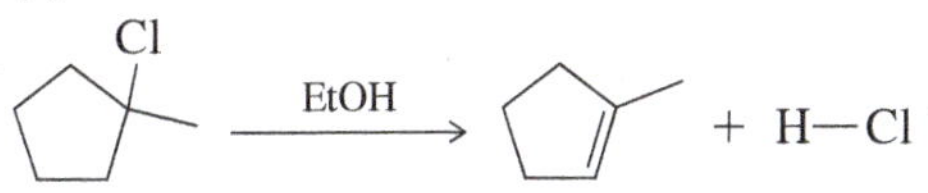

(b)

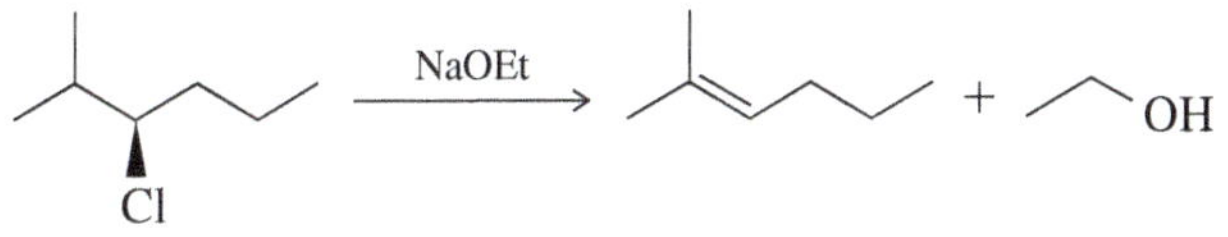

Assessment 12.35 When 2-bromopropane is treated with sodium ethoxide, propene is produced. What molecule is lost from 2-bromopropane in this process?

Assessment 12.36 (a) How would you convert propene to 2-bromopropane (a reaction we learned in Chapter 8)?
(b) How is this reaction from Chapter 8 related to the reaction discussed in Assessment 12.35?

12.5.1 Elimination: two important mechanisms

Similar to substitution, there are just three ways that an elimination can occur. Focusing only on the organic molecule, think about what happens in the reaction. Is it "the C—H bond breaks, ***and then*** the C—C π bond forms, breaking the C—Br σ bond" (possibility 1), "the C—Br bond breaks, ***and then*** the C—C π bond forms by breaking the C—H bond" (possibility 2), or "the C—H bond breaks, the C—C bond forms, and the C—Br bond forms ***at the same time***" (possibility 3)? Electrons can be pushed to illustrate these three possibilities (**Figure 12.50**). [*Note that the statements above with "and then" define the reaction as having two steps.*]

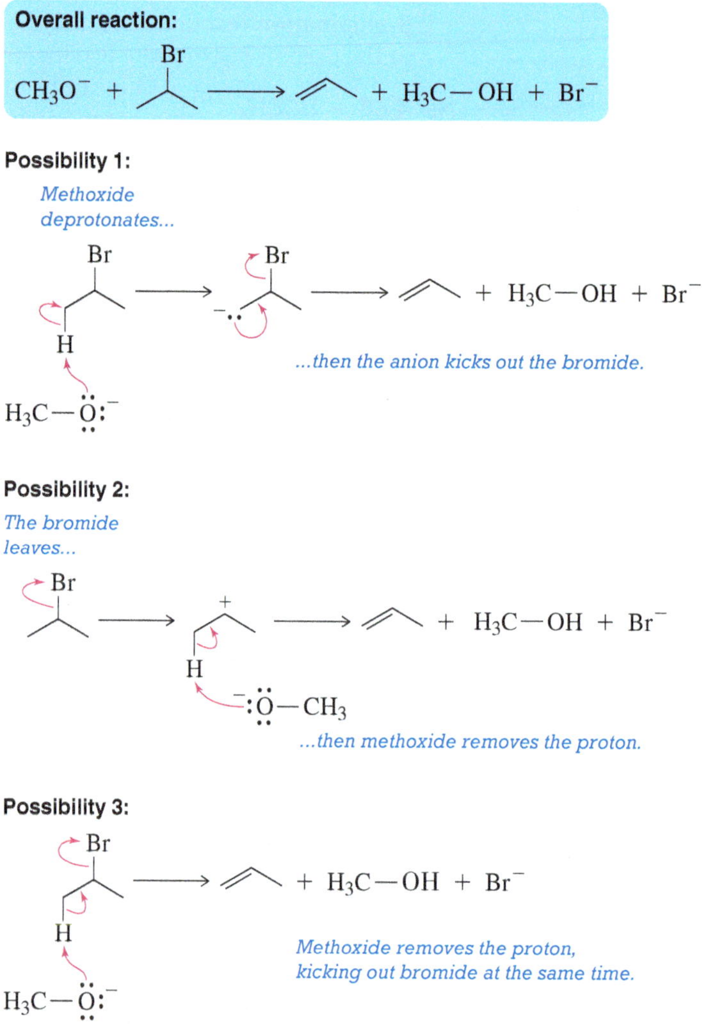

Figure 12.50 Three hypothetical elimination mechanisms.

Possibility 1, referred to as the E1cb mechanism, is examined when we study the chemistry of enolates in Chapter 20. [*This will allow you to focus on the two important elimination mechanisms in this chapter.*]

Possibility 2 is reasonable (**Figure 12.51**). This two-step process features a carbocation intermediate (like the S_N1 reaction). Once formed, in order to become neutral, the carbocation loses a proton to form a new C—C π bond. [*We have seen this before; do you remember when?*] This reaction, the elimination of HBr, proceeds through the opposite mechanism of alkene addition. [*Draw both mechanisms to convince yourself that it's true.*]

Possibility 3 is reminiscent of the S_N2 reaction; everything happens in a single step (**Figure 12.52**). The unstable base grabs a proton, forming a new π bond and a stable leaving group at the same time. Though this is new, our understanding of the S_N2 reaction makes it more plausible.

Possibility 2:

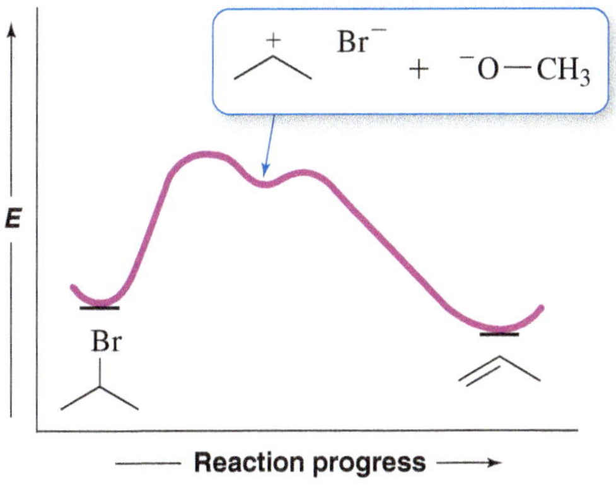

Figure 12.51 Possibility 2 is a reasonable two-step mechanism.

Possibility 3:

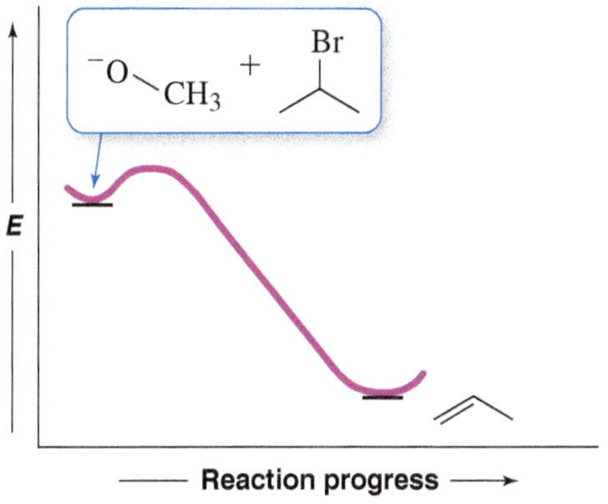

Figure 12.52 Possibility 3 is a reasonable one-step mechanism.

With two reasonable mechanisms, physical organic chemistry can help us figure out which is actually happening. [*This might seem redundant, as similar analyses were done for the two substitution reactions. As such, the analysis here is intentionally briefer. If, however, you need a more complete analysis of any aspect of elimination reactions, you should review the corresponding section on substitution.*]

For this particular reaction, a baseline rate (x) is established at a particular concentration [0.1 M for both reactants (CH_3O^- and 2-bromopropane)] (**Figure 12.53**). Doubling the concentration of 2-bromopropane (control → trial 1) doubles the rate to $2x$. Thus, the reaction is first order with regard to 2-bromopropane; that is, there must be one molecule of 2-bromopropane in the transition state of the rate-determining step. Doubling the concentration of CH_3O^- also doubles the rate. Thus, the reaction is also first order with respect to CH_3O^-; likewise, there must be one molecule of methoxide in the transition state of the rate-determining step. The reaction is second order overall (bimolecular); there are two molecules present in the rate-determining step (Figure 12.53).

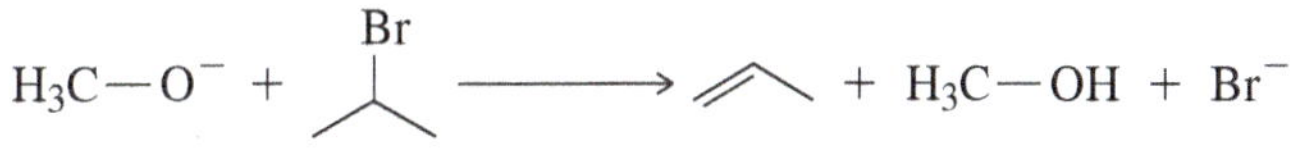

	$[CH_3O^-]$ (M)	[2-Bromopropane] (M)	Rate
Control	0.1	0.1	x
Trial 1	0.1	0.2	$2x$
Trial 2	0.2	0.1	$2x$

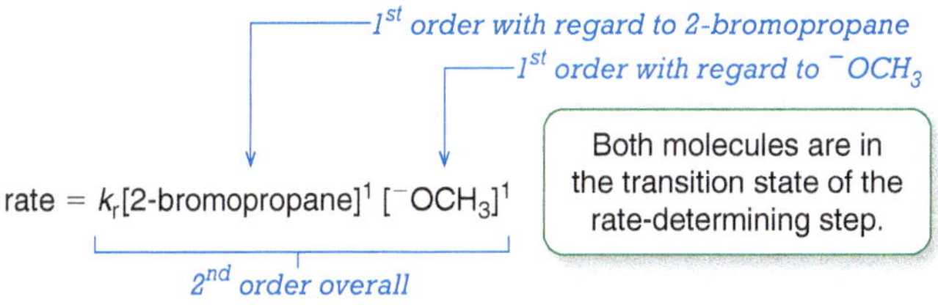

Figure 12.53 The rate law shows that the reaction is 2nd order. There are two molecules in the rate-determining step.

Of the two possibilities (2 and 3), only **possibility 3** has CH_3O^- and 2-bromopropane present in the transition state of the rate-determining step. Thus, this experiment confirms a concerted reaction with a transition state where the new O—H σ bond and C—C π bond are partially formed and the C—H and C—Br σ bonds are partially broken. [*Drawing transition states with several partially broken and formed bonds and partial charges is confusing. Luckily, the arrow-pushing formalism shows the same thing and is perhaps a bit easier to follow.*] This is termed the E2 mechanism, because it is an *Elimination* reaction with a *bimolecular (2 molecules)* rate-determining step (**Figure 12.54**).

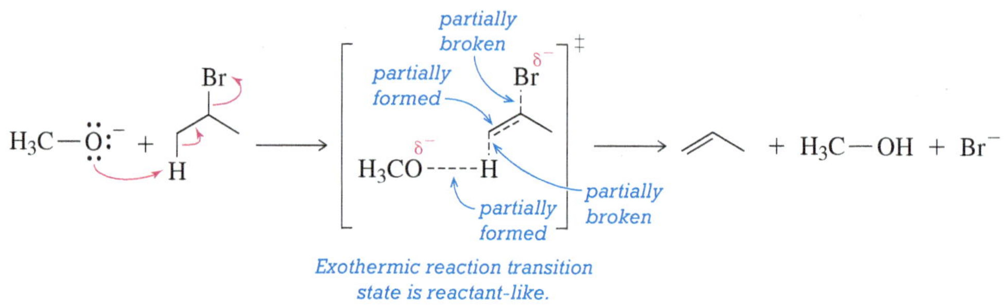

Figure 12.54 This elimination reaction proceeds by possibility 3 (E2 elimination).

If the S_N1 mechanism was an alternative to the S_N2 mechanism, then **possibility 2** is likely an alternative to the E2 mechanism. The formation of 2-methylpropene by reaction of *t*-BuBr with water occurs by this mechanism. The rate data obtained for this reaction suggest that the reaction is first order with regard to *t*-BuBr, zeroth order with regard to H_2O, and first order overall. The obtained rate law reveals that this *Elimination* reaction is *unimolecular (1 molecule in the rate-determining step)*, earning it the label **E1** (**Figure 12.55**).

	[*t*-BuBr] (M)	[H_2O] (M)	Rate
Control	0.1	0.1	*x*
Trial 1	0.2	0.1	2*x*
Trial 2	0.1	0.2	*x*

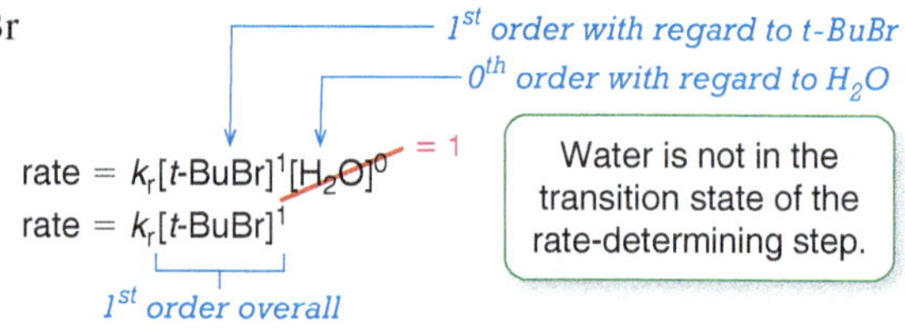

Figure 12.55 The reaction is 1st order. *t*-BuBr is the only molecule in the rate-determining step.

Mechanistically, Br^- leaves in the first (rate-determining) step (**1**), producing a carbocation. [*The transition state of this step involves just one molecule (t-BuBr). Where have we seen this before?*] Then in (**2**), H_2O removes a proton from the carbon next to the highly reactive carbocation, forming the C—C π bond and H_3O^+. The transition state of the E1 rate-determining step is identical to the transition state of the S_N1 rate-determining step (**Figure 12.56**).

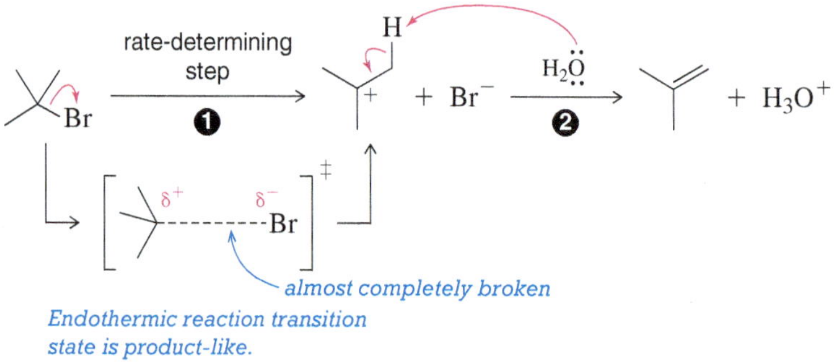

Figure 12.56 This elimination reaction proceeds by possibility 2 (E1 elimination).

[*It bears repeating: The E2/E1 label tells you about the mechanism of the reaction: specifically, the number of molecules in the transition state of the rate-determining step.*]

12.5.1 SUMMARY

E2 reaction: The concerted β-elimination reaction of a haloalkane, proceeding in a single bimolecular step. The base removes a proton from the β-carbon, forming a new C—C π bond while pushing out the leaving group.

E1 reaction: The two-step elimination reaction of a haloalkane, proceeding with a unimolecular rate-determining step. After the leaving group leaves, a proton is removed from the β carbon, resulting in formation of a new C—C π bond.

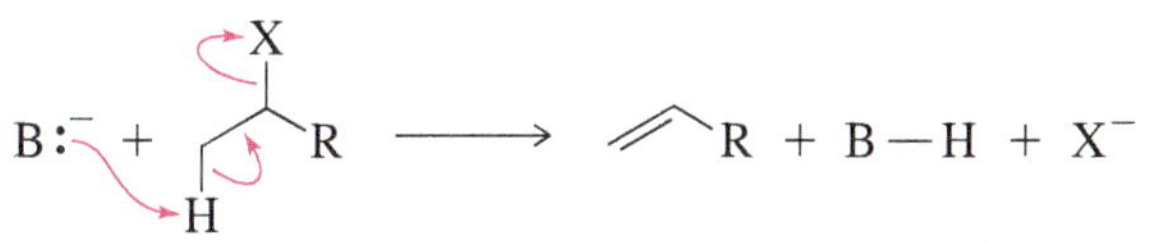

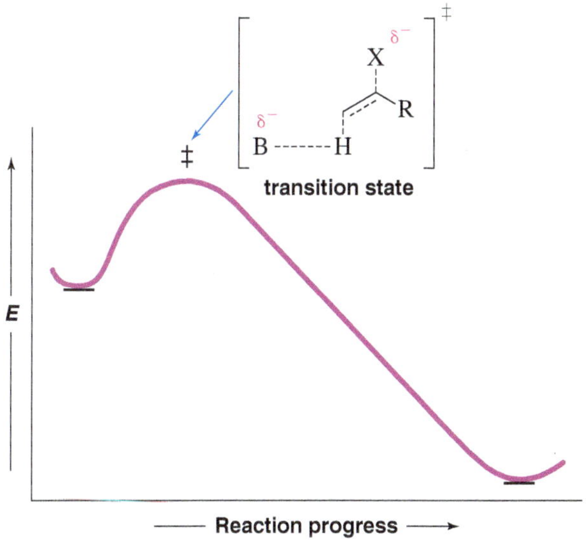

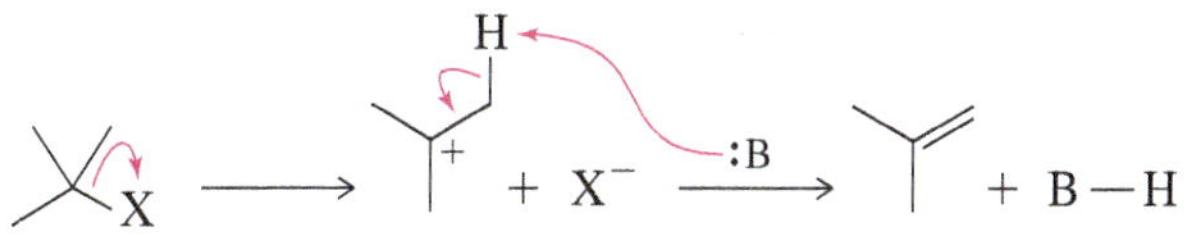

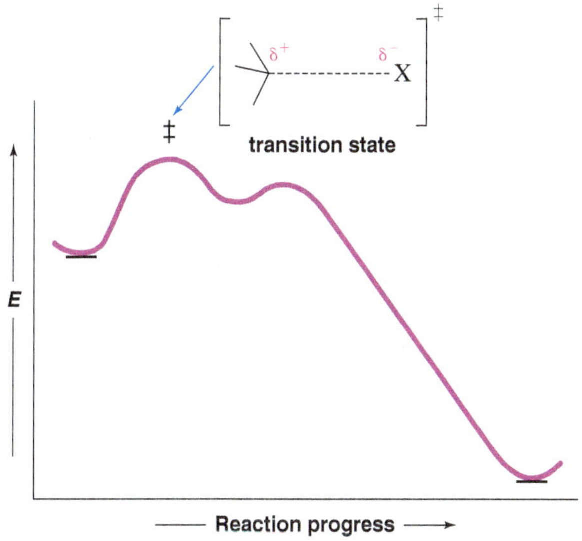

Assessment 12.37 Practice your electron-pushing skills by drawing a mechanism for the following E2 reactions.

(a)

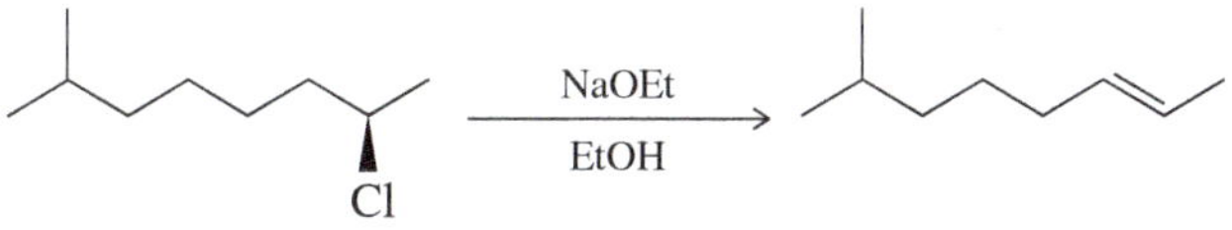

(b)

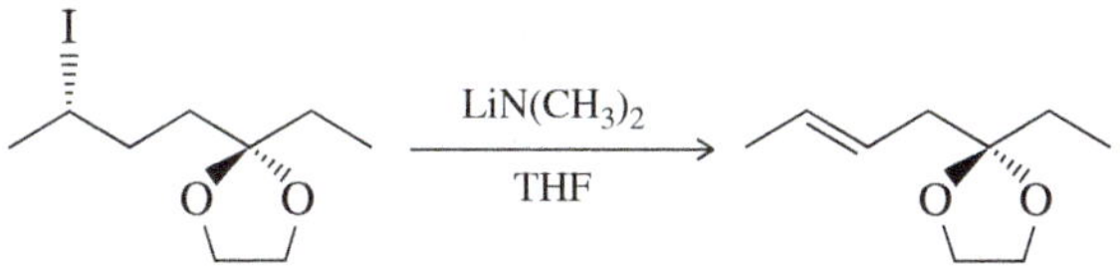

(c)

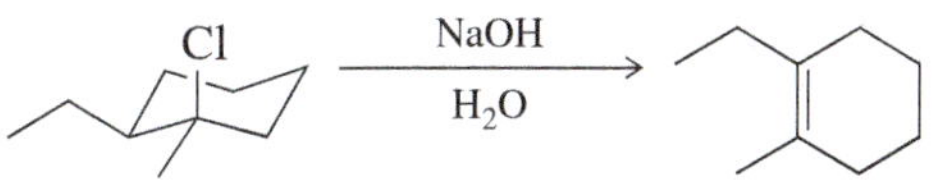

Assessment 12.38 Practice your electron-pushing skills by drawing a mechanism for the following E1 reactions.

(a)

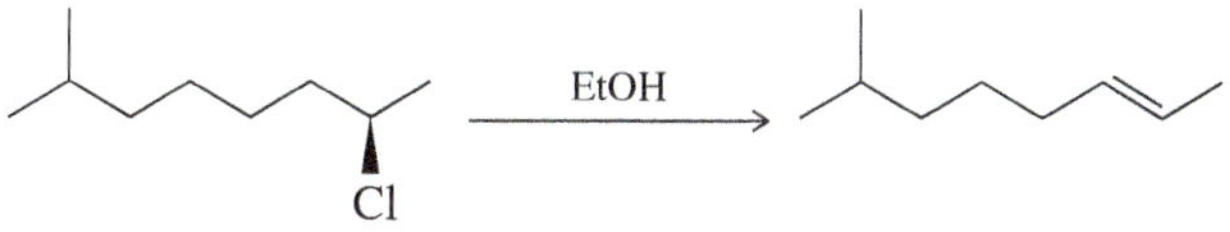

(b)

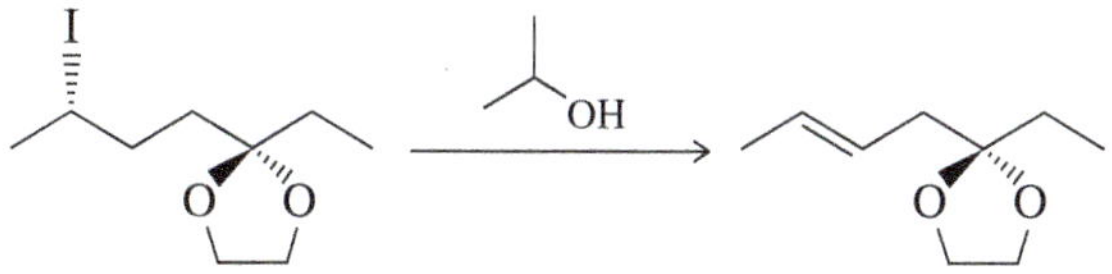

(c)

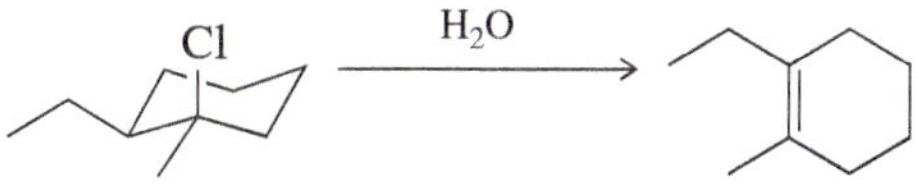

12.5.2 E2 and E1: a spectrum of reactivity

While a reaction always proceeds by one elimination mechanism, it is helpful to think of E1 and E2 mechanisms as occurring along a spectrum of reactivity (**Figure 12.57**), a concept we used when considering S_N1 and S_N2 reactions (see Figure 12.13). Consistent with their effects on S_N1 and S_N2 reactions, the haloalkane substitution, the base, and the solvent in elimination reactions influence the choice in mechanistic pathways (E1 vs. E2). [*You may wonder how to choose between a substitution and elimination pathway. Sections 12.6.1–12.6.5 show how to choose between E2 and E1, and in Section 12.8, we consider all mechanistic possibilities (S_N2, S_N1, E2, E1) at once.*]

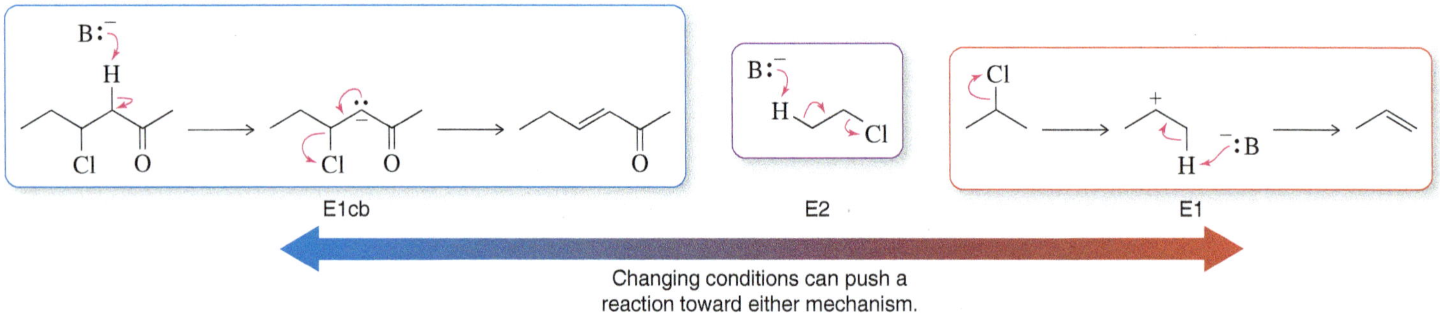

Figure 12.57 All elimination reactions lie on a spectrum between pure E1cb and pure E1, depending on the conditions of the reaction. This chapter focuses on E2 and E1 only.

12.6 Distinguishing between E2 and E1

In this section, our discussion focuses on distinguishing between the two elimination mechanisms. As we discussed for substitution reactions, how can you know, without doing the rate experiment, which elimination mechanism is most likely? And can we predict the mechanism that occurs and the product that forms? [*Absolutely!*] Toward this goal, we will continue to consider the influence each parameter has on making each mechanism (E2 vs. E1) favored or disfavored. Note that the discussion of leaving groups from Section 12.3.4 applies equally to elimination reactions.

12.6.1 E2 vs. E1: where do the electrons go?

We examine the fundamental difference between the E2 and E1 mechanisms the same way we did with substitution reactions. That is, how do the electrons flow and where do they go when forming the new C—C π bond?

E2

In the only step, the strong base removes the β-proton. At the same time, the electron pair collapses to the neighboring carbon, pushing out the leaving group. These electrons, which eventually become the π-bonding electrons, are donated to σ^* of the C—Cl bond (**Figure 12.58**). Just like in the S_N2 reaction, donation to σ^* at a 180° angle from the C—Cl σ bond causes the C—Cl bond to break. [*This is very important when we study the stereochemistry of E2 eliminations in Section 12.7.2.*]

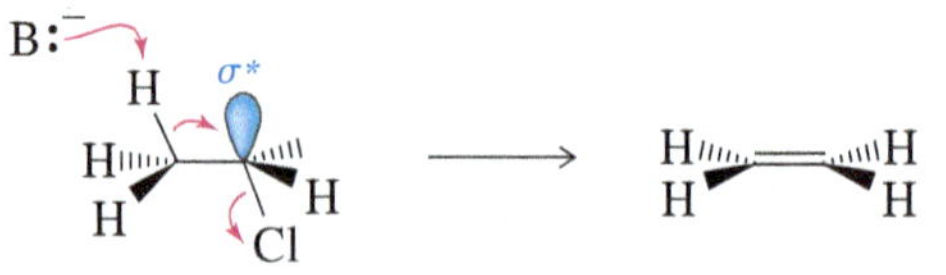

Figure 12.58 Electrons from the C—H σ bond fall into σ^* to break the C—Cl bond.

E1

The leaving group leaves in the first, rate-determining step, resulting in an sp^2-hybridized trigonal planar carbocation (empty *p* orbital). Removal of a proton from the β-carbon causes the electron pair to collapse into the empty *p* orbital, forming the new π bond (**Figure 12.59**).

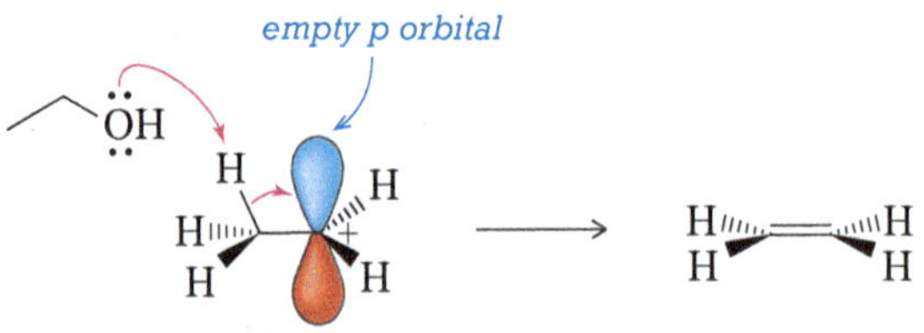

Figure 12.59 Electrons from the C—H σ bond fall into the empty *p* orbital.

As you will see, whether π bond formation occurs through σ^* or an empty p orbital affects the outcome of elimination reactions. For now, let's explore how the reaction conditions predict a particular mechanism.

12.6.2 E2 vs. E1: base

E2 mechanism

A moderately strong base is needed to remove a proton from the β-carbon of a neutral carbon. Thus, negatively charged (anionic) bases tend to react by the E2 mechanism (**Figure 12.60**).

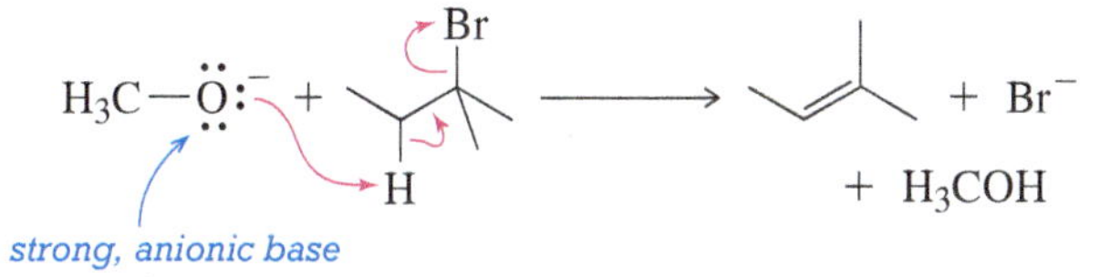

Figure 12.60 Strong bases do E2 eliminations.

It is important to distinguish between anionic species that do S_N2 reactions vs. anionic species that do E2 reactions (**Table 12.11**). Consideration of the pK_a values of the conjugate acids of the anionic species aids this determination. If the pK_a > 11, elimination is favored, whereas a pK_a < 11 favors substitution reactions. Traditionally, stronger bases like hydroxide, methoxide, and amide do E2 reactions, whereas nucleophiles like cyanide, azide, and thiolates are common S_N2 nucleophiles.

Table 12.11 Nucleophiles (S_N2) vs. bases (E2).

S_N2 nucleophiles	E2 bases
CH_3S^- (thiolate)	HO^- (hydroxide)
NC^- (cyanide)	CH_3O^- (methoxide)
N_3^- (azide)	H_2N^- (amide)
AcO^- (acetate)	$(i\text{Pr})_2N^-$ (amide)
PhO^- (phenoxide)	t-BuO^- (t-butoxide)

E1 mechanism

Weak bases favor the E1 mechanism. Molecules like water (H_2O), methanol (CH_3OH), and acetic acid (CH_3CO_2H) are not strong enough bases to abstract a proton from a neutral molecule (**Figure 12.61**). Instead, a carbocation must form prior to their abstracting the acidic β-proton.

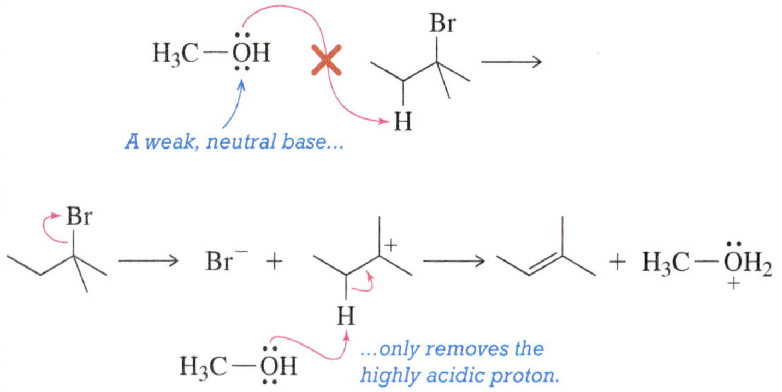

Figure 12.61 Weak bases do E1 eliminations.

The proton next to the carbocation is very acidic. When removed, the conjugate base is the very stable alkene. As discussed in Chapter 4, stable conjugate bases are formed by removing highly acidic protons (**Figure 12.62**).

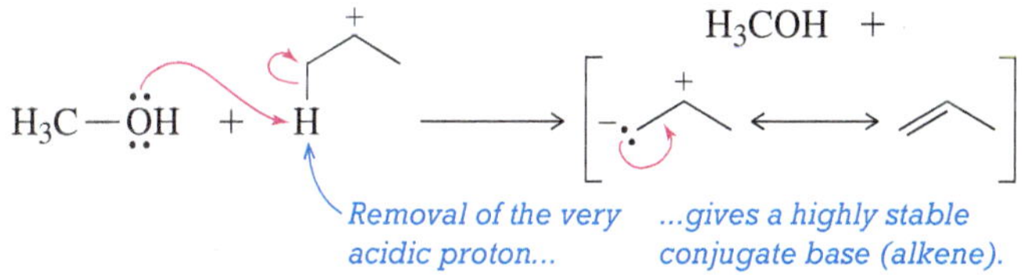

Figure 12.62 Removal of the acidic proton gives a stable conjugate base (the alkene).

12.6.2 SUMMARY

- Strong bases (pK_a of conjugate acid > 11) favor E2 eliminations. Weak, neutral bases allow the carbocation to form first, favoring E1 eliminations.

Assessment 12.39 Would you expect the following bases to favor E1 or E2 elimination?

(a)

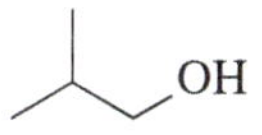

(b)

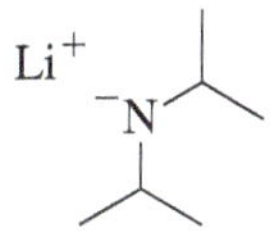

(c)

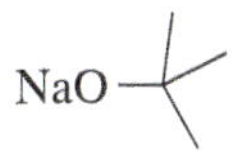

(d) H_2O

Assessment 12.40 Justify on a reaction coordinate diagram the fact that a strong base like sodium amide ($NaNH_2$) results in a faster E2 elimination than does sodium hydroxide (NaOH).

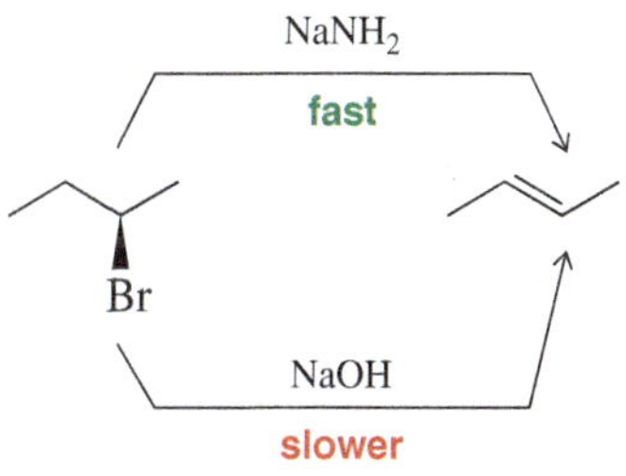

12.6.3 E2 vs. E1: solvent

E2 mechanism

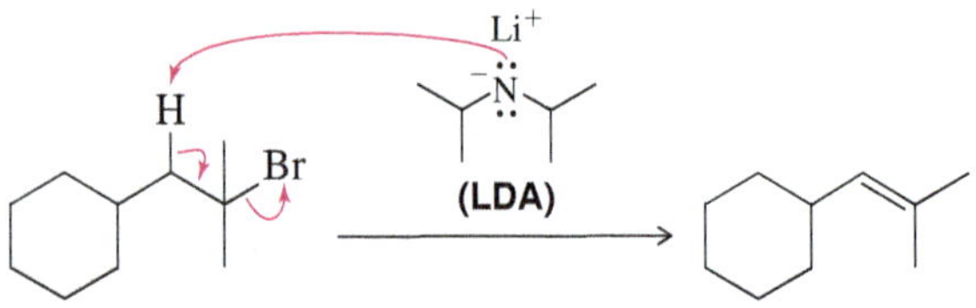

Because strong bases favor E2 reactions, any solvent that increases base strength should also increase the likelihood of an E2 reaction. Polar aprotic solvents solvate only the inorganic cation, increasing the strength of the base, favoring E2 reactions (**Figure 12.63**).

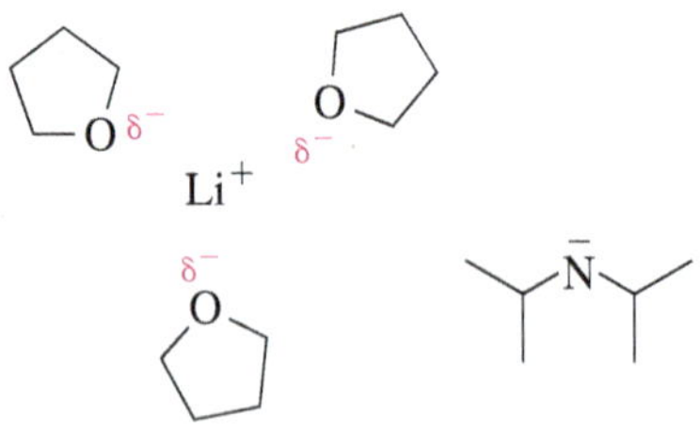

Figure 12.63 Polar aprotic solvents make bases stronger, favoring E2 eliminations.

Polar protic solvents will often be used in combination with strong bases for E2 reactions. This is due to the manner in which strong bases are synthesized. For example, sodium ethoxide is prepared by adding sodium metal to a large excess of ethanol (**Figure 12.65**). This results in a solution of ethoxide in ethanol. When used for an E2 elimination, the haloalkane is added directly to this solution.

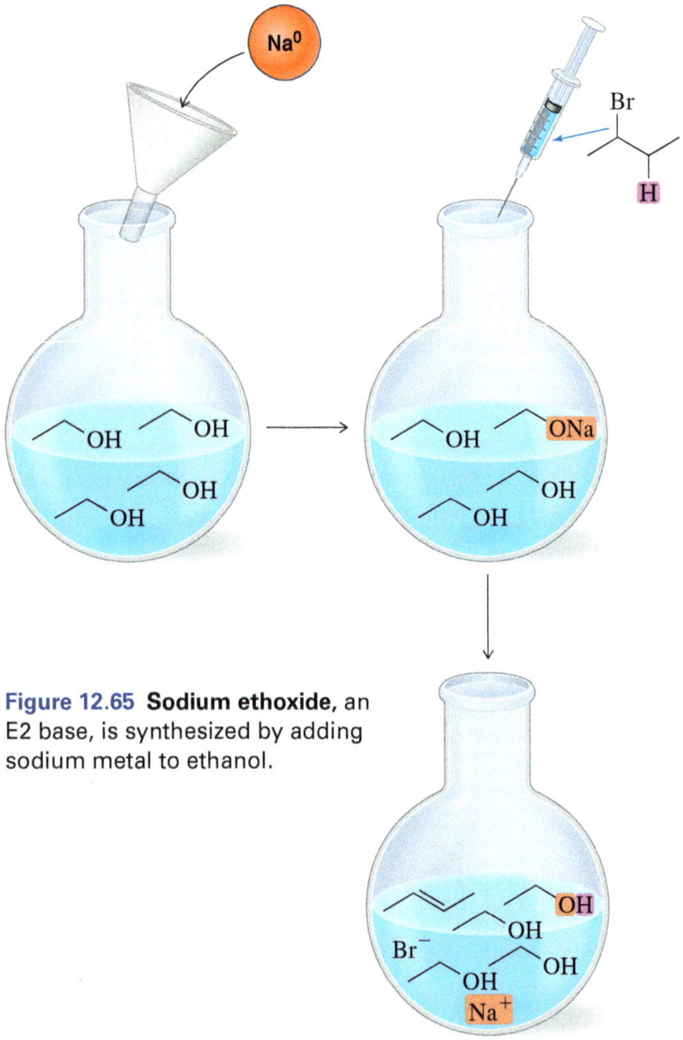

Figure 12.65 Sodium ethoxide, an E2 base, is synthesized by adding sodium metal to ethanol.

E1 mechanism

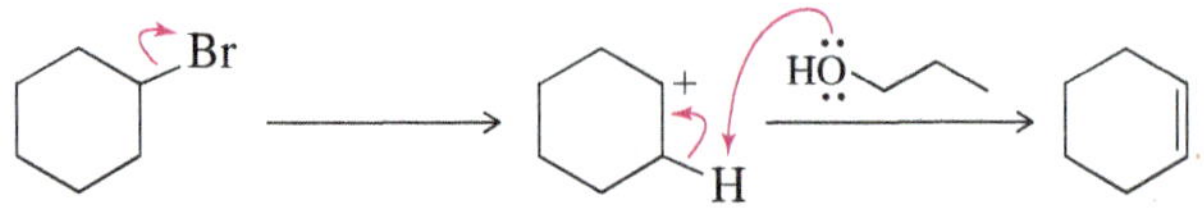

With identical rate-determining steps, anything that makes the S_N1 reaction faster will also favor an E1 mechanism. Polar protic solvents stabilize the leaving group, promoting rapid formation of the carbocation (**Figure 12.64**).

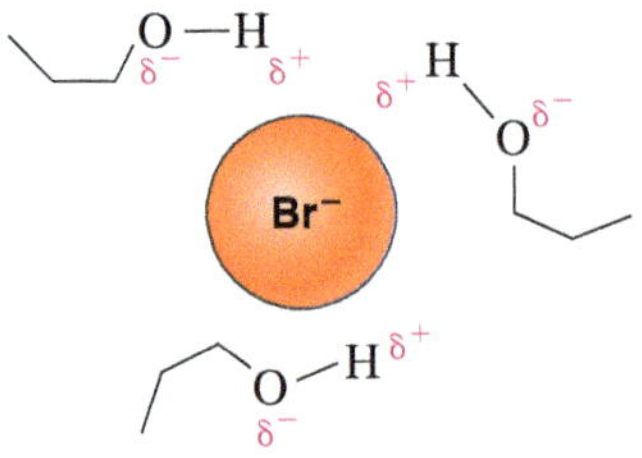

Figure 12.64 Polar protic solvents stabilize leaving groups, favoring E1 eliminations.

The solvated leaving group lowers the energy of the intermediate and the transition state of carbocation formation, resulting in a lower activation energy (**Figure 12.66**).

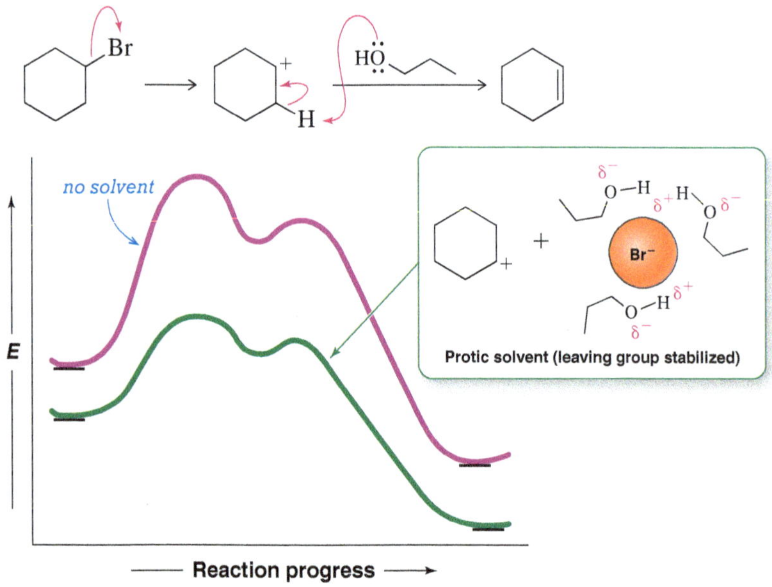

Figure 12.66 By stabilizing the leaving group, the rate of E1 elimination is increased.

12.6.3 SUMMARY

- Polar aprotic solvents make the base stronger, favoring E2 eliminations. Polar protic solvents stabilize the leaving group, favoring E1 eliminations.

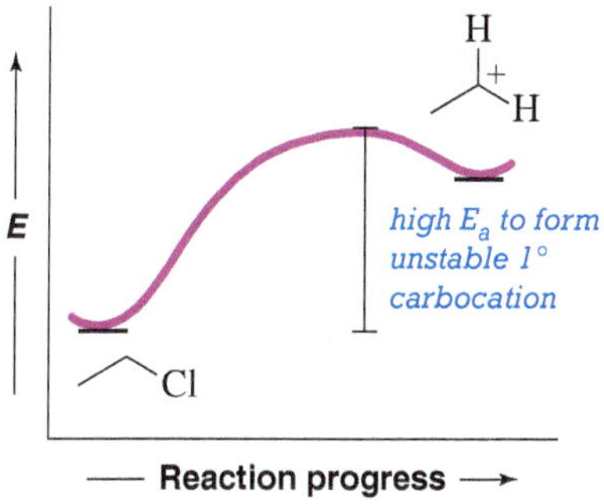

Figure 12.67 1° carbocations never form.

12.6.4 E2 vs. E1: substitution of the haloalkane

Haloalkane substitution is not a discriminating factor between E2 and E1, with a single exception. [*Before exploring the exception, consider why substitution mattered in the S_N2 and S_N1 reactions. Why isn't it as important for β-eliminations?*]

Haloalkane substitution is mechanism determining only when the haloalkane is 1°. Elimination at a 1° carbon is not very common, and it never occurs via the E1 mechanism. Formation of a highly unstable 1° carbocation has an activation energy so high that it simply does not happen (Figure 12.67).

Haloalkane substitution is unimportant for choosing between E1 and E2 for two additional reasons. First, both 2° R—X and 3° R—X allow carbocations to form; both can eliminate through E1 mechanisms. Second, steric hindrance (important in S_N2 reactions) is less important here. The base is removing a proton from the β-carbon. Even with a 3° haloalkane, the β-proton is accessible enough to be removed in the elimination (Figure 12.68).

Figure 12.68 Steric hindrance barely decreases the rate of E2 reactions.

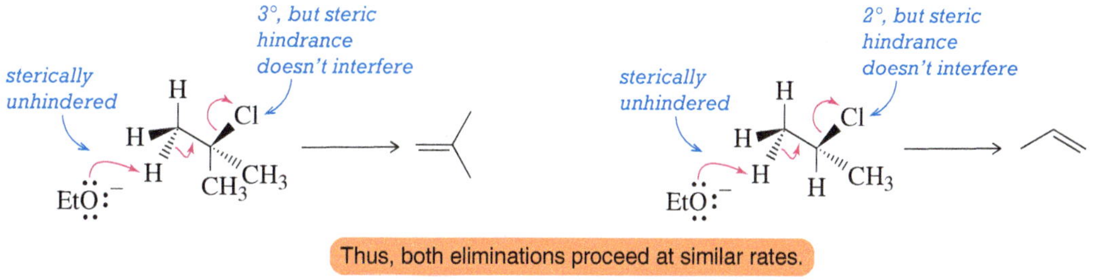

12.6.4 SUMMARY

- 1°, 2°, and 3° haloalkanes are able to undergo E2 eliminations. 2° and 3° haloalkanes can eliminate by the E1 mechanism.

12.6.5 Putting it all together: making a diagnosis

Because the haloalkane substitution is a less important factor in the reaction, the process of distinguishing between E1 and E2 is simpler than that for distinguishing between S_N1 and S_N2. Table 12.12 summarizes the guiding principles discussed so far.

Table 12.12 The particular substitution of the leaving group, base, and solvent influence the mechanism by which elimination reactions occur.

Haloalkane substitution	**1°:** can't form carbocations, but β-proton is accessible	E2 only
	2°: forms relatively stable carbocations; β-proton is accessible	E1 or E2
	3°: forms very stable carbocations; β-proton is accessible	E1 or E2
Base	**Strong:** anionic nucleophiles whose conjugate acids have a $pK_a > 11$	E2 favored
	Weak: neutral bases	E1 favored
Solvent	**Polar aprotic:** solvates inorganic ions, but not anions/nucleophiles or leaving groups; makes bases stronger	E2 favored
	Polar protic: solvates inorganic ions, nucleophiles, and, importantly, leaving groups	E1 favored

Using Table 12.12, the choice of elimination mechanism is made by evaluating each variable and weighing its importance. As you will see in the examples in **Table 12.13**, the choice is made predominantly based on the base strength. Strong bases favor E2, whereas weak bases favor E1. [*Although distinguishing between E1 and E2 reactions is simpler, review the line of reasoning used in Section 12.3.6 to distinguish between* S_N1 *and* S_N2 *reactions. This is especially important when substitution and elimination reactions are considered together in Section 12.8.*]

Table 12.13 The particular substitution of the leaving group, base, and solvent influence the mechanism by which elimination reactions occur.

Reaction	Reaching a diagnosis
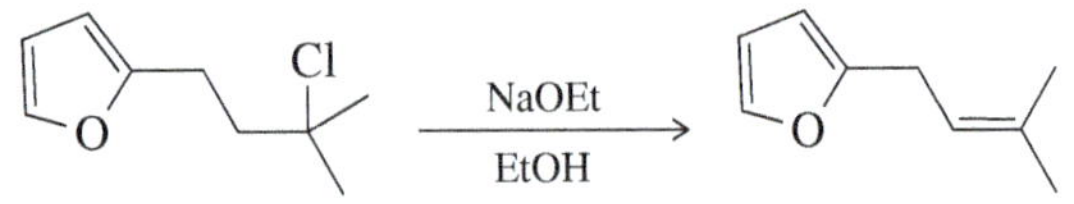	1. ***Substitution***: 3° R—X; can be either E2 or E1 2. ***Base***: alkoxide . . . strong base (pK_a of conjugate acid = 16) → favors E2 3. ***Solvent***: polar protic; favors E1 ***Diagnosis***: In a split decision, the strong base determines the mechanism. The ethoxide anion is too unstable to wait for a carbocation to form before reacting. Thus, this elimination proceeds by the E2 mechanism.
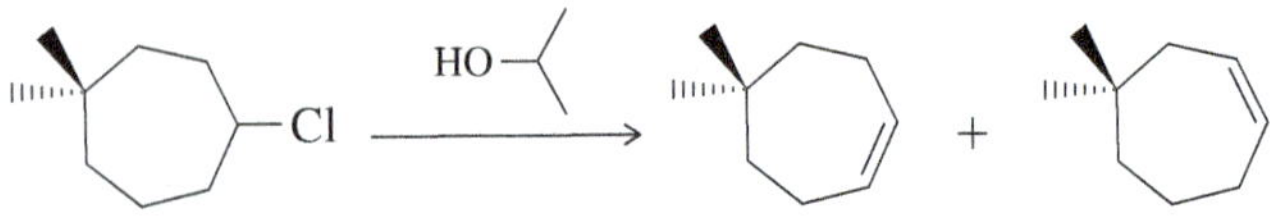	1. ***Substitution***: 2° R—X; can be either E2 or E1 2. ***Base***: neutral . . . weak base, won't remove a proton from a neutral molecule → favors E1 3. ***Solvent***: polar protic, stabilizes leaving group; favors E1 ***Diagnosis***: The neutral base and the protic solvent together suggest that this elimination proceeds by the E1 mechanism.

[*It is never too early to begin thinking about the conditions under which elimination or substitution might be favored. Assessments 12.41 and 12.42 cover only elimination reactions, but you should keep in mind why these conditions may produce substitution reactions instead.*]

Assessment 12.41 Given the reactants shown, what type of elimination would you expect to occur?

(a)

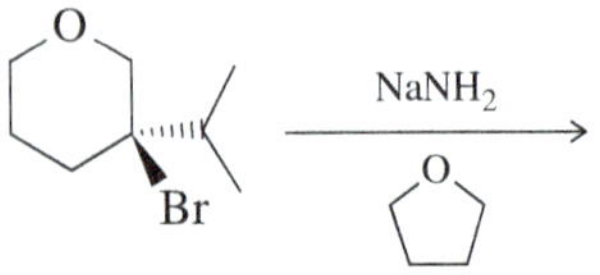

(b)

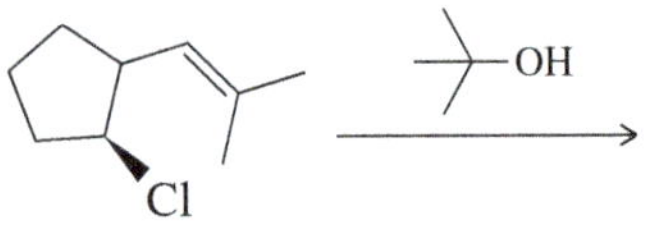

(c)

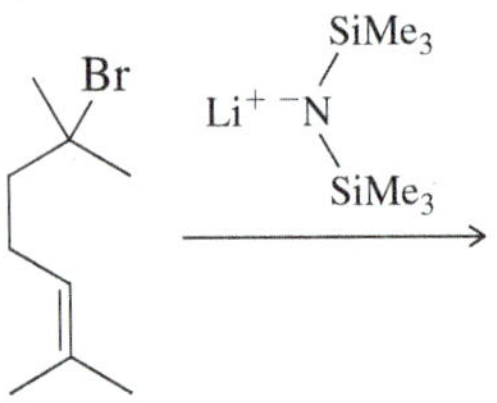

(d)

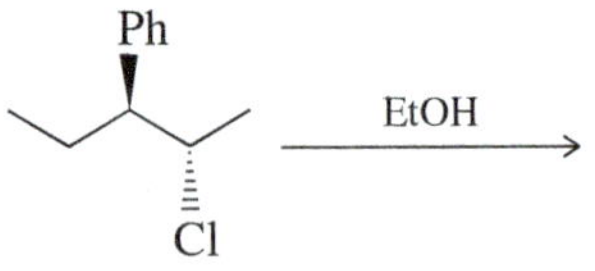

Assessment 12.42 Show a mechanism for the following elimination reactions. Label the mechanism as E1 or E2.

(a)

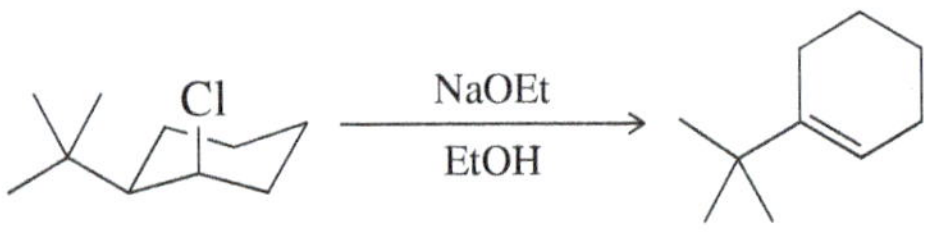

(b)

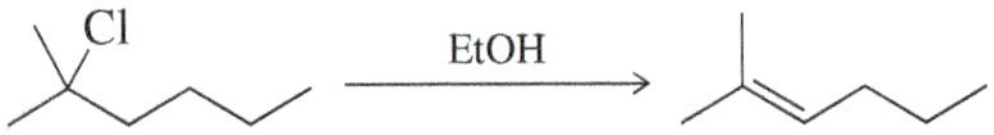

(c)

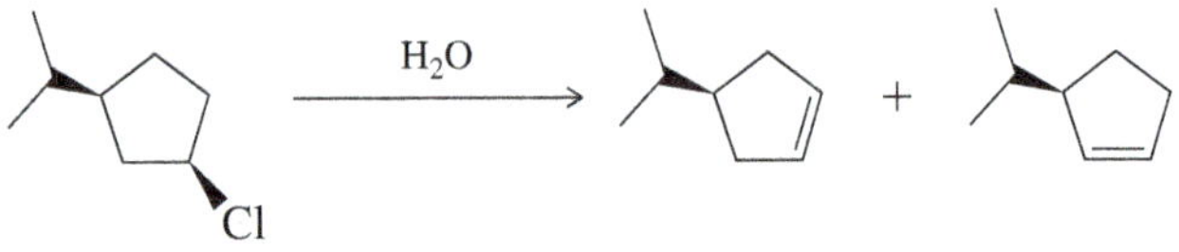

(d)

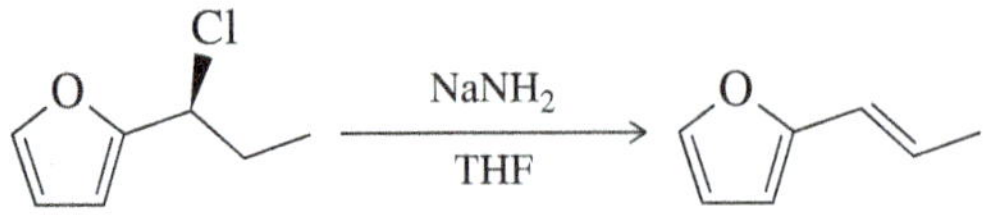

12.7 E2 vs. E1 outcomes

Once more, the choice of E2 vs. E1 may not seem important. And yet, like in S_N2 and S_N1 reactions, the outcome of the reaction depends on the mechanism by which the elimination reaction proceeds. As summarized in **Figure 12.69**, in each case it boils down to the manner in which the π bond is formed. For E2, the electrons from the breaking C—H σ bond must fall into the σ^* antibonding orbital, whereas for E1, the electrons from the breaking C—H σ bond fall into the empty p orbital of the carbocation.

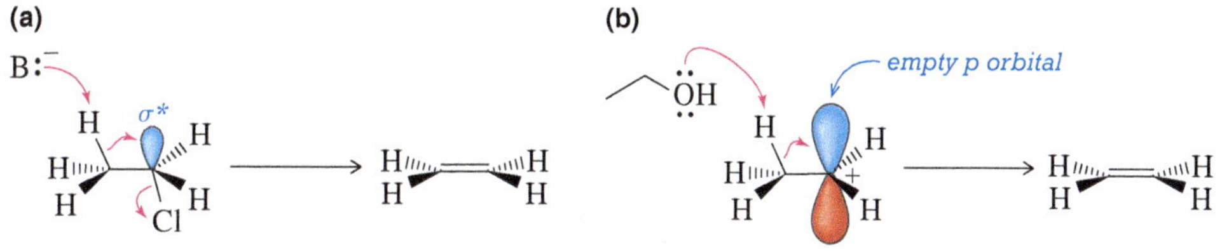

Figure 12.69 The regioselectivity, stereoselectivity, and whether or not rearrangement can occur depend on how the C—C π bond forms. (a) In E2, the electrons from the C—H σ bond fall into σ^* of the neighboring C—Cl bond. (b) In E1, the electrons from the C—H σ bond fall into an empty p orbital

12.7.1 E2 vs. E1 outcomes: regioselectivity

The elimination reactions covered thus far have shown only one product forming. What happens when two different alkenes can be formed through an E2 or E1 elimination? In either reaction, according to **Zaitsev's rule**, the most substituted alkene is formed preferentially [**Figure 12.70(a)**]. This makes sense, given that the more substituted (Zaitsev) alkene is also the more stable alkene (Section 9.2.1). The enhanced stability of more substituted alkenes is a result of enhanced hyperconjugation between neighboring C—H/C—C σ bonds and π^* of the more substituted alkene. In addition, the more substituted alkene minimizes steric strain between the methyl and hydrogen at C_4 [**Figure 12.70(b)**].

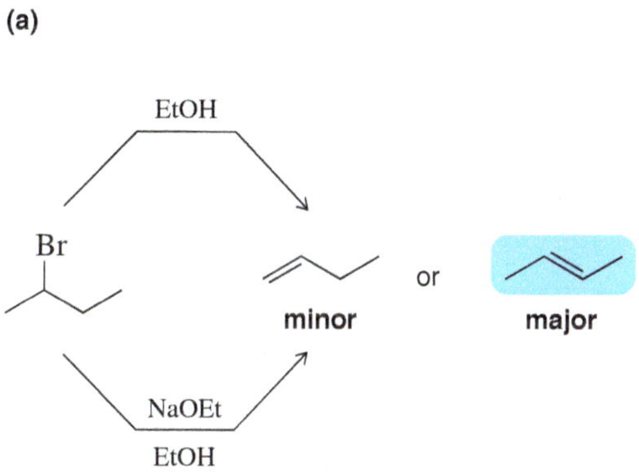

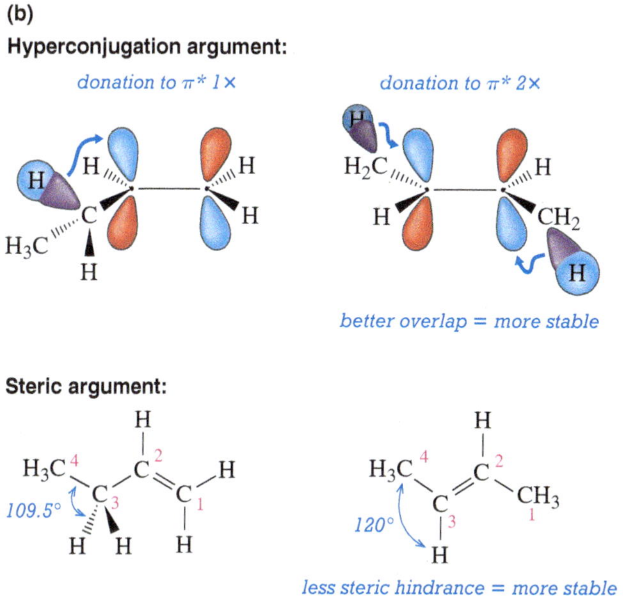

Figure 12.70 Selectivity in elimination. (a) When mechanistically possible, the more substituted alkene forms in either E1 or E2 reactions because (b) it is more stable.

12.7.1 SUMMARY

- In elimination reactions with a choice, the most stable (most substituted) alkene forms preferentially, as long as it is allowed by the mechanism.

There are situations where the anti-Zaitsev (less substituted) alkene can be formed as the major product. [*Note that in every case we have seen so far, the major product has been the more substituted alkene. Even when a minor product wasn't shown, there was likely a small amount of it in the product mixture.*] Formation of the anti-Zaitsev alkene takes advantage of the fact that the less substituted alkene is formed by removal of the less sterically hindered proton [Figure 12.71(a)]. The use of large bases, such as those in Figure 12.71(b), provides convenient access to the less stable alkene products. We present another way to make the anti-Zaitsev alkene when we study the chemistry of ammonium salts in Chapter 25.

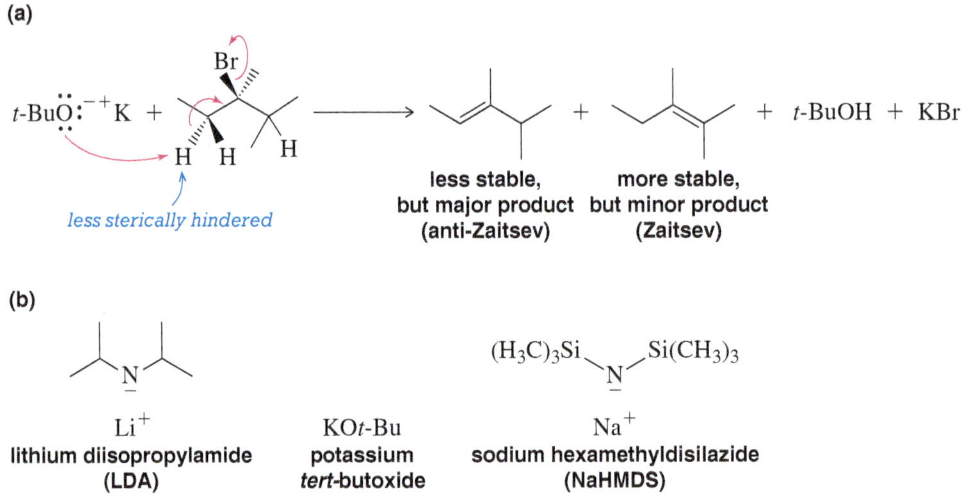

Figure 12.71 Sterically hindered bases give the less stable anti-Zaitsev alkene by (a) removing the least sterically hindered proton. (b) Some anti-Zaitsev bases.

Assessment 12.43 Suggest an appropriate base to synthesize the alkene as the major product from the starting haloalkane.

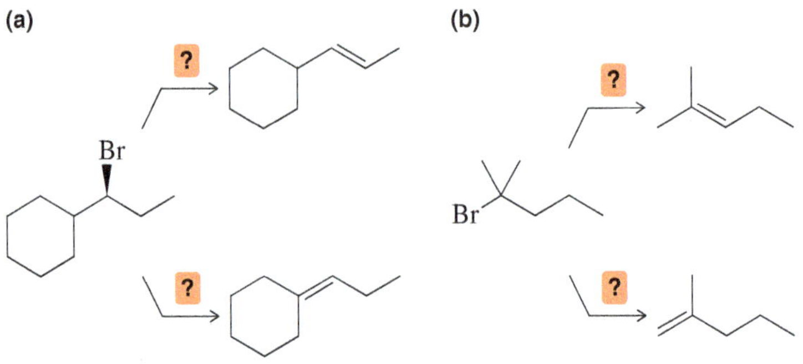

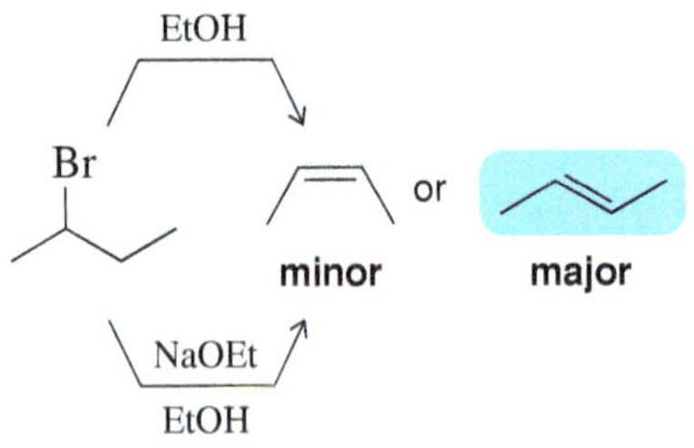

Figure 12.72 When mechanistically possible, the more stable alkene stereoisomer forms in either E1 or E2 reactions.

12.7.2 E2 vs. E1 outcomes: stereochemistry

Which alkene was formed was not the only issue to consider with 2-bromobutane. As you might have noticed, there are two possible stereoisomers of but-2-ene. Because forming the more stable alkene is important (Zaitsev's rule), it is not surprising that the most stable alkene stereoisomer (trans) is the major product (Figure 12.72).

But it's not as simple as just forming the more stable stereoisomer, especially in E2 reactions. When the C—H bond is broken by a strong base, the electrons immediately fall into σ* to break the C—Br bond (Section 12.6.1). As a result, there is a geometric arrangement necessary for the E2 reaction to happen (Figure 12.73). [*Another way of thinking about this is that the empty σ* orbital begins to tug on the electrons in the C—H bond, making the proton more acidic.*] The proton to be removed and the leaving group need to be opposite each other at a 180° dihedral angle. This relationship is described as **anti** (opposite sides) and **coplanar** (same plane) and is best represented using Newman projections to show the various conformations. [*Now is a good time to play with model kits, so you can see this arrangement in 3-D.*]

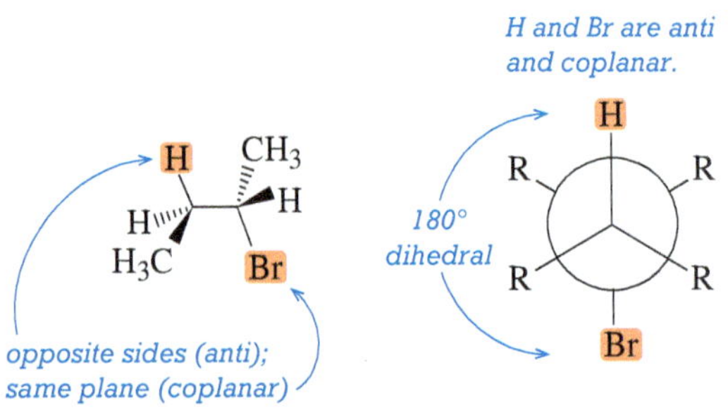

Figure 12.73 The anti and coplanar arrangement allows electrons from the C—H bond to fall into σ* of the C—Br bond and is necessary for E2 reactions to occur.

For 2-bromobutane, there are three important staggered Newman projections (Figure 12.74). In only two of them are there protons (H_a and H_b) anti and coplanar with the Br. As a result, elimination is only possible for these two. As the π bond forms, the groups on the same side (right or left) in the Newman projection remain on the same side in the product. Removing H_a generates the cis-product (CH_3 groups on the same side). Removing H_b gives the trans-product. Although both are mechanistically possible, the more stable trans-alkene is the major

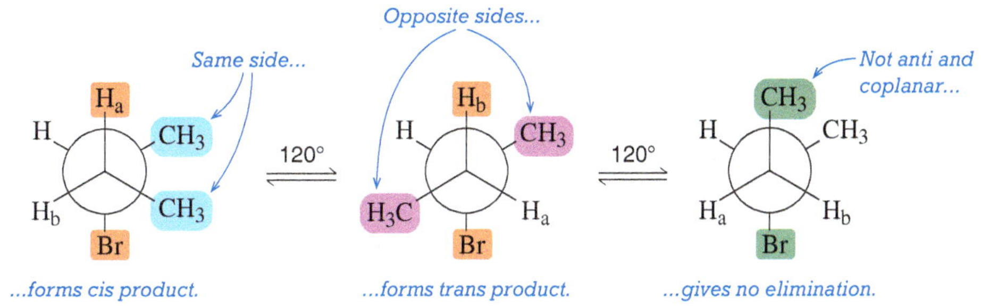

Figure 12.74 E2 elimination only proceeds through the anti and coplanar transition state.

product. [*Drawing Newman projections might be something you haven't done in a while; go back to Section 3.4 and practice some if you need to.*]

The anti and coplanar requirement can force the formation of the less stable alkene stereoisomer (**Figure 12.75**). This is true for elimination in (3*S*,4*R*)-3-bromo-3,4-dimethylhexane under E2 conditions. Of the Newman projections around the $C_3—C_4$ bond, only one has the anti and coplanar arrangement. In this projection, the bulky ethyl groups are on the same side when the proton is removed, so the less stable alkene isomer forms.

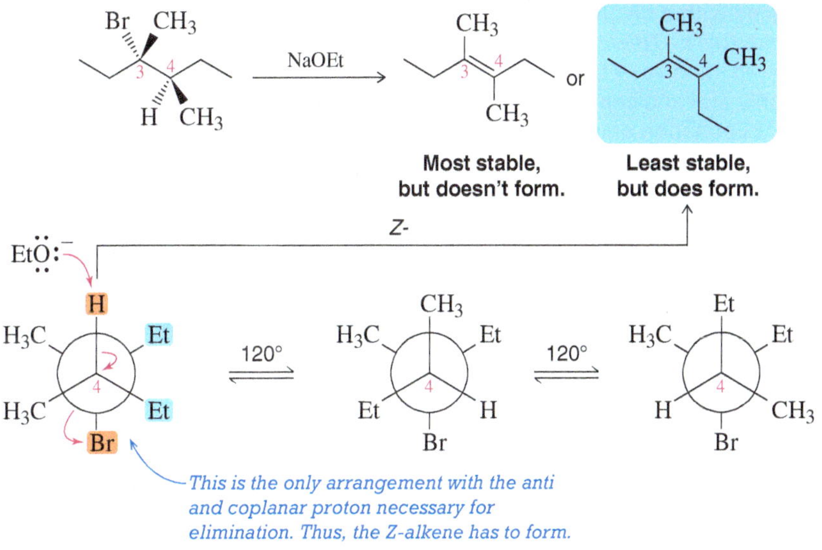

Figure 12.75 The stereochemical outcome of E2 reactions depends on the anti and coplanar arrangement in the transition state.

There is no anti and coplanar requirement for E1 reactions. For the same haloalkane used in Figure 12.75, the E1 reaction proceeds through a carbocation intermediate. When the proton is removed, the electrons from the C—H σ bond can fall into the empty *p* orbital from either side. With both isomers possible, the more stable *E*-alkene forms predominantly (**Figure 12.76**).

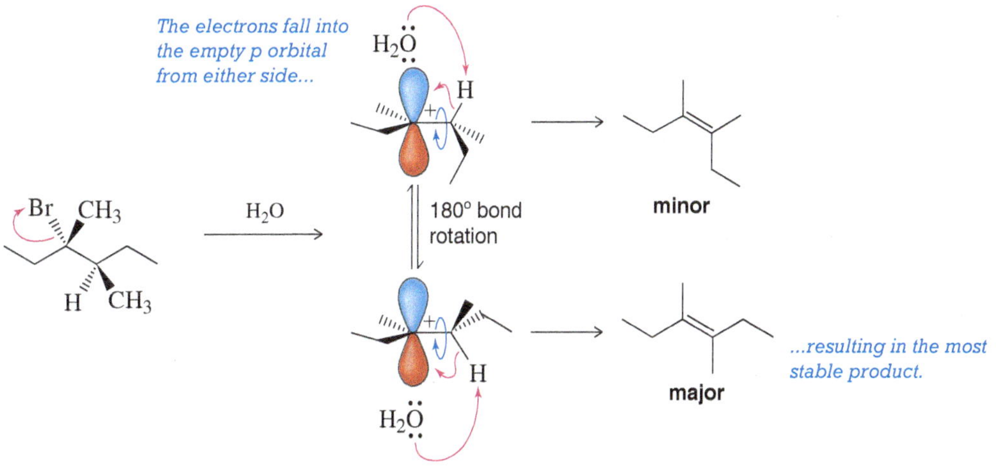

Figure 12.76 In E1 reactions, the more stable alkene stereoisomer is always produced.

12.7.2 SUMMARY

- In the E1 reaction, the most stable stereoisomer is the major product. For E2 reactions, the most stable stereoisomer results, as long as it forms through the anti and coplanar conformation.

12.7.3 E2 vs. E1 outcomes: regiochemistry in cyclic systems

Because the anti and coplanar requirement can change the regiochemical outcome of E2 reactions in cyclic systems, special mention is necessary here. Consider that the elimination reaction of *trans*-1-chloro-2-methyl shown in **Figure 12.77** does not give the expected Zaitsev product. To achieve the anti and coplanar arrangement, the leaving group and proton being removed must be in axial positions on adjacent carbons. Thus, a ring-flip from the most stable conformation is required before elimination can take place [Figure 12.77(a)]. The relevant Newman projections show that only the proton at C_d is anti and coplanar to the leaving chloride [Figure 12.77(b)]. Thus, only the anti-Zaitsev product forms.

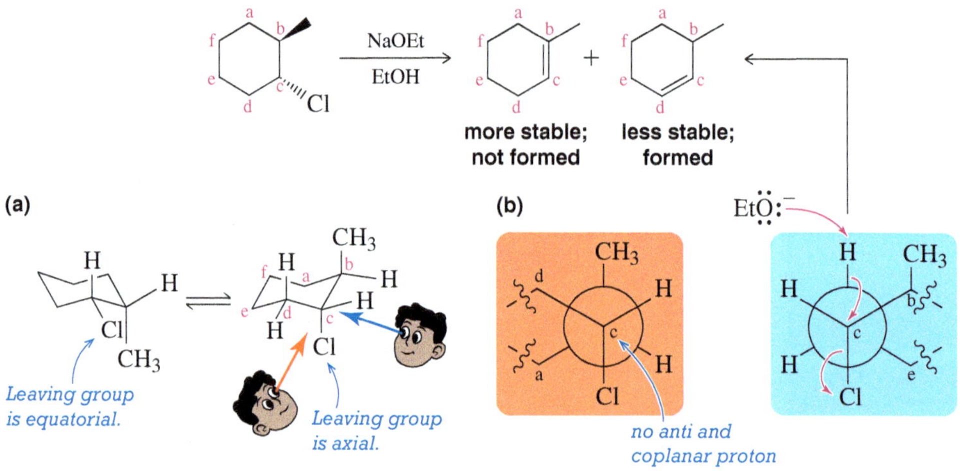

Figure 12.77 Anti and coplanar requirement in cyclic systems. (a) The leaving group must be axial so that it is aligned with a neighboring C—H σ bond. (b) The Newman projections, (c → d) and (c → b), show that the only anti and coplanar proton is on C_d. Elimination only results in the less stable alkene.

E1 reactions, lacking the specific anti and coplanar requirement, always produce the most stable alkene, even in cyclic systems.

Assessment 12.44 Paying close attention to the stereochemical outcome, predict the product of these elimination reactions.

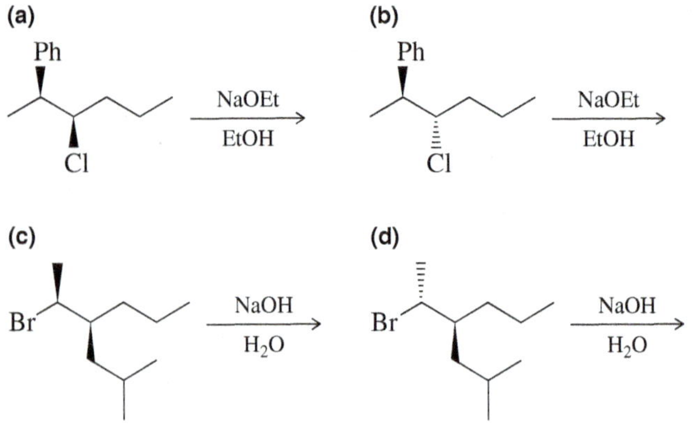

Assessment 12.45 The cis-diastereomer undergoes E2 elimination 500 times faster than the trans form. Explain.

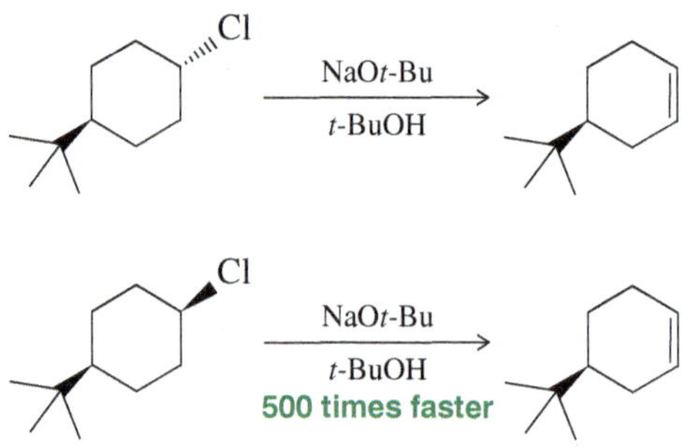

12.7.4 E2 vs. E1 outcomes: rearrangement

As in the S_N1 reaction, carbocation rearrangement is possible for E1 reactions prior to formation of the new alkene. An example of this is shown in **Figure 12.78**. In this case, the

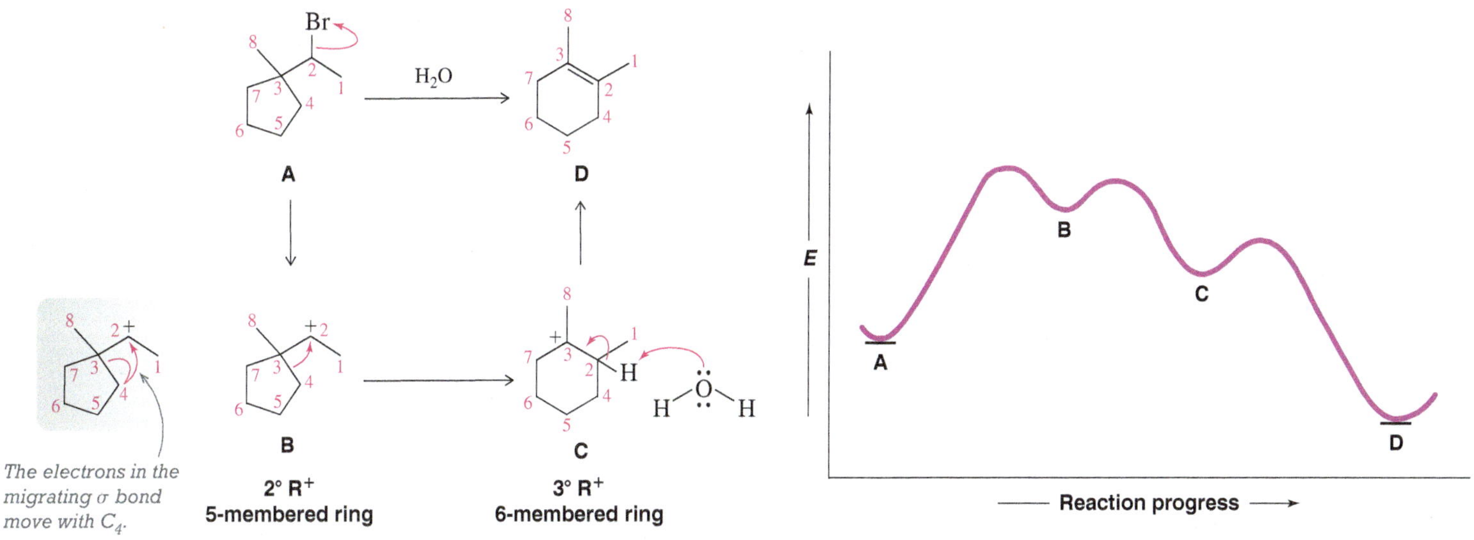

Figure 12.78 Rearrangement can occur in E1 elimination reactions.

rearrangement from a 2° to 3° carbocation also results in ring expansion to form a more stable six-membered ring. Following rearrangement, the resulting carbocation loses a proton from the β-carbon to form the alkene.

E2 reactions, like their concerted S_N2 counterpart, do not undergo rearrangements because they do not proceed through carbocation intermediates.

12.7.4 SUMMARY

- E1 reactions, proceeding through carbocation intermediates, undergo rearrangement if there is an opportunity to form a more stable carbocation. E2 eliminations are not subject to rearrangement.

Assessment 12.46 Provide a mechanism for the following E1 reactions.

(a)

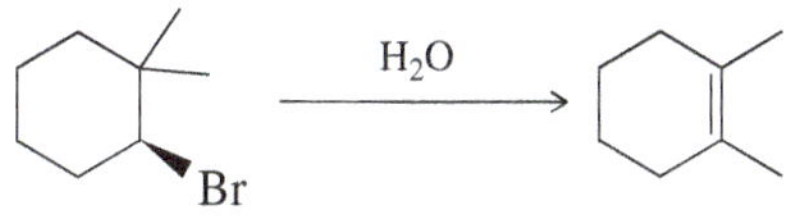

(b)

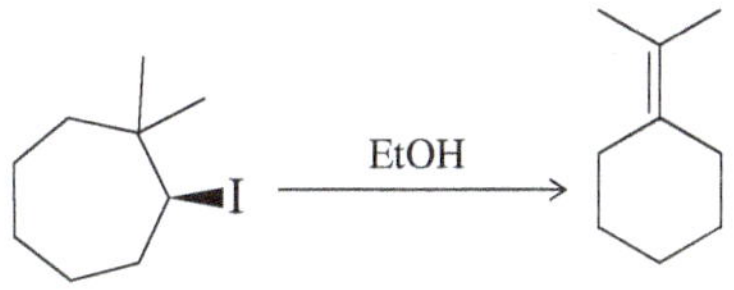

(c)

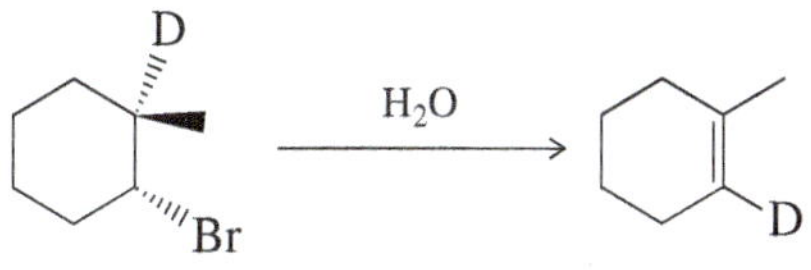

D is an isotope of H.

Assessment 12.47 Predict the product of the following rearrangement-prone E1 eliminations.

(a)

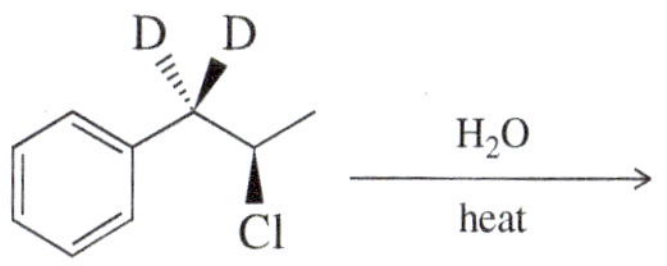

(b)

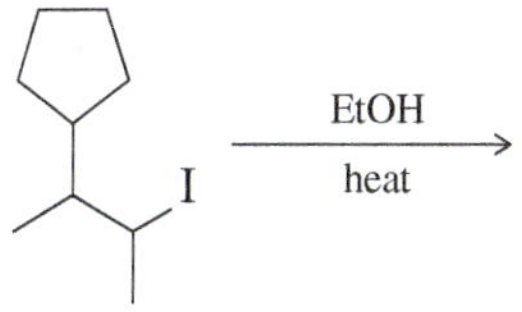

(c)

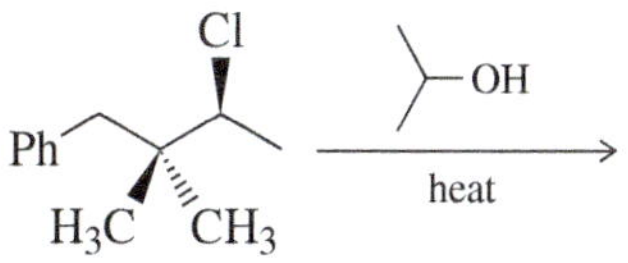

12-C CHEMISTRY IN CONTEXT: Medicinal Chemistry

From chemical weapon to chemotherapy

Following the use of mustard gas as a chemical weapon in World War I, the world feared its deployment in future wars. Thus, scientists undertook studies with these poisonous agents in hopes of finding antidotes. Autopsies of WWI mustard gas victims revealed that stricken soldiers suffered from severe hair loss, shedding of the stomach lining, and a substantially diminished number of immune cells. Because the side effects of mustard gases are localized to rapidly dividing cells, it was thought that mustards (as a general class of compounds) might have utility in killing cancer cells (another class of rapidly dividing cells).

DNA represents a significant part of the machinery for cell division in both healthy and tumor cells. A molecule that can covalently attach to DNA should inhibit the cell division shown in Figure 12.79. Based on this logic, in 1942 doctors at Yale University used a nitrogen mustard as a chemotherapeutic agent in a patient with advanced lymphoma. Though the patient died within 6 months, he showed great improvement over the course of treatments and was more comfortable and in less pain as a result of this new therapy.

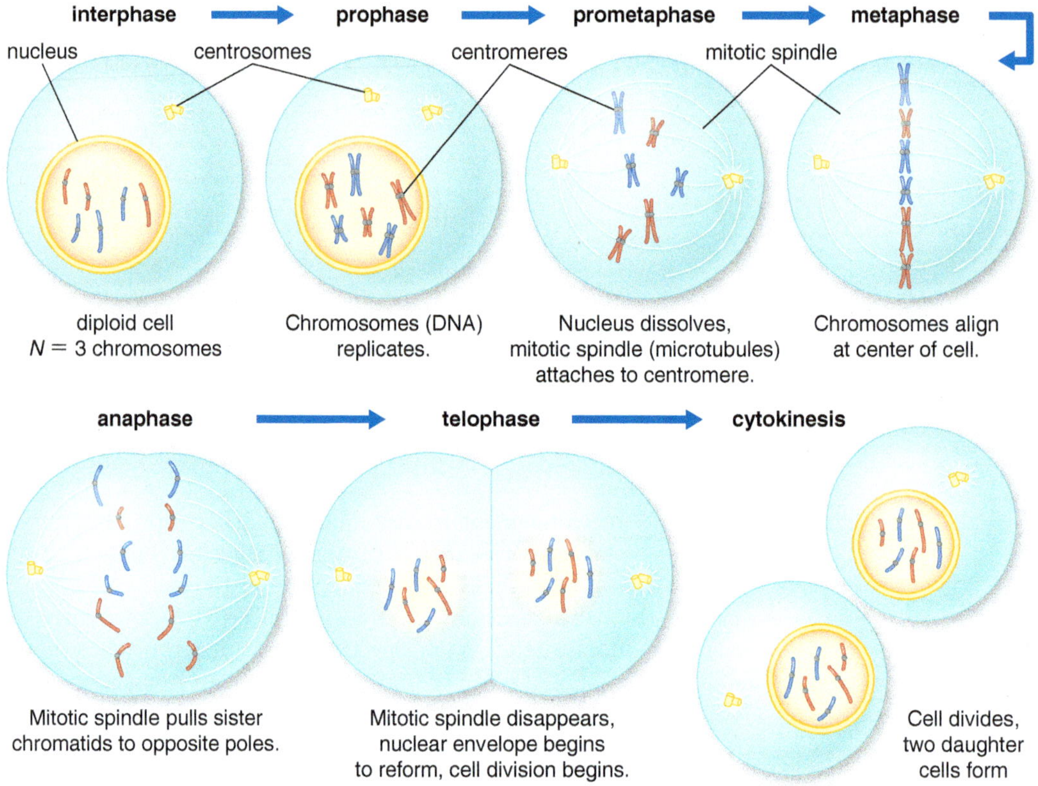

Figure 12.79 Cell division via mitosis.

12.8 Substitution vs. elimination

Given the similarity in conditions and the rules governing mechanistic outcomes, we encountered many reactions in Sections 12.6 and 12.7 where elimination *and* substitution were possible. By considering how different factors (haloalkane substitution, nucleophile, and solvent) favor S_N1 over S_N2 and E1 over E2, it is possible to evaluate all four potential outcomes at the same time. A few general considerations will help us as we delve into these types of problems.

12.8.1 Substitution vs. elimination: role of entropy

Deciding between substitution and elimination boils down to picking the more favorable of two favored processes. For the sake of discussion, imagine an equilibrium between S_N1 and E1 products. [*Assuming they are nearly equal in terms of ΔH, what is the major difference between the two equilibrium processes?*] In the elimination reaction, two molecules react to produce three ($\Delta S_{elim} > 0$). Alternatively, substitution is entropically neutral ($\Delta S \approx 0$; 2 molecules $\rightarrow$ 2 molecules). As a result, running a reaction at higher temperatures favors elimination over substitution (Figure 12.80).

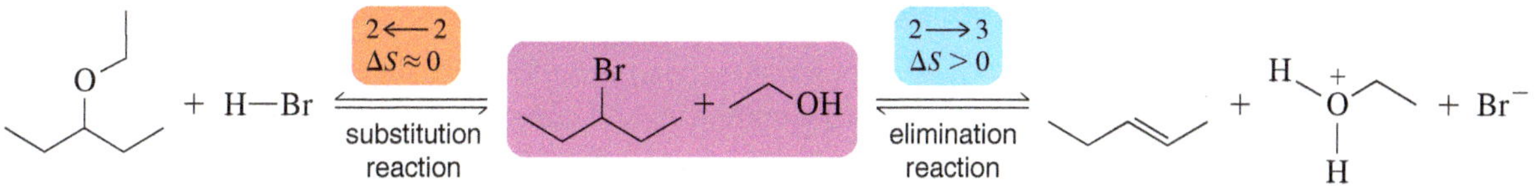

Figure 12.80 Elimination is favored by entropy over substitution. High temperature favors elimination.

12.8.1 SUMMARY

- At high temperatures, the entropy term dominates, making elimination favored over substitution.

Assessment 12.48 Which reaction, E2 or S_N2, would you expect to be more favorable at higher temperatures?

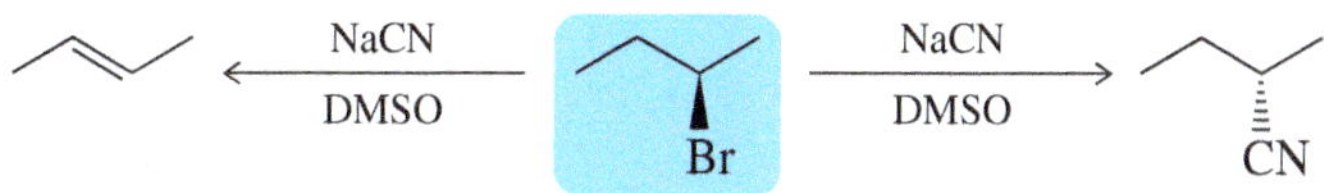

Assessment 12.49 For the following reactions we have not seen yet, which side, if either, would be favored by increasing the temperature?

(a)

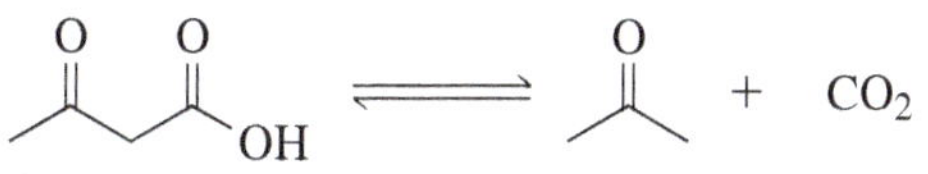

(b)

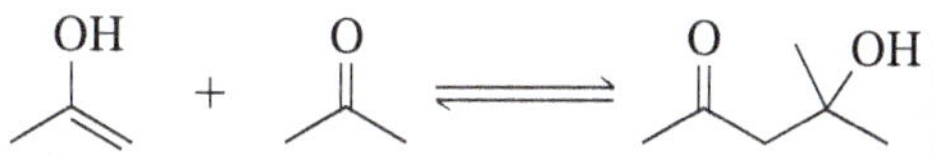

(c)

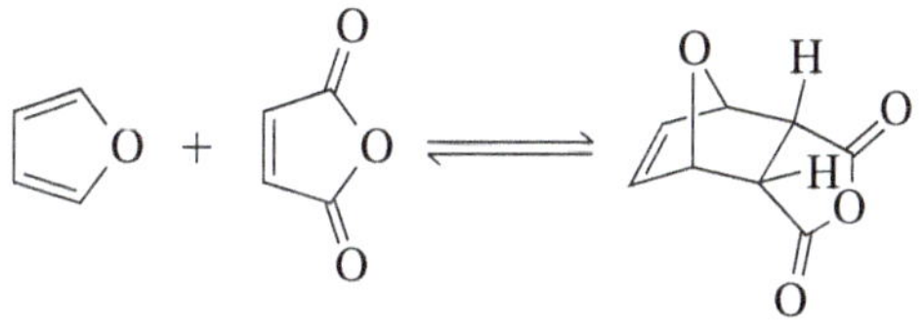

(d)

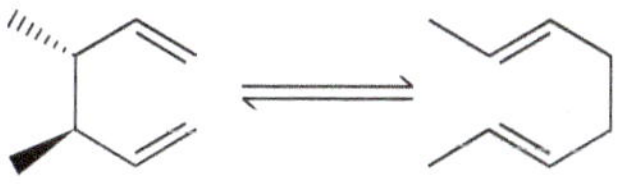

12.8.2 Substitution vs. elimination: E1 and S_N1

E1 and S_N1 reactions proceed through identical first steps (carbocation formation). Because they share the rate-determining step, the reaction outcome (E1 or S_N1) is determined in the second, fast, and non-rate-determining step (**Figure 12.81**). This exothermic step has a transition state similar in structure to the carbocation intermediate, regardless of which direction the reaction is going (E1 or S_N1). As a result, it is difficult to predict which will occur for competing S_N1/E1 reactions. Experimentally, substitution is slightly favored when using weak bases, but higher temperatures favor E1. Usually, a mixture of both is obtained.

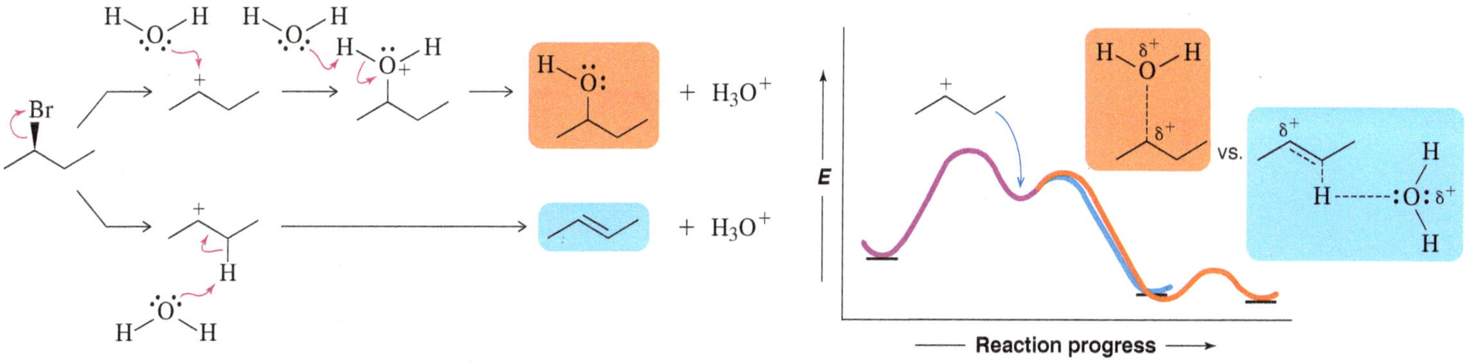

Figure 12.81 Under S_N1/E1 conditions, a mixture of substitution and elimination products is generated.

12.8.2 SUMMARY

- When the reaction is under E1/S_N1 conditions, the product will often include a mixture of both elimination and substitution products.

12.8.3 Substitution vs. elimination: E2 vs. S_N2

Choosing between S_N2 and E2 is a bit more straightforward. For the most part, it boils down to steric hindrance, as seen in the following generalizations (Table 12.14). [*This table should make sense in the context of the mechanism of each reaction studied thus far.*]

Table 12.14 Distinguishing between E2 and S_N2 reactions is largely a function of steric hindrance and base strength.

Entry	Characteristic	Observation	Example
1	1° haloalkane	With very little steric hindrance blocking σ*, substitution is favored.	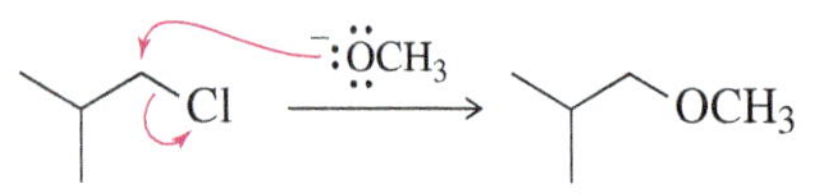
2	3° haloalkane	With σ* very hindered, elimination is the only possibility.	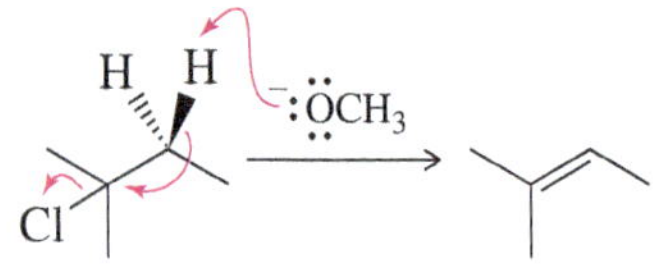
3	Bulky base/nucleophile	Because substitution requires more space to accommodate approach to σ*, bulky bases favor elimination reactions.	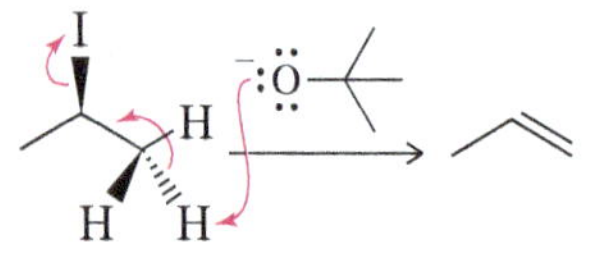
4	Small base/nucleophile	Small nucleophiles can easily access σ*; thus, substitution is favored.	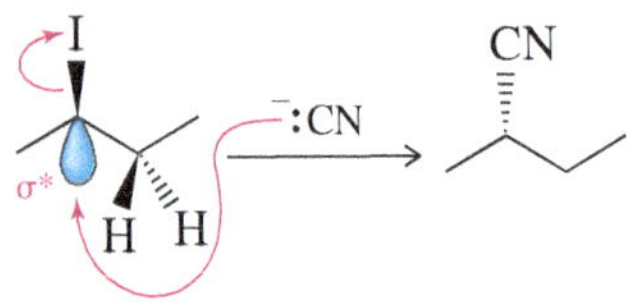
5	Branching near leaving group	Branching creates more steric hindrance around σ* and makes elimination more likely.	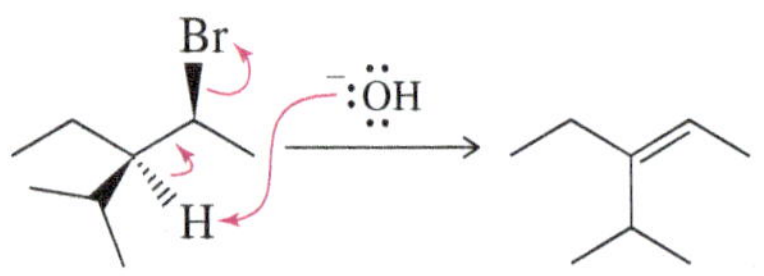
6	2° haloalkane/weak base	There is some steric hindrance, slowing down S_N2, but not enough to prevent it completely.	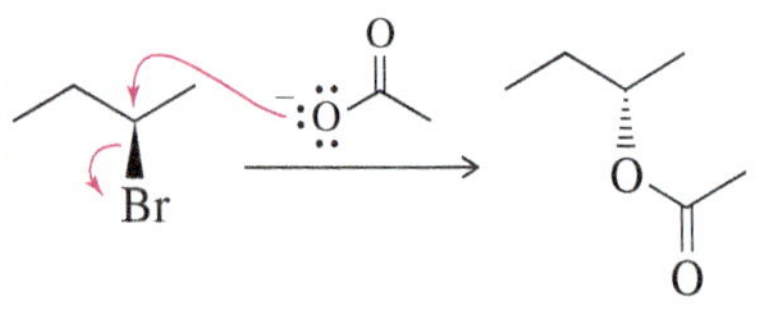

The last entry in Table 12.14 deserves special mention. At a 2° carbon, there is steric hindrance around σ*, but not enough to completely prevent substitution. Thus, you might expect both S_N2 and E2 to occur. Indeed, both are possible, but with slight changes in conditions. The use of a strong base (conjugate acid $pK_a > 11$) in this situation favors elimination [Figure 12.82(a)]. [*This makes sense, right? Strong bases want to grab protons, resulting in elimination.*] If the pK_a of the conjugate acid of the anionic nucleophile is less than 11 [*that is, the anion has no interest in a proton*], then substitution is favored [Figure 12.82(b)].

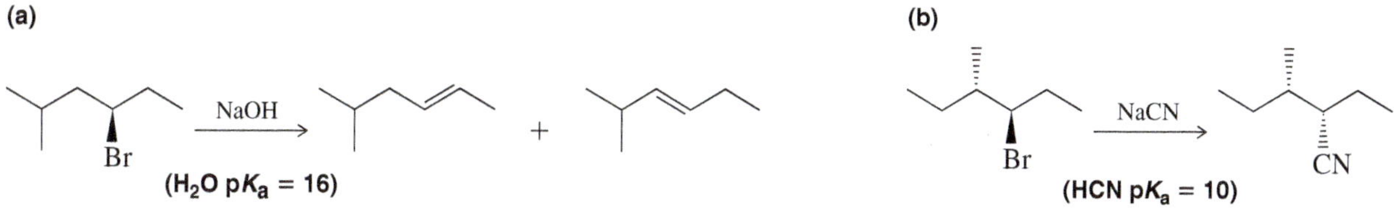

Figure 12.82 **E2 vs. S_N2 at 2° carbons.** (a) Stronger bases perform E2 reactions. (b) Weaker bases perform S_N2 reactions.

Finally, to elaborate on entry 3 in Table 12.14, the bulky bases shown in **Figure 12.83** are too sterically hindered to attack σ*. Thus, they exclusively do E2 reactions.

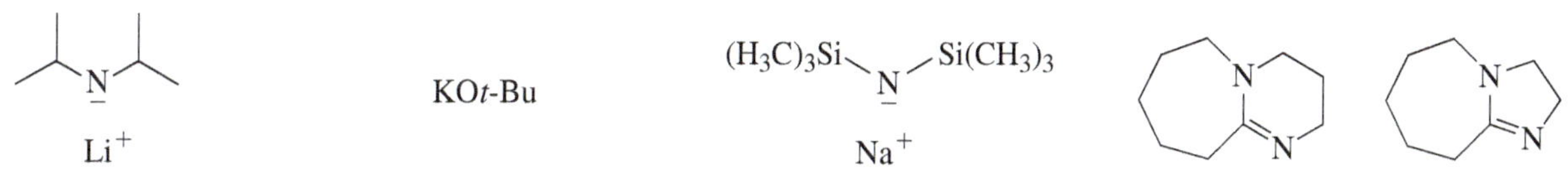

lithium diisopropylamide (LDA) **potassium *tert*-butoxide** **sodium hexamethyldisilazide (NaHMDS)** **DBU** **DBN**

Figure 12.83 **Bulky, non-nucleophilic bases preferentially result in E2 reactions.**

12.8.3 SUMMARY

- Sterically hindered conditions (3° R—X, bulky bases, branching) favor E2. Sterically less demanding conditions (1° R—X, small bases, less branching) favor S_N2. For 2° R—X, strong anionic bases favor elimination; weak anionic nucleophiles favor substitution.

Assessment 12.50 For which of the following reactions would you expect elimination to be more favored than substitution?

(a)

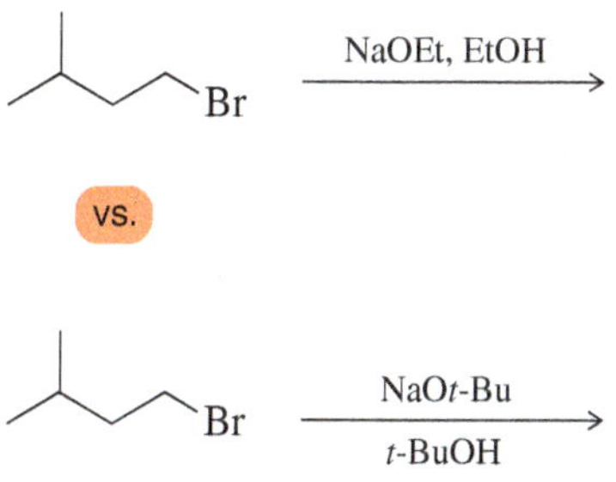

(b)

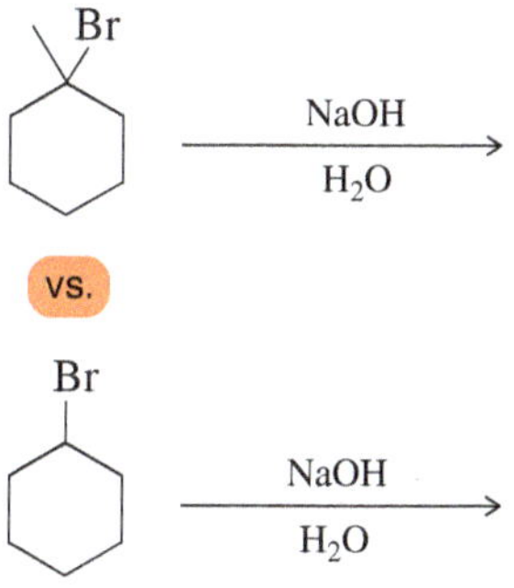

(c)

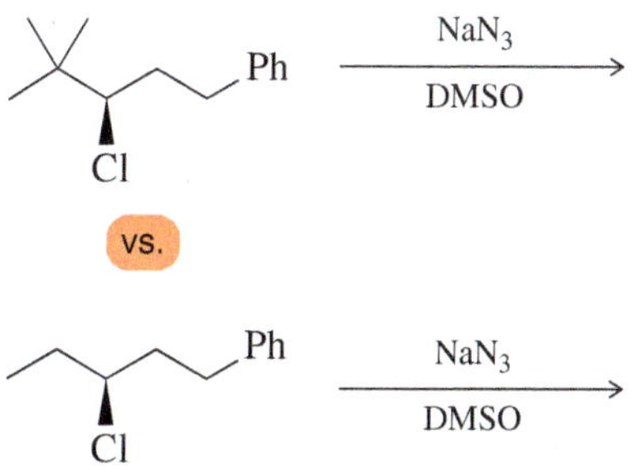

(d)

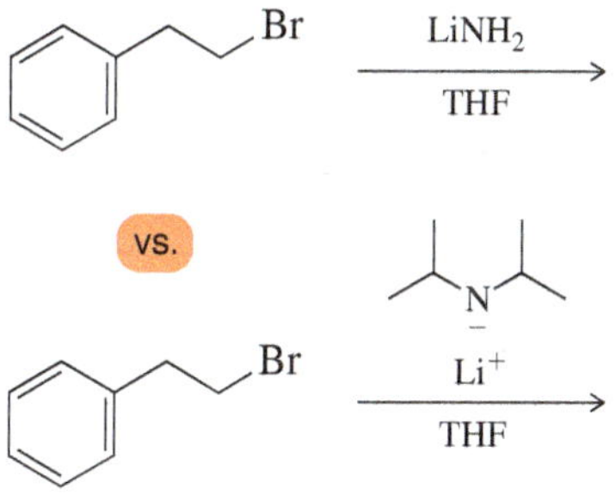

(e)

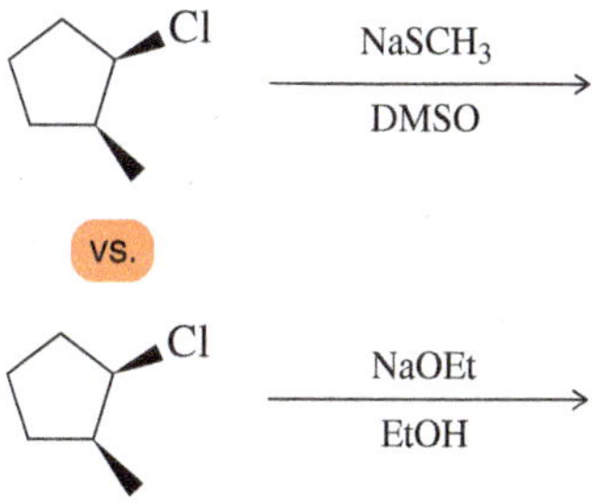

(f)

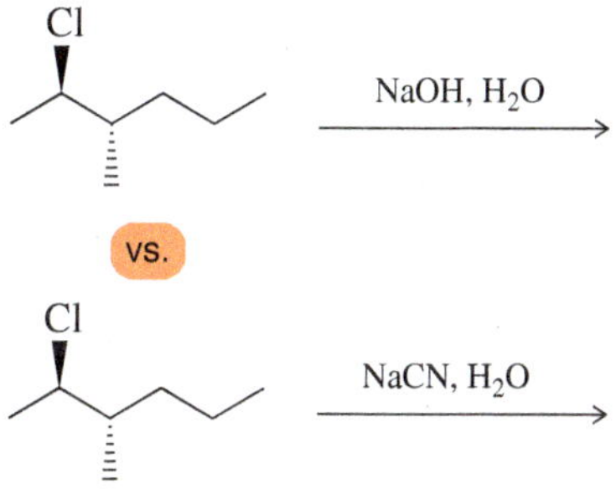

As you work more and more assessments, you will find yourself encountering tricky situations and "just knowing" what is going to happen. This is called chemical intuition and is exactly what we are trying to achieve.

12.8.4 Substitution vs. elimination: worked examples

In the same way that a physician becomes an expert at diagnosing illnesses, the only way to become an expert at diagnosing reaction outcomes is to practice. Some exercises will be worked for you (Table 12.15), but in Assessment 12.51 and those at the end of the chapter, you are encouraged to work as many as possible yourself.

Table 12.15 Examples of substitution and elimination reactions and the mechanism by which they proceed.

Reaction	Reaching a diagnosis
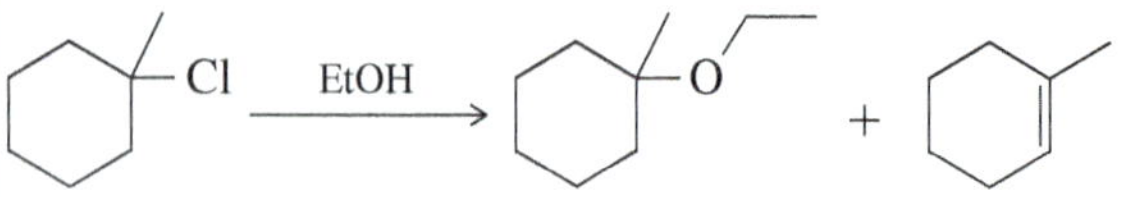	1. ***Substitution***: 3° R—X; can only be S_N1 or E1 2. ***Nucleophile***: neutral alcohol . . . weak nucleophile → favors S_N1 or E1 3. ***Solvent***: polar protic; favors S_N1 or E1 ***Diagnosis***: All categories suggest S_N1 or E1. Because the solvent is the nucleophile, this is a standard solvolysis reaction, proceeding by the S_N1/E1 mechanism to give both substitution and elimination.
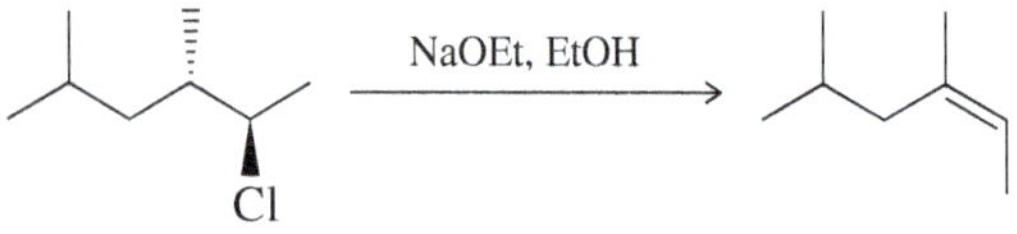	1. ***Substitution***: 2° R—X; can be S_N1, S_N2, E1, or E2 2. ***Nucleophile***: anionic . . . strong nucleophile/base → favors S_N2/E2 3. ***Solvent***: polar protic; favors S_N1 and E1 ***Diagnosis***: Though the categories are mixed, the negatively charged anion does not wait for a carbocation. Thus, this reaction occurs by an E2 or S_N2 mechanism. Because the haloalkane is 2°, the choice between S_N2 and E2 comes down to base strength. Because the pK_a of EtOH, the conjugate acid of EtO^-, is greater than 11, this is an E2 reaction. Note that there are stereochemical issues to consider with regard to the alkene formed.
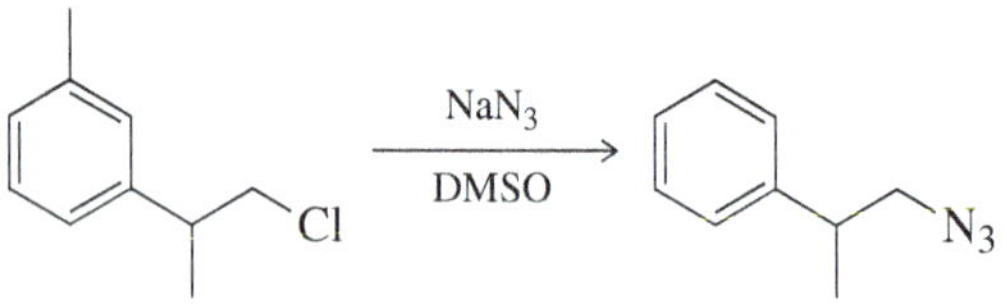	1. ***Substitution***: 1° R—X; can only be S_N2 or E2 (rarely) 2. ***Nucleophile***: anionic . . . strong nucleophile → favors S_N2/E2 3. ***Solvent***: polar aprotic; favors S_N2/E2 ***Diagnosis***: A carbocation cannot form at the 1° carbon, so this reaction must be S_N2 or E2. With a small nucleophile and a sterically unhindered 1° carbon, substitution results.
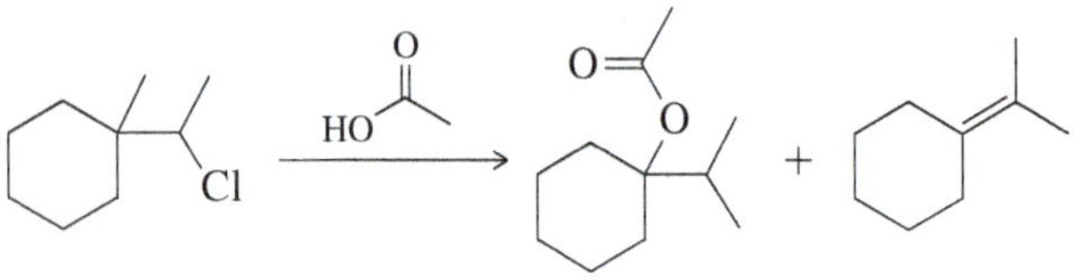	1. ***Substitution***: 2° R—X; can be S_N1, S_N2, E1, or E2 2. ***Nucleophile***: neutral . . . weak nucleophile → favors S_N1/E1 3. ***Solvent***: polar protic; favors S_N1/E1 ***Diagnosis***: Here, everything leans toward E1 or S_N1, suggesting that both will occur and involve rearrangement.
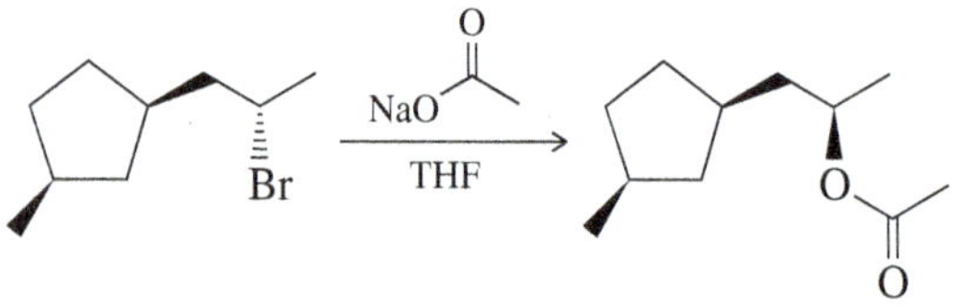	1. ***Substitution***: 2° R—X; can be S_N1, S_N2, E1, or E2 2. ***Nucleophile***: anionic . . . strong nucleophile/base → favors S_N2/E2 3. ***Solvent***: polar aprotic; favors S_N2 and E2 ***Diagnosis***: Based on the anionic nucleophile, this reaction will proceed by an E2 or S_N2 mechanism. The haloalkane is 2°, so the choice between S_N2 and E2 comes down to base strength. Because the pK_a of acetic acid, the conjugate acid of the base, is 5 (less than 11), this reaction proceeds by the S_N2 mechanism. Note that the stereocenter is inverted via backside displacement.

Assessment 12.51 Predict the product of the following reactions.

(a)

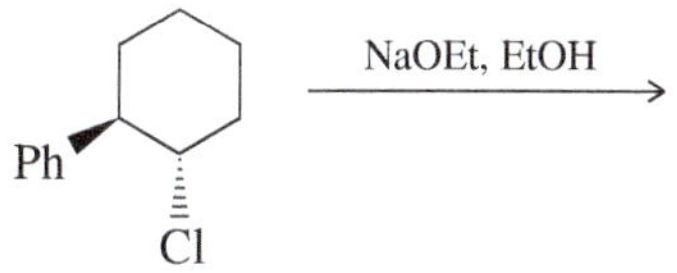

(b)

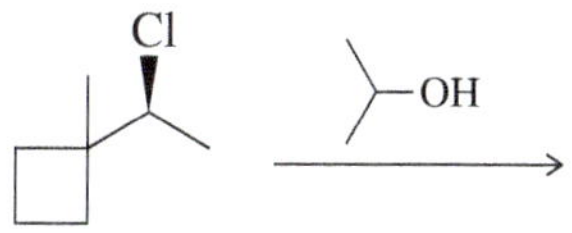

(c)

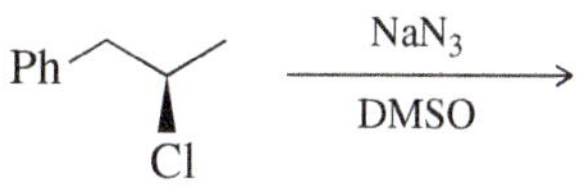

(d)

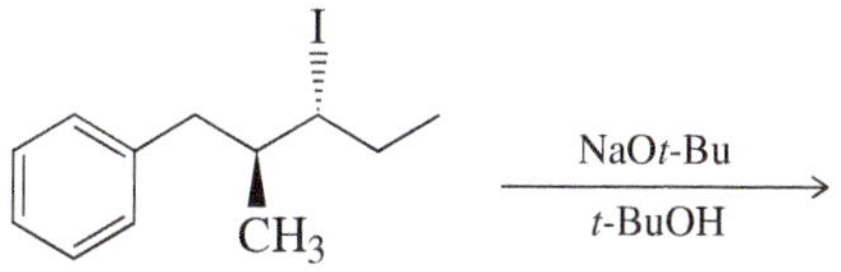

(e)

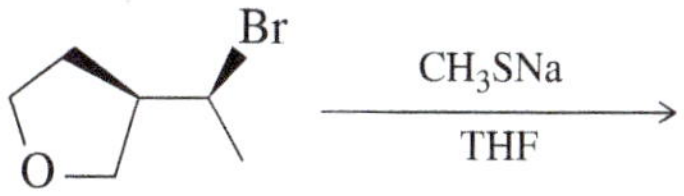

(f)

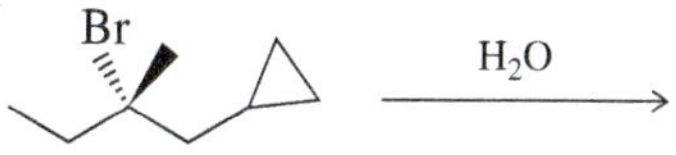

Concluding thoughts: DNA interactive agents as chemotherapy

Sulfur mustards, the cause of such pain and suffering during military conflicts of the early 20th century (and sadly, wars of today), became targets of drug discovery in the 1940s as potential anticancer drugs. The most significant compound that emerged from these efforts was the nitrogen mustard mechlorethamine (Mustargen®). Mechlorethamine works by cross-linking, through alkylation, two guanine bases in DNA (**Figure 12.84**). [*Because you may not have had a biochemistry course, simply focus on the reactions that are occurring; ignore the large molecules.*] Featuring four substitution reactions, this process produces a cross-linked dimer that, through multiple pathways, results in cell death, thereby inhibiting the growth of tumor cells.

The redeeming story of how chemical weapons used in wars between nations became chemical weapons used in the war against cancer reminds us of both the devastation caused by chemistry in the wrong hands and the magnificent outcomes of chemistry when used to improve lives. And the central player in this story, the substitution reaction, is one we continue to explore in Chapter 13. [*Let's go there now!*]

Although mechlorethamine is still used in the treatment of advanced Hodgkin's disease, it has the disadvantage of poor oral bioavailability (Chapter 11). It also reacts very quickly with water, which means that aqueous solutions of the drug must be prepared immediately prior to injection. Because of its instability, mechlorethamine reacts completely within 3 minutes of administration. To mitigate these problems, medicinal chemists attached a benzene ring to the central nitrogen, using resonance stabilization as a means of reducing its nucleophilicity. Unfortunately, compounds of this type were too nonpolar to be administered through injection. Thus, an alkyl chain terminating in a carboxylic acid was attached. This modification resulted in chlorambucil (Leukeran®), an antitumor drug with reduced reactivity and an appropriate level of water solubility (**Figure 12.85**). Leukeran has been used to treat various cancers, including chronic lymphocytic leukemia (CLL), Hodgkin's disease, and breast and testicular cancer. Similar strategies resulted in Alkeran®, an oral chemotherapy used in the treatment of multiple myeloma, a blood cancer.

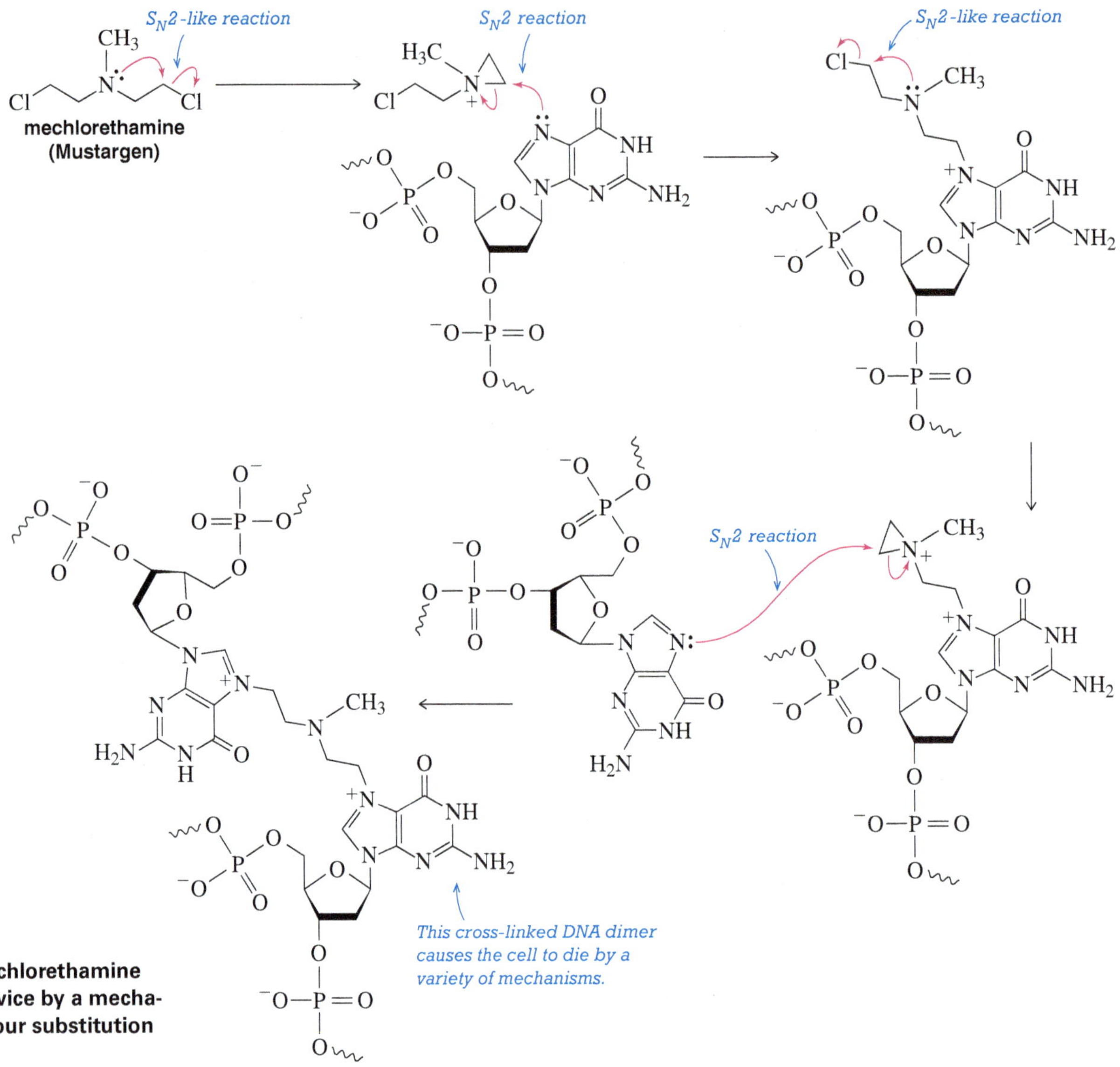

Figure 12.84 Mechlorethamine alkylates DNA twice by a mechanism involving four substitution reactions.

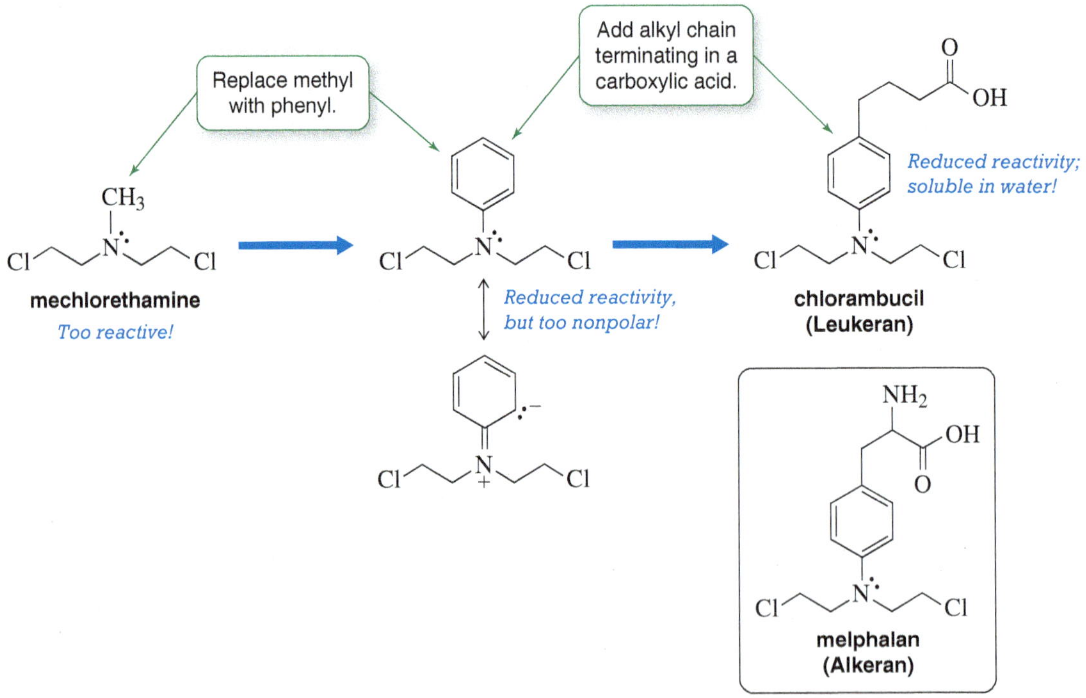

Figure 12.85 The lead compound mechlorethamine resulted in two anticancer drugs.

Important Concepts

[*These are the major concepts covered in each section, excluding reactions.*]

12.1: Substitution and elimination of haloalkanes are examples of double replacement reactions (AB + CD → CB + AD) and decomposition reactions (C → A + B).

12.2: Substitution is a favorable process based on ΔH ($\Delta S \approx 0$). The formation of stronger bonds as well as the formation of stable leaving groups makes substitution favorable. Substitution can proceed through either of two mechanisms. In the S_N2 reaction, the nucleophile displaces the leaving group in a single, concerted, bimolecular step. In the S_N1 reaction, the leaving group leaves to form a carbocation in a unimolecular rate-determining step prior to the carbocation being attacked by the nucleophile.

12.3–12.4: There are several distinguishing characteristics that cause a reaction to proceed by either the S_N1 or S_N2 mechanism.

S_N2 reactions occur via donation of a pair of electrons to σ^*, the antibonding orbital, located at a 180° angle to the leaving group. S_N2 reactions occur most often at non-sterically hindered carbons, with strong, anionic nucleophiles, in polar aprotic solvents. Because of backside attack at σ^*, S_N2 reactions proceed with inversion of stereochemistry. As concerted processes, rearrangement is not observed in S_N2 reactions.

S_N1 reactions occur via donation of a pair of electrons to an empty *p* orbital. S_N1 reactions occur most often at carbons that can become stable carbocations, with weak, neutral nucleophiles, in polar protic solvents. Because bond formation occurs at a planar carbocation, S_N1 reactions proceed with racemization of stereochemistry. Involving carbocations, S_N1 reactions are subject to rearrangement.

12.5: Elimination is a favorable process based on ΔH and ΔS. The formation of stronger bonds, stable leaving groups, and making three molecules from two makes elimination favorable. Elimination can proceed through either of two mechanisms. In the E2 reaction, the base removes the β-proton, causing the electrons to collapse, displacing the leaving group and forming a new π bond in a single, concerted, bimolecular step. In the E1 reaction, the leaving group leaves to form a carbocation in a unimolecular rate-determining step prior to the proton being removed and the new π bond forming.

12.6–12.7: There are several distinguishing characteristics that cause a reaction to proceed by either the E1 or E2 mechanism.

E2 reactions occur via an electron pair collapsing, concurrent with deprotonation, into σ^*, the antibonding orbital, located at a 180° angle to the leaving group. E2 reactions occur most often with strong, bulky, anionic nucleophiles, in polar aprotic solvents. Because of the involvement of σ^*, E2 reactions are stereospecific, requiring the anti and coplanar arrangement of the proton and the leaving group to proceed. As concerted processes, rearrangement is not observed in E2 reactions.

E1 reactions occur via an electron pair collapsing, after deprotonation, into an empty *p* orbital. E1 reactions occur most often at carbons that can become stable carbocations, with weak, neutral bases, in polar protic solvents. Because π bond formation occurs at a planar carbocation, E1 reactions are non-stereospecific, giving the more substituted/more stable alkene (Zaitsev's rule). Involving carbocations, E1 reactions are subject to rearrangement.

12.8: Because they are entropically favored, elimination reactions are favored over substitution reactions at higher temperatures.

When an E1 reaction occurs, it will usually be accompanied by formation of the S_N1 product.

When predicting S_N2 vs. E2 reactions, a couple of generalizations guide the decision. Sterically unhindered environments (1°, 2°, small nucleophiles) favor S_N2 because the nucleophile can easily access σ^*. Sterically hindered environments (2°, 3°, bulky bases) favor E2 because access of the base to σ^* is limited. Strong bases favor E2 (conjugate acid $pK_a > 11$), while weaker bases/nucleophiles favor S_N2 (conjugate acid $pK_a < 11$), an important consideration when the leaving group is on a 2° carbon.

Important Reactions and Mechanisms

[Note that all of these reactions proceed by one of four mechanisms (S_N1, S_N2, E1, or E2). Make sure you recognize the similarities between each of these reactions. Doing so will make learning this material much easier.]

1. S_N2 Reaction (Sections 12.3–12.4): A nucleophilic substitution reaction with a bimolecular rate-determining step. In a concerted step, a good leaving group is displaced by a strong, anionic nucleophile. Because the nucleophile does backside displacement by attacking σ*, this reaction results in inversion of stereochemistry.

General:

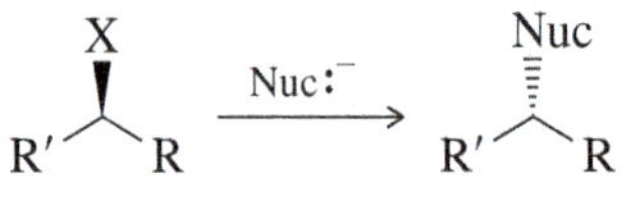

Specific:

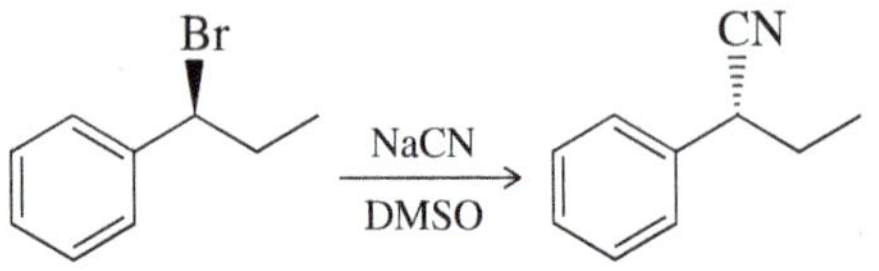

Mechanism:

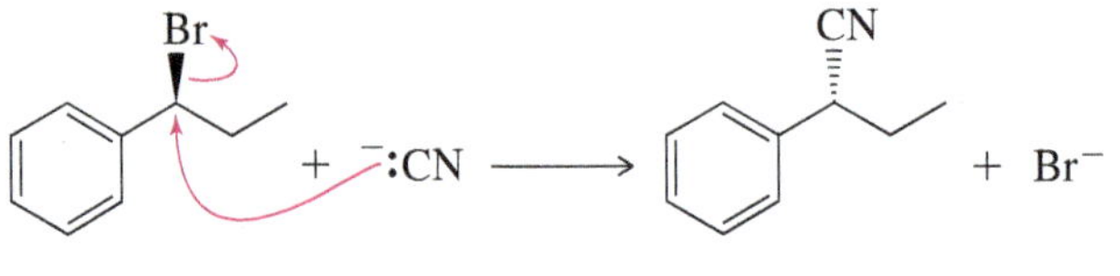

2. S_N1 Reaction (Sections 12.3–12.4): A nucleophilic substitution reaction with a unimolecular rate-determining step. In the first, rate-determining step, a good leaving group on either a secondary or a tertiary haloalkane leaves, forming a carbocation. This carbocation, containing an empty *p* orbital, is attacked by a weak nucleophile in the second step. Because the nucleophile can add to either the front or the back of the carbocation, this reaction forms a racemic mixture of enantiomers.

General:

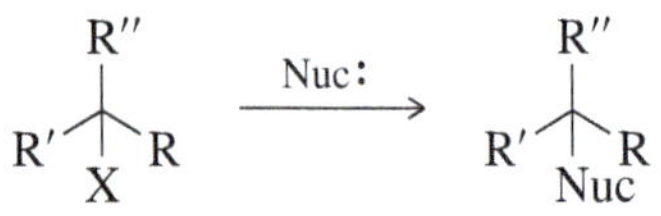

Specific:

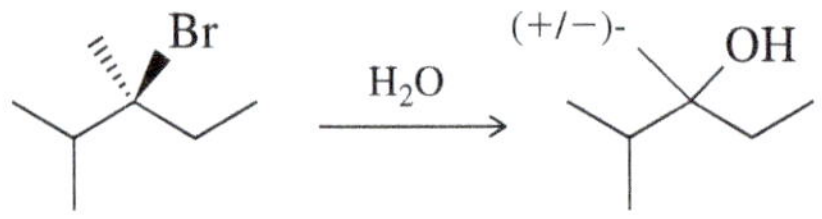

Mechanism:

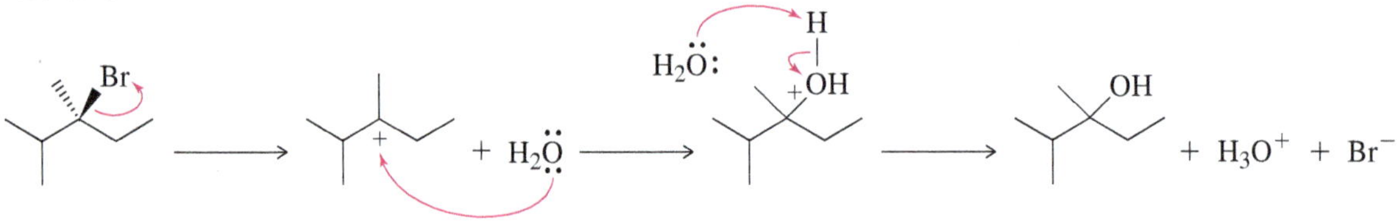

3. E2 Reaction (Sections 12.6–12.7): An elimination reaction with a bimolecular rate-determining step. In a concerted step, a strong base removes a proton from the β-carbon, causing the electrons to collapse into σ*, displacing a good leaving group. The anti and coplanar arrangement of H and the leaving group is required for E2 elimination, making this reaction stereospecific.

General:

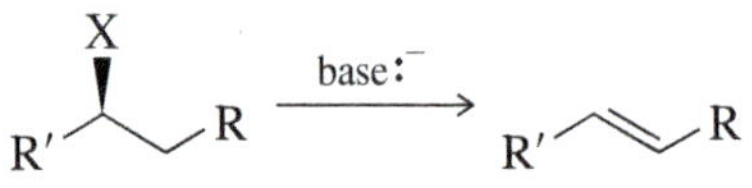

Specific:

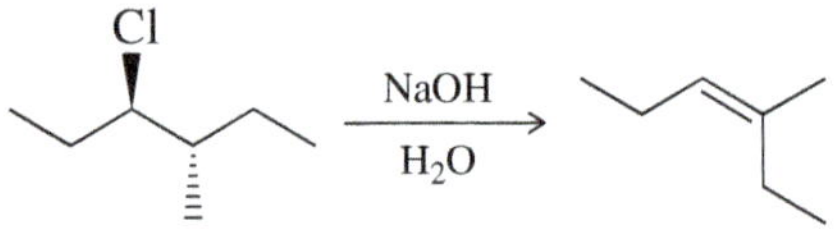

Mechanism:

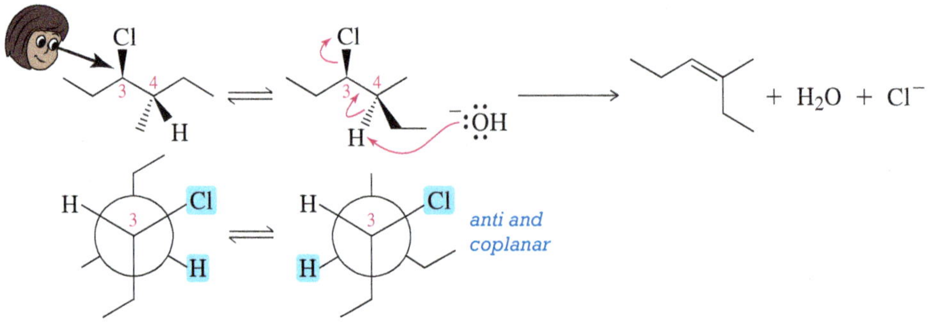

4. **E1 Reaction (Sections 12.6–12.7):** An elimination reaction with a unimolecular rate-determining step. In the first, rate-determining step, a good leaving group on either a 2° or 3° carbon leaves, forming a carbocation. In the second step, a weak base removes a proton from the β carbon forming a new π bond and producing the most stable alkene.

General:

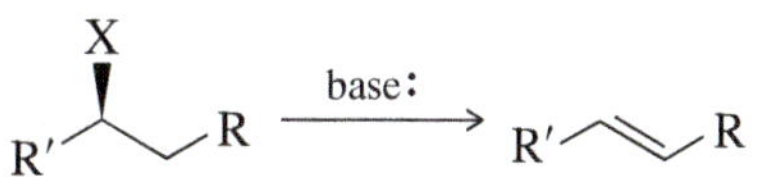

Specific:

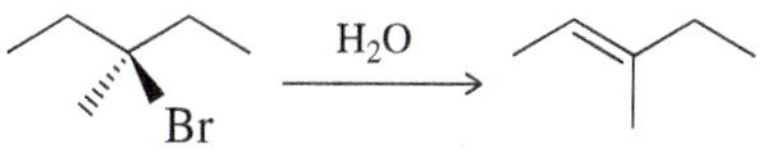

Mechanism:

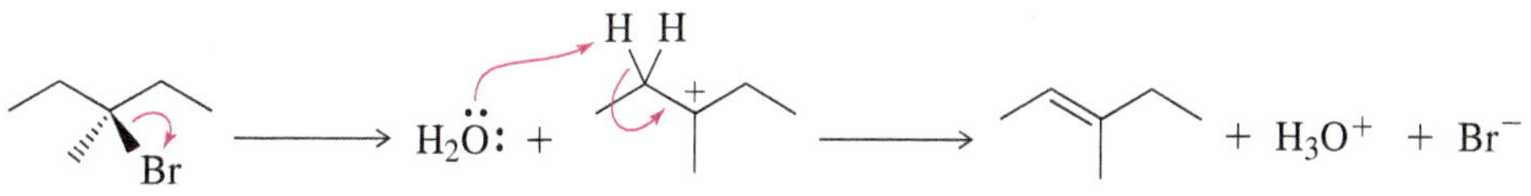

End-of-Chapter Assessments

[Assessments are ranked according to their difficulty; •••• = most difficult.]

12.52 (•) Suggest a mechanism for the following substitution reactions.

(a)

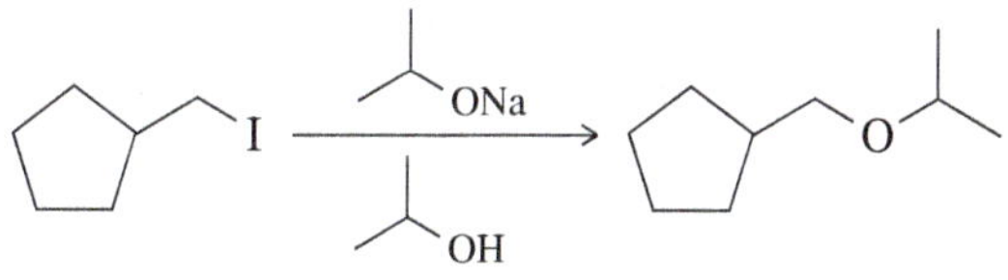

(b)

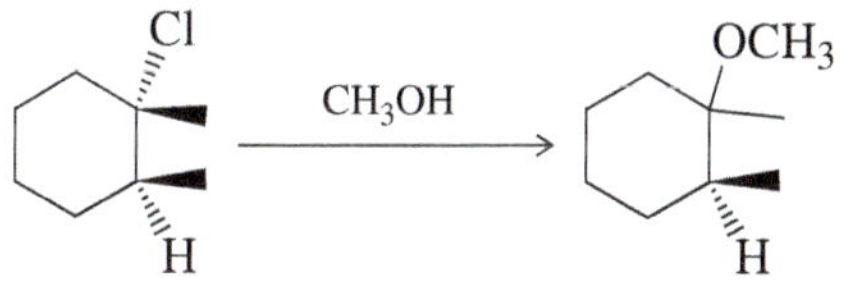

(c)

(d)

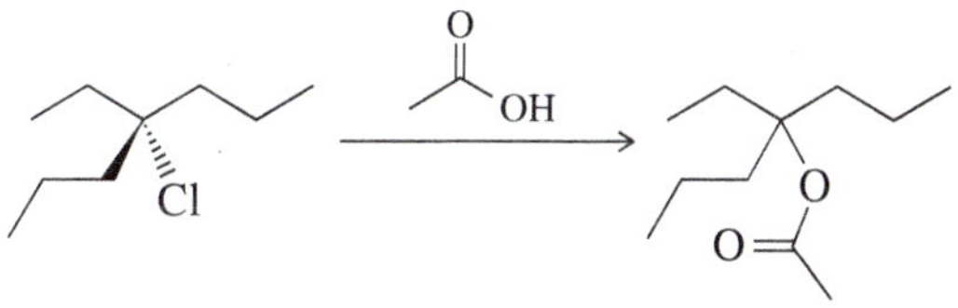

12.53 (•) Suggest a mechanism for the following elimination reactions.

(a)

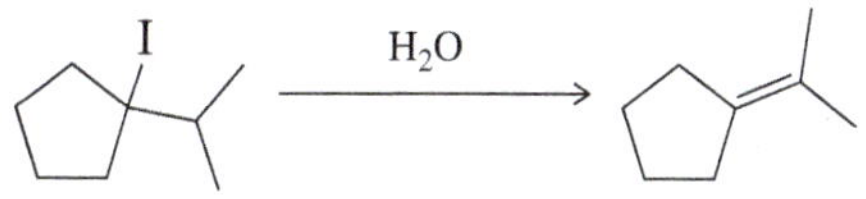

(b)

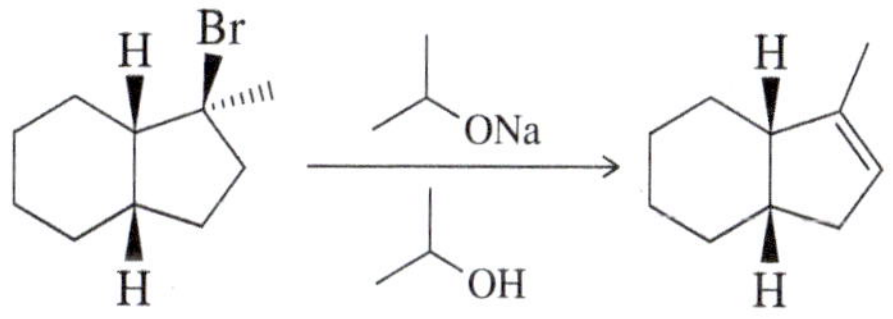

(c)

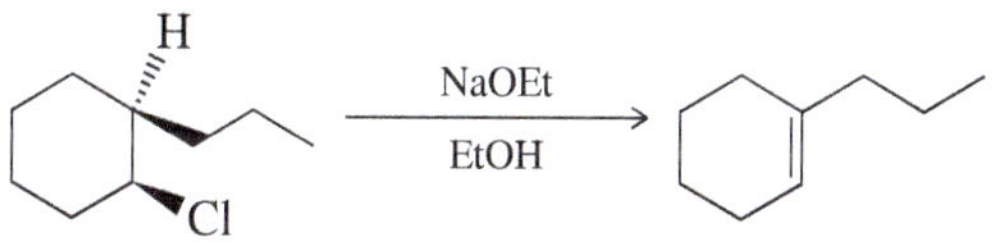

(d)

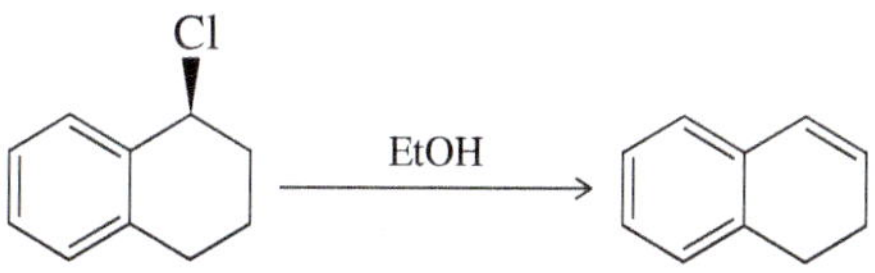

12.54 (••) Predict the product of the following substitution reactions, paying close attention to the stereochemical outcome of the reactions.

(a)

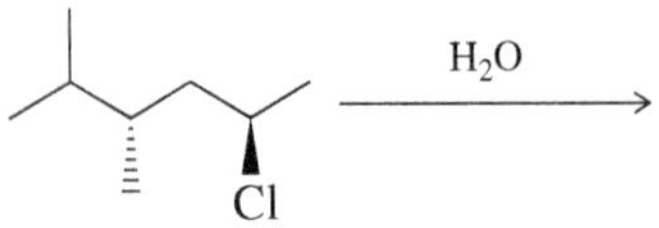

(b)

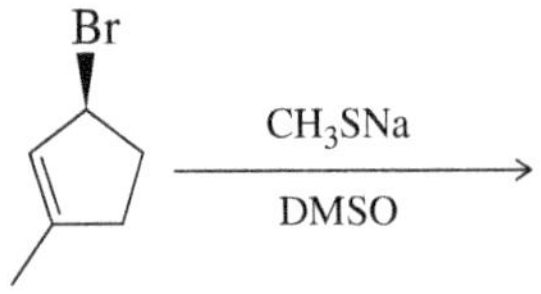

(c)

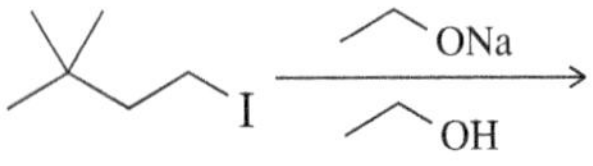

(d)

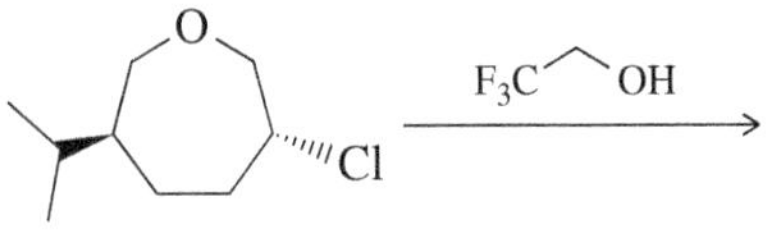

(e)

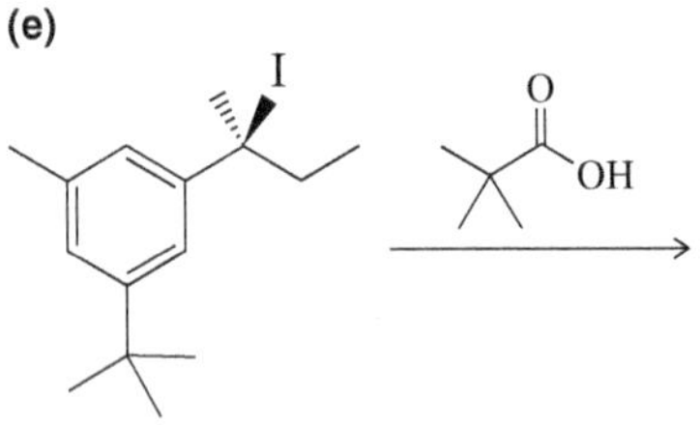

(f)

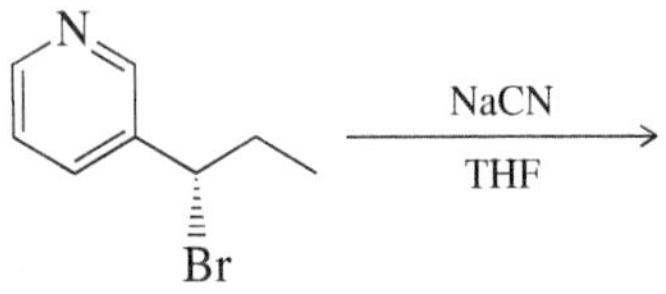

12.55 (••) Predict the major product(s) of the following elimination reactions, paying close attention to the stereochemical outcome of the reactions.

(a)

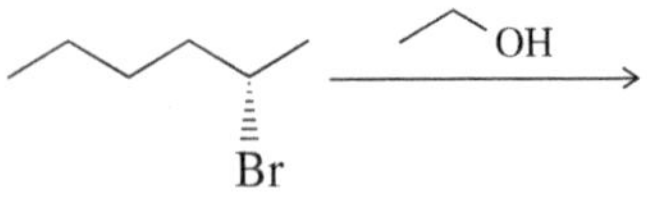

(b)

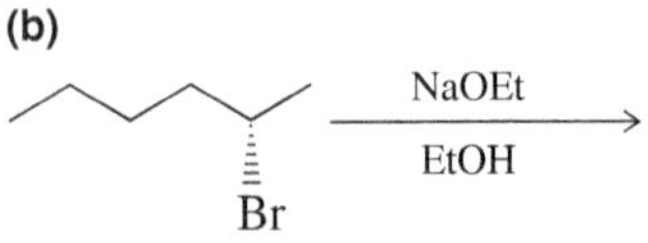

(c)

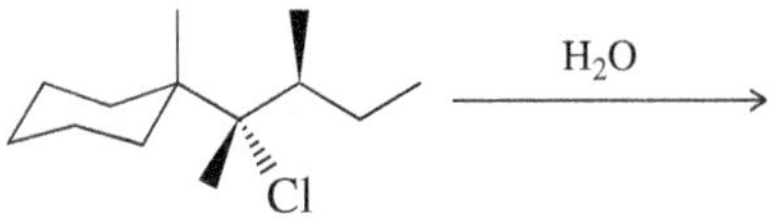

(d)

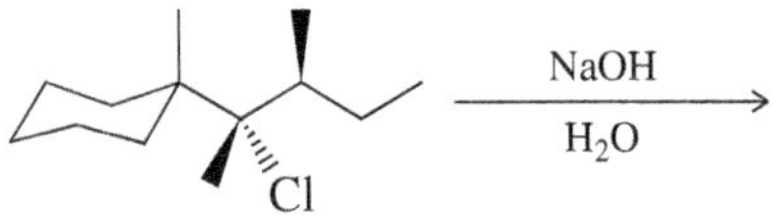

(e)

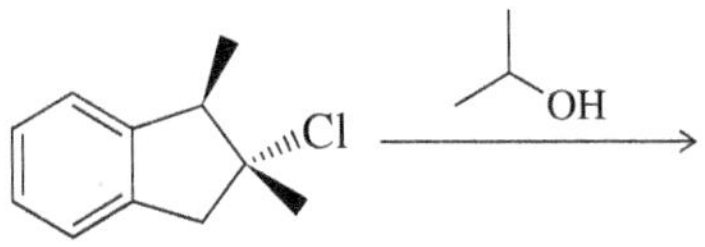

(f)

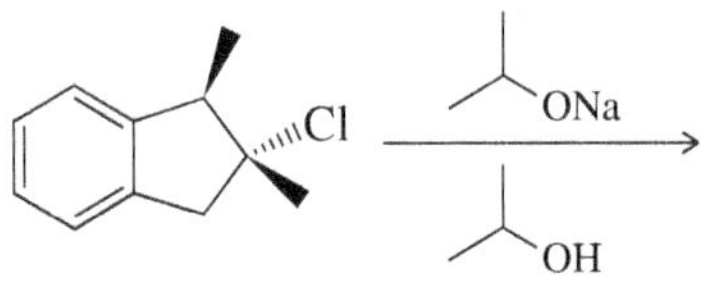

12.56 (•••) Predict the product(s) of the following substitution or elimination reactions, paying close attention to the stereochemical outcome of the reactions.

(a)

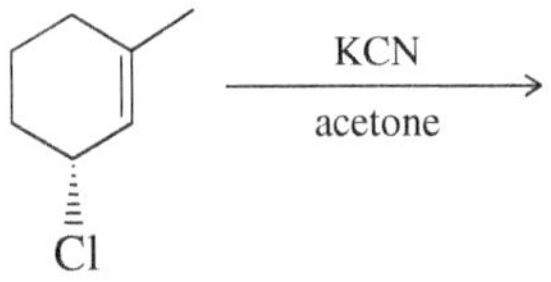

(b)

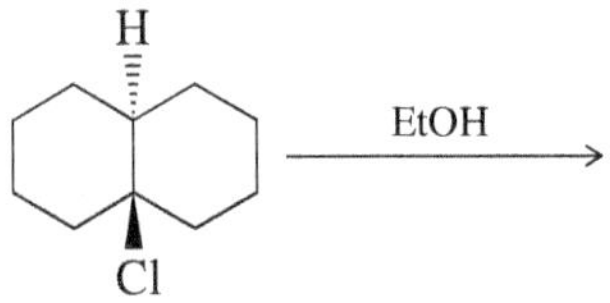

(c)

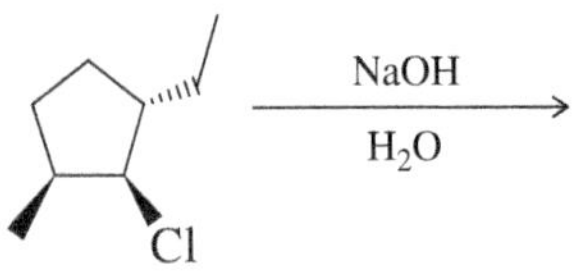

(d)

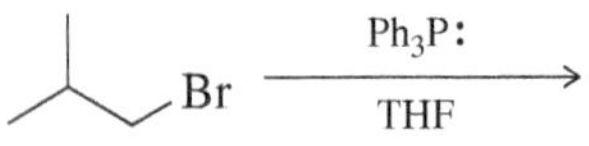

(e)

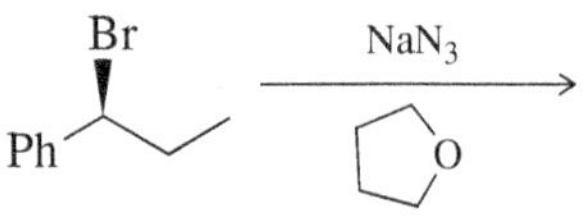

(f)

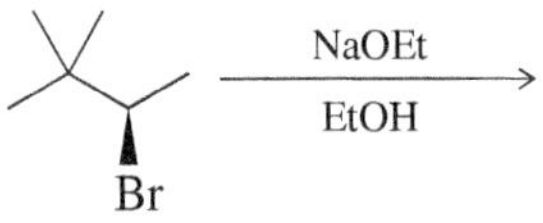

(g)

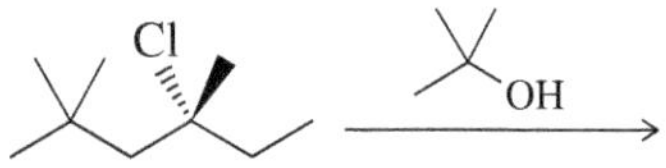

(h)

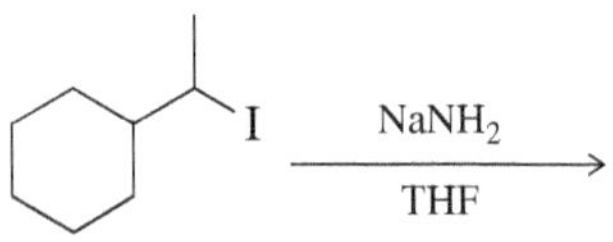

12.57 (•) For each pair, choose the nucleophile that would react most quickly in an S_N2 reaction (assume H_2O is the solvent).

(a)

F^- vs. Br^-

(b)

CH_3O^- vs. 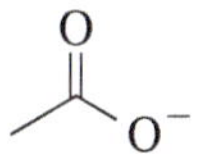

(c)

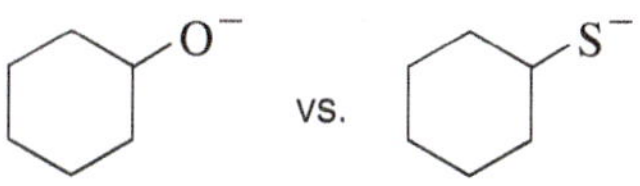

(d)

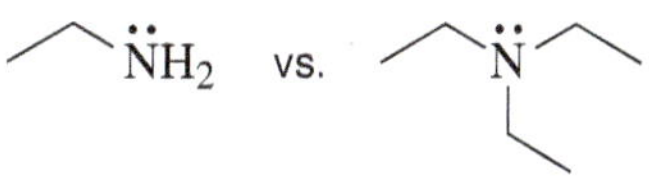

12.58 (••) For each pair, choose the haloalkane that would react most quickly in an S_N2 reaction.

(a)

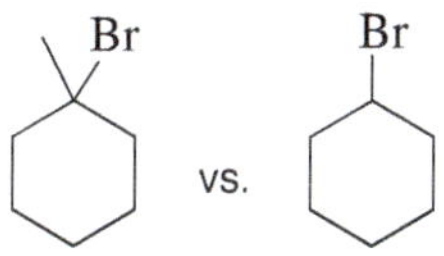

(b)

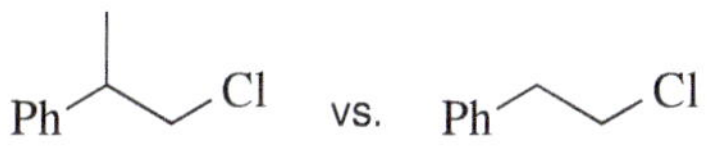

(c)

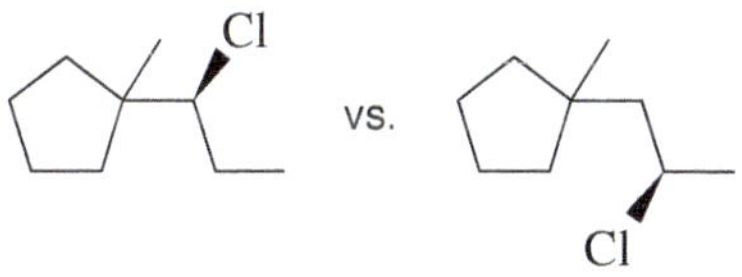

(d)

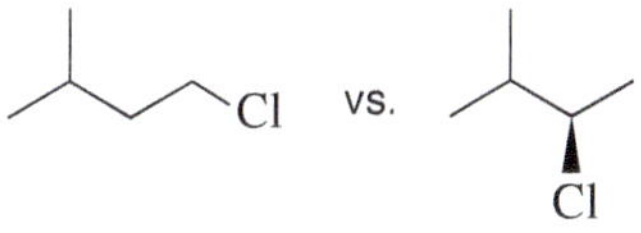

12.59 (••) For each pair, choose the haloalkane that would react most quickly in an S_N1 or E1 reaction.

(a)

(b)

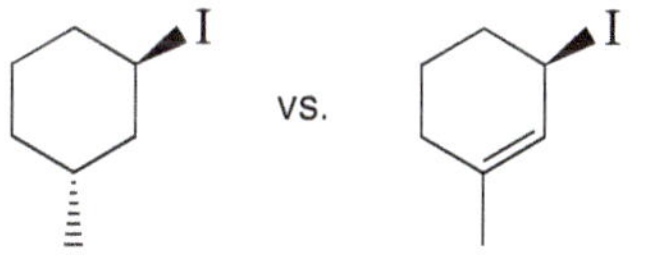

(c)

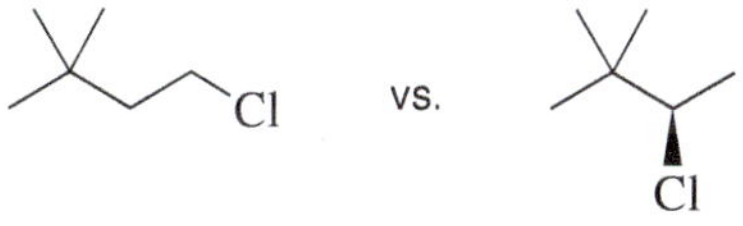

12.60 (•••) Suggest a bromoalkane and the conditions necessary to produce the alkenes shown.

(a)

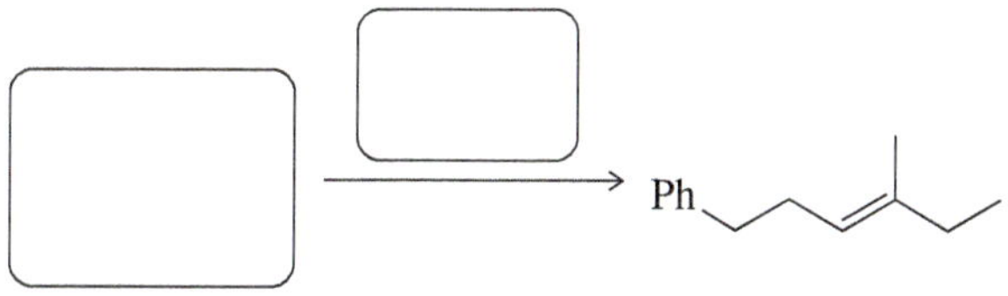

(b)

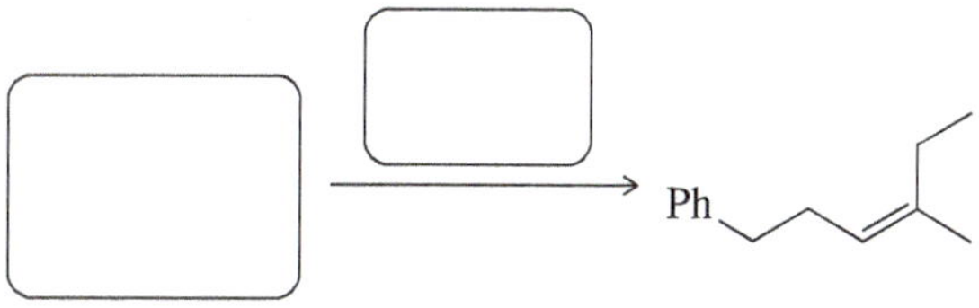

12.61 (••) In Chapter 10, you learned how to make an alkyne by acetylide alkylation with a 1° haloalkane. Suggest a mechanism by which this reaction occurs.

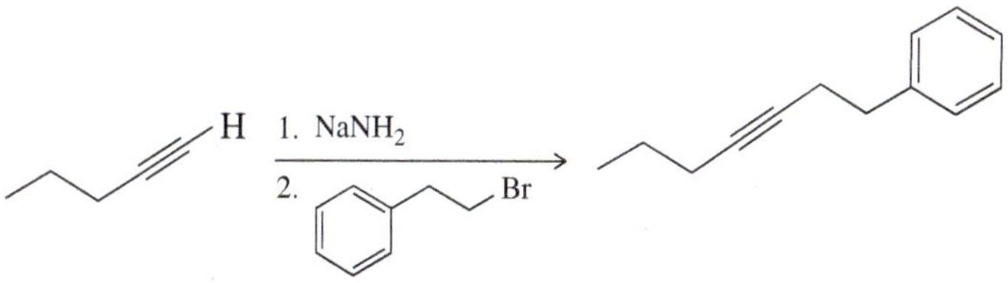

12.62 (•••) Acetylide alkylation, from Assessment 12.61, fails to give the desired product with 2° haloalkanes. Why? What is the actual product of this reaction?

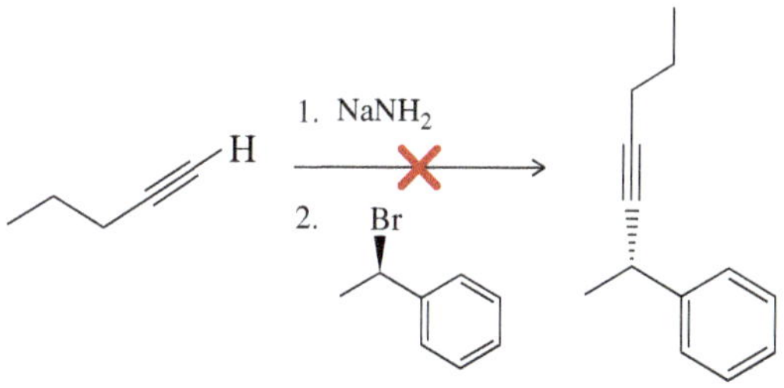

12.63 (••) The introduction of elimination reactions provides a second way to synthesize alkynes in a two-step process starting with an alkene. Suggest a mechanism for both steps of this process.

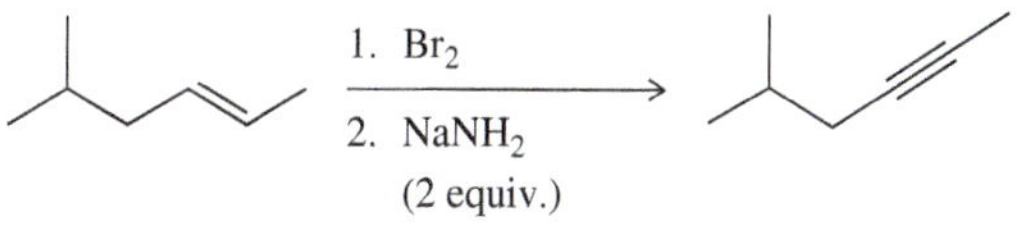

12.64 (•••) When the reaction scheme in Assessment 12.63 is done on a monosubstituted alkene, at least three equivalents of base are needed. Reacting the product of step 2 with D—Cl (D is an isotope of H) incorporates deuterium at the terminal carbon. Explain these two observations.

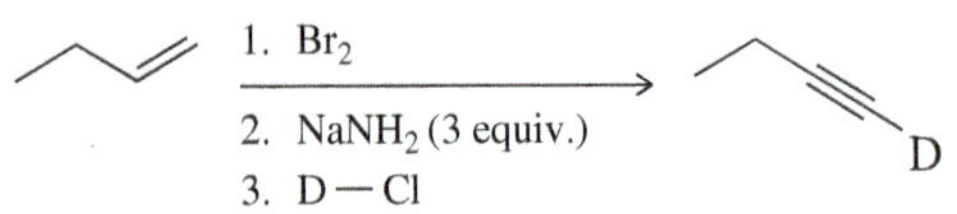

12.65 (••••) Which of the following substitution reactions would you expect to occur more quickly? Explain your answer.

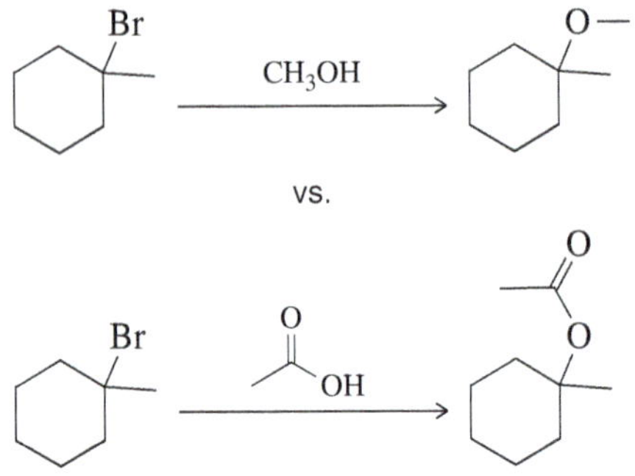

12.66 (•••) Give a mechanism for the following substitution and elimination reactions.

(a)

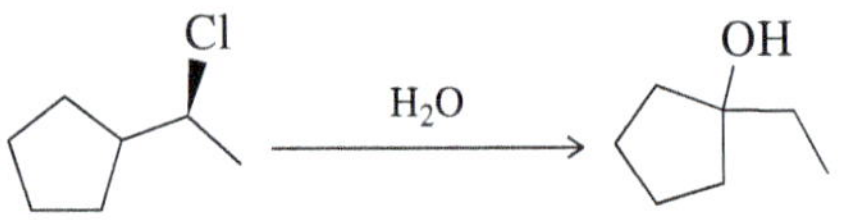

(b)

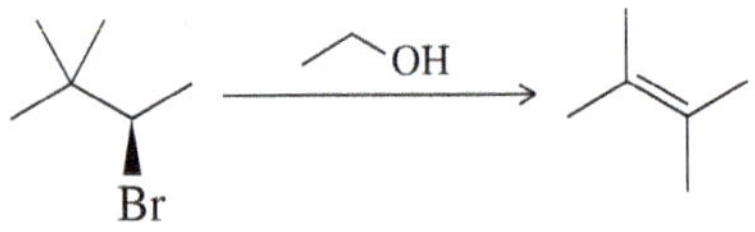

(c)

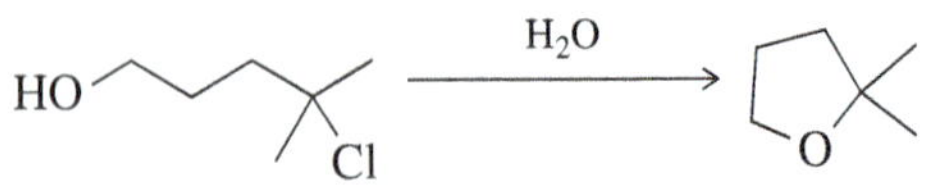

12.67 (•••) In addition to using *m*CPBA (Chapter 9), epoxides can be synthesized from alkenes in the two-step process shown. Give a mechanism for each step of the process.

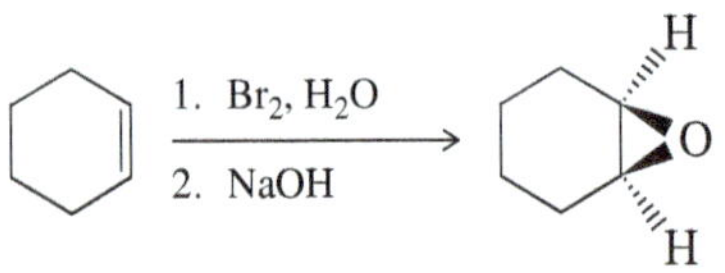

12.68 (••••) The following chlorocyclohexane undergoes neither S_N2 nor E2 under the conditions shown. Why?

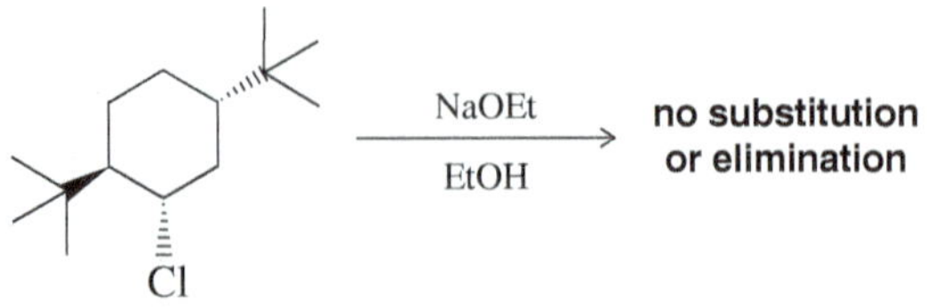

12.69 (••••) When *trans*-4-bromocyclohexanol is treated with base, an intramolecular substitution reaction occurs to give a cyclic ether. This product does not form when *cis*-4-bromocyclohexanol is reacted under the same conditions. Explain these observations.

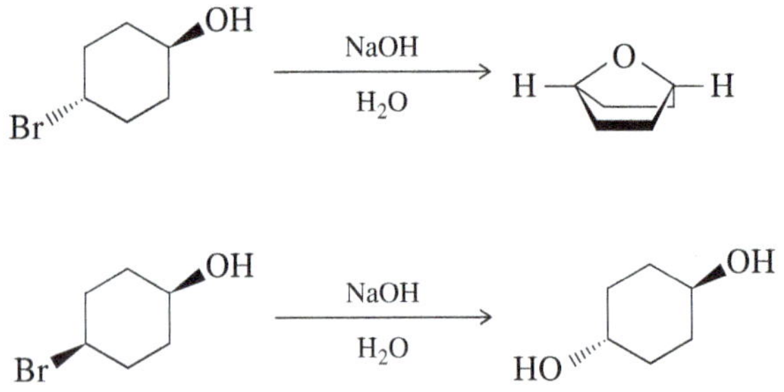

12.70 (••••) The following substitution reaction, between a strong base and a 1° haloalkane, occurs in a single step via backside displacement. Yet it is not technically an S_N2 reaction. Why?

Learning through assessments

Assessments 12.71–12.73 should be answered in order. Each question should be used to help you answer the next. At the end of these assessments you should have learned a new skill.

12.71 (•) Which is a better leaving group, HO^- or H_2O? Explain your answer.

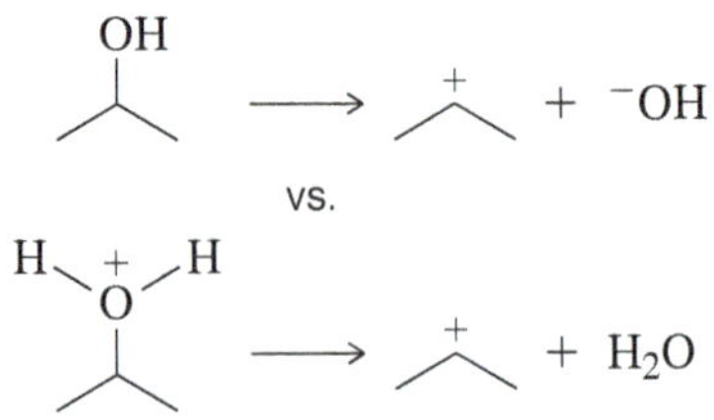

12.72 (••) Using pK_a values, calculate K_{eq} for the following acid–base reaction.

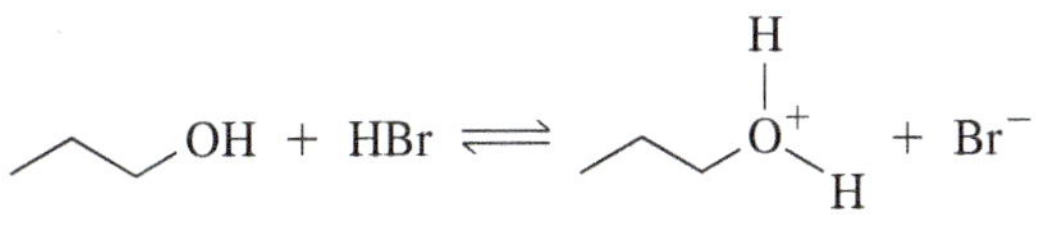

12.73 (••••) Suggest a mechanism for the following reactions.

(a) Substitution:

(b) Substitution:

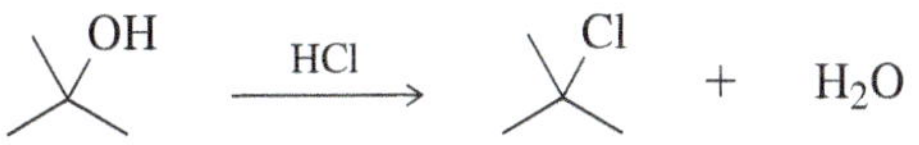

(c) Elimination:

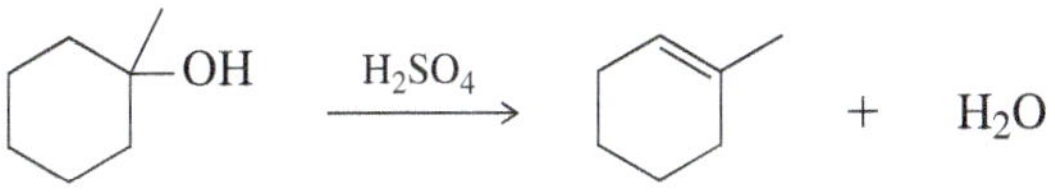

12.74 (••••) Suggest an experiment that conclusively demonstrates that S_N2 reactions at a 1° haloalkane proceed by inversion/backside displacement.

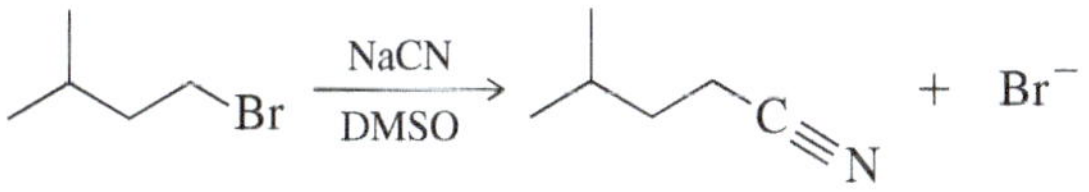

12.75 (••) In this, and previous, chapters, we have seen 1,2-alkyl and 1,2-hydride shifts. If both are possible, as in the carbocation shown, which would you expect to occur? Explain your answer.

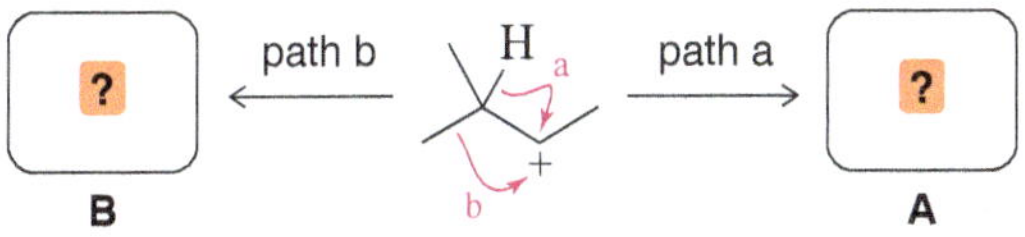

12.76 (•••) In light of your answer to Assessment 12.75, which would you expect to migrate, a methyl group or a hydroxyl group? Explain your answer.

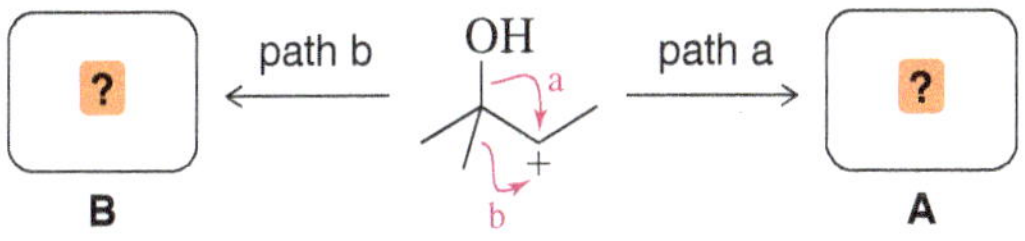

12.77 (••••) The bromoalkanes shown below participate in S_N1 reactions with the relative rates shown. Explain this trend.

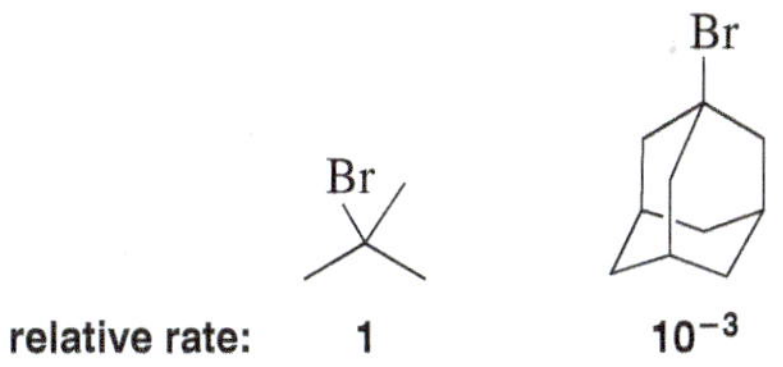

12.78 (••••) **LOOKING AHEAD** Of the three possible elimination mechanisms (Figure 12.50), this chapter focused on two of them (E1 and E2). The third possibility occurs in situations like the one below. What makes this mechanism favored under these conditions?

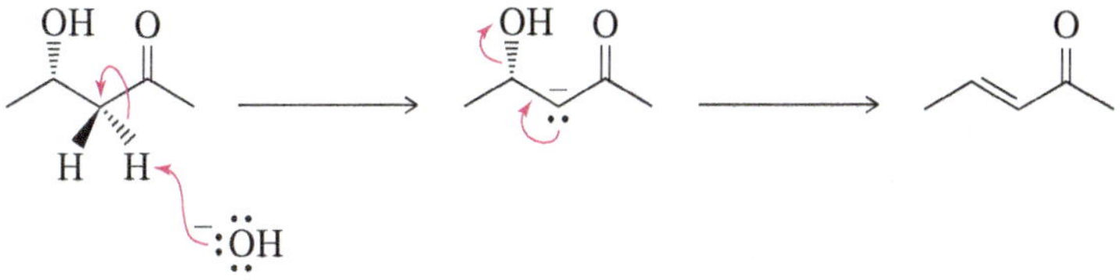

12.79 (••••) **LOOKING BACK** In Chapter 8, we learned about the chemistry of terpenes and the interesting reactions they can undergo. One such reaction is the acid-catalyzed conversion of nerol to terpineol. Suggest a mechanism for this transformation.

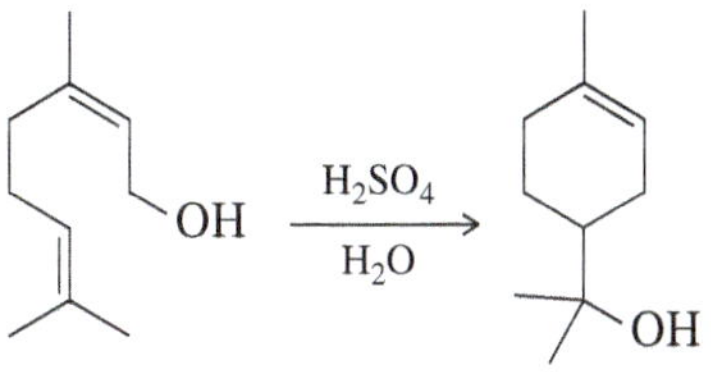

nerol terpineol

12.80 (•••) Crown ethers are able to solvate cations based on their size. Specifically, 15-crown-5 forms stable complexes with sodium. How would the addition of a crown ether change the rate of an S_N2 reaction?

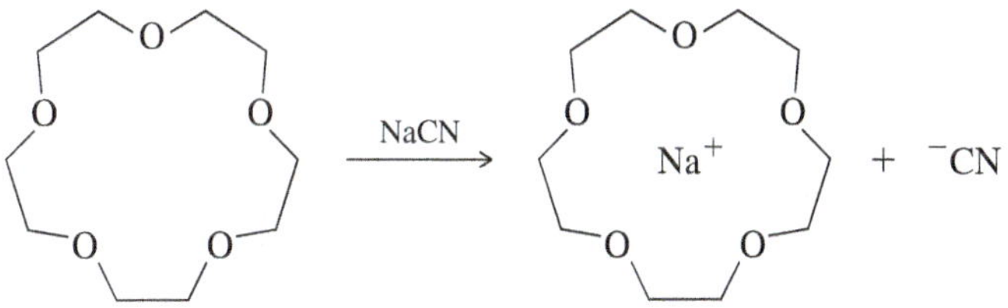

15-crown-5

13

Alcohols, Ethers and Related Compounds: Substitution and Elimination

Learning Objectives

LO 13.0 Relate the name of an alcohol to its structure and vice versa

LO 13.1 Analyze the molecular orbitals of alcohols

LO 13.2 Analyze reactions involving alcohols

LO 13.3 Predict the favorability of reactions involving the substitution of alcohols

LO 13.4 Determine if modifying a hydroxyl group makes it a better or worse leaving group

LO 13.5 Predict the products of an elimination reaction involving an alcohol

LO 13.6 Evaluate oxidation reactions involving alcohols

LO 13.7 Describe the properties of ethers

LO 13.8 Analyze ether synthesis reactions

(continued)

Introduction

In Chapter 7, we introduced the 12 principles of green chemistry and described some examples in which they were implemented. Because you were new to organic chemistry at that time, you may not have fully appreciated the problems caused by traditional organic reagents, solvents, and processes. This chapter on alcohols and ethers is a great opportunity to go further in that discussion [*Or to have that discussion for the first time, if your instructor didn't cover Chapter 7.*]

◦ ◦ ◦ ◦ ◦ ◦ ◦

This chapter marks a subtle change in the approach you need to be successful. Whereas the first reactions we studied were organized largely by their functional group (i.e., alkenes, alkynes, and haloalkanes), we now shift to a more mechanism-centered approach. The functional group approach is still important, but here we directly examine the similarity in mechanisms that the reactions of alcohols and ethers undergo. As a result, alcohols are compared to thiols in the first half of the chapter, and ethers and thioethers are examined concurrently in the second half. Let's get started by assessing prior knowledge that will be helpful in this chapter.

LO 13.9 Predict the products of reactions involving ethers

LO 13.10 Analyze reactions involving epoxides

LO 13.11 Evaluate the use of protecting groups to avoid undesirable side reactions in organic synthesis

"Unintended consequences seem to be a natural outcome of solely reductionist thinking."

—PAUL ANASTAS

Look for opportunities in the text and assessments to connect this material to previously studied concepts. The purple recall boxes will help with this.

Test your understanding

Assessment 13.1 Rank the ability of the following compounds to undergo an S_N1 reaction (1 = fastest, 5 = slowest).

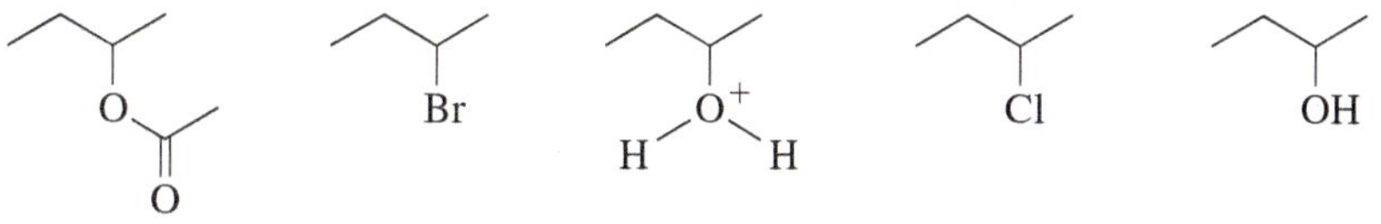

Assessment 13.2 Provide a mechanism for the following S_N1 reaction. What roles (acid/base/nucleophile/electrophile) does ethanol play in the reaction? What functional group is made in this reaction?

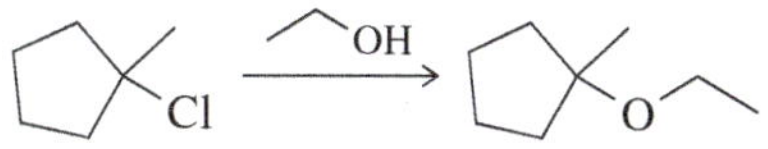

Assessment 13.3 Which of the following carbocations would you expect to undergo rearrangement?

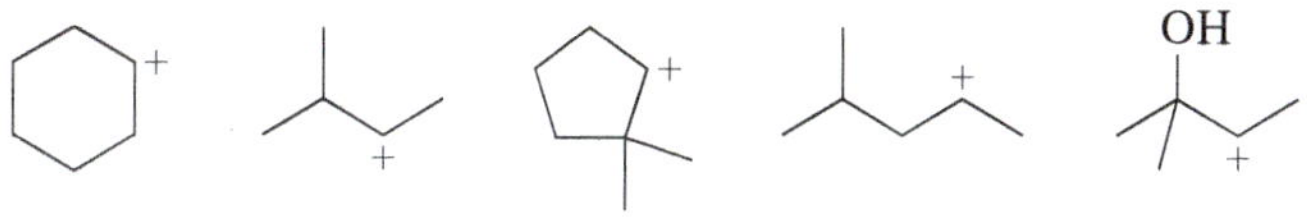

Assessment 13.4 Use oxidation numbers to determine whether the following are redox reactions.

(a)

(b)

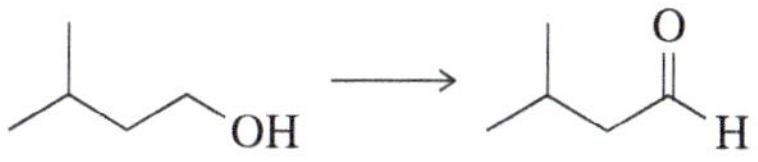

Assessment 13.5 The mechanism of the acid-catalyzed hydration of an alkene to make an alcohol is shown. If this reaction is reversible, draw a mechanism that justifies formation of the alkene from the alcohol under similar conditions. How do you know this mechanism is correct?

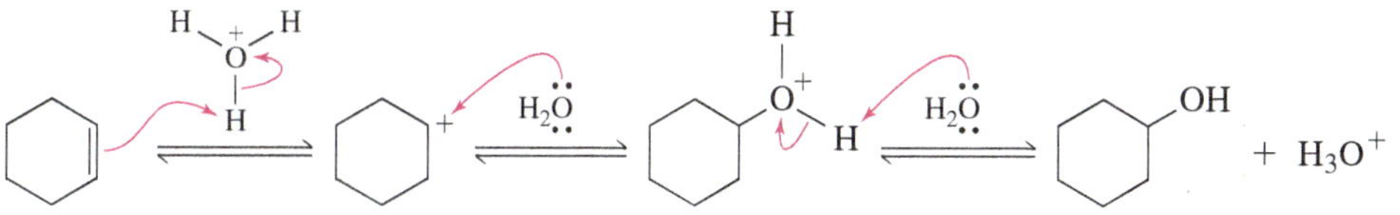

13.1 General structure of alcohols, ethers, thiols and thioethers

Here in Chapter 13, we discuss the chemistry of four related molecule types—namely, alcohols, ethers, thiols, and thioethers (Table 13.1). **Alcohols** represent a general class of compounds in which one hydrogen of a parent molecule has been replaced by a **hydroxyl group** (—OH). An **ether** is a compound that has two alkyl groups attached to a central oxygen. A **thiol** is a molecule where one hydrogen has been replaced by a **sulfhydryl group** (—SH). Because thiols have a high affinity for mercury, they are often referred to as **mercaptans** (from the Latin for "capturing mercury"). Finally, a **thioether** (or **sulfide**) is a compound that has two alkyl groups attached to a central sulfur.

These compounds are all structurally related and thus have similar reactivity. To see how, let's start with water (Figure 13.1). Replacing one of its hydrogens with an alkyl group generates an alcohol. Replacing the other hydrogen with a second alkyl group converts the alcohol into an ether. Replacing the oxygen of an alcohol with a sulfur gives a thiol. Finally, replacing the oxygen of an ether with a sulfur gives a thioether.

Table 13.1 The four molecule classes discussed in Chapter 13.

Molecule class	General structure	Selected examples		
Alcohol	ROH	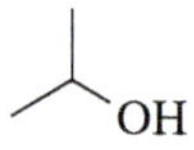 isopropanol	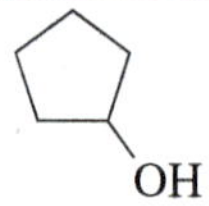 cyclopentanol	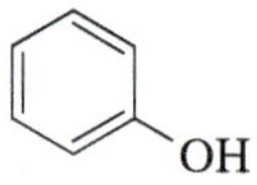 phenol
Thiol	RSH	CH_3SH methanethiol (methyl mercaptan)	 *t*-butylthiol	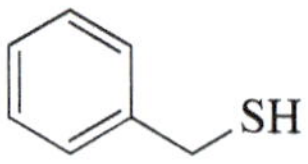 benzyl mercaptan
Ether	ROR ROR′	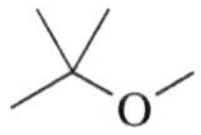 *t*-butyl methyl ether	 ethyl vinyl ether	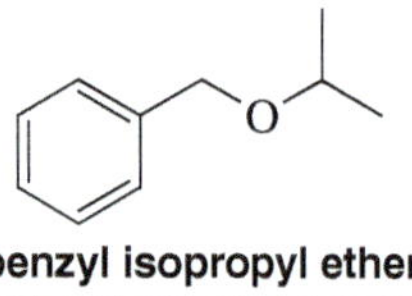 benzyl isopropyl ether
Thioether	RSR RSR′	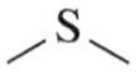 dimethyl sulfide	 1-methylthiopropane	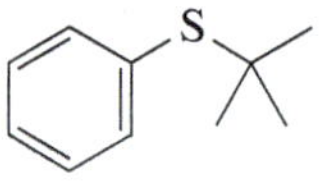 *t*-butyl phenyl sulfide

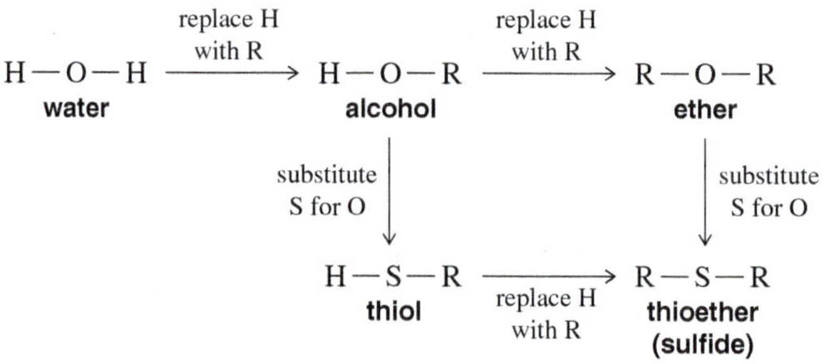

Figure 13.1 The structural relationships between water, alcohols, ethers, thiols, and thioethers help rationalize their similar reactivity patterns. [*These thought experiments suggest how we will do the real transformations.*]

13.2 Nomenclature of Alcohols

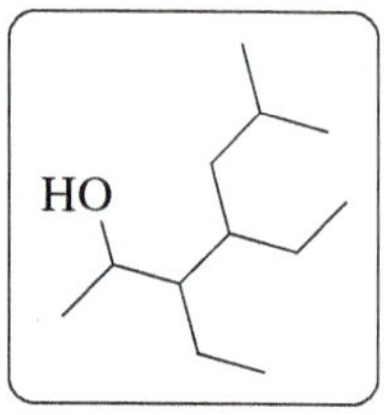

The procedure for naming alcohols according to the IUPAC system is quite similar to the way we name other functional groups—see the rules summarized in Table 13.2. We now use these rules to name the alcohol at right.

Table 13.2 Rules for naming alcohols according to the IUPAC system of nomenclature.

1. Identify the parent chain as the longest carbon chain containing the carbon bearing the hydroxyl functional group. The parent chain will be the root of the name. The suffix -e of the corresponding alkane is replaced with -ol for alcohols.	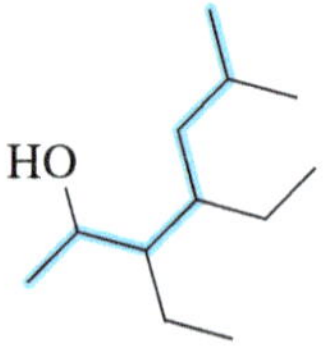 7-carbon parent chain heptanol

(continued)

Table 13.2 ***Continued***

2. Number the carbons of the parent chain such that the hydroxyl carbon has the lowest number. The number of the carbon bearing the hydroxyl group is placed between the parent name and the -ol suffix.	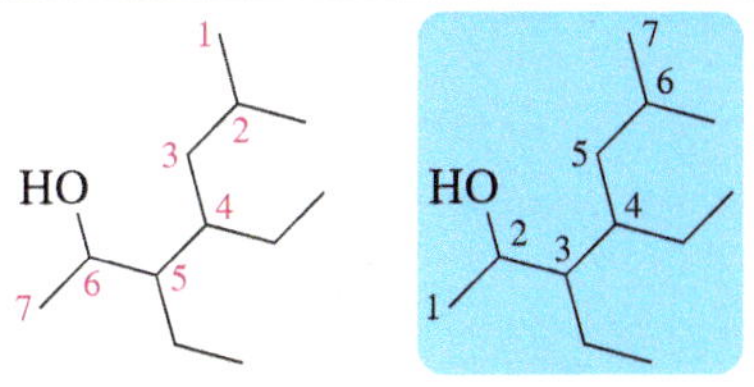 ~~heptan-6-ol~~ heptan-2-ol
3. Substituents on the parent chain are named as described previously.	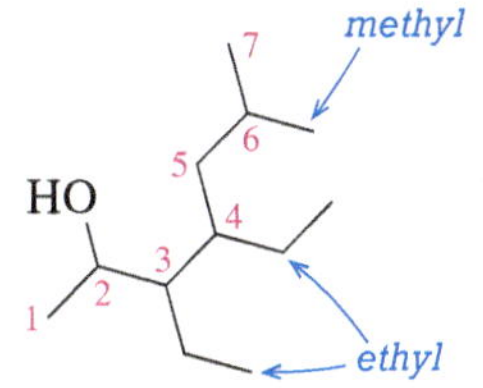
4. With the carbon chain numbered, the name is written following the standard IUPAC alphabetization rules, with numbers used to indicate the location of the substituents and the hydroxyl group.	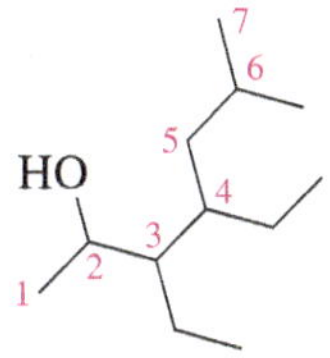 3,4-diethyl-6-methylheptan-2-ol

Table 13.3 **Priority of functional groups in nomenclature.**

Functional group	Priority	Functional group	Priority
Acids	1 (highest)	Ketones	7
Esters	2	Alcohols	8
Acid chlorides	3	Amines, thiols	9
Anhydrides	3	Amines	10
Amides	4	Alkenes, alkynes	11
Nitriles	5	Alkanes	—
Aldehydes	6	Alkyl halides, ethers	— (lowest)

13.2.1 Naming with multiple functional groups

When other functional groups appear in a molecule containing a hydroxyl group, the system established in Chapter 10 is used to assign priority (Table 13.3). A molecule containing a hydroxyl group and any functional group at a lower priority is named as an alcohol. This implies that the carbon containing the hydroxyl group receives the lowest number, the name ends in -ol, and all other functional groups are indicated as substituents (halogens) or by numbers in front of the infix (-en- for alkenes, -yn- for alkynes). For examples of this method, see the three polyfunctional molecules named in Figure 13.2.

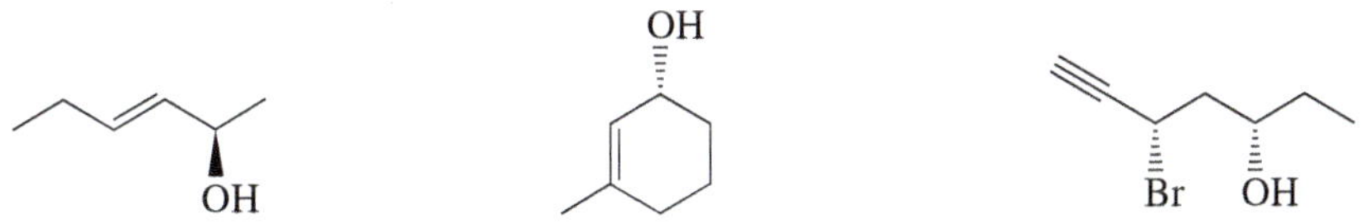

(2*R*,3*E*)-hex-3-en-2-ol (*R*)-3-methylcyclohex-2-enol (3*S*,5*S*)-5-bromohept-6-yn-3-ol

Figure 13.2 **Nomenclature of polyfunctional alcohols.**

(2*S*)-3,3-dimethylbutane-1,2-diol (2*R*)-pentane-1,2,5-triol

Figure 13.3 **Nomenclature of polyols.**

13.2.2 Naming molecules with multiple hydroxyl groups

When naming alcohols containing multiple hydroxyl groups (so-called *polyols*), the suffix is modified to indicate how many hydroxyl groups are present (Figure 13.3). Though the names still end in -ol, indicating the presence of an alcohol, a prefix (di-, tri-, tetra-, etc.) is used to indicate the number of hydroxyl groups.

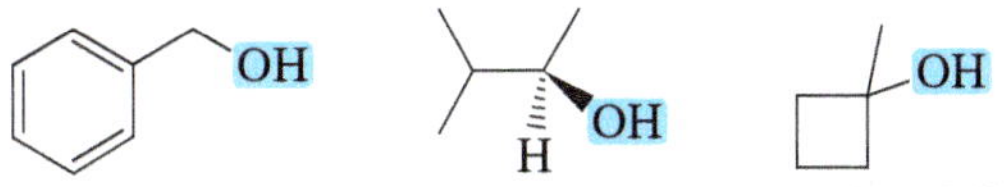

primary (1°) alcohol secondary (2°) alcohol tertiary (3°) alcohol

Figure 13.4 **Substitution of alcohols.**

13.2.3 Designating the substitution of alcohols

Much of the chemistry of alcohols involves the substitution of the hydroxyl group. The substitution may be primary, secondary, or tertiary (Figure 13.4), and it is indicated in the same way it was done with haloalkanes and other leaving groups in Chapters 11 and 12. That is, when the carbon bonded to the hydroxyl group is attached to one carbon, it is primary (1°); to two carbons, secondary (2°); and to three carbons, tertiary (3°).

13.2.4 Naming thiols

Depending on the complexity of the molecule, there are three possible conventions for naming thiols (**Figure 13.5**). For simple thiols, the common name of the alkyl group can be combined with "mercaptan" [Figure 13.5(a)]. For more complex molecules that have a sulfhydryl group as the highest priority functional group, use standard IUPAC rules with the suffix -thiol at the end of the name of the parent chain [Figure 13.5(b)]. When the sulfhydryl group is in a molecule with higher priority functional groups (for example, a hydroxyl group), the sulfhydryl should be named and numbered as a substituent on the parent chain [Figure 13.5(c)].

isopropyl mercaptan **ethyl mercaptan** **(*R*)-5-methylhexane-2-thiol** **(*R*)-3-methylcyclohex-3-enethiol** **(*E*)-6-sulfhydrylhex-3-en-1-ol**

Figure 13.5 The nomenclature of thiols depends on the complexity of the molecule. (a) The term "mercaptan" is used in common names for simple thiols. (b) Standard IUPAC rules place the suffix-thiol at the end of the parent chain. (c) IUPAC rules for thiols in a molecule with at least one higher priority functional group name the sulfhydryl group as a substituent.

13.2 SUMMARY

- Alcohols are named such that the hydroxyl group is afforded the lowest number in the chain. A prefix (di-, tri-, tetra-, etc.) is used to indicate the presence of more than one hydroxyl group in polyols. The substitution of an alcohol is based on the number of carbons attached to the hydroxyl-bearing carbon.

Assessment 13.6 Provide the IUPAC name for the following molecules.

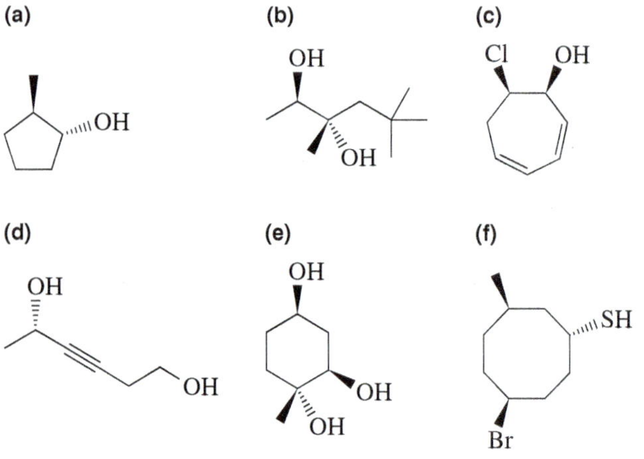

Assessment 13.7 Draw the structure that corresponds to the name provided.

(a) (1*R*,5*S*)-5-methylcyclohex-3-enol
(b) (2*R*,3*R*,4*S*)-heptane-2,3,4-triol
(c) (2*E*,4*S*,6*Z*)-octa-2,6-dien-4-ol
(d) (2*R*,4*Z*)-5-bromopent-4-ene-1,2-diol
(e) (1*S*,4*S*)-4-isopropylcyclopent-2-enethiol

Assessment 13.8 Label each hydroxyl group in Assessment 13.6 as primary (1°), secondary (2°), or tertiary (3°).

Assessment 13.9 Label each sulfhydryl group as primary (1°), secondary (2°), or tertiary (3°).

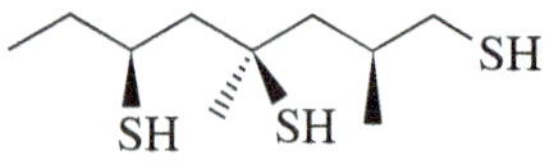

Assessment 13.10 Identify the substitution of carbons (a)–(d) comprising the ether functional group in the following molecules.

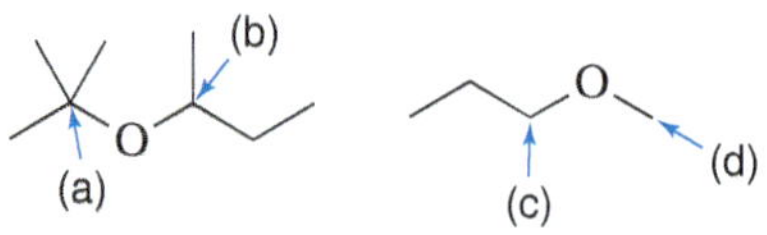

13.3 Physical properties of alcohols

Table 13.4 Dipole moments of select alcohols.

Molecule	μ (debye)
H_2O	1.85
CH_3OH	1.69
CH_3CH_2OH	1.69
$CH_3(CH_2)_7OH$	1.68

Alcohols are polar molecules because they have a bond between carbon and an electronegative oxygen. Their dipole moments (μ) are slightly less than that of water (Table 13.4), and they change very little in a series of larger alcohols.

Because the alcohol group consists of an electronegative oxygen bonded to hydrogen, alcohols can serve as both hydrogen bond donors and hydrogen bond acceptors.

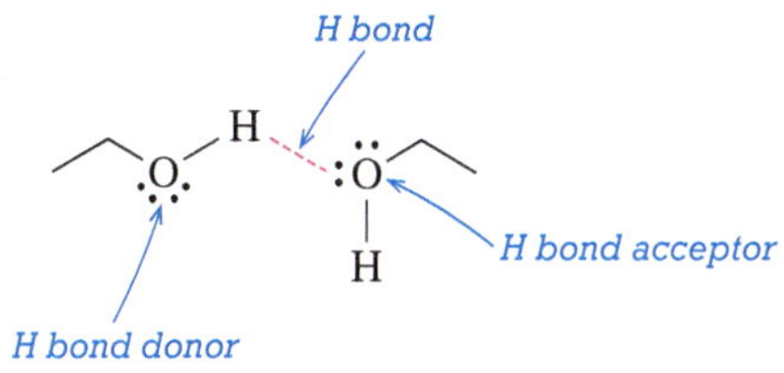

Being polar and having the ability to hydrogen bond greatly affects the physical properties of alcohols.

13.3.1 Boiling points of alcohols

Their polarity and their ability to hydrogen bond means that alcohols have strong intermolecular interactions As a result, alcohols are relatively high boiling compounds, especially compared to non-alcohols of similar structure and molecular weight (Table 13.5).

Similar to the trend seen in the alkane series (Table 3.2), the boiling points and melting points of alcohols increase as the length of the alkyl chain increases due to greater London dispersion forces (Table 13.6). The greater surface area provides more opportunities to form favorable noncovalent interactions with other molecules of itself.

Table 13.5 Comparison of the boiling point of alcohols and non-alcohols.

Compound	bp (°C)
H_2O	100
$CH_3CH_2CH_2CH_2OH$	118
$CH_3CH_2CH_2CH_2Cl$	78
$CH_3CH_2OCH_2CH_3$	35

Table 13.6 Boiling point of alcohols increases with alkyl chain length.

Alcohol	bp (°C)	mp (°C)
CH_3CH_2OH	78	−114
$CH_3CH_2CH_2OH$	97	−126
$CH_3CH_2CH_2CH_2OH$	118	−90
$CH_3CH_2CH_2CH_2CH_2OH$	138	−79

Table 13.7 The water solubility of alcohols decreases as the length of the alkyl chain increases.

Alcohol	Water solubility (mol/100 g H_2O)
Methanol	∞
Ethanol	∞
Propanol	∞
Butan-1-ol	0.11
Pentan-1-ol	0.030
Hexan-1-ol	0.0058
Heptan-1-ol	0.0008
Octan-1-ol	insoluble

13.3.2 Water solubility

Because they can hydrogen bond with water, low-molecular-weight alcohols are highly soluble in water (Table 13.7). Methanol, ethanol, and propanol are infinitely soluble in water. As the length of the alkyl chain increases, the water solubility decreases, with heptan-1-ol being the largest soluble alcohol.

13.3.3 Good solvents

Alcohols are relatively high boiling compounds that can hydrogen bond with reagents, so they make excellent solvents except when using a highly nonpolar reactant. We have already seen them used extensively in substitution and elimination reactions (Chapter 12), where the ability to hydrogen bond made alcohols excellent solvents for S_N1 and E1 reactions (Figure 13.6). By stabilizing the leaving group, alcohols promote carbocation formation, the rate-determining step of these reactions.

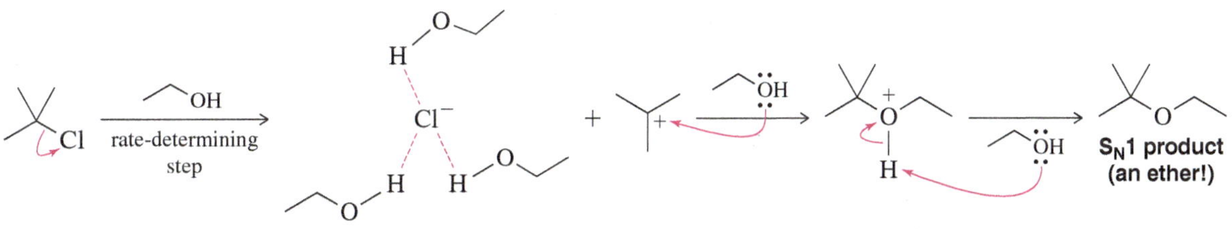

Figure 13.6 Hydrogen bonding in alcohols promotes carbocation formation in S_N1 reactions.

Ethanol is considered to be an environmentally friendly solvent. It is essentially nontoxic, is safe to handle, has a relatively low volatility, and is produced from renewable resources [*like corn*]. Besides water, it may be the greenest solvent.

13.3 SUMMARY

- As hydrogen bond donors and acceptors, alcohols have high boiling points and melting points and are good solvents.

13.4 Molecular orbital picture of alcohols

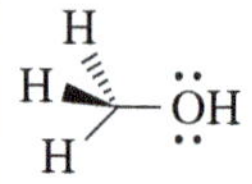

methanol

The molecular orbital (MO) picture of alcohols helps explain their reactivity. [*See if you can draw the molecular orbital picture of methanol (the simplest alcohol) on your own before reading further.*]

The MO picture of methanol is shown in **Figure 13.7**. The carbon and oxygen are both sp^3 hybridized, so the C—O σ bond is formed by the overlap of sp^3-hybridized orbitals. The three C—H σ bonds are formed when the three remaining sp^3-hybridized orbitals of carbon overlap with the 1*s* orbitals of hydrogen. Similarly, the O—H bond is formed when an sp^3-hybridized orbital of oxygen overlaps with the 1*s* orbital of hydrogen. The two lone pairs on oxygen reside in nonbonded sp^3-hybridized atomic orbitals.

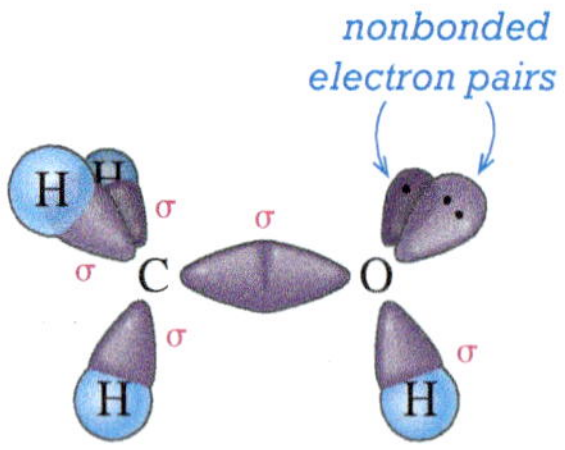

Figure 13.7 Molecular orbital picture of methanol.

Assessment 13.11 Draw the MO picture of 2-methylpropan-2-ol (*t*-butanol).

***t*-butanol**

13.5 Previous alcohol synthesis

Alcohols are often products of alkene addition reactions [*see Chapters 8 and 9*] and substitution reactions [*see Chapter 12*]. These reactions are summarized in **Table 13.8**. [*Test your understanding of these reactions in Assessments 13.12–13.15.*]

Table 13.8 Previously studied alcohol syntheses.

Acid-catalyzed hydration of an alkene (Section 8.4.1)	
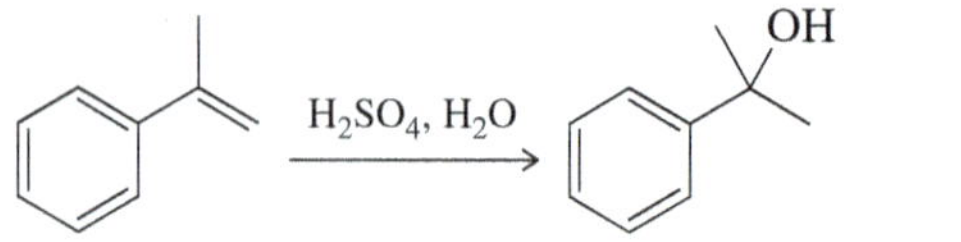	Protonation of the alkene results in the most substituted carbocation that is attacked by water from the front and back to produce a racemic mixture of alcohols. Because it proceeds through a carbocation intermediate, rearrangement is possible for acid-catalyzed hydration.
Oxymercuration–reduction of an alkene (Section 8.4.4)	
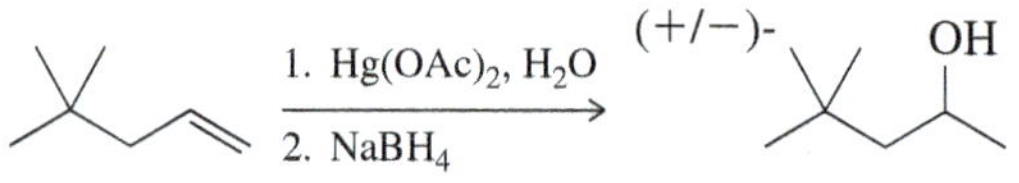	Oxymercuration–reduction of an alkene, proceeding with Markovnikov regioselectivity through anti addition, results in a hydroxyl organomercury compound. Reductive cleavage of the carbon–mercury bond produces the alcohol at the most substituted carbon of what was the alkene.
Hydroboration/oxidation of an alkene (Section 8.4.5)	
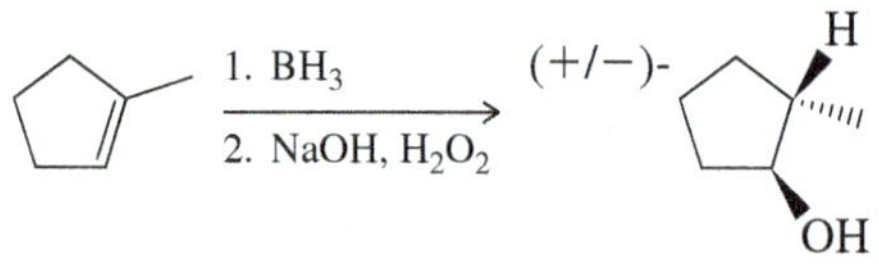	Hydroboration of an alkene, proceeding with anti-Markovnikov regioselectivity and syn stereospecificity, results in an alkyl borane. Oxidation under basic conditions produces an alcohol at what was the least substituted carbon of the alkene.
OsO_4 dihydroxylation (Section 9.1.5)	
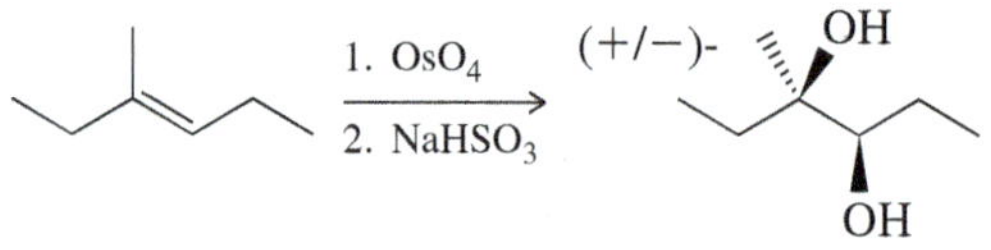	The concerted addition of two oxygens by the formation of an osmate ester is a stereospecific reaction, resulting in syn addition. Reduction of the osmate ester produces the syn diol.

(continued)

Table 13.8 *Continued*

S_N2 reaction with hydroxide (Section 12.3)	
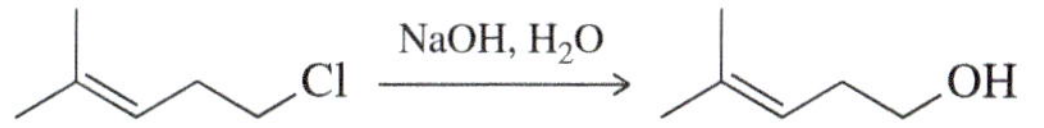	A leaving group at a 1° or methyl carbon can be displaced in a concerted reaction with hydroxide via backside displacement.
S_N1 reaction with water (Section 12.3)	
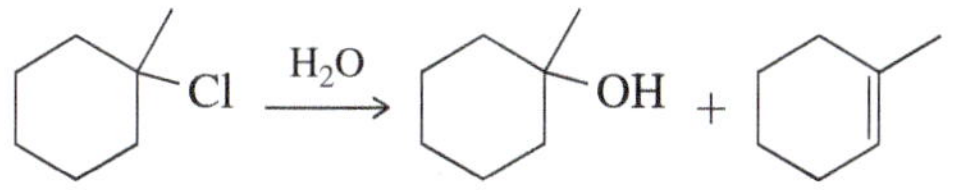	Good leaving groups at 2° or 3° carbons leave to produce a carbocation that is attacked by water from the front and back, often producing a racemic mixture of alcohols. Because it proceeds through a carbocation intermediate, rearrangement is possible for acid-catalyzed hydration.

Assessment 13.12 Show a mechanism for each of the alcohol synthesis reactions in Table 13.8.

Assessment 13.13 Predict the product of each of the following alcohol synthesis reactions.

(a)

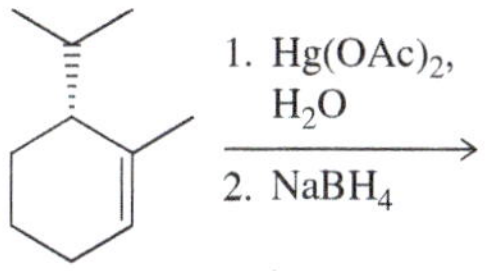

(b)

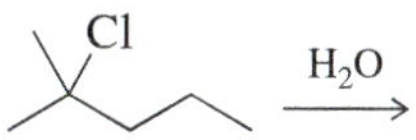

(c)

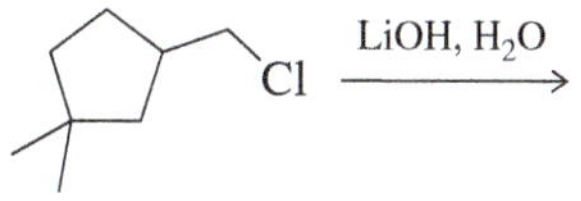

(d)

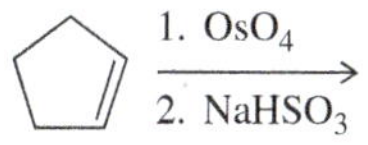

Assessment 13.14 Suggest a reagent to carry out each of the following conversions to an alcohol.

(a)

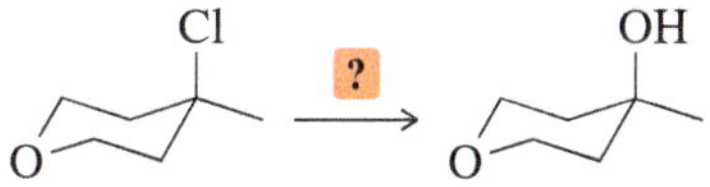

(b)

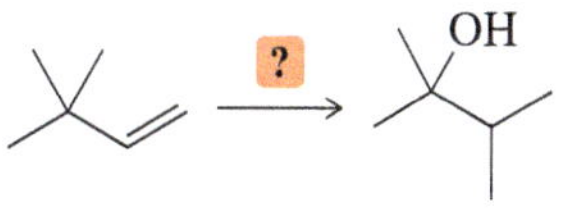

(c)

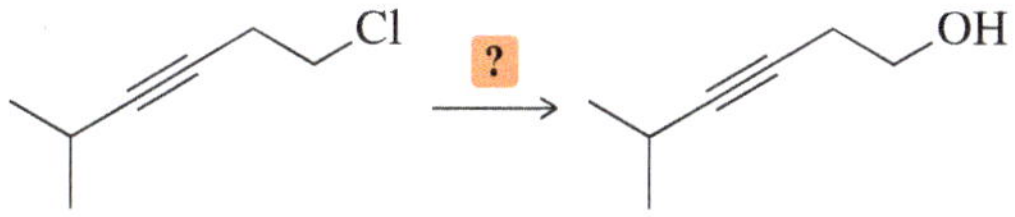

(d)

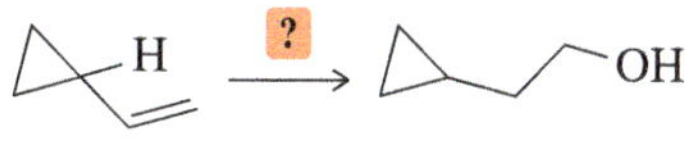

Assessment 13.15 Suggest a reagent and a reactant that could be combined to make each of the following alcohols.

(a)

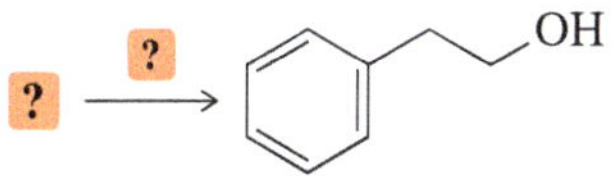

(b)

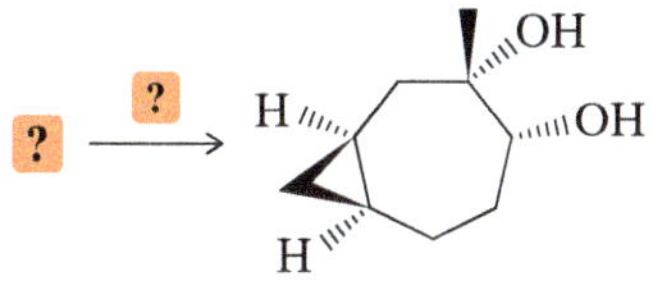

(c)

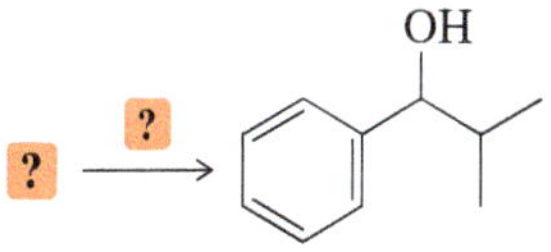

(d)

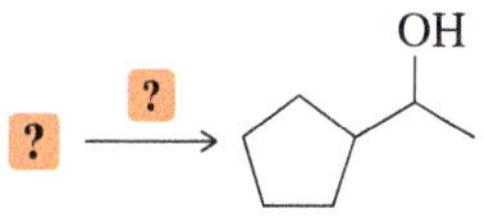

13.6 Reactions of alcohols

Three factors contribute to the chemical reactivity of alcohols: the electron-rich oxygen, the weakly acidic hydrogen, and the polarized C—O bond. [*We'll pay close attention to these as we continue.*]

13.6.1 Alcohols are weak bases

The electron-rich oxygen of the hydroxyl group makes alcohols weakly basic. Like water, alcohols are protonated by very strong acids like HCl and H_2SO_4. Any acid with a pK_a lower than −2 will favorably protonate alcohols [**Figure 13.8(a)**]. The electron-rich oxygen also helps make alcohols weakly Lewis basic [**Figure 13.8(b)**]. Alcohols, like water, react with strong electrophiles in alkene addition (Section 8.4) and S_N1 reactions (Section 12.2), donating a pair of electrons to the empty *p* orbital of carbocations.

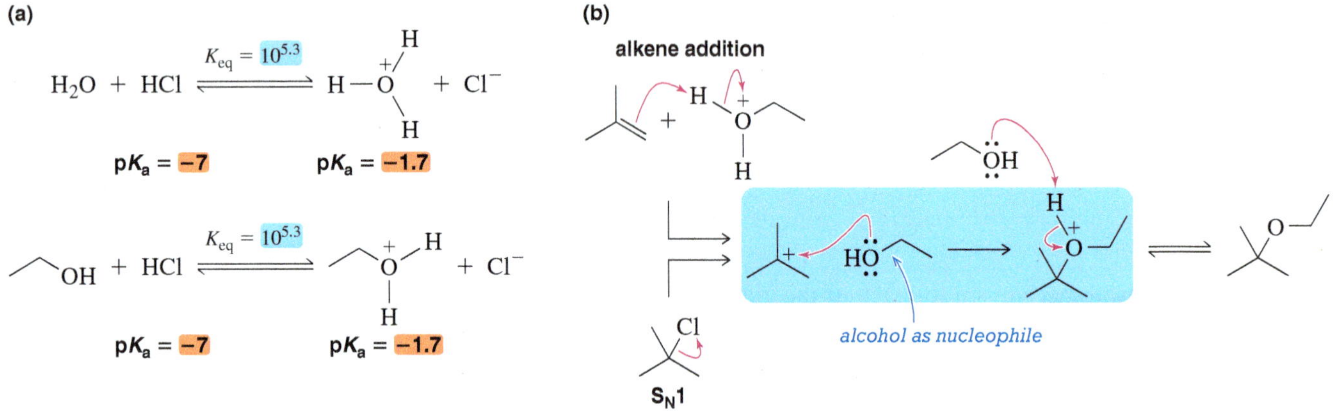

Figure 13.8 Reactivity of alcohols. (a) Alcohols are weak bases. (b) Alcohols act as Lewis bases.

13.6.2 Alcohols are weak acids

Alcohols (pK_a = 16) are relatively weak acids, necessitating strong bases for deprotonation (**Figure 13.9**). Common bases used for this purpose include NaH (conjugate acid, H_2; pK_a = 36) and $NaNH_2$ (conjugate acid, NH_3; pK_a = 38). Thiols, on the other hand, are more easily deprotonated. Sulfur atoms are larger than oxygen atoms, so sulfur is better able to delocalize and stabilize the negative charge in the conjugate base. [*See Section 4.4 for a reminder on how conjugate bases are stabilized.*]

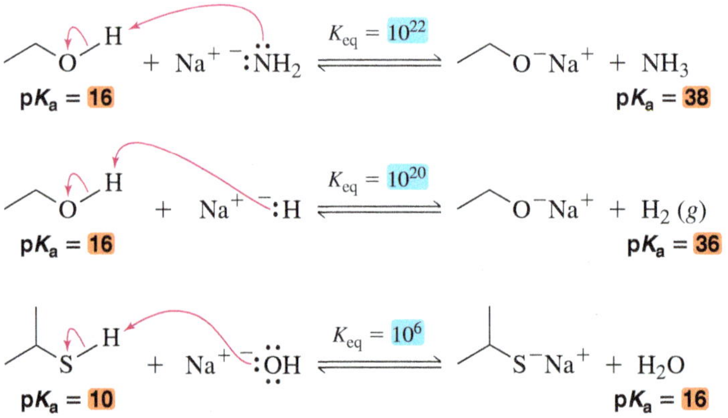

Figure 13.9 Alcohols are weak acids. Thiols are slightly stronger.

Early chemists, lacking readily available strong bases, relied on the reduction of the O—H bond to generate sodium alkoxides. These bases have found utility in reactions studied previously, including the E2 elimination of 2° and 3° alkyl halides (Section 12.5) and the S_N2 reaction of 1° alkyl halides (Section 12.3.2). [*The product of the substitution reaction in **Figure 13.10** is an example of the Williamson ether synthesis, a reaction we present in Section 13.11.1*] Thiols can be similarly deprotonated and alkylated to form thioethers (general form R—S—R).

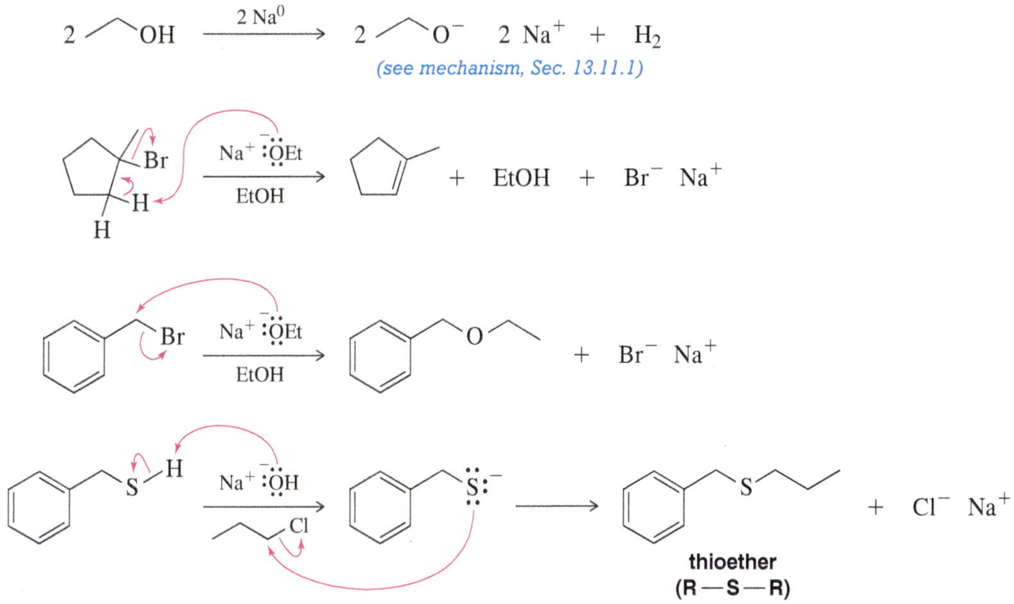

Figure 13.10 **Deprotonated alcohols and thiols** can do substitution and elimination reactions.

Assessment 13.16 Suggest an arrow-pushing mechanism for the following reaction. Is the hydroxyl group acting as a base, acid, Lewis base, or Lewis acid?

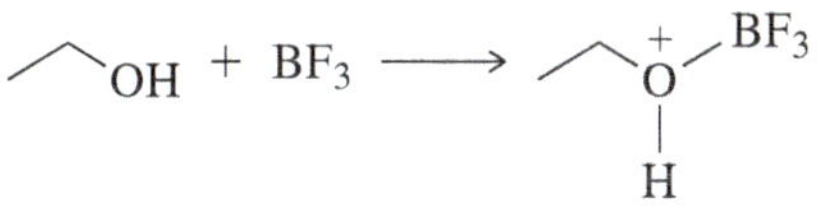

Assessment 13.17 Rationalize the difference in pK_a values for the two hydroxyl groups.

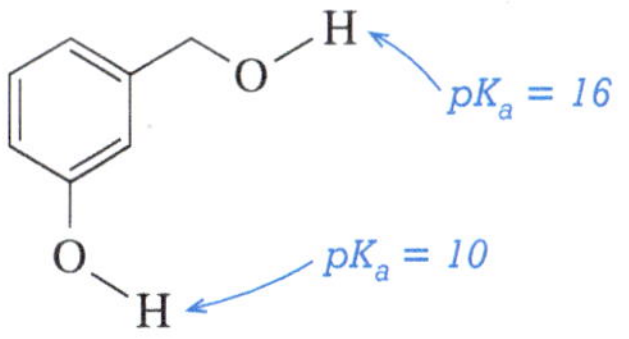

Assessment 13.18 When a secondary haloalkane is treated with sodium ethoxide in ethanol, we predict alkene formation over ether formation. How did we make this determination?

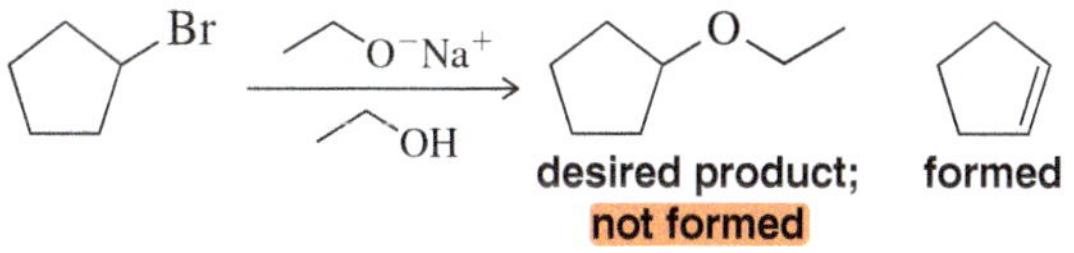

Assessment 13.19 In contrast to the results of Assessment 13.18, when a secondary haloalkane is treated with sodium ethanethiolate, we predict formation of a thioether. How is this rationalized?

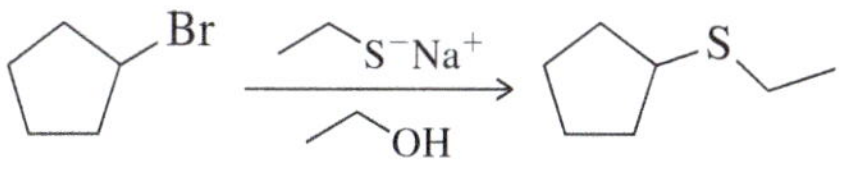

13.7 Hydroxide as a leaving group: substitution

We learned in Chapter 12 that halogens are good leaving groups in substitution reactions. Despite being favorable ($K_{eq} \approx 10^9$) based on pK_a values [**Figure 13.11(a)**], hydroxide is not a good leaving group in S_N2 reactions. Instead, a strong nucleophile like acetylide can also act as a base, deprotonating the alcohol in a similarly favorable reaction [**Figure 13.11(b)**]. When a favorable substitution reaction is competitive with an equally favorable deprotonation reaction, deprotonation is always faster. It is much easier for a nucleophile/base to abstract a sterically unhindered proton than it is to squeeze into the sterically hindered σ* orbital required for an S_N2 reaction [**Figure 13.11(c)**]. [*The idea that deprotonation is faster than substitution or nucleophilic attack is a common theme of later chapters.*]

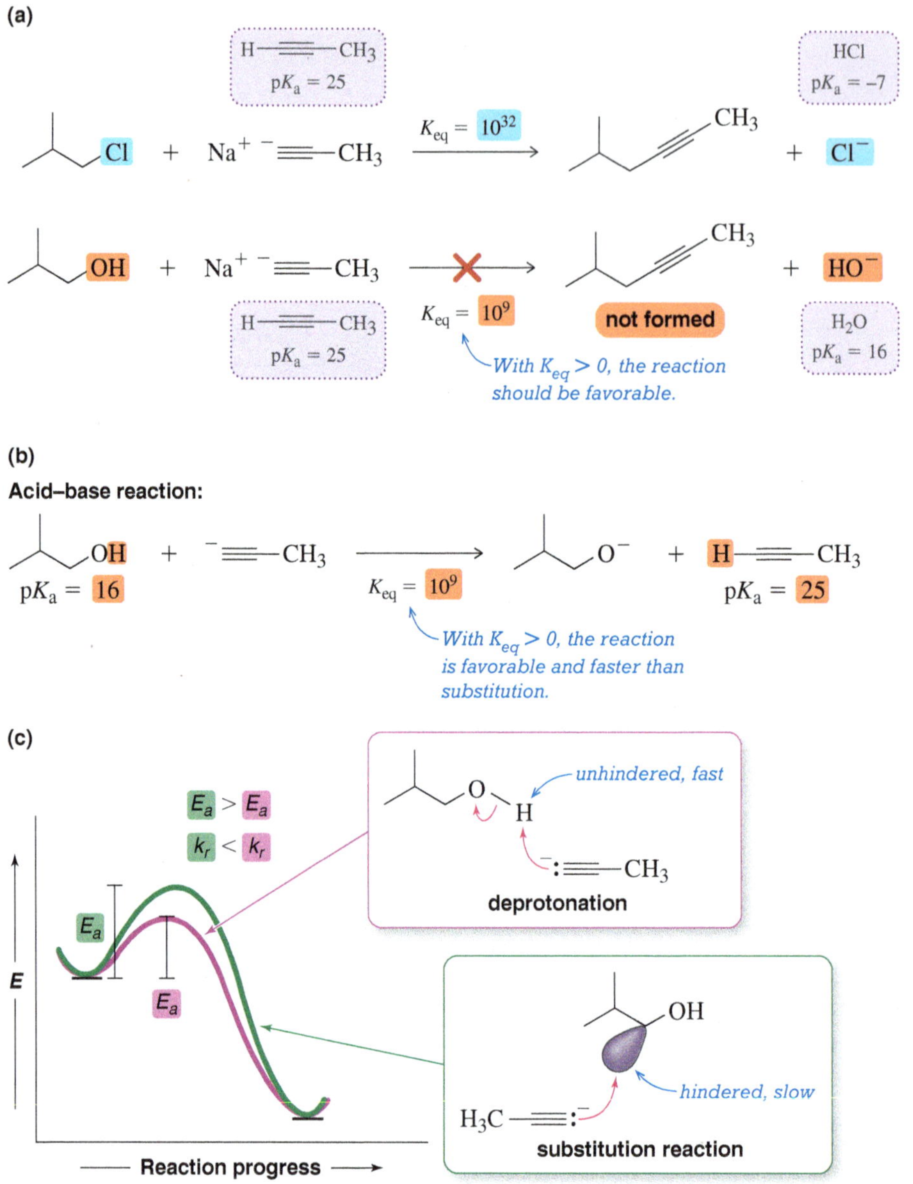

Figure 13.11 **Hydroxide is not a good leaving group.** (a) Substitution should be favorable based on the relative stability of conjugate bases. (b) Deprotonation of alcohols with strong bases is also favorable. (c) Deprotonation is faster than nucleophilic attack because there is less steric hindrance. [*The reactions are approximated to finish at the same energy based on the pK_a values of the bases involved.*]

Assessment 13.20 The intended S_N2 displacement of the 1° chloride by acetylide is unsuccessful for the molecule below. Why?

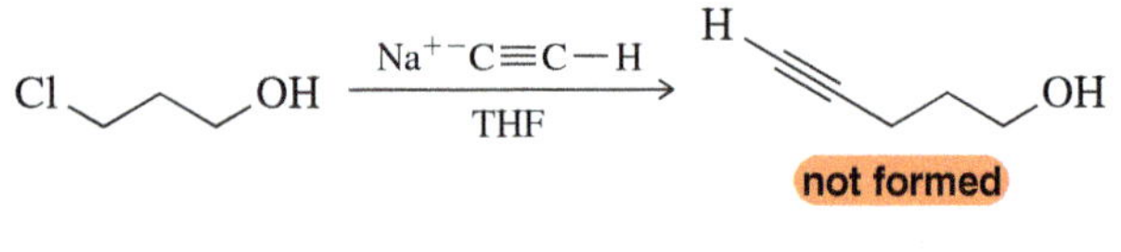

Assessment 13.21 Which of the following bases would favorably deprotonate a hydroxyl group?

(a)

(b) NaOH

(c) NaCN

(d) Et_3N

(e)

(f)
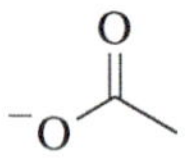

13.7.1 Tosylates are good leaving groups

Rationalizing why hydroxide is a poor leaving group gives insight into how it can be modified for use in substitution reactions. There are two problems with hydroxide as a leaving group:

1. Its conjugate acid (water) has a $pK_a = 16$, so hydroxide is not a very stable anion.
2. Hydroxyl groups have an acidic proton ($pK_a = 16$) that can be removed by strong bases.

[*Pay close attention to how these two problems are addressed in Sections 13.7.1 and 13.7.2 in creating better leaving groups.*]

If the acidic hydrogen is the problem, then we can replace it with something that is not hydrogen (represented by Z in **Figure 13.12(a)**). [*Now we don't have to worry about deprotonation.*] If Z is an electron-withdrawing group (EWG), then it will delocalize the negatively charged oxygen once $^{-}$O—Z leaves, making $^{-}$O—Z a more stable conjugate base/better leaving group (LG). The **sulfonate** functional group solves both problems simultaneously. The most popular sulfonate used for this purpose is **toluenesulfonate,** often referred to as the tosyl functional group. **Tosylates** are synthesized from the reaction of an alcohol and toluenesulfonyl (or tosyl) chloride in the presence of a nonnucleophilic amine base [**Figure 13.12(b)**]. [*Why are tosylates great leaving groups?*]

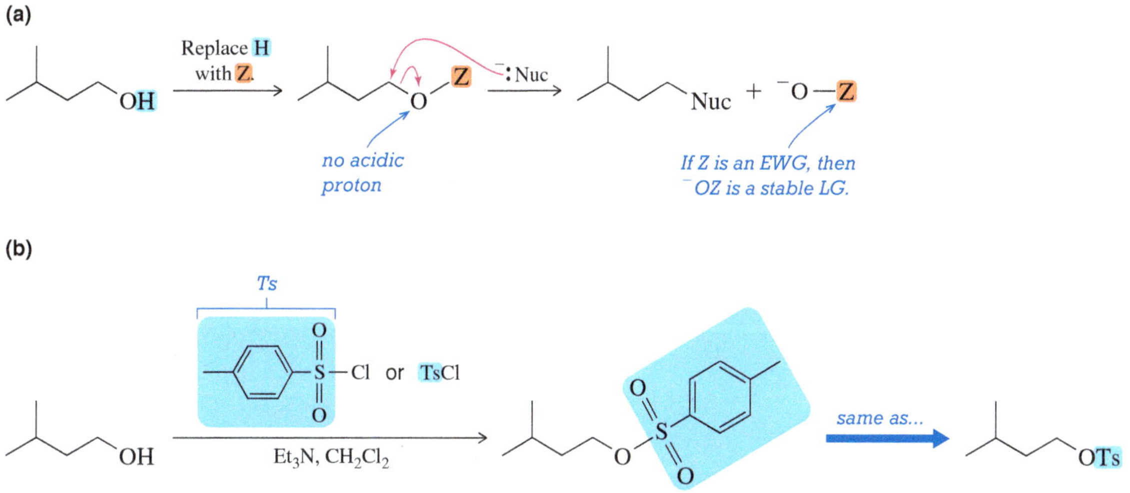

Figure 13.12 Converting hydroxide into a better leaving group. (a) General strategy. (b) Formation of tosylates. [*CH_2Cl_2 in these reactions is just the solvent.*]

We explain in Section 18.7 that substitution at doubly bonded atoms proceeds through a tetrahedral intermediate, but we can ignore this fact for now.

Mechanistically, this reaction looks like the S_N2 reactions studied in Chapter 12, with the nucleophilic hydroxyl group attacking the strongly electrophilic sulfur and displacing the stable chloride ion (**Figure 13.13**). The hydroxyl group, having brought along its proton, is now deprotonated by triethylamine. Triethylamine is not strong enough to deprotonate the alcohol ($pK_a = 16$) itself; it only removes the highly acidic oxonium proton ($pK_a = -2$). The nonnucleophilic base is important in this reaction because it abstracts the highly acidic proton before it has the opportunity to cause a side reaction with any other acid-sensitive functional groups.

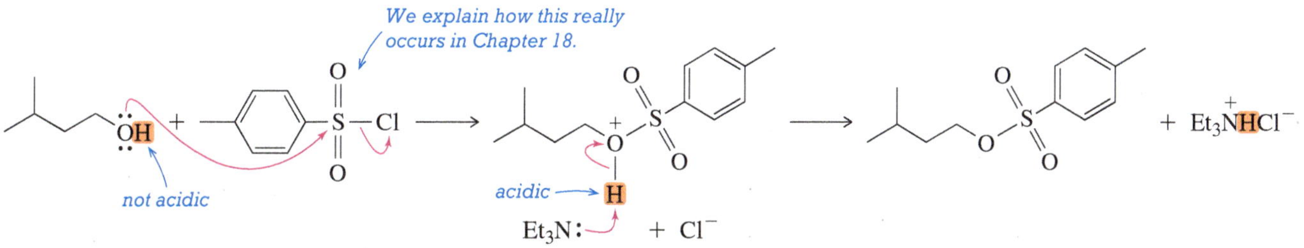

Figure 13.13 Mechanism of tosylate formation. Et_3N is only strong enough to deprotonate the highly acidic proton after the alcohol attacks.

Assessment 13.22 The sulfur atom in toluene sulfonyl chloride (TsCl) is strongly electrophilic. Why?

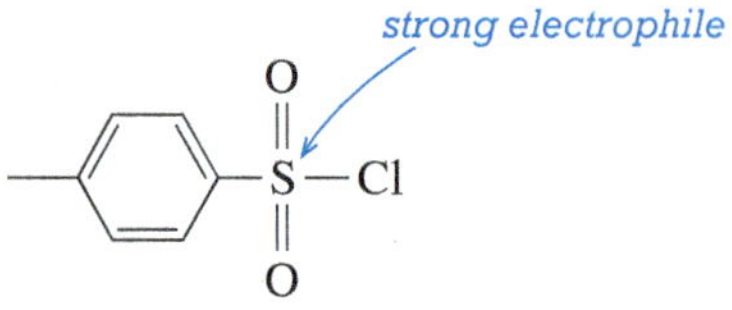

Assessment 13.23 An absent-minded chemist forgot to add triethylamine when attempting to add a tosyl group to the alcohol shown. As a result, the desired product was not isolated. What product might you expect to form? And how could the product's identity be distinguished from the desired product using spectroscopy?

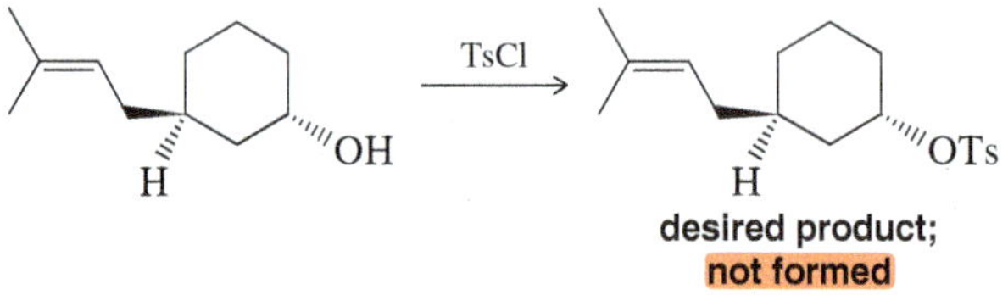

The importance of the amine base in these reactions is demonstrated by working Assessment 13.23 together.

Solved Assessment 13.23

Regardless of whether there is triethylamine present, the reaction begins the same, with the hydroxyl group displacing the chloride from the sulfonyl group (**Figure 13.14**). Having brought along its proton, a strong acid is formed. Triethylamine, if present, would deprotonate this intermediate. The only other species that might be able to abstract the proton is the displaced chloride anion. Chloride is a not a very strong base, and HCl, formed when chloride abstracts the proton, is a very strong acid. We have seen HCl used as a reagent in several reactions thus far. [*Do you see a functional group that the HCl could react with?*]

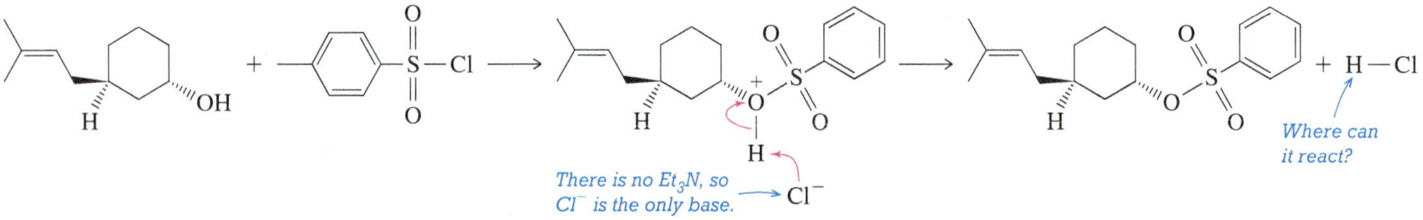

Figure 13.14 **With no Et_3N, HCl forms during tosylate formation.**

In Chapter 8, we studied the addition of strong electrophiles to nucleophilic alkenes. One of the first reactions (Section 8.3) was hydrohalogenation, a reaction where HBr or HCl adds to alkenes. This molecule, containing a trisubstituted alkene, can undergo this reaction (**Figure 13.15**), resulting in the formation of a 3° chloroalkane if triethylamine is left out of the reaction. ■

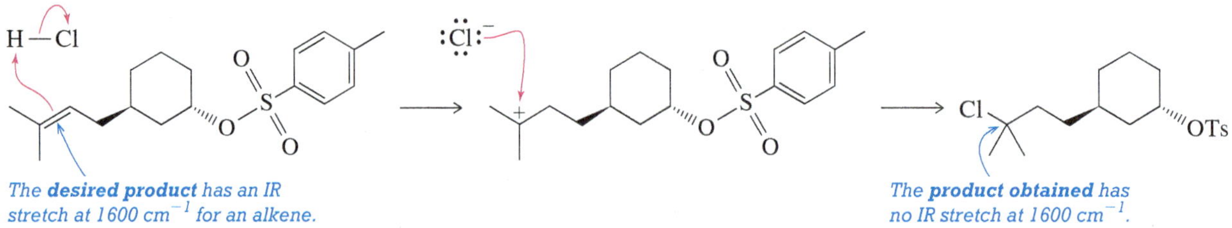

Figure 13.15 **HCl can react with other basic functional groups like alkenes.**

SPECTROSCOPY INSIGHTS

The desired product contains a C=C bond, exhibiting a stretching band in the IR spectrum around 1600 cm^{-1}. This bond is lost in the reaction with HCl. Thus, the obtained product would be missing this band at 1600 cm^{-1}. [*If you haven't covered IR spectroscopy (Chapter 14) yet, you don't need to worry about this.*]

Assessment 13.24 Predict the product of the following sulfonylation reactions.

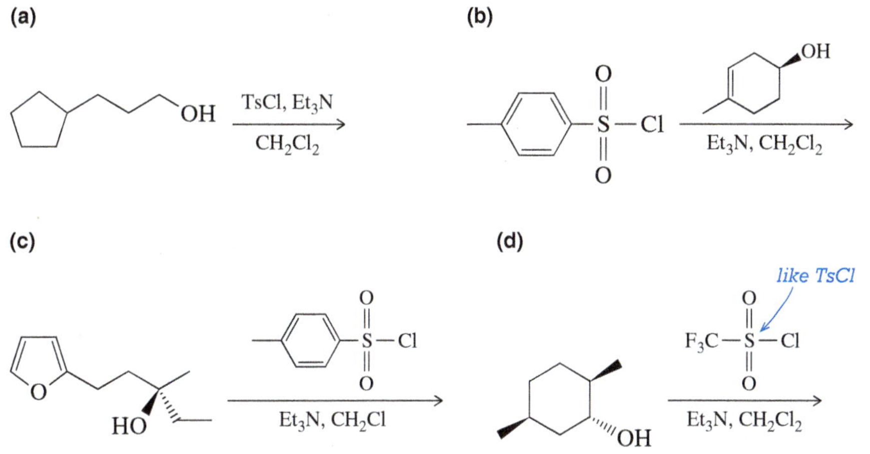

Once formed, tosylates make great leaving groups that can be used in a variety of substitution and elimination reactions. The specific mechanism and outcome depend on the same factors studied in Chapter 12. With a strong nucleophile in an aprotic solvent, S_N2 and E2 reactions are favored. With a weak nucleophile in a protic solvent, S_N1 and E1 reactions are favored. When predicting products of the tosylate reactions, as in **Figure 13.16**, you can imagine the tosylate as just another great leaving group like chloride or bromide. [*For a review of the S_N1 vs. S_N2 vs. E1 vs. E2 rules, see Section 12.8 and practice applying them in Assessment 13.25.*]

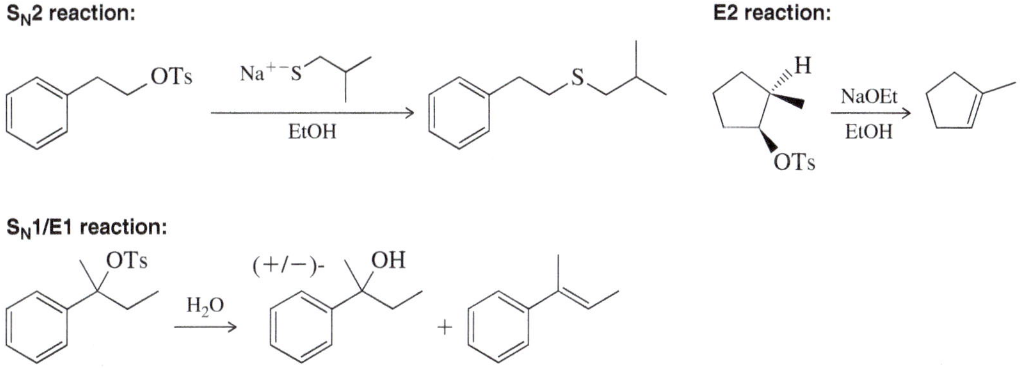

Figure 13.16 Tosylates function like haloalkanes (good leaving groups) in reactions.

Assessment 13.25 Predict the major product(s) of each of the following reactions.

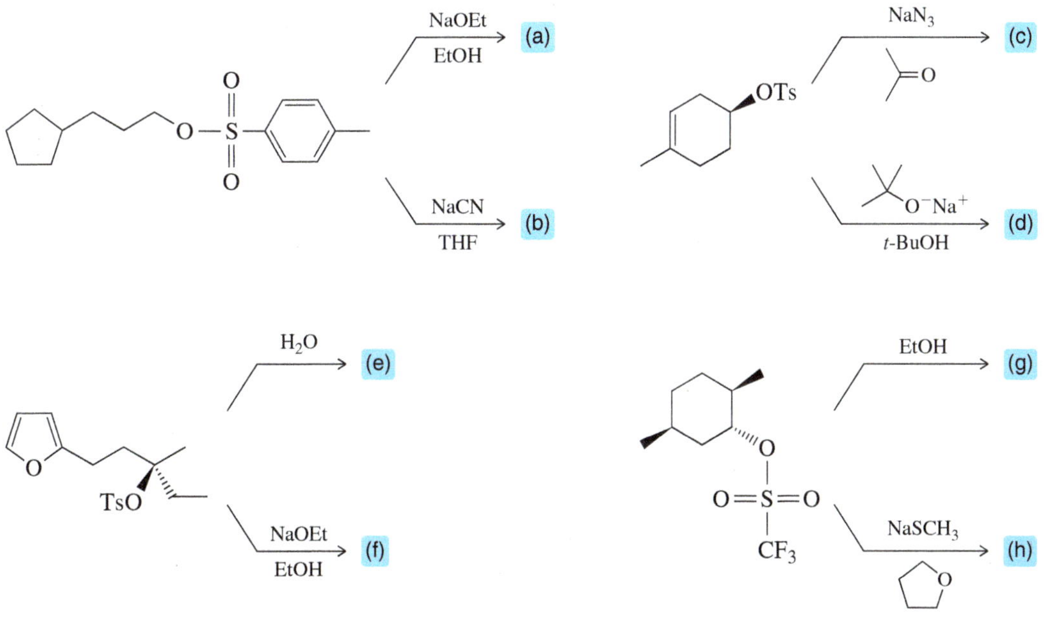

Assessment 13.26 Complete the following multistep syntheses using tosylate formation as one of the steps. The optimum number of steps for each synthesis is shown.

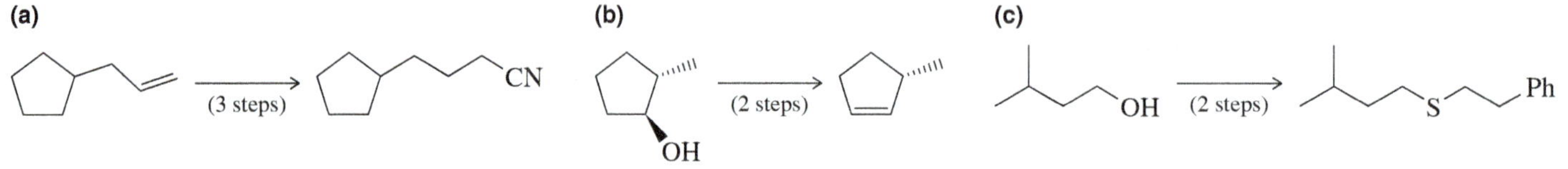

Assessment 13.27 On the reaction coordinate diagram for the disfavored nucleophilic displacement of hydroxide, predict the curve that would demonstrate how using a tosylate makes the substitution favorable.

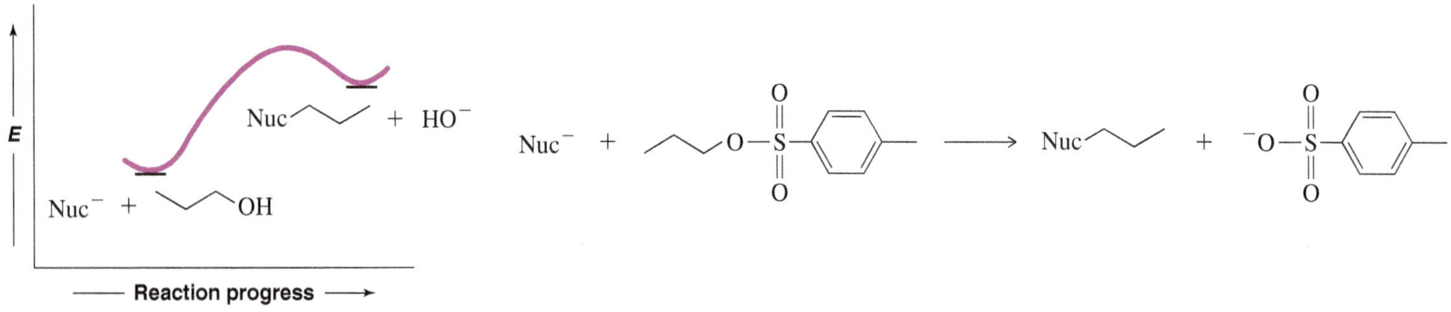

Figure 13.17 Chloride is not a strong enough nucleophile to displace hydroxide, a poor leaving group.

13.7.2 Synthesis of haloalkanes

Because hydroxide is not a good enough leaving group for S_N2 reactions, whereas halides are, the conversion of an alcohol to a haloalkane is a valuable addition to the synthetic chemist's toolbox. Chloride, as a nucleophilic species, is not strong enough to displace the hydroxide of propan-1-ol on its own (**Figure 13.17**). In order to replace a hydroxide with a halide, the hydroxyl of propan-1-ol must first be converted into a good leaving group.

[*Making hydroxide into a better leaving group continues to be the theme of Section 13.7. As you read, note the similarity between the processes. In each, the nucleophilic hydroxyl attacks an electrophile to make a better leaving group.*] While hydroxide is a poor leaving group, water is a very good one. Thus, treatment of an alcohol with a strong acid like HCl [**Figure 13.18(a)**] results in the formation of an oxonium ion. With a high concentration of

Figure 13.18 Activation with acid makes the hydronium ion an excellent leaving group. (a) 1° hydronium ions are displaced by the S_N2 mechanism. (b) 3° hydronium ions are displaced by the S_N1 mechanism.

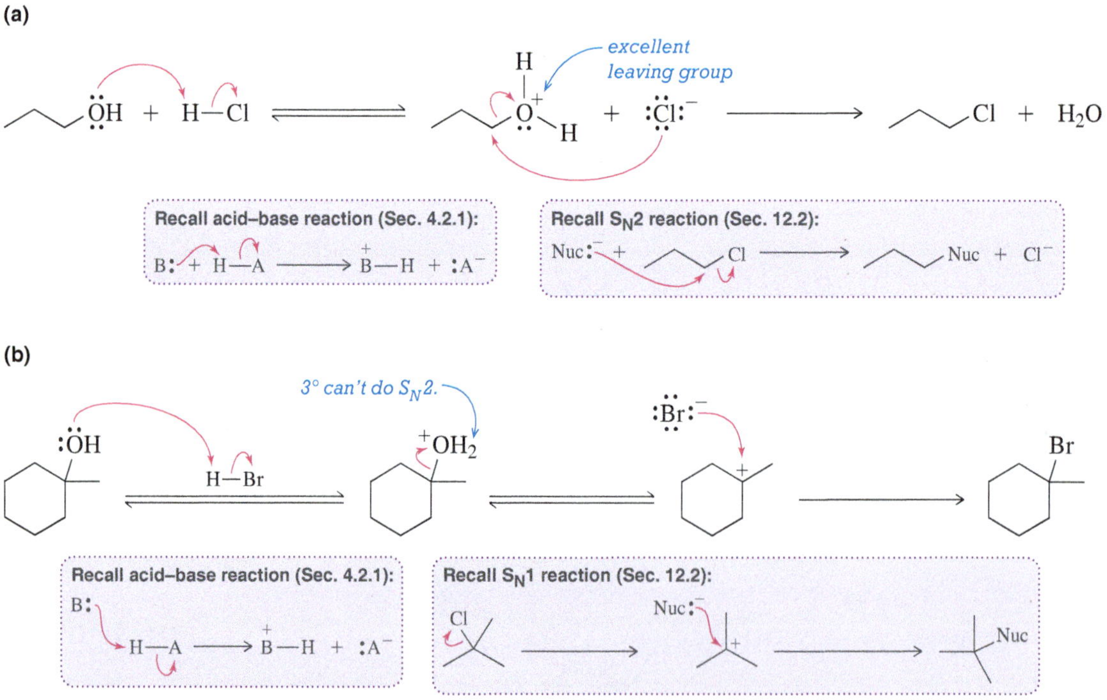

chloride around, substitution occurs to produce the chloropropane. Because the leaving group is 1°, displacement occurs by an S_N2 mechanism, exactly as you would expect based on the rules (anionic nucleophile and little steric hindrance) established in Chapter 12. When a 3° alcohol is used [**Figure 13.18(b)**], the S_N2 reaction is prevented by the steric hindrance around σ*, the antibonding orbital. As a result, conversion to a 3° bromoalkane proceeds through the carbocation intermediate of the S_N1 reaction.

Secondary alcohols can also be converted to haloalkanes by this reaction, also via the S_N1 mechanism. As a result, racemization and rearrangement is possible. That is, if the alcohol being converted is optically pure, then the stereogenic center is racemized. The carbocation, being planar, can be attacked from either side with equal rates (**Figure 13.19**). [*See Section 8.3.1.*]

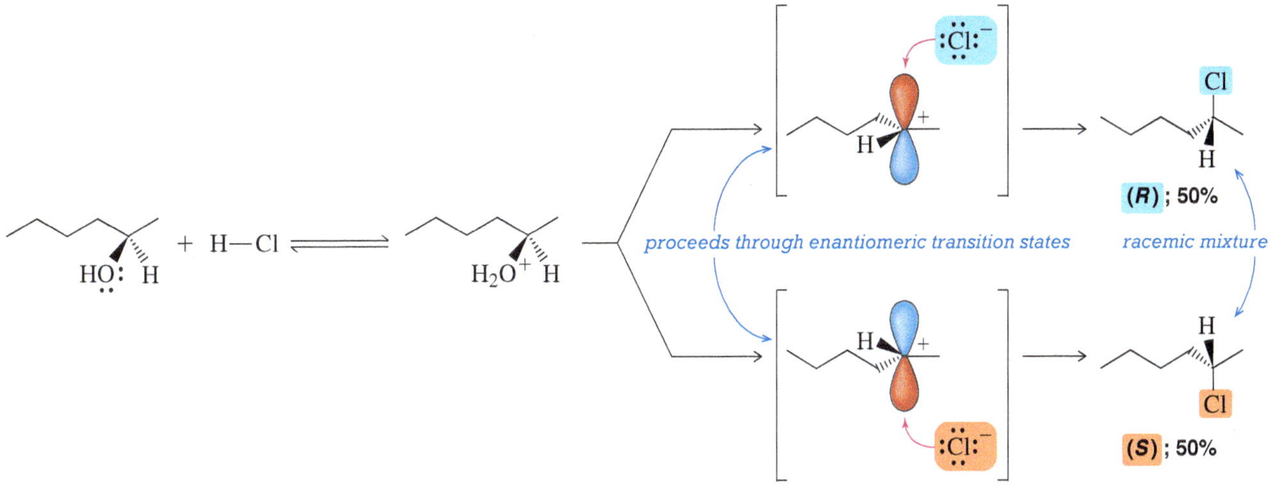

Figure 13.19 Racemization occurs with stereochemically pure alcohols.

Additionally, if the 2° alcohol neighbors a more substituted carbon, rearrangement is possible (**Figure 13.20**) [*just as it is for all carbocation mechanisms; see Sections 8.4.3 and 12.4.2*].

Figure 13.20 Rearrangement occurs if the leaving group is next to a more substituted carbon.

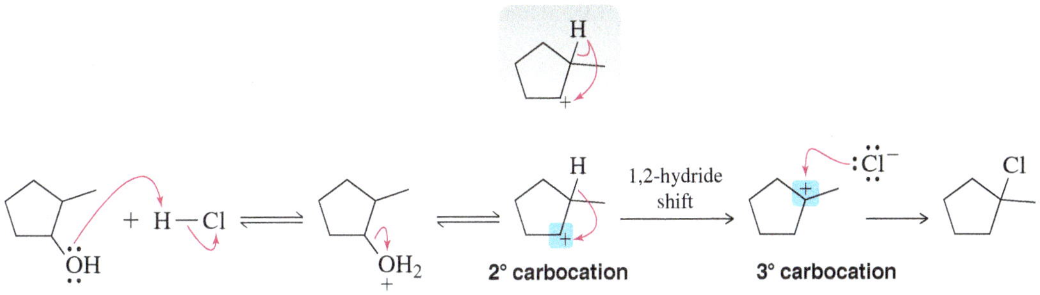

Assessment 13.28 Predict the product of the following reactions.

(a) (b) (c)

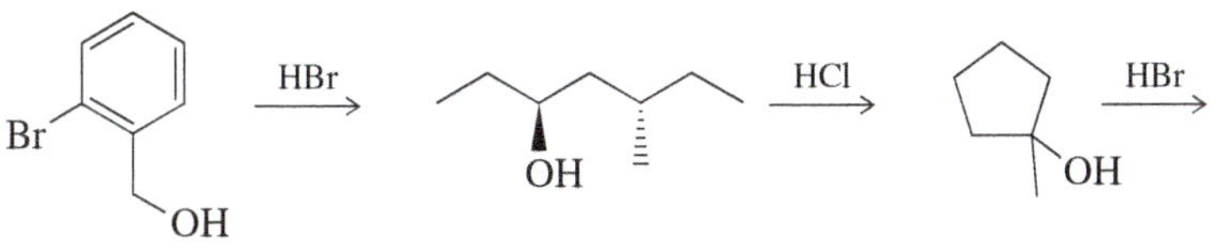

Assessment 13.29 Using an excess of HCl in the following reaction resulted in a product that was not simply the substitution of chlorine for the hydroxyl group. Predict the identity of the product obtained.

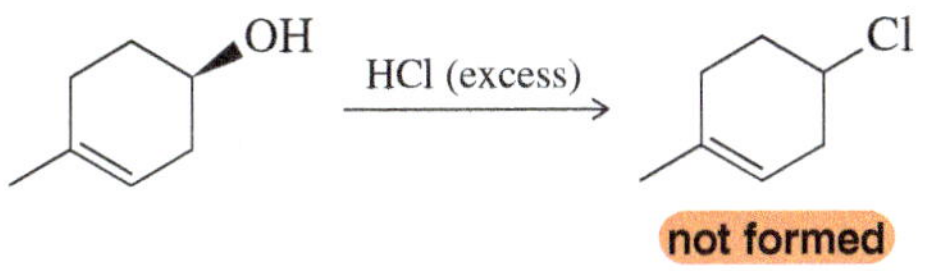

Assessment 13.30 Provide an arrow-pushing mechanism that rationalizes the outcome of the reaction shown.

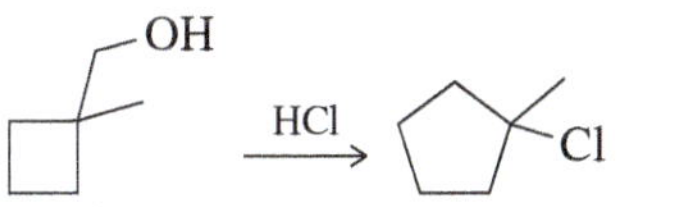

Assessment 13.31 Treatment of the 2° alcohol shown results in two different halogenated products. Rationalize the formation of each. Which do you expect to be the major isomer?

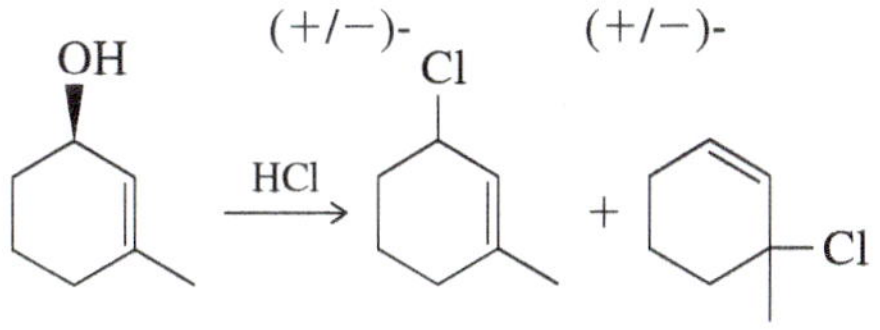

Why do organic chemists need so many ways to make haloalkanes from alcohols? The possibility for racemization, rearrangement, and the fact that many other functional groups are reactive under strongly acidic conditions make necessary the variety of reactions in Sections 13.7.3 and 13.7.4.

13.7.3 Thionyl chloride to make chloroalkanes

An alternative reagent for the synthesis of chloroalkanes is thionyl chloride ($SOCl_2$) in a solvent like anhydrous dichloromethane. **Anhydrous** means that no water is present. Thionyl chloride can be used with or without base resulting in a slightly different mechanism and stereochemical outcome for each set of conditions.

13.7.3.1 In the presence of base

A single equivalent of $SOCl_2$ is enough to convert an alcohol completely (**Figure 13.21**), whereas a large excess of HCl was needed. A nonnucleophilic base such as pyridine prevents the formation of HCl as a by-product. These two factors mean the reaction can be run under milder conditions, reducing the possibility for unwanted reactions with other functional groups [*as occurred in Assessment 13.29*].

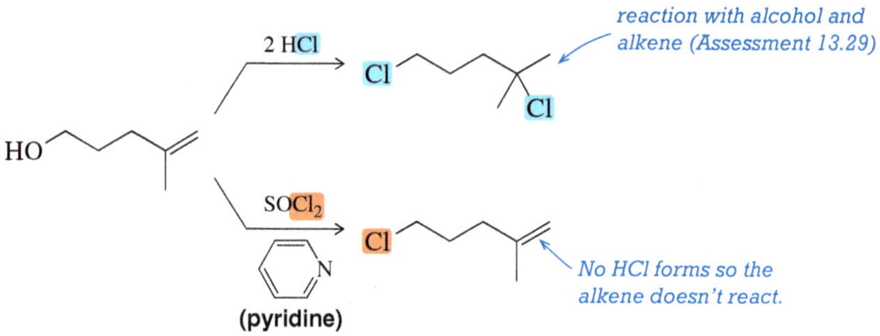

Figure 13.21 **$SOCl_2$ provides options** when HCl doesn't work.

The mechanism for this reaction (**Figure 13.22**) is similar to the one for the reaction between alcohols and HCl. The weakly nucleophilic hydroxyl group attacks the strongly electrophilic sulfur of $SOCl_2$, displacing the leaving group in what initially looks like an S_N2 reaction. The hydroxyl group, having brought along its proton, is now deprotonated by pyridine. Pyridine is not strong enough to deprotonate the alcohol ($pK_a = 16$) itself; it only removes the highly acidic oxonium proton ($pK_a = -2$). Chloride now attacks, displacing the sulfoxide, a significantly improved leaving group compared to hydroxide itself.

Remember, it isn't an S_N2 reaction, but we explain that in Section 18.7.1.

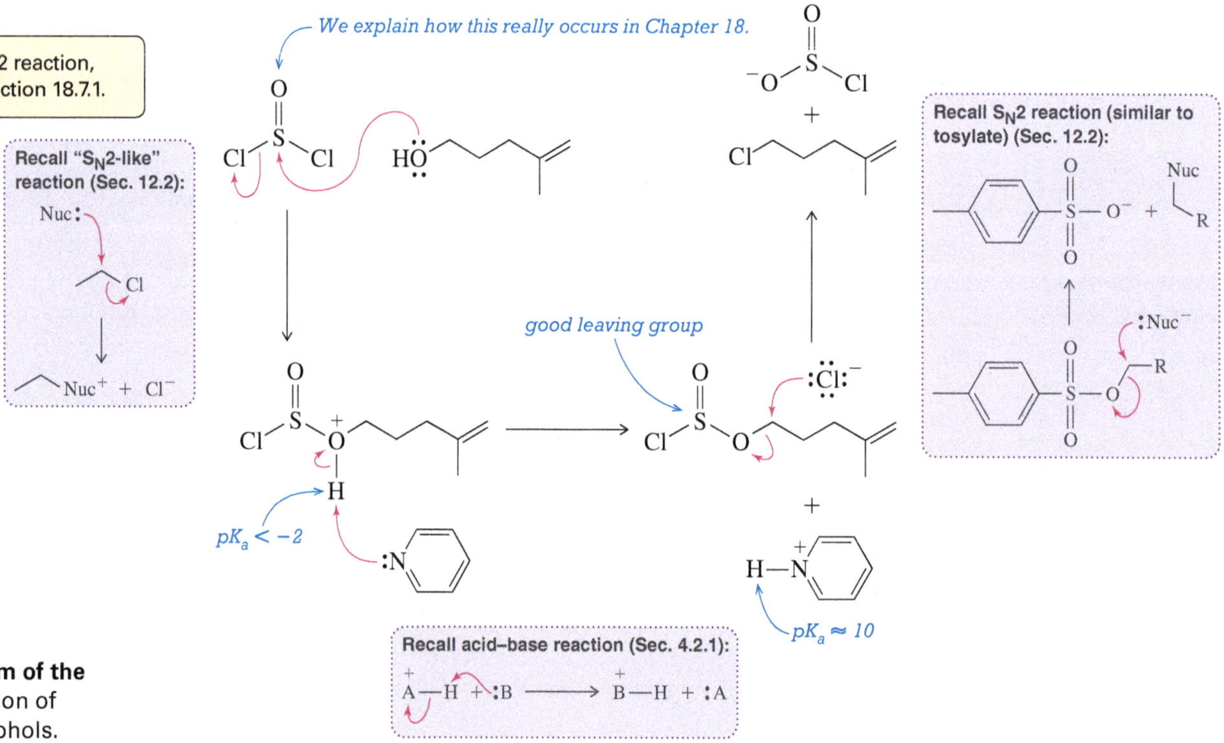

Figure 13.22 **Mechanism of the thionyl chloride** formation of chloroalkanes from alcohols.

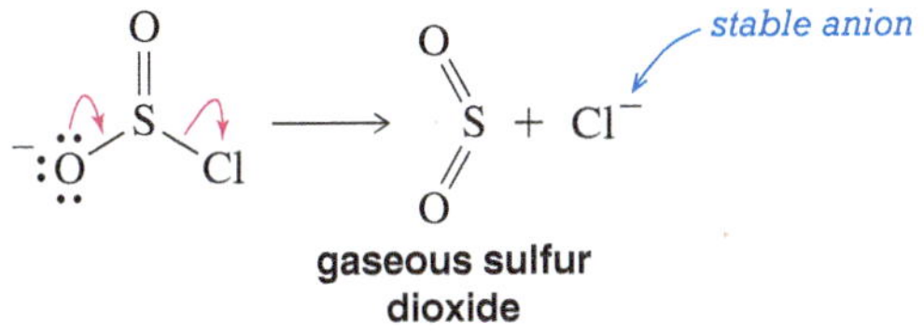

Figure 13.23 The driving force for substitution with $SOCl_2$ is the entropically favored formation of SO_2 gas.

In addition to being an excellent leaving group, SO_2Cl^- decomposes to give gaseous SO_2 and another chloride ion in an entropically favored and irreversible reaction. Formation of these two products represents an important driving force for the overall conversion of an alcohol to a chloroalkane. (**Figure 13.23**).

Using a single enantiomer of a 2° alcohol, the conversion to a chloroalkane proceeds with the inversion of configuration expected of an S_N2 mechanism [**Figure 13.24(a)**]. Rearrangement does *not* occur [**Figure 13.24(b)**] as it would if the reaction proceeded by an S_N1 reaction. Finally, conversion to a chloride via this method can be performed in the presence of acid-sensitive functional groups [**Figure 13.24(c)**].

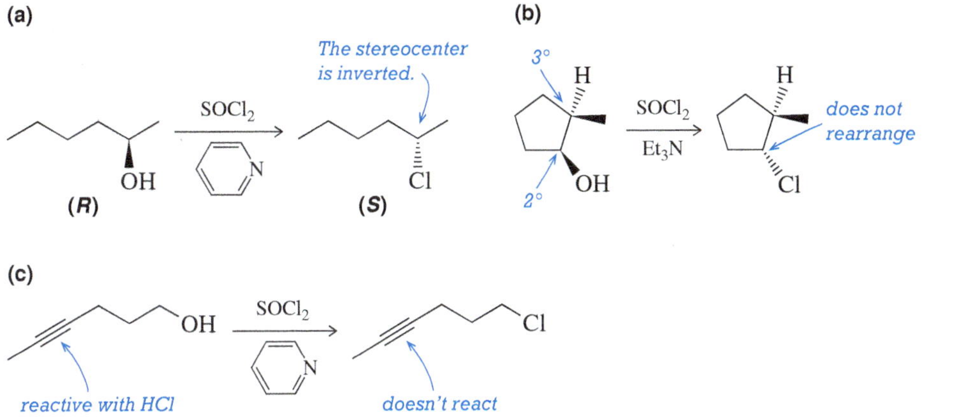

Figure 13.24 Reactions of alcohols with $SOCl_2$ in the presence of pyridine results in (a) inversion of stereochemistry, (b) no rearrangement, and (c) no side reactions of acid-sensitive functional groups.

Using 3° alcohols, where steric hindrance prevents backside displacement, substitution with $SOCl_2$ proceeds via the S_N1 mechanism. (**Figure 13.25**).

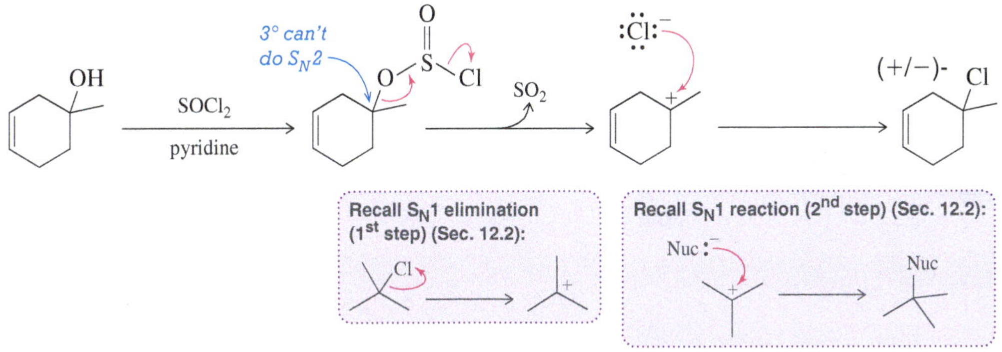

Figure 13.25 Tertiary alcohols react via the S_N1 mechanism with $SOCl_2$.

13.7.3.2 In the absence of base

The reaction of an enantiopure 2° alcohol with $SOCl_2$ in the absence of base (**Figure 13.26**) results in *retention* of configuration. [*Retention of configuration should be surprising because, thus far, in substitution reactions, racemization and inversion of stereochemistry were the only possibilities. Rather than get caught up in an exception to a rule you need to remember, focus on understanding why this is different.*] After forming, the excellent leaving group leaves immediately, in what looks like the first step of an S_N1 reaction. Before it is fully separated from the reactive carbocation, it delivers its chloride to the same face from which it left.

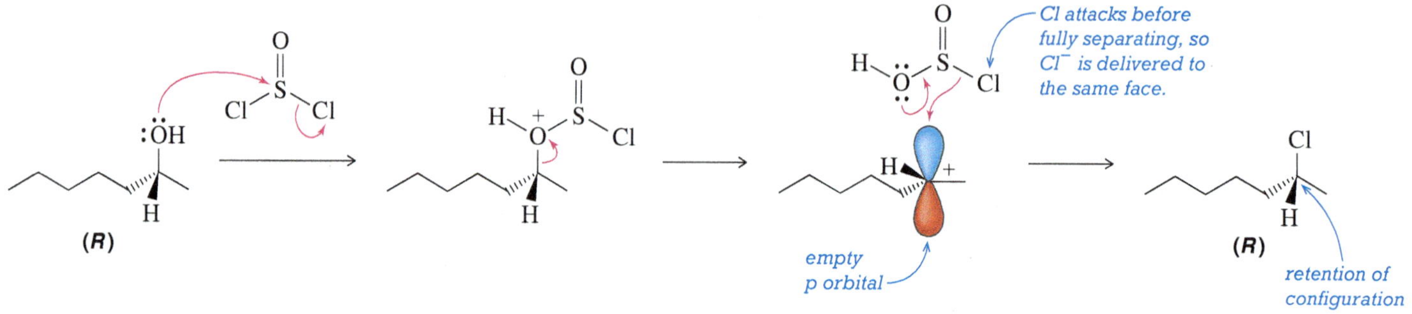

Figure 13.26 Reactions of alcohols with $SOCl_2$ in the absence of pyridine proceed with retention of stereochemistry. Chloride is delivered to the reactive carbocation on the same face from which it left.

What we have just described is consistent with the current mechanistic understanding of these reactions. But the stereochemical and mechanistic rationale regarding $SOCl_2$ is not completely understood, so be sure to consult your instructor regarding how you need to use this information.

Thionyl chloride, then, makes it possible to synthesize both enantiomeric haloalkanes from a single stereochemically pure alcohol (**Figure 13.27**).

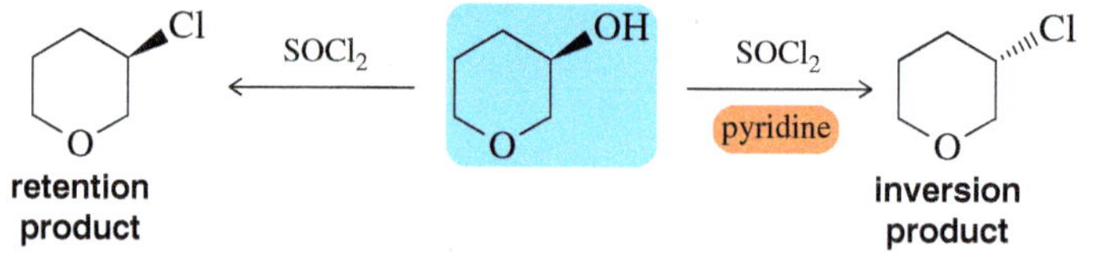

Figure 13.27 Stereodivergent synthesis is possible starting with the same alcohol using $SOCl_2$ with or without pyridine.

Assessment 13.32 When $SOCl_2$ is used in place of HCl, only one product results. Why are these conditions better? [*Compare this to the two products obtained in Assessment 13.31.*]

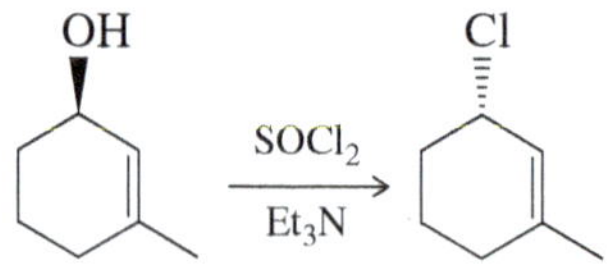

Assessment 13.33 When $SOCl_2$ is used in place of HCl, a single product is formed. Why do these conditions result in the formation of only a single isomer? [*Look back at the products of Assessment 13.29.*]

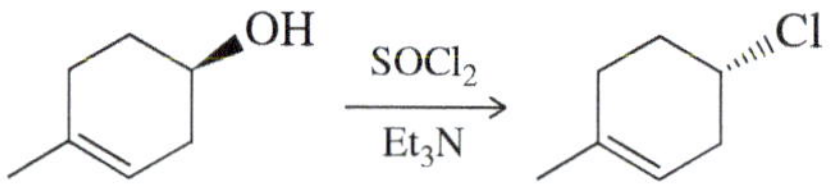

Assessment 13.34 Identify the reagent that should be used to obtain each stereochemical outcome shown.

(a)

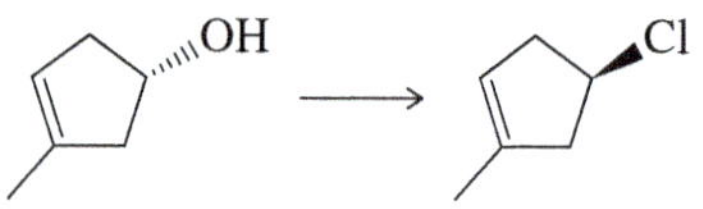

(b)

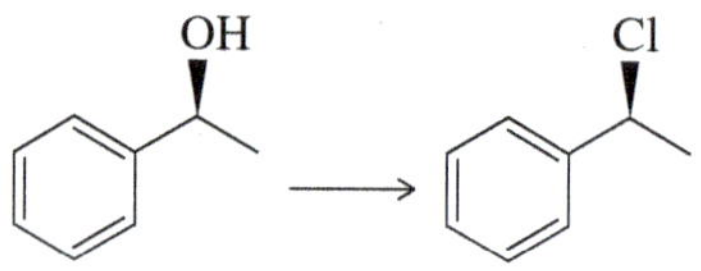

The P—O bond is one of the strongest bonds in organic chemistry. With rare exception, when you encounter a reagent containing phosphorus, you should consider formation of a P—O bond first.

13.7.4 Phosphorus tribromide (PBr_3) to make bromoalkanes

The reaction of alcohols with phosphorus tribromide (PBr_3) proceeds by a similar mechanism, but without the stereochemical inconsistency, to form bromoalkanes. Once more, the weakly basic hydroxyl group attacks the strongly electrophilic phosphorus, displacing a bromide in what can be thought of as an S_N2 reaction. With a positive charge, the oxonium ion thus formed is readily displaced in an S_N2 reaction to give a primary haloalkane (**Figure 13.28**).

Figure 13.28 The PBr_3 substitution mechanism is similar to $SOCl_2$.

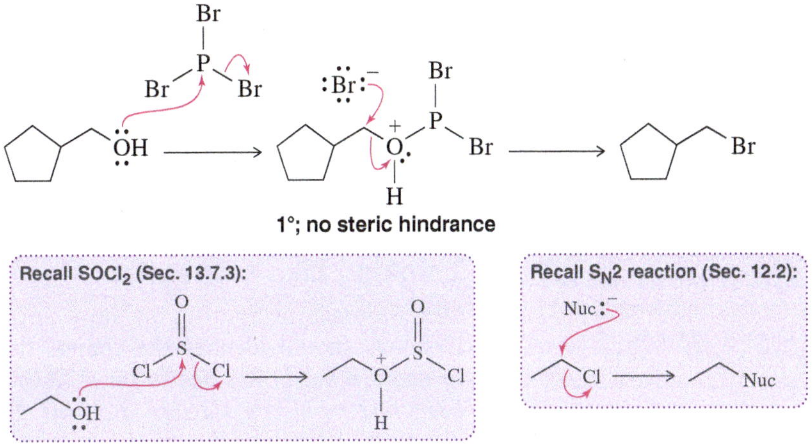

Reaction of a 2° alcohol also proceeds via the S_N2 mechanism, resulting in inversion of stereochemistry without the possibility of rearrangement [Figure 13.29(a)]. As you might expect, a 3° alcohol will be converted to a 3° bromoalkane through an S_N1 reaction because σ* is blocked from nucleophilic attack of the bromide by the steric hindrance of the attached alkyl groups [Figure 13.29(b)].

(a) 2° alcohols undergo inversion:

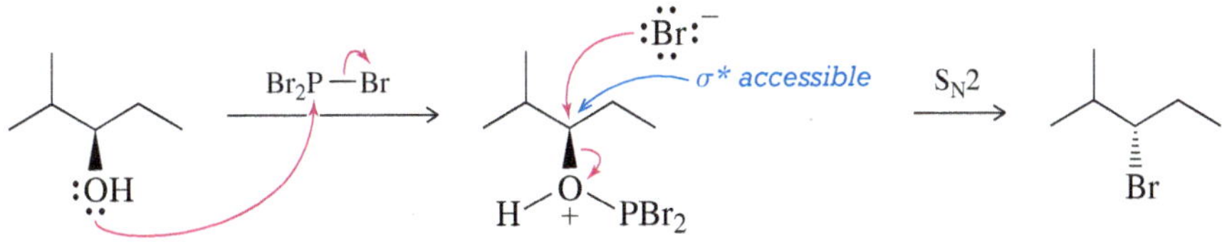

(b) 3° alcohols undergo racemization:

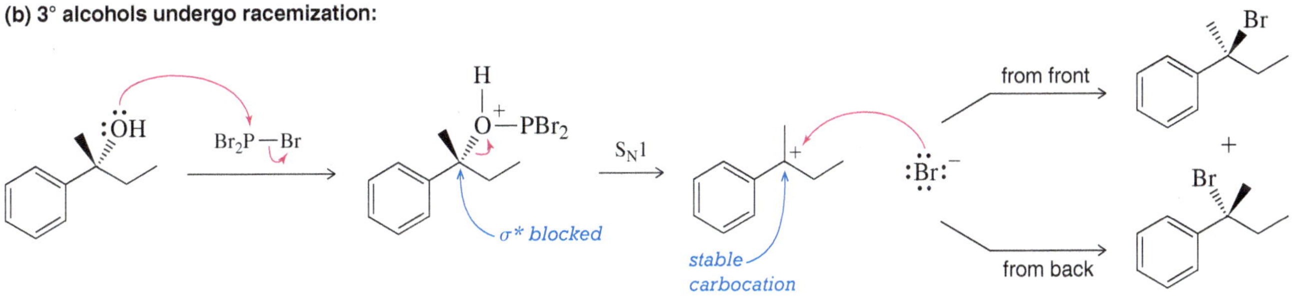

Figure 13.29 The mechanism of PBr_3 with (a) 2° alcohols undergoing inversion and (b) 3° alcohols featuring racemization.

13.7 SUMMARY

- The hydroxyl group is not a good leaving group, making substitution difficult. To do substitution at a hydroxyl-bearing carbon, the hydroxyl group is converted to a tosylate (TsCl, Et_3N), chloroalkane (HCl or $SOCl_2$), or bromoalkane (HBr or PBr_3).

Assessment 13.35 Provide the alcohol that would be used to make the bromoalkanes shown using PBr_3.

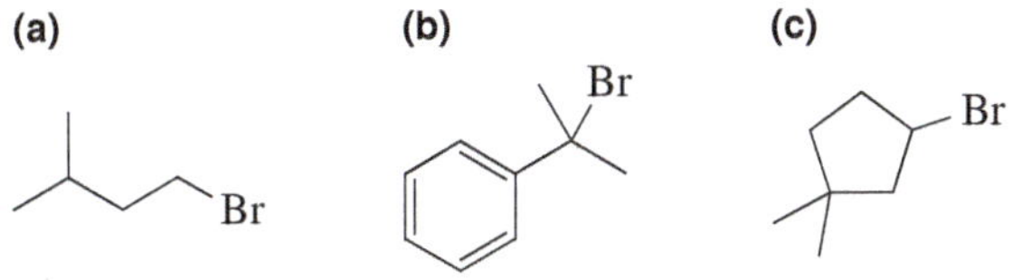

Assessment 13.36 Besides PBr_3 and $SOCl_2$, there are other ways of synthesizing haloalkanes. One such way is shown. Provide an arrow-pushing mechanism that rationalizes formation of the chloroalkane. [*Hint: Dimethyl sulfide is a good nucleophile and Cl_2 is an electrophile. Start by reacting those two together.*]

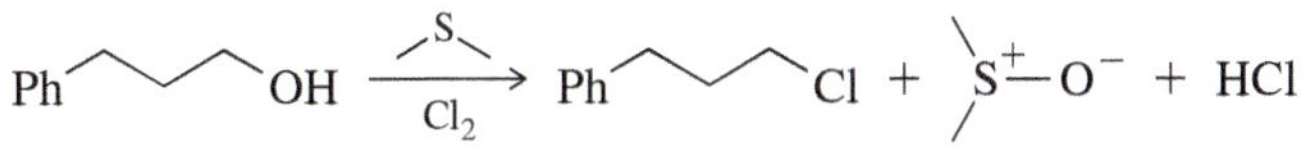

13.8 Hydroxyl as leaving group: elimination

In Section 12.8, when studying the reactions of haloalkanes, substitution was compared to competing elimination reactions. Just as it does for substitution, turning the hydroxyl group into a better leaving group should also make it possible for elimination reactions to form alkenes.

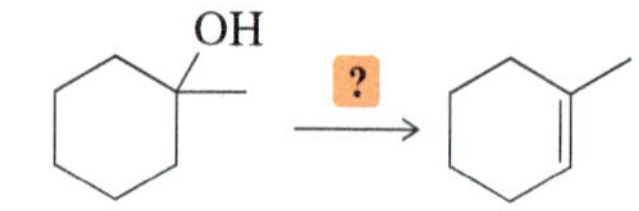

Let's think like a chemist to see how a reagent for this transformation might be developed based on chemistry studied previously.

Thinking Like a Chemist

Our approach to figuring out how to convert an alcohol to an alkene involves two steps. First, we should identify the change that the molecule has undergone. [*What bonds are formed? What bonds are broken? What has been lost from the starting alcohol? Think about this before reading further.*]

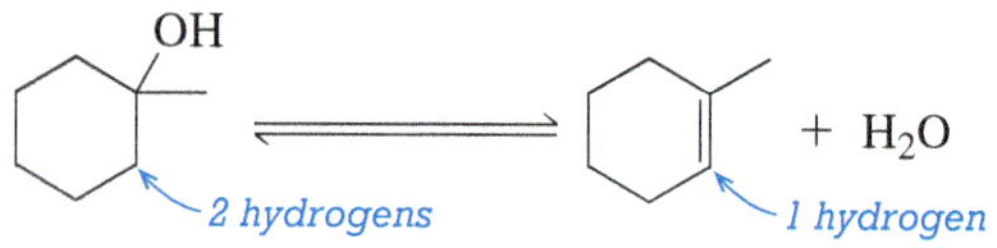

Bonds broken:	**Bonds formed:**
C—O σ	C—C π
C—H σ	O—H σ

Balancing the equation shows that the alcohol has lost a molecule of water in forming the alkene, suggesting that this reaction might be referred to as a **dehydration** reaction. This analysis is confirmed by making a list of bonds formed and broken.

The second step in our approach to this reaction is to identify a related, previously studied reaction. [*Of what does the reaction remind you?*] The flipped reaction in **Figure 13.30** imagines the alkene as the reactant and the alcohol as the product; this is the hydration of an alkene, a reaction we studied in Chapter 8. In essence, then, the dehydration reaction is simply the reverse of the alkene hydration reaction. This means the conditions required (an acid catalyst) for the forward reaction can be used to execute the reverse reaction. Because of the principle of microscopic reversibility, we know that the mechanism of the forward and the reverse reactions will also be exact opposites.

When a reaction is reversible, with an equilibrium constant near $K_{eq} = 1$, as is the case here, then Le Châtelier's principle can be used to push the reaction in one direction or the other. As we saw in Chapter 8, an excess of water drives the reaction to alcohol formation. [*What would push the reaction toward the alkene, the goal of this section?*]

If the presence of water pushes the equilibrium to the alcohol, then the alkene is favored by conditions that specifically exclude water. Hence, we use sulfuric acid (95% H_2SO_4) or phosphoric acid (99% H_3PO_4), two acids that in their pure form are nearly anhydrous. ■

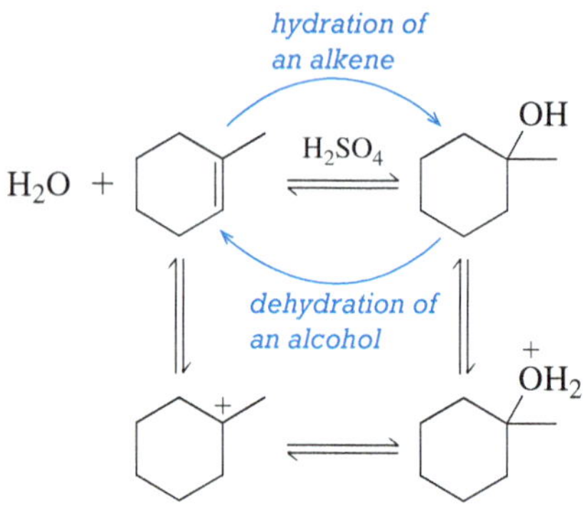

Figure 13.30 Dehydration of an alcohol is the reverse of alkene hydration. Mechanism arrows have been left out intentionally so that you can try to supply them in parts (a) and (b) of Assessment 13.37.

Assessment 13.37 Suggest an arrow-pushing mechanism for the (a) forward and (b) reverse reactions shown in Figure 13.30.

Assessment 13.38 When HCl was used for the attempted dehydration reaction shown, a reaction occurred, but none of the desired product was formed. Suggest the identity of the actual product obtained.

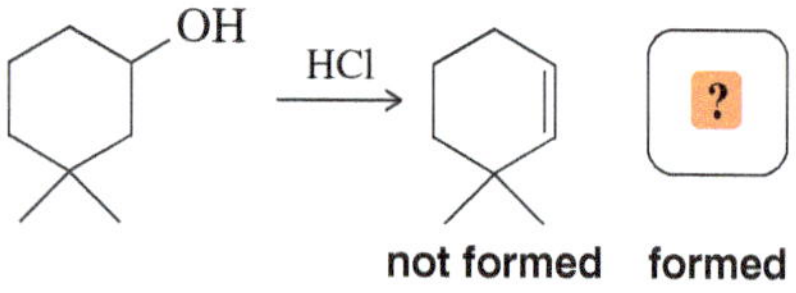

Assessment 13.39 Predict which side of the equilibrium is favored by ΔH, ΔG, and ΔS.

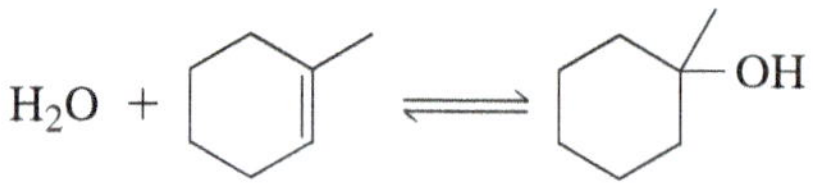

A similar strategy can be used to make alkenes from 3° and 2° alcohols. Proceeding by similar mechanisms, protonation of the hydroxyl group makes a good leaving group (**Figure 13.31**). After the water leaves and a carbocation forms, the neighboring hydrogen is removed with the electrons in the C—H σ bond collapsing into the empty *p* orbital to form the new C—C π bond. In sum, this is just an acid–base reaction followed by an E1 elimination (see Section 12.5.1 for review). As with all previous E1 elimination reactions, if two alkene products are possible, then the most substituted (most stable) alkene will be the major product.

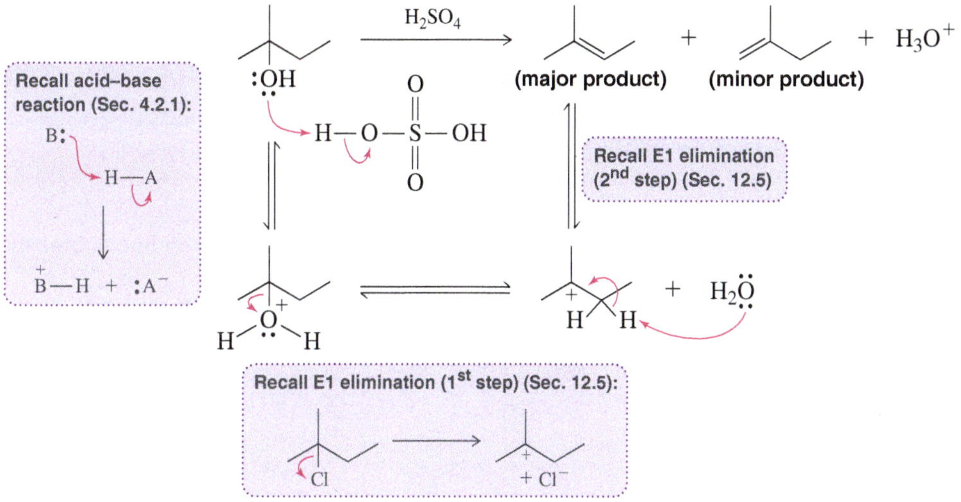

Figure 13.31 **The mechanism of acid-catalyzed dehydration** is simply an acid–base reaction followed by an E1 reaction.

Assessment 13.40 Predict the major product of the following elimination reactions.

(a) (b) (c)

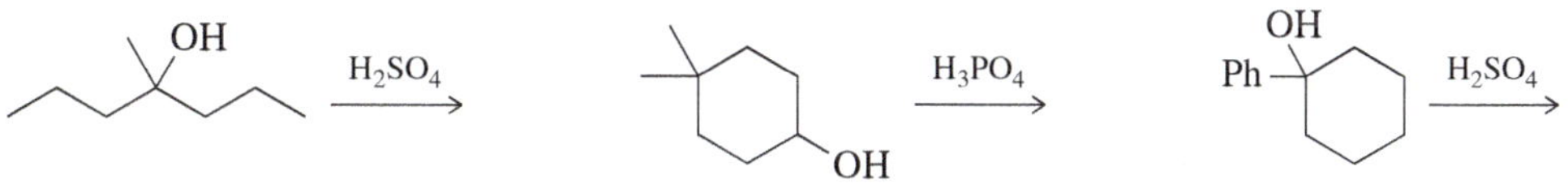

Assessment 13.41 Identify the alcohol(s) that would produce the following alkenes under the given conditions.

(a) (b) (c)

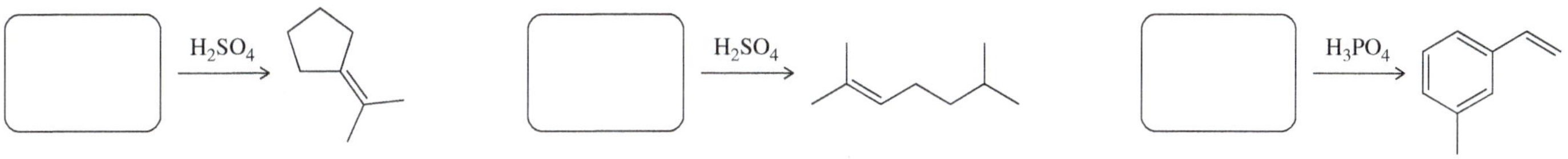

Assessment 13.42 Treatment of the following alcohol was expected to give alkene **A**. Instead, **B** was produced as the major product. Suggest a mechanism by which **B** formed.

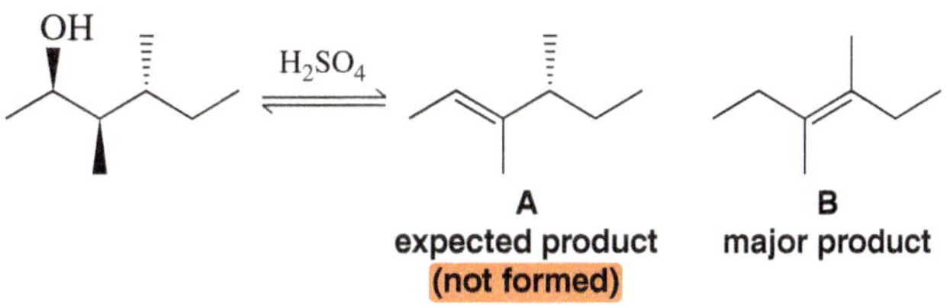

By solving Assessment 13.42 together, we will discover the circumstances that allow rearrangement to occur during dehydration reactions.

13.8.1 Rearrangement in H_2SO_4

The acid-catalyzed dehydration of alcohols is often accompanied by a carbocation rearrangement.

Solved Assessment 13.42

When approaching a complex mechanism, start by numbering the carbons and identifying the bonds formed and broken during the course of the reaction (**Figure 13.32**). [*Don't forget that H_2O is one of the products of the dehydration reaction.*] In addition to the bonds usually formed and broken in the dehydration reaction, we have also broken the $C_3—H$ bond and formed the $C_2—H$ bond. This might be a result of a carbocation undergoing a 1,2-hydride shift. [*Drawing all of the hydrogens may help you notice this.*]

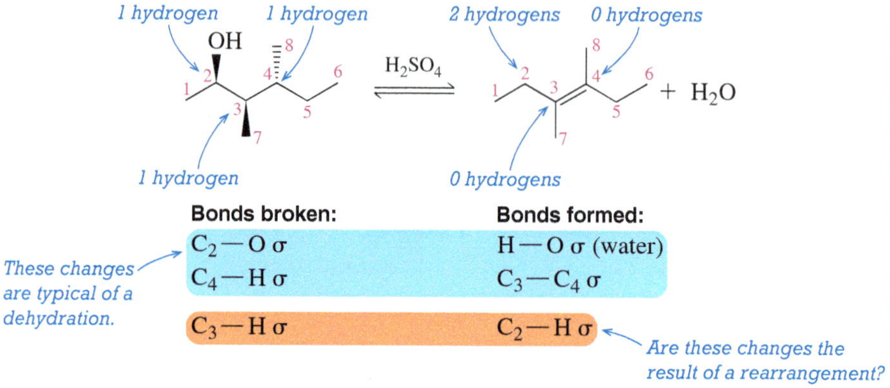

Figure 13.32 Comparing bonds broken to bonds formed shows that a rearrangement has occurred.

The second strategy employed in solving complex mechanisms is to work them in the forward and backward directions (**Figure 13.33**). This allows us to see where they meet in the middle. [*We have used this process in discovering the rearrangement of carbocations in Thinking Like a Chemist, Section 8.4.2.*] In the forward direction, the hydroxyl group is protonated and leaves as water to form a 2° carbocation at C_2. In the reverse direction, the $C_3—C_4$ alkene could only form if the $C_4—H$ bond was eliminated, allowing the electrons to collapse into an empty *p* orbital at C_3.

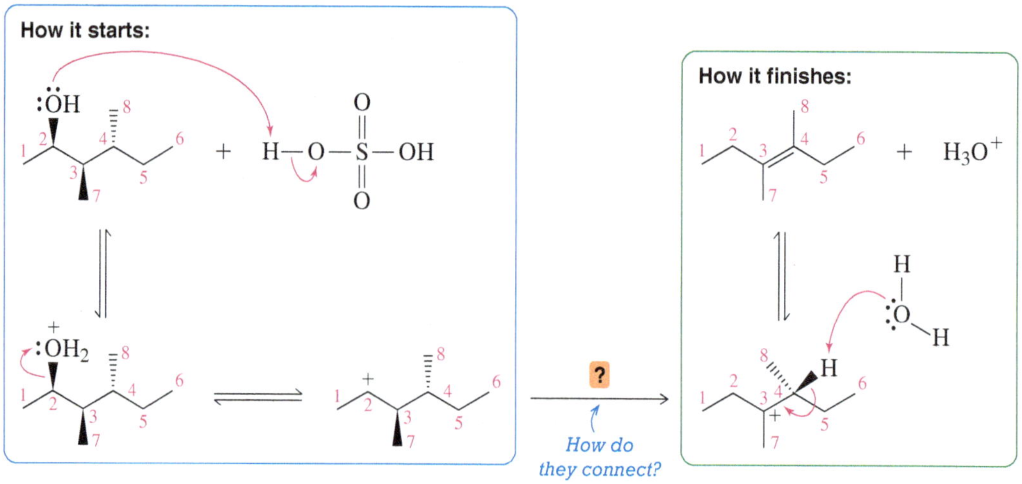

Figure 13.33 Solving the mechanism of a rearrangement.

This analysis shows the mechanism starting and finishing like a standard dehydration reaction but with a carbocation shift from C_2 to C_3 (**Figure 13.34**). We can show the carbocation shift by transferring a hydride from C_3 to C_2 in a concerted step. [*Is this a favorable step?*]

It *is* a favorable step because the carbocation at C_3 is 3°, which is better stabilized by hyperconjugation than the 2° carbocation at C_2. [*Solving the mechanism this way should have been very familiar to you because it was done previously in Table 8.6. Confirm this by comparing the two processes.*] ■

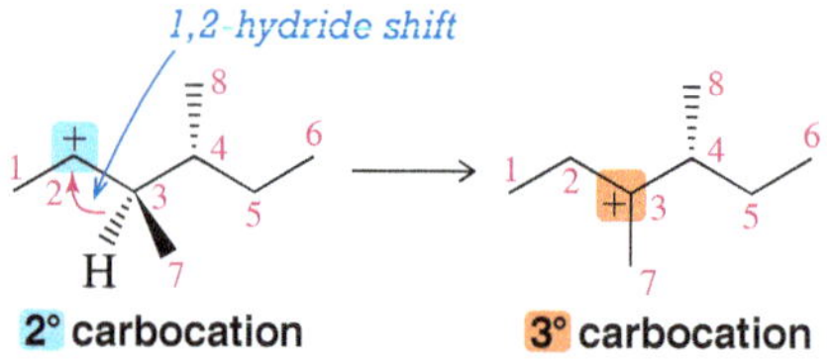

Figure 13.34 A hydride shift converts a 2° carbocation into a more stable 3° carbocation.

Recall from Section 8.4.3 that anytime a carbocation was adjacent to a more substituted carbon, a rearrangement would occur to generate a more stable carbocation. This will always be true, regardless of how the carbocation is formed [*via alkene addition, Chapter 8; via S_N1 or E1, Chapter 12; or via dehydration of an alcohol, Chapter 13*].

Carbocation rearrangement is not limited to the shifting of a hydride. Alkyl shifts occur when a 2° carbocation is adjacent to a fully substituted (4°, quaternary) carbon. Thus, the dehydration of 2,2-dimethylcyclohexanol produces 1,2-dimethylcyclohexene by a mechanism involving a fast 1,2-methyl shift as an intermediate step (**Figure 13.35**).

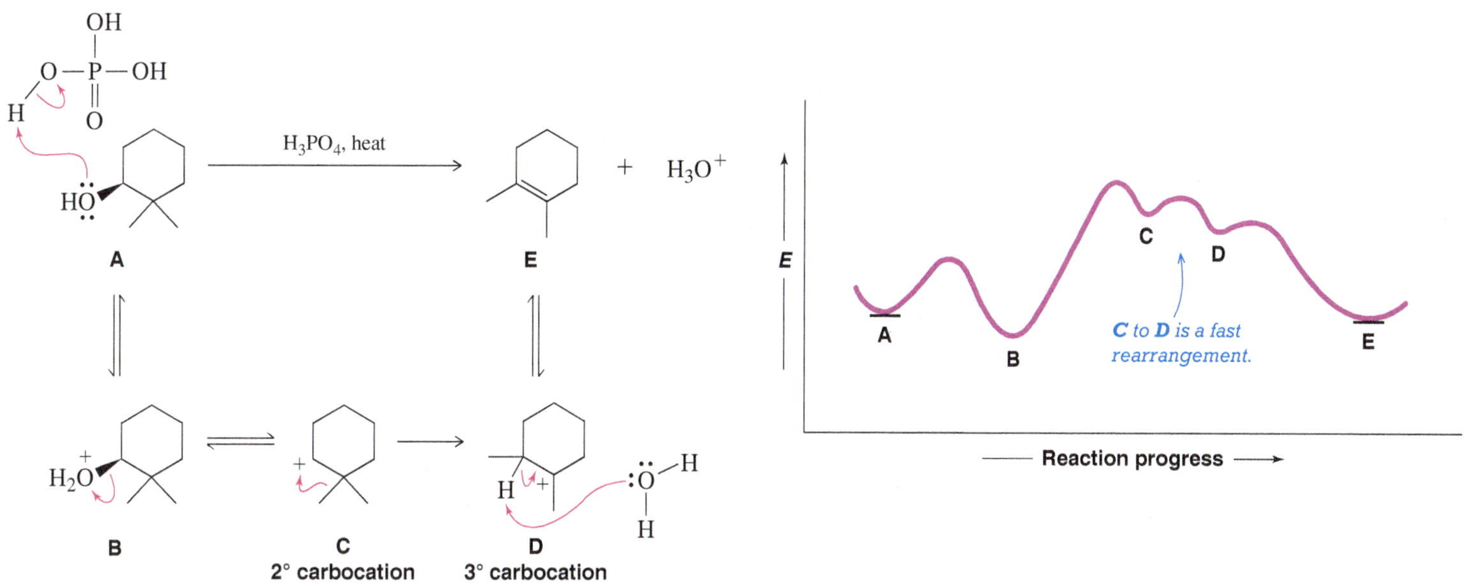

Figure 13.35 Mechanism of acid-catalyzed dehydration with a concerted, fast 1,2-alkyl shift to make a more substituted carbocation.

Assessment 13.43 Predict the product(s) of the reactions shown.

(a)

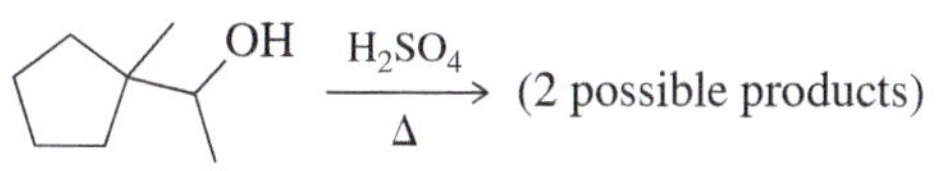

(b)

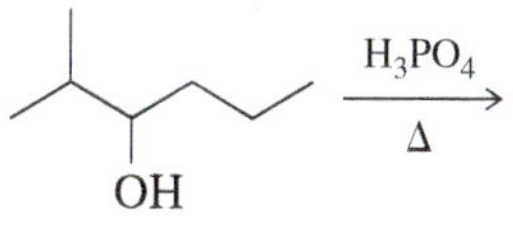

Assessment 13.44 Identify two different alcohols that can be dehydrated (one with rearrangement and one without) to form the alkene shown.

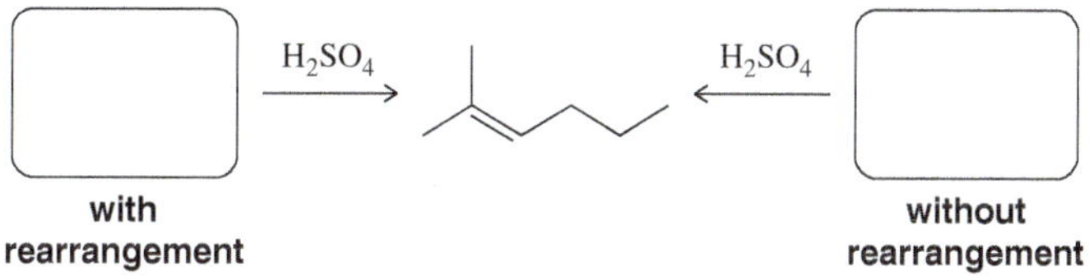

13.8.2 Pinacol rearrangement

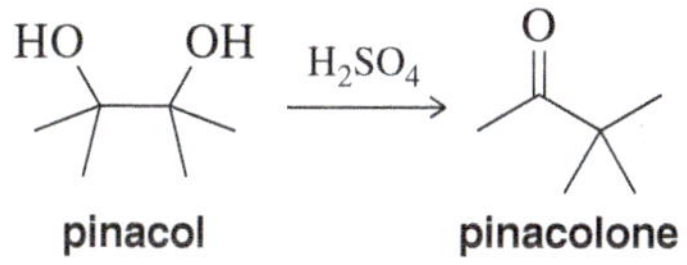

Figure 13.36 The pinacol (to pinacolone) rearrangement. Under dehydration conditions, 1,2-diols produce ketones with skeletal rearrangement.

A special type of rearrangement occurs when the dehydration conditions are applied to 1,2-vicinal diols, such as 2,3-dimethylbutan-2,3-diol (often referred to as pinacol, its common name). [*Recall from Section 9.1.2 that vicinal means the two substituents are on neighboring carbons.*] An alkene might be expected based on what you have read so far here in Section 13.8, but a ketone (pinacolone) is produced when pinacol is treated with sulfuric acid. This reaction, shown in **Figure 13.36**, is referred to as the **pinacol rearrangement**, although it occurs with numerous vicinal diols. [*See if you can figure out the mechanism before reading further.*]

Mechanistically, the reaction begins with protonation of either of the two hydroxyl groups. [*Later on, we'll encounter a situation where the choice of which hydroxyl group matters.*] The hydronium ion is an excellent leaving group, so it leaves to produce a 3° carbocation. At this point in the mechanism, as part of the problem-solving process, it is important to look ahead to where the reaction is going. By numbering the carbons throughout, you can see that one of the methyl groups has migrated from carbon 2 to carbon 3. Given that carbocations allow for alkyl and hydride shifts, it makes sense that the shift occurs at this point in the mechanism (**Figure 13.37**). [*Why would a 3° carbocation rearrange? Aren't 3° carbocations stable?*]

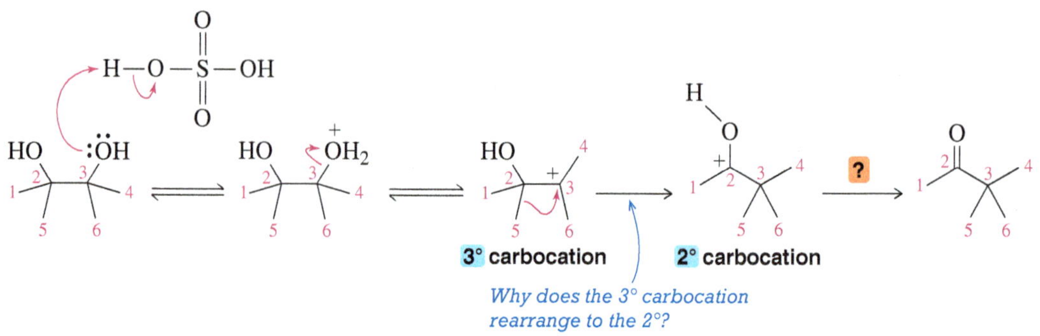

Figure 13.37 Mechanism for the pinacol rearrangement (partially solved).

When solving mechanism problems, we often push electrons thinking only about how to get to the product [*The answer is on the page!*]. It is also important to think about *why* a particular step would happen so we know we are on the correct path. For example, we know that carbocation rearrangement occurs when enhanced hyperconjugation is available at a neighboring carbon. But considering only hyperconjugation isn't enough. Moving from a 3° to a 2° carbocation, based only on hyperconjugation, would be unfavorable. But this particular carbocation has a neighboring oxygen that can donate a pair of electrons through resonance to stabilize the carbocation by delocalizing it over two atoms (**Figure 13.38**). Not only does this help us understand *why* the rearrangement occurs, it also makes clear *what* the next step is, because one of the resonance forms of the carbocation closely resembles a protonated ketone. Thus, the final step is removal of the proton by a water molecule in solution to give the final ketone product.

Whenever one electronic effect is examined, make sure to consider all of them: hyperconjugation, induction, and resonance.

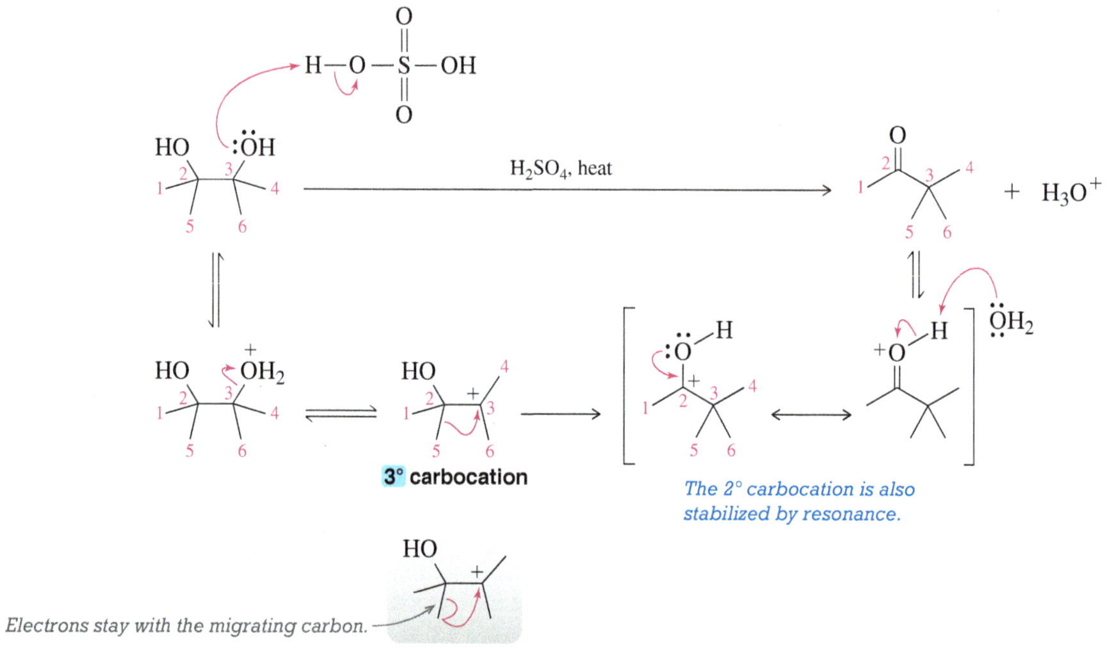

Figure 13.38 Mechanism of the pinacol rearrangement (complete).

In a sense, the pinacol rearrangement is no different than previous dehydration reactions that rearranged. In fact, the steps are identical, as shown in **Figure 13.39** for a different pinacol example that features a hydride shift instead of an alkyl shift. The four steps are as follows: (1) protonate the hydroxyl group, (2) water leaves, (3) rearrange to form a more stable carbocation, and (4) remove a proton.

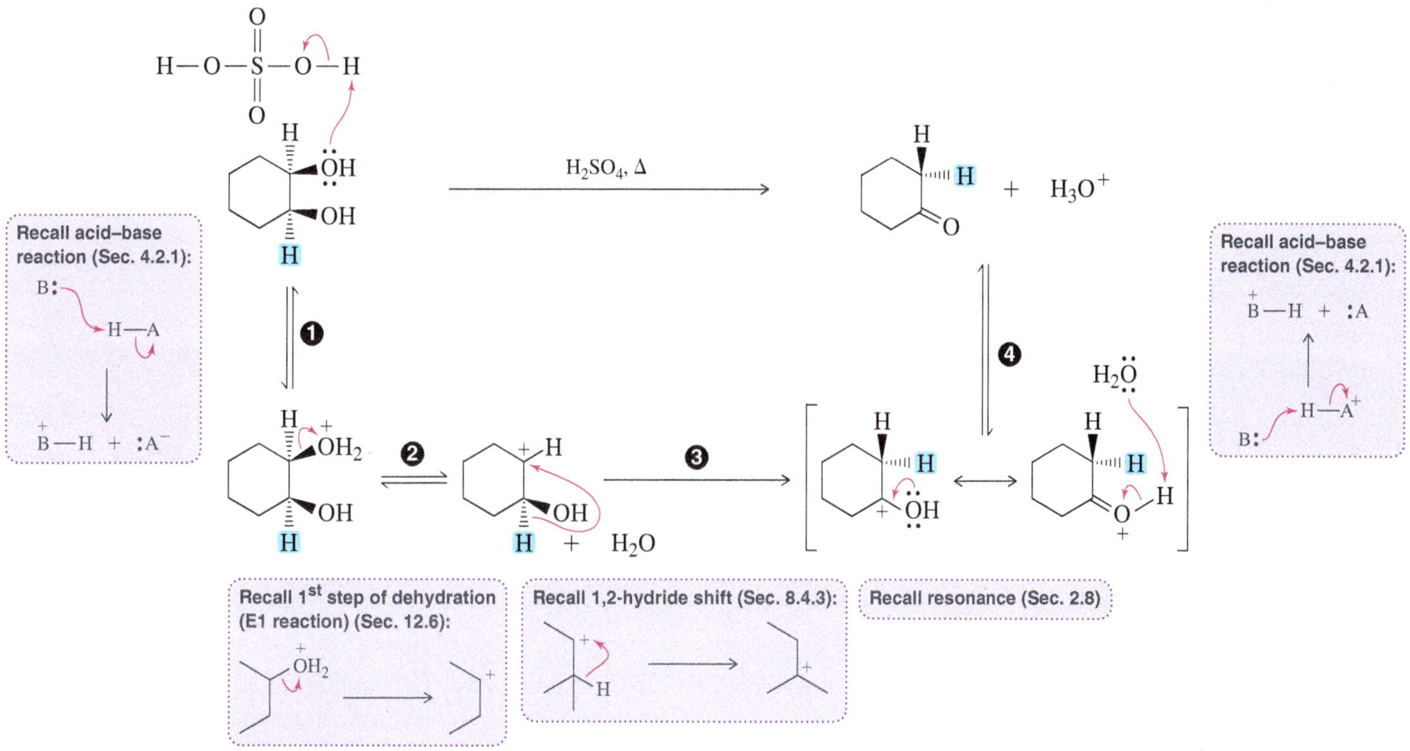

Figure 13.39 Summary of the pinacol rearrangement mechanism.

For pinacol and cyclohexane-1,2-diol, there was no need to choose one particular hydroxyl group to protonate in the first step. When there is a choice, as in the case of the molecule in Figure 13.40, how is this decision made? [*Before proceeding, convince yourself that possibilities A and B give the products shown by drawing a full mechanism for each.*]

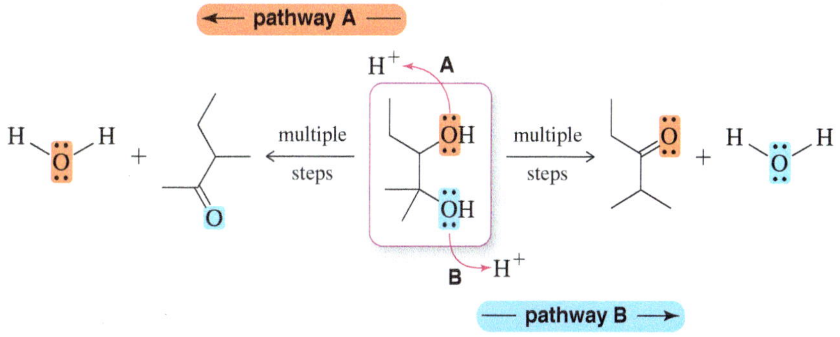

Figure 13.40 Two possible pathways for pinacol rearrangement.

In this situation neither steric nor electronic effects favor protonation of one hydroxyl group over the other. Thus, we must consider an equilibrium where *both* hydroxyl groups can protonate at a nearly equal rate. Once the two hydronium ions are formed, from which will water leave preferentially? The hydronium ion at the more substituted carbon will leave first to form the more substituted, and therefore more stable, carbocation. As shown in Figure 13.41, the 3° carbocation forms preferentially, shifting the equilibrium in that direction via Le Châtelier's principle.

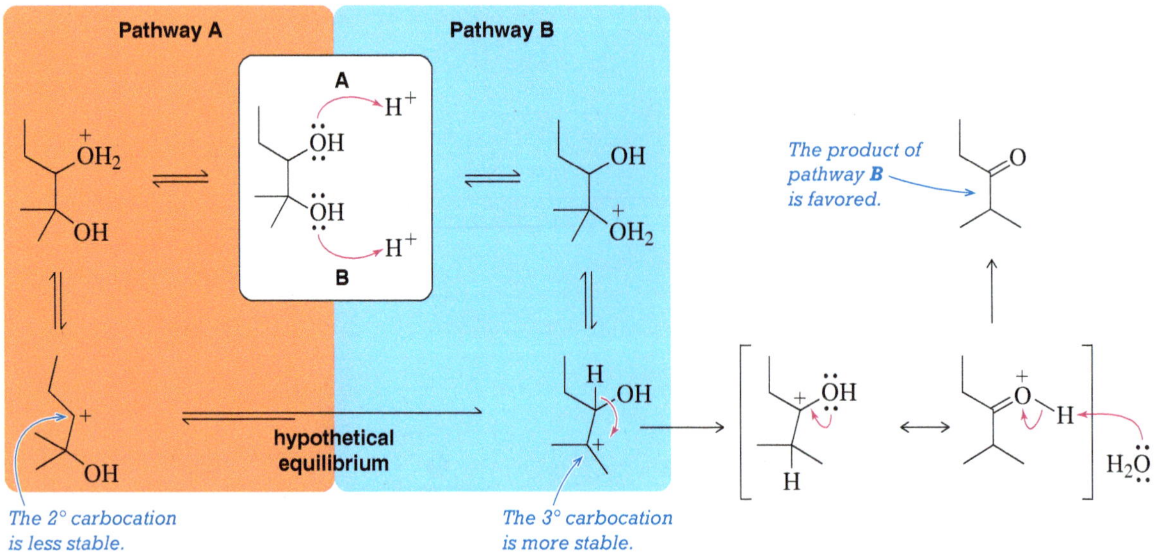

Figure 13.41 Formation of the most stable carbocation determines the outcome of the pinacol rearrangement.

13.8 SUMMARY

- Elimination of water is catalyzed by sulfuric acid. The equilibrium is shifted to the alkene by removal of water from the reaction mixture. The pinacol rearrangement results in elimination of water from a diol to produce a ketone or aldehyde via hydride or alkyl rearrangement.

Assessment 13.45 Show a reaction coordinate diagram for the two processes in Figure 13.41 that rationalizes pathway **B** as the one that gives the major product.

Assessment 13.46 Predict the product of the following pinacol rearrangements.

(a)

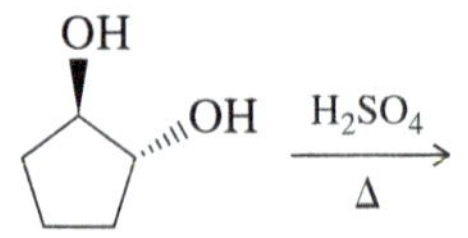

(b)

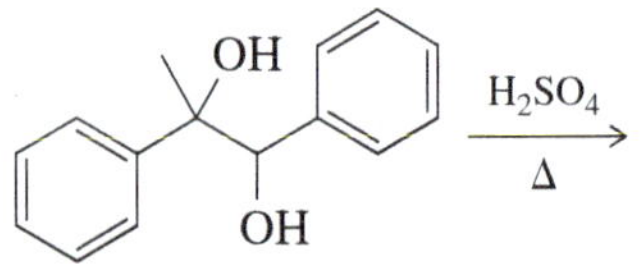

(c)

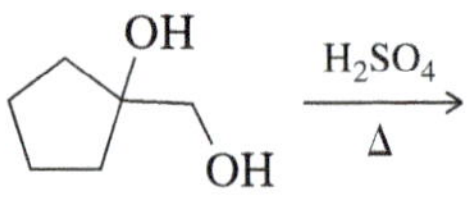

Assessment 13.47 Suggest an alkene that, in two steps, could be converted into each of the following ketones. Each sequence should involve a pinacol rearrangement.

(a)

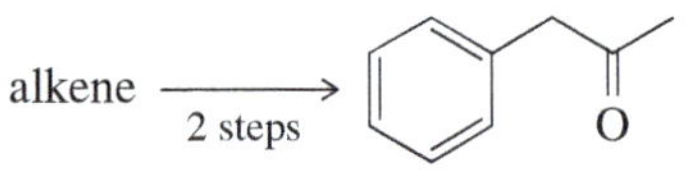

(b)

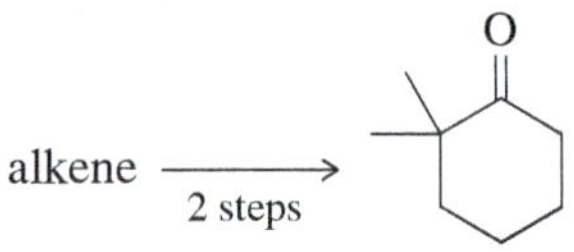

(c)

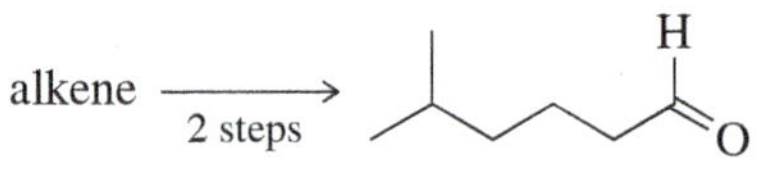

13.9 Oxidation of alcohols

Recall that the highest oxidation state possible for carbon is +4, as in carbon dioxide. [*To review how to calculate oxidation numbers, see Section 9.1.1.*] A 2° alcohol, such as propan-2-ol, with C_2 in the 0 oxidation state, can be oxidized, most often to a ketone (propanone), which has C_2 in the +2 oxidation state (**Figure 13.42**).

Section 13.9.1 explores the oxidation of 2° alcohols to ketones and 1° alcohols to aldehydes and carboxylic acids.

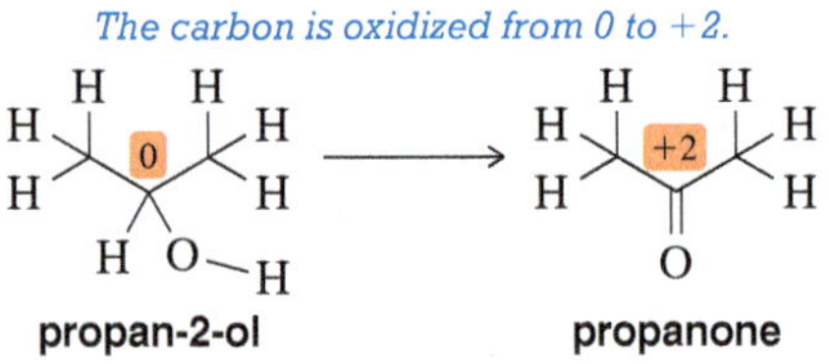

Figure 13.42 Oxidation of propan-2-ol gives propanone.

13.9.1 Oxidation of alcohols: A general mechanism

Though there are many different reagents for oxidizing alcohols to carbonyl compounds, they all have very similar mechanisms.

Let's think like a chemist to "invent" one such reagent, such that this mechanistic commonality becomes self-evident.

Thinking Like a Chemist

Let's first examine the oxidation of propan-2-ol to propanone in **Figure 13.43** by identifying the bonds broken and bonds formed. Can you identify the molecule that is lost as the reaction proceeds from alcohol to ketone? [*It is often helpful to draw in the hydrogens here.*]

Bonds broken:
C—H σ
O—H σ

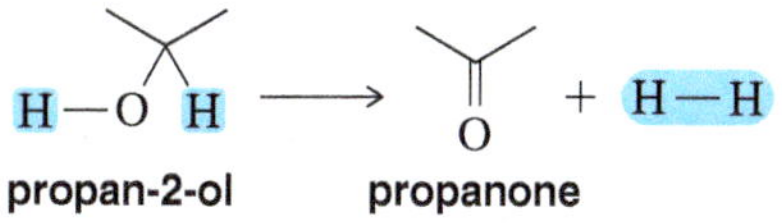

Bonds formed:
C—O π
H—H σ

Figure 13.43 **Identifying the bonds broken and bonds formed** shows that a molecule of hydrogen is lost in the oxidation.

Based on the table of bonds formed and broken in Figure 13.43, the difference between a ketone and an alcohol is two hydrogen atoms or one molecule of hydrogen (H_2). This aligns with our understanding of redox chemistry, where oxidation is thought of as gaining bonds to oxygen and/or losing bonds to hydrogen. With this in mind, we need to come up with a reagent that would remove this H_2. Though there are reagents available that will simply remove H_2 [*those are better talked about in an advanced course*], let's focus on chemistry with which we are familiar. Specifically, can we draw a mechanism where the hydrogen atoms are removed via previously studied elimination mechanisms? [*Draw what this might look like before reading further.*]

Thinking along the lines of an E2 elimination reaction, we might imagine that a base like hydroxide would remove a proton (H^+) from the carbon, pushing the electron pair onto oxygen and kicking off hydride (H^-) as a leaving group. Imagining the reaction this way accomplishes three things. First, we eliminate a molecule of hydrogen ($H^+ + H^- = H_2$). Second, we have formed a C—O π bond. And finally, we have a reasonable mechanism that could allow us to invent a reagent for oxidation, except for one major problem. When we studied E2 reactions in Chapter 12, the leaving group had to be a stable anion—in other words, a good leaving group. Hydride (H^-) is a terrible leaving group, with its conjugate acid (H_2) having a pK_a around 40 (**Figure 13.44**).

A hypothetical, but implausible E2 mechanism:

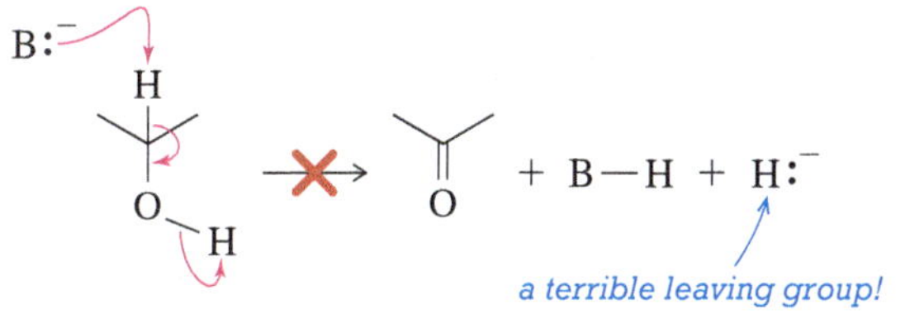

Figure 13.44 **A hypothetical E2 mechanism** cannot occur because hydride is a terrible leaving group.

Although hydride is not a good enough leaving group, the mechanism drawn in **Figure 13.45** would be more plausible if we substituted a good leaving group for it. [*Any ideas what that good leaving group might be?*] Chloride is an excellent leaving group and would make the mechanism significantly more plausible. In this way, HCl would be eliminated, something seen often when studying E2 reactions.

A hypothetical and more plausible E2 mechanism:

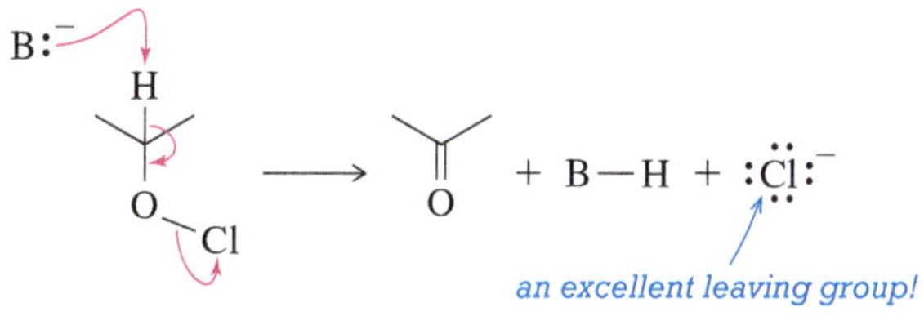

Figure 13.45 **This E2 mechanism is reasonable** because chloride is an excellent leaving group.

To complete the task of inventing a reagent, we need to figure out how to add a chlorine to the oxygen. Because the oxygen of the hydroxyl acts most often as a nucleophile, an electrophilic source of Cl is needed. This is done using the reagent sodium hypochlorite in an acidic solution. [*In acid, NaOCl becomes HOCl.*] **Figure 13.46** shows that protonation of HOCl by acetic acid creates an electrophilic chlorine atom that is attacked by the alcohol, displacing water. Acetate then removes the acidic proton formed when the alcohol attacked. The stage is now set for the E2 reaction in Figure 13.46, with water removing the proton and forming the new C—O π bond. ■

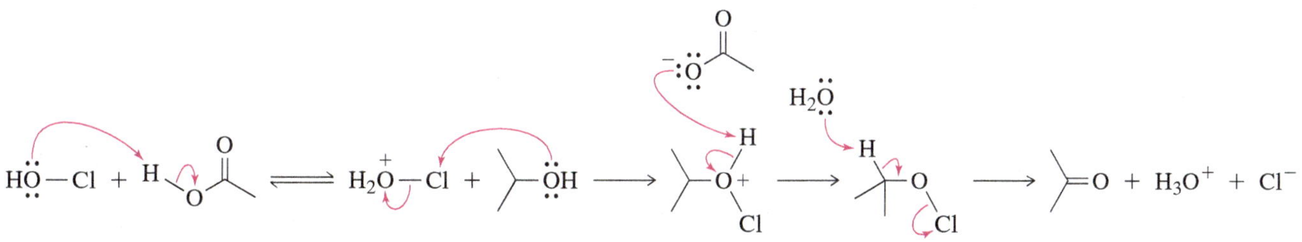

Figure 13.46 **Alcohol oxidation** with hypochlorous acid, from sodium hypochlorite (bleach) in acidic solution.

One purpose of the Thinking Like a Chemist exercise was to identify a reagent for oxidizing alcohols. More important, however, is identification of a common mechanism for alcohol oxidations. Every oxidation reaction we study in Sections 13.9.1–13.9.4 proceeds in a similar fashion (**Figure 13.47**). That is, some good leaving group (X) will be installed at the oxygen, allowing a base in solution to effect the elimination reaction. [*For each oxidation reaction, see if you can identify this common theme.*]

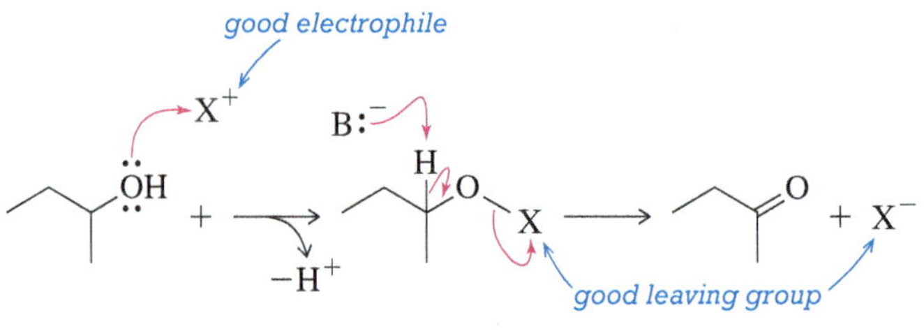

Figure 13.47 The general mechanism of all alcohol oxidations.

Assessment 13.48 The intermediates for the Swern oxidation, a reaction introduced in Section 13.9.4, are shown. Provide the arrow-pushing mechanism that rationalizes the formation of each intermediate and the final product(s).

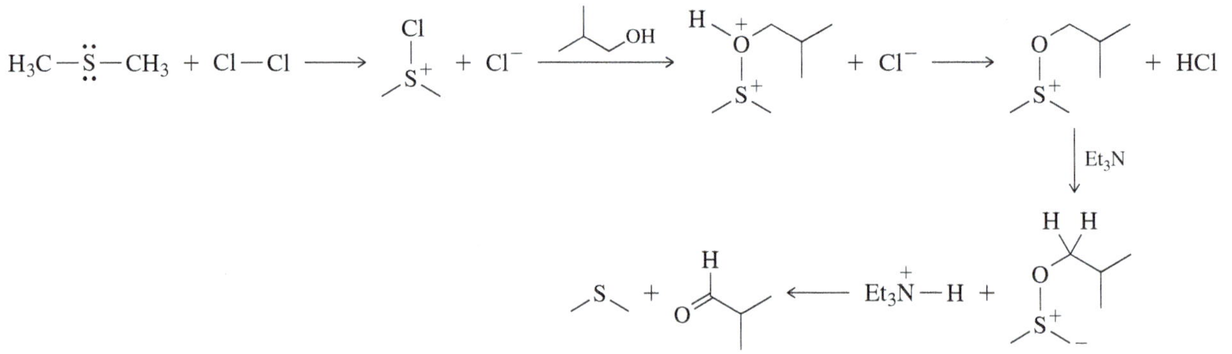

Assessment 13.49 In Assessment 13.48, identify the individual step that represents the general mechanism for all alcohol oxidation reactions.

Assessment 13.50 Every oxidation is accompanied by a reduction. Identify the species that is reduced in the Swern oxidation in Assessment 13.48.

Assessment 13.51 Identify the alcohols that would undergo oxidation to produce the following carbonyl compounds.

(a) (b) (c) (d)

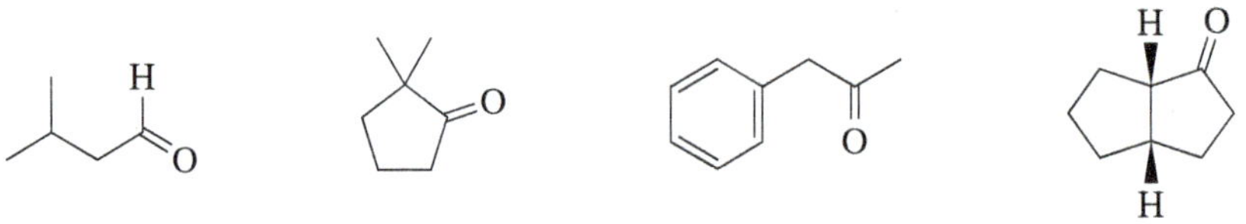

13.9.2 Chromium oxidations of 2° alcohols

A common method for oxidizing alcohols (despite its negative environmental effects) involves the use of chromium reagents. Chromium, a transition metal, can exist in a variety of oxidation states, with +6 and +3 being most stable. In reagents like chromic acid and pyridinium chlorochromate (PCC) (**Figure 13.48**), the chromium has a +6 oxidation state. A high oxidation number indicates that it will serve as an oxidizing agent. The driving force for chromium oxidations is the reduction of chromium(VI) to chromium(IV) and eventually to the more stable chromium(III). The only difference between the two reagents is the solvent used with each. Chromic acid is a highly polar acid, so it is formed and used as an aqueous solution. PCC is a slightly less polar salt, so it is used in dry [*meaning, no water is present*] dichloromethane. [*We will see the consequences of these conditions when we study the oxidation of 1° alcohols.*]

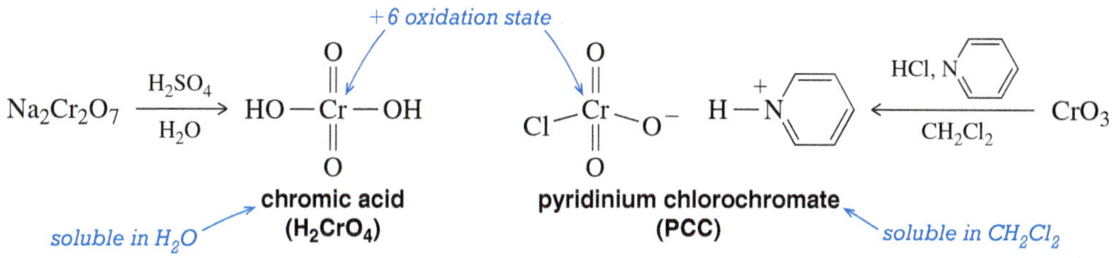

Figure 13.48 Chromic acid is used as an aqueous solution while PCC can be used in dry organic solvents.

Figure 13.49 shows that a ketone is produced with 2° alcohols, regardless of which chromium reagent is used. Mechanistically, the nucleophilic alcohol attacks the electron-deficient chromium of chromic acid, displacing a molecule of hydroxide that quickly deprotonates the hydroxyl proton. [*Forming hydroxide in an acidic solution should normally be avoided. We'll learn later how this reaction really works.*] This chromate ester is the key intermediate for oxidation because it contains an excellent leaving group attached to the oxygen. [*Why is it an excellent leaving group?*] Chromium, in a +6 oxidation state, wants a pair of electrons so it can be reduced to a lower oxidation state. [*Note the similarity between chromium and chloride as good leaving groups.*] With such an excellent leaving group on oxygen, the proton on C_2 is quite acidic, allowing deprotonation by water in an elimination reaction [*likely by E2*] to produce the ketone. This is identical to the elimination reaction using chlorine and hydroxide [*see Thinking Like a Chemist in Section 13.9.1*].

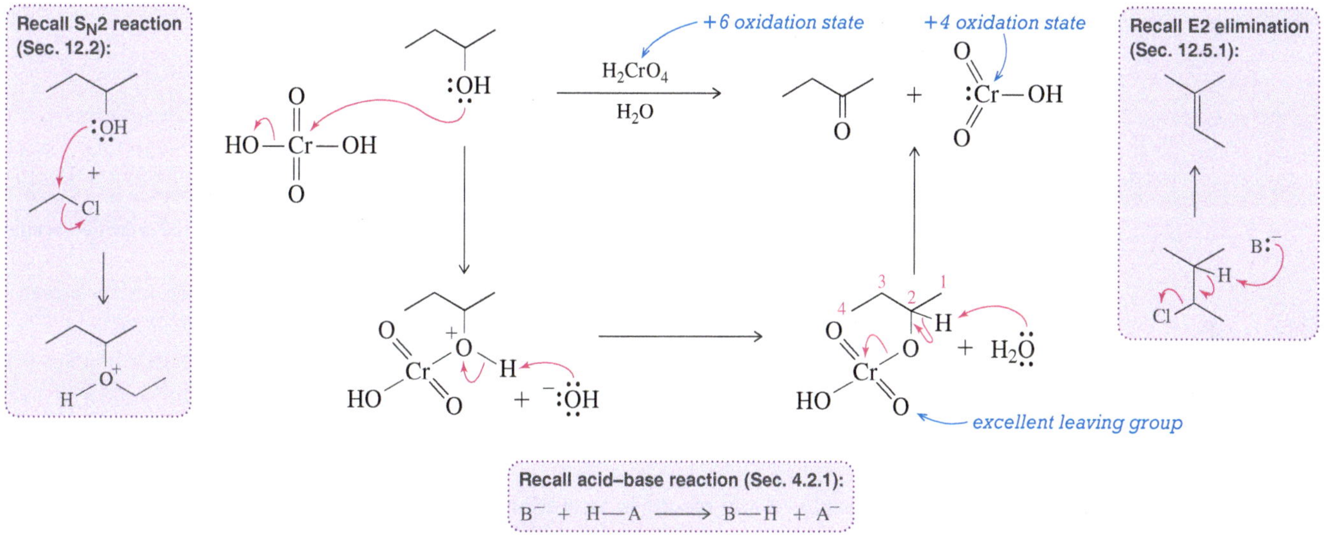

Figure 13.49 The mechanism of the chromic acid oxidation of a 2° alcohol.

Assessment 13.52 Predict the product of the oxidation reactions shown.

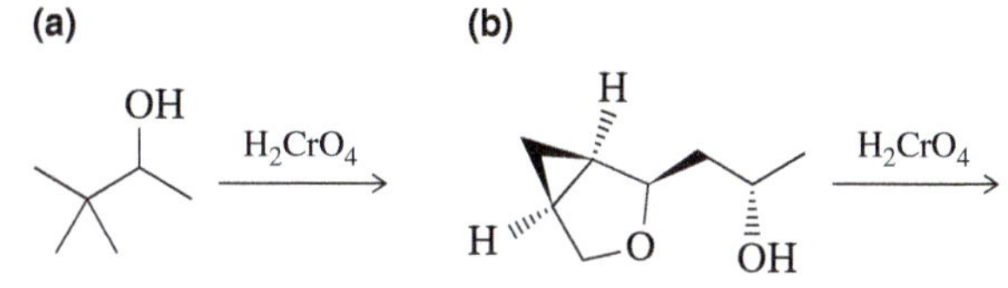

Assessment 13.53 The intermediates for the oxidation of isopropanol to acetone are shown. Calculate oxidation numbers for the indicated atoms, then use them to determine in which step oxidation occurs.

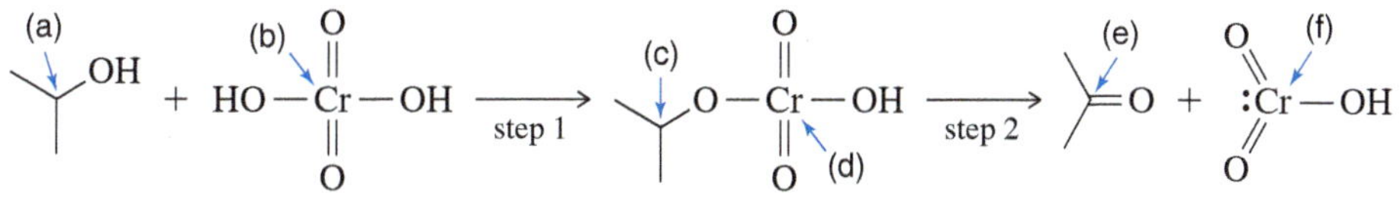

13.9.3 Chromium oxidations of 1° alcohols

By an identical mechanism, 1° alcohols can be oxidized to a carboxylic acid using chromic acid and to an aldehyde using PCC (Figure 13.50). The difference in reaction outcome, often attributed to a difference in oxidizing agent strength, is actually a result of the conditions employed with each.

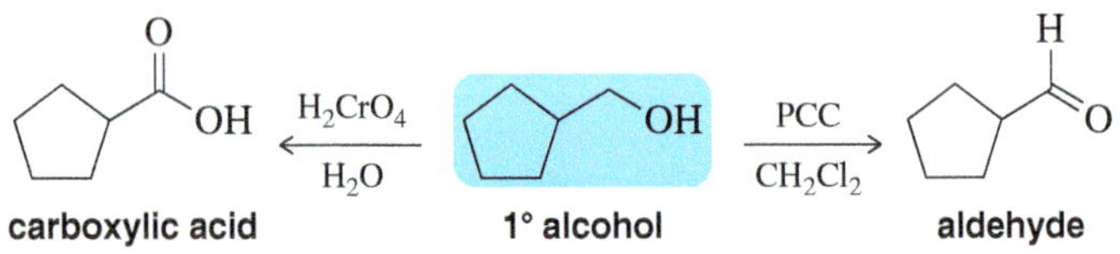

Figure 13.50 **PCC and H_2CrO_4** react with 1° alcohols to give different products.

Chromic acid first oxidizes a 1° alcohol to an aldehyde through the mechanism shown in Figure 13.51. Formation of the chromate ester is followed by elimination to form the C—O π bond. Once formed, in the presence of water, an unfavorable equilibrium is set up between the aldehyde and the hydrate of the aldehyde. Though the equilibrium favors the aldehyde form, there is nevertheless always a small amount of hydrate present. [*This particular reaction is described in detail in Chapter 17. For now, just know that there is some hydrate present when an aldehyde is dissolved in water.*] The hydrate meets the two requirements for oxidation: a nucleophilic hydroxyl group and a hydrogen that can be eliminated. As such, the hydrate is oxidized further, once again by formation of the chromate ester and E2 elimination to form the C—O π bond. By further oxidizing the hydrate of the aldehyde, the aldehyde to hydrate equilibrium is shifted to the hydrate until it has all been converted to the carboxylic acid.

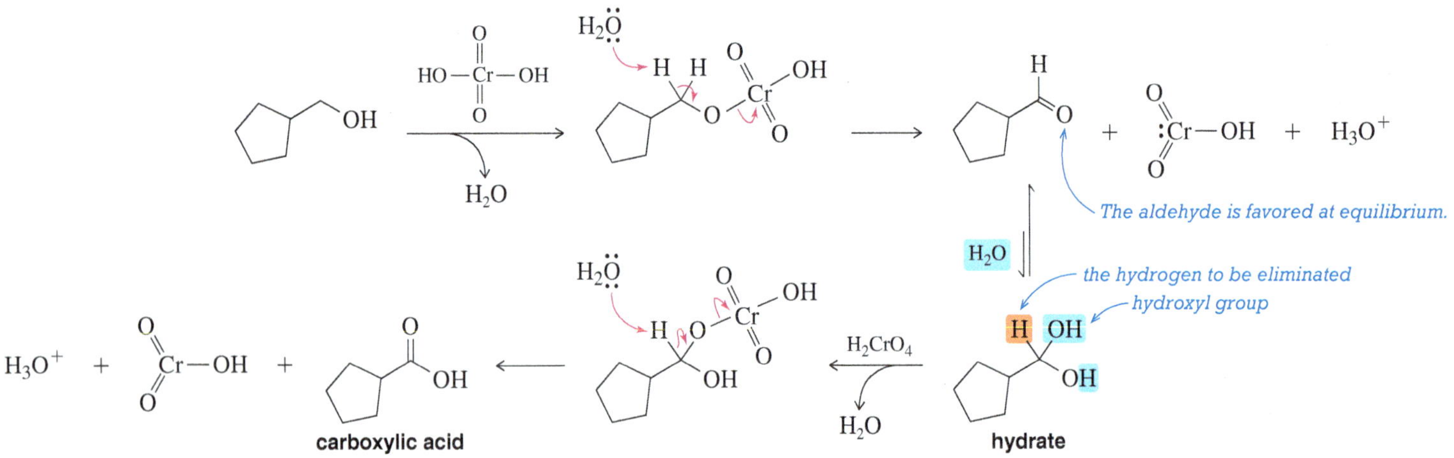

Figure 13.51 **Oxidation of alcohols with chromic acid in water results in formation of a carboxylic acid.**

Assessment 13.54 Tertiary (3°) alcohols are *not* oxidized by chromic acid. Why?

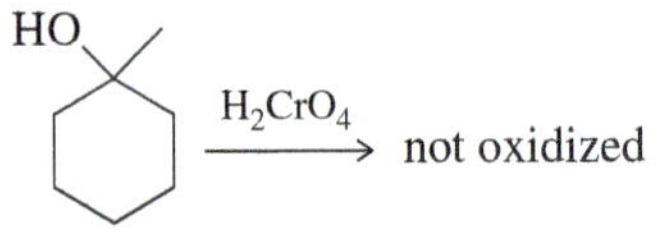

Assessment 13.55 Though ketones, like aldehydes, are in equilibrium with a hydrated form, they cannot be further oxidized. Why?

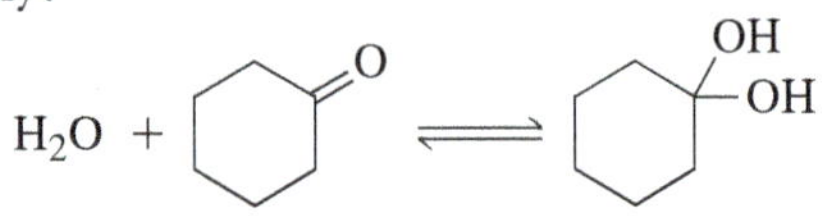

Assessment 13.56 If there is no water present, the hydrate of an aldehyde cannot form. Could an aldehyde itself (not the hydrate) be oxidized to a carboxylic acid? Why?

Chromium has a unique ability to shuffle oxidation states, regenerating the reactive +6 oxidation state.

The ability to synthesize carboxylic acids from 1° alcohols is a valuable synthetic tool. [*Make sure you add this to your reaction sheets (Section 10.7)—namely, "Reactions of alcohols" and "Preparation of carboxylic acids."*] Aldehydes are also important synthetic intermediates, however, so there would be value in being able to stop this reaction at the aldehyde. Unfortunately, this can't be done even if just one equivalent of chromic acid is used.

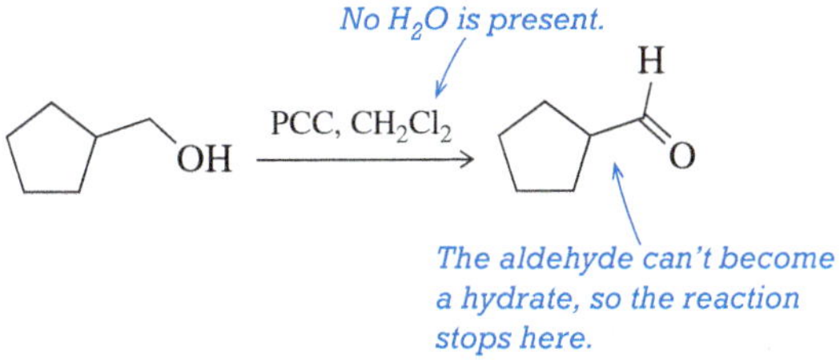

Figure 13.52 **Oxidation of 1° alcohols** stops at the aldehyde when there is no water present.

How, then, can this reaction be stopped at the aldehyde? [*Consider the three requirements mentioned previously before making a guess.*] There will always be an excess of oxidizing agent present after the aldehyde is formed, so the only way to stop the reaction there is to ensure that the hydrate cannot form. The aldehyde itself does not have a nucleophilic hydroxyl group, so it cannot attack the electrophilic oxidizing agent. Hydrate formation is prevented by using an oxidizing agent that does not require water as the solvent. Enter PCC! Because PCC is soluble in dry CH_2Cl_2, no water is present, in which case the hydrate does not form and the reaction stops at the aldehyde (**Figure 13.52**).

Assessment 13.57 Fill in the missing reactant, reagent, or product for each of the following oxidation reactions.

(a)

(b)

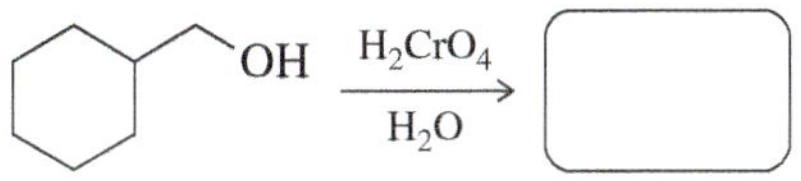

(c)

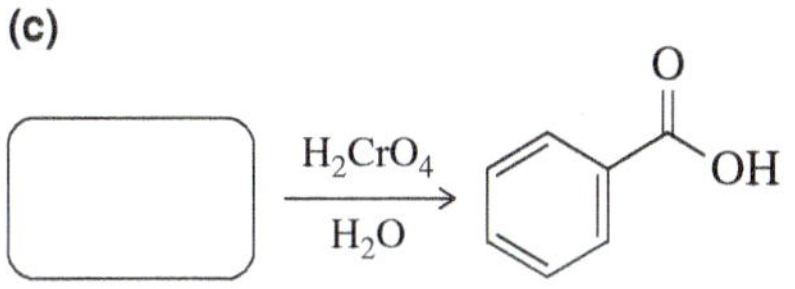

(d)

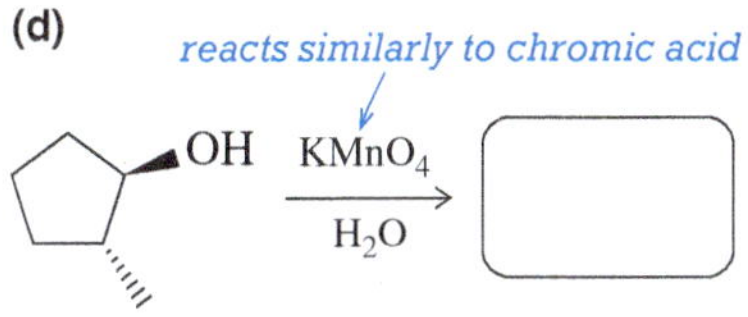

(e)

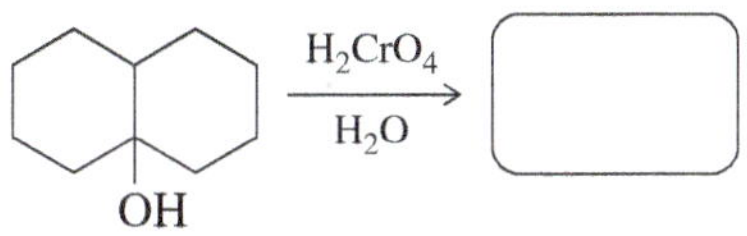

(f)

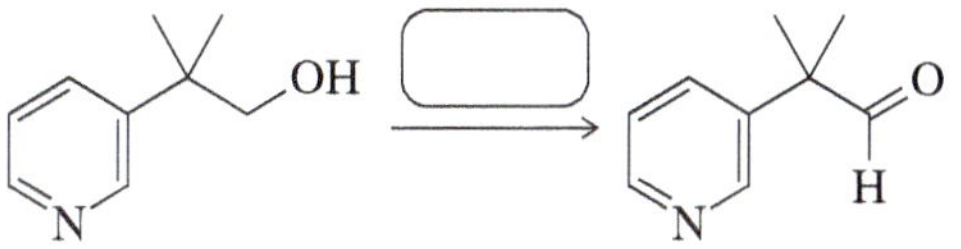

13-A CHEMISTRY IN CONTEXT: Green Chemistry

Chromium(VI) and cancer

Green Chemistry Principle 4: Minimize toxicity by molecular design.

Chromic acid is an example of what is referred to as a hexavalent chromium species because the chromium is in the +6 oxidation state. These compounds have been shown to cause a variety of stomach and intestinal cancers in animals. As depicted in the movie *Erin Brockovich* (2000), Pacific Gas and Electric (PG&E) had to settle various lawsuits because it was found negligent in contaminating the groundwater with this toxic chemical near the town of Hinkley, California. In addition to its use as an alcohol oxidizing agent, hexavalent chromium is used as an anticorrosion additive, which is how it found its way into the Hinkley aquifer.

Because the chromate ion (CrO_4^{2-}) is similar in size and shape to the sulfate (SO_4^{2-}) and phosphate (PO_4^{3-}) ions, it is easily absorbed by cells through nonspecific transport channels. Once in the cell, chromate is reduced to chromium(III) through various redox reactions. In this complex process, reactive oxygen species (ROS) are generated. One such ROS is the hydroxyl radical (•OH), a molecule that can damage DNA strands. This damage can induce mutations in DNA, promoting unregulated cell growth and eventual tumor growth.

As a result of the known link between chromium(VI) and cancer, the use of chromic acid (and PCC) as an oxidizing agent is strongly discouraged, in spite of the efficiency with which it promotes oxidation. This has prompted chemists to search for environmentally benign oxidizing agents like NaOCl (household bleach). The oxidizing agents covered in Section 13.9.4, while not carcinogenic, are by no means green, for reasons that are discussed later.

Assessment 13.58 In Hinkley, the cleanup continues today. In one process PG&E is using to mitigate the chromium(VI) toxicity, ethanol is injected into the soil. How would this reduce the chromium(VI) contamination?

13.9.4 Alternate alcohol oxidation

Because chemists want to make a diversity of molecules, it is important to have a robust synthetic toolbox with multiple ways to achieve the same transformation. The three reactions in this section (Corey–Kim oxidation, Swern oxidation, and the Dess–Martin oxidation) are shown in **Figure 13.53**. They are useful because they proceed efficiently, under mild conditions, and do so with environmentally inert reagents (as compared to the toxic chromium reagents).

Figure 13.53 Alternate alcohol oxidations.

Corey–Kim oxidation:

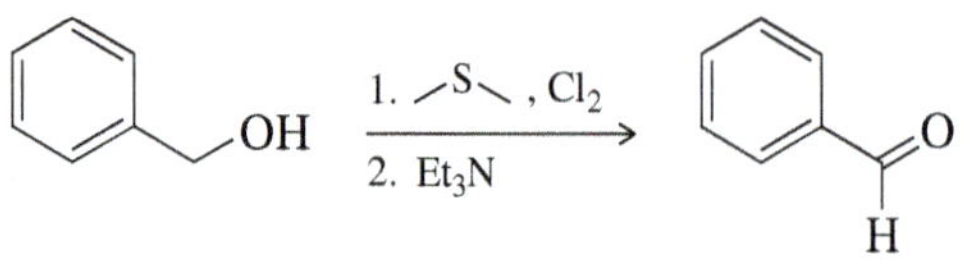

Swern oxidation:

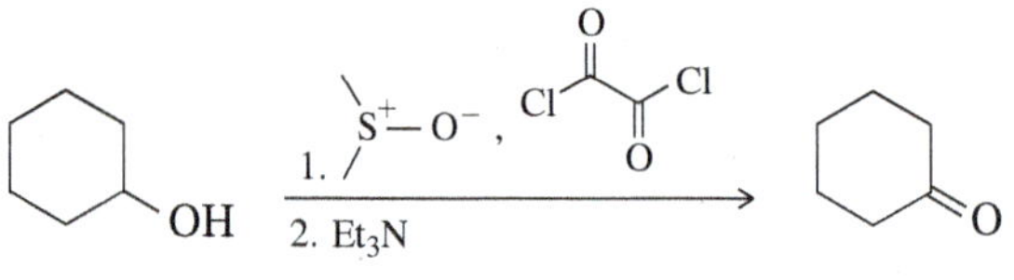

Dess–Martin oxidation:

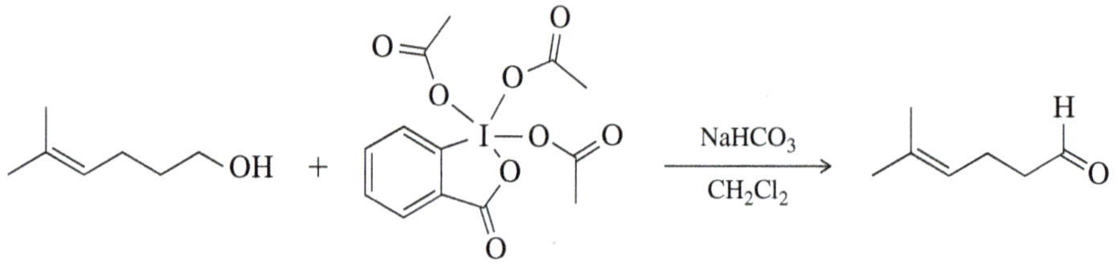

Like the alcohol oxidations studied thus far, these feature the formation of an intermediate containing a good leaving group attached to the hydroxyl oxygen, setting up the elimination reaction to form the C—O π bond (**Table 13.9**). The only difference is the manner in which the hydroxyl group becomes attached to the leaving group. [*It is probably not necessary that you learn all of these mechanisms, so check with your instructor. But you should be able to recognize the common features in all of them.*]

Table 13.9 Alternate oxidation mechanisms have the same key step.

Key step/General mechanism:	
	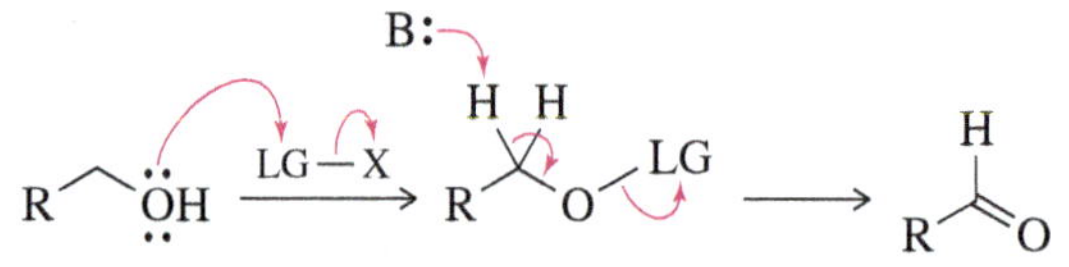
Reaction	**Formation of LG—X (screened in green) and mechanism**
PCC	LG—X is the reagent itself 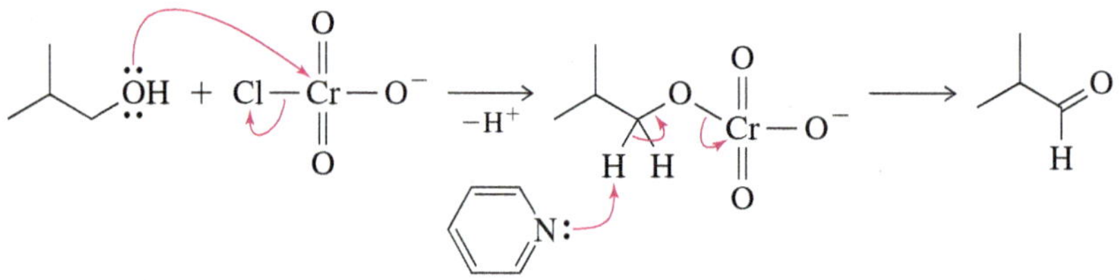
Corey–Kim	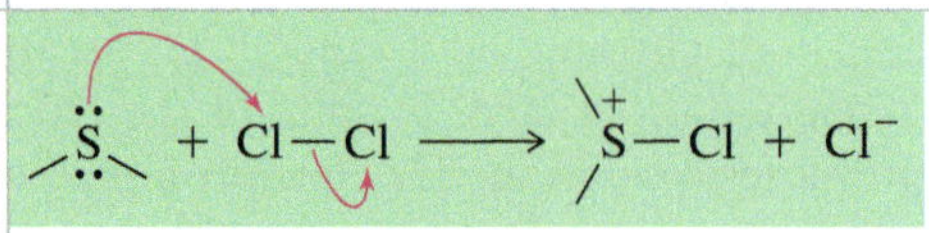 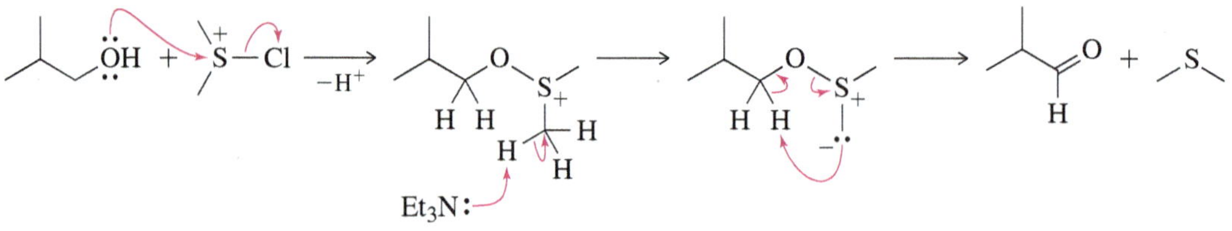

(continued)

Table 13.9 ***Continued***

Reaction	Formation of LG—X (screened in green) and mechanism
Swern	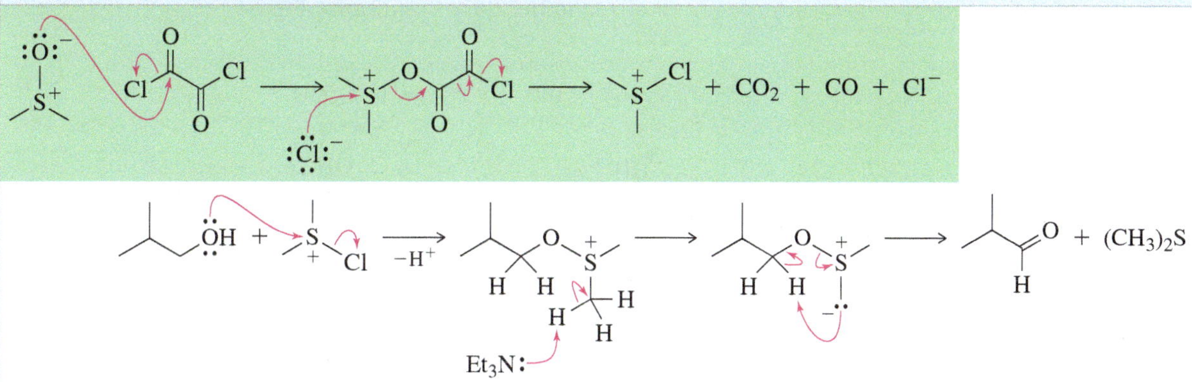
Dess–Martin	LG—X is the reagent itself 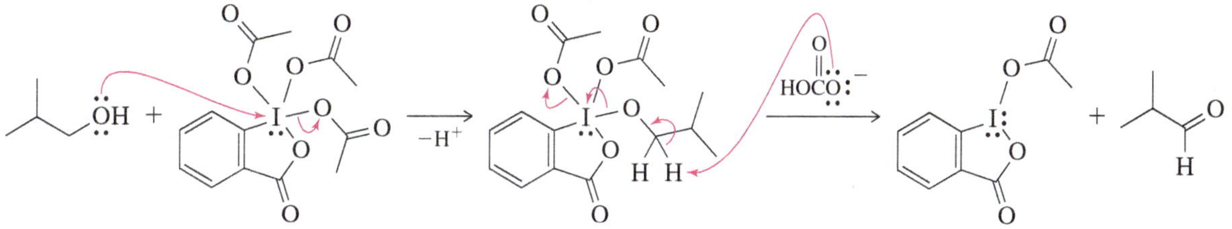

There is no water in these reactions, so 1° alcohols are oxidized to aldehydes under these conditions and 2° alcohols are oxidized to ketones. Tertiary alcohols are not oxidized because they lack the necessary C—H bond.

13.9.1–13.9.4 SUMMARY

- 1° alcohols are oxidized to aldehydes (PCC, CH_2Cl_2) or carboxylic acids (H_2CrO_4, H_2O) depending on the presence or absence of water. 2° alcohols, regardless of reagent used, are oxidized to ketones. The mechanism of all alcohol oxidations is similar, involving formation of an intermediate that undergoes an elimination reaction to form a new C—O π bond.

Assessment 13.59 Predict the product of the following oxidation reactions.

(a)

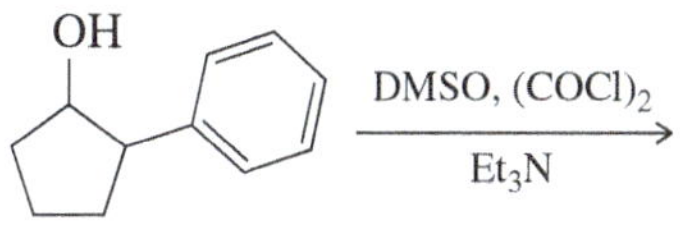

(b)

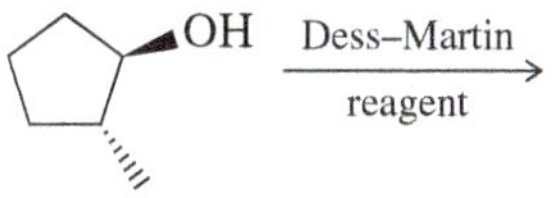

(c)

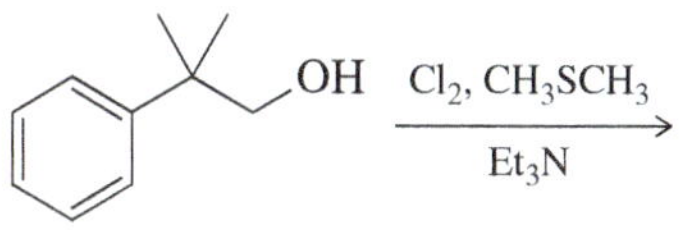

(d)

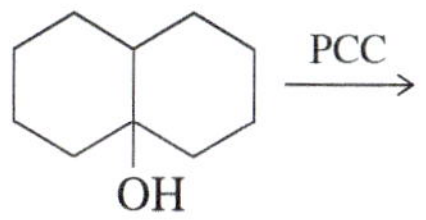

Assessment 13.60 δ-Hydroxyaldehydes are in equilibrium with their hemiacetal form. [*We discuss hemiacetals in Chapter 17.*] Predict the product that would form upon treatment of the hemiacetal with Dess–Martin periodinane.

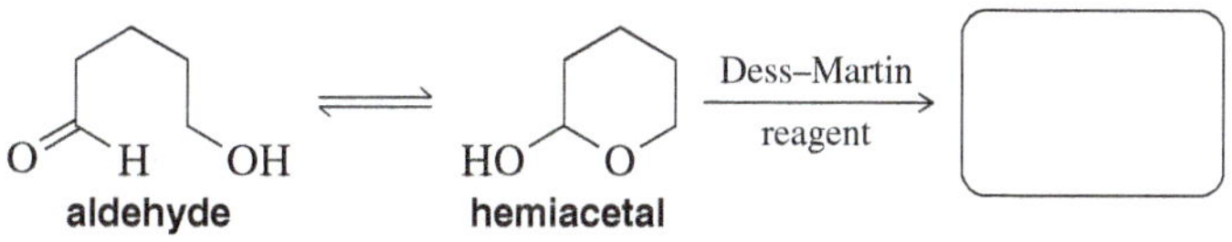

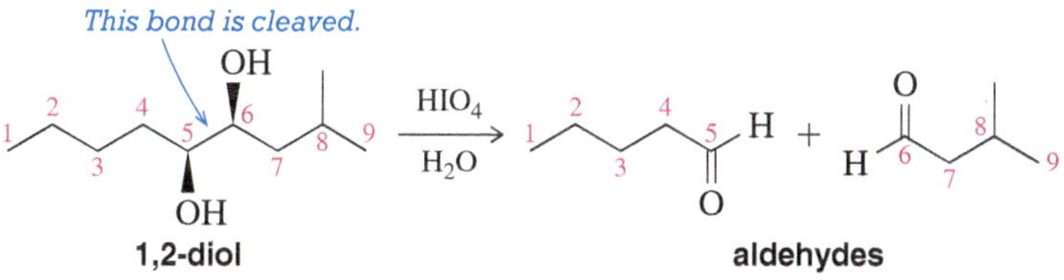

Figure 13.54 **Periodic acid cleaves diols to carbonyl compounds.**

13.9.5 Periodic acid oxidation of diols

1,2-Diols, often referred to as **glycols** (or **vicinal** diols), undergo a special oxidation reaction with periodic acid to produce two aldehydes via an oxidative cleavage reaction (**Figure 13.54**). In essence, the C—C bond between the two hydroxyl groups is cleaved in a reaction analogous to ozonolysis (Section 9.1.7).

When applied to cyclic systems, only syn diols undergo cleavage, which implicates the formation of a cyclic periodate diester. Because trans-diols are *un*able to form this intermediate, they are *not* oxidized under these conditions (**Figure 13.55**).

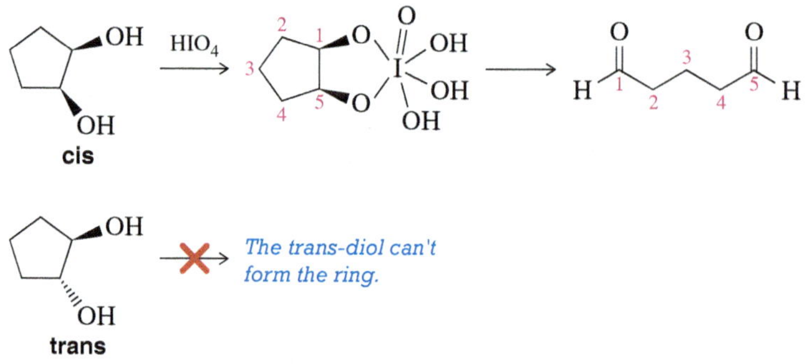

Figure 13.55 Trans-diols on a ring can't be cleaved.

The mechanism is shown in **Figure 13.56**. It begins with nucleophilic addition of the alcohol to the electrophilic iodine, concurrent with breaking of the I—O π bond. [*This reaction is discussed in detail in Chapter 17.*] A proton transfer completes formation of the periodate monoester. A second addition reaction and proton transfer generate the cyclic diester. Cleavage via a 6-electron cyclic transition state results in formation of the dialdehyde. [*The 6-electron cyclic (or aromatic) transition state is common in organic chemistry. In Chapter 22, we explain why it is so stable.*]

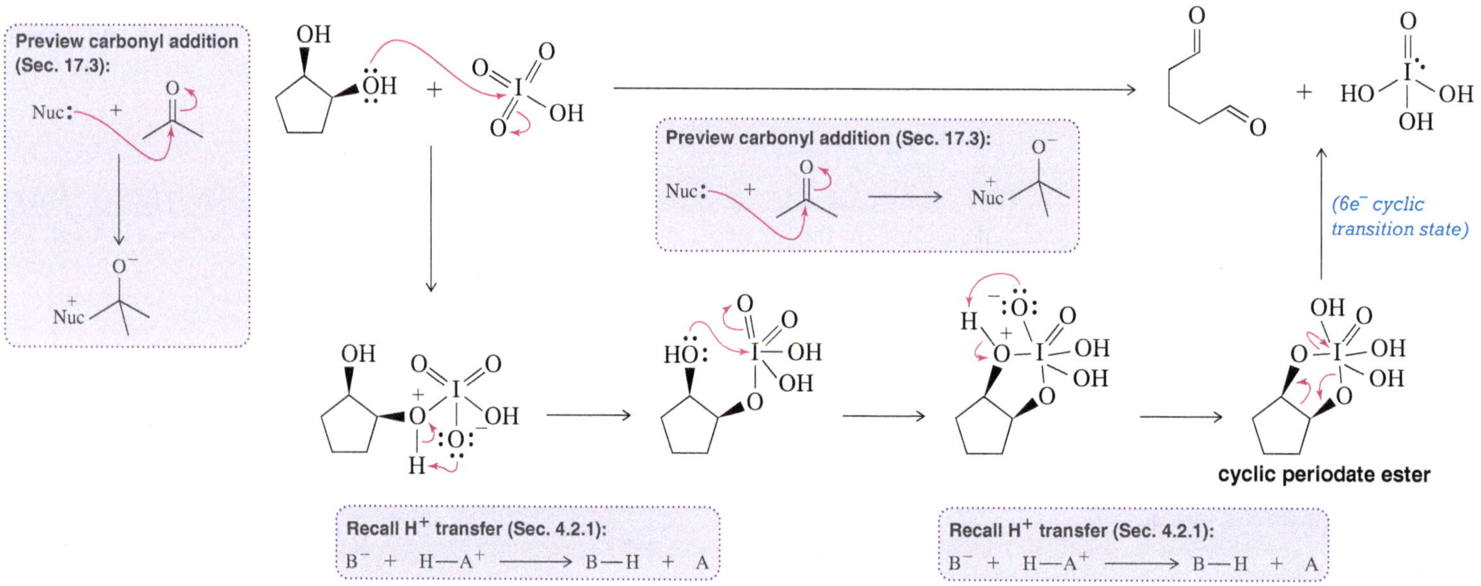

Figure 13.56 The mechanism of periodic acid cleavage.

Used in conjunction with the dihydroxylation of alkenes with OsO_4 (see Section 9.1.5), this reaction provides an alternative to alkene ozonolysis for forming aldehydes and ketones (**Figure 13.57**).

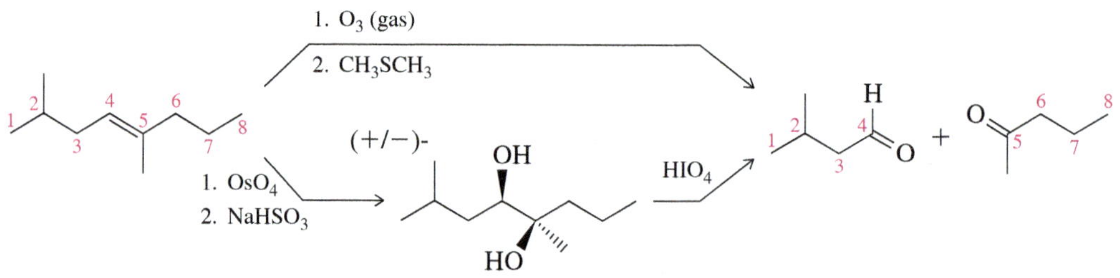

Figure 13.57 The dihydroxylation/periodic acid cleavage sequence with alkenes gives the same products as ozonolysis.

Assessment 13.61 Predict the product of the following reactions.

(a)

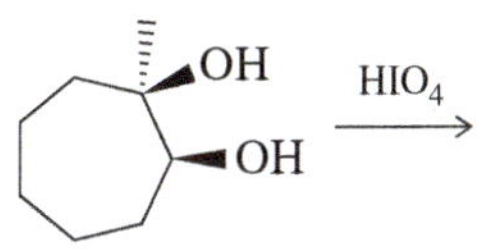

(b)

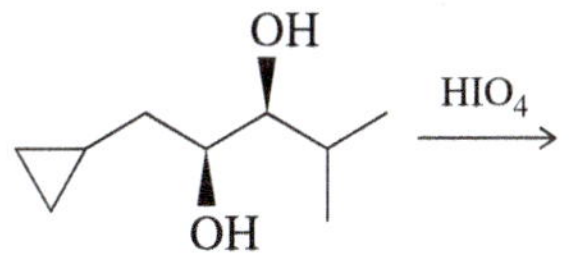

Assessment 13.62 Although trans-diols in rings cannot be cleaved using HIO_4, acyclic trans-diols can. Explain this discrepancy.

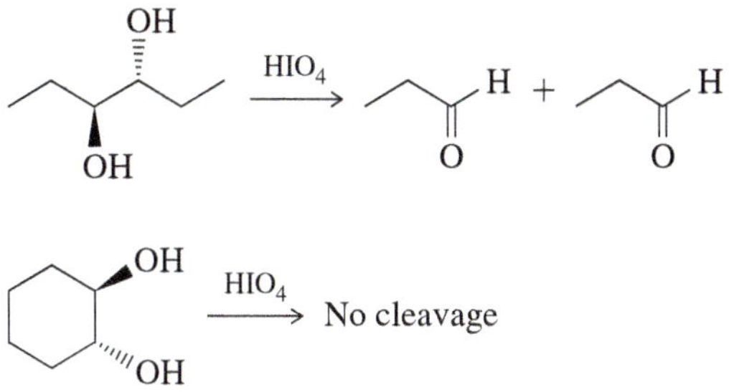

Assessment 13.63 Propose a synthesis of the carbonyl(s) using the (i) ozonolysis and (ii) dihydroxylation/periodic acid cleavage pathways.

(a)

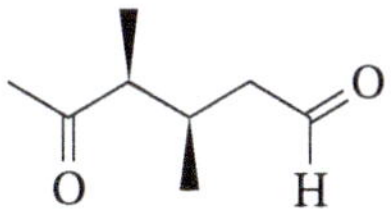

(b)

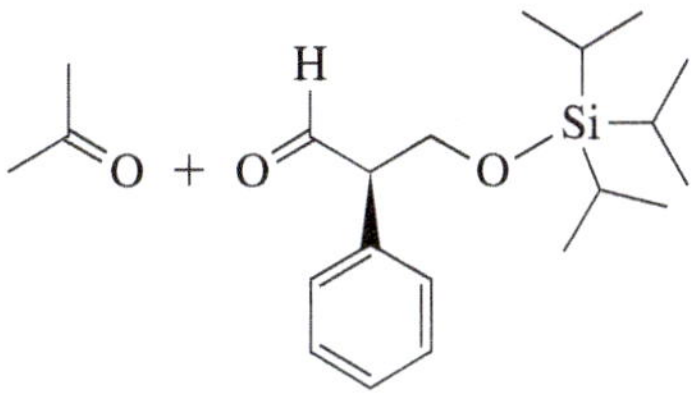

13-B CHEMISTRY IN CONTEXT: Green Chemistry

Catalysis

Green Chemistry Principle 4: Minimize toxicity by molecular design.

Green Chemistry Principle 9: Use catalytic instead of stoichiometric reagents in reactions.

In Section 13.9.5, we discussed the two-step process of alkene dihydroxylation and periodic cleavage to generate new carbonyls. Though very powerful, and often the better alternative to ozonolysis, it is a problematic reaction by the metrics of green chemistry. Osmium tetroxide is one of the most toxic substances known. For comparison, the Occupational Safety and Health Administration (OSHA) sets a permissible exposure limit (PEL) for mercury (an element that strikes fear into all nonchemists) at 1 mg/L. Osmium tetroxide, on the other hand, has a PEL of 0.002 mg/L, suggesting that it is nearly 1000 times more toxic than mercury.

The toxicity of OsO_4, though, is not really an issue for the general public or the environment itself. First, it is extremely expensive and thus is not widely used in large-scale industrial processes. Second, it readily degrades to harmless by-products if released into the environment. The major issue, then, is for the direct user of the chemical—in our case, the chemist running the reaction.

To protect the chemist, processes have been developed that make use of catalytic (sub-stoichiometric) amounts of OsO_4. This is accomplished by using a co-oxidant to continually regenerate OsO_4 for reaction with the alkene, as shown in Figure 13.58. One example of a co-oxidant is *N*-methylmorpholine-*N*-oxide (NMO). NMO, used at stoichiometric levels, is reduced to *N*-methylmorpholine in this process.

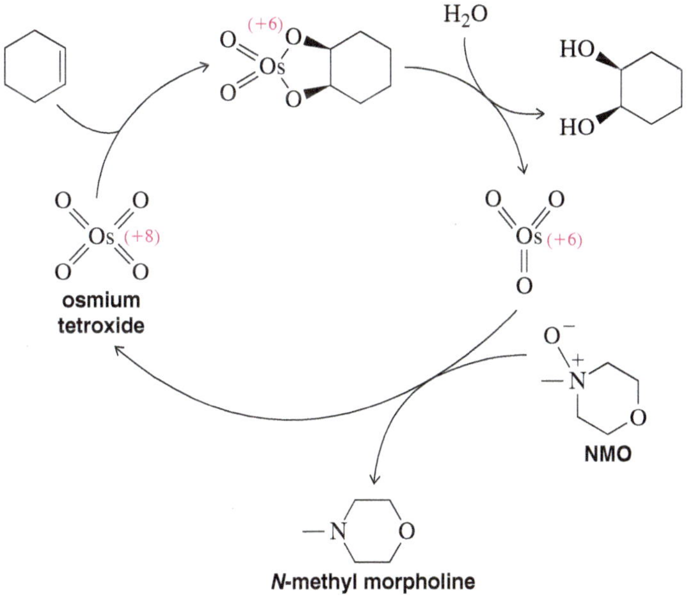

Figure 13.58 Mechanism of catalytic dihydroxylation with a co-oxidant.

Assessment 13.64 Unfortunately, the use of NMO creates an additional green chemistry problem. What is the problem and how might it be solved? [*Hint: See Chemistry in Context 13-E.*]

13.10 Ethers

Closely related to the alcohol functional group is the **ether**, which contains an oxygen bound to two alkyl groups (general structure R—O—R). This structural relationship is shown in Figure 13.59, where replacing one hydrogen of water with an alkyl group makes an alcohol, and replacement of the second hydrogen with an alkyl group makes an ether. [*The arrows in Figure 13.59 aren't meant to represent a synthetic scheme, but they do hint at how you might make an ether.*] The **thioether** functional group can be represented in a similar way, replacing the oxygen of water, an alcohol, and an ether with a sulfur atom to give hydrogen sulfide (H_2S), a thiol (RSH), and a thioether (R—S—R′).

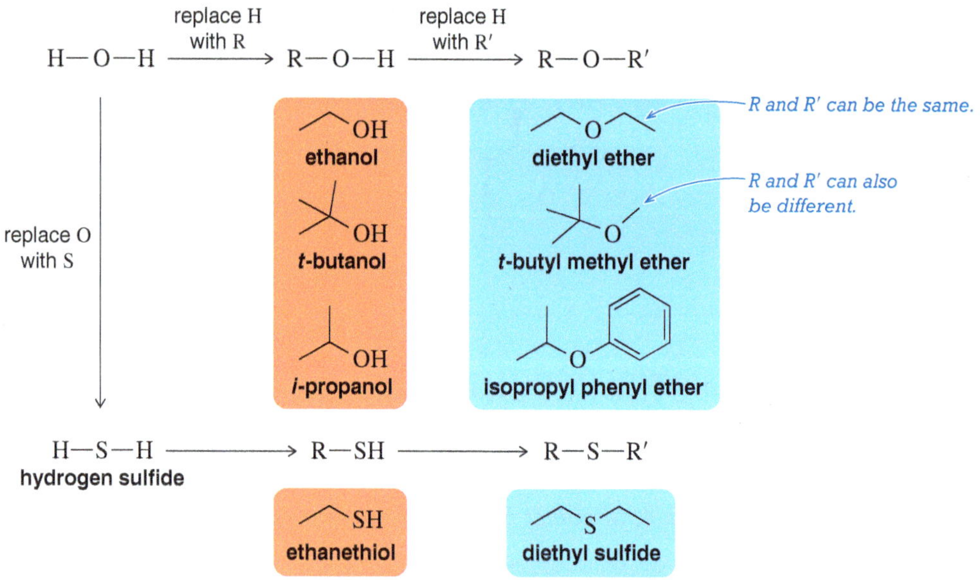

Figure 13.59 The relationship between water (hydrogen sulfide), alcohols (thiols), and ethers (thioethers). R and R′ may be the same, as in diethyl sulfide, or they may be different, as in *t*-butyl methyl ether.

13.10.1 Nomenclature of ethers

The nomenclature of ethers is relatively straightforward. Simple ethers are named according to the formula "[alkyl group 1] [alkyl group 2] ether" with the alkyl groups listed in alphabetical order [Figure 13.60(a)]. When the alkyl groups on either side of oxygen are the same, use the prefix di- to indicate that there are two of the same group. For more complex molecules containing an ether functional group on a large or heavily functionalized parent chain, the smaller —OR group can be named simply as an alkoxy group on the chain [Figure 13.60(b)]. Some other common ether names are shown in Figure 13.60(c).

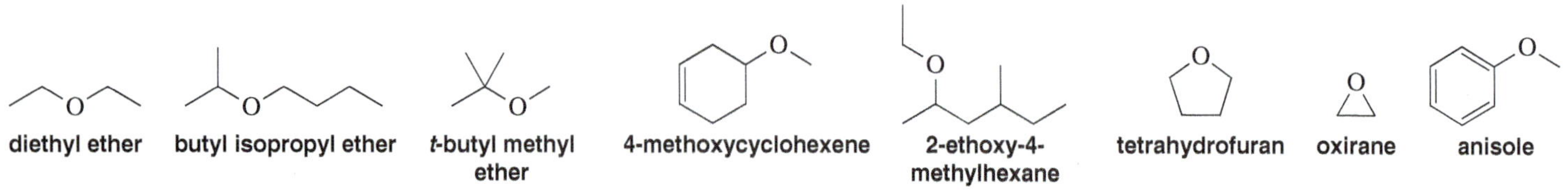

Figure 13.60 Nomenclature of ethers: (a) simple, (b) complex, and (c) common names.

13.10.2 Ethers as good solvents

Epoxides are a specific type of ether that are useful synthetic intermediates (see Section 13.13), but the principal use of ethers in organic chemistry is as reaction solvents. The purpose of a reaction solvent is to bring reactants and reagents together into a single phase so that collisions can occur. Four major factors make ethers excellent solvents.

1. **Intermediate polarity**
Most of the organic molecules that we study are relatively nonpolar. [*Flip through some previous chapters to confirm this.*] And yet many of the reagents, especially the salts used in the substitution reactions of Chapter 12, are polar. Thus, it is important for a solvent to be able to dissolve both polar and nonpolar species. Diethyl ether (or ethyl ether), for example, falls right in the middle of the two extremes (water and benzene) of the dipole moment scale (**Table 13.10**). As such, it makes for an excellent solvent choice. Diethyl ether and tetrahydrofuran (a cyclic ether) are polar aprotic solvents that are often used in substitution reactions. [*See Chapter 12.*]

Table 13.10 Dipole moment of ethers.

Solvent	μ (debye)
Water	1.86
Ethanol	1.69
Tetrahydrofuran	1.63
Dichloromethane	1.6
Diethyl ether	1.15
Benzene	0

2. **Boiling point variability**
Some reactions require higher temperatures to overcome activation energies. The easiest way to heat a reaction to a particular temperature is to heat it to the solvent's boiling point. The temperature at which a solvent boils is often referred to as the **reflux temperature**. The commonly used ether solvents offer a variety of reflux temperatures (**Table 13.11**). As expected, larger and/or more polar ethers boil at higher temperatures.

Table 13.11 Boiling point of ethers.

Solvent	Structure	Boiling point (°C)
Diethyl ether	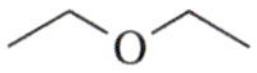	35
t-Butyl methyl ether		55
Tetrahydrofuran		65
DME	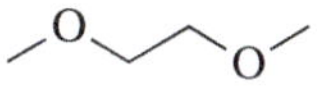	83
1,4-Dioxane	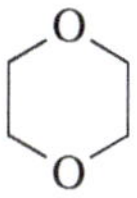	101
Anisole	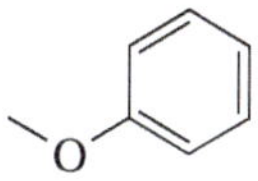	154

3. **Hydrogen bond acceptors**
Ethers are very good hydrogen bond acceptors because they contain an electronegative oxygen. This helps them dissolve ionic salts by sharing an electron pair with metal cations, while also making the anion of the salt more reactive [**Figure 13.61(a)**]. Recall from Chapter 12 that when the metal cation is solvated, the "naked" anion cannot delocalize the charge, resulting in a stronger nucleophile. This is in contrast to the behavior of a salt in a polar protic solvent, where both the cation and the anion are solvated [**Figure 13.61(b)**].

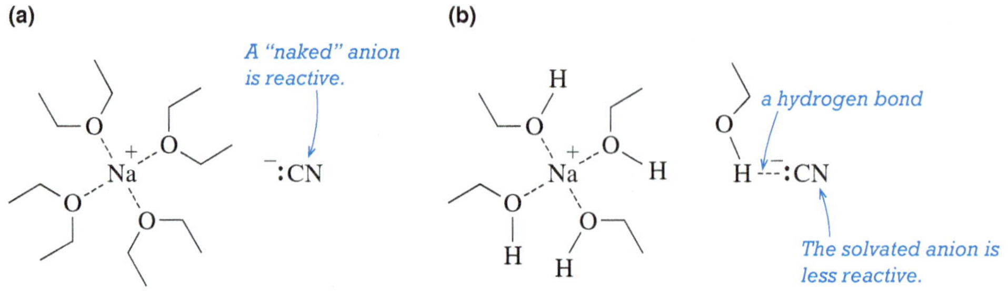

Figure 13.61 Solvent interactions with ionic salts. (a) Ethers can solvate cations but not anions. (b) Protic solvents can solvate cations and anions.

4. **No acidic hydrogen**
   Grignard reagents, to be introduced in Chapter 16, are incredibly strong nucleophiles. Like all nucleophiles, they can also act as strong Brönsted–Lowry bases. Thus, an additional value of ether solvents is that they lack acidic hydrogens, making them very stable (and therefore useful) in reactions featuring strong bases/nucleophiles. Contrast this with a strong base/nucleophile in a protic solvent like ethanol or water. Deprotonation in this environment would happen very quickly (that is, would be very favorable), destroying the nucleophile prior to it reacting as desired (Figure 13.62).

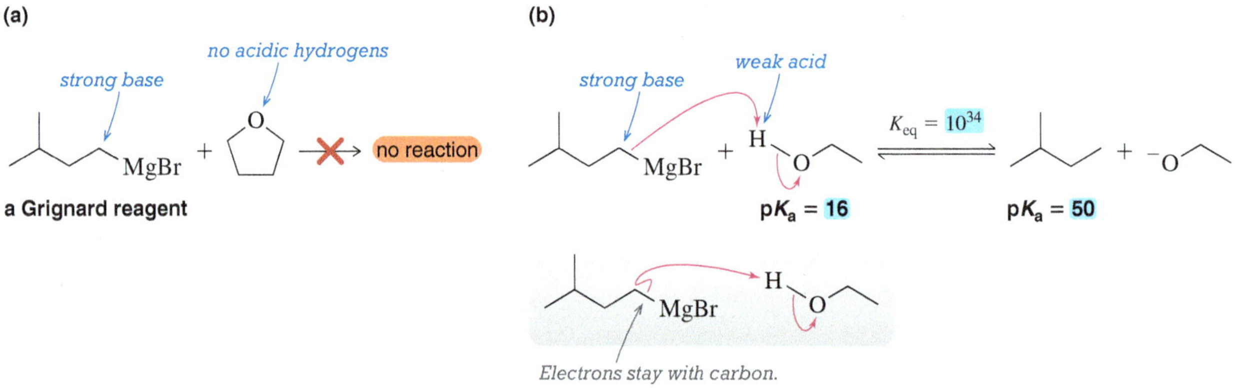

Figure 13.62 Ethers are ideal solvents for strong bases. (a) Ethers lack acidic protons. (b) Strong bases are quenched by the acidic proton in protic solvents.

13.10 SUMMARY

- As polar molecules lacking an acidic hydrogen, ethers are good aprotic solvents, capable of dissolving ionic, polar, and nonpolar compounds.

13-C CHEMISTRY IN CONTEXT: Green Chemistry

Ethers: alternatives to halogenated solvents

Green Chemistry Principle 5: Choose the safest solvent available for any given step.

The pharmaceutical company GlaxoSmithKline (GSK) recently published a solvent selection guide, a modified version of which appears in Table 13.12. This list details which solvents should be used by medicinal chemists to maximize sustainability. Water, as might be expected, tops the list as the greenest of solvents, having an environmental impact (EI) of 10 (1 = worst; 10 = best). Halogenated solvents (EI = 2–6), lauded in Chapter 11 as a result of their ability to dissolve both nonpolar organic compounds and polar reagents, are regarded as having major issues. These issues include the difficulty of disposing them [*halogenated compounds persist in the environment*] and their deleterious effects on human health. They are classified as likely to be carcinogenic and are hepatotoxic (cause chemical-driven liver damage).

(continued)

Ethers, on the other hand, are ranked as suitable replacements for halogenated solvents (EI = 4–9, excluding the highly explosive diisopropyl ether; see Chemistry in Context 13-D). Ethers are generally more easily disposed of and have fewer negative impacts on human health. One drawback, especially for low-molecular-weight ethers, is their high volatility and flammability [*an explosive combination!*]. With this in mind, GSK recommends the use of less volatile ethers, such as *t*-butyl methyl ether, 2-methyltetrahydrofuran, and cyclopentyl methyl ether as alternatives to halogenated solvents.

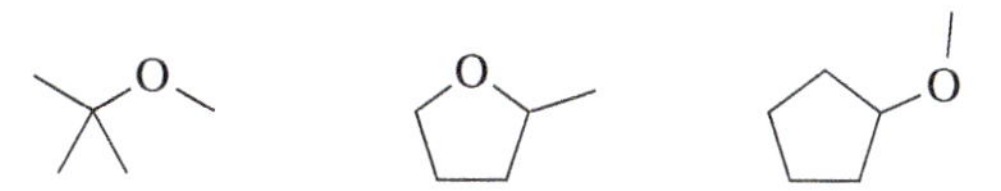

t-butyl methyl ether | **2-methyl-tetrahydrofuran** | **cyclopentyl methyl ether**

Table 13.12 Guide to selecting solvents. [*Boiling points are not given for the solvents with major environmental concerns because these should be avoided.*]

	Minimal environmental concerns (bp)	Moderate environmental concerns (bp)	Major environmental concerns
Water (100 °C)	Water is the greenest option.		
Alcohols	Butan-1-ol (118 °C) Butan-2-ol (100 °C)	Ethanol/IMS (78 °C) *tert*-Butanol (82 °C) Methanol (65 °C) Propan-1-ol (97 °C) Propan-2-ol (82 °C)	2-Methoxyethanol
Esters	*tert*-Butyl acetate (95 °C) Isopropyl acetate (89 °C) Propyl acetate (102 °C) Dimethyl Carbonate (91 °C)	Ethyl acetate (77 °C) Methyl acetate (57 °C)	
Ketones		Methyl isobutyl ketone (117 °C) Acetone (56 °C)	Methyl ethyl ketone (MEK)
Aromatics		*p*-Xylene (138 °C) Toluene (111 °C)	Benzene
Hydrocarbons		Isooctane (99 °C) Cyclohexane (81 °C) Heptane (98 °C)	Petroleum spirit 2-Methylpentane Hexane
Ethers		*tert*-Butyl methyl ether (TBME) (55 °C) 2-Methyl THF (78 °C) Cyclopentyl methyl ether (106 °C)	1,4-Dioxane 1,2-Dimethoxyethane Tetrahydrofuran (THF) Diethyl ether Diisopropyl ether
Dipolar aprotics		Dimethyl sulfoxide (189 °C)	Dimethyl formamide *N*-Methyl pyrrolidone *N*-Methyl formamide Dimethyl acetamide Acetonitrile
Chlorinated	Before using chlorinated solvents, have you considered TBME, isopropyl acetate, ethyl acetate, 2-methyl THF or dimethyl carbonate?		Dichloromethane Carbon tetrachloride Chloroform 1,2-Dichloroethane

Assessment 13.65 Ethers can be converted into radicals, some more easily than others. Which of the following radicals is more stable, and thus, more likely to form?

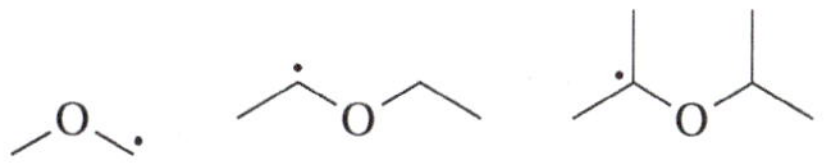

13.11 Synthesis of ethers

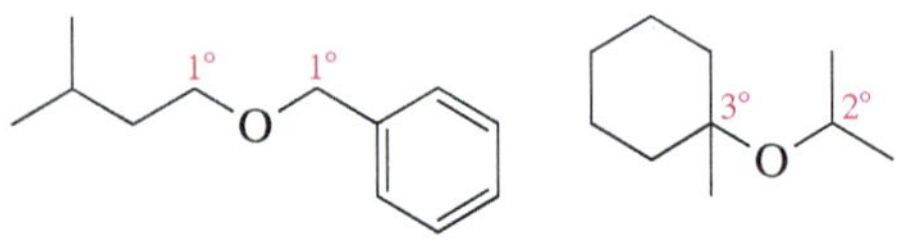

Though it may not be obvious, we have already studied several ways to make ethers. [*Check your reaction sheets.*] In Chapter 8 and 9, when the focus was the reactions of alkenes, the product was, at times, an ether. Similarly, the substitution reactions (S_N1 and S_N2) of haloalkanes discussed in Chapter 12 often resulted in ether products. The choice of which method to use is based on the substitution of the carbons on either side of the ether, as we explain in Sections 13.11.1 and 13.11.2, where the ethers shown at left are synthesized.

13.11.1 Williamson ether synthesis

The Williamson ether synthesis (**Figure 13.63**) is the most popular method for making ethers. [*Just because it has a fancy name doesn't mean it's really a new reaction. You have seen it before.*] Ultimately, it is simply the concerted displacement of a haloalkane by a nucleophilic alkoxide. [*Sound familiar? S_N2!*] The alkoxide can be prepared by either deprotonation (commonly with sodium hydride) or reduction with sodium metal. [*Don't forget that numbering carbons can help you avoid mistakes.*]

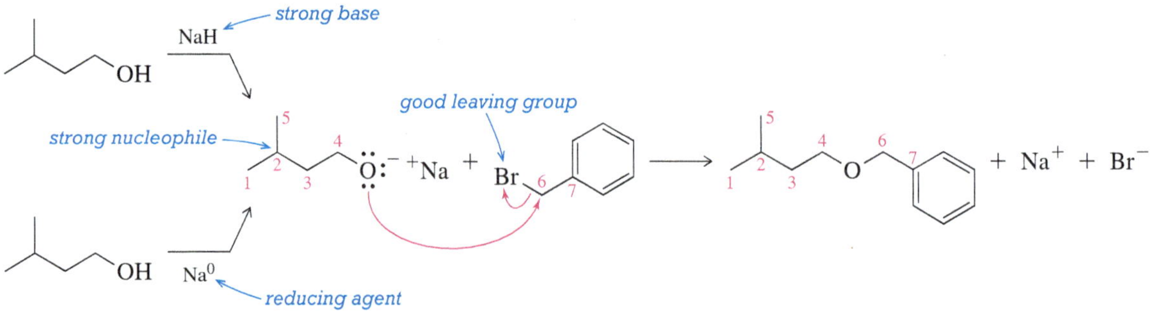

Figure 13.63 The Williamson ether synthesis is the S_N2 reaction between an alkoxide and a haloalkane.

When an ether contains 1° carbons on both sides of the oxygen, two different Williamson reactions can be used to make the same ether because either 1° haloalkane is a viable partner in an S_N2 reaction with a strong base (**Figure 13.64**). [*Why is it important to have a 1° haloalkane in this reaction?*]

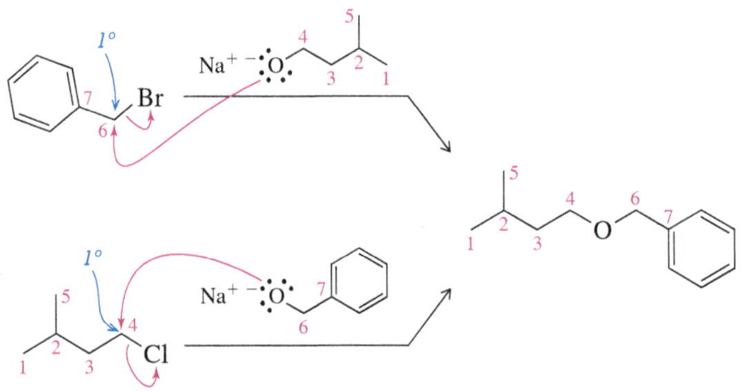

Figure 13.64 Two different Williamson ether syntheses can be used to make the same ether when the carbons on both sides of the oxygen are primary.

When attempting to make a new C—O bond at a 2° carbon in the Williamson ether synthesis, an alternate reaction occurs instead. Recall the process used in Chapter 12 to decide between an S_N2 and E2 reaction. With a strong base that won't wait around for a carbocation to form, at a 2° carbon, the decision was based on the pK_a of the conjugate acid of the nucleophile/base. If the p$K_a \leq 11$, substitution occurs preferentially. If the p$K_a > 11$, elimination occurs preferentially. Thus, the attempted Williamson ether synthesis in **Figure 13.65** would instead result in the elimination product shown.

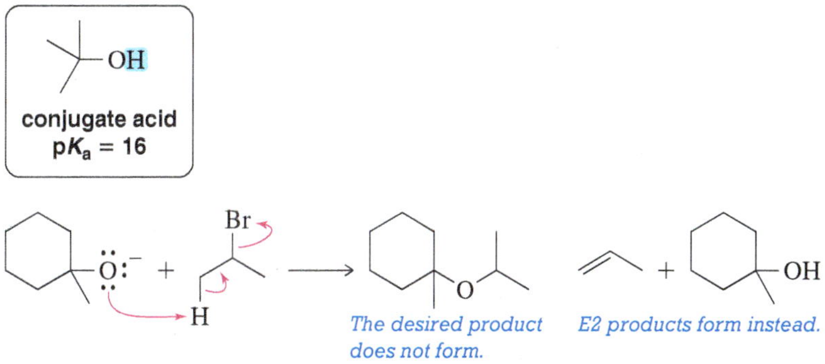

Figure 13.65 The attempted Williamson ether synthesis results in elimination when trying to make a C—O bond at a 2° carbon.

The Williamson ether synthesis, then, can only be used to make ethers with at least one 1° carbon attached to the oxygen. Interestingly, this limitation is not observed when the Williamson strategy is used for the synthesis of thioethers. This is because the conjugate acid of a sulfide (a thiol) has a p$K_a \approx 10$. As a result, a variety of thioethers can be synthesized by this method (**Figure 13.66**).

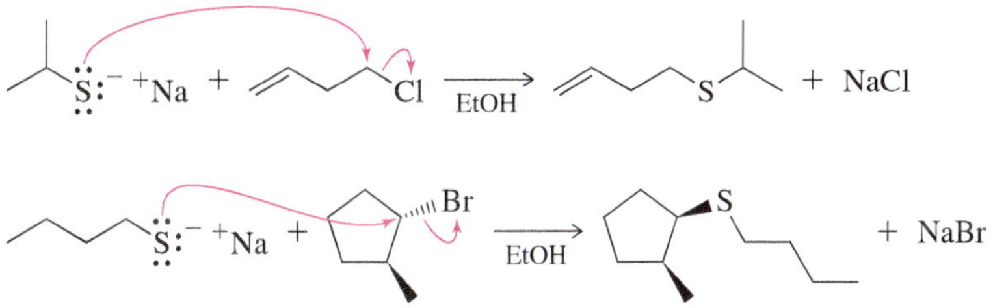

Figure 13.66 Thioether synthesis using the Williamson strategy can make the C—S bond at 1° or 2° carbons.

Assessment 13.66 Predict the product of the following reactions. [*Two of them are Williamson ether syntheses. Why isn't the other?*].

(a)

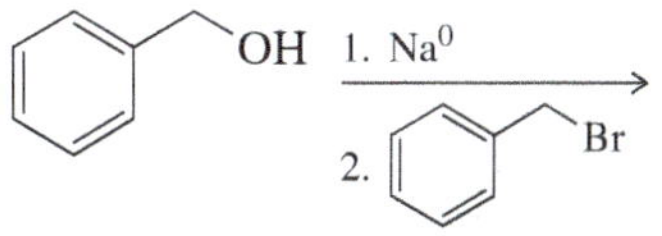

(b)

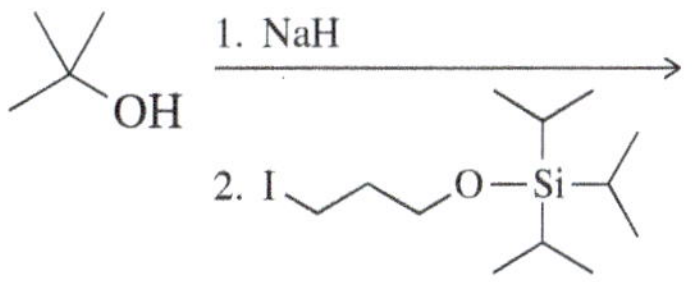

(c)

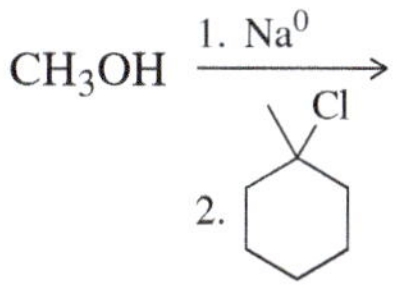

Assessment 13.67 Two different Williamson ether syntheses can be used to make the compound in (a). Show them. The compound in (b), however, can only be made one way. Show it and explain why a second Williamson ether synthesis is not possible.

(a)

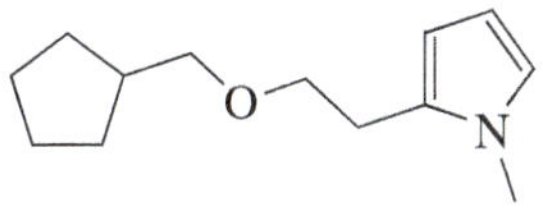

(b)

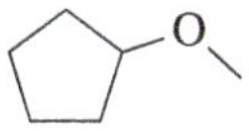

Assessment 13.68 Why is the S_N1 reaction shown an inefficient way of synthesizing ethers?

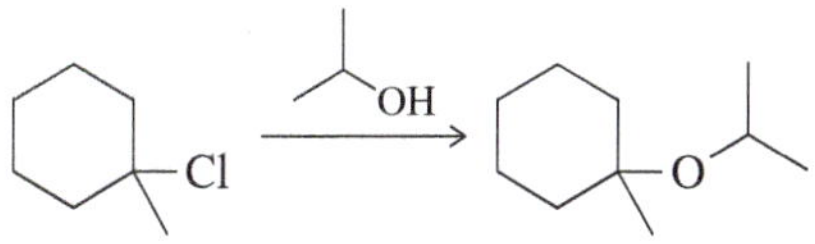

Assessment 13.69 Starting with hydrogen sulfide, suggest a synthesis of the following thioether that makes use of two different haloalkanes.

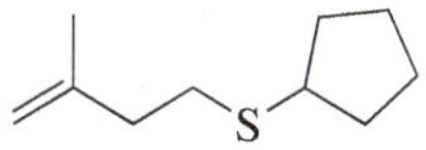

13.11.2 Alkene addition

Fortunately, the limitation of the Williamson ether synthesis can be overcome using reactions introduced in Chapter 8. The electrophilic addition of water, through acid-catalyzed addition or oxymercuration/reduction, results in formation of an alcohol. How might these reactions, shown in **Figure 13.67**, be modified so that an ether is produced?

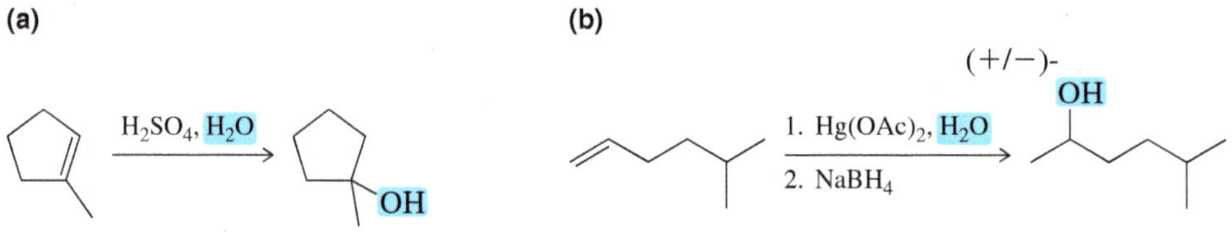

Figure 13.67 Hydration of alkenes by two methods: (a) acid-catalyzed addition of water and (b) oxymercuration/reduction of alkenes.

When the reactions in Figure 13.67 begin, water (HOH) has two hydrogens attached. Water eventually loses one of its protons, with it being replaced with the alkyl group of the alkene to become an alcohol (ROH). If the goal is to make an ether, then, instead of starting with water (two hydrogens), the reaction can be done with an alcohol (ROH) as the reactant. Now when one of the hydrogens of an alcohol is replaced by the R group of the alkene, the product is an ether (ROR). **Figure 13.68** shows how this simple modification (replacing water with an alcohol) turns a reaction we learned previously for synthesizing alcohols into one that produces ethers.

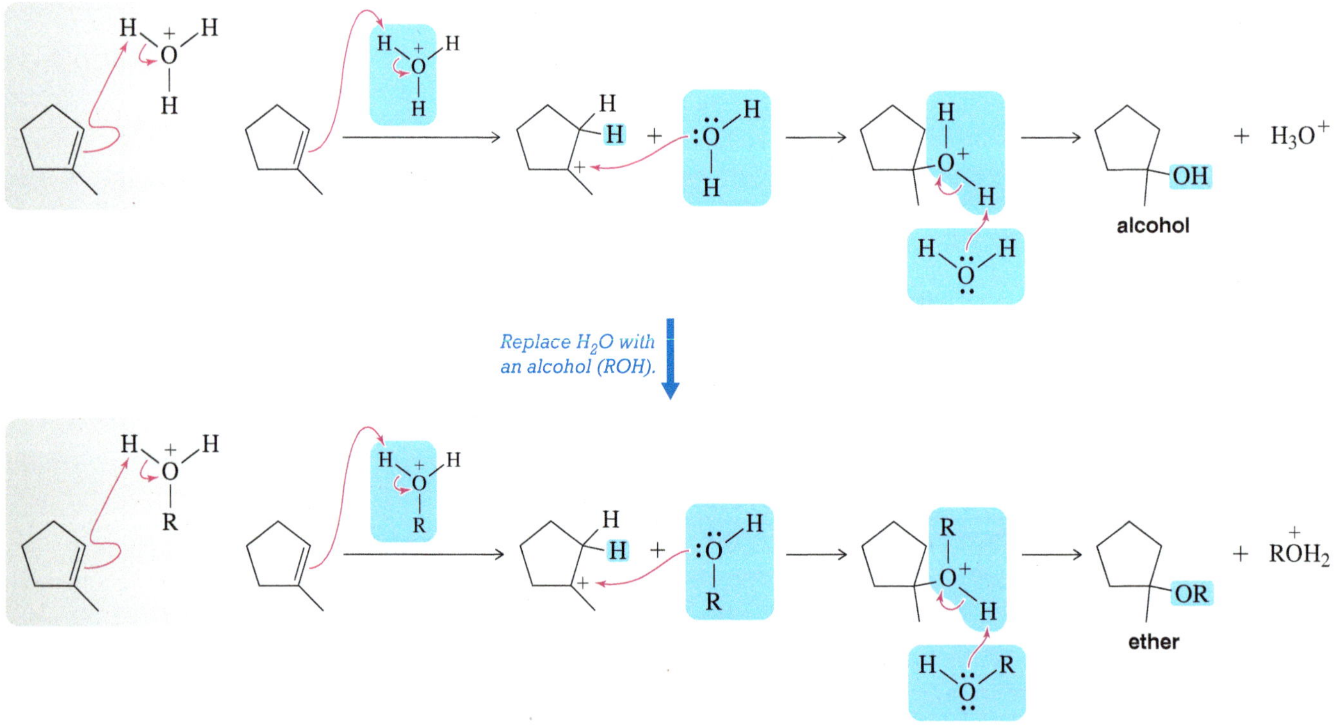

Figure 13.68 Replacing water with an alcohol gives us a way of making ethers. [*H_3O^+ comes from the reaction of H_2SO_4 with H_2O, and ROH_2^+ comes from the reaction of H_2SO_4 with ROH.*]

Because electrophilic addition to alkenes occurs through the most substituted carbocation (Markovnikov's rule), the more highly substituted ether will always be produced when using a nonsymmetrical alkene. The ability to make a new C—O bond at 2° and 3° carbons by electrophilic addition to alkenes makes this method complementary to the Williamson ether synthesis, which can only produce new C—O bonds at 1° carbons. [*Can electrophilic addition to alkenes be used to make new C—O bonds at 1° carbons?*]

As was the case with the Williamson ether synthesis, the ether from Figure 13.65 can be synthesized by two different approaches, using either alkyl group as the alkene or the alcohol (**Figure 13.69**). Moreover, being able to conduct this reaction as an acid-catalyzed addition or as an alkoxymercuration makes this approach highly flexible.

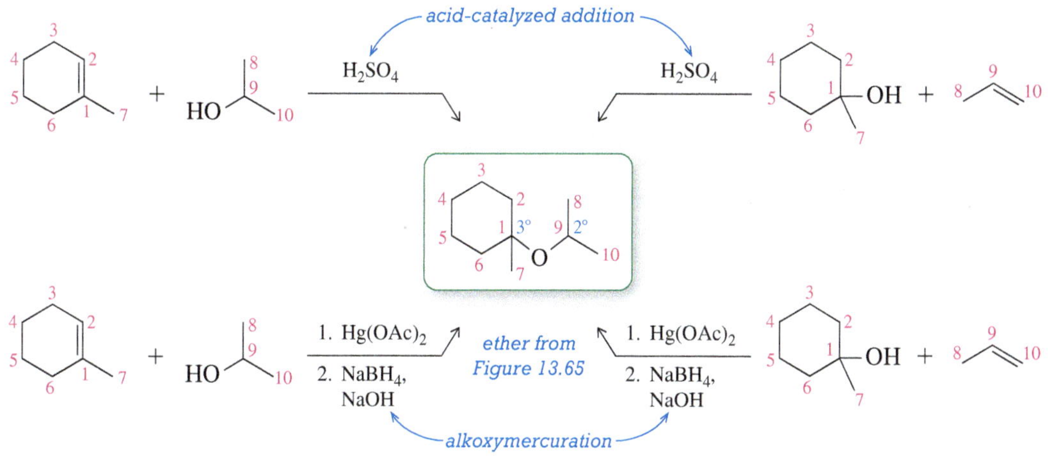

Figure 13.69 Multiple electrophilic additions can be used to make the same ether.

13.11 SUMMARY

- The synthesis of ethers containing at least one 1° carbon can be done using the Williamson ether synthesis, which is an S_N2 reaction of an alkoxide with a 1° haloalkane. The acid-catalyzed addition of an alcohol to an alkene, proceeding through a carbocation intermediate, can be used to make ethers where neither carbon is 1°.

Assessment 13.70 How could the reaction in Figure 13.67(b) be modified to produce the following ether?

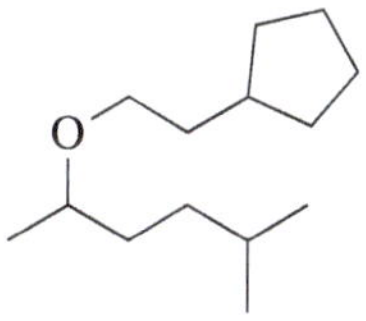

Assessment 13.71 Predict the product and show an arrow-pushing mechanism for the first step of the alkoxymercuration reaction.

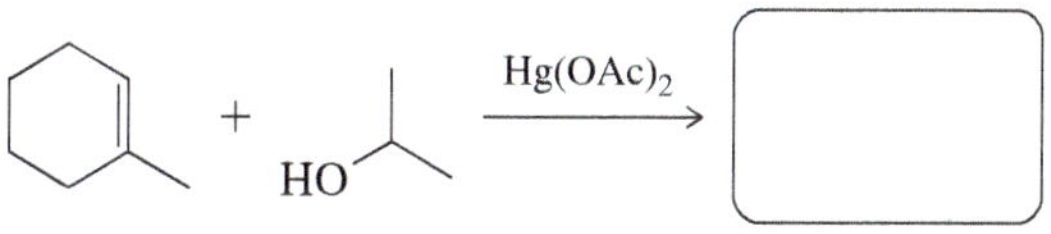

Assessment 13.72 Draw a reaction coordinate diagram for the acid-catalyzed addition of an alcohol to an alkene. Which step is rate-determining?

Assessment 13.73 Identify the alkene and alcohol partners that could be used to make the following ethers.

(a)

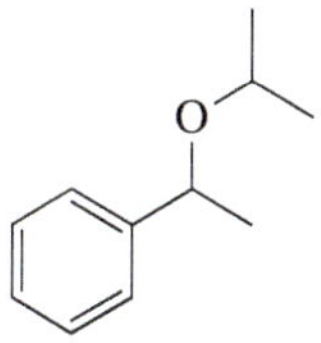

(b)

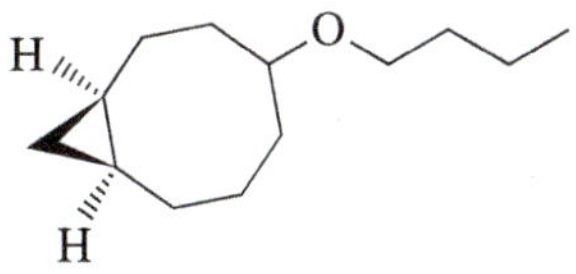

(c)

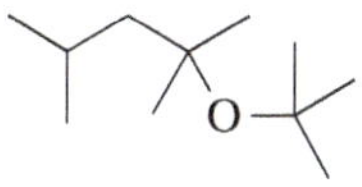

13.12 Reactions of ethers

Figure 13.70 reviews why ethers are so unreactive. Lacking an acidic proton, they are stable in the presence of a strong base. Because the alkoxide (like hydroxide) is a poor leaving group, they also do not readily participate in substitution reactions. For this reason, ethers are inert to both strong bases and strong nucleophiles and are rarely used as synthetic intermediates. [*Studying the reactions in Sections 13.12.1 and 13.12.2, then, is an exercise in relating new material to things studied previously and setting the stage for why epoxides are great synthetic intermediates.*]

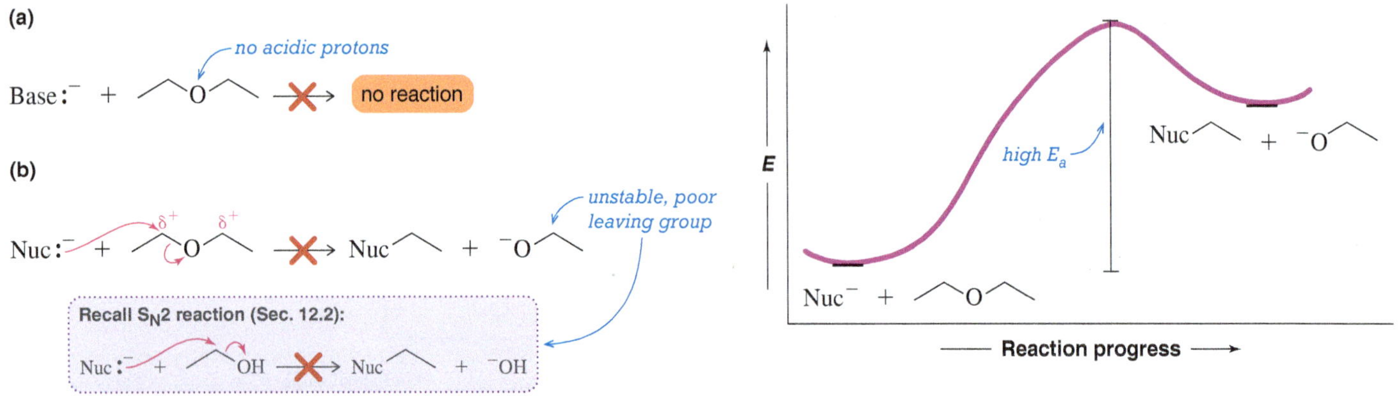

Figure 13.70 Ethers are inert to strong bases and strong nucleophiles because they lack acidic protons, and alkoxides (like hydroxide) are unstable, poor leaving groups.

Ethers and alcohols are similar in their lack of reactivity with strong nucleophiles, which suggests a potential solution. [*How did we get alcohols to react? How did we get hydroxide to be a good enough leaving group?*]

13.12.1 Acidic cleavage of ethers

To cleave ethers, the alkoxide is made into a better leaving group using a strong acid. [*Like alcohols with HCl, yeah?*] These reactions are referred to as cleavage reactions because the ether is being broken down into smaller pieces. [*Not the way we'd want to go if doing synthesis . . . .*] When a hydrohalic acid (HI, HBr, or HCl) is used, the ether is converted to one or two molecules of haloalkane, depending on the number of equivalents of acid used. With a single equivalent, the haloalkane is accompanied by an equivalent of alcohol. The use of two or more equivalents forms two haloalkanes. [*Using the color coding, see if you can figure out how the reactions in* ***Figure 13.71*** *occur.*]

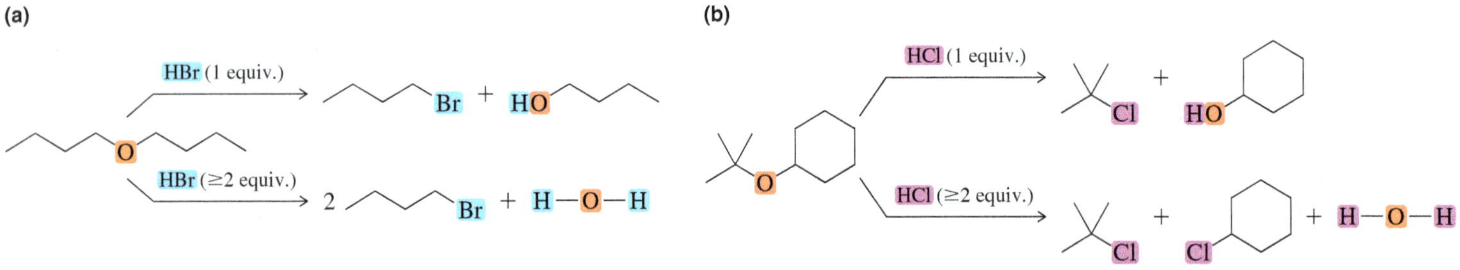

Figure 13.71 Acid cleavage of ethers. (a) HBr cleavage of ethers can make one (with 1 equivalent of HBr) or two equivalents of haloalkane (with ≥ 2 equivalents of HBr). (b) HCl cleavage of ethers can make one (with 1 equivalent of HCl) or two equivalents of haloalkane (with ≥ 2 equivalents of HCl).

The mechanism of the reaction depends on the substitution of the ether carbons in a manner consistent with the process outlined in Chapter 12. [*Imagine how the mechanisms might be different for the reactions in Figure 13.71 before reading further.*] In each case, the reaction begins by protonating the ether oxygen. This converts a poor alkoxide leaving group into an alcohol. [*This is similar to how protonating an alcohol makes water a better leaving group.*] When the carbons attached to the ether oxygen are primary, the halide substitutes for the alcohol in an S_N2 reaction (**Figure 13.72**). With a single equivalent of HBr, the reaction stops here, having formed 1-bromobutane and butan-1-ol. In the presence of an excess of HBr, the hydroxyl group of the alcohol is protonated, making a good leaving group that is displaced in an S_N2 reaction at the 1° carbon. [*See Section 13.7.2.*]

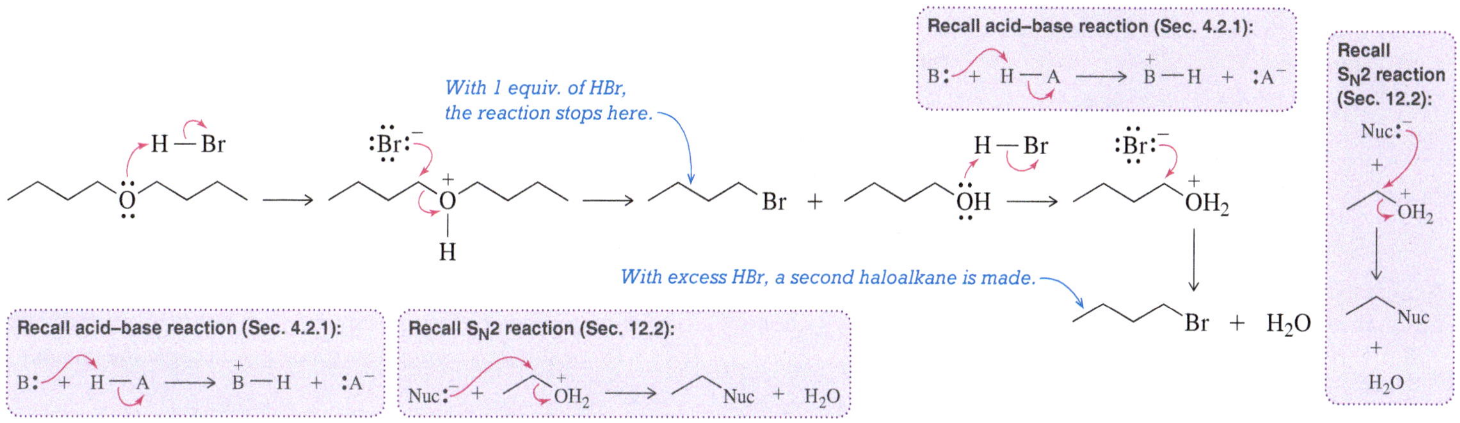

Figure 13.72 HX cleavage of ethers with 1° carbons on each side of the oxygen goes by an S_N2 mechanism.

When the carbons of the ether are not 1°, the substitution no longer proceeds through an S_N2 reaction (**Figure 13.73**). Instead, after protonating the ether oxygen, the chloride displaces the alcohol leaving group in an S_N1 reaction. When a single equivalent of HCl is used, the outcome is dictated by which side of the ether is the more stable carbocation. In this case, the 3° C—O bond breaks to form the more stable 3° carbocation. With an excess of acid, the alcohol is converted to a second haloalkane by an S_N1 reaction.

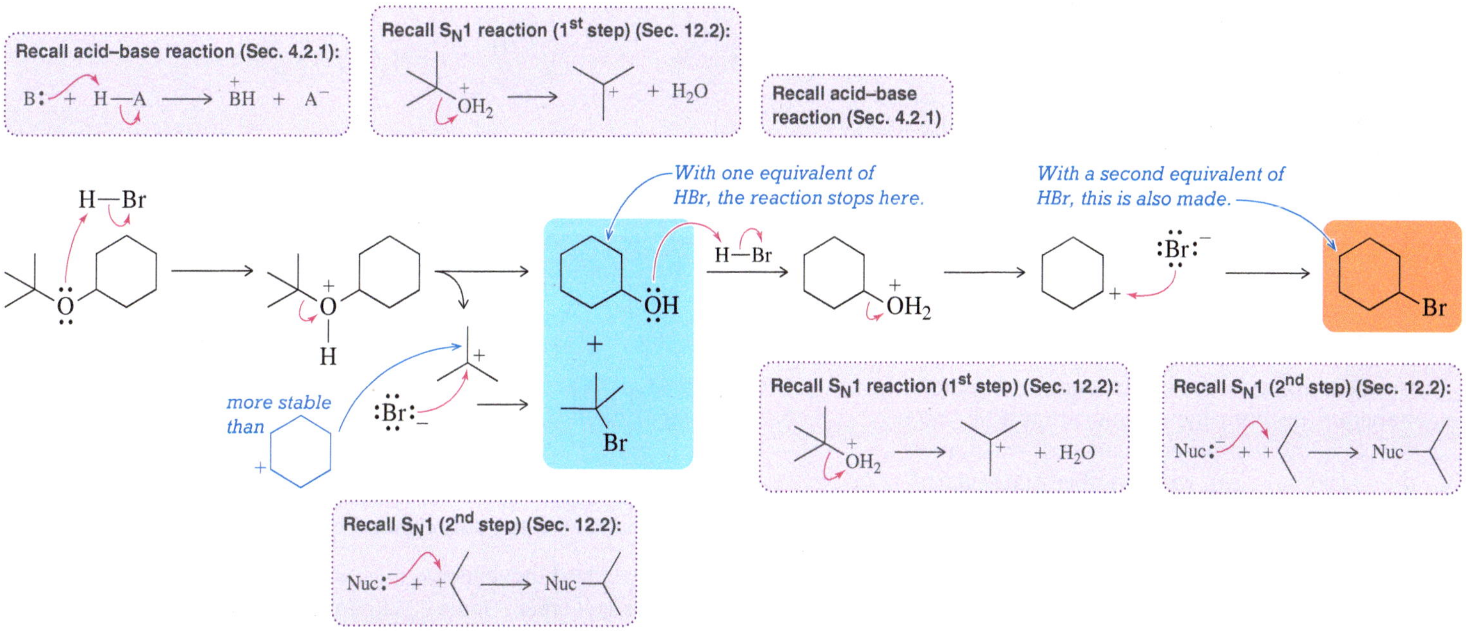

Figure 13.73 HX cleavage of ethers with 2° and 3° carbons on each side of the oxygen occurs by an S_N1 mechanism. [*There's a lot of information here—it might be worth looking at this figure twice.*]

The acid cleavage reaction can also be achieved in the presence of sulfuric acid and water (**Figure 13.74**). With heating, two equivalents of an alcohol are produced.

Figure 13.74 Cleavage of ethers with sulfuric acid in water makes two alcohols.

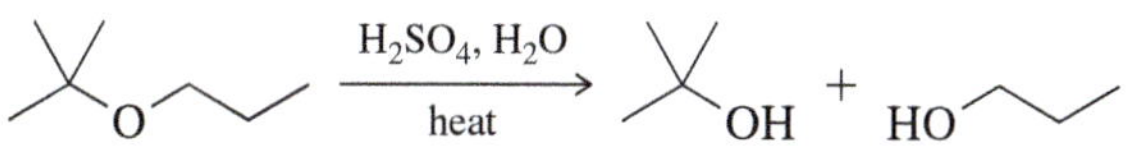

13.12 SUMMARY

- Like alcohols, ethers are nucleophilic at oxygen and deprotonate strong acids. Cleavage of ethers occurs by protonating the oxygen to make a good leaving group that participates in an S_N1 or S_N2 reaction, depending on the substitution of the carbon involved.

Assessment 13.74 Predict the products of the following reactions.

(a) (b) (c)

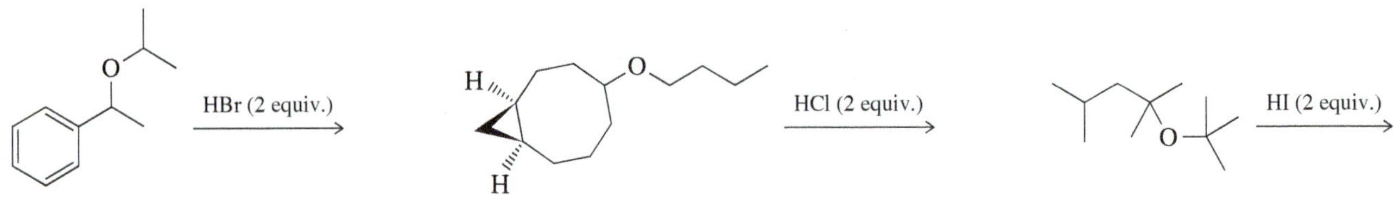

Assessment 13.75 Show an arrow-pushing mechanism that rationalizes formation of the two products. Which C—O bond will break first? Why?

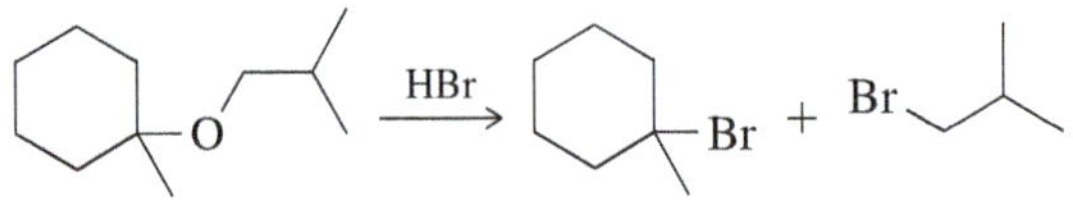

Assessment 13.76 Predict the alcohol and haloalkane that will form upon reaction of the ether shown with one equivalent of HBr. [*Hint: Think carefully about which side will become the halide.*]

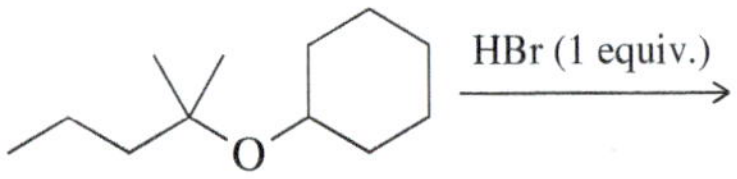

13-D CHEMISTRY IN CONTEXT: Green Chemistry

Explosive peroxides

Green Chemistry Principle 3: Design chemical reactions and synthetic routes to be as safe as possible.

In addition to the volatility and flammability of low-molecular-weight ethers, the use of other ethers is discouraged because of their ability to form explosive peroxides. These peroxides form via autooxidation, an unintentional oxidative process that occurs via reaction with oxygen in air.

Autooxidation (**Figure 13.75**) is a radical process. Radical formation is initiated when the excited state of oxygen abstracts a hydrogen from the ether carbon. The resulting radical is stabilized by both hyperconjugation and resonance. In the first propagation step, reaction of the carbon radical with another molecule of oxygen forms a peroxy radical. This peroxy radical then abstracts a hydrogen from another ether molecule in the second propagation step.

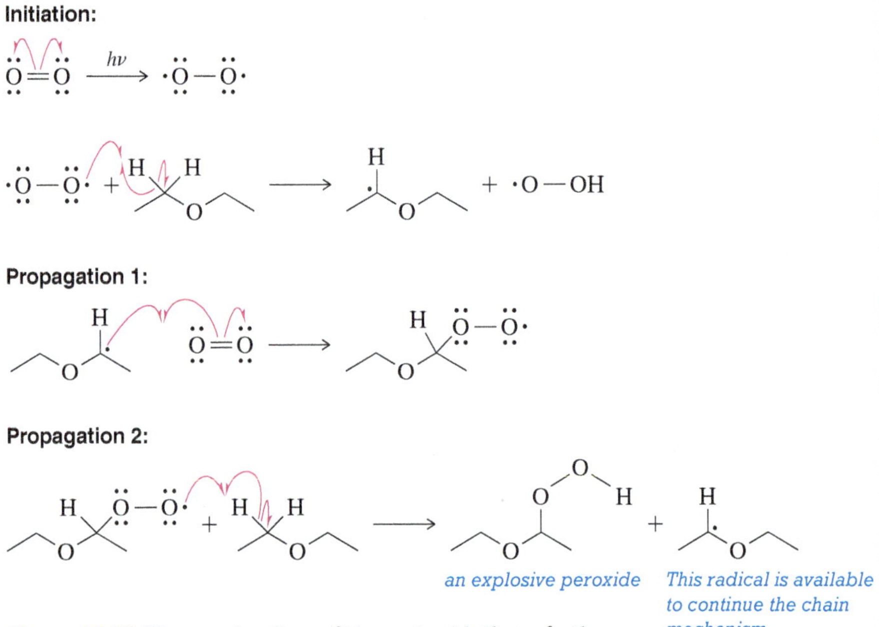

Figure 13.75 The mechanism of the autoxidation of ethers.

Ethers stored under atmospheric conditions develop these peroxides over time. They are extremely unstable molecules that spontaneously degrade in an exothermic reaction, which can become explosive due to the low flash point and high flammability of ethers. To mitigate the risk of peroxide formations, ethers are stored under an inert atmosphere of either argon or nitrogen. Additionally, a small amount of a radical inhibitor like BHT is often added to destroy any alkyl radicals that form before peroxide formation can occur. [*See Chemistry in Context 11-E.*]

In the chemical and pharmaceutical industry, when ethers are needed, specific ethers are chosen that are less prone to the formation of peroxides. As explained in Chapter 11, radicals are stabilized by hyperconjugation. Thus, 2° ethers such as diisopropyl ether (**Figure 13.76**) are much more prone to hydrogen abstraction and eventual peroxide formation than either 1° or methyl ethers.

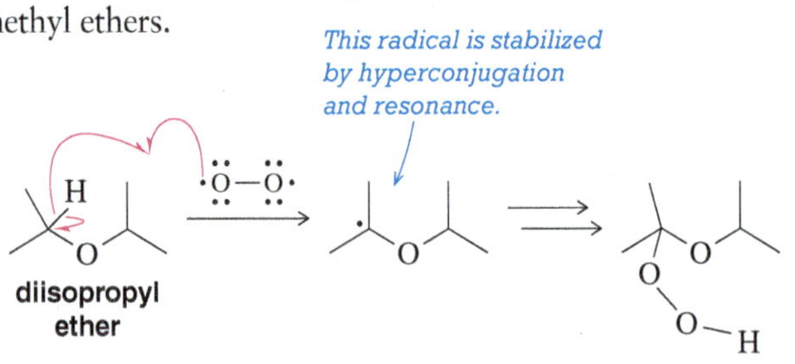

Figure 13.76 Ethers that form stable radicals are more likely to form peroxides.

Assessment 13.77 Methyl *t*-butyl ether (MTBE) is used preferentially over diethyl ether because it is less prone to form peroxides. Explain this observation in terms of the two structures.

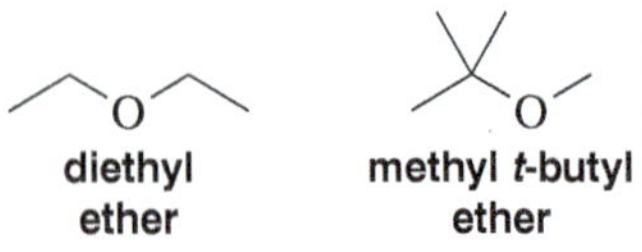

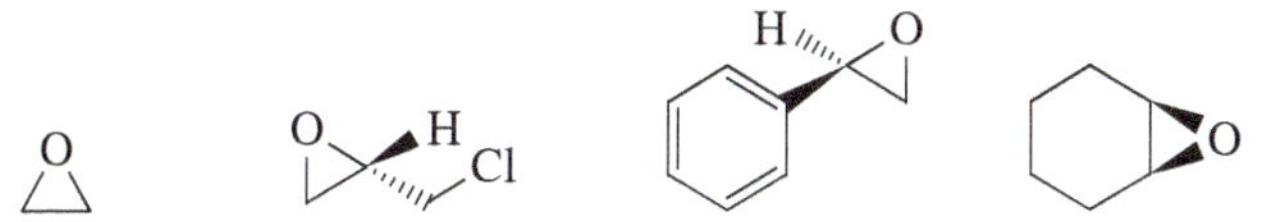

Figure 13.77 **Common epoxides.**

13.13 Epoxides

Epoxides (**Figure 13.77**) are cyclic ethers that are relatively reactive, which allows them to overcome some of the synthetic limitations found with acyclic ethers. Also called **oxiranes**, epoxides are three-membered rings in which one of the atoms in the ring is oxygen.

13.13.1 Synthesis of epoxides—*m*CPBA

Epoxides are most often synthesized by treatment of an alkene with *meta*-chloroperoxybenzoic acid (*m*CPBA). [*We "invented" this reaction in Section 9.1.4 in a Thinking Like a Chemist section.*] Mechanistically (**Figure 13.78**), the terminal oxygen of a peroxyacid is attacked by an electron-rich π bond, triggering a concerted cascade that features the cleavage of a weak O—O σ bond, formation of a protonated three-membered ring, and deprotonation of the three-membered ring to yield the epoxide (oxirane) product. Because the reaction is concerted, epoxide formation happens before there is any opportunity to rotate around a σ bond; as a result, the reaction is stereospecific. This concerted mechanism generates what has been referred to as the butterfly transition state. Because the alkene can attack the peroxyacid from either side with equal rates (enantiomeric transition states), the reaction results in a racemic mixture of products. Several asymmetric versions of this reaction have been developed (Section 9.3.2).

Figure 13.78 **The mechanism of *m*CPBA oxidation** of alkenes.

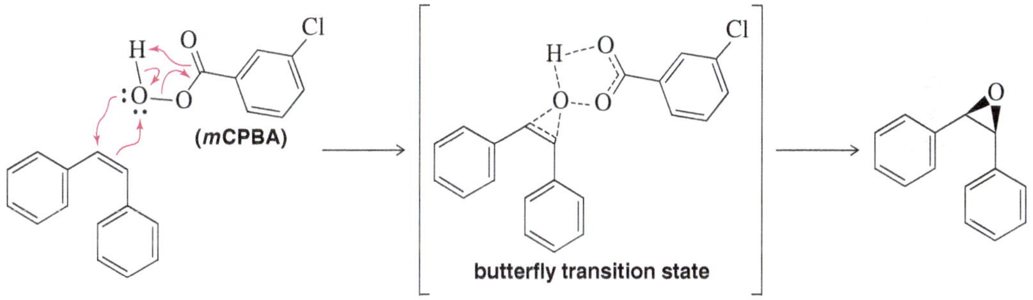

13.13.2 Reactions of epoxides

Epoxides undergo two major types of reactions—those with weak nucleophiles and those with strong nucleophiles. The next two sections explore each of these.

13.13.2.1 Weak nucleophiles add "via S_N1 reaction"

In addition to learning how to make epoxides in Chapter 9, we also learned one reaction that they undergo. Recall that you can dihydroxylate cyclohexene using osmium tetroxide (OsO_4), followed by an oxidative workup to give a syn diol, which results from the concerted addition of the two oxygens to the same face of the molecule. Thinking as synthetic chemists, we needed a way to add hydroxyl groups to the opposite side of the alkene. Epoxides were the key intermediate in this reaction because they allowed us to add the two oxygens in separate steps, ensuring that the second one added on the opposite side (**Figure 13.79**).

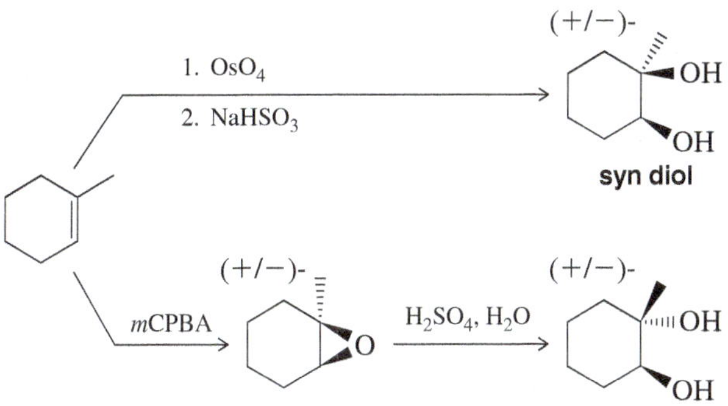

Figure 13.79 **Synthesis of diols** from Chapter 9. The anti diol makes use of an epoxide intermediate.

The mechanism (**Figure 13.80**) begins with protonation of the epoxide. This intermediate should be reminiscent of the bromonium (Section 9.1.2) and mercurinium ions (Section 8.4.4) because all three have three-membered rings in which one of the atoms has a positive charge. Water attacks, opening the epoxide and relieving the ring strain. As was established previously in the bromonium and mercurinium reactions, the oxonium ring is opened by water from the backside, at the carbon that would be the most stable carbocation. In the final step, H^+ is removed from the diol, regenerating H_3O^+, demonstrating that this is an acid-catalyzed process.

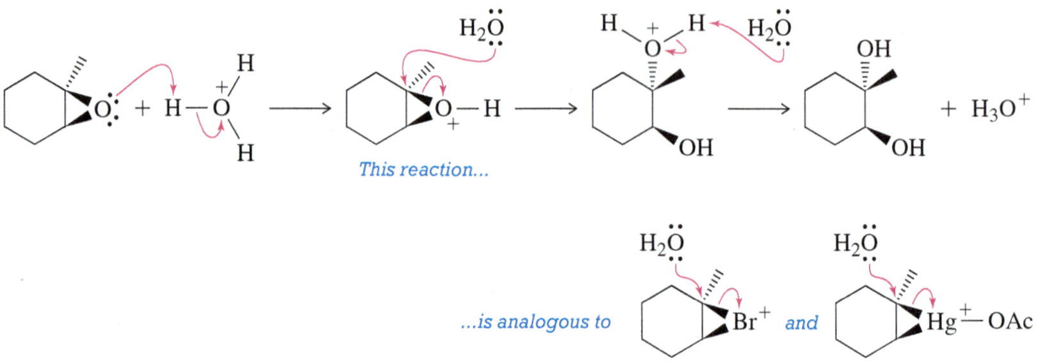

Figure 13.80 Acid-catalyzed addition of water to an epoxide. Compare this figure to Figure 9.36, where this reaction was first studied.

[*The mechanism presented in Figure 13.80 relies on ideas you learned in Chapter 9. However, we have also discussed substitution reactions of haloalkanes in Chapter 12, so it is important to reframe the mechanistic discussion in light of ideas you learned in that chapter.*] When doing substitution reactions with weak nucleophiles like water, S_N1 reactions are favored. The epoxide addition is unique from the substitution reactions of haloalkanes because there are two sites where the reaction can occur. But the defining characteristics are still the same. A weak nucleophile is not that interested in sharing its pair of electrons and will only do so when it can attack a positively charged atom. Fortunately, we can think of the protonated epoxide as equilibrating between two different carbocations [**Figure 13.81(a)**]. The 3° carbocation is more stable than the 2° carbocation, and so there is more charge on the 3° carbon of the protonated epoxide. Water attacks at the site containing the greatest amount of positive charge—the 3° carbon [**Figure 13.81(b)**], but from the backside. Thus, it has characteristics of both the S_N1 (attack at most substituted carbon) and S_N2 reaction (backside attack). [*A similar analysis was done for the nonsymmetrical bromonium ion in Section 9.1.3.*]

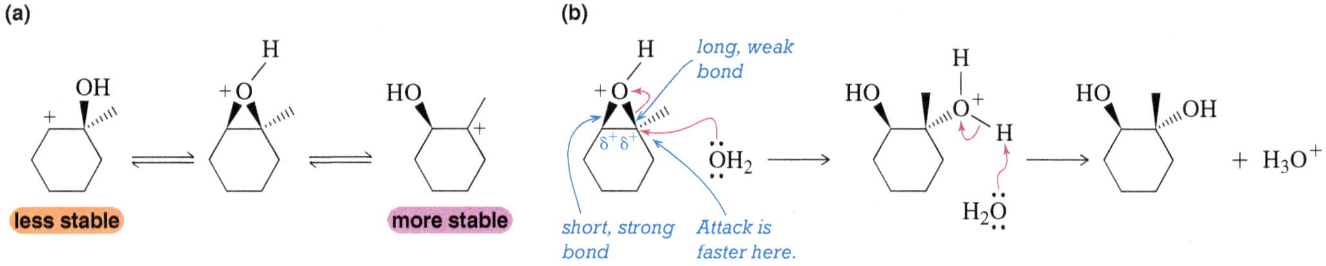

Figure 13.81 Weak nucleophiles open epoxides at (a) the carbon that would be the most stable carbocation. (b) Attack is from the opposite side. [*Recall a similar conversation in Section 8.4.4.*]

From this we can make two generalizations. First, when reacting a weak nucleophile with an epoxide, the epoxide has to be protonated first in order to become electrophilic enough to be attacked. Second, once protonated, regardless of the weak nucleophile being used, attack will be at the more substituted carbon. [*Assessment 13.78 allows you to work some problems with weak nucleophiles that aren't water.*]

Assessment 13.78 Predict the product of the following reactions.

(a)

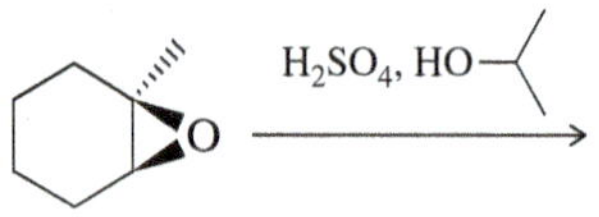

(b)

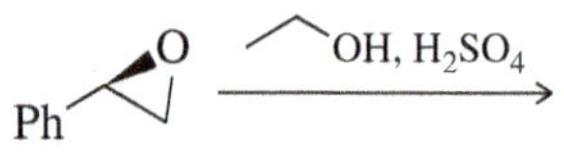

(c)

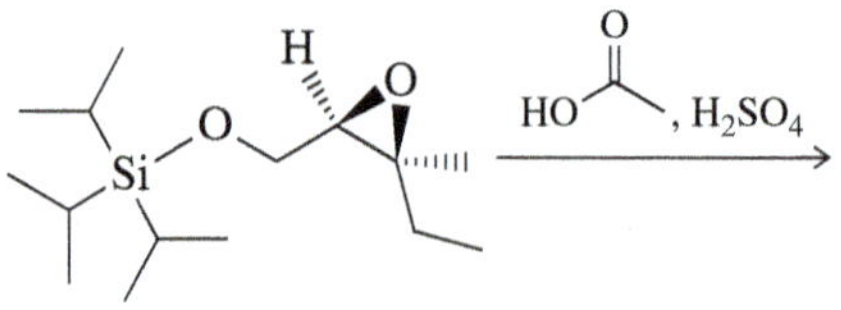

(d)

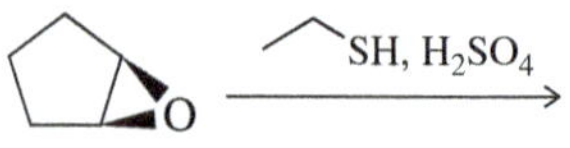

We make a pretty good team . . . I'll help with this one, too. But make sure you are carefully following along.

Assessment 13.79 Suppose you wanted to make the product shown here from an epoxide, necessitating attack at the less substituted carbon. How might you accomplish this?

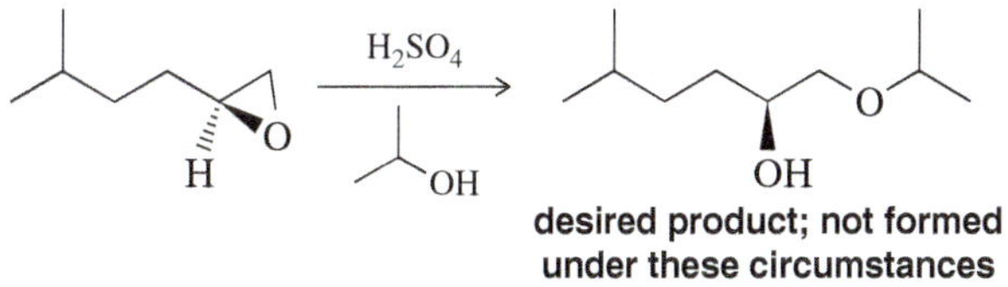

13.13.2.2 Strong nucleophiles add via S_N2 reaction

Assessment 13.79 asked how you might add a nucleophile to the less substituted carbon. The Solved Assessment that follows explains how this is done [*if you haven't figured it out already*].

Solved Assessment 13.79

The reaction of the epoxide in Assessment 13.79 to make the desired compound would require that a nucleophile add to the terminal, less substituted carbon. Unfortunately, the discussion so far has made it clear that weak nucleophiles, because of their propensity to participate in S_N1 reactions, always add to the more substituted carbon of a nonsymmetrical epoxide (**Figure 13.82**).

If S_N1 reactions always add to the more substituted carbon [*because the carbocation is most stable*], is there a substitution reaction that always takes place at the least substituted carbon? [*Yes!*] S_N2 reactions always add to the least substituted carbon because of the diminished steric hindrance. How, then, do you modify the conditions such that an S_N2 reaction will occur? [*Hint: It isn't as easy as whispering to the reaction flask, "Do the S_N2 reaction."*] Whereas weak nucleophiles need a positively charged species to react with, strong nucleophiles can react at neutral carbons, directly displacing even moderate leaving groups. Thus, the answer here is to use *i*-propoxide, the strong-nucleophile version of *i*-propanol (**Figure 13.83**). [*If you had guessed i-propoxide from the beginning, well done!*] ■

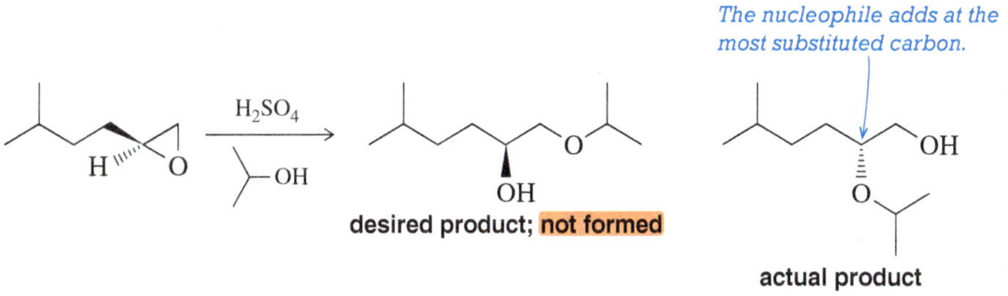

Figure 13.82 **Addition of isopropanol in acid** adds to the more substituted carbon, not forming the desired product.

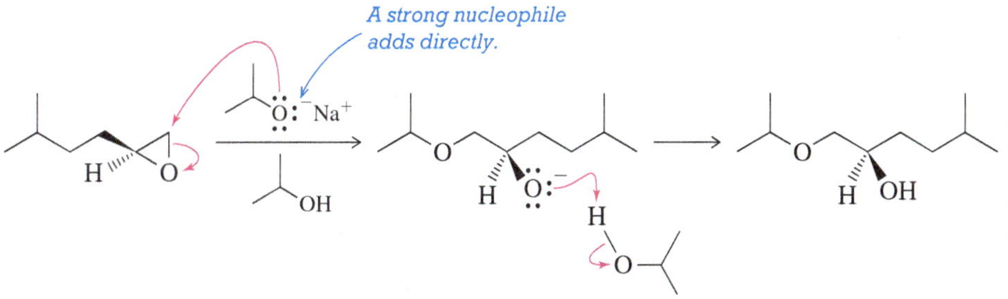

Figure 13.83 **By making a strong nucleophile (isopropoxide),** addition occurs to the 1° carbon via an S_N2 reaction.

If ethers are so unreactive, then why are epoxides (also ethers) reactive with strong nucleophiles as shown in Solved Assessment 13.79? To answer that, look at **Figure 13.84**, which is a modification of Figure 13.70. With an acyclic ether, even a strong nucleophile is unable to displace ethoxide, a poor leaving group. In contrast, reaction between a strong nucleophile and an epoxide is very efficient. [*How does an epoxide, with its three-membered ring, differ from an acyclic ether?*] The epoxide contains substantial angle strain (60° bond angles vs. the desired 109.5° bond angles). As a result, the strained epoxide starts at a much higher energy on the reaction coordinate diagram. This accomplishes two things—namely, the activation energy (E_a) is lowered, and the reaction becomes slightly downhill. This substitution reaction occurs via an S_N2 mechanism, with the electrons of the nucleophile being donated to σ*, resulting in the breaking of the C—O bond of the epoxide.

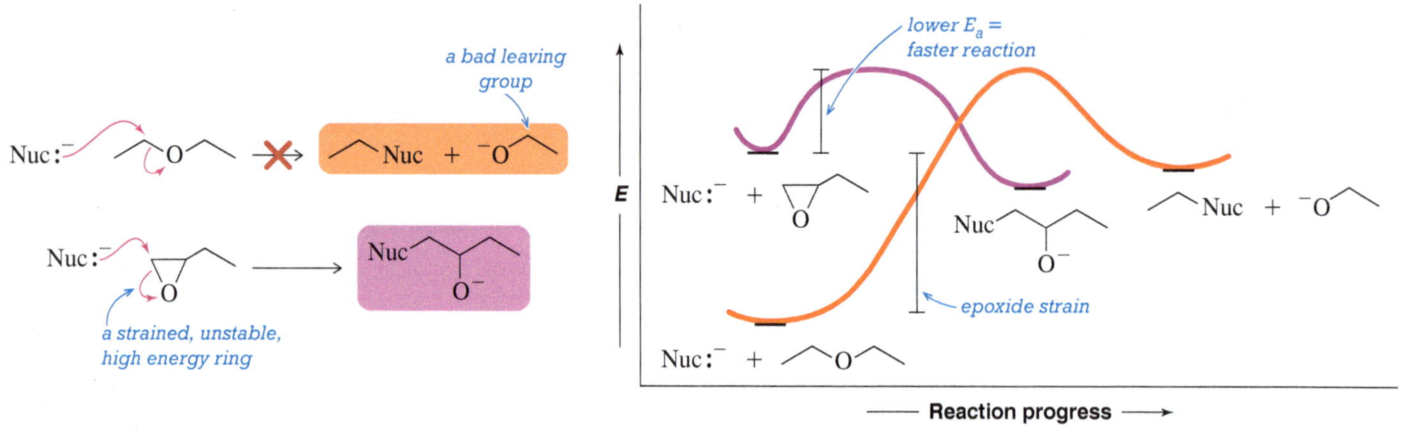

Figure 13.84 The angle strain in the epoxide ring makes it susceptible to nucleophilic attack.

A variety of strong nucleophiles can be used in conjunction with epoxides, including strong nucleophiles seen previously like hydroxide, alkoxides, and acetylides (**Figure 13.85**). Whereas hydroxide and alkoxides already have a proton available in solution (i.e., water and alcohol) for formation of the new O—H bond, this proton has to be administered in a second step for a strong base like acetylide. Acetylide is such a strong base that it reacts with any protic solvents like ethanol or water. As a result, acetylide is typically used in polar aprotic solvents [*like ethers!*] or in liquid ammonia [*though protic, NH_3 has a high pK_a*]. When an acid is added at the end of a reaction to neutralize the product, it is referred to as an **acid quench**. [*Because this will be very common going forward, it is important to understand exactly what is meant by acid quench.*] When you see this, only the negatively charged species in solution are protonated. Though the acids that we use for the quench have the potential to do other reactions, don't overthink the quench step. The idea is simply to turn negative ions, by protonation, into neutral atoms.

Figure 13.85 Strong nucleophiles open epoxides at the least substituted carbon to make alcohols.

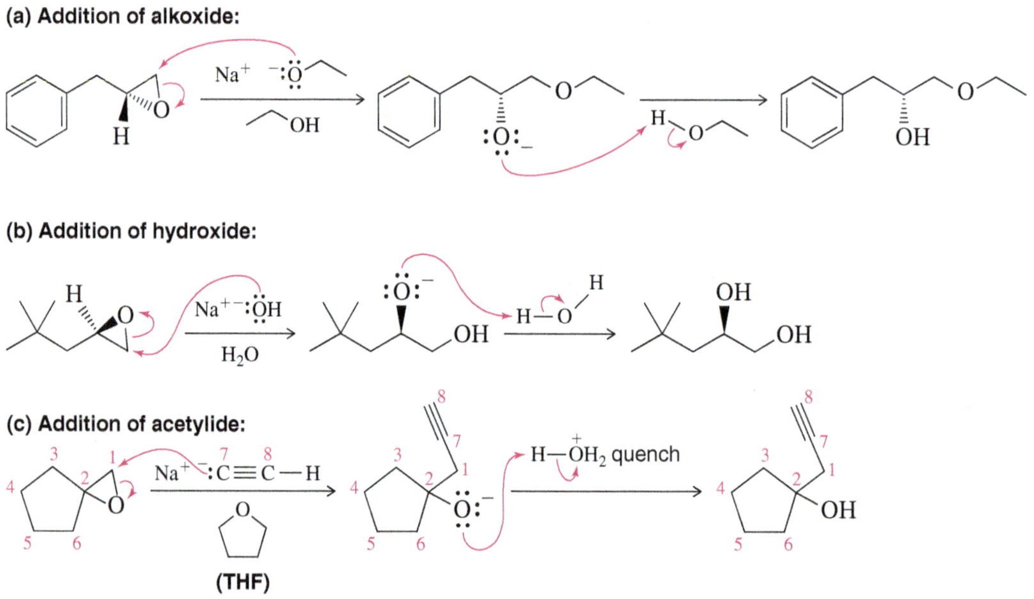

13.13 SUMMARY

- Weak nucleophiles open protonated epoxides on the opposite face at what would be the more substituted carbocation (the more substituted carbon). Strong nucleophiles open epoxides directly by attacking at the least sterically hindered (least substituted) carbon.

Assessment 13.80 Predict the product of the following epoxide addition reactions.

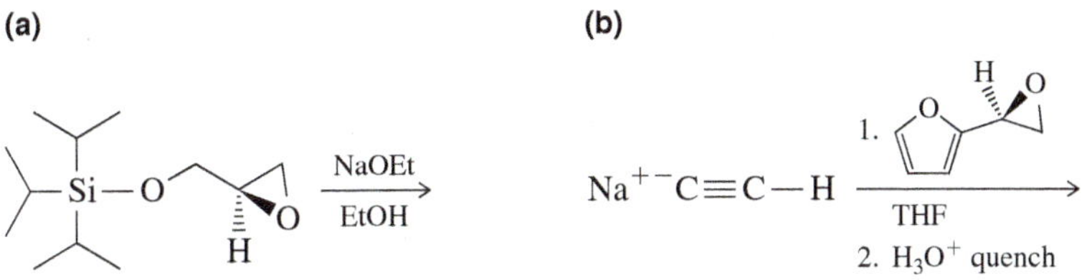

Assessment 13.81 A variety of organometallics, which as strong nucleophiles can react with epoxides, are introduced in Chapter 16. Predict the product of these reactions. [*Hint: Assume the carbon–metal bond in each is ionic, with the carbon possessing the negative charge.*]

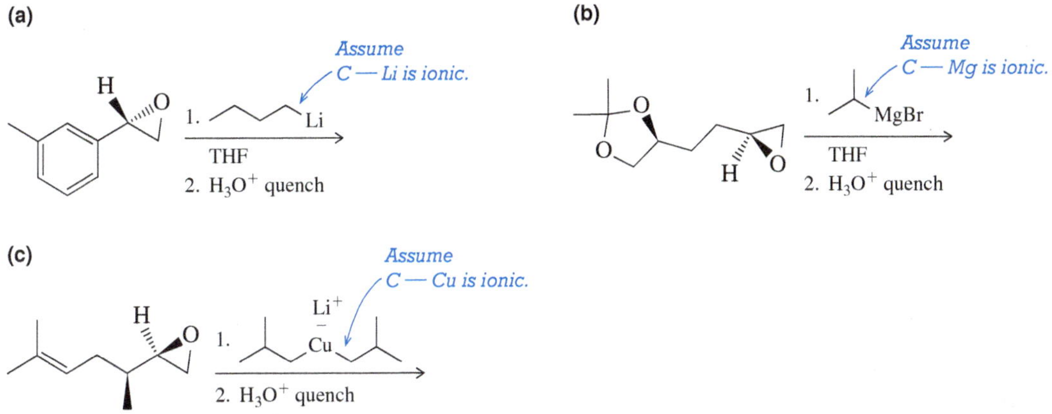

Assessment 13.82 An impatient student worked this problem too quickly and made a common mistake. What was their mistake? How could it be avoided in the future?

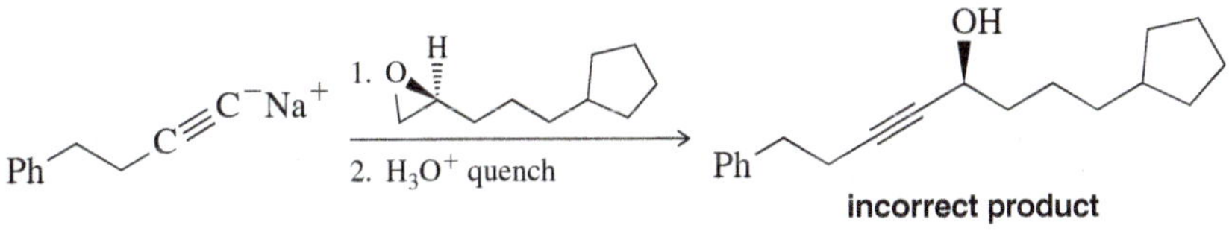

Assessment 13.83 Reaction of the acetylide with the epoxide shown will not form the desired product. What side reaction occurs instead? Why?

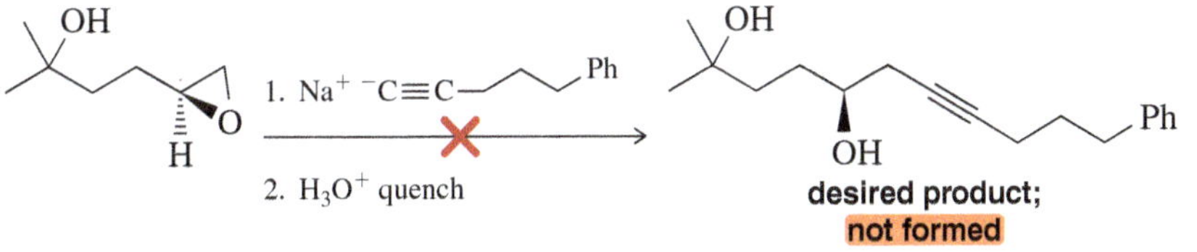

13.14 Protecting groups

Assessment 13.83 presented a situation in which a reaction that looks pretty straightforward ended up not working. [*Go back and work Assessment 13.83 before you read further.*] One way of looking at this reaction, especially in light of the recent discussion on epoxides, is to see only a strong nucleophile and an epoxide. Strong nucleophiles add to the least substituted carbon of the epoxide, opening them efficiently because of the strain on the epoxide. Thus, the mechanism shown in **Figure 13.86** should be plausible.

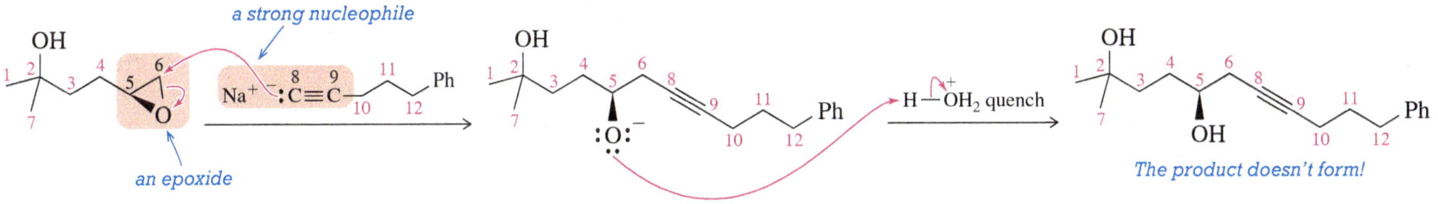

Figure 13.86 An attempted epoxide opening that should work, but doesn't. Why?

And yet the product in Assessment 13.83 doesn't form. So what is wrong here? [*The carbons are numbered in Figure 13.86, so we haven't lost or gained a carbon.*] The problem is that when we looked at the reactants, we focused [*somewhat correctly*] on the functional groups that we have most recently studied. We noticed only that there was an epoxide and a strong nucleophile. But what else is present? [*Look back at Figure 13.86 and see what you find.*] In addition to an epoxide, there is also an alcohol present. And alcohols, having a somewhat acidic proton ($pK_a \approx 16$), readily react with strong bases like acetylide ions. Because there is less steric hindrance around the proton, deprotonation always occurs before nucleophilic addition or substitution. It is much easier to grab a proton than to donate a pair of electrons to σ* of the epoxide. Thus, instead of a substitution reaction at C_6 of the epoxide, the acetylide simply deprotonates the hydroxyl group at C_2 (**Figure 13.87**).

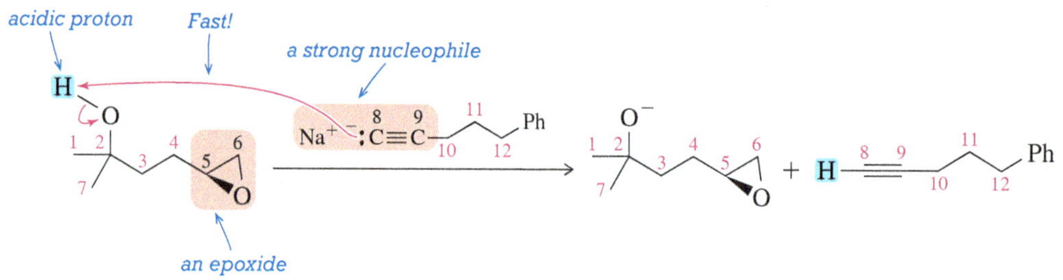

Figure 13.87 The molecule contains an acidic hydroxyl group. The strong nucleophile deprotonates the acidic hydroxyl group faster than it initiates substitution at C_6 of the epoxide.

This situation demonstrates just how important it is to identify all of the functional groups in a reaction and to consider whether there are other reactions that could occur. Perhaps more importantly, is there some modification we can make so that the acetylide attacks the epoxide instead of grabbing the proton? [*I wouldn't ask if there weren't.*]

If the problem in Figure 13.87 is that there was an acidic proton on the hydroxyl group, could we modify the alcohol so that it no longer has a proton? Not coincidentally, we have been studying ethers in this chapter and ethers lack that proton. So maybe we can just convert the hydroxyl group into an ether (i.e., replace H with R; see **Figure 13.88**), making the desired reaction possible. But it's really not that simple because we still wouldn't have the desired product, right? [*See Figure 13.88.*] Only after turning the ether back into a hydroxyl group (i.e., replace R with H) would we have the desired product.

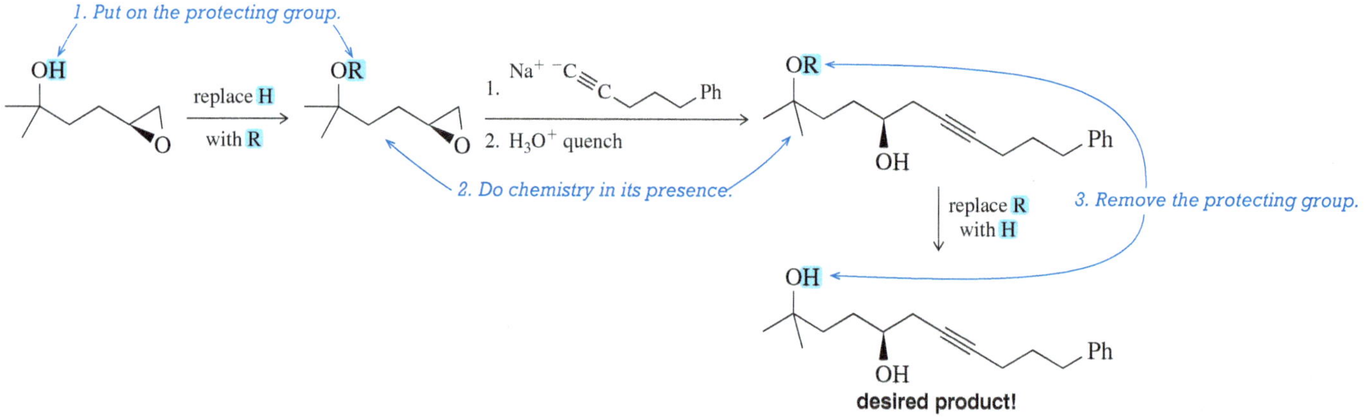

Figure 13.88 A protecting group strategy allows us to avoid the side reaction.

The process shown in Figure 13.88 is called a **protecting group strategy**. In essence, the protecting group, R (undefined at this point), protects the hydroxyl group from being attacked by the acetylide anion. Then, upon removal of the protecting group, the desired product is obtained. As shown in Figure 13.88, there are three requirements for a good protecting group. You must be able to put it on, do chemistry in its presence, and remove it selectively. Ethers are great protecting groups, because, as discussed in Section 13.11, they are inert to a variety of conditions.

13.14.1 Silyl ethers as protecting groups

The biggest challenge of using an ether as a protecting group is removing it. As we discussed in Section 13.12, the only way to cleave ethers is to treat them with very strong acids. Strong acids can react with other functional groups, so these are probably not ideal conditions. Luckily, silyl ethers have a strong affinity for fluoride, allowing them to be cleaved using a fluoride salt. Although there are a wide variety of silyl ethers that can be used as protecting groups, we focus only on the triisopropyl silyl ether (abbreviated as TIPS). In executing the full process (**Figure 13.89**), the silyl ether would be prepared using what looks like a Williamson ether synthesis with triethylamine as the base. The mechanism of the reaction is S_N2-like, with Et_3N serving as a weak base. Once attached, the acetylide can be reacted with the epoxide, which, upon quenching with acid, gives a 2° alcohol. Finally, making use of the strong Si—F bond, tetrabutylammonium fluoride is used to remove the silyl group in another S_N2-like reaction, completing the synthesis of the desired product.

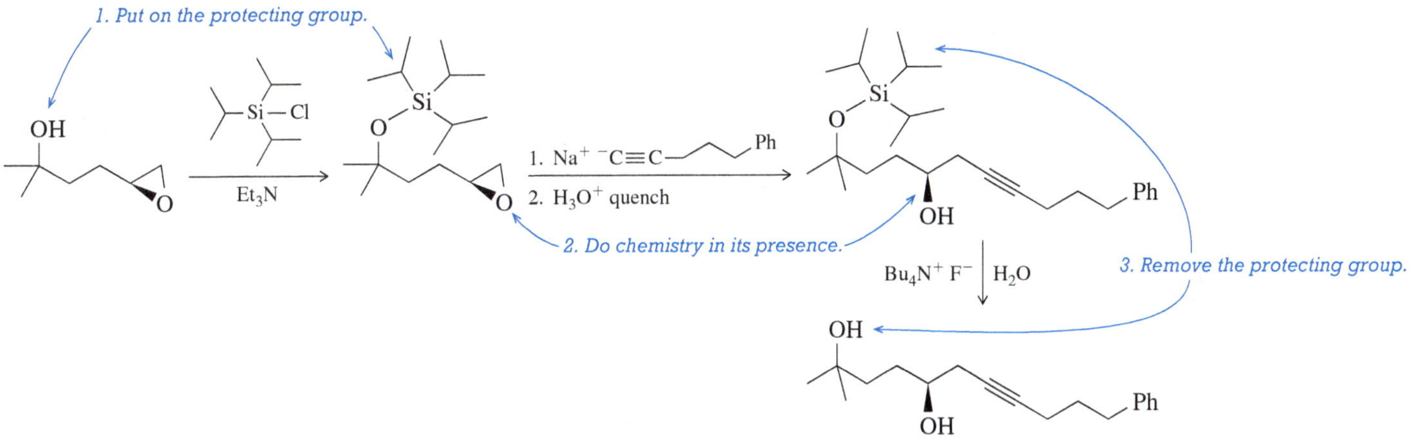

Figure 13.89 Use of a silyl ether protecting group solves the problem.

Other examples of silyl ether protecting groups are shown with their abbreviations in **Figure 13.90**.

In an advanced organic course, you might learn which situation requires the use of one protecting group or another.

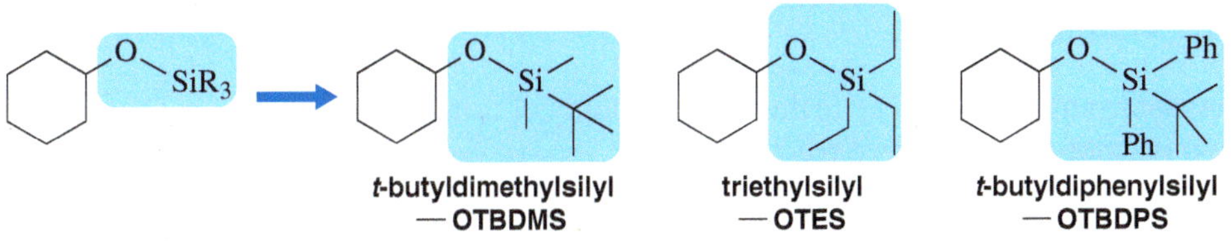

Figure 13.90 Examples of silyl ethers used as protecting groups.

Assessment 13.84 Show how a protecting group might be used to make these reactions successful.

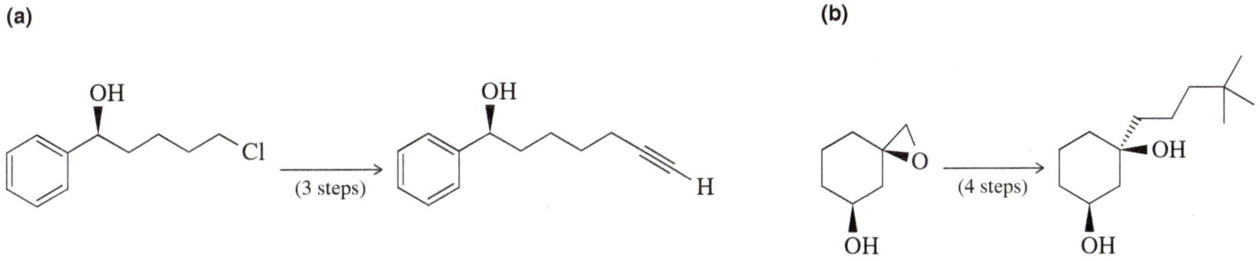

13-E CHEMISTRY IN CONTEXT: Green Chemistry

Protecting groups and atom economy

Green Chemistry Principle 2: Reduce waste by maximizing the number of atoms from all reagents that are incorporated into the final product.

Green Chemistry Principle 8: Minimize the use of temporary derivatives such as protecting groups.

While protecting groups are extremely useful in directing reactivity to a particular functional group, they are extremely wasteful. Recall from Section 7.2.2 that atom economy is a way of measuring how much of the mass contained in the reactants and reagents is preserved in the desired (usually final) product. This can be calculated over a single reaction or a series of steps. To see the impact of protecting groups, the atom economy of a protecting group–free reaction is compared to that of an identical reaction of a protected molecule in **Figure 13.91**.

These calculations imply that 59.2% of the mass that went into making the desired product from only these two steps was wasted. For reactions done on a huge, industrial scale, this amounts to costly waste disposal and potential damage to the environment if the waste is disposed of improperly. This doesn't even take into account the time, energy, and cost of the reagents and solvents that went into putting the protecting group on and taking it off. For these reasons, protecting groups receive special mention in two of the 12 principles of green chemistry.

Though protecting groups continue to be used where absolutely necessary, the design of protecting group–free syntheses has become an important goal of organic and medicinal chemists.

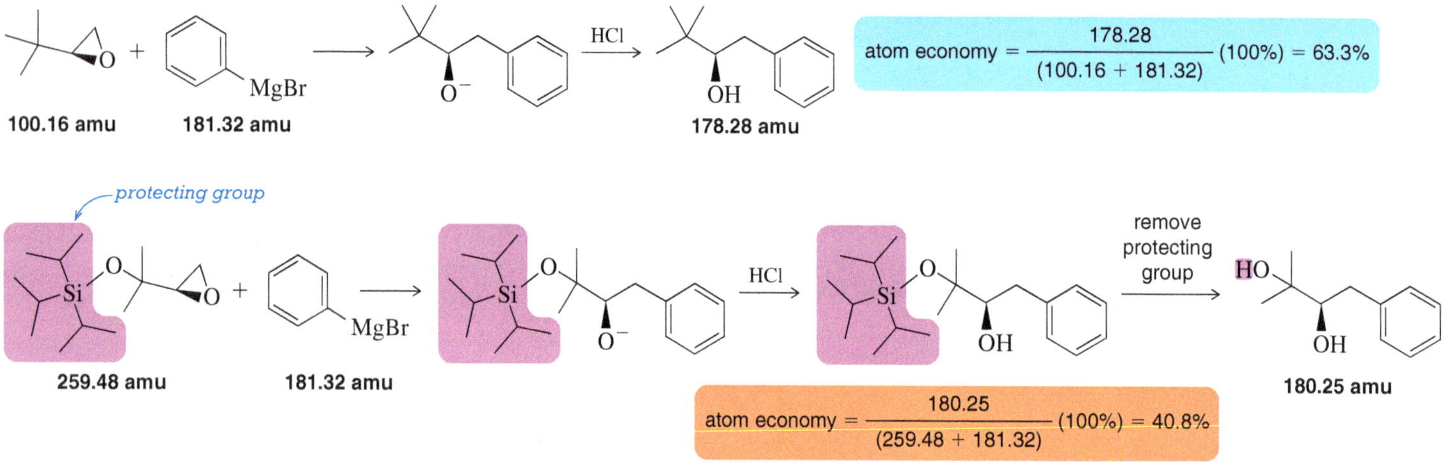

Figure 13.91 **Protecting groups reduce the atom economy of a reaction.**

Assessment 13.85 Calculate the atom economy of using a reagent like the Dess–Martin periodinane in the following oxidation reaction. Is this a green reaction?

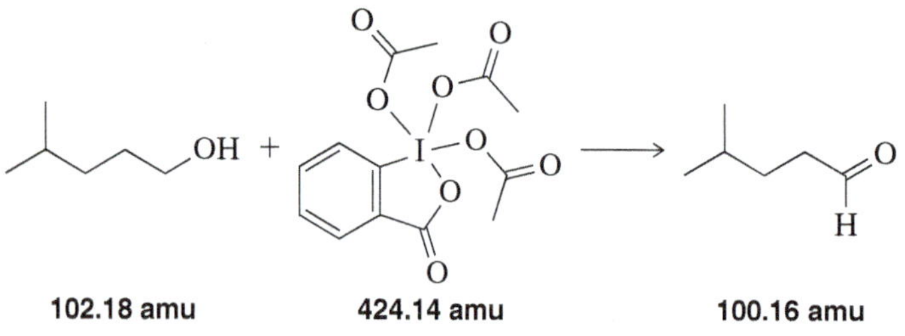

Concluding thoughts: a call to action

Because some of the greenest solutions to problems in organic chemistry have occurred in the area of organometallics, those will be the next reactions we study. But first, we take a two-chapter break where we learn how to determine the structure of the molecules we make and discover using organic spectroscopy.

Normally, this part of the chapter would conclude the story that was threaded through the chapter. Instead, I want to issue a call to action—or maybe two. First, as it relates to your studies of organic chemistry, continue looking for similarities in mechanisms among different functional groups. You won't see much in the way of new reactions going forward; most of what we cover will just be recombining old ideas into new situations. This chapter was a start in that direction. [*Stay on top of your studies, and you will find it getting easier and easier.*]

The second call to action deals with green chemistry. We saw in this chapter a variety of problems caused by classical chemistry and some solutions that have mitigated these problems by offering greener opportunities. But even at the highest levels [*beyond the scope of an organic course*], there are still discoveries to be made. Should you choose to become an organic chemist, always keep the principles of green chemistry in mind as you design reactions, products, and techniques. And if you don't pursue a career in organic chemistry, you can still do your part by demanding that the products you consume originate with companies that behave ethically and with a sustainable future in mind.

End-of-Chapter Materials

Important Concepts

[*These are the major concepts covered in each section, excluding reactions.*]

13.1: Because of their similar structure, water, alcohols, ethers, thiols, and thioethers display similar reactivity.

13.2: Alcohols are named such that the hydroxyl group is afforded the lowest number in the chain. A prefix (di-, tri-, tetra-, etc.) is used to indicate the presence of more than one hydroxyl group in polyols. The substitution of an alcohol is based on the number of carbons attached to the hydroxyl-bearing carbon.

13.3: Alcohols are hydrogen bond donors and acceptors. As a result, they have high boiling points and melting points and are good solvents.

13.4: Two nonbonded lone pairs on the hydroxyl oxygen make alcohols weak nucleophiles.

13.6: Alcohols are weak bases and weak acids.

13.7: The hydroxyl group is not a good leaving group, making substitution difficult. To do substitution at a hydroxyl-bearing carbon, the hydroxyl group should be converted to a tosylate (TsCl, Et_3N), chloroalkane (HCl or $SOCl_2$), or bromoalkane (HBr or PBr_3).

13.8: Elimination of water to make an alkene is the opposite of alkene hydration and can be carried out similarly with sulfuric acid as catalyst. Water is removed from the reaction mixture to shift the equilibrium to the alkene. When a carbocation is involved, rearrangement can occur. The pinacol rearrangement results in elimination of water from a diol to produce a ketone or aldehyde via hydride or alkyl rearrangement.

13.9: Primary (1°) alcohols are oxidized to aldehydes (PCC, CH_2Cl_2) or carboxylic acids (H_2CrO_4, H_2O) depending on the presence or absence of water. Secondary (2°) alcohols, regardless of reagent used, are oxidized to ketones. The mechanism of all alcohol oxidations is similar, involving formation of an intermediate that undergoes an elimination reaction to form a new C—O π bond. Syn diols can be oxidized via cleavage of the C—C σ bond to form two new carbonyls by periodic acid.

13.10: Ethers are polar molecules lacking an acidic hydrogen, so they are good aprotic solvents, capable of dissolving ionic, polar, and nonpolar compounds.

13.11: The synthesis of ethers containing at least one 1° carbon can be done using the Williamson ether synthesis, which is essentially an S_N2 reaction of an alkoxide with a 1° haloalkane. The acid-catalyzed addition of an alcohol to an alkene, proceeding through a carbocation intermediate, can be used to make ethers where neither carbon is 1°.

13.12: Like alcohols, ethers are nucleophilic at oxygen and deprotonate strong acids. Cleavage of ethers occurs by protonating the oxygen to make a good leaving group that participates in an S_N1 or S_N2 reaction, depending on the substitution of the carbon involved.

13.13: Epoxides are strained ethers, so they can be opened with weak or strong nucleophiles. Only after protonation of the epoxide oxygen, weak nucleophiles attack at the more substituted carbon (more stable carbocation) from the opposite side. Strong nucleophiles react directly, opening the epoxide at the least substituted (and least sterically hindered) carbon.

13.14: When multiple functional groups are present in a molecule, it is often necessary to use a protecting group to allow selective reactions. By temporarily blocking one functional group, a reaction is performed with whatever functional group remains. The protected functional group is later deprotected. The silyl ether, formed by the reaction of a chlorosilane and an alcohol, is a commonly used protecting group. Owing to the strength of the Si—F bond, fluoride reagents are used to deprotect silyl ethers, regenerating the original alcohol.

Important Reactions and Mechanisms

1. **Formation of a tosylate (Section 13.7.1):** Nucleophilic attack by the alcohol displaces the chloride, resulting in an acidic proton that is removed by the bulky amine base. The resulting tosylate is used as a good leaving group in substitution and elimination reactions.

General:

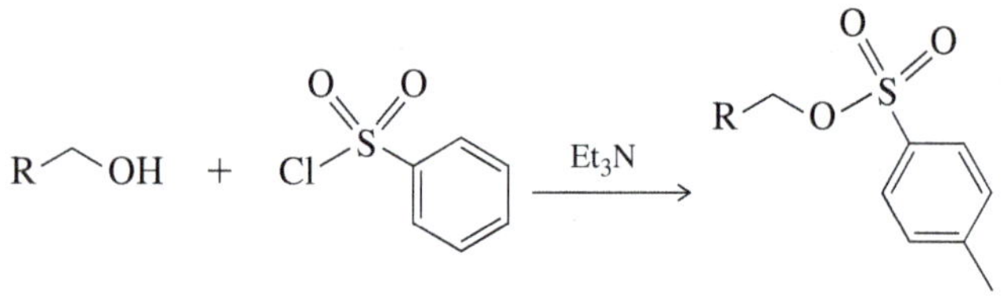

Specific:

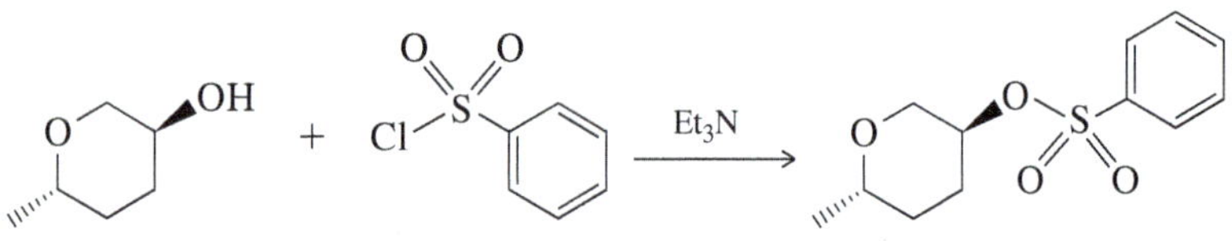

Mechanism:

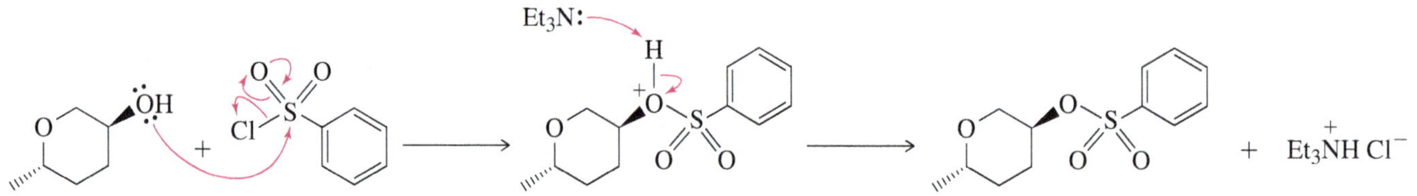

2. **Reaction with HX (Section 13.7.2):** Protonation of an alcohol results in a hydronium ion (a good leaving group). At a 1° carbon, water is displaced by the chloride in an S_N2 reaction. At 2° and 3° carbons, the water leaves to form a carbocation. The resulting carbocation is attacked by the bromide to give an alkyl halide with partial racemization. Like all reactions that involve carbocation intermediates, rearrangement is possible.

General:

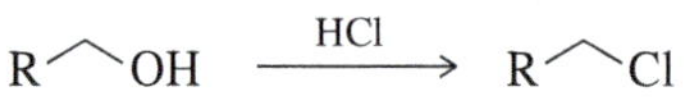

Specific:

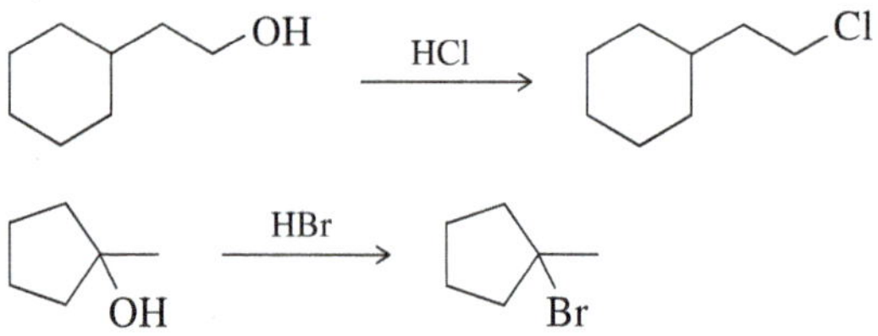

Mechanism:

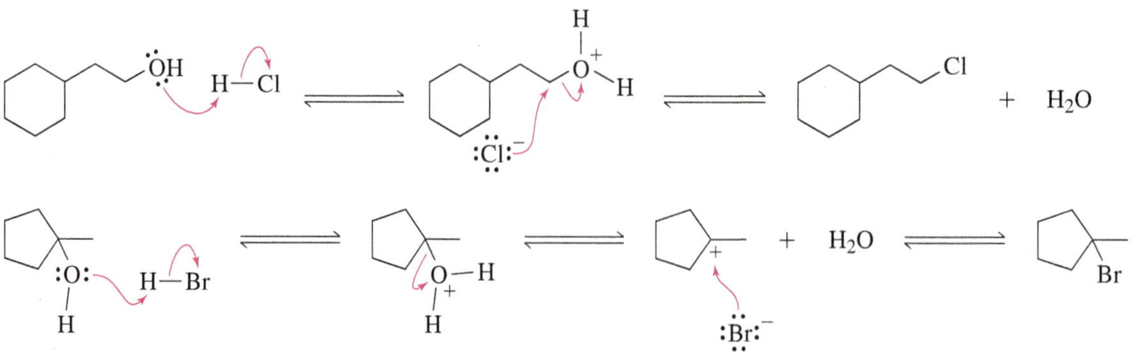

3. Reaction with $SOCl_2$ (Section 13.7.3): The nucleophilic alcohol displaces the chloride in thionyl chloride, resulting in an acidic proton that is removed by pyridine. At a 1° or 2° carbon, chloride then displaces the improved leaving group by an S_N2 reaction. At a 3° carbon, the reaction proceeds through an S_N1 reaction, giving a racemic mixture. In the absence of pyridine, the reaction proceeds similarly, but results in retention of stereochemistry.

General:

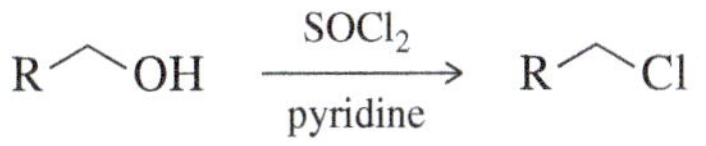

Specific:

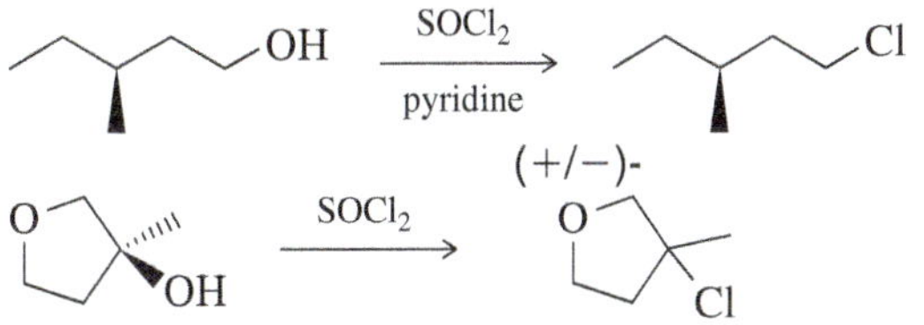

Mechanism:

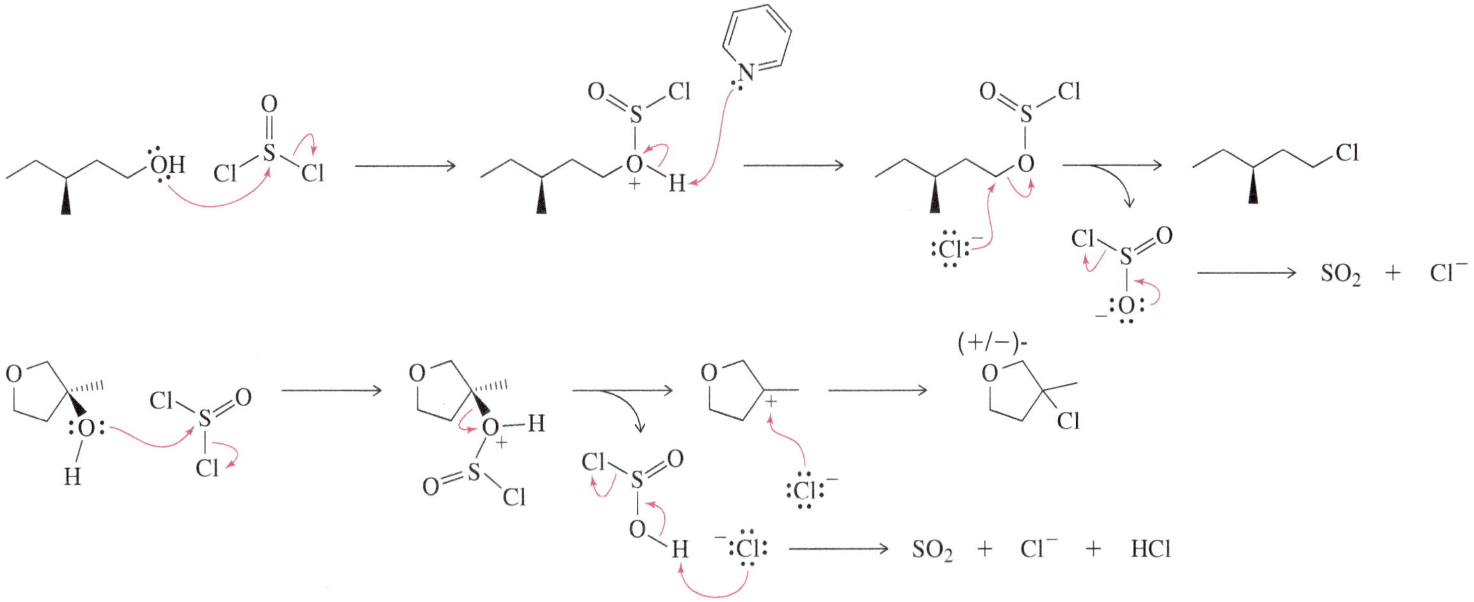

4. Reaction with PBr_3 (Section 13.7.4): The nucleophilic alcohol displaces the bromide in phosphorus tribromide. At a 1° or 2° carbon, bromide then displaces the improved leaving group by an S_N2 reaction. At a 3° carbon, the reaction proceeds through an S_N1 reaction, giving a racemic mixture.

General:

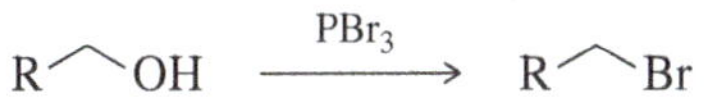

Specific:

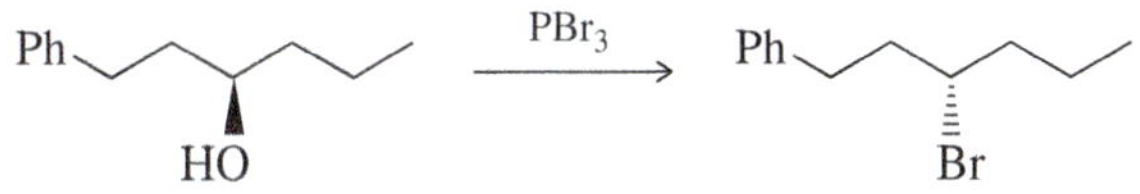

Mechanism:

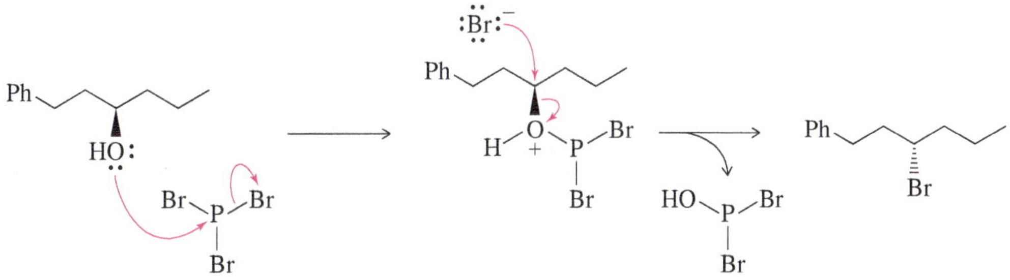

5. **Elimination with H_2SO_4 (Section 13.8):** Protonation of the hydroxyl group produces a hydronium ion (a good leaving group). At a 1° carbon, elimination occurs through an E2 reaction to give an alkene. At a 2° or 3° carbon, elimination occurs through an E1 reaction. Like all reactions that involve carbocation intermediates, rearrangement is possible.

General: **Specific:**

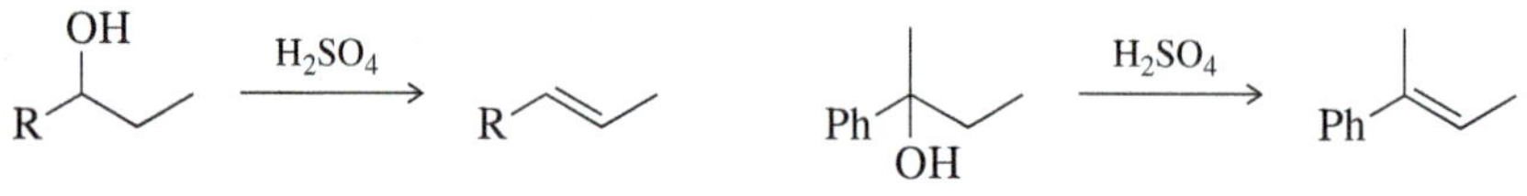

Mechanism:

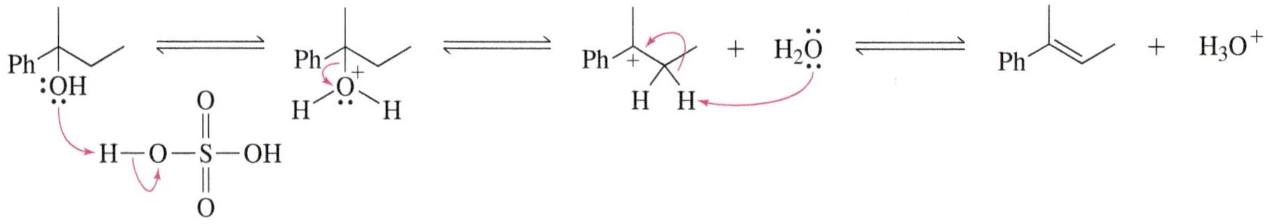

6. **Pinacol rearrangement (Section 13.8.2):** Protonation of one hydroxyl group of a diol produces a hydronium ion (a good leaving group). Water leaves to produce a carbocation that rearranges by a 1,2-alkyl or hydride shift to produce a new carbocation that is further stabilized by resonance. Removal of the acidic proton completes formation of the aldehyde or ketone.

General: **Specific:**

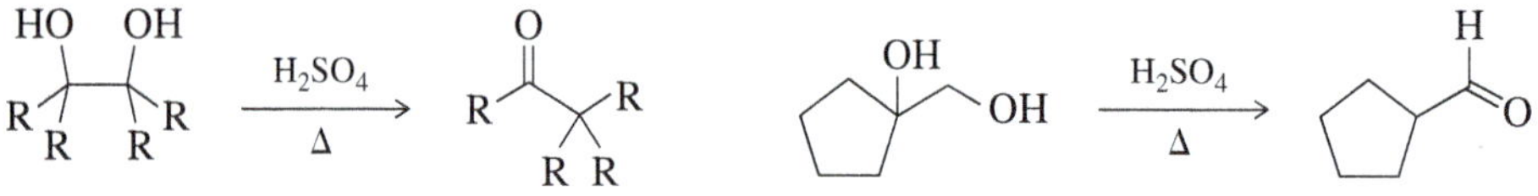

Mechanism:

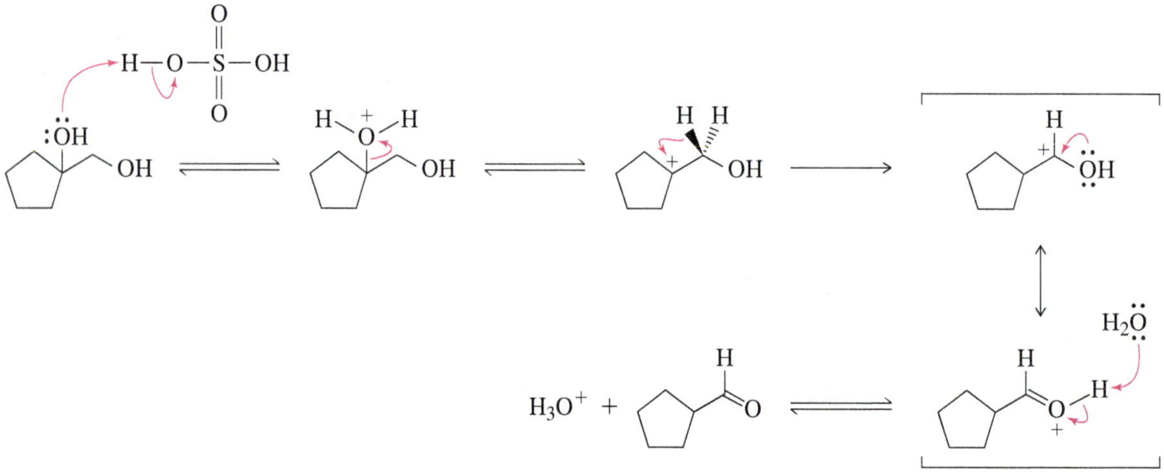

7. **H_2CrO_4 oxidation of 1° alcohols (Section 13.9.3):** Displacement of a hydroxyl group on chromic acid with an alcohol results in formation of a chromate ester. E2 elimination results in reduction of chromium and formation of a new C—O π bond. In the presence of water, the resulting aldehyde equilibrates to a diol that undergoes a second oxidation to give a carboxylic acid. Under similar conditions, a 2° alcohol forms a ketone.

General:

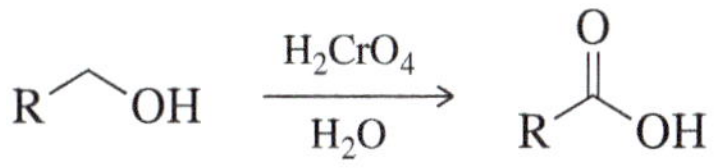

Specific:

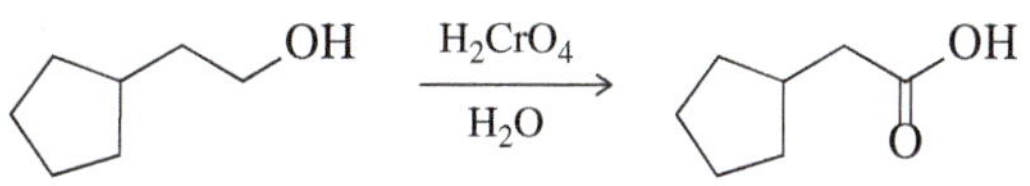

Mechanism:

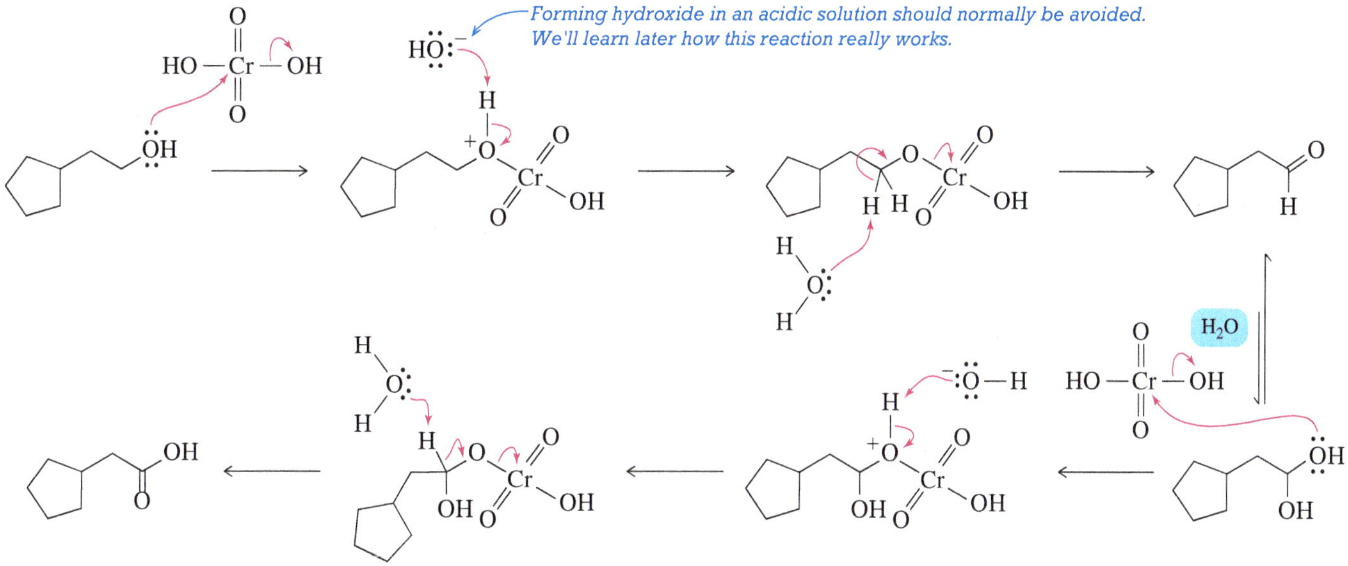

8. **PCC oxidation of 2° alcohols (Section 13.9.2):** Displacement of a chloride from PCC with an alcohol results in formation of a chromate ester. E2 elimination results in reduction of chromium and formation of a new C—O π bond. Because water is not present, the reaction of a 1° alcohol stops at the aldehyde oxidation state.

General:

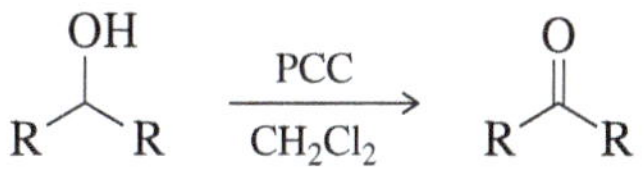

Specific:

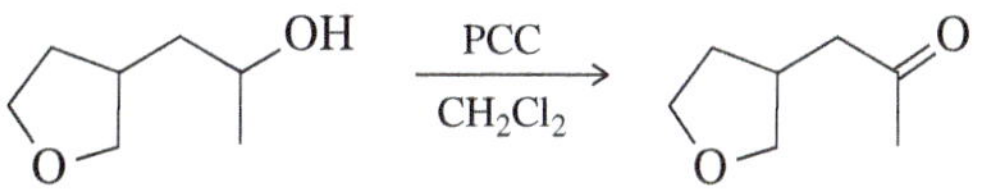

Mechanism:

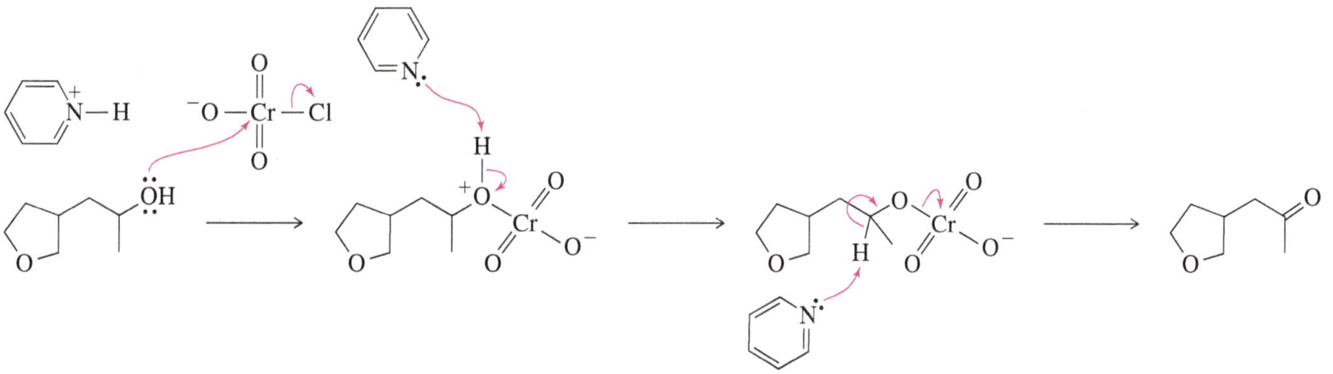

9. HIO_4 cleavage of diols (Section 13.9.5): Formation of a cyclic iodate ester results in cleavage of a C—C σ bond and formation of two carbonyls through a 6-electron cyclic transition state. This reaction only works with syn diols or ones that can become so by rotation around a single bond.

General:

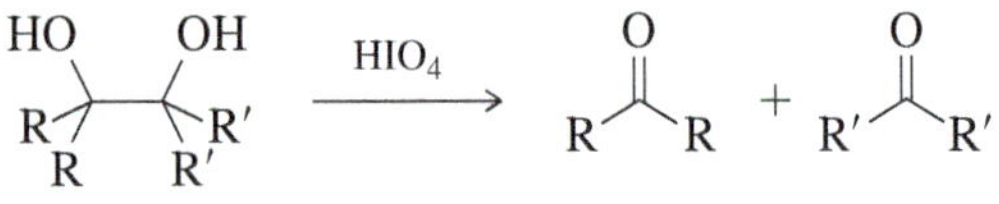

Specific:

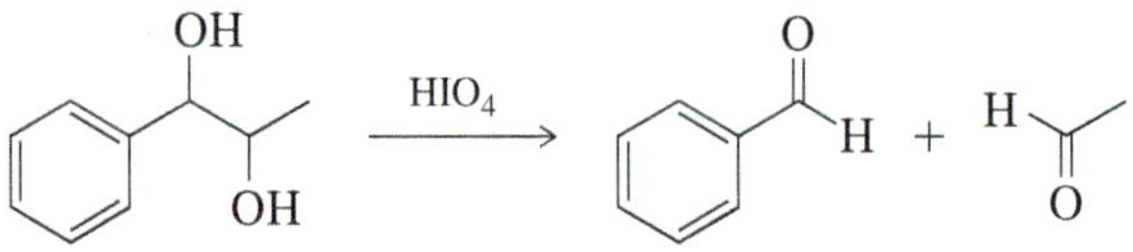

Mechanism:

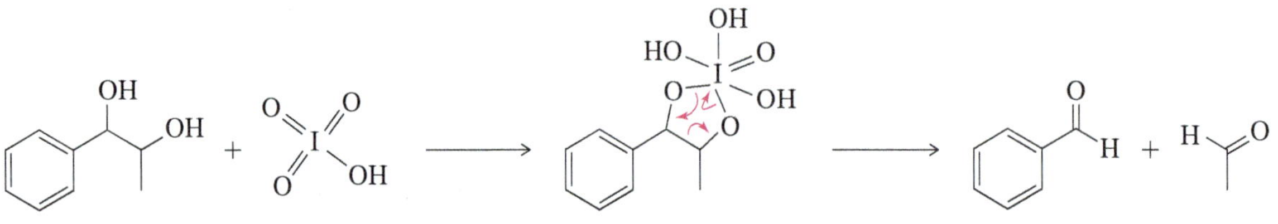

---

10. Williamson ether synthesis (Section 13.11.1): Reduction of an alcohol with sodium or deprotonation of an alcohol gives an alkoxide (strong nucleophile) that displaces leaving groups from 1° carbons in an S_N2 reaction. This reaction does not work at 2° or 3° carbons because elimination by E2 becomes the favored pathway.

General:

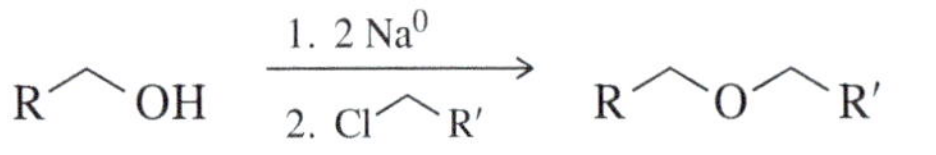

Specific:

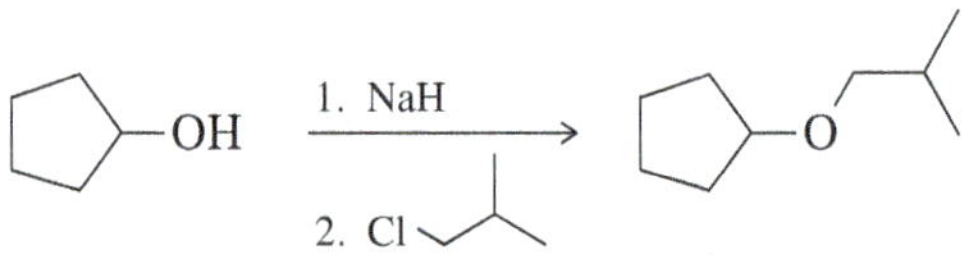

Mechanism:

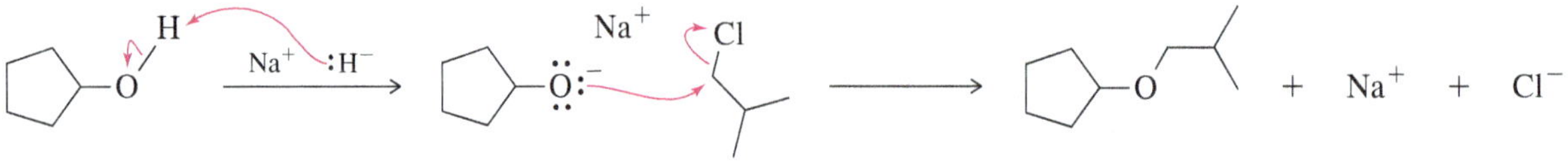

---

11. Oxymercuration/reduction with alcohols (Section 13.11.2): The nucleophilic alkene attacks mercury acetate, forming a bridged mercurinium ion in a concerted process. The alcohol opens the mercurinium ion at the more substituted carbon on the face opposite the mercury to generate the anti hydroxy organomercury compound. In the second step, the carbon mercury bond is reduced by sodium borohydride, replacing the mercury with a hydrogen.

General:

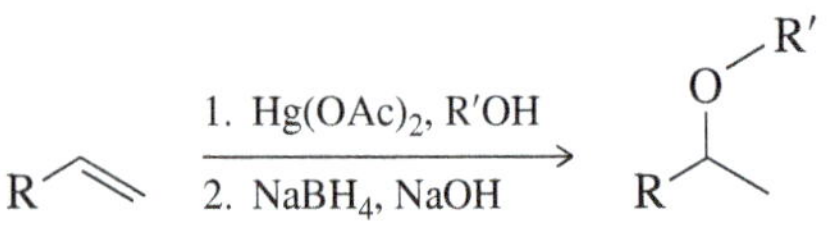

Specific:

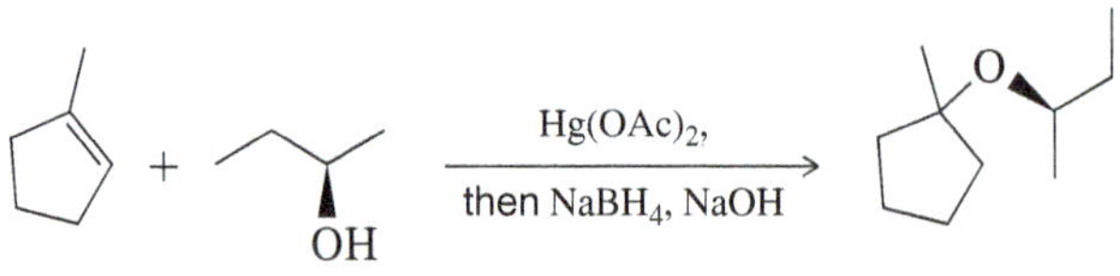

Mechanism:

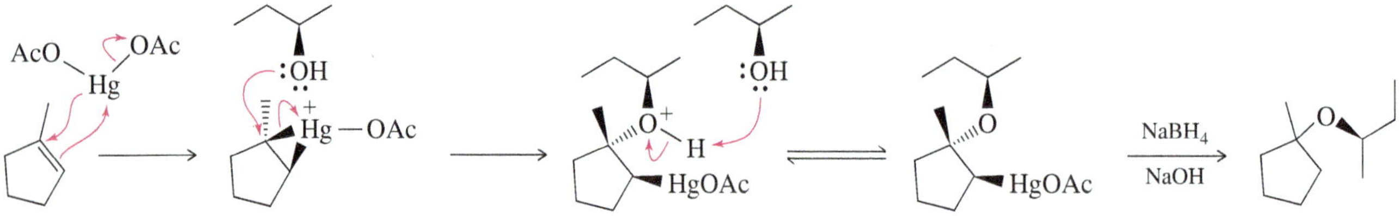

12. HX cleavage of ethers (Section 13.12.1): Protonation of the ether oxygen results in an oxonium ion (a good leaving group). At a 1° carbon, the alcohol is displaced by the halide in an S_N2 reaction. At 2° and 3° carbons, alcohol leaves to form a carbocation. The resulting carbocation is attacked by the halide to give an alkyl halide with partial racemization. The resulting alcohol can undergo a second substitution by protonation with HX and displacement of water according to the mechanism in Section 13.7.2. When this reaction proceeds through a carbocation, rearrangement is possible.

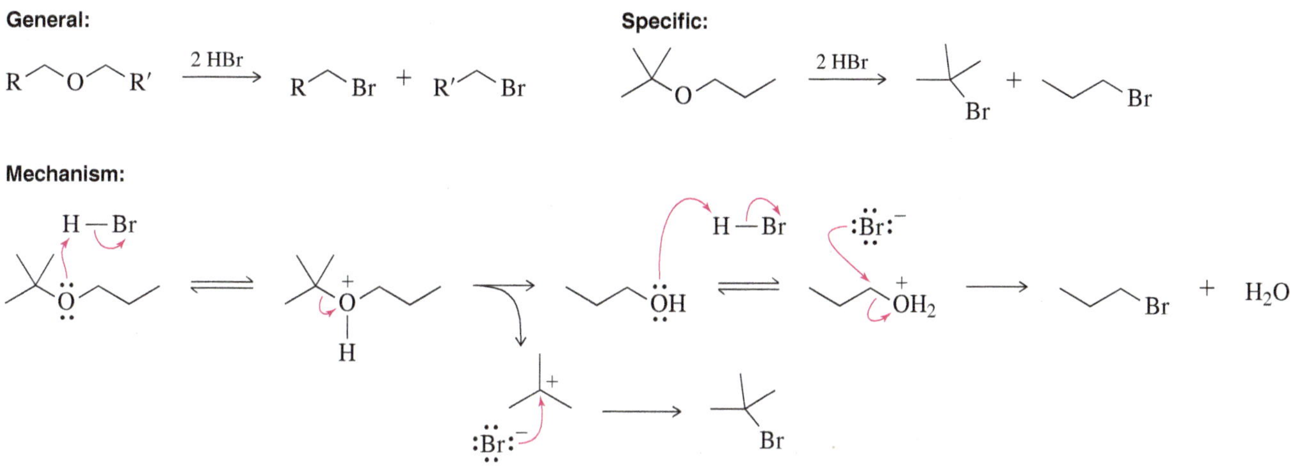

13. Acid-catalyzed hydrolysis of ethers (Section 13.12.1): Protonation of the ether oxygen results in an oxonium ion (a good leaving group). At a 1° carbon, the alcohol is displaced by water in an S_N2 reaction. Because water brought along its proton, a final acid–base reaction occurs to produce an alcohol. At 2° and 3° carbons, alcohol leaves to form a carbocation that is attacked by water in an S_N1 reaction. When proceeding through a carbocation, rearrangement is possible.

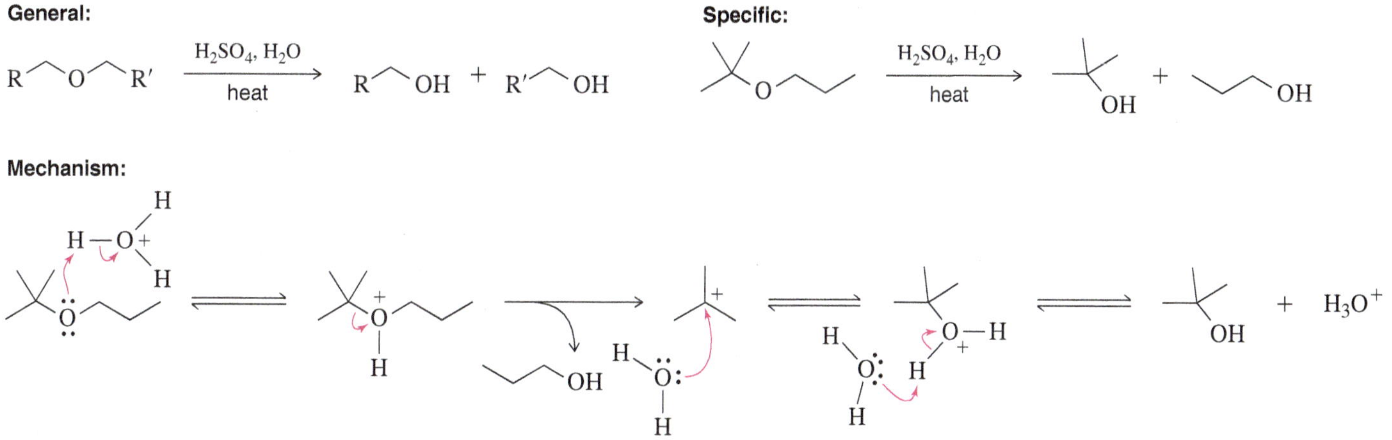

14. Epoxide opening with weak nucleophiles (Section 13.13.2): Protonation of the oxygen of the epoxide ring weakens the C—O bond at the most substituted carbon, initiating nucleophilic attack by water. In an analogous fashion to the chemistry of the bromonium, chloronium, or mercurinium ion, attack occurs at the more substituted carbon from the opposite face of the oxonium ion, resulting in the preparation of the trans diol in a stereospecific reaction.

General:

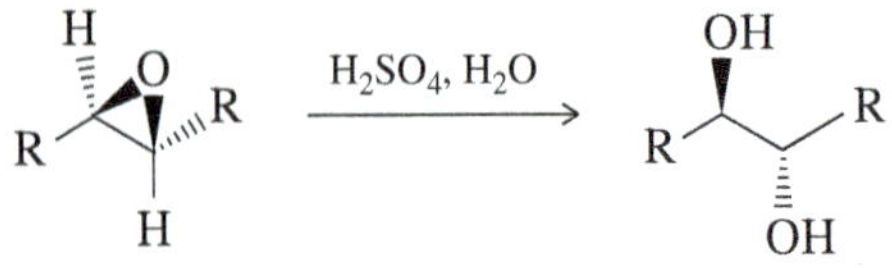

Specific:

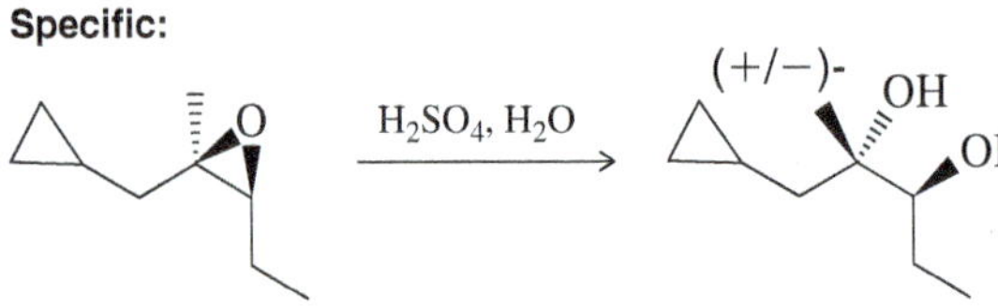

Mechanism:

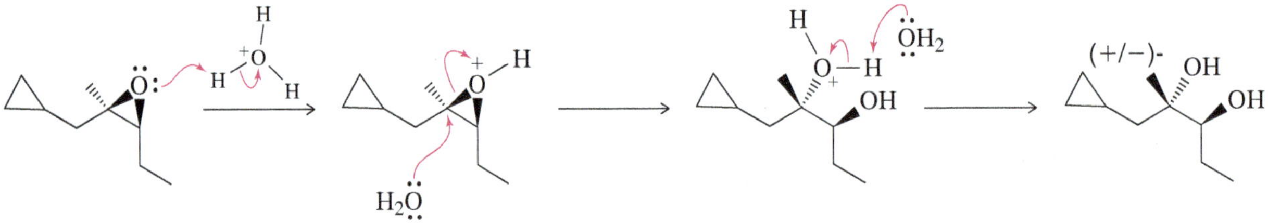

15. Epoxide opening with strong nucleophiles (Section 13.13.2): As we studied in the S_N2 reactions of Chapter 12, strong nucleophiles directly attack epoxides at the least sterically hindered carbon. An acid quench results in formation of an alcohol.

General:

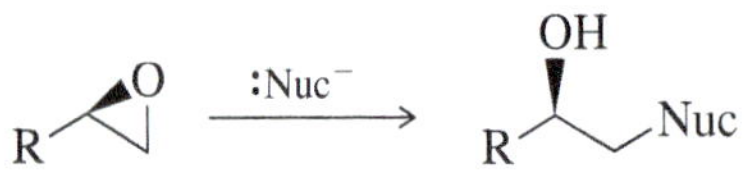

Specific:

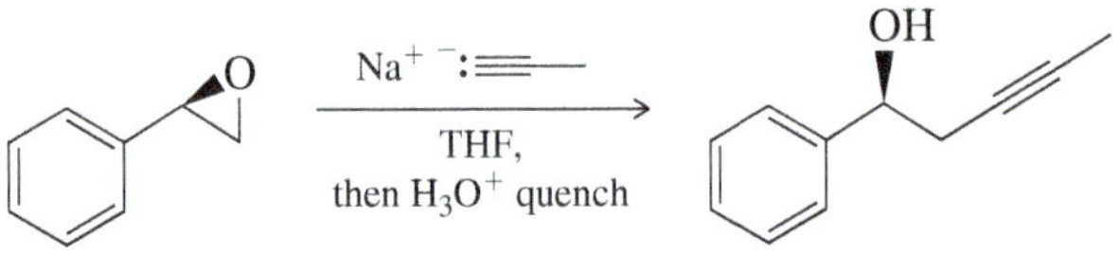

Mechanism:

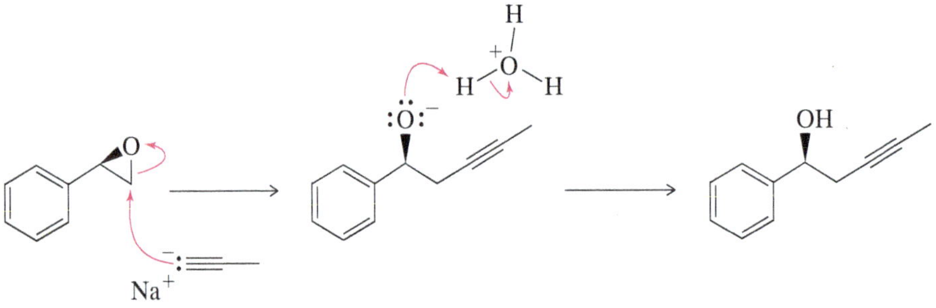

End-of-Chapter Assessments

[*Assessments are ranked according to their difficulty;* •••• = *most difficult.*]

13.86 (••) Using IUPAC rules, name the following molecules.

(a)

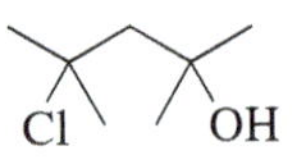

(b)

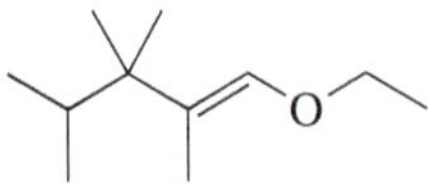

(c)

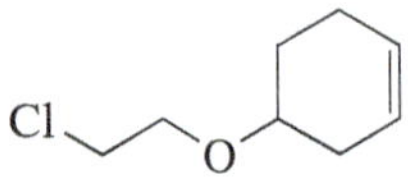

(d)

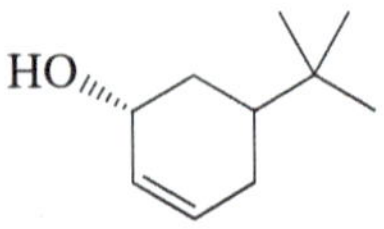

(e)

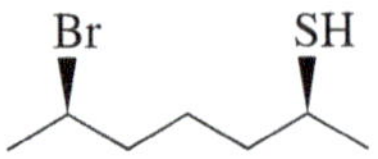

13.87 (••) Draw the correct structure from the following IUPAC names:

(a) (4*R*,2*Z*)-4-methylhex-2-en-1-ol
(b) 4-methoxybut-1-yne
(c) (1*S*,4*R*)-4-bromocyclohex-2-en-1-ol
(d) (2*R*,3*S*)-methoxy-3-methylpentane
(e) (*R*)-2,2-dimethyl-1-phenylpropane-1-thiol
(f) (*S*)-4-isopropoxypent-2-yne.

13.88 (••) Predict the product(s) that would result when the following molecules are allowed to react under the following conditions: (i) 1. BH_3 2. NaOH, H_2O_2; (ii) 1. $Hg(OAc)_2$, H_2O 2. $NaBH_4$; (iii) H_2SO_4, H_2O; (iv) 1. OsO_4 2. $NaHSO_3$; (v) H_2O; (vi) NaOH, H_2O. If there is no reaction, write "no reaction."

Do (i)–(iv) for these molecules.

(a)

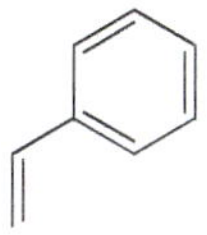

(b)

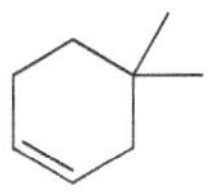

(c)

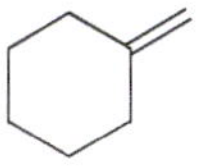

Do (v) and (vi) for these molecules.

(d)

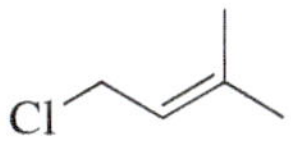

(e)

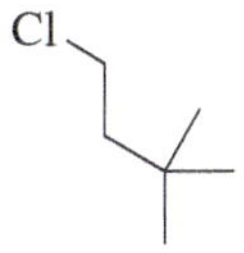

(f)

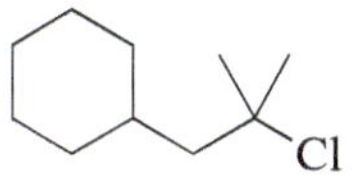

(g)

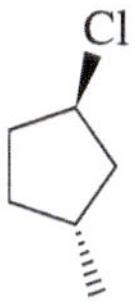

13.89 (••) Predict the product for each of the following reactions.

(a)

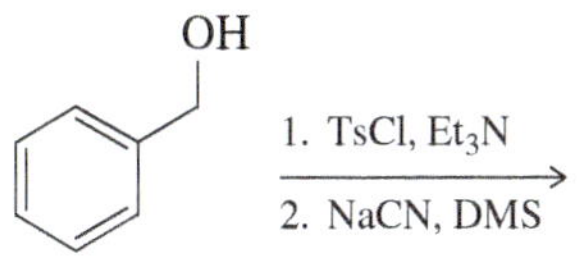

(b)

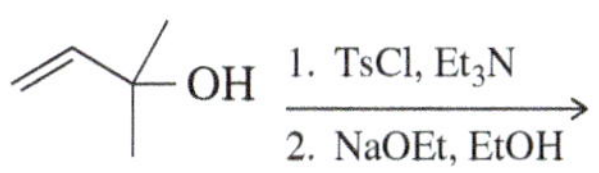

(c)

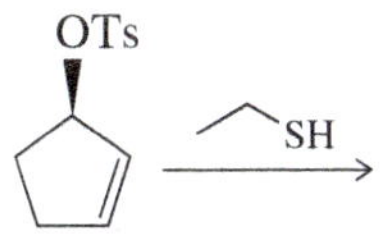

(d)

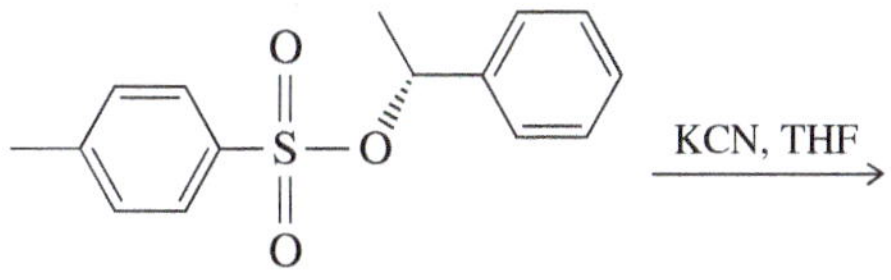

13.90 (••) Identify whether each of the following reactions proceed by an S_N1, S_N2, E1, or E2 mechanism.

(a)

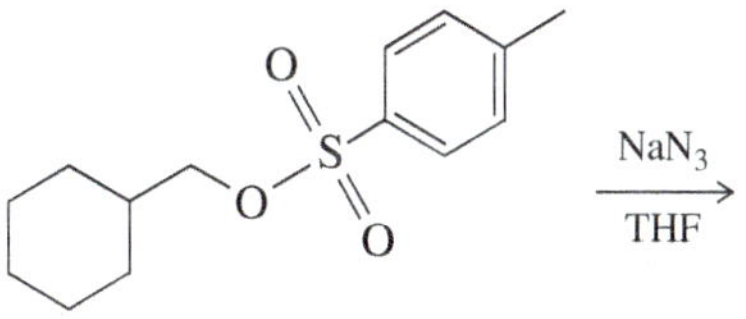

(b)

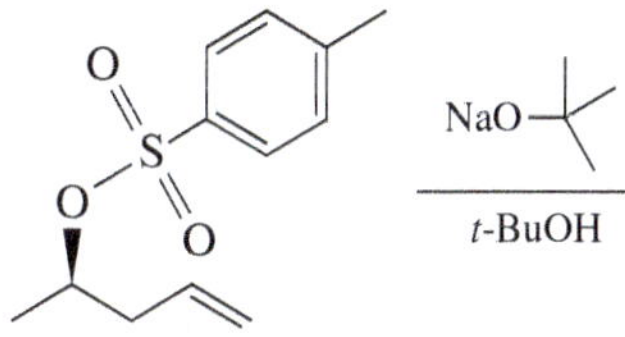

(c)

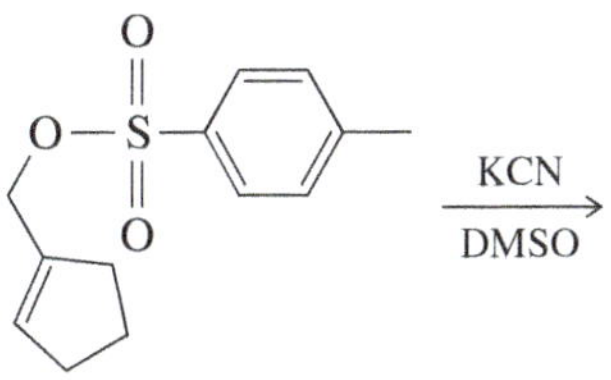

(d)

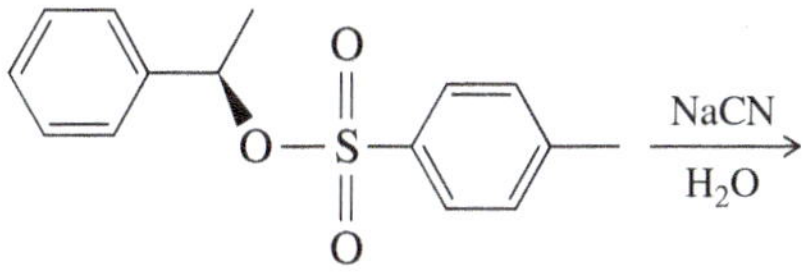

13.91 (•••) A trifluoromethanesulfonate (triflate) can be used in a manner similar to tosylates. Which would you expect to be a better leaving group? Why?

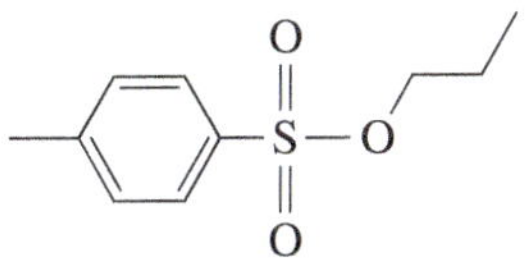

tosylate

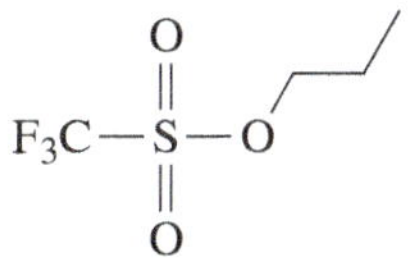

triflate

13.92 (•••) Using an appropriate tosylate intermediate, synthesize the following molecules starting from the appropriate alcohol.

(a)

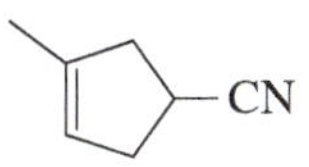

(b)

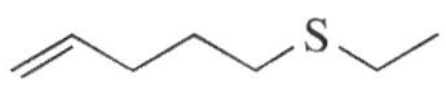

13.93 (••) (a) Show how you would affect the following transformation using a tosylate. (b) Why might this not be the most sustainable method? (c) What reagent might you use instead?

13.94 (•••) Predict the product of the following reactions.

(a)

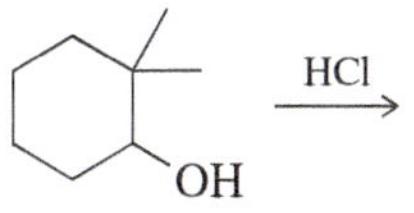

(b)

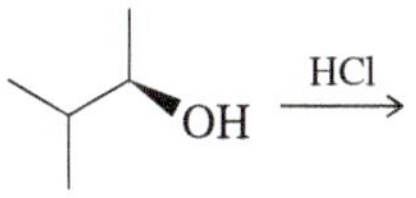

(c)

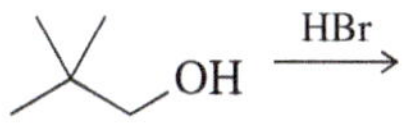

13.95 (••) Draw a reaction coordinate diagram for the reaction of ethanol and HCl. Be sure to label all products, reactants, and intermediates. Calculate the equilibrium constant of the first step.

13.96 (••••) A graduate student attempted the following reaction and did not isolate the expected product. (a) What product did they isolate? (b) What reagent should they have used instead to get their desired product?

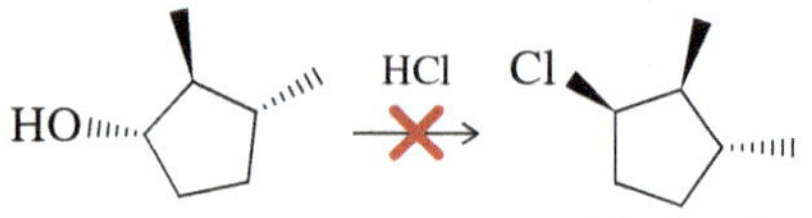

not isolated

13.97 (•••) Suggest the appropriate reagents to carry out the following transformations.

(a)

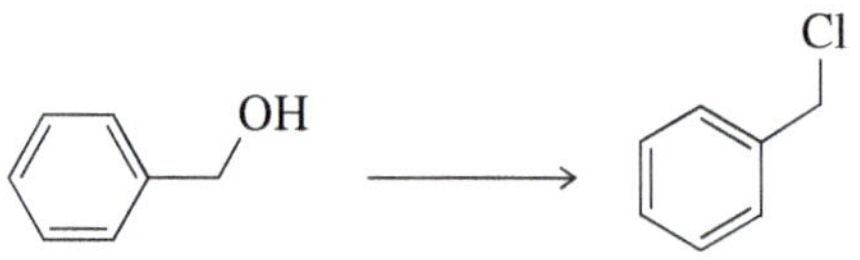

(b)

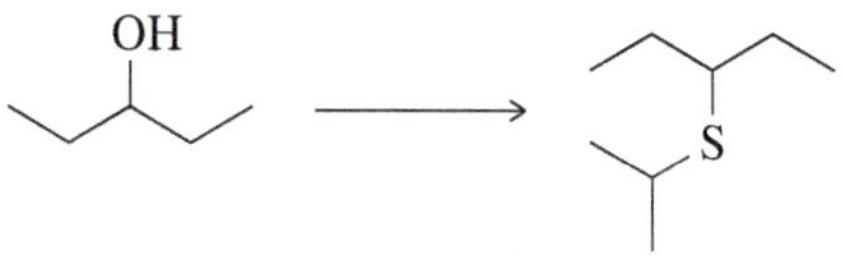

(c)

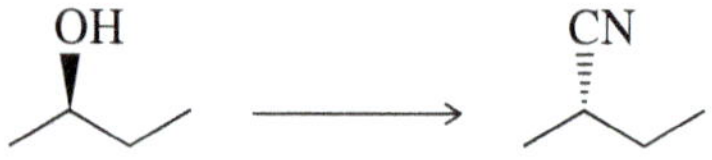

(d)

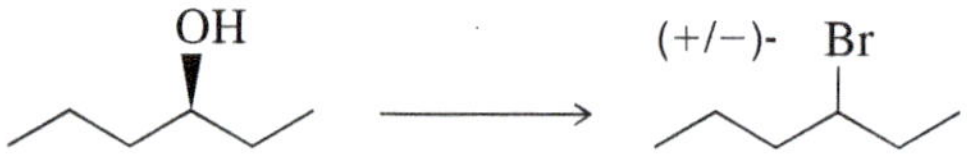

(e)

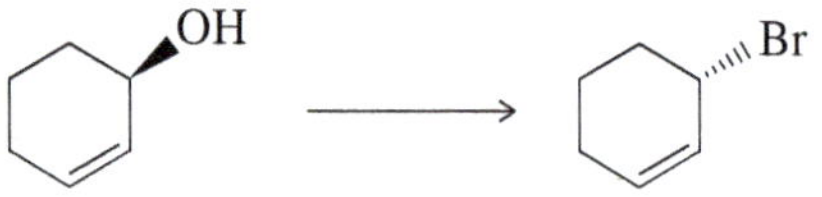

(f)

(g)

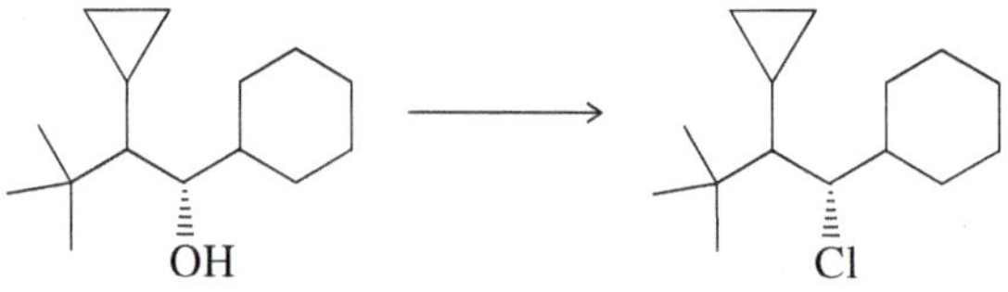

13.98 (•••) Draw a reaction coordinate diagram for the following transformation. Label all product, intermediates, and substrates.

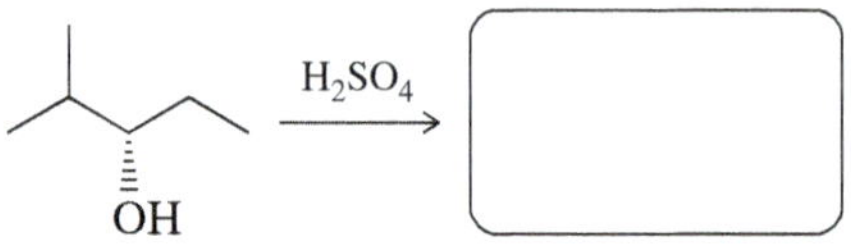

13.99 (•••) Produce a mechanism for the following transformation.

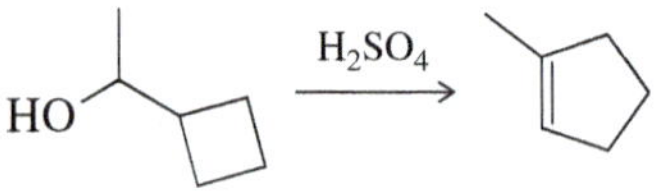

13.100 (•••) Predict the product and draw the mechanism for the acid-catalyzed pinacol rearrangement of the following diols.

(a)

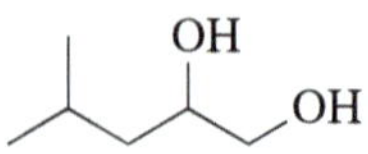

(b)

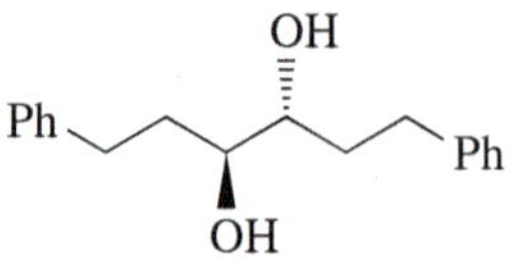

(c)

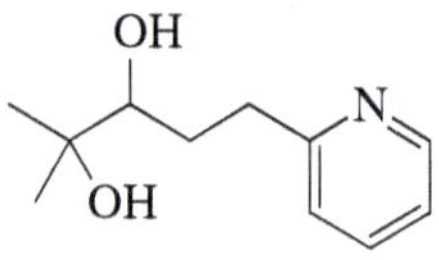

13.101 (•••) Draw a mechanism for the following oxidation reactions.

(a)

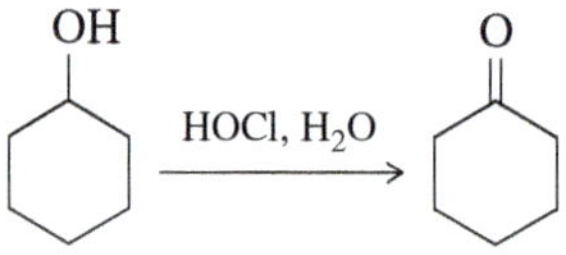

(b)

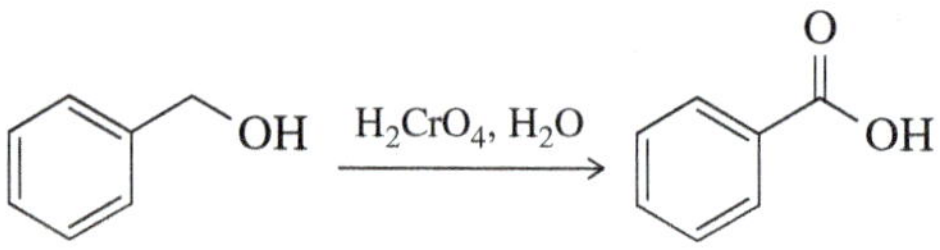

(c)

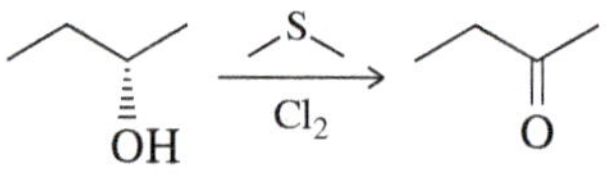

(d)

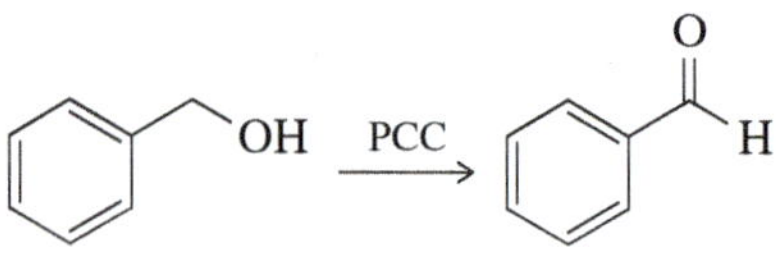

13.102 (••) Identify the two possible combinations of haloalkane and alkoxide that can be used to make the following ether.

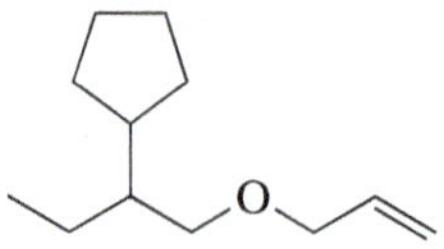

13.103 (••) In contrast to Assessment 13.102, only one combination of haloalkane and alkoxide can be used in the Williamson ether synthesis to make the ether shown. Identify the combination and explain why it is the only combination that works.

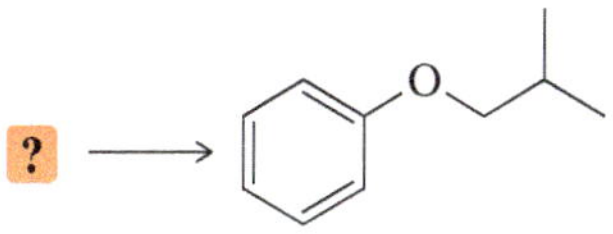

13.104 (••••) Suggest a synthesis for the following molecules beginning with organic molecules containing three or fewer carbons. [*You will need to use a protecting group in these syntheses.*]

(a)

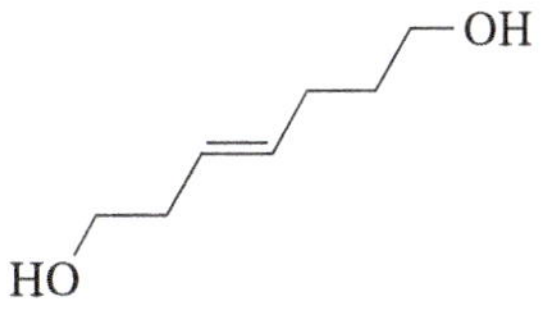

(b)

(c)

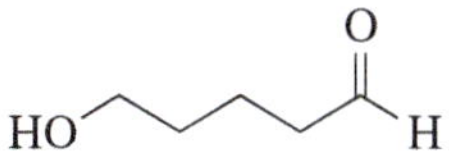

13.105 (•••) The reaction of alkoxides with 2° haloalkanes is not a viable way to form ethers. (a) Why? (b) Why can thioethers be formed by an analogous reaction?

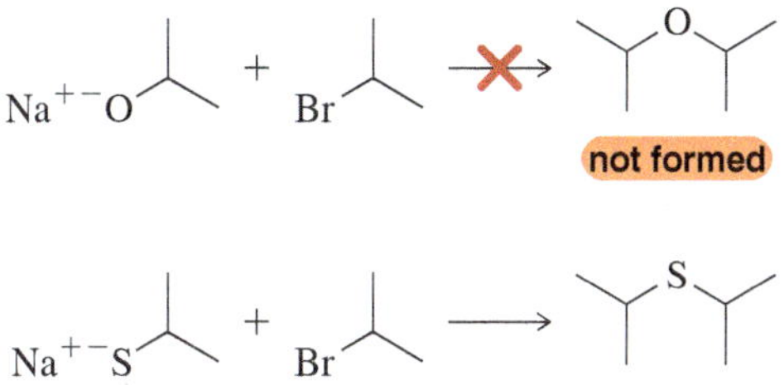

13.106 (••) Predict the product(s) that would result when molecules (a)–(p) are allowed to react under the following conditions: (i) $SOCl_2$; (ii) PBr_3; (iii) $SOCl_2$, NEt_3; (iv) 1. TsCl, Et_3N 2. NaCN; (v) 1. TsCl, Et_3N 2. NaO*t*-Bu; (vi) H_2SO_4; (vii) HCl; (viii) HBr; (ix) PCC; (x) H_2CrO_4, H_2O; (xi) HOCl, H_2O; (xii) HIO_4. If no reaction occurs, write "no reaction."

(a)

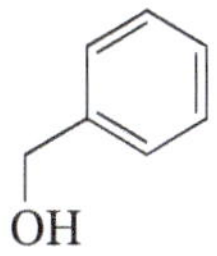

(b)

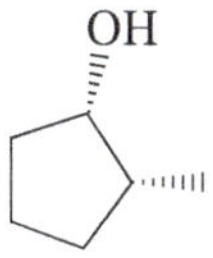

(c)

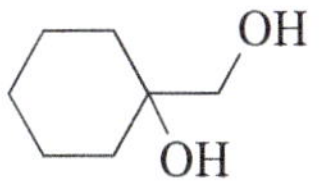

(d)

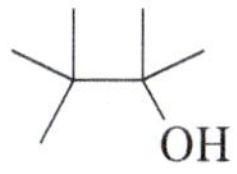

(e)

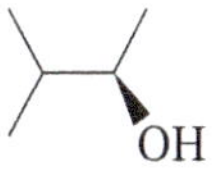

(f)

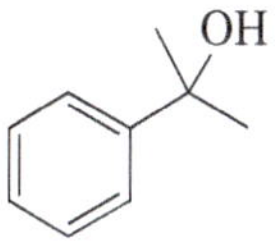

(g)

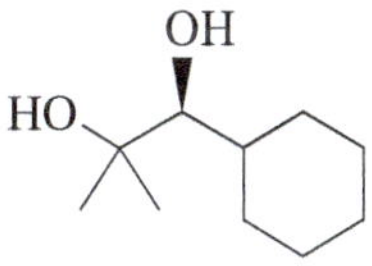

(h)

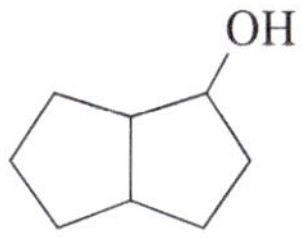

(i)

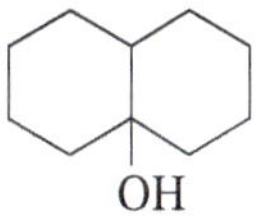

(j)

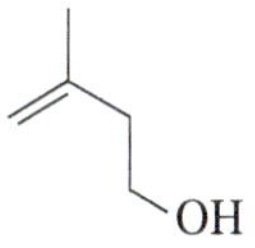

(k)

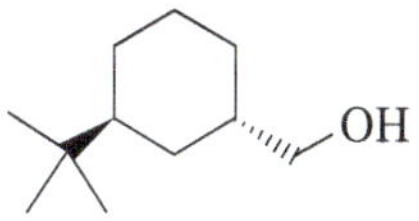

(l)

(m)

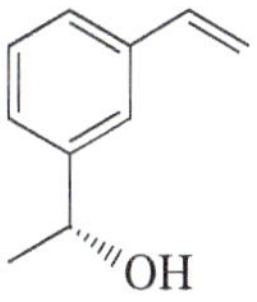

(n)

(o)

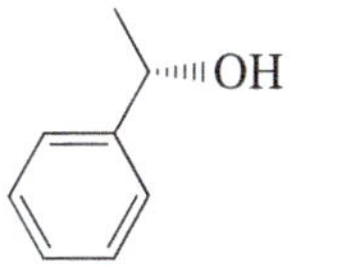

(p)

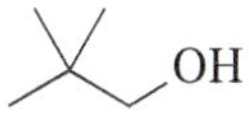

13.107 (••••) **LOOKING BACK** Cleavage of the following ether produces the alcohol and haloalkane only, regardless of how much HBr is used. Thinking about the mechanism of the reaction, explain why bromobenzene is not also a product of this reaction.

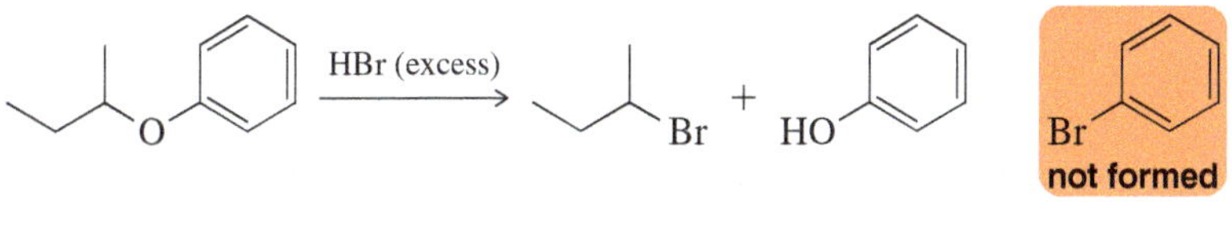

13.108 (••••) Epichlorohydrin reacts twice with strong nucleophiles to form two new C—C bonds. (a) Suggest an arrow-pushing mechanism for each step of the reaction. (b) Despite this reaction beginning with a chiral and nonracemic epichlorohydrin, the product is achiral. Why?

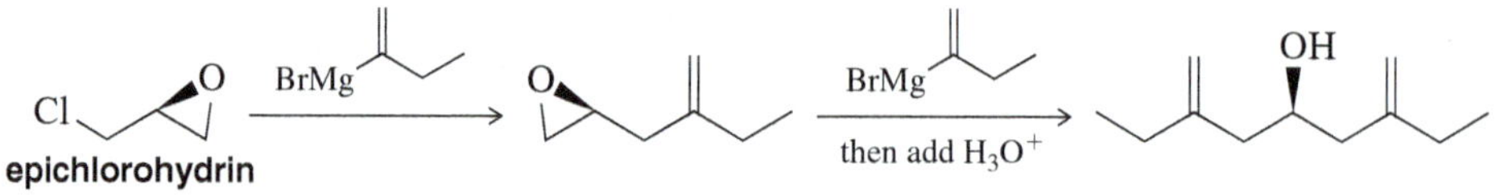

13.109 (••••) Triphenylphosphine and iodine can be used to convert alcohols to iodoalkanes. Suggest a mechanism for this reaction. [*Triphenylphosphine first acts as a nucleophile in this reaction.*]

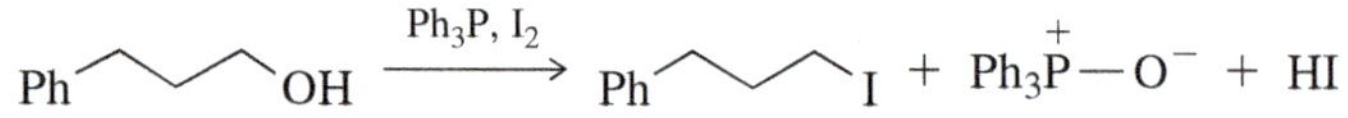

13.110 (••••) Another method for converting alcohols to chloroalkanes makes use of chlorotrimethylsilane (TMSCl) and DMSO. Suggest a mechanism for this reaction to form (a) a 1° chloroalkane and (b) a 3° chloroalkane. [*The reaction begins by the reaction of DMSO and TMSCl and is analogous to the Swern oxidation.*]

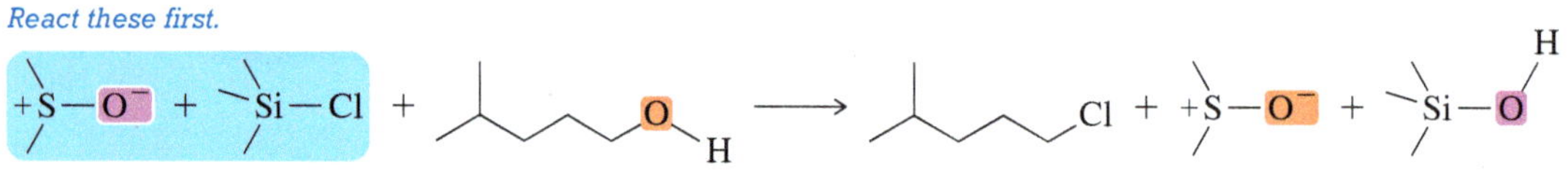

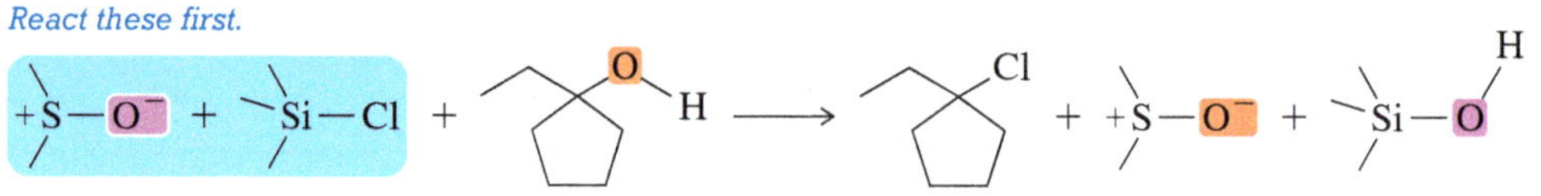

13.111 (••••) **THINKING AHEAD** We explain in Chapter 24 that bisphenols can be oxidized to quinones. (a) Calculate the oxidation numbers of C1 and C_2 in going from reactant to product. (b) Provide a mechanism for this transformation. [*The reaction begins like the alcohol oxidations of Section 13.9.*]

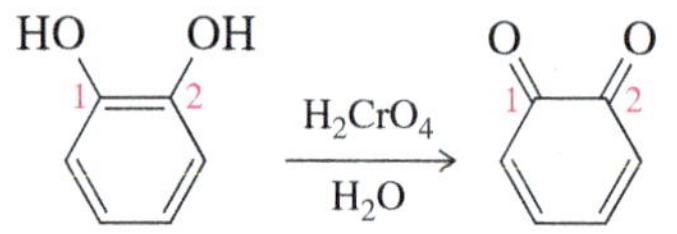

13.112 (••••) **LOOKING BACK** In Chapter 12, we learned that crown ethers were used to increase the rate of S_N2 reactions (Assessment 12.80). Suggest a synthesis of 15-crown-5 using the reactions learned here in Chapter 13.

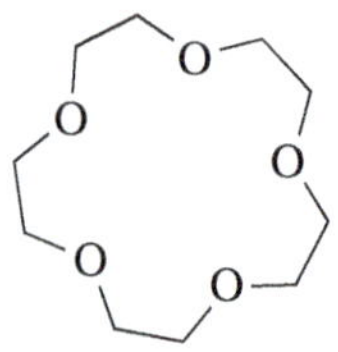

13.113 (•••) **THINKING AHEAD** Thiols are prone to dimerize through the formation of a disulfide bond in a reaction that stylists use to induce permanent curls in hair. [*You'll learn about the importance of this reaction to protein structure in biochemistry.*] What type of reagent would you use to effect this transformation?

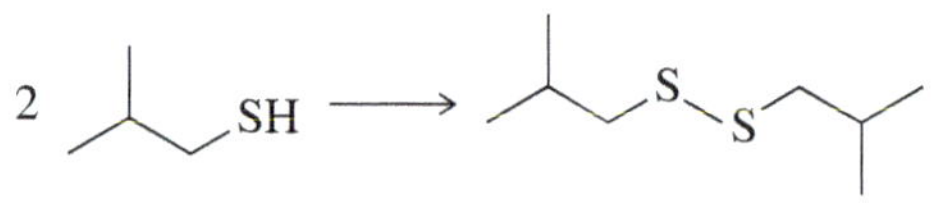

13.114 (•••) **THINKING AHEAD** A chemist attempted the reaction below, one we introduce in Chapter 17, expecting the reaction between a strong nucleophile and a ketone in water to give an alkoxide product. (a) Why did the reaction fail? (b) How could the reaction conditions be changed to give a successful reaction?

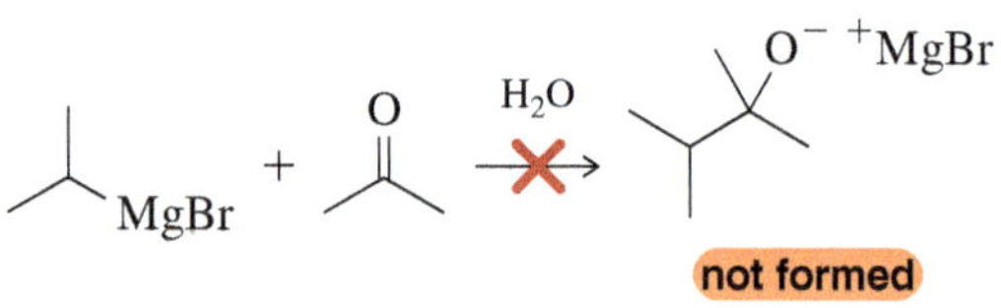

13.115 (•••) **THINKING AHEAD** Benzyl ethers make excellent protecting groups (see Chapter 24) according to the general scheme shown here. (a) How would you protect the 1° alcohol as a benzyl ether in the first step? (b) Suggest a mechanism for step 2. (c) Why type of reaction is step 3? [*Give the reagents for the first step.*]

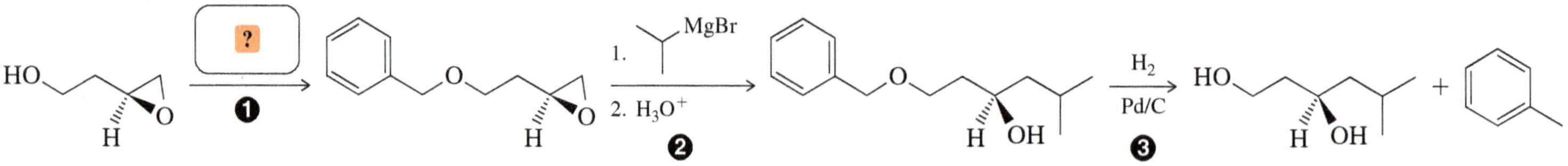

13.116 (•••) Suggest a mechanism for the following substitution reaction. [*We studied a similar reaction of radicals in Assessment 11.53.*]

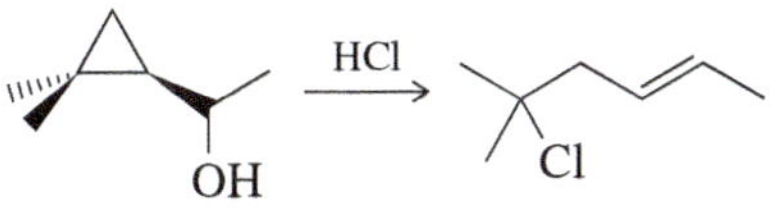

13.117 (•••) LOOKING BACK The acid-catalyzed dehydration we learned in this chapter is reversible, as shown below. [*We studied this reaction in Chapter 8.*]

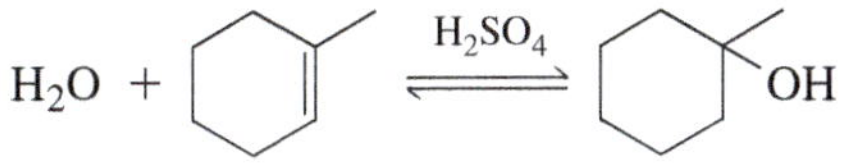

(a) Propose a mechanism for the formation of an alcohol from an alkene.

(b) Without looking back, how do you know this mechanism is correct?

(c) Which side of the reaction would be favored by running the reaction at low temperatures?

(d) How might you shift the equilibrium to the right?

(e) How might you shift the equilibrium to the left?

13.118 (••••) LOOKING BACK Assessment 8.74 revealed that oxymercuration could be used to make cyclic esters. Suggest a likely mechanism for this transformation.

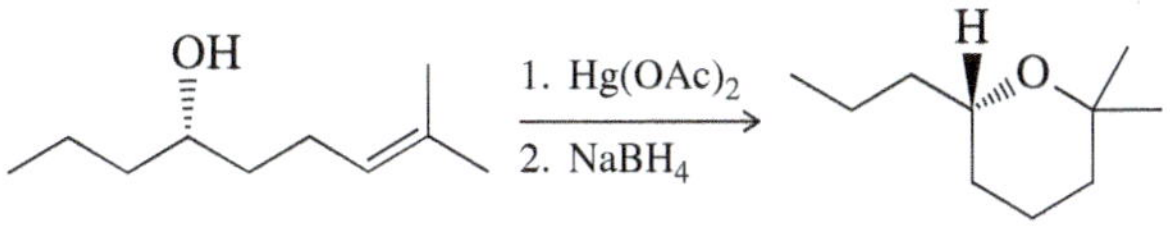

13.119 (••••) LOOKING BACK Chapter 8 discussed the synthesis of cholesterol, which proceeds by a cationic cyclization cascade. Without looking back, suggest a mechanism by which the following reaction occurs. [*The carbons have been numbered for you.*]

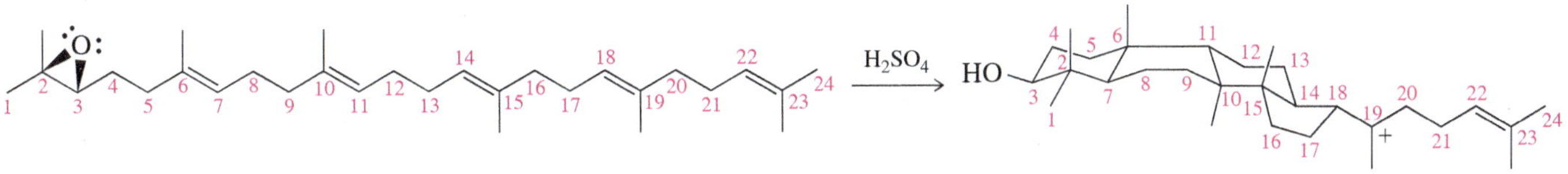

13.120 (••) Suggest reagents to carry out the following transformation. [*One of the answers should be from Chapter 9.*]

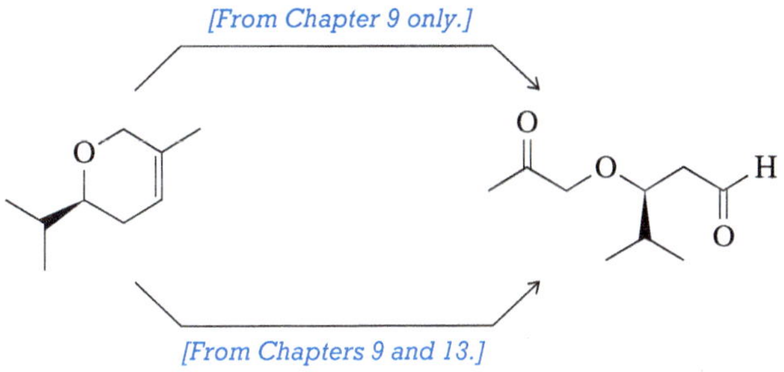

14

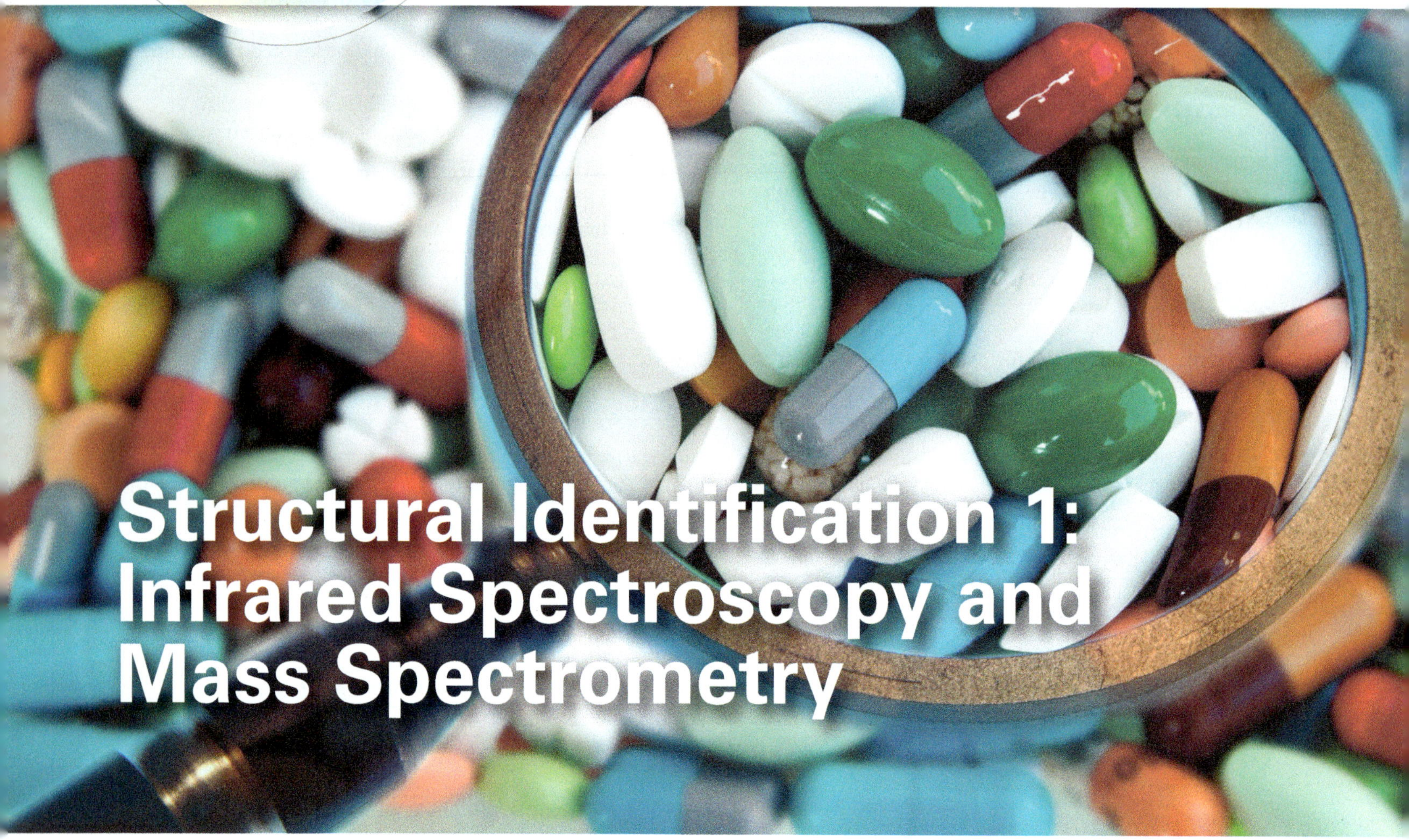

Structural Identification 1: Infrared Spectroscopy and Mass Spectrometry

Learning Objectives

- LO 14.0 Describe the general concepts of spectroscopy
- LO 14.1 Use IR spectroscopy to solve organic chemistry problems
- LO 14.2 Evaluate compounds using IR spectra
- LO 14.3 Determine the functional groups present in a compound based on its IR spectrum
- LO 14.4 Determine the identity of a compound based on its mass spectrum
- LO 14.5 Distinguish between compounds based on the principles of mass spectroscopy
- LO 14.6 Use mass spectrometry to evaluate the fragmentation of compounds
- LO 14.7 Identify a compound based on its IR and mass spectrum

Introduction

In 2011, over 1000 patients at a cardiac clinic in Lahore, Pakistan, became violently ill, experiencing extensive bleeding from the mouth and urinary tract along with plummeting white blood cell counts. Over the course of the next several weeks, more than 200 of these patients died. This tragedy has been attributed to the patients being given heart medicines that, unbeknown to the physicians, were contaminated with antimalarial drugs.

This tragedy illustrates the need for rigorous testing of pharmaceutical agents. Notably, the victims of the Lahore tragedy were using free drugs provided by a government-run clinic. It is estimated that, especially in developing nations, as much as 30% of all medicines are compromised. In 2013, over 100,000 children died as a result of using substandard antimalarial medications, and while some of these may have been a result of intentional forgery, many resulted from the use of expired drugs, substandard production practices, or accidental negligence. No country is immune to this—in fact, the FDA recently identified counterfeit cancer, anticholesterol, and anemia drugs being sold in the United States.

Identifying the structure of medicines, which are typically composed of organic molecules, is important, and the spectroscopic techniques discussed here in Chapter 14 and in Chapter 15 provide convenient ways of doing that. Not only can these techniques be used to confirm the success of a synthetic reaction, but when applied in a cost-effective and simple manner, developing technology can ensure worldwide access to helpful medicines.

"Curiosity—the rover and the concept—is what science is all about: the quest to reveal the unknown."

—AHMED ZEWEIL

As you read through the background theory, be sure to look for connections to previous material, because there are many. Let's start by testing our understanding of some of those concepts.

We have learned many reactions in Chapters 8–13 that can be used for the synthesis of molecules. One aspect not discussed in these chapters is how a chemist knows what they've made when the reaction is finished or how pure that compound is after the synthesis. Thus, while the previous six chapters tell the story of how to make a molecule, Chapters 14 and 15 tell how to know it's been made. There are relatively few reactions to speak of in these chapters, and the material must be thought about in different ways. [*Worry not: this material is mastered simply by working problems, early and often, in the same way you get better at solving puzzles. Don't wait until an exam to find out what you don't know.*]

There are two aspects to spectroscopy: the background theory and the practical application. To the typical organic chemist, the application of spectroscopy is of principal importance. And yet, an understanding of the background creates deeper connections with the material that enhances our ability to solve structural identification problems and strengthens our long-term retention of spectroscopy concepts. To enable both, our study of each technique (infrared spectroscopy and mass spectrometry in Chapter 14 and then nuclear magnetic resonance in Chapter 15) begins with a brief discussion of how it will be used. With a framework of application established, the background theory is used to fill in the gaps.

Test your understanding

Assessment 14.1 Given the molecular formula, calculate the index of hydrogen deficiency.

(a) $C_6H_{14}O$ (b) C_4H_9ON
(c) $C_6H_{10}O$ (d) C_8H_9OBr

Assessment 14.2 Given the molecular formula, (i) what functional groups are possible in each of the following molecules? (ii) Why are your answers for (a) and (b) different?

(a) $C_6H_{14}O$ (b) $C_6H_{10}O$

Assessment 14.3 Rank the following C—H σ bonds in terms of their bond length. Explain your ranking (1 = longest; 3 = shortest).

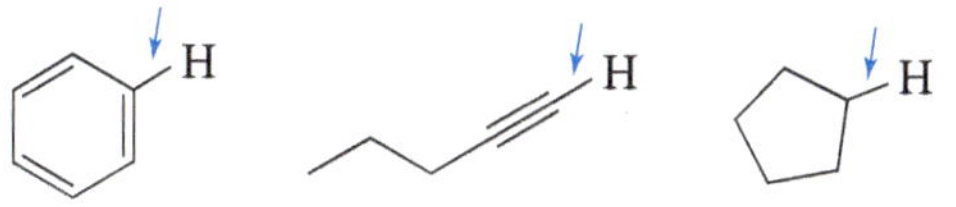

Assessment 14.4 In the periodic table, the number 79.904 appears under the element bromine. What is the significance of this number?

Assessment 14.5 Rank the boiling points of the following molecules. Explain your ranking (1 = highest bp; 3 = lowest bp).

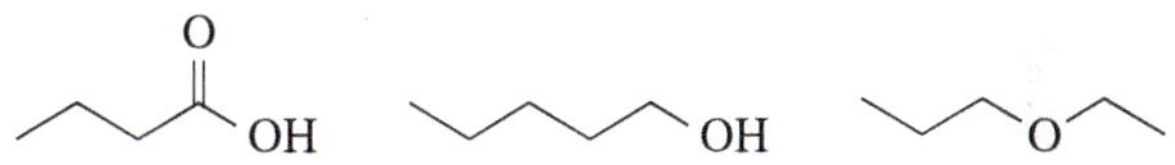

Assessment 14.6 Identify the more stable carbocation in each pair.

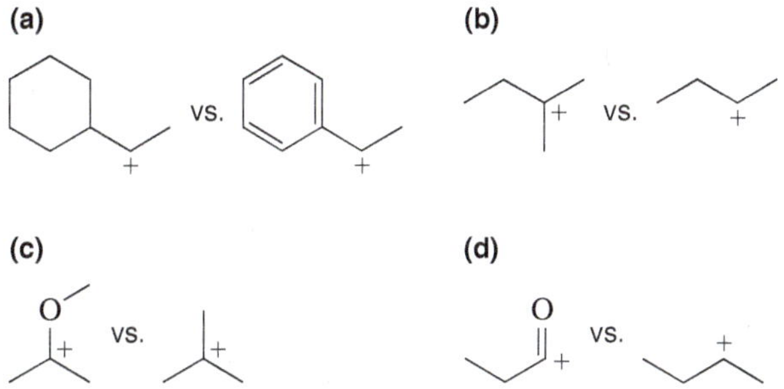

14.1 Structural identification and spectroscopy: an overview

In the lab, a reaction looks significantly different than it does on paper. Consider the alkylation of a terminal alkyne in **Figure 14.1**. The experimental protocol for this reaction involves mixing a colorless liquid with other reagents and a solvent only to isolate and purify another pure liquid that seems unchanged based on physical appearance.

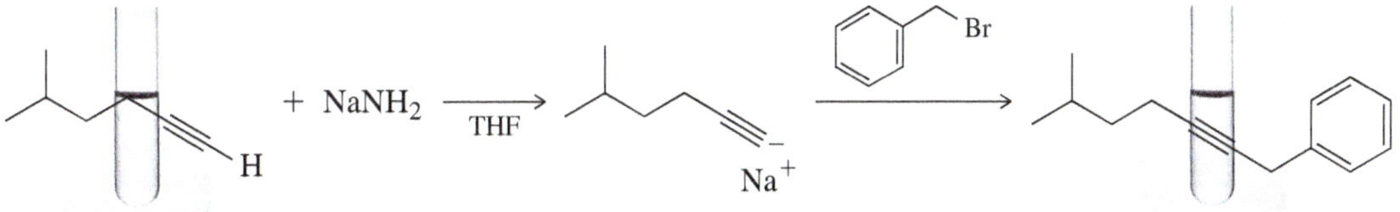

Figure 14.1 Synthesis on paper vs. synthesis in lab. Though it is the same thing, it certainly looks different.

How do we distinguish a viscous oil that is a starting material from a viscous oil that is the desired product or an impurity? How would we identify the structure of a newly discovered compound that shows anticancer properties in hopes of one day synthesizing more of it? How would the FDA identify the structure of a counterfeit drug with the goal of identifying how it got into the United States? These, and other questions, can be solved using spectroscopy.

14.1.1 Properties of light: the electromagnetic spectrum

Organic spectroscopy is the interaction of energy (in the form of light) with organic molecules. This light, referred to as electromagnetic radiation, is simultaneously a particle and a wave. One **photon** (particle) of light possesses a **quantum** of energy (E) that is directly proportional to the **frequency** (v) of light, according to Equation 14.1. The units of frequency are reciprocal seconds, or hertz (s^{-1} or Hz). The proportionality constant between energy and frequency is Planck's constant ($h = 6.63 \times 10^{-34}\ J \cdot s$).

$$E = hv \qquad \text{Equation 14.1}$$

The frequency and wavelength (λ) of light are related by the speed of light ($c = 3.0 \times 10^8$ m/s) through Equation 14.2.

$$c = v\lambda \qquad \text{Equation 14.2}$$

Combining Equations 14.1 and 14.2 by substituting c/λ for v shows that the energy of a photon of light is inversely proportional to the wavelength of light (Equation 14.3).

$$E = hv = h\left(\frac{c}{\lambda}\right) \qquad \text{Equation 14.3}$$

The **electromagnetic spectrum** (Figure 14.2) shows the range of photon energies, and as stated in Equation 14.1, the energy increases as the frequency increases. Within the electromagnetic spectrum is the small range of visible light, corresponding to the frequencies between 7×10^{14} and $4 \times 10^{14}\ s^{-1}$. The take-home message from Figure 14.2 is this: As the frequency of light increases, the wavelength decreases, and the energy increases. Assessment 14.7 helps us relate this fact to our own experiences.

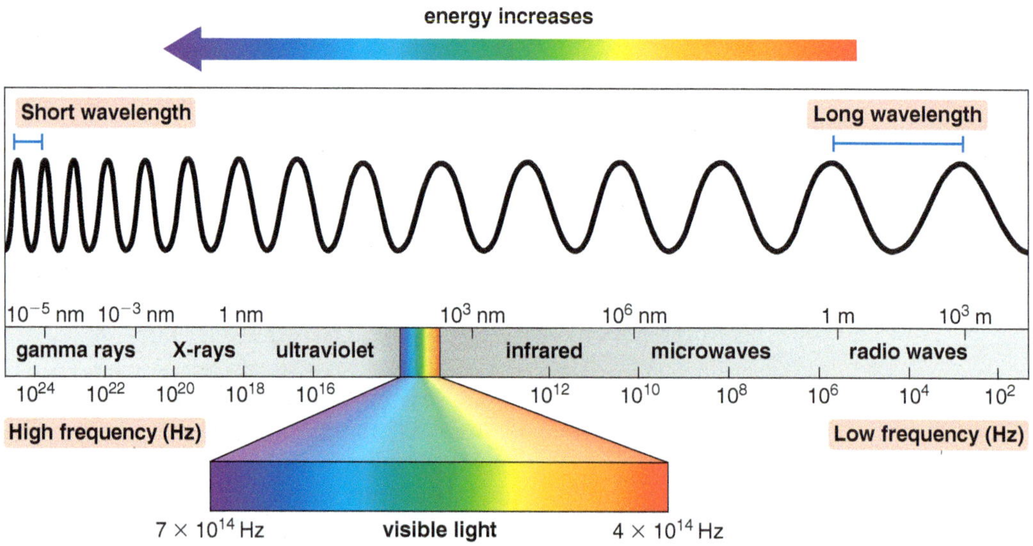

Figure 14.2 The electromagnetic spectrum. As wavelength decreases, frequency increases and energy increases. As a result, infrared and radio waves correspond to low-energy transitions, whereas X-rays and gamma rays correspond to high-energy transitions.

Students [*and myself included*] often struggle with remembering which types of radiation are high energy and which are low energy. By relating the type of radiation to your own experiences, however, you can easily figure this out. You watch TV, open your eyes on sunny days, and comfortably see in color without damaging your eyes; thus, visible light must be of relatively low energy. To the left of the visible region, as drawn in Figure 14.2, is the ultraviolet region (blue, indigo, violet moves into ultraviolet). Ultraviolet radiation might remind you of the UV rays of the sun. You put on sunscreen to protect yourself from these dangerous rays; it makes sense, then, that the ultraviolet region of the spectrum is at a higher energy. And when you get an X-ray, the rest of your body is covered with a lead blanket to protect it from the X-rays. As dangerous radiation, X-rays must also be of high energy. Contrast that with energy to the right of the visible spectrum (red to infrared). At the extreme end are radio waves, which are literally all around you at all times. The fact that you never hear of warnings about traveling among radio waves suggests that these are low-energy waves.

Assessment 14.7

(a) Holding your right hand in front of you, move your index finger up and down slowly (low frequency). Then move your finger up and down as quickly as you can (high frequency). Which took more energy?
(b) When you want to heat up food, do you place it in a microwave or next to a radio?
(c) When getting X-rays in the dentist's chair, why is your body covered with a lead blanket?

Assessment 14.8 How does Assessment 14.7 help you remember that a higher frequency is associated with higher energy?

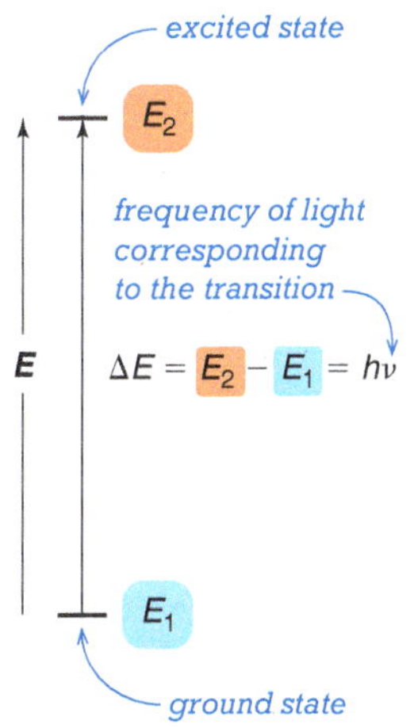

Figure 14.3 The transition from a ground state to an excited state requires a specific quantum of energy.

14.1.2 Quantized energy levels

When we talk about light (electromagnetic radiation) interacting with molecules, we are really discussing how molecules absorb and release light. For example, carrots are orange because β-carotene absorbs visible light corresponding to certain frequencies while reflecting all others. Our eyes process this reflected light as being orange. Much of what we know about the structure of atoms and will soon know about the structure of molecules has to do with the fact that energy levels within molecules are quantized. That is, for a given transition between a ground state and an excited state, there is one particular energy of light that will be absorbed (Figure 14.3). The energy that is absorbed is equal to the difference in energy between the excited state and the ground state. In the case of β-carotene, the transition of an electron from one molecular orbital to another corresponds to the energy associated with a wavelength of 450 nm. Visible light is not absorbed by β-carotene unless it corresponds to the energy of a specific transition.

Assessment 14.9 How much energy is absorbed by β-carotene in the transition that gives carrots their orange color? [*Use Equation 14.3.*]

Assessment 14.10 When the following elements are heated, only discrete wavelengths of light are produced. Why?

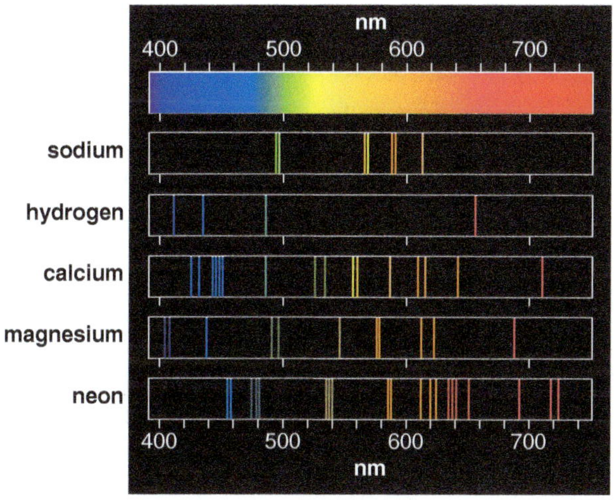

What would happen if energy levels of transitions within molecules weren't quantized? This would mean that a wide variety of wavelengths (radiation energies) would cause the same transitions. In the case of β-carotene, what if all wavelengths of visible light were absorbed? None would be reflected and our eyes would detect carrots as black (or gray, depending on how much light was absorbed; Figure 14.4). In truth, the whole world would be different shades of gray.

Figure 14.4 Quantized energy levels. (a) With them, carrots are orange. (b) Without them, everything would be different shades of gray.

But even worse, chemists would know nothing about the structural features that give rise to the particular transitions. By not knowing which energy caused which transition, we'd have no way of correlating molecular interactions with light with structural features. We wouldn't know of the existence of principal energy levels in atoms. We wouldn't know how much energy is required for rotation around the single bond of ethane. We wouldn't know which functional groups a molecule contains (infrared spectroscopy; Section 14.2). We wouldn't know that colored molecules are usually highly conjugated (UV–vis spectroscopy; Section 21.3.2). And we wouldn't know about the chemical environment of hydrogens and carbons (nuclear magnetic resonance spectroscopy; Chapter 15).

Table 14.1 shows how different frequencies interact differently with molecules, causing the different molecular transitions that allow us to determine structure. The excitation of electrons requires higher energy (UV–vis region) than is required by bond vibration (IR region).

Table 14.1 Different frequencies of electromagnetic radiation cause different molecular transitions.

Region	Wavelengths	Frequencies (Hz)	Molecular transition
X-rays	10^{-3} nm to 10 nm	10^{21} to 10^{18}	Ionization
UV–vis	10 nm to 700 nm	10^{18} to 10^{14}	Electron excitation
Infrared	700 nm to 1 mm	10^{14} to 10^{10}	Molecular vibrations
Microwave	1 mm to 1 m	10^{10} to 10^{6}	Bond rotation
Radio	1 m to 100 km	10^{6} to 10^{2}	Nuclear spin

14.1 SUMMARY

- Spectroscopy is the study of how light interacts with molecules and can give information about their structure. The energy of electromagnetic radiation is directly proportional to the frequency and inversely proportional to the wavelength. Quantized energy levels correlate to specific molecular transitions. Light in the infrared region corresponds to bond vibrations.

14.2 Infrared spectroscopy: practical usage

Infrared (IR) spectroscopy uses light in the infrared region of the electromagnetic spectrum to identify functional groups within a molecule. An IR spectrum is a plot that shows at which wavenumbers (can be thought of as frequencies) the molecule absorbs. Each of these bands is associated with a particular functional group. Table 14.2 shows which bands indicate the presence of which functional groups.

Table 14.2 IR stretching frequencies correlated with functional group.

Type of bond	Wavenumber (cm^{-1})	Intensity
C≡N	2260–2220	Medium
C≡C	2260–2100	Medium to weak
C=C	1680–1600	Medium
C=N	1650–1550	Medium
	~1600 and ~1500–1430	Strong to weak
C=O	1780–1650	Strong
C—O	1250–1050	Strong
C—N	1230–1020	Medium
O—H (alcohol)	3650–3200	Strong, broad
O—H (carboxylic acid)	3300–2500	Strong, very broad
N—H	3500–3300	Medium, broad
C—H	3300–2700	Medium

There are two main ways we'll use IR spectroscopy. Given a molecular formula, you might be asked to examine a spectrum and identify what functional groups are present. Assessment 14.11 gives you an opportunity to do this.

Assessment 14.11 What functional groups are present in a molecule with a molecular formula of $C_4H_{10}O$ and the following IR spectrum? [*Use Table 14.2 for this.*]

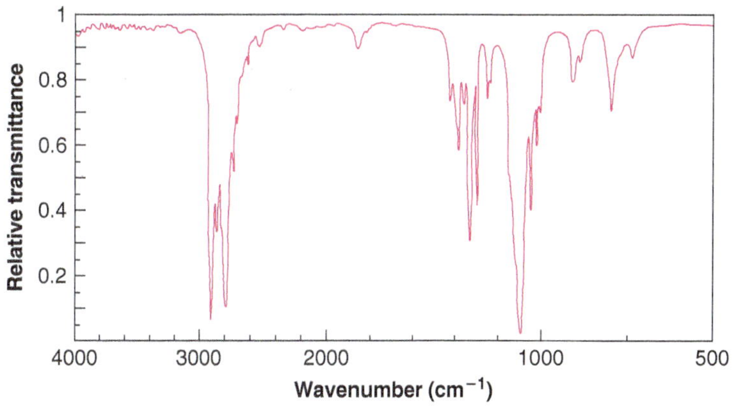

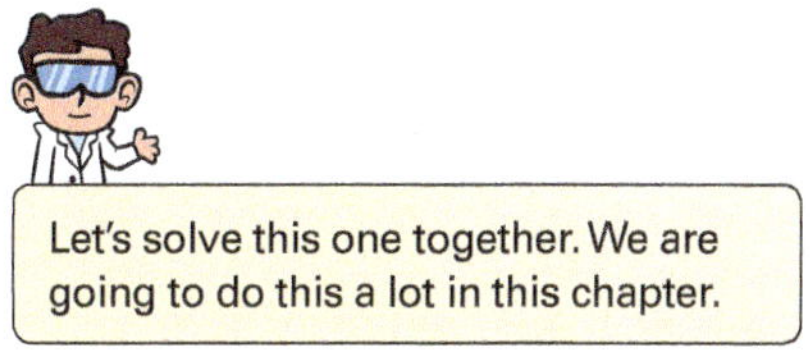

Solved Assessment 14.11

The key to solving IR problems is to use all of the information you have. A big piece of what you know is the molecular formula. Knowing that the molecular formula is $C_4H_{10}O$, what functional groups do you need to look for? Because the only heteroatom (non-carbon/non-hydrogen) is an oxygen, and there is only one of them, this molecule can't possibly be an amine (needs a nitrogen), an alkyl halide (no F, Cl, Br, or I), or a carboxylic acid (needs two oxygens), among many other functional groups. [*What else can't it be?*] The molecular formula actually goes farther than this, telling us how many π bonds or rings it contains. [*Recall our deeper discussion of IHD from Chapter 8.*] By calculating the index of hydrogen deficiency (IHD; Equation 14.4) or degrees of unsaturation, we determine that there are no rings and no double or triple bonds:

$$\text{IHD} = \begin{matrix}\text{number of}\\ \text{carbons}\end{matrix} - \frac{\begin{matrix}\text{number of}\\ \text{hydrogens}\end{matrix}}{2} - \frac{\begin{matrix}\text{number of}\\ \text{halogens}\end{matrix}}{2} + \frac{\begin{matrix}\text{number of}\\ \text{nitrogens}\end{matrix}}{2} + 1 = 4 - \frac{10}{2} - \frac{0}{2} + \frac{0}{2} + 1 = 0 \qquad \text{Equation 14.4}$$

With this information, we can now look at the spectrum with purpose, as in **Figure 14.5**. What functional groups could be present in a fully saturated molecule with only a single oxygen? There are only two possibilities. [*Any ideas?*] Table 14.2 says an alcohol should contain a band around 3300 cm^{-1} for the O—H σ bond and another around 1150 cm^{-1} for the C—O σ bond. There is no band at 3300 cm^{-1}, so this molecule cannot be an alcohol. It *does* have a band around 1150 cm^{-1}, so we know there is a C—O σ bond present. Without any other functional group possibilities, the compound must be an ether. Other techniques (mass spectrometry and nuclear magnetic resonance spectroscopy) would be used to figure out which C_4 ether it must be. ■

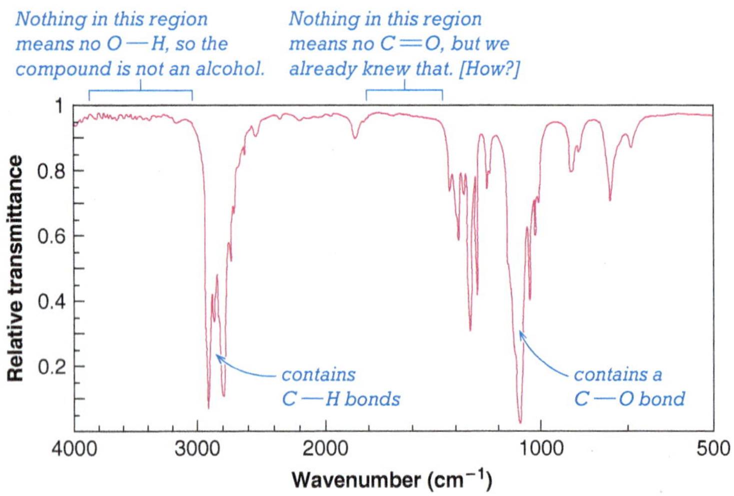

Possible ethers with a molecular formula of $C_4H_{10}O$:

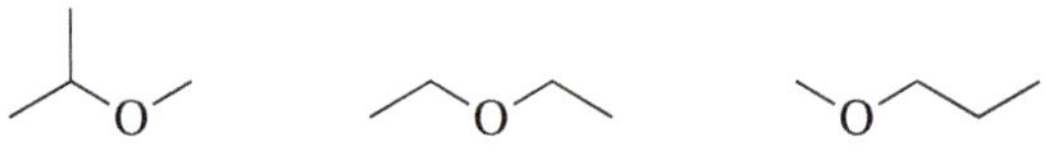

Figure 14.5 Using the IR spectrum to narrow down the structural possibilities to an ether.

A second, and more realistic way that organic chemists use IR spectroscopy is to identify whether a reaction has occurred by looking for the presence or absence of bands in the IR spectrum of the resulting product. Assessment 14.12 gives you practice doing this.

Assessment 14.12 Sodium borohydride reduces hexanal to hexan-1-ol (see Chapter 17). How does the following IR spectrum confirm that an alcohol was produced?

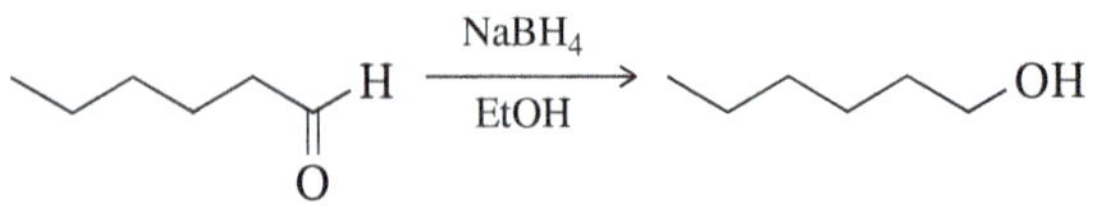

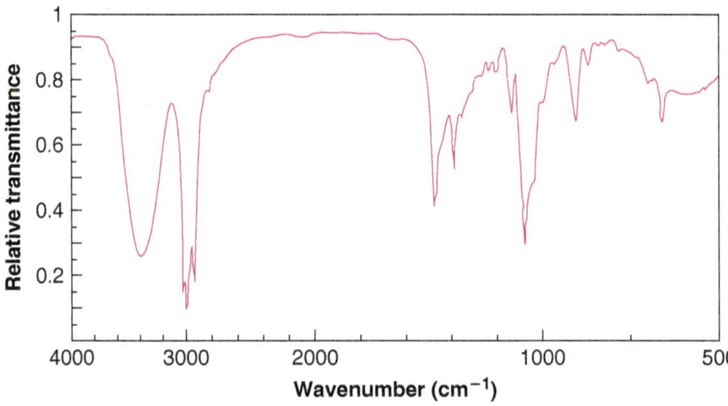

That was fun; how about one more?

Solved Assessment 14.12

If we know what reactant we started with, and if it was characterized previously by IR, then we know specifically what to look for in the spectrum. From the values in Table 14.2, hexanal would contain a band around 1700 cm^{-1} for the aldehyde carbonyl. The spectrum of our product has no such absorbance, so the aldehyde is no longer present. Similarly, we'd expect an alcohol to have a band at 3300 cm^{-1} for the O—H σ bond and a band at 1150 cm^{-1} for the C—O σ bond. Both of these *are* present, as shown in **Figure 14.6**. Based on this analysis, we can be confident that this reaction has proceeded as expected and that hexan-1-ol has been produced. ■

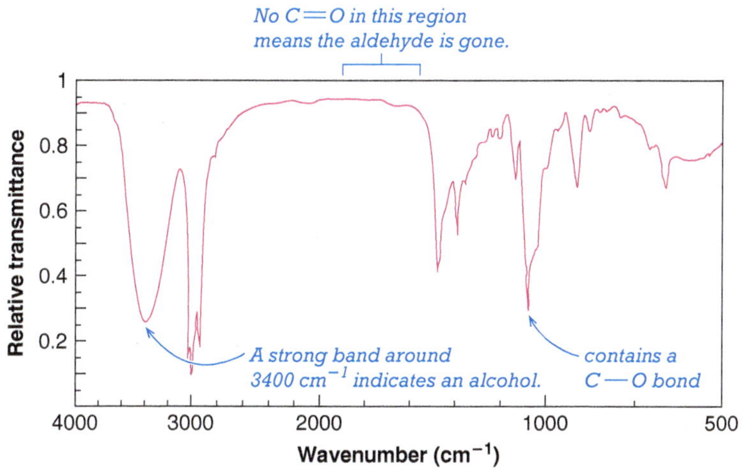

The product must be hexan-1-ol, as expected:

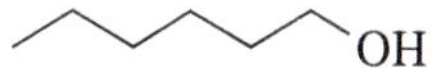

Figure 14.6 **IR spectroscopy confirms the identity of the expected product** and the successful conversion of hexanal to hexan-1-ol.

The purpose of starting with these problems is not to have you stop here, confident that you can now solve any IR problem. Instead, it is to show you the relevance of IR spectroscopy prior to diving deeper into the background theory. That said, it's never too early to start memorizing the frequencies associated with each functional group. [*This is one of the rare times I'll tell you to memorize; flash cards can give you retrieval practice here. So will working assessments.*]

Assessment 14.13 Given the IR spectrum, suggest what functional groups might correspond to the given molecular formula.

(a) C_3H_4O

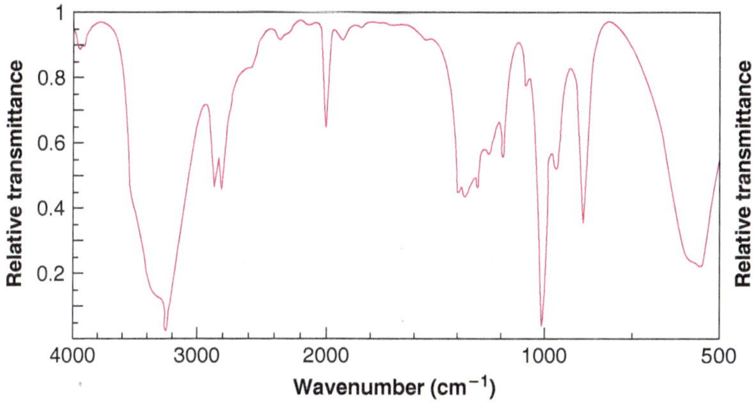

(b) C_3H_7NO

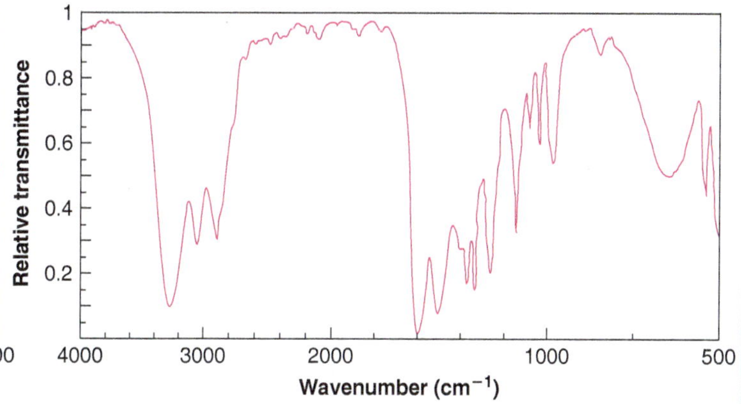

(c) C_4H_8O

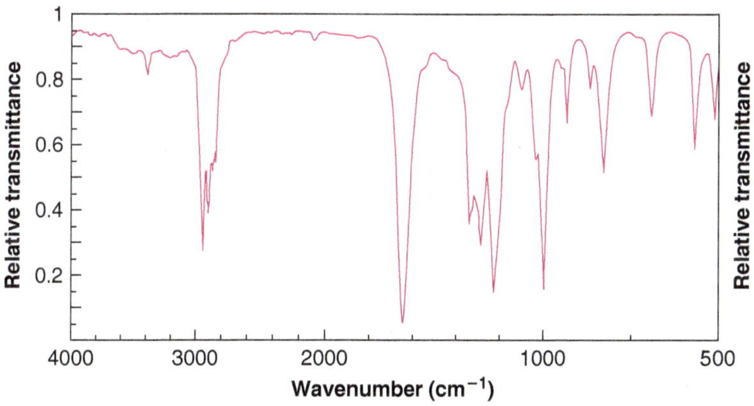

(d) C_7H_6O

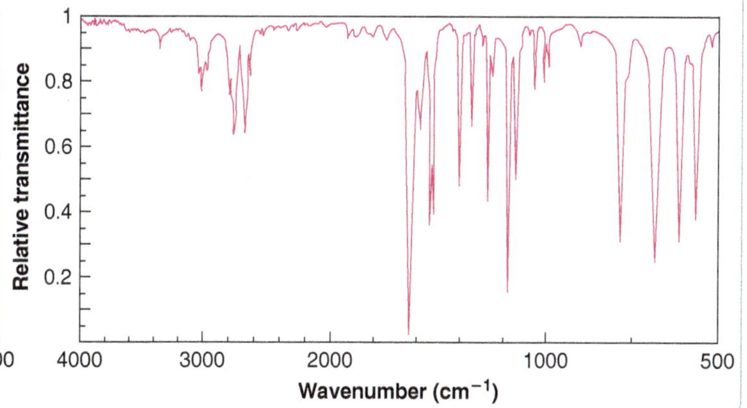

14.3 Infrared spectroscopy: background theory

While many spectroscopic techniques can be used effectively by treating them as a black box and simply interpreting the data that come out, to ensure that you retain this information, it is important to have at least some understanding of how and why the technique works. By doing so, you will make connections between this material and concepts from other classes, strengthening your retention of the information. Section 14.3 discusses some important aspects, before taking you on an infrared tour of the functional groups.

14.3.1 Molecular vibrations

Two atoms connected by a bond constantly vibrate. The electrons in each atom pull the nuclei close. When the two nuclei are too close to each other, they begin to repel. Halfway between is the average bond length shown in **Figure 14.7**. All bonds vibrate, except when the temperature is lowered to 0 K, at which point all motion ceases.

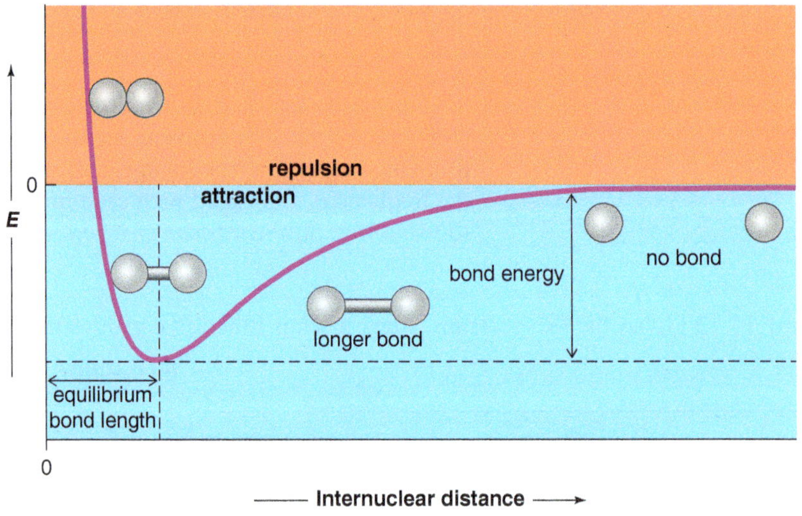

Figure 14.7 The bond length is the equilibrium distance of a vibrating bond.

It is helpful to think about bond vibration as two masses connected by a spring (**Figure 14.8**). As the spring is stretched, the tension begins to pull the masses back together. Over-compressed, the spring now pushes back, forcing the masses apart. In a frictionless environment, this would continue indefinitely with a consistent frequency of vibration. We use this analogy several times throughout Section 14.3, so keep it in mind.

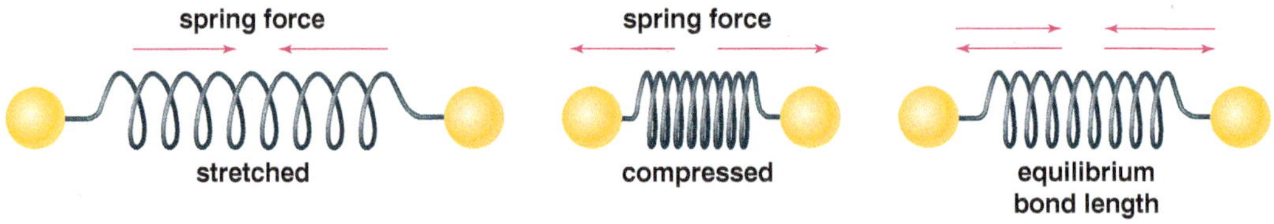

Figure 14.8 A bond vibration is like the vibration of two masses connected by a spring.

In a given molecule containing n atoms, there are $3n - 6$ vibrational modes. In water (**Figure 14.9**), for example, $n = 3$, so water has a total of three vibrational modes. Two of them are stretching modes—one is symmetric, and the other is asymmetric. The symmetric stretching mode has both hydrogens moving toward oxygen and then away at the same time. Alternatively, in the asymmetric stretching mode, one hydrogen comes toward oxygen while the other moves away. The third vibrational mode is a bending mode, called scissoring, where the two hydrogens bend toward each other. We aren't as concerned with the bending modes because they are really low energy vibrations that show up only in the region of the spectrum below 1400 cm^{-1}.

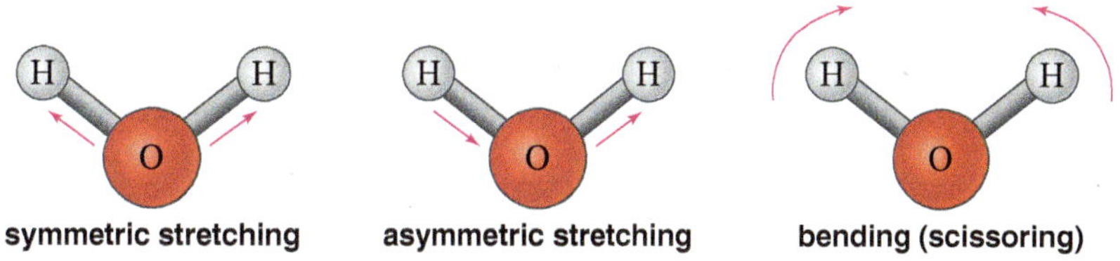

Figure 14.9 **Three vibrational modes of water.**

When a vibration results in a change in the dipole moment, the bond will absorb electromagnetic radiation whose frequency matches the frequency of vibration. Thus, you'd expect the IR spectrum of water to contain three bands, one for each of the vibrational modes. [*It turns out you can only see two of them, but that's for reasons we explain when we discuss alcohols.*]

14.3.2 Measuring a spectrum

Based on these concepts, I believe you can design an infrared spectrometer. Let's think like a chemist and do that.

The following summarizes what we know so far:

- The bonds in molecules vibrate at a particular frequency.
- Electromagnetic radiation at a frequency that matches the vibrating frequency will be absorbed.
- Radiation at all other frequencies will pass through.

Thinking Like a Chemist

Let's start simply with a light source, a sample holder, and a detector, as shown in **Figure 14.10**. When we shine the light on the sample, light at some frequencies will pass through, while light at others will not. [*It's obviously not this simple. What problems do you see with this setup?*]

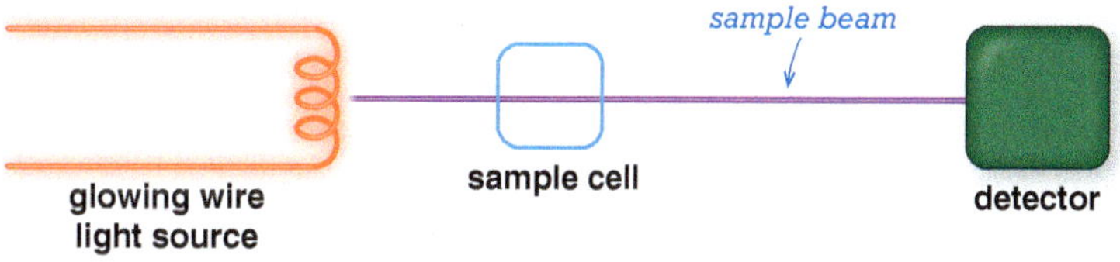

Figure 14.10 **A preliminary IR spectrometer featuring only a light source, a sample, and a detector.**

Most light sources (think of the sun or a lightbulb) cover a wide range of frequencies. If our source does the same, how will we observe whether any light is absorbed? And how could we use that information given that we don't know what frequency, if any, is absorbed? Unless we can identify the frequency that is absorbed, we have no way of translating that to structural information. To mitigate this problem, we can install a **monochromator** after the sample (**Figure 14.11**). As the name implies, a monochromator ("single color") can select and transmit a narrow band of frequencies at a single time. By selecting for each individual frequency, we can see at which frequency the light is absorbed.

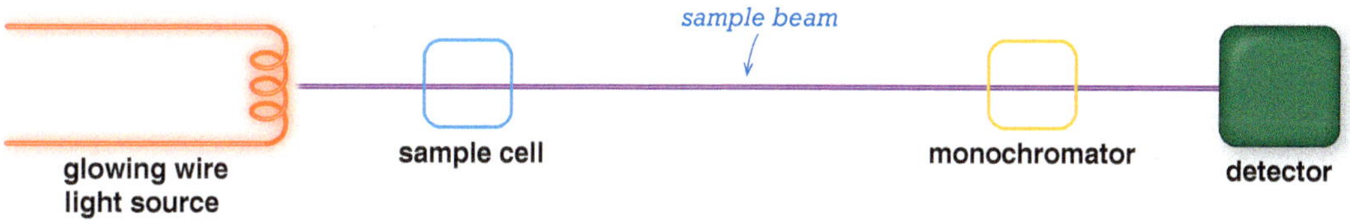

Figure 14.11 **A more evolved IR spectrometer featuring a light source, a sample, a detector, and a monochromator.**

We explain later that some bonds absorb all of the light at a given frequency, while other bonds absorb half, and some absorb very little. With our current setup, if all of the light at a given frequency were absorbed and our detector saw complete darkness, those bonds would be easily detected. But what about those bonds that absorb only a small amount of light. How does a detector recognize that the amount of light coming out of the sample is slightly less than the amount that went in? Does the detector know how much light was shined on the sample? [*No.*] Any ideas on how it could?

A common technique to solve this problem, one you'll study in an instrumental analysis course, is to use a reference, as shown in **Figure 14.12**. In this case, the reference cell is basically an empty sample cartridge that the source light will pass through in parallel with the light that passes through the sample. How much light at each frequency will pass through the blank? [*Thoughts?*] All of the light at every frequency will pass through the reference cell. How much light passes through the sample? At some frequencies, where there is no vibrating bond, all of it passes through. But where there is a bond that matches the frequency of the light, some will be absorbed. Throw in some mirrors to direct the light in appropriate directions and voilà, our detector can compare the light that passes through the sample (the sample beam) to the light that passes through the reference cell (the reference beam), giving the information to be displayed on the computer and then printed on the spectrum. ■

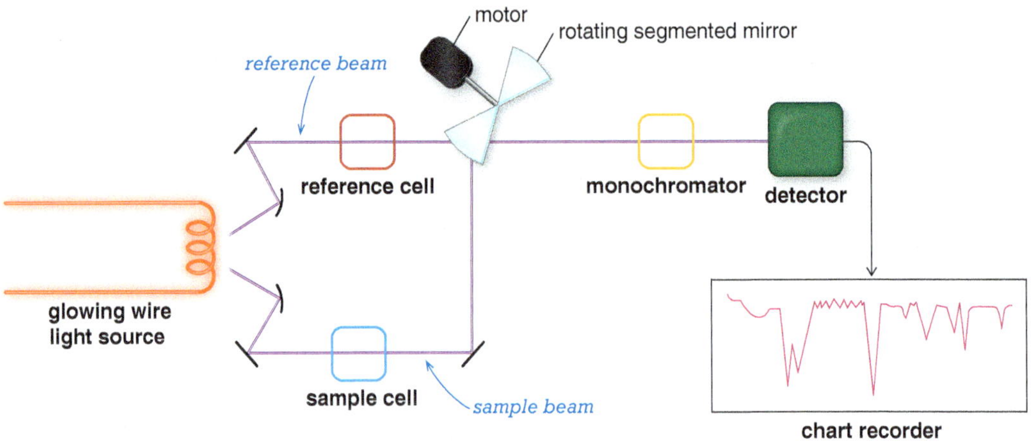

Figure 14.12 Schematic of a dispersive IR spectrometer.

The spectrum can be recorded in two ways. For both, the x axis on the spectrum is in wavenumbers, which is essentially the frequency, ranging from 4000 cm^{-1} (high energy/high frequency) to 400 cm^{-1} (low energy/low frequency). [*We talk more about wavenumbers in Section 14.3.3.*] Most commonly the y axis displays the relative transmittance, ranging from 1 (100%) transmittance at the top to 0 (0%) transmittance at the bottom. When done this way, the detector compares the amount of light that is transmitted through the sample to the light transmitted through the blank. At frequencies where 100% of the light passes through the sample, none is absorbed. Thus, there is no bond corresponding to that frequency (**Figure 14.13**). At frequencies where less than 100% of the light passes through the sample, some is absorbed. There must be some bond that absorbs light at this frequency. These are the values given in Table 14.2.

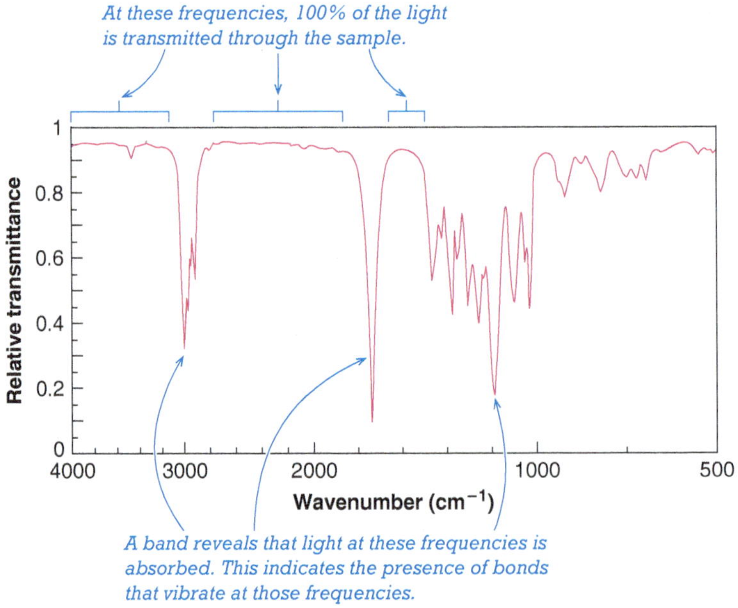

Figure 14.13 This sample IR spectrum shows at which frequencies light is absorbed. Relative transmittance is displayed on the y axis. A relative transmittance of 0 corresponds to a percent transmittance of 0%. Similarly, a relative transmittance of 1 corresponds to a percent transmittance of 100%.

Although relative transmittance is commonly shown on the *y* axis, at times absorption is plotted there instead. The spectrum in **Figure 14.14** is generated this way, and it is basically the flipped image of Figure 14.13. The band, now extending up instead of down, still indicates the frequency at which light is absorbed by the molecule.

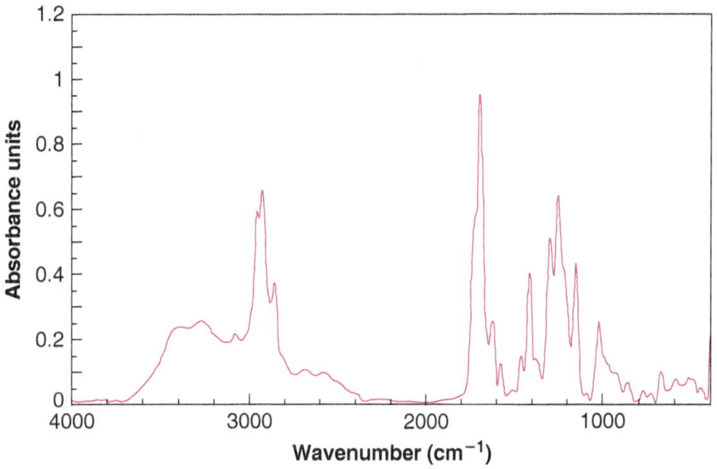

Figure 14.14 This sample IR spectrum shows at which frequencies light is absorbed. Absorbance is displayed on the *y* axis.

The preceding discussion described an older dispersive infrared spectrometer, used as an example here because the technology is more easily understood. Modern labs utilize Fourier transform infrared (FT-IR) spectrometers, a schematic of which is shown in **Figure 14.15**. Using an FT-IR, high-resolution data can be collected over multiple frequencies at the same time. The term *Fourier transform* refers to a complicated mathematical process that the computer uses to analyze the data. A detailed discussion of FT-IR is left to an advanced instrumental analysis course.

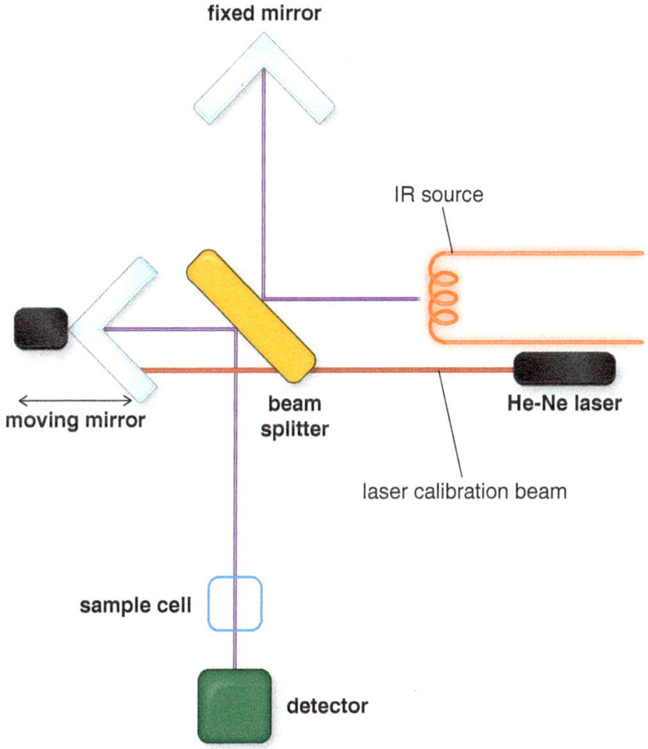

Figure 14.15 Schematic of an FT-IR spectrometer.

14.3.3 The spectrum

We've already seen a few IR spectra, so let's briefly examine a few details. As said previously, IR spectra most commonly have percent transmittance on the *y* axis. Although light is typically discussed in the context of frequency (proportional to energy through Equation 14.1) or wavelength (inversely proportional to energy through Equation 14.3), in IR spectroscopy, we use wavenumbers ($\bar{\nu}$). Whereas wavelength is the distance between the peaks of a propagating wave, the wavenumber is a measure of how many waves there are in a given distance. For IR spectroscopy, the distance is measured in centimeters. As a result, the units of wavenumber are cm^{-1}, or reciprocal centimeters. Wavenumbers, like frequencies, are proportional to energy. Although they are not the same, it is a reasonable approximation to think of wavenumbers as if they are frequencies. That is, bonds at higher wavenumbers vibrate at higher frequencies and are therefore more energetic. The IR spectrum extends from 4000 cm^{-1} to 400 cm^{-1}.

Within the infrared spectrum, there are three regions of interest. The first is the functional group region, spanning from 4000 cm^{-1} to 1400 cm^{-1}. This is the region where the most diagnostic bands indicating the presence of a unique functional group will appear, as shown in **Figure 14.16**. Below 1400 cm^{-1} is the fingerprint region. Bands appearing in this region are unique to molecules (like fingerprints), but there are very few bands we will use to diagnose the presence of a given functional group. Finally, the region from 2000 cm^{-1} to 1600 cm^{-1} is referred to as the aromatic overtone region. A grouping of weakly intense bands in this region [*along with bands at 1600 and 1450 cm^{-1}; see Table 14.4*] indicates the presence of an aromatic ring.

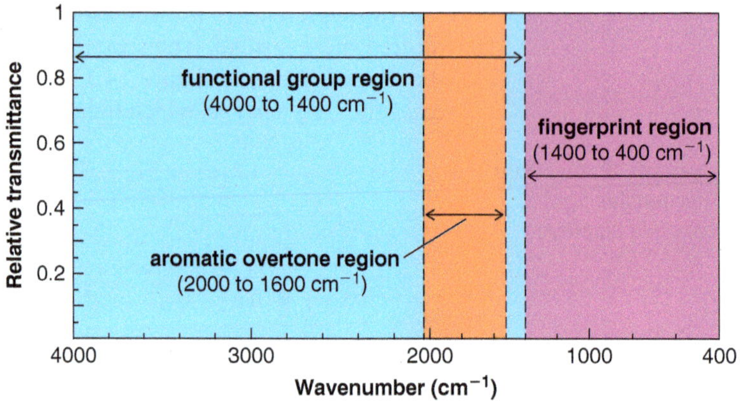

Figure 14.16 Regions of the infrared spectrum. Functional group (4000 cm^{-1} to 1400 cm^{-1}), fingerprint (1400 cm^{-1} to 400 cm^{-1}), and aromatic overtone (2000 cm^{-1} to 1600 cm^{-1}).

14.3.2–14.3.3 SUMMARY

- The infrared spectrometer is designed to identify the frequencies of light that are absorbed by a molecule by the presence of a band in the IR spectrum. These bands, appearing at specific wavenumbers, correlate to specific functional groups.

Assessment 14.14 In Table 14.2, there are two diagnostic bands in the fingerprint region. Which are they, and at what wavenumber do they appear?

14-A SPECTROSCOPY IN CONTEXT: Infrared

A handheld near-infrared spectrometer

The region of the IR spectrum below 1400 cm^{-1} is referred to as the fingerprint region, largely due to its complexity rather than its ability to identify specific functional groups. Each molecule is said to show a unique "fingerprint" in this region. Technology is currently being developed using the idea of a molecular fingerprint to identify counterfeit/substandard drugs, specifically antimalarials.

IR spectroscopy in the 4000 to 400 cm^{-1} region (the so-called mid-IR region) requires complex instrumentation and careful sample preparation, limiting its utility for point-of-care diagnostics. In the near-IR region (4000–13,000 cm^{-1}), however, where light is higher in energy, analysis of bulk materials is possible. In this region, molecules absorb relatively little radiation, but do so with a consistent fingerprint.

Engineers at Global Good, a philanthropic invention company, have developed a handheld device using near-IR spectroscopy that analyzes medicines to ensure they contain the active ingredients (Figure 14.17). The device records the unique spectral pattern of the medicine without damaging the pill. When the device is linked to a smartphone, cloud-based analysis compares the signal produced by the pill with the signal produced by the known active ingredients and conclusively indicates their presence or absence. The device is affordable so that governmental and nongovernmental clinics in developing nations can use this technology, ensuring safe access to needed medicine.

Figure 14.17 With a portable scanner and a smartphone, counterfeit medications can be easily identified.

In proof-of-concept experiments, the near-IR device has been found useful in identifying fake medicines. As of 2020, the device is commercially available.

14.3.4 The bands

Each band in an IR spectrum has two characteristics: its intensity (% transmittance) and its location (wavenumber). All bonds in all molecules vibrate, but not all bonds absorb infrared radiation. In order to absorb infrared radiation, the vibration of the bond needs to cause a periodic change in the dipole moment. The amount of light absorbed is proportional to the intensity of the band. A more polar bond absorbs more light, a less polar bond absorbs less, and a completely nonpolar bond absorbs none (**Figure 14.18**). We don't usually think of a C—H bond as being polar because there is only a slight difference between electronegativity values for C and H. In IR spectroscopy, however, any difference in polarity is enough to cause a vibrating bond to absorb light. The intensities of bands are only discussed qualitatively as being strong, medium, or weak.

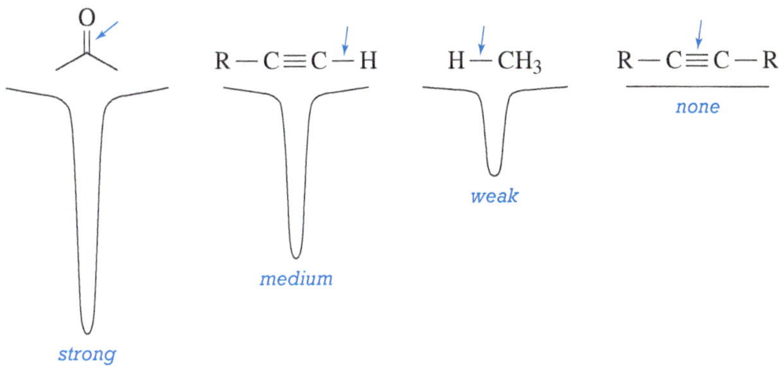

Figure 14.18 The intensity of a band increases as the polarity of the bond increases.

Figure 14.19 The tension on a spring determines the frequency of vibration. The frequency increases as the tension increases.

The wavenumber at which the band appears is based loosely on the strength of the vibrating bond with a few logical tweaks. Recall Figure 14.8, where we compared a bond to two masses connected by a spring. [*Try to visualize the following. By doing so, it will help the location of bands make sense.*] Assuming two equivalent masses (**Figure 14.19**), the frequency of vibration will depend on the tension in the spring. A tightly wound spring, if stretched and released, vibrates at a high frequency, whereas a loosely wound spring (like a Slinky), if stretched and released, vibrates at a low frequency. In this way, the tension of the bond (spring) plays a major role in determining the frequency of vibration and therefore the wavenumber at which it absorbs.

Now let's imagine springs of equal tension but with different masses attached, as shown in **Figure 14.20**. A heavy mass will cause the spring to vibrate slowly (i.e., at a low frequency). A lighter mass would vibrate much more rapidly (i.e., at a higher frequency). As a result, the masses of the atoms connected by a vibrating bond also affect the wavenumber at which the bond absorbs.

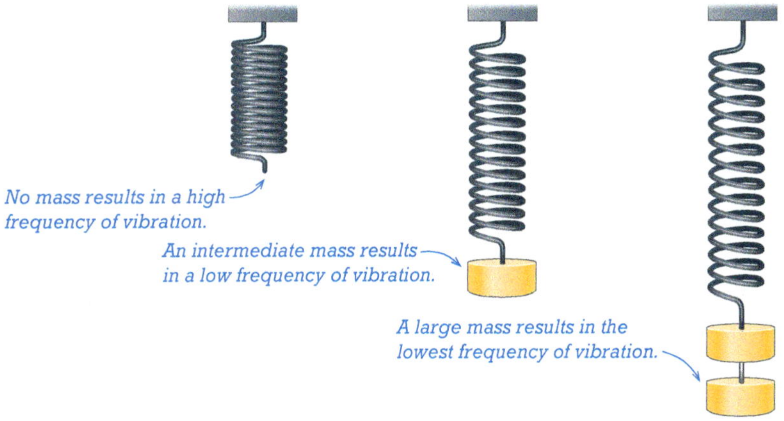

Figure 14.20 The masses connected by the spring determine the frequency of vibration. The frequency decreases as the masses increase.

The frequency (ν) of a vibrating spring with a mass (m) attached at the end can be calculated using Equation 14.5,

$$T = \frac{1}{\nu} = 2\pi\sqrt{\frac{m}{k}} \qquad \text{Equation 14.5}$$

where T is the period of the vibrations of the spring and k is the force constant of the spring. Equation 14.5 is derived from Hooke's law (**Figure 14.21**) and represents a movable mass attached by a spring to an immovable object. [*From physics—you may not have studied it yet, but you will.*] The frequency of vibration depends only on the mass of the object attached to the spring and the tension in the spring (assuming no friction and a massless spring). The tension is related to the force constant (k) of the spring. Equation 14.6, which estimates the wavenumber for a given bond, can be derived from Equation 14.5:

$$\bar{\nu} = \frac{1}{2\pi}\sqrt{\frac{k}{\mu}} \qquad \text{Equation 14.6}$$

where μ is the reduced mass of the atoms attached to the bond (spring) and is given by $\frac{m_1 m_2}{m_1 + m_2}$. Reduced mass is used because, in this case, both masses on the bond (spring) are movable. Once more, the masses attached and the tension in the bond determine the frequency of vibration.

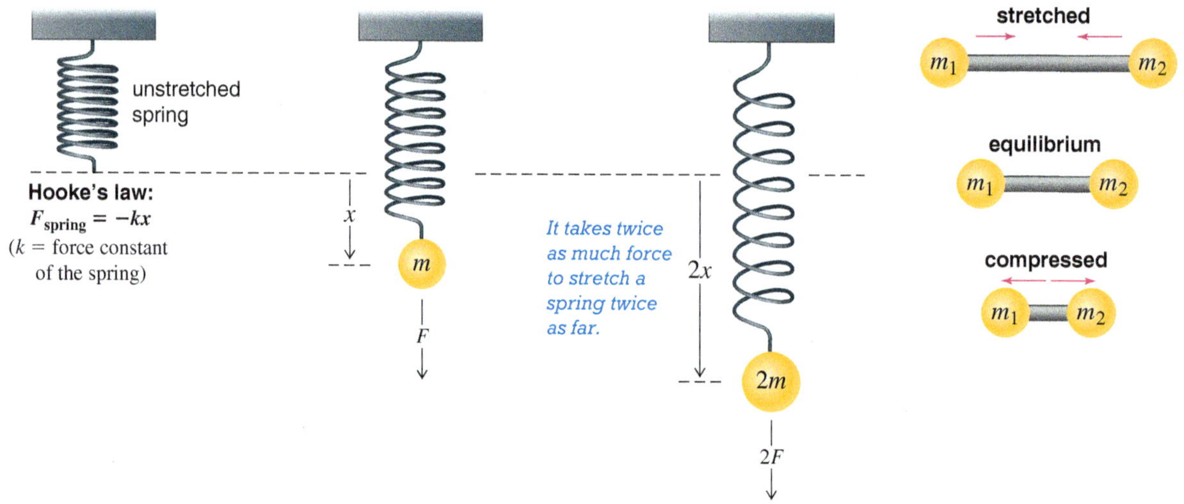

Figure 14.21 Like a mass hanging from a spring, the frequency of a bond vibration depends on the force constant (k; i.e., the strength) of the bond and the masses of the attached atoms, according to a derivative of Hooke's law.

Equation 14.6 helps us rationalize the difference in the bonds shown in **Figure 14.22**. Within the series of C—C bonds (single, double, and triple), triple bonds have a force constant that is triple that of a single bond. [*Recall: C≡C, 200 kcal/mol (Figure 10.7); C—C, ~83 kcal/mol (Table 5.6).*] Thus, they absorb at a much higher wavenumber. The C=C double bond similarly absorbs at a wavenumber nearly twice that of a C—C single bond. The impact of mass can be seen by comparing the C—F, C—Cl, and C—Br bonds. As the mass of the attached atoms increases, the bond vibrates at a lower wavenumber (lower frequency).

(a)

increasing force constant (k)

Bond	Wavenumber ($\bar{\nu}$) (cm^{-1})
$H_3C—CH_3$	995
$H_2C=CH_2$	1827
HC≡CH	2181

(b)

increasing reduced mass (μ)

Bond	Wavenumber ($\bar{\nu}$) (cm^{-1})
C—F	1049
C—Cl	732
C—Br	611

Figure 14.22 Trends in wavenumber: (a) Increasing bond order *increases* the wavenumber and (b) increasing atomic mass *decreases* the wavenumber.

Assessment 14.15 Choose the bond in each pair that you expect to vibrate at the higher wavenumber.

(a) C—N vs. C=N
(b) C—H vs. C—O
(c) C≡C vs. C=C
(d) P—O vs. C—O
(e) C—D vs. C—H
(f) C—Sn vs. C—Si

Assessment 14.16 Calculate the reduced mass for the following bonds.

(a) C—H (b) H—H (c) C—Cl (d) H—Cl

Assessment 14.17 The wavenumbers corresponding to the C—H bonds are shown. According to Hooke's law, these should be the same. Explain the difference. [*This is a Solved Assessment in the next section. Try it on your own before proceeding.*]

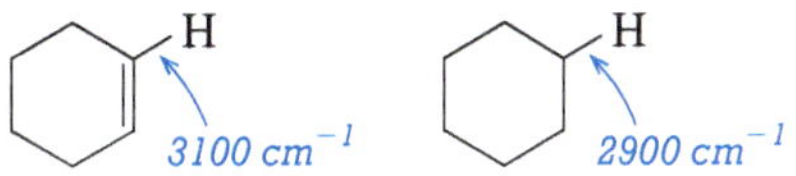

Another great opportunity to work together . . . by solving Assessment 14.17, we'll learn how hybridization impacts the wavenumber of vibration.

14.3.5 Hybridization

Besides the force constant and the reduced masses of the attached atoms, the hybridization of atoms can have an impact on the wavenumber of vibration.

Solved Assessment 14.17

In Assessment 14.17, you were asked to explain the different wavenumber observed for two C—H bonds. According to Hooke's law, the two C—H bonds should have identical force constants because both are σ bonds. Moreover, they both contain the same atoms, so the reduced mass (μ) should be the same for both. Hooke's law then would predict that these two bonds should absorb at the same wavenumber . . . yet they don't.

In a situation like this, we are trained to look for differences. [*Do you see any?*] One of the carbons is sp^3 hybridized, whereas the other is sp^2. This means that one of the C—H σ bonds is formed using an sp^3-hybridized orbital of carbon while the other uses an sp^2-hybridized orbital. How does a change in hybridization change the frequency at which a bond vibrates? [*Where have we discussed hybridization in this way before?*] Recall from Section 2.6.5 that an sp^3-hybridized orbital contains 25% *s* character while an sp^2-hybridized orbital contains 33% *s* character. [*Check out Sections 2.6.5, 4.4.5, and 10.4.4 for a review of this concept.*] More *s* character means the bond is shorter. Shorter bonds are stronger, and stronger bonds vibrate at a higher frequency (**Figure 14.23**). [*Reread this if it didn't make sense the first time.*] This is a general trend we will encounter again. ■

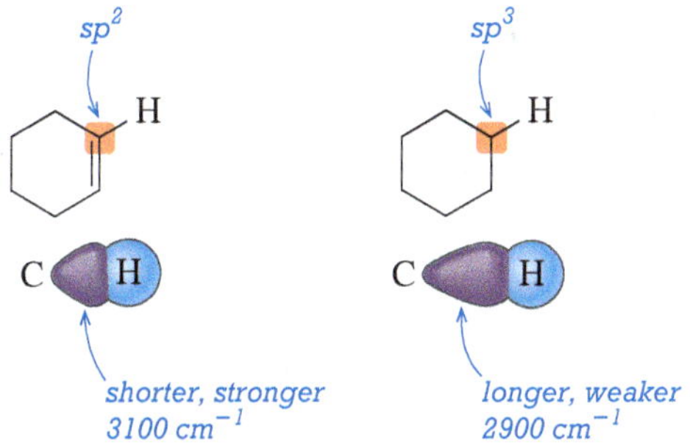

Figure 14.23 Bonds with more *s* character are shorter and stronger and therefore absorb at a higher frequency.

Assessment 14.18 Would you expect an acetylenic C—H to absorb at a higher or lower wavenumber than the C—H in ethene?

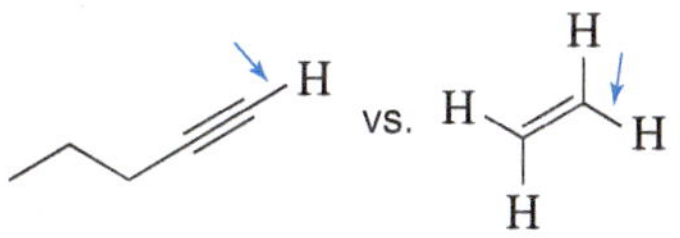

Assessment 14.19 The IR spectrum for anisole contains two C—O stretching bands in the fingerprint region. Match the band to the bond that gives rise to it. Why are these bands so intense?

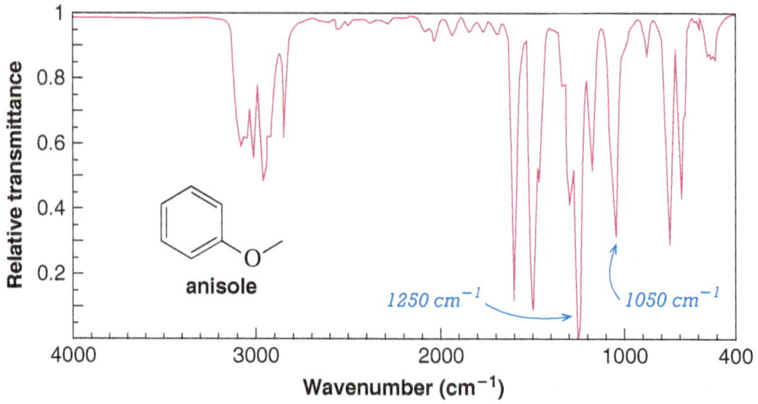

14.3.6 Electronic effects: the carbonyl

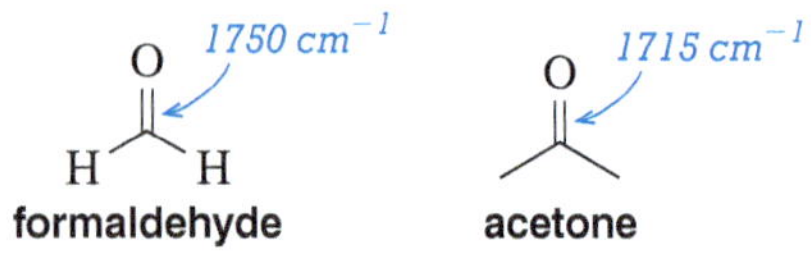

Figure 14.24 **Formaldehyde and acetone absorb at different wavenumbers** despite their similarities.

The two carbonyl groups shown in **Figure 14.24** have IR bands at different wavenumbers despite what might be predicted on the basis of Hooke's law. [*This particular deviation will help us understand the reactivity of carbonyls, giving us a head start on the material in Chapters 17–20.*] Both are C—O double bonds, containing identical reduced masses and nearly identical force constants. And yet they vibrate at different frequencies. [*It will help you to memorize 1715 cm^{-1} for acetone and all other "normal" ketones.*]

What's different here? Whenever we solve problems involving carbonyls, drawing the second-best resonance structure will usually be helpful. The actual structure of any functional group is the combination of all resonance structures, and the second-best resonance structure often tells a lot about the reactivity. This resonance structure, shown in **Figure 14.25**, is referred to as the **zwitterionic** resonance structure because it contains two oppositely charged ions in the same molecule. When this is done for both formaldehyde and acetone, there is a negative charge on oxygen, so no obvious difference there. On the other hand, the carbocation in acetone is 2°, whereas it is unsubstituted in formaldehyde. [*Before you read further, think deeply about the impact of the different carbocations and how they might explain the difference in absorption frequency.*]

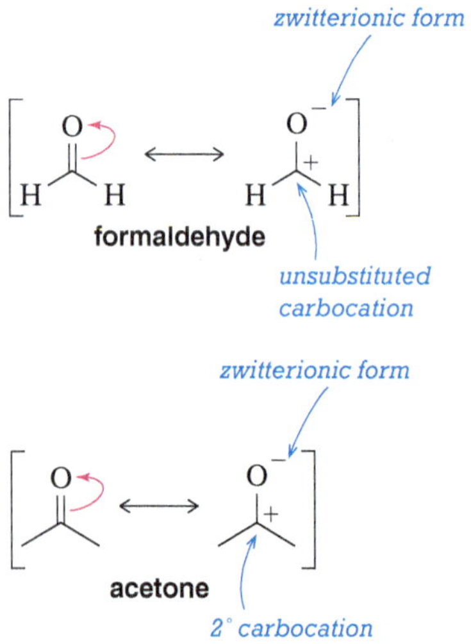

Figure 14.25 **Zwitterionic forms of acetone and formaldehyde.**

As it turns out, the zwitterionic resonance structure explains the difference quite well (**Figure 14.26**). Rather than discussing it in the context of bond strength, which will eventually cause confusion, let's think of a high frequency of vibration arising from a bond's resistance to stretching. [*A spring that resists stretching would vibrate at a high frequency once released, right?*] In formaldehyde's zwitterionic form, there is an unsubstituted carbocation, and unsubstituted carbocations are unstable. An unstable carbocation would hold tightly to a negatively charged oxygen. By holding tightly to the oxygen, the C═O bond in formaldehyde is more resistant to stretching and therefore vibrates at a higher frequency (1750 cm^{-1}). In acetone's zwitterionic form, there is a 2° carbocation, and 2° carbocations are relatively stable. A more stable carbocation will hold *less* tightly to a negatively charged oxygen. By holding *less* tightly to the oxygen, the C═O bond in acetone is *less* resistant to stretching and therefore vibrates at a lower frequency (1715 cm^{-1}). [*Read this again because we will see this trend many times.*]

Figure 14.26 **Carbocation stability in the zwitterionic resonance form rationalizes the difference in absorption frequencies.**

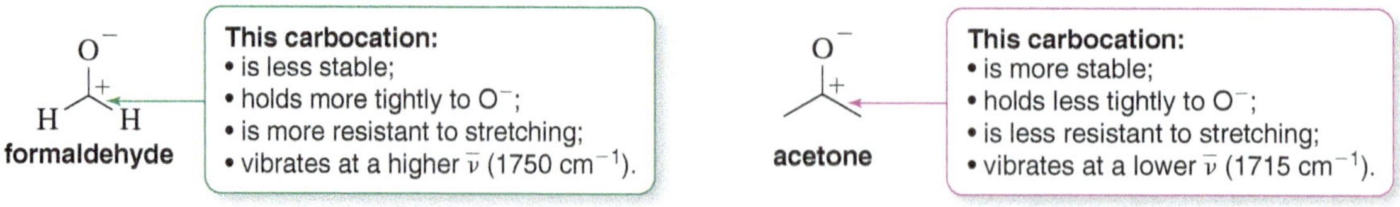

The position of the C═O stretching band for several carbonyls and carboxylic acid derivatives can similarly be predicted based on the stability of the zwitterionic carbocation. **Table 14.3** shows the deviation of the C═O stretching frequency from a normal, saturated ketone ($1715\ cm^{-1}$) based on the attached substituent, using the analysis just presented. For acid chlorides, inductive withdrawing dominates the diminished resonance-donating ability of the highly electronegative chlorine. The net withdrawing effect of the chlorine makes the positive charge on the carbonyl carbon larger, resulting in it holding the negatively charged oxygen more tightly and a higher vibrating frequency. On the other end of the spectrum is the amide. As a less electronegative heteroatom, it demonstrates weaker induction, allowing resonance donation to be the dominant effect. Thus, the central positive charge on the carbonyl carbon in the zwitterionic resonance structure is stabilized. The delocalized/diminished positive charge holds the negatively charged oxygen loosely, resulting in vibration at a lower frequency.

Table 14.3 **C═O stretching frequency** can be rationalized based on inductive and resonance stabilizing effects on the carbocation of the zwitterionic carbonyl structure. [*There are two bands for the anhydride because of symmetric and asymmetric C═O stretching.*]

Functional group	Structural effects	Description	Carbocation stability	Holds negatively charged oxygen	Resulting vibration frequency $\bar{\nu}$ (cm^{-1})
Acid chloride	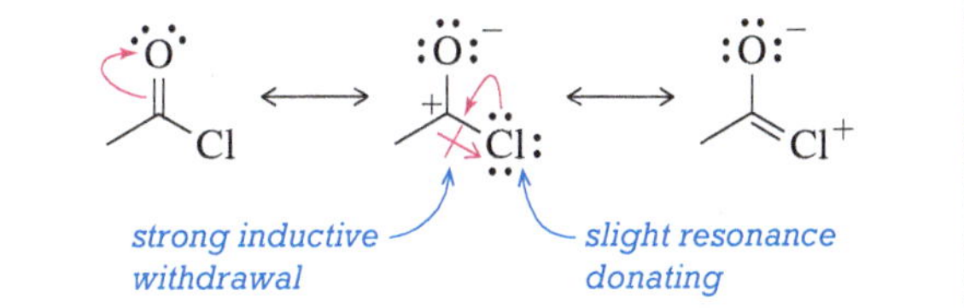	**Strong inductive withdrawal** Weak resonance donating	Destabilized (induction wins)	Most tightly	1785
Anhydride	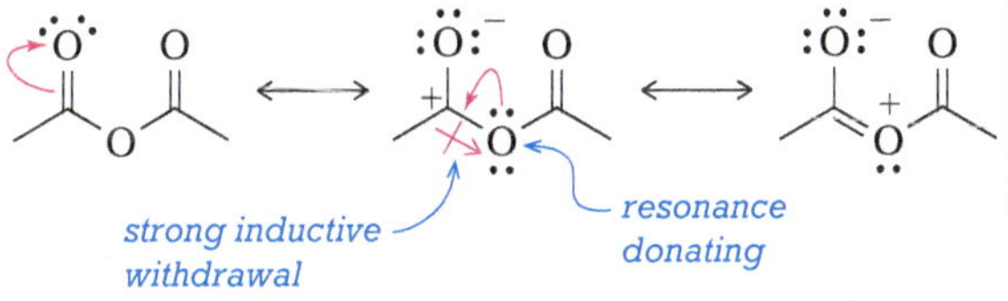	**Strong inductive withdrawal** Intermediate resonance donating	Destabilized (induction wins)	More tightly	1800 and 1740
Ester	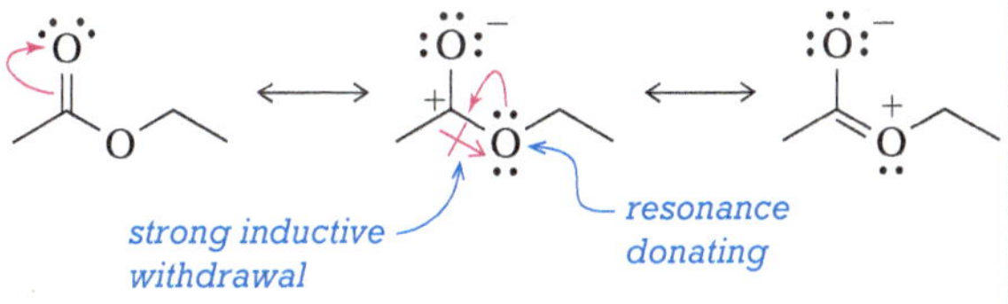	Intermediate inductive withdrawal Intermediate resonance donating	Destabilized (induction wins)	Tightly	1735
Amide	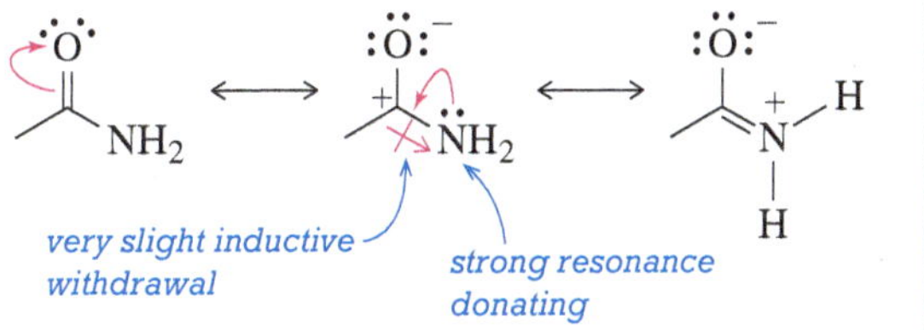	Weak inductive withdrawal **Strong resonance donating**	Stabilized (resonance wins)	Less tightly	1685

Further explanation is required to differentiate between anhydrides and esters (**Figure 14.27**). Both of them have an inductively withdrawing and resonance-donating oxygen attached to the carbonyl carbon, suggesting that they should stretch at the same frequency. The alkoxy oxygen in an ester can only donate an electron pair in one direction, so all of its resonance goes toward stabilizing the positive charge on the zwitterionic carbonyl. On the other hand, the central anhydride oxygen shares an electron pair with both attached carbonyls. The electron pair, delocalized over both carbonyls, is less able to stabilize either carbocation. The resulting diminished resonance donation allows induction to win by a bigger margin for anhydrides than it does for esters.

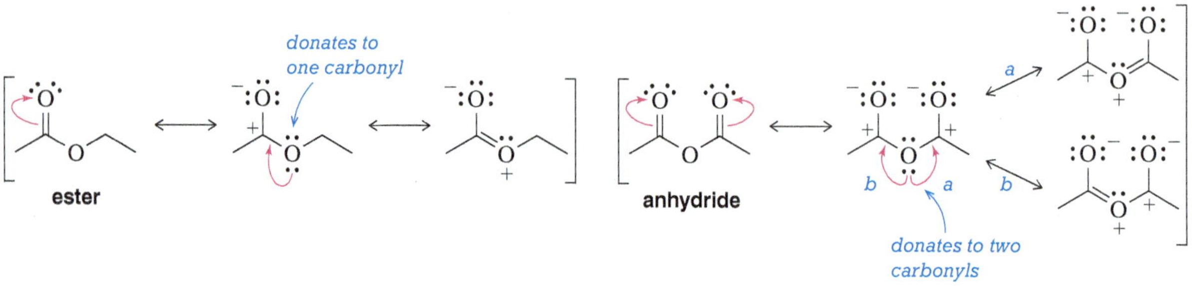

Figure 14.27 **Resonance in anhydrides** is divided equally between the two carbonyls, lessening its frequency-lowering effect.

In the series of cycloalkanones shown in Figure 14.28, the carbonyl stretching band occurs at decreasing wavenumbers moving from the small, strained cyclopropanone to the strain-free cyclohexanone.

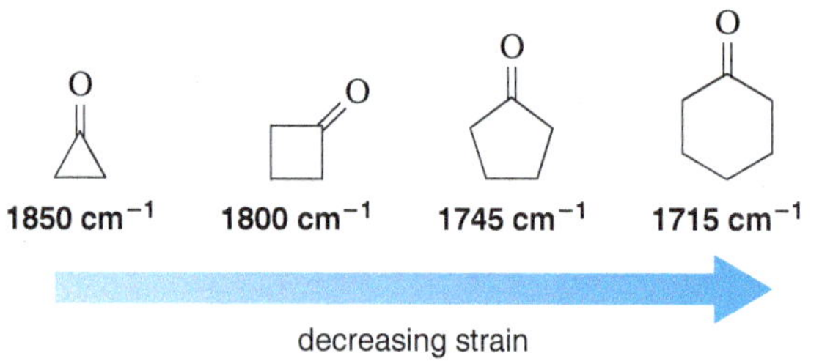

Figure 14.28 **The characteristic carbonyl absorption of cyclic ketones increases as ring size decreases and strain increases.**

14.3.4–14.3.6 SUMMARY

- A more polar bond gives rise to a more intense band. The location of the band in the IR spectrum is a function of the reduced mass of the atoms attached and the force constant of the bond between them. The electrons in bonds with more *s* character are held closer to the nucleus and the bonds vibrate at higher frequencies. The stretching frequency of a carbonyl depends on the ability of attached substituents to stabilize the carbocation in the second-best (zwitterionic) resonance structure.

Assessment 14.20 For each pair, choose the molecule that you expect to have the highest wavenumber for its C=O stretch.

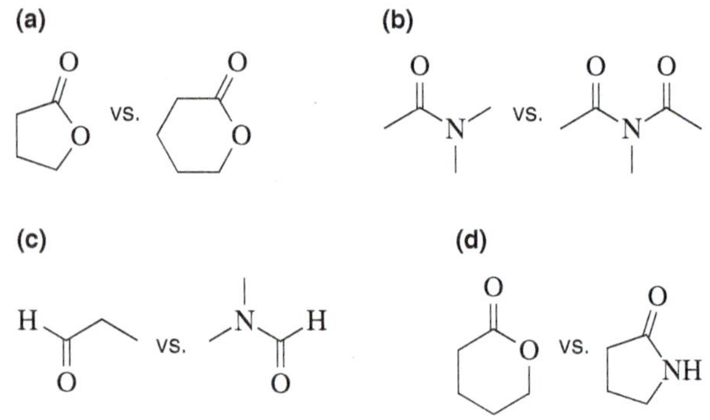

Assessment 14.21 The C=O stretching band for a normal, fully saturated ketone appears at 1715 cm^{-1}. Would you expect the α,β-unsaturated carbonyl to be higher or lower? [*Think about how an alkene might impact the zwitterionic resonance structure.*]

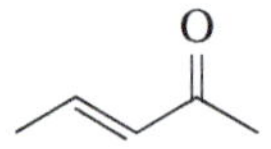

α,β-unsaturated ketone

14.4 Infrared spectroscopy: a tour of functional groups

Now that you understand how IR spectroscopy is used and how it works, we can take a tour of functional groups (Table 14.4). On this tour, we examine the important bands that distinguish each major functional group. Now is the time to familiarize yourself with how these functional groups appear in the IR spectrum. [*At times, the tour bus will stop for a brief discussion of concepts specific to a given functional group.*]

Table 14.4 The important functional groups via infrared spectroscopy: A tour.

Functional group	Description	Representative spectrum
Alkanes	Besides the presence of the C—H stretching bands around 2900 cm^{-1}, the most noteworthy aspect of an alkane spectrum is the lack of bands. There is one band each for the symmetric (2980 cm^{-1}) and asymmetric stretch (2880 cm^{-1}). 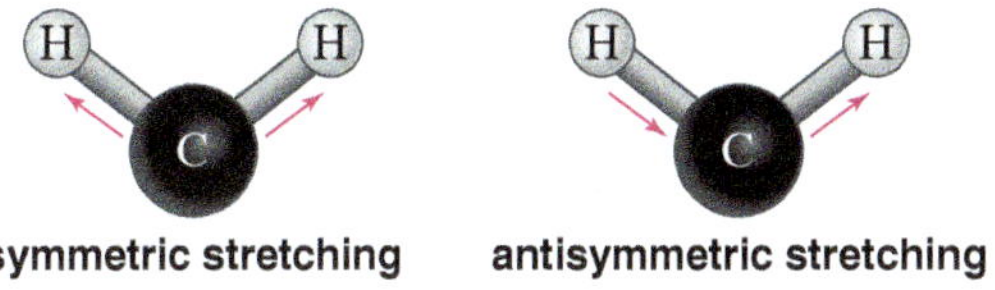	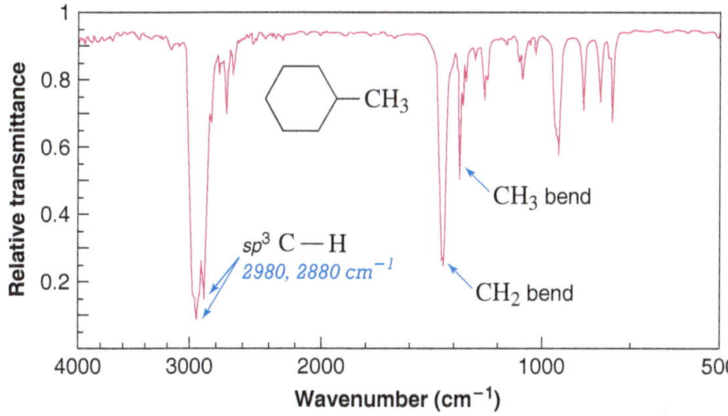
Alkenes	The characteristic bands of an alkene include the weak intensity C═C stretching band at 1660 cm^{-1}. When it contains alkene hydrogens, the characteristic sp^2 C—H appears around 3100 cm^{-1}. Because there are also sp^3-hybridized carbons, as in most organic molecules, there are C—H bands at 2980 and 2880 cm^{-1}.	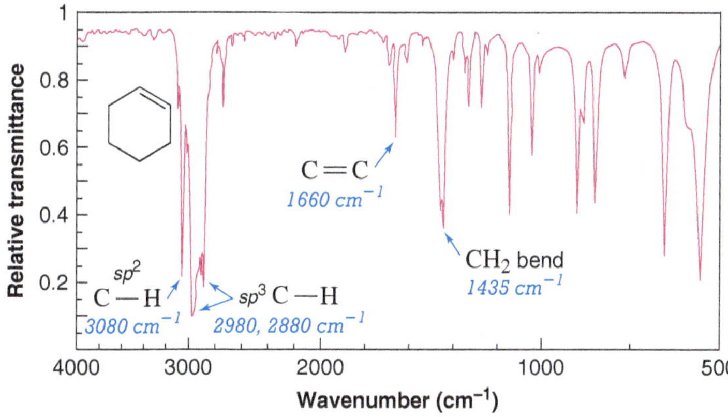
Alkynes	Terminal alkynes contain two distinct bands: ~2100 cm^{-1} for the C≡C stretch and ~3300 cm^{-1} for the *sp*-hybridized C—H stretch.	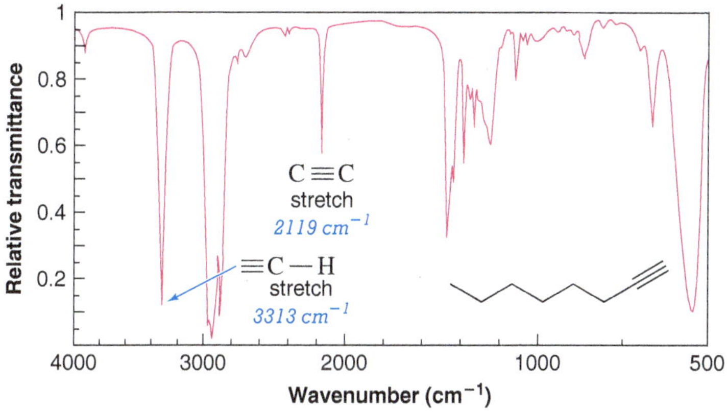
	Internal alkynes, on the other hand, look a lot like alkanes. Lacking a terminal hydrogen, there is no band at 3300 cm^{-1}. To make matters worse, an internal alkyne is a completely nonpolar bond. Thus, even when it does vibrate, it doesn't absorb infrared light.	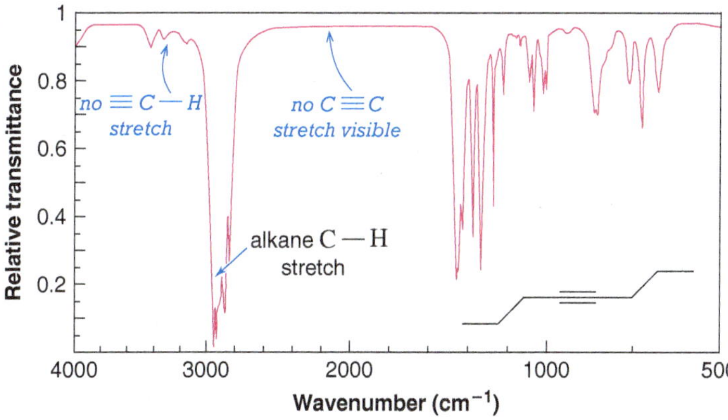

(continued)

Table 14.4 *Continued*

Functional group	Description	Representative spectrum
Alcohols	Alcohols are one of the easiest functional groups to identify as a result of the broad band for the O—H stretch occurring around 3400 cm^{-1}. [*There are few specific facts needed for the MCAT from IR, but the broad stretch at 3400 cm^{-1} is one you should know.*] The C—O stretching frequency around 1050 cm^{-1} will always be there for an alcohol, but is of secondary importance to structural determination.	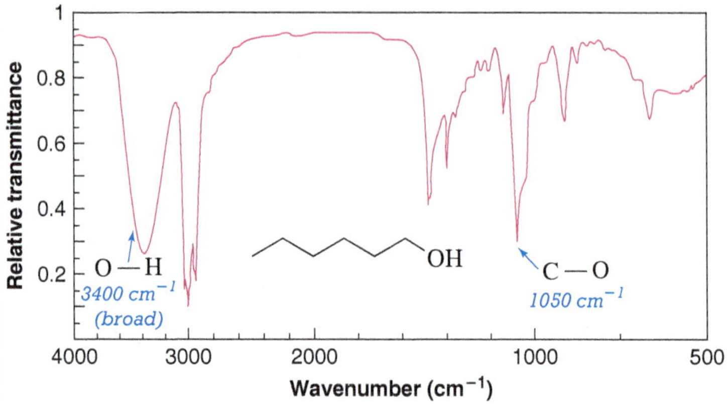

(continued)

The characteristic broad O—H band is a result of hydrogen bonding (**Figure 14.29**). An IR spectrum is typically obtained using a neat sample where hydrogen bonding is possible. [*"Neat" here means it is not a solution but the pure compound itself.*] To understand the broad band, consider first a single molecule of ethanol. With no other molecules around to form hydrogen bonds with it, the O—H bond has a specific bond length. An IR spectrum of an alcohol that cannot engage in hydrogen bonding has a narrow, sharp band. On the other hand, a neat sample of ethanol allows for substantial hydrogen bonding. Within the sample, then, there are a large number of O—H bonds with different bond lengths and different H---O—H bond angles. Furthermore, hydrogen bonding is a dynamic process. On the time scale of IR analysis, the hydrogen bonds exchange among many hydroxyl groups. As a result, the IR spectrum sees all of the different O—H bonds at one time but doesn't resolve them. The result is the characteristic broad O—H stretch.

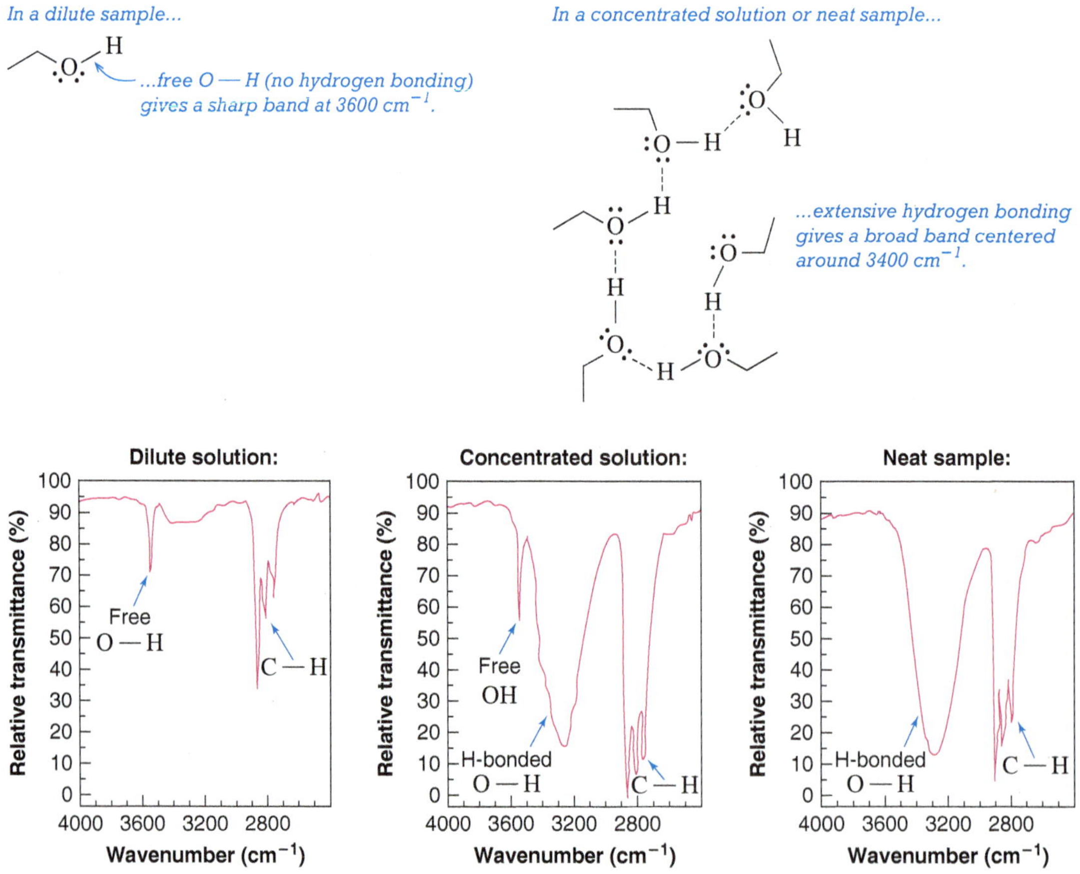

Figure 14.29 Hydrogen bonding causes the shifting and broadening of the O—H stretching band from 3600 cm^{-1} to 3400 cm^{-1}.

Table 14.4 *Continued*

Functional group	Description	Representative spectrum
Ethers	Like alkanes, ethers are noted for their lack of bands, with the exception of the strong C—O stretching band around 1100 cm^{-1}.	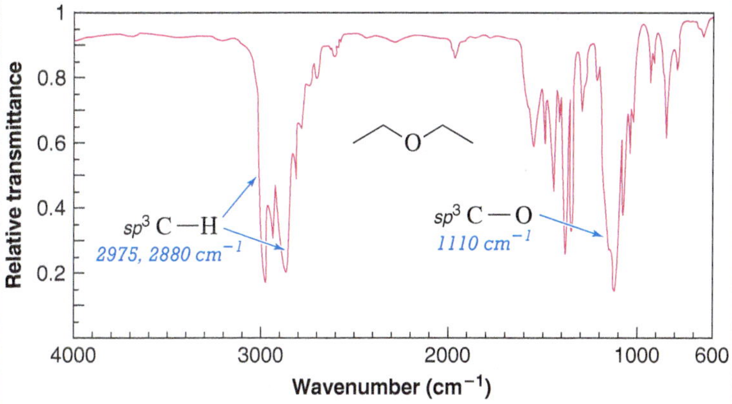
Ketones	Perhaps the most important IR band appears in ketones and other carbonyls. The C═O stretch, discussed in detail in Section 14.3.6, is right around 1715 cm^{-1} for ketones. It is a polar bond, so it is very intense and impossible to miss. [*This is another band that you will need to know on standardized exams.*]	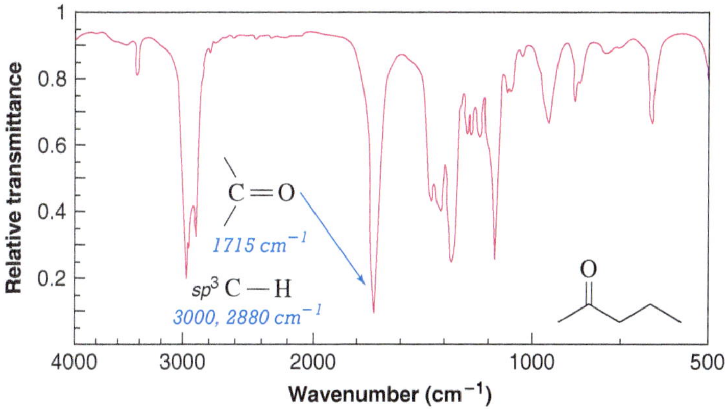
Aldehydes	Aldehydes can be distinguished from ketones in two ways. First, the C═O stretching vibration absorbs at a slightly higher wavenumber (1735 cm^{-1}). Second, there are two distinct aldehydic C—H bands (2840 cm^{-1} and 2710 cm^{-1}) attributed to Fermi resonance, a concept that is beyond the scope of the discussion here.	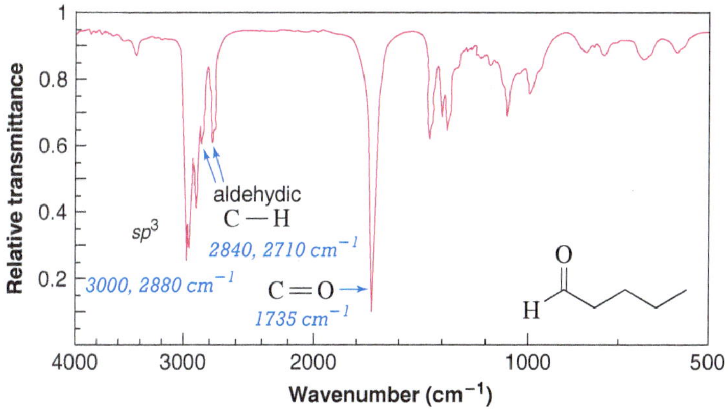
Carboxylic acids	It is almost impossible to miss a carboxylic acid. The broad O—H band extends from 3600 to 2400 cm^{-1}. As with alcohols, the broad band is caused by extensive hydrogen bonding between carboxylic acids. Although other bands aren't usually needed, a carboxylic acid can be diagnosed using the C═O stretch at 1715 cm^{-1} and the sp^2-hybridized C—O stretch at 1250 cm^{-1}.	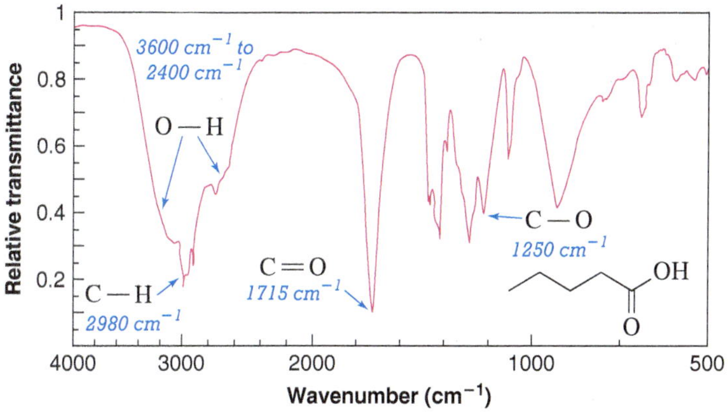

(continued)

Table 14.4 *Continued*

<table>
<tr><th>Functional group</th><th>Description</th><th>Representative spectrum</th></tr>
<tr><td>Esters</td><td>Esters also have the prominent C=O stretching band around 1740 cm^{-1} because of the inductive withdrawal of the electronegative oxygen. Although they are sometimes hidden in the fingerprint region, there are two C—O stretching bands around 1100 cm^{-1}. The sp^2-hybridized C—O bond gives a band around 1200 cm^{-1}, while the sp^3-hybridized C—O bond gives a band around 1000 cm^{-1}. This difference of ~200 cm^{-1} is consistent with the difference seen in Figure 14.23 (Section 14.3.5).</td><td>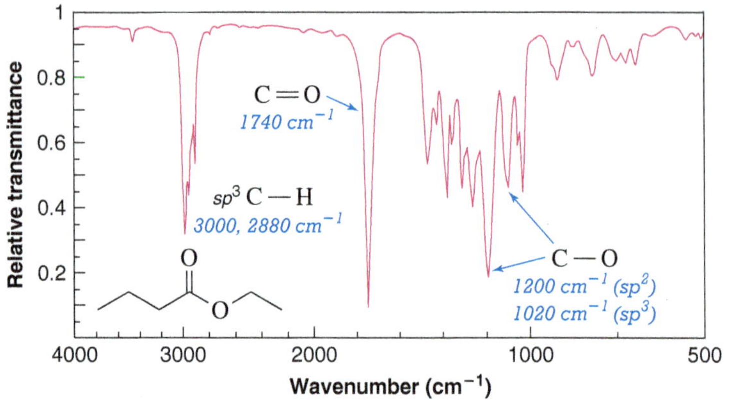
</td></tr>
<tr><td>Amides</td><td>The amide is one of the few carbonyl-containing compounds where the C=O stretch is below 1700 cm^{-1}. This is due to the electron-donating amino group, which stabilizes the zwitterionic form of the carbonyl (Section 14.3.6).</td><td>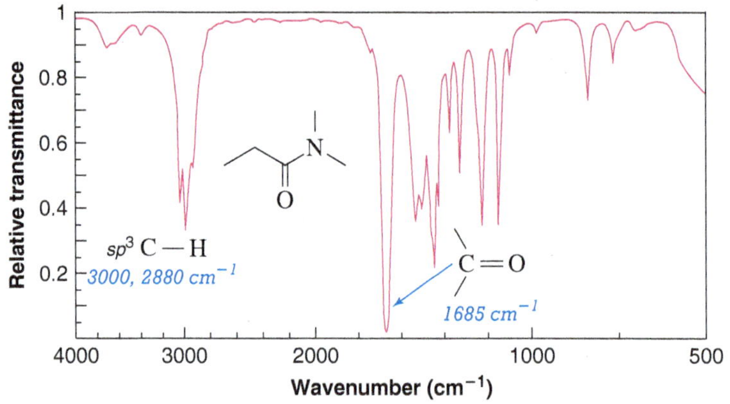
</td></tr>
<tr><td>Aromatics</td><td>The spectrum for an aromatic compound always looks more complex than that for other functional groups. [Compare it to the spectrum for amides above and amines below.] This complexity is a result of sp^2-hybridized C—H bands around 3100 cm^{-1} overlapping with the usual sp^3-hybridized C—H bands. The aromatic overtone region contains several (usually four) low-intensity bands between 2000 and 1600 cm^{-1}. [Note the difference in this region compared to the spectra of other functional groups.] Finally, benzene derivatives usually have distinct bands at 1600 and 1450 cm^{-1}, representing the aromatic ring.</td><td>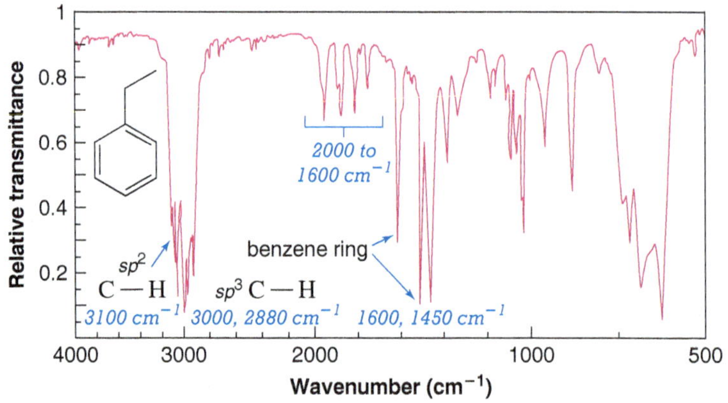
</td></tr>
<tr><td>Amines</td><td>Like alcohols, amines can hydrogen bond. As a result, the band for the N—H is somewhat broad. It is not as broad as the O—H bond, however, consistent with the fact that more electronegative atoms (O > N) are better at hydrogen bonding. 1° amines, which contain an NH_2 group, have a band for both the symmetric and asymmetric stretch. The NH_2 bend around 1600 cm^{-1} is a secondary band, not usually used for diagnosis.</td><td>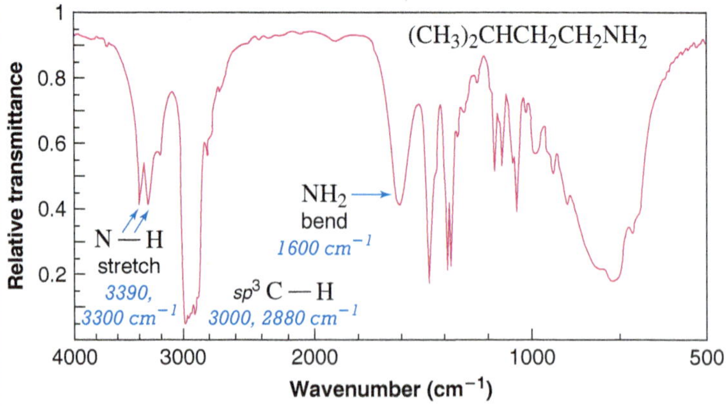
</td></tr>
</table>

(continued)

IR is excellent at distinguishing between 1°, 2°, and 3° amines (**Figure 14.30**). With only one hydrogen on the nitrogen and no opportunity for symmetric and asymmetric stretches, 2° amines have only a single broad band around 3350 cm^{-1}. And 3° amines, with no hydrogens on nitrogen, do not contain a band around 3350 cm^{-1}.

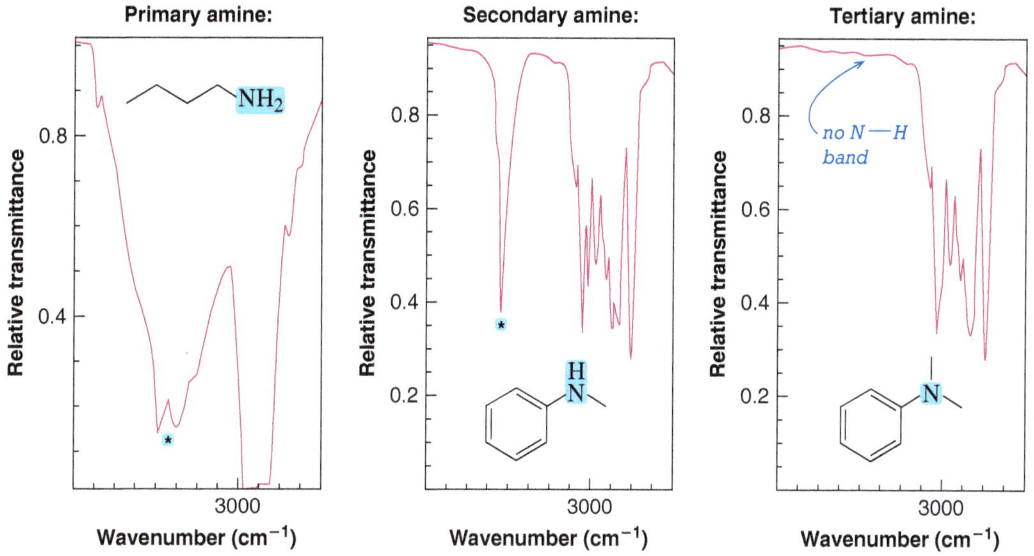

Figure 14.30 **Primary, secondary, and tertiary amines** are easily distinguished by IR spectroscopy.

Table 14.4 ***Continued***

Functional group	Description	Representative spectrum
Nitriles	Nitriles are sometimes confused with alkynes given the sharp band around 2200 cm^{-1} that is predicted by Hooke's law for molecules with a C≡X triple bond. Being more polar than an alkyne, the C≡N band is a bit more intense. [*What's another major difference that would allow you to distinguish between an alkyne and a nitrile? Assessment 14.23 asks this question again.*]	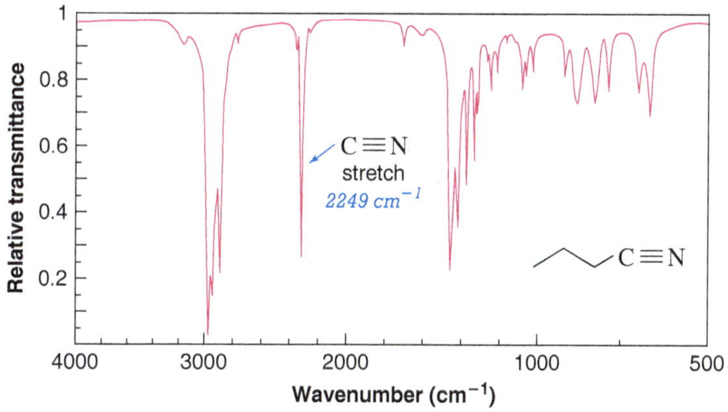

Assessment 14.22 Match each of the four IR spectra to one of the given compounds. [*One of the compounds does not match a spectrum.*]

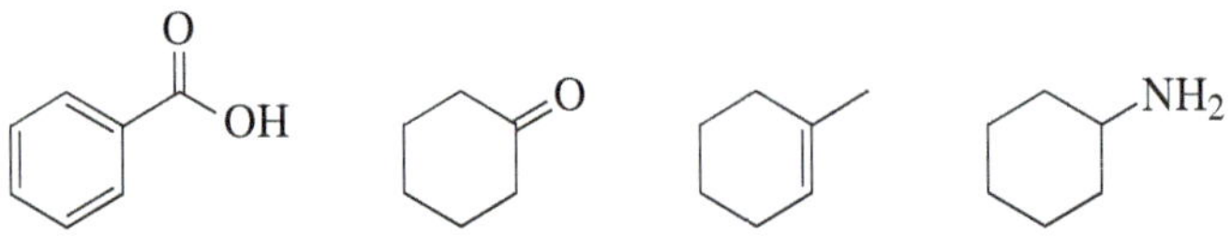

(a)

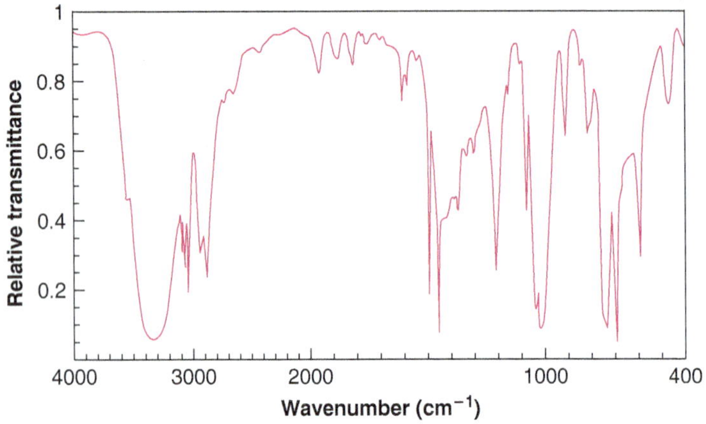

(b)

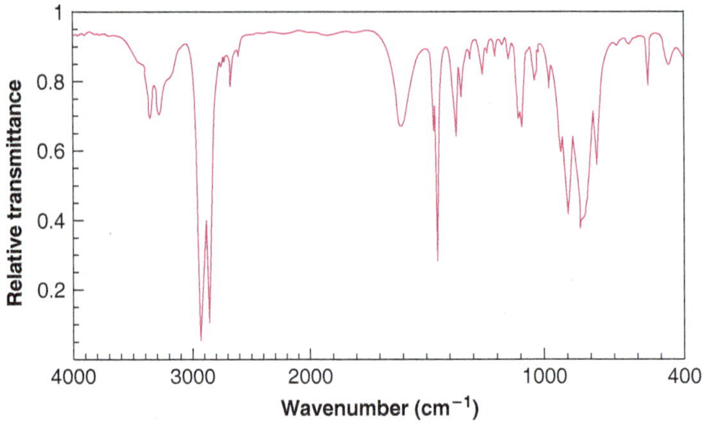

(c)

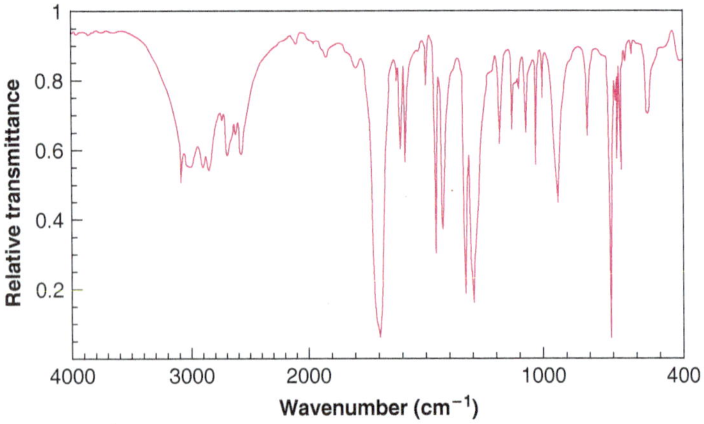

(d)

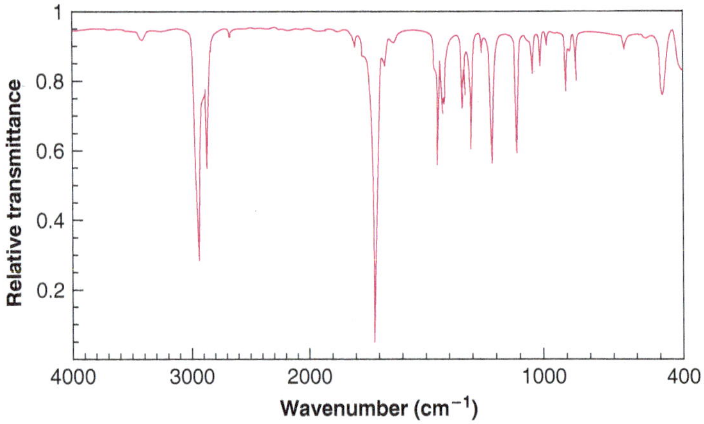

Assessment 14.23 Given that both would show a band around 2200 cm^{-1}, how would you distinguish between hexanenitrile and hept-1-yne on an IR spectrum?

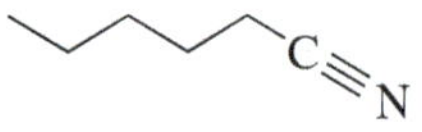

hexanenitrile

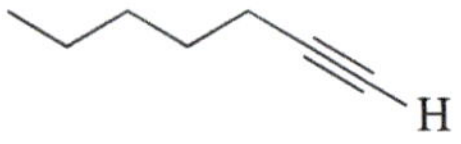

hept-1-yne

Assessment 14.24 Sketch a spectrum containing the important bands for the following compound.

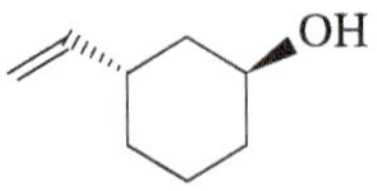

14-B SPECTROSCOPY IN CONTEXT: Raman

Counterfeit antimalarial drugs

A type of spectroscopy that is popular among analytical chemists, though not covered extensively in this chapter, is Raman spectroscopy. It, like infrared spectroscopy, measures the vibrational modes in a molecule, but there are a few key differences. Instead of measuring the absorption of light, Raman relies on the scattering of light for the detection of molecules. When a laser interacts with a molecule, the energy of the photons shifts (termed Raman scattering) based on the molecular vibrations they interact with. As shown in Figure 14.31, a Raman spectrum is quite similar to an IR spectrum, offering a structural fingerprint of the molecule being analyzed. Whereas IR requires a vibrational mode to cause a change in dipole moment to be active, Raman requires bonds to be polarizable. In this way, Raman and IR are complementary techniques.

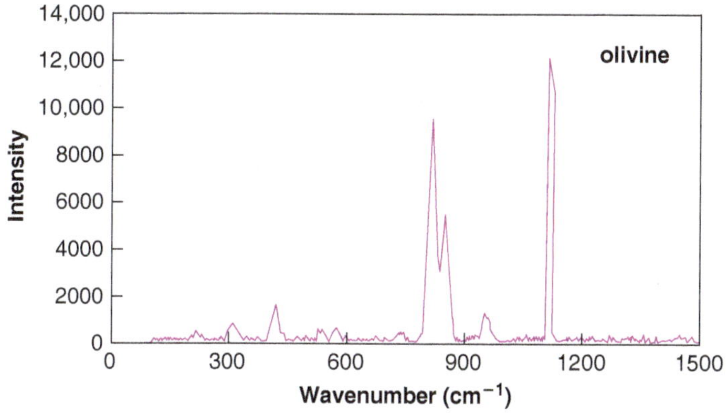

Figure 14.31 Example of a Raman spectrum.

Artesunate is a medication used as part of a combination therapy in the treatment of malaria in Southeast Asia and is a target of counterfeit drug producers. The World Health Organization (WHO) has estimated that as many as 200,000 malaria deaths could be prevented if the available medicines were of good quality. The increase in mortality occurs when the patient believes they are receiving treatment when they are perhaps receiving none of the active ingredient.

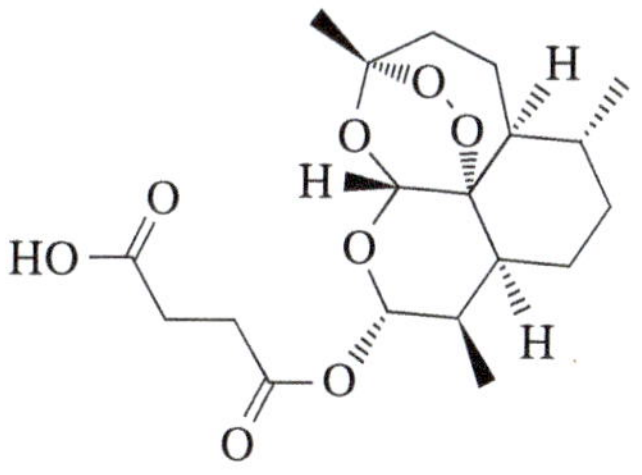

artesunate

In one study of 50 tablets purchased in Southeast Asia, Raman spectroscopy was able to distinguish between authentic and counterfeit artesunate by comparing the structural fingerprint of artesunate to the structural fingerprint of the tablets. Furthermore, Raman spectroscopy was able to identify the composition of the counterfeits. As shown in Figure 14.32, the counterfeit samples consisted principally of starch, calcite, and acetaminophen.

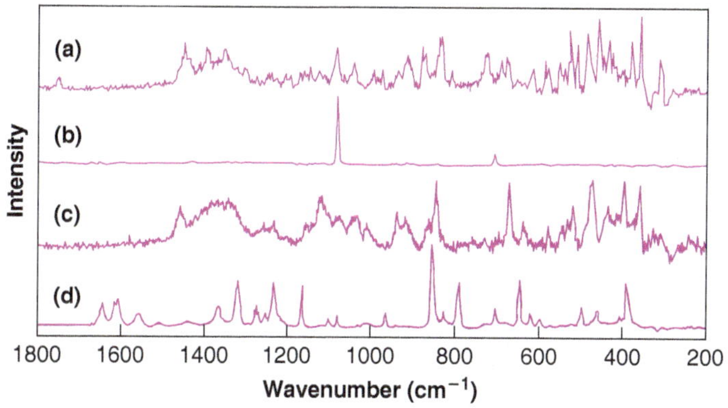

Figure 14.32 Comparison of Raman spectra. (a) Authentic artesunate. Counterfeit pills, alleged to also contain artesunate, contained (b) calcite, (c) starch, and (d) acetaminophen.

14.5 Mass spectrometry: practical usage

In Assessment 14.11, you were given the molecular formula and asked to deduce the structure that corresponded to a given IR spectrum. With the molecular formula, problems of that sort become much simpler than just looking at a spectrum without that information. **Mass spectrometry (MS)** is a technique that gives the molecular mass of an unknown molecule, thus making it possible to determine the molecular formula. Used in this way, it is a great complement to infrared spectroscopy and nuclear magnetic resonance (NMR) spectroscopy (Chapter 15).

As we did for IR, we are first going to work a mass spectrometry problem and then fill in the gaps by learning how and why it works. An example of a mass spectrum is shown in Assessment 14.25. The location of the peak along the x axis gives the molecular mass [*really, the mass to charge ratio, but we explain what that means later*], and the y axis gives the intensity of each peak relative to the tallest peak, which is arbitrarily set at 100%. Each peak represents the molecular mass of a component in the mixture. Even if you start with a pure compound, the conditions of mass spectrometry result in extensive fragmentation of the molecule. As a result, each peak represents the molecular mass of the original compound or the molecular mass of one of the compounds it fragments into. The peak that corresponds to the molecular mass of the original compound is referred to as the molecular ion (M) for reasons that we discuss in Section 14.6. [*Equipped with only this knowledge, you are ready to answer Assessment 14.25. Try it on your own, and then we can work it together.*]

Assessment 14.25 (a) Based on the mass spectrum shown, give the molecular mass of an unknown compound containing only carbon and hydrogen. (b) Based on this molecular mass, suggest a molecular formula for the unknown fully saturated hydrocarbon.

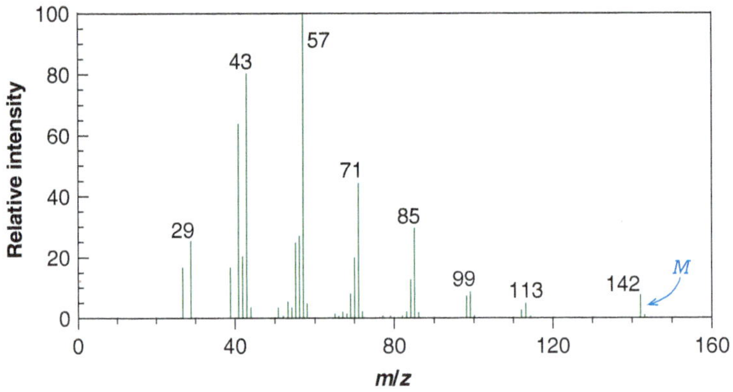

Working assessments before we learned the concepts helped with IR spectroscopy, so let's do the same with mass spectrometry.

Solved Assessment 14.25

The peak corresponding to the molecular mass of the compound being analyzed is the molecular ion (M). Thus, the spectrum shows a molecular mass of 142 atomic mass units for the parent compound. [*Pretty simple so far.*]

If Assessment 14.25 had involved a more complicated spectrum, we almost certainly wouldn't be able to work it right now. But because it concerns a small organic molecule with only carbon and hydrogen, there are only a few possibilities. The general formula of an alkane is C_nH_{2n+2}, so there must be $2n + 2$ hydrogens for every carbon in the molecule. This assumes that the molecule is fully saturated, which was stated in the problem. Table 14.5 shows all the possible molecular masses for saturated hydrocarbons up to the mass range observed in the spectrum. From this analysis, a compound containing 10 carbons and 22 hydrogens would have a mass of 142 amu. Thus, the molecular formula of the parent compound in Assessment 14.25 is $C_{10}H_{22}$. ■

Table 14.5 Possible molecular formulas and corresponding molecular masses of *saturated* hydrocarbons.

Number of carbons	Number of hydrogens	Total molecular mass (amu)
1	4	16
2	6	30
3	8	44
4	10	58
5	12	72
6	14	86
7	16	100
8	18	114
9	20	128
10	22	142
11	24	156
12	26	170

If we'd had an infrared spectrum to go with our mass spectrum, we'd have made our unknown analysis problem simpler. We could calculate the index of hydrogen deficiency and know what functional groups to look for in the IR spectrum. [*This is an important point that is worth repeating often: Each of the spectroscopic techniques we learn in Chapter 14 and 15 is only part of the puzzle. We often need to use them together.*]

Assessment 14.26 For a compound containing a molecular formula of $C_{10}H_{22}$, there are 75 constitutional isomers. To which does the spectrum correspond?

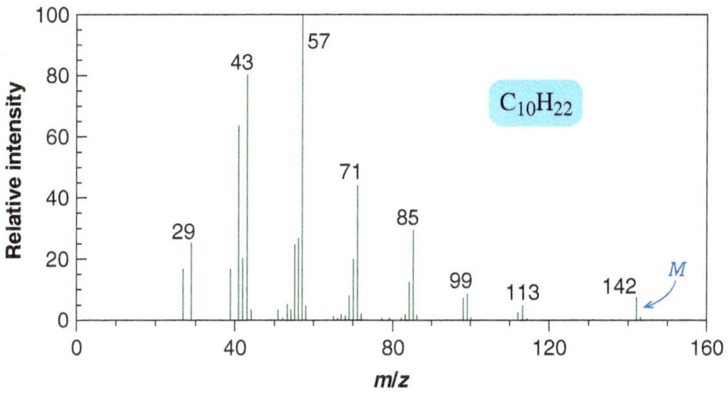

To preview the full power of mass spectrometry, let's do one more.

Solved Assessment 14.26

Given our lack of experience, this problem won't be as easy, but if we think logically, I think you'll find that the solution makes sense. When acquiring a spectrum, the molecule fragments into several smaller pieces. Each of these fragments is represented by a peak in the spectrum. For this Assessment, only focus on the peaks whose molecular masses are labeled.

To use these fragments' molecular mass, we have to think about what they mean. If the parent molecule started with a molecular mass of 142 and fragmented to a molecule with a molecular mass of 113, how much was lost? And, more importantly, what is the molecular formula of the fragment that fell off? **Figure 14.33** shows that a molecular mass of 142 becomes a molecular mass of 113 by losing 29 atomic mass units. The only way to have a molecular mass of 29 is to have lost two carbons and five hydrogens. C_2H_5 can only be an ethyl group. The key: If the fragmentation of a molecule results in the loss of an ethyl group, the molecule must have had an ethyl group to start. [*Read that sentence again because it is the key to using mass spectrometry for structure determination.*]

And, if a fragment falls off in moving from the 142 peak to the 113 peak, then it will likely show up somewhere else in the spectrum. Sure enough, when reading from left to right, the first peak is at 29. [*Continuing left to right, do you see a pattern as you go from labeled peak to labeled peak?*] From 29 to 113, there is a consistent increase of 14 atomic mass units at a time. [*How would you account for 14 amu using only carbon and hydrogen?*] Each increase in 14 amu must represent an additional CH_2. And so, if we were building this molecule from left to right, starting with an ethyl group, we'd add six CH_2 groups to get to a fragment of 113 amu. Add one more ethyl group (C_2H_5; 29 amu) and we have *n*-decane, the straight-chain constitutional isomer of $C_{10}H_{22}$ (**Figure 14.34**). ■

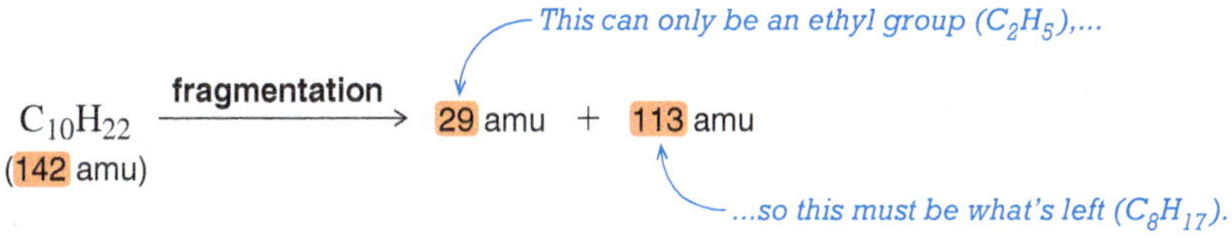

Figure 14.33 Identifying what's lost in fragmentation shows what was present in the original molecule.

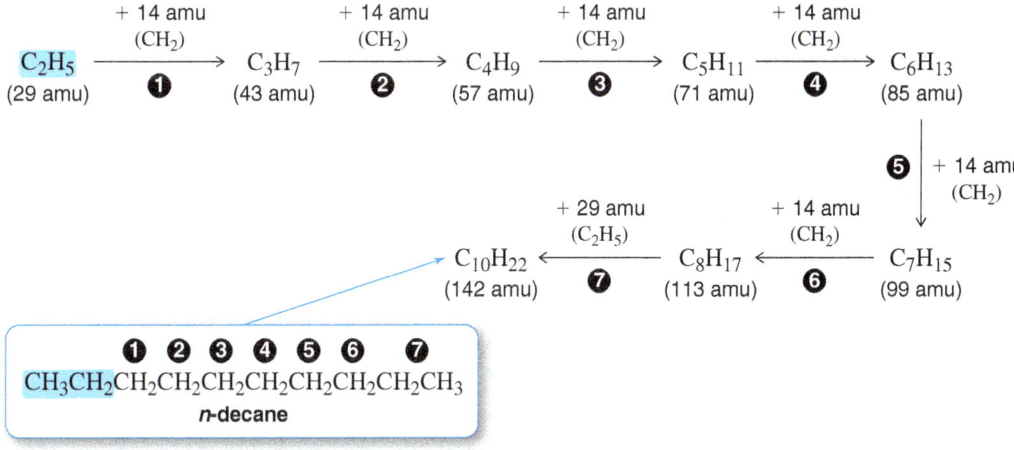

Figure 14.34 By adding up the fragments that were lost, the only solution emerges. [*Most problems won't be as simple as this.*]

Assessments 14.25 and 14.26 represent straightforward and somewhat simple applications of mass spectrometry, but they give us the framework within which we can more deeply examine this incredibly useful structure identification technique.

14.6 Mass spectrometry: background theory

Unlike infrared spectroscopy (Sections 14.2–14.4), NMR spectroscopy (Chapter 15), and UV–vis spectroscopy (Section 21.3.2), mass spectrometry does not involve the interaction of organic molecules with light. Instead, mass spectrometry bombards a molecule with a high-energy beam of electrons, which breaks the molecule into several fragments, similar to how throwing a rock at a window fragments the glass into different pieces. By doing this, we can accurately measure the mass of the molecule and its many fragments. When the molecule is an unknown, knowledge of these molecular masses can quickly lead to structure identification, especially when used in conjunction with other techniques.

There is a second important difference between MS and the other spectroscopic techniques. Whereas a molecule can be recovered after obtaining an IR, NMR, or UV–vis spectrum, the molecule is destroyed in MS. Fortunately, analytical chemists have developed the technique so thoroughly that extremely small sample sizes (on the microgram scale) can be analyzed so that very little material is lost. The following sections highlight the important aspects of mass spectrometry as they apply to organic chemistry. What is covered here barely scratches the surface of its value as an analytical technique. [*You can earn a Ph.D. just studying mass spectrometry.*]

14.6.1 Forming a radical cation

The most common detector used in mass spectrometry can only detect positively charged molecules. As a result, acquisition of a spectrum begins by bombarding a molecule with a high-energy beam of electrons. When an electron collides with a molecule like ethane (**Figure 14.35**), it removes an electron from ethane, forming a radical cation in the process. The radical cation formed by expulsion of one electron from the **parent molecule** is referred to as the **molecular ion (M)**. [*We haven't studied these before because . . . .*] Radical cations are very unstable species. This gives you an idea of how much energy is present in this beam of electrons—namely, enough energy to overcome virtually any reaction barrier. For this reason, you don't want to think about reactions in the mass spectrometer as being typical, or even possible, in the solution phase. [*What happens in the MS stays in the MS.*]

high energy electrons

$$CH_3CH_3 + e^- \longrightarrow [CH_3CH_3]^{\bullet+} + 2e^-$$

radical cation

Figure 14.35 Bombarding ethane with electrons produces a radical cation.

When dealing with reactive intermediates in this text, we have made explicit the atom on which the radical or cation resides. This is not always possible with mass spectrometry because, again, the excess energy in the system makes anything possible. In the case of the radical cation of ethane, we don't know, nor do we really care, which electron is removed. To acknowledge this ambiguity, the molecule is shown in brackets with a radical cation as a superscript [*see Figure 14.35*].

14.6.2 Measuring a spectrum

To understand why the mass spectrometer can only detect positively charged molecules, consider the schematic of a typical instrument (**Figure 14.36**). Injection of the sample results in immediate ionization by the electron gun. After being ionized, the molecular ion and other fragments are accelerated into the flight tube, where they encounter a magnetic field. As charged intermediates, they are attracted to the magnet, with the degree of their attraction being a function of their mass to charge ratio (m/z). Ions that are too heavy continue on a straight path, crashing into the side of the mass spectrometer [*too big . . .*]. Ions that are too light are pulled too strongly, once again crashing into the wall of the flight tube [*. . . too small . . .*]. Ions of the right size continue down the path, making it to the detector [*. . . just right!*]. Although we've talked about how mass spectrometry gives the mass of a molecule, that isn't exactly true. Notice in the spectrum in Assessment 14.26 that it is really the **mass to charge ratio (m/z)** of the ions that is being detected. Fortunately, the charge on almost every fragment is +1, so the mass to charge ratio is effectively the molecular mass. What about uncharged molecules? With no charge, they are unaffected by the magnetic field and therefore go undetected.

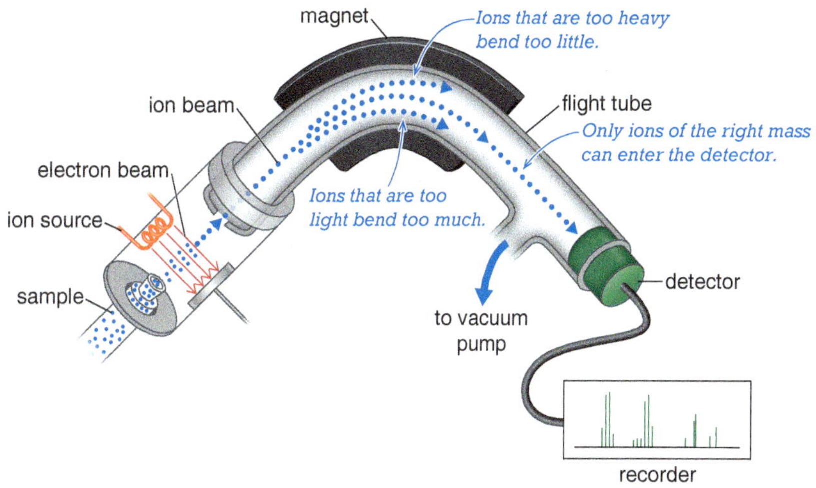

Figure 14.36 **Schematic of a mass spectrometer.**

14.6.3 Isotopes

The mass spectrum in **Figure 14.37** has a molecular ion (peak farthest to the right) of $m/z = 44$, so we know that the molecular mass of the parent molecule is 44 amu. There are multiple possible compounds that have that molecular mass, including acetaldehyde and propane (**Figure 14.38**), the two we consider here. A low-resolution mass spectrometer is unable to distinguish between these two possibilities.

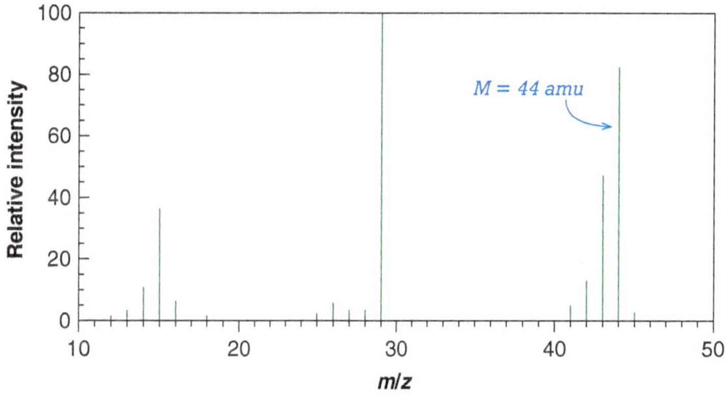

Figure 14.37 **Mass spectrum of an unknown** (M = 44 amu).

acetaldehyde C_2H_4O (44 amu) | propane C_3H_8 (44 amu)

Figure 14.38 **Acetaldehyde and propane** both have a mass of 44 amu.

Fortunately, mass spectrometry can also give extremely high resolution in determining the molecular mass of a molecule/fragment. Whereas a low-resolution instrument gives $m/z = 44$, a high-resolution instrument would give the spectrum shown in **Figure 14.39**. Here, the molecular ion has $m/z = 44.0262$. [*Yes, that's measured to the ten-thousandths place!*] This high-resolution *m*/*z* is referred to as an **exact mass**. Based on the exact masses of acetaldehyde and propane, the molecule shown in the spectrum is clearly acetaldehyde.

Figure 14.39 **The exact mass in a high-resolution mass spectrum** identifies the unknown as acetaldehyde. (a) Data from a high-resolution mass spectrometer. (b) Exact mass calculation of propane and acetaldehyde.

(a)

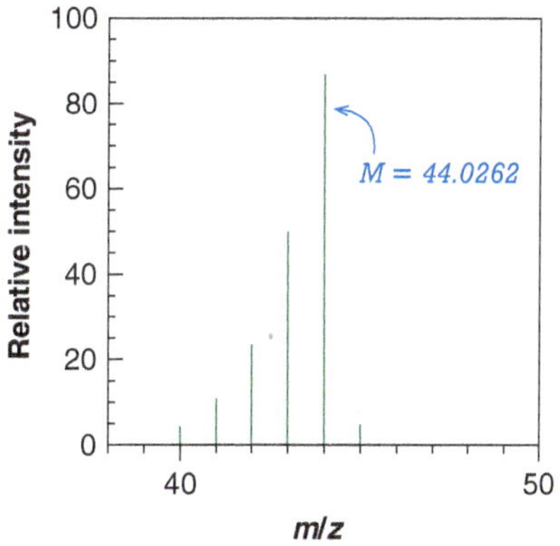

(b)

Propane (C_3H_8)		Acetaldehyde (C_2H_4O)	
3 C	36.0000 amu	**2 C**	24.0000 amu
8 H	8.0626 amu	**4 H**	4.0313 amu
		1 O	15.9949 amu
Total mass	44.0626 amu	**Total mass**	44.0262 amu

How can it be that two molecules with the same low-resolution masses have such different high-resolution masses? When we think of atomic masses, we think of 12 amu for C, 1 amu for H, and 16 amu for O, but these are *rounded* numbers. And, more importantly, the atomic masses given in the periodic table are a *weighted average* of all of the atomic masses of every isotope of a given element. Thus, there is not a single carbon atom in the universe that has an atomic mass of 12.011 amu (the atomic mass of carbon in the periodic table). Instead, 98.9% of all carbon atoms have an atomic mass of 12.0000 (by definition) and 1.1% have an atomic mass of 13.0034 (^{13}C). Table 14.6 shows the isotopic ratios of the commonly encountered elements in organic chemistry.

Table 14.6 Isotopic abundances of common elements in organic chemistry.

Element	M^+		M + 1		M + 2	
Hydrogen	^{1}H	99.9%	^{2}H	0.1%		
Carbon	^{12}C	98.9%	^{13}C	1.1%		
Nitrogen	^{14}N	99.6%	^{15}N	0.4%		
Oxygen	^{16}O	99.8%			^{18}O	0.2%
Sulfur	^{32}S	95.0%	^{33}S	0.8%	^{34}S	4.2%
Chlorine	^{35}Cl	75.5%			^{37}Cl	24.5%
Bromine	^{79}Br	50.7%			^{81}Br	49.3%
Iodine	^{127}I	100.0%				

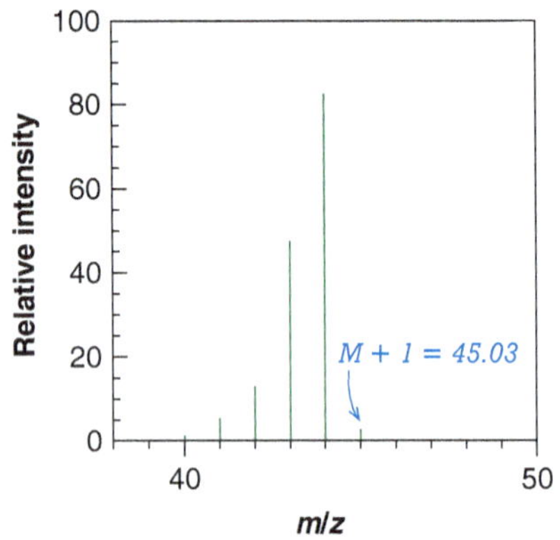

Figure 14.40 The M + 1 peak present in the mass spectrum is a result of a ^{13}C atom being in 1 out of every 50 molecules.

And so when a 1-mg sample of a compound is injected into a mass spectrometer, there are still billions and billions of molecules in the sample. Within this sample, 99 out of every 100 carbons are ^{12}C; the other is ^{13}C. And 999 out of every 1000 hydrogens are ^{1}H; the others is ^{2}H. The mass spectrometer doesn't detect the average atomic mass [*it probably doesn't even know what that is; mass spectrometers have never seen the periodic table.*] Instead, it detects the exact mass of every single molecule that is injected. As shown in Figure 14.39(b), the exact masses of acetaldehyde and propane are different, allowing a mass spectrometer to distinguish between them.

The molecular ion (M) is usually the farthest right peak in the spectrum [. . . *sort of*]. Look closely at the spectrum of what we now know to be acetaldehyde (**Figure 14.40**). Note the tiny, almost invisible peak one atomic mass unit to the right of the molecular ion peak. What is that, and why have we ignored it? Although only one out of every 100 carbon atoms is ^{13}C, if enough molecules are present, it will appear in some. For a molecule containing two carbons, on average, one in 50 will have a ^{13}C atom. And MS is sensitive enough to be able to detect even 1/50th of the molecules in our 1-mg sample. This type of peak, present as a result of the less common isotope of an atom, is referred to as the M + X peak. In the case of the spectrum in Figure 14.40, the mass of the small peak is 1 higher than the molecular ion, so it is called an M + 1 peak (X = 1). More common, and more useful, are the M + 2 peaks we discuss in Section 14.6.4.

Assessment 14.27 Bromine-79 (50.7% abundance) has an atomic mass of 78.918 amu. Bromine-81 (49.3% abundance) has an atomic mass of 80.916 amu. From these data, calculate the average atomic mass of bromine that you would expect to appear in the periodic table.

Assessment 14.28 The ratio of M:M + 1 is given for both decane ($C_{10}H_{22}$) and icosane ($C_{20}H_{42}$). Why is the M + 1 peak more prominent in icosane?

Decane	Icosane
M:M + 1	M:M + 1
10:1	5:1

14.6.4 M + 2 ratios

Assessment 14.29 introduces us to the concept of M + 2 ratios.

Assessment 14.29 The molecule that gave the mass spectrum shown here contains a halogen. Which halogen is present?

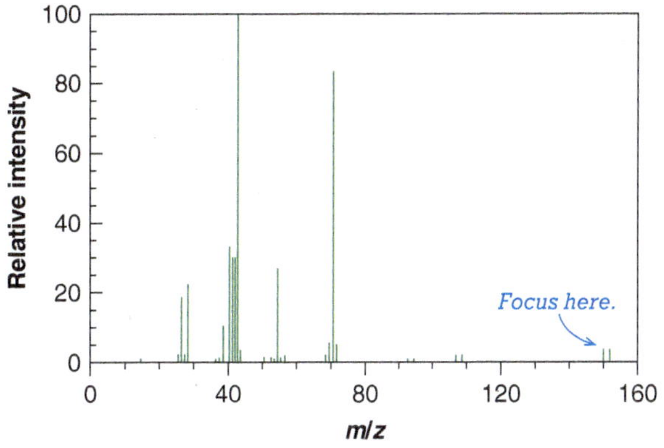

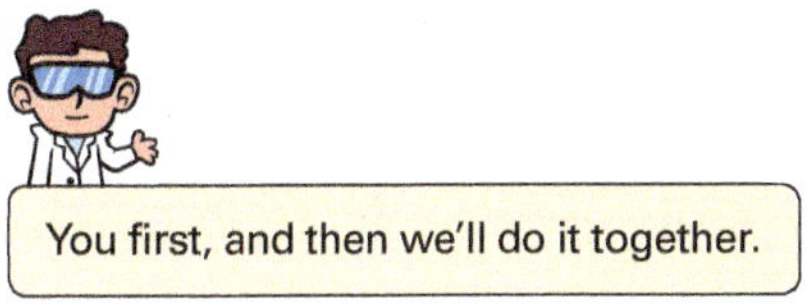

Solved Assessment 14.29

The spectrum in Assessment 14.29 contains two peaks at the far right, giving the appearance of two molecular ion peaks of similar size (**Figure 14.41**). The one on the left, at *m/z* 150, is technically the molecular ion, and the peak at *m/z* 152 is an M + 2 peak because it is 2 amu higher. This implies that the parent molecule has an atom whose slightly less abundant isotope is two atomic mass units higher. [*Look back at Table 14.6. Which element is it?*]

There are only three elements in Table 14.6 that have a significant second-most-abundant isotope that is two atomic mass units higher—namely, sulfur, chlorine, and bromine. To distinguish between these three, you must look at the relative abundance of each isotope. The ratio is ~19:1 for sulfur, 3:1 for chlorine, and ~1:1 for bromine. Based on these ratios, the molecule that gives the spectrum in Assessment 14.29 contains a bromine atom. ■

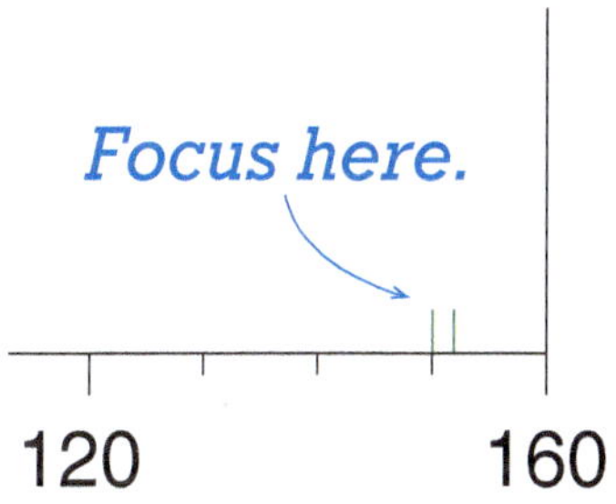

Figure 14.41 A mass spectrum with an M and an M + 2 peak in a nearly 1:1 ratio.

Assessment 14.30 The presence of what atom is indicated by the following mass spectrum?

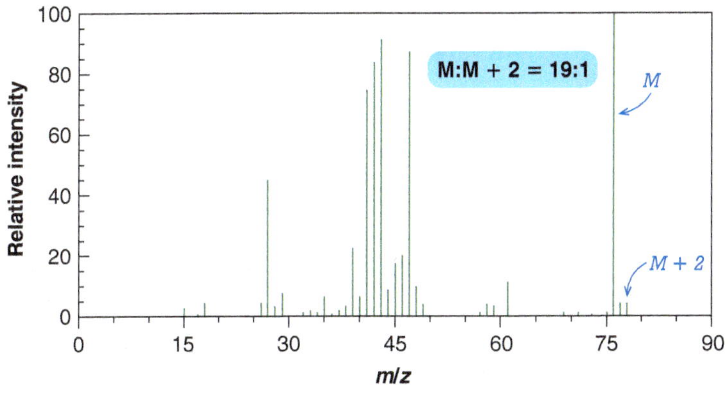

Assessment 14.31 The presence of what atom is indicated by the following mass spectrum?

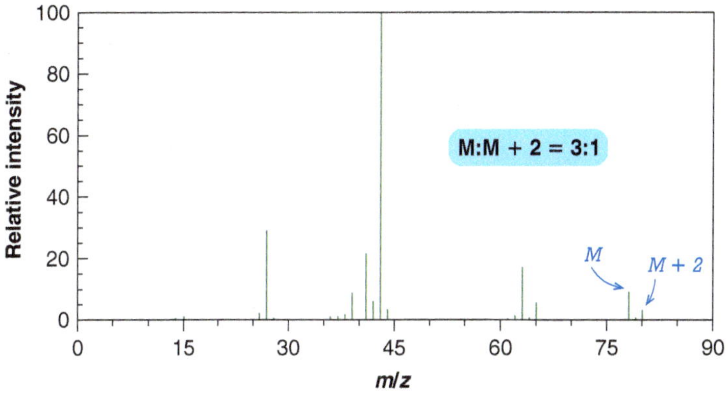

14.6.5 Odd mass/nitrogen rule

Neutral compounds containing an odd number of nitrogens will always have an odd molecular mass, whereas neutral compounds with zero or an even number of nitrogens will have an even molecular mass. [*I've always thought this was an overly broad statement that can't always be true . . . and yet it is.*] Carbon has an even atomic mass (12 amu). It is tetravalent, so there will be $2n + 2$ hydrogens for every carbon in a molecule. Oxygen, with an even atomic mass (16 amu), is also divalent. When you substitute an OH for an H (Figure 14.42), you are replacing an odd mass hydrogen (1 amu) with an odd mass OH (17 amu). Thus, the molecule still has an even molecular mass. Nitrogen, though, has an even atomic mass (14 amu) and is trivalent. When you substitute an NH_2 for an H, you are replacing an odd mass hydrogen (1 amu) with an even mass NH_2 (16 amu). Thus, the formerly even molecular mass is increased by 15, making it odd.

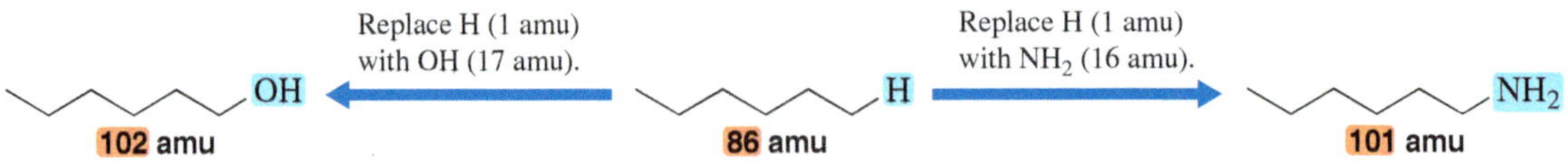

Figure 14.42 Compounds with an odd number of nitrogen atoms have an odd molecular mass due to the trivalency of nitrogen.

If you continue down the list of other elements common to organic molecules, the trend remains. As shown in Table 14.7, elements typically fall into one of two categories: even atomic mass with an even valency or odd atomic mass with an odd valency. The halogens fall into this second category, such that replacing a hydrogen (1 amu) with fluorine (19 amu) keeps the molecular mass even. Nitrogen, however, has an odd valency with an even atomic mass, further accounting for the **odd mass/nitrogen rule.**

Table 14.7 Relationship between an atom's valency and atomic mass to rationalize the odd mass rule.

Atom	Valency	Atomic mass
Hydrogen	Odd (1)	Odd (1 amu)
Carbon	Even (4)	Even (12 amu)
Oxygen	Even (2)	Even (16 amu)
Nitrogen	Odd (3)	Even (14 amu)
Fluorine	Odd (1)	Odd (19 amu)

14.6.6 Fragmentation

The mass spectrometer uses a high-energy beam of electrons to ionize the sample. To form the highly unstable radical cation of the parent molecule, the energy must be extremely high. It turns out that, even in the mass spectrometer, the ball rolls downhill. As it travels through the flight tube (Figure 14.36), then, the parent molecule sometimes fragments into other slightly more stable pieces. The radical cation of ethane, for example, can be represented as having only 1 electron in the σ bond [*again, this is very unstable.*] When this bond breaks (Figure 14.43), the electron goes to one side, resulting in a methyl carbocation and a methyl radical. The resulting carbocation is detected by the mass spectrometer, while the methyl radical is not, because only positively charged molecules are detected.

After fragmentation, we have two unstable fragments. [*Is that really downhill?*] What, then, is the likely driving force for this process? In this case, one molecule is becoming two, so fragmentation is heavily favored by entropy. [*Yes, it is downhill!*]

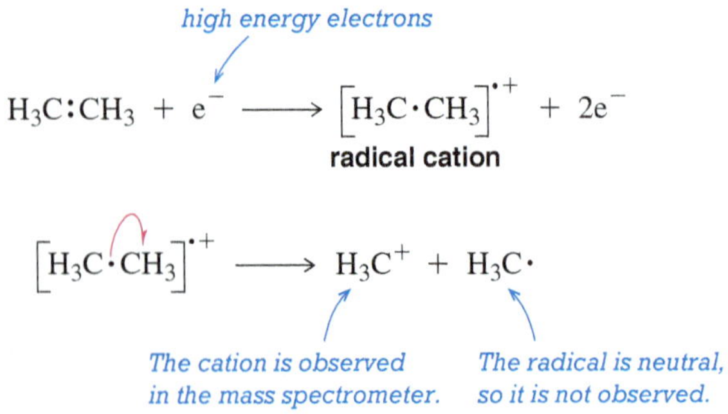

Figure 14.43 Fragmentation of radical cations produces a carbocation (observed in MS) and a radical (not observed).

In Assessment 14.26 at the beginning of Section 14.5, you were asked to reconstruct an alkane using peaks in the mass spectrum that corresponded to various fragments of what turned out to be *n*-decane. We can now understand how this fragmentation might have occurred. Fragmentation can occur at any of the positions indicated in **Figure 14.44** to produce the set of radicals and cations shown. There is no peak in the spectrum corresponding to the loss of a methyl group. This particular fragmentation occurs only rarely because the methyl carbocation and the methyl radical are extremely unstable with no additional alkyl groups to stabilize them through hyperconjugation.

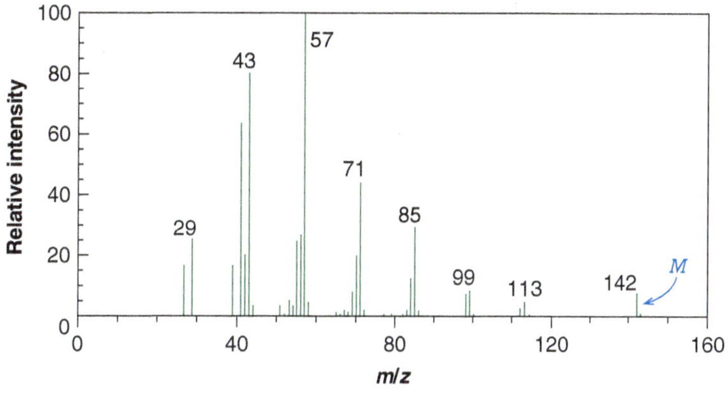

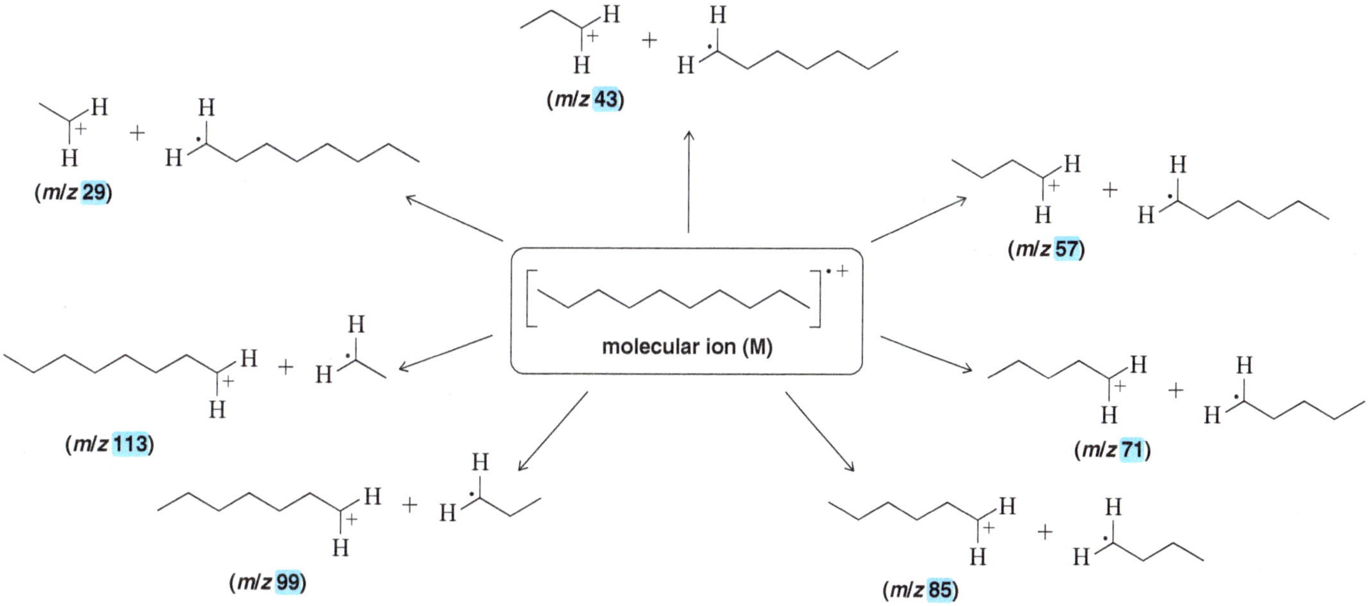

Figure 14.44 Fragmentation of decane to give the mass spectrum shown. Only the carbocations appear in the spectrum. No peak (m/z 15) is present for CH_3^+ because it is too unstable to form. [*None of the radicals are detected because they are neutral.*]

14.6 SUMMARY

- Formed in a high-energy process, the parent ion (M) can fragment into smaller radicals and cations. A plot of the mass to charge ratio (the mass spectrum) identifies the cation fragments and can be used to determine the structure of the parent ion. Isotopic ratios (M:M + 2) can be used to determine the identity of specific atoms (Cl, 3:1; Br, 1:1; and S, 19:1) present in a molecule. The nitrogen rule tells us that molecular ions with an odd mass to charge ratio are known to contain an odd number of nitrogens.

Assessment 14.32 The mass spectrum of 2,2-dimethylbutane is shown. What is the identity of the fragment that gives the peak at *m*/*z* 57?

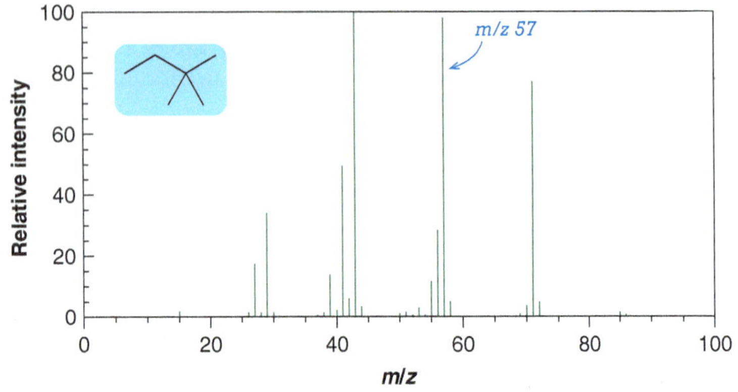

Assessment 14.33 The base peak is the peak of greatest intensity in the mass spectrum, set arbitrarily to 100% relative intensity. Why is the benzyl carbocation (*m*/*z* 91) the base peak in the mass spectrum of *n*-propyl benzene?

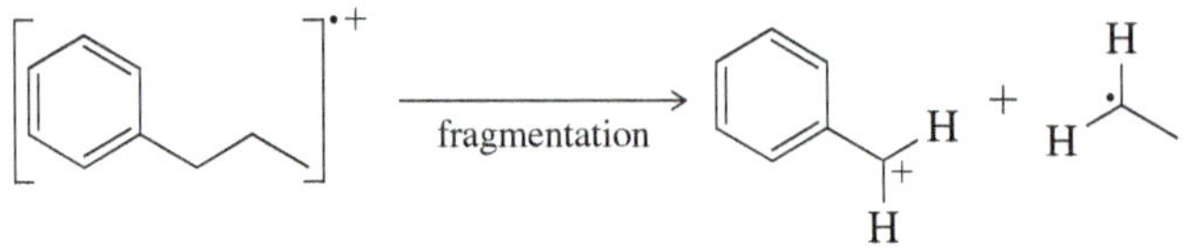

14.7 Mass spectrometry: a tour of fragmentations

Although nuclear magnetic resonance (see Chapter 15) is a simpler technique for complete structure analysis, common fragmentation patterns in mass spectrometry can be used for this purpose, too. Section 14.7 gives a brief overview of some of the most common fragmentations. In each, think carefully about why each fragment forms. Though we don't want to conflate reactivity within the mass spectrometer with reactivity in solution, some of the same principles apply. Finally, when discussing the fragmentation of each functional group, we are only justifying the most common, so-called "primary," fragmentations. In addition to the ones discussed, secondary fragmentations occur to produce a variety of additional peaks in the mass spectrum; we do not justify the formation of these here.

Alkanes: The fragmentation of alkanes was discussed briefly in Section 14.6.6. Recall that alkanes fragment to form the most stable carbocations. Two examples of this are shown in the fragmentations of 2-methylpentane [**Figure 14.45(a)**] and 2,2-dimethylbutane [**Figure 14.45(b)**] that form the 2° (*m*/*z* 43) and 3° carbocations (*m*/*z* 57), respectively.

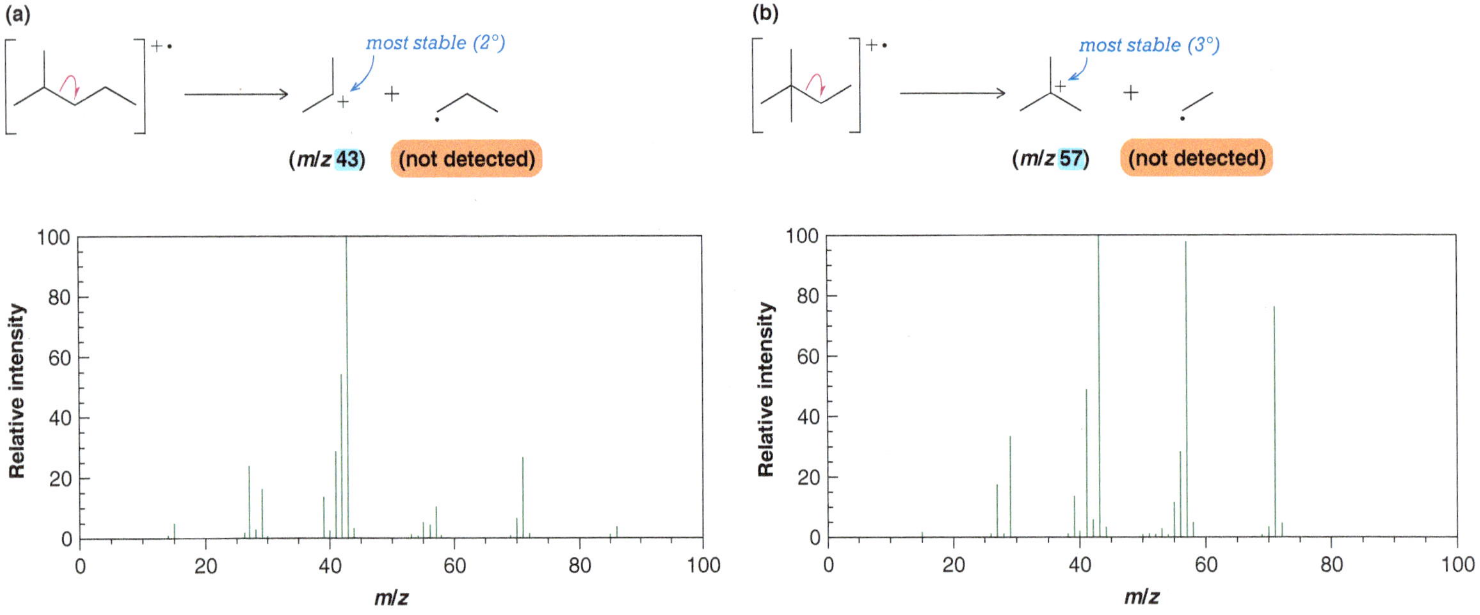

Figure 14.45 Fragmentation of alkanes forms the most stable carbocation: (a) the 2° (*m*/*z* 43) carbocation; (b) the 3° (*m*/*z* 57) carbocation.

Alkenes: When possible, alkenes will fragment via a mechanism similar to that of alkanes. Fragmentation will favor formation of the more stable allylic carbocation. In the case of hex-1-ene, this results in a fragment at *m/z* 41 (**Figure 14.46**).

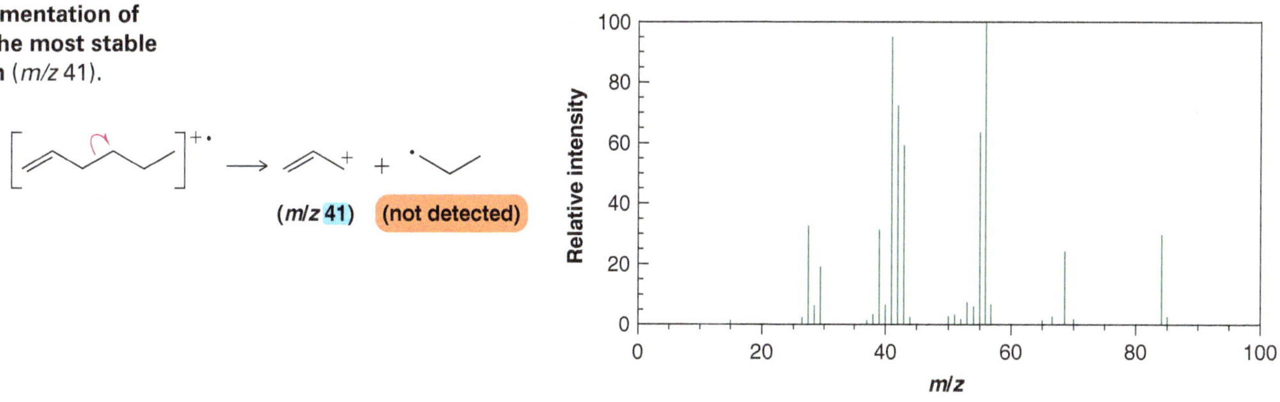

Figure 14.46 Fragmentation of hex-1-ene forms the most stable allylic carbocation (*m/z* 41).

Alkylbenzenes: Alkylbenzenes also fragment to form the more stable carbocation, which in their case is the benzylic carbocation (**Figure 14.47**). In addition to being stabilized by resonance, the benzylic carbocation (*m/z* 91) rearranges to form the more stable tropylium ion. [*We explain why this is so stable in Section 23.2.*]

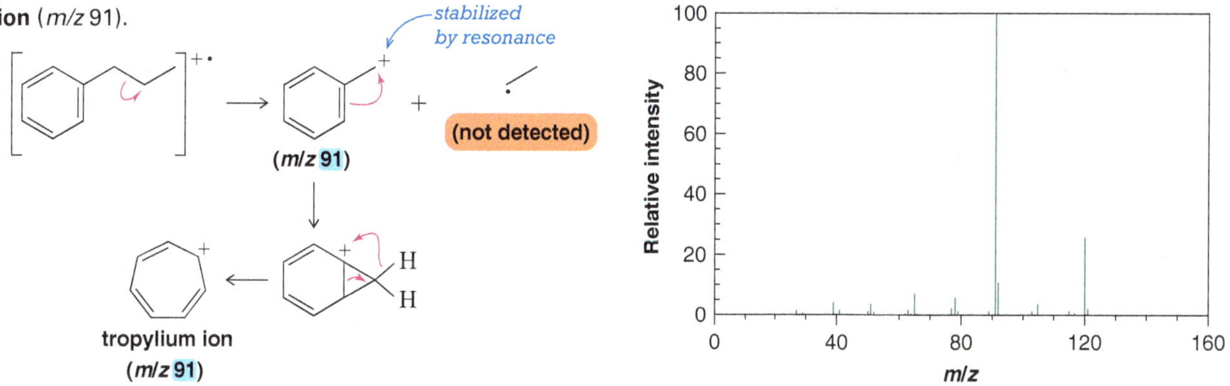

Figure 14.47 Fragmentation of alkyl benzenes is driven by formation of the stable tropylium ion (*m/z* 91).

Haloalkanes: Fragmentation of a haloalkane occurs by cleavage of the C—X bond. In comparison to C—H and C—C σ bonds, C—X bonds are relatively weak. Thus, under the high-energy conditions of the electron beam, these bonds are first to break. In fact, they often break so easily that they fragment immediately, resulting in the complete absence of the parent ion from the spectrum. This is the case for 1-chlorobutane, shown in **Figure 14.48 (a)**. 1-Bromobutane, shown in **Figure 14.48(b)**, has a molecular ion peak for the parent molecule, but its intensity is less than 10% of the base peak.

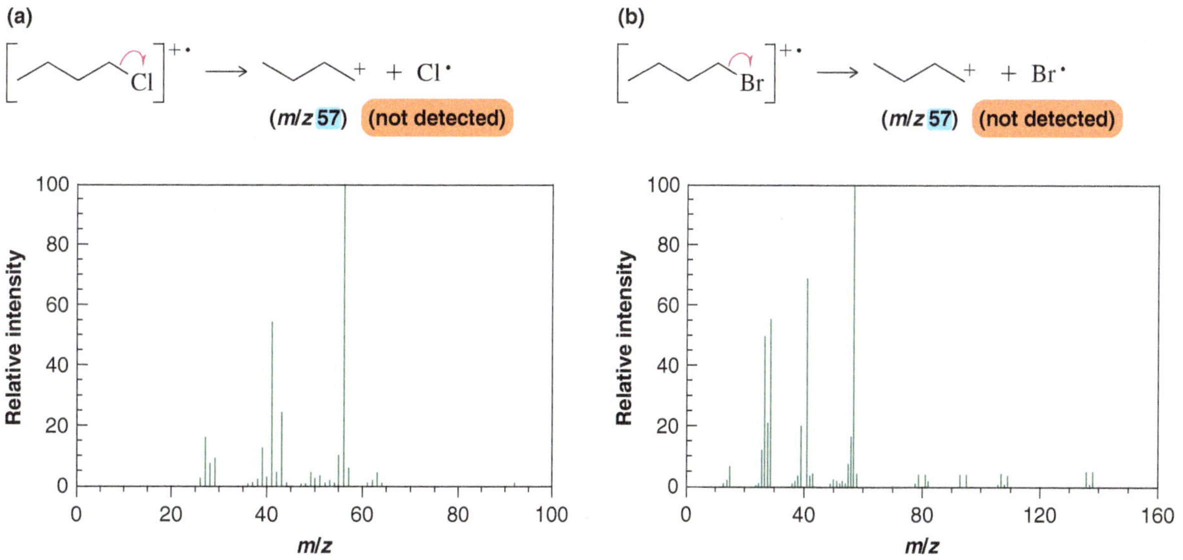

Figure 14.48 Fragmentation of haloalkanes results first in cleavage of the weak C—X bond.
(a) 1-Chlorobutane.
(b) 1-Bromobutane.

Assessment 14.34 The molecular ion (M) is significant in the mass spectrum of benzyl fluoride, but it is barely present in the mass spectrum of benzyl iodide. Why might this be the case?

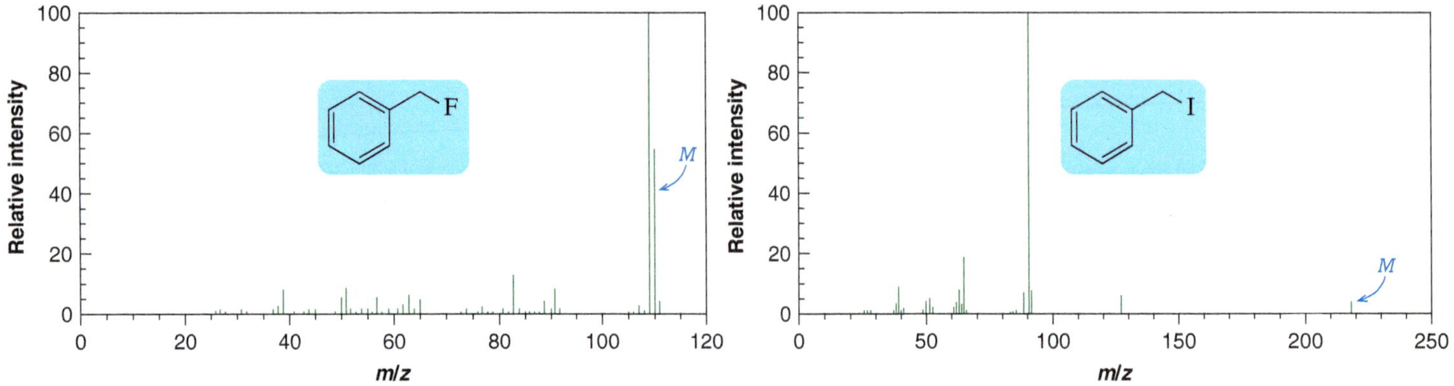

Assessment 14.35 Besides the molecular mass, how do you know that the fragment that gives rise to the peak at m/z 111 in the mass spectrum for 1-bromo-4-chlorobenzene contains only chlorine?

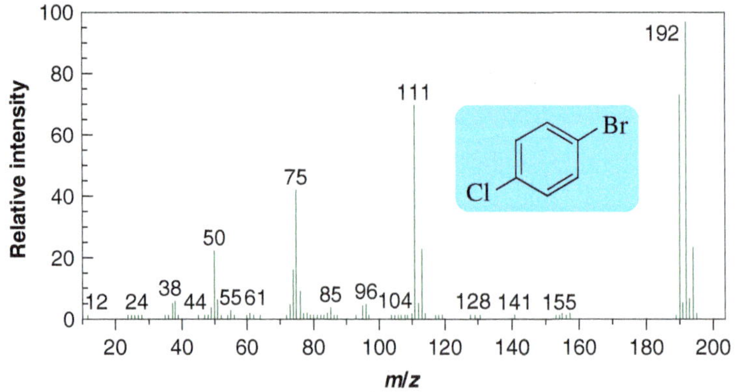

Alcohols: Alcohols fragment by a couple of different mechanisms. First, dehydration (loss of water) occurs within the radical cation to give an alkene, as shown in Figure 14.49. Whereas other fragments result in a cation and a radical, the resulting alkene is still a radical cation. It is detected like any other positively charged fragment.

The second common pathway for the fragmentation of alcohols is referred to as **α-cleavage**. The C—C σ bond of the alcoholic carbon breaks to give an alkyl radical and a carbocation. This carbocation is highly stabilized by resonance with the lone pair of electrons on the

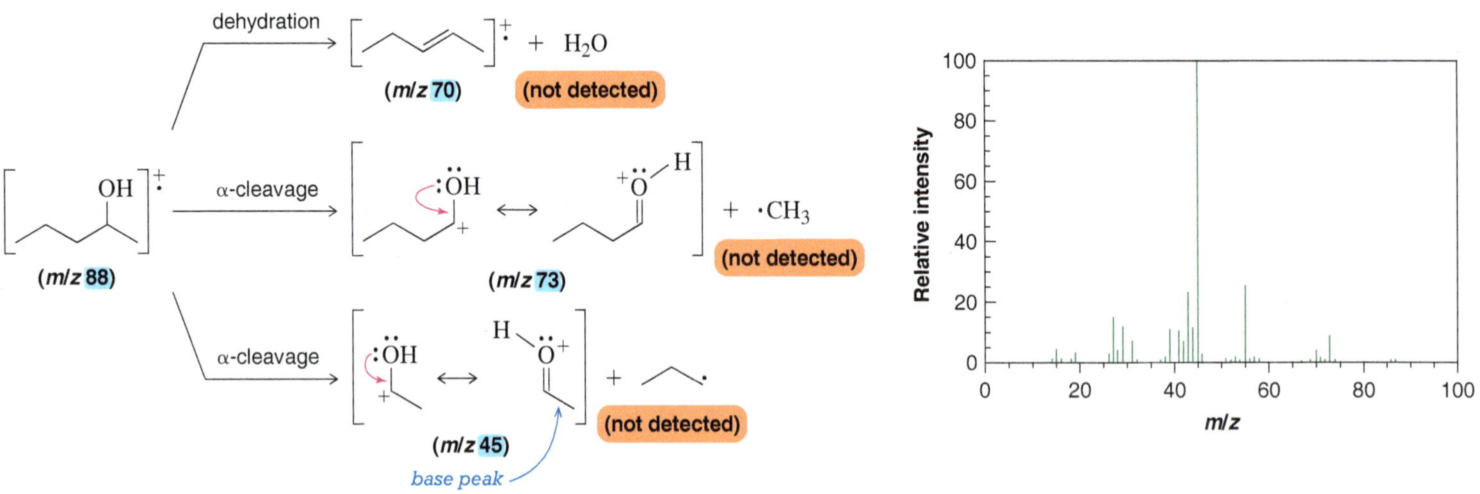

Figure 14.49 Alcohols fragment quickly by dehydration and α-cleavage, so the parent molecule is often absent from the mass spectrum.

hydroxyl group. As a 2° alcohol, pentan-2-ol can undergo two different α-cleavages. These two pathways account for the peaks at *m*/*z* 73 and *m*/*z* 45. The peak at 45 is the base peak, making formation of it the most common pathway for fragmentation.

With three favorable pathways for the fragmentation of alcohols, it is common for the parent molecule to be completely absent from the mass spectrum. [*Luckily, IR (Section 14.4) gives conclusive evidence of a hydroxyl group in a molecule.*]

Assessment 14.36 Propose an arrow-pushing mechanism to account for the α-cleavage fragmentation in Figure 14.49.

Assessment 14.37 In the mass spectrum of hexan-3-ol, identify the fragments arising from two α-cleavage fragmentations and one dehydration fragmentation.

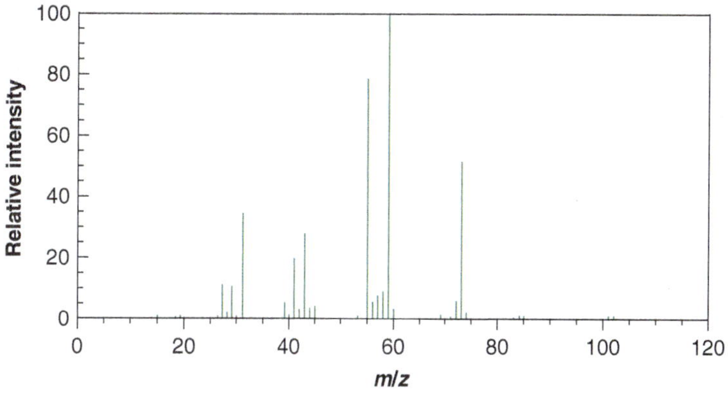

Assessment 14.38 How many ways can a 1° alcohol fragment by α-cleavage?

Amines: Like alcohols, amines also undergo α-cleavage. As shown in the mass spectrum of hexan-1-amine (**Figure 14.50**), the resulting resonance-stabilized carbocation represents the base peak (*m*/*z* 30). Relative to the base peak, the molecular ion of amines is usually small and sometimes missing from the mass spectrum.

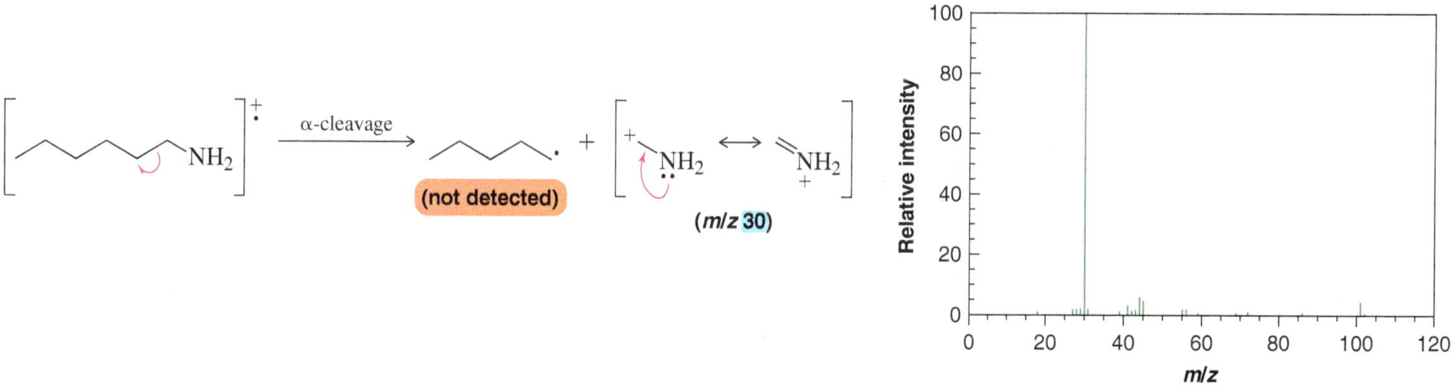

Figure 14.50 Amines fragment quickly by α-cleavage, so the parent molecule is often absent from the mass spectrum.

Assessment 14.39 Suggest three ways that the following tertiary amine can fragment by α-cleavage. At which *m*/*z* would you expect to see peaks corresponding to these fragments?

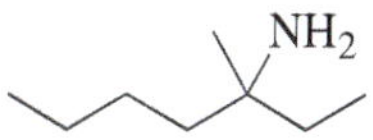

Assessment 14.40 α-Cleavage of the alcohol in Figure 14.49 gave a fragment with an odd mass. Alternatively, α-cleavage of the amine in Figure 14.50 gave a fragment with an even mass. Explain these observations.

Ethers: We said in Chapter 13 that ethers were simply alcohols where a hydrogen had been replaced by an alkyl group. As a result, the reactivities of alcohols and ethers are closely related. This holds for the fragmentation patterns in the mass spectrometer, with α-cleavage being a primary mechanism by which fragmentation of an ether occurs, as shown for diisopropyl ether in **Figure 14.51**. The resulting carbocation (*m/z* 87) is stabilized by resonance.

Ethers can also lose either alkyl group from the central oxygen. Following rearrangement of a hydrogen to the oxygen [*again, high energy . . . anything is possible*], a resonance-stabilized carbocation (*m/z* 59) is produced.

Both possible fragmentations are highly favorable, each producing resonance-stabilized carbocations. For this reason, the parent molecule of ethers is noticeably absent from a spectrum.

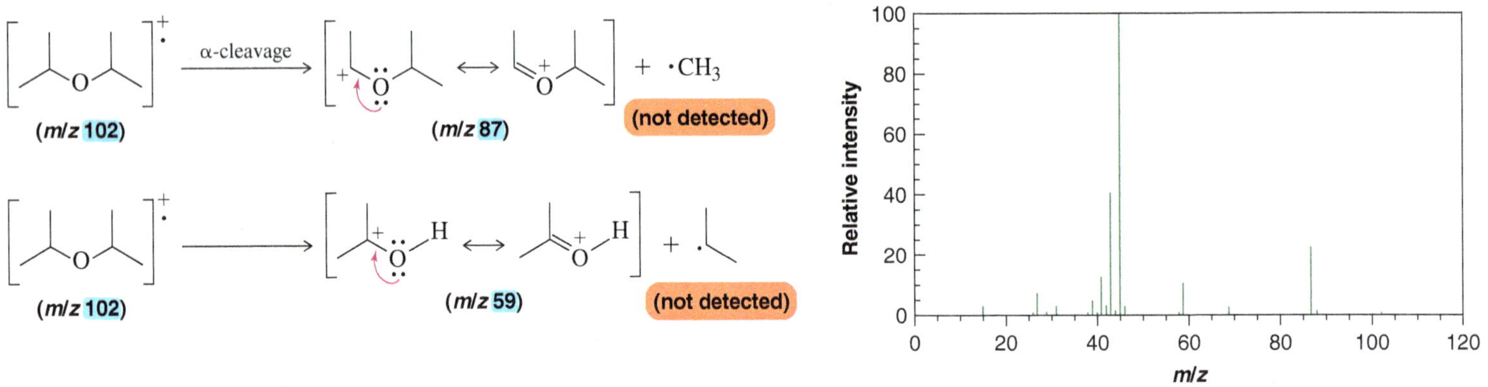

Figure 14.51 Fragmentation of ethers occurs by α-cleavage and cleavage of the C—O bond.

Assessment 14.41 The base peak in Figure 14.51 has *m/z* 45. Suggest a molecular formula and a structure for this fragment.

Assessment 14.42 The parent molecule (*m/z* 108) is very prominent in the mass spectrum of anisole. Why?

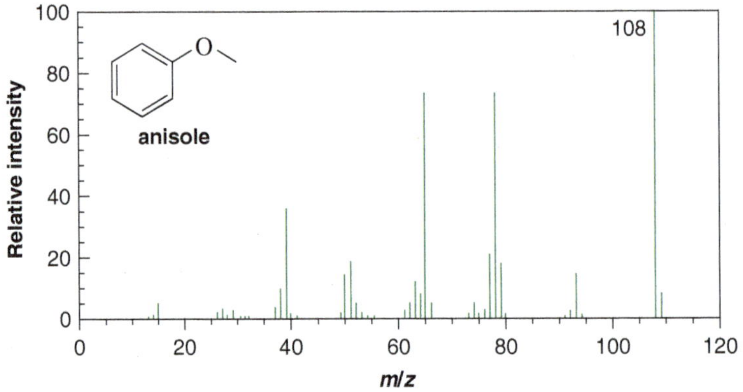

Aldehydes/ketones: There are two major pathways by which aldehydes and ketones fragment. The first, shown in **Figure 14.52**, is identical to α-cleavage. Loss of the propyl radical in butanal gives a carbocation that is stabilized by resonance with a lone pair on the carbonyl oxygen [Figure 14.52(a)]. Hexan-2-one with alkyl groups on either side of the carbonyl, undergoes α-cleavage on both sides to give two resonance-stabilized carbocations (*m/z* 85 and 43) [Figure 14.52(b)].

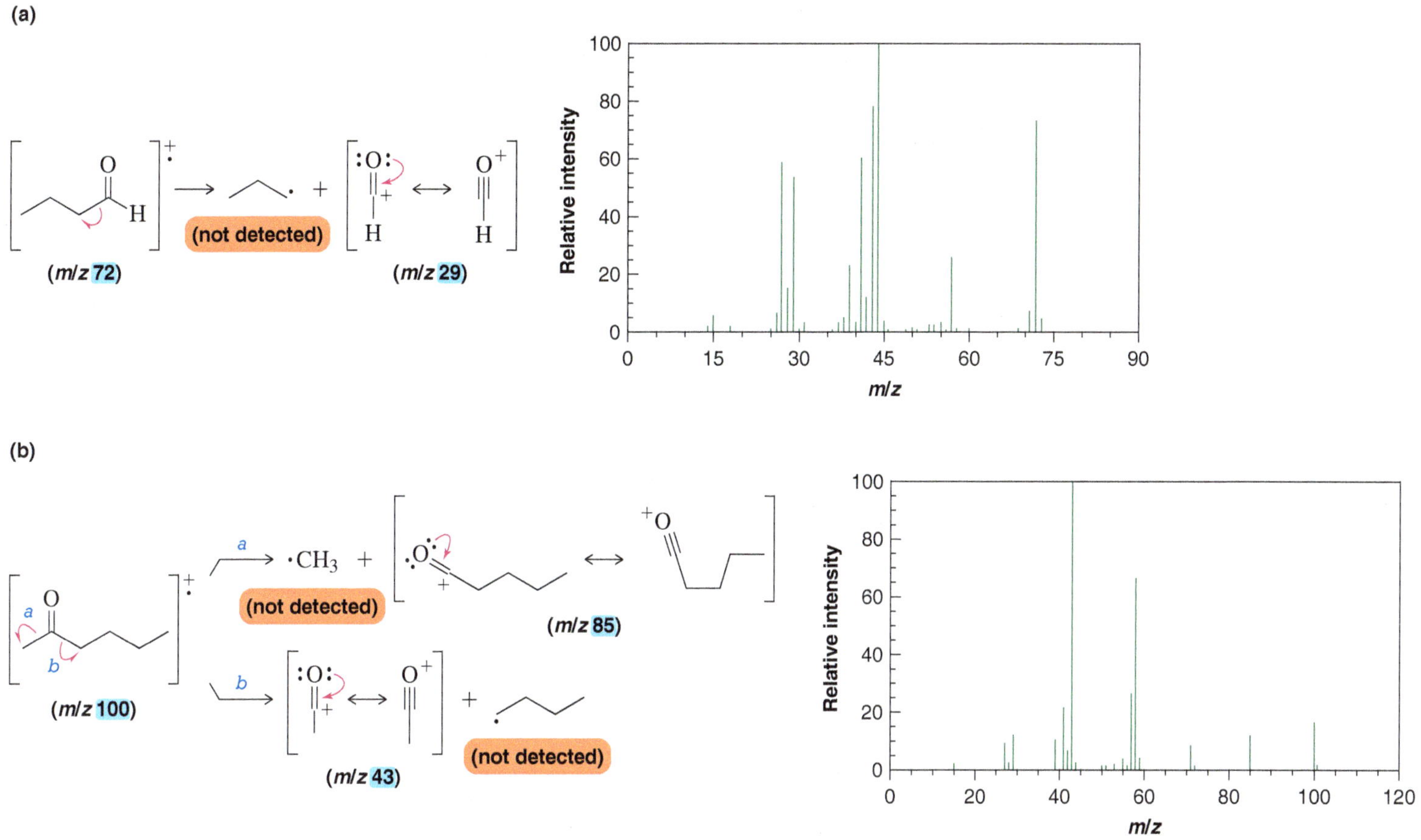

Figure 14.52 Fragmentation of ketones and aldehydes by cleavage at the carbonyl carbon. (a) An aldehyde can cleave by one mechanism. (b) There are two pathways for a ketone to cleave in this way.

A second, somewhat complicated, fragmentation seen in carbonyl compounds is the McLafferty rearrangement (Figure 14.53). After ionization to form the radical cation, fragmentation of hexan-2-one occurs by a mechanism we discuss in more detail in Chapter 23 when we study pericyclic reactions. For now, it can be represented as a 6-electron cyclic transition state in which a hydrogen is transferred from the γ-carbon to the carbonyl while at the same time a new $C_\gamma—C_\beta$ π bond forms and the $C_\alpha—C_\beta$ σ bond breaks. The alkene, being neutral, is not detected. The enol [*recall Chapter 10*], still a radical cation, appears at *m/z* 58. For this mechanism to occur, the carbonyl must be flanked on one side by an alkyl chain of at least three carbons.

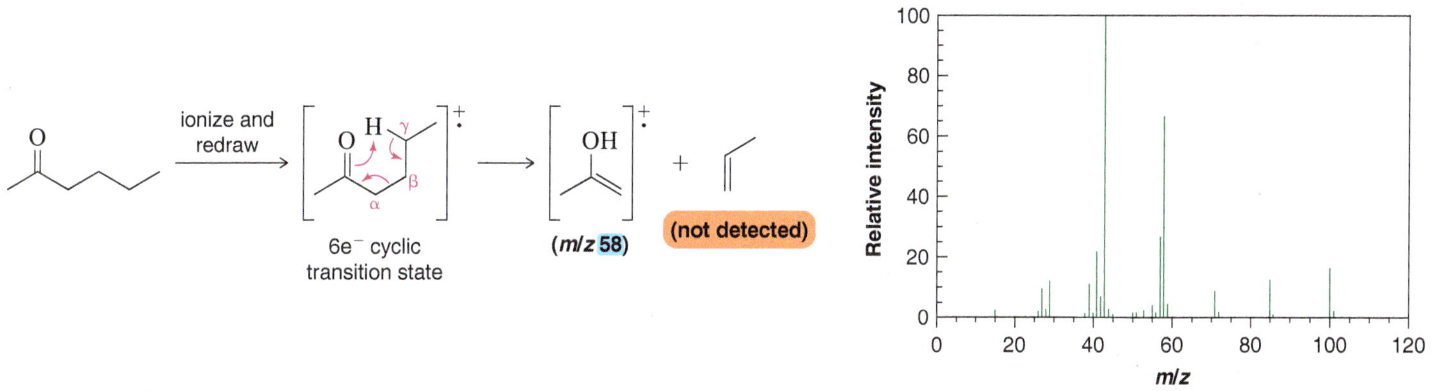

Figure 14.53 The McLafferty rearrangement of ketones.

Assessment 14.43 Identify the peaks in the mass spectrum of octan-4-one that correspond to (a) α-cleavage and (b) the McLafferty rearrangement.

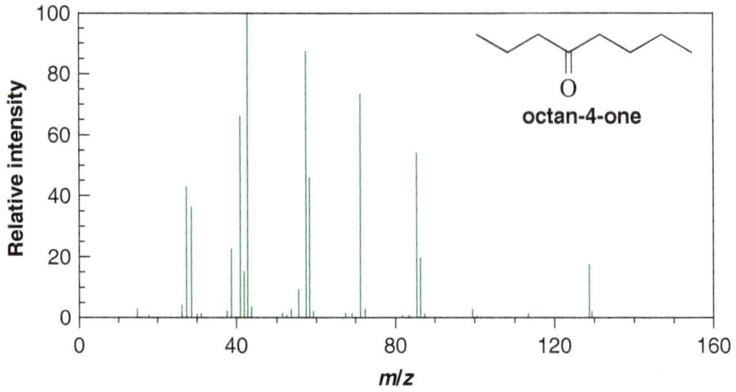

Assessment 14.44 2-Methylnonan-4-one can undergo two different McLafferty rearrangements. Draw the products of each of them.

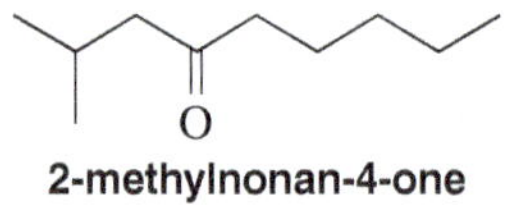

14-C SPECTROSCOPY IN CONTEXT: GC/MS

Traditional Chinese medicines

The power of analytical methods like mass spectrometry is multiplied when used in conjunction with other analytical methods. One example of this is GC/MS, which links gas chromatography with mass spectrometry [Figure 14.54(a)]. In this method, a mixture of compounds is vaporized and injected onto a column that separates the compounds on the basis of polarity. Used alone, the chromatogram generated by GC [Figure 14.54(b)] can provide the number and relative

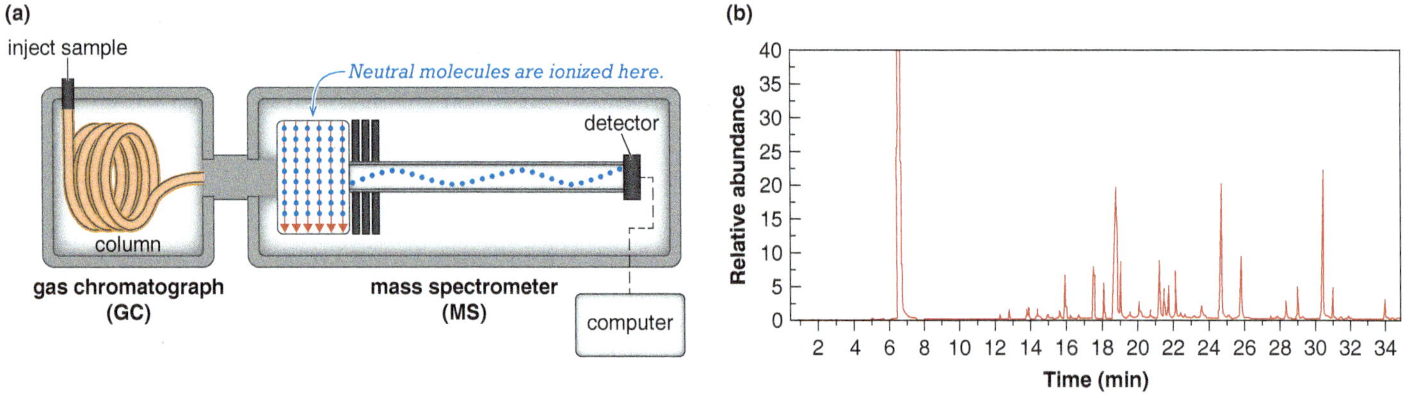

Figure 14.54 Gas chromatography/mass spectrometry. (a) Schematic of a GC/MS. (b) A sample gas chromatogram showing peaks and retention times for compounds as they come off the GC column.

(continued)

concentration of components in the mixture. When a mass spectrometer is placed at the end of the GC column, the molecular mass of each constituent in the mixture can be measured and correlated to the individual peak in the gas chromatogram.

Chinese proprietary medicines refer to natural materials used in traditional Chinese medicine that have been compressed into pills or tablets. Some formulations of these medicines sold in Singapore have been implicated in illnesses and deaths due to impurities. GC/MS was utilized in a study of 41 Chinese proprietary medicines used in the treatment of asthma. As shown in **Figure 14.55**, the peak that came out of the GC at 18.87 min gave a mass spectrum that matched the published mass spectrum of codeine. Codeine, an opioid painkiller and cough suppressant, is not legally sold as a traditional medicine under Singapore law. The results of this study suggest that contamination, misidentification, incorrect labeling, and adulteration are common problems encountered with herbal medicines.

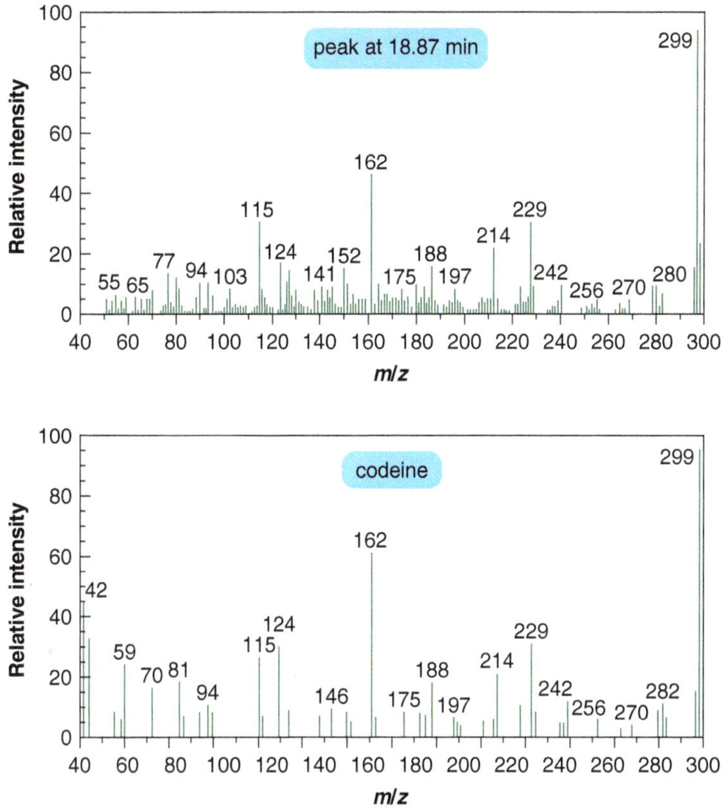

Figure 14.55 **The mass spectrum of an unknown peak at 18.87 min in the GC matches that of codeine.**

14.8 IR and MS combined

Though studied in isolation, all spectroscopic techniques can be used in conjunction with each other to determine the structure of a molecule. Assessment 14.45 shows how this can be done. [*In Chapter 15, we add another technique.*]

Assessment 14.45 Assign the structure based on the mass spectrum and IR spectrum shown.

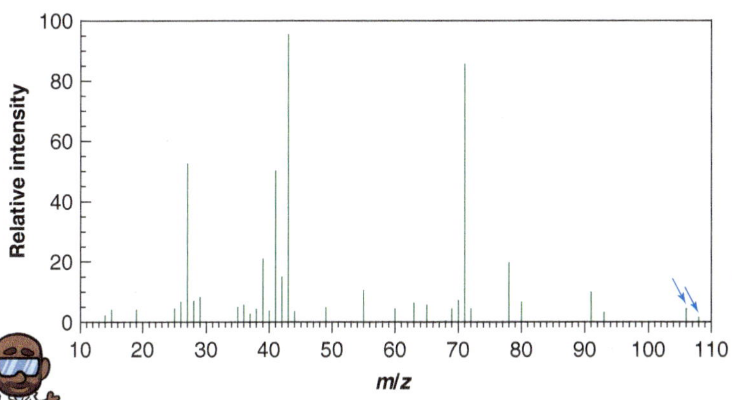

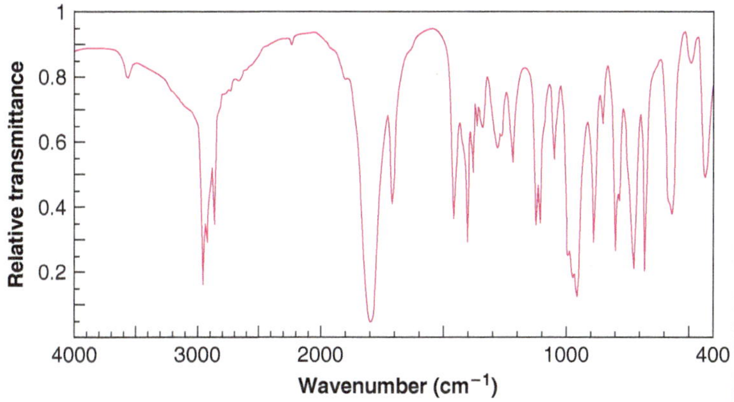

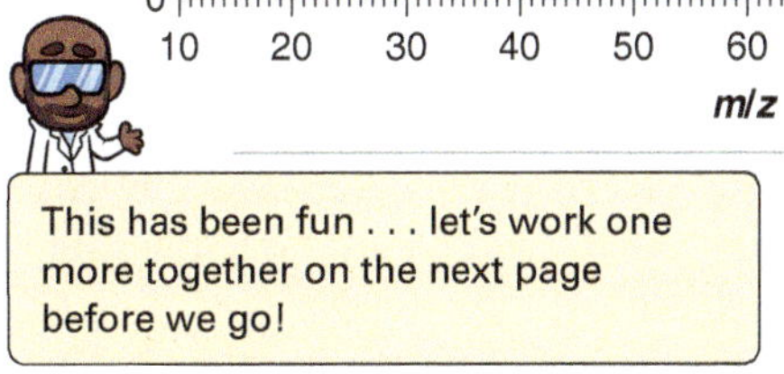

Solved Assessment 14.45

Solving a problem like this requires identifying the pieces of the puzzle that each spectrum provides and then logically putting them together.

The mass spectrum (**Figure 14.56**) conclusively tells us a few things. First, the molecular mass of the molecule is 106 amu. [*It is hard to see this peak, but trust me, it is there.*] This means that there is likely no nitrogen. Because the M:M + 2 peak has a 3:1 ratio, there must be a chlorine present. The signals at 78/80 and 91/93 must also contain a chlorine.

The IR spectrum (**Figure 14.57**) shows a band around 1800 cm^{-1}, suggesting the presence of a carbonyl. Because it is so much higher than 1715 cm^{-1} (a normal ketone) and because the mass spectrum indicated the presence of a chlorine, it is likely that this is an acid chloride (RCOCl). The absence of a C—H band at 3100 cm^{-1} and an alkene band at 1600 cm^{-1} suggests a fully saturated molecule.

At this point, we have to start inferring things based on what might not exactly be described as chemistry logic. The acid chloride (COCl) accounts for 63 amu of the total 106 amu, leaving us to account for the remaining 43 amu. Assuming the presence of only C and H (reasonable given the IR), this 43 amu corresponds to C_3H_7. As an acid chloride containing only a C_3H_7 substituent, the two possibilities are butyryl chloride and isobutyryl chloride. Careful inspection of the MS shows a loss of a methyl group ($106 \rightarrow 91$ amu) and the loss of CH_2CH_2 ($106 \rightarrow 78$). The loss of CH_2CH_2 likely occurs by a McLafferty rearrangement that only occurs in butyryl chloride (**Figure 14.58**). As a result, we conclude that butyryl chloride is the molecule in Assessment 14.45. [*^1H NMR gives us a more conclusive way of differentiating between butyryl and isobutyryl chloride. We explain how in Chapter 15.*] ■

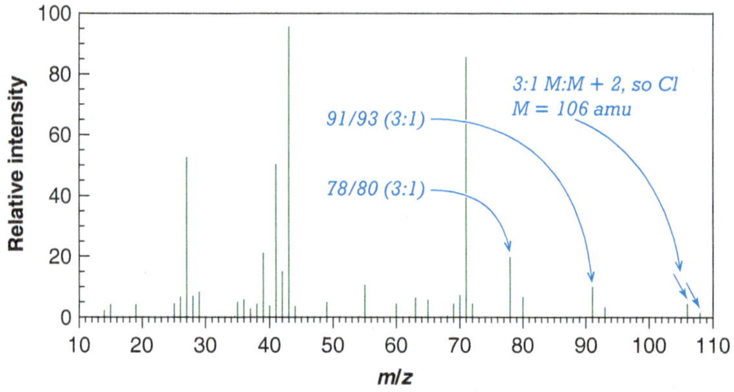

Figure 14.56 **The molecule has an molecular mass of 78 and contains a chlorine.**

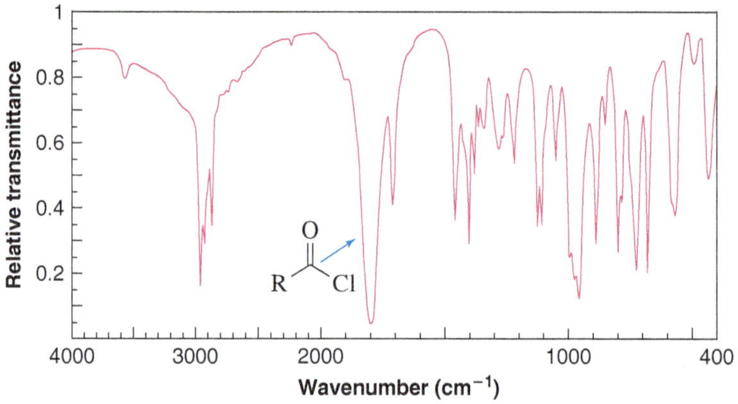

Figure 14.57 **The molecule is an acid chloride.**

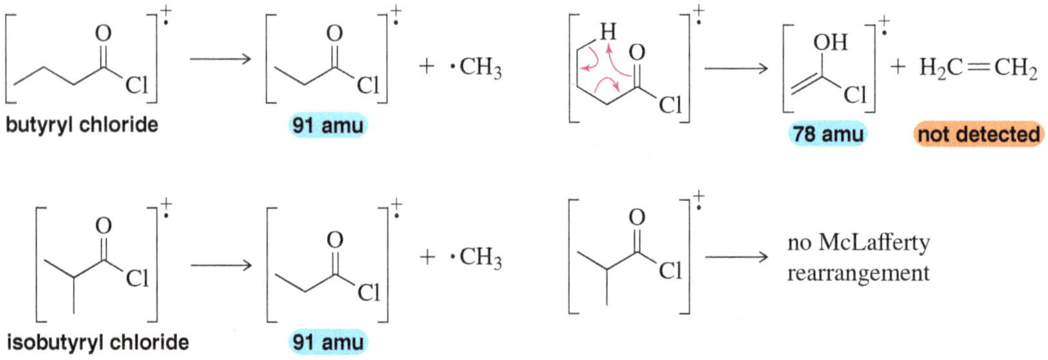

Figure 14.58 **The unknown molecule in Assessment 14.45 is butyryl chloride.**

Concluding thoughts

Infrared spectroscopy and mass spectrometry can help us solve the structure of unknown compounds. These techniques have also identified counterfeit, low-quality, or mislabeled medications throughout the world. Regardless of continued innovation to generate low-cost point-of-care techniques to efficiently identify these counterfeit drugs, there are still situations where this technology is not practical.

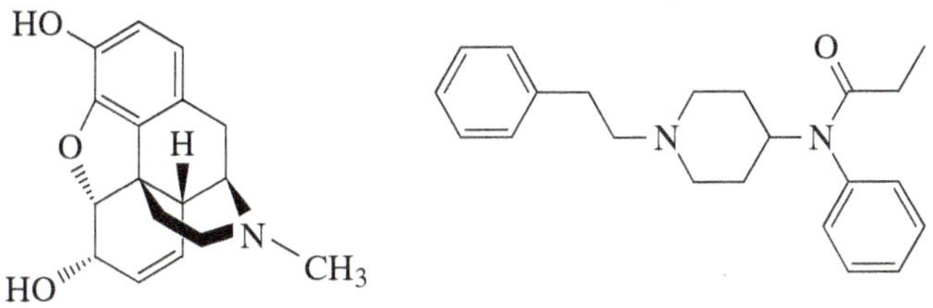

morphine fentanyl

Figure 14.59 Fentanyl has a similar activity to morphine.

Deaths resulting from synthetic opioids and heroin have spiked as a result of heroin being laced with the additive fentanyl. In 2017, there were over 30,000 deaths due to synthetic opioid overdoses. Originally developed as a pain reliever used in the terminally ill, fentanyl has a similar activity to morphine (Figure 14.59) and is highly addictive. When mixed with heroin, the potency of both depressant drugs is increased, resulting in unconsciousness, slowed respiration, and, occasionally, death.

As a public health crisis, the opioid epidemic has been the subject of intense research, both sociologically and scientifically. One potential victory in this battle is the development of inexpensive test strips that can be used to identify the presence of fentanyl in street drugs. Given that spectroscopic methods can't be used on the street, these test strips allow drug users to test their drugs for the presence of fentanyl prior to using them. A large percentage of surveyed drug users have indicated that knowledge of fentanyl contamination would cause a change in their choices regarding drug use.

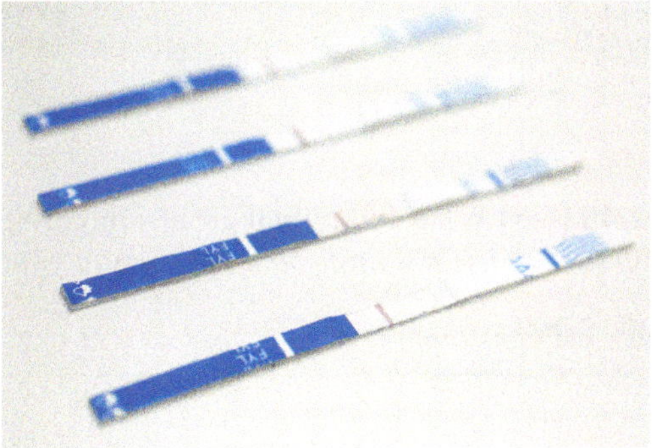

Going forward, we continue our exploration of spectroscopy in Chapter 15 by studying nuclear magnetic resonance, by far the most powerful technique for small molecule structure identification. Meet me in Chapter 15, but only after first working the End-of-Chapter Assessments in Chapter 14.

End-of-Chapter Materials

Important Concepts

[*These are the major concepts covered in each section, excluding reactions.*]

14.1: Spectroscopy is the study of how light interacts with molecules and can give information about their structure. The energy of electromagnetic radiation is directly proportional to the frequency and inversely proportional to the wavelength. Quantized energy levels correlate to specific molecular transitions. Light in the infrared region corresponds to bond vibrations.

14.2: Together with the index of hydrogen deficiency, infrared (IR) spectroscopy is used to identify the functional groups present in a molecule. Specific bond vibrations correspond to bands at certain wavenumbers in the IR spectrum.

14.3: The frequency of bond vibration, analogous to masses connected by a spring, is dependent on the reduced mass of the attached atoms and the strength of the bond between them. The infrared spectrometer is designed to identify the frequencies of light that are absorbed by a molecule by the presence of a band on the IR spectrum. These bands, appearing at specific wavenumbers, correlate to specific functional groups. Electrons in bonds with more *s* character are held closer to the nucleus and the bonds vibrate at higher frequencies. The stretching frequency of a carbonyl is dependent on the ability of attached substituents to stabilize the carbocation in the second-best (zwitterionic) resonance structure of the carbonyl.

14.5: Mass spectrometry is used to determine the molecular mass of a molecule by forming and detecting the mass to charge ratio of a radical cation.

14.6: Formed in a high-energy process, the parent ion (M) can fragment into smaller radicals and cations. A plot of the mass to charge ratio (the mass spectrum) identifies the cation fragments and can be used to determine the structure of the parent ion. Isotopic ratios (M:M + 2) can be used to determine the identify of specific atoms (Cl, 3:1; Br, 1:1; and S, 19:1) present in a molecule. The nitrogen rule tells us that molecular ions with an odd mass to charge ratio are known to contain an odd number of nitrogens.

14.7: Specific fragmentation patterns are indicative of certain functional groups or structural features in the parent molecule.

End-of-Chapter Assessments

[*Assessments are ranked according to their difficulty;* •••• = *most difficult.*]

14.46 (••) Based on Hooke's law, choose the bond in each pair that you expect to vibrate at a higher wavenumber.

(a) C—S vs. C—O
(b) C—F vs. C—Br
(c) C═O vs. C═N
(d) C—S vs. C═S
(e) C═N vs. C≡N
(f) C—S vs. C═O

14.47 (••) Choose the bond in each pair that you expect to have the more intense stretching band.

(a) C—O vs. C—S
(b) C═O vs. C═N
(c) C—H vs. C—F

14.48 (•••) Identify the important bands you would expect to find in an IR spectrum for the following molecules.

(a)

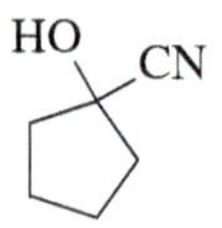

(b)

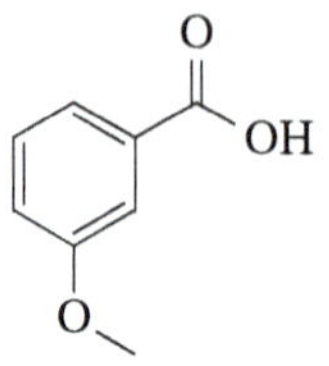

(c)

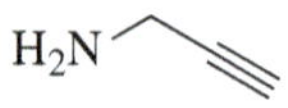

(d)

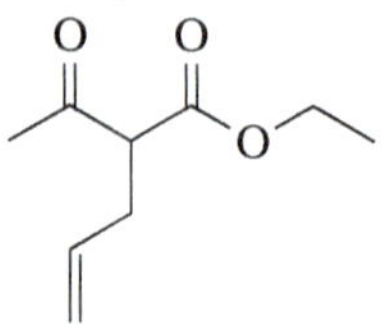

14.49 (•••) What functional groups might be present in the IR spectra for the molecules with the given molecular formulas. [*Be sure to use the molecular formula in your analysis.*]

(a) C_5H_8O

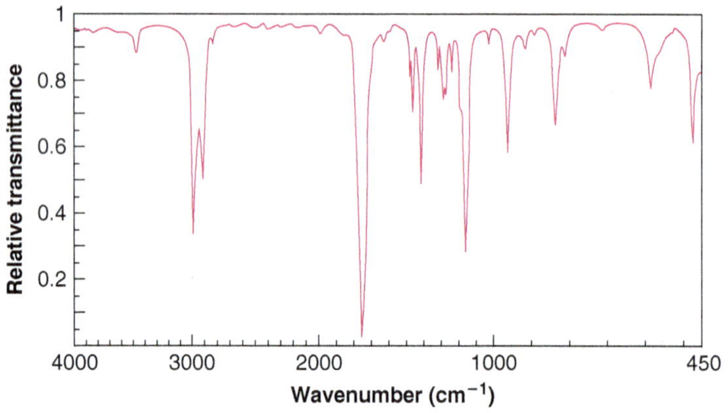

(b) $C_3H_4O_4$

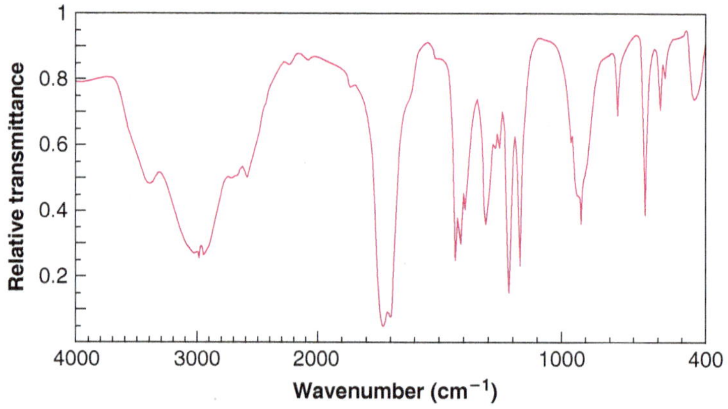

(c) $C_6H_{12}O$

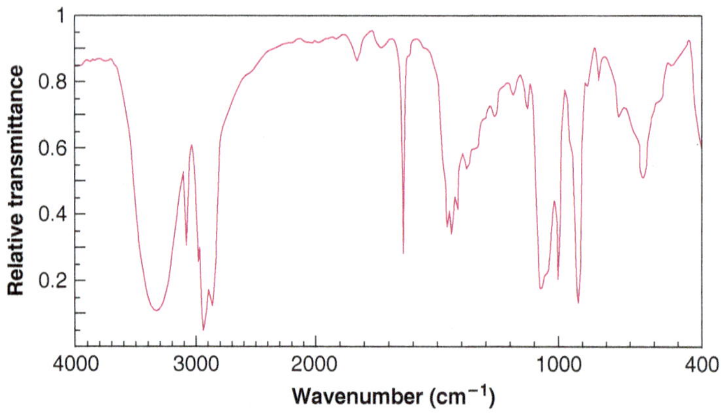

(d) C_6H_7N

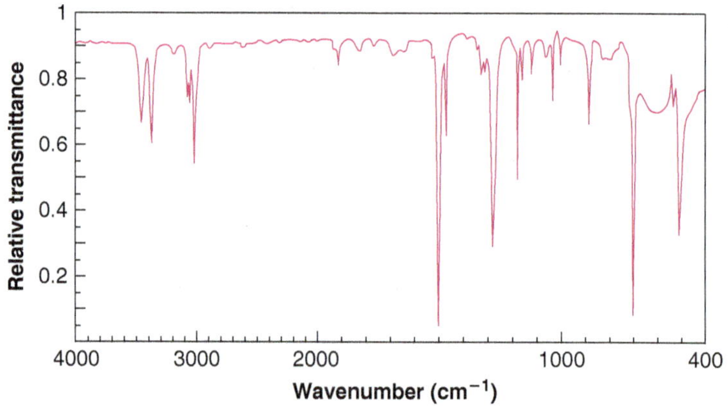

14.50 (•••) Ketones can be reduced to alcohols using the conditions shown (see Chapter 17). Explain how a comparison of an IR spectrum of the reactant and product can be used to determine whether the reaction was successful or not.

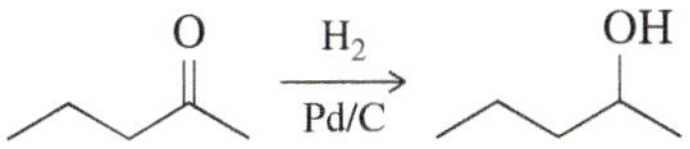

14.51 (•••) A carboxylic acid can be converted to an ester using the conditions shown (see Chapter 18). Explain how a comparison of an IR spectrum of the reactant(s) and product can be used to determine whether the reaction was successful or not.

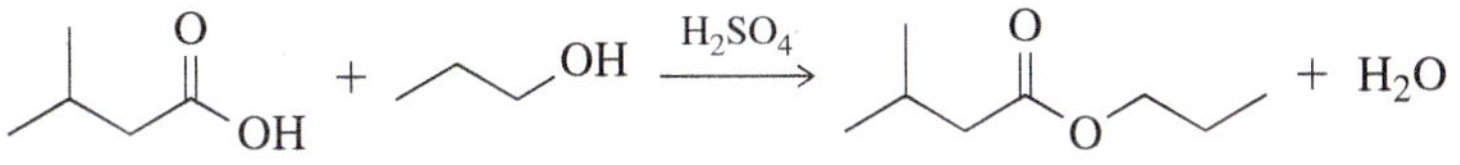

14.52 (••••) Hoping to make the following diene, a chemist treated the diol shown with acid. Based on the IR spectrum, was the reaction successful? If not, what compound was made instead? [*You studied this reaction in Chapter 13.*]

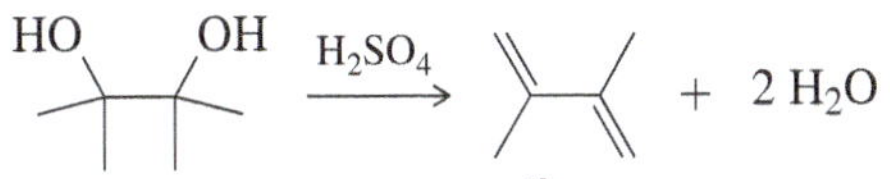

diene

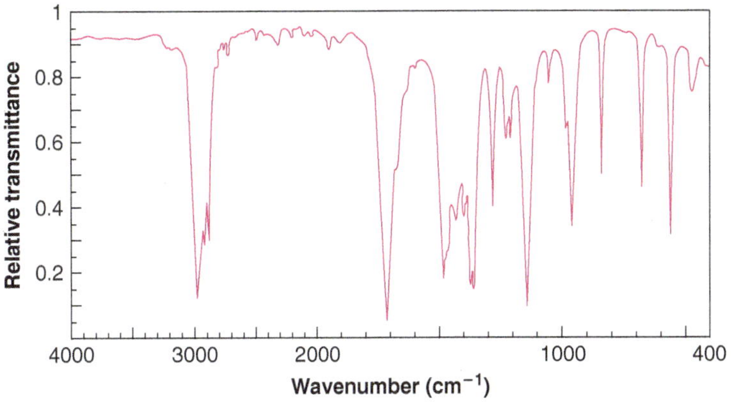

14.53 (••) IR spectroscopy can be used to determine whether acetylide alkylation to make a terminal alkyne is successful, but not when the similar reaction is used to make internal alkynes. Why?

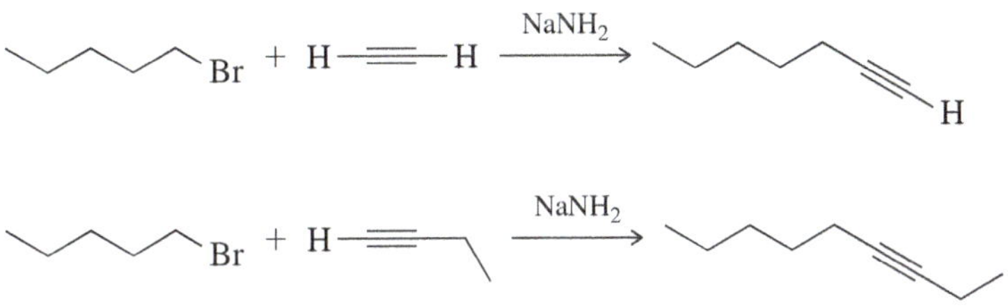

14.54 (••••) Under acidic conditions, alkene **A** can be isomerized to the more stable alkene **B**. How could IR spectroscopy be used to distinguish between **A** and **B**? [*There are a few correct answers.*]

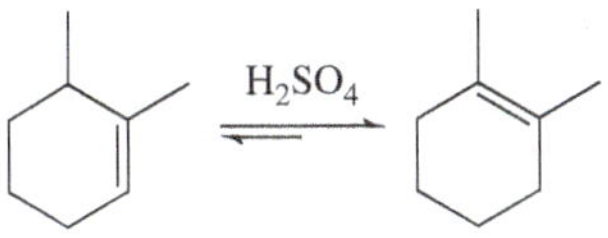

A **B**

LOOKING AHEAD We explain in Chapter 23 that substituents can transmit electronic information through the benzene ring when appropriately positioned. Assessments 14.55–14.59 allow us to use IR spectroscopy to discover this concept and apply it to an example (Assessment 14.60) that we study there.

14.55 (••) Would you expect the stretching band of the carbonyl to appear at a higher frequency for cyclohexanecarbaldehyde or benzaldehyde? Explain.

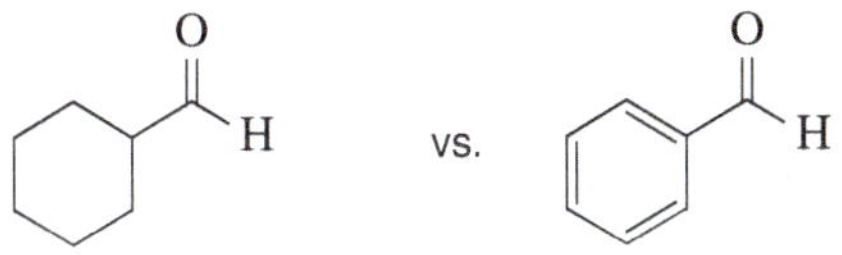

cyclohexanecarbaldehyde **benzaldehyde**

14.56 (•••) Justify the carbonyl stretching frequencies indicated for benzaldehyde and 4-methoxybenzaldehyde.

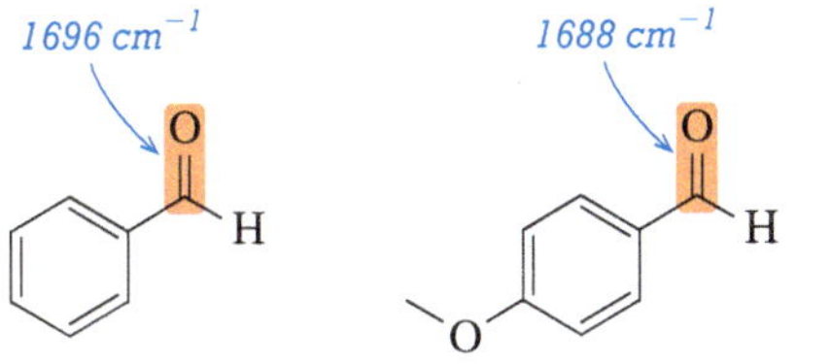

benzaldehyde **4-methoxybenzaldehyde**

14.57 (•••) Justify the carbonyl stretching frequencies for a series of methoxybenzaldehydes. Specifically, why are the 2- and 4-methoxy derivatives similar to each other but different from the 3-methoxy derivative?

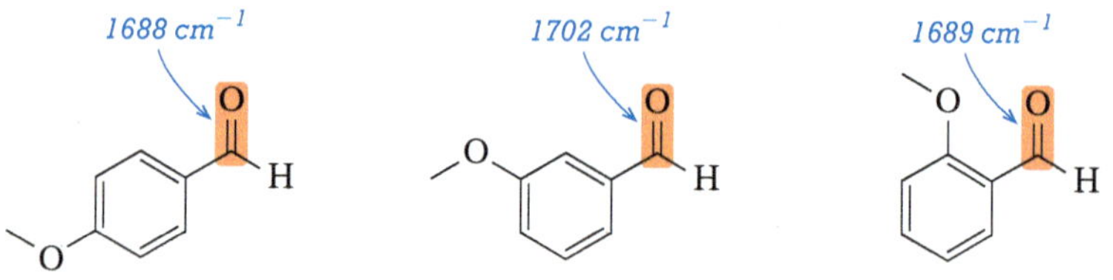

4-methoxybenzaldehyde 3-methoxybenzaldehyde 2-methoxybenzaldehyde

---

14.58 (•••) Justify the carbonyl stretching frequencies indicated for 3-chlorobenzaldehyde and 3-fluorobenzaldehyde.

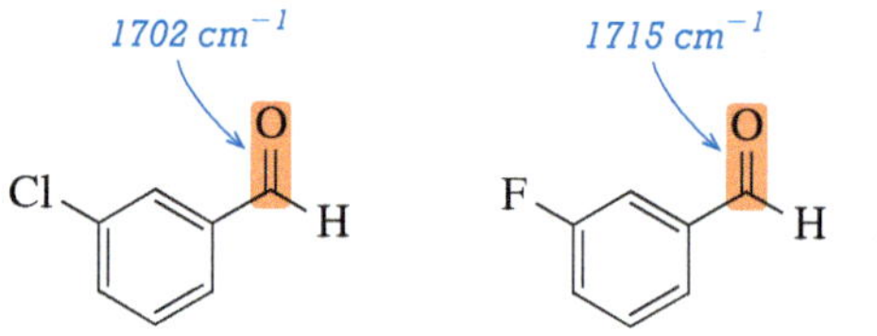

3-chlorobenzaldehyde 3-fluorobenzaldehyde

14.59 (••••) On the basis of your answers to Assessments 14.55–14.58, rank the following acetophenone derivatives on the basis of the carbonyl stretching frequency (1 = highest; 4 = lowest).

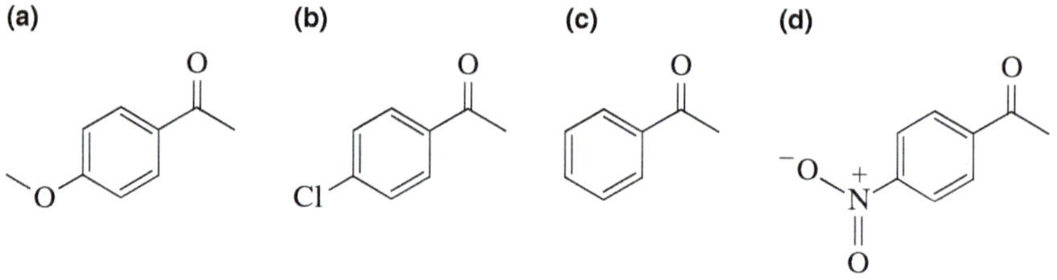

14.60 (••••) The first step of a reaction called electrophilic aromatic substitution (see Chapter 23) is as follows:

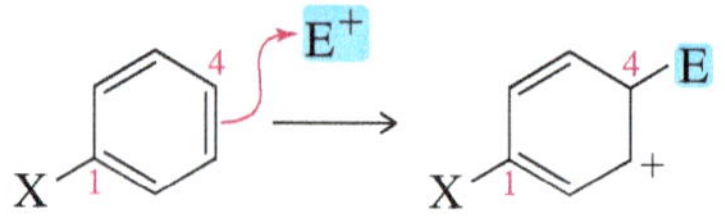

If this step is rate-determining for the overall reaction, which benzene derivative would you expect to react most quickly? Which would react most slowly? [*Use the information you gleaned from working Assessments 14.55–14.59.*]

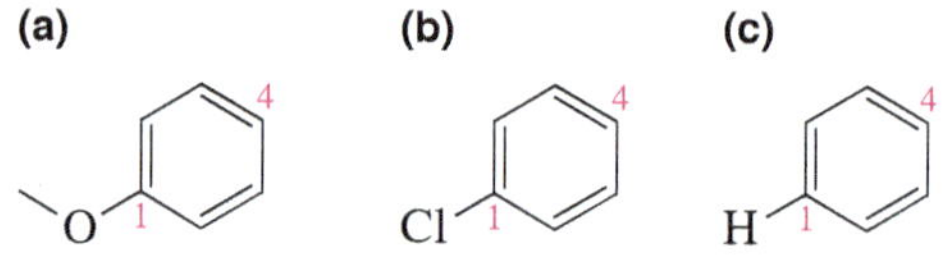

14.61 (••••) Predict the product of the reaction illustrated by the IR spectra.

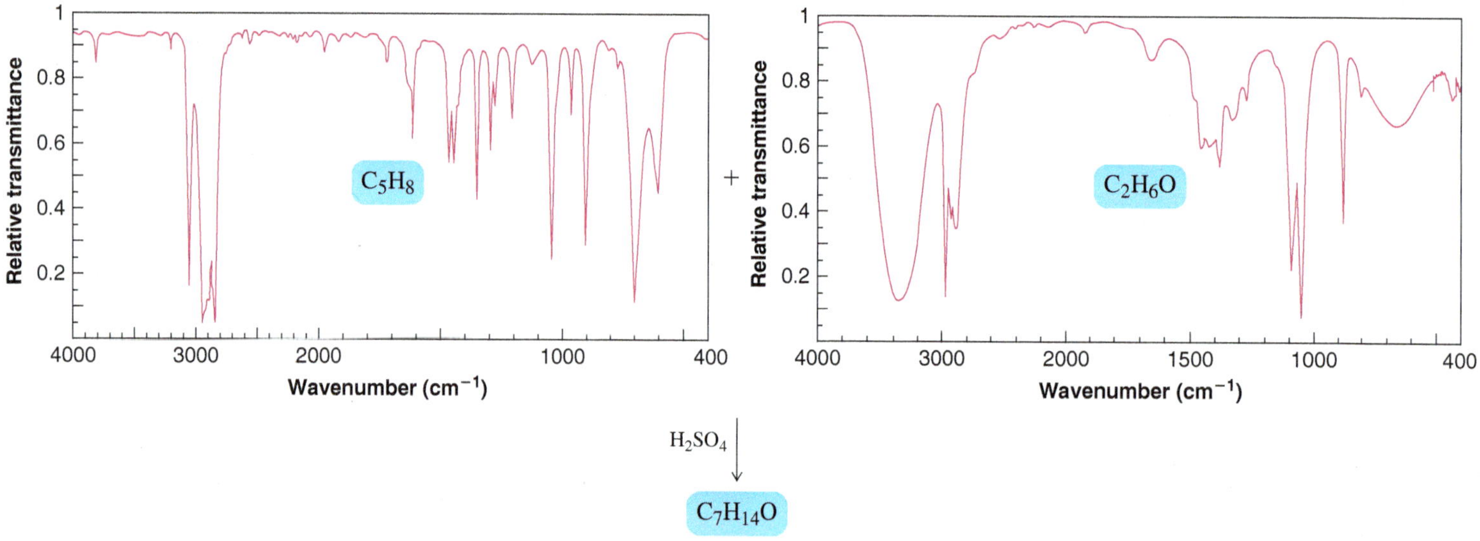

14.62 (••••) For each of the molecules shown, predict the structure of at least one major fragment in the mass spectrum.

(a)

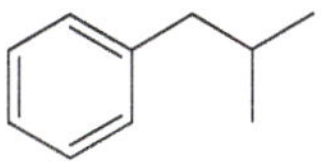

(b)

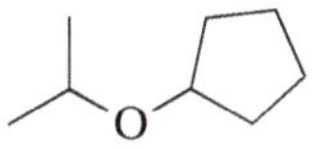

(c)

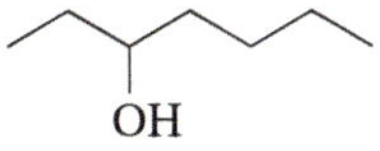

(d)

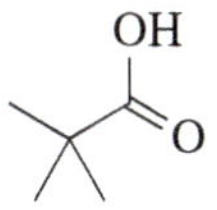

(e)

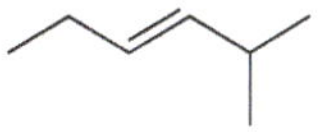

14.63 (•) The following mass spectrum belongs to a straight-chain alkane. Identify the molecule.

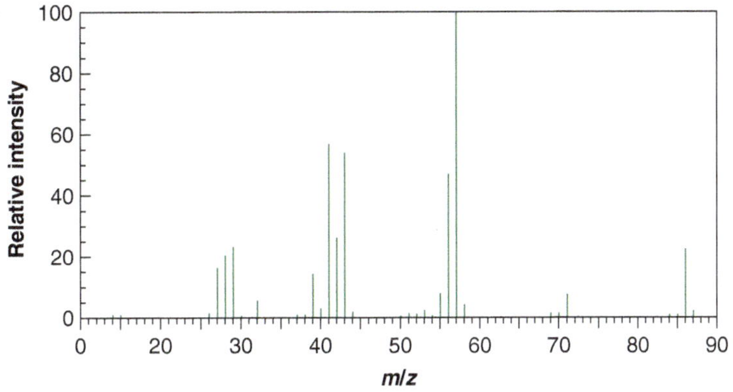

14.64 (•••) Give the structure of the molecule that gives rise to the following IR and mass spectra.

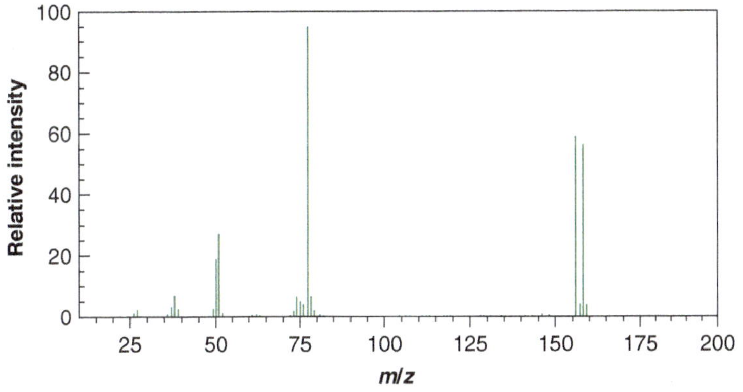

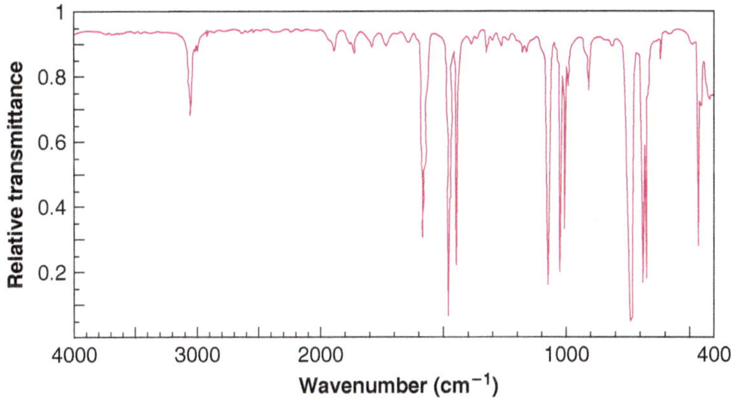

14.65 (•••) Give the structure of the molecule that gives rise to the following IR and mass spectra. [*What's the significance of m/z 59 and the two bands around 3400 cm^{-1} and 3200 cm^{-1}?*]

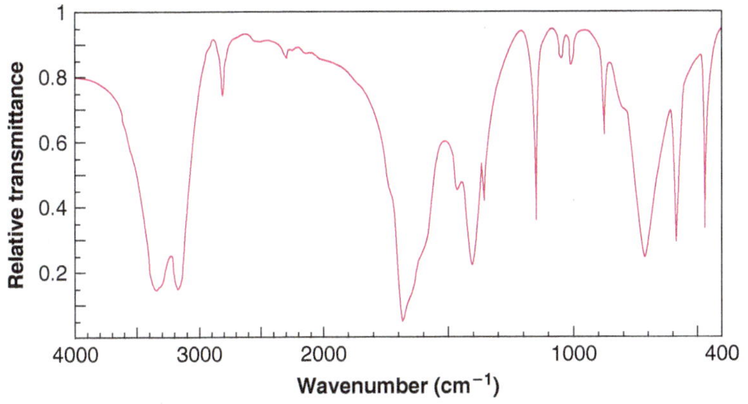

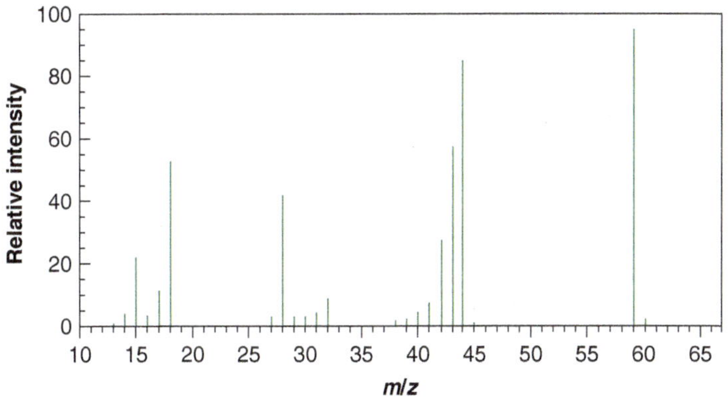

14.66 (•••) Given the structure, draw the fragment that gives rise to each *m/z* peak observed in the mass spectrum. [*The fragment must be positively charged.*]

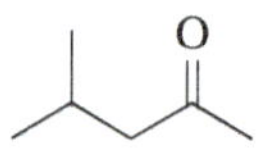

m/z **85, 58, 43**

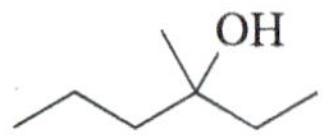

m/z **101, 98, 87, 73**

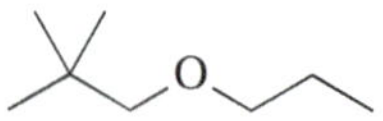

m/z **101, 87, 73, 59**

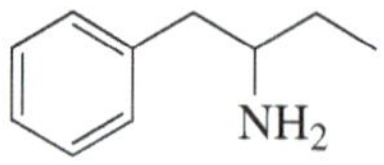

m/z **120, 91, 58**

14.67 (••••) Suggest a mechanism by which each of the fragments in Assessment 14.66 formed.

14.68 (••••) What ratio of M, M + 2, and M + 4 would you expect for 1,6-dibromohexane?

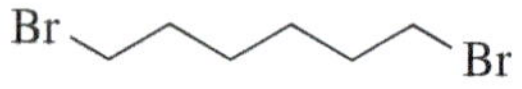

14.69 (••••) The IR and mass spectra for an unknown unsaturated hydrocarbon are shown here. When this molecule is treated with HBr, a mass spectrum with *m/z* 186 and 184 in a 1:1 ratio is produced. Give the structure of (a) the unknown molecule and (b) the product.

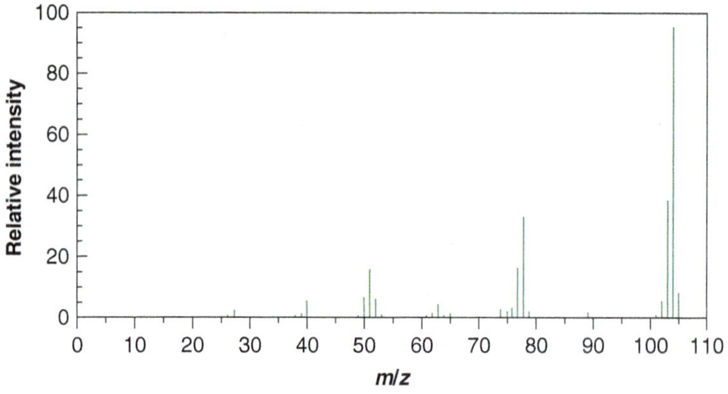

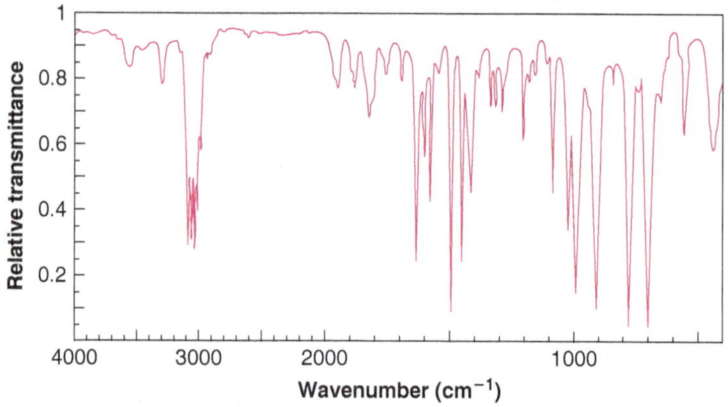

14.70 (•••) A haloalkane that gives the following mass spectrum, when treated with hydroxide, gives what product? [*Which haloalkane gives no significant M + 2 peak?*]

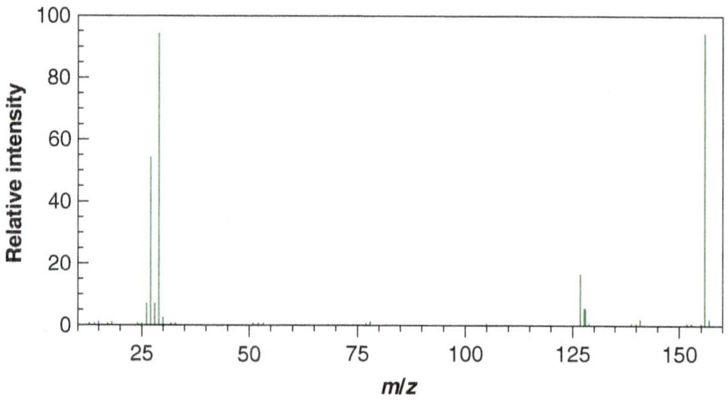

14.71 (••) An alcohol is treated with chromic acid, and the IR and mass spectra shown are obtained. Identify the starting material and the product of the reaction.

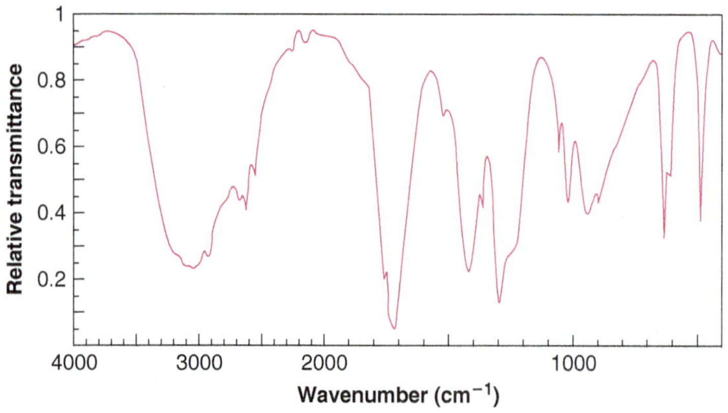

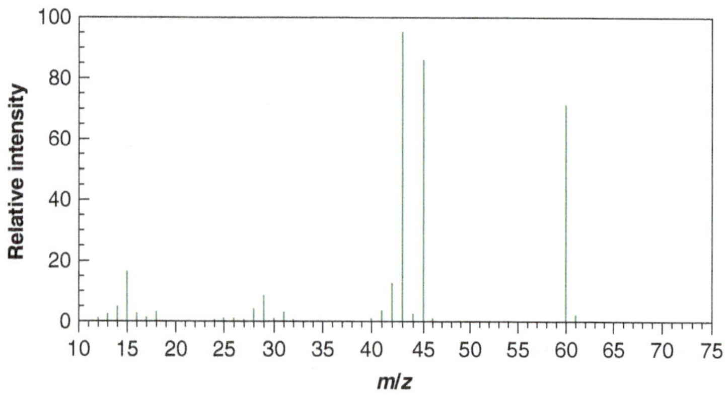

15

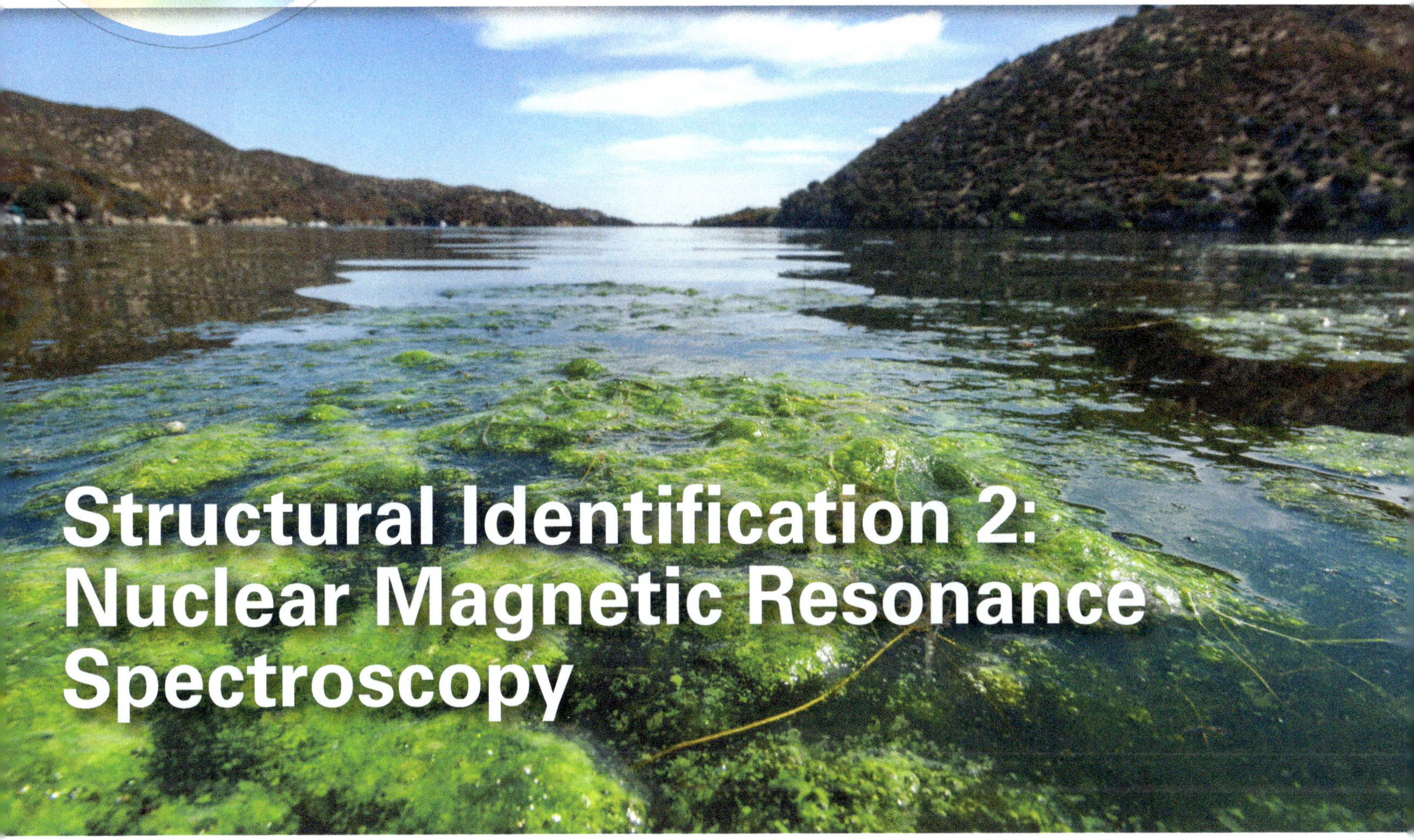

Structural Identification 2: Nuclear Magnetic Resonance Spectroscopy

Introduction

Throughout history, nature has been a prolific source of medicines and important chemical products (**Figure 15.1**). Four thousand years ago, Sumerians recorded the pain-relieving properties of the willow tree, now known to contain salicylic acid as the active ingredient. Ephedrine, a component of the Chinese shrub ephedra (Ma Huang), has been used for thousands of years in traditional Chinese medicine to treat asthma, hay fever, and bronchitis. Foxglove, a plant native to western Europe, was used for centuries as a treatment for dropsy (edema), the buildup of fluids in the lungs, caused by heart failure. In the 18th century, it was discovered to contain the compounds digitoxin and digoxin, known as cardiac glycosides for their ability to cause stronger heart contractions.

As a result of their use in traditional medicines, a variety of plants, fungi, and marine organisms are mined as sources of medicinally important compounds. In addition to being medicines themselves, these isolated **natural products** often serve as lead compounds for further research and the development of clinically useful and safe medicines. A lead compound is a molecule that possesses some useful biological activity but may have unwanted side effects or other issues that prevent its use as a drug. Research is then done to modify the compound to eliminate the unwanted properties while maintaining the useful medicinal properties. The success of these processes is evidenced by the fact that around 70% of new medications in the past 25 years are derived from natural products.

Learning Objectives

LO 15.0 Determine the structure of compounds using NMR spectra

LO 15.1 Understand the background theory of NMR spectroscopy

LO 15.2 Examine the link between sets of equivalent hydrogens and chemical shift

LO 15.3 Analyze the effect of signal splitting

LO 15.4 Determine the location of neighboring hydrogens using coupling constants

LO 15.5 Analyze ^{13}C NMR spectra

LO 15.6 Analyze DEPT spectra

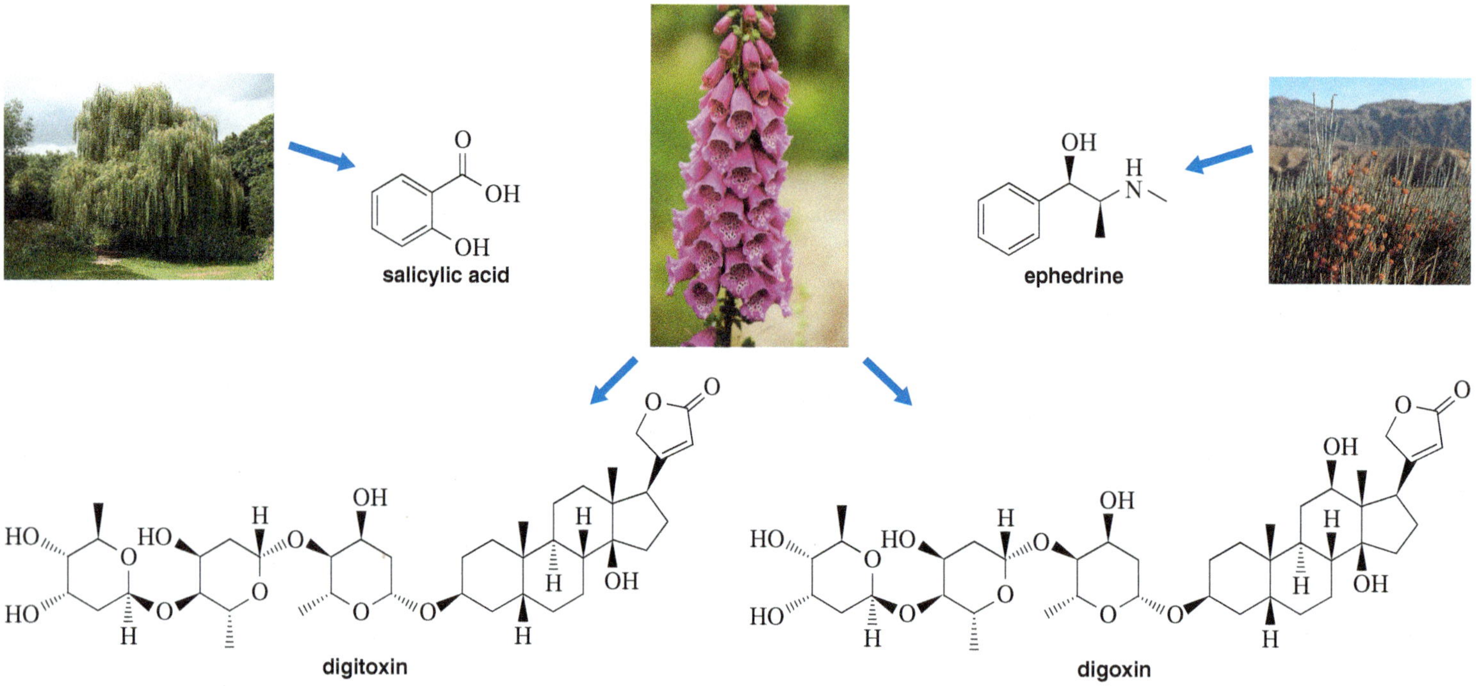

Figure 15.1 **Traditional remedies isolated from natural sources.**

"The art of healing comes from nature, not from the physician. Therefore, the physician must start from nature, with an open mind."

—Paracelsus

The process of natural product isolation is outlined in **Figure 15.2**. Once a natural product is determined to have a desired activity, any further research on it requires elucidation of its structure. This is where infrared spectroscopy and mass spectrometry (Chapter 14) and nuclear magnetic resonance (NMR) spectroscopy (Chapter 15) come in. As we study the principles and application of NMR, a case study on the isolation of curacin A, a potential cancer treatment, is presented.

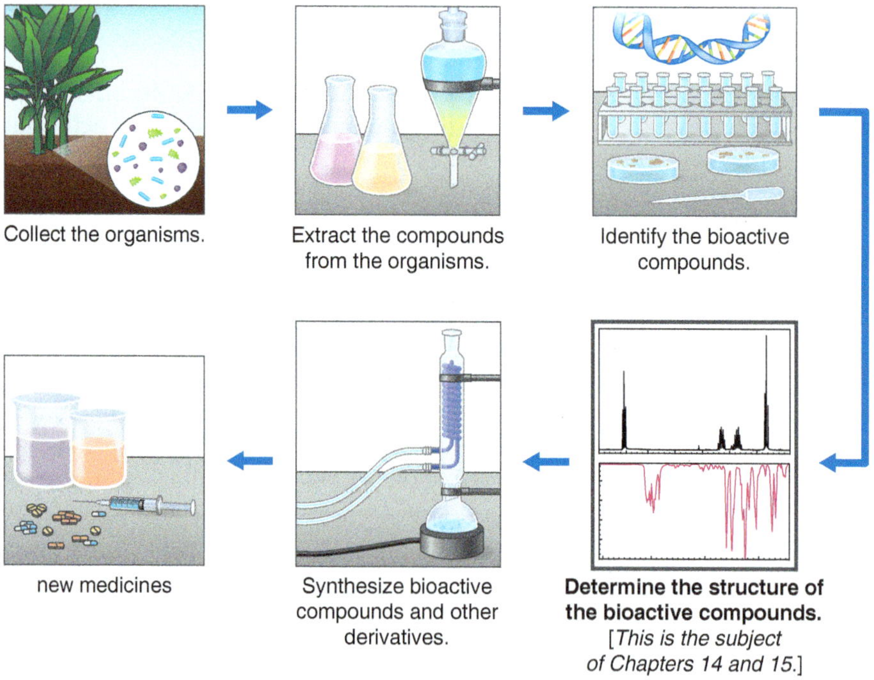

Figure 15.2 **Scheme of natural product isolation leading to a potential medicine.**

Similar to Chapter 14, Chapter 15 focuses on structure identification, rather than looking at reactions and mechanisms. Compared to infrared spectroscopy and mass spectrometry, NMR spectroscopy is far and away the most useful tool for identifying the structures of organic molecules. Many unknown structures can be solved completely just by knowing the molecular formula and obtaining a ^{1}H NMR spectrum. This is a result of the large amount of information contained within a spectrum. [*That's the good news.*]

[*The bad news:*] With all the information available in a spectrum, learning ^{1}H NMR can be overwhelming at times. But for the students who put effort into it right from the beginning of the chapter [*and not a couple nights before an exam*], it becomes easy and [*you'll never believe me, but it's true*] fun! The reason: Once you figure out the patterns and have a handle on the process, solving spectra is as simple as working a jigsaw puzzle. Just as you once had to learn a process for working puzzles (Chemistry in Context 10-A), you must learn a similar process to solve spectra. Using the analogy of a jigsaw puzzle, once the pieces are organized, solving the spectrum is rather simple. As always, there are some important concepts we'll need to use in this chapter. Let's assess our understanding of those now.

Test your understanding

Assessment 15.1 After a day of rowing downstream, it is time to head back to where you began. Is it easier to turn your boat around when the current is strong or when it is weak?

Assessment 15.2 Radical chlorination, a reaction studied in Chapters 5 and 11, replaces a hydrogen with a chlorine. Assuming there is no regioselectivity (all hydrogens can react equally), how many different chloroalkanes are possible upon reaction of each alkane with one equivalent of Cl_2?

(a) (b) (c) (d)

Assessment 15.3 (a) Rank the following bonds in terms of the strength of their bond dipole (1 = weakest, 6 = strongest). (b) Which carbon has the largest δ^{+}?

C—F C—Br C—I C—H C—C C—Cl

Assessment 15.4 Every number in Pascal's triangle is the sum of the two numbers above it. Given this, fill in the missing numbers.

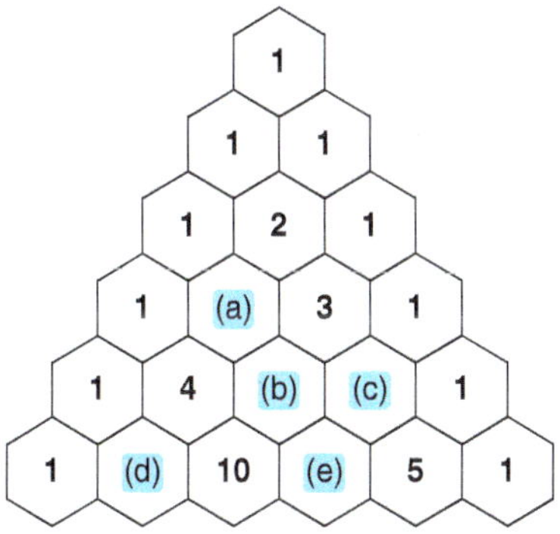

Assessment 15.5 Imagine a reaction where either of the hydrogens at C_4 is replaced with deuterium. Draw the two products that would result. What is the relationship between the two products you've produced in each example?

(a)

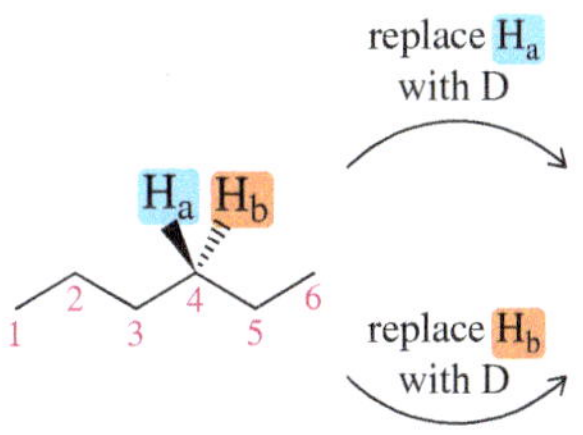

(b)

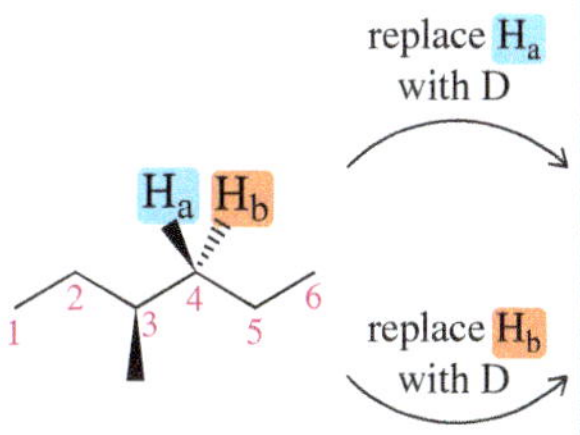

15.1 Nuclear magnetic resonance: an overview

Nuclear magnetic resonance (NMR) is another type of spectroscopy. As we learned in Chapter 14, spectroscopy is simply the interaction of energy (in the form of electromagnetic radiation) with, in our case, organic molecules. As described by Equation 15.1, one **photon** (particle) of light possesses a **quantum** of energy (E) directly proportional to the **frequency**

(v) of light. The units of frequency are reciprocal seconds (s^{-1}). The proportionality constant between energy and frequency is Planck's constant ($h = 6.63 \times 10^{-34}\ J \cdot s$). Because the frequency and wavelength (λ) of light are related by the speed of light ($c = 3.0 \times 10^8\ m/s$) through Equation 15.2, the energy is inversely proportional to the wavelength (Equation 15.3).

$$E = hv \qquad \text{Equation 15.1}$$

$$c = v\lambda \qquad \text{Equation 15.2}$$

$$E = hv = h\left(\frac{c}{\lambda}\right) \qquad \text{Equation 15.3}$$

Nuclear magnetic resonance uses energy in the radio frequency region of the electromagnetic spectrum. This energy corresponds to the energy required to flip nuclear spin. Under conditions in which an NMR spectrum is recorded, this energy depends on the environment of the nucleus. [*We'll talk more about this.*] Being at the long wavelength/low-frequency end of the electromagnetic spectrum (**Figure 15.3**), radio waves are of very low energy. As a result, the radio waves that are all around us pose no danger. This also implies that flipping nuclear spin must not require much energy.

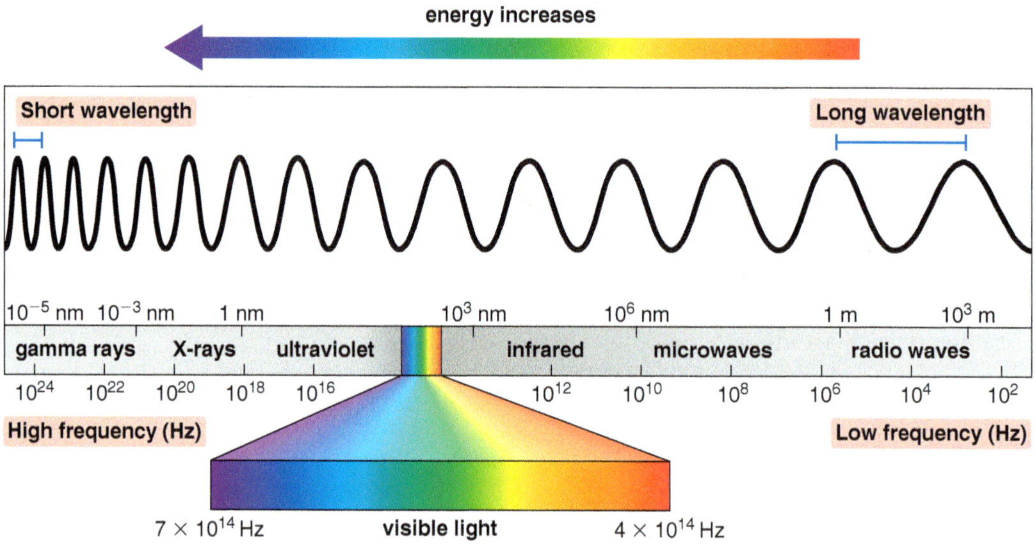

Figure 15.3 The electromagnetic spectrum. As wavelength decreases, frequency increases and energy increases. Radio waves correspond to the flipping of nuclear spin, a low-energy transition.

Table 15.1 shows how different frequencies interact differently with molecules, causing the different molecular transitions that allow us to determine structure. Note that bond vibration (IR) requires more energy than flipping nuclear spin (radio).

Table 15.1 Different frequencies of electromagnetic radiation cause different molecular transitions.

Region	Wavelengths	Frequencies (Hz)	Molecular transition
X-rays	10^{-3} nm to 10 nm	10^{21} to 10^{18}	Ionization
UV–Vis	10 nm to 700 nm	10^{18} to 10^{14}	Electron excitation
Infrared	700 nm to 1 mm	10^{14} to 10^{10}	Molecular vibrations
Microwave	1 mm to 1 m	10^{10} to 10^6	Bond rotation
Radio	1 m to 100 km	10^6 to 10^2	Flipping nuclear spin

15.2 NMR spectroscopy: practical usage

Similar to Chapter 14, we begin by using NMR to determine a structure. By first seeing how it is applied, we will have a framework that will allow us to better understand the background theory later.

NMR spectroscopy uses electromagnetic radiation in the radio region of the electromagnetic spectrum to give us information on the chemical environment of every hydrogen in a given molecule. Armed with this knowledge, complete structure determination is usually possible. An NMR spectrum is a plot showing the frequencies at which electromagnetic radiation is absorbed. The notation of frequency in units of ppm is explained later. On the spectrum in **Figure 15.4**, each spike in the spectrum is a **signal**, representing where electromagnetic radiation was absorbed.

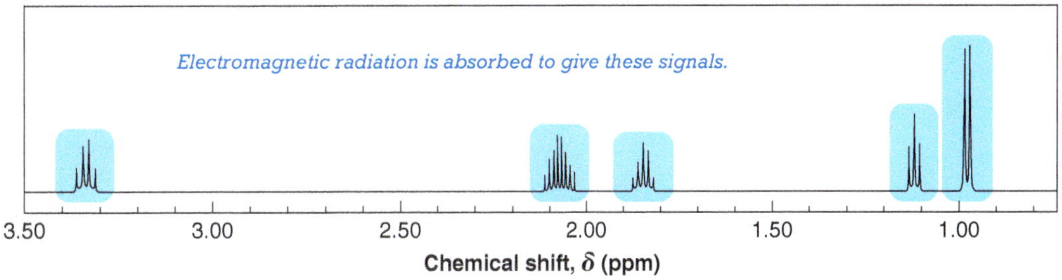

Figure 15.4 A sample ^{1}H NMR spectrum shows where electromagnetic radiation is absorbed. [*Each signal can contain a different number of peaks. For example, the signal at 0.9 is composed of two peaks.*]

15.2.1 The anatomy of a signal

NMR can be used to determine a structure once you know what information is given by an individual signal. There are four important parts of the signal, as summarized in **Figure 15.5**.

1. ***Signal:*** Each signal represents a set of one or more equivalent hydrogens that absorb light at that frequency [expressed as a **chemical shift (δ)** in units of parts per million (ppm)—more on that later]. [*We know there is a* ***hydrogen.***]
2. ***Chemical shift:*** The location of the signal gives us information about the electronic environment of the hydrogen responsible for the signal. It tells us how close the hydrogen is to any functional groups in the molecule. **Table 15.2** lists representative chemical shift values for certain hydrogen types. Based on a chemical shift of 2.4, the signal in Figure 15.5 is likely at the α-carbon of a carbonyl. [*There are other possibilities, but to simplify, we will assume that we now know there is a hydrogen* ***next to a carbonyl.***]
3. ***Integration:*** The integration tells us how many hydrogens are responsible for giving that signal, at least in a relative sense. [*We can assign a value to one integral whose expected value we know to make them absolute numbers.*] That is, if the chemical shift depends on the chemical environment, then every hydrogen that is in the exact same chemical environment will be represented by that signal. With an absolute integration of two, the signal represents two hydrogens. [*We know there are* ***two hydrogens****, likely in the form of a* CH_2 *next to a carbonyl.*]
4. ***Splitting:*** A signal may be made up of one or more peaks within the signal. The splitting of the signal tells us how many hydrogens are on **neighboring carbons**. Neighboring carbons are the carbons directly attached to the carbon bearing the hydrogen represented by the signal. The number of neighbors can be determined by subtracting 1 from the number of peaks the signal is split into. A singlet (one peak) implies that there are zero neighbors, a doublet (two peaks) implies one neighbor, a triplet (three peaks) implies two, and so on. The signal in Figure 15.5 is a triplet, so it is next to two hydrogens, most likely a CH_2. [*We know that there are two hydrogens, likely in the form of a* CH_2 *with a carbonyl on one side* ***and another*** CH_2 ***on the other side.***]

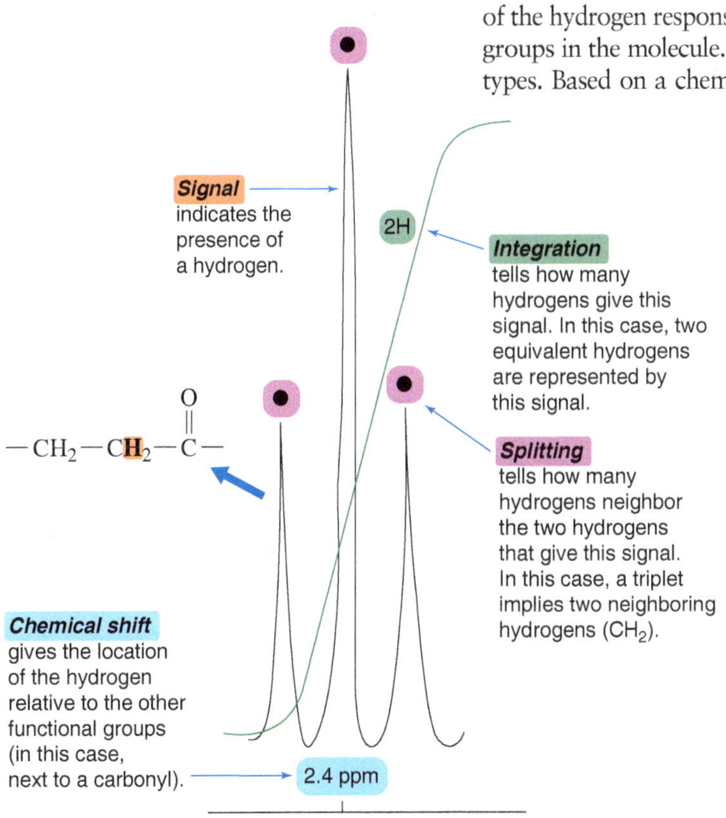

Figure 15.5 The anatomy of an NMR signal.

Table 15.2 Chemical shift values correspond to hydrogens in certain chemical environments.

Type of proton	Approximate δ (ppm)	Type of proton	Approximate δ (ppm)
Alkane ($-CH_3$) **methyl**	0.9	$>C{=}C<$ with CH_3 **allylic**	1.7
Alkane ($-CH_2-$) **methylene**	1.3	$Ph-H$ **aromatic**	7.2
Alkane ($-\underset{\vert}{C}H-$) **methine**	1.4	$Ph-CH_3$ **benzylic**	2.3
$-\overset{O}{\overset{\Vert}{C}}-CH_3$ **methyl ketone**	2.1	$R-\overset{O}{\overset{\Vert}{C}}-H$ **aldehyde**	9–10
$-C\equiv C-H$ **acetylenic**	2.5	$R-\overset{O}{\overset{\Vert}{C}}-OH$ **acid**	10–12
$R-CH_2-X$ (X = halogen, O)	3–4	$R-OH$ **alcohol**	Variable, ~2–5
$>C{=}C<$ with H **vinyl**	5–6	$Ar-OH$ **phenol**	Variable, ~4–7
		$R-NH_2$ **amine**	Variable, ~1.5–4

Note: These values are approximate because all chemical shifts are affected by neighboring substituents. The numbers given here assume that alkyl groups are the only other substituents present.

Were we trying to solve the structure of a spectrum containing the peak in Figure 15.5, we'd now have one piece of the puzzle. Doing the same for every peak in the spectrum would allow us to piece the puzzle together to give the final structure. Let's convince ourselves that we can do this by working Assessment 15.6.

Assessment 15.6 The spectrum shown here is for a molecule with a molecular formula of C_4H_9Cl. Suggest a structure for this molecule.

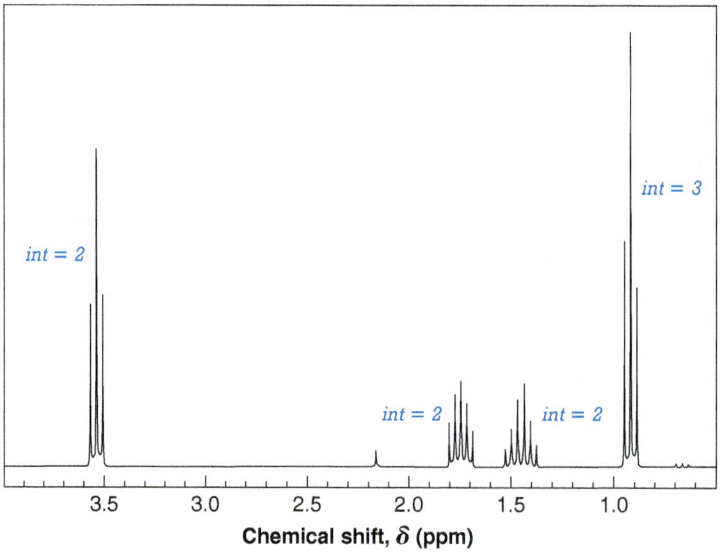

Since you have never solved an NMR problem, let's work this one together.

Solved Assessment 15.6

The key to solving all spectroscopy problems is to use all of the information you have. And a big piece of it is the molecular formula. Knowing that the molecular formula is C_4H_9Cl, what functional groups are possible? Because the only heteroatom (non-carbon/non-hydrogen) is a chlorine, the molecule must be a haloalkane. Using the index of hydrogen deficiency (IHD), calculated to be zero for C_4H_9Cl using Equation 15.4, all molecules containing rings or π bonds are ruled out. This will be helpful to us as we use Table 15.2 to suggest what functional group each hydrogen is near.

$$\text{IHD} = \text{number of carbons} - \frac{\text{number of hydrogens}}{2} - \frac{\text{number of halogens}}{2} + \frac{\text{number of nitrogens}}{2} + 1 = 4 - \frac{9}{2} - \frac{1}{2} + \frac{0}{2} + 1 = 0 \qquad \text{Equation 15.4}$$

When you first look at a spectrum to try to determine a structure, it can be overwhelming. This is like the feeling when you read the words "propose a mechanism for this complex reaction" or "synthesize this big molecule starting from three carbons or fewer." [*You might have this same feeling when you dump out the 1000 pieces of a puzzle you want to put together.*] In each of these cases, we have to trust our problem-solving process; we need to do the same here. [*Lay out the pieces.*] We start by looking at each individual peak to get as much information from each as possible. Laying out this information in a table will then allow us to put the information together to give the structure.

The steps that follow are summarized in **Figure 15.6** and **Table 15.3**. Let's analyze the signals from right to left (a → d).

Signal a appears at 0.9 ppm, integrates to 3, and is a triplet (three peaks), implying that it neighbors two hydrogens. The simplest [*not always correct, but **always** where you should start*] way for three hydrogens to be in the same environment is a CH_3, or methyl group. The simplest way [*again*] to neighbor two hydrogens is to be next to a CH_2.

*Our guess for signal **a**:* A CH_3 attached to a CH_2, or $CH_3{-}CH_2$. The bolded **H** is the one that the signal is telling us about. The others are hydrogens we believe to be present as neighbors. [*Make sure this is clear before continuing.*]

Signal b appears at 1.45 ppm, integrates to 2, and is a sextet (six peaks), implying that it neighbors five hydrogens. The simplest way for two hydrogens to be in the same environment is a CH_2, or a methylene group. The simplest way to have five neighbors is to be between a CH_3 and a CH_2. [*It is impossible to be next to a CH_5 or between a CH_4 and a CH, right?*]

*Our guess for signal **b**:* A CH_2 between a CH_2 and a CH_3, or $CH_2{-}CH_2{-}CH_3$.

Signal c appears at 1.75 ppm, integrates to 2, and is a pentet/quintet (five peaks), implying that it neighbors four hydrogens. The simplest way for two hydrogens to be in the same environment is a CH_2, or a methylene group. The simplest way to have four neighbors is to be between two CH_2 groups.

*Our guess for signal **c**:* A CH_2 between a CH_2 and a CH_2, or $CH_2{-}CH_2{-}CH_2$.

Signal d appears at 3.53 ppm, integrates to 2, and is a triplet (three peaks), implying that it neighbors two hydrogens. The simplest way for two hydrogens to be in the same environment is a CH_2, or a methylene group. The simplest way to have two neighbors is to be next to a CH_2 group. A CH_2 group that only has one neighboring group is missing its fourth bond. The chemical shift of 3.52 helps us here. [*Using the molecular formula and Table 15.2, what else is probably attached to this CH_2?*] Because of its location in the spectrum, the hydrogen that gives this signal is likely on a carbon bearing the chlorine atom.

*Our guess for signal **d**:* A CH_2 between a CH_2 and a Cl, or $Cl{-}CH_2{-}CH_2$.

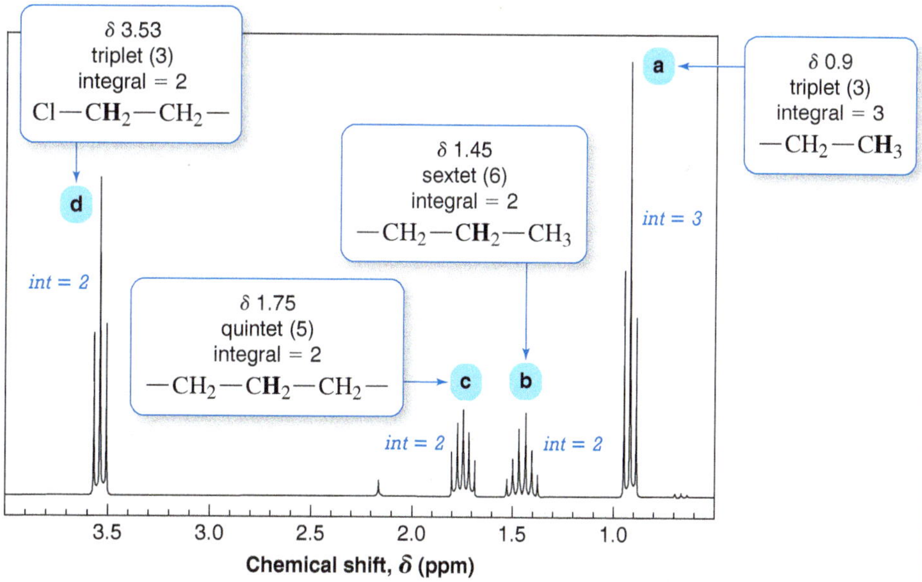

Figure 15.6 Spectrum from Assessment 15.6 (C_4H_9Cl).

Table 15.3 Summarized data from Figure 15.6.

Signal	Chemical shift, δ (ppm)	Integral	Splitting (*N*)	Neighbors (*N* − 1)	Guess
a	0.9	3	Triplet (3 peaks) (t)	2	$CH_3{-}CH_2$
b	1.45	2	Sextet (6 peaks) (sext)	5	$CH_2{-}CH_2{-}CH_3$
c	1.75	2	Quintet (5 peaks) (quint)	4	$CH_2{-}CH_2{-}CH_2$
d	3.53	2	Triplet (3 peaks) (t)	2	$Cl{-}CH_2{-}CH_2$

And with that, we have our pieces assembled in Table 15.3. Don't forget: The bolded hydrogens are the ones each signal provides direct evidence for; the neighboring hydrogens are implied by the splitting pattern of the signal. Before we put them together, NMR problems have a way of making sure we can't be incorrect about the implied neighbors. Consider peak **a**, a CH_3 which we believed to have a CH_2 as a neighbor. Peak **b** happens to be a CH_2 that also has a CH_3 as a neighbor. Thus, it is likely that peak **a** is attached to peak **b** and they are each other's neighbors. Peak **b**'s other neighbor is a CH_2, and peak **c**, not coincidentally, is a CH_2 that also has a CH_2 as a neighbor. Hence, peak **b** is probably attached to peak **c** and they are each other's neighbors. Peak **c**'s other neighbor is a CH_2, and peak **d**, not coincidentally, is a CH_2 that also has a CH_2 as a neighbor. [*No doubt the previous paragraph is confusing—read it again and then continue.*] In summary, every hydrogen that we guess is a neighbor of one peak must also be present in the spectrum. If we correctly believe a hydrogen will have a CH_2 as a neighbor, the spectrum should tell us that there is a CH_2 in the molecule. In this way, we can always check our guesses.

And so, assembling the pieces as shown in **Figure 15.7**, we can be confident that we have a CH_3 attached to a CH_2 attached to a CH_2 attached to a CH_2 attached to a chlorine, or 1-chlorobutane. Note that as a hydrogen moves closer to the functional group, it appears at an increasing chemical shift. [*We aren't done yet though* . . . .] Before finishing a spectroscopy problem, always check your answer by looking at the proposed molecule and imagining what each hydrogen would look like in the spectrum. [*We'll do more of this later on.*] ■

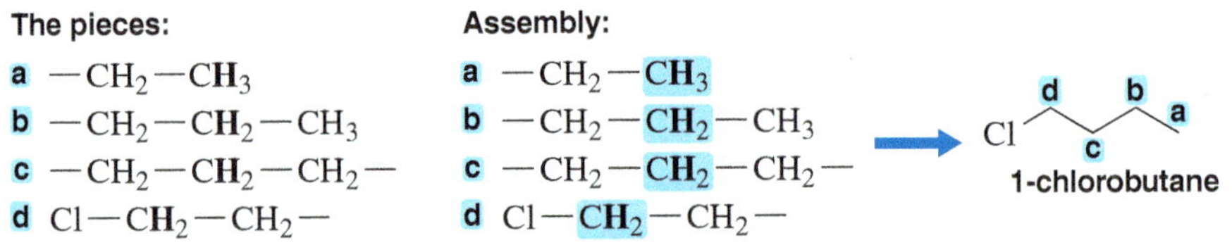

Figure 15.7 The structure of the unknown molecule in Assessment 15.6 is 1-chlorobutane.

15.2.1 SUMMARY

- Using the information gained from each signal in an NMR spectrum (chemical shift, integration, and splitting) and then piecing together the information that is gained from each, a structure, like a puzzle, can be assembled.

Assessment 15.7 The spectrum shown here is for a molecule with a molecular formula of $C_5H_{10}O$. Suggest a structure for this molecule. The integration (int) is indicated on the spectrum.

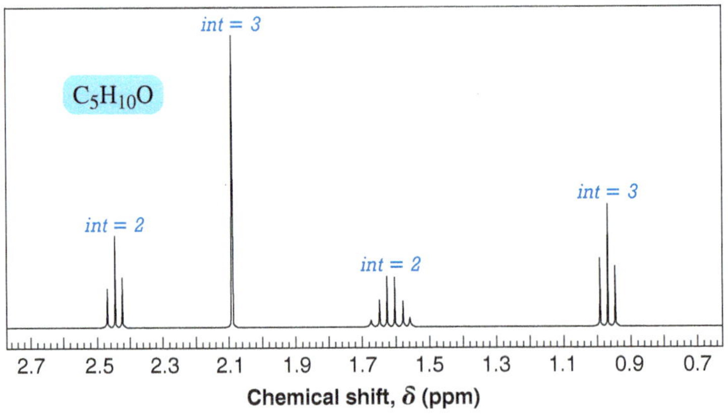

Assessment 15.8 Draw the peak that would correspond to the hydrogens at C_3 in the molecule shown. Be sure to indicate the integration and place it at an appropriate chemical shift.

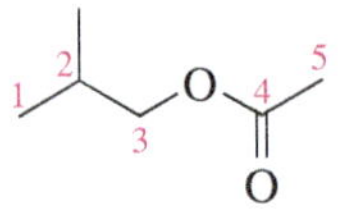

15-A CHEMISTRY IN CONTEXT: Natural Product Isolation

Curacin A

In 1994, it was discovered that extracts of the filamentous cyanobacteria *Lyngbya majuscula* were toxic to brine shrimp. Brine shrimp toxicity can indicate potential anticancer properties, so these extracts were purified by column chromatography to identify the exact compound responsible for the toxicity. One molecule isolated in this study was curacin A, the structure of which was not known at the time of isolation. The name was chosen to recognize that the molecule was isolated in Curaçao. The spectroscopic techniques discussed in Chapter 14 and 15 were used to determine the structure of this natural product.

Following purification, a high-resolution mass spectrum was obtained, giving an exact mass of 373.2439 amu. This exact mass can only arise from a molecular formula of $C_{23}H_{35}NOS$. [*Computer programs can easily calculate this.*] As we learned in Chapter 14, the molecular formula not only tells what atoms are present, but also suggests the presence of functional groups and rings through calculation of its index of hydrogen deficiency. [*Try figuring it out on your own before proceeding to Chemistry in Context 15-B.*]

[*As you read through these Chemistry in Context boxes, understand that only part of the story is told here. The amount of labor that goes into solving a complicated structure like that of curacin A is extensive and far beyond the scope of an undergraduate organic chemistry course. However, taking a glimpse at this story from a novice perspective allows us to appreciate the power of spectroscopy for unknown compound identification.*]

Assessment 15.9 Calculate the index of hydrogen deficiency for curacin A, which has a molecular formula of $C_{23}H_{35}NOS$.

15.3 NMR spectroscopy: background theory

Even though it was a relatively simple problem, being able to solve the structure in Assessment 15.6 shows that NMR can be used effectively by treating it as a black box and simply interpreting the data that comes out. However, to be able to do more complicated problems, it is important to understand how NMR works. By doing so, you will make connections between this and previously studied material, strengthening your retention of this information. Sections 15.3.1–15.3.3 discuss some important aspects of the background theory of NMR before we do a deeper dive into the anatomy of a spectrum in the context of this theory.

15.3.1 Nuclear spin

NMR involves the absorption of energy in the radio frequency region of the electromagnetic spectrum. This energy is absorbed when the nucleus of an atom flips from a low-energy spin state to a high-energy spin state. For reasons that are better suited for a physics class, nuclei are NMR active if they have an odd number of protons and/or neutrons. This combination results in a nuclear magnetic moment and angular moment, often referred to as nuclear spin. Table 15.4 provides a short list of NMR active nuclei. We focus predominantly on ^{1}H and ^{13}C because they are most prevalent in organic chemistry.

Table 15.4 NMR active nuclei have an odd number of protons and/or neutrons.

Nucleus	Number of protons	Number of neutrons	Spin quantum numbers	NMR active?
^{1}H	1	0	$+1/2, -1/2$	Yes
^{13}C	12	13	$+1/2, -1/2$	Yes
^{19}F	9	10	$+1/2, -1/2$	Yes
^{31}P	15	16	$+1/2, -1/2$	Yes
^{12}C	6	6	0	No

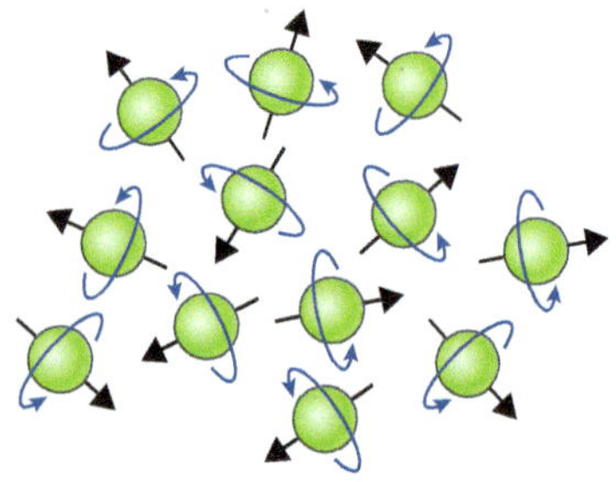

Figure 15.8 Spinning nuclei generate magnetic fields in all directions.

Protons (^{1}H) have two spin states with quantum numbers of $+\frac{1}{2}$ and $-\frac{1}{2}$. [*This is just like the spin states of electrons; Section 2.2.2.*] There is no difference between the energy of these spin states, so nuclei can easily flip between the two. Charged particles, upon spinning, generate tiny magnetic fields, similar to a tiny bar magnet. The direction of the magnetic field can be predicted using the **right-hand rule** typically taught in physics courses. If you curl your right hand in the direction of rotation, the magnetic field will be in the direction of your extended thumb. In a collection of nuclei, there is no energy difference between the spin states, so the nuclei are generating tiny magnetic fields in all directions (Figure 15.8).

Assessment 15.10 When rowing on a lake on a calm day, is it easier to row in any particular direction?

Assessment 15.11 Would you expect the following nuclei to be NMR active?

(a) ^{18}O (b) ^{14}N (c) ^{81}Br (d) ^{16}O (e) ^{2}H

15.3.2 Nuclear spin in a magnetic field

The energies of the two spin states become different when the nuclei are placed into a strong external magnetic field [Figure 15.9(a)]. When a magnetic field is applied, all of the nuclei align in either of two directions, each direction representing one of the two possible spin states. When the magnetic field of the nucleus is aligned *with* the external magnetic field, it is referred to as **α spin**. When the magnetic field of the nucleus is aligned directly *against* the external magnetic field, it is called **β spin** [Figure 15.9(b)]. These are the only two spin states that exist in the presence of an external magnetic field. A spin at any other position will relax to the nearest minimum [Figure 15.9(c)]. ΔE is the difference in energy between these spin states.

Figure 15.9 In an external magnetic field, only two spin states (α and β) are possible. α is with the field; β is against the field.

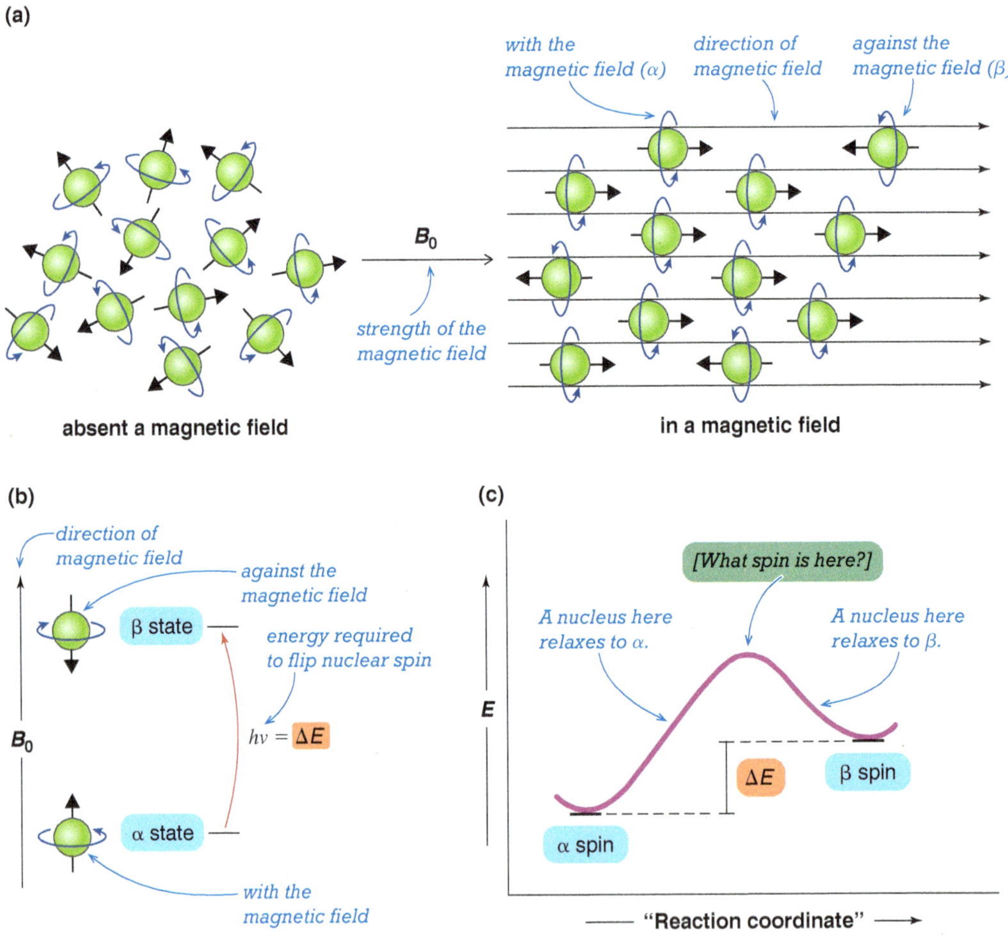

Assessment 15.12 (a) When rowing on a river, is it easier to row directly with or directly against a moderate current? (b) Is it possible to row directly perpendicular to the current?

Assessment 15.13 Answer the question in Figure 15.9(c).

Like rowing on a river with the moderate current, α spin is the more stable, lower-energy spin state. It is harder, but still possible, for a nucleus to be in the β spin state. The difference between these energies depends on the strength of the magnetic field. [*Does the strength of the current make it more difficult to row against it? Yes!*] **Figure 15.10** shows the relationship between ΔE, the difference in energy between α and β spin, as a function of the magnetic field felt by the spinning nuclei. A typical magnetic field strength (B_0) is 4.7 tesla (T). At this field strength, just over half of the nuclei are α and slightly less than half are β, implying that the energy difference is very small [2×10^{-5} kcal/mol (8×10^{-5} kJ/mol)]. This small energy difference falls within the radio frequency region of the electromagnetic spectrum. If the principle of NMR could be boiled down to one figure, Figure 15.10 is it. [*Make sure you understand what it's saying and plan on referring to it throughout the chapter.*]

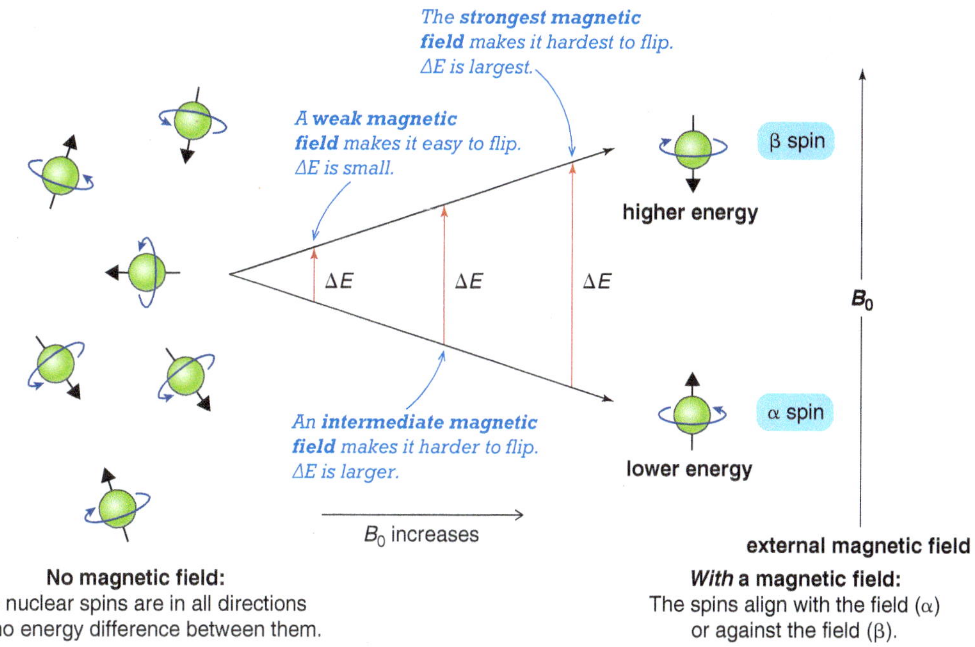

Figure 15.10 The difference in energy between α and β spin states depends on the strength of the magnetic field.

Thus far, we have talked about placing nuclei in a magnetic field, so the first two letters of NMR make sense. What about resonance? [*It is not what organic chemists usually think about.*] When the axis of nuclear rotation is changed, referred to as **precession**, in a magnetic field, it does so in a similar fashion to a spinning top, as shown in **Figure 15.11**. The absorption of energy only occurs when the frequency of energy matches the frequency of precession. When this occurs, it is said that the precession frequency and frequency of electromagnetic radiation are in resonance. This absorption of energy causes the nuclear spin to flip. From now on, the term **resonance frequency** will be used to describe the frequency that causes a nucleus to flip.

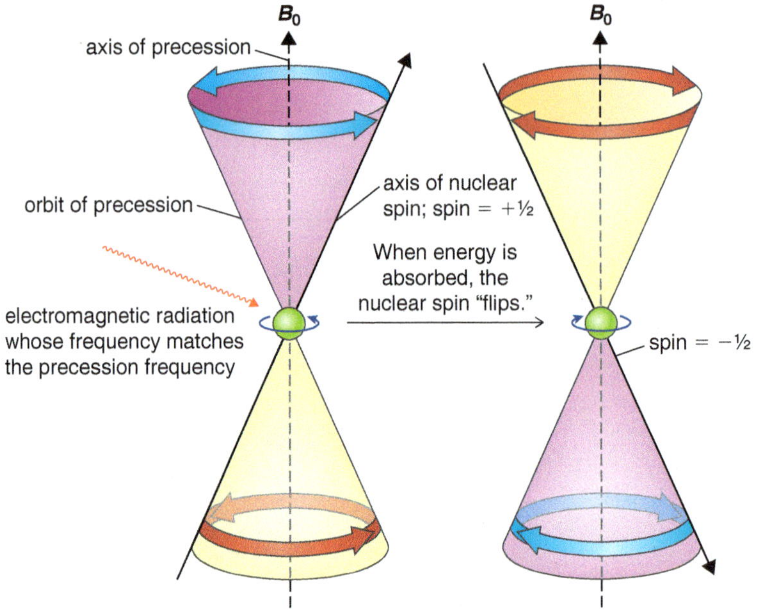

Figure 15.11 **Absorption of energy occurs when the frequency of precession matches the frequency of electromagnetic radiation.** [*Resonance.*]

15.3.2 SUMMARY

- The difference in energy between α and β spin is directly proportional to the strength of the magnetic field.

15.3.3 Acquiring a spectrum

The important components of an NMR spectrometer are a magnet and a radiofrequency (RF) transmitter (**Figure 15.12**). To acquire a spectrum, the sample is dissolved in a solvent without hydrogens, such as $CDCl_3$ or C_6D_6, and placed in the spectrometer. The sample is pulsed with a quick, intense burst of RF energy; energy that is absorbed by the sample results in nuclei flipping from α to β. With all nuclei at the less stable β spin state, slightly more than half of them will relax back to their original state to reestablish the equilibrium between α and β. The nuclei relax at different rates for reasons we discuss in Section 15.4, giving off energy that is received by the RF detector and transformed to data that are stored in the computer. After several scans, a Fourier transform converts the data to a spectrum like the ones we have studied.

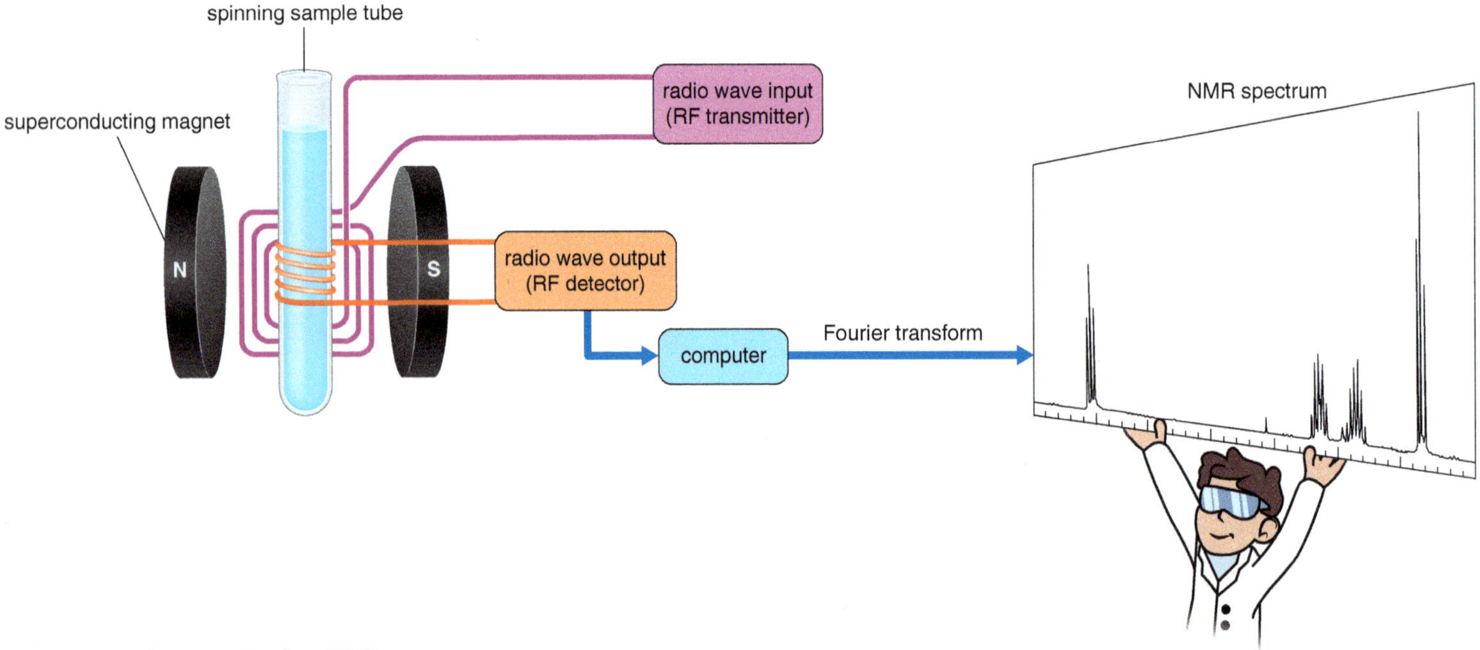

Figure 15.12 **Schematic of an NMR spectrometer.**

15.4 ^{1}H NMR spectroscopy: theory to practice

Consider a collection of noninteracting hydrogen nuclei. At a constant magnetic field like 4.7 T, the difference between α and β for each of them would be constant. That is, when flipped from α to β with a pulse of RF energy, all nuclei would relax at the exact same rate and would generate one signal. If this situation were representative of what happens in a molecule, every NMR spectrum would look like that shown in **Figure 15.13(a)**. And yet we have already seen several spectra like that shown in **Figure 15.13(b)**. How do we get a spectrum with so many signals and such variety in signal shapes? [*What determines the difference in energy between α and β spin states?*]

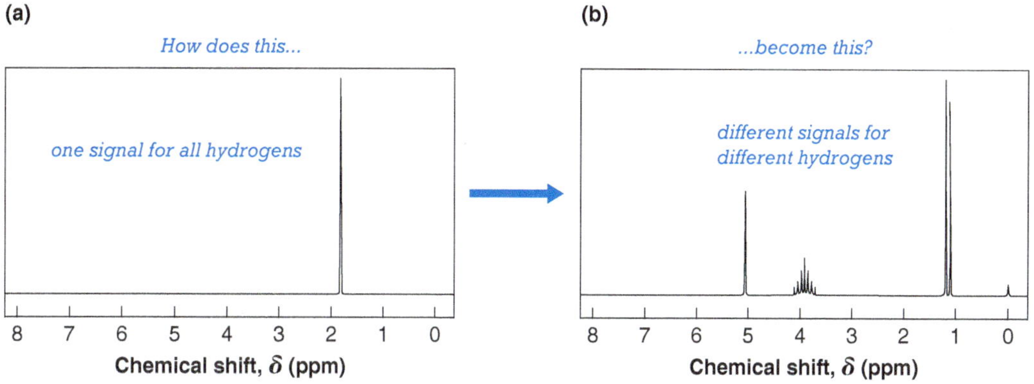

Figure 15.13 NMR spectra. (a) If all nuclei had the same difference in energy between α and β, spectra would look like this. (b) But this is what they really look like. Why?

The gap between α and β spin states is a function of the magnetic field felt by the individual nuclei. Thus, if an NMR spectrum shows multiple signals representing different hydrogens, it must be that each hydrogen feels an ever-so-slightly different magnetic field due to being in a different chemical environment. Essentially, different nuclei are at different positions on our favorite figure, Figure 15.10 (redrawn in **Figure 15.14**), implying that each requires a different energy to flip the spin. How the magnetic field is modulated by the chemical environment and how we can use that information to elucidate structure is the topic of Sections 15.4.1–15.4.4.

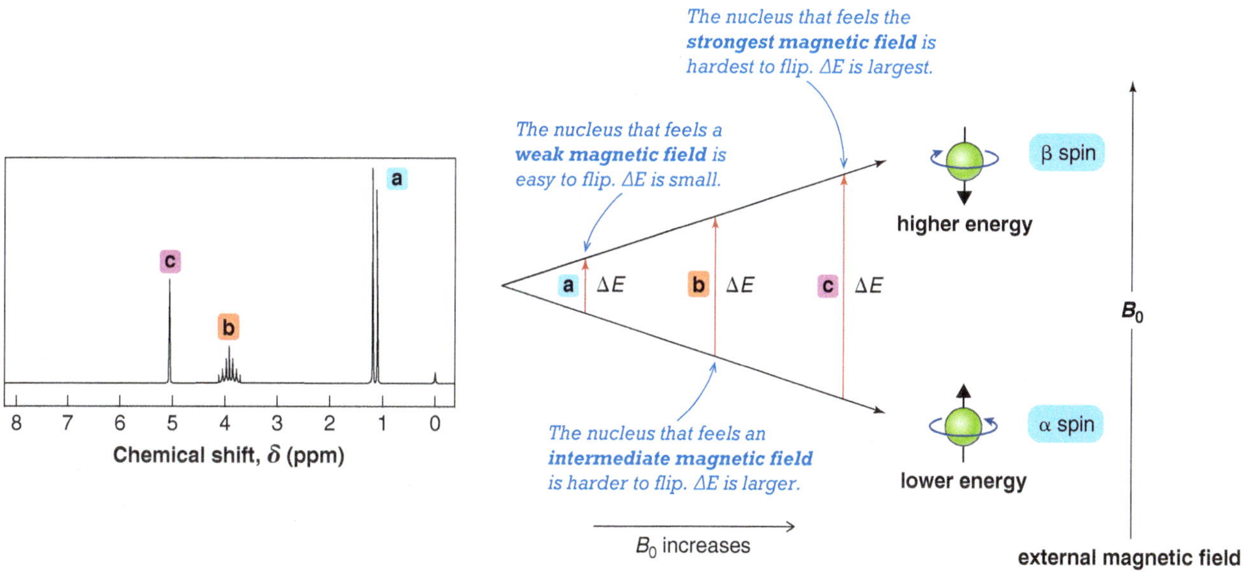

Figure 15.14 Slight differences in the magnetic field result in slightly different Δ*E* values between α and β.

The following sections will be reminiscent of Section 15.2.1 (the anatomy of a signal), even following the same format. The difference is that here we will see how the background theory results in this anatomy. In this way, we will take a deeper dive into the anatomy of a signal, filling in the knowledge gaps created by Section 15.2.1. This will give us other ways of solving problems in the event that we forget the steps we might have tried to simply memorize. Equipped with a better understanding of the pieces that make up a given spectrum, we will work more complicated examples in Section 15.4.5.

15.4.1 How many sets of equivalent hydrogens: the signals

If we accept that any difference in chemical environment, no matter how slight, will cause a difference in the magnetic field felt by nuclei, then we should expect all equivalent hydrogens to be represented by the same signal, and we should expect a different signal for each set of equivalent hydrogens in a molecule. The 3 hydrogens in a CH_3 are equivalent and that set is represented by the same signal, but a CH_3 will always be different than a CH_2 or a CH. A hydrogen close to a functional group will be different than a hydrogen farther away. A CH_2 next to a CH_3 will be different than a CH_2 next to another CH_2. Again, regardless of how slight the difference, it is enough to make each set of equivalent hydrogens distinct from each other. And each set of equivalent hydrogens will have a distinct signal in the NMR spectrum. **Figure 15.15** shows some example molecules and the sets of equivalent hydrogens in each.

4 sets of equivalent hydrogens (**4** signals) | **3** sets of equivalent hydrogens (**3** signals) | **4** sets of equivalent hydrogens (**4** signals) | **10** sets of equivalent hydrogens (**10** signals)

Figure 15.15 Unique sets of equivalent hydrogens give distinct signals in the NMR spectrum.

To better understand what it means to be a distinct set of equivalent hydrogens, let's look at examples of equivalent hydrogens. Consider the ester, with a molecular formula of $C_4H_8O_2$, shown in Figure 15.15. It has eight hydrogens, and yet it has only three sets of equivalent hydrogens. This is because hydrogens on the same carbon are usually equivalent, especially when there is free rotation around bonds and when there is no stereocenter. So the three hydrogens of a CH_3 and the two hydrogens of a CH_2 are in the same chemical environment, feel the same magnetic field, and are therefore equivalent. The second way for hydrogens to be equivalent is via symmetry within the molecule. Especially with achiral molecules, hydrogens on opposite sides of a symmetry plane are equivalent, belong to the same set of equivalent hydrogens, and will all be represented by the same signal. **Figure 15.16** shows examples of molecules containing equivalent hydrogens as a result of symmetry.

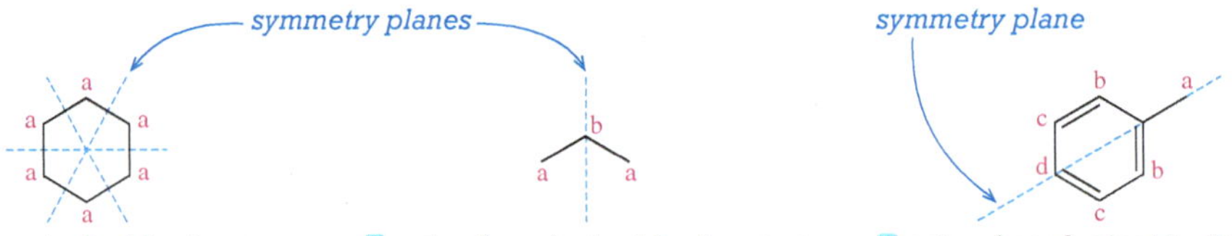

1 set of equivalent hydrogens (**1** signal) | **2** sets of equivalent hydrogens (**2** signals) | **4** sets of equivalent hydrogens (**4** signals)

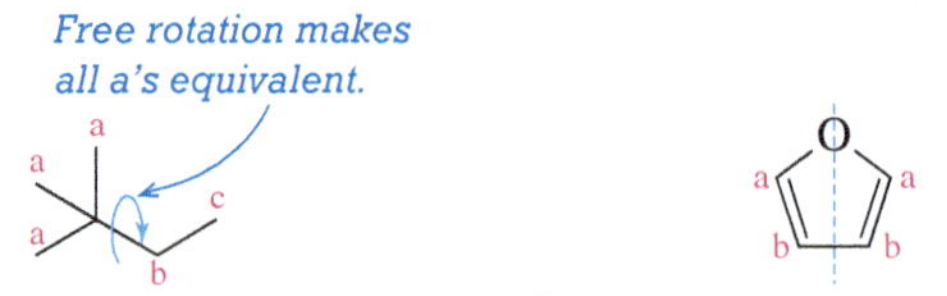

3 sets of equivalent hydrogens (**3** signals) | **2** sets of equivalent hydrogens (**2** signals)

Figure 15.16 Symmetrical molecules have fewer sets of equivalent hydrogens.

Assessment 15.14 How many sets of equivalent hydrogens are in each molecule shown?

(a)

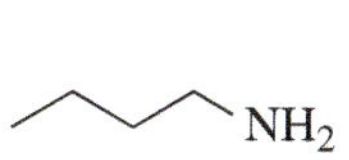

(b)

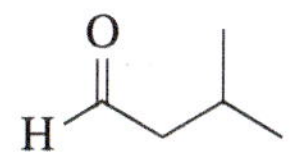

(c)

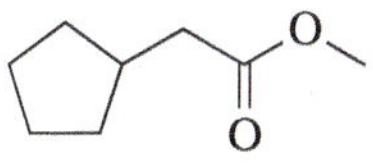

Assessment 15.15 How many unique ^{1}H NMR signals would you expect in an NMR spectrum for the following molecules?

(a)

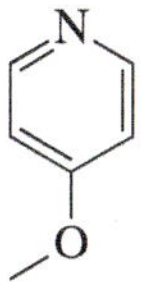

(b)

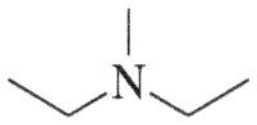

(c)

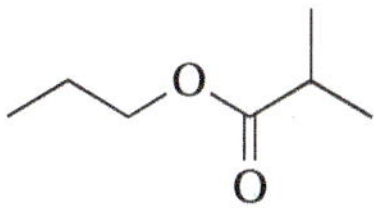

Assessment 15.16 How many sets of equivalent hydrogens are present in the molecule that resulted in this NMR spectrum? [*Recall that some signals can be split into multiple peaks—they are still just one signal.*]

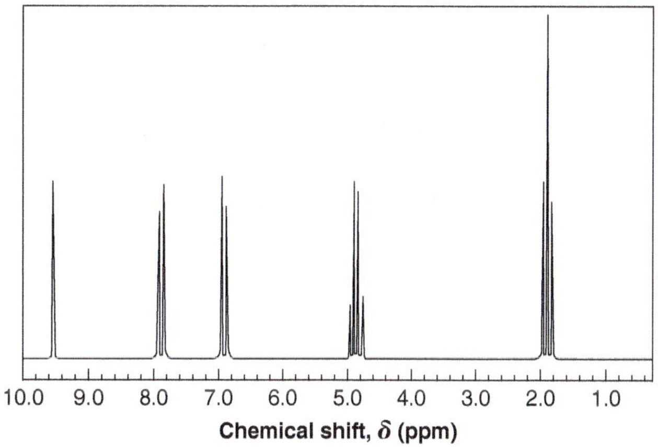

Assessment 15.17 (**LOOKING AHEAD**) If rotation is restricted, as in the case of the molecule shown, the hydrogens labeled a and b are nonequivalent. Why?

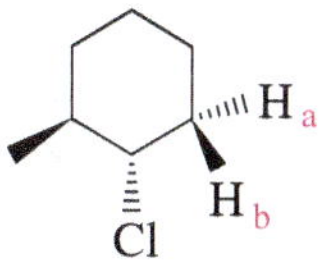

Assessment 15.18 Without worrying about the relative location of the signals (i.e., the chemical shift) or the splitting patterns, draw a spectrum of the following molecule. Be sure to label each signal based on the set of equivalent hydrogens to which it corresponds. [*We expand on this question in future assessments.*]

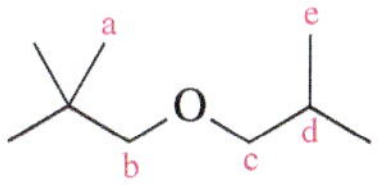

15.4.2 How many hydrogens in each set: the integration

As mentioned previously, the ester in Figure 15.15 has a molecular formula of $C_4H_8O_2$ but exhibits only three signals in its NMR spectrum. The NMR detects one quantum of energy for each individual hydrogen as it relaxes from β to α. Because the three hydrogens of the C_a methyl group are in the same chemical environment, there will be one quantum of energy detected for each hydrogen. This results in a signal that is three times more intense than a signal for just one hydrogen atom. Within the spectrum, this intensity is displayed as the integral (**Figure 15.17**) and is set relative to the other peaks in the spectrum. [*Note: Because the computer doesn't know what molecule is in the spectrometer, these integral values will often be relative, not absolute.*] The integration tells you how many hydrogens are in each set of equivalent hydrogens.

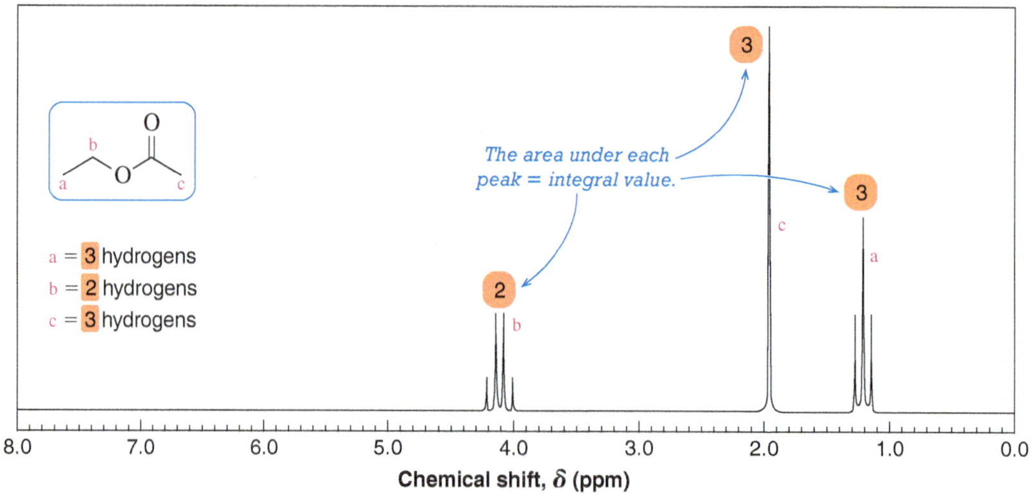

Figure 15.17 The integration tells the relative number of hydrogens represented by each signal.

Assessment 15.19 For the molecules shown, give the number of signals expected and the relative ratio of the signal integrations.

(a) (b)

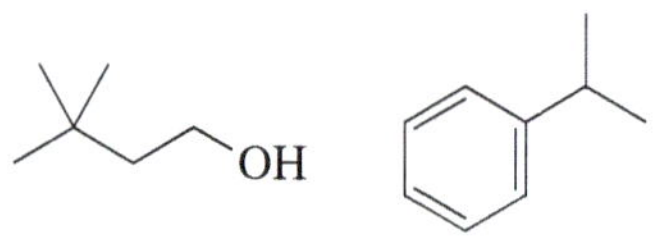

Assessment 15.20 Without worrying about the relative location of the signals (i.e., the chemical shift) or the splitting patterns, draw a spectrum of the following molecule, being sure to indicate the integration of each peak. Label each signal based on the set of equivalent hydrogens to which it corresponds. [*We expand on this question in future assessments.*]

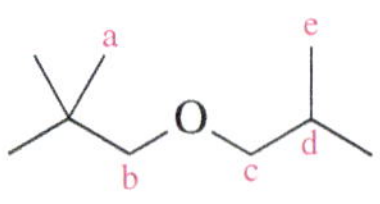

15.4.3 What functional group it is close to: the location

Until now, we have simply accepted that hydrogens in different chemical environments feel a different magnetic field. Finally, we can discuss why this is. A collection of nuclei that don't interact with each other "feel" only the external magnetic field. However, nuclei in a molecule are surrounded by electrons that, as spinning, charged particles, also create their own magnetic field. Depending on the type of electrons, these induced magnetic fields can work with or against the external magnetic field. If the spinning electrons work *against* the magnetic field, nearby nuclei will feel a smaller magnetic field, moving them to the left in our favorite figure (Figure 15.10). This results in a smaller ΔE between α and β and a lower resonance frequency. Stated another way, it will require less energy to flip the nuclear spin (**Figure 15.18**). On the other hand, spinning electrons working *with* the magnetic field will create a larger magnetic field for nearby nuclei, moving them to the right and creating a larger ΔE between α and β and a higher resonance frequency. It will therefore require more energy to flip the nuclear spin.

1. *Chemical shift:* As you saw in previous spectra, the position of the signals in the spectrum is referred to as the **chemical shift (δ)** and is expressed in units of parts per million (ppm). The chemical shift is a simplified way of telling how different the resonance frequency of a given nucleus is relative to the resonance frequency of a standard. The most common standard used in NMR is tetramethylsilane (TMS), which is defined to have a resonance frequency of 0 MHz. Deuterated chloroform ($CDCl_3$), the most common NMR solvent, is often spiked with 1% TMS to ensure that the signal for TMS is present in all spectra. Equation 15.5 shows how the chemical shift is calculated.

$$\text{chemical shift, } \delta \text{ (ppm)} = \frac{\text{shift from TMS (Hz)}}{\text{spectrometer frequency (MHz)}} \qquad \text{Equation 15.5}$$

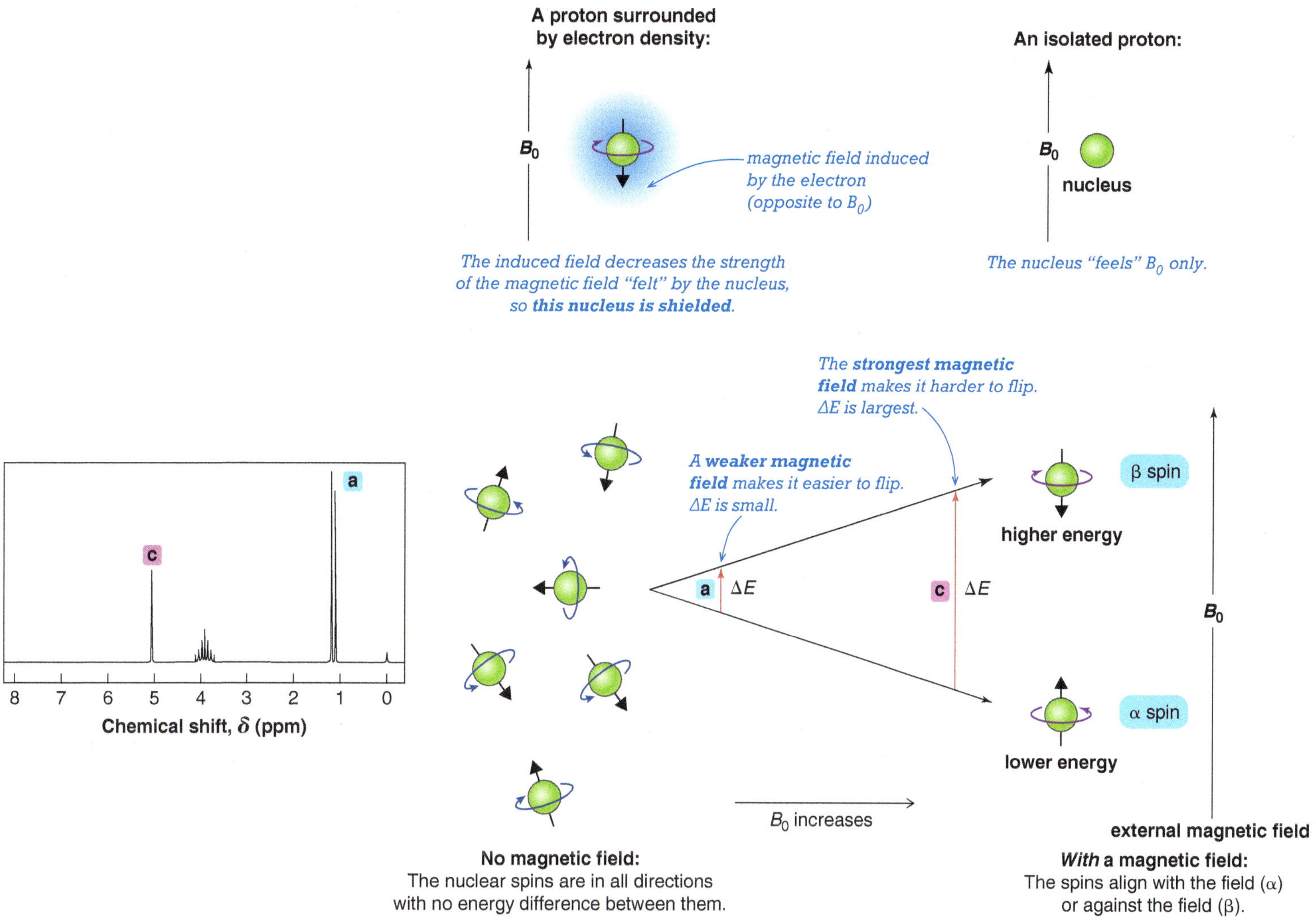

Figure 15.18 The influence of neighboring electrons on nuclei. Electrons that reduce the magnetic field make it easier to flip nearby nuclei.

$$\text{Chemical shift, } \delta \text{ (in ppm)} = \frac{\text{shift downfield from TMS (in Hz)}}{\text{spectrometer frequency (in MHz)}}$$

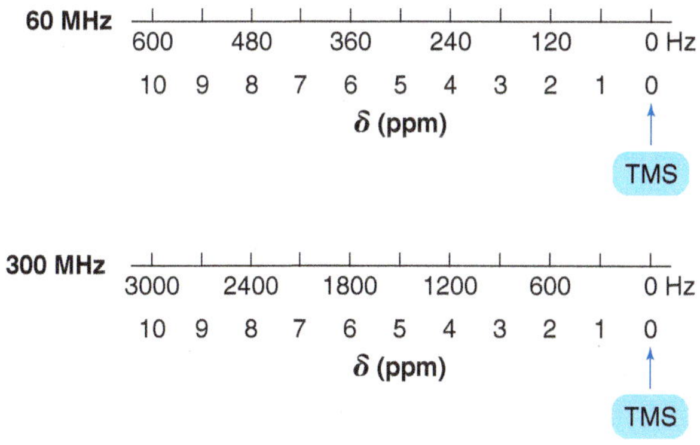

Figure 15.19 Regardless of spectrometer frequency, signals for identical hydrogens will have identical chemical shifts.

NMR spectrometers come in a variety of frequencies, ranging from 60 MHz to 1 GHz (1000 MHz), with the most common instruments being either 300 MHz or 500 MHz. The major difference between the lower and higher frequency instruments is the resolution and the time required to acquire a spectrum. Figure 15.19 shows that, regardless of the spectrometer frequency, the same chemical shift value will be measured. For example, a hydrogen with a resonance frequency 60 Hz higher than TMS will appear at 1 ppm because 60 Hz is $1/10^6$ of 60 MHz. When this same signal is recorded on a 300-MHz frequency, it will appear 300 Hz higher than TMS. The signal will also appear at 1 ppm because 300 Hz is $1/10^6$ of 300 MHz. Recall, from Table 15.2, the chemical shift values for the common functional groups. [*Go back and look at Table 15.2 again.*]

Assessment 15.21 (a) Calculate the resonance frequency of an aldehydic proton (δ 9.3 ppm) if it is detected on a 60-MHz NMR spectrometer. (b) What if it were detected on a 300-MHz instrument?

2. *Effects of electronegativity on chemical shift:* Electrons in σ bonds work against the magnetic field, making it easier for nearby nuclei to flip. Stated another way, we say that σ-bonded electrons **shield** nuclei from the external magnetic field. When nuclei are shielded or **deshielded** to a different degree, they will have different resonance frequencies. This results in the signals for the nuclei appearing at different chemical shifts.

Table 15.5 lists chemical shift values for the methyl hydrogens in a variety of related compounds, including tetramethylsilane, the standard against which all other compounds are measured. Recall that σ-bonded electrons shield a nucleus from the external magnetic field. It makes sense, then, that a highly electronegative atom, polarizing the σ bond by withdrawing electrons, will deshield the nucleus. In doing so, the nucleus feels a greater portion of the external magnetic field and becomes harder to flip, requiring a higher resonance frequency. The terms "shielded" and "deshielded" are often used to indicate relative location in the NMR spectrum. Deshielded signals are at higher chemical shifts, whereas shielded signals are at lower chemical shifts. In older terminology [*that you should be familiar with*], **downfield** is synonymous with deshielded, whereas **upfield** is synonymous with shielded. [*These terms are used interchangeably in this text to get you used to hearing both.*]

Table 15.5 Chemical shift differences as a result of electronegativity.

	CH_3—F	CH_3—OH	CH_3—Cl	CH_3—Br	CH_3—I	CH_3—CH_3	CH_3—SiR_3
Electronegativity of X	4.0	3.4	3.2	3.0	2.7	2.6	1.9
Chemical shift of methyl hydrogens (ppm)	δ 4.3	δ 3.4	δ 3.0	δ 2.7	δ 2.2	δ 0.9	δ 0.0 (defined)

Table 15.5 helps explain why TMS is used as the internal standard. The C—Si bond is one of the few C—X bonds where the carbon is more electronegative. As a result, the C—Si bond is polarized in the direction of carbon, making it partially negative [Figure 15.20(a)]. With more electron density around the carbon, the methyl hydrogens of TMS are highly shielded from the external magnetic field. An additional benefit, as shown in Figure 15.20(b), is that the TMS peak, a singlet, is in a region where few other hydrogens absorb. As a result, the standard is never confused as arising from the molecule being analyzed.

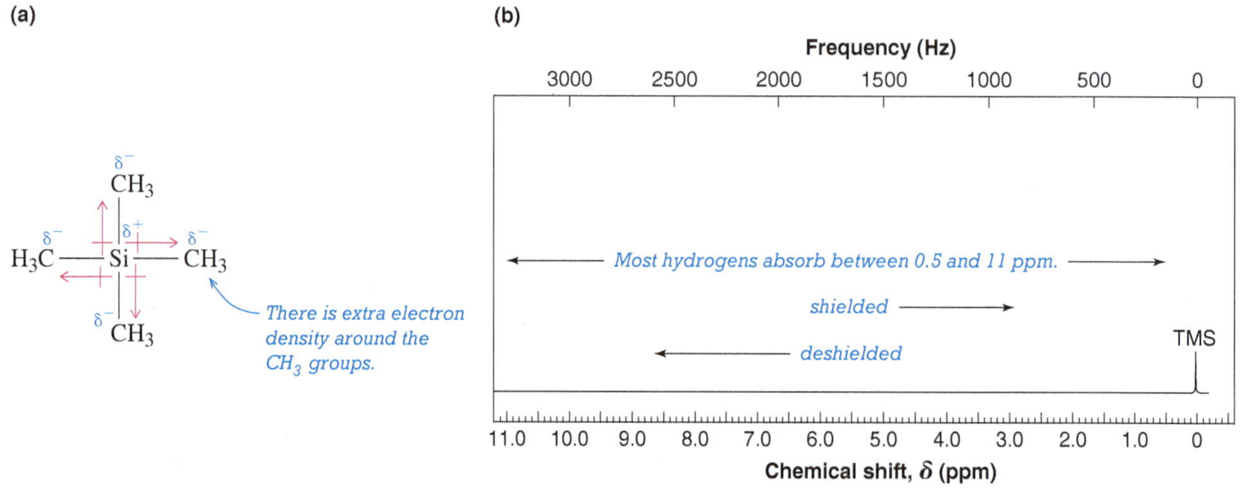

Figure 15.20 TMS is the internal standard (a) because silicon releases electrons to the methyl groups and (b) TMS appears in an area of the spectrum where few hydrogens absorb.

The effect of an electronegative atom can be clearly shown in the ^1H NMR spectrum of 1-chlorobutane (Figure 15.21). The hydrogens on carbon **d**, closest to the electronegative chlorine, are highly deshielded and show up around 3.6 ppm. As the hydrogens get farther away from the electronegative atom, the chemical shift decreases. The methyl hydrogens at **a** barely feel any effect from the chlorine, appearing in the range expected for an alkyl hydrogen. This is consistent with the effects of induction dropping off with distance, as discussed in Section 4.4.4.

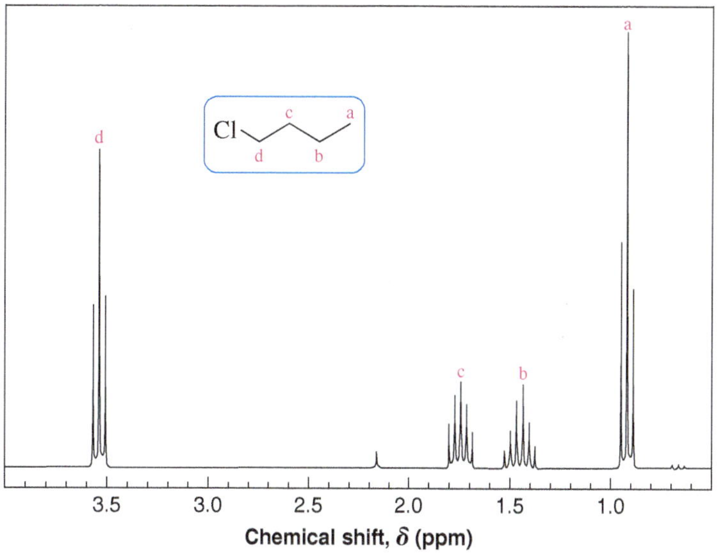

Figure 15.21 Hydrogens closer to electronegative atoms are more deshielded.

3. ***Chemical shift of aromatic, alkenyl, and aldehydic hydrogens:*** Table 15.2 shows that aromatic hydrogens (δ 7–8 ppm) are significantly deshielded compared to the corresponding alkyl hydrogens (~δ 0.9 ppm). This suggests that something other than electronegativity is operating here. When benzene, or any aromatic ring, is placed in an external magnetic field, a ring current (**Figure 15.22**) is induced for the six circulating π electrons. Like spinning electrons, this ring current produces its own magnetic field. At the center of the benzene ring, the induced magnetic field opposes the external magnetic field. On the outer edges of benzene, however, the magnetic field has turned in the opposite direction. As a result, the hydrogens of benzene are in an environment where the induced field is working with the external magnetic field. This results in deshielding of the hydrogens, causing them to resonate at a higher frequency.

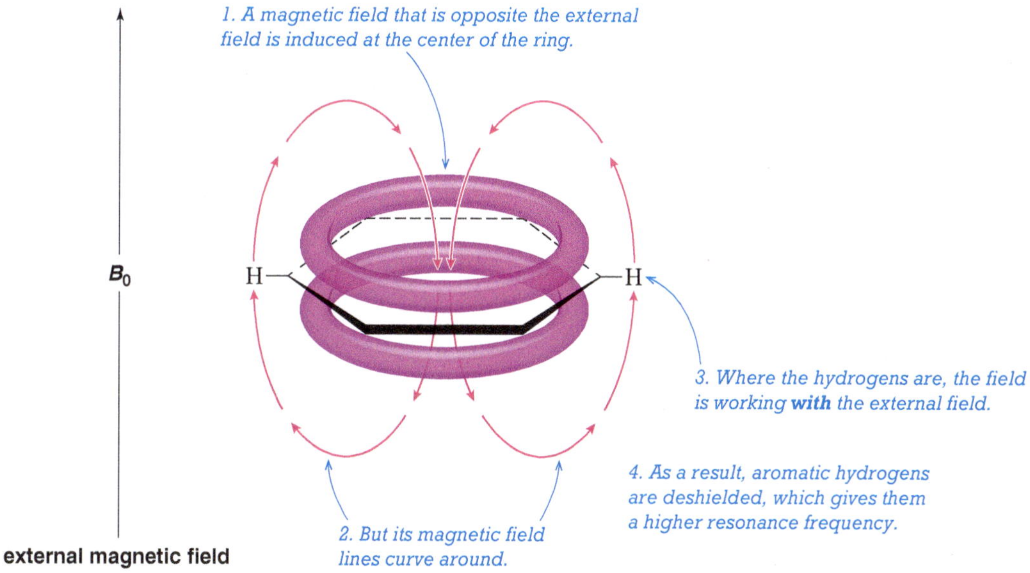

Figure 15.22 The induced magnetic field in benzene works with the external magnetic field to deshield the hydrogens on an aromatic ring.

For similar reasons, alkenyl hydrogens (δ 5–6 ppm) are deshielded relative to alkyl hydrogens. The induced magnetic field of the π electrons works against the magnetic field at the center of the π bond, but reinforces it on the edges where the alkenyl hydrogens are [**Figure 15.23(a)**]. Once more, this deshields alkenyl hydrogens, thereby increasing their resonance frequency. Given the electronegativity of oxygen, we might expect aldehydic hydrogens to be similar to hydrogens on sp^3-hybridized C—X. And yet the aldehydic proton, as shown in Table 15.2, is between δ 9 and 10 ppm. Like the C—C π bond, the external magnetic field induces a magnetic field by circulation of the C—O π bond. The heavily deshielded aldehydic hydrogen requires a much higher than expected resonance frequency.

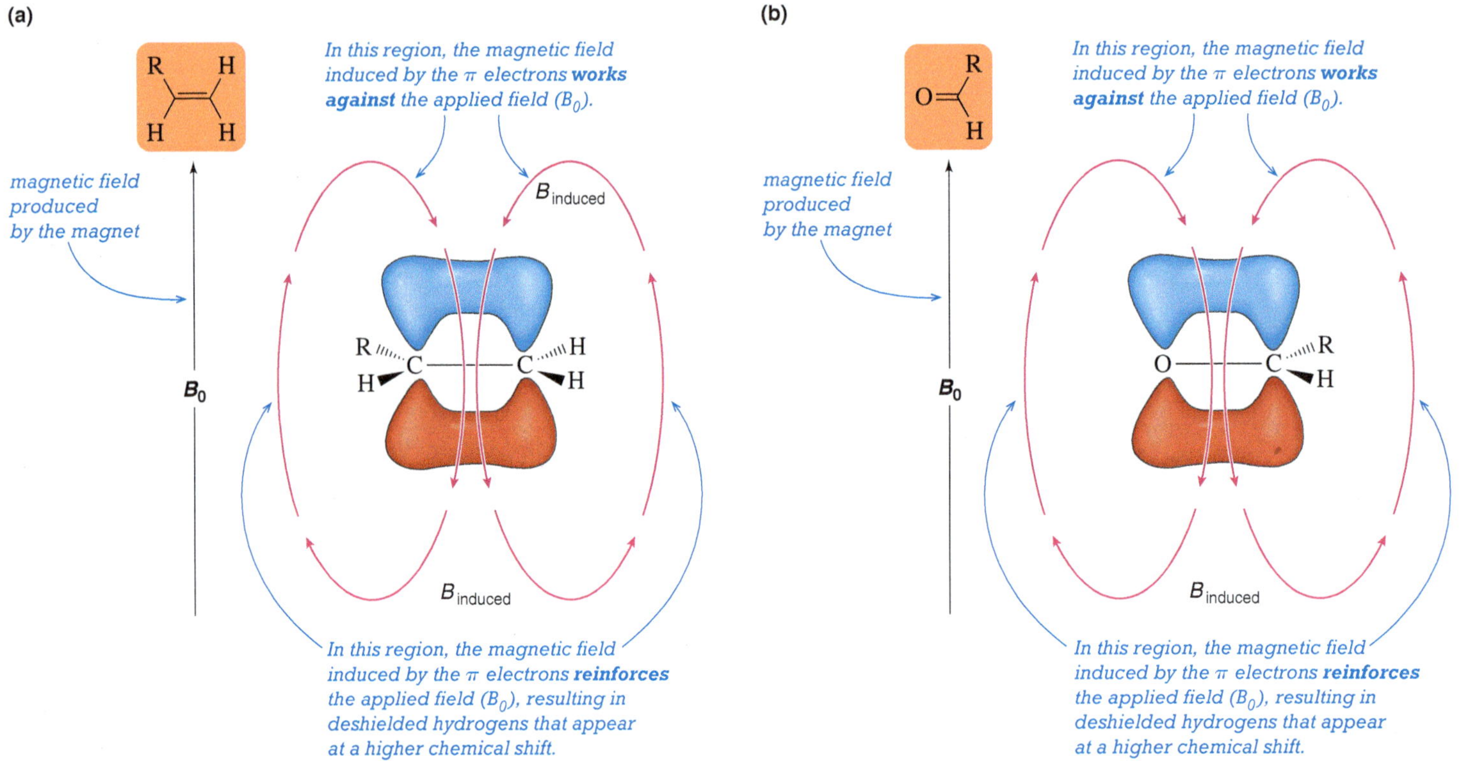

Figure 15.23 The induced magnetic field works with the external magnetic field to deshield (a) alkenyl hydrogens and (b) aldehydic hydrogens.

Assessment 15.22 [18]-Annulene, like benzene in Figure 15.22, is an aromatic ring (see Chapter 23 for more on this). The hydrogens at the center of the ring appear at a chemical shift of −1.8 ppm, whereas the outer hydrogens appear at a chemical shift of 8.9 ppm. Why?

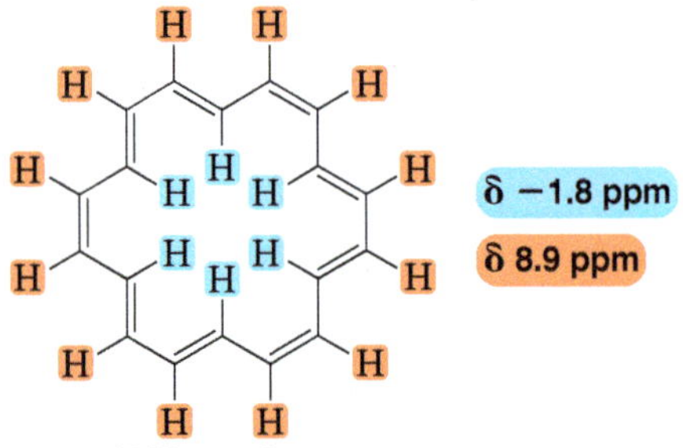

[18]-annulene

Assessment 15.23 The methyl hydrogens in propane appear at a chemical shift of 0.9 ppm, whereas the methyl hydrogens of propene appear around 2.5 ppm. Explain.

4. ***Chemical shift of acetylenic hydrogens:*** Compared to aromatic, alkenyl, and aldehydic protons, the acetylenic hydrogen (the hydrogen attached to an alkyne carbon) is more shielded, appearing farther upfield (~δ 2.5 ppm). The principle is similar to what happens with aromatic rings, alkenes, and aldehydes in spite of it seeming to be the opposite.

The external magnetic field induces a magnetic field that opposes it along an axis running through the C≡C bond. On the outer edges of the alkyne, however, the magnetic field has turned in the opposite direction. Unlike the earlier examples, the acetylenic hydrogen is in the region where the induced magnetic field works *against* the external field (**Figure 15.24**). As a result, the hydrogen is *shielded* and experiences a *weaker* magnetic field. The weaker magnetic field reduces the energy between the α and β spin states, lowering the resonance frequency of acetylenic hydrogens relative to other functional groups.

Figure 15.24 The induced magnetic field in acetylene works against the external magnetic field to shield the acetylenic hydrogen.

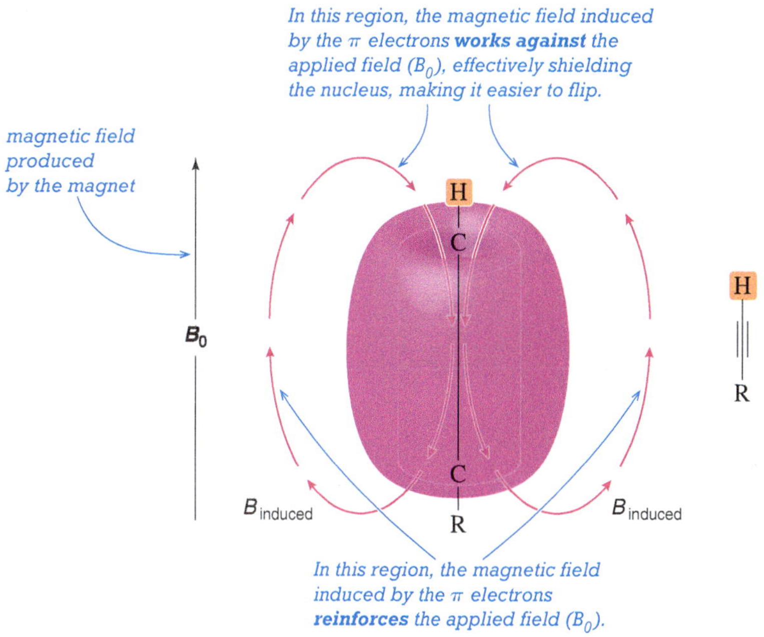

5. ***Chemical shift of amines, alcohols, carboxylic acids:*** These three functional groups are unique in their chemical shifts. [*Where and why have we seen these grouped together? See Section 14.4.*] Hydroxyl, amino, and carboxyl protons can all form hydrogen bonds. Just as hydrogen bonding made the O—H and N—H stretching bands broad in an IR spectrum, the ability to hydrogen bond results in highly variable chemical shifts for these protons. As seen in Table 15.2, hydroxyl protons appear as a broad signal between 2 and 6 ppm. Given the variability of this signal, it is difficult to assign the hydroxyl group to an unknown molecule in the absence of additional experiments. [*What other technique could you use to determine whether an alcohol is present in your unknown?*]

Consider an unknown that gives the IR and ^{1}H NMR spectra shown in **Figure 15.25**. The IR spectrum provides conclusive evidence that there is a hydroxyl group present, but it is difficult to know which peak in the ^{1}H NMR spectrum corresponds to the hydroxyl hydrogen.

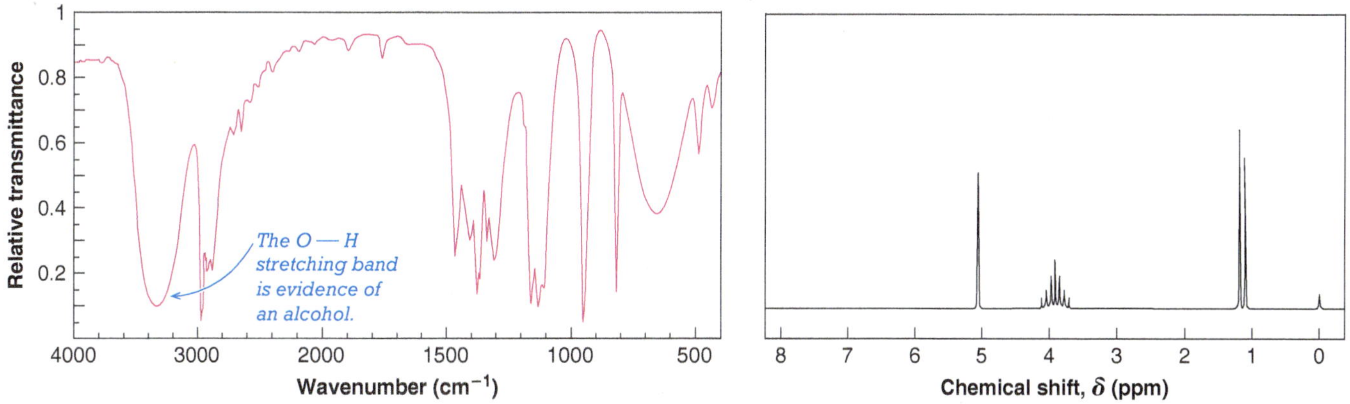

Figure 15.25 The IR and ^{1}H NMR spectra for an unknown alcohol. Which peak in the ^{1}H NMR spectrum corresponds to the hydroxyl proton?

Let's think like a chemist and design an experiment that allows us to determine which peak corresponds to the hydroxyl proton.

Thinking Like a Chemist

Knowing that there is a hydroxyl group in the unknown molecule, we can isolate which peak corresponds to it by modifying the molecule such that the peak corresponding to the hydroxyl proton disappears. How might we do this? To solve a problem like this, we need to figure out how a hydroxyl proton is different from hydrogens bound to carbon and manipulate it appropriately. Given that the proton is slightly acidic, it is reasonable to think about deprotonating it to form an alkoxide (**Figure 15.26**). And that would likely work, but we don't usually take NMR spectra of charged molecules. Furthermore, deprotonating it would change the chemical environment of nearby protons, likely causing many peaks in the spectrum to move.

Besides being deprotonated, what else does a hydroxyl proton do that an alkyl hydrogen does not? Hydroxyl groups are good hydrogen-bond donors and acceptors. Recall from Chapter 14 that hydrogen bonding is not really a static process. This implies that alcohols in solution will pass their protons around through exchange of hydrogen bonds, as shown in **Figure 15.27**. Notably, the fact that this exchange happens faster than can be detected by the NMR spectrometer results in the broadening of the peak and its varied location.

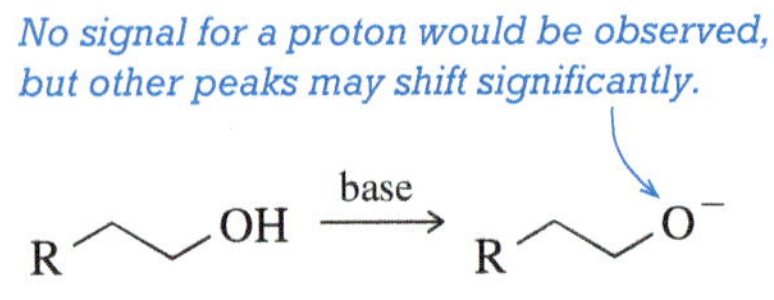

Figure 15.26 Deprotonation of the unknown alcohol would cause the hydroxyl signal to disappear, but would likely move other peaks, too.

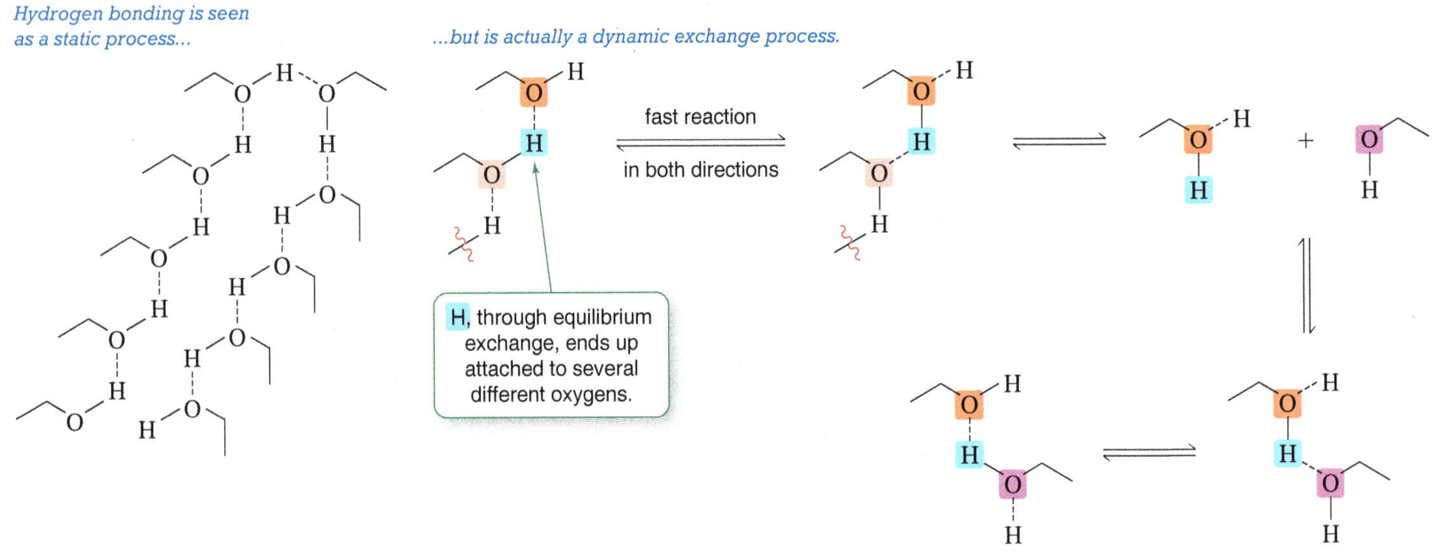

Figure 15.27 Hydrogen bonding is an exchange process.

If alcohols hydrogen bond in an exchange process, is there something we could exchange "protons" with that would make the signal disappear? Do you know of an atom that acts like hydrogen in every way except that it doesn't show up in a ^{1}H NMR spectrum? [*Think about what makes an atom NMR active from Section 15.3.1.*] Deuterium is used specifically in NMR solvents because it doesn't show up in the ^{1}H NMR spectrum. By adding a drop of D_2O, the alcohol can hydrogen bond with it, exchanging an H for a D (**Figure 15.28**). Thus,

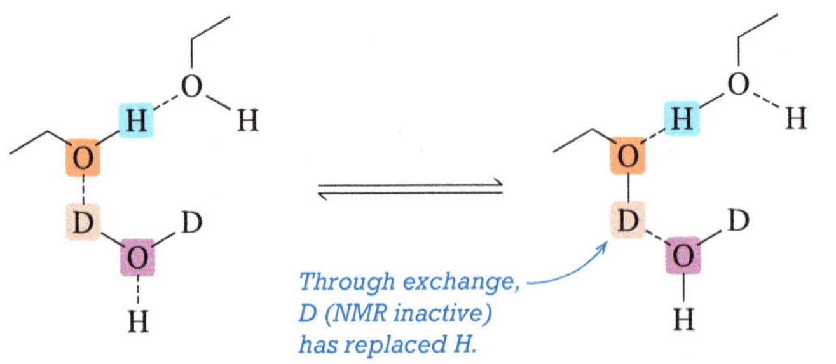

Figure 15.28 Deuterium replaces hydrogen through a hydrogen-bonding exchange process.

any signal for a hydrogen-bonding hydrogen will disappear. Using this experiment, the spectrum in **Figure 15.29(a)** becomes the spectrum in **Figure 15.29(b)**. The peak at 5.1 ppm disappears, suggesting that it was probably the signal for the hydroxyl proton. ■

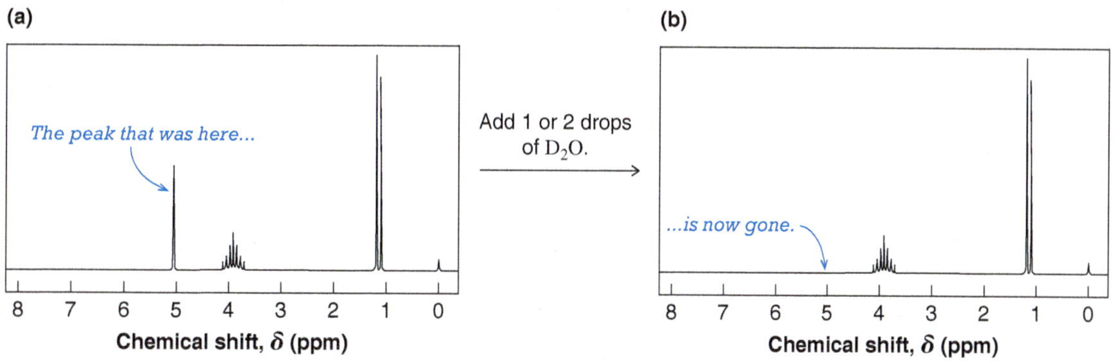

Figure 15.29 Adding D_2O to an NMR solution causes the exchange of a deuterium atom for any hydrogen that can hydrogen bond. (a) The peak at 5.1 ppm is present. (b) The peak at 5.1 ppm is now gone. Conclusion: The signal at 5.1 ppm was the signal for the hydroxyl proton.

Because nitrogen is less electronegative, amino protons appear over a slightly narrower range (1.5 to 4 ppm), which you would expect based on lesser hydrogen bonding relative to hydroxyl groups [**Figure 15.30(a)**]. The signal for a carboxylic acid shows up around 11 ppm, but because it hydrogen bonds so extensively, it is often so broad that it blends in with the baseline, making it easy to miss unless you are looking for it [**Figure 15.30(b)**]. [*Fortunately, the presence of a carboxylic acid can be conclusively determined with an IR spectrum.*]

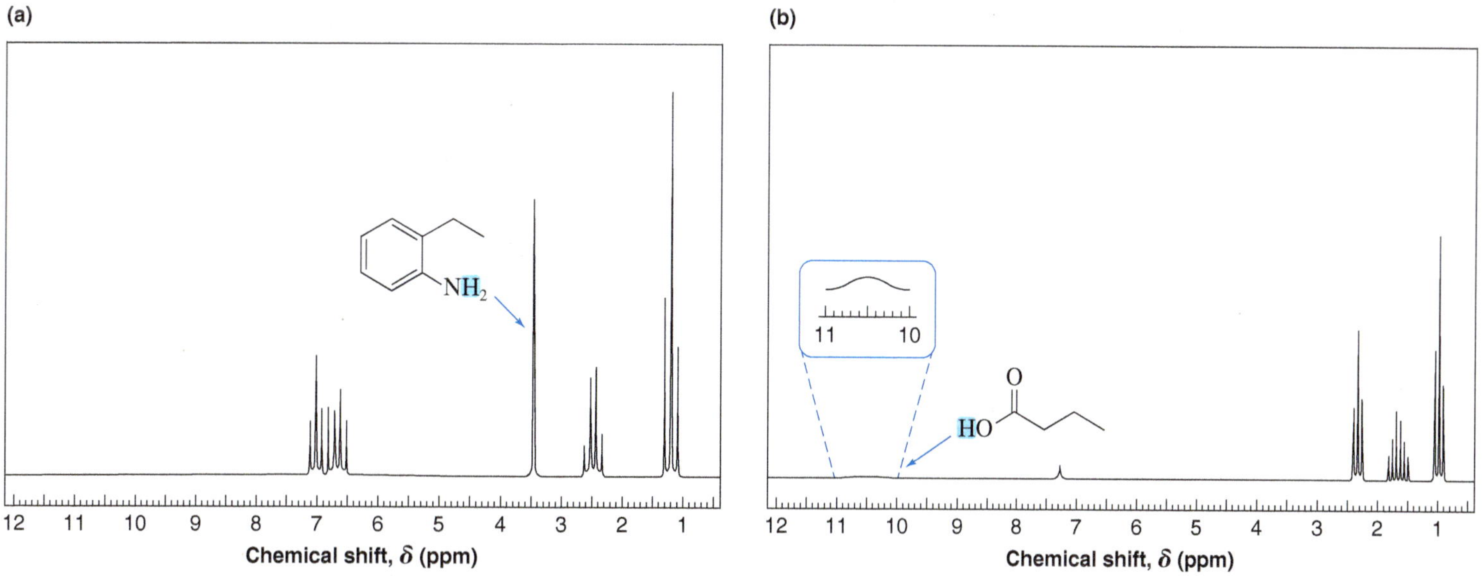

Figure 15.30 Other hydrogen-bonding functional groups. (a) The amino hydrogen is between 1.5 and 4 ppm. (b) The carboxylic acid proton often blends with the baseline around 11 ppm.

15.4.3 SUMMARY

- The chemical shift of a hydrogen depends on the strength of the magnetic field felt by that hydrogen and is influenced by neighboring functional groups. As a result, the chemical shift tells us how close the hydrogen is to a particular functional group.

Assessment 15.24 Assuming that the hydrogen in fluoromethane feels the magnetic field indicated by the red arrow, show where an (a) aldehydic, (b) alkenyl, (c) aromatic, and (d) alkynyl hydrogen would appear on a modified Figure 15.10. [*Would the magnetic field felt by these other hydrogens be stronger or weaker than the one felt by CH_3F?*]

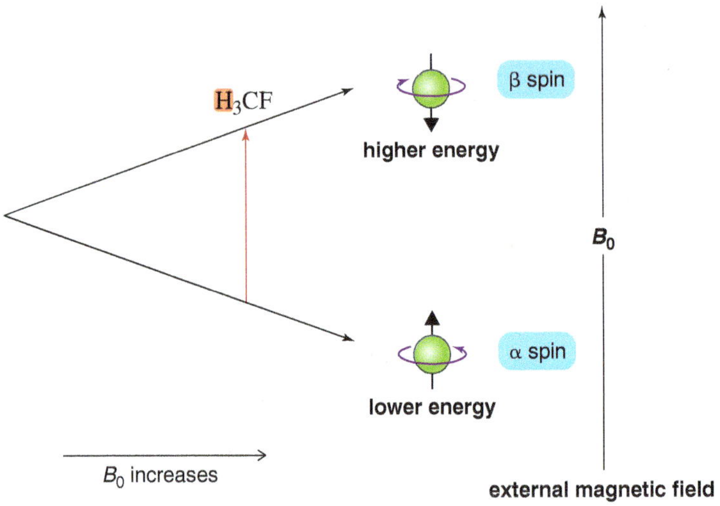

Assessment 15.25 Rank hydrogens (a)–(j) in terms of their chemical shifts (1 = lowest, 10 = highest). If a hydrogen appears at a variable chemical shift, you can exclude those from the ranking. [*There are lots of correct answers here, so make your best guess.*]

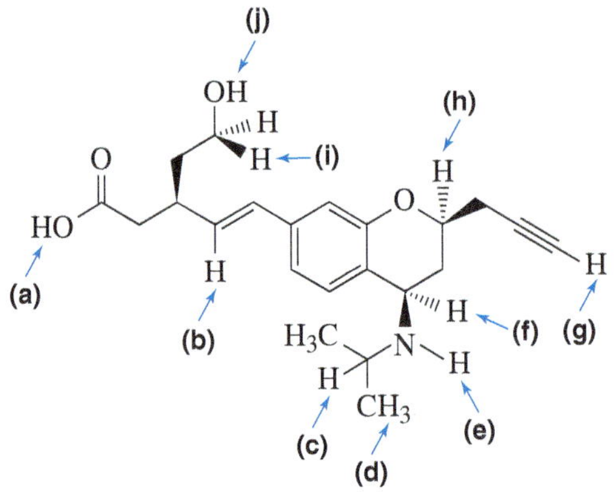

Assessment 15.26 Without worrying about the splitting patterns, draw a spectrum of the following molecule, being sure to indicate the integration and chemical shift of each peak. Label each signal based on the set of equivalent hydrogens to which it corresponds. [*We expand on this question in future assessments.*]

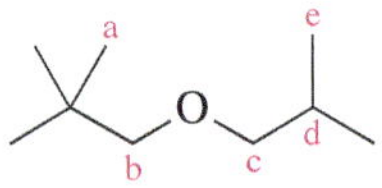

Assessment 15.27 An unknown compound (C_7H_6O) gives the IR spectrum shown here. At what chemical shifts would you expect to see signals in the ^{1}H NMR spectrum?

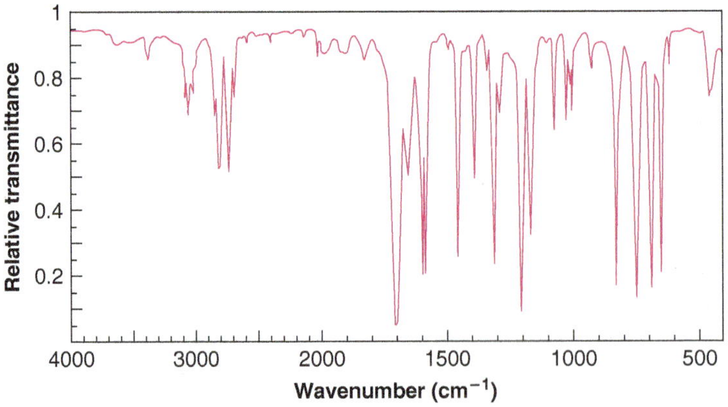

15-B CHEMISTRY IN CONTEXT: Natural Product Isolation

^{1}H NMR of curacin A

Continuing our study of curacin A from Chemistry in Context 15-A, we can now look at its ^{1}H NMR spectrum. From the index of hydrogen deficiency of 7 calculated in Assessment 15.9, we know to look for hydrogen signals that indicate the presence of double bonds, triple bonds, and rings. The structure of curacin A is shown in **Figure 15.31**, but when chemists were analyzing the spectrum, they would not have had access to this information. [*In other words, spectral analysis is easier in hindsight.*] Analysis of the ^{1}H NMR spectrum shows that there are nine hydrogens in the alkene region. [*We would discover later that the signal at 5.1 ppm is not an alkene hydrogen.*] From this, we could expect curacin A to be a highly unsaturated molecule containing as many as four alkenes.

Lots of other information can be gained just by analyzing the chemical shifts of the peaks. There are at least five hydrogens on carbons bound to electronegative atoms, with the peak at 3.2 ppm representing a methyl group. In the range of 1.7 to 2.4 ppm, there are several allylic hydrogens (on carbons attached to an alkene). This gives more confirmation of the presence of alkenes in curacin A. Finally, the three-hydrogen doublet at 1.2 ppm indicates a methyl group with a single neighboring hydrogen. [*Read on to see how we know this.*]

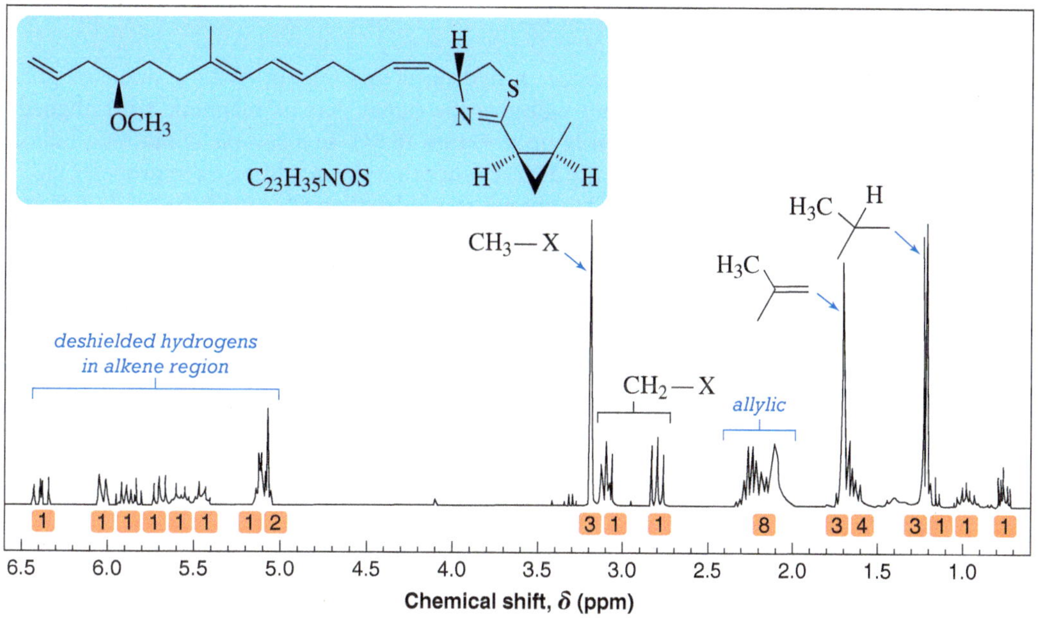

Figure 15.31 **^{1}H NMR spectrum of curacin A.**

15.4.4 How many neighboring hydrogens: the splitting

So far, we have discussed the integration and chemical shift of a signal. This allows us to know what pieces of the puzzle are contained within a spectrum. But until now, there has been no way for us to link these pieces. [*Imagine puzzle pieces lacking an image: We know what pieces we have, but connecting them would be a guessing game.*] The splitting, or **multiplicity**, of a signal allows us to link the pieces because it takes the information we have about one signal and connects it to another signal by illustrating the number of hydrogen neighbors each hydrogen has.

Formally, the **$N + 1$ rule** says that a hydrogen with N neighbors will appear in the spectrum as an $N + 1$ multiplet (M), as shown in Equation 15.6.

$$N + 1 = M \qquad \text{Equation 15.6}$$

The statement of this rule begins with a known structure and predicts the appearance of the signal [**Figure 15.32(a)**]. More often, however, we will look at a signal and determine how many neighbors it has by subtracting 1 from the number of peaks in the signal. In this way, we simply rearrange Equation 15.6 [**Figure 15.32(b)**].

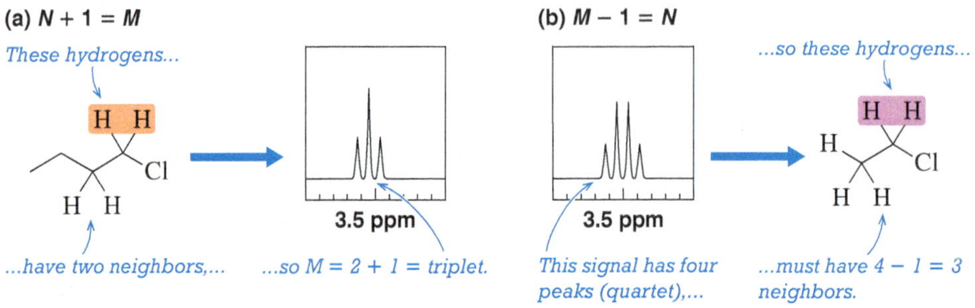

Figure 15.32 The multiplicity of a signal indicates the number of neighboring hydrogens. (a) Using $N + 1 = M$ to predict how a signal will appear based on the structure. (b) Using $M - 1 = N$ to predict a structure based on the multiplicity of the signal.

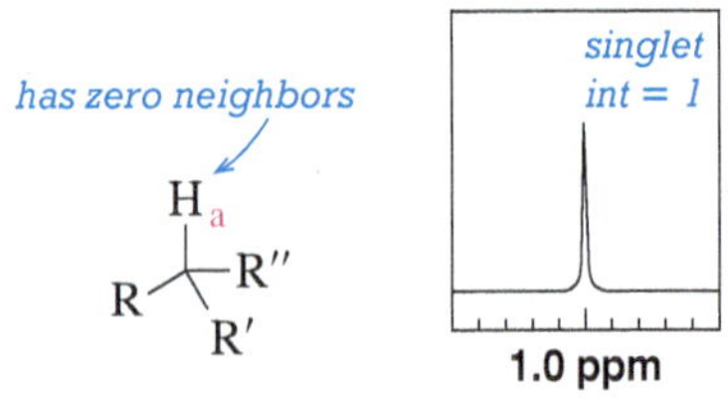

Figure 15.33 With zero neighbors, a hydrogen feels only the external magnetic field. It appears as a singlet in the spectrum.

Like everything here in Chapter 15, signal splitting can be explained using Figure 15.10 (our favorite figure). Just as spinning electrons create a magnetic field that is felt by nearby nuclei, other spinning nuclei create magnetic fields felt by nearby nuclei. First consider a hydrogen (H_a) with no neighboring hydrogens (**Figure 15.33**). Because it feels only the external magnetic field, every one of these hydrogens in a sample containing billions of molecules will resonate at the same chemical shift, let's say δ 1.0 ppm to make it simple. With zero neighbors ($N = 0$), $N + 1 = 1$.

Now, consider a hydrogen (δ 1.0 ppm) with a single neighbor and assume that H_a, the hydrogen that will give the signal, has its magnetic field aligned with the external magnetic field (arbitrarily up in **Figure 15.34**). In a sample containing billions and billions of molecules, there is a nearly 50% chance that the neighboring hydrogen (H_b) will be aligned with the field (up) and a 50% chance it will be aligned against the field (down). When the spins are aligned, the external magnetic field is enhanced, making H_a resonate at a higher frequency (requiring more energy to flip). When the spins are in opposite directions, the magnetic field is diminished, making H_a resonate at a lower frequency (requiring less energy to flip). And so, instead of getting a singlet at δ 1.0, a doublet will be produced with the height of the doublet peaks being in a 1:1 ratio. The 1:1 ratio is very important here because it represents the proportion of times the signal will appear at this frequency. With one neighbor ($N = 1$), $N + 1 = 2$.

Figure 15.34 With one neighbor, a hydrogen is influenced by the magnetic field of the neighboring hydrogen. It appears as a doublet (d) in the spectrum.

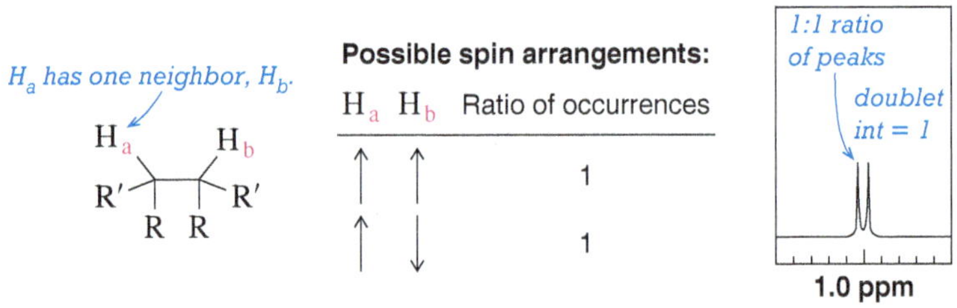

Assessment 15.28 In Figure 15.34, H_a was assumed to be "up." How does the analysis change if we assume instead that H_a is down?

Assessment 15.29 Though Figure 15.34 was concerned with the appearance of H_a, how would H_b appear in the spectrum?

Before we proceed, let's elaborate on the interaction between spin states. We can actually quantify this interaction, or **coupling** between neighboring nuclei. [*And we explain in Section 15.5.1 that the degree to which two nuclei couple depends on several factors.*] We can assume that every time the magnetic fields of two nuclei are aligned, the chemical shift increases by some value we'll call x (as a change in chemical shift, δ, we'll call it Δδ). When the magnetic fields of two nuclei are in opposite directions, the chemical shift will decrease by the same value ($\Delta\delta = x$). **Table 15.6** shows this analysis for a hydrogen with up to three neighbors. In every case, the distance between the peaks of the multiplets is $2x$. This value of $2x$ is defined as the **coupling constant** and is a measure of the interaction between the magnetic fields of neighboring hydrogens.

Table 15.6 Multiplicity as a function of neighboring hydrogens.

Number of neighbors (*N*)	Structure	H_a	H_b	$\Delta\delta$	Occurrences	Peak appearance
0		↑	No H_b	0	1	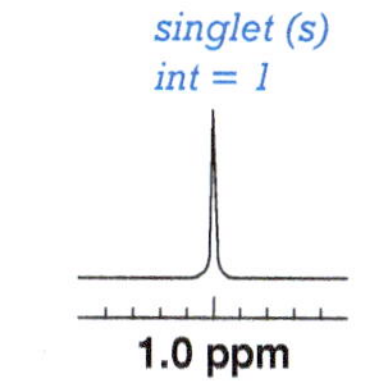
1	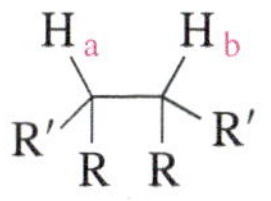	↑	↑	$+x$	1	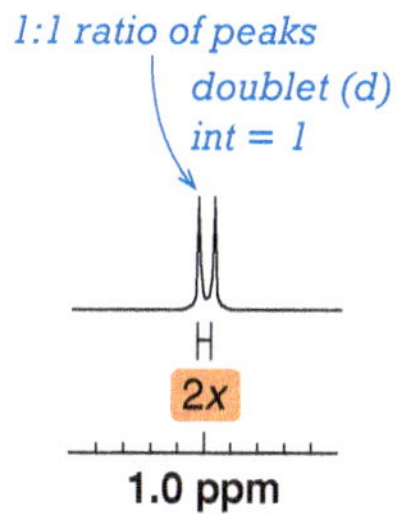
		↑	↓	$-x$	1	
2	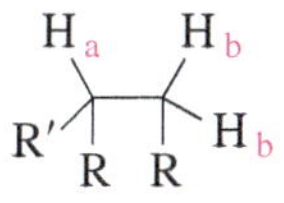	↑	↑ ↑	$+x + x = +2x$	1	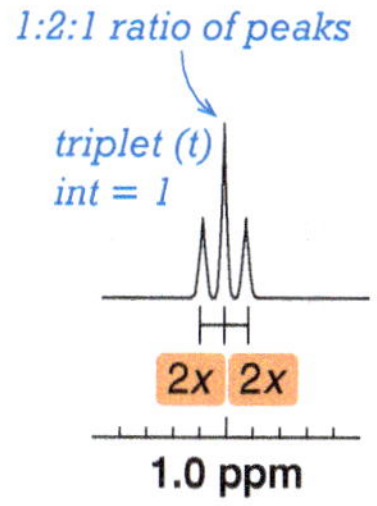
		↑	↑ ↓	$+x - x = 0$	2	
		↑	↓ ↑	$-x + x = 0$		
		↑	↓ ↓	$-x - x = -2x$	1	
3	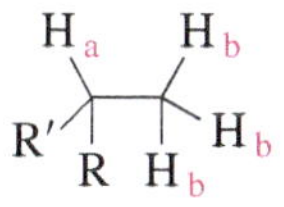	↑	↑ ↑ ↑	$+x + x + x = +3x$	1	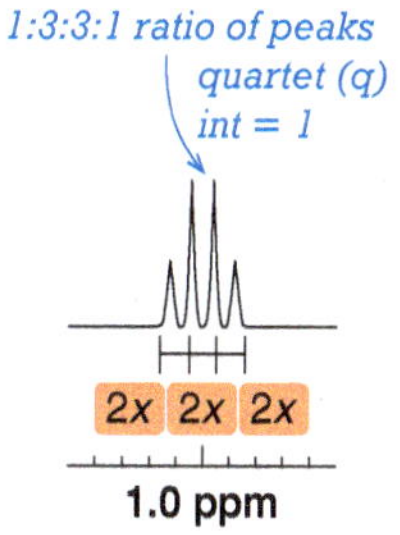
		↑	↑ ↑ ↓	$+x + x - x = x$	3	
		↑	↑ ↓ ↑	$+x - x + x = x$		
		↑	↓ ↑ ↑	$-x + x + x = x$		
		↑	↑ ↓ ↓	$+x - x - x = -x$	3	
		↑	↓ ↑ ↓	$-x + x - x = -x$		
		↑	↓ ↓ ↑	$-x - x + x = -x$		
		↑	↓ ↓ ↓	$-x - x - x = -3x$	1	

You may have noticed a pattern for the peak heights. As you increase the number of neighboring hydrogens, you are essentially moving to the next row in Pascal's triangle (**Figure 15.35**).

singlet (s)	1	0 neighbors
doublet (d)	1 1	1 neighbors
triplet (t)	1 2 1	2 neighbors
quartet (q)	1 3 3 1	3 neighbors
quintet (quint)	1 4 6 4 1	4 neighbors
sextet (sext)	1 5 10 10 5 1	5 neighbors
septet (sept)	1 6 15 20 15 6 1	6 neighbors
octet (oct)	1 7 21 35 35 21 7 1	7 neighbors

Figure 15.35 Pascal's triangle depicts the relative peak heights for the number of neighbors.

Assessment 15.30 (a) Draw all of the possible spin states to explain why a hydrogen with four neighbors appears as a quintet (quint, five peaks). (b) How is it possible that a hydrogen could have four neighbors?

Assessment 15.31 The spectrum provided has signals representing five sets of equivalent hydrogens. How many neighbors does each hydrogen have?

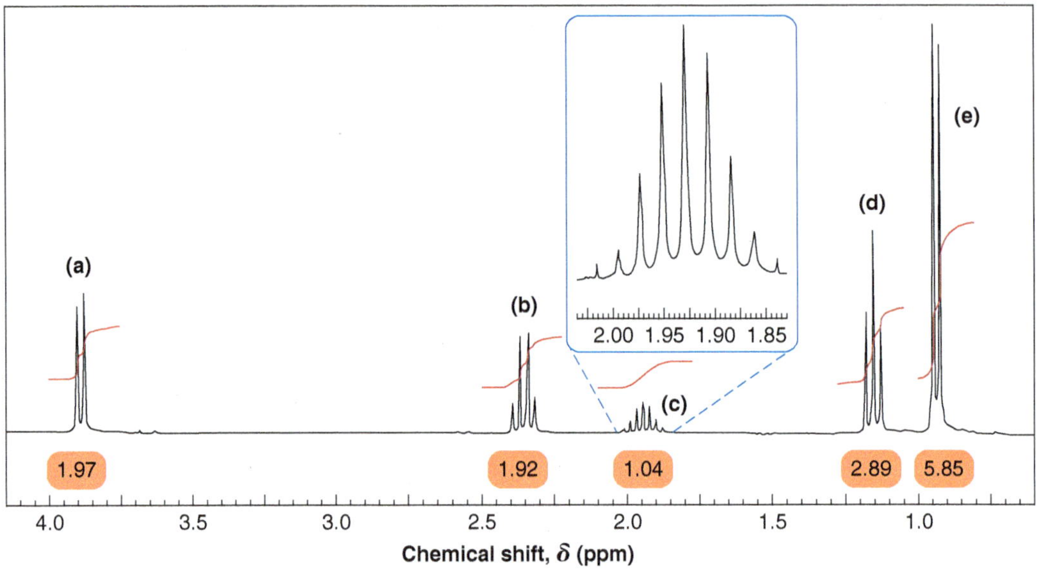

Assessment 15.32 Draw a spectrum of the following molecule, being sure to indicate the multiplicity, integration, and chemical shift of each peak. Label each signal based on the set of equivalent hydrogens to which it corresponds.

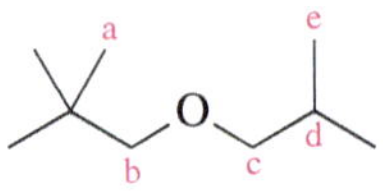

Assessment 15.33 Being able to recognize patterns of integration and multiplicity for common functional groups makes structure identification more efficient. Draw the pattern of integration and multiplicity you'd expect to see for each common alkyl group.

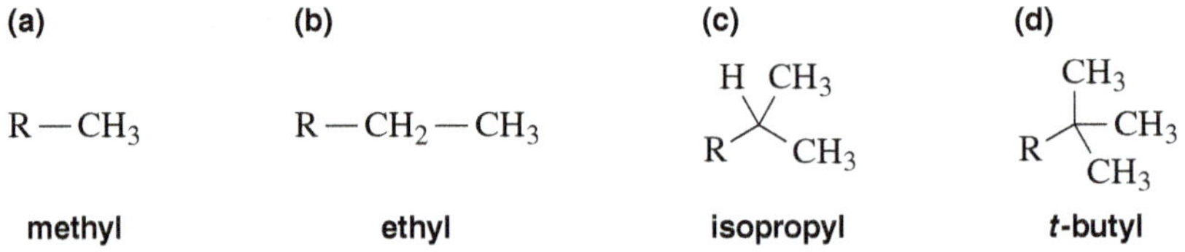

The common substituents we encounter in organic chemistry will make it easier to solve structures. Let's identify these together.

Assessment 15.34 The spectrum shown has three peaks in close proximity. How do you know this is not a triplet (t)?

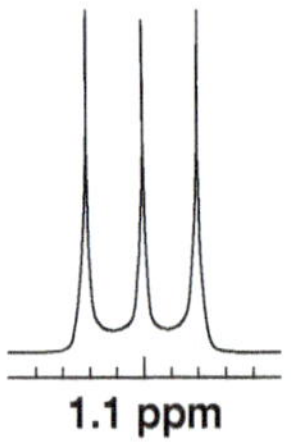

Solved Assessment 15.33

To answer this question, label each hydrogen and make a table. [*We are essentially going in the opposite direction we'll work when solving spectra.*] By including the integration value and expected multiplicity in the table, the spectrum for the fragments can be easily drawn. The results of this process are shown in **Table 15.7**. ■

Table 15.7 Common splitting patterns.

Fragment	Structure	Signals	Integration	Number of neighbors (*N*)	Multiplicity (*M*)	Appearance in NMR spectrum
Methyl	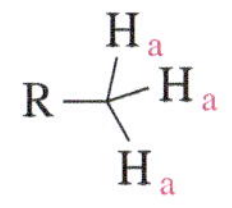	H_a	3	0	Singlet (s)	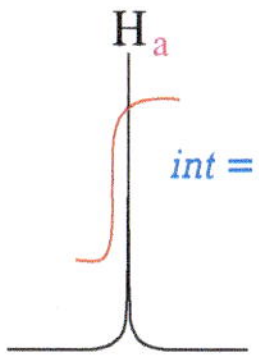
Ethyl	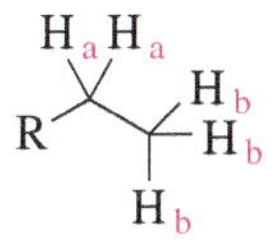	H_a H_b	2 3	3 2	Quartet (q) Triplet (t)	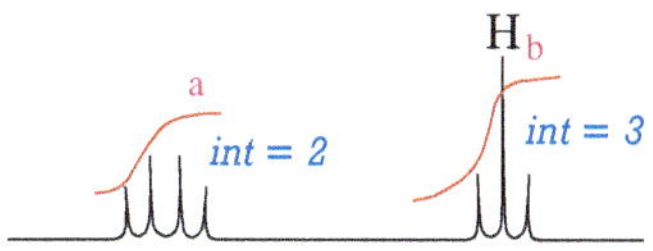
i-Propyl	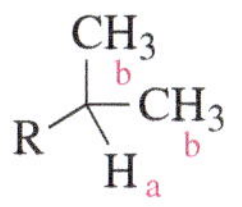	H_a H_b	1 6	6 1	Septet (sept) Doublet (d)	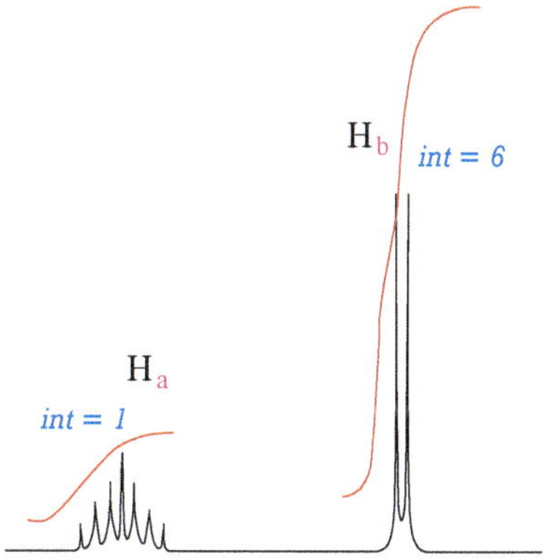
t-Butyl	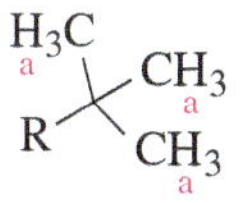	H_a	9	0	Singlet (s)	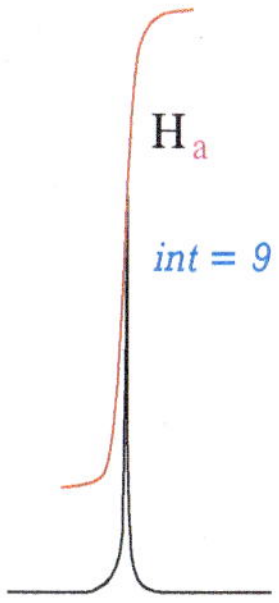

Be on the lookout for these common patterns when solving spectra. It can save you lots of time on an exam—time that can be spent on more difficult questions or just checking your answers.

15.4.4 SUMMARY

- The magnetic field generated by spinning hydrogen nuclei splits the signal of neighboring hydrogens. The $N + 1$ rule says that a hydrogen with N neighbors will appear as an $N + 1$ multiplet.

Assessment 15.35 Draw a spectrum for each of the following molecules, being sure to indicate the multiplicity, integration, and chemical shift of each peak. Label each signal based on the set of equivalent hydrogens to which it corresponds.

(a)

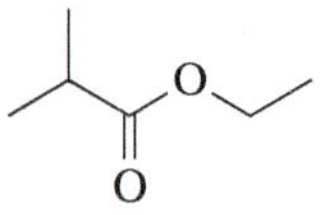

(b)

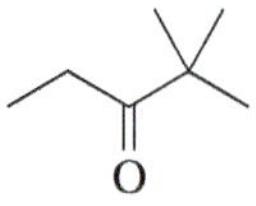

(c)

(d)

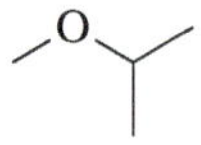

15.4.5 Putting it all together: structure determination

Having fully "dissected" the anatomy of a signal [*Ha!*], we can now use what we know to practice solving ^{1}H NMR problems. Provide a structure for the four ^{1}H NMR spectra in Assessment 15.36 and then we'll work them together. Different strategies will be used in each example, but common to all of them is starting by making a table and thinking logically. As you become more practiced at solving problems like this, you'll be able to start taking shortcuts. And in many cases, you'll see a spectrum and just "have a hunch" that, after careful analysis, will end up being correct.

Assessment 15.36 Given the molecular formula, determine the structure that corresponds to the spectrum shown.

(a) $C_6H_{13}Br$

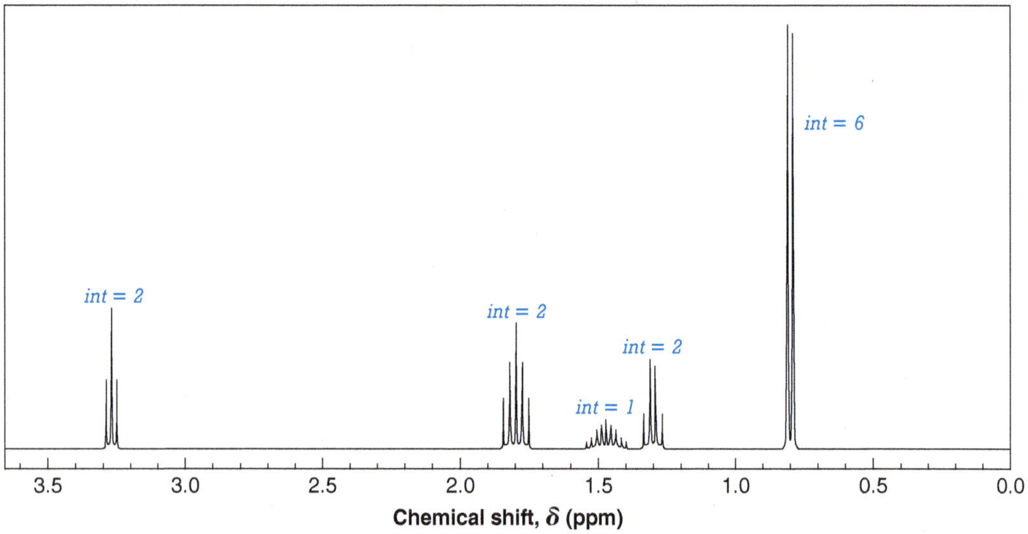

(b) $C_6H_{12}O_2$

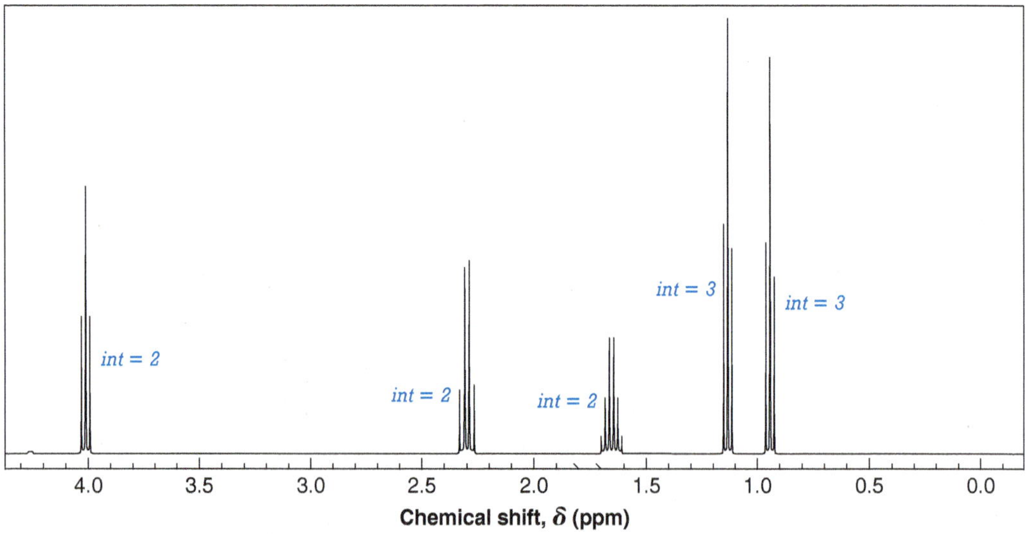

(c) $C_7H_{14}O_2$; IR: 3400 cm^{-1} (broad band), 1715 cm^{-1}.
^{1}H NMR spectrum: 1.23 (9H, s); 2.60 (2H, t); 3.10 (1H, broad s); 3.50 (2H, t).

(d) $C_8H_8O_2$

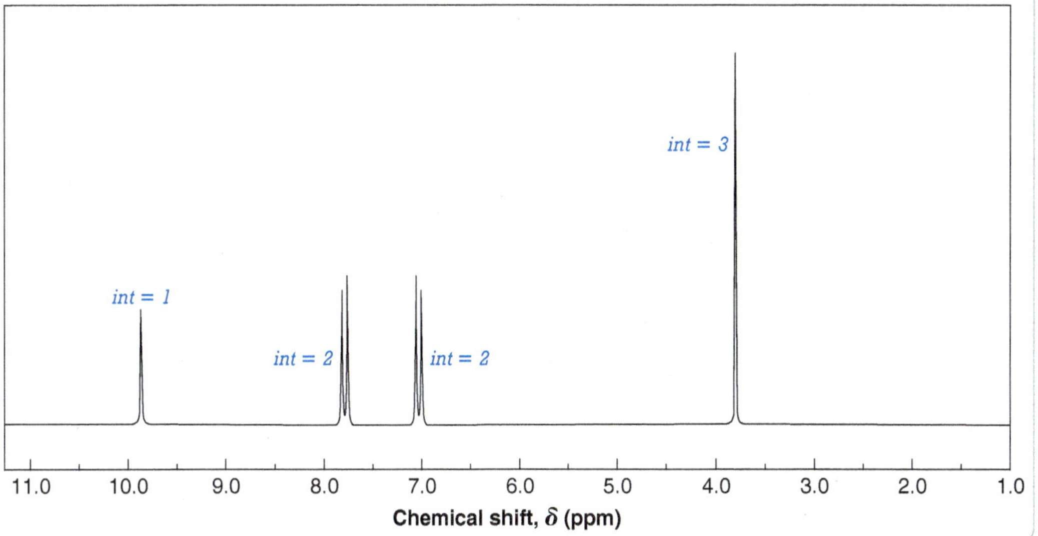

We'll work these together, but only after you have attempted them on your own.

Solved Assessment 15.36

(a) $C_6H_{13}Br$

Always start by calculating the index of hydrogen deficiency.

$$\text{IHD} = \text{number of carbons} - \frac{\text{number of hydrogens}}{2} - \frac{\text{number of halogens}}{2} + \frac{\text{number of nitrogens}}{2} + 1 = 6 - \frac{13}{2} - \frac{1}{2} + \frac{0}{2} + 1 = 0$$

Based on the IHD, there are no π bonds or rings in the molecule. From this we know it is just an alkyl halide. We can now label the peaks as shown in **Figure 15.36** in order to construct **Table 15.8**.

Figure 15.36 Label the peaks in the NMR spectrum.

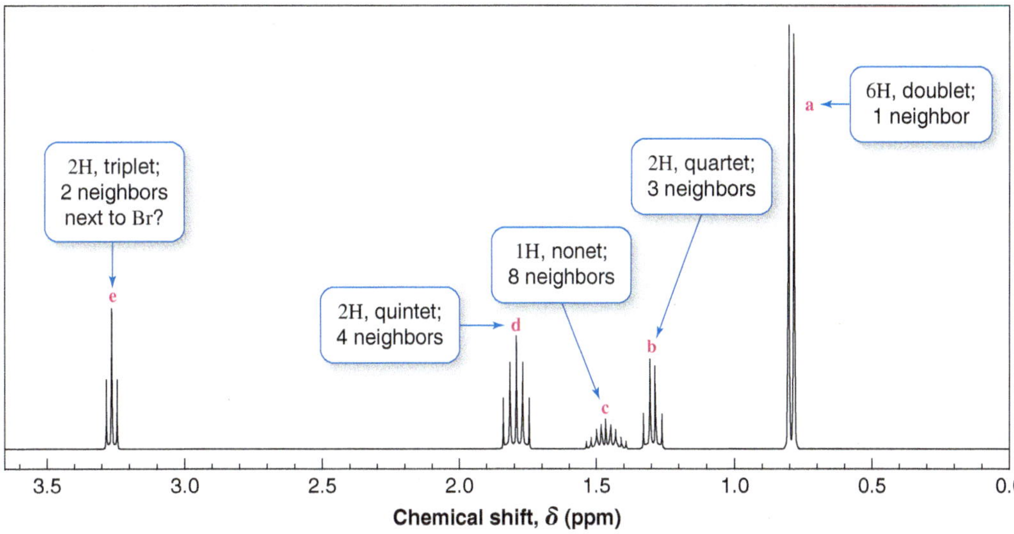

Table 15.8 Summarized data from Figure 15.36.

Signal	δ (ppm)	Integral	Multiplicity (*M*)	Number of neighbors ($N = M - 1$)	Guess
a	0.8	6	Doublet (d, 2 peaks)	1	$CH_3-CH-CH_3$
b	1.3	2	Quartet (q, 4 peaks)	3	$-CH_2-CH_3$ or $-CH-CH_2-CH_2$
c	1.48	1	Nonet (non, 9 peaks)	8	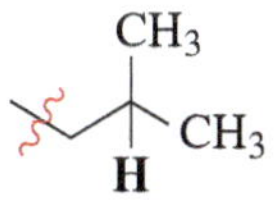
d	1.78	2	Quintet (quint, 5 peaks)	4	$-CH_2-CH_2-CH_2-$
e	3.25	2	Triplet (t, 3 peaks)	2	$Br-CH_2-CH_2-$

Brief analysis: The only tricky assignment is the signal at 0.7 ppm (signal **b**). With three neighbors, the simplest guess would be for it to have a single methyl group as a neighbor. This possibility is ruled out because there is no peak that integrates to three in the spectrum. Thus, it must be that the three neighbors of signal **b** are a CH_2 and a CH. This is confirmed by the CH of signal **c**.

To put the molecule together, we can trace the carbon framework through our guesses: **a**→**c**→**b**→**d**→**e**→Br. The methyl groups of signal **a** are attached to the CH of signal **c**. The CH of signal **c** is attached to the CH_2 of **b**, which is attached to another CH_2 (signal **d**), which is itself attached to the CH_2 of signal **e**, which is attached to the bromine atom. By mapping the puzzle pieces in this way, the structure that results is 1-bromo-4-methylpentane (**Figure 15.37**).

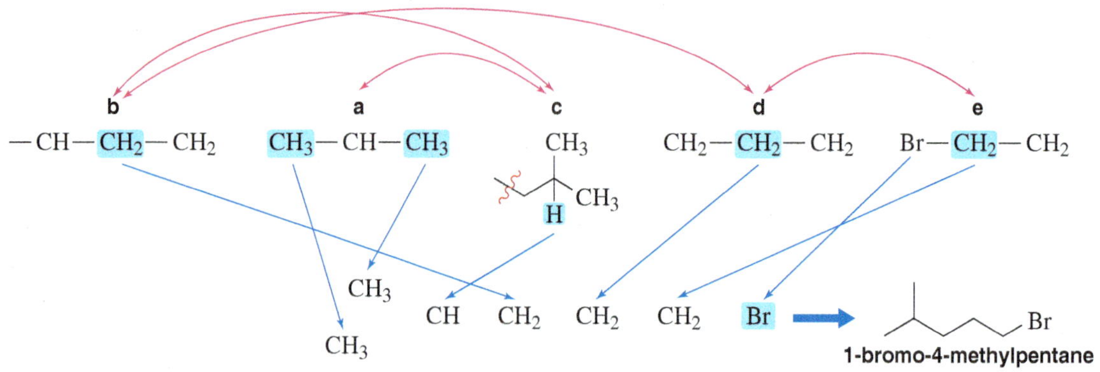

Figure 15.37 The structure is revealed through mapping the bond connections.

(b) $C_6H_{12}O_2$
Calculate the IHD:

$$\text{IHD} = \text{number of carbons} - \frac{\text{number of hydrogens}}{2} - \frac{\text{number of halogens}}{2} + \frac{\text{number of nitrogens}}{2} + 1 = 6 - \frac{12}{2} - \frac{0}{2} + \frac{0}{2} + 1 = 1$$

An IHD = 1 suggests the presence of a single double bond equivalent, which could be a ring or a π bond. With no peaks between 5 and 6 ppm (**Figure 15.38**), it is unlikely to be an alkene. The presence of two methyl groups, indicating a carbon chain with two ends, suggests, but doesn't completely confirm, the absence of a ring. Finally, the chemical shift of 2.3 ppm for peak **d** is in the range of being next to a carbonyl. This carbonyl would represent the IHD, making a ring impossible. [*An IR would make the presence of a carbonyl clear.*]

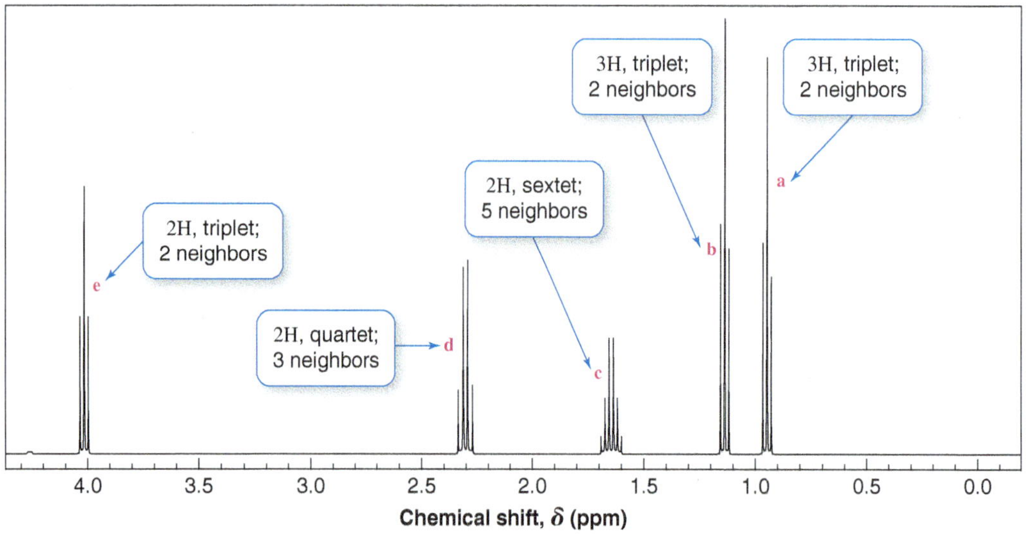

Figure 15.38 Label the peaks in the NMR spectrum.

Brief analysis: The peak at 2.3 ppm, as stated previously, likely represents a hydrogen next to a carbonyl. Because there is only one peak in this range (2–3 ppm; **Table 15.9**), there must be only one type of hydrogen next to the carbonyl. Thus, a ketone is unlikely. The peak at 4.0 ppm is for a hydrogen on a carbon directly attached to an electronegative atom, and the only possible electronegative atom is oxygen. Once more, with only one peak in the range of 3–5 ppm, there is not a hydrogen on the "other side" of the oxygen. Thus, an ether is unlikely. Having ruled out a ketone and an ether, the most likely way to have two oxygens in the same molecule is an ester functional group (RCO_2R).

Table 15.9 Summarized data from Figure 15.38.

Signal	δ (ppm)	Integral	Multiplicity (*M*)	Number of neighbors ($N = M - 1$)	Guess
a	0.9	3	Triplet (t)	2	$CH_3{-}CH_2{-}$
b	1.1	3	Triplet (t)	2	$CH_3{-}CH_2{-}$
c	1.6	2	Sextet (sext)	5	${-}CH_2{-}CH_2{-}CH_3$
d	2.3	2	Quartet (q)	3	$O{=}C{-}CH_2{-}CH_3$
e	4.0	2	Triplet (t)	2	$O{-}CH_2{-}CH_2{-}$

As we did previously, we can trace the carbon framework through our guesses. Note that the carbon framework is actually in two pieces, or three, depending on whether we count the carbonyl itself as a fragment. Thus, we see **b** → **d** → C=O and **a** → **c** → **e** → O. In one fragment, the methyl group of signal **b** is attached to the CH_2 of signal **d**, which based on its chemical shift is attached to the carbonyl (C=O). In the second fragment, the methyl group of signal **a** is attached to the CH_2 of signal **c**. The CH_2 of signal **c** is attached to the CH_2 of signal **e**, which is attached to an oxygen. By mapping the puzzle pieces in this way, the structure that results is propyl propanoate (Figure 15.39).

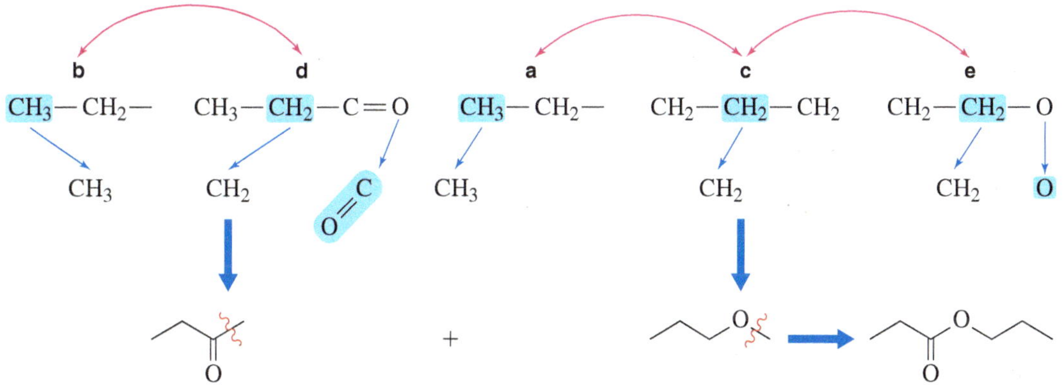

propyl propanoate

Figure 15.39 The structure is revealed through mapping the bond connections.

(c) $C_7H_{14}O_2$

Always start by calculating the index of hydrogen deficiency.

$$\text{IHD} = \text{number of carbons} - \frac{\text{number of hydrogens}}{2} - \frac{\text{number of halogens}}{2} + \frac{\text{number of nitrogens}}{2} + 1 = 7 - \frac{14}{2} - \frac{0}{2} + \frac{0}{2} + 1 = 1$$

In this case, we have IR data that tell us that the IHD comes from a carbonyl, as represented by the band at 1715 cm^{-1}. [*Always use the relevant information you are given.*] Similarly, the band at 3400 cm^{-1} indicates the presence of an alcohol. Because the information was already given to us in table form, we need only complete Table 15.10.

Table 15.10 Summarized data from Assessment 15.36(c).

Signal	δ (ppm)	Integral	Multiplicity (*M*)	Number of neighbors ($N = M - 1$)	Guess
a	1.23	9	Singlet (s)	0	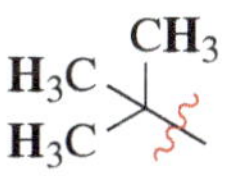
b	2.60	2	Triplet (t)	2	$CH_2{-}CH_2{-}$
c	3.10	1	Broad singlet (br s)	0	${-}OH$
d	3.50	2	Triplet (t)	2	${-}CH_2{-}CH_2{-}$

Brief analysis: Given the knowledge obtained from the IR spectrum, assignment of peak **c** becomes clearer. The broad singlet suggests that the hydrogen that gives that signal is exchanging (hydrogen bonding; Section 15.4.3). [*How could you prove that it was the OH signal?*] Peak **b** at 2.60 ppm represents a hydrogen likely next to the carbonyl. And peak **d** is for a hydrogen next to an electronegative atom. Thus, it must be next to the hydroxyl group. Finally, the nine-hydrogen singlet is a *t*-butyl group, one of the common patterns from Table 15.7.

With all of this information, the pieces of the puzzle can be assembled, as shown in **Figure 15.40**. The CH_2 of signal **b** is attached to the CH_2 of signal **d** and the carbonyl (C=O). The CH_2 of signal **d** is attached to the hydroxyl group, based on its chemical shift. By mapping the puzzle pieces in this way, the structure that results is 1-hydroxy-4,4-dimethylpentan-3-one.

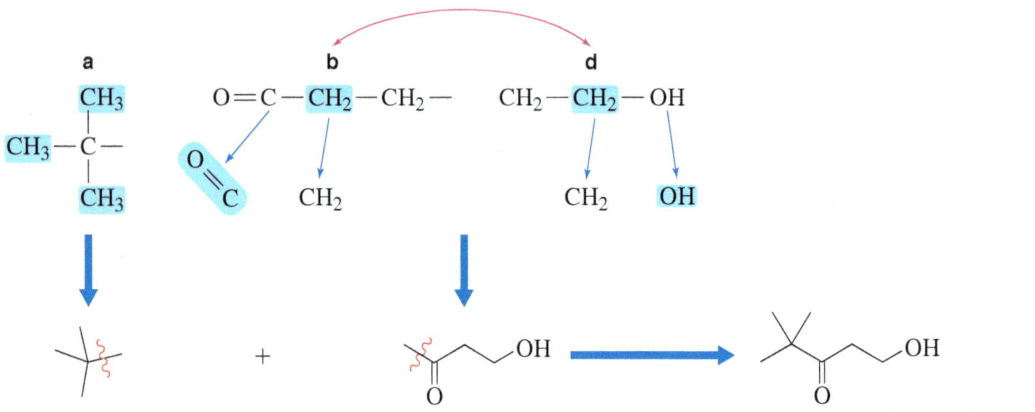

Figure 15.40 The structure is revealed through mapping the bond connections.

(d) $C_8H_8O_2$
Calculate the IHD:

$$\text{IHD} = \text{number of carbons} - \frac{\text{number of hydrogens}}{2} - \frac{\text{number of halogens}}{2} + \frac{\text{number of nitrogens}}{2} + 1 = 8 - \frac{8}{2} - \frac{0}{2} + \frac{0}{2} + 1 = 5$$

When the index of hydrogen deficiency is 4 or more, you should immediately look for evidence of an aromatic ring in the spectral data you have. The labeled spectrum in **Figure 15.41** shows peaks between 7 and 8 ppm that have a total integration of 4. This not only confirms the presence of an aromatic ring, but also tells us that it must be disubstituted. [*With four hydrogens on the benzene ring, how many non-hydrogen groups are attached?*] The remaining IHD is likely accounted for by the aldehyde indicated by the chemical shift of 9.8 for peak **d** (Table 15.2).

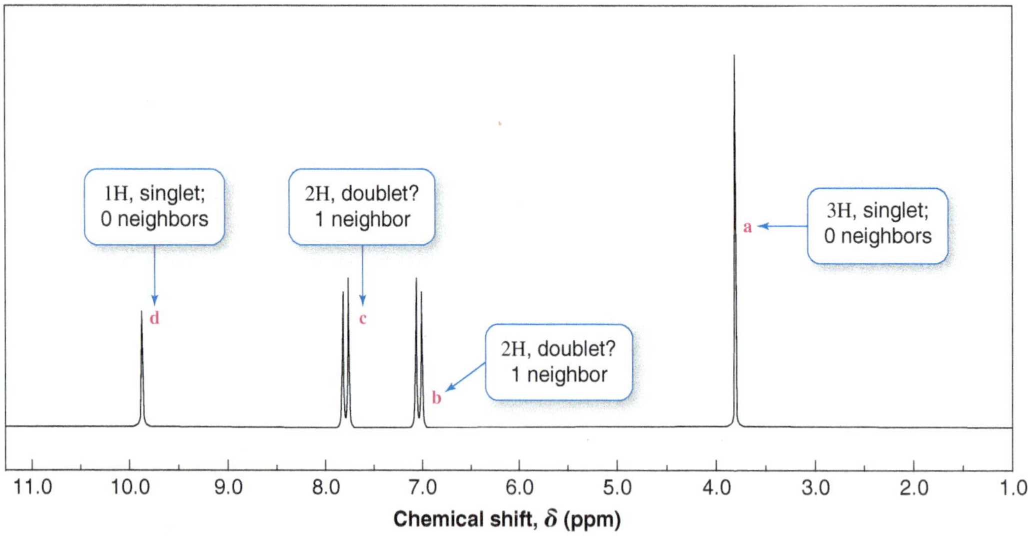

Figure 15.41 Label the peaks in the NMR spectrum.

Given the information in the spectrum itself, there is really no need to create a table for this one. Let's consider how this might be true (**Figure 15.42**). First, peak **a** at 3.8 ppm is indicative of a methyl group attached to an electronegative oxygen. Thus, one of the pieces is a methoxy group. Peaks **b** and **c** together represent a total of four aromatic hydrogens, suggesting we have one of three possible disubstituted aromatic structures. [*Why are there three and only three?*] The final piece is the aldehyde indicated by peak **d**.

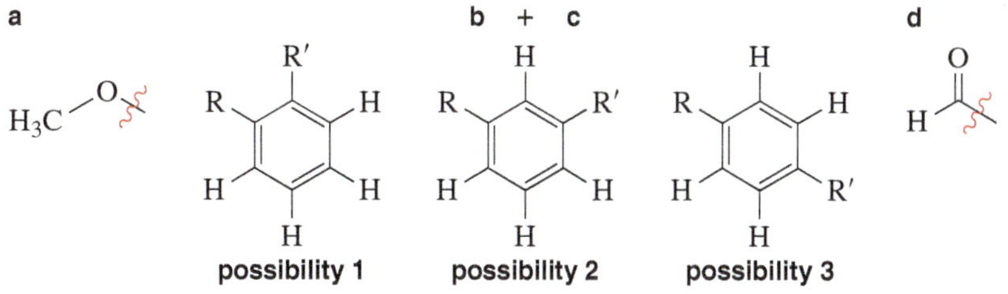

Figure 15.42 The pieces of the puzzle.

With a disubstituted ring and two substituents, the pieces should be easy to assemble. The only challenge is determining which of the three possible aromatic structures is represented by peaks **b** and **c**. To do this, let's consider how each of the combinations might appear, while keeping symmetry in mind (**Figure 15.43**). Both possibilities 1 and 2, because there is no symmetry, should have a signal for each of the four hydrogens (**a–d**). Possibility 3 has a symmetry plane that cuts through the middle of the ring, bisecting the substituents. As a result, there are only two types of hydrogens (**a** and **b**). We'd expect this substitution to give a pattern of two signals, each integrating to 2. With one neighbor each, these signals should appear as a doublet. [*Confirm the pattern in Figure 15.41.*] And, thus, the ring in Assessment 15.36(d) must be 1,4-disubstituted. Combining the 1,4-disubstituted ring with the aldehyde and methoxy group gives the molecule anisaldehyde. ■

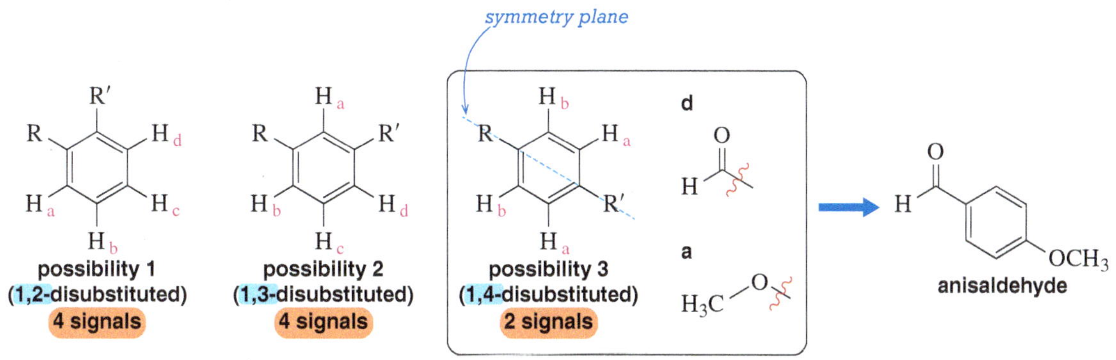

Figure 15.43 The pattern in the aromatic region of the ^{1}H NMR spectrum can only be for a 1,4-disubstituted ring.

15.5 Advanced ^{1}H NMR topics

In the preceding sections, molecules were specifically chosen that, as a result of their structure, give relatively simple NMR spectra. Here in Section 15.5, however, the spectra aren't always as straightforward as the ones you have encountered up to now.

If the spectra in this section aren't as simple, let's figure out what makes them complex!

15.5.1 Coupling constants

In Section 15.4.4, we defined the coupling constant as being a measure of the interaction of the magnetic fields between two nuclei. The coupling constant, or J value, usually reported in units of frequency (Hz), can be measured as the distance between two peaks within a multiplet (**Figure 15.44**).

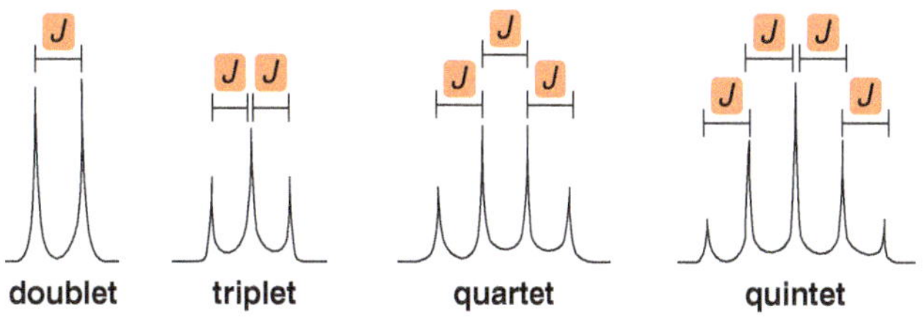

Figure 15.44 The coupling constant (*J* value) is the distance between peaks in a multiplet and is measured in Hz.

The value of the coupling constant is, in fact, not constant but depends on the orientation between the two hydrogens that are coupling to each other. Specifically, it depends on the dihedral angle between the two hydrogens. The graph in **Figure 15.45**, referred to as a Karplus diagram, shows the J value as a function of the dihedral angle. It is not important to remember the specific values, because there will be slight deviations from these values. But a couple of important points emerge. First, the largest coupling constant is when two hydrogens are anti to each other ($\theta = 180°$, $J_{ab} = 15$ Hz). The smallest coupling constant is when two hydrogens have a 60° dihedral angle ($J_{ab} = 4$ Hz). Finally, with a range of coupling constants between 3 and 15, the average coupling constant is around 7 Hz. [*This last one is why previous spectra had such well-defined splitting patterns.*]

Figure 15.45 The coupling constant (J_{ab}) depends on the dihedral angle as shown in the Karplus diagram.

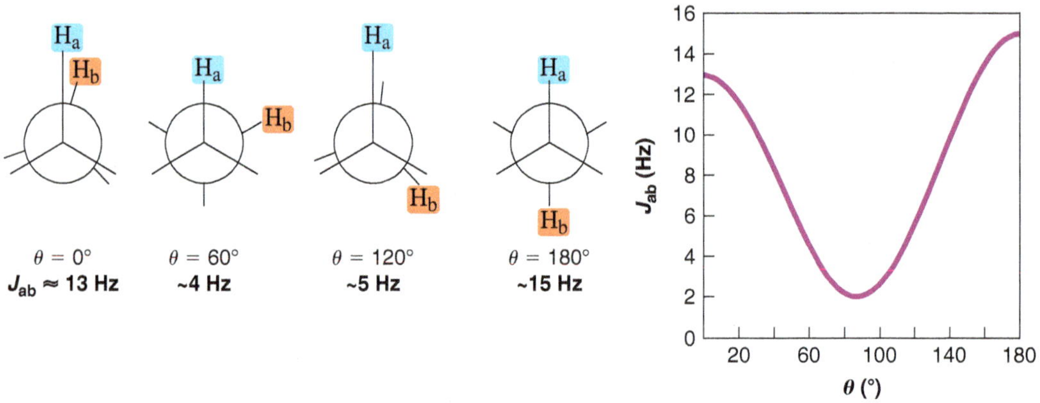

The coupling constants predicted in Figure 15.45 result in the varied coupling constants shown in **Figure 15.46**. These values show that the coupling constant can be used to distinguish between cis ($J = 15$ Hz) and trans ($J = 10$ Hz) alkenes. Importantly, the coupling constant between hydrogens connected to the sp^3-hybridized carbons of a freely rotating alkyl group is the average of the coupling constants for all possible conformations ($J = 7$ Hz). This is because the timescale of NMR analysis is slower than the time it takes to rotate around a single bond. All that NMR can detect is the average of all coupling constants. We explain why this is important when we consider the implications of different coupling constants between hydrogens on neighboring carbons.

Figure 15.46 Coupling constants for specific molecules.

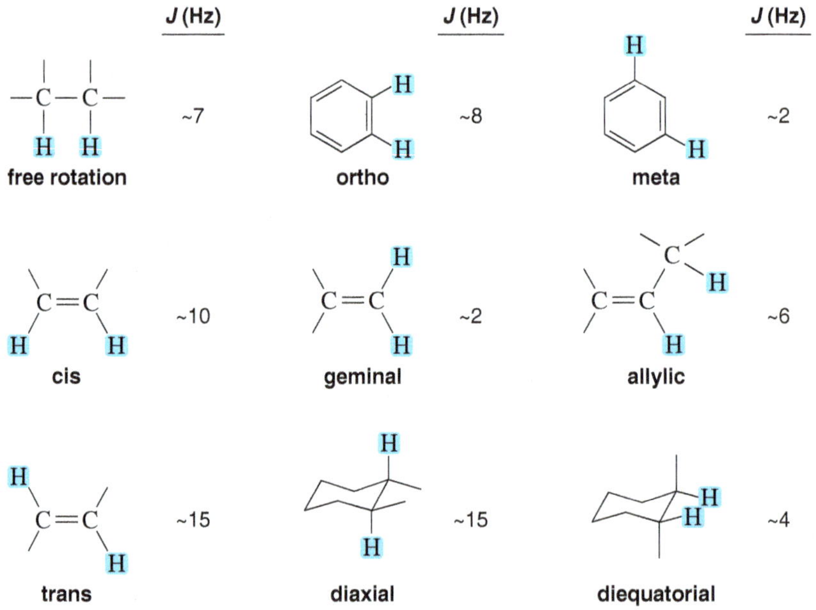

Figure 15.47 shows how coupling constants can be used to distinguish between two diastereomers of ethyl cinnamate. In Figure 15.47(a), a coupling constant of 16 Hz between alkene hydrogens is indicative of a trans-alkene. The spectrum in Figure 15.47(b) shows a

coupling constant of 9.5 Hz, indicating the alkene is cis. Importantly, and this will always be the case, the coupling constant for two hydrogens will be the same for each signal. Thus, if H_a splits H_b by a certain amount (J_{ab} = 16 Hz), H_b will *always* split H_a by the exact same amount (J_{ba} = 16 Hz). Comparing coupling constants across a spectrum can be useful in determining which hydrogens are next to which.

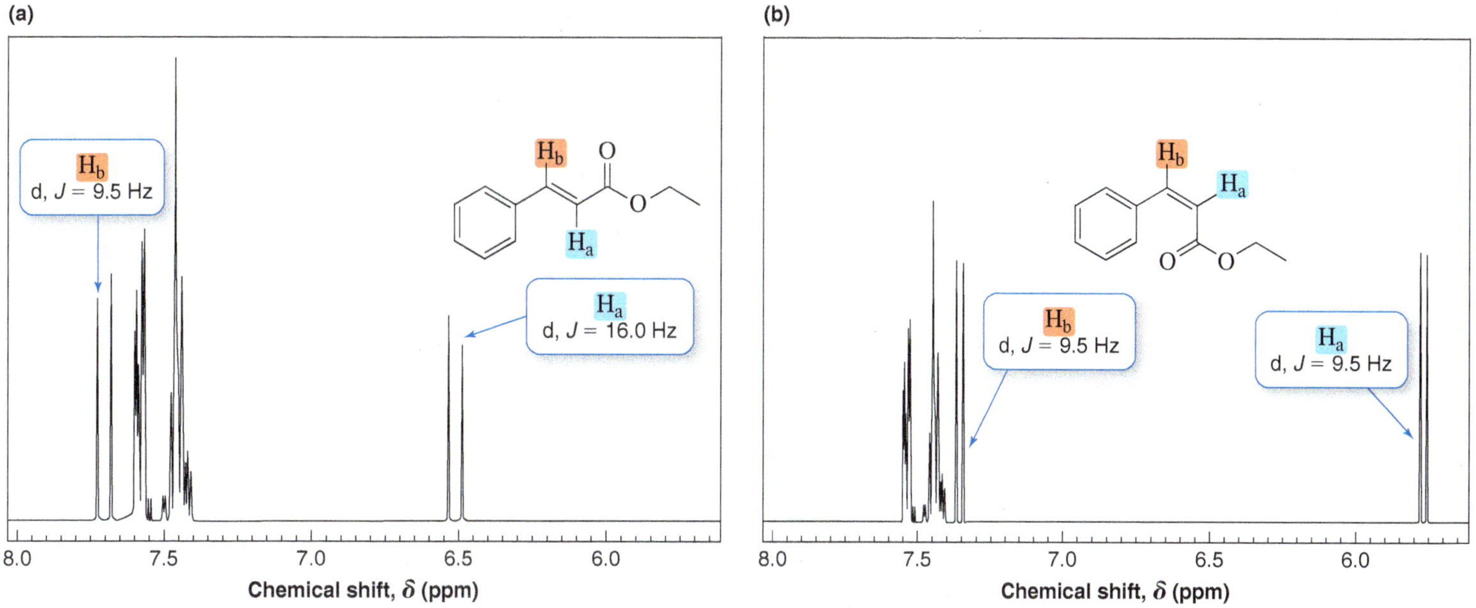

Figure 15.47 Alkene isomers can be distinguished by the coupling constants of the alkenyl hydrogens. (a) *trans*-Ethyl cinnamate. (b) *cis*-Ethyl cinnamate.

Assessment 15.37 A chemist made what was thought to be compound **A** or **B**. How could coupling constants between H_a and H_b be used to distinguish between the two isomers? [*Hint: Draw a chair conformation of each.*]

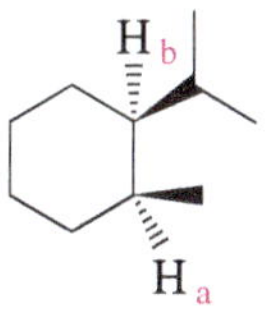

A

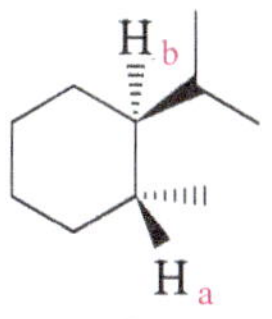

B

Assessment 15.38 What coupling constant would you expect between H_a and H_b in cyclopent-2-en-1-one?

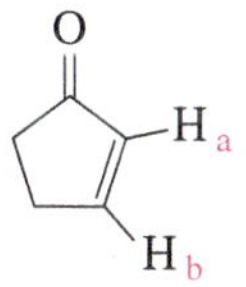

Assessment 15.39 Alkene hydrogens usually appear at similar chemical shifts between 5 and 6 ppm. The alkene hydrogens in the structure shown are very different. Rationalize the difference in chemical shifts between H_a and H_b in the structure shown.

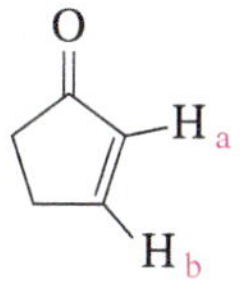

15.5.2 Splitting trees

Consider the expanded signal for H_a in *trans*-but-2-enoic acid shown in **Figure 15.48**. If you were analyzing this spectrum with the goal of structure determination, it would be difficult to characterize the signal for H_b as a standard multiplet (singlet, doublet, triplet, etc.).

Figure 15.48 **1H NMR spectrum of *trans*-but-2-enoic acid.** [*Remember that you often won't see a signal for the carboxylic acid hydrogen. And the humps on the sides of the middle two peaks are buried peaks themselves.*]

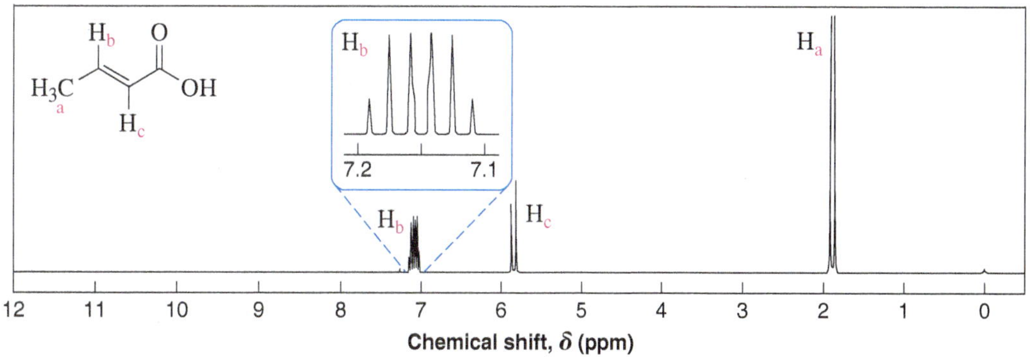

Thinking Like a Chemist

Let's think like a chemist and see how we can rationalize the appearance of this signal.

It turns out that in Section 15.4.4, we weren't telling the entire story when it came to hydrogens with neighbors on more than one adjacent carbon. Back then, when we looked at a CH_2 like the one in **Figure 15.49**, we said it had four neighbors and appeared as a quintet. While this is mostly true, it is better to discuss this situation as if the CH_2 has three neighbors (CH_3) and then another one neighbor (CH). Said another way, H_a splits H_b into a quartet and H_c splits H_b into a doublet. And yet it shows up as a quintet. [*We'll see why later, but return to Figure 15.48 for now.*]

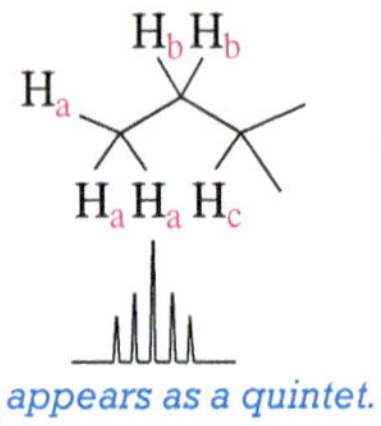

Figure 15.49 Hydrogens with four neighbors appears as a quintet.

If we analyze H_a of *trans*-but-2-enoic acid (**Figure 15.50**) in the same way, we can say it has three neighbors (a CH_3) and one neighbor (a CH) instead of four neighbors (a CH_3 and a CH). This is a subtle but important difference. Discussed this way, we can consider the situation as if only H_a were a neighbor to H_b. In this case, H_b would appear as a *quartet* ($J_{ab} \approx 7$ Hz), because of free rotation around the C—C σ bond. If, on the other hand, only H_c were a neighbor to H_b, H_b would appear as a *doublet* ($J_{bc} \approx 15$ Hz). Now look back at the blown-up image of the signal for H_b in Figure 15.48. [*What does it look like?*] On careful analysis, H_b looks like two quartets, or a *doublet of quartets*. [*It may be hard to tell, but the two quartets are overlapping. We'll see why soon.*]

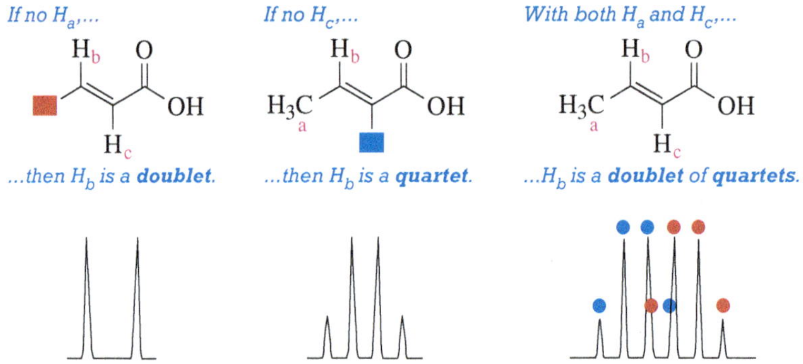

Figure 15.50 **H_b is a doublet of quartets.** H_a makes it a quartet, whereas H_c makes it a doublet. [*The red and blue dots show the two overlapping quartets.*]

So why does it appear this way as opposed to the quintet for the CH_2 in Figure 15.49? We can draw a **splitting tree** to see why. A splitting tree does exactly what we did in Figure 15.50 by combining the coupling constants arising from each hydrogen to predict what the signal will look like. [*You'll need a ruler if you want to do this with me.*] First, split H_b into a doublet due to the influence of H_c (**Figure 15.51**). The distance between the two lines needs to be 15 mm, representing the coupling constant (J_{bc} = 15 Hz) of the resulting doublet. The influence of H_a is illustrated by splitting each of the resulting doublet lines into a quartet. Once more, the distance between each line should be 7 mm, accounting for the coupling constant (J_{ab} = 7 Hz) of each resulting quartet. At this point, we have a sense of where the lines will be relative to each other. By drawing a spectrum underneath the eight lines, keeping in mind the relative peak heights, you'll get a signal that looks like the *doublet of quartets* in Figure 15.51. ■

Figure 15.51 Using a splitting tree to predict the appearance of a multiplet.

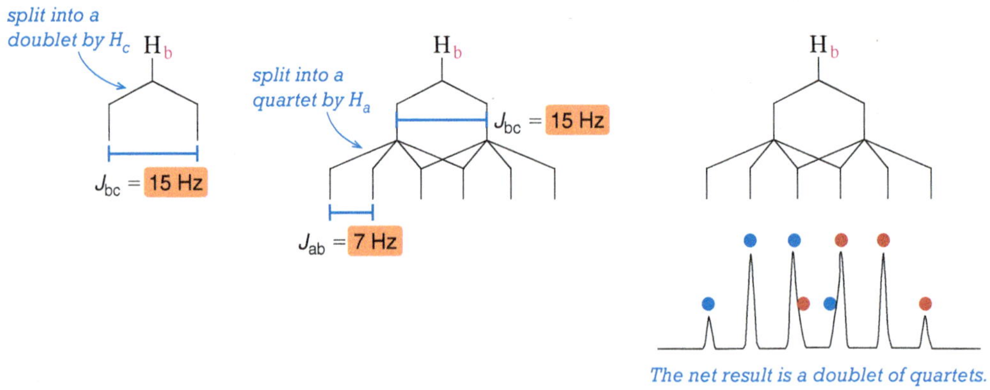

Figure 15.52 shows a splitting tree drawn in the same way for the situation of the alkyl hydrogens in Figure 15.49. Because the alkyl group has free rotation, both coupling constants are equal to the average ($J_{ab} = J_{bc} = 7$ Hz). First, split H_b into a doublet due to the influence of H_c. The distance between the two lines needs to be 7 mm, representing the coupling constant ($J_{bc} = 7$ Hz) of the resulting doublet. The influence of H_a is illustrated by splitting each of the resulting doublet lines into a quartet. Once more, the distance between each line should be 7 mm, accounting for the coupling constant ($J_{ab} = 7$ Hz) of each resulting quartet. [*Look at what happens.*] At this point, we see that there is overlap between several lines of the quartets. This occurs because the coupling constants are the same. By drawing a spectrum underneath the five lines, keeping in mind the relative peak heights, you'll get a signal that is essentially the quintet we'd have predicted based on the discussion in Section 15.4.4. [*Counting the overlapping lines will also tell you the multiplicity.*] And so what could be called a doublet of quartets appears as a quintet.

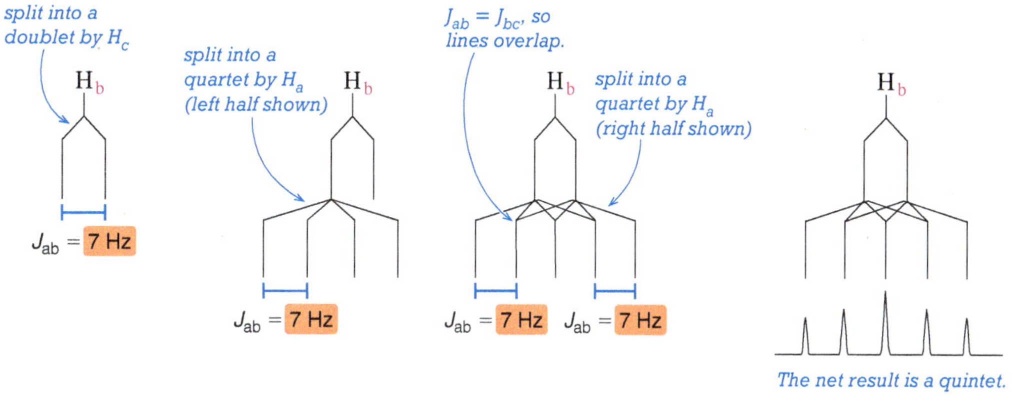

Figure 15.52 A splitting tree shows with identical coupling constants. Lines overlap to give the splitting pattern predicted by the $N + 1$ rule.

15.5.1–15.5.2 SUMMARY

- When a hydrogen is split by two nonequivalent hydrogens with identical coupling constants, the signal will appear as predicted by the $N + 1$ rule. When a hydrogen is split by two nonequivalent hydrogens with different coupling constants, the signal will not adhere to the $N + 1$ rule.

Assessment 15.40 Draw the expected signal for a hydrogen with the following coupling constants.

(a) H_a: δ 5.34 ($J_{ac} = 12, J_{ab} = 2$)

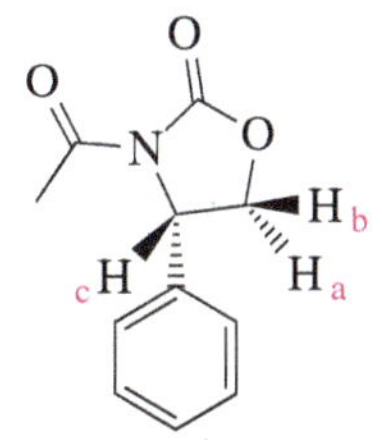

(b) H_a: δ 3.34 ($J_{ac} = 9, J_{ab} = 4$)

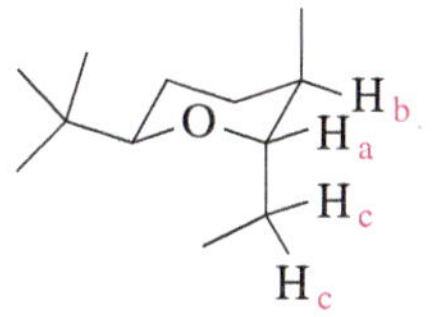

Assessment 15.41 Though it wasn't discussed, what coupling constant would you expect for H_a and H_c in the spectrum of *trans*-but-2-enoic acid in Figure 15.50? Justify your answer.

15.5.3 Diastereotopic hydrogens

In Section 15.4, we treated every CH_2 group as if the two hydrogens were equivalent. This was possible because there was usually free rotation around the C—C σ bonds of the acyclic alkanes and the two hydrogens were in the same chemical environment. This won't always be the case. Consider the labeled CH_2 in the oxazolidinone in **Figure 15.53**.

[*This is the molecule used in Assessment 15.40(a).*] We can test for chemical equivalence by replacing each hydrogen with a deuterium atom. If the exact same molecule is produced by each substitution, then the two hydrogens are said to be isotopic. Doing this experiment for the oxazolidinone, however, results in two diastereomers. In this case, the two methylene hydrogens are said to be diastereotopic and chemically nonequivalent. As a result, there will be a different signal for each.

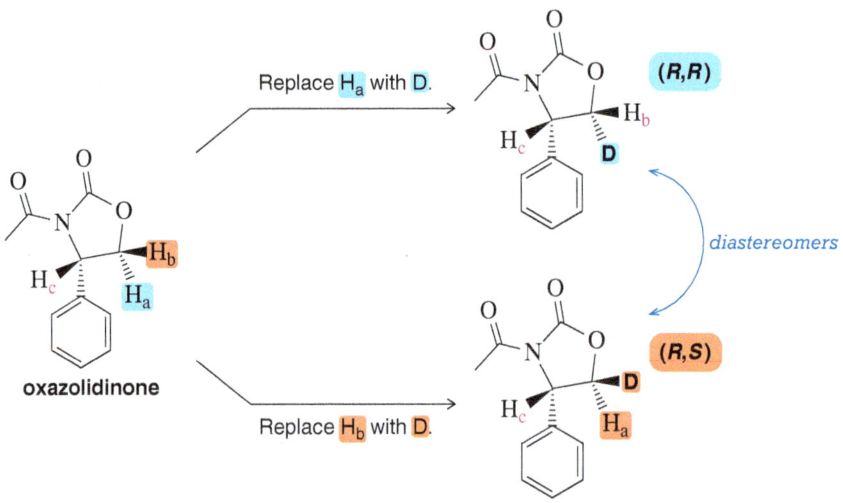

Figure 15.53 Diastereotopic hydrogens are not chemically equivalent.

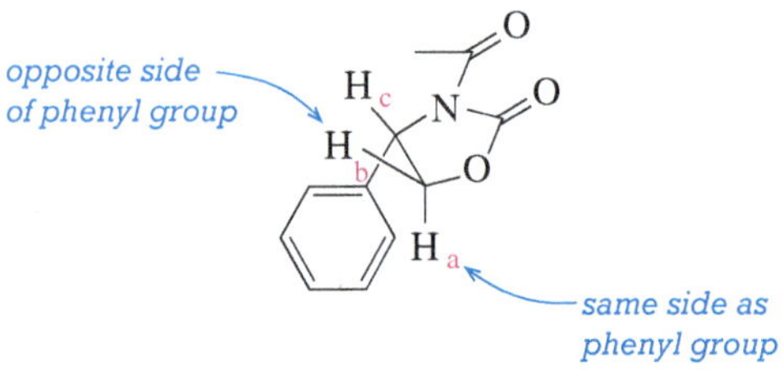

Figure 15.54 Diastereotopic hydrogens are not chemically equivalent because they are in different chemical environments.

By drawing out the 3-D structure of the oxazolidinone, it becomes clearer that H_a and H_b are in different chemical environments. H_b is on the opposite side of the ring relative to the phenyl group, whereas H_a is on the same side (**Figure 15.54**). This slight difference in environment is enough to give them different resonance frequencies.

Figure 15.55 shows the spectrum of the oxazolidinone. Note specifically the appearance of H_a and H_b. Because they are nonequivalent, they can also couple to each other. Thus, H_a appears as a doublet of doublets, coupling to H_b and H_c. H_b appears as a doublet of doublets, too, coupling to H_a and H_c. H_c, also shown in Figure 15.55, appears as a triplet because the coupling constants (J_{ac} and J_{bc}) are apparently equal. [*The large chemical shift difference between H_a and H_b is atypical but is due to H_a being in the deshielded region relative to the benzene ring. See Figure 15.22.*]

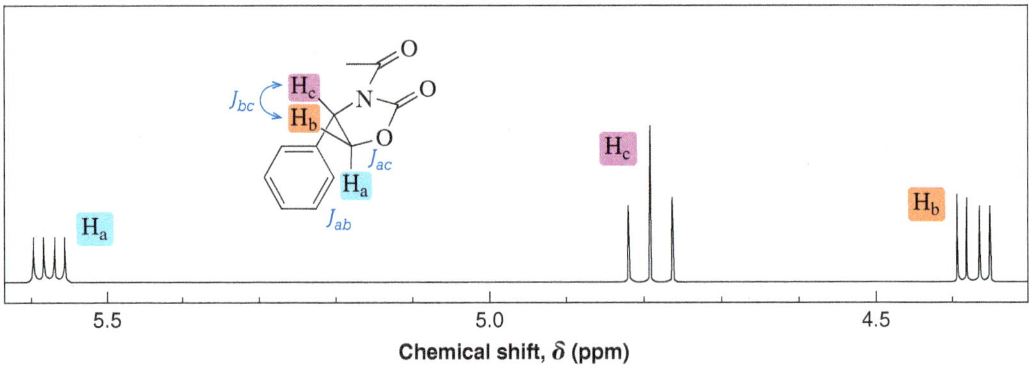

Figure 15.55 Diastereotopic hydrogens give complicated signals.

Diastereotopicity and the resulting complex spectra are not limited to molecules in rings. **Figure 15.56** shows two examples of diastereotopic methylenes, confirmed by replacement of each with deuterium. The spectra show there is a separate signal for each hydrogen of the methylene groups.

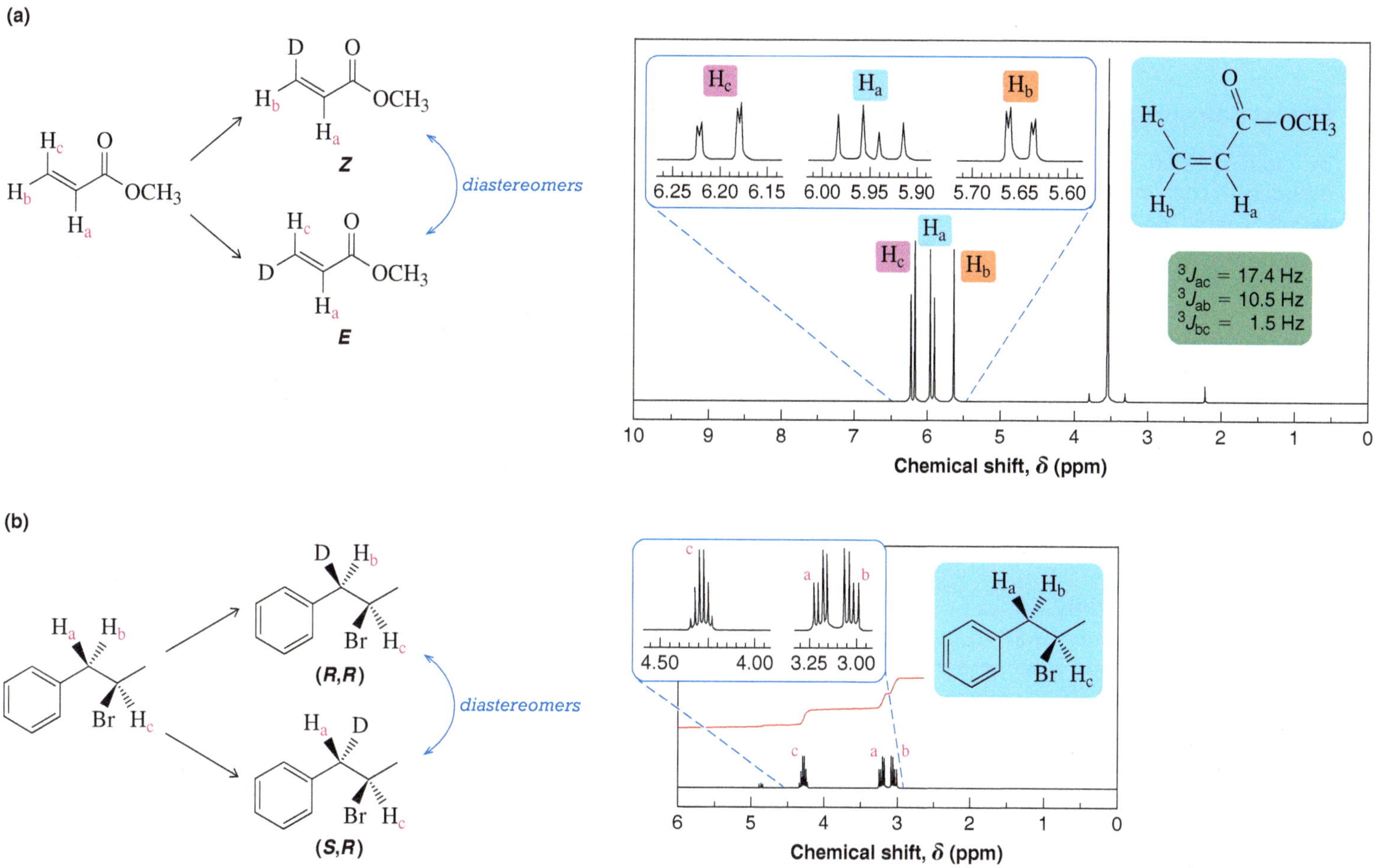

Figure 15.56 Diastereotopicity in acyclic molecules.

Assessment 15.42 Replace H_a and H_b in hexane with a deuterium. What is the relationship between the two products you obtain? Based on this, would you expect the two hydrogens to give one or two signals in the ^{1}H NMR spectrum?

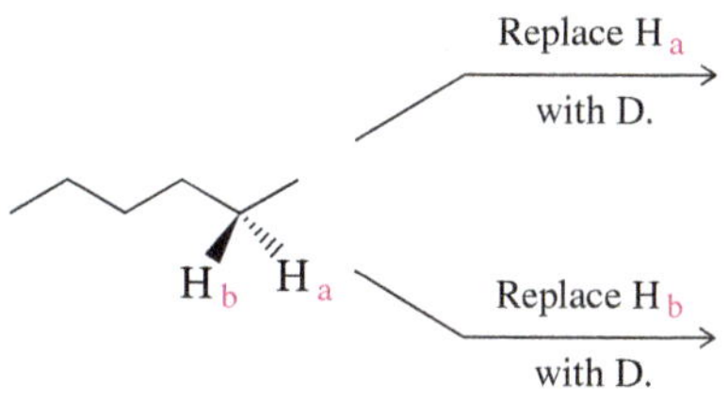

Assessment 15.43 Replace H_a, H_b, and H_c in methyl benzene with a deuterium. What is the relationship between the three products you obtain? Based on this, how many signals would you expect for these hydrogens in the ^{1}H NMR spectrum?

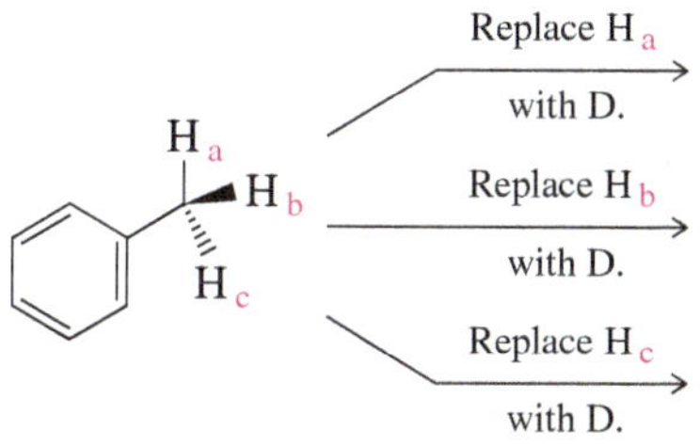

Assessment 15.44 Replace H_a and H_b in acetone with a deuterium. What is the relationship between the two products you obtain? Based on this, would you expect the two hydrogens to give one or two signals in the ^{1}H NMR spectrum?

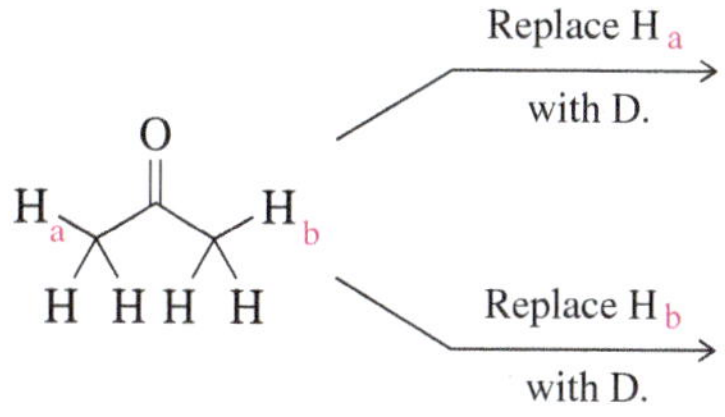

Assessment 15.45 Sketch the signals you would expect to see for H_a, H_b, and H_c in the molecule shown. The important coupling constants are given.

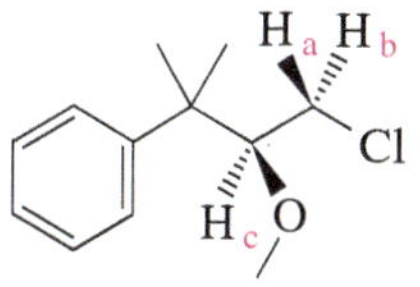

J_{ab} = 2 Hz
J_{ac} = 7 Hz
J_{bc} = 5 Hz

15-C CHEMISTRY IN CONTEXT: Natural Product Identification

Using coupling constants

By zooming in on individual peaks, we can see the utility of coupling constants in structure elucidation. Figure 15.57 shows the signal for H_a, which integrates to 1 H. Because this hydrogen is a doublet of doublets, it must have two hydrogens as neighbors. And because the coupling constants are different, these two neighbors must be in different chemical environments. [*What does the 15-Hz coupling constant indicate?*]

Similarly, the signal for H_b is a doublet of triplets (Figure 15.58). The 15-Hz coupling constant suggests that H_b is a neighbor of H_a. H_a makes H_b a doublet. The two H_c make H_b a triplet. The average coupling constant of 7 Hz indicates free rotation.

Expanding out slightly, we can infer from the information in Figures 15.57 and 15.58 that the neighboring H_a and H_b are in a trans relationship. This is due to the common coupling constant shared between the two (J_{ab} = 15 Hz) (Figure 15.59).

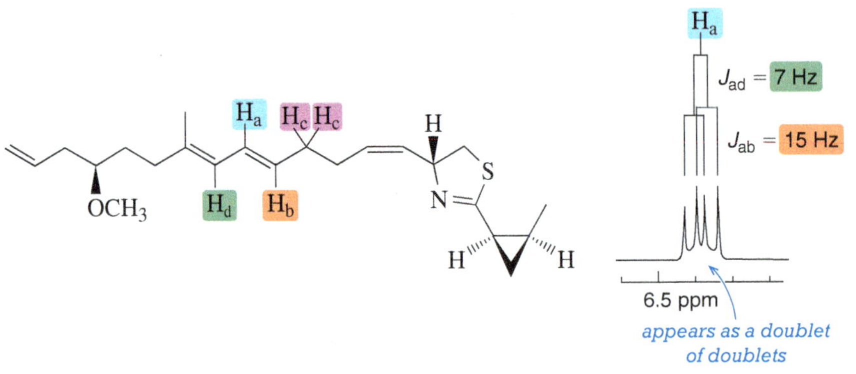

Figure 15.57 **The coupling constants of H_a suggest two neighboring C—H groups.**

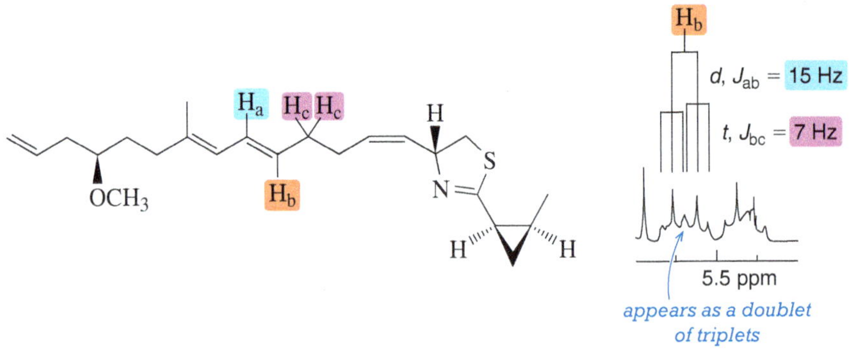

Figure 15.58 **The coupling constants of H_b suggest neighboring CH and CH_2 groups.**

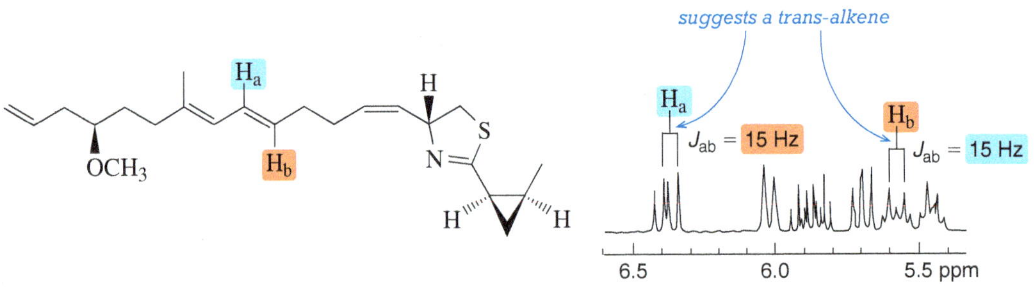

Figure 15.59 **A coupling constant of 15 Hz indicates a trans relationship between alkene hydrogens.**

15.6 ^{13}C NMR spectroscopy: an overview

In addition to ^{1}H NMR spectra, ^{13}C NMR is also a useful technique for structure identification. Specifically, it allows us to identify functional groups that don't contain hydrogens (e.g., a carbonyl or internal alkyne) along with carbon atoms with no hydrogens. Whereas we may have inferred the presence of these using ^{1}H NMR spectroscopy, ^{13}C NMR makes it possible to identify them directly. The background theory for ^{13}C NMR is covered in less detail than it was for ^{1}H NMR, simply because many of the principles are the same. Instead, we focus on the differences between the two techniques.

Signal to noise: Whereas almost every hydrogen is NMR active, only 1% of carbon atoms (^{13}C) can be detected using NMR. This fact alone accounts for many of the differences discussed here. The low abundance of ^{13}C results in spectra with a lower signal-to-noise ratio (Figure 15.60).

To achieve spectra containing useful information, more scans or a more concentrated sample are required. The advent of Fourier transform (FT) NMR has enabled the acquisition of useful spectra in spite of this complication. Even in a highly resolved ^{13}C NMR spectrum, the choppy baseline is contrasted with the perfectly smooth baseline of ^{1}H NMR spectra.

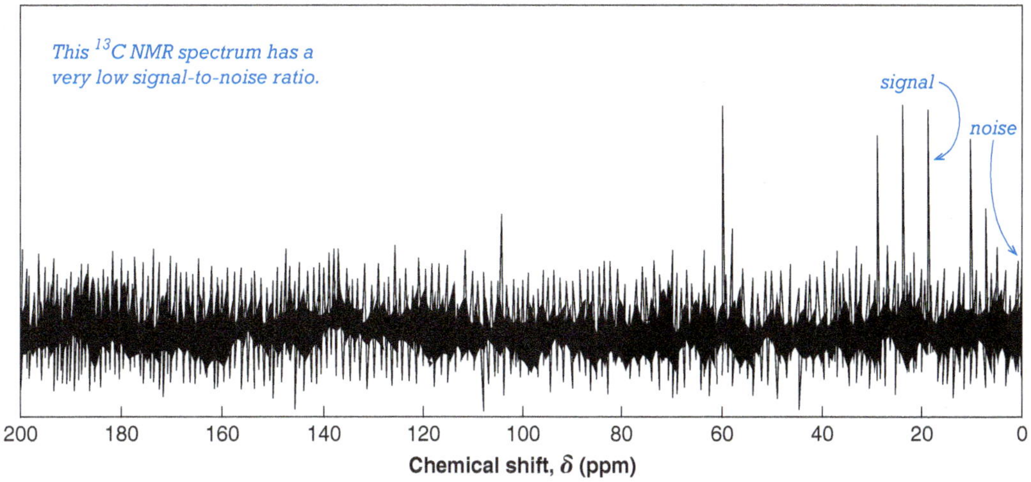

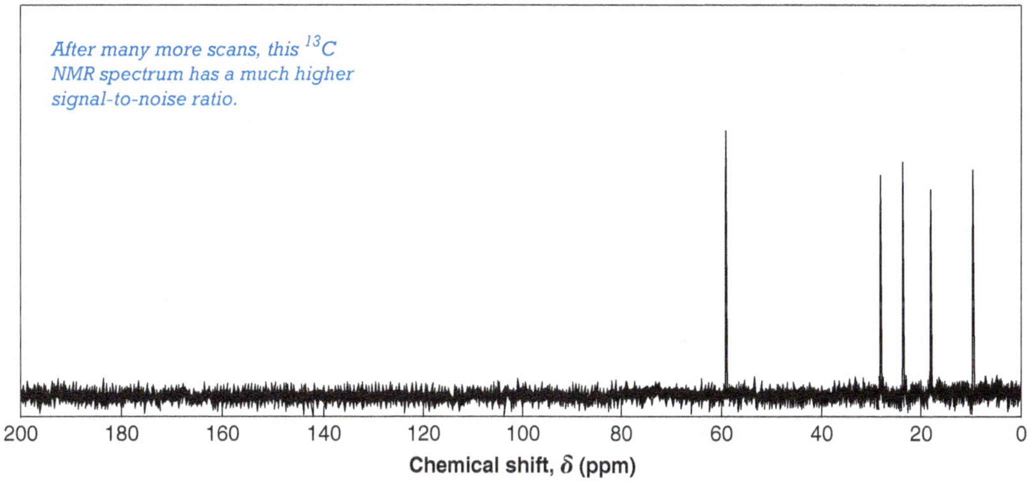

Figure 15.60 ^{13}C NMR requires more scans to give a useful spectrum. (a) One scan. (b) 200 scans.

The spectral range: As seen in **Figure 15.61**, the spectral width in ^{13}C NMR runs from 0 to 250 ppm. As in ^{1}H NMR, the resonance frequency of ^{13}C atoms is based on their chemical environment, with highly deshielded carbons giving a signal at a higher chemical shift. The carbon of tetramethylsilane (TMS) is highly shielded, so it is set to 0 ppm and used as a standard for ^{13}C NMR just as it is for ^{1}H NMR.

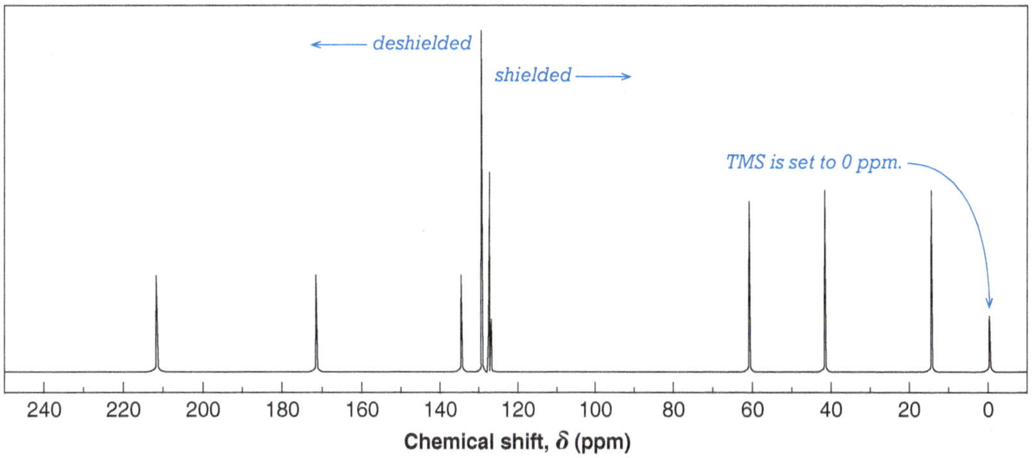

Figure 15.61 The spectral range of ^{13}C NMR is considerably larger than it is of ^{1}H NMR.

Integrals: In contrast to ^{1}H NMR, the signal heights for a single carbon atom vary due to a number of factors, including their chemical environment and the number of attached hydrogens. As a result, it is impossible to compare the relative peak heights to assign an integral value for the individual signals. This limits the utility of ^{13}C NMR to identify the sets of equivalent carbons, because it gives no information about how many of each are present.

Splitting: A ^{13}C atom would be split by its ^{13}C atom neighbor, if it ever had any. Because of the low abundance of ^{13}C, it hardly ever happens that two ^{13}C atoms are bonded to each other. As a result, a ^{13}C spectrum does not show ^{13}C—^{13}C coupling. ^{13}C atoms are bonded, however, to hydrogen atoms that generate their own magnetic field. Given the poor resolution of ^{13}C spectra already, the splitting of the individual signals into doublets, triplets, or quartets by attached hydrogens makes ^{13}C spectra more complex. To avoid having to analyze these complex spectra, and because ^{1}H NMR provides most of this information anyway, ^{13}C NMR spectra are obtained in **proton-decoupled** mode. Using this method, the hydrogens are rendered invisible, making it appear as if each carbon has zero attached hydrogens. Thus, as seen in Figure 15.61, a standard ^{13}C NMR spectrum shows only singlets.

15.7 ^{13}C NMR spectroscopy: analysis of a spectrum

Because of the differences described in Section 15.6, analyzing a ^{13}C NMR spectrum is greatly simplified compared to a ^{1}H NMR spectrum. The only information given in a standard ^{13}C NMR spectrum is the number of sets of equivalent carbons and where they appear in proximity to certain functional groups.

Equivalence is established in the same way it was for hydrogens in ^{1}H NMR. **Figure 15.62** shows how many signals you'd expect in the ^{13}C NMR spectrum of the molecules shown. Symmetry and rotation around a C—C single bond make carbons equivalent.

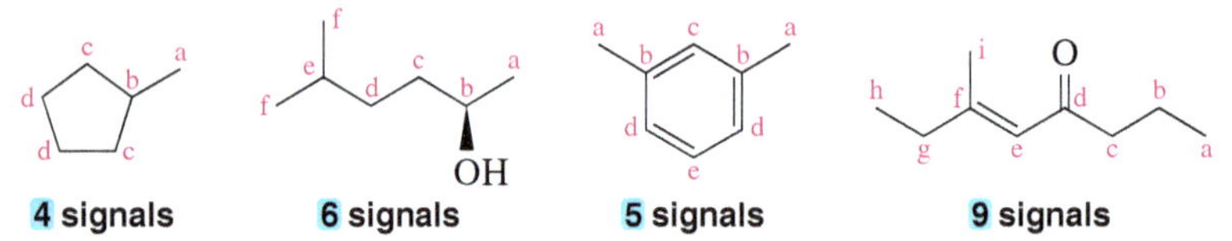

Figure 15.62 Nonequivalent carbons give distinct signals in the ^{13}C NMR spectrum.

With no integration and splitting, there is still value in a spectrum that contains only a singlet for each set of equivalent carbons present. One value is being able to quickly assess symmetry in the molecule. **Figure 15.63(a)** shows a ^{13}C NMR spectrum for a molecule with a molecular formula of $C_5H_{10}O$. With five carbons in the molecular formula, you should immediately look for five signals in the ^{13}C NMR spectrum. Figure 15.63(a) has exactly this many. Because carbons are made different by symmetry, there must be no symmetry in the molecule. Alternatively, the spectrum in **Figure 15.63(b)** for $C_6H_{10}O_2$ has just four signals, so there is probably a plane of symmetry in the molecule. While this type of analysis won't always get you to a structure, it reinforces the need to "use all the information each spectrum gives you."

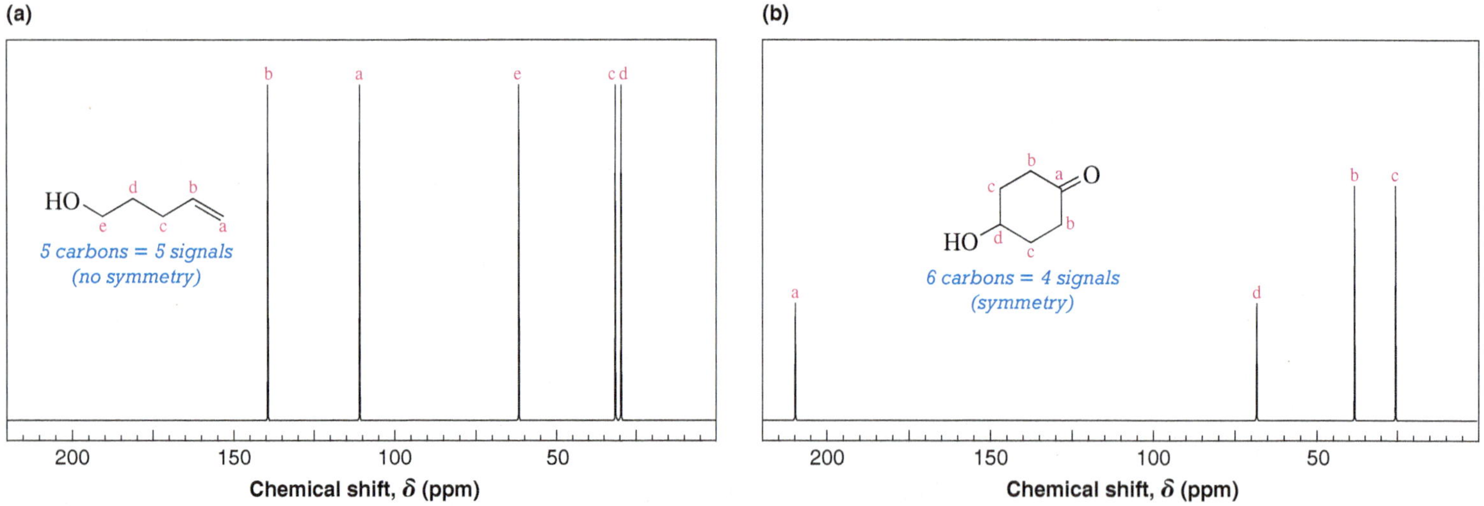

Figure 15.63 ^{13}C NMR can be used to show symmetry based on the molecular formula and the expected number of carbon signals. (a) This $C_5H_{10}O$ compound is *not* symmetrical, whereas (b) this $C_6H_{10}O_2$ compound *is* symmetrical.

Assessment 15.46 How many signals would you expect in the ^{13}C NMR spectrum of each molecule shown?

(a) (b) (c) (d)

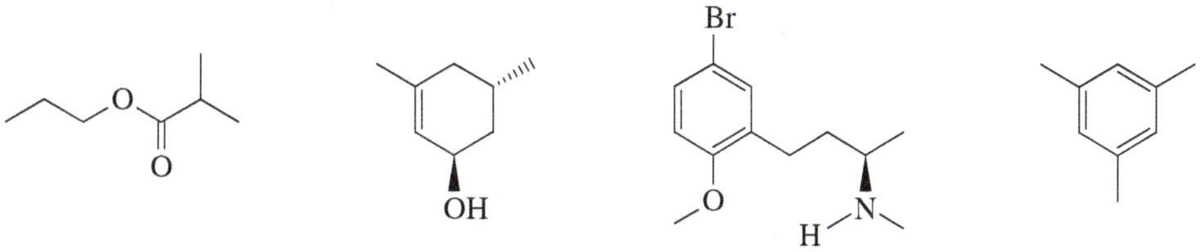

Assessment 15.47 Based on the ^{13}C NMR spectrum shown, does the compound with molecular formula $C_{13}H_{28}O$ have symmetry?

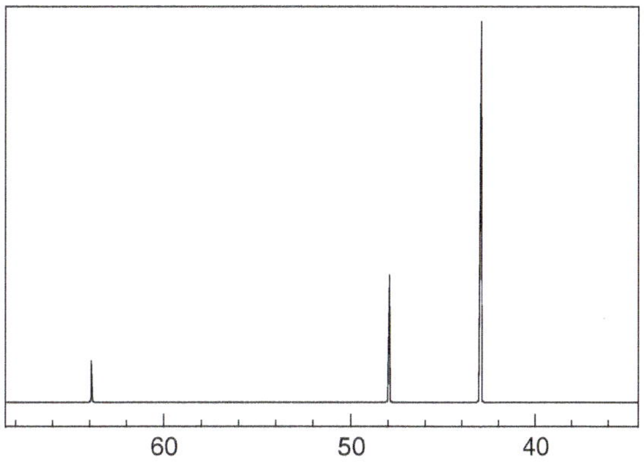

The location of the signals gives us information on the chemical environment in which the carbon resides. Similar to hydrogens, the presence of σ-bonded electrons shields carbons from the external magnetic field, making it easier for a ^{13}C nucleus to flip. In the presence of electron-withdrawing groups, the nuclei are deshielded, so they feel the external magnetic field more strongly and flip at a higher frequency (higher chemical shift). The chemical shift trends in ^{13}C NMR parallel the trends seen in ^{1}H NMR, just at a greater magnitude given the wider spectral range in ^{13}C NMR. **Figure 15.64(a)** shows the range of chemical shift values you'd expect for different functional groups. **Figure 15.64(b)** shows this same information, but further simplified.

(a)

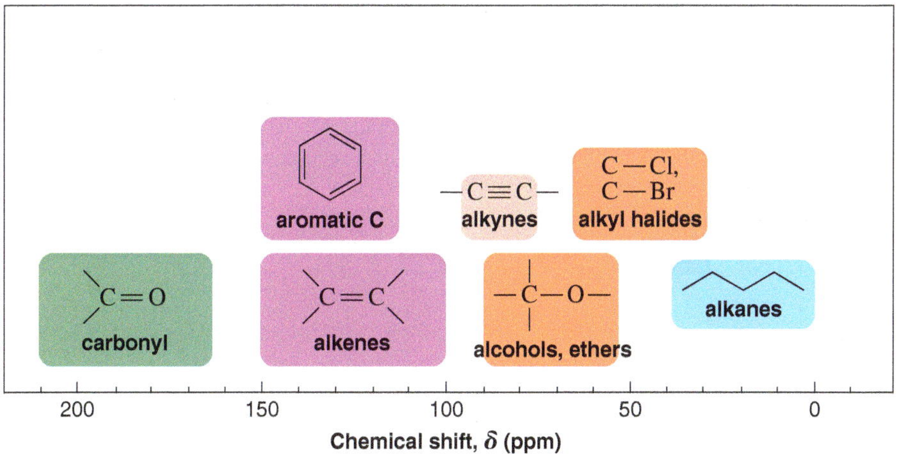

(b)

Carbon type	Chemical shift (ppm)
sp^3 C—C	0–40
sp^3 C—X (X = electronegative atom)	40–100
C≡C	80–100
C=C, aromatic	100–150
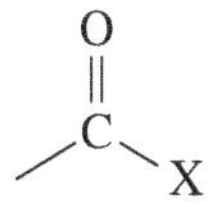	150–200
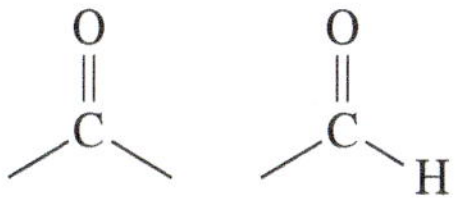	>200

Figure 15.64 Chemical shift values in ^{13}C NMR. (a) Expanded ranges. (b) Simple table.

Using the information in Figure 15.64, the signals in the ^{13}C NMR spectra of the molecules in **Table 15.11** can be assigned.

Table 15.11 **Assignment of ^{13}C NMR spectra.**

Molecule and spectrum	Explanation of assignment
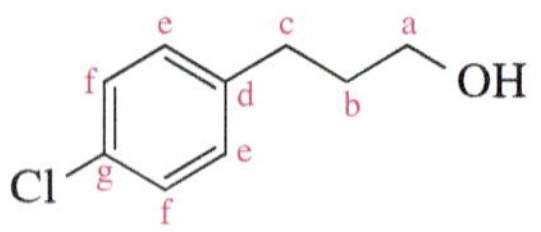	Carbons **b** and **c** are sp^3 hybridized and are not directly attached to any functional groups: 0–50 ppm. Carbon **a** is sp^3 hybridized and is connected to an electronegative oxygen: 50–100 ppm. Carbons **d–g** are sp^2 hybridized and aromatic: 100–150 ppm.
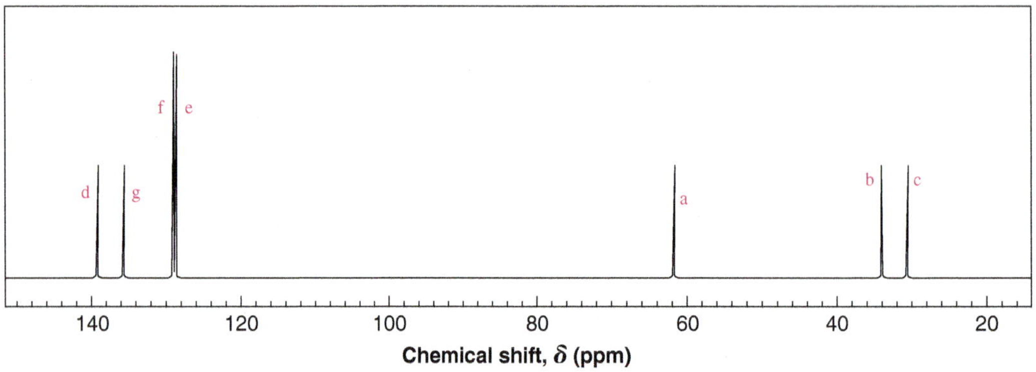	
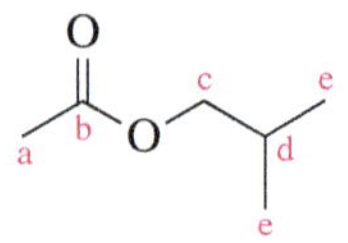	Carbons **a, d,** and **e** are sp^3 hybridized and are not directly attached to any functional groups: 0–50 ppm. Carbon **c** is sp^3 hybridized and is connected to an electronegative oxygen: 50–100 ppm. Carbon **b** is an sp^2-hybridized carbon of an ester: 150–200 ppm.
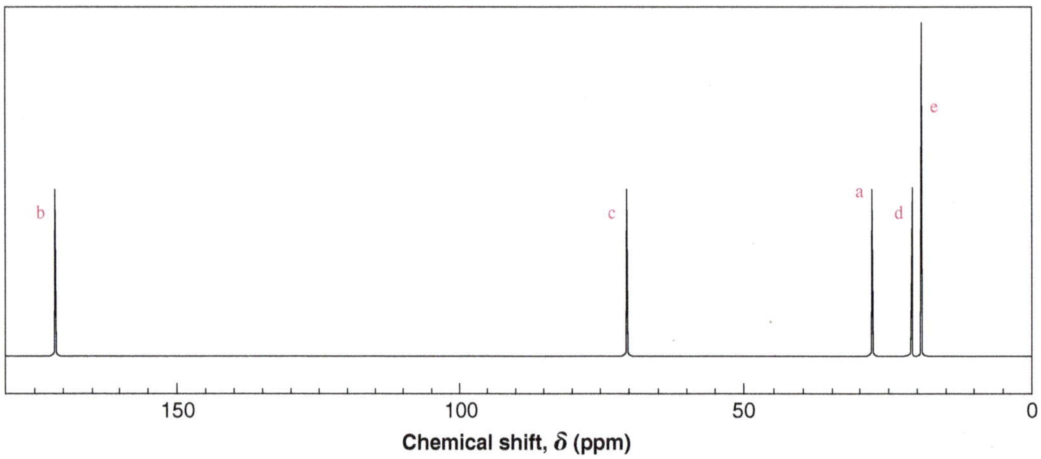	
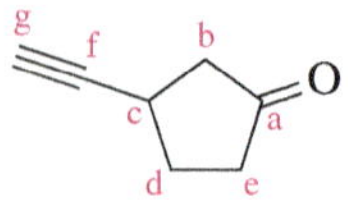	Carbons **b, c, d,** and **e** are sp^3 hybridized and are not directly attached to any functional groups: 0–50 ppm. Carbons **f** and **g** are the *sp*-hybridized carbons of an alkyne: 70–100 ppm. Carbon **a** is an sp^2-hybridized carbon of a ketone: >200 ppm.
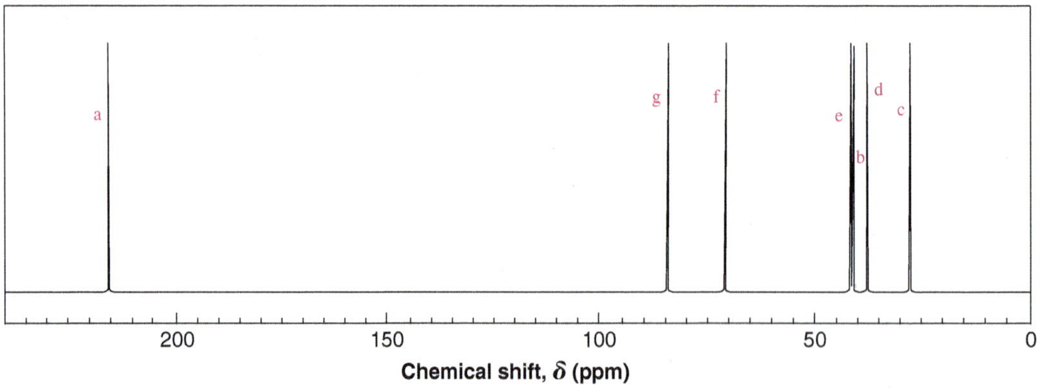	

Assessment 15.48 Suggest a reasonable chemical shift value for each set of equivalent carbons in the molecules shown. [*Use a number in the middle of the ranges in Figure 15.64.*]

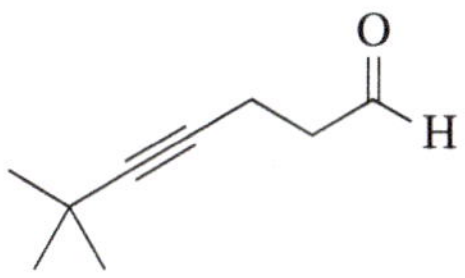

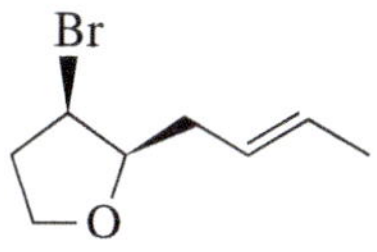

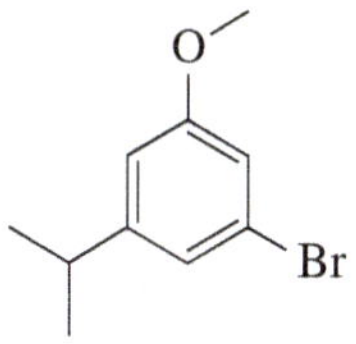

Assessment 15.49 Draw the ^{13}C NMR spectrum of each molecule in Assessment 15.48.

Assessment 15.50 Assign each signal in the ^{13}C NMR spectra to the molecule shown.

(a)

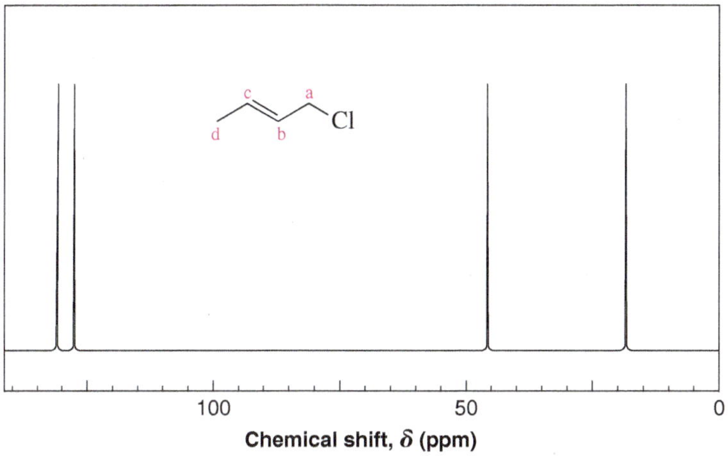

(b)

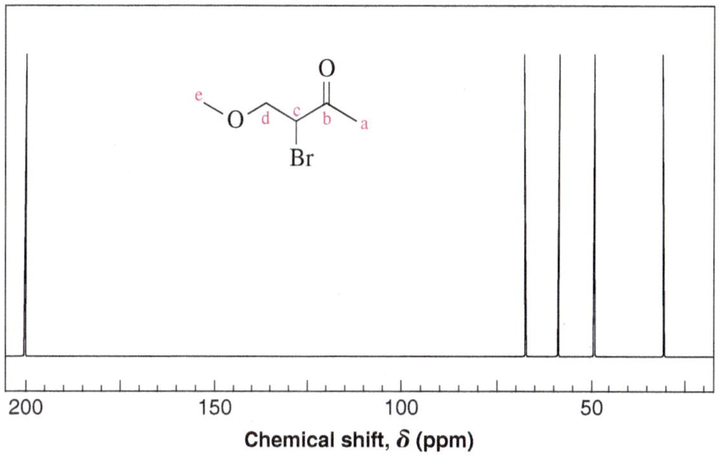

15.7.1 Putting it all together: structure determination

As you might expect, given the lack of information in a ^{13}C NMR spectrum, structure elucidation by ^{13}C NMR is difficult and is rarely used this way. To see why it's complicated, work Assessment 15.51 on your own, and then we'll work it together. [*This exercise will also help us understand the need for the advanced methods discussed in Section 15.8.*]

Assessment 15.51 Give the structure of a molecule (molecular formula = $C_6H_{14}O$) that corresponds to the ^{13}C NMR spectrum shown.

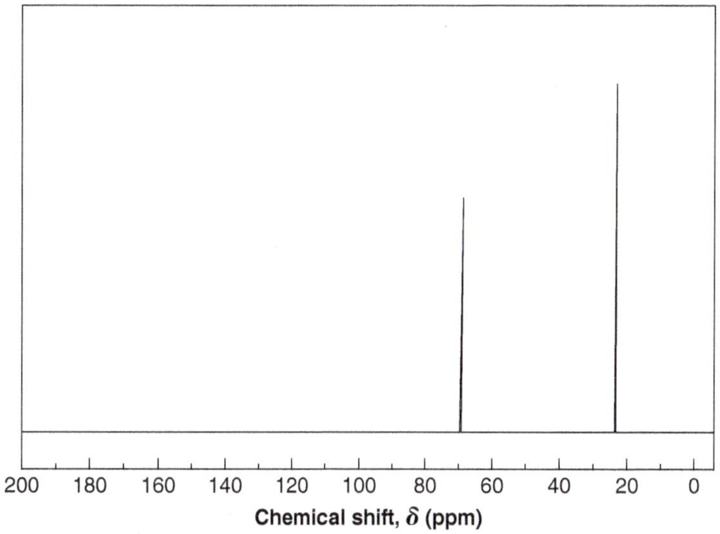

Solved Assessment 15.51

Always start by calculating the index of hydrogen deficiency.

$$\text{IHD} = \text{number of carbons} - \frac{\text{number of hydrogens}}{2} - \frac{\text{number of halogens}}{2} + \frac{\text{number of nitrogens}}{2} + 1 = 6 - \frac{14}{2} - \frac{0}{2} + \frac{0}{2} + 1 = 0$$

The IHD tells us that there are no π bonds or rings in the molecule. Knowing that there are no π bonds, the oxygen must be part of an ether or an alcohol.

There are only two peaks in the ^{13}C spectrum (**Figure 15.65**). With six carbons, there must be symmetry in the molecule. Using the chemical shift ranges above, the carbon at 63 ppm (peak **a**) must have an electronegative atom attached, while the carbon at 12 ppm (peak **b**) must be sp^3 hybridized and attached only to other carbons and hydrogens.

With only this information, we are left to logically figure out the structure using skills that aren't really NMR skills. Instead, what are all the possible ways that two peaks can represent six carbons? The ratios of peak **a** to peak **b** could only be 1:5, 2:4, 3:3, 4:2, or 5:1. [*Wouldn't it be great if we had integrals here?*] We know that peak **a** represents a carbon attached to an oxygen, and there is only one oxygen in the molecule, and oxygen can only have two carbons attached, so peak **a** can't represent any more than two carbons (**Figure 15.66**). This immediately rules out three possibilities, leaving the only two possible **a**:**b** ratios as 1:5 and 2:4.

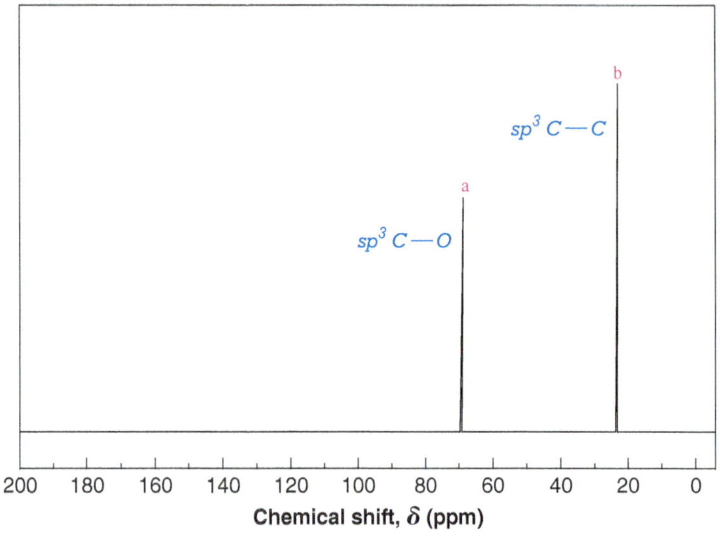

Figure 15.65 Label the peaks in the ^{13}C NMR spectrum.

Possibilities for C_a:

C—OH
*if peak **a** corresponds to 1 carbon*

C—O—C
*if peak **a** corresponds to 2 carbon*

C_a can't possibly be these. Why?

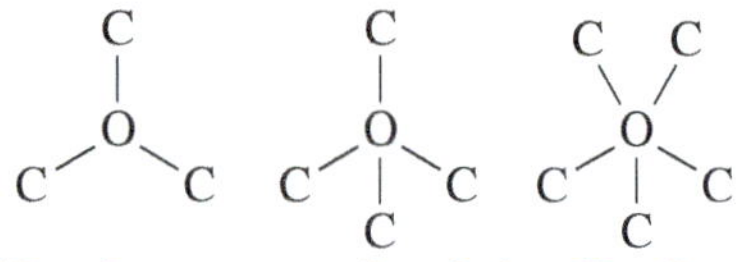

*if peak **a** corresponds to 3, 4, or 5 carbons*

Figure 15.66 Narrowing down the possibilities for peak **a**.

If peak **a** represents just one carbon, then the molecule must be an alcohol. If we were to start sketching molecules out, we'd start by drawing a C—OH, as shown in **Figure 15.67**. Our challenge now becomes including the five remaining carbons, where every one of the five is identical, and we have to do it without rings or π bonds (IHD = 0). Some of the molecules you might have drawn are in Figure 15.67. [*When you are doing a spectroscopy problem incorrectly, you'll know it. There will be a feeling of desperation where you are making blind guesses, hoping that your pencil will come up with a structure that you look back on and realize—hey, that's correct. That's how I feel right now and you should, too. Unfortunately, none of the molecules in Figure 15.67 are correct.*]

Possibilities for C_a:

C—OH

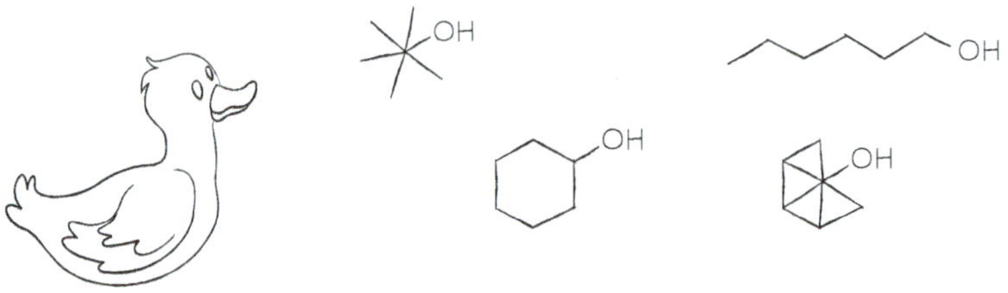

Figure 15.67 Some "possible" structures . . . but not even close.

If peak **a** represents two carbons, then the molecule must be an ether because both carbons are attached to an oxygen. Now, we can start by drawing C—O—C, as shown in **Figure 15.68**. At this point, it becomes a bit clearer that the symmetry plane will cut through the oxygen. And so we need to add two carbons to each side; there are only two ways to do this. The resulting molecules are di-*n*-propyl ether and di-*i*-propyl ether, both of which are much more reasonable than any of the structures we drew in Figure 15.67. Di-*n*-propyl ether should have three peaks in the ^{13}C NMR spectrum, leaving di-*i*-propyl ether as the last remaining possibility and the structure to which the spectrum corresponds. ■

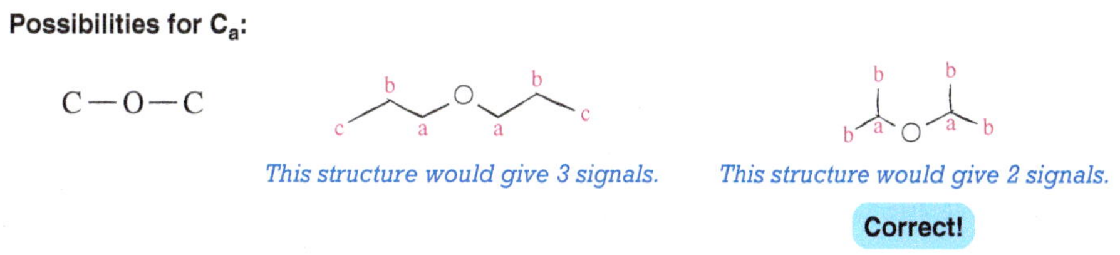

Figure 15.68 The only two reasonable possibilities lead to the correct answer.

There were two purposes to Solved Assessment 15.51. First, to show that ^{13}C NMR isn't very useful as a stand-alone technique. In addition to the molecular formula and a ^{13}C NMR, what if we had a mass spectrum, an IR spectrum, and a ^{1}H NMR spectrum to go along with the IHD? **Figure 15.69** shows how simple this would have been. From the IHD, it has to be an alcohol or ether. IR shows no O—H band, so it's an ether. ^{13}C NMR shows that there is symmetry. ^{1}H NMR has the common splitting pattern associated with *i*-propyl (1H, septet; 6H, doublet). MS confirms an isopropyl fragment (*m*/*z* 43). [*Boom! Structure determined! Pizza for everyone!*]

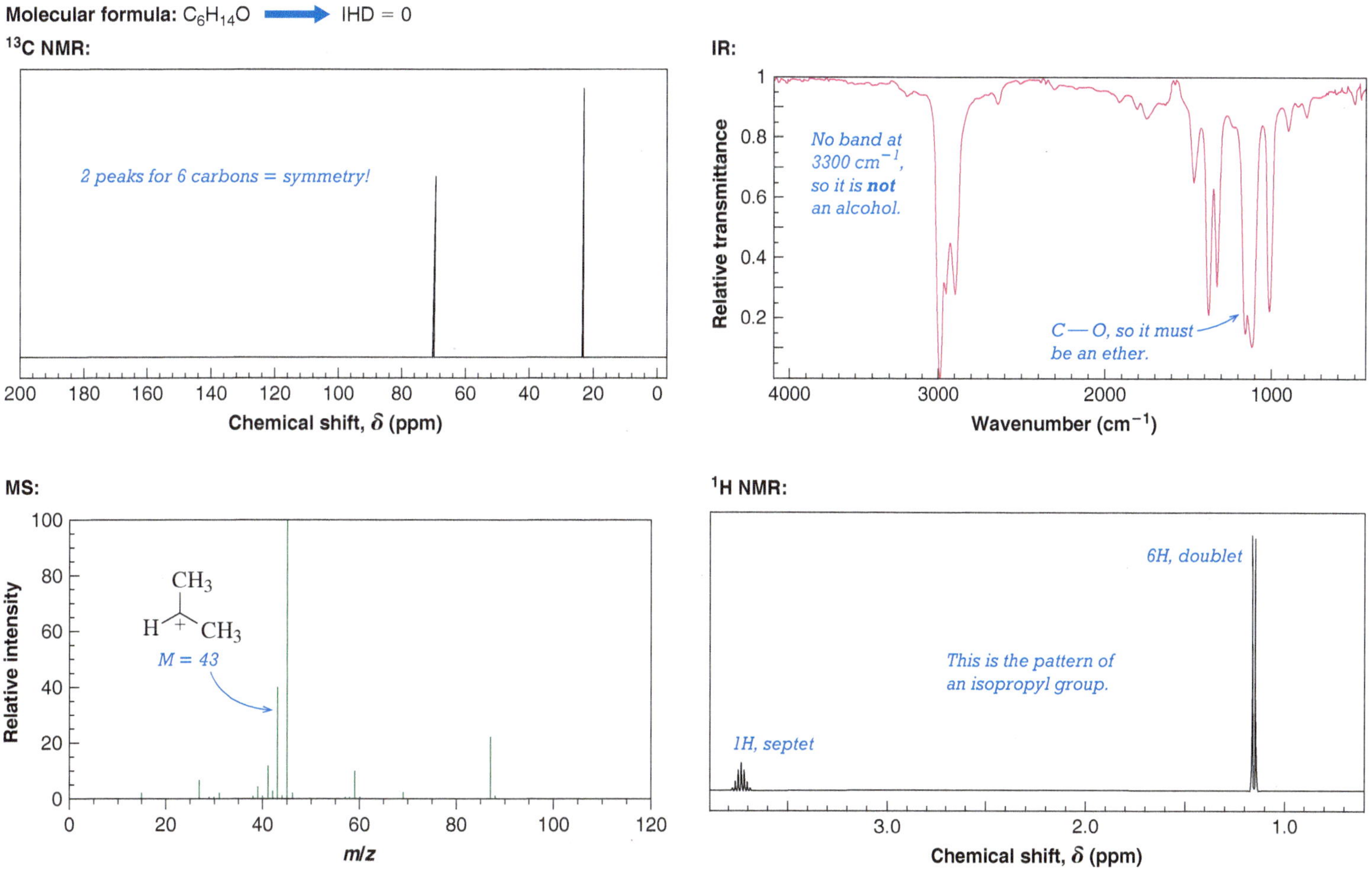

Figure 15.69 With more information, structure determination is greatly simplified.

The second purpose of Assessment 15.51 is to show that logic can get you to the right answer, even if the technique itself isn't adequate. And finally, by seeing how inadequate ^{13}C NMR is as a stand-alone technique, we can wonder, "How could ^{13}C NMR be reinvented to make it more useful?" [*Integrals, splitting—it turns out to be even better than that. Read on.*]

15.7.1 SUMMARY

- Because the signal for a carbon shows up only as a singlet, it is difficult to fully elucidate the structure of a molecule using only a ^{13}C NMR spectrum.

15-D CHEMISTRY IN CONTEXT: Natural Product Identification

Assigning the ^{13}C NMR spectrum of curacin A

^{13}C NMR on its own is not great for structure elucidation, but it can make clearer some of the information gleaned from other data. Let's go back to the ^{13}C NMR spectrum of curacin A, which is shown in **Figure 15.70**. First of all, we see 23 peaks, one for each carbon. [*The three tallest peaks around 128 are for the C_6D_6 solvent, if you're counting. And the peak nearest 132 is actually two peaks.*] Recall from Chemistry in Context 15-A that we had an index of hydrogen deficiency of 7. The 100–150 ppm region confirms that there are eight alkene carbons for a total of four alkenes. The peak at 168 ppm is normally indicative of a carbonyl carbon, but in this case it represents an imine (C=N). Regardless, the double bond indicated by this peak is the fifth degree of unsaturation. The remaining IHD of 2 can only be attributed to the two rings of curacin A. Finally, the peaks between 40 and 100 ppm are for carbons bound directly to an electronegative atom (O, O, N, and S). The remaining peaks are aliphatic sp^3-hybridized carbons.

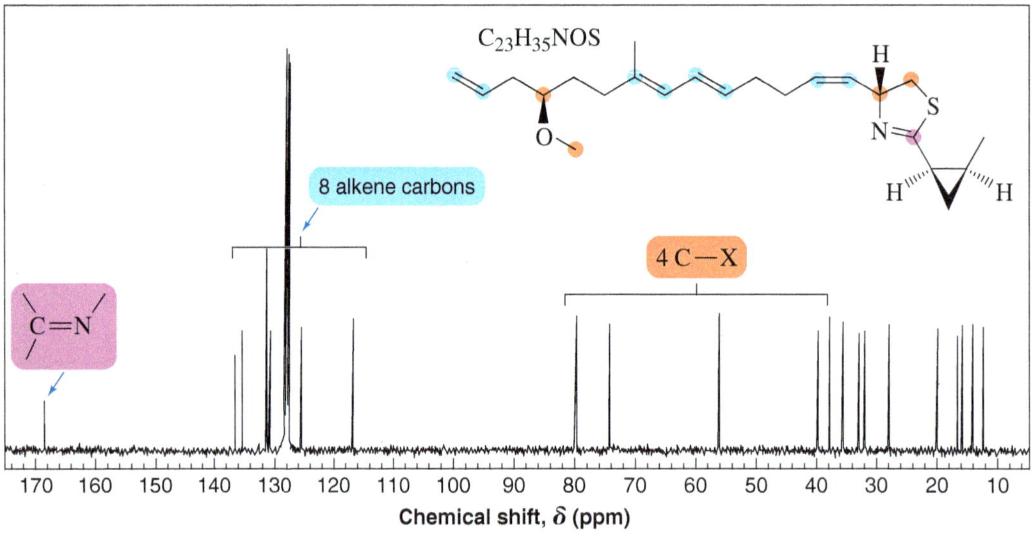

Figure 15.70 ^{13}C NMR of curacin A.

Assessment 15.52 An IHD > 4 often suggests a benzene ring, whose carbon signals will also fall in the range of 100 to 150 ppm. Using the ^{1}H NMR of curacin A (Chemistry in Context 15-B), how do you know there is no benzene ring in curacin A?

15.8 Advanced ^{13}C NMR topics

When a spectrum is obtained in proton-decoupled mode, it is simpler to read, but much information is lost. For simple molecules, by altering the pulse sequence, a spectrum can be run with **off-resonance decoupling**. Off-resonance decoupling generates a spectrum where each carbon is split only by attached hydrogens, according to the $N + 1$ rule. In this mode, a methyl group appears as a quartet (CH_3), a methylene (CH_2) as a triplet, a methine (CH) as a doublet, and a carbon with no hydrogens as a singlet. **Figure 15.71** shows both the simulated off-resonance decoupled and proton-decoupled spectrum of δ-valerolactone.

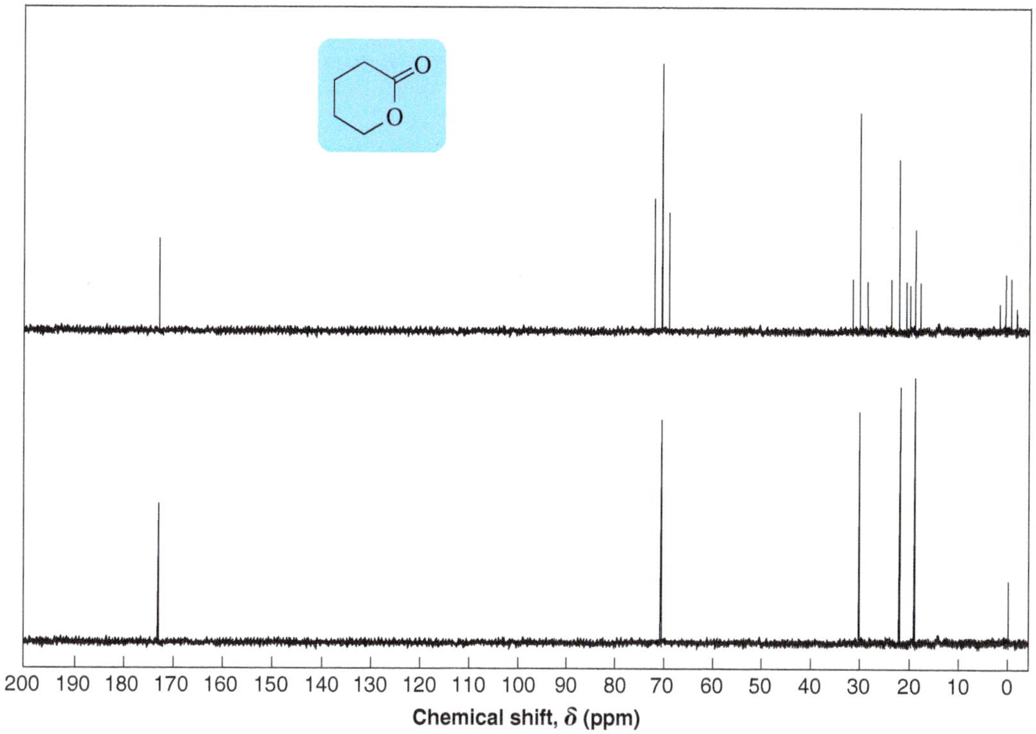

Figure 15.71 Simulated off-resonance decoupled and proton-decoupled spectrum of δ-valerolactone.

The simulated off-resonance decoupled spectrum of δ-valerolactone makes this seem like a useful method. And it can be for simple molecules. However, what Figure 15.71 does not approximate is the extremely large C—H coupling constants. Whereas coupling constants for neighboring hydrogens are in the 2–15 Hz range, C—H coupling constants are in the 100–300 Hz range. These larger coupling constants are a result of the closer proximity of the interacting magnetic fields between atoms directly attached to each other. **Figure 15.72** shows an off-resonance decoupled spectrum of phenyl acetate. There are only three carbons between 15 and 75 ppm, but the large coupling constants result in the peaks nearly covering this entire range. [*What would it look like if there were ten carbons in this range?*] Additionally, the benzene ring should contain three doublets and one singlet for its three methines (CH) and one carbon with zero hydrogens. Unless you knew what to look for, though, it would be almost impossible to identify these doublets around 125–145 ppm.

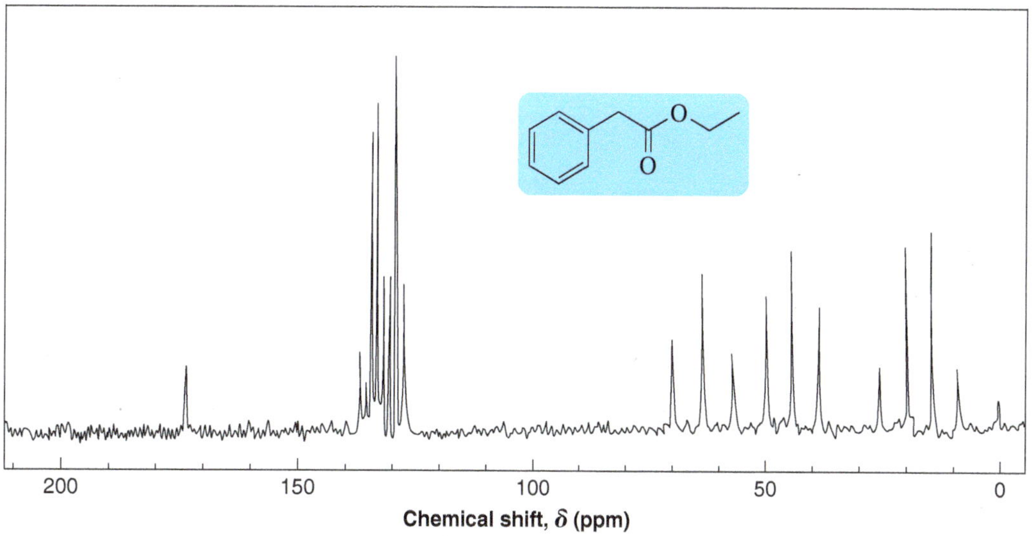

Figure 15.72 Off-resonance decoupled spectrum of a larger molecule.

Assessment 15.53 The C—H coupling constant depends on the carbon's hybridization as shown. Explain this observation.

Molecule	J_{C-H} (Hz)
CH_4	125
$H_2C{=}CH_2$	156
$HC{\equiv}CH$	249

Modern instruments have the capability to combine the simplicity of a proton-decoupled spectrum with the information of an off-resonance decoupled spectrum. In an experiment referred to as **DEPT** (distortionless enhanced polarization transfer), every signal is still a well-resolved and well-separated singlet, while telling whether each is a C, CH, CH_2, or CH_3. This is done by acquiring three different spectra using slightly different pulse sequences: normal, DEPT-90 and DEPT-135. [*This is the second time pulse sequences have been mentioned; don't worry about what that means. It has only to do with how and for how much time the radiofrequency is transmitted to the nuclei.*] Table 15.12 shows how the three sequences distinguish between carbon types. The normal ^{13}C NMR spectrum shows a signal (singlet) for every carbon in the molecule. The DEPT-90 spectrum shows a signal (a singlet) for only methine (CH) carbons. The DEPT-135 spectrum shows a signal for any carbon containing a hydrogen, but the signal for a methylene (CH_2) is a negative peak, extending below the baseline. From this information, the type of carbon present can be logically deduced.

Table 15.12 Summary of DEPT spectra.

	Type of ^{13}C	Number of protons	Proton-decoupled ^{13}C NMR	DEPT-90	DEPT-135
Quaternary	—C— (with bonds above and below)	0			
Methine	—C—H (with bonds above and below)	1			
Methylene	—CH_2 (with bond above)	2			
Methyl	—CH_3	3			

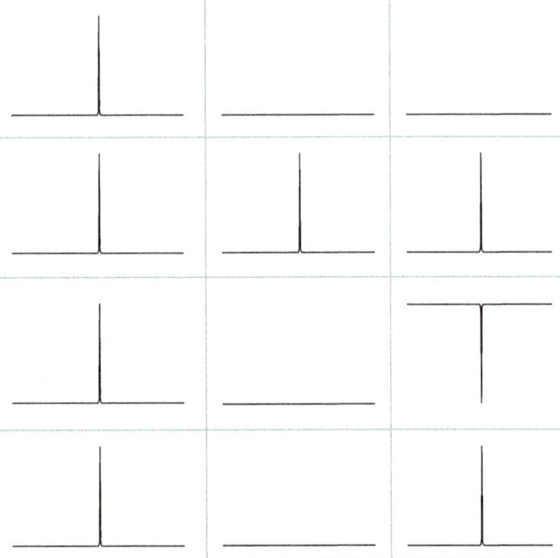

Figure 15.73 shows how these data can be interpreted for but-3-en-2-one. Figure 15.73(a) is the normal spectrum, showing four singlets, one for each carbon in the molecule. Figure 15.73(b), the DEPT-90 spectrum, shows only one signal at δ 137 ppm, indicating that this peak represents the CH at carbon **c**. Figure 15.73(c), the DEPT-135 spectrum, shows three signals. The negative peak at δ 129 ppm directly implicates this as representing the CH_2 of carbon **d**. The signal at δ 26 ppm, for which there is no positive peak in the DEPT-90, must represent the C_a methyl group. Finally, the peak at δ 199 ppm, appearing in neither the DEPT-90 nor DEPT-135, must have no hydrogens attached, making it carbon **b**.

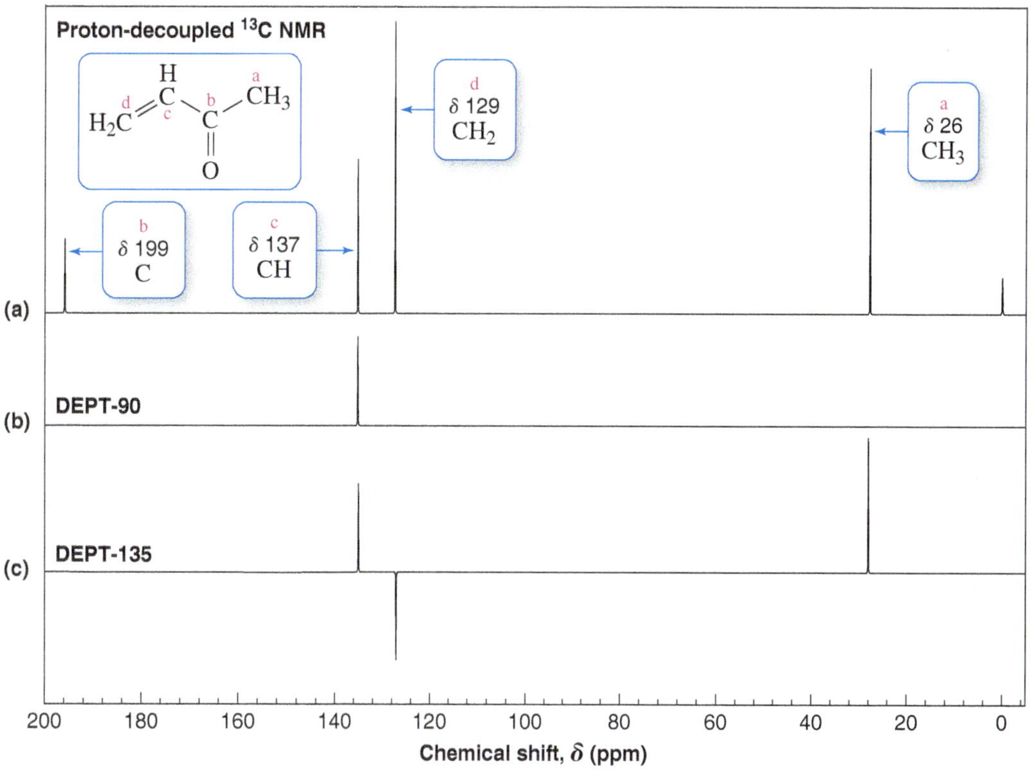

Figure 15.73 DEPT spectrum of but-3-en-2-one. (a) The normal proton-decoupled ^{13}C NMR spectrum. (b) The DEPT-90 gives a signal for all CH groups. (b) The DEPT-135 gives a positive signal for CH and CH_3 groups and a negative signal for CH_2 groups.

Figure 15.73 shows how useful the DEPT experiment is for assigning a spectrum to a known compound. To see how it is used for structure identification, the DEPT spectra for the unknown of Assessment 15.51 are shown in **Figure 15.74**. [*Before doing the analysis here, go back and look at the struggle that we went through in Solved Assessment 15.51.*]

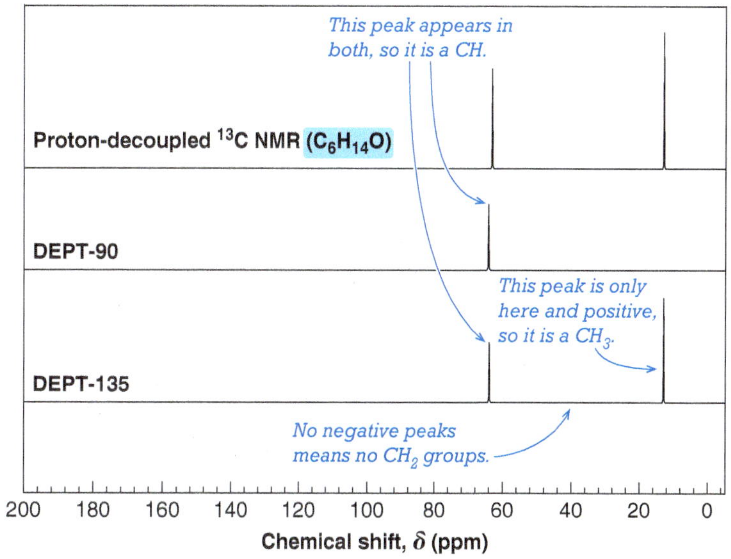

Figure 15.74 The DEPT spectra of unknown $C_6H_{14}O$ make structure identification straightforward.

The analysis of this spectrum (**Figure 15.75**) is as simple as (a) calculating the index of hydrogen deficiency from the molecular formula and (b) recognizing that there is symmetry from the ^{13}C NMR spectrum. (c) The DEPT-90 spectrum shows that the carbon bound to oxygen is a methine (CH). Bound to an oxygen and a hydrogen, this carbon has two additional groups attached (to be determined). (d) The shielded carbon δ 12 ppm is a methyl group, representing the only other carbons in the molecule. As a result, the two additional groups must be methyl groups. At this point we know we have at least an oxygen bound to an isopropyl group. (e) Based on the symmetry and the molecular formula, there must be a second isopropyl group attached to the oxygen, making the unknown molecule diisopropyl ether.

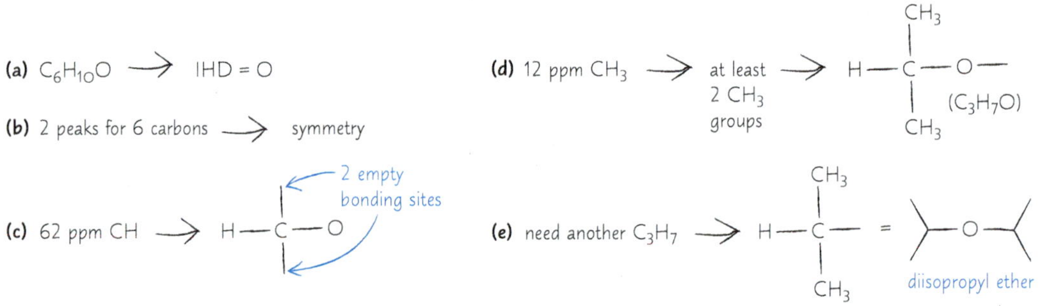

Figure 15.75 The structure is identified as diisopropyl ether.

Assessment 15.54 Draw the expected results of a DEPT sequence for the molecules shown.

(a)

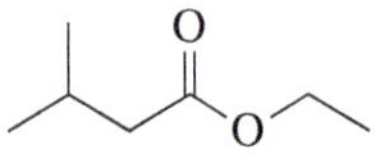

(b)

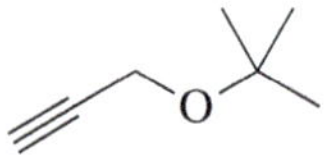

Assessment 15.55 Provide a structure that corresponds to the given DEPT spectra and the molecular formula.

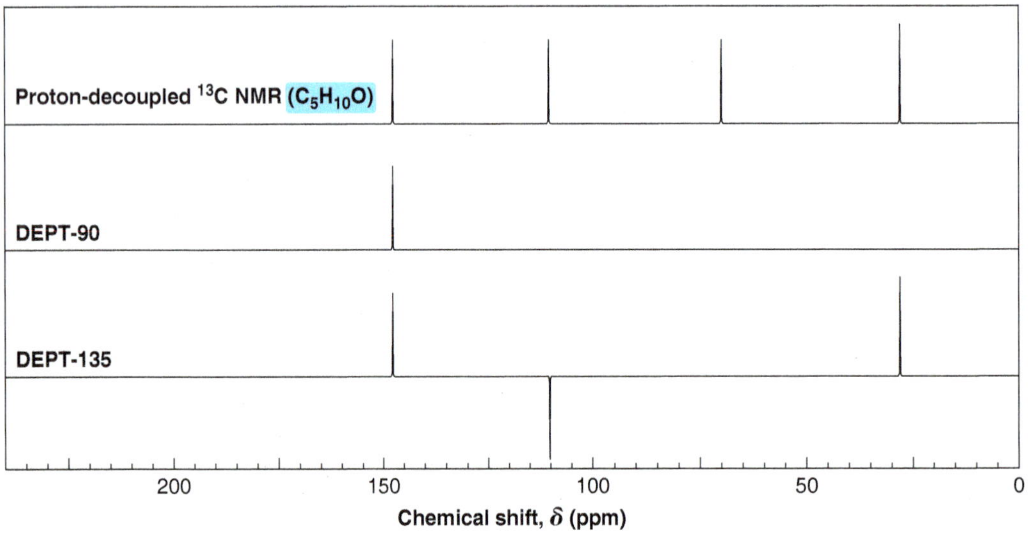

Assessment 15.56 With information from the ^{1}H NMR and the ^{13}C DEPT spectra, structure elucidation becomes even easier. Provide the structure that corresponds to the following data. [*The identity of the carbons comes from the DEPT experiment.*]

$C_6H_{11}BrO_2$

IR: 1745 cm^{-1}

^{1}H NMR: δ 1.25 (t, 3H), 2.18 (quint, 2H), 2.58 (t, 2H), 3.46 (t, 2H), 4.15 (q, 2H)

^{13}C NMR: δ 14.2 (CH_3), 27.8 (CH_2), 32.5 (CH_2), 32.6 (CH_2), 60.5 (CH_2), 172.4 (C)

Concluding thoughts

Given curacin A's potential as an anticancer drug, its structural identification is only the beginning of the story. Because isolation from a natural source produces only a tiny amount, to more fully explore its biological activity, it is necessary to synthesize it. Let's do that in Chapter 16!

The structural elucidation of unknown natural products is a tedious and time-consuming task. In conjunction with isolating the biologically important molecules themselves, structure elucidation can be the topic of an entire Ph.D. project. The snippets about curacin A threaded throughout this chapter tell only a tiny part of the story. And as you might expect, just looking at the two spectra shown [^{1}H NMR (Chemistry in Context 15-B) and ^{13}C (Chemistry in Content 15-D)] is not nearly enough to complete the task of determining the structure for a molecule of this complexity. There are other NMR experiments [*not discussed in this chapter*] that played a key role in this work: two-dimensional NMR experiments that can tell you which hydrogens are near each other [*a COSY experiment*], which hydrogens are attached to which carbons [*a HETCOR experiment*], and even which carbons are two or three carbons away from which hydrogens [*an HMBC experiment*]. The combination of these experiments created more and more specific pieces of the puzzle, ultimately leading to the curacin A structure. [*Do these 2-D experiments sound interesting? If so, perhaps you are a future natural products chemist.*]

End-of-Chapter Materials

Important Concepts

[*These are the major concepts covered in each section, excluding reactions.*]

15.1: The energy of a photon of light is proportional to the frequency and inversely proportional to the wavelength. Electromagnetic radiation in the radio wave region of the electromagnetic spectrum corresponds to the energy required to flip a nucleus.

15.2: For each unique set of equivalent hydrogens in a molecule, a signal is generated in an NMR spectrum. The chemical shift of the signal gives information about the electronic environment of the hydrogen responsible for it. The integration of the signal tells how many of that type of hydrogen are present in the molecule. The splitting pattern of the signal tells how many hydrogens are on neighboring carbons. Using the information contained within each signal in the NMR spectrum, a structure can be generated.

15.3: Nuclear spin can be aligned with (α) or against (β) an applied magnetic field. The energy required to flip a nucleus from α to β depends on the strength of the magnetic field felt by the nucleus.

15.4: The influence of neighboring atoms, π bonds, and other hydrogens on the magnetic field experienced by individual hydrogens is responsible for the chemical shift and multiplicity that result for each signal. When a nucleus is shielded from the magnetic field, it requires less energy to flip from α to β, resulting in a lower chemical shift for that nucleus. A deshielded nucleus requires more energy to flip from α to β, resulting in a higher chemical shift for that nucleus. Hydrogens that can participate in hydrogen bonding often give broad, sometimes nonexistent, signals across a wide range of chemical shifts in an NMR spectrum. The $N + 1$ rule says that the signal for a hydrogen with N neighbors will appear in the spectrum as a multiplet with $N + 1$ peaks.

15.5: The coupling constant (J value) is a measure of the interaction between the magnetic fields of neighboring nuclei and can be measured as the distance between two peaks of a multiplet. The J value is a function of the dihedral angle and, so used, can give information about the relative stereochemistry of neighboring hydrogens. The average coupling constant is 7 Hz and most commonly appears when there is free rotation around the bond that separates neighboring hydrogens. A splitting tree can be constructed to rationalize the appearance of nonstandard multiplicities. Diastereotopic hydrogens are chemically nonequivalent and therefore split each other, while appearing at different chemical shifts.

15.6: ^{13}C NMR spectroscopy is similar in theory to ^{1}H NMR spectroscopy with electromagnetic radiation in the radio wave region required to flip ^{13}C nuclei from α to β. Because ^{13}C is of limited abundance relative to ^{12}C in organic molecules, a ^{13}C NMR spectrum is acquired using more scans than ^{1}H NMR. A ^{13}C spectrum is obtained in proton-decoupled mode so that each carbon is represented as a singlet.

15.7: A ^{13}C NMR spectrum shows one singlet for each unique set of equivalent carbons contained in a molecule. The chemical shift range is 0–250 ppm. ^{13}C NMR isn't usually amenable to full structure determination except in simple molecules.

15.8: When run in off-resonance decoupling mode, each signal in a ^{13}C NMR is split by the number of attached hydrogens, making this a useful technique for structure identification of simple molecules. Similar information is available using a DEPT experiment, an experiment that is amenable to more complex molecules.

End-of-Chapter Assessments

[*Assessments are ranked according to their difficulty; •••• = most difficult.*]

Be sure to take breaks while doing these to maximize your retention of the material. See Section 1.4 for why this is important.

^{1}H NMR boot camp

Assessments 15.57–15.64 steadily increase in difficulty while offering a repetitive experience that will help you interpret ^{1}H NMR spectra.

15.57 (•) How many distinct signals would you expect to see in the ^{1}H NMR spectrum of the following molecules? [*Ignore diastereotopic hydrogens for the sake of this assessment.*]

(a) (b) (c) (d) (e) (f)

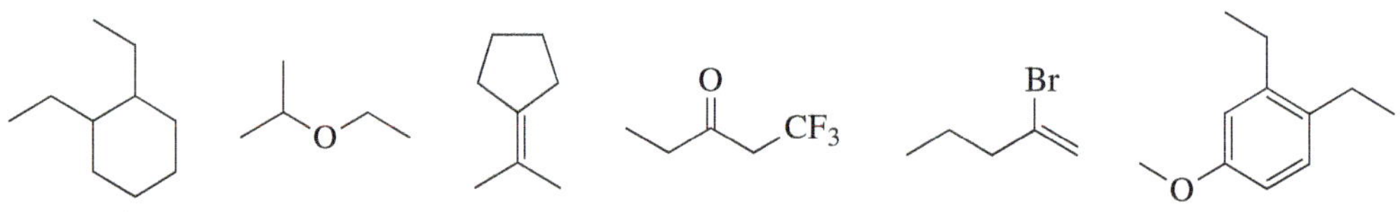

15.58 (•) For the following molecules, give the integration you would expect for the signal associated with the hydrogens at the labeled carbons. [*Pay attention to the symmetry, or lack of symmetry, in the molecules.*]

(a) (b) (c) (d) (e) (f)

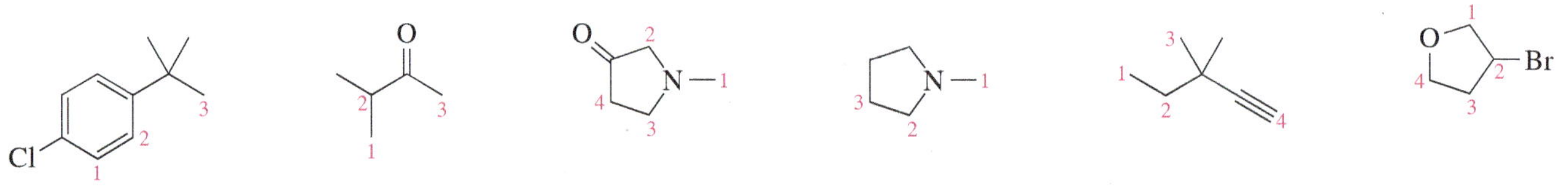

15.59 (•) For the following molecules, give the chemical shift for each indicated hydrogen.

(a) (b) (c) (d) (e) (f)

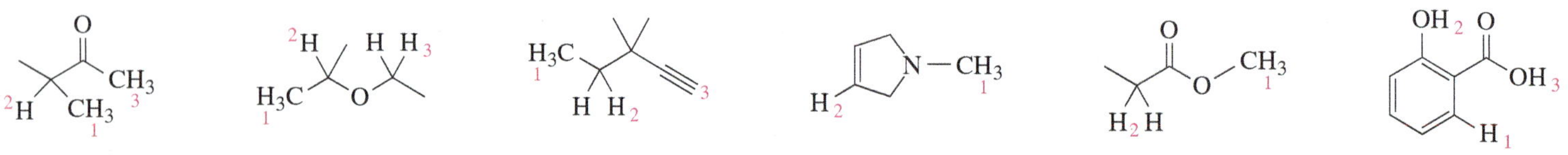

15.60 (•) Predict the splitting pattern for each of the indicated hydrogens in Assessment 15.59.

15.61 (•) Draw the signal for the following multiplicities. What is the ratio of peaks within each signal?

(a) doublet (c) quartet (e) sextet
(b) triplet (d) quintet (f) septet

15.62 (••) For the hydrogen(s) screened in blue, draw the signal you would expect to see in a ^{1}H NMR spectrum. At which chemical shift would the signal appear?

(a) (b) (c) (d) (e)

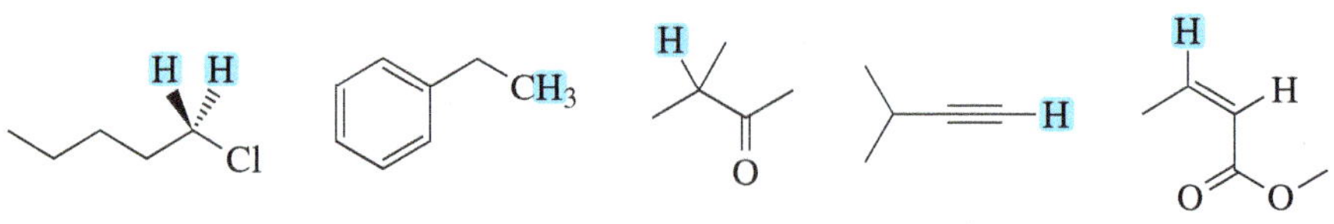

15.63 (•••) Complete the table of ^{1}H NMR data you'd generate for each of the following molecules.

(a) (b) (c) (d)

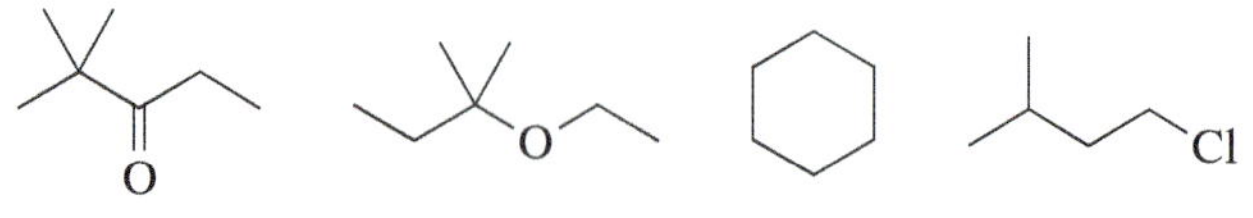

15.64 (•••) Draw the ^{1}H NMR spectrum you would expect to see for each of the molecules in Assessment 15.63.

^{13}C NMR boot camp (Assessments 15.65–15.68)

15.65 (•) How many distinct signals would you expect to see in the ^{13}C NMR spectrum of the molecules shown in Assessment 15.57?

15.66 (•) For the molecules in Assessment 15.58, give an approximate chemical shift for each indicated carbon. [*The range of correct answers is large here.*]

15.67 (•••) Draw the ^{13}C NMR spectrum you would expect to see for each of the molecules shown.

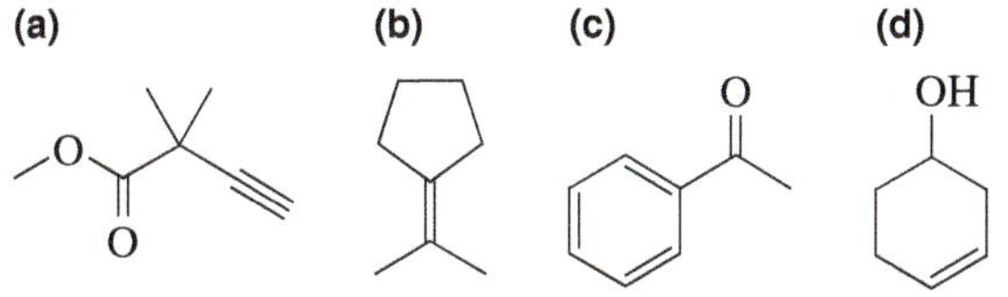

15.68 (•••) Why does it take more time to acquire a ^{13}C NMR spectrum than it does to obtain a ^{1}H NMR spectrum?

15.69 (•••) Assign the peaks in the ^{1}H NMR spectrum for the molecule shown.

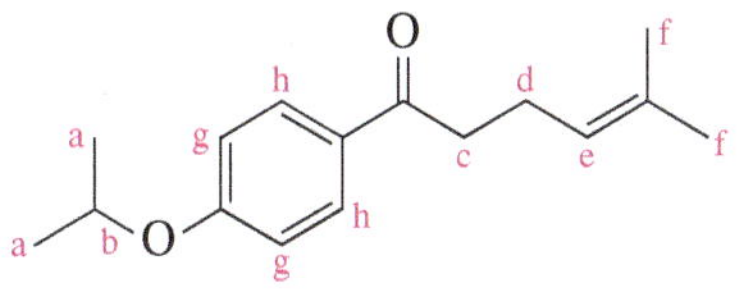

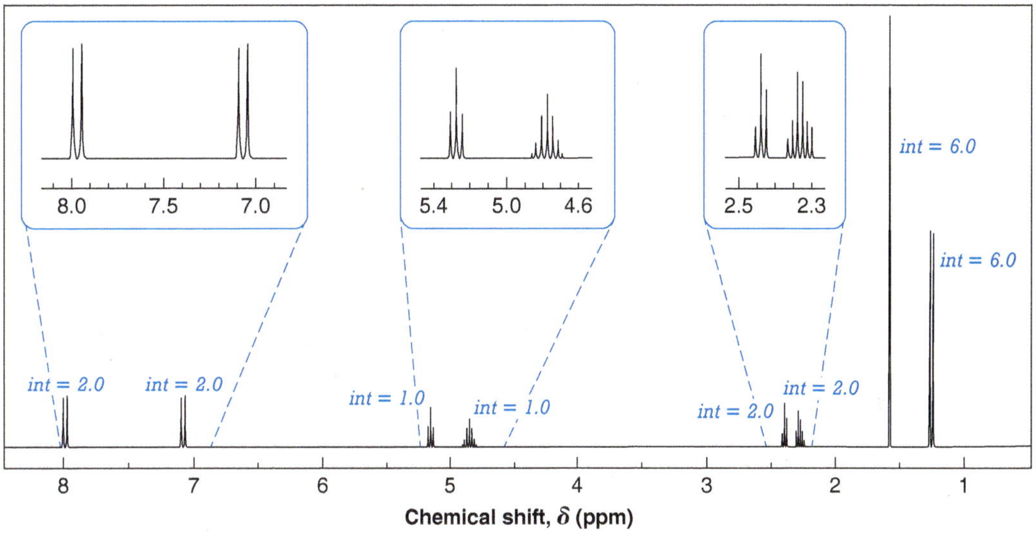

15.70 (••••) Using the spectra provided, identify the unknown compound with the molecular formula $C_9H_{11}NO_2$.

IR:

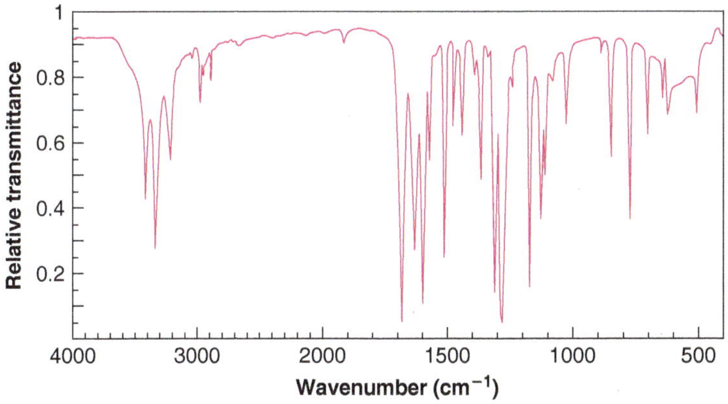

^{13}C NMR:

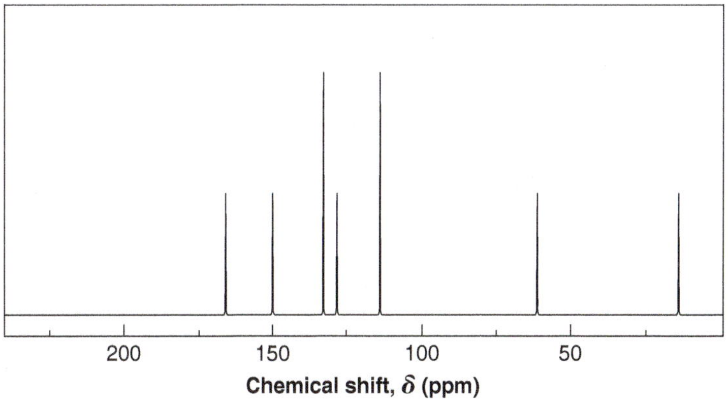

^{1}H NMR:

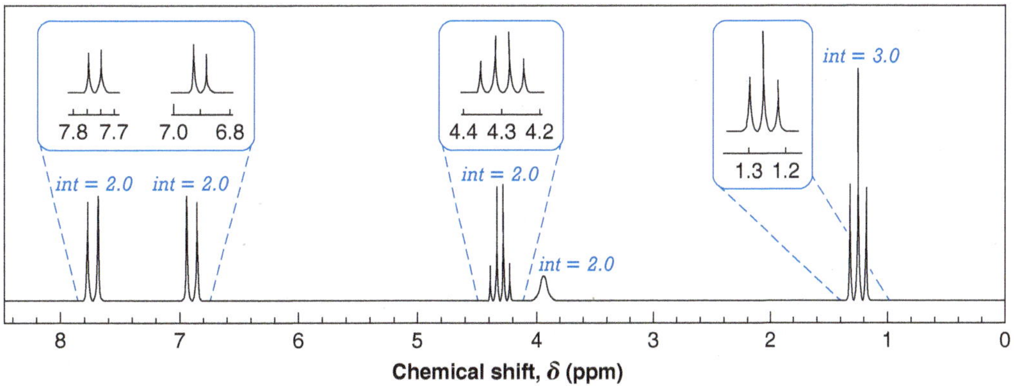

15.71 (••••) Using the spectra provided, identify the unknown compound with the molecular formula $C_9H_8O_4$.

IR:

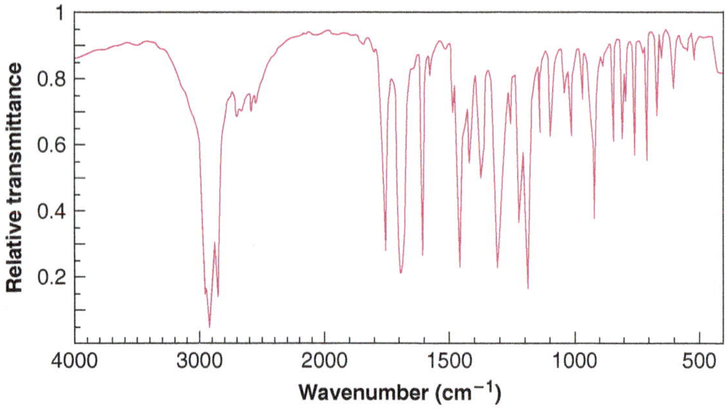

^{13}C NMR:

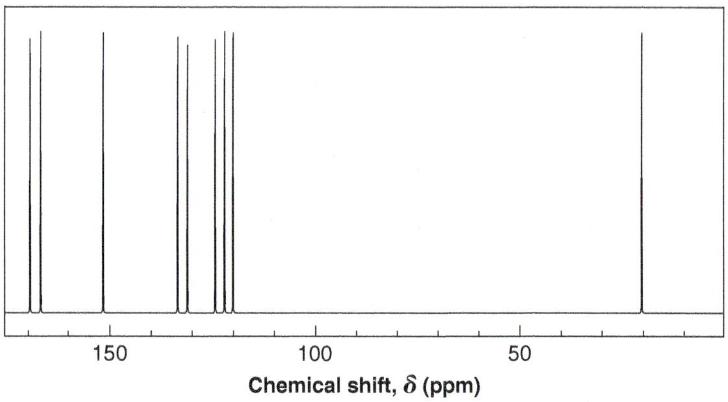

^{1}H NMR:

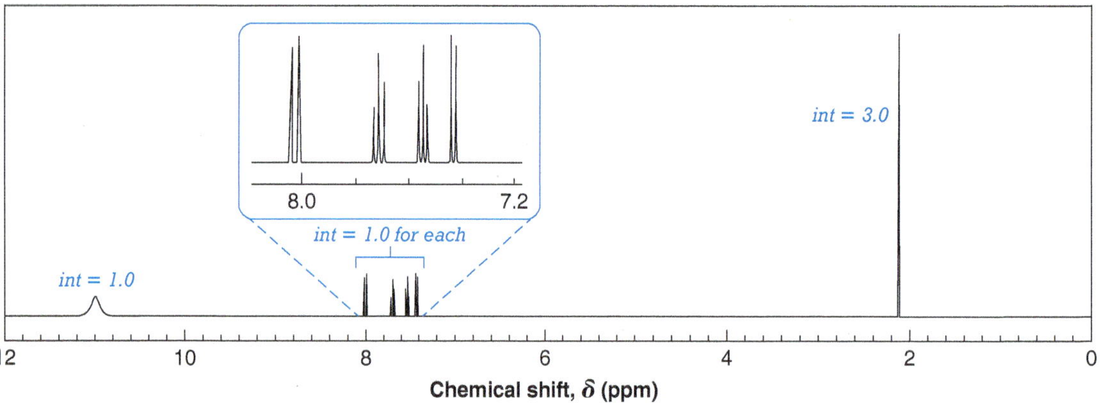

15.72 (•) Can NMR distinguish between the following two molecules? Why or why not? [*Recall Chapter 6.*]

15.73 (••) Esterification of the alcohols in Assessment 15.72 with Mosher's acid will allow them to be distinguished. Draw the ester products and tell why NMR can now be used to differentiate between them.

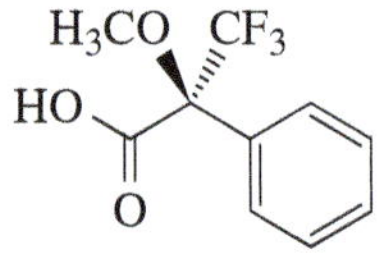

Mosher's acid

15.74 (•••) A graduate student ran a reaction that produced a mixture of the following two compounds. After painstaking purification, she is able to separate the two compounds. Using ^{1}H NMR, how can she determine which diastereomer is in which separated sample?

15.75 (•••) The following compound is an antifungal medication. Sketch its ^{1}H NMR spectrum.

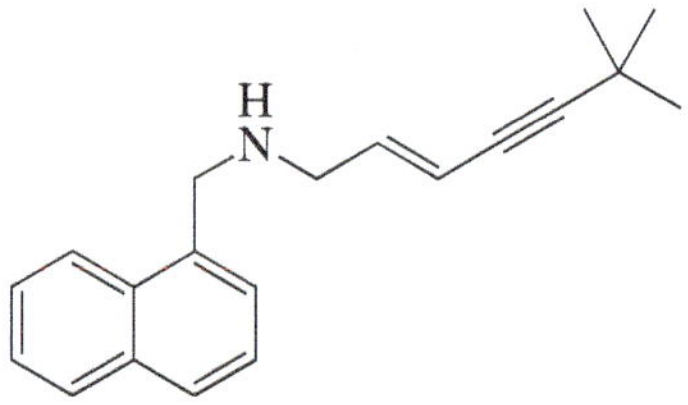

15.76 (••••) For the compound shown, produce a table of the shift, integration, and multiplicity of each peak you would expect to see in a ^{1}H NMR spectrum.

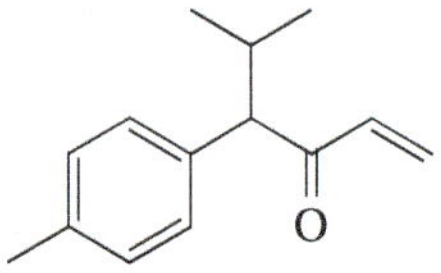

15.77 (•••) In the lab, ^{1}H NMR is often used to verify that a reaction has worked as expected by comparing the product spectrum with what is expected. Given the ^{1}H NMR of the reactant shown, draw the spectrum you'd expect to see of the product that results.

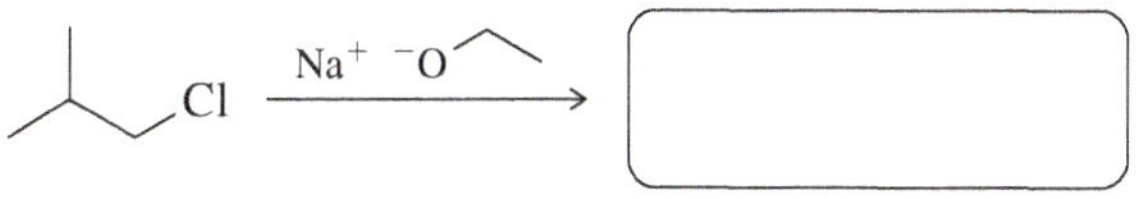

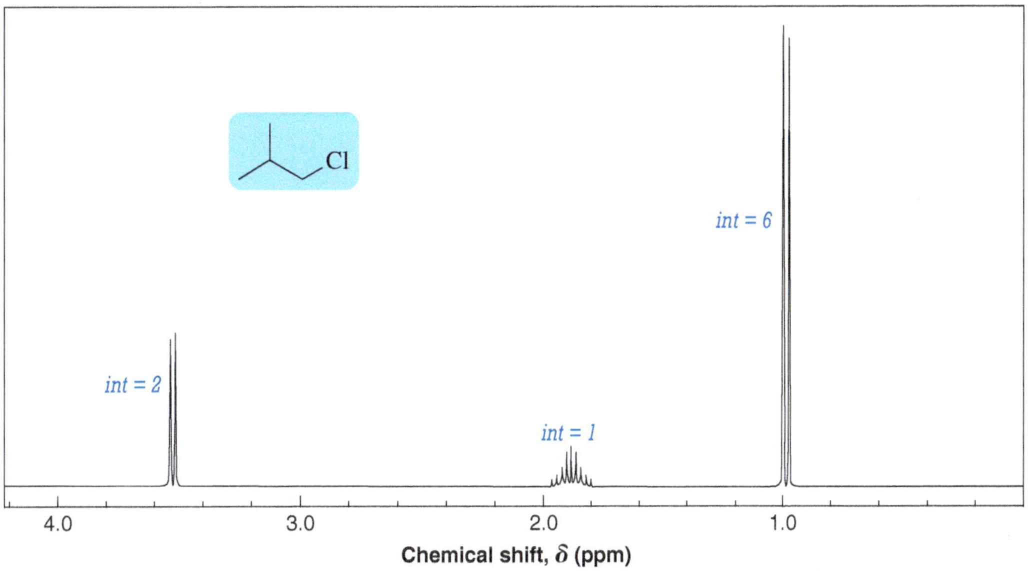

15.78 (•••) Given the ^{1}H NMR spectrum and the molecular formula, suggest a structure for each molecule. [*The IR spectrum suggests the presence of two C=O bonds.*]

(a) Important IR bands (cm^{-1}): 1728, 1708
[Note: The pK$_a$ of this molecule is around 10.]

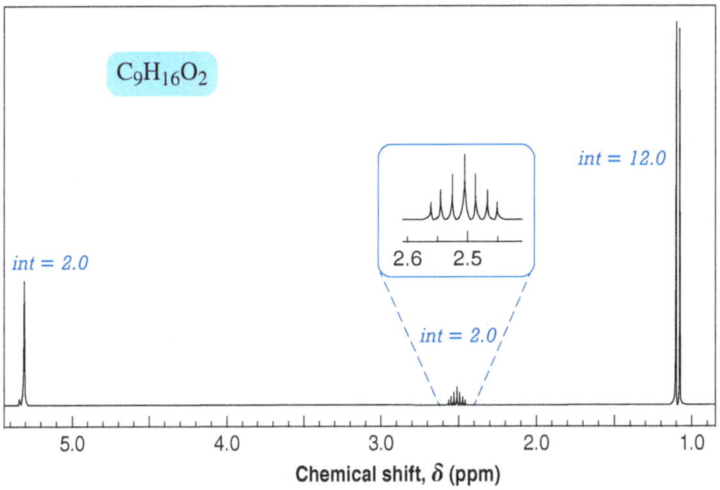

(b)

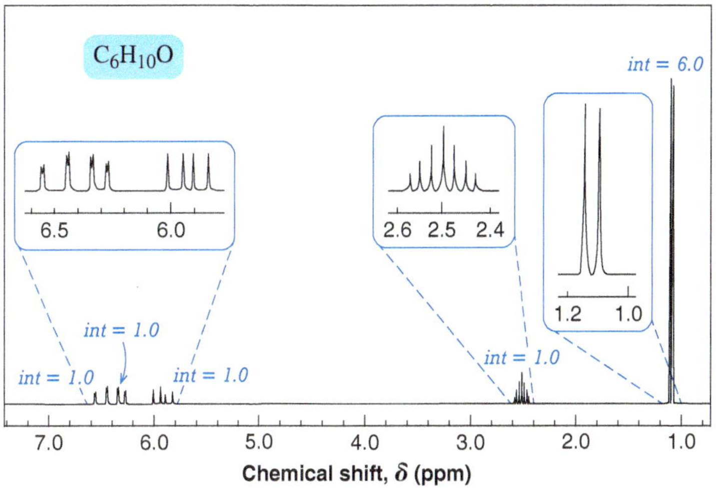

(c) Important IR bands (cm^{-1}): 3400–2500, 1610, 1450

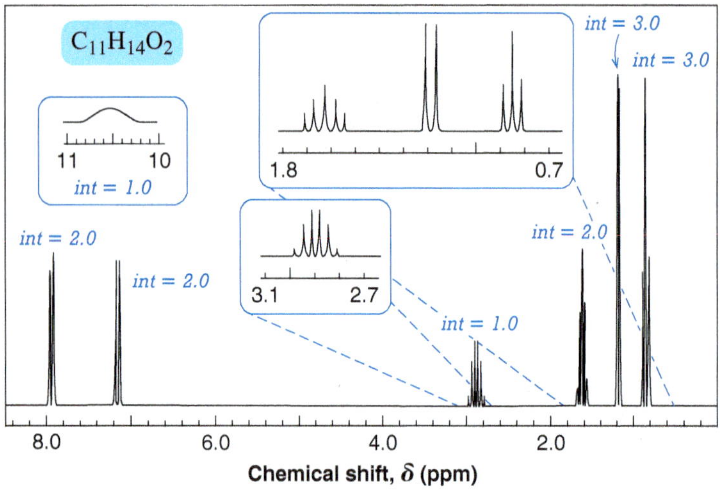

15.79 (•••) (a) Using the ^{1}H NMR spectra, identify the reactants and product for this reaction. (b) Provide an arrow-pushing mechanism for the reaction.

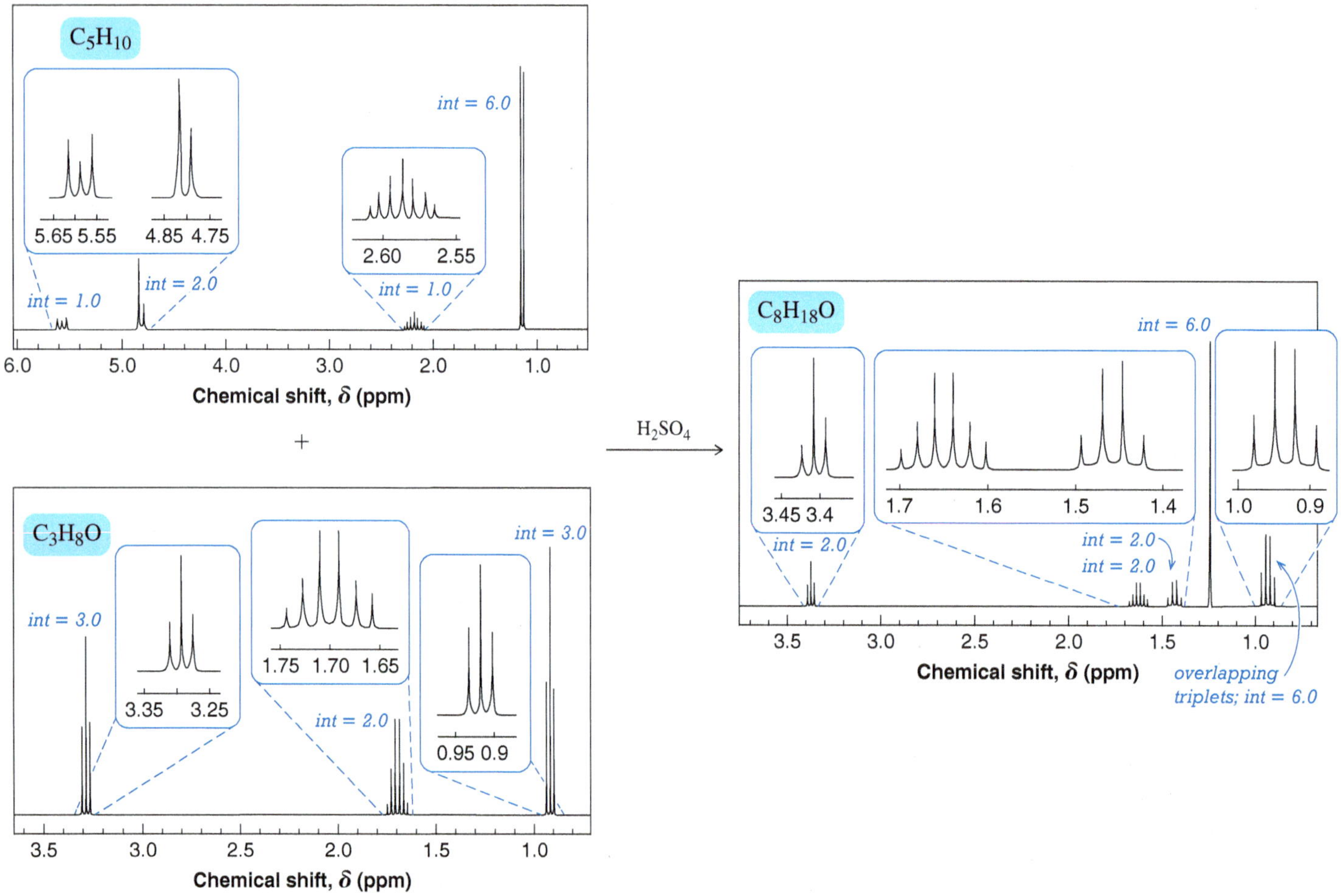

15.80 (•••) The spectral data below are presented in a manner similar to what you would find in a chemistry research journal. Identify the structure for each set of data.

(a) $C_4H_6O_2$: ^{1}H NMR: δ 2.14 (2H, pentet), 2.45 (2H, triplet), 4.38 (2H, triplet); IR (cm^{-1}): 1745, 1200

(b) $C_5H_{13}N$: ^{1}H NMR: δ 1.17 (6H, d, J = 6.6 Hz), 2.35 (6H, s), 2.72 (1H, sept, J = 6.6 Hz)

(c) $C_7H_{14}O_2$: ^{1}H NMR: δ 0.91 (3H, t, J = 7.0 Hz), 1.11 (6H, d, J = 7.0 Hz), 1.82 (2H, sextet), 2.40 (1H, sept, J = 7.0 Hz), 4.14 (2H, t, J = 7.1 Hz); IR (cm^{-1}): 1745, 1200

15.81 (•••) (a) Using the ^{1}H NMR spectra, identify the reactants and product for this reaction. (b) Provide an arrow-pushing mechanism for the reaction.

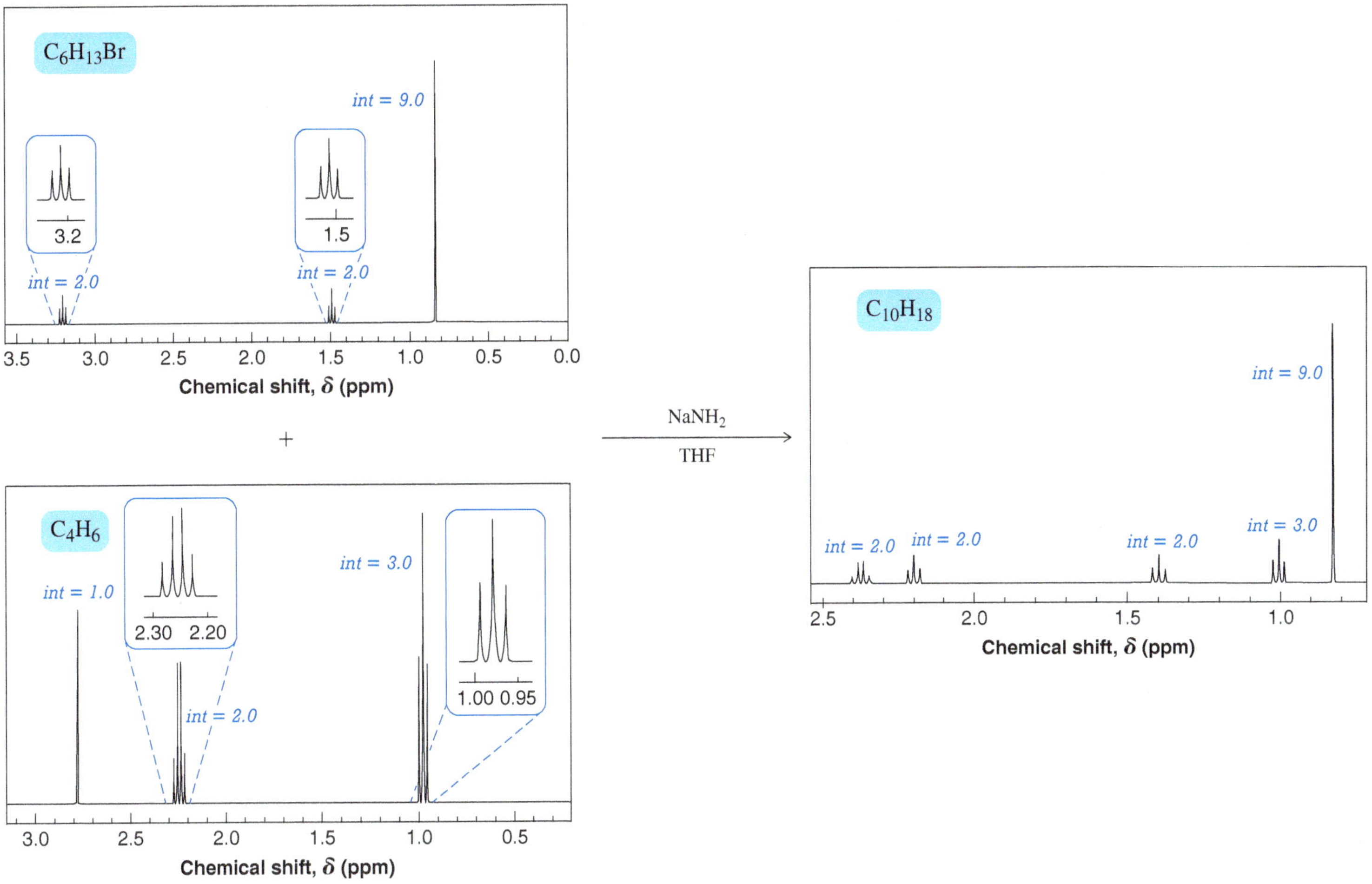

15.82 (•••) (a) Using the ¹H NMR spectra, identify the reactants and product for this reaction. (b) Provide an arrow-pushing mechanism for the reaction.

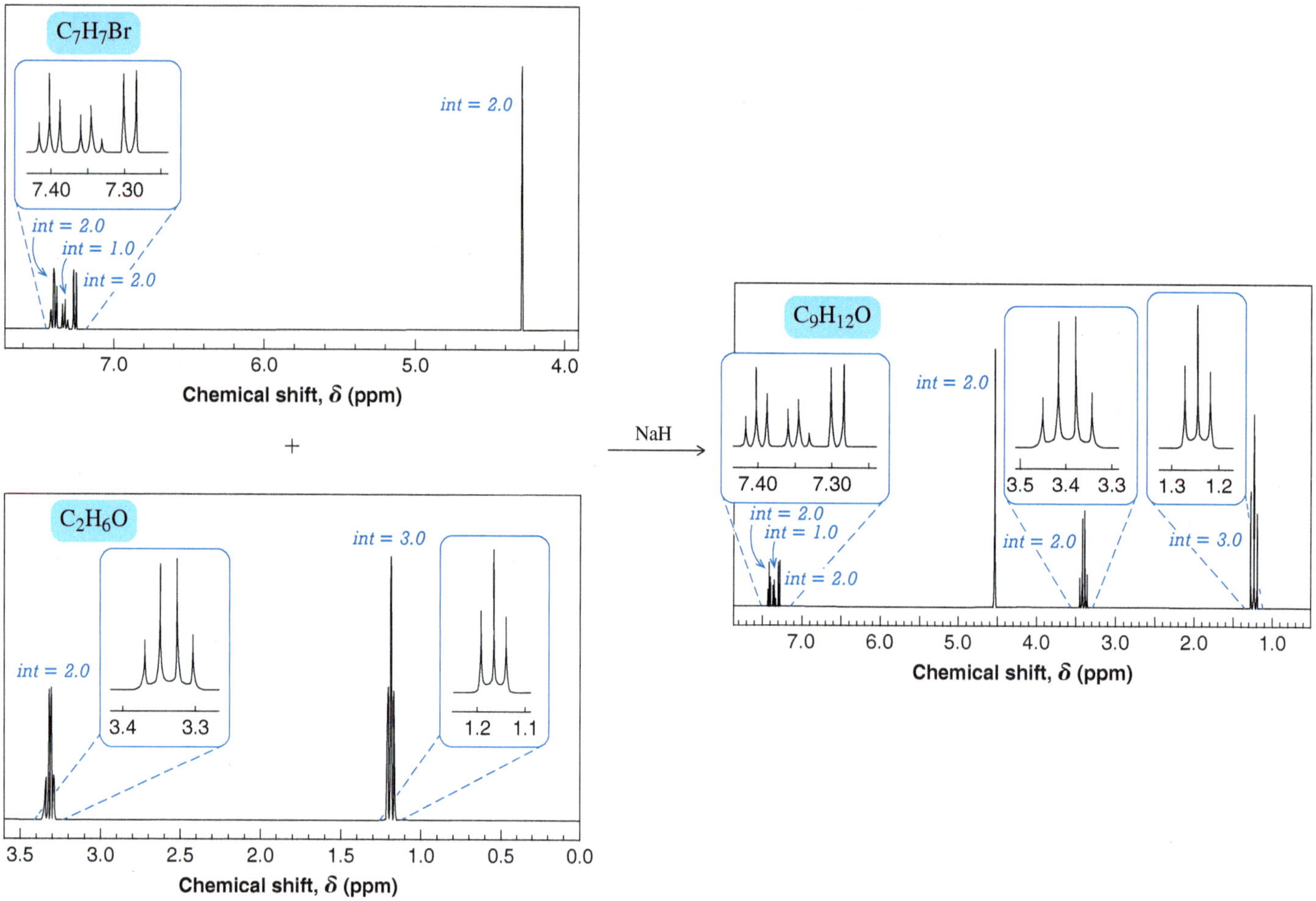

15.83 (•••) Give the structure using only the ¹³C NMR and the mass spectrum shown.

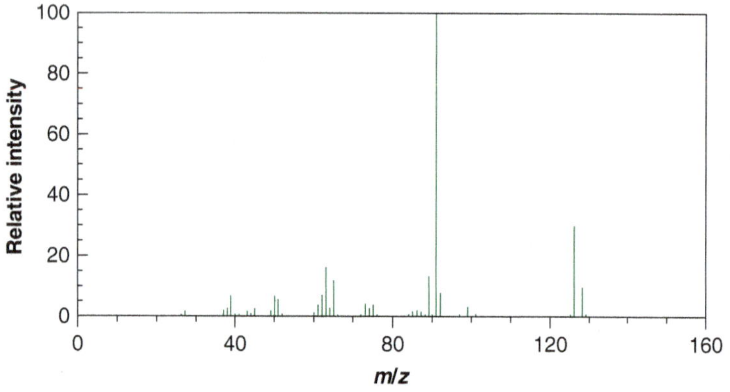

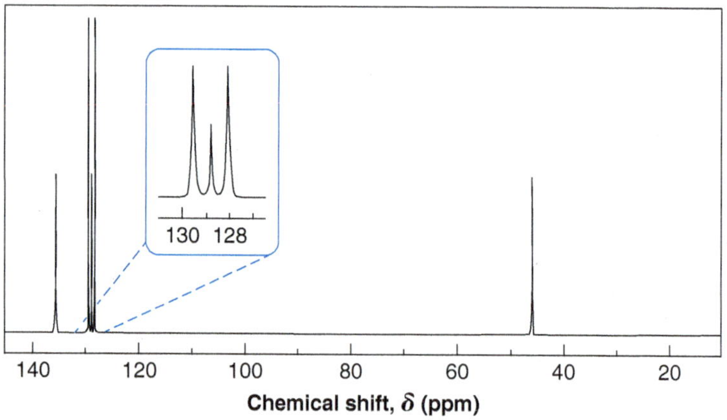

16

Metals in Organic Chemistry

Introduction

In Chapter 10, we began thinking about the synthesis of larger molecules from smaller ones. Critically important to organic synthesis is a retrosynthetic strategy and a toolbox full of reactions allowing us to form carbon–carbon bonds. The use of metals in organic chemistry is indispensable because they allow for the selective formation of new C—C bonds using highly functionalized substrates.

In Chapters 14 and 15, we discussed using analytical techniques (IR, NMR, and MS) to identify unknown compounds. One complicated example of structure identification dealt with the natural product curacin A (Chemistry in Context 15-B). Curacin A was isolated from the filamentous cyanobacteria *Lyngbya majuscula*, a prevalent source of marine natural products. The term "natural product," used in this way, refers to secondary metabolites (meaning not essential for life, like proteins, DNA, or RNA) produced by organisms for a variety of reasons (self-defense, waste disposal, by accident, etc.). In initial assays, curacin A was found to be cytotoxic, possessing antiproliferative activity against a variety of cell lines, making it a potential anticancer compound.

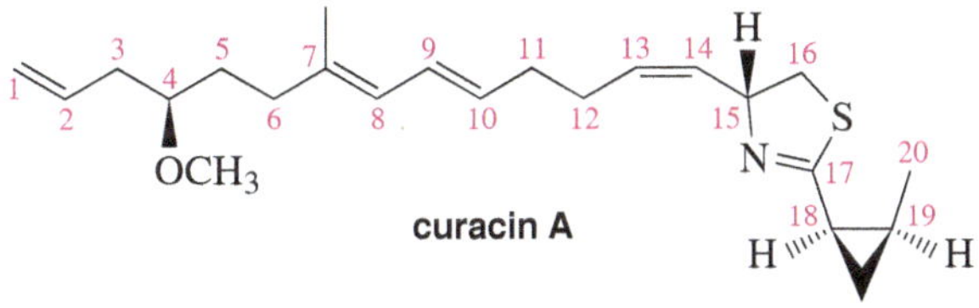

curacin A

Learning Objectives

LO 16.0 Determine the products of reactions with Grignard and organolithium reagents

LO 16.1 Evaluate carbene reactions

LO 16.2 Compare and contrast Suzuki, Sonogashira, Stille, and Negishi coupling reactions

LO 16.3 Analyze cross-coupling reactions

LO 16.4 Predict the product of metathesis reactions

LO 16.5 Evaluate the use of palladium as a catalyst

"The synthesis of substances occurring in Nature, perhaps in greater measure than activities in any other area of organic chemistry, provides a measure of the conditions and powers of science."

—Robert Burns Woodward, 1965 Nobel Prize in Chemistry

An important area of organic chemistry research [*definitely my favorite*] is natural product synthesis. Following isolation and identification of curacin A, this molecule became a target for synthesis by a number of natural product chemists. Why would a molecule produced naturally need to be synthesized in the laboratory? We answer that question here in Chapter 16, and in the process, we demonstrate the power of organometallic reactions for the synthesis of molecules like curacin A.

○ ○ ○ ○ ○ ○ ○

As defined previously, organic chemistry is the study of compounds containing mainly carbon, hydrogen, oxygen, and nitrogen. You might be surprised to learn, then, that the use of *inorganic* reagents is fundamental to the field of organic synthesis. Specifically, organometallic chemistry has enabled myriad selective, high-yielding, and green methods for synthesizing new carbon–carbon bonds. The mechanisms in this chapter are different from those seen previously and, in many cases, will not resonate with reactions you'll see later. However, to appreciate the modern state of organic synthesis, a general understanding of organometallic chemistry has a place in a first-year organic chemistry text. Before making this case, let's assess our current understanding of the material necessary for success in Chapter 16.

Test your understanding

Assessment 16.1 (i) Indicate the direction of the bond dipole for each of the following bonds based on the Pauling electronegativity values in Figure 2.20. (ii) Which atom is partially positive? (iii) Which is partially negative?

(a)	(b)	(c)	(d)	(e)	(f)	(g)	(h)	(i)
C—Cl	C—Mg	C—O	C—Li	C—Sn	C—F	C—B	C—S	C—Zn

Assessment 16.2 Using pK_a values, calculate the equilibrium constants for the following acid–base reactions.

(a) (acetylenic proton: the terminal H of $H-C\equiv C-H$)

$$H-C\equiv C-H + {}^{-}NH_2 \rightleftharpoons H-C\equiv C^{-} + NH_3$$

pK_a = 25; pK_a = 38

(b)

$$CH_4 + H^{-} \rightleftharpoons {}^{-}CH_3 + H_2$$

pK_a = 50; pK_a = 36

Assessment 16.3 Suggest a reagent for the transformation of a 1° alcohol to a 1° alkyl halide.

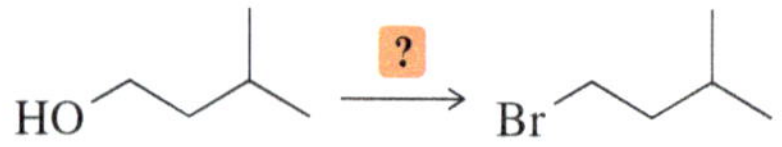

Assessment 16.4 Give the oxidation state of the palladium in each of the following forms.

(a) Pd/C
(b) AcO—Pd—OAc
(c) $Pd(PPh_3)_4$
(d) H_3C—Pd—Br

16.1 Organometallics: an overview

To appreciate the need for organometallic reagents, we have to recognize that most bonds are formed when an electron-rich nucleophile attacks an electron-poor electrophile. Aside from alkenes and alkynes, which are weak nucleophiles that attack using π electrons on neutral carbons, carbon is most often an electrophile. Consider, for example, that the functional groups studied thus far feature an electronegative atom (F, Cl, Br, O, or N) attached to carbon. The attached electronegative atom polarizes the C—X σ bond such that the less electronegative carbon is partially positive, or electrophilic (**Figure 16.1**).

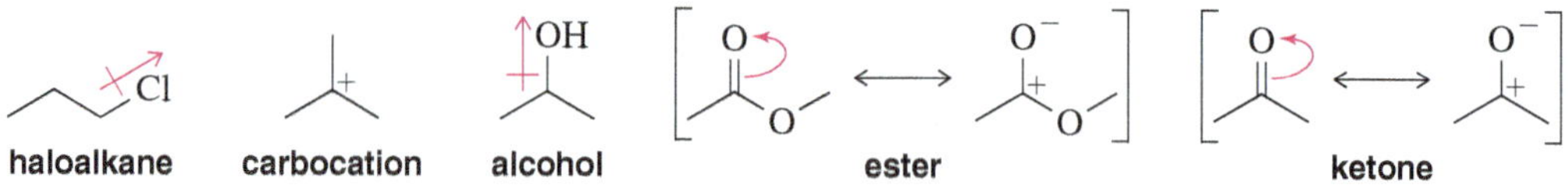

Figure 16.1 Functional groups featuring an electrophilic carbon.

The fact that carbon is most often electrophilic makes it difficult to form new carbon–carbon bonds [*the most important goal of organic synthesis*] unless nucleophilic carbon species can be generated to react with these electrophiles. The organometallic reagents we study in this chapter, shown in **Figure 16.2**, will enable us to do just that. With a metal (M) attached to carbon, the C—M bond is polarized such that carbon, being more electronegative than the metal, now has a partial negative charge, making it nucleophilic. Going forward, one way to simplify our approach to organometallic reactions is to consider the C—M bond to be ionic, with the carbon possessing a full negative charge. [*We do this despite the fact that electronegativity differences suggest that some of these C—M bonds are polar covalent. Recall Section 2.3.4.*]

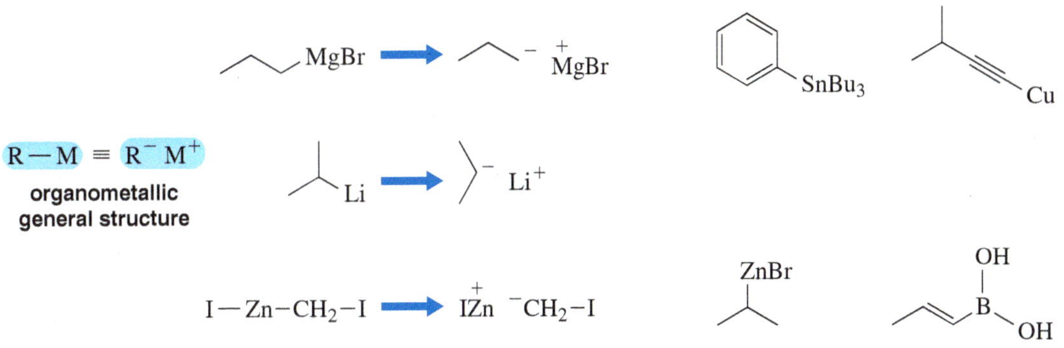

Figure 16.2 Organometallic reagents where carbon is nucleophilic. (The symbol "≡" means "is equivalent to" or "is identical to.")

16.1 SUMMARY

- The carbon attached to the metal in organometallic reagents (R—M) is nucleophilic.

16.2 Reduction of C—X

We encountered our first organometallic reagent in Chapter 10, when studying the reaction of acetylide ions. By deprotonating the acetylenic proton ($pK_a = 25$), as shown in **Figure 16.3**, the resulting sodium acetylide is technically an organometallic. Though it reacts as if the C—Na bond is ionic, it can also be drawn as a covalent bond. In the covalent form, it matches the general formula (R—M) shown in Figure 16.2 for organometallics, where R represents a hydrocarbon and M represents the metal.

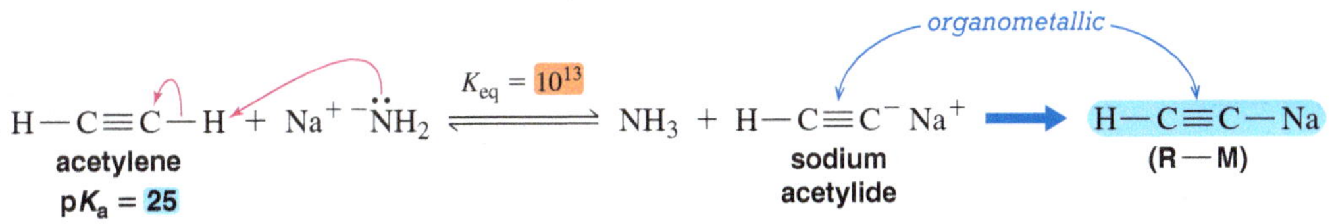

Figure 16.3 Deprotonation of acetylide makes an organometallic reagent.

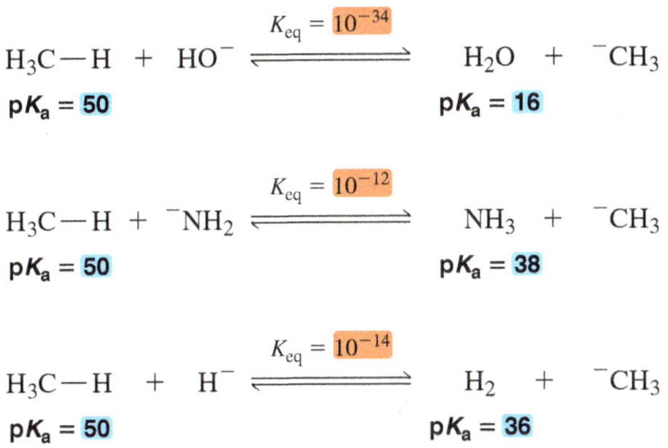

Figure 16.4 No base is strong enough to favorably deprotonate methane.

Preparation of the acetylide ion by deprotonation is only possible because the pK_a of the acetylenic proton is 25. Thus, strong bases like sodium amide ($NaNH_2$) or sodium hydride (NaH) can deprotonate it. The requirement of a low pK_a severely limits organometallic formation through acid–base reactions. Consider an attempt at making sodium methylide by this reaction (**Figure 16.4**). Because the pK_a of methane is 50, neither hydroxide, nor sodium amide, nor sodium hydride is strong enough to deprotonate methane.

Because deprotonation of an alkane is *not* possible for making a carbanion, we have to look for another way. In the Williamson ether synthesis (introduced in Section 13.11.1, but shown again in **Figure 16.5**), there were two ways of making the necessary alkoxide. The first way uses a strong base (NaH) to deprotonate an alcohol. The second utilizes a strong reducing agent (most often sodium metal) to reduce the polarizable O—H bond.

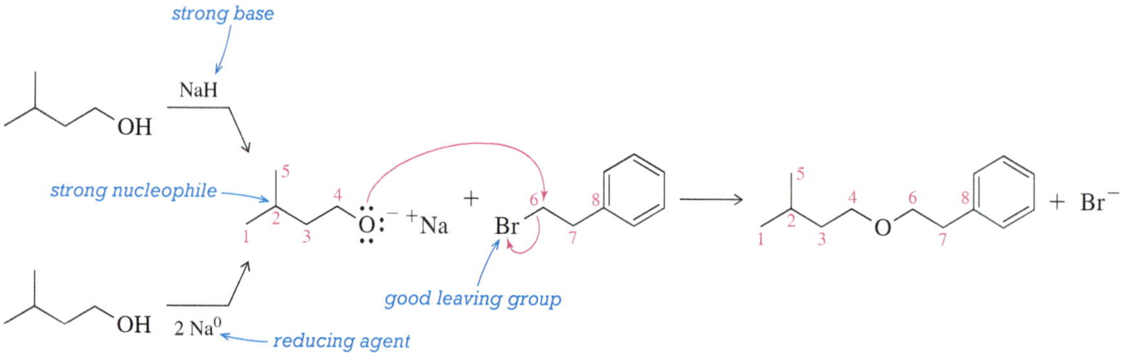

Figure 16.5 The Williamson ether synthesis is the S_N2 reaction between an alkoxide and a haloalkane.

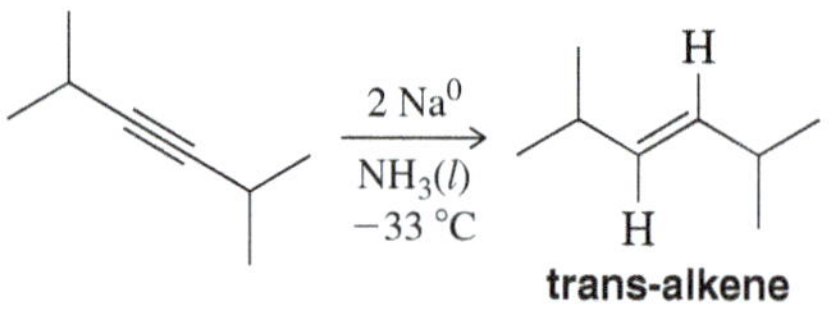

Figure 16.6 Dissolving-metal reduction of alkynes gives a trans-alkene.

Sodium metal is used in a similar way to reduce alkynes to the trans-alkene, as shown in Figure 16.6. This reaction was discussed in detail in Section 10.6.3.

The reactions in Figures 16.5 and 16.6 are united in ways that provide insight into how we might be able to use them to make carbanions. First, they occur by single-electron transfer—reactions that we can illustrate using fishhook arrows in the electron-pushing formalism. [*See radical reactions in Chapter 5 and 11.*] Second, each begins with the addition of an electron to a reducible bond (O—H σ or C—C π). Once formed, the resulting anion is protonated while the electron-poor free radical is attacked by a second electron to give an alkoxide [Figure 16.7(a)] or a carbanion [Figure 16.7(b)]. While the formation of the alkoxide stops here, the carbanion is protonated to give the trans-alkene.

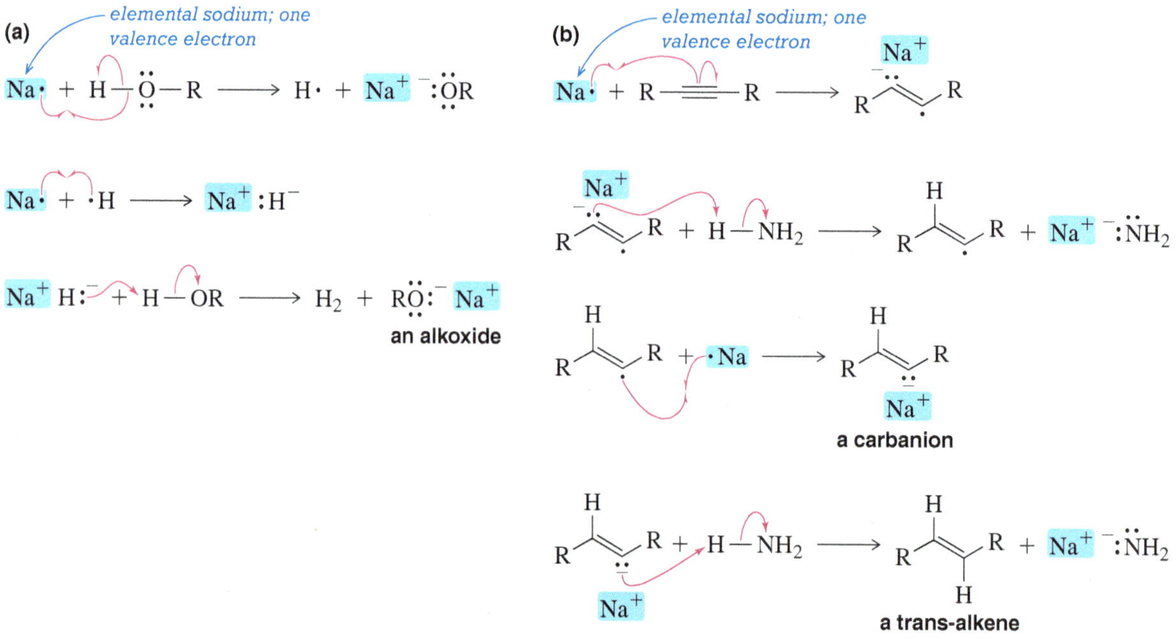

Figure 16.7 The mechanism of dissolving-metal reduction. (a) Cleavage of the O—H bond gives a stable alkoxide. (b) Reduction of the weak C—C π bond gives a radical anion initially, terminating in a trans-alkene.

In the initial reduction of the O—H σ bond (Figure 16.7), the electron pair ultimately resides on an electronegative atom, making a stable anion. It is this resulting alkoxide that makes the O—H σ bond reducible. This teaches us that a carbanion can be made by the reduction of a C—X bond, where the resulting X^- anion would be stable. What could X be that would make X^- stable? When X is a halogen like chlorine, bromine, or iodine, the C—X bond can be reduced by a reactive metal (Mg or Li being most common; Figure 16.8). We explore each of these in more detail in the following sections. The result of these reactions is a C—M bond—that is, an organometallic.

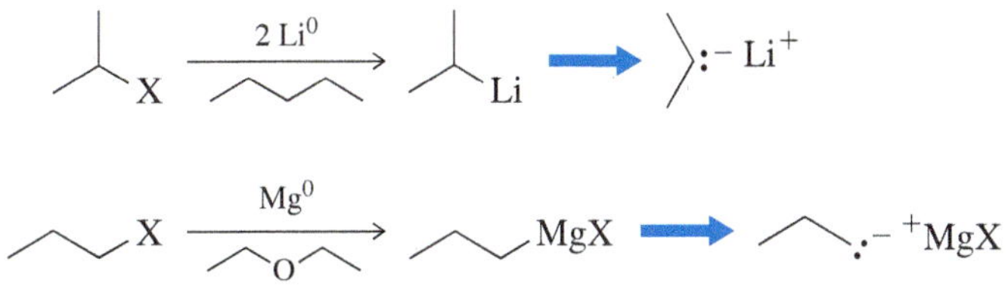

Figure 16.8 Reduction of a C—X bond with magnesium or lithium gives an organometallic reagent.

16.2.1 Organolithium reagents

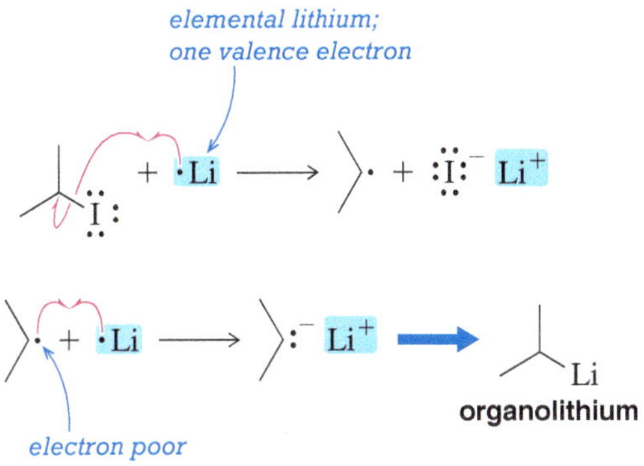

Figure 16.9 Reduction of a C—I bond with lithium gives an organolithium reagent.

Organolithium reagents are typically made by the reaction of iodoalkanes with two equivalents of lithium, as in **Figure 16.9**. Lithium is right above sodium in the periodic table, so Li is a strong reducing agent. Desperate to gain a noble gas configuration, each lithium atom rapidly ejects an electron to become Li^+. The first electron attacks the iodine, homolytically breaking the C—I bond to form a carbon radical and iodide (I^-). Iodide, as a result of its large size and its noble gas configuration, is an extremely stable anion, and forming it is the driving force for this reaction. The resulting carbon radical is attacked by the second electron to give a carbanion. Though the lithium countercations won't be important when discussing reactions of organolithium reagents, one of the lithium ions ends up associated with the carbanion while the other associates with the iodide.

Because the resulting organolithium is an extremely strong base (Section 16.2.3), these reagents are formed in inert, aprotic solvents, like pentane.

16.2.2 Grignard reagents

The formation of organomagnesium halides is almost identical to the formation of organolithium reagents (**Figure 16.10**). Organomagnesium halides are called **Grignard reagents** in honor of their inventor, the French chemist Viktor Grignard. The only difference between them and organolithium reagents is that magnesium, being in group 2, loses 2 electrons to gain a noble gas configuration. As a result, only a single magnesium atom is required. In the case of reduction of a C—Br bond with magnesium, the first electron attacks the bromine, homolytically breaking the C—Br bond to form a carbon radical and bromide (Br^-). Bromide, as a result of its large size and its noble gas configuration, is an extremely stable anion, and forming it is the driving force for this reaction. The resulting carbon radical is attacked by the second electron to give a carbanion. Though the magnesium countercation won't be important when discussing reactions of Grignard reagents, the Mg^{2+} ion associates with the negatively charged carbanion and bromide ions.

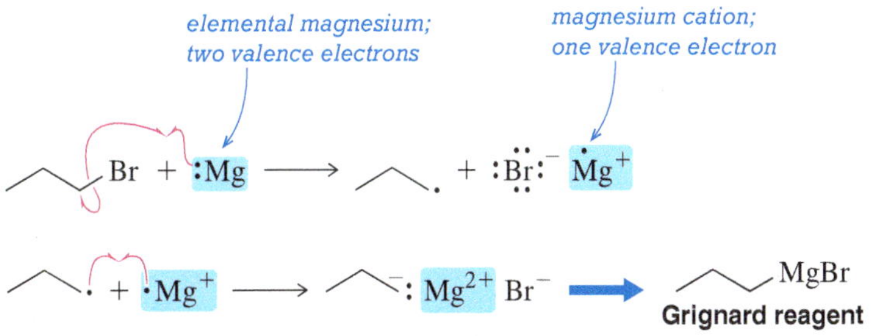

Figure 16.10 Reduction of a C—Br bond with magnesium gives an organomagnesium halide (a Grignard reagent).

Because the resulting Grignard reagent is an extremely strong base (Section 16.2.3), these reagents are formed in inert, anhydrous aprotic solvents like diethyl ether and tetrahydrofuran.

16.2–16.2.2 SUMMARY

- Reduction of alkyl, alkenyl, and aryl halides is used to make organolithium and Grignard reagents.

16.2.3 Reactions of Grignard and organolithium reagents

Organolithium and Grignard reagents can be thought of as possessing an ionic bond between carbon and the corresponding metal. The metals are less electronegative than carbon, so the carbon is negatively charged and the metal is positively charged in this ionic bond. Because they are carbanions, these reagents are extremely reactive, reacting as strong bases or strong nucleophiles. Differing only in the countercation (Li^+ or Mg^{2+}), organolithium and Grignard reagents react identically in virtually every situation. Often the choice between the two organometallics is based on the ease of preparation or the commercial availability of the reagents. Typically, Grignard reagents, using less reactive magnesium metal, are prepared more easily and safely, but a wider range of simple organolithium reagents are available for purchase.

16.2.3.1 Reactions with acid

Upon forming an organolithium or Grignard reagent, we have succeeded in generating a carbanion. Consider the formation of methyllithium from methyl iodide. Methylide (CH_3^-) is an extremely strong base, with its conjugate acid methane (CH_4) having a $pK_a = 50$. In fact, the alkyl anions are the strongest bases studied in Chapter 4. [*See Table 4.5 to confirm this.*] As strong bases, these reagents react quickly and favorably with even weakly acidic protons (**Figure 16.11**). It is for this reason that typical solvents like water ($pK_a = 16$) and ethanol ($pK_a = 16$) cannot be used in the formation or reaction of organolithium or Grignard reagents. Instead, pentane and ethers are used, respectively.

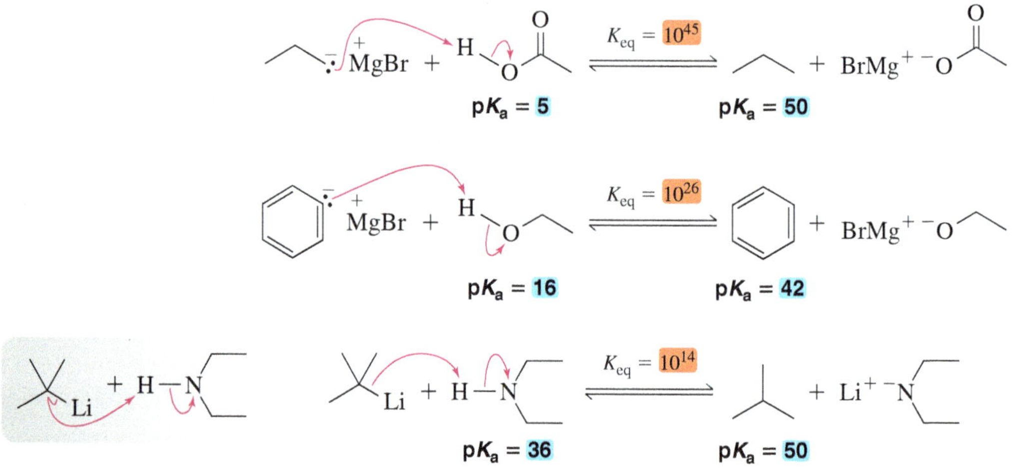

Figure 16.11 Organolithium and Grignard reagents favorably deprotonate strong and weak acids.

Assessment 16.5 Calculate K_{eq} for the following acid–base reactions.

(a)

(b)

Assessment 16.6 Predict the product that would form when a Grignard reagent is prepared in the presence of deuterated water.

$$CH_3{-}Br + Mg^0 \xrightarrow[D_2O]{} ?$$

deuterated water

Assessment 16.7 Predict the product that would result from the reaction of an organolithium reagent with a ketone when a hydroxyl group is present in the ketone substrate. [*We discuss the reaction of organolithium reagents with ketones in Section 16.2.3.3 and in Chapter 17.*]

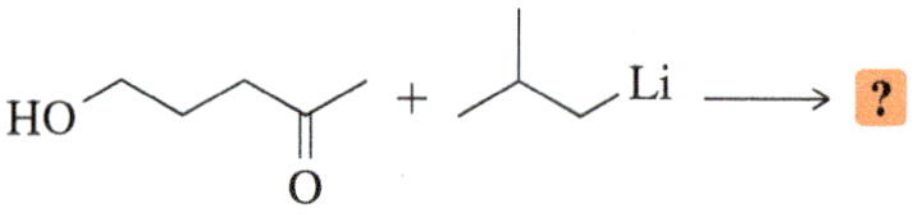

16.2.3.2 Reactions with epoxides

Epoxides (Section 13.13) are a special kind of ether that, as a result of being constrained in a three-membered ring, react with strong nucleophiles. As a result, organolithium and Grignard reagents, which are strong nucleophiles, open epoxides at the least sterically hindered carbon in what looks like an S_N2 reaction. Several examples are shown in **Figure 16.12**. Once more, because they are strong bases, these reactions must be run in the absence of acidic protons. As a result, diethyl ether, *t*-butylmethyl ether, and tetrahydrofuran are common solvents for these reactions. The alkoxide product that results from epoxide opening must be quenched with an acid to complete the formation of an alcohol. [*The addition of an organolithium/Grignard reagent is a key reaction for your reaction sheets. Why is it so special?*]

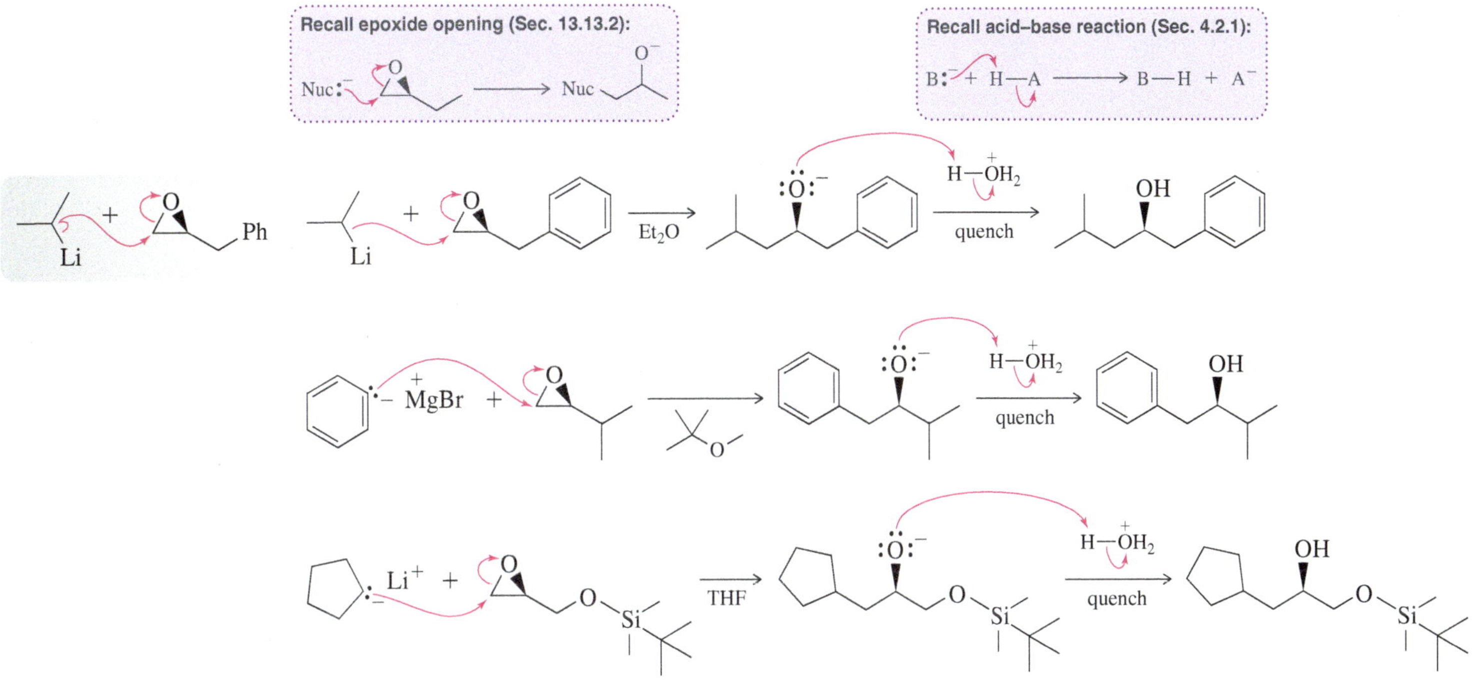

Figure 16.12 Organolithium and Grignard reagents favorably react with epoxides.

One of the important goals of organic synthesis, as discussed in Chapter 10, is the formation of new C—C σ bonds. Without reactions that do this, it is impossible to combine two small organic molecules to make larger all-carbon skeletons like the alcohol in **Figure 16.13**.

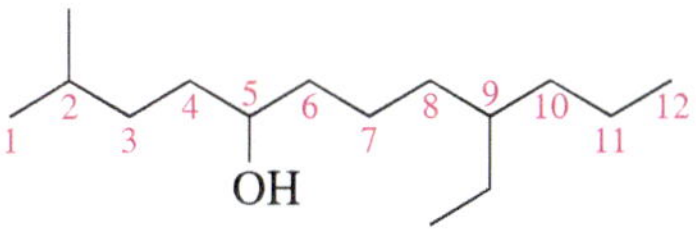

9-ethyl-2-methyl-dodecan-5-ol

Figure 16.13 The alcohol to be synthesized while Thinking Like a Chemist.

Let's think like a chemist to see how the reaction just learned could be used in multiple ways to make 9-ethyl-2-methyl-dodecan-5-ol.

Thinking Like a Chemist

When strong nucleophiles add to epoxides, opening the three-membered ring, the oxygen of the epoxide ends up attached to the epoxide carbon that was not attacked. The new bond forms at what was formerly the other epoxide carbon. [*Confirm that this makes sense by looking back at the reactions in Figure 16.12.*] This concept is illustrated in **Figure 16.14** with a general nucleophile (Nuc⁻). When a nucleophile adds to C_1 of the epoxide, forming the new bond there, the oxygen, in the form of an alkoxide initially, ends up attached to C_2.

For an alcohol in the middle of a long alkyl chain, there are usually two different epoxide/organolithium combinations that will produce it. Using the carbon numbers shown in Figure 16.13 for 9-ethyl-2-methyl-dodecan-5-ol, can you come up with these two combinations? [*Give it a shot before looking*

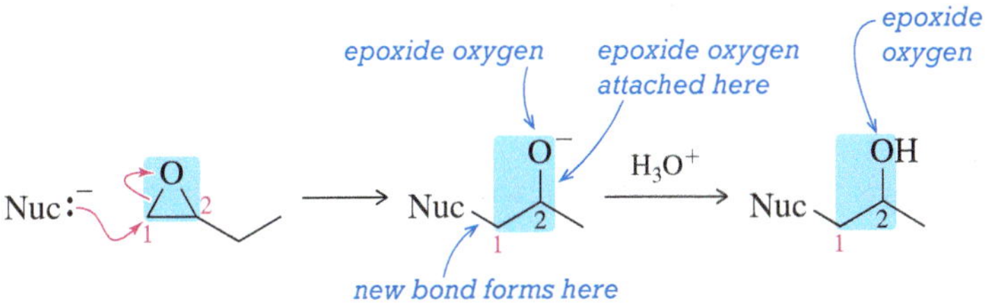

Figure 16.14 Nucleophilic addition to an epoxide occurs at the least substituted carbon of an epoxide, leaving the oxygen attached to the other epoxide carbon.

at ***Figure 16.15***.] Regardless of which epoxide we use, the epoxide oxygen must end up at C_5. This implies that the organolithium reagent should attack C_4 if using the epoxide in Figure 16.15(a) to form a new $C_3—C_4$ bond. Alternatively, the organolithium reagent should attack C_6 if using the epoxide in Figure 16.15(b) to form a new $C_6—C_7$ bond. ■

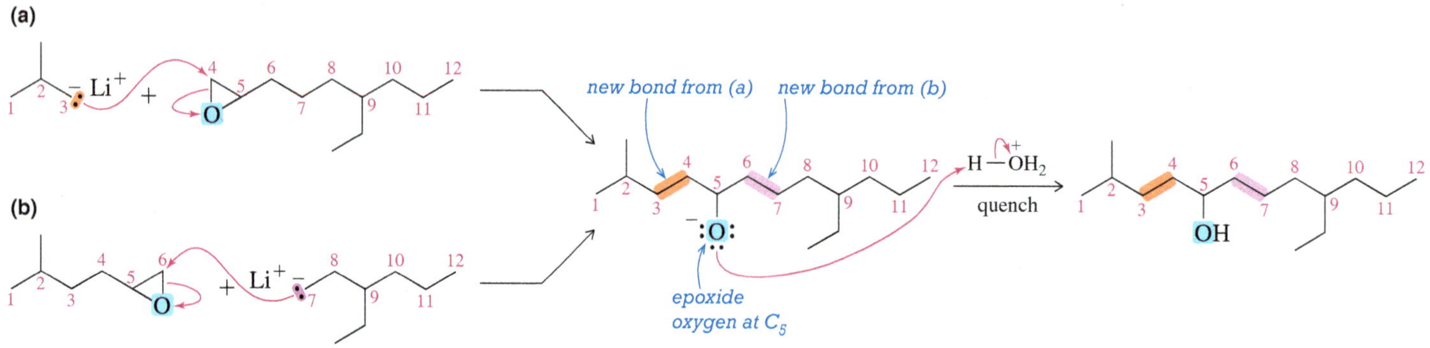

Figure 16.15 Two epoxide/organolithium combinations to make the same alcohol.

Assessment 16.8 Predict the product of the following epoxide addition reactions.

(a)

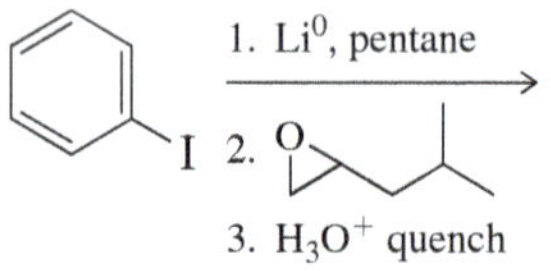

(b)

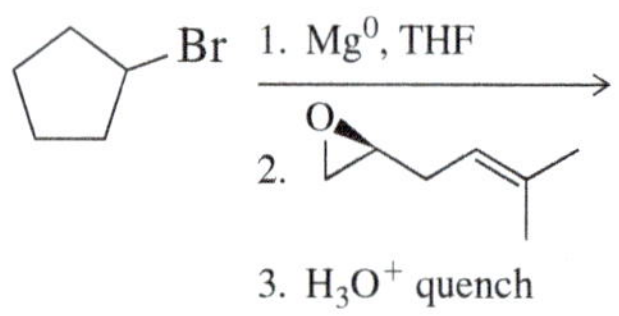

Assessment 16.9 Addition to an epoxide occurs via an S_N2 reaction, but the stereochemistry of the epoxide is retained in the following reaction. Why?

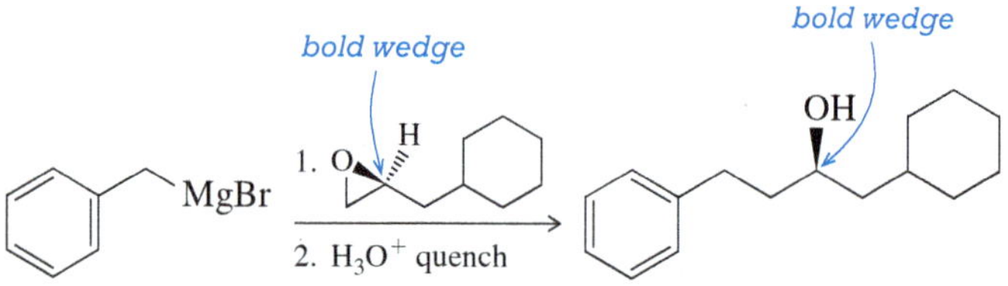

Assessment 16.10 Working backward, design a synthesis of the following alcohol using two different epoxide/Grignard reagent combinations.

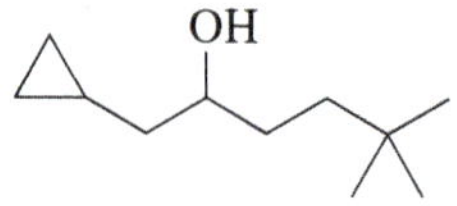

Assessment 16.11 Using the epoxide shown, addition of an organolithium reagent, when followed by an acid quench, produces only the starting epoxy alcohol. Why? How could the reaction be modified to produce the desired molecule? [*Hint: Look back at Section 13.14.*]

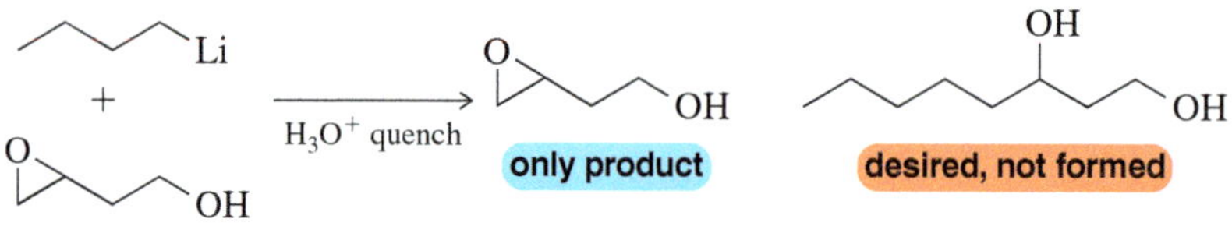

16.2.3.3 Reactions with ketones and aldehydes

We study the reactivity of ketones and aldehydes in detail in Chapter 17. Because addition to ketones and aldehydes is one of the most important reactions of organolithium and Grignard reagents, we mention it here as well. [*Confirm with your instructor whether you need to learn this now.*] Because we already know that organolithium and Grignard reagents are strong nucleophiles, let's first explore why aldehydes and ketones are electrophilic. Because of the electronegative oxygen, the carbonyl (C=O) group is polarized in the direction of oxygen, giving the carbon a partial positive charge. Furthermore, a second resonance structure can be drawn for the carbonyl in which the oxygen has a negative charge and the carbon has a positive charge (**Figure 16.16**). This valid and contributing resonance structure is referred to as the zwitterionic resonance structure. It forms the basis of all carbonyl reactivity and is discussed in detail in Chapter 17. The fact that the second resonance structure of a given functional group often determines its reactivity can be confirmed by looking at the electron density map of acetone shown in Figure 16.16. The blue color means that the central carbon is highly electron-deficient.

Figure 16.16 The central carbon of the carbonyl (C=O) group is electrophilic.

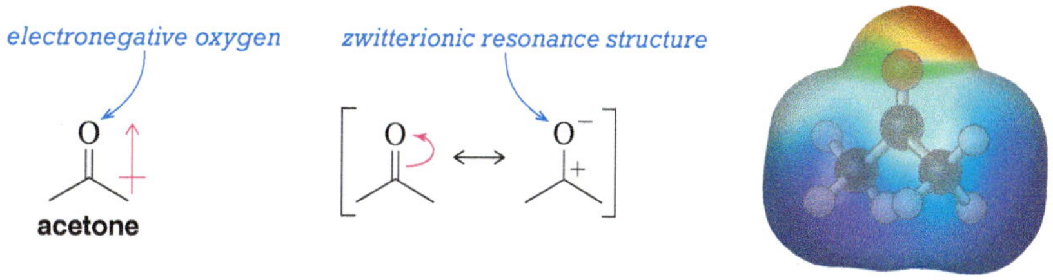

In solution, a strongly nucleophilic Grignard reagent (essentially a carbanion) will readily attack the central carbon of the carbonyl group in an aldehyde, as shown in **Figure 16.17**. Upon attacking, the π electrons in the C=O bond are pushed onto the electronegative oxygen to give an alkoxide as the initial product. Protonation of this alkoxide results in formation of an alcohol. The zwitterionic structure is important in predicting the reactivity of the carbonyl. As such, attack by the nucleophile on this resonance form can be shown to result in the same product. Because the structure of the carbonyl is the combination of both the neutral and zwitterionic forms, these two ways of showing attack are identical.

Figure 16.17 Attack of a Grignard reagent on a ketone, via either resonance form, gives an alcohol after acid quench.

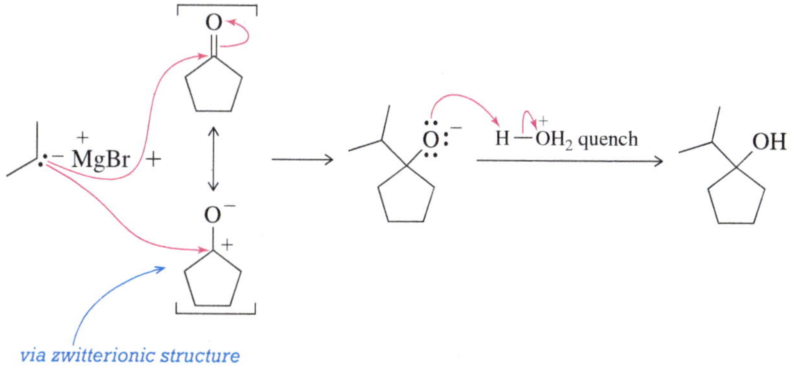

Although the reactions of carbonyls are explored more deeply in Chapter 17, recognize for now that the formation of C—C bonds using organolithium and Grignard reagents with ketones and aldehydes is a very useful tool for organic synthesis. More examples are shown in **Figure 16.18**.

Figure 16.18 Examples of organolithium and Grignard reagents reacting with aldehydes and ketones.

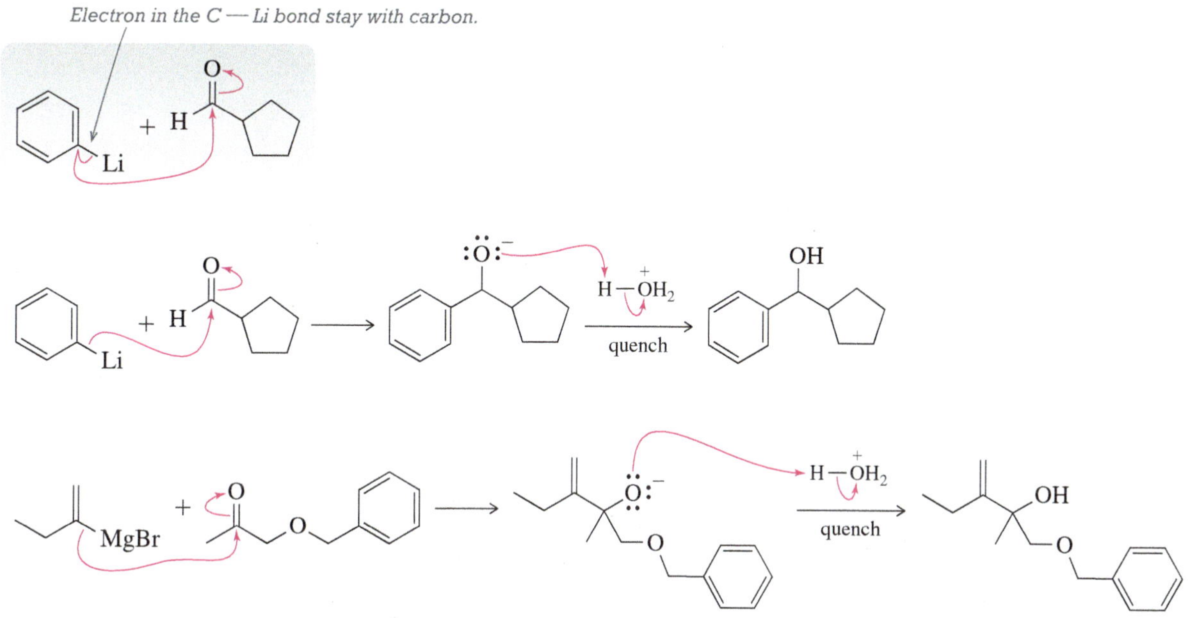

16.2.3 SUMMARY

- Organolithium and Grignard reagents have similar reactivity, reacting as strong bases and as nucleophiles with epoxides, aldehydes, and ketones.

Assessment 16.12 Predict the product of the following aldehyde and ketone addition reactions.

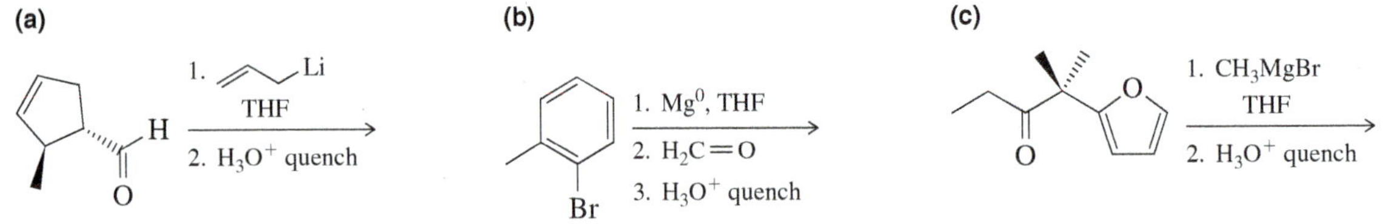

Assessment 16.13 The addition of organometallic reagents to aldehydes and ketones is a versatile C—C bond-forming reaction. What type of alcohol (1°, 2°, or 3°) is produced in each of the reactions in Assessment 16.12?

Assessment 16.14 Choose a ketone and a Grignard reagent that will give the following alcohol. Is there more than one way to do this?

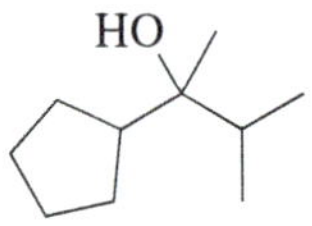

16-A CHEMISTRY IN CONTEXT: Natural Product Synthesis

Grignard reaction

When curacin A was first discovered, it was found to have significant biological activity, including antiproliferative properties that suggested its potential as an anticancer therapy. It was discovered to inhibit cell growth by binding to microtubules and interfering with normal mitotic spindle formation. This mechanism of action was found to be similar to a structurally unrelated compound, prompting an interest in obtaining more of the compound for further biological testing. Because *Lyngbya majuscula*, even at its highest production level, produces only a small amount of curacin A, it was necessary for synthetic chemists to provide the material for biological assays and, if successful, ultimately produce sufficient material for manufacturing a medicine. [*Reason #1 for natural product synthesis: to secure more material for biological studies.*]

The Grignard reaction, one of the most important C—C bond-forming reactions we will study [*especially in Chapter 17*], was utilized to form the C_3—C_4 bond of curacin A (**Figure 16.19**). In the presence of a chiral molecule, a single enantiomer of the 2° alcohol was produced. Recall from Chapter 6 that a chiral environment makes attack from one face proceed more rapidly, resulting in an optically active product.

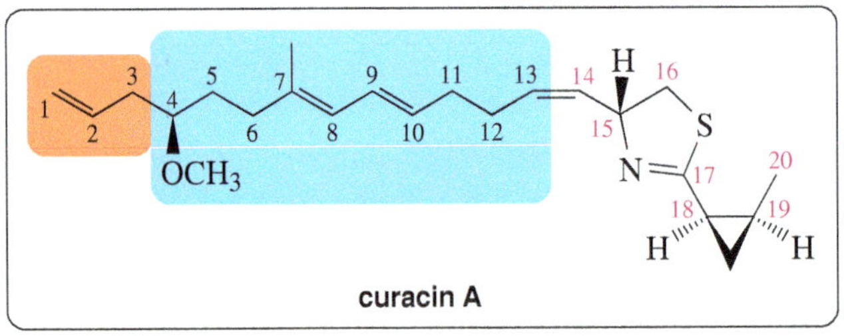

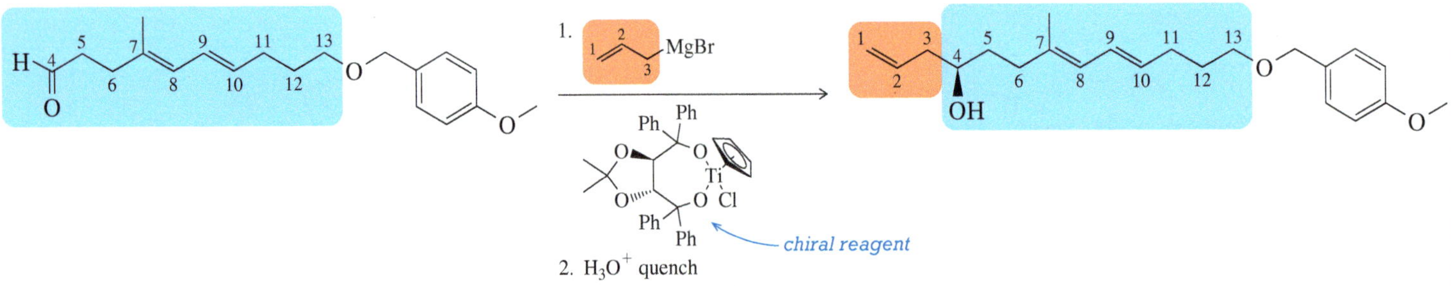

Figure 16.19 Grignard addition to an aldehyde in a chiral environment stereoselectively forms the C_3—C_4 bond in curacin A.

Assessment 16.15 Ignoring stereochemistry, provide a mechanism of the reaction in Figure 16.19.

16.3 Carbenes

The overarching theme of Chapters 8 and 9 was electrophilic addition to alkenes. One such electrophile that adds to alkenes is the carbene, first introduced in Assessment 8.79 (see the end-of-chapter materials in Chapter 8).

As a review of what carbenes do, work Assessment 16.16 on your own. Then we'll work it together.

Assessment 16.16 [*formerly Assessment 8.79(e)*] Provide an arrow-pushing mechanism for the cyclopropanation of cyclohexene with methylene carbene. Rationalize the outcome.

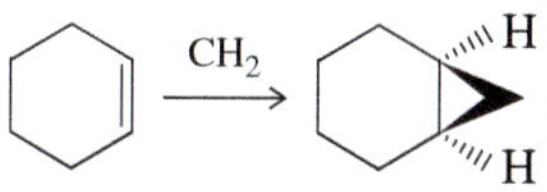

Solved Assessment 16.16

To solve a mechanism problem using reagents we haven't used before, it is first important to identify the inherent reactivity of the alkene and the carbene. The alkene, based on what we learned in Chapters 8 and 9, is likely a nucleophile. This implies that the carbene is an electrophile, but this should be confirmed before proceeding. **Figure 16.20** shows the Lewis structure of CH_2. With two hydrogens attached, there must also be a single pair of electrons on the central carbon. Analysis of this structure suggests that the carbon is electron-deficient, sp^2-hybridized, and trigonal planar and has an empty *p* orbital. In other words, the carbene is an electrophile, despite having a lone pair of electrons. This particular carbene is referred to as a **singlet carbene** because the 2 electrons are paired in a single orbital.

Figure 16.20 Analysis of carbene reveals that it is a strong electrophile.

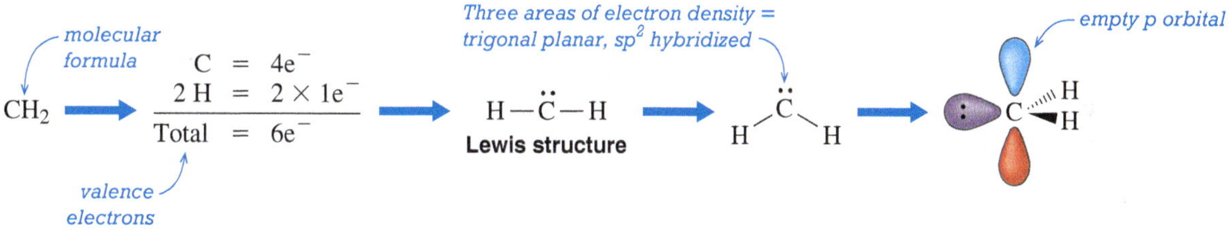

Having established that this reaction will be like the alkene addition reactions from Chapters 8 and 9, the first step must be to have the π electrons of the alkene attack the empty *p* orbital of the carbene. [*Does the reaction stop there?*] The intermediate that would result from only alkene attack, shown in **Figure 16.21**, possesses a carbocation and a carbanion, two highly reactive species, in close proximity. The lone pair of electrons on the carbanion then attacks the empty *p* orbital of the carbocation to form the second C—C σ bond of the cyclopropane.

Recall from Chapters 8 and 9 that when a slow first step is followed by a fast, intramolecular step, the overall reaction is concerted. Thus, the correct mechanism (**Figure 16.22**) shows the π bond at C_1 attacking the carbene while the lone pair on the carbene attacks C_2. ■

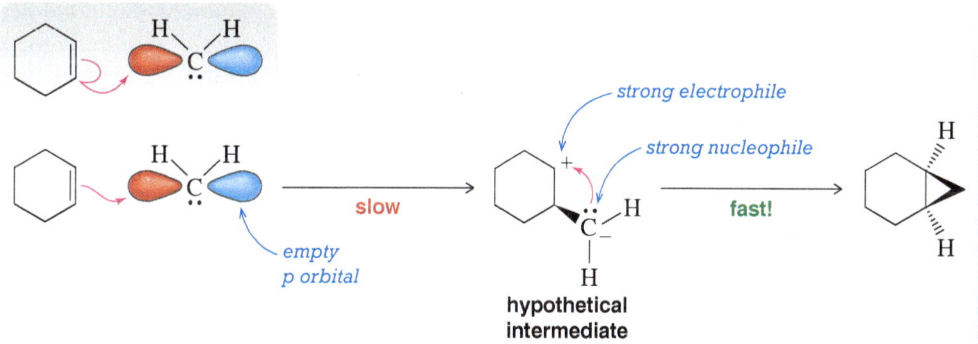

Figure 16.21 The nonconcerted mechanism of carbene addition.

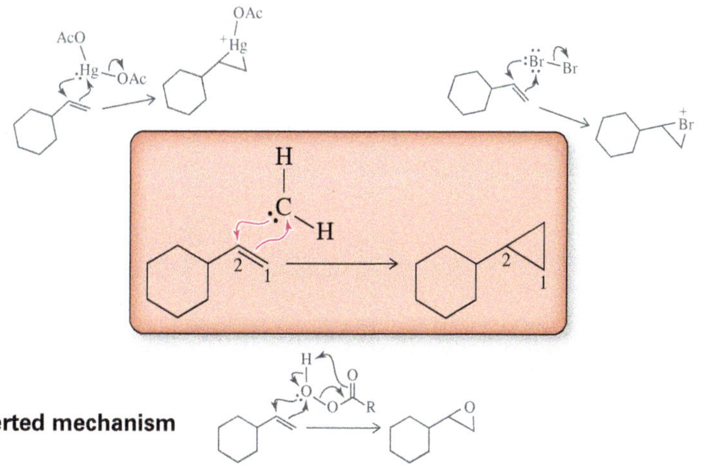

Figure 16.22 The concerted mechanism of carbene addition.

There are three common ways to form carbenes, all of which are related to each other. By examining two nonorganometallic methods (Sections 16.3.1 and 16.3.2), our understanding of the Simmons–Smith reaction (an organometallic reaction like the Grignard formation; Section 16.3.3) will be enhanced.

16.3.1 Carbene formation: α-elimination

The first carbene formation reaction utilizes the α-elimination of chloroform under basic conditions, as shown in Figure 16.23.

$$\underset{\text{chloroform}}{HCCl_3} + KO\textit{t}\text{-Bu} \longrightarrow \underset{\text{dichloro carbene}}{:CCl_2} + KCl + \textit{t}\text{-BuOH}$$

Figure 16.23 Formation of a carbene via α-elimination.

α-Elimination is a reaction we haven't seen previously, but it is analogous to the more common β-elimination studied in Chapter 12 (E2 reaction). Figure 16.24 and Figure 16.25 show these reactions in close proximity, so we can see what they have in common. Though the E2 elimination of haloalkanes is a concerted reaction, we can imagine it as the stepwise removal of a proton from C_1, followed by the halide leaving spontaneously from the neighboring C_2 (also referred to as the β-carbon; hence, β-elimination). If only these two arrows were drawn, the product would be a molecule containing a C_1 carbanion and a C_2 carbocation. The more contributing resonance structure of this odd-looking molecule is an alkene. Analyzed in this way, the driving force for the halide leaving in β-elimination is for the negatively charged molecule to become neutral. [*This is **not** the mechanism of E2 elimination. It is just a way of viewing it that helps α-elimination make more sense.*]

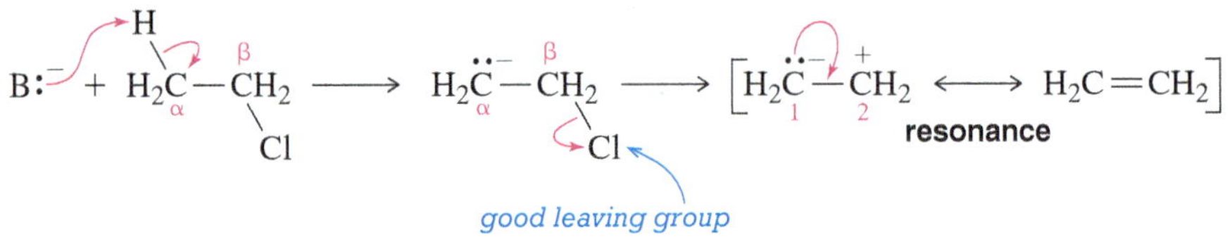

Figure 16.24 An alternate view of E2 elimination.

If we think of β-elimination as the initial deprotonation to give a carbanion followed by a halide leaving to make a neutral molecule, then α-elimination is almost identical. The pK_a of chloroform ($CHCl_3$) is around 15, meaning it can be deprotonated to give a carbanion stabilized by the three electron-withdrawing chlorine atoms. This intermediate is analogous to the negatively charged intermediate shown in Figure 16.24. If the driving force of β-elimination was for the carbanion to become neutral, then this negatively charged intermediate can become neutral in a similar way. The only difference is that the intermediate lacks a leaving group at a neighboring carbon, so the halide leaves from the carbon bearing the negative charge. Because the leaving group leaves from the same carbon, this is referred to as α-elimination.

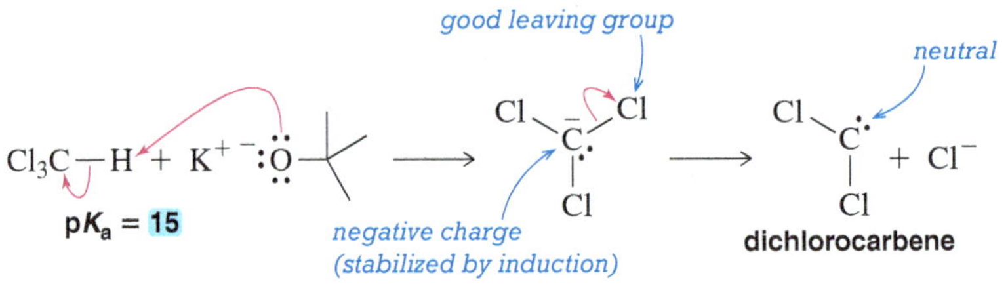

Figure 16.25 The mechanism of α-elimination.

Assessment 16.17 Estimate the value of entropy ($\Delta S > 0$ or $\Delta S < 0$) for the elimination step shown.

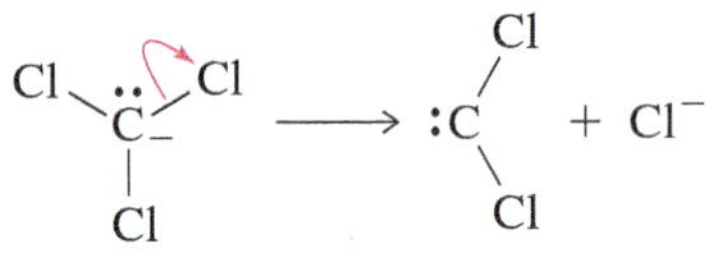

16.3.2 Carbene formation: diazomethane

A second method of forming a carbene utilizes the thermal decomposition of diazomethane (CH_2N_2) as shown in **Figure 16.26**. The carbene forms in an identical fashion to the α-elimination in Section 16.3.1. Although the most contributing resonance structure of diazomethane has a negative charge on nitrogen, a second resonance structure can be drawn where the negative charge is delocalized onto carbon. In this resonance structure, a negative charge is on the same carbon as N_2—perhaps the best leaving group we have studied to this point. With slight heating, nitrogen gas leaves via α-elimination to give a carbene in an entropically favored step.

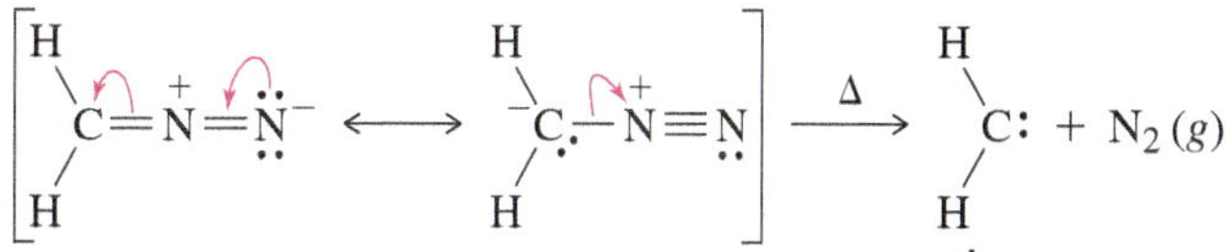

Figure 16.26 Thermal decomposition of diazomethane gives carbene and nitrogen gas.

16.3.3 Carbene formation: Simmons–Smith reaction

This brings us to the Simmons–Smith reaction. Carbene formation via α-elimination of an anion following deprotonation (Section 16.3.1) requires an acidic proton. How can a carbanion be generated in the absence of an acidic proton? [*Think about what we did in in Section 16.2.*] Using a dissolving metal, in this case zinc, reduction of a C—I bond in diiodomethane occurs to give what is referred to as the Simmons–Smith reagent. Mechanistically similar to the formation of organolithium and Grignard reagents, the reaction begins with zinc expelling two electrons to become Zn^{2+} (**Figure 16.27**). The first electron attacks the iodine, homolytically breaking the C—I bond, forming iodide and a carbon radical. The carbon radical is attacked by the second electron to give a carbanion. In solution, the carbanion, the Zn^{2+} cation, and iodide aggregate to form ICH_2ZnI, the Simmons–Smith reagent.

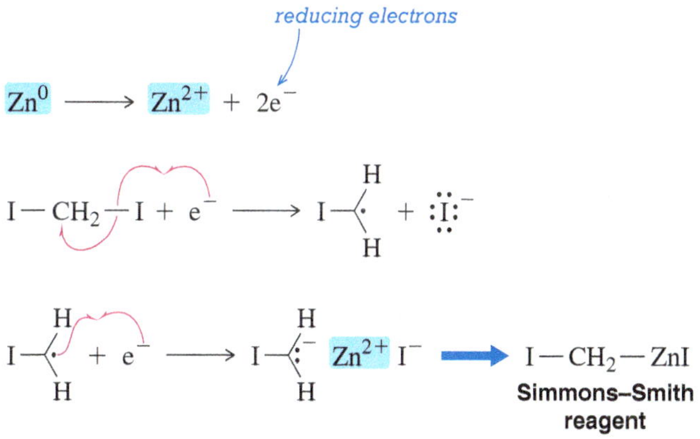

Figure 16.27 The mechanism of formation for the Simmons–Smith reagent is similar to the formation of organolithium and Grignard reagents.

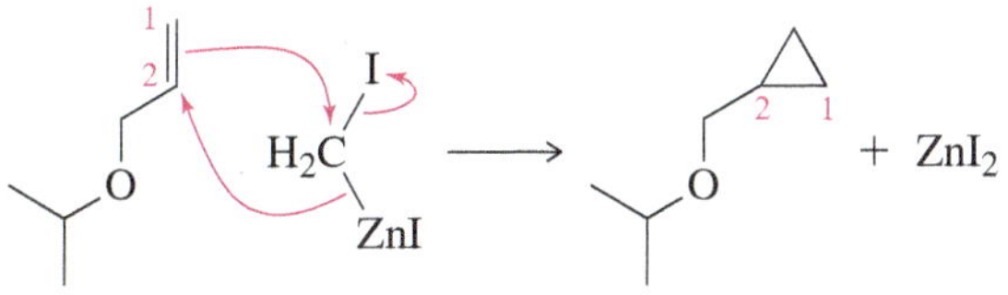

Figure 16.28 The mechanism of the Simmons–Smith cyclopropanation reaction.

Even though it could decompose spontaneously to the carbene, the Simmons–Smith reagent is stable. It only reacts in the presence of an alkene. Mechanistically similar to the carbene reaction studied in Assessment 16.16, the π bond at C_1 attacks the carbon of the Simmons–Smith reagent, kicking out iodide, while at the same time, the electrons in the C—Zn bond attack C_2 to form the cyclopropane (**Figure 16.28**).

16.3 SUMMARY

- Carbenes can be generated by α-elimination, decomposition of diazomethane, and formation of the Simmons–Smith reagent. They react with alkenes to form cyclopropanes.

Assessment 16.18 Cyclopropanation using any of the reagents discussed here is stereospecific. (a) What does this say about the mechanism? (b) Draw a reaction coordinate diagram for cyclopropanation.

Assessment 16.19 Work backward to show how the cyclopropane would be synthesized from the chloroalkane shown.

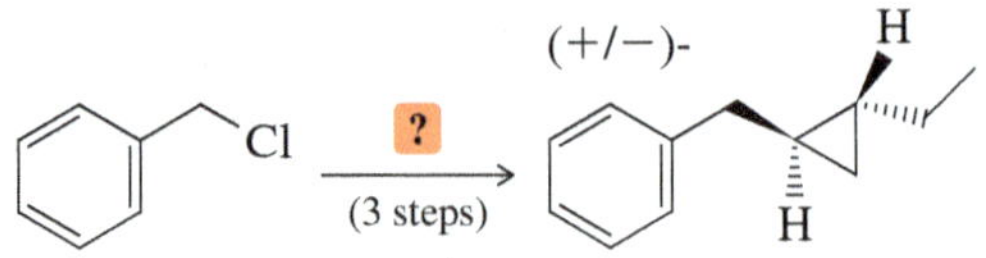

Assessment 16.20 Predict the product of the following reactions.

(a)

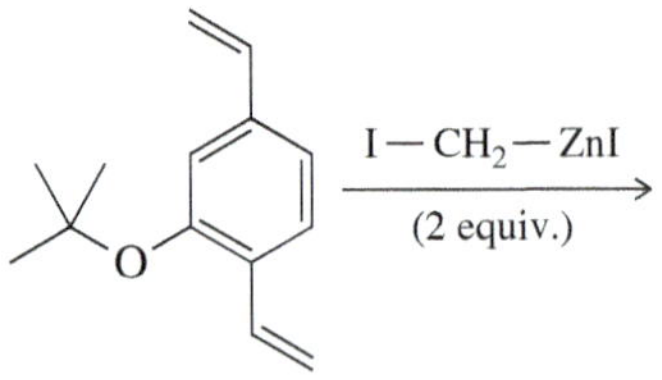

(b)

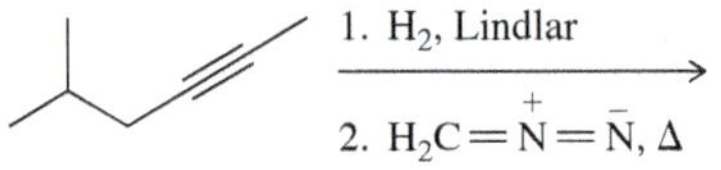

(c)

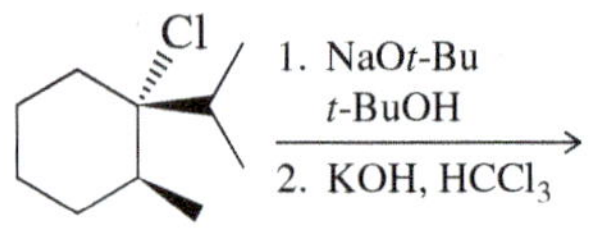

16-B CHEMISTRY IN CONTEXT: Natural Product Synthesis

Cyclopropanation

When curacin A was first isolated, its structure was deduced by a variety of advanced spectroscopic methods. [*See Chemistry in Context 15-A through 15-D.*] While these methods are excellent for identifying the functional groups present and the attachment order of the carbon skeleton, determining the relative and absolute stereochemistry is more difficult. With seven stereocenters in the natural product (**Figure 16.29**), there are $2^7 = 128$ possible stereoisomers that would give the same carbon skeleton, but only one of these compounds represents the actual structure of curacin A. Because of its notable biological activity, it was important to conclusively determine both the relative and absolute stereochemistry of curacin A. The laboratory synthesis of curacin A confirmed the structure as being the one shown in Figure 16.29. [*Reason #2 for natural product synthesis: conclusive structure determination.*]

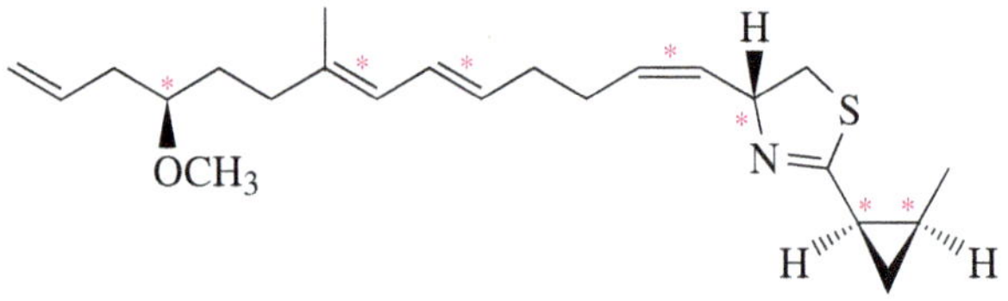

curacin A

Figure 16.29 A compound with seven stereocenters has 128 possible stereoisomers. Curacin A is the exact diastereomer shown. (Each stereocenter is indicated with an asterisk.)

The C_{17}—C_{20} cyclopropane fragment of curacin A was synthesized using a Simmons–Smith cyclopropanation. A chiral additive was used to deliver the Simmons–Smith reagent to the top face of the alkene, ensuring that only a single enantiomer of the cyclopropyl alcohol was formed (**Figure 16.30**).

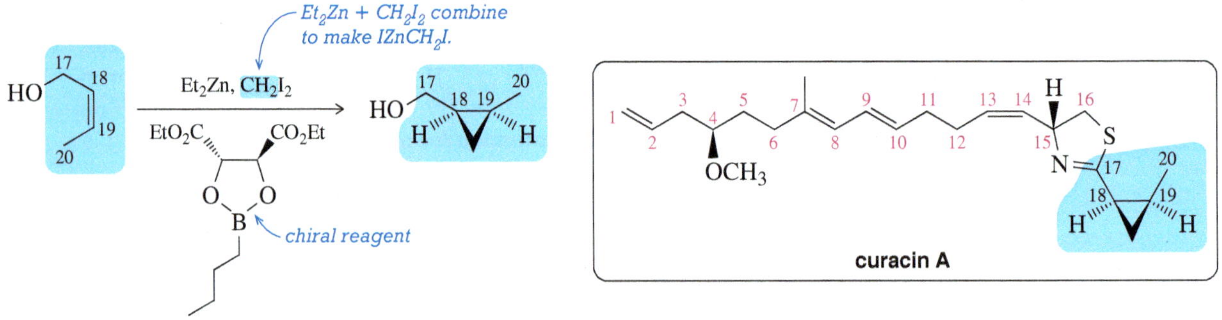

Figure 16.30 Cyclopropanation in a chiral environment produces a single enantiomer.

Assessment 16.21 Suggest a mechanism for the reaction shown in Figure 16.30 using $IZnCH_2I$ as the cyclopropanating reagent.

Assessment 16.22 To determine the stereochemistry of curacin A by synthesis, it would have been necessary to prepare all stereoisomers of the C_{17}—C_{20} cyclopropane fragment. How would the reaction in Figure 16.30 be modified to produce the other stereoisomers shown here? [*Think about the concepts of stereoselectivity (Chapter 6) and stereospecificity (Chapter 8).*]

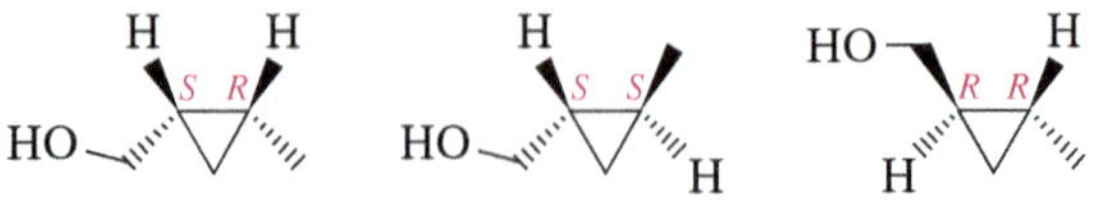

16.4 Transition metal–catalyzed coupling

Arrow-pushing hasn't been used historically by inorganic and organometallic chemists, but it can enhance our understanding of these reactions, even if a "real" organometallic chemist might look at what is drawn and say, "Well, not exactly . . . ."

There are few ways that are as reliable for making new C—C bonds as coupling reactions catalyzed by transition metals. Using an organometallic reagent and a halogenated compound in the presence of a transition metal catalyst, virtually any C—C bond can be made. It is difficult to find a recent natural product synthesis [*see Chemistry in Context 16-C*] where transition metal catalysis does not feature prominently.

Unfortunately, transition metal catalysis is complicated by a need for understanding the chemistry of transition metals, which, with filled *d* orbitals and empty *f* orbitals, can seem weird. At times, an inability to understand causes students to instead choose rote memorization to get through the material. [*What's the point of that?*]

To avoid simply memorizing, Sections 16.4.1–16.4.6 simplify the reactions in a way that makes them understandable. This is done by showing you the real mechanism of the reactions, but then showing that same mechanism using our arrow-pushing formalism in a way that relates it to reactions studied previously. Although many transition metals can be used to catalyze these coupling reactions, we focus on palladium.

16.4.1 The catalytic cycle: general considerations

The reactions here in Section 16.4.1 have multiple things in common. [*Look for these commonalities as a way of better understanding the new reactions.*] First, they are called coupling reactions because two organic fragments are coupled together to make larger molecules. In all of the reactions (Negishi, Stille, Suzuki, and Sonogashira), an organohalide (R—X) is being coupled to an organometallic compound (R—M, where R = organic molecule; M = metal). Finally, these reactions are all catalyzed by palladium, typically in the form of "palladium tetrakis" [$Pd(PPh_3)_4$] [*for short . . . it's actually tetrakis(triphenylphosphine)palladium.*] The general catalytic cycle for these coupling reactions is shown in **Figure 16.31**. Note that Pd^0 is used to represent palladium tetrakis in order to simplify the picture.

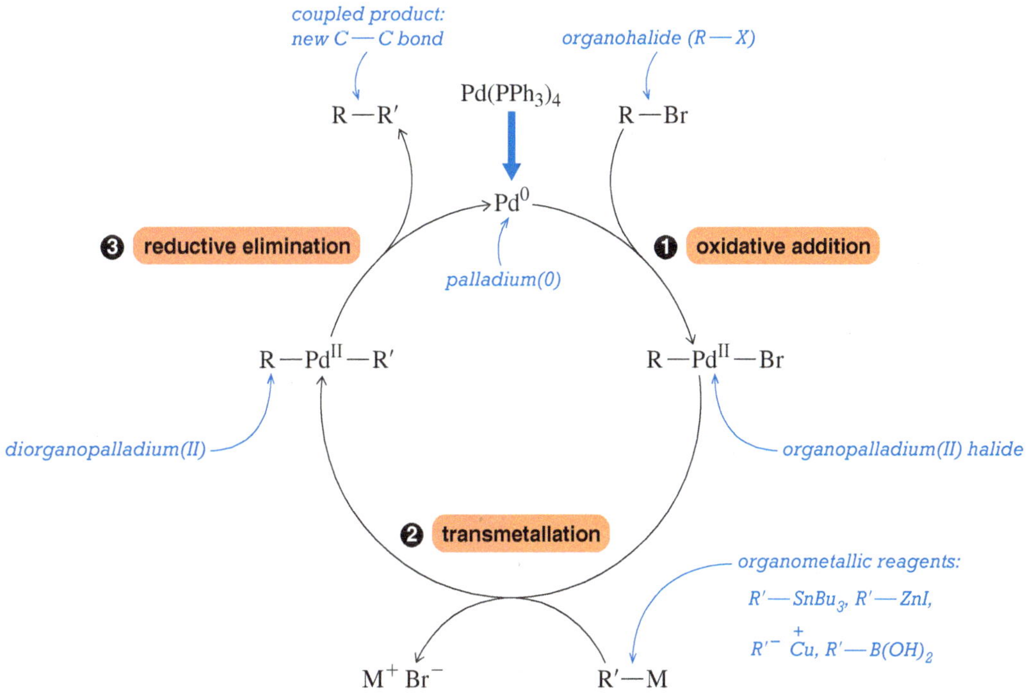

Figure 16.31 The catalytic cycle of organometallic/organohalide coupling reactions.

Regardless of the organometallic reagent (R—M) and the halogenated compound (R—X) being used, these reactions proceed by the mechanism shown in Figure 16.31. ***Oxidative addition*** of the organohalide to palladium(0) gives an organopalladium halide (step 1). At this point, ***transmetallation*** with the organometallic reagent (R—M) produces a diorganopalladium (step 2). In the final step (step 3), ***reductive elimination*** occurs to give

Note the absence of mechanistic arrows in Figure 16.31. This is how the catalytic cycle would be shown by an organometallic chemist. And while the three steps (oxidative addition, transmetallation, and reductive elimination) sound simple enough, for a novice to learn them this way would be to simply memorize—perhaps worthwhile, but definitely not satisfying.

the coupled product with a new C—C σ bond and, importantly, regenerates palladium(0) so that the cycle can be repeated again. And that's it. Just oxidative addition, transmetallation, and reductive elimination [*lather, rinse, and repeat.*]

Let's expand upon each step, relating it mechanistically and conceptually to how we talk about chemistry, even at the risk of oversimplification. [*In other words, let's translate this to the language of a first-year organic chemist.*]

1. *Oxidative addition of R—X to palladium*: Palladium is a transition metal, so it can do interesting chemistry. Pertinent to what we are studying here, it has several easily accessible oxidation states. [*Recall chromium(VI) and chromium(III) from Chemistry in Context 13-A.*] The two important oxidation states are palladium(0) and palladium(II). In palladium tetrakis, palladium is at the zero oxidation state.

   There are two ways that this reaction can be represented, but neither tells the whole story, so both are just proposals. However, by framing this reaction in ways that are analogous to reactions seen previously, we can visualize how it *might* work.

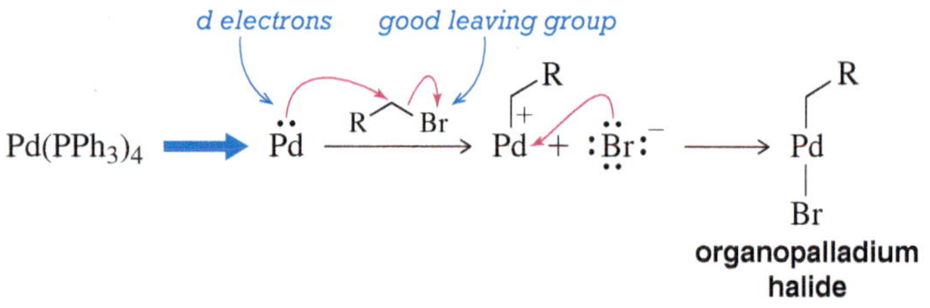

Figure 16.32 **Oxidative addition depicted as an "S_N2 reaction."**

   The case for S_N2: Palladium, like mercury [*see oxymercuration in Chapter 8*], has *d* electrons that it can use like "lone pairs." If the palladium in palladium tetrakis (**Figure 16.32**) were to use a lone pair to do an S_N2-like reaction on a haloalkane, the product would be a cationic organopalladium species and a negatively charged halide. This is reasonable because bromide is an excellent leaving group. Following displacement of the halide, we'd expect it to attack the positively charged palladium [*negatives attack positives*]. The result? An organopalladium halide where palladium now has a +2 oxidation state. The halide can attack because palladium has the empty *d* orbitals to accommodate extra electron pairs that second row elements do not.

   The case for radicals: Palladium has *d* electrons that it could use to do single-electron, or free radical, chemistry. The C—Br bond is relatively weak and participates in free radical reactions. If a single electron on palladium attacks the bromine, the C—Br bond could homolytically cleave, forming a new Pd—Br bond and a carbon radical. The carbon radical could then combine with a "palladium radical" to complete formation of the organopalladium(II) halide (**Figure 16.33**).

Figure 16.33 **Oxidative addition depicted as a "radical reaction."**

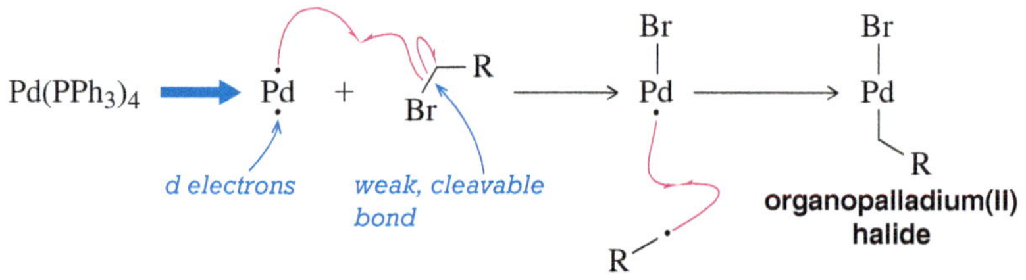

   Depending on the organohalide used, either of these representations of oxidative addition could be correct—alternatively, maybe both are completely off-base. Maybe it's some concerted version of one or a combination of the two . . . who knows? [*There is nothing wrong with not knowing; that's how research questions arise.*] But—and this is the important part—both possibilities should look familiar to you because they are rooted in mechanisms presented previously. And if that helps us understand what is going on, then the goal is accomplished.

2. *Transmetallation:* Looking back at Figure 16.31, the transmetallation step should be easier to relate to previous studies. When studying organolithium and Grignard reagents, we mentioned that carbon–metal bonds, despite having varying degrees of ionic/polar covalency, can be thought of as having a negatively charged carbon attached to a positively charged metal. And we know that carbanions are strong nucleophiles. In essence, then, the transmetallation step is simply the displacement of a halide from an atom by a strong nucleophile. [*Sound familiar?*] That description sounds a lot like an S_N2 reaction (**Figure 16.34**).

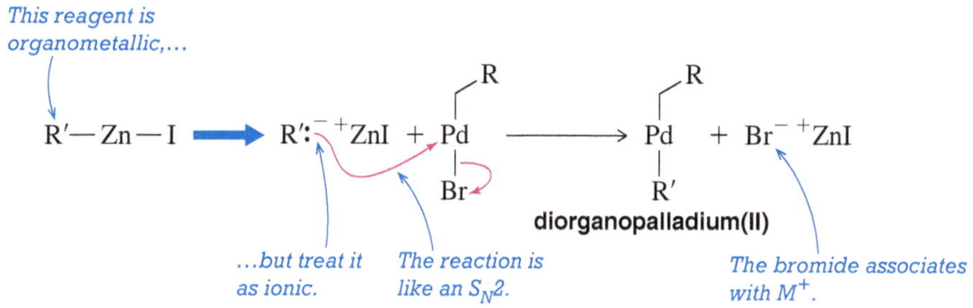

Figure 16.34 **Transmetallation depicted as an S_N2 reaction.**

3. *Reductive elimination:* The best way to understand reductive elimination is to view it as the reverse of oxidative elimination. [*That certainly makes sense, based on the name.*] And so the diorganopalladium can lose both alkyl groups by homolytic cleavage to produce two carbon free radicals while regenerating palladium tetrakis, as shown in **Figure 16.35**. The two carbon radicals would then combine to make a new C—C σ bond. In this process, the oxidation number of palladium has been reduced from +2 to 0—reductive elimination!

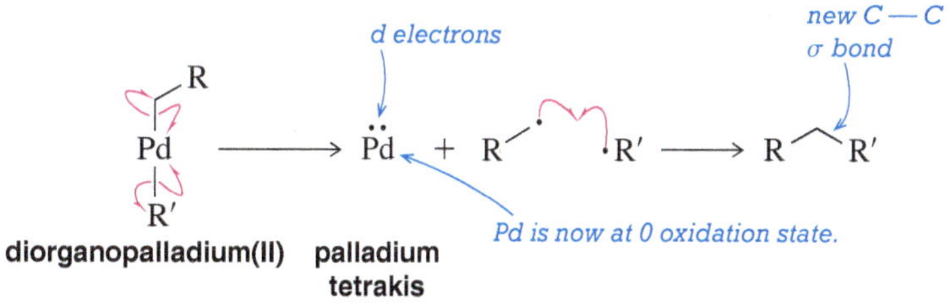

Figure 16.35 **Reductive elimination depicted as a stepwise radical reaction.**

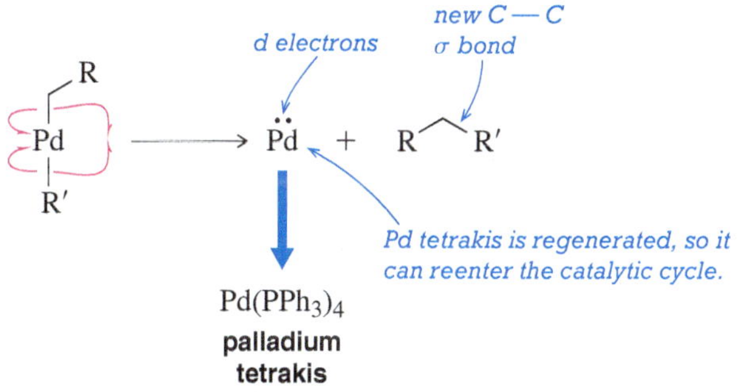

Figure 16.36 **Reductive elimination depicted as a concerted "radical" reaction.**

When RBr reacts with R′M, only R—R′ is produced. If the two radicals separated as in Figure 16.35, then it would be possible for the radicals to combine such that R—R and R′—R′ would result. Because these compounds are never observed, a more likely electron-pushing rationale would involve a concerted combination of radicals to form the new C—C σ bond (**Figure 16.36**).

If reductive elimination is the reverse of oxidative addition, then why aren't steps 1 and 3 reversible? It turns out that because of the stability of Br•, Br^-, R^+, and R•, and the relative weakness of the R—Br bond, step 1 probably is (**Figure 16.37**). But once the irreversible transmetallation occurs, step 3 results in the formation of a very stable C—C σ bond. Because of the stability of this bond, step 3 is probably irreversible.

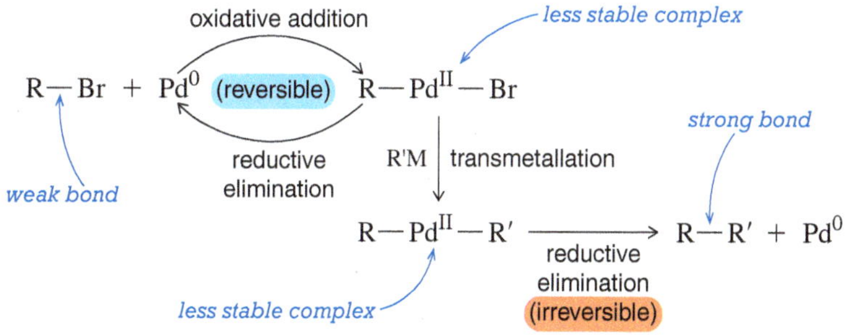

Figure 16.37 **The reversibility of oxidative addition and reductive elimination.**

The purpose of Figures 16.32–16.37 and the interleaved dialogue was not necessarily to tell you how these reactions occur, much of which is unknown. Instead, these figures are meant to show you that even for complicated reactions, you are equipped to understand the underlying chemical reactivity that drives these processes. As a way of concluding, **Figure 16.38** is a reimagining of Figure 16.31 using the "language" of organic chemistry.

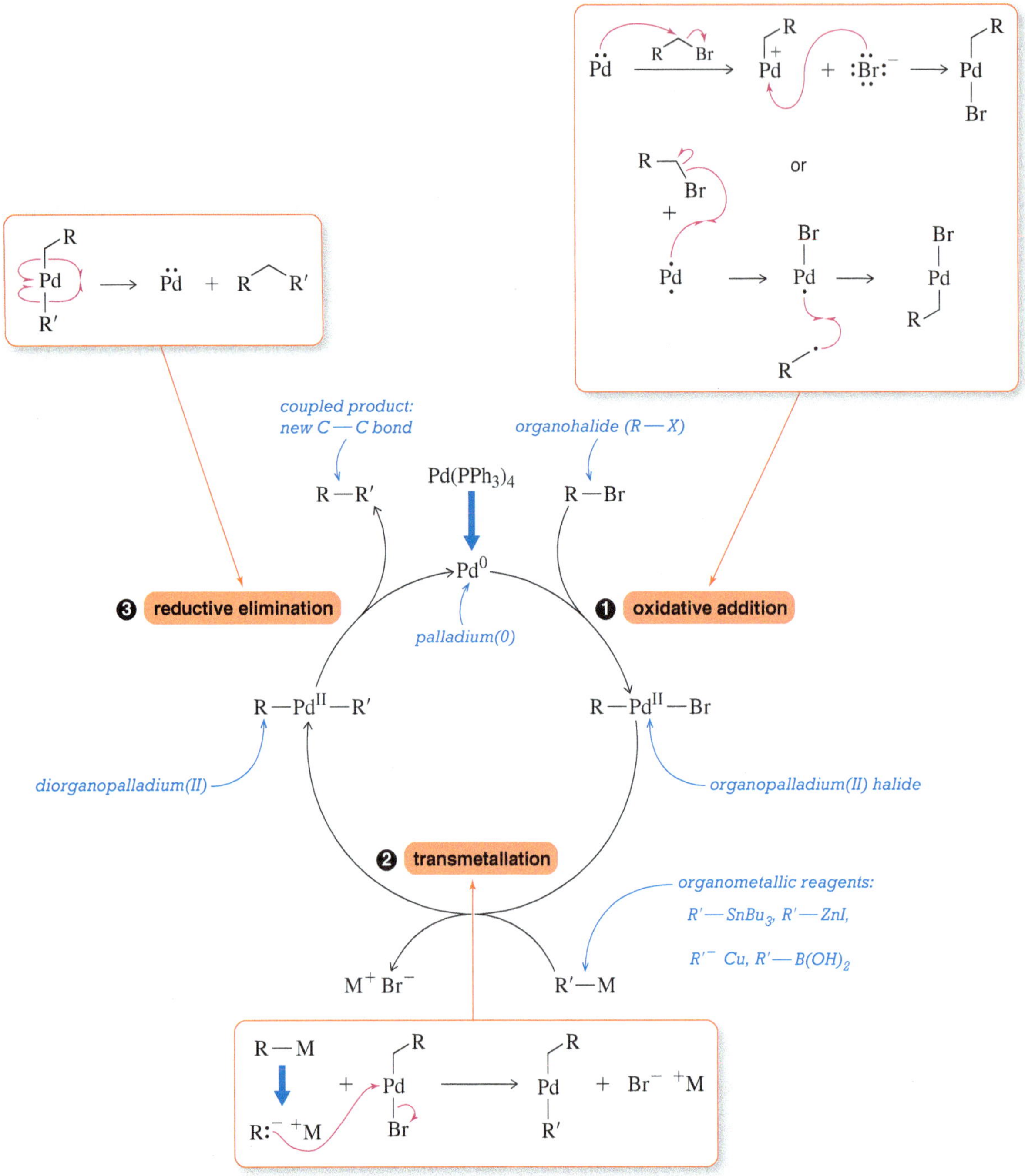

Figure 16.38 **Translating the catalytic cycle of transition metal–catalyzed coupling reactions.** [*This simplified catalytic cycle ignores the fact that the use of different metals and different ligands can dramatically change the outcome of these reactions.*]

Organic chemists use the names of the discoverers of reactions as a shorthand way of referring to the reactions. Don't get hung up on the names here. Instead, focus on the mechanism and the fact that it is basically the same for the next four sections.

16.4.1 SUMMARY

- The catalytic cycle of organometallic/organohalide coupling reactions always contains the following three steps: oxidative addition, transmetallation, and reductive elimination. Because these reactions involve organometallic reagents, our arrow-pushing formalism only approximates the electron flow in these reactions.

16.4.2 Negishi coupling

The Negishi coupling is the transition metal–catalyzed reaction between an organohalide and an organozinc halide. As shown in **Figure 16.39**, the reaction follows a path identical to that described in Section 16.4.1. Following *oxidative addition* of the C—Br σ bond to Pd, an organopalladium halide is produced where palladium is now in the +2 oxidation state. *Transmetallation* then occurs with the organozinc halide to give a diorganopalladium species. *Reductive elimination* results in the coupling of the two organic fragments to produce a new C—C σ bond, while regenerating $Pd(PPh_3)_4$ (Pd in the 0 oxidation state) for continued use in the catalytic cycle.

Figure 16.39 The catalytic cycle for the Negishi coupling. [*PPh_3 ligands have been removed for clarity.*]

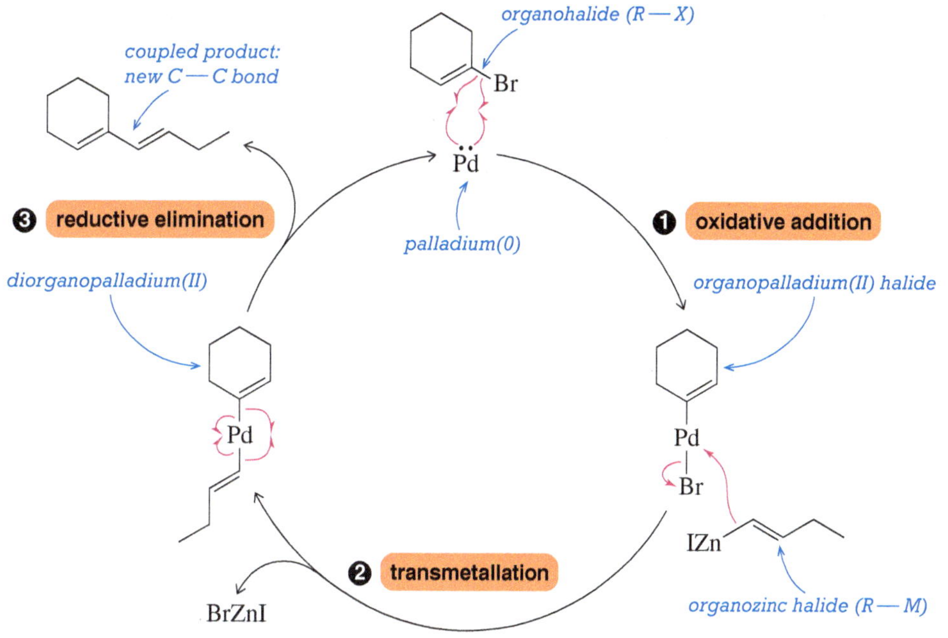

Several examples of the Negishi coupling are shown in **Figure 16.40**. The shorter, stronger C—F bond does not readily participate in palladium coupling reactions common to the other halides [Figure 16.40(a)]. Using an ambiphilic molecule (PTS), the Negishi reaction can be conducted in the presence of water to form a new sp^2–sp^3 C—C bond [Figure 16.40(b)] [*Recall the ambiphilic molecule used in Section 7.2.5.*]. Other good leaving groups can be used in place of halides in palladium coupling reactions. Figure 16.40(c) shows the use of a tosylate [*see Section 13.7.1*] as a polarizable C—X bond. Importantly, these coupling reactions are stereospecific with the *Z* stereochemistry retained from reactant to product. This observation is suggestive of a concerted mechanism for all palladium coupling reactions where there is no opportunity to isomerize to a more stable product.

Figure 16.40 Examples of the Negishi coupling between haloalkanes and organozinc halides.

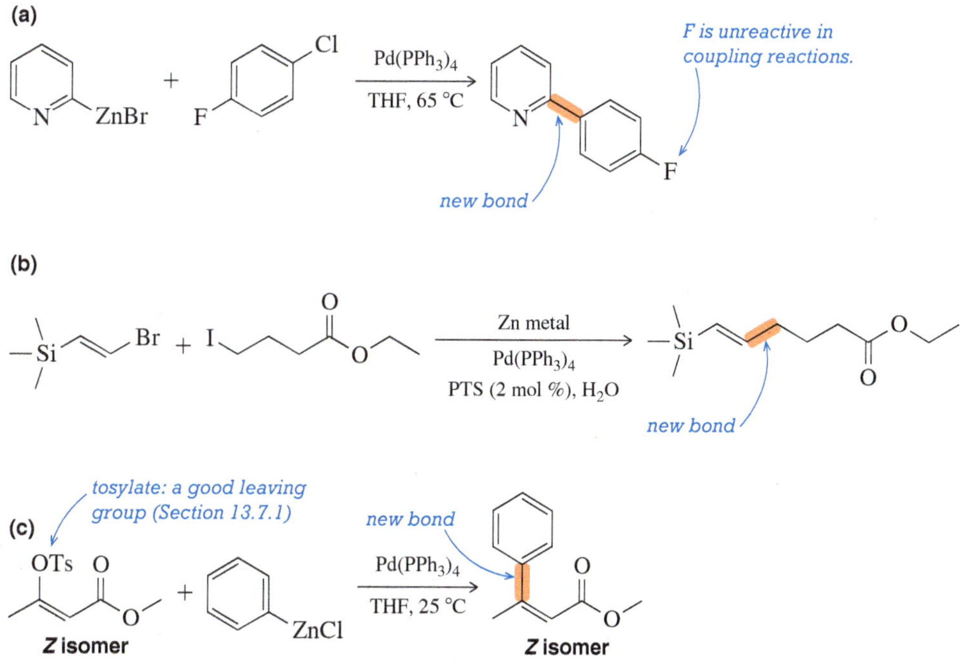

Assessment 16.23 Predict the product of the following Negishi coupling reactions.

(a) (b) (c)

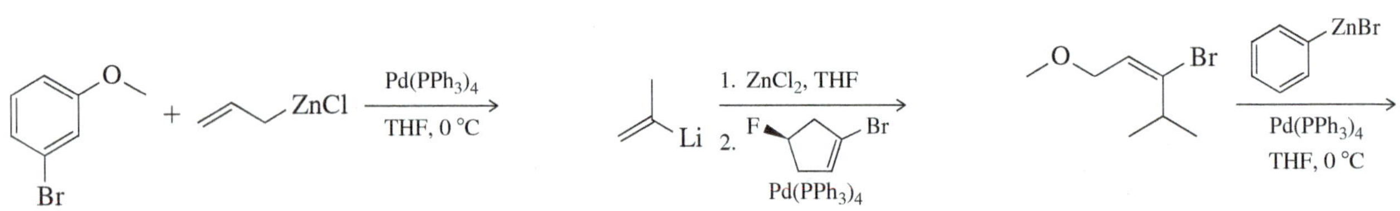

Assessment 16.24 Work backward to an appropriate organozinc halide and organohalide to make the bonds indicated by the blue arrows. There may be two possibilities for each.

(a) (b)

16.4.3 Stille coupling

The Stille coupling is the transition metal–catalyzed reaction between an organohalide and an organotin compound (also referred to as a stannane). As shown in **Figure 16.41**, the reaction follows a path identical to that described in Section 16.4.1. Following *oxidative addition* of the C—Br σ bond to Pd, an organopalladium halide is produced where palladium is now in the +2 oxidation state. *Transmetallation* then occurs with the stannane to give a diorganopalladium species. *Reductive elimination* results in the coupling of the two organic fragments to produce a new C—C σ bond, while regenerating $Pd(PPh_3)_4$ (Pd in the 0 oxidation state) for continued use in the catalytic cycle.

This should seem familiar, as we just covered it in Section 16.4.2.

Figure 16.41 The catalytic cycle for the Stille coupling. [*PPh_3 ligands have been removed for clarity.*]

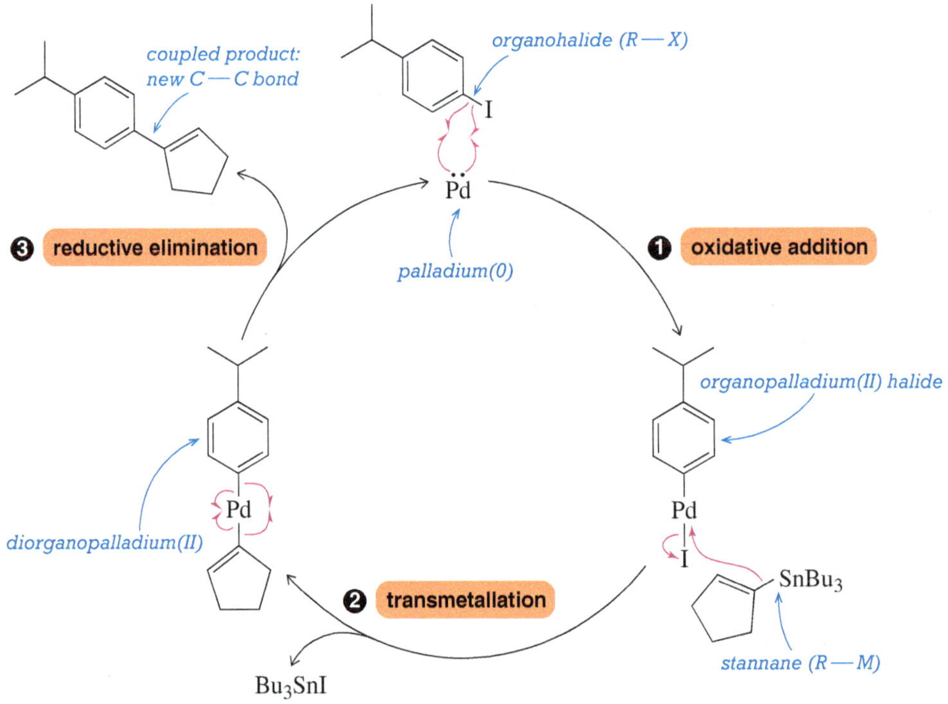

Several examples of the Stille coupling are shown in **Figure 16.42**. Like the Negishi coupling, the Stille coupling results in retention of stereochemistry when conducted on a stereochemically pure alkene [Figures 16.42(a) and 16.42(b)]. A triflate ($CF_3SO_3^-$), like a tosylate, is an excellent leaving group that can be used in place of a halide in palladium coupling reactions. Owing to the enhanced stability of the C—Sn bond, as compared to the more ionic C—Zn bond, intramolecular Stille coupling reactions are efficient [Figure 16.42(c)]. Trimethyl tin can be substituted for tributyl tin as the metal (M) component of Stille couplings [Figure 16.42(c)]. From a green chemistry perspective, the reaction has the advantage of being catalytic, but the use of toxic organotin compounds limits the utility of the Stille coupling reaction on an industrial scale.

Figure 16.42 Examples of the Stille coupling between haloalkanes and organozinc halides.

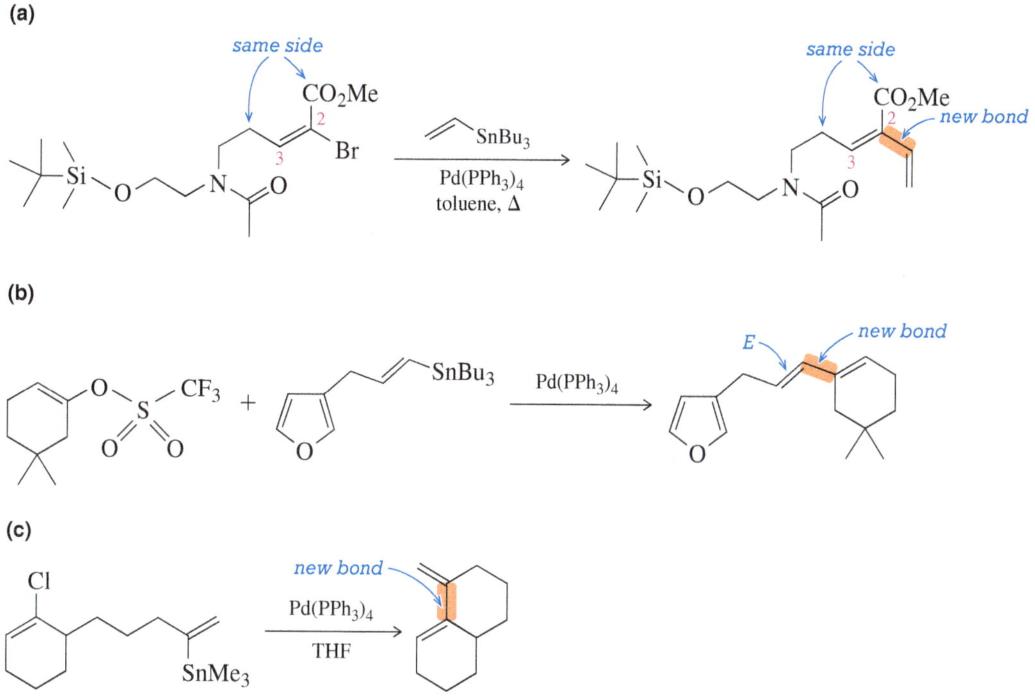

Assessment 16.25 Predict the product of the following Stille coupling reactions.

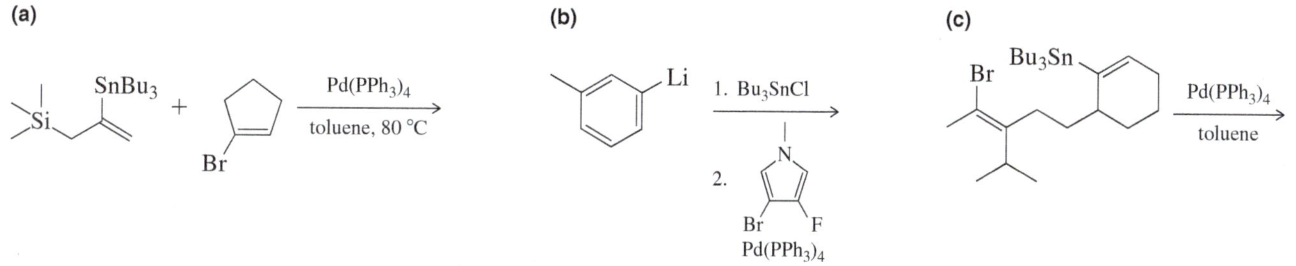

Assessment 16.26 Work backward to an appropriate organotin and organohalide to make the bonds indicated by blue arrows. There may be two possibilities for each.

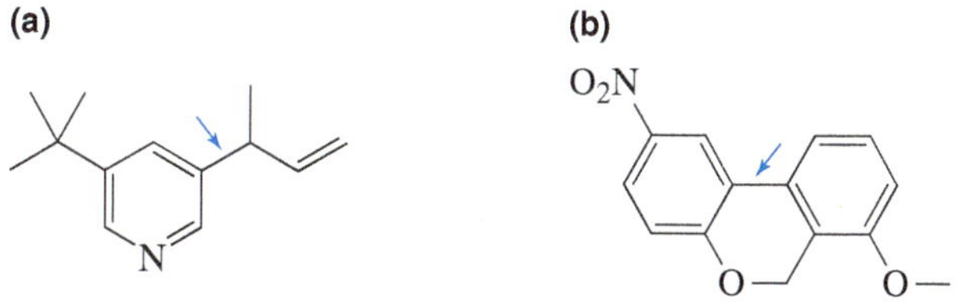

Assessment 16.27 In Figure 16.42(a), the C_2—C_3 alkene goes from *E* to *Z* in the reaction. Why is this still considered to be retention of stereochemistry?

16.4.4 Suzuki coupling

The Suzuki coupling is the transition metal–catalyzed reaction between an organohalide and a boronic acid. As shown in **Figure 16.43**, the reaction follows a path nearly identical to that described in Section 16.4.1, diverting only because the Suzuki reaction requires a base. Following *oxidative addition* of the C—Br σ bond to Pd, an organopalladium halide is produced where palladium is now in the +2 oxidation state. The bromide is displaced by *t*-butoxide. *Transmetallation* then occurs with a boronate species derived from the boronic acid to give a diorganopalladium species. *Reductive elimination* results in the coupling of the two organic fragments to produce a new C—C σ bond, while regenerating $Pd(PPh_3)_4$ (Pd in the 0 oxidation state) for continued use in the catalytic cycle.

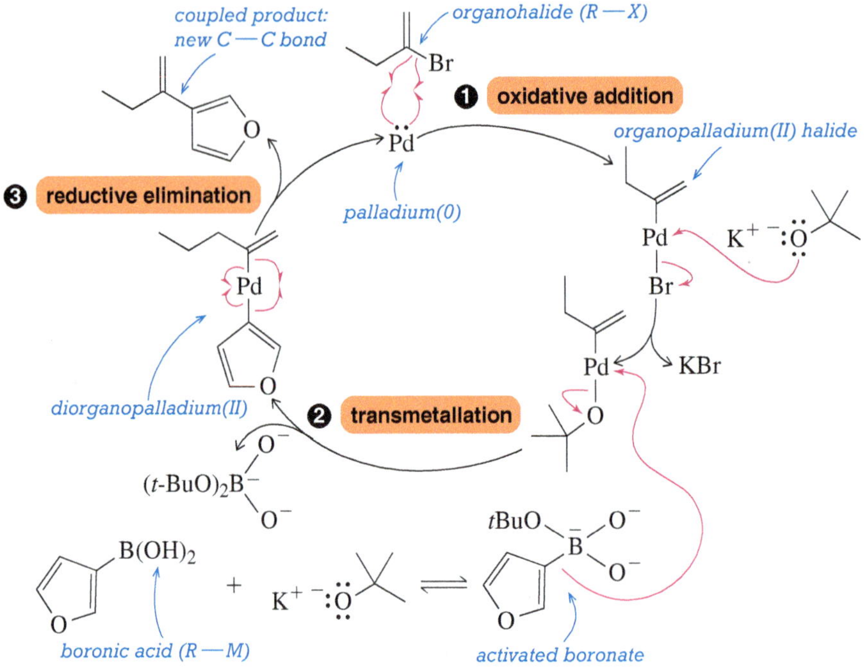

Figure 16.43 The catalytic cycle for the Suzuki coupling. [*PPh_3 ligands have been removed for clarity.*]

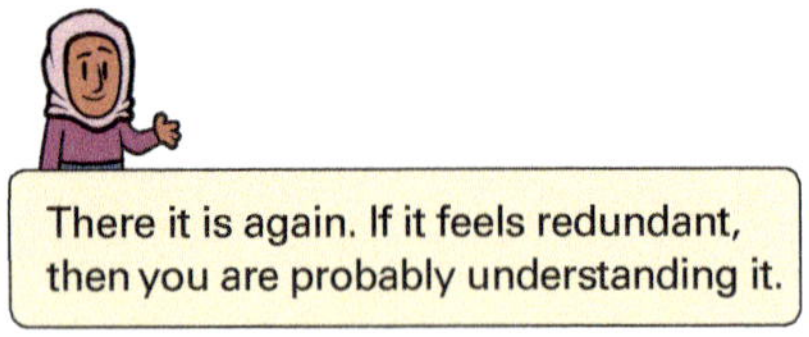

Several examples of the Suzuki coupling are shown in **Figure 16.44**. The Suzuki coupling is especially useful compared to the Stille coupling because boronic acids are environmentally benign. Thus, it is more amenable to scaling up for industrial syntheses. Because of the lack of reactivity of the C—B bond, Suzuki couplings can be run in water, requiring an equivalent of base present to activate the boronic acid. Using the Suzuki coupling, new

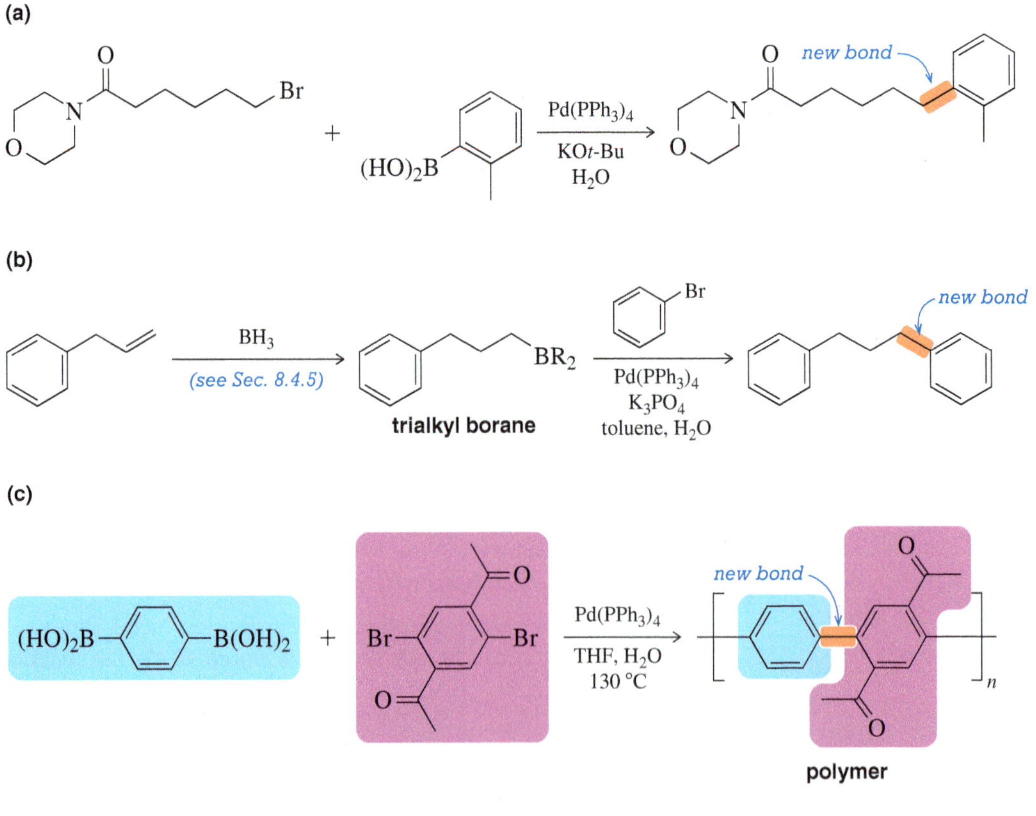

Figure 16.44 Examples of the Suzuki coupling between haloalkanes and alkylboron species.

C—C bonds can be made between sp^2 and sp^3 carbons [Figures 16.44(a) and 16.44(b)]. Showcasing the way previous chemistry is incorporated into this chapter, the hydroboration of an alkene [*see Section 8.4.5*] gives a trialkyl borane. The trialkyl borane can be used in place of a boronic acid in the Suzuki coupling reaction [Figure 16.44(b)]. Finally, the Suzuki coupling reaction can be used in polymerization reactions [Figure 16.44(c)]. [*We present polymerization in more detail in Chapter 28.*]

Assessment 16.28 Predict the product of the following Suzuki coupling reactions.

(a)

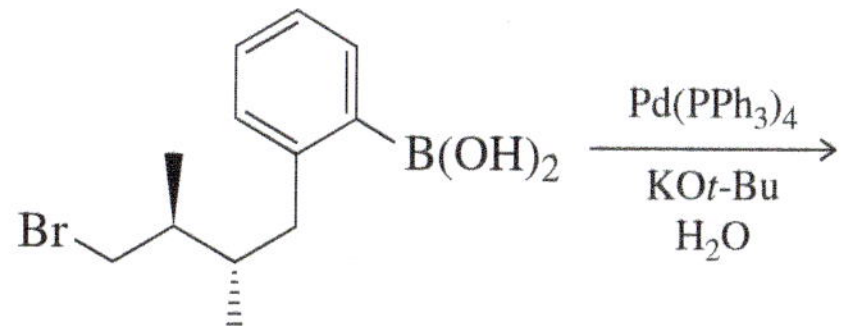

(b)

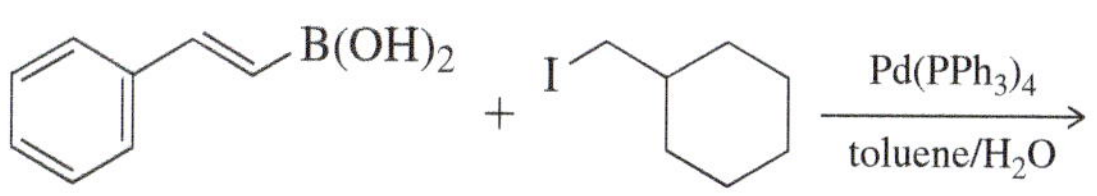

Assessment 16.29 Suggest a synthesis of the molecule on the right using the starting materials on the left. [*Use chemistry from Chapters 8 and 10 and a Suzuki coupling in your answer.*]

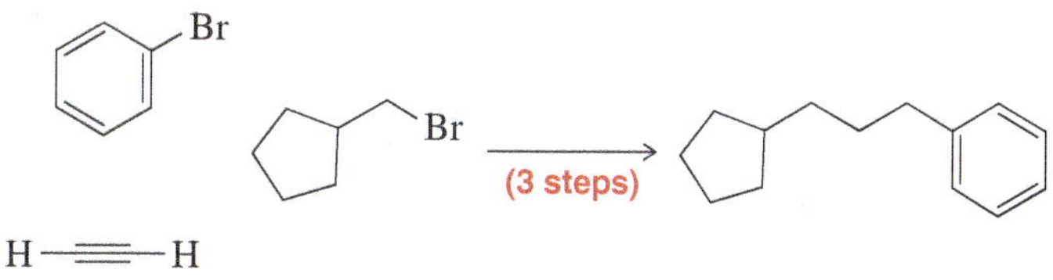

16-C CHEMISTRY IN CONTEXT: Natural Product Synthesis

Suzuki coupling

Although the synthesis of a natural product is a worthy goal in its own right, sometimes it is the *means* to the end—the novel organic synthesis that produces the molecule—that provides the greatest value. An English proverb says, "Necessity is the mother of invention," meaning that it is only the need for a particular product/solution that drives the innovation required to create it. A new reaction for forming a particular bond is only invented when a molecule calls for that bond to be formed. This is the greatest value of natural product synthesis. If chemists only synthesized molecules for which available technology existed, new reactions would never be invented and an infinite number of compounds would never be made. By letting nature provide us with the target molecules (natural products), chemists are forced to invent new reactions. [*Reason #3 for natural product synthesis: a template for developing new and useful reactions.*]

The C_1—C_{13} fragment of curacin A, representing the majority of the carbon skeleton, was assembled using the Suzuki coupling as a key reaction (**Figure 16.45**). Using a palladium catalyst, an alkenyl boronate was coupled to an alkenyl iodide to make the C_8—C_9 σ bond. A boronate was used in place of the usual boronic acid, but the reaction is unchanged compared to what was discussed in Section 16.4.4.

Figure 16.45 Suzuki coupling to make the C_8—C_9 bond of curacin A.

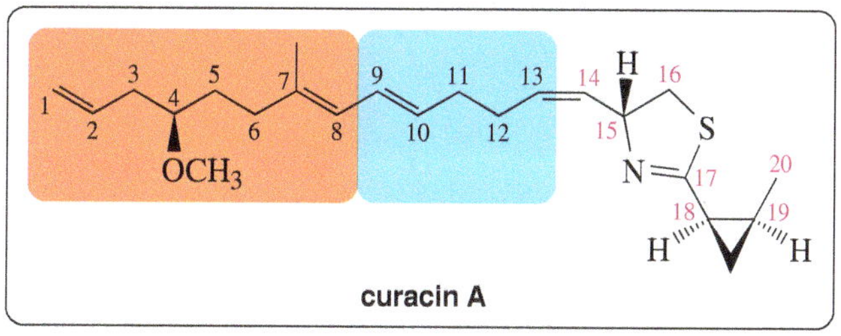

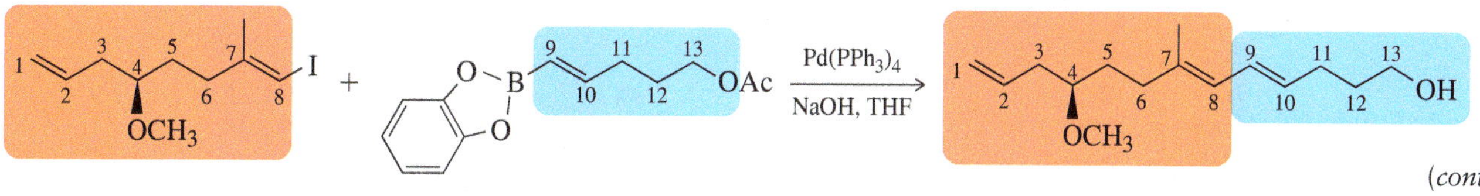

(continued)

While the study of curacin has advanced our understanding of cancer therapies and the mechanism of antiproliferative drugs, it has not, to this point, led to an actual drug. Some would argue, incorrectly, that there is little value in synthetic efforts that don't lead directly to a usable product.

The inherent value of natural product synthesis cannot be accurately judged without looking at the downstream effects. Consider losartan, an angiotensin II receptor antagonist (Chemistry in Context 26-B), marketed under the trade name Cozaar® as a high blood pressure medication. The synthesis of losartan involved the Suzuki coupling shown in **Figure 16.46**. This reaction, which may have found its "necessity for invention" in a natural product synthesis, has also found applicability in the production of a drug that, as a generic medication, was prescribed over 51 million times in 2017. In other words, natural product synthesis continues to be an important field of study because reactions developed in these contexts today may find extremely important uses tomorrow.

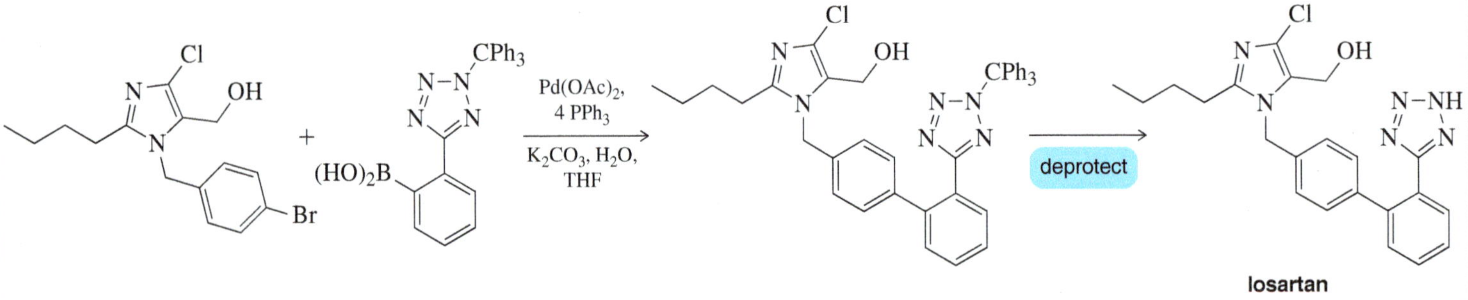

Figure 16.46 The synthesis of losartan utilized a Suzuki coupling.

Assessment 16.30 Provide a mechanism for the Suzuki coupling shown in Figure 16.45.

Last time: If it sounds like you've read this before, it's because it is nearly identical to what was presented in Sections 16.4.2–16.4.4.

16.4.5 Sonogashira coupling

The Sonogashira coupling is the transition metal–catalyzed reaction between an organohalide and a copper acetylide. As shown in **Figure 16.47**, the reaction follows a path identical to that described in Section 16.4.1. Following *oxidative addition* of the C—Br σ bond to Pd, an organopalladium halide is produced where palladium is now in the +2 oxidation state. *Transmetallation* then occurs with the copper acetylide to give a diorganopalladium species. *Reductive elimination* results in the coupling of the two organic fragments to produce a new C—C σ bond, while regenerating $Pd(PPh_3)_4$ (Pd in the 0 oxidation state) for continued use in the catalytic cycle.

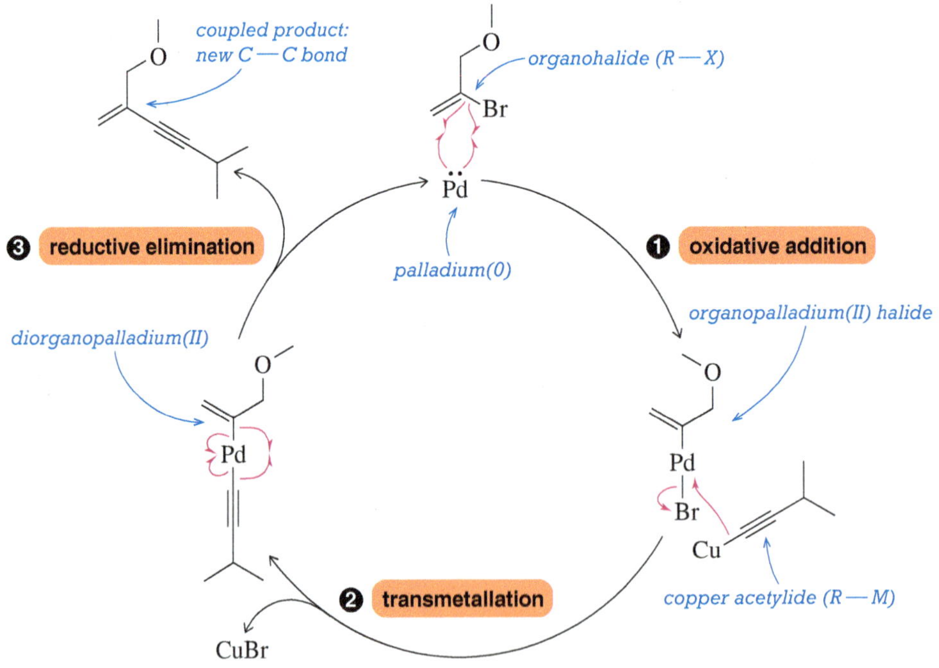

Figure 16.47 The catalytic cycle for the Sonogashira coupling. [*PPh_3 ligands have been removed for clarity.*]

Experimentally, the copper acetylide is generated by the reaction of a terminal acetylene with diisopropyl amine in the presence of a catalytic amount of copper(I) halide like CuI by a reaction shown in **Figure 16.48**. The diisopropyl amine is not strong enough to fully deprotonate the acetylene based on pK_a values [*see Chapter 10*], but enough of the copper acetylide is generated to allow the reaction to be driven to completion by Le Châtelier's principle. In the course of the catalytic cycle shown in Figure 16.47, the copper(I) salt is regenerated to react with the next acetylide ion.

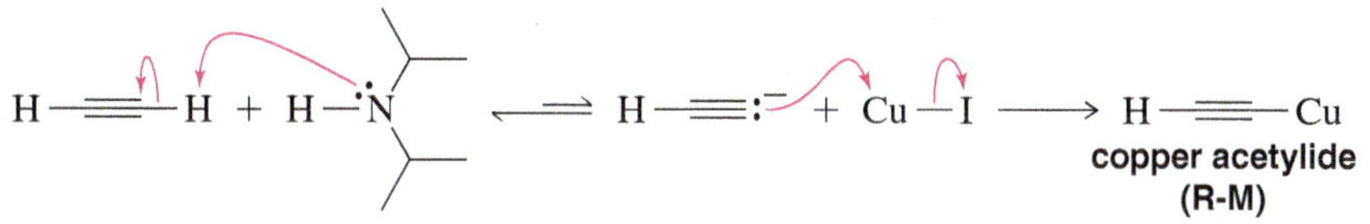

Figure 16.48 The formation of the copper acetylide by deprotonation.

Two examples of the Sonogashira coupling are shown in **Figure 16.49**. Because all Sonogashira reactions involve a copper acetylide, this reaction is limited in scope as compared to the palladium coupling reactions presented in Sections 16.4.2–16.4.4. Like those reactions, however, the Sonogashira is stereospecific when coupling to stereochemically pure alkenes [Figure 16.49(a)].

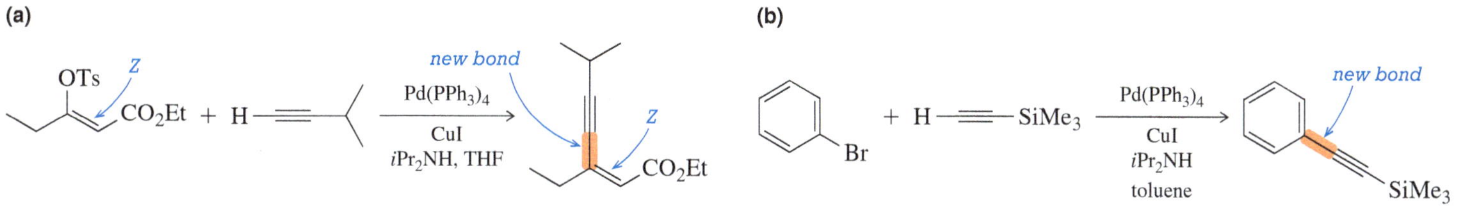

Figure 16.49 Examples of the Sonogashira coupling between haloalkanes and copper acetylides.

Assessment 16.31 Predict the product of the following Sonogashira coupling reactions.

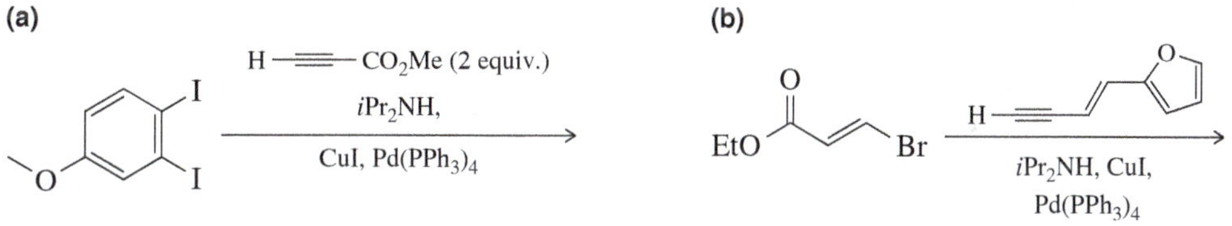

Assessment 16.32 The trimethylsilyl (TMS) group, used as a protecting group for alcohols, can also be used a protecting group for terminal alkynes. Show how TMS-acetylene could be used to link together two aryl halides using the Sonogashira reaction. [*Hint: Deprotection of the TMS-acetylene can be done using KF in H_2O.*]

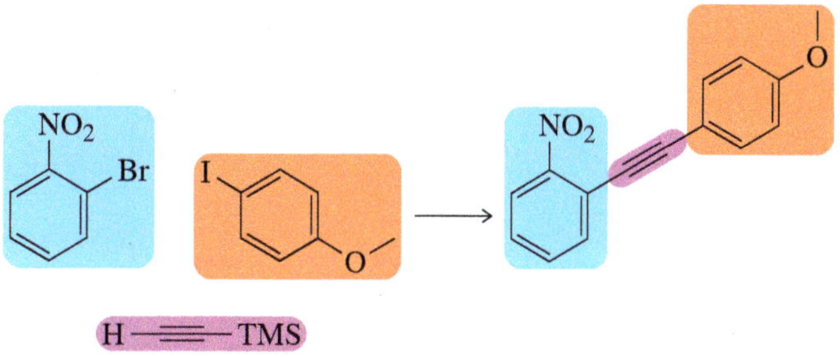

16.4.6 Transition metal–catalyzed coupling: concluding thoughts

While some of the transition metal–catalyzed coupling reactions can be used interchangeably, the choice will often depend on the availability of starting materials or the ease of preparation of reagents. Nevertheless, **Table 16.1** summarizes these reactions so they can be directly compared.

Table 16.1 Summary of transition metal–catalyzed coupling reactions. $Pd(PPh_3)_4$ is used in each.

Reaction	Organometallic	Aryl halide	Product
Negishi	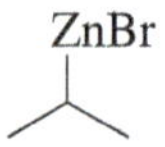 organozinc	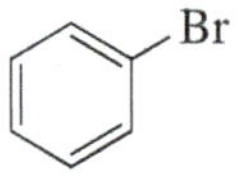	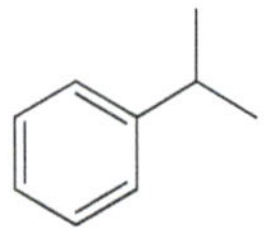
Stille	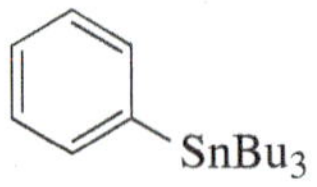 organotin	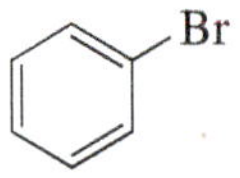	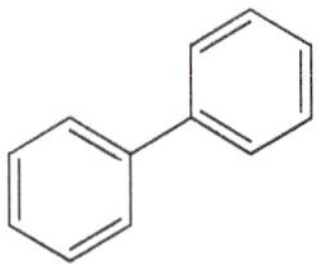
Suzuki	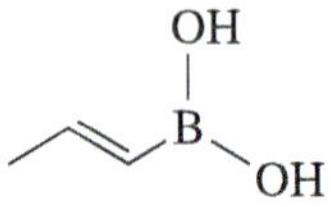 boronic acid	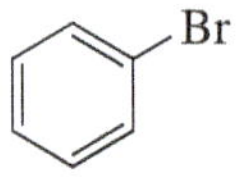	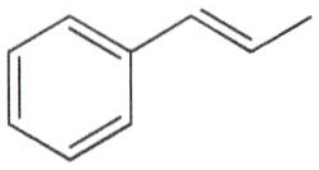
Sonogashira	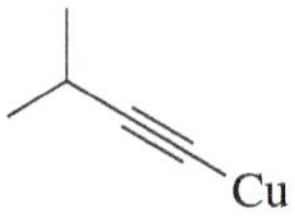 copper acetylide	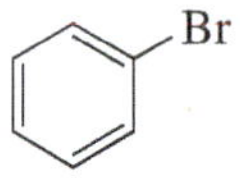	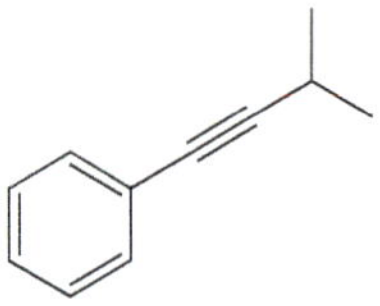

16.4.2–16.4.6 SUMMARY

- The Negishi (RZnBr), Stille ($RSnBu_3$), Suzuki [$RB(OH)_2$], and Sonogashira (R—C≡C—Cu) reactions occur by the same general mechanism in palladium-catalyzed coupling.

16.5 Gilman reagents/organocuprates

In the mid-1950s, Henry Gilman, appropriately deemed a pioneer in the development of modern organometallic chemistry, discovered lithium diorganocuprates, reagents that now bear his name. **Gilman reagents** are prepared by the reaction of two equivalents of an organolithium reagent with a copper(I) salt, most often CuI (Figure 16.50). Bearing a formal negative charge on copper, cuprates are more reactive than the corresponding neutral organocopper reagents. Gilman reagents can be similarly formed using the corresponding Grignard reagent, with the only difference being the countercation of the resulting organocuprate salt.

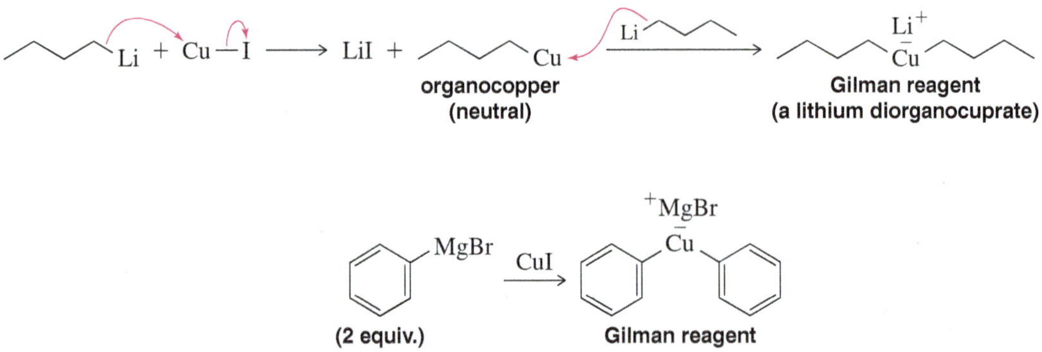

Figure 16.50 **Two equivalents of an organolithium or Grignard reagent and CuI produce a Gilman reagent.**

Gilman reagents are incredibly useful for the formation of C—C bonds. Throughout the text, we present four reactions of diorganocuprates. Among the four listed in Table 16.2, we discuss cross-coupling and epoxidation here in Chapter 16. Substitution of acid chlorides is presented in detail in Section 19.9.2, while conjugate addition is encountered in Section 21.5.4.

Table 16.2 Reactions of Gilman reagents (dioorganocuprates)

Reaction	Example	To be studied in:
Cross-coupling	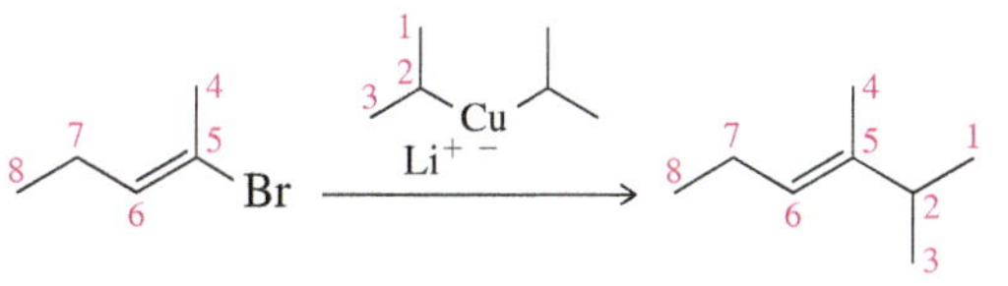	Section 16.5.1
Epoxide opening	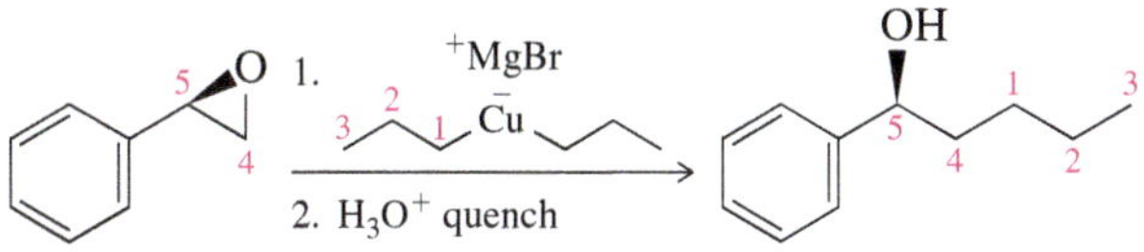	Section 16.5.2
Substitution of acid chlorides	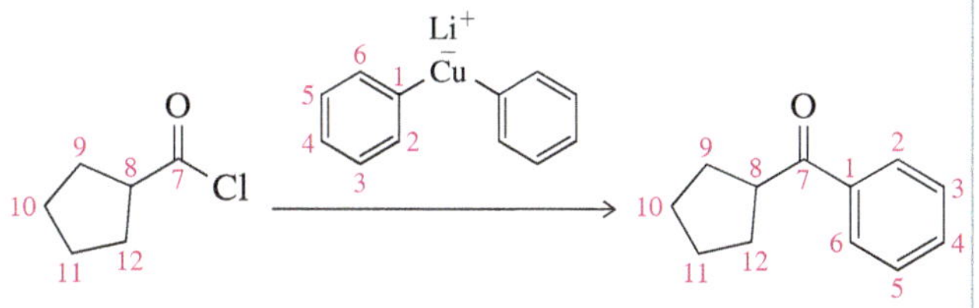	Section 19.9.2
Conjugate addition to α,β-unsaturated carbonyls	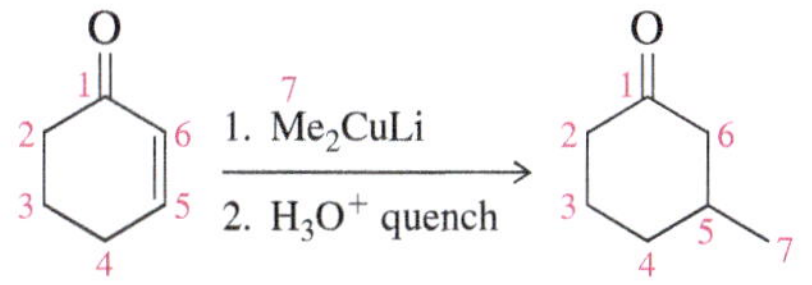	Section 21.5.4

16.5.1 Cross-coupling of Gilman reagents

The cross-coupling of Gilman reagents with alkyl, alkenyl, and aryl halides is very similar to the palladium-catalyzed coupling reactions in Section 16.4, with the exception that palladium is *not* required. An example of each of the three cross-couplings is shown in **Figure 16.51**.

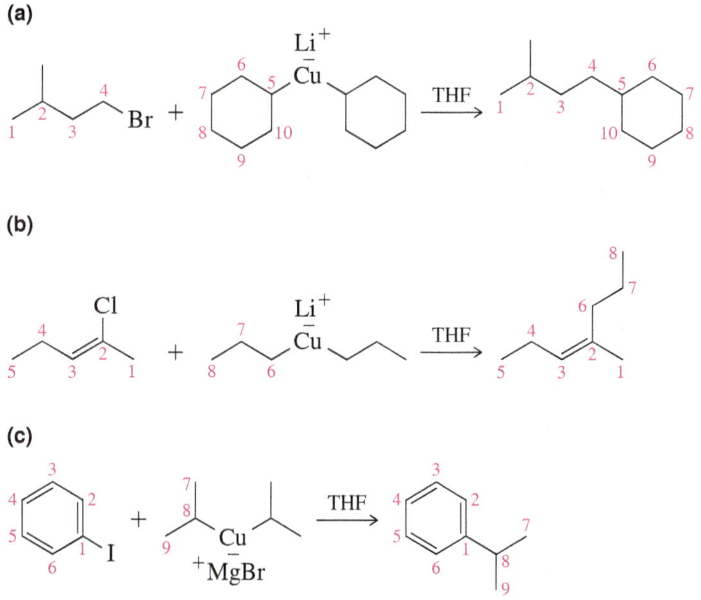

Figure 16.51 Cross-coupling of organocuprates with (a) alkyl halides, (b) alkenyl halides, and (c) aryl halides.

The mechanism of diorganocuprate cross-coupling is not straightforward. Not only are the mechanisms substrate dependent, but they are also still subject to debate. Regardless, and this is a point worth elaborating upon, the mechanistic differences we note here are important for our later understanding of what cuprates will and won't react with. [*Take a quick look back at Figure 16.51 and see if you can imagine a mechanism for the three reactions shown.*]

Consider the coupling reaction of a cuprate with a 1° alkyl halide to make a new C—C σ bond. By our previous rationale of considering a C—M bond as being ionic, with a negative charge on carbon, this reaction has all of the characteristics of an S_N2 reaction. And for this particular substrate, the research suggests that it just might proceed by an S_N2 reaction, as shown in **Figure 16.52**.

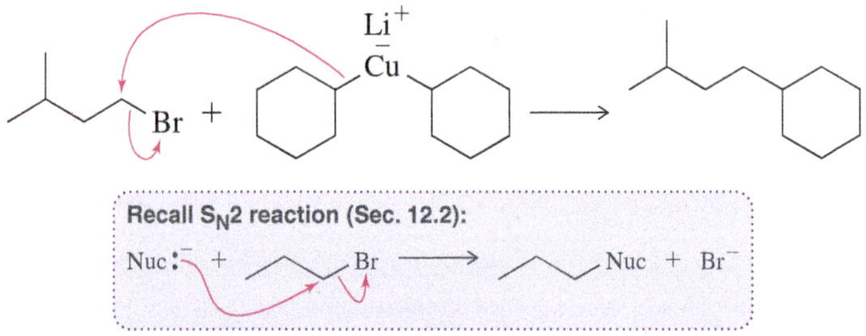

Figure 16.52 Cross-coupling with 1° alkyl halides proceeds by an S_N2 reaction.

What about cross-coupling with an alkenyl or aryl halide? We know from Chapter 12 that S_N2 reactions only occur at sp^3-hybridized carbons. As such, the sp^2 C—X bonds of alkenyl and aryl halides cannot undergo the reaction shown in Figure 16.52. Instead, these reactions are presumed to occur through a process of oxidative addition and reductive elimination, similar to what was discussed in Section 16.4. The cuprate oxidatively inserts into the C—I bond, going from Cu^+ to Cu^{3+}, as shown in **Figure 16.53(a)**. Oxidative insertion is followed by a concerted reductive elimination to produce the new C—C bond, at the same time generating an unreactive organocopper reagent. **Figure 16.53(b)** shows the implications of a concerted reductive elimination, with the reaction being stereospecific in retaining (not inverting) the stereochemistry of the starting alkenyl halide. Importantly, the inability of organolithium and Grignard reagents to do this type of chemistry makes them unable to do cross-coupling reactions. [*This is the first, but not the last, time we will see different reactivity of these seemingly related organometallic reagents.*]

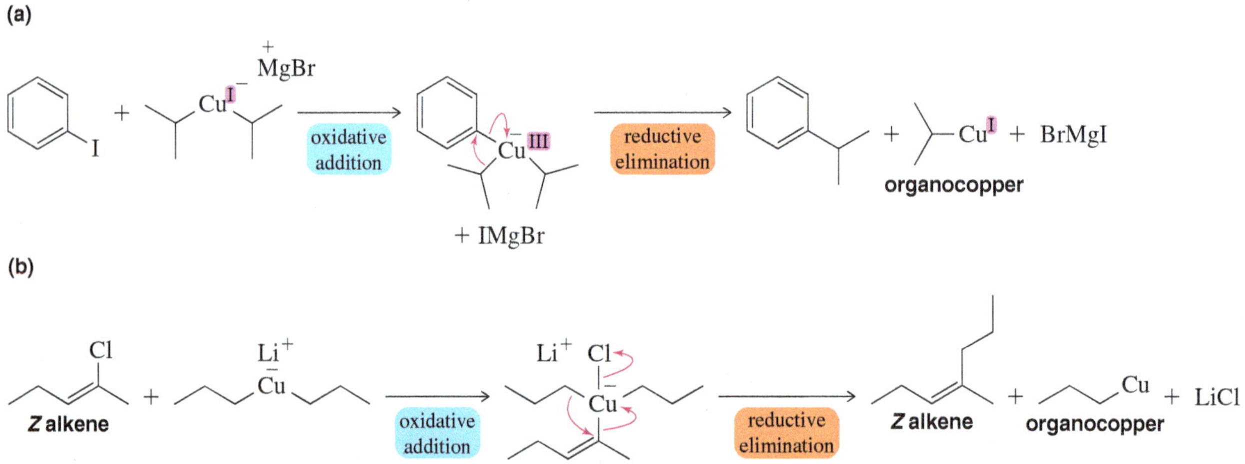

Figure 16.53 Cross-coupling of diorganocuprates with (a) aryl and (b) alkenyl halides proceeds by an oxidative insertion/reductive elimination process.

Back in Figure 16.50, the synthesis of diorganocuprates was shown to require two equivalents of an organolithium or Grignard reagent. And yet in each of the examples we have seen thus far, only one of the alkyl groups was transferred in the cross-coupling reaction. When the alkyl group being transferred is small and inexpensive, the problem with this reaction is limited to one of diminished atom economy. What about a case like that in **Figure 16.54**, where a new C—C bond is being formed between two large fragments? This problem is solved by using **dummy ligands** like cyanide. A dummy ligand in this context is a group that binds so tightly with copper that it cannot transfer in cross-coupling reactions.

Now, only a single equivalent of the organolithium is required to make the diorganocuprate with CuCN, saving time and materials while eliminating waste [*Section 7.2.1*] and maximizing atom economy [*Section 7.2.2*]. [*Going forward, look for RCuCNLi and R_2CuLi to be used interchangeably.*]

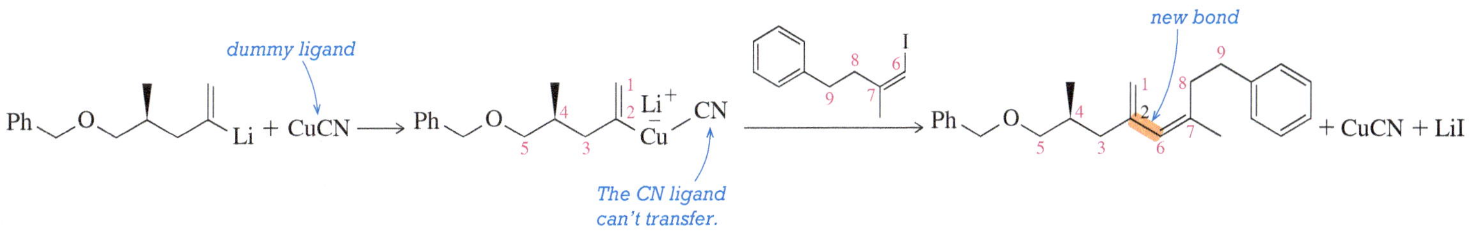

Figure 16.54 The use of a dummy ligand like cyanide eliminates the need for a second equivalent of organolithium.

Assessment 16.33 Predict the product of the diorganocuprate cross-coupling reactions shown.

(a)

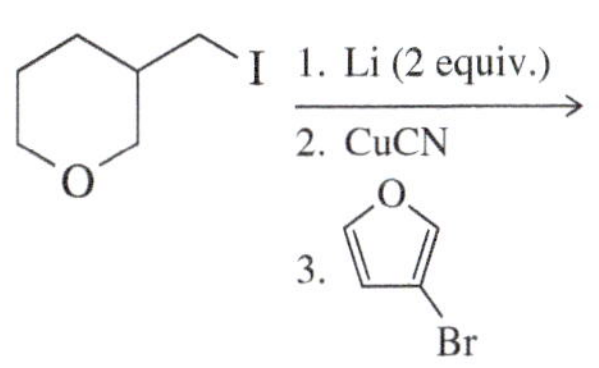

(b)

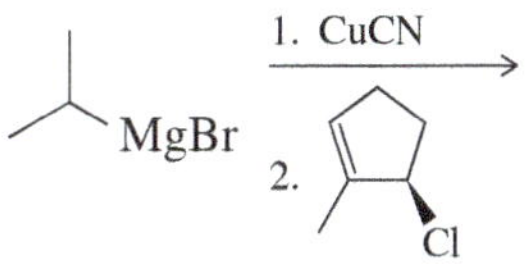

(c)

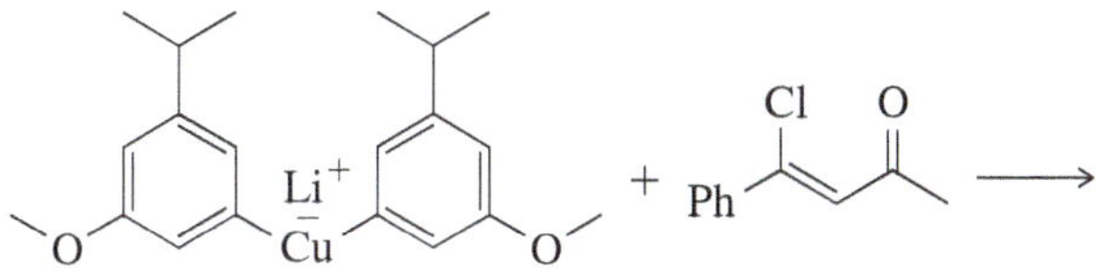

Assessment 16.34 Based on the stereochemical result alone, how can you tell that this reaction does *not* proceed by an S_N2 mechanism?

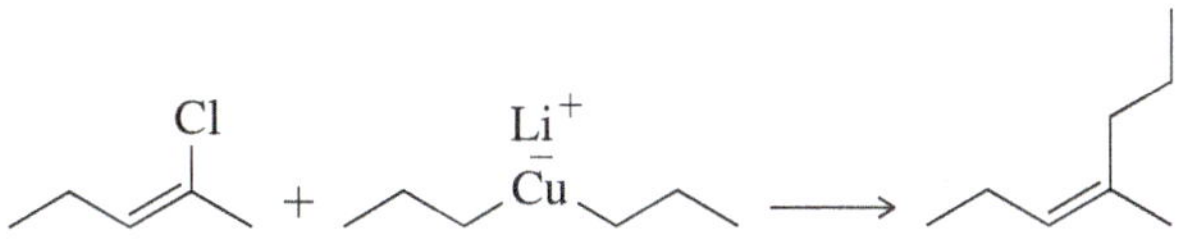

Assessment 16.35 Suggest a synthesis of the following molecule starting with the reagents shown, using cuprate cross-coupling as the key step. [*You will need a reaction from Chapter 13 to complete this synthesis.*]

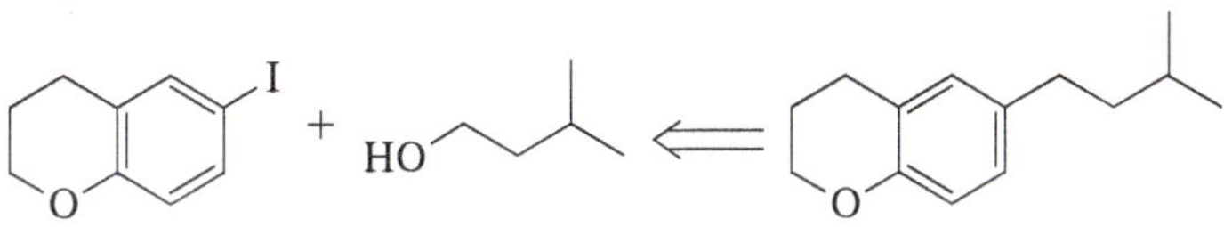

16.5.2 Epoxide opening with Gilman reagents

In Chapter 13, and again in Section 16.2.3, we studied the ring opening of epoxides by organolithium and Grignard reagents. [*Look back at Figure 16.12 to see these reactions.*] Recall that epoxides have a significant amount of strain that allows them to react, unlike their acyclic ether counterparts. Gilman reagents react similarly to organolithium and Grignard reagents, opening epoxides at the least substituted carbon, as shown in **Figure 16.55**. Mechanistically, the reaction proceeds in a manner similar to an S_N2 reaction, with the nucleophilic carbon donating a pair of electrons to the least sterically hindered carbon. Though this reaction has an identical mechanism, regardless of whether a Grignard, organolithium, or Gilman reagent is used to open the epoxide, this is the only reaction where this will be true. [*Again, Grignard and organolithiums proceed by one mechanism, and (except in this case) Gilman reagents proceed by another.*] So if all three do the same thing by the same mechanism, then why, or when, would a student choose to open an epoxide with an organolithium, Grignard, or cuprate? This decision usually relates to the ease of purchase or preparation of a particular reagent. In other words, that decision is one you will be equipped to make when you are working in the lab. [*For now, there is rarely a reason to choose between one or another, though Assessment 16.37 gives one of those situations.*]

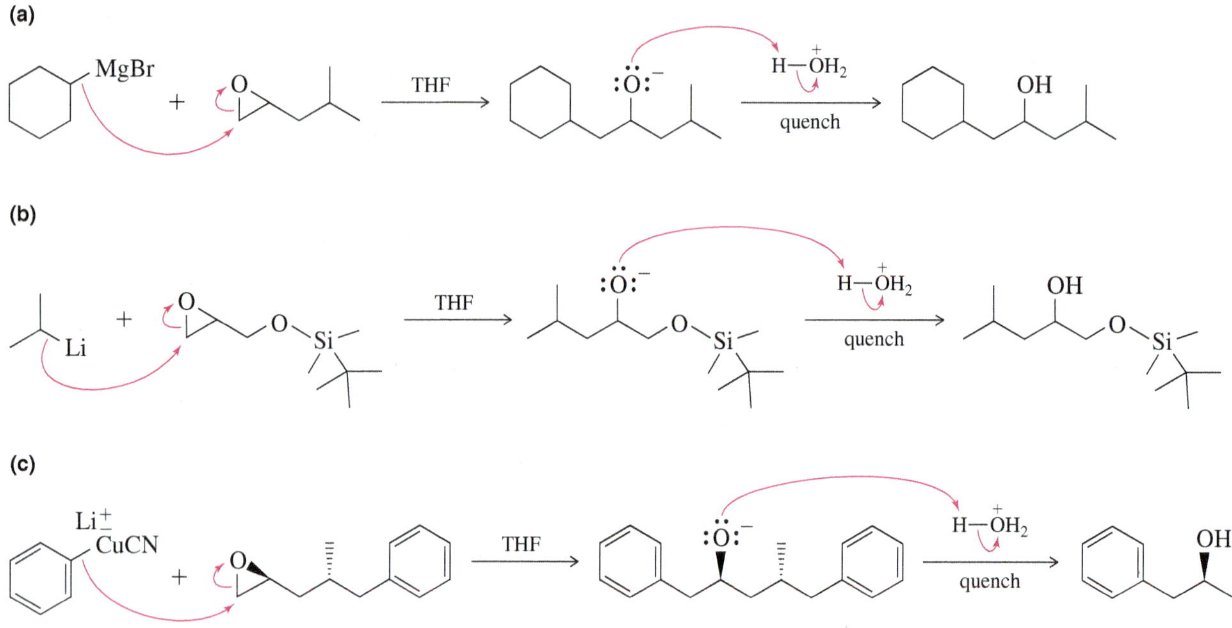

Figure 16.55 **Epoxide opening by a diorganocuprate occurs via an S_N2 mechanism.**

In summary, and to set the stage for later, cuprates react with C—X bonds in cross-coupling reactions and with C—O bonds when highly strained by a three-membered ring. Each of these bonds, C—X and the strained C—O, are relatively weak, polarizable bonds (**Figure 16.56**). Going forward, that's going to be an important indicator of whether a cuprate will react with a particular functional group. Cuprates are unreactive with polar functional groups containing strong bonds (aldehydes, ketones, and esters). Alternatively, weak, polarized bonds (as in α,β-unsaturated carbonyls or acid chlorides) are easily broken by reaction with cuprates. [*Don't worry about these reactions yet; we'll come to them eventually.*]

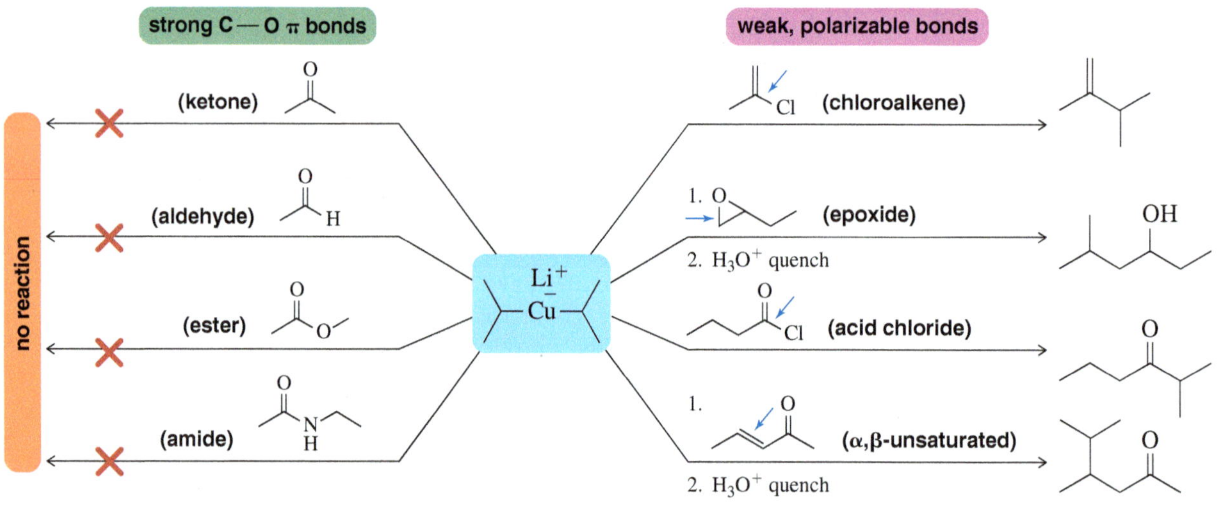

Figure 16.56 **Cuprates only react with weaker, polarized bonds** (as in acid chlorides and α,β-unsaturated carbonyls), but not with aldehydes and ketones.

16.5 SUMMARY

- Diorganocuprates (Gilman reagents) react with organohalides in cross-coupling reactions and as nucleophiles that open epoxides.

Assessment 16.36 Predict the product of the following epoxide opening reactions.

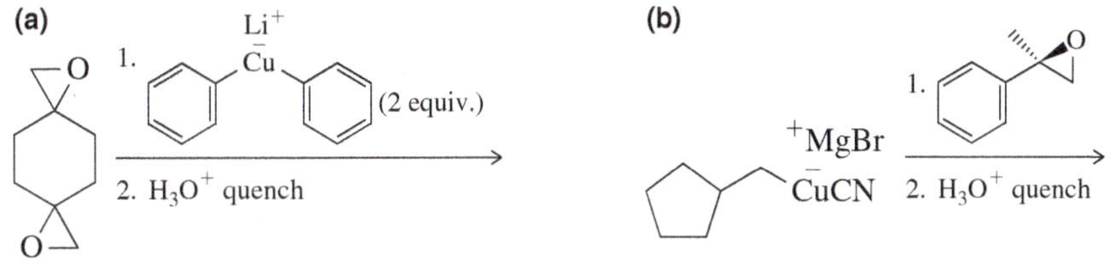

Assessment 16.37 The situation shown here is an example where a cuprate is the only organometallic that will allow the product alcohol to be obtained. What is the problem with using a Grignard or an organolithium to attempt the same reaction?

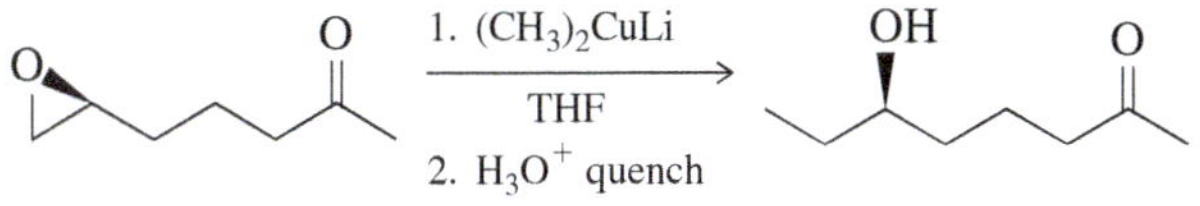

16.6 Olefin metathesis

Olefin (another word for alkene) **metathesis** is an important C=C bond forming reaction, resulting from the cleavage of two C=C bonds, while reforming two others. [*Recall double displacement, or metathesis, reactions from general chemistry, which take the form of AB + CD → AD + BC.*] This reaction is catalyzed by transition metal carbene catalysts. The two most common catalysts are the Grubbs ruthenium-based catalyst and the Schrock molybdenum-based catalyst, both shown in **Figure 16.57**. These catalysts are named for their founders, Robert Grubbs and Richard Schrock, for which they shared the 2005 Nobel Prize in Chemistry with Yves Chauvin.

Figure 16.57 Grubbs and Schrock olefin metathesis catalysts.

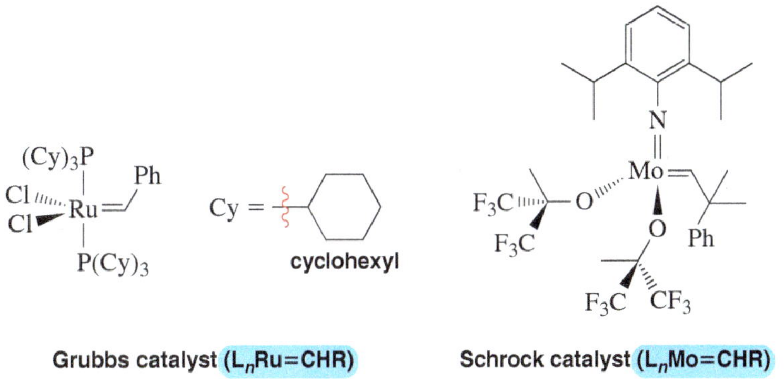

Figure 16.58 shows an example of a cross-metathesis, in which two different alkenes are connected by one of the new C—C double bonds. To simplify our understanding of this reaction, we consider only an abbreviated form of the Grubbs catalyst ($L_nRu{=}CHR$), with all ligands excluded. [*This is a complicated reaction, so we need to slowly and carefully analyze it by relating it to reactions studied previously. For now, use the color coding in Figure 16.58 to track the location of the new bonds.*]

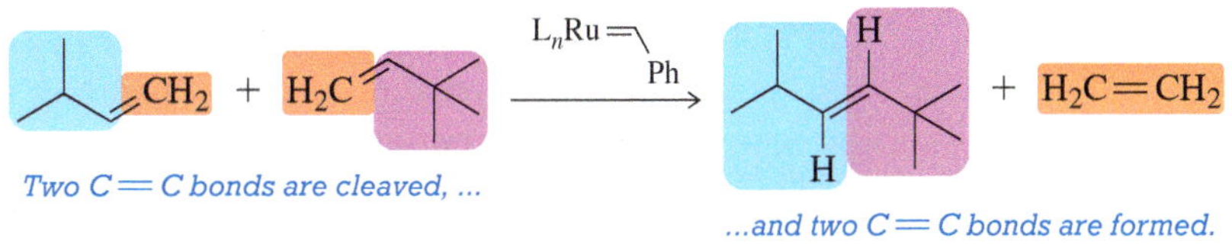

Figure 16.58 A cross-metathesis reaction using the Grubbs carbenoid metathesis catalyst.

To start our analysis, we have to first understand what is meant by a carbenoid. Essentially, a carbenoid is a molecule that is bound to a metal but exhibits carbene-like reactivity. In the same way that the Simmons–Smith reagent (a carbenoid) reacts like a carbene, the ruthenium carbenoid of the Grubbs catalyst will show similar reactivity (**Figure 16.59**). Thus, we can expect concerted reactivity, similar to what was seen in Section 16.3.

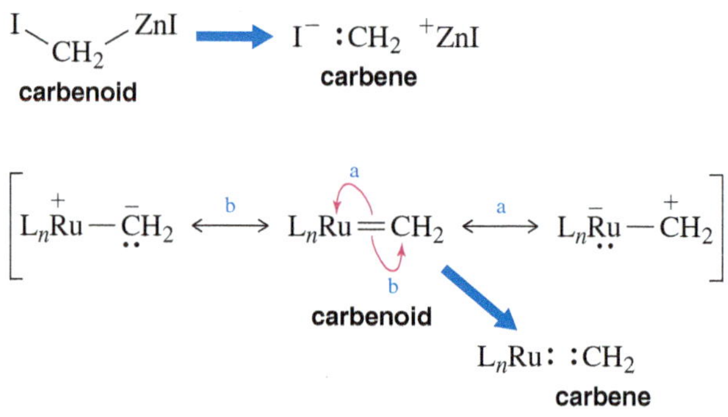

Figure 16.59 The reactivity of ruthenium carbenoids is similar to that of the Simmons–Smith reagent.

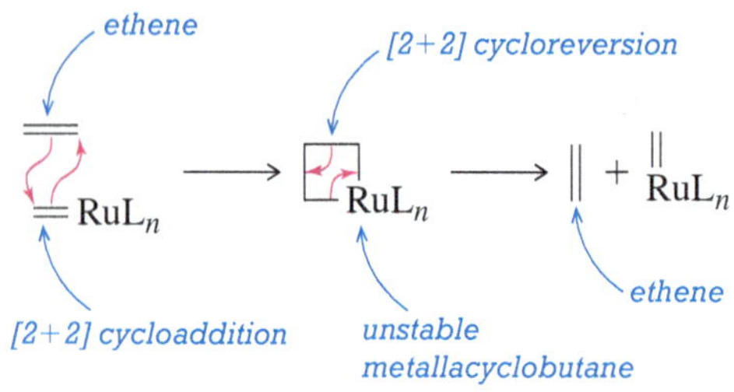

Figure 16.60 Alkenes react reversibly with ruthenium carbenoids via [2+2] cycloaddition reactions.

Like carbenes, carbenoids have some electrophilic character. As a result, ruthenium carbenoids react with alkenes. In this case, however, the reaction follows a pathway not seen previously. Collision between an alkene and the Grubbs catalyst results in a [2+2] cycloaddition to give a new four-membered ring, called a metallacyclobutane. [*We discuss cycloaddition reactions in Chapter 22.*] The designation of [2+2] tells that there are two sets of 2 electrons combining in the concerted reaction between ethene and the Grubbs catalyst shown in **Figure 16.60**. Because four-membered rings are unstable, the reverse reaction, a cycloreversion, occurs to regenerate the starting materials.

Because [2+2] reactions are reversible, metathesis reactions occur under thermodynamic control, equilibrating through a series of metallacyclobutanes to give the most stable product. In **Figure 16.61**, for example, we start with two monosubstituted alkenes and produced a disubstituted alkene and ethene. The formation of ethene, a gas, is entropically favored, giving the reaction as written a negative ΔG. In addition, ethene bubbles out of solution, driving the reaction to the product side according to Le Châtelier's principle.

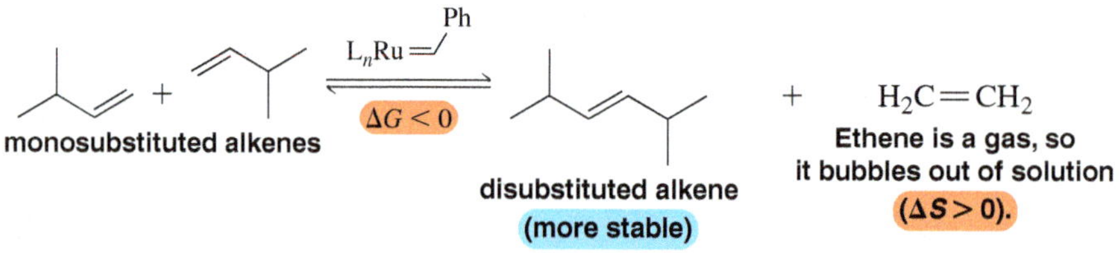

Figure 16.61 The formation of a disubstituted alkene and a gas thermodynamically favors the product side of the reaction.

The overall mechanism is shown in **Figure 16.62**. The reaction begins with an initiation step in which an alkene reacts with the catalyst to generate the catalytically active ruthenium carbenoid. A reversible [2+2] cycloaddition with one of the alkenes is then followed by a [2+2] cycloreversion. The result of this is the incorporation of the ruthenium into one of the alkene starting materials, generating the ethene by-product at the same time. A second [2+2] cycloaddition occurs with a second alkene and is, once more, followed by [2+2] cycloreversion to give the more stable disubstituted alkene, the product of the reaction, while regenerating the catalyst to re-enter the catalytic cycle.

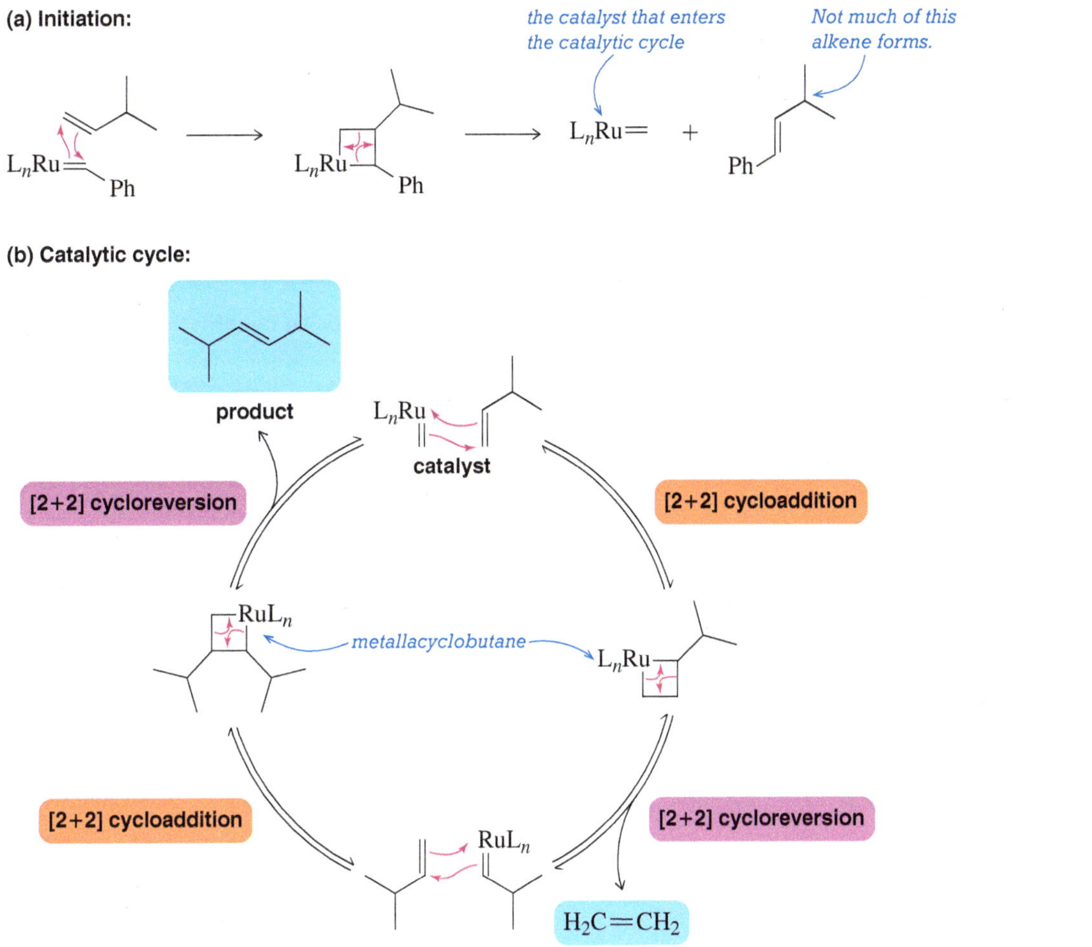

Figure 16.62 The catalytic cycle of alkene metathesis. (a) The initiation step forms the active catalyst. (b) The catalytic cycle produces the more stable product under thermodynamically controlled conditions.

Recent advances have allowed cross-metathesis to be efficient and high yielding, but that is beyond the scope of what is being discussed here.

As a reversible reaction, where only one possible disubstituted alkene can form, olefin metathesis is quite reliable. In **Figure 16.63**, however, two different alkenes are used in the reaction. A reaction of this type is referred to as a cross-metathesis because two different alkenes are being stitched together. If a stoichiometric ratio of alkenes is used, because none of the three possible disubstituted alkenes is particularly more stable than the others, a statistical ratio of alkene products will result.

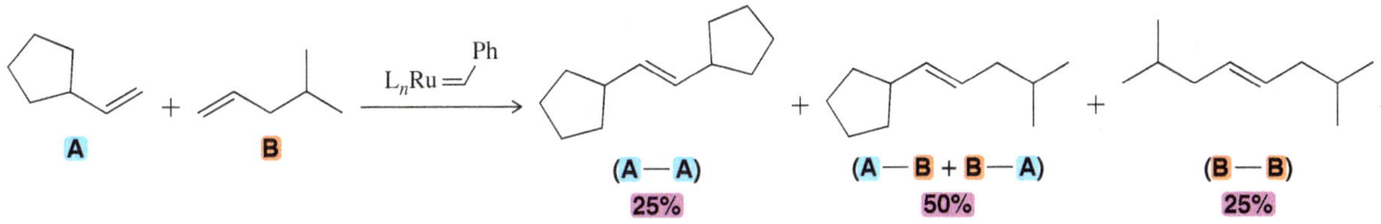

Figure 16.63 Cross-metathesis produces a statistical mixture of similarly stable disubstituted alkenes.

Perhaps the most impressive utility of metathesis reactions is in synthesizing new rings. Ring-closing metathesis, as it is called, takes terminal alkenes within the same molecule and stitches them together to make rings of various sizes. An example is shown in **Figure 16.64**. Still under thermodynamic control, ring-closing metathesis is even more favorable than the metathesis reaction in Figure 16.61. Entropically, the reaction begins with one molecule and makes two, one of which is a gas. And enthalpically, a new disubstituted alkene is formed at the expense of two monosubstituted alkenes. The combination of these two factors gives a very negative ΔG and a highly favorable reaction ($K_{eq} > 1$).

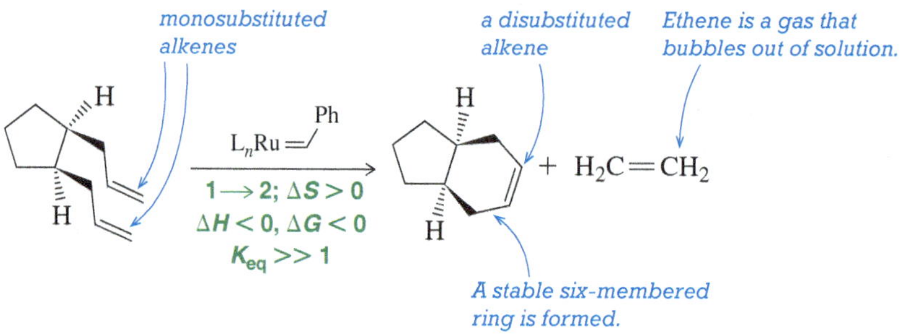

Figure 16.64 Ring-closing metathesis.

16.6 SUMMARY

- Alkene metathesis is the coupling of two alkenes to make a new C=C bond with extrusion of a smaller alkene. Either the Grubbs or the Schrock catalyst can be used in alkene cross-metathesis and ring-closing metathesis.

Assessment 16.38 Predict the product of the following metathesis reactions. If two different alkenes are used, only predict the major, cross-metathesis product.

(a)

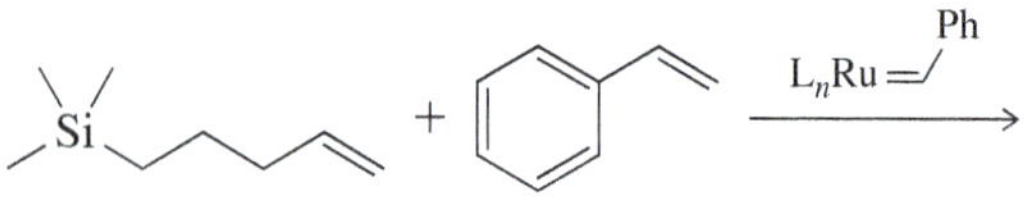

(b)

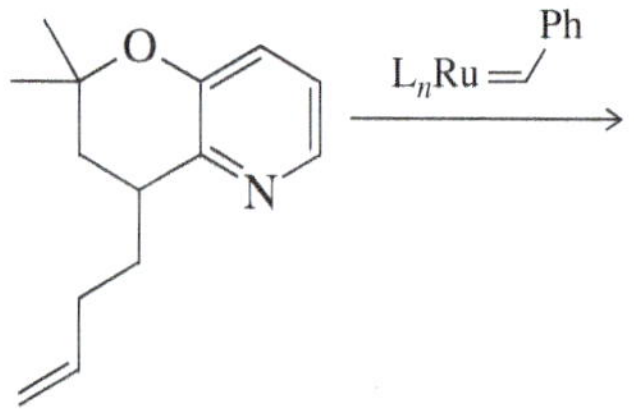

(c)

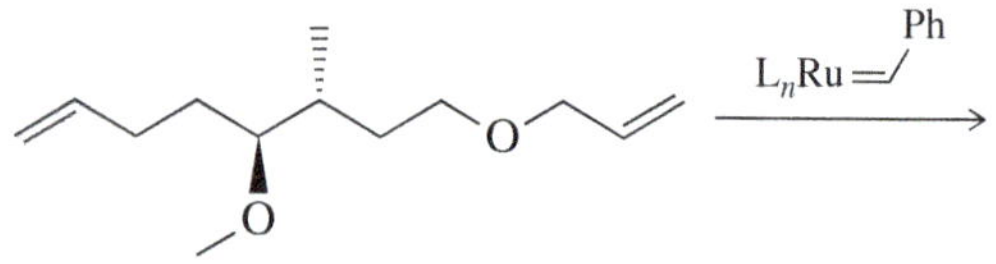

Assessment 16.39 Suggest the alkene partners that might be used in a metathesis reaction to produce the following products.

(a)

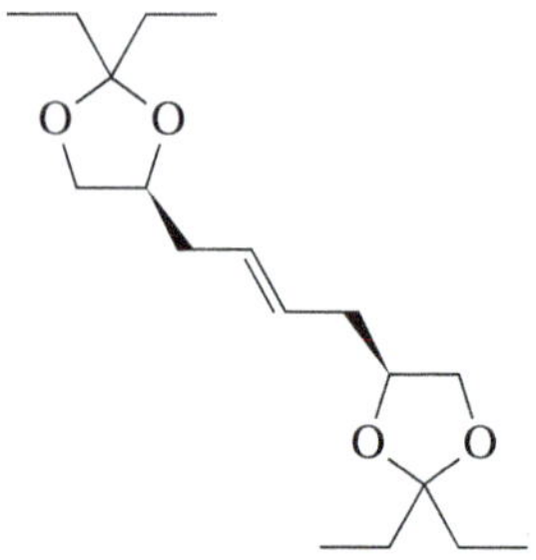

(b)

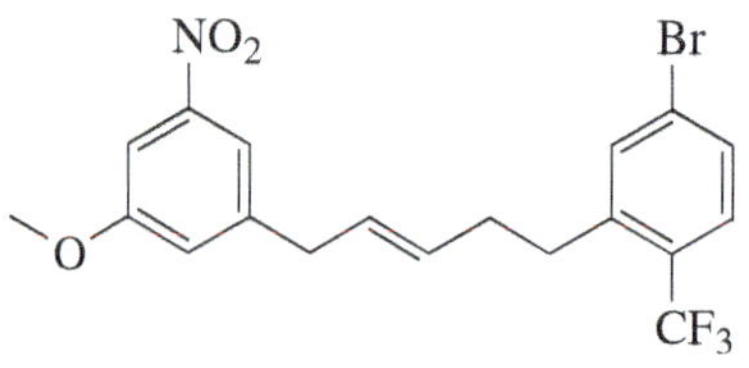

(c)

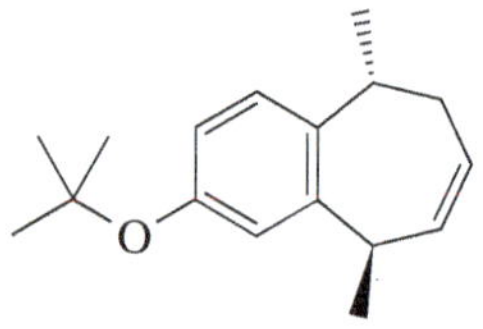

16-D CHEMISTRY IN CONTEXT: Green Chemistry

Petroleum products from renewable feedstocks

In Section 7.2.7 (part of the green chemistry chapter), the importance of using renewable feedstocks as a source for organic molecules was discussed. One popular renewable feedstock is natural plant and vegetable oil. Unfortunately, the composition of these oils often limits their use to just a few applications. The conversion of animal fats to biodiesel fuel is an example of a modification of a renewable feedstock into usable chemicals (**Figure 16.65**).

Seeking a way to diversify the compounds that can be produced from renewable plant oils, Elevance Renewable Sciences has developed a catalytic (GCP 9), low-pressure, low-temperature (GCP 6) process employing metathesis

(continued)

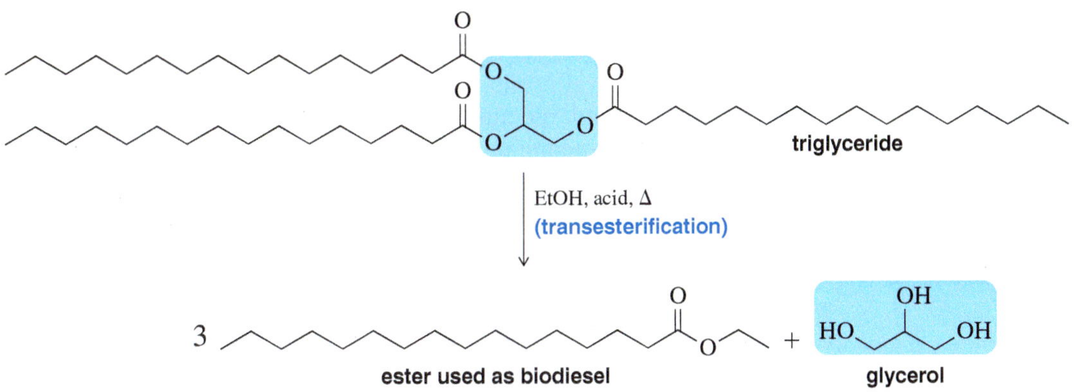

Figure 16.65 Transesterification of animal fats and vegetable oils results in long-chain esters, which can be used as biodiesel fuel, and glycerol.

technology to convert unsaturated plant oils into useful olefins and specialty chemicals with a reduced environmental footprint. As shown in **Figure 16.66**, the cross-metathesis of plant oils with but-1-ene results in chain-shortened oils that, after transesterification (Section 19.7.2), produce novel C_{10} and C_{12} unsaturated esters. Additionally, olefins, usually obtained from petroleum feedstocks, are generated in the cross-metathesis step as useful co-products of this technology. Elevance won a Presidential Green Chemistry Challenge Award for this technology in 2012.

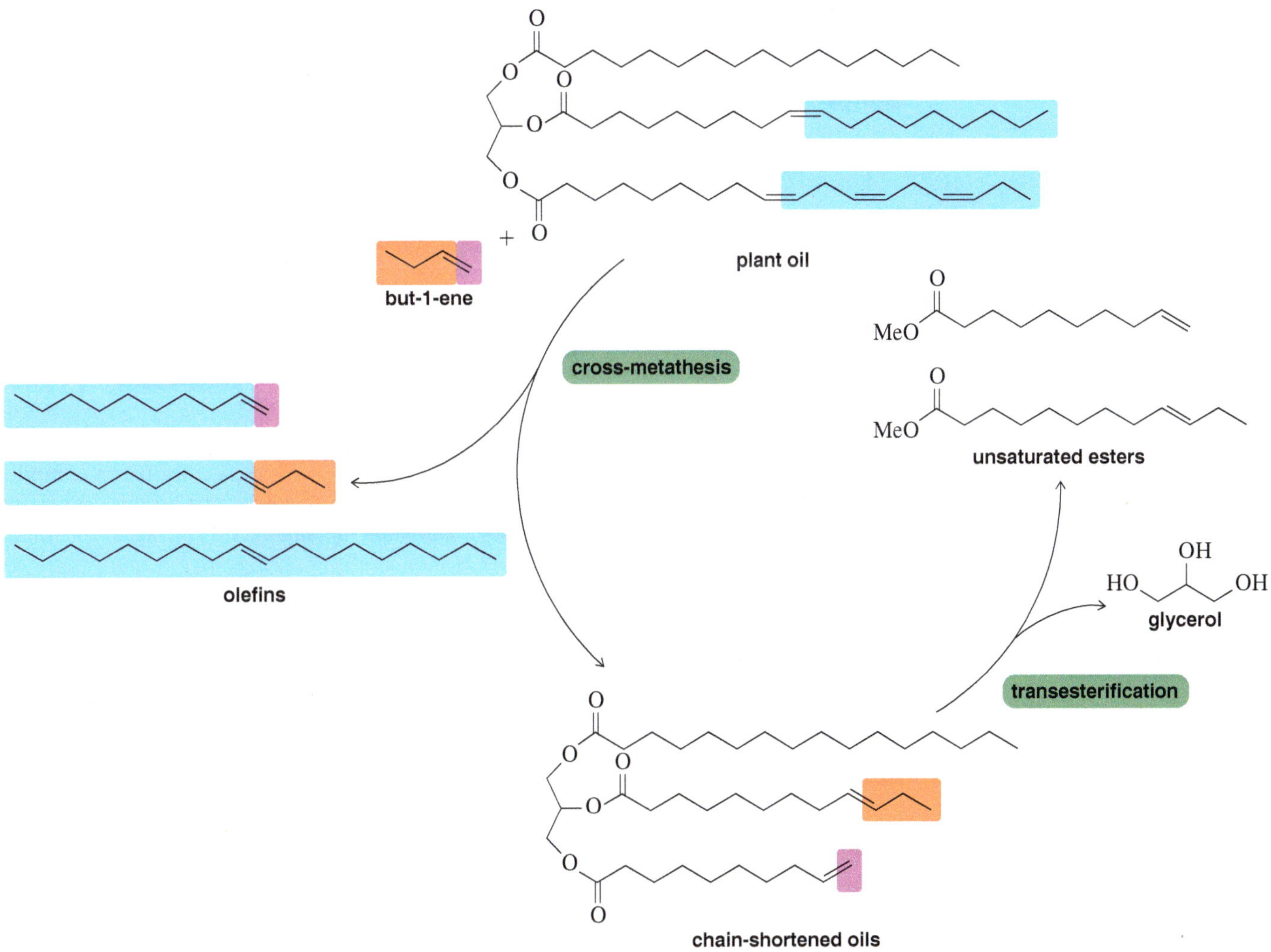

Figure 16.66 Cross-metathesis of plant oils produces novel esters from renewable feedstocks.

Concluding thoughts

The synthesis of natural products has spurred a great deal of innovation in organic chemistry research. In fact, a large number of the reactions we have studied and will study in this book were likely invented in efforts to synthesize natural products. The use of organometallic reactions for the synthesis of curacin A is only one example.

A final thought: Many consider natural product synthesis to be as much an art as it is a science. Sure, scientific principles go into the invention of reactions, but there is an artistic side. In the same way that a beautiful building might inspire an architect to pursue novel ways of constructing it, a beautiful molecule [*yes, there are beautiful molecules*] can inspire a chemist to pursue novel ways of constructing it. An example of one of these overwhelmingly complex and beautiful molecules is maitotoxin (**Figure 16.67**), a potent toxin produced by the dinoflagellate *Gambierdiscus toxicus*. [*Devise a synthesis of it starting from molecules containing three carbons or fewer . . .*]

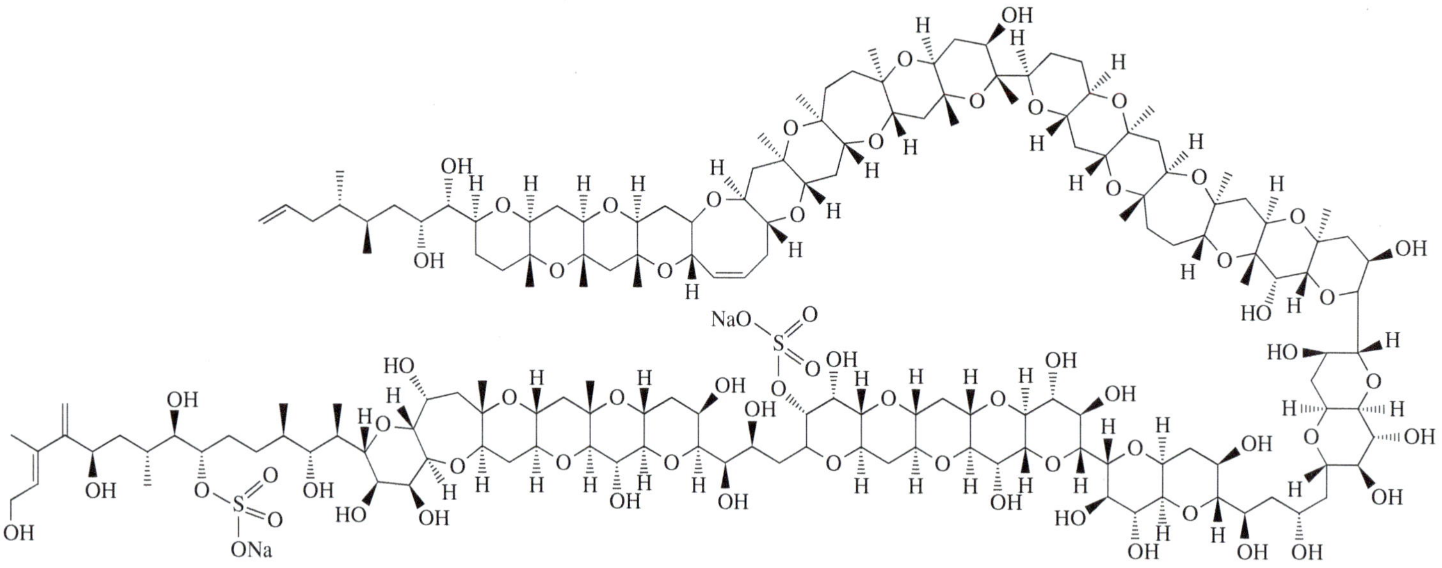

Figure 16.67 The (impressive) structure of maitotoxin, a potently toxic natural product.

Inspired or not, we present a number of important reactions with applicability to natural product synthesis in Chapters 17–20. Let's go there now.

Considering that nature makes these beautifully complex molecules using complicated cellular machinery unavailable to a synthetic chemist, the challenge to produce molecules in equally efficient ways is formidable. Yet in the end, the satisfaction of accomplishing just that may be its own reward.

Important Concepts

[These are the major concepts covered in each section, excluding reactions.]

16.1: Organometallics (R—M) are useful because they confer a partial negative charge on carbon, making them useful as nucleophiles in C—C bond forming reactions.

16.2: The reduction of C—X bonds by lithium and magnesium allows formation of organolithium and Grignard (organomagnesium) reagents, respectively. Both organolithium and Grignard reagents are strongly basic and react as strong nucleophiles with epoxides, ketones, and aldehydes. The strain in the epoxide makes it susceptible to nucleophilic attack, while a partial positive charge on carbon in ketones and aldehydes makes them good electrophiles.

16.3: The carbene is an intermediate that has characteristics of a nucleophile (a nonbonded electron pair) and an electrophile (an empty *p* orbital). Carbenes can be produced by α-elimination, decomposition of diazomethane, and treatment of geminal diiodoalkanes with metallic zinc. Carbenes react with alkenes to form cyclopropanes by a mechanism first studied in Chapter 8.

16.4: Transition metal–catalyzed coupling is used to form C—C bonds between alkenyl or aryl halides with a variety of organometallic reagents. While all of the reactions are catalyzed by palladium, they are named based on the organometallic reagent used: Negishi (RZnBr), Stille ($RSnBu_3$), Suzuki ($RB(OH)_2$), and Sonogashira (R—C≡C—Cu). The catalytic cycles of these reactions are nearly identical, involving the oxidative addition of the C—X bond to Pd, transmetallation with the organometallic reagent, and then reductive elimination to create the new C—C bond and regenerate the palladium(0) catalyst.

16.5: Gilman reagents (diorganocuprates) are formed by the reaction of organolithium or Grignard reagents with copper(I) salts. They undergo cross-coupling reactions with alkyl, aryl, or alkenylhalides and open epoxides at the least sterically hindered carbon.

16.6: Alkene metathesis is the coupling of two alkenes to produce two new alkenes. When used productively, a larger alkene is generated from two smaller ones. The Grubbs ($L_nRu{=}CHR$) and Schrock ($L_nMo{=}CHR$) catalysts can be used for cross-metathesis or to form rings via ring-closing metathesis.

Important Reactions and Mechanisms

1. **Formation of an organolithium (Section 16.2.1):** A lithium atom donates an electron to iodine, forming iodide, a carbon radical, and a lithium cation. A second lithium atom donates an electron to the electron-deficient radical, thus forming a carbanion.

General:

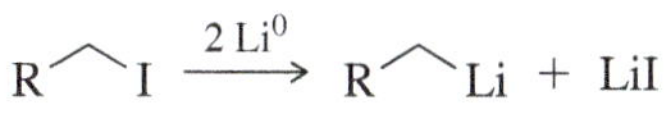

Specific:

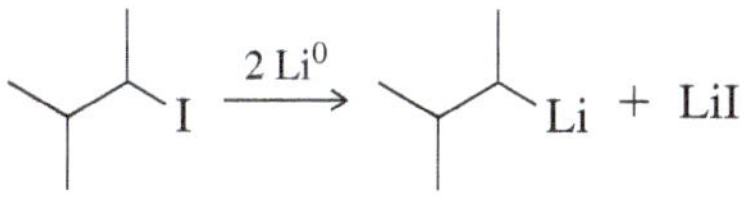

Mechanism:

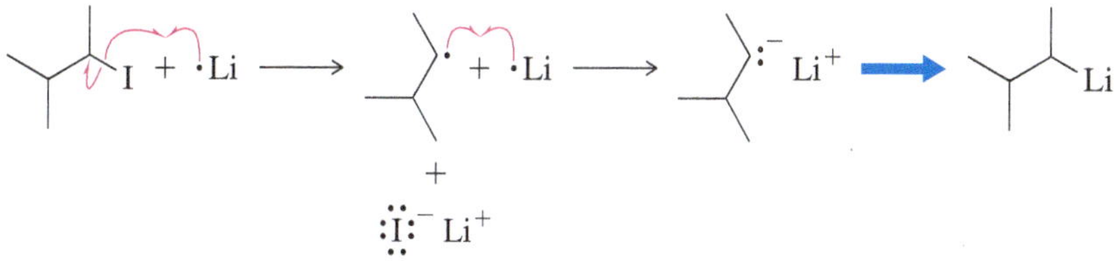

2. **Formation of a Grignard reagent (Section 16.2.2):** A magnesium atom donates an electron to bromide, forming bromide, a carbon radical, and a magnesium bromide ($BrMg^+$) cation. The magnesium cation donates a second electron to the electron-deficient radical, thus forming a carbanion.

General:

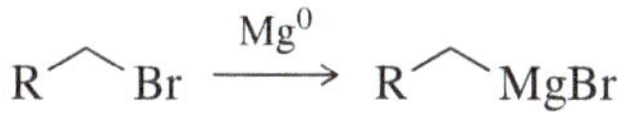

Specific:

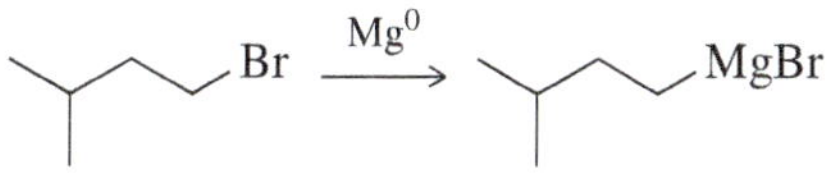

Mechanism:

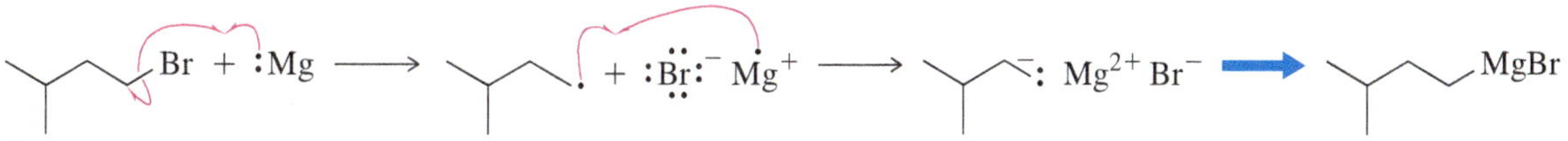

3. **Organolithium with epoxide (Section 16.2.3):** As we studied in the S_N2 reactions of Chapter 12, strong nucleophiles directly attack epoxides at the least sterically hindered carbon. An acid quench results in formation of an alcohol. Replacing the organolithium with a Grignard reagent results in the exact same reaction.

General:

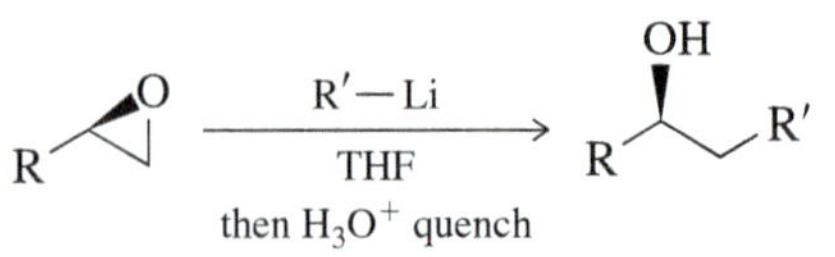

Specific:

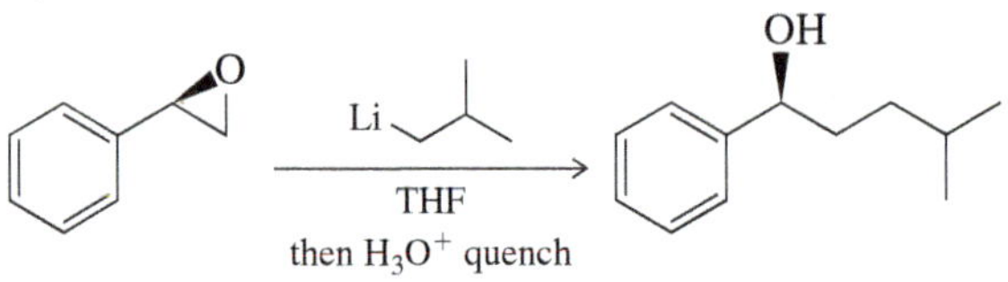

Mechanism:

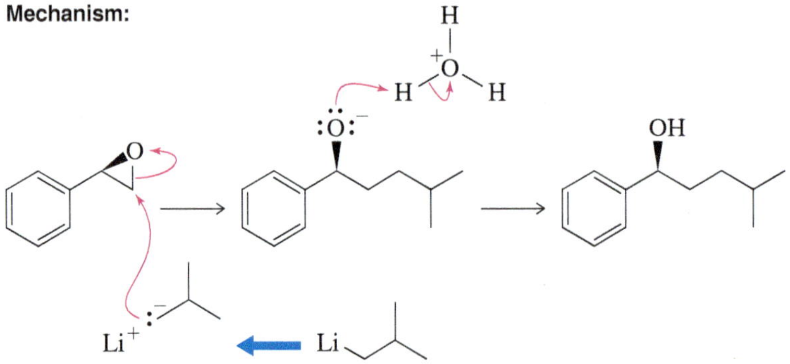

4. **Grignard reagent with aldehyde/ketone (Section 16.2.3):** The strong nucleophile attacks the electrophilic carbonyl carbon, pushing the π electrons onto the oxygen. This forms an alkoxide, which is subsequently quenched in a second step to give an alcohol. Replacing the Grignard reagent with an organolithium results in the exact same reaction.

General:

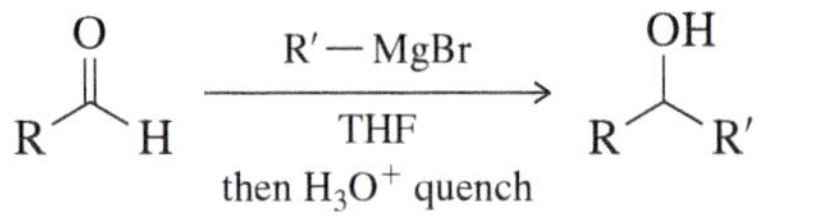

Specific:

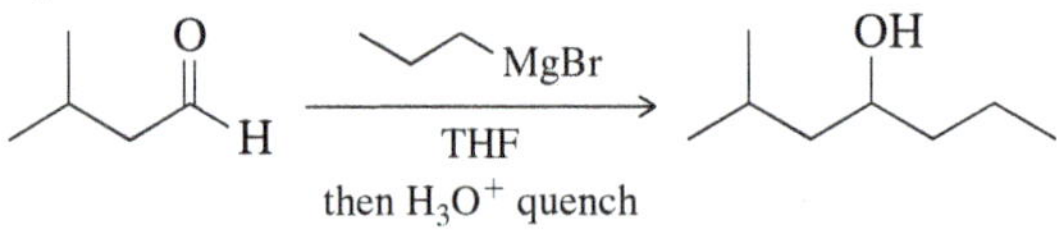

Mechanism:

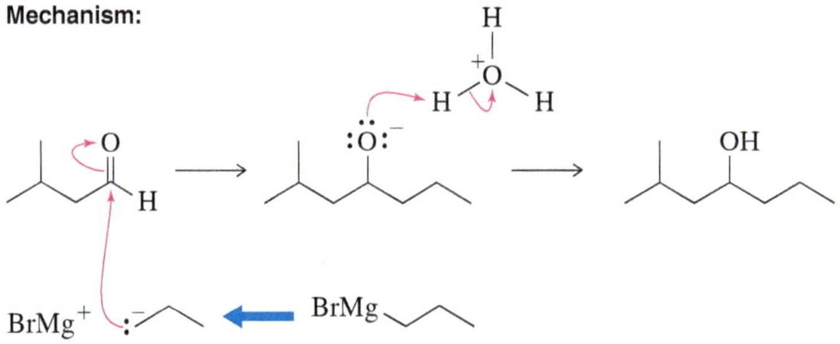

5. **Carbene with alkene (Section 16.3):** As species that are both nucleophiles and electrophiles, carbenes react with alkenes to give cyclopropanes. The π electrons of the alkene attack the carbene while at the same time the lone pair on the carbene attacks the developing carbocation. [*This mechanism is analogous to those studied in Chapters 8 and 9.*]

General:

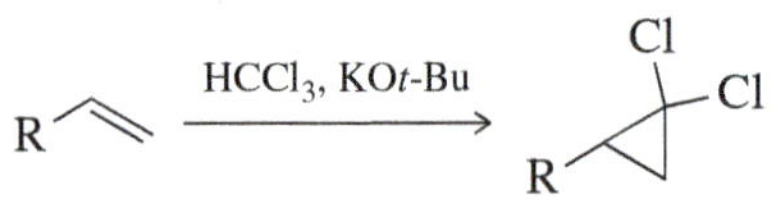

Specific:

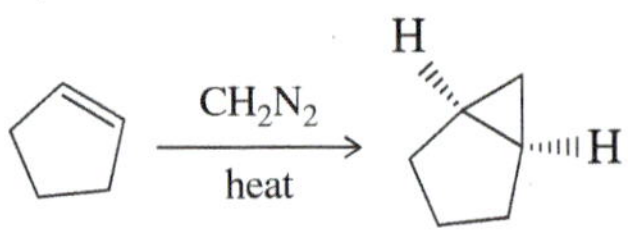

Mechanism:

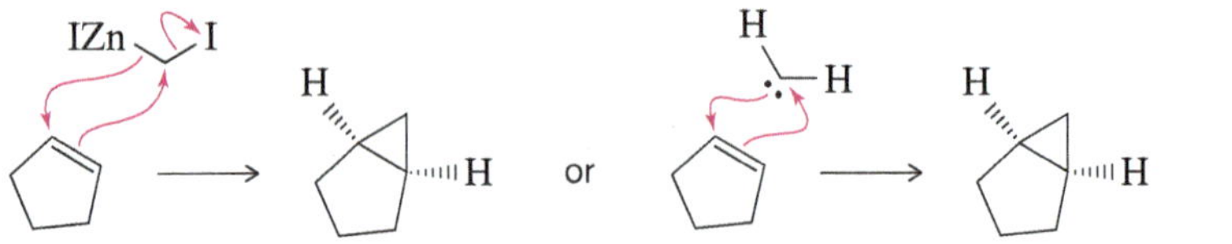

[In Simmons–Smith] *[With other carbenes]*

6. **Formation of carbenes (Section 16.3):** Carbenes are formed by α-elimination reactions in which a good leaving group leaves from a carbon containing a negative charge. In the Simmons–Smith reaction, the anion is generated by zinc donating a pair of electrons for the diiodoalkane in a reaction that is analogous to Grignard formation.

Mechanism:

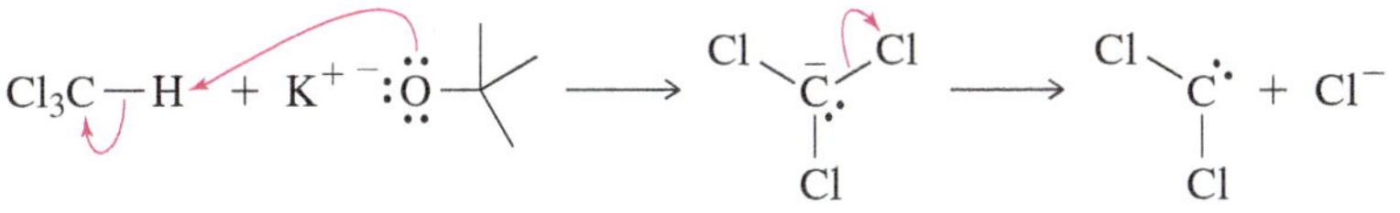

Mechanism:

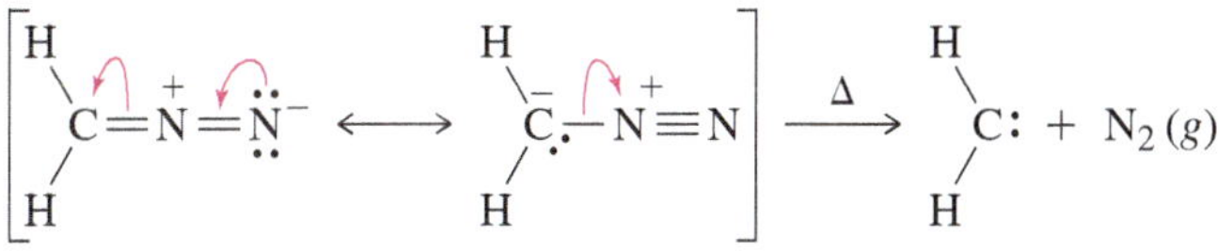

Mechanism:

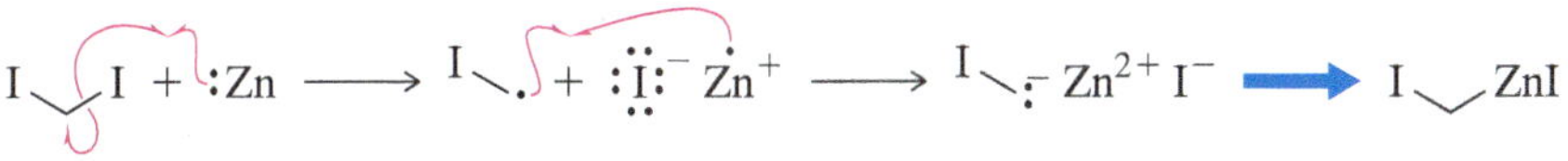

[*Simmons–Smith*]

7. **Transition metal–catalyzed coupling (Section 16.4):** Organotin (Stille), organozinc (Negishi), alkylboronic acids (Suzuki), and copper acetylides (Sonogashira) are coupled to organohalides using palladium tetrakis as the catalyst. All four reactions proceed by a catalytic cycle involving oxidative addition, transmetallation, and reductive elimination.

General:

$$R—M \quad + \quad R'—X \xrightarrow{Pd(PPh_3)_4} R—R' \quad + \quad MX$$

Specific:

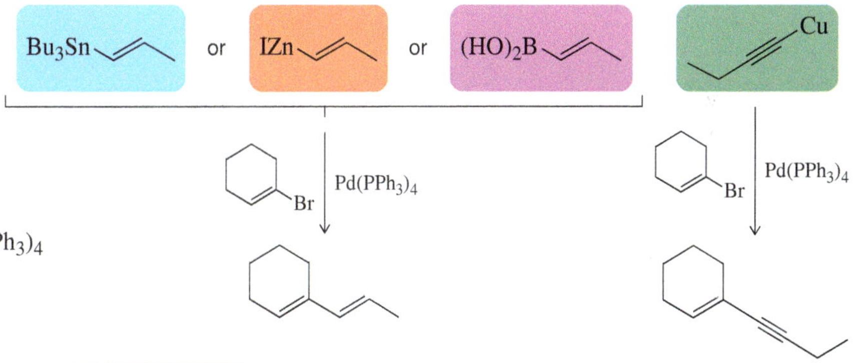

Mechanism:

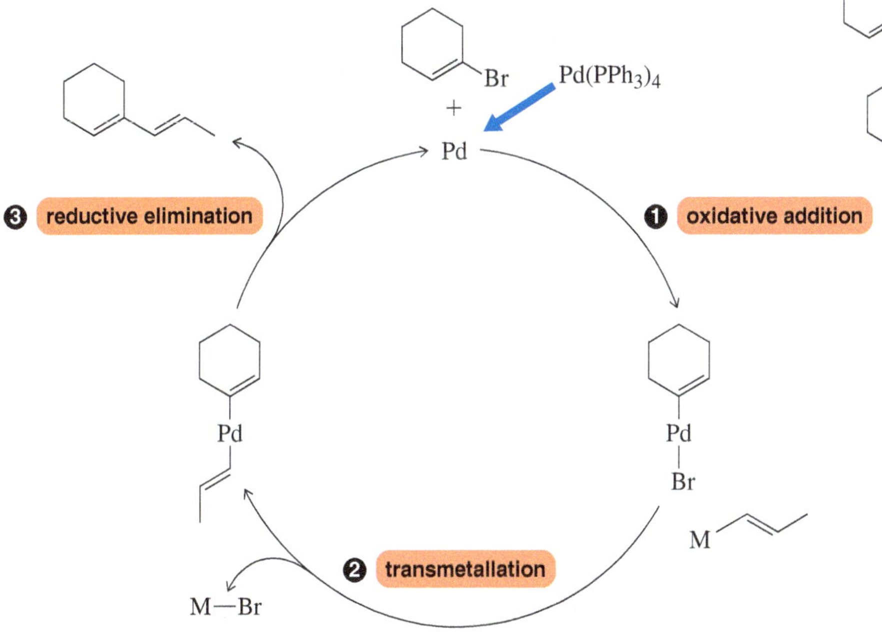

8. **Olefin metathesis (Section 16.6):** Using a ruthenium or molybdenum carbenoid, alkenes can be coupled to form a new C=C bond. The catalytic cycle involves a series of reversible [2+2] cycloaddition and [2+2] cycloreversion reactions, ultimately proceeding to make the most stable alkene.

General:

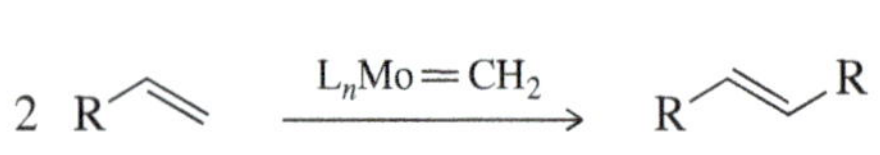

Specific:

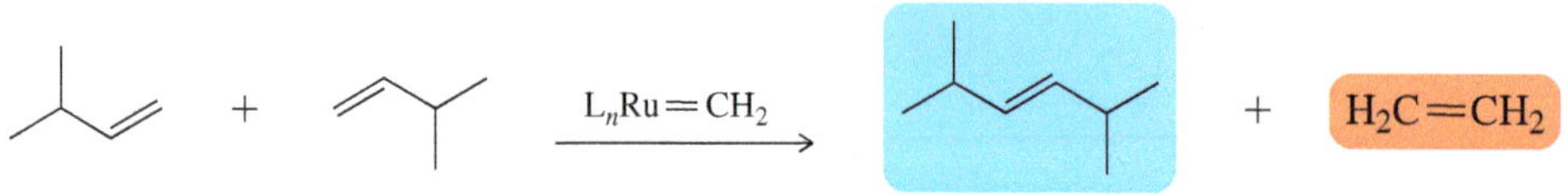

Mechanism:

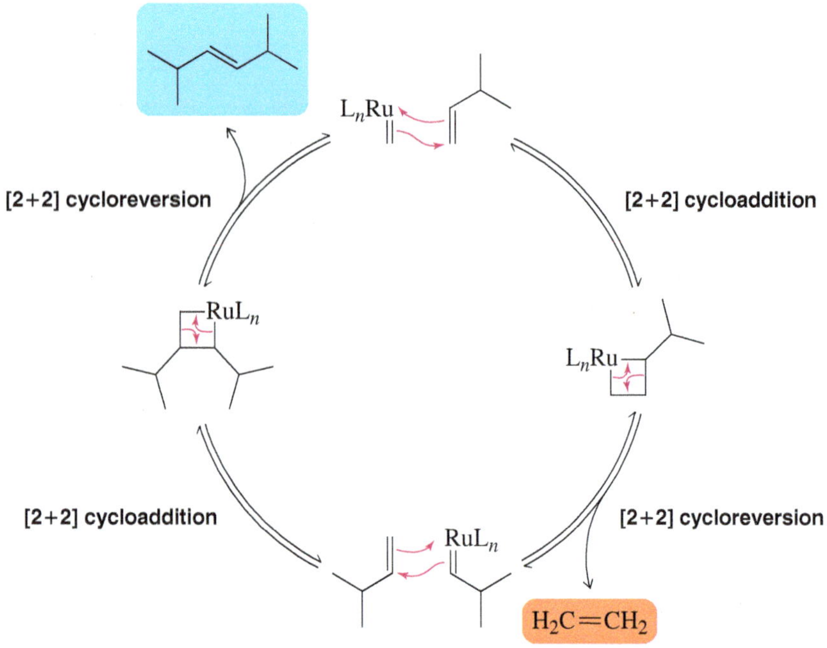

End-of-Chapter Assessments

[*Assessments are ranked according to their difficulty; •••• = most difficult.*]

16.40 (•) Predict the product(s) that would result when the alkyl halides shown are allowed to react under the following conditions: (i) PhZnBr, $Pd(PPh_3)_4$; (ii) $PhB(OH)_2$, $Pd(PPh_3)_4$; (iii) $PhSnBu_3$, $Pd(PPh_3)_4$; (iv) $Cu{-}C{\equiv}C{-}CH_3$, $Pd(PPh_3)_4$; (v) 1. Mg^0 2. PhCHO 3. H_3O^+ quench; (vi) 1. 2 Li^0 2. Oxirane 3. H_3O^+ quench; (vii) 1. $H_2C{=}CHSnBu_3$, $Pd(PPh_3)_4$ 2. ICH_2ZnI; (viii) 1. 2 Li^0 2. CuI (0.5 equiv.) 3. $BrCH{=}CH_2$.

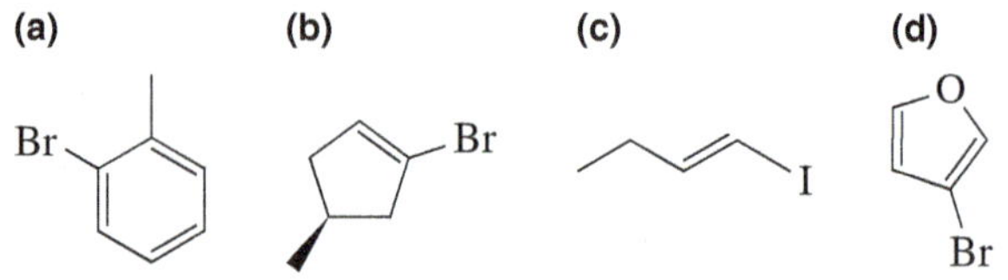

16.41 (••) The molecules shown were synthesized in the pharmaceutical industry using cross-coupling reactions. Suggest the precursors that could be used to make the bonds indicated by blue arrows. [*There are several correct answers.*]

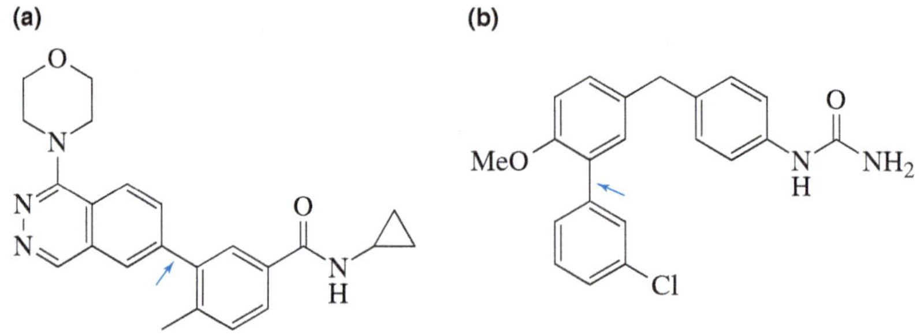

16.42 (••) Which of the following solvents are reasonable choices for a Grignard reaction? Justify your choices. [*Hint: Carbonyls are good electrophiles (see Chapter 17).*]

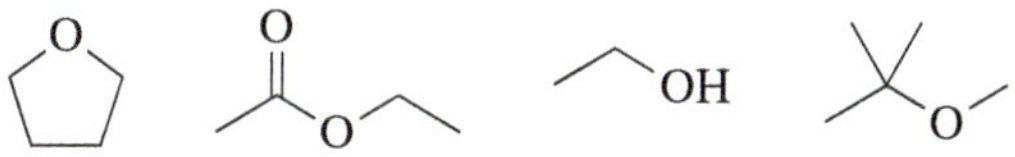

16.43 (•••) LOOKING BACK A chemist failed to generate the diol using the reaction shown here. (a) Suggest a reason why this reaction did not work as written. (b) How could the reaction conditions be modified to allow formation of the diol? [*It may require more than one step.*]

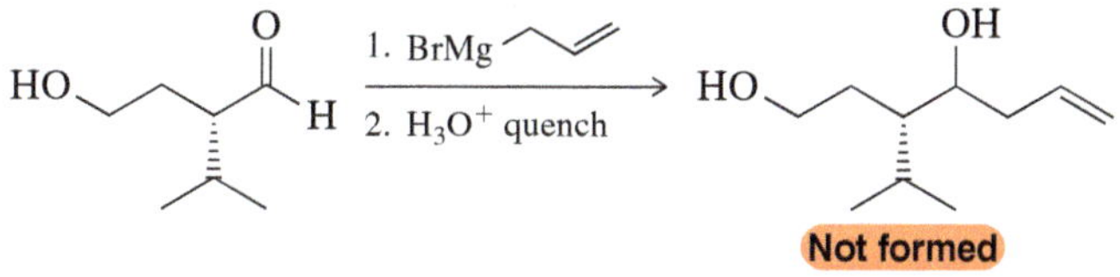

16.44 (•••) LOOKING AHEAD A chemist failed to generate the alcohol using the reaction shown here. (a) Suggest a reason why this reaction did not work as written. (b) How could the reaction conditions be modified to allow formation of the product? [*Answer in general terms; we explain how to fix this in Chapter 17.*]

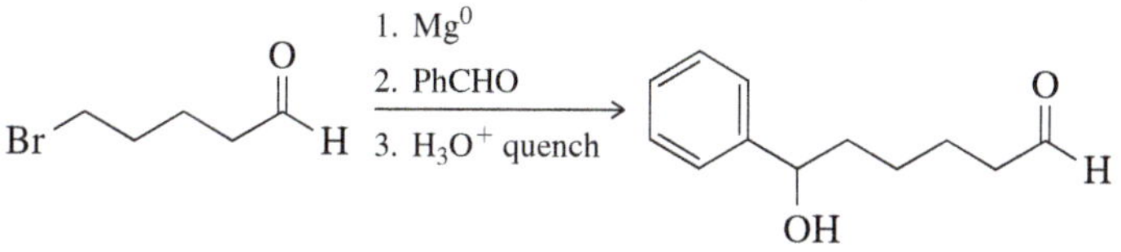

16.45 (•••) LOOKING AHEAD Palladium coupling reactions can be used for the synthesis of amines according to the following catalytic cycle:

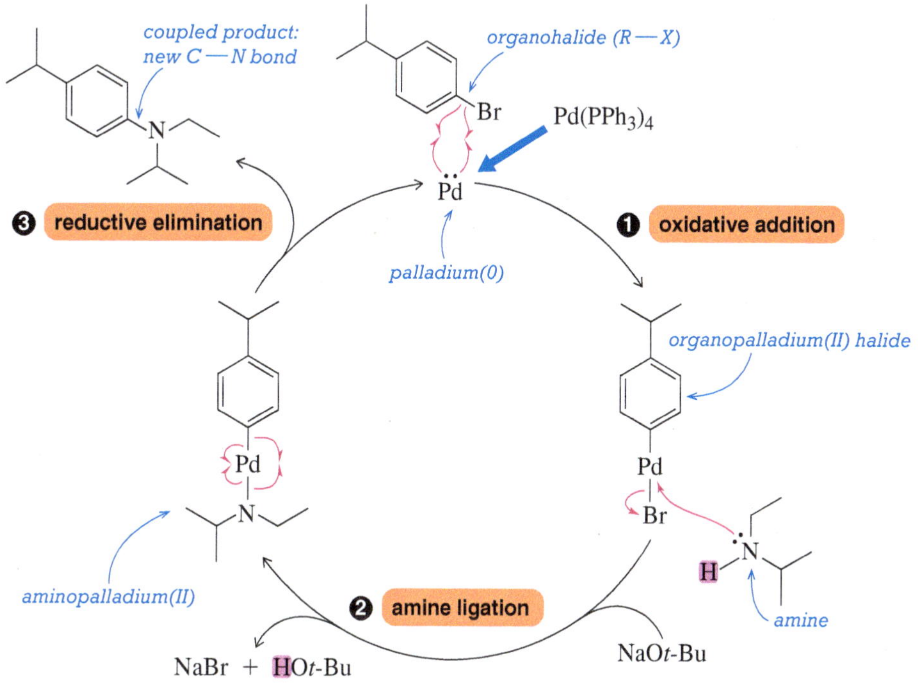

Based on this catalytic cycle, predict the products of the following reactions. [*We describe this reaction in more detail in Chapter 25.*]

(a)

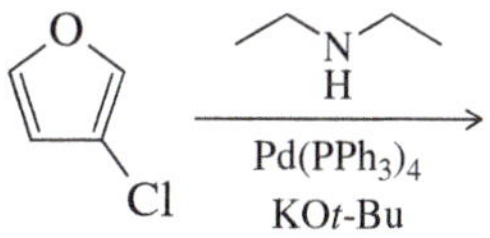

(b)

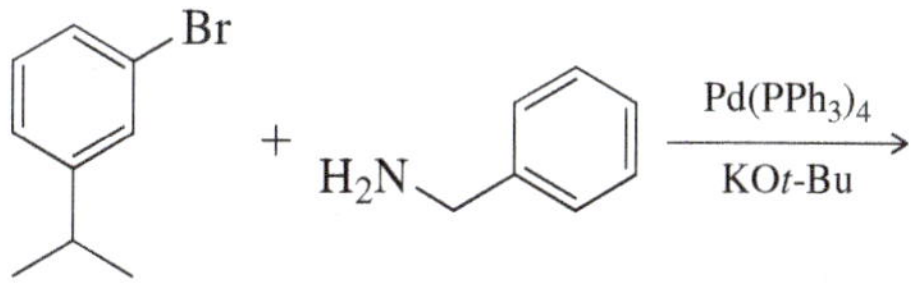

16.46 (••••) Several syntheses of laulimalide, an antiproliferative natural product, have featured the use of a scheme similar to that shown here. Provide the product for each synthetic step shown. Based on your answer to (c), identify which portion of laulimalide has been synthesized. [*Two of the reactions are from this chapter, whereas the other is from Chapter 13.*]

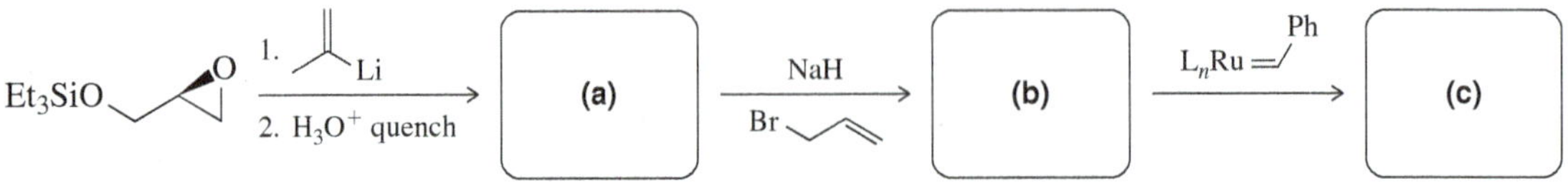

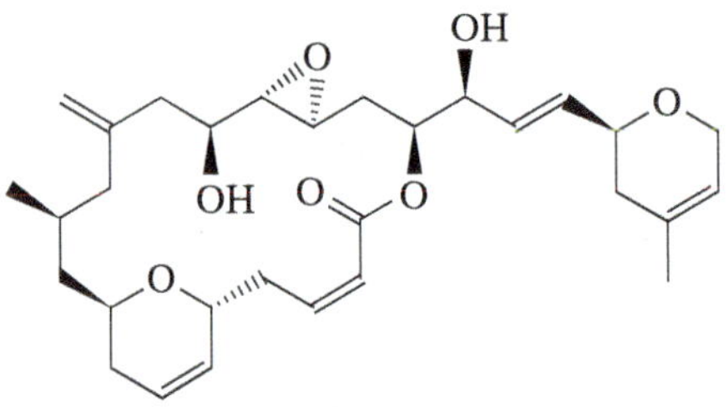

laulimalide

LOOKING AHEAD: In Chapter 28, we study how the reactions here in Chapter 16 can be used to make long-chain polymers. This is often done using monomers containing functional groups that can react in two directions and is enabled by the fact that reagents from Chapter 16 are often used in catalytic quantities. Assessments 16.47–16.49 provide a preview of the chemistry we study there.

16.47 (••) Suggest a mechanism by which the following polymer is synthesized using repetitive Stille coupling reactions. [*You only need to show the sequence until three monomer units are incorporated.*]

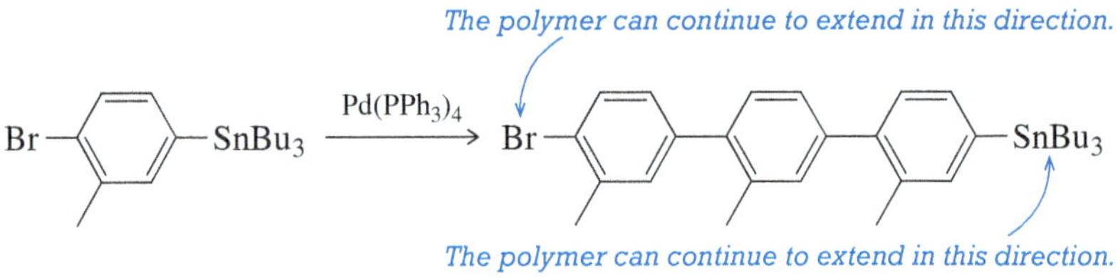

16.48 (••) Predict the polymer that would result from the following Suzuki coupling reaction. [*Your product should incorporate a maximum of three of each of the monomers.*]

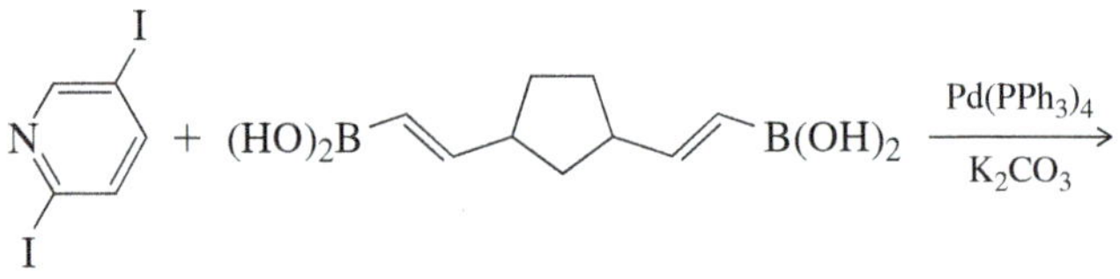

16.49 (•••) Polymers can be made by ring-opening metathesis polymerization (ROMP), proceeding by a mechanism similar to that for alkene metathesis. Suggest a mechanism for the following polymerization reaction.

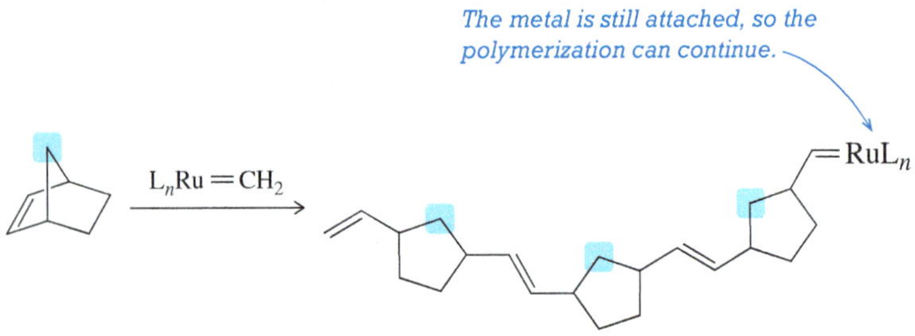

LOOKING AHEAD: The Heck reaction is a common C—C bond-forming reaction that makes use of the same palladium catalysts studied here in Chapter 16. By working Assessments 16.50–16.54, we will learn the mechanism of the Heck reaction.

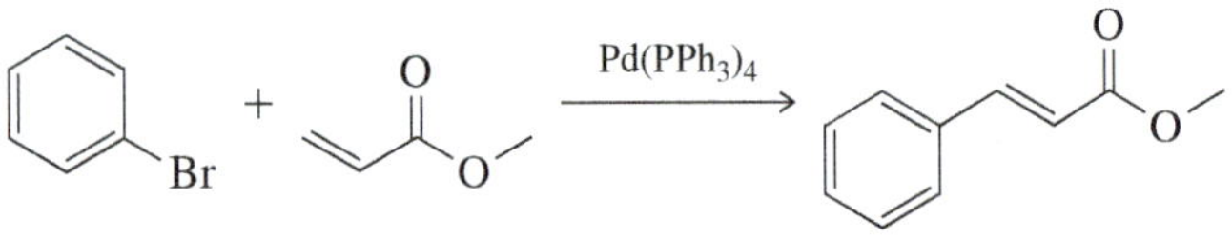

16.50 (•) Oxidative insertion of Pd^0 into an aryl bromide creates a palladium(II) intermediate. Would you expect palladium in this species to be a nucleophile or an electrophile?

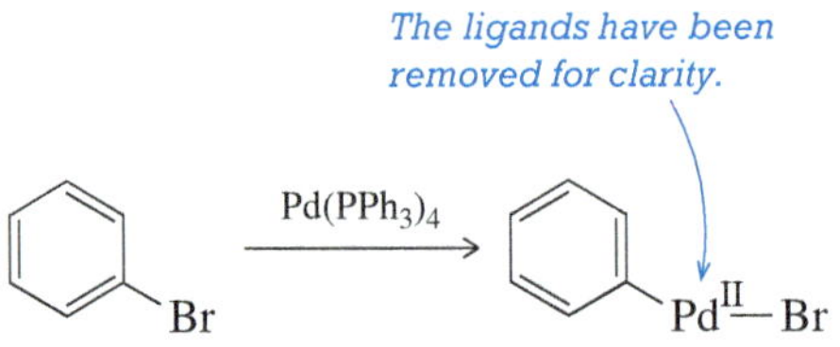

16.51 (••) (a) Suggest a one-step, concerted, arrow-pushing mechanism that rationalizes formation of the new C—C σ bond and the C—Pd σ bond. (b) While the mechanism is similar to electrophilic alkene addition reactions, rationalize the regiochemical outcome.

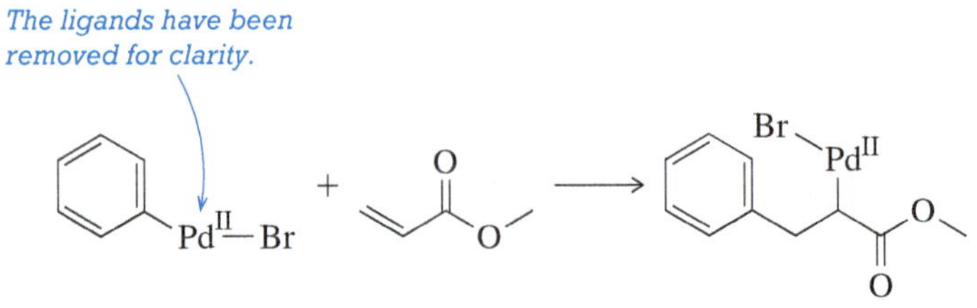

16.52 (•••) Alkyl palladium species can undergo a reaction called a β-hydride elimination, illustrated by the following electron movement arrows. Predict the product(s) that you would obtain in this reaction.

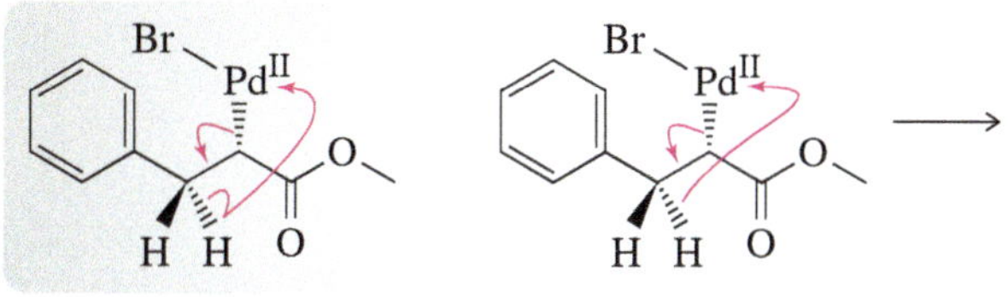

16.53 (•••) Palladium(0) is regenerated using a base added to the Heck reaction mixture by a mechanism analogous to carbene formation, discussed in Section 16.3. Suggest a mechanism for this reaction.

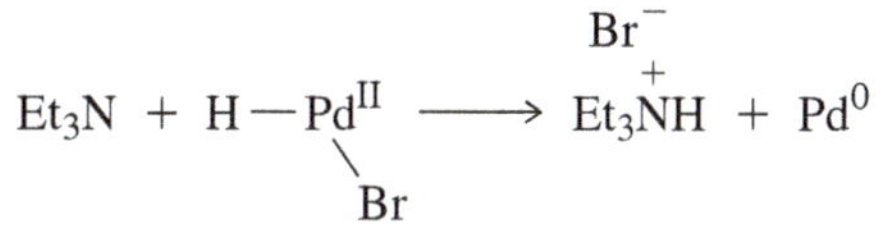

16.54 (•••) Predict the product of the following Heck reactions.

(a)

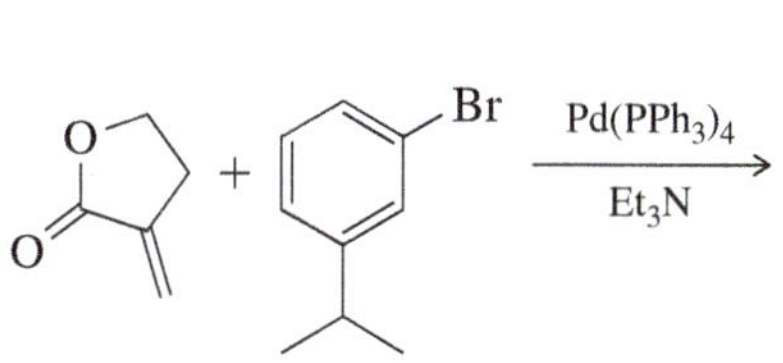

(b)

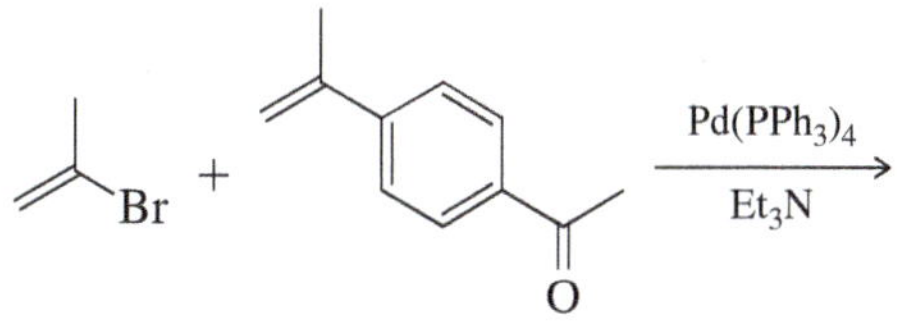

(c)

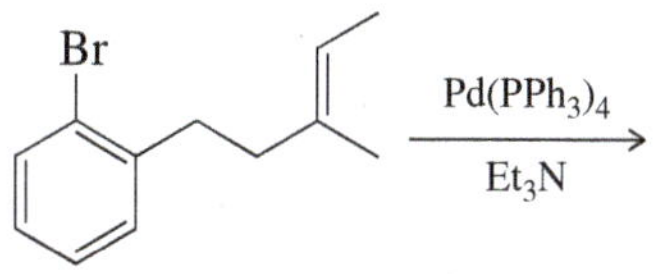

16.55 (••••) LOOKING AHEAD In Chapter 23, we show that cuprates add to the β-carbon of α,β-unsaturated ketones by a mechanism similar to the Heck reaction in Assessment 16.54. Predict the product of this so-called conjugate addition reaction.

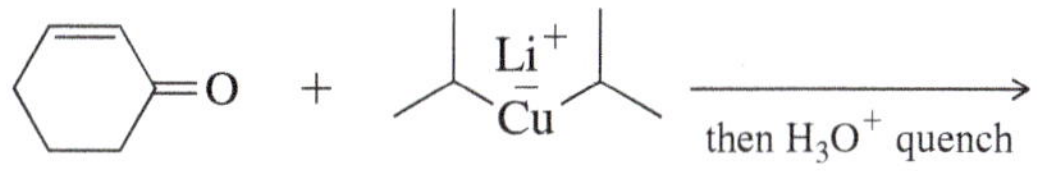

α,β-unsaturated ketone

16.56 (•••) A Suzuki coupling reaction was used to make losartan, a medicine used to treat high blood pressure. Predict the product of this reaction.

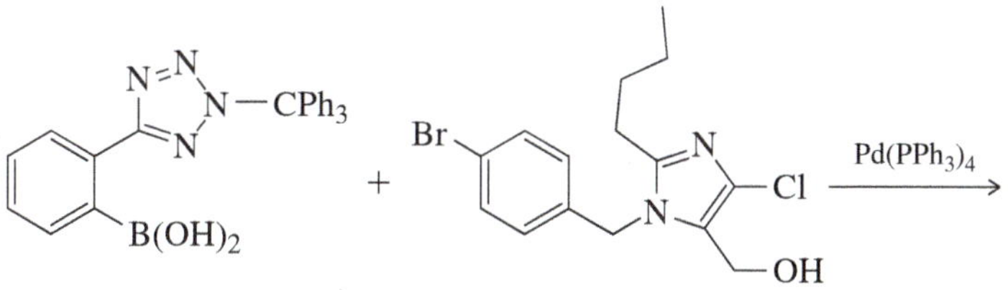

16.57 (••••) FROM THE CHEMICAL LITERATURE The synthesis of the following intermediate was recently reported using an organolithium reagent. Identify a likely pair of starting materials that could have been used (*Org. Lett.* 14, 2012, 4666).

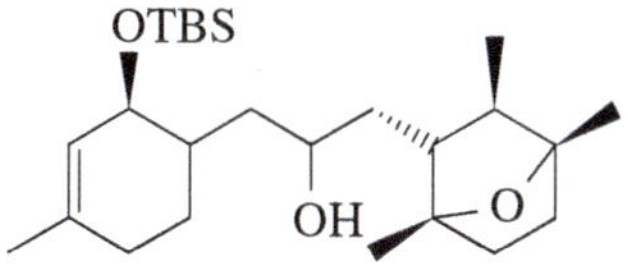

16.58 (••) The following transformation is used as an early step in the synthesis of cilastatin, a drug used to counter antibiotic resistance. Identify a possible reagent for this step.

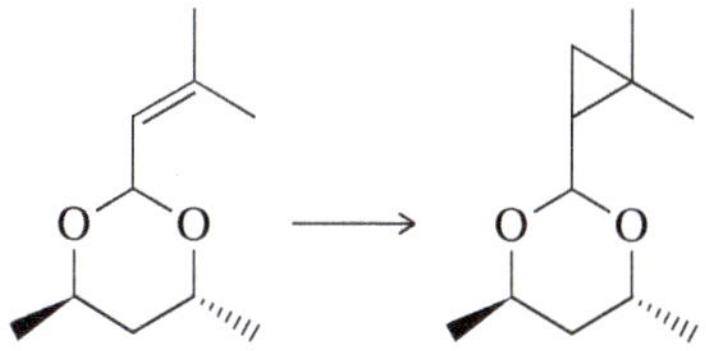

16.59 (•••) Nitrogen-containing heterocycles form particularly stable carbenes and are commonly used as ligands in organometallic chemistry. (a) Why is the carbene shown particularly stable? [*It may be helpful to draw the molecular orbital picture.*]

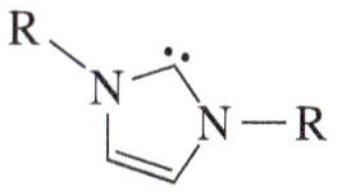

16.60 (••••) A graduate student attempted the following transformation but got a poor yield of the desired product. What side product was produced?

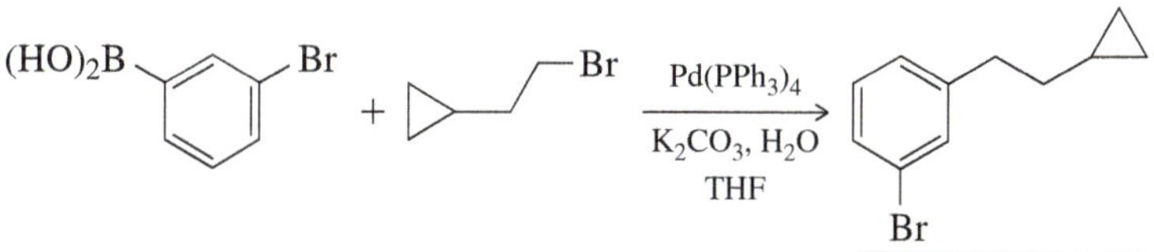

formed in low yield

16.61 (•••) FROM THE CHEMICAL LITERATURE The following transformation was used in the synthesis of (−)-exiguolide. Suggest the name of this reaction and the reagents used (*Org. Lett.* **17**, 2015, 4651).

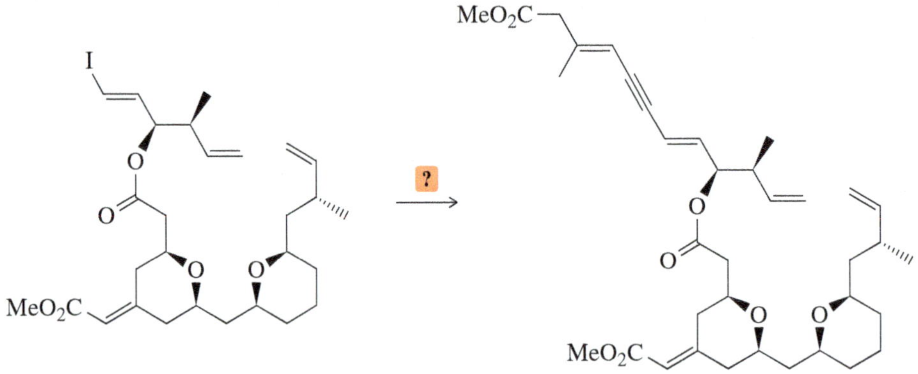

16.62 (••••) FROM THE CHEMICAL LITERATURE The synthesis of (−)-exiguolide was completed from the intermediate in Assessment 16.61 using a two-step process. Suggest reagents to complete the synthesis (*Org. Lett.* **17**, 2015, 4651). [*One reaction is from this chapter, whereas the other is from Chapter 10.*]

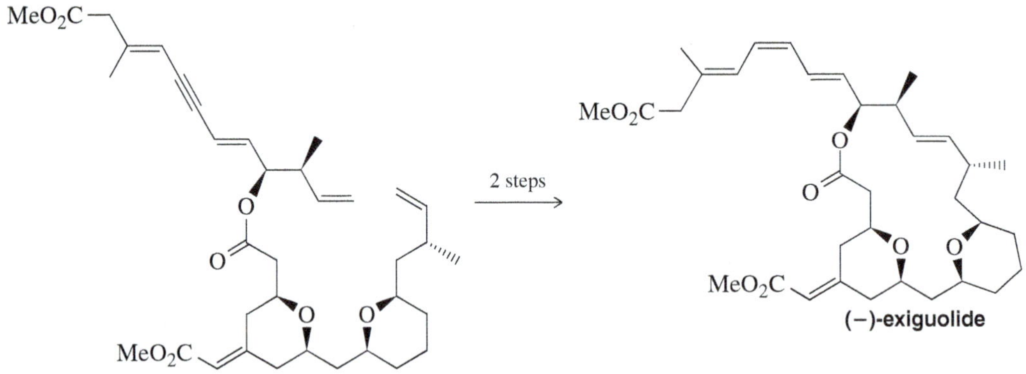

17

Carbonyl Addition Reactions: Aldehydes and Ketones

Introduction: NAD^+ cofactors

The chemistry of carbonyls is of central importance in biological systems. The oxidation and reduction of aldehydes and ketones, for example, play a vital role in a number of metabolic pathways. Dehydrogenase enzymes catalyze these reactions by stabilizing the transition state and lowering the activation energy of reaction, thus increasing the rate of reaction.

Enzymes often make use of cofactors to carry out these vital biochemical reactions. Many cofactors, nonprotein molecules essential for the activity of the enzyme, are small organic molecules. One of the more important cofactors is nicotinamide adenine dinucleotide (NAD^+), synthesized biochemically using the vitamin niacin (**Figure 17.1**). The dehydrogenases use NAD^+ as a cofactor in catalyzing the removal or addition of hydrogen (H_2) from substrates.

As both an oxidizing and a reducing agent, this cofactor has two reactive forms, NAD^+ and NADH. The chemistry of NAD^+ is based on the interconversion between these two forms, depending on the substrate undergoing reaction.

○ ○ ○ ○ ○ ○ ○

Here in Chapter 17 we study the chemistry of aldehyde and ketone reactivity and, in the process, explore the activity of *alcohol dehydrogenase* (a cause of hangovers associated with alcohol consumption), *aldehyde dehydrogenase* (which reduces the hangover effects),

Learning Objectives

LO 17.0 Determine the IUPAC names for aldehydes and ketones

LO 17.1 Analyze synthesis reactions of alkenes and ketones

LO 17.2 Compare the reactivity of aldehydes and ketones

LO 17.3 Assess carbonyl addition reactions (including Grignard, organolithium, and cyanide reagent reactions)

LO 17.4 Evaluate the stereochemical outcomes of carbonyl addition reactions

LO 17.5 Predict the products of Wittig reactions

LO 17.6 Analyze acetal formation reactions

LO 17.7 Evaluate the addition of nucleophiles to carbonyls

LO 17.8 Determine the outcomes of carbonyl reduction reactions

LO 17.9 Determine the products of carbonyl oxidation reactions

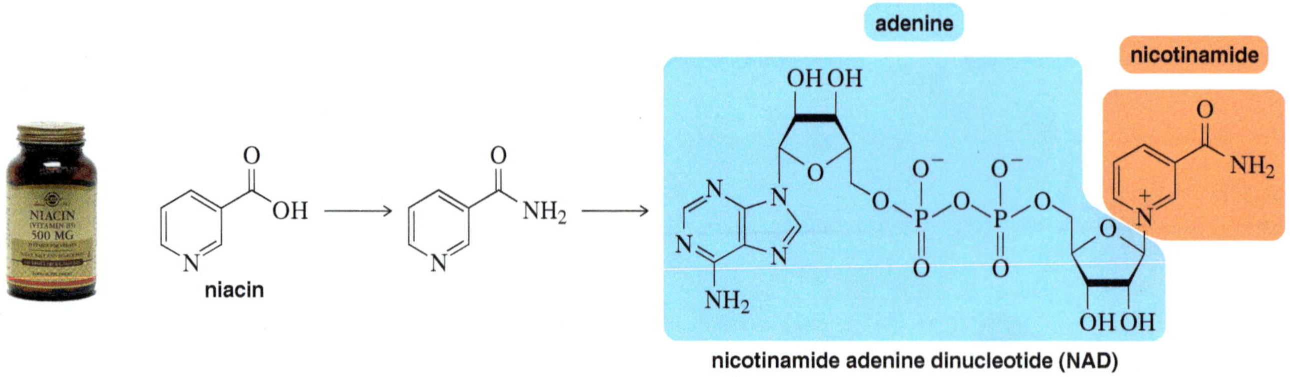

Figure 17.1 **NAD^+ is an important cofactor used for biological redox reactions.**

> *"If you can't fly, then run, if you can't run, then walk, if you can't walk, then crawl, but whatever you do, you have to keep moving forward."*
>
> —Martin Luther King Jr.

glutamate dehydrogenase (which helps cells generate energy), and *lactate dehydrogenase* (which is suggested to result in muscle pain from vigorous exercise and is now a target for cancer chemotherapy).

At this point in your journey through organic chemistry, you may want to take note of how far you have come and how far you have left to go. A reaction coordinate diagram of your progress (**Figure 17.2**) shows your journey is about 75% complete. [*That's surely a good feeling.*]

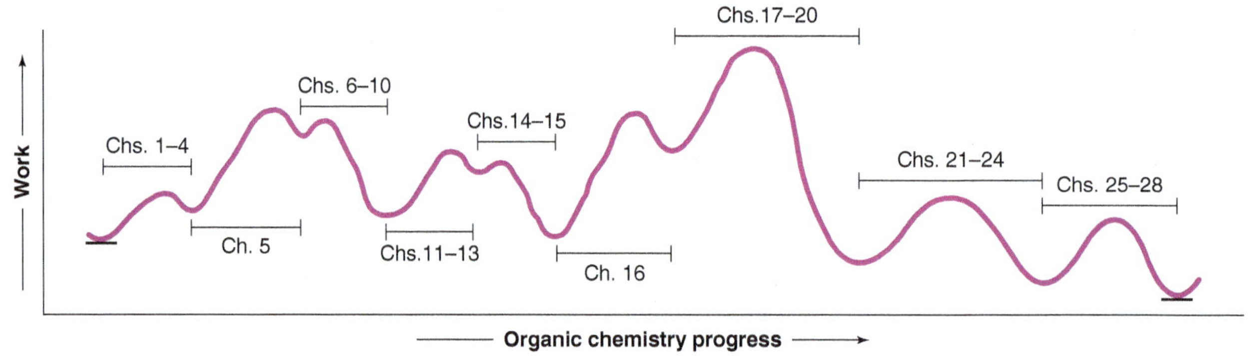

Figure 17.2 **Assessing your progress in organic chemistry with a reaction coordinate diagram.** Chapters 17–20 are the hill with the highest activation energy, but you will finish in a much more stable place.

Chapters 17–20 also give you a head start on your next likely chemistry endeavor—biochemistry. The chemistry of biological systems is heavily influenced by the chemistry of carbonyls. With that in mind, let's check your understanding of previously learned concepts by working Assessments 17.1–17.5.

Test your understanding

Assessment 17.1 In each case, circle the stronger nucleophile.

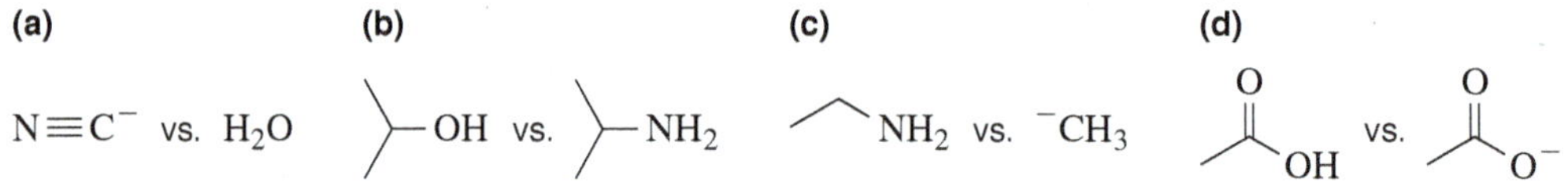

Assessment 17.2 Identify the following reactions as oxidation or reduction (based on what happens to the organic molecule).

(a)

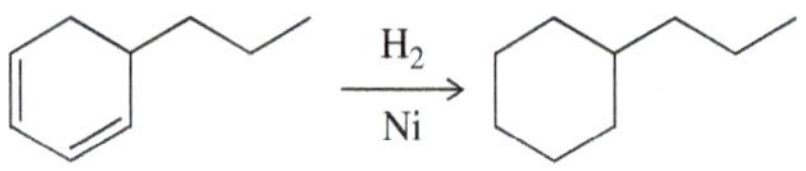

(b)

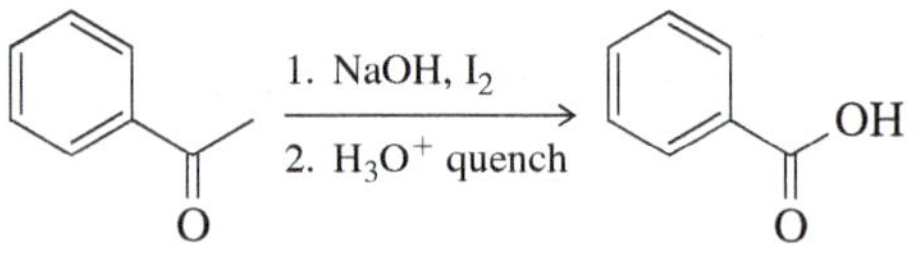

(c)

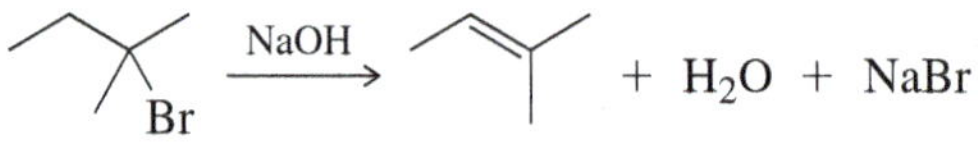

(d)

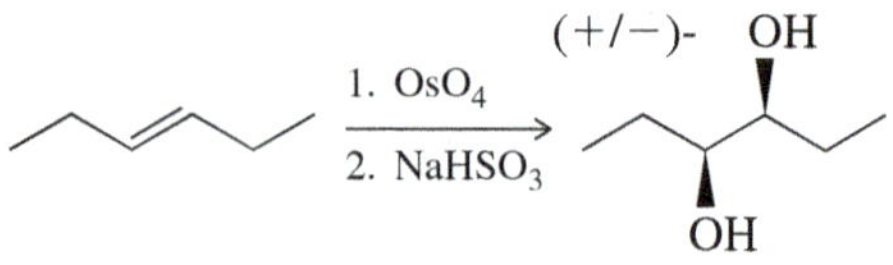

Assessment 17.3 Which carbonyl would you expect to vibrate at a higher wavenumber? Explain your reasoning.

(a) (b) (c) (d)

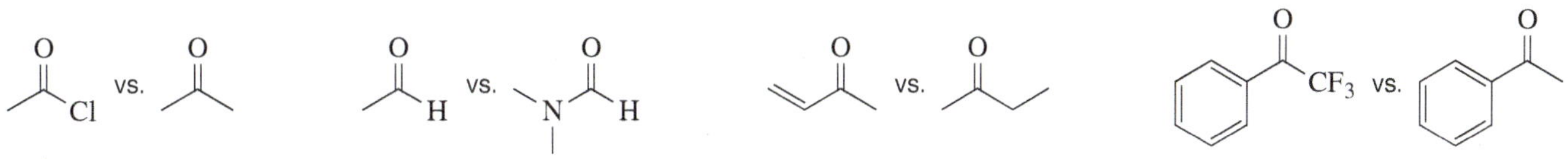

Assessment 17.4 Based on the analysis you used in Assessment 17.3, which carbonyl would you expect to react most quickly with a nucleophile?

(a) (b) (c) (d)

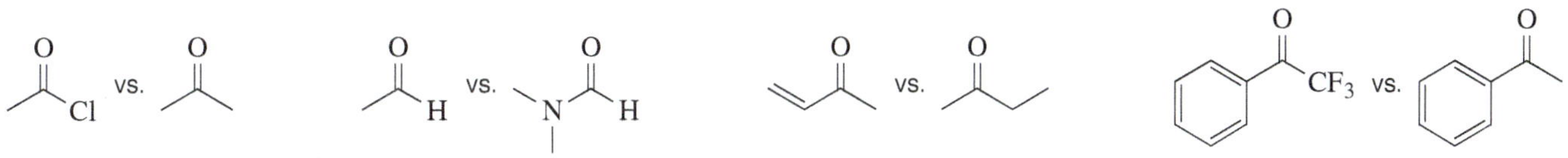

Assessment 17.5 Show all possible resonance structures of acetophenone.

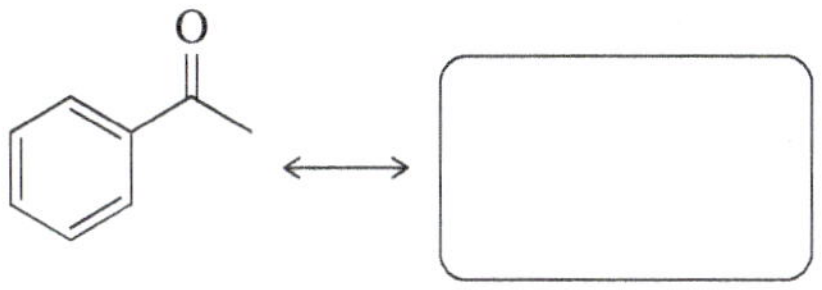

multiple structures

17.1 Introduction to aldehydes and ketones

Aldehydes and ketones are two classes of unsaturated molecules containing the carbonyl functional group, which is composed of a carbon atom double-bonded to an oxygen atom (**Table 17.1**). [*We discuss other carbonyl compounds in Chapter 18.*]

Table 17.1 Common classes of carbonyl compounds.

Class	General formula	Class	General formula
Ketones	R—C(=O)—R′	Aldehydes	R—C(=O)—H
Carboxylic acids	R—C(=O)—OH	Acid chlorides	R—C(=O)—Cl
Esters	R—C(=O)—O—R′	Amides	R—C(=O)—NH_2

If the carbonyl group is between two carbons, regardless of their hybridization, the molecule is referred to as a **ketone** (represented as RCOR) (**Figure 17.3**). **Aldehydes** (represented as RCHO) have the carbonyl group flanked by a carbon and a hydrogen. The simplest carbonyl compound, formaldehyde, is substituted with two hydrogens.

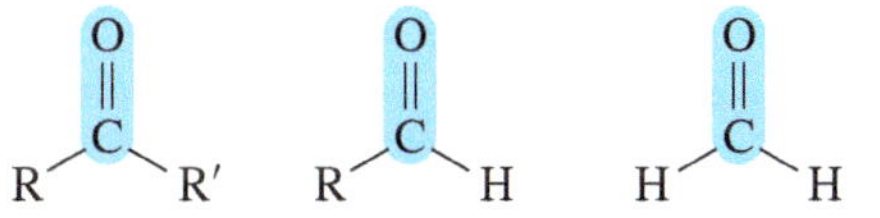

ketone RCOR′ aldehyde RCHO formaldehyde CH_2O

Figure 17.3 The relationship between ketones and aldehydes. [*The carbonyl group is screened in blue.*]

17.1.1 Physical properties

The melting and boiling points of aldehydes and ketones fall between those of alkanes and alcohols of similar molecular weight (Table 17.2). Aldehydes and ketones have higher melting and boiling points than the corresponding alkanes due to enhanced dipole–dipole interactions of the highly polar C=O bond. Aldehydes and ketones, however, have lower melting and boiling points than the corresponding alcohols. This is due to the fact that alcohols are able to accept *and* donate hydrogen bonds, whereas aldehydes and ketones can only *accept* hydrogen bonds.

Table 17.2 The boiling and melting points of aldehydes and ketones are intermediate between those of alcohols and alkanes of similar molecular weight.

Compound (class)	Pentane (alkane)	Butanal (aldehyde)	Butan-2-one (ketone)	Butan-1-ol (alcohol)
Structure		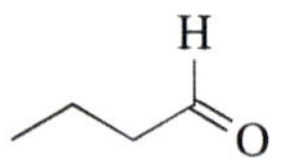	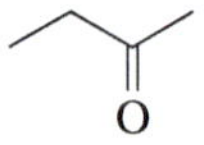	
Molecular weight (g/mol)	72.15	72.11	72.11	74.12
Boiling point (°C)	36.1	74.8	79.6	117.7
Melting point (°C)	−129.8	−96.9	−86	−89.8

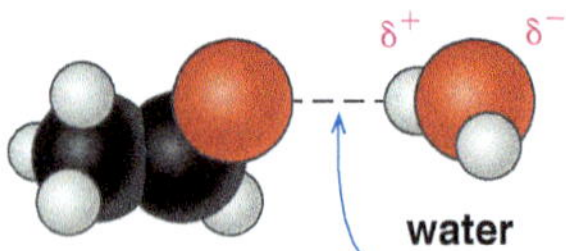

a hydrogen bond between a slightly positive hydrogen and a lone pair on the oxygen in the carbonyl group

Because aldehydes and ketones can accept hydrogen bonds from water, low-molecular-weight aldehydes and ketones are soluble in water. However, the nonpolar carbon chains of larger aldehydes and ketones disrupt this hydrogen bonding. As a result, higher-molecular-weight aldehydes and ketones are *in*soluble in water. The smallest insoluble aldehyde is octanal (Table 17.3).

Table 17.3 The solubility of aldehydes of increasing molecular weight: higher-molecular-weight aldehydes are insoluble in water.

Aldehyde	**Structure**	**Molecular weight (g/mol)**	**Water solubility (g/100 g H_2O)**
Formaldehyde/methanal	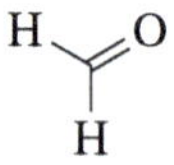	30.0	Miscible
Acetaldehyde/ethanal	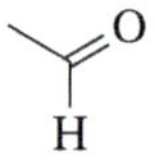	44.1	Miscible
Propanal	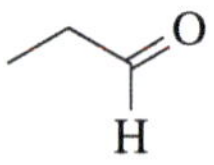	58.1	16
Butanal	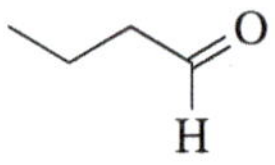	72.1	7
Pentanal	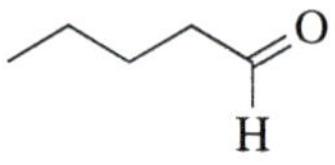	86.1	4
Hexanal	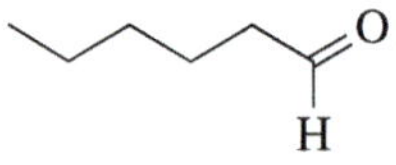	100.2	1
Heptanal	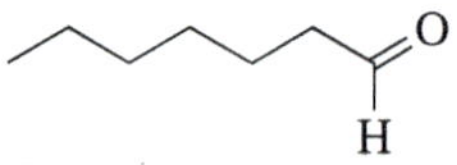	114.2	0.1
Octanal	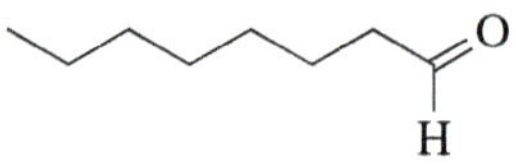	128.2	Insoluble

17.1.2 Nomenclature of aldehydes and ketones

The IUPAC rules for naming aldehydes and ketones are summarized in Table 17.4. Stepping through these rules, we name the following two compounds.

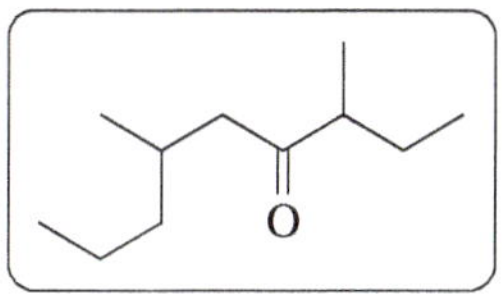

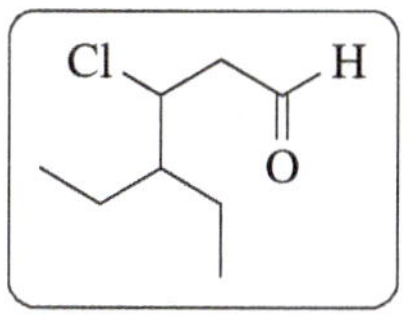

Table 17.4 IUPAC rules for naming aldehydes and ketones.

1. The parent chain (the longest carbon chain containing the carbonyl carbon) forms the root of the name. If the molecule is fully saturated, the infix -an- is used, and the suffix –e is replaced with –one for ketones and –al for aldehydes.	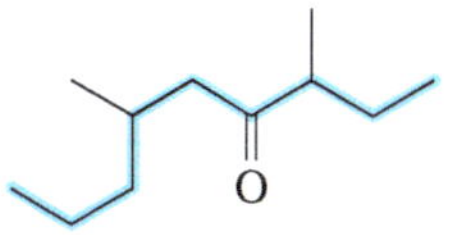 **9-carbon parent chain** nonanone	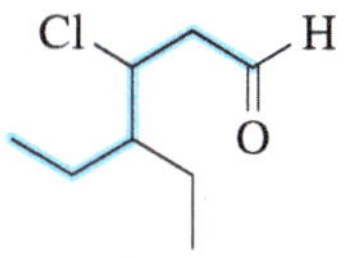 **6-carbon parent chain** hexanal
2. The parent chain is numbered such that the carbonyl carbon gets the lowest number. This means that the aldehyde carbon is always carbon 1 and therefore does not need to be specified in the name.	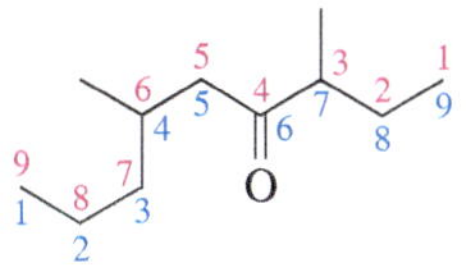 ~~nonan-6-one~~ nonan-4-one	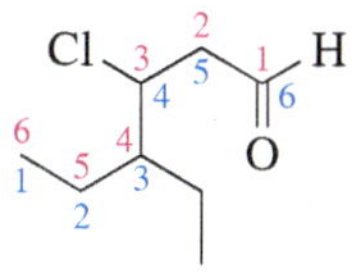 *unnecessary* → ~~6-hexanal~~ 1-hexanal hexanal
3. Substituents on the parent chain are named as was done previously.	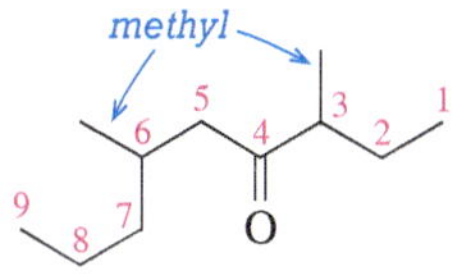	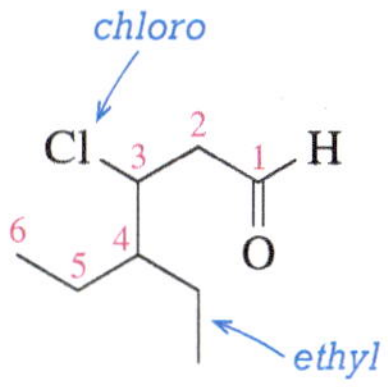
4. The name is written following IUPAC alphabetization rules, with numbers used to indicate the location of substituents and the carbonyl (only in the case of ketones).	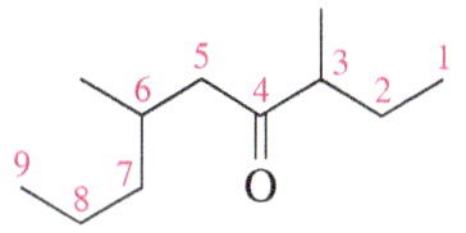 3,6-dimethylnonan-4-one	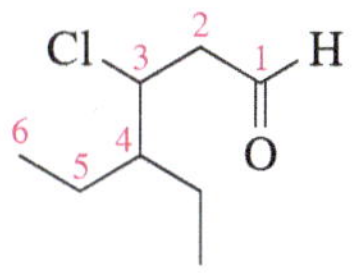 3-chloro-4-ethylhexanal

Table 17.5 IUPAC Functional group naming priority.

Functional group	Priority
Acids	1 (highest)
Esters	2
Acid chlorides	3
Anhydrides	3
Amides	4
Nitriles	5
Aldehydes	6
Ketones	7
Alcohols	8
Amines	9
Thiols	10
Alkenes, alkynes	11
Alkanes	—
Alkyl halides, ethers	(lowest)

Many aldehydes and ketones contain other functional groups. Under the IUPAC system, when numbering the chains, only carboxylic acids and carboxylic acid derivatives receive higher priority than aldehydes and ketones (Table 17.5).

The name of a polyfunctional molecule containing an aldehyde ends in -al, indicating that the molecule is an aldehyde, while all other functional groups are named as substituents. With an aldehyde and a ketone in the same molecule, the substituent name oxo- shows the location of the ketone. Figure 17.4 gives examples.

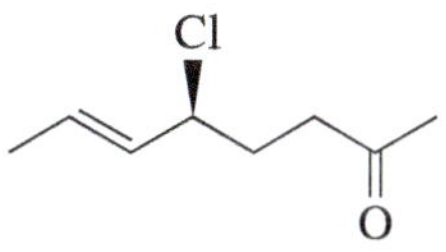

(5*S*,6*E*)-5-chlorooct-6-en-2-one

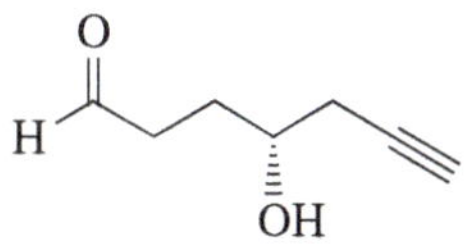

(4*R*)-4-hydroxyhept-6-ynal

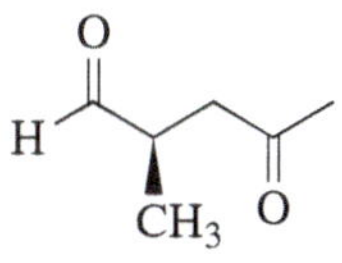

(2*R*)-4-oxo-2-methylpentanal

Figure 17.4 IUPAC nomenclature for some polyfunctional aldehydes and ketones.

Assessment 17.6 Name the following aldehydes and ketones.

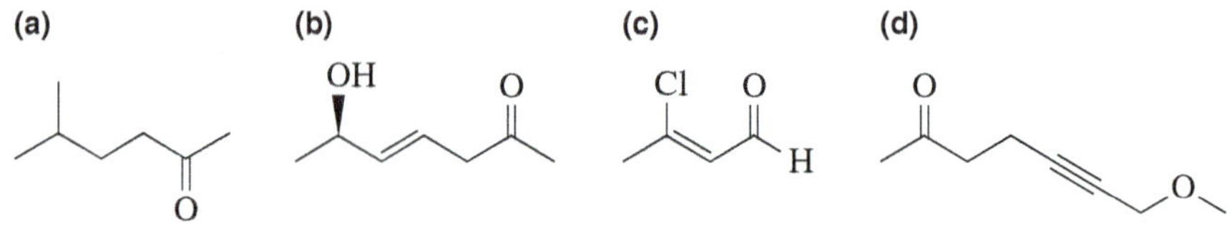

Assessment 17.7 Give a structure that corresponds to the name provided.

(a) (*R*)-5-chlorocyclohex-2-enone
(b) (2*E*,6*R*)-6,7-dihydroxy-7-methyloct-2-en-4-one
(c) (*S*)-1-bromo-3-methylhex-5-yn-2-one
(d) *E*-4-oxopent-2-enal

17.1.3 Synthesis of aldehydes and ketones

Although we have not yet studied reactions of aldehydes and ketones, **Table 17.6** reminds us of reactions used for the synthesis of aldehydes and ketones that we learned previously. These reactions will be important for any synthesis we do involving aldehydes and ketones [*and should be on the reaction sheets you began creating in Chapter 10.*]

Table 17.6 Methods for synthesizing aldehydes and ketones.

Hydroboration/oxidation of an alkyne (Section 10.9.5)	
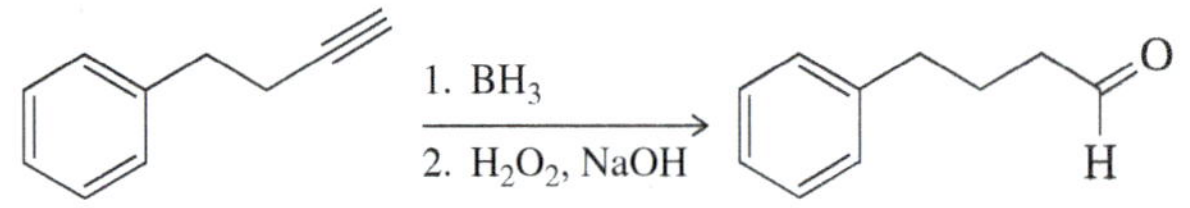	Hydroboration of an alkyne, proceeding with anti-Markovnikov regioselectivity, results in a terminal alkenyl borane. Oxidation of the alkenyl borane, under basic conditions, produces an enol that tautomerizes to the more favorable aldehyde form. Hydroboration/oxidation is an efficient method for making ketones or aldehydes using symmetrical, internal alkynes or terminal alkynes, respectively.
Oxymercuration of an alkyne (Section 10.9.4)	
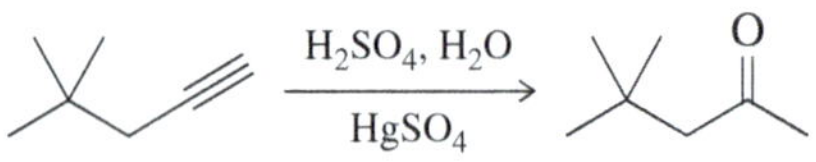	Oxymercuration of an alkyne, proceeding with Markovnikov regioselectivity, results in an enol, which tautomerizes to the thermodynamically more stable ketone. This reaction is an efficient method for making ketones, using either symmetrical, internal alkynes or terminal alkynes.
Oxidation of a primary alcohol (Section 13.9)	
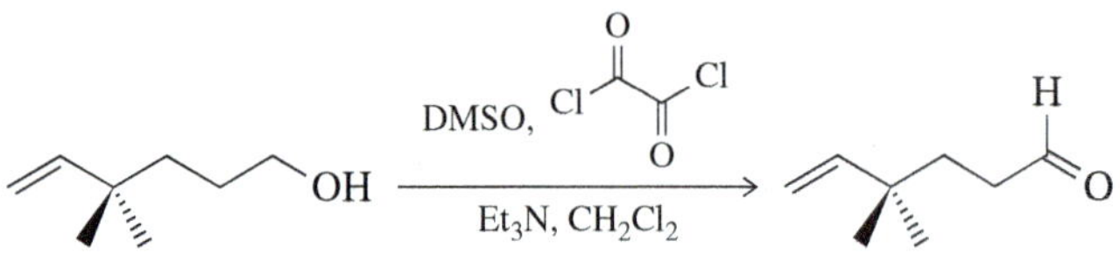	The Swern oxidation of primary alcohols is an efficient method for synthesizing aldehydes.
Oxidation of a secondary alcohol (Section 13.9)	
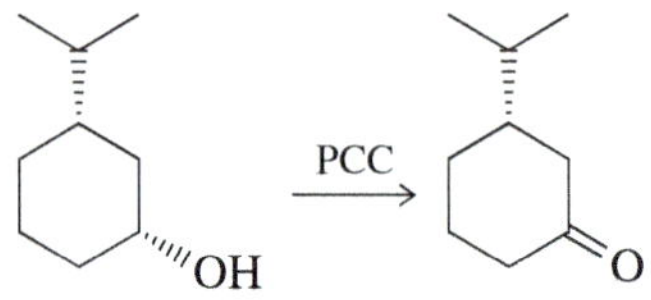	Oxidation of a secondary alcohol with pyridinium chlorochromate (PCC) is an efficient, though potentially toxic, method for the synthesis of ketones. Other oxidations can be used with similar results.
Ozonolysis of an alkene (Section 9.1.7)	
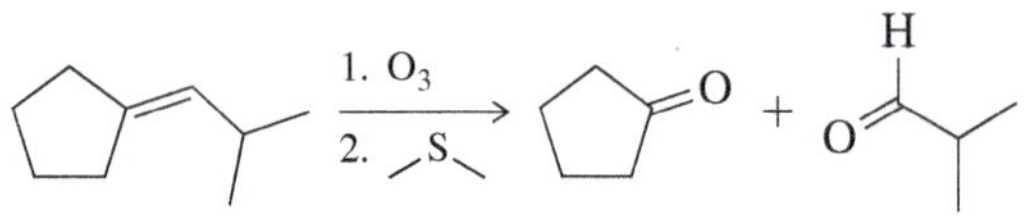	The reaction of an alkene with the highly oxidizing gas ozone results in formation of an ozonide intermediate. Reduction of this intermediate generates two new carbonyls (an aldehyde and/or a ketone), with the net result being the attachment of an oxygen to each carbon of the original alkene.

Assessment 17.8 Predict the product of the following aldehyde/ketone syntheses.

(a)

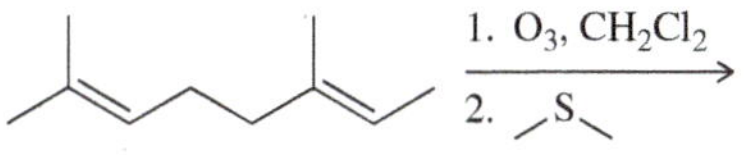

(b)

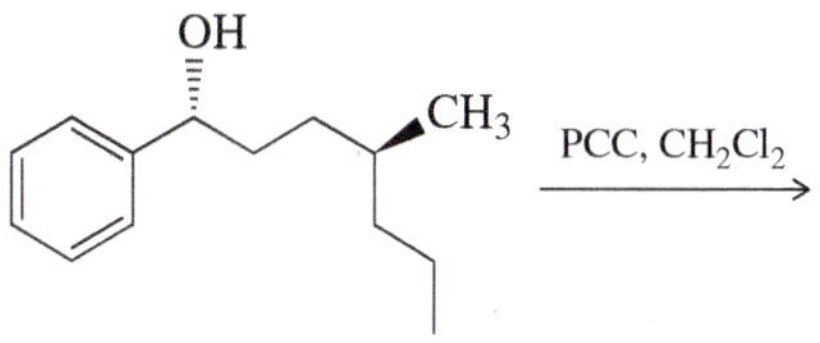

(c)

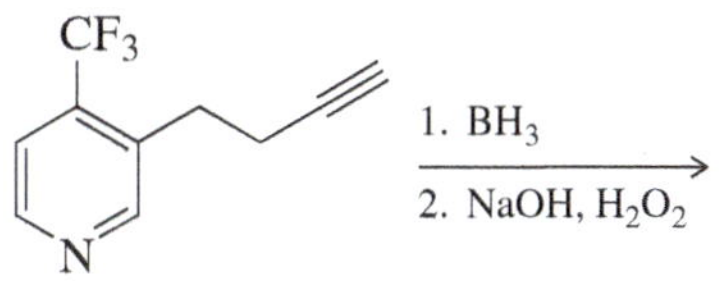

(d)

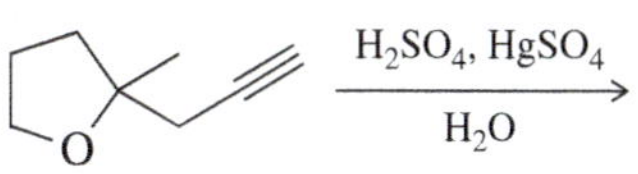

Assessment 17.9 Predict the reagents or reactant(s) necessary to complete the following syntheses.

(a)

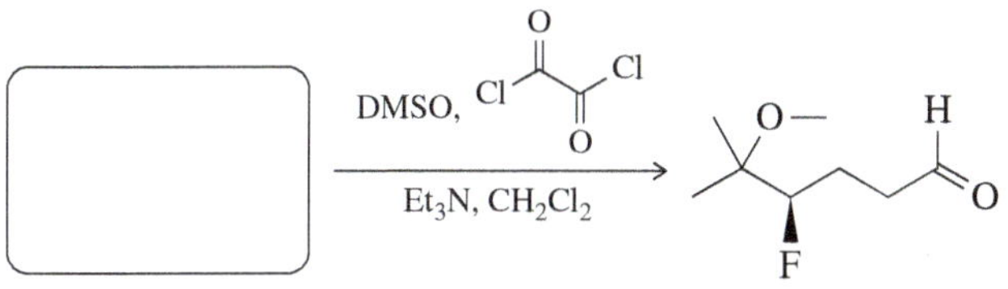

(b)

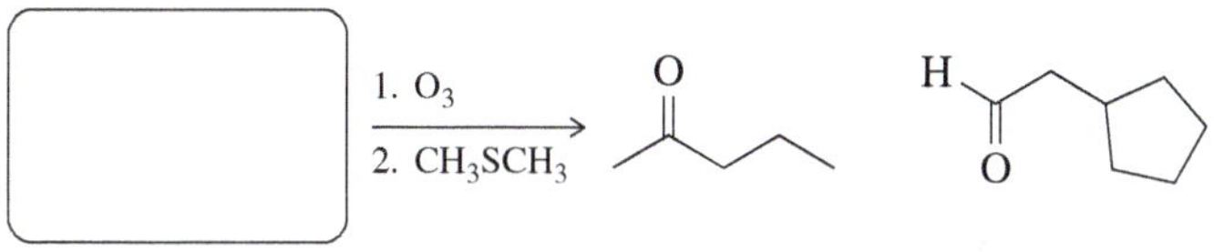

(c)

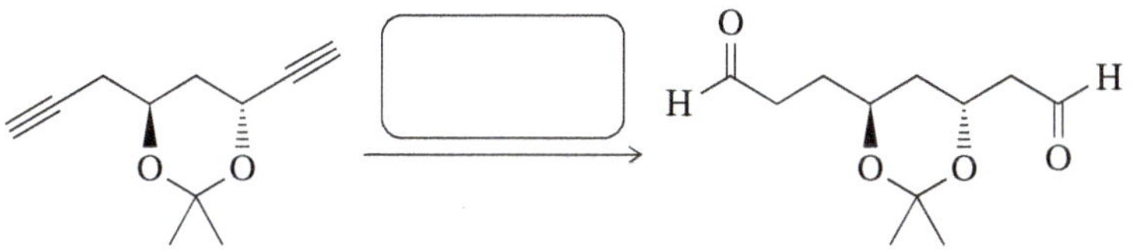

17.1.4 Molecular orbitals of the carbonyl

The molecular orbital picture of the carbonyl helps us understand the chemistry of aldehydes and ketones. The C=O double bond features a σ bond and a π bond between carbon and oxygen atoms that are each sp^2 hybridized (**Figure 17.5**). The lone pairs on oxygen are also in sp^2-hybridized atomic orbitals. The carbonyl C—H σ bonds form via the combination of carbon's sp^2 orbital and hydrogen's 1*s* orbital.

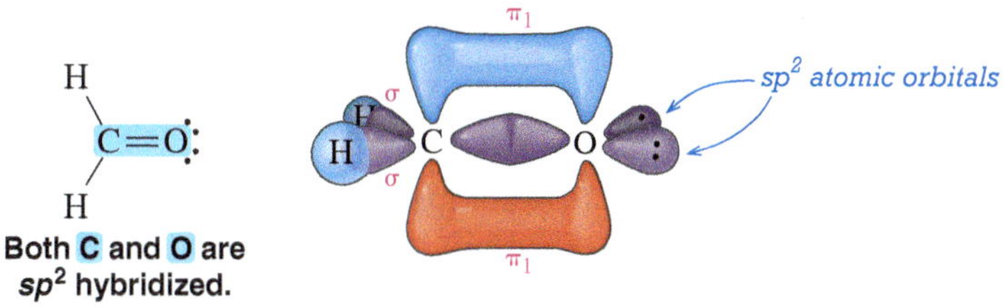

Figure 17.5 Molecular orbital picture of formaldehyde.

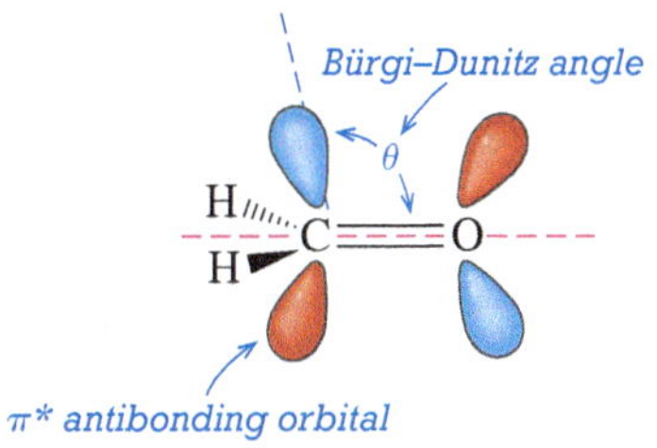

Figure 17.6 The location of the π* antibonding orbital rationalizes the direction of nucleophilic attack of carbonyls.

The π* antibonding orbital is located within the same plane as the C—O π bond at an angle of 107° (the **Bürgi–Dunitz angle**) relative to an axis slicing through the π bond (**Figure 17.6**). The location of π* is used by advanced organic chemists to rationalize the manner in which nucleophiles react with carbonyls. [*Recall that nucleophiles also add to antibonding orbitals during* S_N2 *reactions (Section 12.3.1).*]

17.2 Resonance structure of the carbonyl group

The carbonyl group is best represented as a combination of two resonance structures (**Figure 17.7**). Moving the π electrons onto the electronegative oxygen generates a resonance structure with a negatively charged oxygen and a positively charged carbon. This is referred to as the **zwitterion** resonance structure because it is a neutral molecule with two opposite formal charges. The electrons should not be moved in the opposite direction because the less electronegative carbon is *not* a suitable location to which electrons can be pushed. [*To review this concept, see the rules for drawing resonance structures in Section 2.8.2.*]

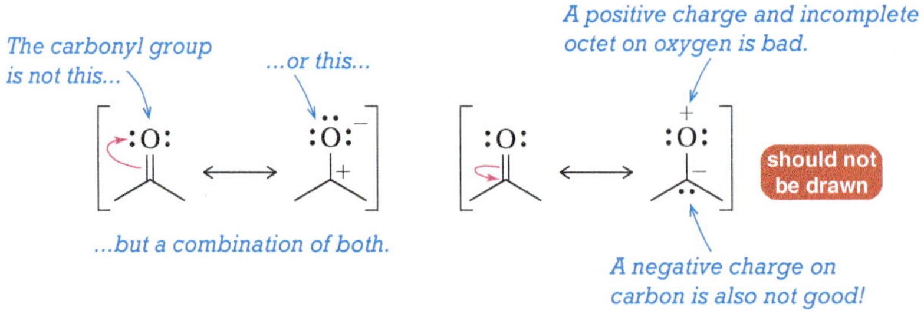

Figure 17.7 The zwitterionic resonance structure of the carbonyl has a negative charge on oxygen and a positive charge on carbon.

Because it has the maximum number of covalent bonds and full octets [**Figure 17.8(a)**], while minimizing charge separation, the neutral resonance structure is considered the major contributor (Section 2.8.3). The zwitterion structure has one fewer covalent bond, separates charges, and leaves carbon with an incomplete octet. As the minor contributor, the zwitterion resonance structure is still very important, helping explain the highly polarized C—O π bond. An electron density map of the carbonyl [**Figure 17.8(b)**] clearly shows unequal sharing of electrons between carbon and oxygen (similar to what is predicted by the zwitterion structure).

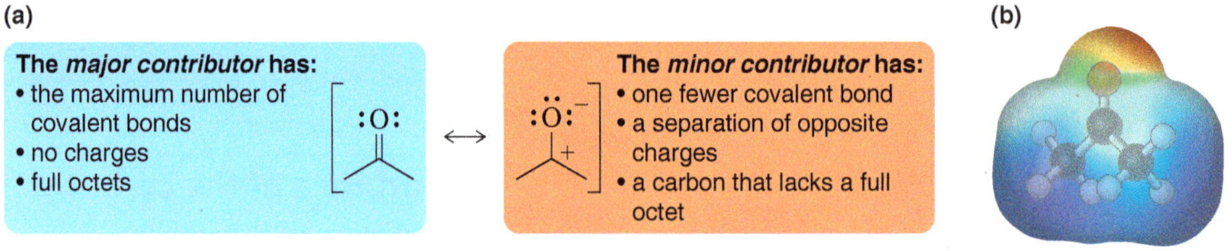

Figure 17.8 Contributing resonance structures. (a) The minor contributor (zwitterion) predicts the electron distribution in the (b) electron density map of the carbonyl.

17.2.1 Importance of resonance structures: deeper analysis

The carbonyl C═O double bond is a highly polar, short, and very strong bond. In our studies of alkenes, it was observed that the π bond between two atoms is usually weaker than the σ bond between those same atoms. This is not the case for the carbonyl group. The C—O π bond energy is estimated as being between 85 and 95 kcal/mol, while the average C—O σ bond-dissociation energy is near 88 kcal/mol. [*Recall that we estimated the C—C π bond at 65 kcal/mol.*] At the low end of the estimated range, the C—O is about as strong as the σ bond; at the high end, the C—O π bond is slightly stronger than the C—O σ bond.

A comparison of the C═C and C═O double bonds further illustrates the increased strength of the carbonyl (**Figure 17.9**). As expected, the stronger C═O bond is significantly shorter (0.11 Å).

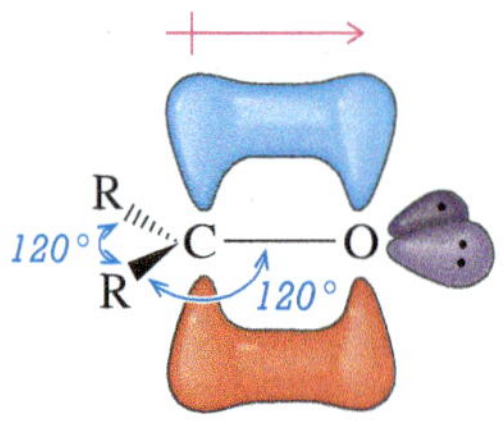

	Length	Energy
Ketone C═O bond	1.23 Å	178 kcal/mol (745 kJ/mol)
Alkene C═C bond	1.34 Å	146 kcal/mol (611 kJ/mol)

Figure 17.9 The carbonyl double bond is shorter and stronger than the alkene double bond.

Other chemical properties give us insights into the nature of the C=O bond. The dipole moment (μ) of the polar formaldehyde molecule can be measured and compared to similar molecules such as ethene, methanol, and chloromethane [**Figure 17.10(a)**]. Although Cl and O are similar in electronegativity, formaldehyde is significantly more polar than chloromethane. Finally, consider the ^{13}C NMR data for the central carbon of the structurally related molecules acetone (207 ppm) and 2-methylpropene (142 ppm) [**Figure 17.10(b)**]. When compared to the *t*-butyl carbocation (320 ppm) and a resonance-stabilized benzylic carbocation (244 ppm), it becomes clear that the carbonyl carbon is more similar to a resonance-stabilized carbocation than the central carbon of 2-methylpropene. [*Once finished with this section, you can forget all of these chemical shift numbers, as long as you understand the general point they help make.*]

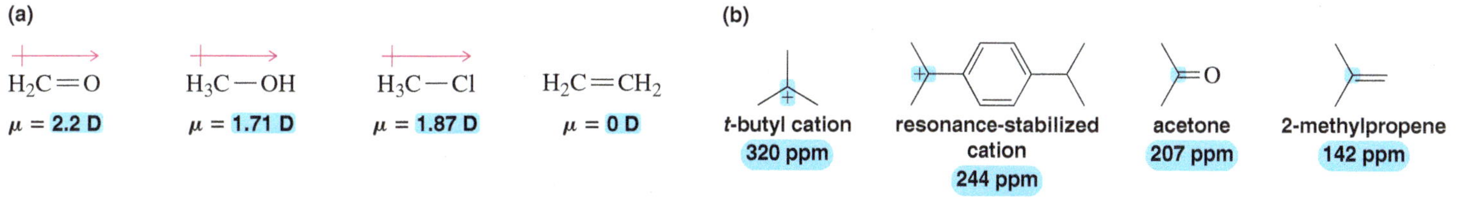

Figure 17.10 Carbonyl properties rationalized by a zwitterionic structure. (a) Formaldehyde is more polar than expected. (b) The ^{13}C NMR of a ketone is more deshielded than expected.

So, why consider these data? In virtually every chemistry book you might read—including this one—the carbonyl is presented as a combination of two resonance structures, where the neutral resonance structure is the major contributor. And yet the data suggest that the carbonyl has properties inconsistent with a simple covalent bond between carbon and an electronegative atom. In sum, the bond is stronger, shorter, and more polar (and the carbon more deshielded) than expected based on the covalently bonded resonance form. [*Here's where it gets a bit crazy, so stay with me.*] What if the covalent resonance structure is not the major structure? Instead, perhaps the zwitterion resonance form is the major contributor and the carbonyl is best thought of as a resonance-stabilized carbocation. Maybe the C—O π bond is actually an ionic π bond. [*What's an ionic π bond, you ask?*] According to this outside-the-box idea, the carbonyl is made up of a covalent σ bond, but the π bond is actually an **ionic π bond** between a negatively charged oxygen and a positively charged carbon. This resonance form doesn't adhere to Lewis's octet rule, and truthfully, drawing the carbonyl as the zwitterion feels awkward, but this idea can help us understand the properties and the reactivity better. [*You might be wondering what is gained by thinking of the carbonyl as a zwitterion? Read on . . . .*]

17.2.2 Zwitterion resonance structure and reactivity

Almost every carbonyl reaction is more easily understood in the context of the zwitterion resonance form. To see why, let's go back to the second rule of organic chemistry—namely, negatives attack positives. Thus, nucleophiles attack electrophiles, Lewis bases attack Lewis acids, and so on. The challenge in applying this rule is identifying the charges, because they might be in empty *p* orbitals or antibonding orbitals, or they might exist as partial charges or formal charges, and so forth.

Figure 17.11 Predicting carbonyl reactivity with the neutral resonance contributor.

Based on what you know so far, how might acetone react with a nucleophile (like hydroxide) or an electrophile (like H^+)? Think about this first using the covalent, neutral resonance structure of acetone (**Figure 17.11**). [*Don't read ahead before thinking about this.*]

Now, do the same using the zwitterion resonance form (**Figure 17.12**). [*Even as a first-day organic student, you could have told me that the negative hydroxide will bond with the positively charged carbon, while the proton would bond with the negatively charged oxygen.*]

Figure 17.12 Predicting carbonyl reactivity [*is easier?*] with the zwitterionic resonance contributor.

Put simply, the zwitterion resonance form, as the major or minor contributor, simplifies the prediction of carbonyl reactivity. At some point [*on an organic exam, or maybe the DAT or the MCAT*], you will be confronted by the reaction of a carbonyl with an unknown reagent. My advice? Draw the zwitterion resonance form, *then* think about what goes where. [*For many students, just drawing the zwitterion causes the light bulb to come on.*] In identifying the properties explained by the zwitterion form and demonstrating its problem-solving utility, it is clear that the zwitterion structure is too important to be regarded as just a "minor resonance contributor."

Assessment 17.10 Another way the preceding might be presented is to say that the minor resonance structure(s) reveal the reactivity of a molecule. Show how the minor resonance structure(s) can help us understand that the following reactions occur with the regioselectivity shown.

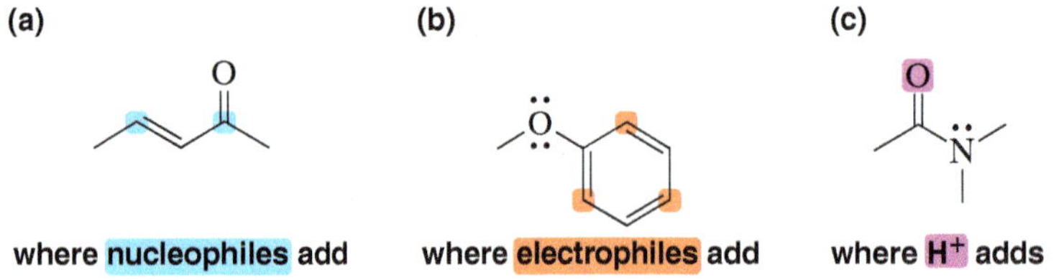

17-A CHEMISTRY IN CONTEXT: Biochemistry

The chemistry of a hangover (part 1)

The human body can rid itself of ethanol and other toxic alcohols using NAD^+ and the enzyme alcohol dehydrogenase (ADH), which catalyzes the oxidation of alcohol to form acetaldehyde. ADH is an enzyme that likely evolved to metabolize toxic alcohols present in partially fermented foods or generated by gut bacteria. For consumers of alcoholic beverages, ADH in the liver spends a fair amount of its time oxidizing ethanol to acetaldehyde.

Mechanistically, ADH uses the NAD^+ cofactor as part of the oxidative process (Figure 17.13). The enzymatic mechanism begins with ethanol and NAD^+ binding within the active site of the enzyme. Next, an active-site base deprotonates the hydroxyl group of ethanol, with the electrons collapsing to form a new C—O π bond while kicking out a hydride ion. The hydride ion is immediately donated to the nearby NAD^+ cofactor and the result is NADH, a reduced form of NAD^+. If NAD^+ is an oxidizing agent, does that mean NADH is a reducing agent? [*We explore its chemistry later on in this chapter.*]

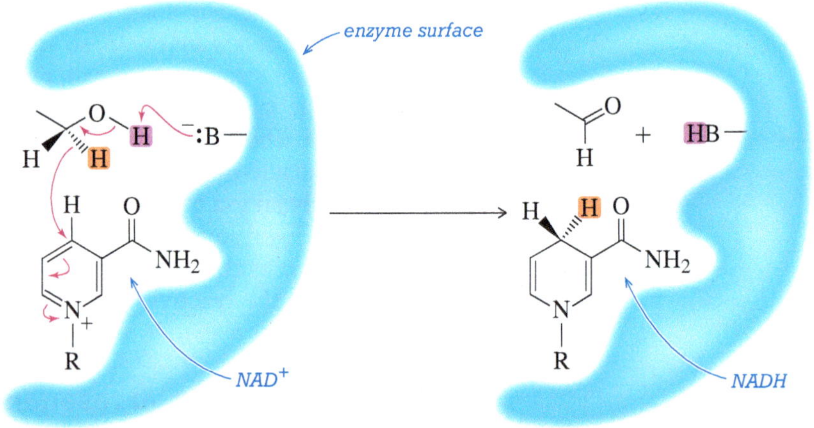

Figure 17.13 The mechanism of enzymatic ethanol oxidation.

Comparing the chromium oxidations we studied in Chapter 13 helps us understand this new biochemical mechanism (Figure 17.14). In the presence of a chromium reagent, oxidation occurs by first forming the chromate ester, with the concomitant loss of the hydroxyl proton. At this point water (or another base in solution) removes the hydrogen from carbon, forming the π bond, kicking out the chromium species as a leaving group, and generating a reduced form of chromium.

The benefits of removing ethanol in this way are short-lived, however, because acetaldehyde is actually more toxic to humans than ethanol alone.

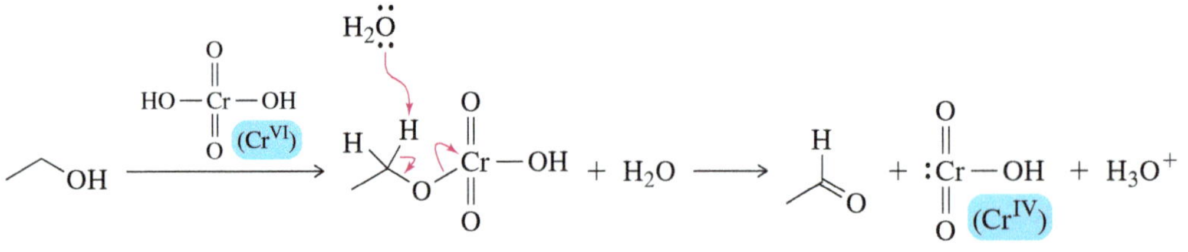

Figure 17.14 The mechanism of chromic acid alcohol oxidation is similar to the mechanism in Figure 17.13.

17.3 Nucleophilic addition reactions of carbonyls

Nucleophilic addition reactions involving carbonyl compounds are a major topic of this chapter. Although these reactions are new to you, it turns out that they have much in common with the addition reactions we've already studied. For example, let's compare an addition reaction we studied in Section 8.4.1 (the hydration of propene, an alkene, to form propan-2-ol) to a nucleophilic addition reaction involving an aldehyde (the hydration of ethanal to form ethane-1,1-diol). Both reactions result in the addition of water, the formation of C—H and C—O bonds, and the breaking of an O—H σ bond and a C—X π bond (**Figure 17.15**).

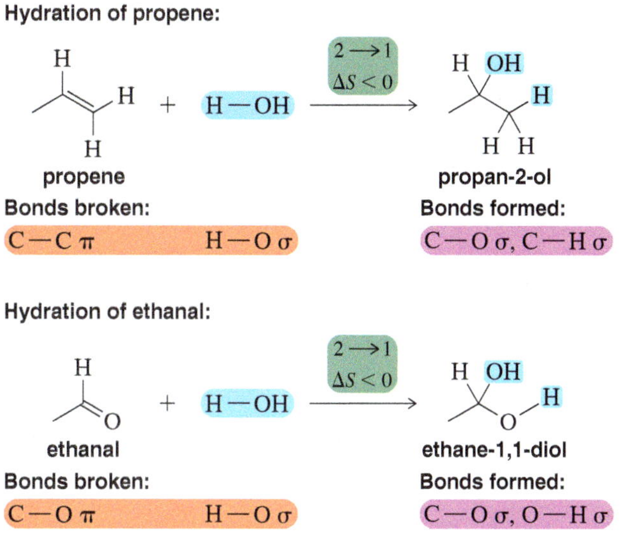

Figure 17.15 The hydrations of propene and ethanal are similar.

Assessment 17.11 Identify the bonds broken and formed in the following addition reaction. (a) Would you expect this reaction to be favored based on entropy? (b) Based on enthalpy [*qualitatively*]? (c) Overall?

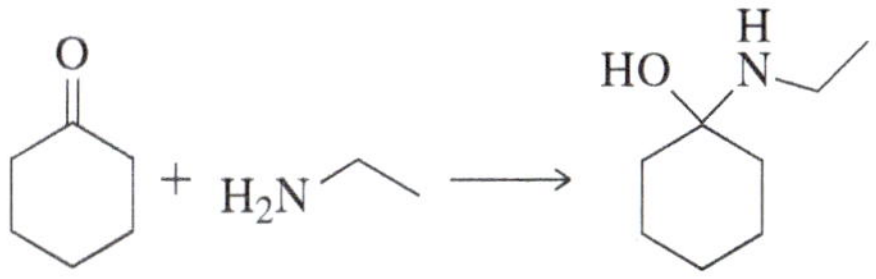

17.3.1 Carbonyl reactions with strong and weak nucleophiles

The challenge in this chapter, more so than in previous chapters, lies in the large number of new reactions to be studied. The key to your success is learning how the reactions are connected *conceptually*. We have already begun this process, learning to recognize simple carbonyl addition reactions. Furthermore, every reaction in this chapter can be categorized into one of two groups:

1. reactions with strong nucleophiles
2. reactions with weak nucleophiles

We have seen this theme previously in Chapter 12 (S_N1 vs. S_N2 and E1 vs. E2), and it is one that comes up again in Chapter 18. Although the products vary from reaction to reaction, each starts the same way. All that is left to understand is *why* each reaction finishes the way it does.

17.3.1.1 Reactions with Strong Nucleophiles (Basic Conditions): Mechanism 17A

The first reaction type we explore is the addition of a strong nucleophile with an aldehyde or a ketone. Strong nucleophiles react quickly, attacking atoms with only a partially positive charge. That is, they don't wait around for positively charged intermediates prior to reacting. [*Recall that we used the same reasoning for the S_N2 mechanism.*] As a result, the nucleophile adds to the carbonyl, pushing the π electrons onto the oxygen and forming an alkoxide intermediate. Typically, this alkoxide intermediate is protonated in an acid quench step. The addition of a Grignard reagent to a carbonyl shown in **Figure 17.16** is representative of the reaction mechanism for strong nucleophiles.

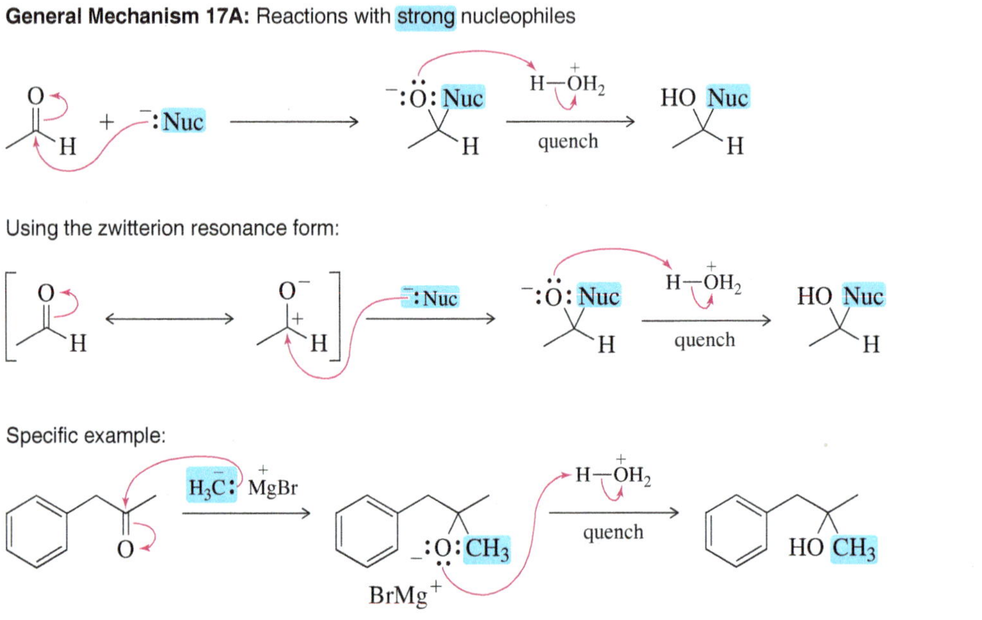

Figure 17.16 **Mechanism 17A: the addition of strong nucleophiles** proceeds quickly with the neutral carbonyl. In general, strong nucleophiles add first, then the alkoxide intermediate is protonated.

The process in Figure 17.16 is referred to throughout this chapter as General Mechanism 17A.

17.3.1.2 Reactions with Weak Nucleophiles (Acidic Conditions): Mechanism 17B

The reaction of an aldehyde or ketone with a weak nucleophile is slower than its reaction with a strong nucleophile. A nucleophile is considered weak when the atom donating the lone pair of electrons is neutral (Section 17.3.2). These types of nucleophiles (H_2O, CH_3OH, etc.) do *not* react directly (S_N2) with neutral molecules in substitution reactions. Instead, they only react by the S_N1 mechanism. That is, a full carbocation (a more reactive intermediate) must form prior to reacting. When weak nucleophiles react with the carbonyl group, a protonation step is necessary to create a reactive carbocation that is ultimately attacked by the weak nucleophile (**Figure 17.17**). The addition of water represents the general reaction of weak nucleophiles. With weak nucleophiles, a final deprotonation step is required to give the neutral product.

The process in Figure 17.17 is referred to throughout this chapter as General Mechanism 17B.

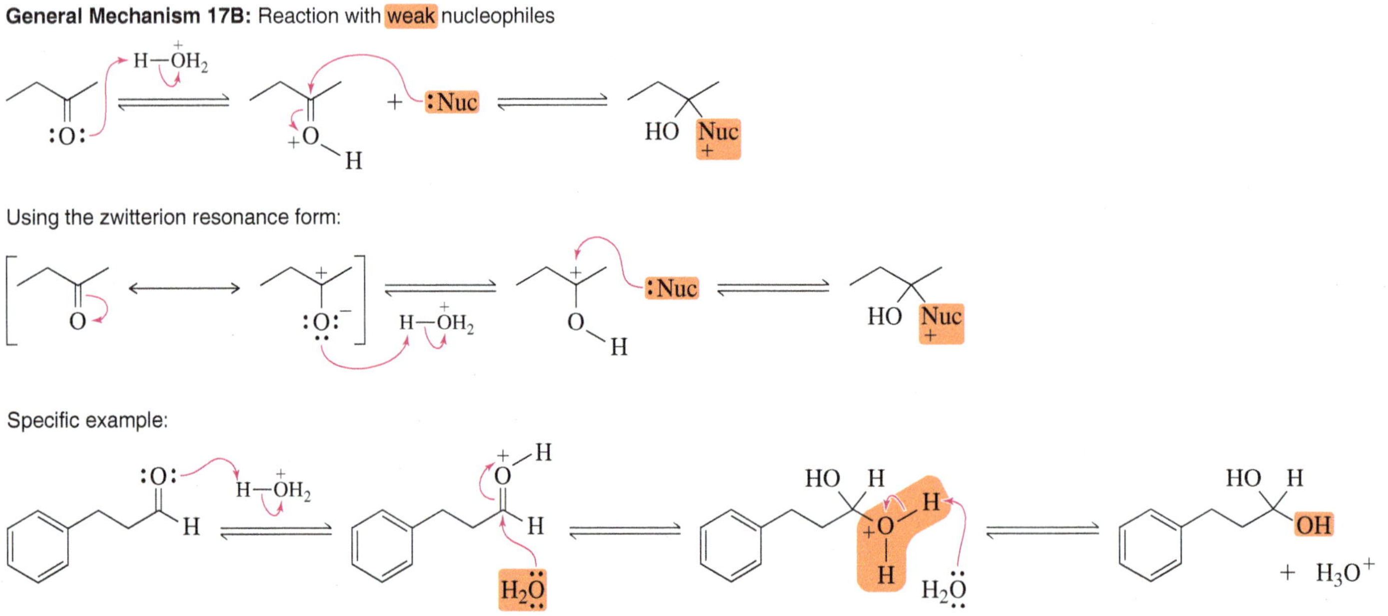

Figure 17.17 **Mechanism 17B: the addition of weak nucleophiles** proceeds slowly with a neutral carbonyl. In general, the carbonyl protonates first with weak nucleophiles, then the nucleophile adds.

17.3.2 Relative nucleophilic strength

Because the mechanism of carbonyl addition reactions depends on the strength of the attacking nucleophile, we need a method to assess nucleophilic strength. Using pK_a values [*as we have in the past*] allows us to assess the relative strength of three important nucleophiles: carbanions (CH_3^-), alcohols (CH_3OH), and amines (CH_3NH_2).

To examine relative nucleophilic strength, let's assume that all three will react with the carbonyl to give an alkoxide (conjugate acid has a pK_a = 16). From there, we can determine whether the product of the first step is more or less stable than the reactants. Although we are using pK_a values here, these are *not* acid–base reactions. As we have done many times previously, we can use the pK_a value of the conjugate acid of the base on each side to tell us about the direction of favorability. [*Recall that a base whose conjugate acid has a higher* pK_a *value is less stable.*]

In the case of the methyl anion (a carbanion) shown in **Figure 17.18**, the reaction proceeds from a very unstable base (conjugate acid CH_4; pK_a = 50) to a relatively stable one (conjugate acid *i*PrOH; pK_a = 16). Ignoring entropy, this is a very rapid, downhill reaction. Thus, the carbanion is a very strong nucleophile—one of the strongest—and adds to aldehydes and ketones by General Mechanism 17A.

Figure 17.18 The reactive methyl anion adds to the carbonyl in a fast and favorable reaction.

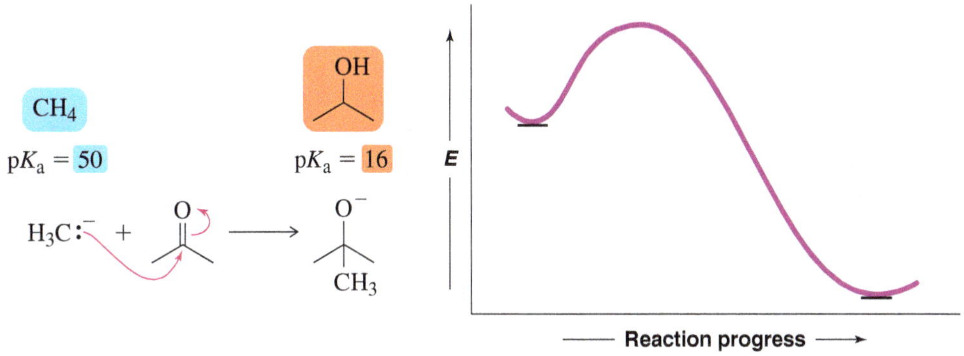

For methanol (conjugate acid $CH_3OH_2^+$; pK_a = −1.7), the reaction is highly unfavorable [*uphill*] (**Figure 17.19**). This makes sense qualitatively as well, because a negatively charged alkoxide (conjugate acid *i*PrOH; pK_a = 16) is less stable than the neutral oxygen of methanol. Thus, a neutral oxygen is considered a weak nucleophile, reacting only by General Mechanism 17B.

Figure 17.19 The weak, neutral nucleophile methanol adds slowly in an unfavorable reaction.

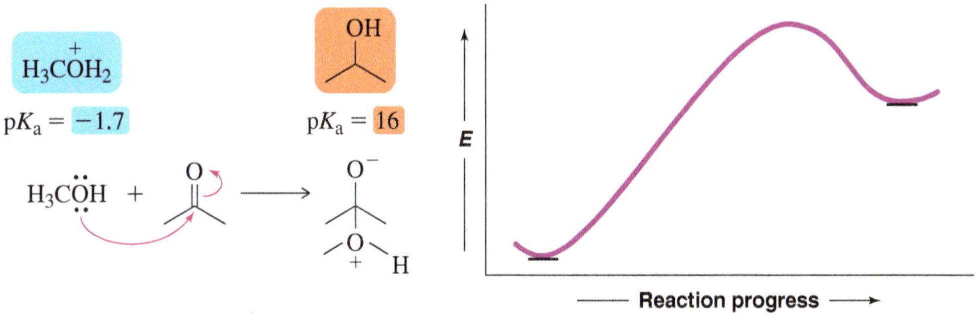

Lastly, let's look at the nucleophilic strength of a typical amine. The nucleophilicity of methyl amine, containing nitrogen, is intermediate between a neutral oxygen and a carbanion (**Figure 17.20**). This is illustrated by the pK_a values as the reaction goes from a base (conjugate acid $CH_3NH_3^+$; pK_a = 10) to the alkoxide (conjugate acid *i*PrOH; pK_a = 16). We explain in Section 17.8.4 that, because direct addition is only slightly disfavored [*slightly uphill*], nitrogen-based nucleophiles can add by General Mechanism 17A but add more quickly via General Mechanism 17B when acid is present. [*It seems like there is always one reaction that fits neither category. At least* pK_a *values help us understand why.*]

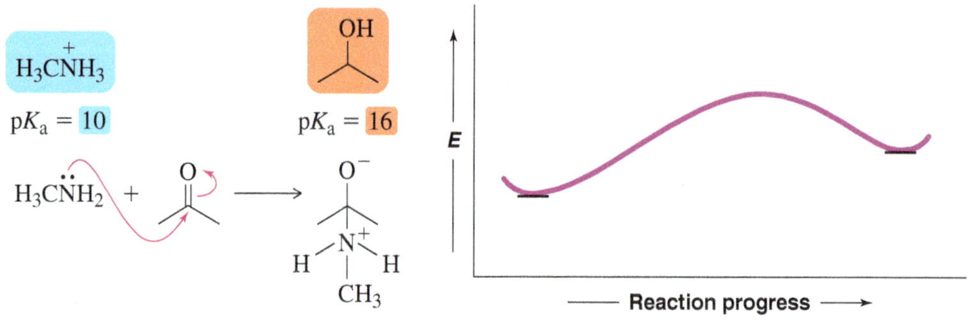

Figure 17.20 Amines are intermediate-strength nucleophiles that, while neutral, can attack a carbonyl directly in a slightly disfavored reaction.

17.3.2 SUMMARY

- Relative nucleophilic strength can be assessed using the pK_a of the conjugate acid of a given nucleophile. In general, nucleophiles that donate a lone pair from negatively charged atoms are strong, whereas nucleophiles that donate a lone pair from a neutral atom are weak. Because nitrogen is slightly electronegative, readily sharing electrons, amines are considered to be of intermediate strength.

Assessment 17.12 Classify the following nucleophiles as strong, weak, or intermediate. Would you expect each to add to a carbonyl directly or wait for a carbocation to form?

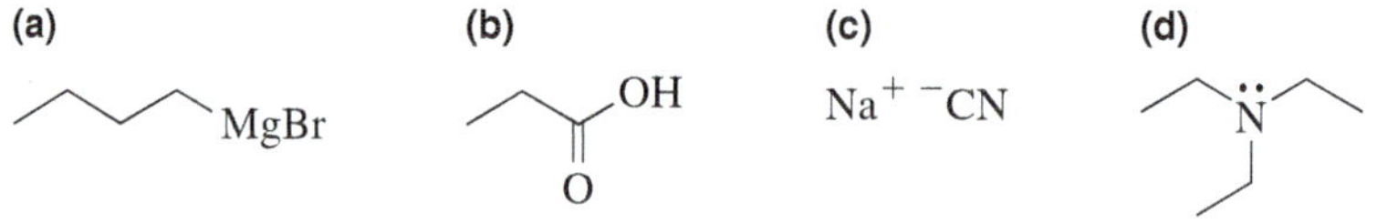

17.4 Comparing aldehyde and ketone reactivity

The reactions we study in this chapter can occur with both aldehyde and ketones, albeit often at different rates. Would you expect an aldehyde or a ketone to react more quickly as an electrophile (meaning with nucleophiles)? Would you expect an aldehyde or a ketone to be a stronger base (grab a proton) more rapidly? The following sections provide the answers to these questions.

17.4.1 Reactivity as an electrophile: steric and electronic factors

Assessing relative reactivity requires us to consider both steric and electronic factors. As shown in **Figure 17.21**, ketones are more sterically hindered than aldehydes due to the presence of two large groups attached to the carbonyl carbon. As a result, nucleophiles have better access to the carbonyl carbon of an aldehyde, making aldehydes more reactive than ketones. Formaldehyde, lacking non-hydrogen groups, is the most reactive carbonyl.

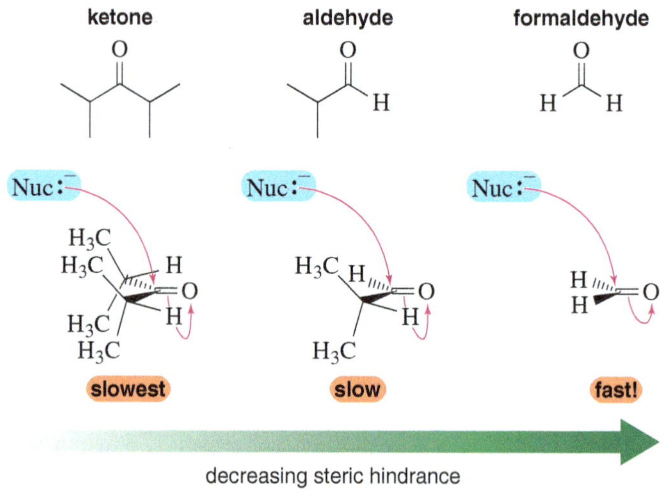

Figure 17.21 As steric hindrance decreases, nucleophilic addition to the carbonyl becomes faster.

Electronic factors can be best rationalized by examining the zwitterion resonance structures. In its zwitterion resonance form, the carbocation of the ketone is better stabilized by hyperconjugation from the two attached alkyl groups. In contrast, the aldehyde carbocation is stabilized by only one alkyl group (**Figure 17.22**). Thus, the aldehyde is the stronger electrophile, reacting more quickly with nucleophiles. Formaldehyde, with only hydrogens attached, has no stabilizing hyperconjugative effects and therefore reacts most quickly with nucleophiles.

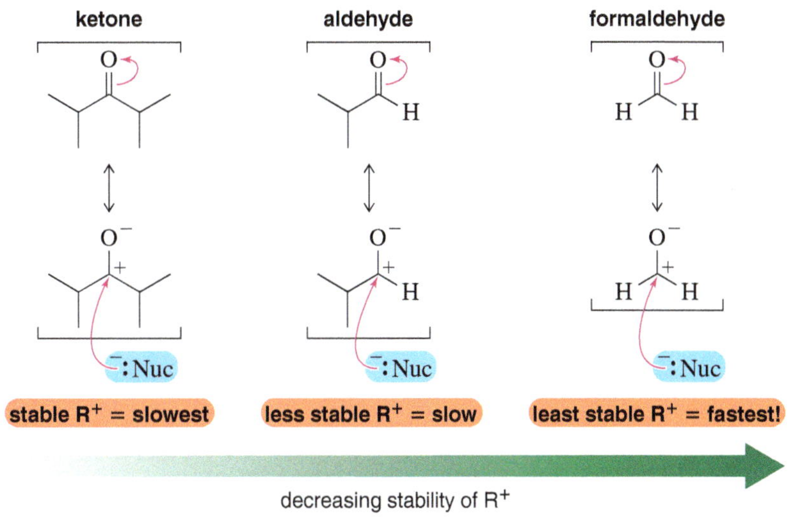

Figure 17.22 As the stability of the carbocation in the zwitterion resonance structure decreases, the rate of nucleophilic addition increases.

17.4.1 SUMMARY

- Aldehydes, compared to ketones, are less sterically hindered and thus more easily attacked by nucleophiles. In addition, with one fewer alkyl group, the zwitterion form of an aldehyde is less stable and more reactive than the corresponding zwitterion form of a ketone, once more making aldehydes more easily attacked by nucleophiles.

Assessment 17.13 Choose the carbonyl in each pair you would expect to react most quickly with a strong nucleophile.

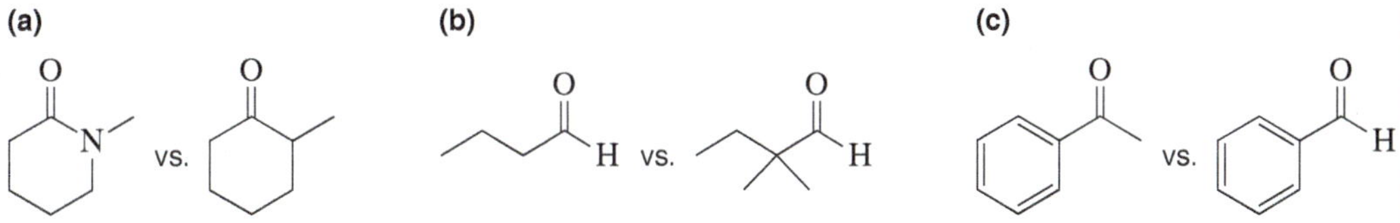

17.4.2 Reactivity as a base: steric and electronic factors

Because aldehydes and ketones protonate away from the groups at the central carbon, there is little difference in reactivity due to steric considerations. However, the effect of electronic factors on reactivity is significant. The basicity of the oxygen can be assessed by comparing the stability of the resulting acids. The protonated ketone can be stabilized by hyperconjugation from two alkyl groups, whereas a protonated aldehyde is stabilized by one alkyl group (**Figure 17.23**). Because the conjugate acid of a ketone is more stable, it follows that ketones are more nucleophilic (basic) at oxygen.

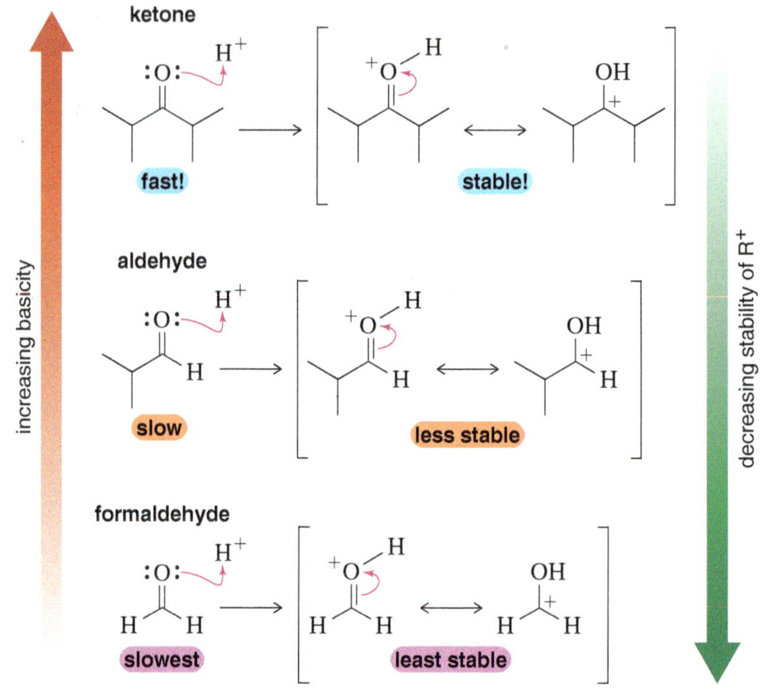

Figure 17.23 The carbonyl that forms the most stable carbocation (conjugate acid) after protonation is the strongest base.

Assessment 17.14 Choose the carbonyl from each pair you would expect to react most quickly with H^+. Explain your choice.

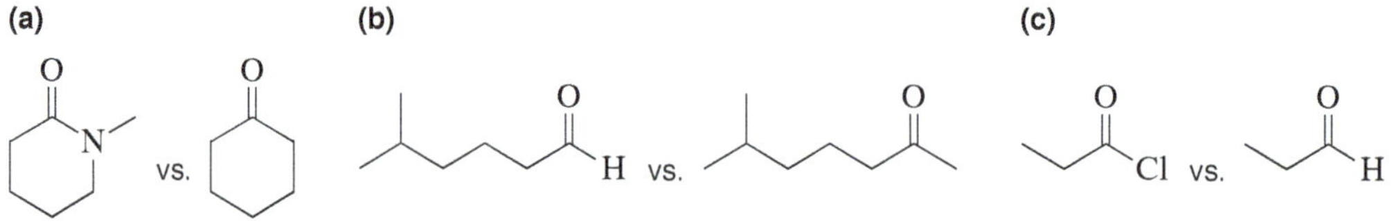

Assessment 17.15 The two ketones shown here protonate at relatively equal rates. Explain why this is true, given their obvious steric differences.

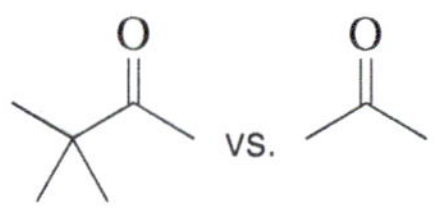

17.5 Addition of carbon nucleophiles to aldehydes and ketones

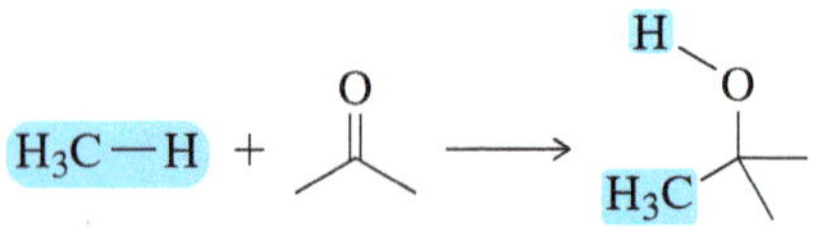

Figure 17.24 Addition of methane to acetone is favored on the basis of ΔH, but because of incompatible reactivity, the reaction does *not* occur.

Before we study the addition of carbon nucleophiles, the equilibrium constant (K_{eq}) for adding methane to acetone, a representative ketone, can be estimated by estimating ΔG (**Figure 17.24**). Based on ΔH and ΔS for the reaction, what value would you expect for ΔG?

By analogy to alkene addition reactions (Section 8.2), the addition of methane to acetone should be a favorable process. Based on the bond-dissociation energies of the formed and broken bonds, the reaction is exothermic. And yet, because carbonyls are typically electrophiles and methane is not at all acidic, this reaction will not proceed as written. Instead, to form C—C bonds using an aldehyde or ketone, a strong nucleophile is needed to react with the electrophilic carbonyl.

The addition of carbon nucleophiles to aldehydes and ketones follows General Mechanism 17A (**Figure 17.25**). The unstable carbanion adds directly to the neutral carbonyl, forming an alkoxide. Quenching of the alkoxide with acid gives an alcohol as product (see Figure 17.24). In each of the examples of carbon nucleophile addition that follow, the mechanism is basically the same. The only difference is the manner in which the nucleophile is prepared, prior to reacting with the carbonyl.

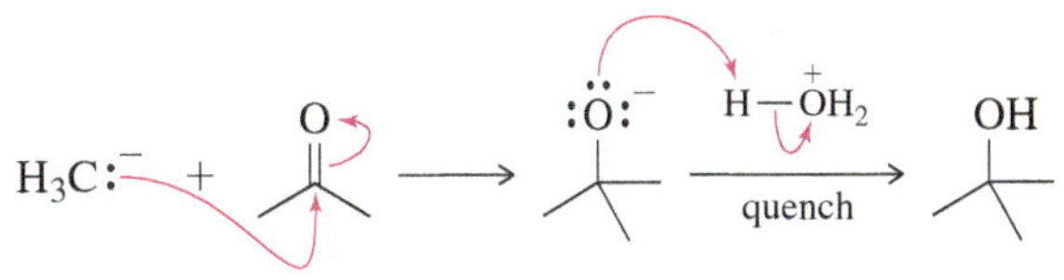

Figure 17.25 **The addition of carbon nucleophiles proceeds by General Mechanism 17A.**

When studying the carbon nucleophiles that add to a carbonyl, you should focus on two key aspects.

- Learn how the reagents are prepared and reacted. You'll notice that the same theme is repeated over and over—namely, strong nucleophiles add to the carbonyl, and the resulting alkoxide is protonated.
- The reactions allow C—C σ bond formation in a large number of diverse products. For each nucleophile studied, we can make primary, secondary, and tertiary alcohols, depending on the carbonyl used.

We now take a closer look at the following carbon-based nucleophiles: Grignard reagents, organolithium reagents, acetylide anions, and cyanide ions.

17.5.1 Grignard reagents

The most logical way to make a carbanion is to use an acid–base reaction (**Figure 17.26**). However, alkanes, like methane, are simply too weakly acidic to be deprotonated. Looking back at the pK_a values in Table 4.5, there is no base strong enough to deprotonate alkanes. Recall from Section 16.2, however, that carbanions can be formed in the reaction of organohalides with magnesium to make Grignard reagents via the reduction of the C—X bond.

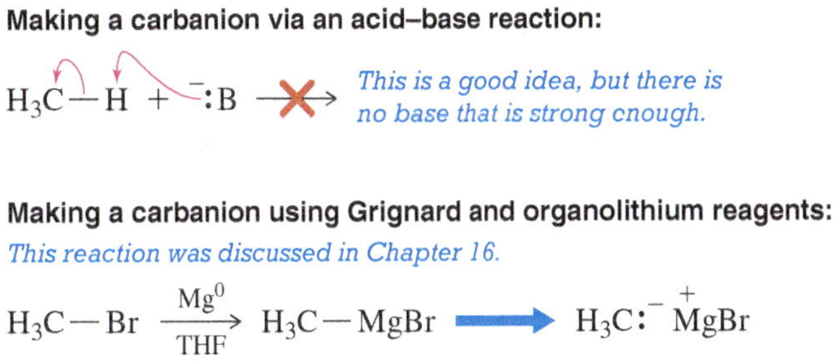

Figure 17.26 **Grignard reagents are formed by the reduction of C—X bonds.**

Grignard reagents are strong nucleophiles that rapidly add to carbonyls, attacking directly via General Mechanism 17A (**Figure 17.27**). [*We can show the mechanism using the zwitterion form, too.*] After addition, the magnesium bromide cation associates with the alkoxide. Protonation of the alkoxide forms the alcohol.

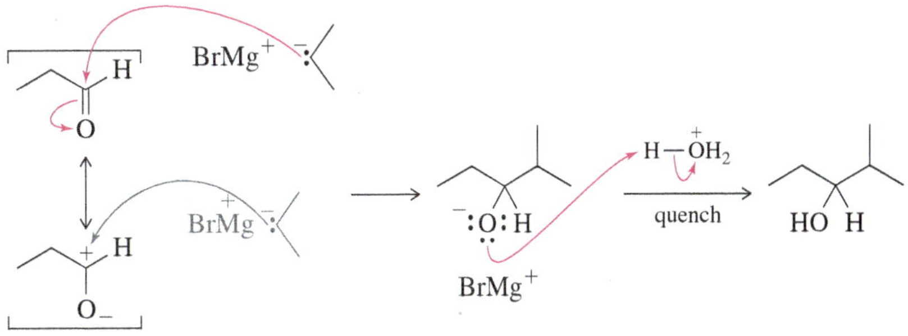

Figure 17.27 **Grignard reagents are strong nucleophiles, so they react by General Mechanism 17A.**

The diversity of structures available by the reaction of Grignard reagents and aldehydes/ketones (**Table 17.7**) makes this an important tool in the synthetic chemist's toolbox. [*Recall our reaction sheets from Section 10.7.*]

Table 17.7 **The reaction between Grignard reagents and aldehydes/ketones** provides access to a variety of different alcohols while forming new carbon–carbon bonds.

Product	Reaction	Description
1° alcohols	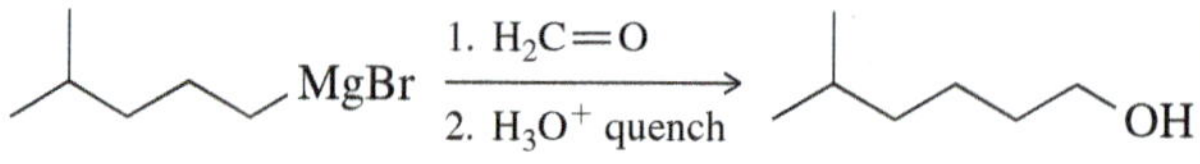	When the Grignard reagent adds to formaldehyde, a primary alcohol is produced.
2° alcohols	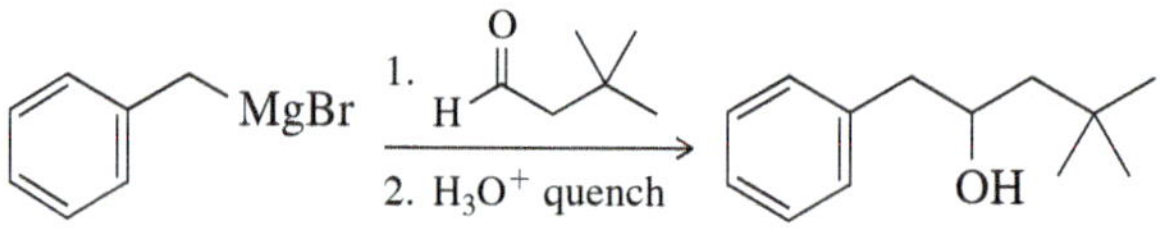	Reaction between an aldehyde and a Grignard reagent gives a secondary alcohol.
3° alcohols	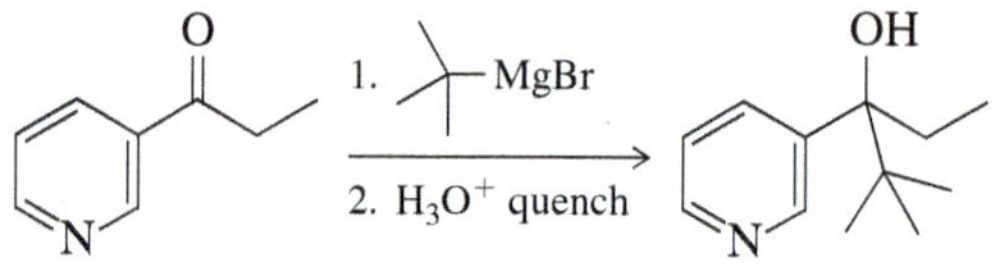	Although ketones react more slowly than aldehydes, Grignard reagents still add quickly to form tertiary alcohols.

Assessment 17.16 Predict the product of the following Grignard addition reactions.

(a)

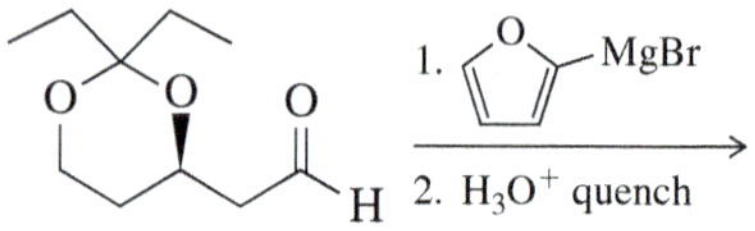

(b)

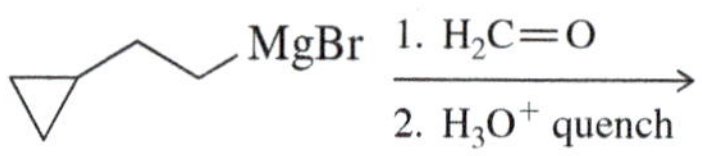

(c)

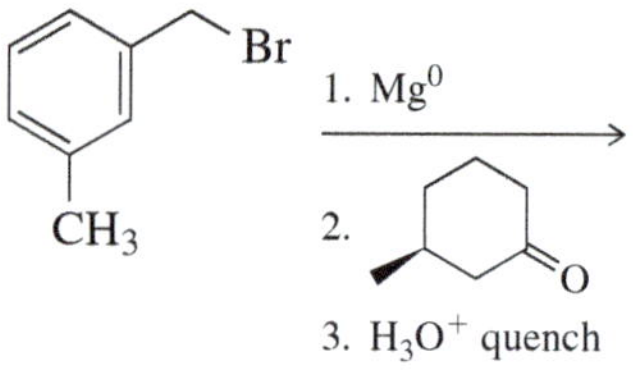

Assessment 17.17 A student, when solving the following "predict-the-product" question, made a common mistake by writing the answer shown here. Explain why this reaction would not work as written.

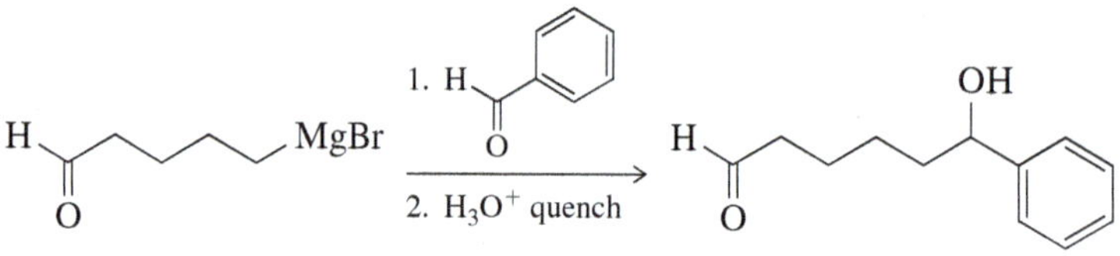

17.5.2 Organolithium reagents

Organolithium reagents, also from Section 16.2, react similarly to Grignard reagents, meaning that identical products are produced by the same mechanism regardless of which reagent is chosen (Table 17.8). [*The choice of which reagent to use is based on the ease of obtaining*

Table 17.8 **The reaction between organolithium reagents and aldehydes/ketones** provides access to a variety of different alcohols while forming new carbon–carbon bonds.

Product	Reaction	Description
1° alcohols	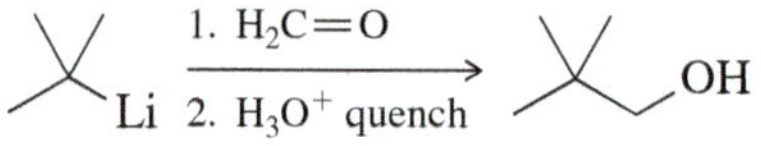	Organolithium reagents react with formaldehyde to make 1° alcohols.
2° alcohols	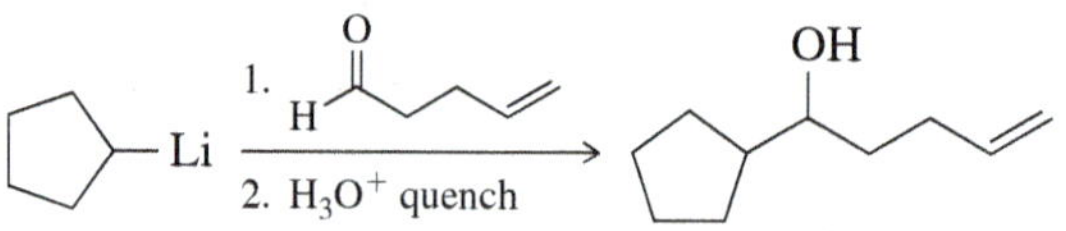	Aldehydes and organolithium reagents react to produce 2° alcohols.
3° alcohols	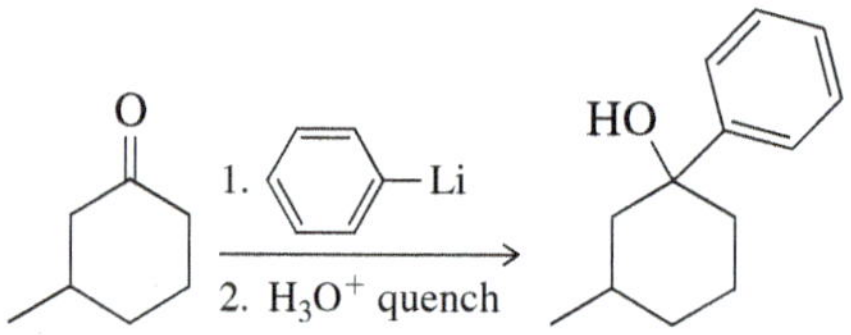	A ketone reacts more slowly than an aldehyde, but organolithium reagents still add quickly to make 3° alcohols.

the organometallic reagent.] As a strong nucleophile, the reaction follows General Mechanism 17A, and as shown in Figure 17.28, nucleophilic addition to the carbonyl is followed by an acid quench.

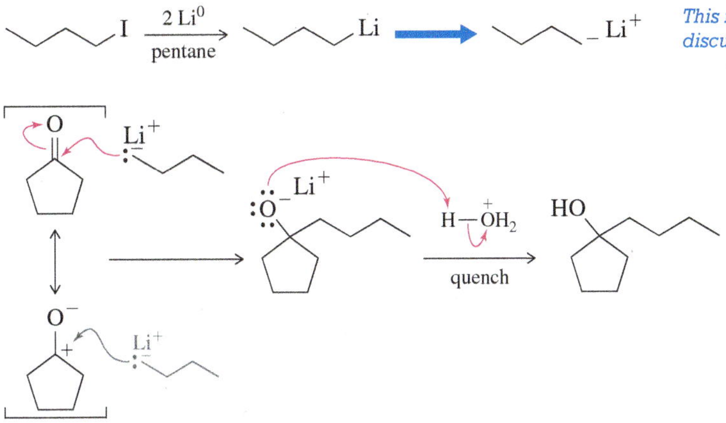

This reaction was discussed in Chapter 16.

Figure 17.28 Organolithium reagents are strong nucleophiles, so they react by General Mechanism 17A.

Assessment 17.18 Suggest an organolithium reagent to allow a complete synthesis of the molecule shown.

(a)

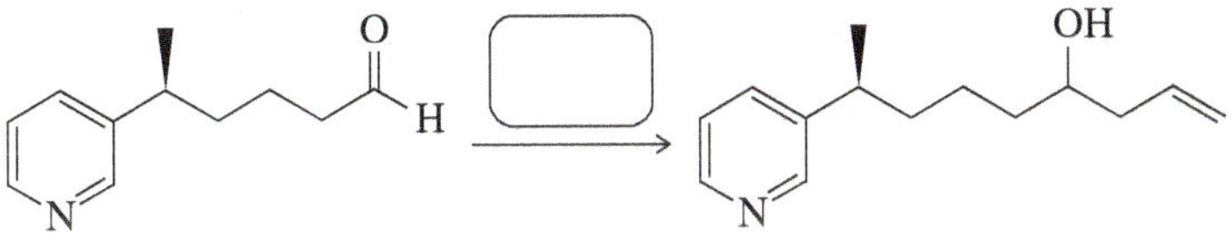

(b)

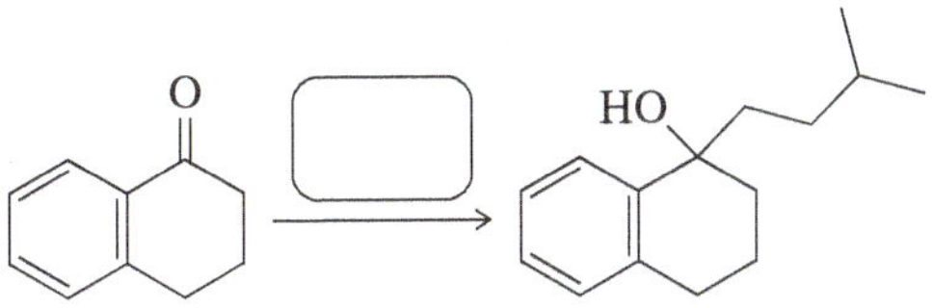

(c)

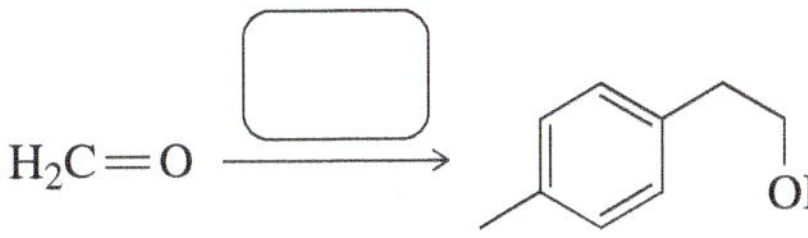

Assessment 17.19 Suggest a carbonyl and an organolithium reagent that might be used to synthesize each alcohol shown. [*There are multiple correct answers.*]

(a)

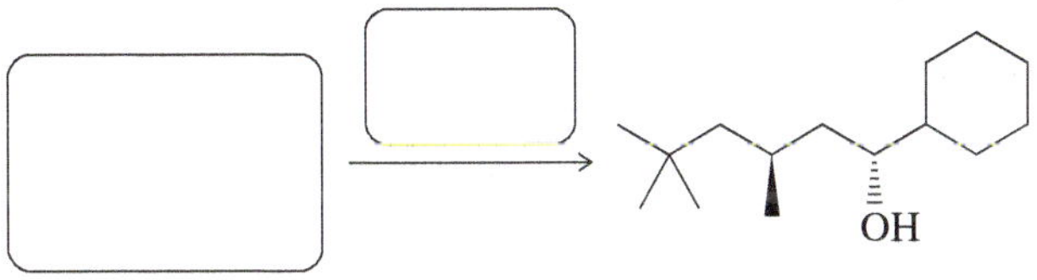

(b)

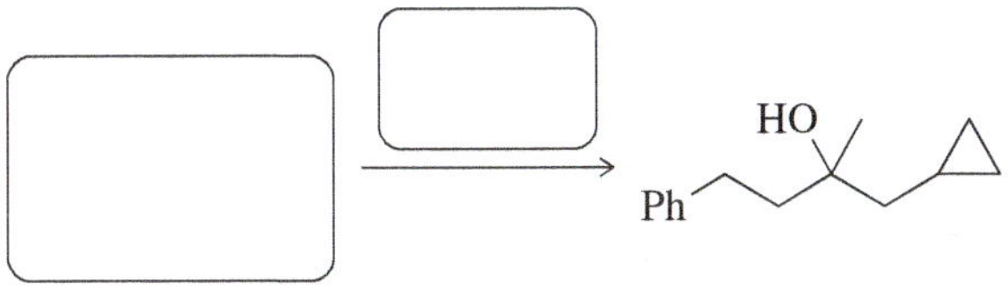

(c)

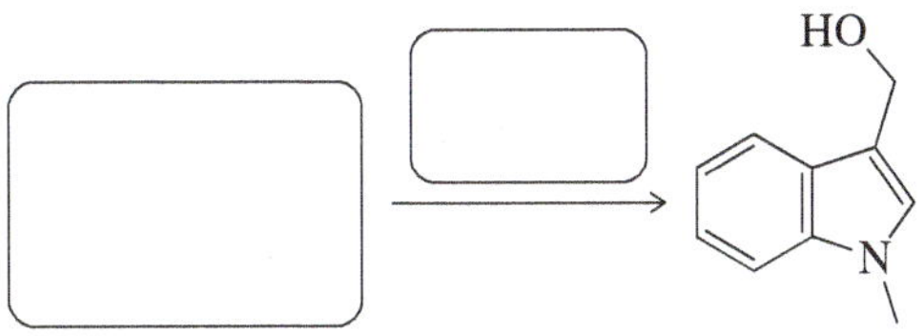

17.5.3 Acetylide anions

Recall that acetylide ions used in S_N2 reactions in Section 10.5 made it possible to form carbon–carbon bonds. Generated by deprotonation of the corresponding alkyne [Figure 17.29(a)], acetylide ions are not as reactive as Grignard reagents. Yet, based on the relative pK_a values of the conjugate acids of the bases involved [*an analysis we have done many times before*], the addition of acetylide ions is a favorable reaction [Figure 17.29(b)].

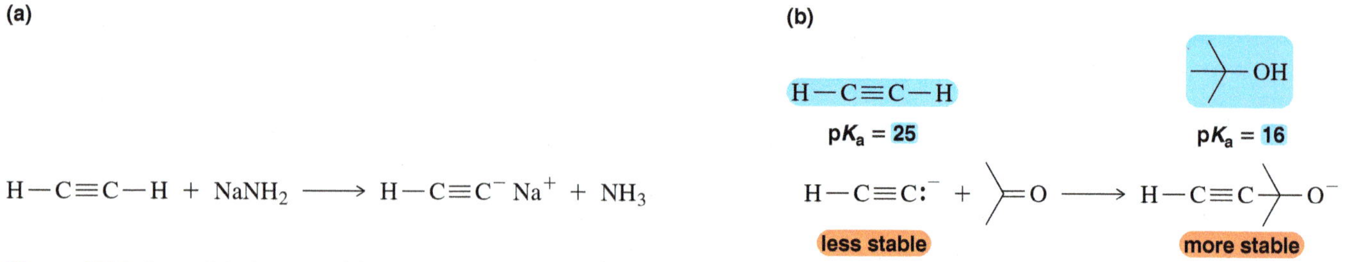

Figure 17.29 Acetylide ions are (a) generated by deprotonation of terminal alkynes and are (b) reactive enough to attack carbonyls.

As a strong nucleophile, the acetylide ion directly adds by General Mechanism 17A (Figure 17.30). Because these reactions are just slight variations on the same theme, the mechanisms and synthesis products shown in Table 17.9 should look familiar by now. [*If not, go back and confirm that we are simply showing the same thing over and over again.*]

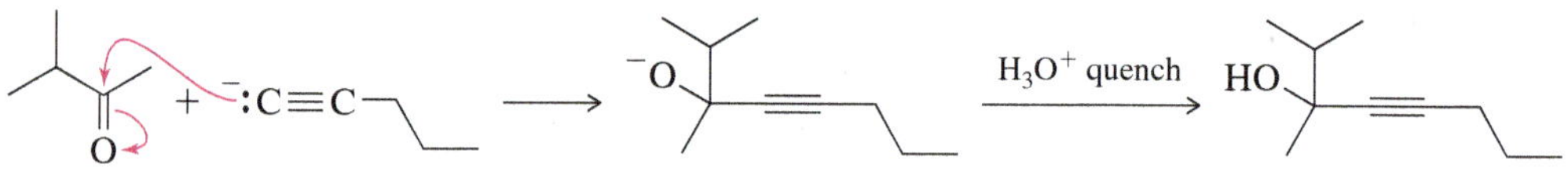

Figure 17.30 Acetylide ions are strong nucleophiles, so they react by General Mechanism 17A.

Table 17.9 The reaction between acetylide ions and aldehydes/ketones provides access to a variety of different alcohols while forming new carbon–carbon bonds.

Product	Reaction	Description
1° alcohols	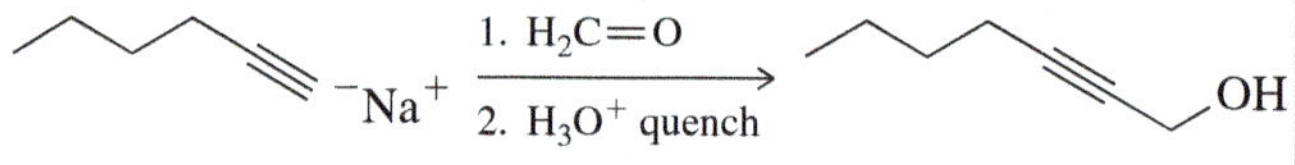	The acetylide anion adds to formaldehyde to produce a 1° alcohol.
2° alcohols	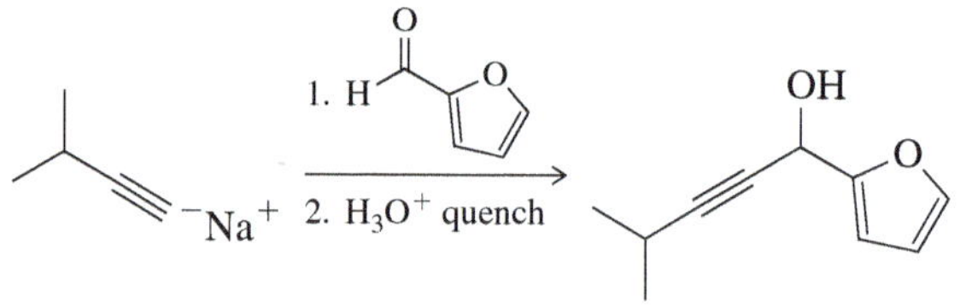	The reaction between an aldehyde and an acetylide ion gives a 2° alcohol.
3° alcohols	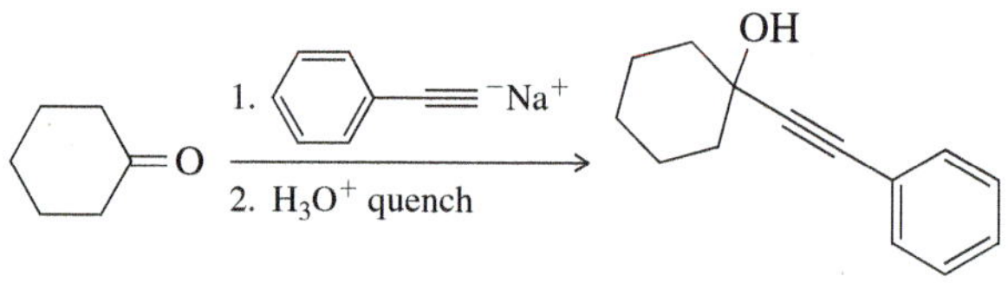	The reaction of a ketone with an acetylide ion generates a 3° alcohol.

Assessment 17.20 Suggest an acetylide ion and a carbonyl that might be used to make the following products.

(a) oct-4-yn-3-ol
(b) 2,6-dimethylhept-3-yn-2-ol
(c) 5-phenylhex-2-yn-1-ol

17.5.4 Cyanide ions

Table 17.10 shows, in terms of stability (based on conjugate acid pK_a values), that cyanide is the weakest nucleophile of those studied thus far. What are the consequences of using a weaker nucleophile? Is an addition reaction still possible?

Table 17.10 pK_a values for the conjugate acids of several nucleophiles. Cyanide is a weaker nucleophile than the carbon nucleophiles studied previously.

Reagent	Nucleophile	Conjugate acid	pK_a
$H_3C-MgBr$	H_3C^-	CH_4	~50
H_3C-Li	H_3C^-	CH_4	~50
$Na^+ \; ^-C\equiv CH$	$^-C\equiv CH$	$H-C\equiv CH$	25
NaCN	^-CN	HCN	10

Imagine a reaction similar to the addition of Grignard reagents; that is, imagine that the nucleophile is added first and the product is then quenched with acid (**Figure 17.31**). Using pK_a values of the conjugate acids shows us that this reaction starts with a stable cyanide anion and, after addition, generates a less stable alkoxide. Thus, this reaction is disfavored as performed.

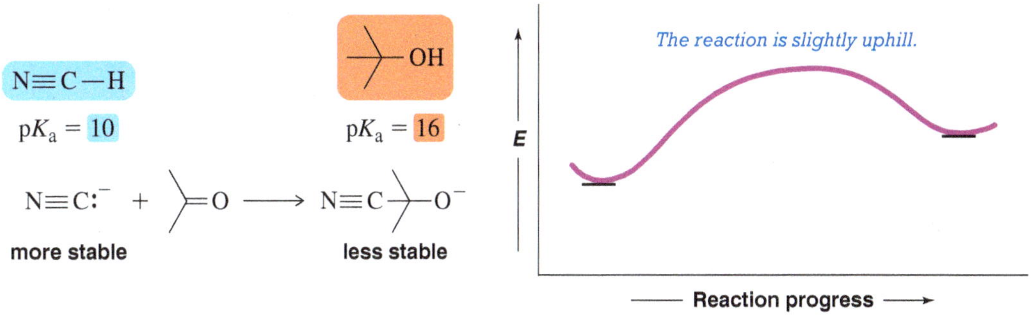

Figure 17.31 The addition of cyanide is uphill/disfavored.

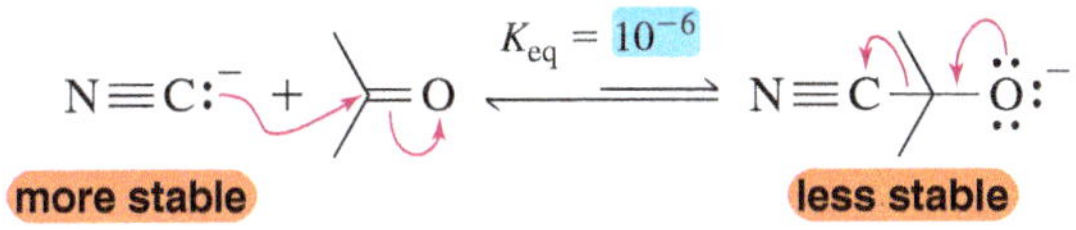

Figure 17.32 The reverse reaction/elimination of cyanide is favored.

The small 6 pK_a unit difference means that addition (while unfavorable) reversibly occurs to produce some alkoxide. In the reverse direction, the lone pair of electrons on the alkoxide collapses down to kick out the more stable cyanide ion (**Figure 17.32**). [*This type of reaction is studied in Chapter 18.*]

Can the equilibrium of this slightly unfavorable reversible reaction be shifted to the product side? According to Le Châtelier's principle, the equilibrium can be shifted toward the product side by either adding more reactants or removing the products. Thus, the addition of cyanide is accomplished using a **substoichiometric** amount of cyanide and a full equivalent of hydrogen cyanide. In this way a second equilibrium reaction occurs in which the alkoxide rapidly deprotonates the hydrogen cyanide (**Figure 17.33**). Removal of the alkoxide from the equilibrium shifts the first reaction to the right, making the overall reaction favorable for forming a **cyanohydrin**.

Figure 17.33 Cyanide reacts by a modified form of General Mechanism 17A. As a weaker, but still negatively charged nucleophile, the proton needs to be in solution already to help shift the equilibrium to the product side.

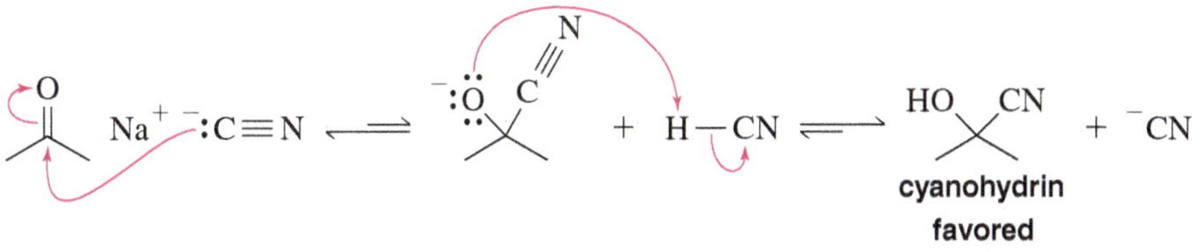

Table 17.11 again reminds us of the variety of cyanohydrins that can be prepared by this reaction. This reaction illustrates, for the first time, the impact of a weaker nucleophile. Although still like General Mechanism 17A, the reaction conditions are changed, shifting the equilibrium toward the desired product. [*This rationale explains why weaker nucleophiles like alcohols react via General Mechanism 17B—stay tuned!*]

Table 17.11 The reaction between NaCN/HCN and aldehydes/ketones provides access to a variety of different cyanohydrins while forming new carbon–carbon bonds.

Product	Reaction	Description
1° cyanohydrin	 glycolonitrile	The industrially important intermediate glycolonitrile (a precursor to EDTA) forms when formaldehyde is reacted with NaCN/HCN.
2° cyanohydrin	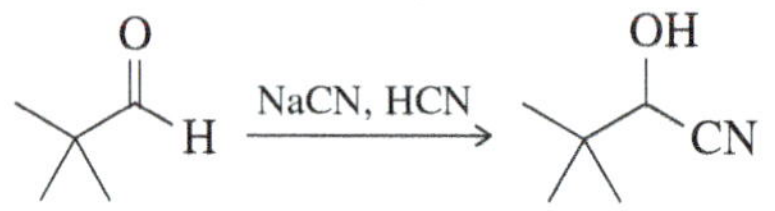	Addition of cyanide to aldehydes results in the formation of a 2° cyanohydrin.
3° cyanohydrin	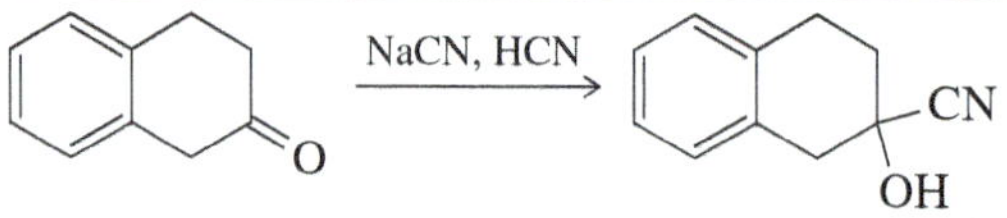	The reaction of a ketone with NaCN/HCN generates a 3° cyanohydrin.

Assessment 17.21 Predict the product of the following cyanide addition reactions.

(a)

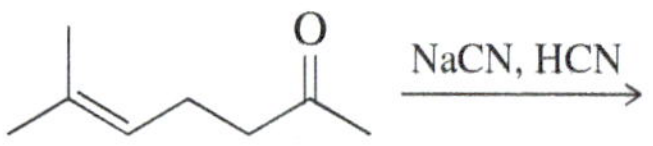

(b)

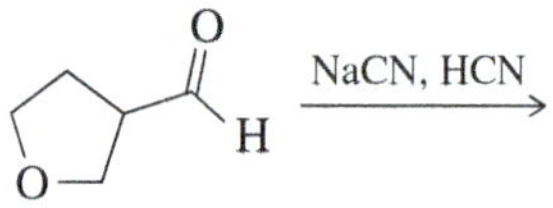

Assessment 17.22 Suggest a carbonyl to react with NaCN/HCN to produce the following cyanohydrins.

(a)

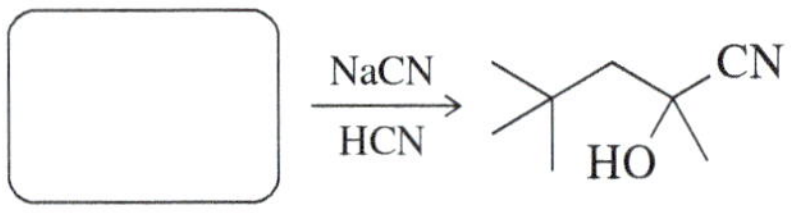

(b)

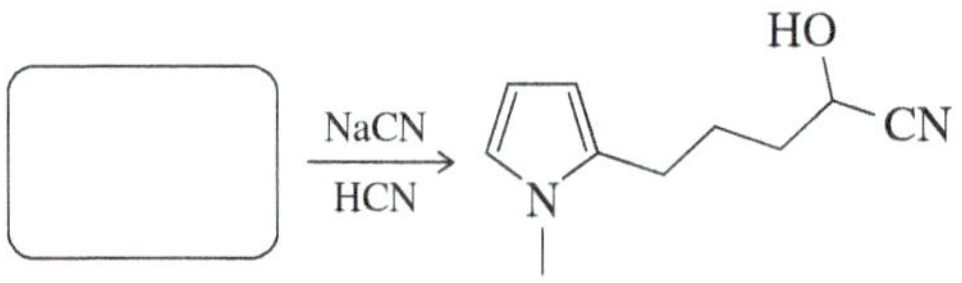

(c)

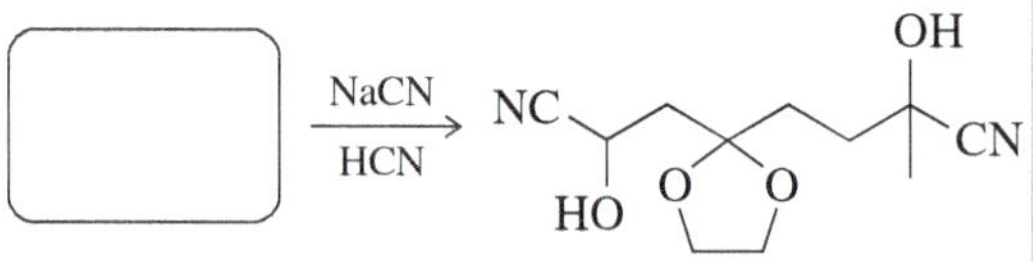

Assessment 17.23 For each of the following carbonyl addition reactions, would you expect a racemic mixture or a mixture enriched in one stereoisomer? In each case, draw both possible stereoisomeric products.

(a)

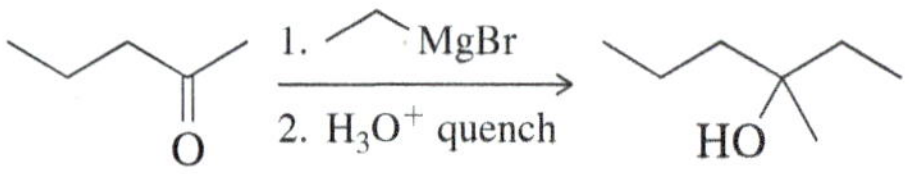

(b)

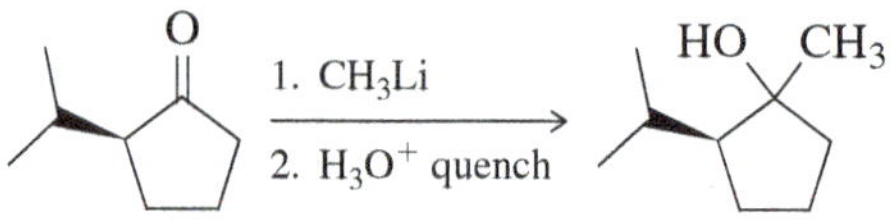

You might have anticipated the stereochemical outcome of these reactions, but let's find out by working Assessment 17.23 together.

17.5.5 The stereochemistry of addition reactions

As was the case when we studied the stereochemistry of S_N2 reactions, the location of the antibonding orbital will feature prominently in dictating the stereochemical outcome of carbonyl addition reactions.

Solved Assessment 17.23

When you are asked to predict the stereochemical outcome of an addition reaction, you need to examine the shape of the reacting functional group and determine whether the transition state favors attack from one side over the other. We know that these reactions proceed by General Mechanism 17A, so they feature direct addition to the carbonyl carbon. This carbon is sp^2 hybridized and is therefore planar. When the nucleophile attacks the carbonyl, the electrons are adding to π^*, resulting in the following trajectory (Bürgi–Dunitz angle) (Figure 17.34).

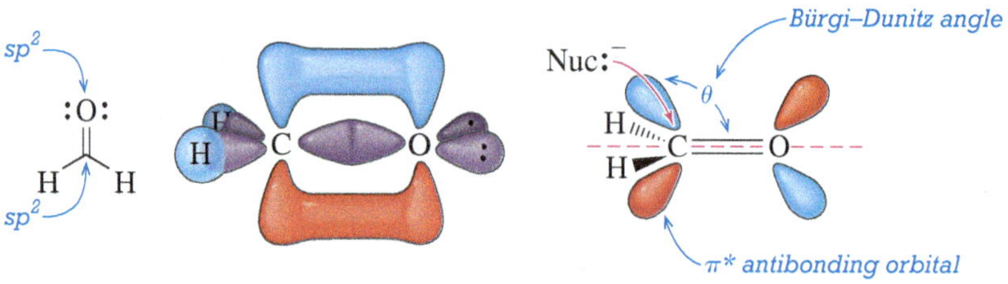

Figure 17.34 Nucleophiles add to π* along the Bürgi–Dunitz angle.

In Assessment 17.23(a), attack from the top gives the (*S*) alcohol, whereas attack from the bottom gives the (*R*) alcohol. As shown in **Figure 17.35**, close examination reveals that these are enantiomeric transition states. Enantiomeric transition states are equal in stability (Section 6.6), resulting in equal activation energies, equal rates of reaction, and the production of a racemic mixture containing (*R*)- and (*S*)-3-methylhexan-3-ol.

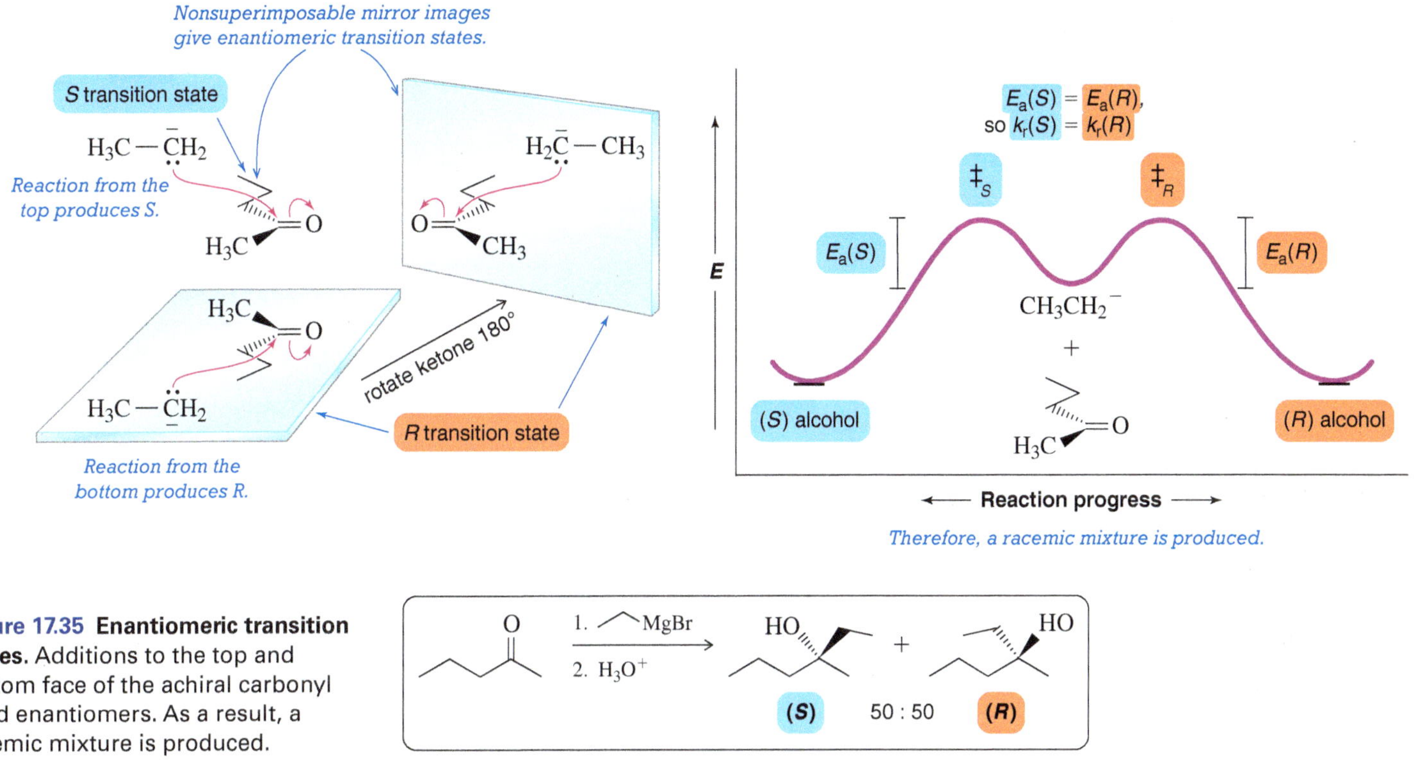

Figure 17.35 Enantiomeric transition states. Additions to the top and bottom face of the achiral carbonyl yield enantiomers. As a result, a racemic mixture is produced.

In Assessment 17.23(b) (**Figure 17.36**), addition from the top of the planar ketone gives the (1*R*,2*S*) alcohol. Attack from the opposite side is represented as the (1*S*,2*S*) transition state, yielding the (1*S*,2*S*) alcohol. [*Convince yourself of this.*] These are diastereomeric transition states, so they have unequal energies and occur with unequal rates. As a result, a mixture enriched in one diastereomer is produced. ■

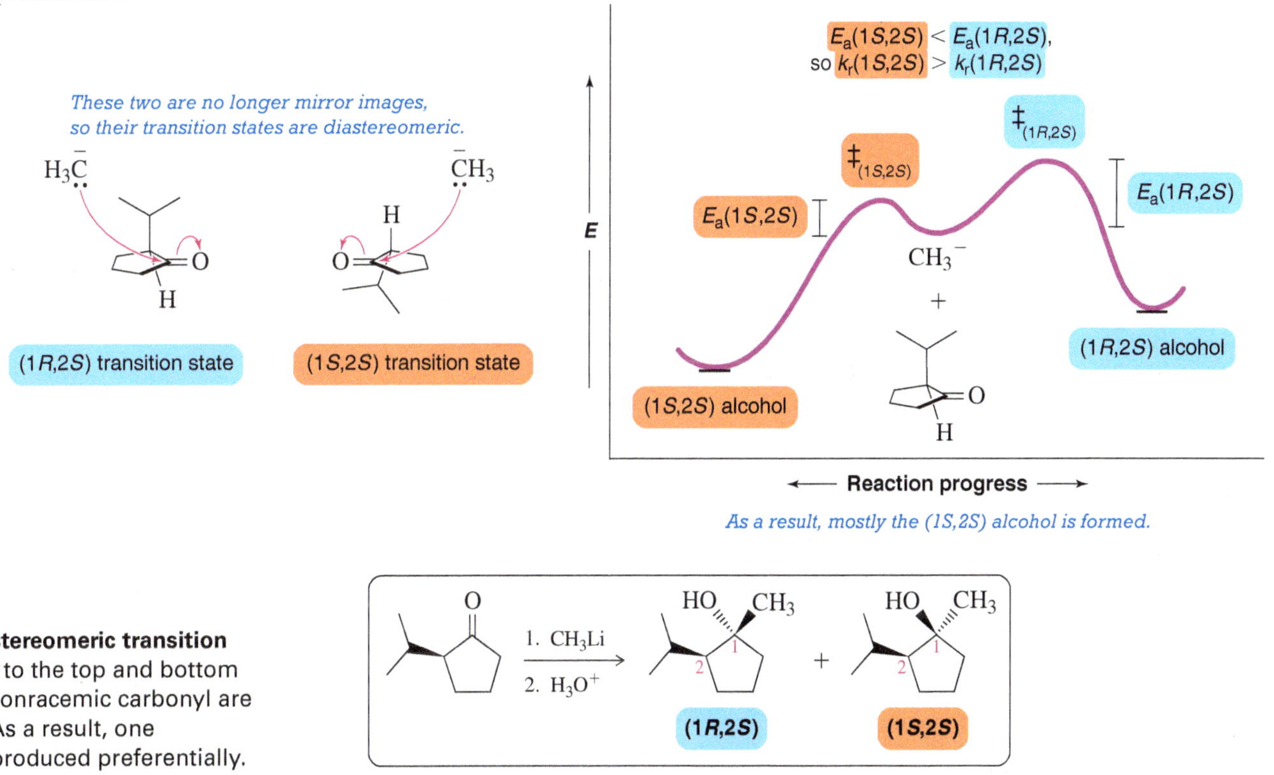

Figure 17.36 Diastereomeric transition states. Additions to the top and bottom face of a chiral, nonracemic carbonyl are diastereomeric. As a result, one diastereomer is produced preferentially.

17.5.5 SUMMARY

- Addition to aldehydes and ketones occurs from either side at equal rates only when the two possible transition states are enantiomers.

Assessment 17.24 For the following carbonyl addition reactions, would you expect a racemic mixture or a mixture enriched in one stereoisomer? In each, draw both possible stereoisomeric products.

(a)

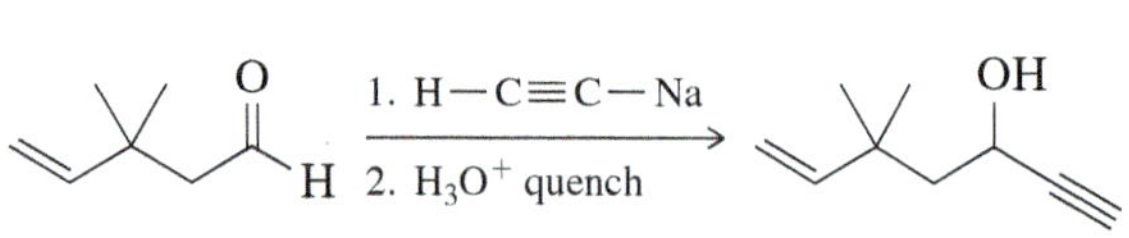

(b)

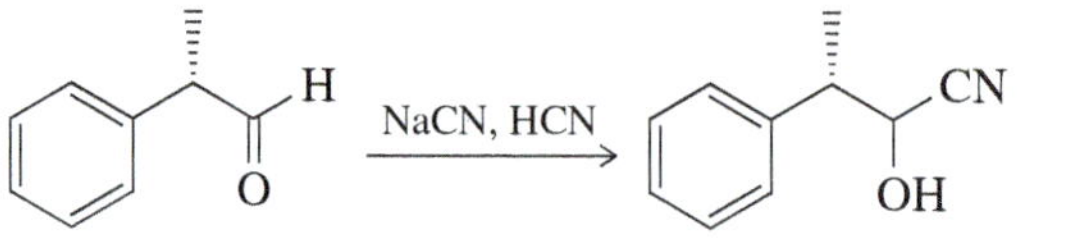

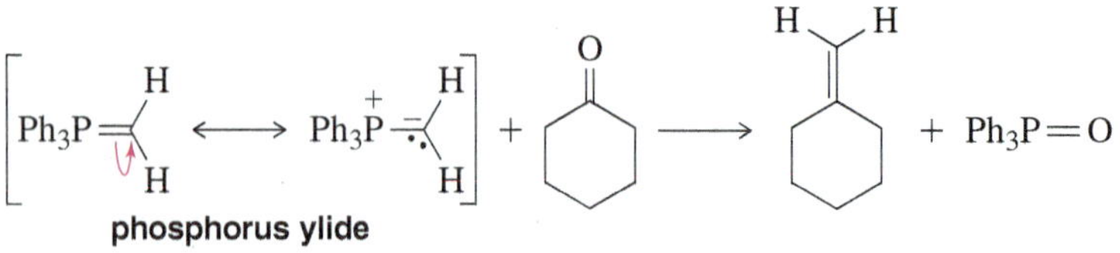

Figure 17.37 **The overall Wittig reaction.**

17.6 The Wittig reaction

The Wittig reaction is a versatile synthesis tool that has a different [*and somewhat complicated*] mechanism compared to what we've studied so far (**Figure 17.37**). The reaction is essentially the double displacement reaction between a phosphorus **ylide**, a molecule containing opposite charges on adjacent atoms, and an aldehyde or ketone to produce an alkene. The reaction forms two new carbon–carbon bonds: 1 σ bond and 1 π bond.

Instead of memorizing the new mechanism, **Table 17.12** compares it, step by step, to reactions we already understand well. [*By approaching it this way, with each step making sense, its mechanism will not seem so complicated.*]

Table 17.12 **A step-by-step mechanism of the Wittig reaction.**

Step	Reaction
1. **Strong nucleophiles add directly to the electrophilic carbon of a carbonyl (Section 17.3).** Although often represented as a neutral molecule containing a C=P double bond, the phosphorus ylide is better represented as a zwitterion. Note the negatively charged carbon and the adjacent positively charged phosphorus of the ylide. The ylide is a strong nucleophile, using a lone pair to attack the carbonyl carbon, breaking the C—O π bond via General Mechanism 17A.	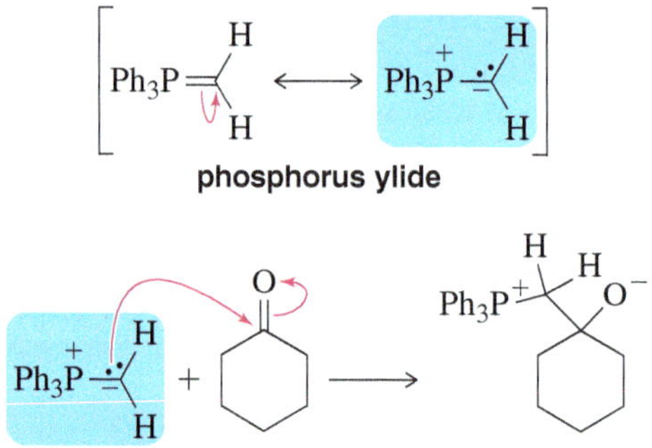
2. **Formation of the strong P—O single bond is the driving force for many reactions involving phosphorus (Section 13.7.4).** The resulting alkoxide attacks the positively charged phosphorus to form a four-membered ring containing a P—O bond, an intermediate known as an **oxaphosphetane.** Phosphorus, being a third-row element, possesses empty *d* orbitals that can accommodate the electrons donated by the oxygen. The favorable formation of the relatively unstable four-membered ring shows just how strong the new P—O bond is. Being intramolecular, this step is much faster than the preceding nucleophilic addition.	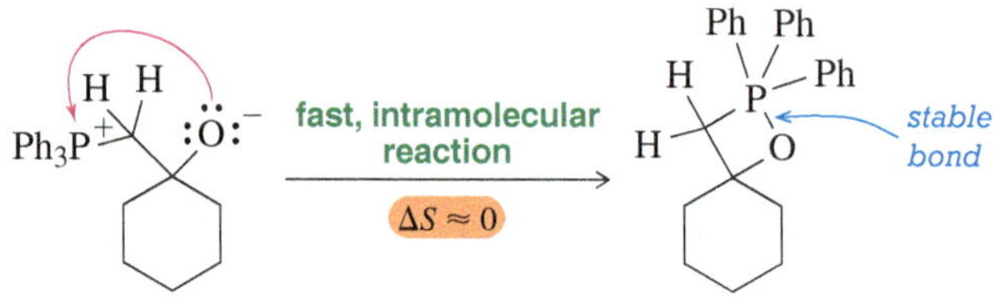
3. **Four-membered rings are unstable (Section 3.6).** The unstable ring decomposes in an entropically favorable step to give two new molecules. The mechanism here is similar to one we observed in ring-closing metathesis (Section 16.6). 4. **The only bond stronger than the P—O single bond is the P=O double bond (Section 13.7.4).** Replacing two σ bonds with two π bonds isn't typically favored, but this is a favored process when one is a P—O π bond. In sum, favorable enthalpy, relief of ring strain, and an increase in entropy drive this final step.	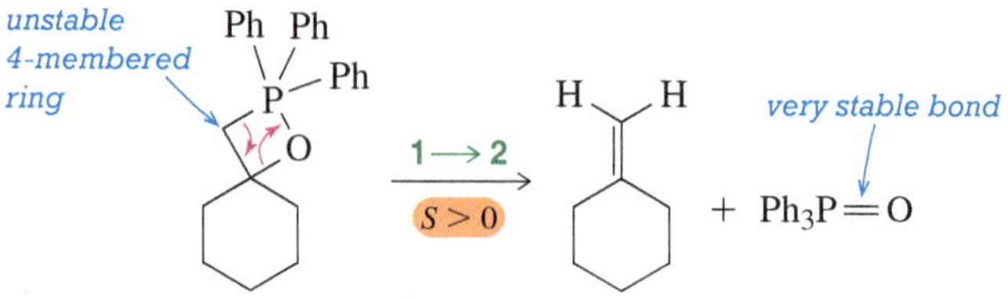

As a double replacement reaction [*a term from general chemistry*] (**Figure 17.38**), C_1 of the carbonyl exchanges its oxygen for the phosphorus-bound alkyl group, forming a new C=C double bond in the process. You might have noticed that the formation of the new C=C double bond during the Wittig reaction is visually reminiscent of the reverse of ozonolysis (Section 9.1.7). [*Cues like this can be helpful when predicting Wittig products.*]

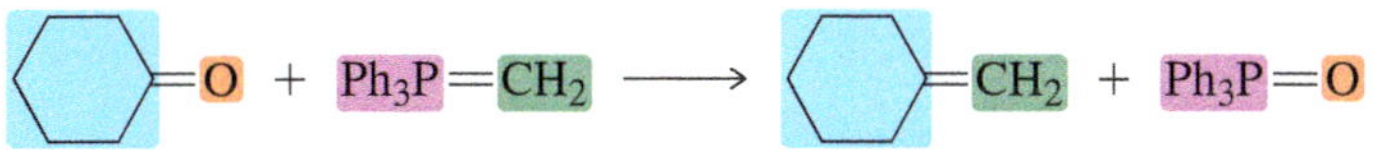

Figure 17.38 The overall Wittig reaction is essentially a double replacement reaction.

Assessment 17.25 Predict the product of the following Wittig reactions.

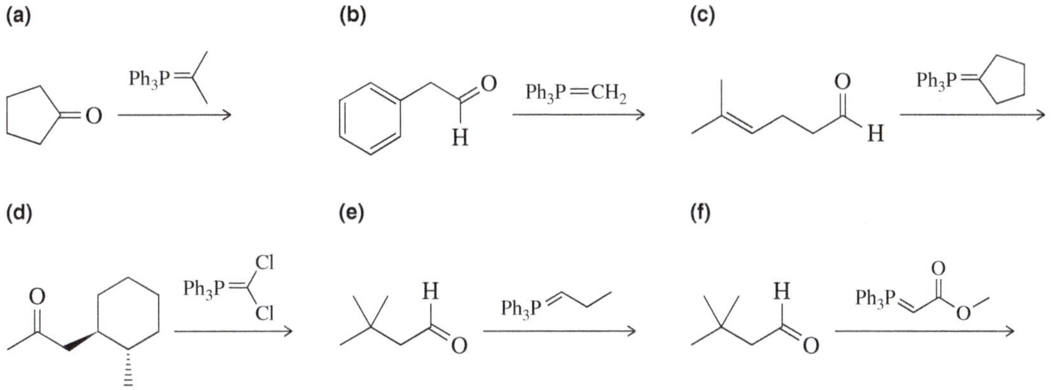

17.6.1 The Wittig reaction and stereoselectivity

When reacting an aldehyde (or a nonsymmetrical ketone) and an ylide with different substituents at C_1, two diastereomeric alkene products are possible. And yet, in these cases, there is usually a single major product. For example, consider the reactions in Assessment 17.25(e) and (f) (**Figure 17.39**). The Wittig reaction in (e) gives almost exclusively the less stable cis or *Z* alkene, whereas the Wittig reaction in (f) gives the more stable trans or *E* alkene as the exclusive product. These different stereochemical outcomes, both involving the same aldehyde, must be due to differences in the ylides.

Assessment 17.25(e):

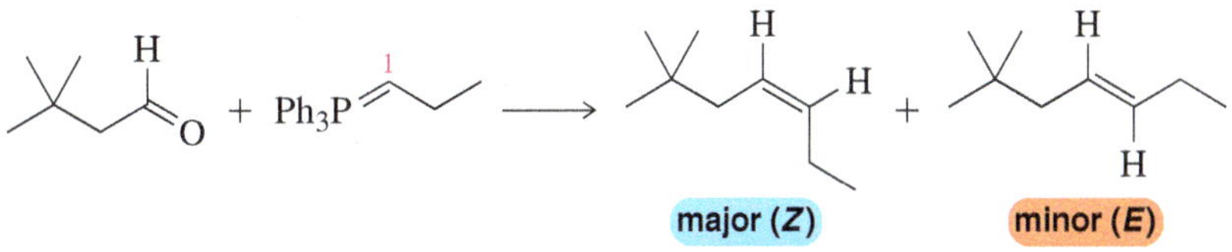

Assessment 17.25(f):

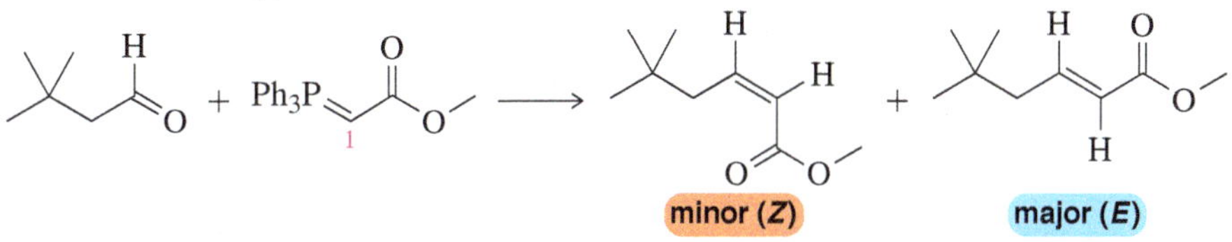

Figure 17.39 Stereoselectivity in the Wittig reaction.

Differences in the stability of the ylides in (e) and (f) determine the stereochemical outcome of these reactions. In its zwitterion form, the negative charge on the ylide in (f) is stabilized by resonance (**Figure 17.40**). These **stabilized ylides** react more slowly to give *E* alkenes as products. Ylides not stabilized by resonance, **nonstabilized ylides**, are more powerful nucleophiles. These ylides react quickly in (e) to give the less stable *Z* alkenes. The reason for this difference is not well understood and is a subject of current organic chemistry research. At the very minimum, though, we have a general rule that helps us predict Wittig outcomes: *Stable* ylides produce *stable* alkenes, whereas *unstable* (nonstabilized) ylides produce *unstable* (less stable) alkenes.

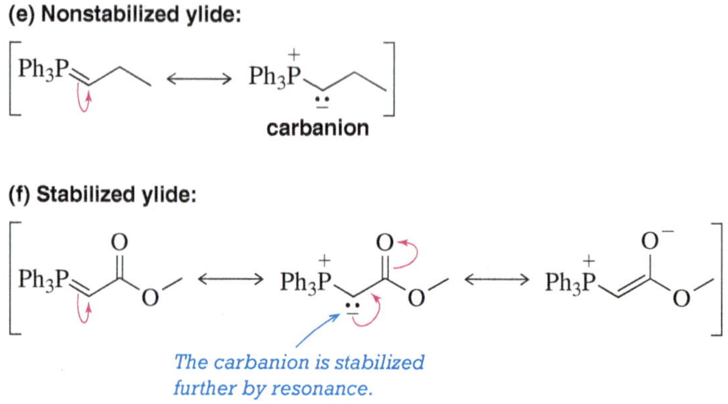

Figure 17.40 A stabilized ylide is one whose anion can be stabilized by resonance.

Assessment 17.26 Suggest the appropriate carbonyl and Wittig reagent to make the following alkenes.

(a) (*E*)-7-methylnon-4-en-3-one
(b) (*Z*)-2-phenylhex-3-ene
(c) *cis*-hept-2-ene

Assessment 17.27 Stabilized ylides, such as the one shown here, fail to react with ketones. Suggest why this might be true.

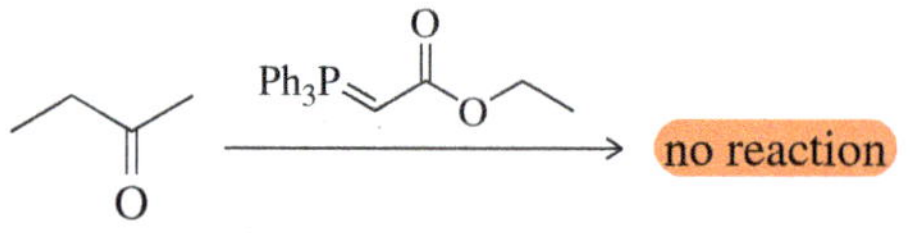

Assessment 17.28 Predict the product of the following Wittig reactions, being sure to specify the stereochemical outcomes.

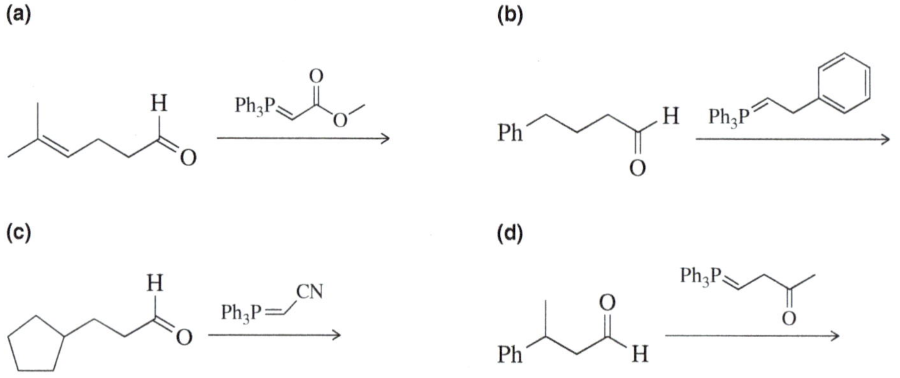

17-B CHEMISTRY IN CONTEXT: Green Chemistry

Wittig variants

The Wittig reaction, which won Georg Wittig the 1979 Nobel Prize in Chemistry, is a very important tool in the synthetic chemist's toolbox. Enabling the stereoselective synthesis of a variety of alkenes, this reaction has featured prominently in a number of important syntheses.

One drawback of the reaction is its poor atom economy; it generates a substantial amount of triphenylphosphine oxide waste. As shown in **Figure 17.41**, 440.48 atomic mass units of starting materials are required to make 162.19 atomic mass units of the desired product. The remaining 278.29 atomic mass units are simply waste.

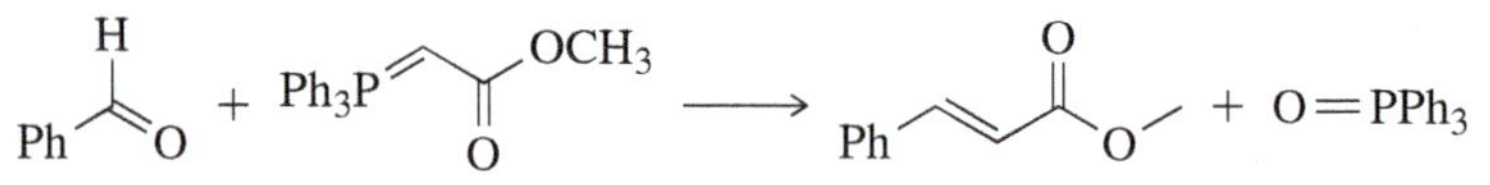

Chemical formula:	C_7H_6O	$C_{21}H_{19}O_2P$	$C_{10}H_{10}O_2$	$C_{18}H_{15}OP$
Molecular weight:	106.12	334.35	162.19	278.29

440.48

(162.19/440.48)(100%) = 37%

Figure 17.41 The Wittig reaction has a poor atom economy.

(continued)

If the reaction makes 907 kg (1 ton) of the desired product (standard in the pharmaceutical industry), then 1556 kg of triphenylphosphine oxide waste would be produced. Thus, Wittig variants that reduce the environmental impact of the reaction are sorely needed.

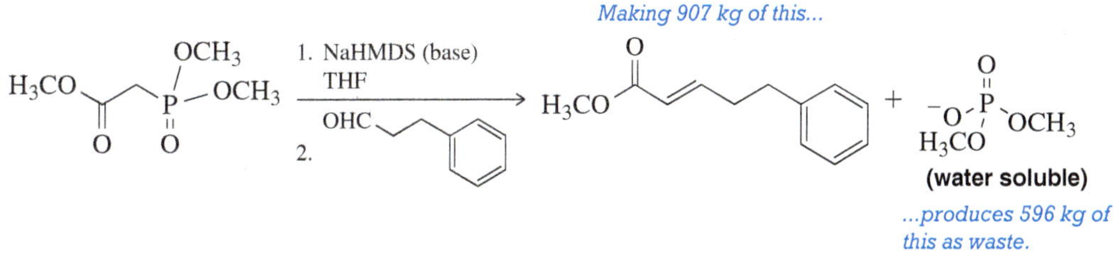

Figure 17.42 The Wittig–Horner variant has a slightly better atom economy.

The Wittig–Horner variant (**Figure 17.42**) utilizes a phosphonate in place of the triphenyl phosphonium. A deprotonated phosphonate reacts with the aldehyde by a mechanism similar to the Wittig reaction, producing the more stable *E* alkene in the process. The water-soluble phosphonate byproduct (smaller atomic mass than PPh_3) is removed by washing with water.

Another unique approach uses triphenylphosphine in substoichiometric quantities, regenerating it as the reaction proceeds. Diphenylsilane reduces the modified phosphine oxide to the phosphine used in the Wittig reaction, according to the catalytic cycle in **Figure 17.43**. The waste associated with the use of stoichiometric silane creates a separate problem, but it is at least a step in the green direction.

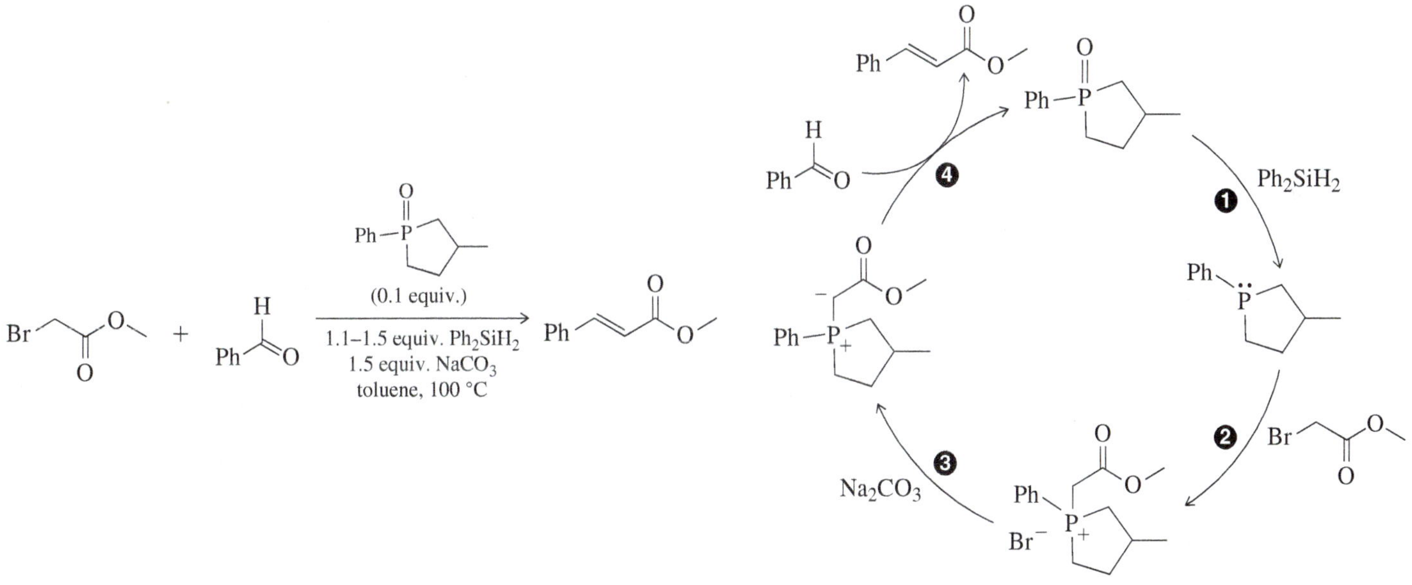

Figure 17.43 A greener Wittig reaction and its catalytic cycle.

17.7 Addition of oxygen nucleophiles to aldehydes and ketones

The addition of oxygen nucleophiles has been discussed previously (Section 13.9) in the context of alcohol oxidation. Recall that when using chromic acid, oxidation of 1° alcohols went all the way to the carboxylic acid. Often attributed to the strength of chromic acid as an oxidizing agent, it is actually because a hydrate forms when water is present in the reaction mixture (**Figure 17.44**).

Section 17.7 explores the addition of weak, oxygen-based nucleophiles to aldehydes and ketones, with an emphasis on the mechanism and the driving force for these thermodynamically controlled reactions. [*For example, the hydrate formation in Figure 17.44.*]

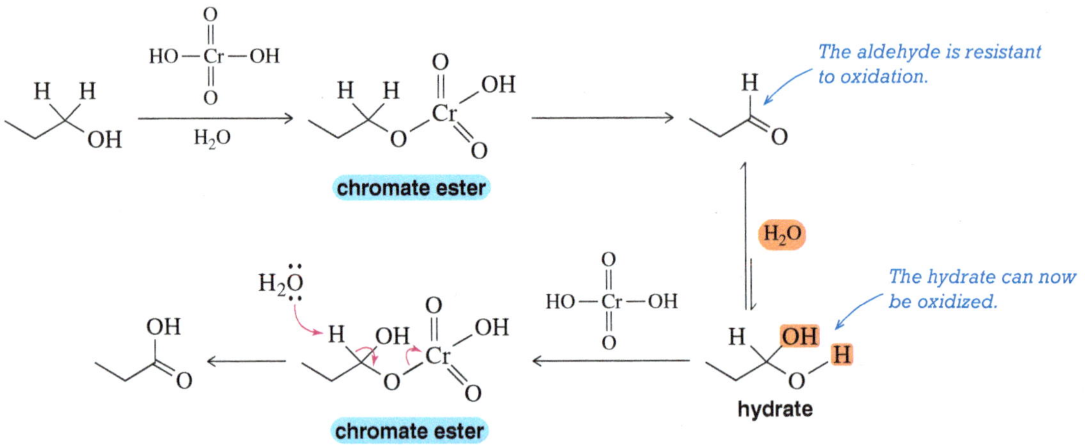

Figure 17.44 **Only the hydrate form of an aldehyde can be oxidized using chromic acid.** [*Recall Section 13.8.*]

17.7.1 Addition of water: hydrate formation

As made clear in Figure 17.44, hydrate formation is simply the addition of one molecule of water to an aldehyde or ketone. Thus, this reaction can be examined in a manner similar to our examination of previous carbonyl additions (**Figure 17.45**). As always, K_{eq}, as a function of ΔG, is estimated by considering ΔS and ΔH. As in all addition reactions, the production of one molecule from two is entropically disfavored ($\Delta S < 0$). To estimate ΔH, the bonds broken and formed are identified, approximating the energy of each. In a qualitative sense, assume that the O—H σ bond strengths are similar regardless of molecule type and that the C—O π bond is slightly weaker than the C—O σ bond. Using these assumptions, this reaction is slightly favored to the right ($\Delta H < 0$). Experimentally, an equilibrium mixture of aldehyde and water consists of 99% aldehyde and 1% hydrate. This confirms that the entropy term is more important in favoring the left side of the reaction. [*Reread these last two sentences. The hydrate is disfavored, yet hydrate formation allows further oxidation. How can this be?*]

Figure 17.45 **Addition of water to the aldehyde is *not* favored enough by ΔH to make it an overall favored reaction.**

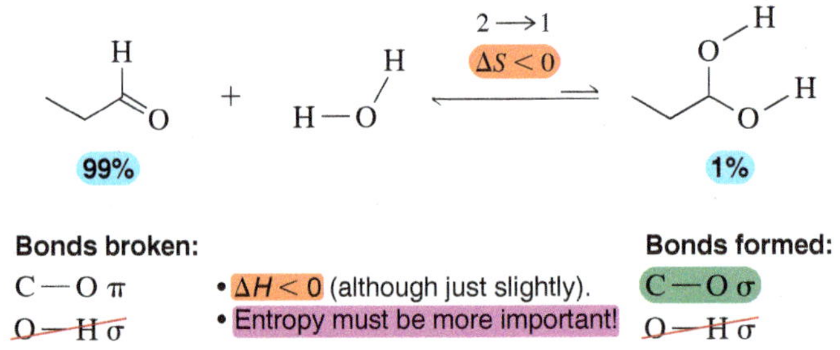

This is an example of Le Châtelier's principle (**Figure 17.46**). As long as some hydrate molecules are present, formation of the chromate ester removes hydrate molecules from the equilibrium. This shifts hydrate formation to the right until all of the aldehyde has been oxidized.

Figure 17.46 **Le Châtelier's principle allows the unfavorable addition of water to proceed in oxidation reactions.**

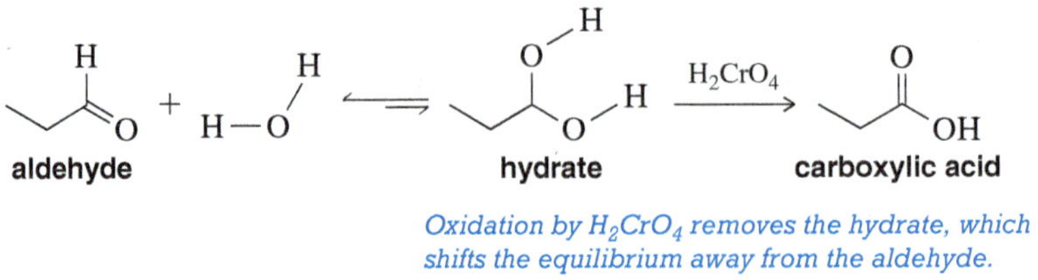

Assessment 17.29 Explain why hydrate formation is more favorable for the compound on the right in each pair.

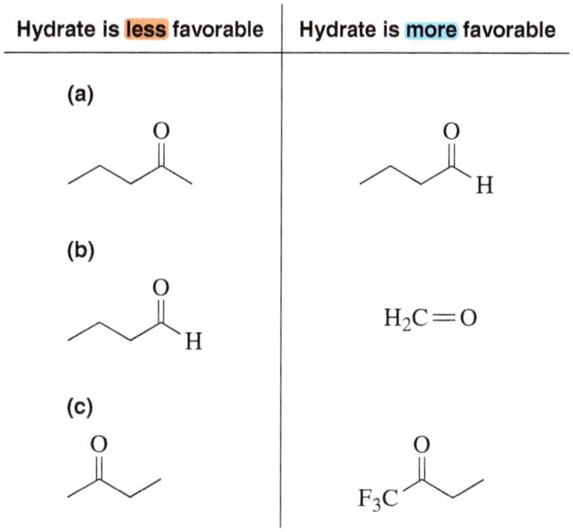

The nucleophilic addition of water to a carbonyl does not proceed by General Mechanism 17A for a variety of reasons. Direct addition of water, a weak nucleophile, to the carbonyl would be very slow. Reaction of two neutral molecules to give one molecule containing opposite charges is disfavored by both entropy and enthalpy. In terms of the nucleophiles involved, the reaction begins with water (conjugate acid H_3O^+; $pK_a = -1.7$) and finishes with an alkoxide (conjugate acid ROH; $pK_a = 16$). Thus, this is a very slow reaction (**Figure 17.47**).

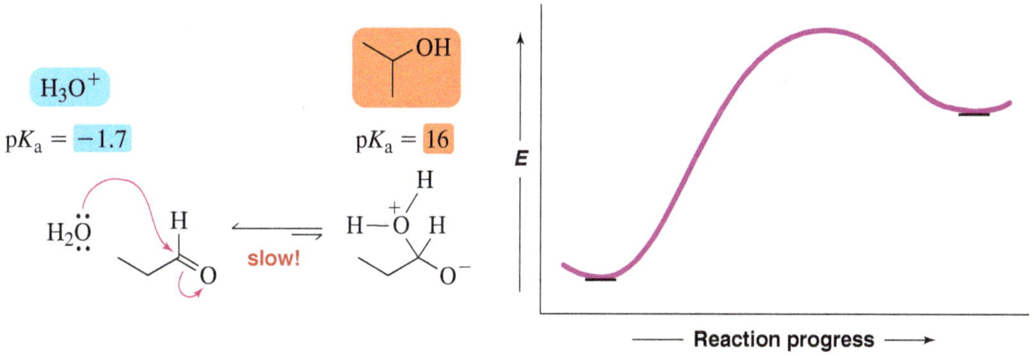

Figure 17.47 Addition of water to a neutral carbonyl is a slow, disfavored reaction.

The analysis in Figure 17.47 explains why protonation of the carbonyl is required prior to attack by weak nucleophiles. Water, having little interest in sharing an electron pair, only attacks very strong electrophiles. Protonation of the carbonyl generates an oxonium ion containing a resonance structure with a positive charge on the carbonyl carbon. Water attacks the strong electrophile, bringing along its protons. In a final step, an acidic proton is removed, producing the hydrate. The regeneration of H_3O^+ demonstrates that this is an acid-catalyzed reaction. [*Recall that a catalyst increases the rate, yet is recovered unchanged from the reaction.*] **Figure 17.48** contrasts the reaction coordinate diagrams of the catalyzed and uncatalyzed reactions, showing how acid increases the rate of the reaction.

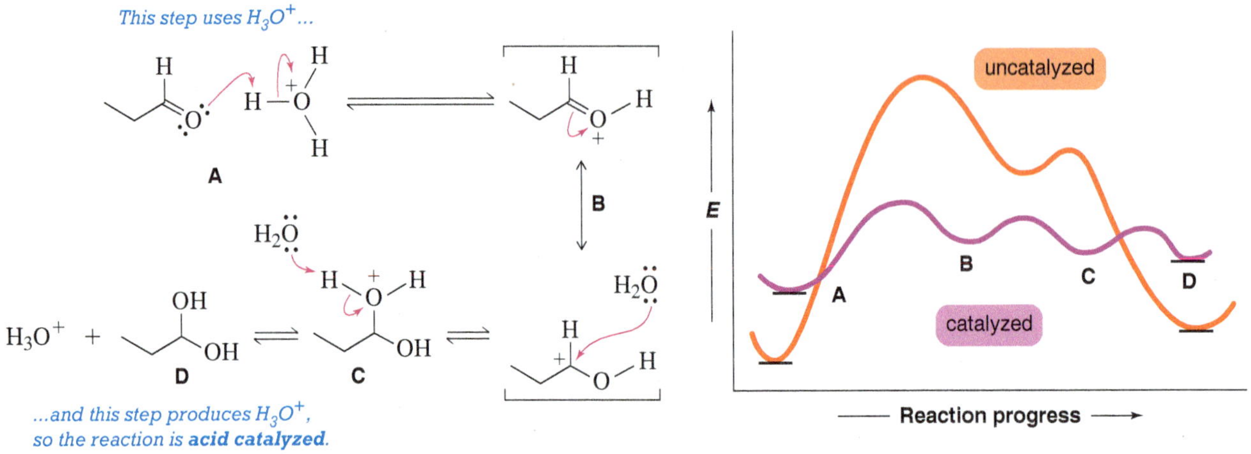

Figure 17.48 The rate of addition of a weak nucleophile (like water) to a neutral carbonyl can be increased by using an acid catalyst.

Assessment 17.30 Hydrate formation can also be catalyzed by hydroxide.

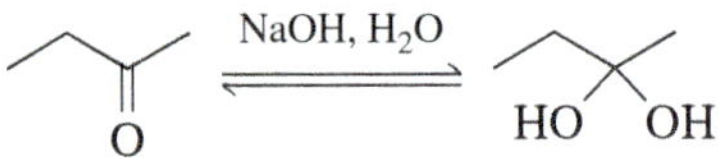

(a) Suggest a mechanism for this reaction.
(b) Draw a reaction coordinate diagram for the mechanism you generated.
(c) Does this occur by General Mechanism 17A or 17B?
(d) Do you expect a larger K_{eq} for the acid- or base-catalyzed hydrate formation reaction?

17.7.2 Addition of alcohols: hemiacetal formation

As we just learned, water adds only after protonation of the carbonyl to give the hydrate, a minor constituent in the equilibrium. Further reaction of the hydrate can shift the equilibrium, but because of its instability, the hydrate cannot be isolated as a product. As we proceed, the analysis seems to always return to the same issue—namely, why is the hydrate less favored?

The answer in this case is that entropy (disfavored) wins over enthalpy (slightly favored). To ensure you understand this concept, consider what happens when water is replaced with methanol in the acid-catalyzed addition reaction shown in Figure 17.49. [*In other words, do the analysis for the reaction in Figure 17.49 before reading further.*]

Figure 17.49 The addition of an alcohol to a ketone is favored by ΔH, but not by ΔS.

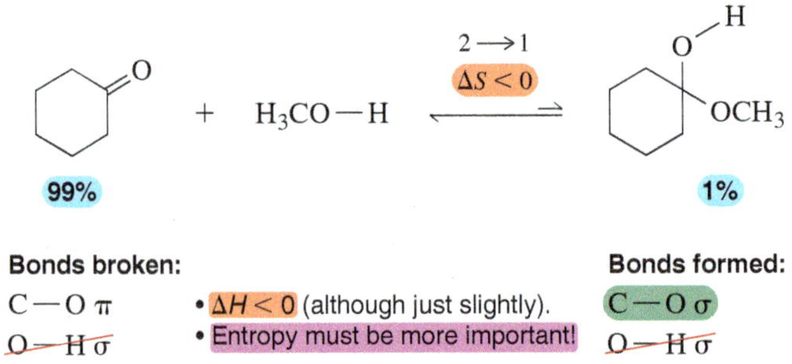

As you might have guessed, there is essentially no change. Throughout this text, whatever water does in a reaction, alcohols do as well; this case is no different. The reaction is still entropically disfavored, like all addition reactions. Once more, it is only slightly favored by ΔH. The net result, confirmed experimentally with ΔS winning again, favors the ketone side of the equilibrium.

> Hemiacetals are missed by students so often that it is important to practice it right away. When you see what you think is an alcohol, ask yourself, "Or is it a hemiacetal?" This will pay off for you eventually.

The minor constituent in the equilibrium mixture of methanol and butan-2-one has a special name. With a hydroxyl group, it looks like an *alcohol*. The fact that it contains an oxygen with two carbons attached suggests it is an *ether*. It is neither. Instead, a molecule containing a hydroxyl group and an alkoxy group bound to the same carbon is called a **hemiacetal.**

Nomenclature is not often stressed this way, but it is important that you recognize the difference. Alcohols and ethers react very differently than hemiacetals. Thus, it is very important to quickly identify the presence of a hemiacetal. The general formula RO—C—OH, where R represents any alkyl group, provides a visual cue that the molecule is a hemiacetal.

Assessment 17.31 Identify the hemiacetal functional group in each of the following molecules. These molecules may not be stable enough to be the favorable product in an equilibrium reaction.

(a) (b) (c) (d) (e)

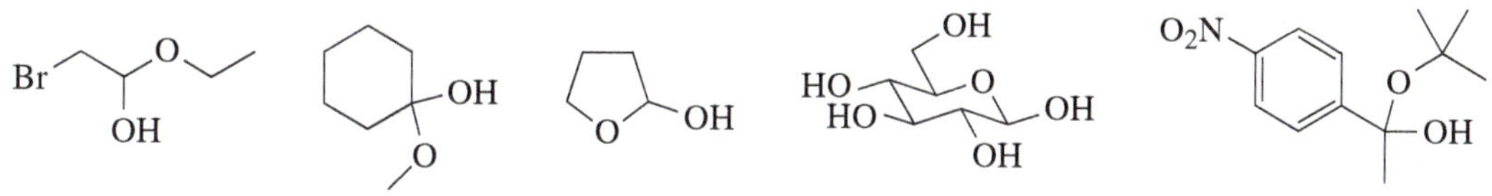

Figure 17.50 shows that hemiacetal formation proceeds by the same mechanism as the addition of water. In step 1, protonation activates the carbonyl group for attack by methanol, the weak nucleophile, in step 2. In step 3, removal of a proton produces the hemiacetal, regenerating the acid catalyst. Once again, the hemiacetal is still disfavored (even though it is formed more quickly under acidic conditions), due to the greater importance of ΔS. [*Remember, catalysts increase the reaction rate but do not shift the equilibrium.*]

$$CH_3OH + H_2SO_4 \rightleftharpoons CH_3\overset{+}{O}H_2 + HSO_4^-$$

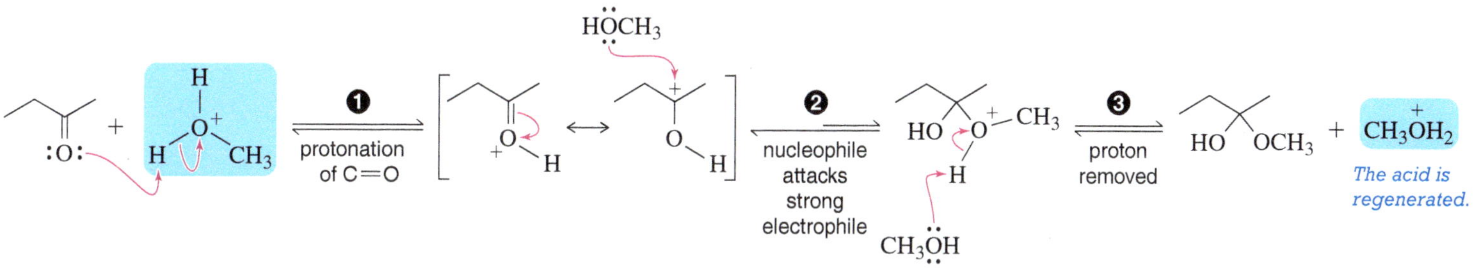

Figure 17.50 Like the hydration of ketones, acid catalyzes the formation of hemiacetals according to General Mechanism 17B.

Assessment 17.32 Provide a mechanism for the formation of the hemiacetals shown. [*Only (c) is favored as written.*]

(a)

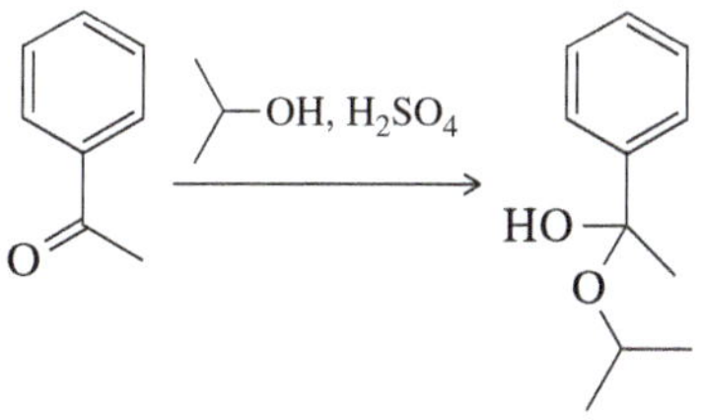

(b)

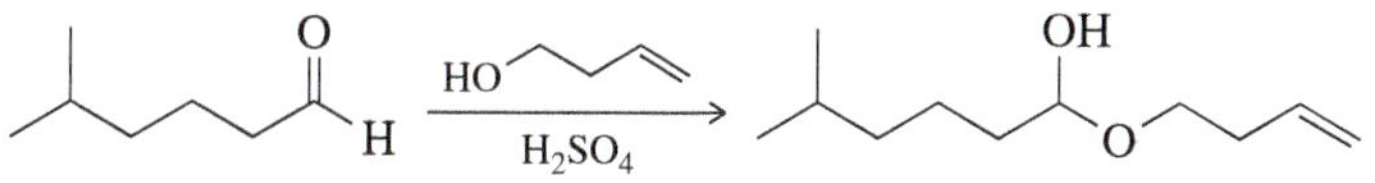

(c)

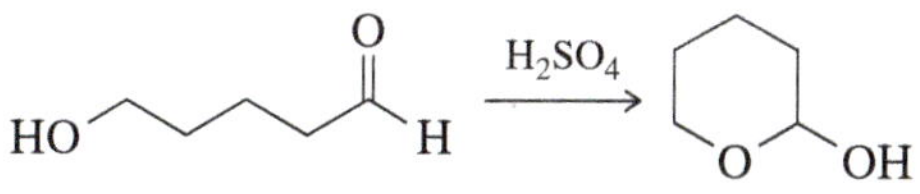

The first two additions of oxygen nucleophiles that we have looked at, CH_3OH and H_2O, have been disfavored reactions [*and thus need not be on your reaction sheets.*] Is there value in learning mechanisms of reactions that don't actually proceed? Actually, stable hemiacetals exist, and many have biological importance. For example, glucose equilibrates between its open-chain and ring-closed forms (Figure 17.51). [*What type of reaction occurs to close the ring? What functional groups are present in the open-chain form? In the ring?*] This reaction is simply the intramolecular addition of an alcohol to an aldehyde to form a hemiacetal. You could draw the mechanism of hemiacetal formation—in fact, you did just that in Assessment 17.32(c)! And, unlike any case studied previously, here the hemiacetal is the favored side of the equilibrium. [*Any ideas why?*]

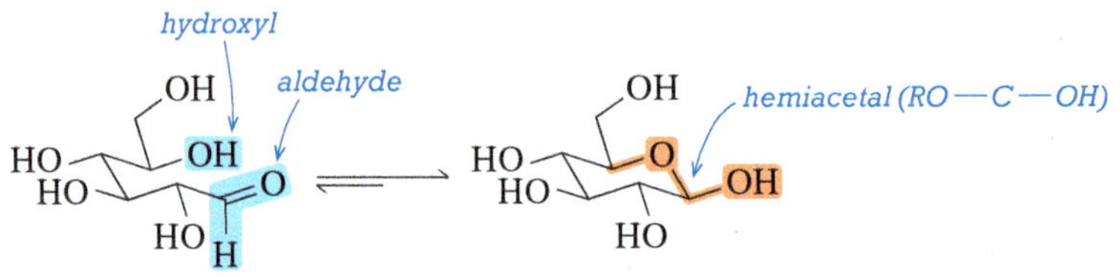

Figure 17.51 Sugars exist in the ring-closed hemiacetal form.

Excising the hydroxyl groups makes analysis of this reaction simpler. Like all previous hemiacetal-forming reactions, $\Delta H < 0$. But here one molecule reacts to give one molecule. As an intramolecular reaction, $\Delta S \approx 0$, in which case enthalpy finally wins, and the hemiacetal is favored (Figure 17.52). The hemiacetal has less freedom to rotate than the open-chain form. This loss of **rotational entropy** is significantly less important than the change in molecules, or **translational entropy**. As a result, rotational entropy can often be ignored.

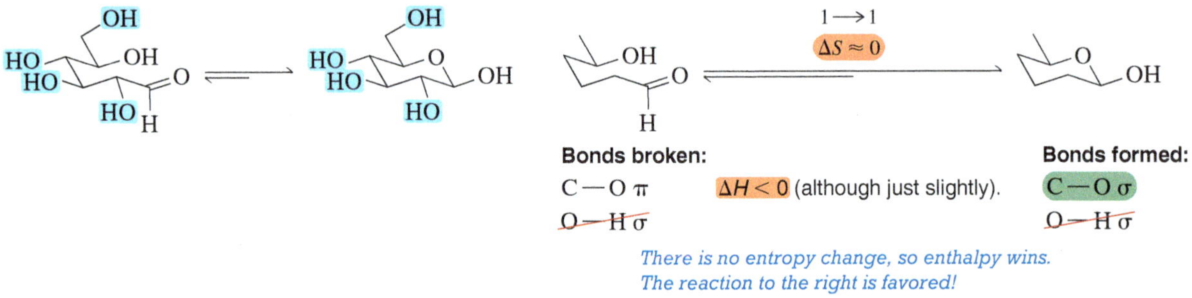

Figure 17.52 Intramolecular formation of a hemiacetal has $\Delta S \approx 0$.

Assessment 17.33 Which of the following cyclic hemiacetals would you expect to have the highest K_{eq} for their formation? Explain your answer.

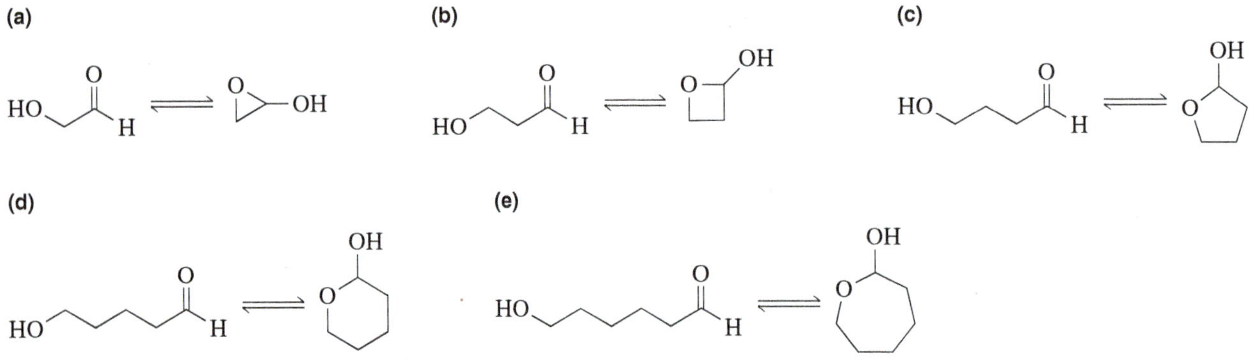

Assessment 17.34 Which of the following cyclic hemiacetals would you expect to have the higher K_{eq} for their formation? Explain your answer.

17.7.3 Addition of alcohols: acetal formation

Even though acyclic hemiacetals cannot be isolated, they are intermediates on the pathway from ketones or aldehydes to **acetals** (**Figure 17.53**). The acetal is a functional group often misread by organic students as an ether. Acetals, with two alkoxy groups coming off the same carbon (RO—C—OR), show very different reactivity as compared to ethers. [*Finally, we see the origin of the name **hemi**acetal—namely, it is halfway between a ketone and an acetal.*]

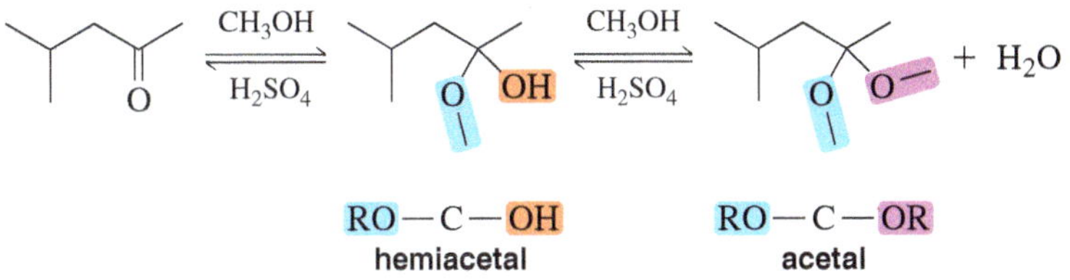

Figure 17.53 **Alcohols and carbonyls react to favorably make acetals.**

Unlike hemiacetals and hydrates, acetals can be isolated. [*Consider the competition between entropy and enthalpy in thinking about why this might be true.*] Owing to the greater stability of the C—O σ bond as compared to the C—O π bond, this reaction has a favorable ΔH (**Figure 17.54**). Additionally, the O—H bonds in water are especially stable. Entropically, this reaction results in the production of two molecules from three, a key difference from hemiacetal formation. Making two molecules from three, while disfavored, is *less* disfavored than making one from two. Thus, ΔS is not as important here, ΔH wins, and acetal formation is slightly favored.

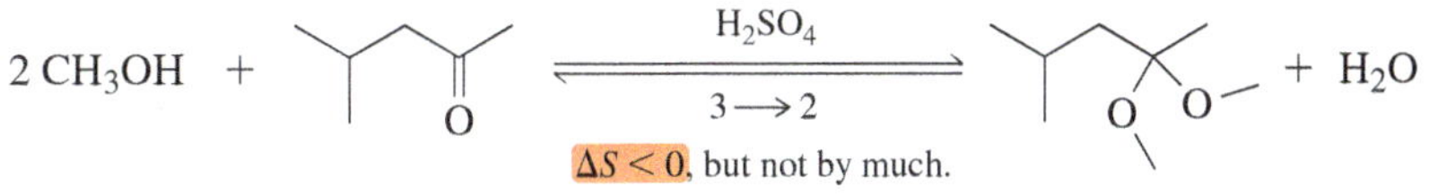

Bonds broken:
O—H σ
O—H σ
C—O σ
C—O π

- $\Delta H < 0$
- ΔH is more negative and ΔS is less important here, so $\Delta G < 0$.
- $K_{eq} > 1$, so the right side is favored!

Bonds formed:
C—O σ
C—O σ
H—O σ
H—O σ
These bonds in water are especially strong.

Figure 17.54 **When entropy is not as negative, ΔH is most important, making the formation of acetals a favorable reaction.**

In addition to the subtle thermodynamic differences, there is another reason this reaction is synthetically useful. The production of water as a by-product allows the direction of equilibrium to be manipulated using Le Châtelier's principle. As shown in **Figure 17.55**, the addition of methanol, or the removal of water, drives the reaction to the right. We should similarly be able to drive the reaction to the formation of the ketone by removing methanol or adding water. [*How might you ensure an excess of methanol in the reaction mixture? How might you remove water?*]

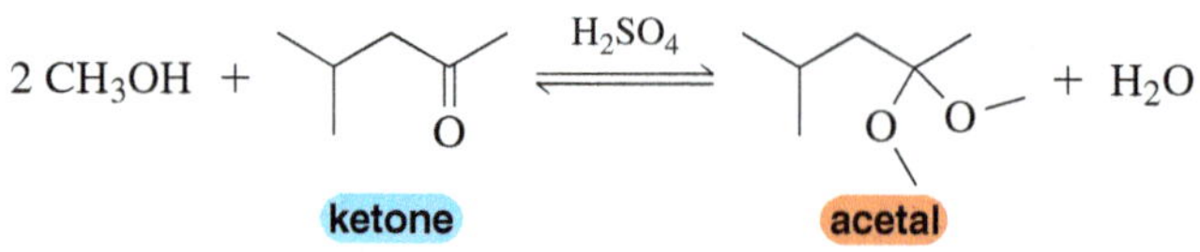

Figure 17.55 The formation of acetals can be shifted in either direction, depending on the desired product.

To shift the reaction to the **acetal** side:
- increase the concentration of methanol
- remove water from the reaction

To shift the reaction to the **ketone** side:
- increase the concentration of water
- remove methanol from the reaction

Experimentally, the ketone is dissolved in methanol as the solvent, ensuring a large excess. Shifting the equilibrium by removing water is more complicated. With a boiling point of 64 °C, methanol boils before water. Thus, distillation cannot be used to remove water selectively. Adding molecular sieves to the reaction mixture allows for the removal of water from the methanol/water mixture. A molecular sieve is a bead-like material containing pores that are the perfect size to trap water.

Removing water from a mixture containing an immiscible alcohol is done with a **Dean–Stark apparatus** (**Figure 17.56**). The alcohol and water form a low-boiling **azeotrope**, a mixture that boils at a lower temperature than each of the compounds (water and the alcohol) would individually. As a result, the vapor contains both water and alcohol. Cooling the vapor condenses it into a separate trap. Within the trap, water (the more dense liquid) settles to the bottom, allowing the alcohol to return to the reaction flask. By removing water, the equilibrium is shifted to the acetal.

Figure 17.56 The Dean–Stark apparatus facilitates the removal of water from reaction mixtures.

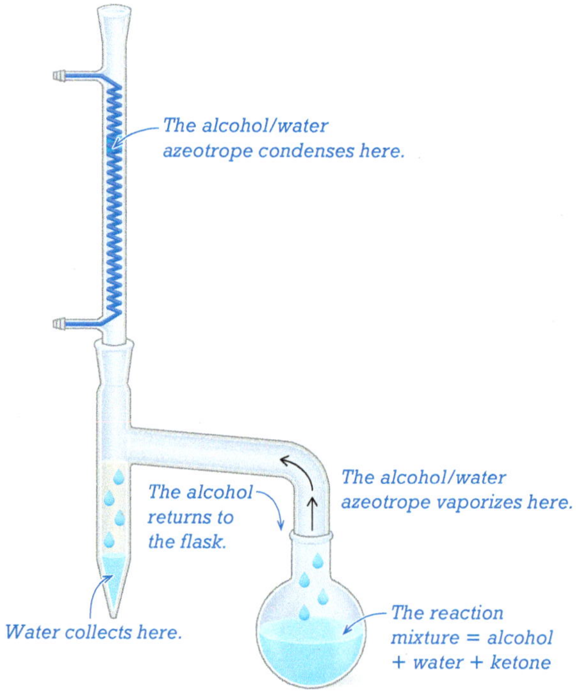

Let's think together to see how a chemist might have figured out the acetal mechanism.

Thinking Like a Chemist

The reasoning that led chemists to understand complex reaction mechanisms can be lost when learning from a textbook. The mechanisms discussed here make sense based on things studied previously, but this wasn't always the case. For example, acetal formation was at one time not well understood. Experiments were designed to elucidate these mechanisms. To see how this might work, use a separate sheet of paper to write out a mechanism for the acid-catalyzed formation of the dimethyl acetal (**Figure 17.57**). [*We already know how a hemiacetal forms, so you are already halfway there.*]

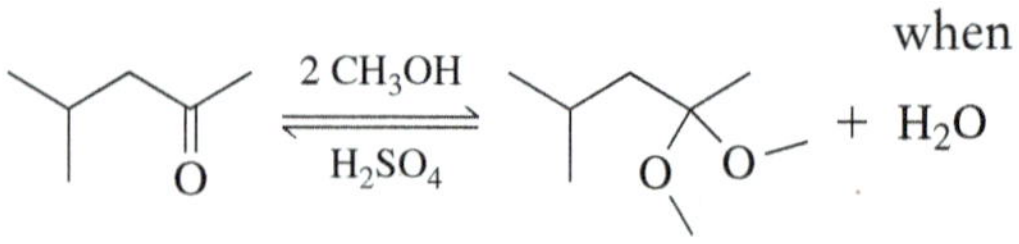

Figure 17.57 Propose a mechanism of acetal formation before proceeding.

[*Did you write down your idea? To get the most out of this exercise, you should try to come up with something.*]

Because we know how hemiacetals are formed, we begin there. Based on the chemistry discussed previously, there are two conceivable pathways by which the acetal forms—namely, an S_N2 reaction or an S_N1 reaction.

Possibility 1: S_N2 reaction

In an S_N2 reaction (Chapter 12), the hemiacetal hydroxyl group could displace the protonated methanol hydroxyl group. Loss of a proton would then generate the hemiacetal and the acid catalyst (**Figure 17.58**).

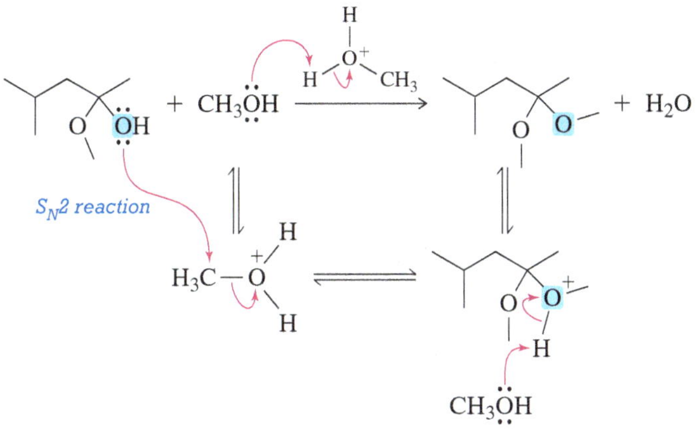

Figure 17.58 Hemiacetal to acetal by an S_N2 mechanism.

Possibility 2: S_N1 reaction

Alternatively, the protonated hemiacetal hydroxyl group could leave as water, producing a carbocation. Attack by methanol on the carbocation would then be followed by removal of a proton to generate the acetal (**Figure 17.59**).

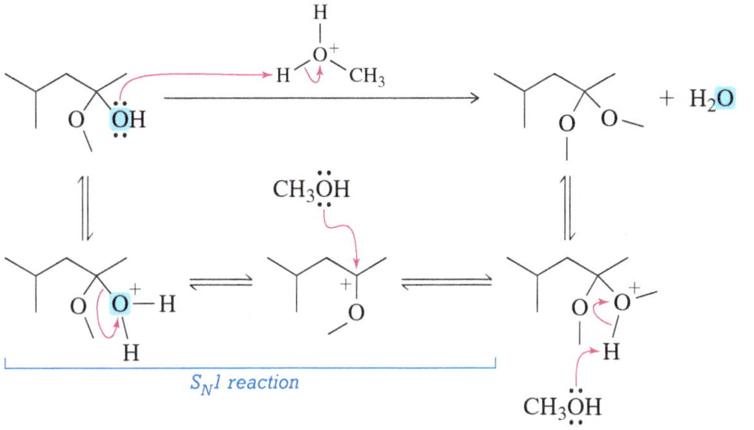

Figure 17.59 Hemiacetal to acetal by an S_N1 mechanism.

Given both of these mechanisms, which do you think is correct? Recall that reactions follow the lowest energy pathway [*and there can only be one such pathway*]. To elucidate the correct pathway, chemists might label the ketone oxygen with ^{18}O. Although the label does not change the reaction mechanism, ^{18}O can be detected and distinguished from ^{16}O. When performed with the labeled ketone in unlabeled methanol, the label is found only in the water after the reaction has gone to completion (**Figure 17.60**). How does this experimental observation impact your choice of mechanistic pathways?

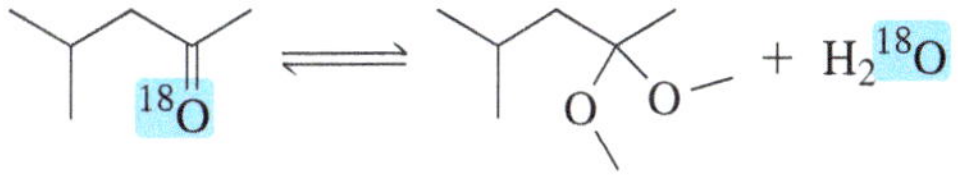

Figure 17.60 With ^{18}O-labeled ketone, ^{18}O- labeled water is produced.

Looking back at the two mechanisms, only possibility 2 would produce ^{18}O-labeled water. You have thus learned one way by which chemists might determine a reaction mechanism. And, equally important, you now know the mechanism of acetal formation. [*More to come on this . . . .*] ■

17.7.4 A detailed look at the acetal mechanism

The experiment described in Thinking Like a Chemist says nothing about *why* possibility 2 was best. In solution, either hydroxyl (methanol or the hemiacetal) can be protonated with equal favorability. The proton is probably moving from base to base in a random manner. The favored pathway is then a result of which "next" reaction is faster. The S_N2 reaction requires two molecules to come together in the transition state. The S_N1 reaction, which requires only a leaving group to leave, forms two molecules from one. Typically, a collision between two molecules is slower than the expulsion of a leaving group, especially when a highly stable carbocation is produced. As shown in **Figure 17.61**, the carbocation formed by water leaving is stabilized by resonance from the attached oxygen.

Figure 17.61 The formation of a resonance-stabilized carbocation is faster than an S_N2 reaction.

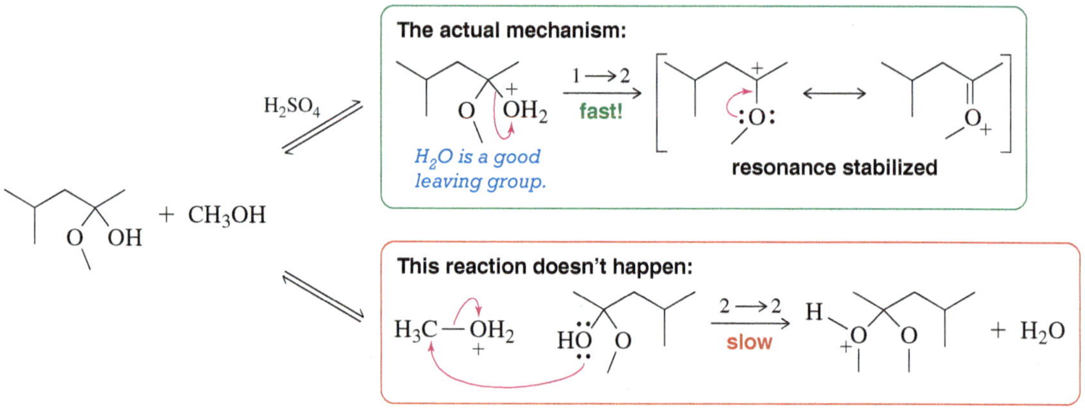

In sum, the mechanism of acetal formation is shown in **Figure 17.62**. The mechanism for the reverse reaction is shown in **Figure 17.63**. According to the principle of microscopic reversibility, a reaction proceeds by the same mechanism in either direction, and these two figures show that.

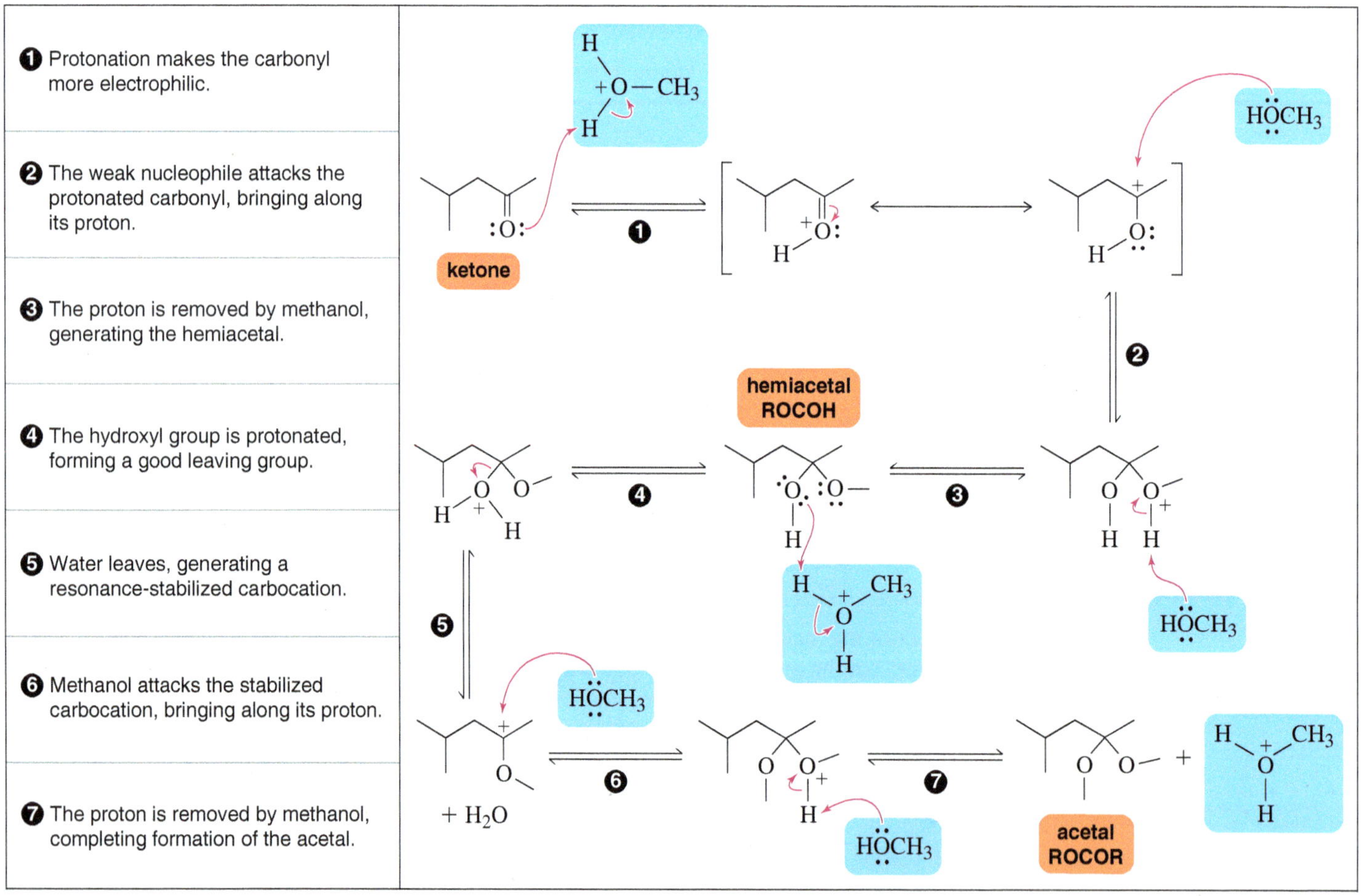

Figure 17.62 General mechanism of acetal formation. According to the principle of microscopic reversibility, acetal formation and acetal hydrolysis have equal, but opposite, mechanisms. The direction of the reaction depends on the conditions used and relies on Le Châtelier's principle to drive the reaction in a particular direction.

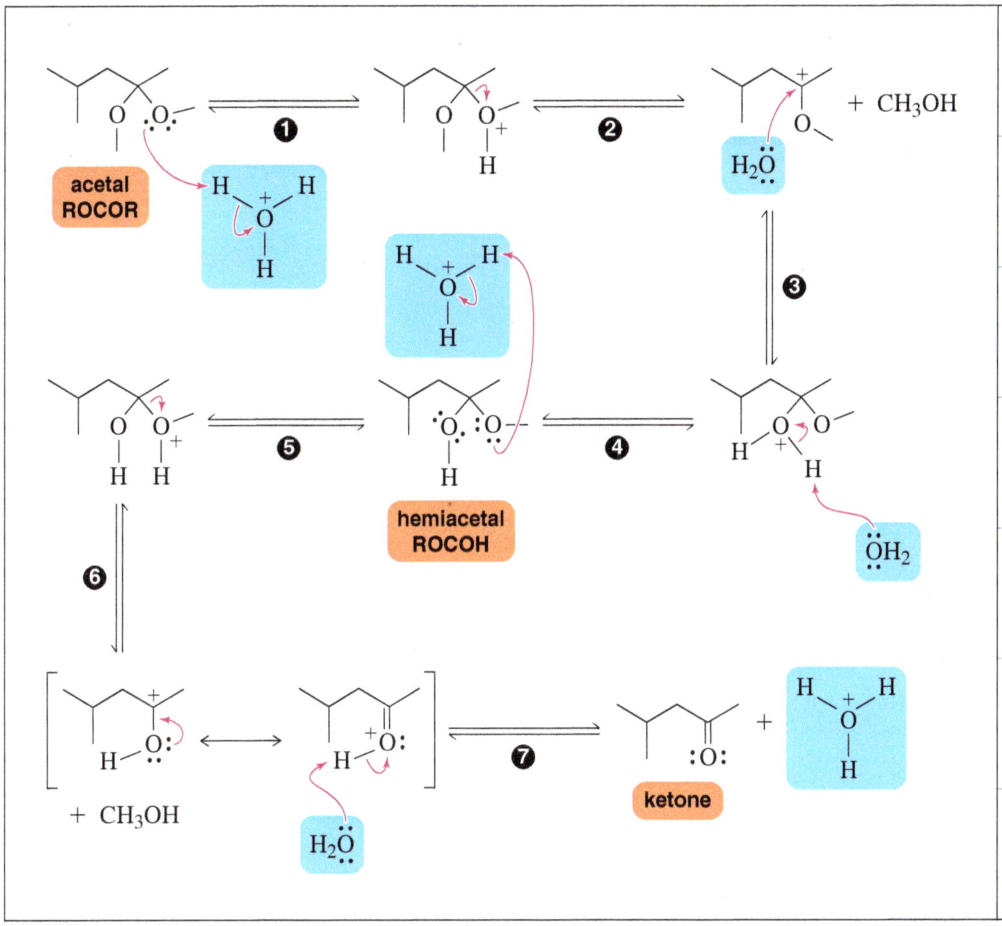

❶ Protonation of the oxygen of the acetal makes a good leaving group.

❷ Methanol leaves, producing a resonance-stabilized carbocation.

❸ Water attacks the carbocation, bringing along its proton.

❹ The acidic proton is removed by water, generating the hemiacetal.

❺ The alkoxy group is protonated, forming a good leaving group.

❻ Methanol leaves, generating a resonance-stabilized carbonyl.

❼ Removal of the proton generates the ketone.

Figure 17.63 General mechanism of acetal hydrolysis. According to the principle of microscopic reversibility, acetal formation and acetal hydrolysis have equal, but opposite, mechanisms. The direction of the reaction depends on the conditions used and relies on Le Châtelier's principle to drive the reaction in a particular direction.

Assessment 17.35 Predict the acetal that would be formed under the following conditions.

(a)

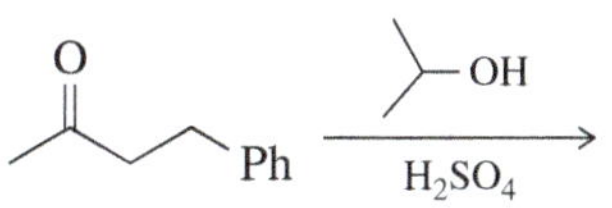

(b)

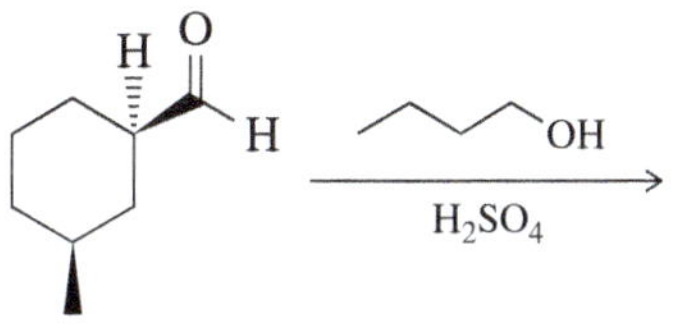

(c)

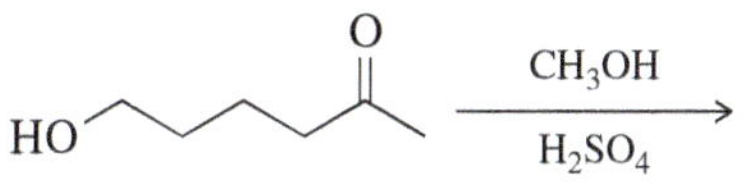

Assessment 17.36 Suggest a mechanism that rationalizes the product(s) you predicted in 17.35(a)–(c) above.

Assessment 17.37 A chemist ran the following acetal formation using ^{18}O-labeled ethanol. Where would you expect the label to be at the end of the reaction?

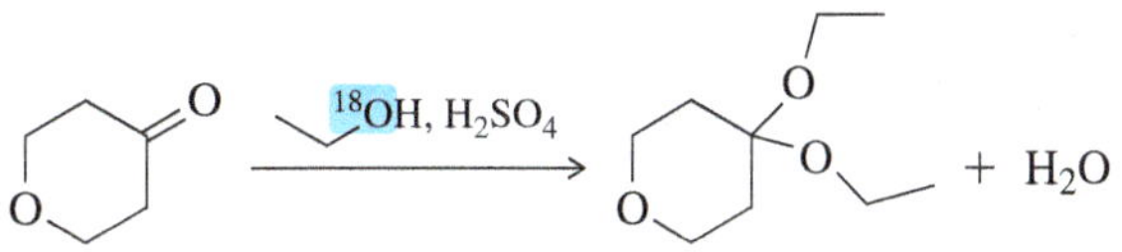

17.7.5 Acetals as protecting groups

One of the more important uses of acetals is as protecting groups for aldehydes and ketones. Protecting groups are necessary when a chemist wants to do a reaction with one functional group but in the presence of another functional group that exhibits a similar reactivity pattern. [*Recall our studies of alcohol protecting groups in Section 13.14.*]

Solved Assessment 17.17

In Assessment 17.17 (Section 17.5.1), you were asked to explain why a Grignard reaction did not form the expected product via nucleophilic addition to a carbonyl. [*Look at that assessment again before continuing.*]

Most of the molecules studied so far have contained one functional group. We may tend, especially if we memorized some of the reactions, to zoom in on one functional group per molecule. As a result, the reaction in **Figure 17.64** becomes simplified in our minds to be the reaction of a Grignard reagent with an aldehyde, with no other functional groups present. Making too many assumptions about the lack of reactivity of other functional groups causes students to overlook obvious reactivity. "Seeing" what is important amidst other unreactive functional groups is a valuable skill.

With that in mind, zoom back out and see what other functional group is present in the Grignard reagent (**Figure 17.65**). This molecule contains both a Grignard reagent and an electrophilic aldehyde. The intramolecular reaction is faster than the intermolecular reaction, rendering the molecule useless as a Grignard reagent in the reaction shown.

Before making the Grignard reagent, the aldehyde needs to be converted into an unreactive functional group (we need a protecting group). After Grignard addition to the carbonyl, this functional group should be converted back to the aldehyde (the protecting group must be removed). Is there a reaction that reversibly forms an unreactive functional group, starting with an aldehyde? Yes, there is! ■

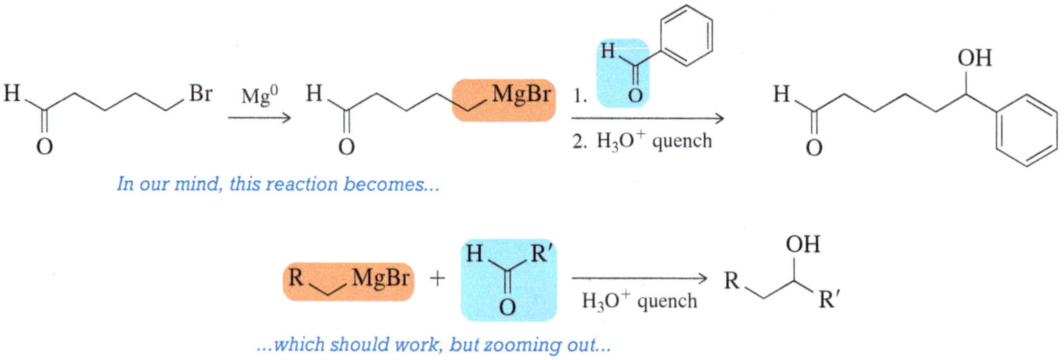

Figure 17.64 A Grignard reaction which, at first glance, seems like it should produce the alcohol shown.

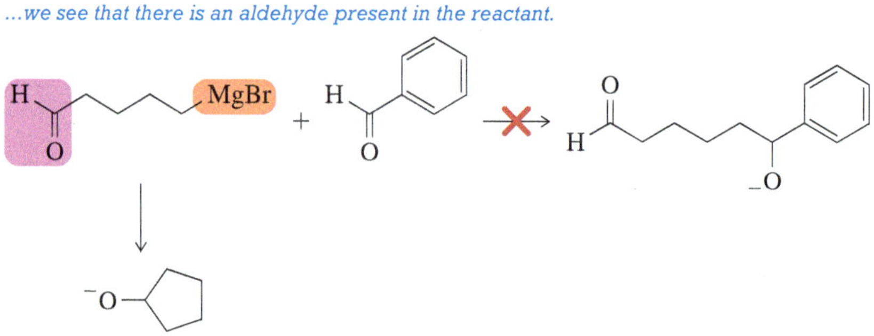

Figure 17.65 Seeing the entire reaction (zooming out) offers a different perspective.

Reversibility makes the acetal an ideal carbonyl-protecting group. Acetals are also unreactive under strongly basic conditions and are easily removed through hydrolysis. Typically, ethylene glycol is used to protect aldehydes and ketones (**Figure 17.66**). The reaction to form the acetal from ethylene glycol has no significant contribution from entropy (2 molecules → 2 molecules, so $\Delta S \approx 0$).

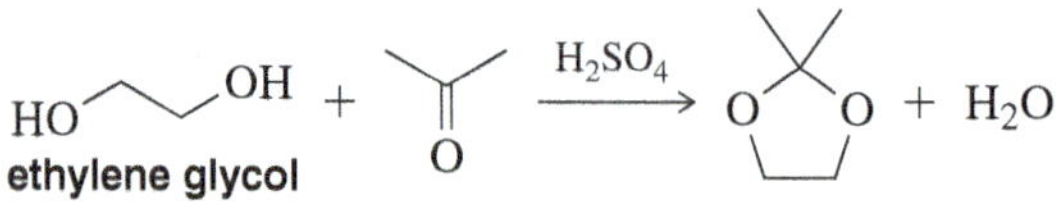

Figure 17.66 Ethylene glycol protects aldehydes by favorably forming an acetal.

Mechanistically, the reaction proceeds in a manner identical to that in Figure 17.62. The lone difference is the final alcohol addition step, which is an *inter*molecular reaction in Figure 17.62 but is an *intra*molecular reaction to generate the five-membered ring in **Figure 17.67**.

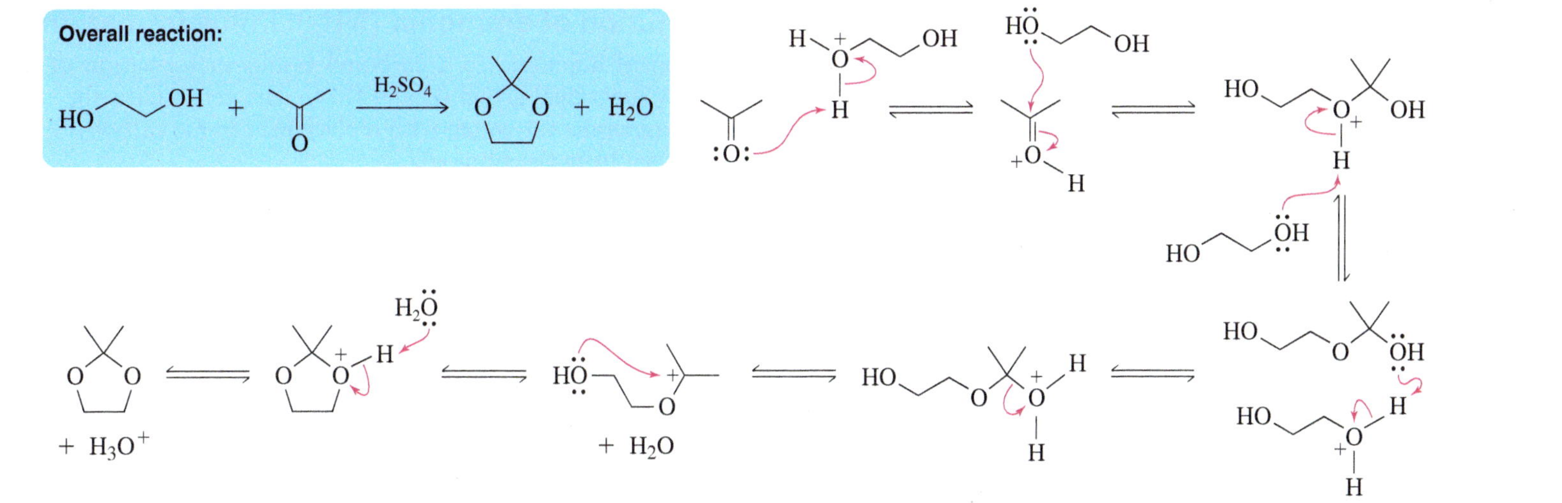

Figure 17.67 The mechanism for the formation of an acetal using ethylene glycol. It is similar to the mechanism in Figure 17.62.

The overall synthetic process (**Figure 17.68**) involves protecting the bromoaldehyde as the acetal before making the Grignard reagent. Nucleophilic addition is followed by an acid quench. Deprotection under acid-catalyzed conditions ultimately produces the aldehyde.

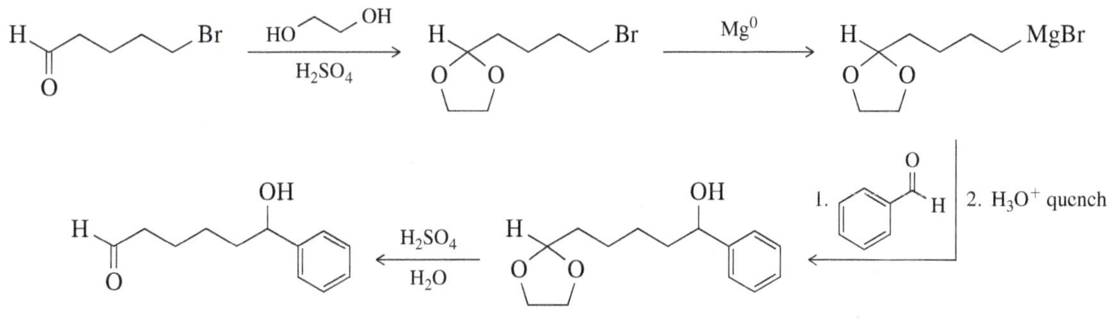

Figure 17.68 The overall synthetic process of using a protecting group.

Assessment 17.38 A chemist attempted unsuccessfully to generate the alkynyl ketone shown here using an acetylide alkylation strategy. How might this scheme be modified to produce the desired product?

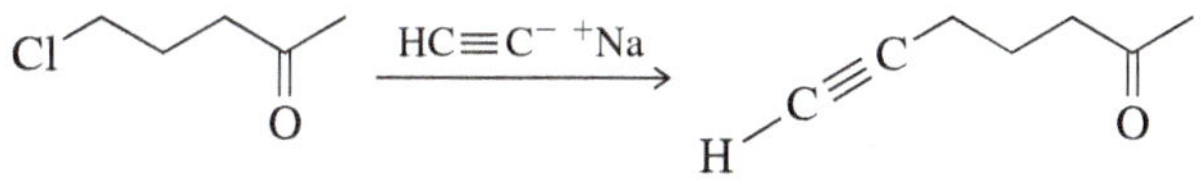

desired, but not obtained

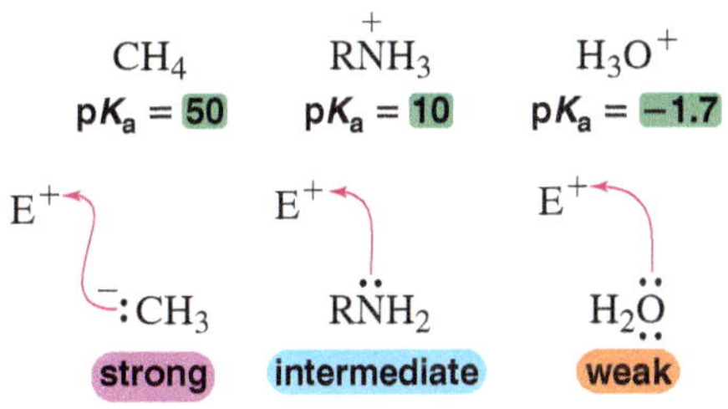

Figure 17.69 Neutral nitrogen is of intermediate strength between negative carbon and neutral oxygen nucleophiles.

17.8 Addition of nitrogen nucleophiles to aldehydes and ketones

The rate of addition of neutral nitrogen nucleophiles falls between that of a negatively charged carbon and a neutral oxygen. Being neutral, they are not as prone to donate an electron pair as carbanions. Because nitrogen is less electronegative than oxygen, the nitrogen nucleophiles are more willing to donate an electron pair than oxygen nucleophiles (**Figure 17.69**). This creates an interesting mechanistic situation, with nitrogen nucleophiles adding by either General Mechanism 17A or 17B.

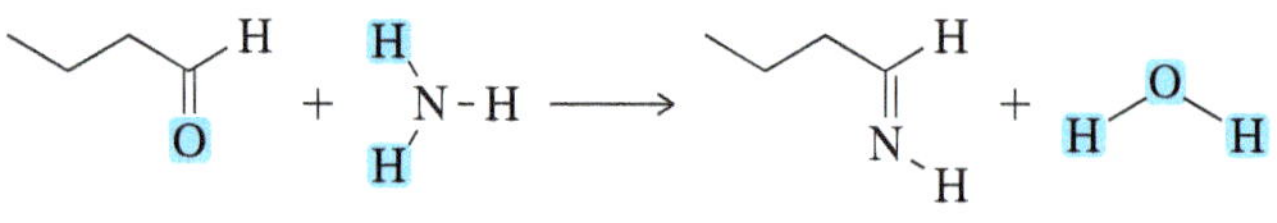

Figure 17.70 The addition of ammonia forms an imine.

17.8.1 Addition of ammonia (NH_3): imine formation

The addition of ammonia to a carbonyl results in formation of an imine, a new functional group (**Figure 17.70**). The net result is the substitution of ammonia for water, with replacement of the carbonyl oxygen by nitrogen from the ammonia.

Mechanistically, in the presence of acid, the reaction begins by protonation of the carbonyl oxygen, giving an activated electrophile (**Figure 17.71**). Ammonia donates an electron pair to the protonated carbonyl, bringing along its three hydrogens. The acidic proton is next removed from nitrogen by excess ammonia. The intermediate thus formed is analogous to the hydrate formed in the earlier addition of water. Just as the hydrate is unstable (converting back to the ketone), this analogous intermediate is also unstable. In this case, however, it can proceed productively to generate a new compound.

Figure 17.71 Ammonia reacts with aldehydes/ketones in a manner analogous to hydrate formation.

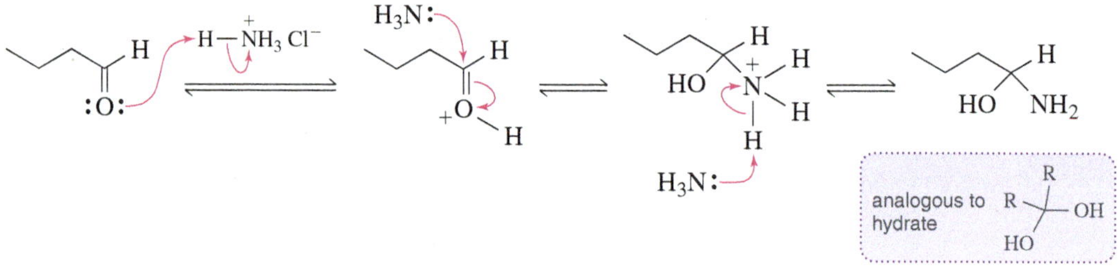

Protonation of the hydroxyl group results in the formation of water. After water leaves, a resonance-stabilized carbocation is formed. At this point, the two potential reactions shown in **Figure 17.72** can occur:

- Possibility 1: Nucleophilic addition of a second molecule of ammonia.
- Possibility 2: Deprotonation to give the neutral imine.

[*Possibility 2 is favored, but why? Why does* **DEPROTONATION** *to make the imine occur instead of a second* **ADDITION**?]

Figure 17.72 Formation of the resonance-stabilized carbocation makes nucleophilic attack or deprotonation possible.

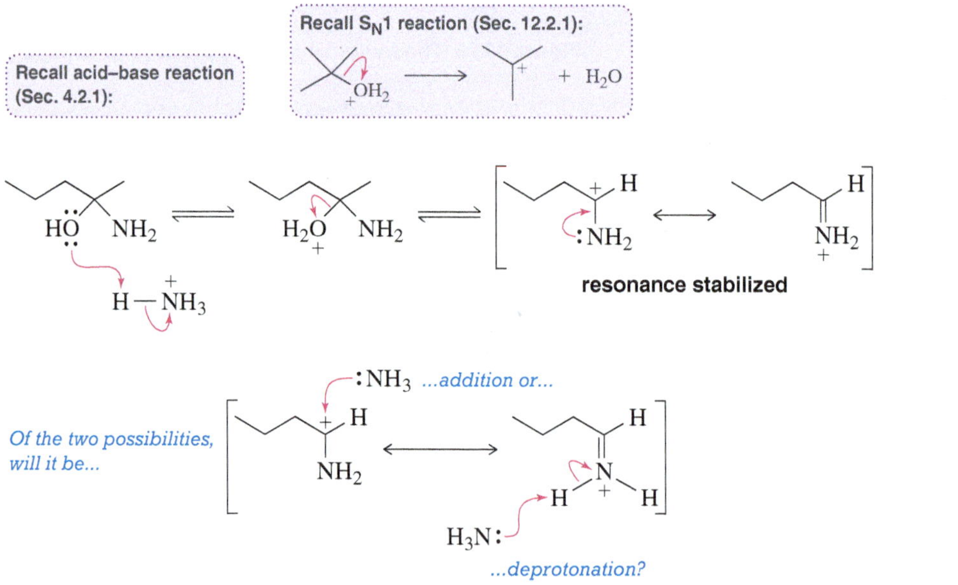

The best way to understand the answer is by comparing the two processes. We have done this previously (see Section 13.7) and determined that deprotonation is the faster reaction (lower E_a). It is also thermodynamically favored on the basis of $\Delta S \approx 0$ (2 molecules → 2 molecules). As such, the imine is produced in a high-yielding reaction (**Figure 17.73**).

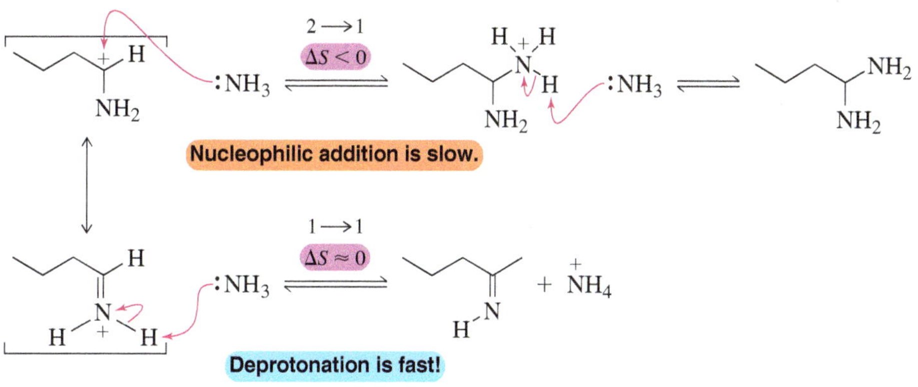

Figure 17.73 Deprotonation is faster and more favored than nucleophilic attack.

Assessment 17.39 Predict the product of the following amine addition reactions.

(a) (b) (c)

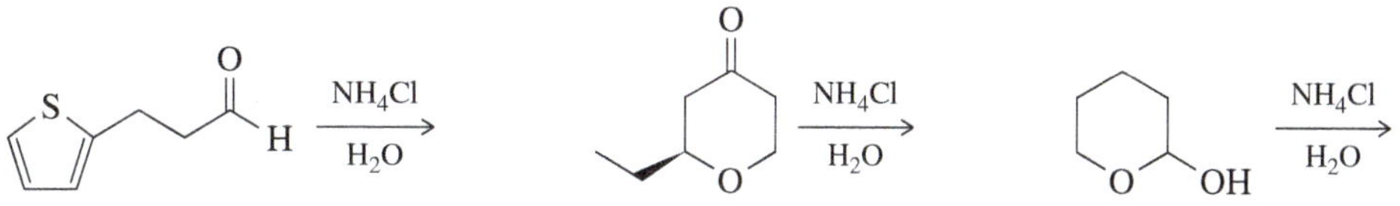

Assessment 17.40 The formation of imines is reversible. How might the equilibrium be shifted toward the ketone side of the equilibrium?

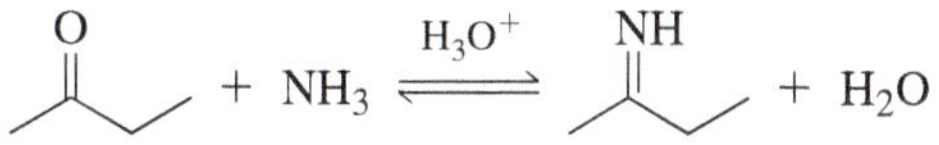

17-C CHEMISTRY IN CONTEXT: Biochemistry

Producing energy from amino acids

The Krebs (or citric acid) cycle is a nine-step pathway used by cells to generate energy (**Figure 17.74**). Many of the molecules in the pathway get hijacked by other biological processes, resulting in the need for a secondary method for replenishing intermediates. One such intermediate that enters the cycle through a secondary pathway is α-ketoglutarate.

α-Ketoglutarate, normally produced by a decarboxylation reaction in the Krebs cycle, is also generated by the deamination reaction of glutamate (**Figure 17.75**). The deamination reaction makes use of NAD^+ as a cofactor.

The process starts with the binding of glutamate and NAD^+ in the active site of glutamate dehydrogenase (similar to that in Chemistry in Context 17-A). As shown in **Figure 17.76**, deprotonation of a hydrogen from the amino group causes the electrons to collapse down, forming a C—N π bond, while donating a hydride ion to the neighboring NAD^+ cofactor.

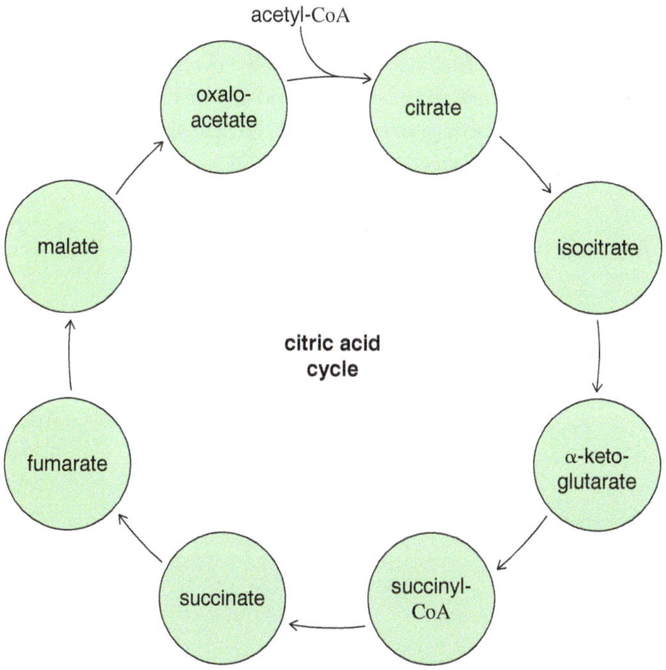

Figure 17.74 The citric acid cycle.

(continued)

Figure 17.75 Deamination of glutamate to make α-ketoglutarate.

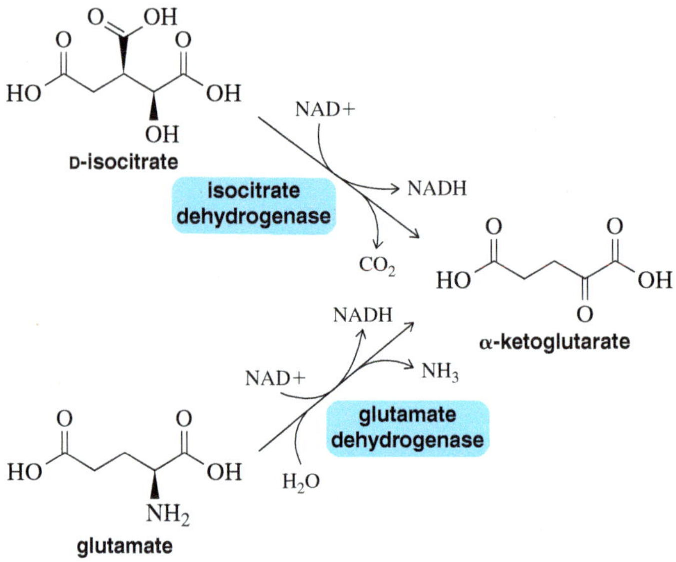

Figure 17.76 The mechanism of enzymatic amine oxidation.

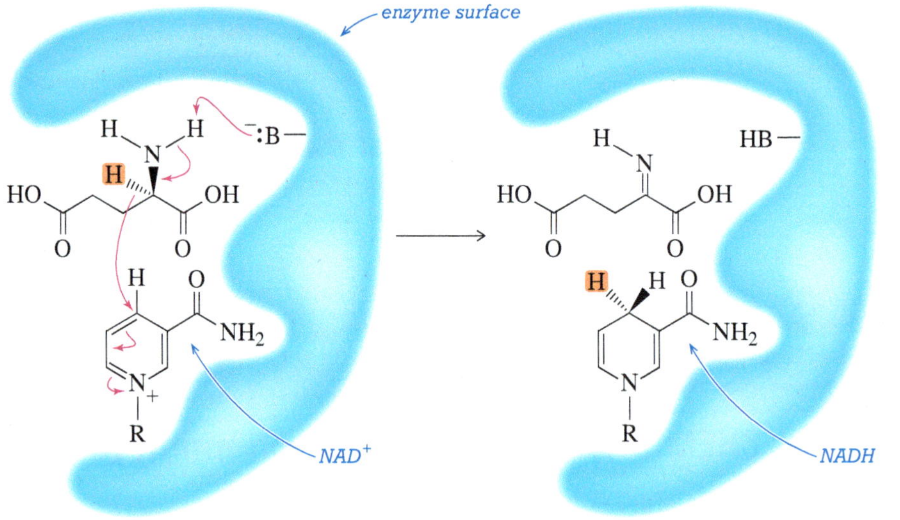

The imine, in an aqueous environment, equilibrates to the α-ketoglutarate for use in the Krebs cycle. Imine hydrolysis (Figure 17.77) proceeds by the opposite mechanism [*recall the principle of microscopic reversibility*] discussed in Section 17.8.1 for imine formation.

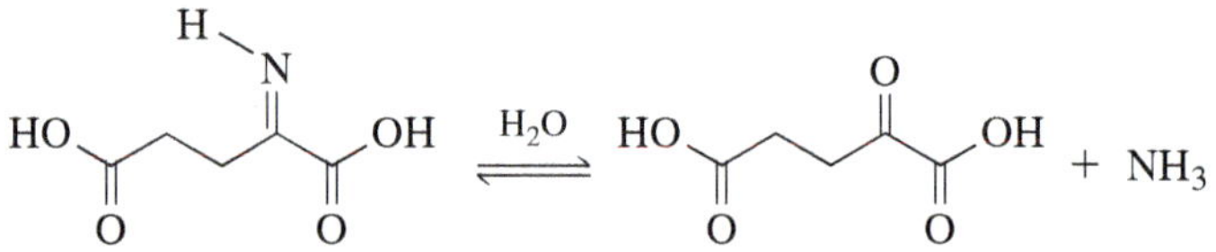

Figure 17.77 Hydrolysis of the imine.

Assessment 17.41 Provide a mechanism for the hydrolysis of the imine shown in Figure 17.77.

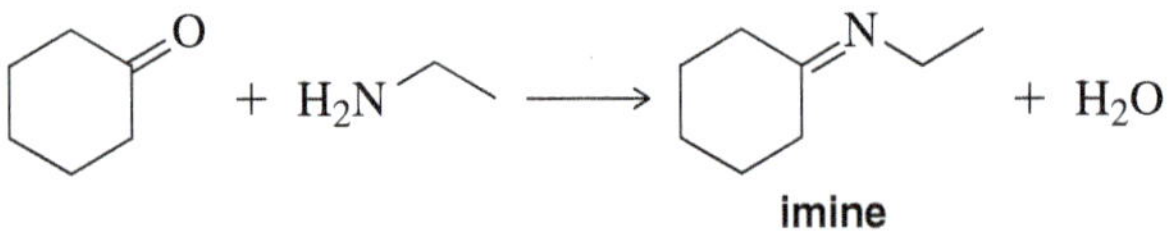

Recall acetal formation (Sec. 17.7.3):

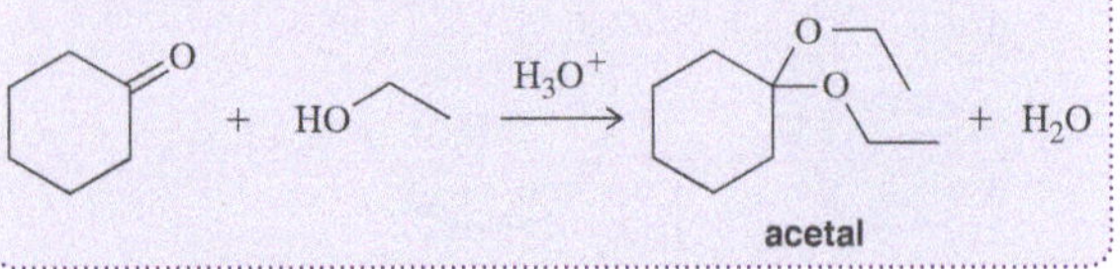

Figure 17.78 Primary (1°) amines react with ketones to give imines.

17.8.2 Addition of a primary amine (RNH_2): imine formation

Addition of a primary amine to an aldehyde or a ketone resembles the addition of ethanol (Figure 17.78). Whereas ethanol addition produces an acetal, here an imine is produced. [*If you could pinpoint one difference between ethanol and ethylamine that leads to this different reactivity, what would it be?*]

In the absence of acid—remember, we can do these reactions with or without it—the nucleophilic nitrogen directly attacks the electrophilic carbon of the carbonyl (General Mechanism 17A). This intermediate, being both acidic ($R_2NH_2^+$) and basic (RO—), undergoes a series of proton transfers to give a heminal, a molecule that is similar to the hemiacetals seen previously (Figure 17.79).

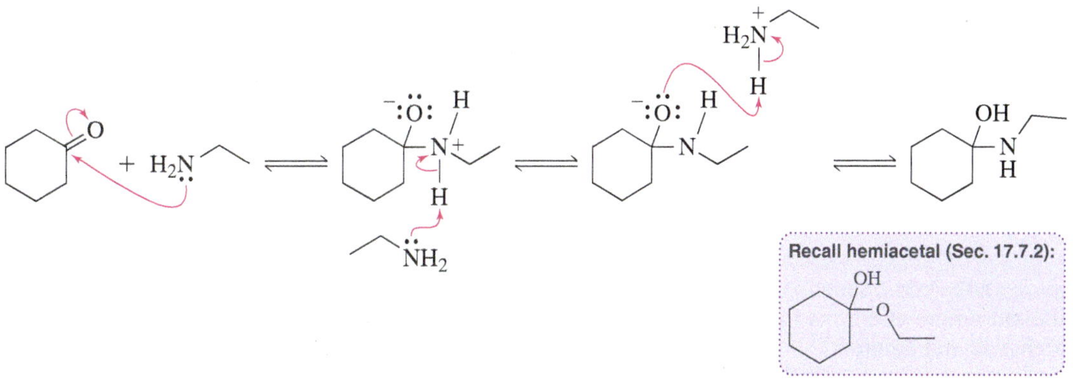

Figure 17.79 The attack of an amine on a carbonyl is similar to the formation of a hemiacetal.

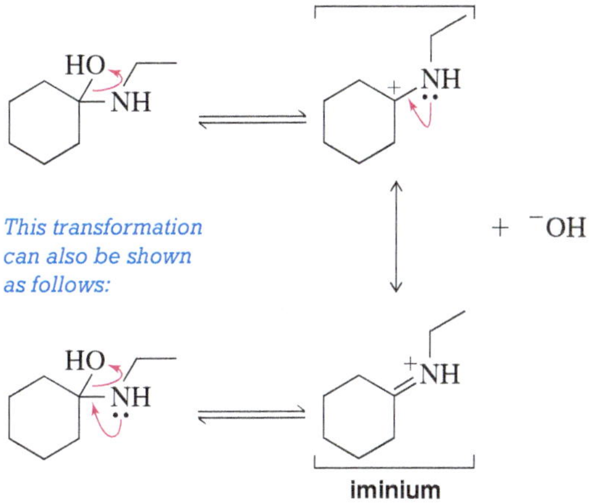

Figure 17.80 Hydroxide is a not a great leaving group, unless it is being forced out.

[*Alert: To this point, the addition of ethylamine should make sense based on previous information. The next step, however, violates one of our previous "rules."*]

This hemiacetal-like intermediate is relatively unstable, decomposing by the loss of hydroxide. [*But isn't hydroxide a poor leaving group?*] Even though hydroxide is a poor leaving group, look closely at what is accomplished by the loss of hydroxide (**Figure 17.80**). That is, a highly resonance-stabilized carbocation is formed. As a result, instead of the water leaving on its own, it is actually forced out by the donation of nitrogen's lone pair.

Deprotonation of this intermediate is less sterically hindered and entropically neutral compared to nucleophilic attack, so deprotonation is once again preferable to nucleophilic attack. Hydroxide, as the strongest base in solution, likely deprotonates the iminium, generating the imine (**Figure 17.81**).

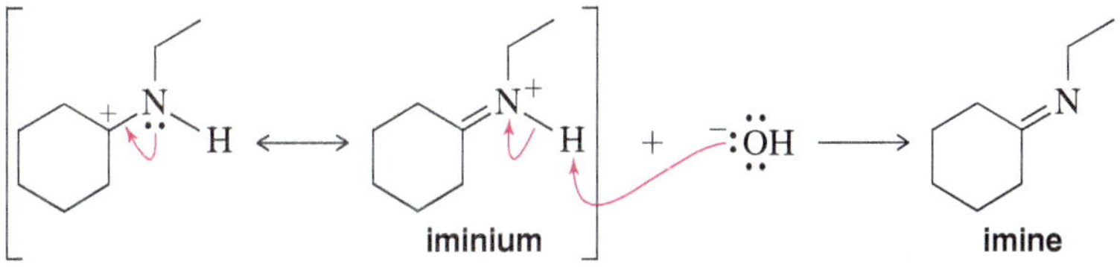

Figure 17.81 After leaving, hydroxide quickly removes a proton to form the imine.

What about hydroxide as a leaving group? Previously (Sections 13.7.2 and 17.7), hydroxide left only after it was protonated (i.e., turned into water, which is a better leaving group). Hydroxide was coaxed to leave, making for a fast and favorable reaction. Here, hydroxide is being kicked out by the electron-donating nitrogen. Although this may be a slow reaction, the immediate protonation of hydroxide makes this a favorable process.

Assessment 17.42 Suggest an amine and an aldehyde/ketone that could be used to make the following imines.

(a)

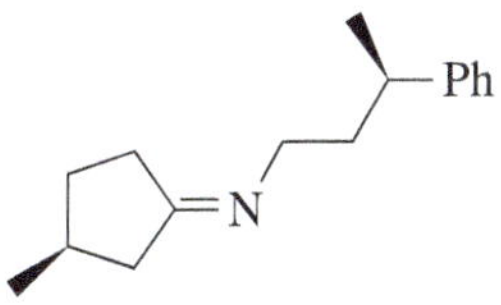

(b)

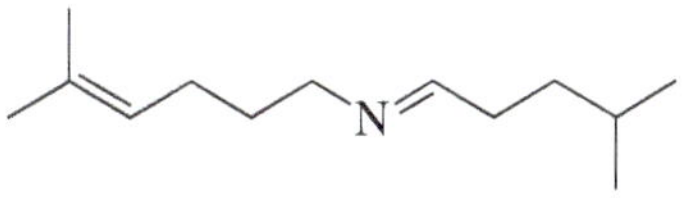

(c)

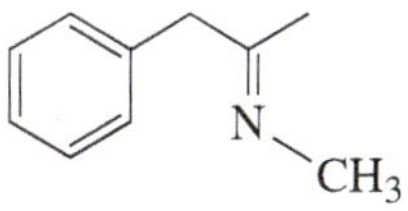

(d)

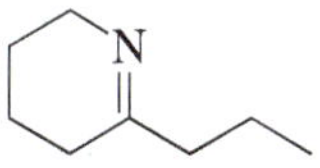

Assessment 17.43 If the imine formation reaction is conducted using an amine whose hydrogens have been replaced with deuterium atoms, then where would the deuterium end up in the product?

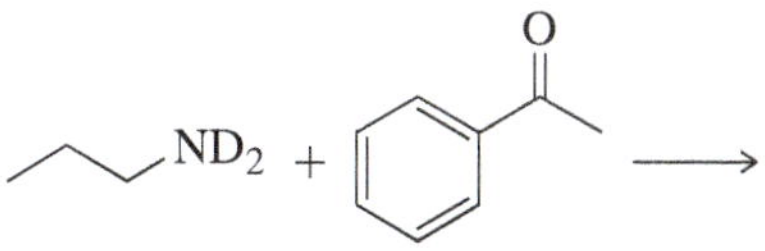

Assessment 17.44 Imine formation is reversible and can be driven to the ketone side by adding water [*Le Châtelier's principle*]. Show a mechanism for the acid-catalyzed version of this reaction.

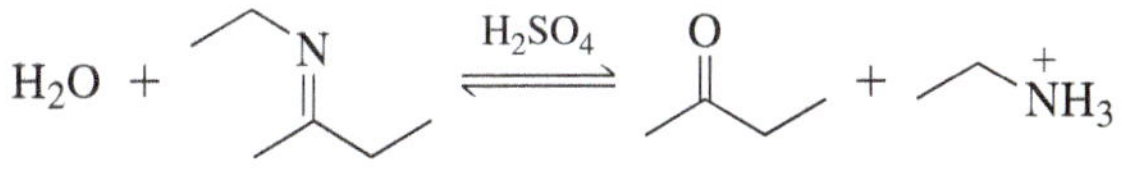

17.8.3 Addition of a secondary amine (R_2NH): enamine formation

In both the addition of ammonia and the addition of a 1° amine, oxygen ultimately forms two new O—H bonds, forming a molecule of water in the process. In the net reaction (ignoring the catalyst), these hydrogens originated on the amine. [*Verify that this is true.*] A secondary amine contains only one N—H bond, so a different reaction outcome is expected. The reaction of a 2° amine and a carbonyl forms an **enamine**, a functional group featuring a neutral nitrogen singly bonded to the carbon of an alkene. Figure 17.82 compares these three reactions.

Figure 17.82 A comparison of different amine additions to aldehydes and ketones.

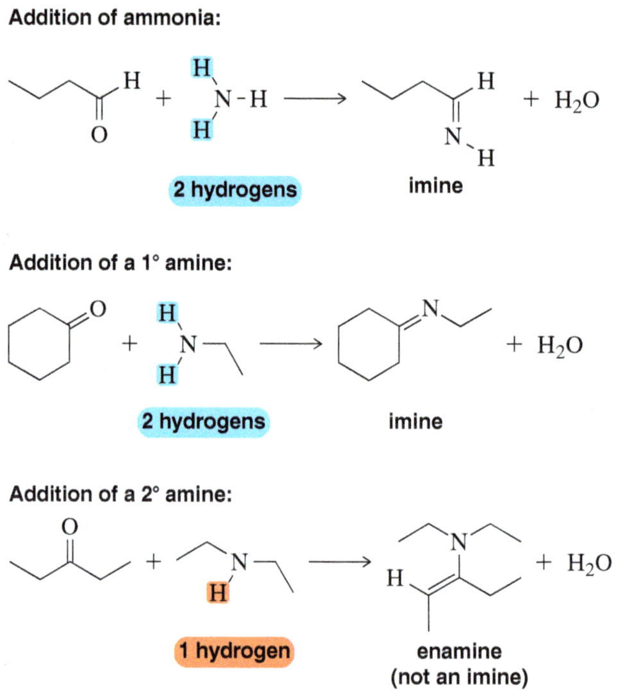

Figure 17.83 shows that the addition of a 2° amine is mechanistically similar to the addition of a 1° amine. The nucleophilic nitrogen first attacks the carbon of the carbonyl. Consecutive acid-base reactions give the neutral hemiacetal-like intermediate. Once again, nitrogen donates an electron pair, kicking out hydroxide.

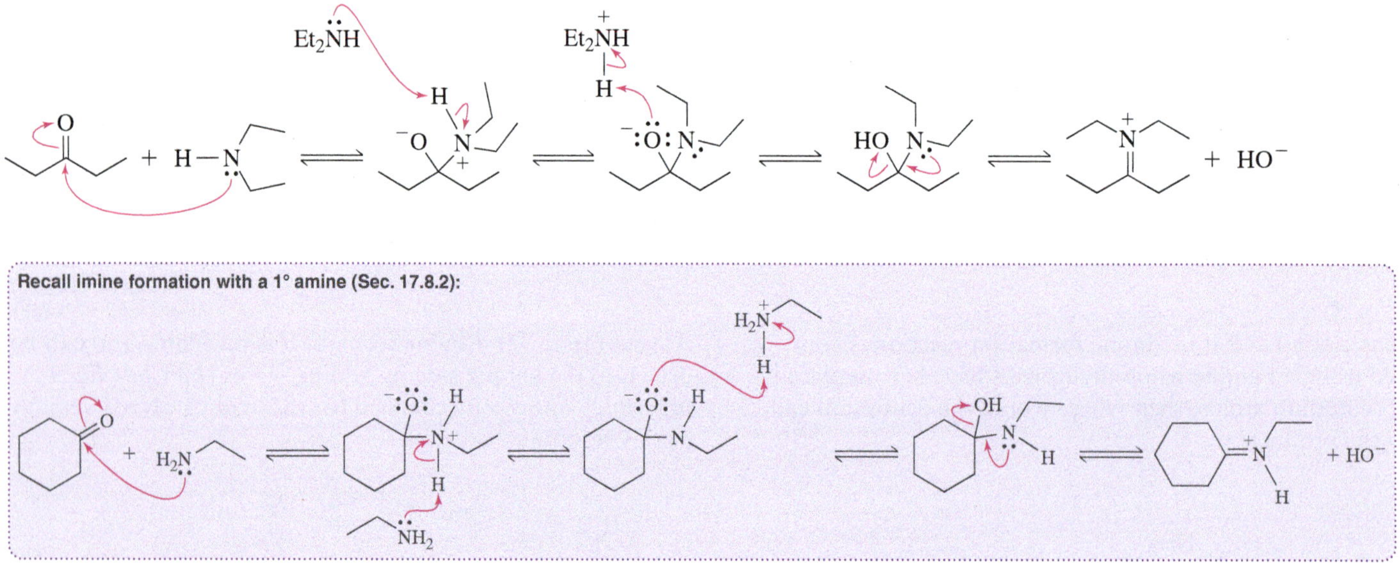

Figure 17.83 A comparison of the addition of a 2° amine to the previously studied 1° amine.

So far, the reaction is identical to the reaction of ammonia and a 1° amine. In those reactions, the next step would involve deprotonation of the positively charged nitrogen. The nitrogen here, though positively charged, has no proton to be removed (**Figure 17.84**).

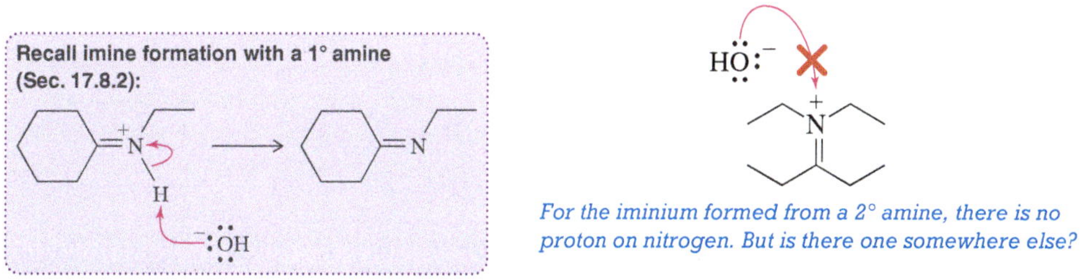

Figure 17.84 With a secondary imine, the resulting iminium does not have an acidic proton on nitrogen.

There are several outcomes possible here (**Figure 17.85**).

1. The reaction could stop at the **iminium ion**. [*Charged molecules are rarely isolated, so this is unlikely.*]
2. The hydroxide could attack the carbon of the iminium ion, returning the electron pair to nitrogen. [*This is the reverse of the previous step and is also unlikely.*]
3. The hydroxide removes a proton from a carbon neighboring the iminium ion. [*This is the step that occurs.*]

Deprotonation of the α-**carbon** results in a carbanion stabilized via a neutral resonance structure. By far the largest contributor to the resonance hybrid, the neutral structure represents the enamine functional group.

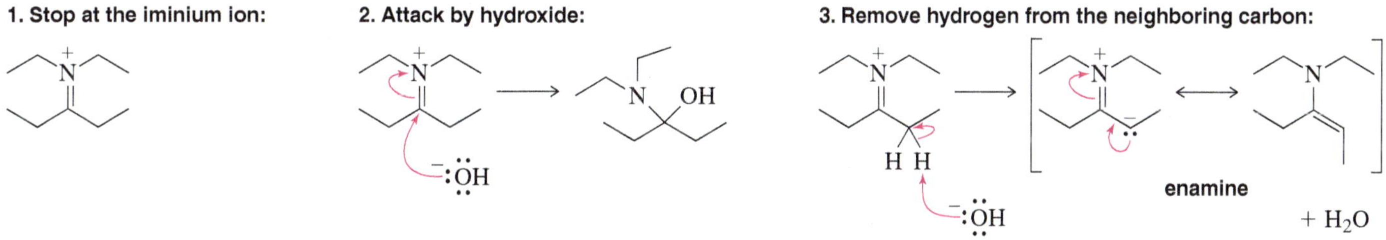

Figure 17.85 Three possible fates of the iminium ion.

Many secondary amines form enamines with carbonyls. When two alkenes, or alkene diastereomers, can form, this reaction, under thermodynamic control, always gives the more stable enamine (**Figure 17.86**).

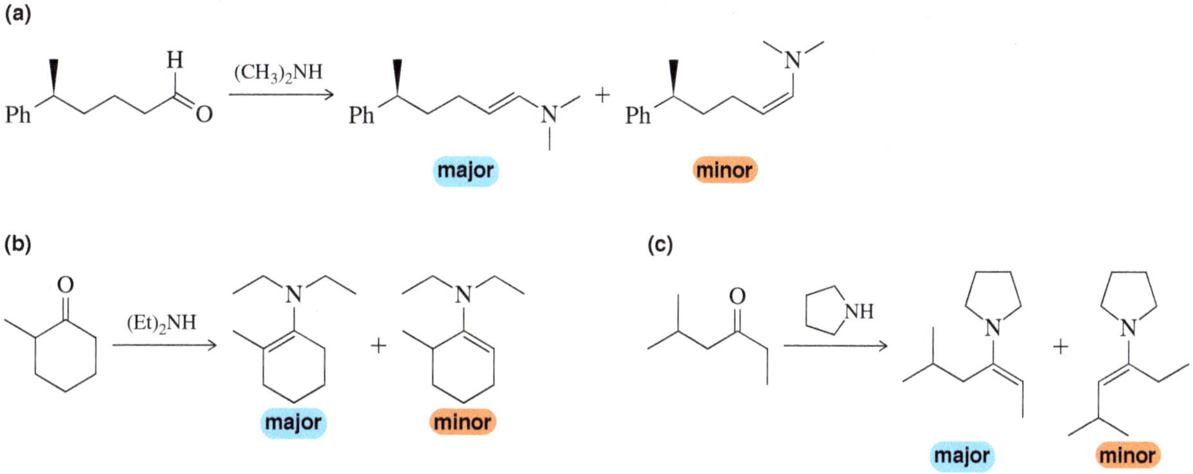

Figure 17.86 Enamine formation is under thermodynamic control.

Assessment 17.45 Predict the enamine that will form upon treatment of the amine and carbonyl shown.

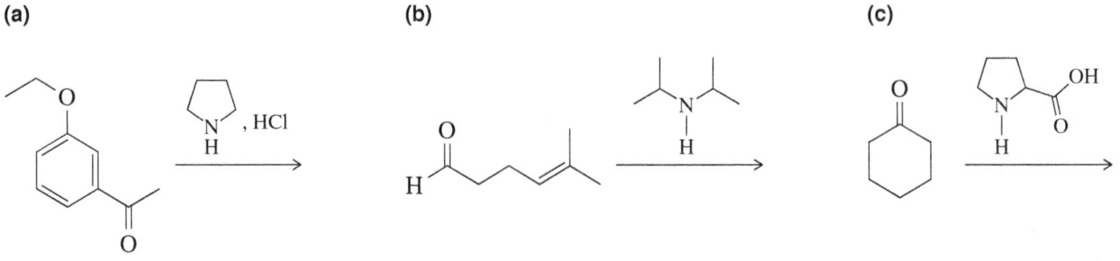

Assessment 17.46 Rationalize the major and minor products for each reaction in Figure 17.86.

17.8.4 pH dependence of amine addition

Although not necessary, the presence of acid can speed up amine addition. But to what extent? And what can this tell us about the mechanism? If some acid increases the rate, does more acid increase the rate more?

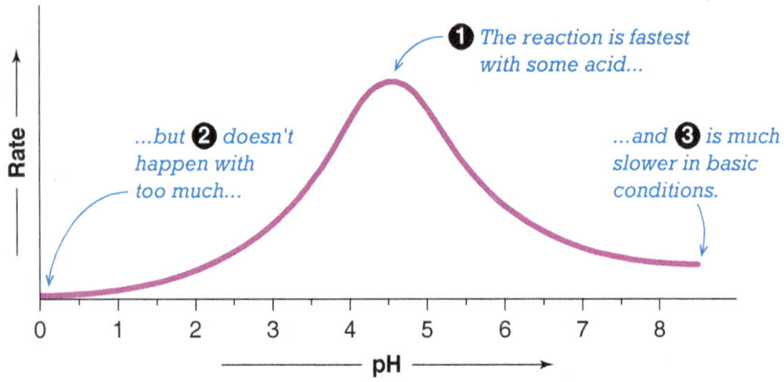

Figure 17.87 Amine addition is slow at low pH, stops at high pH, and is fastest around pH = 4.

The rate of methyl amine addition has been examined at several acid concentrations (**Figure 17.87**). As the pH decreases from 8 to 4 (i.e., as the acid concentration increases), the reaction rate increases. It turns out, though, that adding additional acid beyond pH 4 begins to *slow* the reaction. Why might this be? [*Consider the mechanistic steps from previous sections. What parts would be sped up by acid? What steps would be slowed?*]

To simplify, think of imine formation as the combination of amine addition and dehydration. For each step, consider the observations made at different points of the curve in Figure 17.87. For the first step, the reaction occurs in slightly basic solutions (pH = 8). Because it is a strong enough nucleophile, methylamine attacks a neutral carbonyl, but slowly. At pH 4, the carbonyl can be protonated, resulting in a more reactive electrophile. At this pH, the rate of the first step increases. At higher acid concentrations, both the carbonyl and the amine are protonated, leaving no nucleophile to attack the carbonyl. Thus, at pH = 1, the reaction basically stops (**Figure 17.88**).

Figure 17.88 Protonation of the carbonyl, as long as not all amine is protonated, increases the rate of the first step.

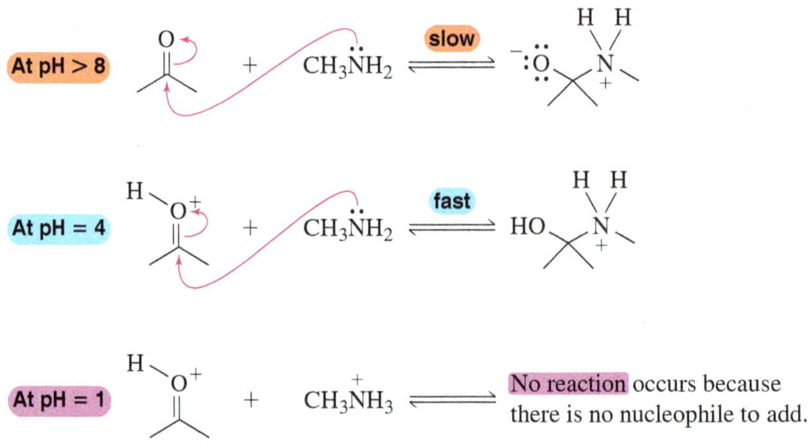

In the second step of imine formation (**Figure 17.89**), our concern is the hydroxyl group leaving. In a mildly acidic solution (pH = 4), the hydroxyl can be protonated, making it a better leaving group. In a basic solution (pH > 8), there are no protons around to protonate the hydroxyl group and help it leave. As such, it leaves as hydroxide more slowly.

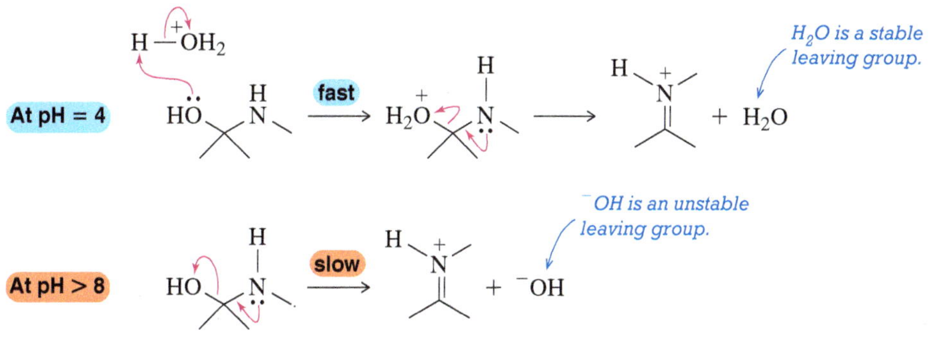

Figure 17.89 Protonation of the hydroxyl group increases the rate of the second step.

17.8.4 SUMMARY

- Imine formation is fastest around a pH of 4. This pH ensures that there are enough protons to assist the hydroxyl group leaving, but not so many that all of the amines are protonated and rendered nonnucleophilic.

17.8.5 Bringing it all together: summary of amine addition

It is helpful to look at nucleophilic substitution of the three types of amines (NH_3, 1°, and 2°) together, as well as to consider the substitution of a 3° amine [*the only type of amine we haven't looked at yet*]. In each case, $\Delta S \approx 0$, so the driving force is formation of a π bond (C=N or C=C). Table 17.13 shows how these reactions are related, each beginning with the addition of the amine.

Table 17.13 Summary of amine additions and why different products are produced with each.

Type of addition	Reaction
Addition of NH_3 to an aldehyde or ketone gives an imine.	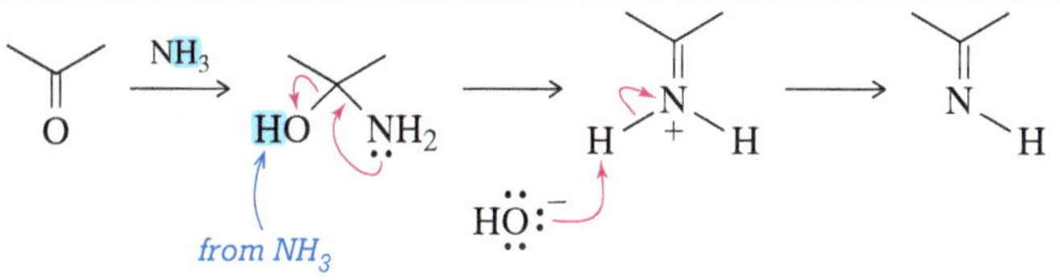
Addition of $EtNH_2$ (1° amine) to an aldehyde or a ketone gives an imine.	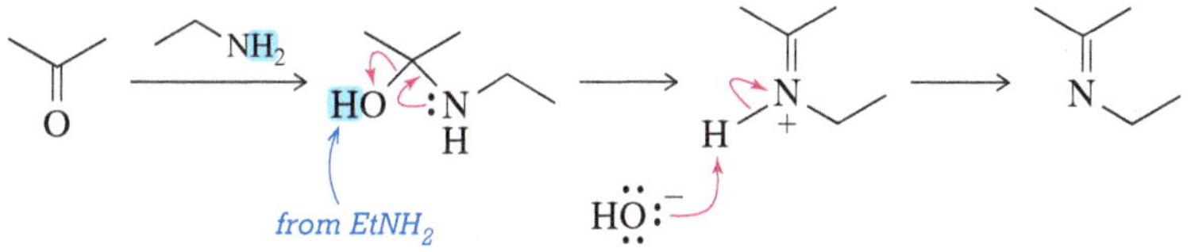
Addition of Et_2NH (2° amine) to an aldehyde or a ketone gives an enamine because there is only one hydrogen on the nitrogen.	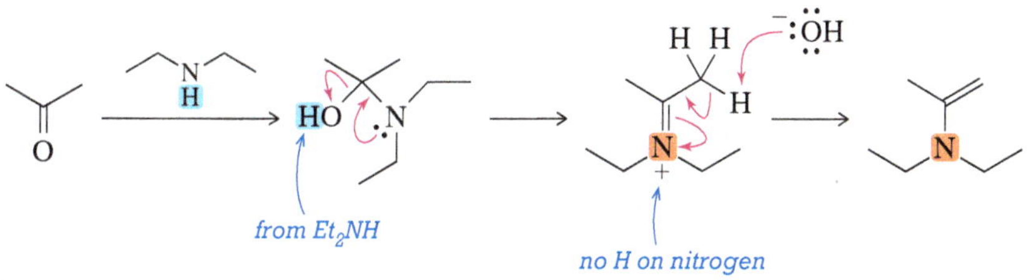
Addition of Et_3N (3° amine) does not react with aldehydes or ketones because there is no hydrogen on nitrogen.	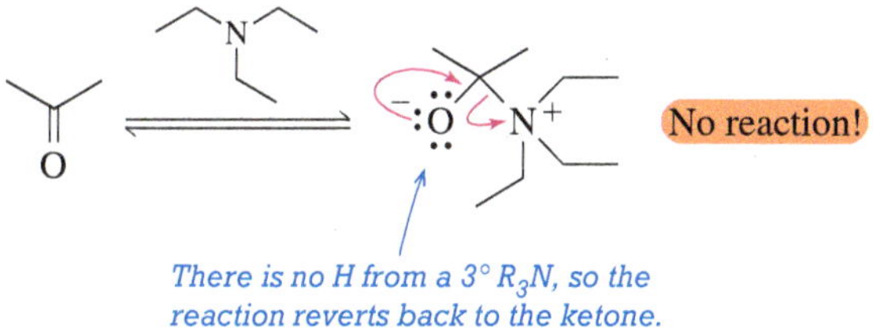

Based on Table 17.13, any amine containing at least one hydrogen can react with a carbonyl. For the most part, addition is independent of the group attached to the nitrogen. **Table 17.14** shows examples of amine derivatives that are used to form imines. These reactions were once used to characterize unknown compounds, but modern spectroscopic methods have made them less useful in this regard.

Table 17.14 Amines can be used to derivatize ketones and aldehydes by imine formation.

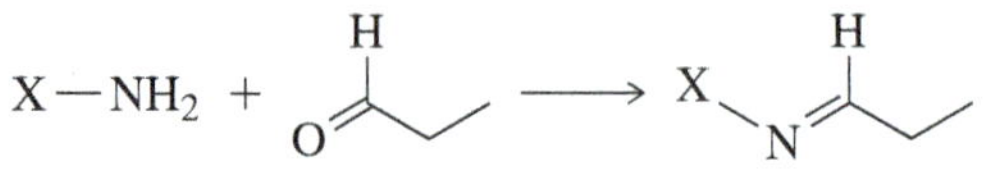

Reagent name	Reagent	Product name	Product
Hydrazine	$H_2N—NH_2$	Hydrazone	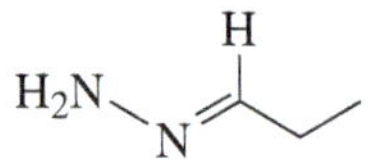
Hydroxylamine	$HO—NH_2$	Oxime	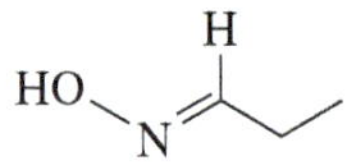
Phenylhydrazine	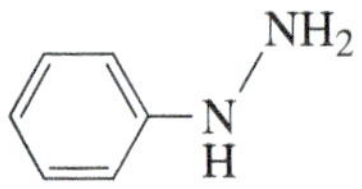	Phenylhydrazone	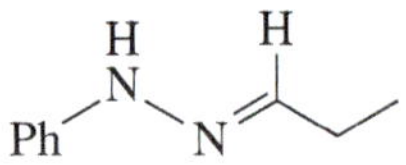
2,4-Dinitro-phenylhydrazine	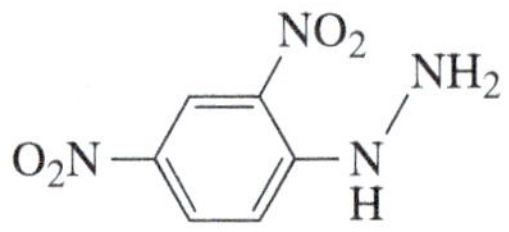	2,4-Dinitro-phenylhydrazone	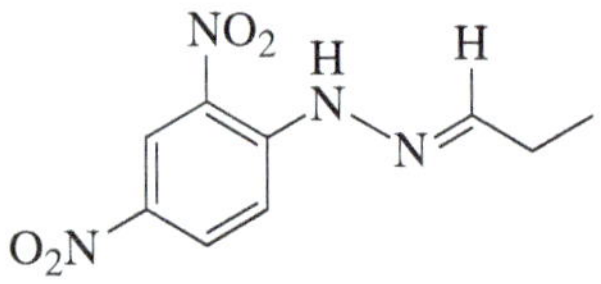
Semicarbazide	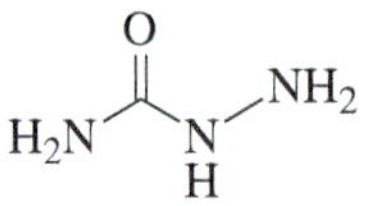	Semicarbazone	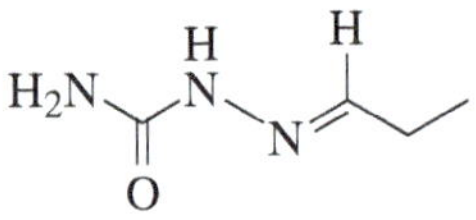

Assessment 17.47 Predict the aldehyde/ketone derivative that would be produced in the following reactions.

(a)

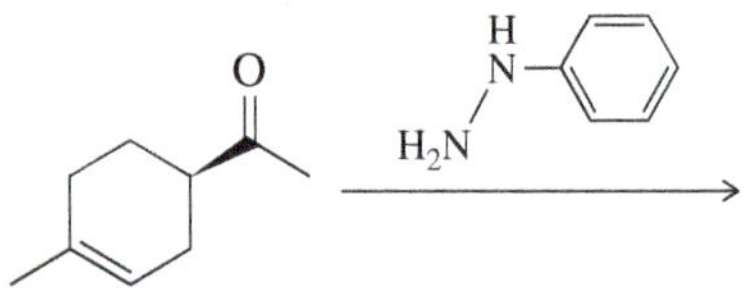

(b)

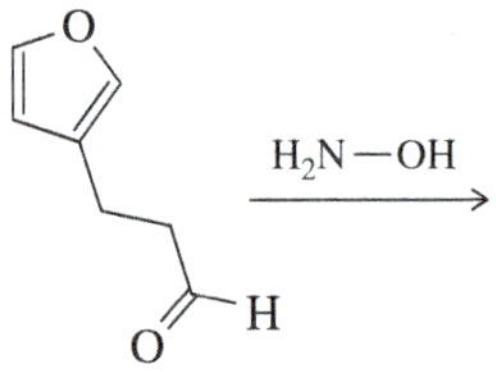

(c)

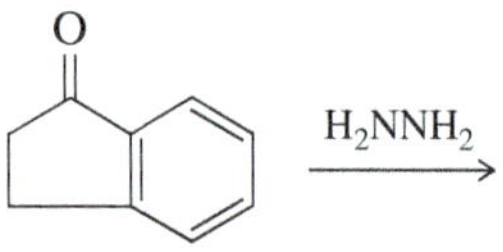

Assessment 17.48 The semicarbazide reagent contains three nitrogen atoms, yet only one of them attacks the carbonyl in the reaction shown. Why do the other two nitrogen atoms *not* participate in this reaction?

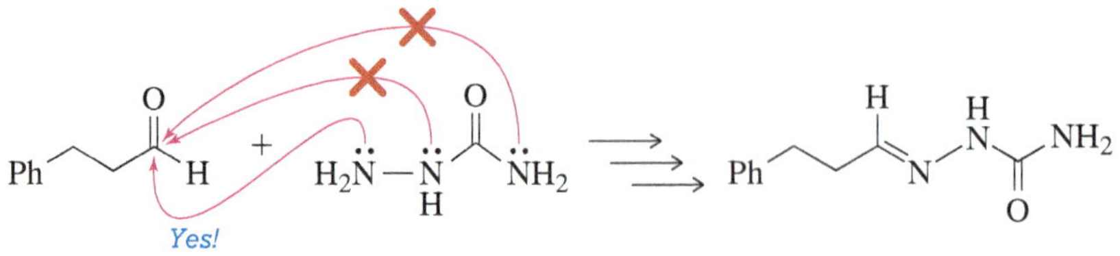

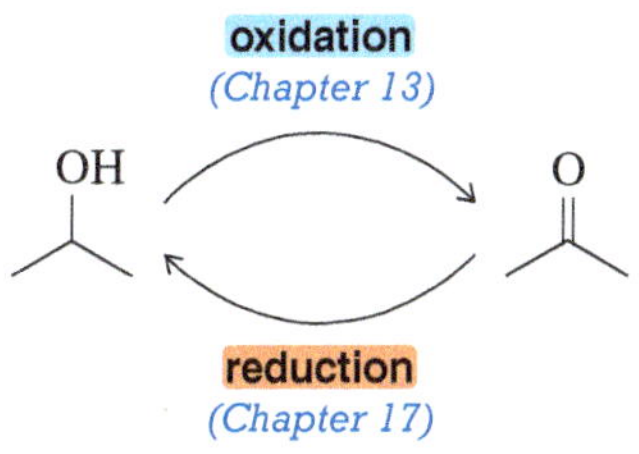

Figure 17.90 **Alcohols and ketones/aldehydes are at different oxidations states.**

17.9 Reduction

Section 13.9 featured the formation of ketones and aldehydes by oxidation of the corresponding alcohols. There are circumstances where the opposite reaction is required, in which an alcohol is formed by reduction of a carbonyl (**Figure 17.90**). This section explores this useful synthetic tool, examining in detail the use of hydrogenation, hydride reducing agents, and, if complete removal of the carbonyl is desired, the Wolff–Kishner reaction.

17.9.1 Hydrogenation

Reduction is often described as the net addition of H_2 to a molecule. Recall from Section 9.2 that we studied the reduction of a C—C π bond using H_2 and a palladium catalyst. This reaction works for aldehydes and ketones, too, because we are forming two new σ bonds at the expense of one σ bond and one π bond (**Table 17.15**). Like alkene hydrogenation, concerted addition of hydrogen occurs from one face of the transition metal catalyst to the carbonyl (**Figure 17.91**). [*Compared to alkenes, is addition of hydrogen as favorable?*]

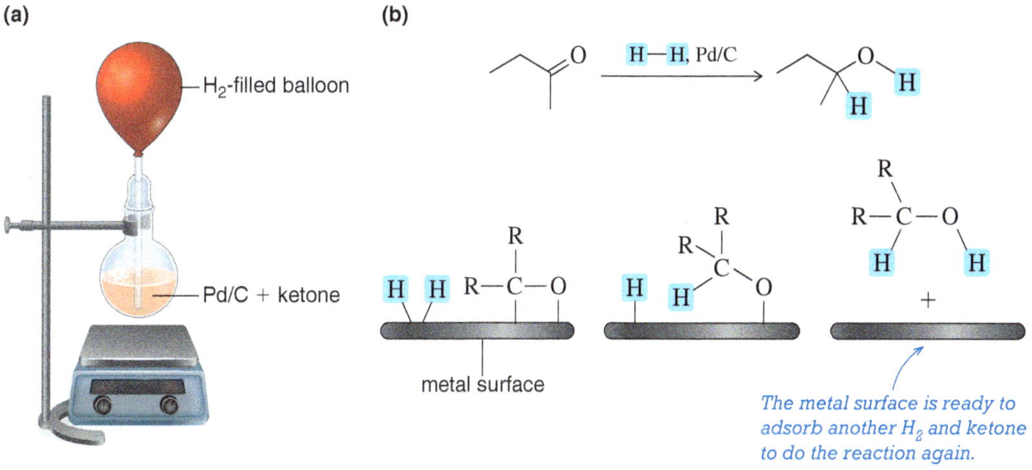

Figure 17.91 **The hydrogenation of a ketone** is similar in mechanism to the hydrogenation of an alkene.

We can assess the relative favorability of alkene and carbonyl hydrogenations experimentally. From Section 9.2, the ΔH of alkene hydrogenation was approximately −27 kcal/mol. This value is based on the sum of the bonds broken and formed (**Figure 17.92**). Given the relative stability of the C—O π bond, is reduction of the carbonyl more or less favorable than alkene hydrogenation? [*Consider this question before moving on.*]

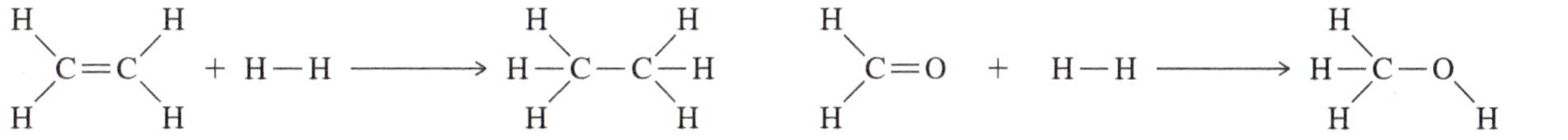

Bonds broken: C—C π ~65 kcal/mol; H—H σ 104 kcal/mol
Bonds formed: 2 C—H σ 98 kcal/mol

$\Delta H = (65 + 104) - (98 + 98) = -27$ kcal/mol

Bonds broken: C—O π ~85 kcal/mol; H—H σ 104 kcal/mol
Bonds formed: C—H σ and O—H σ 98 kcal/mol 105 kcal/mol

$\Delta H = (85 + 104) - (98 + 105) = -14$ kcal/mol

Figure 17.92 **The hydrogenation of a ketone is *not* as favorable as the hydrogenation of an alkene.**

The C—O π bond is stronger than the C—C π bond. Thus, it takes more energy to break it, resulting in a less negative ΔH. In other words, carbonyl hydrogenation is less favorable than alkene hydrogenation. For this reason, reduction of a carbonyl requires higher H_2 pressures and elevated temperatures. We see the consequences of the reduced reactivity in Section 17.9.2.

Table 17.15 The hydrogenation of an aldehyde or a ketone makes it possible to synthesize primary and secondary alcohols, respectively.

Product	Reaction	Description
1° alcohols	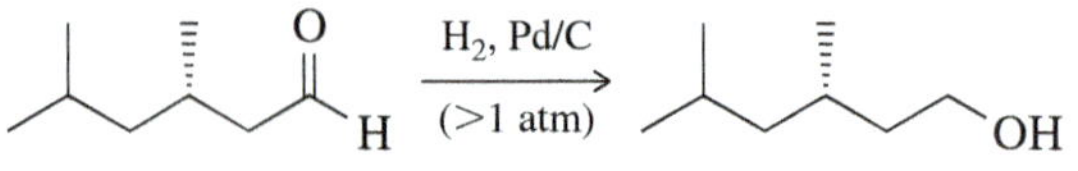	Hydrogenation of an aldehyde with elevated H_2 pressures produces a 1° alcohol.
2° alcohols	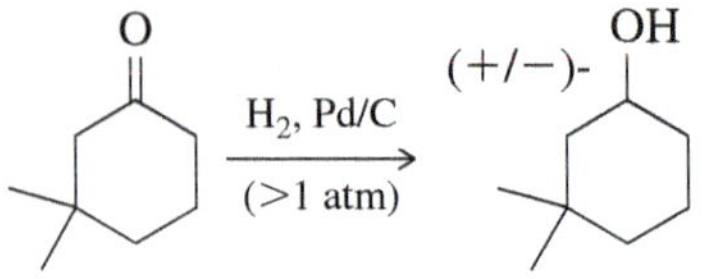	2° alcohols result from hydrogenation of ketones. In the absence of a chiral environment, a racemic mixture forms.

Assessment 17.49 Predict the product of the hydrogenation reactions shown.

(a)

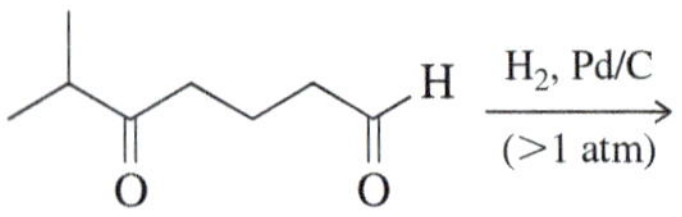

(b)

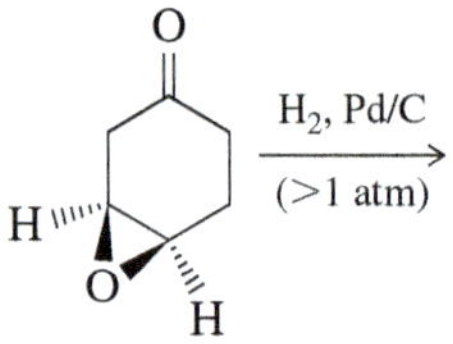

(c)

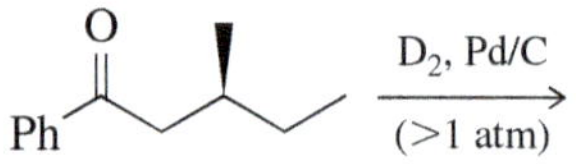

(d)

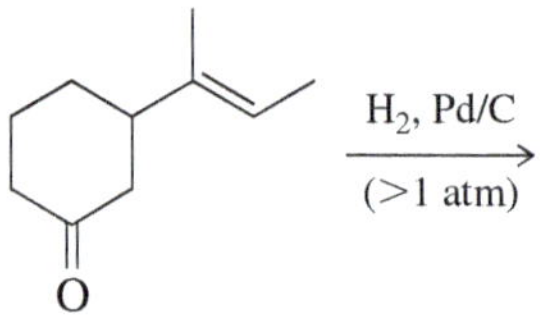

17-D CHEMISTRY IN CONTEXT: Asymmetric Synthesis

Enantioselective hydrogenation

In the absence of a chiral environment, the hydrogenation of ketones gives a racemic mixture of products.

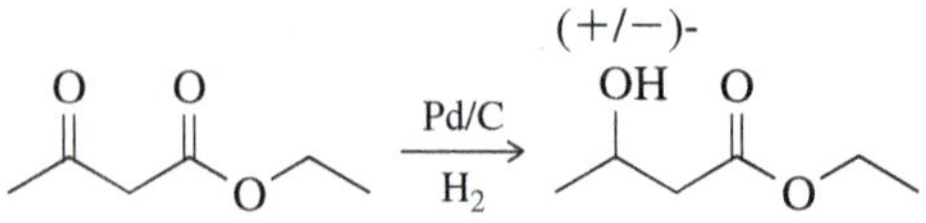

As discussed in Section 6.6, addition of hydrogen to the top face of the planar carbonyl is the enantiomer of addition to the bottom face (**Figure 17.93**). These enantiomeric transition states have identical activation energies and therefore proceed with equal rates to produce a racemic mixture.

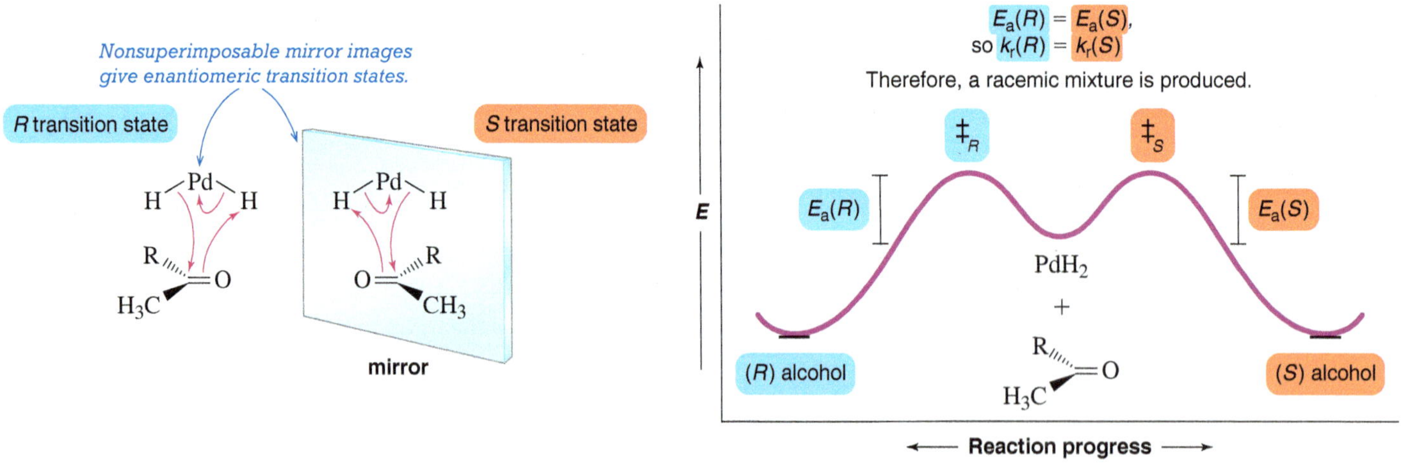

Figure 17.93 Reaction in an achiral environment gives enantiomeric transition states and therefore equal rates.

(continued)

Using a chiral catalyst developed by Nobel Prize winner Professor Ryoji Noyori, the situation changes (**Figure 17.94**). The enantiomerically pure catalyst results in transition states that are diastereomeric. The rate of formation of each enantiomer is now different, and enantioselectivity is achieved.

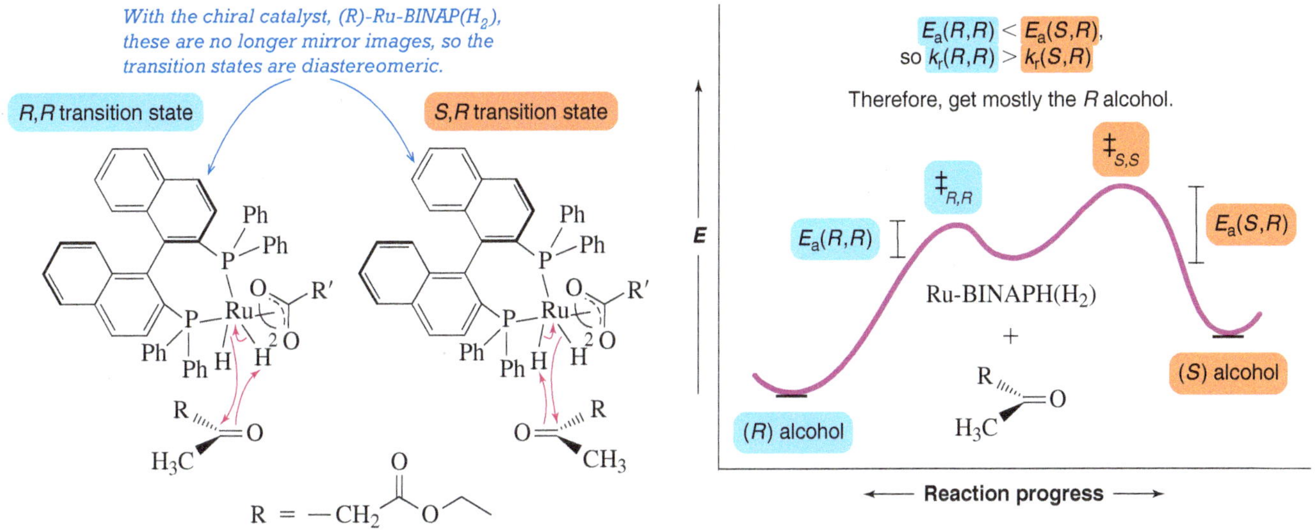

Figure 17.94 **Reaction in a chiral environment gives diastereomeric transition states and therefore unequal rates.**

The versatility of this and similar systems lies in the fact that either enantiomeric alcohol can be produced, depending on the catalyst enantiomer chosen (**Figure 17.95**).

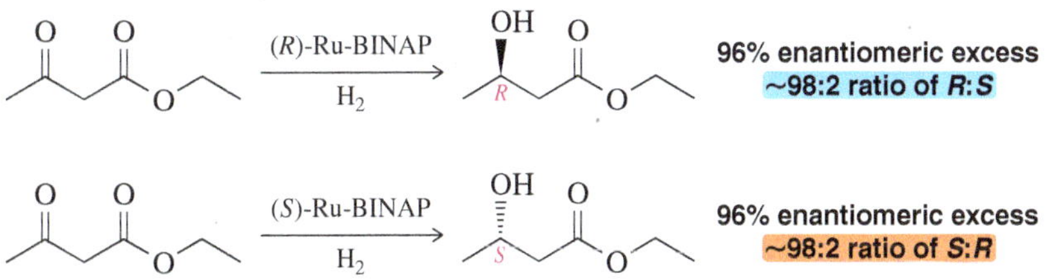

Figure 17.95 **Different catalyst enantiomers produce opposite alcohol enantiomers.**

17.9.2 Reduction with hydride reagents

Carbonyls can also be reduced using what are referred to as hydride reagents.

Let's think like a chemist to see how a reagent like this may have been conceived.

Thinking Like a Chemist

Assessment 17.49(d) asked you to predict the product of a ketone hydrogenation reaction. In this case, there was an alkene present in the same molecule. The use of higher H_2 pressures for ketone hydrogenation results, in this particular case, in the reduction of both π bonds [**Figure 17.96(a)**]. Given that C—C π bonds are more easily reduced than C—O π bonds, we can only selectively reduce the alkene in the presence of the ketone [**Figure 17.96(b)**].

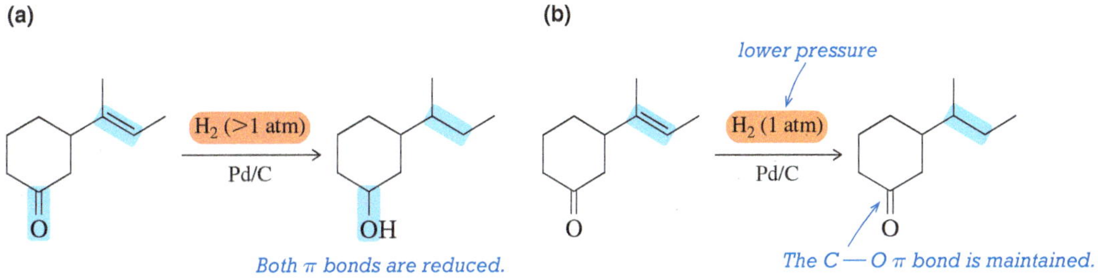

Figure 17.96 **Selective ketone hydrogenation is *not* possible.** (a) At high pressures, both the alkene and the carbonyl are reduced. (b) At low pressures, only the alkene is reduced.

As synthetic chemists, we might want a way to reduce the ketone in the presence of the alkene. [*What if a potentially useful pesticide being synthesized required an alcohol and an alkene to be present?*] To design a new reagent for this, we need to recognize the inherent reactivity differences between carbonyls (most often electrophiles; this chapter) and alkenes (most often nucleophiles; Chapters 8 and 9) (**Figure 17.97**).

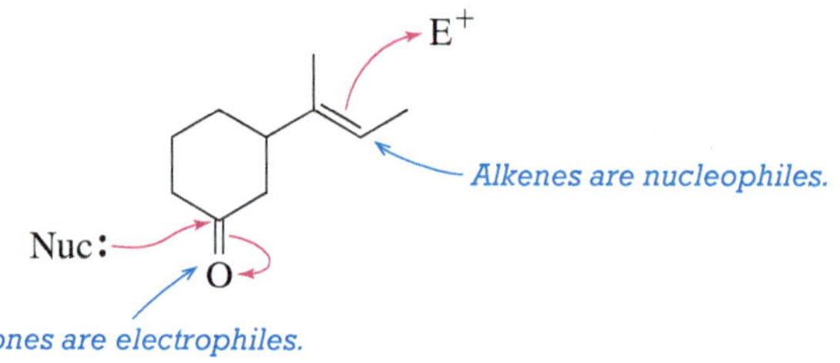

Figure 17.97 A nucleophilic reducing agent should selectively reduce the electrophilic ketone.

The net reduction reaction results in addition of hydrogens to the carbon and oxygen of the ketone. Following General Mechanism 17A, a nucleophilic hydride (H^-) should react selectively with the carbonyl (**Figure 17.98**), pushing the C—O π electrons onto oxygen. Quenching of this alkoxide with acid would generate a 2° alcohol. ■

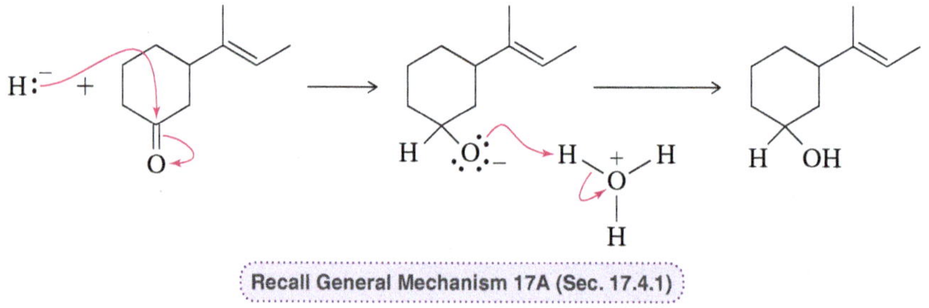

Figure 17.98 A hypothetical selective reduction of a carbonyl in the presence of an alkene using a nucleophilic reducing agent. This is only hypothetical because H^- usually reacts as a base and not a nucleophile.

Though sodium hydride (NaH) might seems like a good option for this reaction, it reacts only as a *base*. To reduce carbonyls under nucleophilic conditions, sodium borohydride ($NaBH_4$) and lithium aluminum hydride ($LiAlH_4$) are used.

Recall that boron and aluminum are exceptions to the octet rule because they are stable with six valence shell electrons [*because they are neutral*], and they are also stable with a full octet [*because every element (except H and He) is stable with eight valence electrons*]. As a result, sodium borohydride and lithium aluminum hydride can be thought of as equilibrating between the negative and neutral forms (i.e., borane and alane) in **Figure 17.99(a)**. A simplified mechanism of carbonyl reduction involves BH_4^- donating a hydride to a carbonyl, resulting in an alkoxide that is protonated by ethanol [**Figure 17.99(b)**]. [*This is General Mechanism 17A.*]

In reality, each molecule of BH_4^- reduces four carbonyls [**Figure 17.99(c)**]. After giving up a hydride ion, BH_4^- becomes BH_3 (borane). With six valence electrons, sp^2 hybridization, and an empty *p* orbital, borane is a good Lewis acid. Thus, the alkoxide reacts with it instead of grabbing a proton. [*This will be a key step in the design of future selective reducing agents.*] When this happens, a new hydride reagent is generated ($ROBH_3^-$). This reagent can reduce a second equivalent of aldehyde. The process continues until a single molecule of borohydride has reduced as many as four carbonyls.

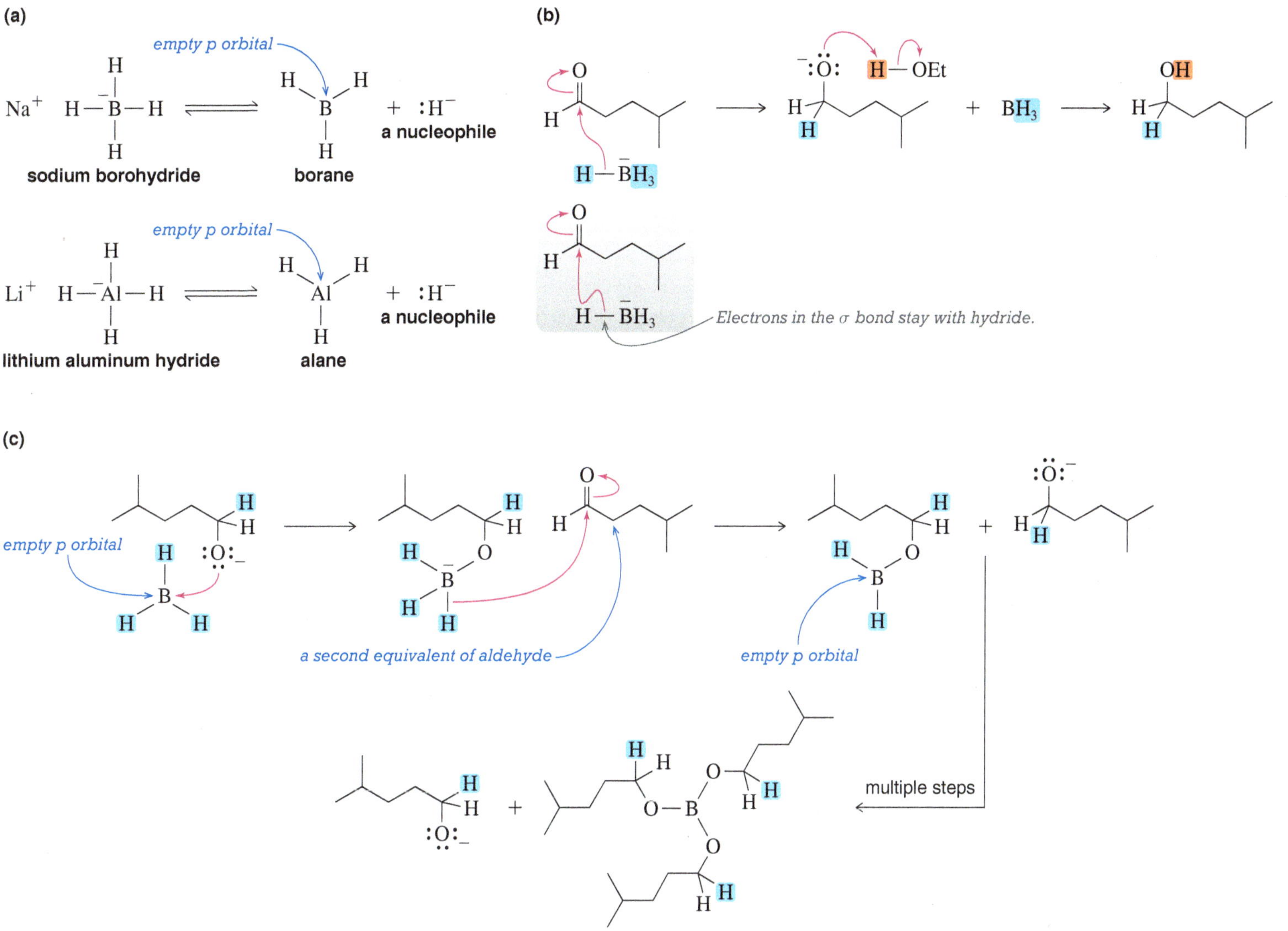

Figure 17.99 Hydride reduction of carbonyls. (a) Lithium aluminum hydride and sodium borohydride are hydride donors. (b) The mechanism of reduction can be thought of as proceeding by General Mechanism 17A with ethanol as the source of the acidic proton. (c) Each molecule of $NaBH_4$ can reduce four carbonyls because a borohydride is continually regenerated. [*Recall the addition of borane to alkenes in Section 8.4.5.*]

Lithium aluminum hydride reacts similarly but is much more reactive (**Figure 17.100**). As a result, reductions are conducted at lower temperatures (−78 °C) in aprotic solvents. With no solvent hydrogens available, an acid quench is required to generate the hydroxyl group.

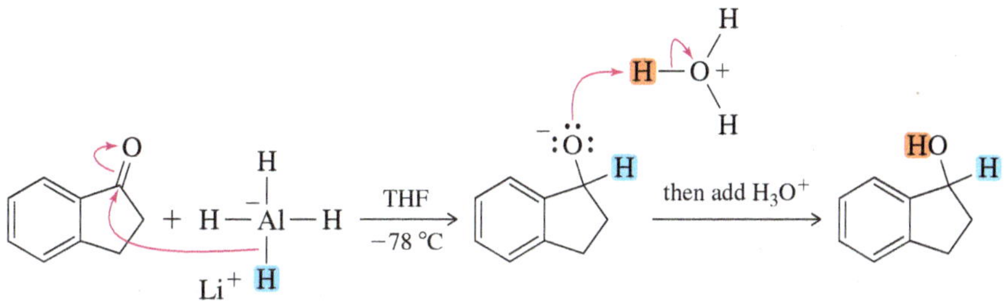

Figure 17.100 The reduction of a ketone with lithium aluminum hydride is done at low temperature. [*This is General Mechanism 17A.*]

17.9.2 SUMMARY

- Sodium borohydride and lithium aluminum hydride donate a hydride to the carbonyl carbon, producing 1° or 2° alcohols.

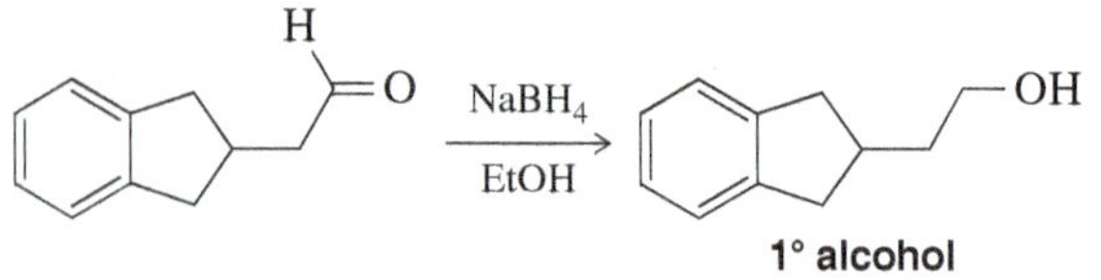

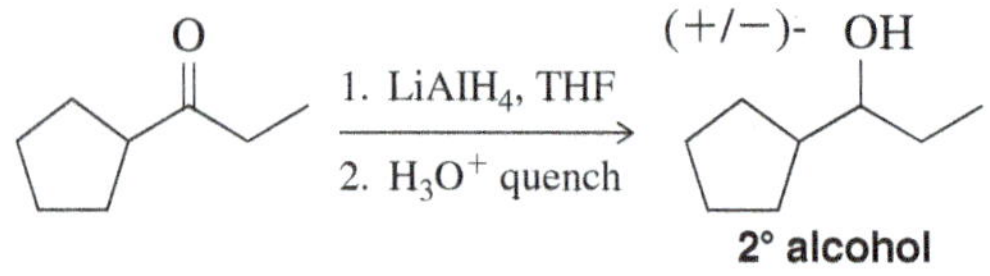

Assessment 17.50 Predict the product of the following reactions. If a new stereocenter is generated, indicate whether or not you would expect an equal mixture of stereoisomers to be produced.

(a)

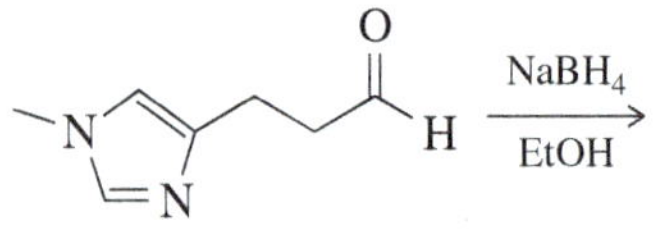

(b)

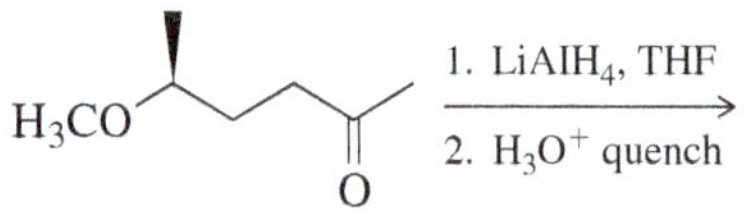

(c)

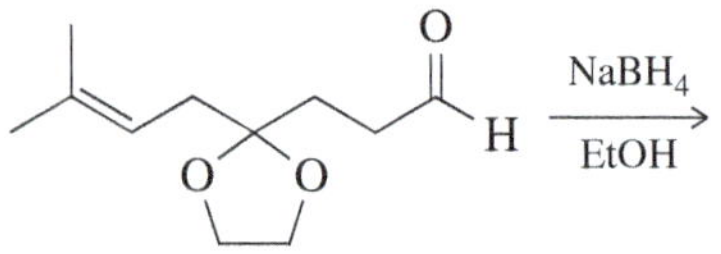

17-E CHEMISTRY IN CONTEXT: Green Chemistry

High-speed ball milling

Most reactions are facilitated by a solvent that can dissolve both the reactants and reagents, allowing them to collide and form new products. Because the solvent is not part of the usable final product, it is typically discarded. To avoid contaminating the environment with potentially toxic solvent waste, chemists are seeking solutions to this problem. Potential solutions include recyclable solvents, solvents derived from renewable resources, or developing solvent-free reaction conditions.

Without a solvent, the number of collisions between molecules is decreased. One emerging technology, called high-speed ball milling (HSBM), facilitates these collisions under solvent-free conditions. In HSBM (**Figure 17.101**), metal ball bearings are placed inside a vessel containing the organic reactants, which is then shaken at high speeds. This creates enough force to turn even solid reagents into an amorphous mixture, allowing the reaction to occur.

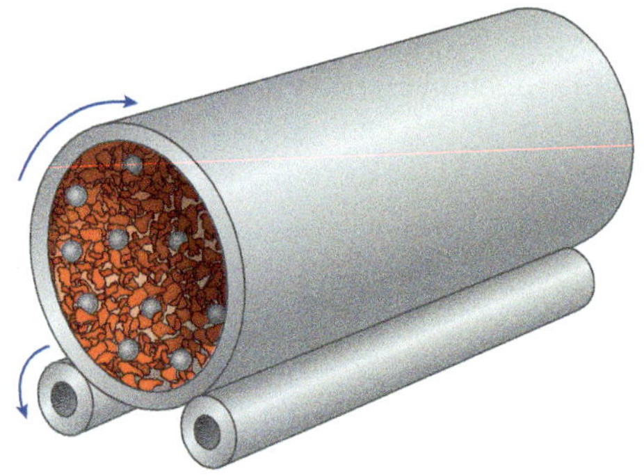

Figure 17.101 Schematic of high-speed ball milling reactions.

Although early in its development, HSBM has found utility in the sodium borohydride reduction of carbonyls. The yields are similar to those obtained in solvent-run reactions.

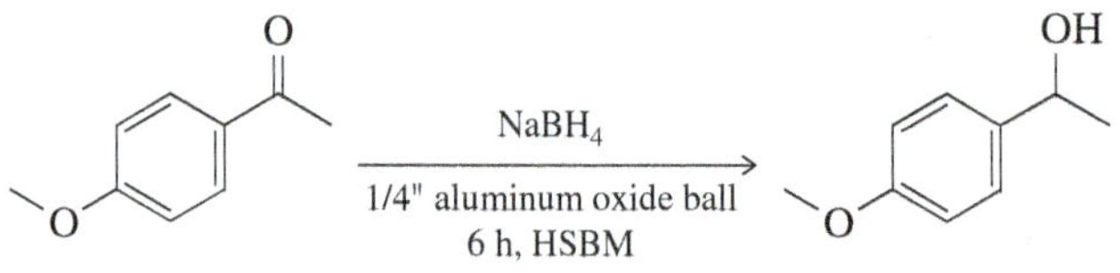

17-F CHEMISTRY IN CONTEXT: Biochemistry

Nature's reducing agent

Chapter 5 showed that enzymes catalyze reactions in either direction, proceeding by equal but opposite mechanisms (see the principle of microscopic reversibility in Section 5.2.5). Having seen a couple of examples of dehydrogenases as oxidizing agents, it should not be surprising that they can also serve as reducing agents.

Lactate dehydrogenase (LDH) is an example of a reducing enzyme. It catalyzes the reduction of pyruvate to lactate, which is an important reaction for anaerobic respiration (**Figure 17.102**). The soreness associated with strenuous exercise has been attributed to the buildup of lactic acid in overworked muscles.

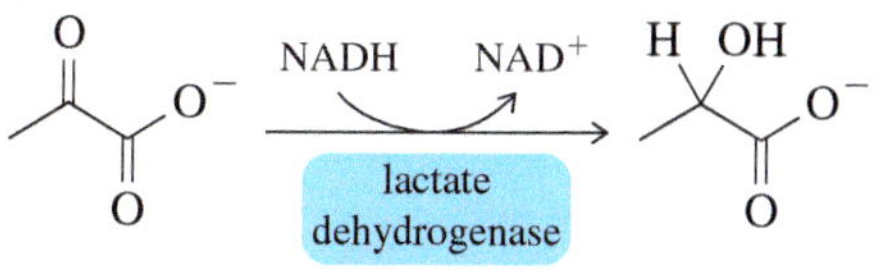

Figure 17.102 Lactate dehydrogenase can reduce ketones to alcohols.

Mechanistically, LDH uses NADH (the reduced form of NAD^+) as cofactor (**Figure 17.103**). Upon binding pyruvate and NADH in the active site, NADH transfers a hydride anion to the carbonyl carbon. Protonation of the alkoxide results in the alcohol. NAD^+ is regenerated in this process. [*Note the similarity between this reaction and the other hydride reducing agents. This reaction is the exact reverse of the dehydrogenase oxidation mechanism (Chemistry in Context 17-A).*]

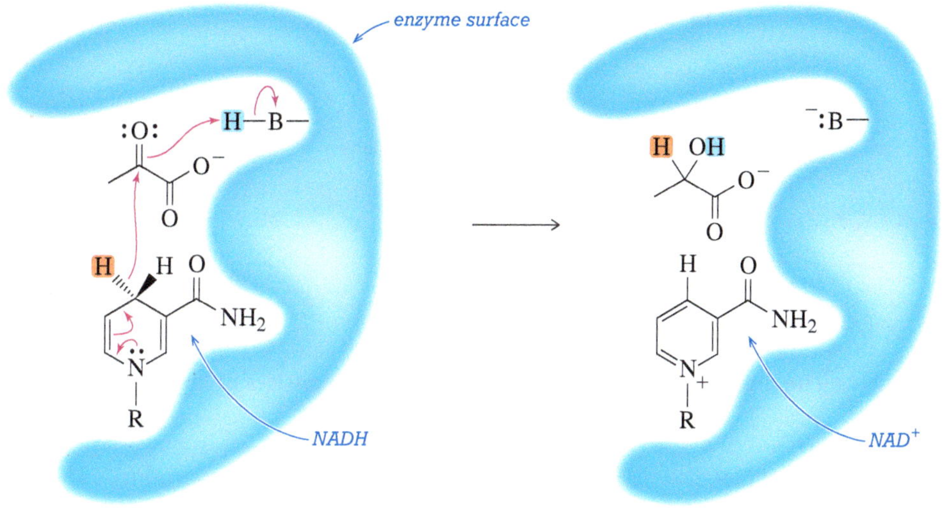

Figure 17.103 The mechanism of enzymatic reduction.

17.9.3 Exhaustive reduction: Wolff–Kishner reaction

The major purpose of developing new reactions is to use them as tools for synthesis. Situations will arise where a carbonyl needs to be completely removed from a molecule. The conversion of aryl ketones to alkyl benzenes is one such case—a process that is effected by the Wolff–Kishner reaction.

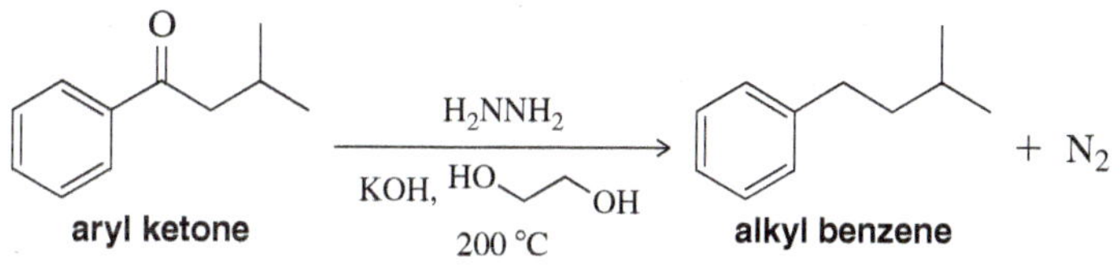

Complete removal of the stable carbonyl requires harsh conditions (high temperatures, high pressures, concentrated bases) that allow any activation barrier to be overcome. As a result, the mechanism presented here is unlike any other. [*Your instructor will clarify whether you need to be able to reproduce it.*] And yet the reaction proceeds through four major steps (**Figure 17.104**), two of which are familiar.

1. Initially, condensation of hydrazine and a ketone occurs to form a hydrazone via a standard imine-forming mechanism. [*Recall that hydrazone is one of the special imines in Table 17.14.*]
2. Initiated by a strong base and extensive heating, tautomerization occurs. In addition to the double bond switching from the C=N to N=N, the hydrogen that was on the terminal nitrogen is now on carbon. [*Recall the keto–enol tautomerization reaction from Section 10.8.3.*]

3. Finally, in the new step, the N≡N triple bond forms, kicking out a carbanion in the process. Although carbanions are rarely leaving groups, this reaction is similar to an elimination reaction [*E2; Section 12.5*]. The driving force for this reaction is the formation of the stable N_2 along with the positive entropy associated with forming a gas.
4. The reaction finishes with a simple acid–base reaction.

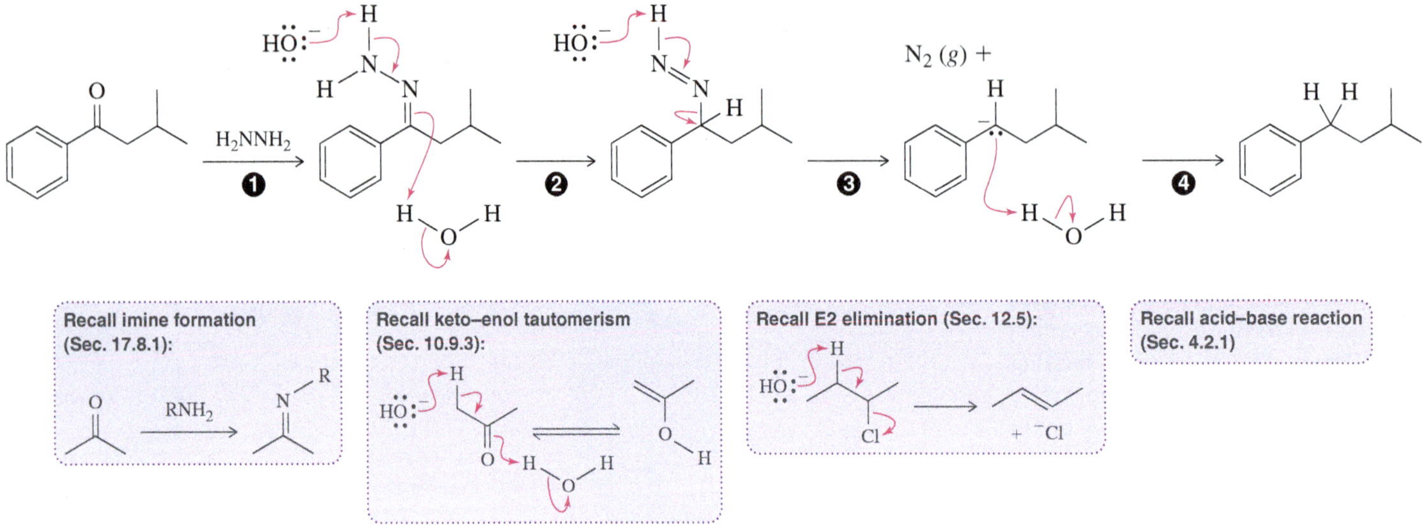

Figure 17.104 The mechanism of the Wolff–Kishner reaction consists of four steps, two of which we have seen in other contexts.

Assessment 17.51 Show the product expected by the Wolff–Kishner reduction of the following aldehydes/ketones.

(a)

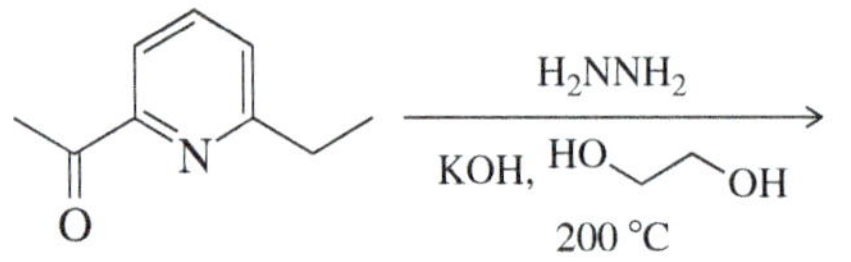

(b)

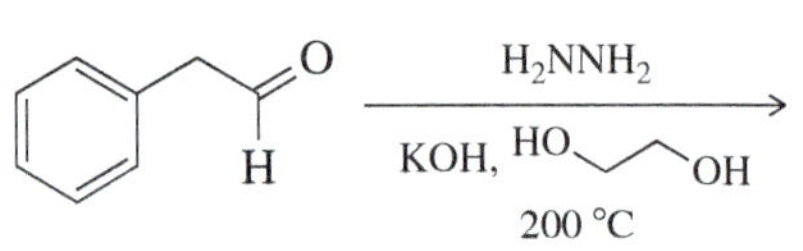

Assessment 17.52 Show the product you would expect if the Wolff–Kishner reaction were done using deuterated water and deuterated ethylene glycol.

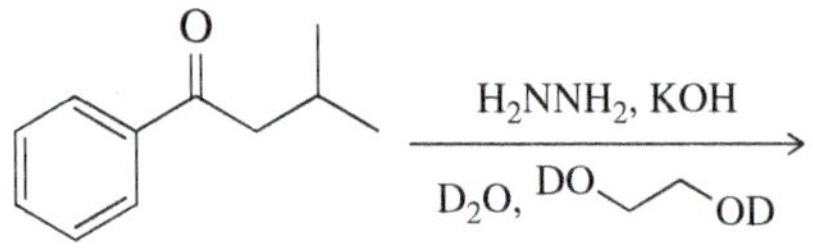

17.10 Oxidation of aldehydes and ketones

Ketones can be further oxidized, though not easily. With no hydrogen attached to the carbon of a ketone (a requirement for easy oxidation discussed in Chapter 13), ketone oxidation requires significantly harsher conditions. The production of adipic acid from cyclohexanone is an industrially useful ketone oxidation (**Figure 17.105**). Adipic acid is a precursor to the polymer nylon, so it is produced by this reaction at a rate of 2.5 billion kilograms per year.

The rest of this section reviews what we have learned previously about aldehyde oxidation.

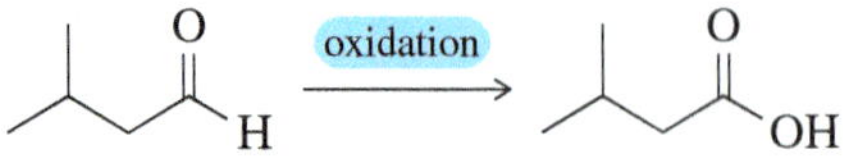

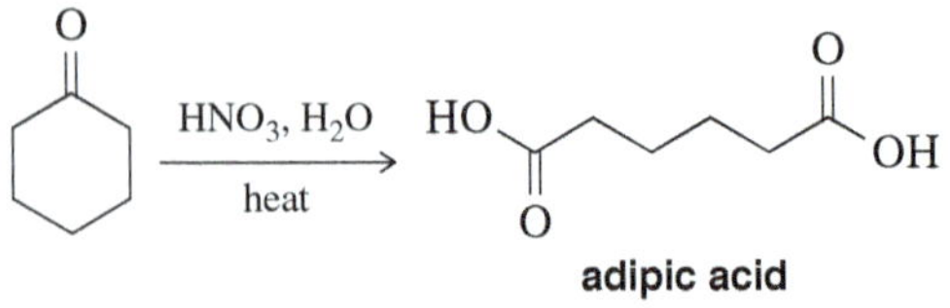

Figure 17.105 The industrial synthesis of adipic acid uses nitric acid as the oxidizing agent.

Chromic acid (H_2CrO_4) is the simplest method for oxidizing aldehydes (**Figure 17.106**). Because chromic acid is used in an aqueous solution, there will always be a small amount of the hydrate form of the aldehyde present at equilibrium (step 1; see Section 17.7.1). This hydrate reacts with chromic acid (step 2) to form a chromate ester, elimination of which forms the new C—O π bond (step 3).

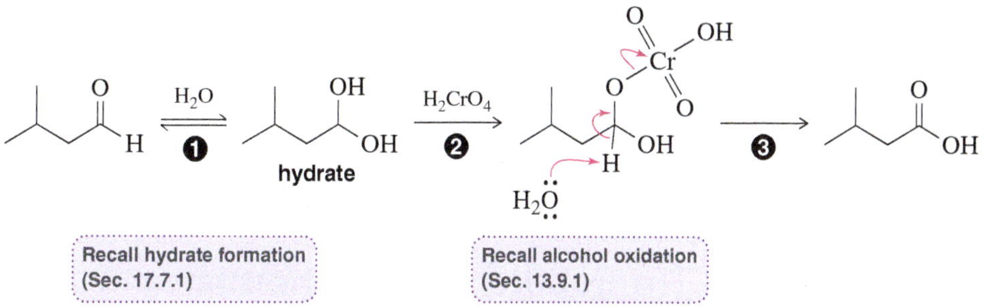

Figure 17.106 Chromic acid oxidation of aldehydes.

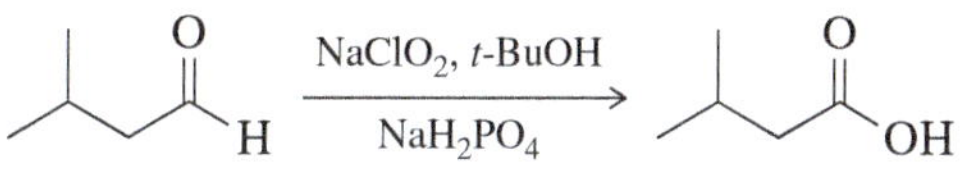

Figure 17.107 The Pinnick oxidation avoids the use of toxic chromium reagents.

Because it is highly toxic, chemists rarely use H_2CrO_4 to oxidize aldehydes to carboxylic acids. Instead, the most common reagent system used for the mild oxidation of aldehydes is a buffered solution of chlorous acid ($HClO_2$). The oxidation of aldehydes under these conditions is referred to as the Pinnick oxidation (**Figure 17.107**).

Although the Pinnick oxidation mechanism differs from that for chromic acid, each step is analogous to reactions seen in other contexts (**Figure 17.108**). Protonation of the aldehyde results in a more electrophilic carbonyl (step 1, General Mechanism 17B; Section 17.3.1). The chlorite anion now reacts as a nucleophile, attacking the strong electrophile (step 2, General Mechanism 17B; Section 17.3.1). Finally, the original aldehydic hydrogen is removed in a concerted, 6-electron, cyclic transition state to give the new C—O π bond and hypochlorous acid (step 3—similar to step 3 in Figure 17.106). [*These favorable concerted 6-electron cyclic transition states are explained in Chapter 20.*]

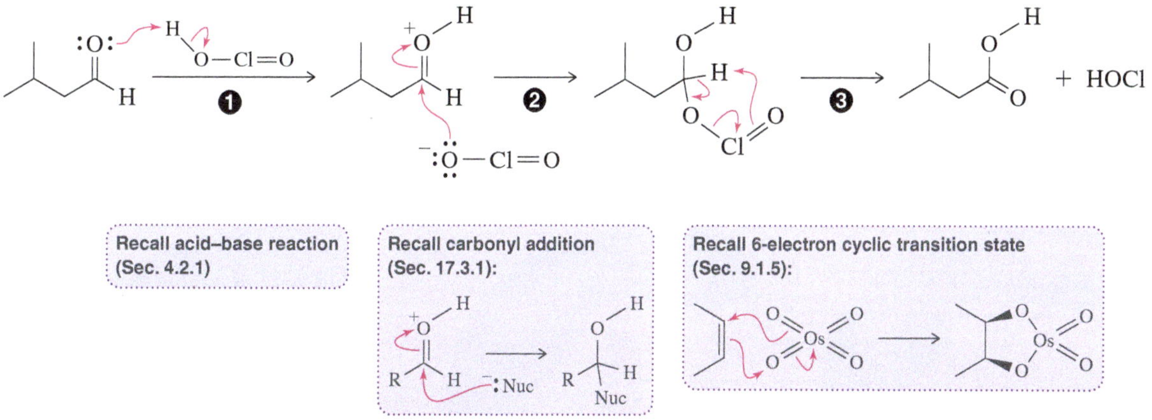

Figure 17.108 The use of hypochlorous acid is environmentally safer than the use of carcinogenic chromium oxidizing agents.

These mild reaction conditions allow highly functionalized aldehydes to be efficiently oxidized to the carboxylic acids with no observable side reactions (**Figure 17.109**).

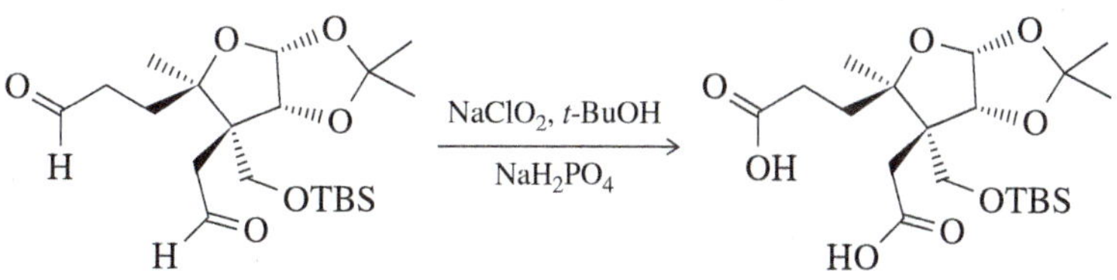

Figure 17.109 The Pinnick oxidation is mild and selective for aldehydes over other functional groups.

Assessment 17.53 Predict the product of the following Pinnick oxidations.

(a)

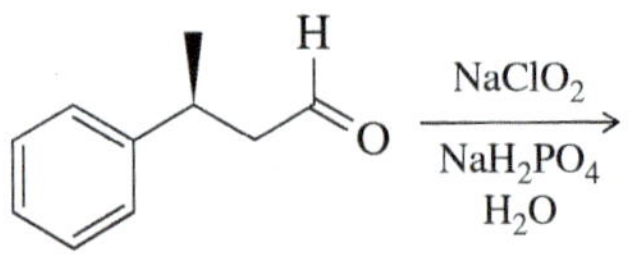

(b)

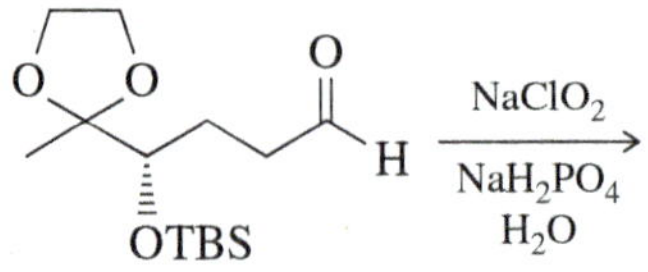

(c)

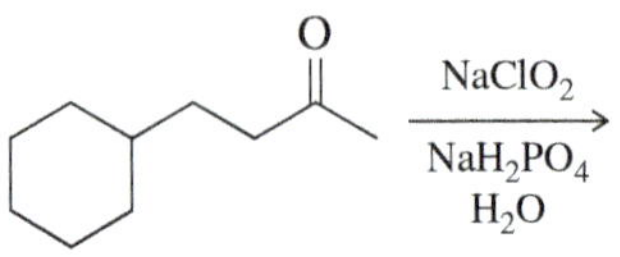

Assessment 17.54 During an oxidation reaction, there must also be a reduction. What is reduced in the Pinnick oxidation?

17-G CHEMISTRY IN CONTEXT: Biochemistry

The chemistry of a hangover (part 2)

Chemistry in Context 17-A showed that the body removes ethanol using alcohol dehydrogenase (ADH). ADH oxidizes it to acetaldehyde, a substance that is more poisonous than ethanol itself. Additional enzymes are necessary to remove acetaldehyde and other toxic aldehydes. The enzymes that do these reactions are called aldehyde dehydrogenases (ALDH). Once more, the NAD^+ cofactor assists in the oxidation process.

Oxidation to acetic acid begins with binding of acetaldehyde and NAD^+ at the active site (**Figure 17.110**). Nucleophilic attack by a side chain sulfur results in a covalently bound complex [*like the hydrate necessary for aldehyde oxidation in solution*]. By the same pathway used by alcohol dehydrogenases (Chemistry in Context 17-A), an active site base deprotonates the hydroxyl group. Electrons now collapse down to form a new C—O π bond, donating a hydride ion to the nearby NAD^+ cofactor. The newly oxidized, enzyme-bound thioester is cleaved by a reaction presented in Chapter 18. Acetic acid, which is nontoxic and water soluble, is easily expelled from the body.

Physiologically, it is the accumulation of acetaldehyde in the brain that makes a person feel unwell after consuming too much alcohol. The liver more quickly and efficiently oxidizes ethanol to acetaldehyde than it oxidizes acetaldehyde to acetic acid. When too much ethanol is consumed, such that the aldehyde dehydrogenase cannot keep up, the effects of a hangover persist.

Acetaldehyde can also cause a change in complexion during alcohol consumption. In some populations, large numbers of individuals possess a gene encoding for an inactive form of ALDH. These individuals often become flushed (red-faced) when consuming even small quantities of alcohol. This is a result of not being able to oxidize any acetaldehyde produced by ethanol oxidation.

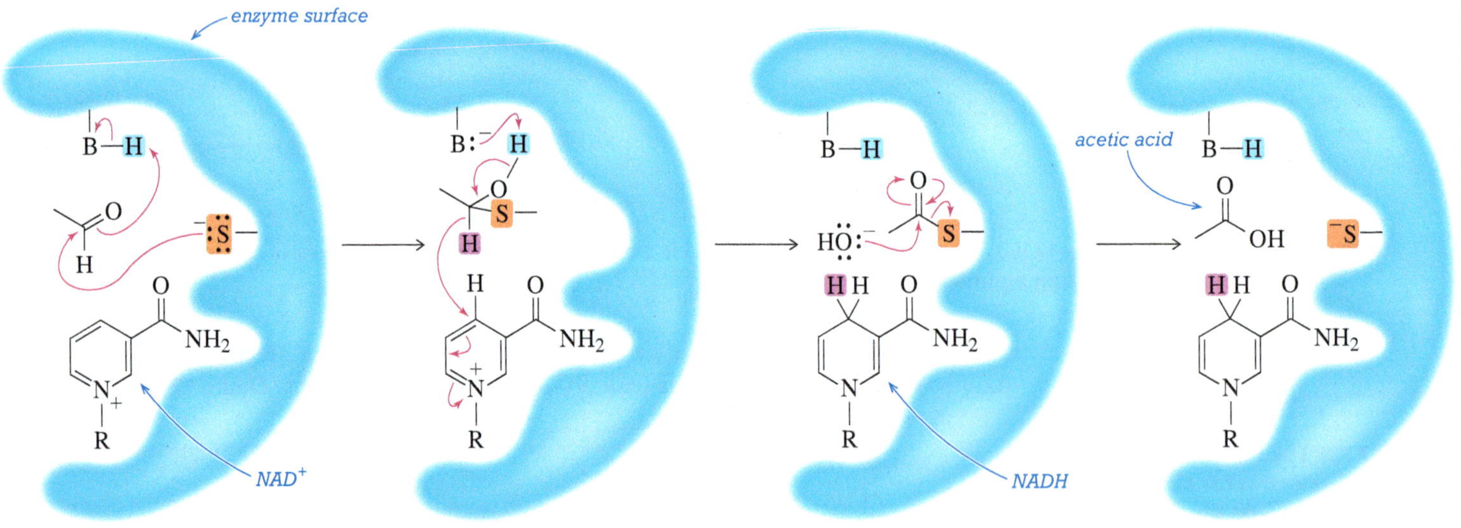

Figure 17.110 **The biological oxidation of acetaldehyde to acetic acid.**

Concluding thoughts

The chemistry of carbonyls has tremendous synthetic utility, thus expanding the diversity of molecules we can make. Equally important, a fundamental understanding of carbonyl chemistry provides a foundation upon which your future biochemistry class will build. For example, many of the concepts important to carbonyl chemistry play a role in the chemistry of dehydrogenases that use the NAD^+ cofactor. This understanding has led to the development of a number of medicines that inhibit these enzymes.

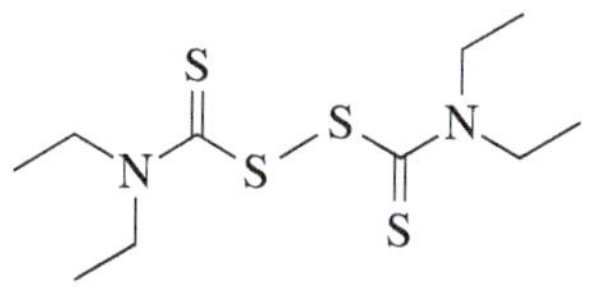

disulfiram

An example of a drug that inhibits a hydrogenase is disulfiram, a treatment for chronic alcoholism. By inhibiting aldehyde dehydrogenase (ALDH) and preventing the oxidation of acetaldehyde, a small amount of alcohol makes the patient feel lousy. The first drink brings about the onset of hangover symptoms. By providing immediate negative feedback, the patient is discouraged from taking that first drink. Studies suggest a 50% success rate (complete abstinence) for patients on disulfiram.

Similarly, lactate dehydrogenase has emerged as a target for anticancer drugs, based on the observation that cancer cells generate energy differently than healthy cells. Known as the Warburg effect, tumor cells generate energy anaerobically even when oxygen levels are sufficient for aerobic respiration. Because lactate dehydrogenase plays a key role in anaerobic respiration (Chemistry in Context 17-F), its inhibition has potential as a target for future chemotherapies. Although it is very early in the process, dichloroacetic acid is one molecule that has made it to clinical trials based on this mechanism of cell growth inhibition.

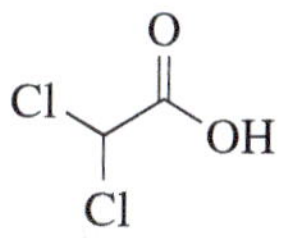

dichloroacetic acid

End-of-Chapter Materials

Important Concepts

[*These are the major concepts covered in each section, excluding reactions.*]

17.1: Aldehydes (RCHO) and ketones (RCOR) are unsaturated compounds containing a C=O double bond. They boil and melt at intermediate temperatures because they are polar, hydrogen bond acceptors, but not hydrogen bond donors. Aldehydes and ketones are trigonal planar, resulting from the sp^2 hybridization of the carbonyl carbon. The C—O π bond is formed by overlap between a p orbital on carbon and a p orbital on oxygen.

17.2: There are two important structures that contribute to the resonance hybrid of aldehydes and ketones. The neutral form is assumed to contribute more than the zwitterionic form, even though the chemical reactivity, and several other properties, are best explained by the zwitterionic form.

17.3: There are two important mechanisms for nucleophilic addition to aldehydes and ketones. Represented in this chapter as General Mechanisms 17A and 17B, the choice of mechanism depends on the conditions under which the reaction is run. In General Mechanism 17A, typical of strongly basic conditions, the strong nucleophile first adds to the carbonyl to generate an alkoxide, which is then protonated to generate an alcohol product. In General Mechanism 17B, typically with weak nucleophiles under acidic conditions, the carbonyl is protonated first, initiating nucleophilic attack.

17.4: Aldehydes are more reactive in nucleophilic addition reactions because they are less sterically hindered than are ketones. Additionally, the carbocation in the zwitterion resonance structure of an aldehyde is less stable than the zwitterion resonance structure of a ketone. For similar reasons, ketones protonate more rapidly at the carbonyl oxygen than do aldehydes.

17.5: Carbon nucleophiles, including Grignard reagents, organolithium reagents, acetylide ions, and cyanide ions, add by General Mechanism 17A. Because the carbonyl is planar, attack from the top and the bottom proceeds at equal rates, resulting in the formation of a racemic mixture when a new stereocenter is generated.

17.6: The Wittig reaction is the reaction of phosphorus ylides with aldehydes and ketones to form a new carbon–carbon double bond. It is driven by the formation of the exceptionally stable phosphorus–oxygen double bond. When using a stabilized ylide, one that is stabilized by resonance, the *E* alkene is produced preferentially. An unstabilized ylide reacts to give exclusively the *Z* alkene.

17.7: Neutral, oxygen-based nucleophiles add to aldehydes and ketones only under acidic conditions by General Mechanism 17B. Acetals, formed by the addition of two equivalents of an alcohol, can be formed reversibly, driven by Le Châtelier's principle. Owing to the reversibility of their formation, acetals are excellent protecting groups for aldehydes and ketones.

17.8: The addition of ammonia and primary amines with aldehydes and ketones produces imines. Because amines are intermediate-strength nucleophiles, they can add by either General Mechanism 17A or 17B. These reactions are fastest at pH = 4, where there is enough acid to increase the rate, but not so much acid that all of the nucleophilic amine is protonated. Secondary amines react with aldehydes and ketones to give enamines, through the intermediacy of the iminium ion.

17.9: Aldehydes and ketones can be reduced by hydrogenation, requiring higher pressure of hydrogen gas than the corresponding alkenes, owing to the increased strength of the C—O π bond. Carbonyls can also be reduced using sodium borohydride and lithium aluminum hydride, each of which donate a hydride ion to the electrophilic carbon of the carbonyl by General Mechanism 17A. The carbonyl can be completely reduced to the alkane under harsh reducing conditions like the Wolff–Kishner reaction.

17.10: Aldehydes can be oxidized to carboxylic acids only when the hydrate form of the aldehyde is present. Most commonly done using toxic chromium reagents, the use of hypochlorous acid represents an environmentally safer oxidation reaction.

Important Reactions and Mechanisms

1. **Grignard addition (Section 17.5.1):** A Grignard reagent (a strong nucleophile) reacts by General Mechanism 17A. The carbanion attacks the electrophilic carbonyl carbon, pushing the π electrons onto the oxygen, forming an alkoxide, which is subsequently quenched in a second step to give an alcohol. Grignard reactions can be used for the synthesis of 1°, 2°, or 3° alcohols.

General:

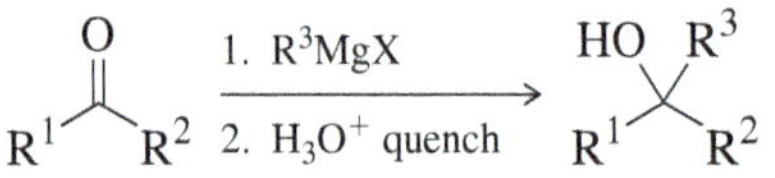

Specific:

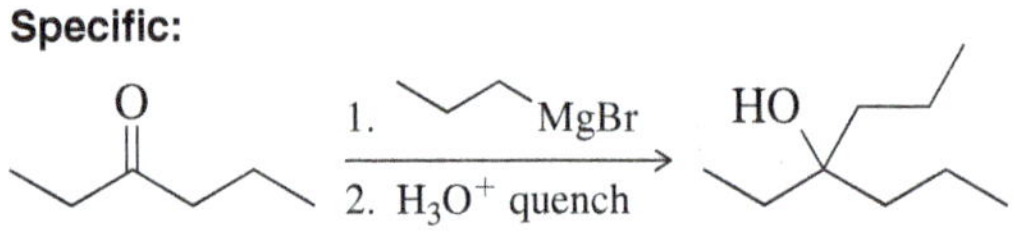

Mechanism:

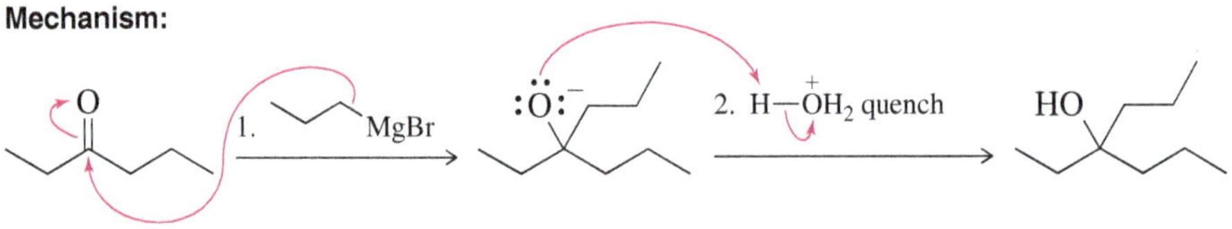

2. **Organolithium addition (Section 17.5.2):** Like a Grignard reagent, organolithium reagents are strong nucleophiles that add by General Mechanism 17A. The carbanion attacks the electrophilic carbonyl carbon, pushing the π electrons onto the oxygen, forming an alkoxide, which is subsequently quenched in a second step to give an alcohol. Organolithium reagents can be used for the synthesis of 1°, 2°, or 3° alcohols.

General:

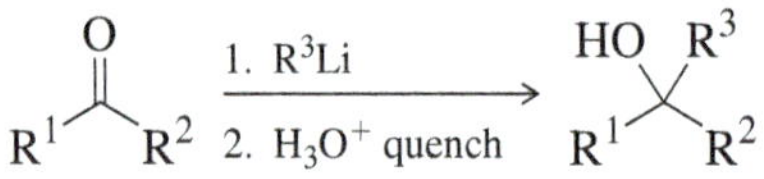

Specific:

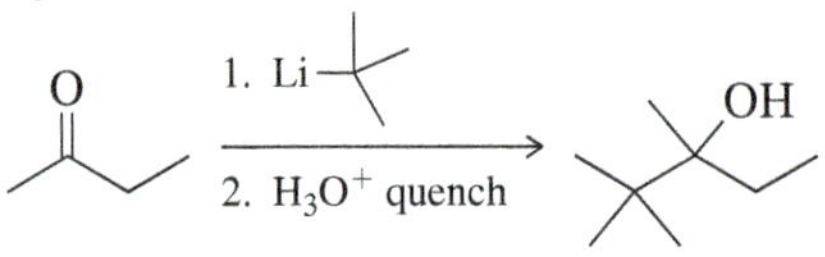

Mechanism:

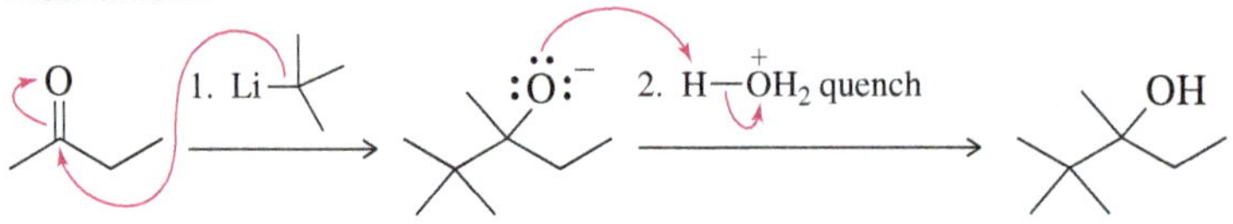

3. **Acetylide addition (Section 17.5.3):** Acetylide anions, formed by deprotonation of terminal alkynes, add as strong nucleophiles by General Mechanism 17A. The acetylide anion attacks the electrophilic carbonyl carbon, pushing the π electrons onto the oxygen, forming an alkoxide, which is subsequently quenched in a second step to give an alcohol. Acetylide addition can be used for the synthesis of 1°, 2°, or 3° alcohols.

General: **Specific:**

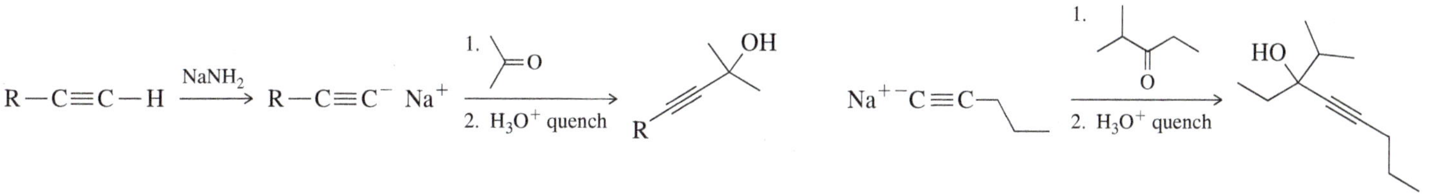

Mechanism:

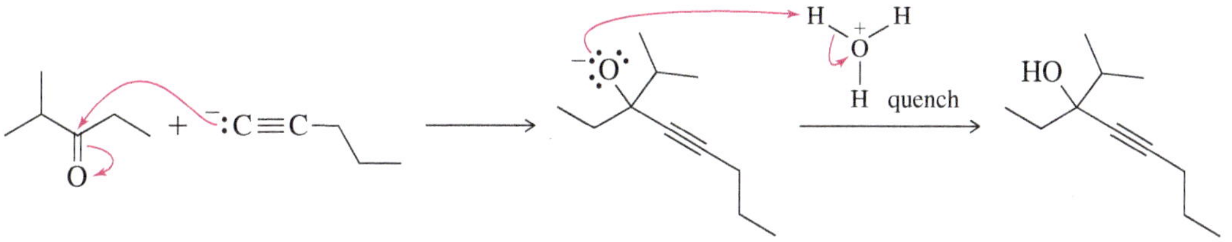

4. **Cyanide addition (Section 17.5.4):** Cyanide (a weaker, but negatively charged anion) adds by General Mechanism 17A. Because the formation of the alkoxide is disfavored, the reaction is driven to completion by the presence of hydrocyanic acid, which protonates the alkoxide immediately, resulting in the favorable formation of a cyanohydrin.

General: **Specific:**

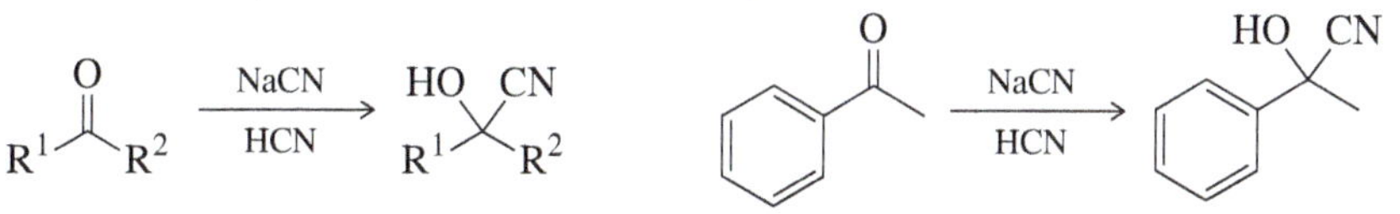

Mechanism:

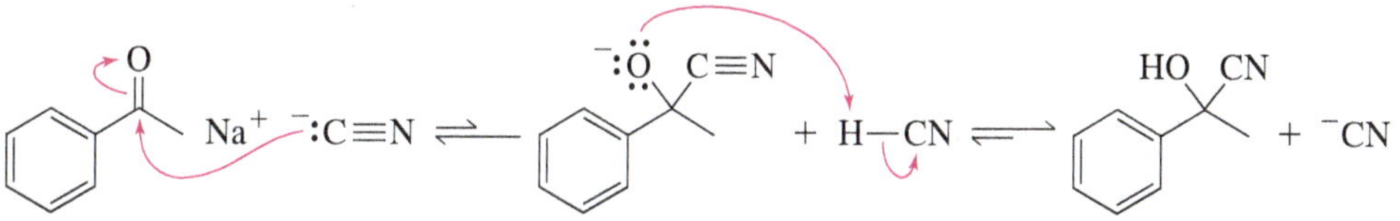

5. **Wittig reaction (Section 17.6):** Driven by the formation of the very stable P=O double bond, the Wittig reaction forms an alkene by reacting a phosphorus ylide with an aldehyde or a ketone. The carbanion of the ylide attacks the electrophilic carbonyl carbon, pushing the π electrons onto the oxygen, forming an alkoxide, which subsequently attacks the positively charged phosphorus atom to form a four-membered ring. Decomposition of this ring results in the new alkene and triphenyl phosphine oxide. Stabilized ylides give *E* alkenes preferentially, whereas unstable ylides predominantly produce *Z* alkenes.

General: **Specific:**

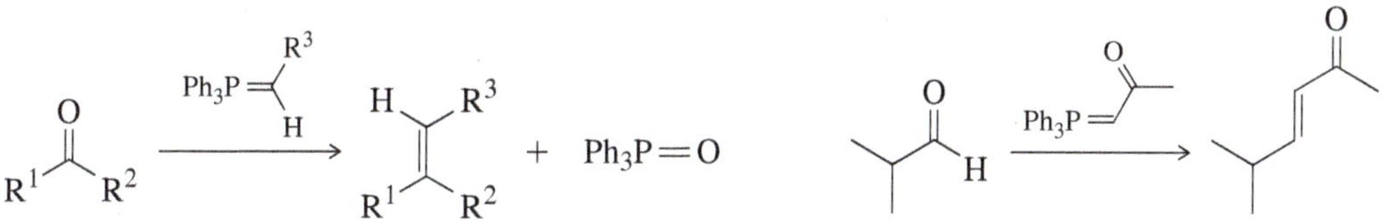

Mechanism:

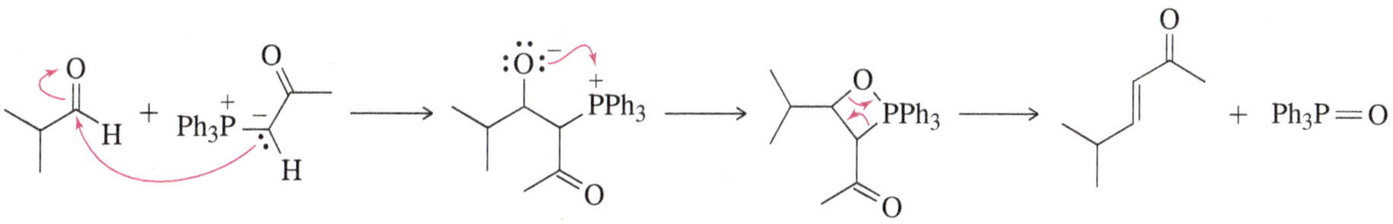

6. **Acetal formation (Section 17.7.1):** Alcohols (weak nucleophiles) add to carbonyls by General Mechanism 17B, which requires that the carbonyl be protonated prior to attack. Attack of the alcohol on the protonated carbonyl results eventually in the hemiacetal. In what can be thought of as an S_N1 reaction, the hemiacetal reacts under acidic conditions to produce the acetal.

General:

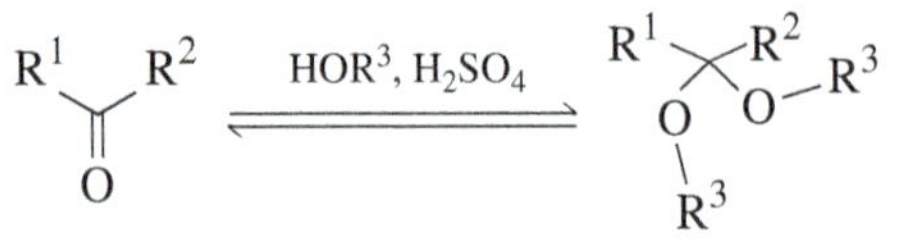

Specific:

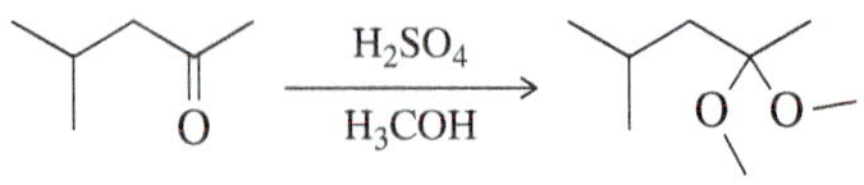

Mechanism:

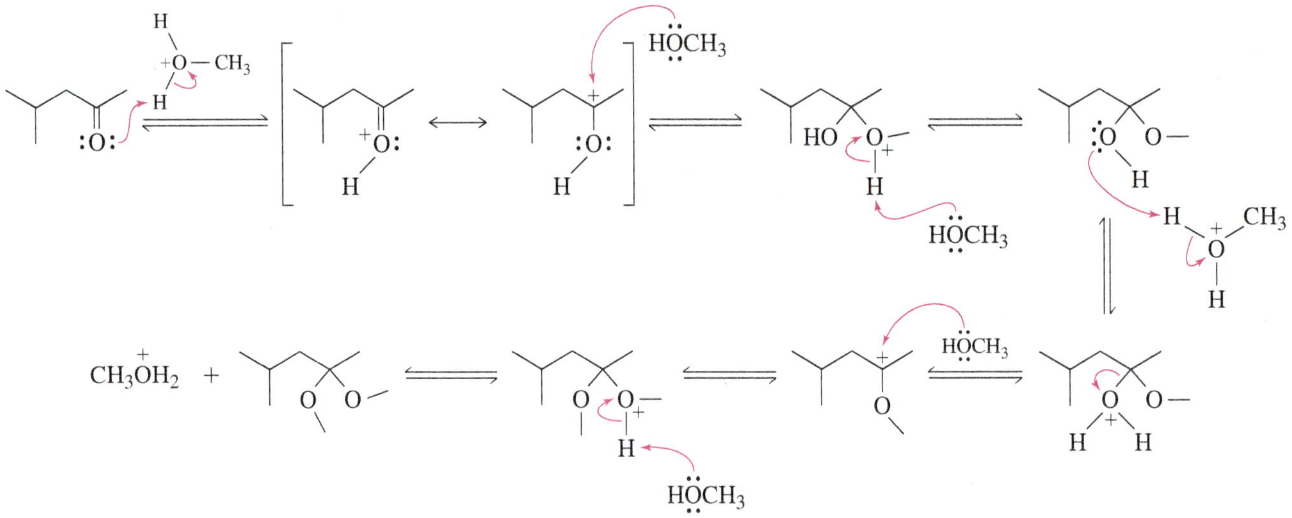

7. **Imine formation (Sections 17.8.1, 17.8.2):** Amines (intermediate-strength nucleophiles) can add by General Mechanism 17A or 17B. Nucleophilic attack by nitrogen on the electrophilic carbonyl carbon pushes the π electrons onto oxygen. A series of proton transfers results in formation of the heminal. Pushed by the collapsing lone pair of the amine, hydroxide is expelled, resulting in formation of the C=N double bond of the iminium ion. Deprotonation of the iminium by the expelled hydroxide results in the neutral imine. Ammonia and primary amines can react with aldehydes and ketones to produce imines.

General:

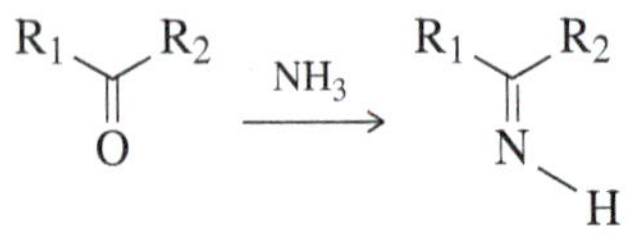

Specific:

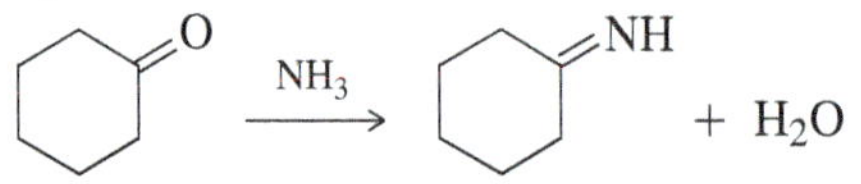

Mechanism:

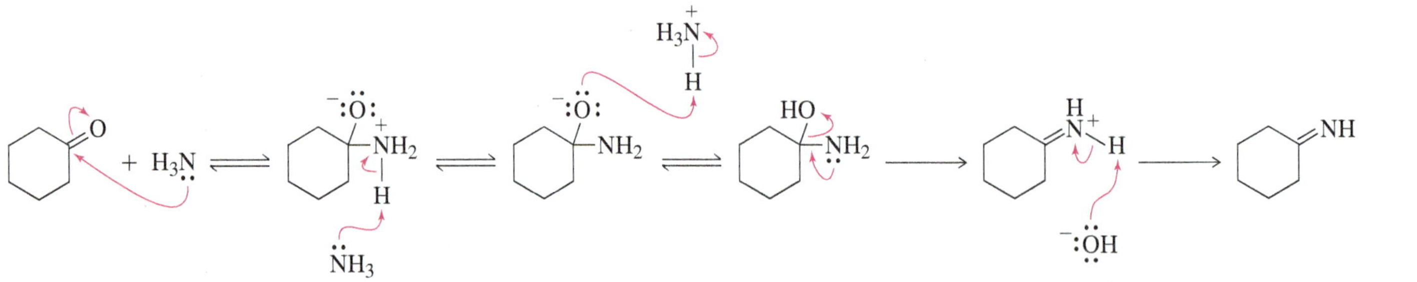

8. Enamine formation (Section 17.8.3): Secondary amines (intermediate-strength nucleophiles) can add by General Mechanism 17A or 17B. Proceeding by a pathway identical to that of primary amines, this reaction diverges at the point of forming the iminium ion. Because there is no acidic proton attached to the nitrogen, the expelled hydroxide instead removes a proton from the carbon α to the iminium, resulting in formation of the enamine.

General:

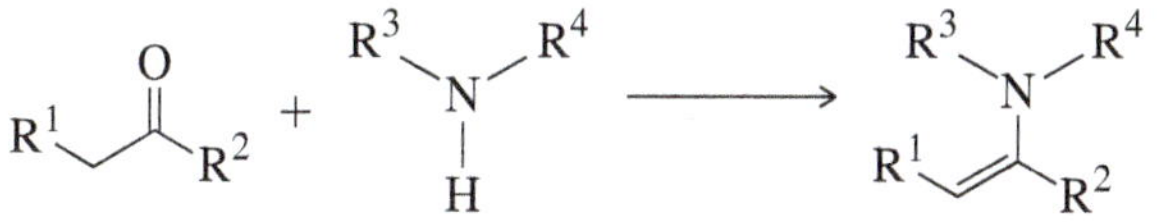

Specific:

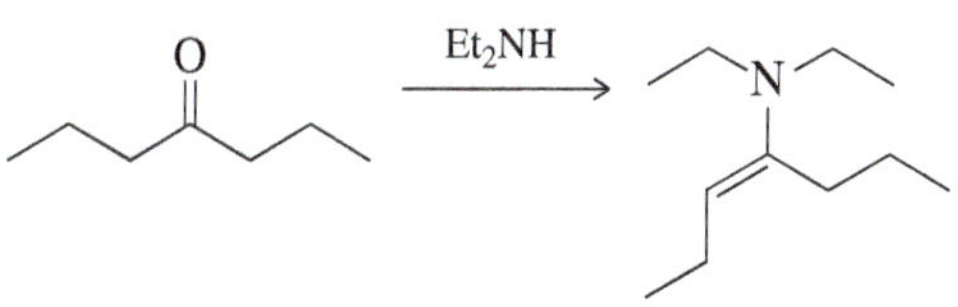

Mechanism:

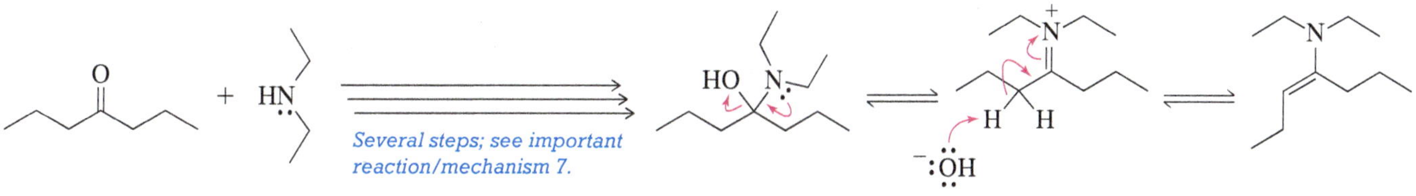

9. Hydrogenation (Section 17.9.1): Aldehydes and ketones can be reduced by the concerted syn addition of H_2 across the C—O π bond. Although it is mechanistically similar to the hydrogenation of alkenes, the C—O π bond is stronger than the C—C π bond, so higher pressures of hydrogen are required to reduce carbonyls via this process.

General:

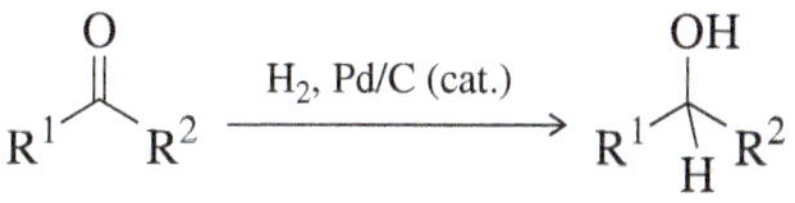

Specific:

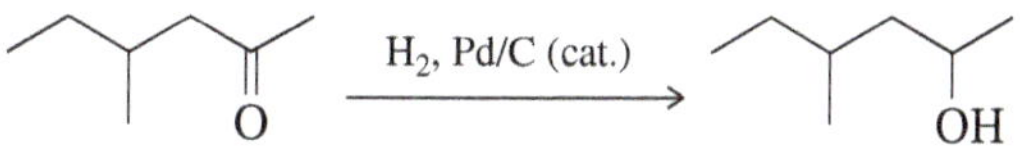

Mechanism:

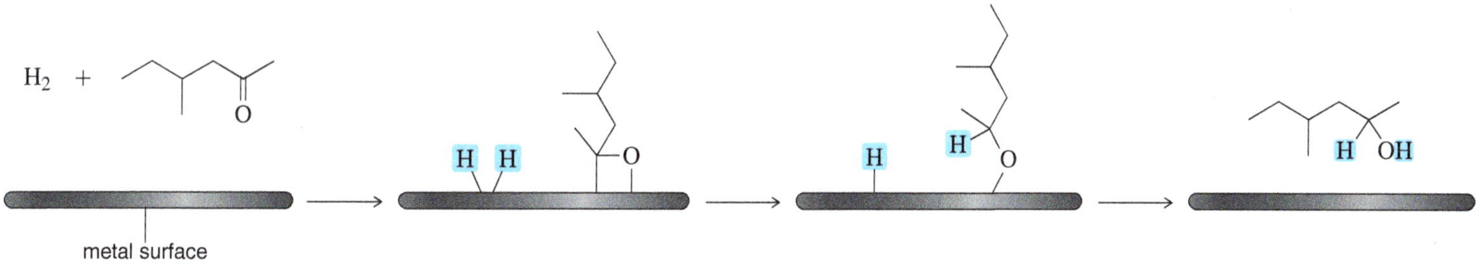

10. Hydride addition (Section 17.9.2): Carbonyls can be reduced using the nucleophilic reducing agents sodium borohydride ($NaBH_4$) and lithium aluminum hydride ($LiAlH_4$). These reducing agents donate a hydride anion to the electrophilic carbonyl carbon, pushing the π electrons onto the oxygen, forming an alkoxide, which is subsequently quenched in a second step to give an alcohol. While they accomplish the same transformation, $LiAlH_4$ is a stronger reducing agent than $NaBH_4$.

General:

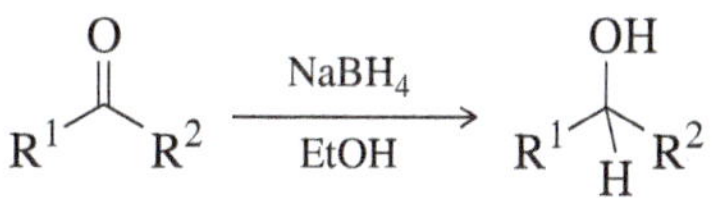

Specific:

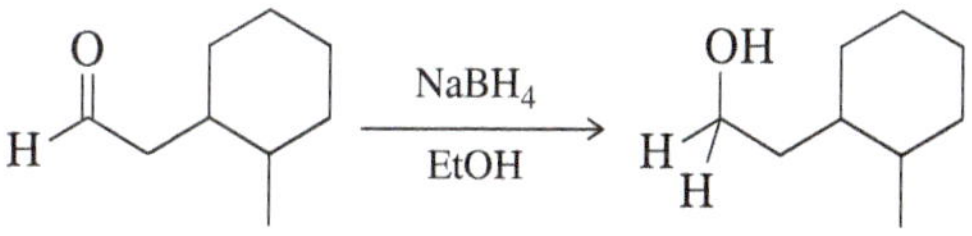

Mechanism:

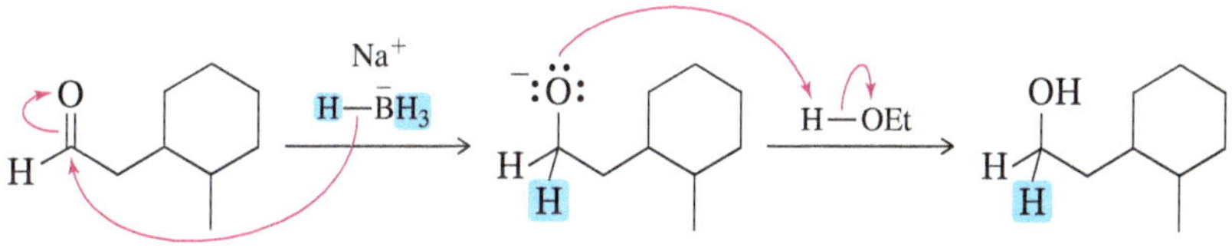

11. Wolff–Kishner reaction (Section 17.9.3): The Wolff–Kishner reaction is the exhaustive reduction of a carbonyl to an alkane. This complicated mechanism starts by formation of an imine with hydrazine. Tautomerization forms one of the new C—H σ bonds. Elimination of nitrogen gas results in an anion that is quenched to form the second C—H σ bond.

General:

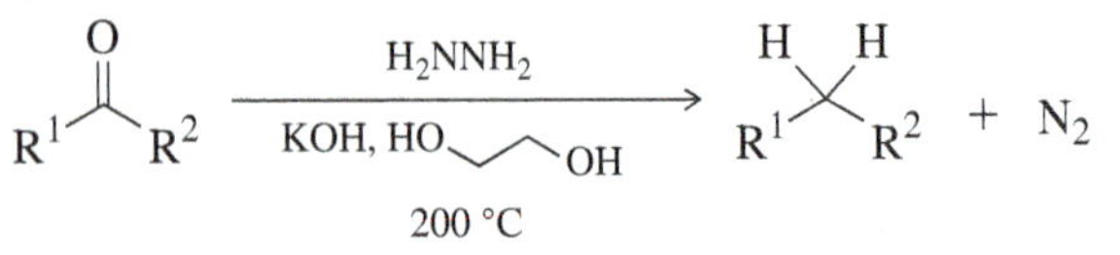

Specific:

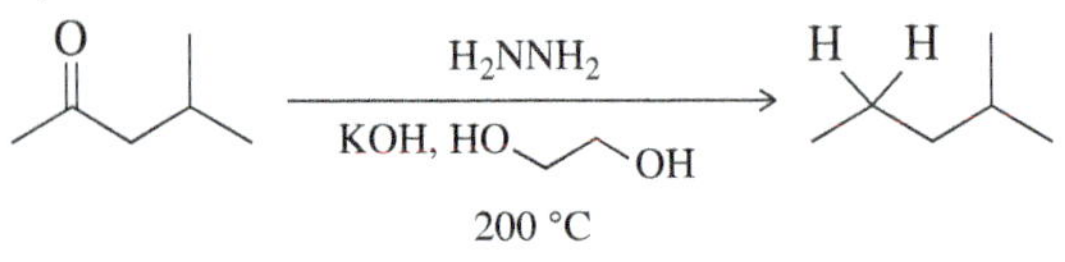

Mechanism:

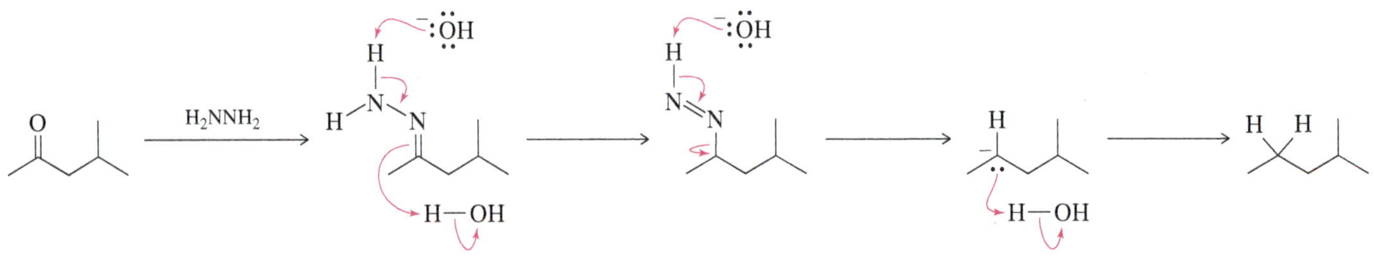

End-of-Chapter Assessments

[*Assessments are ranked according to their difficulty;* •••• = *most difficult.*]

17.55 (•) Name the following aldehydes and ketones according to the IUPAC rules of nomenclature.

(a) (b) (c) (d)

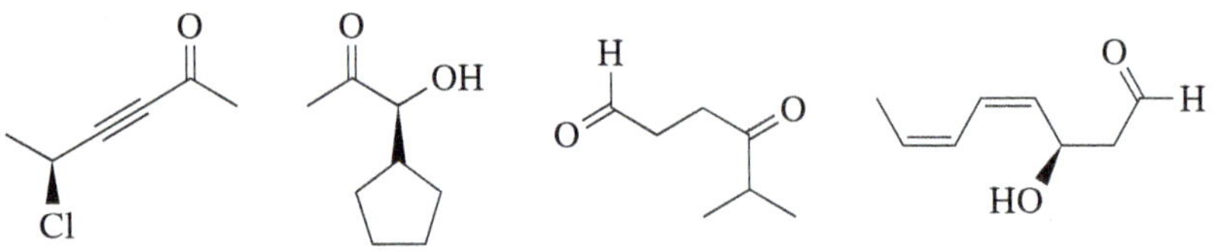

17.56 (•) The following molecules were named incorrectly according to IUPAC nomenclature. Give the correct name of these compounds.

(a) 3-oxo-5-methylhexan-2-ol
(b) (*R*)-6-bromo-7-oxoheptan-2-one
(c) 1-chloro-3-methylheptan-7-one
(d) (*E*)-2-methylhex-2-en-6-al

17.57 (•) Predict the product(s) that would result when the aldehydes and ketones shown are allowed to react under the following conditions: (i) 1. PhMgBr, THF 2. H_3O^+ quench; (ii) 1. EtLi, pentane 2. H_3O^+ quench; (iii) NaCN/HCN; (iv) $Ph_3P{=}CHCH_2CH_3$; (v) $Ph_3P{=}CHCO_2Et$; (vi) EtOH, H_2SO_4; (vii) $HOCH_2CH_2OH$, H_2SO_4; (viii) NH_3; (ix) $CH_3CH_2CH_2NH_2$, HCl; (x) $(CH_3CH_2)_2NH$; (xi) $NaBH_4$, EtOH; (xii) 1. $LiAlH_4$, THF 2. H_3O^+ quench; (xiii) H_2 (>1 atm), Pd/C, heat; (xiv) H_2NNH_2, KOH, $HOCH_2CH_2OH$.

(a) (b) (c) (d)

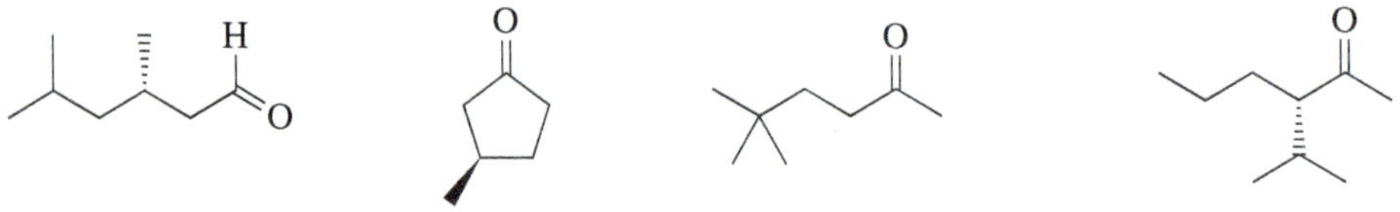

(e) (f) (g) (h)

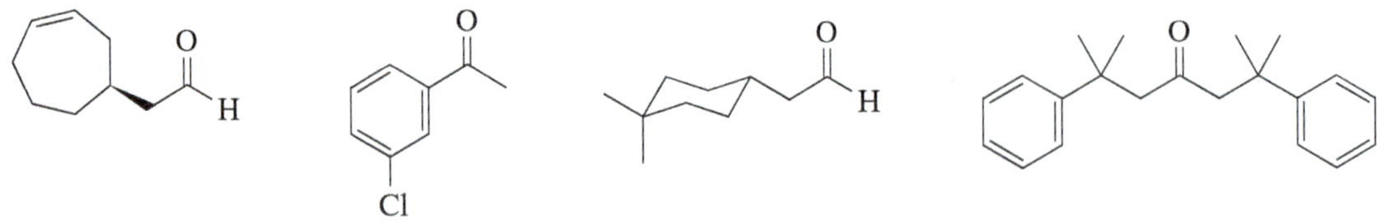

17.58 (•) Fill in the blank with the reagents necessary to complete the synthesis shown.

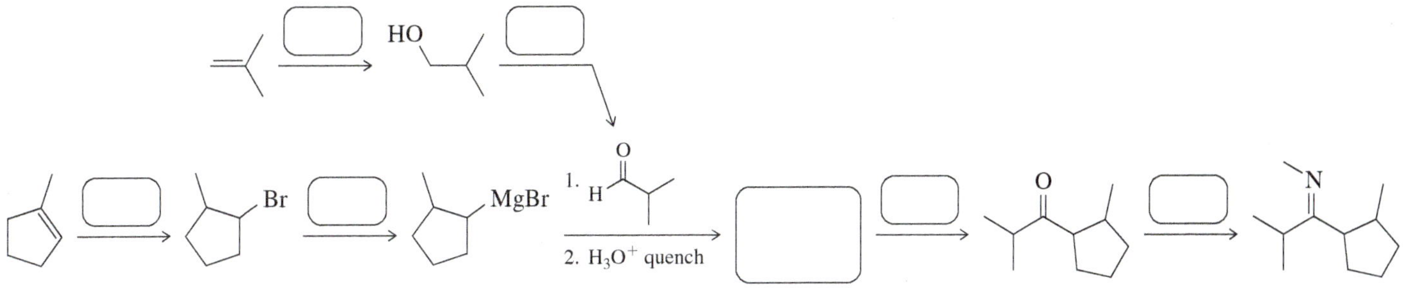

17.59 (•) Suggest an arrow-pushing mechanism for molecule (a) reacting with each set of reagents (i–xiv) shown in Assessment 17.57.

17.60 (••) Draw the molecular orbital picture of but-3-en-2-one.

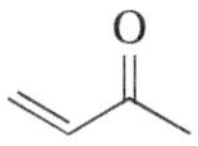

17.61 (•••) Provide a mechanism for the following cyclization to make a spiroketal.

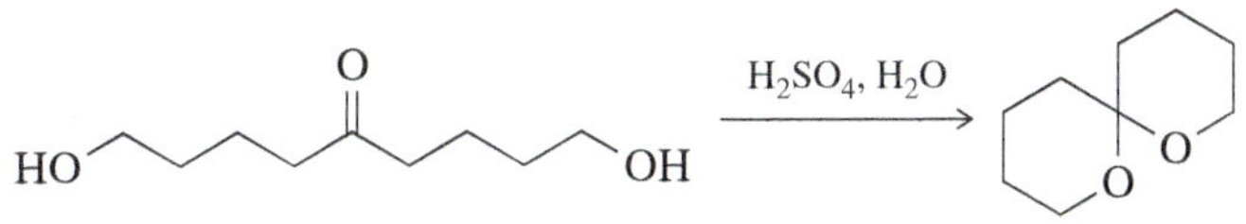

spiroketal

17.62 (•••) Starting with formaldehyde and the alkyl halides shown as the only sources of carbon, provide a synthesis of the following alcohol.

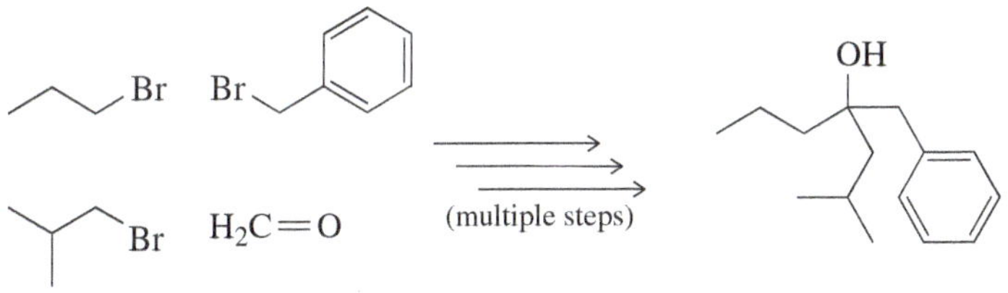

17.63 (•••) Imagine a situation where, seeking approval, you proudly walk into your advisor's office to show her the alkyl halide (shown below) that you synthesized. Unfortunately, she points out that you made a bromoalkane one carbon shorter than desired. Propose a synthesis you could use to rectify this unlikely scenario.

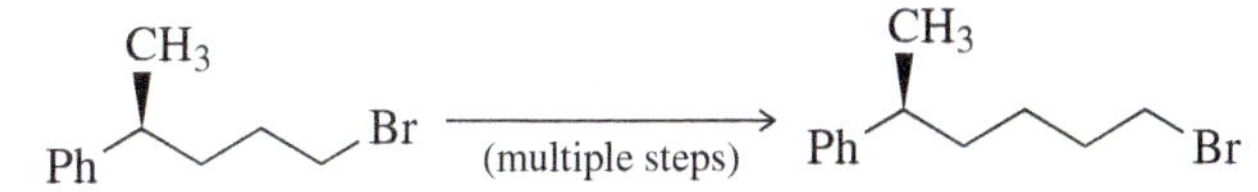

the compound you made *the compound you need*

17.64 (••) In lieu of quenching the product of Grignard addition to a carbonyl with acid, alkyl halides can be added directly to generate ether products. Predict the product and show the mechanism of this process.

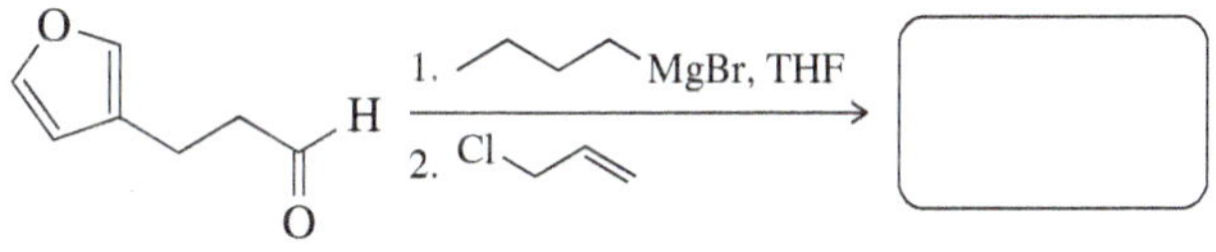

17.65 (••) Fill in the blanks with the reagents necessary to complete the synthesis shown.

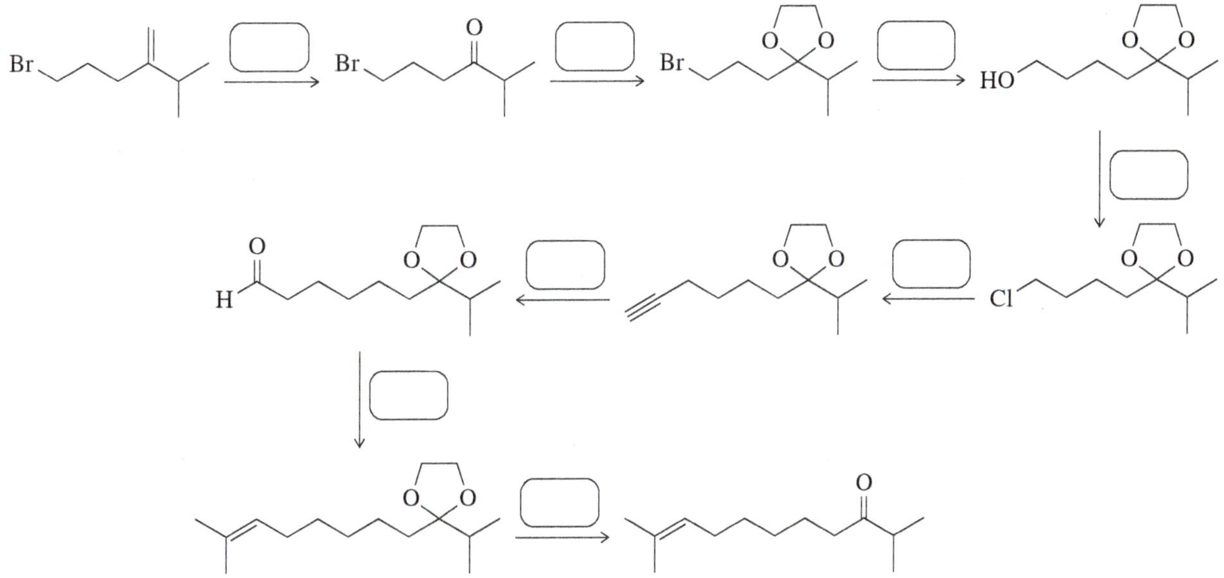

17.66 (•••) Suggest a synthetic scheme, involving a protecting group, to generate the molecule shown starting with the molecule at the left.

(a)

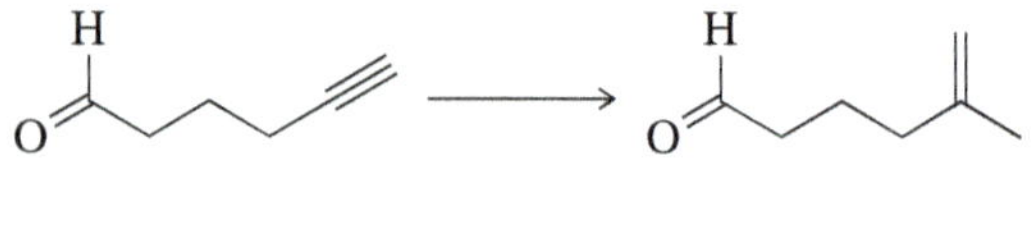

(b)

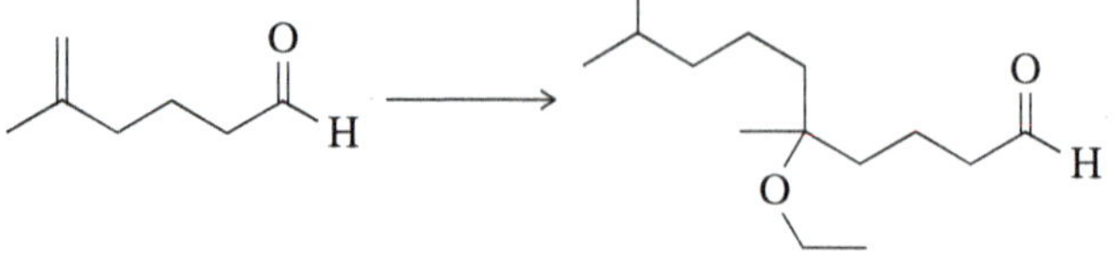

(c)

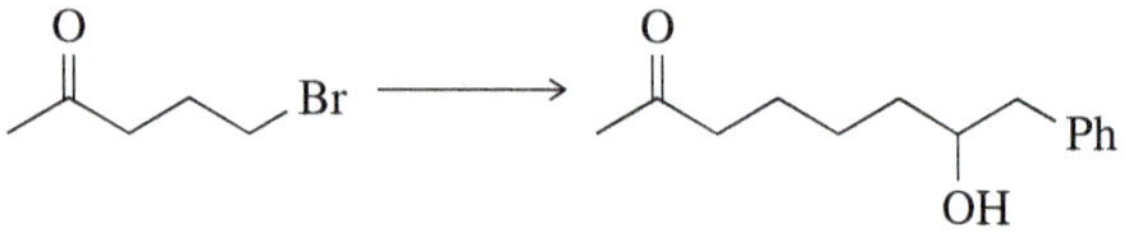

(d)

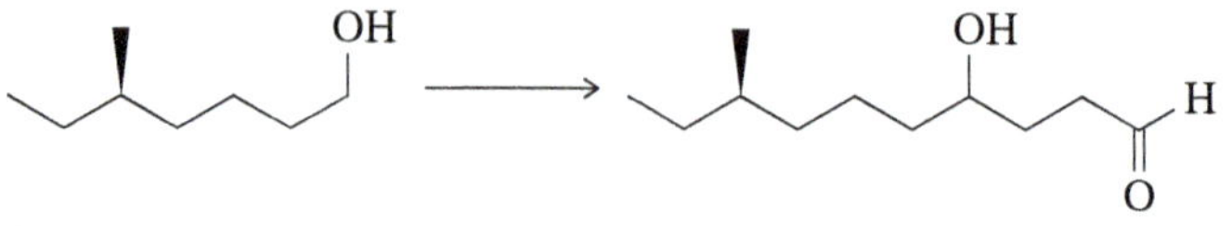

17.67 (••••) As discussed in Section 17.9.1, alkenes can be hydrogenated selectively in the presence of ketones. Suppose that was not the case, and suggest how you might use a protecting group strategy to generate **A** from **B**.

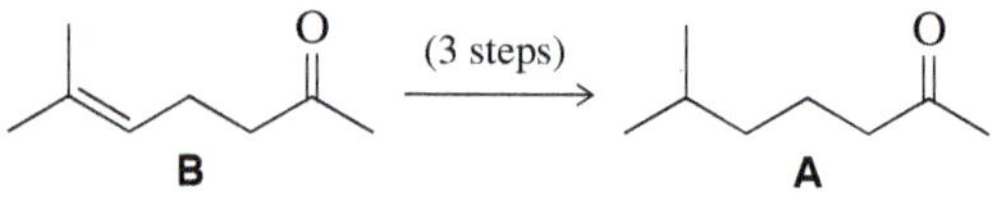

17.68 (•••) We learned in Section 17.7.5 that ethylene glycol was a good protecting group for ketones and aldehydes. For similar reasons, a ketone is often used as a protecting group for 1,2-diols. Predict a product and suggest a mechanism for this protecting-group strategy. Also suggest a reaction you might use to remove the protecting group.

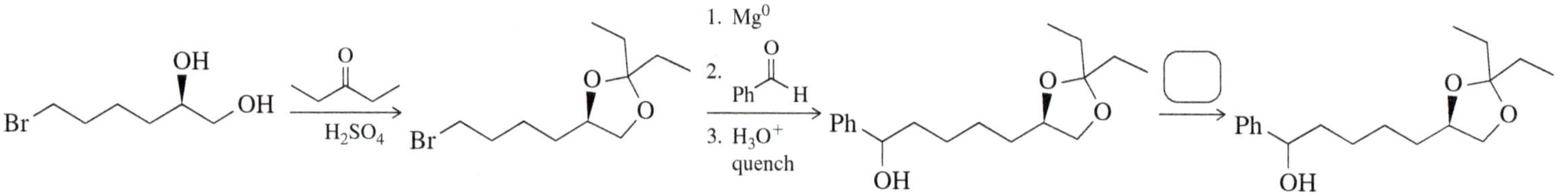

17.69 (•••) Synthesize the molecule on the right, beginning with the molecule on the left.

(a)

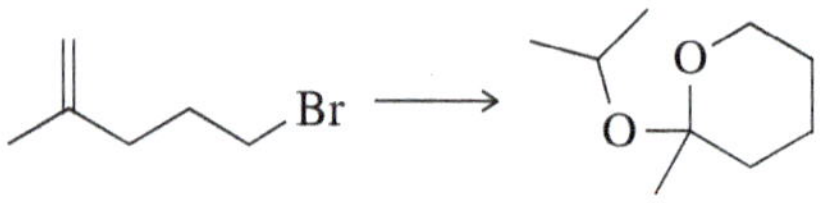

(b)

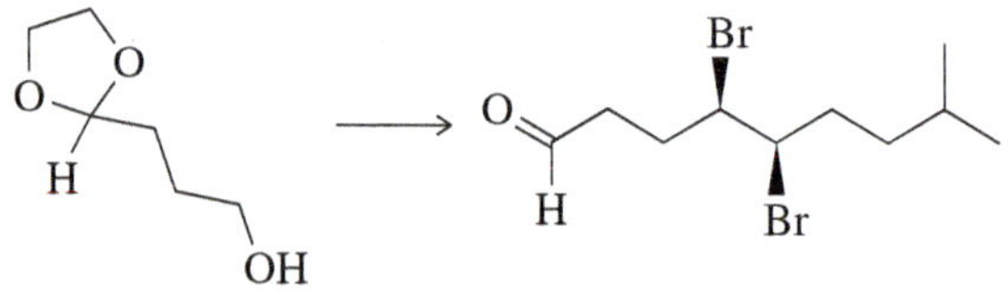

(c)

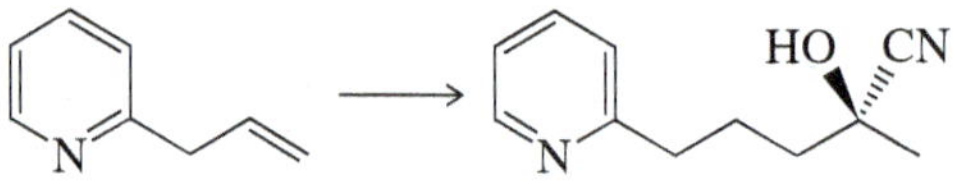

17.70 (••••) When a ketone is dissolved in ^{18}O-labeled water, the ^{18}O label is incorporated into the ketone. Suggest a mechanism that explains this observation.

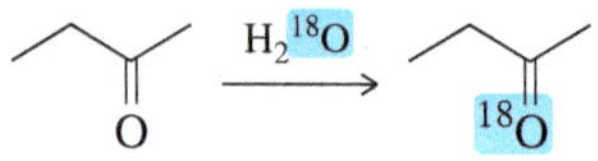

17.71 (••••) A common (but not always wise) practice for cleaning up spills in the organic lab is to rinse with acetone and wipe with paper towels. **True story:** While weighing out lithium aluminum hydride for a large-scale reduction, an undergraduate student spilled a sizable amount on the lab bench. The student began to clean up the spill with acetone and paper towels, but a fire started immediately, reaching almost to the ceiling of the lab. Why did this fire start under what are pretty typical cleanup conditions? [*The fire was extinguished right away and caused no damage. And a valuable, but in hindsight, obvious, lesson was learned by all.*]

17.72 (••••) LOOKING AHEAD In Chapter 18, we study in detail the reaction of nucleophiles with carboxylic acid derivatives. One such derivative, the acid chloride, is shown here reacting with a Grignard reagent to give a tertiary alcohol. Based on the manner in which Grignard reagents react with aldehydes and ketones, suggest a mechanism of this multistep reaction. [*Hint: The reaction to make A is a two-step reaction (not a direct displacement) of the chloride ion.*]

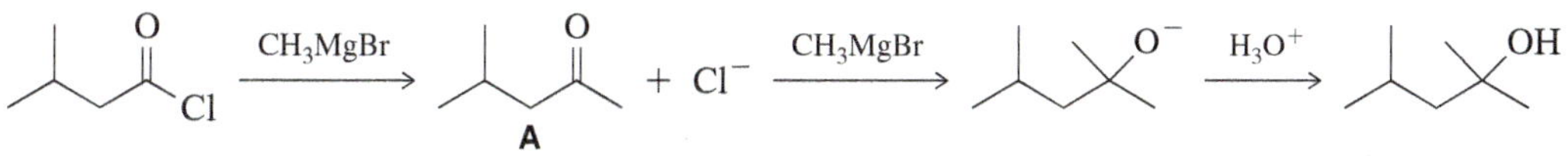

17.73 (•••) LOOKING AHEAD The second best resonance structure is often predictive of how a molecule's functional group will react. With that in mind, how would you expect an enamine to react with a good electrophile like $CH_3—Cl$?

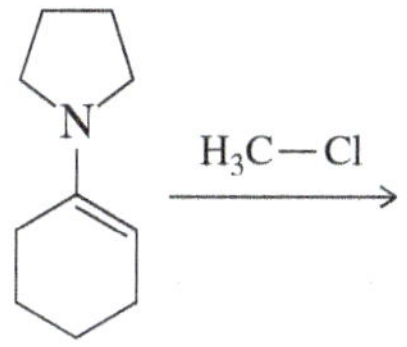

17.74 (•••) LOOKING AHEAD How would the product you obtained in Assessment 17.73 react if H_2O was added to the reaction mixture? Suggest a mechanism of this reaction.

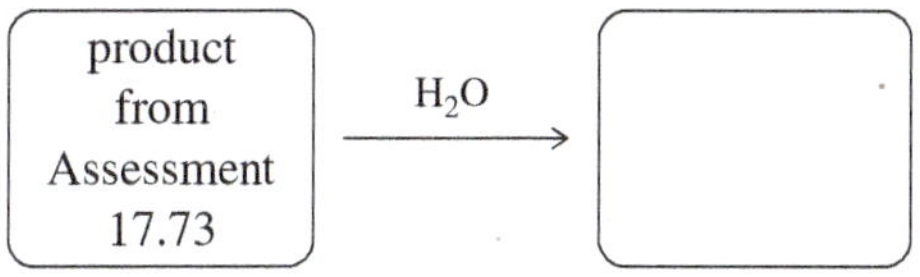

17.75 (•••) One of the drawbacks of the Wolff–Kishner reaction is the harsh conditions needed to completely remove both C—O bonds. The Mozingo reaction was developed as a mild, but foul-smelling, version of this reaction. Beginning with a ketone and 1,3-propanedithiol, a thioketal is produced. The thioketal contains relatively weak C—S σ bonds that undergo hydrogenolysis using a nickel catalyst. Suggest a mechanism by which the thioketal might form.

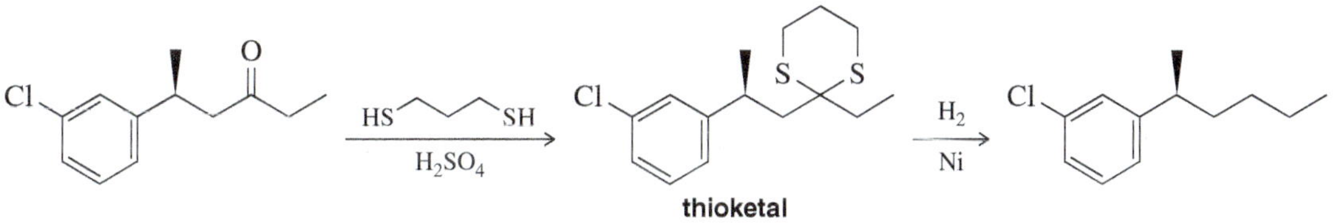

17.76 (•••) Here in Chapter 17, we saw the carbon of the carbonyl function exclusively as an electrophile. The use of 1,3-propanedithiol (see Assessment 17.75) gives us a method by which the polarity of the carbonyl can be reversed and the carbonyl carbon instead participates in reactions as a nucleophile. Suggest a mechanism for each step in this process.

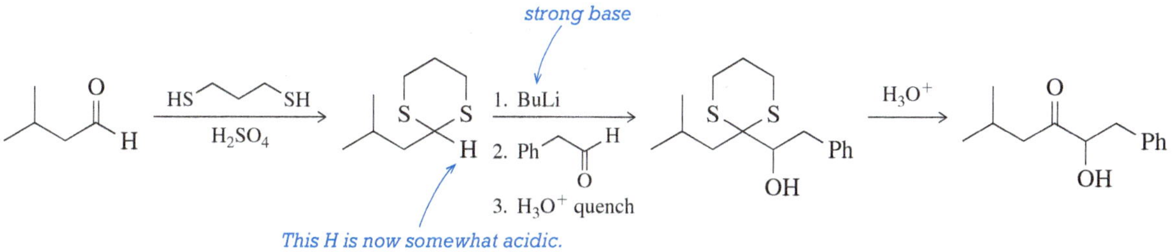

17.77 (•••) LOOKING AHEAD Reductive amination is a commonly used procedure for synthesizing amines using two reactions we studied in this chapter. Give a mechanism for each step of the two-step process. [*$NaCNBH_3$ is just like $NaBH_4$.*]

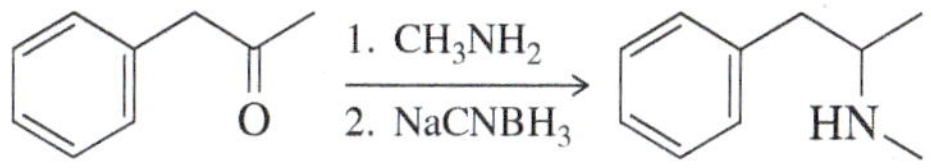

17.78 (•••) LOOKING AHEAD We show in subsequent chapters that the combination of a secondary amine and formaldehyde in the presence of a nucleophile is a method for forming a new C—C bond. Suggest a mechanism for this general process.

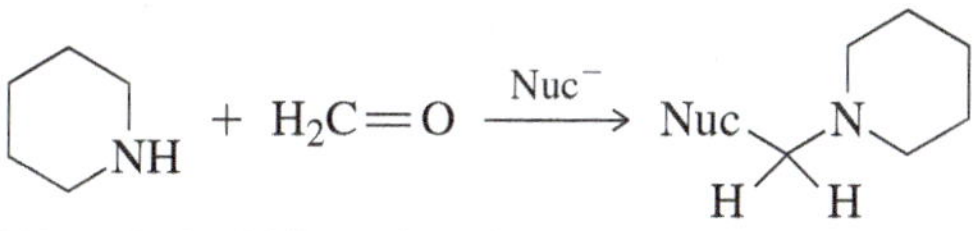

17.79 (••) LOOKING AHEAD We show in Chapter 20 that α,β-unsaturated ketones are electrophilic at C_4. Rationalize this observation.

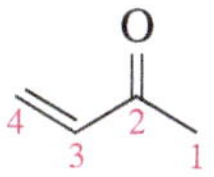

18

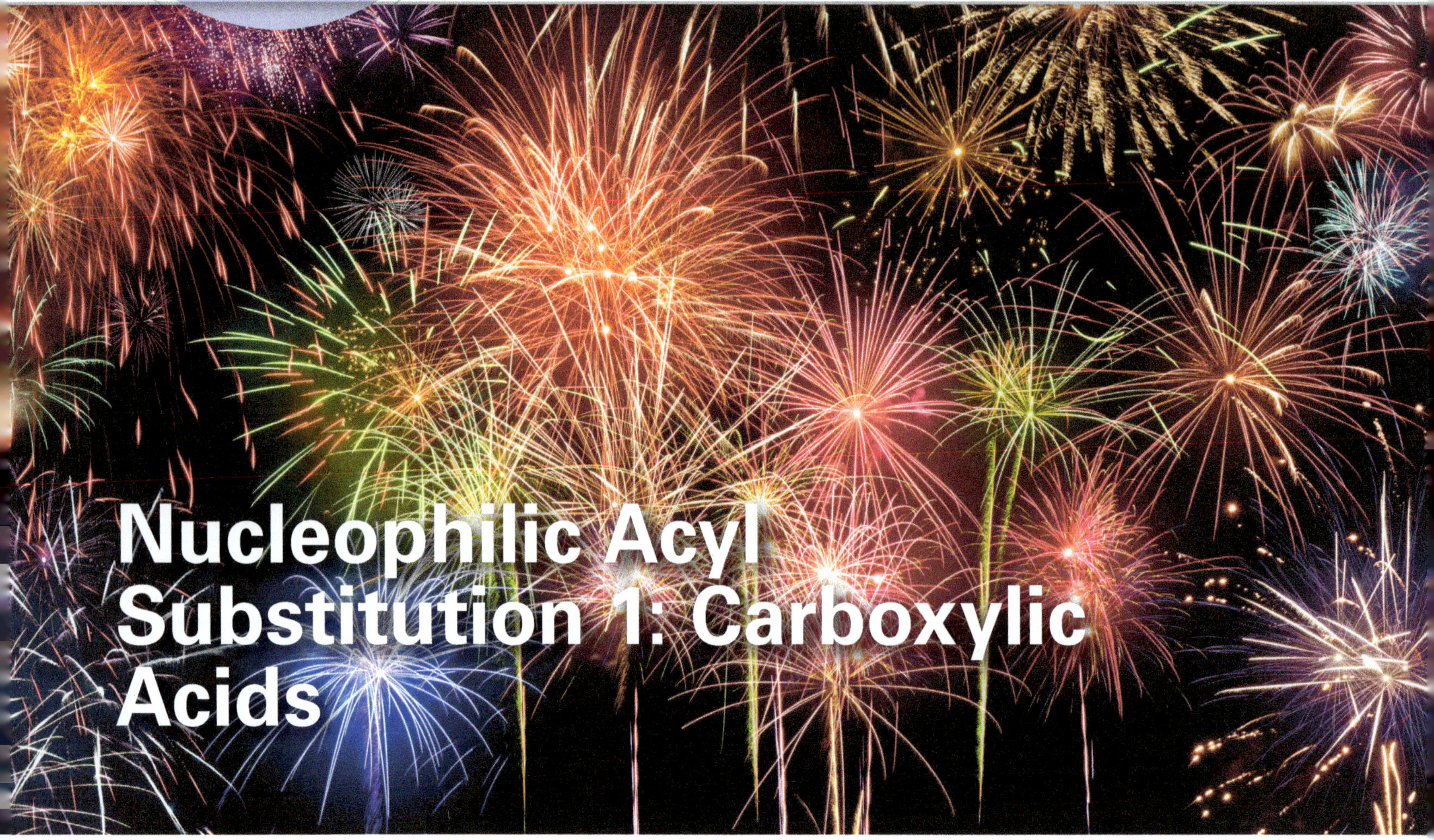

Nucleophilic Acyl Substitution 1: Carboxylic Acids

Learning Objectives

LO 18.0 Relate the name of a carboxylic acid to its structure and vice versa

LO 18.1 Analyze the molecular orbitals of carboxylic acids

LO 18.2 Compare the physical properties of carboxylic acids

LO 18.3 Evaluate carboxylic acid synthesis reactions

LO 18.4 Determine a method for separating carboxylic acids from nonpolar, nonacidic compounds

LO 18.5 Predict the products of nucleophilic acyl substitution reactions

LO 18.6 Evaluate reactions of carboxylic acids

LO 18.7 Examine decarboxylation reactions

Introduction

Recently, a popular restaurant chain ran an advertisement in which they boasted, "We put Sodium Benzoate back where it belongs. Into 200 lbs of fireworks." Sodium benzoate is the conjugate base of benzoic acid, a carboxylic acid, the functional group of interest here in Chapter 18. It is a food preservative that has been used for over 100 years, most often in acidic foods like sodas and salad dressings. Under acidic conditions it is converted to benzoic acid. By acting as an antifungal agent, sodium benzoate increases the shelf life of foods, making them safer to eat for longer periods. This chapter tells the story of sodium benzoate and why a marketing strategy centered on its removal might be misguided.

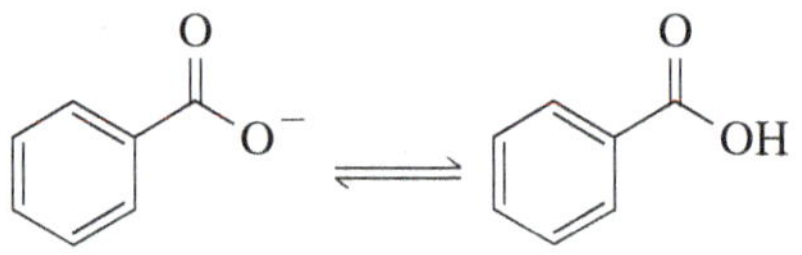

benzoate benzoic acid

"He invited people to sign a petition that demanded either strict control of, or a total ban on, dihydrogen monoxide . . . . Yes, 86 percent of the passersby voted to ban water (H_2O) from the environment. Maybe that's what really happened to all the water on Mars."

—NEIL DEGRASSE TYSON

As we transition from Chapter 17, the chemistry of carbonyl addition reactions, to Chapters 18 and 19, the chemistry of carbonyl substitution reactions, keep in mind that everything we have learned previously is important to what we are learning now. In all three of these chapters, the carbonyl (C=O) is the key functional group. So what is the difference between addition and substitution reactions? That difference is *leaving groups*. In essence, Chapters 18 and 19 combine the language of Chapter 12 (which discusses S_N1/S_N2 reactions) with that of Chapter 17 (carbonyl addition reactions). Recognizing this will help you understand this material. [*It is important to remind ourselves of what we already know, so let's do that now by working Assessments 18.1–18.6.*]

Test your understanding

Assessment 18.1 Draw two more resonance structures for each of the following molecules.

(a) (b) (c)

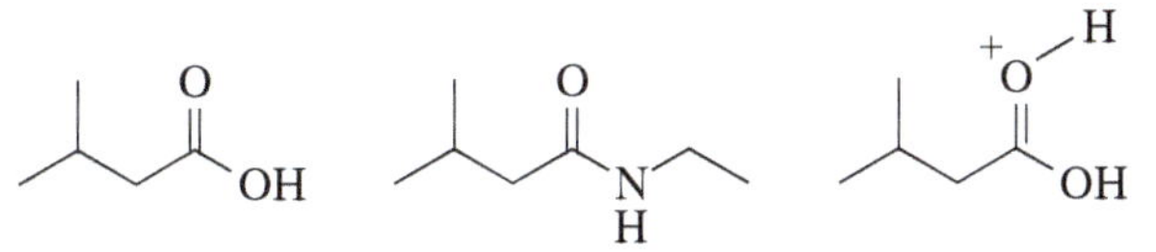

Assessment 18.2 Order these molecules from lowest C=O IR stretching frequency to highest.

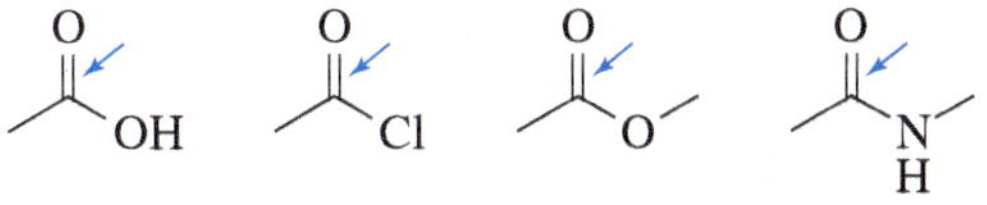

Assessment 18.3 Would you expect ΔS to be greater than, less than, or equal to zero in the following reactions?

(a) (b) (c)

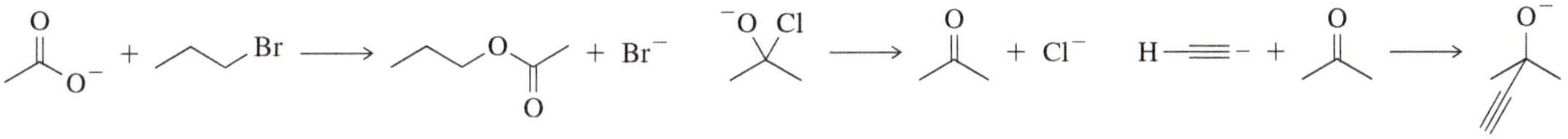

Assessment 18.4 Calculate K_{eq} for the acid–base reaction shown. Which side is favored and why?

Assessment 18.5 Calculate the oxidation numbers for the indicated atoms.

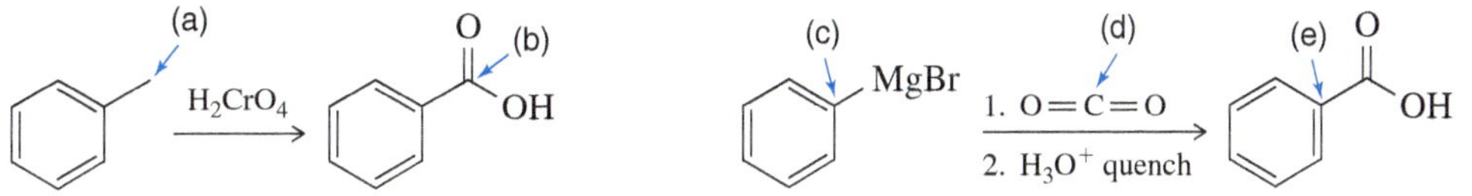

Assessment 18.6 How does knowing the mechanism of a reversible reaction in one direction tell you the mechanism in the opposite direction? Use a reaction coordinate diagram to explain.

18.1 General structure of carboxylic acids

A **carboxylic acid** (RCO_2H) is a molecule containing the carboxyl functional group attached to either an alkyl group (as in acetic acid) or a hydrogen (as in formic acid). The carboxyl (**carb**onyl + hydr**oxyl**) functional group is the combination of a C=O double bond (a carbonyl) and an OH (hydroxyl) bound to the same carbon (**Figure 18.1**).

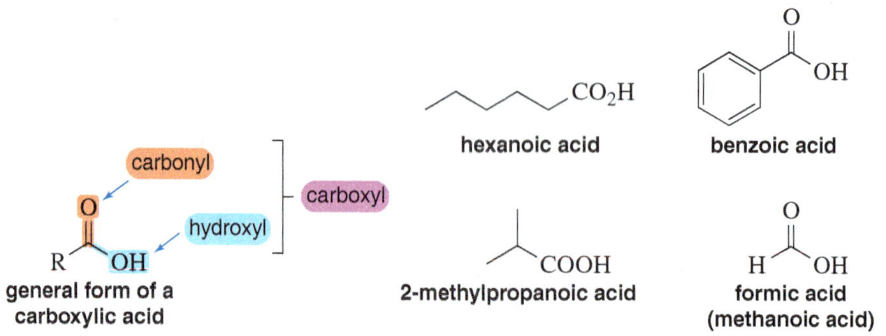

Figure 18.1 Representative structures of carboxylic acids.

18.1.1 Structure of carboxylic acid derivatives

Related to carboxylic acids are **carboxylic acid derivatives**, shown in **Figure 18.2**. These molecules are at the same oxidation state as carboxylic acids and can be thought of as the products of condensation reactions of carboxylic acids with the respective reagents. [*These are not actual reactions, just a way of categorizing them that is described in greater detail in Chapter 19.*] The **acid chloride** is the hypothetical condensation product of a carboxylic acid and HCl. **Esters** result from the hypothetical condensation of acids and alcohols, **anhydrides** from the condensation of two carboxylic acids, and **amides** from the condensation of a carboxylic acid and an amine. **Nitriles** are essentially dehydrated 1° amides. All but the nitrile feature a carbonyl group attached to an alkyl group and a substituent more electronegative than carbon.

The chemistry of these derivatives is closely related to the chemistry of carboxylic acids, which is introduced in Section 18.7 and is covered in detail in Chapter 19.

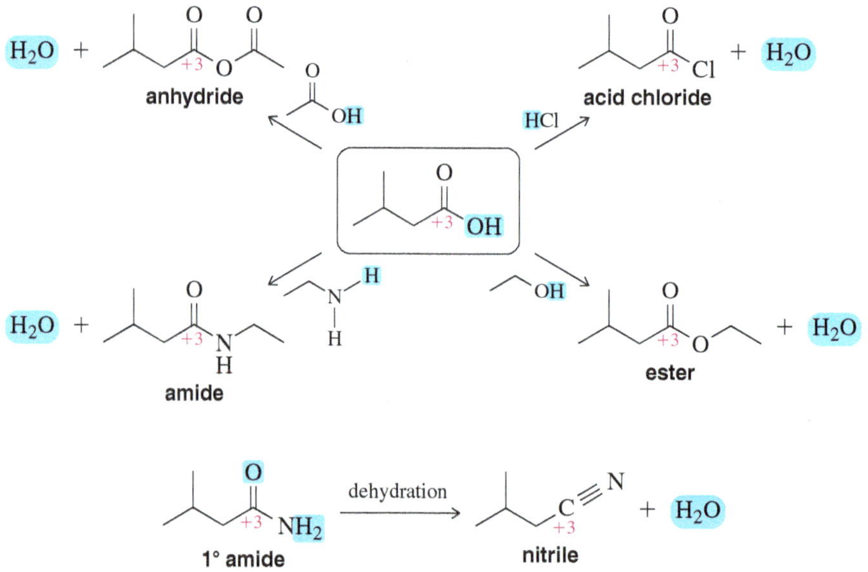

Figure 18.2 Carboxylic acid derivatives as the hypothetical products of condensation reactions. Like carboxylic acids, the carbonyl carbons of all carboxylic acid derivatives have an oxidation state of +3. [*These are not actual reactions.*]

18.2 Nomenclature of carboxylic acids

The IUPAC rules for naming carboxylic acids are summarized in **Table 18.1**. Following these rules, the two carboxylic acids shown here can be named.

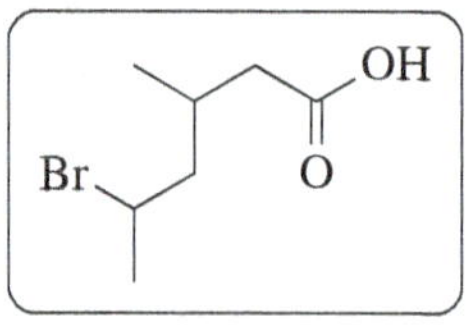

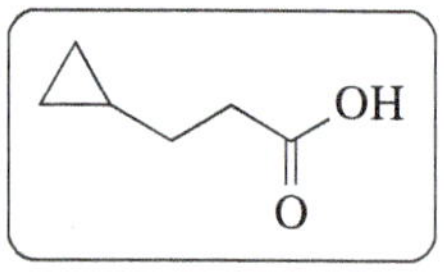

Table 18.1 Rules for naming carboxylic acids according to the IUPAC system of nomenclature

1. The parent chain is the longest carbon chain containing the carboxyl carbon and forms the root of the name. If the molecule is fully saturated, the infix -an- is used, and the suffix -e is replaced with -oic acid.	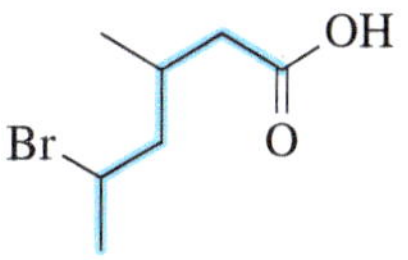 **6-carbon parent chain** **hexanoic acid**	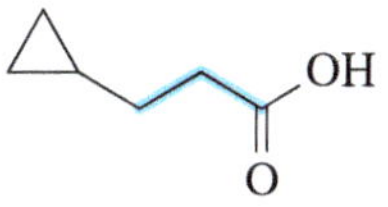 **3-carbon parent chain** **propanoic acid**
2. The parent chain is numbered such that the carboxyl carbon gets the lowest number. Because the carboxyl carbon is always carbon 1, its position is not specified in the name.	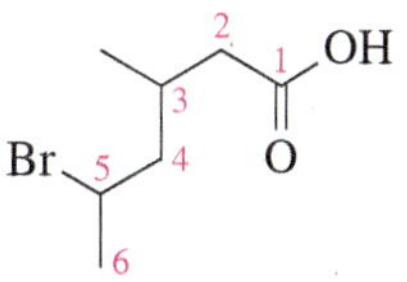 ~~1-hexanoic acid~~ **hexanoic acid**	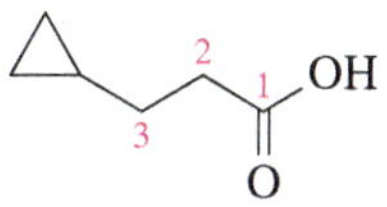 ~~1-propanoic acid~~ **propanoic acid**
3. Substituents on the parent chain are named as done previously. Small rings are named as substituents on the parent chain.	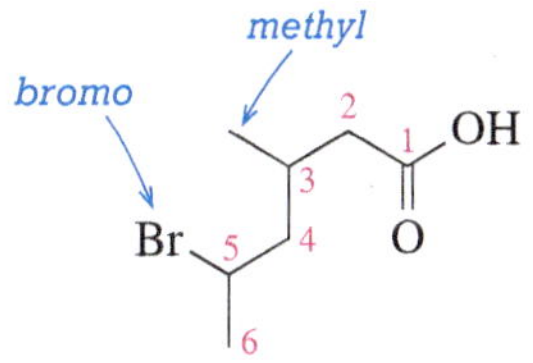	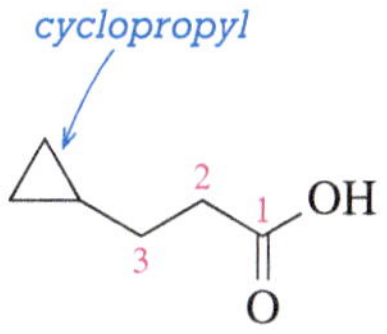
4. The name is then written following standard IUPAC alphabetization rules, with numbers used to indicate the location of substituents.	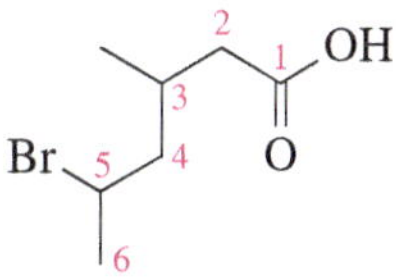 **5-bromo-3-methylhexanoic acid**	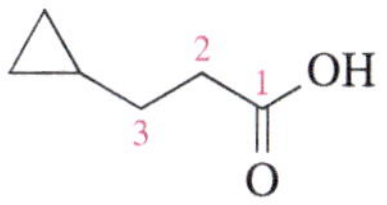 **3-cyclopropylpropanoic acid**

Table 18.2 IUPAC Functional group naming priority.

Functional group	Priority
Acids	1 (highest)
Esters	2
Acid chlorides	3
Anhydrides	3
Amides	4
Nitriles	5
Aldehydes	6
Ketones	7
Alcohols	8
Amines	9
Thiols	10
Alkenes, alkynes	11
Alkanes	—
Alkyl halides, ethers	(lowest)

18.2.1 Naming with rings

Naming carboxylic acids on rings deserves special mention. The ring is named as a cycloalkane with the term "carboxylic acid" at the end (the name of the cycloalkane + "carboxylic" is all one word). The ring carbon bearing the carboxylic acid is always designated as carbon 1 when numbering substituents (**Figure 18.3**).

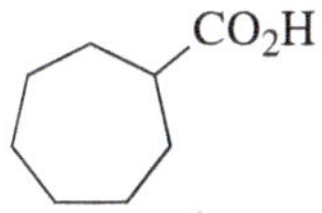

cycloheptanecarboxylic acid

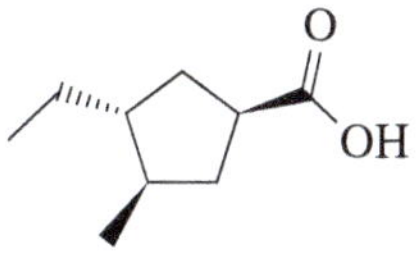

(1*S*,3*R*,4*R*)-3-ethyl-4-methylcyclopentanecarboxylic acid

Figure 18.3 IUPAC nomenclature for carboxylic acids on rings.

18.2.2 Naming with multiple functional groups

Under the IUPAC system, carboxylic acids receive the highest priority of all functional groups (**Table 18.2**). The name of a polyfunctional carboxylic acid ends in -oic acid, indicating that the molecule is a carboxylic acid. All other functional groups are named as substituents, as shown in **Figure 18.4**.

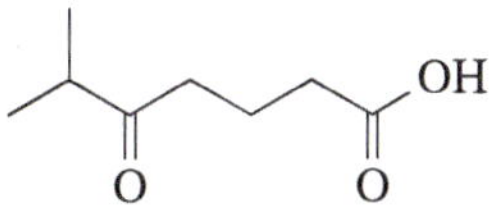

6-methyl-5-oxoheptanoic acid

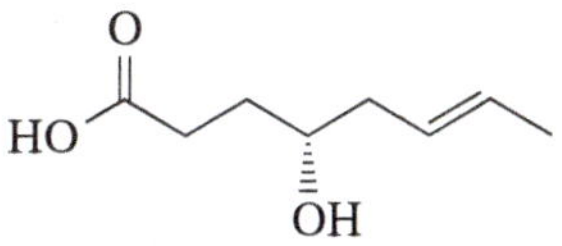

(4*R*,6*E*)-4-hydroxyoct-6-enoic acid

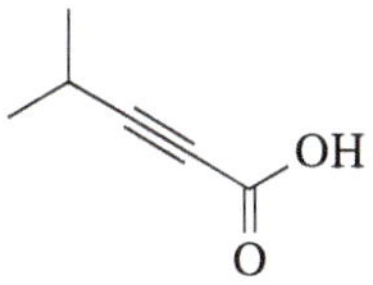

4-methylpent-2-ynoic acid

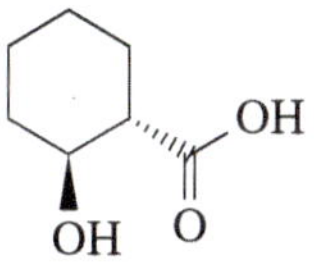

(1*S*,2*S*)-2-hydroxycyclohexanecarboxylic acid

Figure 18.4 IUPAC nomenclature for some polyfunctional carboxylic acids.

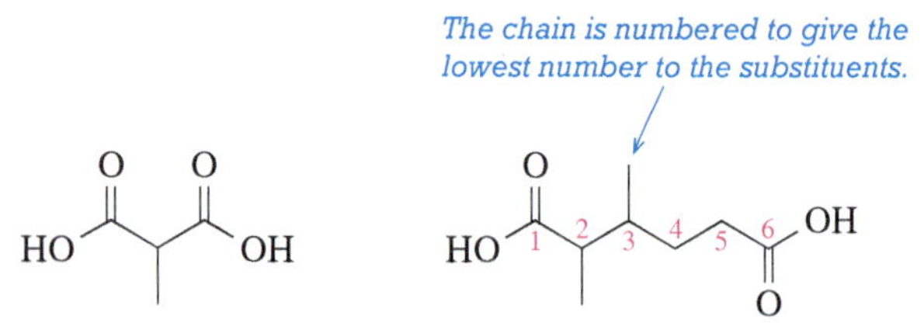

2-methylpropanedioic acid 2,3-dimethylhexanedioic acid

Figure 18.5 Nomenclature of polyacids.

18.2.3 Naming polyacids

Molecules with more than one carboxyl functional group are named using a prefix (di-, tri-, tetra-, etc.) with "-oic acid" to indicate the number of carboxyl groups (**Figure 18.5**). Because both carboxylic acids will occupy an end of the parent chain, there is no need to indicate the location of the carboxylic acid groups.

18.2.4 Common names of acids

Many carboxylic acids are referred to by their common name. Some examples are shown in **Figure 18.6**.

Figure 18.6 Common names of carboxylic acids.

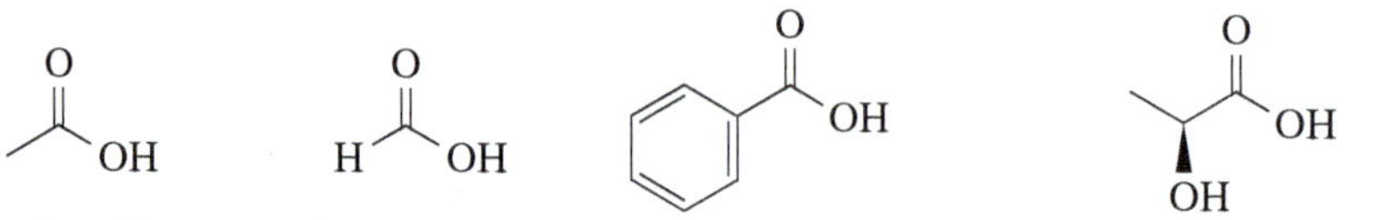

acetic acid (ethanoic acid) | formic acid (methanoic acid) | benzoic acid | lactic acid [(*S*)-2-hydroxypropanoic acid]

Assessment 18.7 Name the following carboxylic acids.

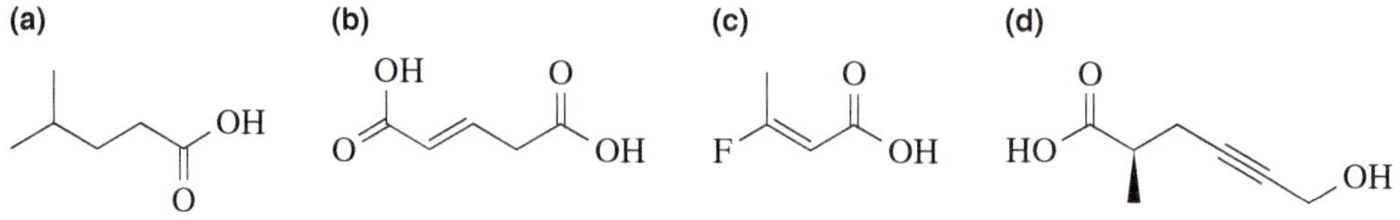

Assessment 18.8 Draw the structure that corresponds to the given name.

(a) (2*E*,5*R*)-3,5-dimethylhept-2-en-6-ynoic acid
(b) (1*S*,3*R*)-3-methylcyclohexanecarboxylic acid
(c) (2*R*,3*R*)-2,3-dihydroxybutanedioc acid
(d) (9*Z*,11*E*)-octadeca-9,11-dienoic acid

18.3 Molecular orbital picture of carboxylic acids

Similar to other carbonyls, the C═O bond of the carboxylic acid features a σ and π bond between sp^2-hybridized carbon and oxygen atoms (Figure 17.5). The C—OH bond is formed between the sp^2-hybridized central carbon and an sp^2-hybridized oxygen. The bond angles around the central carbon are all near 120° (**Figure 18.7**).

The carboxylic acid can be represented by three valid resonance structures (**Figure 18.8**). The nonbonded electron pair on O(2), delocalized throughout the π system (as indicated by resonance), must sit in a p orbital aligned with the neighboring C—O π bond. As a result, both oxygens are sp^2 hybridized. These resonance structures account for carboxylic acid reactivity.

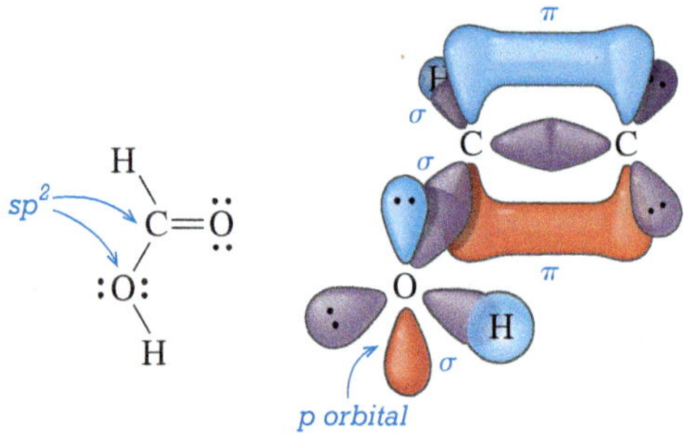

Figure 18.7 Molecular orbital picture of formic acid.

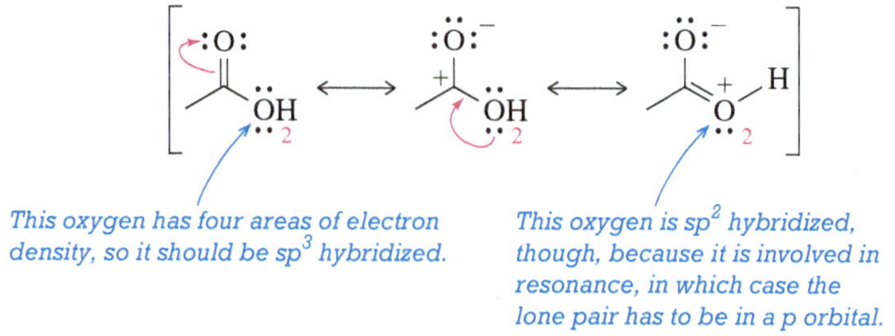

Figure 18.8 Because the carboxylic acid exhibits resonance, both oxygens are sp^2 hybridized.

Assessment 18.9 Draw the molecular orbital picture of ethanoic acid.

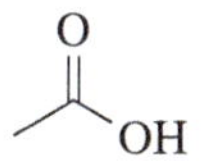

ethanoic acid

Assessment 18.10 Give the hybridization of each of the indicated nitrogen atoms.

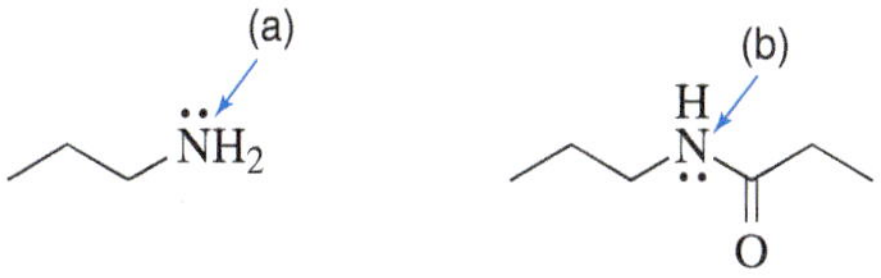

18.4 Physical properties of carboxylic acids

Carboxylic acids are highly polar and are both hydrogen bond donors and acceptors. Although they contain a hydroxyl and a carbonyl group, a carboxylic acid is markedly different from the analogous aldehyde and alcohol. Carboxylic acids are more water soluble because of extensive hydrogen bonding with water. Because pure carboxylic acids also form extensive hydrogen bonds with each other, they boil and melt at higher temperatures (**Table 18.3**).

Table 18.3 A comparison between the properties of alcohols, aldehydes, and carboxylic acids.

Compound	Melting point (°C)	Boiling point (°C)	H_2O solubility (g/L H_2O)
 butan-1-ol	−89.8	117.7	73
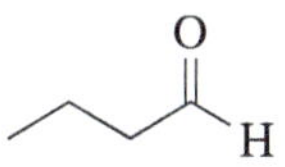 butanal	−96.9	74.8	7.6
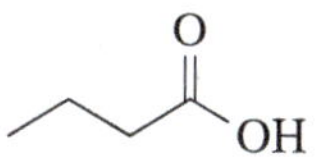 butanoic acid	−5.1	163.8	Completely miscible

The water solubility of carboxylic acids decreases with the addition of more carbons to the alkyl chain, becoming immeasurable for dodecanoic acid (C_{12}). At this point, the stability gained by hydrogen bonding of the carboxyl group with water is not enough to overcome the unfavorable entropy of having a large nonpolar chain dissolved in water. [*Recall Sections 11.3.3 and 13.3.2 on water solubility.*] Additionally, large nonpolar chains sterically impede hydrogen bonding between the carboxyl group and water (**Table 18.4**).

Table 18.4 Larger carboxylic acids have stronger intermolecular forces [higher melting point (mp) and boiling point (bp)] and are less soluble in H_2O.

IUPAC name	Common name	Formula	mp (°C)	bp (°C)	Solubility (g/100 g H_2O)
Methanoic	Formic	$HCOOH$	8	101	∞ (miscible)
Ethanoic	Acetic	CH_3COOH	17	118	∞
Propanoic	Propionic	CH_3CH_2COOH	−21	141	∞
Butanoic	Butyric	$CH_3(CH_2)_2COOH$	−6	163	∞
Pentanoic	Valeric	$CH_3(CH_2)_3COOH$	−34	186	3.7
Hexanoic	Caproic	$CH_3(CH_2)_4COOH$	−4	206	1.0
Octanoic	Caprylic	$CH_3(CH_2)_6COOH$	16	240	0.7
Decanoic	Capric	$CH_3(CH_2)_8COOH$	31	269	0.2
Dodecanoic	Lauric	$CH_3(CH_2)_{10}COOH$	44	—	i (insoluble)
Tetradecanoic	Myristic	$CH_3(CH_2)_{12}COOH$	54	—	i
Hexadecanoic	Palmitic	$CH_3(CH_2)_{14}COOH$	63	—	i
Octadecanoic	Stearic	$CH_3(CH_2)_{16}COOH$	72	—	i

18.4.1 Spectroscopy of carboxylic acids

The IR spectra of carboxylic acids are characterized by a broad O—H stretching band spanning from 3400 to 2500 cm^{-1} (**Figure 18.9**). As in alcohols, this is a result of extensive inter- and intramolecular hydrogen bonding. The C=O stretching band characteristic of all carbonyls and carboxylic acid derivatives falls around 1710 cm^{-1}. Relative to the band for a normal, aliphatic ketone (1715 cm^{-1}) this value is a result of the cancelling of resonance (donating) and inductive (withdrawing) effects of the hydroxyl group. [*See Section 14.3.6 for a review.*]

Figure 18.9 IR spectrum of acetic acid.

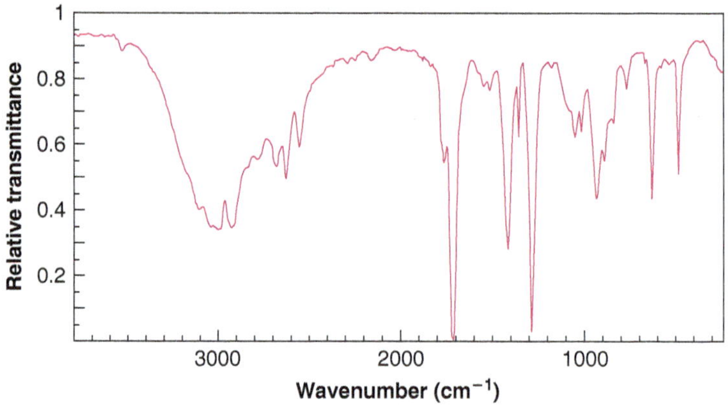

The ^{1}H NMR spectrum of a carboxylic acid displays a broad singlet around 11 ppm, corresponding to the carboxyl hydrogen (**Figure 18.10**). This signal is often so broad that it blends in with the baseline, making conclusive characterization by NMR less reliable. Hydrogens α to the carbonyl show up between 2 and 3 ppm as a result of the inductive effect of the carbonyl and the circulation of electrons in the C—O π bond. The carboxyl carbon shows up around 175 ppm, consistent with all carboxylic acid derivatives and slightly shielded relative to aldehydes and ketones (200 ppm).

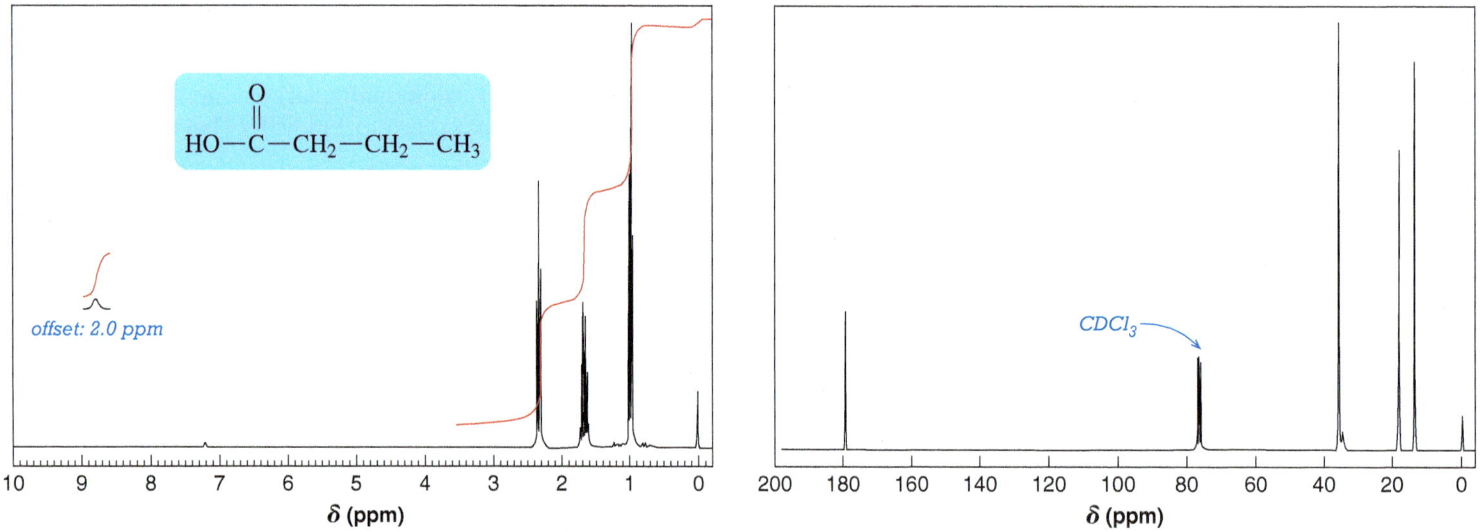

Figure 18.10 The ^{1}H and ^{13}C NMR spectra of butanoic acid.

Aliphatic carboxylic acids show two distinct fragmentation patterns in the mass spectrum. They undergo the McLafferty rearrangement to give a fragment at *m*/*z* 60, which is often the base peak in the spectrum (**Figure 18.11**). Longer chain carboxylic acids will commonly lose the alkyl group from the β-carbon to give a resonance-stabilized fragment at *m*/*z* 73.

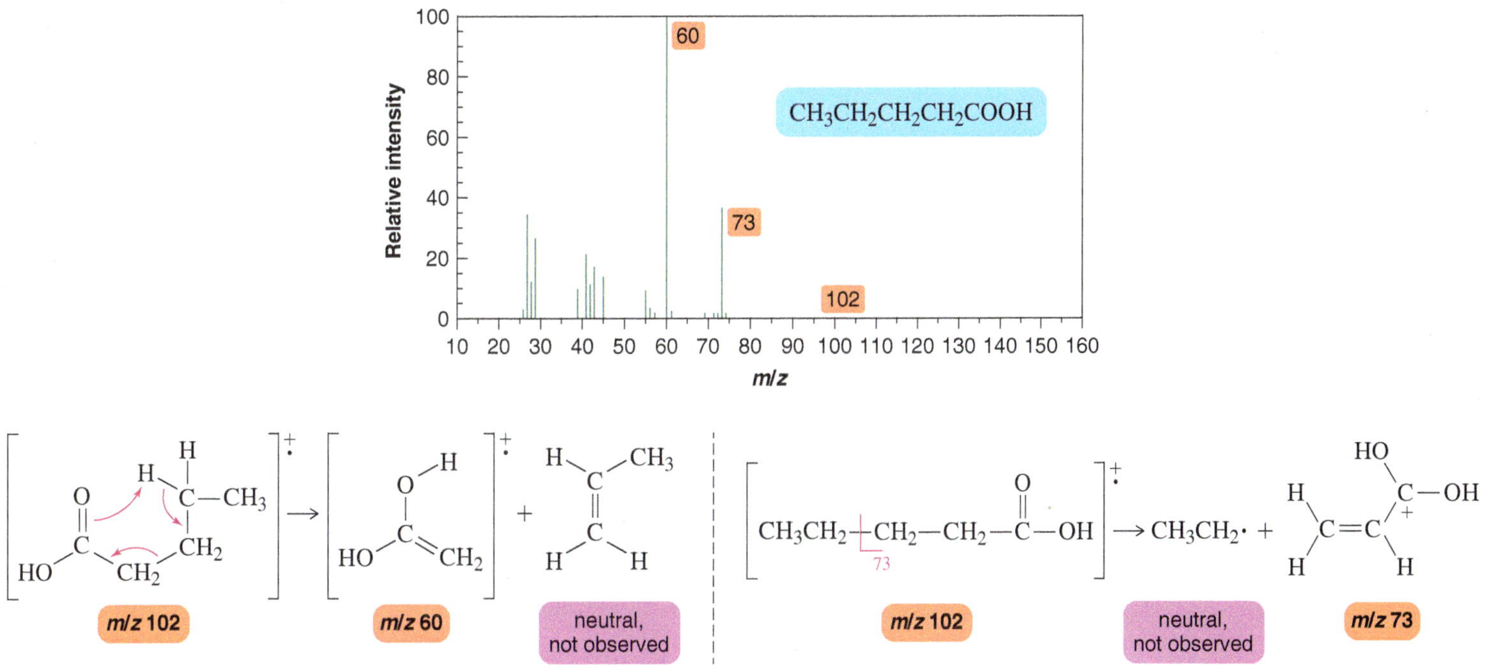

Figure 18.11 **The mass spectrum of pentanoic acid.**

18.4.2 Acidity of carboxylic acids

Carboxylic acids, as the name implies, are acidic. Acetic acid has a pK_a value of 4.7, meaning that even with weak bases, the **carboxylate anion** makes up a substantial component of the equilibrium mixture (**Figure 18.12**).

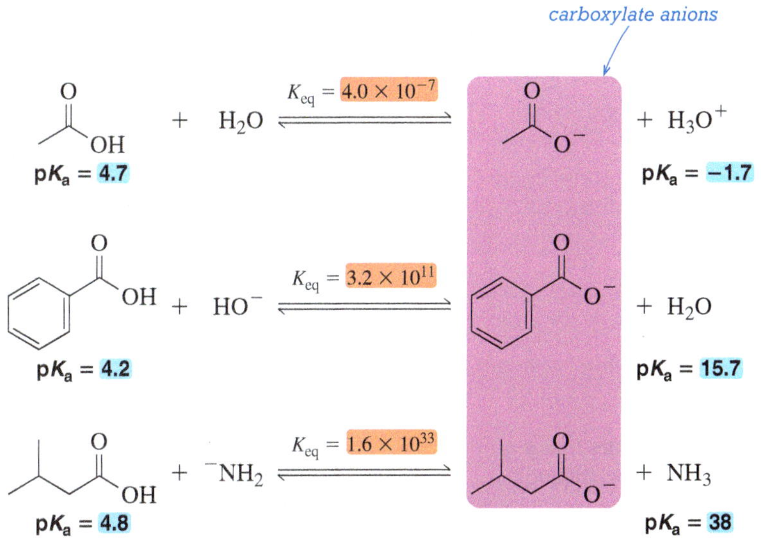

Figure 18.12 **Carboxylic acids can be deprotonated by strong and weak bases.** [*See Section 4.3.3 for calculating K_{eq} values.*]

Carboxylic acids are ~10^{11} times more acidic than alcohols. Deprotonation of an alcohol (pK_a ≈ 16) gives an alkoxide with a negative charge localized on a single oxygen. The negative charge on a carboxylate ion is delocalized over two oxygens through resonance. This delocalization results in enhanced stabilization in the conjugate base of a carboxylic acid (pK_a ≈ 5). As demonstrated on the reaction coordinate diagram in **Figure 18.13**, deprotonation of a carboxylic acid is more favorable than deprotonation of an alcohol.

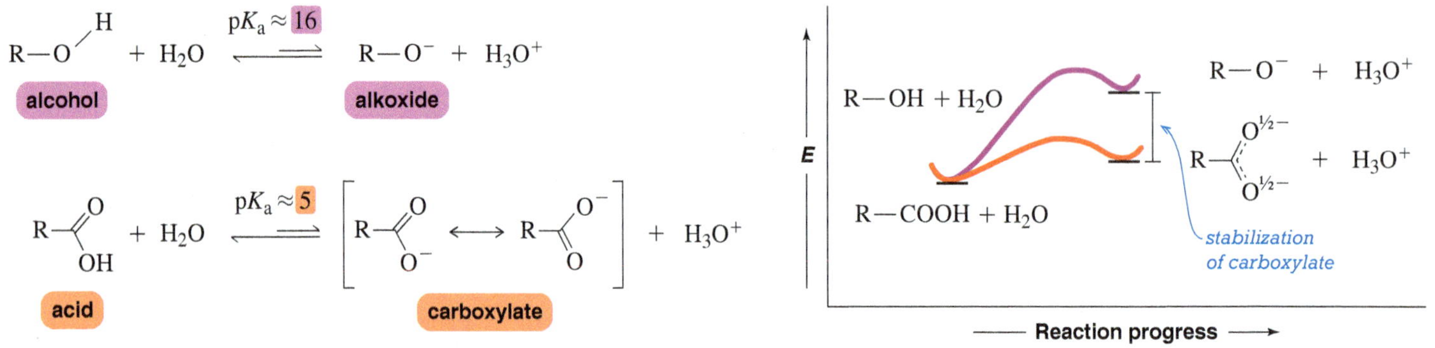

Figure 18.13 Deprotonation of a carboxylic acid is "less uphill"—requires less energy—on a reaction coordinate diagram than deprotonation of an alcohol.

Carboxylic acids are also acidic because of the inductively withdrawing carbonyl neighboring the acidic hydroxyl group (**Figure 18.14**). The electron-deficient carbonyl carbon pulls electron density toward it, creating a significant partial positive charge on the acidic hydrogen, as illustrated in the electrostatic potential map. This same concept accounts for the strong hydrogen bonds formed within carboxylic acid dimers (**Figure 18.15**).

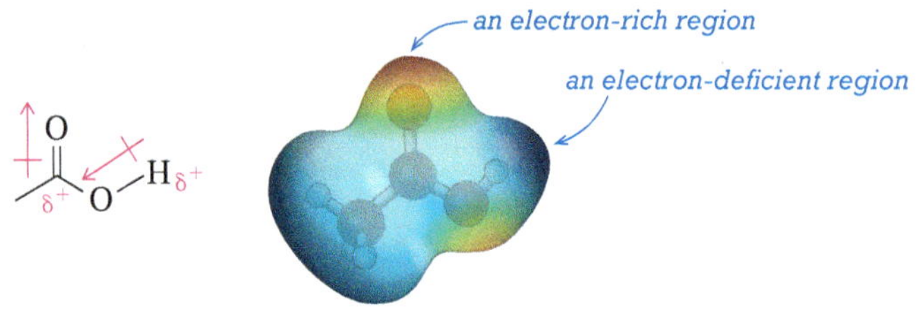

Figure 18.14 The strong induction of the carbonyl makes the carboxylic acid proton electron-deficient and, therefore, very acidic.

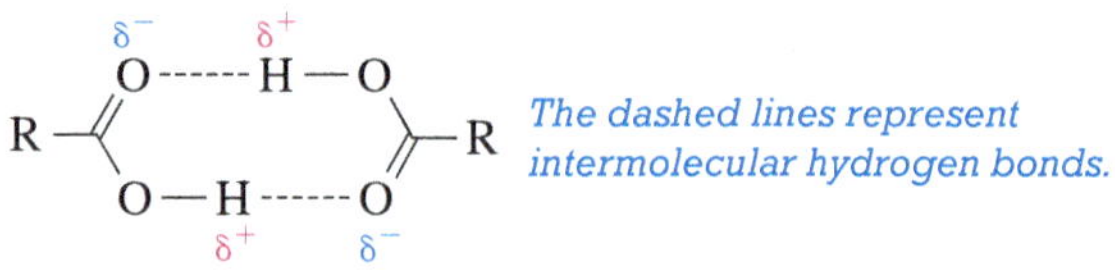

Figure 18.15 Carboxylic acids form dimers through intermolecular hydrogen bonding.

Induction can be used to explain deviations from the standard pK_a value (~5) among a series of carboxylic acids. A nearby electron-withdrawing group induces a partial positive charge near the carboxylate (**Figure 18.16**). This electrostatic interaction stabilizes the negative charge in the conjugate base, making the original acid more acidic.

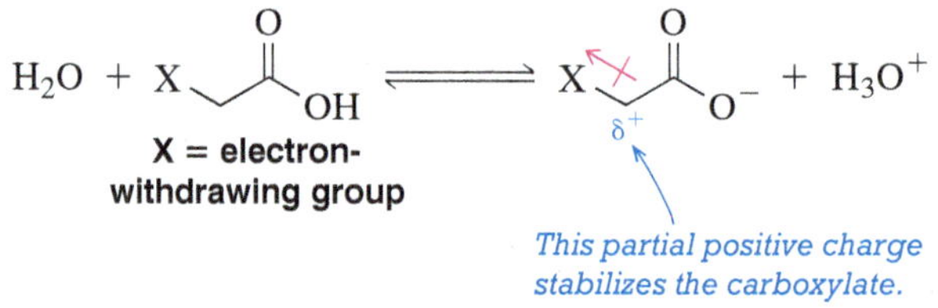

The pK_a lowering effect of induction is enhanced by adding more electron-withdrawing groups or having them more proximal to the carboxylic acid.

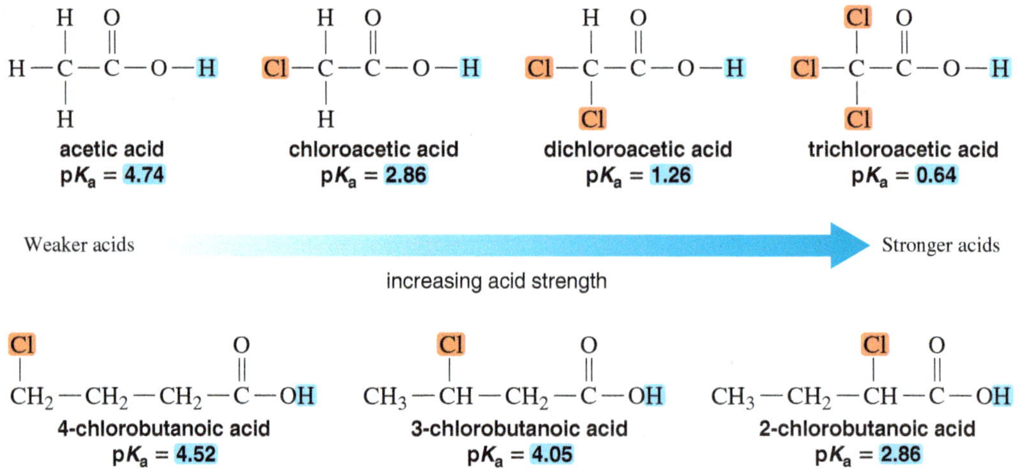

Figure 18.16 Neighboring electron-withdrawing groups make carboxylic acids more acidic through induction.

The impact of electron-withdrawing groups, as illustrated in **Table 18.5**, can result in substituted carboxylic acids ranging in pK_a from 0 to 5. Note that a stronger electron-withdrawing group (NO_2 vs. F) increases acidity (pK_a = 1.68 vs. 2.59).

Table 18.5 Substituted carboxylic acids have wide-ranging pK_a values.

Acid	K_a	pK_a
F_3CCOOH	5.9×10^{-1}	0.23
Cl_3CCOOH	2.3×10^{-1}	0.64
$Cl_2CHCOOH$	5.5×10^{-2}	1.26
$O_2N{-}CH_2COOH$	2.1×10^{-2}	1.68
$NCCH_2COOH$	3.4×10^{-3}	2.46
FCH_2COOH	2.6×10^{-3}	2.59
$ClCH_2COOH$	1.4×10^{-3}	2.86
$CH_3CH_2CHClCOOH$	1.4×10^{-3}	2.86
$BrCH_2COOH$	1.3×10^{-3}	2.90
ICH_2COOH	6.7×10^{-4}	3.18
CH_3OCH_2COOH	2.9×10^{-4}	3.54
$HOCH_2COOH$	1.5×10^{-4}	3.83
$CH_3CHClCH_2COOH$	8.9×10^{-5}	4.05
PhCOOH	6.46×10^{-5}	4.19
$PhCH_2COOH$	4.9×10^{-5}	4.31
$ClCH_2CH_2CH_2COOH$	3.0×10^{-5}	4.52
CH_3COOH	1.8×10^{-5}	4.74
$CH_3CH_2CH_2COOH$	1.5×10^{-5}	4.82

Stronger acids

Weaker acids

Assessment 18.11 Show how the equilibrium constants are calculated in Figure 18.12.

Assessment 18.12 Choose the more acidic carboxylic acid in each pair. Justify your choice.

(a)

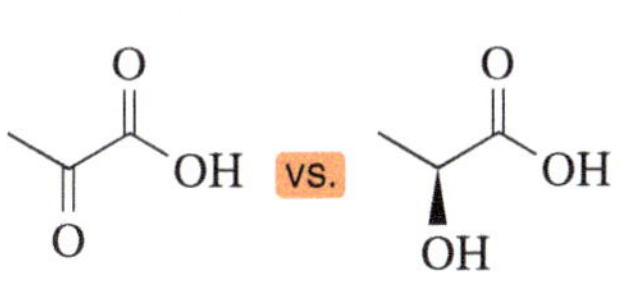

(b)

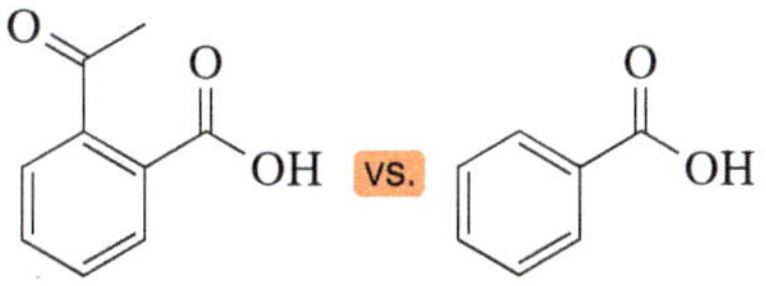

(c)

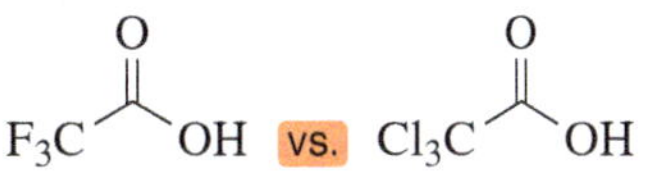

(d)

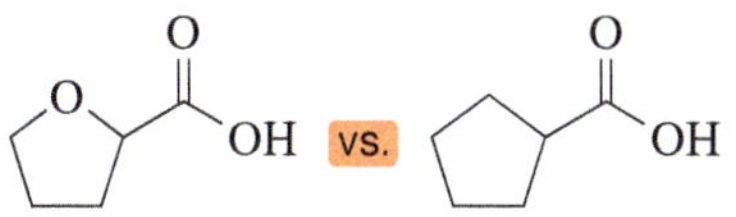

Assessment 18.13 Both an amide and an amine contain a nitrogen with a lone pair of electrons. Which nitrogen would you expect to be the stronger base?

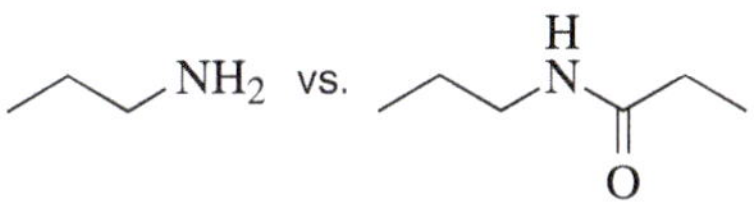

Assessment 18.14 Which oxygen of a benzoic acid would you expect to protonate first when exposed to a very strong acid?

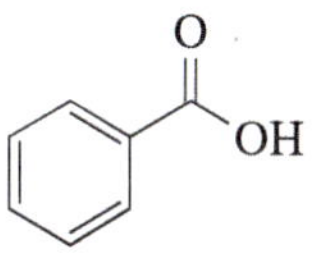

benzoic acid

Assessment 18.15 Which would you expect to be more soluble in H_2O: a carboxylic acid or a carboxylate? Why?

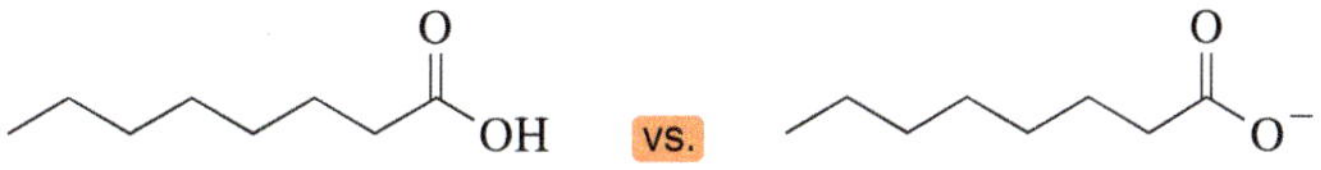

The synthesis of carboxylic acids is divided into four sections. Section 18.5.1 contains previously studied reactions in which carboxylic acids are products. Section 18.5.2 consists of reactions seen before that, with slight modifications, now result in carboxylic acids. A new carboxylic acid synthesis is discussed in Section 18.5.3, while Section 18.5.4 previews carboxylic acid synthesis reactions we study in subsequent chapters. Make sure all of these reactions make it onto your reaction sheets.

18.5 Synthesis of carboxylic acids

18.5.1 Previous syntheses of carboxylic acids

The synthesis of carboxylic acids has been discussed previously. [*Confirm this using the "Preparation of carboxylic acids" reaction sheet you created in Chapter 10.*] The oxidation reactions for this purpose are shown in **Table 18.6**. The oxidations of 1° alcohols and aldehydes are the most useful of these reactions.

Table 18.6 Previous syntheses of carboxylic acids

Oxidation of a 1° alcohol (Section 13.9.3)	
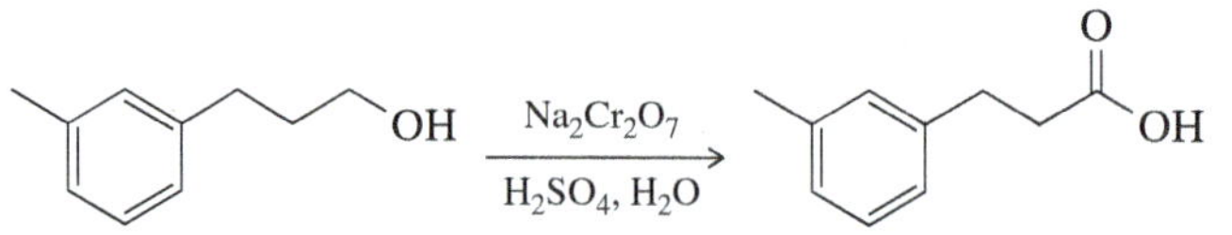	In the presence of water, the oxidation of a 1° alcohol results in a carboxylic acid, proceeding through an intermediate aldehyde.
Oxidation of an aldehyde (Section 17.10)	
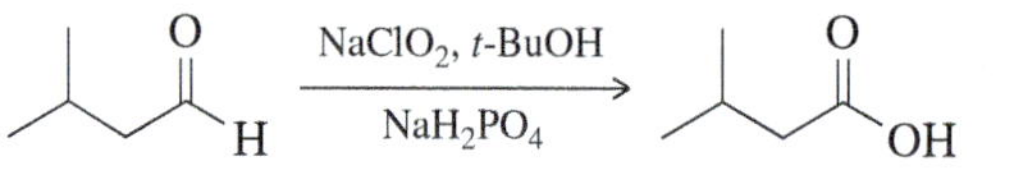	A mild, buffered, selective oxidation of an aldehyde with sodium chlorite gives a carboxylic acid under environmentally friendly conditions.
Exhaustive oxidation of a ketone (Section 17.10)	
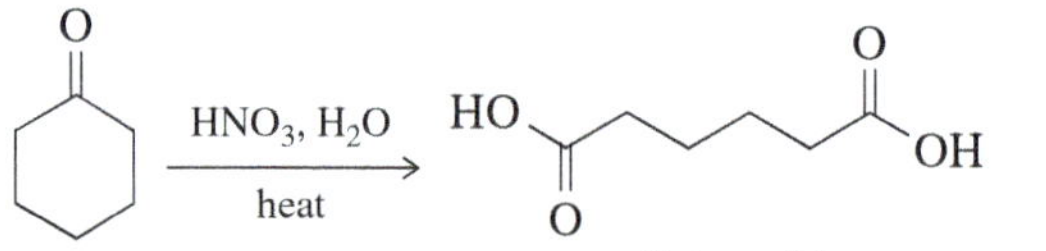 adipic acid	Though only effective with simple substrates, oxidation of a ketone with nitric acid is useful for the industrial synthesis of adipic acid, a precursor to the polymer nylon.

18.5.2 Modification of previous reactions

There are several reactions studied previously that allow for the synthesis of carboxylic acids when applied to different substrates or when run under conditions modified slightly from what you saw before. Those are discussed in this section.

18.5.2.1 $KMnO_4$ oxidation of an alkene or an alkyne

Previously (Section 9.1.5):	To make carboxylic acids:
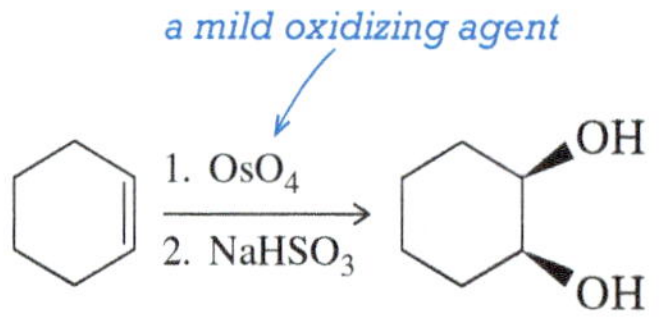	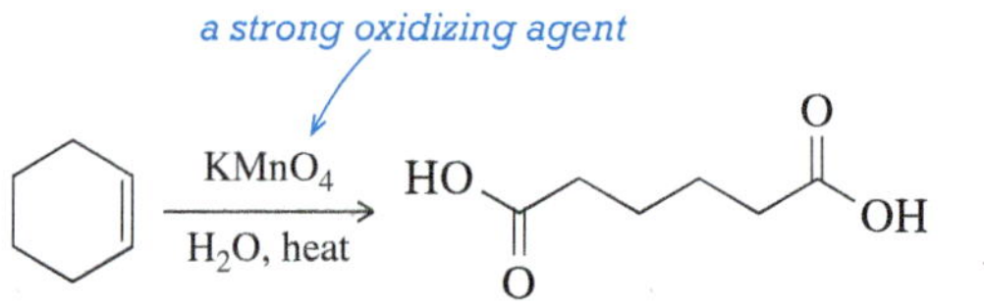

Despite producing different outcomes, $KMnO_4$, a strong oxidizing agent, is similar in structure to osmium tetroxide (**Figure 18.17**). Thus, the mechanism for potassium permanganate oxidation is similar. Formation of a cyclic manganate ester occurs through the identical concerted 6-electron cyclic transition state. At low temperature, the diol can be isolated but in lower yield due to overoxidation compared to the diol produced by the reaction of OsO_4. At higher temperatures, the cyclic product breaks down to form two aldehydes through another 6-electron cyclic transition state. The resulting aldehydes, in the presence of water and a strong oxidizing agent, are converted to carboxylic acids via the hydrate. If ketones were produced instead, then further oxidation would *not* occur. [*This final oxidation is identical to the chromic acid oxidations from Chapters 13 and 17.*]

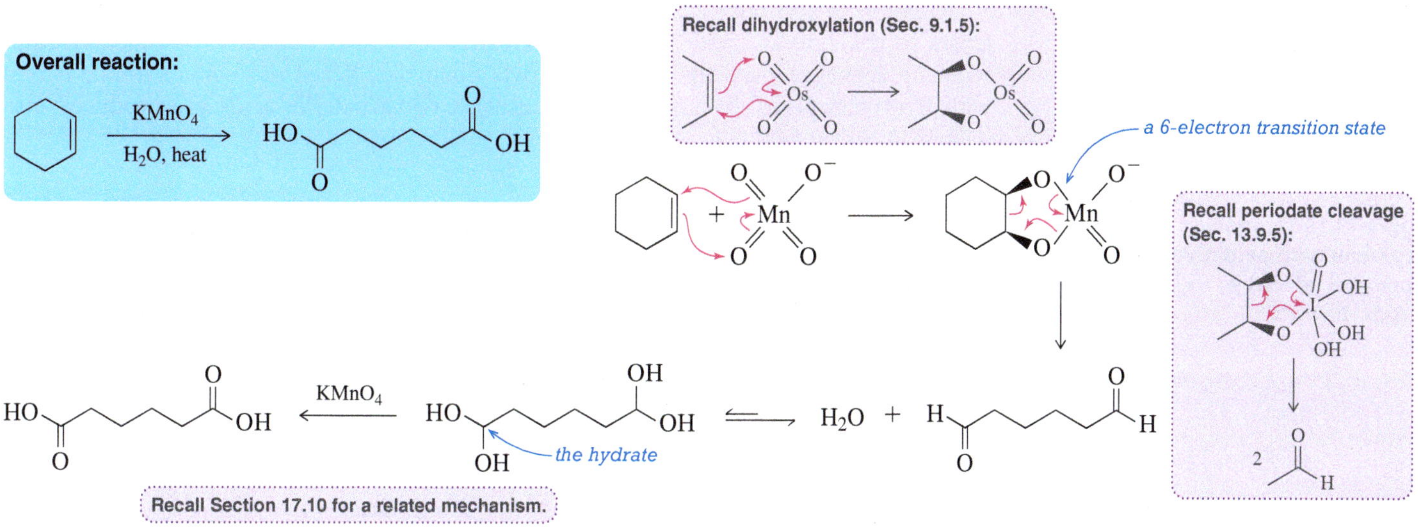

Figure 18.17 **The $KMnO_4$ cleavage of alkenes** is similar to reactions studied previously, including dihydroxylation with OsO_4 and the oxidation of aldehydes by H_2CrO_4.

Alkynes, which begin at a higher oxidation state than alkenes, can be similarly cleaved to give carboxylic acids through a diketone intermediate (**Figure 18.18**). The hydrated form of the diketone is presumably cleaved further via a mechanism analogous to the periodate cleavage reaction from Section 13.9.5.

Figure 18.18 **The $KMnO_4$ cleavage of alkynes** to give carboxylic acids proceeds by a mechanism similar to that shown for the cleavage of alkenes (Figure 18.17). $KMnO_4$ behaves similarly to how IO_4^- did in Section 13.9.5.

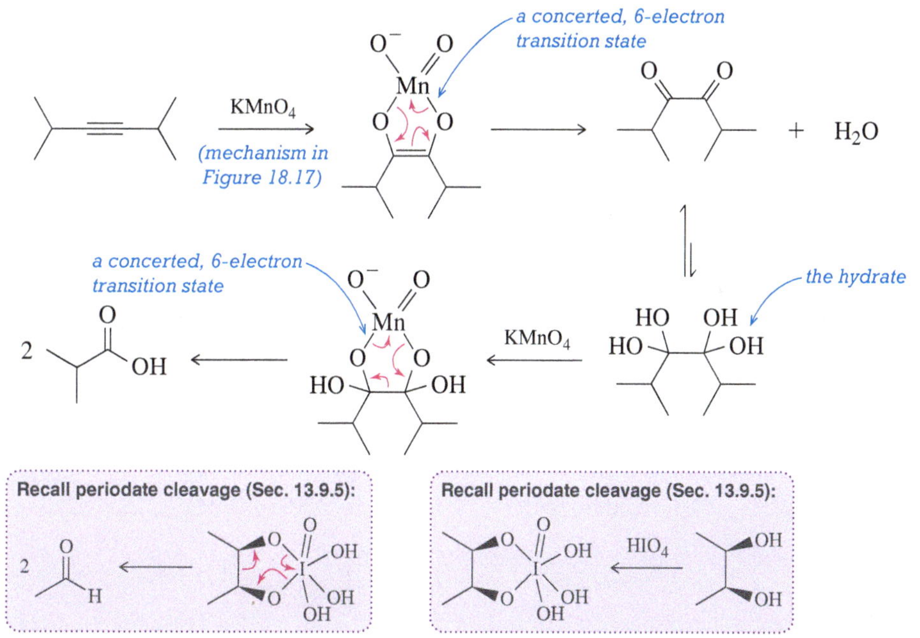

Although poorly understood, the mechanisms shown in Figures 18.17 and 18.18 are reasonable, based on reactions studied previously.

18.5.2.2 Ozonolysis of an alkene (with oxidative work up)

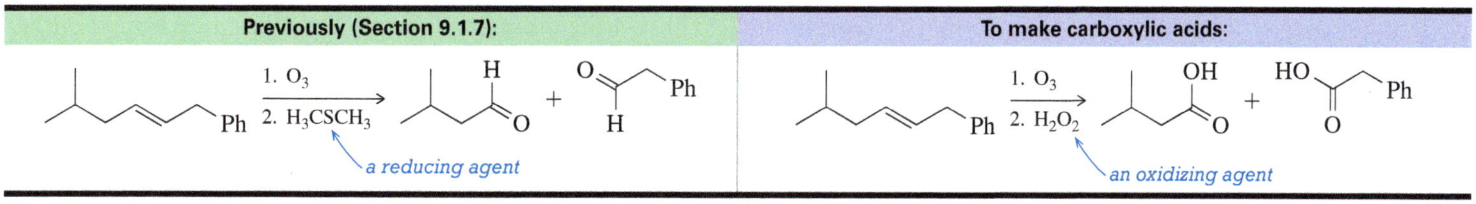

Recall from Section 9.1.7 that ozonolysis of an alkene proceeded by a complicated mechanism to generate an unstable ozonide. Treatment of this ozonide with dimethylsulfide (a reducing agent) reduced the ozonide to two aldehydes. When this same ozonide is treated with hydrogen peroxide (an oxidizing agent), two carboxylic acids are produced instead (**Figure 18.19**).

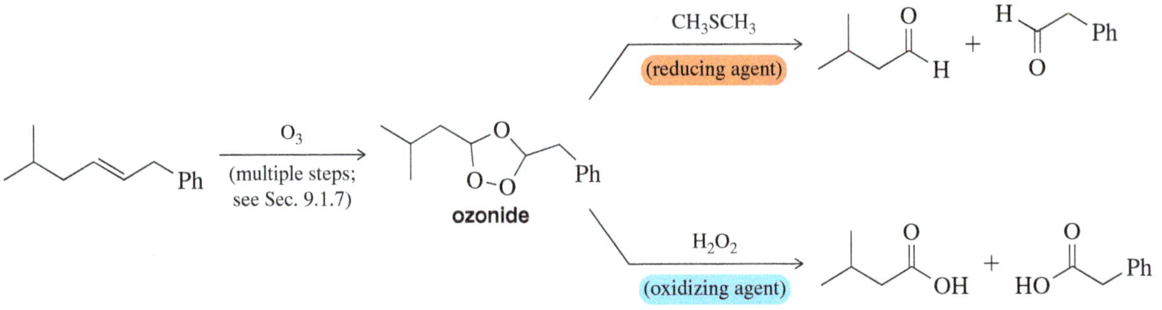

Figure 18.19 Ozonolysis of an alkene with oxidative workup produces a carboxylic acid.

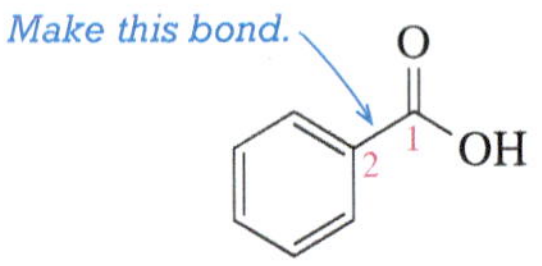

18.5.3 Grignard addition to CO_2

The oxidation reactions in Table 18.6 [*at the beginning of Section 18.5*] are useful for synthesizing carboxylic acids when the full carbon skeleton is already in place. There are circumstances, however, when a molecule needs to be extended by a single carbon with a carboxylic acid as the desired product.

With the goal of making the $C_1—C_2$ bond of benzoic acid, let's *think like a chemist* and see how reactions we saw in Chapter 17 help us do that.

Thinking Like a Chemist

There are two ways to make a new bond between two groups R_1 and R_2. R_1 could be a nucleophile that attacks an electrophile R_2, or R_2 could be a nucleophile that attacks R_1. [*This excludes radical possibilities, despite some studies suggesting they are reasonable also.*] Applied to our case here, we could imagine a hypothetical C_2 phenyl carbocation being attacked by a carbanion on the carbonyl carbon, C_1 (this is possibility 1 in **Figure 18.20**). Alternatively, a phenyl anion at C_2 could attack a carbocation on C_1 (this is possibility 2 in Figure 18.20). [*Between these two possibilities, which seems more reasonable?*]

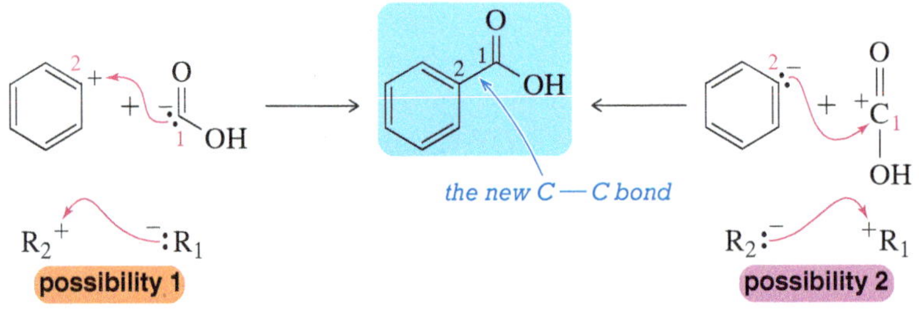

Figure 18.20 Two possible ways to make a new carbon–carbon bond (excluding radicals).

The process in Figure 18.20 is meant only to get our brains working toward a solution, but the hypothetical intermediates we came up with are not equivalent—one is better. In possibility 1, there is a carbocation on the benzene. Back in Chapter 12, we learned that alkenyl halides cannot undergo S_N1 reactions because alkenyl carbocations are unstable. Furthermore, carbonyl anions are also unlikely, representing a reversal of normal carbonyl polarity (see Assessment 17.76). On the other hand, possibility 2 has a carbanion on benzene, which is not necessarily stable, but is similar to the carbanion of a Grignard reagent. [*What about the carbocation in possibility 2?*] Possibility 2 is more reasonable when you consider a resonance structure of the carbocation electrophile. By pushing a pair of electrons, the molecule resembles a protonated form of carbon dioxide (CO_2), a structure that could exist (**Figure 18.21**).

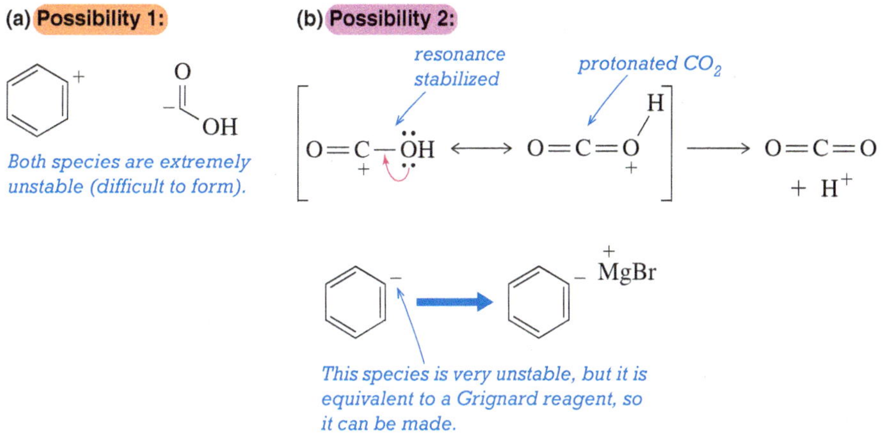

Figure 18.21 Comparison of the stability of intermediates in the hypothetical formation of a carbon–carbon bond. (a) Possibility 1 is unlikely because of the instability of the two intermediates. (b) Possibility 2 is reasonable because the intermediates are stable enough to be made.

This thought process leads us to a way that we could carry out this transformation (**Figure 18.22**). We can't protonate the carbonyl first and then have the Grignard reagent attack. [*Why? Because Grignard reagents are quenched by strong acids.*] But because Grignard reagents are such strong nucleophiles, they can easily attack neutral carbonyls [*which can be shown in zwitterionic form—see Chapter 17*]. Upon attacking, electrons in the C—O π bond are pushed onto the oxygen to give a carboxylate. Quenching the carboxylate with acid forms benzoic acid, having extended the molecule by an extra carbon. [*And we made a new carbon–carbon bond, which is always an awesome thing to be able to do.*] ■

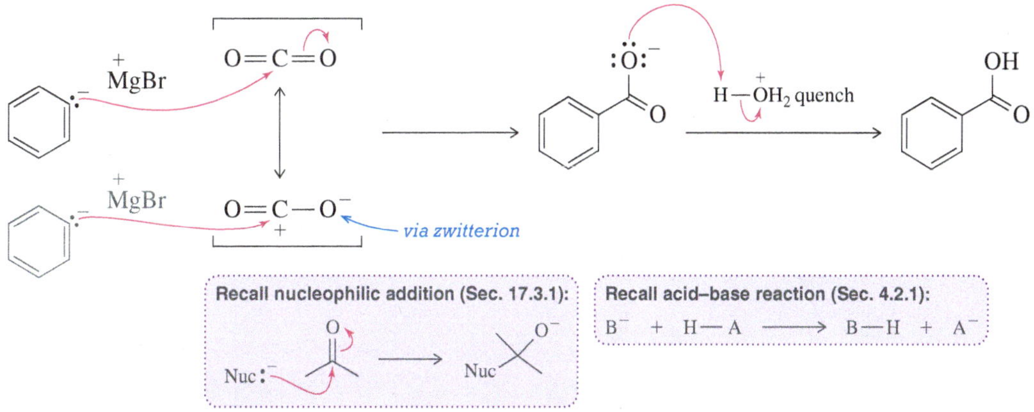

Figure 18.22 Grignard addition to CO_2 produces a chain-extended carboxylic acid.

Experimentally, preparation of the Grignard reagent is done in a polar aprotic solvent like THF or ether. Once it is formed, carbon dioxide (often by way of dry ice) is bubbled into the solution, where addition occurs prior to quenching with acid. This reaction can be used with a variety of Grignard reagents (**Figure 18.23**).

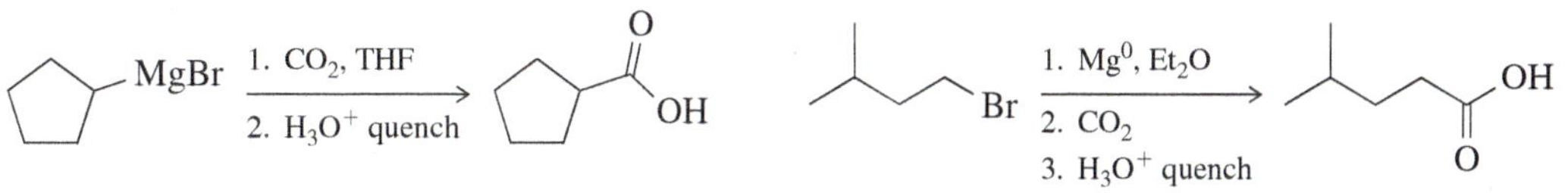

Figure 18.23 Selected examples of Grignard addition to CO_2.

Assessment 18.16 Predict the product of the following reactions.

(a)

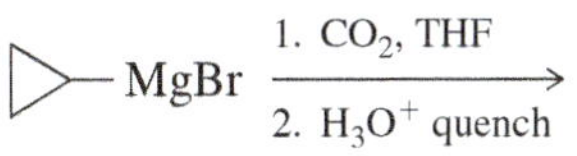

(b)

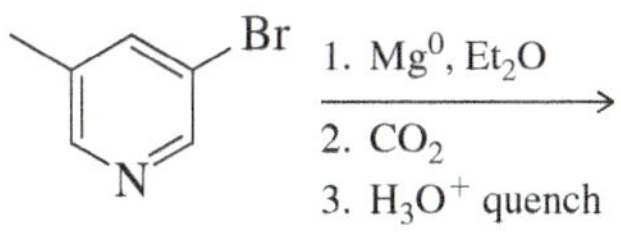

Assessment 18.17 Beginning with the alkene shown, synthesize the carboxylic acid on the right.

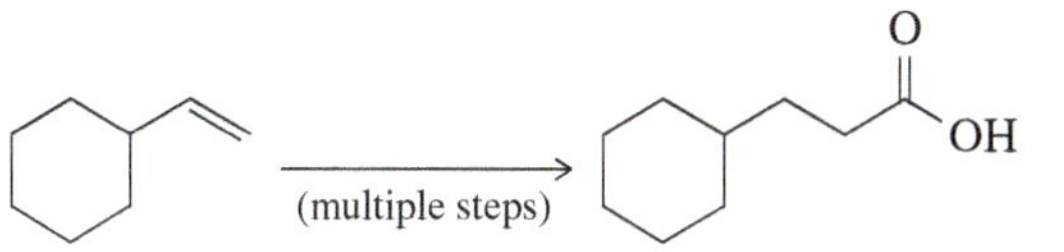

18.5.4 Chromic acid oxidation of alkyl benzenes

The side chain of alkyl benzenes can be oxidized to a carboxylic acid by treatment with hot aqueous chromic acid or basic potassium permanganate. For example, the oxidation of toluene results in benzoic acid (**Figure 18.24**).

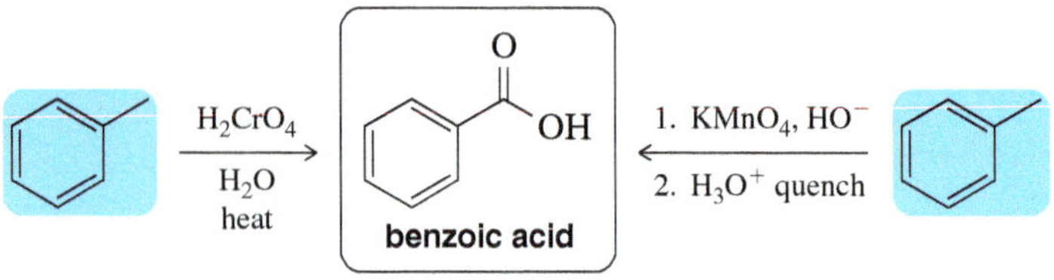

Figure 18.24 The oxidation of toluene to benzoic acid.

The mechanism of this reaction is not well understood: For reasons we explain in Chapter 24, it likely proceeds through the formation of a radical on the benzylic carbon. Any alkyl group with a C—H bond that neighbors the benzene ring is oxidized under these conditions (**Figure 18.25**). [*Assessments 18.18 and 18.19 cover the material in all of Section 18.5.*]

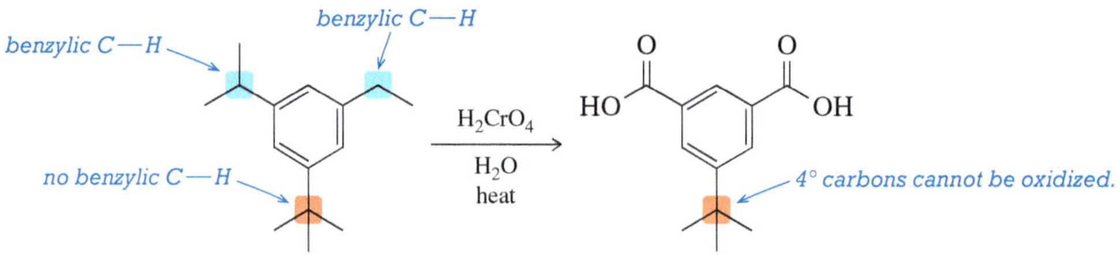

Figure 18.25 The oxidation of alkyl benzenes only occurs if there is a hydrogen at the benzylic carbon.

Assessment 18.18 Identify the best reagent to carry out each of the following transformations.

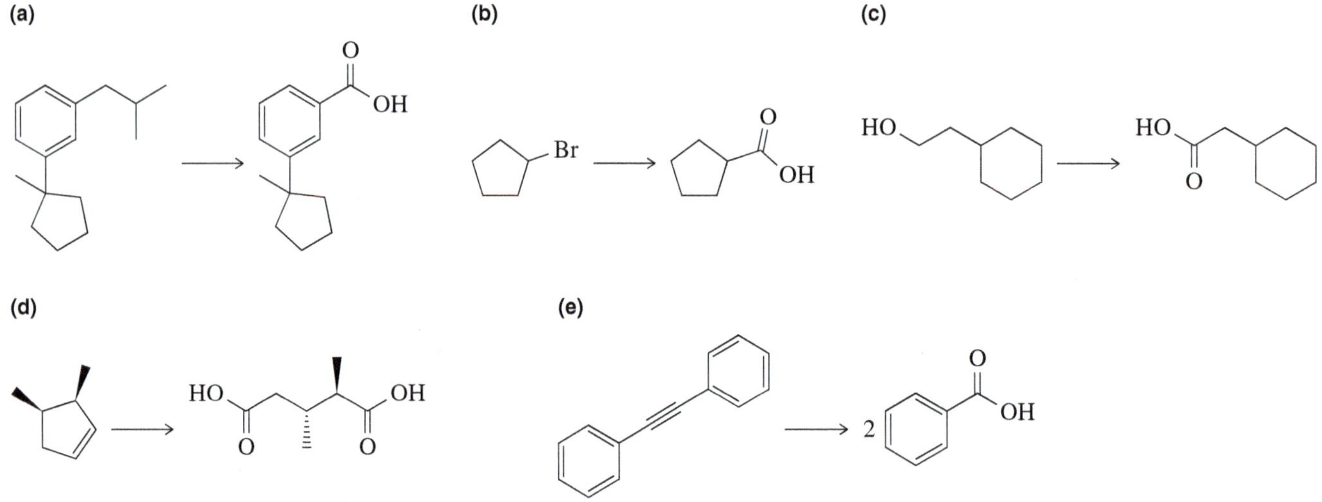

Assessment 18.19 Using the reactions from Section 18.5, suggest five different methods by which the following carboxylic acid could be made.

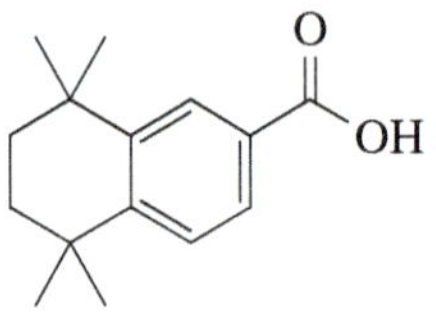

18-A CHEMISTRY IN CONTEXT: Green Chemistry

The biosynthesis and industrial synthesis of benzoic acid

Benzoic acid is produced by many fruit-bearing plants through complicated biochemical pathways. One pathway begins with shikimate (**Figure 18.26**) and involves 16 enzyme-mediated steps that occur in the plastid, cytosol, and mitochondria. The final conversion of benzaldehyde to benzoic acid is similar to oxidation reactions described in Sections 17.10 and 18.5. It is catalyzed in the mitochondria by benzaldehyde dehydrogenase, an enzyme from the family of oxidoreductases. We have seen these previously in the discussion of the oxidation of alcohols by NAD^+ cofactor-dependent enzymes. [*See Chemistry in Context 17-A.*]

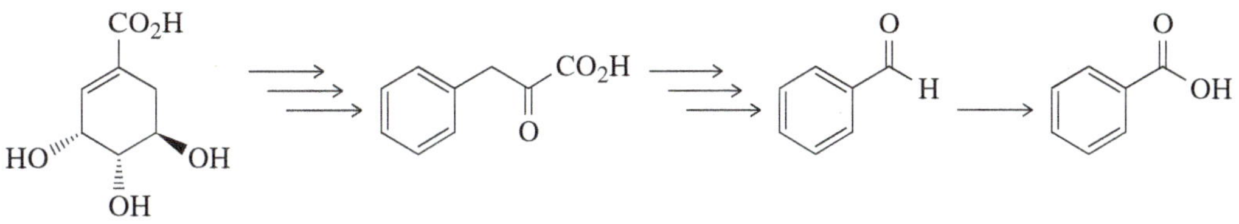

Figure 18.26 The biosynthesis of benzoic acid is mediated, in the final step, by the NAD^+-dependent, enzymatic oxidation of benzaldehyde in the mitochondria.

The modern synthesis of benzoic acid demonstrates the power of green chemistry. As shown in **Figure 18.27**, the synthesis could occur by the reaction of toluene with hot chromic acid (see Section 18.5.4). Chromic acid is especially toxic, however, and is known to cause cancer [*Chemistry in Context 13-A*]. The reaction of trichlorobenzene with calcium hydroxide in the presence of catalytic iron salts is a slightly greener option, but it produces benzoic acid contaminated by chlorinated benzenes—molecules with major environmental issues [*see Chemistry in Context 11-C*]. The current synthesis uses molecular oxygen [*environmentally harmless and renewable*] along with catalytic [*good!*] manganese salts [*nontoxic*] to oxidize toluene. Probably the only way to green the process further would be to discover a renewable source of toluene [*get to work on that!*], which is currently produced from petroleum feedstocks [*see Chemistry in Context 3-B*].

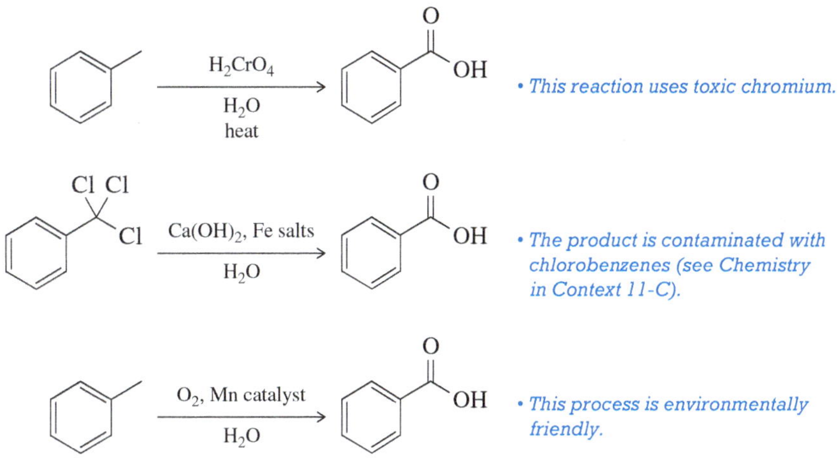

Figure 18.27 The oxidation of toluene with molecular oxygen and manganese catalysts is a green process used for the industrial preparation of benzoic acid and sodium benzoate.

18.6 Reactions of carboxylic acids: acid–base reactions

Recall from Section 18.4.2 that carboxylic acids are moderately acidic ($pK_a = 5$). The most common reaction of carboxylic acids is the simple acid–base reaction to make carboxylate ions. These proceed like all acid–base reactions, with the electron-rich base grabbing the proton and pushing electrons onto the electronegative oxygen. Carboxylates obtained in this way can be used in a variety of reactions, including the S_N2 reaction with a 1° or 2° haloalkane (Section 18.8.2).

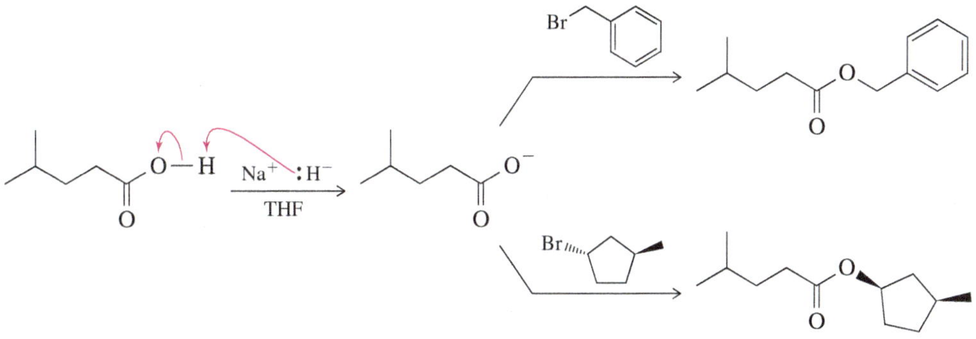

18.6.1 Extraction of carboxylic acids

The acidity of carboxylic acids allows for the design of experimental protocols that permit carboxylic acids to be separated from nonpolar, nonacidic compounds. As might be expected, insoluble carboxylic acids can be made more soluble in water by deprotonation. Consider decanoic acid, which has a solubility of 0.2 g/100 g H_2O (**Figure 18.28**). After being deprotonated by sodium hydroxide, the resulting sodium decanoate is 50 times more soluble in water (10 g/100 g H_2O). The newly formed sodium carboxylate is now a polar ionic salt.

Figure 18.28 Deprotonation of carboxylic acids increases their solubility in water.

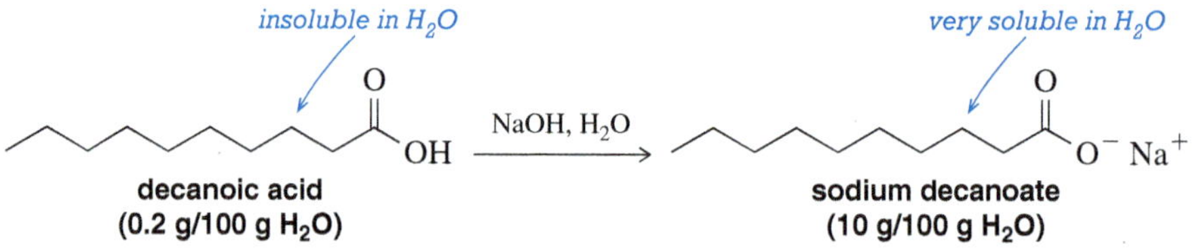

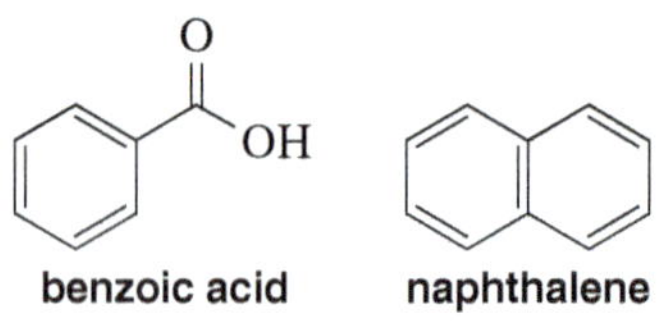

Consider then how you might separate a mixture of the carboxylic acid benzoic acid and the neutral, nonpolar compound naphthalene (labeled as "organic impurities" in **Figure 18.29**). [*This is a common experiment that you might have already performed in lab.*] When the carboxylic acid is neutral, both compounds are very soluble in organic solvents (diethyl ether). By extracting the diethyl ether with an aqueous solution of sodium hydroxide, the deprotonated carboxylate salt moves preferentially to the denser aqueous layer. Separation of the layers leaves two solutions: One contains naphthalene in ether and the other contains the benzoate anion in water. Acidification of the basic aqueous layer regenerates the benzoic acid (insoluble in water). Addition of diethyl ether allows the benzoic acid to move into the diethyl ether layer. Separation of the layers and evaporation of ether result in pure benzoic acid.

Figure 18.29 Separation of carboxylic acids from neutral compounds via acid–base extraction.

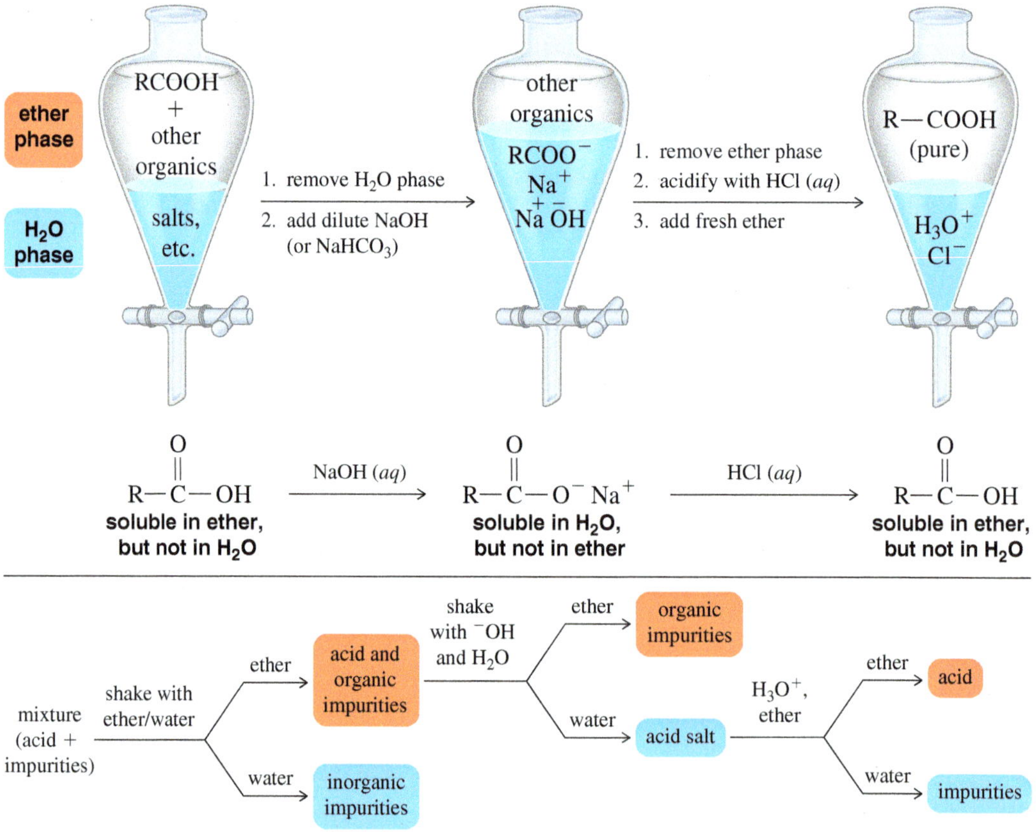

Assessment 18.20 The acid-catalyzed hydrolysis of an ester (Sections 18.7.2 and 19.7.1) results in the formation of an equal amount of carboxylic acid and alcohol. (a) Design a flow chart for separating the two products. (b) Once separated, how could you distinguish between the carboxylic acid and alcohol using IR spectroscopy?

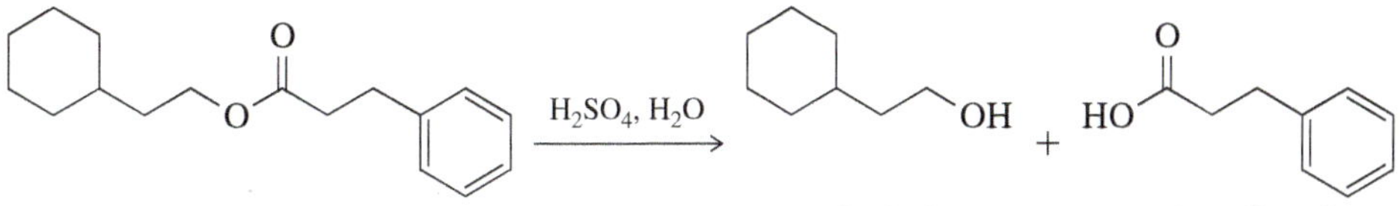

Assessment 18.21 At the end of a $LiAlH_4$ reduction (Section 19.8.1), there was found to be a small amount of unreacted amide along with the amine product dissolved in diethyl ether. How might you remove the amine from the ether?

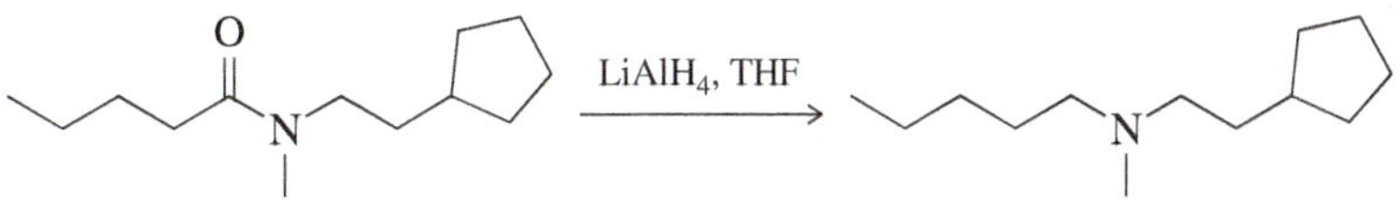

18-B CHEMISTRY IN CONTEXT: Food Chemistry

Sodium benzoate preservative

Sodium benzoate, found in plants and fruits, has been used since 1909 as a food preservative. Food preservatives extend the shelf life of foods by minimizing the growth of bacteria and fungi, which decreases the cost to consumers by reducing preparation and shipping costs.

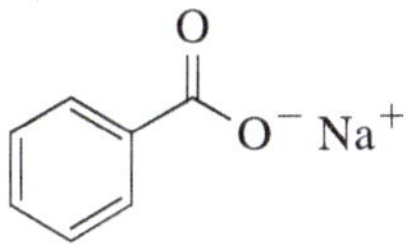

sodium benzoate

Although not completely understood, the mechanism of sodium benzoate as a preservative was recently proposed to be a result of the acid–base equilibrium between benzoate and benzoic acid (**Figure 18.30**). It is commonly used in acidic foods where sodium benzoate exists in its protonated form (benzoic acid).

Bacteria are unable to absorb ionized benzoate. Only under acidic conditions (pH < 4), when the equilibrium is shifted to benzoic acid, can it be absorbed. Once in the cytosol, where the pH is between 6 and 7, the equilibrium shifts back to ionized benzoate. Bacteria are unable to remove ionized benzoate, resulting in a buildup of benzoate that triggers acidification of the cytosol (**Figure 18.31**). This acidification presumably aids in the removal of benzoate from the cytosol and is believed to be responsible for the slowed bacterial growth. This hypothesis, albeit intriguing, has been called into question by recent research and requires further study.

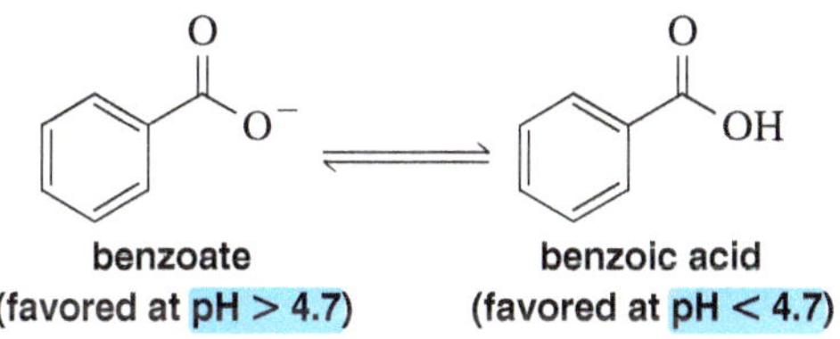

Figure 18.30 The position of equilibrium between benzoate and benzoic acid depends on the pH of the solution.

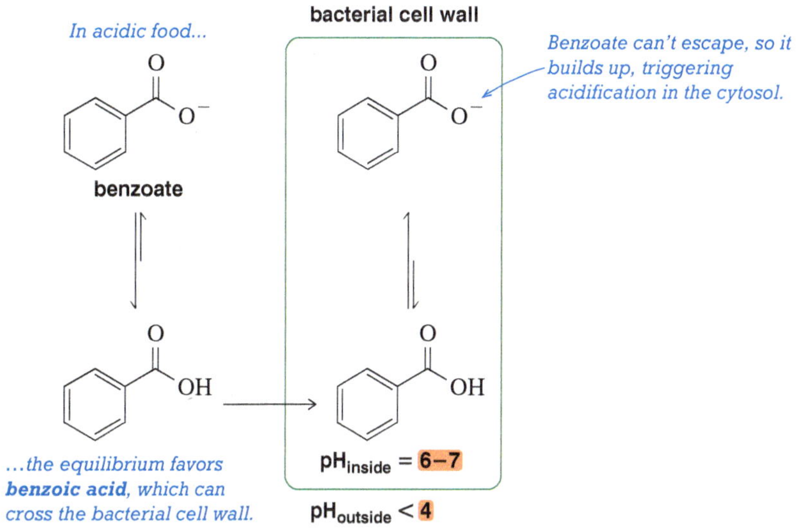

Figure 18.31 According to one hypothesis, accumulation of benzoate in the cytosol of bacteria and the resulting acidification of the cytosol slow bacterial growth.

18.7 Reactions of carboxylic acids—nucleophilic acyl substitution

An important reaction of carboxylic acids is nucleophilic acyl substitution. Put simply, it is the *substitution* by a *nucleophile* at an *acyl* (or carbonyl) carbon. Applied to carboxylic acids, a nucleophile will displace the hydroxyl group that is attached to the carbonyl carbon. This reaction is not limited to carboxylic acids and is more important in Chapter 19, where the reactions of carboxylic acid derivatives are discussed.

In other words, a solid understanding of the nucleophilic acyl substitution mechanism will help you do well in Chapter 19.

As illustrated in **Figure 18.32**, the net result of nucleophilic acyl substitution is displacement of a leaving group (X) by a nucleophile. [*We described* S_N2 *reactions this way.*] We learned in Chapter 17 that nucleophiles attack the carbonyl carbon of aldehydes and ketones, pushing the π electrons onto oxygen. Given that the leaving group in carboxylic acids and carboxylic acid derivatives (see Chapter 19) is attached to a carbonyl carbon, these reactions demonstrate characteristics of both S_N2 and carbonyl addition reactions.

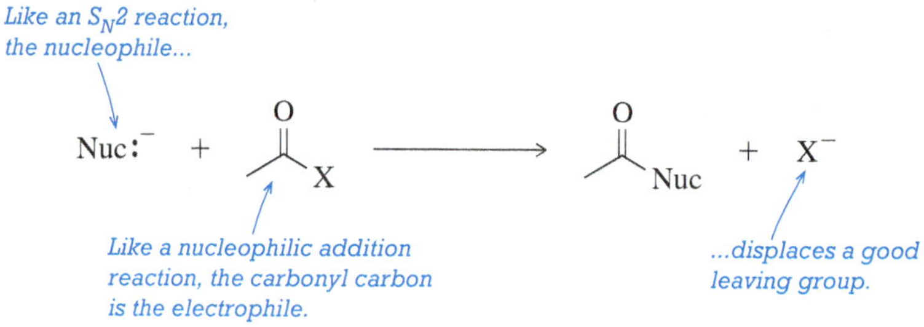

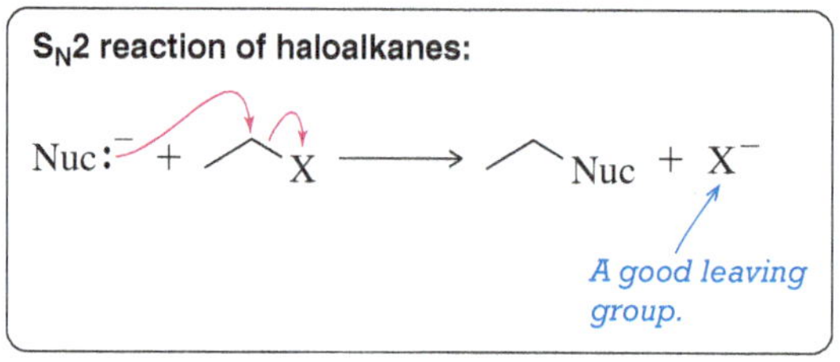

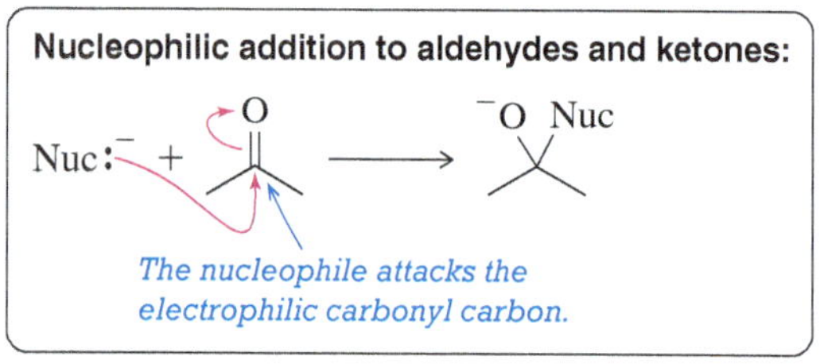

Figure 18.32 Nucleophilic acyl substitution of carboxylic acids and carboxylic acid derivatives has characteristics of S_N2 substitution and carbonyl addition.

18.7.1 General mechanism of nucleophilic acyl substitution

We have previously separated reactions into two broad categories: (1) reactions with strong nucleophiles and (2) reactions with weak nucleophiles. This was the case when studying the S_N1 versus S_N2 and E1 versus E2 mechanisms (Chapter 12) and again when studying nucleophilic addition to aldehydes and ketones (Chapter 17). The reactions of carboxylic acids and carboxylic acid derivatives fall into similar categories. Using designations similar to those from Chapter 17, this section explores how the general mechanism of nucleophilic acyl substitution differs based on the strength of the reacting nucleophile.

18.7.1.1 General Mechanism 18A: reaction with strong nucleophiles

The first reaction type is the substitution of a strong nucleophile. In **Figure 18.33**, the general leaving group X is used to represent the various leaving groups of carboxylic acids (OH) and carboxylic acid derivatives. [*While the rate will change with each carboxylic acid derivative, the course of the reaction is identical. We discuss these rate differences in Chapter 19.*] As in the S_N2 reaction, strong nucleophiles react quickly, attacking atoms with a partial positive charge. They don't wait around for fully cationic intermediates. As a result, the nucleophile adds to the carbonyl, pushing the π electrons onto oxygen and generating an alkoxide, referred to as the **tetrahedral intermediate.** This short-lived intermediate degrades by collapse of an oxygen lone pair to reform the C—O π bond, displacing a leaving group. Named for the two steps in the overall reaction, this is referred to as the **addition/elimination** mechanism. The substitution reaction between ethoxide and an acid chloride (a reaction we present in Chapter 19) is representative of this mechanism. [*For reasons we'll see later, carboxylic acids don't typically react by this mechanism with strong nucleophiles.*]

General Mechanism 18A: Reactions with strong nucleophiles

Specific example:

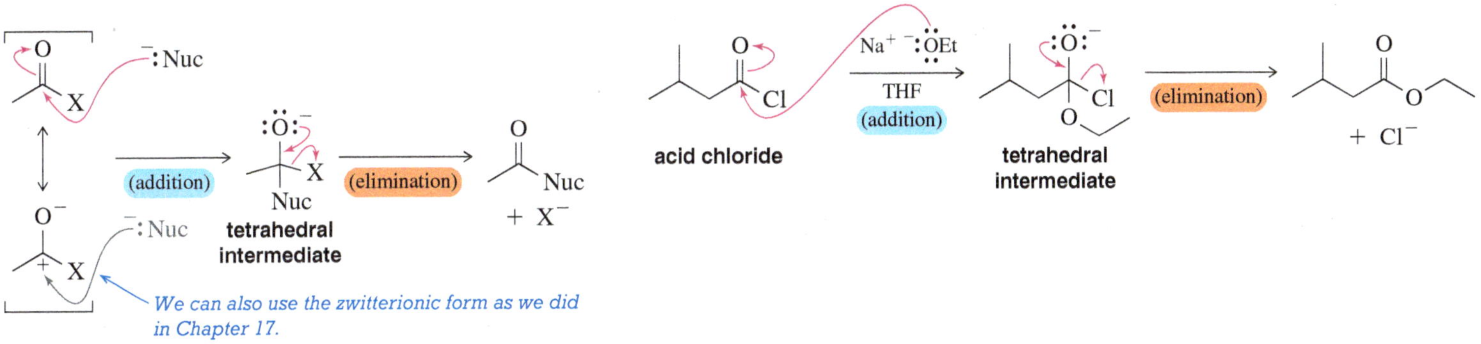

Figure 18.33 The addition of strong nucleophiles proceeds quickly with the neutral carboxylic acid derivatives. Strong nucleophiles add to make the tetrahedral intermediate, which collapses to give the new product. This is referred to throughout Chapters 18 and 19 as General Mechanism 18A.

Before continuing to the reactions with weak nucleophiles [*I bet you know where the conversation is going*], a comparison should be made between this and the related reaction of aldehydes and ketones. Two questions need to be answered. First, we didn't see collapse of an intermediate with aldehydes and ketones (Chapter 17), so what makes that step favorable now? And why wasn't it favorable then?

These two questions are inexorably linked; by answering the first, the answer to the second will become clear. Consider the tetrahedral intermediate formed by addition of ethoxide to an acid chloride (**Figure 18.34**). When the lone pair on oxygen collapses down, three good things happen in terms of the energetics of the reaction [*ball rolling downhill*]. The strong C—O π bond is reformed. [*Good.*] Chloride, a stable leaving group, gets to leave. [*Better.*] And one molecule becomes two, meaning $\Delta S > 0$. [*Best!*]

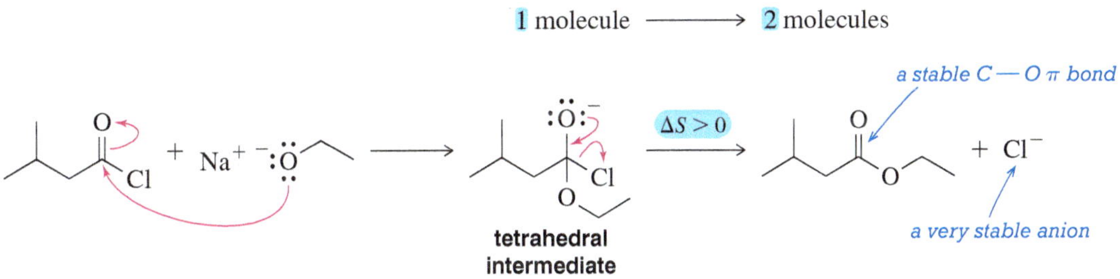

Figure 18.34 Collapse of the tetrahedral intermediate is favorable due to stable bond formation (C—O π bond), a stable leaving group (Cl^-), and overall favorable entropy ($1 \rightarrow 2$, so $\Delta S > 0$)

When a nucleophile adds to a ketone or an aldehyde, we don't speak of a "tetrahedral intermediate" because the reaction stops there (**Figure 18.35**). This is because an alkyl anion (a terrible leaving group) is the only possible leaving group following nucleophilic attack on a ketone. Similarly, a hydride anion (also a terrible leaving group) is the only possible leaving group following nucleophilic attack on an aldehyde. [*These terrible leaving groups never leave!*] Regardless of the potential gains of reforming the C—O π bond and the favorable entropy, it simply cannot happen.

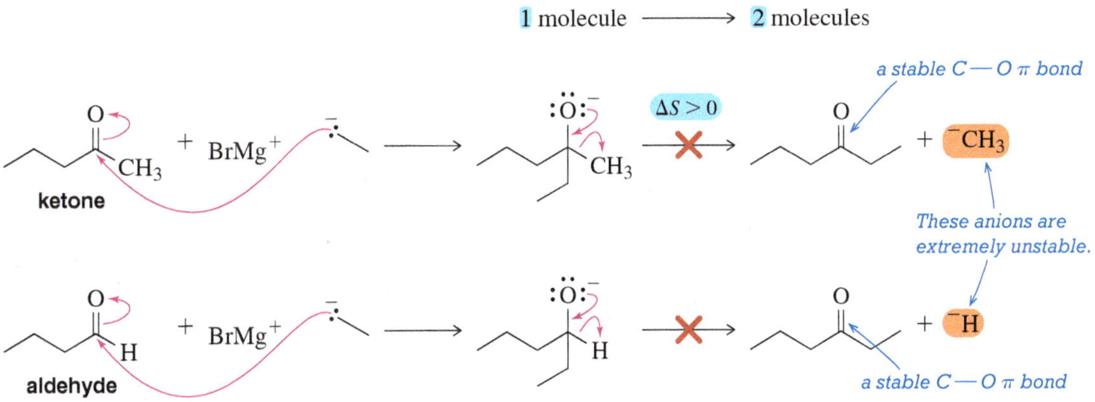

Figure 18.35 Collapse of the tetrahedral intermediate is favorable due to stable bond formation (C—O π bond) and overall favorable entropy ($1 \rightarrow 2$; $\Delta S > 0$), but it can't happen without a good leaving group.

18.7.1.2 General Mechanism 18B: reaction with weak nucleophiles/acidic conditions

Carbonyls react more slowly with weak nucleophiles than strong nucleophiles. As in previous chapters, a nucleophile is considered weak when the atom donating the lone pair of electrons is neutral (Section 17.3.2). [*Except for alkyl amines (RNH_2, R_2NH, etc.), which are considered strong, neutral nucleophiles . . . recall the S_N2 reaction of amines.*] This type of nucleophile (H_2O, CH_3OH, etc.) does not react directly with neutral carboxylic acids or carboxylic acid derivatives. Instead, a full carbocation (a more reactive intermediate) formed by protonation initiates attack by the nucleophile (**Figure 18.36**). Because the weak nucleophile brings along a proton, a deprotonation step occurs to give the neutral tetrahedral intermediate. Under acidic conditions, the leaving group is protonated prior to leaving to give a resonance-stabilized carbocation. Deprotonation generates the new carboxylic acid derivative. The Fischer esterification (a reaction we discuss in Section 18.7.2) is representative of this mechanism.

General Mechanism 18B: Reaction with weak nucleophiles

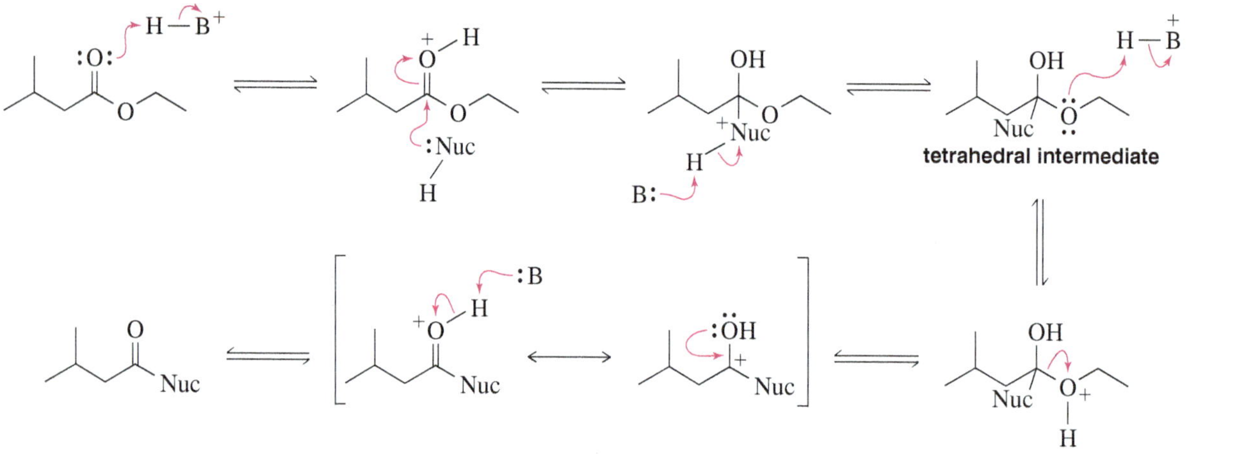

Specific example:

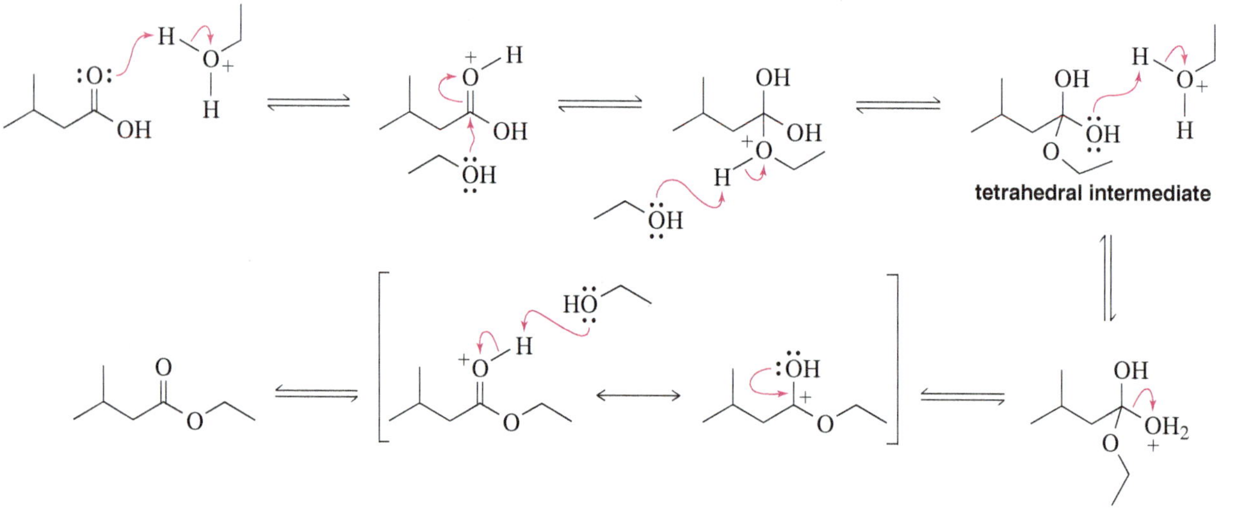

Figure 18.36 The substitution of weak nucleophiles necessitates protonation of the carbonyl to make a reactive intermediate. With weak nucleophiles, the carbonyl protonates first, then the nucleophile adds. Similarly, the leaving group is protonated prior to leaving. This is referred to throughout Chapters 18 and 19 as General Mechanism 18B.

Although General Mechanism 18B as shown has six steps, is more time-consuming, and seems significantly more complicated than General Mechanism 18A, it really isn't (**Figure 18.37**). The addition/elimination in General Mechanism 18A can be summarized as "*nucleophilic addition* to tetrahedral intermediate, *elimination* to product." The corresponding addition/elimination in General Mechanism 18B would be summarized as "protonate, *nucleophilic addition*, deprotonate to tetrahedral intermediate, protonate, *elimination*, deprotonation to product." In other words, the four extra steps in General Mechanism 18B are simply acid–base reactions. This observation reveals the second important point: Most reactions occurring by General Mechanism 18B are acid catalyzed because an acidic proton is usually generated in the last step. Finally, because the nucleophiles are weak, these reactions are reversible, being pushed in one direction by Le Châtelier's principle. [*This is one of many similarities you'll see between acetal formation (Section 17.7.4) and reactions occurring by General Mechanism 18B.*]

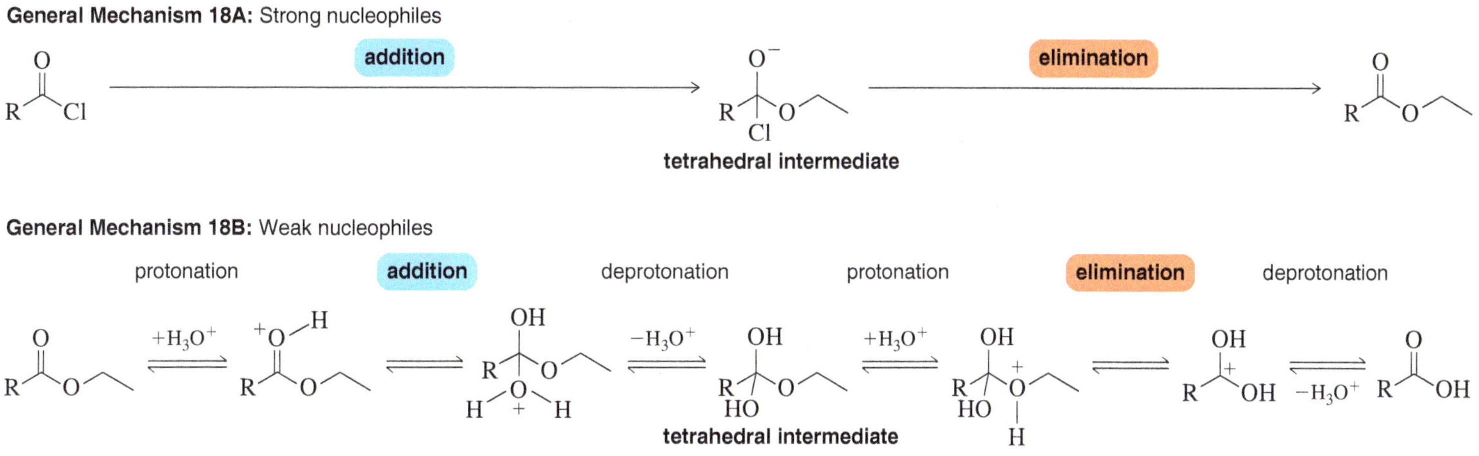

Figure 18.37 A comparison of General Mechanisms 18A and 18B shows that they differ only by the addition of four extra protonation/deprotonation steps in General Mechanism 18B.

18.7.2 Fischer esterification

The first nucleophilic acyl substitution we will study is the Fischer esterification. Discovered in 1895 by Emil Fischer, in this reaction a carboxylic acid is converted to an ester by treating it with an alcohol and an acid catalyst. [*The name tells exactly what the reaction does, right? This will help you remember it when you need it for synthesis.*]

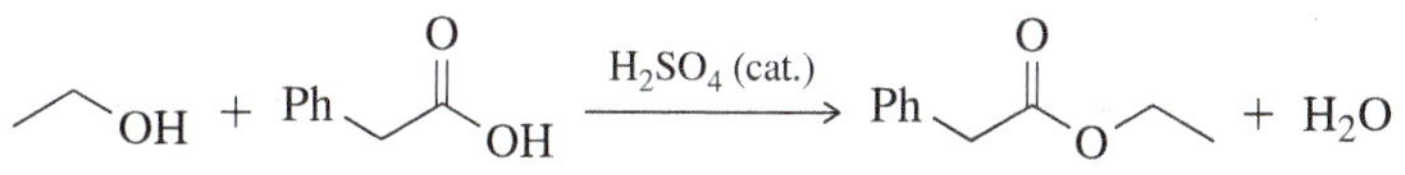

As you should expect for a reaction with a weak nucleophile, the Fischer esterification occurs via General Mechanism 18B (**Figure 18.38**). Protonation of the carbonyl occurs to give a positively charged [*and more reactive*] electrophile. Ethanol adds to the carbocation, bringing along its proton. This proton, now acidic, is removed to complete formation of the tetrahedral intermediate. Protonation of a hydroxyl group creates an oxonium ion, which is a good leaving group. [*What do good leaving groups do? They leave.*] Departure of the water produces a carbocation that is then deprotonated to complete formation of the ester, regenerating the acid catalyst.

The analysis of Figure 18.38 is closely related to how we described acetal formation in Section 17.7.4. [*Look back at Figure 17.62 for its mechanism.*] **Figure 18.39** (with mechanistic arrows removed) shows that acetal formation and Fischer esterification share the first five steps, differing only by the methyl group (acetal formation) and hydroxyl group (esterification) attached to the carbonyl carbon. Once the carbocation is formed in step five, the two pathways diverge.

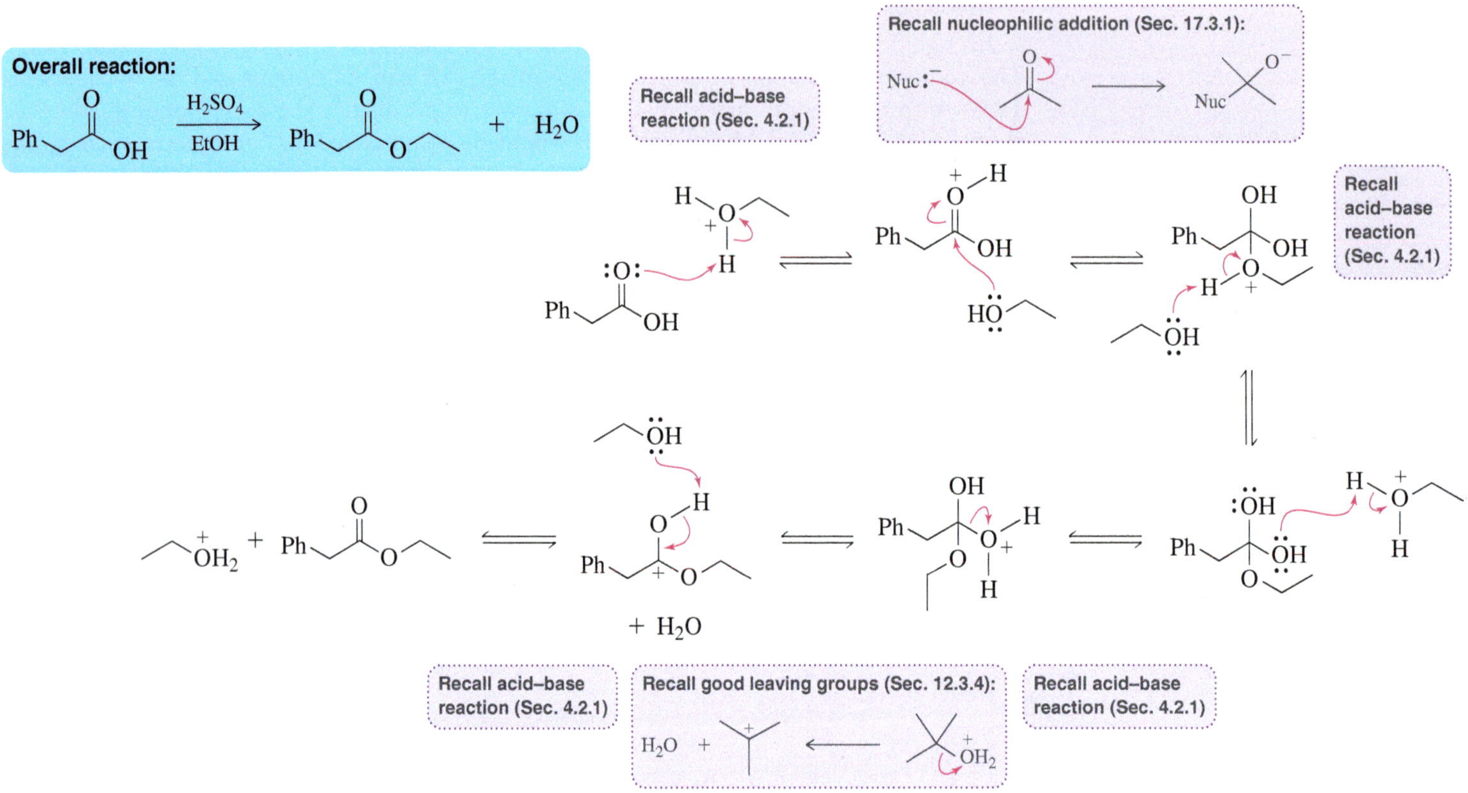

Figure 18.38 The mechanism of Fischer esterification.

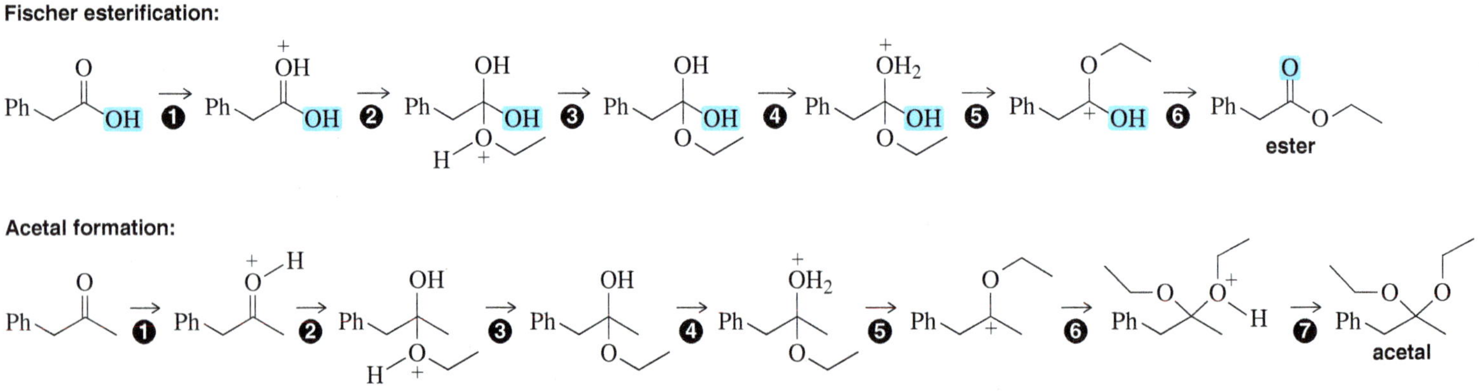

Figure 18.39 A direct comparison of the reaction pathways of Fischer esterification and acetal formation (Section 17.7.3).

For an example of where we have seen this before, look back at the discussion in Section 17.8 of why amines plus ketones give imines instead of the corresponding acetal products.

Why do these pathways diverge after step 5 (Figure 18.39)? [*Is there a difference between the intermediates?*] This is an example of how entropy affects the outcome of a reaction. For esterification, the carbocation intermediate has an acidic proton that can be removed to neutralize it. Like all acid–base reactions, this is entropically neutral (favored) and sterically unhindered (fast). On the pathway to acetal formation, there is no such acidic proton. Thus, addition is the only way to stabilize the carbocation, resulting ultimately in acetal formation (**Figure 18.40**).

Two aspects of the mechanism are worth mentioning because they relate to concepts discussed throughout Chapters 18 and 19. First, Fischer esterification is the substitution of an alkoxy group for a hydroxyl group, so why not simply protonate the hydroxyl group [*as we did when converting alcohols into better leaving groups*]? Carboxylic acids and carboxylic acid derivatives always protonate at the carbonyl oxygen because the resulting conjugate acid from carbonyl protonation is stabilized by resonance. A protonated hydroxyl group, in comparison, has no additional means of stabilization (**Figure 18.41**).

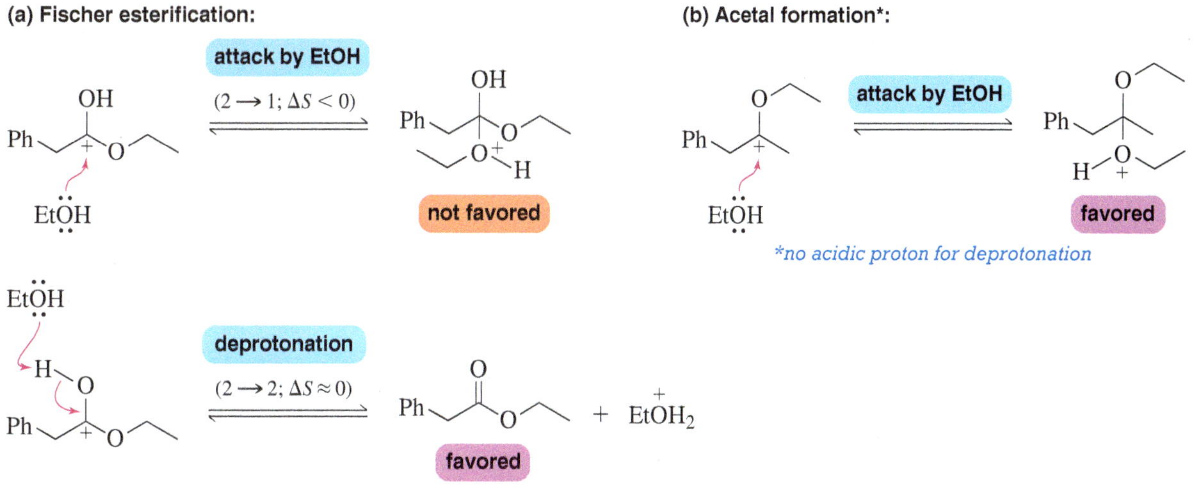

Figure 18.40 The conditions that favor ester or acetal formation after step 5 in Figure 18.39. (a) When deprotonation is favored over nucleophilic addition, the result is ester formation. (b) When there is no proton that can be removed, nucleophilic addition is favored, resulting in acetal formation.

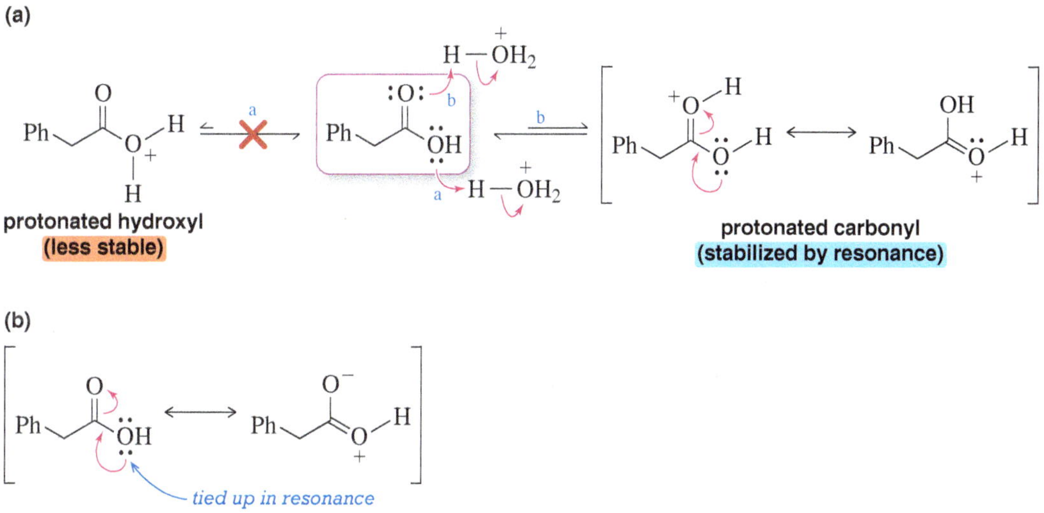

Figure 18.41 Protonation occurs at the carbonyl oxygen preferentially because (a) the conjugate acid is better stabilized and (b) the lone pair on the hydroxyl group is tied up in resonance.

The chemical reasoning and analysis described in the last two paragraphs are revisited multiple times in Chapter 19, so make sure you understand it well.

SPECTROSCOPY INSIGHTS

The product of this equilibrium (either direction) can be easily confirmed by the presence (carboxylic acid) or absence (ester) of the broad O—H stretching band (3300–2700 cm^{-1}) in the infrared spectrum.

Second, after forming the tetrahedral intermediate, which hydroxyl group is protonated? The resulting conjugate acids from protonating each are similarly stable. In the presence of a strong acid, all sites will, at some point, be protonated (**Figure 18.42**). [*Through random collisions, the proton will be transferred millions of times.*] Thus, this step is driven by the favored side of equilibrium or whichever side the experimental conditions are designed to push the reaction (Le Châtelier's principle). In the case of the example we have been discussing, protonation of the alkoxy oxygen would result in a return to the carboxylic acid. Protonation of either hydroxyl group results in ester formation, because either water molecule can leave.

In sum, the Fischer esterification reaction is reversible, with the outcome determined by the favored side of equilibrium. For esterification or ester hydrolysis (the reverse), both useful reactions, we can use the same strategy we used in driving acetal formation in a particular direction (see Section 17.7.4). Using Le Châtelier's principle, adding alcohol or removing water drives the reaction toward the ester. Water can be removed using molecular sieves [*Section 17.7.3*]. When the ester is being formed using an alcohol that is immiscible with water, the Dean–Stark apparatus from Figure 17.56 can be used. Alternatively, the reaction can be driven back toward the carboxylic acid by the addition of water or the removal of the (usually) more volatile alcohol (**Figure 18.43**).

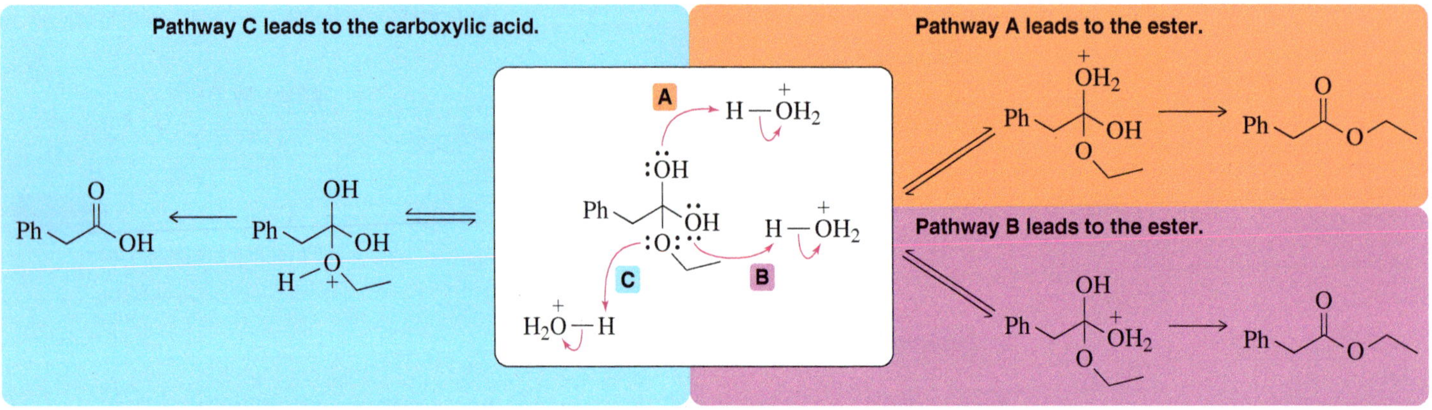

Figure 18.42 Protonation at either site results in similarly stable conjugate acids. Thus, the site of protonation is determined by the favored side of the equilibrium. For each of pathways A–C, see Figure 18.38 to complete the mechanism.

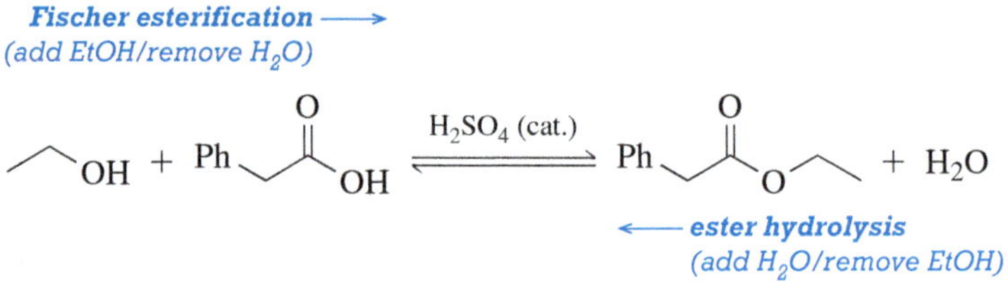

Figure 18.43 Using Le Châtelier's principle can make the Fischer esterification/ester hydrolysis synthetically useful in either direction.

18.7.2 SUMMARY

- Fischer esterification is the acid-catalyzed conversion of carboxylic acids to esters. As a reaction with a weak nucleophile, it proceeds by General Mechanism 18B and can be driven in either direction using Le Châtelier's principle.

Assessment 18.22 Predict the product of the following reactions.

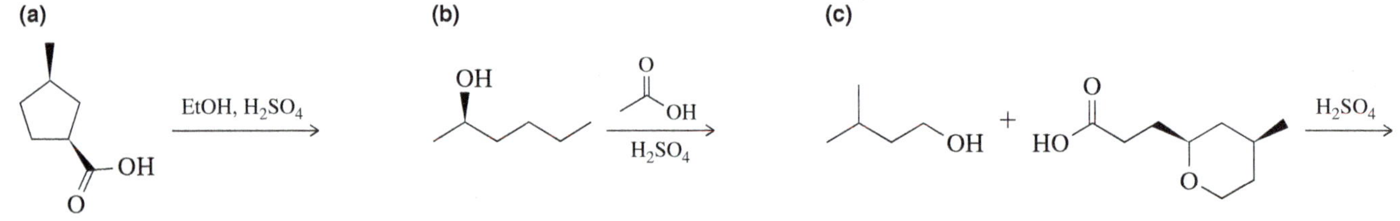

Assessment 18.23 Draw a reaction coordinate diagram, labeling all of the intermediates in the mechanism shown in Figure 18.38.

Assessment 18.24 Show a mechanism for the ester hydrolysis reaction in Figure 18.43. How do you know that this is the correct mechanism?

Assessment 18.25 A chemist attempted to produce the methyl ester shown using the Fischer esterification. Although a reaction occurred, including the Fischer esterification, the desired product was *not* isolated. Predict the product that was isolated.

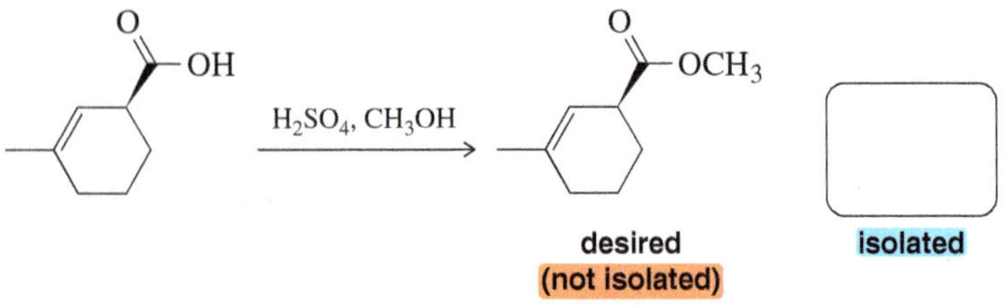

Assessment 18.26 A chemist replaced only the carbonyl oxygen (assume this is possible) with ^{18}O before running the Fischer esterification reaction shown. Half of the ^{18}O label ended up in the water. What does this tell you about the mechanism of the reaction?

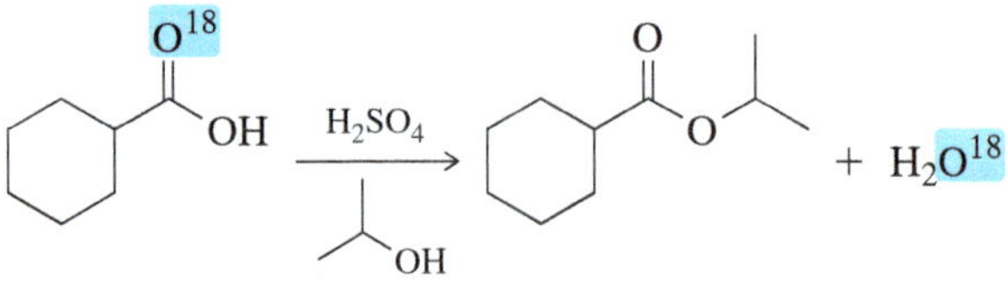

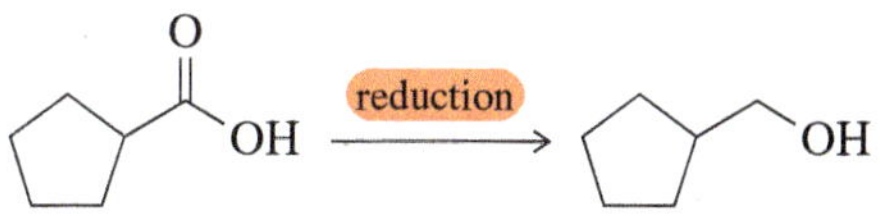

18.7.3 Reduction with $LiAlH_4$

The central carbon of a carboxylic acid is in the +3 oxidation state. As a result, it is a good substrate for reduction to 1° alcohols (−1 oxidation state). In Chapter 17, you were introduced to two nucleophilic reducing agents, sodium borohydride ($NaBH_4$) and lithium aluminum hydride ($LiAlH_4$), only one of which reduces carboxylic acids. [*Any ideas which?*]

To choose a reagent, the relative reactivity of carboxylic acids must be assessed. In Chapter 17, the reactivity of aldehydes and ketones was compared via the zwitterionic resonance structure and the degree to which the carbocation was stabilized/destabilized. Ketones, with an extra stabilizing alkyl group (by hyperconjugation), are less reactive than aldehydes. Similar analysis in **Figure 18.44** with the carboxylic acid reveals a carbocation that can be stabilized by resonance from the attached oxygen. Thus, carboxylic acids are relatively poor electrophiles; a strong reducing agent is needed. [*Which is stronger, $NaBH_4$ or $LiAlH_4$?*]

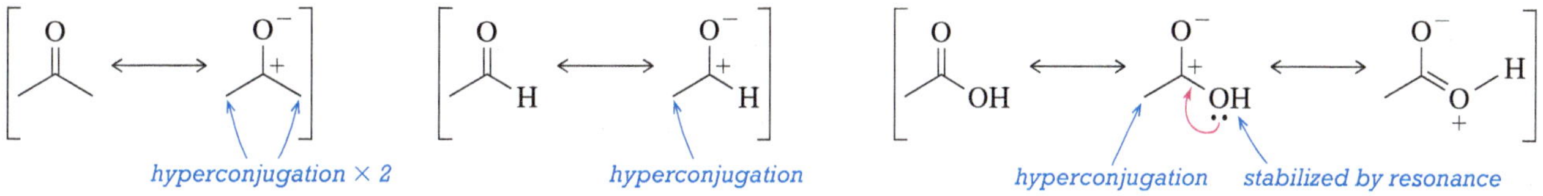

Figure 18.44 Carboxylic acids are relatively weak electrophiles because the carbocation in the zwitterionic form is stabilized by resonance donation of a lone pair by the hydroxyl oxygen.

Sodium borohydride is not strong enough to reduce carboxylic acids, so lithium aluminum hydride must be used. The enhanced nucleophilicity that enables reduction of the carboxylic acid also complicates matters in terms of the mechanism of the reaction. With a focus on the big picture, this mechanism is essentially General Mechanism 18A—what you should expect for an acyl substitution reaction with a *strong* nucleophile. [*This mechanism is shown one step at a time to make it easier to digest.*]

Usually when we show a mechanism, we ignore the countercation and other inorganic species. This is not the case now, and **Figure 18.45** shows us the importance of these other species in the reaction mechanism. Because strong nucleophiles are also strong bases, $LiAlH_4$ first reacts as a base, forming the carboxylate ion (Figure 18.45). [*$LiAlH_4$ is always used in polar aprotic solvents like THF for this reason.*]

Figure 18.45 Lithium aluminum hydride acts as a base to form a carboxylate.

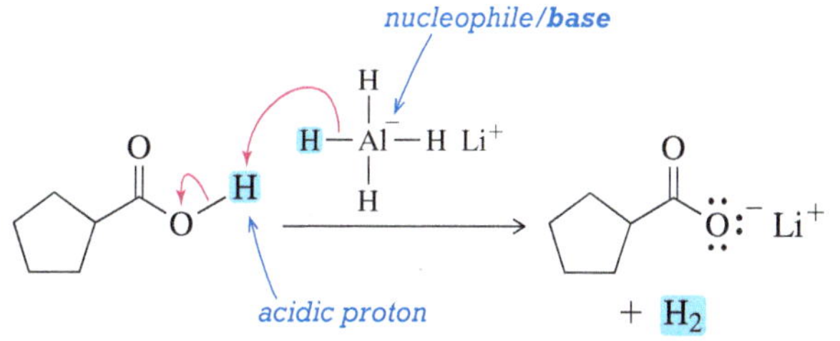

After forming the carboxylate, one proposed mechanism has alane donating a hydride ion to the carbonyl carbon at the same time a strong bond forms between oxygen and aluminum (**Figure 18.46**). This step, reminiscent of when borane adds to an alkene (Section 8.4.5), represents the **addition** of a nucleophile to form the tetrahedral intermediate. A lone pair on oxygen then collapses back down, reforming the π bond, and kicking out the leaving group [***elimination***] to form an aldehyde. These two steps are identical to General Mechanism 18A, nucleophilic acyl substitution with a strong nucleophile.

Figure 18.46 Reduction occurs by addition/elimination of the strongly nucleophilic hydride ion (General Mechanism 18A).

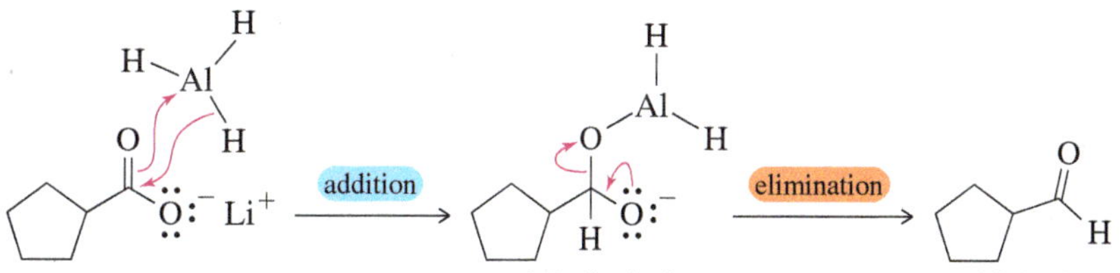

The resulting aldehyde is reduced by $LiAlH_4$ by General Mechanism 17A (carbonyl addition of strong nucleophiles). Hydride is transferred to the carbonyl carbon, pushing the π electrons to oxygen (**Figure 18.47**). The alkoxide is then quenched by addition of a strong acid. [*Look back at Section 17.9.2 if you have forgotten this reaction.*]

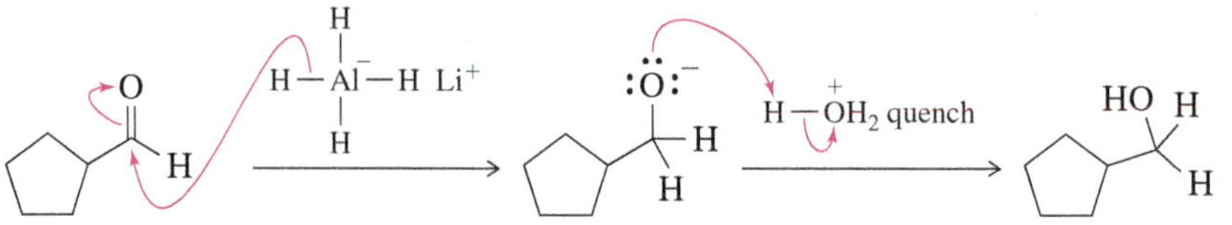

Figure 18.47 The aldehyde is further reduced to a 1° alcohol by General Mechanism 17A (see Section 17.9.2).

SPECTROSCOPY INSIGHTS

Stopping at the aldehyde would give a new ^1H NMR signal between 9 and 10 ppm, whereas the 1° alcohol will have a new signal between 3 and 4 ppm.

The overall mechanism is shown in **Figure 18.48**. While complicated, it is essentially the addition/elimination of a strong nucleophile with a carboxylic acid derivative (General Mechanism 18A) preceded by an acid–base reaction and followed by nucleophilic addition to a carbonyl (General Mechanism 17A).

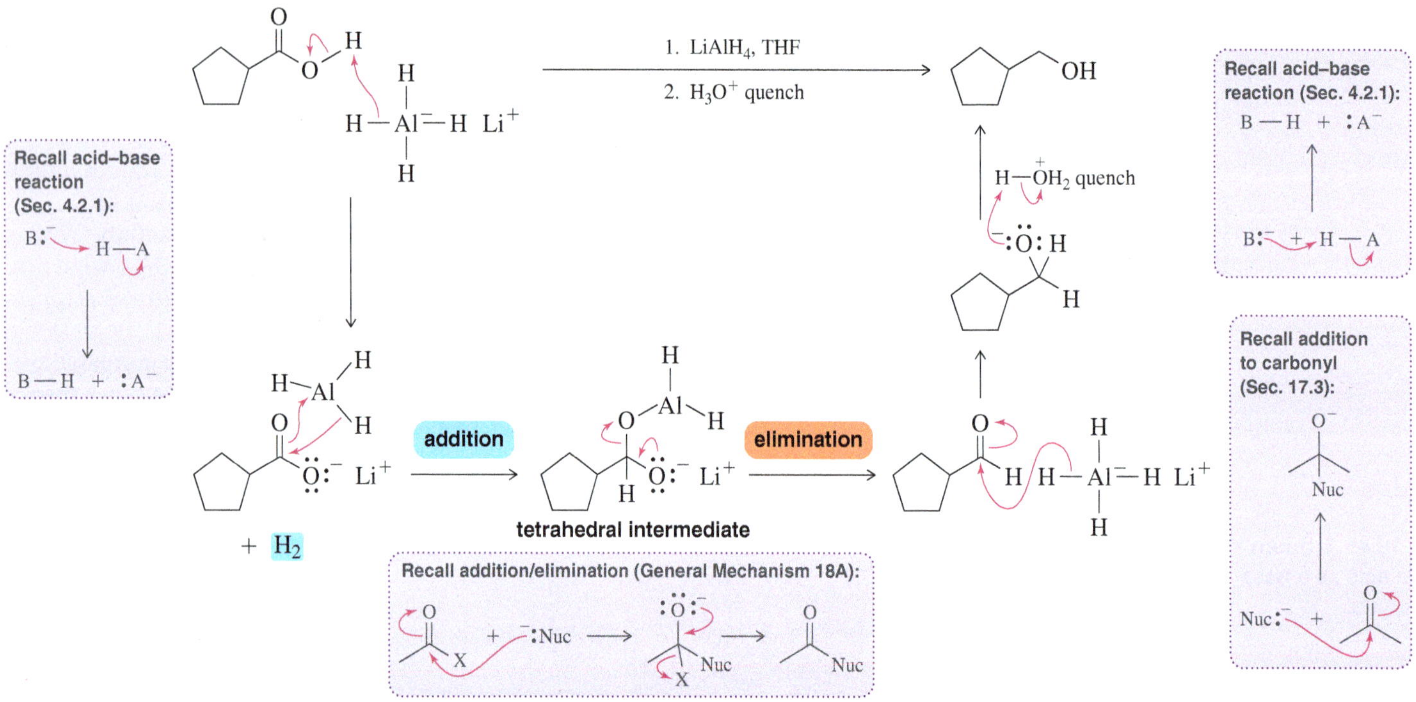

Figure 18.48 Summary of the mechanism of $LiAlH_4$ reduction of a carboxylic acid.

Because it is an intermediate on the way to an alcohol, you might wonder whether this reaction ever stops at the aldehyde. It's a question of "what happens next?" when a reactive functional group (aldehyde) forms in the presence of another reactive functional group (carboxylic acid). Despite how we represent them, reactions involve collisions between many billions of molecules. After one molecule of carboxylic acid is reduced, the solution would contain many molecules of unreacted carboxylic acid and a single molecule of the aldehyde. Because aldehydes are more reactive than carboxylic acids (**Figure 18.49**), they will be reduced preferentially. As a result, reduction of a carboxylic acid with $LiAlH_4$ never stops at the aldehyde (Figure 18.49). [*We encounter this same issue in Side Conversation 5 in Section 19.8.1.*] If the goal is to make a 1° alcohol, this isn't a problem, but what if, as a synthetic chemist, you wanted to make an aldehyde using this type of reaction? [*Stick around for Chapter 19 for a cool solution to this problem.*]

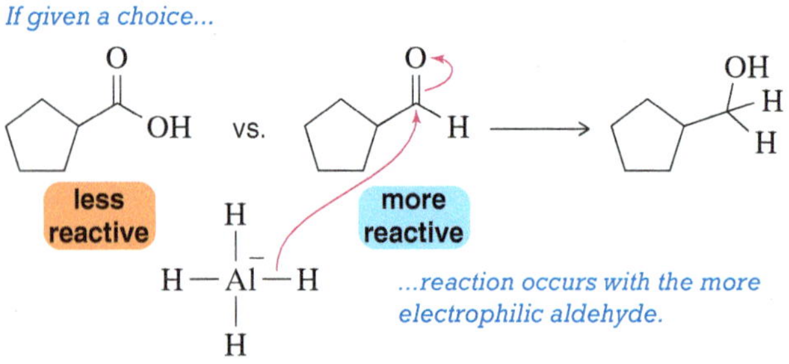

Figure 18.49 Because aldehydes are more reactive, reduction of carboxylic acids never stops at the aldehyde.

18.7.3 SUMMARY

- $LiAlH_4$ is an especially powerful reducing agent that converts carboxylic acids to 1° alcohols by the addition of two hydride ions to the central carbon of the carboxylic acid. The reaction proceeds by a modification of General Mechanism 18A.

Assessment 18.27 Predict the product of the following reactions.

(a)

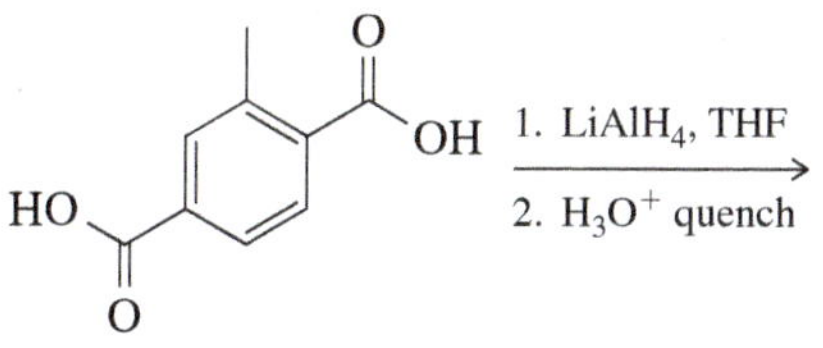

(b)

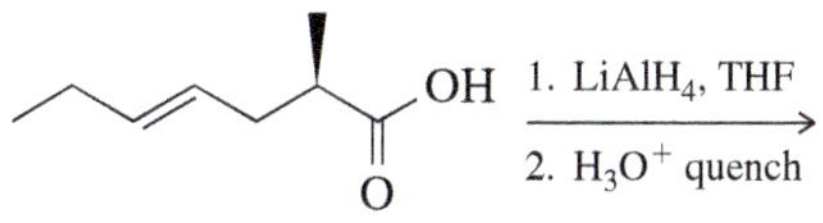

(c)

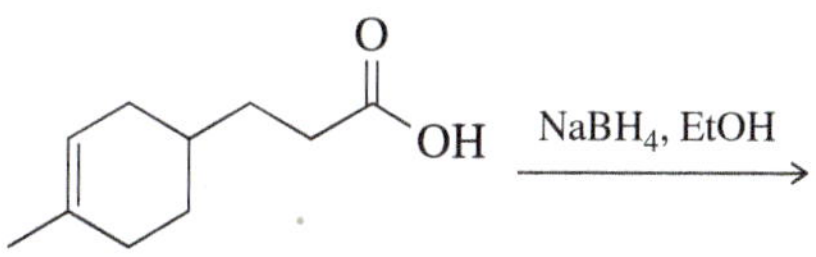

Assessment 18.28 Identify the carboxylic acids that would be reacted with $LiAlH_4$ to make the 1° alcohols shown.

(a)

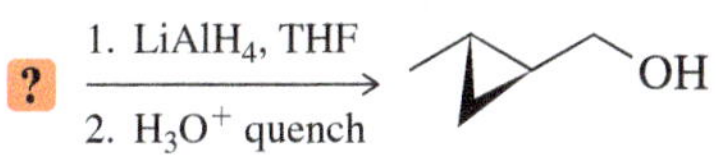

(b)

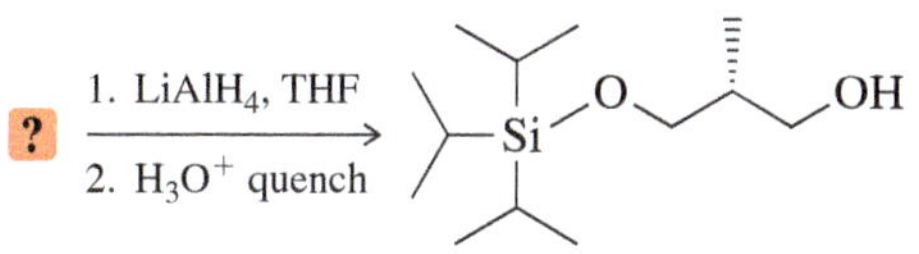

(c)

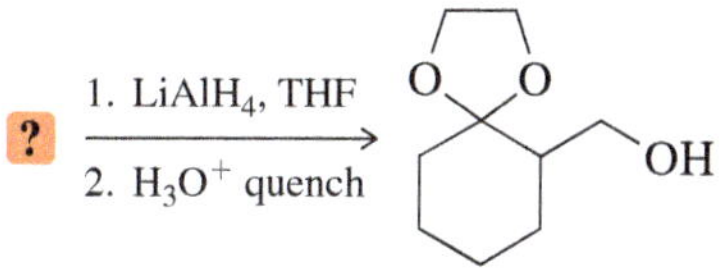

18.7.3.1 Selective reduction

Thus far, you have learned three major carbonyl reducing agents: $LiAlH_4$, $NaBH_4$ and H_2/Pd (Section 17.9). These reducing agents have been used to reduce aldehydes, ketones, carboxylic acids, alkenes, and alkynes. Table 18.7 summarizes the functional groups reduced by each reagent. We are now able to think about selective reduction, or reducing one functional group in the presence of another.

Table 18.7 Selected reducing agents and the functional groups they will reduce.

Reducing agent	Functional groups reduced
$LiAlH_4$	Aldehyde, ketone, carboxylic acid
$NaBH_4$	Aldehyde, ketone
H_2/Pd (1 atm)	Alkene, alkyne
H_2/Pd (>1 atm)	Alkene, alkyne, aldehyde, ketone

Let's work Assessment 18.29 together.

Assessment 18.29 How would you reduce the alkene in the presence of a carboxylic acid?

Solved Assessment 18.29

Assessment 18.29 asks us to reduce an alkene without reducing the carboxylic acid. According to Table 18.7, $LiAlH_4$ is the only reagent that reduces carboxylic acids. Moreover, only hydrogenation (and not $NaBH_4$ or $LiAlH_4$) reduces alkenes. As a result, selective reduction can be achieved using H_2/Pd as the reducing agent (**Figure 18.50**). ■

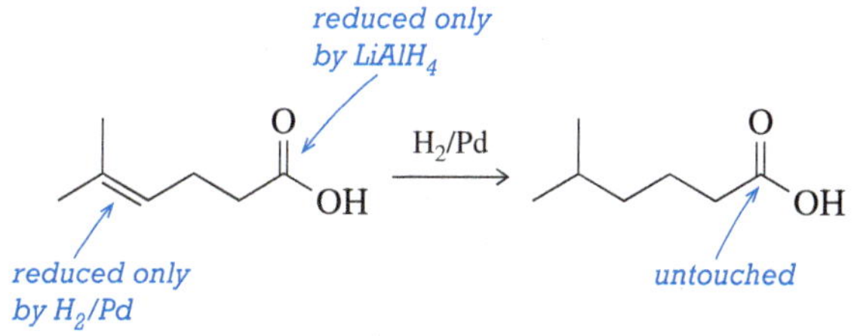

Figure 18.50 Alkenes can be reduced in the presence of carboxylic acids by hydrogenation with a palladium catalyst.

Figure 18.51 shows how selective reduction can be used to reduce a single functional group in the presence of others (ketone, alkyne, and carboxylic acid). Sodium borohydride selectively reduces the ketone because a ketone is more reactive than a carboxylic acid. Hydrogenation at atmospheric pressure is strong enough to reduce only the alkyne. High-pressure hydrogenation reduces alkynes and ketones. Similarly, $LiAlH_4$, the stronger reducing agent, reduces both ketones and carboxylic acids.

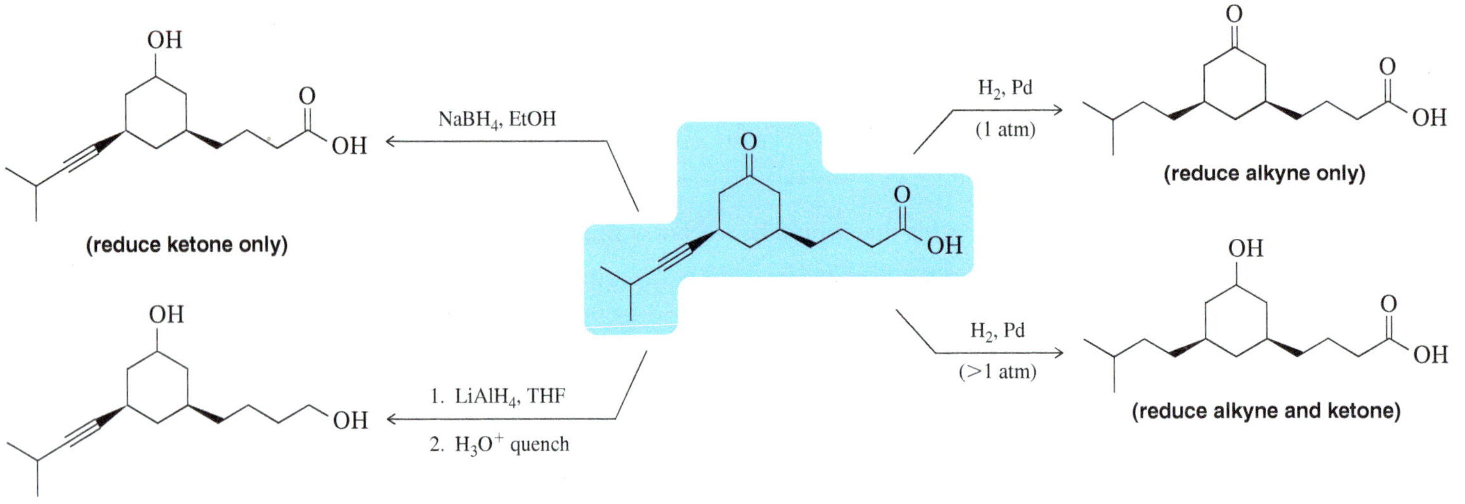

Figure 18.51 Selective reduction of an alkyne, ketone, and carboxylic acid.

Assessment 18.30 Suggest the reagent(s) required to convert the reactant on the left to the products on the right.

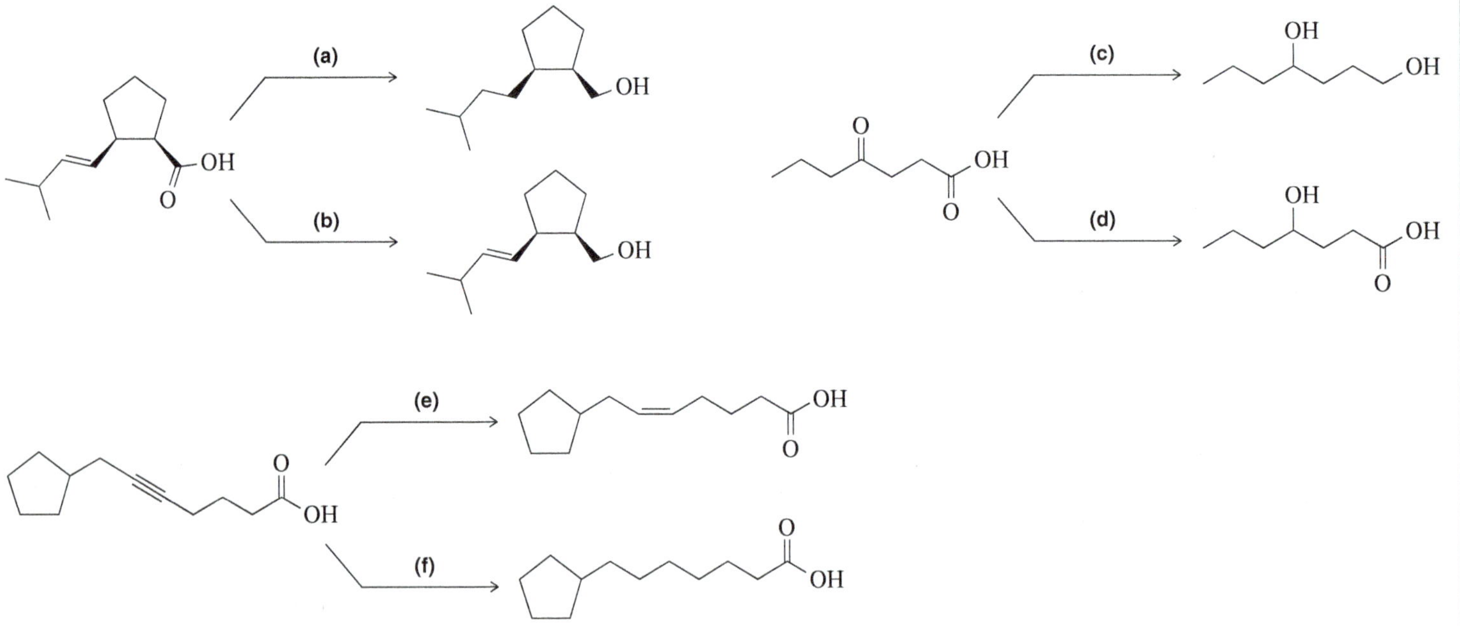

Assessment 18.31 In Figure 18.51, there was no method given for reducing a carboxylic acid in the presence of an aldehyde/ketone. Given that aldehydes are always reduced before carboxylic acids, suggest a multistep strategy for accomplishing the following transformation.

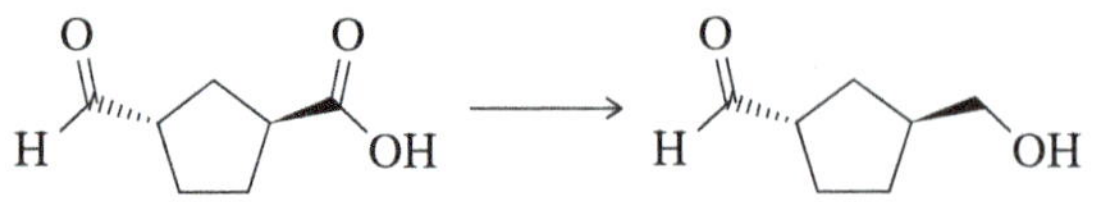

18.7.4 Organolithium addition

The third nucleophilic acyl substitution reaction in this section is the reaction of organolithium reagents with carboxylic acids. This particular reaction suffers from a problem that is overcome by using two equivalents of the organolithium reagent per equivalent of carboxylic acid. The reaction of two equivalents of phenyllithium with butyric acid in **Figure 18.52** is a representative example of this substitution reaction.

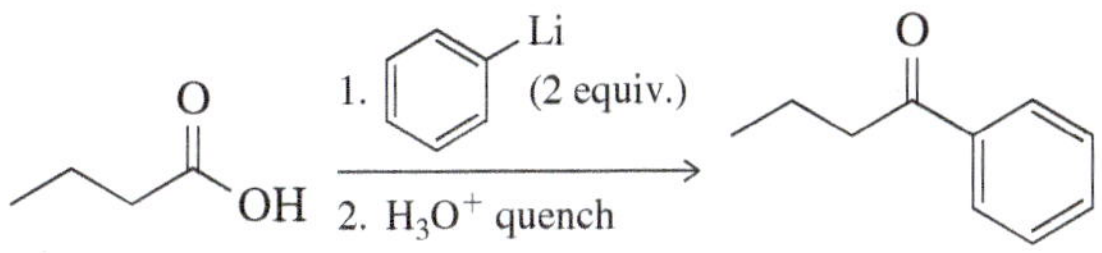

Figure 18.52 The reaction of organolithiums with carboxylic acids gives ketones.

Appearing as simply the replacement of a hydroxyl group with an alkyl group, it is not that straightforward. It is another example of an addition/elimination mechanism; you must look harder to find it. Like $LiAlH_4$, organolithium reagents are extremely strong bases. Deprotonation is faster than nucleophilic addition, so deprotonation of the carboxylic acid by the organolithium occurs first to make a carboxylate (**Figure 18.53**). [*That's one equivalent of organolithium that won't be incorporated into the product.*] Carboxylate ions are negatively charged, so they are bad electrophiles. But organolithium reagents are extremely powerful nucleophiles. And so, with no protons to be removed, nucleophilic addition occurs at the carbonyl carbon. [*Note: **Addition!***] This results in the formation of a dianion, likely stabilized by coordination with lithium cations.

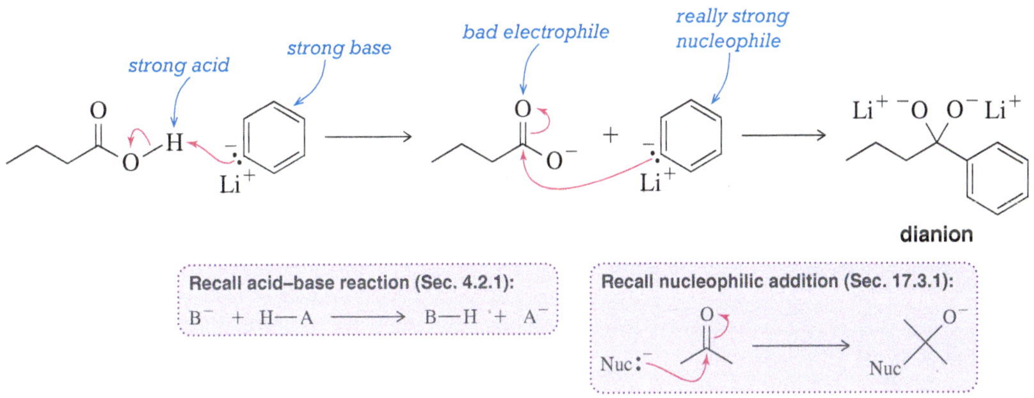

Figure 18.53 After deprotonation, addition of phenyllithium to the carboxylate gives a dianion.

Under normal circumstances, addition to a carboxylic acid or carboxylic acid derivative is followed by collapse of the tetrahedral intermediate. For this to happen in the dianion, O^{2-} would have to be eliminated, which is a preposterous idea. Instead, the dianion persists in solution until the reaction is quenched with acid (**Figure 18.54**) to produce a compound with two hydroxyl groups attached to the same carbon. [*Where have you seen this before?*] This is a ketone hydrate, the less favored side of an equilibrium reaction with the ketone. [*See Section 17.7.1 for a review.*] In forming the ketone from the hydrate, the elimination step of the overall addition/elimination mechanism occurs. [*See if you can draw the mechanism of ketone formation from the hydrate under acidic conditions.*]

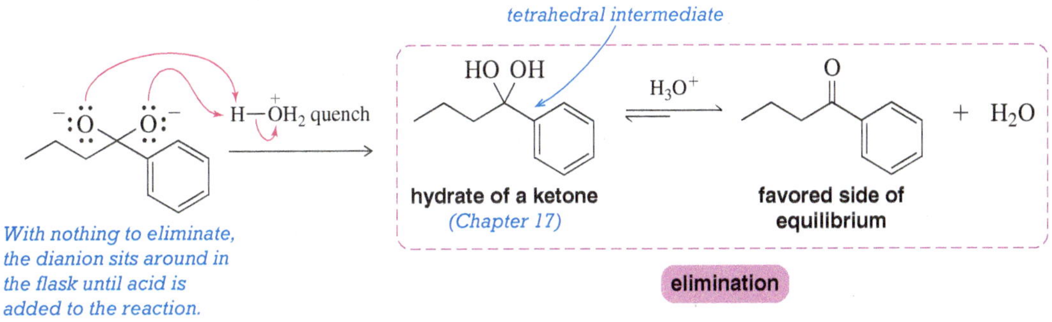

Figure 18.54 Protonation of the dianion makes a hydrate that equilibrates to the more stable ketone. This is the elimination step of the addition/elimination mechanism.

In sum, this reaction can be thought of as following the standard addition/elimination reaction mechanism represented by either General Mechanisms 18A or 18B, but with other steps intermingled. You might consider this as a deprotonation/***nucleophilic addition***/quench/***elimination*** sequence. [*Look back at Figures 18.53 and 18.54 and convince yourself that this is a valid way of describing the sequence.*]

Make sure to number the carbons so an avoidable mistake isn't made.

Because a full equivalent of the organolithium is wasted in this reaction, it fails the "green" test due to poor atom economy. [*In Chapter 19, this problem is solved with a new approach.*] Despite these limitations, the reaction offers synthetic flexibility, with two ways to make an individual ketone. As shown in **Figure 18.55**, two combinations of carboxylic acid and organolithium reagent give the same product.

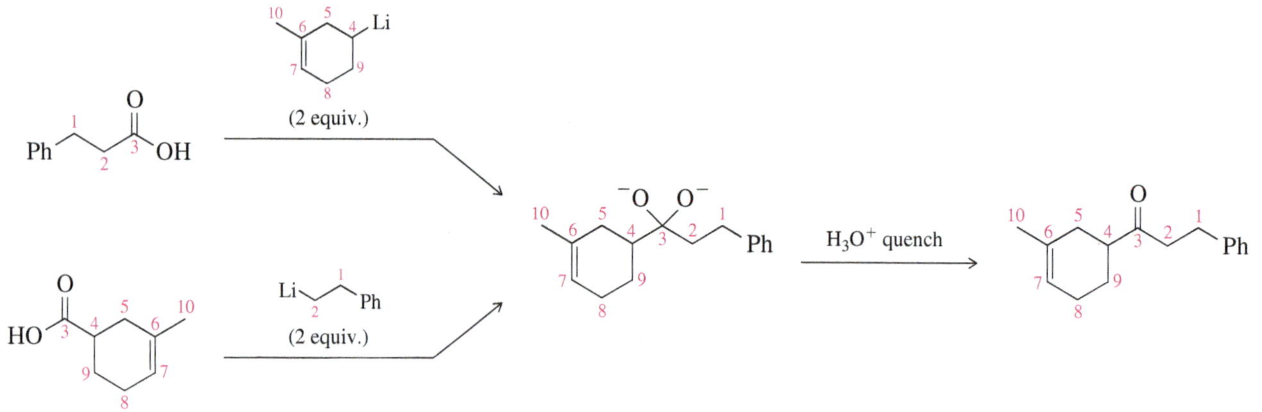

Figure 18.55 Ketone synthesis by organolithium addition offers flexibility in providing two ways to make each ketone.

18.7.4 SUMMARY

- The use of two equivalents of an organolithium reagent overcomes the diminished reactivity of carboxylates. After deprotonation by the first organolithium, the second adds to the carbon of the carboxylate. Acidification and elimination of water give a ketone.

Assessment 18.32 Predict the product of the following reactions.

(a)

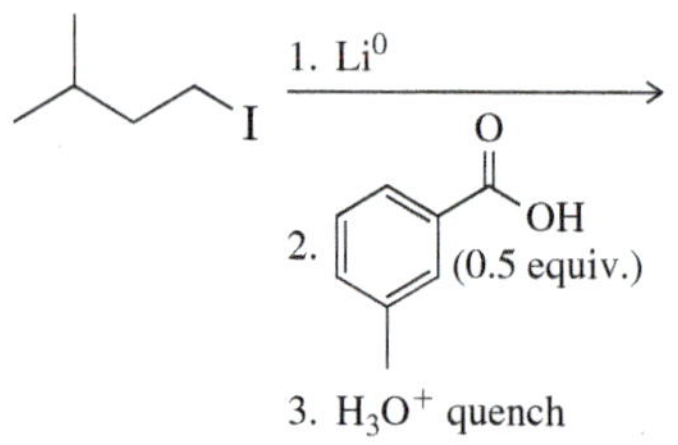

(b)

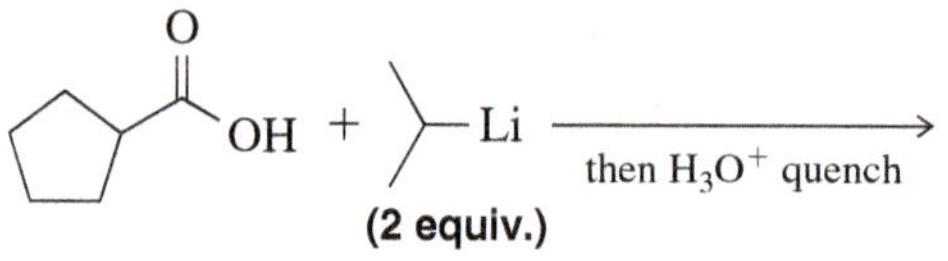

Assessment 18.33 Identify the organolithium and carboxylic acid that could be used to make each ketone. There are two answers for each.

(a)

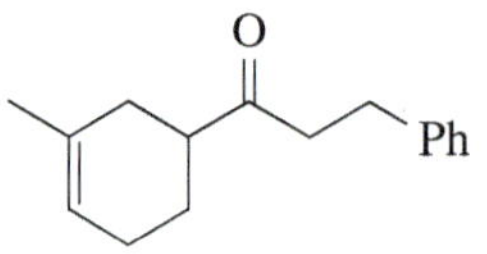

(b)

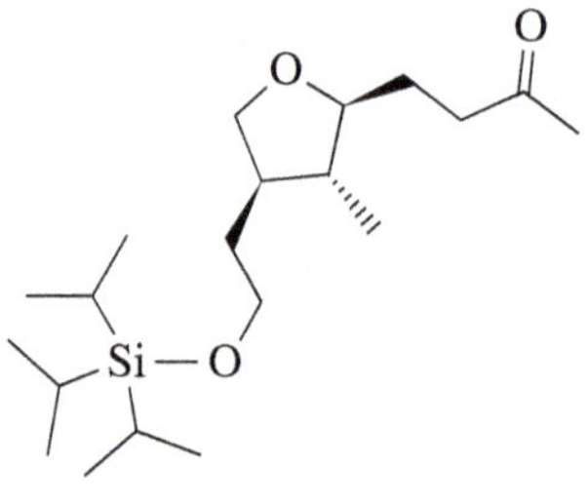

(c)

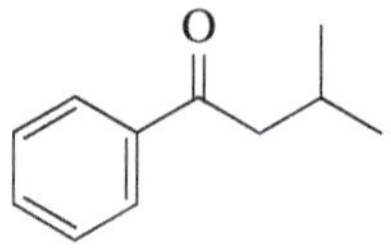

Assessment 18.34 Of the two ways you proposed to make the ketone in Assessment 18.33(b), which is best? Why?

Assessment 18.35 While working a "predict-the-product(s)" question, a student gave the incorrect answer shown. What mistake did this student make?

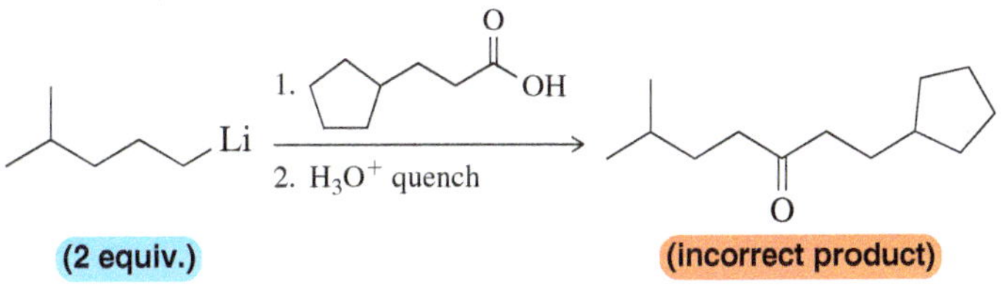

18.7.5 Formation of acid chlorides

[*These next few sentences are going to sound like something you've heard before—pay close attention and see if you can figure out where you've heard it.*] The last two sections revealed problems with using carboxylic acids for substitution reactions. In short, strong nucleophiles deprotonate them, and the resulting carboxylate ion is not a very good electrophile (**Figure 18.56**). Furthermore, hydroxide is a poor leaving group in attempted carboxylic acid substitution reactions. [*Recall the conversation from Chapter 13.*]

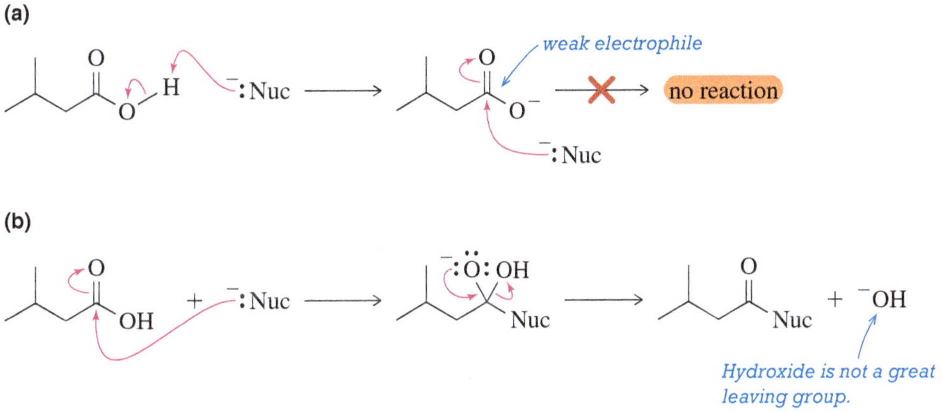

Figure 18.56 Carboxylic acids are not great substrates for substitution because (a) strong nucleophiles deprotonate them and (b) hydroxide is not a good leaving group.

These two problems are going to complicate the synthesis of carboxylic acid derivatives in Chapter 19. As such, we should think like a chemist and see if we can solve both problems before we get there.

Thinking Like a Chemist

The solution lies in finding a group (X) that does not contain the limitations of the OH group. It needs to be a good leaving group and have no acidic protons. [*Do you remember this conversation from Chapter 13?*] In **Figure 18.57**, an X group (or atom) that meets these requirements should allow weak nucleophiles to do substitution on what was once a carboxylic acid. [*What could X be?*]

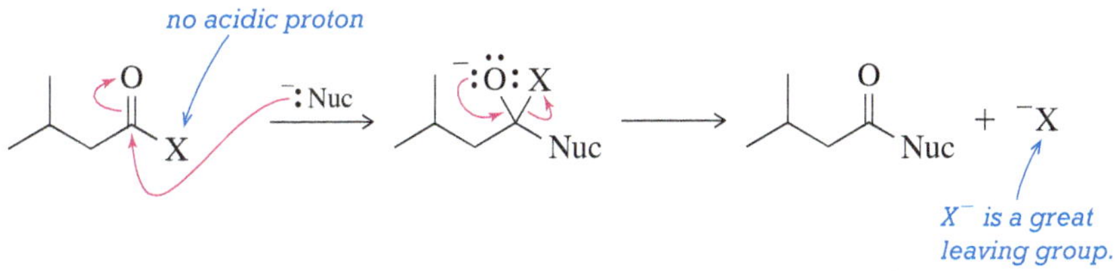

Figure 18.57 Substitution should be fast and favorable if X has no acidic protons and is a great leaving group.

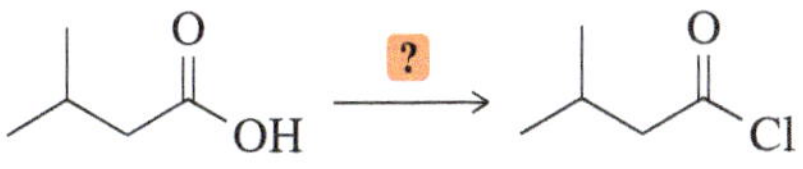

Figure 18.58 Conversion of a carboxylic acid to an acid chloride would be a useful reaction.

In Chapter 13, conversion of an alcohol to a chloroalkane made better substrates for both substitution and elimination reactions. Chloride is an excellent leaving group, and it has no protons to donate to strong bases. In this way, it is the ideal leaving group. This "new" functional group, shown in **Figure 18.58**, is called an acid chloride. [*The big question: How do we make it?*]

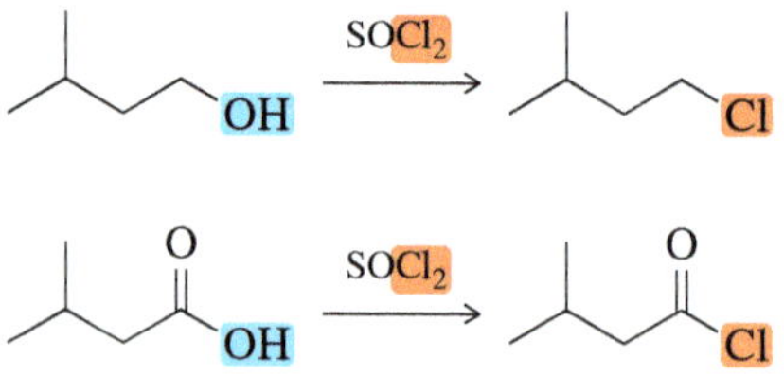

Figure 18.59 Thionyl chloride converts alcohols to haloalkanes (Section 13.7.3) and carboxylic acids to acid chlorides.

[*It is important to recognize the connections to previous material, especially Chapter 13. Let's start there with a potential solution.*] Recall that we used thionyl chloride ($SOCl_2$) in Chapter 13 to convert an alcohol to a chloroalkane. In a research lab, if you mixed $SOCl_2$ and a carboxylic acid, you'd find out that thionyl chloride is indeed the reagent of choice here as well (**Figure 18.59**). ■

Assessment 18.36 The absence of what band in the IR spectrum of the product in Figure 18.59 would be consistent with full conversion of the carboxylic acid to an acid chloride?

The formation of acid chlorides goes through two different addition/elimination mechanisms (**Figure 18.60**). It starts with the carbonyl oxygen attacking the electrophilic sulfur atom, breaking the S—O π bond to form the sulfoxide version of the tetrahedral intermediate. [*Imagine that the S═O bond behaves like the C═O bond.*] Collapse of the electron pair reforms the S—O π bond, eliminating a chloride. The carboxylic acid proton, now even more acidic, is removed by the chloride ion.

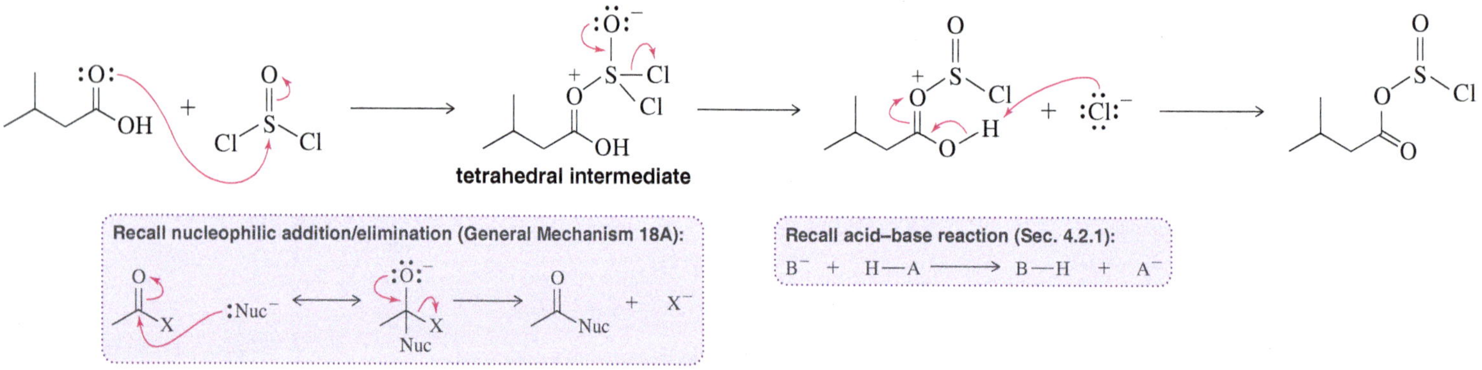

Figure 18.60 The first half of the thionyl chloride reaction with a carboxylic acid features an addition/elimination reaction of a sulfonyl group.

Analysis of the structure produced during the first half of the reaction (Figure 18.60) shows an electrophilic carbonyl with a very good leaving group attached. As a result, a chloride ion attacks the carbonyl in the second addition step of the overall mechanism (**Figure 18.61**). Collapse of the resulting tetrahedral intermediate kicks out the sulfoxide leaving group, which engages in an entropically favored elimination reaction to generate gaseous sulfur dioxide and another very stable chloride ion. The formation of these two products is the driving force for the overall reaction. [*Note the similarities between this mechanism and the one in Figure 13.22.*]

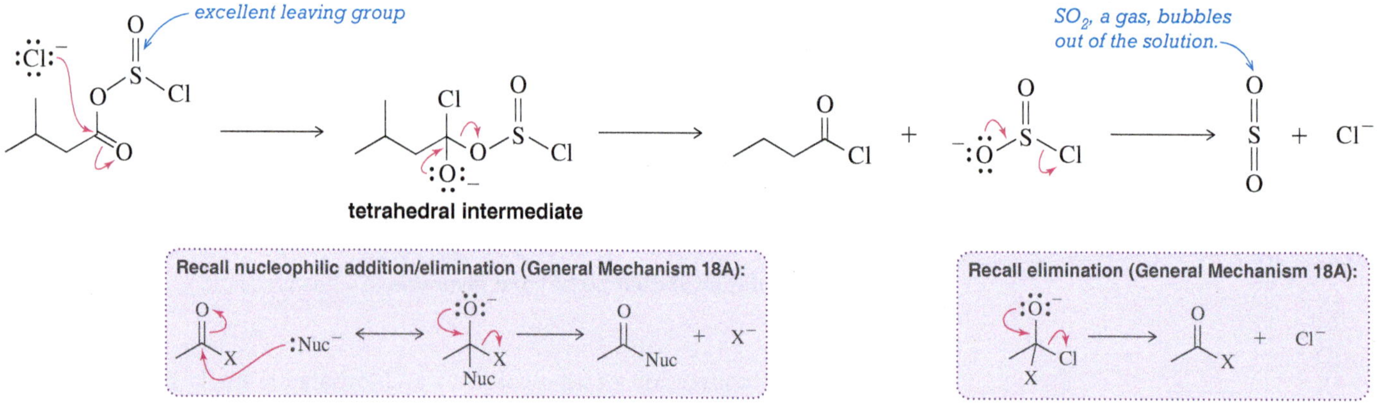

Figure 18.61 The second half of the thionyl chloride reaction with a carboxylic acid features an addition/elimination reaction at the carbonyl and an entropically driven formation of gaseous SO_2, which is the overall driving force of this reaction.

The focus of Chapter 19 is the interconversion of carboxylic acid derivatives via nucleophilic acyl substitution reactions (strong and weak nucleophiles; General Mechanisms 18A and 18B). All carboxylic acid derivatives can be converted into a carboxylic acid by hydrolysis, and all carboxylic acid derivatives can be synthesized from an acid chloride. Thus, the conversion of an acid to an acid chloride (**Figure 18.62**) is the lynchpin reaction in the chemistry of Chapter 19. [*You don't need to worry about how these reactions work yet.*]

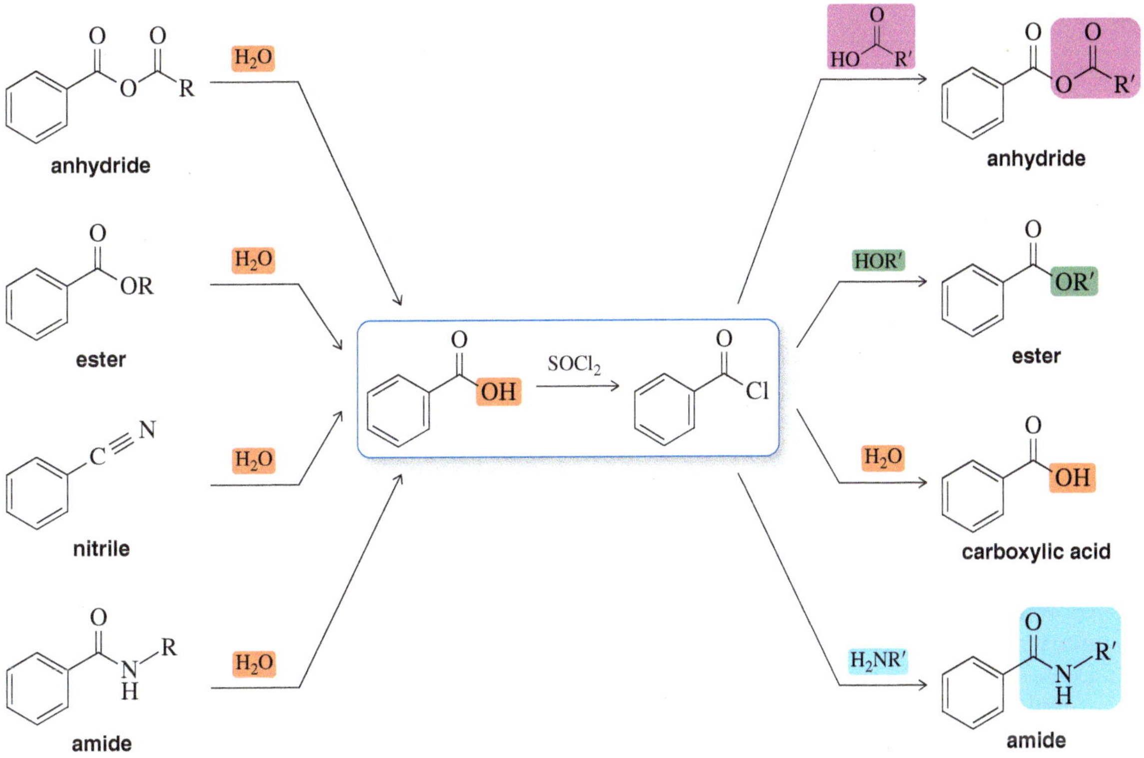

Figure 18.62 Formation of an acid chloride makes it possible to interconvert *all* of the carboxylic acid derivatives.

Assessment 18.37 Identify the carboxylic acid and the reagents needed for synthesizing the acid chlorides shown.

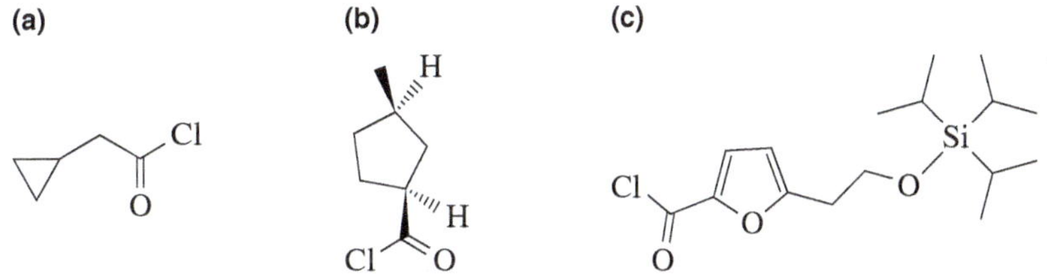

Assessment 18.38 What reagent do you think would be useful for synthesizing an acid bromide from a carboxylic acid? [*Chapter 13 might help here.*]

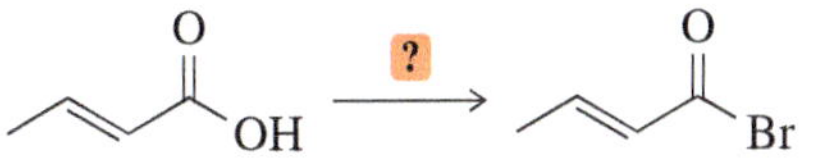

Assessment 18.39 In three steps, synthesize the acid chloride on the right from the alkene on the left.

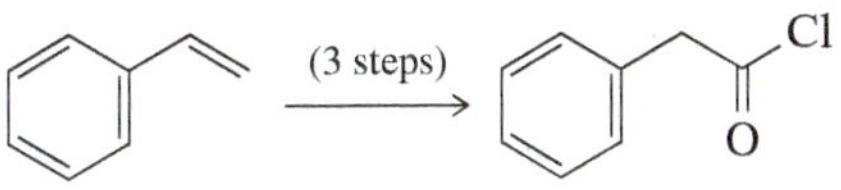

18-C CHEMISTRY IN CONTEXT: Biochemistry

Metabolism of sodium benzoate

Because not all compounds we digest are healthy, useful, or safe, our bodies have a way of detoxifying and removing them from our system, usually through biochemical reactions that occur in the liver.

Sodium benzoate and its conjugate acid, benzoic acid, are seen as foreign substances, so the body has an efficient mechanism for eliminating them. The process involves enzymatic attachment (referred to as conjugation) of benzoate to a molecule of glycine to make hippuric acid (**Figure 18.63**). This reaction has a lot in common with the reactions studied in this chapter. First, sodium benzoate is phosphorylated by ATP [*a big molecule*]. Attachment of the phosphate converts the oxygen of benzoate into a much better leaving group. [*Sound familiar?*] The phosphate is replaced by attachment of the sulfur of HSCoA [*another huge molecule*] to give a thioester, which is simply an ester with a sulfur in place of oxygen. The —SCoA [*another good leaving group*] is replaced by substitution with the nitrogen of the amino acid glycine. The resulting compound is hippuric acid, a water-soluble metabolite that is easily excreted in urine.

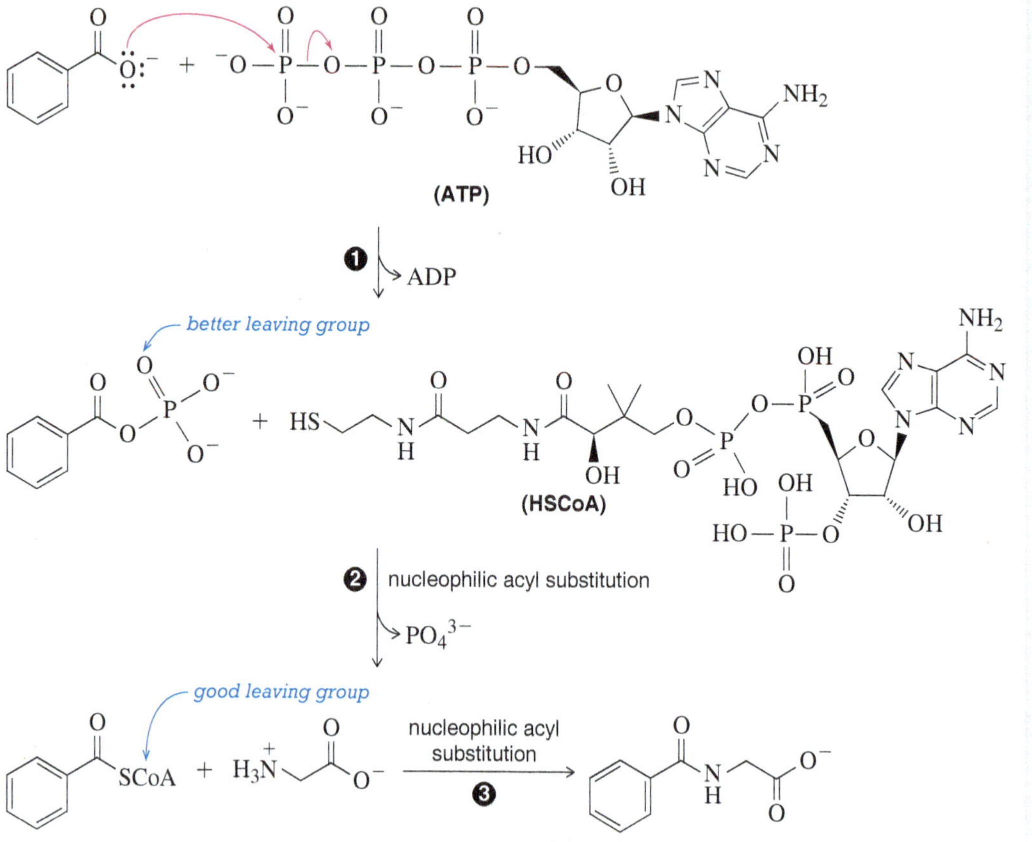

Figure 18.63 **The enzymatic synthesis of hippuric acid** features activation of a carboxylic acid followed by two nucleophilic acyl substitution reactions.

The size of the molecules may make the mechanism shown in Figure 18.63 seem quite daunting, but this is all chemistry you have seen before. Converting a carboxylic acid to an acid chloride makes it better at undergoing substitution. Similarly, phosphorylating an acid and converting it into a thioester also activates it for substitution. [*In Chapter 19, we look specifically at how we might synthesize hippuric acid from benzoic acid in the lab.*]

Assessment 18.40 Suggest a mechanism for steps 2 and 3 in Figure 18.63.

To understand why alternative pathways are necessary for synthesis, let's review Assessment 18.25 from earlier. Go back and check over your attempt at this one.

18.8 Other reactions of carboxylic acids

18.8.1 Ester synthesis with diazomethane

In learning new reactions, there's always that lingering question of "why?" [*Why do we need another way to make an ester?*]

Solved Assessment 18.25

In Assessment 18.25, we were asked why, in spite of esterification occurring, the expected product was *not* the major product isolated (**Figure 18.64**).

This is another example where tunnel vision might cause us to miss other reactive parts of the molecule. Certainly, carboxylic acids react with alcohols in the presence of acid to form esters. [*But what **other** functional group is present?*] There is also an alkene, which, as we learned in Chapters 8 and 13, reacts with strong acids and alcohols to make ethers (**Figure 18.65**). [*It is important to analyze a molecule for all possible reactive sites.*]

From here, predicting the product is as simple as writing a mechanism for how the alkene will react. The final product, after two reactions, contains both an ester and an ether (**Figure 18.66**). ■

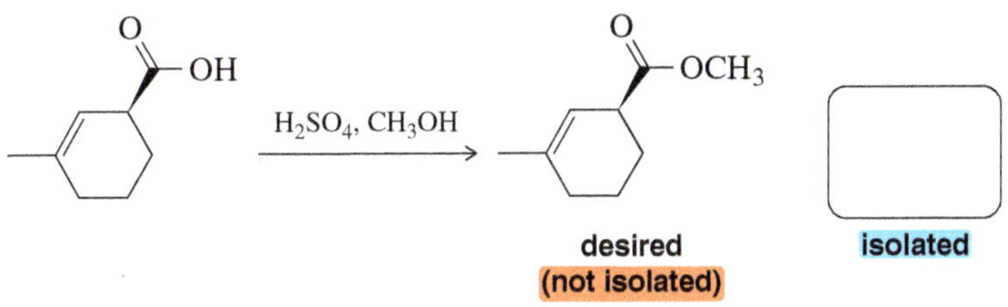

Figure 18.64 What product forms instead of the desired product?

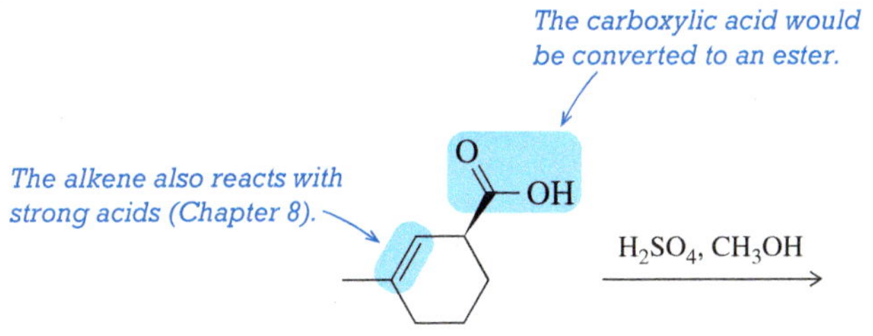

Figure 18.65 Analyze all functional groups; alkenes also react with strong acids.

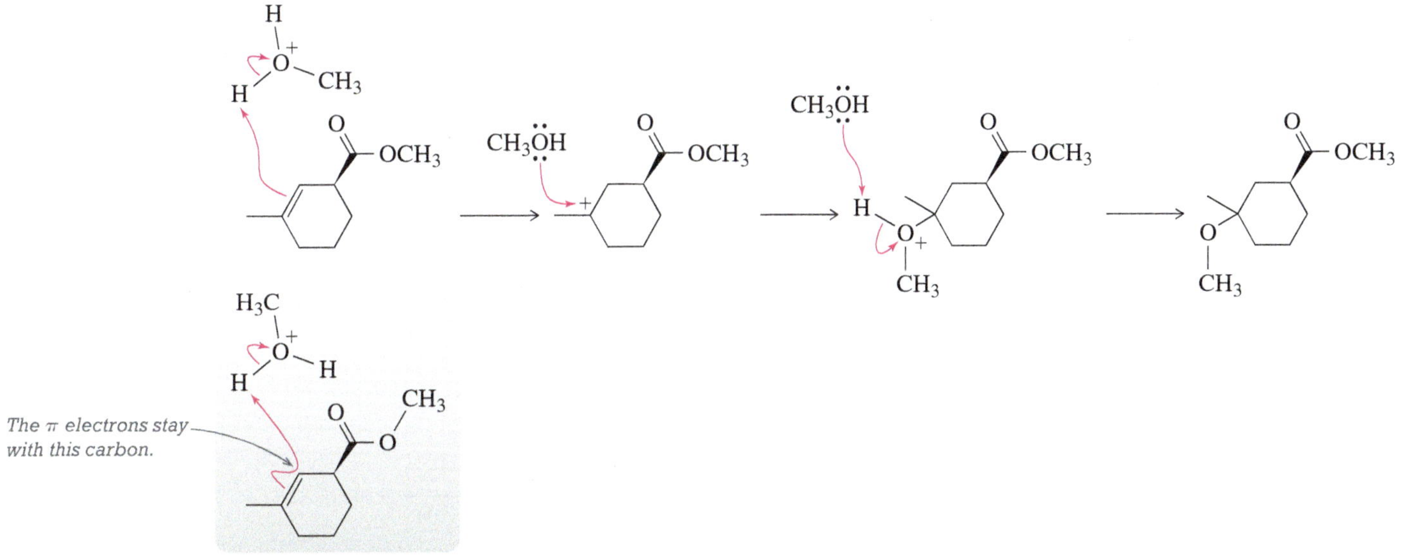

Figure 18.66 The product contains both an ester (from the carboxylic acid) and an ether (from the alkene).

The outcome of Assessment 18.25 shows a need for ester-forming reactions that occur under neutral, or even basic, conditions. Most of our focus on ester formation has been about replacing the entire OH group with an OR group. Another viable strategy, given the acidity of the carboxylic acid, would be to replace the H with an R (**Figure 18.67**).

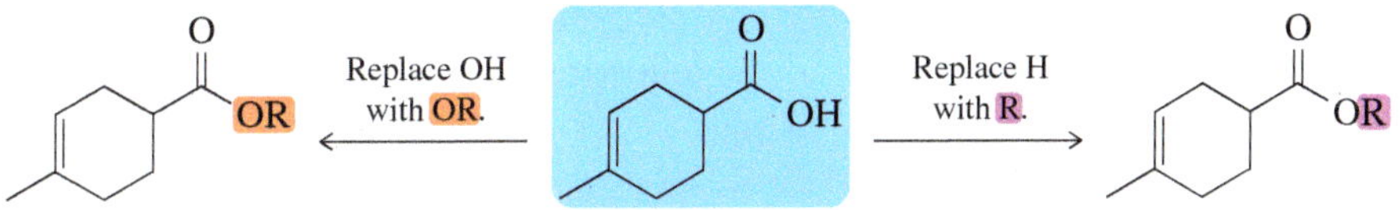

Figure 18.67 Replacing the hydrogen with an alkyl group should be a viable strategy for ester formation.

Diazomethane (CH_2N_2) is a reagent that accomplishes this transformation. The reaction begins by deprotonating the acidic proton to make a carboxylate ion, while forming a methyl diazonium compound (**Figure 18.68**). With a sterically unhindered methyl group and a great leaving group (N_2), methyl diazonium is an ideal S_N2 substrate. Substitution by the carboxylate in an S_N2 reaction is also entropically favorable because the nitrogen gas produced immediately leaves the reaction vessel.

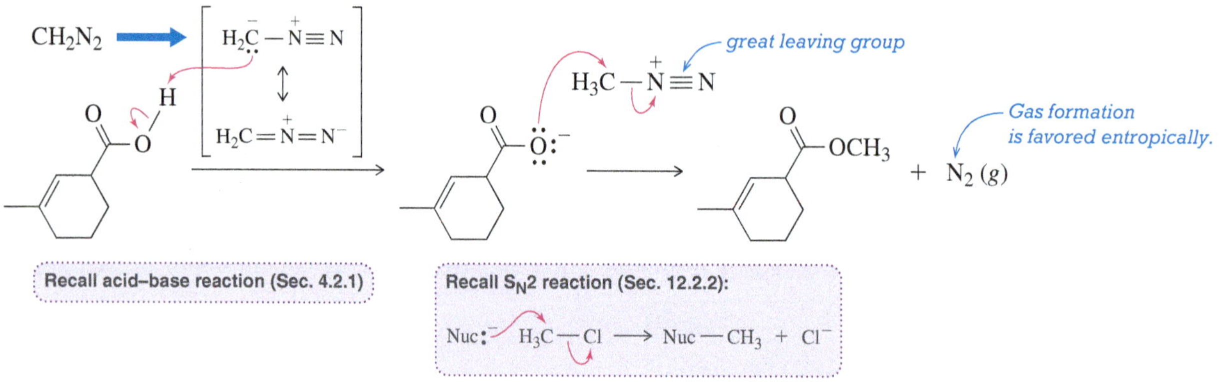

Figure 18.68 Diazomethane converts carboxylic acids to esters under neutral conditions. An acid–base reaction is followed by an S_N2 reaction.

Unfortunately, this reaction is limited to formation of methyl esters because larger diazo compounds enable other reactions to occur instead (see Assessment 18.42).

18.8.2 Ester synthesis by S_N2 reactions of carboxylate ions

Related to the reaction of diazomethane is the substitution reaction of carboxylates with haloalkanes or other good leaving groups. We saw these reactions in Chapter 12 while studying S_N2 reactions. The carboxylate ion, as a negatively charged (or strong) nucleophile, donates a pair of electrons directly to σ^*, displacing the halide leaving group in a concerted reaction. In a polar aprotic solvent, this reaction is especially efficient with 1° and methyl halides. Because the pK_a of the conjugate acid of a carboxylate ion is less than 11, these nucleophiles do substitution over elimination at 2° carbons as well. When the leaving group is at a stereocenter, the reaction proceeds with inversion of stereochemistry (**Figure 18.69**).

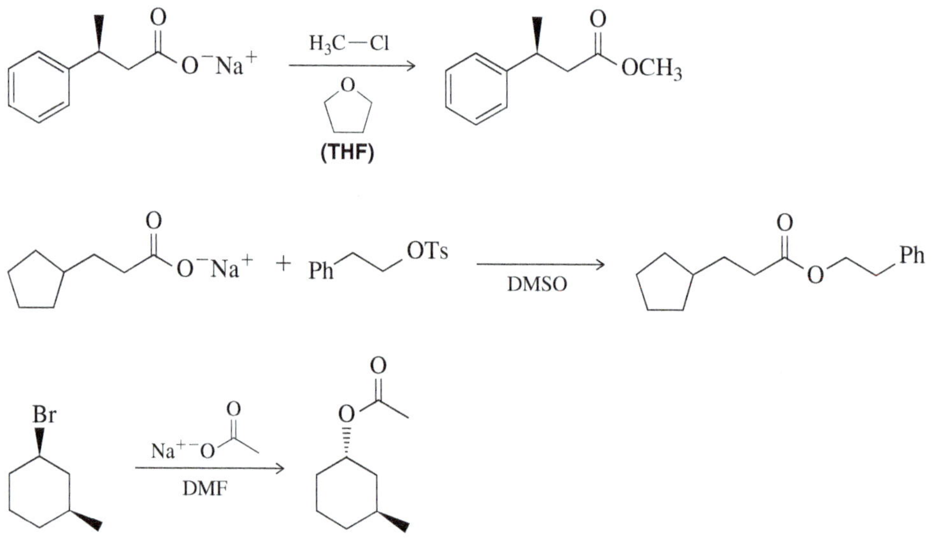

Figure 18.69 S_N2 reactions of carboxylate ions occur at carbons with methyl-, 1°-, and 2°-substituted leaving groups (halides and tosylates).

Assessment 18.41 Predict the product of the following reactions.

(a)

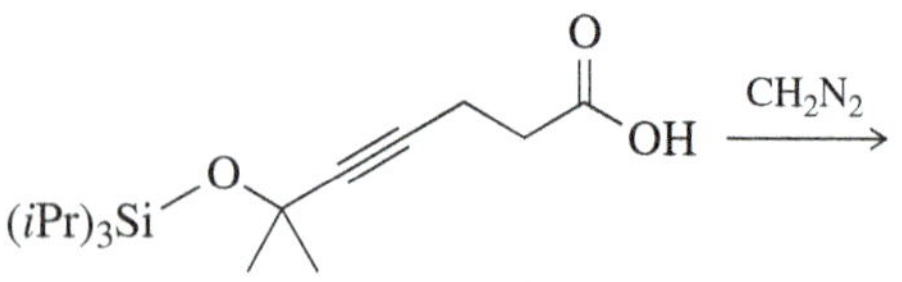

(b)

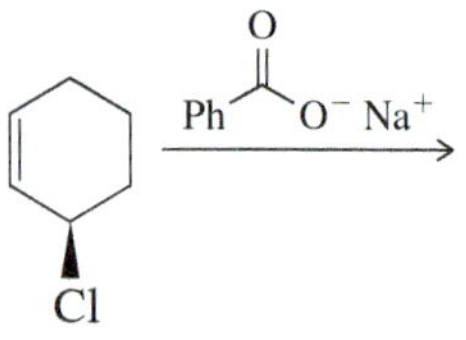

(c)

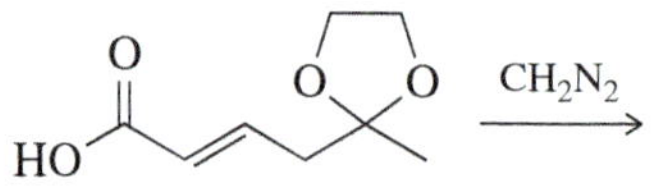

(d)

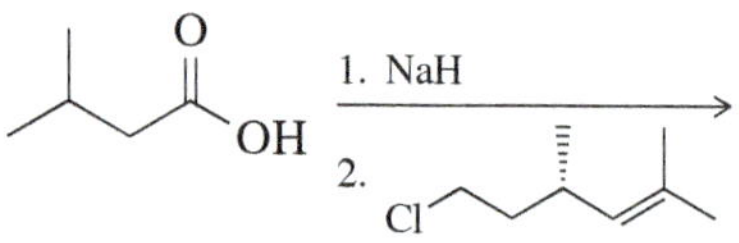

Assessment 18.42 A chemist hoped to make an ethyl ester using diazoethane. At the end of the reaction, only the original carboxylic acid, ethene, and nitrogen were produced. Suggest an arrow-pushing mechanism that accounts for these products.

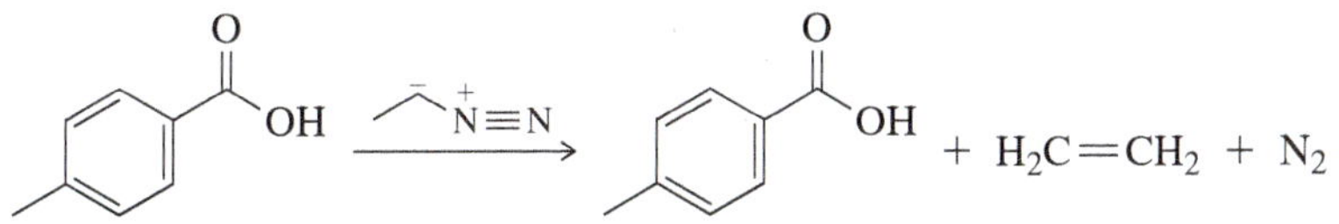

18.8.3 Ester synthesis by alkene addition

In Sections 18.8.1 and 18.8.2, we saw that the oxygen of a carboxylic acid can, after deprotonation, participate in an S_N2 reaction to make a new ester via C—O σ bond formation. [*Look back at Figure 18.69 for a reminder.*] By now we are conditioned, when developing new reactions, to reason, "If I can do it using an S_N2 mechanism, can I also do it as an S_N1 reaction?" The answer is yes, especially in the general sense of a carboxylic acid reacting with a carbocation, as in **Figure 18.70**.

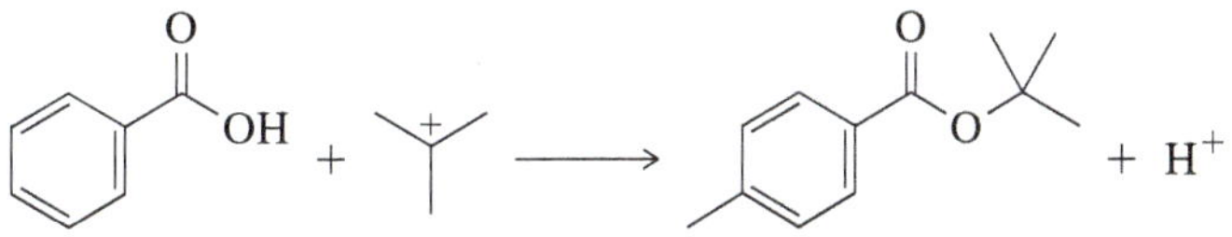

Figure 18.70 **A carboxylic acid plus a carbocation** should produce an ester via formation of a new C—O σ bond.

In practice, this reaction is not often done using the S_N1 mechanism specifically, but it does make use of a carbocation intermediate. The carbocation is instead generated by reaction of an alkene with a strong acid (see Section 8.4). This carbocation is attacked by the carbonyl oxygen. The newly formed intermediate, possessing an acidic proton, is deprotonated, likely by another molecule of alkene, allowing the next molecule of acid to be esterified. Because the reaction begins with an acidic proton and produces an acidic proton, it is acid-catalyzed (**Figure 18.71**).

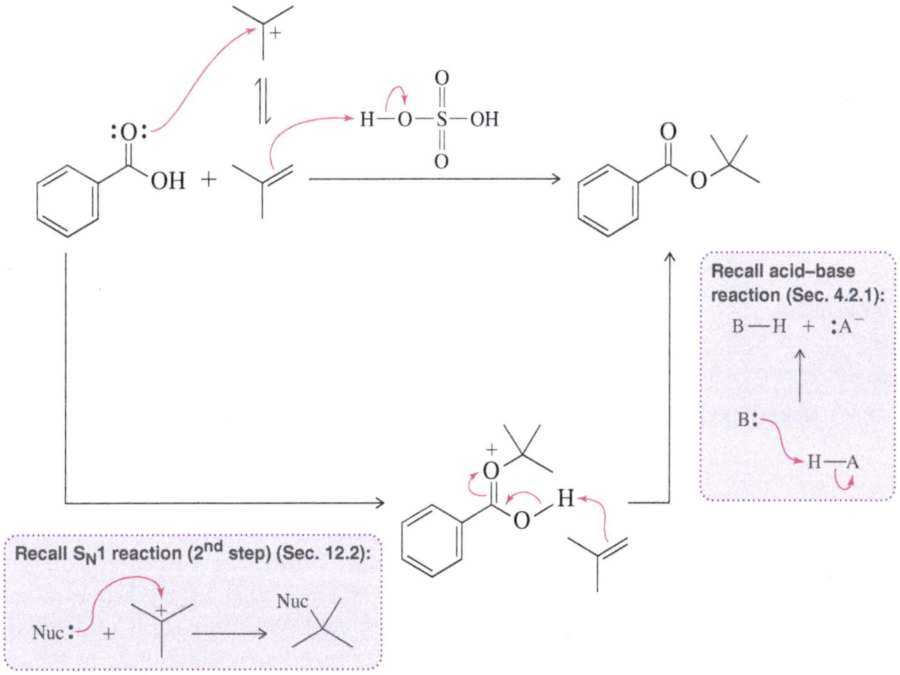

Figure 18.71 **Alkene addition** (from Chapter 8) with a carboxylic acid as the nucleophile produces an ester.

Assessment 18.43 Predict the product of the reactions shown.

(a)

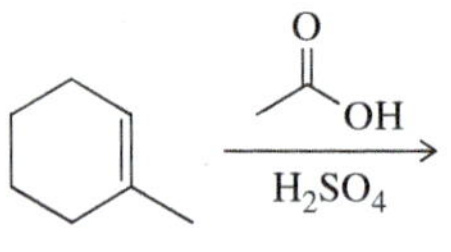

(b)

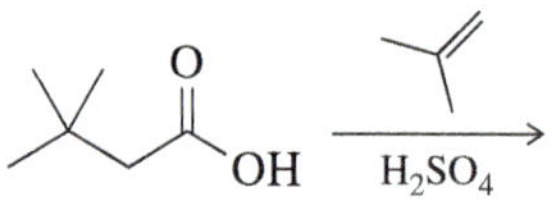

(c)

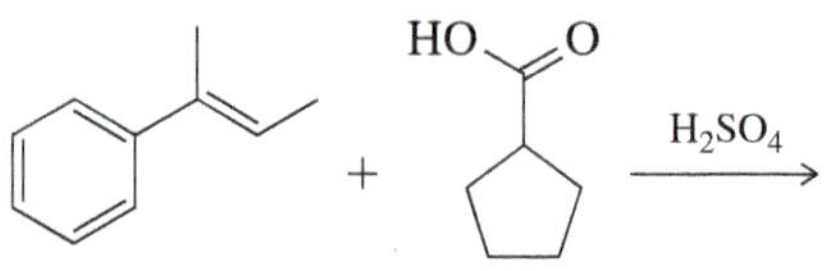

Assessment 18.44 Suggest an acid and alkene that could be combined to make the esters shown.

(a)

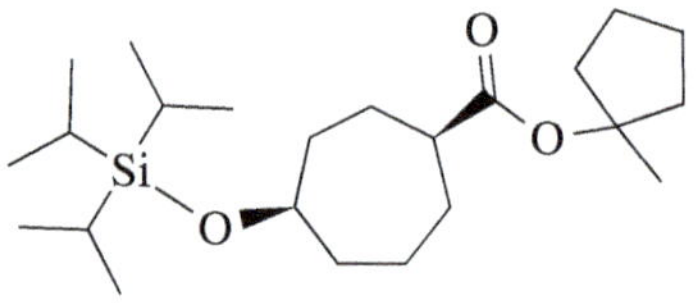

(b)

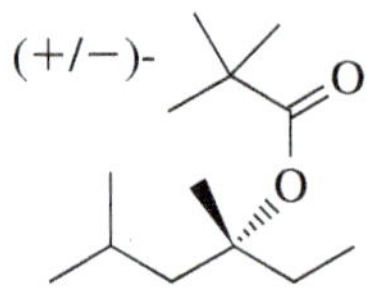

Assessment 18.45 Show a reaction coordinate diagram for the mechanism in Figure 18.71. What is the rate-determining step?

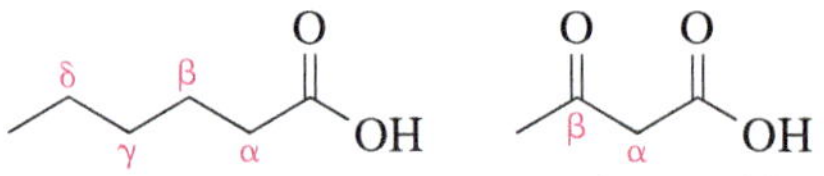

β-keto acid

18.8.4 Decarboxylation of β-keto acids

The last reaction we discuss in Chapter 18 is the thermal decarboxylation of **β-keto acids**. These are named using a convention in which a carbon's distance from a particular functional group is designated by a Greek letter. For a carboxylic acid, the carbon directly attached to the carboxylic acid carbon is α, the next is β, and so on. Thus, a β-keto acid has a ketone at the second carbon from the carboxylic acid. The carboxylic acid carbon, which is carbon 1 in IUPAC nomenclature, does not receive a Greek letter designation in this naming convention.

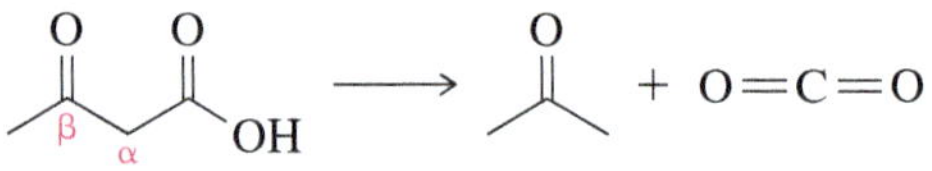

Figure 18.72 Decarboxylation of β-keto acids.

The decarboxylation of β-keto acids in **Figure 18.72** is a complicated reaction. For one, it doesn't really fit into the general categories of reactions we have studied in this chapter. And second, it occurs by a mechanism we haven't seen very often. But, as we explain in Chapter 20, it is a very important reaction for your synthetic toolbox. So, as we study it, focus on what it has in common with things seen before. We'll start with what makes it favorable. [*For that, go back to the basics . . . any ideas?*]

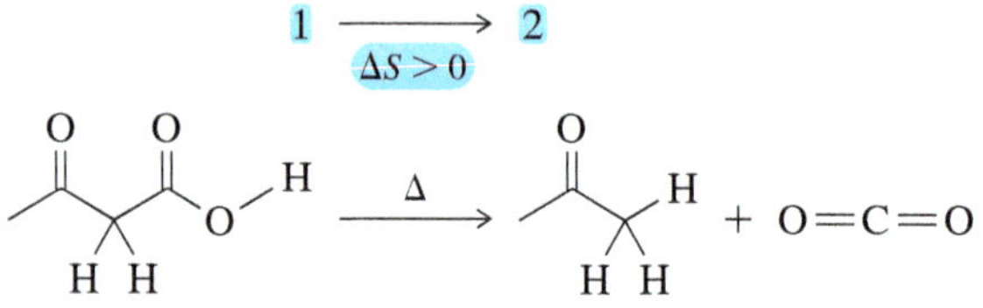

Bonds broken:
C—C σ
O—H σ

Bonds formed:
C—H σ
C—O π

Figure 18.73 Decarboxylation is heavily favored by entropy.

The favorability of a reaction is determined by ΔG, which is composed of ΔH and ΔS [*recall Chapter 5*]. To begin, the decarboxylation reaction forms two molecules from one, and one of those is a gas. Thus, it is highly favored by entropy ($\Delta S > 0$). By showing the hydrogens, as in **Figure 18.73**, we see that ΔH is a result of breaking two σ bonds and forming a σ and a π bond. While π bonds are usually substantially weaker than σ bonds, this is a C—O π bond (~85 kcal/mol), which is nearly as strong as a C—C σ bond (~88 kcal/mol). In conjunction with the entropic driving force, this ends up being a favorable reaction, albeit with a high activation energy (which is why it needs to be heated).

Although β-keto acids are typically drawn with the two carbonyls pointed in the same direction [*just a convention*], they exist more often as the conformation in which the OH of the carboxylic acid is near the oxygen of the ketone (**Figure 18.74**). This conformation is favored by the hydrogen bond that can form between the acidic proton and the ketone. From this conformation, the reaction occurs by a concerted 6-electron, cyclic transition state (also referred to as an **aromatic transition state**) to give carbon dioxide (CO_2) and an enol. We study enols in detail in Chapter 20, but have also seen them before as molecules that immediately equilibrate to the more stable ketone. [*Check out Section 10.8.3 for a review of enols.*]

Chapter 21 explains why aromatic transition states have low activation energies.

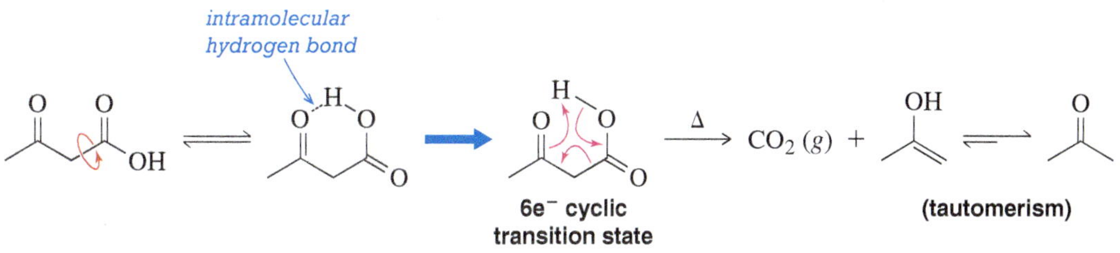

Figure 18.74 The mechanism of the thermal decarboxylation of β-keto acids involves a concerted reaction with an aromatic transition state.

As we have seen previously, concerted reactions can sometimes be more easily understood by imagining them as stepwise reactions (**Figure 18.75**). This allows us to rationalize the movement of each individual pair of electrons. But make no mistake, the decarboxylation reaction of β-keto acids is concerted, as shown in Figure 18.74. [*What is shown in Figure 18.75 is only meant to help us understand Figure 18.74. Always, always show the mechanism as concerted.*] Our hypothetical mechanism begins with transfer of the proton from the carboxyl group to the ketone. A resonance structure can be drawn for the protonated ketone, where the positive charge is shown on carbon. [*We did this a lot in Chapter 17.*] This zwitterionic structure now has an electron-poor carbon proximal to the electron-rich carboxylate. We can imagine the carbocation pulling on the electrons in the C—C σ bond, eventually breaking the bond to give the enol and what is essentially a resonance structure of CO_2.

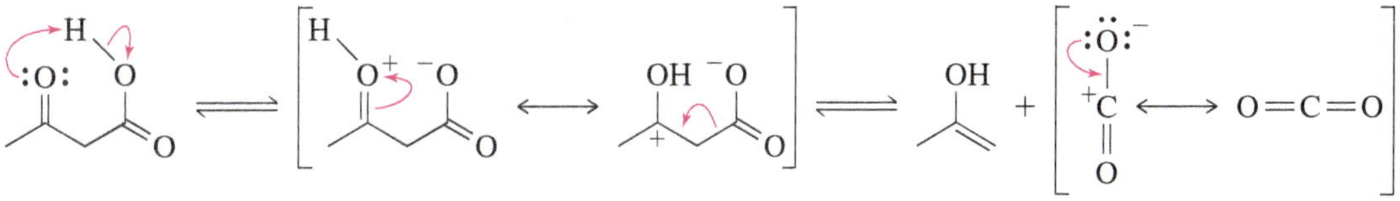

Figure 18.75 A hypothetical stepwise mechanism of decarboxylation that helps us understand the concerted mechanism in Figure 18.74. [*To be clear, the mechanism is concerted.*]

There are two takeaways from studying this reaction now. First, keep it in your synthetic toolbox [*reaction sheets*] because you will need it later. And second, these reactions are only possible when the concerted, 6-electron cyclic transition state (the aromatic transition state) is possible.

18.8.3 SUMMARY

- β-Keto acids undergo thermal decarboxylation by a mechanism involving a 6-electron cyclic transition state. This reaction will be especially useful for synthesis in Chapter 20.

Assessment 18.46 Predict the product of the reactions shown.

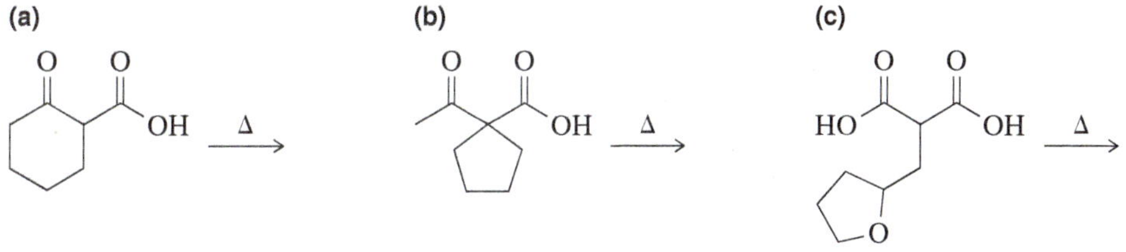

Assessment 18.47 Decarboxylation of an α-keto acid and a γ-keto acid is not possible. Why?

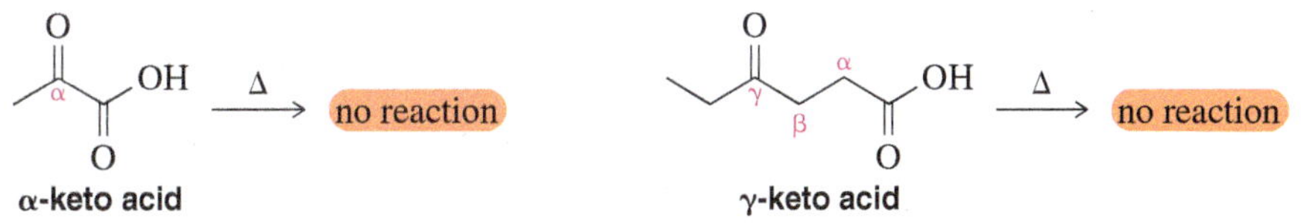

18-D CHEMISTRY IN CONTEXT: Food Chemistry

A precursor to benzene?

In 1993, it was confirmed that the combination of ascorbic acid (vitamin C) and benzoic acid in the presence of copper salts can result in the radical decarboxylation of benzoic acid and production of benzene. Although this reaction is different than the decarboxylation studied in Section 18.8.4, we can draw an arrow-pushing mechanism (**Figure 18.76**). Abstraction of the carboxylic acid hydrogen results in the formation of a carboxyl radical. The carboxyl radical then collapses to produce carbon dioxide and a benzene radical. The benzene radical abstracts a hydrogen from solution to complete the synthesis of benzene.

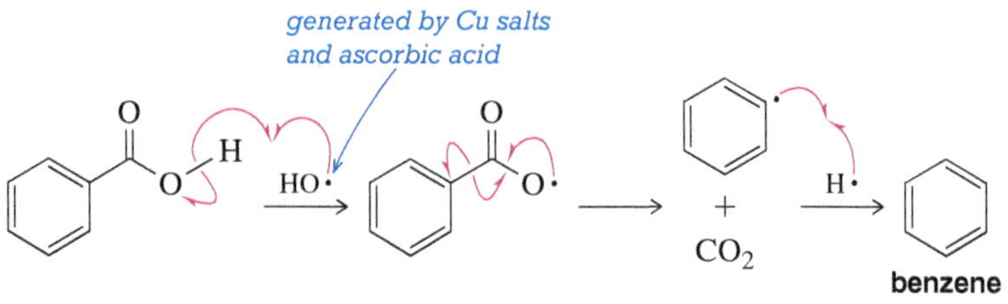

Figure 18.76 The radical-mediated decarboxylation of benzoic acid is catalyzed by ascorbic acid and copper salts.

This reaction is relatively slow at low temperatures and in the absence of light, conditions that mimic those present in a typical can of soda on a refrigerator shelf. Nonetheless, small amounts of benzene have, at times, been discovered in sodas. This is problematic because benzene is a known carcinogen, and these findings have prompted calls for the removal of sodium benzoate as a food preservative in sodas.

Concluding thoughts

Back in Chapter 13, I made an appeal to you, as students of science and chemistry, to keep the principles of green chemistry in your mind at all times. Whether you are a research chemist, a physician, or simply a consumer who can hold businesses, retailers, and restaurants accountable, this remains the case. However, let's review sodium benzoate in light of the advertisement mentioned in the introduction to this chapter. Since the start of the 20th century, sodium benzoate was confirmed to be harmless in human feeding studies conducted by the U.S. Department of Agriculture. The U.S. Food and Drug Administration has continually conferred upon sodium benzoate GRAS status, reaffirming this in 1973. GRAS status means the compound is "generally recognized as safe" as a food additive. Quoting the Select Committee on GRAS Substances (SCOGS) directly, "There is no evidence in the available information to show that benzoic acid and sodium benzoate as food ingredients constitute a hazard to the general public when used at levels that are now current . . . ." Continuing, "The highest no effect level . . . is approximately 90-fold the amount that would be consumed if an individual's diet [consisted] only of those foods containing the greatest amounts of sodium benzoate in current usage." In short, sodium benzoate, while found in fireworks, seems to have virtually no risk associated with its use at current levels.

But what about the formation of benzene, a carcinogen, from sodium benzoate? That is something that happens, but not at levels that cause any harm. A recent study of 451 beverages found an average concentration of benzene below the European Union drinking water limit of 1 μg/L. Only infant-ready carrot juice had unacceptable levels, attributed to the higher heating necessary to remove microorganisms. The point is, *it's the dose that makes the poison*. And because of how little is present from the conversion of sodium benzoate to benzene, multiple research studies have concluded that "food is not an important pathway for benzene exposure." Your benzene exposure is far greater when standing beside your car as you fill it with gasoline or when walking down a crowded city street.

So let's review: Sodium benzoate is found naturally in fruits, berries, milks, cheeses, and spices. [*The restaurant that advertises that their food contains no sodium benzoate serves all of these things.*] Via the conversion to hippuric acid, our bodies have a way of rapidly excreting the sodium benzoate/benzoic acid that we consume. The U.S. FDA considers it safe, as does the most recent research into its potential toxicity. So why would a restaurant use sodium benzoate as an advertising tagline? It's simple, really: to scare consumers into believing that their food is better, cleaner, and purer; their food is "chemical-free." [*Think about that one for a minute.*]

Stepping down from the soapbox, let's get ready to see all the different derivatives we can convert carboxylic acids into in Chapter 19. By the way, I haven't forgotten my promise to show you how chemists make hippuric acid.

This type of advertising is a danger to the environmental, sustainability, and green chemistry movements. Overstating the danger of food preservatives is as dangerous as suggesting that hexavalent chromium in the drinking water is safe (Chapter 13), or arguing that vaccinations cause autism, or that humans have no role in climate change. It is extremely important that we as scientists use our science knowledge to stand against businesses, governments, and individuals who are damaging our environment in pursuit of profits. But we should similarly stand against businesses that use fear-mongering tactics and ignore real science in pursuit of those same profits.

End-of-Chapter Materials

Important Concepts

[*These are the major concepts covered in each section, excluding reactions.*]

18.1: Carboxylic acids consist of a carboxyl (a carbonyl and a hydroxyl) bound to a hydrogen or an alkyl group. Carboxylic acid derivatives are at the same oxidation state as carboxylic acids and can be thought of as forming by the condensation reaction between carboxylic acids and other reagents.

18.2: When naming acyclic carboxylic acids, the carboxyl functional group always receives the lowest number, even in the presence of other functional groups.

18.3: The central carbon of a carboxylic acid is sp^2 hybridized. Because resonance is possible in a carboxylic acid, both oxygens are sp^2 hybridized.

18.4: Carboxylic acids are excellent hydrogen bond donors and acceptors, often existing as dimers. As a result, they are highly water soluble and have high melting and boiling points. Carboxylic acids are identified by a broad band between 3400 and 2600 cm^{-1} in their IR spectrum and a broadened peak representing the carboxylic acid hydrogen around 11 pm in the ^{1}H NMR spectrum (so broad it is often invisible). Carboxylic acids undergo the McLafferty rearrangement in the mass spectrometer. As the name implies, carboxylic acids are acidic, with a $pK_a \approx 5$ for acetic acid. Inductive effects can result in some carboxylic acids having a pK_a value as low as 0.2.

18.5: Carboxylic acids can be synthesized using modifications of previously studied reactions, including the oxidation ($KMnO_4$ and O_3) of alkenes and alkynes, reaction of a Grignard reagent with CO_2, and by the chromic acid oxidation of alkylbenzenes.

18.6: Deprotonation of a carboxylic acid results in the formation of a good nucleophile that can be used effectively in substitution reactions. The extraction of carboxylic acids from a mixture containing neutral organic compounds can be accomplished by deprotonation to make a carboxylate ion that is more soluble in water.

18.7–18.7.1: There are two important mechanisms for nucleophilic acyl substitution with carboxylic acids and carboxylic acid derivatives. Represented in this chapter as General Mechanisms 18A and 18B, the choice of mechanism depends on the conditions under which the reaction is run. In General Mechanism 18A, typical of strongly basic conditions, the strong nucleophile first adds to the carbonyl to generate a tetrahedral alkoxide intermediate that collapses to expel a good leaving group and reform the C—O π bond. In General Mechanism 18B, typically with weak nucleophiles under acidic conditions, the carbonyl is protonated first, initiating nucleophilic attack to form a tetrahedral intermediate after a second proton transfer reaction. A second sequence featuring protonation, elimination, and a final deprotonation results in the reformed carbonyl.

18.7.2–18.7.5: Through variations on nucleophilic acyl substitution reactions, carboxylic acids can be converted to esters (Fischer esterification; H_2SO_4, ROH), reduced to 1° alcohols ($LiAlH_4$), converted to ketones (two equivalents of RLi), and transformed into acid chlorides ($SOCl_2$).

18.8: When the Fischer esterification isn't possible, carboxylic acids can be converted into esters via their reaction with diazomethane or via the S_N2 reaction of carboxylate ions. The decarboxylation of β-keto acids results in the formation of CO_2 and a ketone via a 6-electron cyclic transition state.

Important Reactions and Mechanisms

1. **$KMnO_4$ oxidation of an alkene, 18.5.2:** Oxidation of an alkene occurs by a mechanism similar to the reaction of alkenes and OsO_4. Because of the harsh conditions, the reaction continues through decomposition of the permanganate ester by a 6-electron cyclic transition state to give two carbonyls. Further oxidation results in the formation of carboxylic acids.

General:

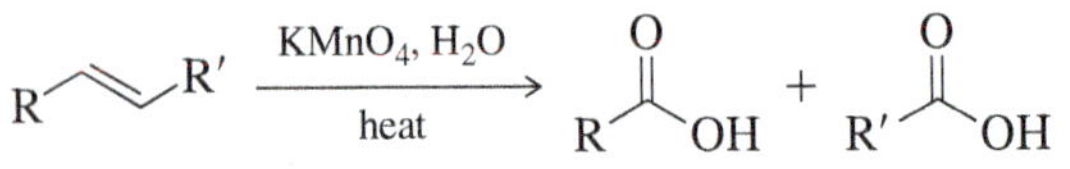

Specific:

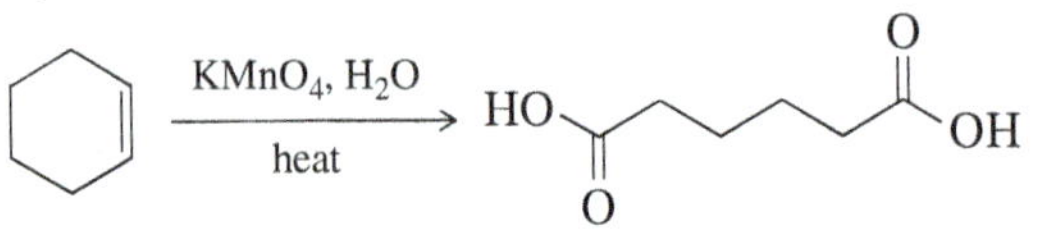

Mechanism:

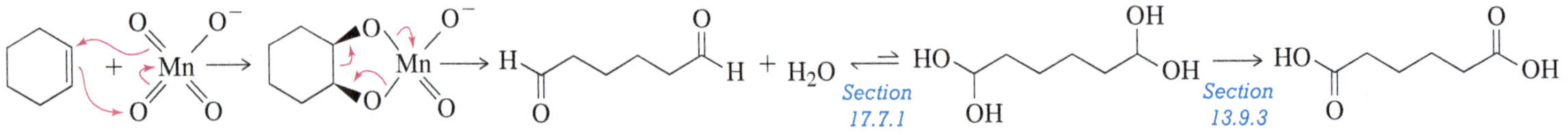

2. **$KMnO_4$ oxidation of an alkyne, 18.5.2:** Oxidation of an alkyne occurs by a mechanism similar to the reaction of alkenes and OsO_4. Because of the harsh conditions, the reaction continues through decomposition of the permanganate ester by a 6-electron cyclic transition state to give a diketone. Hydration of the diketone gives a diol that reacts to form a second permanganate ester. Decomposition of the permanganate ester results in the formation of two carboxylic acids.

General:

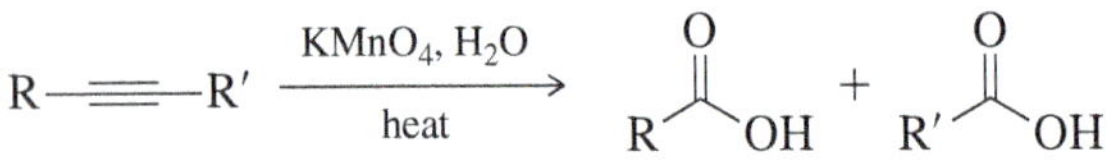

Specific:

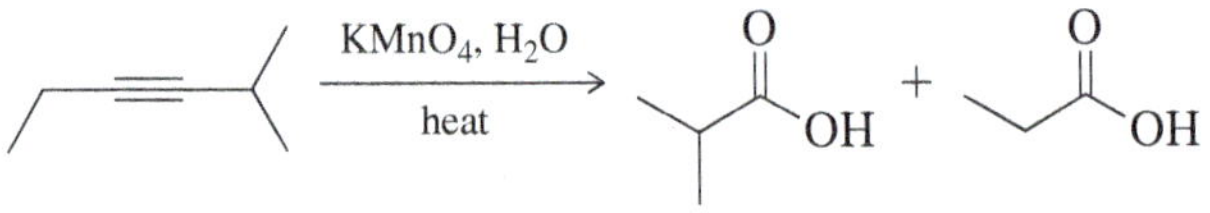

Mechanism:

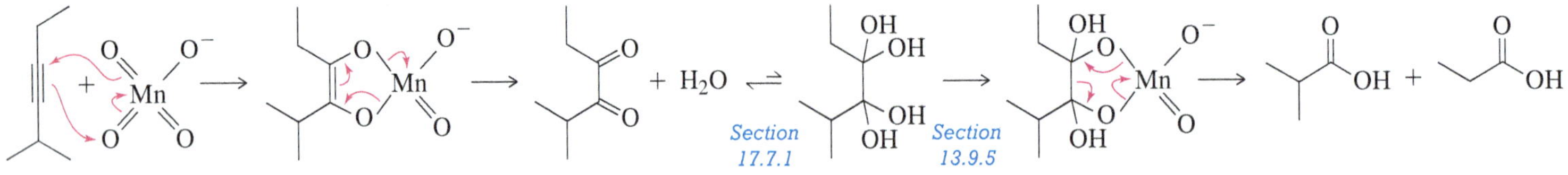

3. **Ozonolysis of alkenes, oxidizing workup, 18.5.2:** The highly reactive oxidizing gas ozone oxidatively cleaves both the σ and π bonds of a 1,2-disubstituted alkene in a three-step process to initially generate a five-membered ozonide. This reactive ozonide is oxidized on exposure to hydrogen peroxide, producing a carboxylic acid at each of the original carbons of the 1,2-disubstituted alkene carbons. [*The ozonolysis mechanism is found in Section 9.1.7.*]

General:

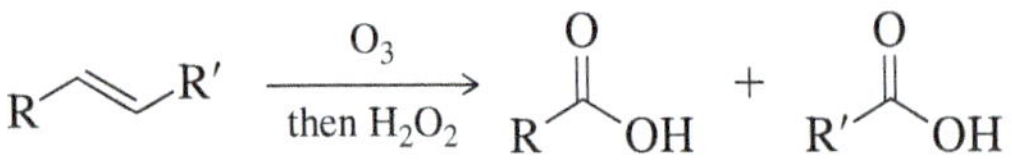

Specific:

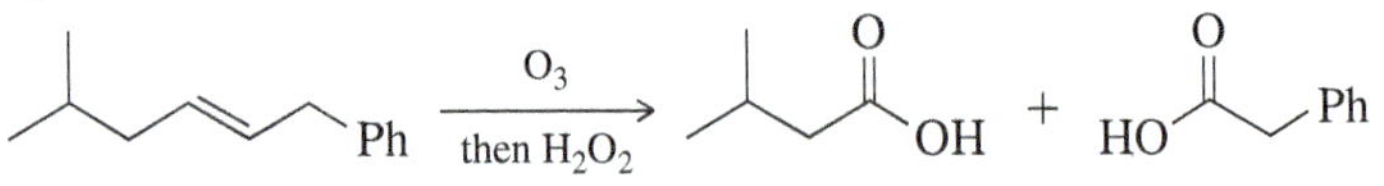

Intermediate steps:

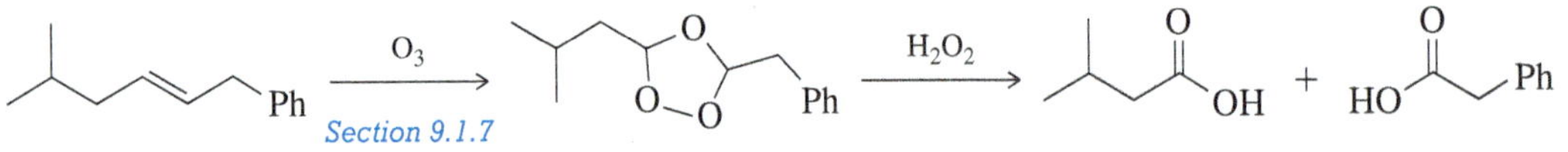

4. **Grignard addition to CO_2, 18.5.3:** A Grignard reagent (a strong nucleophile) reacts by General Mechanism 17A. The carbanion attacks the electrophilic carbon of CO_2, pushing the π electrons onto the oxygen. The resulting carboxylate is quenched with acid in a second step to give a carboxylic acid.

General:

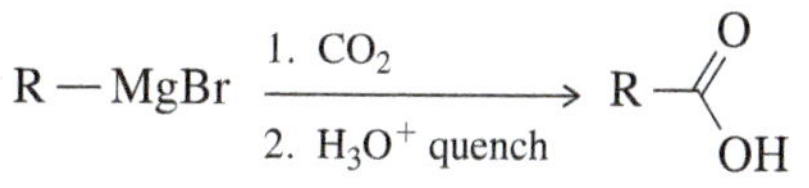

Specific:

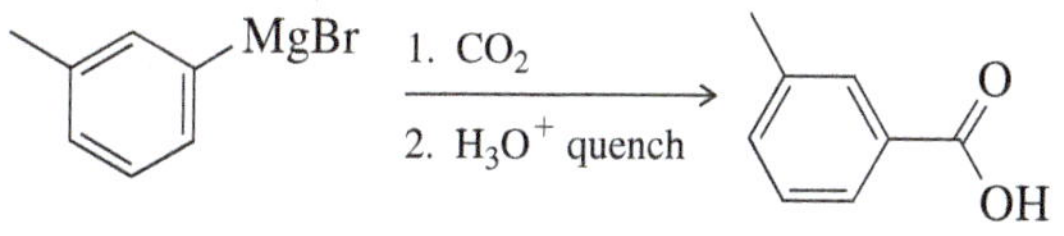

Mechanism:

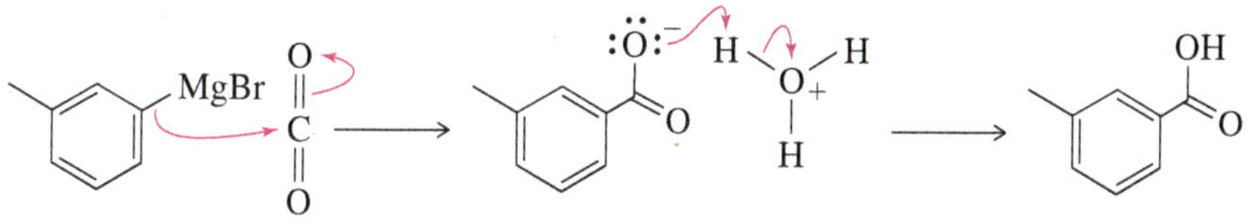

5. **Chromic acid oxidation of alkyl benzenes, 18.5.4:** Benzylic groups, as long as there are hydrogens at the benzylic carbon, are oxidized to carboxylic acids by a mechanism that is not well understood.

General:

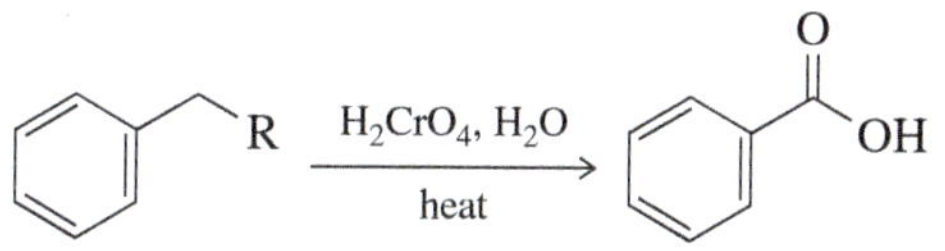

Specific:

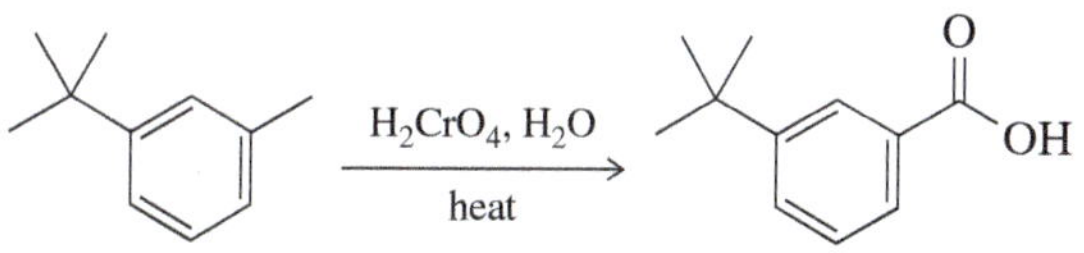

6. **Acid–base reactions of carboxylic acids, 18.6:** Carboxylic acids are deprotonated favorably by bases whose conjugate acids have a pK_a > 5.

General:

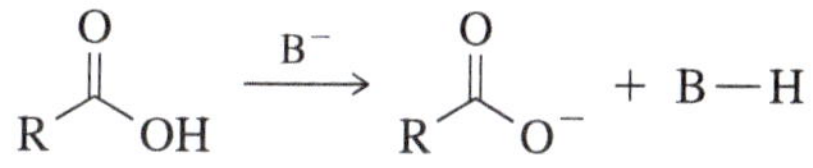

Specific:

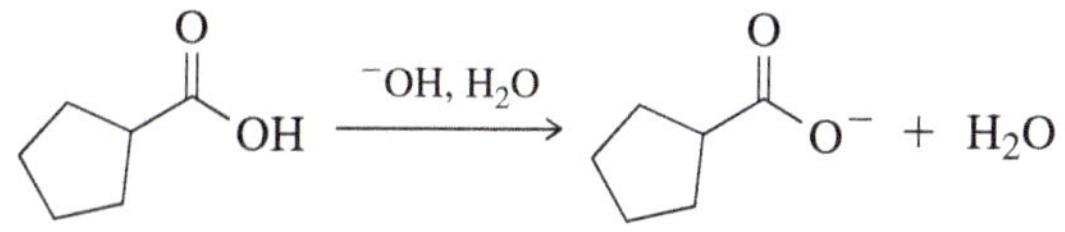

Mechanism:

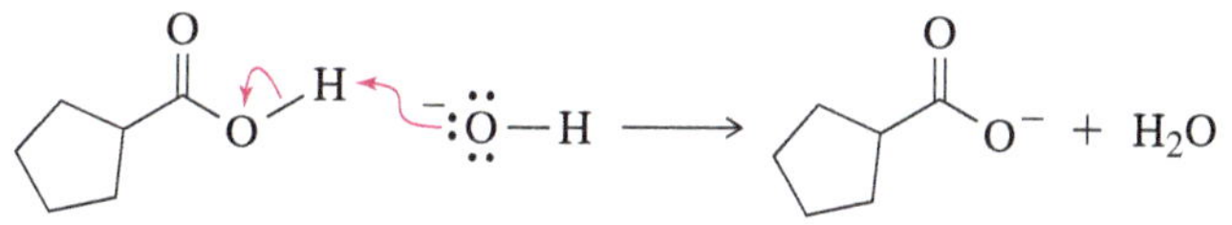

7. **Fischer esterification, 18.7.2:** The Fischer esterification is the acid-catalyzed conversion of a carboxylic acid to an ester. The reaction proceeds by General Mechanism 18B, where protonation of the carboxylic acid precedes attack of the alcohol. Following two proton transfer reactions, water leaves to give a resonance-stabilized carbocation. A final deprotonation step produces the neutral ester.

General:

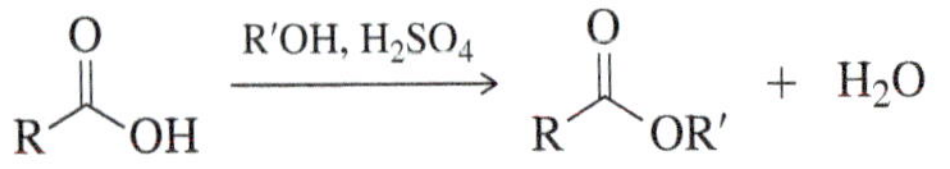

Specific:

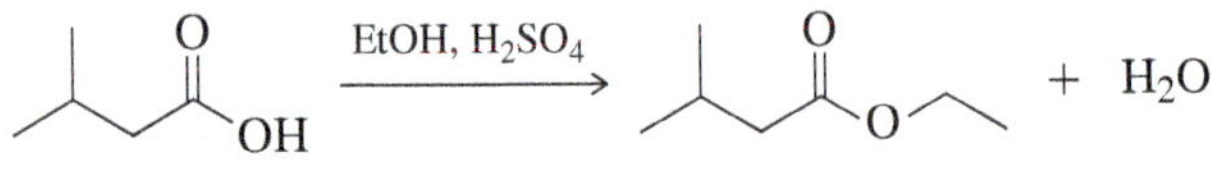

Mechanism:

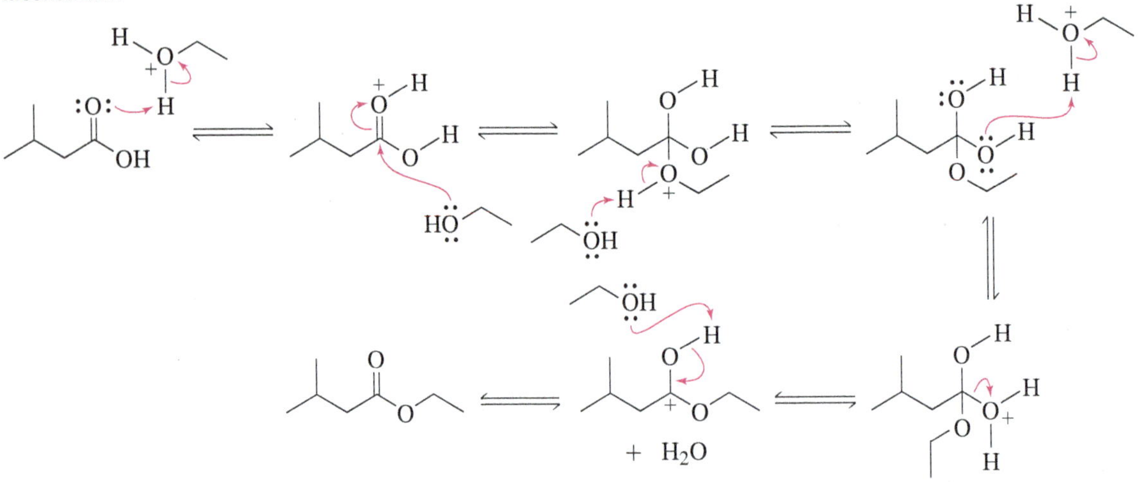

8. **$LiAlH_4$ reduction of carboxylic acids, 18.7.3:** The reduction of acids with $LiAlH_4$ begins by deprotonation of the acid with a hydride ion to form hydrogen gas. The resulting carboxylate reacts with AlH_3 to give a tetrahedral intermediate that collapses to give an aldehyde. The aldehyde undergoes further reduction by General Mechanism 17A, which, after protonation, results in a 1° alcohol.

General:

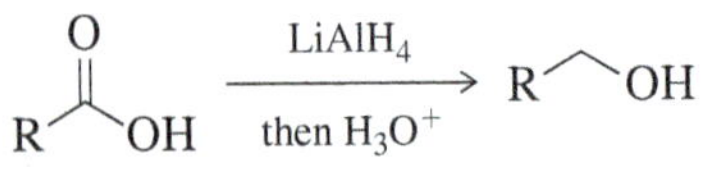

Specific:

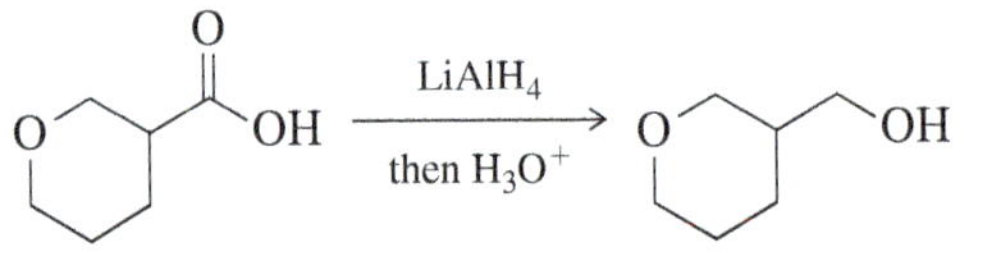

Mechanism:

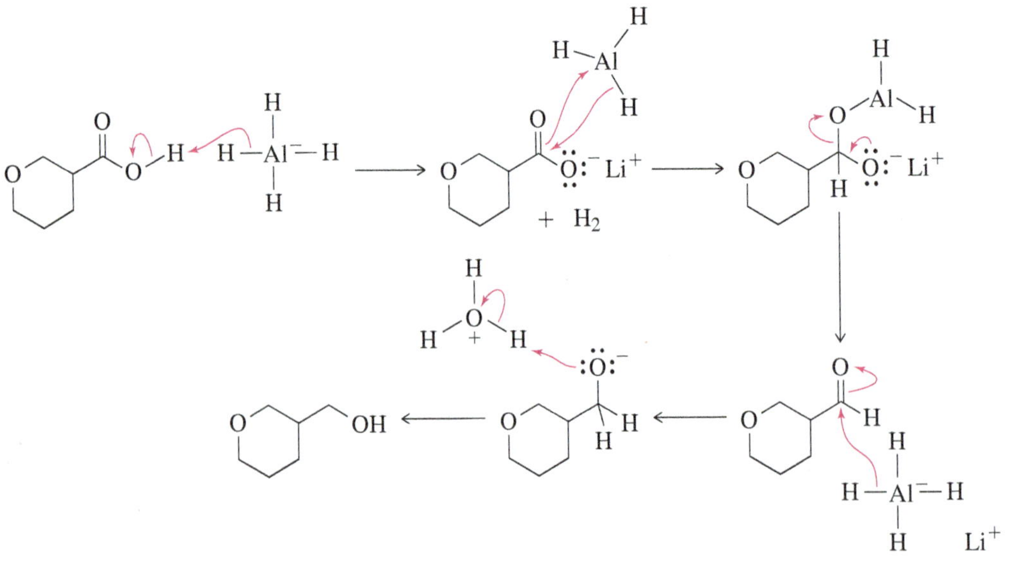

9. **Organolithium addition to carboxylic acids, 18.7.4:** The first equivalent of an organolithium reagent, as a strong base, deprotonates the carboxylic acid to form a carboxylate ion. The carboxylate ion, despite being a weak electrophile, is attacked by the strongly nucleophilic organolithium to form a dianion. Protonation of the dianion results in the elimination of water to form a ketone.

General:

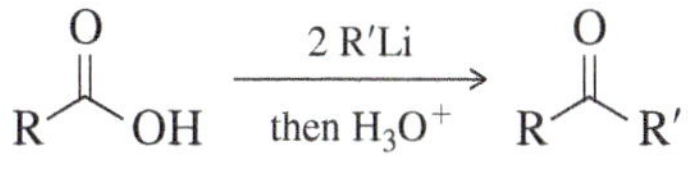

Specific:

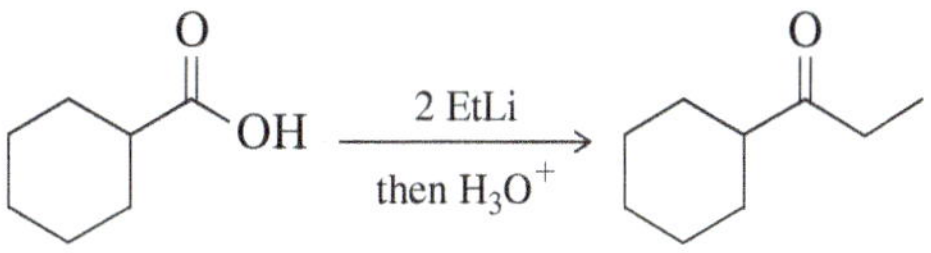

Mechanism:

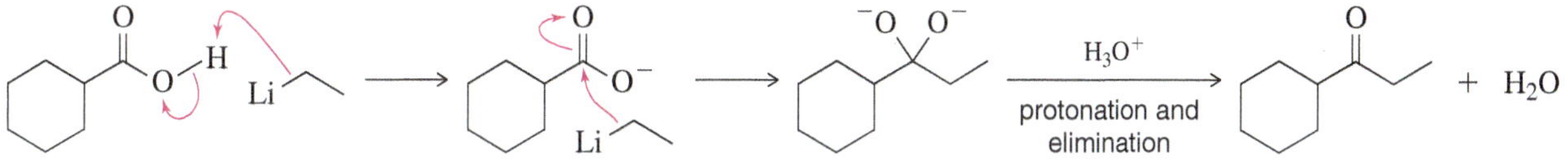

10. **Thionyl chloride with carboxylic acids, 18.7.5:** The strongly electrophilic $SOCl_2$ is attacked by the carbonyl oxygen of the carboxylic acid to produce a chloride ion. The chloride ion deprotonates the acidic proton to complete the formation of an activated species. The chloride ion attacks the central carbonyl (General Mechanism 18A) to give a tetrahedral intermediate. The tetrahedral intermediate collapses, reforming the C—O π bond (the acid chloride) and kicking out a good leaving group.

General:

Specific:

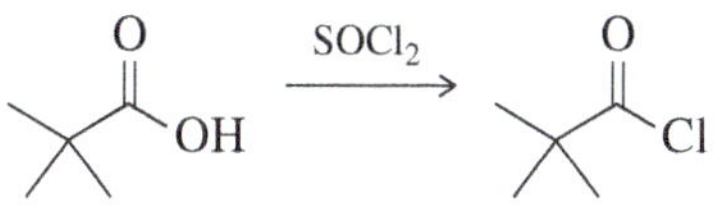

Mechanism:

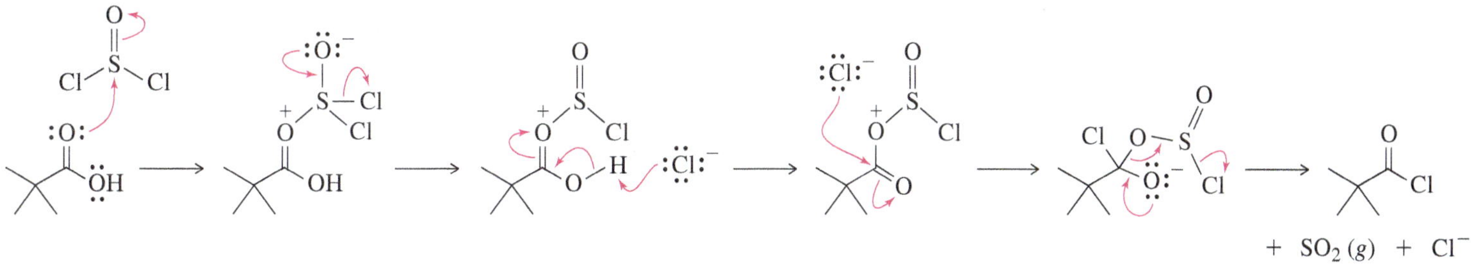

11. **Ester formation with diazomethane, 18.8.1:** Diazomethane deprotonates a carboxylic acid to produce a nucleophilic carboxylate. The nucleophilic carboxylate attacks the methyl diazonium in an S_N2 reaction to produce a new C—O σ bond and N_2 (an excellent leaving group).

General:

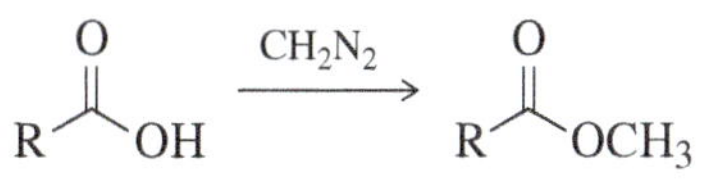

Specific:

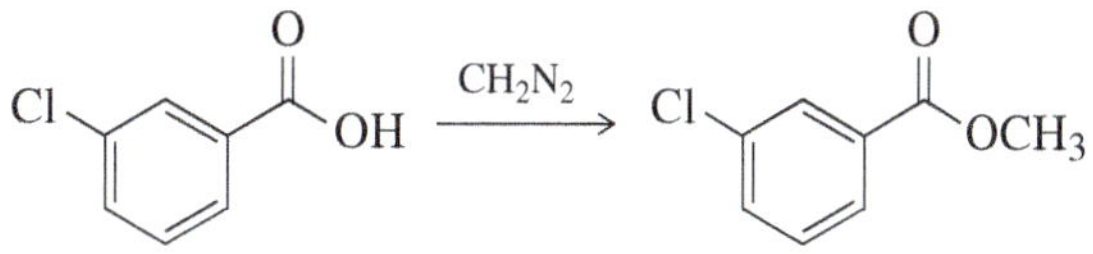

Mechanism:

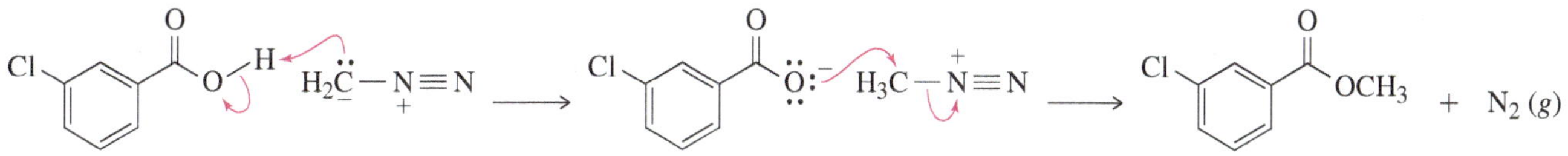

12. Acid-catalyzed addition of carboxylic acids to alkenes, 18.8.2: Protonation of an alkene with a strong acid results in the formation of a carbocation. A carboxylic acid attacks the carbocation, bringing along its proton. The acidic proton is removed in a final step to regenerate the catalyst and produce the ester.

General:

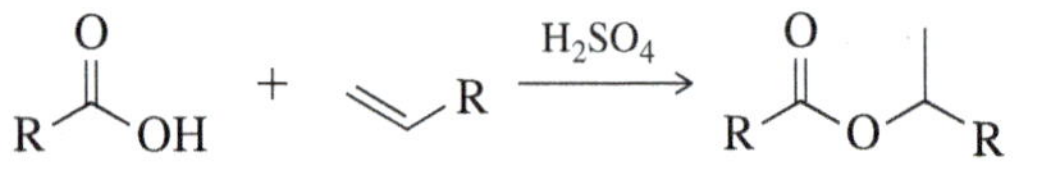

Specific:

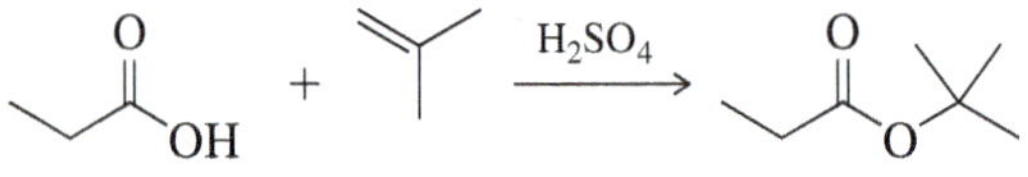

Mechanism:

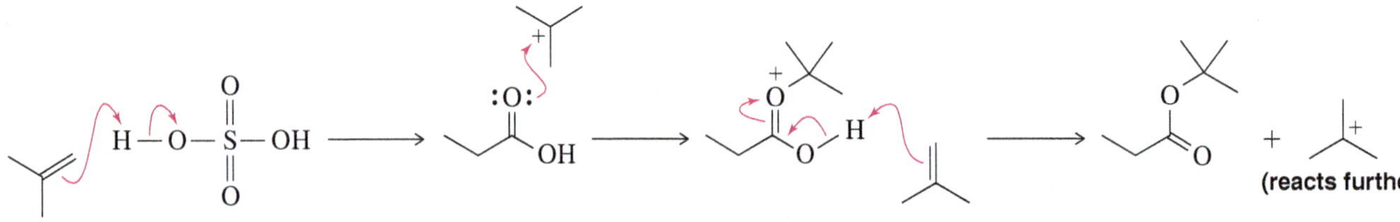

13. Decarboxylation of β-keto acids, 18.8.3 Proceeding through a 6-electron cyclic transition state, β-keto acids undergo thermal decarboxylation to give an enol and carbon dioxide. The resulting enol tautomerizes to the thermodynamically more favorable ketone.

General:

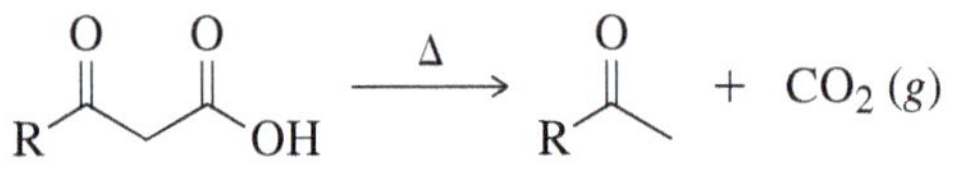

Specific:

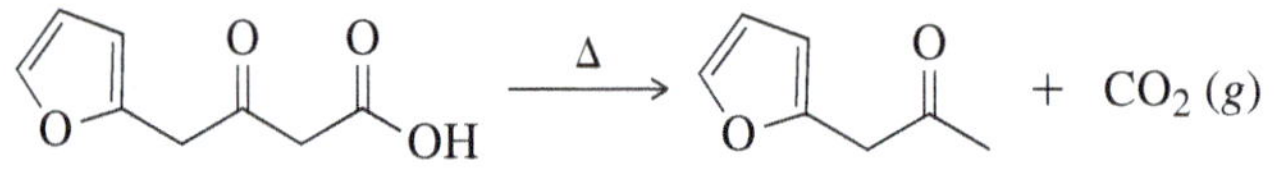

Mechanism:

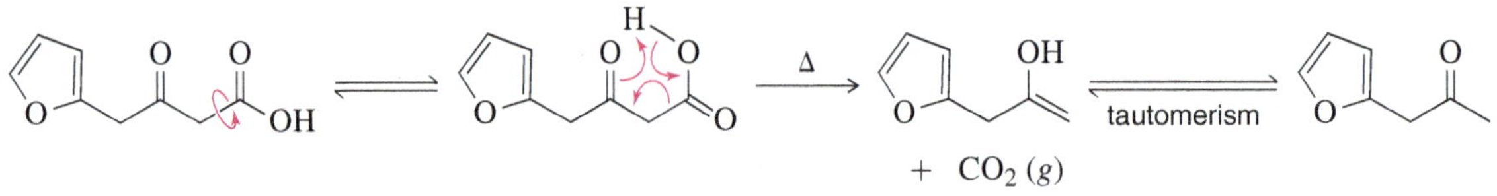

End-of-Chapter Assessments

[*Assessments are ranked according to their difficulty;* •••• = *most difficult.*]

18.48 (•) Name the following carboxylic acids.

(a)

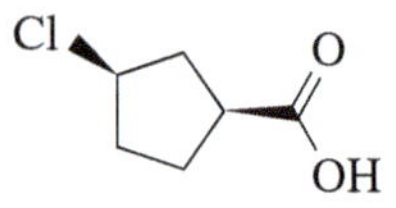

(b)

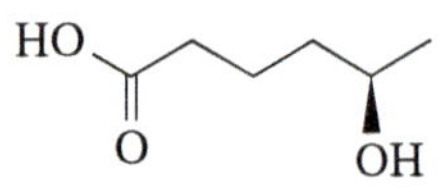

(c)

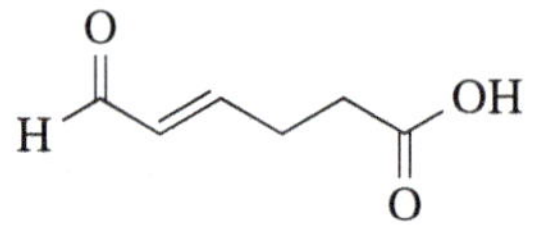

(d)

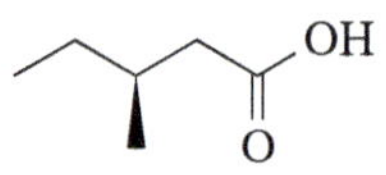

18.49 (••) The following carboxylic acids were named incorrectly. Provide the correct name.

(a) 2-heptanecarboxylic acid
(b) 6-bromocyclohexane carboxylic acid
(c) 1-carboxylic acid,5-hexanone

18.50 (••) Predict the product(s) that would result when the carboxylic acids shown are allowed to react under the following conditions: (i) EtOH, H_2SO_4; (ii) 1. $LiAlH_4$, THF 2. H_3O^+ quench; (iii) 1. EtLi (2 equivalents), pentane 2. H_3O^+ quench; (iv) $SOCl_2$; (v) CH_2N_2; (vi) 1-methylcyclohexene, H_2SO_4.

(a)

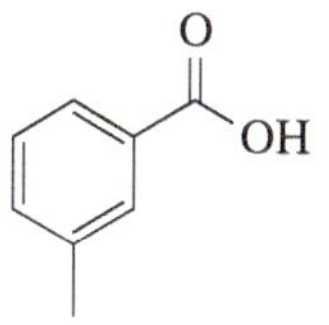

(b)

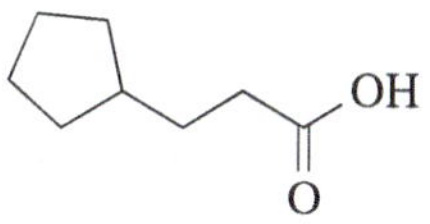

(c)

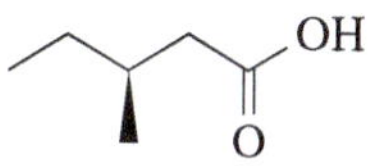

(d)

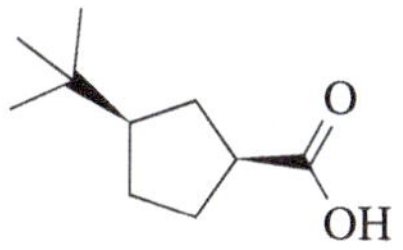

18.51 (••) Fill in the blanks with the reagents necessary to complete the syntheses shown.

(a)

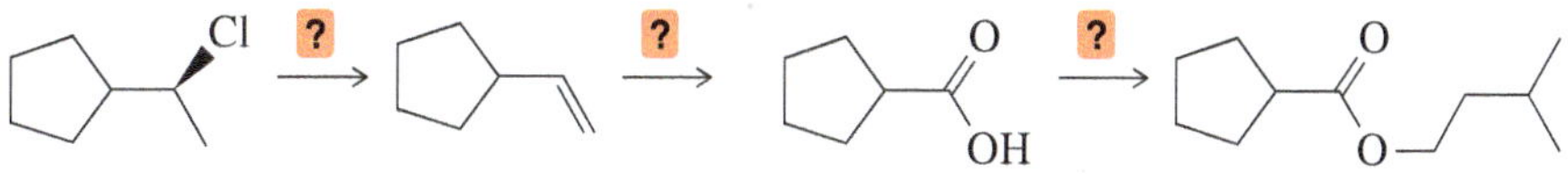

(b)

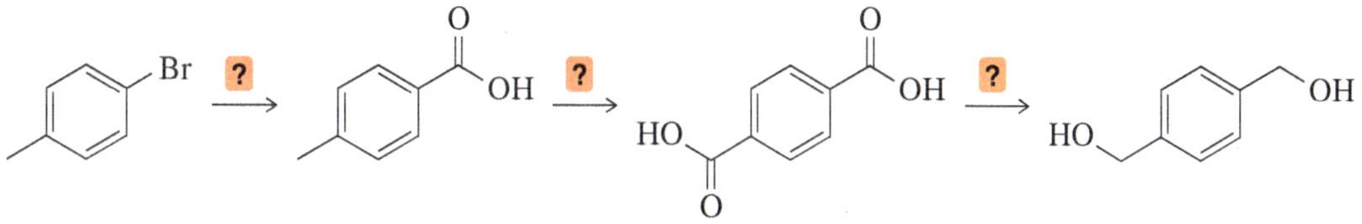

18.52 (•••) When ^{18}O-labeled ethanol is used in the Fischer esterification reaction, which oxygen in the product—a or b—ends up with the label?

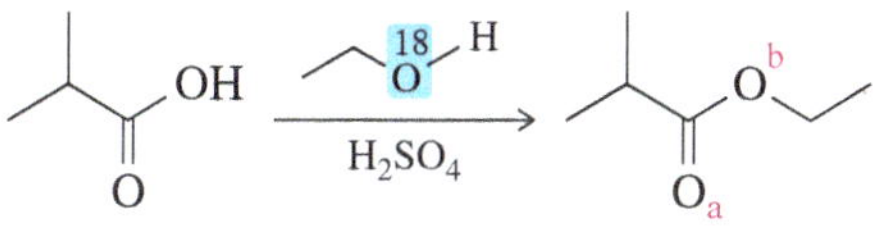

18.53 (••••) Dissolving a carboxylic acid in ^{18}O-labeled water over time results in complete incorporation of the ^{18}O label into the carboxylic acid. In acid, this process happens more quickly. Rationalize these observations.

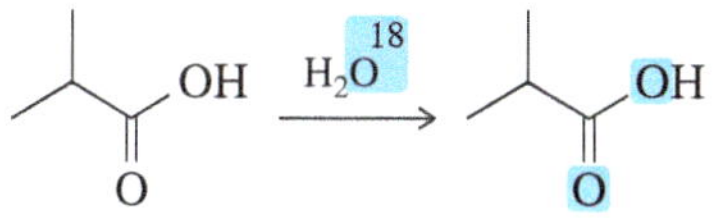

18.54 (•••) **LOOKING BACK** In Chapter 13, toluene sulfonyl chloride was used to convert alcohols into better leaving groups. Suggest a reagent that might be used to convert a sulfonic acid to a sulfonic acid chloride.

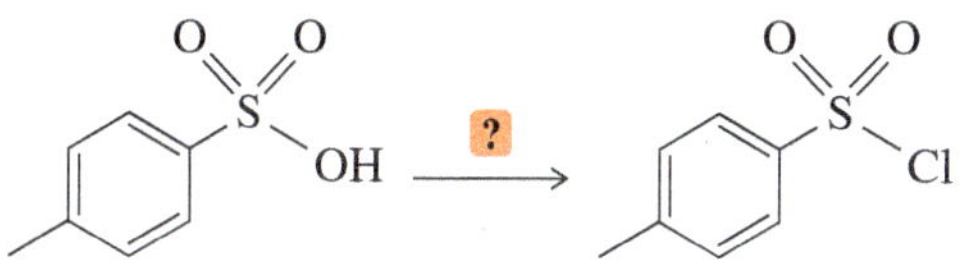

18.55 (•••) If you used $LiAlD_4$ to reduce a carboxylic acid, how many deuterium atoms would be incorporated into the resulting alcohol? What does this tell you about the mechanism?

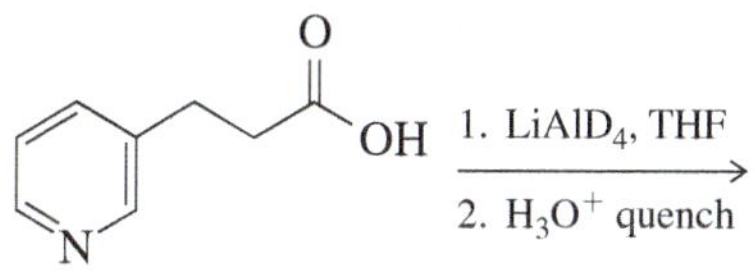

18.56 (•••) Only one of the carboxylic acid–forming reactions also allowed for the formation of a new C—C bond. Which is it?

18.57 (•••) Of the three reactions we learned for forming an ester, only one can be used to synthesize each of the molecules below. For each, which is it? Justify your answers.

(a)

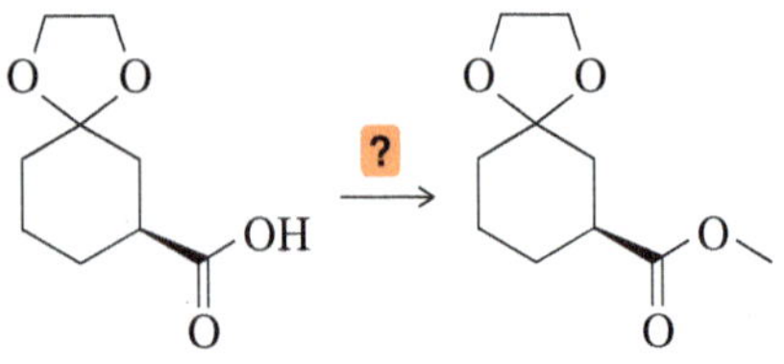

(b)

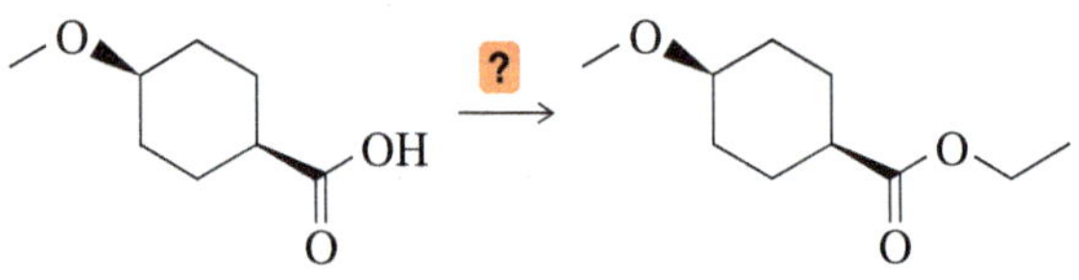

(c)

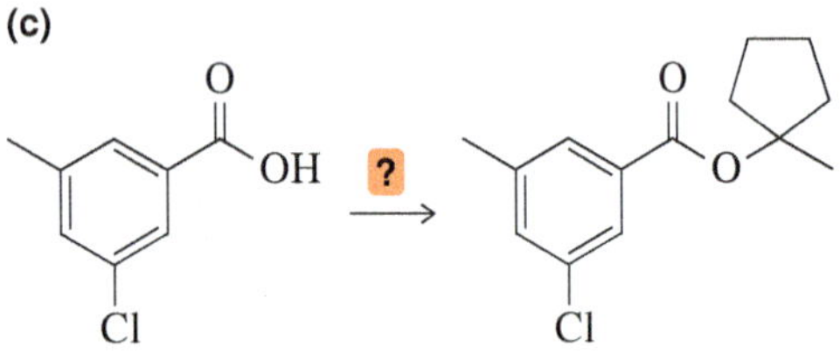

18.58 (••) Predict the product formed when each of the following are subjected to the following reduction conditions: (i) $LiAlH_4$, then H_3O^+; (ii) $NaBH_4$, EtOH; (iii) H_2, Pd/C (1 atm); (iv) H_2, Pd/C (5 atm); (v) H_2, Lindlar catalyst.

(a)

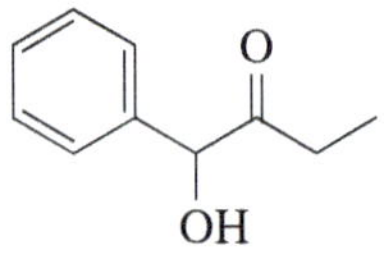

(b)

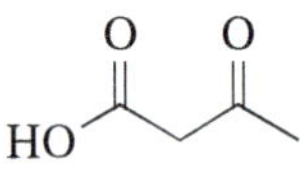

(c)

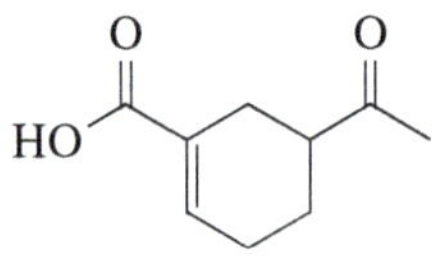

(d)

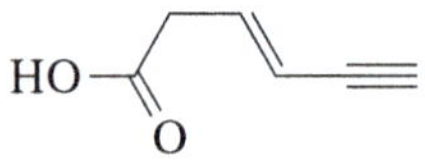

18.59 (••••) When a chemist attempted the following reaction sequence, the desired product was not formed. (a) Why? (b) Suggest a solution to the problem. [*Think about chemistry from the end of Chapter 13.*]

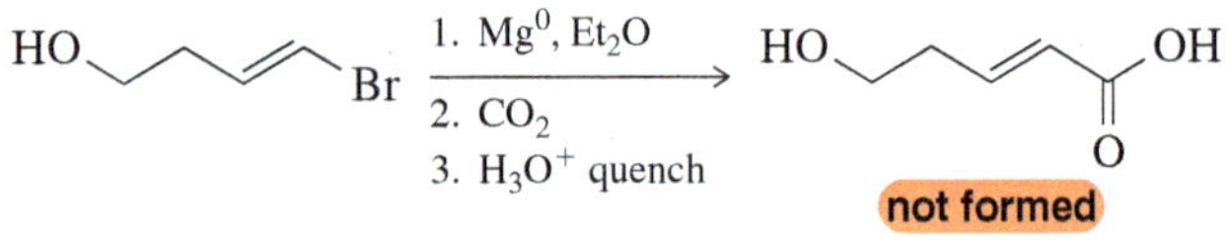

18.60 (••••) PET, a polymer used in clothing and food storage containers, has the oligomer structure below. Suggest two molecules and reaction conditions that could be used for its synthesis.

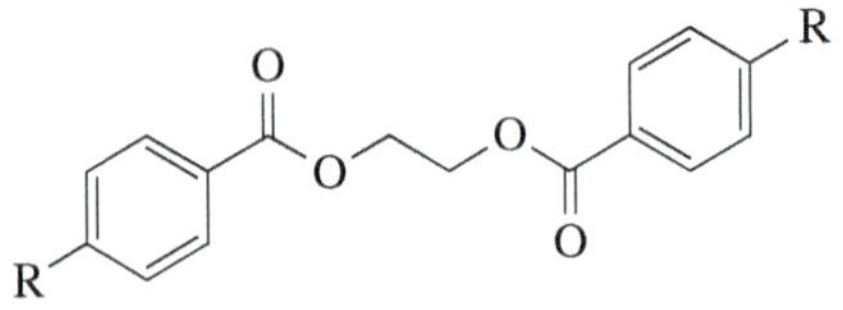

Assessments 18.61–18.63 should be worked sequentially as part of learning a way to convert carboxylic acids to amides, a functional group we will study in detail in Chapter 19.

18.61 (•) Identify the products of an acid–base reaction between benzoic acid and dicyclohexylcarbodiimide (DCC). [*What is the most basic atom in DCC?*]

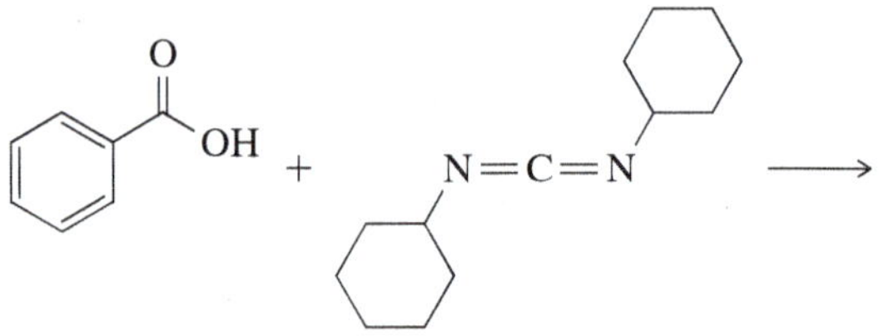

DCC

18.62 (••) Suggest an arrow-pushing mechanism that rationalizes the formation of the intermediate shown from the products obtained in Assessment 18.61.

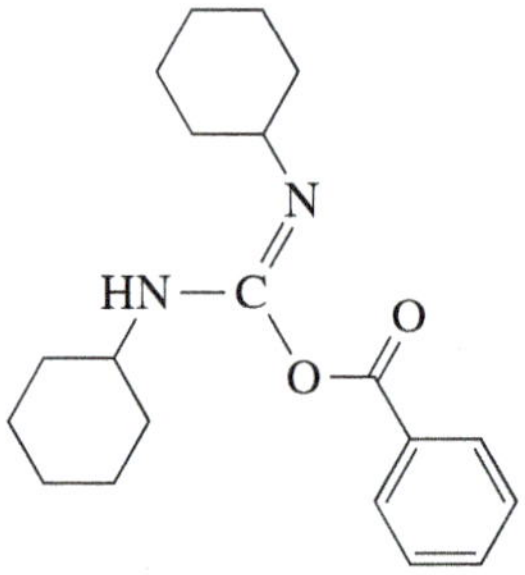

18.63 (••••) The intermediate shown in Assessment 18.62 is considered an activated carboxylic acid, much like the acid chlorides we observed briefly in Chapter 18 and will study in detail in Chapter 19. Reaction of this intermediate with a nucleophilic amine results in the formation of the amide shown. Suggest a mechanism of this reaction. [*This reaction is a nucleophilic acyl substitution reaction where the amine is a strong nucleophile.*]

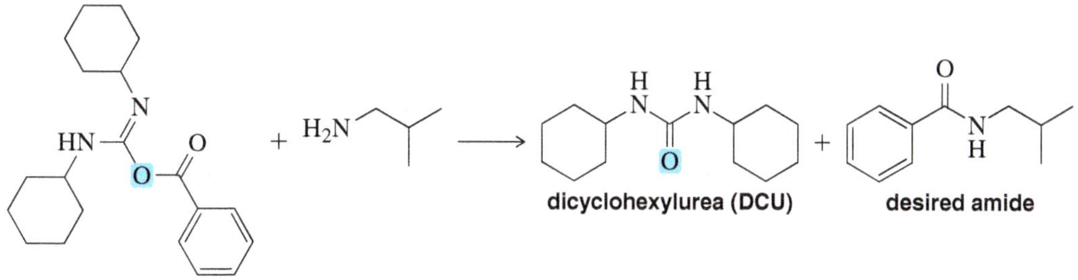

18.64 (••) **GREEN CHEMISTRY** Calculate the overall atom economy of the reaction that is featured in Assessments 18.61 to 18.63. [*It is shown in total below.*] On the basis of this calculation, do you believe this to be a "green" reaction?

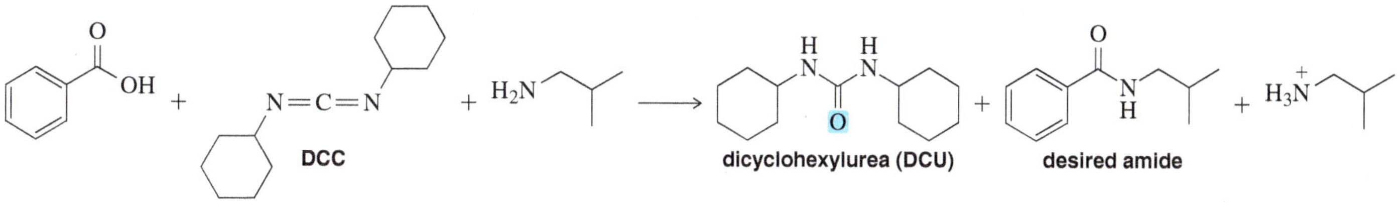

18.65 (••••) The Fischer esterification can be done intramolecularly. (a) Predict a mechanism of the reaction shown. (b) Why is the intramolecular Fischer esterification more favorable than the intermolecular version?

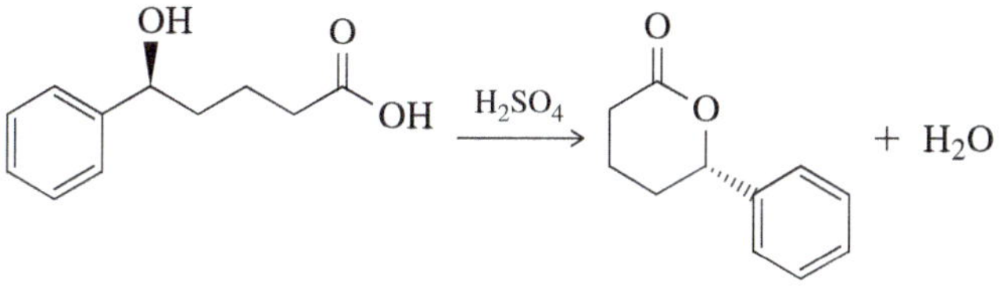

18.66 (••••) An erstwhile graduate student attempts the following reaction, but isolates only starting material. Explain why esterification cannot occur under these conditions.

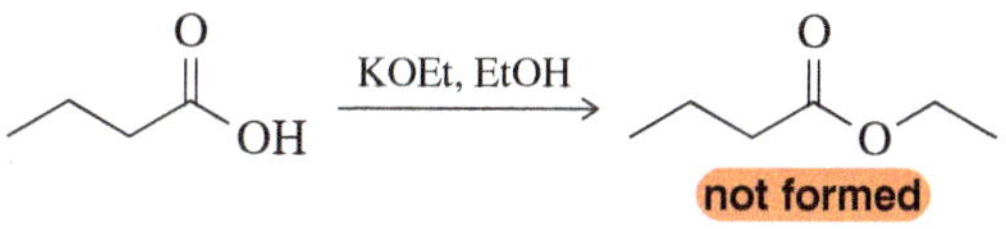

18.67 (••) Predict the identity of the β-keto acid that undergoes decarboxylation to give the following products.

(a)

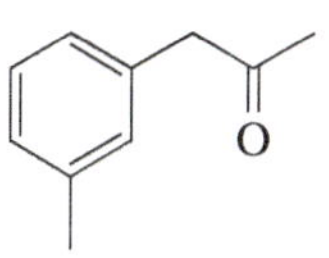

(b)

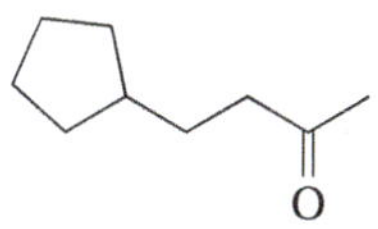

(c)

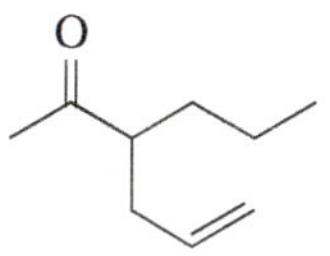

18.68 (••••) **LOOKING AHEAD** In Chapter 19, we will learn about the hydrolysis of *t*-butyl esters. In the reaction below, the hydrolysis is coupled to the decarboxylation reaction learned in this chapter. Suggest a mechanism for this reaction. [*Hint: The formation of t-butanol proceeds by an* S_N1 *reaction.*]

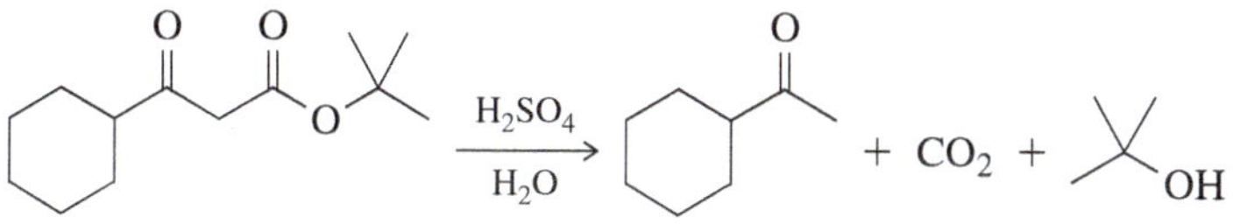

18.69 (••••) **LOOKING AHEAD** While acidic conditions were used in Assessment 18.68, decarboxylation of esters can also be conducted under basic conditions to give, at least temporarily, the enolate product shown. [*We'll learn more about the chemistry of enolates in Chapter 20.*] Suggest a mechanism of this reaction. [*Hint: The formation of chloromethane proceeds by an* S_N2 *reaction.*]

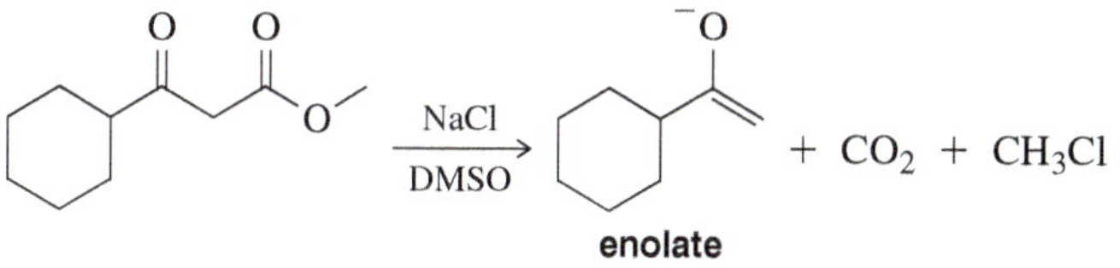

18.70 (••••) FROM THE LITERATURE Suggest the reagents needed for this transformation that was used in the total synthesis of halofuginone, a veterinary medicine used in farm animals (*Org. Process Res. Dev.* **23**, 2019, 990).

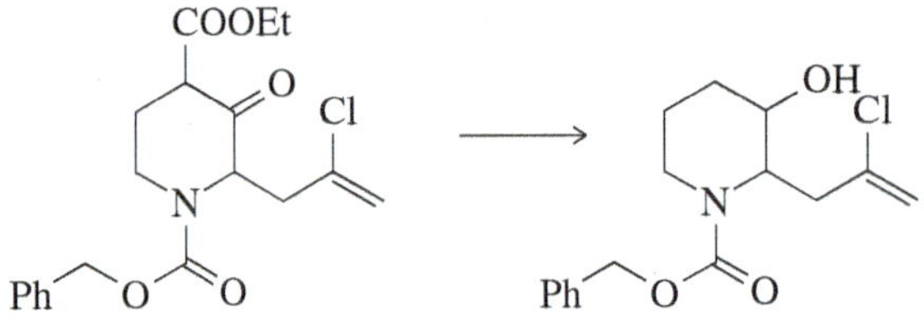

18.71 (••••) LOOKING BACK Decarboxylation can also be used to make bromoalkanes by the heat-induced cleavage of acyl hypobromites. Suggest a mechanism of both steps of the reaction shown. [*The decarboxylation proceeds by a radical mechanism. Make sure your mechanism includes initiation and propagation steps.*]

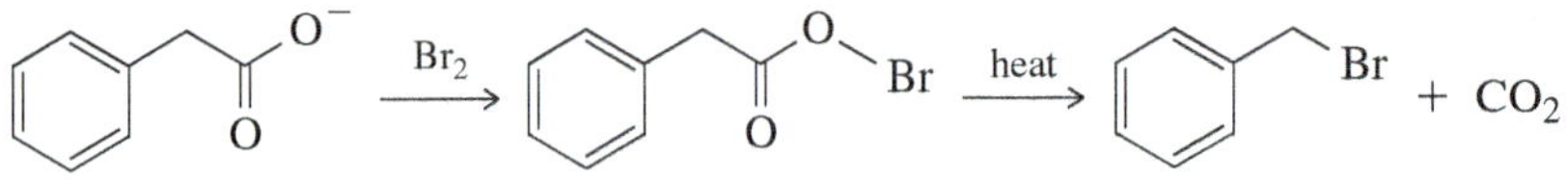

acyl hypobromite

18.72 (••••) (a) Predict the product of the intramolecular Fischer esterification shown below. (b) Suggest a mechanism for the reaction.

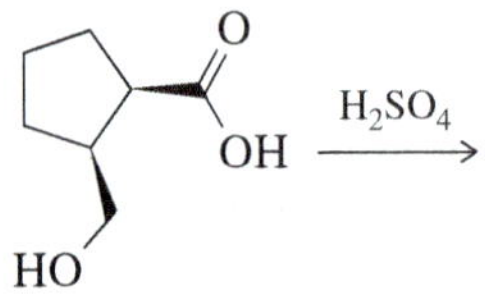

18.73 (••) LOOKING BACK Now that we have learned the mechanism of nucleophilic acyl substitution (General Mechanisms 18A and 18B), the mechanism of the below reactions from Chapter 13 can be drawn in a better, more correct way. Draw the mechanism for the formation of the chloroalkane and tosylate shown.

(a)

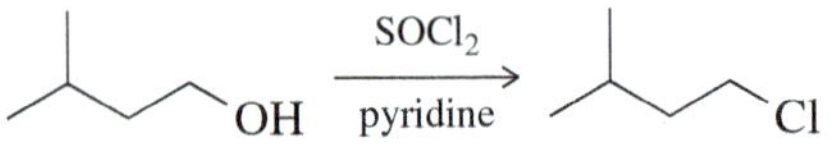

(b)

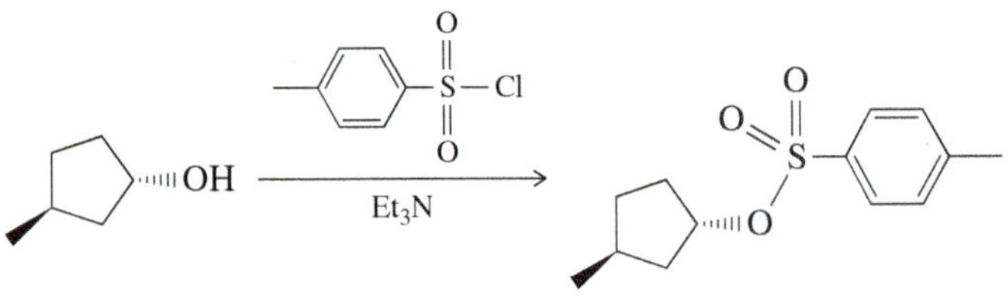

18.74 (•••) FROM THE LITERATURE (a) Suggest reagents for the following two-step transformation that is used for the synthesis of dihydropyridomycins, potential treatments for tuberculosis. (b) How do you think the chiral center at C_2 was synthesized as a single enantiomer (*ACS Med. Chem. Lett.* **4**, 2013, 264)?

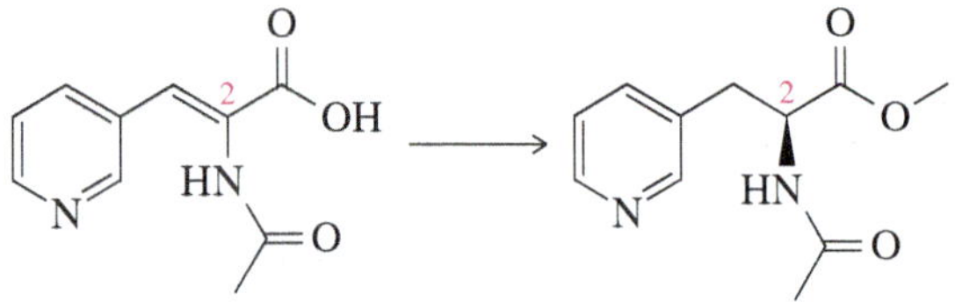

18.75 (••••) FROM THE LITERATURE Suggest an arrow-pushing mechanism of the following reaction used in the synthesis of potential drug candidates to treat chikungunya and dengue fevers (*Org. Lett.* **19**, 2017, 1156).

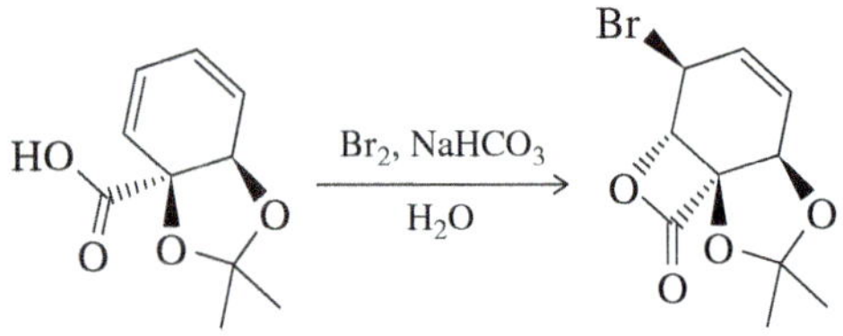

19

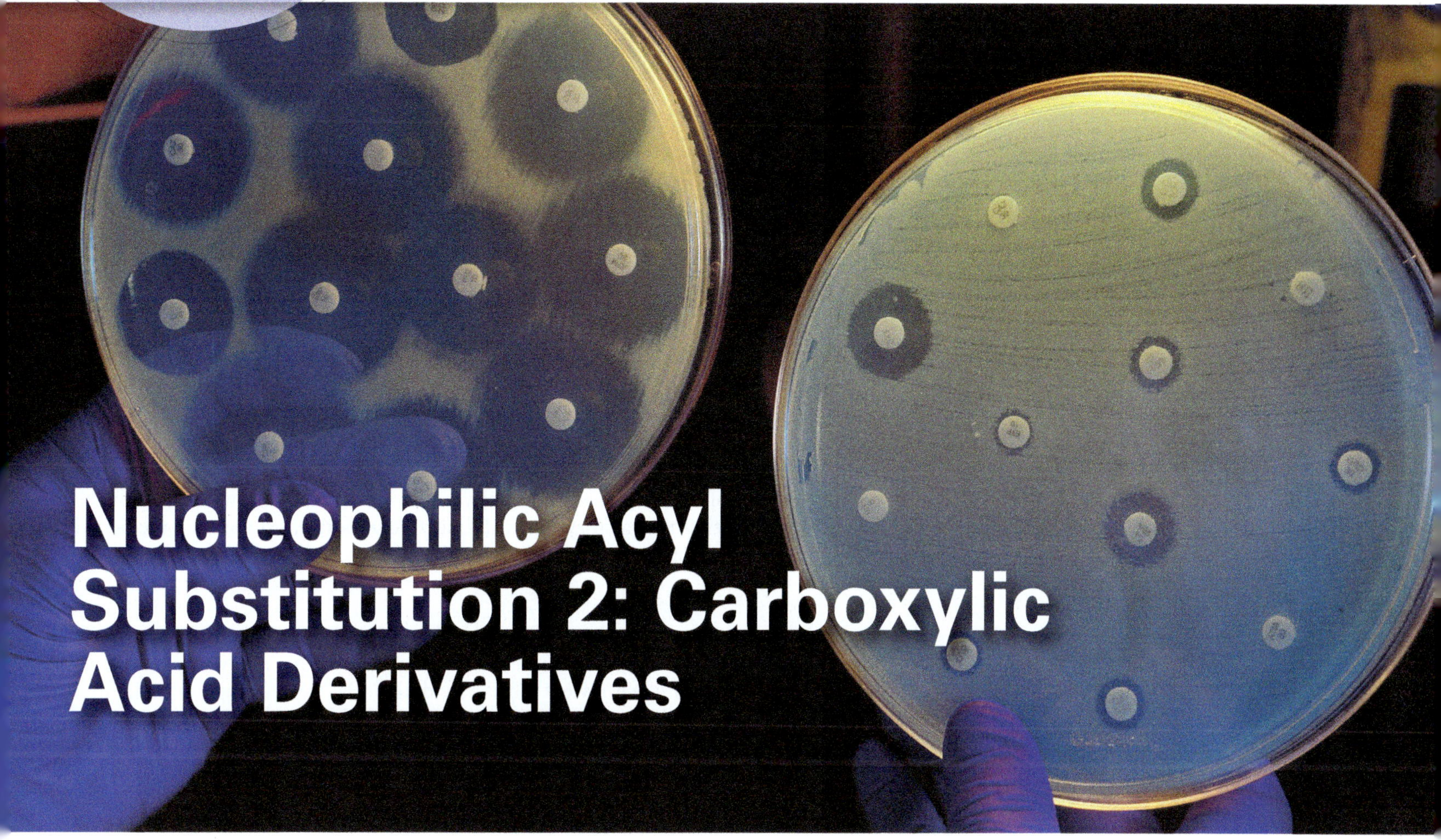

Nucleophilic Acyl Substitution 2: Carboxylic Acid Derivatives

Introduction

Humans and bacteria have an incredibly close relationship. Recent research has shown that there are more bacterial cells than there are human cells in our bodies, an astonishing situation made possible by the fact that bacterial cells are much, much smaller than human cells. Bacteria's prevalence is not a bad thing, given that bacteria form our first line of defense against infection-causing pathogens, aid digestion of carbohydrates and fatty acids, and even modulate our immune response against foreign invaders. Some bacteria, however, are pathogenic, and they cause diseases such as pneumonia, meningitis, and tuberculosis.

Though the term "antibiotic" was not officially coined until 1942, during ancient times moldy bread (which hosts microscopic fungi with antibacterial properties) was used to treat wounds and infections. In 1640, the use of mold was prescribed for treatment in a pharmacology text. Today, antibiotics are classified either as bacteriostatic or bacteriocidal depending on their mode of action. Bacteriostatic antibiotics halt the growth of bacteria, allowing the human immune system to "catch up" and defeat them. For example, sulfa drugs (discussed in Chapter 23) act by inhibiting bacterial synthesis of folate, a nutrient necessary for bacterial DNA synthesis and reproduction, and thus are bacteriostatic.

Bacteriocidal antibiotics, discussed here in Chapter 19, directly kill bacteria. The β-lactam antibiotics, first discovered accidentally by Alexander Fleming, are bacteriocidal and include the penicillins, cephalosporins, and cephamycins. They are named for the four-membered amide-containing ring referred to as a β-lactam, which is integral to their structures (**Figure 19.1**). These antibiotics act by inactivating peptidoglycan transpeptidase, an enzyme responsible for forming the bacterial cell wall.

Learning Objectives

LO 19.0 Compare the structures of acid chlorides, anhydrides, esters, amides, and nitriles

LO 19.1 Explain the physical properties and spectral characteristics of carboxylic acid derivatives

LO 19.2 Analyze the reactions of carboxylic acid derivatives

LO 19.3 Analyze functional group conversion reactions

LO 19.4 Analyze reduction reactions of carboxylic acid derivatives

LO 19.5 Predict the products of carboxylic acid derivative reactions with organometallics

LO 19.6 Analyze nitriles as carboxylic acid derivatives and their reactions

Figure 19.1 **β-Lactam antibiotics.**

Penicillins:

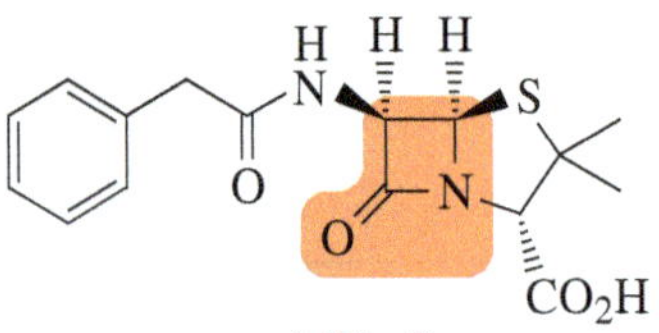

penicillin G

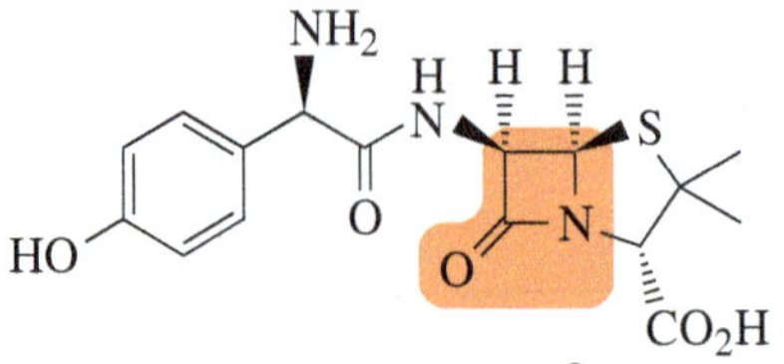

amoxicillin (Amoxil®)

Cephalosporins/cephamycins:

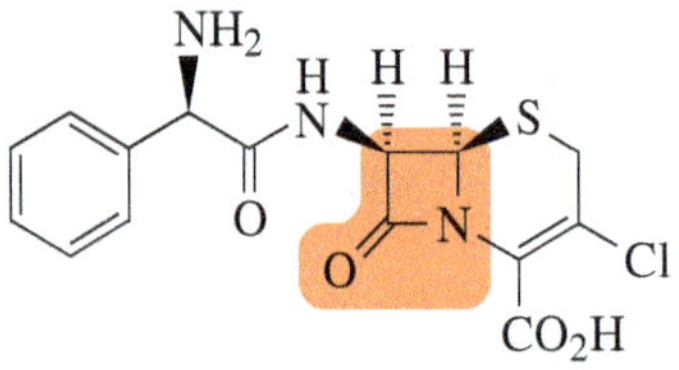

cefaclor (Ceclor®)

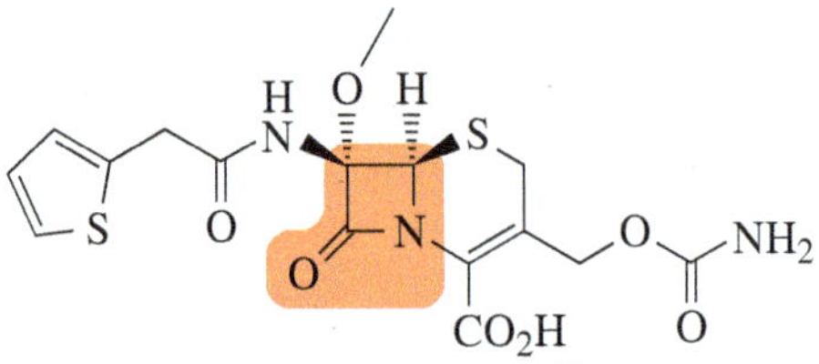

cefoxitin (Mefoxin®)

"The history of [bacterial] liberty is a history of [antibiotic] resistance."

—Woodrow Wilson (modified)

Through our study of carboxylic acid derivatives and their reactivity, we will learn the story of the β-lactam antibiotics, a story that will likely be important to you as a medicinal chemist, a doctor, or simply an informed citizen.

○ ○ ○ ○ ○ ○ ○

In Chapter 19 (part three in our four-part study of carbonyls), we study the nucleophilic acyl substitution process introduced in Chapter 18. Although featuring many new reactions, much of this chapter will feel like a review. This is by design. By "re-learning" material in a new context, you are more likely to recall this information when it is necessary to do so. Speaking of which, the information in Chapter 19 will appear again in a variety of situations, including a biochemistry course and later on standardized exams like the MCAT. As a result, you need to learn this material well, not only in the context of organic chemistry, but also in the context of what it will mean going forward. [*The story of β-lactam antibiotics should help with this.*]

We can't learn new material unless we have a solid foundation, so let's start by reviewing what we learned previously.

Test your understanding

Assessment 19.1 Rank the following anions based on their stability as potential leaving groups. Explain your reasoning (1 = most stable; 5 = least stable).

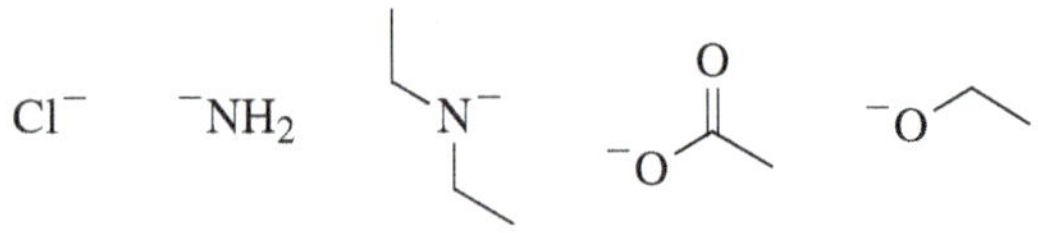

Assessment 19.2 Would you expect a ketone or an ester to be more reactive with a strong nucleophile? Justify your answer.

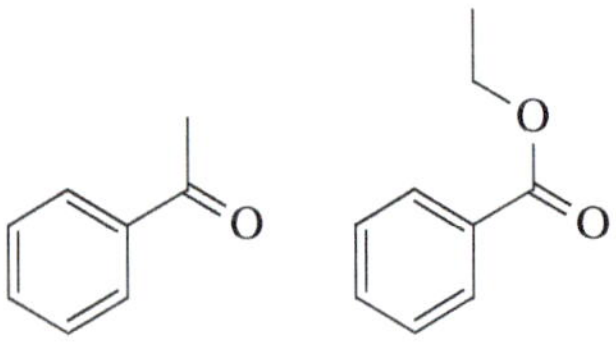

Assessment 19.3 Identify the bonds formed and bonds broken in the following transesterification reaction. [*It has been color-coded to assist you.*] Which would you expect to be the favored side of equilibrium, based on this analysis?

Assessment 19.4 The following reaction is reversible. How would you drive it to the right? To the left?

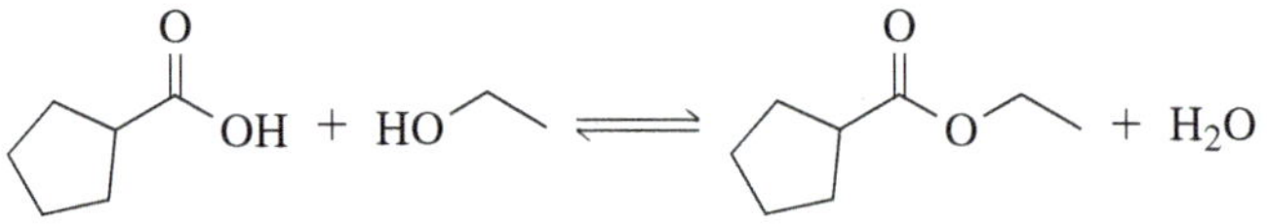

Assessment 19.5 In Chapter 17, reactions were characterized as occurring by General Mechanisms 17A and 17B. In Chapter 18, reactions were characterized as occurring by General Mechanisms 18A and 18B. (a) What is the difference between General Mechanisms 17A and 17B? (b) Between General Mechanisms 18A and 18B? (c) What do General Mechanisms 17A and 18A have in common? (d) What do General Mechanisms 17B and 18B?

Assessment 19.6 Which basic atom, a or b, would you expect to protonate most favorably? Justify your answer.

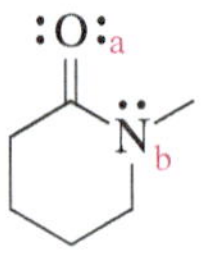

19.1 General structure of carboxylic acid derivatives

We study the five carboxylic acid derivatives shown in Figure 19.2 in this chapter.

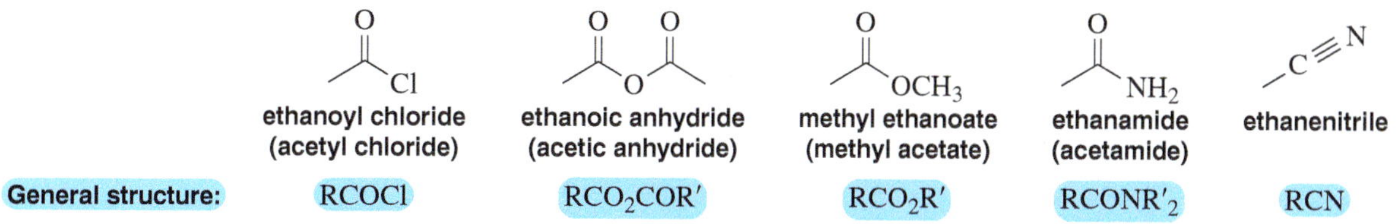

Figure 19.2 Carboxylic acid derivatives.

Carboxylic acid derivatives are, as the name implies, closely related to the carboxylic acids of Chapter 18. These molecules have the same oxidation state as carboxylic acids and can be thought of as the products of various carboxylic acid condensation reactions. Figure 19.3 shows several hypothetical condensation reactions between carboxylic acids and different reagents, which are identified on top of the respective arrows. For example, the **acid chloride** is the hypothetical condensation product of a carboxylic acid and HCl. **Nitriles** are essentially dehydrated amides. All but the nitrile feature a carbonyl group attached to an alkyl group and an electronegative substituent. [*Figure 19.3 doesn't show actual reactions, just a way of categorizing them that is elaborated upon in this chapter.*]

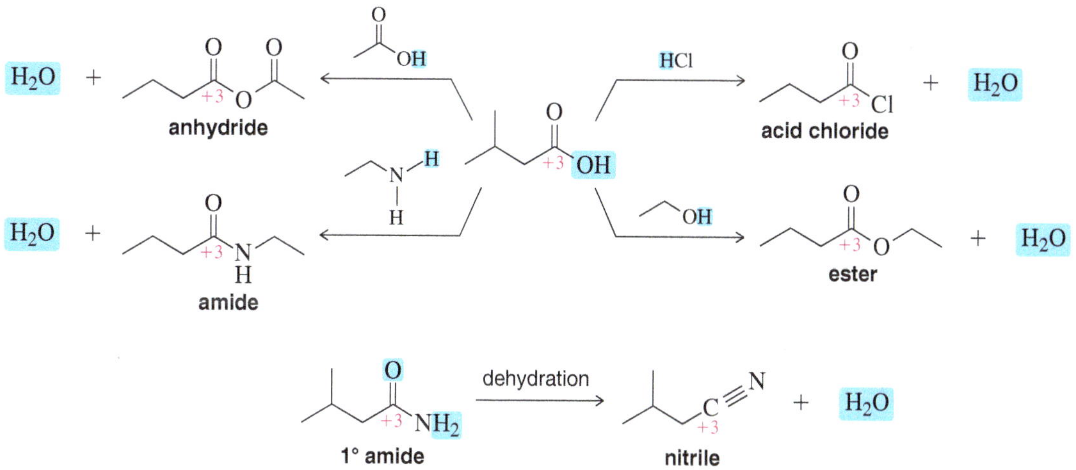

Figure 19.3 Carboxylic acid derivatives as the hypothetical products of condensation reactions. Like carboxylic acids, all have an oxidation state of +3.

A cyclic ester is referred to as a **lactone**, while a cyclic amide is a **lactam** (Figure 19.4).

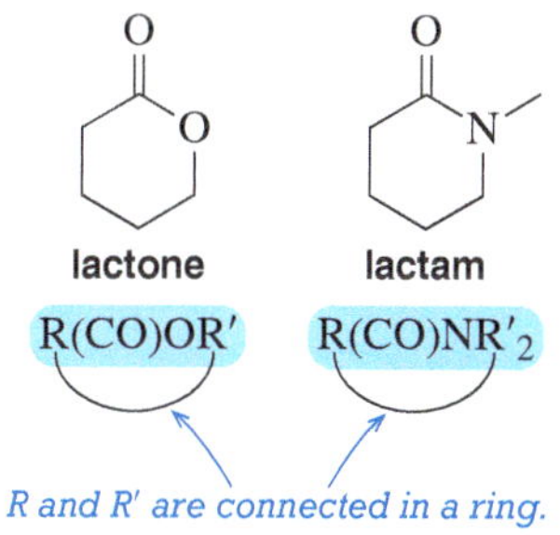

Figure 19.4 Cyclic esters and amides are called lactones and lactams, respectively.

19.1 SUMMARY

- Acid chlorides, anhydrides, esters, amides, and nitriles, which have the same oxidation state as carboxylic acids, are considered to be carboxylic acid derivatives.

Assessment 19.7 Which of the following are considered carboxylic acid derivatives?

(a) (b) (c) (d) (e) (f)

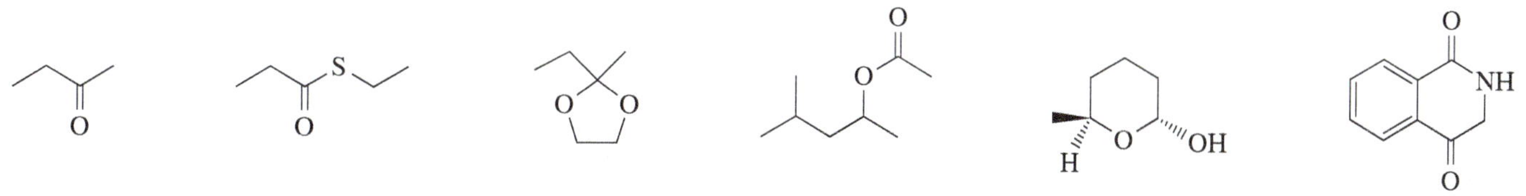

Assessment 19.8 In Chapter 13, you were introduced to the chemistry of toluene sulfonyl chloride. A sulfonyl chloride is a derivative of what functional group?

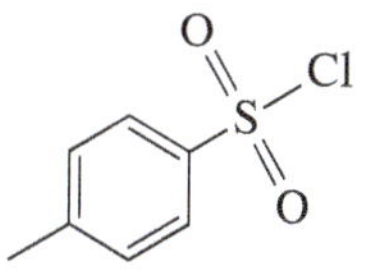

toluene sulfonyl chloride

19.2 Nomenclature of carboxylic acid derivatives

The nomenclature of carboxylic acid derivatives begins with naming the molecule as if it were a carboxylic acid and then modifying that name based on the particular carboxylic acid derivative. Table 19.1 shows how this is done.

Table 19.1 Nomenclature of carboxylic acid derivatives requires first naming as carboxylic acids and then modifying as appropriate.

Functional group	Rules	Example
Acid chloride	Name as a carboxylic acid and then replace "-oic acid" with "-oyl chloride."	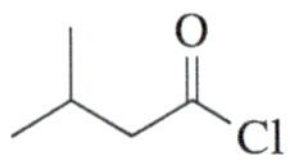 **3-methylbutanoyl chloride** 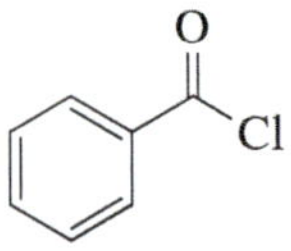 **benzoyl chloride** 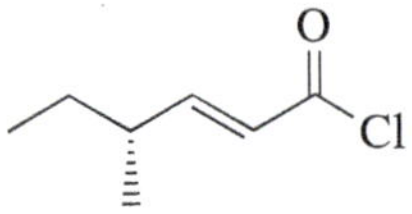 **(2*E*,4*R*)-4-methylhex-2-enoyl chloride**
Anhydride	If symmetrical, replace "-oic acid" with "-oic anhydride." When unsymmetrical, name each side of the anhydride as if it were a carboxylic acid, drop "acid" from each, "anhydride," and order the names alphabetically.	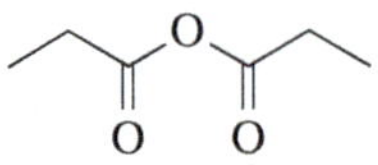 **propanoic anhydride** 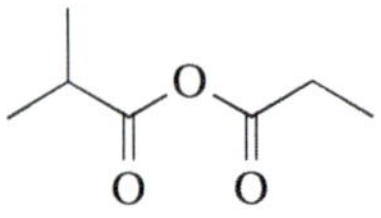 **isobutanoic propanoic anhydride** 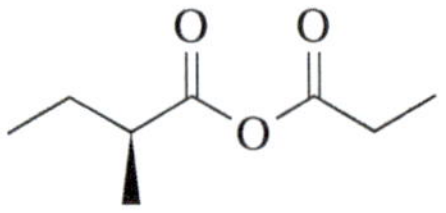 **(*S*)-2-methylbutanoic propanoic anhydride**
Ester	1. Name the group attached to the oxygen as a substituent ending in "-yl." 2. Name the carbonyl side as a carboxylic acid, replacing "-oic acid" with "-oate." 3. Combine the names so it's written as alk*yl* alkan*oate*.	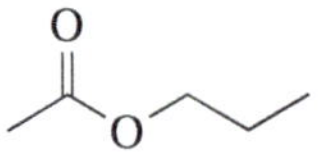 **propyl ethanoate** 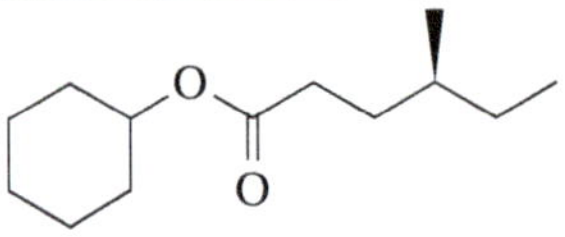 **cyclohexyl (*S*)-4-methyl hexanoate** 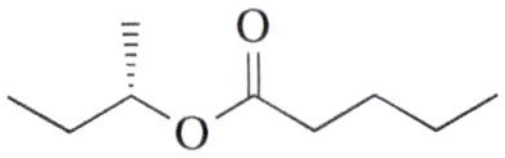 **(*S*)-1-methylpropyl pentanoate**
Amide	Name as a carboxylic acid and then replace "-oic acid" with "-amide." When there are substituents on the nitrogen, these should be named as alkyl groups and placed after an italicized "*N*-" to indicate they are attached to the nitrogen.	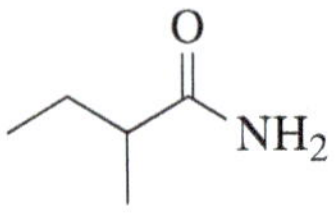 **2-methylbutanamide** 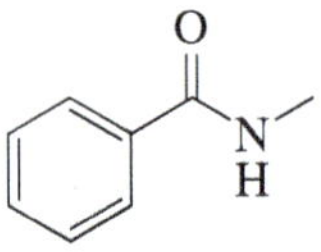 ***N*-methylbenzamide** 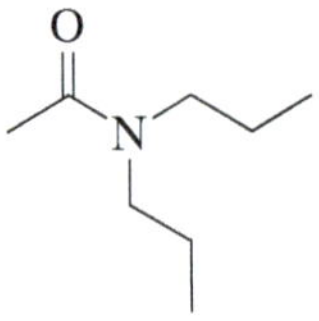 ***N*,*N*-dipropylethanamide**
Nitrile	Name as a carboxylic acid and then replace "-oic acid" with "-e nitrile."	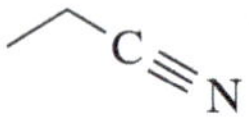 **propanenitrile** 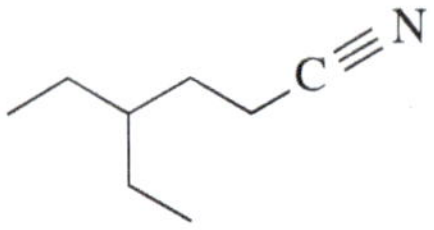 **4-ethylhexanenitrile** 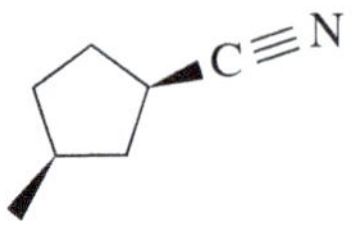 **(1*R*,3*S*)-3-methyl-cyclopentanenitrile**

19.2 SUMMARY

- The nomenclature of carboxylic acid derivatives begins by naming the molecule as a carboxylic acid and making the appropriate modification based on the particular derivative.

Assessment 19.9 Provide the IUPAC name for the following molecules.

(a)

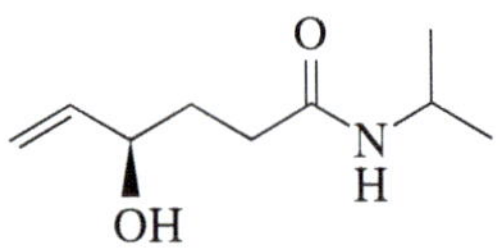

(b)

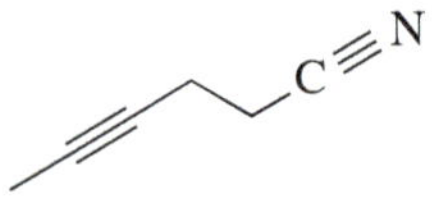

(c)

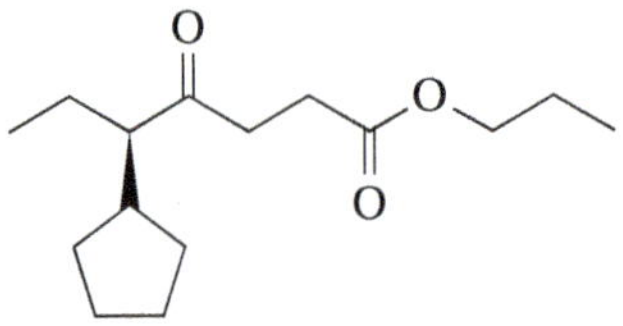

(d)

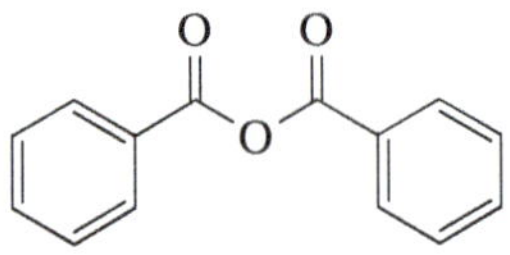

(e)

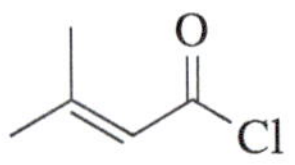

Assessment 19.10 Provide a molecular structure that corresponds to the given name.

(a) benzoic propanoic anhydride
(b) (*S*)-4-phenyloctanoyl chloride
(c) (3*R*,5*R*)-5-chloro-3-isopropylcycloheptyl-1-enenitrile
(d) cyclopropyl (3*S*,5*E*)-3,5-dimethylhept-4-enoate
(e) (*R*)-*N*,*N*-diethyl 5-cyclohexyl-3-methoxypentanamide

19.3 Molecular orbital picture of carboxylic acid derivatives

Figure 19.5 shows the molecular orbital picture for formamide, which we treat as representative of all carboxylic acid derivatives. The C═O bond of the amide features a σ and π bond between an sp^2-hybridized carbon and oxygen. The C—NH_2 bond is formed between the sp^2-hybridized central carbon and an sp^2-hybridized nitrogen. The bond angles around the central carbon are all near 120°.

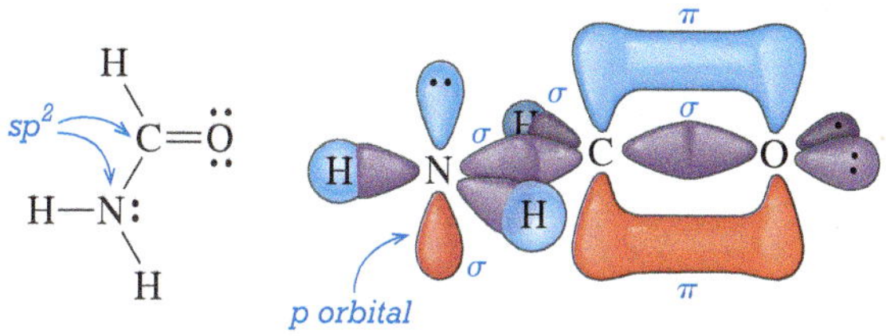

Figure 19.5 Molecular orbital picture of formamide.

Carboxylic acid derivatives can be represented by two valid resonance structures, as shown in **Figure 19.6** for the ester and amide. One of the resonance structures involves a lone pair on the atom attached to the carbonyl carbon, so this atom is sp^2 hybridized, with one lone pair in a *p* orbital. Where there is a second or third lone pair, these reside in sp^2 -hybridized orbitals. The enhanced resonance stability accounts for the difference in the hybridization expected based on valence shell electron pair repulsion (VSEPR) theory. These resonance structures also account for the reactivity we explore throughout this chapter.

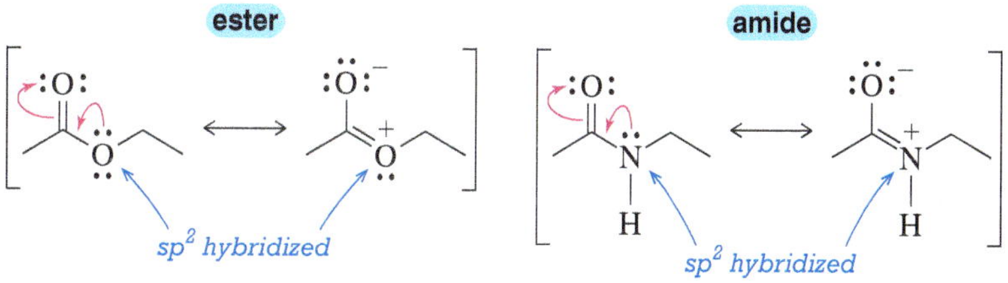

Figure 19.6 Because resonance structures can be drawn for carboxylic acid derivatives, we know that the atom attached to the carbonyl is sp^2 hybridized.

19.3 SUMMARY

- The electronegative atom attached to the carbonyl of a carboxylic acid derivative is sp^2 hybridized because it participates in resonance with the carbonyl.

Assessment 19.11 Predict the hybridization of the indicated atoms.

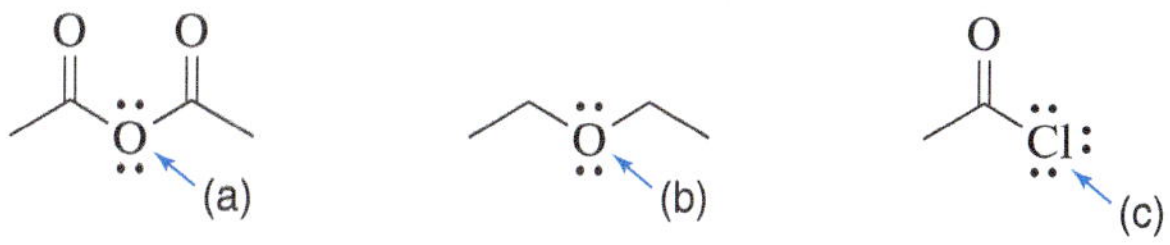

19.4 Physical properties of carboxylic acid derivatives

The physical properties of carboxylic acid derivatives and carboxylic acids are compared in **Table 19.2**. Unlike carboxylic acids, which are capable of hydrogen bonding as both hydrogen bond donors *and* acceptors, acid chlorides, anhydrides, esters, and nitriles are *only* hydrogen bond acceptors. As a result, they are less soluble in water and boil/melt at lower temperatures. In comparison, unsubstituted amides have two hydrogens available to form hydrogen bonds. This makes them boil and melt at higher temperatures than the corresponding carboxylic acid.

Table 19.2 Physical properties of carboxylic acid derivatives compared to the corresponding carboxylic acid.

Compound	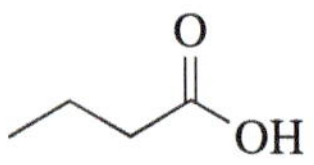 butanoic acid	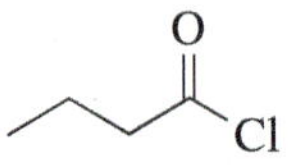 butanoyl chloride	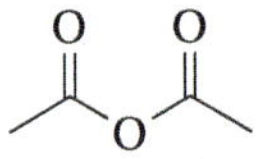 acetic anhydride	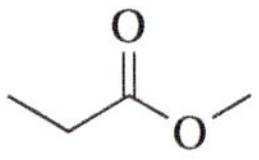 methyl propanoate	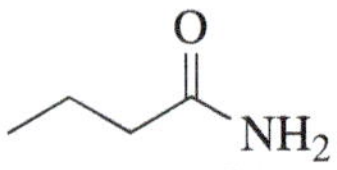 butanamide	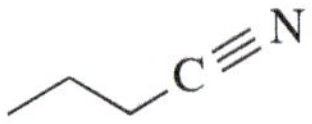 butanenitrile
Melting point (°C)	−5.1	−89	−73.1	−88	115	−112
Boiling point (°C)	163.8	102	139.8	80	216	118
Water solubility (g/L H_2O)	Completely miscible	Decomposes	26, decomposes	72	163	0.33

The physical properties of amides depend on the number of alkyl groups attached to the nitrogen. The addition of each nonpolar alkyl group to nitrogen makes the molecule more lipophilic, and eliminates one hydrogen that can participate in hydrogen bonding. Table 19.3 illustrates this concept. The addition of the first methyl group only raises the boiling point slightly, despite increasing the molecular weight and size of the molecule. Adding a second methyl group, making the amide only a hydrogen bond acceptor, causes a significant decrease in boiling point.

Table 19.3 Hydrogen bonding in amides contributes to its high boiling and melting points.

Amide	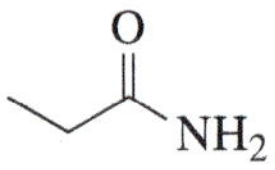 propanamide	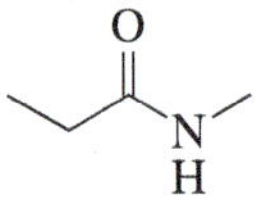 *N*-methylpropanamide	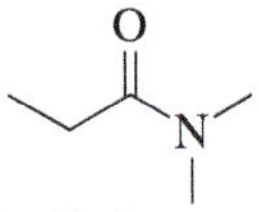 *N,N*-dimethylpropanamide
Molecular weight (amu)	73.1	87.1	101.5
Boiling point (°C)	213	217	174

19.4.1 Spectroscopy of carboxylic acid derivatives

The IR spectra of carboxylic acid derivatives are characterized by a strong C=O stretch around 1715 cm^{-1} for all but the nitrile, which shows a characteristic C≡N stretch around 2200 cm^{-1}. Table 19.4 lists the characteristic IR bands for the simplest member of each carboxylic acid derivative.

Table 19.4 Important IR bands for carboxylic acid derivatives.

Compound	Wavenumber (cm^{-1})	Bonds
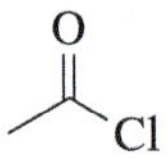 acetyl chloride	~700 ~1785	$C_{(sp^2)}$—Cl C=O
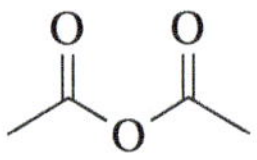 acetic anhydride	~1050 ~1800 ~1740	C—O C=O (symmetrical) C=O (asymmetrical)
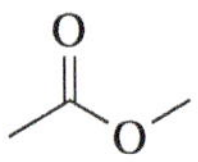 methyl acetate	~1050 ~1250 ~1735	$C_{(sp^3)}$—O $C_{(sp^2)}$—O C=O
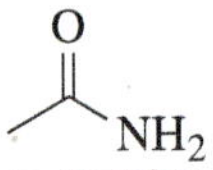 acetamide	~1000 ~3300 ~1685	$C_{(sp^2)}$—N N—H C=O
 acetonitrile	~2200	C≡N

The position of the C=O stretch can be predicted based on resonance and induction (**Table 19.5**). [*This is a review of what we did when first studying IR spectroscopy in Chapter 14.*] Recall from Chapters 14 and 17 that the zwitterionic structure of the carbonyl explains its reactivity and physical properties. For acid chlorides, inductive withdrawing dominates. Because the C—Cl bond is longer and reduces orbital overlap, the resonance-donating ability of chlorine is diminished. The net withdrawing effect of the chlorine makes the positive charge on the carbonyl larger, causing it to hold the negatively charged oxygen more tightly, which results in a higher vibrating frequency. On the other end of the spectrum [*Ha!*] is the amide. The less electronegative nitrogen demonstrates weaker induction, allowing resonance to win. The central positive charge on the carbonyl in the zwitterionic resonance structure is stabilized, the oxygen is loosely held, and the vibration frequency is lowered.

Table 19.5 **C=O stretching frequency** can be rationalized based on inductive and resonance-stabilizing effects on the carbocation of the zwitterionic carbonyl structure.

Functional group	Structural effects	Description	Resulting wavenumber $\bar{\nu}$ (cm^{-1})
Acid chloride	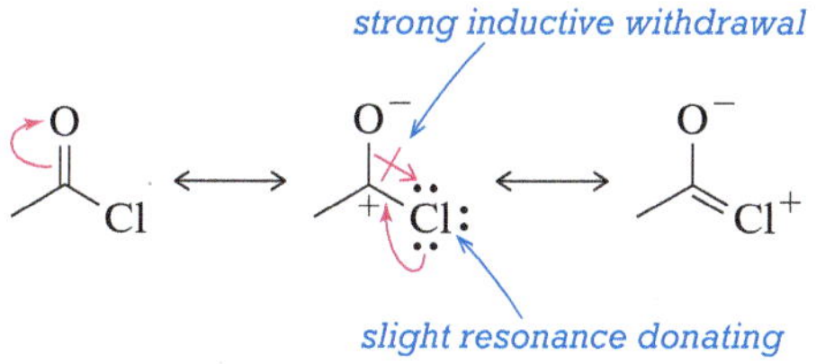	**Strong inductive withdrawal (increases $\bar{\nu}$)** Slight resonance donating (decreases $\bar{\nu}$)	1785
Anhydride	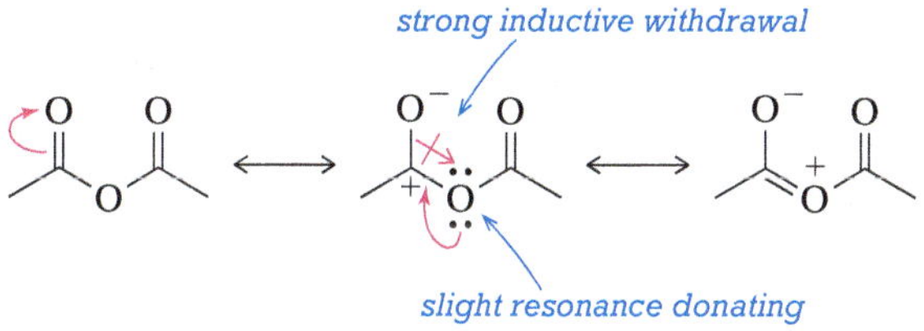	**Strong inductive withdrawal (increases $\bar{\nu}$)** Slight resonance donating (decreases $\bar{\nu}$)	1800 and 1740
Ester	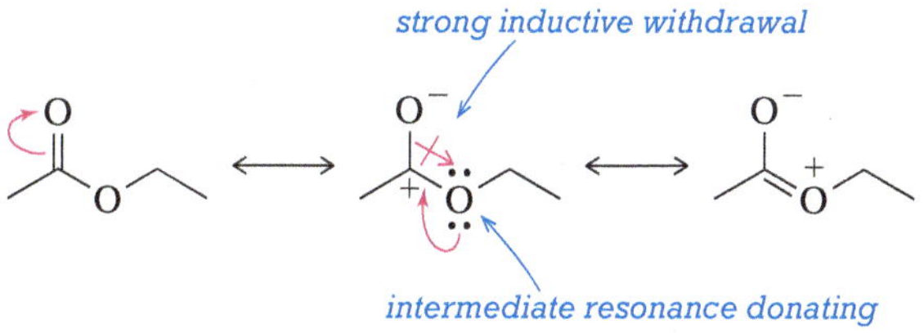	Strong inductive withdrawal (increases $\bar{\nu}$) Intermediate resonance donating (decreases $\bar{\nu}$)	1735
Amide	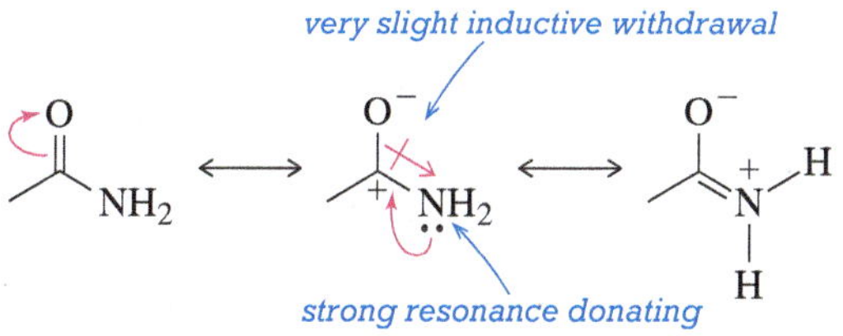	Very slight inductive withdrawal (increases $\bar{\nu}$) **Strong resonance donating (decreases $\bar{\nu}$)**	1685

How can we differentiate between anhydrides and esters (**Figure 19.7**)? Both have an inductively withdrawing and resonance-donating oxygen attached to the central carbonyl. Do they then stretch at the same frequency? No: The oxygen in an ester can only donate an electron pair in one direction, toward the positive charge in the zwitterionic resonance structure. On the other hand, the central anhydride oxygen can donate an electron pair in two directions. This cuts the resonance donation in half, allowing induction to win by a bigger margin for anhydrides than it does for esters. [*How might this impact the reactivity of anhydrides and esters?*]

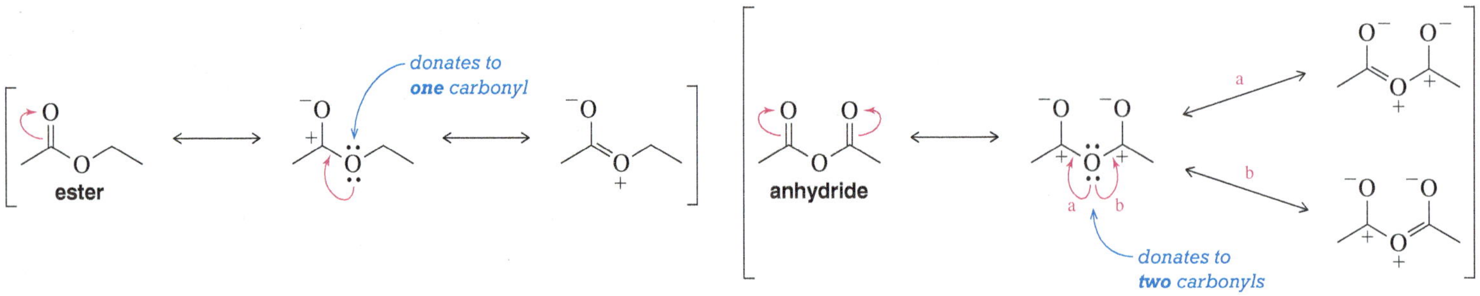

Figure 19.7 **Resonance in anhydrides** is divided equally between the two carbonyls, lessening its frequency-lowering effect.

In the ^{1}H NMR, characteristic signals of all carboxylic acid derivatives appear between 2 and 3 ppm (hydrogens at the α carbon) (**Figure 19.8**). Esters also contain a signal around 4.2 ppm, corresponding to the hydrogen of the alkoxy substituent. This hydrogen is deshielded compared to an alcohol or ether (3 to 4 ppm) because of resonance withdrawing by the attached carbonyl. The carbonyl carbons exhibit a ^{13}C signal around 175 ppm—lower than aldehydes and ketones (~200 ppm). The nitrile carbon, shielded by the C≡N triple bond, gives a signal around 118 ppm.

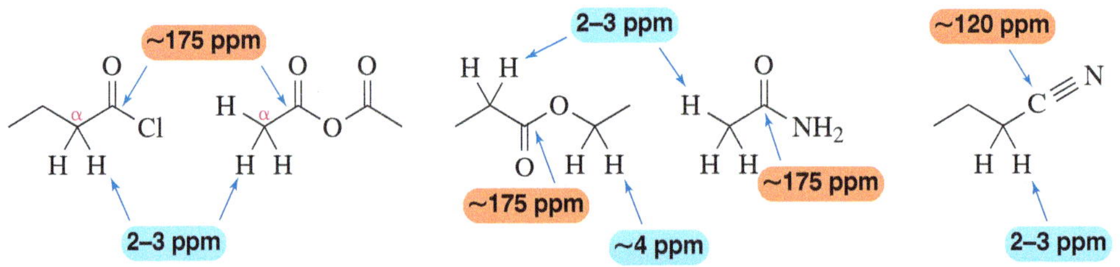

Figure 19.8 **Characteristic ^{1}H (screened in blue) and ^{13}C signals (screened in orange)** in the carboxylic derivatives.

19.4 SUMMARY

- Carboxylic acid derivatives, lacking the hydrogen-bonding ability of carboxylic acids, are generally less water soluble and have weaker intermolecular interactions.
- The IR stretching frequency of carboxylic acid derivatives decreases from 1785 to 1685 cm^{-1} (acid chlorides to amides) based on the cumulative effects of induction and resonance.

Assessment 19.12 Would you expect the C=O in acid chlorides or acid bromides to absorb at a higher wavenumber? Explain your answer.

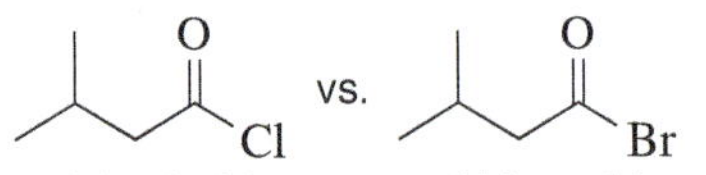

acid chloride **acid bromide**

Assessment 19.13 Would you expect the C=O in esters or thioesters to absorb at a higher wavenumber? Explain your answer.

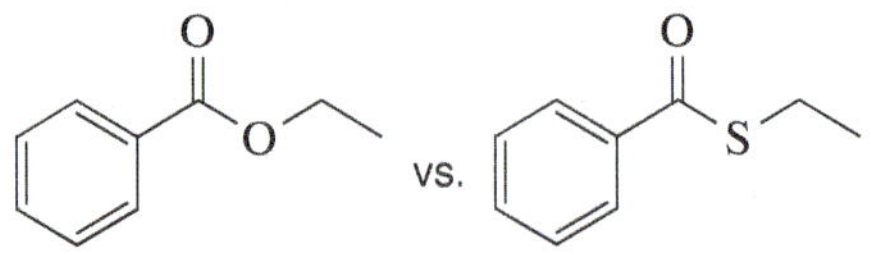

ester **thioester**

Assessment 19.14 How would you expect the IR and ^{1}H NMR spectra for propanamide and *N*,*N*-diethylpropanamide to differ?

propanamide ***N*,*N*-diethylpropanamide**

19.5 Synthesis of carboxylic acid derivatives: general considerations

The synthesis of carboxylic acid derivatives is best understood by studying their reactions because one carboxylic acid derivative is often used to make another. **Figure 19.9** shows an example of how each carboxylic acid derivative can be prepared. Note the ranking of reactivity and that there are more ways to go downhill than there are ways to go uphill. [*Makes sense, right?*]

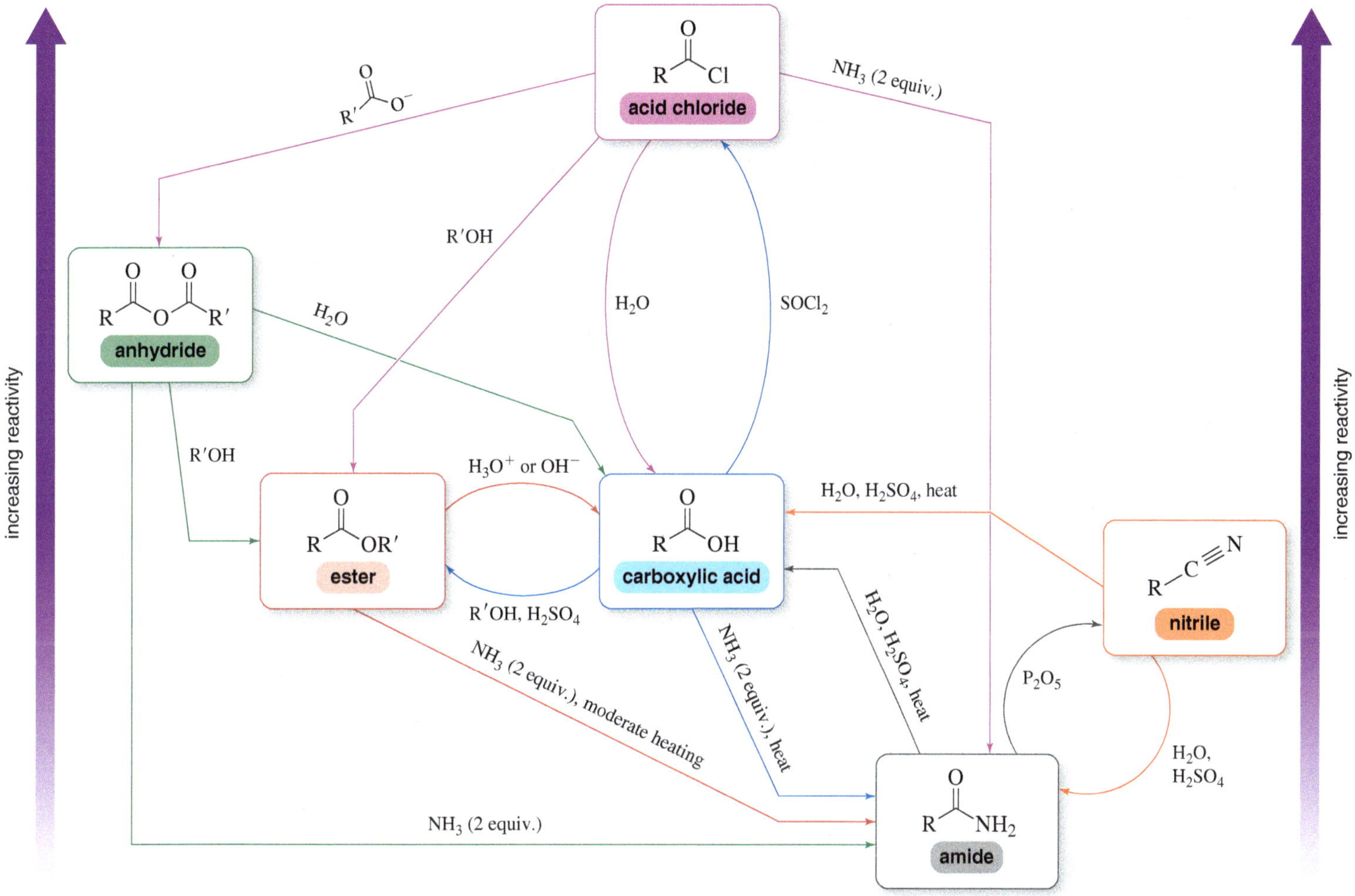

Figure 19.9 **Synthesis of carboxylic acid derivatives** by conversion from other carboxylic acid derivatives.

Figure 19.9 may appear overwhelming at first because there are myriad ways to go from one derivative to another. Take a few minutes to study the picture. A version of this figure introduces each class of reactions we consider in this chapter, and after working through the next several sections, Figure 19.9 should be much less imposing.

19.5 SUMMARY

- It is more efficient to synthesize less reactive (more stable) derivatives from more reactive (less stable) derivatives (i.e., make amides from acid chlorides).

19.6 Reactions of carboxylic acid derivatives—nucleophilic acyl substitution

Previously we have organized reactions into two broad categories: (1) reactions with strong nucleophiles and (2) reactions with weak nucleophiles. This was the case when studying the S_N1 versus S_N2 and E1 versus E2 reactions (Chapter 12), the addition of nucleophiles to aldehydes and ketones (Chapter 17), and then again with the reactions of carboxylic acids (Chapter 18). The reactions of carboxylic acid derivatives, which differ only by the presence of a leaving group attached to the carbonyl, should fall into similar categories. Adopting the same naming conventions used in Chapter 18 (General Mechanism 18A and General Mechanism 18B), the next section examines how the mechanisms of nucleophilic acyl substitution differ based on the strength of the reacting nucleophile. [*It never hurts to review material a second time.*]

19.6.1 General Mechanism 18A: reaction with strong nucleophiles

The first reaction type is the addition of a strong nucleophile to a carboxylic acid derivative. In **Figure 19.10**, the general leaving group X is used to represent an arbitrary leaving group. [*While the rate will change with each carboxylic acid derivative, the mechanism is identical. We study these rate differences in Section 19.6.3.*] Strong nucleophiles react quickly, attacking atoms with a partially positive charge. Here, the nucleophile adds directly to the carbonyl, pushing the π electrons onto oxygen and generating a short-lived **tetrahedral intermediate**. The lone pair on oxygen quickly collapses to reform the C—O π bond and displace the good leaving group. This is often referred to as the **addition/elimination** mechanism. [*Why?*] The substitution of ethoxide onto an acid chloride (a reaction we study in Section 19.7.2) is representative of this mechanism.

Figure 19.10 The addition of strong nucleophiles proceeds quickly with the neutral carboxylic acid/carboxylic acid derivative. In sum, strong nucleophiles add first to make the tetrahedral intermediate, which collapses to give the product. This is referred to throughout Chapters 18 and 19 as General Mechanism 18A.

General Mechanism 18A: Reaction with strong nucleophiles

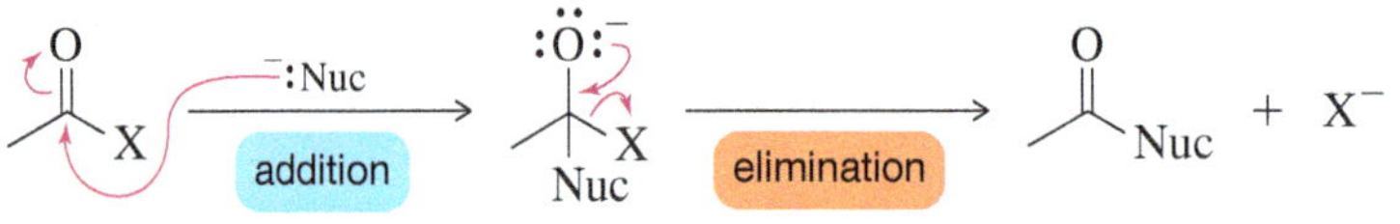

Specific example:

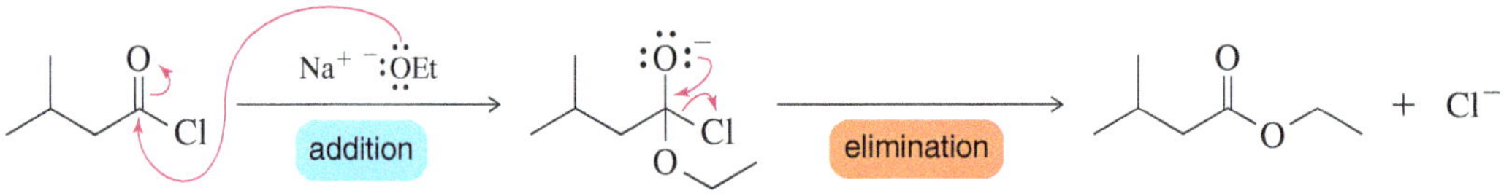

Assessment 19.15 You might expect that aldehydes and ketones (Chapter 18) could undergo the addition/elimination mechanism. With strong nucleophiles, however, nucleophilic addition is the only outcome. Why?

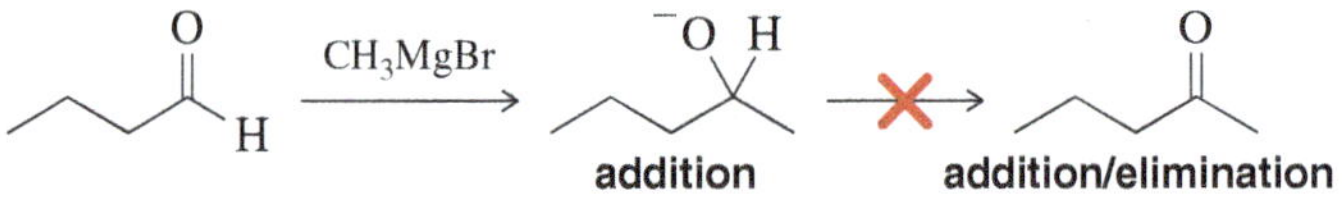

19.6.2 General Mechanism 18B: reaction with weak nucleophiles/ acidic conditions

Weak nucleophiles such as H_2O, CH_3OH, etc., in which the atom donating the electron pair is neutral (Section 17.3.2), react more slowly than strong nucleophiles and do not react quickly with neutral carboxylic acid derivatives. Instead, a full cation (a more reactive intermediate), formed by protonation, initiates attack by the nucleophile (**Figure 19.11**). Because the weak nucleophile brings along its proton, a deprotonation step occurs to give the neutral tetrahedral intermediate. Under acidic conditions, the leaving group is protonated and then leaves, forming a cation. Deprotonation generates the new carboxylic acid derivative. The addition of water to an ester (a reaction we study in Section 19.7.1) is representative of this mechanism.

In comparing General Mechanisms 18A (strong nucleophiles) and 18B (weak nucleophiles) (**Figure 19.12**), a couple of important points should be mentioned. With six steps, General Mechanism 18B seems much more complicated, but it really isn't. The four extra steps in General Mechanism 18B are simply acid–base reactions [*convince yourself of this*]. This observation reveals the second important point: Most reactions occurring by General Mechanism 18B are acid catalyzed because the acidic proton required to protonate the carbonyl is regenerated in the last step. Finally, because the nucleophiles are weak, these reactions are easily reversible, in which case the reaction often needs to be pushed in one direction using Le Châtelier's principle. [*One of the reactions we studied in Chapter 18 is the exact reverse of a reaction we study here in Chapter 19. Any ideas which one it is?*]

General Mechanism 18B: Reaction with weak nucleophiles

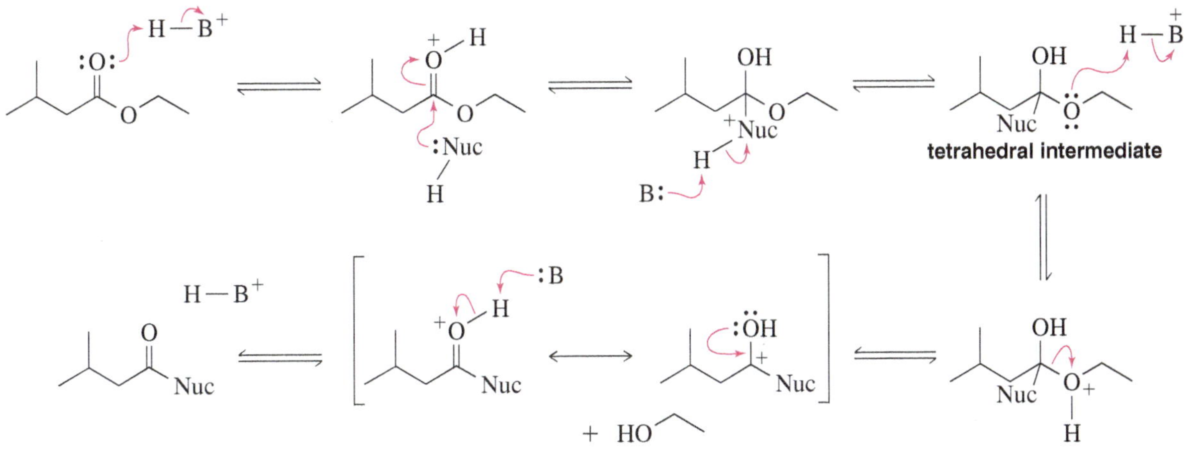

Specific example:

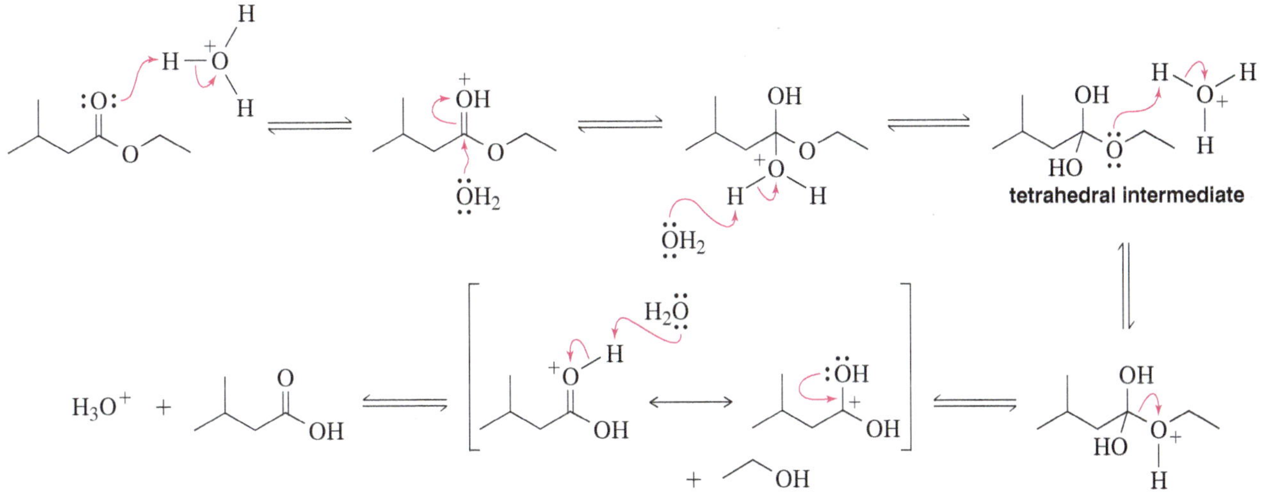

Figure 19.11 The substitution of weak nucleophiles necessitates protonation of the carbonyl to make a reactive intermediate. In sum, with weak nucleophiles, the carbonyl protonates first, then the nucleophile adds. Similarly, the leaving group is protonated prior to leaving. This is referred to throughout Chapters 18 and 19 as General Mechanism 18B.

General Mechanism 18A: Strong nucleophiles

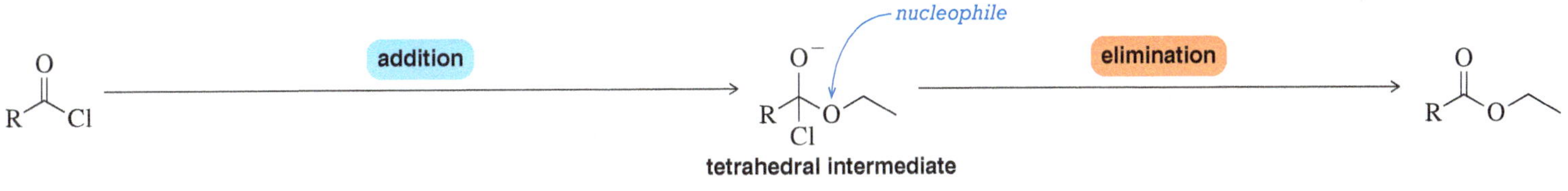

General Mechanism 18B: Weak nucleophiles

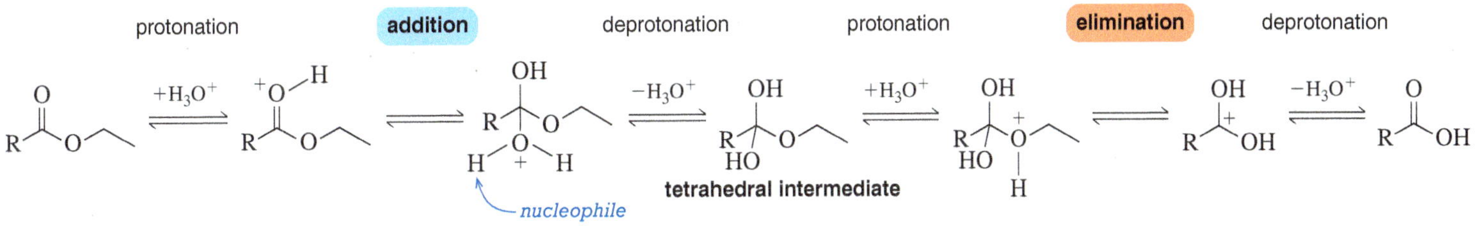

Figure 19.12 A comparison of General Mechanism 18A and 18B shows that 18B differs only by four extra protonation/deprotonation steps.

19.6.3 Relative reactivity of carboxylic acid derivatives

In Chapter 18, the rate of reaction of the carboxylic acid functional group depended only on the strength of the nucleophile. In Chapter 19, the relative electrophilicity of the carboxylic acid derivatives equally impacts the rate. Therefore, a nucleophile that is considered weak when reacting with one carboxylic acid derivative might be considered strong when reacting with another. [*More is said about this later.*]

There are two ways to rationalize the relative reactivity of carboxylic acid derivatives, depending on which of the two steps—addition or elimination—is rate determining. Luckily, both descriptions give the same relative reactivity.

When the first step is rate determining, the rate of substitution will be determined by the electrophilicity of the carbonyl carbon. As such, stabilization of the zwitterionic form of the carbonyl by the attached group rationalizes the relative reactivity (Table 19.6). The chlorine of the acid chloride withdraws electron density by induction, giving the carbonyl carbon a larger positive charge and making it the most reactive carboxylic acid derivative. Anhydrides, with the induction of the oxygen atom, and the half resonance-donating ability of the central oxygen (see Section 19.4.1; Figure 19.7), are the second most reactive. Esters, because of the induction and full resonance-donating ability of the oxygen, come in third. And, finally, the strong electron-donating ability of nitrogen makes amides the least reactive carboxylic acid derivative. [*This should remind you of how we rationalized C═O stretching frequency in the IR spectra of carboxylic acid derivatives.*]

Table 19.6 Carboxylic acid derivative reactivity can be rationalized based on inductive and resonance-stabilizing effects on the carbocation of the zwitterionic carbonyl structure.

Functional group	Structural effects	Description	Resulting vibration frequency $\bar{\upsilon}$ (cm^{-1})
Acid chloride	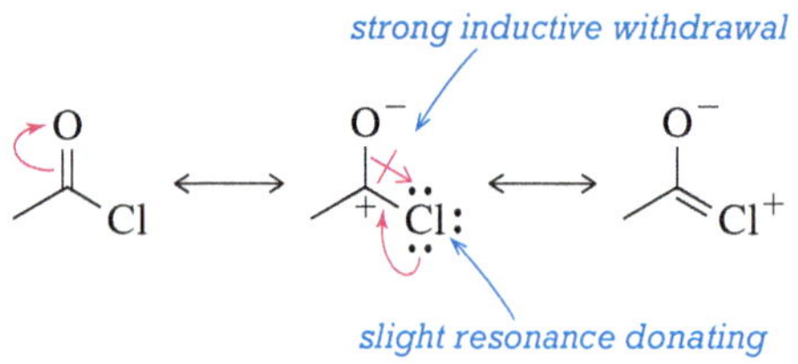	**Strong inductive withdrawal (increases reactivity)** Slight resonance donating (decreases reactivity)	1785
Anhydride	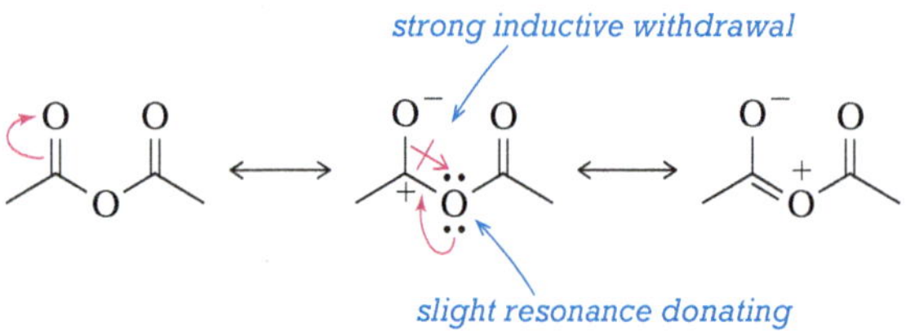	**Strong inductive withdrawal (increases reactivity)** Slight resonance donating (decreases reactivity)	1800 and 1740
Ester	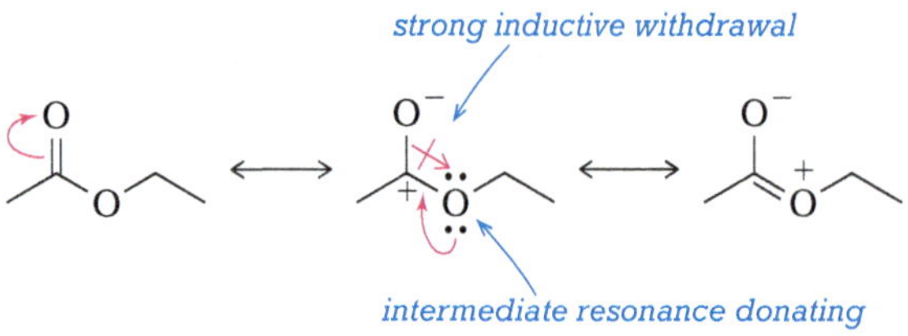	Strong inductive withdrawal (increases reactivity) Intermediate resonance donating (decreases reactivity)	1735
Amide	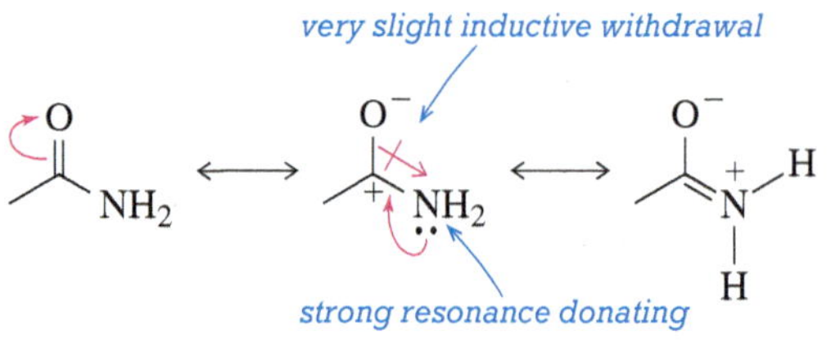	Very slight inductive withdrawal (increases reactivity) **Strong resonance donating (decreases reactivity)**	1685

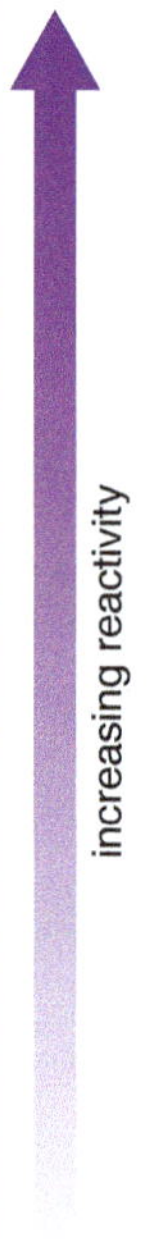

A second way to understand relative reactivity is to compare the derivatives by leaving group ability (**Figure 19.13**). [*Which step determines the rate of a reaction?*] Comparison of the pK_a values of the conjugate acids of the leaving groups shows that the order of stability is $Cl^- > RCOO^- > RO^- > H_2N^-$. This analysis would technically only apply to the rare case in which the second step is the slow, rate-determining step because that is when the elimination occurs. Nonetheless, it parallels the reactivity analysis in Table 19.6 and thus provides a second way to remember the relative reactivity of carboxylic acid derivatives.

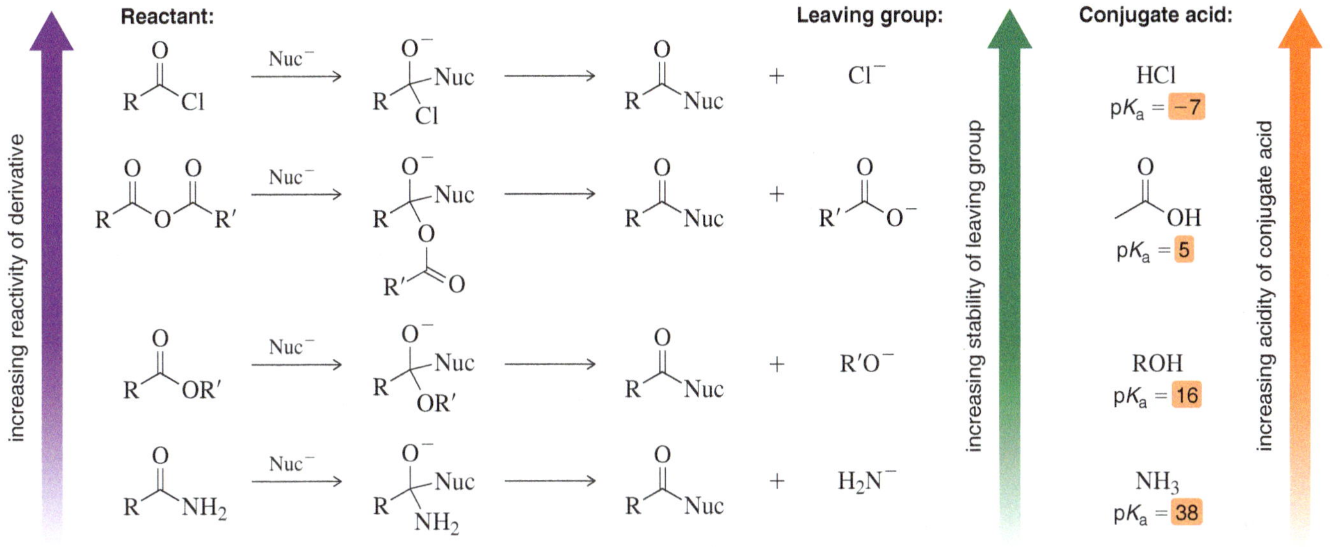

Figure 19.13 **Carboxylic acid derivative reactivity** can be rationalized by considering the stability of the leaving group.

19.6 SUMMARY

- With strong nucleophiles, nucleophilic acyl substitution proceeds by direct attack to form a tetrahedral intermediate (General Mechanism 18A). The tetrahedral intermediate collapses to reform the carbonyl, expelling a good to moderate leaving group.
- With weak nucleophiles, protonation of the carboxylic acid derivative is first required to initiate nucleophilic acyl substitution (General Mechanism 18B). A series of proton transfers is all that differentiates General Mechanisms 18A and 18B.
- The relative reactivity of carboxylic acid derivatives is a function of the cumulative effects of induction and resonance on the carbonyl carbon. The order of reactivity, from most reactive to least reactive, is acid chlorides > anhydrides > esters > amides.

Assessment 19.16 In addition to considering the rate of a reaction, substituent effects can be used to discuss the relative favorability/spontaneity of a reaction. Of the two ways to rationalize reactivity discussed here, which most appropriately justifies the statement, "acid chlorides react more spontaneously with water than amides."

Assessment 19.17 Would you expect esters or thioesters to be more reactive in a nucleophilic acyl substitution reaction? Explain your answer.

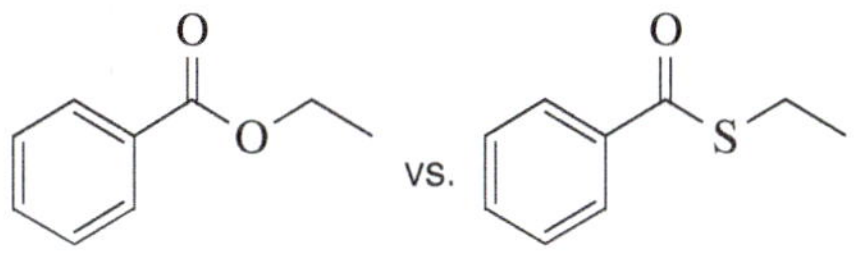

ester thioester

Assessment 19.18 Imides are closely related to amides. Which would you expect to be more reactive in a nucleophilic acyl substitution reaction? Why?

imide amide

19-A CHEMISTRY IN CONTEXT: Biochemistry

The peptide bond

A peptide consists of two or more amino acids linked in a chain. In biochemistry, the C—N bond of an amide is often called a peptide bond because it is the bond that links two amino acids. Large peptides, consisting of more than 50 amino acids, are called proteins. One important role played by proteins in living organisms is that of enzymes, which are catalysts for biochemical reactions (**Figure 19.14**).

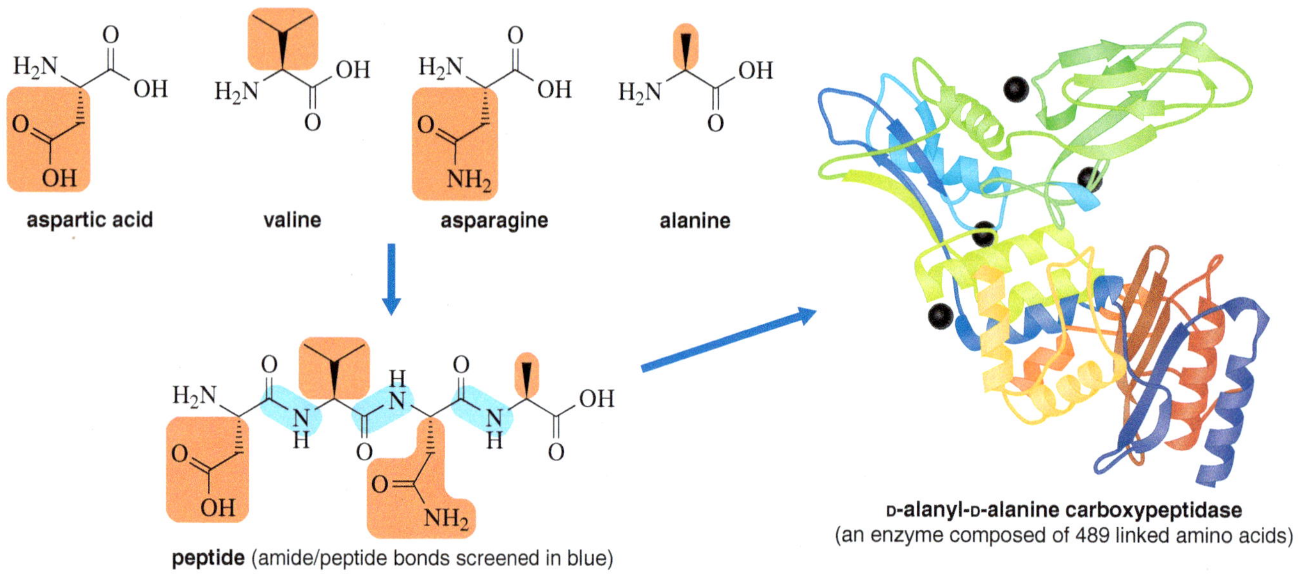

Figure 19.14 Though dynamic, enzymes have a somewhat fixed 3-D structure because of the stability of their peptide bonds.

Peptide bonds are also important for imparting rigidity and stability to bacterial cell walls by forming cross-links between polysaccharide chains and creating structural molecules called peptidoglycans. **Figure 19.15** shows the alternating *N*-acetylmuramic acid (MurNAc) and *N*-acetylglucosamine (GlcNAc) residues that make up the polysaccharide chains connected by the peptide cross-linker.

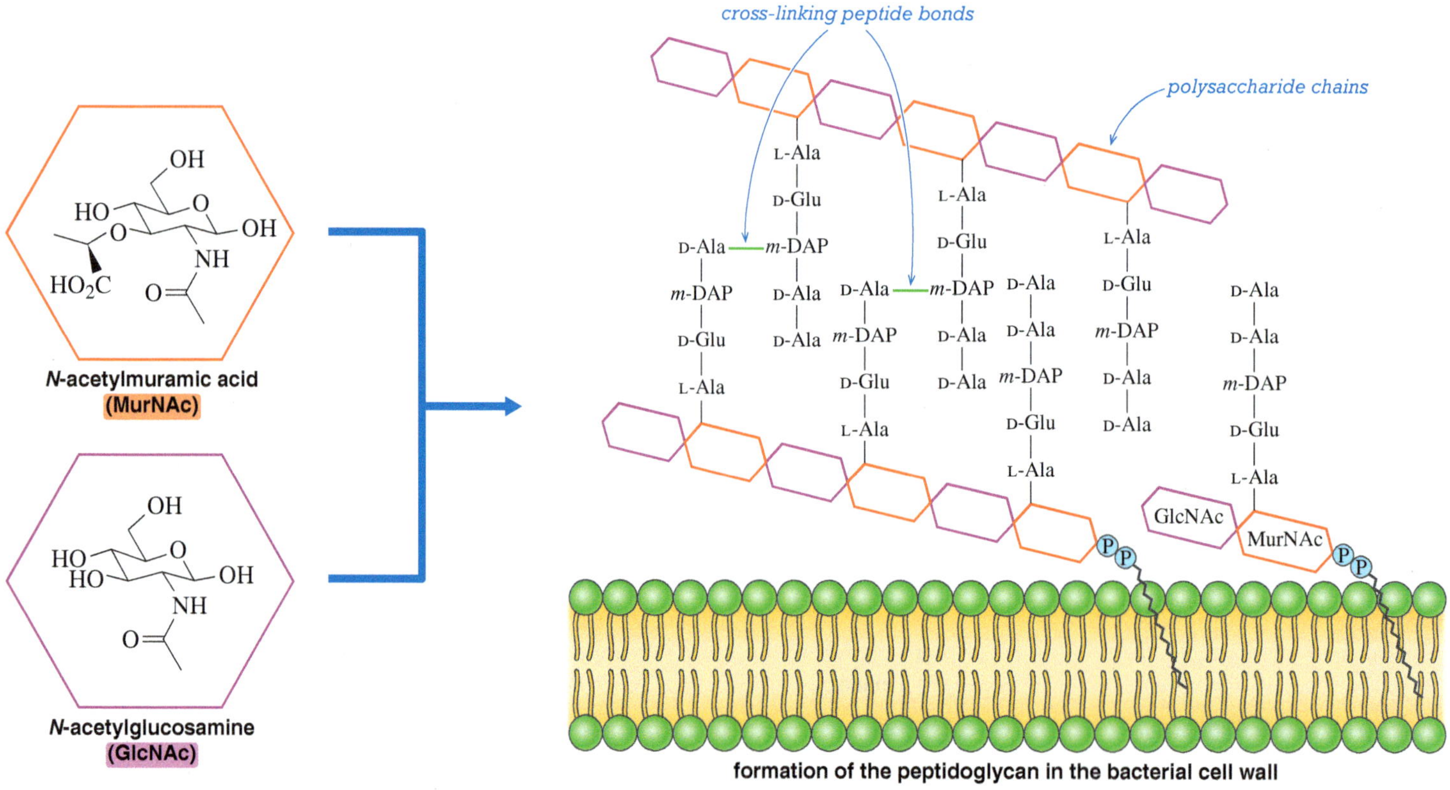

Figure 19.15 The bacterial cell wall is made rigid by peptide bonds.

(continued)

From an evolutionary perspective, the amide (or peptide) bond is well suited for these important roles for two reasons. Because nitrogen is strongly electron donating in forming the resonance structure containing the C═N bond, it is a very strong bond. It is the strength of this bond that keeps enzymes together. This resonance structure also restricts rotation around the C—N bond. Recall that the π bond requires overlapping *p* orbitals. There are only two conformations [*which two?*] of the amide bond that maintain this required alignment (**Figure 19.16**). Rotation between the two must overcome a high activation barrier because the π character of the C—N bond must be temporarily broken to allow it to occur. This restricted rotation helps enzymes maintain their shape, which in turn restricts catalytic activity to specific substrates.

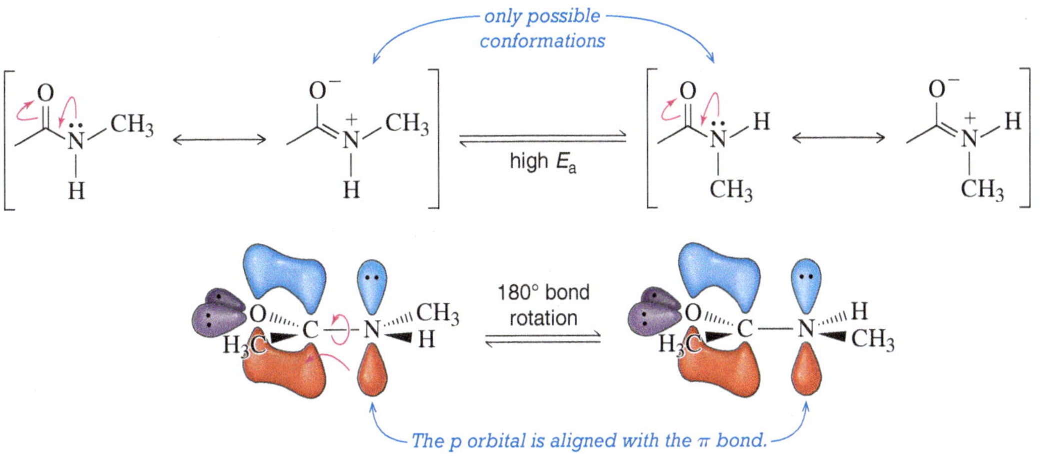

Figure 19.16 **Resonance in the amide** makes the peptide bond strong and restricts rotation around the C—N bond.

19.7 Functional group interconversion

As stated previously, and as shown in Figure 19.9, the synthesis of carboxylic acids/carboxylic acid derivatives often arises from functional group interconversion from other carboxylic acids/carboxylic acid derivatives. Before proceeding through this section, though, which features as many as 16 [*but really only two*] new reactions, familiarize yourself with the following three disclaimers:

1. There are two main ways that the material in Sections 19.7.1–19.7.4 can be organized. To reinforce the relative ranking of the reactivity of the derivatives, it has been organized based on the reagent that is attacking and the product that is formed. [*You can and should also organize this material according to the reactions of each derivative. Use your reaction sheets for this.*]
2. These reactions occur by either General Mechanism 18A or 18B. While it is important that you understand *why* which mechanism occurs, showing them every time can become space-consuming and redundant. Thus, mechanisms are shown only when it is necessary to illustrate some concept. [*And I'll ask you to draw some in the text and assessments.*]
3. The new concepts covered in these sections are the **relative reactivity of derivatives, solubility effects on reactivity**, and **autocatalysis**, among others. We highlight these concepts in the form of **SIDE CONVERSATIONS**, selecting one example in the chapter to illustrate each concept.

That these are side conversations does not diminish their importance. Setting them apart this way is to highlight them as being general concepts with application beyond just where they are discussed. When you've learned a new concept, try to find the other places it applies.

Finally, the chemistry of nitriles is discussed in a separate section because, despite being carboxylic acid derivatives, their reactivity is more like that of aldehydes and ketones.

19.7.1 Reactions with water: synthesis of carboxylic acids

The reaction of water with each carboxylic acid derivative forms a carboxylic acid (**Figure 19.17**). Water (a weak nucleophile) does not react directly with all carboxylic acid derivatives. This section will help us figure out how to modify reaction conditions so that two otherwise unreactive molecules react.

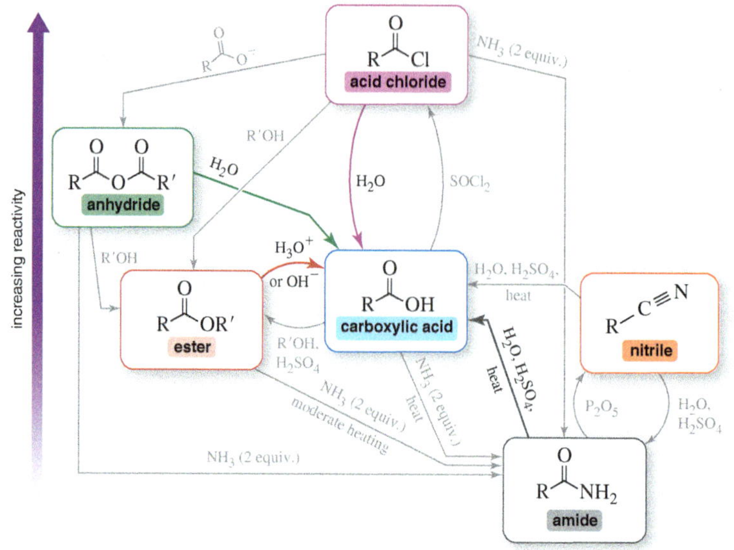

Figure 19.17 **The reaction of carboxylic acid derivatives** with water produces carboxylic acids. [*Nitriles are in Section 19.10.*]

19.7.1.1 H_2O with acid chlorides

As shown in **Figure 19.18**, low-molecular-weight acid chlorides react with water by General Mechanism 18A. Direct attack by water on the carbonyl carbon results in formation of the tetrahedral intermediate. The tetrahedral intermediate collapses to displace the chloride (a better leaving group than water). Because water brought along its proton in the attack, it is now acidic. Thus, a final acid–base reaction occurs to give the carboxylic acid.

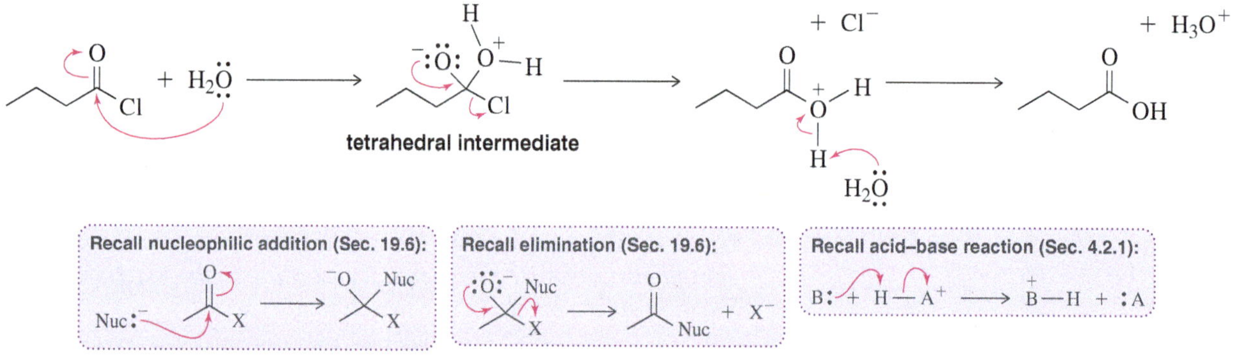

Figure 19.18 **Water reacts with small acid chlorides** by General Mechanism 18A.

[*Read the previous paragraph again.*] Two issues should arise. First, isn't water a weak nucleophile that should react via General Mechanism 18B? And second, why are *low*-molecular-weight acid chlorides specified? What does molecular weight have to do with reactivity? [*I'm glad you asked!*]

SIDE CONVERSATION 1 **Relative strength of nucleophiles** We considered nucleophilic strength in Chapter 12 when we compared S_N1 versus S_N2 reactions and in Chapter 17 when we examined the relative reactivity of aldehydes and ketones. In these cases, nucleophiles were categorized as strong or weak, almost in an absolute sense. But reactivity is always relative. For example, water is too weak to attack a neutral ketone. However, protonate the ketone, and water willingly shares its electron pair. By making the electrophile stronger, water is itself strong enough to attack. The example in **Figure 19.19** illustrates a concept that comes up throughout this section—namely, if the electrophile is strong enough, even weak nucleophiles react by General Mechanism 18A. Acid chlorides are the most reactive carboxylic acid derivative, so they are reactive enough that water reacts as if it is a strong nucleophile. ■

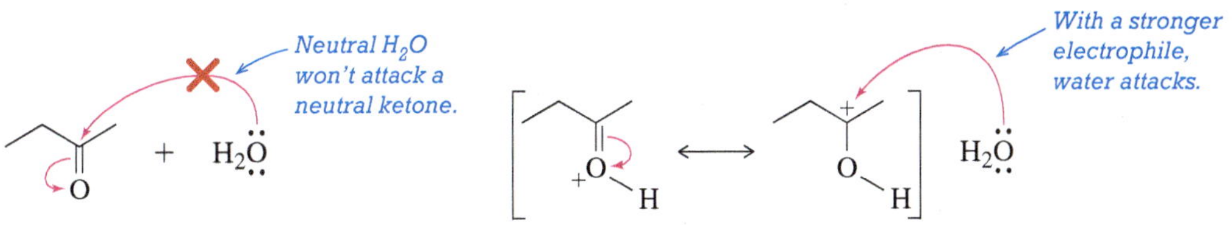

Figure 19.19 **The strength of a nucleophile** depends on the electrophile with which it reacts.

SIDE CONVERSATION 2 **Reactivity and solubility** When studying functional group reactivity, students often mistakenly ignore the rest of the molecule. [*Recall the conversation about protecting groups in Section 13.14.*] Why is it that a small acid chloride reacts in water but a larger, nonpolar one does not? [*Ideas?*] As a molecule gets bigger and more nonpolar, its water solubility decreases; eventually, the aqueous and organic phases separate. Now collisions can occur only at the phase boundary. Even though the functional group is unchanged, collisions occur far less frequently (**Figure 19.20**). ■

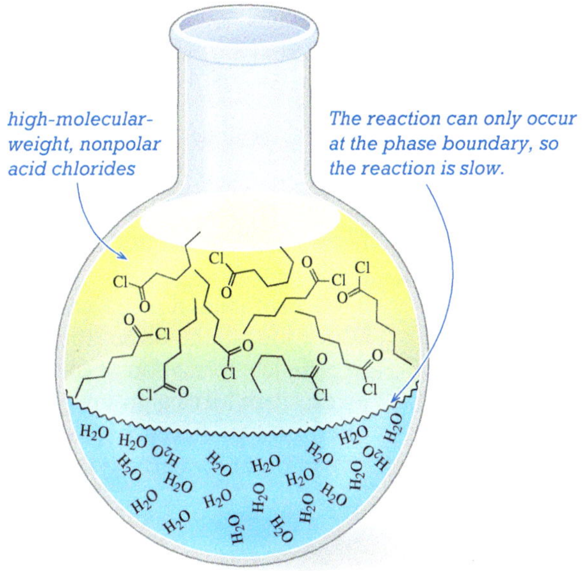

Figure 19.20 **High-molecular-weight, nonpolar acid chlorides** react slowly because they are insoluble in water.

To convert larger acid chlorides to carboxylic acids more quickly, hydroxide (a stronger nucleophile) can be used (**Figure 19.21**). Collisions are still restricted to the phase boundary, but now each collision is more likely to result in a reaction. [*Remember A, the frequency factor, in the Arrhenius equation?*] Notice in Figure 19.21 that an acid quench is needed if you use hydroxide. [*Why is this?*]

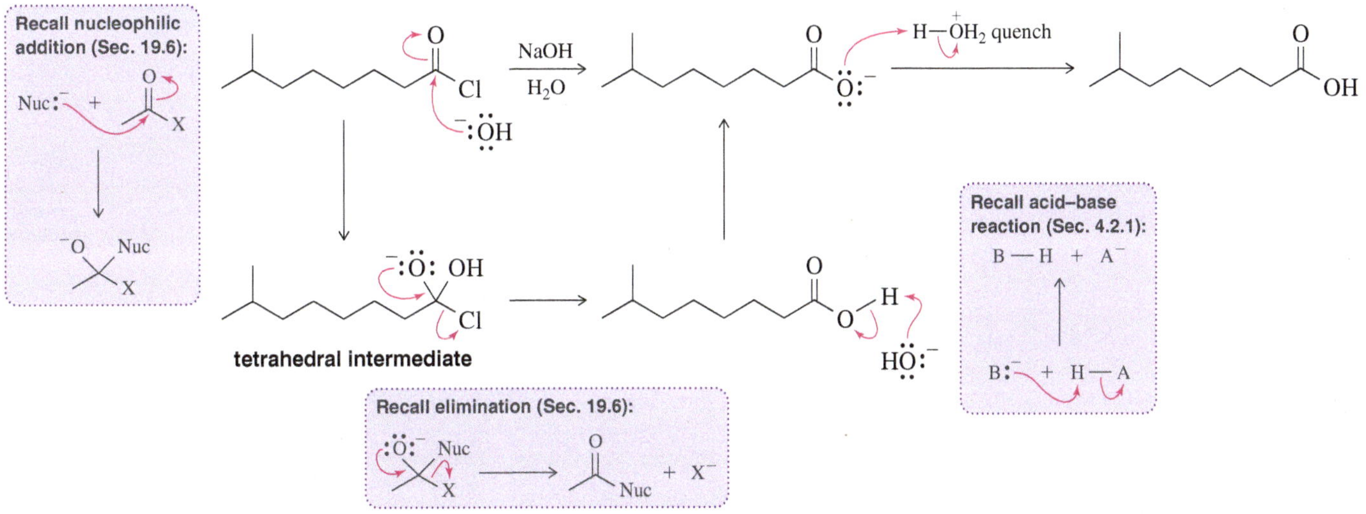

Figure 19.21 **Hydroxide is a stronger base,** so it reacts more quickly with acid chlorides.

19.7.1.2 H_2O with anhydrides

Anhydrides react like acid chlorides. Thus, low-molecular-weight anhydrides are soluble in water and react immediately [Figure 19.22(a)]. For higher molecular weight anhydrides, the reaction is faster under basic or acidic conditions [Figure 19.22(b)].

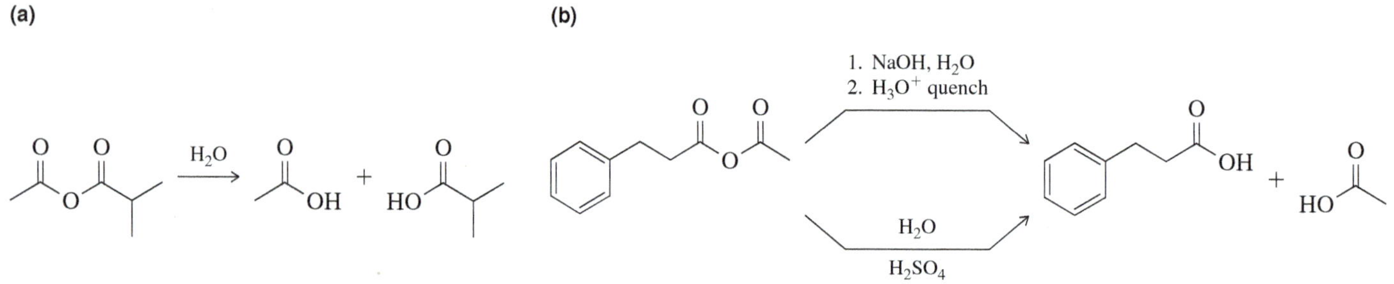

Figure 19.22 A comparison of experimental conditions for low- and high-molecular-weight, nonpolar anhydrides. (a) Low-molecular-weight anhydrides react directly (General Mechanism 18A) with water. (b) Larger, more lipophilic anhydrides react more quickly under basic (General Mechanism 18A) and acidic conditions (General Mechanism 18B).

19.7.1.3 H_2O with esters

Esters are too stable to react with water (a weak nucleophile). Instead, ester hydrolysis occurs in the presence of acid via General Mechanism 18B. [*Where have you seen this reaction before? Hint: Look at it in both directions.*] Ester hydrolysis (Figure 19.23) is the reverse of the Fischer esterification. [*Go back and compare it to the Fischer esterification in Figure 18.38.*] Thus, protonation of the carbonyl gives a positively charged [*and more reactive*] electrophile. Water adds to the carbocation, bringing along its proton. This proton, now acidic, is removed to complete formation of the tetrahedral intermediate. Protonation of an alkoxy group creates an excellent leaving group. [*What do excellent leaving groups do? They leave.*] The departure of ethanol produces a carbocation-like intermediate that is itself deprotonated to complete formation of the carboxylic acid and regenerate the acid catalyst.

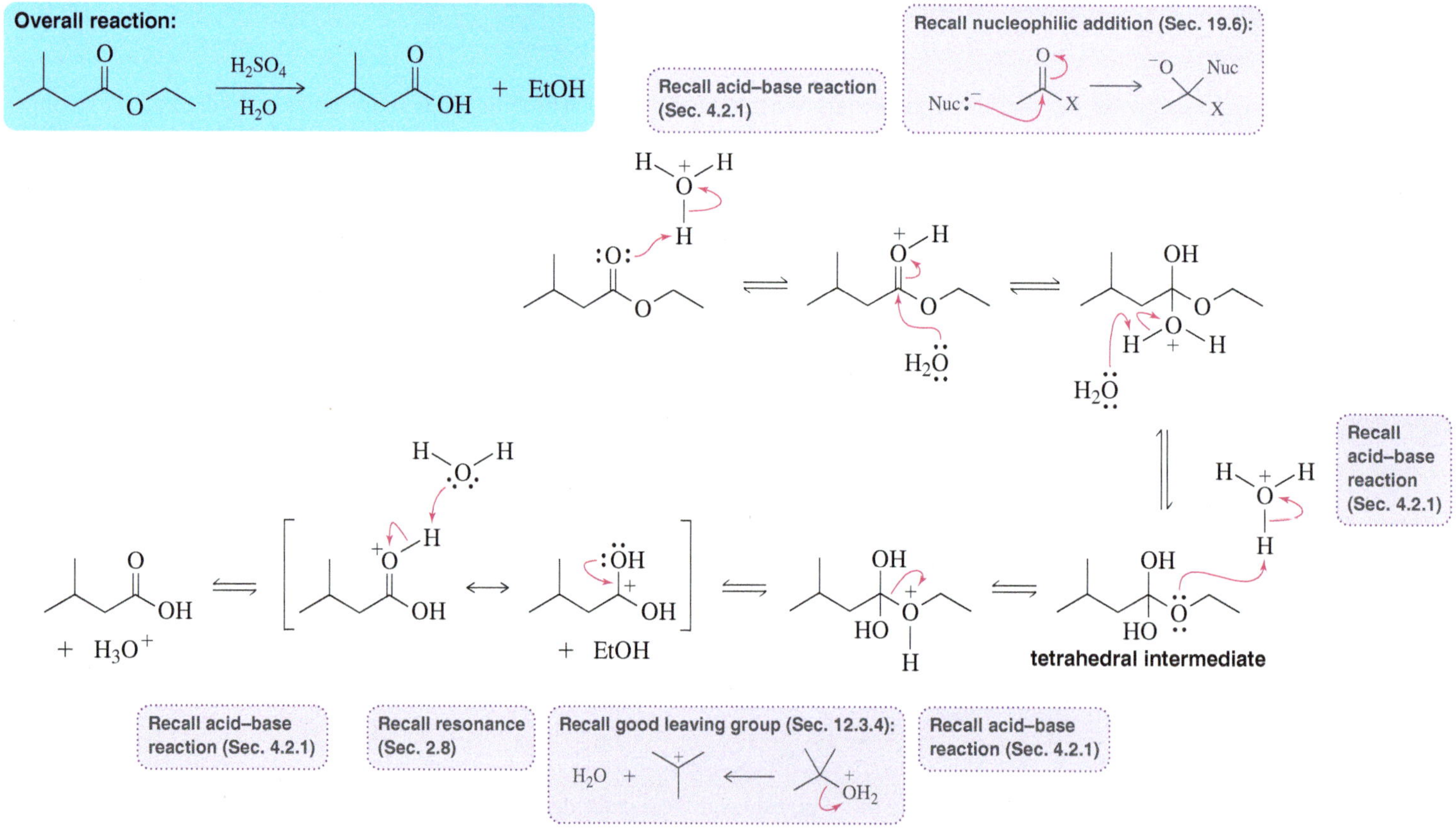

Figure 19.23 The mechanism of acid-catalyzed ester hydrolysis is the opposite of Fischer esterification (Figure 18.38).

An acid catalyst makes the ester more electrophilic, allowing the weak nucleophile water to attack. Is there another way ester hydrolysis can be accomplished? Let's discover one together.

Thinking Like a Chemist

The basic problem is that esters are too stable to react with water. Protonating the ester enhanced its electrophilicity. We can show this on a reaction coordinate diagram for the hypothetical first step of the reaction. A neutral ester and water have a high activation energy of reaction, and protonation of the ester raises its energy level, making the activation barrier smaller. The reaction *starts* higher on the hill, so attack by water is faster (**Figure 19.24**).

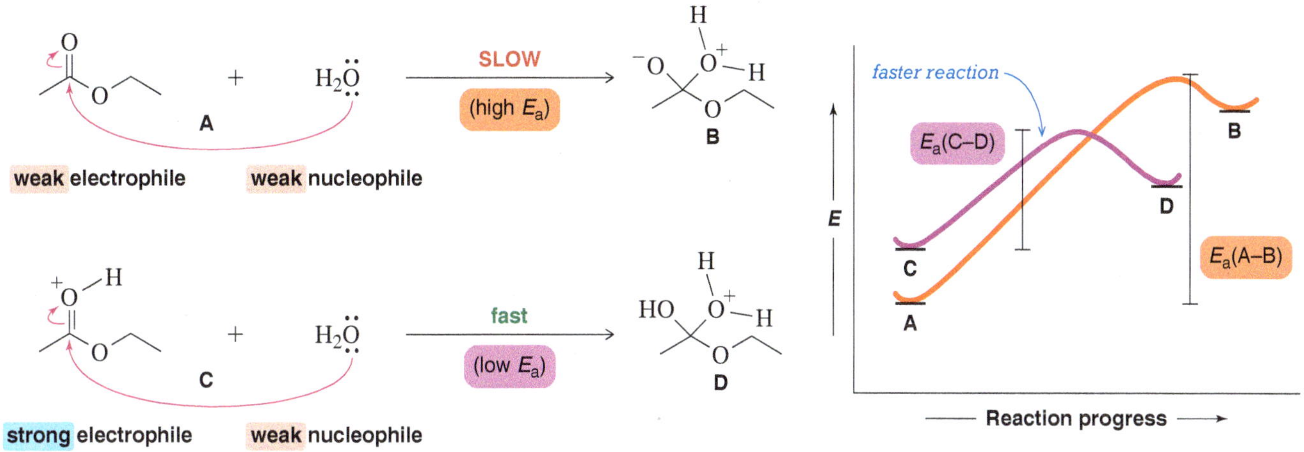

Figure 19.24 **Protonation of the ester** makes it a stronger electrophile, increasing the rate of reaction with water.

Modifying the electrophile increased the rate. What about modifying the nucleophile? [*What would make water a stronger nucleophile? Consider this before continuing.*] Neutral nucleophiles are weak, but negatively charged nucleophiles are generally stronger. Hydroxide (a much stronger nucleophile) can react directly with the neutral ester by General Mechanism 18A. Once more, the less stable hydroxide starts higher on the hill, reducing the activation energy (**Figure 19.25**).

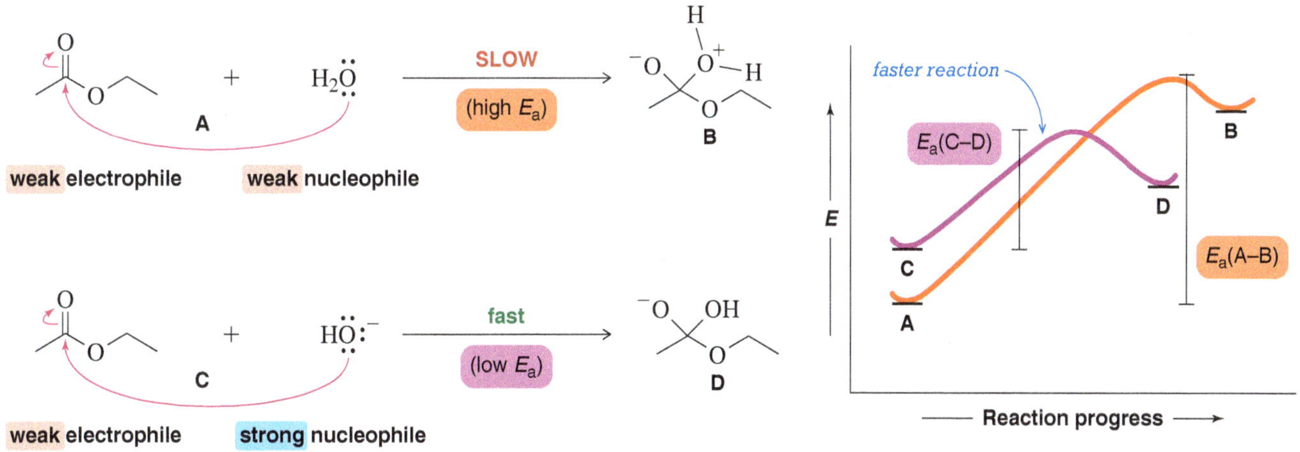

Figure 19.25 **Deprotonation of water** makes it a stronger nucleophile, increasing the rate of reaction with the ester.

The hydroxide attacks and forms the tetrahedral intermediate, which collapses to reform the C—O π bond and expel the alkoxide (**Figure 19.26**). Because the product of the reaction is a carboxylic acid ($pK_a = 5$) and because the solution is basic, the carboxylic acid is immediately deprotonated. Thus, to make a carboxylic acid in this manner, a final acid quench step is required. This overall process, the base-promoted hydrolysis of an ester, is referred to as **saponification**.

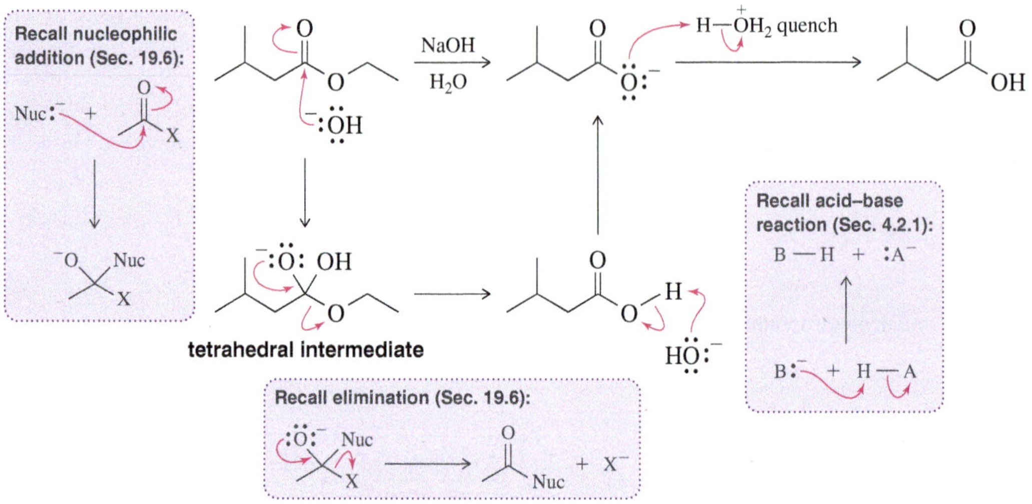

Figure 19.26 **The mechanism of basic ester hydrolysis** (saponification).

[If adding acid (H^+) and adding base (HO^-) both make hydrolysis faster, what about adding both? Unfortunately, they'd just combine to make water. But enzymes can do exactly this—see Chemistry in Context 19-C.] ■

Let's look more closely at the fate of the tetrahedral intermediate in **Figure 19.27**. When the electron pair collapses, there are two possible leaving groups—namely, hydroxide (HO^-) and ethoxide (EtO^-). The conjugate acids of these (H_2O and EtOH, respectively) each have a $pK_a \approx 16$. Thus, they are equally good leaving groups. When hydroxide leaves, the reaction reverts to the starting ester, but when ethoxide leaves, a carboxylic acid is formed. This carboxylic acid is [*quickly and favorably*] deprotonated. This deprotonation drives the reaction to completion, making it functionally irreversible. [*This same principle occurs in the Claisen condensation in Chapter 20.*]

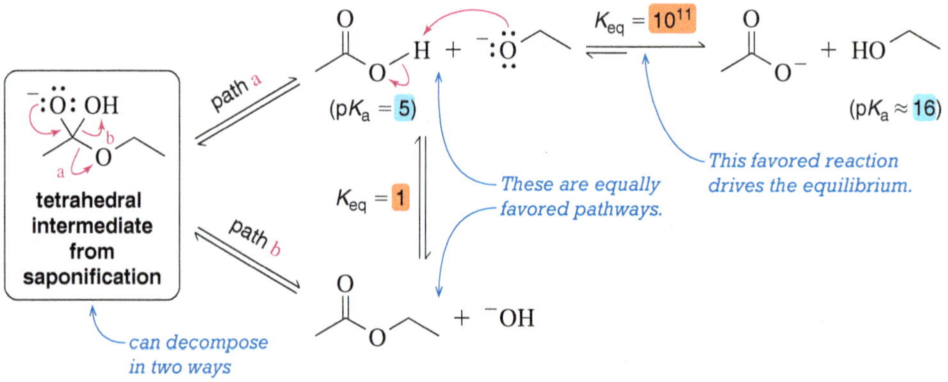

Figure 19.27 **Favorable deprotonation of the resulting carboxylic acid** drives the reaction to completion.

19.7.1.4 H_2O with amides

As the least reactive carboxylic acid derivative, amide hydrolysis requires harsher conditions. In addition to adding acid or base, the reaction is typically heated to force it to completion. Once more, when base is used, a final acid quench step is required to produce the carboxylic acid (**Figure 19.28**).

Figure 19.28 The hydrolysis of amides with (a) acid or (b) base requires heating. [*Try showing a mechanism of these reactions on your own.*]

(a)

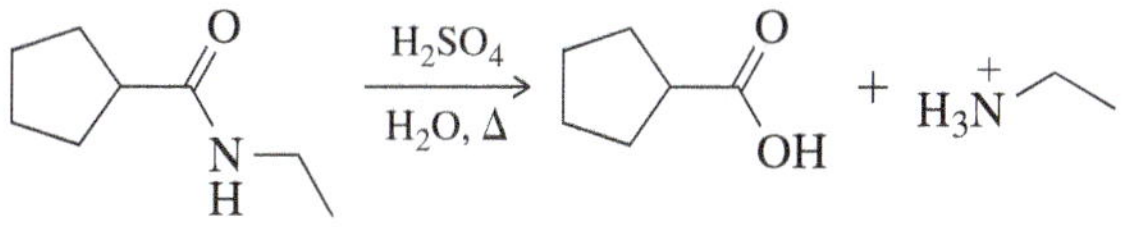

(b)

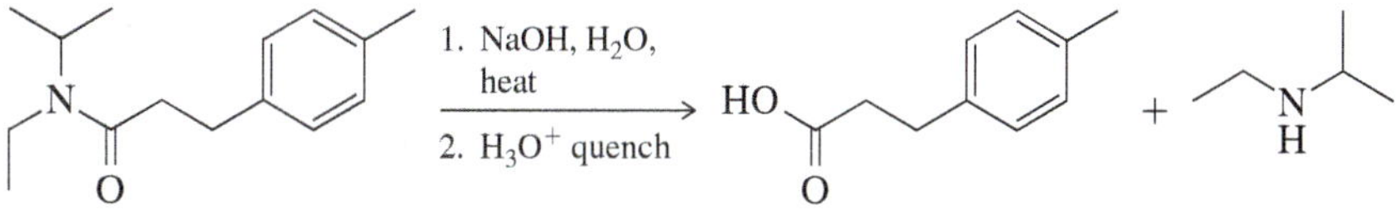

19.7.1 SUMMARY

- The reaction of water with carboxylic acid derivatives produces carboxylic acids.
- Water-soluble acid chlorides and anhydrides are electrophilic enough to react directly with pH neutral water. Water-insoluble acid chlorides and anhydrides, along with esters and amides, require acidic or basic conditions for hydrolysis to occur with reasonable efficiency.

Assessment 19.19 Given the reversibility of the acid-catalyzed hydrolysis/Fischer esterification shown, how would you modify the reaction to make (a) a carboxylic acid or (b) an ester?

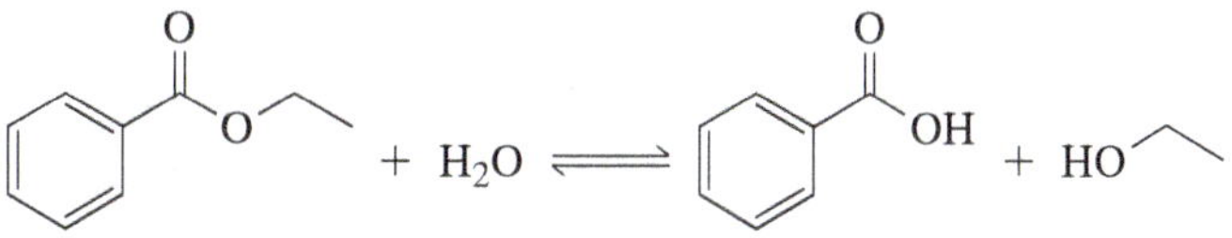

Assessment 19.20 Show a mechanism of each of the following hydrolysis reactions. Pay careful attention to whether acid, base, or neither is present.

(a)

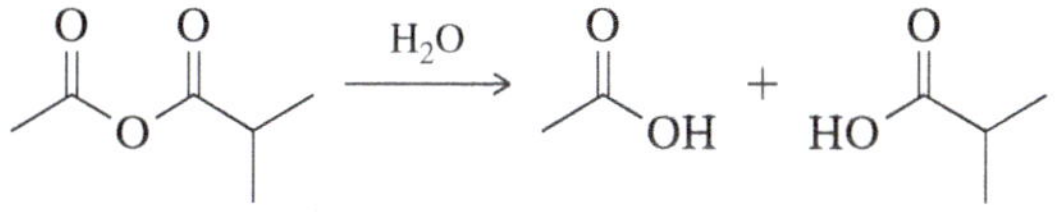

(b)

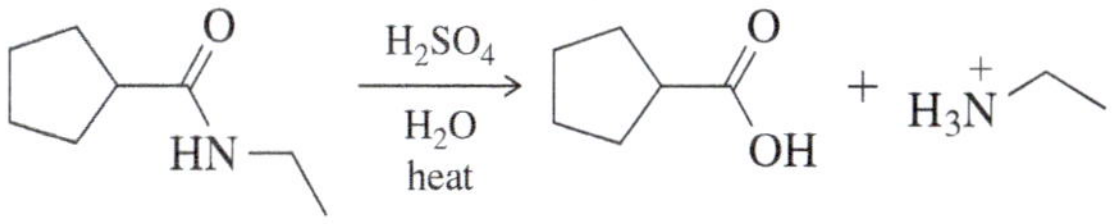

(c)

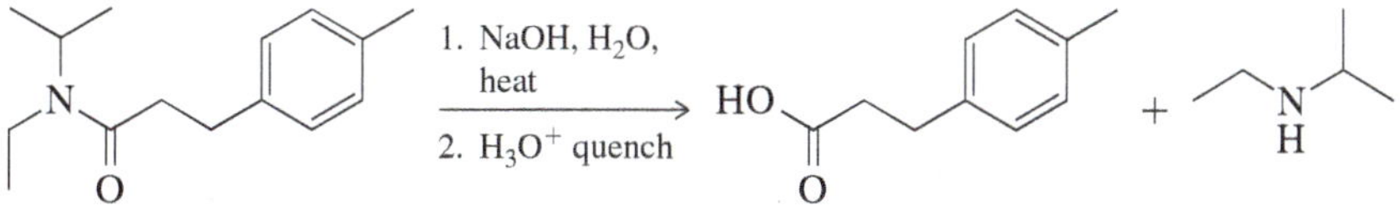

Assessment 19.21 Use a reaction coordinate diagram to rationalize the importance of deprotonation in determining the outcome of the saponification reactions as shown in Figure 19.27.

Assessment 19.22 Saponification gets its name from a process used for centuries to make soap. Reaction of lye (NaOH or KOH leached from wood ashes) with animal fats produced soaps. Predict the structure of these soaps.

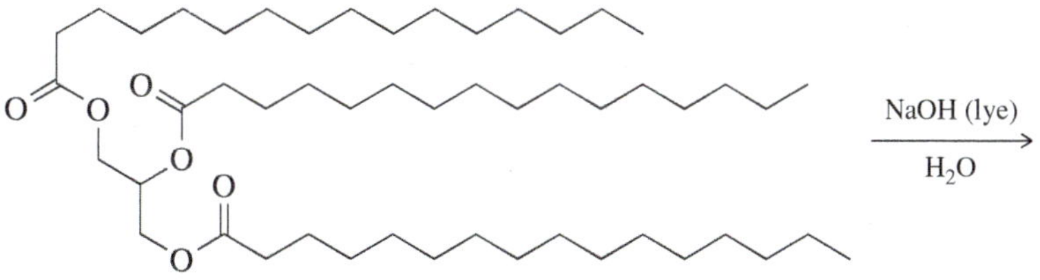

19.7.2 Reactions with alcohols: synthesis of esters

Because water and alcohols differ only by the attachment of an alkyl group in place of one of the hydrogens, the reactions of alcohols with carboxylic acid derivatives resemble those in the previous section. These reactions, when they occur, form an ester (**Figure 19.29**).

Figure 19.29 The reaction of carboxylic acid derivatives with alcohols produces esters.

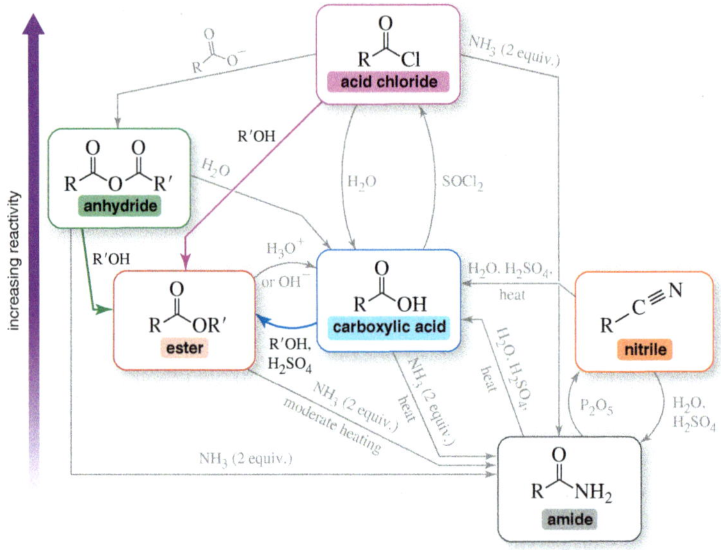

19.7.2.1 ROH with acid chlorides

Like water, ethanol is a weak nucleophile that reacts with highly electrophilic acid chlorides by General Mechanism 18A. [*See* SIDE CONVERSATION 1.] Because alcohol solvents are less polar than water, the solubility issues encountered in water are not relevant here, and there is never a need to add acid or base to promote product formation, though addition of either would make the reaction faster. In the mechanism shown in **Figure 19.30**, acid is generated at the end of the reaction.

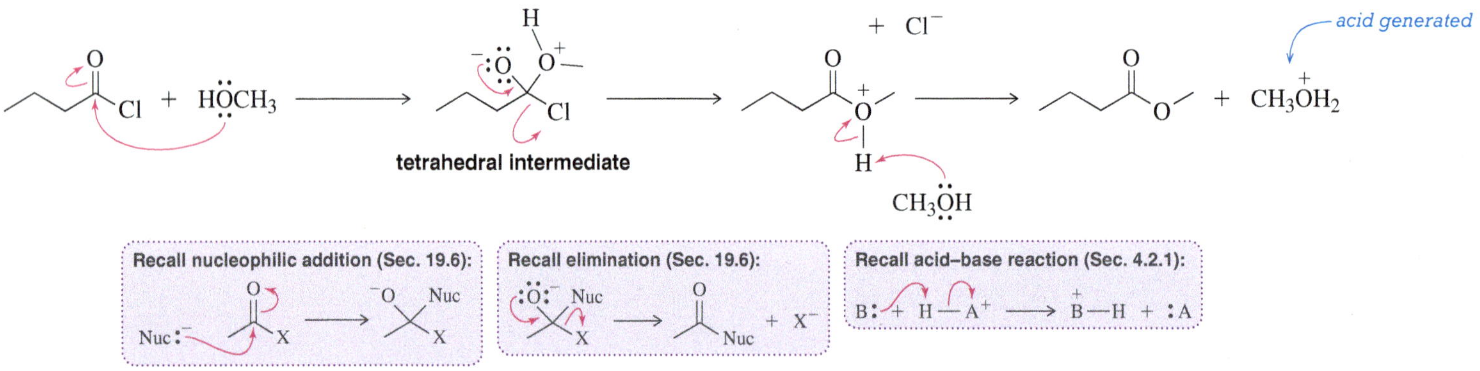

Figure 19.30 Ethanol (and all alcohols) react with acid chlorides by General Mechanism 18A.

SIDE CONVERSATION 3 **Autocatalysis** Acid is formed in the final step of esterification from acid chlorides (see Figure 19.30). Each time the reaction happens, the pH is lowered slightly. Although acid is not needed to promote the reaction of acid chlorides and alcohols, it can catalyze the reaction by protonating the acid chloride, making it more electrophilic. With acid present, the reaction occurs by General Mechanism 18B [**Figure 19.31(a)**]. For each proton that is used to initiate the reaction, two protons are produced, further lowering the pH [**Figure 19.31(b)**]. Each successive reaction produces more catalyst and speeds up the reaction [**Figure 19.31(c)**]. This reaction demonstrates **autocatalysis**, wherein the product of a reaction (in this case, a pair of protons) catalyzes the reaction itself. ■

Figure 19.31 The reaction of alcohols with acid chlorides results in autocatalysis. (a) Two protons are produced for each proton that starts a reaction. As the reaction progresses, (b) the pH decreases and (c) the rate of reaction increases.

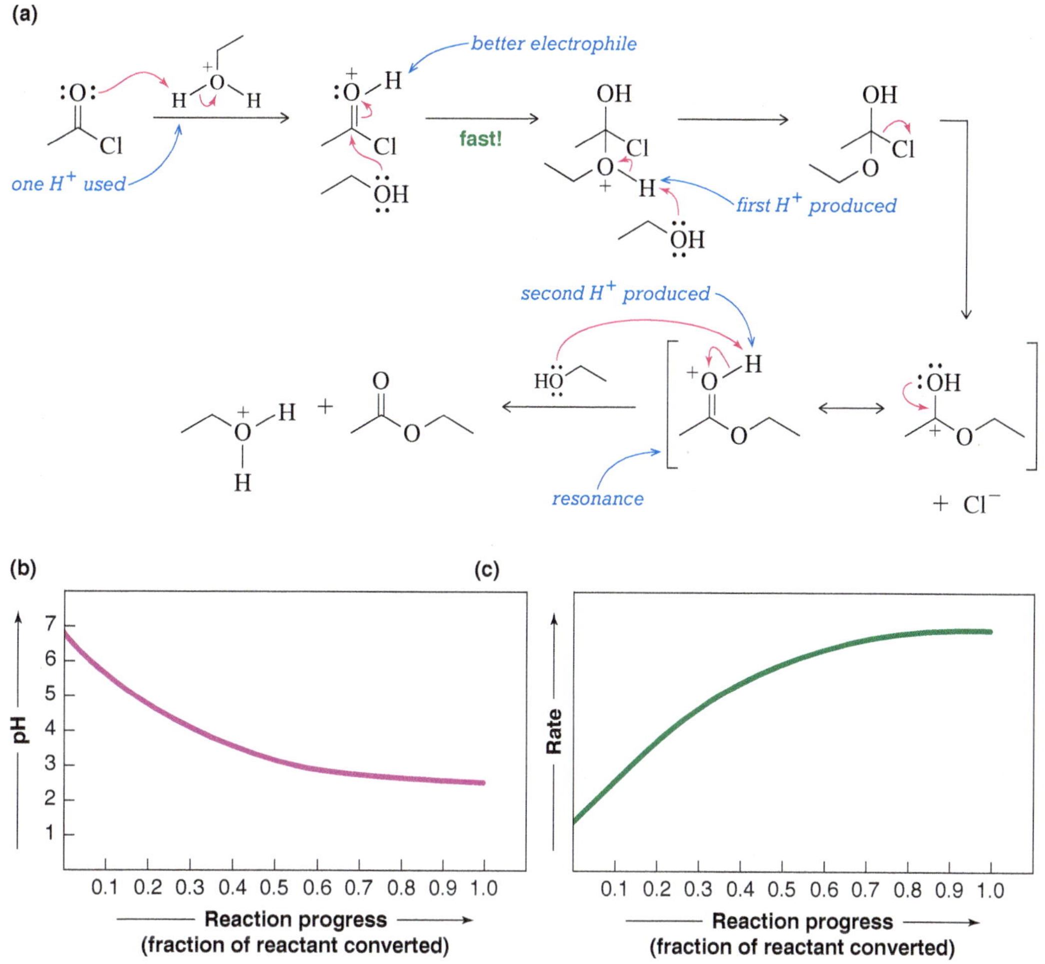

SIDE CONVERSATION 4 **Avoiding side reactions** In addition to increasing the reaction rate, a strong acid can cause side reactions in molecules containing acid-sensitive functional groups (**Figure 19.32**). [*You encountered a similar scenario in Solved Assessment 13.23. How did we solve this problem then?*] To avoid acid-promoted side reactions, alcohols are often reacted with acid chlorides in the presence of a neutral amine base. Because the base is fairly weak, the proton is only removed after the alcohol attacks the acid chloride. ■

Figure 19.32 A neutral amine base is used to prevent acid-promoted side reactions.

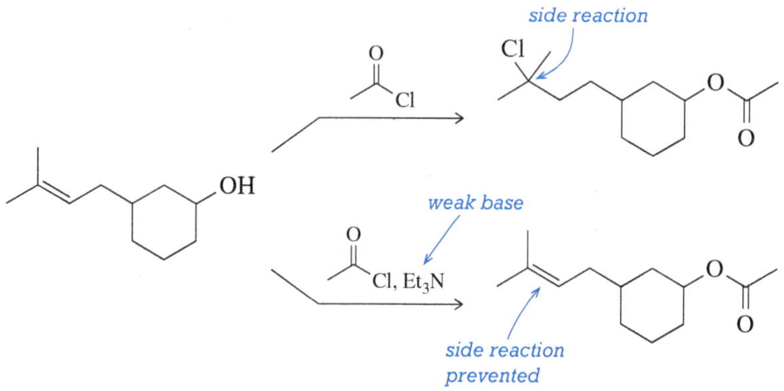

19.7.2.2 ROH with anhydrides

Anhydrides, again, react similarly to acid chlorides. A neutral amine base is used to prevent acid-promoted side reactions. Acetic anhydride is used to protect alcohols, as shown in **Figure 19.33**. Protection of the 1° alcohol as an ester is favored over the 2° because of steric hinderance.

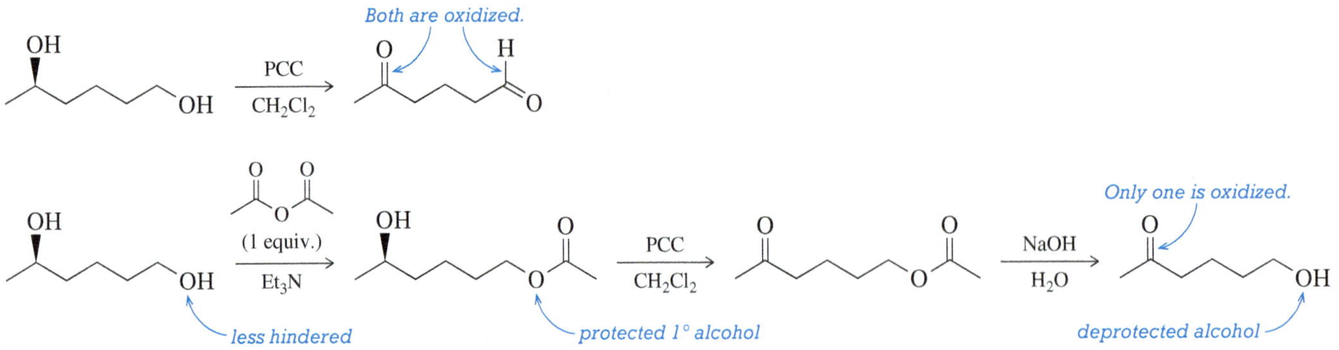

Figure 19.33 Acetic anhydride is often used to protect alcohols as esters.

Although they react similarly, chemists generally use anhydrides instead of acid chlorides because anhydrides are slightly more stable (Figure 19.9), so they are more easily handled and stored.

19.7.2.3 ROH with esters

The nucleophilic acyl substitution of an ester with an alcohol simply produces a different ester (**Figure 19.34**). Referred to as **transesterification,** this reaction has $\Delta G \approx 0$ ($K_{eq} \approx 1$). The equilibrium is shifted using Le Châtelier's principle. Most often, this is possible when the starting alcohol boils at a higher temperature than the product alcohol. The lower-boiling alcohol evaporates and drives the equilibrium to the product side. Either acidic or basic conditions can be used. [*In truth, these reactions are not useful unless the ester is itself the final product in a synthetic sequence. This is because the alkoxy group doesn't dramatically change the reactivity of an ester. Assessments 19.40 and 19.44 further illustrate this principle.*]

Figure 19.34 Ester substitution with an alcohol needs to be driven by Le Châtelier's principle because neither side is favored by ΔG.

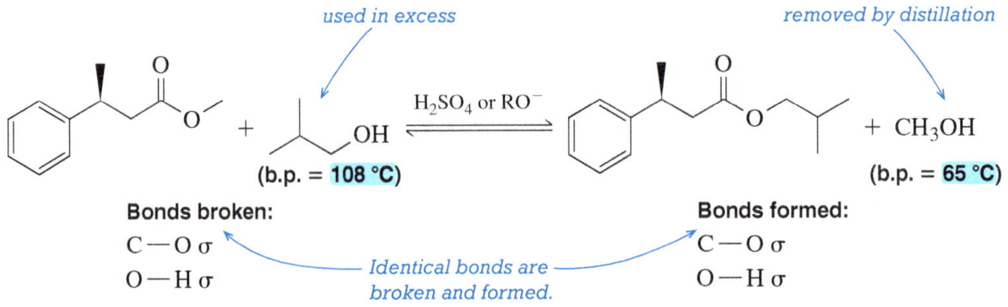

19.7.2.4 ROH with amides

Amides are typically unreactive with alcohols, so multiple steps are usually required if you want to prepare an ester from an amide. [*Any thoughts on how to do the synthesis in* ***Figure 19.35****? It comes up again in Section 19.7.5.*]

Figure 19.35 How do you convert an amide to an ester? Think about it, then see Section 19.7.5.

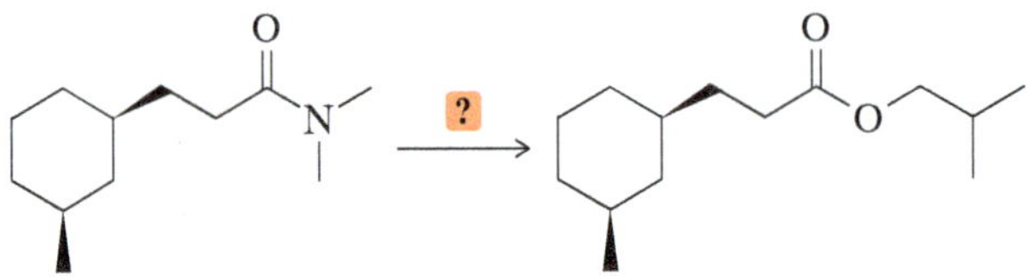

19.7.2 SUMMARY

- The reaction of alcohols with carboxylic acid derivatives produces esters.
- Acid chlorides and anhydrides are electrophilic enough to react directly with pH neutral alcohols but require a weak base to prevent the occurrence of acid-promoted side reactions. Transesterification requires acidic or basic conditions. The reaction of alcohols with amides is not a practically useful reaction.

Assessment 19.23 For each pair, identify the faster reaction.

(a)

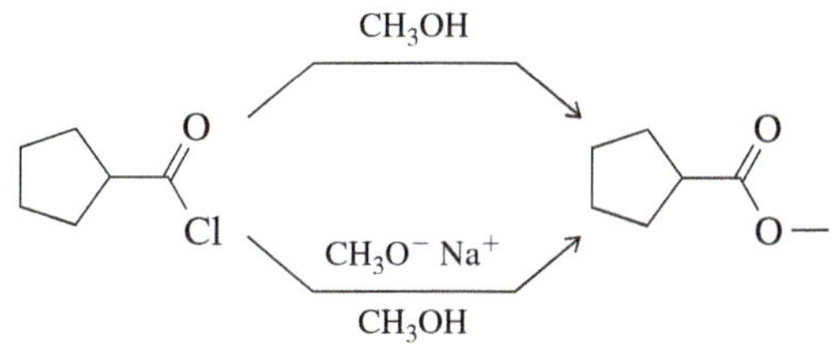

(b)

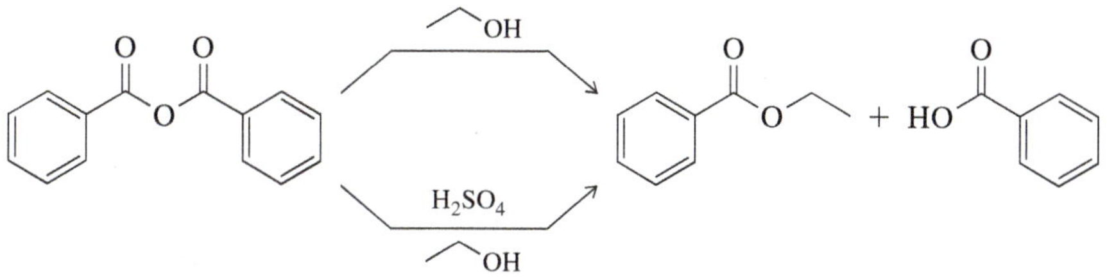

Assessment 19.24 A chemist unsuccessfully attempted to produce the 1,4-cyclohexanediol by hydration of the cyclohexene shown. (a) Provide a mechanism for the formation of the actual product. (b) Suggest a pathway using acetyl chloride as a protecting group that will allow for the formation of the desired product.

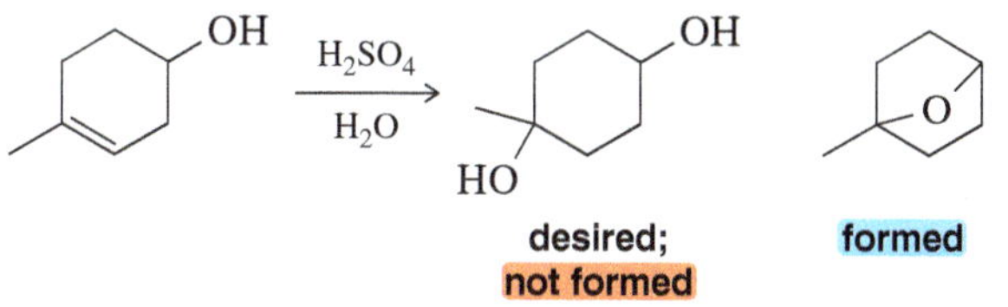

Assessment 19.25 Tertiary amines (like triethyl amine) and pyridine are typically used as a weak base in nucleophilic acyl substitution reactions. Why aren't 2° or 1° amines used?

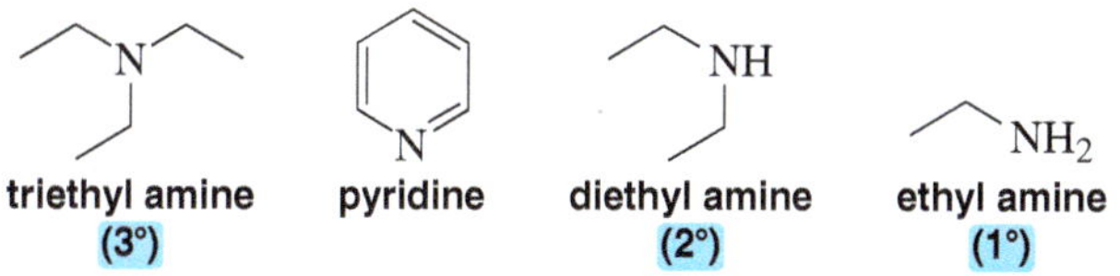

Assessment 19.26 The transesterification reaction shown cannot be performed efficiently. Why? Suggest a two- and three-step synthesis that will enable the synthesis of ester **B** from ester **A**.

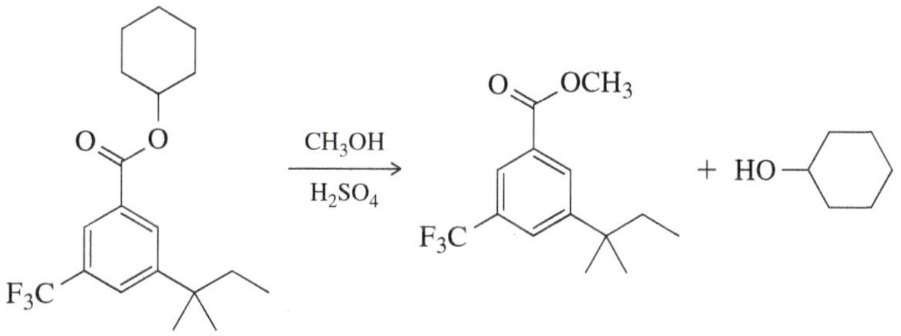

A **B**

Assessment 19.27 Predict the product of the following reactions. [*The nucleophile and the electrophile have been placed in different spots about the reaction arrow, something that commonly trips students up.*]

(a)

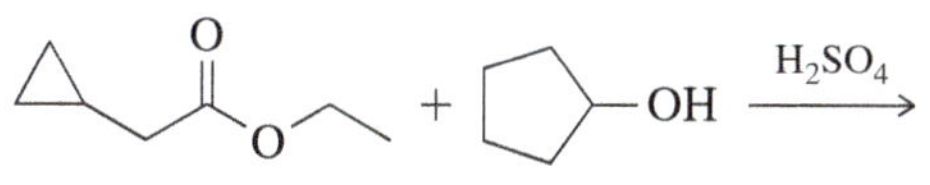

(b)

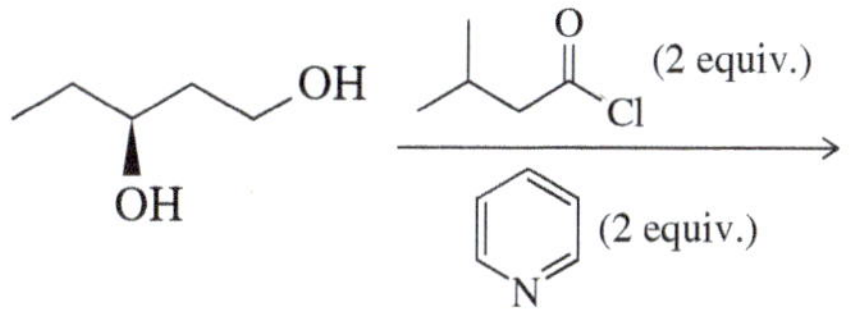

(c)

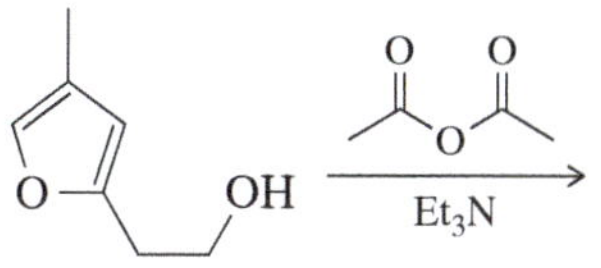

(d)

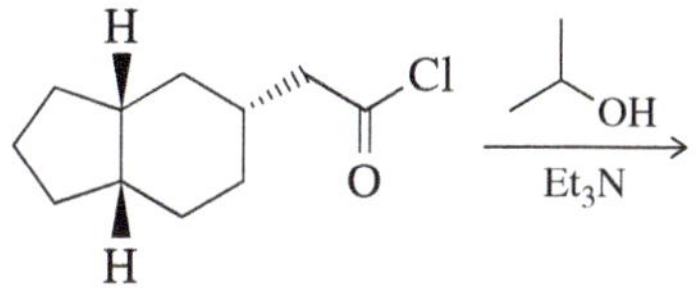

19.7.3 Reactions with amines: the synthesis of amides

Amines are stronger nucleophiles than water and alcohols, so amines readily react with carboxylic acid derivatives to give amides (**Figure 19.36**).

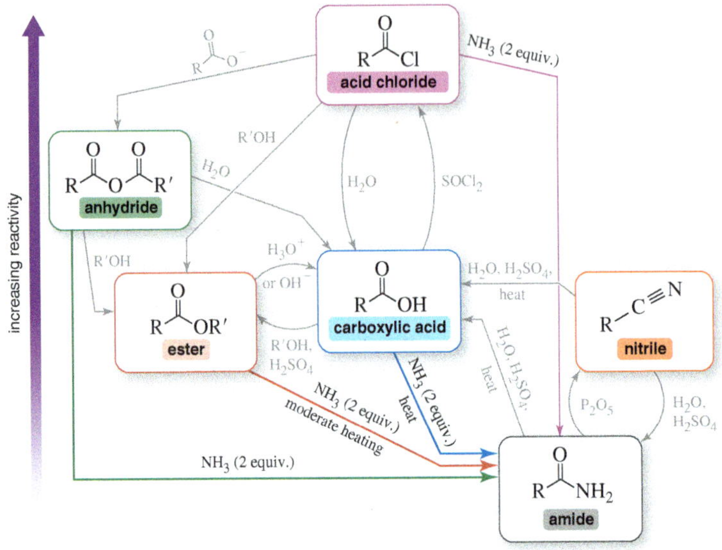

Figure 19.36 The reaction of carboxylic acid derivatives with amines produces amides.

19.7.3.1 Amines with acid chlorides

Unlike water and alcohols, amines are moderately strong nucleophiles. [*Recall the reaction of amines with aldehydes and ketones in Section 17.8.*] As a result, amines react more readily with highly electrophilic acid chlorides by General Mechanism 18A. Acid is generated by the reaction (**Figure 19.37**), and this acid is immediately deprotonated by unreacted amine in solution. Because one equivalent does the substitution and one does the deprotonation, two equivalents of amine are required. The second equivalent is usually discarded as waste.

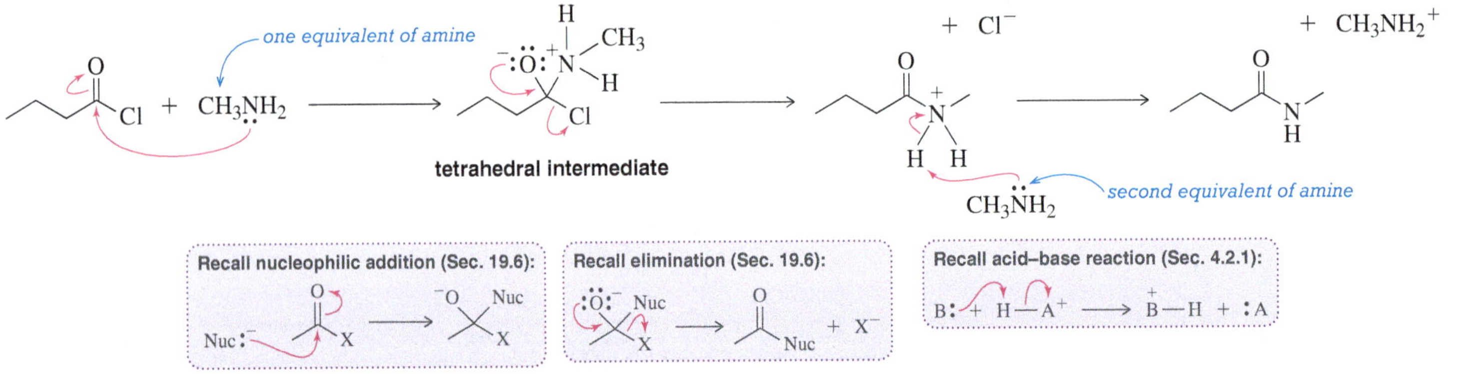

Figure 19.37 Amines react with acid chlorides by General Mechanism 18A. Two equivalents of amine are required to fully convert the acid chloride to the amide.

Ammonia, 1°, and 2° amines can form amides with acid chlorides (**Figure 19.38**). Tertiary amines do not, however, because there is no proton to remove. [*Does this make sense, considering your answer to Assessment 19.25?*]

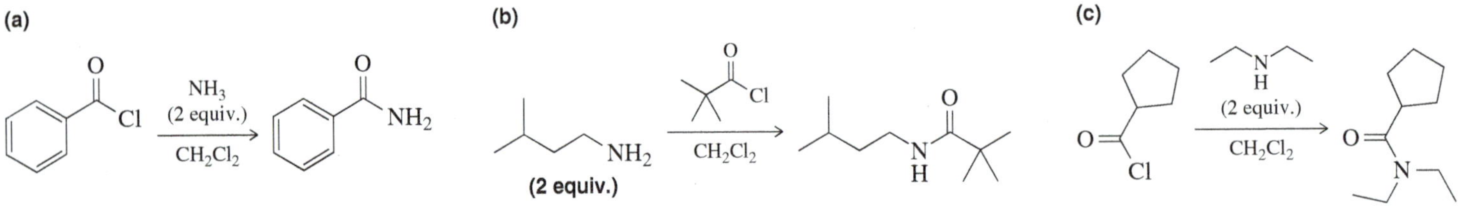

Figure 19.38 (a) Ammonia, (b) 1°, and (c) 2° amines react with acid chlorides to form amides.

19-B CHEMISTRY IN CONTEXT: Green Chemistry

Avoiding waste

Using two equivalents of amine is fine when the amine is small, inexpensive, and commercially available. But what if it's not? What if the synthesis of the amine in **Figure 19.39** required 14 synthetic steps and two years of graduate school? Would you discard half of this precious material as waste? [*Trust me, you don't want to lose any late-stage synthesis material. Ever.*] But we still need a base to remove the proton. [*Any ideas where we might get a nonnucleophilic base that won't predictably attack the acid chloride but will deprotonate the acidic hydrogen as it forms?*] Triethyl amine is perfect for this. Indeed, the reaction of amines with acid chlorides is often done with a neutral amine base acting as the second equivalent of amine.

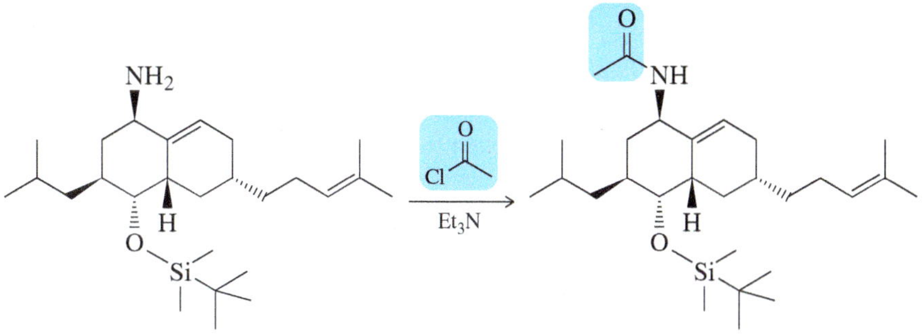

Figure 19.39 The use of a weak amine base (Et_3N) means that only one equivalent of the nucleophilic amine is consumed. Et_3N serves as the second equivalent of amine.

19.7.3.2 Amines with anhydrides

Anhydrides, again, react like acid chlorides. A neutral amine base is used to prevent the waste of one equivalent of amine. Acetic anhydride is commonly used to protect amines, as shown in **Figure 19.40**.

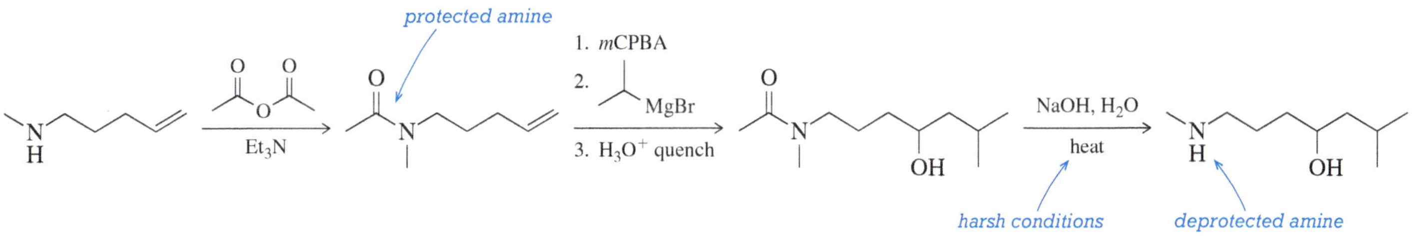

Figure 19.40 Acetic anhydride is used as a protecting group for amines. Removal of the resulting amide requires harsher conditions.

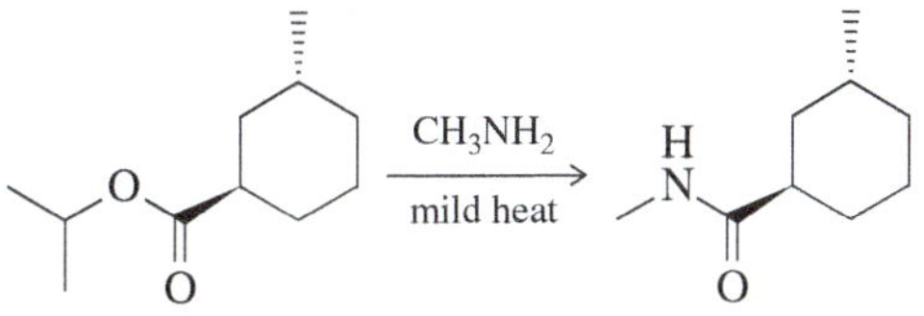

Figure 19.41 Amines react with esters only with mild heating.

19.7.3.3 Amines with esters

Esters are generally too stable to react with amines at ambient temperature (25 °C). With moderate heating, substitution occurs through General Mechanism 18A (**Figure 19.41**). You can't add an acid catalyst because added acid will simply protonate the basic amine. Because of the difficulty of this reaction, the conversion of esters to amides typically occurs through a multistep pathway. [*See Section 19.7.5.*]

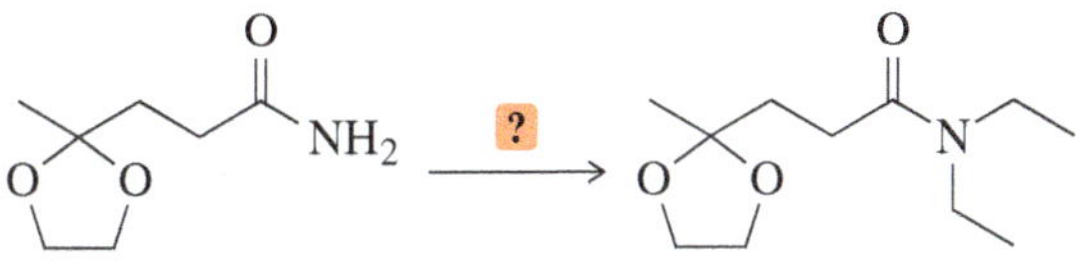

Figure 19.42 How do you convert an amide to a different amide? Think about it, then see Section 19.7.5.

19.7.3.4 Amines with amides

The reaction of amines with amides is not possible because of the diminished reactivity of amides. If you want to prepare an amide from a different amide, multiple steps are required. [*How might you carry out the conversion in **Figure 19.42**? We study this reaction in Section 19.7.5.*]

19.7.3 SUMMARY

- The reaction of amines with carboxylic acid derivatives produces amides.
- Acid chlorides and anhydrides are electrophilic enough to react directly with amines but require a second equivalent of amine to drive the reaction to completion. Moderate heating is required to convert esters to amides. The conversion of one amide to another is not practically useful.

Assessment 19.28 Because amines are moderately basic, acid catalysts are not useful for increasing the rate of acyl substitution reactions. Calculate K_{eq} of an amine reacting with sulfuric acid.

Assessment 19.29 $\Delta G = 0$ for transesterification, so Le Châtelier's principle could be used to drive the reaction in one direction or the other (Figure 19.34). $\Delta G = 0$ for transamidation, too, yet amides can't be converted directly to other amides (Figure 19.42). What reaction parameter is preventing this reaction from occurring?

Assessment 19.30 Azides can be reduced to give amines using triphenyl phosphine in a reaction we discuss in Chapter 25. The amine, once formed, cyclizes to give the cyclic amide (lactam) shown, even without heating. (a) Show a mechanism for the second step. (b) Why is the reaction so efficient, as compared to other amine reactions with esters?

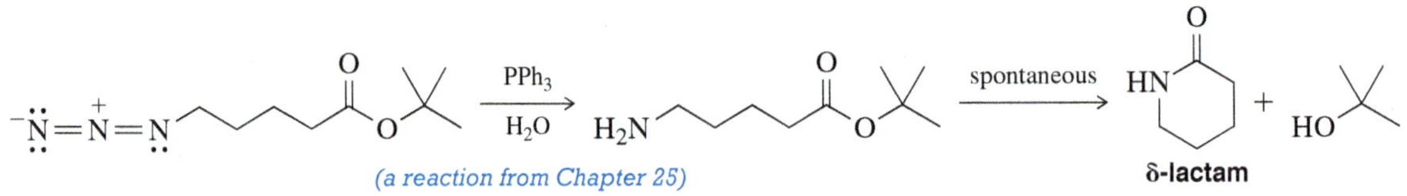

Assessment 19.31 Predict the product of the following reactions. [*The nucleophile and the electrophile have been placed in different spots about the reaction arrow, which can trip students up.*]

(a)

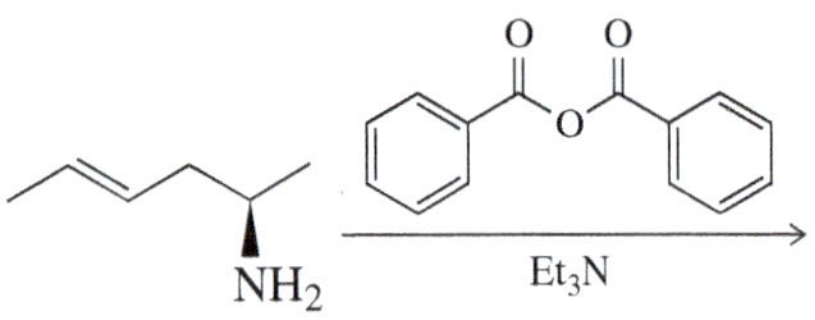

(b)

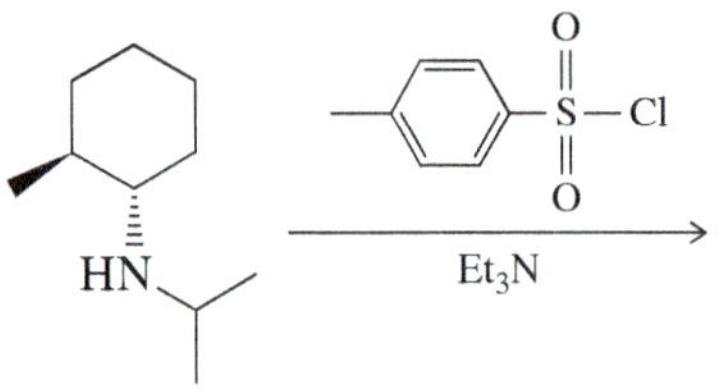

(c)

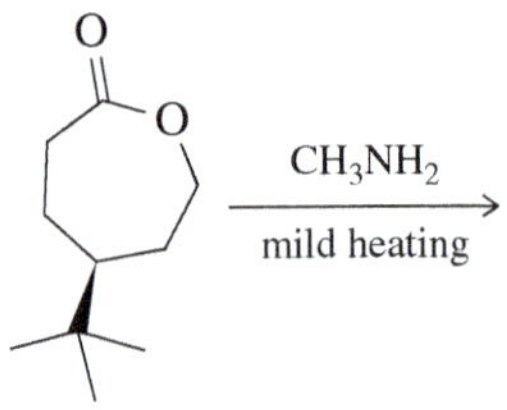

(d)

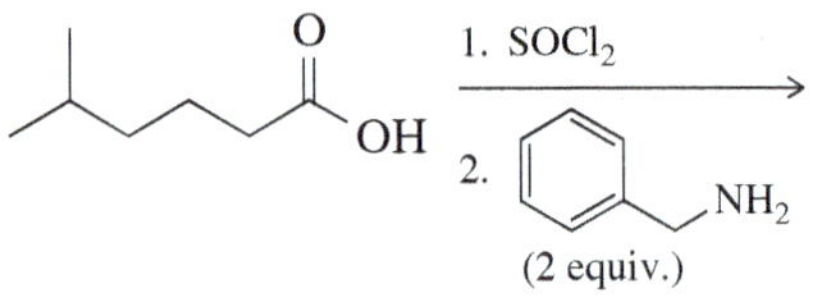

19-C CHEMISTRY IN CONTEXT: Biochemistry

Building the bacterial cell wall

The bacterial cell wall is rigid as the result of cross-links between the peptide regions of two different peptidoglycans. The overall reaction (**Figure 19.43**) is a transamidation, with a terminal amine group of peptidoglycan 2 (PG2 in the figure) displacing an amino acid (alanine) from the terminal end of peptidoglycan 1 (PG1). Amides are poor electrophiles and amines are bad leaving groups, so this reaction is extremely slow in solution. Bacteria do this with the enzyme D-alanyl-D-alanine carboxypeptidase, which is known as a transpeptidase because it exchanges one peptide (amide) bond for another.

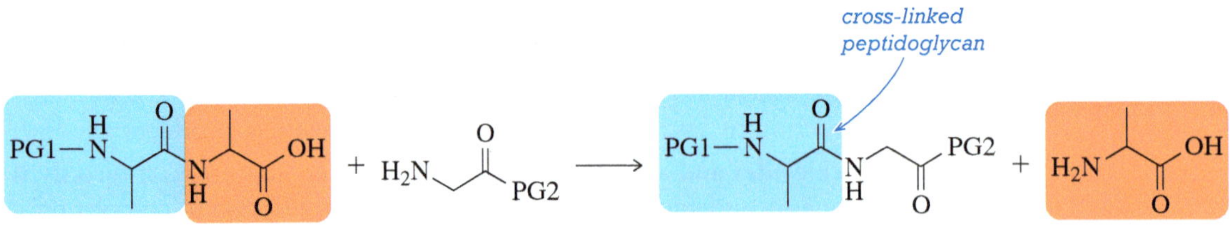

Figure 19.43 The overall reaction catalyzed by D-alanyl-D-alanine carboxypeptidase is a transamidation.

(*continued*)

Let's break the mechanism down into two halves. In the first half (**Figure 19.44**), the relatively unreactive amide is converted to an ester, which is a more reactive carboxylic acid derivative. The amide of peptidoglycan 1 interacts with a positively charged active site residue and becomes more electrophilic. At the same time, a base converts an alcohol into a stronger alkoxide nucleophile. Finally, the tetrahedral intermediate collapses to expel an amide leaving group after protonation. The result of these two steps is an enzyme-bound ester (peptidoglycan 1) and release of the free amino acid alanine.

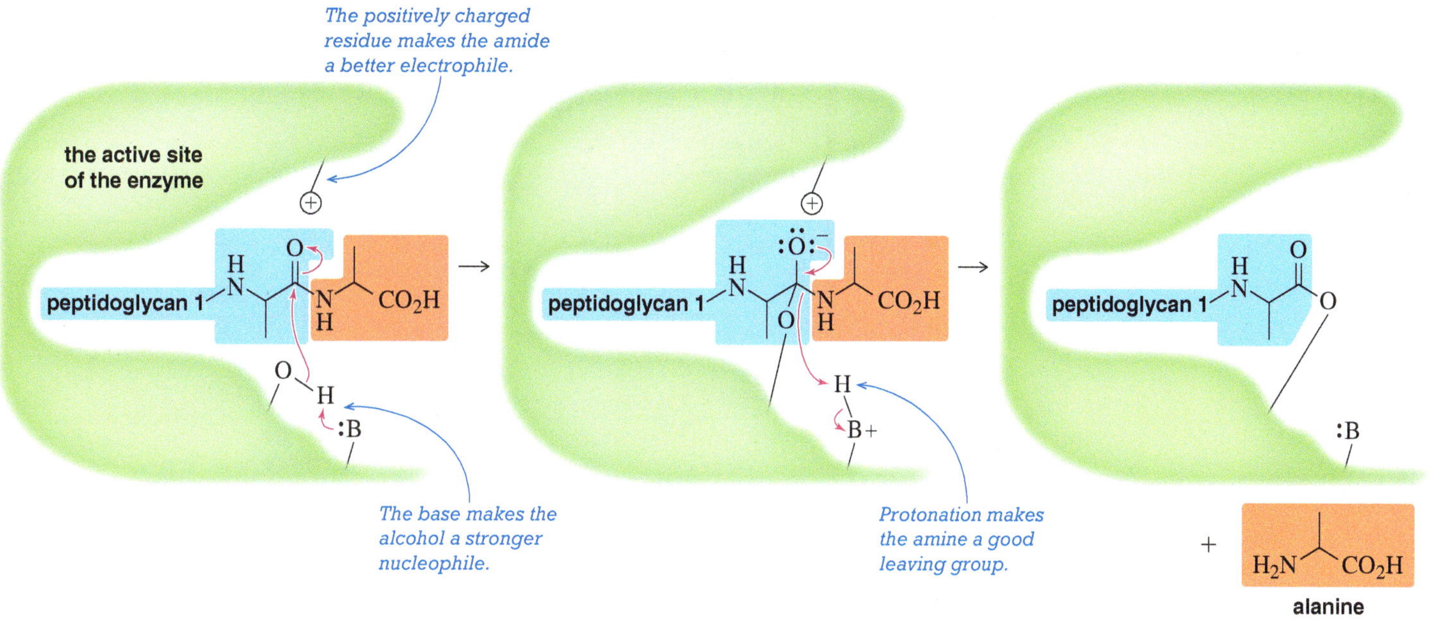

Figure 19.44 The first half of the peptidoglycan cross-linking catalyzed by D-alanyl-D-alanine carboxypeptidase.

The second half of the mechanism (**Figure 19.45**) attaches peptidoglycan 2 to peptidoglycan 1. The amine attacks the enzyme-bound ester, forming a tetrahedral intermediate. The ester is again activated by the positively charged active site residue, while the neighboring active site base deprotonates the resulting ammonium. Collapse of the tetrahedral intermediate forms the cross-linked peptidoglycan, which is released from the active site, freeing it up to do the conversion again.

[*This is a complicated biochemical mechanism. You certainly don't need to know all of the steps shown. Focus, instead, on the steps you recognize, such as nucleophilic attack, the tetrahedral intermediate, and elimination of a leaving group. And don't forget to be in awe of how enzymes make slow reactions fast!*]

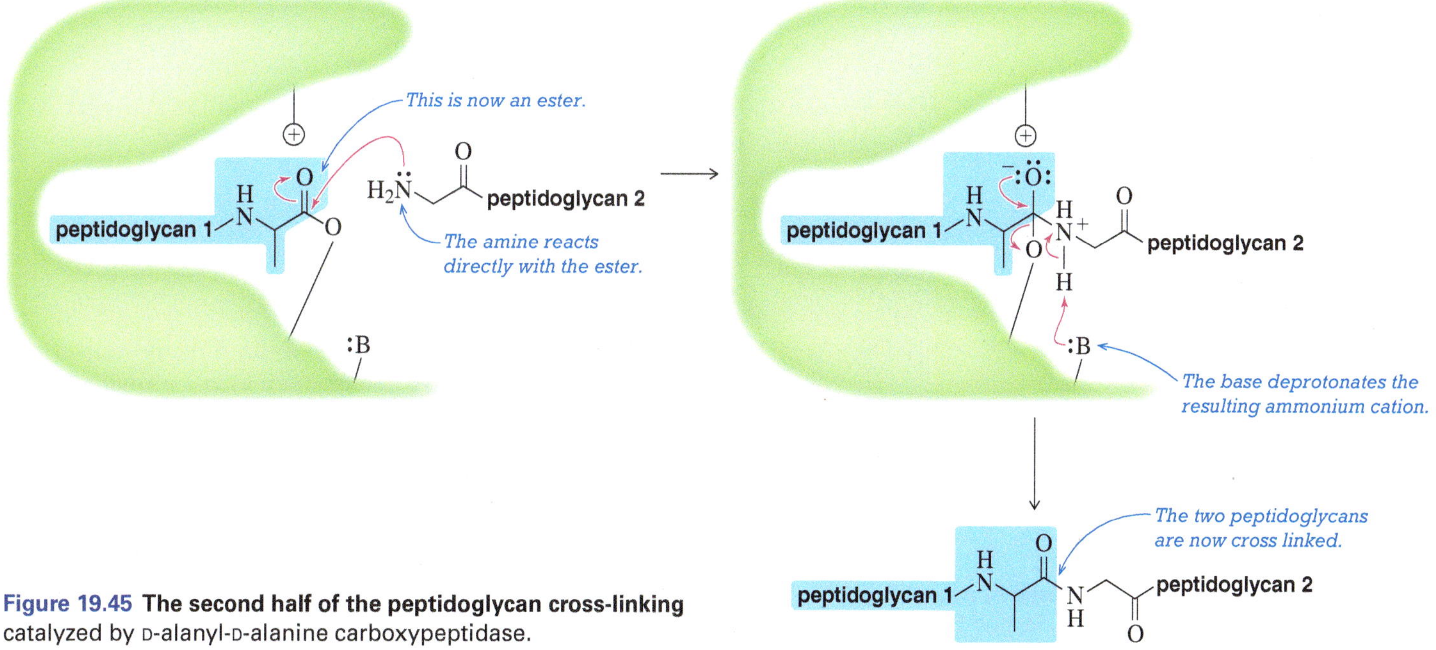

Figure 19.45 The second half of the peptidoglycan cross-linking catalyzed by D-alanyl-D-alanine carboxypeptidase.

19.7.4 Reactions with carboxylates: the synthesis of anhydrides

The synthesis of anhydrides is performed most efficiently by reacting acid chlorides with carboxylate ions (**Figure 19.46**). [*Think about why this must be before reading further.*]

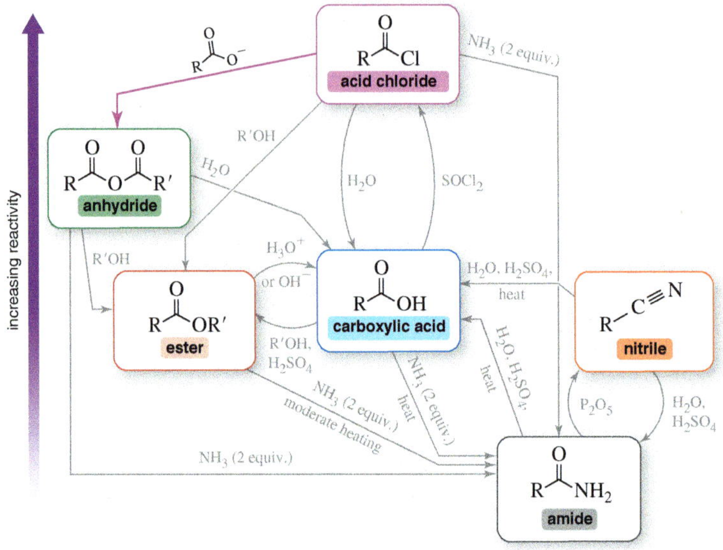

Figure 19.46 The reaction of carboxylic acid derivatives with acid chlorides produces anhydrides.

Take a moment to review Figure 19.46. Acid chlorides are the only carboxylic acid derivative more reactive than anhydrides, so you can easily make an anhydride from an acid chloride. Esters and amides, on the other hand, are more stable than anhydrides. To go from an ester or an amide to an anhydride would be uphill in potential energy. [*The ball rolls downhill, remember.*] Additionally, the resonance-stabilized carboxylate ion, though negatively charged, is a weak nucleophile—it is only powerful enough to displace a chloride ion. This reaction (**Figure 19.47**) proceeds by General Mechanism 18A.

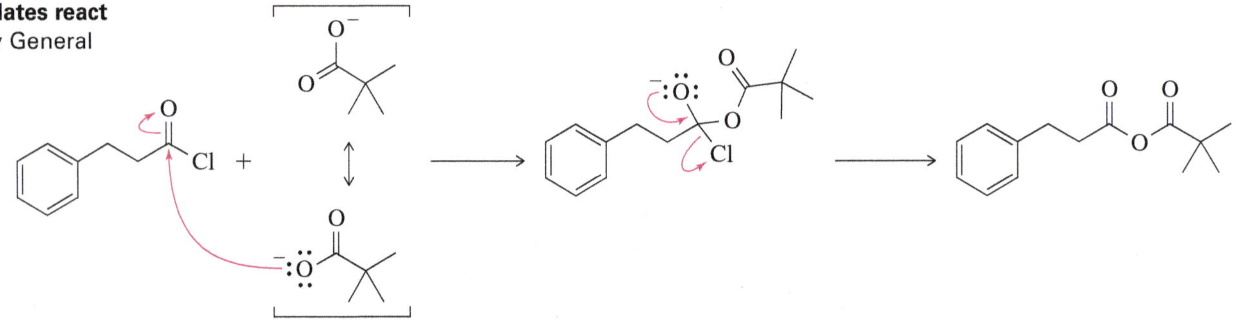

Figure 19.47 Carboxylates react with acid chlorides by General Mechanism 18A.

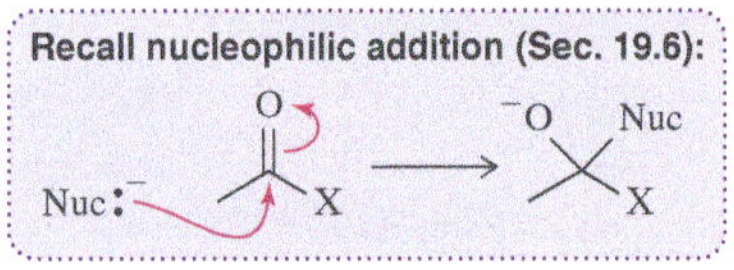

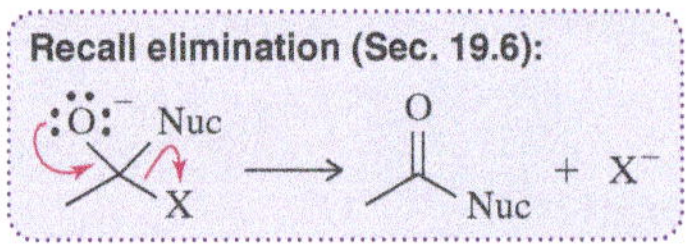

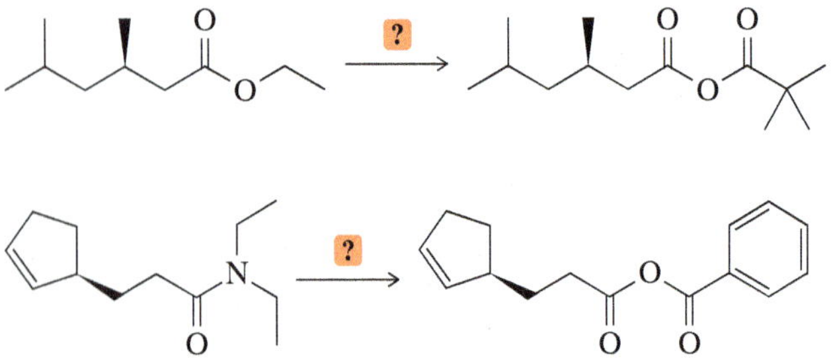

Figure 19.48 How do you convert an ester or an amide to an anhydride? Think about it, then see Section 19.7.5.

To reiterate, esters and amides cannot be directly converted to anhydrides. To prepare an anhydride from an ester or amide, multiple steps are usually required. [*Any thoughts on how you might carry out the conversions in* ***Figure 19.48****? We look at these in Section 19.7.5.*]

19.7.4 SUMMARY

- The reaction of a carboxylate ion with an acid chloride gives an anhydride. Acid chlorides are the only carboxylic acid derivative that react with weakly reactive resonance-stabilized carboxylate nucleophiles.

Assessment 19.32 Predict the product of the following reactions.

(a)

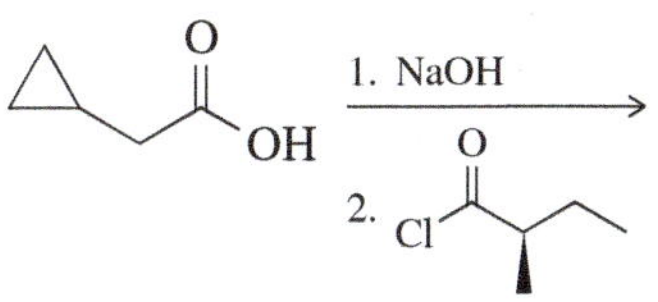

(b)

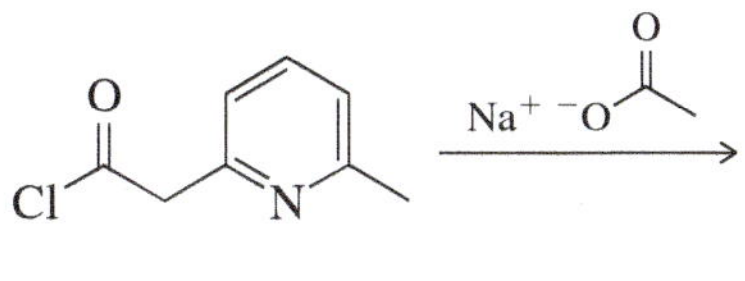

(c)

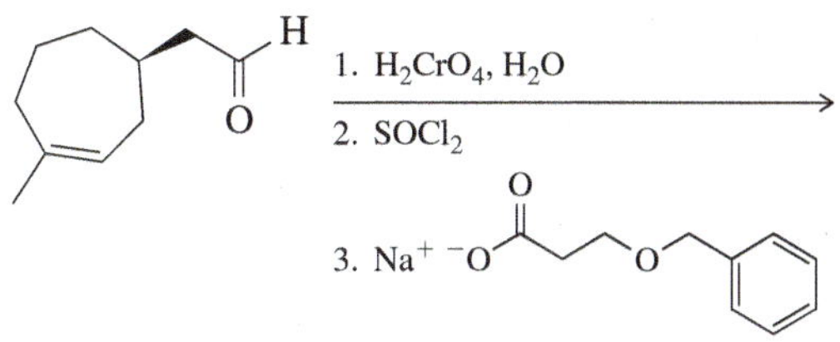

19.7.5 The synthesis of carboxylic acid derivatives

Finally, we have arrived at Section 19.7.5. You were directed here on four previous occasions in order to check your guesses on the more difficult syntheses. These "difficult" syntheses are summarized in **Figure 19.49**.

(a) Amide to amide (from Figure 19.42):

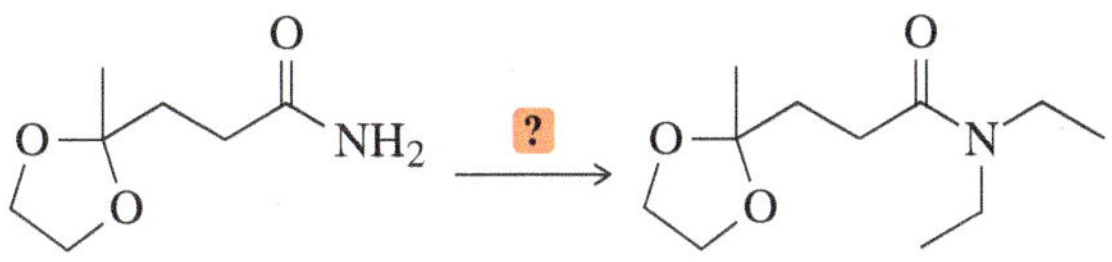

(b) Amide to ester (from Figure 19.35):

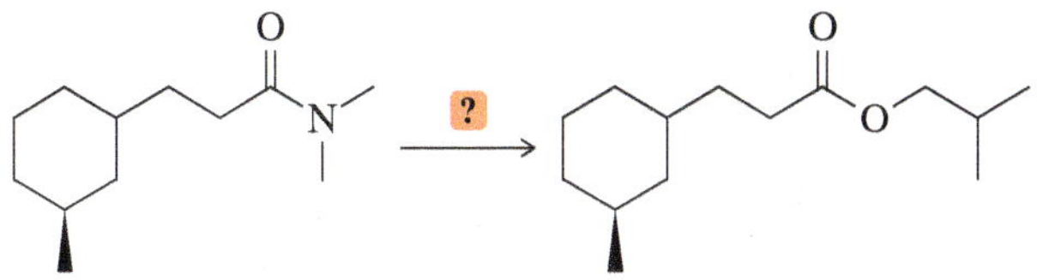

(c) Ester to anhydride (from Figure 19.48):

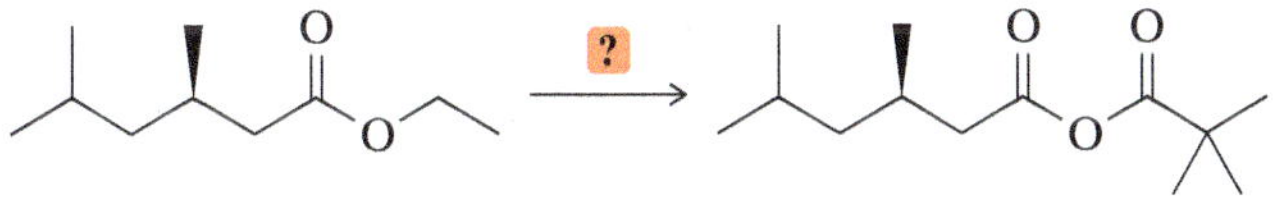

(d) Amide to anhydride (from Figure 19.48):

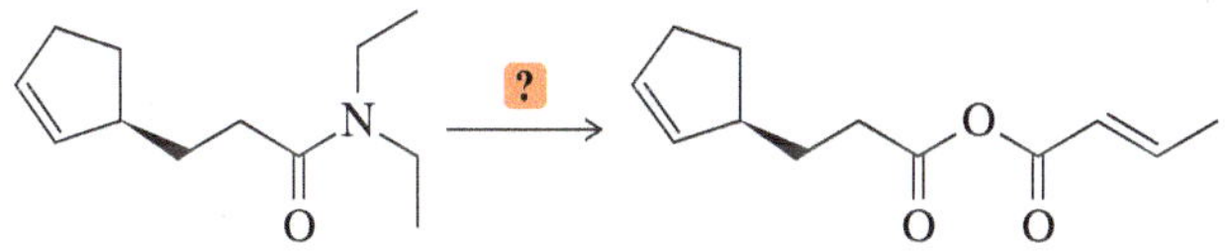

Figure 19.49 Syntheses previously deemed challenging and requiring multiple steps.

Let's think like a chemist to see if a general strategy can be developed for the synthesis of these and other carboxylic acid derivatives.

Thinking Like a Chemist

The primary issue with doing the syntheses in Figure 19.49 in a single step is the diminished reactivity of the reactants. Recall that reactivity decreases in the order of acid chlorides, anhydrides, esters, and amides. The reactions in Sections 19.7.1 to 19.7.4 that were most efficient were predominantly going downhill in reactivity, from the least stable to the more stable carboxylic acid derivative.

The idea that reactions proceed more efficiently when products are less reactive is certainly not new. [*We learned it in Chapter 1.*] Now it gives us a clue how to do the syntheses in Figure 19.49. We need to get to the "top" of the reactivity hill—to the acid chloride. [*Is there a reaction that allows us to do this?*] Recall from Chapter 18 that carboxylic acids react with thionyl chloride ($SOCl_2$) to make acid chlorides (**Figure 19.50**). [*I promised this reaction would be important to Chapter 19; now we see why.*] ■

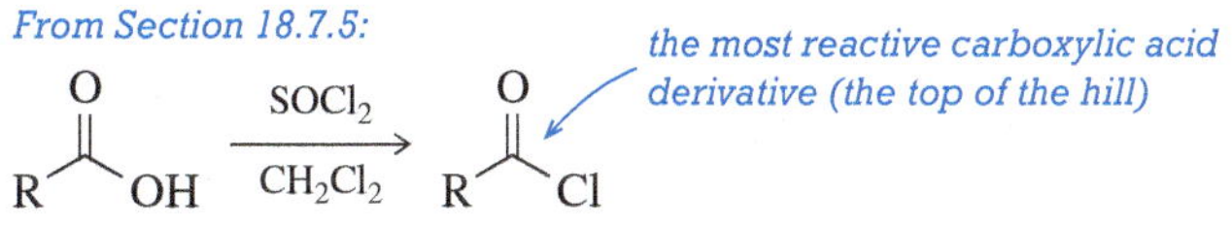

Figure 19.50 Carboxylic acids are converted to acid chlorides with $SOCl_2$.

Fortunately, Section 19.7.1 showed us that every carboxylic acid derivative can be hydrolyzed to a carboxylic acid, even if some require harsher conditions (e.g., acid, base, and/or heating). Thus, the solution here lies in an initial conversion to a carboxylic acid, which is then converted to the acid chloride. And now, being at the top of the hill, any other derivative can be made with the appropriate reagent. [*Look carefully at the examples in* **Table 19.7**, *then practice some on your own in Assessment 19.33.*]

Table 19.7 Multistep synthesis requires getting to the top of the reactivity hill to make more-reactive derivatives from less-reactive starting materials.

Transformation	Example
Amide to amide	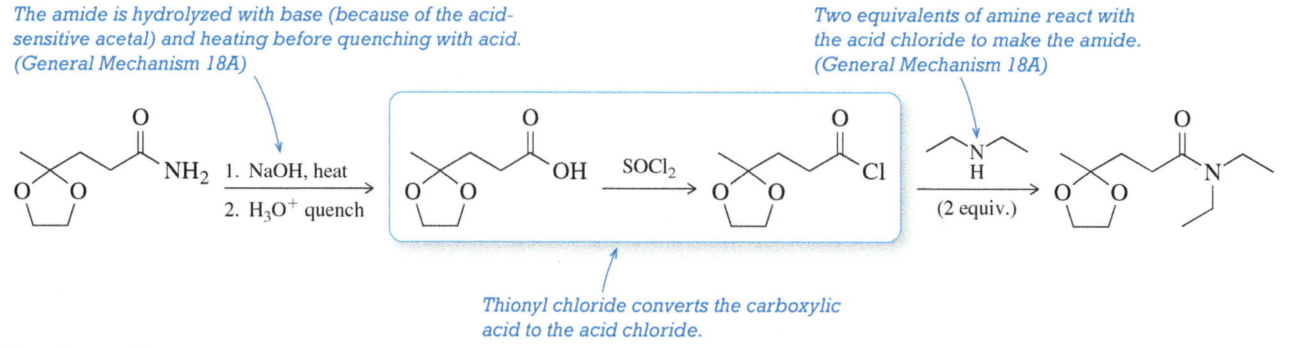
Amide to ester	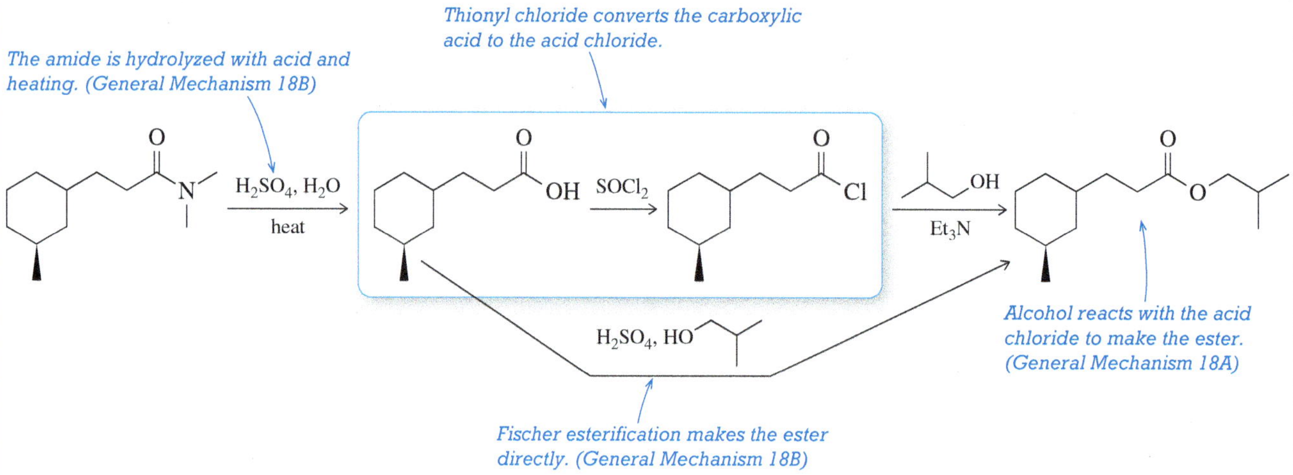
Ester to anhydride	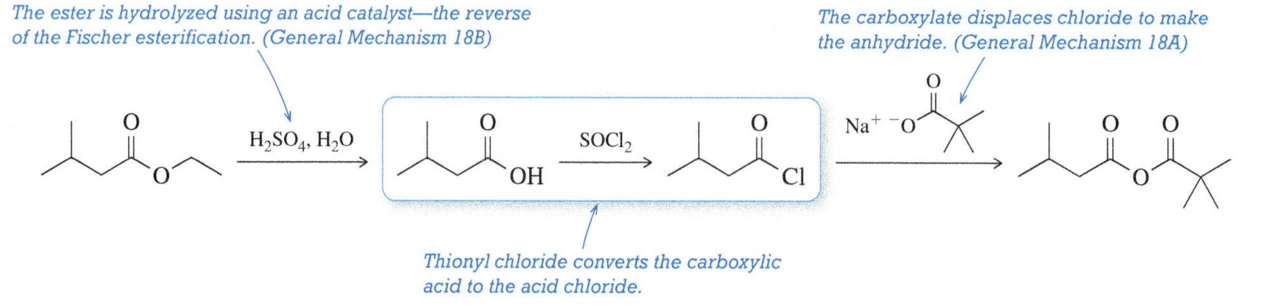
Amide to anhydride	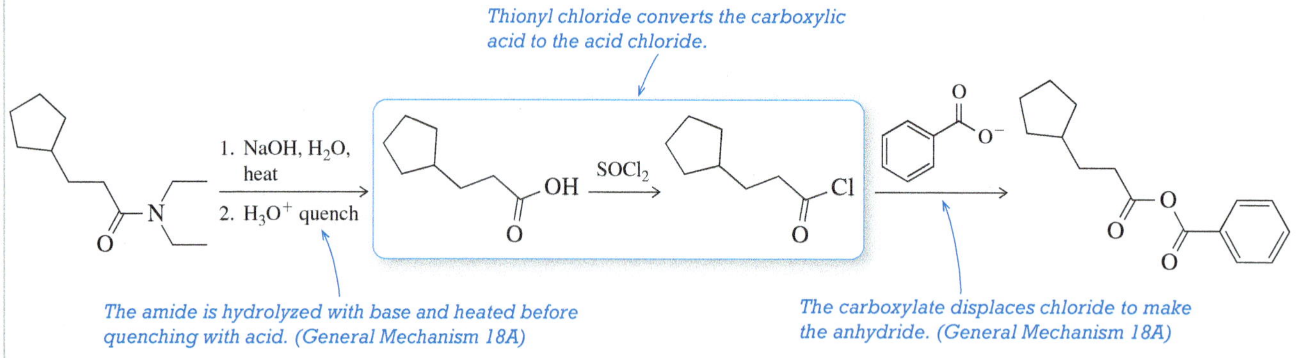

19.7.5 SUMMARY

- In general, carboxylic acid derivative synthesis requires going up the reactivity hill so that all future reactions are downhill. The conversion of a carboxylic acid to an acid chloride using $SOCl_2$ is the most important reaction in this regard.

Assessment 19.33 Starting with the molecule on the left, synthesize the carboxylic acid derivative on the right.

(a)

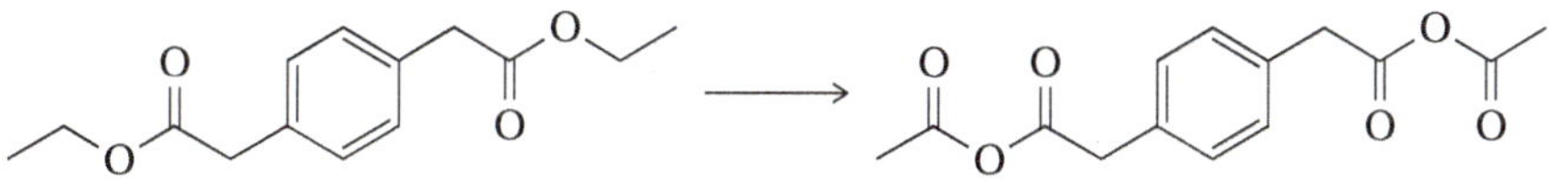

(b)

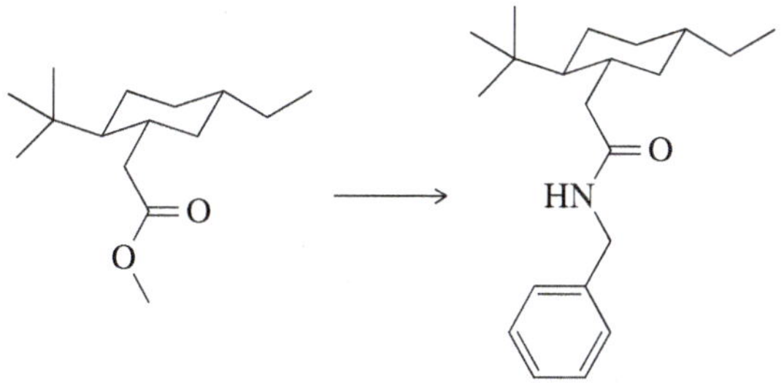

(c)

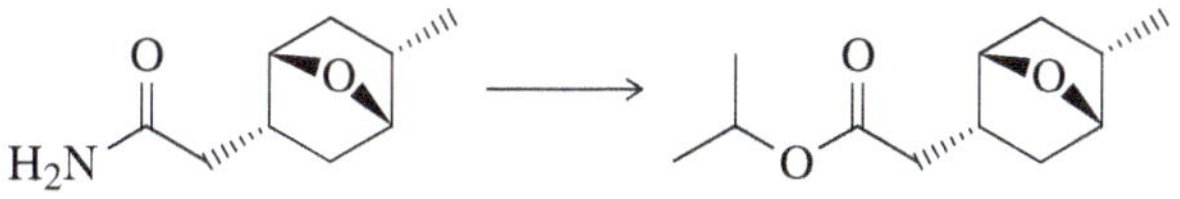

Assessment 19.34 Synthesis of carboxylic acid derivatives doesn't always have to start with another carboxylic acid derivative. Synthesize the amide on the right from the alkene on the left.

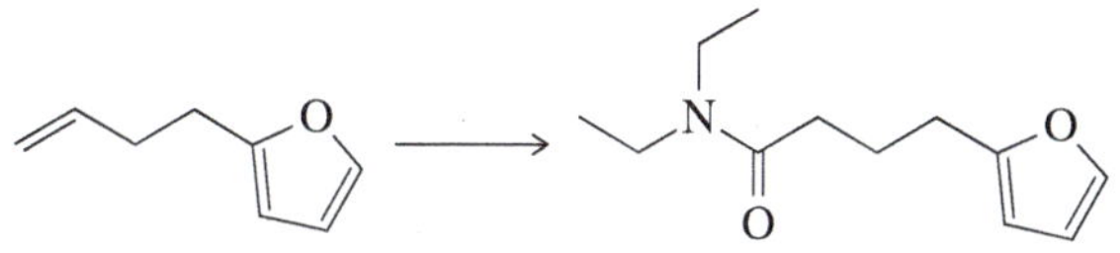

19-D CHEMISTRY IN CONTEXT: Biochemistry

Irreversible inhibition of transpeptidase

The penicillins and cephalosporins/cephamycins are called β-lactam antibiotics because of the four-membered amide ring at the core of their structure. When carboxylic acid derivatives are contained within a ring, Greek letters are used to indicate the size of the ring. A four-membered ring is named β, a five-membered ring is γ, and a six-membered is δ (**Figure 19.51**).

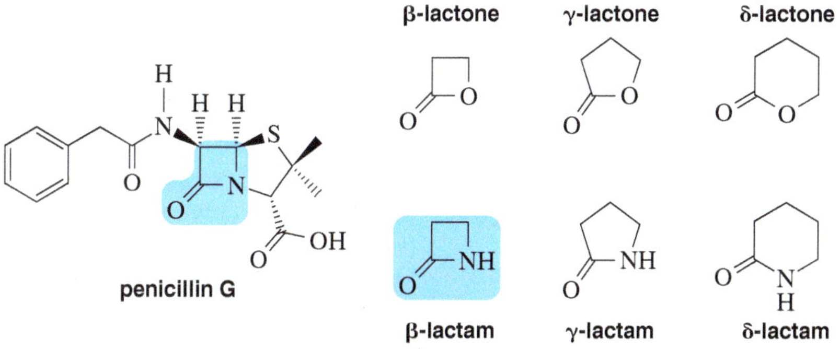

Figure 19.51 The use of Greek prefixes indicates the size of lactone and lactam rings.

The β-lactam antibiotics have structural similarities with the peptidoglycans in bacterial cell walls, so they are excellent inhibitors of the peptidoglycan cross-linking enzyme D-alanyl-D-alanine carboxypeptidase (**Figure 19.52**). [*See Chemistry in Context 19-C.*] Recall that enzyme active sites possess a somewhat rigid structure due to the restricted rotation of the amide (peptide) bond. [*See Chemistry in Context 19-A.*] The active site usually binds only the substrate of reactions it catalyzes. Drugs that look like the substrate are therefore good inhibitors.

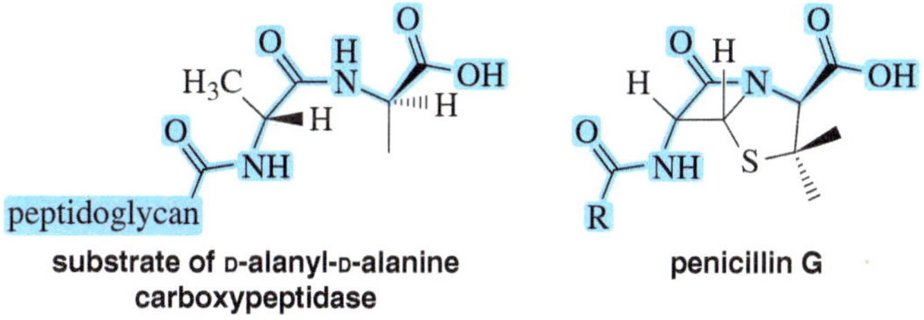

Figure 19.52 Penicillin G and all penicillins and cephalosporins/cephamycins are similar in structure to the peptidoglycan substrate of D-alanyl-D-alanine carboxypeptidase.

(continued)

While many drugs inhibit enzymes reversibly, β-lactam antibiotics act by irreversibly binding with D-alanyl-D-alanine carboxypeptidase (**Figure 19.53**). In the active site, an alcohol reacts with the lactam ring in the characteristic nucleophilic acyl substitution reaction that is the focus of this chapter. Although amides are typically unreactive, opening the strained four-membered ring is a driving force for this reaction. Once the drug is bound, the enzyme is no longer able to catalyze further cross-linking of the peptidoglycans, the cell wall can no longer be stabilized, and the bacterium ends up dying.

Because human cells do not make cell walls, we don't possess the enzyme D-alanyl-D-alanine carboxypeptidase. Thus, the β-lactams are drugs that never target anything but the cellular machinery of bacteria. [*Why is this a good thing in a drug? Fewer side effects!*]

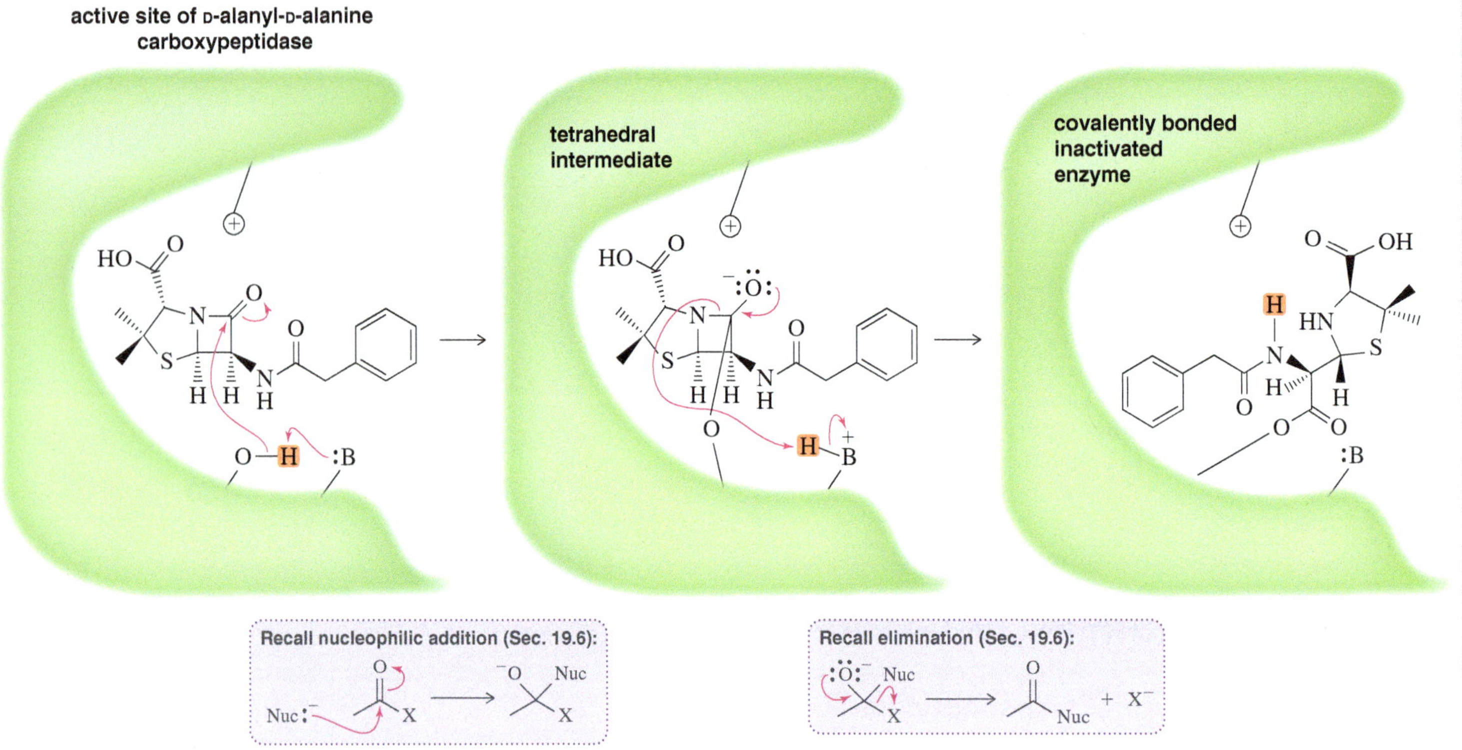

Figure 19.53 Irreversible inhibition of D-alanyl-D-alanine carboxypeptidase by penicillin G.

19.8 Reduction of carboxylic acid derivatives

Carboxylic acids contain a carbonyl carbon at the +3 oxidation state. As a result, they are good substrates for reduction to 1° alcohols and amines (in the case of amides). This section focuses mostly on the reduction of esters and amides because these reactions are the most synthetically useful.

19.8.1 Reduction to the alcohol oxidation state: $LiAlH_4$

Lithium aluminum hydride reduces all of the carboxylic acid derivatives. This section treats each carboxylic acid derivative individually but recognizes their shared similarities, particularly the mechanism. [*Look for the* **SIDE CONVERSATIONS**.]

19.8.1.1 $LiAlH_4$ with esters

$LiAlH_4$ reduces aldehydes and ketones by General Mechanism 17A (**Figure 19.54**). When an ester replaces the aldehyde, as shown in Figure 19.54, the alkoxide formed initially is now a tetrahedral intermediate, which immediately collapses to give an aldehyde, expelling the ethoxide as a leaving group. [*But the reaction doesn't stop here, because?*]

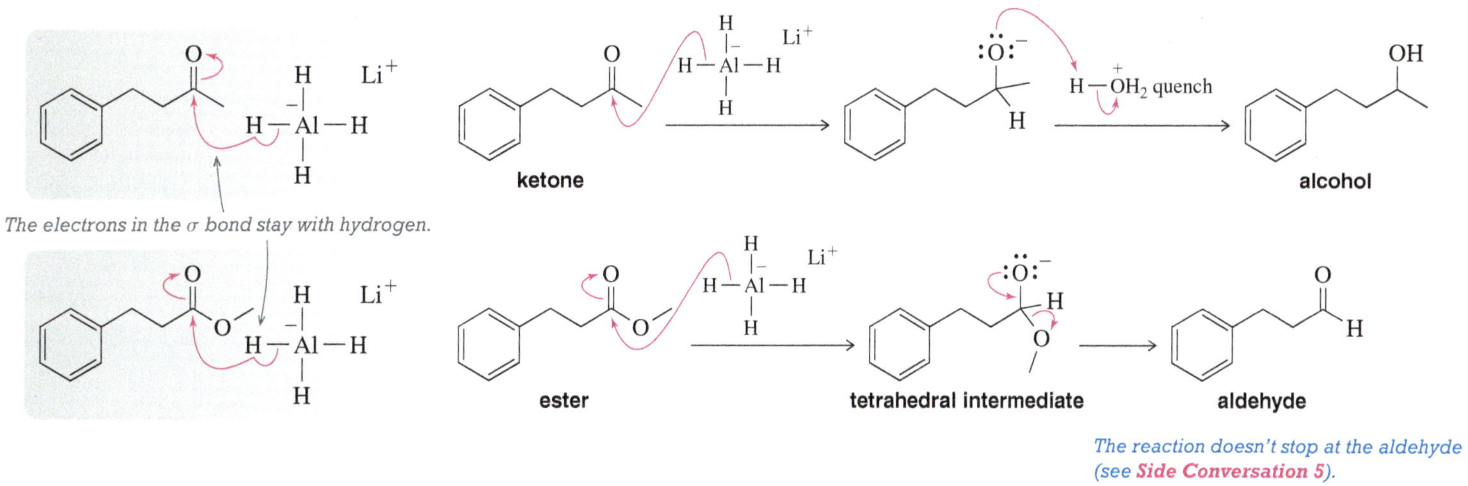

Figure 19.54 The reduction of an ester occurs by General Mechanism 18A, but the reaction doesn't stop at the aldehyde.

SIDE CONVERSATION 5 The product is more reactive than the reactant To understand what happens next, think about how reactions occur. When we draw a reaction, we show one molecule reacting with one other molecule. In solution, there are countless billions of molecules of each reactant. So let's imagine that a single molecule of the ester reacts with a single molecule of $LiAlH_4$ to form one aldehyde. [*What would the solution look like now?*] The result would be lots of unreacted ester, lots of unreacted $LiAlH_4$, and one aldehyde. [*What reaction will happen next?*] Our intention was to use $LiAlH_4$ to reduce the ester, but that's not how chemistry works. Because aldehydes (the product) are more reactive than esters (the reactant), the aldehyde will be reduced preferentially (**Figure 19.55**). Mechanistically, the reduction of the aldehyde occurs as it was shown in Section 17.9.2 (by General Mechanism 17A). ■

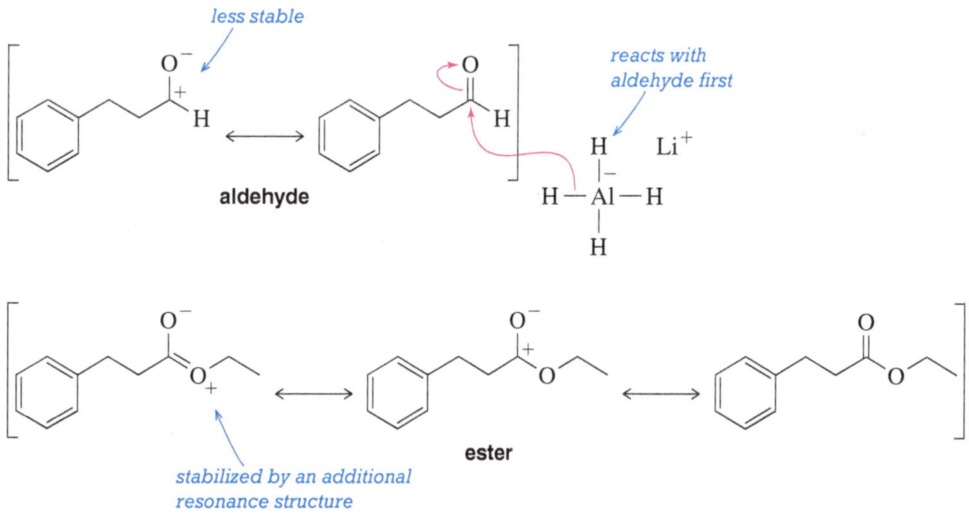

Figure 19.55 The more reactive aldehyde will always be reduced before the resonance-stabilized ester.

And so the reduction of esters cannot stop at the aldehyde because every aldehyde that forms will be more reactive than the starting ester. This idea of the product of a reaction being more reactive than the reactant will come up again. [*Soon!*] The overall mechanism of ester reduction is shown in **Figure 19.56**. It is simply the combination of nucleophilic acyl substitution (General Mechanism 18A) followed by nucleophilic addition (General Mechanism 17A).

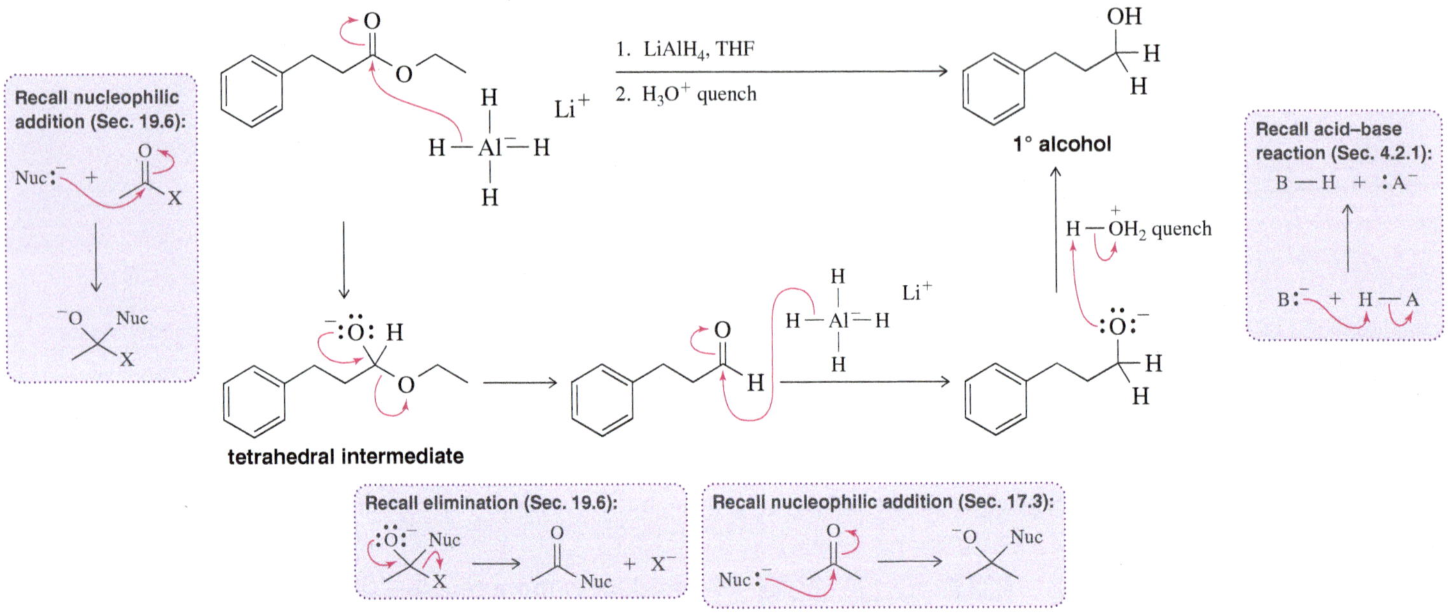

Figure 19.56 **The overall mechanism of ester reduction with $LiAlH_4$.**

19.8.1.2 $LiAlH_4$ with acid chlorides and anhydrides

The reduction of acid chlorides to 1° alcohols with lithium aluminum hydride isn't important enough to discuss in detail. Generally, carboxylic acids and esters are cheaper and easier to make than acid chlorides. To make an alcohol from an acid chloride you usually would have started with a carboxylic acid. Thus, it is better to reduce the carboxylic acid directly. **Figure 19.57** shows an example of an acid chloride reacting by the same mechanism by which esters are reduced (see Figure 19.56 to review). The reduction of anhydrides, also of little synthetic utility for similar reasons, gives primary alcohols by an identical mechanism.

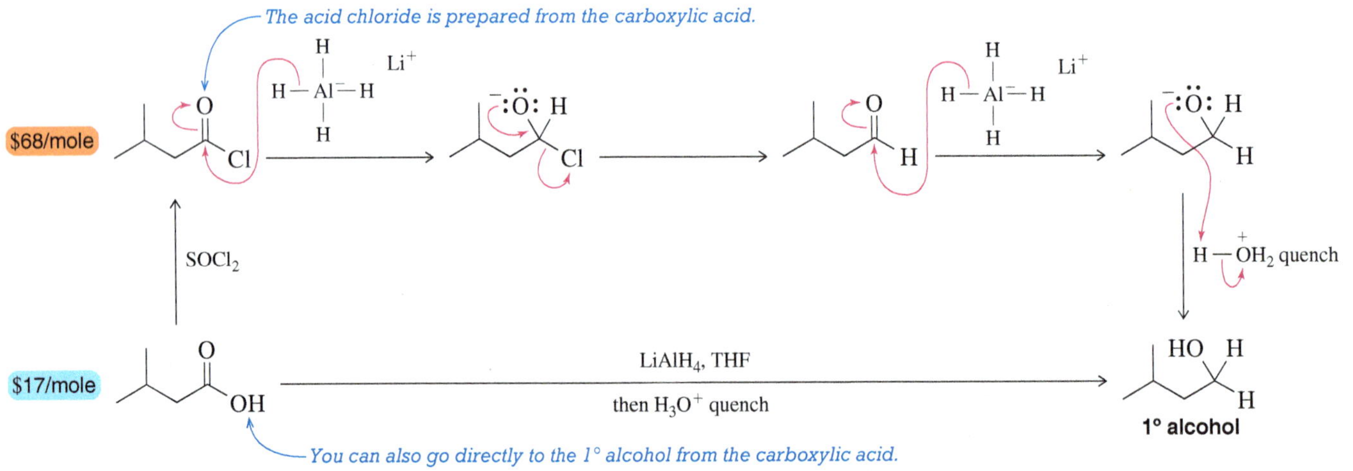

Figure 19.57 **Acid chlorides can be reduced to 1° alcohols,** although it usually makes more (sustainable and financial) sense to reduce carboxylic acids directly.

19.8.1.3 $LiAlH_4$ with amides

The reduction of amides with $LiAlH_4$ goes by a slightly different pathway compared to acid chlorides because amides (R_2N^-) are poor leaving groups (**Figure 19.58**). Like the other reactions here in Chapter 19, the nucleophilic hydride ion adds to the carbonyl to form the tetrahedral intermediate. Typically, the tetrahedral intermediate would collapse to reform the C—O π bond, but the nitrogen anion is a terrible leaving group, making this a very slow process. Instead, the tetrahedral intermediate sticks around and reacts with the Lewis acid alane (AlH_3). Alane was formed when AlH_4^- donated a hydride ion (H^-) to the amide. [*AlH_3 closely resembles BH_3, a Lewis acid we studied in Chapter 8.*]

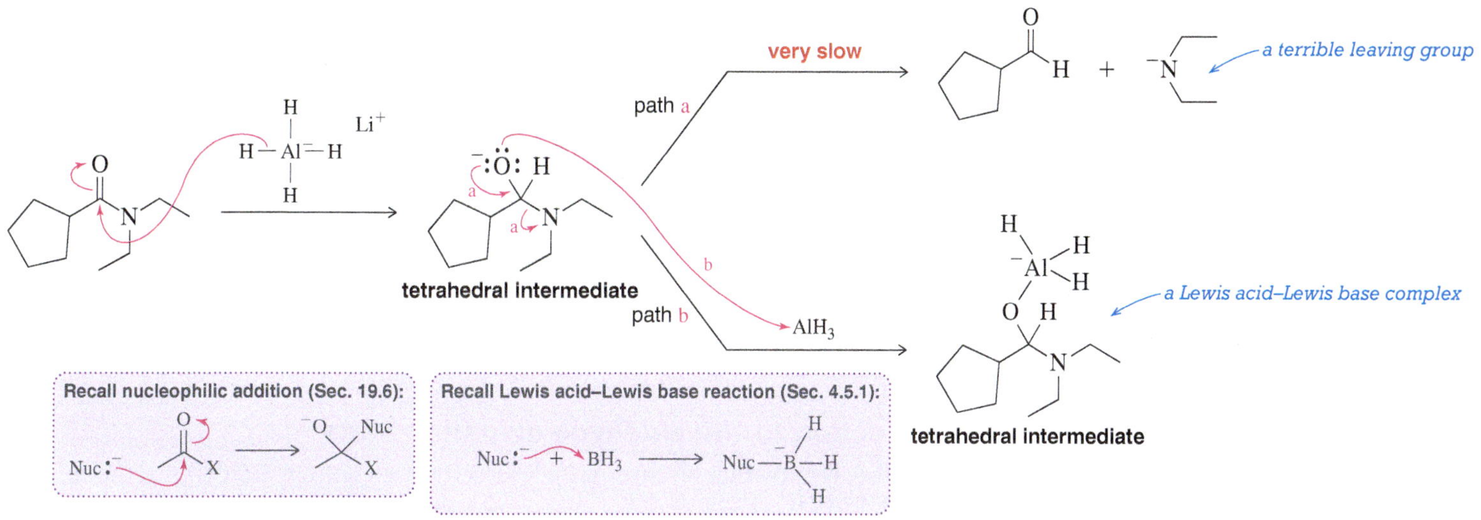

Figure 19.58 Hydride addition to an amide gives a stable tetrahedral intermediate that, instead of collapsing, reacts with alane to form a Lewis acid–Lewis base complex.

This Lewis acid–Lewis base complex undergoes a reaction seen in Section 17.8 when studying imine and enamine formation. The nitrogen's lone pair kicks out the oxyaluminate leaving group, forming a C—N π bond. The leaving group possesses a negatively charged aluminum with three hydrogens attached, so it is a strong reducing agent. [*It looks a lot like AlH_4^-, right?*] It donates a hydride ion to the iminium carbon, pushing the π electrons onto the nitrogen of the resulting amine product (**Figure 19.59**).

Figure 19.59 The formation of the iminium ion is followed by reduction of the iminium ion to the amine.

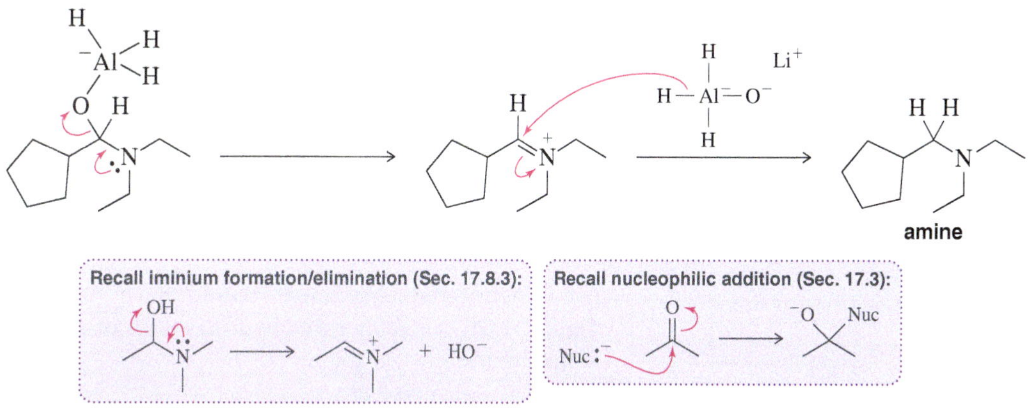

19.8.1 SUMMARY

- Hydride reduction ($LiAlH_4$ or $NaBH_4$) of acid chlorides, anhydrides, and esters (only by $LiAlH_4$) results in the formation of alcohols. Amides are reduced to amines by $LiAlH_4$. The choice of hydride reducing agent is based on the relative electrophilicity of the carboxylic acid derivatives.

Assessment 19.35 Formation of the following alcohol can occur by reduction of four different functional groups, all of which contain carbon at the +3 oxidation state. Show them.

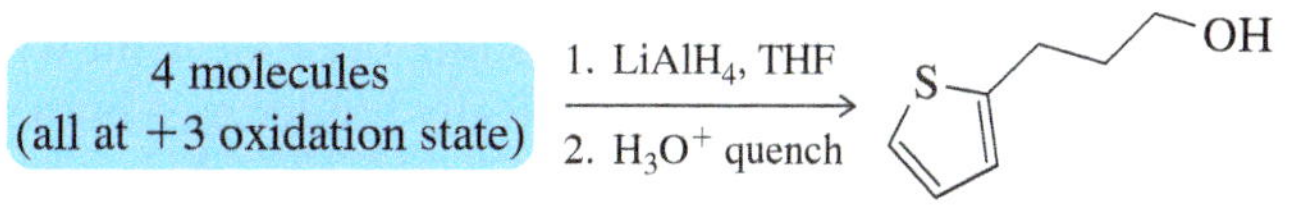

Assessment 19.36 Predict the product of the following reductions.

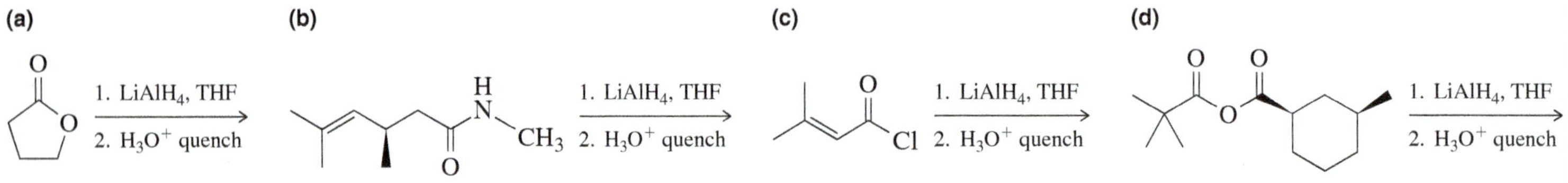

Assessment 19.37 Show a mechanism for the lithium aluminum hydride reduction of benzoic anhydride.

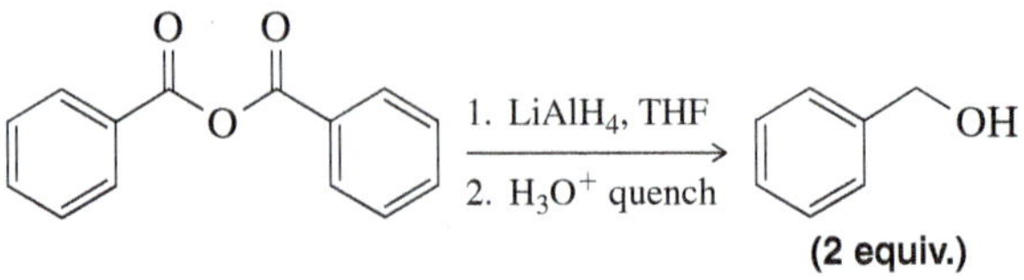

Assessment 19.38 Predict the product of the multistep synthesis reaction shown.

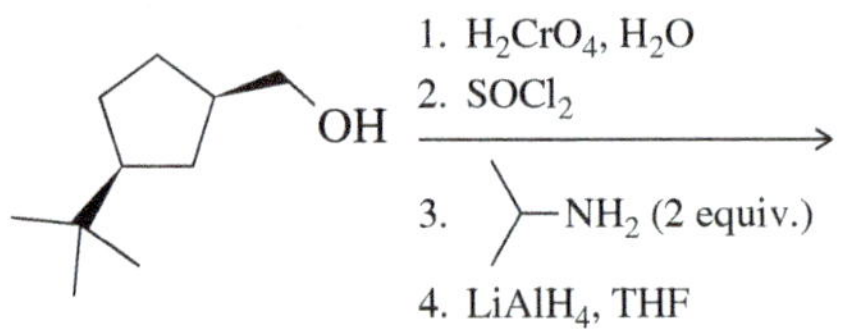

19.8.2 Reduction to the aldehyde oxidation state

In Section 19.8.1, we saw that all carboxylic acid derivatives (except amides) react to form a 1° alcohol with $LiAlH_4$.

What if we wanted to stop at the aldehyde intermediate? Can you imagine a way? Let's think like a chemist and develop a method for doing this.

Thinking Like a Chemist

What causes $LiAlH_4$ to reduce an ester all the way to an alcohol (**Figure 19.60**)? First, notice that $LiAlH_4$ has four hydrides to donate. In essence, even when using a 1:1 stoichiometric ratio, there will be an excess of hydride. Second, $LiAlH_4$ is a very strong reducing agent. And finally, the ester is significantly less reactive than the aldehyde. As a result, the aldehyde, once formed, is reduced selectively over the starting ester. [*Think about how each of these problems might be solved before reading further.*]

SPECTROSCOPY INSIGHTS

A chemist who discovered this over-reduction would have noticed the presence of a broad band in the IR spectrum around 3300 cm^{-1} for the O—H stretch (in the alcohol) and the absence of a sharp band around 1730 cm^{-1} for the aldehyde C=O stretch.

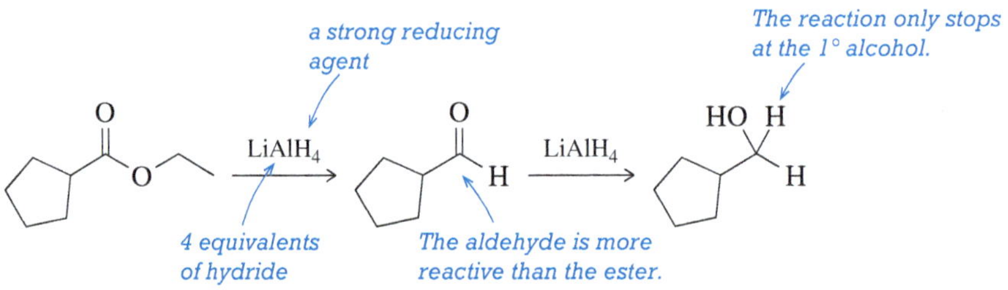

Figure 19.60 Why reduction of an ester is difficult to stop at the aldehyde.

A better reducing agent would have only one hydride ion present. As shown in **Figure 19.61**, replacing three hydrides with other groups gives a reducing agent containing a single hydride.

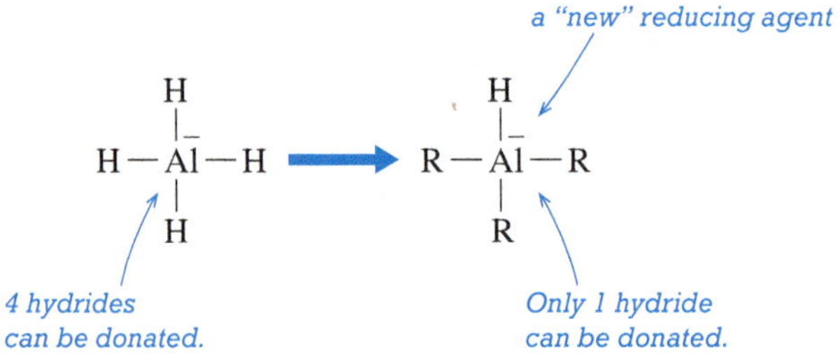

Figure 19.61 Replacing three hydrides with different groups allows only one hydride to be donated per our "new" reducing agent.

To reduce the reactivity of $LiAlH_4$, we have to think about why it is so reactive. The AlH_4^- ion has a very unstable negative charge. [*Why?*] It wants to donate a hydride ion in order to become neutral AlH_3. Adding three electron-withdrawing substituents should stabilize the negative charge. The sterically hindering *t*-butoxy groups in **Figure 19.62**, containing an electron-withdrawing oxygen, are used for this purpose.

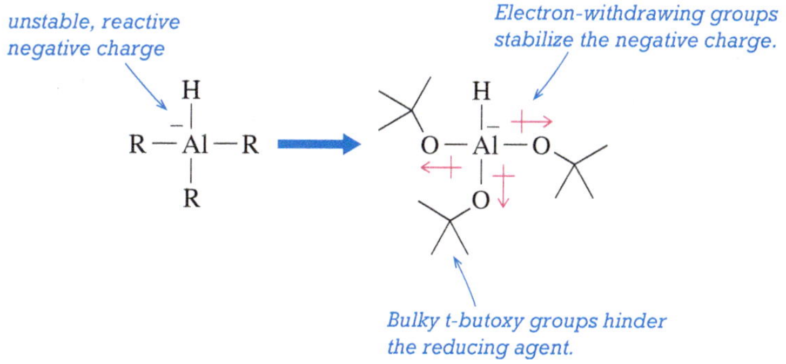

Figure 19.62 Incorporating three *t*-butoxy groups into our "new" reducing agent stabilizes the negative charge and provides steric hindrance, making a less reactive reducing agent.

Finally, the aldehyde is significantly more reactive than the ester. [*Is there a different carboxylic acid derivative we could start with, for which this isn't true?*] The acid chloride, the most reactive carboxylic acid derivative, is much more reactive than an aldehyde (**Figure 19.63**). Thus, as the reaction progresses and both functional groups are present, the acid chloride will be reduced first. ■

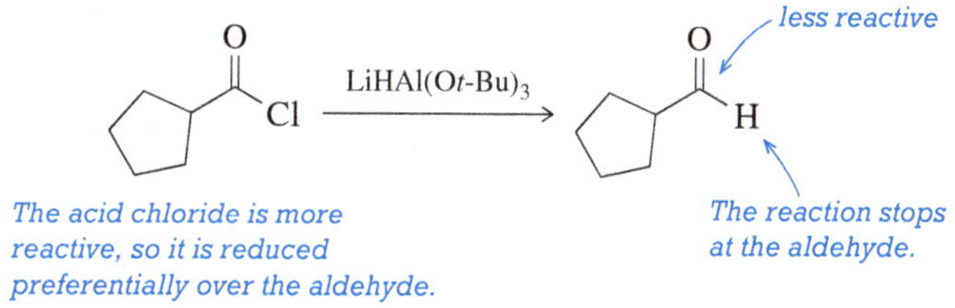

Figure 19.63 The acid chloride is more reactive than the aldehyde, allowing the reaction to stop at the aldehyde.

In conclusion, we can reduce a carboxylic acid derivative to an aldehyde by reacting a more stable, less reactive, more sterically hindered reducing agent containing a single hydride ion with a highly reactive acid chloride. However, what if we wanted to transform the ester itself to an aldehyde? Using our method, the ester must first be converted to an acid chloride, usually through the two-step sequence of hydrolysis to the acid and treatment with thionyl chloride (**Figure 19.64**).

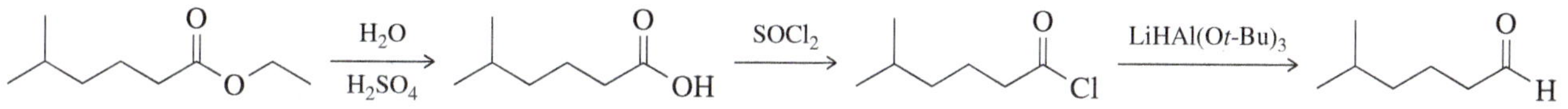

Figure 19.64 Conversion of an ester to an aldehyde first requires conversion to the acid chloride.

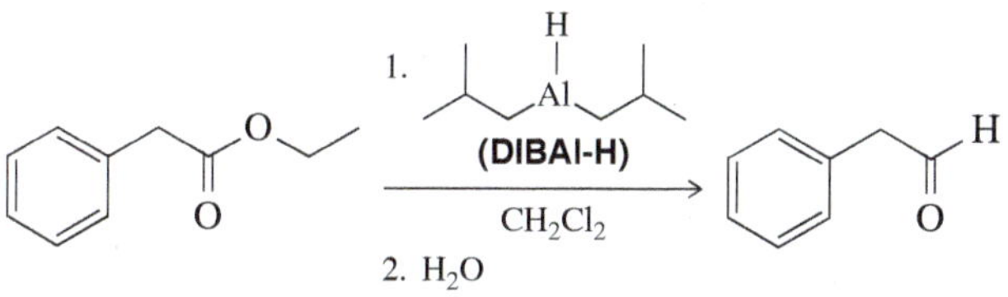

Figure 19.65 Direct conversion of an ester to an aldehyde with DIBAl-H.

It turns out that direct conversion of an ester to an aldehyde is possible using [*in one of the coolest reactions you'll ever see*] diisobutylaluminum hydride (DIBAl-H) in dichloromethane at low temperature, after aqueous workup (**Figure 19.65**).

Full appreciation of this reaction starts with recognizing that DIBAl-H (**Figure 19.66**) isn't like $NaBH_4$ and $LiAlH_4$, the other hydride reducing agents. With no negative charge, DIBAl-H has no real need to donate a hydride ion, at least not yet. [*If DIBAl-H isn't a hydride donor, what is it? Look carefully at the structure in Figure 19.66 and think about its expected reactivity.*] Neutral aluminum with three bonds is 2 electrons short of a full octet. The aluminum is electron-deficient, is sp^2 hybridized, and contains an empty *p* orbital. The closest analog that we have studied is borane (BH_3) or alane (AlH_3), two molecules we have seen behave as strong Lewis acids.

Figure 19.66 DIBAl-H is more of a Lewis acid than a reducing agent.

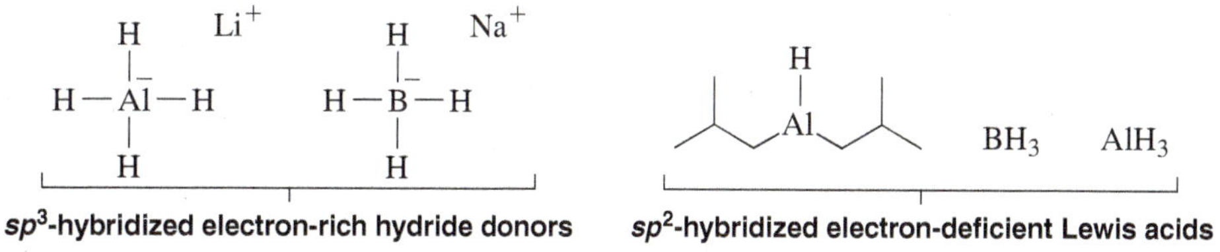

And so when the ester first encounters DIBAl-H, it does so as part of a Lewis acid–Lewis base reaction. The carbonyl oxygen donates a pair of electrons to the empty *p* orbital on aluminum, forming a Lewis acid–Lewis base complex (**Figure 19.67**). [*Look at the complex in Figure 19.67: What should happen next?*]

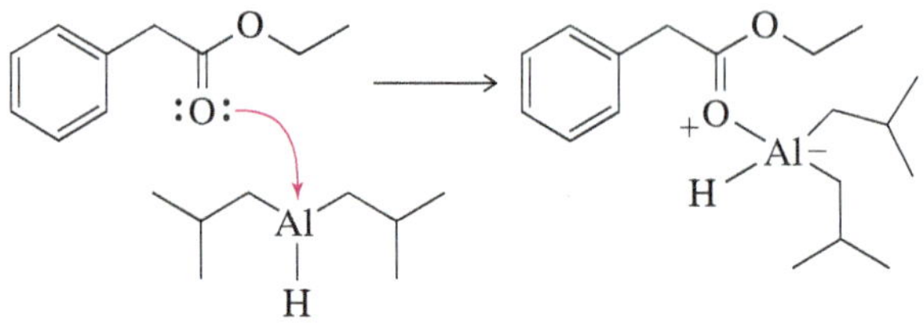

Figure 19.67 DIBAl-H and the ester form a Lewis acid–Lewis base complex.

[*Here's where it gets exciting!*] By forming this Lewis acid–Lewis base complex, we have converted a largely unreactive ester into a very strong electrophile, as demonstrated by the resonance structure in **Figure 19.68**. Similarly, aluminum now has a negative charge, resulting in a more nucleophilic source of hydride. With a nucleophilic hydride source so close to an electrophilic carbon, an intramolecular hydride transfer occurs to give what might be thought of as a tetrahedral intermediate.

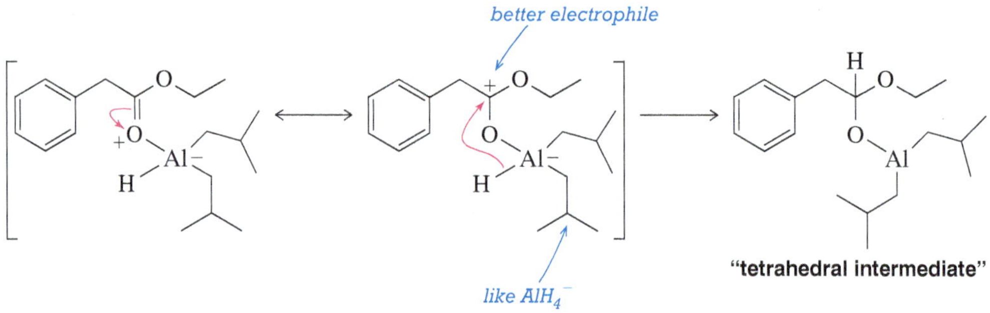

Figure 19.68 Hydride transfer within the Lewis acid–Lewis base complex reduces the carbonyl carbon.

It turns out that the "tetrahedral intermediate" in Figure 19.68 is actually very stable, especially at low temperature. In fact, it remains intact while the remaining ester molecules are converted into it. [*What functional group does it resemble? Hint: It's one you were trained in Chapter 17 to look for.*] The carbon with two oxygens attached most closely resembles an acetal, a generally stable functional group, except when placed in water. Although mechanistically different than previous acetal hydrolysis (because of the aluminum), this intermediate is hydrolyzed when water is added to the reaction after all of the ester has been reduced (**Figure 19.69**). Although DIBAl-H also reduces aldehydes, we only added one equivalent of DIBAl-H. There is none left to reduce the new aldehyde.

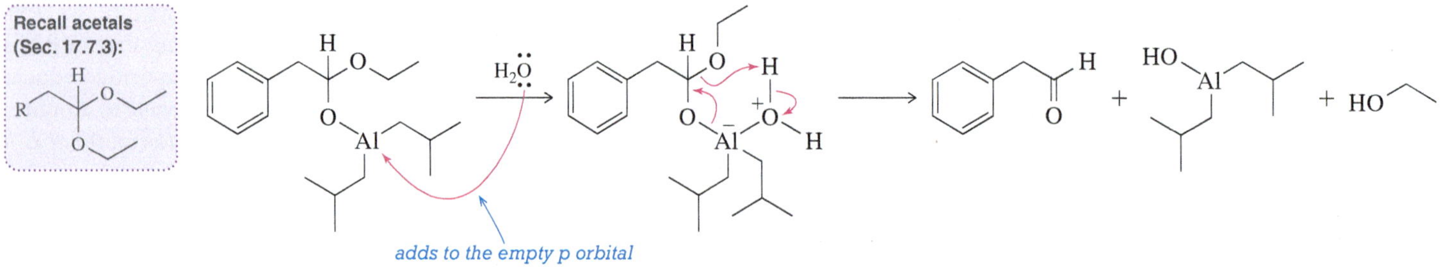

Figure 19.69 Hydrolysis of the complex, after all of the DIBAl-H is consumed, produces the aldehyde as the final product.

19.8.2 SUMMARY

- Stopping a reduction at the more reactive aldehyde oxidation state requires manipulation of the reaction conditions. A single equivalent of hydride donated by a less reactive hydride reducing agent [$LiHAl(Ot\text{-}Bu)_3$] allows the conversion of acid chlorides to aldehydes. The use of DIBAl-H to form a stable complex prior to aqueous workup allows esters to be converted to aldehydes.

Assessment 19.39 Predict the product of the following reaction sequences.

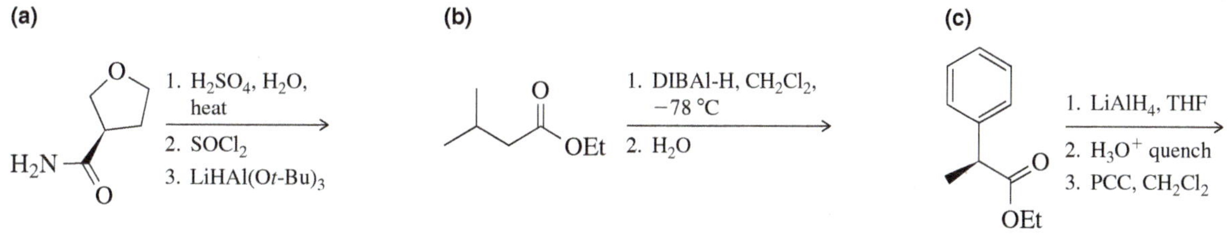

Assessment 19.40 The esters shown differ only by the alkoxy group. (i) Predict the product(s) obtained when these react with DIBAl-H. (ii) Based on your answer, in a sequence like this, would there ever be a need to convert from one ester to another?

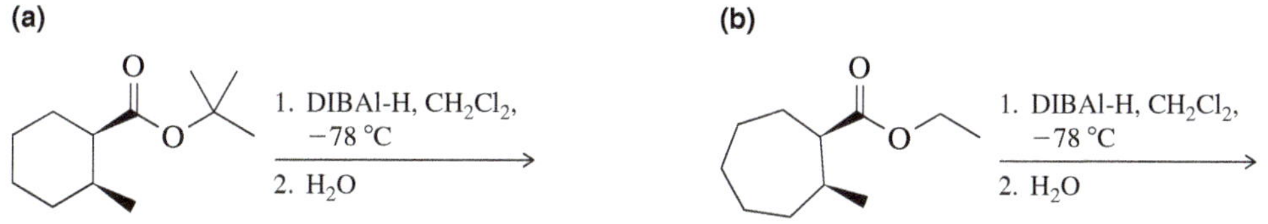

Assessment 19.41 When the ester attacked the aluminum of DIBAl-H, why did the carbonyl oxygen attack preferentially over the alkoxy oxygen?

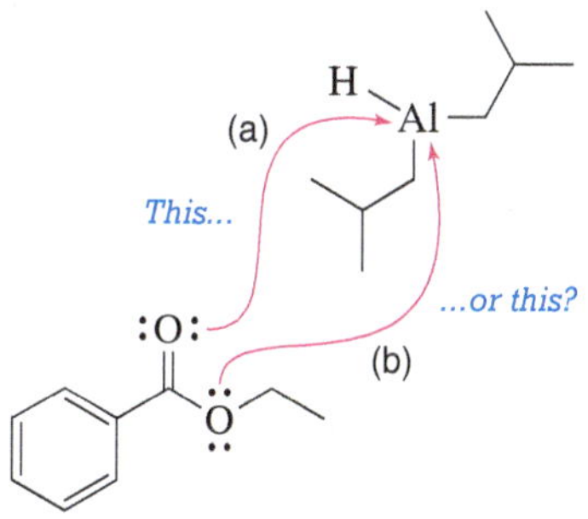

19.9 Organometallics with carboxylic acid derivatives

Carboxylic acid derivatives react with organometallics like Grignard and organolithium reagents. But before we go there, work Assessment 19.42.

Assessment 19.42 A chemist futilely attempted to synthesize the ketone shown by reacting an ester with phenyl magnesium bromide. (a) Predict the identity of the actual product(s) obtained. (b) How might IR spectroscopy be used to confirm a successful reaction?

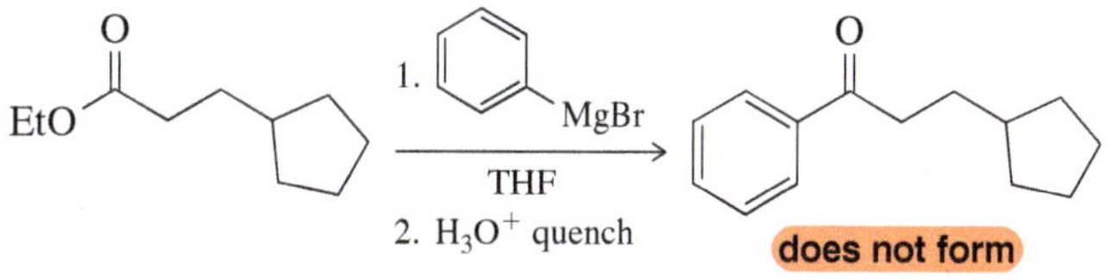

The outcome of Grignard reagents reacting with esters is discussed in the solution to Assessment 19.42(a). Make sure you have thought about it to get the most out of what follows.

19.9.1 Reaction to alcohols: ester plus Grignard reagent

Solved Assessment 19.42(a)

Grignard reagents are strong enough nucleophiles to react with an ester. Because they are so strong, we'd expect them to react by General Mechanism 18A, attacking the carbonyl carbon directly [*as they did in Chapter 17*] to form a tetrahedral intermediate (**Figure 19.70**). Collapse of the tetrahedral intermediate should expel the ethoxide leaving group, producing a ketone. But Assessment 19.42 makes clear that the ketone is *not* the product of the reaction. So what happens instead of, or in addition to, what has been shown here? [*If you aren't sure, briefly reread Section 19.8.1 for a hint.*]

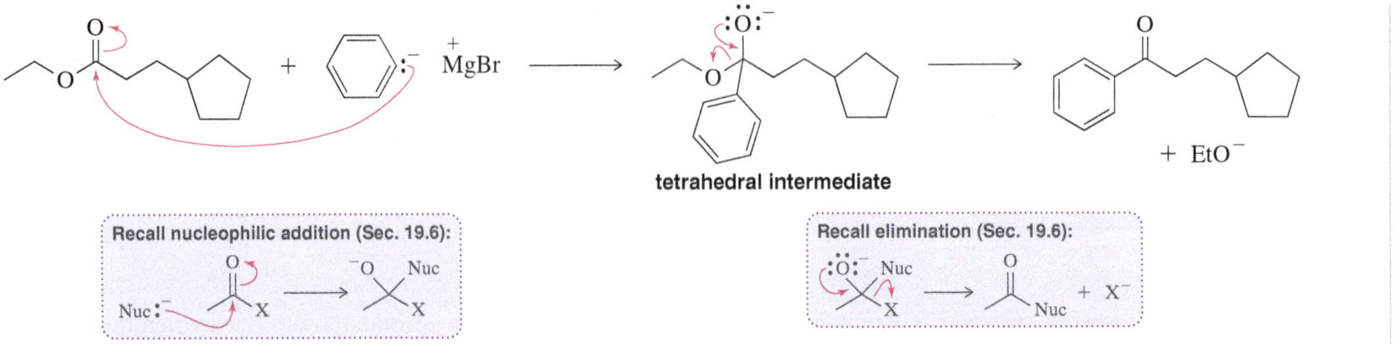

Figure 19.70 Grignard reagents do substitution with esters by General Mechanism 18A.

Consider again how reactions occur. What has been shown thus far is a single molecule of Grignard reagent reacting with a single molecule of ester. When the first molecule of ketone has been produced, there are still untold billions of Grignard molecules in solution. In a competition between a ketone and an ester, the ketone is more electrophilic, and so a Grignard reagent will immediately attack the ketone, this time by General Mechanism 17A, to give an alkoxide (**Figure 19.71**). [*Does this sound familiar? See* **SIDE CONVERSATION 5**: *Section 19.8.1.*] And so for every ester molecule that participates in the reaction, two molecules of the Grignard reagent are consumed and one molecule of alkoxide is produced. After this reaction, the alkoxide is quenched to give a 3° alcohol.

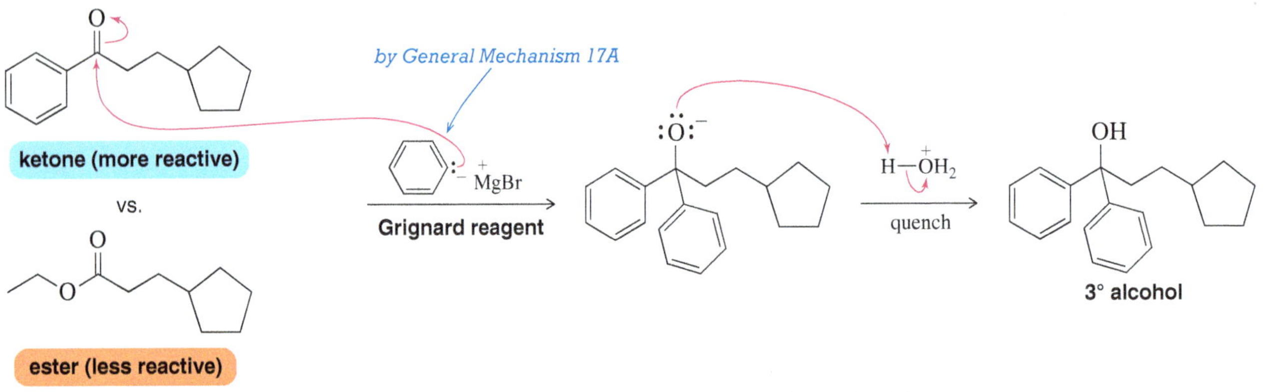

Figure 19.71 The more reactive ketone reacts before a second molecule of ester is able to.

If the ester and the Grignard reagent are used in a 1:1 stoichiometric ratio, then half of the ester will be converted to the 3° alcohol while the other half will remain unreacted. This is, once again, because each ester molecule consumes two molecules of the Grignard reagent. Because synthetic chemists don't enjoy reactions where the best possible yield is 50%, this reaction is usually done with two equivalents of the Grignard reagent. This ensures that 100% of the ester is consumed and the 3° alcohol can be produced in a 100% yield (**Figure 19.72**). ■

Figure 19.72 Esters are usually reacted with *two* equivalents of Grignard reagent to ensure maximum yield of the 3° alcohol.

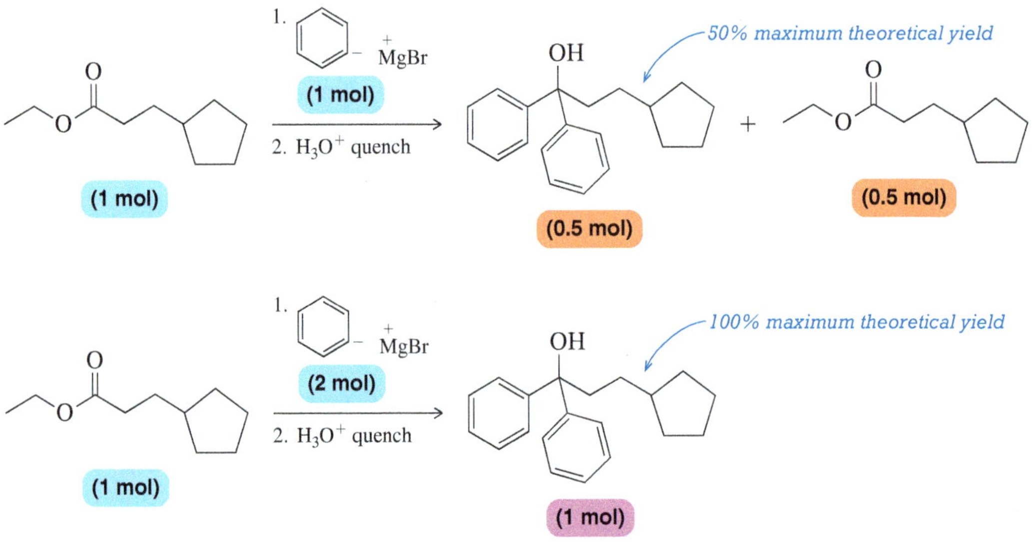

Assessment 19.43 Predict the product of the following reactions.

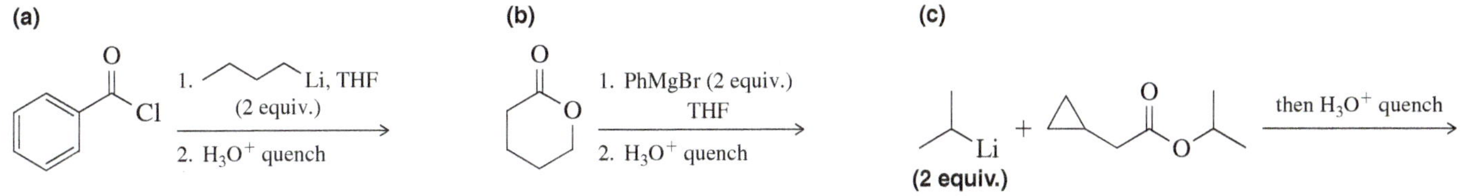

Assessment 19.44 The esters shown differ only by the alkoxy group. (i) Predict the product(s) obtained when these react with a Grignard reagent. (ii) Based on your answer, in a sequence like this, would there ever be a need to convert from one ester to another?

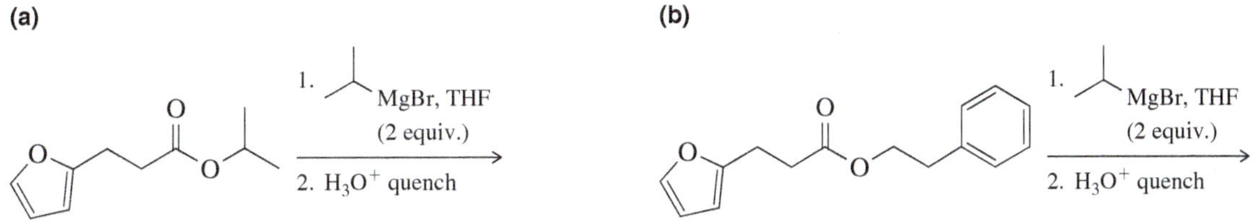

Assessment 19.45 Suggest a multistep synthesis to make the 3° alcohol from the alkene on the left.

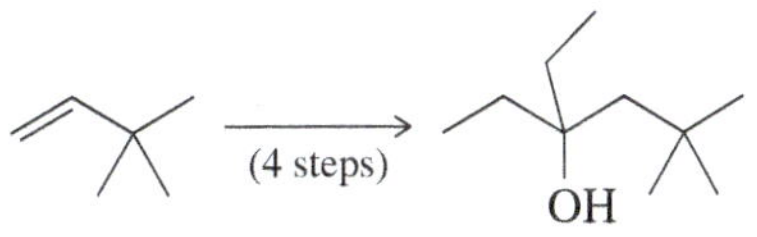

Assessment 19.46 Why is a ketone more reactive/electrophilic than an ester?

19.9.2 Reaction to form a ketone: cuprate plus acid chloride

As we learned in Section 19.9.1, the reaction between an ester and an organometallic doesn't stop at a ketone. But what if a synthetic chemist wants to do that? It turns out that lithium diorganocuprates (R_2CuLi; Gilman reagents) react with acid chlorides to give ketones. Recall from Section 16.5 that **Gilman reagents** only react with molecules containing an easily reducible bond. Most common among these are C—X bonds, where X is a halogen. The mechanism in Figure 19.73 makes it possible to predict the product of the reaction (and relates it to most of the reactions in this chapter).

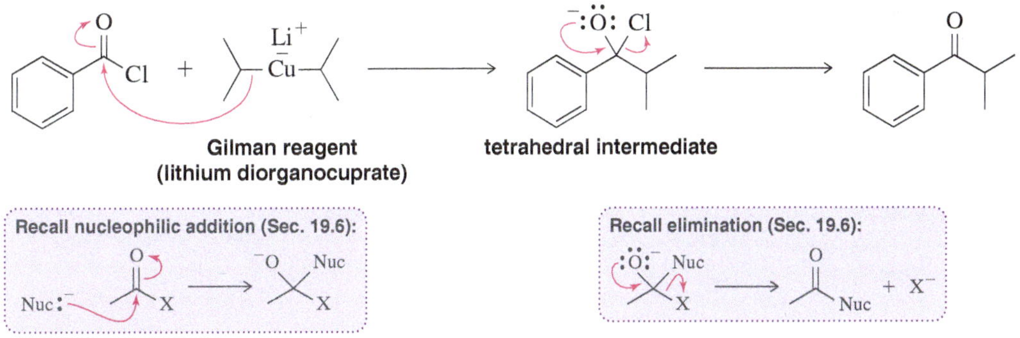

Figure 19.73 The mechanism of Gilman reagents reacting with acid chlorides. This approximate mechanism relates this reaction to those studied previously, while making it possible to predict the product.

Although the mechanism in Figure 19.73 is suitable for the purposes of this course, to understand why the reaction stops at the ketone requires a quick review of how Gilman reagents are believed to react. [*See Section 16.5 for a full treatment.*] Gilman reagents are organometallics like Grignard and organolithium reagents, but they are not simply strong nucleophiles that donate a pair of electrons to electron-deficient sites. Instead, they react only when there are easily reducible bonds present. The presumed mechanism involves oxidative addition of the cuprate to the C—Cl bond. This complex reductively eliminates to form the new C—C bond in the ketone. Because neither a ketone nor an aldehyde contain an easily reducible bond (conjugated C=C or C—X), cuprates will not react with them. As a result, the mechanism in Figure 19.74 shows why the reaction of Gilman reagents with acid chlorides stops at the ketone.

SPECTROSCOPY INSIGHTS

The conversion of an acid chloride to a ketone can be confirmed by a shift of the ^{13}C NMR signal for the carbonyl from ~175 to ~200 ppm.

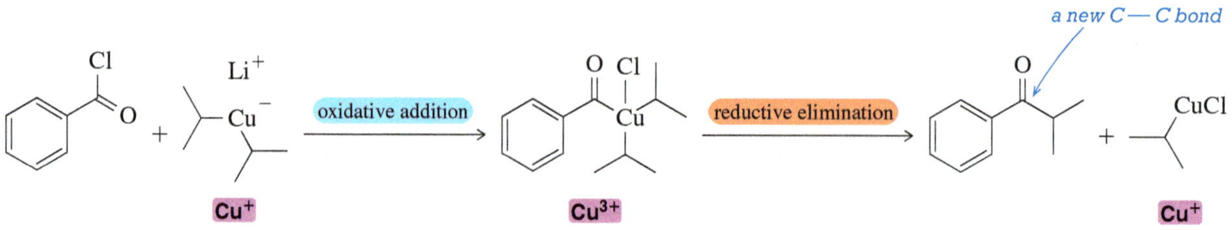

Figure 19.74 **The organometallic mechanism of Gilman reagents with acid chlorides.** This mechanism explains why Gilman reagents react with acid chlorides to make ketones.

19.9 SUMMARY

- Grignard and organolithium reagents are unable to stop at the more reactive ketone, reacting twice with esters to produce 3° alcohols. The most effective way to stop at a ketone is to react an acid chloride with a cuprate.

Assessment 19.47 Predict the product of the following reactions.

(a)

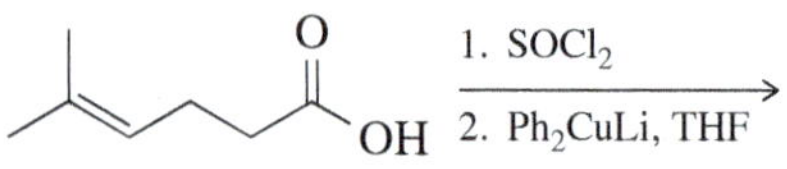

(b)

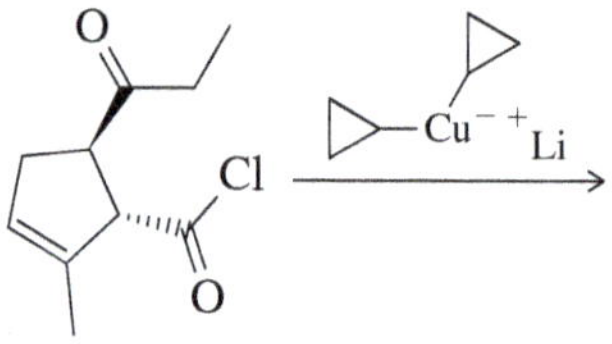

(c)

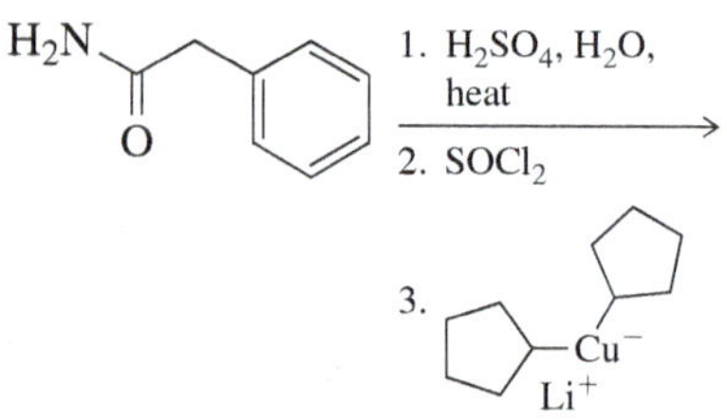

Assessment 19.48 Suggest a series of steps involving a cuprate reagent that would convert the reactant on the left to the product on the right. The ideal number of steps is shown.

(a)

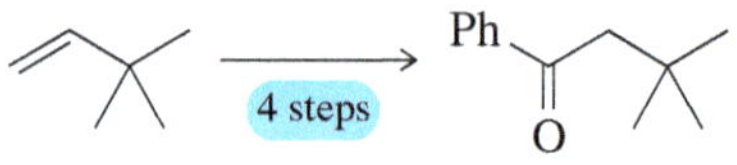

(b)

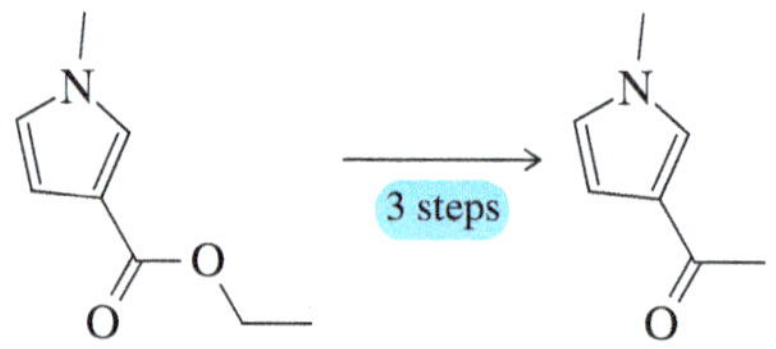

(c)

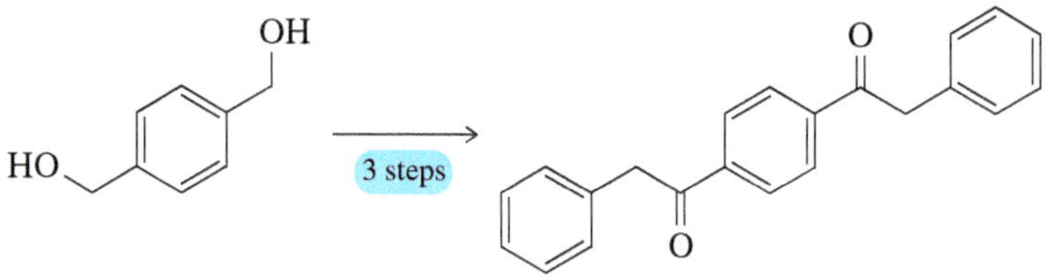

19.10 Reactions of nitriles

The central carbon of a nitrile has the same +3 oxidation state as a carboxylic acid. Therefore, it is considered to be a carboxylic acid derivative. There is no leaving group, however, so nitriles undergo nucleophilic addition reactions rather than substitution. Thus, the reactions in this section are closely related to the nucleophilic addition reactions we studied in Chapter 17 (**Figure 19.75**). The major difference between the two is that an imine is produced when the nitrile addition reaction is quenched, not an alcohol.

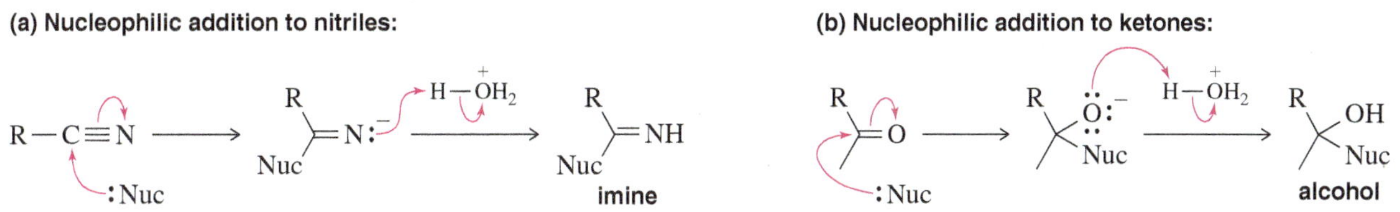

Figure 19.75 Nucleophilic addition to nitriles is analogous to nucleophilic addition to aldehydes and ketones.

This imine rarely remains in the product. Instead, it is hydrolyzed (often in the presence of acid) to give a ketone. The mechanism shown in **Figure 19.76** is the reverse of the imine formation reaction studied in Section 17.8. [*Compare the mechanism in Figure 19.76 to Section 17.8.1 to confirm that these are, in fact, the reverse of each other.*]

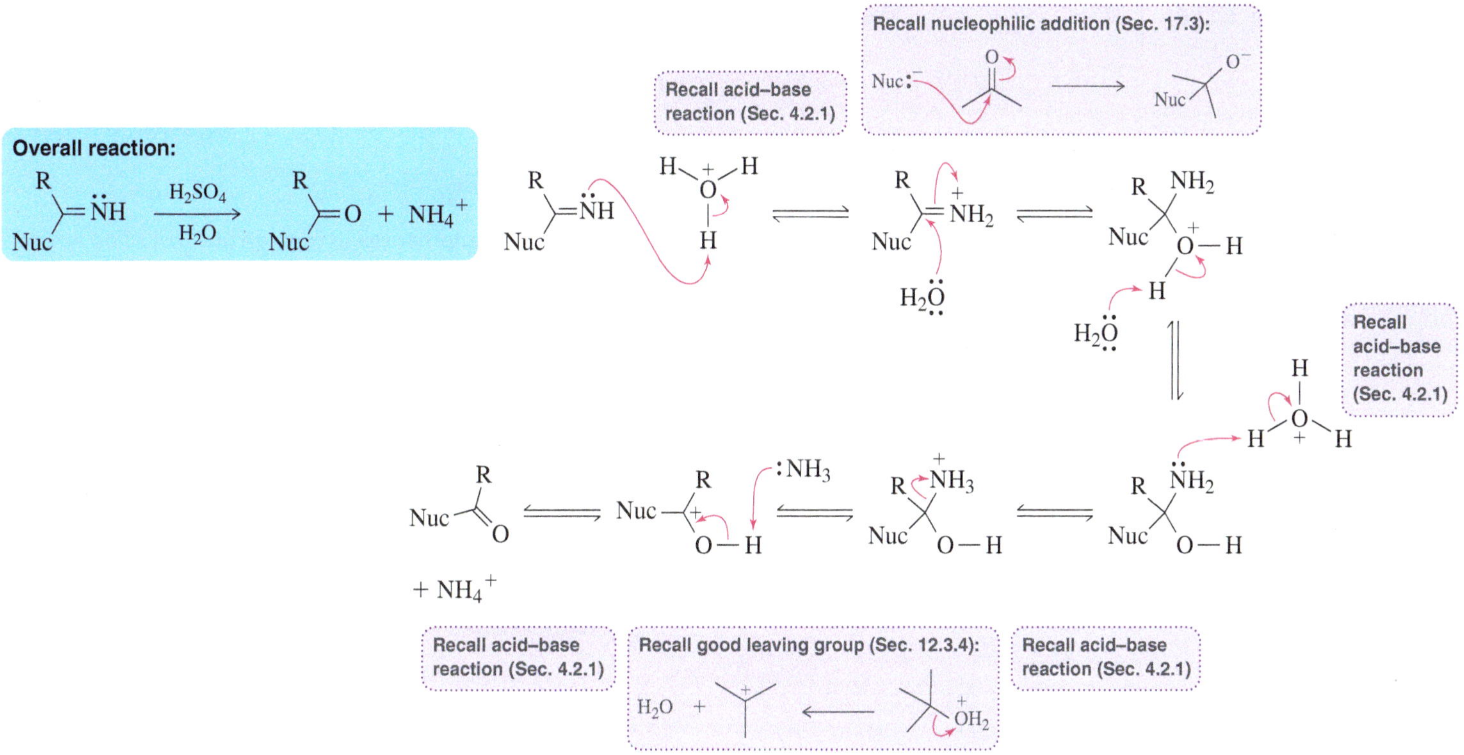

Figure 19.76 The hydrolysis of an imine results in a ketone.

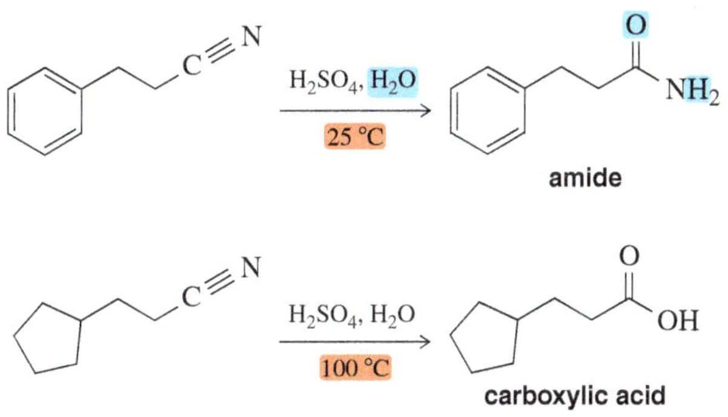

Figure 19.77 The hydration of nitriles at room temperature produces an unsubstituted amide. Hydrolysis of nitriles at higher temperatures produces a carboxylic acid.

19.10.1 Hydration/hydrolysis

The hydration/hydrolysis of nitriles, depending on the conditions of the reaction, results in either amides (hydration at room temperature) or carboxylic acids (hydrolysis with heating) (**Figure 19.77**).

Compared to aldehydes and ketones, nitriles are relatively weak electrophiles. Thus, protonation is required for them to react with water, a weak nucleophile (**Figure 19.78**). After protonation, water attacks, breaking the C≡N π bond while bringing along its proton. Deprotonation of the resulting acidic proton gives an intermediate containing an OH group attached to a doubly bonded carbon. [*Where have we seen an intermediate like this before?*] Like an enol, this tautomerizes to give the more stable C=O bond. Mechanistically, protonation of the nitrogen is followed by deprotonation of the alcohol to form the unsubstituted amide.

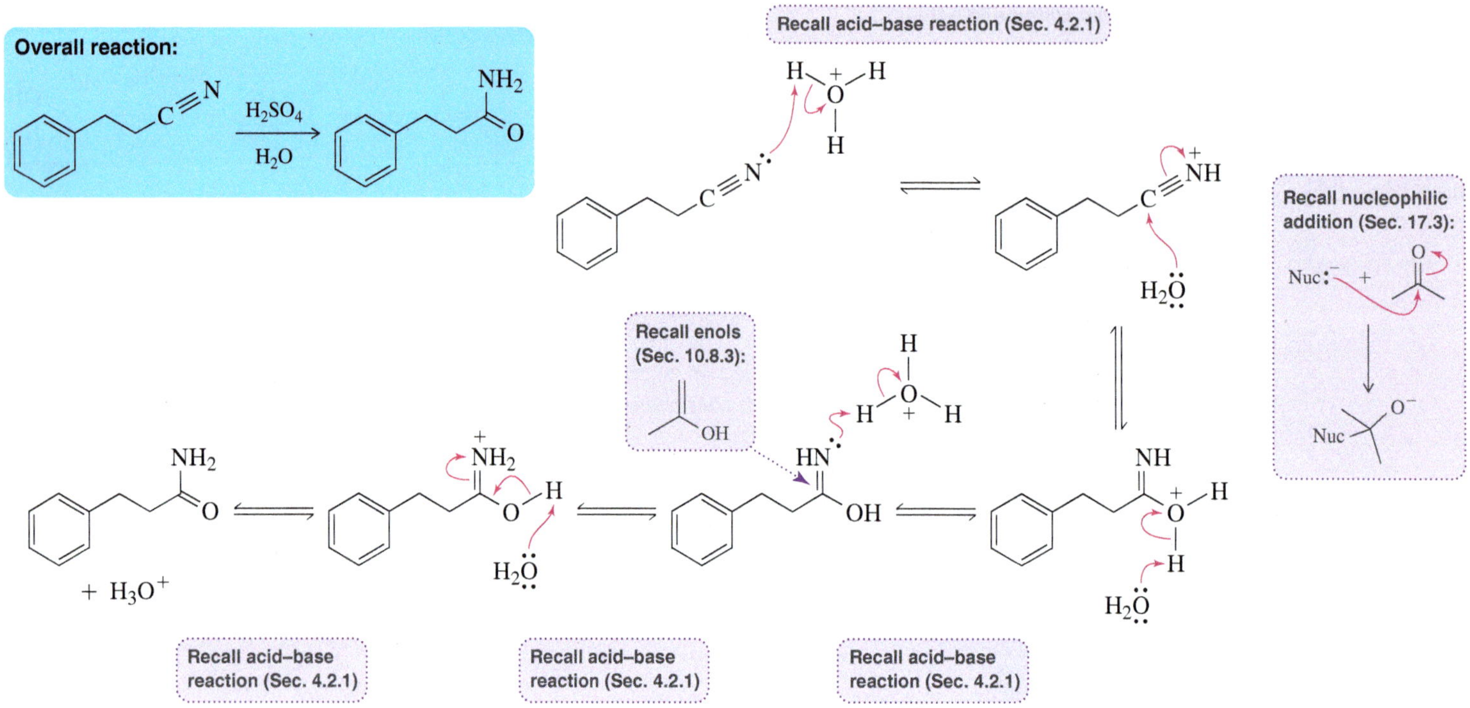

Figure 19.78 **The hydration of nitriles** initially gives an intermediate similar to an enol, which tautomerizes to form an unsubstituted amide.

Amides, the product in Figure 19.78, can undergo hydrolysis to an acid. Also, recall from Section 19.7.1 that, owing to the stability of amides, their conversion to carboxylic acids required heating. As shown in **Figure 19.79**, the mechanism begins with protonation of the carbonyl oxygen, making the amide more electrophilic. Attack by water is followed by deprotonation to give a tetrahedral intermediate. Protonation of the amine allows it to leave. Removal of a proton from the resulting cation results in a carboxylic acid.

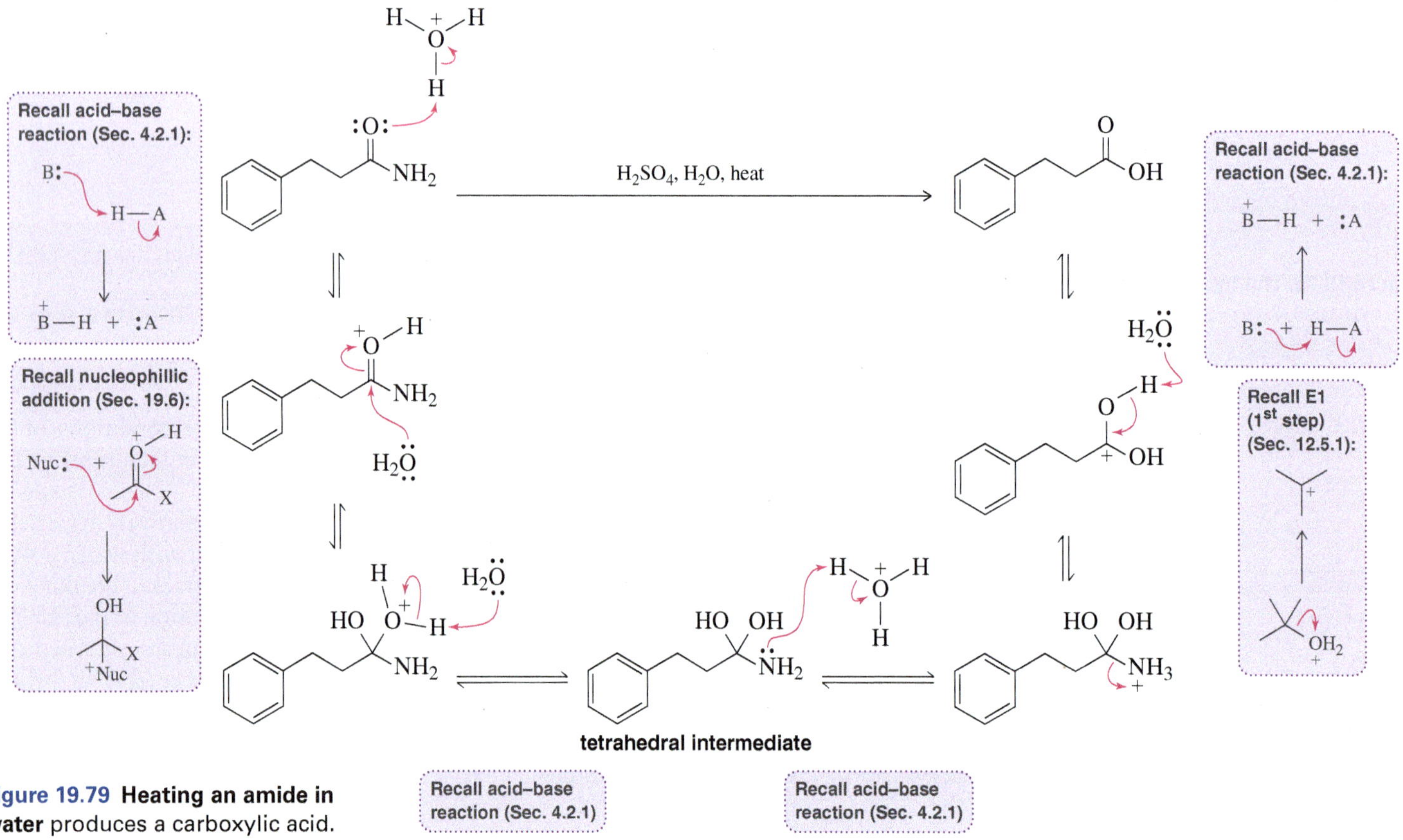

Figure 19.79 **Heating an amide in water** produces a carboxylic acid.

Assessment 19.49 How would you synthesize a nitrile from a haloalkane? (See Chapter 12 for a review.)

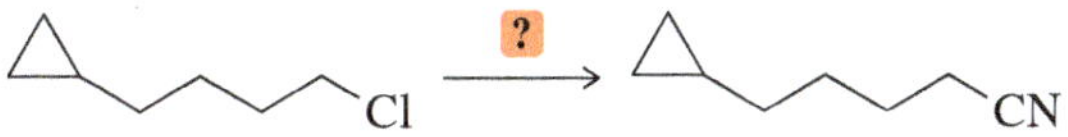

Assessment 19.50 How would you synthesize the cyanohydrin shown? (See Chapter 17 for a review.)

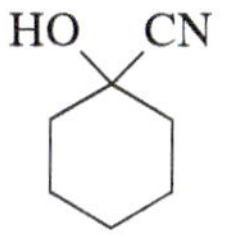

Assessment 19.51 The hydration of a nitrile can be reversed to prepare a nitrile from an amide. This is typically done using P_2O_5, a dehydrating agent, with a base and heating. Suggest a mechanism for this reaction.

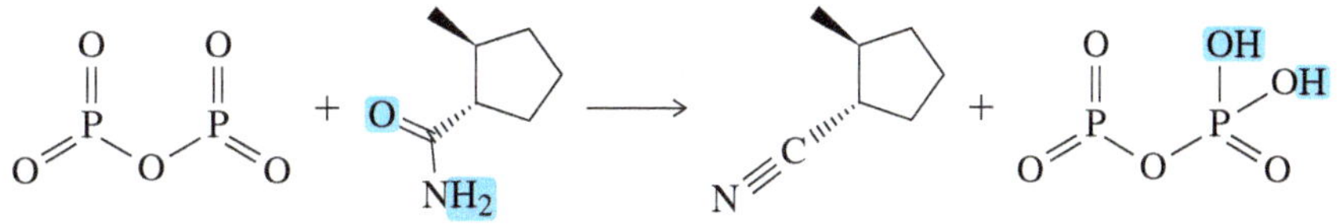

Assessment 19.52 Predict the product of the following reaction sequences.

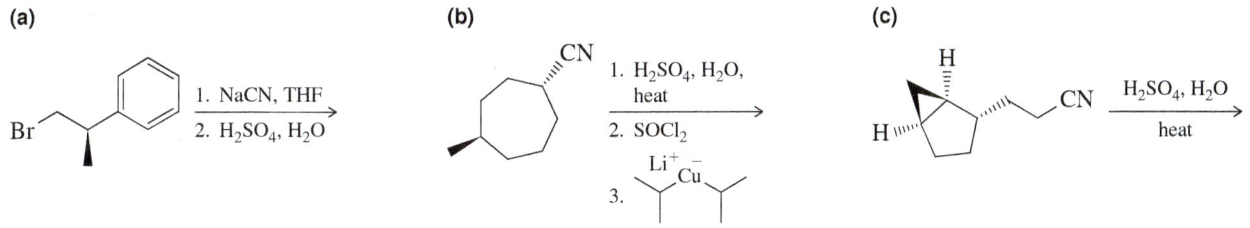

19.10.2 The addition of organolithium and Grignard reagents

Unlike weak nucleophiles like water, which only add after protonation, Grignard and organolithium reagents add directly to nitriles, producing an imine after quenching the imide with acid (**Figure 19.80**). Mechanistically, this reaction is most closely related to General Mechanism 17A, which was observed when studying nucleophilic addition to aldehydes and ketones. In an aqueous workup, the imine is hydrolyzed to produce a ketone. This reaction lets us form C—C bonds and gives an additional way to make ketones directly from carboxylic acid derivatives.

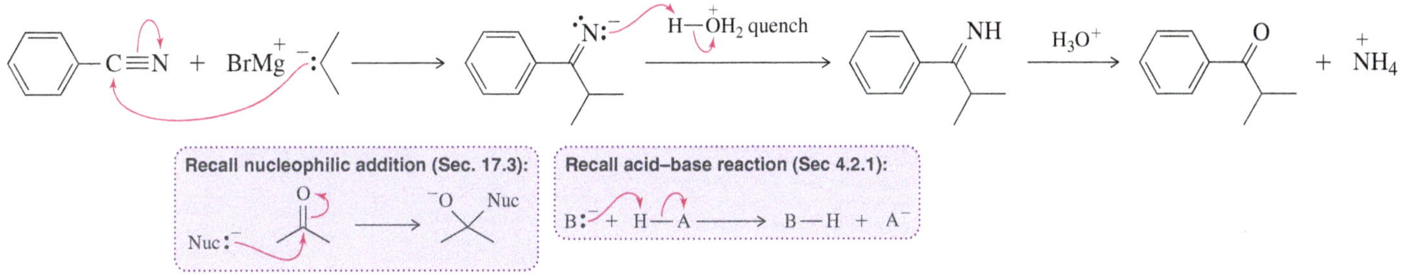

Figure 19.80 **Ketone synthesis by the addition of a Grignard reagent to nitriles.** The initial imine intermediate is hydrolyzed to a ketone by the mechanism in Figure 19.76.

Assessment 19.53 Predict the product of the following reaction sequences.

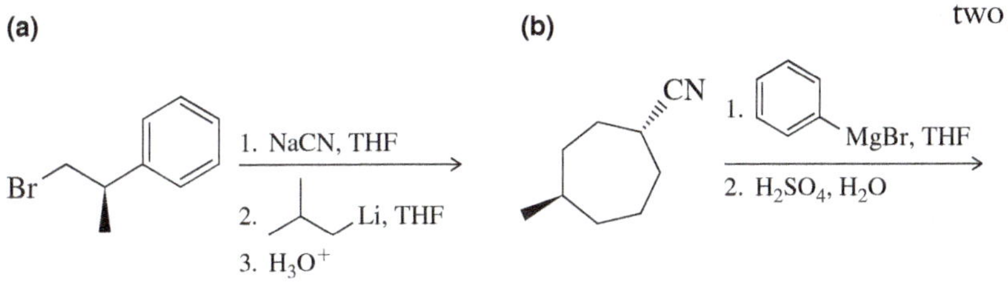

Assessment 19.54 Beginning with an amide, two different pathways produce the same compound. Predict the product of these two pathways.

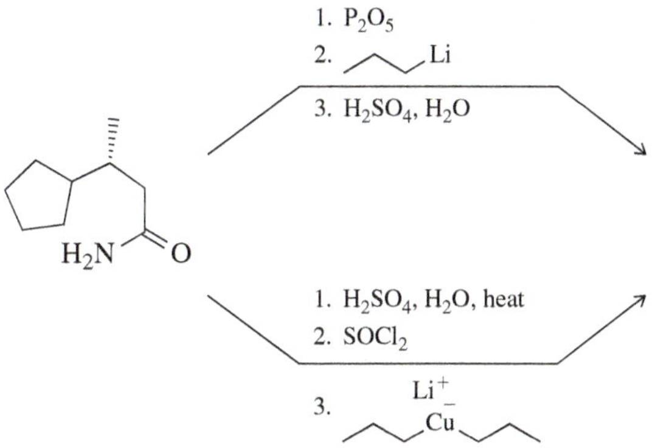

Assessment 19.55 Give the nitrile and organometallic reagent you would use to make the following ketones. There are two possible answers for both.

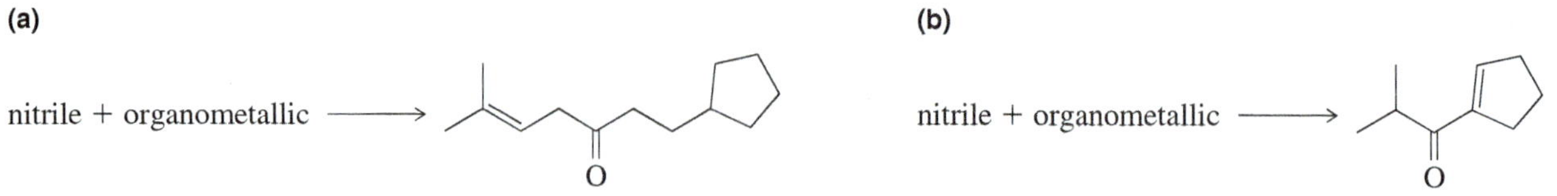

19.10.3 Reduction to the aldehyde

Nitriles can be reduced to aldehydes using DIBAl-H (**Figure 19.81**). Mechanistically similar to ester reduction (Section 19.8.2), the reaction begins with nitrogen donating a lone pair of electrons to the empty *p* orbital of aluminum. [*Recall that DIBAl-H does reduction reactions, but is itself a Lewis acid.*] Formation of the Lewis acid–Lewis base complex is another example [*see Section 19.8.2*] of a reaction that makes the electrophile stronger while at the same time creating a reducing agent. Intramolecular hydride transfer from aluminum breaks one of the C—N π bonds, forming an iminoaluminum species. Aqueous workup of this intermediate hydrolyzes the N—Al bond, before the resulting imine is hydrolyzed to an aldehyde by the mechanism shown in Figure 19.76 (Section 19.10).

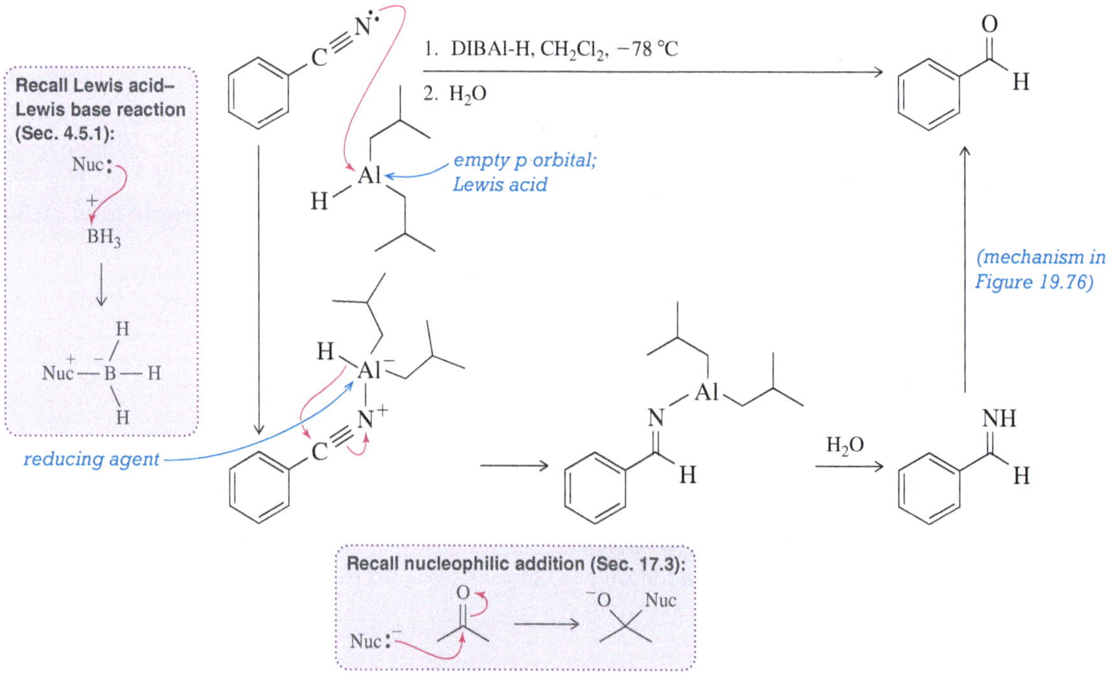

Figure 19.81 The reduction of a nitrile with DIBAl-H is similar to the reduction of an ester with DIBAl-H.

Assessment 19.56 Predict the product of the following reaction sequences.

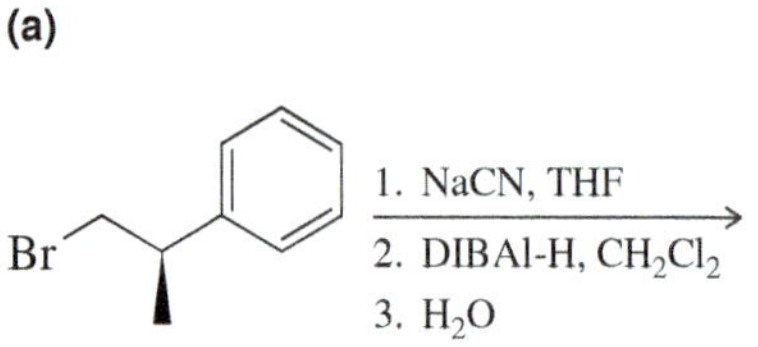

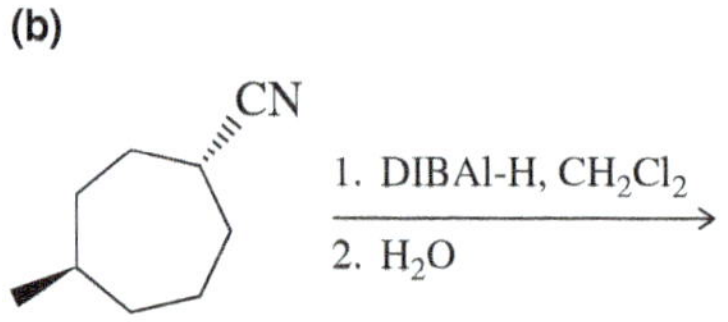

Assessment 19.57 Beginning with an amide, three different pathways produce the same compound. Predict the product of these three pathways.

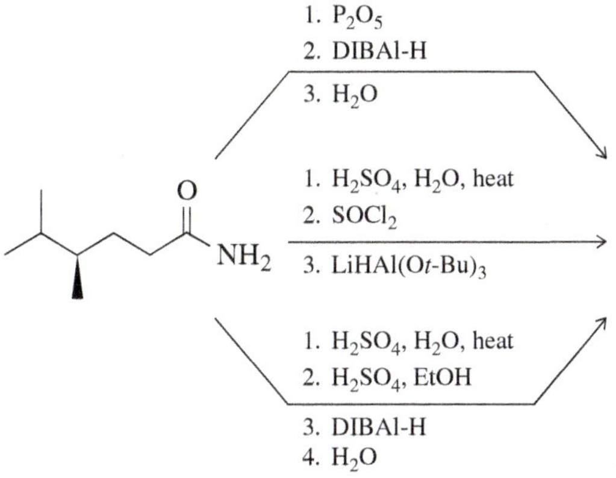

Assessment 19.58 Give the nitrile you would react with DIBAl-H to make the following aldehydes.

(a)

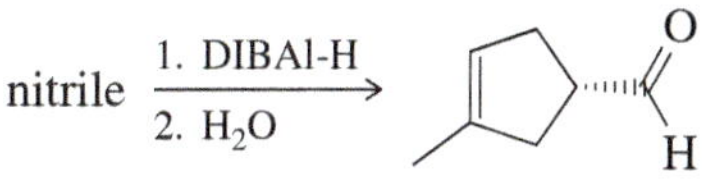

(b)

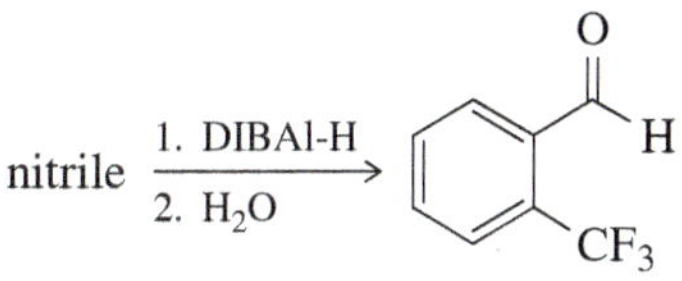

19.10.4 Reduction to the amine

With $LiAlH_4$ (a strong reducing agent), nitriles are reduced to amines (**Figure 19.82**). Although the actual mechanism is somewhat complicated, it is essentially the addition of two equivalents of hydride to the nitrile carbon, breaking each of the two C—N π bonds. The amine hydrogens are installed during the acid quench step. In this way, the reaction is a lot like the reduction of ketones and aldehydes, occurring by General Mechanism 17A, but occurring two times.

Resembling General Mechanism 17A, a hydride ion is donated directly to the nitrile carbon, breaking the C—N π bond while forming alane (AlH_3). The AlH_3, a Lewis acid, is important here because the negatively charged nitrogen is a strong Lewis base. Formation of a Lewis acid–Lewis base complex by donation of a pair of electrons to an empty *p* orbital [*similar to that seen for ester and nitrile reduction by DIBAl-H; Sections 19.8.2 and 19.10.3*] occurs quickly. Intramolecular hydride transfer from aluminum to carbon breaks the second C—N π bond, once more resembling General Mechanism 17A. Quenching of the resulting amide is followed by an aqueous workup to replace aluminum with a proton (**Figure 19.83**).

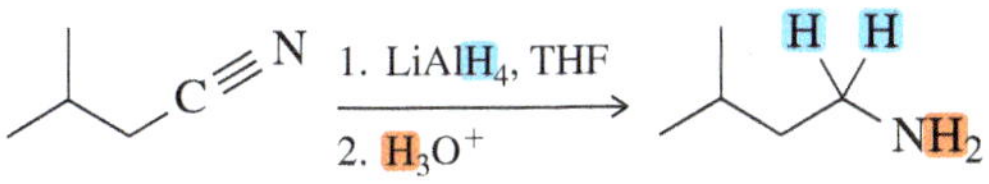

Figure 19.82 Nitrile reduction with $LiAlH_4$ occurs via two nucleophilic additions (General Mechanism 17A).

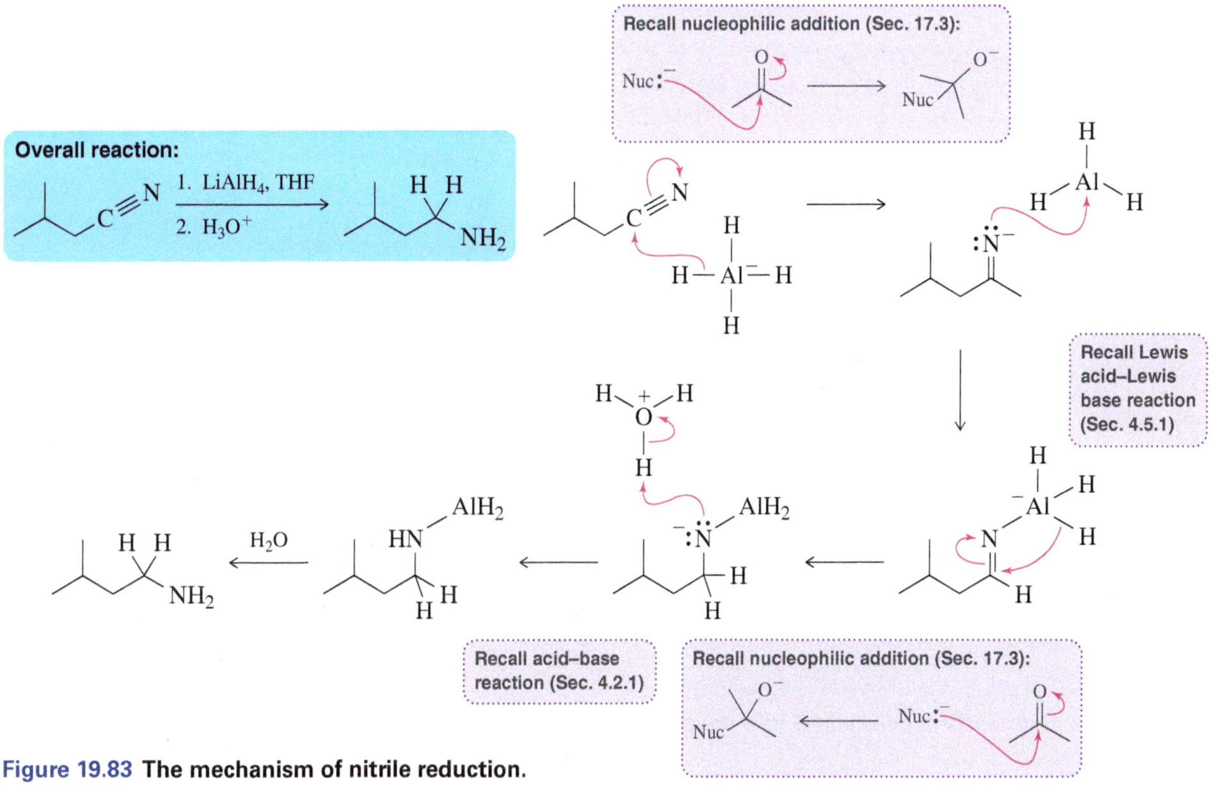

Figure 19.83 The mechanism of nitrile reduction.

19.10 SUMMARY

- Although they are carboxylic acid derivatives, nitriles lack a leaving group, so they do addition reactions like aldehydes and ketones. Among the synthetically useful transformations are the hydrolysis of nitriles to amides and carboxylic acids (H_3O^+, with or without heat), the conversion to ketones (RMgBr or RLi), the reduction to aldehydes (DIBAl-H), and the reduction to amines ($LiAlH_4$).

Assessment 19.59 Predict the product of the following reaction sequences.

(a)

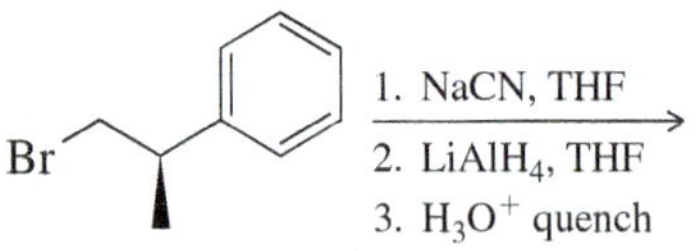

(b)

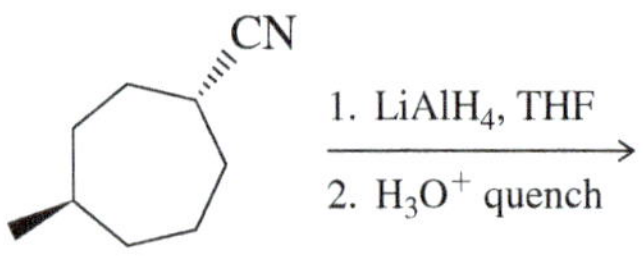

Assessment 19.60 Reduction of two different carboxylic acid derivatives with $LiAlH_4$ would give the amine shown. Identify these carboxylic acid derivatives.

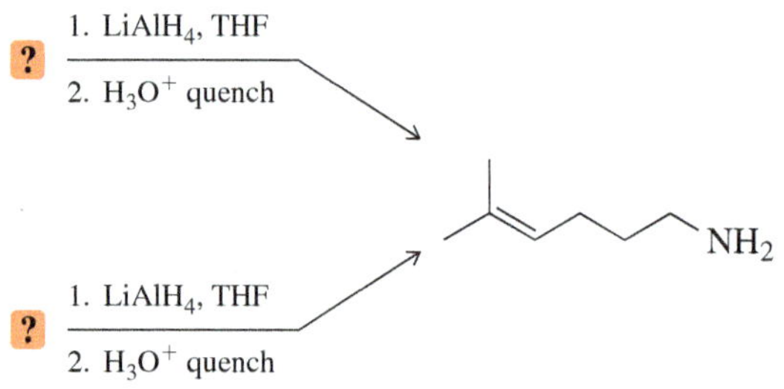

19-E CHEMISTRY IN CONTEXT: Biochemistry

Bacteria fight back!

If something were hindering your ability to survive, you might try to destroy it. In a similar way, bacteria living today have developed resistance to many β-lactam antibiotics, evolving enzymes (β-lactamases) that destroy β-lactam antibiotics before they reach the target enzyme, D-alanyl-D-alanine carboxypeptidase. β-Lactamases destroy the drugs by the same mechanism we've studied in this chapter. As shown in Figure 19.84, hydrolysis of penicillin G proceeds through the formation of a covalently bonded ester

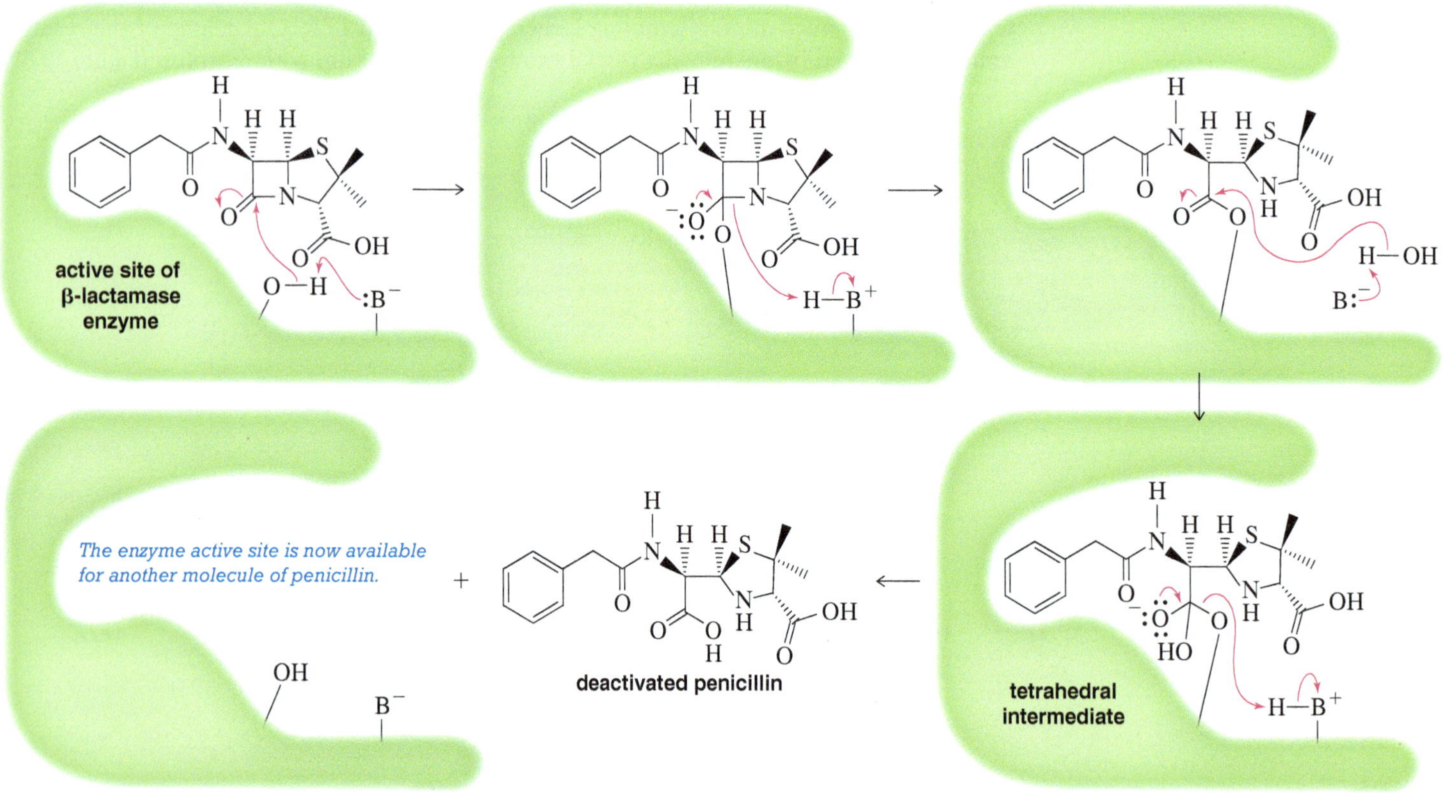

Figure 19.84 Hydrolysis of penicillin G by a drug-destroying β-lactamase.

(continued)

reminiscent of the mechanism by which D-alanyl-D-alanine carboxypeptidase is inhibited. [*See Chemistry in Context 19-C.*] In this case, however, the covalently bonded ester is hydrolyzed to a carboxylic acid, freeing up the enzyme to deactivate another molecule of the penicillin.

[*Once more, don't get caught up in the complexity of Figure 19.84. Instead, focus on the application of the chemistry we have studied here in Chapter 19.*]

In the "war on bacteria," medicinal chemists fired the first shot by developing molecules (β-lactam antibiotics) that inhibit D-alanyl-D-alanine carboxypeptidase. Bacteria responded by creating β-lactamases to destroy the β-lactam antibiotics. [*As a medicinal chemist, what would you do next?*] In turn, medicinal chemists invented inhibitors of the drug-destroying β-lactamases and co-administered them with β-lactam antibiotics so that the antibiotics would reach their target. One example of a β-lactamase inhibitor, shown in **Figure 19.85**, is clavulanic acid, which is co-administered with amoxicillin under the trade name Augmentin®. Clavulanic acid does not inhibit D-alanyl-D-alanine carboxypeptidase. It only inhibits the destruction of amoxicillin. [*The battle continues . . . your move, bacteria!*]

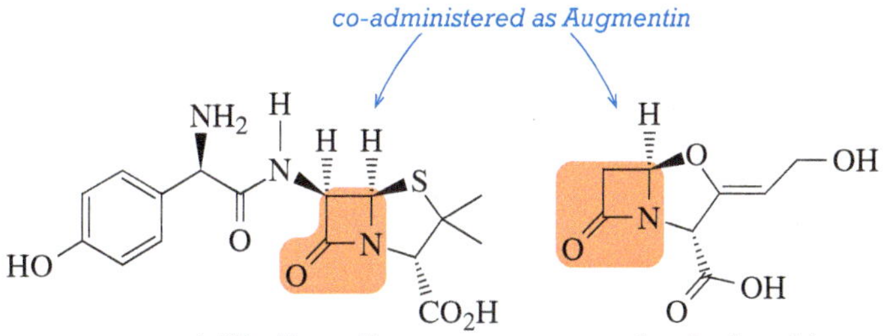

amoxicillin (Amoxil)
(transpeptidase inhibitor)

clavulanic acid
(a β-lactamase inhibitor)

Figure 19.85 Co-administration of amoxicillin with clavulanic acid prevents degradation of amoxicillin, leaving it to inhibit cell wall biosynthesis.

Concluding thoughts

In the mid-20th century, Sir Frank MacFarlane Burnet, co-winner of the 1960 Nobel Prize in Physiology or Medicine, imagined the elimination of infectious disease as a result of the "golden age" of antibiotics. Then bacteria developed antibiotic resistance. As more and more antibiotics are used, sometimes carelessly, bacteria have become resistant to even the strongest medicines. All it takes for antibiotic resistance to become established in a population is for one bacterium to survive as a result of a random mutation. This surviving bacterium, resistant to whatever drug was intended to kill it, very quickly produces offspring that, in turn, pass on this mutation in their DNA.

And so, instead of vanquishing bacterial disease, we find ourselves in a dangerous spot where we may run out of options for effective antibiotics. This is a problem that requires solutions from many sources, including [*one day*] some of you. Private and public research funding needs to be devoted to the development [*by organic chemists!*] of new antibiotics with novel mechanisms. Physicians need to resist overprescribing antibiotics, even when patients insist. The agriculture industry needs to avoid the use of antibiotics in farm animals when possible. And, finally, ordinary citizens can contribute by fully taking the prescribed dose and properly disposing of unused antibiotics.

But before we are in a position to invent or prescribe antibiotics, we need to complete our journey through the reactions of carbonyls by studying the nucleophilic addition and substitution reactions of enols and enolates. Let's go there now!

Important Concepts

[*These are the major concepts covered in each section, excluding reactions.*]

19.1–19.3: Carboxylic acid derivatives are at the same oxidation state as carboxylic acids and can be synthesized from carboxylic acids by a hypothetical condensation reaction. The important carboxylic acid derivatives are acid chlorides, anhydrides, esters, amides, and nitriles. Nomenclature of each carboxylic acid derivative begins with the name of the analogous carboxylic acid prior to the appropriate modification. Within the molecular orbital picture, the heteroatom attached to the carbonyl is sp^2 hybridized so that an electron pair can be delocalized via resonance onto the carbonyl oxygen.

19.4: With the exception of 1° and 2° amides, carboxylic acid derivatives, because they are unable to serve as hydrogen bond donors, have weaker intermolecular interactions than the corresponding carboxylic acid. The carbonyl stretching frequency of carboxylic acid derivatives is a function of the degree to which the carbocation in the zwitterionic resonance structure is stabilized or destabilized by the attached substituents. The order by decreasing wavenumber is as follows: acid chloride ($\sim$1785 cm^{-1}), anhydride ($\sim$1800/1745 cm^{-1}), ester ($\sim$1745 cm^{-1}), and amide ($\sim$1685 cm^{-1}).

19.5–19.6: The carboxylic acid derivatives can be interconverted, most efficiently going from least stable (high energy) to most stable (low energy). The most common reaction of carboxylic acid derivatives is nucleophilic acyl substitution, proceeding by either General Mechanism 18A or 18B. With strong nucleophiles, direct formation of a tetrahedral intermediate is followed by expulsion of a leaving group to reform the carbonyl (General Mechanism 18A). With weak nucleophiles, a series of acid–base reactions is necessary to allow formation of the tetrahedral intermediate and, after expelling a good leaving group, the new carboxylic acid derivative (General Mechanism 18B). The reactivity of carboxylic acid derivatives decreases in the following order: acid chloride, anhydride, ester, and amide.

19.7: Carboxylic acid derivatives are hydrolyzed (i.e., react with water) to form carboxylic acids. They react with alcohols to form esters and with amines to form amides. Acid chlorides react with carboxylates to form anhydrides. As the reactivity of the carboxylic acid derivative decreases, harsher conditions (e.g., acid, base, and/or heat) are needed to ensure complete reactivity. The synthesis of acid chlorides with $SOCl_2$ allows a return to the top of the reactivity hill so that all of the more stable derivatives can be easily prepared.

19.8: The reduction of carboxylic acid derivatives utilizes four reagents: $NaBH_4$, $LiAlH_4$, $LiHAl(Ot\text{-}Bu)_3$, and DIBAl-H. With the exception of amides, which produce amines, $LiAlH_4$ reacts with all carboxylic acid derivatives to make alcohols. DIBAl-H reduces esters to amides by formation of a stable complex. $LiHAl(Ot\text{-}Bu)_3$ reduces highly reactive acid chlorides to aldehydes by providing a single equivalent of hydride.

19.9: Grignard and organolithium reagents add twice to esters to make 3° alcohols. Synthesizing ketones from carboxylic acid derivatives requires the reaction of an acid chloride with a cuprate. Cuprates react by a different mechanism than Grignard and organolithium reagents and do not react with the resulting ketone.

19.10: Nitriles are the fifth carboxylic acid derivative. They lack a good leaving group, so nitrile reactivity is more similar to the reaction of aldehydes and ketones. Using addition reactions, nitriles can be hydrated to amides, hydrolyzed to carboxylic acids, reduced to amines or aldehydes, and converted to ketones by the reaction of Grignard or organolithium reagents.

Important Reactions and Mechanisms

1. **Nucleophilic acyl substitution—General Mechanism 18A (Sections 19.6–19.7):** The substitution of a strong nucleophile onto a carboxylic acid derivative. In the first step, the nucleophile attacks to form a tetrahedral intermediate with a negative charge on the carbonyl oxygen. The tetrahedral intermediate decomposes by collapse of an electron pair on oxygen to reform the carbonyl, while expelling a good leaving group.

General:

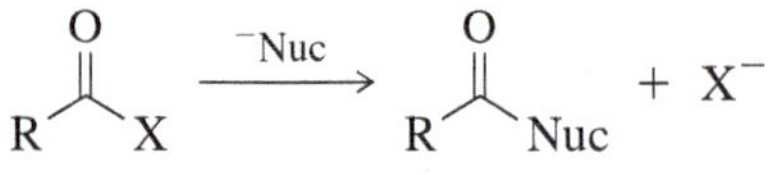

Specific:

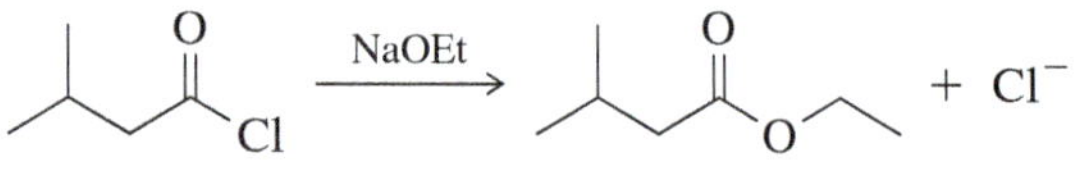

Mechanism:

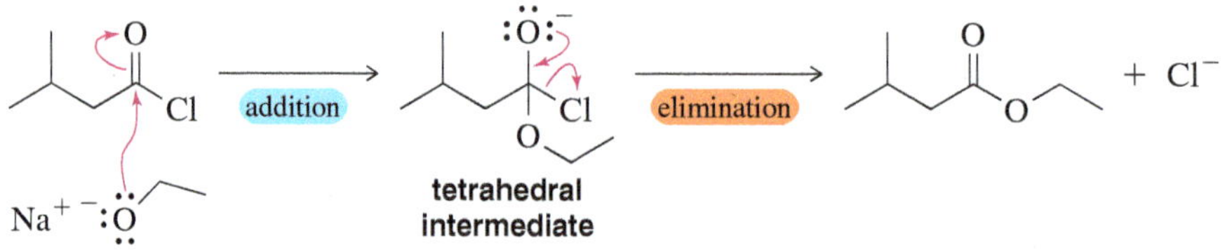

2. **Nucleophilic acyl substitution—General Mechanism 18B (Sections 19.6–19.7):** The substitution of a weak nucleophile onto a carboxylic acid derivative. Protonation of the carbonyl activates it for nucleophilic attack. Deprotonation of the resulting product gives a neutral tetrahedral intermediate. Protonation gives a good leaving group that is finally expelled by collapse of an electron pair to reform a protonated carbonyl intermediate. A final deprotonation completes the reversible mechanism, often driven to completion using Le Châtelier's principle.

General:

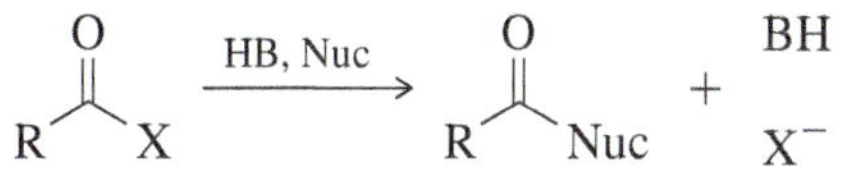

Specific:

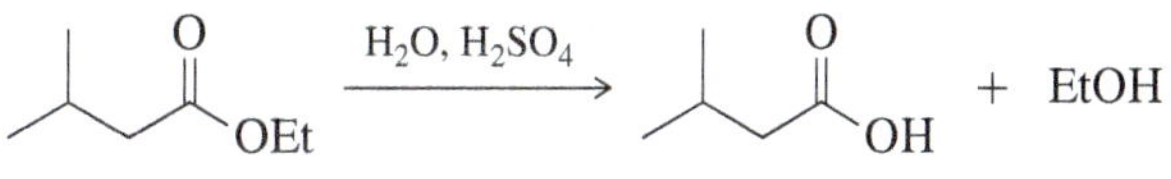

Mechanism:

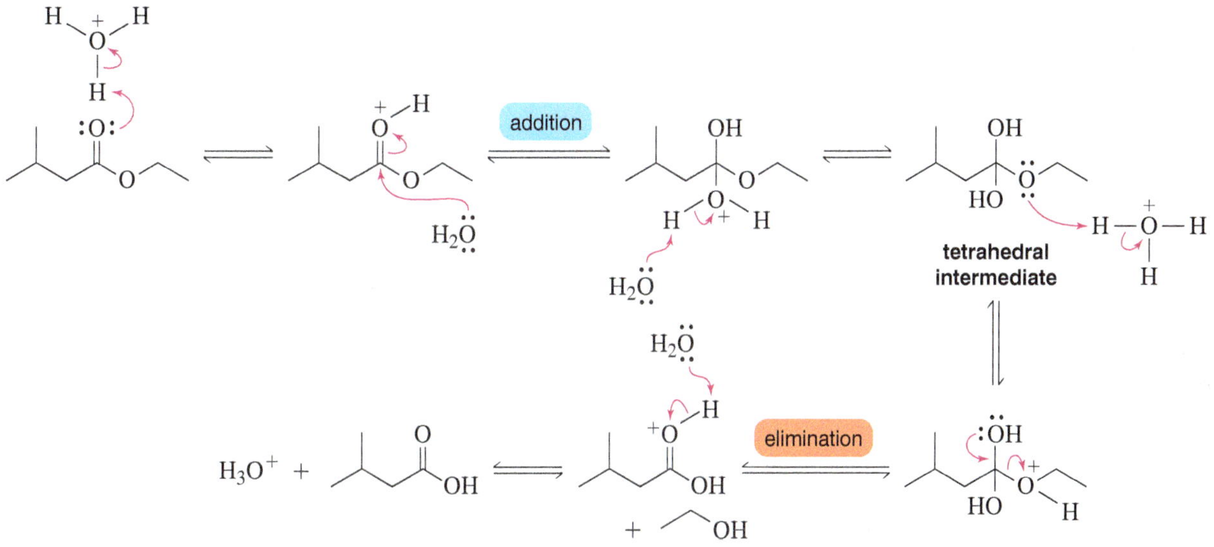

[*Note that General Mechanism 18B is simply General Mechanism 18A but with four acid–base steps needed to make stronger electrophiles and better leaving groups. Several of the reactions that follow proceed by identical mechanisms.*]

---

3. **Reduction to an alcohol—$LiAlH_4$ (Section 19.8.1):** Two equivalents of hydride add to the carbonyl carbon in separate steps to reduce all of the carboxylic acid derivatives except amides and nitriles to 1° alcohols. Proceeding by General Mechanism 18A, addition of hydride results in a tetrahedral intermediate that collapses with expulsion of a leaving group to produce an aldehyde. The aldehyde is subsequently reduced by addition of the second hydride ion. An acid quench results in the 1° alcohol product.

General:

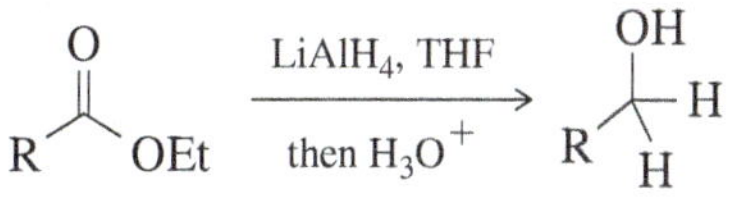

Specific:

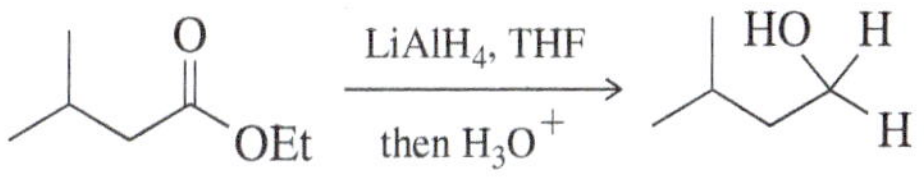

Mechanism:

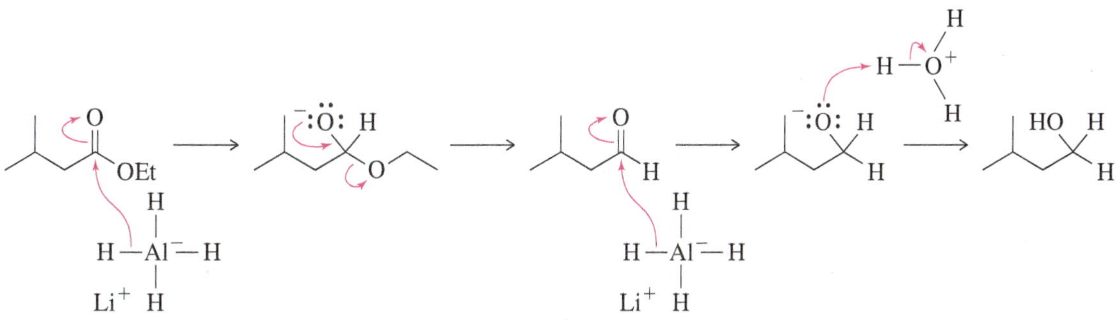

4. **Reduction to an amine—$LiAlH_4$ (Section 19.8.1):** Two equivalents of hydride add to the carbonyl carbon in separate steps to reduce amides to amines. Proceeding by General Mechanism 18A, addition of hydride results in a stable tetrahedral intermediate. The alkoxide donates a pair of electrons to the Lewis acid AlH_3 (alane), making a good leaving group. The oxyaluminate is expelled with assistance from the lone pair on nitrogen, thus producing an iminium ion. The iminium ion is subsequently reduced with the second equivalent of hydride, resulting in the amine product.

General:

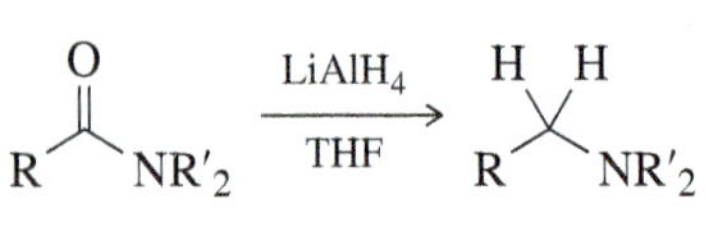

Specific:

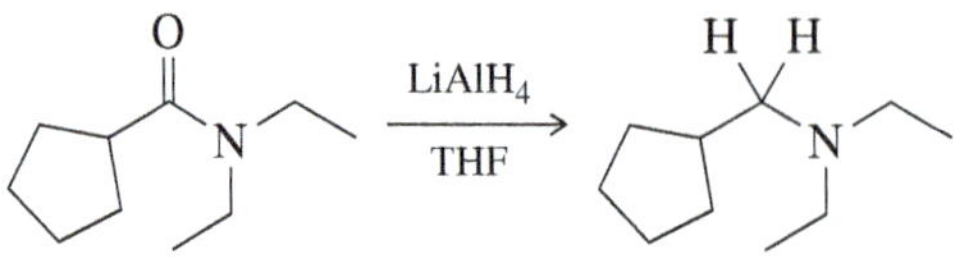

Mechanism:

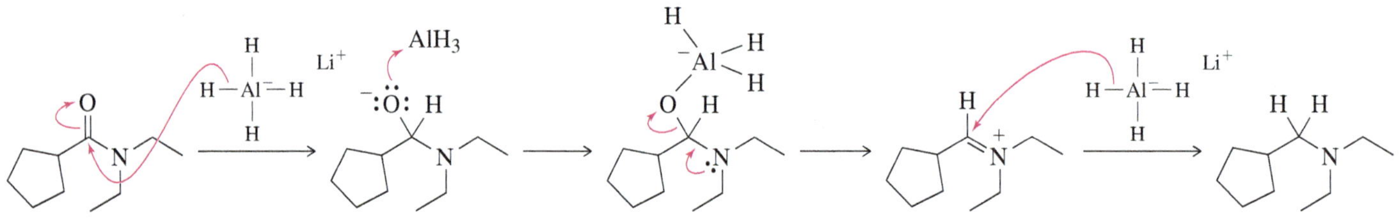

5. **Reduction to an aldehyde—LiHAl(O*t*-Bu)$_3$ (Section 19.8.2):** Useful only with acid chlorides, a single equivalent of hydride ion reacts via General Mechanism 18A. Formation of the tetrahedral intermediate is followed by reformation of the carbonyl and expulsion of the chloride leaving group to give the aldehyde. A second reduction does *not* occur because the reactivity of the aldehyde is reduced compared to the acid chloride, coupled with the fact that LiHAl(O*t*-Bu)$_3$ contains only a single hydride equivalent.

General:

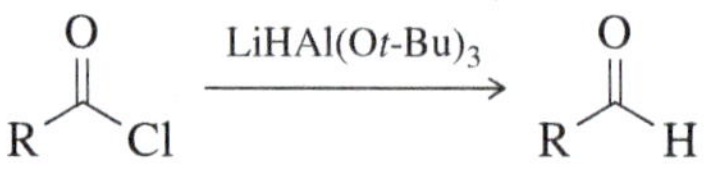

Specific:

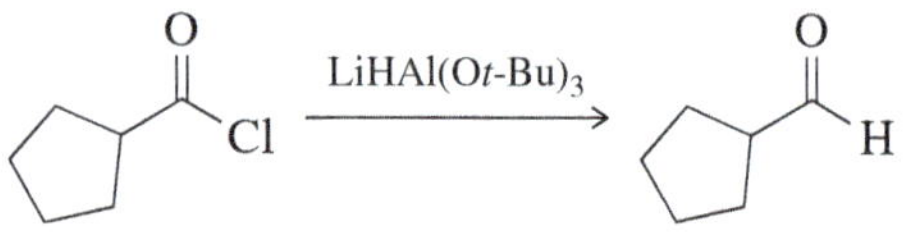

Mechanism:

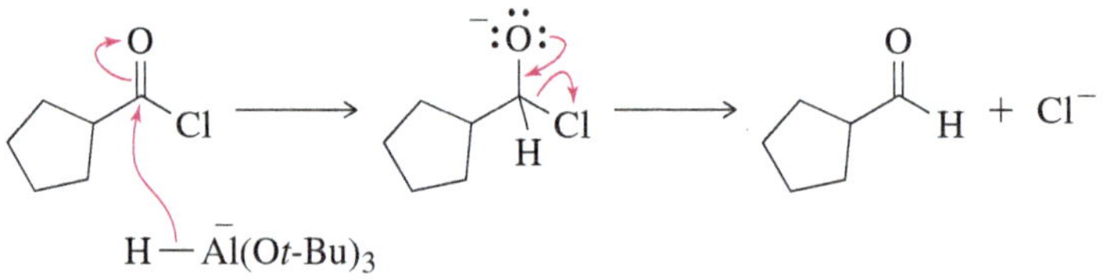

6. **Reduction to an aldehyde—DIBAl-H (Section 19.8.2):** With esters, the Lewis acid–promoted reaction occurs via a modified General Mechanism 18B. A Lewis acid–base reaction between the carbonyl oxygen and DIBAl-H makes the ester more electrophilic while at the same time placing a formal negative charge on aluminum. A resulting hydride transfer from aluminum to the carbonyl carbon gives a stable intermediate. This intermediate, when reacted with water, collapses to form an aldehyde. Subsequent aldehyde reduction is not possible at this point because all of the DIBAl-H has been consumed.

General:

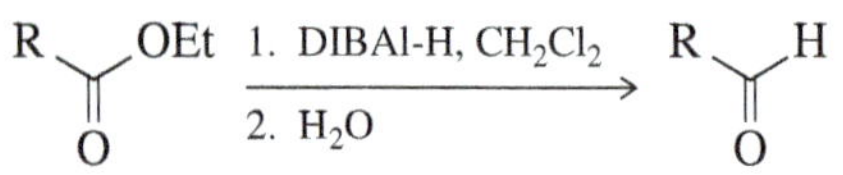

Specific:

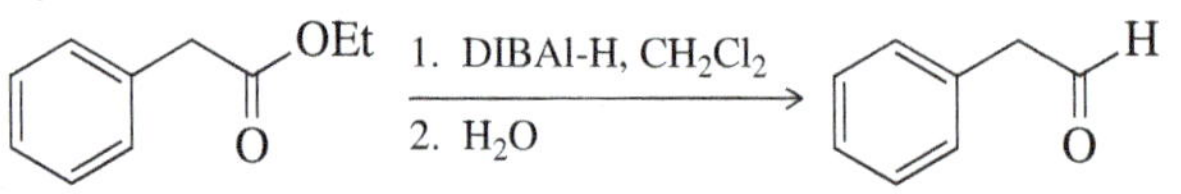

Mechanism:

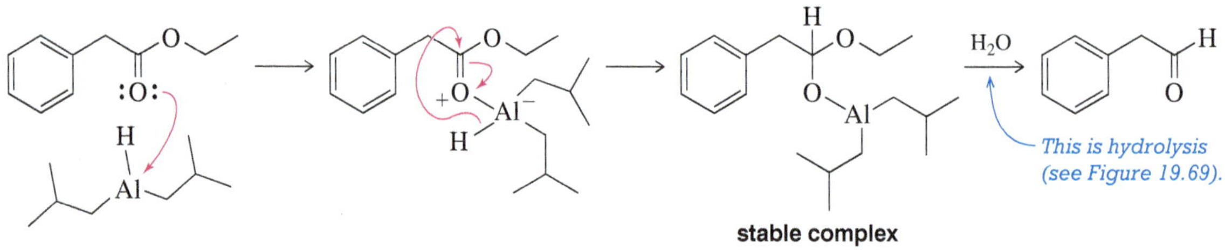

7. **Grignard reagents with esters (Section 19.9.1):** Proceeding by General Mechanism 18A, addition of the strong nucleophile results in the tetrahedral intermediate. Collapse of the tetrahedral intermediate with expulsion of the alkoxide produces a ketone. The ketone is more reactive than the starting ester, so it immediately reacts with the Grignard reagent to give a 3° alkoxide. Protonation of the alkoxide gives the 3° alcohol product. The reaction works identically with organolithium reagents.

General:

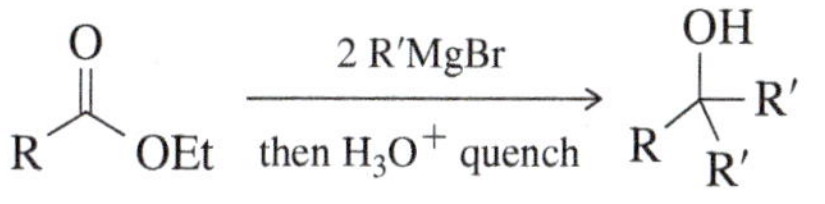

Specific:

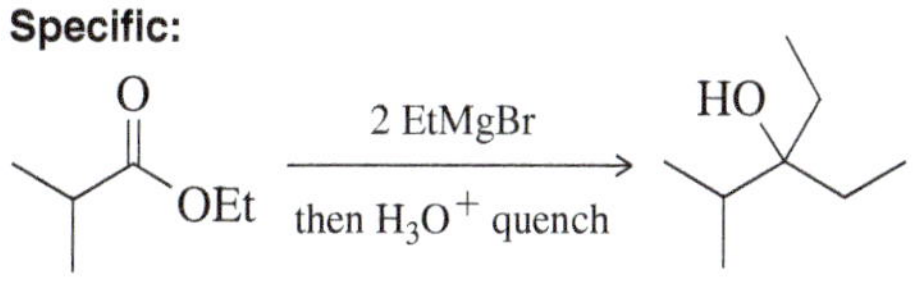

Mechanism:

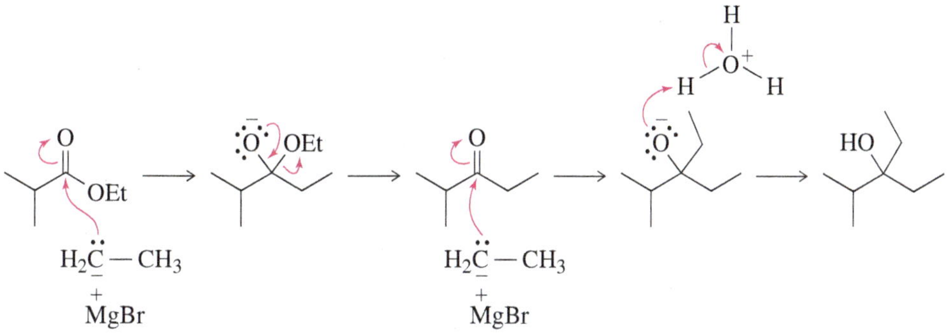

---

8. **Cuprates with acid chlorides (Section 19.9.2):** To enable direct synthesis of ketones from carboxylic acid derivatives, acid chlorides react with cuprates by an altogether different mechanism. Oxidative insertion of the cuprate into the C—Cl bond is followed by reductive elimination to give the new C—C bond. Analogous mechanisms are discussed in Chapter 16. It can be formally represented as occurring by General Mechanism 18A.

General:

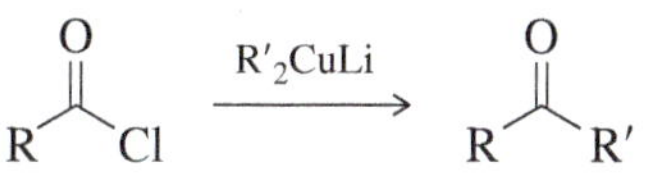

Specific:

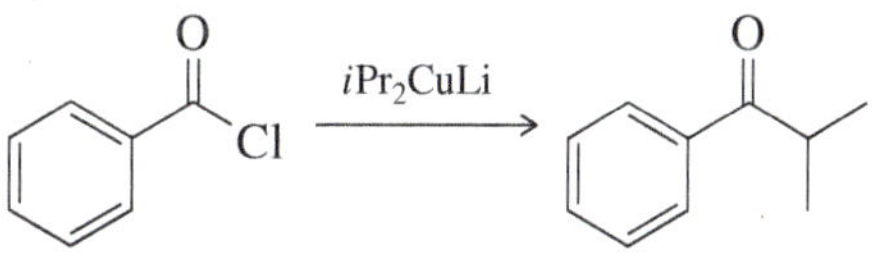

Mechanism:

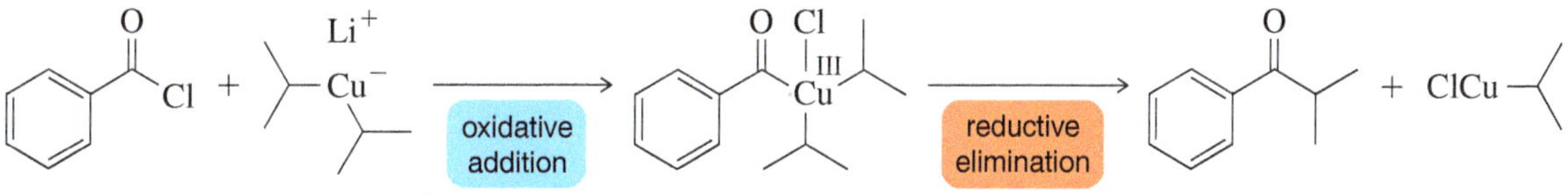

Formal mechanism:

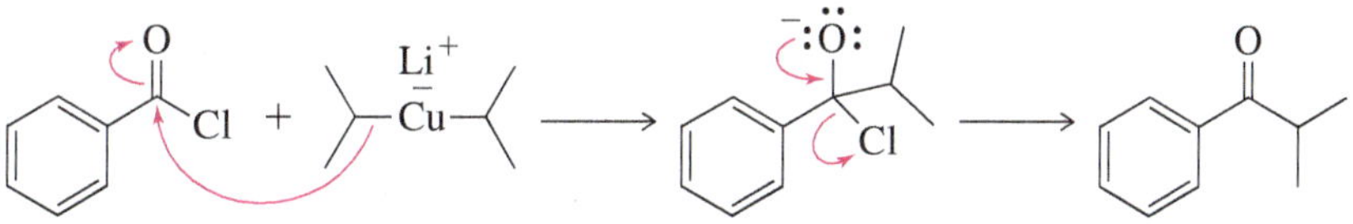

9. Hydration of nitriles (Section 19.10.1): Reacting like aldehydes and ketones, the hydration of nitriles occurs by an addition reaction. Protonation of the nitrile nitrogen is required to allow water (a weak nucleophile) to attack the nitrile carbon. Transfer of a proton from oxygen to nitrogen over two steps results in amide formation. If this reaction is heated, hydrolysis of the resulting amide occurs to give a carboxylic acid. This mechanism is covered in Section 19.7.1.

General:

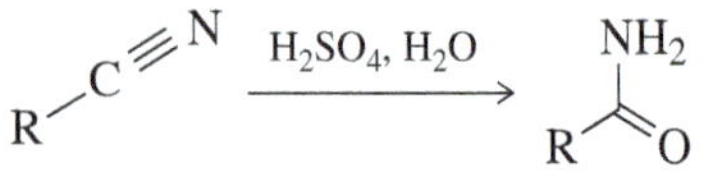

Specific:

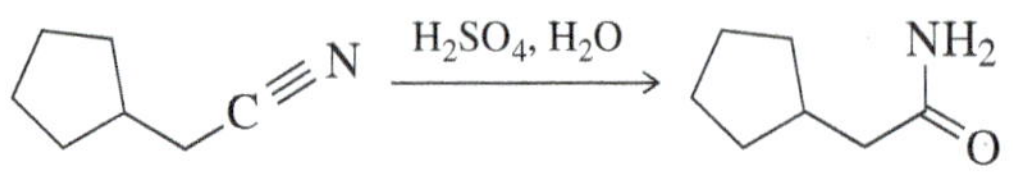

Mechanism:

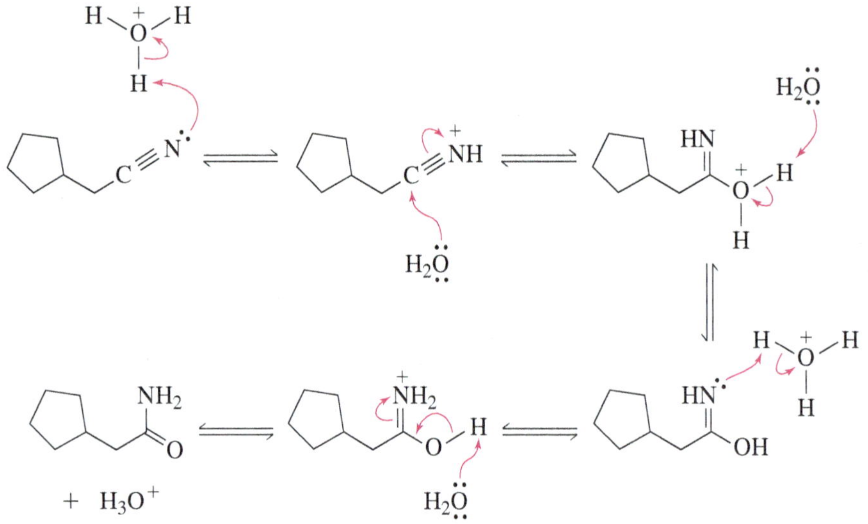

10. Grignard reagents with nitriles (Section 19.10.2): As strong nucleophiles, Grignard reagents add directly to nitriles. Following an acid quench, the neutral imine is hydrolyzed to give a ketone.

General:

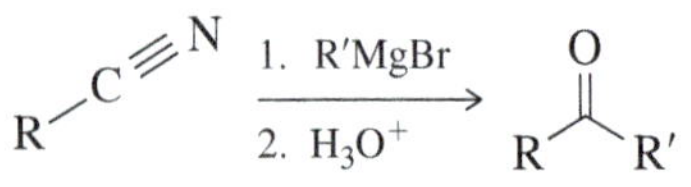

Specific:

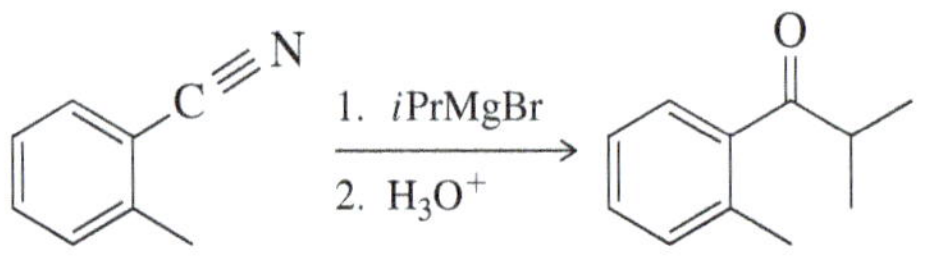

Mechanism:

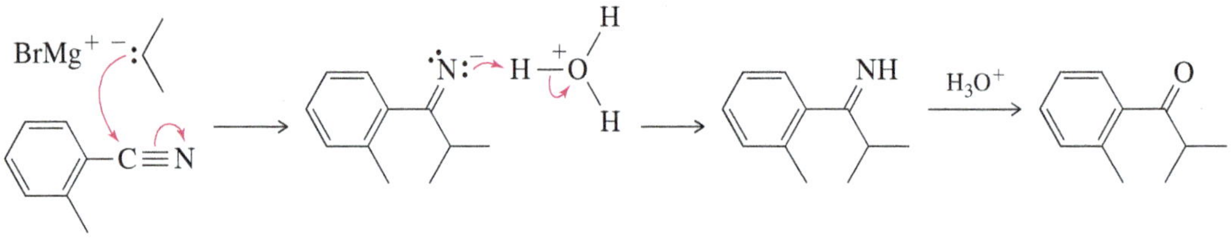

11. Reduction of nitriles—DIBAl-H (Section 19.10.3): A Lewis acid–Lewis base reaction between the nitrile nitrogen and DIBAl-H makes the nitrile more electrophilic, while at the same time placing a formal negative charge on aluminum. A resulting hydride transfer from aluminum to the nitrile carbon gives a stable intermediate. This intermediate, when reacted with water, forms an imine that is subsequently hydrolyzed to an aldehyde.

General:

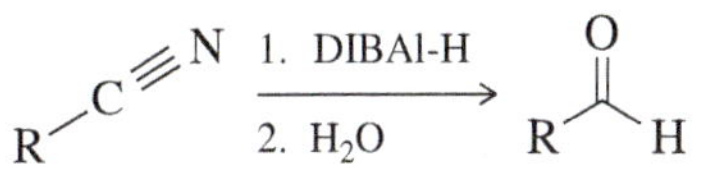

Specific:

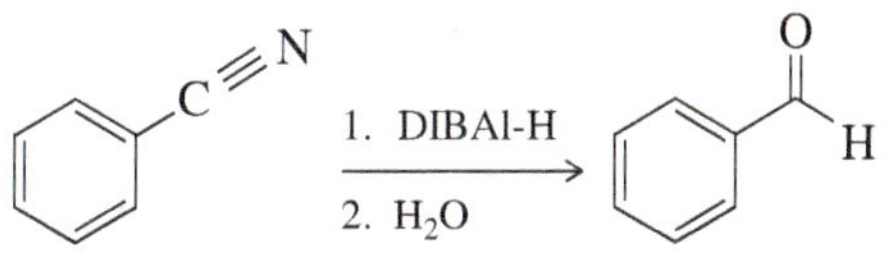

Mechanism:

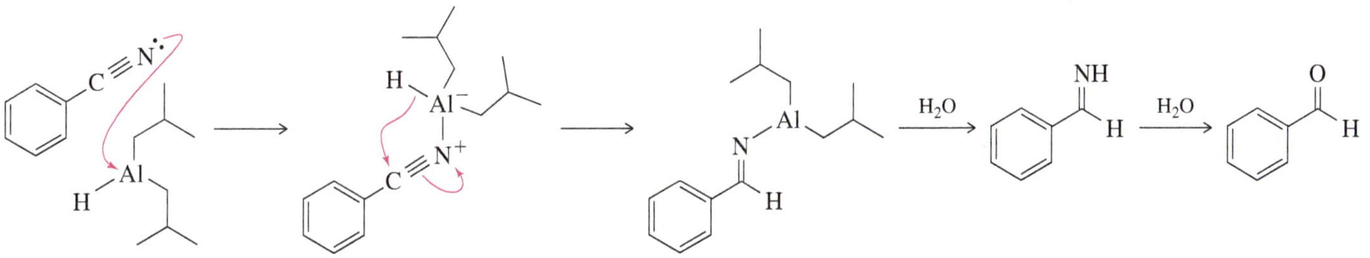

12. Reduction of nitriles—LiAlH$_4$ (Section 19.10.4): Addition of a strongly nucleophilic hydride ion is followed by donation of a pair of electrons on nitrogen to the Lewis acid AlH$_3$ (alane). Hydride transfer from aluminum to the imine carbon produces an amide that is protonated in an acid quench step. Hydrolysis of the resulting complex gives a 1° amine.

General:

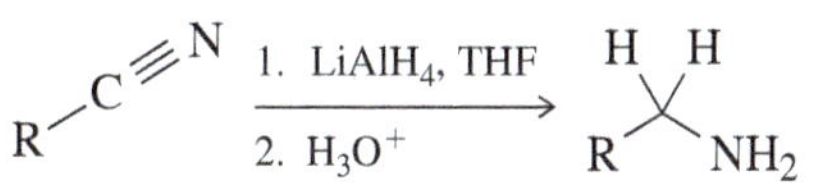

Specific:

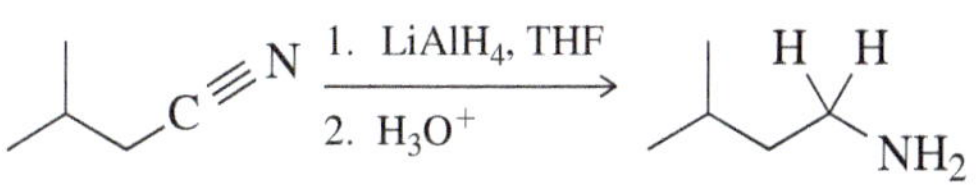

Mechanism:

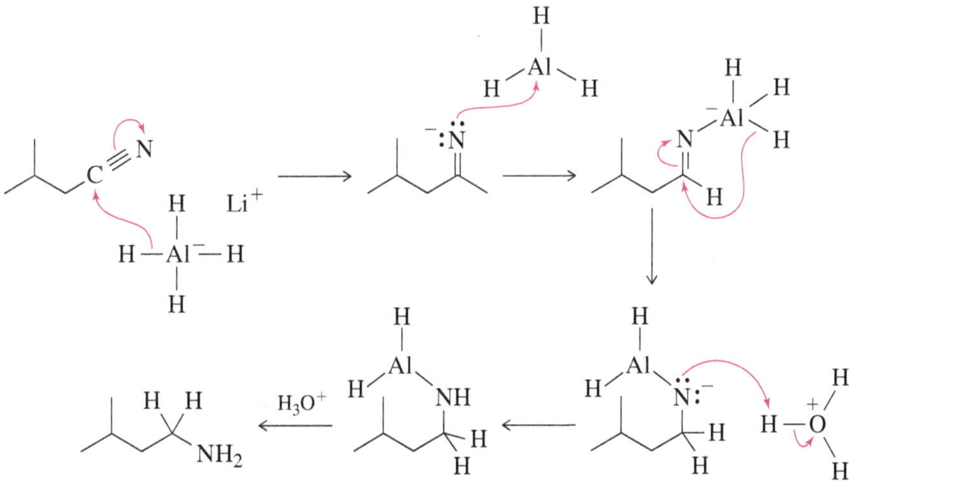

End-of-Chapter Assessments

[*Assessments are ranked according to their difficulty; •••• = most difficult.*]

19.61 (••) Recall from Chapter 18 that our bodies eliminate sodium benzoate by conversion to hippuric acid. How might an organic chemist carry out this transformation? [*What functional group is present in hippuric acid?*]

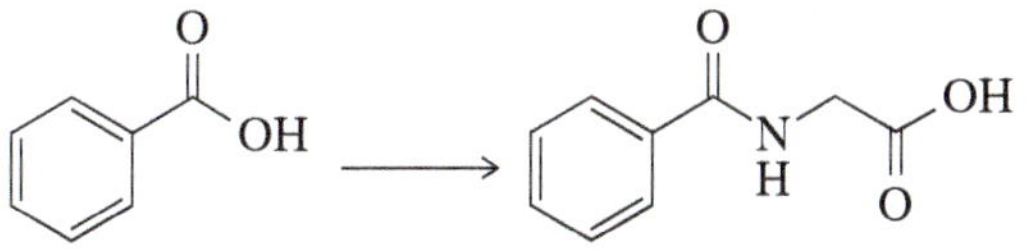

hippuric acid

19.62 (••) **LOOKING BACK** In Chapter 9, we used *m*CPBA to form epoxides from alkenes. How might you synthesize *m*CPBA from a carboxylic acid? [*Hint: Start by making an acid chloride.*]

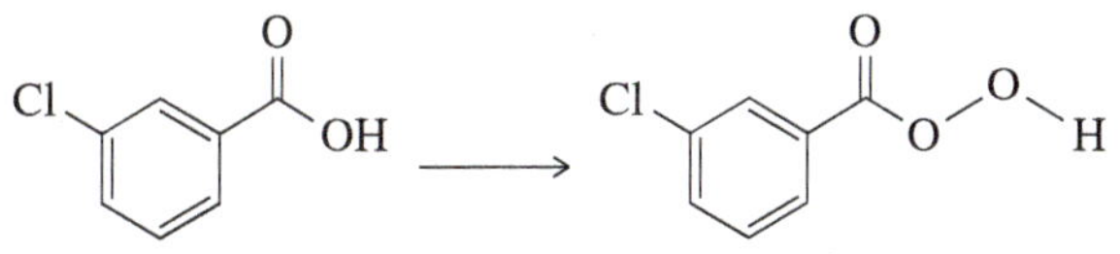

***m*CPBA**

19.63 (••) The reduction of an acid chloride isn't very useful. (a) Given how acid chlorides are made, why isn't it a useful reaction? (b) Show a mechanism for how the reaction would occur. [*It will look very similar to what you have seen with $LiAlH_4$ and esters.*]

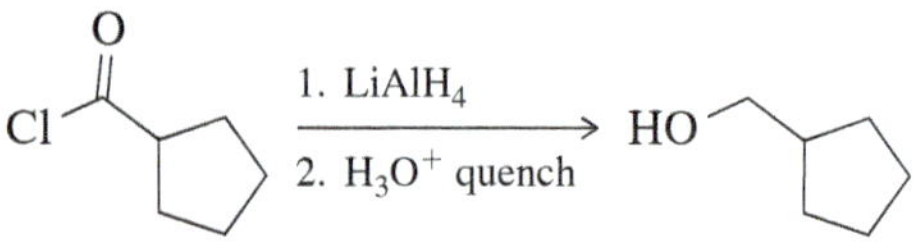

19.64 (••) Throughout the chapter, we saw the formation of carboxylates that necessitated an acid quench step. According to the definition given in the chapter, carboxylates are technically carboxylic acid derivatives. How might you modify Figure 19.9 to include carboxylate reactivity?

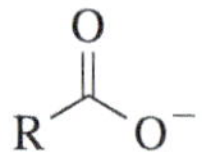

carboxylate

19.65 (•••) LOOKING AHEAD In Chapter 23, we show that amino groups donate electron density through resonance to an attached benzene ring, as in aniline. When aniline is reacted with acetic anhydride, the resulting functional group, while still electron donating, does so to a lesser extent. (a) Show the product of the reaction and (b) explain the diminished electron-donating ability.

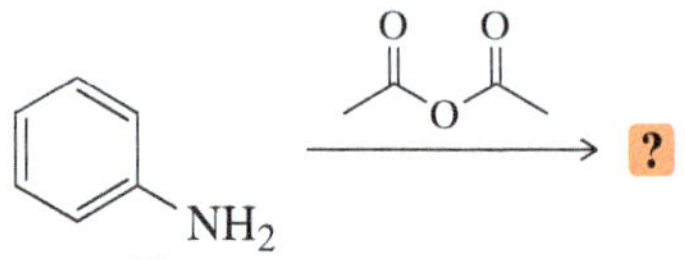

aniline

19.66 (•) LOOKING AHEAD In Chapter 25, we show that phthalimide can be deprotonated by hydroxide to make a good S_N2 nucleophile, as shown in the following reaction. Why is the proton in phthalimide so acidic?

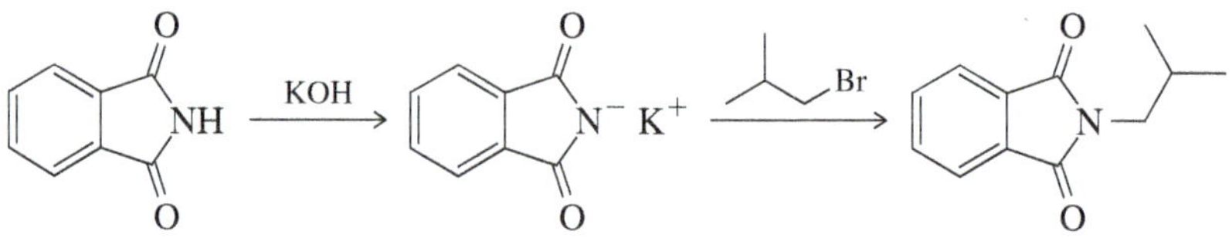

phthalimide

19.67 (•••) LOOKING AHEAD In Chapter 23, we discuss the sulfa antibiotics, one of which is sulfanilamide (it contains the sulfonamide functional group). Suggest a two-step reaction by which a sulfonamide could be synthesized from a sulfonic acid. [*Note how these functional group names compare to ones studied in Chapters 18 and 19.*]

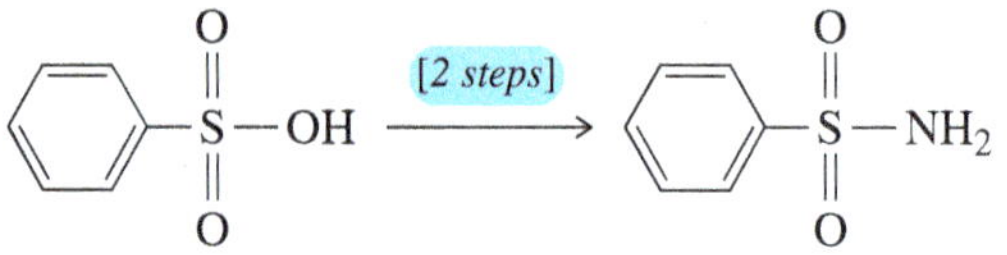

19.68 (•••) A chemist mixed up vials containing *N*,*N*-dimethylacetamide and *N*,*N*-dimethylformamide. How might IR spectroscopy be used to conclusively determine the identity of the compound in each vial? [*Think about the difference between the IR spectrum of an aldehyde and a ketone (Chapter 14).*]

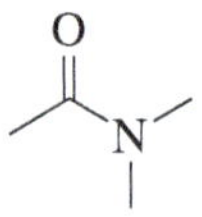

***N*,*N*-dimethylacetamide**

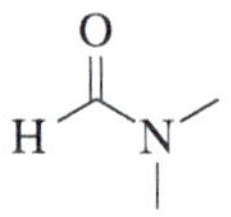

***N*,*N*-dimethylformamide**

19.69 (••) On a day in which the laboratory air conditioning stopped working, a chemist hypothesized that the atmosphere would provide enough ambient heat to drive the nitrile hydrolysis reaction to the carboxylic acid. Based on the IR of the resulting product, was this hypothesis correct?

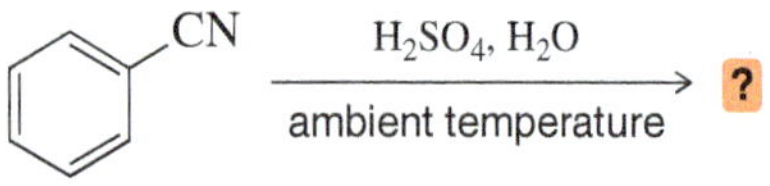

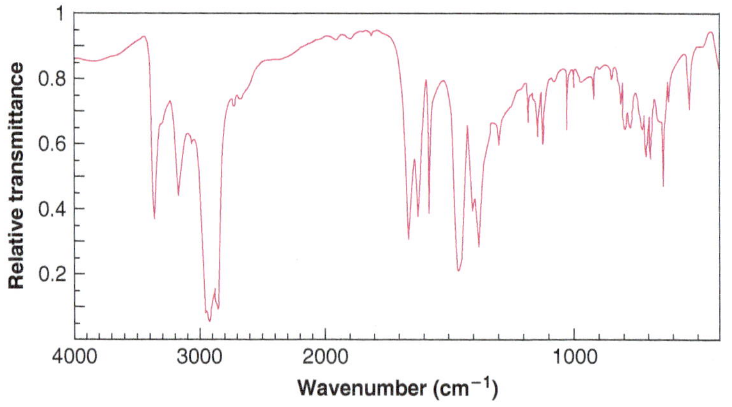

19.70 (••••) The following scheme has replaced the organic products with spectra of the compound produced by the reaction shown. Based on your knowledge of spectroscopy and reactivity, identify molecules (b), (d), and (f) and the reagents for (a), (c), and (e).

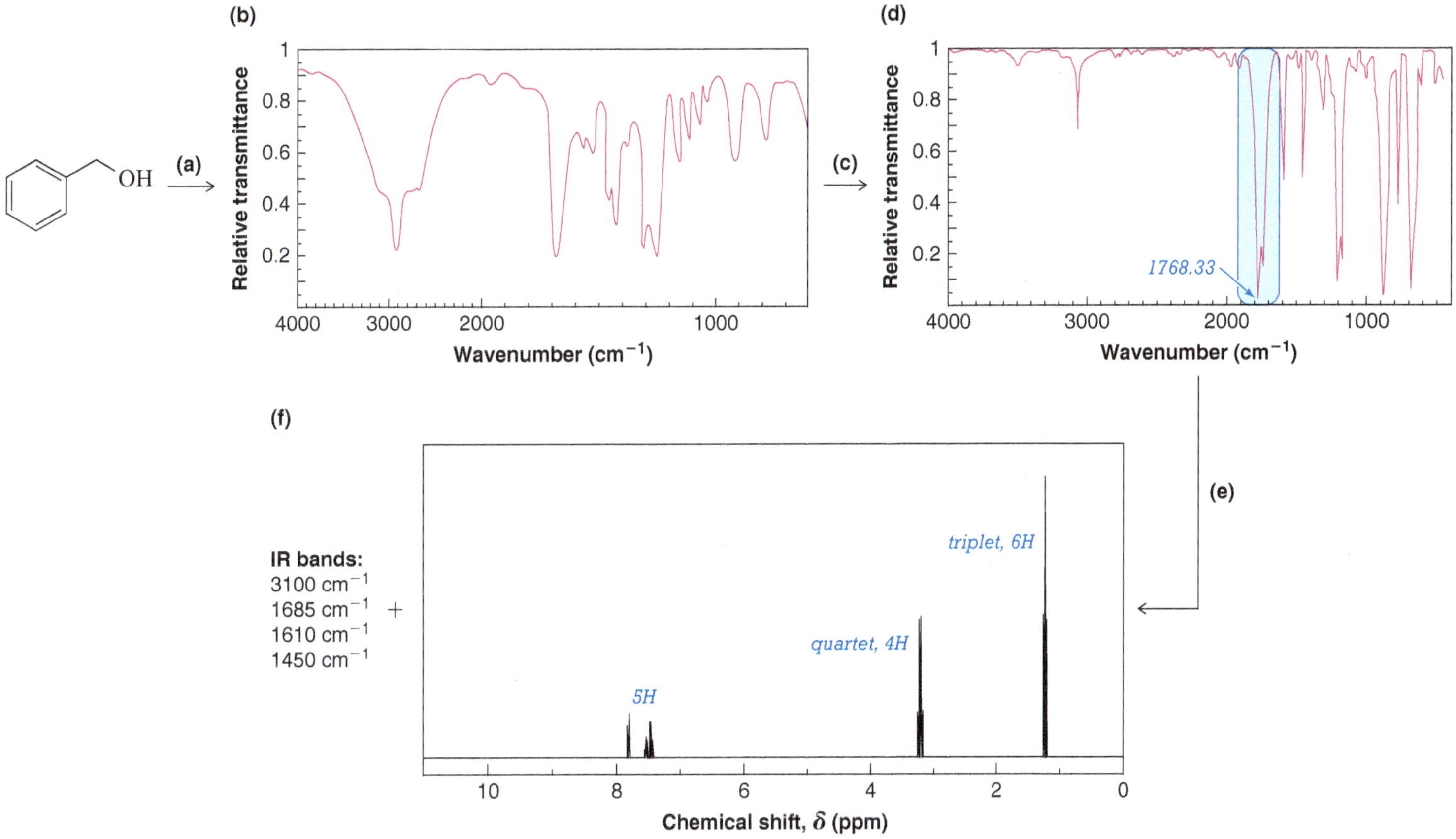

19.71 (•••) A chemist performed the following deprotection, expecting to isolate the 1° alcohol. Instead, a new lactone was produced. (a) Show a mechanism for the overall reaction and (b) rationalize the production of the larger lactone.

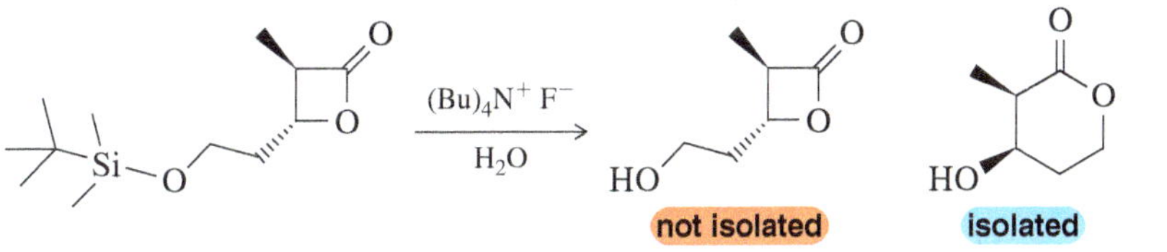

19.72 (•••) Suggest a synthesis of the molecule on the right from the molecule on the left. [*Each of these requires only the functional group transformations covered here in Chapter 19. The ideal number steps is shown, but there may be more than one possibility.*]

(a)

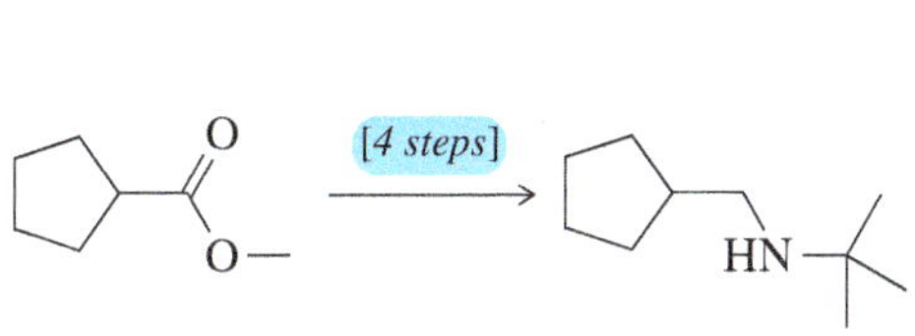

(b)

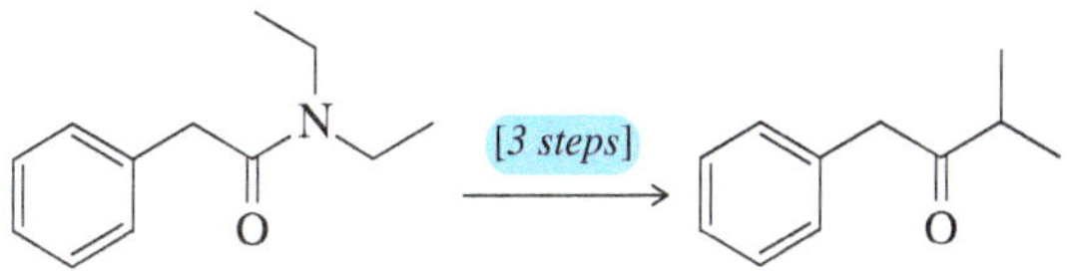

(c)

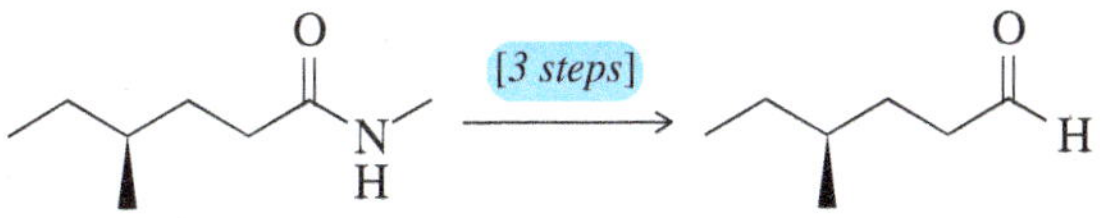

19.73 (••••) Suggest a synthesis of the molecule on the right from the molecule on the left. [*Unlike Assessment 19.72, these syntheses require reactions from previous chapters; be sure to number the carbons.*]

(a)

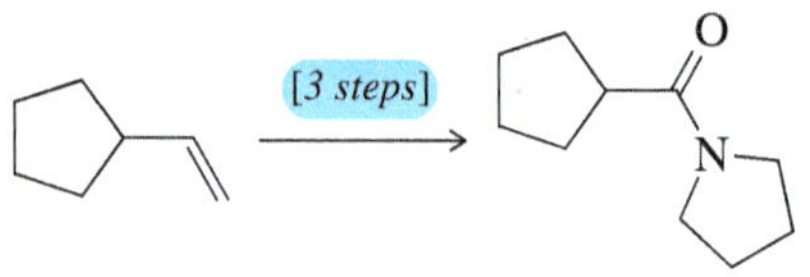

(b)

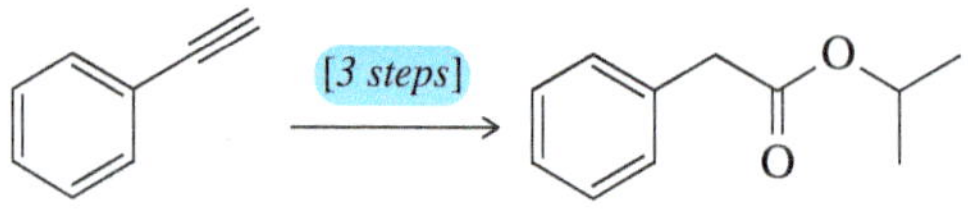

(c)

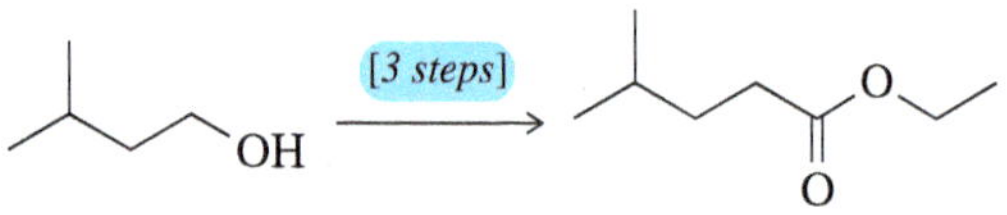

(d)

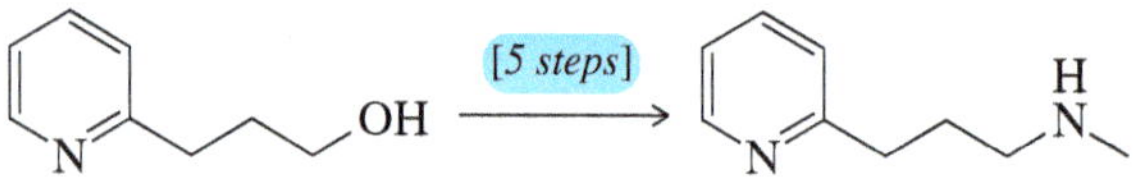

19.74 (••••) **LOOKING AHEAD** The conversion of an acid chloride to an amide is followed by reduction to give an amine, as shown here. By analogy to this reaction, how might you synthesize an amine from a ketone? [*You will need to consider a reaction learned in Chapter 17.*]

19.75 (••••) When DIBAl-H is used to reduce an ester with THF as the solvent, the reduction does not stop at the aldehyde. Explain this result. [*Think about how THF might interfere with the stable complex.*]

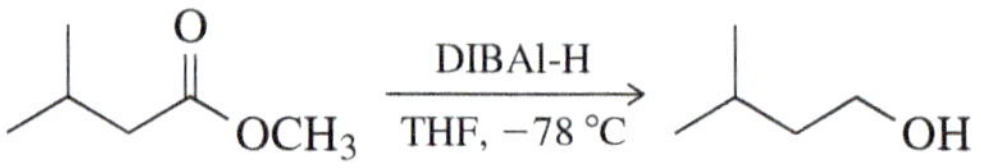

19.76 (•••) In addition to reducing esters, DIBAl-H will reduce aldehydes when exposed to them directly. Suggest a mechanism for how this might occur.

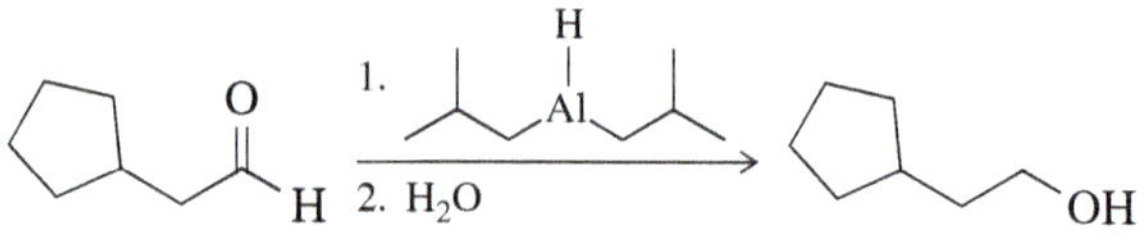

19.77 (•••) Reduction of a cyclic anhydride with lithium aluminum anhydride is followed by an aqueous acidic workup to give a cyclic ester (lactone). Suggest a mechanism for this reaction. [*Hint: Predict the reduction product first, then consider the esterification process.*]

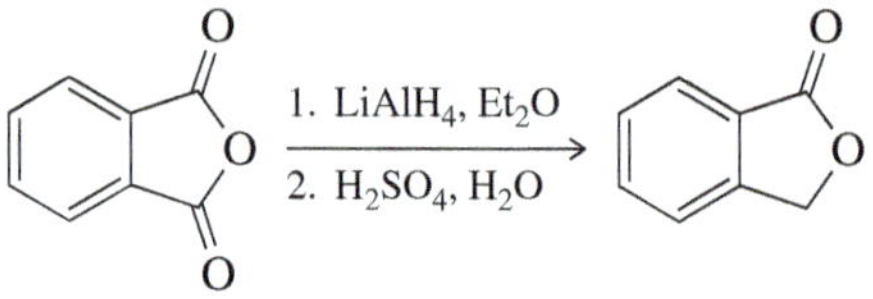

19.78 (••••) In light of your answer to Assessment 19.77, predict the product you'd expect if the anhydride was treated instead with two equivalents of MeMgBr, followed by a similar acidic workup.

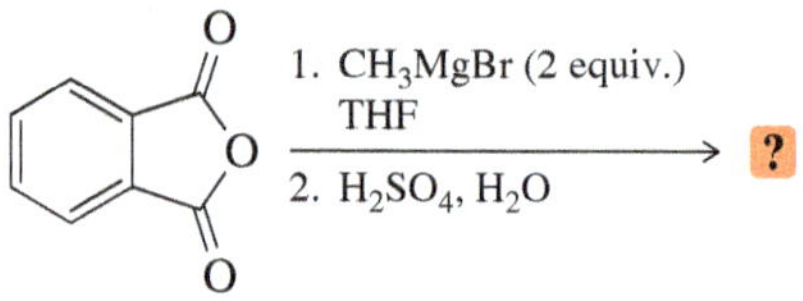

19.79 (••••) In Section 19.10.1, the hydration reaction of nitriles was shown to produce 1° amides. In the opposite direction, dehydration of 1° amides results in the synthesis of nitriles. Suggest a mechanism of a reaction that employs ethyl dichlorophosphate and DBU (a base) for this process. [*Recall from Chapter 17 that phosphorus has a strong affinity for oxygen.*]

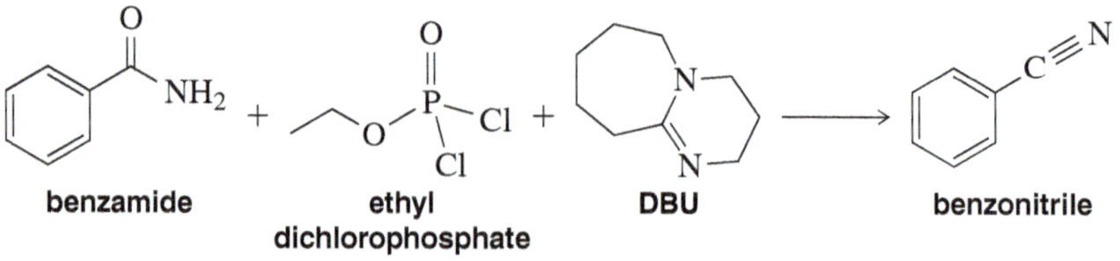

19.80 (•••) **LOOKING BACK** The use of DCC as a coupling agent for carboxylic acids and amines to make amides was shown in Assessments 18.61–18.63. (a) Suggest a mechanism for this reaction. (b) How is using this process analogous to the conversion of a carboxylic acid to an acid chloride prior to further derivatization?

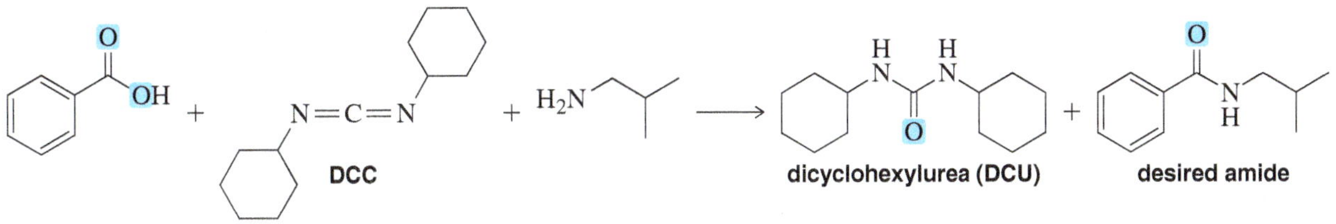

19.81 (•••) DCC can be used to synthesize amides and esters from carboxylic acids. Predict the product of the following reactions that make use of this technique.

(a)

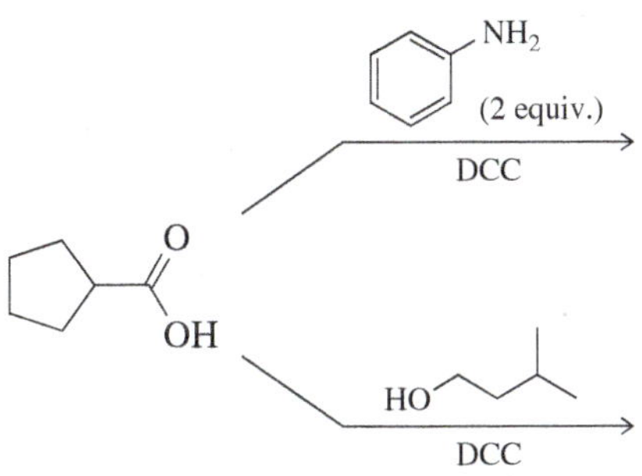

(b)

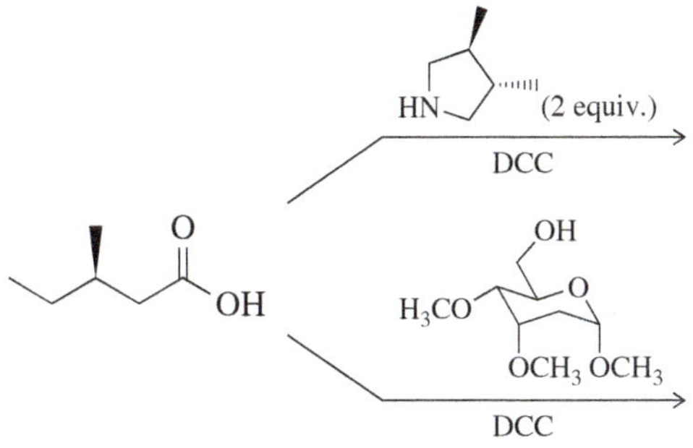

19.82 (•••) **LOOKING AHEAD** We show in Chapter 26 that DCC can also be used to attach an amino acid monomer to a growing peptide chain bound to a polymer resin. Predict the product of the following reaction. [*Ignore the unfamiliar parts and focus on the fact that this is simply the reaction of an amine and a carboxylic acid.*]

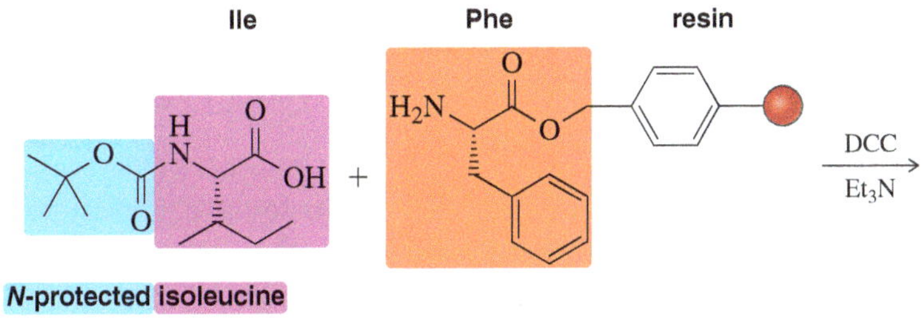

19.83 (•••) **LOOKING AHEAD** In Chapter 25, we show that aldehydes undergo a reductive amination reaction that features aspects of imine formation from Chapter 17 and the reduction of amides from this chapter. Suggest a mechanism for reductive amination. [*$NaCNBH_3$ is just a special form of $NaBH_4$.*]

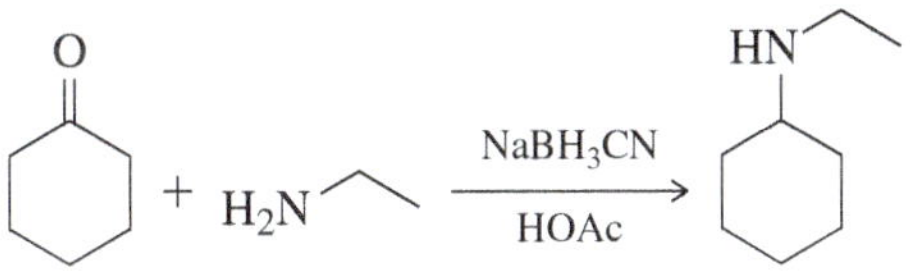

19.84 (•••) **LOOKING AHEAD** In Chapter 28, we show that nucleophilic acyl substitution can be used to synthesize the polymer Kevlar, which is used to make bulletproof vests because it is one of the strongest materials known. Suggest a mechanism for the formation of Kevlar.

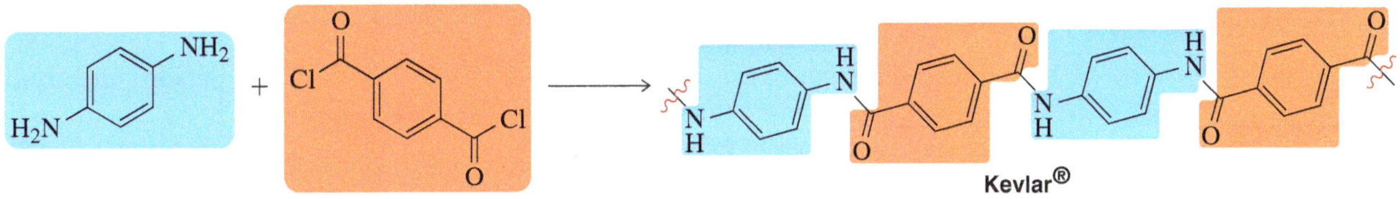

19.85 (•••) **LOOKING AHEAD** In Chapter 28, we show that phosgene, similar to an acid chloride, is used to convert bisphenol A to a polycarbonate polymer. Suggest a mechanism for the coupling of two bisphenol A molecules as shown below.

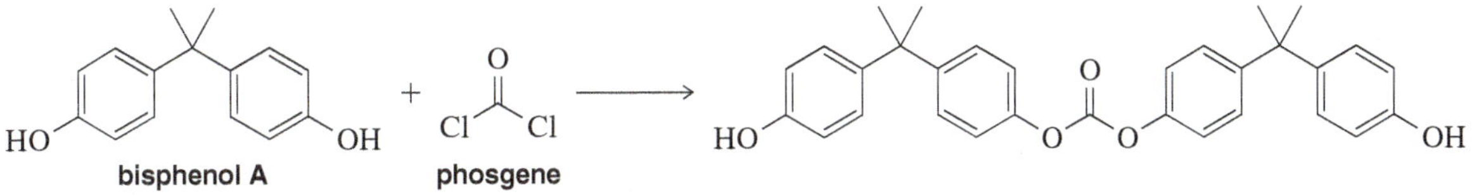

20

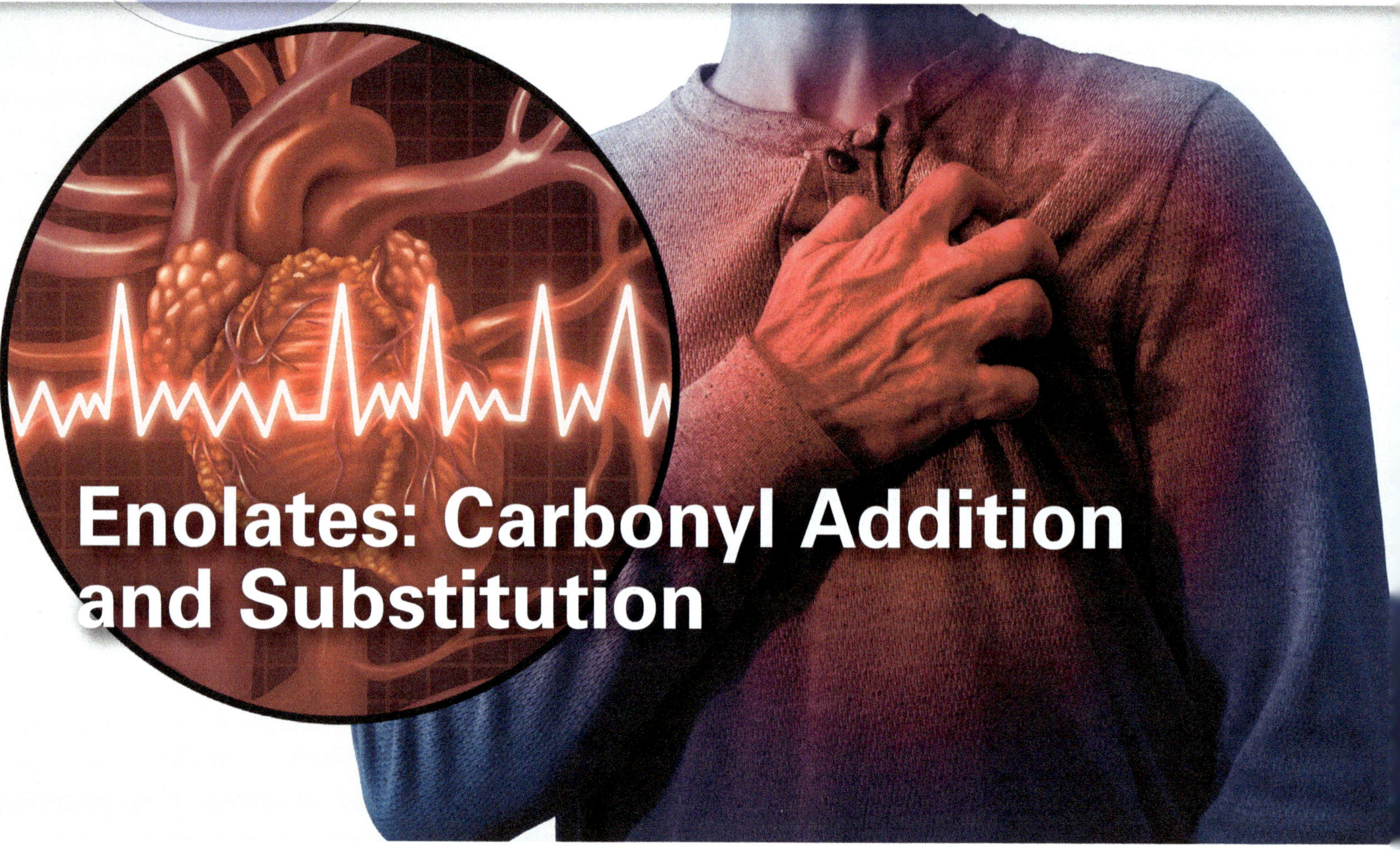

Enolates: Carbonyl Addition and Substitution

Learning Objectives

LO 20.0 Compare enols to alkenes

LO 20.1 Analyze the general mechanism of aldol reactions

LO 20.2 Analyze aldol reactions

LO 20.3 Compare crossed aldol condensation under classical and modern conditions

LO 20.4 Analyze intramolecular aldol reactions

LO 20.5 Analyze Claisen reactions and their variations

LO 20.6 Analyze acetoacetic ester and malonic ester synthesis reactions

LO 20.7 Describe the mechanism of enamine synthesis reactions

Introduction

According to the Centers for Disease Control, heart disease accounts for one in every four deaths and is the leading cause of death for men, women, and most ethnic groups. And it's not just a burden on the individuals and families who suffer loss; heart disease is also a burden on the U.S. economy, costing over $100 billion each year in treatments, medicines, and lost productivity. Cholesterol is a key player in the development of coronary heart disease. As LDL cholesterol builds up on the inner walls of heart arteries, white blood cells accumulate in those areas to remedy the situation. The continual buildup of cholesterol and the immune system response result in plaques on the arterial walls that restrict blood flow. If the plaque ruptures, blood clots can form rapidly, ultimately blocking the coronary artery, resulting in myocardial infarction (i.e., a heart attack).

Although high cholesterol levels are usually attributed to a person's dietary habits and exercise patterns, more than half of the cholesterol in the human body is made in the liver by the pathway shown in **Figure 20.1**. Because diet and exercise are not always enough, medicines can be used to control cholesterol levels. Fortunately, the outcome-determining step occurs very early in the 20-step biosynthesis leading to cholesterol. By inhibiting the enzymatic reduction of β-hydroxy-β-methylglutaryl-CoA (HMG-CoA), statins have been successful at reducing the amount of cholesterol produced by the body.

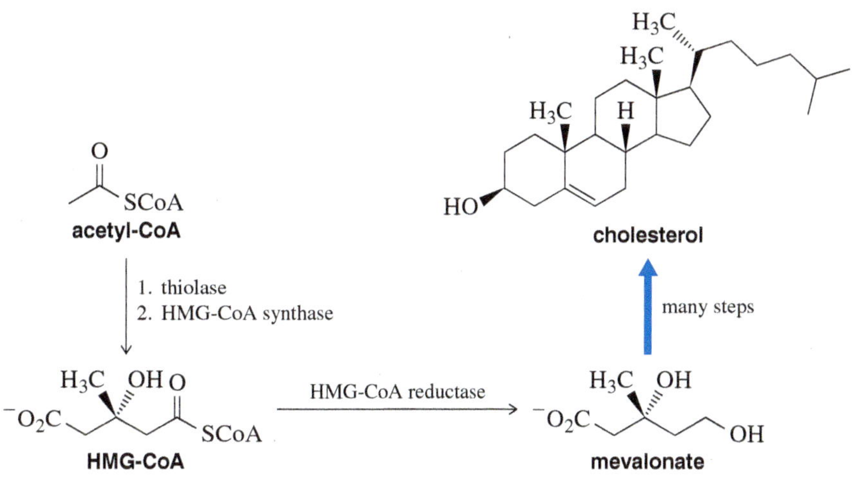

Figure 20.1 Cholesterol is a major contributor to heart disease. The reduction of HMG-CoA is a key step in the biosynthesis of cholesterol, so a drug that interferes with the operation of HMG-CoA reductase should reduce the amount of cholesterol in the bloodstream. This should limit, in turn, the buildup of the LDL cholesterol plaques that can lead to heart attack.

"To imagine is human; to invent, divine."

—Chemically modified proverb

Enolate reactions are useful for the synthesis of complex molecules like cholesterol and are prevalent in biological systems (see Chemistry in Context 20-B). In addition to accomplishing our organic chemistry–related goals in Chapter 20, we will see how knowledge of biochemical mechanisms inspire the synthesis of drugs that positively impact society.

Modern organic chemists work under the assumptions that there is a reagent for every transformation, any molecule can be synthesized, *and* molecules can always be bent to the will of the organic chemist. [*This isn't true.*] We have seen many situations in this book where molecules don't react how we want them to. [*The ball rolls downhill, but it doesn't always end up where we want it.*] This is especially true when thermodynamically controlled reactions find their way to the lowest potential energy, regardless of the chemist's desires. When there are several states of similar low potential energy, a complex mixture of products results. The earliest-studied reactions of enols and enolates with carbonyls were especially difficult to control because optimal reagents were unavailable. Instead, early chemists had to "trick" the system to make the lowest-potential-energy product the desired molecule. This chapter gives us a unique opportunity to look at how early organic chemists were able to manipulate reactions, coaxing them to do what they wanted them to do. Our access to modern reagents and a modern understanding of chemical principles inspires great admiration for what early organic chemists accomplished without them. Finally, reflecting on how they solved problems helps us develop our own problem-solving strategies, ones with application in the organic chemistry lab, the physician's office, or elsewhere.

As synthetic chemists, we always look forward to learning new carbon–carbon bond-forming reactions. [*In case you needed the motivation to continue reading . . . .*] We'll learn several ways to do this in Chapter 20. As the fourth and final chapter on the carbonyl, this chapter examines the chemical reactions enabled by the presence of a carbonyl group. Specifically, we look at reactions that take place at the α-carbon, the carbon directly attached to a carbonyl (**Figure 20.2**). Before we get started looking at these reactions, let's test our understanding of some concepts that will aid our understanding.

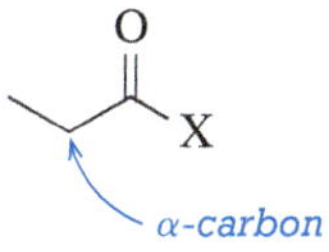

Figure 20.2 The α-carbon is the carbon directly attached to the carbonyl. [*You'll soon know why it deserves such special recognition.*]

Test your understanding

Assessment 20.1 Beginning with acetylene and benzyl bromide and using any other inorganic reagents, propose a synthesis of the alkene shown here.

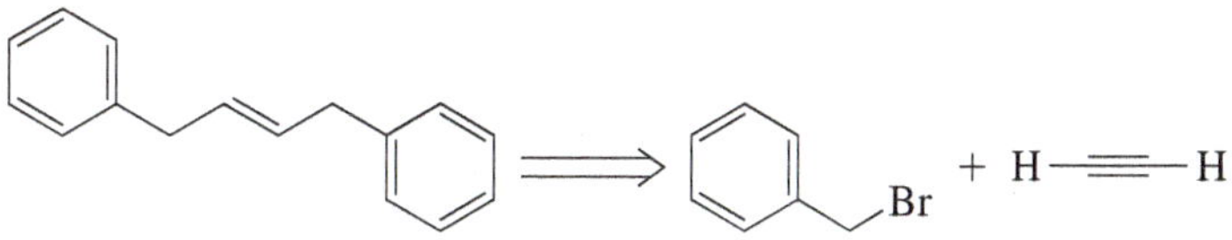

Assessment 20.2 Predict the product of the following series of reactions.

(a) (b) (c)

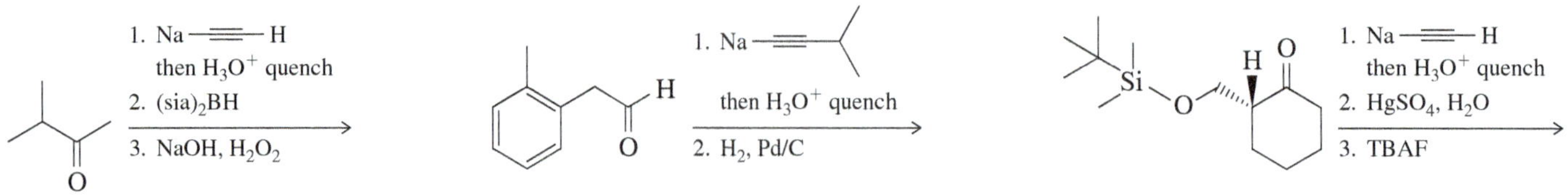

Assessment 20.3 Suggest reagents and conditions to effect the following transformations.

(a) (b)

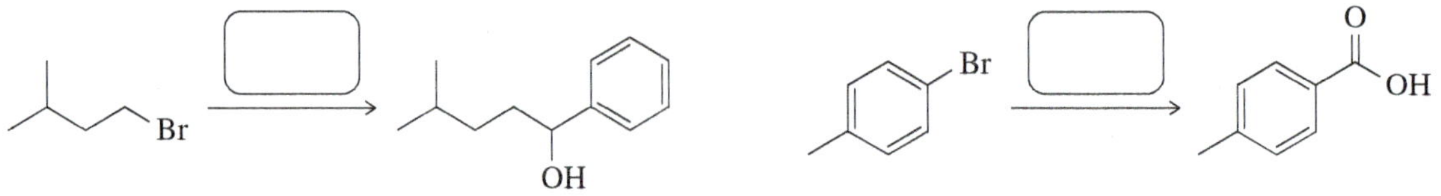

Assessment 20.4 Provide an arrow-pushing mechanism for the **keto–enol tautomerization** under acidic and basic conditions.

(a) (b)

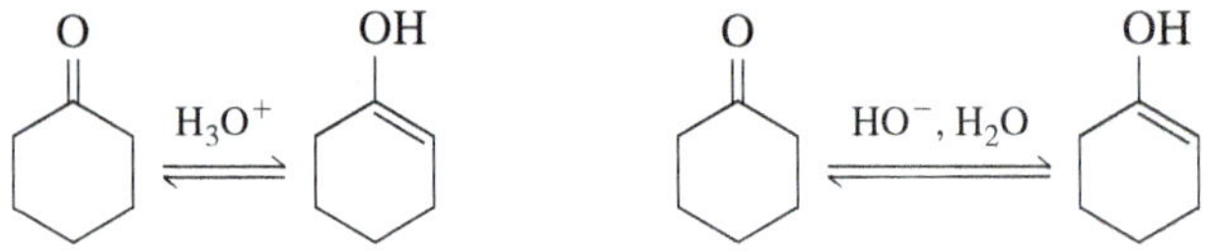

Assessment 20.5 Although the keto form is typically more stable than the enol form, the enol form of the molecule shown here is actually more stable. Explain this observation.

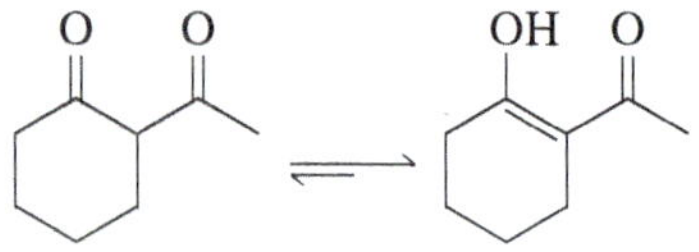

Assessment 20.6 Predict the product of the following reactions.

(a) (b)

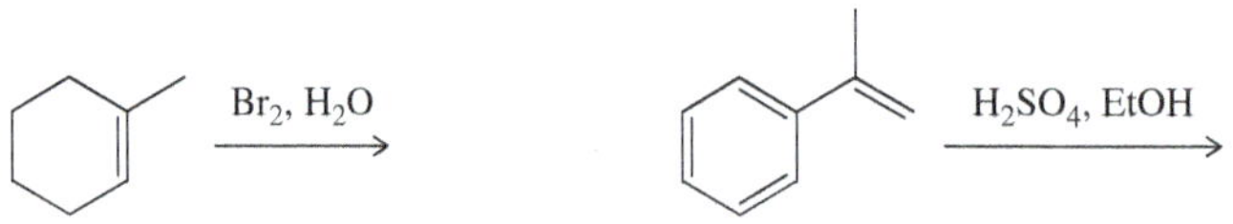

20.1 Chemistry at the α-carbon

The focus here in Chapter 20 is on functionalizing the α-carbon of carbonyls, as represented by the myriad possibilities shown in Figure 20.3(a). In each of these transformations, a hydrogen is replaced by the group that is added. This is a first clue that these reactions occur through the intermediacy of enols and enolates, two species that are nucleophilic at the α-carbon [Figure 20.3(b)]. [*To form either the enol or the enolate, an α-hydrogen must be removed.*]

(a) (b)

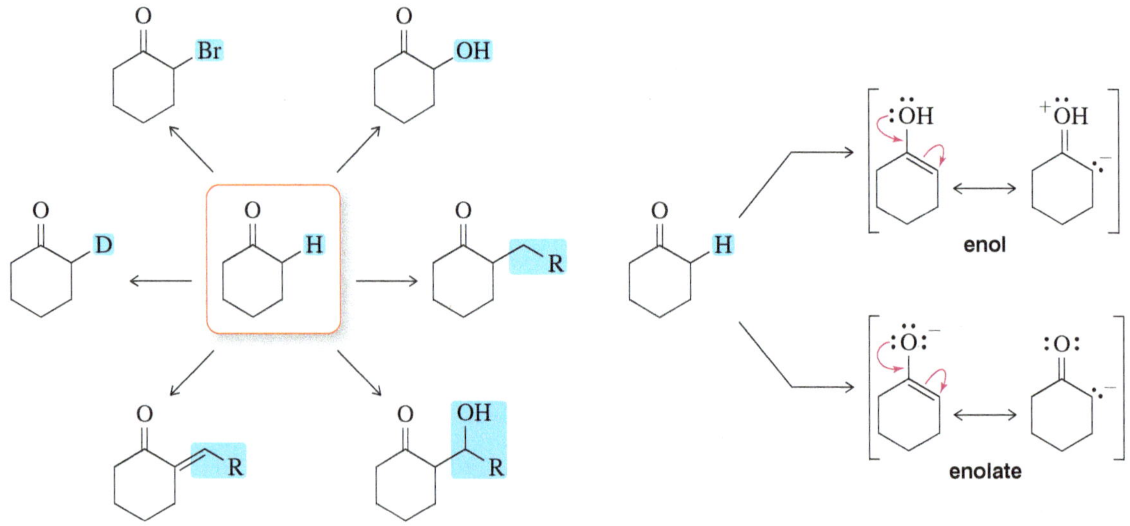

Figure 20.3 Reactions at the α-carbon (a) enable the formation of various new molecules and (b) occur through the intermediacy of enols and enolates.

20.2 Enols

As a new functional group, we begin our exploration of enols by comparing them to alkenes. [*See Chapters 8 and 9.*] Alkenes react predominantly as nucleophiles, donating the C—C π electrons to strong electrophiles. An enol can be thought of as an alkene with a hydroxyl group attached. A comparison of alkene and enol reactions with HBr is shown in **Figure 20.4**. The first notable difference is that HBr addition to an enol results in a carbocation stabilized by resonance and hyperconjugation. The carbocation from alkene addition is only stabilized by hyperconjugation. For this reason, enols are stronger nucleophiles. In the second step of alkene addition, the bromide ion reacts with the carbocation to form a bromoalkane. In the second step of enol addition, bromide can attack the carbocation to form 1-bromocyclopentanol, or the bromide can act as a base, removing the acidic proton to form cyclopentanone. Deprotonation is usually faster than nucleophilic attack and that is the case here. Furthermore, this deprotonation is favored by entropy and by formation of the stable C—O π bond. [*You have seen this mechanism previously in Section 10.8.3 and Assessment 20.4(a).*]

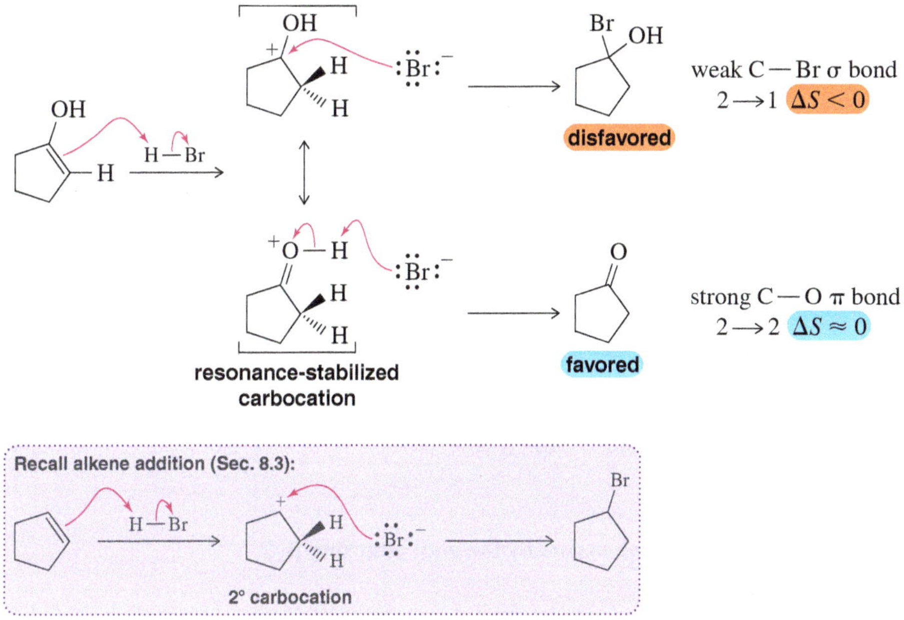

Figure 20.4 Enols are stronger nucleophiles than alkenes because a resonance-stabilized carbocation is formed. The ketone forms preferentially because deprotonation is faster and more favorable than nucleophilic attack.

Assessment 20.7 A chemist attempted to add phenylmagnesium bromide to a cyclohexanone derivative to produce the tertiary alcohol shown here. Unfortunately, the desired product was not formed. What was the product of the reaction? Why did it form instead of the desired product?

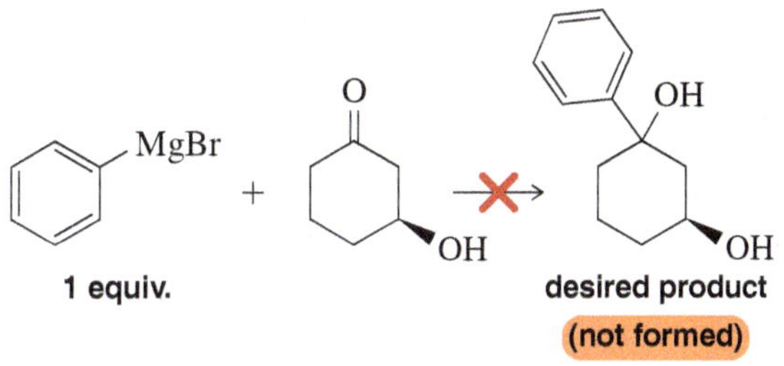

Assessment 20.8 Rationalize the difference in K_{eq} for the following reactions. Be sure to account for both ΔS and ΔH.

20.2.1 Reactions of enols

Recall from Section 10.8.3 on alkyne hydration that ketones are the favored side of the equilibrium in the keto–enol tautomerism. In spite of there being less of it at equilibrium, **α-substitution** of ketones/aldehydes proceeds through the enol form. Once formed by the mechanism you showed for Assessment 20.4, the enol reacts with whatever electrophiles are present.

20.2.1.1 Deuterium exchange

Figure 20.5 shows the preparation of d^6-acetone, a useful NMR solvent, through deuterium exchange. The keto–enol equilibration reaction occurs in the presence of DCl and D_2O to replace **α-hydrogens** with deuterium, a heavy isotope of hydrogen. During equilibration, the predominant electrophile in solution is D_3O^+. Thus, the enol is statistically more likely to attack D_3O^+ than H_3O^+, continuing until all α-hydrogens have been replaced with deuterium.

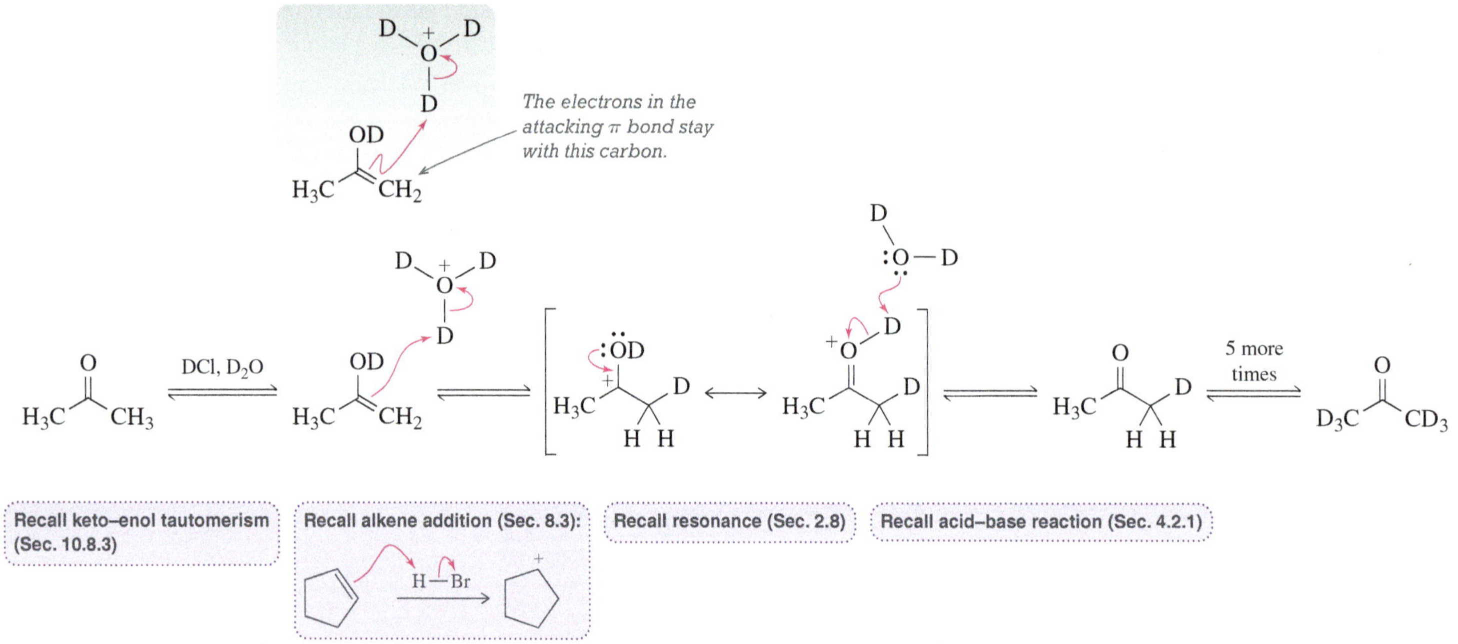

Figure 20.5 **Deuterium exchange occurs by the enol attacking D_3O^+.**

20.2.1.2 α-Bromination

The reaction of enols can be used for the synthesis of α-bromo ketones (Figure 20.6). Following the acetic acid–catalyzed equilibration to the enol, the π bond of the enol attacks bromine to generate the resonance-stabilized carbocation. When an alkene reacts with Br_2, a bromonium ion forms because it is more stable than the 2° carbocation. Resonance stabilization of the carbocation obviates the need for the bromonium ion forming in the enol reaction. Once more, deprotonation by bromide is faster and more favored than nucleophilic attack.

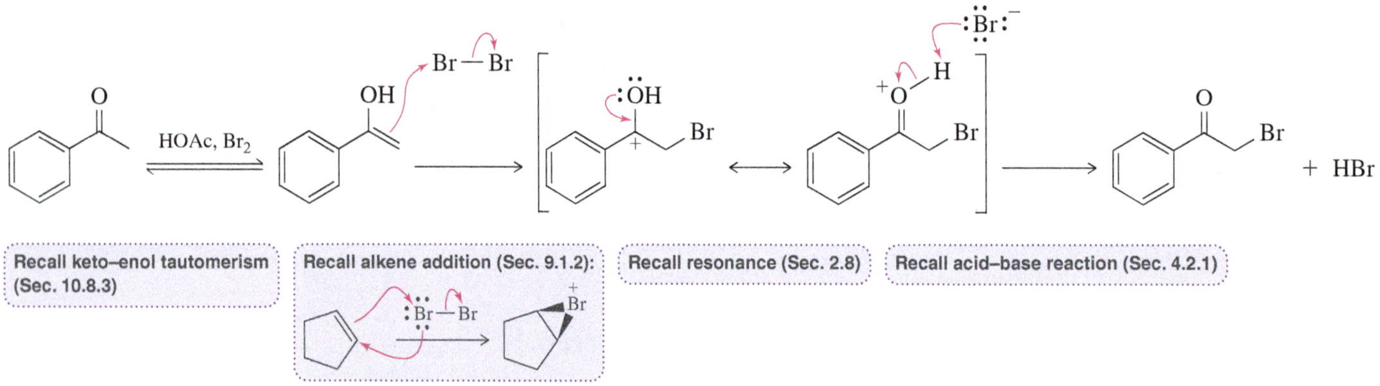

Figure 20.6 **α-Bromination occurs by the enol attacking Br_2.**

20.2.1.3 Racemization

Not all reactions of enols are productive. Imagine that we had produced (*S*)-2-methylcyclopentanone as a single enantiomer (**Figure 20.7**). Although the keto form is more stable than the enol, the presence of acid catalyzes its formation. Once formed, the stereochemistry of the starting material is lost. In reforming the ketone, the enol grabs a proton from solution on either face of the planar alkene at equal rates to produce a racemic mixture through enantiomeric transition states. [*This is simply the acid-catalyzed keto–enol tautomerization mechanism we learned in Section 10.8.3 with negative consequences for our stereocenter.*]

Figure 20.7 Keto–enol tautomerism can cause a loss of stereochemical purity.

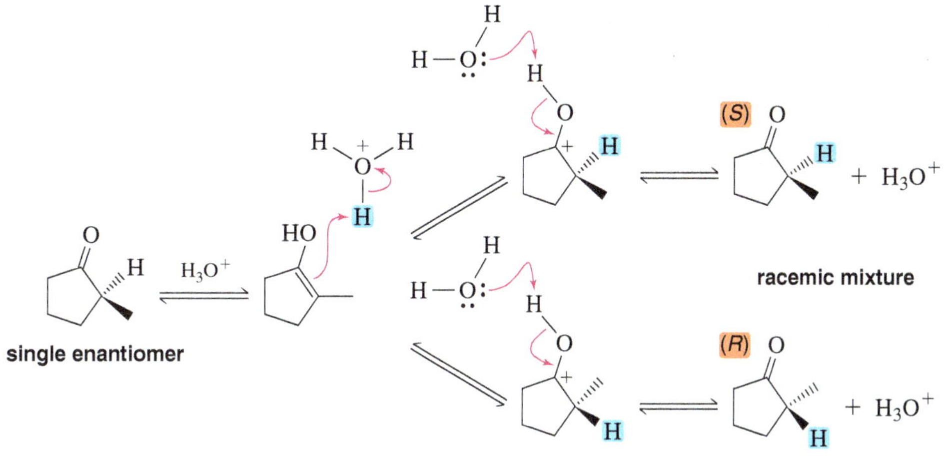

20.2.1.4 α-Hydroxylation

Structurally related to enols, **silyl enol ethers** can be oxidized with *m*CPBA to generate α-hydroxy carbonyl compounds, which have interesting biological properties (**Figure 20.8**). In a process analogous to alkene epoxidation, the alkene of the silyl enol ether attacks the electrophilic oxygen, kicking out the carboxylate ion (a good leaving group). The carboxylate attacks the silicon, cleaving the Si—O bond in the resonance-stabilized carbocation and generating the ketone.

Figure 20.8 Oxidation of silyl enol ethers gives α-hydroxy carbonyls. [*There is evidence that this occurs through an epoxide intermediate.*]

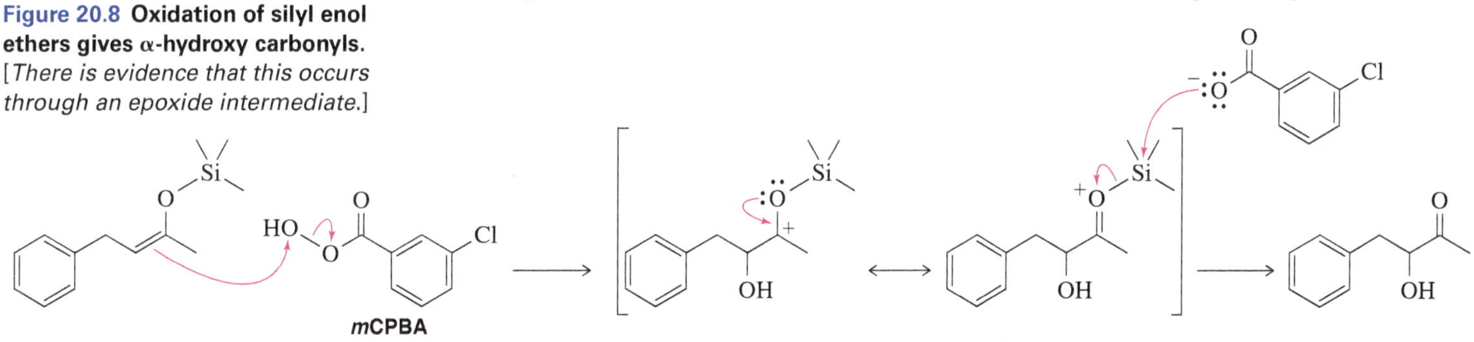

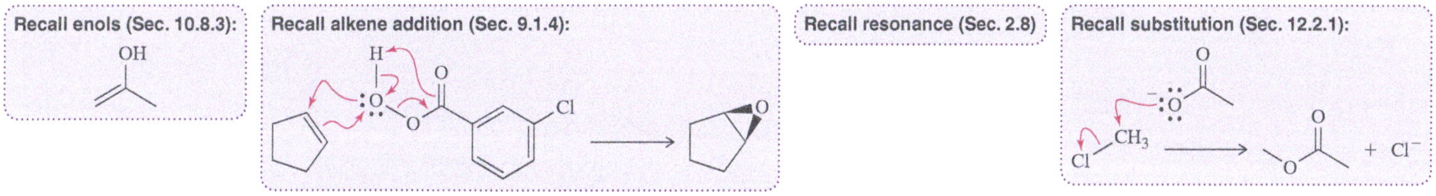

Assessment 20.9 Predict the major product of the following reactions.

(a)

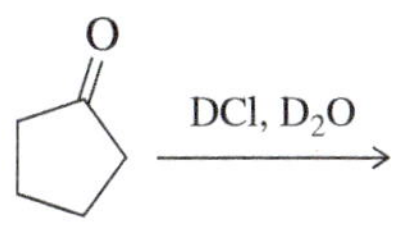

(b)

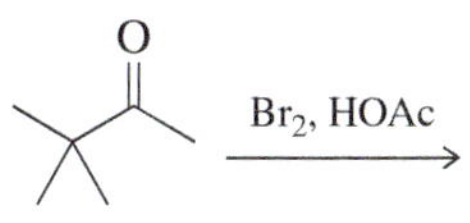

(c)

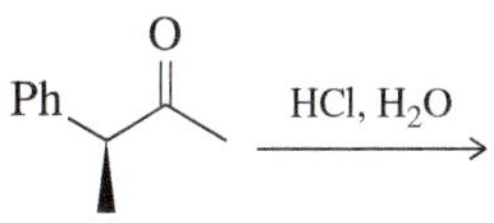

(d)

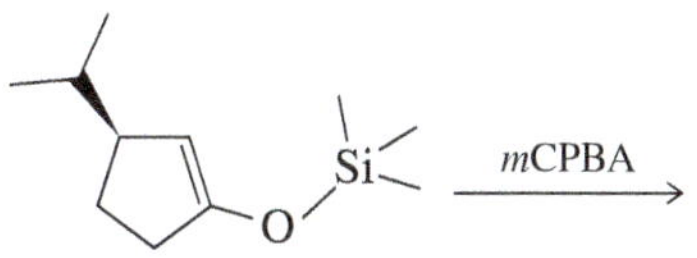

Figure 20.9 has been color-coded because tracking carbons is important for recognizing where bonds form in these reactions.

20.2.2 Acid-catalyzed aldol reaction

In the cases presented in Assessment 20.9, the enol acted as a nucleophile, but only with very strong electrophiles (strong acids, bromine, or *m*CPBA). The reaction of an enol with an aldehyde, shown in **Figure 20.9**, does not occur. The enol, like the weak nucleophiles that react with lone pairs on neutral atoms [$S_N1/E1$; *Chapter 12*], is not a strong enough nucleophile. In addition, the aldehyde is not a strong enough electrophile due to the instability of the alkoxide resulting from nucleophilic attack.

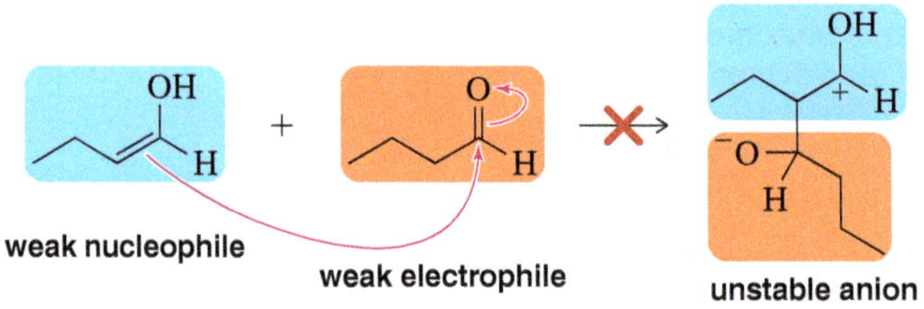

Figure 20.9 An enol is not a strong enough nucleophile to attack an aldehyde directly.

If a nucleophile is too weak, the reaction can be made faster by enhancing the reactivity of the electrophile. [*Could we also make the nucleophile stronger? We do this in Section 20.3.1.*] This is done by adding acid to the reaction (**Figure 20.10**). [*In Chapter 17, we protonated carbonyls to encourage reaction with water and alcohols, which also are weak nucleophiles.*] The acid plays two roles in this reaction: It catalyzes formation of the enol, and it protonates

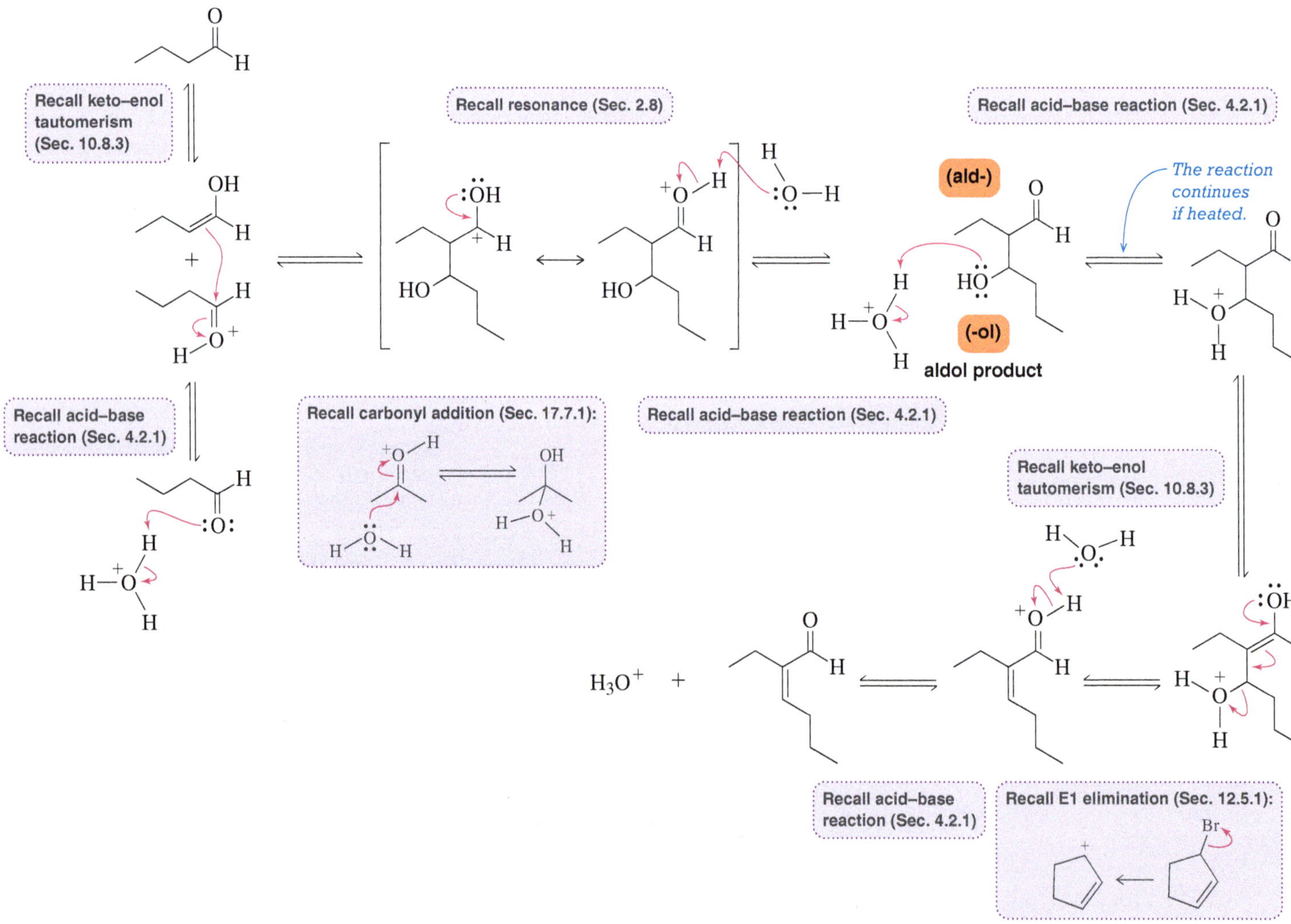

Figure 20.10 The mechanism of the acid-catalyzed aldol reaction. [*You have seen all of these reactions before.*]

the aldehyde to make it a better electrophile. As seen in the previous enol reactions, nucleophilic attack by the π bond results in a resonance-stabilized carbocation that is subsequently deprotonated to give a β-hydroxy aldehyde. This reaction began with acid, was made faster by acid, and produced acid as a product; it is acid-catalyzed. The resulting β-hydroxy aldehyde was historically referred to as an **aldol** product because it possesses an **ald**ehyde and alcoh**ol**. We now refer to reactions of this and related types as **aldol reactions**. If the reaction is allowed to continue with heating, protonation of the hydroxyl group is followed by elimination through the enol to give an α,β-unsaturated aldehyde. [*Although it looks complicated, with the exception of using an enol, there is no reaction in Figure 20.10 that you have not seen before. Convince yourself of this before moving on.*]

Table 20.1 briefly analyzes the energetics of each step to illustrate an important point. [*Spoiler alert!*] This reaction is under thermodynamic control, equilibrating to give the most stable product. Almost every reaction from Figure 20.10 is unfavorable with the exception of the final two. [*As you study Table 20.1, think about how a reaction with so many unfavorable steps proceeds to completion.*]

Table 20.1 The energetics of the acid-catalyzed aldol reaction. Few steps are favorable, but the overall reaction is, especially when heated to form the α,β-unsaturated aldehyde.

Step of the acid-catalyzed aldol	Analysis
1. Formation of the enol 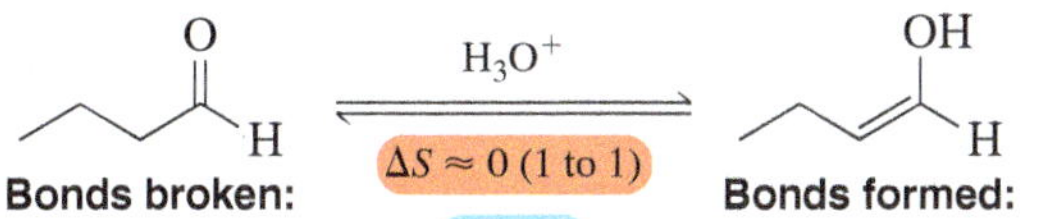 $\Delta S \approx 0$ (1 to 1) $\Delta H > 0$ $\Delta G > 0$; $K_{eq} < 1$ **Bonds broken:** C—H σ, C—O π **Bonds formed:** O—H σ, C—C π	Formation of the enol, as we learned in Chapter 13, is unfavorable because the C—C π bond is less stable than the C—O π bond.
2. Protonation of the aldehyde 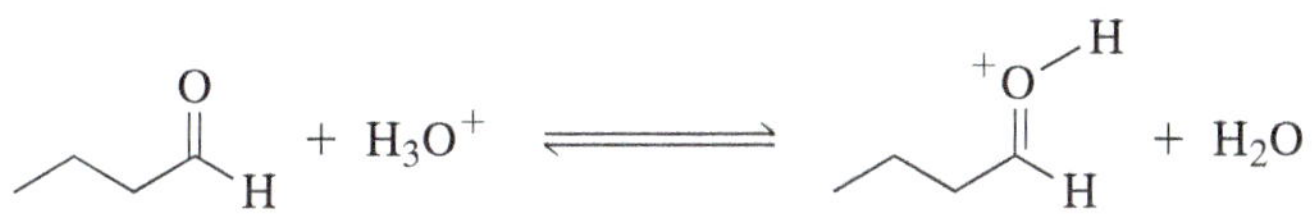 pK_a = −1.7 (H$_3$O$^+$); pK_a = −10 (protonated aldehyde) $K_{eq} = 10^{-8.3} = 5.0 \times 10^{-9}$ $\Delta G > 0$	Protonation of the aldehyde with hydronium ion is disfavored. Using pK_a values for H_3O^+ (−1.7) and $RCHOH^+$ (−10), we calculate $K_{eq} < 1$.
3. Enol attacks the protonated ketone 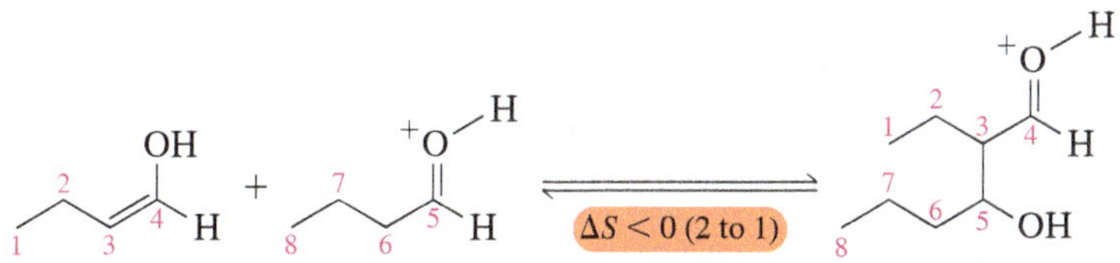 $\Delta S < 0$ (2 to 1) $\Delta H < 0$ $\Delta G \approx 0$; $K_{eq} \approx 1$ **Bonds broken:** C_3—C_4 π, C_5—O π **Bonds formed:** C_3—C_5 σ, C_4—O π	In this case, entropy and enthalpy work against each other. The reaction is favored by enthalpy ($\Delta H < 0$) because a C—C σ bond forms at the expense of a C—C π bond. Because two molecules are reacting to make one, $\Delta S < 0$. Without knowing specific numbers, we can estimate that ΔG is only slightly favorable, if at all. [*Pay attention to the numbering of carbons in this step.*]
4. Deprotonation of the product 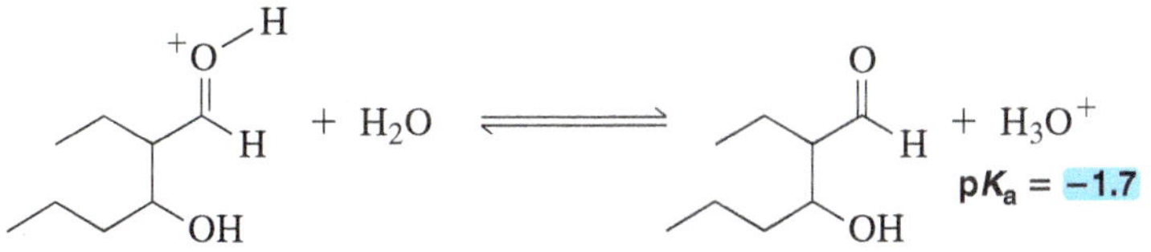 pK_a = −10 (protonated product); pK_a = −1.7 (H$_3$O$^+$) $K_{eq} = 10^{8.3} = 2.0 \times 10^{8}$ $\Delta G < 0$	Deprotonation of the aldehyde with hydronium ion is favored. Using pK_a values for H_3O^+ (−1.7) and $RCHOH^+$ (−10), we calculate $K_{eq} > 1$. [*This is the opposite of the acid–base reaction in step 2.*]

(continued)

Table 20.1 ***Continued***

Step of the acid-catalyzed aldol	Analysis
Overall reaction 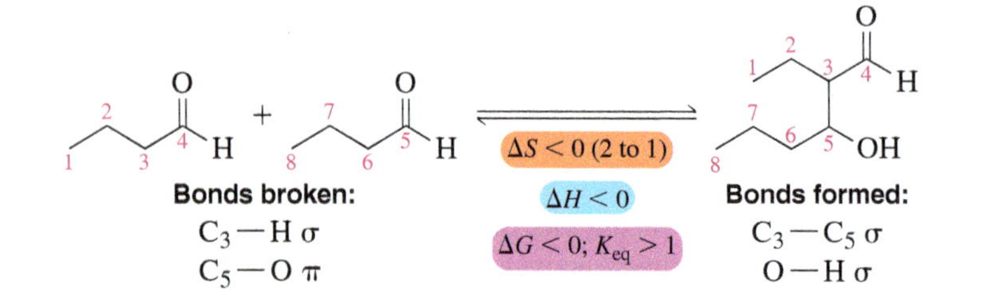	The reaction forms one molecule from two, so it is disfavored by entropy. It is favored by enthalpy because two σ bonds are formed at the expense of one σ and one π bond. Because we know that the reaction proceeds in the forward direction, it must be true that the enthalpy term is slightly more important than the entropy term for this reaction.
Overall reaction (with heating) 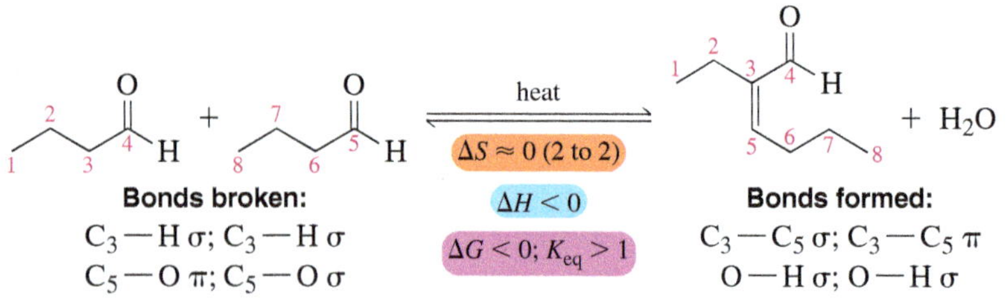	When heated, the elimination of water means the overall reaction now goes from 2 molecules to 2 molecules ($\Delta S \approx 0$). The reaction remains favorable in terms of enthalpy and is therefore favorable overall ($\Delta G < 0$).

Having analyzed each individual step of the process, we see that none but step 4 is very favorable and some are heavily disfavored. [*The elimination to make an α,β-unsaturated aldehyde is also favored.*] Unfavorable protonation of the aldehyde, unfavorable enol formation, and unfavorable enol addition are all driven by formation of the final product. Recall Le Châtelier's principle, which says that as you remove a molecule from the product side of the equilibrium, the equilibrium shifts toward the products. Even if the equilibrium of step 1 results in just a few molecules of enol, when one of these collides with a protonated aldehyde, the enol is consumed. This causes the equilibrium of step 1 to shift, thereby producing another molecule of enol, a process that continues until the aldehyde is fully converted to the aldol product. A final point: this reaction is under **thermodynamic control**, meaning the product distribution is a function of product stability. [*We explain why soon, but it is more difficult to control reactions under thermodynamic control.*]

Before we move on, there are actually two possible alkene regioisomers that can form during the elimination step. **Figure 20.11** shows that elimination occurs only to produce the **α,β-unsaturated aldehyde**—none of the β,γ-unsaturated aldehyde is formed. The name implies that the molecule is unsaturated at the α and β carbons. As we discuss later, conjugated alkenes, or ones where resonance structures can be drawn, are more stable. [*Read ahead about this in Section 21.1.2.*] Once more, the reaction is under thermodynamic control.

The analytical approach that provides this insight is fundamental to understanding reactions you will see later. So, go back and read over this section more carefully before moving on to the next section. You should be able to solve the following assessments before proceeding.

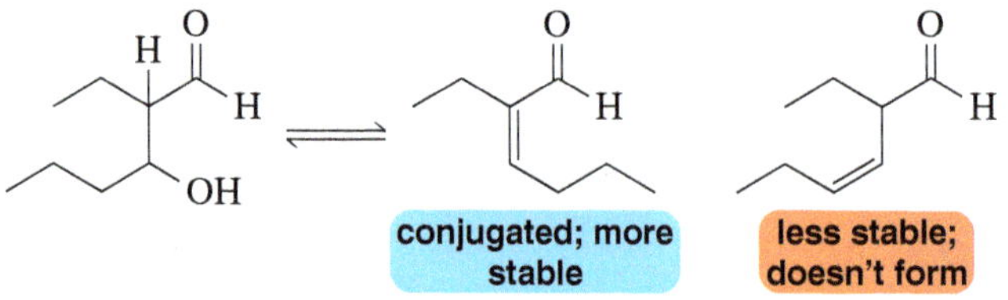

Figure 20.11 Because it is more stable, only the α,β-unsaturated aldehyde is formed.

Assessment 20.10 Predict the major product of the following reactions.

(a)

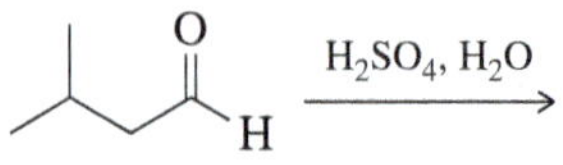

(b)

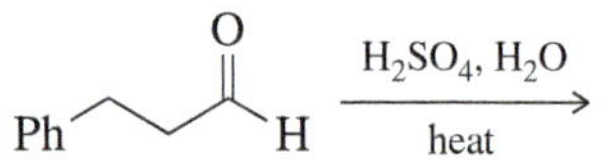

Assessment 20.11 Provide an arrow-pushing mechanism for the following acid-catalyzed aldol reaction.

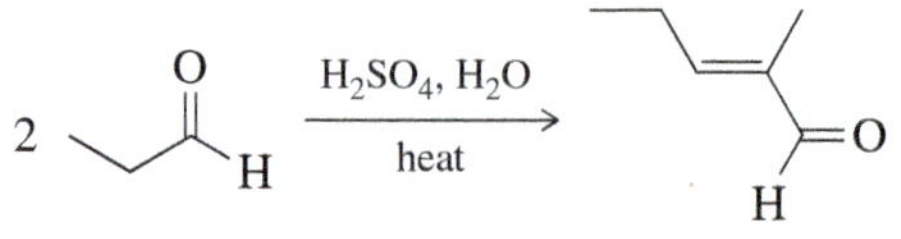

Assessment 20.12 For the following retro-aldol reaction, justify why the product side of the reaction is favored, especially at high temperatures.

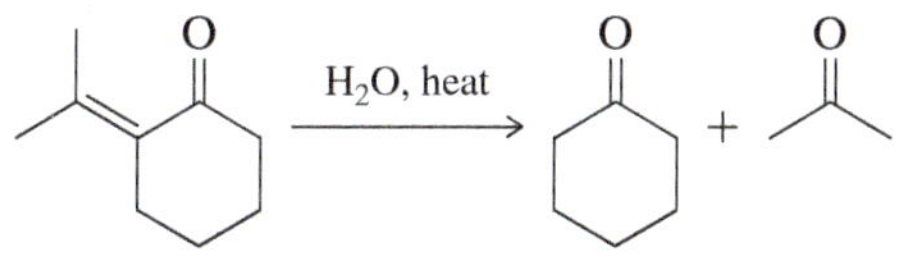

Assessment 20.13 In the acid-catalyzed aldol reaction of 3,3-dimethylbutanal, two alkene stereoisomers (**A** and **B**) are possible. Which do you expect to be the major product? Why?

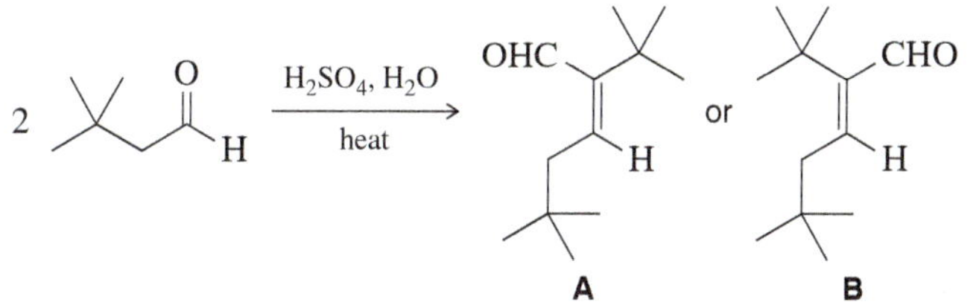

Assessment 20.14 Draw a reaction coordinate diagram for the acid-catalyzed aldol reaction. Make sure that each of the steps in Table 20.1 is represented.

20.3 Enolates

Enols are not strong enough nucleophiles to attack neutral carbonyls. For enols to attack, it is necessary to protonate the carbonyl, making them stronger electrophiles. Assessment 20.15 gives us another idea of how enols can be made to react. [*Focus on how the nucleophile changes in each.*]

Assessment 20.15 Which of the following S_N2 and E2 reactions, respectively, is faster? Justify your choice.

(a)

(b)

Assessment 20.15 makes clear that negatively charged nucleophiles/bases are stronger than neutral ones. By deprotonating an enol, or more typically, the α-carbon of a carbonyl ($pK_a \approx 20$), an enolate is generated (**Figure 20.12**). This is accomplished using a strong base like lithium diisopropyl amide (LDA). The resulting lithio species are technically tautomers of each other because they differ only in the position of a single atom. Viewed without the lithium cation, the **enolate** or **enolate ion** is a hybrid of two resonance structures. While it's clear that an enolate is more reactive than an enol, it is less clear whether we should expect enolates to react preferentially at oxygen or carbon. [*Spoiler: They react preferentially at carbon.*]

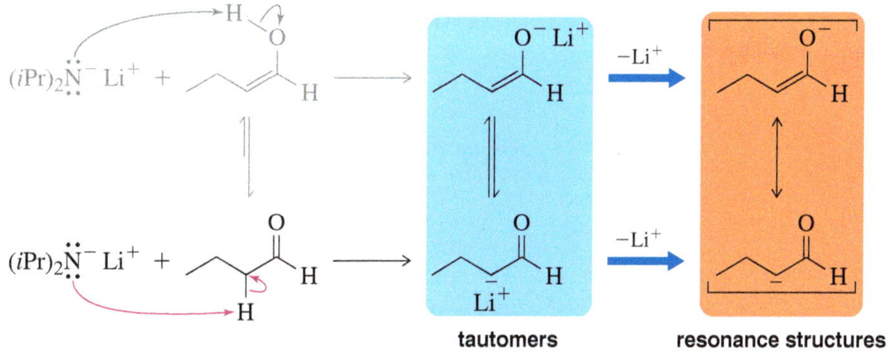

Figure 20.12 Deprotonation of the α-carbon gives enolates. Enolates are stronger nucleophiles than enols.

Assessment 20.16 Which structure contributes more to the resonance hybrid of an enolate?

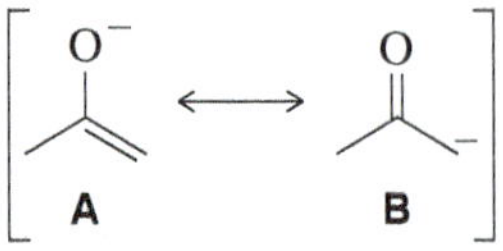

Let's work 20.16 together.

Solved Assessment 20.16

When comparing two resonance structures, it is better to have the negative charge on the more electronegative atom. Thus, we'd expect **A** to contribute more, a fact backed up by an electron density map showing that most of the negative charge is centered around oxygen (**Figure 20.13**). ■

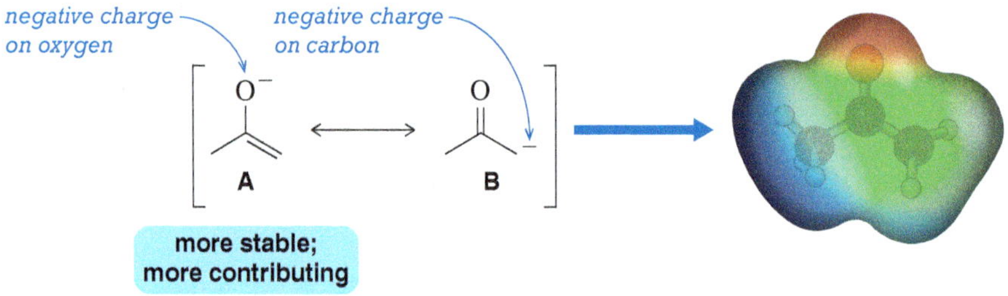

Figure 20.13 The electron density map shows that the majority of negative charge resides on the oxygen of an enolate.

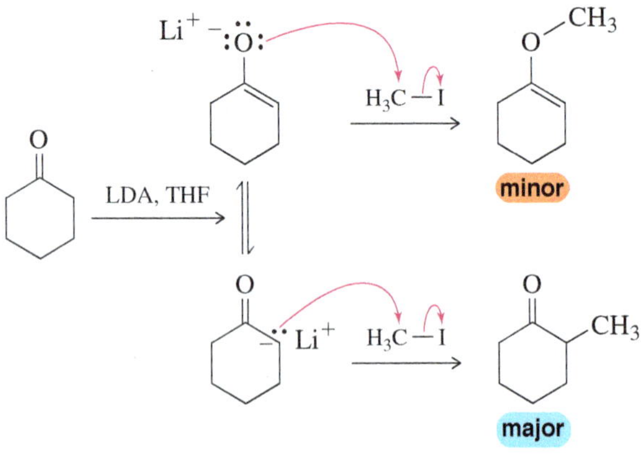

Figure 20.14 Enolates react predominantly at carbon.

Based on the electron distribution in Figure 20.13, you might expect enolates to react with electrophiles at the oxygen atom (**O-alkylation**). And yet the S_N2 reaction between an enolate and chloromethane results in **C-alkylation** to form 2-methylcyclohexanone (**Figure 20.14**). [*Think about why this might be true before continuing.*]

Figure 20.15 makes a thermodynamic argument for the regioselectivity by comparing the stability of the resulting products. As is true when comparing keto and enol tautomers [*Section 10.8.3*], the C—O π bond is significantly stronger than the C—C π bond. If the reaction were under thermodynamic control (i.e., if it were allowed to equilibrate to the more stable product), then the C-alkylation to form 2-methylcyclohexanone would be driven by maintaining the C—O π bond. This reaction isn't reversible, however, so it *isn't* under thermodynamic control. Instead, it is under **kinetic control**, in which case relative rates dictate the regioselectivity. [*Thus, we need another reason.*]

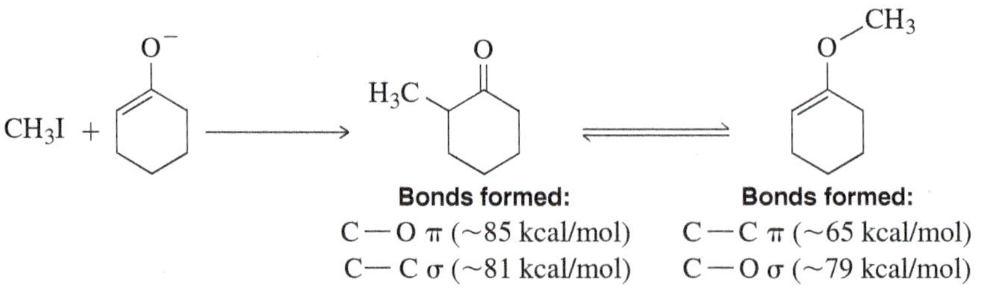

Figure 20.15 Alkylation at carbon is thermodynamically favored by maintaining the C—O π bond. [*This is a hypothetical equilibrium only. Enolate alkylation is irreversible.*]

The kinetic arguments begin with the structure of the enolate shown in **Figure 20.16(a)**. Although we have often ignored the inorganic counterion, it plays an important role in this reaction. With the lithium cation included, the respective enolate structures are tautomers of each other. [*Recall Figure 20.12.*] Whereas the keto tautomer is more stable than the enol isomer because of the strength of the C—O π bond, the O—Li enolate structure is favored over the C—Li enolate structure because the O—Li bond is much stronger than the C—Li bond. This is consistent with **hard and soft Lewis acid and base (HSAB) theory**.

Finally, polar aprotic solvents like THF form coordinate bonds with inorganic cations. [*Recall solvent effects in the* S_N2 *reaction; Section 12.3.5.*] Aggregation of the THF molecules around the lithium cation sterically crowds the oxygen, making *C*-alkylation the faster reaction [**Figure 20.16(b)**].

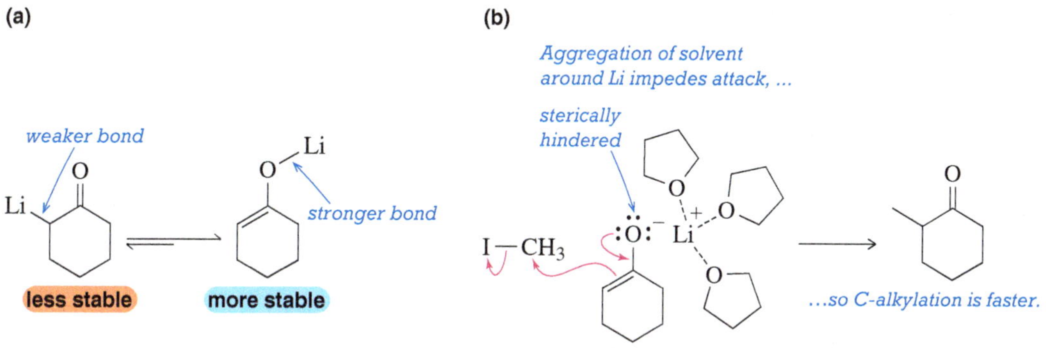

Figure 20.16 The aggregation of solvent around lithium sterically impedes *O*-alkylation. *C*-alkylation is faster.

SIDE CONVERSATION HSAB theory HSAB theory starts by defining Lewis acids and bases as being hard or soft on a relative scale. To simplify, smaller, less polarizable, and higher-charged (oxidation state) species are considered to be harder. Larger, more polarizable, and lesser-charged (oxidation state) species are considered to be softer. **Table 20.2** lists some representative species. Once identified as hard or soft, HSAB theory suggests that hard Lewis acids will react preferentially with hard Lewis bases. The opposite is also true—namely, soft Lewis acids will react preferentially with soft Lewis bases. The implication for enolates is that hard oxygen prefers to bond with hard lithium. ■

Table 20.2 A list of hard and soft acids and bases. [*You only need to understand the relative hardness of acids and bases.*]

Hard acids	Borderline	Soft acids	Hard bases	Borderline	Soft bases
H^+, Li^+, Na^+, K^+			F^-, Cl^-	Br^-	H^-, I^-
Be^{2+}, Mg^{2+}, Ca^{2+}			H_2O, OH^-, O^{2-}		H_2S, HS^-, S^{2-}
BF_3, BCl_3, $B(OR)_3$	BBr_3, $B(CH_3)_3$	BH_3, Tl^+, $Tl(CH_3)_3$	ROH, RO^-, R_2O, $CH_3CO_2^-$		RSH, RS^-, R_2S
Al^{3+}, $Al(CH_3)_3$, $AlCl_3$, AlH_3			NO_3^-, ClO_4^-	NO_2^-, N_3^-, N_2	SCN^-, CN^-, RNC, CO
Cr^{3+}, Mn^{2+}, Fe^{3+}, Co^{3+}	Fe^{2+}, Co^{2+}, Ni^{2+}	Cu^+, Ag^+, Au^+	CO_3^{2-}, SO_4^{2-}, PO_4^{3-}	SO_3^{2-}	$S_2O_3^{2-}$
	Cu^{2+}, Zn^{2+}, Rh^{3+}	Cd^{2+}, $Hg_2^{\ 2+}$	NH_3, RNH_2	$C_6H_5NH_2$	R_3P, C_6H_6
	Ir^{3+}, Ru^{3+}, Os^{2+}	Hg^{2+}, Pd^{2+}, Pt^{2+}		pyridine	
SO_3	SO_2	Pt^{4+}			

There is one more factor involved in the regioselectivity of enolates, and it also involves HSAB theory. Because iodomethane is a neutral electrophile, it is considered to be a soft Lewis acid, in which case we would expect it to react at the "softer" carbon of the enolate.

20.3 SUMMARY

- Enolates, though consisting of two major resonance contributors, react predominantly at the carbon.

Assessment 20.17 Specify which in each pair is the harder Lewis base. Justify your choice beyond just looking at Table 20.2.

(a) CH_3S^- vs. CH_3O^- **(b)** CH_3O^- vs. 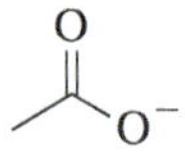 **(c)** I^- vs. F^-

Assessment 20.18 Specify which in each pair is the harder Lewis acid. Justify your choice beyond just looking at Table 20.2.

(a) Al^{3+} vs. B^{3+} **(b)** Li^+ vs. Na^+ **(c)** Mg^{2+} vs. Na^+

Assessment 20.19 Predict the major product of the following reactions.

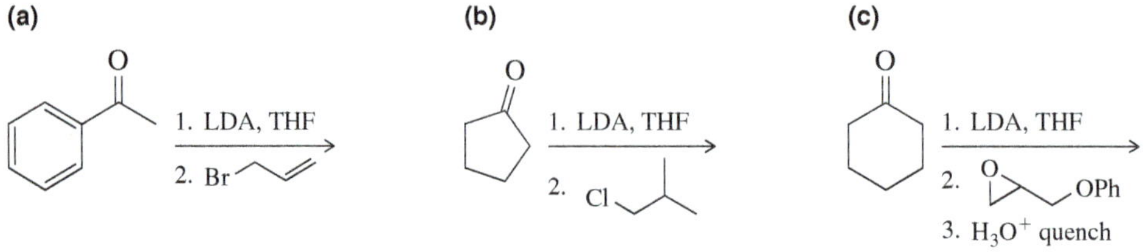

Assessment 20.20 Epichlorohydrin, utilized in the following reaction, is a common building block in natural product synthesis. Give an arrow-pushing mechanism for the reaction shown.

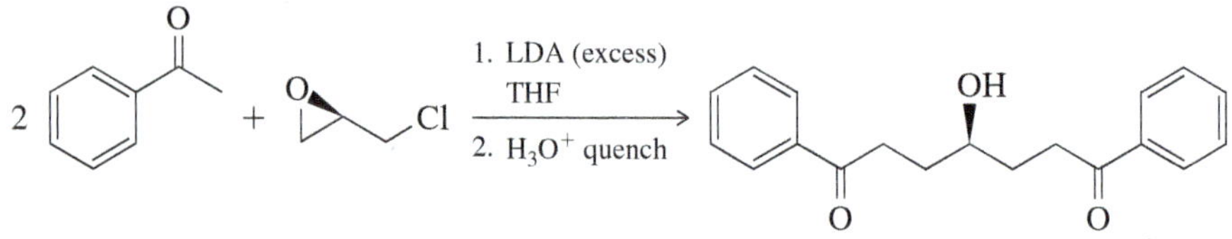

Assessment 20.21 LDA can be used to form enolates on esters and nitriles. Predict the product of these alkylation reactions.

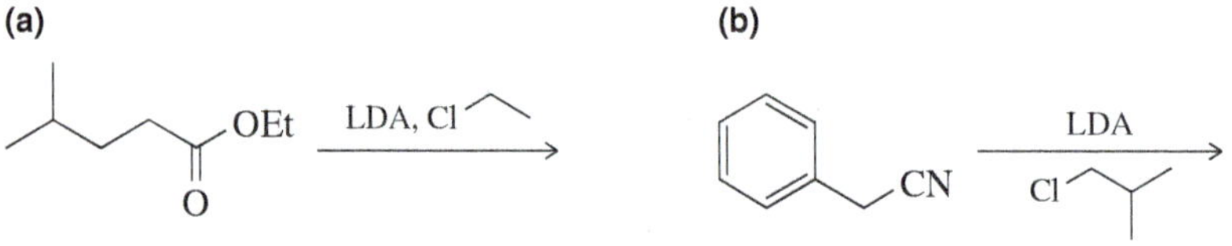

20.3.1 Modern aldol reaction

We will eventually return to the classical base-catalyzed aldol reaction, so here in Section 20.3.1 we frame that discussion in the context of the modern aldol reaction. Enolates are formed by the removal of an α-hydrogen by LDA in a highly favored, essentially irreversible step (**Figure 20.17**). LDA is an ideal base because it is sterically hindered and not prone to act as a nucleophile. Following formation of the enolate, the aldehyde is added to the reaction mixture, allowing nucleophilic attack to occur. [*This is simply General Mechanism 17-A, but with a different nucleophile.*] An acid quench completes formation of the aldol product. A racemic mixture of aldol products results because the enolate attacks equally from the front or back of the aldehyde.

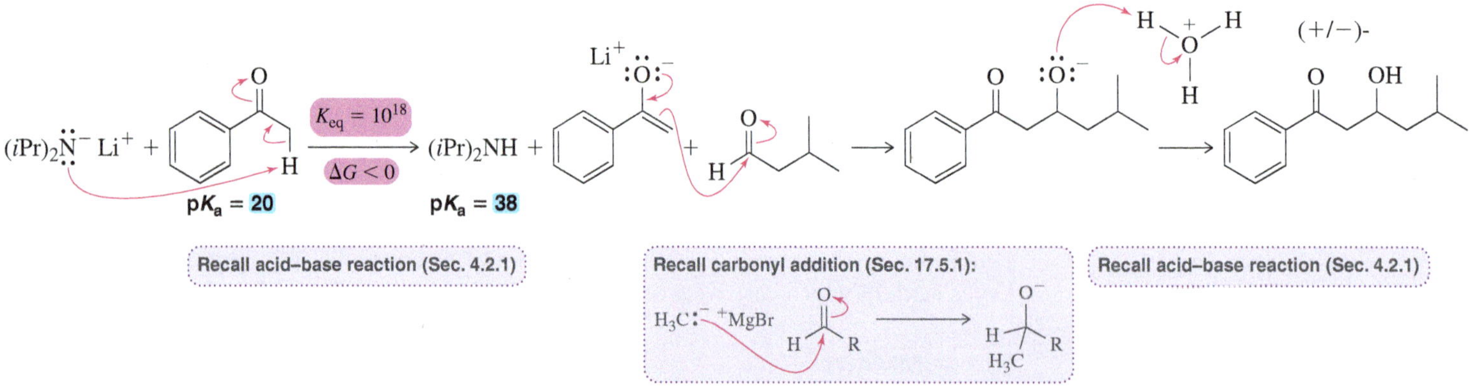

Figure 20.17 The mechanism of the modern aldol reaction. Because a strong base is used, the ketone is completely converted to the enolate prior to adding the aldehyde to solution. [*The importance of the experimental setup will be made clear in future sections.*]

Assessment 20.22 Predict the major product of the following reactions.

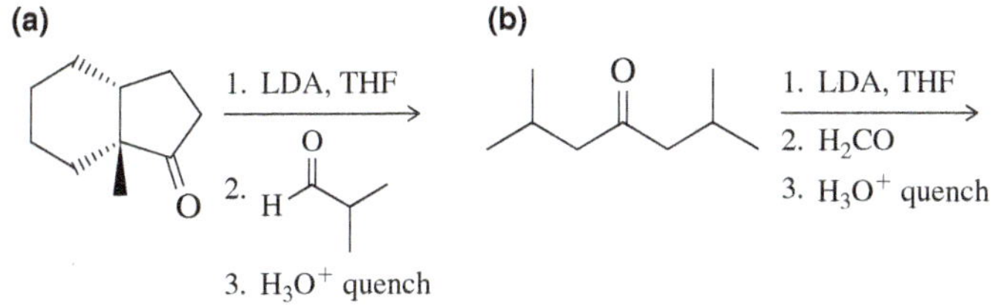

Assessment 20.23 Working backward, suggest an aldol reaction that could be used to make the following. Ignore stereochemistry.

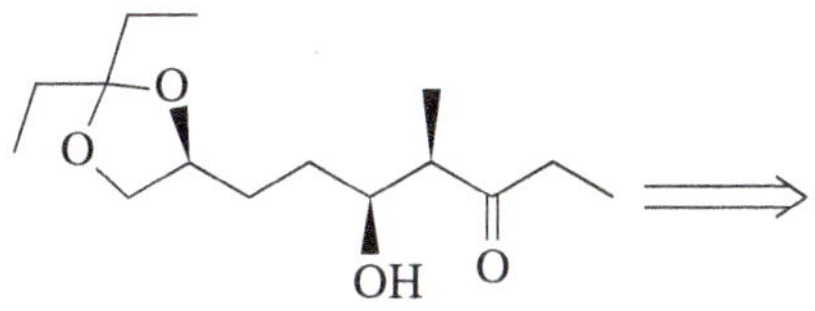

Assessment 20.24 Show an arrow-pushing mechanism for each step of the following reaction.

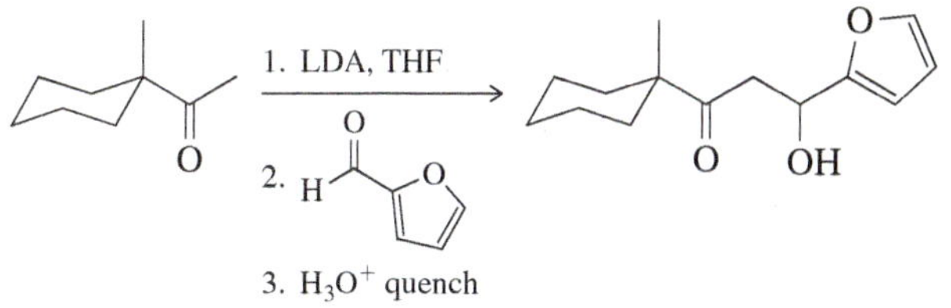

Assessment 20.25 Most of the time (using LDA), the ketone and the aldehyde are added in separate steps. What if both were present in the reaction mixture when the base was added? What product(s) might result?

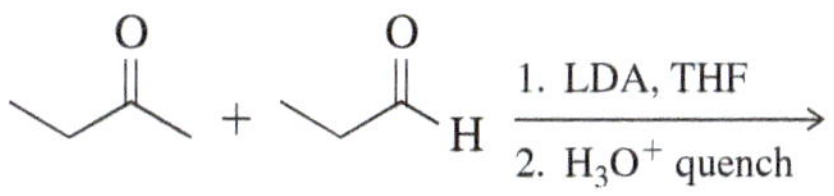

20.3.2 Classical base-catalyzed aldol

Originally discovered around 1870, the aldol reaction is conducted under basic conditions with hydroxide as the catalyst. The mechanism (**Figure 20.18**) is similar to the mechanism of the modern aldol reaction (Figure 20.17). The enolate, formed by deprotonation, reacts directly with the aldehyde in a carbonyl addition reaction to form an alkoxide. Protonation of the alkoxide completes formation of the β-hydroxyaldehyde. Although the reaction can stop here, an elimination reaction occurs with heating to give an α,β-unsaturated carbonyl. Hydroxide is a reactant in steps 1 and 4 and a product in steps 3 and 5; hence, it is a catalyst. [*Note the similarity between this and the acid-catalyzed reaction in Figure 20.10.*]

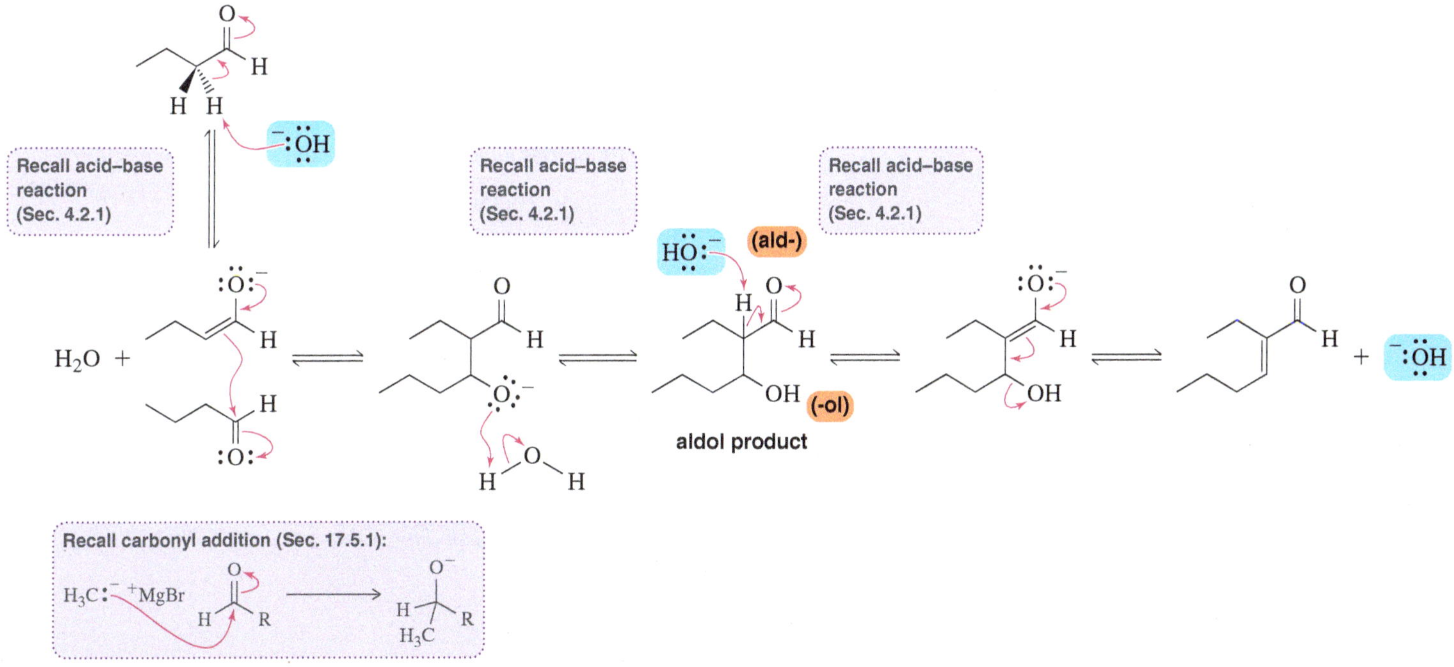

Figure 20.18 The mechanism of the base-catalyzed aldol reaction. [*Perhaps with the exception of the elimination, all of these reactions should be very familiar to you.*]

As was done for the acid-catalyzed aldol (Table 20.1), **Table 20.3** briefly analyzes the energetics of each step to illustrate the same important point. This reaction is under thermodynamic control, so it equilibrates to give the most stable product. [*The modern aldol reaction is kinetically controlled. The classical aldol is thermodynamically controlled. Try to figure out the difference as you read through Table 20.3.*]

Table 20.3 The energetics of the base-catalyzed aldol reaction. Few steps are favorable, but the overall reaction is, especially when heated to form the α,β-unsaturated aldehyde.

Step of the base-catalyzed aldol	Analysis
1. Formation of the enolate 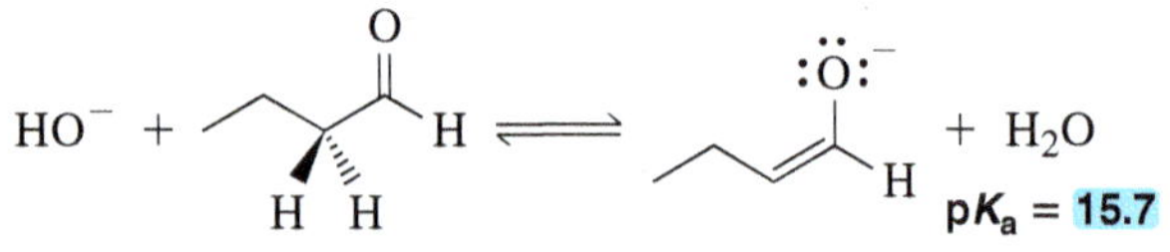 pK_a = 20; pK_a = 15.7 $K_{eq} = 10^{-4.3} = 5.0 \times 10^{-5}$ $\Delta G > 0$	Deprotonation of the α-carbon with hydroxide is unfavorable. Using pK_a values for H_2O (15.7) and the α-proton (20), we calculate $K_{eq} < 1$.
2. Enolate attacks the aldehyde 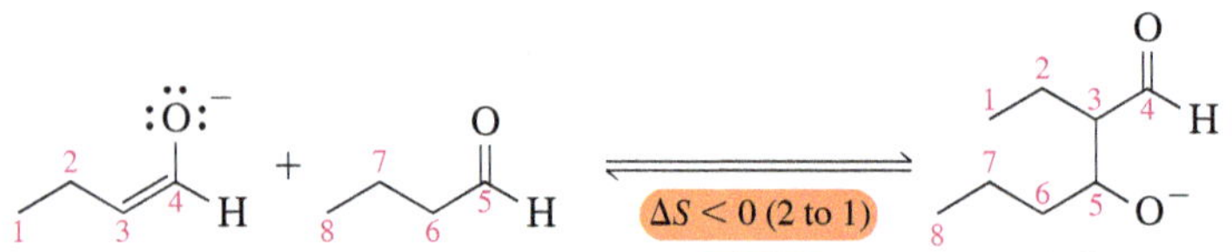 $\Delta S < 0$ (2 to 1); $\Delta H < 0$; $\Delta G \approx 0$; $K_{eq} \approx 1$ **Bonds broken:** $C_3—C_4$ π; $C_5—O$ π **Bonds formed:** $C_3—C_5$ σ; $C_4—O$ π	The reaction is favored by enthalpy ($\Delta H < 0$) because a C—C σ bond forms at the expense of a C—C π bond. Two molecules are reacting to make one, so $\Delta S < 0$. We estimate that ΔG is slightly favorable. [*Pay attention to the numbering of carbons in this step.*]
3. Protonation of the alkoxide 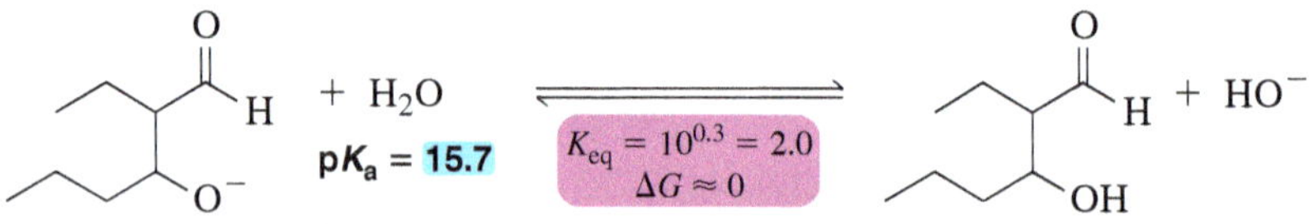 pK_a = 15.7; pK_a = 16 $K_{eq} = 10^{0.3} = 2.0$ $\Delta G \approx 0$	Protonation of the alkoxide with water is barely favored. Using pK_a values for H_2O (15.7) and ROH (16), we calculate $K_{eq} \approx 2$.
Overall reaction 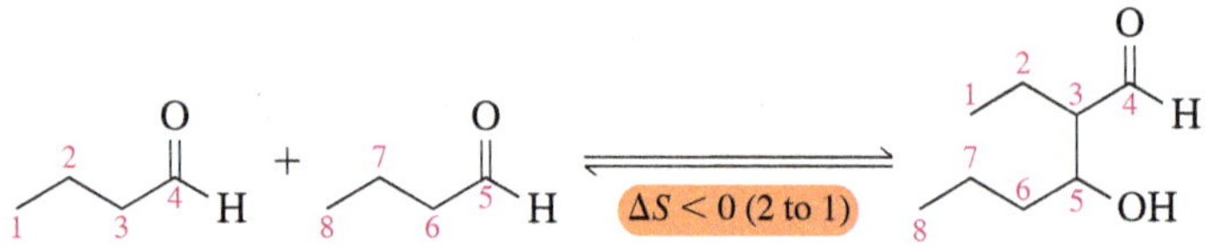 $\Delta S < 0$ (2 to 1); $\Delta H < 0$; $\Delta G < 0$; $K_{eq} > 1$ **Bonds broken:** $C_3—H$ σ; $C_5—O$ π **Bonds formed:** $C_3—C_5$ σ; O—H σ	The overall reaction to make the β-hydroxyaldehyde is disfavored by entropy and favored by enthalpy. Because the reaction is favorable, the enthalpy term must be more important than the entropy term. [*These conclusions are identical to those in Table 20.1 because the change in catalyst has no impact on thermodynamic values.*]
Overall reaction (with heating) 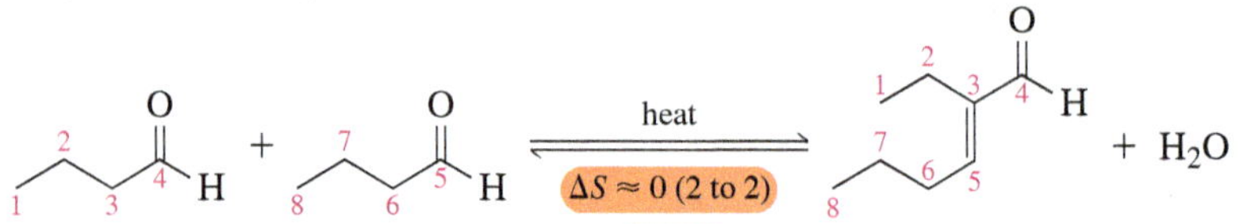 heat; $\Delta S \approx 0$ (2 to 2); $\Delta H < 0$; $\Delta G < 0$; $K_{eq} > 1$ **Bonds broken:** $C_3—H$ σ; $C_3—H$ σ; $C_5—O$ π; $C_5—O$ σ **Bonds formed:** $C_3—C_5$ σ; $C_3—C_5$ π; O—H σ; O—H σ	When heated, the elimination of water means the overall reaction now goes from 2 molecules to 2 molecules ($\Delta S \approx 0$). By forming a stable conjugated alkene, the reaction remains favorable in terms of enthalpy and is therefore favorable overall ($\Delta G < 0$). [*Again, this is the same as in Table 20.1.*]

The elimination step in the aldol reaction proceeds by neither the E1 nor the E2 mechanism. Instead, it proceeds by a mechanism briefly mentioned in Chapter 12 and compared to the E2 reaction in **Figure 20.19**. Because the α-proton is somewhat acidic and hydroxide is not a great leaving group, deprotonation results in a resonance-stabilized intermediate that collapses in a second step to eliminate hydroxide. Slightly modifying the language of Section 12.5.1, "the base removes the proton, ***and then*** the leaving group leaves."

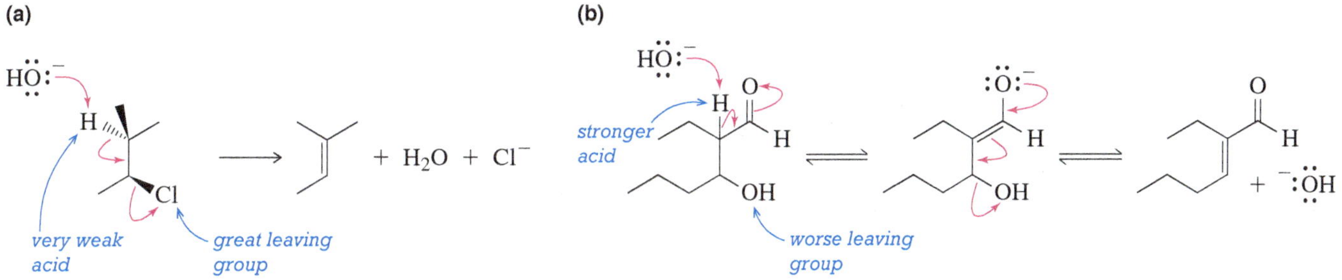

Figure 20.19 A comparison of elimination reactions. (a) A concerted E2 elimination. (b) The E1cb mechanism operates when the proton is slightly acidic and the leaving group is poor.

Although enolate formation in step 1 (Table 20.3) is unfavorable using hydroxide as the base, it is very favorable when using LDA in the modern aldol reaction. The base-catalyzed aldol reaction proceeds to completion as a result of Le Châtelier's principle, but the use of a stronger base would have made the reaction more efficient. Unfortunately, classical chemists didn't have access to bases like LDA. [*These didn't come into being until the advent of organolithium and Grignard reagents between 1900 and 1930.*] The strongest bases available to the classical chemist were those that could be obtained by mixing elemental sodium with water or alcohols. [*As you proceed, we're going to examine a number of problems caused by the use of these weaker bases. The solutions to these problems show the genius of the early chemists that invented them.*]

Assessment 20.26 Predict the major product of the following reactions.

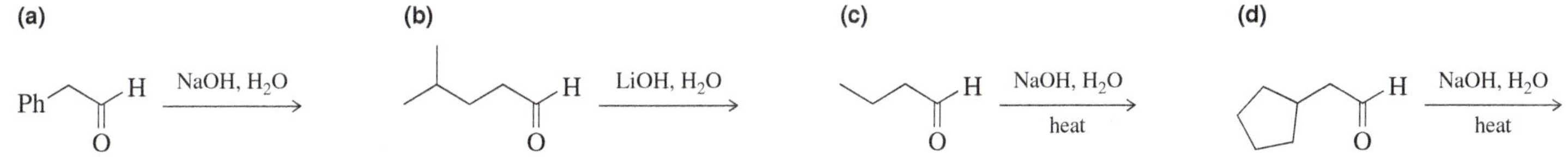

Assessment 20.27 Show a mechanism of the following base-catalyzed aldol reaction. Make sure your mechanism demonstrates that hydroxide is a catalyst.

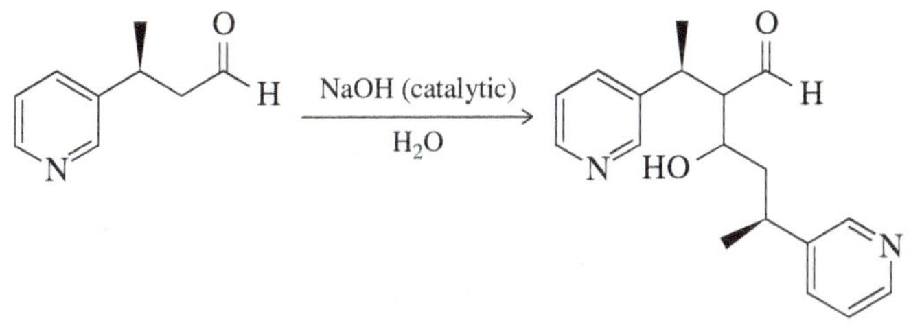

Assessment 20.28 Draw a reaction coordinate diagram for the base-catalyzed aldol reaction. Make sure that each of the steps in Table 20.3 is represented.

Assessment 20.29 Thus far, all of the classical aldol reactions we have shown have involved the condensation of aldehydes. Ketones give varying results in aldol reactions under similar conditions. When heated, for example, the aldol reaction proceeds smoothly to produce the α,β-unsaturated ketone. In the absence of heat, however, only the reactants are isolated from the reaction mixture. Rationalize these results.

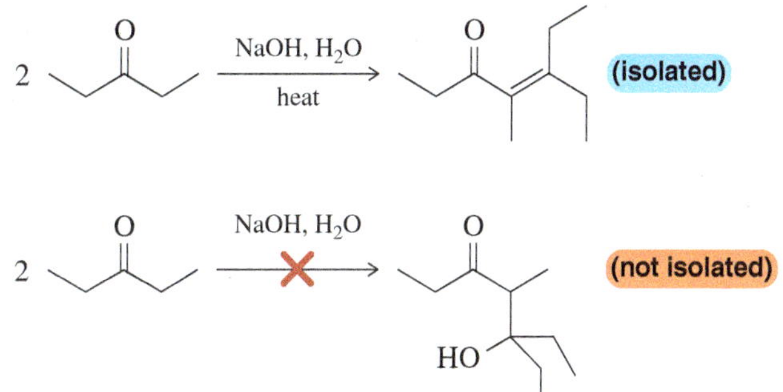

Let's make sure we understand what is happening in Assessment 20.29 by working it together.

Solved Assessment 20.29

The aldol reaction with aldehydes proceeds to make either β-hydroxyaldehydes (no heat) or α,β-unsaturated aldehydes (with heat) depending on the conditions. The aldol reaction of ketones, however, only works if the reaction is driven to the α,β-unsaturated ketone. Because this reaction is under thermodynamic control, formation of the β-hydroxyketone must be an uphill process. [*Make sure you understand this before reading further.*]

Figure 20.20 directly compares the aldol reaction of aldehydes and the aldol reaction of ketones. [*In which step would you expect a difference?*] Because the enolate formation is unfavorable anyway, ketone enolates seem as reasonable as aldehyde enolates. We also know that if a β-hydroxyketone forms, it favorably eliminates. [*Assessment 20.29 tells us this.*] Thus, the difference must be in the step where the enolate attacks the ketone. Recall from Section 17.4 that aldehydes are more electrophilic than ketones. As a result, breaking the C—O π bond in a ketone is a net uphill process, with the reaction being "downhill" only if heat is applied to allow formation of the α,β-unsaturated ketone. [*The activation energies have been removed from Figure 20.20 so the relative energies can be directly compared.*] ■

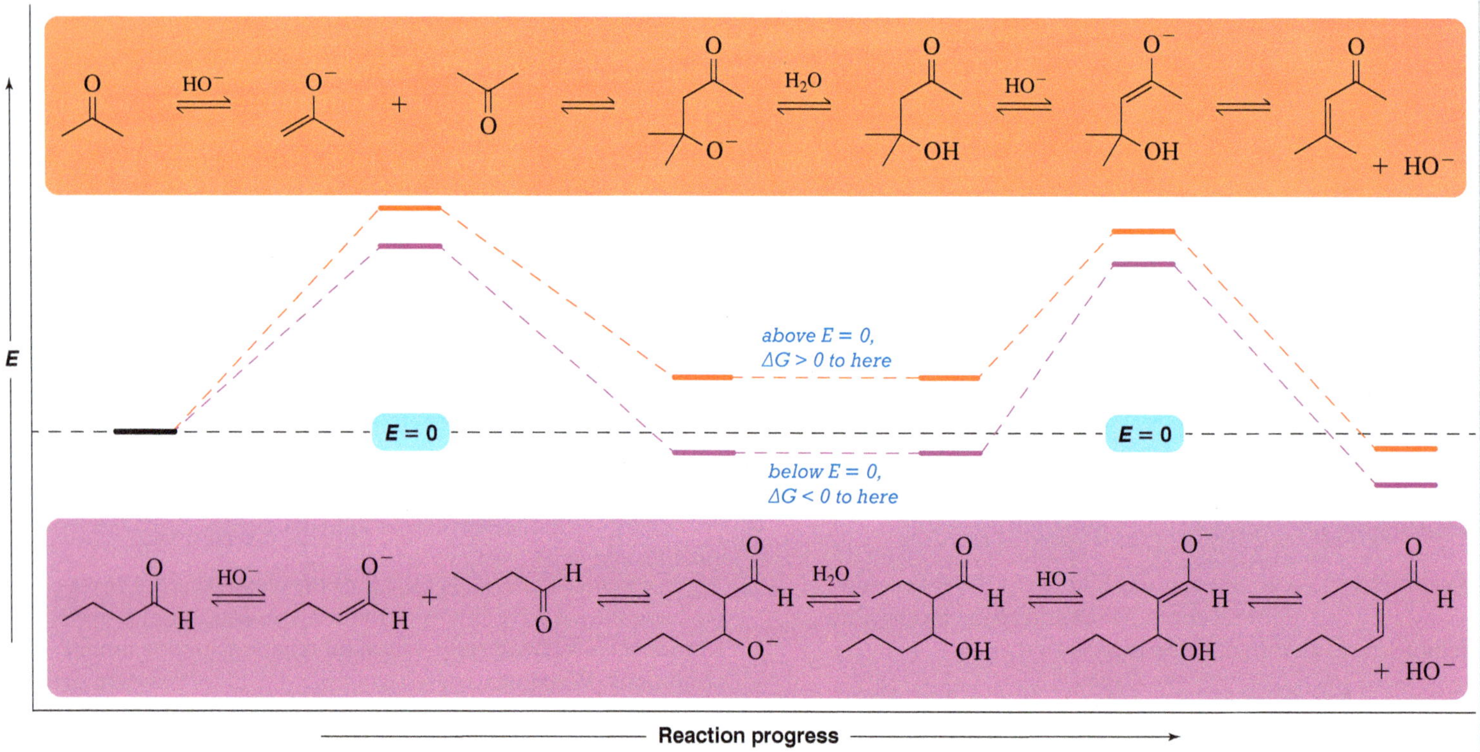

Figure 20.20 **Ketones only do aldol reactions under conditions that favor elimination.** Because ketones are less reactive, nucleophilic addition is unfavorable ($\Delta G > 0$).

To summarize, ketones are not good electrophiles for enolates. [*Don't worry; this works in our favor in Section 20.3.3.*]

Assessment 20.30 Suggest reagents and conditions to prepare the following aldol products.

(a) **(b)** **(c)**

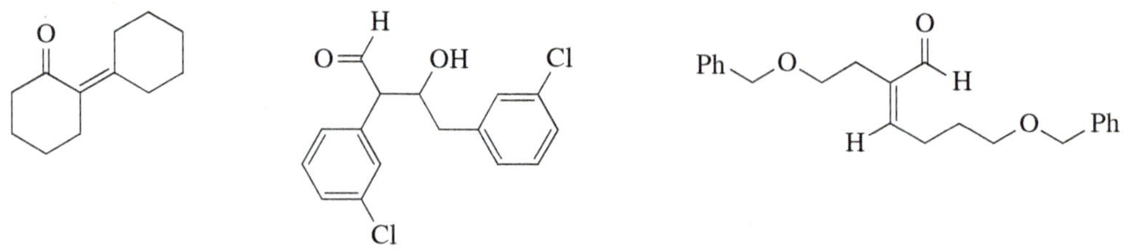

Assessment 20.31 Predict the product of the following reactions.

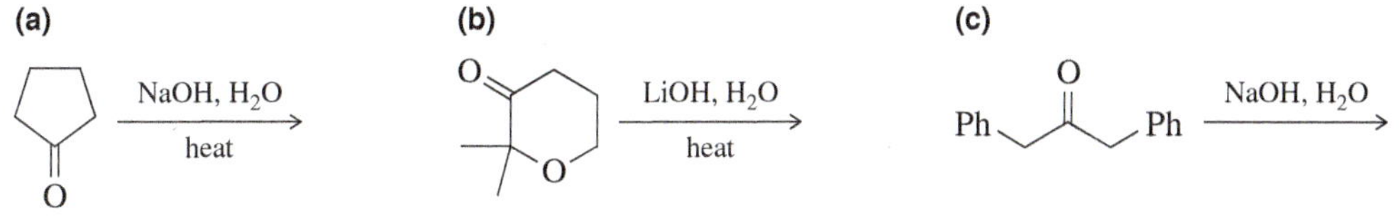

Assessment 20.32 Rank the reactivity of the following carbonyls with nucleophiles, from least reactive to most reactive.

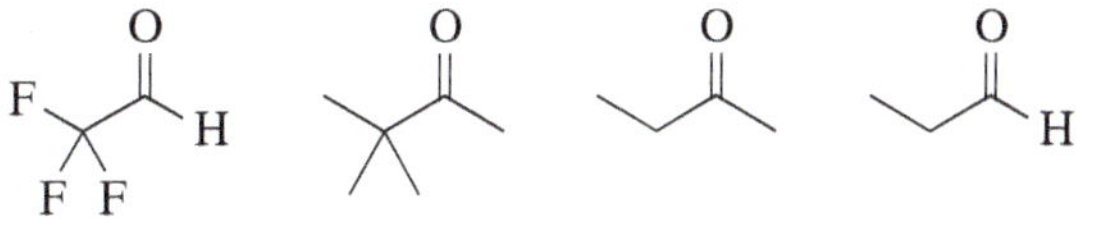

20.3.3 Crossed aldol condensation

Thus far, we have only studied the base-catalyzed **self-condensation**, the reaction of an enolate with the aldehyde from which it derived. **Figure 20.21** shows the retrosynthesis for making an aldol product by reacting an enolate and an aldehyde different from which it derived.

Figure 20.21 Retrosynthesis of a crossed aldol product.

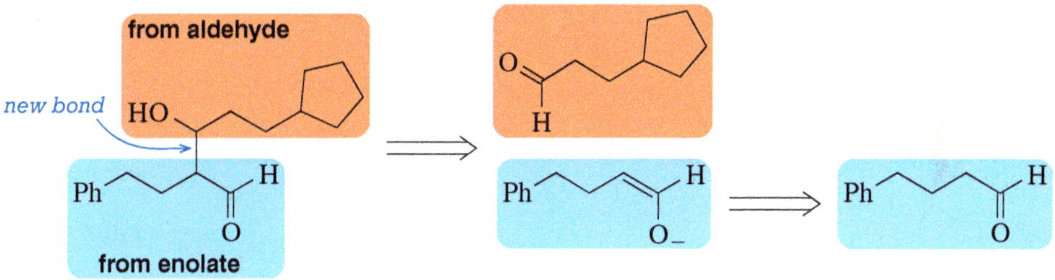

When the reaction suggested by Figure 20.21 is run under classical conditions [i.e., NaOH and H_2O; see **Figure 20.22(a)**], the desired crossed aldol product is obtained as one of four aldol products in nearly equal amounts [**Figure 20.22(b)**]. [*A synthesis that generates four different products in equal yields has very little utility.*] The failure of this reaction is a result

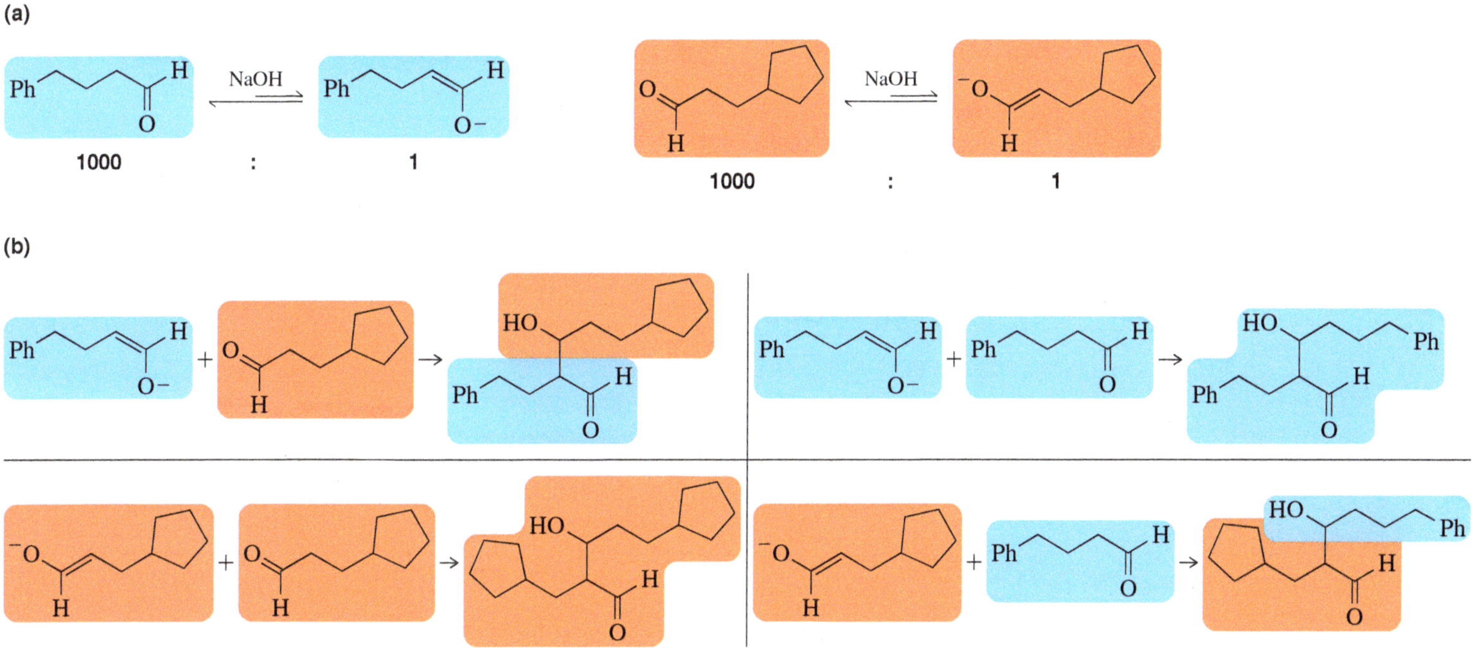

Figure 20.22 The crossed aldol reaction under classical conditions produces four products in nearly equal yields. (a) Unfavorable enolate formation keeps a large concentration of both aldehydes present. (b) Random collisions between enolates and aldehydes give four products.

of the reaction being under thermodynamic control. Given that enolate formation is unfavorable, you have 1000 molecules of each aldehyde for every molecule of each enolate [Figure 20.22(a)]. Thus, both aldehydes are available to be attacked by either enolate, and all four products are equally stable. [*Remember, reactions are random collisions.*] Thus, you end up with two self-condensation products and two crossed aldol products.

Assessment 20.33 Draw an arrow-pushing mechanism that justifies the formation of each of the products in Figure 20.22.

Assessment 20.34 Predict the four products in each of the following reactions. What ratio of products would you expect from the reactions?

(a)

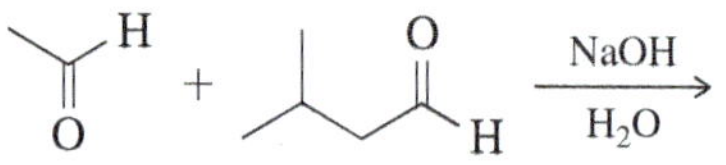

(b)

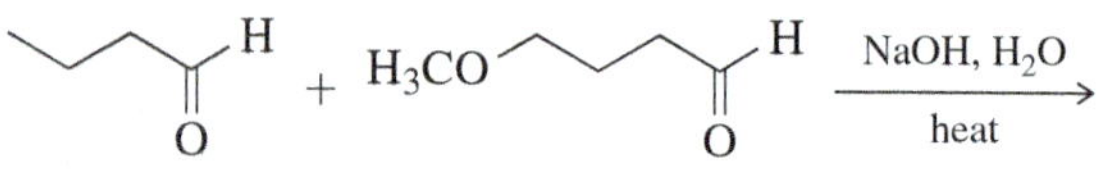

Modern organic chemists can easily perform **crossed aldol reactions** to make any desired product (**Figure 20.23**). By completely deprotonating an aldehyde with a strong base and then adding the aldehyde electrophile, only one aldol reaction occurs.

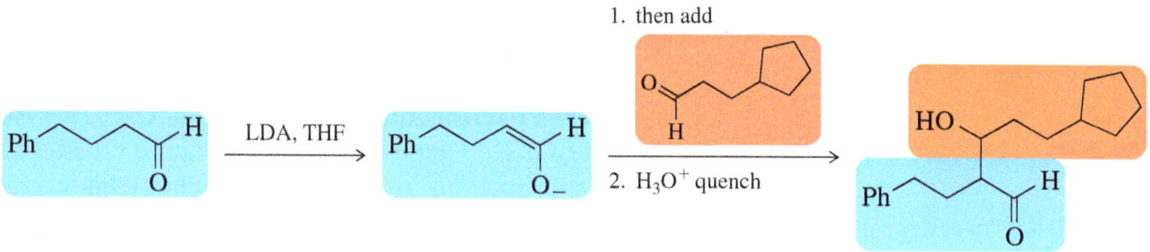

Figure 20.23 Crossed aldol reactions are simple when using strong bases. A strong base completely forms the enolate.

Assessment 20.35 Predict the product of the following modern aldol reactions. Compare your answers to those you obtained in Assessment 20.34.

(a)

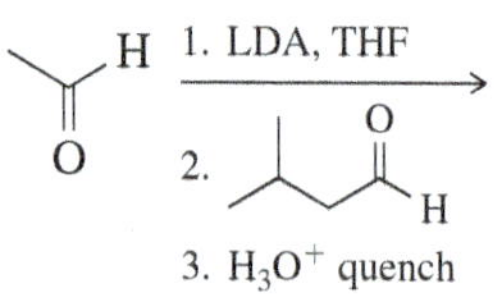

(b)

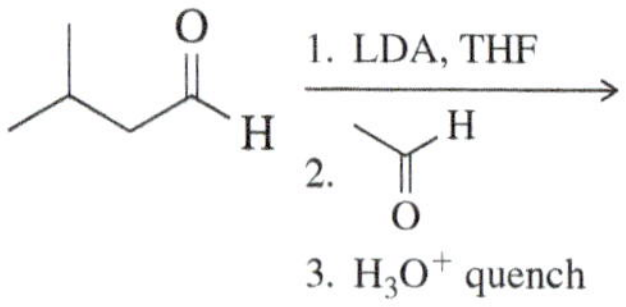

(c)

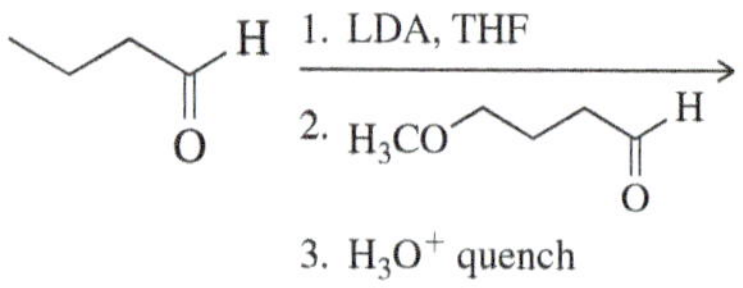

(d)

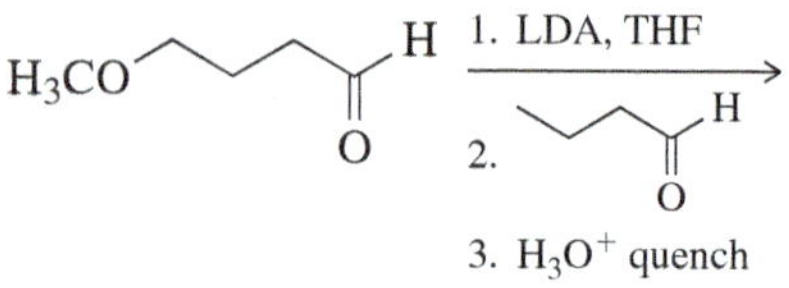

20-A CHEMISTRY IN CONTEXT: Green Chemistry

Atom economy in the aldol reaction

The classical version of the aldol reaction is one of the more sustainable reactions that we study. It is usually high-yielding, takes place at room temperature (requiring no energy to heat or cool the reaction mixture), uses a catalytic amount of base, is done in a nontoxic solvent (water), and proceeds with 100% atom economy (**Figure 20.24**).

Most reactions can always be improved, as demonstrated by the aldol reaction in **Figure 20.25**, which is done in the absence of solvent, reducing the aqueous waste associated with the reaction, while also conserving water.

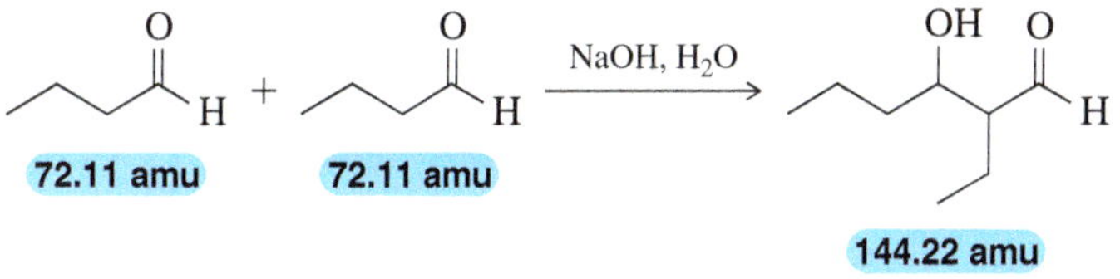

atom economy = [(72.11 + 72.11)/144.22] × 100 = **100%**

Figure 20.24 **The aldol is a very green reaction, occurring with 100% atom economy.**

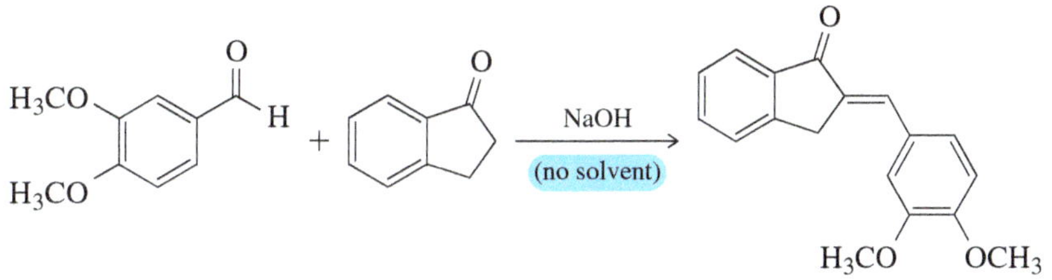

Figure 20.25 **Solvent-free reactions are more sustainable, conserving water and eliminating aqueous waste streams.**

Assessment 20.36 The modern aldol reaction offers a number of advantages over the classical aldol reaction in terms of regio- and stereoselectivity. Compare its efficiency to that of the classical aldol. Is it a greener reaction, too?

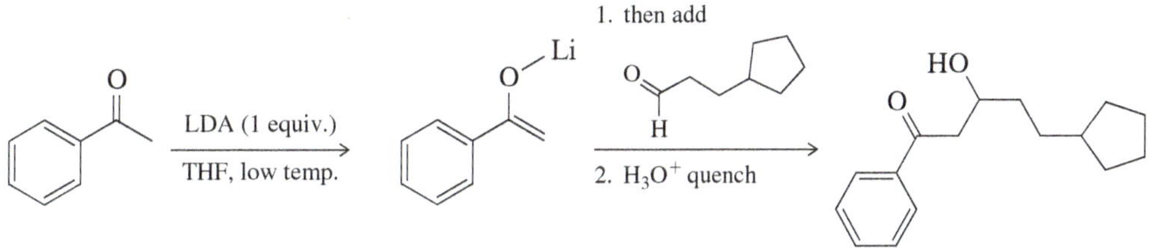

Without access to the bases we have today, classical organic chemists needed creativity and problem solving to enable the crossed aldol reaction. Two facts that we know now perhaps guided them to a solution. First, there must be an α-hydrogen to form an enolate. And second, ketones can become enolates but cannot act as electrophiles in the aldol reaction [*see Section 20.3.2*]. [*With these two facts in mind, think about a selective crossed aldol you might design before finishing this paragraph.*] The **Claisen–Schmidt reaction** is the result of this analysis (**Figure 20.26**). Using an aldehyde without α-hydrogens ensures that it cannot become an enolate. Reacting it with the enolate derived from a ketone ensures that the aldehyde is the only electrophile. By heating the reaction, the Claisen–Schmidt reaction selectively produces α,β-unsaturated ketones.

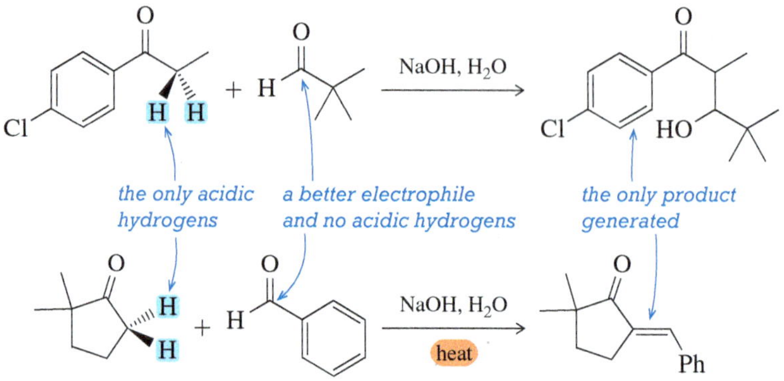

Figure 20.26 The classical crossed aldol reaction. By imposing restrictions on the reaction partners, the crossed aldol reaction is efficient.

20.3.3 SUMMARY

- In the crossed aldol reaction, the enolate of a ketone reacts with an aldehyde that contains no α-hydrogens. The modern aldol reaction is still better because it has fewer restrictions.

Assessment 20.37 Predict the product of the following crossed aldol reactions.

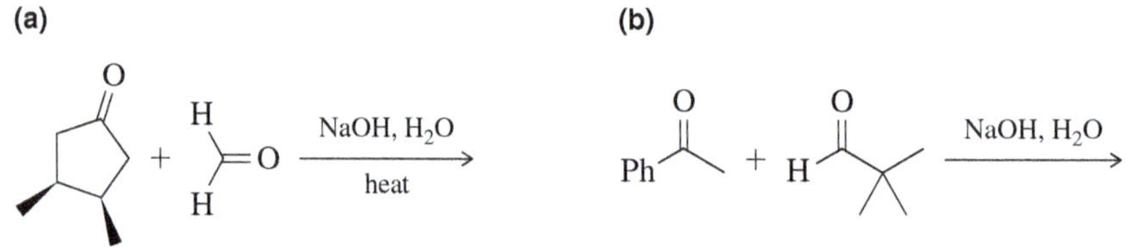

Assessment 20.38 Provide the reagents and conditions you might use to make the following α,β-unsaturated ketones via the Claisen–Schmidt reaction. [*Don't get distracted by the other stuff. Simply make the new bonds next to the carbonyl.*]

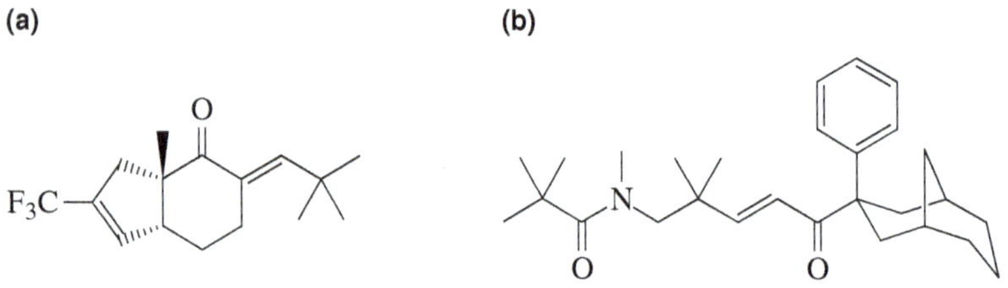

Assessment 20.39 The crossed aldol reaction can be used to synthesize amino alcohols, a structural motif that is prevalent in a number of medicinally important compounds. In this process, the aldol reaction between nitromethane and an aldehyde is followed by reduction of the nitro group to generate the 1,2-aminoalcohol. Show a mechanism for the first reaction and explain why the nitro group is able to facilitate enolate formation in the crossed aldol reaction.

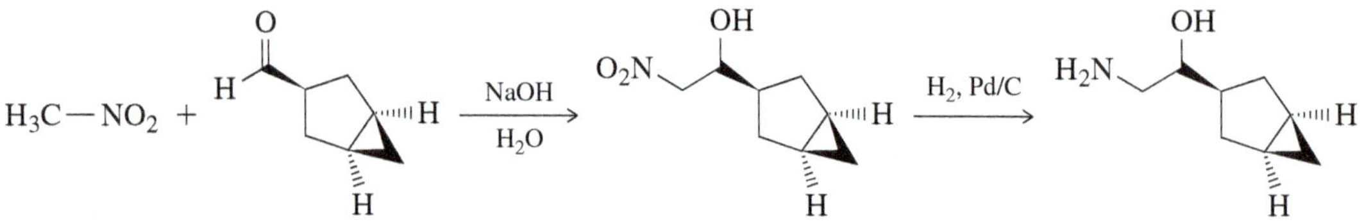

20-B CHEMISTRY IN CONTEXT: Biochemistry

Synthesis of sucrose by the aldol reaction

The aldol reaction is prevalent in biological systems, for example, in the biosynthesis of sucrose by plants. In the cytosol of plant cells, the reaction of dihydroxyacetone phosphate and glyceraldehyde 3-phosphate is simply an enzyme-catalyzed aldol reaction (**Figure 20.27**). Following C—C bond formation, the product spontaneously cyclizes to give the five-membered furanose ring, a hemiacetal.

The aldol reaction is reversible in biological systems as well. Enzymes do not alter the equilibrium of a reaction. As catalysts, they increase the rate by which equilibrium is achieved by stabilizing the transition state between reactants and products. As such, they also catalyze the reverse reaction. The aldolase enzyme does the exact reverse of the process depicted in Figure 20.27 as part of glycolysis in mammalian cells, a process that generates energy for cells. The direction that the enzyme pushes the reaction is determined by the relative concentrations of reactants and products [*Le Châtelier's principle*].

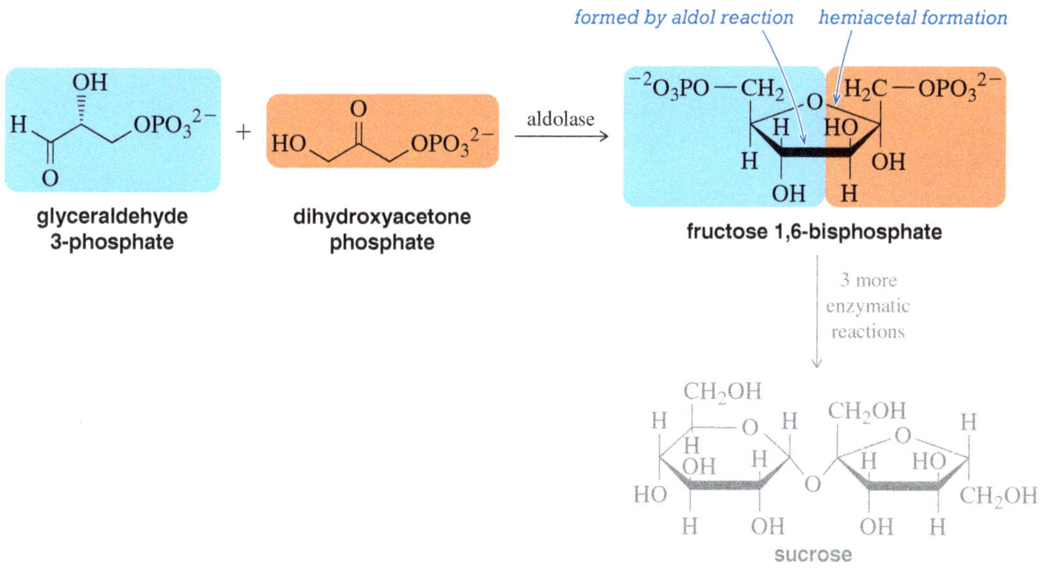

Figure 20.27 **Biosynthesis of sucrose by an enzyme-catalyzed aldol reaction.**

Assessment 20.40 Using active-site general bases (B^-) and acids (B—H) as needed, draw an arrow-pushing mechanism for the aldolase-catalyzed production of fructose 1,6-bisphosphate from dihydroxyacetone phosphate and glyceraldehyde 3-phosphate.

Assessment 20.41 The aldolase-catalyzed process of converting fructose 1,6-bisphosphate to dihydroxyacetone phosphate and glyceraldehyde 3-phosphate is referred to as a retro-aldol reaction. Show an arrow-pushing mechanism for this process. [*As a result of the principle of microscopic reversibility, this reaction should be the exact opposite of the mechanism you drew in Assessment 20.40.*]

20.3.4 Intramolecular aldol reaction

The aldol reaction can be conducted intramolecularly in an entropically favored process when the reaction is driven to formation of the α,β-unsaturated aldehyde.

Let's learn the mechanism of the intramolecular aldol reaction by working Assessment 20.42, first on your own, and then together.

Assessment 20.42 Suggest a mechanism for the intramolecular aldol reaction. [*Hint: All of the steps are identical to those in the intermolecular version.*]

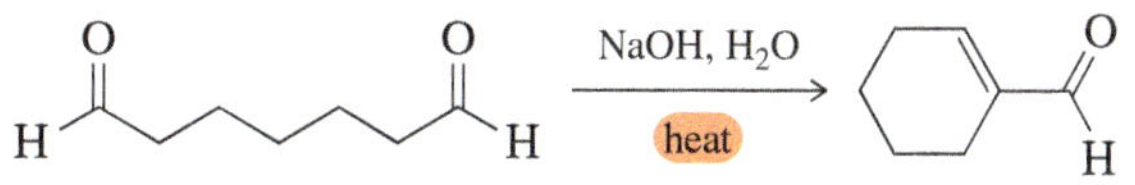

Solved Assessment 20.42

As we work through the mechanism, keep in mind everything we know already about the aldol reaction. Coupling this knowledge to our general organic problem-solving process, a clear picture of the intramolecular aldol reaction should quickly emerge. To start, we number the carbons and identify the bonds formed and broken in **Figure 20.28**.

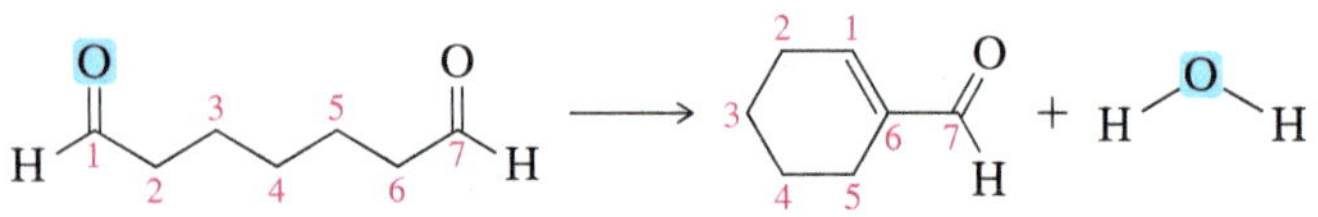

Figure 20.28 Number the carbons and identify the bonds formed and broken.

Check your numbering by verifying that, for example, carbons 2–5 are CH_2 groups in both the reactant and the product and that C_7 is still an aldehyde.

Bonds broken: C_1—O π; C_1—O σ; C_6—H σ; C_6—H σ

The number of bonds broken will be equal to the number of bonds formed in a given reaction.

Bonds formed: C_6—C_1 σ; C_6—C_1 π; O—H σ; O—H σ

The oxygen in H_2O originates at C_1.

This is where our knowledge of the aldol reaction is useful. Focusing on the C_1—C_6 σ and π bonds, we can analyze how the new bonds might form (**Figure 20.29**). C_1, being an aldehydic carbon, is most often electrophilic. C_6, moreover, can become nucleophilic when turned into an enolate. Recognizing this makes it clearer how the reaction will proceed.

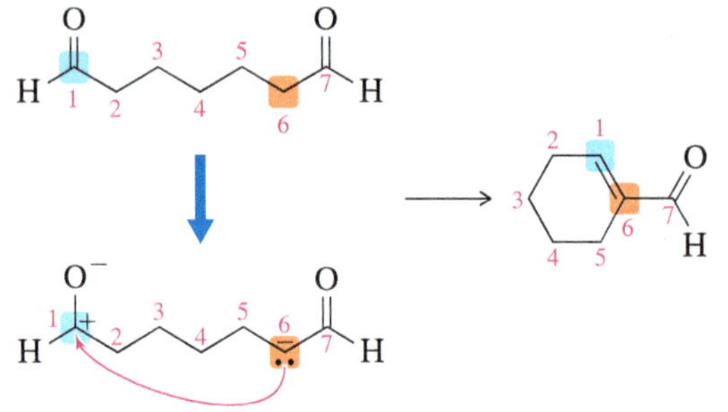

Reactions occur between nucleophiles and electrophiles.

C_1 is an aldehyde, which is most often electrophilic because of the zwitterionic resonance structure.

C_6 can become an enolate; enolates are good nucleophiles.

C_6 can now attack C_1.

Figure 20.29 The inherent reactivity of functional groups suggests how the new C—C bond might form.

From this brief analysis, we can apply our knowledge of the aldol reaction to give the mechanism shown in **Figure 20.30**. The enolate formed by deprotonation of C_6 attacks the aldehyde at C_1. Protonation of the resulting alkoxide gives the β-hydroxyaldehyde, which, with heating, eliminates through the E1cb mechanism to give the α,β-unsaturated aldehyde. ■

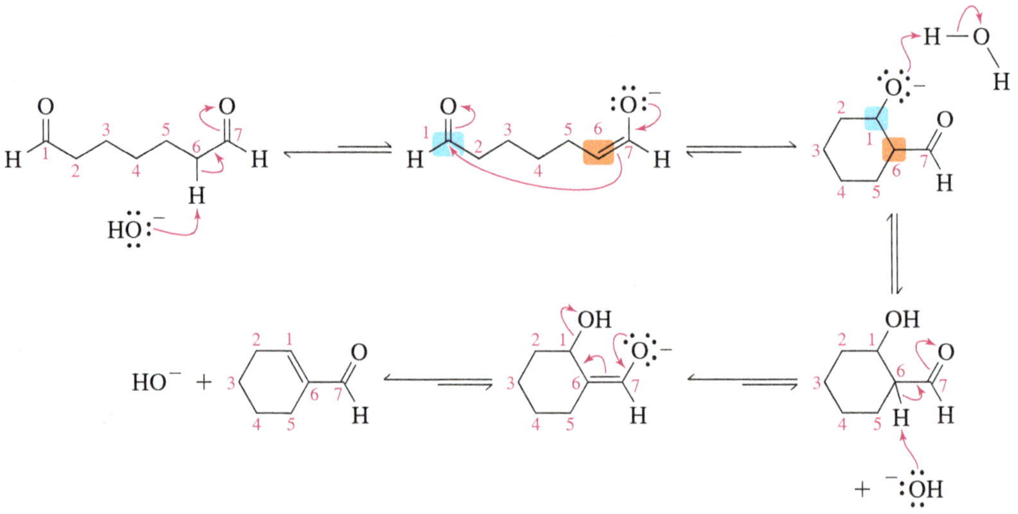

Figure 20.30 The mechanism of the intramolecular aldol reaction.

Diketones can similarly participate in intramolecular aldol reactions, as shown in the mechanism in **Figure 20.31**. [*The mechanism is nearly identical to the one in Figure 20.30.*]

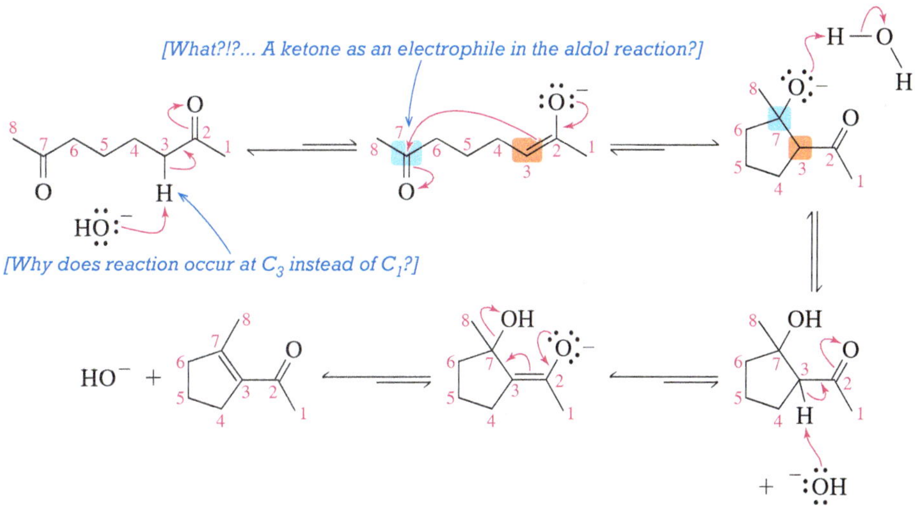

Figure 20.31 The mechanism of the intramolecular aldol reaction with a diketone. [*Consider the questions posed in the figure before reading further.*]

Let's answer the two questions posed in Figure 20.31. First, the enolate forms at C_3 instead of C_1 because the reaction is under thermodynamic control. The pK_a values of the C_3 and C_1 protons are nearly identical, and the resulting enolates are almost equally stable. However, the enolate at C_3 attacks the C_7 ketone to form a stable five-membered ring (**Figure 20.32**). The C_1 enolate attacks the C_7 ketone to form a seven-membered ring. In solution, both of these things happen, but the reaction equilibrates through the more stable five-membered ring. Le Châtelier's principle then drives the reaction to form exclusively the product resulting from deprotonation of C_3.

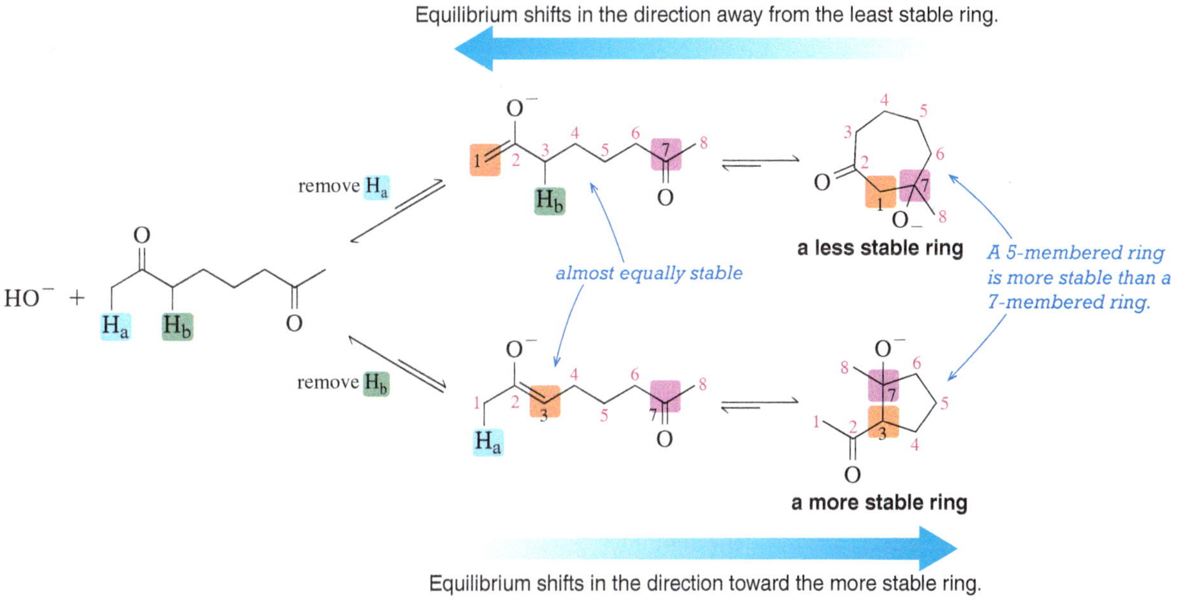

Figure 20.32 Formation of the more stable ring dictates the regioselectivity of enolate formation.

The second question asks why the ketone is an electrophile in the aldol reaction. Figure 20.20 showed that this was unfavorable in an intermolecular aldol reaction. An intermolecular and an intramolecular aldol reaction are compared in **Figure 20.33**. Whereas the intermolecular aldol reaction has to overcome unfavorable entropy, intramolecular aldol reactions are entropically neutral (1 molecule to 1 molecule; $\Delta S \approx 0$). [*Full disclosure: The entropy term is still slightly negative, owing to a loss of rotational entropy that results from ring formation. This has a very small impact on the reaction.*] As a result, the intramolecular reaction is driven by ΔH and is therefore favorable. [*Heating the reaction drives it to form the α,β-unsaturated ketone anyway.*]

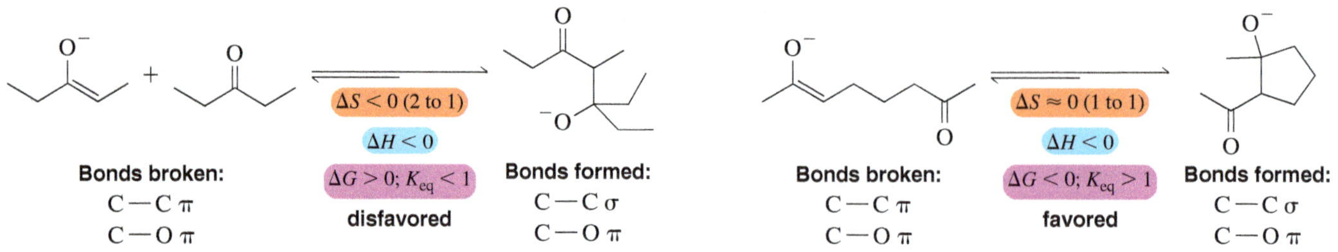

Figure 20.33 Ketones can be the electrophile in intramolecular aldol reactions because $\Delta S \approx 0$.

20.3.4 SUMMARY

- The **intra**molecular aldol reaction proceeds efficiently because it lacks the disfavorable entropy change of the **inter**molecular aldol while maintaining the favorable change in enthalpy.

Assessment 20.43 Predict the product and show a mechanism for the following reaction.

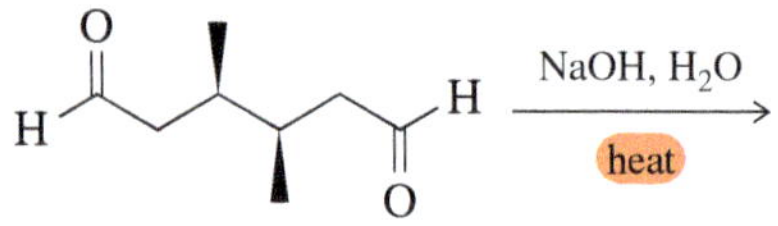

Assessment 20.44 Unlike in any aldol reaction we have studied previously, the following step is favored to the right. Explain this observation.

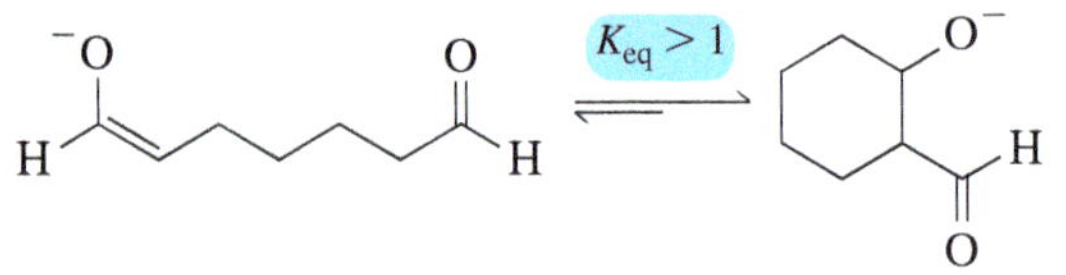

Assessment 20.45 The intramolecular aldol reaction can also be performed using acid catalysis, as shown in the synthesis of the following seven-membered ring. Show a mechanism for this reaction.

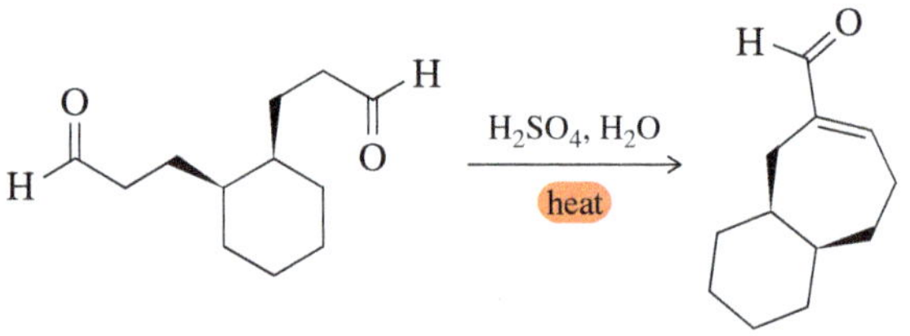

Assessment 20.46 The following diketone can do an intramolecular aldol reaction to form a four- or a six-membered ring. (a) Predict both products and (b) show the mechanism by which both are formed. (c) Which will be the predominant product? Justify your choice.

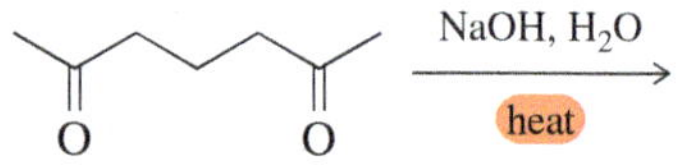

Strategy for cyclohexanone synthesis

1. ***Enolates to make ketones: Claisen***
2. *Intramolecular Claisen: Dieckmann*
3. *Removing the ester (TLaC)*
4. *Adding additional alkyl groups*
5. *Adding additional acyl groups*

This exercise spans multiple sections, so the "strategy" sidebar provides a reminder of where we are in the process of thinking like a chemist. We begin by considering how the aldol reaction might be modified such that a β-dicarbonyl can be formed. Section 20.4.2 circles back to making cyclohexanones specifically.

20.4 Claisen condensation

The remainder of this chapter discusses how organic chemists develop new reactions. Our goal is to develop a synthesis of cyclohexanone using enolate reactions of acyclic molecules. Although somewhat contrived, this exercise allows us to "reinvent" reactions using thought processes similar to those used by early organic chemists who first considered these problems.

Thinking Like a Chemist

Figure 20.34 summarizes a thought process that would ultimately result in the Claisen condensation. To start, how could we modify the aldol reaction so that a ketone is produced? Hydride is a terrible leaving group, so replacing it with a better leaving group (**X**) should allow the electron pair in the tetrahedral intermediate to collapse back down and reform the C—O π bond (step 1, Figure 20.34). In essence, the reaction would be between an enolate and a carboxylic acid derivative, and the mechanism would be a nucleophilic acyl substitution from Chapter 19. [*In the aldol reaction, enolates added to aldehydes and ketones (Chapter 17). In the Claisen condensation, enolates do substitution reactions (Chapter 19). Again, no new chemistry is happening here.*] Unfortunately, using a weak base like hydroxide means that enolate formation is unfavorable and hydroxide is known to react with each of the carboxylic acid derivatives we studied in Chapter 19 (step 2, Figure 20.34). Luckily, classical chemists could add sodium to alcohols to make alkoxide bases (step 3, Figure 20.34). Using an alkoxide base, the choice of an ester as the reactive functional group ensures that if substitution happens, it simply regenerates the starting ester (step 4, Figure 20.34). Eventually, an ester enolate forms, attacking another molecule of ester to produce a β-keto ester. [*A ketone is produced.*] ■

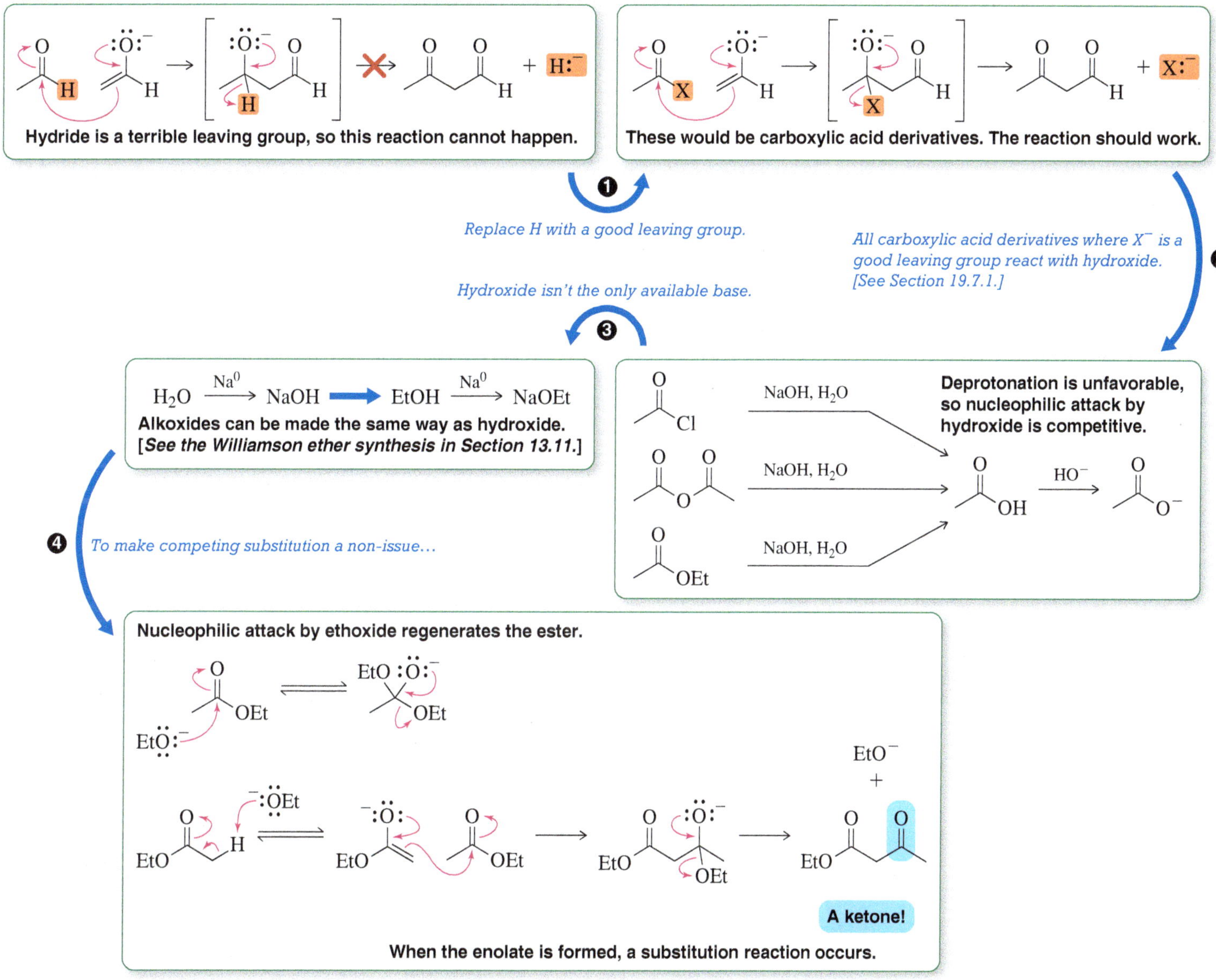

Figure 20.34 **The aldol reaction becomes the Claisen condensation, giving a way to make ketones.**

Unlike the aldol reaction, the Claisen condensation requires a stoichiometric amount of base to proceed. Before looking at other examples of the Claisen reaction, let's try to understand why. To do this, we must consider the mechanism (Table 20.4) in the same careful way that we did for the aldol reaction (see Tables 20.1 and 20.3).

Table 20.4 The energetics of the Claisen condensation. The highly favorable deprotonation step is the driving force for this reaction.

Step of the Claisen condensation	Analysis
1. Formation of the enolate 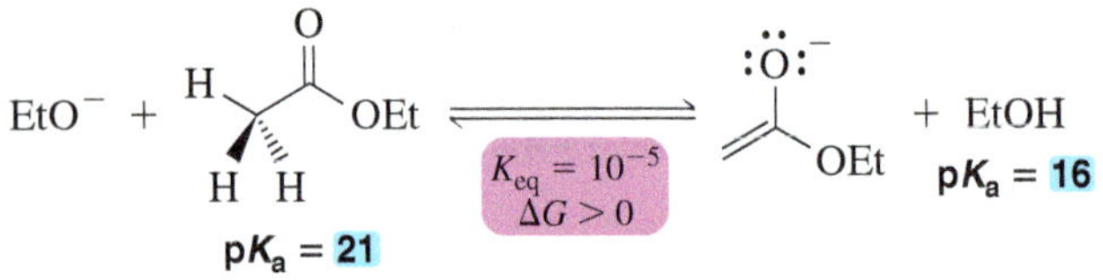	Deprotonation of the α-carbon with ethoxide is unfavorable. Using pK_a values for ethanol (16) and the α-proton of an ester (21), we calculate $K_{eq} < 1$.
2. Enolate attacks the ester 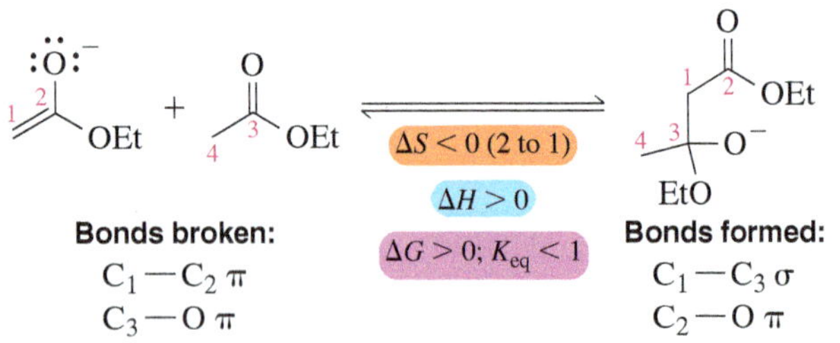	Because esters are worse electrophiles than ketones, this reaction is disfavored by enthalpy ($\Delta H > 0$) despite a C—C σ bond forming at the expense of a C—C π bond. Two molecules are reacting to make one, so $\Delta S < 0$. As a result, this C—C bond-forming step is unfavorable ($\Delta G > 0$). [*If unfavorable is followed by unfavorable, then why does the Claisen condensation even happen?*]
3. Elimination of ethoxide 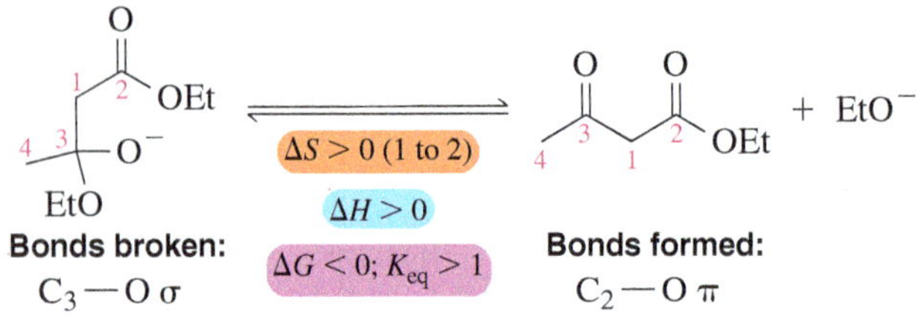	Because one molecule forms two, the second step of nucleophilic acyl substitution is entropically favored ($\Delta S > 0$). Forming a π bond at the expense of a σ bond is slightly disfavored by enthalpy. Overall, this is a slightly favorable step ($\Delta G < 0$). [*The reaction doesn't stop here.*]
Despite the product of step 3 being a β-keto ester, the Claisen reaction wouldn't work if that were the final step. When you consider that the first two steps were disfavored and the third step is only slightly favored, then the overall reaction to this point would have $\Delta G > 0$. The driving force for the Claisen reaction is step 4.	
4. Deprotonation of the product 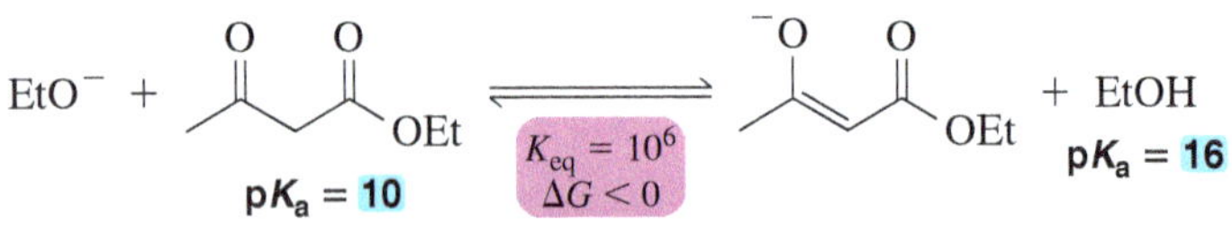	Deprotonation of the resulting β-keto ester is highly favored. Using pK_a values for ethanol (16) and the α-proton between two carbonyls (10), we calculate $K_{eq} > 1$. [*Continue reading for an explanation of the lower pK_a value.*]

The favorability of step 4 is a result of resonance stabilization of the conjugate base—namely, the carbanion is delocalized over two carbonyls (**Figure 20.35**). A CH_2 group between two electron-withdrawing substituents is said to be an activated methylene and has a lower pK_a value than a proton α to just one carbonyl.

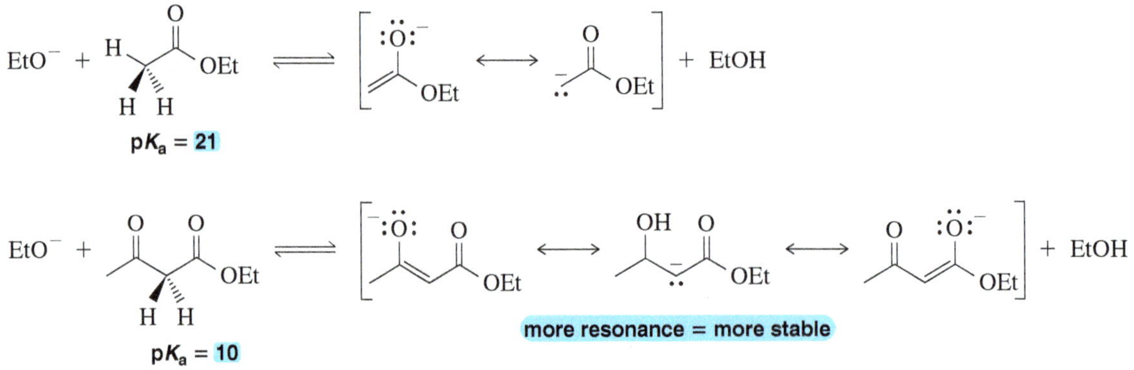

Figure 20.35 Activated methylenes have two electron-withdrawing groups to stabilize anions by resonance. [*See if you can push electrons in the resonance structures.*]

There are a couple of implications of this final deprotonation. First, the deprotonation is the thermodynamic driving force of the reaction—see the reaction coordinate diagram in **Figure 20.36**. Without it, the Claisen condensation doesn't happen. Because ethoxide is consumed in the final step, this reaction cannot be done catalytically; a full equivalent of base is required. For esters to participate in the Claisen reaction, there must be at least two α-hydrogens. [*Assessment 20.49 asks you to think about why this is.*]

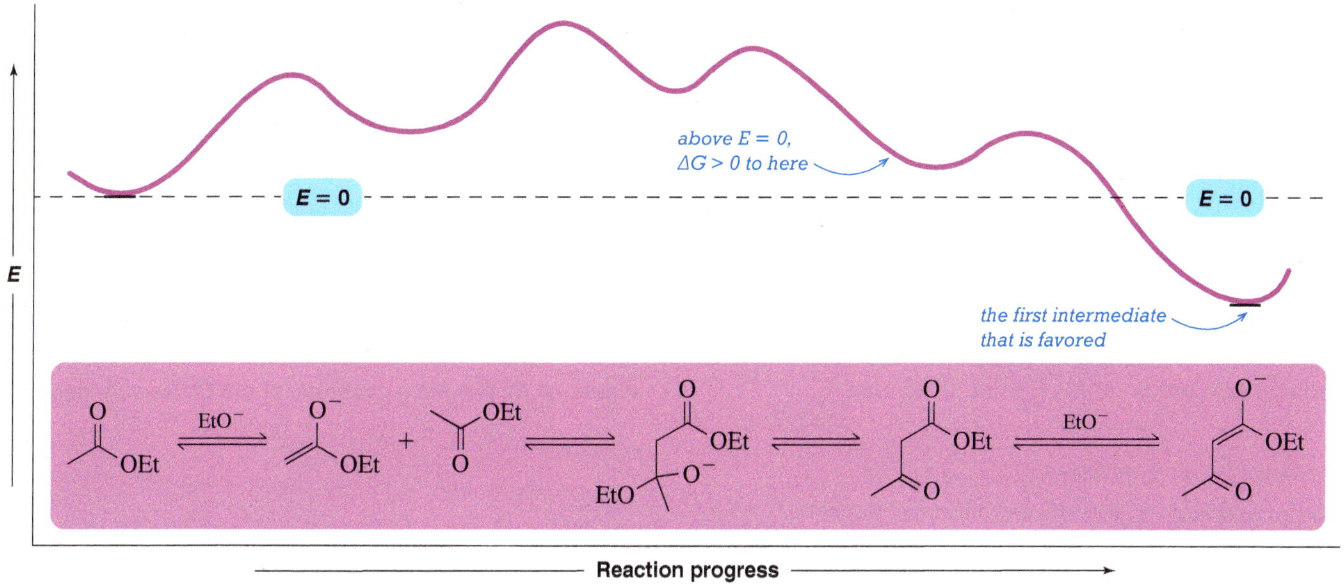

Figure 20.36 Reaction coordinate diagram for the Claisen condensation. It is only favored because of the final deprotonation step.

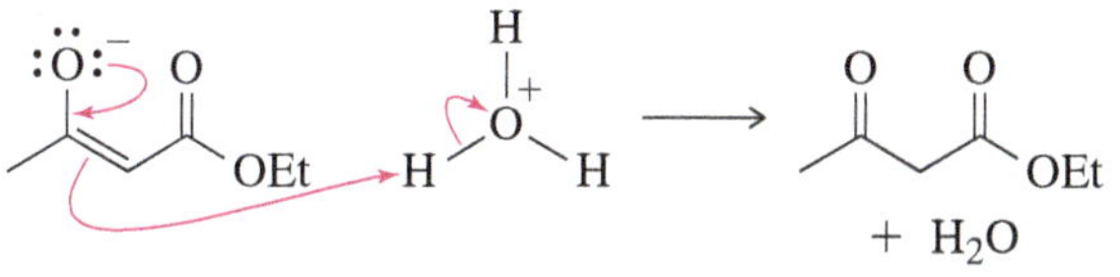

Figure 20.37 The Claisen reaction finishes with an acid quench.

Finally, to generate the desired ketone product, the resulting enolate is quenched with acid, as shown in Figure 20.37. [*Don't be confused by the fact that the quenched enolate gives the same product as that of step 3. Forming the enolate happened spontaneously and is the only reason the Claisen reaction exists.*]

20.4 SUMMARY

- The Claisen condensation is the C—C bond forming reaction between two esters to form a β-keto ester.

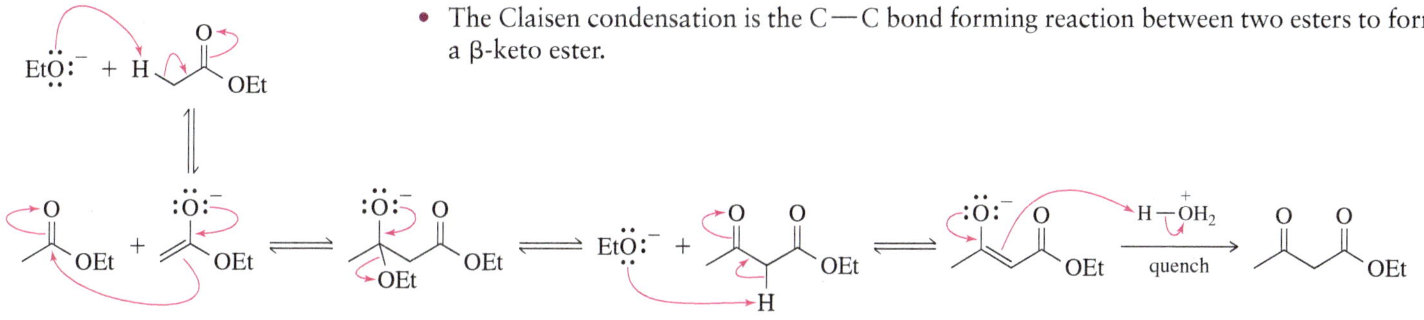

Assessment 20.47 Predict the product of the following reactions.

(a)

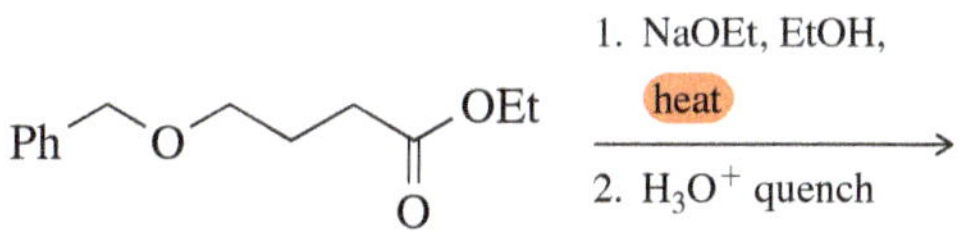

(b)

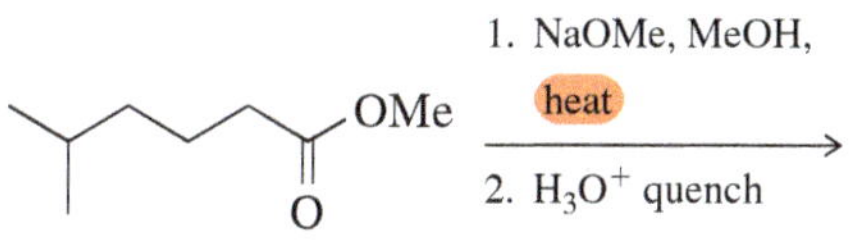

Assessment 20.48 A chemist chose to use isopropanol and sodium isopropoxide instead of ethoxide and ethanol in the Claisen reaction shown. The desired product was not isolated. Identify the product that formed and rationalize its formation.

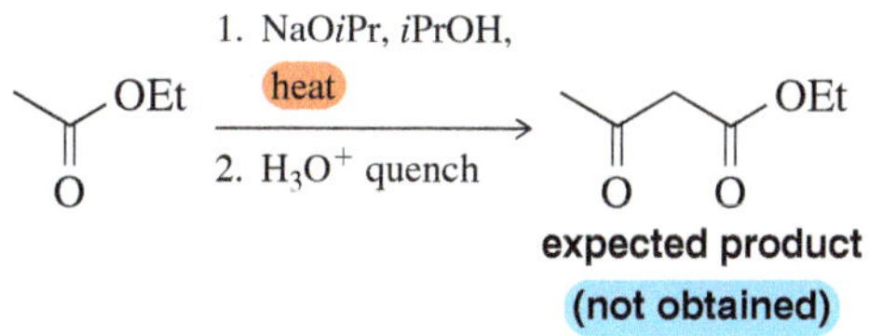

Assessment 20.49 A different chemist attempted to do the Claisen reaction of ethyl 2-methylpropanoate. Explain why this reaction failed in the context of the Claisen mechanism.

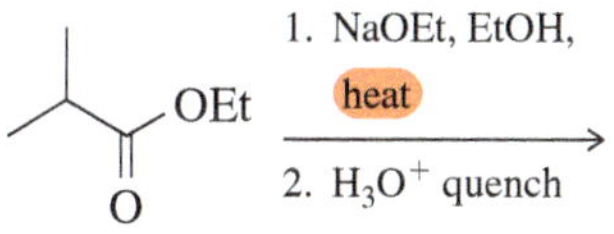

20-C CHEMISTRY IN CONTEXT: Green Chemistry

Claisen reaction

Although not as efficient as the aldol reaction, the Claisen condensation is relatively green when compared to other reactions we've studied (**Figure 20.38**). The disadvantages of the Claisen condensation compared to the aldol reaction include its requirement for a full equivalent of base, elevated temperatures, and a quench step, and its significantly lower atom economy. Fortunately, the use of ethoxide as the base allows for ethanol, a safe and nontoxic solvent, to be used.

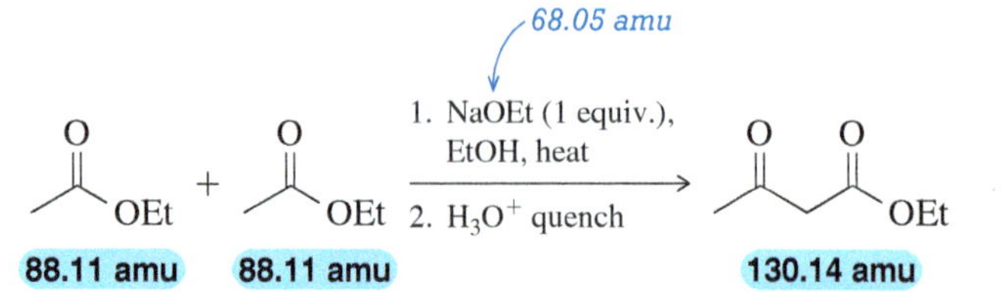

atom economy = [130.14)/(88.11 + 88.11 + 68.05)] × 100 = **53%**

Figure 20.38 The atom economy of the Claisen condensation.

Assessment 20.50 If the Claisen condensation is run using methyl acetate, the atom economy is improved. (a) Does that mean the new process is more efficient? (b) Fill in the missing reactants and solvents, calculate the new atom economy, and evaluate the efficiency of the new process. (c) Is it greener than the one in Figure 20.38?

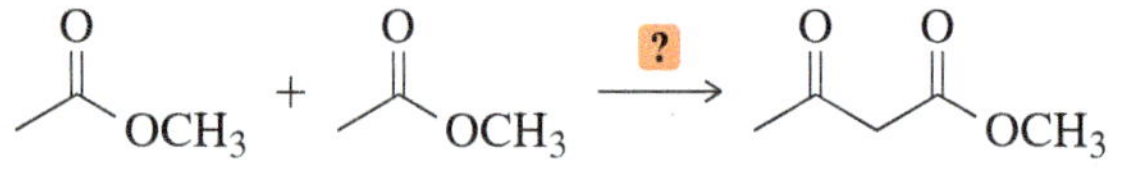

20-D CHEMISTRY IN CONTEXT: Biochemistry

Claisen in biological systems

The Claisen condensation of acetyl-CoA, catalyzed by the enzyme thiolase, demonstrates the important role this reaction plays in biological systems. Although SCoA is a complex molecule, it is simply the thiol portion of a thioester. As a result, acetyl-CoA reacts similarly to the esters we have seen in this chapter to form acetoacetyl-CoA (**Figure 20.39**), the beginning step of the production of mevalonate and the biosynthesis of cholesterol.

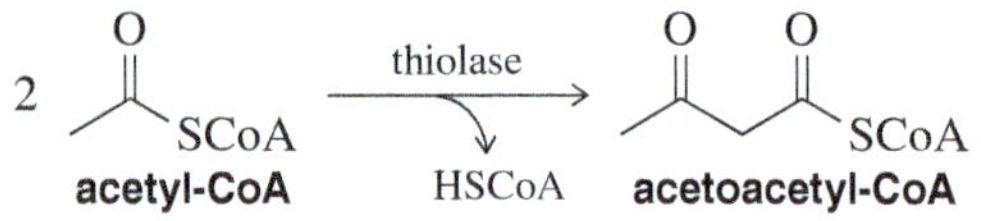

Figure 20.39 The formation of acetoacetyl-CoA is the beginning of many biological processes.

The reverse reaction, the retro-Claisen reaction, is another example. Used in the β-oxidation of saturated fatty acids, this reaction provides a source of energy for mammalian cells. In the example shown in **Figure 20.40**, palmitic acid is oxidized at the β-carbon to give a β-keto ester, which is a substrate of the thiolase enzyme. Following the retro-Claisen reaction, the resulting myristoyl-CoA can undergo a similar process, eventually degrading the fatty acid two carbons at a time while producing a total of seven acetyl-CoA molecules.

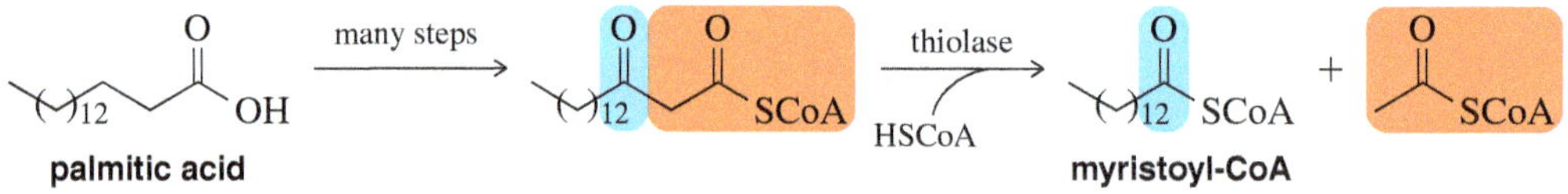

Figure 20.40 A retro-aldol reaction is repeated in the degradation of fatty acids.

Assessment 20.51 Draw an arrow-pushing mechanism for the thiolase-catalyzed retro-Claisen reaction to produce myristoyl-CoA. [*You can use active-site bases (B^-) and acids (B—H) in your mechanism.*]

20.4.1 Crossed Claisen condensation

The **crossed Claisen condensation** suffers from the same problems that made the crossed aldol reaction difficult to accomplish. [*The successful crossed aldol reaction utilized an aldehyde without acidic hydrogens and the enolate of a ketone.*] Because both of the Claisen components are esters, a similar strategy only gets us halfway to a workable crossed Claisen condensation (**Figure 20.41**).

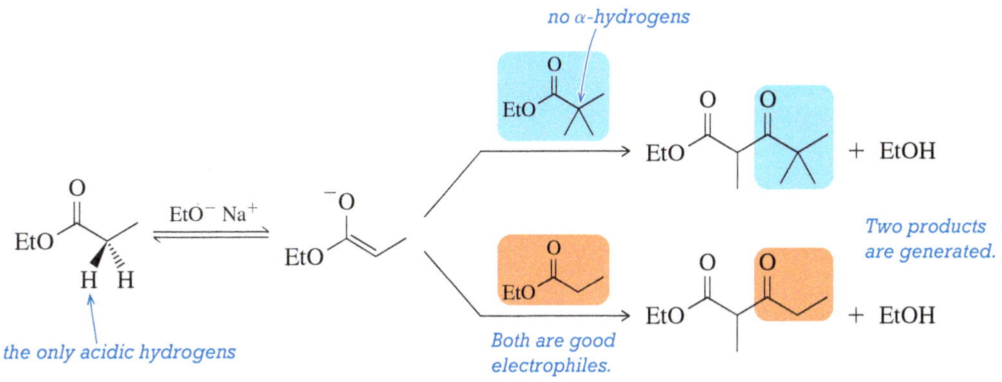

Figure 20.41 Because both esters are good electrophiles, the crossed Claisen condensation gives two products.

The solution to this problem is in the experimental details. In the aldol reaction, the ketone and the aldehyde are placed in the flask together and the base is added, because the bases are weak and we're relying on a thermodynamic driving force to give the desired product. As shown in Figure 20.41, executing the crossed Claisen in this manner gives multiple products. To obtain the desired selectivity in the crossed Claisen reaction, chemists developed a creative solution wherein the ester lacking α-hydrogens was stirred with ethoxide, while the desired electrophile was added slowly to the reaction vessel. **Figure 20.42** shows what the actual solutions look like. Because it has no acidic hydrogens, no reaction occurs between ethoxide and ester **B** in the round-bottomed flask. Upon addition of one drop of ester **A**, deprotonation occurs to give enolate, $\mathbf{A}^-$. Because only a small amount of ester **A** was added, the enolate $\mathbf{A}^-$ is much more likely to collide with ester **B**. This process continues until all of ester **A** is used up. **A** is continually added one drop at a time, allowing only a single product to form.

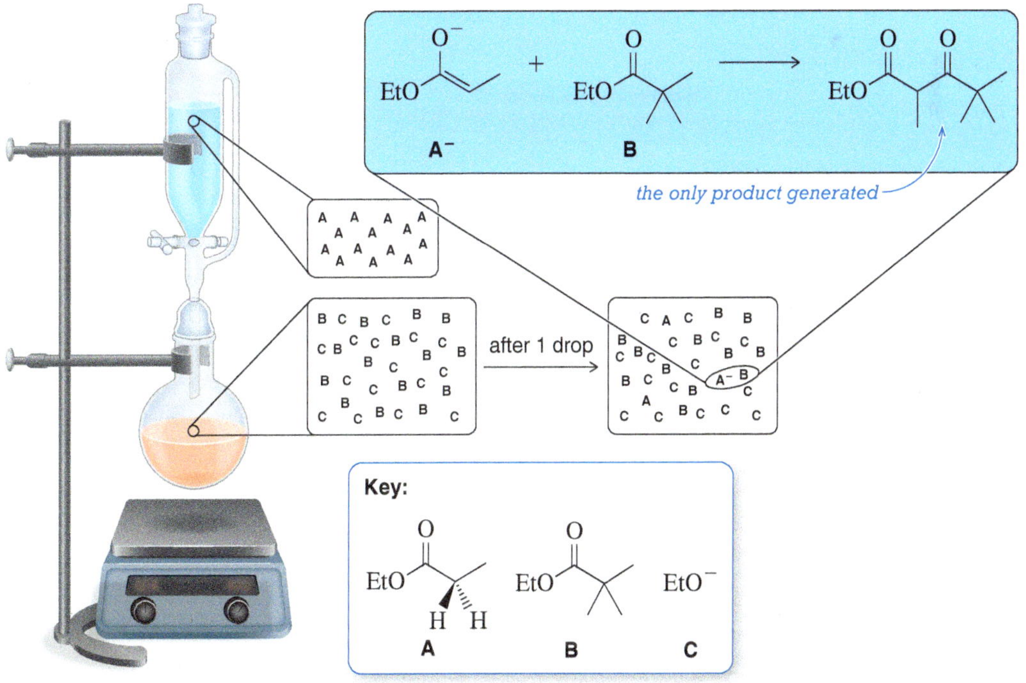

Figure 20.42 The experimental setup for the crossed Claisen condensation. Deprotonation of the product is still the thermodynamic driving force (Section 20.4), and an acid quench is needed to complete the reaction.

20.4.1 SUMMARY

- The crossed Claisen reaction is successful when utilizing one ester that lacks α-hydrogens, while carefully manipulating the experimental conditions.

Assessment 20.52 Suggest a synthesis of the following crossed Claisen products. Be sure to specify the order of addition and the experimental setup.

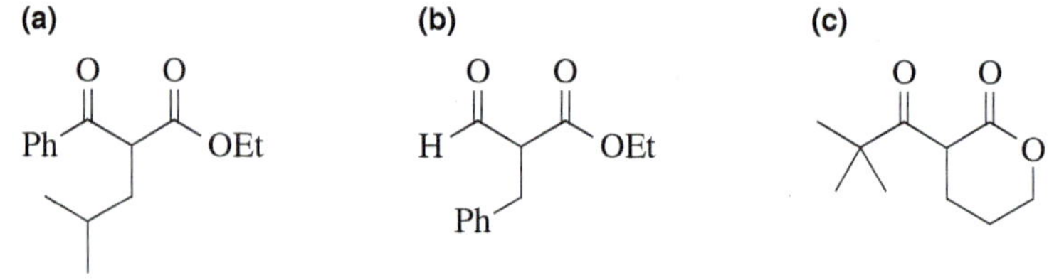

Strategy for cyclohexanone synthesis

1. *Enolates to make ketones: Claisen*
2. ***Intramolecular Claisen: Dieckmann***
3. *Removing the ester (TLaC)*
4. *Adding additional alkyl groups*
5. *Adding additional acyl groups*

20.4.2 Dieckmann condensation (intramolecular Claisen)

To apply the Claisen condensation to our synthesis of a cyclohexanone, the reaction must be conducted intramolecularly in a reaction first reported by Walter Dieckmann a decade after the discovery of the Claisen condensation. [*It is therefore called the Dieckmann condensation.*]

As in the Claisen mechanism, the Dieckmann condensation begins with enolate formation followed by intramolecular attack of the enolate on the ester to give a tetrahedral intermediate (**Figure 20.43**). Being intramolecular, enolate attack is more favored than the same step in the Claisen condensation (1 molecule to 1 molecule; $\Delta S \approx 0$). Collapse of the tetrahedral intermediate is followed immediately by removal of the proton from the more acidic dicarbonyl compound. [*As in the Claisen condensation, this is the driving force of the reaction.*] An acid quench produces the final product.

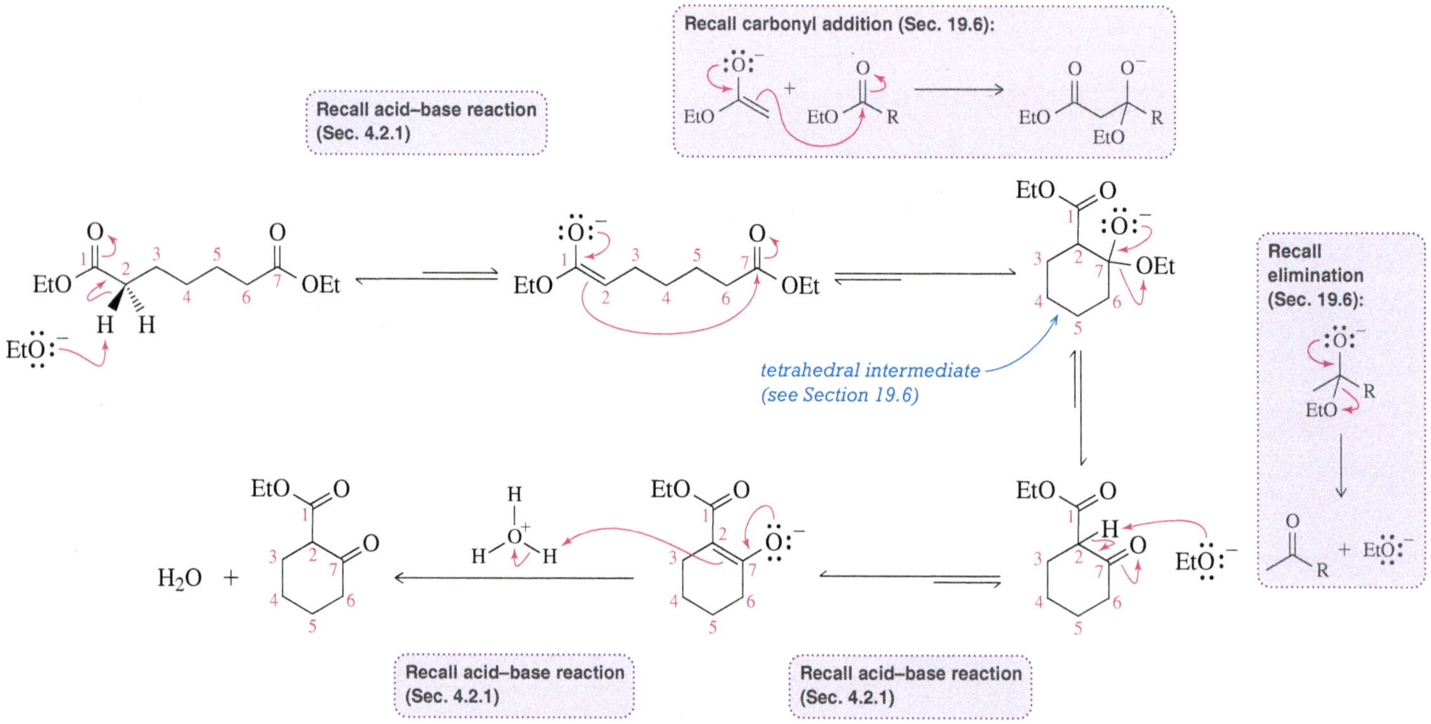

Figure 20.43 The mechanism of the Dieckmann condensation. It is identical to the Claisen condensation, except that it is more favored because it is intramolecular.

Assessment 20.53 The Dieckmann condensation shown in Figure 20.43 is illustrated with equilibrium arrows that appropriately represent which side of each reaction is favored. Justify the position of equilibrium for each reaction.

Assessment 20.54 The Dieckmann product can exist as the enol isomer. Why might it be favored over the keto isomer in this case?

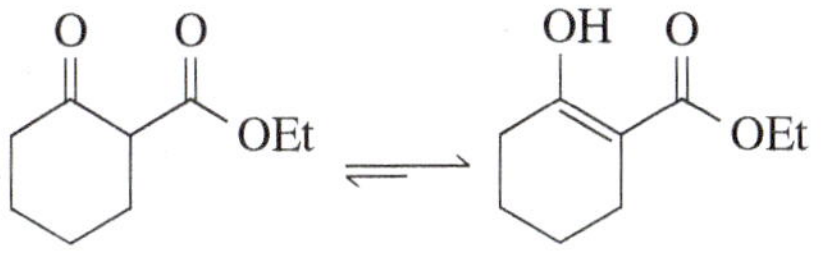

Assessment 20.55 Predict the product of the following Dieckmann condensation reactions.

(a)

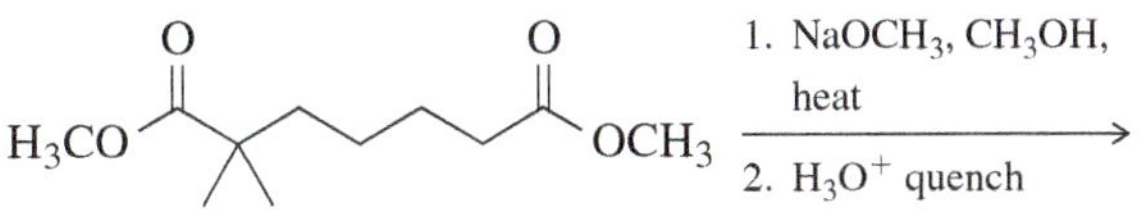

(b)

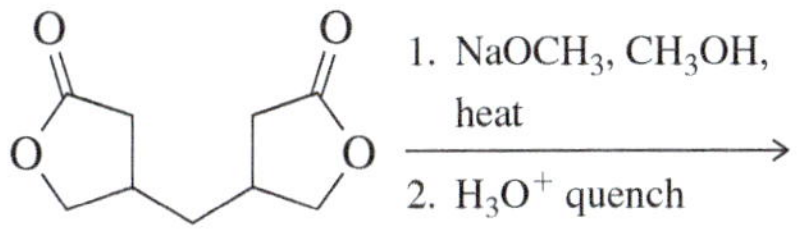

Assessment 20.56 In Assessment 20.55(b) an alkoxide is used that doesn't match the alkyl group of the ester but will still get the desired product. Why? How is this different from the result obtained in Assessment 20.48?

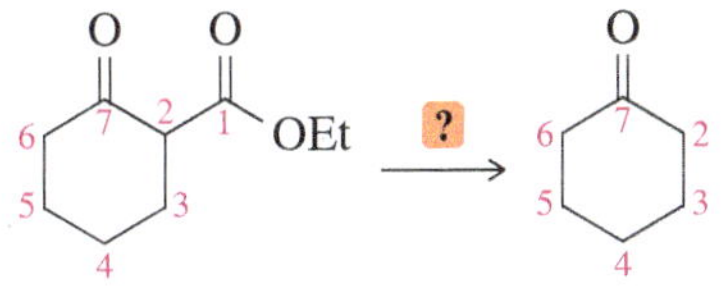

Figure 20.44 Removal of the C_1 ester to complete the synthesis of cyclohexanone.

Although we have succeeded in making a six-membered ring, we haven't technically synthesized cyclohexanone (**Figure 20.44**).

Thinking Like a Chemist

If we number the carbons as in Figure 20.44, then we need to break the $C_1—C_2$ bond and replace it with a hydrogen at C_2. Limiting ourselves to 2-electron processes, **Figure 20.45** imagines the $C_1—C_2$ bond breaking in two hypothetical ways. The electrons can move to either C_2 or C_1. [*Which seems more reasonable?*] If the $C_1—C_2$ bond breaks and the electrons move to C_2, two resonance-stabilized intermediates are produced, one of which is essentially an enolate. For this reason, the $C_1—C_2$ bond breaks in this way.

To complete the synthesis, we need a way of removing the C_1 ester attached at C_2. Let's figure out how to do this by thinking like a chemist. By recalling reactions studied previously, we can learn how to convert the β-keto ester into cyclohexanone while also learning strategies that might help us solve future problems.

Strategy for cyclohexanone synthesis

1. *Enolates to make ketones: Claisen*
2. *Intramolecular Claisen: Dieckmann*
3. ***Removing the ester (TLaC)***
4. *Adding additional alkyl groups*
5. *Adding additional acyl groups*

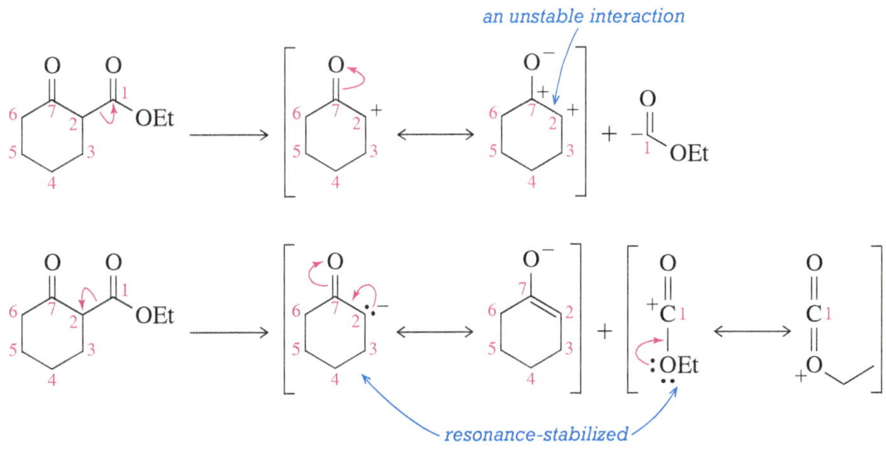

Figure 20.45 Hypothetical $C_1—C_2$ bond breaking to give the more stable, resonance-stabilized intermediate.

Bond breaking in this way also helps us understand the next step. [*Again, this is all hypothetical and part of the problem-solving process that will eventually lead to a set of reactions that can be used to carry out this transformation.*] With an anion at C_2, protonation would be expected to give the ketone product (**Figure 20.46**). Furthermore, the resonance-stabilized C_1 carbocation is reminiscent of carbon dioxide but with an ethyl group attached to one of the oxygens.

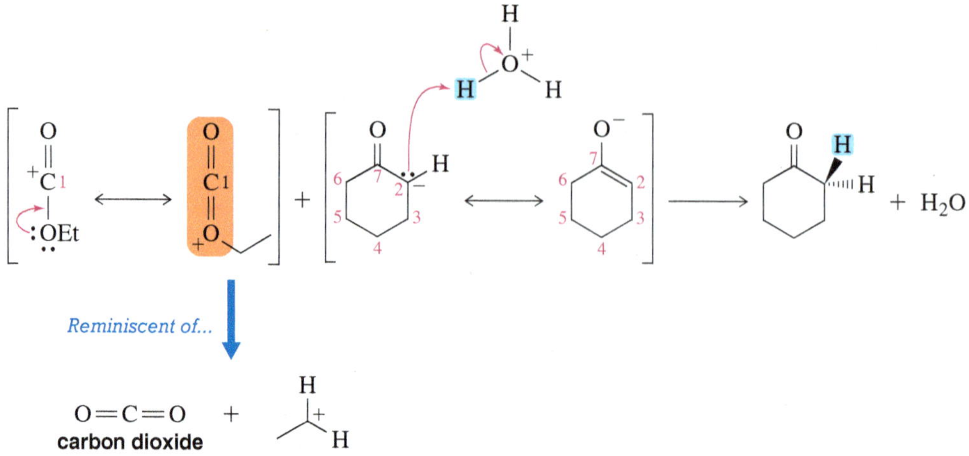

Figure 20.46 Our hypothetical $C_1—C_2$ bond breaking is more reasonable. One intermediate looks like carbon dioxide. The second, if protonated, gives the desired ketone. [*Do you know a reaction where CO_2 is completely removed from a molecule?*]

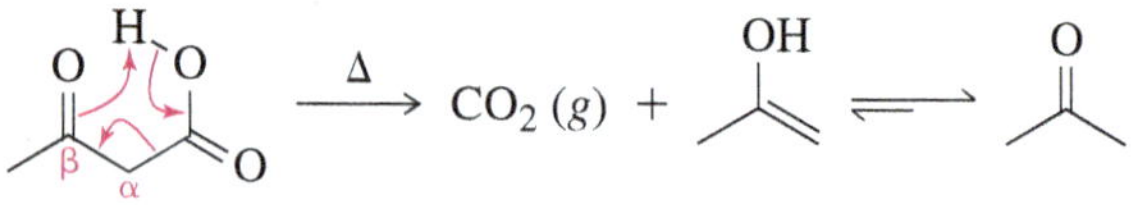

Figure 20.47 The thermal decarboxylation of β-ketoacids. We first studied this reaction in Chapter 18.

This analysis, which suggests the formation of a carbon dioxide–like molecule, is reminiscent of the decarboxylation reaction studied in Section 18.8.4. In this reaction, shown in **Figure 20.47**, a thermal decarboxylation occurs when a β-ketoacid is heated. Proceeding by a 6-electron cyclic transition state, decarboxylation initially forms an enol that tautomerizes to a ketone.

And so, working in the forward direction, the preceding analysis suggests that our β-keto ester can be first converted into a carboxylic acid using a saponification reaction followed by quenching with acid. [*We learned this sequence in Section 19.7.1.*] By heating the resulting β-keto acid, the thermal decarboxylation reaction will occur to give us the desired target molecule, cyclohexanone (**Figure 20.48**). ■

Figure 20.48 Removal of the ester by saponification, acidification, and thermal decarboxylation generates cyclohexanone.

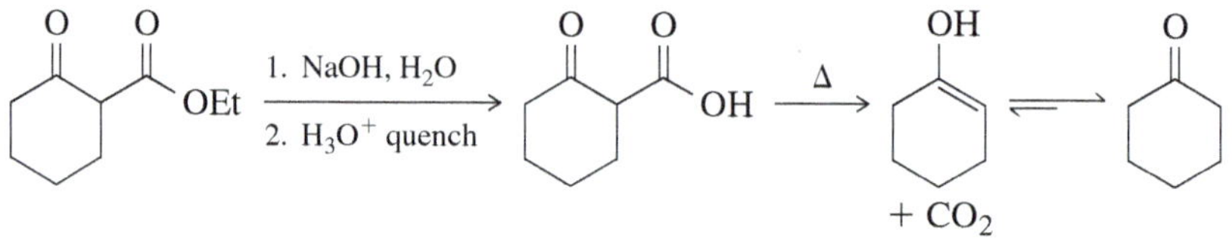

Assessment 20.57 Show how diethyl pimelate, a common Dieckmann starting material, might be synthesized from the following starting materials.

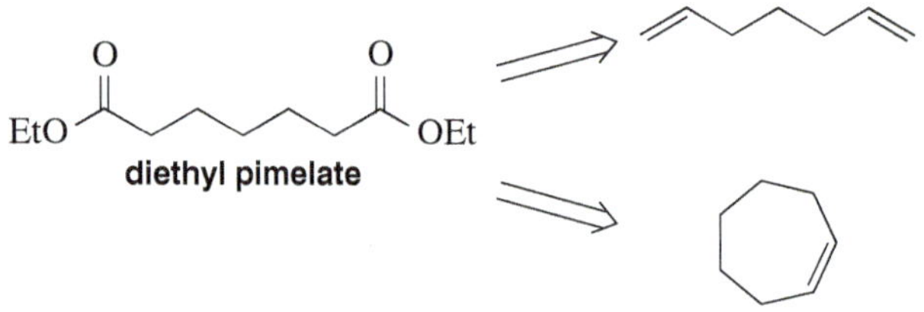

Assessment 20.58 (a) Would you expect ΔG to be greater than or less than 0 in the following thermal decarboxylation reaction? (b) Why does this reaction always proceed to completion?

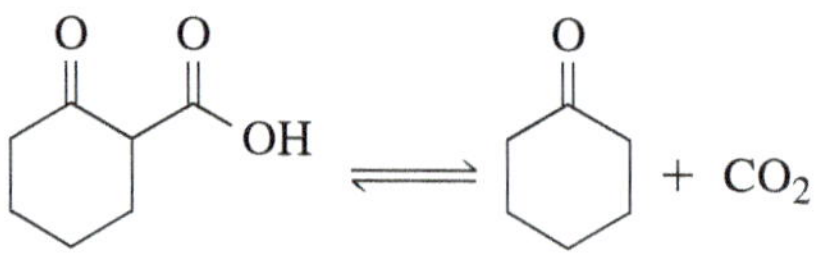

Assessment 20.59 Esters can be directly decarboxylated using the Krapcho reaction with diesters and β-keto esters. Suggest a mechanism for this reaction. [*Number the carbons.*]

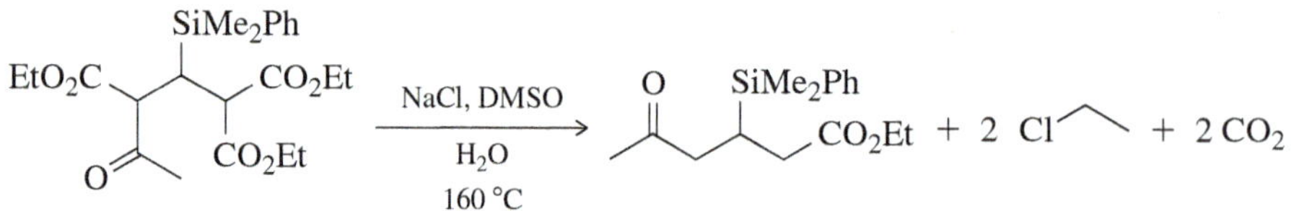

Strategy for cyclohexanone synthesis

1. *Enolates to make ketones: Claisen*
2. *Intramolecular Claisen: Dieckmann*
3. *Removing the ester (TLaC)*
4. ***Adding additional alkyl groups***
5. *Adding additional acyl groups*

20.5 Alkylation of activated methylenes

Continuing with our cyclohexanone synthesis, we can now attempt to add an alkyl substituent. The idea is to form the enolate of cyclohexanone and react it with an alkyl halide in an S_N2 reaction. Using modern conditions, the enolate could be formed using a full equivalent of LDA to completely deprotonate cyclohexanone [**Figure 20.49(a)**]. Unfortunately, classical chemists didn't have access to strong bases and would have unsuccessfully attempted this reaction using a base like sodium methoxide [**Figure 20.49(b)**]. [*Any ideas why it failed?*]

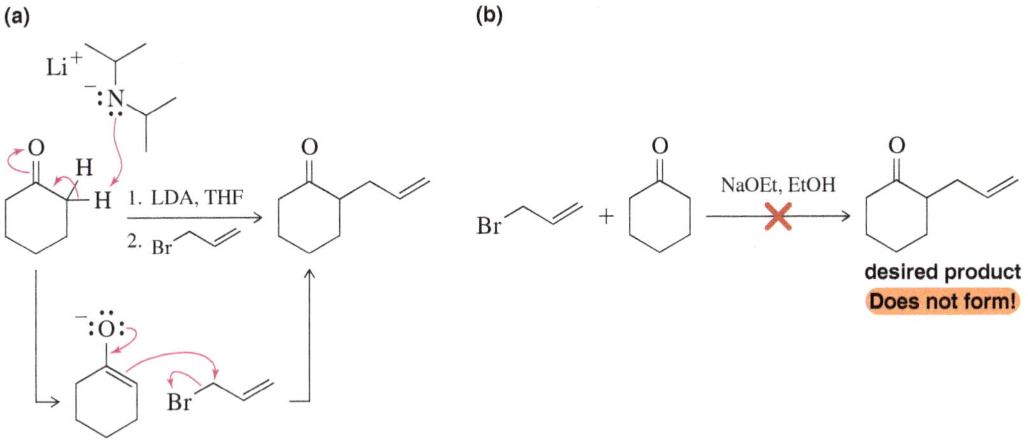

Figure 20.49 Alkylation of an enolate. (a) With strong bases under modern conditions, the reaction is effective. (b) Using the classical, weaker bases, the reaction fails.

The reaction in Figure 20.49(b) fails because the base is too weak. Ethoxide is unable to favorably deprotonate the cyclohexanone, so there is mostly ethoxide present at equilibrium (**Figure 20.50**). Ethoxide, as we learned in Chapter 13, is a great S_N2 nucleophile that reacts with alkyl halides to make ethers [*recall the Williamson ether synthesis*].

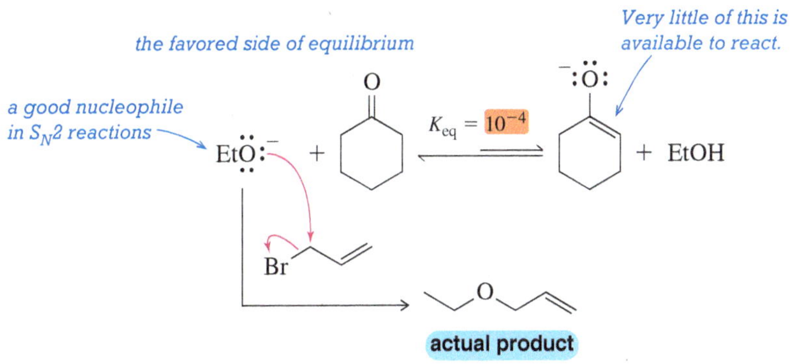

Figure 20.50 Ethoxide is not strong enough to remove the α-hydrogen, so it does an S_N2 reaction instead.

While modern chemists solved this problem by inventing stronger bases, classical chemists did the opposite—that is, they made the proton to be removed more acidic. [*An acid–base reaction can be made more favorable by strengthening either the base or the acid.*] For an acid to be stronger, its conjugate base must be more stable (i.e., weaker). Drawing lessons from the last step of the Claisen and Dieckmann reactions [*Section 20.4*], an ester at the α-carbon is used to stabilize the conjugate base by allowing greater resonance stabilization (**Figure 20.51**). The presence of the ester lowers the pK_a of the α-proton from 20 to 10. As the predominant nucleophile in solution, alkylation proceeds efficiently via an S_N2 reaction.

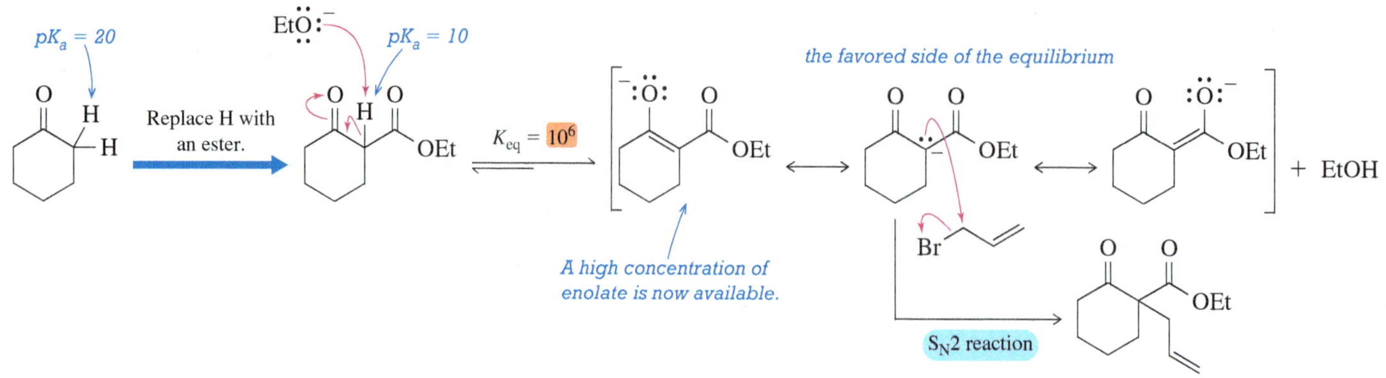

Figure 20.51 The presence of an ester at the α-carbon makes the proton more acidic. Ethoxide, a classical base, is now strong enough.

There are two advantages to this approach. First, the β-keto ester in Figure 20.51 is the product of the Dieckmann condensation, so it would have been readily available for classical chemists. Second, because deprotonation is faster than nucleophilic attack and is heavily favored, there is no chance for ethoxide to do the S_N2 reaction with the haloalkane. The only nucleophile left to attack after the deprotonation is complete is the enolate itself. Because the ultimate goal was to make the alkyl-substituted cyclohexanone, all that remains is to saponify, acidify, and thermally decarboxylate the resulting β-ketoacid to give 2-allylcyclohexanone (**Figure 20.52**). [*Recall Section 20.4.2.*]

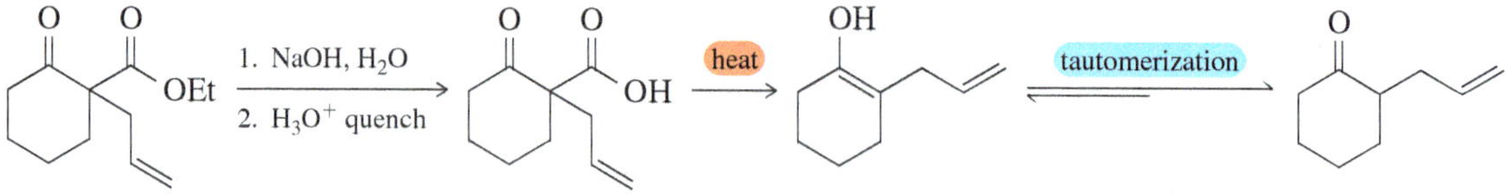

Figure 20.52 Following alkylation, saponification, acidification, and thermal decarboxylation, the α-substituted ketone is obtained.

20.5 SUMMARY

- Alkylation of β-keto esters is followed by removal of the ester group to generate alkylated ketones.

Assessment 20.60 Predict the product of the following reactions. In part (b), NaH is simply a more powerful base that can be employed when you need one that cannot act as a nucleophile.

(a)

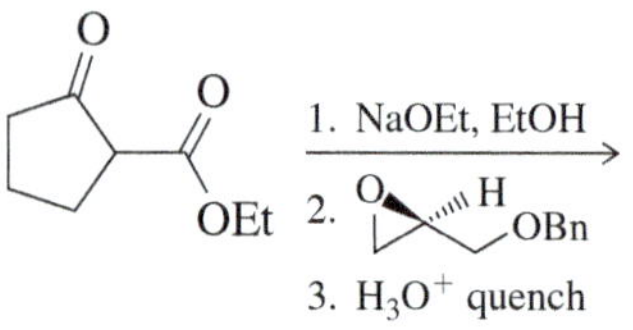

(b)

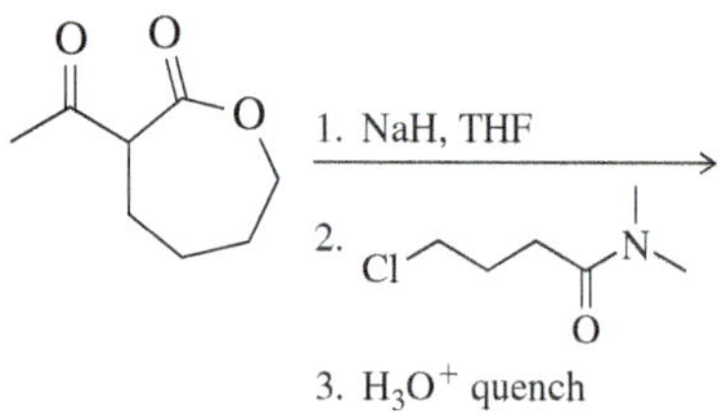

Assessment 20.61 Using the sequence discussed in the preceding section, show how you might synthesize the following substituted cyclopentanone starting with hexane-1,6-diol.

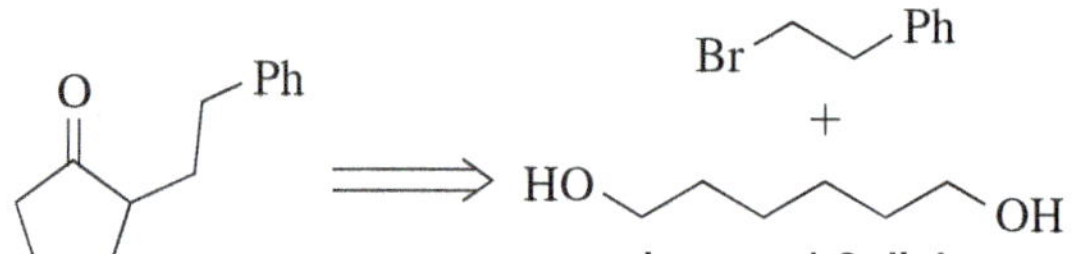

hexane-1,6-diol

20.5.1 Acetoacetic and malonic ester synthesis

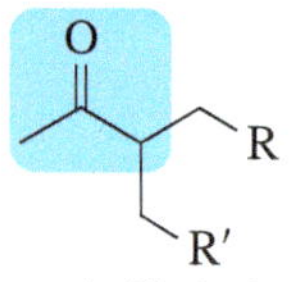

substituted acetone derivative

The strategy used for cyclohexanone synthesis is also used to form substituted acetone derivatives. Because acetone is the simplest ketone, chemists have long been concerned with ways of derivatizing it using enolate chemistry. As before, the use of a classical base, like ethoxide, fails because it isn't strong enough to produce a meaningful amount of enolate. Using ethyl acetoacetate—essentially acetone with a removable, electron-withdrawing ester group—makes the proton more acidic, rendering ethoxide sufficiently basic to completely deprotonate the α-carbon [**Figure 20.53(a)**].

This strategy, called the **acetoacetic ester synthesis**, is especially convenient for the synthesis of acetone derivatives because ethyl acetoacetate is readily available as the product of the Claisen condensation of ethyl acetate [**Figure 20.53(b)**].

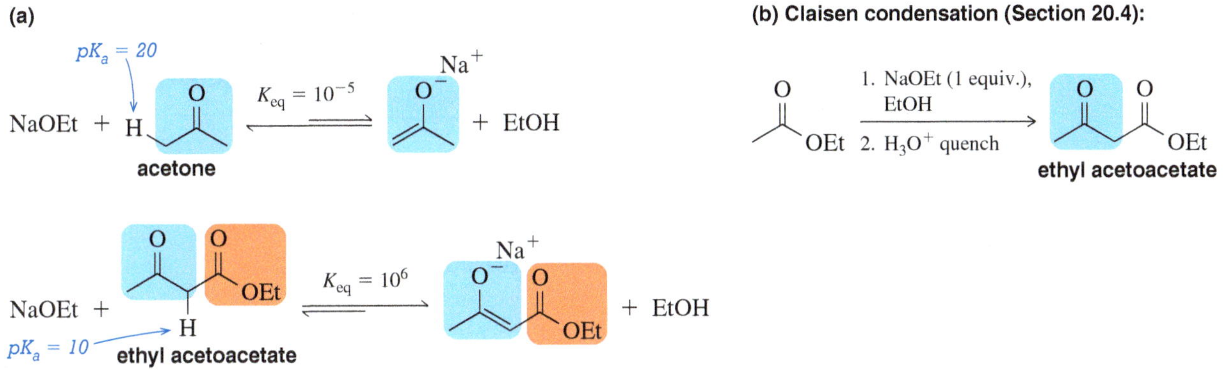

Figure 20.53 While acetone isn't acidic enough, ethyl acetoacetate is. (a) Enolate formation is heavily favored. (b) Ethyl acetoacetate is synthesized by the Claisen condensation.

Let's think about how we might construct a molecule using the acetoacetic ester synthesis by looking at it retrosynthetically (**Figure 20.54**). The first goal is to find the acetone with one or two groups [*never three*] coming off the α-carbon. With acetone identified, we can imagine the presence of the ester to facilitate enolate formation, giving us ethyl acetoacetate. Our final disconnection reveals the two alkyl halides that will be used to alkylate the enolate of ethyl acetoacetate. There are two acidic hydrogens on ethyl acetoacetate, so it is possible to add two alkyl groups. [*Numbering carbons helps with this process.*]

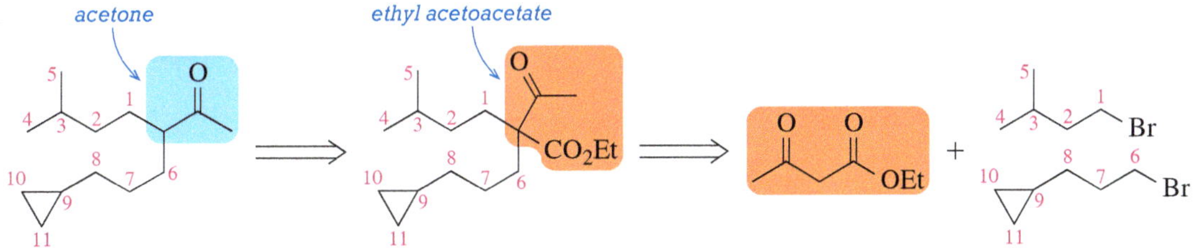

Figure 20.54 Retrosynthetic analysis of the acetoacetic ester synthesis.

Assessment 20.62 Why can't you add a third alkyl group using this strategy?

In the forward direction (**Figure 20.55**), this process is similar to the synthesis of α-substituted cyclohexanones [*Check out Section 20.5*]. Deprotonation of ethyl acetoacetate with ethoxide gives an enolate that is alkylated in an S_N2 reaction [*recall Section 12.2.1*]. The process is repeated with the other alkyl halide to give the disubstituted ethyl acetoacetate. Having served its purpose, the ester is removed via the three-step process of saponification [*Section 19.7.1*], acidification, and thermal decarboxylation [*Section 18.8.4*].

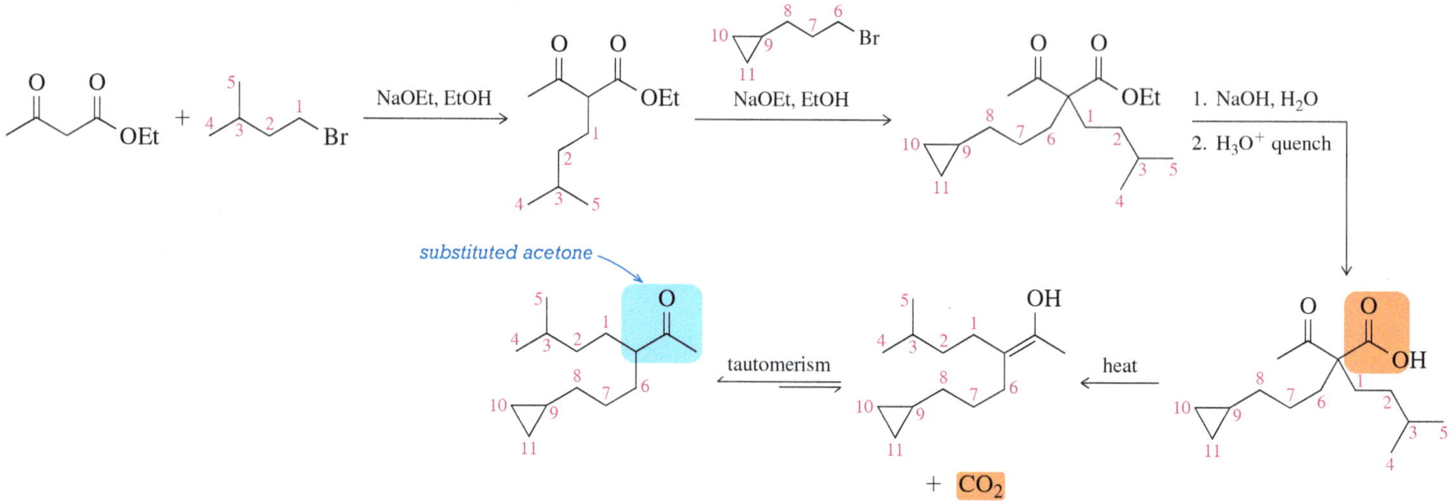

Figure 20.55 Acetoacetic ester synthesis to prepare a disubstituted acetone derivative.

Assessment 20.63 Provide a mechanism for each step shown in Figure 20.55.

Assessment 20.64 It's not a coincidence that only primary alkyl halides have been shown as substrates for the acetoacetic ester synthesis. Why might secondary alkyl halides be less efficient in this process?

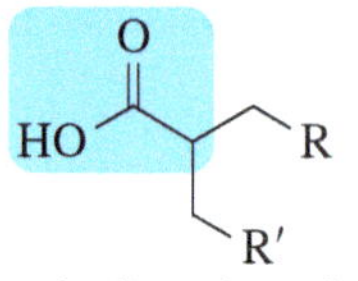

substituted acetic acid derivative

Let's think like a chemist and see how the acetoacetic ester synthesis can be adapted to the synthesis of substituted acetic acids.

In addition to making substituted acetones, the synthesis of substituted acetic acids is also a valuable reaction.

Thinking Like a Chemist

Retrosynthetic analysis starts by identifying the acetic acid (**Figure 20.56**). In a manner analogous to that in Figure 20.55, an attached ester would make it possible to add two haloalkanes via a stabilized enolate. A final disconnection yields a haloalkane and a bifunctional molecule containing a carboxylic acid and an ester.

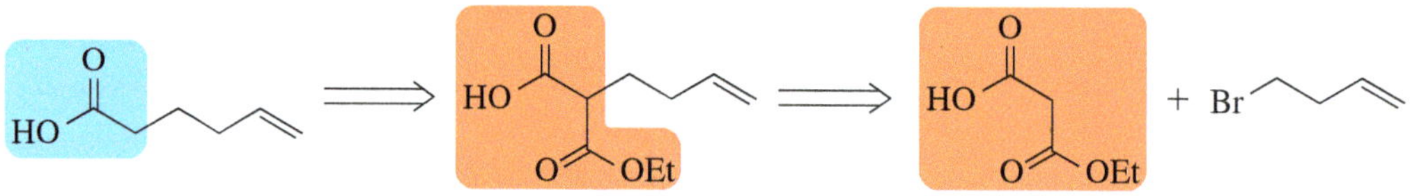

Figure 20.56 Retrosynthetic analysis to make a substituted acetic acid derivative.

In the forward direction (**Figure 20.57**), we encounter a problem with our choice of starting material. Attempted deprotonation with ethoxide would exclusively form the carboxylate instead of the enolate. Luckily, the carboxylic acid can be esterified to give a diester with an acidic central carbon ($pK_a = 10$). This molecule, called diethyl malonate, is deprotonated to favorably form an enolate that is alkylated in an S_N2 reaction. Our choice of starting material is suitable, then, because both esters can be saponified in a single step. [*This is an example of an initial plan not working out, but leading to the correct solution—it happens all the time.*] Acidification gives us a β-diacid, a compound that undergoes thermal decarboxylation to give the substituted acetic acid derivative.

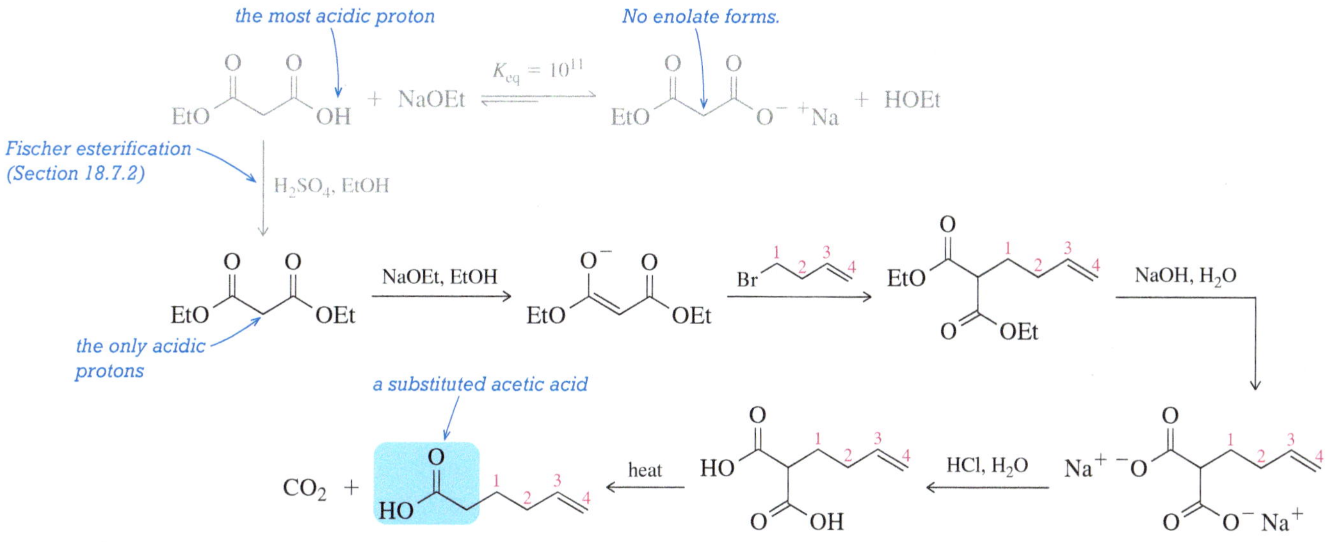

Figure 20.57 The malonic ester synthesis prepares a substituted acetic acid.

The starting material for this sequence is also readily available as the Fischer esterification product of malonic acid (**Figure 20.58**). As a result, when the scheme presented in Figure 20.57 is used for the synthesis of acetic acid derivatives, it is referred to as the **malonic ester synthesis.** ■

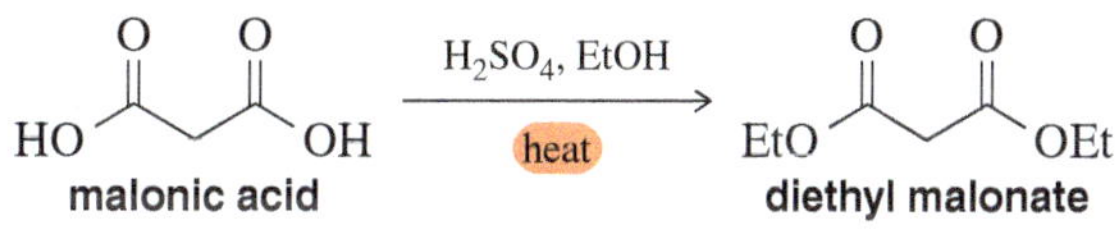

Figure 20.58 Fischer esterification of malonic acid gives diethyl malonate, the starting material for the malonic ester synthesis.

Assessment 20.65 Show how to synthesize the following molecules using the acetoacetic ester or malonic ester synthesis. [*In part (d), Bn is a benzyl group, $PhCH_2$—.*]

(a) (b) (c) (d)

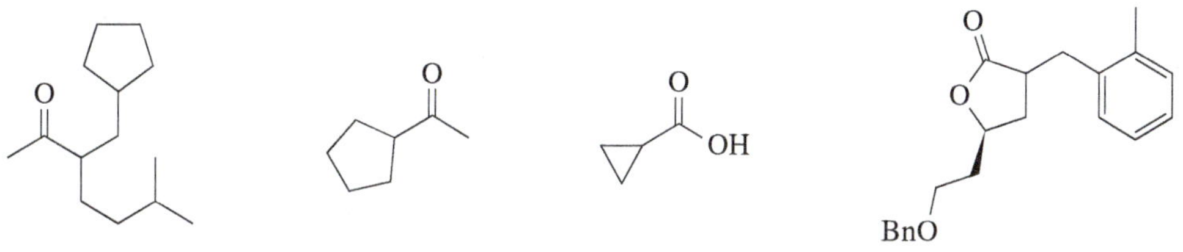

Assessment 20.66 Acylation using either the acetoacetic ester or malonic ester synthesis is *not* effective. Why? [*Think about what's in solution and consider the reversibility of the reaction.*]

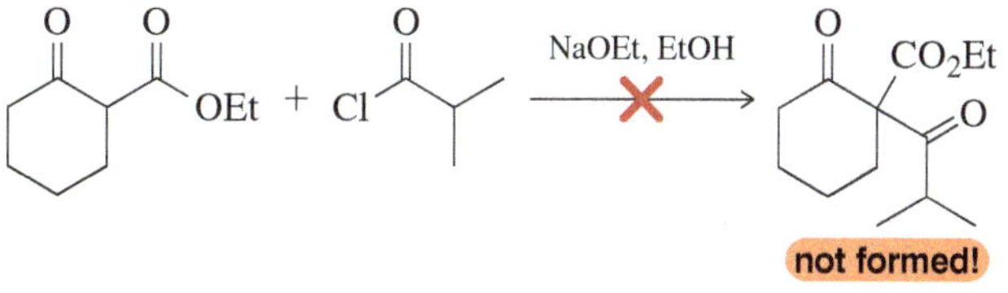

Assessment 20.66 shows that it isn't possible to add an acyl group by forming a stabilized enolate. Let's work together to figure out why. In doing so, we'll be a step closer to figuring how to make compounds like the one in Assessment 20.66.

20.6 Enamines

We learned how to make enamines in Chapter 17. Finally, we learn their importance to organic synthesis.

Solved Assessment 20.66

Because ethoxide is a strong enough base to form a stabilized enolate and because acid chlorides are reactive electrophiles, **Figure 20.59** suggests a mechanism by which the reaction could work. Reactions, however, are random collisions, and recall from Section 19.7.2 that alkoxides and acid chlorides react quickly to form esters. This is one reason why the acylation is difficult. There is a second reason, too, but it is a bit more subtle. If a tricarbonyl product did form, ethoxide, which is still present in solution, could attack the newly attached ketone. If that occurs, we'd end up with a tetrahedral intermediate, which could expel the ethoxide that just added or the enolate of the original β-keto ester. On the basis of pK_a values, we expect the enolate to be the better leaving group. For this reason, tricarbonyl compounds are unstable under these conditions. ■

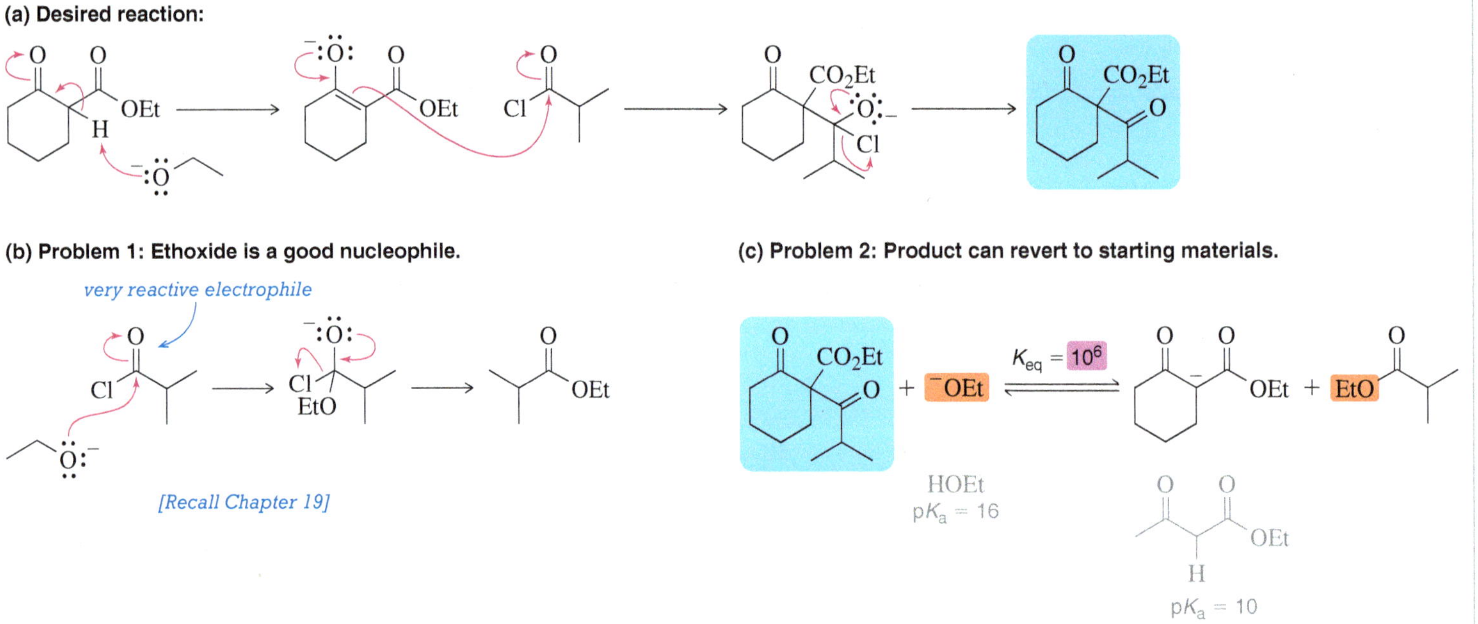

Figure 20.59 **Acylation of stabilized enolates is unsuccessful for two reasons.** (a) Desired reaction. (b) Ethoxide can act as a nucleophile instead of a base. (c) The tricarbonyl product and ethoxide are less stable than the enolate and ethyl ester.

Strategy for cyclohexanone synthesis

1. *Enolates to make ketones: Claisen*
2. *Intramolecular Claisen: Dieckmann*
3. *Removing the ester (TLaC)*
4. *Adding additional alkyl groups*
5. ***Adding additional acyl groups***

A solution to this problem is based on the similarity between enolates, enols, and **enamines**, compounds we made in Section 17.8.3. **Figure 20.60** shows that each contains a lone pair of electrons on an electronegative atom attached to an alkene. A resonance structure can be drawn where each has a negative charge on the terminal carbon. Thus, if enols and enolates are nucleophiles, enamines should be as well. In fact, the reactivity of enamines is intermediate between enols and enolates, owing to the reduced electronegativity of nitrogen. Because nitrogen is a good electron-donating group, enamines are nucleophilic enough to attack acid chlorides, alkyl halides, and other good electrophiles.

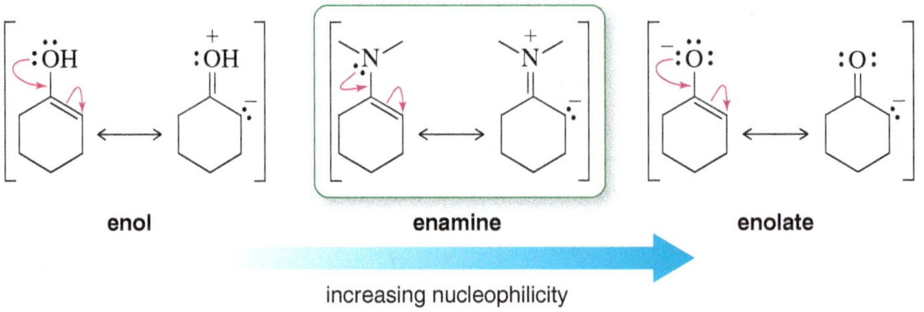

Figure 20.60 **The relative reactivity of nucleophilic species studied in this chapter.**

Figure 20.61 shows an example of the **Stork enamine synthesis**, which can be used for synthesis of the desired cyclohexanone derivative. [*This also represents a general procedure for adding acyl groups to any ketone or aldehyde.*] In Chapter 17, we learned that enamines are synthesized by the reaction of a secondary amine with a ketone. Unlike enolates and enols, enamines are relatively stable molecules that can be isolated, purified, and even stored for a brief time. In this way, the synthesis of the enamine essentially shifts the equilibrium completely to the side of the "enolate" (in the form of the enamine) even in the absence of the electrophile. Upon mixing the enamine and the acid chloride, the enamine attacks the carbon of the acid chloride to give a tetrahedral intermediate. [*See Section 19.6.*] The tetrahedral intermediate collapses, kicking out chloride (a good leaving group). The resulting iminium is then hydrolyzed by the addition of water to give a ketone. [*The hydrolysis mechanism is the opposite of the imine formation mechanism learned in Section 17.8.1. Also see Figure 17.77.*]

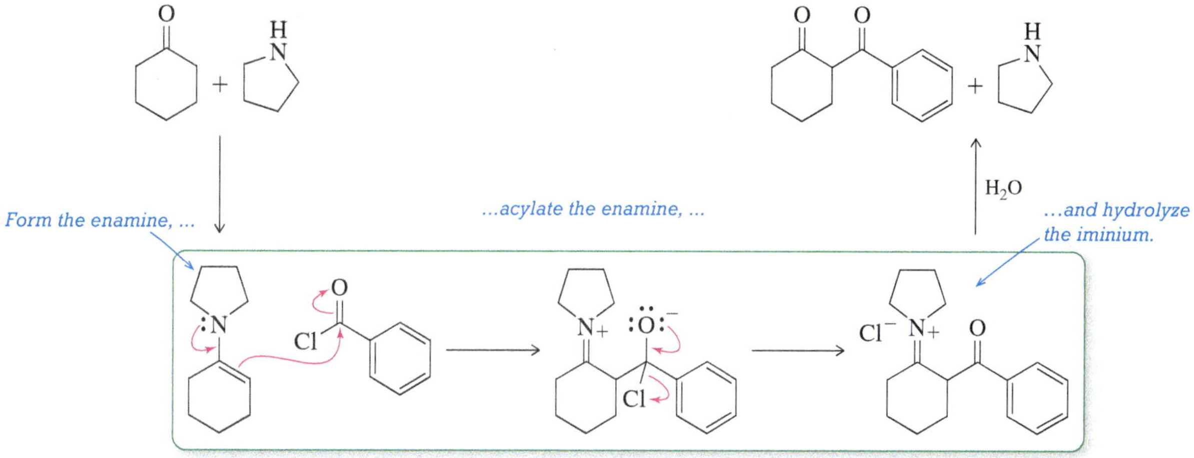

Figure 20.61 The acylation of a ketone in the Stork enamine synthesis.

This concludes the pages dedicated to the synthesis of cyclohexanone. The reactions introduced here are general, so they can be utilized to make different ring sizes as well as open-chain compounds.

The Stork enamine synthesis can also be used with haloalkanes to give α-alkyl substituted ketones, as shown in **Figure 20.62**.

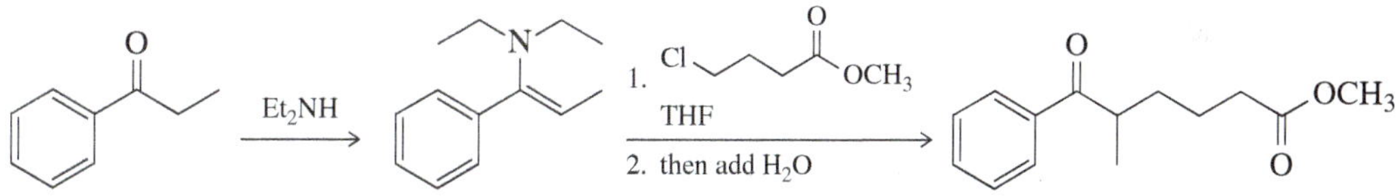

Figure 20.62 The alkylation of a ketone using the Stork enamine synthesis.

20.6 SUMMARY

- The Stork enamine synthesis, named in honor of Gilbert Stork, is used to acylate or alkylate the α-carbon of ketones and aldehydes.

Assessment 20.67 Show how the Stork enamine synthesis could be used to synthesize the following molecules.

(a)

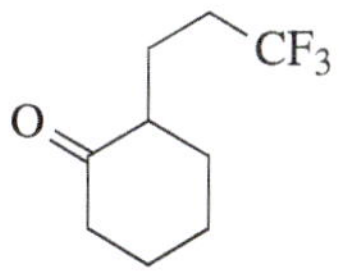

(b)

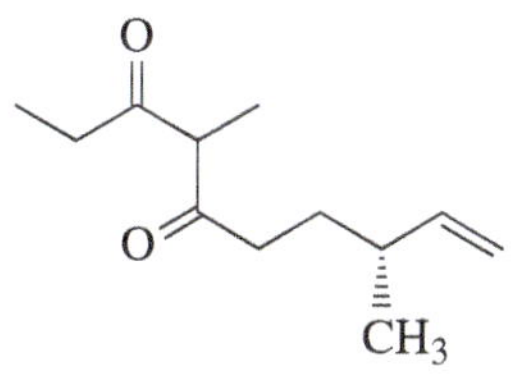

Assessment 20.68 Propose a synthesis of the following molecule starting from ethyl 3-methylbutanoate and using a Claisen reaction and an enamine reaction. [*Numbering carbons will help you here.*]

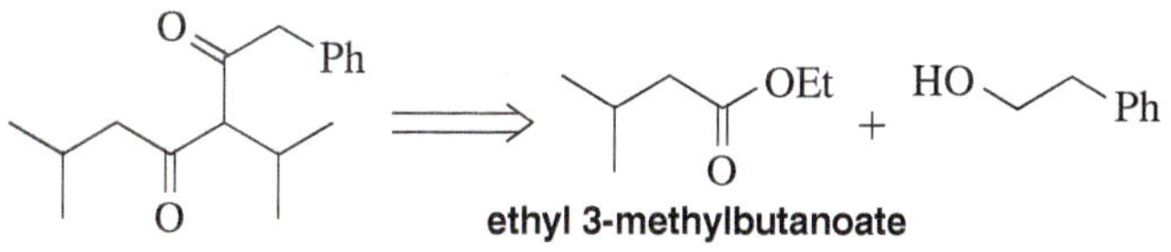

20-E CHEMISTRY IN CONTEXT: Green Chemistry

Organocatalysis

The Stork enamine synthesis is not very efficient because the enamine is usually isolated before reacting with the alkyl halide. Furthermore, the atom economy is usually pretty poor, especially when a small molecule is being made. Figure 20.63 shows a problem with evaluating reactions on the basis of percent atom economy when different sized molecules are being made. In this instance, the same reaction varies from 44 to 64% atom economy. As a result, it is more important to evaluate individual processes rather than generalize about the reaction itself.

One approach to improving this reaction is illustrated in the asymmetric organocatalytic aldol reaction shown in Figure 20.64. Not only does it employ a catalytic amount of a readily available amino acid, but in the process, it produces a molecule that is of high enantiomeric purity. Moreover, the reaction has 100% atom economy if you ignore the catalyst.

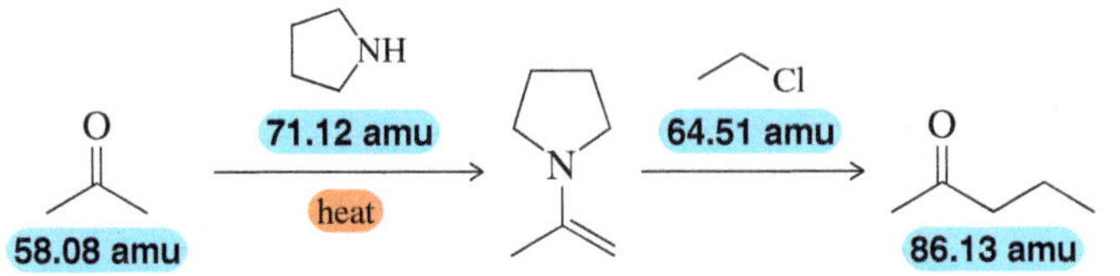

atom economy = [86.13/(58.08 + 71.12 + 64.51)] × 100 = **44%**

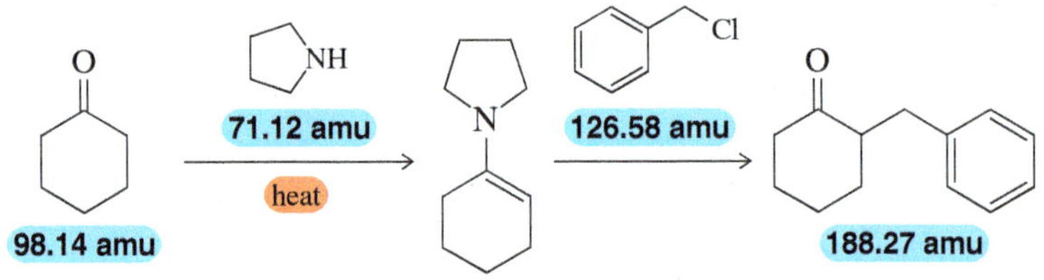

atom economy = [188.27/(98.14 + 71.12 + 126.58)] × 100 = **64%**

Figure 20.63 The atom economy for the Stork enamine synthesis varies based on the size of the molecule being synthesized.

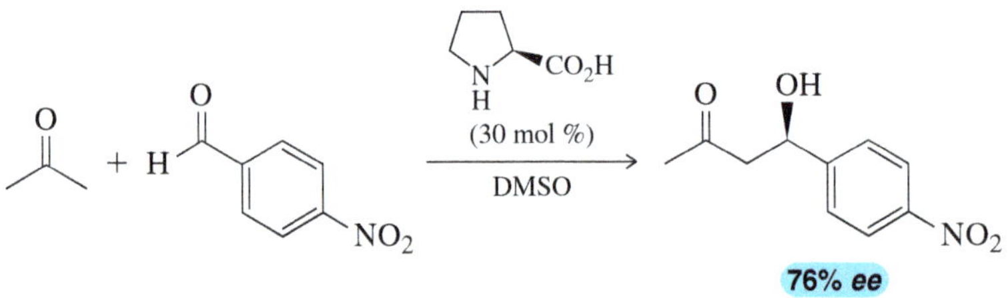

Figure 20.64 Proline-catalyzed aldol reaction.

Assessment 20.69 Provide a mechanism for the proline-catalyzed aldol reaction in Figure 20.64. [*Hint: Use the amino group in proline to make an enamine, then regenerate the proline because it is a catalyst.*]

20-F CHEMISTRY IN CONTEXT: Biochemistry

Claisen/aldol reactions

Recall that enolates are used in the synthesis of sucrose (Chemistry in Context 20-B) and mevalonate (Chemistry in Context 20-D). A third reaction of enolates important in biological systems is the reaction of the enolate of an ester with a ketone. One of its more important uses is in the citric acid cycle, which oxidizes acetyl-CoA to generate energy in the form of NADH and $FADH_2$. [*You'll study this in biochemistry.*] The cycle begins with the reaction of acetyl-CoA with oxaloacetate catalyzed by citrate synthase to produce citrate, as shown in Figure 20.65. Once generated, citrate is immediately dehydrated by aconitase, where it continues in the citric acid cycle.

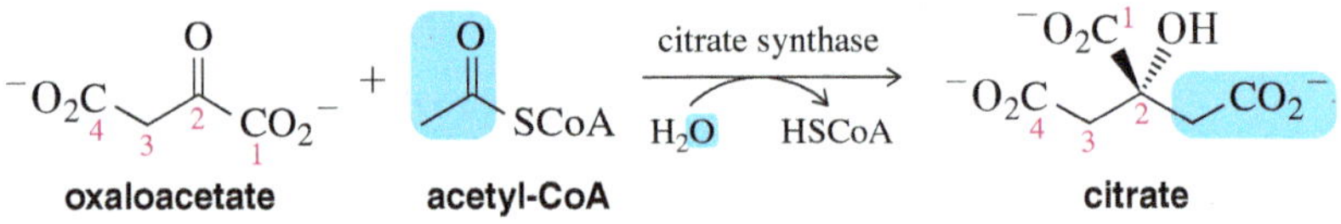

Figure 20.65 Synthesis of citrate by an enolate reaction.

Assessment 20.70 The process catalyzed by citrate synthase really consists of two steps. The first is the aldol-like process, and the second is the hydrolysis of the thioester to generate citrate. Show an arrow-pushing mechanism for these two reactions. As we have mentioned previously, enzymes typically have active site bases (B^-) and acids (B—H) present whenever they need them, so use these as necessary to produce your mechanism.

20.7 Special topic: reactions of enolates in the synthesis of atorvastatin

Before wrapping up this chapter, let's look back at the problem we outlined in the beginning—that is, the synthesis of a molecule that might help lower cholesterol levels, thereby mitigating the risk of heart disease. To fully understand what we are discussing here, you should review the material covered in Chemistry in Context 20-A, 20-B, and 20-C. This discussion is centered on the three reactions shown in **Figure 20.66**, two of which involve reactions of enolates. The first is the thiolase-catalyzed Claisen reaction of acetyl-CoA to give acetoacetyl-CoA. The second is the HMG-CoA synthase–catalyzed enolate addition of a third molecule of acetyl-CoA to give HMG-CoA. The third reaction is the reduction of HMG-CoA by HMG-CoA reductase.

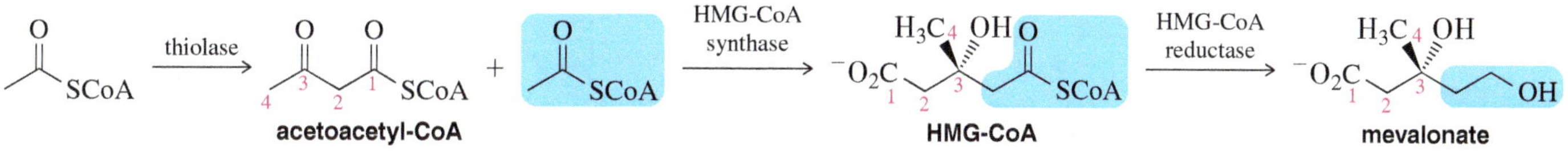

Figure 20.66 Important early reactions toward the biosynthesis of cholesterol.

Because mevalonate is used by the body for just one purpose, inhibiting its synthesis interferes only with cholesterol biosynthesis. Thus, the ideal enzyme to inhibit is HMG-CoA reductase. Careful analysis of the HMG-CoA reductase reaction in **Figure 20.67** reveals a reaction that is similar to aldehyde and ester hydride reductions. [*See Sections 17.9.2 and 19.8.1.*] The reaction takes place at the active site of an enzyme and utilizes cofactors instead of hydride reducing agents, but familiar intermediates are involved. Reduction of the thioester first produces a tetrahedral intermediate, while the second intermediate is the aldehyde that is further reduced to mevalonate.

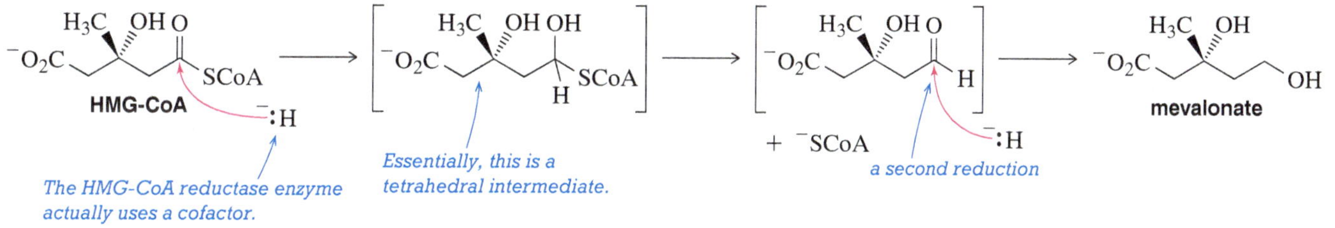

Figure 20.67 Intermediates involved in the HMG-CoA reductase–catalyzed formation of mevalonate, a cholesterol precursor. [*The hydride ion is actually donated by a cofactor in the active site of the enzyme.*]

HMG-CoA reductase has an active site that binds tightly to HMG-CoA, mevalonate, and the two intermediates in the scheme in Figure 20.67. Typically, enzyme inhibitors are molecules that mimic the substrate, the product, the intermediates, and/or the transition states of enzymatic reactions. Similar in structure to the HMG-CoA reductase intermediate, atorvastatin (**Figure 20.68**) is an outstanding inhibitor of HMG-CoA reductase.

Figure 20.68 Atorvastatin is an excellent inhibitor of HMG-CoA reductase. [*Note the similarities between the fragments screened in blue.*]

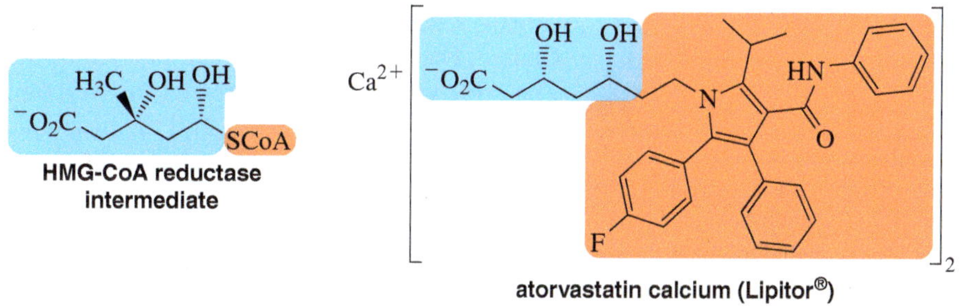

The synthesis of atorvastatin and other members of the statin family utilizes aldol and related reactions. One of the early syntheses of atorvastatin at Pfizer (formerly Parke-Davis) utilized the scheme shown in **Figure 20.69**. [*Several details have been omitted for simplification.*] Ignoring the stereochemical outcome, your familiarity with enolate reactions should allow you to follow along with the synthesis. Simply focus on the functional groups that are reacting and ignore those that are just along for the ride. [*You can demonstrate your understanding in Assessment 20.71.*]

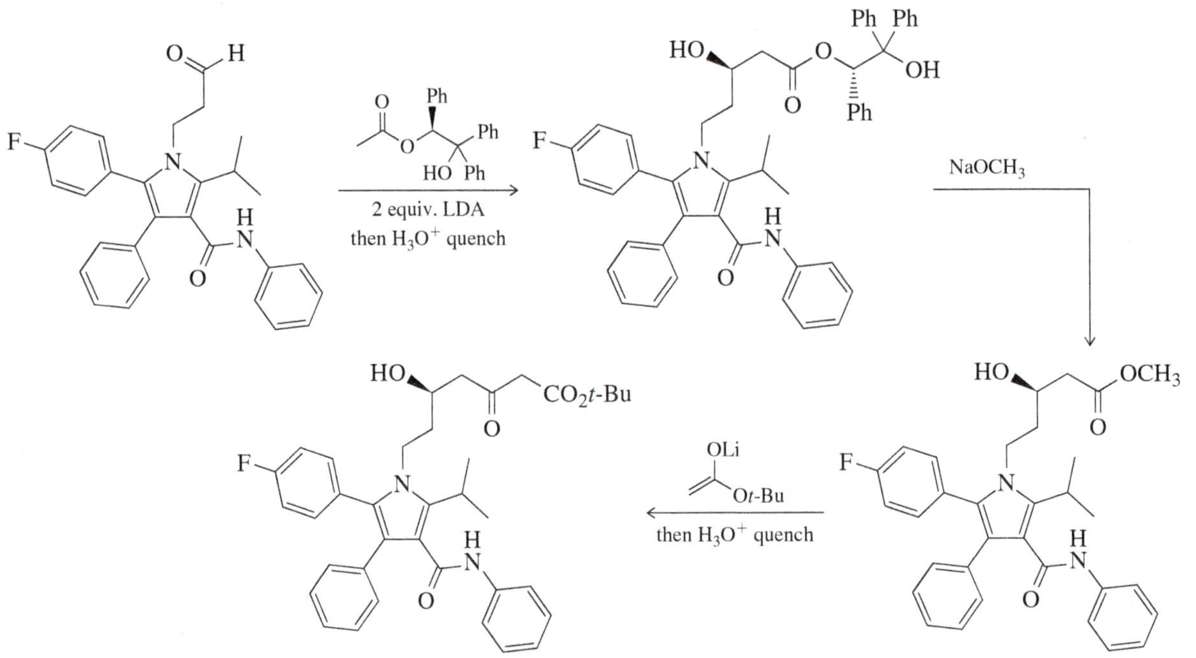

Figure 20.69 The synthesis of atorvastatin involved an aldol and Claisen reaction.

Atorvastatin and the other statin drugs have proven very successful in lowering cholesterol levels and greatly diminishing the incidence of cholesterol-related heart disease. While the effectiveness of statins in preventing first cardiac events is subject to debate, their effectiveness in preventing cardiac events in patients with pre-existing cardiovascular disease is certain. Taken as prescribed, statins have been claimed to lower LDL cholesterol levels by about 30% and have been estimated (in an average of studies) to reduce the risk of cardiac events by about 60% and the incidence of stroke by 17% after long-term treatment.

Assessment 20.71 Provide an arrow-pushing mechanism for the three reactions shown in Figure 20.69 for the atorvastatin synthesis scheme.

Concluding thoughts

There are few reactions that have impacted the field of organic synthesis as significantly as enolate reactions. And yet it was the ingenuity of the classical chemists (Wurtz, Borodin, Claisen, Dieckmann, and others) that led to the reactions being utilized as broadly as they are. The creativity they demonstrated in overcoming limitations should inspire even the novice organic chemist. Their work, and the mechanistic insights it provided, significantly influenced those who later played vital roles in developing and extending the modern version of these reactions. More importantly to you as an organic chemistry student, the process by which they approached these limitations and overcame the obstacles they faced provides insight into how organic chemists solve problems. It teaches us how to solve problems and inspires us to do the same, regardless of the career where we ultimately apply those skills.

For now, let's apply these problem-solving skills to the end-of-chapter assessments and, soon thereafter, to Chapter 21.

Important Concepts

[*These are the major concepts covered in each section, excluding reactions.*]

20.2: Enols, tautomers of ketones and aldehydes, react with strong electrophiles to give substituted carbonyls in an entropically neutral reaction ($\Delta S \approx 0$).

20.2.1: Through equilibration to enols, the α-hydrogen of a carbonyl can be replaced with deuterium, bromine, or a hydroxyl group.

20.2.2: In the acid-catalyzed aldol reaction, we see that, because of Le Châtelier's principle, a reaction consisting of a series of slightly unfavorable steps can be driven to completion by a single favorable step. Reactions like this are said to be under thermodynamic control—that is, the reaction outcome is determined by the stability of the products.

20.3: The proton α to a carbonyl is somewhat acidic ($pK_a = 20$) and can be removed by strong bases such as LDA. The resulting enolates are stronger nucleophiles than the corresponding enols and can therefore participate in S_N2 reactions. As explained by hard and soft Lewis acid and base (HSAB) theory, enolates preferentially react at the α-carbon.

20.3.1: LDA is a strong, bulky base. It completely (and kinetically) deprotonates a ketone to form an enolate. This allows the chemist to choose which carbonyl will serve as the enolate and which will serve as the electrophile in the modern aldol reaction.

20.3.2: Under classical conditions, the use of sodium hydroxide as base prevents complete formation of enolates. As a result, the classical aldol reaction is under thermodynamic control. Once again, Le Châtelier's principle drives the reaction to completion.

20.3.3: The classical conditions make it challenging to achieve a selective aldol reaction between two different carbonyls. By making an enolate of a ketone (not a good electrophile in the aldol reaction) and reacting it with an aldehyde with no α-hydrogens (can't become an enolate), a selective crossed aldol reaction is possible.

20.3.4: Because it does not have a negative entropy term ($\Delta S \approx 0$), the intramolecular aldol is a much more efficient process, with ketones also serving as electrophiles in this process. The ring size formed, if there are multiple possibilities, will be determined by which is most stable, because this reaction is still under thermodynamic control. As a general rule, five- and six-membered rings will form most often, because these have the least amount of strain.

20.4: The Claisen condensation is another example of a series of unfavorable steps being driven forward by one very favorable step—in this case, the formation of the stabilized enolate.

20.4.1: To do the crossed Claisen condensation, the experimental conditions are manipulated in such a way to allow only the desired reaction to occur (Figure 20.42).

20.4.2: The intramolecular Claisen condensation, referred to as the Dieckmann condensation, is best suited for the preparation of five-membered rings or larger. With a neutral entropy term, it is more efficient than the intermolecular version. It can be followed by the saponification, acidification, and thermal decarboxylation sequence to produce cycloalkanones.

20.5: If one carbonyl makes an α-hydrogen acidic, then a second electron-withdrawing group will make it more acidic. Activated methylenes can be reacted with moderately strong bases like alkoxides to fully form enolates for use in S_N2 reactions. Used in conjunction with the saponification, acidification, and thermal decarboxylation sequence, substituted carbonyls can now be made.

20.5.1: The acetoacetic ester synthesis is a sequence of reactions involving enolate alkylation, ester saponification, acidification, and thermal decarboxylation. It makes it possible to synthesize substituted acetone derivatives. A similar process (the malonic ester synthesis) can be used to make substituted acetic acid derivatives.

20.6: Secondary amines can be used in reactions with carbonyls to make enamines, which can serve as functional equivalents of enolates in reactions with alkyl halides and acid chlorides.

Important Reactions and Mechanisms

Major Category I: Form an enol, react with an electrophile

1. **Deuterium exchange (Section 20.2.1.1):** Acid-catalyzed keto–enol tautomerization in the presence of D_3O^+ results in the replacement of all α-hydrogens with deuterium atoms.

General:

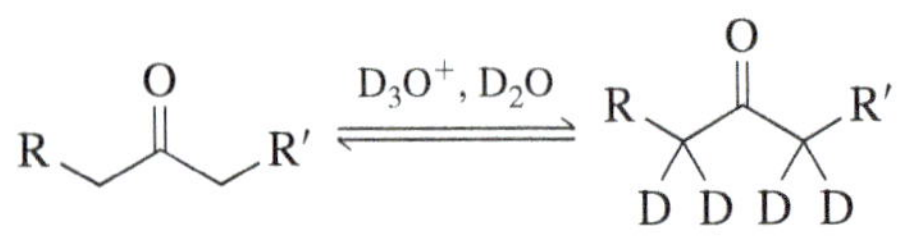

Specific:

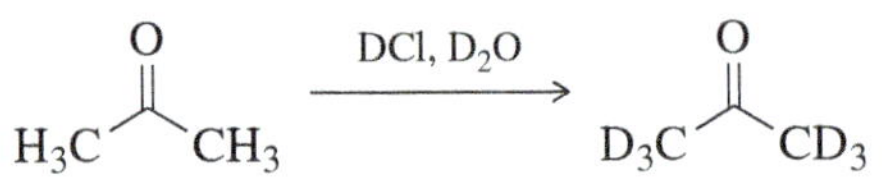

Mechanism:

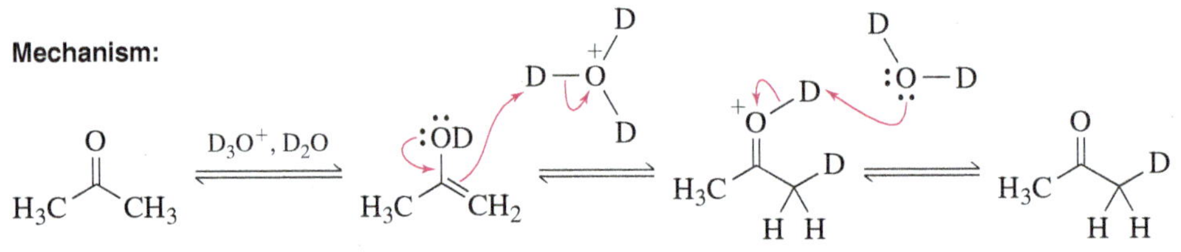

2. **α-Bromination (Section 20.2.1.2):** Acetic acid–catalyzed tautomerization results in an enol that attacks the strongly electrophilic bromine. The reaction can stop at the addition of a single bromine atom.

General:

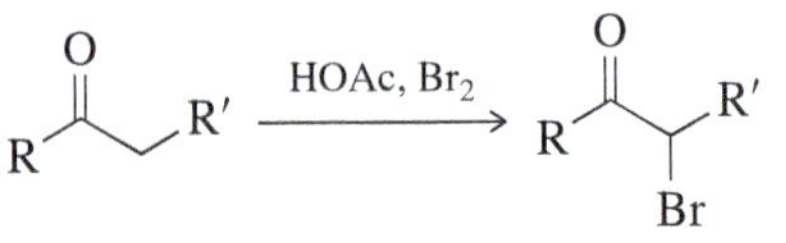

Specific:

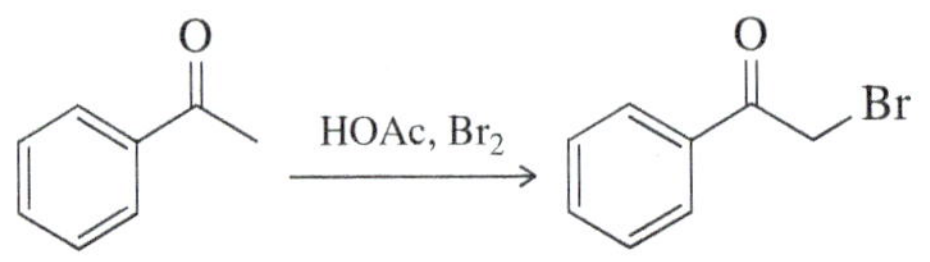

Mechanism:

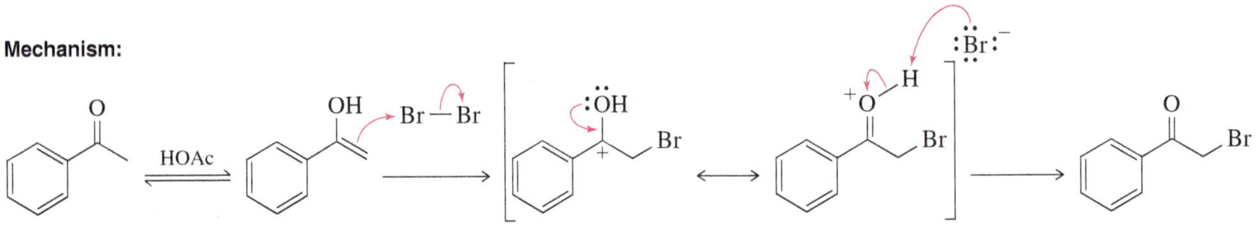

3. **α-Hydroxylation of silyl enol ethers (Section 20.2.1.4):** Assisted by the lone pair on oxygen, the alkene attacks *m*CPBA, kicking out a carboxylate anion and generating an oxonium. The carboxylate attacks the silicon, returning a pair of electrons to the carbonyl oxygen. [*There is evidence that this occurs through an epoxide intermediate.*]

General:

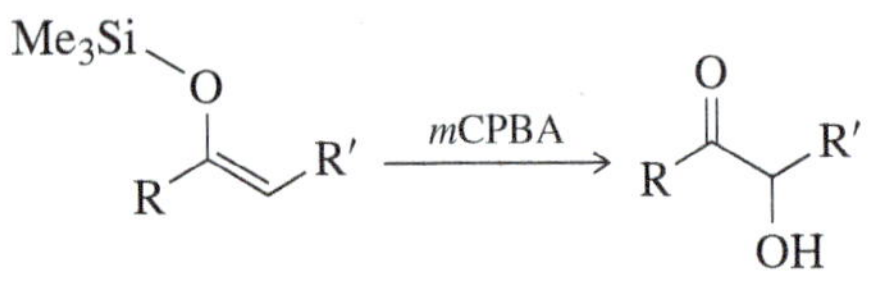

Specific:

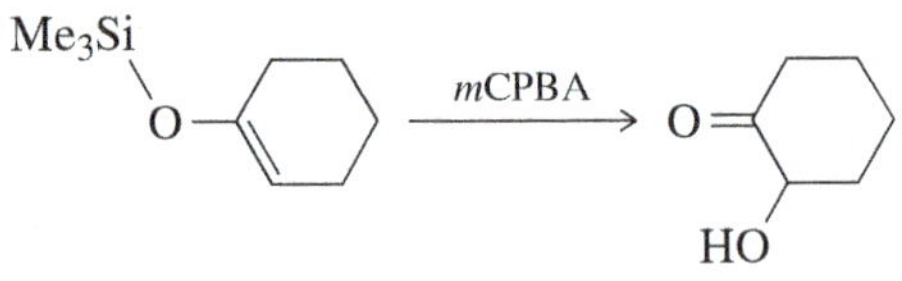

Mechanism:

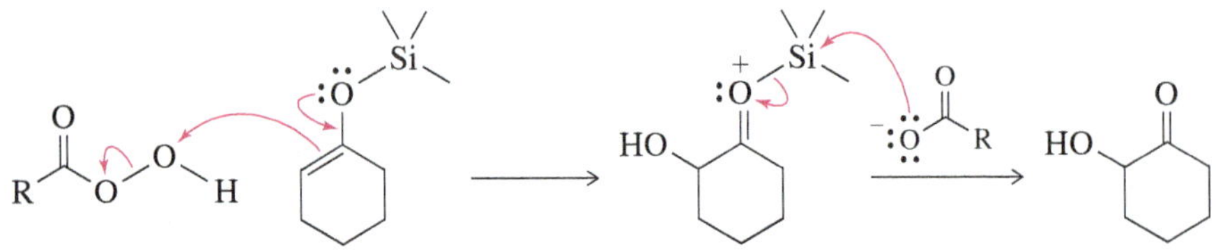

4. Acid-catalyzed aldol reaction (Section 20.2.2): The enol of one aldehyde attacks the electrophilic protonated aldehyde. Removal of a proton results in the β-hydroxyaldehyde. If the reaction is heated, protonation of the hydroxyl group is followed by elimination through the enol to give an α,β-unsaturated carbonyl. This reaction can also be intramolecular.

General:

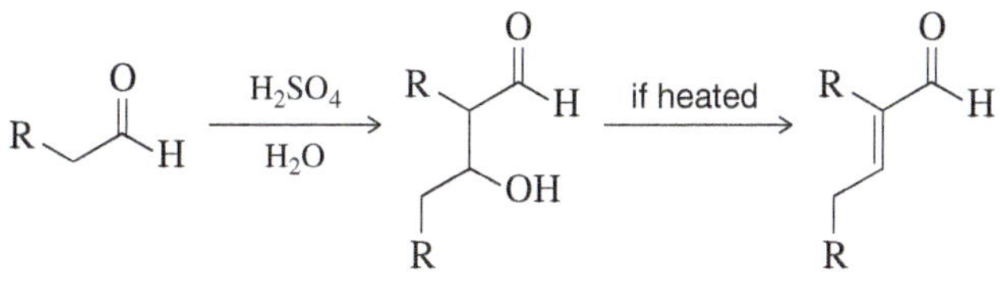

Specific:

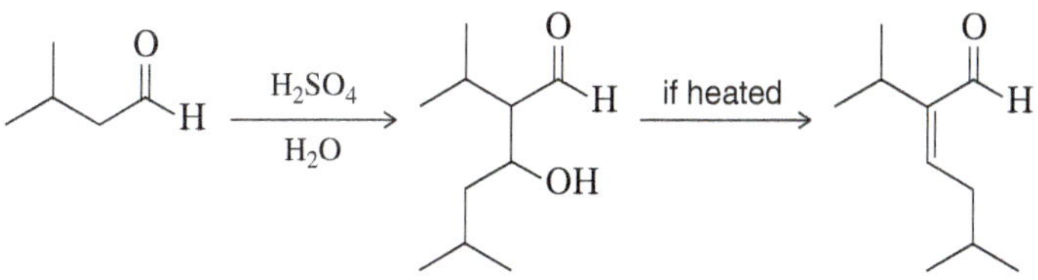

Mechanism:

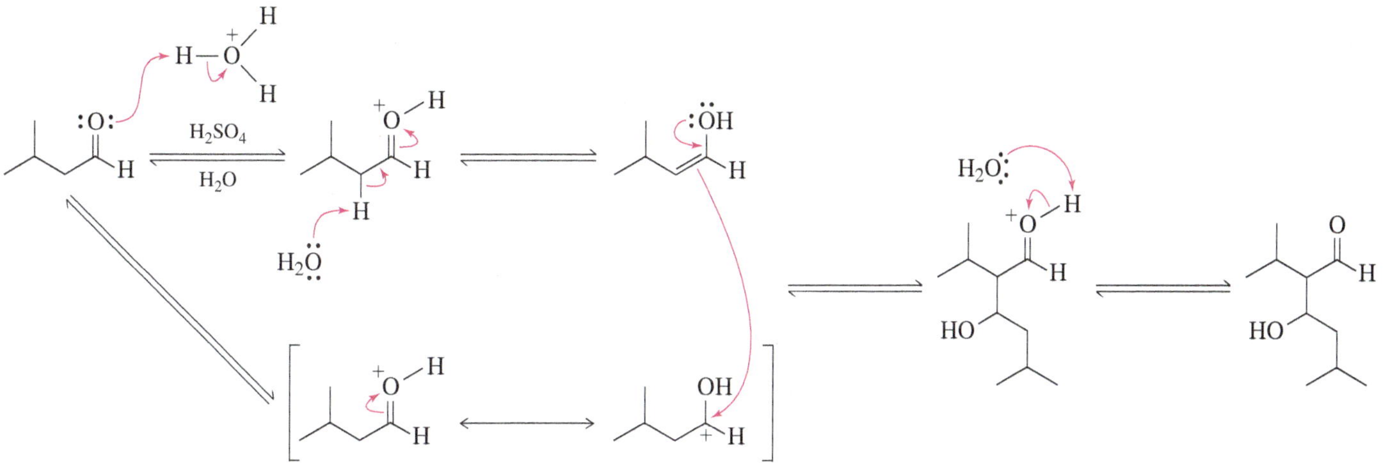

With heating...

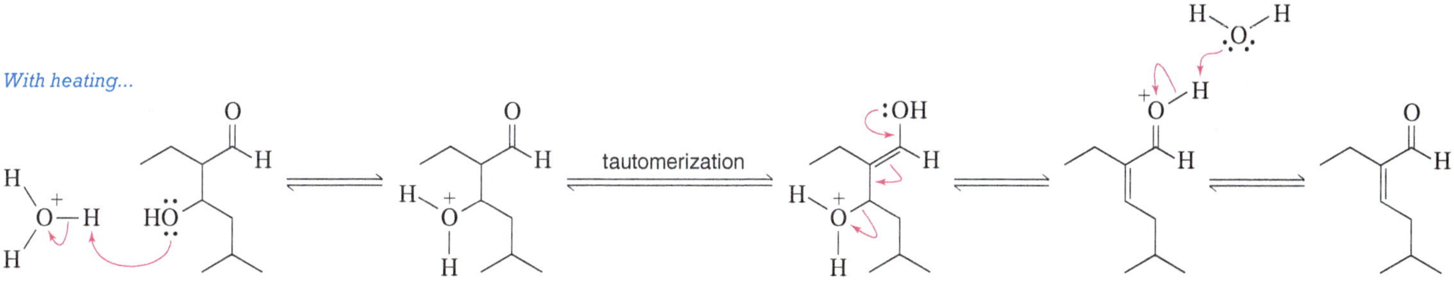

Major Category II: Form an enolate, react with an electrophile

5. **Base-catalyzed aldol reaction (Sections 20.3.2–20.3.4):** Removal of an α-hydrogen results in an enolate that, as a nucleophile, adds to the aldehyde. The alkoxide is quenched with water to give the β-hydroxyaldehyde. With heating, elimination of water produces an α,β-unsaturated carbonyl. This reaction can also be intramolecular. It can be done with different carbonyls in a crossed aldol if reacting an enolate of a ketone and an aldehyde with no α-hydrogens.

General:

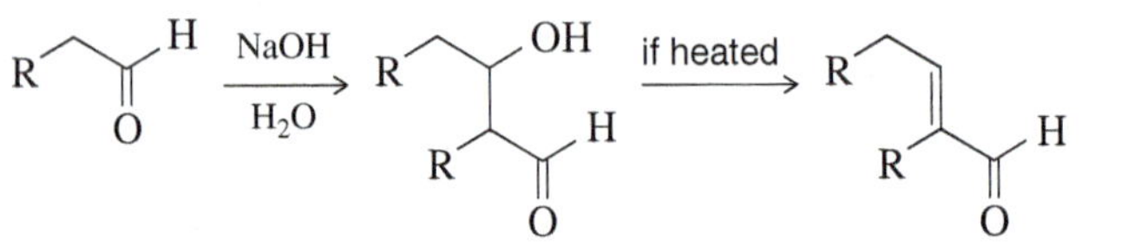

Specific:

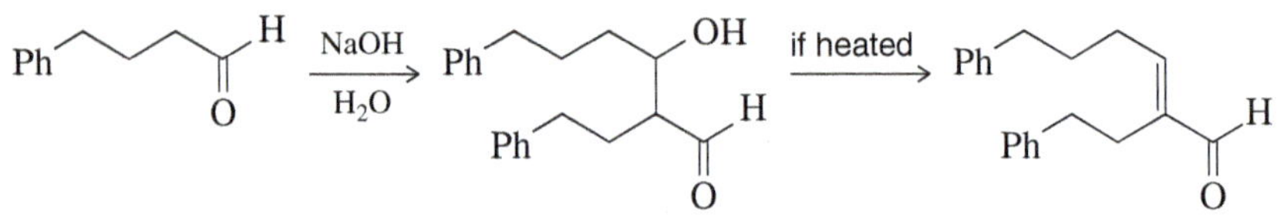

Mechanism:

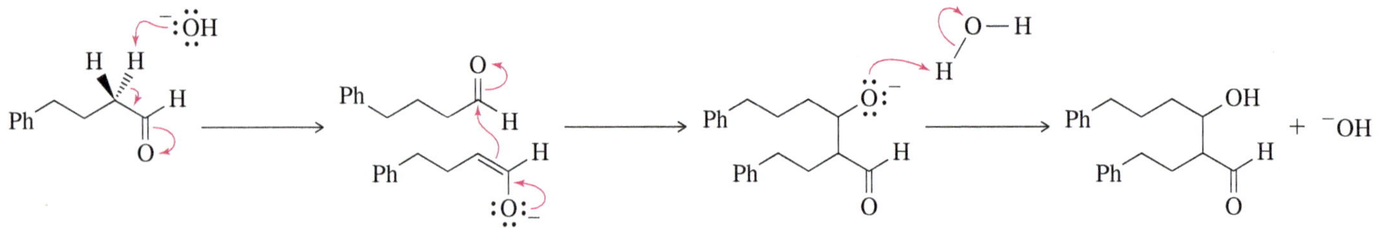

With heating...

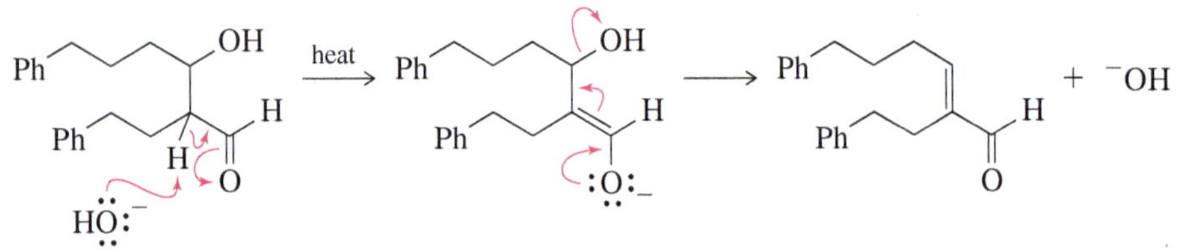

6. **Modern (kinetic) aldol reaction (Section 20.3.1):** Using a strong base (most commonly lithium diisopropylamide, LDA), an enolate is formed at low temperature in the absence of the electrophilic reaction partner. Upon slow addition of the aldehyde, the enolate attacks the carbonyl, thus producing an alkoxide intermediate. Most commonly, the reaction is quenched at this point to generate the α-hydroxy ketone.

General:

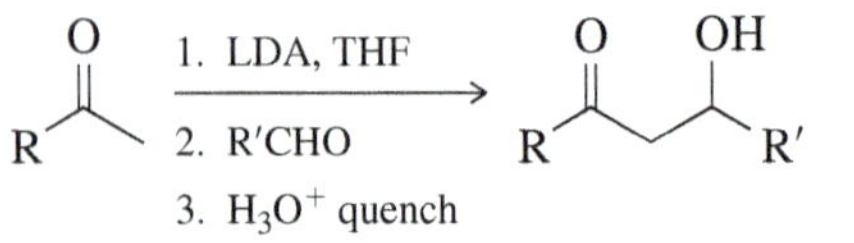

Specific:

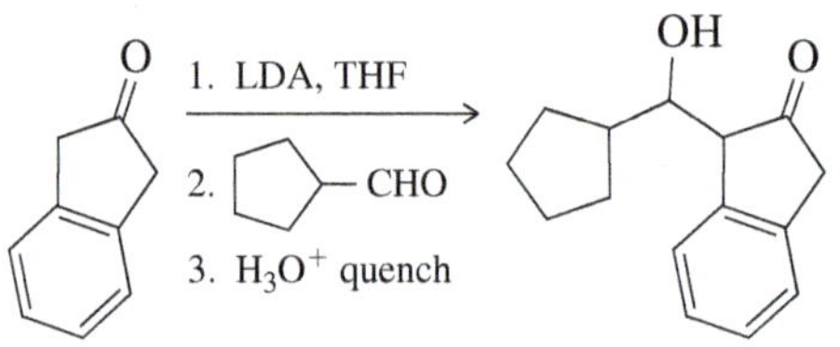

Mechanism:

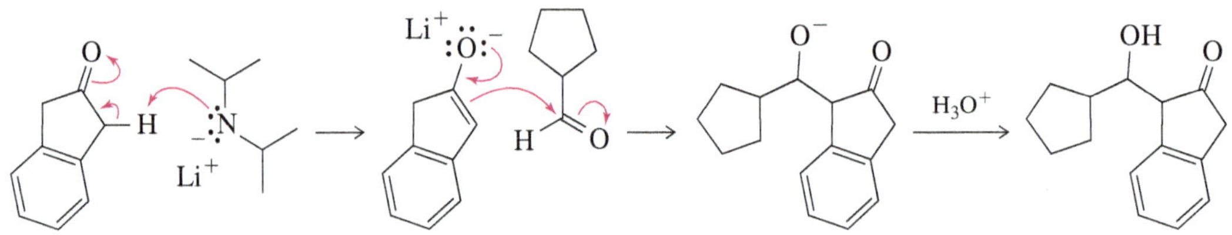

7. Modern enolate alkylation (Section 20.3): Using a strong base (most commonly lithium diisopropylamide, LDA), an enolate is formed at low temperature in the absence of the electrophilic reaction partner. Upon slow addition of a primary alkyl halide, the enolate does an S_N2 reaction, thus displacing the halide leaving group.

General:

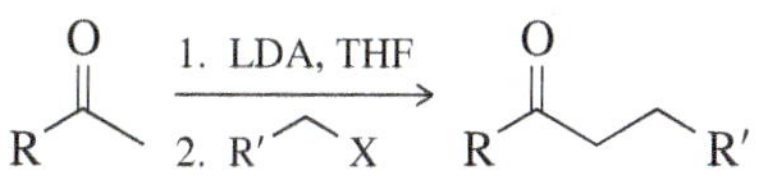

Specific:

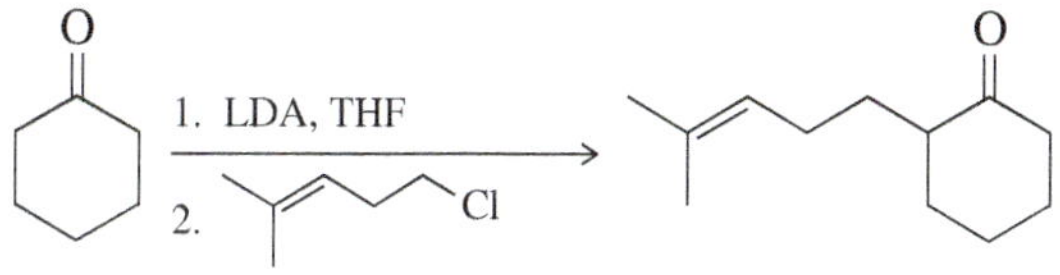

Mechanism:

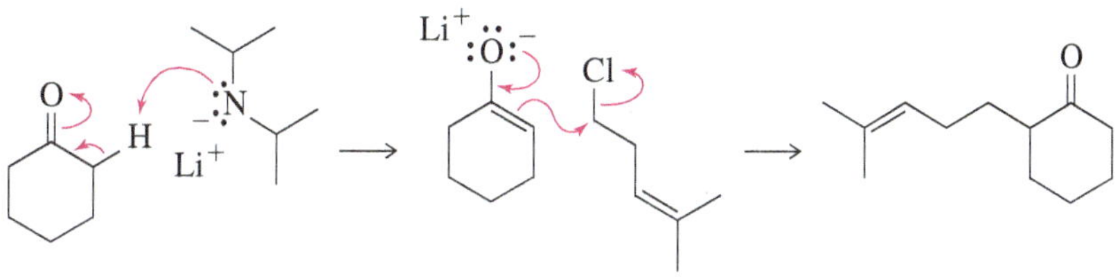

8. Claisen/Dieckmann condensation (Section 20.4): The enolate of one ester attacks another ester to give a tetrahedral intermediate. Collapse of the tetrahedral intermediate generates a β-keto ester with a very acidic proton, which is immediately removed by the ethoxide base. Quenching of the stabilized enolate with acid generates the β-keto ester product. It can also be done between two different esters with the correct experimental modification. The intramolecular version is referred to as the Dieckmann condensation.

General:

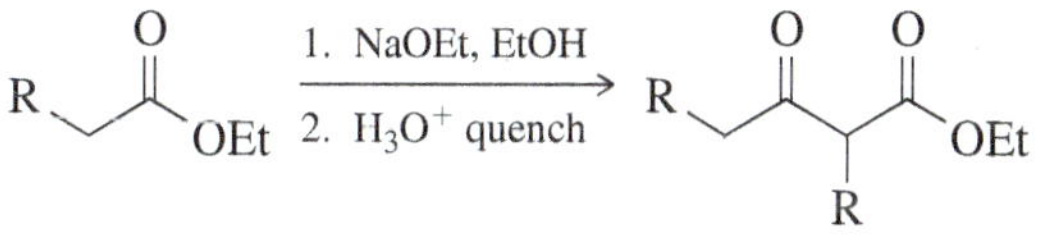

Specific:

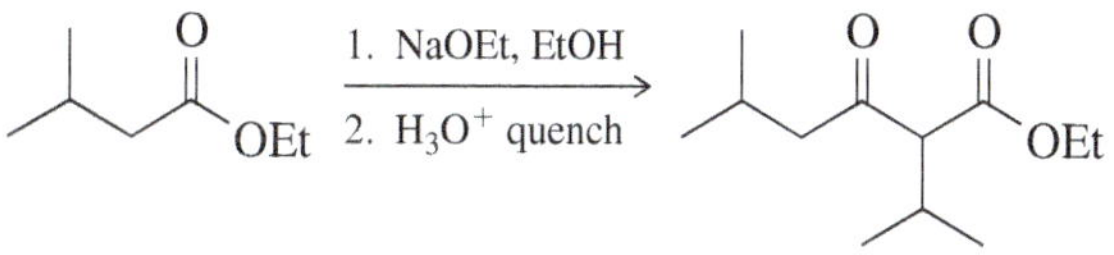

Mechanism:

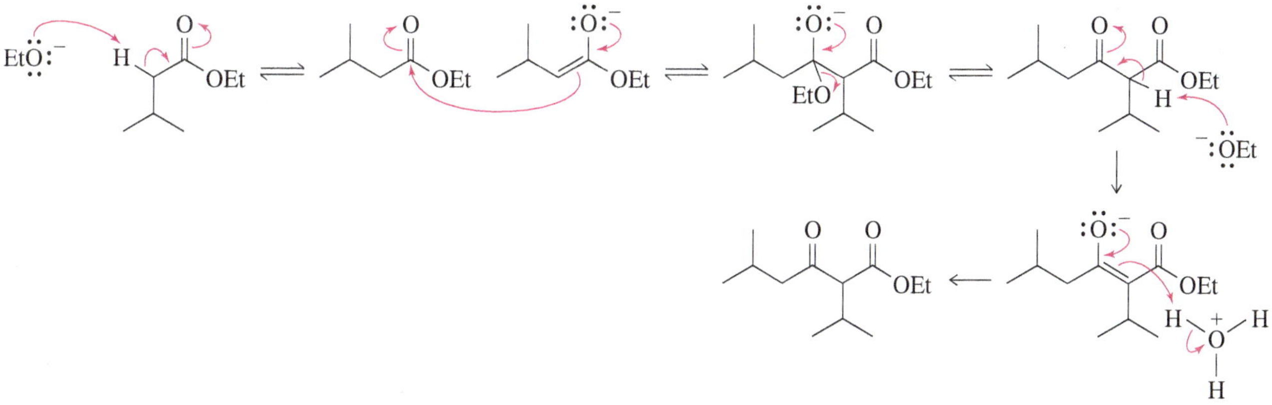

9. Alkylation of an activated methylene (Section 20.5): A methylene between two carbonyls is somewhat acidic ($pK_a \approx 10$), resulting in a resonance-stabilized anion after deprotonation. Once formed, the anion can do S_N2 substitution reactions with primary alkyl halides.

General:

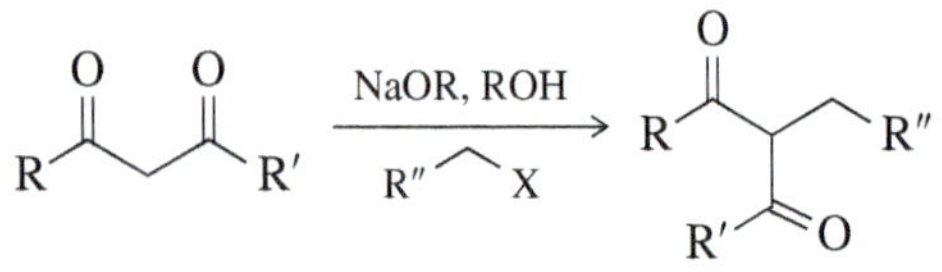

Specific:

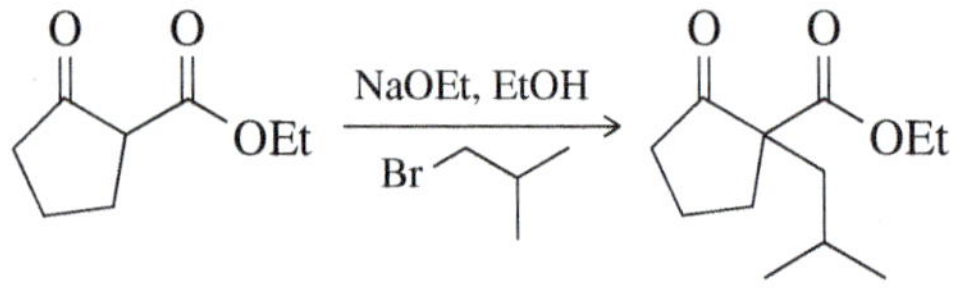

Mechanism:

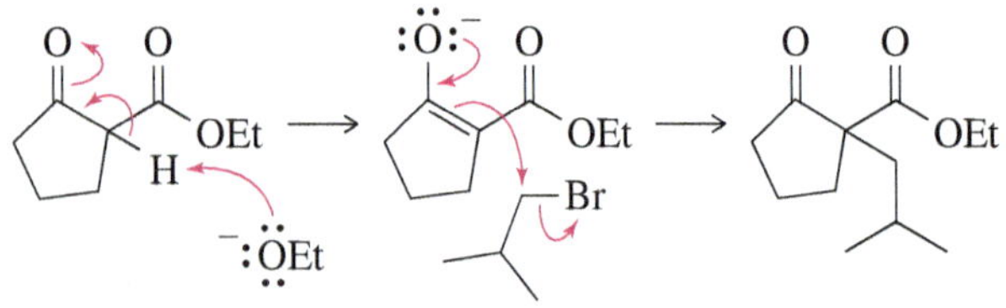

---

10. Acetoacetic ester synthesis (Section 20.5.1): This is a sequence of steps used to prepare derivatives of acetone. Alkylation of the activated methylene of an alkyl acetoacetate is followed by the sequential steps of saponification (Section 19.7.1), acidification, and thermal decarboxylation (Section 18.8.4).

General:

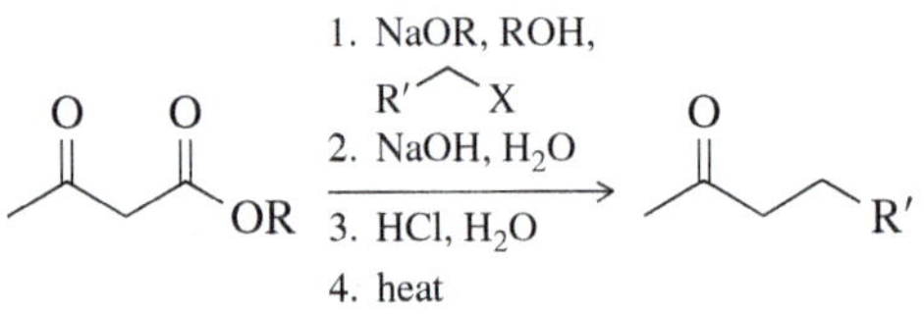

Overall scheme:

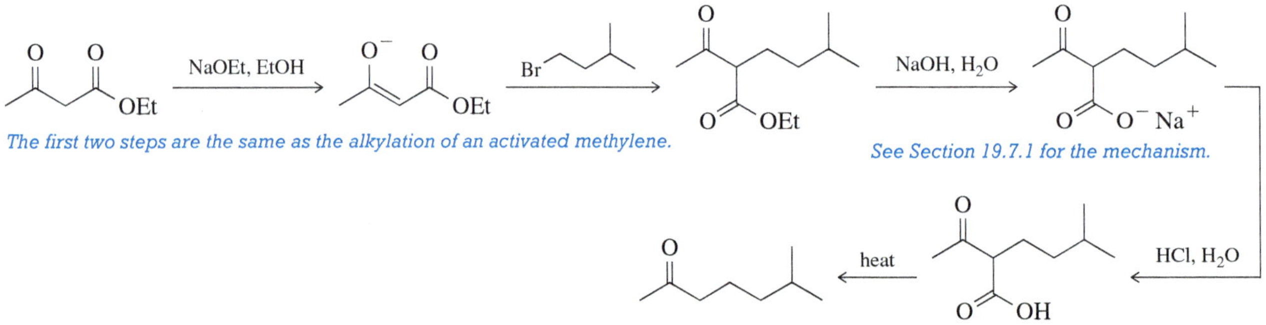

The first two steps are the same as the alkylation of an activated methylene.

See Section 19.7.1 for the mechanism.

See Section 18.8.4 for the mechanism.

11. Malonic ester synthesis (Section 20.5.1): This is a sequence of steps used to prepare derivatives of acetic acid. Alkylation of the activated methylene of a dialkyl malonate is followed by the sequential steps of saponification (Section 19.7.1), acidification, and thermal decarboxylation (Section 18.8.4).

General:

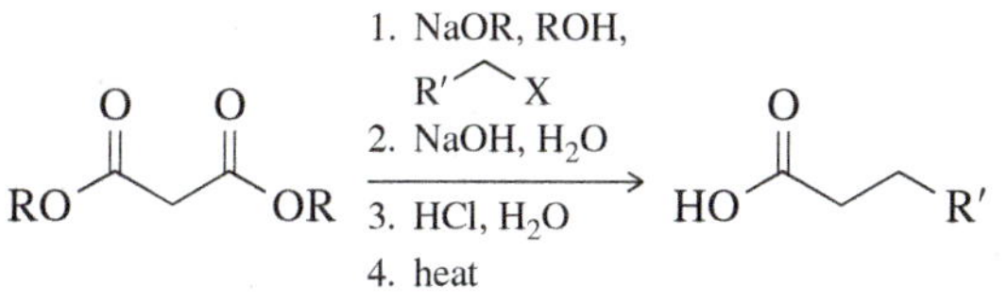

Overall scheme:

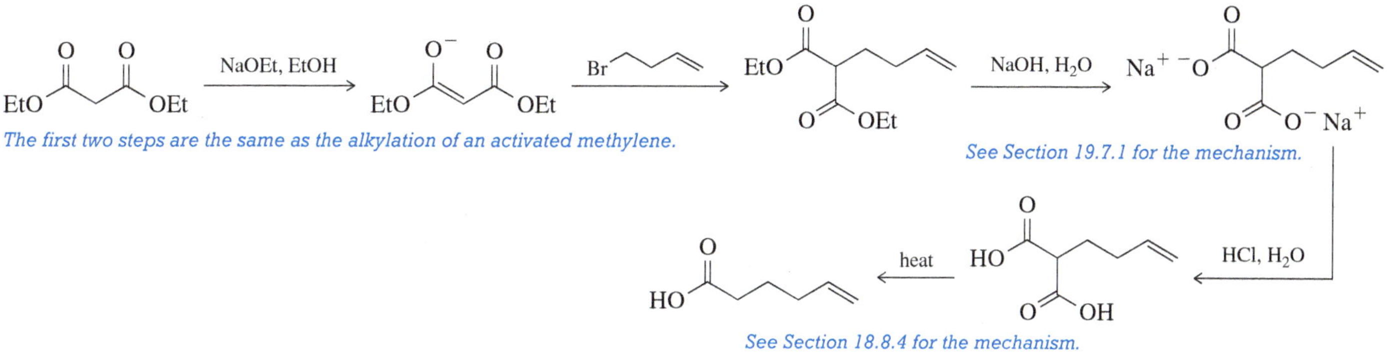

12. Stork enamine synthesis (Section 20.6): A secondary amine reacts with a ketone to give an enamine (see Section 17.8.3). The enamine donates a pair of electrons to an electrophile with assistance from the lone pair on nitrogen. Following the substitution reaction, hydrolysis of the iminium regenerates the ketone (see Section 17.8). Electrophiles in this reaction are typically primary alkyl halides and acid chlorides.

General:

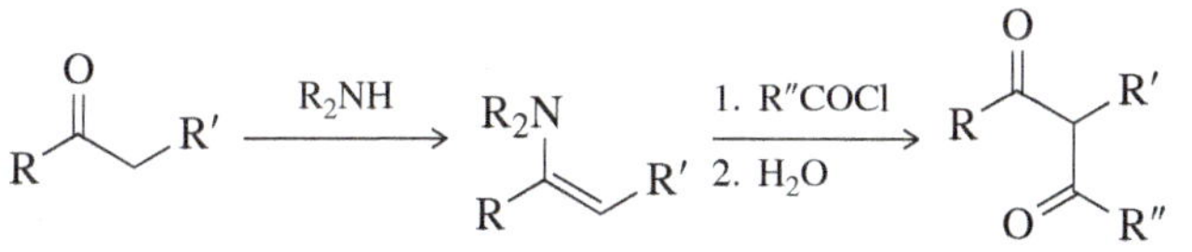

Specific:

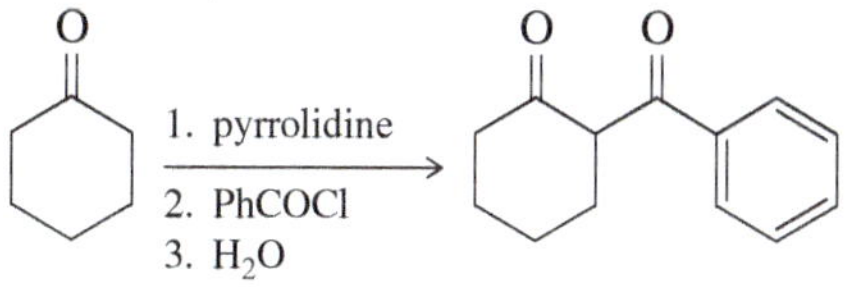

Mechanism:

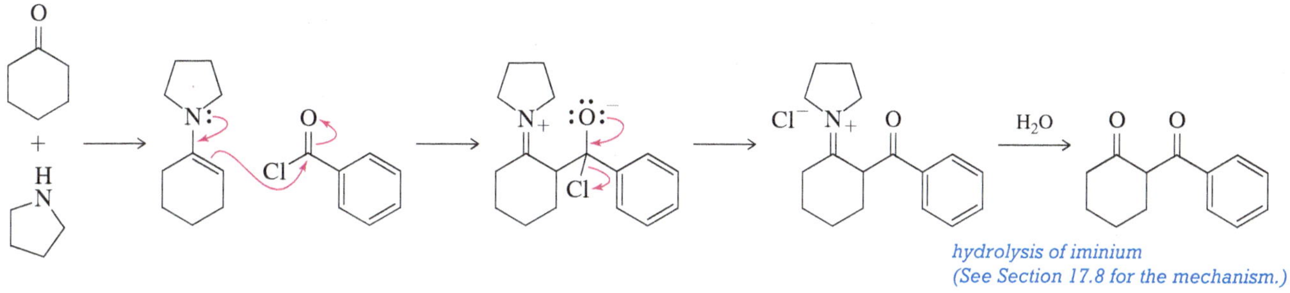

End-of-Chapter Assessments

[*Assessments are ranked according to their difficulty;* •••• = *most difficult.*]

20.72 (••) Identify the enolate(s) that would form on treatment of each of the following carbonyls with base. [*When there are two possibilities, draw both.*]

(a) **(b)** **(c)** **(d)**

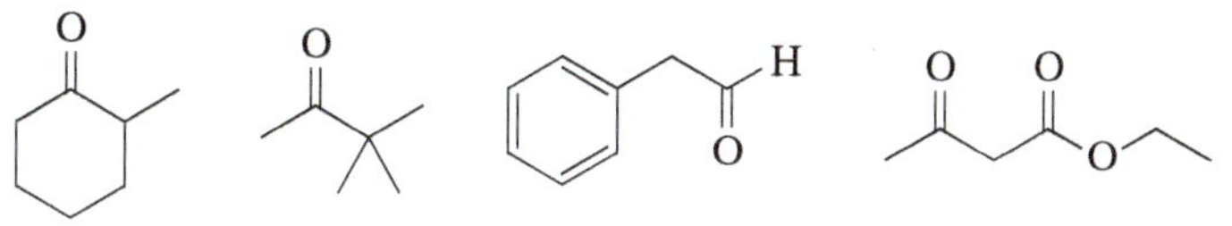

20.73 (••) Predict the non-aldol product that would result when the aldehydes/ketones shown here are allowed to react under the following conditions: (i) D_2SO_4, D_2O; (ii) Br_2, HOAc, (iii) HCl, H_2O. [*Some of these will do no reaction.*]

(a) **(b)** **(c)**

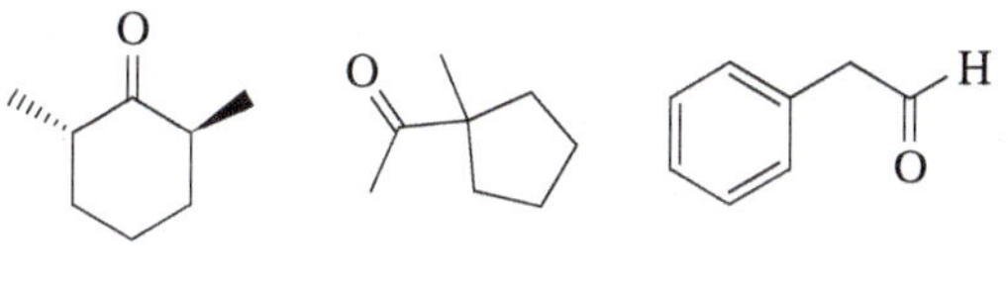

20.74 (•••) Predict the aldol product that would result when the aldehydes/ketones shown here are allowed to react under the following conditions: (i) NaOH, H_2O; (ii) NaOH, H_2O, heat; (iii) HCl, H_2O; (iv) HCl, H_2O, heat.

(a)

(b)
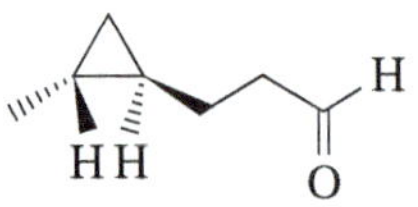
(c)
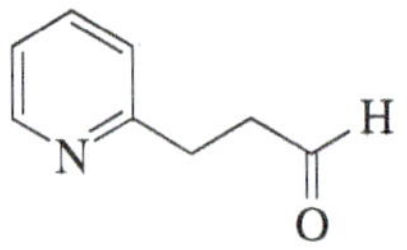
(d)
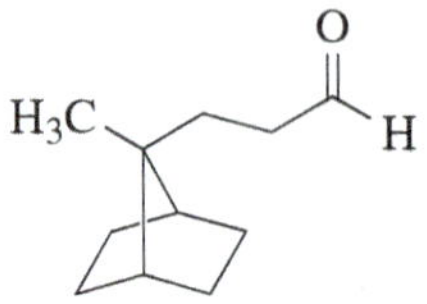
(e)
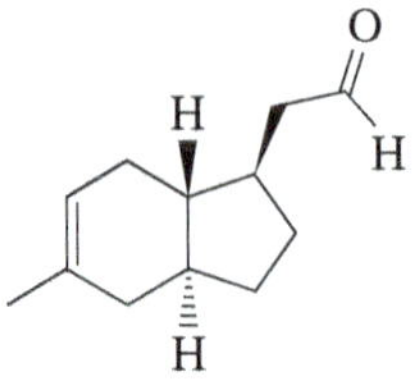
(f)
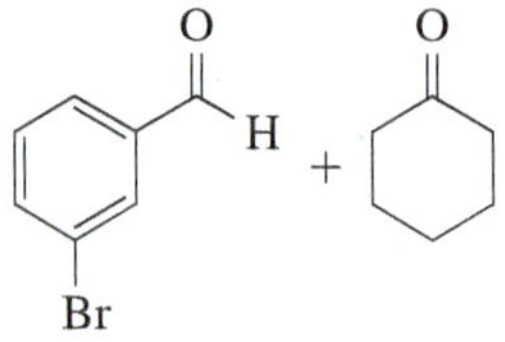
(g)
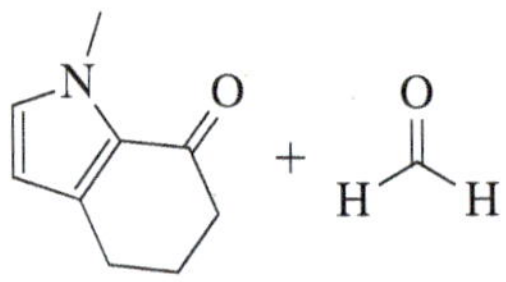
(h)
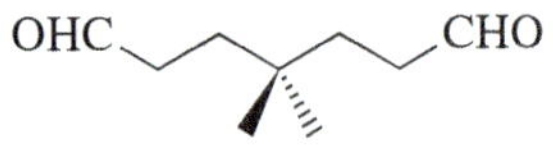
(i)
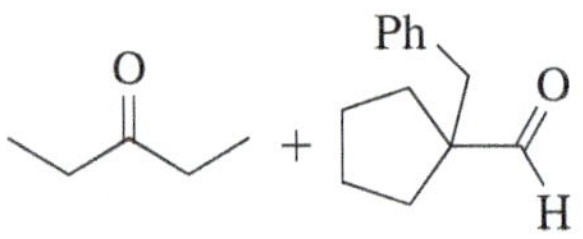
(j)
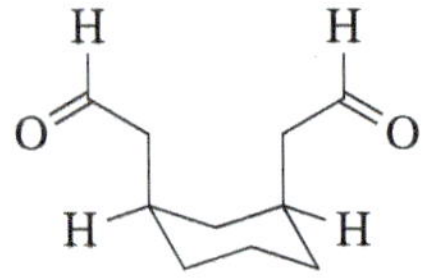
(k)
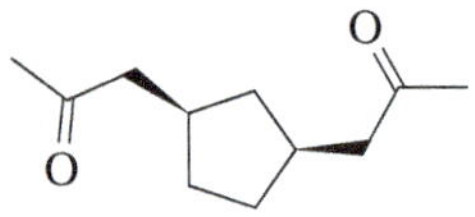

20.75 (•••) Predict the product that would result when each of the following esters is stirred with sodium ethoxide in ethanol until the reactions are quenched with acid.

(a)
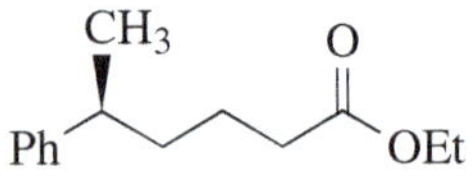
(b)
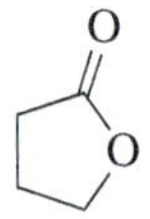
(c)
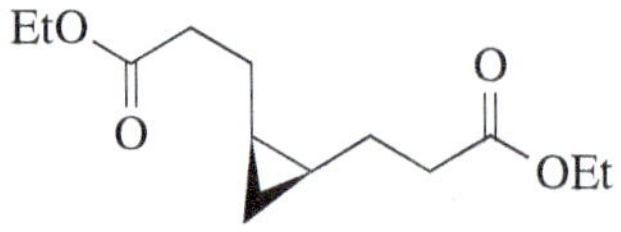
(d)
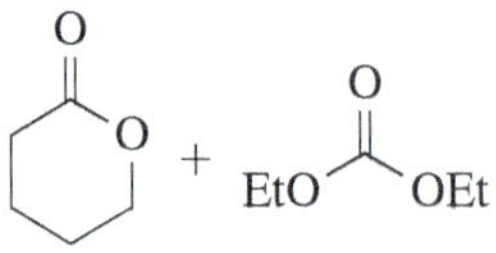
(e) *[Assume the set-up from Figure 20.42 is used to generate one product here.]*
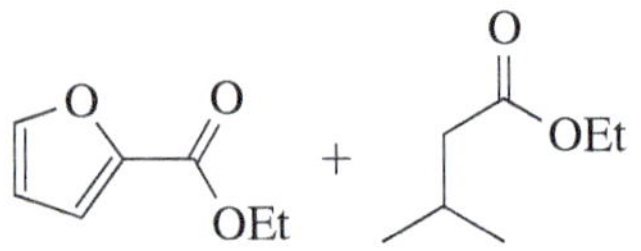

20.76 (•••) Predict the product that would result when each of the following carbonyls is treated with LDA in THF, followed by the electrophile. If the electrophile is an aldehyde, finish the reaction with an acid quench.

(a)
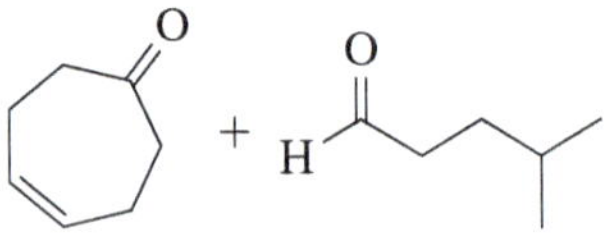
(b)
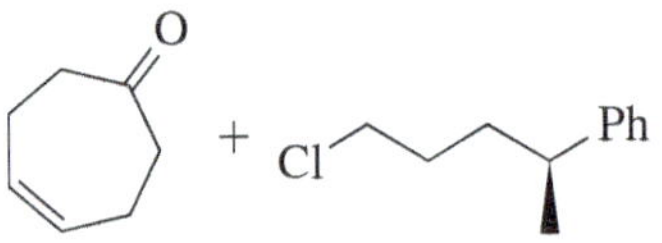
(c)
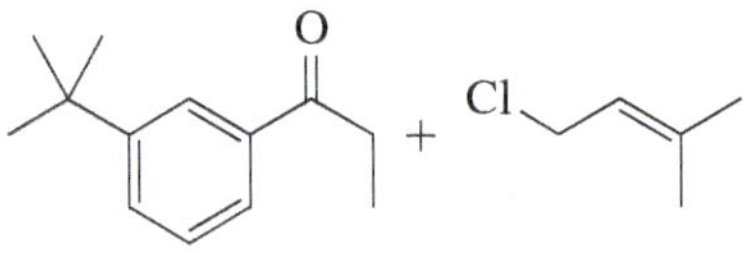
(d)
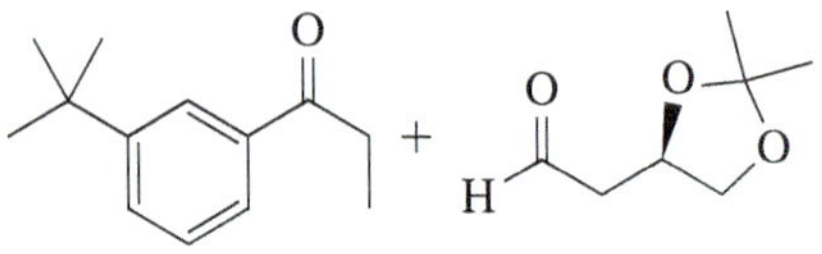
(e)
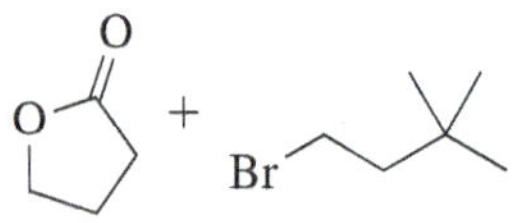
(f)
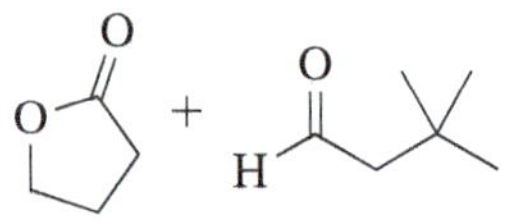

20.77 (•••) Beginning with ethyl acetoacetate or diethyl malonate, synthesize the following molecules.

(a)
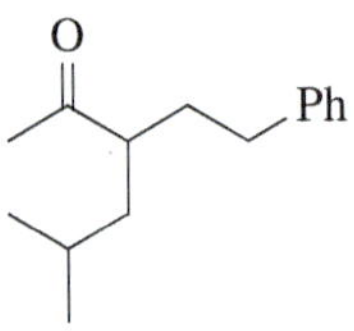
(b)
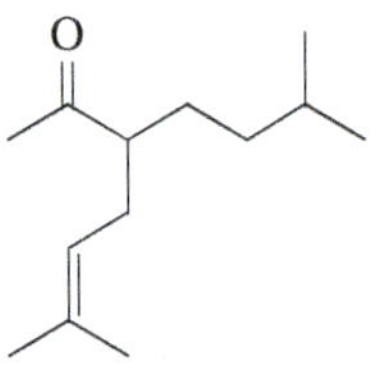
(c)
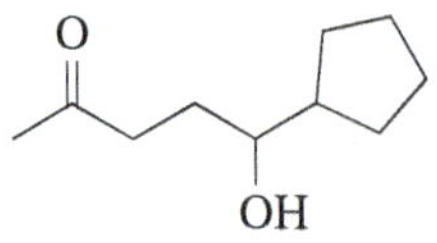
(d)
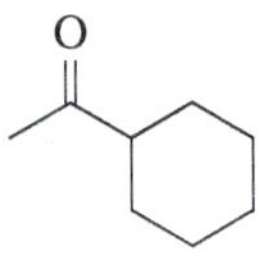
(e)
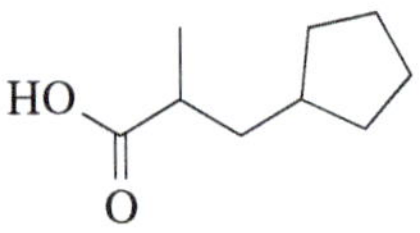
(f)
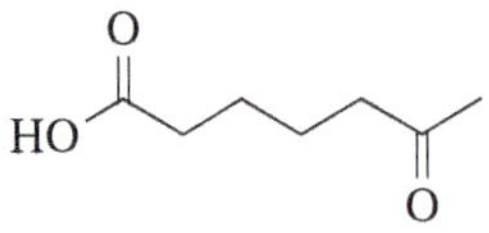
(g)
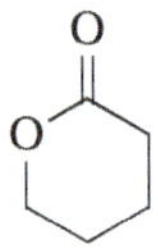
(h)
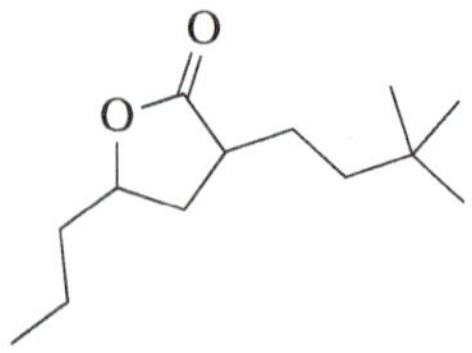

20.78 (••) Predict the product that would result when each of the following aldehydes or ketones is treated with pyrrolidine, followed by the alkyl halide or acid halide shown.

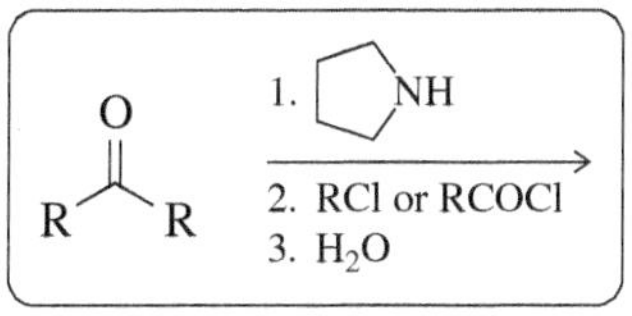

(a)

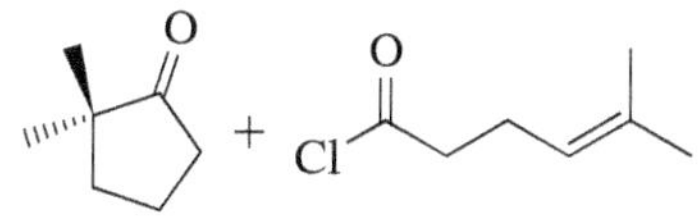

(b)

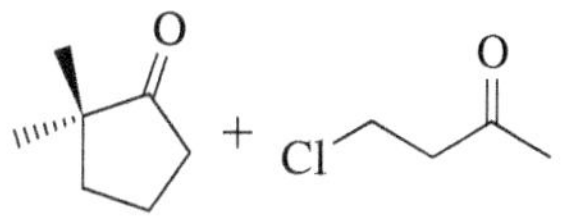

(c)

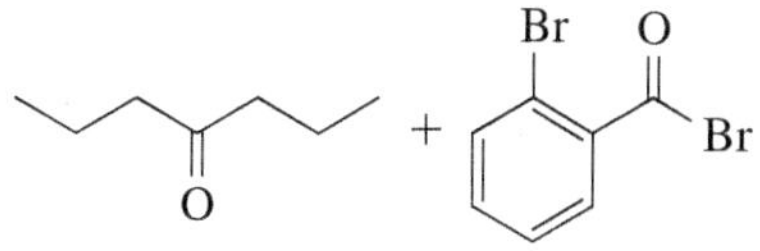

(d)

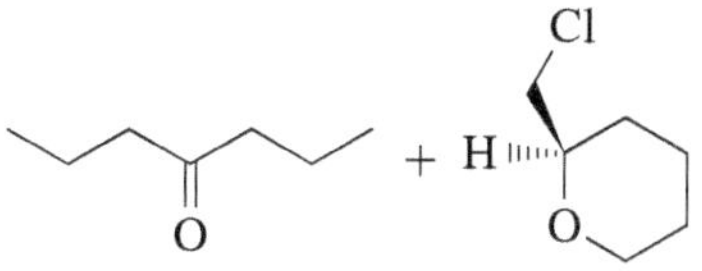

(e)

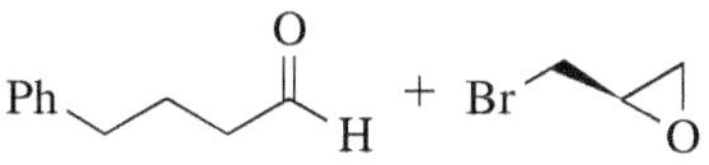

(f)

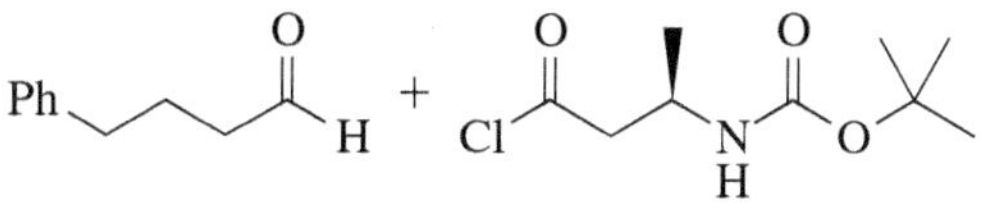

20.79 (••) Calculate the equilibrium constant for the acid–base reaction for each of the compounds below with (i) NaOH; (ii) NaOEt; (iii) LDA.

(a)

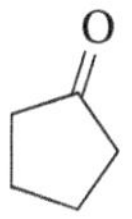

(b)

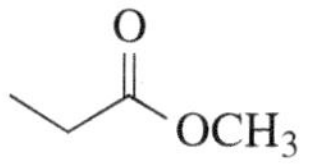

(c)

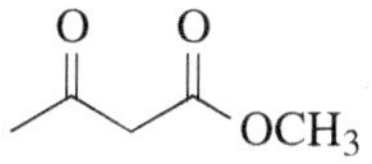

20.80 (••••) For unsymmetrical ketones, there are two possible enolates that can form. With regard to the deprotonation of 2-methylcyclohexanone using the bulky, non-nucleophilic base lithium diisopropyl amide (LDA), explain the following observations.

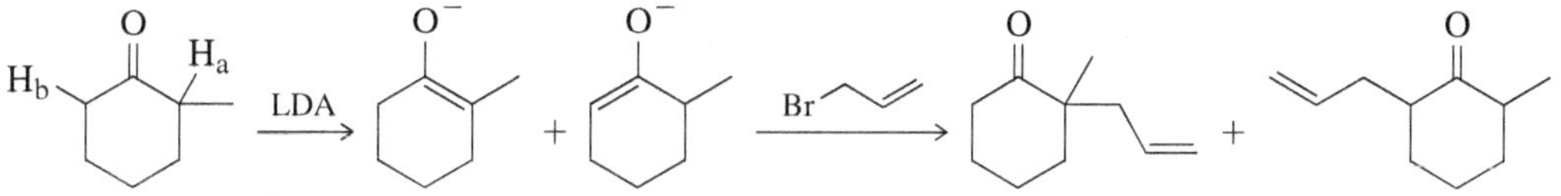

A **B** **C** **D**

(a) The kinetic enolate (formed the fastest) is enolate **B**.
(b) The thermodynamic enolate (most stable) is enolate **A**.
(c) If you run this reaction using 1.0 equivalents of LDA, you end up with exclusively product **D**.
(d) If you run this reaction using 0.9 equivalents of LDA, you end up with exclusively product **C**.
(e) If you run this reaction with 1.0 equivalents of LDA plus 1 drop of methanol, you end up with exclusively product **C**.

20.81 (•••) Draw a reaction coordinate diagram that accounts for the fact that **A** is more stable but **B** forms more quickly in Assessment 20.80.

20.82 (•••) Making sure to include at least one reaction of enolates, synthesize the following starting with molecules containing three or fewer carbons.

(a)

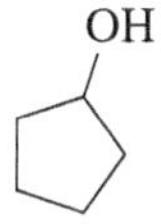

(b)

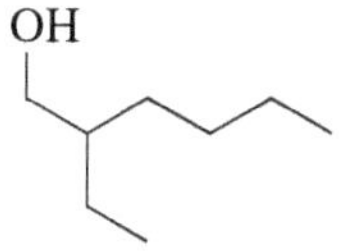

(c)

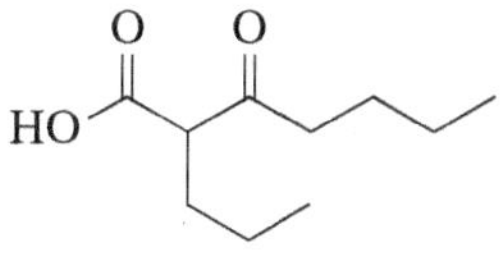

20.83 (•••) Ethyl acetoacetate can be deprotonated twice with LDA to form the dienolate shown. (a) Suggest an arrow-pushing mechanism by which the dienolate forms. (b) Which proton, H_a or H_b is likely removed first? Why? (c) Finally, predict the product that would result when the dienolate is allowed to react with benzaldehyde. [*The last enolate to form is the one that will react first.*]

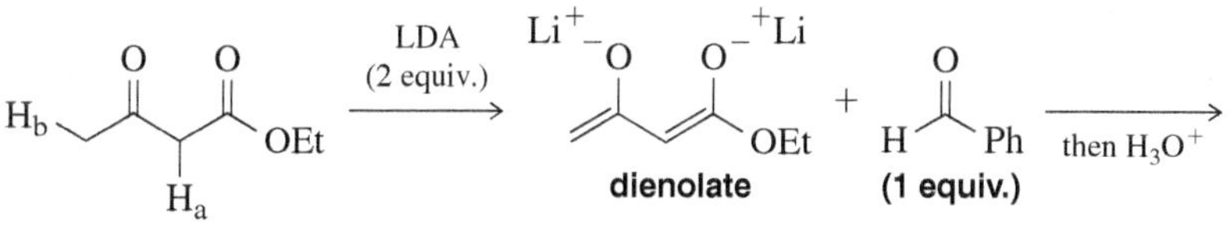

20.84 (•••) Based on your answer to Assessment 20.83, predict the product of the reaction shown.

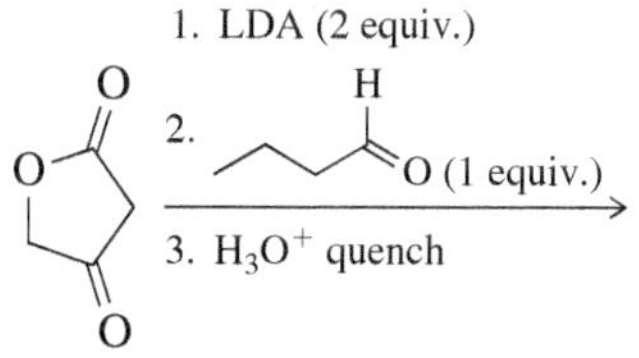

20.85 (••••) **FROM THE LITERATURE** The synthesis of homopropargylic alcohols has recently been achieved via the decomposition of 5,6-dihydro-2-pyrone triflates, as shown in the following synthetic scheme. Suggest a mechanism for each step of the scheme (*J. Am. Chem. Soc.* 2008, *130*, 5050–5051).

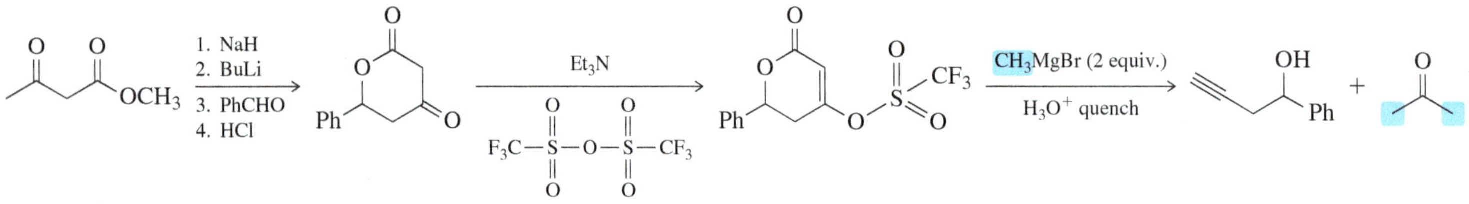

20.86 (••) **FROM THE LITERATURE** Curcumin is a naturally occurring compound (isolated from an herb, *Curcuma longa* L.), which has shown several promising medicinal properties, including anti-HIV and antimalarial activity. Compound **A** was recently synthesized as an intermediate toward some derivatives of curcumin. Beginning with 4-ethylcyclohexanone, suggest a synthesis of compound **A** (*Eur. J. Org. Chem.* 2013, *4*, 146–148).

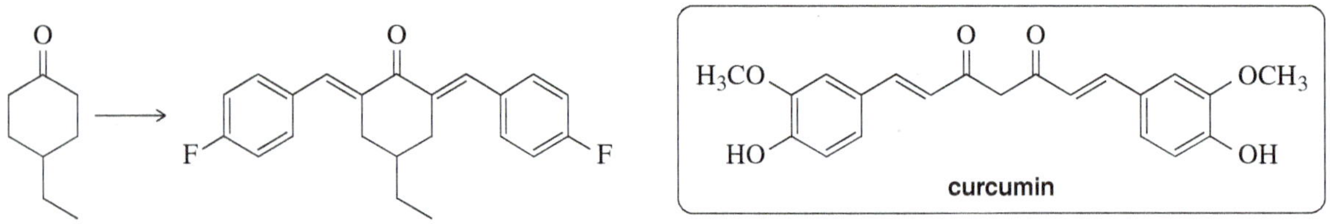

A

20.87 (•••) **FROM THE LITERATURE** One limitation of the Dieckmann condensation is encountered when you attempt to do the reaction on an unsymmetrical diester. For the following scheme, identify the two regioisomeric products you would expect for the Dieckmann condensation of molecule **A**. Which of the two do you think will be favored? Why? (*Russ. J. Org. Chem.* 2014, *50*, 105–109)

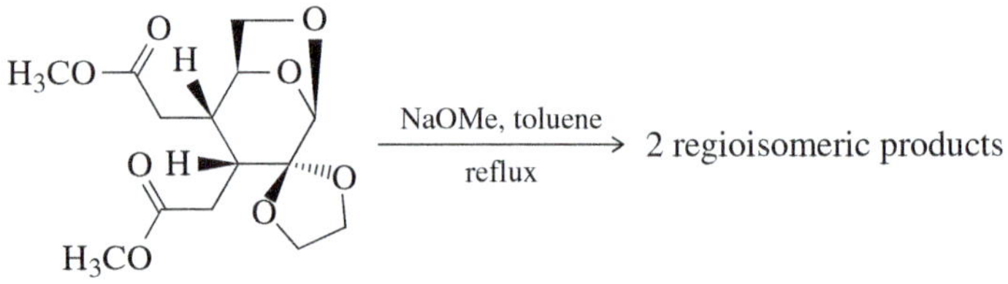

A

20.88 (•••) **FROM THE LITERATURE** Cholesterol is known to play a unique role in maintaining the integrity of eukaryotic cells by providing order to the lipid membranes. Of the many sterol intermediates in the biosynthetic pathway to cholesterol, none is as good as cholesterol at fulfilling this role. To better understand the manner in which cholesterol is able to organize the membrane, chemists have recently synthesized a cholesterol derivative lacking the C18 and C19 methyl groups. A key reaction in their scheme was a highly regioselective Dieckmann condensation of **A**. Predict the two regioisomers you would expect to form and indicate which you believe to be the predominant isomer (*J. Org. Chem.* 2014, *79*, 5636–5643).

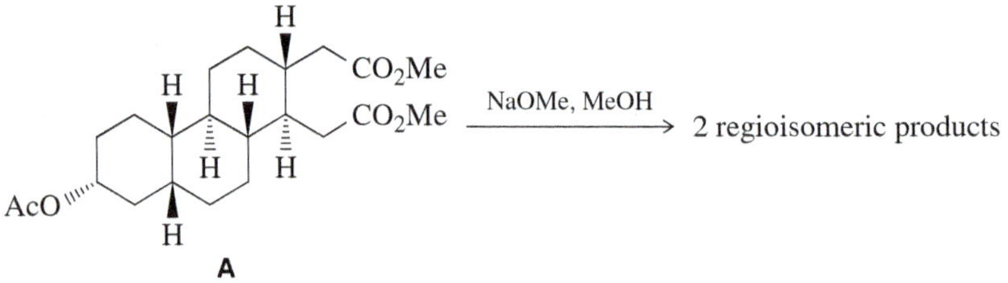

A

20.89 (•••) The reaction of sodium hydroxide, a methyl ketone, and iodine was once used to indicate the presence of a methyl ketone by formation of iodoform as a precipitate in the reaction. Suggest a mechanism for each step. [*Step 1 involves the chemistry of Chapter 20; step 2 is from Chapter 19.*]

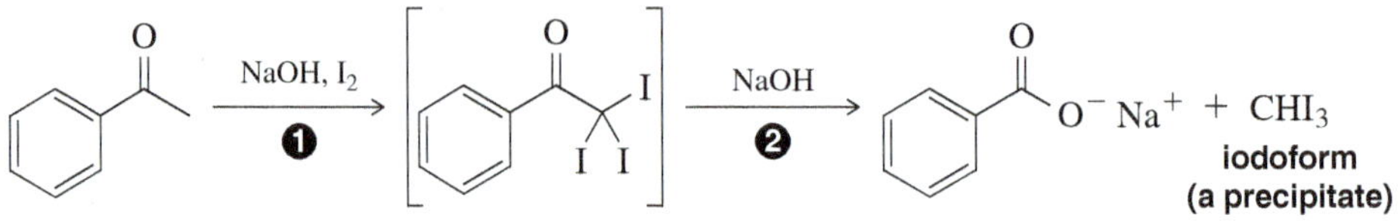

20.90 (•••) Which of the following substrates would you expect to form a precipitate in the iodoform reaction? [*See Assessment 20.89.*]

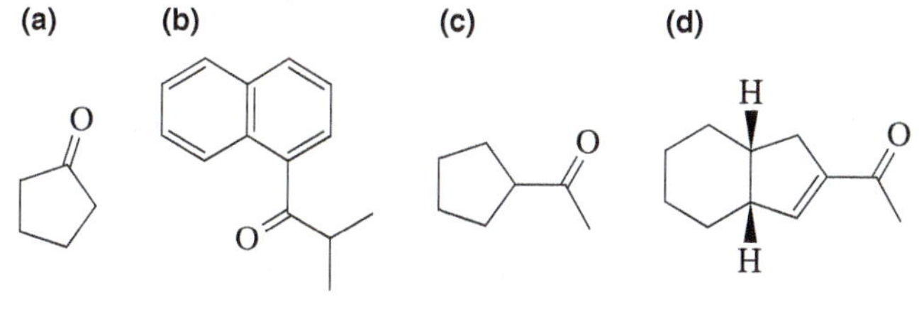

20.91 (••••) Bis(2-ethylhexyl)phthalate (DEHP) is a plasticizer used to increase the flexibility of polyvinylchloride (PVC), a plastic used in the production of many everyday products, including personal care products and children's toys. DEHP is one of a number of phthalates that has been shown to possess endocrine-disrupting properties. Provide a synthesis of DEHP using but-1-ene and phthalic anhydride as your only sources of carbon.

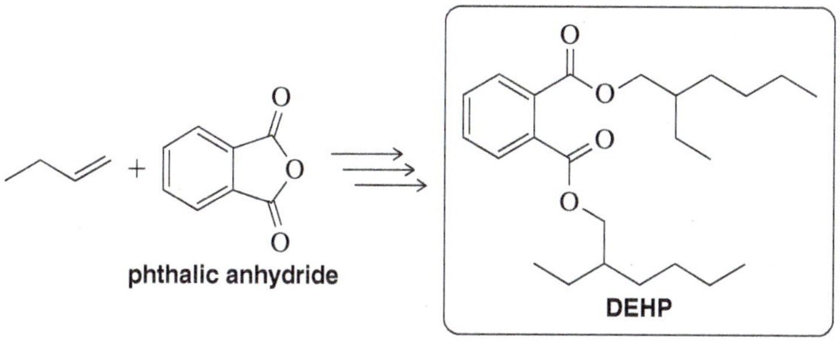

20.92 (•••) **FROM THE LITERATURE** As part of their efforts toward synthesizing and evaluating the tumor activity of 4-hydroxycoumarin derivatives, chemists employed the Reformatsky reaction to form a key carbon–carbon bond. The Reformatsky reaction, related to reactions presented here in Chapter 20, is shown here. Provide an arrow-pushing mechanism for the reaction (*Bioorg. Med. Chem. Lett.* 2004, *14,* 5527–5531).

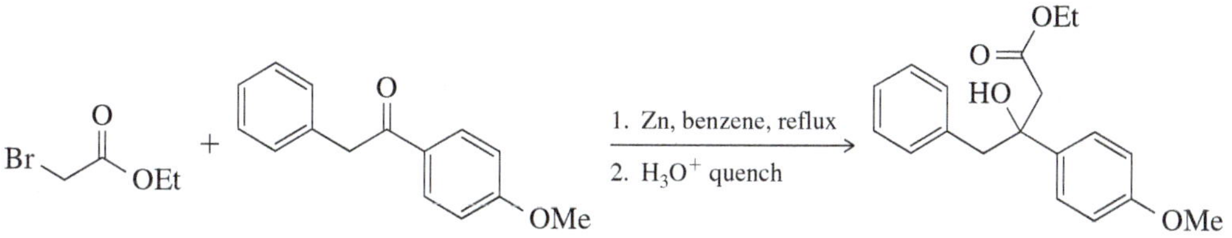

20.93 (••••) The Knoevenagel condensation is related to the reactions presented here in Chapter 20 and is shown here. Suggest an arrow-pushing mechanism that rationalizes the formation of the products. Your mechanism should account for the fact that only a catalytic amount of base is used in the reaction.

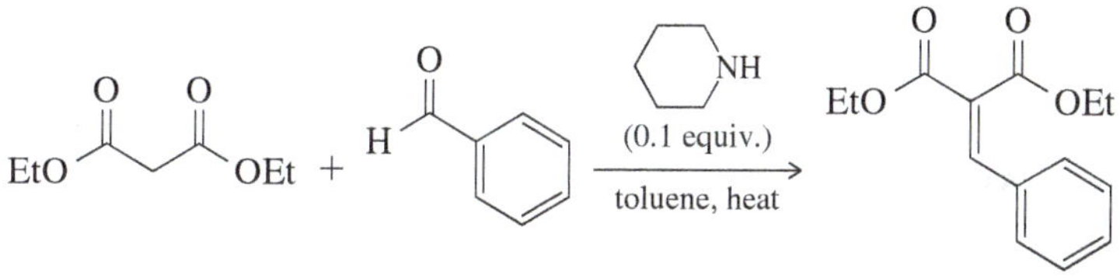

20.94 (••••) The Doebner modification is related to the reaction in Assessment 20.93. Under similar conditions, it results in a decarboxylated product. Rationalize the product formation using an arrow-pushing mechanism. Pyridine is both a base and the solvent in this reaction.

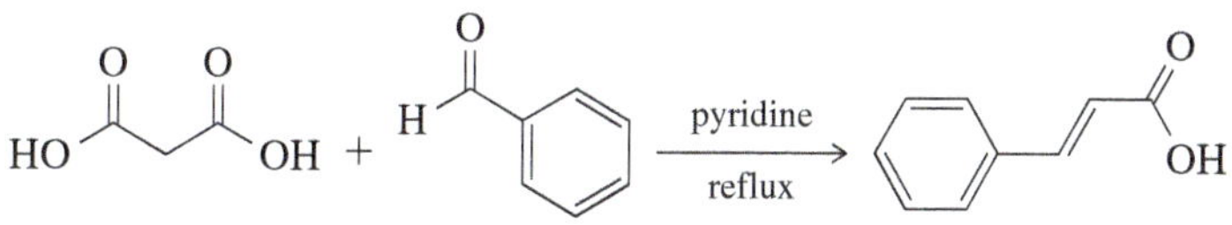

20.95 (••••) The Henry reaction is related to reactions presented here in Chapter 20. Provide an arrow-pushing mechanism that accounts for the formation of the following product (ignore stereochemistry). Sodium hydride is a strong base.

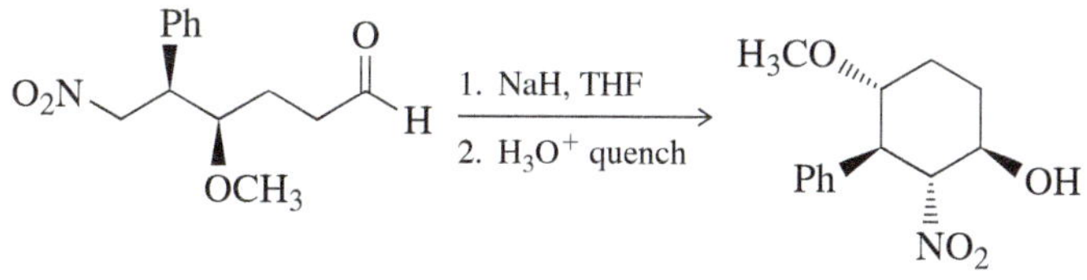

20.96 (••••) (a) Suggest a mechanism for the following reaction used in the synthesis of indinavir (a molecule we introduced in Chapter 9). (b) Why does this reaction selectively produce the (*R*) configuration at the indicated stereocenter?

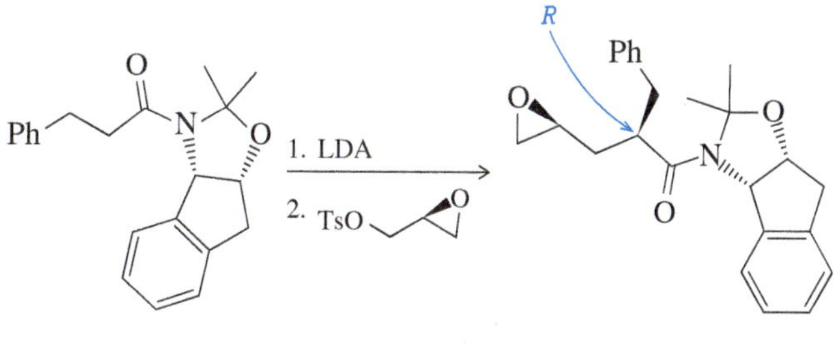

21

Conjugated Systems 1: Stability and Addition Reactions

Learning Objectives

LO 21.0 Distinguish between conjugated and nonconjugated systems

LO 21.1 Analyze the effects of conjugation on molecular properties

LO 21.2 Describe the molecular orbitals of a conjugated molecule

LO 21.3 Correlate the color of a species to wavelengths absorbed, HOMO–LUMO energy gap, and molecular structure

LO 21.4 Assess the chemical properties of sunscreens

LO 21.5 Analyze the stereochemical outcomes of addition reactions in conjugated systems

LO 21.6 Analyze nucleophilic addition reactions to conjugated species

Introduction

Nutritionists recommend that people should consume 5–13 servings of fruits and vegetables each day, depending on their total caloric intake. These fruits and vegetables have several potential health benefits, including lowered risk of diabetes and heart disease, reduced blood pressure, and decreased risk of some cancers. Some of these effects, especially among brightly colored fruits and vegetables, are attributed to the presence of highly conjugated antioxidants like β-carotene, petunidin (an anthocyanin), and lycopene, present in carrots, blueberries, and tomatoes, respectively.

The evolution of plants and flowers to produce these colored compounds suggests they have multiple benefits for the plants themselves. It has been speculated that the bright colors of flowers help attract pollinators. The bright color of fruits might distract herbivores from consuming leaves instead, thus providing a level of protection to the plant. And finally, by appearing attractive and tasty to animals, the seeds in seed-bearing fruits can be spread via consumption and digestion.

The benefits of these fruits and vegetables to plants, animals, and insects alike are due in large part to the presence of highly conjugated molecules. Here in Chapter 21 we learn how these molecules produce color and how their reactivity has the potential to mitigate disease.

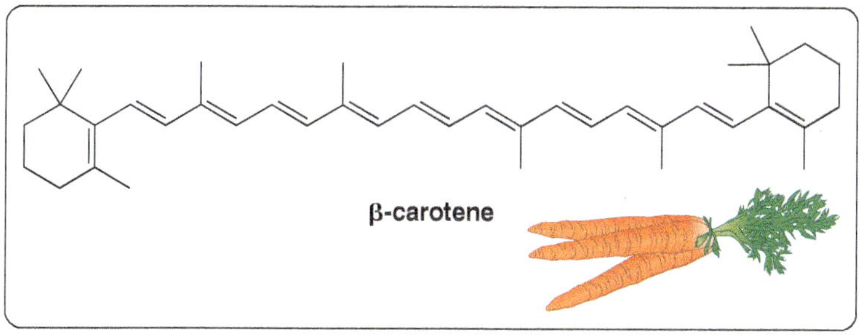

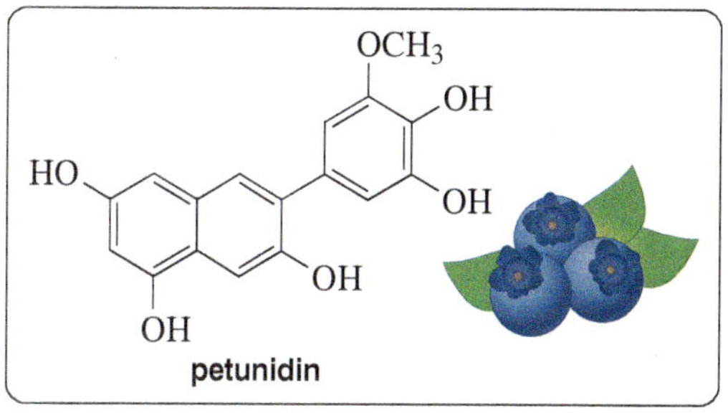

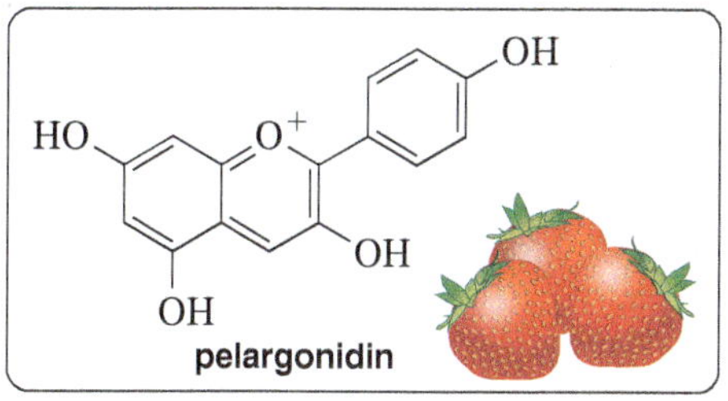

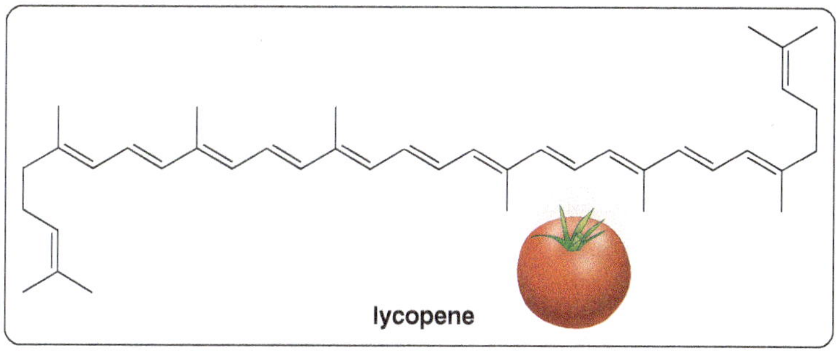

"A good person dyes events with his own color . . . and turns whatever happens to his own benefit."

—Seneca the Younger

At this point, the journey through carbonyls is complete. [*At least until you take a biochemistry class.*] Whereas Chapters 17–20 were united under the theme of carbonyl addition and substitution, the theme for Chapters 21–24 is conjugated systems. Chapter 21 discusses the stability and addition reactions of acyclic conjugated systems, while increasing our reliance on molecular orbital pictures. In Chapter 22, molecular orbital pictures are used to explain the reactivity of systems that can't be directly described by "negatives attack positives." Finally, in Chapters 23 and 24, the focus switches to benzene, one of the most stable organic compounds known. As you study these chapters, it is important to reflect on, and deliberately connect, the information contained in each of them. Let's look back at what we've learned previously to set the stage.

Test your understanding

Assessment 21.1 Draw the molecular orbital picture of *trans*-but-2-ene. Be sure to label all σ and π bonds. Is there free rotation around the $C_2—C_3$ bond? Why or why not?

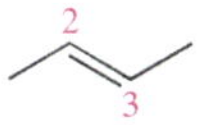

***trans*-but-2-ene**

Assessment 21.2 Draw all possible resonance structures for the reactive intermediates shown.

(a)

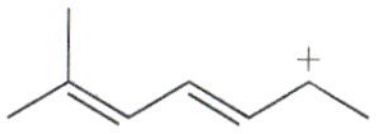

(b)

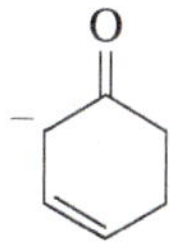

(c)

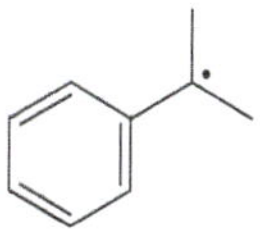

Assessment 21.3 Rank the following carbocations in order of stability (1 = most stable; 5 = least stable). Explain your order.

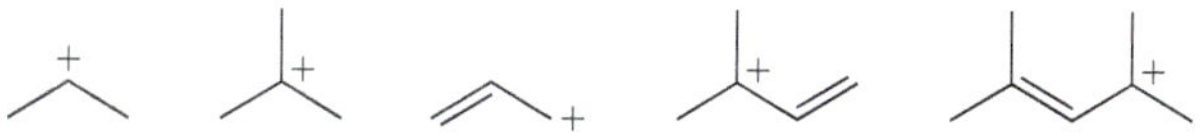

Assessment 21.4 Rank the following alkenes in order of stability (1 = most stable; 5 = least stable). Explain your order.

Assessment 21.5 Predict the product and provide a mechanism for the reaction of 1-methylcyclohexene with HBr.

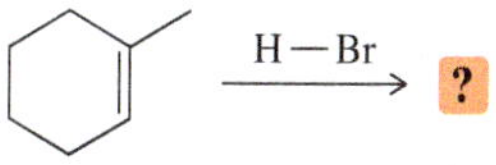

21.1 Conjugated systems: an overview

Molecules are said to be conjugated when they contain a series of alternating single and multiple bonds that result in three or more consecutive atoms containing *p* orbitals (**Figure 21.1**). These neighboring *p* orbitals, overlapping via resonance, delocalize electrons throughout the π system. This delocalization results in enhanced stability (Section 21.2) and altered reactivity patterns. Three types of conjugated systems—namely, alkenes, carbonyls, and reactive intermediates—are discussed next in Sections 21.1.1–21.1.3.

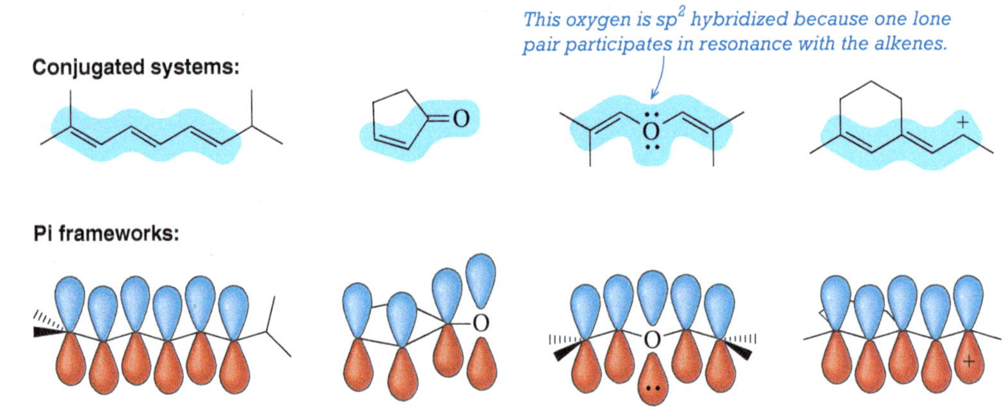

Figure 21.1 Every atom in a conjugated system contains a *p* orbital, which allows the electron density to be delocalized.

21.1.1 Conjugated alkenes

Alkenes (Chapters 8 and 9) contain a π bond between two sp^2-hybridized carbon atoms. Most of the ones studied so far were **isolated** (or **nonconjugated**) **alkenes** (**Figure 21.2a**). [*We never called them that, but that's because we hadn't talked much about conjugation yet.*] An isolated alkene is surrounded on all sides by sp^3-hybridized atoms, which limits the overlapping system of *p* orbitals to the C—C π bond itself. Alternatively, **conjugated alkenes** are separated by one C—C single bond. This arrangement allows for an extended π system where at least four consecutive atoms utilize a *p* orbital for bonding [*Think resonance!*]. **Figure 21.2b** shows a conjugated diene and triene. Two alkenes that share a central carbon are referred to as **cumulated alkenes** (**Figure 21.2c**).

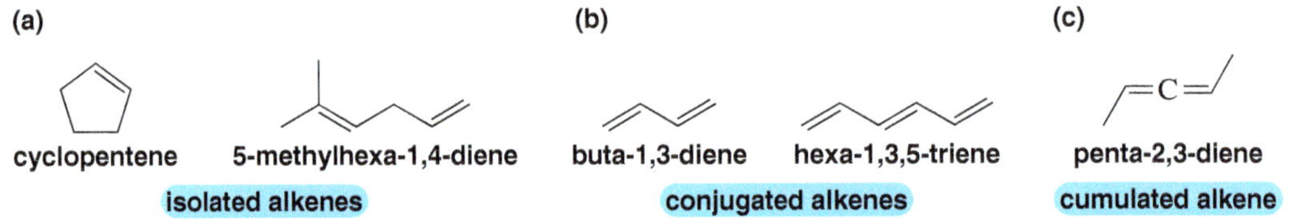

Figure 21.2 Alkenes can be classified as (a) isolated (or nonconjugated), (b) conjugated, or (c) cumulated.

21.1.2 Conjugated carbonyls

Alkenes can also be conjugated to carbonyls, with the alkene and the carbonyl separated by one C—C single bond (**Figure 21.3**). Molecules of this type are referred to as α,β-unsaturated. This nomenclature system uses Greek letters to indicate the distance of each carbon from the carbonyl group. The carbon directly attached to the carbonyl carbon is α, the next carbon is β, and so on. In this way, an α,β-unsaturated ketone, because of the alkene (unsaturation), has fewer than the maximum possible number of hydrogens at the α and β carbons. Recall from Chapter 20 that α,β-unsaturated carbonyls are products of the aldol reaction performed at elevated temperature. We show in Section 21.5 that alkenes become electrophilic when conjugated to electron-withdrawing carbonyl groups. [*See if you can draw a resonance structure that shows this.*]

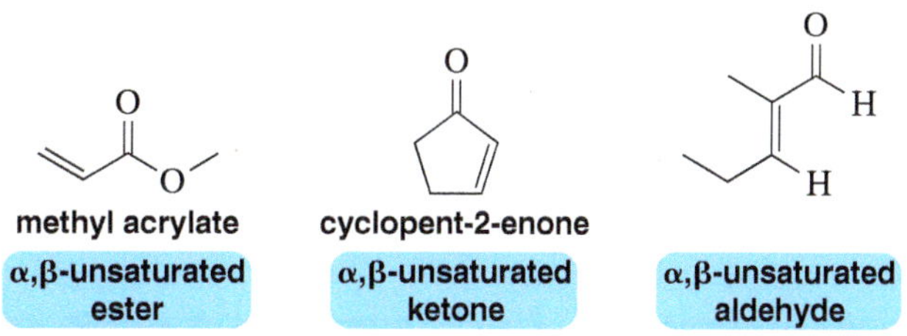

Figure 21.3 α,β-Unsaturated carbonyl compounds are conjugated molecules.

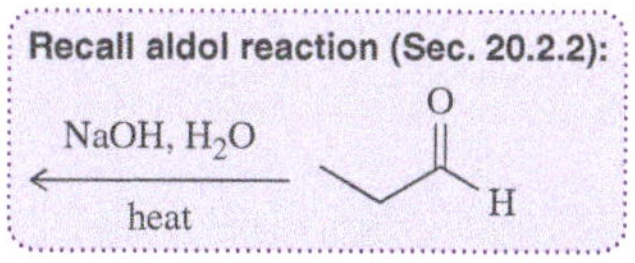

21.1.3 Conjugated reactive intermediates

Reactive intermediates are conjugated when the carbocation, carbanion, or radical neighbors a double or triple bond (**Figure 21.4**). We have studied molecules of this nature throughout the text. In Chapter 11, allylic radicals were intermediates in allylic halogenation. S_N1 and E1 reactions of allylic carbocations were important topics in Chapter 12. And enolates (carbanions neighbored by an electron-withdrawing carbonyl) were of central importance in Chapter 20. In each of these situations, the stability of the reactive intermediate was altered as a result of the neighboring π system. [*Why?*]

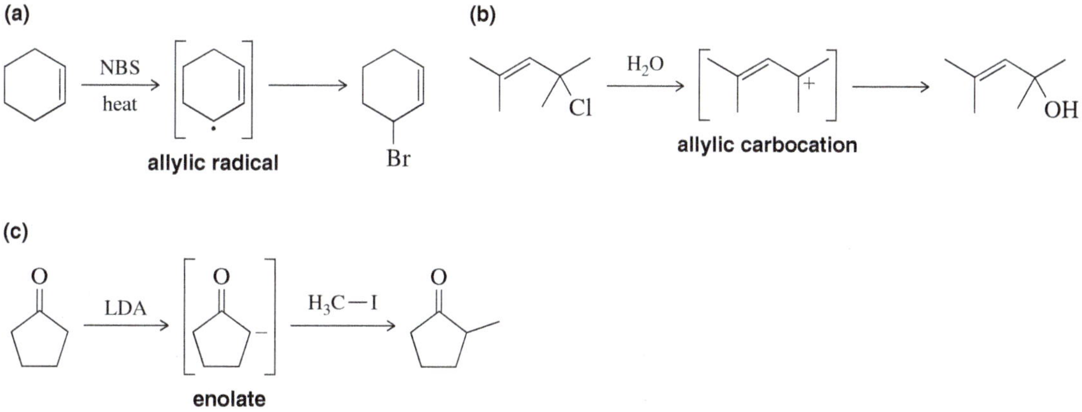

Figure 21.4 Conjugated (a) radicals, (b) carbocations, and (c) carbanions display altered reactivity.

Assessment 21.6 Identify each of the molecules shown as conjugated or nonconjugated.

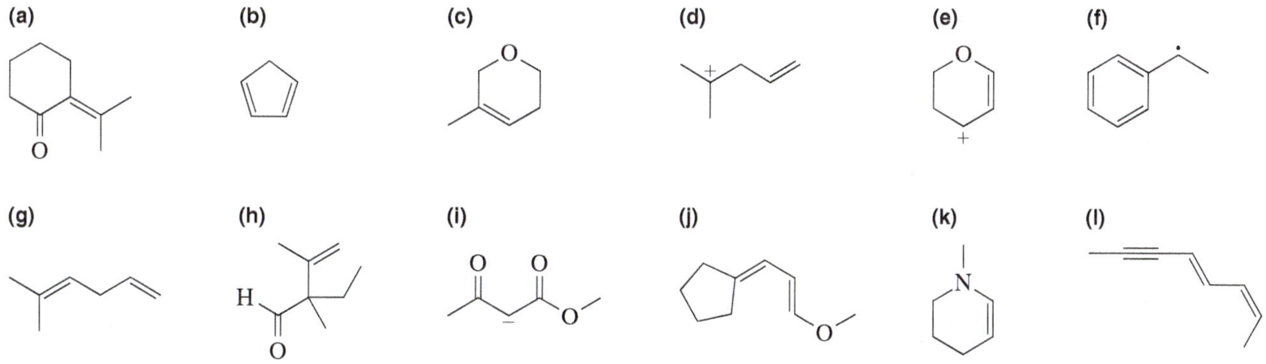

21.2 Enhanced stability in conjugated systems

One way of identifying the conjugated molecules in Assessment 21.6 is to look for places where resonance structures can be drawn. After all, extended π systems of delocalized electrons have, to this point in the text, been illustrated using resonance structures. With the correlation between conjugation and resonance delocalization, it should come as no surprise that conjugated systems show enhanced stability [*Section 4.4.3*]. The next three sections discuss how this enhanced stability has been observed as well as shed light on the degree to which conjugation increases stability.

21.2.1 Conjugated alkenes

In Section 9.2.1, we learned that the heat of hydrogenation ($\Delta H°_{hydr}$) can be used to determine alkene stability. [*See Assessment 9.38.*] **Figure 21.5** shows the $\Delta H°_{hydr}$ of cyclohexa-1,4-diene and cyclohexa-1,3-diene. Because two alkenes are reduced in both compounds and both reactions give the same final product, these numbers can be directly compared. For the nonconjugated cyclohexa-1,4-diene, $\Delta H°_{hydr}$ is as expected, releasing 28.6 kcal/mol (120 kJ/mol) for each alkene reduced. On the other hand, hydrogenation of the conjugated cyclohexa-1,3-diene is about 2 kcal/mol (9 kJ/mol) *less* exothermic, meaning less heat is released.

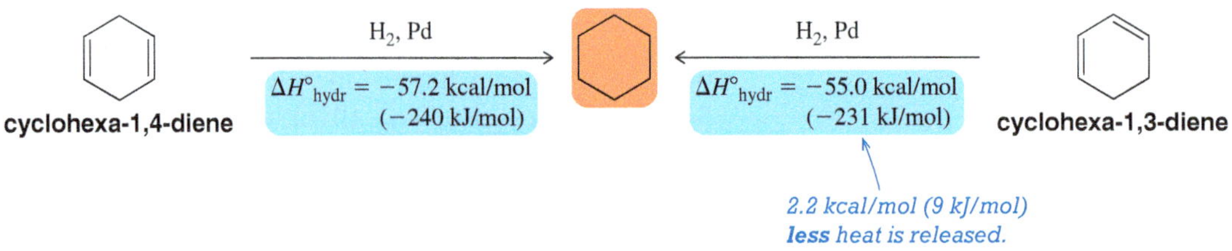

Figure 21.5 Reduction of a conjugated diene releases *less* heat than a nonconjugated diene.

To understand what this means, let's consider the energies associated with each of the two steps of the overall reductions to cyclohexane. The reductions of both cyclohexa-1,4-diene and cyclohexa-1,3-diene pass through cyclohexene. When cyclohexa-1,4-diene is reduced to cyclohexene, 28.6 kcal/mol are released. Similarly, when cyclohexene is reduced to cyclohexane, 28.6 kcal/mol are released, giving a total $\Delta H°_{hydr} = -57.2$ kcal/mol. For cyclohexa-1,3-diene, we know that 28.6 kcal/mol are released when cyclohexene is reduced to cyclohexane (second step). If the total $\Delta H°_{hydr} = -55$ kcal/mol, what is $\Delta H°_{hydr}$ for the reduction of cyclohexa-1,3-diene to cyclohexene? Simple subtraction shows that 26.4 kcal/mol are released in this first step. The first steps of the reduction of each can be compared on a reaction coordinate diagram (**Figure 21.6**). Reduction of a conjugated alkene only releases less energy if it starts lower on the hill; it must be 2.2 kcal/mol more stable!

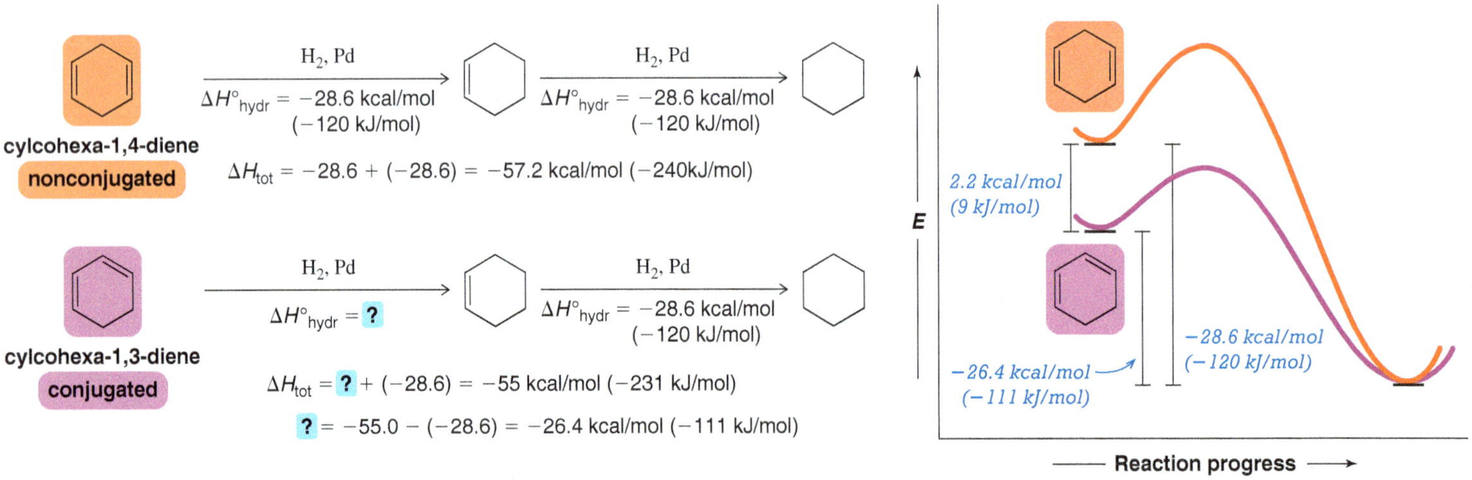

Figure 21.6 Conjugated alkenes start lower on the reaction coordinate diagram and are therefore more stable than nonconjugated alkenes.

21.2.2 Conjugated carbonyls

Recall from Section 20.3.2 that the aldol reaction is a condensation reaction between an enolate and an aldehyde [*or a ketone*]. When heated, in the final step, water is eliminated from a β-hydroxyaldehyde to make an α,β-unsaturated aldehyde. Although it wasn't discussed in Chapter 20, another alkene could have formed via a mechanism shown in **Figure 21.7**. However, the nonconjugated alkene is *never* obtained as a product in the aldol reaction, suggesting that it is substantially less stable than the conjugated alkene.

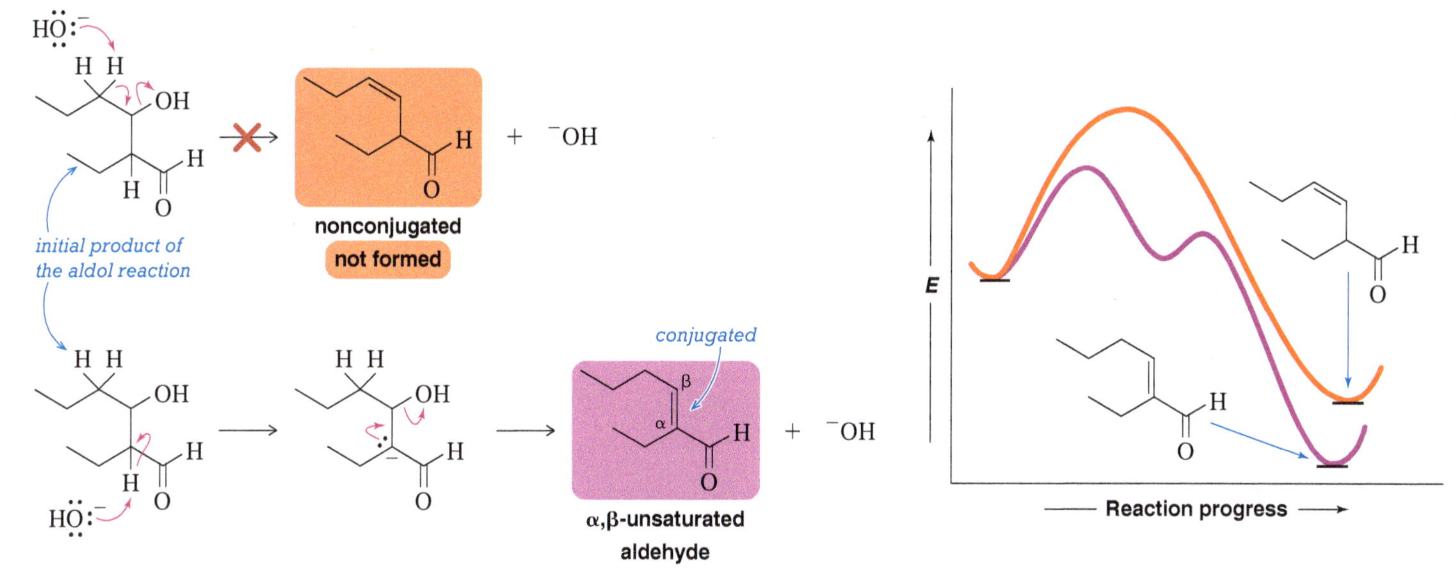

Figure 21.7 α,β-Unsaturated aldehydes (conjugated) form preferentially over nonconjugated aldehydes in aldol reactions.

Recall $\Delta G = -RT \ln K_{eq}$; see Section 5.1.3.

The enhanced stability from being conjugated can be shown by imagining the hypothetical equilibrium reaction in **Figure 21.8** between the α,β-unsaturated aldehyde and the nonconjugated aldehyde. The calculated ΔG for this equilibrium is around −9 kcal/mol. This results in a calculated equilibrium constant (K_{eq}) of 3.3×10^6 at room temperature. In sum, being conjugated (stabilized by resonance) dramatically influences the outcome of the aldol reaction.

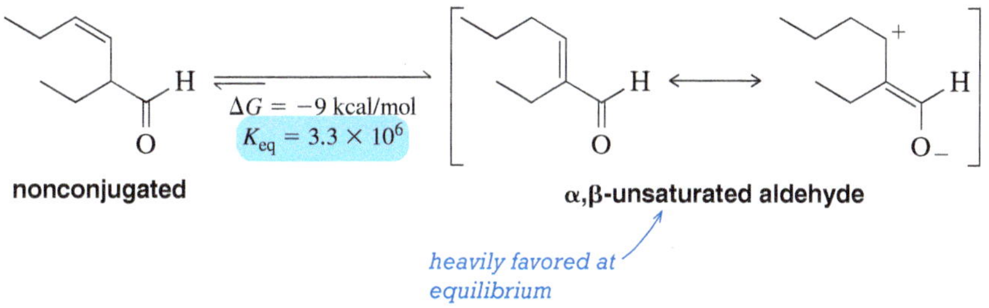

Figure 21.8 The more stable, resonance-stabilized (conjugated) alkene is favored in an equilibrium reaction.

21.2.3 Conjugated reactive intermediates

The stability of allylic radicals was discussed in Section 11.5.3 while introducing halogenation with *N*-bromosuccinimide (NBS). The relative stability of radicals is reflected in the bond-dissociation energy for the bond that would homolytically break to form the radical. Shown in **Figure 21.9**, the bond-dissociation energy for a 2° C—H bond to form a nonconjugated radical is 95 kcal/mol. When the 2° C—H bond is allylic (next to an alkene), the bond-dissociation energy drops to 87 kcal/mol. This 8 kcal/mol difference reflects the increased stability the radical gets from delocalization via resonance. Preferential formation of the more stable allylic radical drove the high selectivity in the allylic halogenation reactions.

In Section 12.2.1, we learned that the rate-determining step of an S_N1 reaction is formation of the carbocation in the first step. The stability of the (conjugated) allylic carbocation is demonstrated by comparing the relative rates of the solvolysis reactions of a 3° haloalkane and a

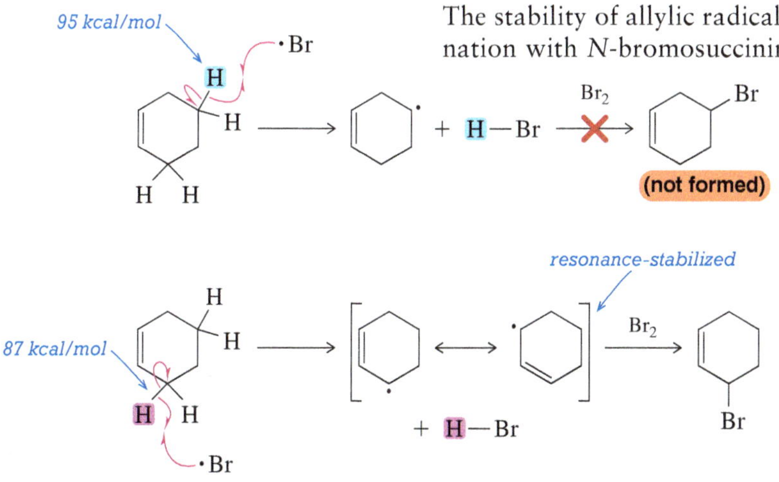

Figure 21.9 Allylic radicals (conjugated) are stabilized by resonance and thus form more quickly.

3° allylic haloalkane (**Figure 21.10**). Formation of the resonance-stabilized allylic carbocation occurs over 100 times faster than formation of the hyperconjugation-only-stabilized 3° carbocation.

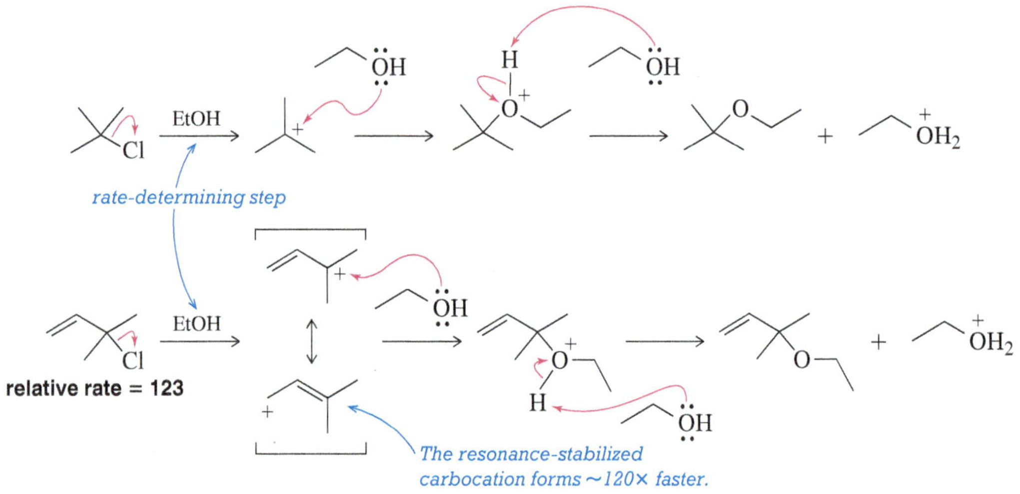

Figure 21.10 **Allylic carbocations (conjugated) are stabilized by resonance** and thus form more quickly.

Recall from Chapter 4 that a high pK_a value for an acid means that it is weak—it does not give up a proton easily and become the conjugate base. Thus, the conjugate base is unstable. In Chapter 16, the invention of Grignard and organolithium reagents was discussed as a way to form carbanions because deprotonation of unfunctionalized alkanes is impossible (pK_a = 50 for propane). However, for a hydrogen on a carbon next to a carbonyl (i.e., the α-carbon; Section 20.3), the pK_a drops to 20. An activated methylene, between two carbonyls, has a pK_a = 10 (Section 20.4). These (conjugated) enolate anions, with the negative charge delocalized by resonance onto the carbonyl oxygens, form more rapidly and more favorably than unfunctionalized carbanions (**Figure 21.11**).

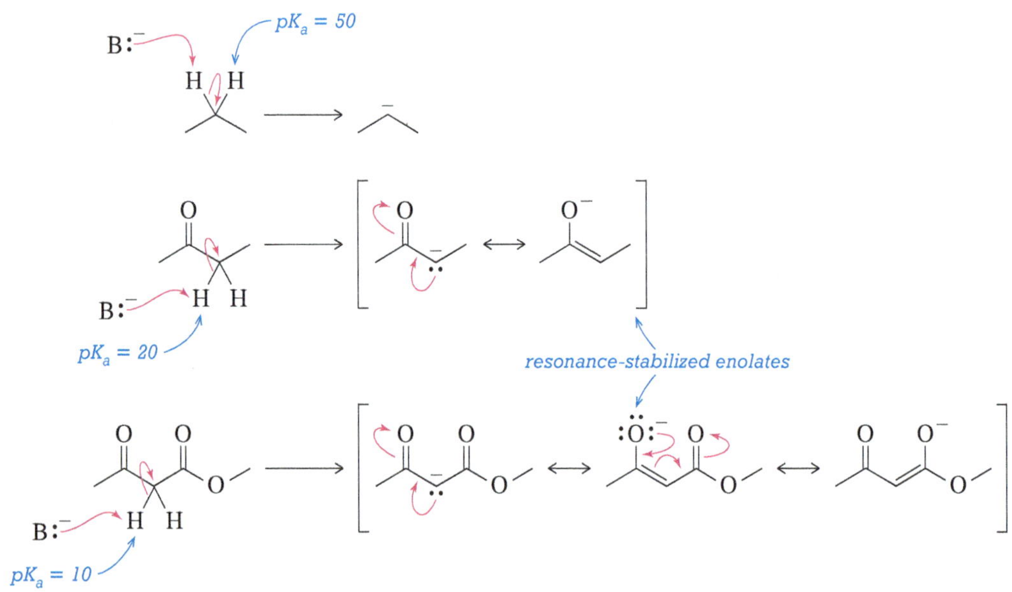

Figure 21.11 **Enolates (conjugated) are stabilized by resonance** and thus form more favorably.

21.2 SUMMARY

- Conjugated systems, by delocalizing electron density through resonance, are more stable than their nonconjugated counterparts. Conjugated reactants react more slowly, while conjugated intermediates form more quickly.

Assessment 21.7 To this point, hydrogenation has always been an exothermic process. Using the numbers from Figure 21.6, calculate ΔH°_{hydr} for each step of the reduction of benzene. [*We study benzene and the value obtained for step 1 in Section 23.2.1.*]

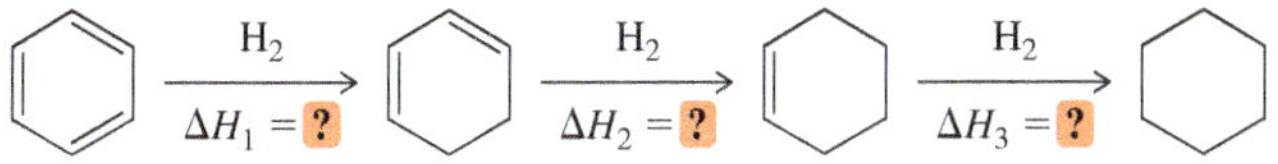

$$\Delta H_1 + \Delta H_2 + \Delta H_3 = \Delta H_{tot} = -49.5 \text{ kcal/mol } (-208 \text{ kJ/mol})$$

Assessment 21.8 Which carbonyl would you expect to absorb infrared radiation at a higher frequency? Explain.

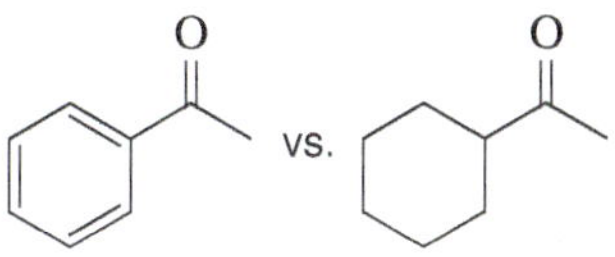

Assessment 21.9 There were actually two possible products in the solvolysis reaction from Figure 21.10. Show both products. Which would you expect to be more stable? Why? [*This concept will be important in Section 21.4.1.*]

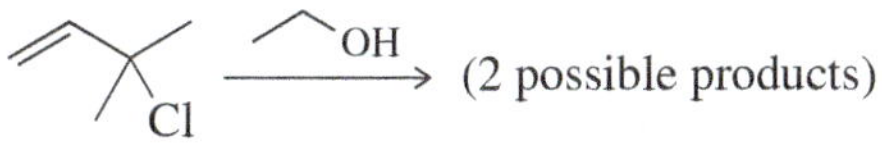

(from Figure 21.10)

Assessment 21.10 In Chapter 12, we learned that ethoxide does elimination preferentially at 2° haloalkanes, whereas acetate does substitution. Explain this in the context of conjugation.

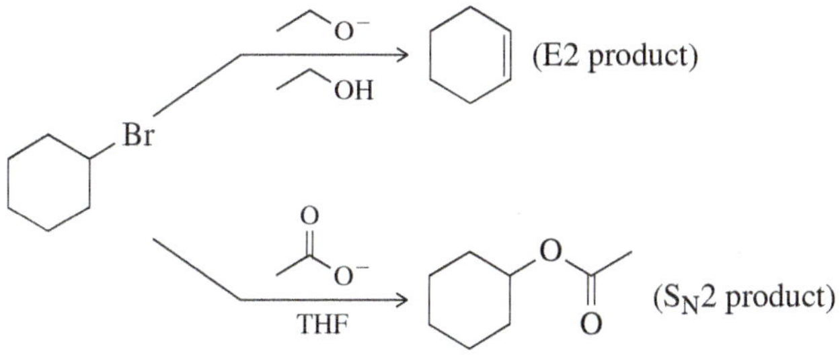

Assessment 21.11 *p*-Nitrophenol ($pK_a = 7.2$) is ten times more acidic than *m*-nitrophenol ($pK_a = 8.4$). Explain. [*This concept is the subject of Section 24.2.1 in the chapter on benzene.*]

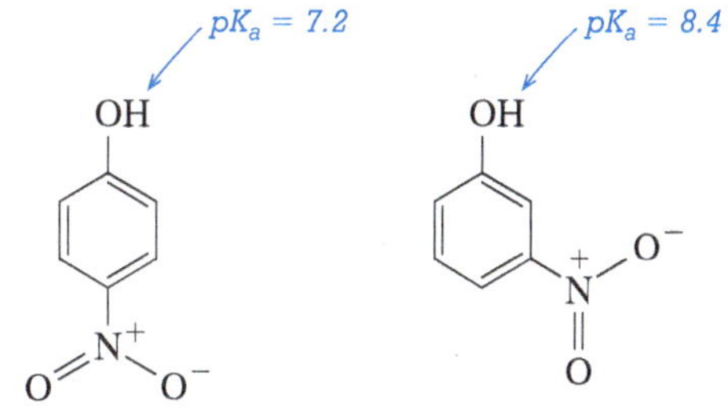

***p*-nitrophenol** ***m*-nitrophenol**

21.3 Molecular orbital picture of conjugated alkenes

The stability, reactivity, and as we explain in Section 21.3.2, the fact that conjugated alkenes absorb ultraviolet (UV) and visible light can be explained by examining the molecular orbital picture of conjugated alkenes. Specifically, we focus in this section on the system of π molecular orbitals. [*The molecular orbital diagrams you will be drawing are derived from complicated mathematical calculations more suitable to a physical chemistry course and beyond the scope of this book. Our discussion here won't require math.*]

A simple way to think about molecular orbital theory is as the **linear combination of atomic orbitals** (**LCAO**). This term, LCAO, in many ways encapsulates everything we need to know about forming molecular orbitals. As discussed in Chapter 2, molecular orbitals are formed by the overlap/combination (the C in LCAO) of two atomic orbitals (the AO in LCAO). **Figures 21.12(a)** and **21.12(b)** show that when the orbitals that are overlapping have *s* character (e.g., *s* orbitals or sp^x hybridized), the molecular orbitals formed are σ orbitals. Overlapping *p* orbitals form π molecular orbitals [**Figure 21.12(c)**].

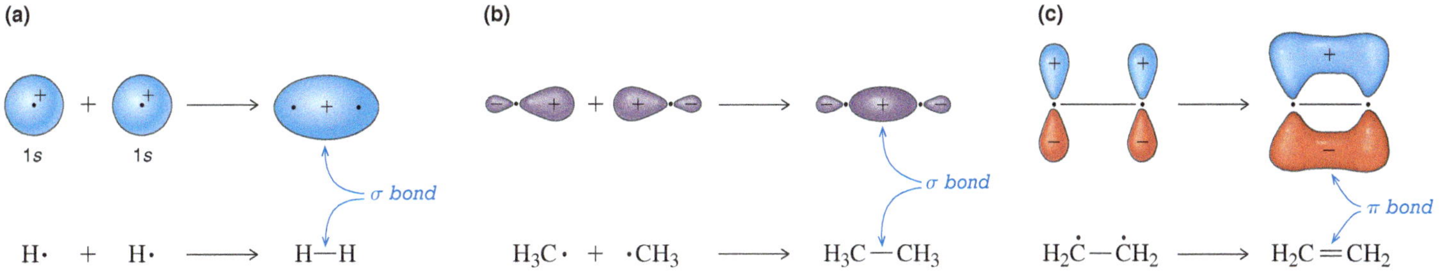

Figure 21.12 Overlapping atomic orbitals produce molecular orbitals. (a) Overlap of the *s* orbitals of two hydrogen atoms forms the σ bond in H_2. (b) Overlap of sp^3-hybridized orbitals makes the σ bond in ethane. (c) Overlap of two *p* orbitals makes the π bond in ethene.

To summarize what has so far been a review of the molecular orbital theory discussed in Chapter 2 and throughout the book, the key message is this: For every atomic orbital that is combined, one molecular orbital is produced. If two atomic orbitals combine, two molecular orbitals result. If four, then four. Put a different way, the combination of n atomic orbitals will produce n molecular orbitals.

This brings us to the L in LCAO. Atomic orbitals themselves are actually wavefunctions, with bonding between them representing the combination of these atomic orbital wavefunctions. As wavefunctions, there are two ways for them to combine. They can be added in phase (a constructive interaction), or they can be added out of phase (a destructive interaction). In-phase addition occurs when lobes of like sign (+ with +; − with −) combine. Out-of-phase addition occurs when lobes of opposite sign (+ with −) combine. The + sign cancels the − sign, so this creates an area between two nuclei where the molecular orbital wavefunction equals zero. The result is a node—a region of zero overlap, or bonding, in that particular molecular orbital. [*For this reason, out-of-phase addition can be thought of as subtraction.*] Thus, for the combination of any two atomic orbitals, two molecular orbitals are produced. It's this last statement that explains the L in LCAO, deriving from a mathematical linear expansion expression for molecular orbitals. [*That's the last you'll hear about that, for now.*]

Figure 21.13(a) shows the in-phase addition (constructive interaction; + with +) of the wavefunctions of two 1*s* orbitals to make a σ bonding molecular orbital (the σ bond). **Figure 21.13(b)** shows out-of-phase addition (i.e., subtraction/destructive interaction; + with −) to make a σ* antibonding molecular orbital. The σ* orbital has a node between the two nuclei. This is the σ* where nucleophiles donate electron density during the S_N2 reactions of haloalkanes (Chapter 12). **Figure 21.13(c)** shows both the

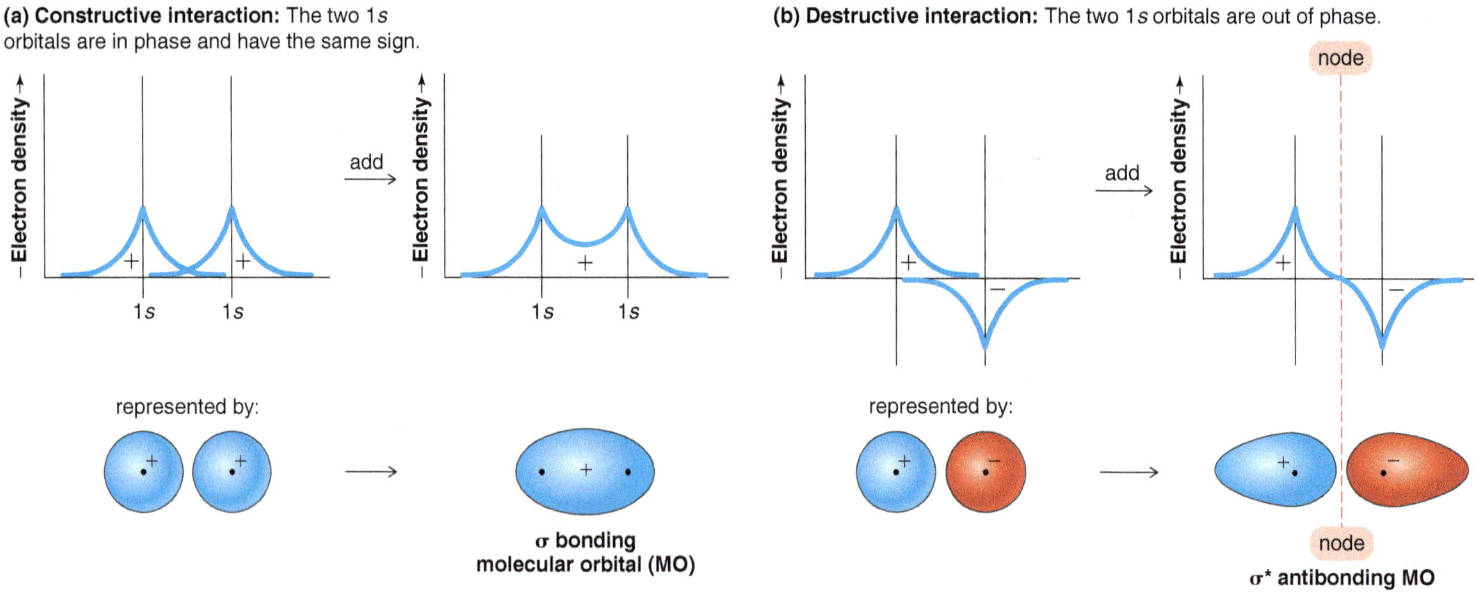

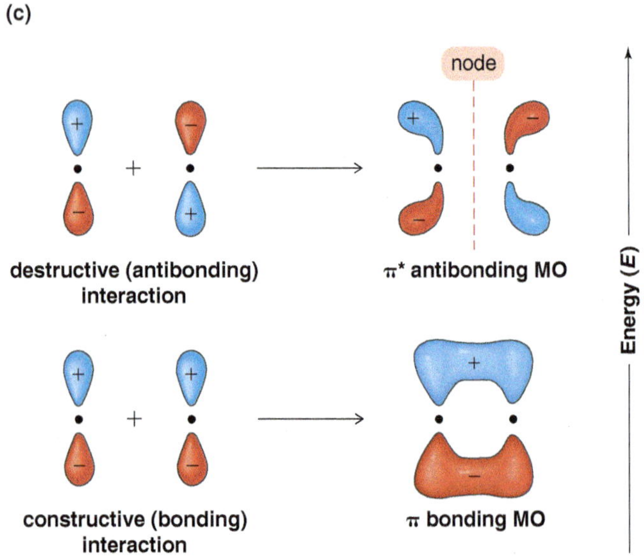

Figure 21.13 The combination of two atomic orbitals produces two molecular orbitals. (a) In-phase addition of *s* orbitals produces a σ bond. (b) Out-of-phase addition of *s* orbitals produces σ*, the antibonding orbital. (c) In-phase addition of *p* orbitals produces a π bond, while out-of-phase addition produces the π* antibonding orbital.

in-phase and out-of-phase additions of *p* atomic orbitals to make the π bonding molecular orbital (the π bond, full overlap, no node) and the π^* antibonding molecular orbital (no overlap between *p* orbitals, 1 node). The π^* orbital is where nucleophiles add to carbonyls, as defined by the Bürgi–Dunitz angle (Section 17.1.4). To reiterate, bonding orbitals contain greater overlap and are lower energy/more stable orbitals. Antibonding orbitals always contain a node—an area where there is no overlap—and are therefore less stable.

Assessment 21.12 Suppose a molecule is formed by the overlap of 16 atomic orbitals. (a) How many molecular orbitals will be present? (b) How many will be bonding? (c) How many will be antibonding?

Assessment 21.13 Ethane has seven bonding orbitals (all σ bonds). How many antibonding orbitals are present?

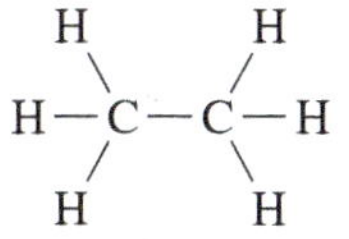

ethane

Assessment 21.14 Hexa-1,3,5-triene uses six *p* orbitals, each containing a single electron, to make its three π bonds. How many total molecular orbitals are made by the six *p* orbitals?

hexa-1,3,5-triene

Assessment 21.15 The molecular orbital picture of H_2 can be represented by the following diagram. Label σ and σ^* on the diagram—that is, which is (a) and which is (b)? Which is lower in energy? Why?

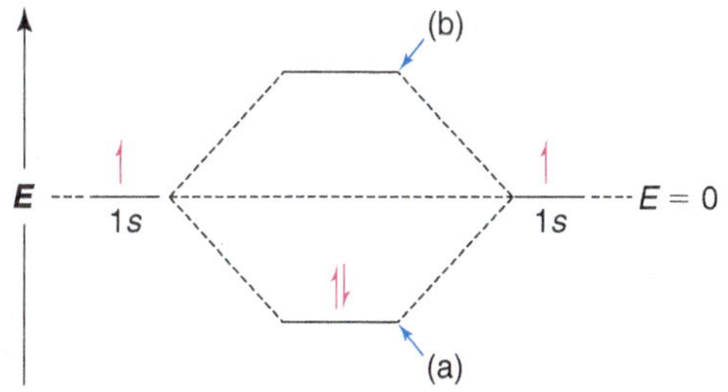

[*Here's where it gets interesting.*] **Figure 21.14(a)** shows what a π molecular orbital (π bond) would look like for the isolated (nonconjugated) alkene ethene. In this case, you can confidently say that the 2 electrons in the π bond are located above and below the σ bond. Alternatively, consider the case of buta-1,3-diene, a conjugated alkene. As was apparent in Section 21.2, conjugation implies that resonance structures can be drawn. Look closely at the π electrons between C_1 and C_2 in the first resonance structure of **Figure 21.14(b)**. Depending on which resonance structure you consider, these electrons could be forming the π bond between C_1 and C_2, as a lone pair on C_1, or forming a π bond between C_2 and C_3. So where are they really? Recall that the true electronic structure is a hybrid of all valid resonance structures, so electrons are shared between the *p* orbitals of all four carbons.

Figure 21.14 illustrates the key difference between conjugated and nonconjugated alkenes as it relates to the molecular orbital picture. Instead of a conjugated diene being represented by two sets of two overlapping *p* orbitals to make two isolated π bonds (and two π^* antibonding orbitals), they are better represented by one set of four overlapping *p* orbitals to make two π bonding orbitals (and two π^* antibonding orbitals). [*Four atomic orbitals combine to make how many molecular orbitals? Four!*] This is because the neighboring *p* orbitals that interact via resonance structures belong not to a single π bond but, instead, to the entire π system.

Figure 21.14 Location of π electrons in a bonding orbital. (a) The electrons in an isolated π bond are only between the two connected carbons. (b) The electrons in a conjugated π bond are delocalized throughout the π system.

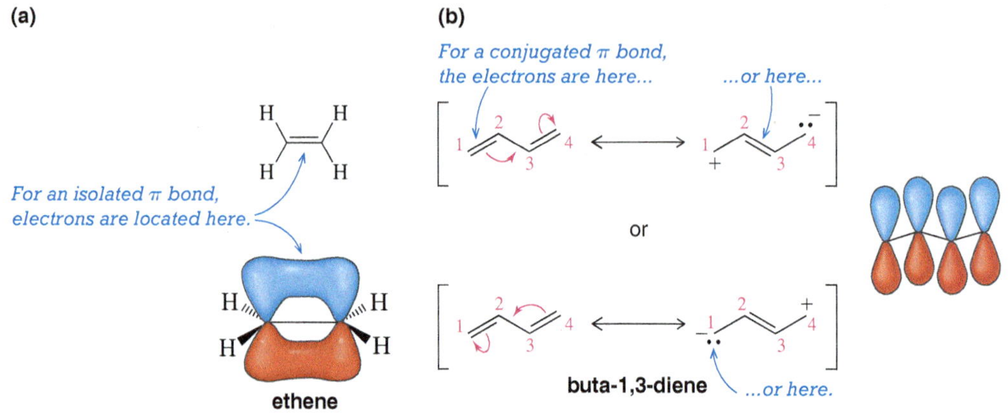

Drawing molecular orbital pictures can be complicated, but is simplified by following a few rules. [*These rules arise from the math.*]

1. Combining n atomic orbitals gives n molecular orbitals, each designated as ψ_n.
2. The most stable/lowest energy molecular orbital has maximum overlap between lobes of like sign. [*The signs of the lobes are not charges. They are the sign of the wavefunction.*] Like signs as neighbors indicates overlap, which is good!
3. The number of nodes—areas with no overlap (lobes of opposite sign as neighbors) in consecutive molecular orbitals—increases by one as you move from low energy (stable) to high energy (less stable).
4. Each molecular orbital should have symmetry around the node(s).
5. There is one p electron for each p orbital. These are placed in molecular orbitals (2 electrons in each), beginning with the lowest energy MO and working up. [*This is the aufbau principle.*]
6. Molecular orbitals lower in energy than nonbonded atomic orbitals ($E = 0$) are bonding; molecular orbitals higher in energy than nonbonded atomic orbitals ($E = 0$) are antibonding. Filled orbitals contain electrons; unfilled orbitals do not.

Figure 21.15 shows how these rules would be applied to buta-1,3-diene, which has four p orbitals, each contributing 1 electron for a total of 4 π electrons. The arrangement of orbitals that produces maximum overlap (ψ_1, pronounced "psi-one") has all lobes with like signs as neighbors. [*Rule 2. Say it with me: "plus-plus-plus-plus" across the top and "minus-minus-minus-minus" across the bottom.*] With the maximum possible overlap, ψ_1 is the most stable molecular orbital. The next orbital (ψ_2) should have one node [*rule 3—one more node than* ψ_1]. To ensure symmetry, the order goes [*say it!*] "plus-plus-minus-minus" across the top and "minus-minus-plus-plus" across the bottom. [*Why not "plus-plus-plus-minus?" It has one node, right? See Assessment 21.16.*] Continuing in this pattern gives ψ_3 [*"plus-minus-minus-plus," which gives two nodes*] and ψ_4 [*"plus-minus-plus-minus," which gives three nodes*]. Note that ψ_4 has no overlap; as such, it is the least stable MO of buta-1,3-diene. With the orbitals constructed, 2 electrons each (a total of 4 electrons in the π system of butadiene) are placed in ψ_1 and ψ_2 [*rule 5: The lowest energy MOs are filled first*]. These are filled bonding orbitals [*rule 6*]. There are no remaining electrons for ψ_3 or ψ_4. These are unfilled antibonding orbitals [*also rule 6*].

In addition to using the ψ_n designation, certain orbitals are identified differently. In Figure 21.15, ψ_2 is referred to as the **HOMO**, or the **highest occupied molecular orbital.** Exactly as the name implies, this orbital is the highest energy/most reactive/least stable molecular orbital that contains electrons. It is the highest energy bonding orbital. Alternatively, ψ_3 is referred to as the **LUMO**, or the **lowest unoccupied molecular orbital.** The LUMO is the lowest energy/least reactive/most stable molecular orbital that does *not* contain electrons.

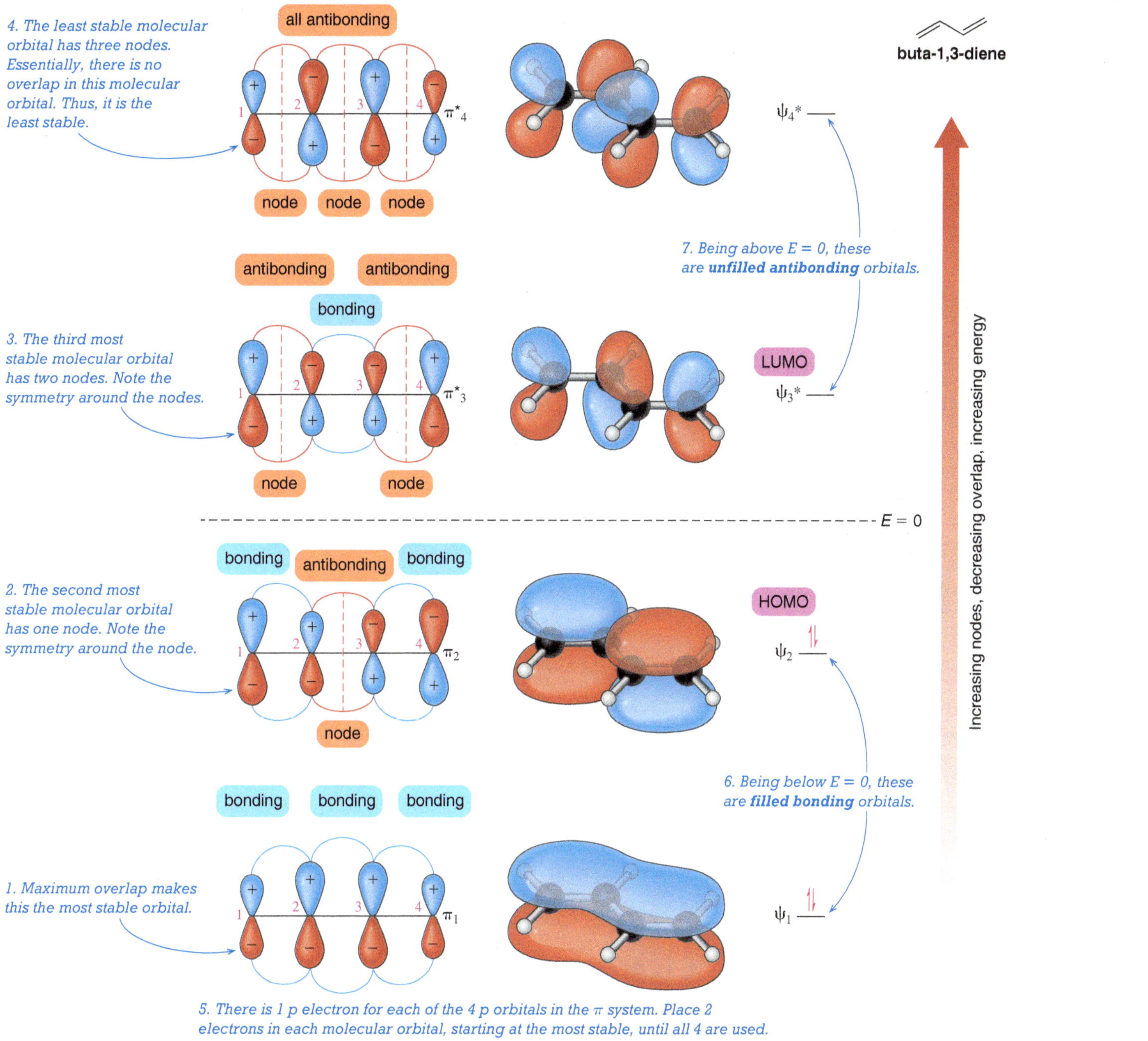

Figure 21.15 Drawing the molecular orbital picture of buta-1,3-diene. Bonding orbitals are more stable than the nonbonded atomic orbitals. Antibonding orbitals are less stable than nonbonded atomic orbitals.

It is the lowest energy antibonding orbital. [*Confirm the identity of the HOMO and LUMO by looking back at Figure 21.15.*] The HOMO–LUMO designations become important when we consider UV and visible light spectroscopy (Section 21.3.2) and reactions of conjugated systems in Chapter 22.

21.3 SUMMARY

- The molecular orbital picture of conjugated systems involves a linear combination of atomic orbitals to make as many π MOs as there are *p* atomic orbitals contributing.

Assessment 21.16 Why is this *not* a viable representation of the ψ_2 molecular orbital of buta-1,3-diene?

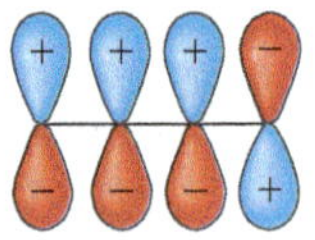

Assessment 21.17 The ψ_2 molecular orbital for octa-1,3,5,7-tetraene is shown. Draw ψ_3.

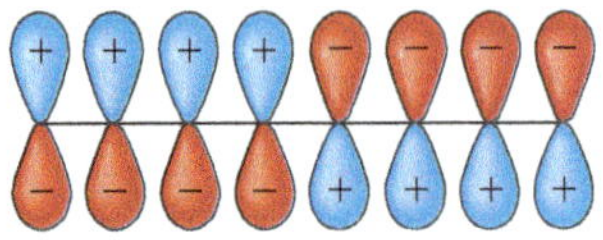

Assessment 21.18 Which of the following would you expect to be a more stable molecular orbital? Why?

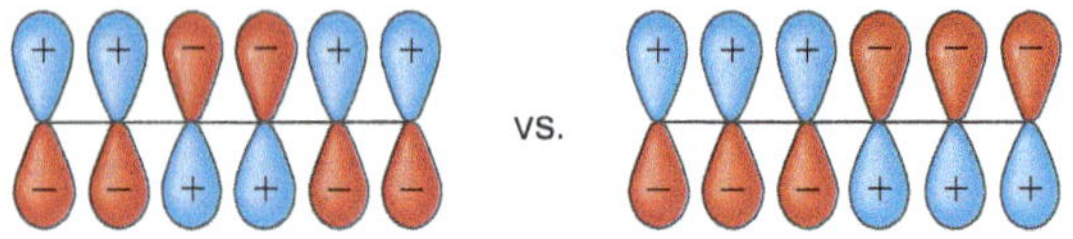

Assessment 21.19 Which of the following would you expect to be a more stable molecular orbital? Why?

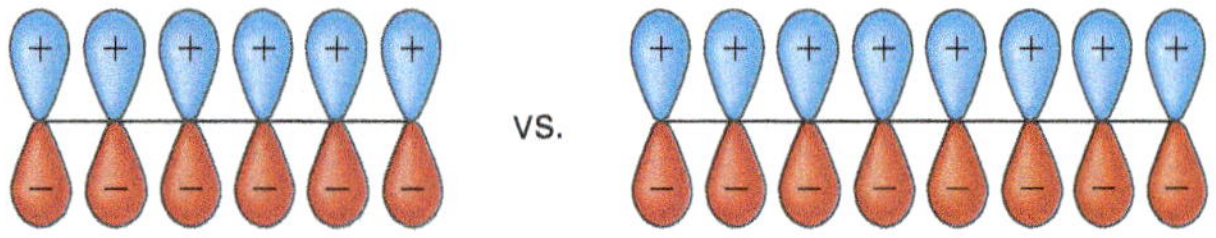

Assessment 21.20 Using the rules described in Section 21.3, draw the molecular orbital picture of hexa-1,3,5-triene. Label the HOMO and LUMO.

hexa-1,3,5-triene

Let's solve Assessment 21.19 to see how conjugation increases stability in ways that parallel our understanding of resonance.

21.3.1 Implications for stability

The molecular orbital picture of conjugated systems has implications for the stability of these systems.

Solved Assessment 21.19

Assessment 21.19 asks us to identify the more stable molecular orbital. Although it isn't stated in the problem, two things are clear about these molecular orbitals. First, they are from different molecules, because one has six overlapping *p* orbitals (hexa-1,3,5-triene) and the other has eight (octa-1,3,5,7-tetraene) (**Figure 21.16**). Second, because both have maximum overlap, they are the most stable molecular orbital in each of the respective molecules.

Being able to draw more resonance structures makes a molecule more stable because there are more atoms on which to delocalize electrons. Similarly, a molecular orbital that is delocalized over more atoms is more stable. Thus, the molecular orbital of octa-1,3,5,7-tetraene is more stable (**Figure 21.17**). ■

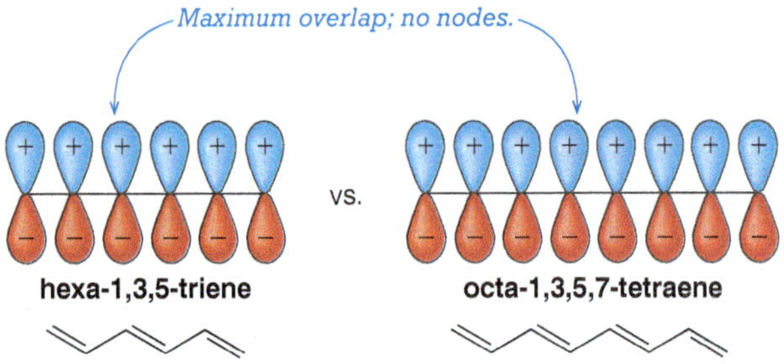

Figure 21.16 Maximum overlap occurs in the most stable molecular orbital.

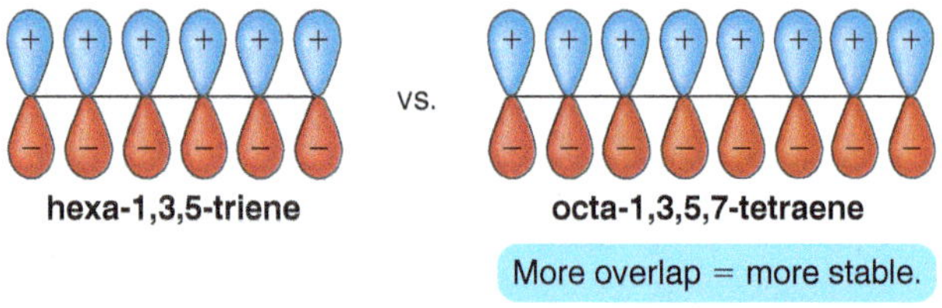

Figure 21.17 More overlapping orbitals/more conjugation makes for a more stable molecular orbital.

The trend illustrated in Solved Assessment 21.19 has significant implications. Basically, an increase in conjugation allows for more *p* orbitals to overlap, increasing the amount of delocalization possible in the π system. Because ψ_1 for conjugated molecules is at a lower energy than a nonconjugated π molecular orbital (ψ_1 for ethene), conjugated molecules are more stable than nonconjugated ones. And with increased conjugation comes enhanced stability. A comparison of the MO pictures of ethene, buta-1,3-diene, and hexa-1,3,5-triene in **Figure 21.18** makes this clear. Note that while the specific orbital overlap isn't shown for

Make sure you rotate the book 90° clockwise to see this figure as intended.

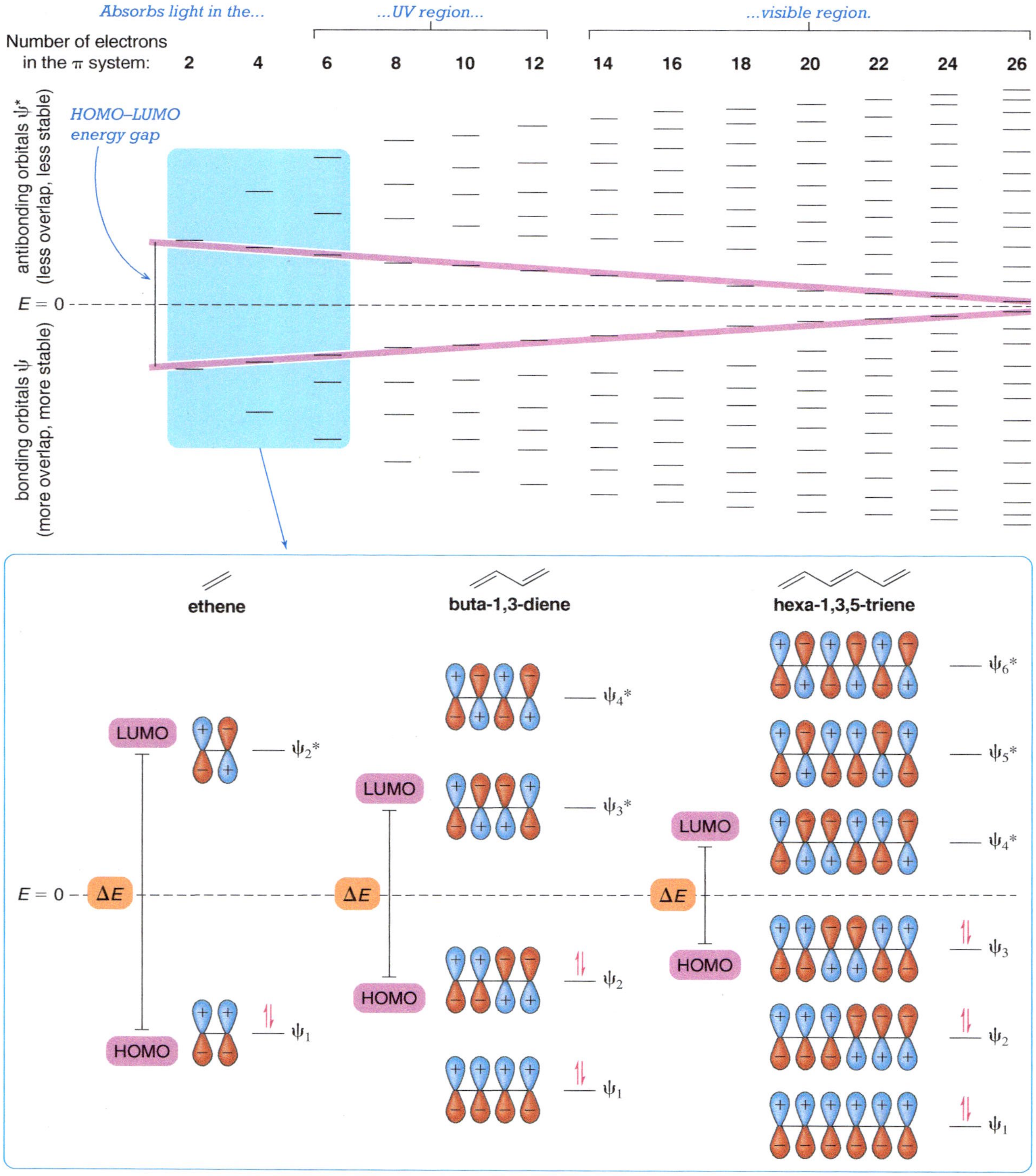

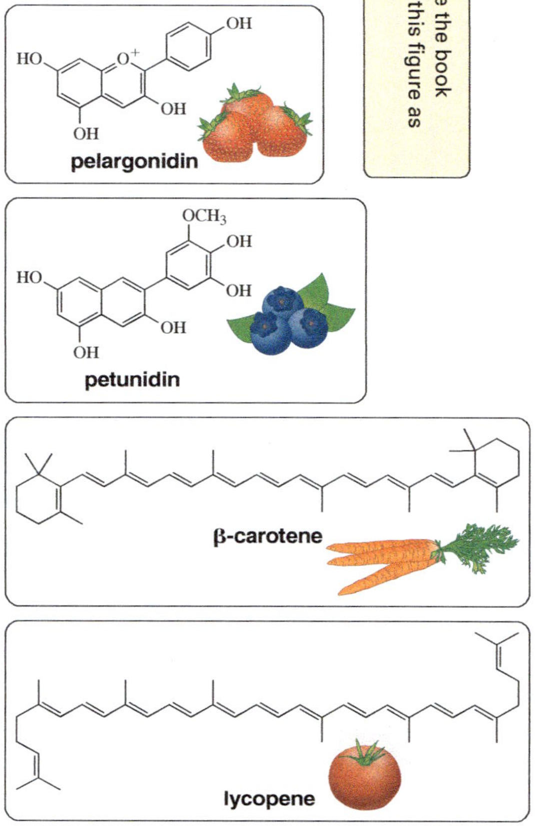

From $n = 2$ to $n = 26$:

1. ψ_1 becomes more stable.
2. HOMO–LUMO energy gap shrinks.
3. Energy of transition corresponds to light in the ultraviolet, then visible region.
4. Number of π molecular orbitals = number of overlapping π electrons.

Figure 21.18 Increasing conjugation results in better delocalization of electrons, greater orbital overlap, more stable molecules, and a smaller HOMO–LUMO energy gap. The number of electrons in each π system is indicated at the top of the figure and where they go is shown in the expansion of the π systems shaded in blue.

larger conjugated systems, the trend continues with the most stable molecular orbital (ψ_1) in each larger polyene being more stable than the previous. [*Figure 21.18 is one of those pictures with a ton of information. Look at it carefully, but don't be overwhelmed by it. We will return to it several times going forward.*]

21.3.2 Implications for UV–vis spectroscopy

Most of the standard compounds you've worked with in lab, like acetone, benzoic acid, and diethyl ether, are colorless (or appear white, if solid), implying that these molecules do *not* absorb or reflect light. We know from studying spectroscopy (Chapters 14 and 15) that this is not the case, because all organic molecules absorb electromagnetic radiation in the infrared (corresponding to bond vibration) and radio frequency (corresponding to flipping of ^{1}H and ^{13}C nuclei) regions. As seen in the electromagnetic spectrum in **Figure 21.19**, both the IR and radio frequencies consist of very low energy radiation.

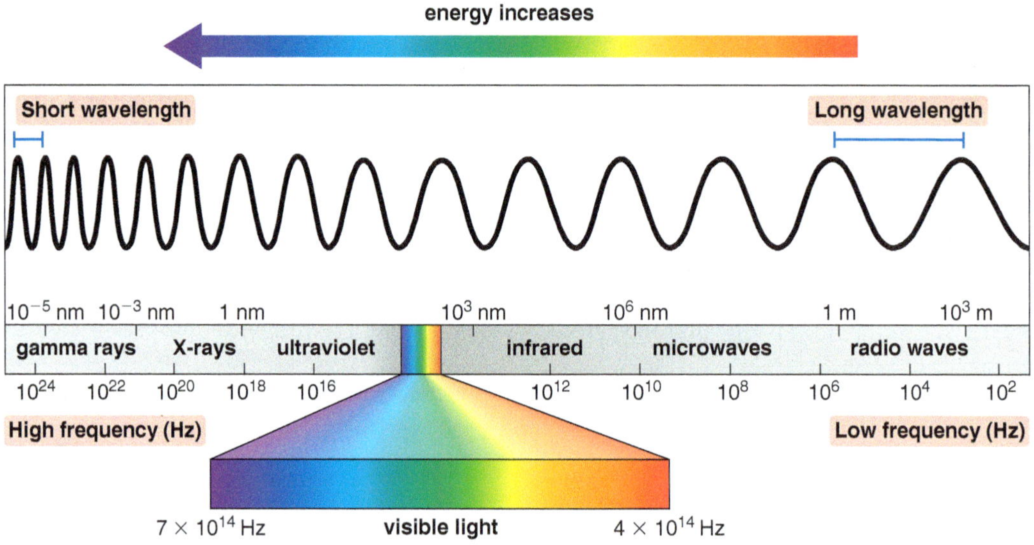

Figure 21.19 The electromagnetic spectrum. As wavelength decreases, frequency increases and energy increases. Infrared and radio waves correspond to low-energy transitions.

The fact that most molecules are colorless implies that there is no energy transition that corresponds to frequencies in the visible region. In addition to nuclei flipping (NMR spectroscopy) and bond vibration (infrared spectroscopy), molecules can also undergo electronic transitions. These electronic transitions are a result of electron excitation from occupied to unoccupied orbitals—for example, from the highest occupied molecular orbital (HOMO) to the lowest unoccupied molecular orbital (LUMO). Because the HOMO is the last bonding π orbital and the LUMO is the first antibonding π orbital, this excitation is referred to as a $\pi \rightarrow \pi^*$ transition. In ethene, the $\pi \rightarrow \pi^*$ transition requires 173 kcal/mol, corresponding to a wavelength (λ_{max}) of 165 nm, according to Equation 21.1, which relates wavelength to energy.

$$\lambda_{max} = (N_A \times h \times c)/\Delta E, \qquad \text{Equation 21.1}$$

where h (Planck's constant) = 6.63×10^{-34} J·s, N_A (Avogadro's number) = 6.02×10^{23} mol^{-1}, and c (speed of light) = 3.00×10^{17} nm/s.

These wavelengths are termed λ_{max} because they represent the wavelength of maximum absorption—that is, the wavelength that corresponds to the exact energy of the $\pi \rightarrow \pi^*$ transition. Note in Figure 21.18 and **Figure 21.20** that with increasing conjugation, the HOMO–LUMO energy gap decreases. For this reason, the $\pi \rightarrow \pi^*$ transitions for buta-1,3-diene and hexa-1,3,5-triene correspond to λ_{max} = 217 nm (132 kcal/mol) and λ_{max} = 258 nm

(111 kcal/mol). The ultraviolet region of the electromagnetic spectrum ranges from 10 to 400 nm, so buta-1,3-diene and hexa-1,3,5-triene absorb UV light and are said to be UV active. As such, they can be detected and analyzed using a UV–vis spectrometer. [*UV–vis spectroscopy was once used for the identification of organic compounds. Now it is primarily used for quantitative analysis of organic compounds.*]

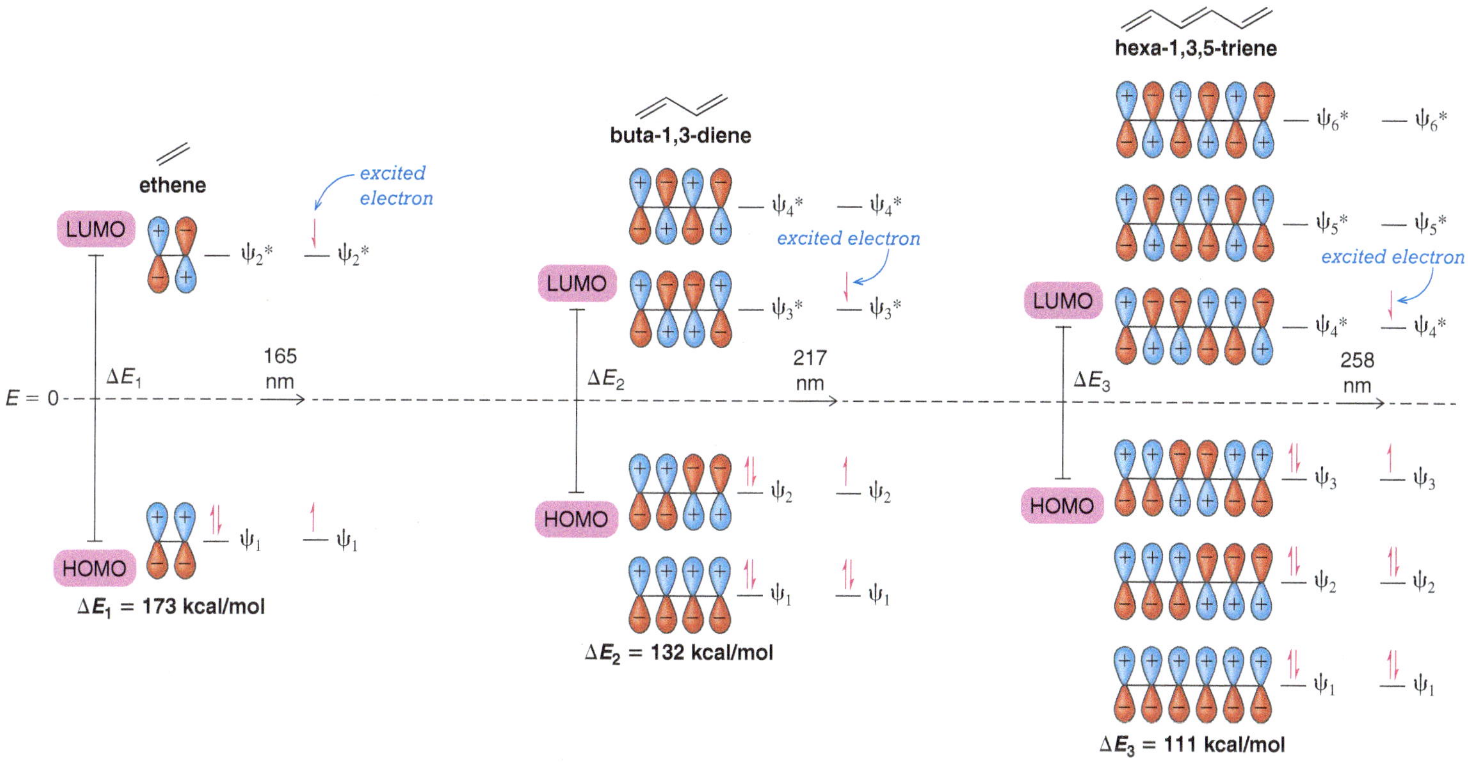

Figure 21.20 Irradiation with light causes excitation of an electron from the HOMO to the LUMO ($\pi \rightarrow \pi^*$). As conjugation increases, the HOMO–LUMO energy gap decreases, causing the irradiation frequency to move into the UV and visible regions of the electromagnetic spectrum.

Table 21.1 shows that with increasing conjugation, λ_{max} increases, moving farther into the UV region of the electromagnetic spectrum. Although all of the molecules would continue to be clear or colorless to the human eye, they would be visible under UV light. As conjugation increases beyond the molecules in Table 21.1, λ_{max} continues to increase, eventually moving into the visible region of the electromagnetic spectrum (390 to 700 nm). For example, lycopene and β-carotene in Figure 21.21 are highly conjugated and colored molecules. Specifically, lycopene absorbs electromagnetic radiation at λ_{max} = 472 nm, which corresponds to the color blue. Thus, when light (of all visible wavelengths) is shined on lycopene, the blue is absorbed, and all others are reflected. As a result, lycopene appears to our eye as red/orange. Similarly, carotene absorbs at λ_{max} = 450 nm, which corresponds to blue/violet and appears to the human eye as orange/yellow.

Table 21.1 Increasing conjugation decreases the HOMO–LUMO energy gap and raises λ_{max}.

Conjugated polyene	λ_{max} (nm)
ethene	165
buta-1,3-diene	217
hexa-1,3,5-triene	258
octa-1,3,5,7-tetraene	290
deca-1,3,5,7,9-pentaene	334
dodeca-1,3,5,7,9,11-hexaene	364

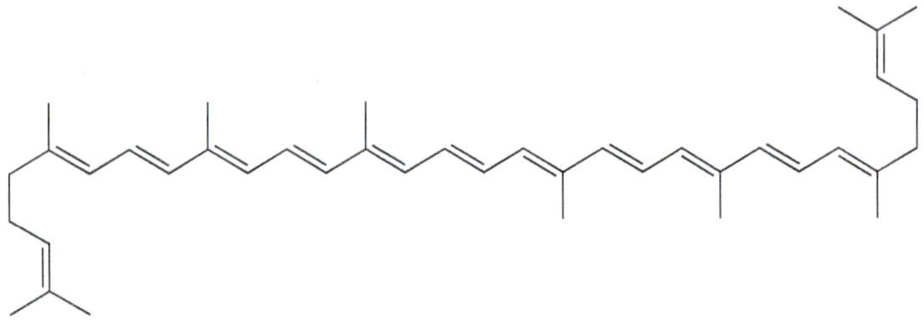

lycopene
λ_{max} = 472 nm

Observed color of compound	Color of light absorbed	Approximate wavelength of light absorbed (nm)
green	red	700
blue-green	orange-red	600
violet	yellow	550
red-violet	yellow-green	530
red	blue-green	500
orange	blue	450
yellow	violet	400

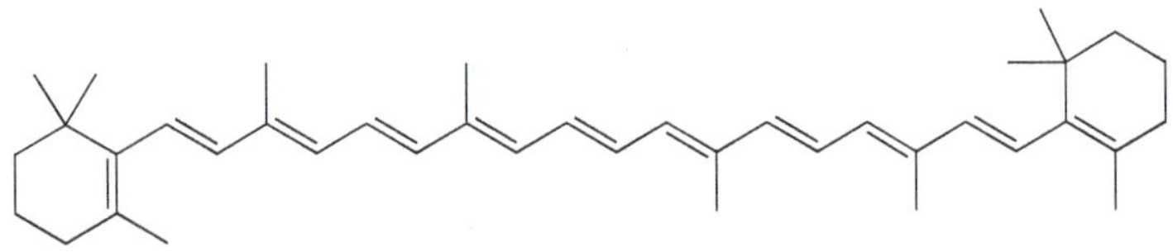

β-carotene
λ_{max} = 450 nm

Figure 21.21 Lycopene and β-carotene, because of their extensive conjugation and small HOMO–LUMO gap, absorb light in the visible region. As a result, they appear to us as colored molecules.

21.3.2 SUMMARY

- A small HOMO–LUMO gap allows a conjugated molecule to absorb light in the UV–vis region. It is why many vegetables and fruits are so colorful.

Assessment 21.21 Which of the following molecules would you expect to absorb in the visible region of the electromagnetic spectrum?

(a)

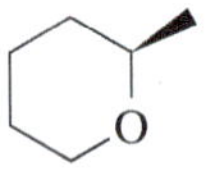

(b)

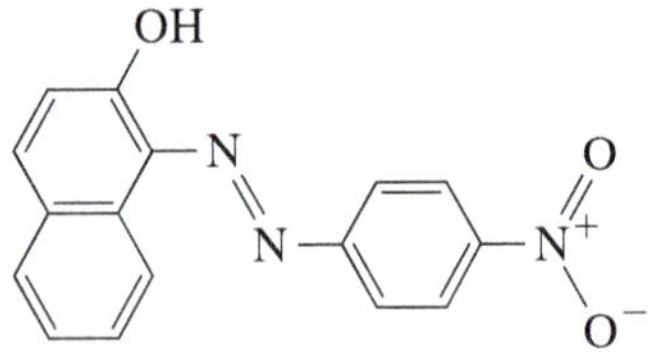

(c)

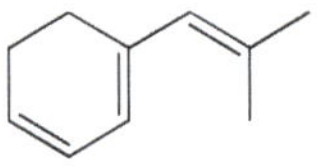

(d)

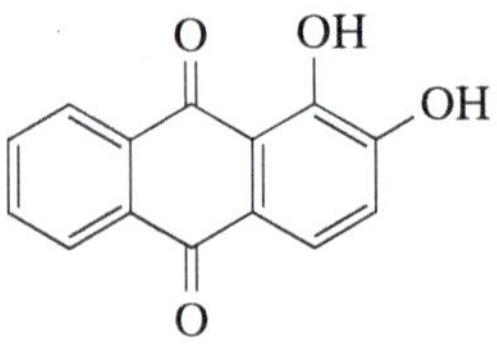

(e)

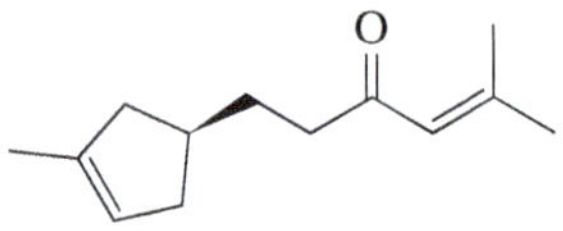

Assessment 21.22 Petunidin gives blueberries their blue color. Estimate λ_{max} for petunidin based on it appearing as blue.

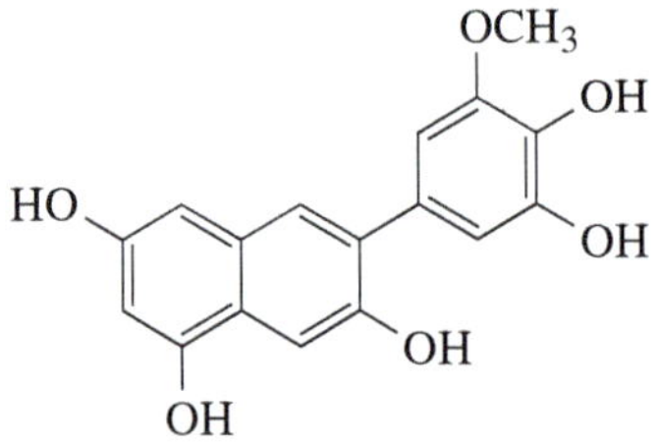

petunidin

Assessment 21.23 The following molecule, discovered in 1856 by William Perkin, was the first synthetic organic dye. It has a λ_{max} = 540 nm. What color is it?

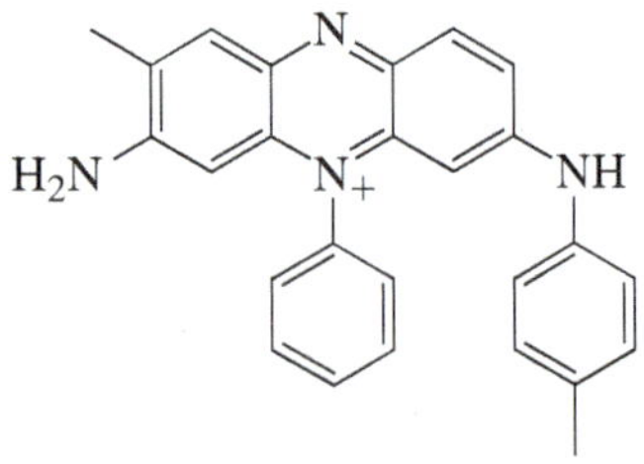

Assessment 21.24 Suppose you wanted to design a green organic dye. What would be the HOMO–LUMO energy gap for this green dye? [*1 kcal* = *4.184 kJ*]

21.3.3 Molecular orbitals of allylic systems

In Section 21.2.3, we considered the stability imparted to carbocations, radicals, and anions by resonance with a neighboring alkene. The stability of these allylic systems can also be rationalized by drawing the molecular orbital picture of these systems. The rules from Figure 21.15 are used here, with only a slight change, as a result of having an odd number of *p* orbitals. The most stable molecular orbital (ψ_1) has zero nodes, and the least stable molecular orbital (ψ_3) has two nodes. As a result, the intermediate molecular orbital (ψ_2) must have one node. To maintain symmetry, the node is located at the central carbon of the allylic system. [*This is the change compared to butadiene.*] As demonstrated in **Figure 21.22**, electrons are added to the lowest energy molecular orbitals (2 for the carbocation, 3 for the radical, and 4 for the anion). ψ_2 for the allylic radical is technically a singly occupied molecular orbital (SOMO). [*We'll see the reactivity of a SOMO in Section 22.4.1.*]

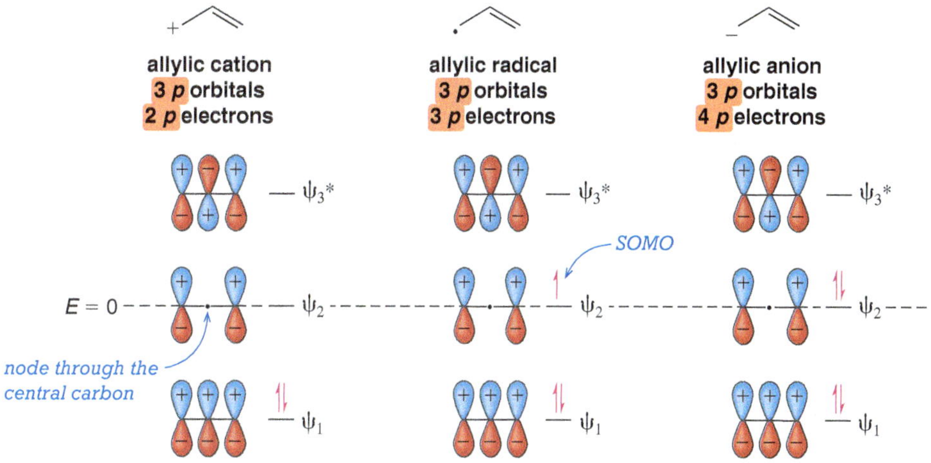

Figure 21.22 Molecular orbital diagrams of allylic systems. A low-energy ψ_1 explains the enhanced stability of these systems.

Assessment 21.25 Identify the HOMO and LUMO of the allylic cation and the allylic anion shown in Figure 21.22.

21-A CHEMISTRY IN CONTEXT: Chemistry at the Beach

Sunscreen

One of the [*annoying, yet practically mandatory*] rituals of doing outdoor activities in the summer is lathering every uncovered part of skin with sunscreen. The purpose of sunscreen is to prevent harmful, high-energy ultraviolet (UV) rays from causing damage to our skin. Damage caused by chemical reactions can eventually lead to the development of skin cancer. There are two kinds of ingredients used in sunscreens. The first kind includes inorganic materials like titanium oxide and zinc oxide, which scatter or reflect harmful rays away from skin. The second kind absorbs UV radiation, dissipating the contained energy as heat. What type of molecules would be useful for absorbing UV radiation? The conjugated molecules shown in **Figure 21.23** are examples of some that do.

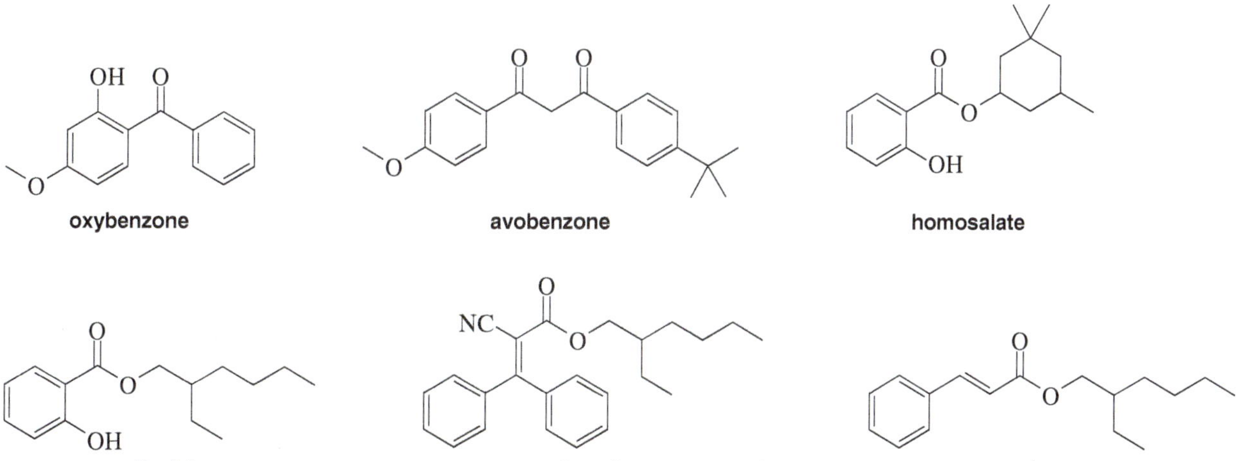

Figure 21.23 Common ingredients in sunscreen.

(continued)

The ingredients in Figure 21.23 work in the same way that causes conjugated molecules to absorb visible light. Excitation of an electron from the HOMO to the LUMO of the conjugated molecules requires a wavelength of radiation that falls in the UV range. Once excited, the energy is released as heat that doesn't cause damage to the skin when the electron returns to the HOMO.

Humans seem to be the only critters at the beach that worry about harmful UV rays. Coral, for example, spends its entire life in the shallow ocean being bombarded by UV rays from the sun, and yet these UV rays do no apparent damage. It is believed that algae living within the coral produce chemicals that are transported throughout the coral and serve as sunscreen. This idea of getting sun protection from within the organism led to researchers imagining the possibility of a sunscreen pill. It was reported that a sunscreen pill could be available in five years, but that was in 2011, and I know of nothing available yet.

[*Two things: First, some things are marketed online as "sunscreen pills." Read the reviews and talk to a doctor before using them. Second, my wife thinks that because I'm an "organic chemist" (yes, she uses air quotes when she calls me this), I should have invented this pill by now. So help me out. Invent a sunscreen pill, and send me some samples.*]

Assessment 21.26 The ingredients in Figure 21.23 are all relatively nonpolar organic molecules. Why is this an attractive feature of sunscreens? What is a disadvantage of this feature?

21.4 Electrophilic addition reactions of conjugated systems

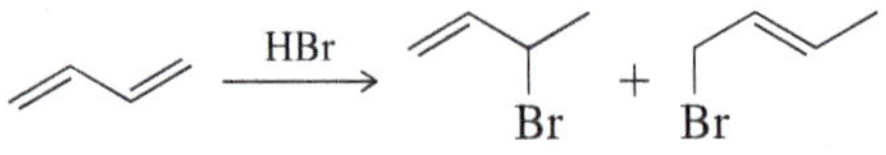

Figure 21.24 The addition of HBr to buta-1,3-diene produces two products.

In addition to changing properties like stability and the absorption of light, conjugation also changes the way molecules react. For example, the reaction of buta-1,3-diene with HBr shown in **Figure 21.24** is similar to the alkene addition reactions studied in Chapter 8. Unlike those alkene addition reactions, however, two products result from this reaction. [*See if you can figure out a mechanism that accounts for the formation of both products before reading further.*]

21.4.1 1,2- versus 1,4-addition to butadiene

Mechanistically, this reaction looks like the alkene addition reactions in Section 8.3. The nucleophilic alkene abstracts a proton from HBr to make a resonance-stabilized carbocation with the new hydrogen attached to C_1. In the second step, bromide attacks the carbocation at C_2 to make the **1,2-addition** product or at C_4 to make the **1,4-addition** product. These terms are used throughout the chapter, so take a minute to look at the products in **Figure 21.25** to make sure you understand why these are named as such. Assuming that the hydrogen always adds to the "1" carbon, the nucleophile can either add to carbon "2" or carbon "4." [*Why can't it add to carbon "3"?*]

Figure 21.25 The mechanism of 1,2- and 1,4-addition of HBr to a diene.

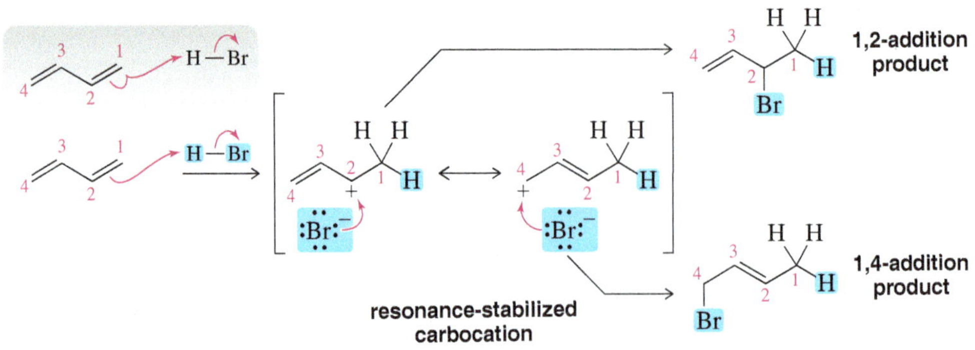

Although two products are possible in this reaction, conditions can be manipulated to give one selectively (Table 21.2).

Table 21.2 Experimental details of HBr addition to buta-1,3-diene.

Condition	Example
1. When the reaction is run at 0 °C, only the 1,2-addition product is produced.	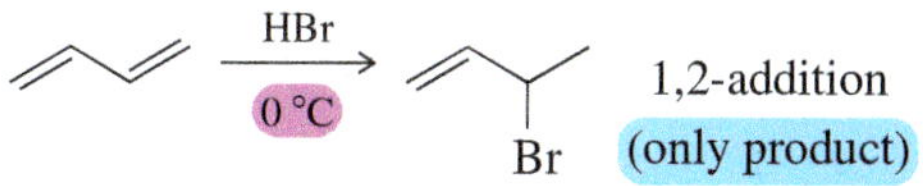 HBr, 0 °C; 1,2-addition (only product)
2. When the reaction is run at 100 °C, only the 1,4-addition product is produced.	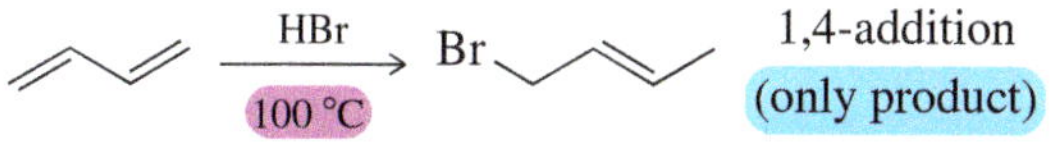 HBr, 100 °C; 1,4-addition (only product)
3. When the reaction is run at 50 °C, both products are produced in nearly equal amounts.	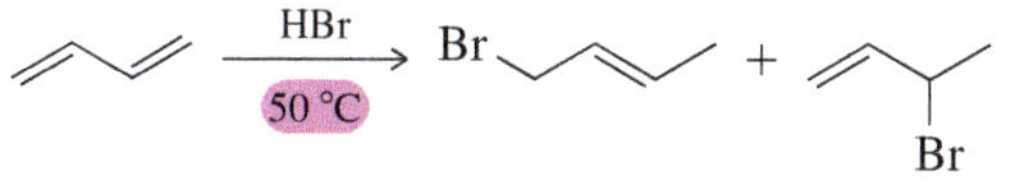 HBr, 50 °C; 1,4-addition + 1,2-addition
4. If the 1,2-addition product is heated to 100 °C, it is converted completely to the 1,4-addition product.	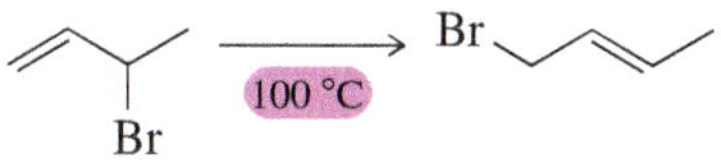 100 °C
5. If the 1,4-addition product is cooled to 0 °C or heated to 100 °C, it does not change.	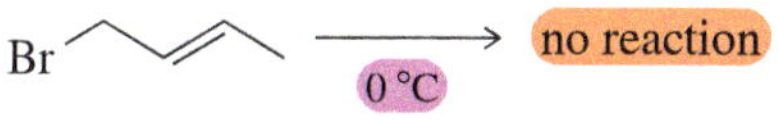 0 °C; no reaction

Let's consider the experimental details and think like a chemist to see if we can figure out what's going on here.

Thinking Like a Chemist

The best place to start is to compare the two products of the reaction and then think about what the data mean with regard to it. Let's start with the stability of the two products. Is there anything that makes the 1,2- or 1,4-product more stable? [*Look back at Figure 21.25 and decide before continuing.*]

To make this determination, we can consider the hypothetical equilibrium reaction in Figure 21.26 and estimate ΔG. Entropically, neither side is particularly favored because there is a single molecule on each side. The same is true of ΔH, where our analysis shows identical bonds formed and bonds broken, except for one small detail. [*Do you see it? Look back at Section 9.2.1 to see if that jogs your memory.*] The 1,4-product has an alkene that is disubstituted, whereas the 1,2-product has a monosubstituted alkene. Disubstituted alkenes are about 2.7 kcal/mol more stable, so the 1,4-product would clearly be the favored side of the equilibrium. And so our first observation based on the experiments described in Table 21.2: *At high temperature, the more stable product is produced.*

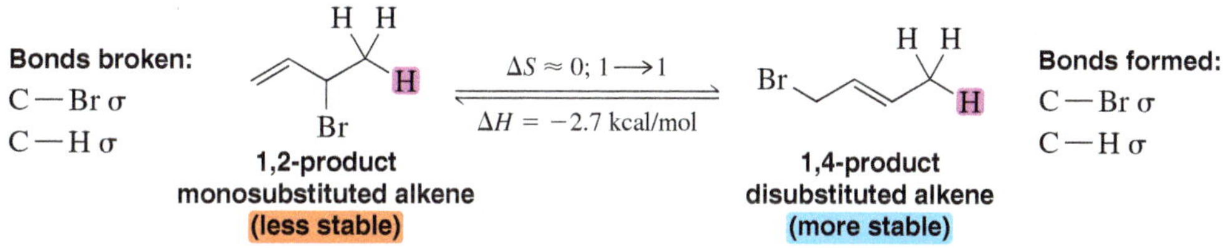

Figure 21.26 The 1,4-product is more stable than the 1,2-product because it has a more substituted alkene.

If the *most* stable product is forming at *high* temperature, then the *least* stable product must be forming at *low* temperature. Thus, besides considering the *thermodynamic* stability of the product [*the ball rolls downhill*], there must be a second factor that determines the outcome of reactions. [*We studied this in Section 5.1.7—* ***slowly*** *or* ***quickly*** *look back.*] As explained in Chapter 5, reaction outcome is governed by either thermodynamic control (product stability) or kinetic control (reaction rate). The less stable product can form only if it forms more quickly (Figure 21.27). It follows from this statement that the activation energy (E_a) for forming the 1,2-product is lower than that for forming the 1,4-product. This brings us to our second observation: *At low temperature, the fastest-forming product is favored.* [*Why might this be?*] ■

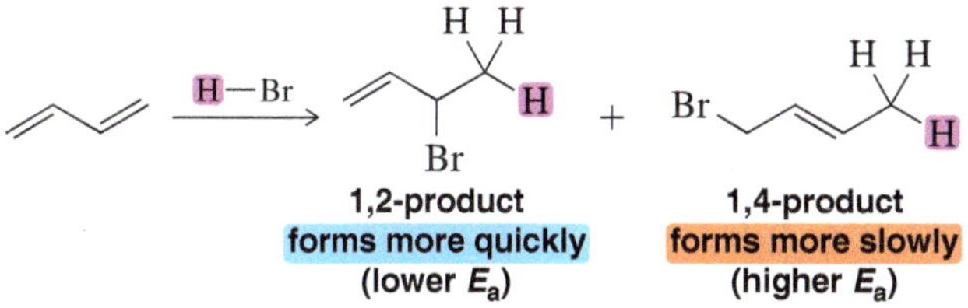

Figure 21.27 The 1,2-product is favored at low temperature because it forms more quickly than the 1,4-product.

This reaction represents an interesting situation that pops up occasionally in organic reactions. It is a situation where the product that forms fastest, with a lower activation energy, is not the most stable product. A reaction whose outcome is governed by reaction rate is said to be under **kinetic control**. When the stability of the product determines the reaction outcome, the reaction is said to be under **thermodynamic control**. [*In most of the reactions covered in Chapters 1–20, the more stable (thermodynamic) product was also the fastest-forming (kinetic) product.*] Because both of these products form from the same intermediate in the first step, we need to examine this reaction in a way that was only done before in Section 13.7.3 when we studied the reaction of alcohols with $SOCl_2$.

Immediately after the alkene removes the proton from HBr [*as in, immediately!*], what does the solution look like? Is the bromide closer to the carbocation at C_2 or C_4? Or is it far removed? [*We don't distinguish between these two when we draw out reaction mechanisms, but it certainly matters in solution.*] Immediately [*in the first femtosecond*] after deprotonation, the bromide is very close to the carbocation that formed when it donated the proton (**Figure 21.28**). And this will be true regardless of the temperature at which the reaction is run. [*How can Br^- move away from the carbocation?*]

Figure 21.28 When HBr adds to the alkene, the bromide (Br^-) and the C_2 carbocation (R^+) are initially in close proximity.

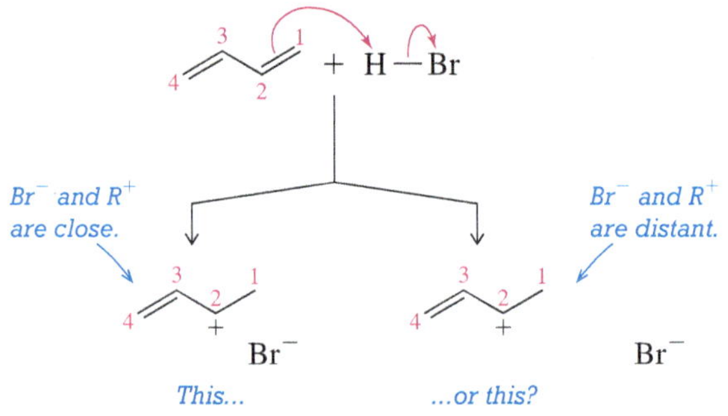

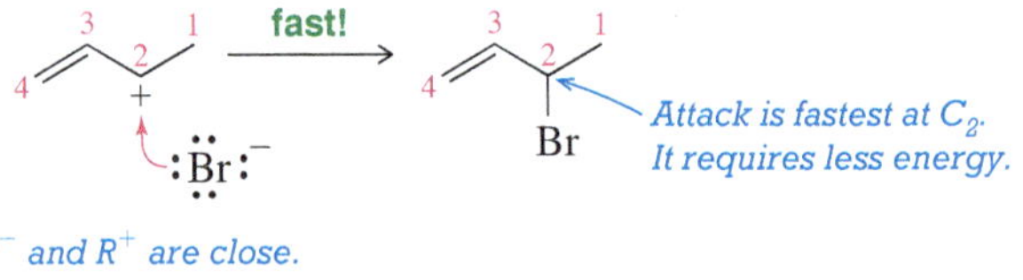

Figure 21.29 At low temperature, bromide reacts more quickly with the closer carbocation at C_2.

If the bromide in close proximity to C_2 is going to attack the carbocation, is it faster to attack at C_2 or C_4? Donating a pair of electrons to C_2 would require the least movement and is therefore a much faster process (**Figure 21.29**). In this way, bromide never has the option to react at C_4. And at low temperature, this is what drives the reaction. [*What changes at higher temperature?*]

The reactants in high-temperature reactions have more kinetic energy, so the molecules move faster. This energy allows the bromide to move away from C_2 before reacting. And yet Br^- still wants to share its electron pair. Eventually, a collision between Br^- and the carbocation will occur, but it does so with a preference for attack at C_4 (**Figure 21.30**). Not only does attack at C_4 leave the more substituted alkene, but attack at C_4 is also less sterically hindered.

Figure 21.30 At high temperature, bromide attacks from a distance. Product stability makes attack at C_4 more favorable.

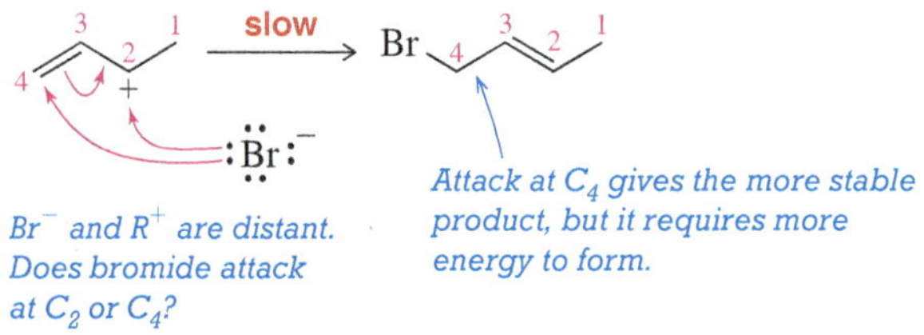

In conclusion, the reaction is under kinetic control at low temperature, where the easiest, fastest path is the one in which bromide moves a short distance to attack the carbocation at C_2. At high temperature, the reaction is under thermodynamic control, where the extra energy in the system allows the bromide to move away from the carbocation before returning to attack at C_4, leaving behind the more stable alkene. These two outcomes can be demonstrated on a reaction coordinate diagram (**Figure 21.31**), where 1,2-addition has a lower E_a but produces a less stable product. 1,4-Addition, with a higher E_a [*that's why the high temperature is needed*], produces a more stable product.

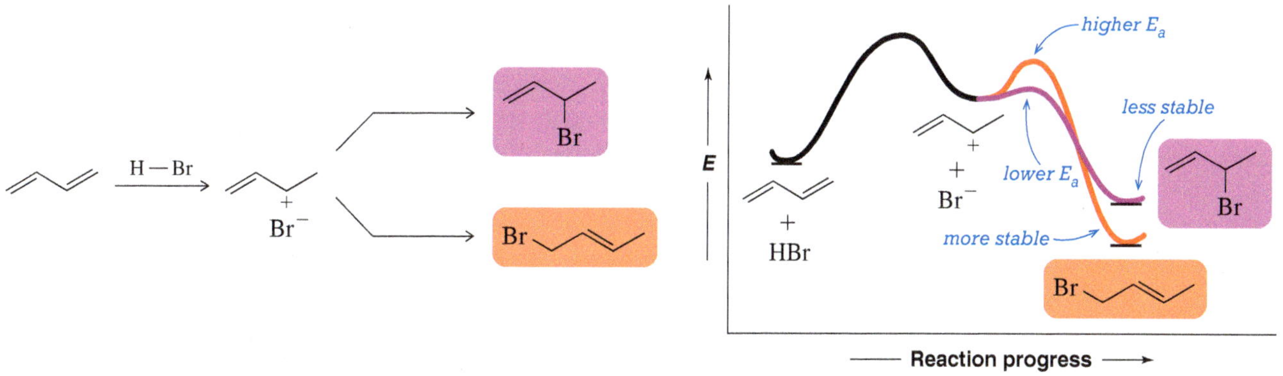

Figure 21.31 Kinetic control versus thermodynamic control on a reaction coordinate diagram. Reactions under kinetic control have a low activation energy and occur at low temperature. Reactions under thermodynamic control occur at high temperature and produce the more stable product.

All that remains is to discuss conditions 4 and 5 from Table 21.2. Specifically, in condition 4, heating the 1,2-product results in conversion to the 1,4-product. This makes sense in the context of **Figure 21.32**, where adding energy breaks the C_2—Br bond. [*Remember the first step of the S_N1 reaction? See Section 12.2.1.*] Once Br^- leaves, it can return at either C_2 or C_4 because there is lots of energy in the system. Once again, attack at C_4 makes the more stable product. In condition 5, where cooling the 1,4-product results in no change, the C_4—Br bond never breaks because there is no energy in the system, so no alternate product can ever form.

Figure 21.32 Heating the 1,2-product allows equilibration to the more stable 1,4-product.

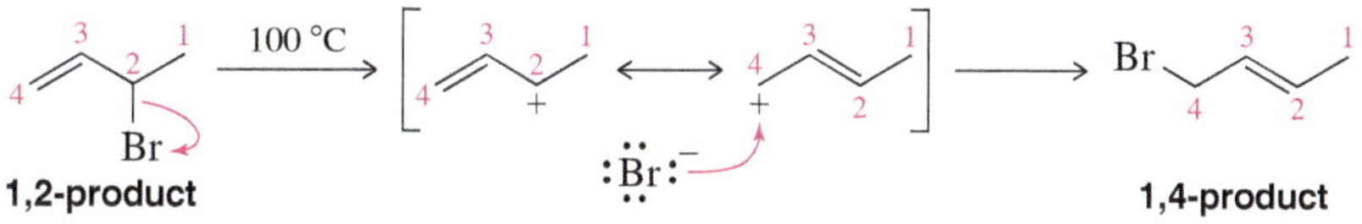

A variety of electrophiles can react with dienes under conditions of kinetic or thermodynamic control. When an unsymmetrical diene undergoes these reactions, the alkene that can form the most stable carbocation is always the one that reacts preferentially in the rate-determining first step, regardless of temperature (**Figure 21.33**).

Figure 21.33 The electrophilic addition to unsymmetrical dienes. The alkene that forms the most stable carbocation reacts preferentially.

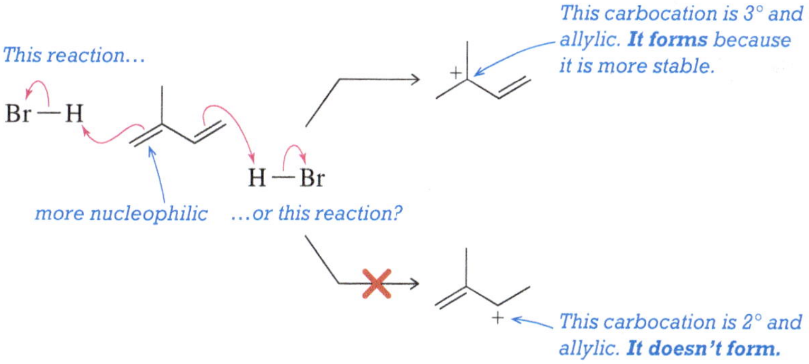

21.4.1 SUMMARY

- For electrophilic addition to buta-1,3-diene, kinetically controlled conditions (low temperature) favor 1,2-addition. Under thermodynamically controlled conditions (high temperature), 1,4-addition is favored, giving the product containing the most stable/most substituted alkene.

Assessment 21.27 Predict the product of the following addition reactions to dienes.

(a)

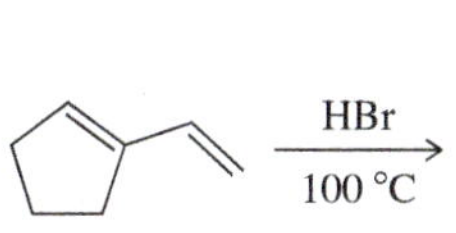

(b)

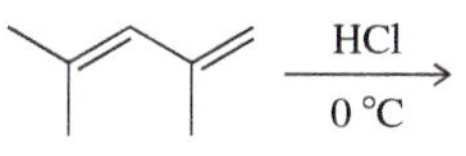

(c)

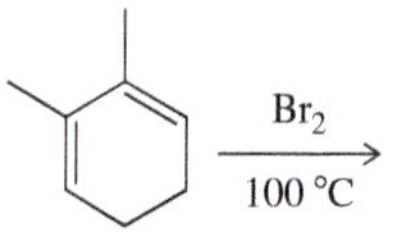

(d)

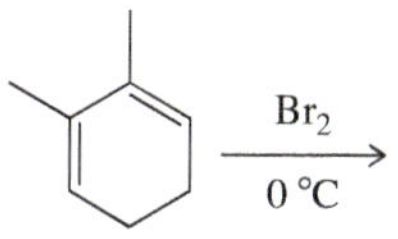

Assessment 21.28 This section discusses 1,2- versus 1,4-addition to dienes. Why is there no such thing as 1,3-addition to a diene?

Assessment 21.29 The following diene gives the same product regardless of whether the reaction is run under conditions of kinetic (0 °C) or thermodynamic (100 °C) control. Predict the product and explain this observation.

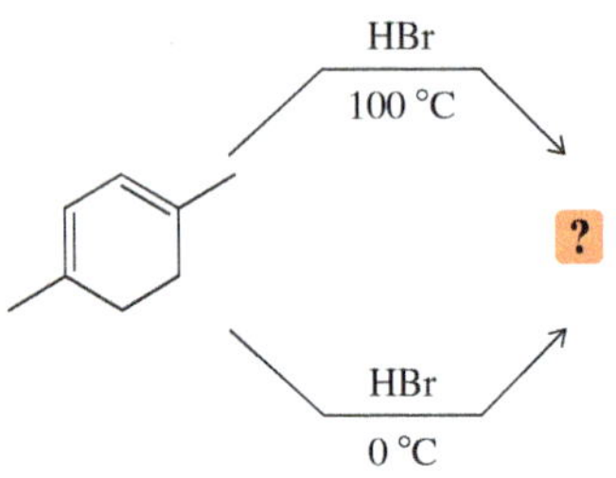

Assessment 21.30 In all of the 1,4-addition reactions, it was possible to form a cis-alkene and yet its formation was never shown. Why does a cis-alkene *not* form favorably under conditions of thermodynamic control?

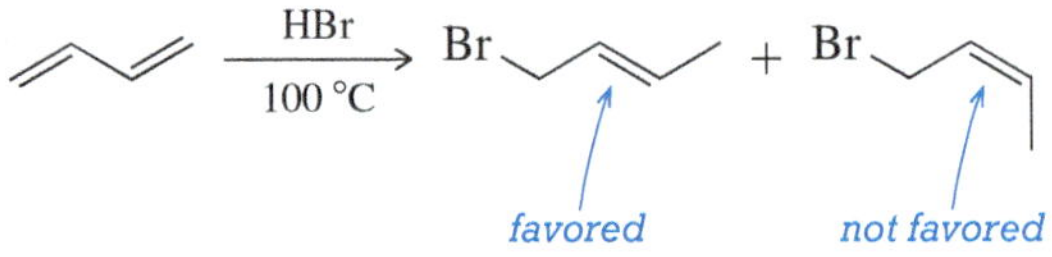

21-B CHEMISTRY IN CONTEXT: Laundry Room Chemistry

A world in color

In 1856, William Perkin, while working to synthesize quinine from impure aniline, was cleaning up black tar after another failed reaction. While doing so, he noticed that part of the tar was a bright purple color. Perkin isolated, purified, and characterized this impurity, which he called mauveine, realizing it could be used as a synthetic dye (**Figure 21.34**). Within a year of mauveine's discovery/invention, Perkin had patented its structure and opened a factory to mass produce it. This discovery is credited with creating the synthetic dye industry, allowing our world to be brightened by the colorful clothes we wear. [*I wish my failed reactions changed the world.*]

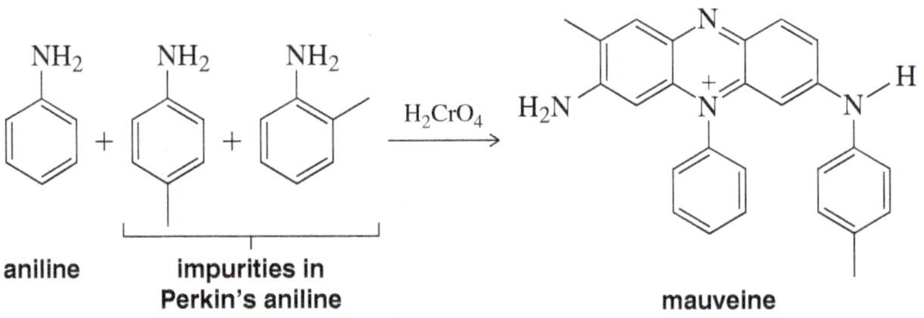

Figure 21.34 The likely reaction that resulted in Perkin's mauveine.

Probably the only negative of our colorful clothes is that we have to separate them from white clothing prior to washing. [*Why might this be?*] For one reason, washing colors with bleach (used on whites) causes the color to fade or turn completely white. Bleach, or sodium hypochlorite (NaOCl), is a molecule we have encountered before. In Chapter 9, it was the molecule that we "created" when thinking like a chemist about how to make epoxides from alkenes. In Chapter 13, we "created" it as a way of oxidizing alcohols to ketones. As an electrophilic oxidizing agent, NaOCl reacts with alkenes. When these alkenes are oxidized, the conjugated π system is disrupted. This causes the HOMO–LUMO energy gap to increase, resulting in molecules that don't absorb light in the visible region. With no visible light absorbed, all colors are reflected back equally, and our eyes see white. **Figure 21.35** shows a hypothetical example of oxidizing indigo with bleach. [*In reality, the harsh conditions of bleach oxidize indigo to many different, nonconjugated, and colorless compounds.*]

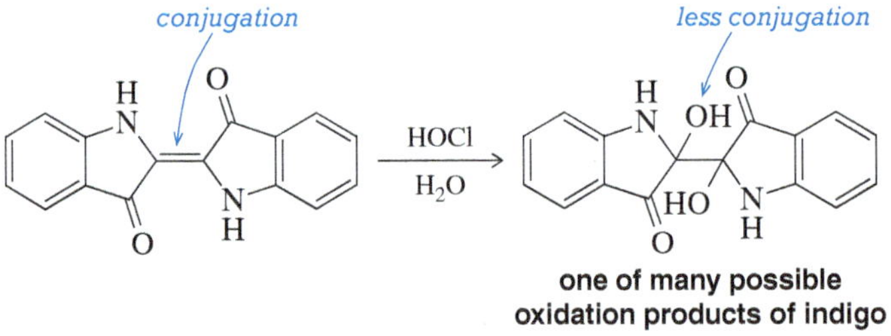

Figure 21.35 The oxidation of indigo removes the conjugation, turning it white.

Assessment 21.31 Show a mechanism of the reaction in Figure 21.34. [*Only joking, but maybe we can give it a shot in Chapter 23 when we study aromatic reactions.*]

21.5 Nucleophilic addition to conjugated carbonyls

In Chapters 8 and 9, and again in Section 21.4, alkenes have predominantly reacted as nucleophiles in addition reactions with relatively strong electrophiles (HBr, Br_2, H_3O^+, etc.). When conjugated to carbonyls or other strong electron-withdrawing groups, alkenes can also react as electrophiles.

Assessment 21.32 allows you to consider why this is on your own first, and then we'll work it together.

Assessment 21.32 Nucleophilic addition to the α,β-unsaturated ketone shown can occur at either C_2 or C_4. Why?

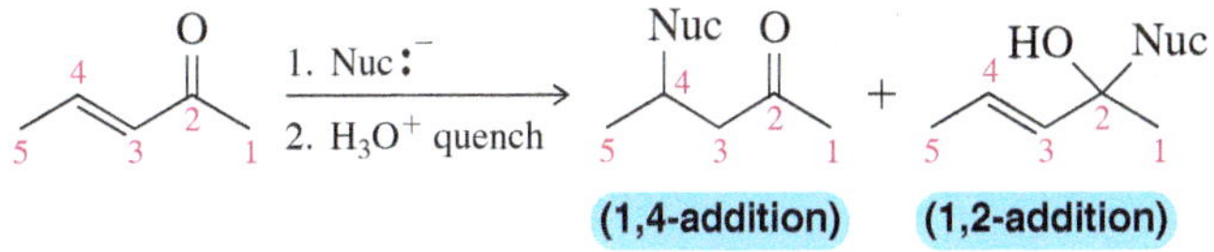

Solved Assessment 21.32

Because our discussion of conjugated systems has thus far centered on the alkene, this problem is best approached by looking at the carbonyl first. Recall from Section 17.2.2 that any difficult problem involving carbonyls can be more easily solved by looking at the zwitterion resonance structure. First, we know that nucleophiles add to carbonyl carbons because a resonance structure can be drawn in which the π electrons of the carbonyl are moved onto the oxygen. This zwitterionic resonance structure has a positive charge on carbon and a negative charge on oxygen. [*What does the conjugated alkene allow us to do now that we couldn't do with nonconjugated carbonyls?*] Neighboring the carbocation at C_2 is a π bond that allows another resonance structure to be drawn where C_4 becomes a carbocation. The electron deficiency at C_2 and C_4, as showcased by the resonance hybrid of the α,β-unsaturated ketone, accounts for the electrophilicity of the alkene (**Figure 21.36**). ■

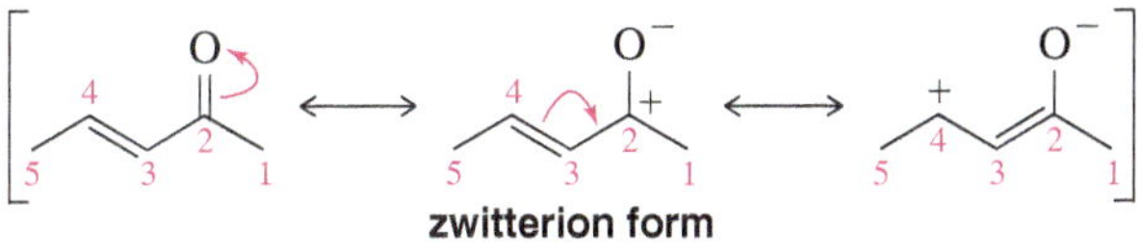

Figure 21.36 **Electron delocalization by resonance** makes C_2 and C_4 electrophilic.

Assessment 21.33 (a) Could a nucleophile ever add to C_3? (b) Why or why not?

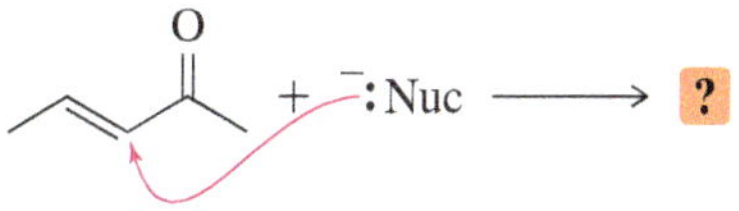

21.5.1 1,2- versus 1,4-addition to α,β-unsaturated carbonyls

The terminology of 1,2- and 1,4-addition used in Section 21.4 can be used to describe nucleophilic addition to α,β-unsaturated carbonyls. Consider the addition of a nucleophile that is immediately followed by a proton quench. Thus, the two groups that will be adding via 1,2- or 1,4-addition are Nuc and H (**Figure 21.37**). When the nucleophile adds directly to the carbonyl (as we saw throughout Chapter 17), an alkoxide is produced. Quenching this alkoxide with acid forms an alcohol. In this result, the hydrogen adds to atom 1 of the π system, whereas the nucleophile adds to atom 2 [*1,2-addition!*]. When the nucleophile adds instead to the β-carbon of the conjugated system, an enolate (see Chapter 20) is produced. Protonation of the enolate on oxygen results in an intermediate that has a hydrogen added to atom 1 of the π system and the nucleophile added to atom 4 [*1,4-addition!*]. The confusion arises because the enol product of 1,4-addition quickly, and favorably, tautomerizes to the ketone [*see Sections 10.8.3 and 20.2*]. Thus, naming the 1,4-addition based only on the outcome of the reaction is confusing. [*Luckily, we don't learn organic reactions without paying close to attention to the mechanism.*]

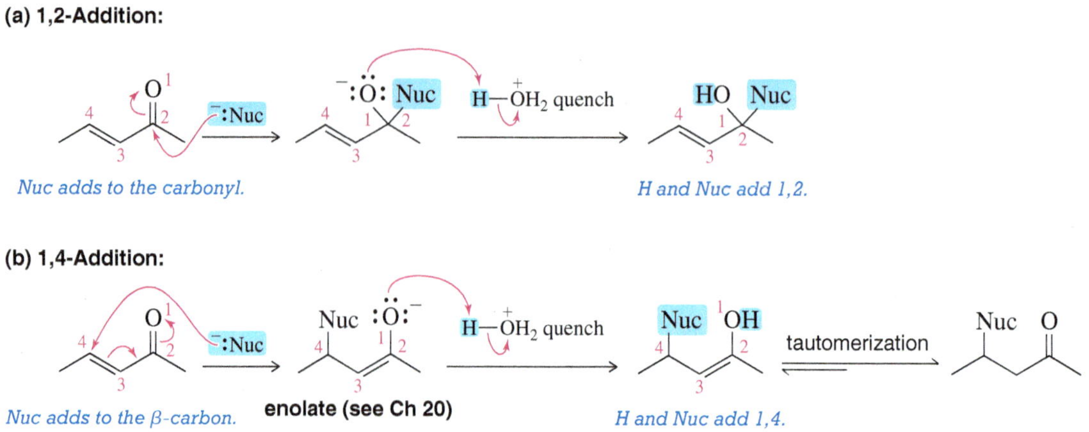

Figure 21.37 1,4-Addition versus 1,2-addition. (a) 1,2-Addition describes direct nucleophilic addition to the carbonyl, as studied in Chapter 17. (b) 1,4-Addition describes nucleophilic addition to the β-carbon of an α,β-unsaturated carbonyl.

Because 1,2- and 1,4-addition are both possible, it is important to determine which occurs when. For reasons similar to those encountered in Section 21.4.1, the selectivity is based on whether a reaction is under kinetic or thermodynamic control. Instead of being temperature-dependent, however, it is based on the strength of the nucleophile that is adding. Before we see why, let's figure out which product—namely, that of 1,2- or 1,4-addition—is more stable. [*Look at Figure 21.37 and make your guess before proceeding.*]

To answer this question, we can consider the hypothetical equilibrium reaction between the 1,2- and 1,4-addition products in **Figure 21.38**, analyzing it based on ΔG, ΔH, and ΔS. Each side is just one molecule, so neither is favored based on ΔS. In terms of bonds formed and broken, O—H σ and C—H σ are pretty similar, as are the C—C σ bonds. Ultimately, the 1,4-addition product is more stable because it contains a C—O π bond in place of the C—C π bond of the 1,2-product. This alone makes the 1,4-product the more stable, or thermodynamic, product.

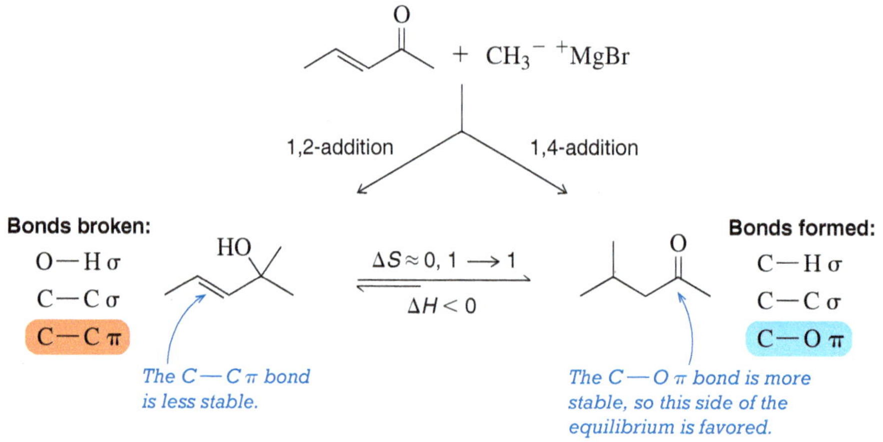

Figure 21.38 The 1,4-product is more stable than the 1,2-product because it has a C—O π bond in place of a C—C π bond.

The kinetic product arises from 1,2-addition because the carbonyl carbon has the greatest degree of carbocation character. This should not be surprising given what we know about the zwitterionic resonance structure from Chapter 17 (**Figure 21.39**).

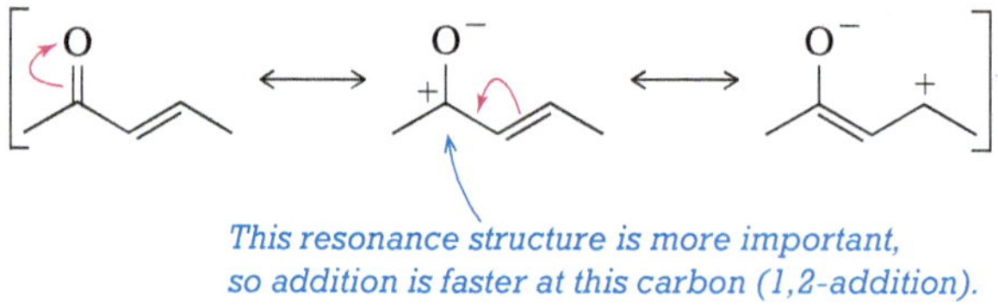

Figure 21.39 The carbonyl carbon is more electrophilic than the β-carbon of an α,β-unsaturated ketone. Thus, 1,2-addition is faster than 1,4-addition.

And so when thinking about whether a reaction will produce the 1,2- or 1,4-product, it comes down to this: With irreversible addition, equilibration is *not* possible, so the faster (kinetically controlled) 1,2-addition predominates. When a nucleophile adds reversibly, equilibration is possible, so the more stable (thermodynamically controlled) 1,4-product is produced. [*What type of nucleophiles add irreversibly? Reversibly? Think about* S_N2 *vs.* S_N1*, E2 vs. E1, General Mechanism 17A vs. 17B, and General Mechanism 18A vs. 18B.*]

For a reaction to be reversible, the nucleophile needs to add and then be able to get kicked back out. In other words, it needs to be a reasonably stable leaving group. Once added, strong nucleophiles like Grignard and organolithium reagents do not leave. Thus, strong nucleophiles do 1,2-addition with no chance of equilibration (**Figure 21.40**).

Figure 21.40 Strong nucleophiles add quickly and irreversibly under kinetic control to give the 1,2-product.

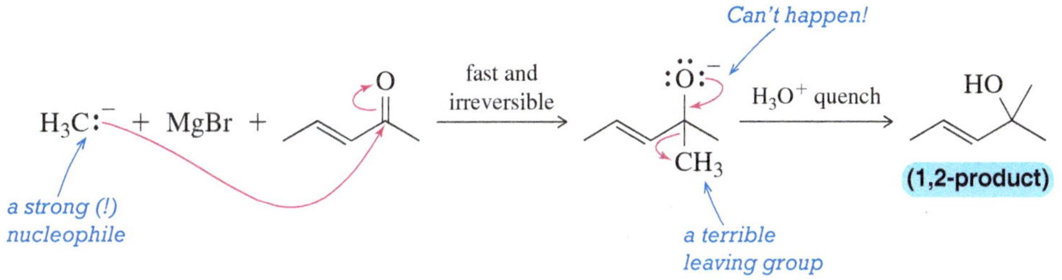

The conjugate acid of sodium acetoacetate has pK_a = 10, so sodium acetoacetate is clearly a weaker nucleophile than the Grignard reagent. [*Recall it doing* S_N2 *reactions at both 1° and 2° carbons in Section 20.5.*] Nearly every time sodium acetoacetate encounters an α,β-unsaturated ketone, the faster 1,2-addition occurs. However, because the acetoacetate anion is stable, it can be kicked out, allowing the reaction to reverse. Every once in a while, then, the acetoacetate anion does the slower 1,4-addition reaction. After protonation by a protic solvent (like CH_3OH), the more stable 1,4-product results (**Figure 21.41**).

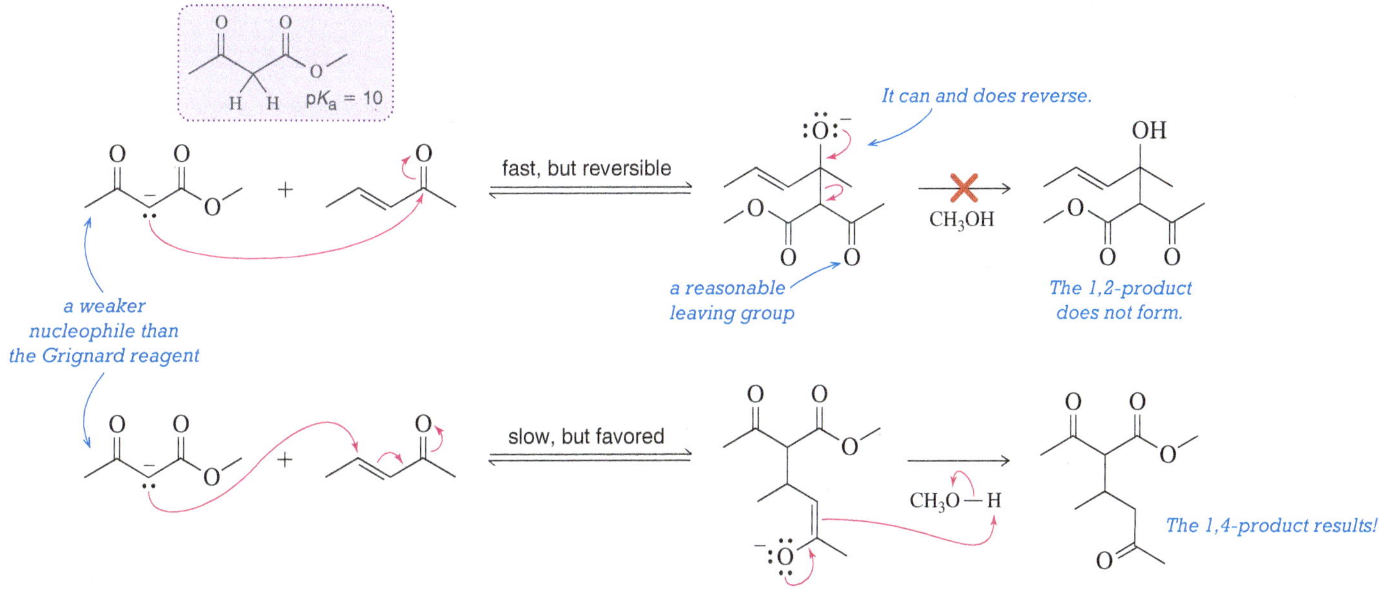

Figure 21.41 Weak nucleophiles add reversibly under thermodynamic control to give the more stable product.

What, then, does it mean to be a strong nucleophile in this context? Here, it isn't as clear as it has been in the past. But of the nucleophiles studied so far, Grignard reagents, organolithium reagents, and acetylide anions are strong and do 1,2-addition (**Table 21.3**—left column). Cyanide, stabilized enolates (malonates/β-keto esters), alkoxides, and thiolates are weaker nucleophiles and do 1,4-addition (Table 21.3—right column). If the conjugate acid of the nucleophile has $pK_a \geq 20$, 1,2-addition predominates. If the conjugate acid of the nucleophile has $pK_a < 20$, 1,4-addition predominates.

Table 21.3 Classification of nucleophiles that do 1,2- or 1,4-addition.

1,2-Addition occurs with these nucleophiles	1,4-Addition occurs with these nucleophiles
Grignard reagents (RMgBr): 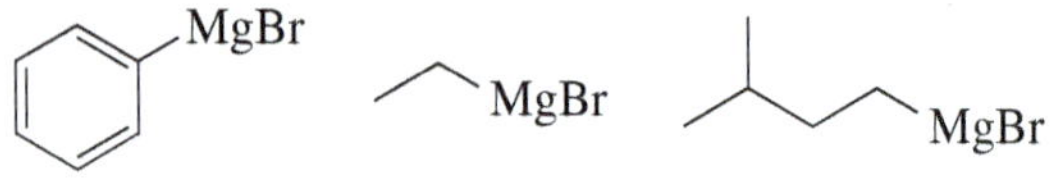	**Cyanide:** Na^+ ^-CN
Organolithium reagents (RLi): 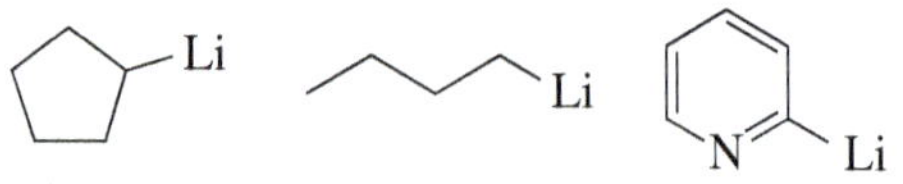	**Alkoxides/thiolates:** 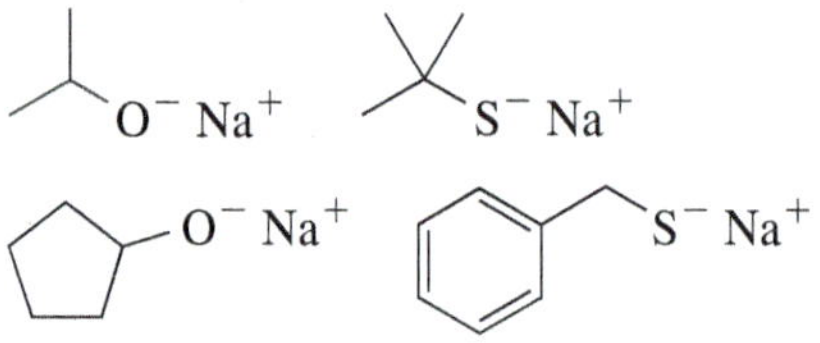
Acetylide anions (R—C≡C⁻): 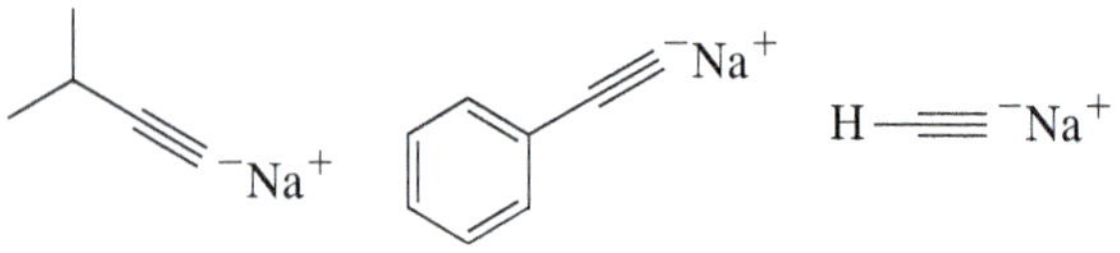	**Stabilized enolates:** 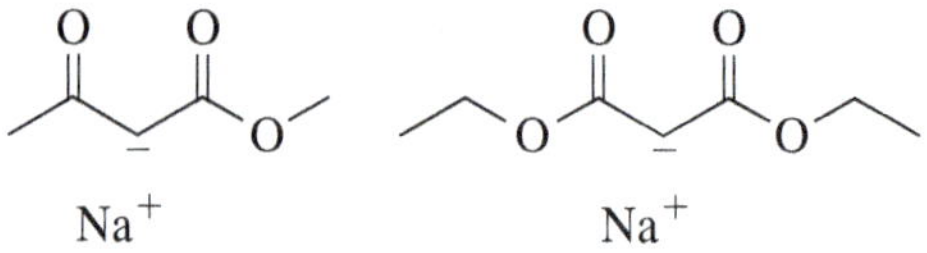

21.5.1 SUMMARY

- In nucleophilic addition to α,β-unsaturated carbonyls, kinetically controlled conditions (strong nucleophiles) favor 1,2-addition. Under thermodynamically controlled conditions (weak nucleophiles), 1,4-addition is favored, giving the product that maintains the strong C—O π bond.

Assessment 21.34 In Section 21.4, a reaction coordinate diagram was shown comparing 1,2- and 1,4-addition to dienes. Draw a similar reaction coordinate diagram that rationalizes the outcome of 1,2- and 1,4-addition reactions to α,β-unsaturated carbonyls.

Assessment 21.35 Cyanide, listed as a nucleophile that does 1,4-addition in Table 21.3, does the following 1,2-addition. Why does it do 1,2- instead of 1,4-addition here?

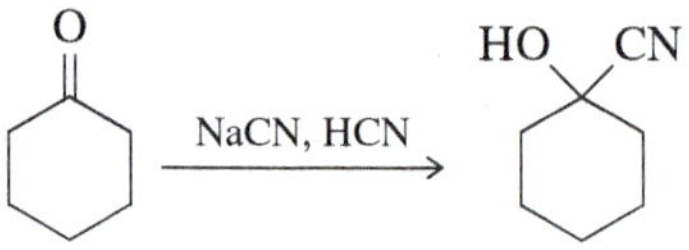

(1,2-addition)

Assessment 21.36 Would you expect the following nucleophiles to do 1,2- or 1,4-addition?

(a) 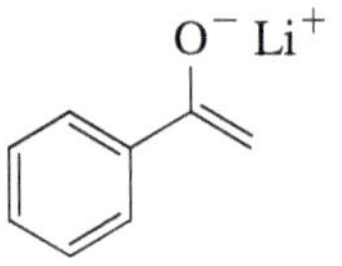

(b) 

(c) HO^-

(d) 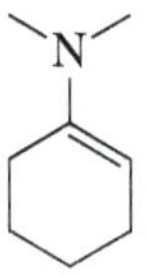

(e) 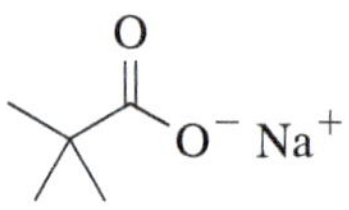

(f) 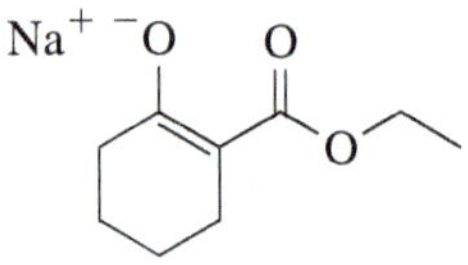

(g) 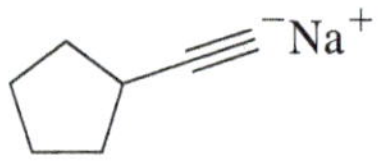

(h) 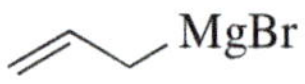

(i) 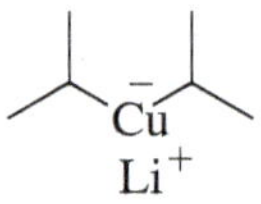

21.5.2 Michael reaction

Owing to the importance of this reaction, any conjugated carbonyl that can undergo a 1,4-conjugate addition is referred to as a Michael acceptor.

The 1,4-addition (sometimes called conjugate addition) of stabilized enolates is referred to as the Michael reaction, named for its inventor, one of the ingenious classical chemists we studied in Chapter 20. The original Michael reaction, discovered in 1897 and shown in **Figure 21.42**, was between diethyl malonate and ethyl cinnamate under basic conditions. Mechanistically, ethoxide deprotonates the acidic proton ($pK_a = 10$) to make the stabilized enolate. As a weak nucleophile, 1,4-addition occurs to give an enolate that deprotonates ethanol to produce the 1,4-product, while forming a new C—C bond!

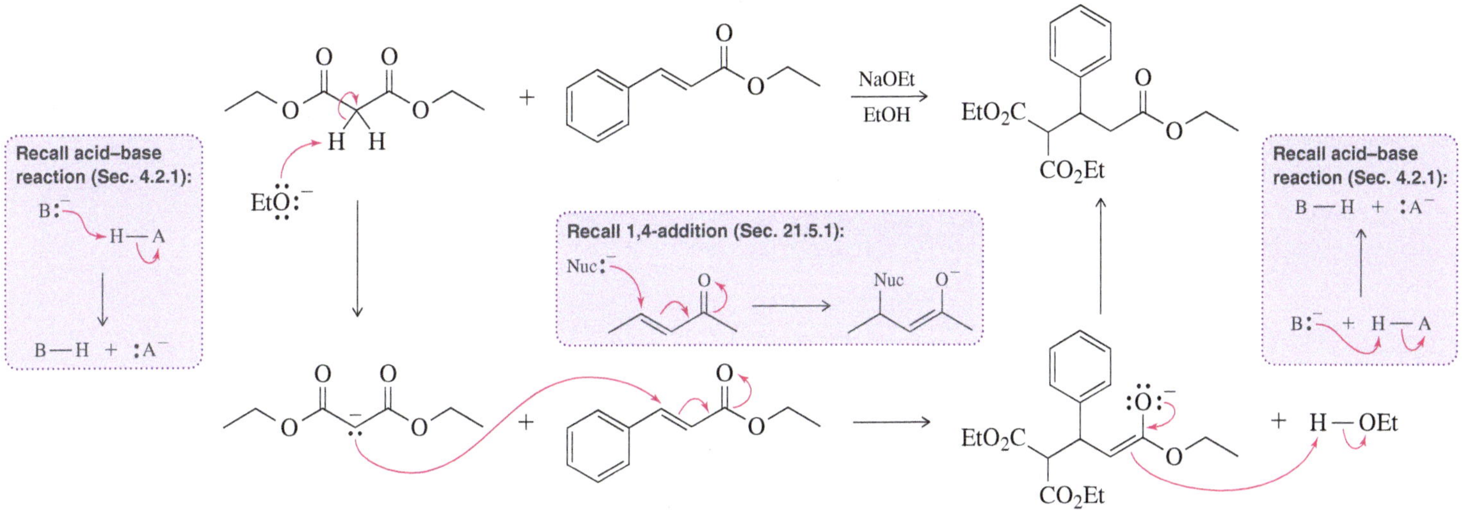

Figure 21.42 The original Michael reaction.

The Michael reaction, as shown in the two examples in **Figure 21.43**, is diverse and can be used for C—C bond formation in a number of situations.

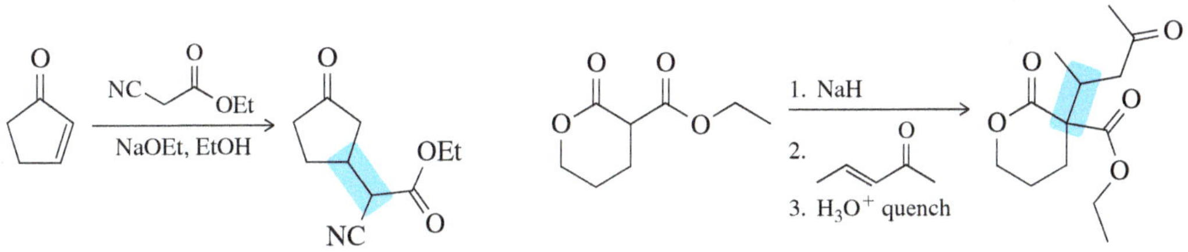

Figure 21.43 Representative Michael reactions. [*The newly formed C—C bond is screened in blue.*]

Assessment 21.37 Whereas stabilized enolates do 1,4-addition, unstabilized (normal) enolates can do both 1,2- and 1,4-addition depending on the situation. Why might this be?

Assessment 21.38 Rationalize the outcome of the following reactions in light of your answer to Assessment 21.37.

(a)

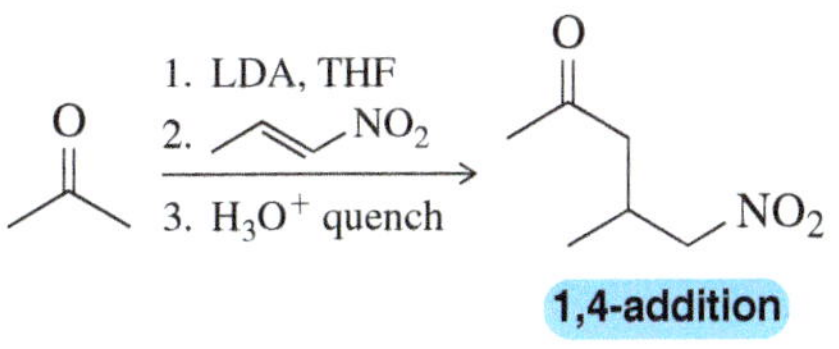

(b)

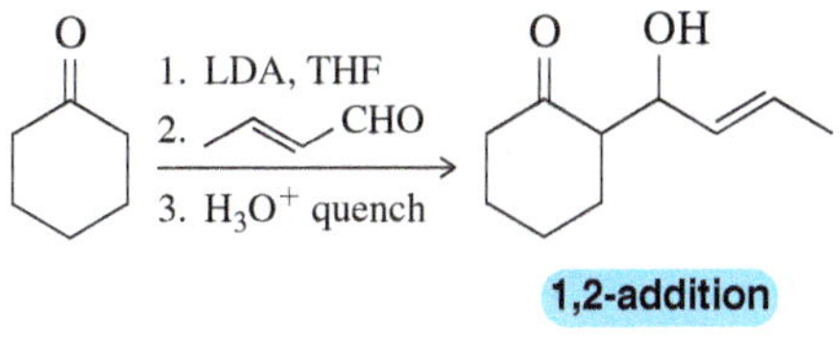

Assessment 21.39 Identify the reagent and reactant(s) that would give the products shown through 1,4-addition.

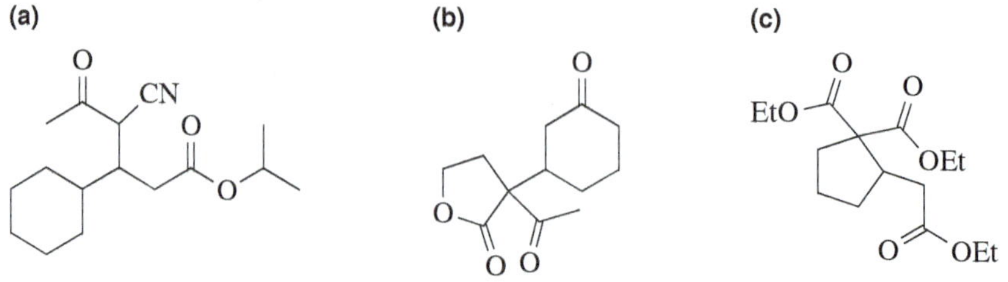

Assessment 21.40 Molecules with oxygenation at C_1 and C_3 can be synthesized using a Claisen reaction. A molecule with oxygenation at which carbons would be best synthesized by a Michael addition? Explain your answer.

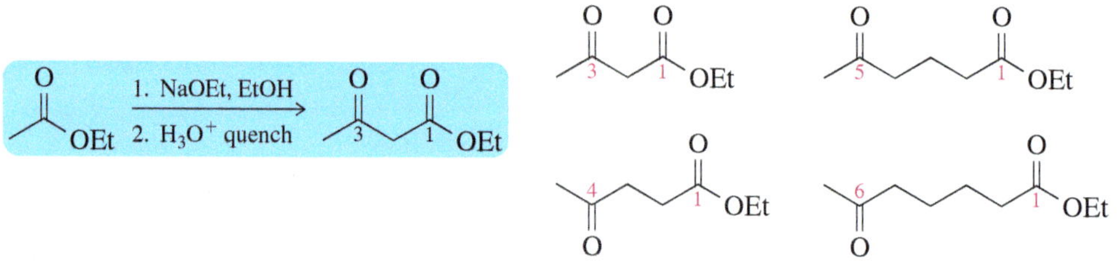

21.5.3 Robinson annulation

One of the more famous 1,4-addition reactions, once again from the classical period discussed in Chapter 20, is the **Robinson annulation** (annulation = ring-forming reaction).

I'm a big fan of Nobel Prize winners. And because it is most fun to learn this reaction in the context of solving complicated mechanisms, work Assessment 21.41 before reading further, and then we'll solve it together.

Assessment 21.41 Suggest a mechanism for the original Robinson annulation, as reported by 1947 Nobel Laureate Sir Robert Robinson in 1935.

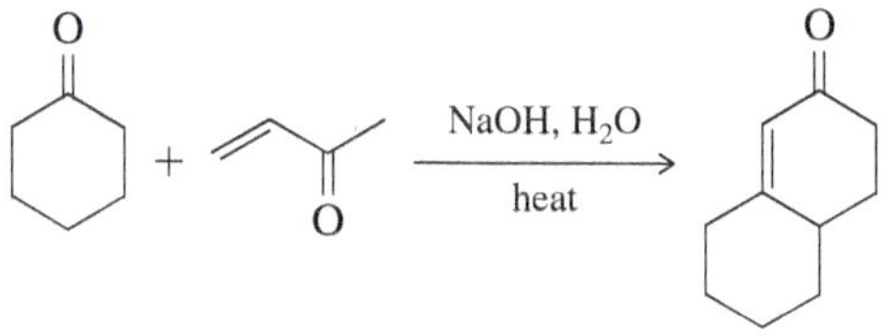

Solved Assessment 21.41

When solving mechanism problems, start by identifying the bonds broken and formed in the reaction (**Figure 21.44**). [*Remember: The answer is on the page, so use it to guide you.*] Numbering the carbons and drawing the hydrogens will help with this. Doing this analysis reveals two things. First, there are several bonds broken and formed in this reaction. Second, many carbons are gaining and losing hydrogens. Given that the reaction is in base, it is likely that the movement of hydrogens is a result of acid–base reactions. Thus, let's focus first on the key bonds that are forming and think about how those might happen. [*Try this on your own, keeping in mind reactions studied previously.*]

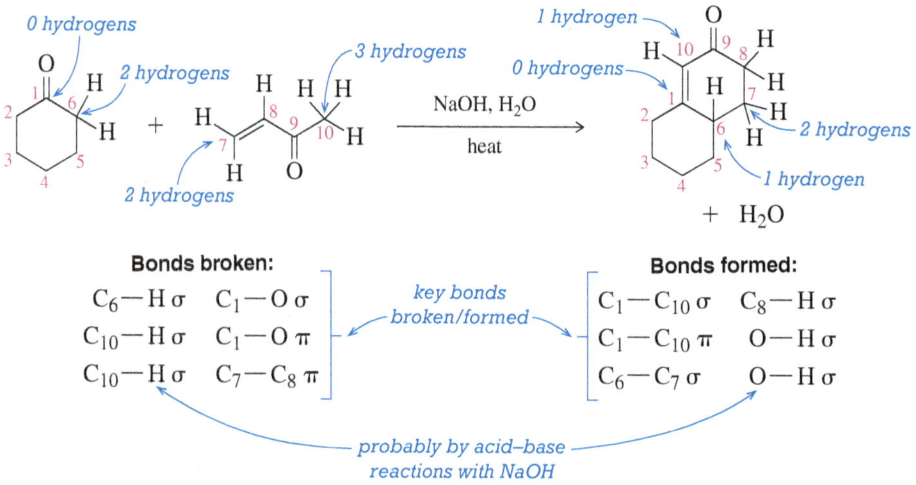

Figure 21.44 Identifying the bonds broken and bonds formed is the first step in solving a mechanism problem.

Let's look first at the bond that forms between C_6 and C_7 (Figure 21.44). Can you imagine a way that this bond might form? C_7 is the β-carbon of an α,β-unsaturated ketone. [*Is this carbon typically an electrophile or a nucleophile?*] The β-carbon is typically electrophilic, so a nucleophile at C_6 could reasonably attack C_7 (**Figure 21.45**). A carbanion at C_6 would be an enolate, meaning the $C_6—C_7$ bond forms in a Michael reaction. How about the $C_1—C_{10}$ σ bond? C_1 is a carbonyl carbon—an electrophile in the reactions discussed in Chapters 17–20. If an anion at C_{10} (again an enolate) attacked it, this would be . . . [*your answer?*] . . . an aldol reaction. Finally, once an aldol reaction occurs to form the $C_1—C_{10}$ bond, elimination of water, promoted by heating an aldol reaction, would form the key $C_1—C_{10}$ π bond. [*We show this in Figure 21.48 if it is confusing right now.*]

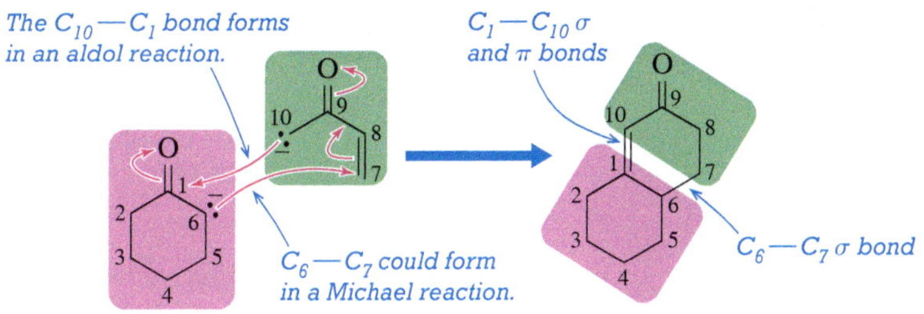

Figure 21.45 A hypothetical way bonds could break and form based on the numbering of the carbons.

Putting these two observations (Michael and aldol reactions) together, we can begin drawing a mechanism. In base, deprotonation of the cyclohexanone would give an enolate that can do a 1,4-addition to produce an enolate at C_8. This step forms the key $C_6—C_7$ bond. Turning our attention to the aldol reaction, an enolate at C_{10} does an intramolecular aldol reaction to form the $C_{10}—C_1$ bond. Protonation of the alkoxide completes this reaction, thus forming the β-hydroxy ketone. [*Confirm that what we have shown here in* **Figure 21.46** *comes directly from Figure 21.45.*]

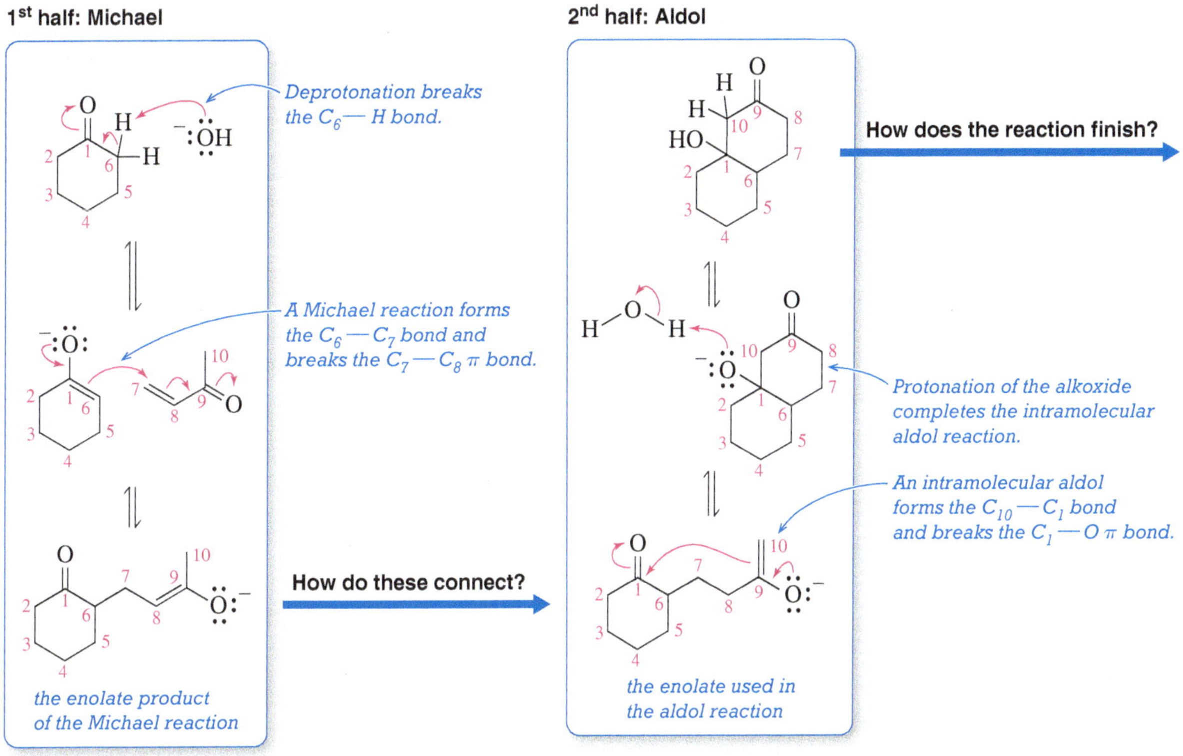

Figure 21.46 Solving the first and second halves of the Robinson annulation mechanism.

There are still two issues to address in the mechanism. First, Figure 21.46 shows that the enolate product of the conjugate addition is *not* the enolate in the aldol reaction. Recall from our discussion of intramolecular aldol reactions in Section 20.3.4 that the choice of which enolate reacts is based less on acidity of the proton and more upon the ring size that will form. In this way, the enolate product of the Michael reaction would form an unstable four-membered ring (**Figure 21.47**). Thus, it cannot do the intramolecular aldol reaction. Instead, it is protonated by water at C_8. At this point, a proton at C_{10} is removed to make the enolate that is able to form the six-membered ring. All of the reactions in Figure 21.46 are reversible, implying that the reaction is driven to completion by forming the most stable product: the six-membered ring.

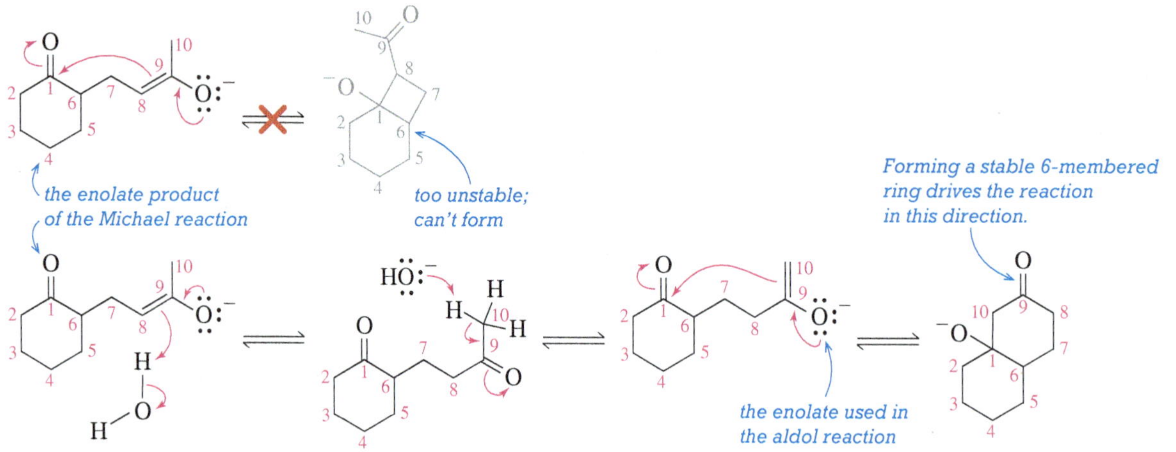

Figure 21.47 Connecting the two halves of the Robinson annulation. Reversible enolate exchange occurs to make an enolate that can form a stable six-membered ring.

The final step in the mechanism is to eliminate the β-hydroxy group of the aldol product. This entropically favored step results from heating the intramolecular aldol reaction, as we did in Section 20.3.4. Occurring by the rarely seen E1cB mechanism (Section 12.5.1) in **Figure 21.48**, deprotonation to make a stable enolate is followed by elimination of hydroxide and formation of the α,β-unsaturated ketone. The E1cB mechanism only occurs when the intermediate anion is stabilized by an electron-withdrawing group—in this case the carbonyl that allows for enolate formation. ■

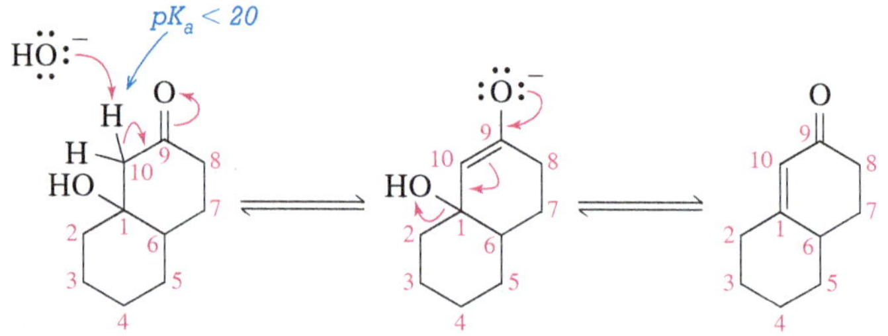

Figure 21.48 The E1cB elimination of hydroxide completes the Robinson annulation mechanism.

Assessment 21.42 The mechanism of the Robinson annulation was discussed across Figures 21.46–21.48. It can be helpful when studying for exams to have this mechanism in one place. Write the overall mechanism for the Robinson annulation. [*Hint: Michael → enolate exchange → aldol → elimination.*]

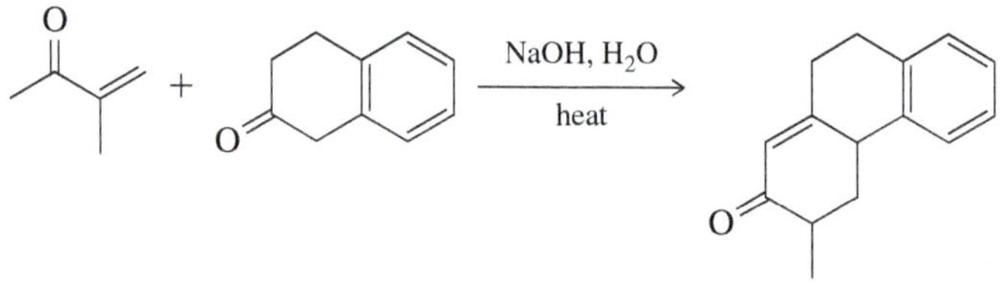

Assessment 21.43 Predict the product of the following reactions.

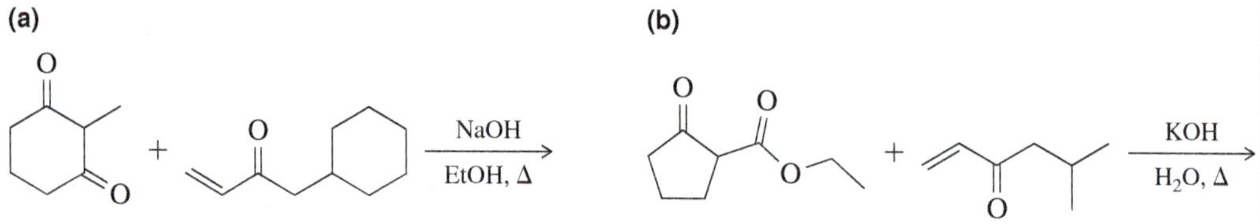

Assessment 21.44 Show how the following molecule might be synthesized using the Robinson annulation.

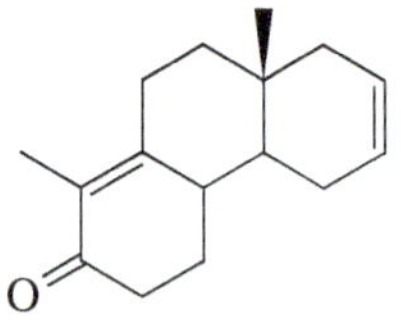

21.5.4 Cuprate addition

Remember, Grignard and organolithium reagents only do 1,2-addition, even if 1,4-addition is possible.

The final 1,4-addition we present is the addition of lithium diorganocuprates (**Gilman reagents**) to α,β-unsaturated carbonyls. Recall from Chapter 16 that Gilman reagents only react with molecules containing an easily reducible bond. While the reducible bond in Section 19.9.2 was the C—Cl bond of an acid chloride, the alkene of an α,β-unsaturated carbonyl is similarly reducible. Although not exactly correct, the mechanism in Figure 21.49 makes it possible to predict the product of the reaction (and relates it to most of the reactions in this chapter). The mechanism is similar to how we might draw it if we were reacting a Grignard or organolithium reagent.

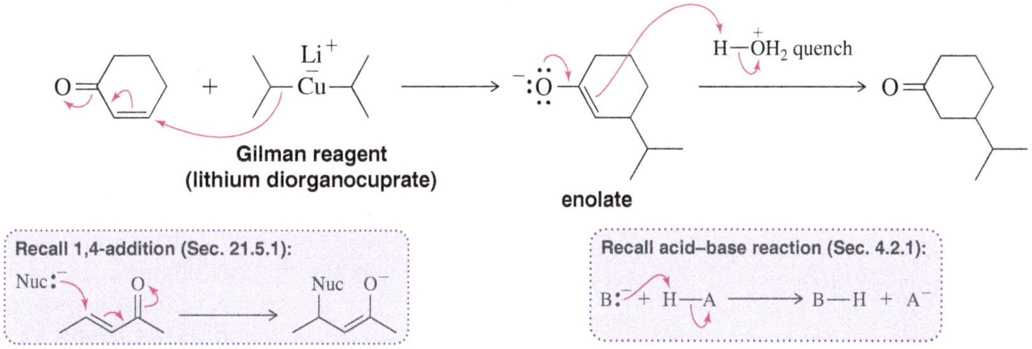

Figure 21.49 The mechanism of Gilman reagents with α,β-unsaturated ketones. This approximate mechanism relates this reaction to those studied previously, while allowing easy prediction of the product.

SPECTROSCOPY INSIGHTS

Conjugate addition can be confirmed by the absence of the C═C stretching band around 1600 cm^{-1} in the IR spectrum. Moreover, the resonance-stabilized carbonyl band at 1685 cm^{-1} will shift to 1715 cm^{-1}.

Because the mechanism drawn in Figure 21.49 implies that Grignard reagents would do 1,4-addition, it is important to quickly review how Gilman reagents are believed to react in these systems. [*See Section 16.5 for a full treatment.*] Gilman reagents are organometallics like Grignard and organolithium reagents, but they are not simply strong nucleophiles that donate a pair of electrons to electron deficient sites. Instead, Gilman reagents react only when there are easily reducible bonds present. The presumed mechanism (Figure 21.50) involves an initial complexation between the copper and the π electrons of the alkene. Oxidative addition then forms a new Cu—C bond and an enolate. Reductive elimination forms the new C—C σ bond. Quenching of the enolate produces the ketone after tautomerization.

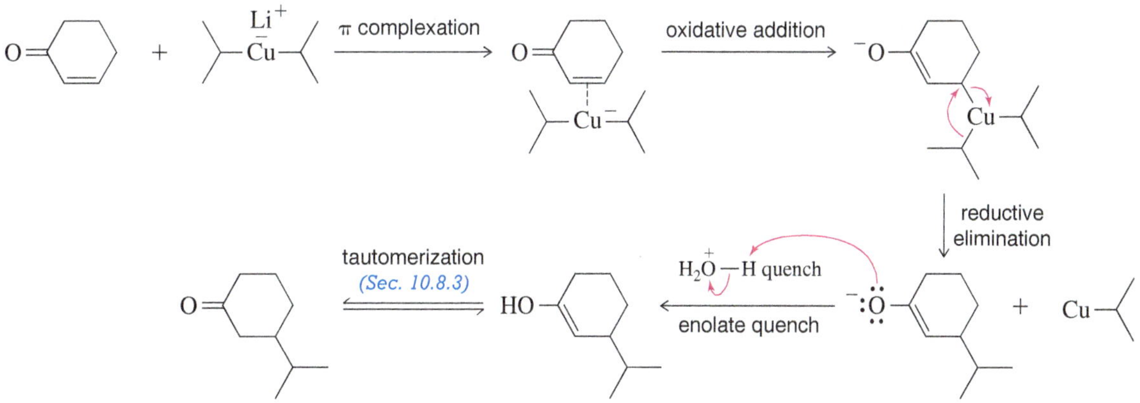

Figure 21.50 The organometallic mechanism of Gilman reagents with α,β-unsaturated carbonyls. This mechanism explains why Gilman reagents do 1,4- as opposed to 1,2-addition.

Assessment 21.45 Predict the product of the following reactions.

(a)

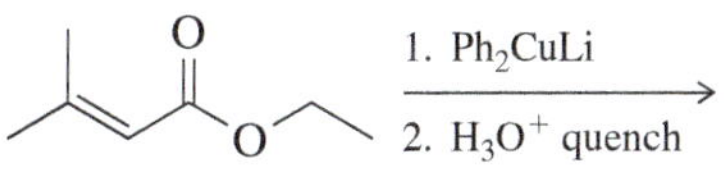

(b)

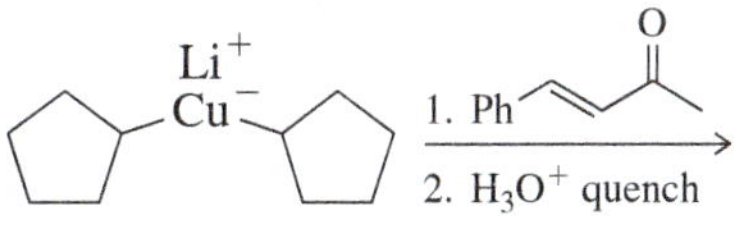

(c)

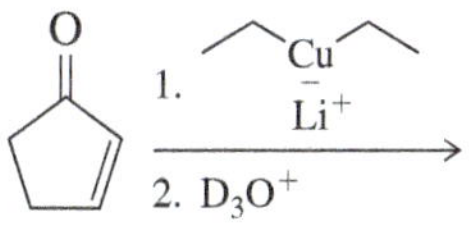

Assessment 21.46 Suggest a reagent and a reactant that could be used to form the following molecules by conjugate addition of a cuprate. There can be multiple possibilities.

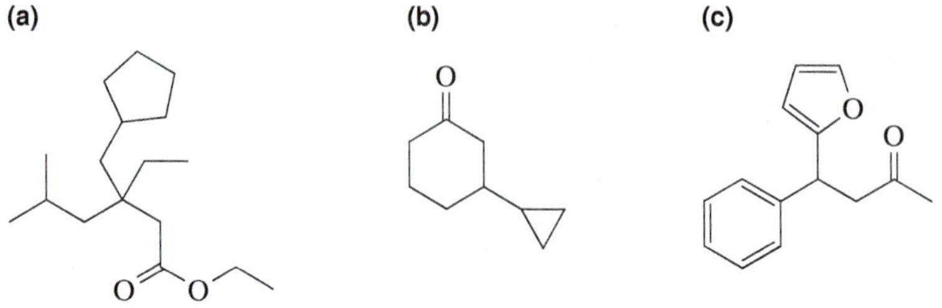

Assessment 21.47 To this point, we have seen four reactions that can be done by Gilman reagents. What are they? What do they have in common?

Concluding thoughts

To recap, conjugation is not only a structural feature, it directly influences the properties and reactivity of a molecule. Several diseases, including cancer and cellular aging, have been attributed to the presence of reactive oxygen species. These reactive oxygen species are essentially free radicals of the type studied in Chapter 11. β-Carotene, one of the highly conjugated and brightly colored molecules discussed in Section 21.3.2, is understood to have a number of health benefits. In addition to it being converted to vitamin A, some studies have suggested that in healthy individuals, a diet rich in β-carotene can mitigate cancer risk, although further investigation is needed to confirm this. One suggested mechanism by which β-carotene and other antioxidants might improve health is by reacting with reactive oxygen species like the peroxy radical, seen reacting with vitamin E in Section 11.7. [*Hence the term antioxidants.*] Recall from Chapters 5 and 11 that free radicals are 1 electron short of a full octet, so they are electron-deficient. In Section 21.4.1, conjugated systems were shown to react efficiently with HBr and other electrophiles. Similarly, the dyes used on our clothes can be oxidized by similar electrophilic reactions with bleach (Chemistry in Context 21-B). From these two examples, we can rationalize the manner in which β-carotene might react with a peroxy radical. Addition of the peroxy radical [*using fishhook arrows—see Chapter 11*] to one of the alkenes in β-carotene creates a highly conjugated, delocalized, and therefore highly stable free radical. Through this process, the highly reactive peroxy radical is quenched, reducing the cellular damage it might otherwise have caused.

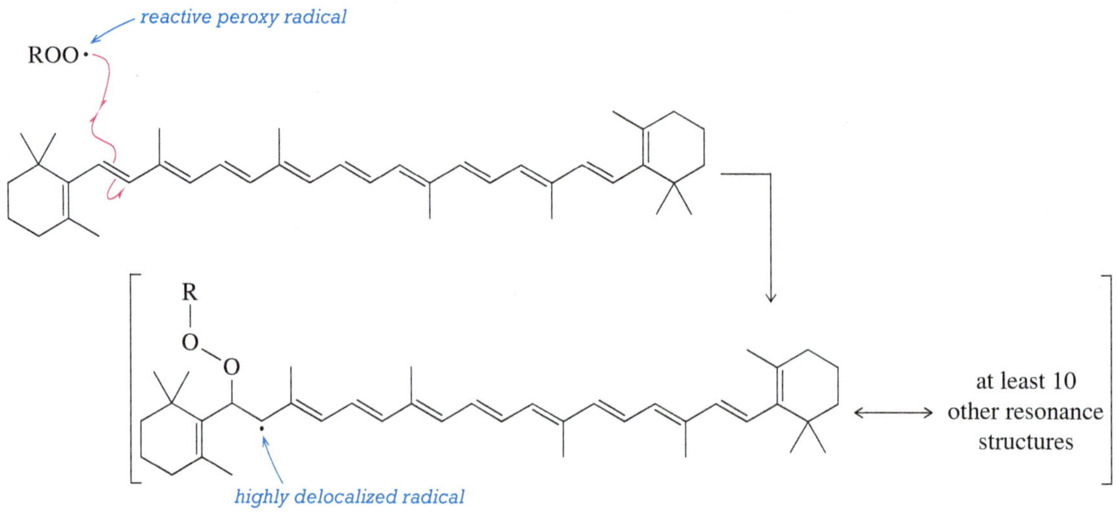

End-of-Chapter Materials

Important Concepts

[These are the major concepts covered in each section, excluding reactions.]

21.1: Molecules are said to be conjugated when they contain a series of alternating single and multiple bonds resulting in a contiguous system of *p* orbitals spread out over more than three atoms.

21.2: Consistent with the impact of resonance on stability, conjugated π systems are more stable than nonconjugated systems. Conjugated reactants react more slowly; conjugated products/intermediates form more quickly.

21.3: The stability of conjugated systems is a result of the delocalization of electron density over multiple atoms. Using MO theory, a conjugated π system is represented as a linear combination of *p* atomic orbitals. The highest occupied molecular orbital (HOMO) is the highest energy molecular orbital that is occupied. Usually, this is a bonding molecular orbital (π) because it is below the energy level of a nonbonded atomic orbital. The lowest unoccupied molecular orbital (LUMO) is the lowest energy molecular orbital that is not occupied. Usually, this is an antibonding molecular orbital (π*) because it is above the energy level of a nonbonded atomic orbital. Excitation of an electron from π to π* is useful for UV–vis spectroscopy and accounts for the color of highly conjugated molecules. Light absorbed for highly conjugated molecules falls in the visible region because the HOMO–LUMO gap is small.

21.4: Electrophilic addition can give products that result from kinetic or thermodynamic control. 1,2-Addition proceeds preferentially at low temperatures because there is not enough energy to overcome the activation barrier required for 1,4-addition. Alternatively, 1,4-addition occurs at high temperature. With multiple outcomes possible, the reaction proceeds to make the most stable product. Usually, this product has the most substituted alkene.

21.5: In addition to the 1,2-addition discussed in Chapters 17–20, nucleophiles can do 1,4-addition to α,β-unsaturated carbonyls. Strong nucleophiles usually do 1,2-addition, while 1,4-addition occurs with weak nucleophiles that allow equilibration to a product that maintains the stronger C—O π bond. The Michael reaction is the addition of enolates to these systems, while the Robinson annulation is a ring-forming reaction that features an initial Michael reaction followed by an intramolecular aldol reaction. Finally, lithium diorganocuprates do 1,4-addition by a mechanism involving oxidative addition and reductive elimination.

Important Reactions and Mechanisms

1. **Addition to dienes at low temperature: kinetic control (Section 21.4.1):** A π bond attacks HBr in an acid–base reaction to form the most substituted, resonance-stabilized carbocation. At low temperature, the resulting bromide immediately attacks the carbocation, resulting in 1,2-addition.

General:

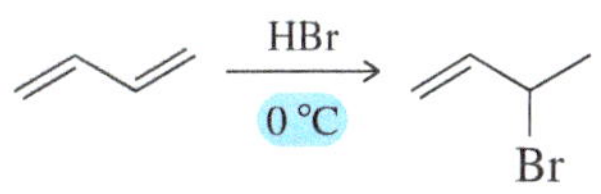

Specific:

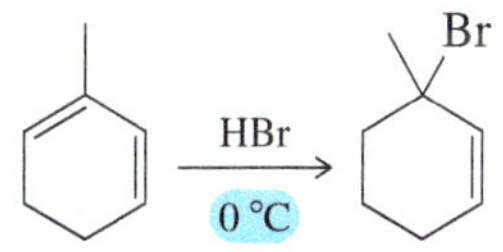

Mechanism:

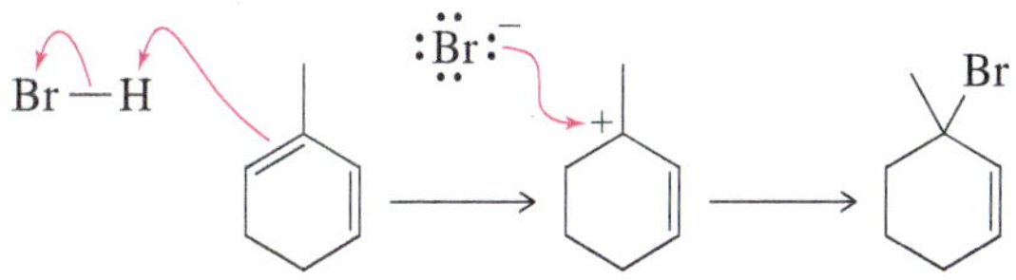

2. **Addition to dienes at high temperature: thermodynamic control (Section 21.4.1):** A π bond attacks HBr in an acid–base reaction to form the most substituted, resonance-stabilized carbocation. At high temperature, the resulting bromide attacks at the carbon of the allylic cation to give the most substituted alkene via 1,4-addition.

General:

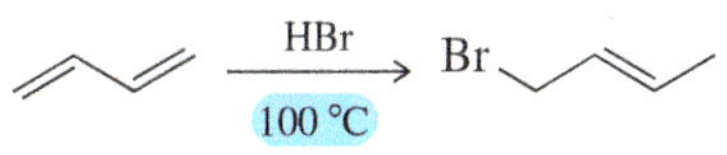

Specific:

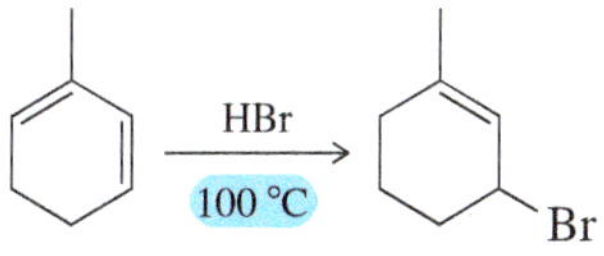

Mechanism:

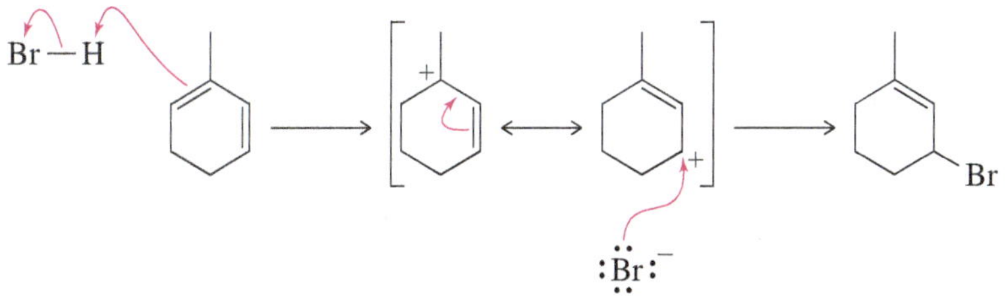

3. **Michael reaction (1,4-conjugate addition) (Section 21.5.2):** A weak nucleophile, such as a stabilized enolate, adds to the β-carbon of an α,β-unsaturated carbonyl to produce an enolate. Quenching of the enolate results in formation of a ketone that is substituted at the β-carbon. 1,4-Conjugate addition is favored by weak nucleophiles (thermodynamic control), whereas direct 1,2-addition to the carbonyl is favored by strong nucleophiles (kinetic control).

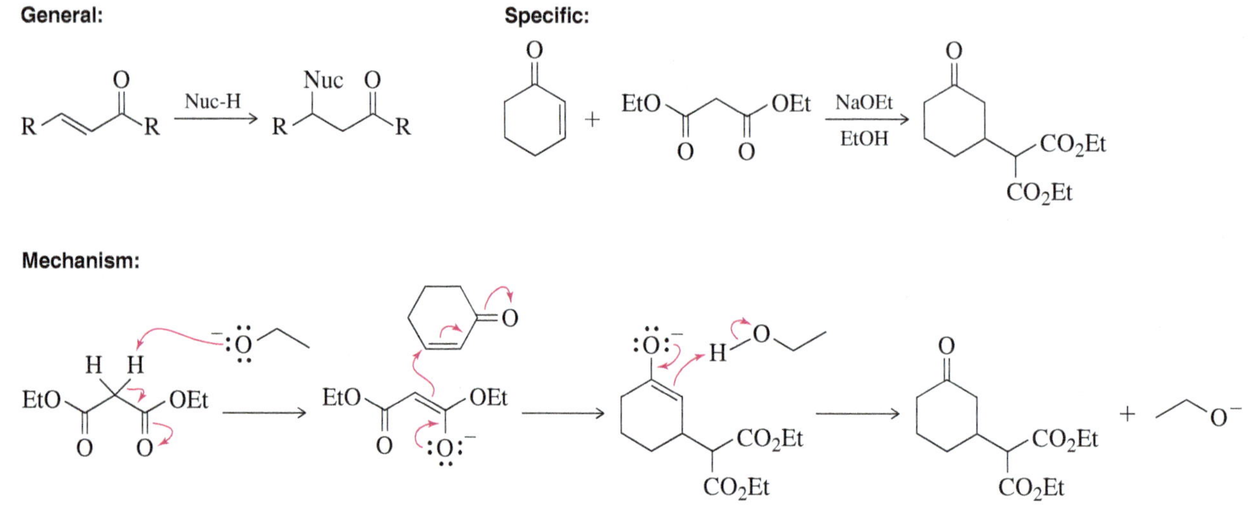

4. **Robinson annulation (Section 21.5.3):** Deprotonation of an α-proton forms an enolate that does 1,4-conjugate addition. The resulting enolate is protonated, revealing a new ketone. Deprotonation forms a second enolate that does an intramolecular aldol reaction. Protonation of the resulting alkoxide generates a β-hydroxyketone. The α,β-unsaturated ketone is formed via deprotonation to form a new enolate and subsequent elimination of the hydroxyl group (E1cb mechanism). This reaction is under thermodynamic control, proceeding with preferential formation of a six-membered ring.

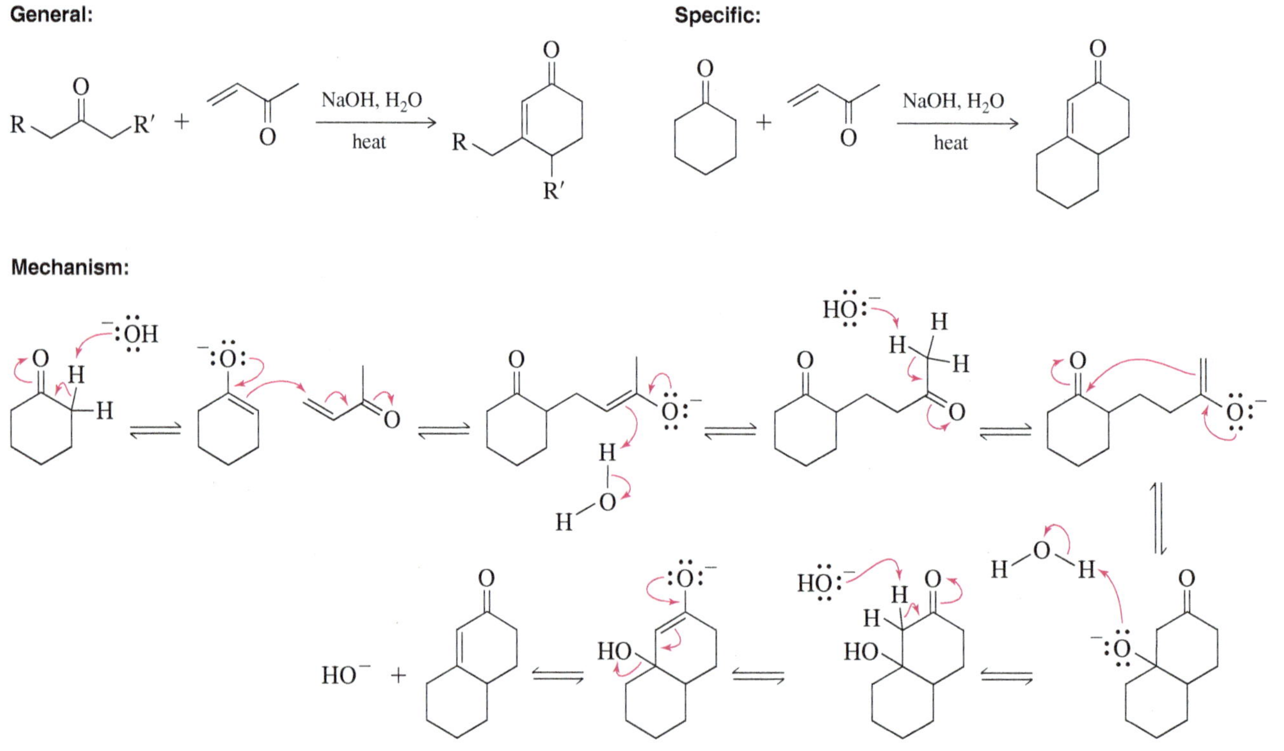

5. **Organocuprate addition (Section 21.5.4):** Formation of an initial complex between Cu and the π bond is followed by oxidative addition to give a C—Cu σ bond and an enolate. Reductive elimination results in formation of a new C—C σ bond at the β-carbon. Quenching of the enolate generates the ketone product. An alternative mechanism can be drawn that reliably predicts the product of these reactions and better relates it to standard nucleophilic addition reactions studied to this point.

General:

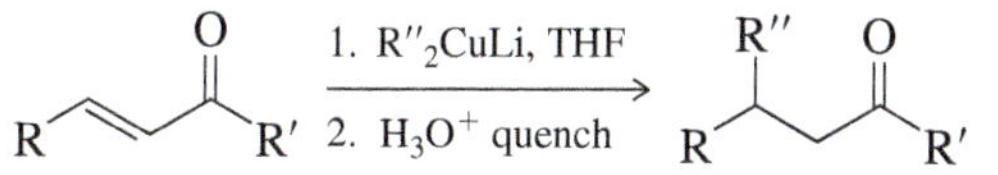

Specific:

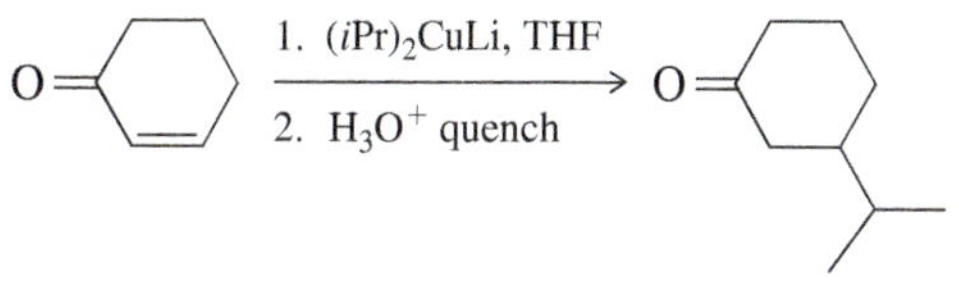

Mechanism:

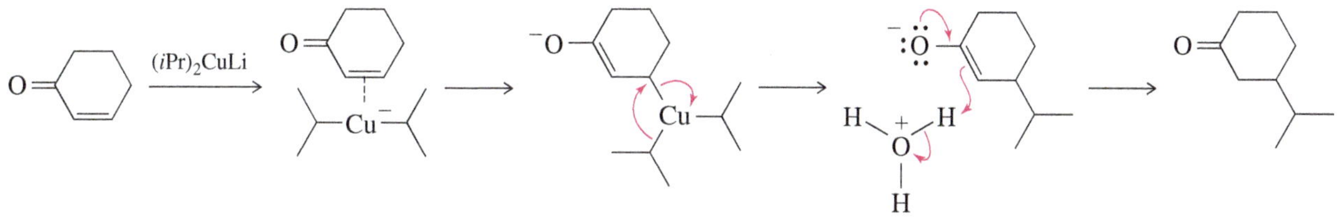

Mechanism (alternate):

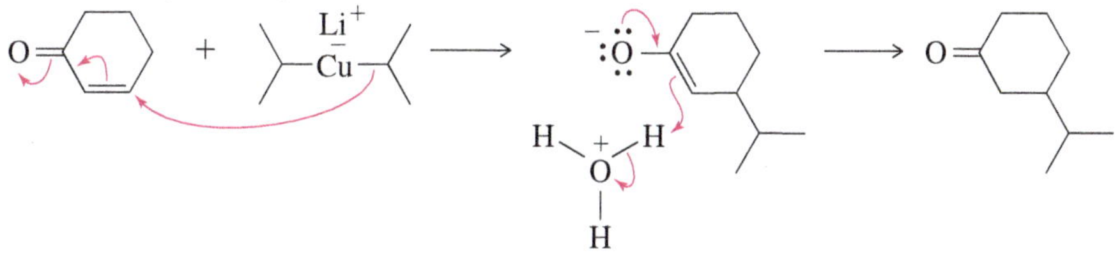

End-of-Chapter Assessments

[Assessments are ranked according to their difficulty; •••• = most difficult.]

21.48 (•) Would you expect the following molecules to absorb in the UV–visible region of the electromagnetic spectrum?

(a)

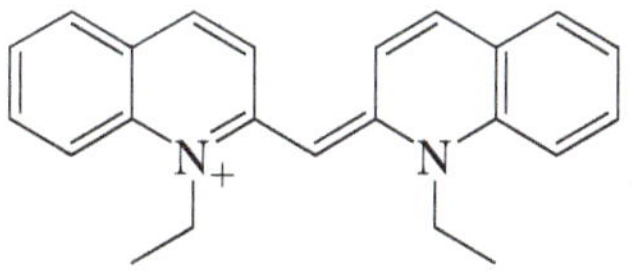

(b)

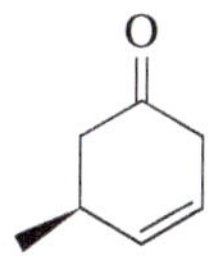

(c)

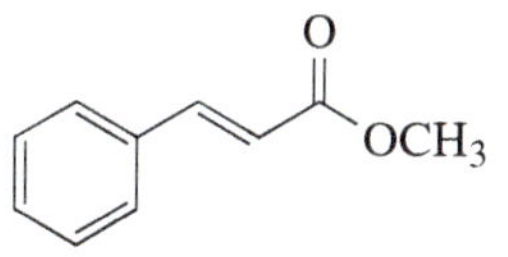

(d)

(e)

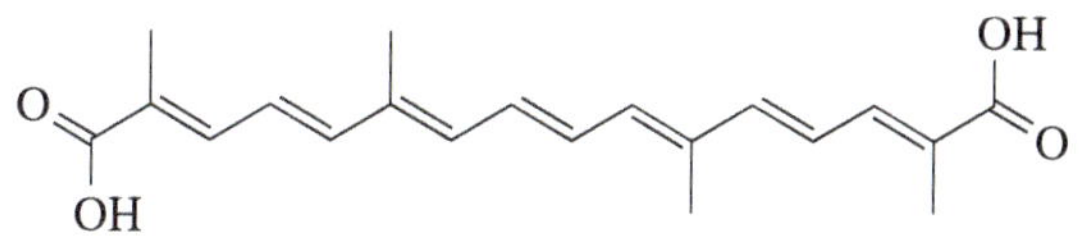

[What type of molecule is this? See Chapter 8.]

21.49 (•) For each of the following hypothetical equilibrium reactions, predict which side would be favored.

(a)

(b)

(c)

21.50 (•) On a per alkene basis, which would have the more negative heat of hydrogenation?

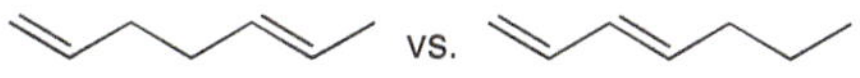

21.51 (•) How many π electrons are involved in the conjugated π system for each of the following molecules?

(a)

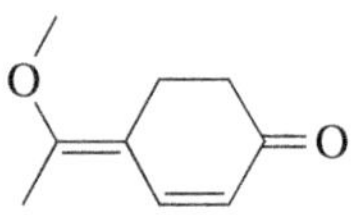

(b)

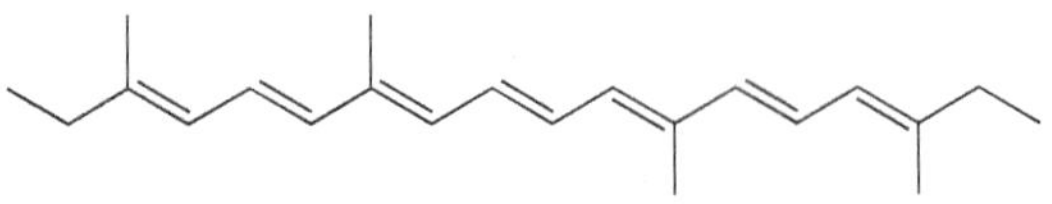

(c)

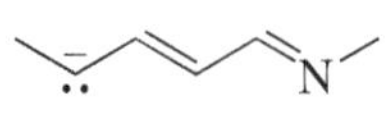

21.52 (•) How many p orbitals are involved in the conjugated π system for each of the molecules in Assessment 21.51?

21.53 (•) Draw the molecular orbital picture of octa-1,3,5,7-tetraene, indicating which is the HOMO and which is the LUMO.

21.54 (•) Which of the following reactions would you expect to be faster/more favorable in each pair? Why?

(a)

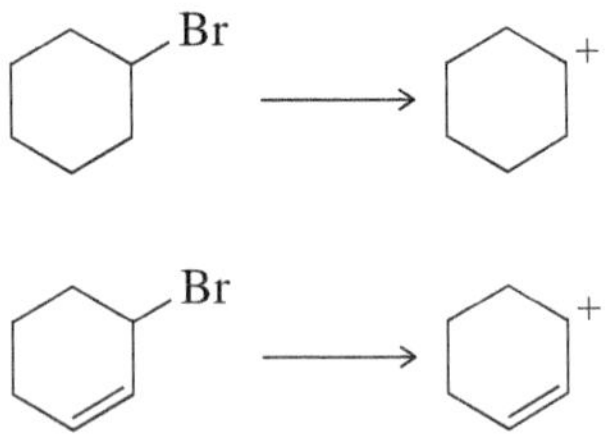

(b)

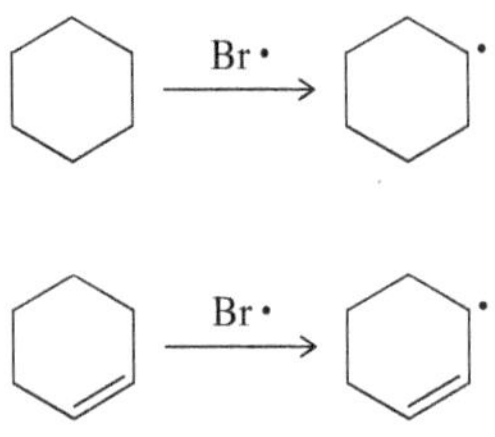

(c)

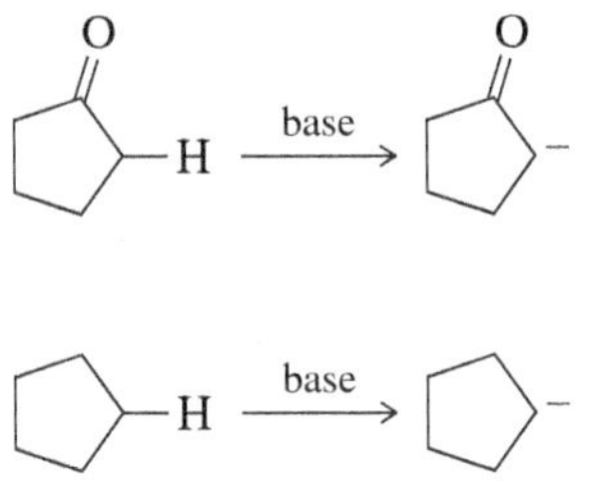

(d)

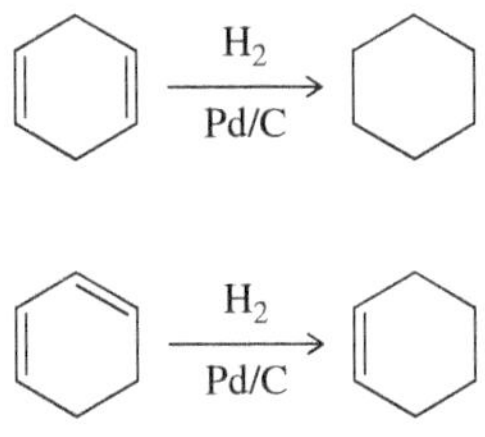

21.55 (••) Molecule **A** will equilibrate to a conjugated system under basic conditions, but molecule **B** will not. Explain why and draw the mechanism to form the conjugated system for molecule **A**.

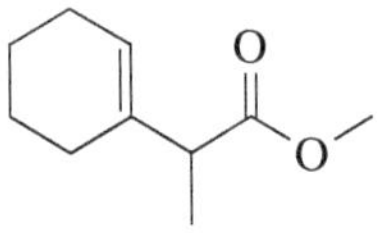

A

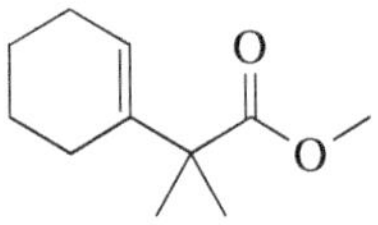

B

21.56 (••) The dye alizarin normally forms an orange solution when dissolved. However, if KOEt is added to the solution, it turns blue very rapidly. Rationalize this result. [*Hint: We'll learn in Chapter 24 that hydroxyl groups bonded to a benzene ring are more acidic than normal alcohols.*]

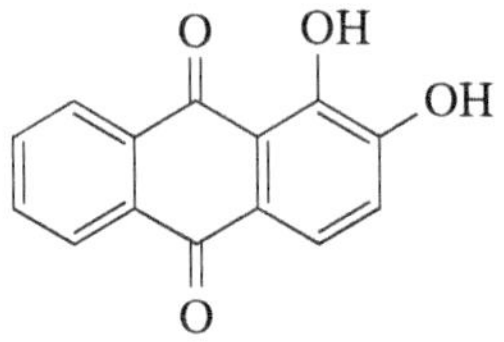

21.57 (••) Predict the product of the following reactions.

(a)

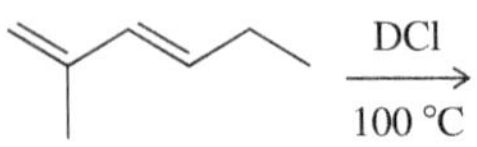

(b)

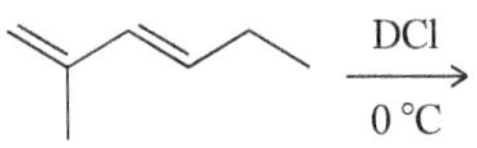

(c)

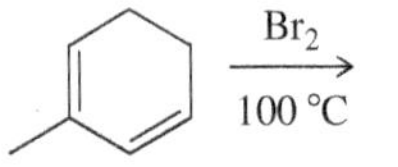

(d)

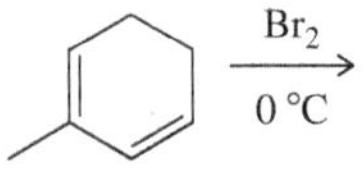

(e)

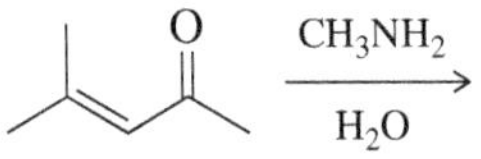

(f)

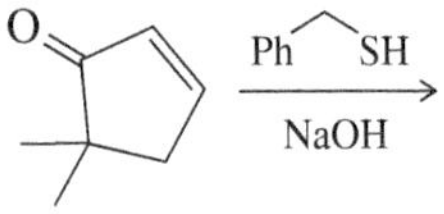

(g)

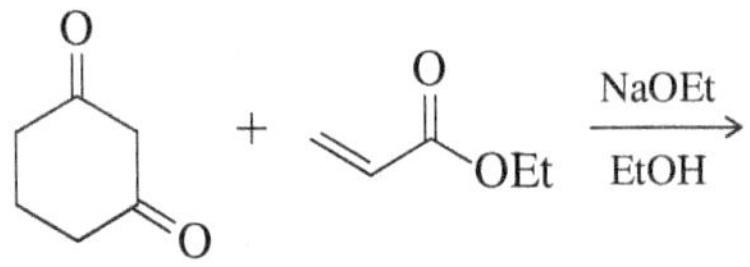

(h)

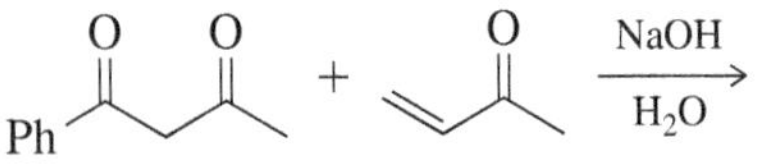

(i)

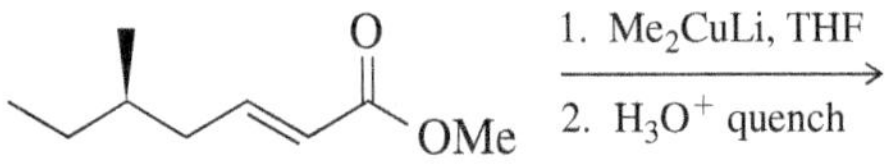

(j)

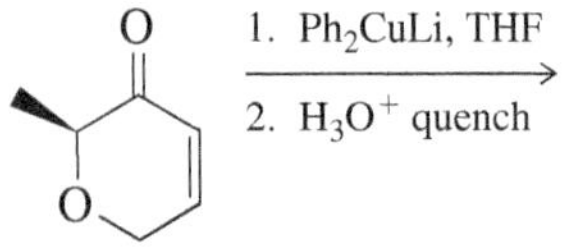

21.58 (•••) In Chapter 9, electron-rich alkenes were oxidized under acidic conditions with *m*CPBA. Conjugated alkenes can be oxidized using the same reagent, but under basic conditions. Suggest a mechanism for this reaction. [*Think about what is electron-rich and what is electron-poor in the reaction. Also, identify the bonds formed and broken.*]

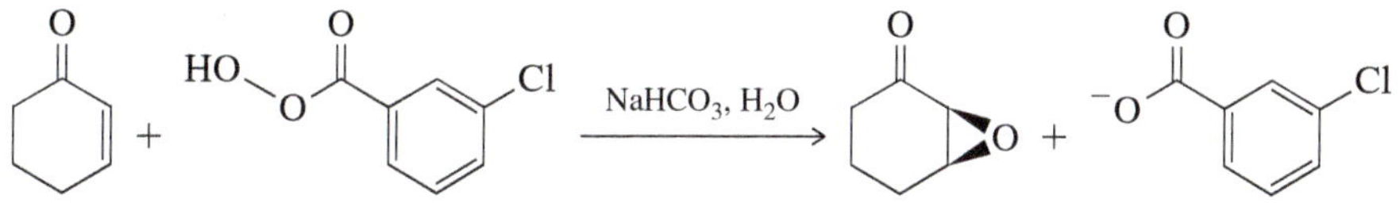

21.59 (••) In Chapter 16, electron-rich alkenes were converted to cyclopropanes using carbenes. A similar reaction occurs for conjugated alkenes under basic conditions. Suggest a mechanism for this reaction. [*The most acidic proton is on the indicated carbon.*]

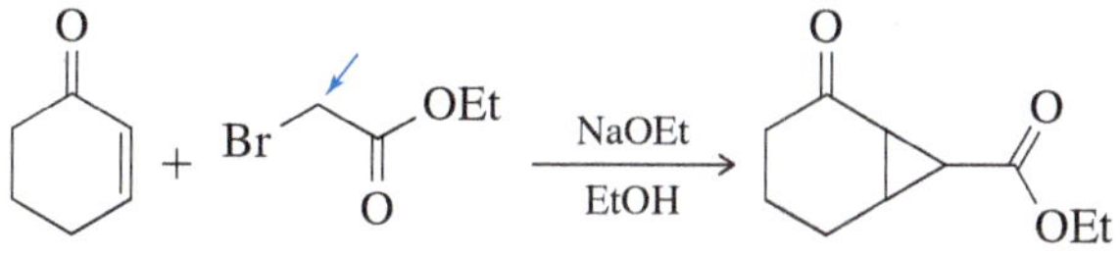

21.60 (•••) The reactivity of cyclopropanes often mimics that of alkenes. (a) On the basis of this, suggest a mechanism for the following reaction. (b) Besides opening the three-membered ring, what is the driving force for this reaction?

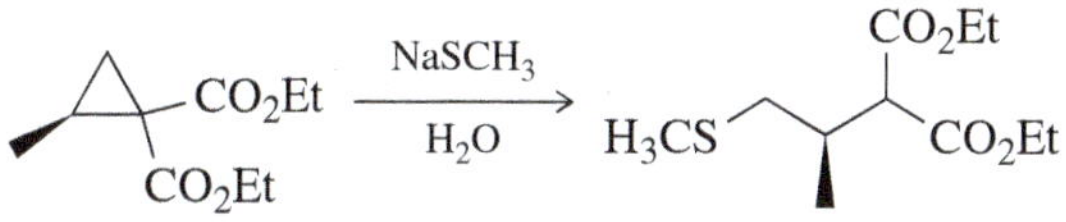

21.61 (••) In Chapter 11, when studying the autooxidation of fatty acids, product **A** was formed to the exclusion of product **B**, even though both are mechanistically possible. Why is **A** favored?

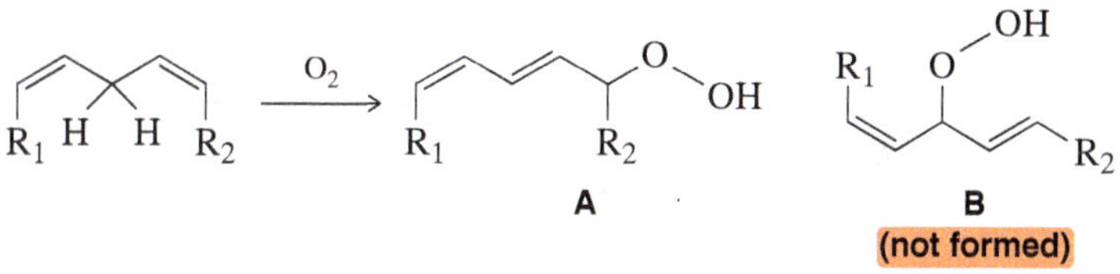

21.62 (•••) FROM THE LITERATURE Using a Robinson annulation, the intermediate shown was made in an early step in the total synthesis of the steroid pregnanolone. Predict the starting materials and reagents used (*Org. Lett.* 2018, *20*, 946–949).

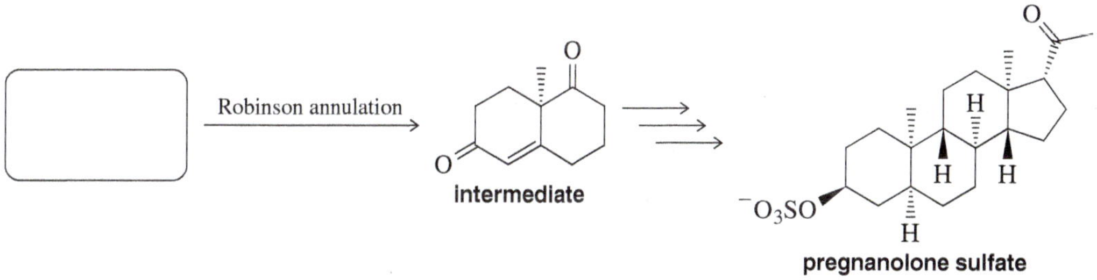

21.63 (•••) FROM THE LITERATURE The following is a later step in the synthesis from Assessment 21.62. Suggest a reagent that would allow synthesis of the product shown (*Org. Lett.* 2018, *20*, 946–949).

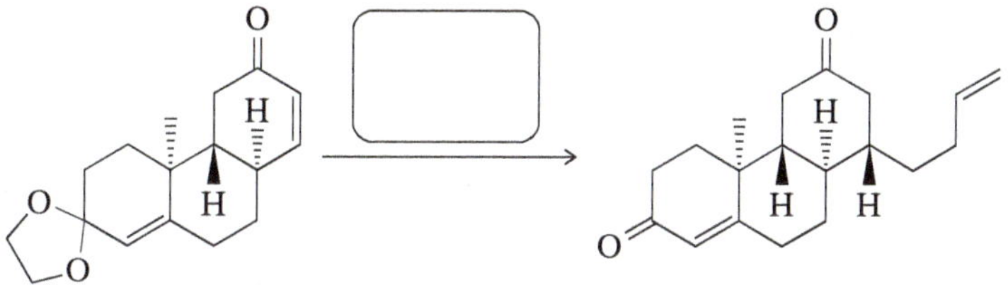

21.64 (•••) The enolates formed by deprotonation of a nitroalkane undergo Michael reactions preferentially. (a) Suggest a mechanism for this reaction. (b) What does the fact that 1,4-addition is preferred tell you about the pK_a of nitroalkanes?

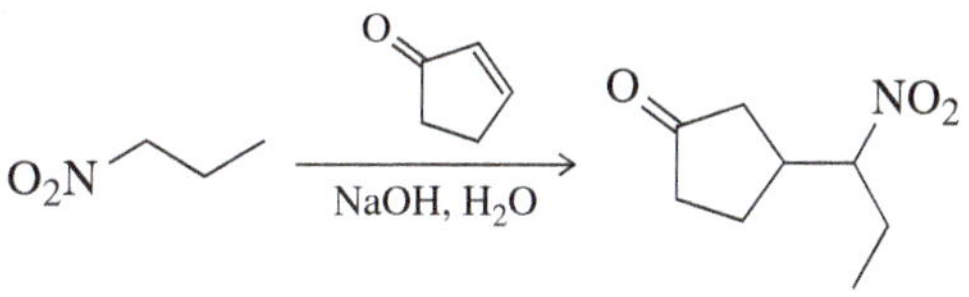

21.65 (•••) In addition to the nitro group making a hydrogen acidic (Assessment 21.64), it can also be used to make alkenes electrophiles and allow for conjugate addition. Predict the product of the following conjugate addition reactions.

(a)

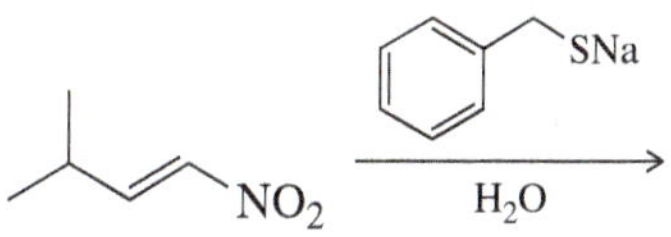

(b)

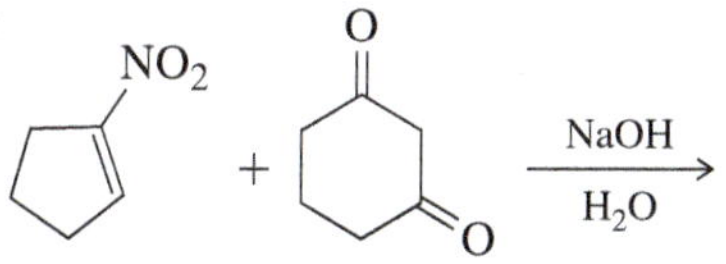

21.66 (•••) FROM THE LITERATURE The following were the first several steps in a synthesis of estrone. Predict the reagents used in each step (*J. Am. Chem. Soc.* **1986**, *108*, 1239–1244). [*Hint: The first step is a Michael reaction.*]

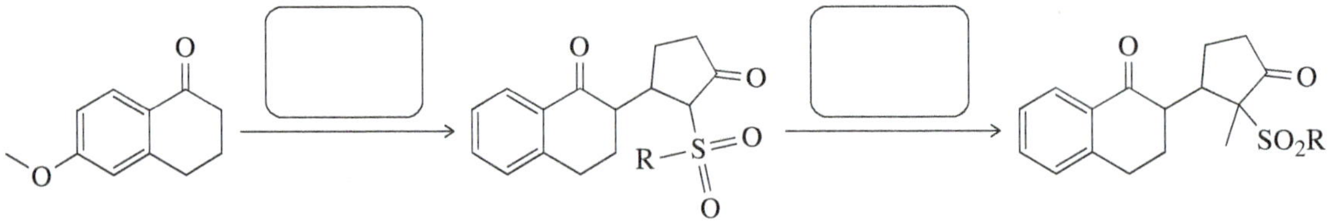

21.67 (•••) Conjugate addition with a cuprate, followed by addition of a 1° haloalkane, can be used to form two new C—C bonds in a single reaction. Suggest a mechanism for this reaction.

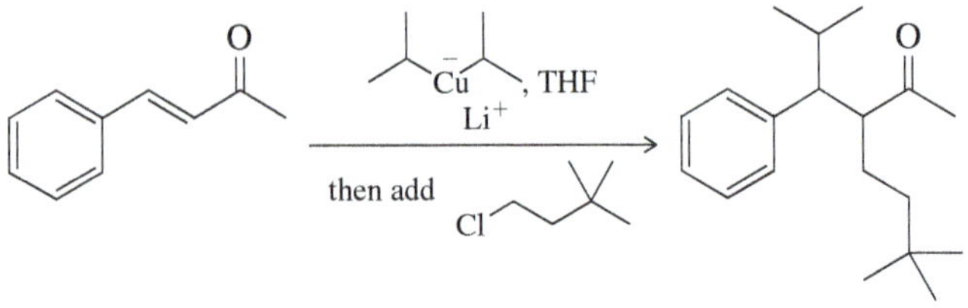

21.68 (•••) FROM THE LITERATURE The following transformation was found to occur in areas with large NO_2 emissions. Suggest a mechanism for the reaction (*J. Phys. Chem.* **2013**, *117*, 14132–14140). [*Hint: Use the fishhook arrows associated with radical reactions.*]

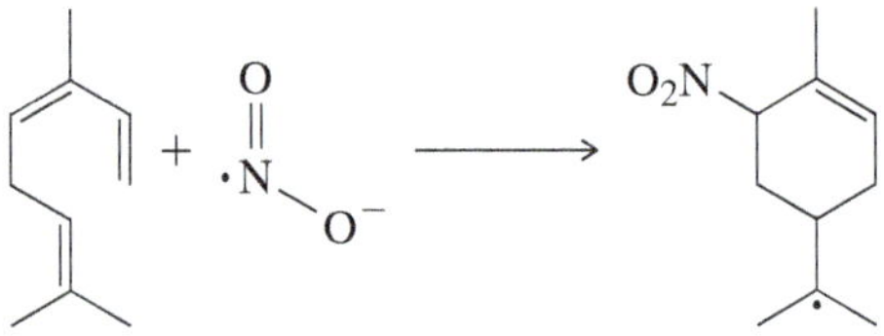

21.69 (••••) LOOKING AHEAD In Chapter 23, we introduce aromatic and antiaromatic molecules. The molecular orbital picture of an aromatic and antiaromatic molecule are shown here. Based on these, which would you expect to be more stable? [*Think carefully about the energy that would be released upon bond formation. Review Section 21.3.1 if you need help.*]

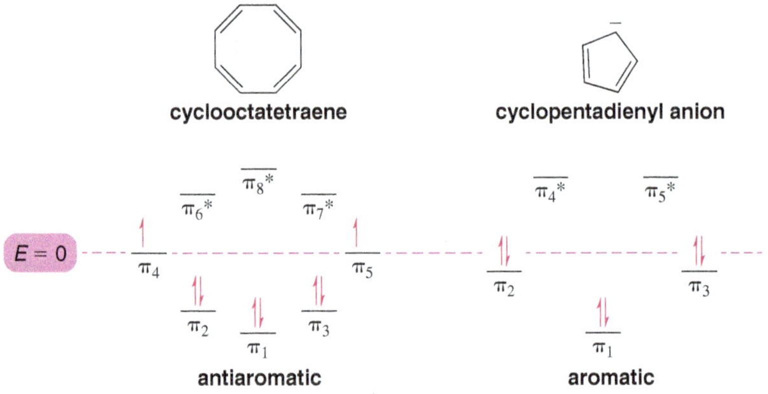

21.70 (••••) LOOKING BACK Hydroboration, an electrophilic addition reaction like those studied in Section 21.4, only gives 1,2-addition to buta-1,3-diene, regardless of temperature. Why?

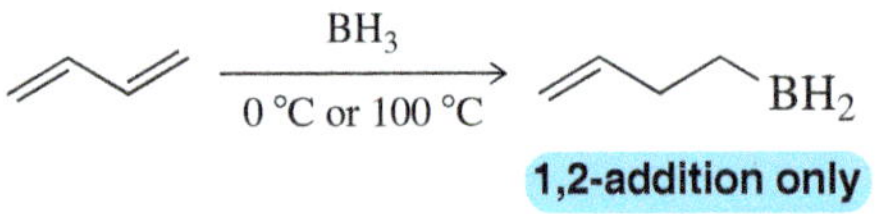

21.71 (•••) FROM THE LITERATURE Indium metal can be used to reduce electron-deficient alkenes in a reaction that is analogous to the reduction of alkynes with sodium and ammonia. Suggest a mechanism for the reaction that accounts for how the alkene being conjugated to an electron-withdrawing group makes it possible (*Org. Lett.* **2001**, *3*, 2603–2605). [*Indium typically ionizes to form In^{3+}.*]

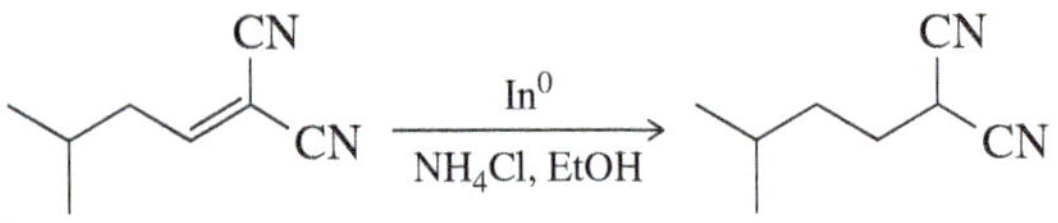

21.72 (•••) FROM THE LITERATURE The following reaction was used in the synthesis of aculeatin A, a natural product that is active against KB cell lines. Although it only worked under acidic conditions, a mechanism can be drawn where the reaction might proceed under basic conditions. Suggest this mechanism (*J. Org. Chem.* **2014**, *79*, 1498–1504). [*The most acidic proton is indicated . . . and number your carbons!*]

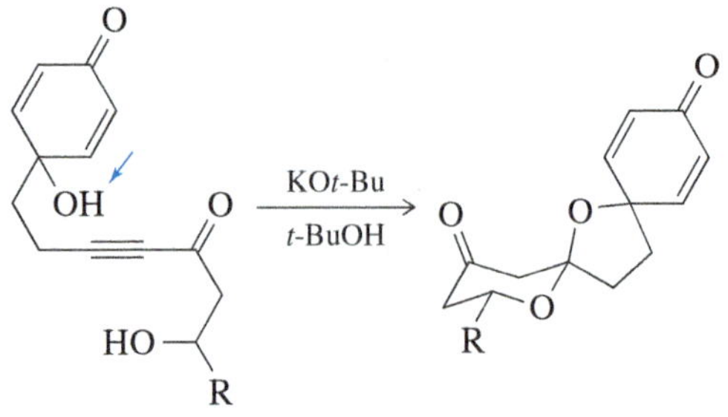

21.73 (•••) FROM THE LITERATURE The following covalent inhibitor blocks function in a protease found in the porcine epidemic diarrhea virus by reacting with a cysteine amino acid residue (shown below) in the active site. Draw the expected complex that forms between the inhibitor and the enzyme active site (*J. Med. Chem.* 2017, *60*, 3212–3216.) [*Assume the presence of active site bases if you need them.*]

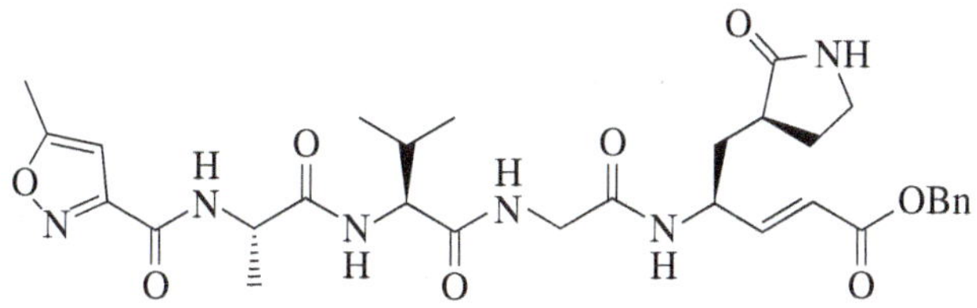
covalent inhibitor

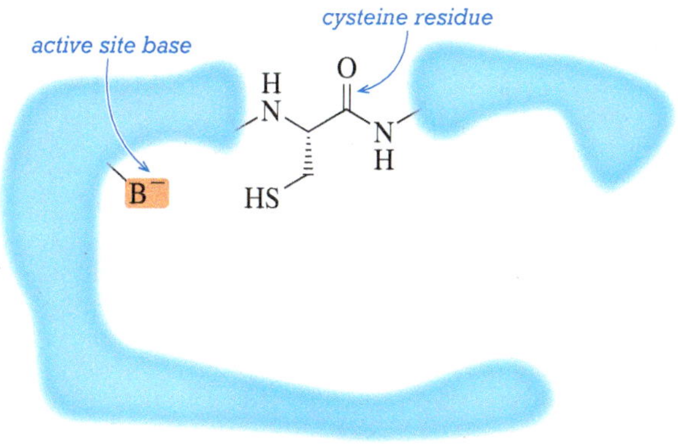

enzyme active site

21.74 (•••) LOOKING AHEAD In Chapter 24, we learn that benzylic alkyl halides undergo faster S_N2 reactions. Thinking about the concept of conjugation, how might the presence of neighboring π bonds lower the energy of the S_N2 transition state?

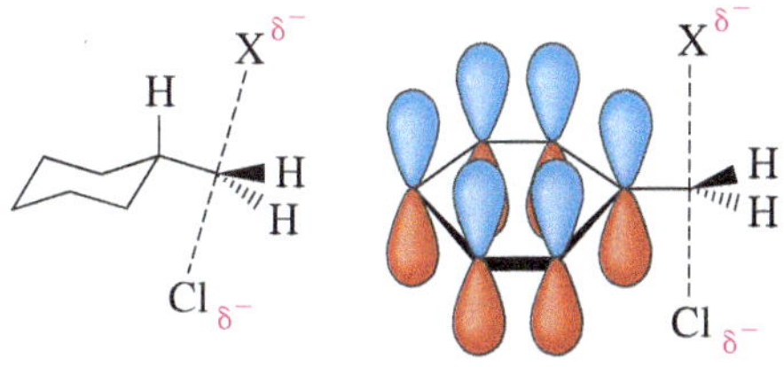

21.75 (•••) Leaves make their "food" using chlorophyll, the highly conjugated molecule responsible for the green color in leaves. Leaves also contain molecules like cryptoxanthin, similar in structure to β-carotene in Figure 21.21. Why do leaves change colors from green to orange/yellow during the fall?

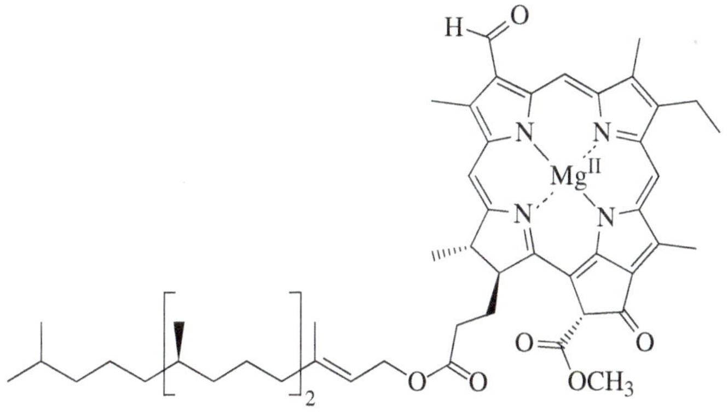
chlorophyll

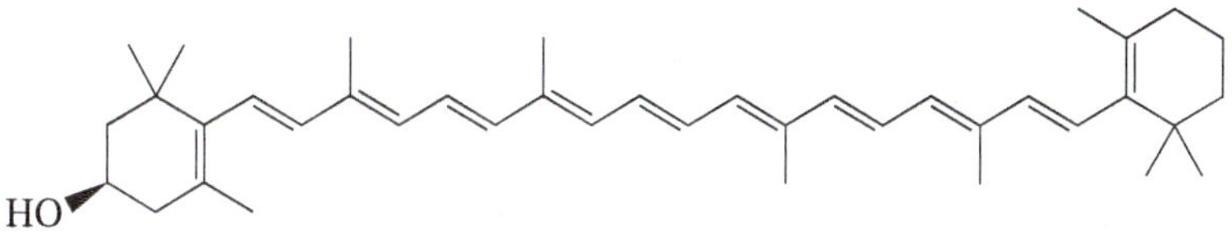
cryptoxanthin

22

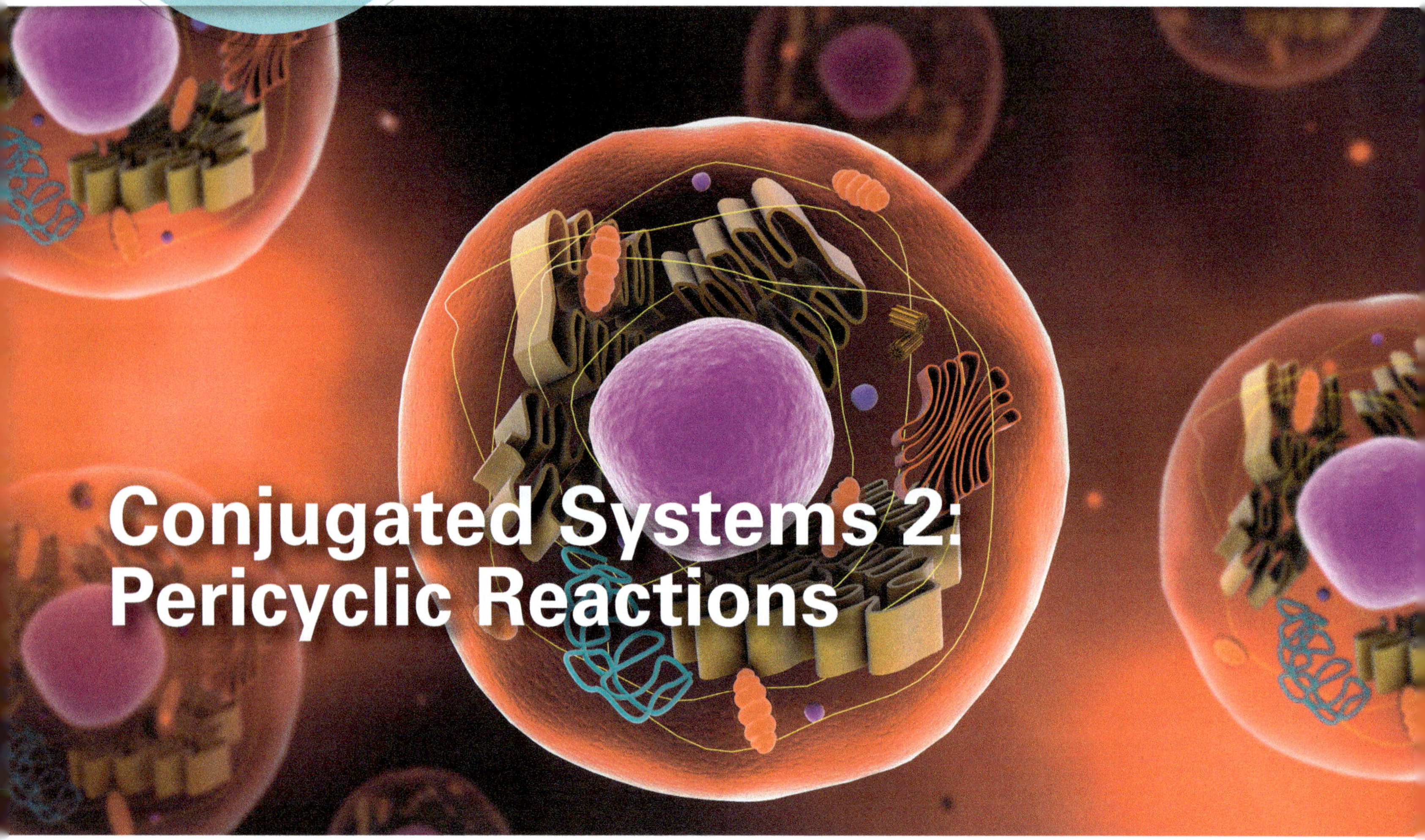

Conjugated Systems 2: Pericyclic Reactions

Learning Objectives

LO 22.0 Compare pericyclic reaction mechanisms

LO 22.1 Analyze Diels–Alder reactions

LO 22.2 Predict the products of pericyclic reactions

Introduction

As the only scientific discipline that makes molecules, chemistry (particularly organic chemistry) has never been more important regarding our modern understanding of biological processes and the treatment of disease. [*Yes, I'm biased, but this is an organic chemistry book.*] There is a symbiotic relationship between biology and synthetic organic chemistry, because many of the reactions that we have studied in this course have a similar enzyme-catalyzed variant in nature. In Chemistry in Context 17-A, for example, we learned about the NAD-dependent oxidation of alcohols by alcohol dehydrogenase. The aldol reaction from Chapter 20 is catalyzed by HMG-CoA synthase to make cholesterol in a pathway that is inhibited by statins to prevent heart disease. The efficiency of enzyme-catalyzed reactions has led organic chemists to successfully explore whether enzymes could be used outside of organisms as catalysts for laboratory-scale synthesis. Enzymes offer a few advantages over traditional catalysts. First, because they provide a chiral environment, high stereoselectivity is possible in enzyme-catalyzed reactions. Second, enzymes occur in nature, so reactions that use them are greener than those that use toxic metal catalysts.

Of the reactions introduced here in Chapter 22, the Diels–Alder reaction has probably had the most impact on the field of organic chemistry. Especially during the middle to late 20th century, it has enabled numerous syntheses that weren't possible previously. As a testament to its impact, its inventors (Otto Diels and Kurt Alder) won the Nobel Prize in Chemistry in 1950. Many of the syntheses that the Diels–Alder reaction enabled were of natural products (complex molecules made by nature). Synthetic chemists often design their syntheses based on how nature might make the same molecule, so these chemists

"First they said my [cyclol] structure [of proteins] couldn't exist. Then when it was found in Nature they said it couldn't be synthesized in a laboratory. Then when it was synthesized they said it wasn't important in any way."

—DOROTHY M. WRINCH

began to wonder whether nature had its own enzyme for doing the Diels–Alder reaction—a so-called Diels–Alderase. [*Even Diels and Alder suggested there might be such a thing.*] Besides satisfying their intellectual curiosity (often enough of a motivation), the discovery and study of a natural Diels–Alderase would enable greener and more efficient catalyst design that could have far-reaching impacts in synthetic endeavors. This chapter discusses the search for and the discovery of a Diels–Alderase enzyme and the creation of an antibody that catalyzes the same reaction.

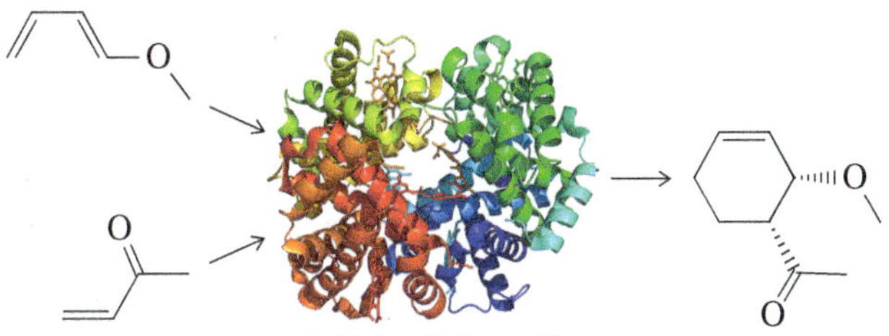

A Diels–Alderase?

○ ○ ○ ○ ○ ○ ○

"Negatives attack positives" has been repeatedly used to explain the how and why of organic chemistry mechanisms. Similarly, resonance structures have helped us to rationalize the stability and reactivity of organic molecules. Both of these concepts are ways that give us a sense of what is happening, but they are simply models constructed to simplify our understanding of difficult concepts. At the root of all reactivity is the interaction of molecular orbitals. This chapter, with mostly concerted reactions that can't be directly explained by "negatives attack positives," instead relies on our understanding of the molecular orbital picture of conjugated systems. In these systems, the driving force is "HOMOs attack LUMOs." It turns out that this is not that much different than "negatives attack positives" because we have really been talking about the interaction of atomic and molecular orbitals all along. And the ball still always rolls downhill. [*Let's prepare for these modified thought processes by testing our understanding of some important concepts.*]

Test your understanding

Assessment 22.1 Draw two important resonance structures involving the lone pair on oxygen for the molecule shown. Which carbons are most likely to act as nucleophiles?

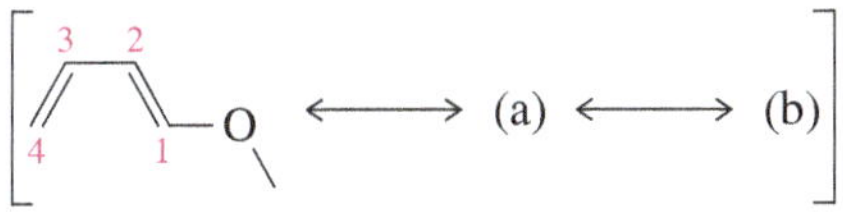

Assessment 22.2 Draw two important resonance structures involving the C—O π bond for the molecule shown. To which carbons would you expect a nucleophile to add?

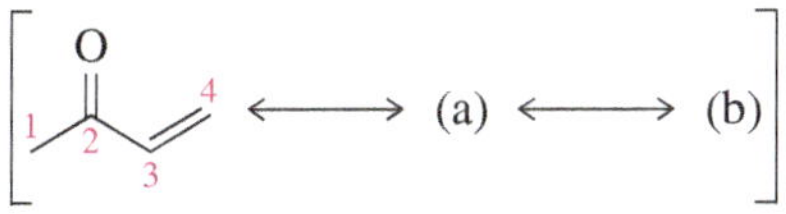

Assessment 22.3 Draw the molecular orbital picture for (a) ethene and (b) buta-1,3-diene. Label the HOMO and LUMO of each.

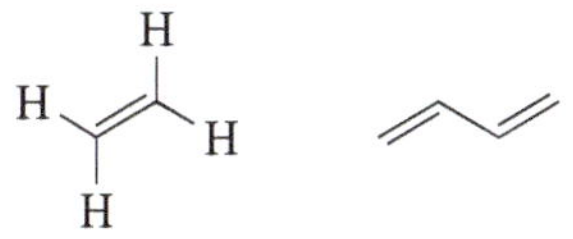

ethene buta-1,3-diene

Assessment 22.4 What frequency of light would be required to excite an electron if the HOMO–LUMO energy gap was 33.9 kcal/mol (142 kJ/mol)?

Assessment 22.5 The S_N2 reaction is the concerted, backside displacement of a good leaving group by a nucleophile [*see Chapter 12*]. Why do nucleophiles attack from the back in S_N2 reactions?

Assessment 22.6 Identify the bonds formed and the bonds broken in the following reaction. Would you expect this reaction to be favored based on ΔH? Based on ΔS? Based on ΔG?

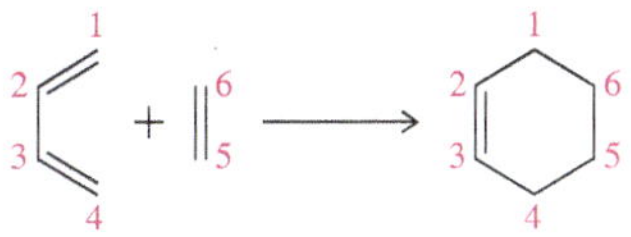

22.1 Pericyclic reactions: an overview

Although they will be studied for the first time in detail here, **pericyclic reactions** have been discussed, though not by that name, throughout the book. Pericyclic reactions (*peri* = around) are concerted, lack charged intermediates, and feature the cyclic flow of electrons. Most common among these are the 6-electron cyclic transition states whose stability is explained in Chapter 23. The decarboxylation of a β-keto acid from Section 18.8.4 is representative of a pericyclic reaction, as are several steps in the ozonolysis reaction from Section 9.1.7 (**Figure 22.1**).

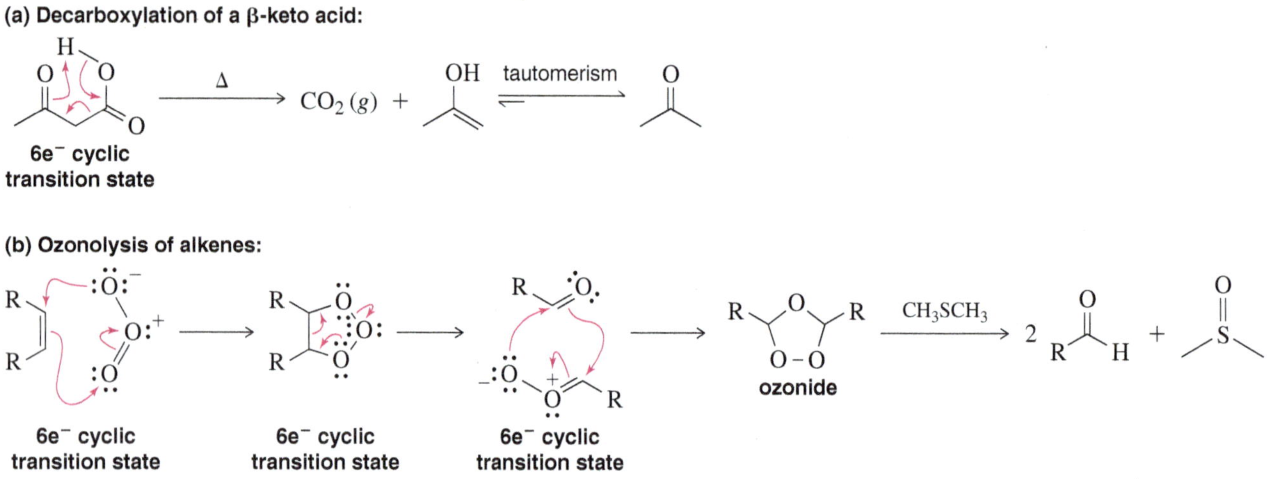

Figure 22.1 Pericyclic reactions studied previously. (a) Decarboxylation of a β-keto acid is a pericyclic reaction (Section 18.8.4). (b) The mechanism of ozonolysis contains several pericyclic reactions (Section 9.1.7).

Assessment 22.7 We have studied a number of pericyclic reactions previously. Draw the mechanism of the steps shown. The section number where this material was first studied is given for your review.

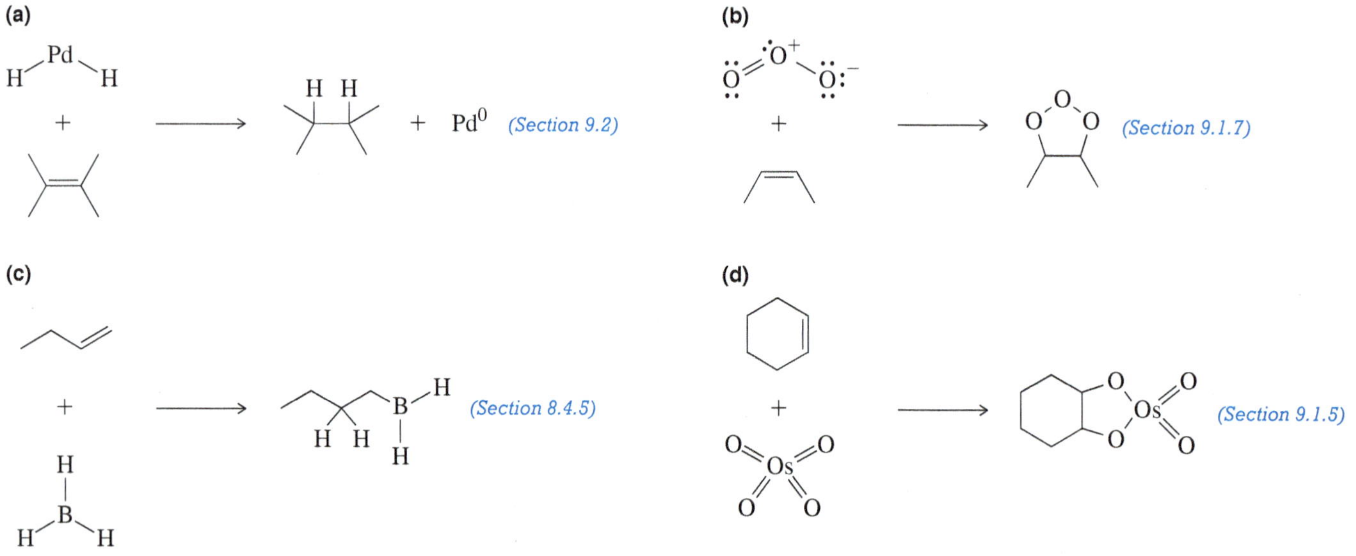

The three pericyclic reactions we discuss are cycloadditions, electrocyclic reactions, and sigmatropic rearrangements. The challenge in these reactions is their concertedness and explaining how bonds form between two neutral, nonpolar molecules/atoms. It turns out that "negatives attack positives" [*rule 2 of organic chemistry*] is buried deep within the molecular orbital analysis. Thus, before examining these new reactions, let's revisit two reactions we know well in the context of frontier molecular orbital theory. This will clarify why "HOMOs attack LUMOs" is the new "negatives attack positives."

22.2 Frontier molecular orbital theory

Developed in the 1950s by Nobel Laureate Kenichi Fukui, **frontier molecular orbital (FMO) theory** states that the outcome of a reaction is determined by the interaction of only the highest occupied molecular orbital (HOMO) and the lowest unoccupied molecular orbital (LUMO). [*You may want to review HOMOs and LUMOs in Section 21.3.*] That is, even when we can draw multiple bonding and antibonding molecular orbitals for a particular molecule, it is only the ones on the "frontiers" that matter—namely, the occupied orbital of highest energy (HOMO) and the unoccupied orbital of lowest energy (LUMO).

While this is based on the complicated math associated with molecular orbital theory, it is not so different from things we learned even as general chemistry students. For example, why do carbon and silicon exhibit such similar reactivity that they were placed in the same column of the periodic table? Chemical reactivity of individual elements is based on the valence electrons (**Figure 22.2**), regardless of how many electrons are at lower energy levels. Those valence electrons are the outermost electrons or, in the language of this chapter, the frontier electrons.

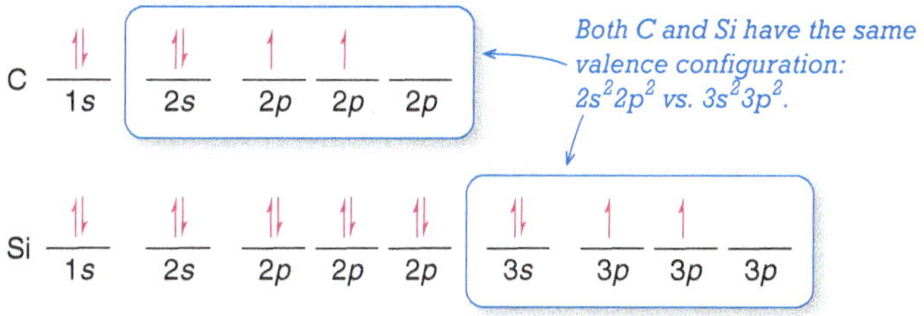

Figure 22.2 Silicon and carbon display similar reactivity because they have the same valence shell (frontier) electron configuration.

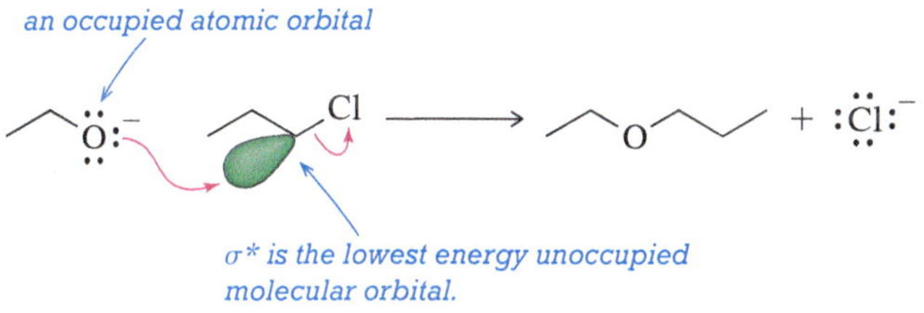

Figure 22.3 The S_N2 reaction of ethoxide and a haloalkane occurs because an occupied atomic orbital of ethoxide reacts with the lowest unoccupied molecular orbital (LUMO) of 1-chloropropane.

Although the electrons in carbon and silicon are in atomic orbitals, the fact that all of the lower energy atomic orbitals are ignored when discussing reactivity makes it a great analogy for frontier molecular orbital theory. Let's look briefly at two reactions we have studied previously to show that we have always known about frontier molecular orbital theory. The first is the S_N2 reaction between ethoxide and 1-chloropropane (Section 12.2 and **Figure 22.3**). Ethoxide and 1-chloropropane each contain a number of σ and σ* molecular orbitals, but this reaction is based on a lone pair on oxygen being donated to the C—Cl antibonding orbital (σ*). In orbital terms, this is an occupied atomic orbital (sp^3 of oxygen) reacting with an unoccupied molecular orbital (σ*) of the haloalkane. That sounds a lot like the HOMO reacting with the LUMO.

The second reaction is the addition of isopropyl magnesium bromide (a Grignard reagent) to acetone in **Figure 22.4**. Once more, each of these compounds contains a number of molecular orbitals. In Section 17.1.4, the angle of attack of a nucleophile on the carbonyl carbon (the Bürgi–Dunitz angle; see Figure 17.6) was a result of the nucleophile donating a lone pair of electrons (from an occupied sp^3 atomic orbital) to an unoccupied molecular orbital (π*) of the ketone. Once more, the HOMO reacts with the LUMO.

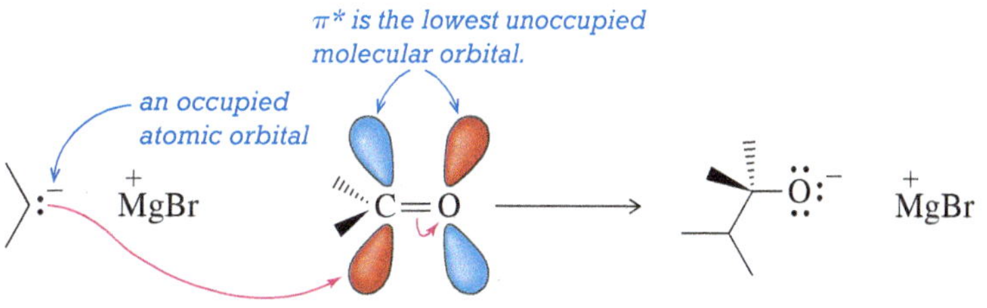

Figure 22.4 Carbonyl addition of a Grignard reagent is really the reaction of an occupied atomic orbital of the Grignard reagent reacting with the lowest unoccupied molecular orbital (LUMO) of acetone.

Early in Chapter 1, organic chemistry was simplified as "negatives attack positives." But we also said that this would be discussed in several different ways—namely, electron-rich with electron-poor and nucleophiles with electrophiles. *HOMOs attack LUMOs* is just one more way. Note in Figures 22.3 and 22.4 that the nucleophile uses an occupied orbital and the electrophile uses the lowest unoccupied molecular orbital. Correlating nucleophiles with HOMOs and electrophiles with LUMOs will help us understand the properties of the reactions in the sections that follow.

Why don't HOMOs react with HOMOs or LUMOs react with LUMOs to make a new bond? [*Any ideas?*] Two filled orbitals (HOMOs) can't overlap because there would be too many electrons, whereas a bond between two LUMOs would have zero electrons and would therefore not be a bond at all. Whenever two orbitals overlap to make a bond, there needs to be a total of exactly 2 electrons in the overlapping orbitals (**Figure 22.5**). Most commonly, this is 2 electrons in the HOMO and zero in the LUMO. [*What about 1 electron in each? Have we seen that before? See Section 11.5.1.1.*]

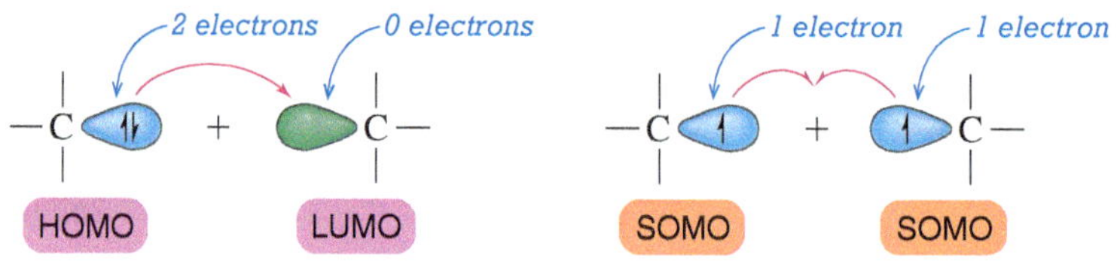

Figure 22.5 Bond formation between two orbitals must contain a total of exactly 2 electrons. (SOMO = singly occupied molecular orbital.)

22.3 Diels–Alder cycloaddition

The **Diels–Alder reaction** is the concerted reaction between a **diene** and a **dienophile** to make a cyclohexene. It is a cycloaddition reaction because, as shown in **Figure 22.6**, ethene *adds* to buta-1,3-diene to make a *cyclo*hexene. The terminology of dienophile [*loves dienes*] is analogous to the similar term electrophile [*loves electrons*]. The concerted mechanism is represented by the simultaneous movement of 6 electrons. This 6-electron cyclic (aromatic) transition state has a low activation energy for reasons that are discussed in Chapter 23. Disfavored by entropy (2 molecules → 1 molecule; $\Delta S < 0$), the Diels–Alder reaction is heavily favored by enthalpy with the net formation of two σ bonds, compared to the breaking of two π bonds.

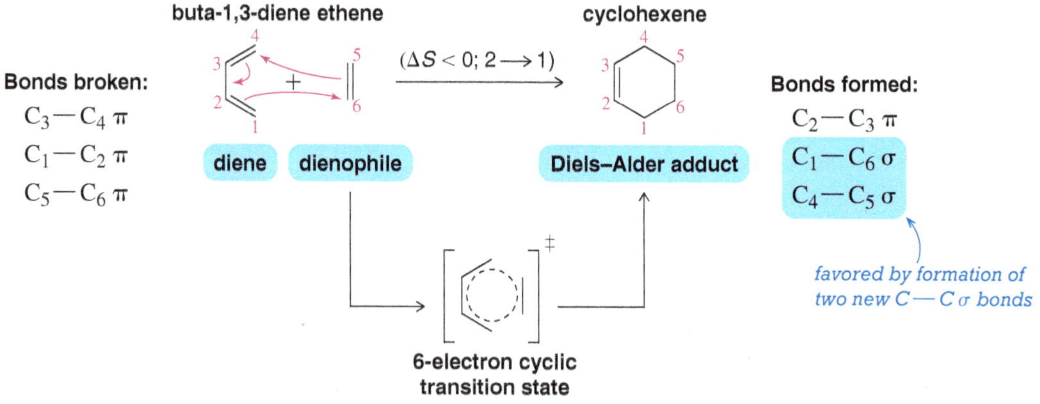

Figure 22.6 The Diels–Alder reaction is a concerted cycloaddition, which occurs through an aromatic transition state and is enthalpically favored by the formation of two new C—C σ bonds.

This reaction occurs between two nonpolar, uncharged molecules, so it can't be understood as "negatives attack positives," but it can be as "HOMOs attack LUMOs." [*My graduate school advisor taught me to think of the Diels–Alder reaction as the diene and the dienophile clicking their π systems together like Legos®.*] With this in mind, there are two possible ways that the reaction could occur. It could occur with the HOMO of the diene reacting with the LUMO of the dienophile [**Figure 22.7(a)**]. Alternatively, the HOMO of the dienophile could react with the LUMO of the diene [**Figure 22.7(b)**]. In both of these cases, the signs of the lobes of the

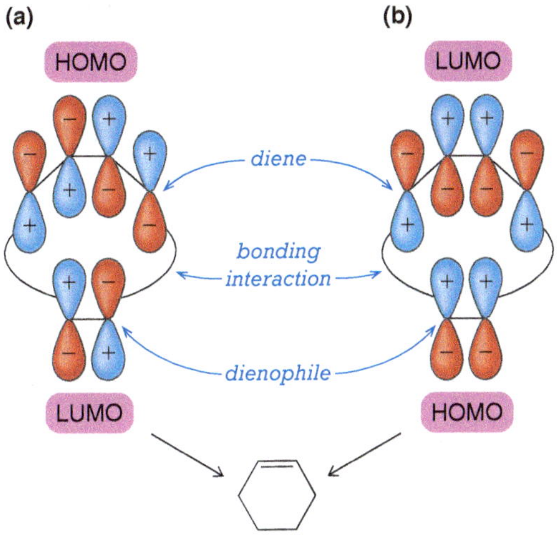

Either combination forms cyclohexene.

Figure 22.7 Molecular orbital overlap in the Diels–Alder reaction occurs when (a) the HOMO of the diene reacts with the LUMO of the dienophile or (b) the HOMO of the dienophile reacts with the LUMO of the diene.

molecular orbitals match up to form the new C_1—C_6 and C_4—C_5 bonds. Where like signs overlap, electrons are delocalized and bonding occurs. [*For drawing molecular orbital pictures of conjugated dienes, see the rules in Section 21.3.*]

To better understand the influence of the frontier molecular orbitals in the Diels–Alder reaction, let's briefly look at a reaction that doesn't occur. Although a similar concerted mechanism could be drawn in which two molecules of ethene react to form a cyclobutane, the orbital overlap is not right for this to happen. Using the HOMO of one ethene molecule and the LUMO of another, overlap to form a new bond is only possible on one side (**Figure 22.8**). The other side, where + overlaps with −, creates a new antibonding orbital with a lobe between the two nuclei, thus preventing formation of the cyclobutane. [*Compare this with Figures 22.7(a) and 22.7(b), where overlap of like coefficients creates new bonding orbitals.*] We explain in Section 22.4 that cyclobutane can be made using this type of reaction with a slight twist. [*Any thoughts on how this might be possible?*]

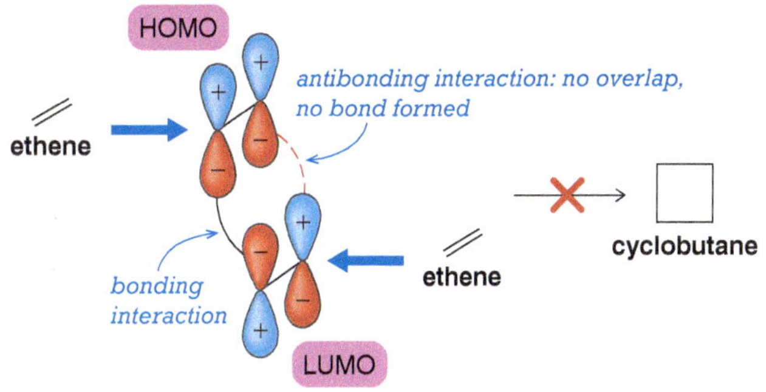

Figure 22.8 Molecular orbital overlap in the cycloaddition of two ethene molecules cannot occur because of the mismatched orbital coefficients that create a new antibonding interaction. Cyclobutane cannot be made under these conditions.

Although a full understanding of this would require a deeper look at frontier molecular orbital theory than is relevant to this course, we can think like a chemist to see how experimental results have demonstrated the favorability of this orbital arrangement.

It turns out that the Diels–Alder reaction occurs predominantly through overlap of $HOMO_{diene}$ with $LUMO_{dienophile}$.

Thinking Like a Chemist

This is where the correlation of HOMOs and LUMOs with nucleophiles and electrophiles becomes pertinent. In trying to determine whether the diene uses the HOMO or the LUMO, we are really asking whether it is the "nucleophile" or "electrophile" in the reaction. [*Note the quotes: These aren't really nucleophiles or electrophiles.*] We know that increased electron density makes nucleophiles more reactive and electrophiles less reactive, whereas decreased electron density makes nucleophiles less reactive and electrophiles more reactive. In a similar way (**Figure 22.9**), increased electron density should make HOMOs more reactive

Figure 22.9 The modulated reactivity of HOMOs and LUMOs (a) Electron donating makes HOMOs more reactive (and LUMOs less reactive). (b) Electron withdrawing makes LUMOs more reactive (and HOMOs less reactive).

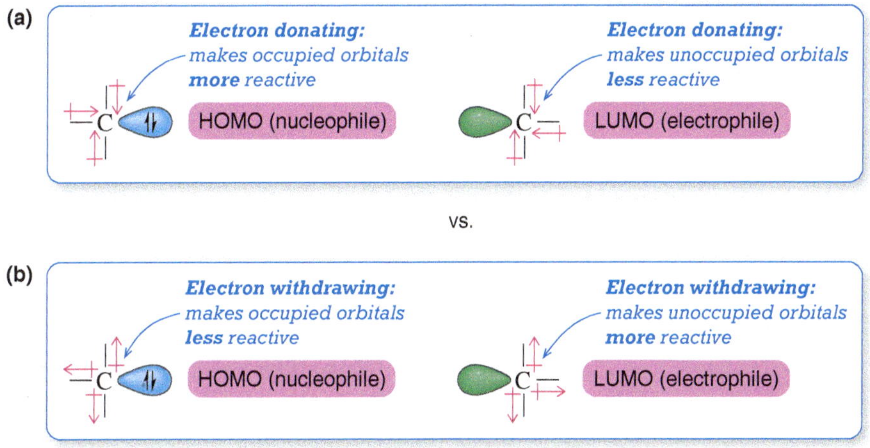

and LUMOs less reactive, whereas decreased electron density should make HOMOs less reactive and LUMOs more reactive. [*Carefully reread this paragraph and examine Figure 22.9 to be sure it makes sense before reading further.*]

And so our experiment becomes this: Place electron-donating and electron-withdrawing groups on each constituent (diene and dienophile) and see how the reaction rate changes. The constituent for which the rate increases with electron-donating groups attached must be using the HOMO (because it is the "nucleophile"). The constituent for which the rate increases with electron-withdrawing groups attached must be using the LUMO (because it is the "electrophile"). A methoxy group will be used as the electron-donating group and a nitro group as the electron-withdrawing group. Both of these groups exhibit their respective effects through resonance. These two will be compared to a hydrogen, which is neither electron-donating nor electron-withdrawing.

The results of this experiment are summarized in **Figure 22.10**. The rate of the Diels–Alder reaction with ethene is increased when an electron-donating group is attached to the diene and decreased when an electron-withdrawing group is attached to the diene [Figure 22.10(a)]. Inversely, the rate of the Diels–Alder reaction with buta-1,3-diene is increased when an electron-withdrawing group is attached to ethene and decreased when an electron-donating group is attached to ethene [Figure 22.10(b)].

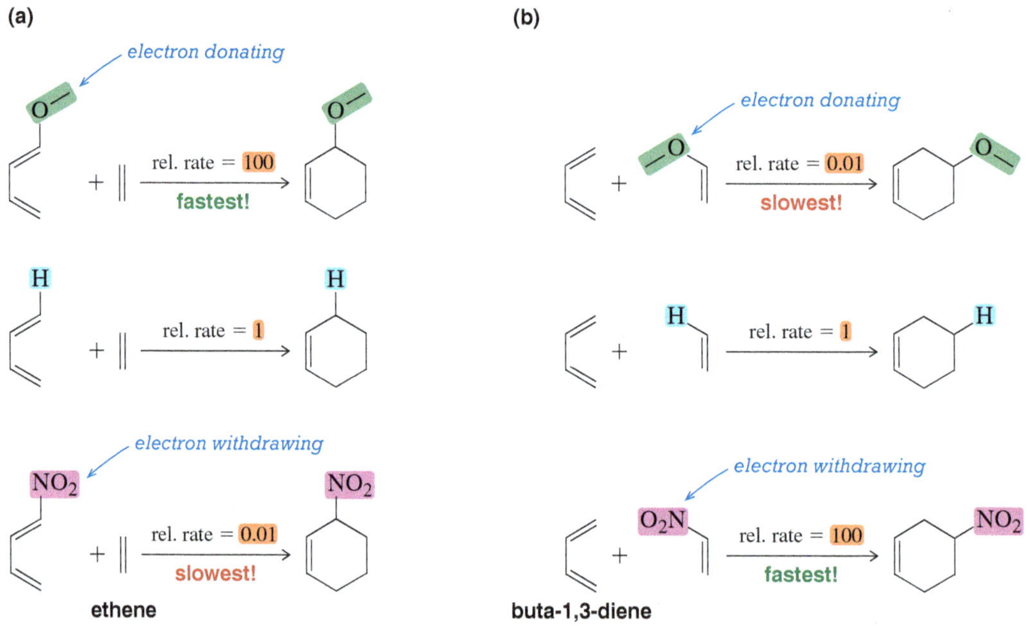

Figure 22.10 Substituent effects on the rate of the Diels–Alder reaction. (a) Electron-donating groups on the diene make the Diels–Alder reaction faster, whereas electron-withdrawing groups on it make the reaction slower. (b) Conversely, electron-withdrawing groups on the dienophile make the Diels–Alder reaction faster, whereas electron-donating groups on it make the reaction slower. [*The trend is real but the relative rates were chosen to illustrate the point.*]

The data in Figure 22.10 are consistent with dienes using their HOMO and dienophiles using their LUMO in normal Diels–Alder reactions. Furthermore, the fastest and most efficient Diels–Alder reactions are achieved between electron-rich dienes and electron-poor dienophiles. [*These are often referred to as normal electron demand Diels–Alder reactions.*] ■

22.3 SUMMARY

- The Diels–Alder reaction is a cycloaddition where a diene reacts with a dienophile in a concerted reaction featuring a 6-electron cyclic transition state. In normal electron demand Diels–Alder reactions, the HOMO of the diene reacts with the LUMO of the dienophile. As a result, electron-rich dienes and electron-poor dienophiles give the fastest Diels–Alder reactions.

22-A CHEMISTRY IN CONTEXT: Biochemistry

Putative Diels–Alderase?

In 1997, studies of the biosynthesis of macrophomic acid by the fungus *Macrophoma commelinae* suggested that it involved a Diels–Alder reaction (**Figure 22.11**). Decarboxylation of oxaloacetate gives an electron-rich dienophile that was proposed to cyclize with an electron-poor diene. [*This is the opposite of the normal electron distribution in the Diels–Alder reaction. Thus, we'd call this an inverse electron demand Diels–Alder.*] If this is the pathway that the enzyme catalyzes, this would have represented an early example of a Diels–Alderase enzyme.

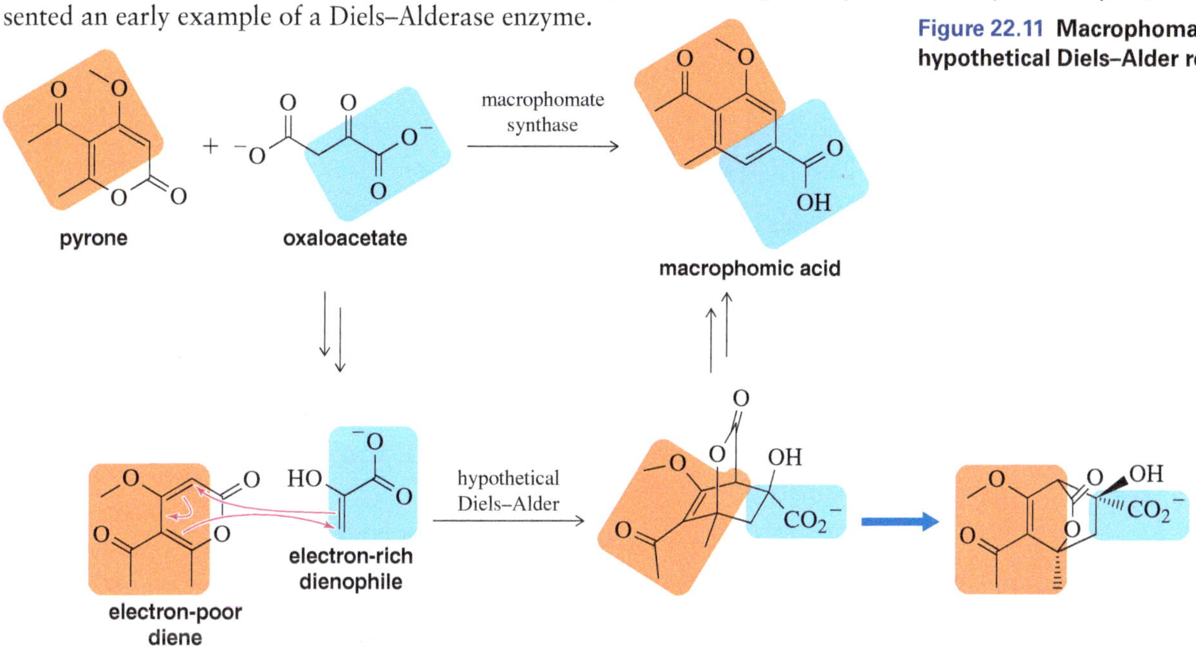

Figure 22.11 Macrophomate synthesis by a hypothetical Diels–Alder reaction.

Further mechanistic studies revealed an alternate possibility for macrophomic acid formation (**Figure 22.12**). Instead of a Diels–Alder reaction, the synthesis could occur by a stepwise process where a Michael addition (Section 21.5.2) is followed by an intramolecular aldol with an extended enolate (Section 20.3.4). Whenever there are two possible pathways, a reaction only occurs by the one with the lowest activation energy. Later computer modeling of the two pathways revealed that the Michael/aldol pathway has a much lower activation energy. Thus, it appears that macrophomate synthase is, in fact, not a Diels–Alderase. [*Cue the sad music . . . .*]

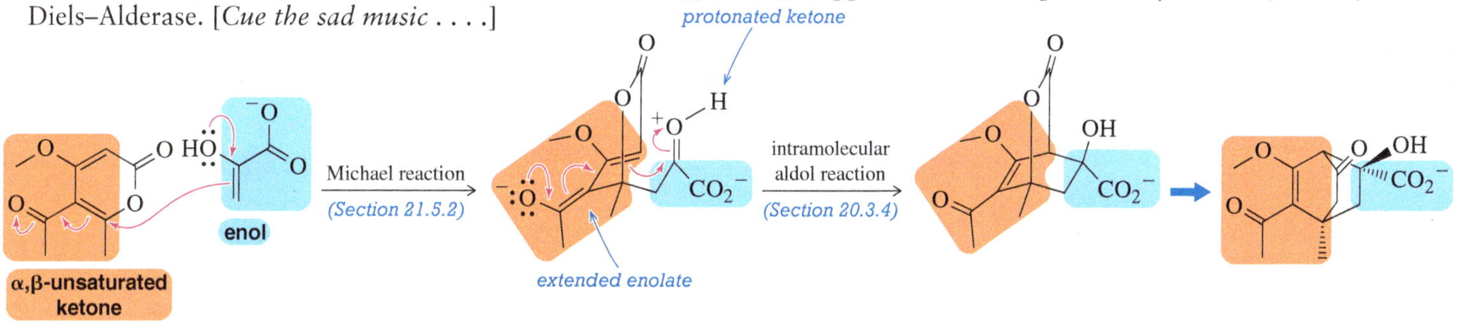

Figure 22.12 Macrophomate synthesis by an alternate Michael reaction/intramolecular aldol sequence.

Assessment 22.8 The hypothetical Diels–Alder product from Figure 22.11 undergoes two more reactions to make macrophomic acid. Step 1 is referred to as a retro-Diels–Alder. (a) Show an arrow-pushing mechanism for this concerted transformation. (b) Show a mechanism for step 2. (c) What is the driving force for step 2? [*Hint: We explain in Chapter 23.*]

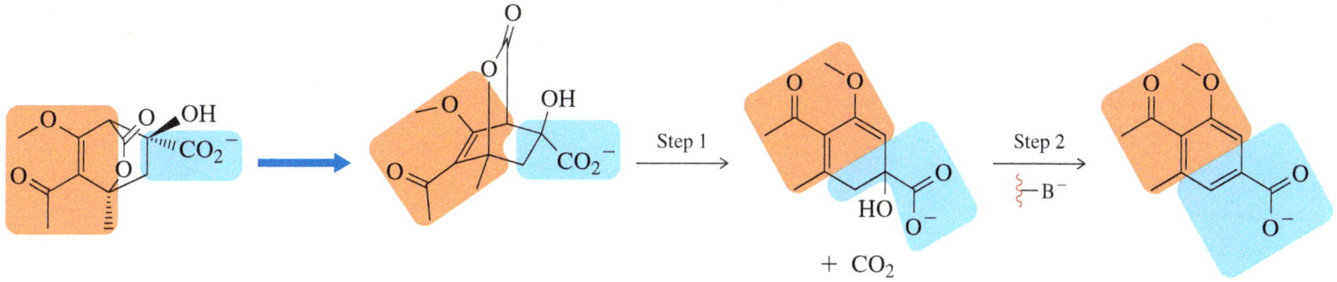

Disclaimer: The Diels–Alder reaction is not easy to understand, at least at first. When the diene and dienophile in Figure 22.13 react, there are four possible diastereomers that can form, each of which is a racemic mixture. Thus, there are eight possible products. And yet only one pair of enantiomers is produced. To understand why this lone racemic product is produced and how we can ourselves predict Diels–Alder products, we have to consider several aspects of the reactions independently. These sections are united under the title "implications of concertedness." By learning each one individually and assessing your knowledge of each before moving on, you can gradually accumulate the knowledge that will allow you to predict the product from the beginning. This may feel like a long and meandering journey, but the assessments are like checkpoints on a map. [*Thus, it is imperative that you do the assessments as you go. Keep a molecular model kit close. It will come in handy here.*]

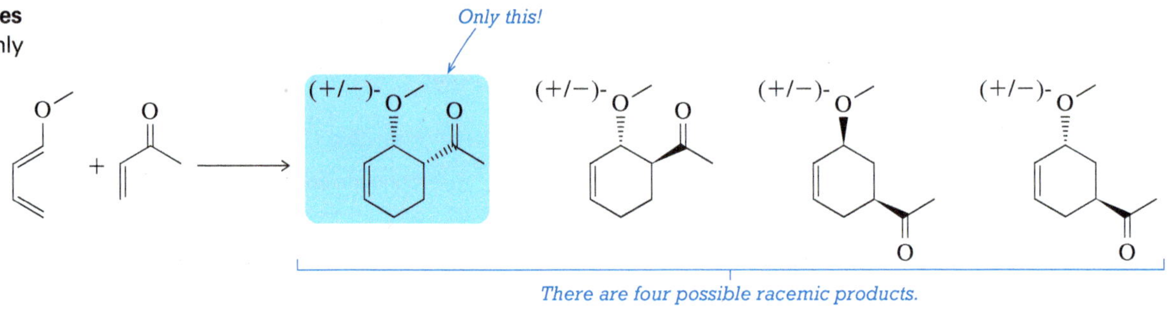

Figure 22.13 Possible outcomes of the Diels–Alder reaction. Only one is produced—read on!

22.3.1 Implications of concertedness #1: the *s*-cis conformation is required

The Diels–Alder reaction is concerted, so in order for both ends of the diene to interact with the dienophile at the same time, the diene must be able to access the *s*-cis conformation (Figure 22.14). The terminology *s*-cis versus *s*-trans should be new to you, but it is similar to the way cis and trans have been used in the past. The only difference is that the *s* refers to the conformation around the single bond. [*You will only see this conformation discussed in the context of the Diels–Alder, so don't worry about the what as much as the why.*] When the two ends of the diene (C_1 and C_4) are pointing in the same direction in the same plane, they are said to be cis (on the same side) as it relates to the σ bond. When the two ends of

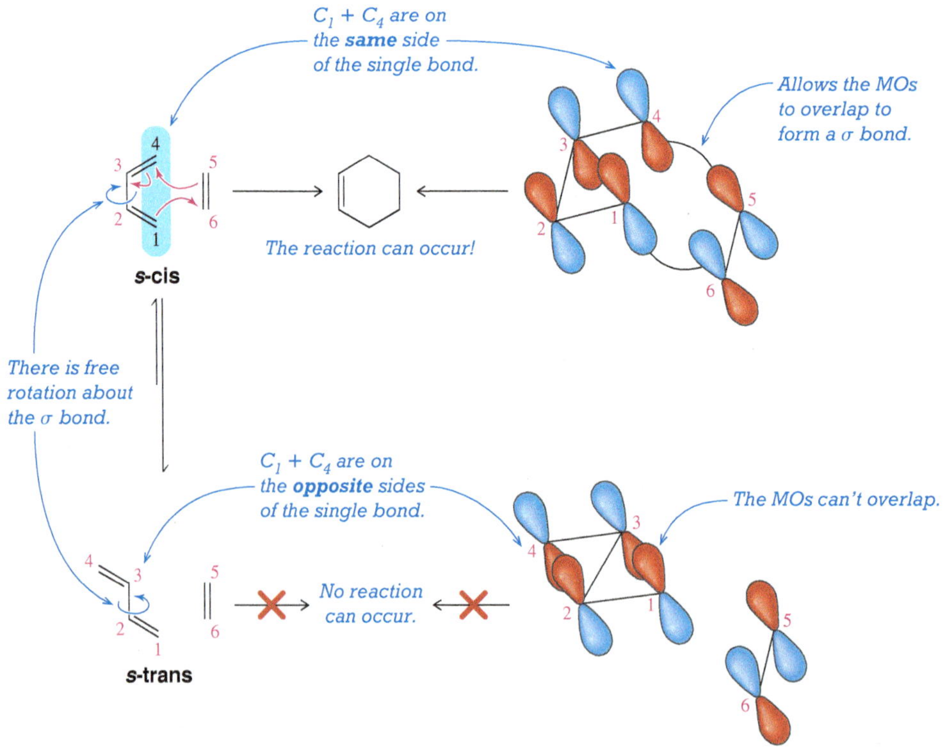

Figure 22.14 The *s*-cis conformation is required for the Diels–Alder reaction to occur.

the diene (C_1 and C_4) are pointing in opposite directions in the same plane, they are said to be trans (on different sides) as it relates to the σ bond. There is free rotation around this single bond, so in the absence of some conformational restraint, the *s*-cis conformation is always accessible, even though it is less stable.

Assessment 22.9 Maleic anhydride, with two electron-withdrawing groups attached to the alkene, is one of the most reactive dienophiles known. And yet, with the diene shown, a Diels–Alder reaction will never occur. Why?

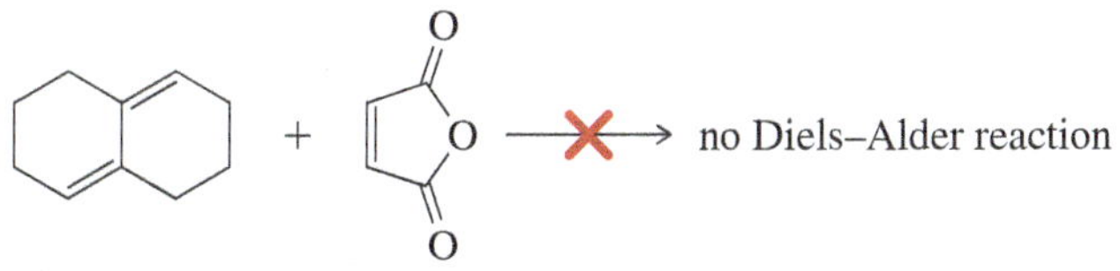

diene
dienophile (maleic anhydride)

Assessment 22.10 Diene **A** participates in a fast and efficient Diels–Alder reaction with maleic anhydride, the powerful dienophile from Assessment 22.9. However, the related diene **B** does *not* undergo a Diels–Alder reaction. Why?

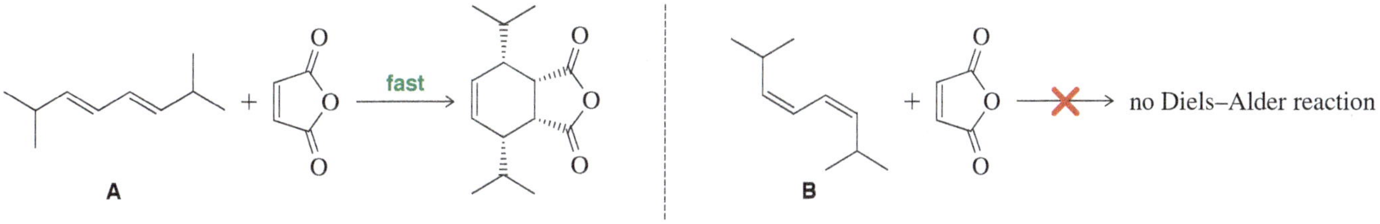

A **B**

Assessment 22.11 Diene **A** undergoes a much faster Diels–Alder reaction with maleic anhydride as compared to diene **B**. Why might this be true?

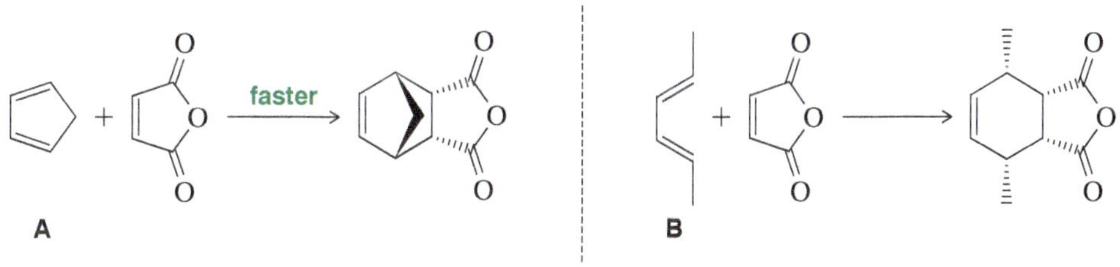

A **B**

The next three sections deal with the stereochemical outcome of the Diels–Alder reaction. In the absence of a chiral environment, the planar diene and dienophile can come together in two different ways—namely, diene attacks dienophile from the top or diene attacks dienophile from the bottom. Because these two ways are energetically equivalent, a racemic mixture results from any Diels–Alder reaction that produces new chiral centers (**Figure 22.15**).

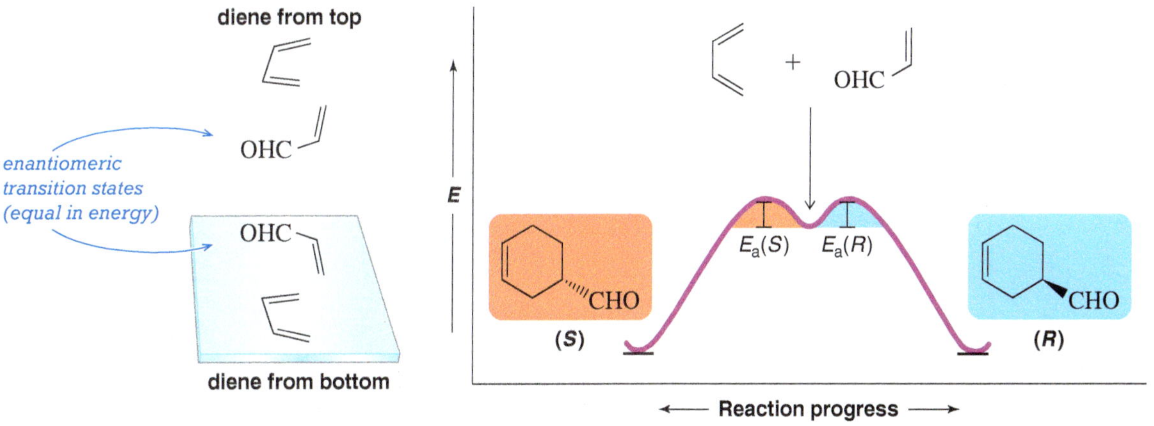

Figure 22.15 The Diels–Alder reaction gives a mixture of stereoisomers. The diene approaching the dienophile from the top is the enantiomer of the diene approaching the dienophile from the bottom.

- $E_a(S) = E_a(R)$
- Therefore, $k_r(S) = k_r(R)$
- As a result, a racemic mixture is obtained.

To simplify matters in the next three sections, only the product that arises from the diene attacking the dienophile from the top is shown. I recommend that you do the same just so your pictures will match mine. The assessments will remind us that a racemic mixture is produced.

22.3.2 Implications of concertedness #2: the stereospecificity of the dienophile

Because of the concertedness of the Diels–Alder reaction, all of the bonds form and break at the same time. As a result, the Diels–Alder reaction is stereospecific with regard to the substituents on the dienophile (**Figure 22.16**). [*Recall that stereospecific implies that only a single diastereomeric product can be produced as a result of the mechanism of the reaction.*] Just like we saw in the concerted, stereospecific reactions of Chapters 8 and 9, there is no chance for rotation around bonds in the Diels–Alder reaction. Thus, if the dienophile is a cis-alkene, then the product will have the two substituents on the same side—a cis-cyclohexene. If the dienophile is a trans-alkene, then the product will have the two substituents on opposite sides—a trans-cyclohexene.

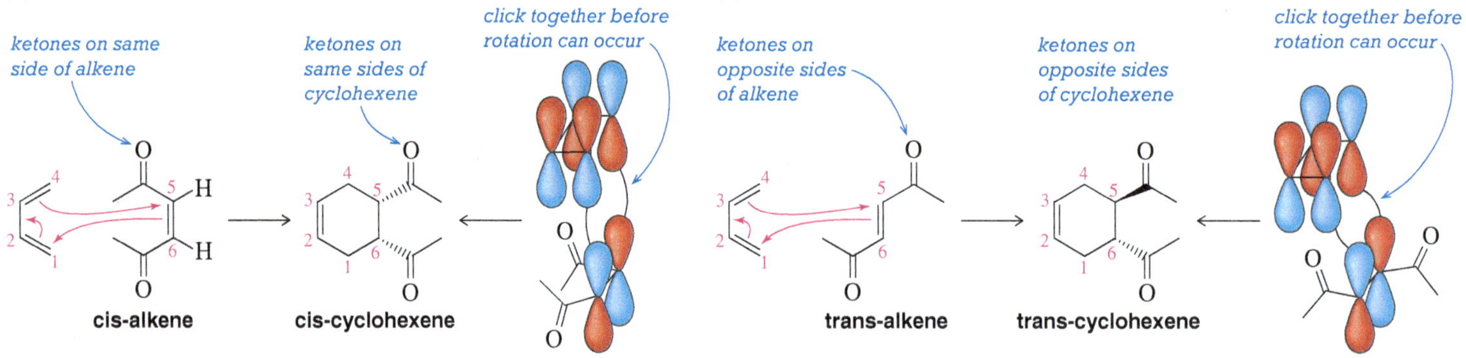

Figure 22.16 The Diels–Alder reaction is stereospecific with regard to the dienophile. Because there is no chance for rotation, (a) a cis-dienophile gives a cis-cyclohexene, while (b) a trans-dienophile gives a trans-cyclohexene.

Assessment 22.12 Reduction of an alkyne using the Lindlar catalyst, a reaction presented in Section 10.6.2, produces only the cis-alkene. Why?

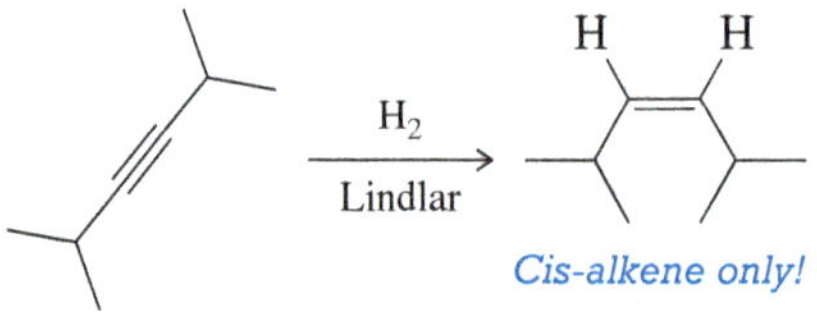

Assessment 22.13 In each Diels–Alder reaction shown, predict the product that will result.

(a)

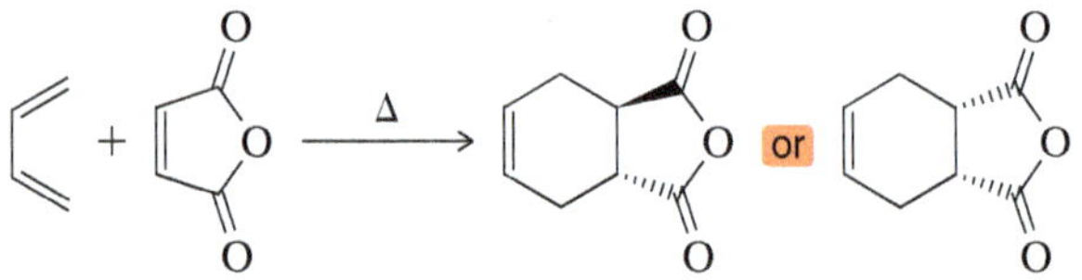

(b)

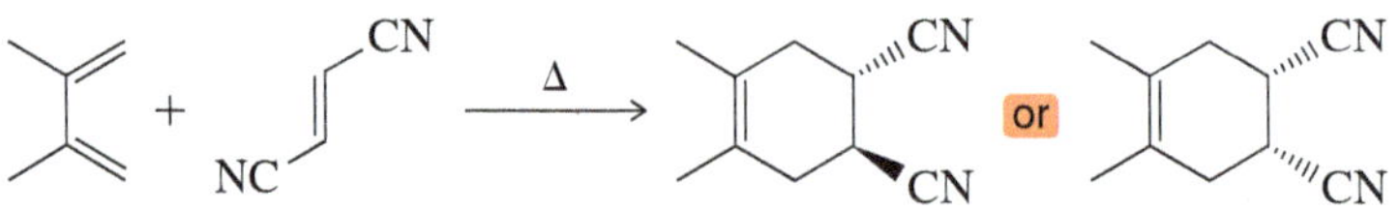

(c)

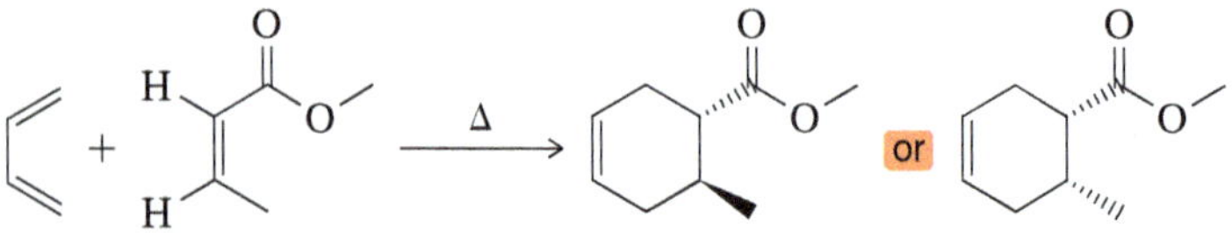

Assessment 22.14 Figure 22.16(a) shows a unique example where the Diels–Alder reaction gives a single product (no enantiomers) regardless of whether the diene attacks the dienophile from the top or bottom. (a) Show the product of the diene attacking from the bottom and confirm that the same product is obtained. (b) What is special about this product that makes this true?

22.3.3 Implications of concertedness #3: the stereospecificity of the diene

Because it is concerted, the Diels–Alder reaction is stereospecific with regard to the substituents on the diene. Thus, substituents on the inside of the *s*-cis-diene will end up on the same side in the product. If the diene attacks the dienophile from the top, these inner substituents (labeled with i in **Figure 22.17**) will be on the top face of the cyclohexene product. Substituents on the outside of the *s*-cis-diene will end up on the same side in the product. If the diene attacks the dienophile from the top, these outer substituents (labeled with o in Figure 22.17) will be on the bottom face of the cyclohexene product.

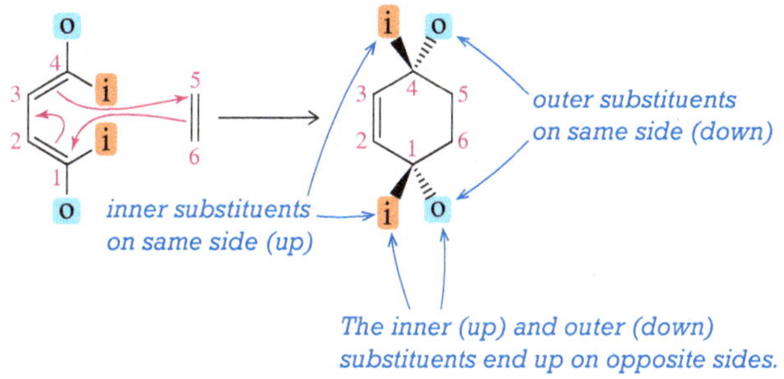

Figure 22.17 The Diels–Alder reaction is stereospecific with regard to the diene. Because there is no chance for rotation, inner substituents (i) on the diene end up on top of the cyclohexene product, while outer substituents (o) end up on the bottom of the cyclohexene product.

[*A set of molecular models will make this section easier to follow.*] While it may seem straightforward to remember that inner substituents end up on top and outer substituents end up on the bottom, what isn't as clear is why. Keep in mind that this is only when the diene attacks the dienophile from the top. If it attacked from the bottom, these would be reversed. When the diene reacts with the dienophile, C_1 and C_4 rehybridize from sp^2 in the diene to sp^3 in the cyclohexene. This rehybridization occurs in such a way that minimizes steric clash between the inner substituents and the approaching dienophile. When the dienophile pushes up into the diene, it pushes the inner substituents (the ones closest to the approaching dienophile) upward. This forces the preferred rotation shown in **Figure 22.18(a)**, with the inner substituents being on top of the resulting cyclohexene. Molecular models show the steric clash in the disfavored rehybridization [**Figure 22.18(b)**] much more clearly.

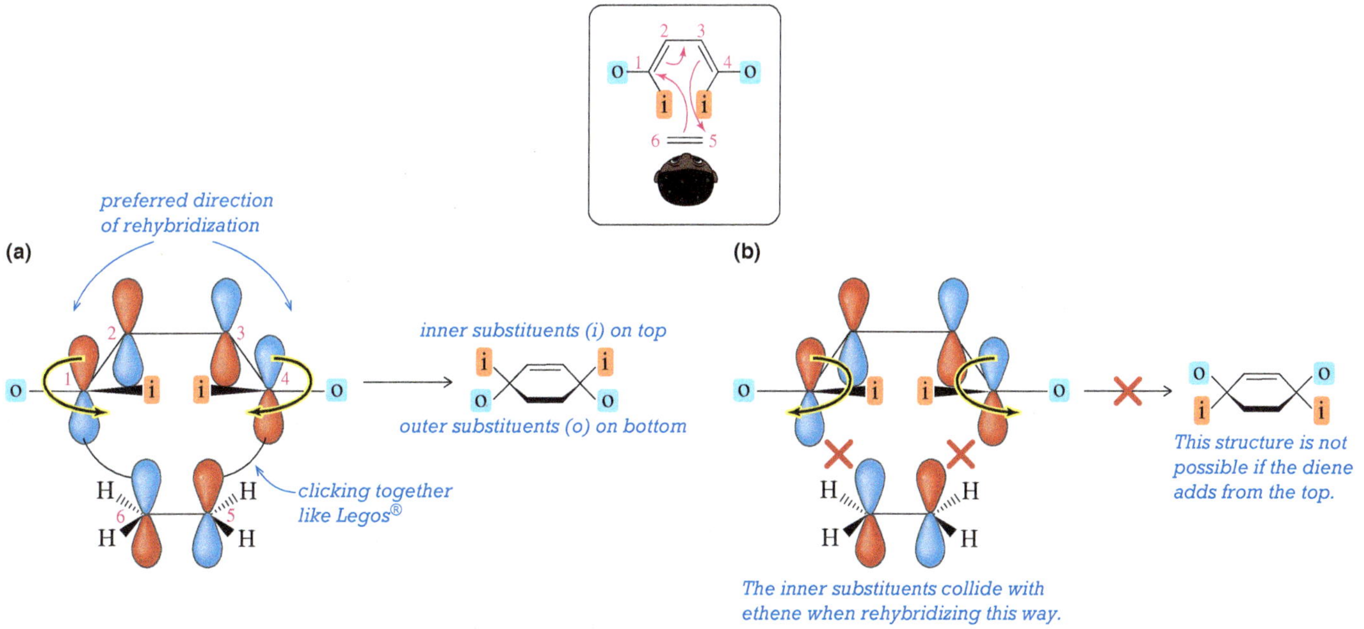

Figure 22.18 Rehybridization away from the approaching dienophile places the inner substituents of the diene on top of the cyclohexene. (a) Favored reaction. (b) Disfavored reaction.

Assessment 22.15 Assuming the diene approaches the dienophile from the top, predict the product of the following Diels–Alder reactions.

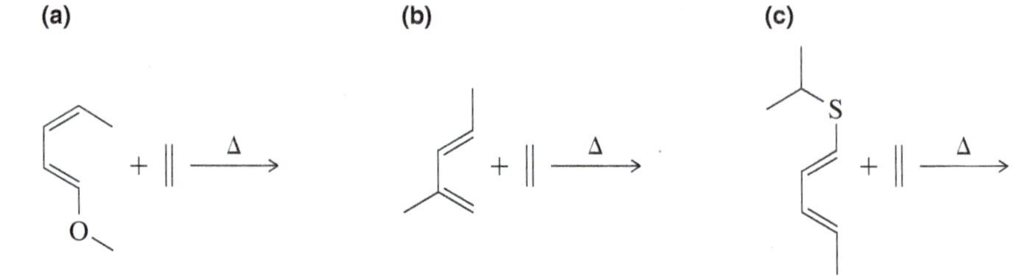

Assessment 22.16 Furan (a) and cyclopentadiene (b) are cyclic dienes that are excellent Diels–Alder substrates. Predict the stereochemical outcome for each of the Diels–Alder reactions shown.

Assessment 22.17 Show how your answers to Assessment 22.15 would change if the diene approaches the dienophile from the bottom. Why?

22.3.4 Implications of concertedness #4: relative stereochemistry of the diene and dienophile

[*Take a breath and stay with me—this gets tricky!*] What we know thus far about the diene is that inner substituents end up on top of the cyclohexene, whereas outer substituents end up on the bottom of the cyclohexene (Section 22.3.3). For the dienophile, substituents on the same side end up cis in the cyclohexene, whereas substituents on opposite sides end up trans in the cyclohexene (Section 22.3.2). But what is their relative stereochemistry? Consider the Diels–Alder reaction in **Figure 22.19**. Even following the rules set out by implications 2 and 3, we still have two possible products, assuming the diene adds from the top. [*If the diene adds from the bottom, we'd have two other products to consider. See Assessment 22.18.*] These two products differ by the relative stereochemistry of the diene and dienophile substituents. [*Which diastereomer is more stable,* **A** *or* **B***?*]

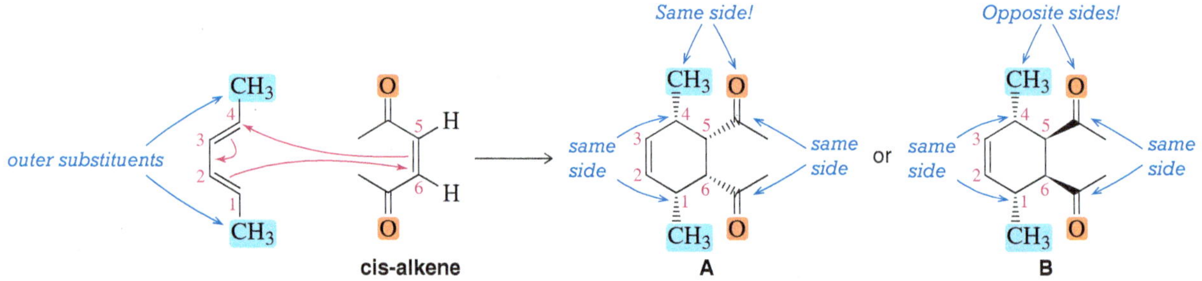

Figure 22.19 Two diastereomeric products are possible if the diene approaches the dienophile from the top. [*With substituents on opposite sides of the ring, you'd expect product **B** to be more stable than **A**, yet **A** is the favored product.*]

The difference between the two products in Figure 22.19 arises as a result of two possible transition states, called endo and exo. As shown in **Figure 22.20**, the **endo transition state** is the one that places the substituents of the dienophile *under* the approaching diene. The **exo transition state**, on the other hand, places the substituents of the dienophile *away* from the approaching diene. The exo transition state has less steric hindrance, yet the endo transition state is of lower energy. A reaction whose outcome is determined by the transition state energy as opposed to product stability is under kinetic control [*recall Section 21.4.1*].

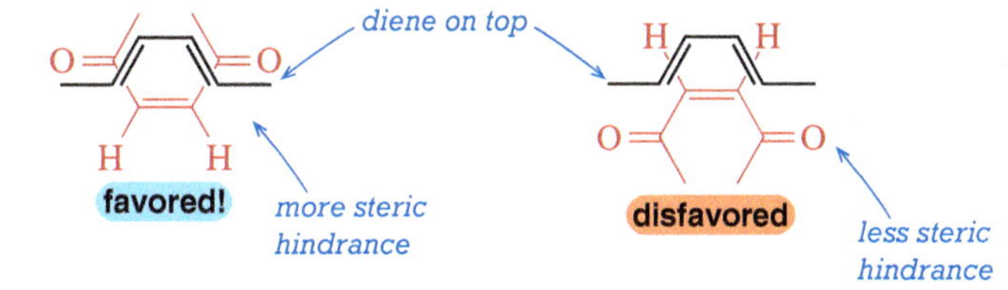

Figure 22.20 The endo transition state is favored, despite being more sterically hindered than the exo transition state.

Why the endo transition state is favored is still being debated. The most accepted hypothesis has to do with what is referred to as secondary orbital overlap (**Figure 22.21**). In the endo transition state, the two sp^2-hybridized carbonyl carbons of the planar dienophile are tucked under the planar diene. These are both planar, so there isn't much steric hindrance to destabilize the transition state. More important, however, is the nonbonding, yet still stabilizing, interaction of the C—O π bonds and the C_2—C_3 π bond of the forming alkene. This secondary orbital interaction makes the endo transition state favored in the same way that forming new bonds makes a reaction more exothermic, although to a lesser degree. [*Hence the reason it is called secondary (instead of primary) orbital overlap.*] This is very difficult to see in a two-dimensional representation but is easily seen with molecular models.

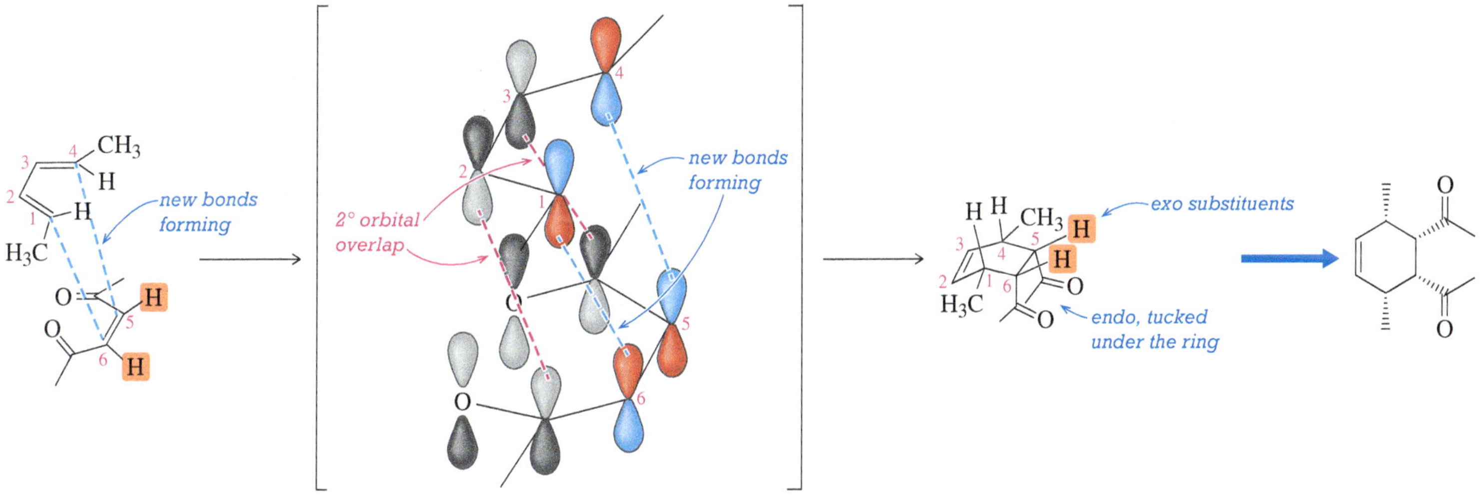

Figure 22.21 Secondary orbital overlap. Interaction between the dienophile substituents and the forming C_2—C_3 π bond makes the endo transition state favored.

In conclusion, when the diene approaches the dienophile [*from the top*], the dienophile substituents are always tucked underneath the ring to maximize secondary orbital overlap. This is true regardless of the size of the substituents. In the event that the dienophile is a trans-alkene, the substituent with *p* orbitals/π bonds should be placed under the ring preferentially.

Assessment 22.18 Predict the product of the following Diels–Alder reactions. Where a racemic mixture is produced, show both enantiomers and explain how each is formed.

(a)

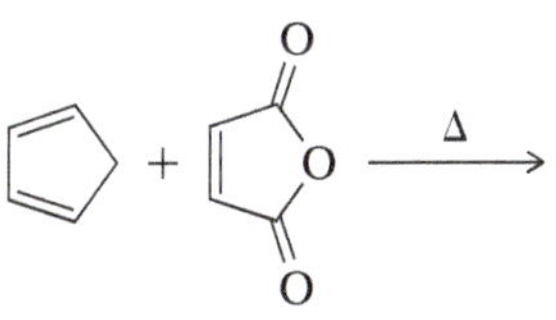

(b)

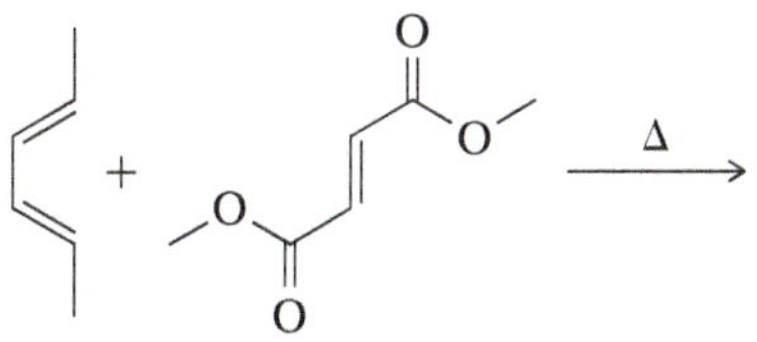

(c)

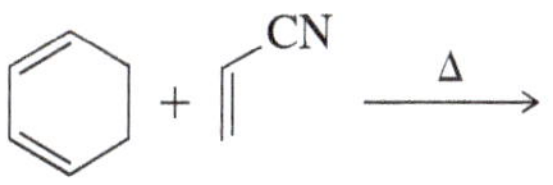

(d)

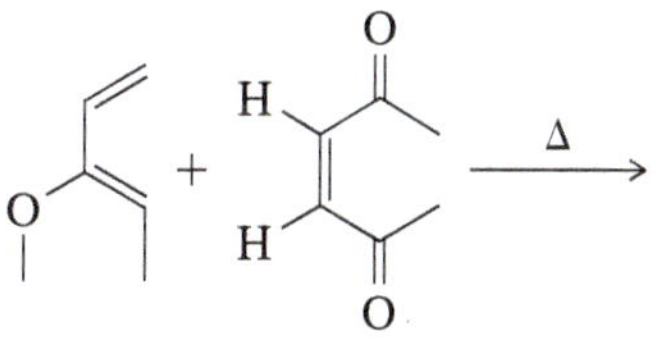

22-B CHEMISTRY IN CONTEXT: Biochemistry

Finally, a Diels–Alderase

[*Cue the happy music.*] In 2000, researchers confirmed a long-held hypothesis that lovastatin (**Figure 22.22**), a fungal metabolite of *Aspergillus terreus*, was synthesized via a Diels–Alder reaction catalyzed by the enzyme lovastatin nonaketide synthase (LNKS). [*Lovastatin is a cholesterol-lowering drug like atorvastatin, which we studied in Chapter 20.*] LNKS is the first naturally occurring Diels–Alderase enzyme to be purified.

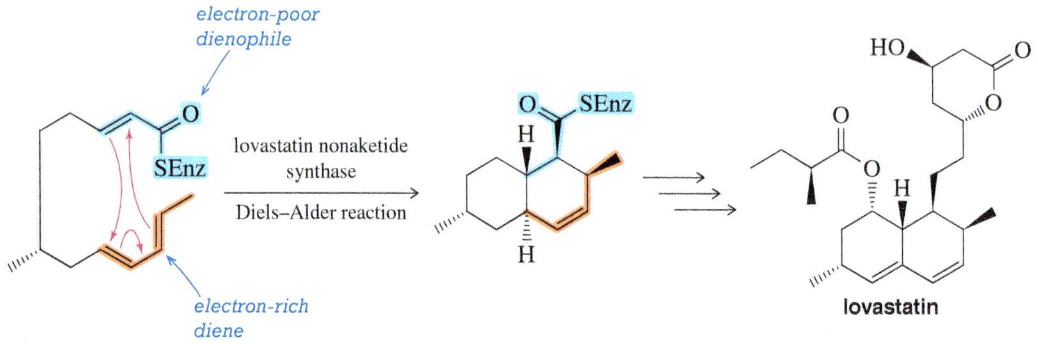

Figure 22.22 **The biosynthesis of lovastatin is presumed to occur through a Diels–Alder reaction.**

Abyssomicin C is a spirotetronate antibiotic isolated from the marine actinomycete *Verrucosispora maris*. As a result of its antibacterial activity, synthetic chemists sought its synthesis on a larger scale for biological studies. One such synthesis, completed in 2005, utilized an intramolecular Diels–Alder reaction like the one shown in **Figure 22.23**. Illustrating the interplay [*and importance*] of natural product synthesis, this successful synthesis led investigators to consider whether the Diels–Alder reaction might be a key step in the biosynthesis of abyssomicin. As hypothesized, in 2016 spirotetronate cyclase was confirmed to act as a stand-alone Diels–Alderase enzyme. Notably, it has also been shown to catalyze Diels–Alder reactions for other substrates, suggesting its potential utility as an industrial-scale biocatalyst for use in future synthesis. [*This is a great example of how organic synthesis helps us understand biology, and biology, in turn, has the potential to make organic synthesis easier.*]

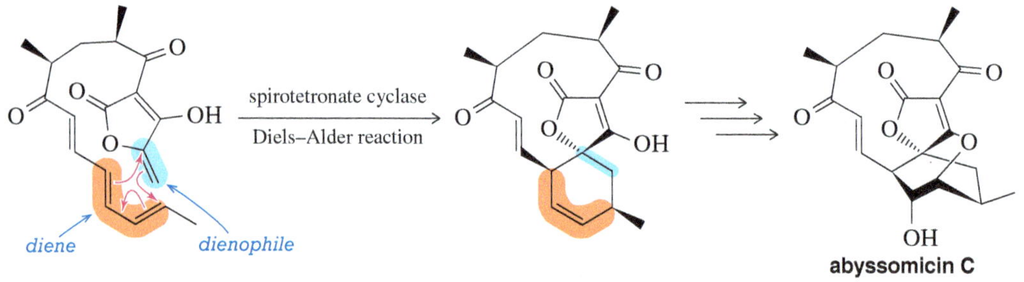

Figure 22.23 **A Diels–Alderase is involved in the biosynthesis of the antibiotic abyssomicin C.**

22.3.5 Implications of concertedness #5: regioselectivity

Thus far, most of the Diels–Alder reactions shown have involved symmetrical dienes and dienophiles. Assessment 22.19 provides one where a nonsymmetrical, electron-rich diene reacts with a nonsymmetrical electron-poor dienophile [*the most efficient Diels–Alder reaction*].

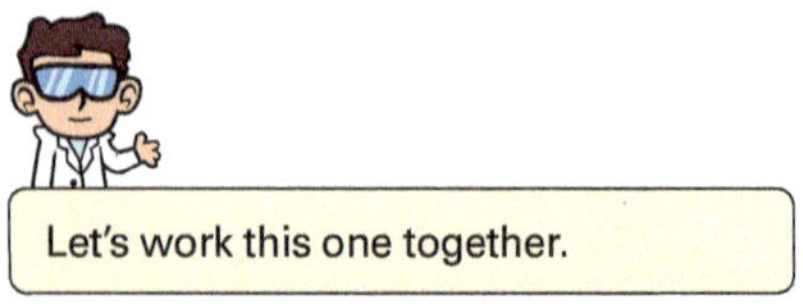

Assessment 22.19 Predict the product of the Diels–Alder reaction shown when the diene attacks from the top. This is the Diels–Alder reaction from Figure 22.13.

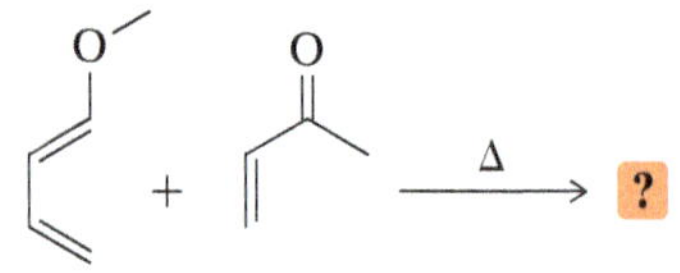

Solved Assessment 22.19

Ignoring stereochemistry for now, this Diels–Alder reaction can result in one of two regioisomers. Using the numbering shown in **Figure 22.24**, the two new σ bonds can be between C_1—C_6 and C_4—C_5 or between C_1—C_5 and C_4—C_6. The reaction is concerted, so we might expect that both products would form in equal amounts. But we'd be wrong . . . again. [*It seems like every time there are two possibilities in the Diels–Alder reaction, only one actually happens. What a cool reaction!*]

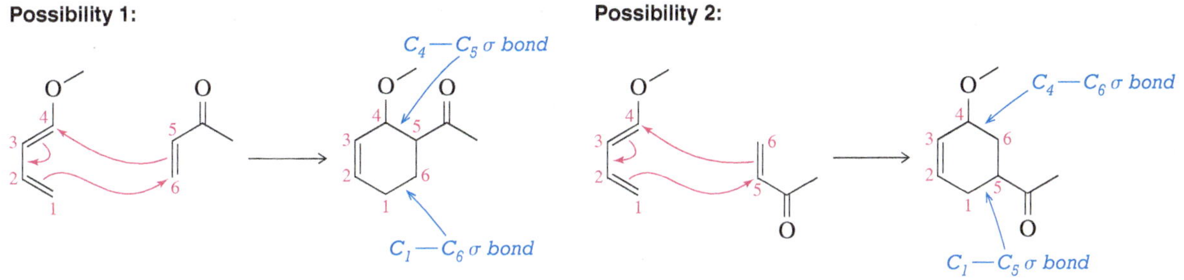

Figure 22.24 Two possible regiochemical outcomes of the Diels–Alder reaction in Assessment 22.19.

Although it's concerted, the Diels–Alder reaction is most efficient when an electron-rich diene using its HOMO reacts with an electron-poor dienophile using its LUMO. So to predict the regiochemical outcome, we need to find where the "electron-richness" lies in the diene and where the "electron-poorness" lies in the dienophile. [*How can we find out where electron density is or isn't in a neutral molecule that has a conjugated π system?*] Drawing the resonance structures (**Figure 22.25**) shows a negative charge on C_3 and C_1 of the diene and a positive charge on C_6 of the dienophile. In frontier molecular orbital terms, we say that the coefficient of the HOMO is largest on C_1 and C_3, and the coefficient of the LUMO is largest on C_6 of the dienophile. In "negative attacks positive" terms, the partially negative C_1 preferentially binds with the partially positive C_6. In this way, a single regioisomer is created.

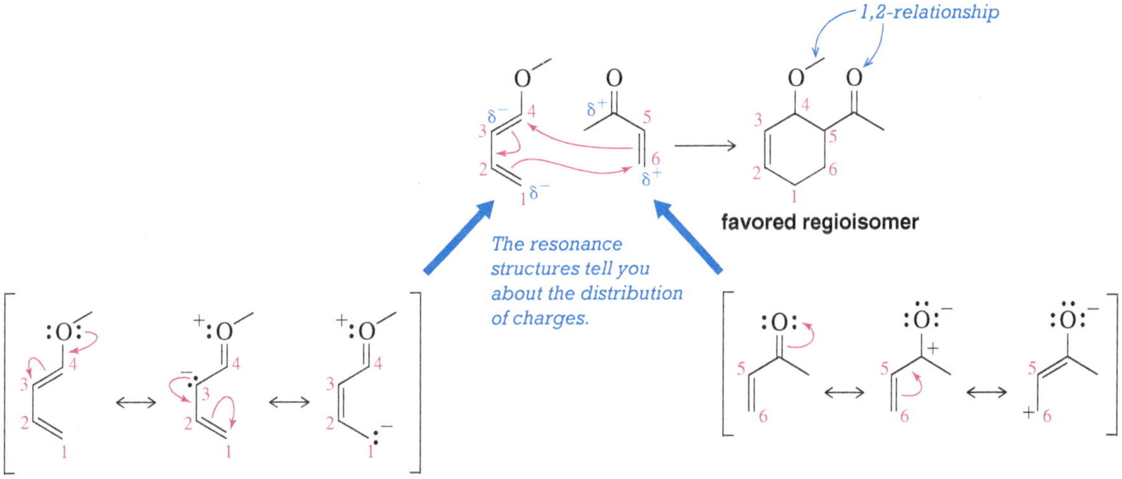

Figure 22.25 The regiochemical outcome of the Diels–Alder reaction is determined by the electron distribution, as predicted by resonance structures of the diene and dienophile.

Completing the problem, we can now address the stereochemical issues. With the diene attacking from the top, the dienophile tucks the ketone under the diene—the endo transition state (**Figure 22.26**). [*Imagine lifting the diene off the page and sliding it over the diene.*] The Legos click together. C_4 rehybridizes so that the methoxy group (an outer substituent) is down. C_5, via the endo transition state, rehybridizes so that the ketone is also down. And there you have it—except for one small detail. This is just one enantiomer of a racemic mixture that is produced in this reaction. [*How is the other enantiomer produced? See Assessment 22.18.*] ■

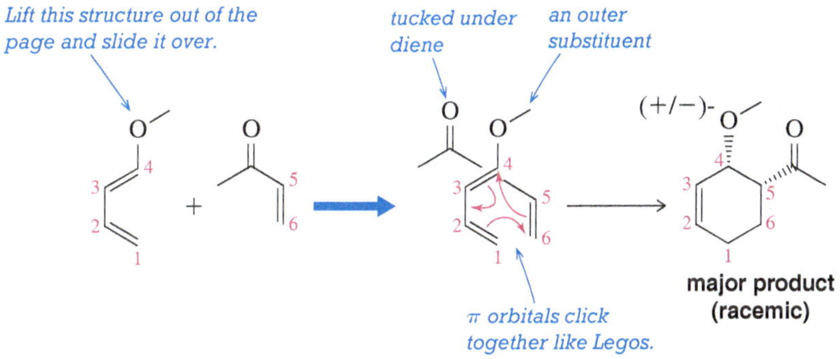

Figure 22.26 A racemic mixture of one diastereomer is produced in the Diels–Alder reaction of Assessment 22.19.

A similar process can be used when there is an electron-donating group at the 2-position of the diene (**Figure 22.27**). Drawing the resonance structure of the diene shows that there is more electron density on C_1. Thus, C_1 will more likely form a bond with C_6 of the dienophile.

The combined results of Figures 22.25 and 22.27 reveal that the major product of Diels–Alder reactions will always have a 1,2- or 1,4-relationship between the electron-donating group on the diene and the electron-withdrawing group on the dienophile.

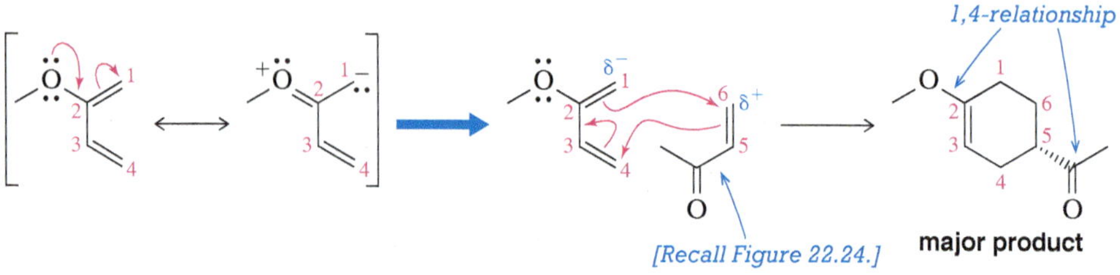

Figure 22.27 Regiochemical outcome with an electron-donating group at C_2.

22.3.1–22.3.5 SUMMARY

- The Diels–Alder reaction, being concerted, requires the *s*-cis conformation of the diene and is highly stereospecific, proceeding through the more stable endo transition state. The regioselectivity is dictated by the relative position of the electron-donating and electron-withdrawing substituents on the diene and dienophile, respectively.

Assessment 22.20 In Assessment 22.19, the diene attacked the dienophile from the top. Show the product you'd obtain if the diene attacked the dienophile from the bottom.

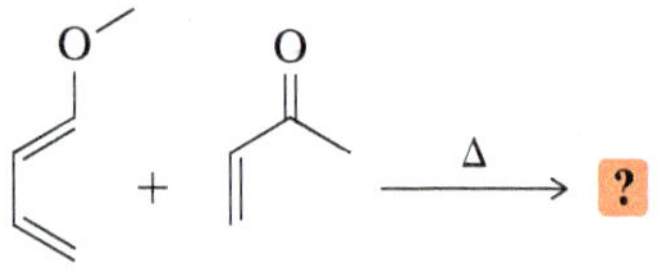

Assessment 22.21 Although the methyl group does not donate by resonance, the same regiochemistry is observed in the following reaction. [*Compare it to the product of Figure 22.26.*] Why might you expect this outcome?

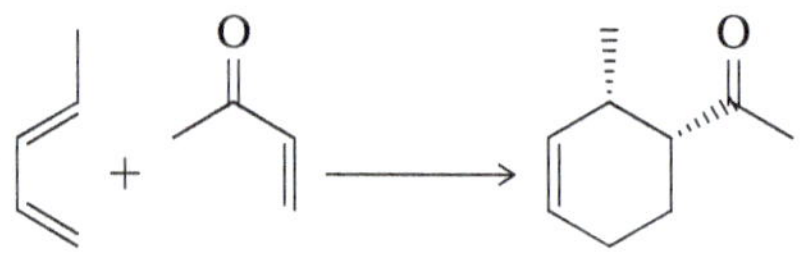

Assessment 22.22 Predict the product of the following Diels–Alder reactions. Where a racemic mixture is produced, show both enantiomers and explain how each is formed. [*Hint: Do you have a 1,2- or 1,4-relationship between the electron-donating and electron-withdrawing groups?*]

(a)

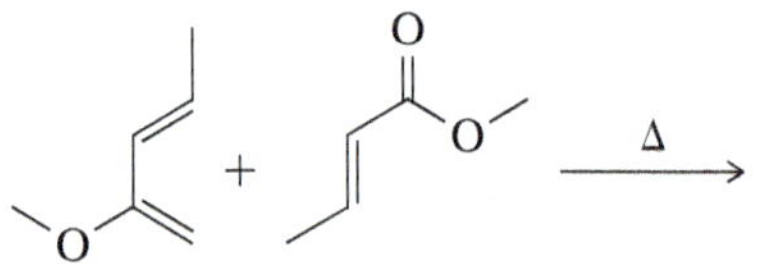

(b)

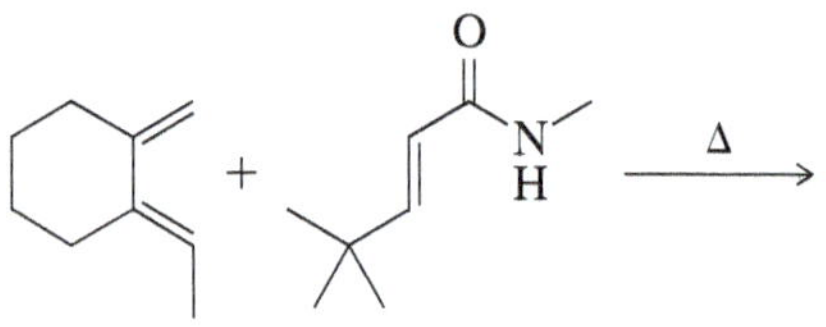

Assessment 22.23 Give the diene and dienophile that could be used to make the following Diels–Alder products.

(a)

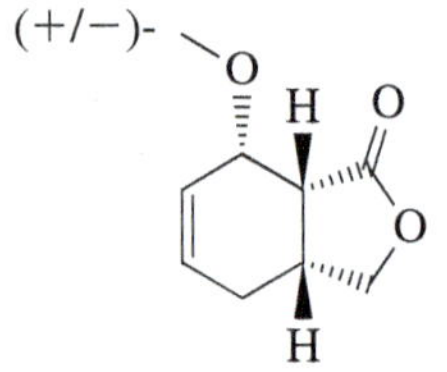

(b)

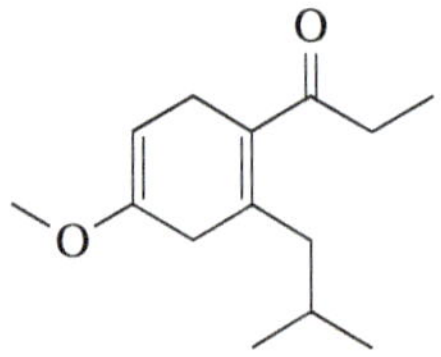

(c)

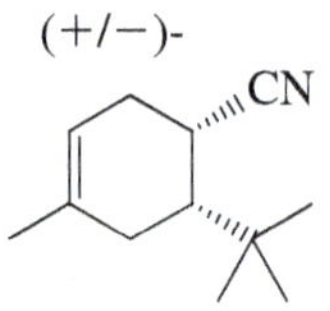

(d)

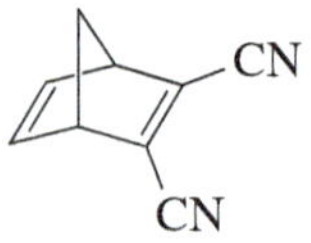

Assessment 22.24 Based on your reading of Section 22.3, is the Diels–Alder reaction under kinetic or thermodynamic control? How do you know?

22-C CHEMISTRY IN CONTEXT: Green Chemistry

Enantioselective Diels–Alder

The Diels–Alder reaction is one of the greenest reactions known because it has 100% atom economy, meaning 100% of the atoms in the reactants end up in the final product. These reactions are even greener when done in solvent-free conditions or in environmentally benign solvents like water. Probably the only drawback of the Diels–Alder reactions seen thus far is the lack of stereoselectivity. For example, each Diels–Alder reaction in Section 22.3 resulted in a racemic mixture. There has been a great deal of research done on performing catalytic, enantioselective Diels–Alder reactions. One such example involves iminium organocatalysis using the amine shown in **Figure 22.28**. The term organocatalysis is meant to highlight the use of an organic molecule as a catalyst in lieu of traditional metal (inorganic) catalysts.

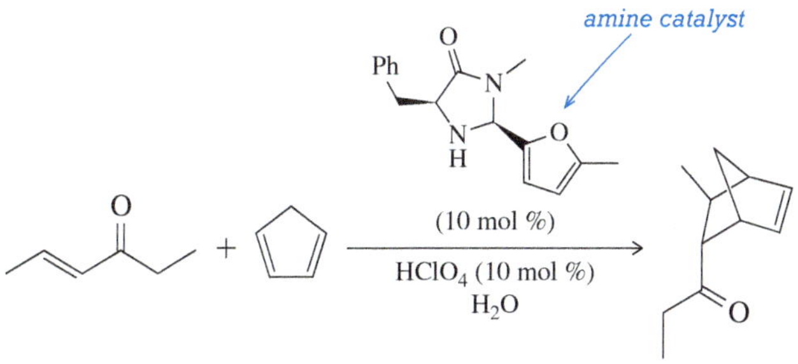

Figure 22.28 Enantioselective organocatalysis of the Diels–Alder reaction.

This type of catalysis takes advantage of the fact that electron-poor dienophiles do fast Diels–Alder reactions. While the ketone in **Figure 22.29** is electron-deficient, when converted to an iminium by a mechanism covered in Section 17.8.3 it becomes even more electron-deficient due to conjugation with the cationic nitrogen. With the catalyst providing a chiral environment, the Diels–Alder reaction proceeds selectively from one face of the dienophile. [*See Sections 6.6.2 and 9.3 for a discussion of how chiral environments make reactions enantioselective.*] In the presence of water as (the greenest!) solvent, the iminium is hydrolyzed back to ketone, releasing the Diels–Alder product and regenerating the catalyst to perform the next Diels–Alder reaction.

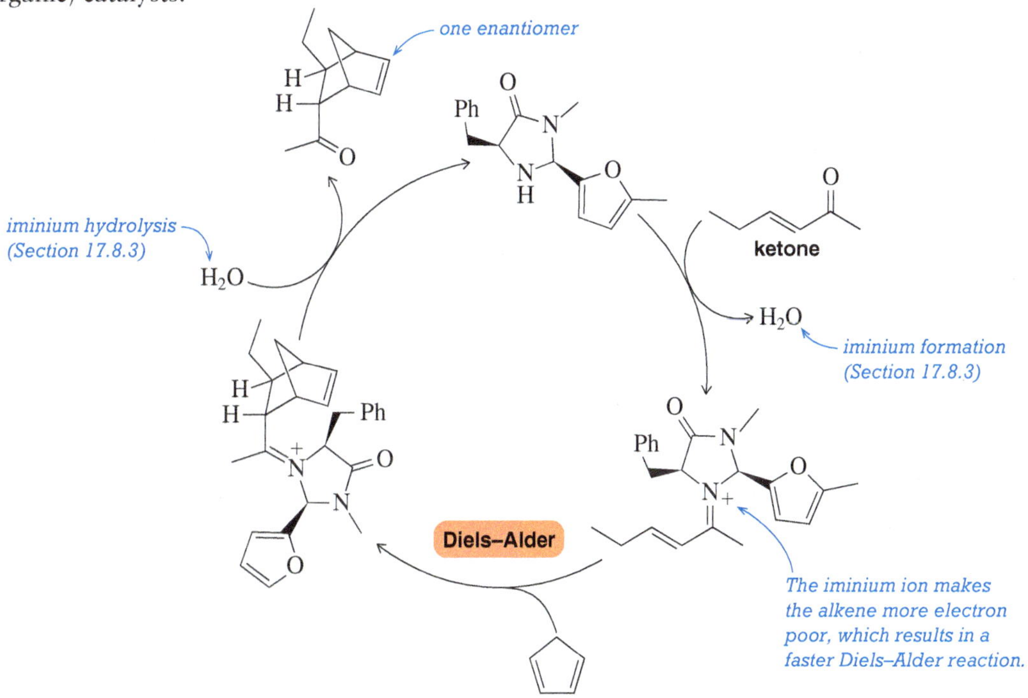

Figure 22.29 The catalytic cycle of the organocatalyzed Diels–Alder reaction in water.

Assessment 22.25 Show a mechanism for the reaction in Figure 22.29 in which the ketone and amine combine to make the iminium ion. [*See Section 17.8.3 if you need help.*]

Assessment 22.26 What stereochemical result would you expect if the Diels–Alder reaction with the free α,β-unsaturated ketone was faster than reaction with the iminium-bound alkene?

22.4 Other pericyclic reactions

The Diels–Alder may be the most important pericyclic reaction, but there are others that are extremely useful for making new carbon–carbon bonds. The following sections briefly cover a few of these. These reactions are based on the same frontier molecular orbital theory at the heart of the Diels–Alder reaction.

22.4.1 [2 + 2] Cycloaddition

The Diels–Alder reaction is a concerted cycloaddition between a diene and a dienophile. Another way of characterizing that cyclization is based on the number of π electrons in the diene that interact with the π electrons in the dienophile. In this way, the Diels–Alder reaction is referred to as a [4 + 2] cycloaddition [**Figure 22.30(a)**]. Could a similar [2 + 2] cycloaddition occur between two molecules of ethene to make a cyclobutane under similar conditions? [*The Diels–Alder reaction is always done with heating, making it a thermal cycloaddition.*] Although an arrow-pushing mechanism could be drawn that rationalizes the outcome [**Figure 22.30(b)**], the [2 + 2] cycloaddition isn't possible with heating. To understand why, we have to look at the orbitals involved. Whereas the Diels–Alder uses the HOMO and LUMO of the diene and dienophile, respectively, the HOMO and LUMO of the two ethene molecules do not overlap properly.

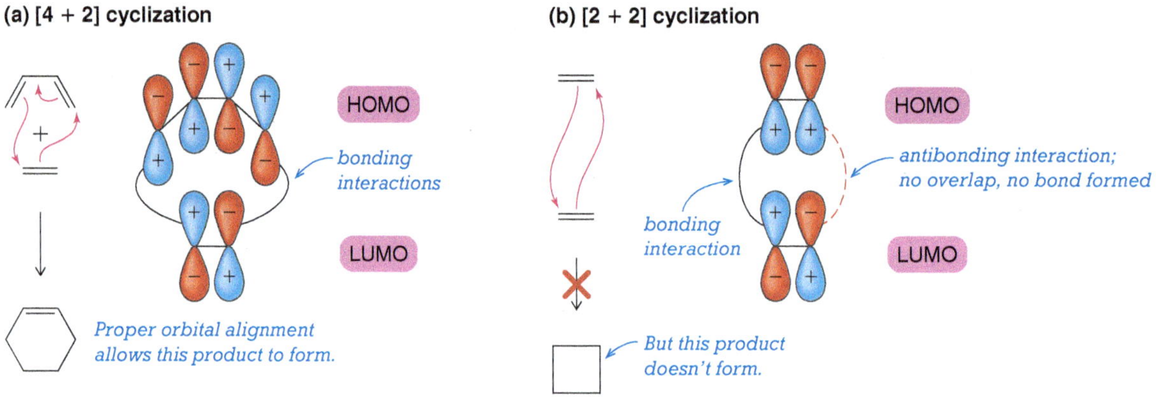

Figure 22.30 A comparison of the molecular orbital overlap in the [4 + 2] and [2 + 2] cycloaddition reactions. (a) The thermal [4 + 2] cycloaddition forms product because the orbitals of the HOMO and LUMO align to form bonding interactions. (b) The thermal [2 + 2] cycloaddition of two ethene molecules cannot occur because the mismatched orbital coefficients create a new antibonding interaction. Cyclobutane cannot be made under thermal conditions (with heating).

Despite having considerable ring strain, the four-membered ring is still stable enough to form. The ring doesn't form solely because proper overlap isn't achieved using the HOMO of one ethene and the LUMO of the other. [*Are there other orbitals that might be able to overlap properly?*] It turns out that there would be proper overlap between two HOMOs [**Figure 22.31(a)**], but then we'd be trying to make a new orbital that holds 4 electrons (maximum possible is 2). [*So that won't work.*] Alternatively, proper overlap is possible between two LUMOs [**Figure 22.31(b)**], but then our new molecular orbital would have 0 electrons. [*So that won't work either.*]

Figure 22.31 Possible orbital overlaps that could produce cyclobutane from a [2 + 2] cycloaddition. (a) HOMO with HOMO would have too many electrons (4). (b) LUMO with LUMO would have too few electrons (0).

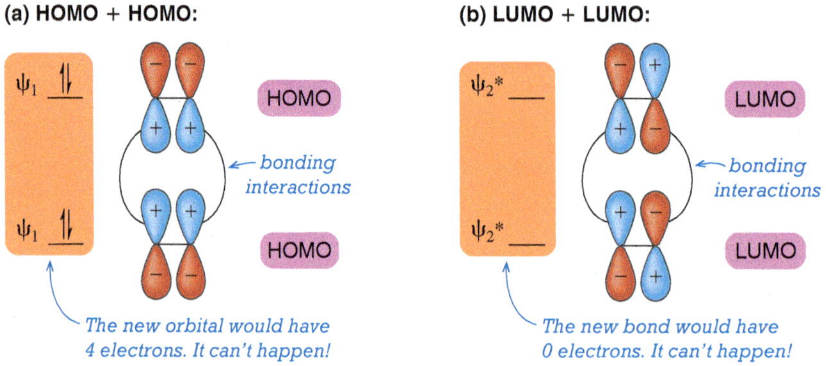

In order for two molecules of ethene to make cyclobutane, we need proper overlap between orbitals that contain more than 0 and fewer than 4 electrons between them. The solution lies in Section 22.2 and Figure 22.9. [*It also lies in why carrots are orange (Section 21.3.2) and how sunscreen protects our skin from damaging UV rays (Chemistry in Context 21-A).*]

The only thing missing here is the proper bookkeeping of electrons but that is beyond the scope of an introductory organic text. If you are interested in learning about this, perhaps a career in physical organic chemistry is in your future.

Instead of heating the solution, it can be irradiated with a wavelength of light that corresponds to the HOMO–LUMO energy gap of ethene (~175 nm). This results in excitation of a single electron from the HOMO to the LUMO (**Figure 22.32**). With 1 electron, what was formerly the LUMO is now a **singly occupied molecular orbital (SOMO)**. The SOMO of one ethene molecule now has the proper overlap for bonding with the original LUMO of another ethene molecule. With that, we can now make four-membered rings. [*Pretty cool, huh?*]

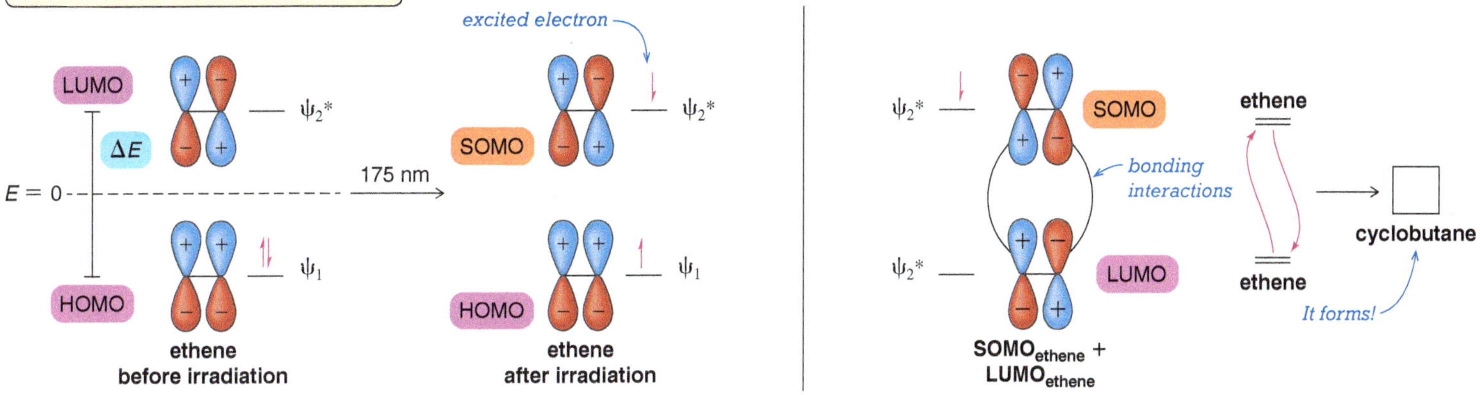

Figure 22.32 Excitation of an electron allows a [2 + 2] cycloaddition. SOMO with LUMO has proper orbital overlap.

Assessment 22.27 Predict the product of the following [2 + 2] cycloadditions.

(a)

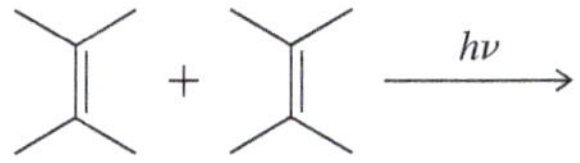

(b)

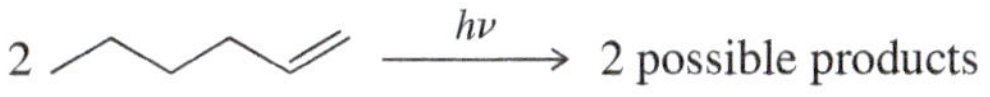

Assessment 22.28 What wavelength of light should be shined on buta-1,3-diene if cycloocta-1,4-diene is the desired product?

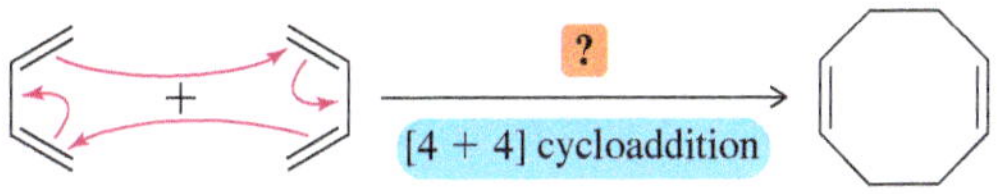

22.4.2 Electrocyclic reactions

Electrocyclic reactions are defined as the conversion of acyclic conjugated molecules to ring structures through the movement of π electrons. The formation of the new σ bond can be promoted thermally (with heat) or photochemically (with light) as the energy source. Although these reactions are reversible, the cyclization of hexa-1,3,5-triene in **Figure 22.33** is typically favored by the net formation of a σ bond at the expense of one π bond. The concerted mechanism is represented by the cyclic movement of 6 electrons. [*We've seen that before, right?*]

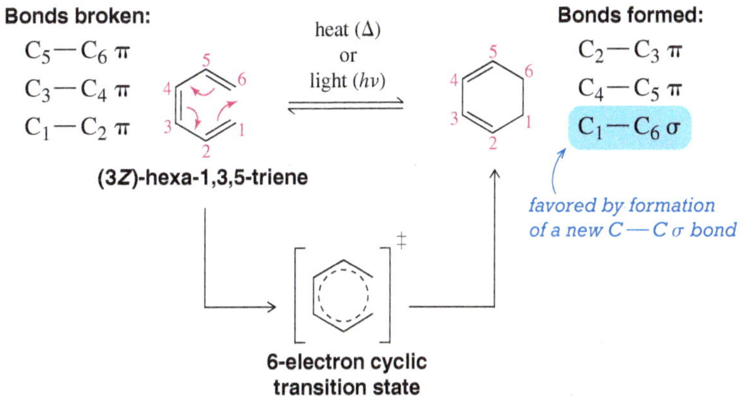

Figure 22.33 Electrocyclic reactions are concerted. They occur through aromatic transition states, and although they are reversible, they are often enthalpically favored by the formation of a new C—C σ bond.

Formation of the new $C_1—C_6$ σ bond in Figure 22.33 occurs when the *p* orbitals at C_1 and C_6 rotate 90 degrees so that there is overlap. The mechanism and the stereochemical outcome of these reactions are explained by the Woodward–Hoffman rules listed in Table 22.1. The 90-degree rotation of the two *p* orbitals can happen in two ways. **Conrotatory** ring closure implies that the C_1 and C_6 *p* orbitals rotate in the same direction (either clockwise or counterclockwise). **Disrotatory** ring closure implies that the C_1 and C_6 *p* orbitals rotate in opposite directions (one goes clockwise, the other goes counterclockwise).

Table 22.1 Woodward–Hoffman rules for electrocyclic reactions.

Electrons in the π system	Heat (Δ) or light (*hν*)?	Motion of *p* orbitals . . .	. . . illustrated	Rotation?
4*n*	Heat (Δ)	Conrotatory	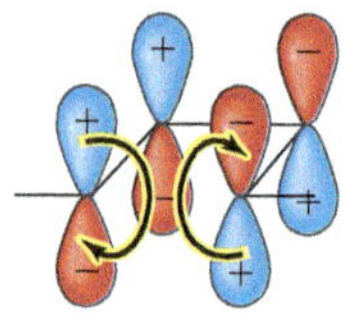	Both rotate clockwise (or both rotate counterclockwise).
	Light (*hν*)	Disrotatory	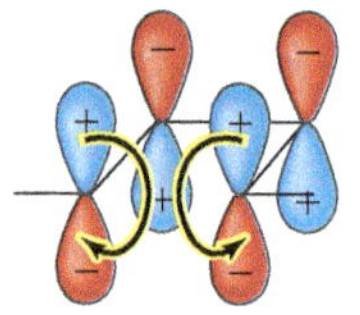	One rotates clockwise, whereas the other rotates counterclockwise.
4*n* + 2	Heat (Δ)	Disrotatory	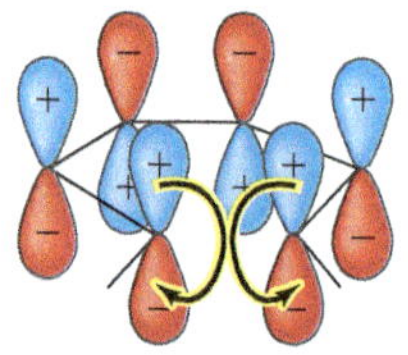	One rotates clockwise, whereas the other rotates counterclockwise.
	Light (*hν*)	Conrotatory	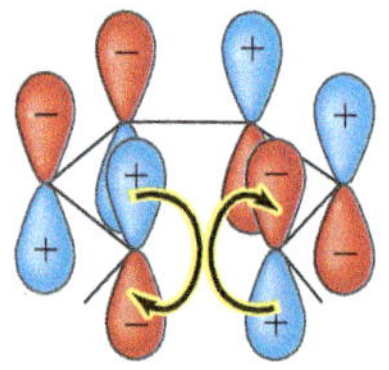	Both rotate clockwise (or both rotate counterclockwise).

Table 22.1 tells us which *p* orbital rotation is (symmetry) allowed depending on the number of electrons present in the π system and whether or not the reaction is conducted thermally (heat; Δ) or photochemically (light; *hν*). There are two types of π systems to be considered, $4n$ π and $4n + 2$ π. In this context, a π system is considered a $4n$ π system if $n = 1, 2, 3$, etc. when $4n$ is set equal to the number of electrons in the π system. A π system is considered a $4n + 2$ π system if $n = 0, 1, 2$, etc. when $4n + 2$ is set equal to the number of electrons in the π system. [*In other words, 4, 8, 12, etc. are 4n π systems, whereas 2, 6, 10, etc. are 4n + 2 π systems. Let's work a few to make sure we get this before moving on, especially since we'll be doing more of this in Chapter 23.*]

Assessment 22.29 Classify the following conjugated systems as having $4n$ or $4n + 2$ π electrons.

(a)
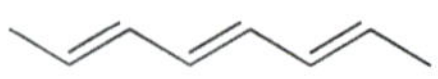

(b)
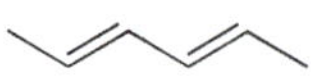

(c)

(d)
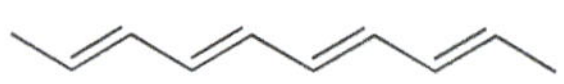

(e)

(f)
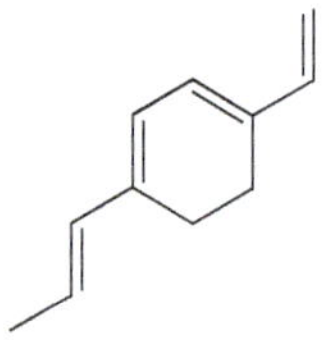

For π systems that are substituted at the terminal carbons, the Woodward–Hoffman rules predict the stereochemical outcome. An example of this is the thermal cyclization of the triene shown in **Figure 22.34**. A triene contains three double bonds, so it has a total of 6 π electrons. Because $n = 1$ when $4n + 2 = 6$, this is considered a $4n + 2$ π system. Rotation of the *p* orbitals and, by extension, the terminal methyl groups, will occur in a disrotatory fashion under thermal conditions [*4n + 2, heat, disrotatory; see Table 22.1*]. Thus, one methyl group rotates clockwise, whereas the other rotates counterclockwise. The two methyl groups in the resulting product are cis.

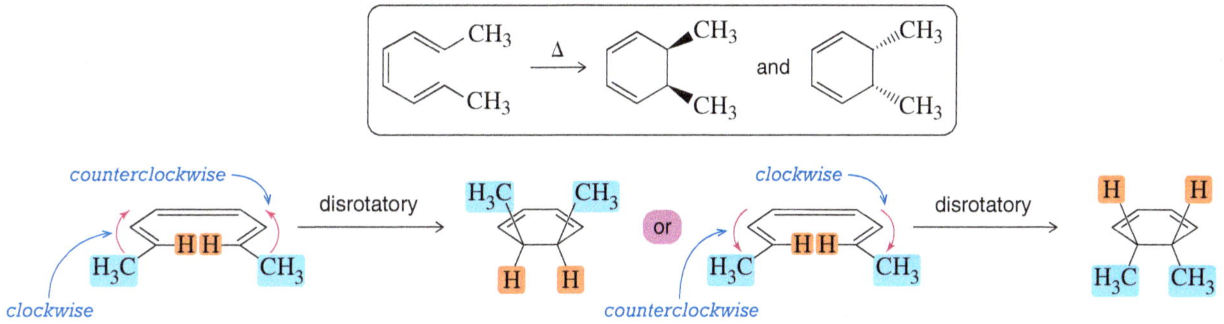

Figure 22.34 Disrotatory motion in the thermal 4*n* + 2 cyclization gives cis methyl groups.

The Woodward–Hoffman rules can be memorized. [*Even 17 years after graduate school, the words "four-heat-con" roll off the tongue like I learned it yesterday.*] But even easier than memorizing them is understanding their origin. Let's look at some molecular orbitals to do this. The rotation of the substituents during the electrocyclic reaction is necessary to align like coefficients of the *p* orbitals to create the overlap that forms the new σ bond. In the thermal cyclization of butadiene [**Figure 22.35(a)**], a $4n$ π system, the important molecular orbital is the HOMO. Thus, conrotatory rotation of the *p* orbitals [counterclockwise and counterclockwise in **Figure 22.35(b)**; clockwise and clockwise in **Figure 22.35(c)**] forms the new σ bond. [*From Table 22.1, this would be a 4n, heat, conrotatory cyclization.*]

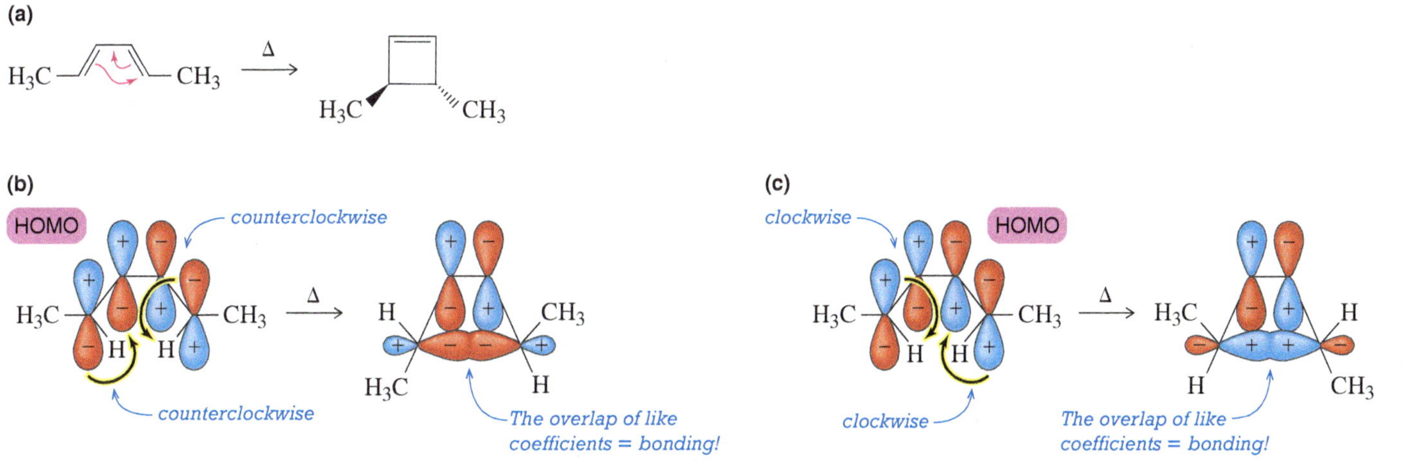

Figure 22.35 Molecular orbital explanation of Woodward–Hoffman rules. (a) Thermal cyclization of a butadiene. (b) Conrotatory rotation (counterclockwise for each) gives a trans product. (c) Conrotatory rotation (clockwise for each) gives the enantiomeric trans product.

What happens if we excite the same butadiene from Figure 22.35 as shown in **Figure 22.36(a)**? With an electron excited into the LUMO, cyclization now occurs from the SOMO. [*That's the important difference: Thermal cyclizations use the HOMO, whereas photochemical cyclizations* *use the LUMO/SOMO.*] In order to get the coefficients of the terminal *p* orbitals to overlap and form the σ bond, disrotatory movement is required [clockwise and counterclockwise in **Figure 22.36(b)**]. As a result, the two methyl groups end up on the same side of the ring. [*From Table 22.1, this would be a 4n, light, disrotatory cyclization.*]

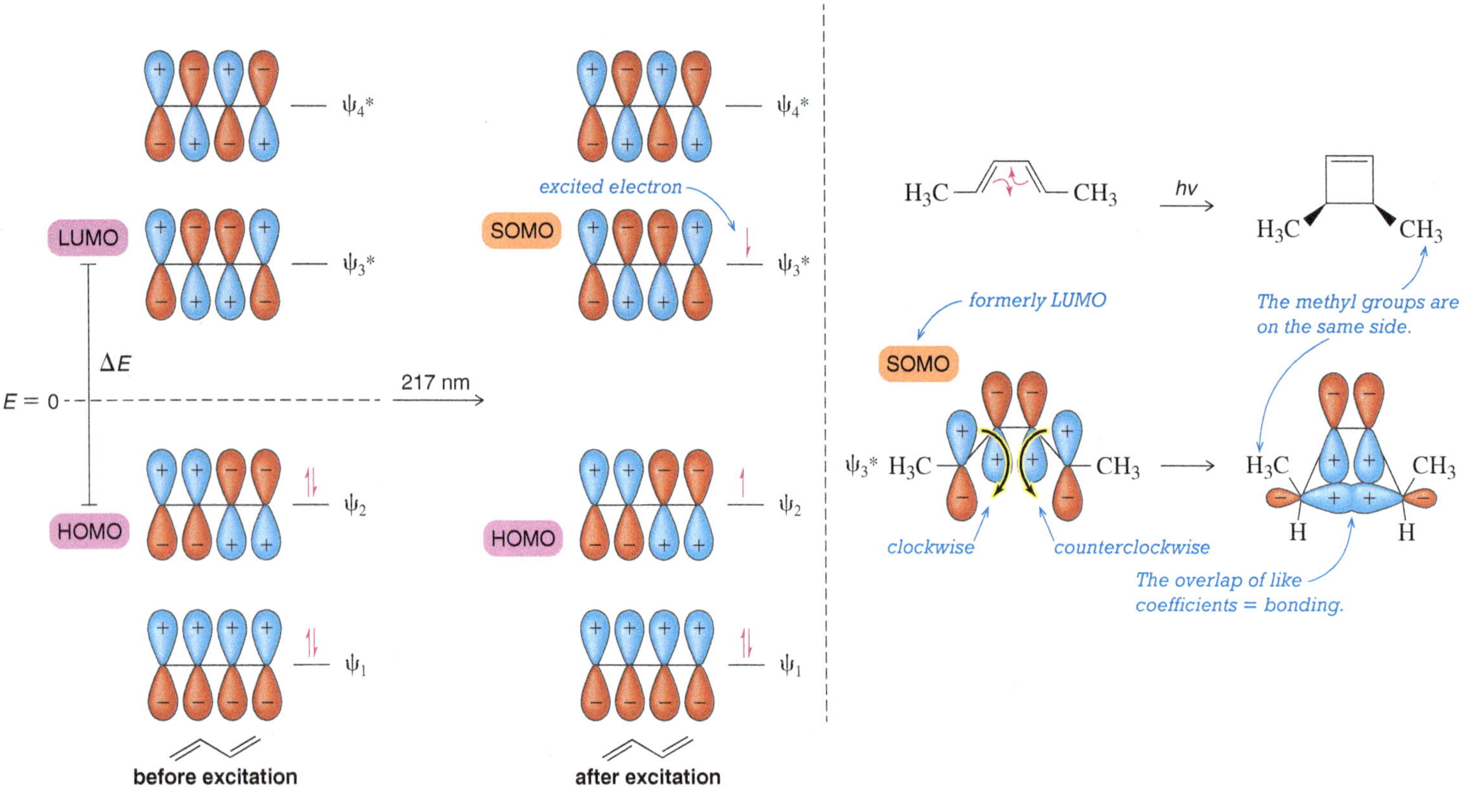

Figure 22.36 Photochemical electrocyclization of a 4*n* π system. Cyclization through the singly occupied molecular orbital requires disrotatory motion to align like coefficients.

As mentioned previously, electrocyclic reactions are reversible. The ring opening of the cyclobutene in **Figure 22.37** under thermal conditions is a 4-electron system (2 electrons in the π bond and 2 in the σ bond), so we expect conrotatory motion [*4, heat, con; see Table 22.1*]. In a nonsymmetrical system, the two possible conrotatory motions will give two different products. Moving both clockwise gives a cis,cis-diene with two bulky groups trying to share

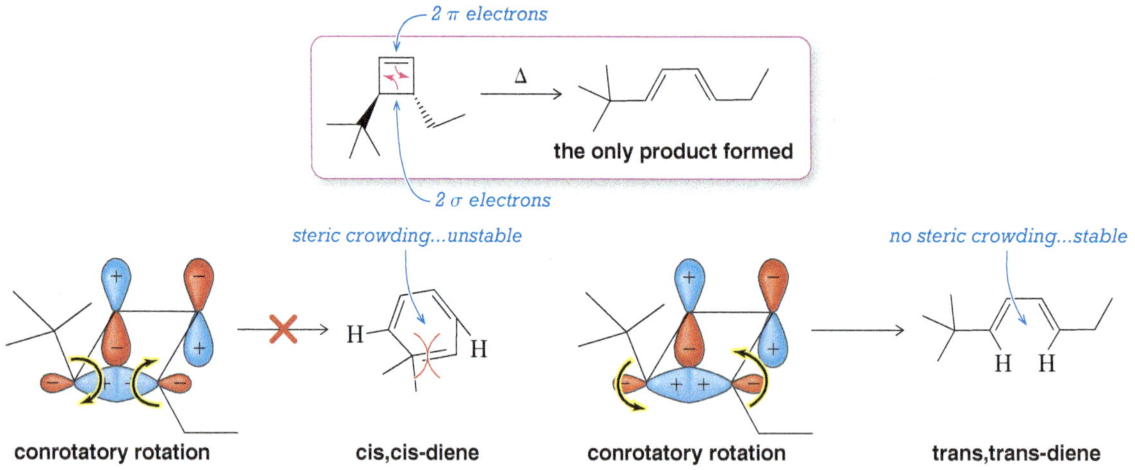

Figure 22.37 Thermal ring opening of a cyclobutene. The trans,trans-diene, formed by conrotatory ring opening (both counterclockwise), forms preferentially. The cis,cis-diene does not form due to steric crowding.

the same space, making this a very unstable product. Alternatively, moving both groups counterclockwise gives the trans,trans-diene, which minimizes steric crowding. As a result, this is the only product of this reaction. This brings up an important point for all electrocyclic reactions: When two products are possible, the more stable one is produced preferentially.

Assessment 22.30 Predict the product of the following electrocyclic reactions.

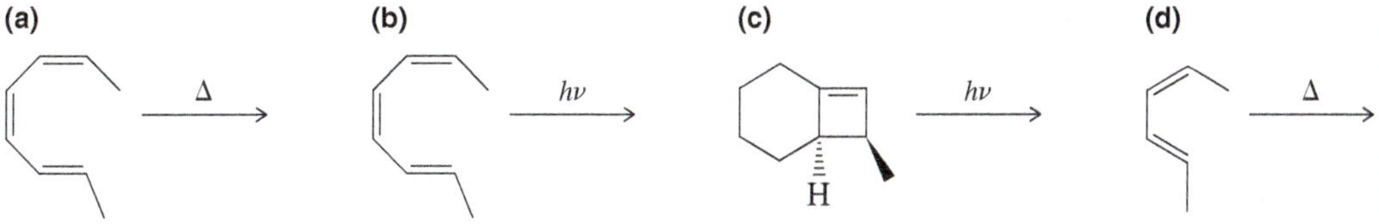

Assessment 22.31 Electrocyclic reactions favor the ring-closed form. (a) What is the reason for this? It is especially true for conjugated trienes, but not so much for conjugated dienes, whose equilibrium favors the ring-opened form. (b) Why might this be true?

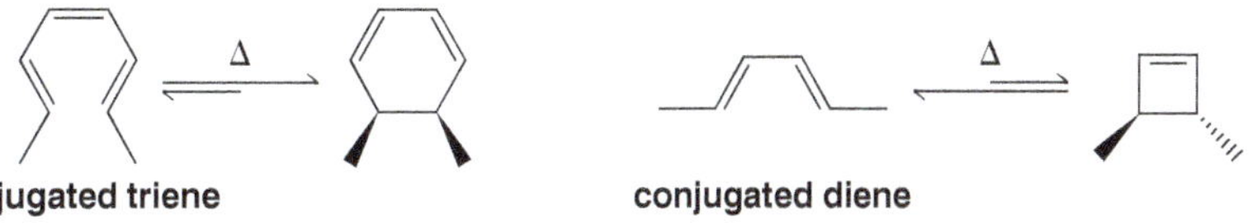

22.4.3 Sigmatropic rearrangement (Cope/Claisen)

The third and final pericyclic reaction type we discuss is sigmatropic rearrangement. Sigmatropic rearrangements are so named because they involve a concerted reorganization of the π system that results in the breaking of one σ bond and the formation of another. Although there are multiple types, we focus only on [3,3] shifts. The designation of [3,3] indicates the number of atoms between the bond being formed and the bond being broken. A representative example of a [3,3]-sigmatropic rearrangement is the Cope rearrangement shown in **Figure 22.38**. When heated, the 1,5-diene undergoes rearrangement through a concerted, cyclic, 6-electron transition state. [*There's that thing again.*] Exchange of the $C_3—C_4$ σ bond for the new $C_1—C_6$ σ bond is what makes this a sigmatropic rearrangement [*tropic = Greek for turn*]. The $C_4—C_5—C_6$ fragment (three atoms) detaches and reattaches to the $C_1—C_2—C_3$ fragment (three atoms), making this a [3,3] rearrangement. Given that the product is also a 1,5-diene, this reaction is reversible. Thus, the ultimate result of a Cope rearrangement is dictated by product stability. [*Between the two, which is more stable?*]

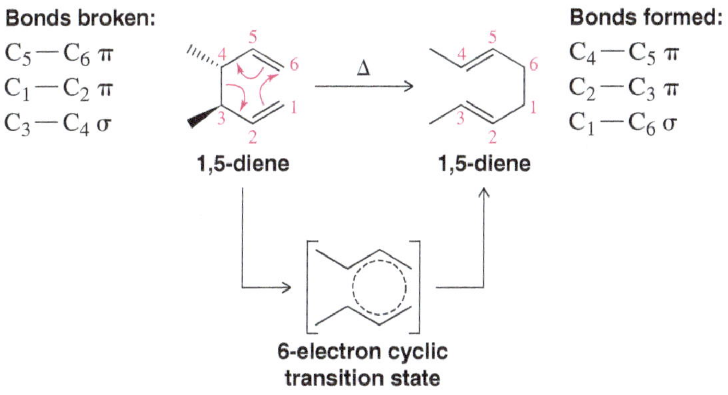

Figure 22.38 The Cope rearrangement is a [3,3]-sigmatropic rearrangement. It is concerted, occurs through an aromatic transition state, and favors the more stable side of the equilibrium.

There are two interesting aspects of the Cope rearrangement. First, it is a reversible reaction, so the favored side of the equilibrium is dictated by product stability. [*The ball still rolls downhill.*] But, as seen in Figure 22.38, the bonds formed and broken are identical, so at first glance, this reaction looks to have $\Delta H = 0$ [*And $\Delta S \approx 0$, too, because one molecule becomes one molecule.*] Closer inspection reveals a slight difference. [*Do you see it? Look at the alkenes.*] The product on the right has two disubstituted alkenes, whereas the reactant has two monosubstituted alkenes. Recall from Chapter 8 that disubstituted alkenes are more stable. Thus, forming the more stable alkenes is the driving force for this Cope rearrangement (**Figure 22.39**).

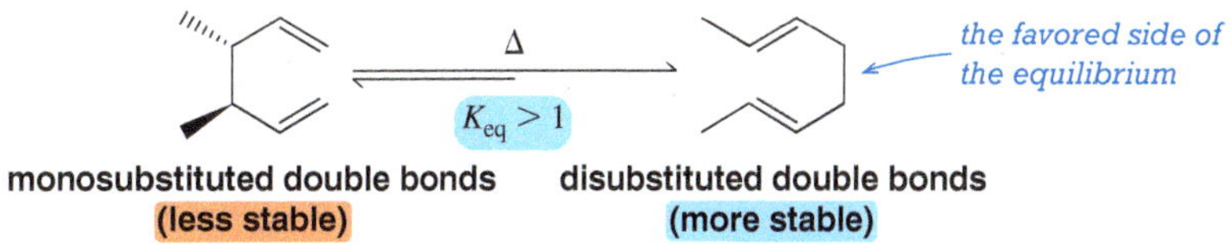

Figure 22.39 Formation of the more stable alkene is the driving force of the Cope rearrangement.

The second interesting aspect deals with the formation of the two trans-alkenes. Although cis-alkenes could hypothetically be possible, this concerted reaction is stereospecific for formation of the trans. Recall that the chair conformation of six-membered rings is the most stable ring conformation. With six carbons involved in the Cope rearrangement, this reaction, and all [3,3]-sigmatropic rearrangements, occur through the chair-like transition state shown in **Figure 22.40**. This conformation results in the *p* orbitals of the two alkenes pointing at each other and aligned with the σ-bonding orbital between C_3 and C_4, enabling the eventual overlap/bond formation. Being concerted, there is no chance for rotation around the σ and π bonds. Thus, the methyl groups maintain their relationship.

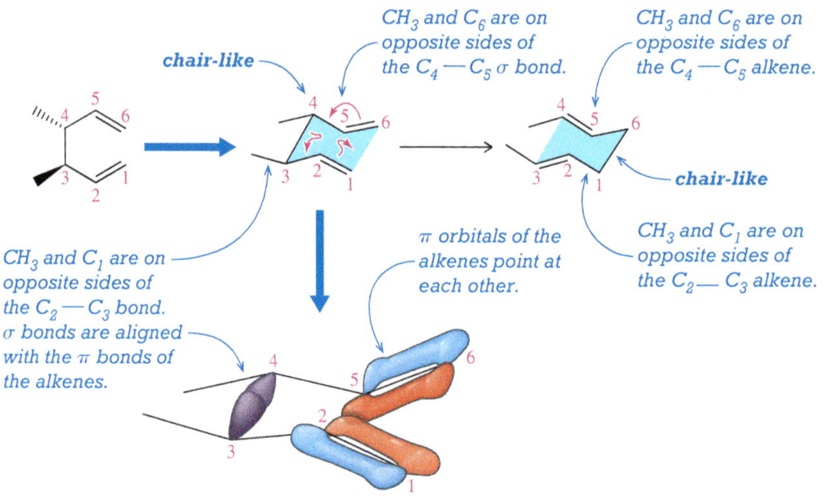

Figure 22.40 Rearrangement through a chair-like transition state results in the stereospecificity of the Cope rearrangement.

And so predicting the stereochemical outcome of the Cope rearrangement requires placing the molecule in the most stable chair-like transition state and then preserving the relationship of the methyl groups through the reaction. Two more examples of this are shown in **Figure 22.41**. In the chair-like conformation of Figure 22.41(a), the C_5—C_6 alkene is cis, so the methyl at C_6 has to occupy the less stable, axial position. For similar reasons, both methyls (at C_6 and C_1) are in the axial positions of Figure 22.41(b). With only one axial methyl group, the chair-like transition in Figure 22.41(a) is more stable and more easily attained. Thus, the Cope rearrangement in Figure 22.41(a) will be faster than the Cope rearrangement in Figure 22.41(b).

Figure 22.41 Stereospecific Cope rearrangement of a 1,5-diene. The orientation of the methyl groups is maintained through the concerted rearrangement. (a) Reaction of a cis,trans-diene. (b) Reaction of a cis,cis-diene.

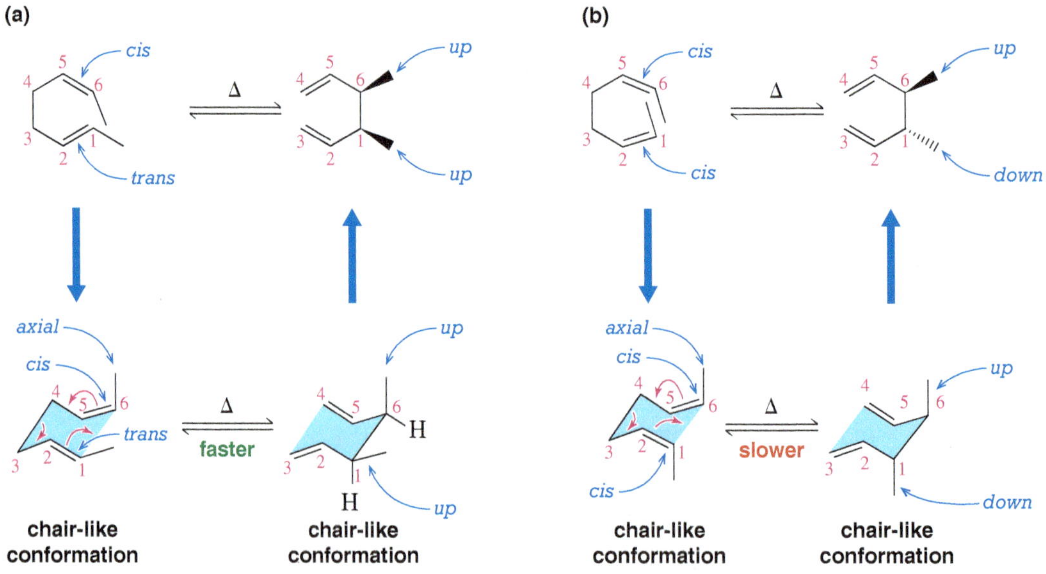

The Claisen rearrangement of allyl vinyl ethers (**Figure 22.42**) is another example of [3,3]-sigmatropic rearrangement. In essence, this is the Cope rearrangement but with O_4 in place of a carbon. [*Confirm this by comparing the reactions in Figures 22.42 and 22.38.*] The only notable difference between it and the Cope rearrangement is that it is more favorable in a particular direction. [*Which direction do you think is favored? And why?*] On the reactant side, there are two C—C π bonds and a C—O σ bond. The product has a C—C π bond, a C—O π bond, and a C—C σ bond. The main difference is the C—O π bond, which is about 20 kcal/mol more stable than a C—C π bond. Thus, formation of this bond makes the carbonyl-containing side favored in Claisen reactions.

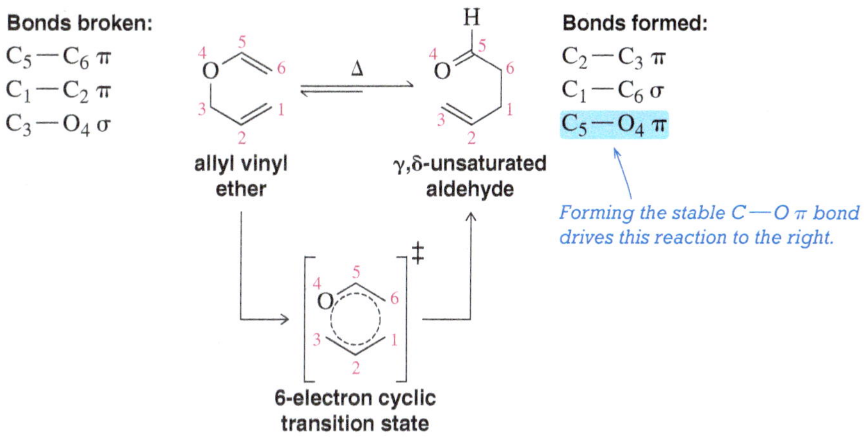

Figure 22.42 **The Claisen rearrangement of allyl vinyl ethers** to give an unsaturated carbonyl. It is concerted, occurs through an aromatic transition state, and favors the more stable side of the equilibrium.

Like the Cope rearrangement, the Claisen rearrangement occurs through a chair-like transition state, making it stereospecific for formation of a single stereoisomer, as shown in **Figure 22.43**.

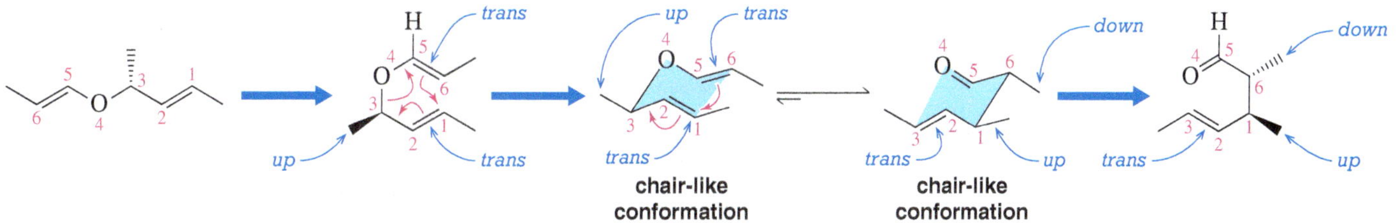

Figure 22.43 **Rearrangement through a chair-like transition state** results in the stereospecificity of the Claisen rearrangement.

Assessment 22.32 Predict the product of the following sigmatropic rearrangements. Be sure to rationalize the stereochemical outcome with a chair-like transition state.

(a)

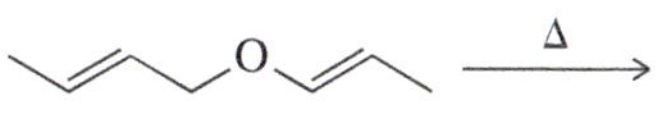

(b)

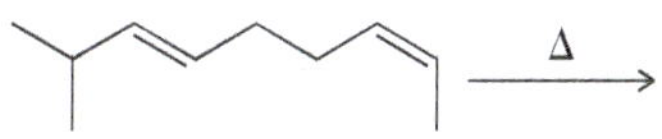

(c)

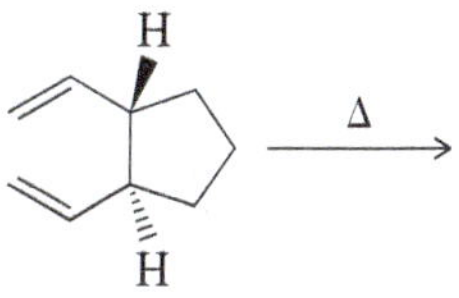

(d)

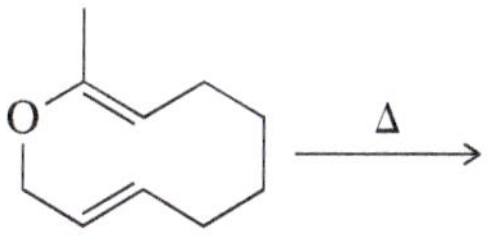

Assessment 22.33 The oxy-Cope reaction is heavily favored to the right because the product can equilibrate to an even more stable compound **A**. (a) Show an arrow-pushing mechanism for the first half of this reaction. (b) Draw compound **A** and explain why it makes the oxy-Cope reaction so favorable.

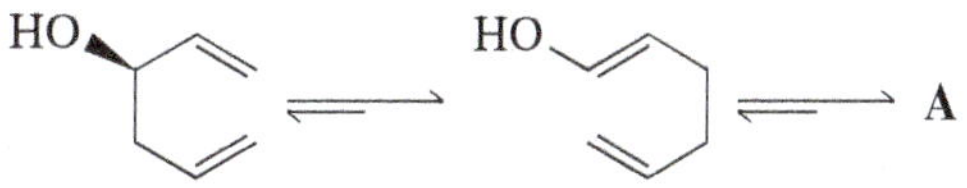

Concluding thoughts

Even Figure 22.44 has an example of a 6-electron cyclic transition state, and yet you still haven't been told why that's a good thing. I think you're ready . . . do you? On to Chapter 23!

The story of the search for a Diels–Alderase illustrates the importance of organic chemistry to biology and vice versa. But the interplay in the case of a Diels–Alderase actually goes much deeper. It was observed that although there were over 1500 known enzyme-catalyzed reactions, there was not an enzyme that had been conclusively determined to catalyze a Diels–Alder reaction. As a result of this, chemists set out to make their own. Specifically, an antibody was prepared to bind selectively to the molecule in **Figure 22.44(a)** because it was similar in structure to the transition state of the desired Diels–Alder reaction. This antibody, in the presence of the diene and dienophile shown in **Figure 22.44(b)**, effected a faster Diels–Alder cyclization reaction. Catalysis was accomplished by lowering the energy of the transition state. With this advance, the concept of Diels–Alder biocatalysis was proven a decade before a Diels–Alderase enzyme was confirmed to exist. [*In some ways, the development of catalytic antibodies for the Diels–Alder affirmed that a Diels–Alderase was out there . . . again, the symbiosis of biology and organic chemistry.*]

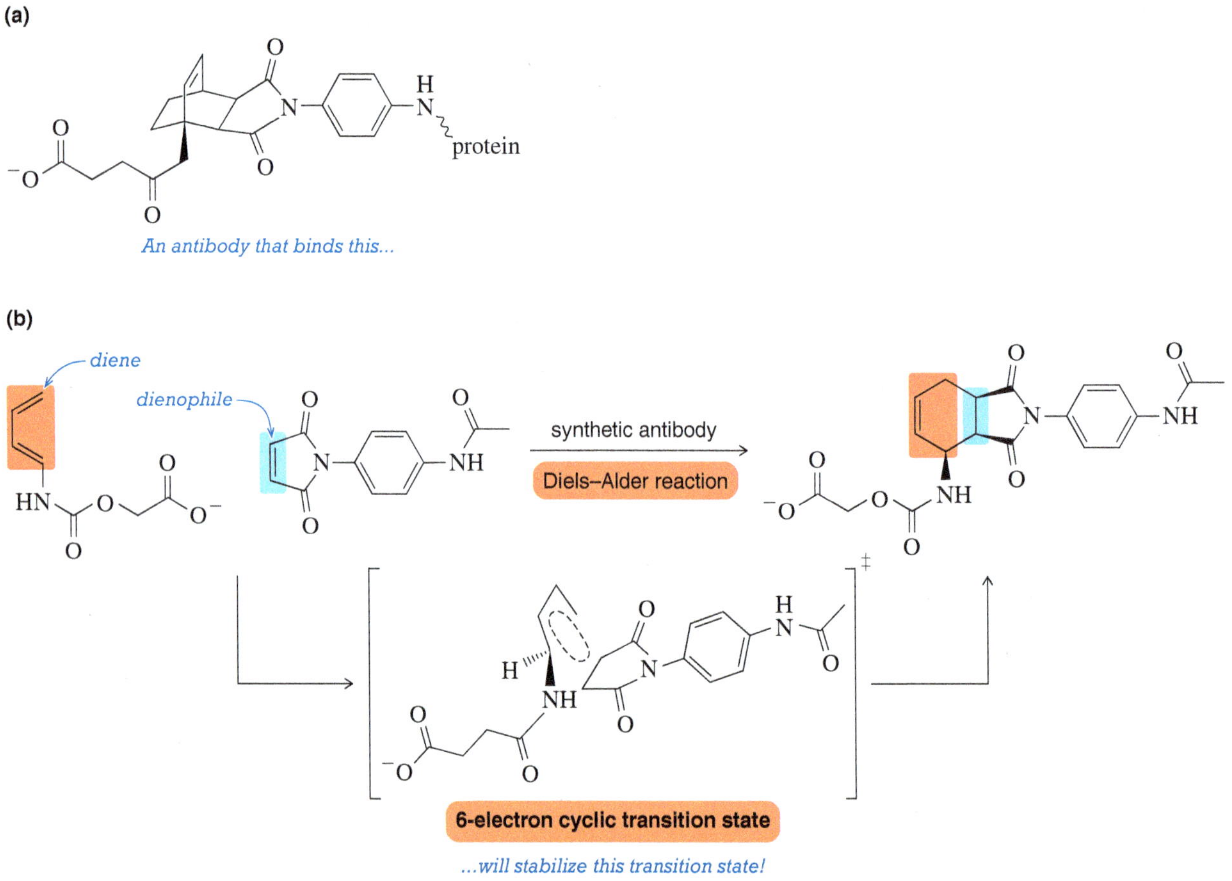

Figure 22.44 The development of a catalytic antibody for the Diels–Alder reaction. (a) Creation of an antibody that binds a transition state analog of a Diels–Alder reaction results in (b) a catalytic antibody that stabilizes the transition state and increases the rate of a Diels–Alder reaction.

End-of-Chapter Materials

Important Concepts

[These are the major concepts covered in each section, excluding reactions.]

22.1: Pericyclic reactions are concerted reactions that involve a cyclic flow of electrons. This chapter focuses on cycloaddition, electrocyclic, and sigmatropic rearrangements. The outcomes of these reactions are explained by examining the molecular orbital pictures of the reactants.

22.2: Pericyclic reactions can be explained using frontier molecular orbital theory. Simply summarized, the highest occupied (HOMO) and lowest unoccupied molecular orbitals (LUMO), lying at the frontier of energy levels, dictate the reactivity in these systems.

22.3: The Diels–Alder reaction is the concerted cycloaddition between a diene and a dienophile. In a normal Diels–Alder reaction, the HOMO of the diene reacts with the LUMO of the dienophile. As a result, electron-rich dienes and electron-poor dienophiles give the fastest Diels–Alder reactions. Being concerted reactions, the *s*-cis conformation is required, and the Diels–Alder is stereospecific and regioselective.

22.4.1: A [2 + 2] cycloaddition is made possible by exciting 1 electron from the HOMO of ethene to the LUMO.

22.4.2: Electrocyclic reactions result in the reversible conversion of acyclic conjugated molecules to ring structures through the movement of π electrons. Under thermal conditions, the structure of the HOMO determines the outcome. Alternatively, under photochemical conditions, an electron is excited from the HOMO to the LUMO, and the LUMO (now the SOMO) determines the outcome. These different outcomes are summarized by the Woodward–Hoffman rules.

22.4.3: The Cope and the Claisen rearrangements are [3,3]-sigmatropic rearrangements. The Cope rearrangement is the interconversion of 1,5-dienes proceeding through a concerted, chair-like, 6-electron cyclic transition state, favoring the more stable side of the equilibrium. The Claisen rearrangement is the reaction of allyl vinyl ethers proceeding through a concerted, chair-like, 6-electron cyclic transition state, favoring the side of equilibrium that maintains the stable C—O π bond.

Important Reactions and Mechanisms

1. **Diels–Alder reaction (Section 22.3):** The cycloaddition reaction of a diene and a dienophile gives a cyclohexene in a concerted step featuring a 6-electron cyclic transition state. Three π bonds break to form one π bond and two σ bonds. It is fastest when an electron-rich diene reacts with an electron-poor dienophile. Being concerted, it is stereospecific, reacting primarily through the endo transition state. It is highly regioselective based on matching the largest coefficient of the HOMO of the diene with the largest coefficient of the LUMO of the dienophile.

General:

Specific:

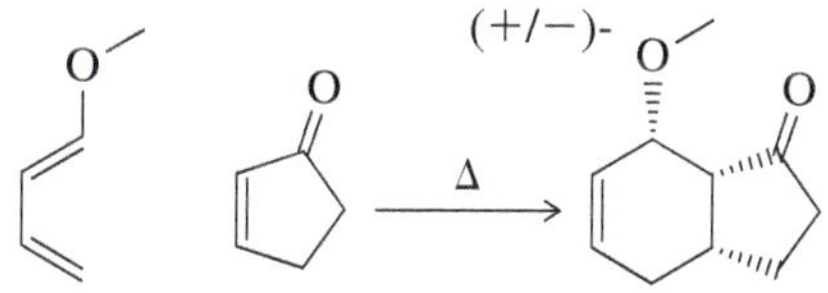

Mechanism:

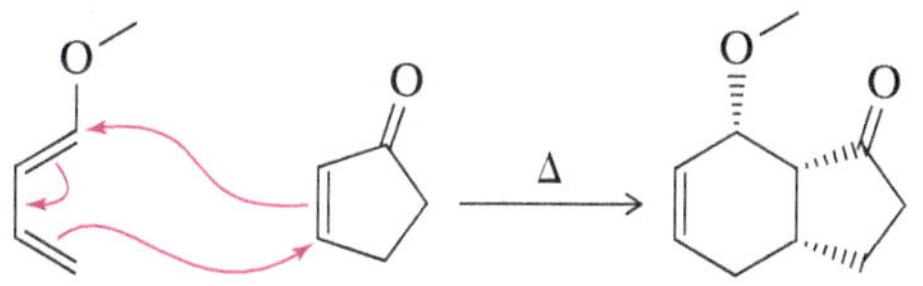

2. **Electrocyclic reactions (Section 22.4.2):** The reversible conversion of acyclic conjugated molecules to ring structures through the movement of π electrons can be initiated under thermal or photochemical conditions. Two π bonds break to form one π bond and one σ bond in the ring-closing direction. The orientation of substituents in the product is explained by the Woodward–Hoffman rules.

General:

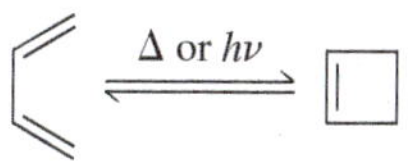

Specific:

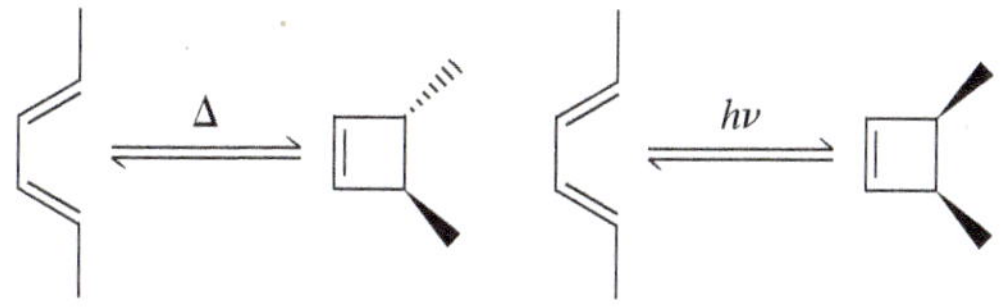

Mechanism:

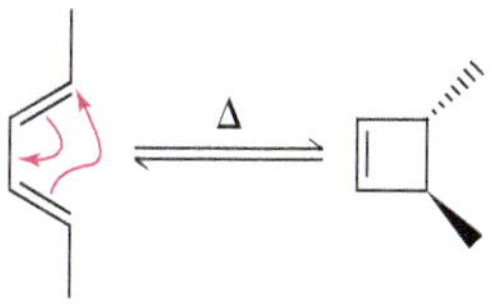

conrotatory based on Woodward–Hoffman rules

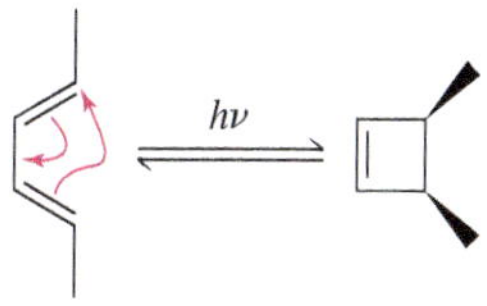

disrotatory based on Woodward–Hoffman rules

3. **Cope rearrangement (Section 22.4.3):** A sigmatropic rearrangement in which a 1,5-diene interconverts via a 6-electron cyclic transition state. Two π bonds and one σ bond break to form two new π bonds and one new σ bond. The favored side of the equilibrium is based on the relative stability of the reactant and product. The reaction is stereospecific through a chair-like transition state.

General:

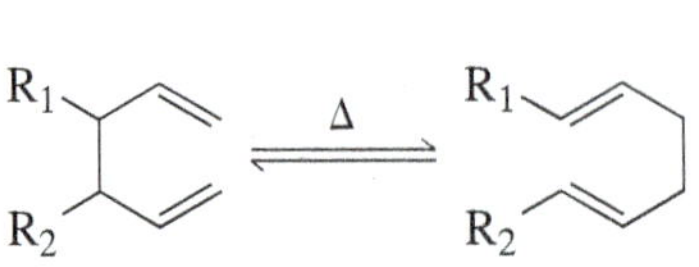

Specific:

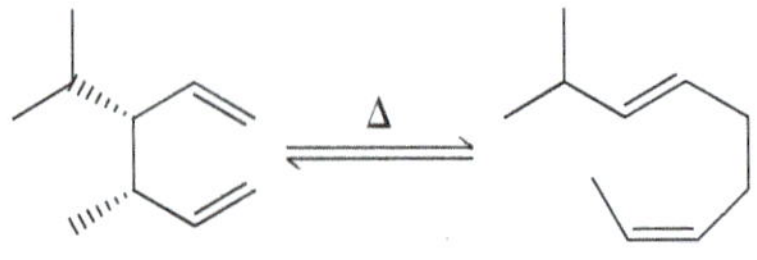

Mechanism:

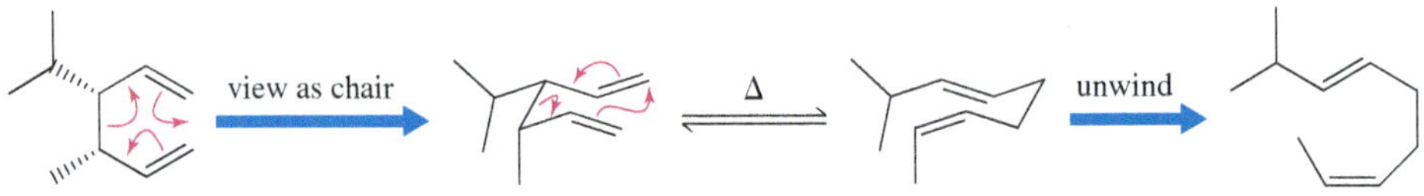

4. **Claisen rearrangement (Section 22.4.3):** A sigmatropic rearrangement in which an allyl vinyl ether is converted to a γ,δ-unsaturated carbonyl through a 6-electron cyclic transition state. Two C—C π bonds and one C—O σ bond break to form a new C—C π bond, a new C—O π bond, and one new C—C σ bond. The favored side of the equilibrium is based on the enhanced stability of the C—O π bond. The reaction is stereospecific through a chair-like transition state.

General:

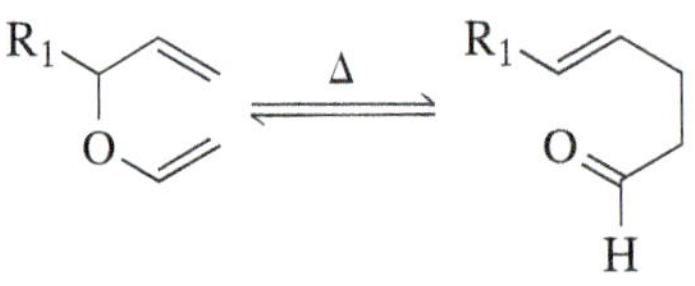

Specific:

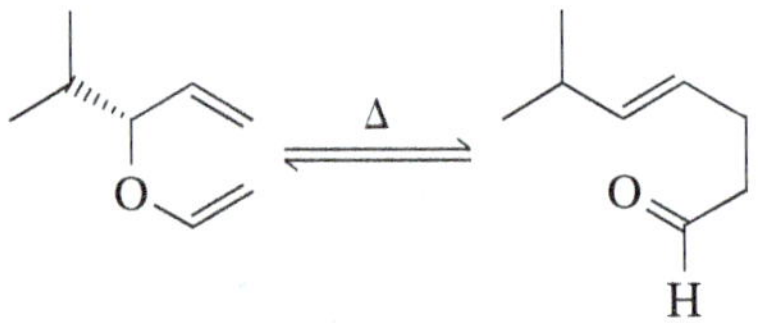

Mechanism:

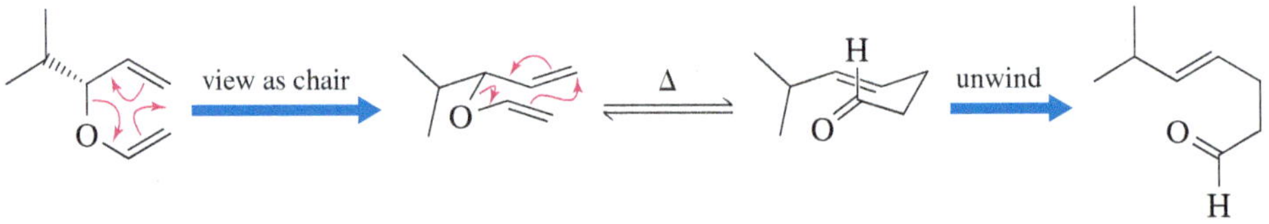

End-of-Chapter Assessments

[*Assessments are ranked according to their difficulty;* •••• = *most difficult.*]

22.34 (•) Identify the following dienes as being in the *s*-cis or *s*-trans conformation. If they are in the *s*-trans conformation, draw them in the *s*-cis conformation. [*It may not always be possible.*]

(a)

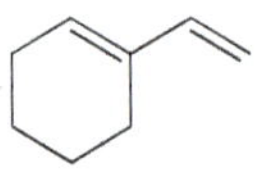

(b)

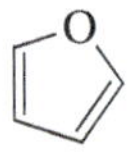

(c)

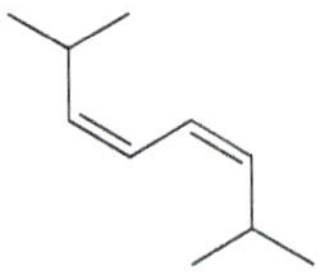

(d)

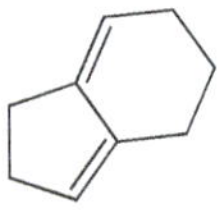

(e)

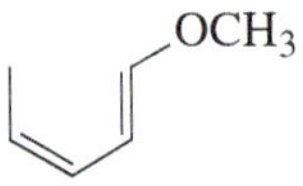

22.35 (•) Provide a mechanism of the Diels–Alder reaction shown and predict the regioisomer that will form.

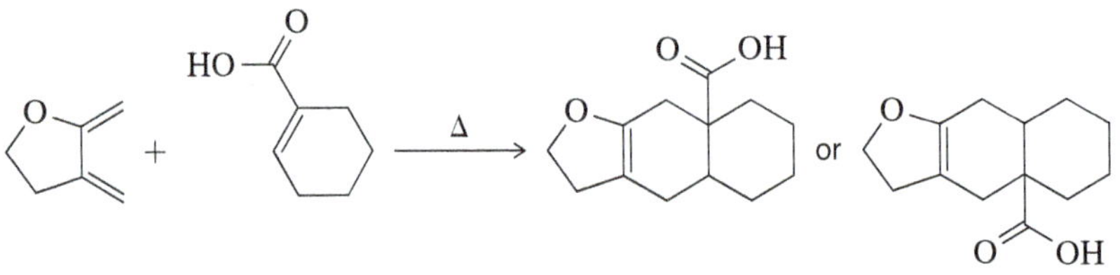

22.36 (••) Predict the product of the Diels–Alder reactions shown.

(a)

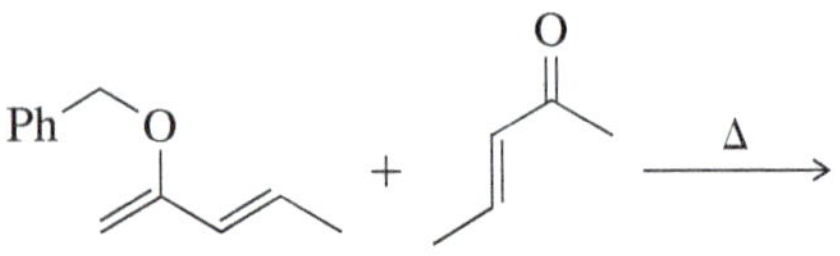

(b)

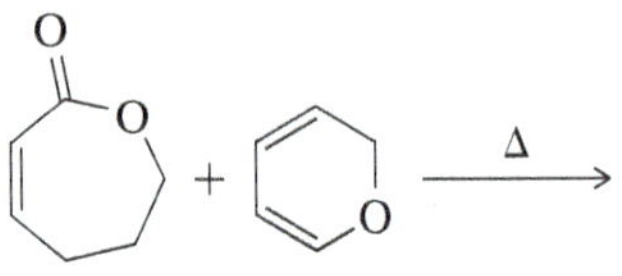

(c)

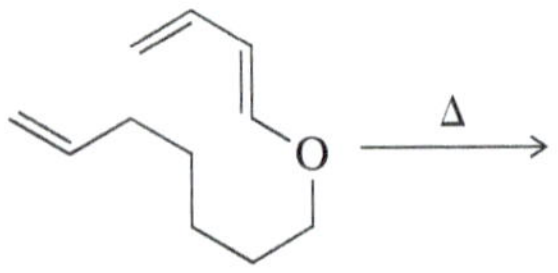

22.37 (•) For the following electrocyclic reactions, did the substituents move in a conrotatory or disrotatory direction? Would you use heat or light to cause movement in this direction?

(a)

(b)

(c)

22.38 (••) Given the conditions, would you expect conrotatory or disrotatory ring closing/opening? Justify this on the basis of the molecular orbital picture.

(a)

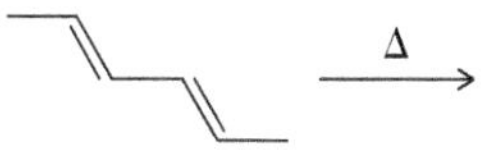

(b)

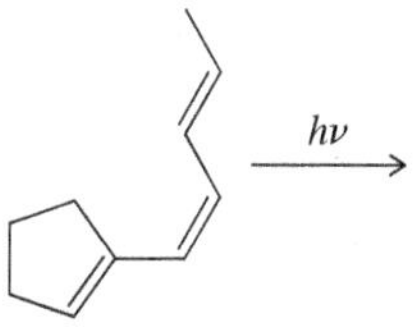

(c)

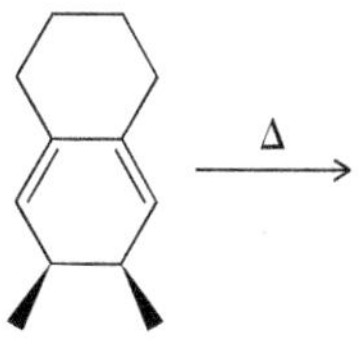

22.39 (••) Draw the products you'd expect to form in Assessment 22.38.

22.40 (•) Predict the product of the Cope reactions shown.

(a)

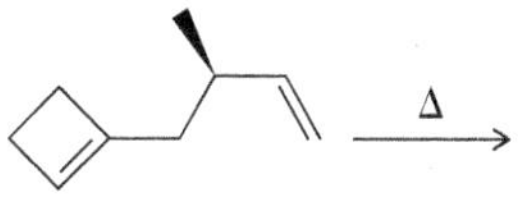

(b)

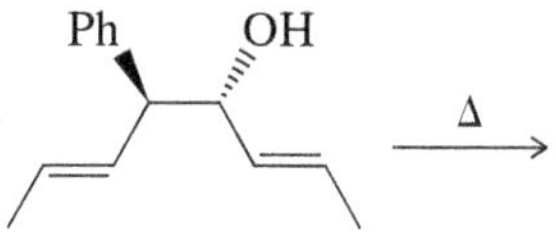

(c)

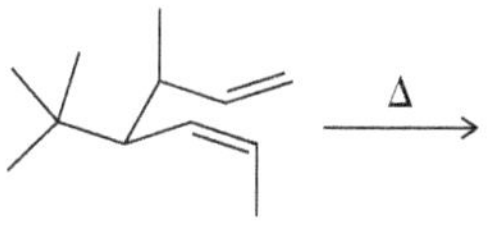

22.41 (•) Predict the product of the Claisen reactions shown.

(a)

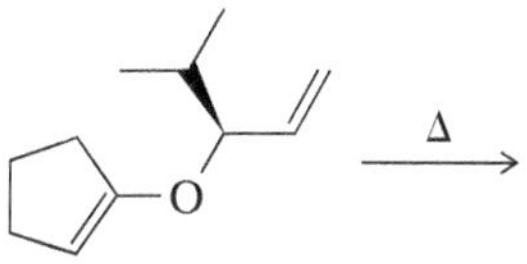

(b)

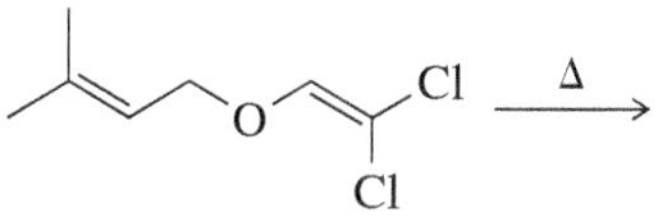

(c)

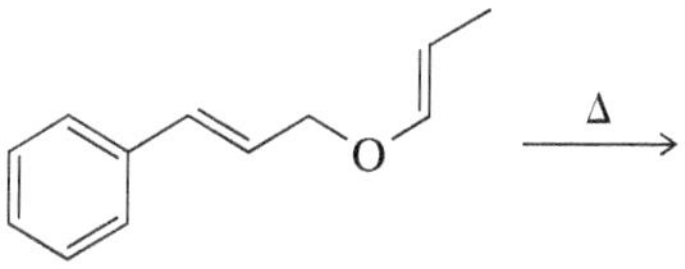

22.42 (••) Based on the bonds that form and break in the pericyclic reactions studied in this chapter, how can you distinguish between (a) Diels–Alder, (b) a 6 π electrocyclic, and (c) Cope/Claisen reactions?

22.43 (••) Predict the product of the following reactions. [*When all of the reactions from this chapter are shown together, you must first decide which type of reaction each is. Is it a Diels–Alder, an electrocyclic, or a sigmatropic rearrangement? Drawing the product will be easier once this determination is made.*]

(a)

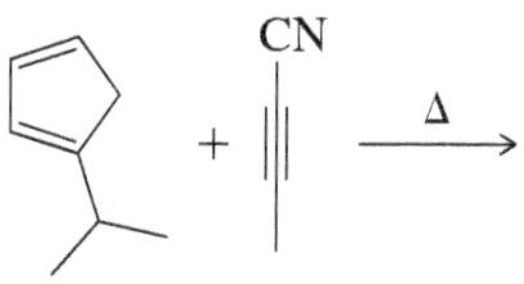

(b)

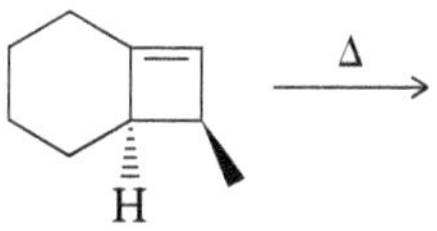

(c)

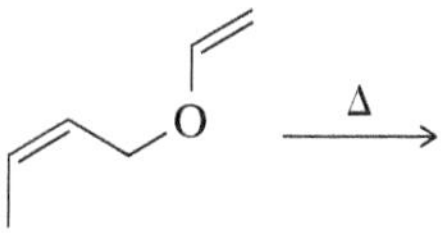

(d)

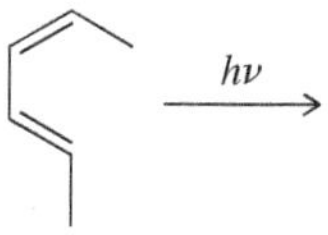

(e)

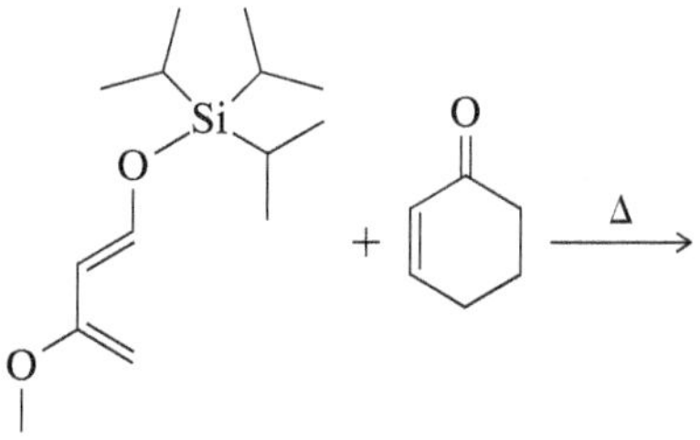

(f)

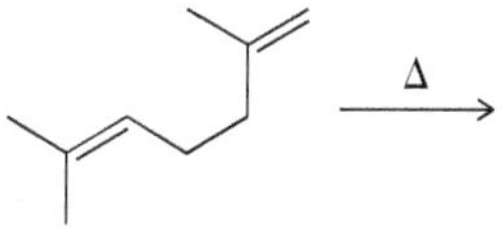

22.44 (••) Suggest a diene and a dienophile that would give the following products.

(a) **(b)** **(c)**

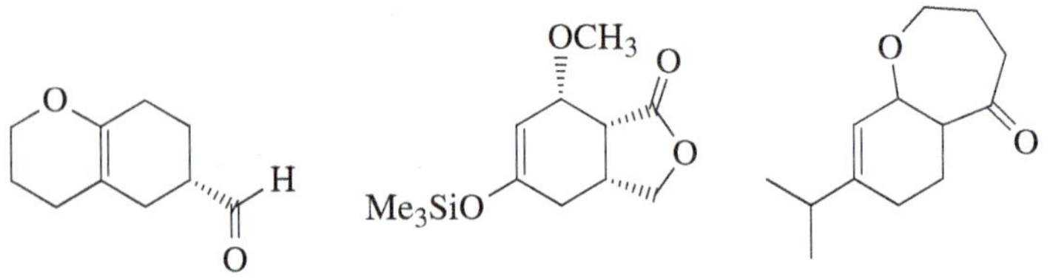

22.45 (••••) Starting with the molecules on the left, suggest a synthesis of the product at right.

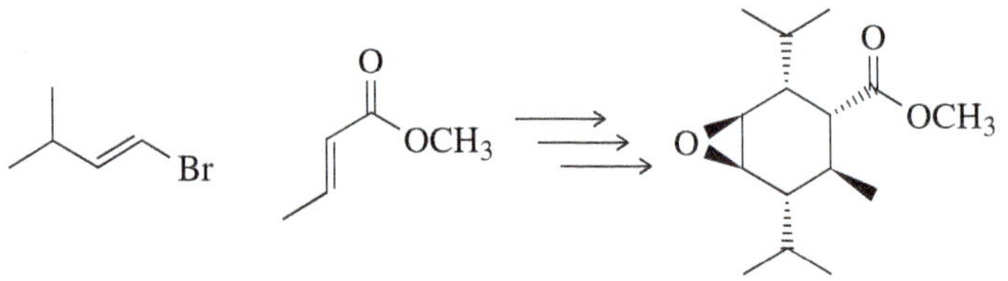

22.46 (•••) By forming a Lewis acid–Lewis base complex with the dienophile, Lewis acids are able to increase the rate of Diels–Alder reactions. Why might this be true?

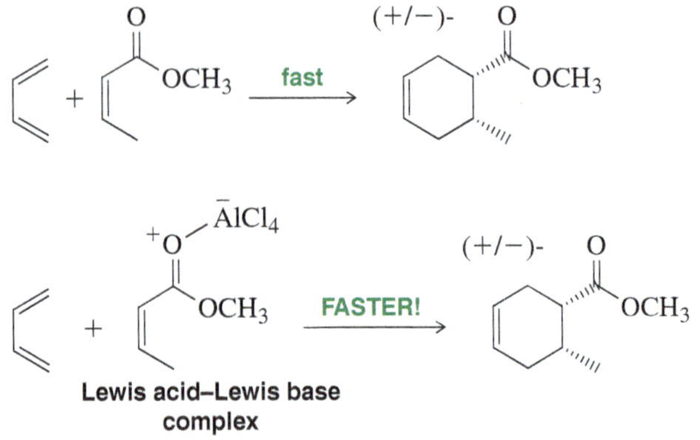

22.47 (••) Describe how the following factors would affect the rate and/or equilibrium of the reaction shown: (a) increasing the concentration of the reactants; (b) removing the product from the reaction mixture; (c) increasing the temperature; (d) adding a Lewis acid to the solution; (e) adding an electron-donating group to the diene; (f) adding an electron-donating group to the dienophile; (g) adding an electron-withdrawing group to the diene; (h) adding an electron-withdrawing group to the dienophile.

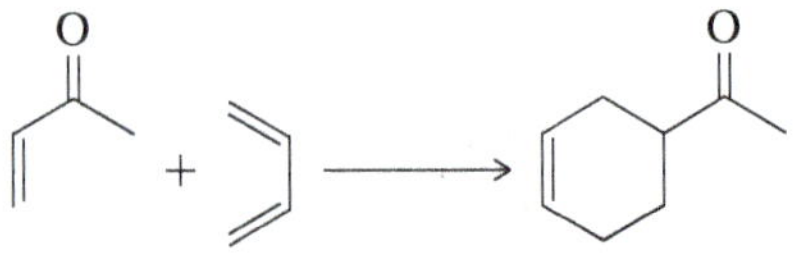

22.48 (•••) Whereas the Diels–Alder reaction typically gives a racemic mixture, the following Diels–Alder reaction proceeds to give a single stereoisomer as product. Why?

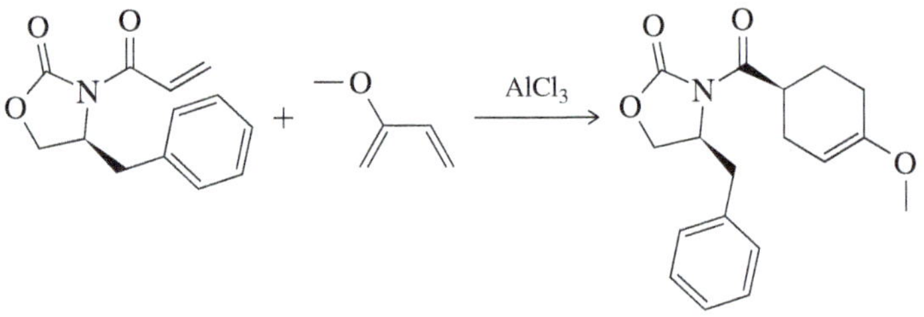

22.49 (•••) **FROM THE LITERATURE** The synthesis of bolivianine involved the step shown. Suggest a reagent that would enable synthesis of the intermediate shown (*J. Am. Chem. Soc.* 2013, *135*, 9291–9294.).

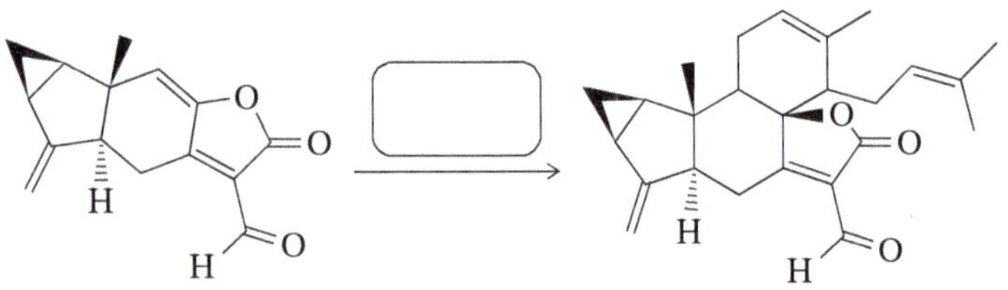

22.50 (•••) **FROM THE LITERATURE** The [2 + 2] cyclization is especially useful when done intramolecularly. (a) What makes intramolecular reactions more favorable? (b) Predict the product of the following [2 + 2] cyclization (slight modification of a reaction from *Angew. Chem.* 2011, *50*, 5149).

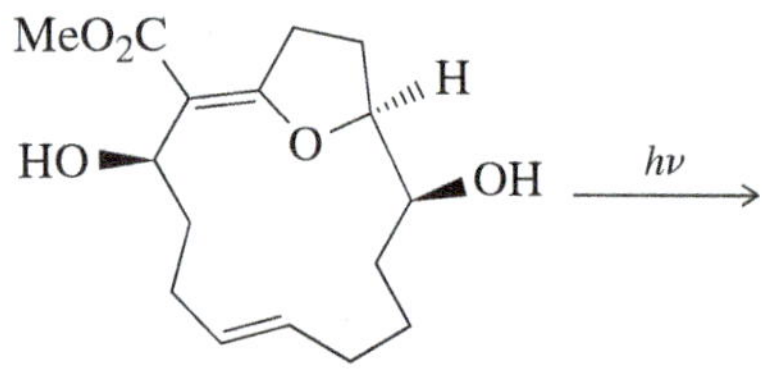

22.51 (••) The Heck reaction, reminiscent of the catalytic coupling reactions of Chapter 16, has been used to prepare the triene below. Predict the products that result when the triene is (a) heated or (b) exposed to light. [*Table 22.1 will help you solve these.*]

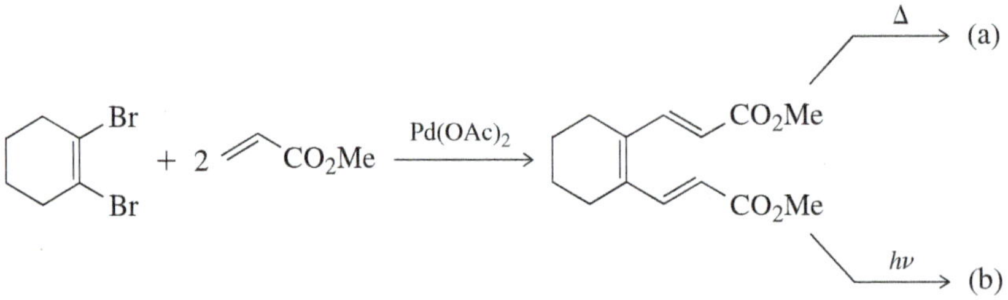

22.52 (••) LOOKING AHEAD The Claisen rearrangement can be used to attach allyl groups to benzene via initial alkylation of a phenol (Chapter 24). (a) Predict the identity of **A** and **B** and (b) provide an arrow-pushing mechanism for each step. [*The phenol alkylation is simply a Williamson ether synthesis reaction (Chapter 13).*]

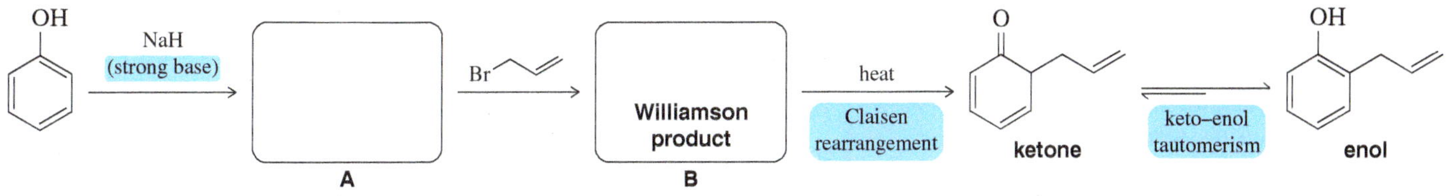

22.53 (•••) The Ireland–Claisen reaction is a variant of the Claisen reaction studied in this chapter. (a) Suggest a mechanism for the first step. (b) Predict the product (**B**) that would result from heating intermediate **A**. (c) Hydrolysis of **B** gives a carboxylic acid. Suggest a mechanism for this reaction.

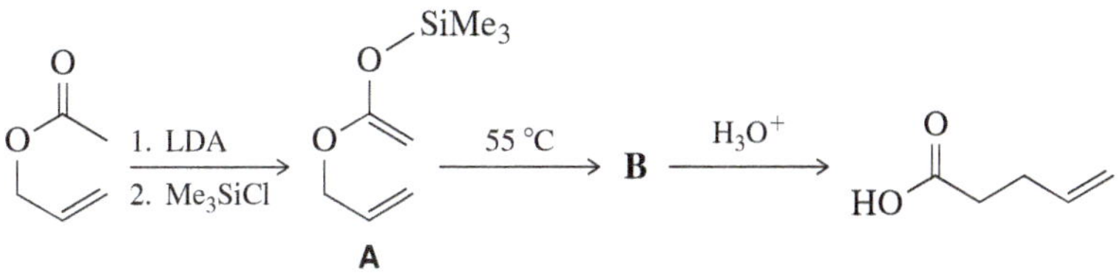

22.54 (•••) Only after working Assessment 22.53, predict the product of the following reactions.

(a)

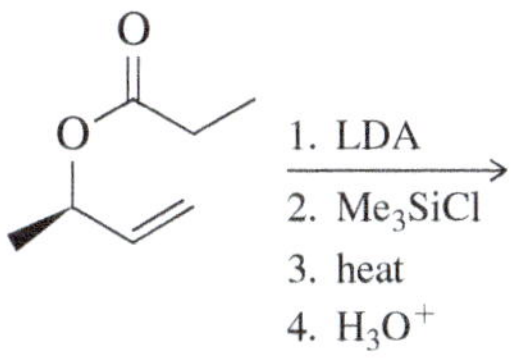

(b)

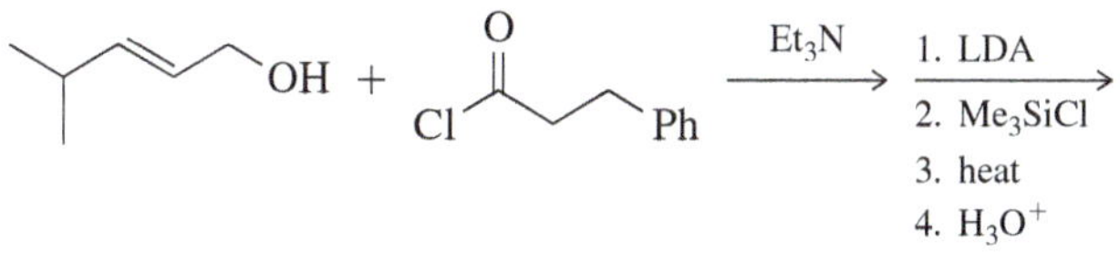

22.55 (•••) The Johnson–Claisen reaction is a variant of the Claisen reaction studied in this chapter. (a) Suggest a mechanism for the first step. (b) Predict the product (**B**) that would result from heating intermediate **A**.

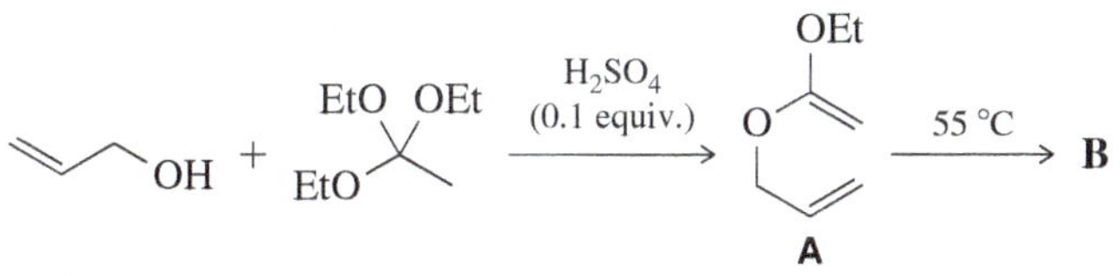

22.56 (•••) The Eschenmoser–Claisen reaction is a variant of the Claisen reaction studied in this chapter. Based on your answers to Assessments 22.53–22.55, predict a product of this reaction.

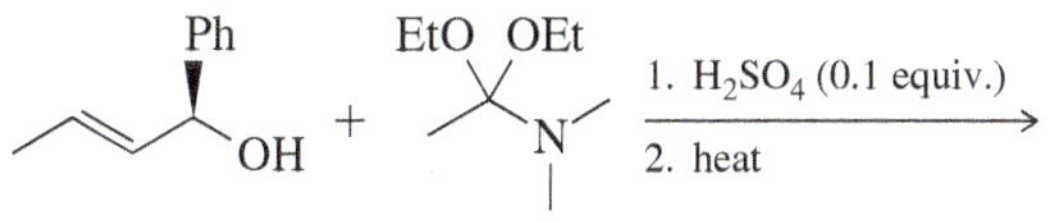

22.57 (••) Organic chemists often invent reactions that look very similar to each other, as was demonstrated in Assessments 22.53–22.56. What is the value of having these three Claisen variants?

22.58 (•••) LOOKING BACK In Chapter 20, we studied the aldol reaction. Although not discussed at the time, this reaction is stereospecific, proceeding through the Zimmerman–Traxler transition state shown here. (a) Show an arrow-pushing mechanism for this concerted reaction. (b) Why is this a favorable mechanism?

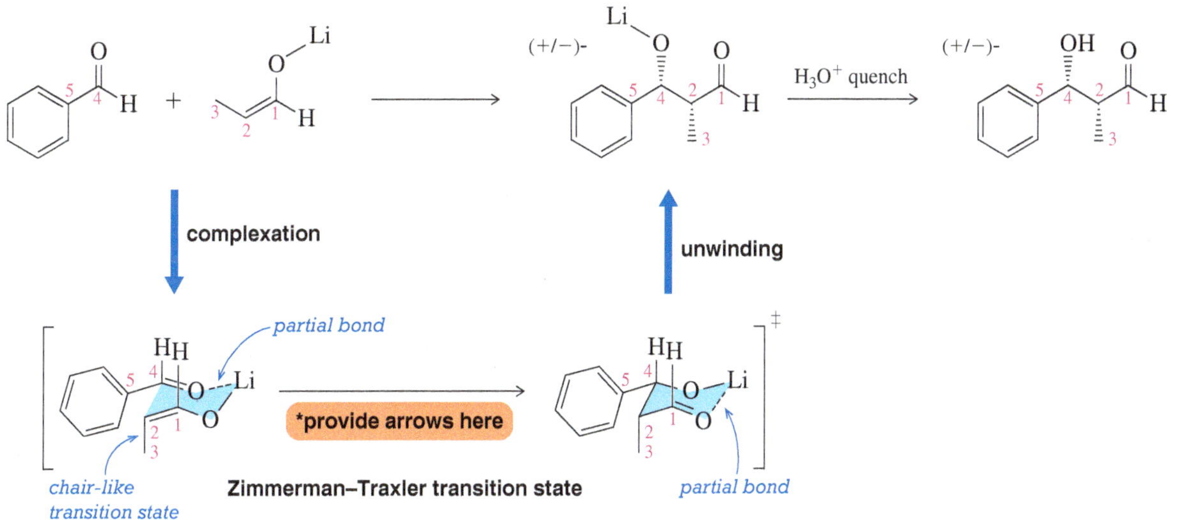

22.59 (••••) LOOKING BACK The enolate in Assessment 21.58 is referred to as the Z(O)-enolate. Predict the product of the reaction in Assessment 21.58 if an E(O)-enolate were used instead. The chair-like transition state has been drawn for you. [*This is a difficult problem. Push electrons, then redraw a chair-like product. Number the carbons as they were in Assessment 21.58. Then, using those same numbers, unwind the chair to give the straight-chain product.*]

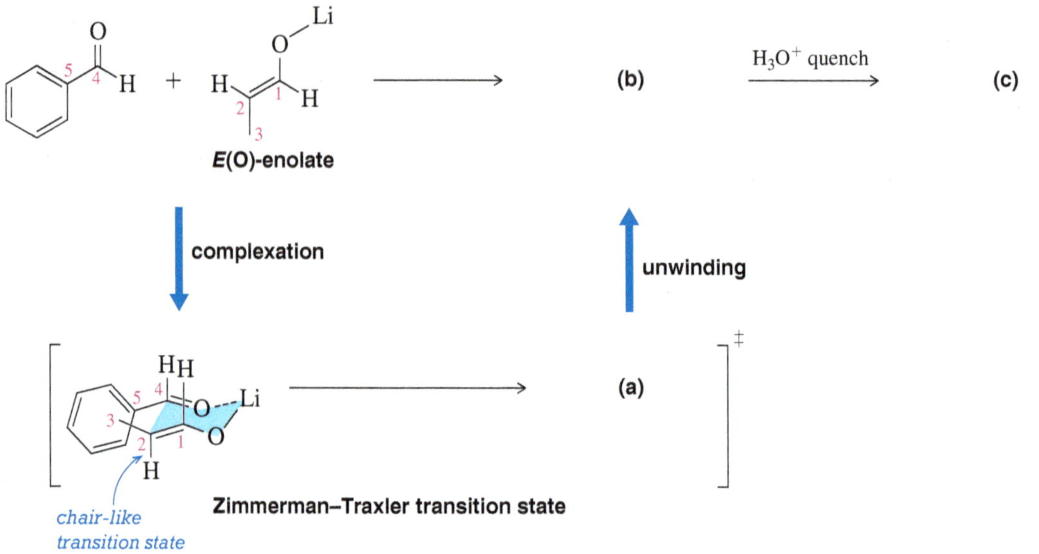

22.60 (••••) While not covered explicitly in this chapter, the ene reaction occurs similarly to the Diels–Alder reaction but replaces the electrons from one π bond in the diene with the electrons in a C—H bond. Draw the mechanism for the following reaction. [*Number the carbons and draw in the hydrogens of the product. And, of course, make a note of bonds formed and bonds broken.*]

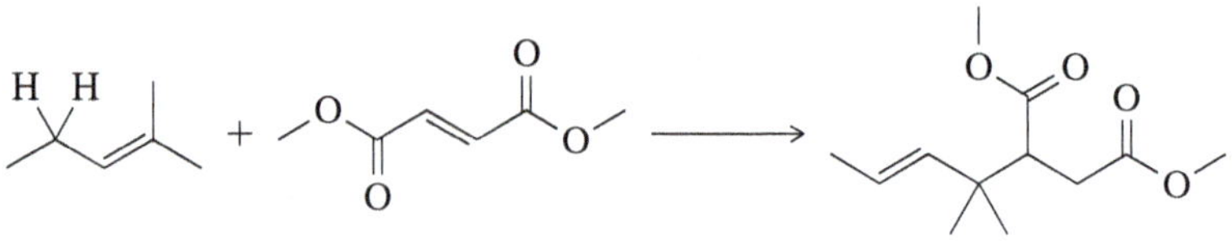

22.61 (•••) FROM THE LITERATURE The following Diels–Alder reaction occurs more rapidly in the presence of an electric field. Suggest a diene and a dienophile that would make the product shown (*Nature* 2016, *531*, 88–91).

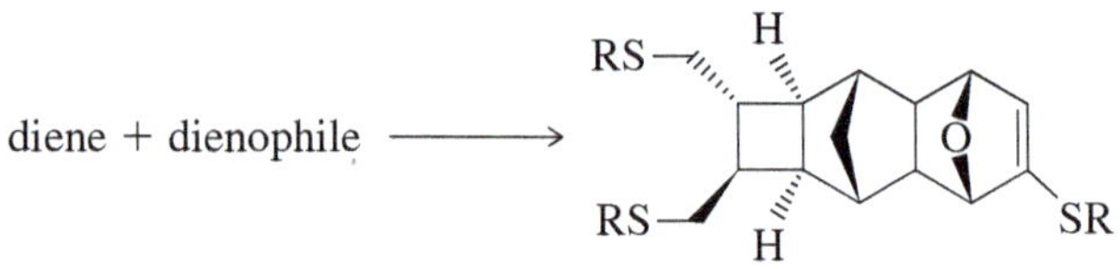

22.62 (••••) FROM THE LITERATURE The product of the Stille coupling reaction (**A**) tautomerizes in a basic solution to give compound **B**. **B** spontaneously converts to **C**. (a) Propose a structure for **A**. (b) Suggest a mechanism for the conversion of **A** to **B**. (c) Suggest a mechanism for the conversion of **B** to **C** (modified from *Org. Lett.* 2001, *3*, 3875–3878).

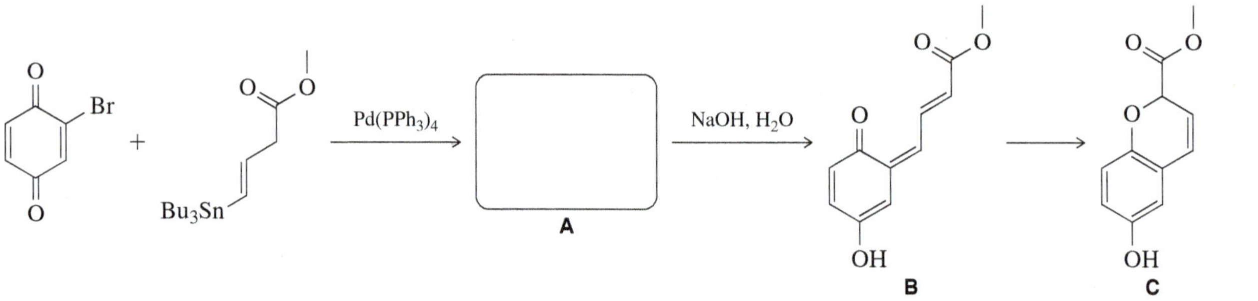

23

Benzene 1: Aromatic Stability and Substitution Reactions

Introduction

In Chapter 19, while studying nucleophilic acyl substitution reactions of carboxylic acid derivatives, we were introduced to bacteriocidal antibiotics. The penicillins, cephalosporins, and cephamycins, which make up the β-lactam antibiotic family, kill bacteria by disrupting their ability to synthesize a cell wall. A second class of antibiotics is referred to as bacteriostatic. These act by slowing the growth of bacteria, thus allowing the immune system to "catch up" and defeat them.

The story of sulfa drugs, a family of bacteriostatic antibiotics, is one that begins with the formation of an early conglomerate of chemical companies in pre-World War II Germany. In 1927, Gerhard Domagk was hired by the IG Farben conglomerate to investigate the potential use of dyes as treatments for human disease. This work was an extension of work by Paul Ehrlich, a German physician who believed that certain "dyes" could stain bacterial cells. One such drug that came out of Domagk's work was Prontosil, originally developed as a red dye. Legend has it that, prior to Prontosil's approval as a drug, Domagk gave it to his daughter because she was dying from a streptococcal infection. She fully recovered, but her skin was permanently stained red as a result of the drug. This experiment, which would never be allowed today, confirmed the viability of Prontosil as an

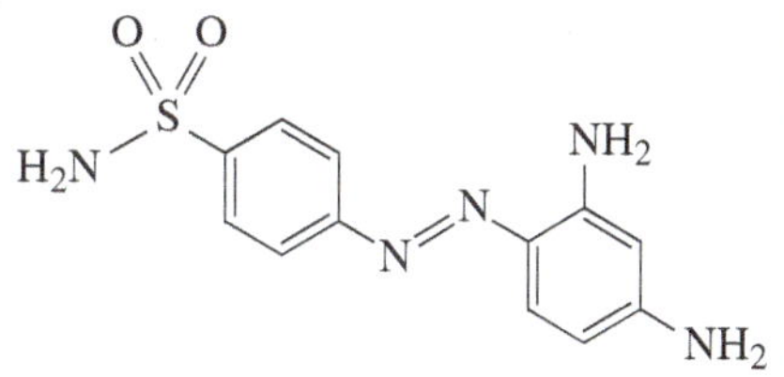

Prontosil

Learning Objectives

- **LO 23.0** Explain the stability of benzene
- **LO 23.1** Apply the key concepts of aromaticity
- **LO 23.2** Account for differences between aromatic and antiaromatic molecules
- **LO 23.3** Analyze molecules using Frost circles
- **LO 23.4** Name benzene derivatives according to the IUPAC system
- **LO 23.5** Analyze the general mechanism of aromatic substitution reactions
- **LO 23.6** Analyze the regioselectivity and stereochemistry of electrophilic aromatic substitution reactions
- **LO 23.7** Analyze nucleophilic aromatic substitution reactions
- **LO 23.8** Analyze aromatic substitution reactions involving diazonium or palladium

"But look! What was that? One of the snakes had seized hold of its own tail, and the form whirled mockingly before my eyes. As if by a flash of lightning I awoke; and this time also I spent the rest of the night in working out the consequences of the hypothesis."

—Friedrich August Kekulé

antibiotic. As we study aromatic rings, we will learn the story of the sulfa antibiotics—molecules that are credited with ushering in the modern era of medicinal chemistry.

○ ○ ○ ○ ○ ○ ○

Chapter 23 begins a run of chapters (23–25) where most of what is presented are things you have seen before. In Chapter 23, for example, the chemistry of benzene also includes concepts like resonance, alkene addition, and the Hammond postulate. There are two benefits to this. First, this is a great review for what is likely a fast-approaching comprehensive final exam. And second, because of how closely the material in Chapter 23 relates to previously studied material, it tends to be more easily comprehended than previous chapters. So be confident as you read further, and have fun—many students actually enjoy this chapter [*or so they say*]. But first, let's test our understanding of some things we've learned previously.

Test your understanding

Assessment 23.1 Calculate the index of hydrogen deficiency for a molecule having the molecular formula C_6H_6.

Assessment 23.2 Draw resonance structures to identify which of the following is the more stable carbocation.

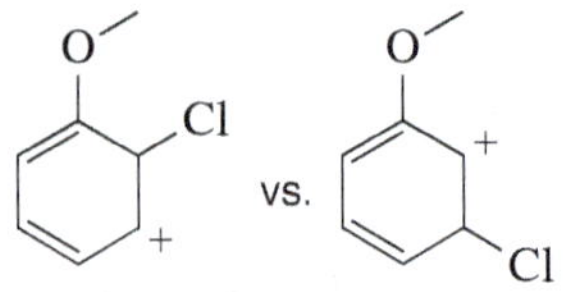

Assessment 23.3 Is the transition state for the reaction shown more reactant-like or more product-like? Justify your answer.

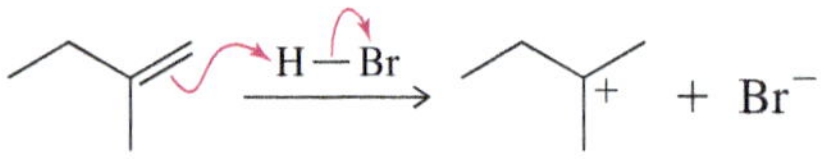

Assessment 23.4 Br_2 is such a good electrophile that it was initially used in addition reactions to alkenes (Section 9.1.2). Which is a better electrophile, Br_2 or Br^+?

Assessment 23.5 Suggest an arrow-pushing mechanism that accounts for the formation of the following Lewis acid–Lewis base complexes. Label the Lewis acid and Lewis base in each.

(a) $Cl{-}Cl + FeCl_3 \longrightarrow Cl{-}\overset{+}{Cl}{-}\overset{-}{Fe}Cl_3$

(b)

(c) $AlBr_3 + Br^- \longrightarrow \overset{-}{Al}Br_4$

Assessment 23.6 Assign the hybridization of the indicated nitrogens in the molecules shown.

(a)

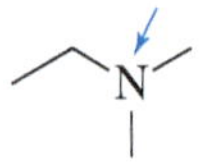

(b)

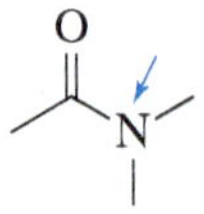

23.1 The discovery of benzene: a very stable molecule

By the mid-19th century, benzene had been discovered independently on three occasions. Michael Faraday (1822) discovered it by isolating an oily residue produced from gas lamps used at the time. Eilhard Mitscherlich (1833) distilled benzoic acid in the presence of lime. And Charles Mansfield and August Wilhelm von Hofmann (1845) isolated benzene from coal tar. The development of combustion analysis in 1822 meant that the molecular formula of benzene (C_6H_6) was known early on; however, determination of its structure lagged behind. The various spectroscopies that we use today (NMR, IR, and UV–vis), as well as mass spectrometry, had not yet been invented, so chemists had to use various other means to confirm the structure of newly discovered molecules. On the basis of its molecular formula, the structures in **Figure 23.1** were proposed for benzene.

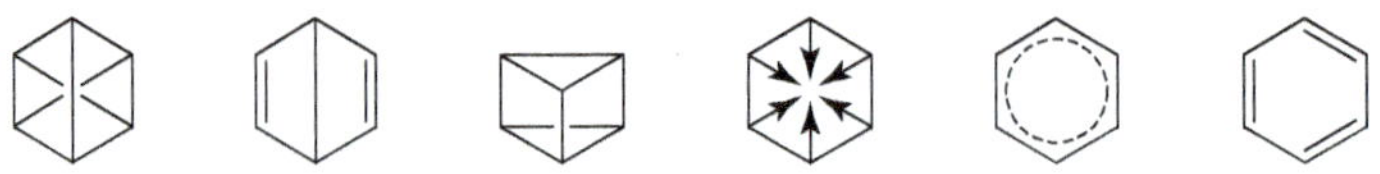

A B C D E F

Figure 23.1 Proposed structures of benzene in the mid-19th century. The structure proposed by Kekulé (**F**) is the actual structure.

To a modern chemist, or even a student of chemistry, some of the structures in Figure 23.1 look unlikely at best, and absurd at worst. Even the most highly regarded scientific minds of the 1800s had barely a fraction of the understanding that a second-semester organic chemistry student has today. [*We stand on the shoulders of giants.*] What these chemists did know is that a molecule with a molecular formula of C_6H_6 had an index of unsaturation of 4. Furthermore, they understood that other unsaturated molecules, like the alkene in **Figure 23.2**, reacted with a solution of bromine, turning it from orange to colorless. Benzene, however, showed no such reactivity. The assumption, then, was that benzene's unsaturation must come from something other than C—C double bonds. Thus, structures like **A**, **C**, and **D** in Figure 23.1 couldn't be ruled out.

Figure 23.2 Unsaturated molecules, like alkenes, react with bromine. Benzene does not.

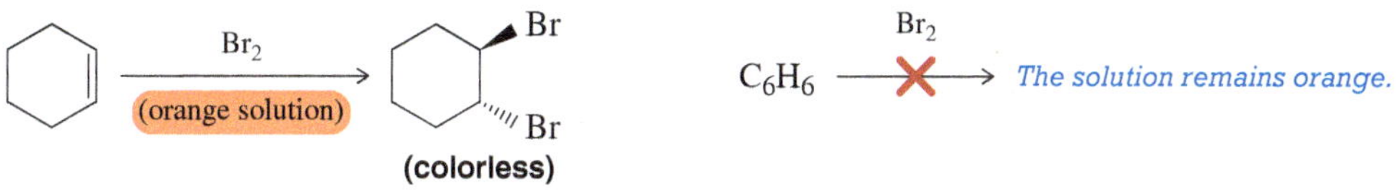

In 1865, Friedrich August Kekulé suggested that benzene was a six-membered hexagonal ring with alternating single and double bonds. [*It is said that this picture of benzene came to him in a dream where he observed a snake biting its own tail.*] Because it was known that there were only three possible isomers of disubstituted benzene, Kekulé posited that benzene existed as the rapid equilibrium between two structures (**Figure 23.3**). This hypothesis underwent a slight, but significant, revision with the discovery of resonance, initially by Johannes Thiele in 1899 and with further development by Linus Pauling in the early 1930s. Instead of being composed of two equilibrating structures, the two Kekulé structures of benzene are now known to be equally contributing resonance structures. As a result, benzene is not either of the Kekulé structures individually, but is instead a combination of the two at all times. For this reason, benzene is sometimes represented as a hexagon with a circle inside.

Figure 23.3 Structure of benzene. Originally proposed as an equilibrating structure, the advent of resonance gave the modern representation of benzene.

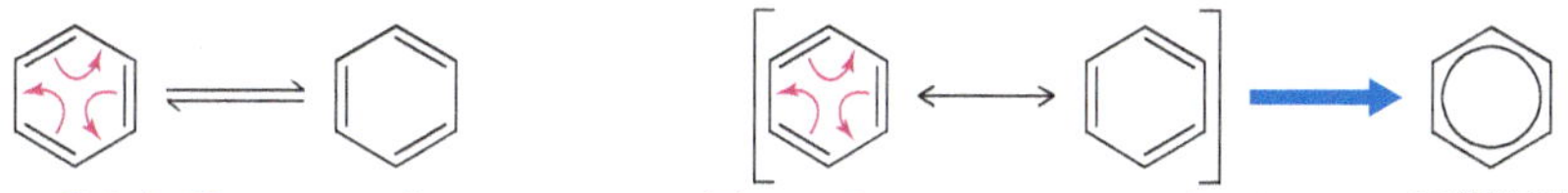

Assessment 23.7 The average C—C single bond length is 1.53 Å, and the average C=C double bond length is 1.31 Å. All of the C—C bonds in benzene are the same length (1.42 Å). Explain.

Assessment 23.8 If Kekulé's original hypothesis had been correct and benzene was really an equilibrium between two structures, how many distinct isomers would exist for 1,2-dichlorobenzene?

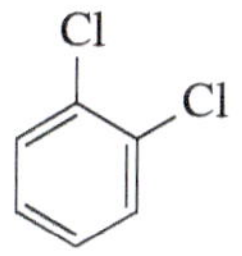

1,2-dichlorobenzene (one isomer)

23.2 Aromaticity

A principal objection to the Kekulé structures of benzene was the presence of double bonds. Alkenes were known at the time to do addition reactions with bromine, but benzene did not. Thus, there was something particularly stable about benzene, even compared to the conjugated alkenes of Chapters 21 and 22. The concept of the 6 π electrons delocalized throughout the ring, making them unreactive with nucleophiles, was one possible explanation. Furthermore, the concept of resonance shows that neither C—C bond in benzene is really a double bond. Instead, they are all halfway between a single and a double bond [*see Assessment 23.7*]. While reasonable, neither of these is sufficient to explain just how much more stable benzene is than would be expected. The following sections describe just how stable benzene is and provide rules for when a molecule is expected to be that stable.

23.2.1 Stability

In Section 9.2.1, and again in Section 21.2.1, we learned that the heat of hydrogenation ($\Delta H°_{hydr}$) can be used to determine alkene stability. [*Review Assessment 21.7.*] To see how stable benzene is, we can treat the hydrogenation of benzene as equivalent to the hydrogenation of three molecules of cyclohexene. The $\Delta H°_{hydr}$ of cyclohexene is -28.5 kcal/mol (-120 kJ/mol). Thus, ignoring the effects of conjugation from Section 21.2.1, you'd expect a $\Delta H°_{hydr}$ of benzene to be $3 \times (-28.5 \text{ kcal/mol}) = -86.7$ kcal mol (-360 kJ/mol). And yet (**Figure 23.4**), this value is actually -49.5 kcal/mol (-208 kJ/mol). The difference between these two values, 36.2 kcal/mol (152 kJ/mol), is referred to as the resonance energy of benzene and represents how much less heat is released when benzene is reduced. [*What does this mean?*]

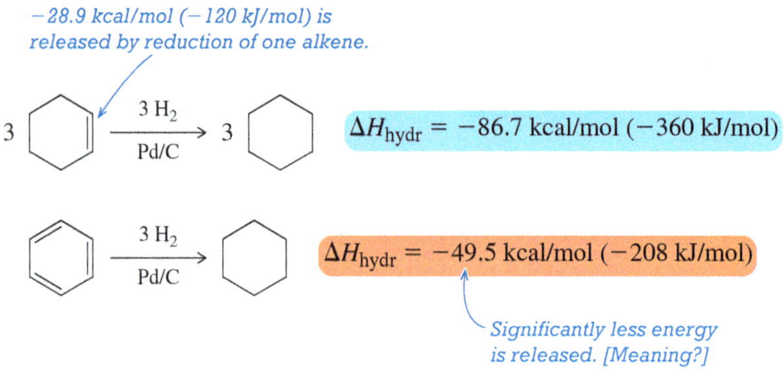

Figure 23.4 **Based on a comparison to three molecules of cyclohexene,** benzene should have $\Delta H°_{hydr} = -86.7$ kcal/mol.

The comparison in Figure 23.4 makes an assumption that ignores the effects of conjugation on stability that were presented in Section 21.2.1. If we treat the stepwise reduction of benzene as the reduction of a conjugated triene, followed by a conjugated diene, then cyclohexene, we'd expect $\Delta H°_{hydr} = -80.4$ kcal/mol (-338 kJ/mol) for benzene (**Figure 23.5**). This number is closer than -86.7 kcal/mol to -49.5 kcal/mol, but it is still 31 kcal/mol (7.4 kJ/mol) away from the actual heat of hydrogenation of benzene.

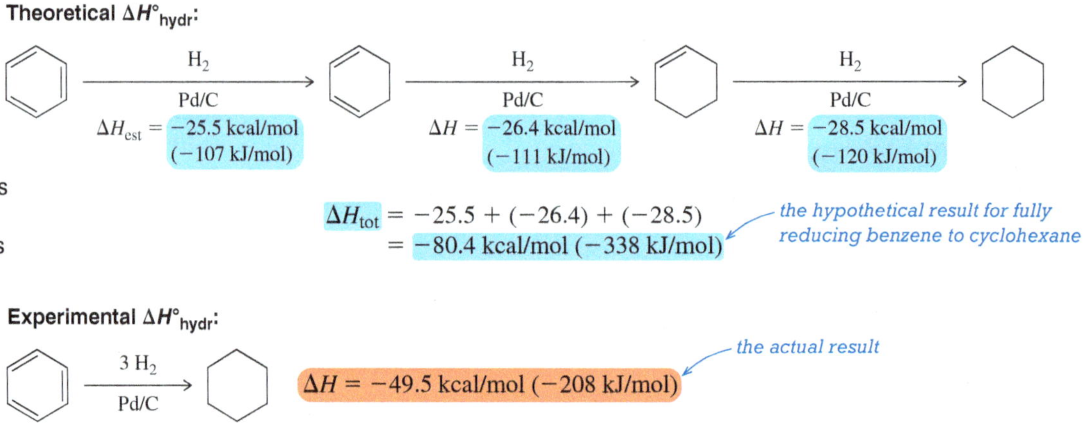

Figure 23.5 **Treating benzene as a conjugated triene** still does not account for its experimental $\Delta H°_{hydr}$. $\Delta H_{est} = -25.5$ kcal/mol is the value estimated for reducing one of the double bonds of a conjugated triene. $\Delta H = -26.4$ kcal/mol is the value obtained from Figure 21.6 for reducing one of the double bonds of a conjugated diene.

We can calculate $\Delta H°_{hydr}$ for the first step using the known values for the reduction of cyclohexadiene and comparing them with $\Delta H°_{hydr}$ of benzene. Using the math shown in **Figure 23.6**, the only way that benzene has $\Delta H°_{hydr} = -49.5$ kcal/mol (-208 kJ/mol) is if the first step (reduction of the triene to diene) has $\Delta H°_{hydr} = +5.4$ kcal/mol ($+23$ kJ/mol). [*That's not a typo.*] Hydrogenation of an alkene, one of the more exothermic reactions we have studied, is *actually endothermic* for the first double bond of benzene. Plotted on a reaction coordinate diagram in comparison to cyclohexene further illustrates just how stable benzene must be in order for the first reduction to be an uphill process. This stability arises because benzene is an **aromatic** molecule. All aromatic molecules show the enhanced stability of benzene.

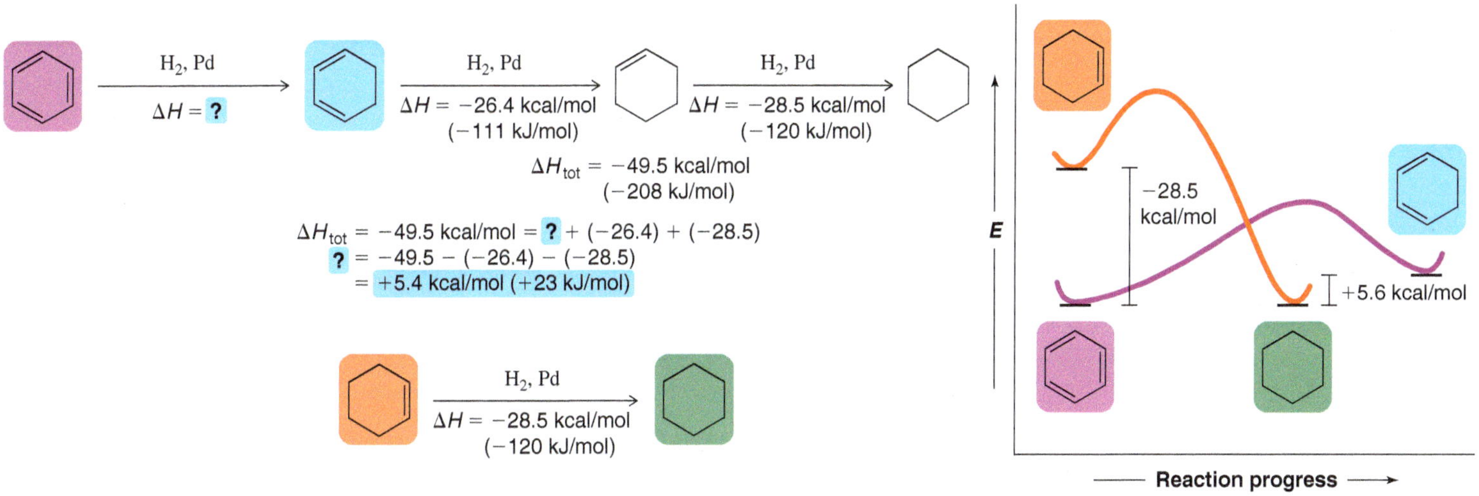

Figure 23.6 Benzene starts much lower on the reaction coordinate diagram and is therefore significantly more stable than other alkenes.

23.2.2 Hückel's rules for aromaticity

Table 23.1 Hückel's rules for aromaticity.

A molecule is aromatic if:
1. It is planar.
2. It is cyclic.
3. Every atom in the ring has a filled or unfilled *p* orbital.
4. It contains $4n + 2$ electrons in the cyclic, conjugated π system.

The enhanced stability of benzene explains why early chemists had trouble determining its structure. The discovery of other molecules with a similar stability profile led German chemist Erich Hückel to develop rules for aromaticity. These rules have successfully predicted which molecules will show enhanced stability beyond what is predicted by normal valence bond theory (Table 23.1).

To understand these rules, we can apply them first to benzene (Figure 23.7). Each carbon is sp^2 hybridized and has a trigonal planar shape. Thus, benzene is a planar molecule (rule 1). It is a ring (rule 2), too. Each carbon atom has a *p* orbital used for forming the three C—C π bonds (rule 3). With 6 electrons in the π cloud, setting $4n + 2 = 6$ and solving for n gives an integer value of $n = 1$. Thus, rule 4 is satisfied. Having satisfied all four rules, benzene is an aromatic molecule and can be expected to have enhanced stability.

It is said that Hückel's inability to translate mathematical principles into common, understandable language hampered the general acceptance of his ideas. This is why it is so important for scientists to have good written and oral communication skills.

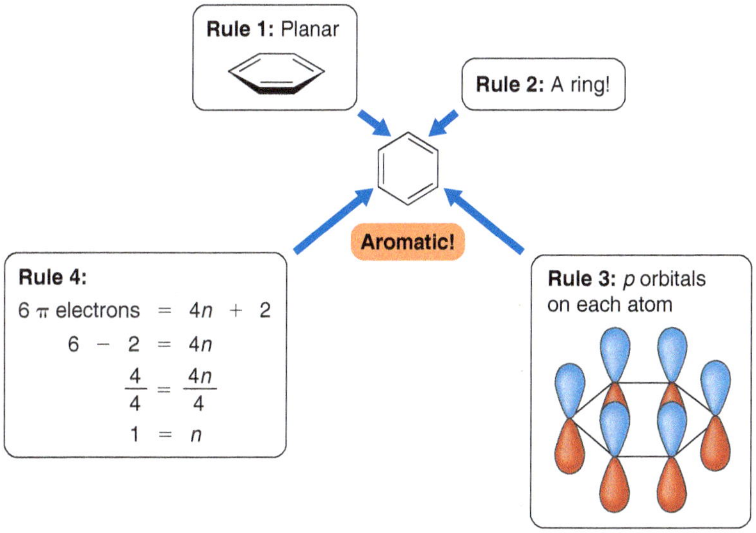

Figure 23.7 The application of Hückel's rules to benzene.

The most challenging part of identifying aromatic molecules is the $4n + 2$ rule. Students consistently ask what n is. It really has no relevance, except that when the total number of electrons is set equal to $4n + 2$ and n is an integer, the molecule is aromatic [*assuming all other conditions are met*]. Table 23.2 shows this calculation for several aromatic molecules that meet rules 1–3 of aromaticity.

Table 23.2 Calculation of 4n + 2 for potentially aromatic molecules.

(a)	(b)	(c)	(d)
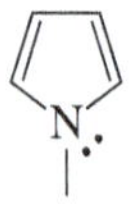	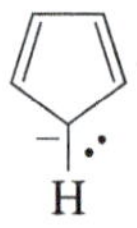	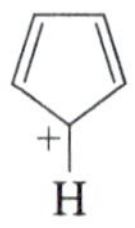	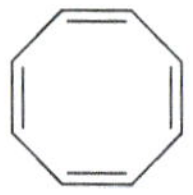
2 π bonds = 4 e^- 1 lone pair = 2 e^-	2 π bonds = 4 e^- 1 lone pair = 2 e^-	2 π bonds = 4 e^-	4 π bonds = 8 e^-
6 e^- = 4n + 2 n = 1	6 e^- = 4n + 2 n = 1	4 e^- = 4n + 2 n = 1/2	8 e^- = 4n + 2 n = 3/2
Aromatic!	Aromatic!	Not aromatic	Not aromatic

[*We find out in Section 23.3 that the cyclopentadienyl cation (c) and octatetraene (d), if planar, are actually antiaromatic . . . stay tuned!*]

23.2.3 Tour of aromatic molecules

In addition to benzene, several other molecules show the same enhanced stability. The purpose of this section is twofold. First, it introduces the types of aromatic ring systems that exist, along with their common names. Additionally, each type of aromatic molecule encountered will provide more examples of how Hückel's rules are applied.

Though they are still aromatic, adding more rings reduces the resonance energy as compared to benzene itself.

23.2.3.1 Hydrocarbons

The other known aromatic hydrocarbons basically consist of a varying number of benzene rings (**Figure 23.8**). Naphthalene, for example, is composed of two fused benzene rings, making it aromatic. Additionally, when you treat it as one large, ten-membered ring, there is a total of 10 π electrons (five π bonds). Ten π electrons give a value of n = 2 when set equal to 4n + 2. A similar process can be used to show that phenanthrene and anthracene [*two polycyclic aromatic hydrocarbons; see Chemistry in Context 23-A*] are also aromatic.

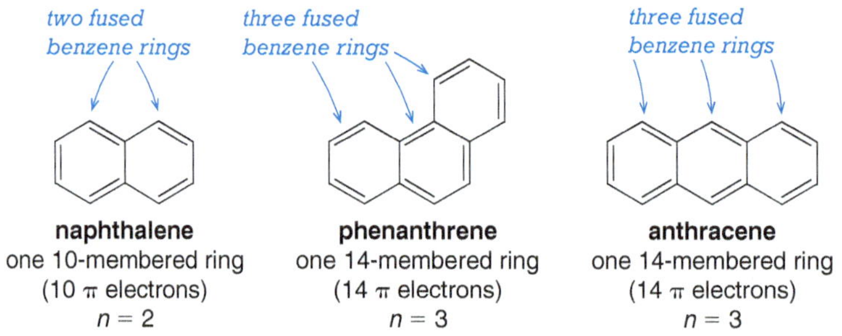

Figure 23.8 The aromatic hydrocarbons naphthalene, phenanthracene, and anthracene.

According to Hückel's rules, larger rings can also be aromatic, as long as they exist in a planar conformation. The ten-membered ring in **Figure 23.9** is *not* aromatic because it cannot be planar. Because of the trans configuration of the central double bonds, there is a substantial steric clash between two hydrogens trying to exist in the same space in the planar conformation. As a result, the molecule twists itself out of a planar conformation. This twisting prevents the p orbitals from aligning, making the molecule nonaromatic. When C_1 and C_6 are connected by a carbon bridge, the π system is nearly planar, allowing this molecule to possess some aromatic character.

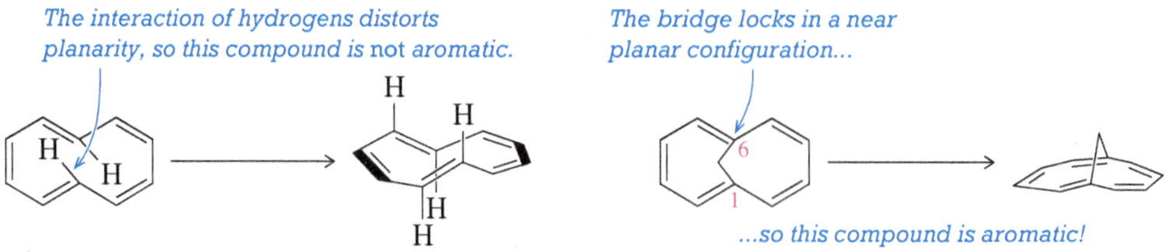

Figure 23.9 A nonplanar π system is not aromatic, but can become aromatic if it is constrained in a nearly planar conformation.

23-A CHEMISTRY IN CONTEXT: Green Chemistry

Polycyclic aromatic hydrocarbons

Polycyclic aromatic hydrocarbons (PAHs) (**Figure 23.10**) consist of multiple fused benzene rings. Although they occur naturally as a result of forest fires and volcano eruptions, a primary source of PAHs in the environment is the coal and oil industry. PAHs are found in wood smoke, cigarette smoke, coal tar used for roofing, and asphalt used on roads. Being somewhat inert and highly lipophilic, these compounds are classified as persistent organic pollutants (POPs).

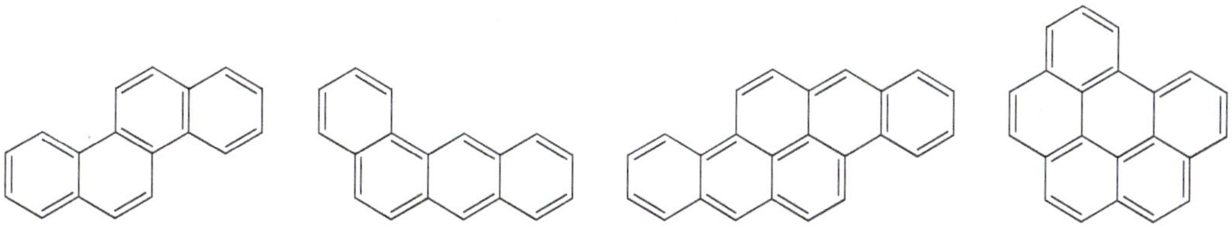

Figure 23.10 PAHs are suspected to be cancer-causing agents.

As early as 1775, when chimney sweeps were found to have an abnormally high occurrence of testicular cancer, the compounds contained within soot were suspected carcinogens. PAHs were later identified as chemicals present in soot. They have been conclusively linked to various cancers, resulting from intercalation of the PAHs between DNA base pairs. The danger posed by PAHs in the environment made it necessary to develop techniques to remove them, especially in sites of heavy contamination. Four remediation technologies have been used to clean PAH-contaminated soils. These are listed in **Table 23.3**.

Table 23.3 Sample methods for cleaning PAH-contaminated soil.

Classification	Example technology
Physical–chemical	Extraction from soil with vegetable oil
Biological	Phytoremediation: use of plants to absorb the contaminants
Chemical	Chemical oxidation with $KMnO_4$ or H_2O_2 to increase polarity/water solubility
Thermal	Incineration to complete combustion

23.2.3.2 Heteroaromatics

Heteroaromatics, still aromatic, also have a lower resonance energy as compared to benzene.

Heteroaromatic rings are aromatic rings that contain at least one non-carbon atom. Heteroatoms like nitrogen, oxygen, and sulfur each carry at least one lone pair that may or may not be a part of the aromatic ring. In the case of pyridine [**Figure 23.11(a)**], the lone pair on nitrogen is placed in an sp^2-hybridized orbital because the C—N π bond is already a part of the 6-electron $(4n + 2)$ π system. Alternatively, the lone pair on nitrogen in pyrrole [**Figure 23.11(b)**] is required to complete a 6-electron $(4n + 2)$ π system.

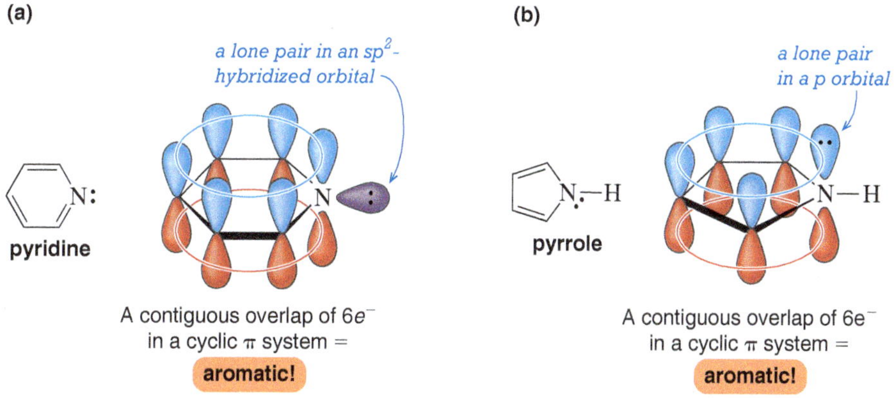

Figure 23.11 Heteroaromatic rings. (a) Pyridine's three C—C π bonds account for the 6 $(4n + 2)\pi$ electrons. (b) Pyrrole uses two C—C π bonds and one lone pair for its 6 $(4n + 2)\pi$ electrons.

Using pyridine and pyrrole as a guide, when does a lone pair count in the π system and when does it *not* count? All molecules hybridize in such a way to be as stable as possible. [*The ball rolls downhill.*] In Chapter 2, when first studying resonance, we showed that

VSEPR-predicted hybridization would be avoided if an alternate hybridization would allow delocalization by resonance. The same principle is at work here. If the sulfur in thiophene (Figure 23.12), for example, exists in the VSEPR-predicted sp^3 hybridization, there is no contiguous system of p orbitals in the ring. Alternatively, by hybridizing as sp^2, one lone pair sits in a p orbital, making a contiguous system of 6 $(4n + 2)$ π electrons, thus making thiophene aromatic.

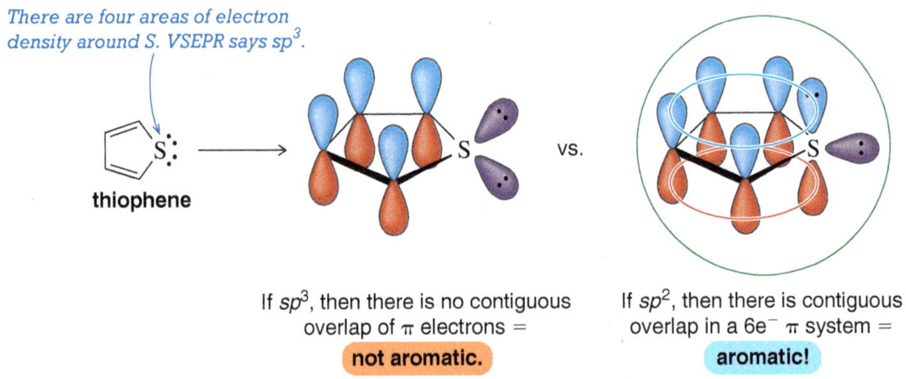

Figure 23.12 Thiophene is aromatic because the sulfur is sp^2 hybridized with a lone pair in a p orbital.

Assessment 23.9 Imidazole is a heteroaromatic base. Which nitrogen, a or b, is most basic?

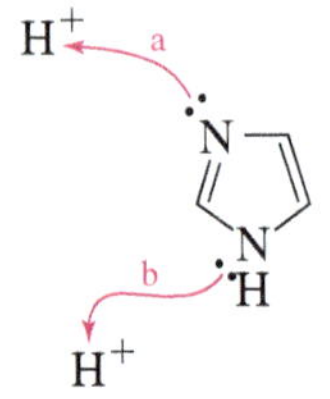

Assessment 23.10 Between pyrrole and pyrrolidine, which nitrogen would be most nucleophilic? Why?

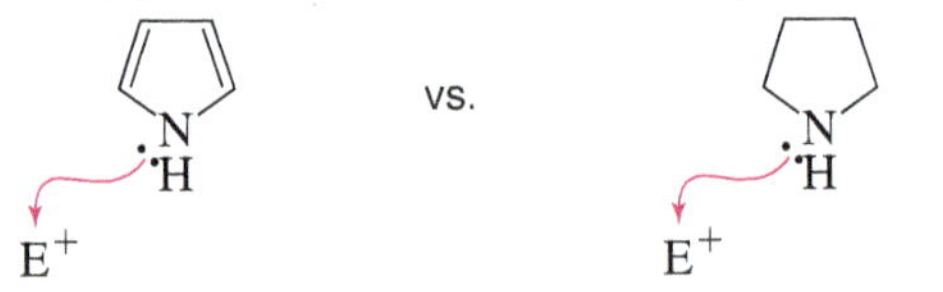

23.2.3.3 Charged aromatics

Aromaticity is not limited to neutral molecules. Several anionic aromatic molecules are shown in Figure 23.13(a). For these compounds, the anionic carbon hybridizes in a way that places a lone pair in a p orbital, allowing it to be part of the $4n + 2$ electron π system. Alternatively, when a carbocation is part of an aromatic ring [Figure 23.13(b)], it hybridizes so that the empty p orbital is part of the $4n + 2$ electron π system, even though it contributes 0 electrons to that π system.

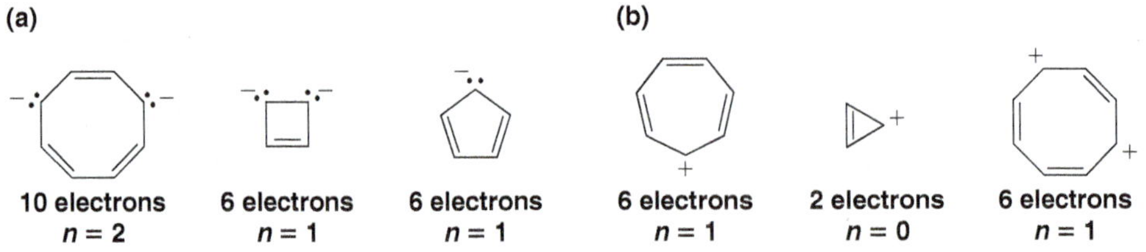

Figure 23.13 Charged aromatic rings. (a) When in a p orbital, the lone pair of a carbanion contributes 2 electrons to the $4n + 2$ π system. (b) Carbocations, with empty p orbitals, help complete the contiguous π system but contribute 0 electrons to the $4n + 2$ π system.

23.2.2–23.2.3 SUMMARY

- Planar, cyclic molecules featuring a contiguous π system of $4n + 2$ electrons are aromatic.

Assessment 23.11 (i) Is the molecule aromatic, yes or no? (ii) For the aromatic molecules, solve for n in Hückel's rule. For the other molecules, explain which of the rules of aromaticity is being broken.

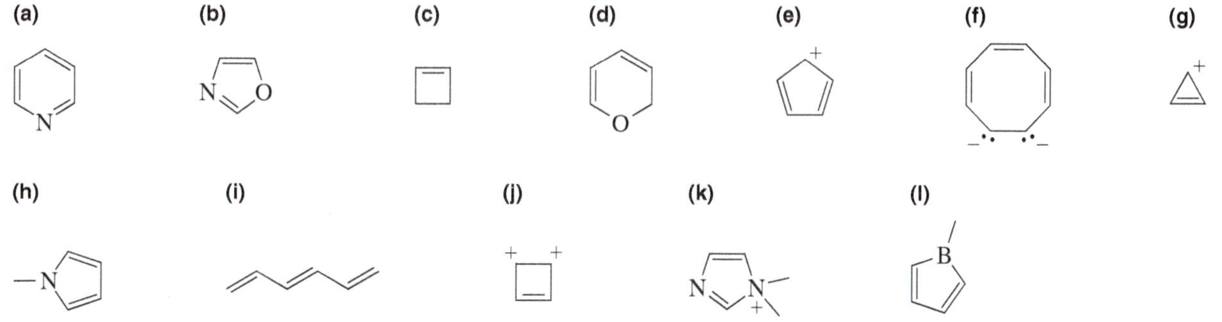

Assessment 23.12 Which of the following concerted reactions would have a more stable transition state? Why?

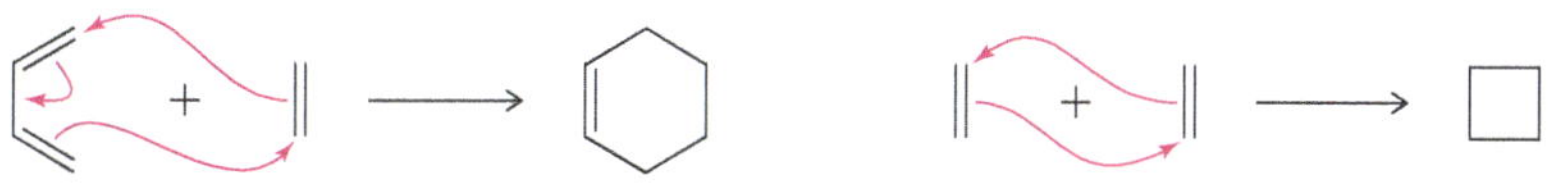

Assessment 23.13 Cyclopentane has pK_a = 50, whereas cyclopentadiene has pK_a = 16. Explain this difference.

cyclopentane pK_a = 50   cyclopentadiene pK_a = 16

Assessment 23.14 Which resonance structure, **A** or **B**, is most contributing?

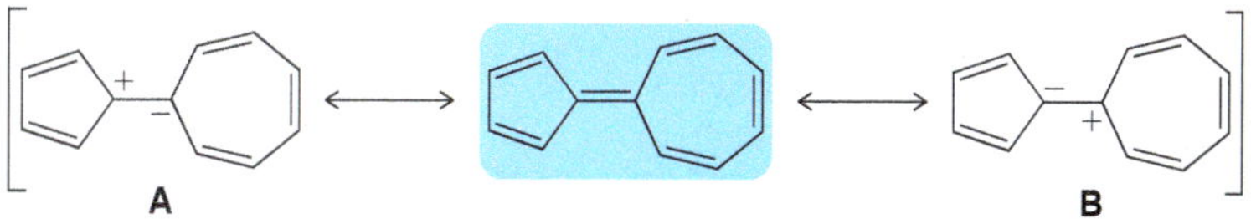

A   **B**

Assessment 23.15 What is the hybridization of the indicated nitrogen? Why does it hybridize in this way?

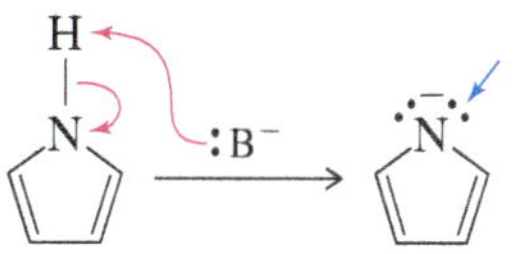

23.3 Antiaromaticity

Figure 23.14 shows three classes of molecules on a relative stability scale. Thus far, we have studied the reactivity of two of these classes—namely, aromatic and nonaromatic molecules. Aromatic molecules, the ones highlighted in Section 23.2, are extremely stable—*way more* stable than we'd have expected. The rest of the molecules in the book have been nonaromatic. They do not adhere to Hückel's rules, and they are exactly as stable as we expect based on valence bond theory, angle strain, bond-dissociation energies, and so on. Antiaromatic molecules, the third class of molecules, are *way less* stable than we'd expect them to be. [*Why might an effect that makes molecules extremely unstable have been discovered much later than aromaticity?*]

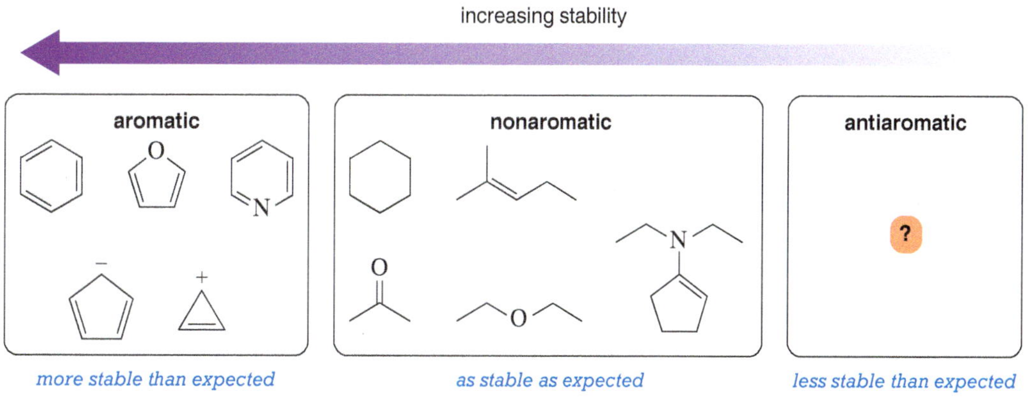

Figure 23.14 Aromatic, nonaromatic, and antiaromatic molecules on a spectrum of stability.

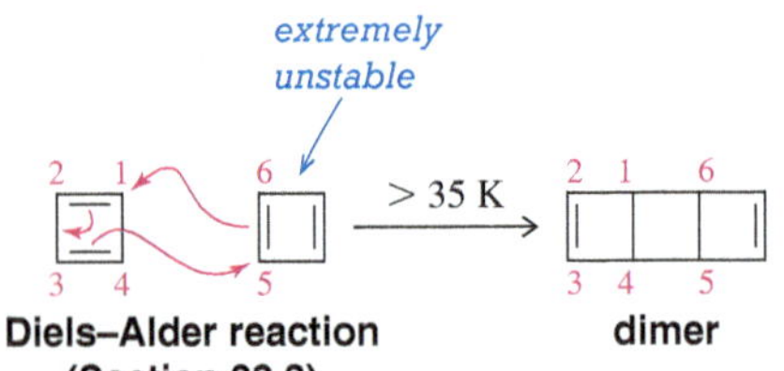

Figure 23.15 Cyclobutadiene, an antiaromatic molecule, rapidly dimerizes by a Diels–Alder reaction due to its instability.

23.3.1 Instability

The instability of antiaromatic molecules is best illustrated by looking at cyclobutadiene. [*How unstable is it?*] At temperatures above 35 K (−228 °C), it dimerizes through a Diels–Alder reaction (Section 22.3) to give the molecule shown in Figure 23.15. That's right—cyclobutadiene on its own, with the angle strain of a four-membered ring, is so unstable that it would prefer to be a dimer containing three of those angle strain–inducing four-membered rings. [*In other words, it's extremely unstable.*] Recent studies have called into question this traditional interpretation and instead have suggested that angle strain, torsional strain, and repulsion strain account for its instability instead of antiaromaticity. So let's consider some other examples.

The cyclopentadienyl carbocation in Figure 23.16(a) is also more unstable than it should be as a result of being antiaromatic. In Chapter 21, the ability to draw multiple resonance structures for carbocations conjugated to alkenes made them quite stable. As such, it is surprising that the cyclopentadienyl carbocation is so unstable. In Figure 23.16(b), the pK_a of propene is around 43, with the conjugate base stabilized by resonance with a neighboring alkene. It is surprising, then, that the pK_a of cyclopropene is 61—even higher than the pK_a of methane itself. This stands in stark contrast to cyclopentadiene, where the aromatic conjugate base makes the removal of a proton much more favorable (Assessment 23.13). [*Do you see a difference, besides ring strain, that might account for the difference in stability between the cyclopentadienyl anion and the cyclopropenyl anion?*]

Figure 23.16 Examples of antiaromaticity. (a) The cyclopentadienyl carbocation is very unstable, despite having a resonance-stabilized carbocation. (b) The conjugate base of cyclopropene is extremely unstable, too, despite being stabilized by resonance (as in propene).

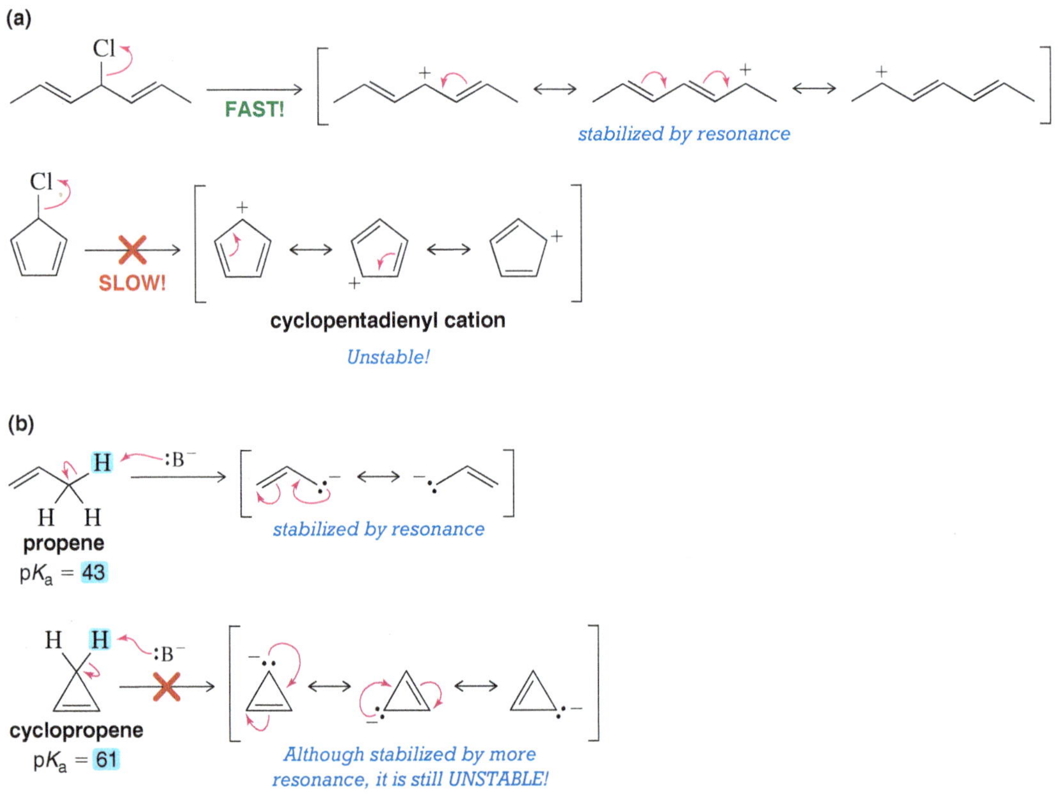

Recall (Section 23.2) that aromatic molecules have electrons in a contiguous, cyclic π system. Rings with $4n + 2$ π electrons (i.e., 2, 6, 10, . . . π electrons, when $n = 0, 1, 2, \ldots$) are aromatic [*assuming the other Hückel rules are adhered to.*] Figures 23.15 and 23.16 show, on the other hand, that antiaromatic molecules have 4 π electrons. Thus, the $4n + 2$ rule does not apply to these molecules.

Think about that for a minute. Delocalization by resonance, in virtually every situation in the book, makes molecules more stable. In antiaromatic systems, however, resonance makes molecules significantly less stable.

23.3.2 Hückel's/Breslow's rules for antiaromaticity

In 1967, chemist Ronald Breslow amended Hückel's rules for aromaticity to account for systems in which delocalization of electrons made a molecule less stable. These modified

Table 23.4 Hückel's/Breslow's rules for antiaromaticity.

A molecule is antiaromatic if:
1. It is planar.
2. It is cyclic.
3. Every atom in the ring has a filled or unfilled p orbital.
4. It contains $4n$ electrons in the cyclic, conjugated π system.

rules (Table 23.4) have successfully predicted which molecules will show diminished stability beyond what is predicted by normal valence bond theory. In essence, they predict which molecules shouldn't exist. The only difference between the rules for antiaromaticity and the rules for aromaticity (Table 23.1) is in rule 4.

To understand these rules, let's first apply them to the cyclopentadienyl carbocation, as shown in Figure 23.17. Each carbon is sp^2 hybridized and has a trigonal planar shape. Thus, it is a planar molecule (rule 1). It is a ring (rule 2). Each carbon atom has a p orbital used for forming the two C—C π bonds and the carbocation (rule 3). With 4 electrons in the π cloud, setting $4n = 4$ and solving for n gives an integer value of $n = 1$. Thus, rule 4 is satisfied. Having satisfied all four rules, the cyclopentadienyl carbocation is an antiaromatic molecule and can be expected to be very unstable.

Figure 23.17 The application of Hückel's/Breslow's rules to the cyclopentadienyl cation.

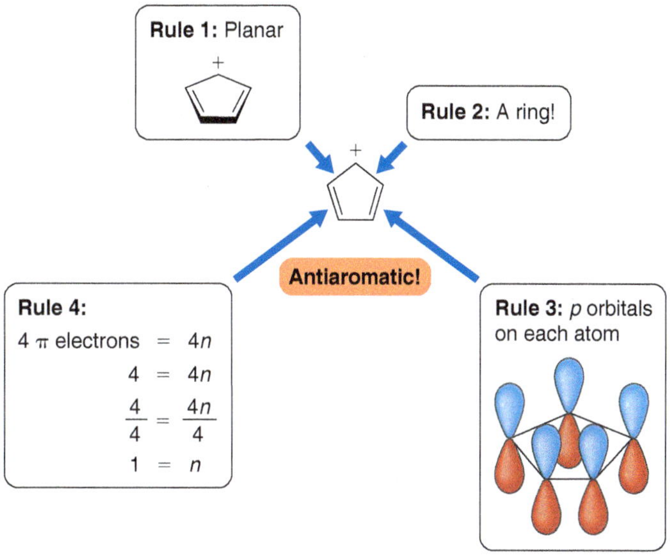

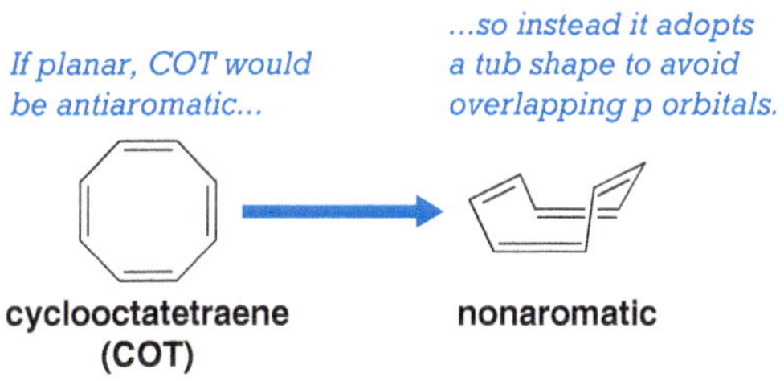

Figure 23.18 Cyclooctatetraene (COT) adopts a tub-shaped conformation to avoid being antiaromatic.

There is still a substantial amount of debate surrounding whether molecules can truly be antiaromatic. One reason is that they are extremely hard to prepare and study. [*Does it make sense why?*] Anything that is added to the molecule to make it antiaromatic gives it the chance to modify itself in some way to avoid adhering to the rules for antiaromaticity in Table 23.4. Cyclooctatetraene (COT), for example, has a contiguous system of 8 π electrons ($n = 2$ when $4n = 8$), is a ring, and has a p orbital on each carbon (Figure 23.18). And although we draw it in a way that makes it look like it is planar, it is in fact tub-shaped. [*Why?*] If it were planar, it would be antiaromatic, highly unstable, and nonexistent. So it adopts the tub-shaped conformation to be as stable as possible. [*The ball is still rolling downhill.*]

23.3 SUMMARY

- Planar, cyclic molecules featuring a contiguous π system of $4n$ electrons are antiaromatic and highly unstable.

Assessment 23.16 (i) Classify the following molecules as aromatic, nonaromatic, or antiaromatic. (ii) For the aromatic and antiaromatic molecules, solve for n in Hückel's/Breslow's rule. For the other molecules, explain which of the rules of aromaticity is being broken.

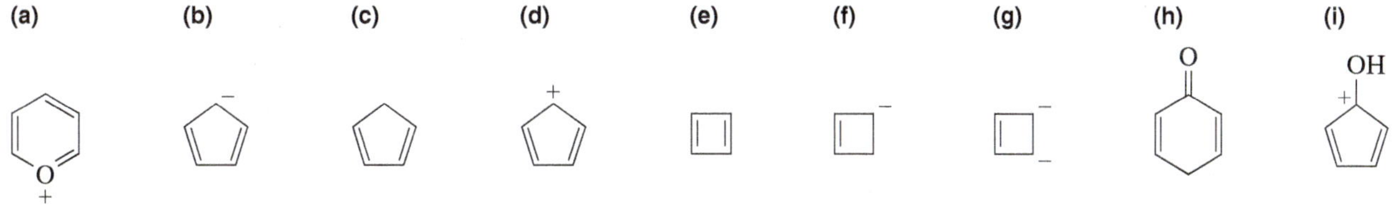

Assessment 23.17 If cyclobutadiene is nonexistent at temperatures greater than −238 °C, why can the cyclobutadiene dication be observed at temperatures as high as −10 °C?

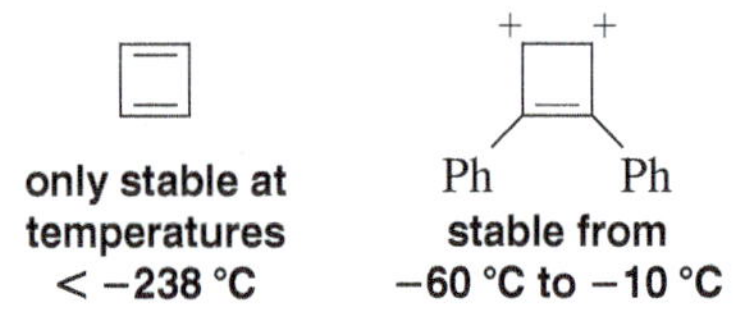

Assessment 23.18 Why were aromatic molecules discovered before antiaromatic molecules?

Assessment 23.19 In benzene, all of the C—C bonds are the same length (~1.42 Å) [*see Assessment 23.7*]. Recent studies have shown that cyclobutadiene has two distinct bond lengths. Account for this fact.

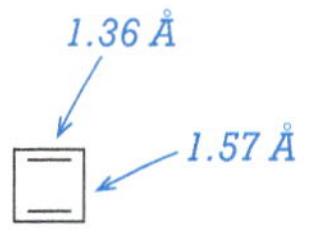

Assessment 23.20 In Sections 22.3 and 22.4.1, we learned that [4 + 2] cycloadditions occur under thermal conditions, but [2 + 2] cycloadditions require photochemical conditions. Explain this in the context of Sections 23.2 and 23.3.

Table 23.5 Summary of the rules for drawing molecular orbital pictures of conjugated π systems.

1. Combining *n* atomic orbitals gives *n* molecular orbitals.
2. More overlap gives more stable orbitals.
3. The number of nodes increases by one as energy increases.
4. Each molecular orbital has symmetry around the node(s).
5. The delocalized π electrons are added to each molecular orbital (2 electrons in each), beginning with the lowest energy MO.
6. Molecular orbitals lower in energy than nonbonded atomic orbitals ($E = 0$) are bonding; molecular orbitals higher in energy than nonbonded atomic orbitals ($E = 0$) are antibonding.

23.4 Molecular orbitals of aromatics and antiaromatics

We first covered the rules for drawing the molecular orbital picture for conjugated π systems in Section 21.3. [*For a complete review, look back at this section before reading further.*] These rules are summarized in **Table 23.5**. The most stable molecular orbital in benzene is the one with maximum overlap, which also has zero nodes (**Figure 23.19**). The least stable molecular orbital has no overlap, due to six nodes. The only [*somewhat confusing*] change when these rules are applied to

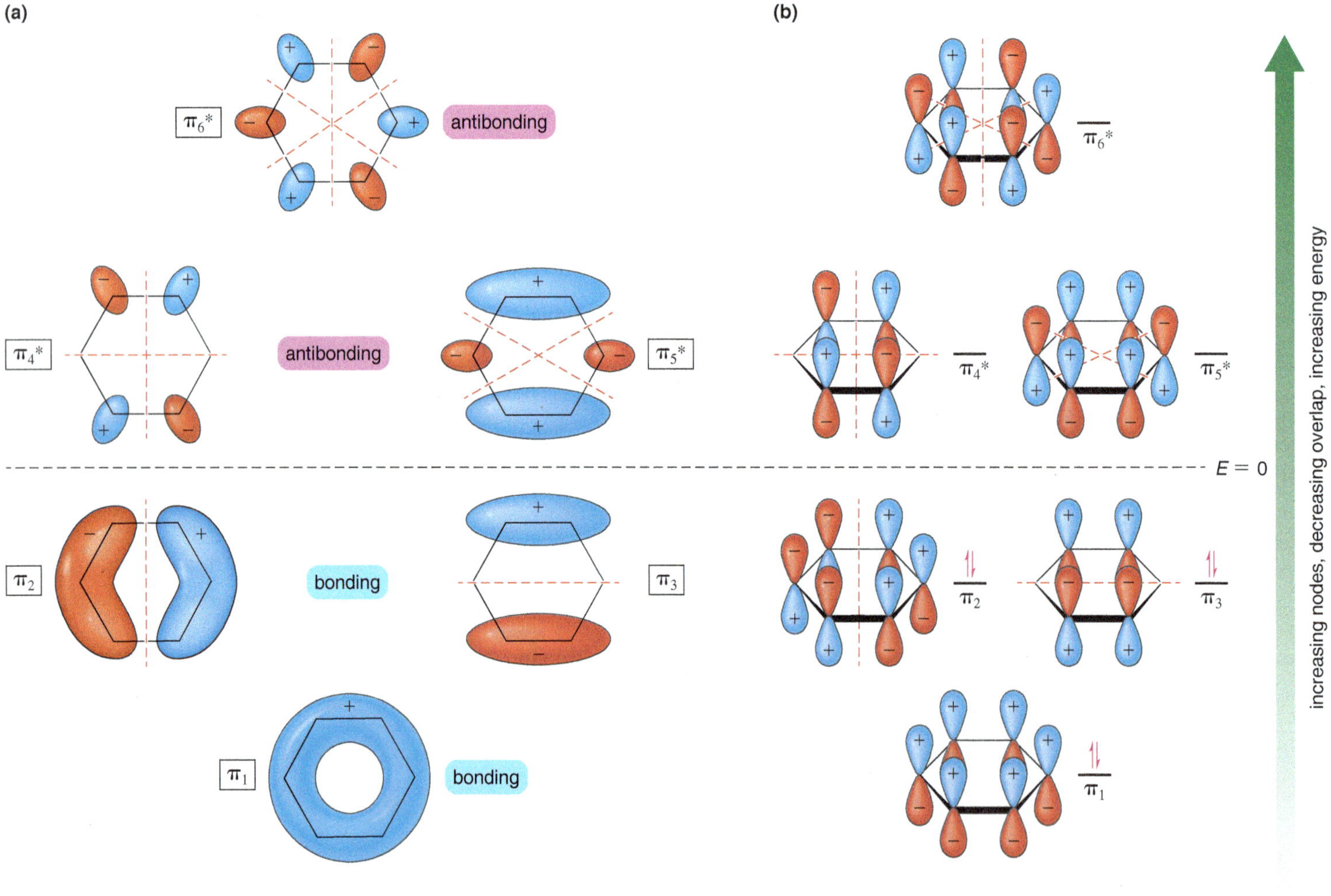

Figure 23.19 The molecular orbital picture of benzene. (a) Top view showing an increase in the number of nodal planes as the energy increases. (b) Side view showing molecular orbitals filled starting with the lowest energy orbitals. [*The red dashed lines represent nodes.*]

benzene deals with the nodal planes. Because benzene is a ring, each nodal plane cuts through the molecular orbitals in two places. [*Thus,* π_6* *uses three nodal planes to create six total nodes.*] As a result, one nodal plane creates two areas of no overlap, as shown in π_2 and π_3. The second complication is that there are now two ways [*symmetry maintained in each*] to draw each nodal plane. The first way, as in π_2, cuts through bonds on opposite sides of the ring. The second, as in π_3, cuts through two atoms on opposite sides of the ring. This creates two **degenerate orbitals**, another term for orbitals of equal energy.

In similar fashion, increasing the number of nodal planes to two in π_4* and π_5* gives four areas where there is no overlap. Once more, the two ways of cutting through the ring give orbitals of equal energy (π_4* and π_5*). The molecular orbitals are filled starting from the lowest energy, with π_1, π_2, and π_3 being bonding orbitals. The unfilled orbitals (π_4*, π_5*, and π_6*) are antibonding orbitals because the energies are higher than the nonbonding *p* atomic orbitals. Having three molecular orbitals below the nonbonding level ($E = 0$) explains the enhanced stability of benzene. [*We compare this to cyclobutadiene next.*]

For cyclobutadiene (**Figure 23.20**), just as for benzene, the most stable molecular orbital (π_1) has maximum overlap, and the least stable molecular orbital (π_4) has no overlap. There are, once more, two ways to draw one nodal plane cutting through the ring, creating two orbitals of equal energy (π_2 and π_3). These two molecular orbitals are equidistant between π_1 (bonding, stable) and π_4* (antibonding, unstable), so they sit right at $E = 0$ (nonbonding). Furthermore, they are degenerate, so when we place the 4 electrons (from the four participating *p* orbitals), π_2 and π_3 each get 1 electron. From this picture, cyclobutadiene can be better thought of as having a pair of nonbonded free radicals. Consistent with its antiaromaticity, cyclobutadiene is a highly unstable molecule.

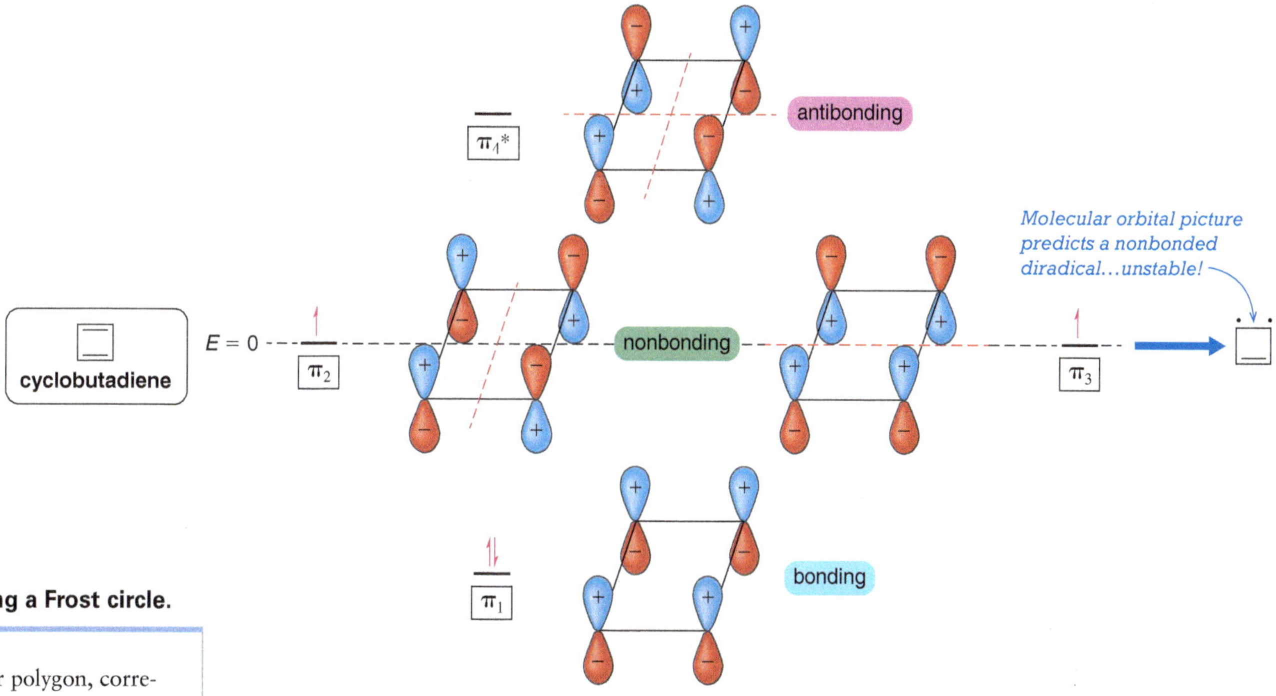

Figure 23.20 The molecular orbital picture of cyclobutadiene. Placing 1 electron in each of the degenerate orbitals (π_2 and π_3) creates nonbonding orbitals, that, like a diradical, are extremely unstable. [*The red dashed lines represent nodes.*]

Table 23.6 Drawing a Frost circle.

1. Draw a circle.
2. Inscribe a regular polygon, corresponding to the ring size of the molecule of interest, in the circle with one vertex pointing down.
3. Each intersection of the polygon with the circle represents a molecular orbital.
4. $E = 0$ bisects the circle.
5. Fill the molecular orbitals starting with the lowest energy (most stable) molecular orbital.

Although there is a lot of math that goes into drawing the molecular orbital pictures in Figures 23.19 and 23.20, a simple way of predicting the relative energies of aromatic and antiaromatic molecular orbitals, called **Frost circles**, was devised by A. A. Frost in 1953. The rules for drawing Frost circles are listed in **Table 23.6**, and benzene is shown as a representative example in **Figure 23.21**.

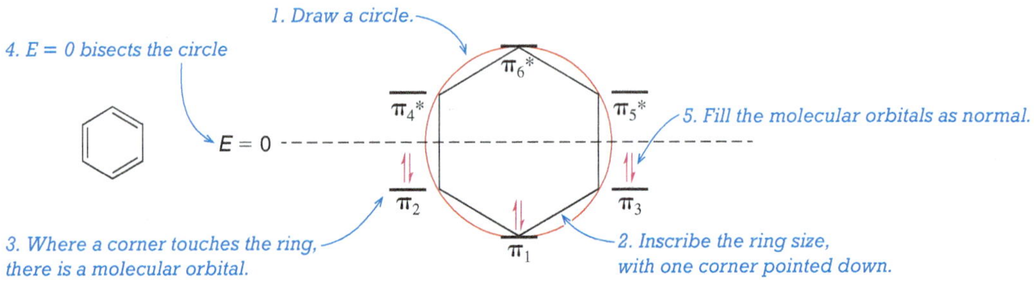

Figure 23.21 The Frost circle for benzene predicts the relative energies of its molecular orbitals.

Using the rules in Table 23.6, the relative energies of molecular orbitals can be quickly drawn for several molecules, as shown in **Figure 23.22**. In each (and all) antiaromatic species, there will be two degenerate molecular orbitals that each have a single electron (nonbonded MOs). This is a function of having $4n$ π electrons instead of $4n + 2$ π electrons. [*Cyclooctatetraene is really nonaromatic because it is nonplanar. Figure 23.22 shows that attaining nonplanarity is important because it avoids antiaromaticity.*] For all aromatic species, all orbitals below $E = 0$ will be full, bonding MOs.

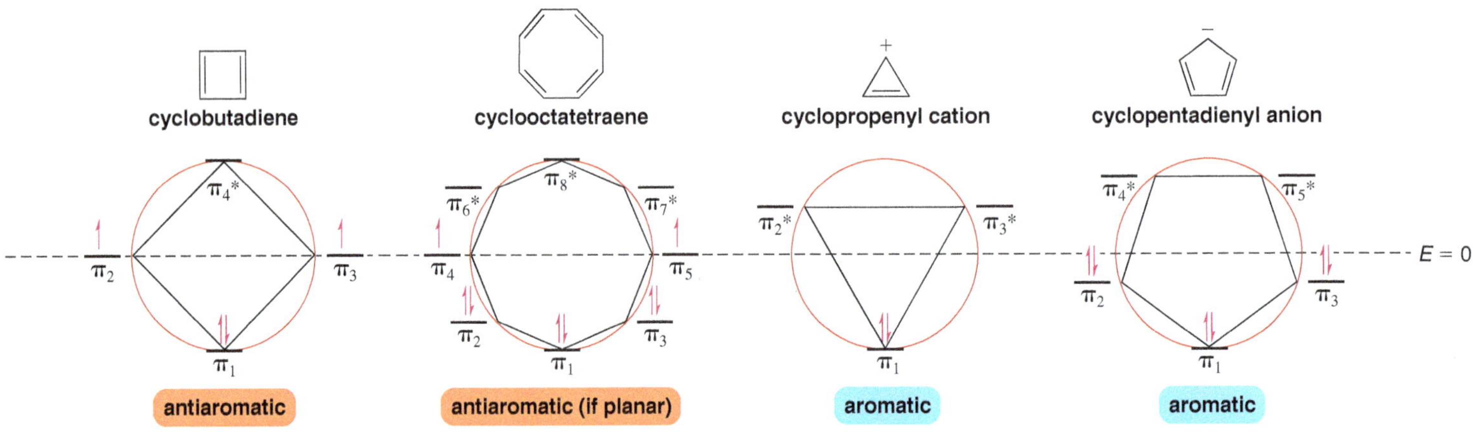

Figure 23.22 Frost circles to illustrate aromaticity and antiaromaticity.

Assessment 23.21 Draw a Frost circle for the cyclopropenyl *anion* and compare it to the Frost circle for the cyclopropenyl *cation* (Figure 23.22). What has changed?

Assessment 23.22 Draw a Frost circle for the cyclopentadienyl *cation* and compare it to the Frost circle for the cyclopentadienyl *anion* (Figure 23.22). What has changed?

Assessment 23.23 An incomplete Frost diagram for a seven-membered ring is shown. Include the necessary electrons to make it represent an *aromatic* molecule. Then the draw the structure to which it would correspond.

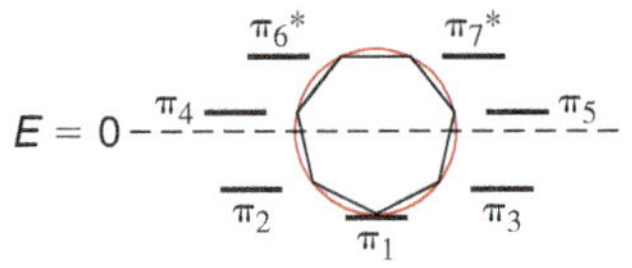

Assessment 23.24 An incomplete Frost diagram for a seven-membered ring is shown. Include the necessary electrons to make it represent an *antiaromatic* molecule. Then the draw the structure to which it would correspond.

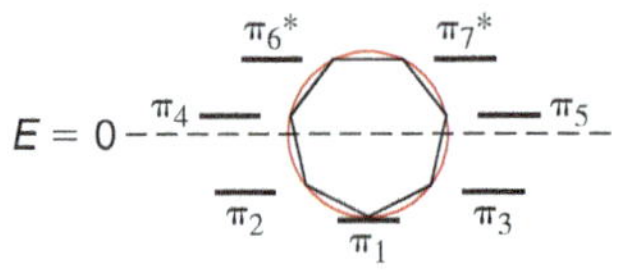

23.5 The nomenclature of benzene

Molecules containing benzene have diverse structures, and there are several ways of correctly naming benzene derivatives. This is a result of the IUPAC allowing a mixture of common names of benzene derivatives to be used within their system. The sections that follow are meant to provide only a brief introduction to benzene nomenclature.

23.5.1 Common names of some aromatics

Because they are so ubiquitous, a variety of benzene derivatives are known by their common name (**Figure 23.23**). When these molecules are further substituted, common names are often incorporated into a systematic IUPAC name. [*This is technically only allowed for phenol, benzaldehyde, and benzoic acid. In other cases, a "close-enough" name will be generated.*]

Figure 23.23 Common names of some benzene derivatives.

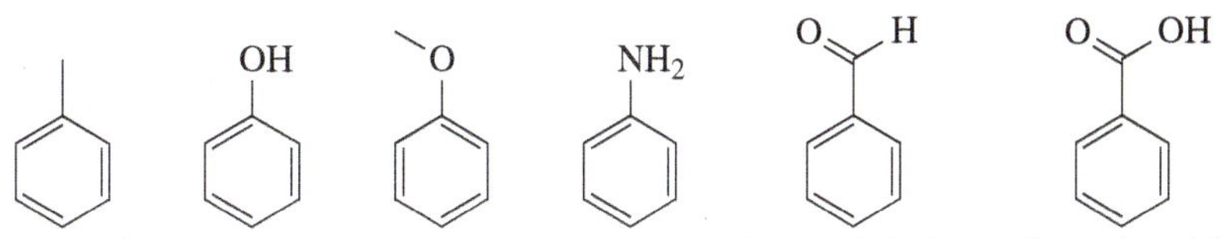

toluene phenol anisole aniline benzaldehyde benzoic acid

23.5.2 IUPAC nomenclature

The systematic method of IUPAC nomenclature involves identifying the substituents on the ring and numbering them such that the lowest combination of numbers is used [*see Section 3.5.1 for naming cycloalkanes*]. The substituents are then alphabetized and placed in front of "benzene" to complete the name. If there is only one substituent, there is no need to number it. Several examples are shown in **Figure 23.24**.

Figure 23.24 Systematic nomenclature of benzene derivatives.

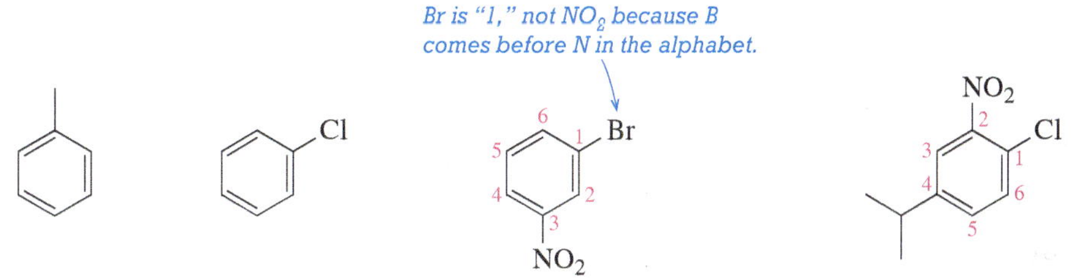

methylbenzene chlorobenzene 1-bromo-3-nitrobenzene 1-chloro-4-isopropyl-2-nitrobenzene

Compounds containing a phenol, benzaldehyde, or benzoic acid as a base structure can be named by incorporating those common names into the IUPAC name (**Figure 23.25**). When this is done, the base functional group (hydroxyl, aldehyde, or carboxylic acid) is always carbon 1 and the rest of the ring is numbered so that the lowest combination of numbers is used. [*See Section 13.2 for naming cyclic alcohols.*]

Figure 23.25 Properly employing common names in systematic nomenclature.

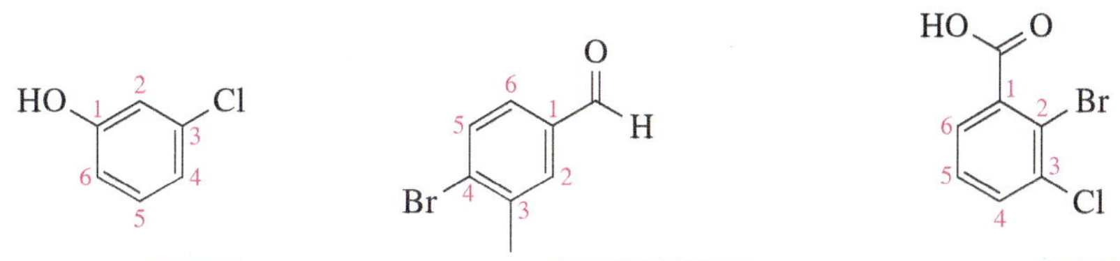

3-chlorophenol 4-bromo-3-methylbenzaldehyde 2-bromo-3-chlorobenzoic acid

The incorporation of common names into IUPAC names creates confusion for some other benzene derivatives because IUPAC doesn't officially recognize their use. If *toluene* is recognized as the base compound of the structures in **Figure 23.26**, then names might be generated that are technically incorrect according to IUPAC.

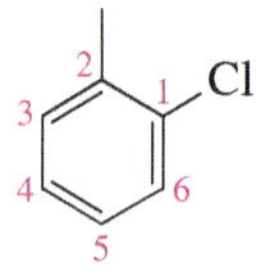

1-chloro-2-methylbenzene

2-chlorotoluene

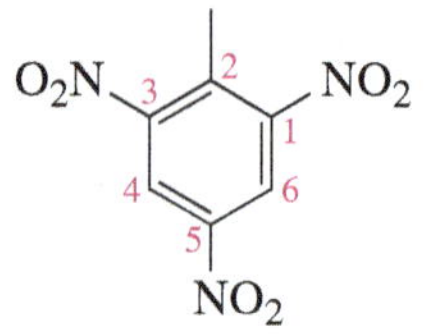

2-methyl-1,3,5-trinitrobenzene

2,4,6-trinitrotoluene

Figure 23.26 Improper nomenclature employing common names. The correct names have a blue screen, whereas the incorrect ones have an orange screen. [*While incorrect by IUPAC rules, they are still commonly used.*]

23.5.3 Positions on the aromatic ring

An alternative system is used to refer to positions on the ring specifically for disubstituted benzene derivatives. These become extremely important when we study the reactions of benzene derivatives, but they can be used for naming as well. Figure 23.27 illustrates the use of ortho (o-), meta (m-), and para (p-) to identify one position on the ring relative to another position on the ring. This is a key distinction—whereas the use of numbers to identify positions is absolute, the use of ortho, meta, and para is relative. Ortho implies a 1,2-relationship, meta a 1,3-relationship, and para a 1,4-relationship. As a result, there are two ortho positions, two meta positions, and one para position.

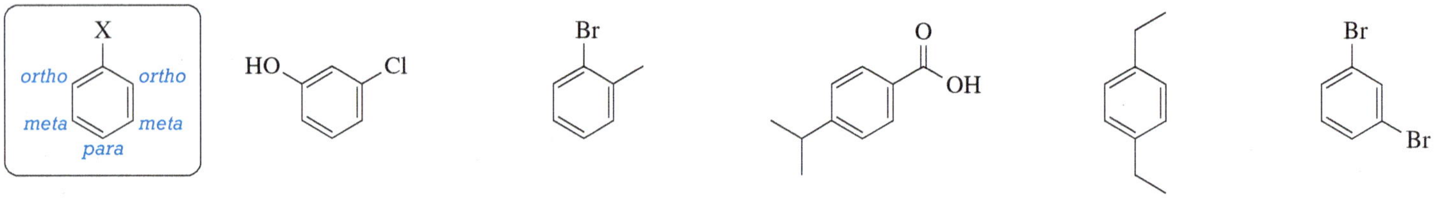

m-chlorophenol
3-chlorophenol

o-bromotoluene
2-bromotoluene
2-bromo-1-methylbenzene

p-isopropylbenzoic acid
4-isopropylbenzoic acid

p-diethylbenzene
1,4-diethylbenzene

m-dibromobenzene
1,3-dibromobenzene

Figure 23.27 Nomenclature using ortho, meta, and para to designate relative positions on the ring. These terms (and their abbreviations, o-, m-, and p-) are italicized when used in a chemical name.

Assessment 23.25 Name the following benzene derivatives according to IUPAC rules.

(a)

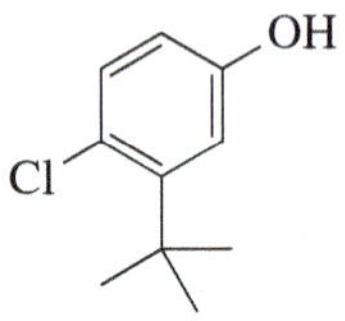

(b)

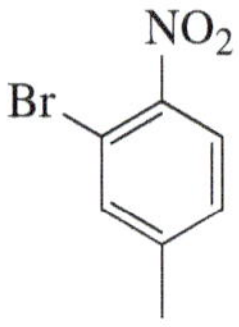

(c)

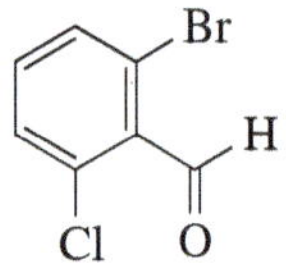

Assessment 23.26 Name the following benzene derivatives using ortho, meta, and para to identify the relative position of substituents.

(a)

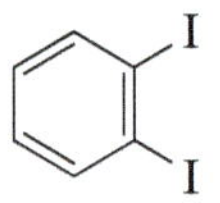

(b)

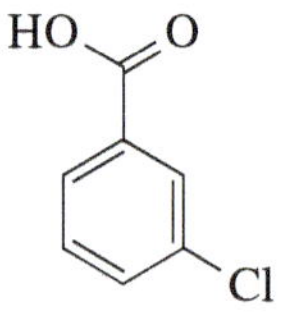

(c)

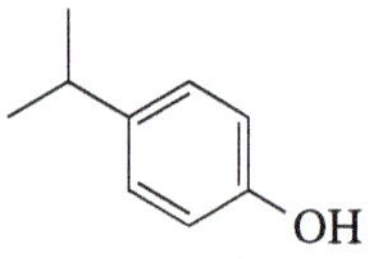

Assessment 23.27 Convert the given name into the corresponding IUPAC name.

(a)

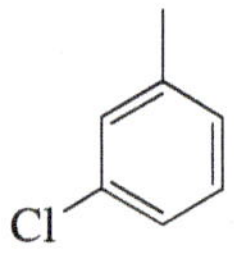

m-chlorotoluene

(b)

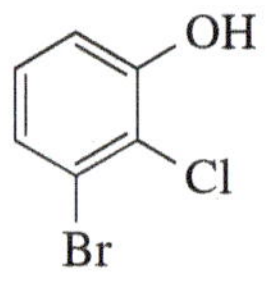

m-bromo-*o*-chlorophenol

(c)

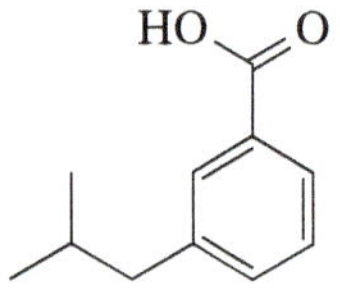

m-isobutylbenzoic acid

Assessment 23.28 The substitution reactions of benzene are introduced in Section 23.6. One such reaction is shown here. Given the product of the reaction, would you say the major reaction resulted in ortho, meta, or para substitution?

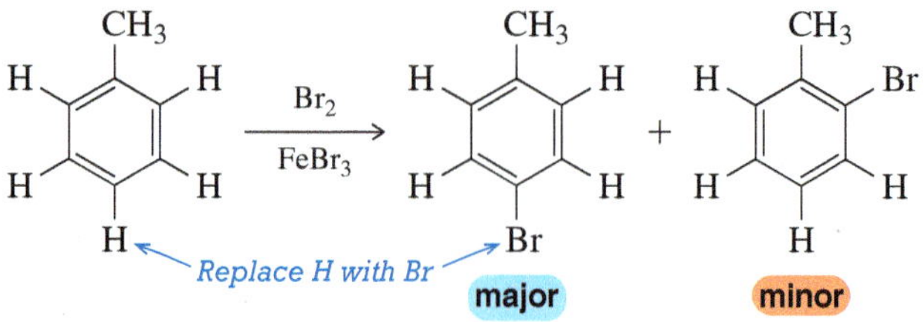

23-B CHEMISTRY IN CONTEXT: Medicinal Chemistry

In vitro to *in vivo*

Early studies on Prontosil at IG Farben showed very little activity against bacteria cultures. In the modern pharmaceutical industry, this might cause Prontosil to be abandoned as a drug target. However, Gerhard Domagk believed that drugs worked synergistically with an organism's immune system. Thus, Prontosil was tested against bacterial infections in mice, where it showed moderate activity. The lack of activity in bacterial cultures can be explained by the later discovery that Prontosil is not itself a drug, but is instead a prodrug. A prodrug is a molecule that is converted by an organism into the active component. In the case of Prontosil, reduction of the diazo functional group occurs in the liver to produce sulfanilamide, the first sulfa (sulfonamide) drug (**Figure 23.28**).

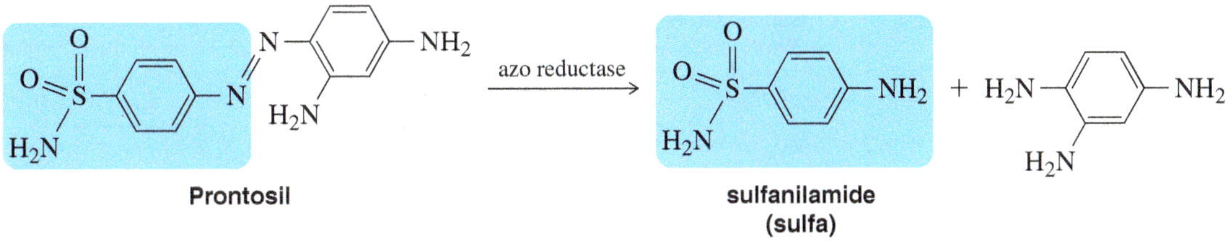

Figure 23.28 The reduction of Prontosil in the body produces sulfanilamide, the active antibiotic.

Sulfanilamide was later discovered to be a bacteriostatic antibiotic. This means it doesn't kill bacteria—it only halts their replication. Its mechanism of action is to irreversibly inhibit bacterial synthesis of folate, which is crucial to DNA synthesis in both bacteria and humans. The enzyme dihydropteroate synthase catalyzes the alkylation of *p*-aminobenzoic acid as part of the biosynthesis of folic acid (**Figure 23.29**). Sulfanilamide resembles the structure of *p*-aminobenzoic acid, so it is a competitive inhibitor of dihydropteroate synthesis. This prevents folic acid synthesis in bacteria, which prevents them from synthesizing the DNA necessary for replication.

Fortunately, we humans do not make our own folic acid. Instead, we get it from the foods we eat. Bacteria can't do this, so sulfa drugs target only bacterial, and not human, systems.

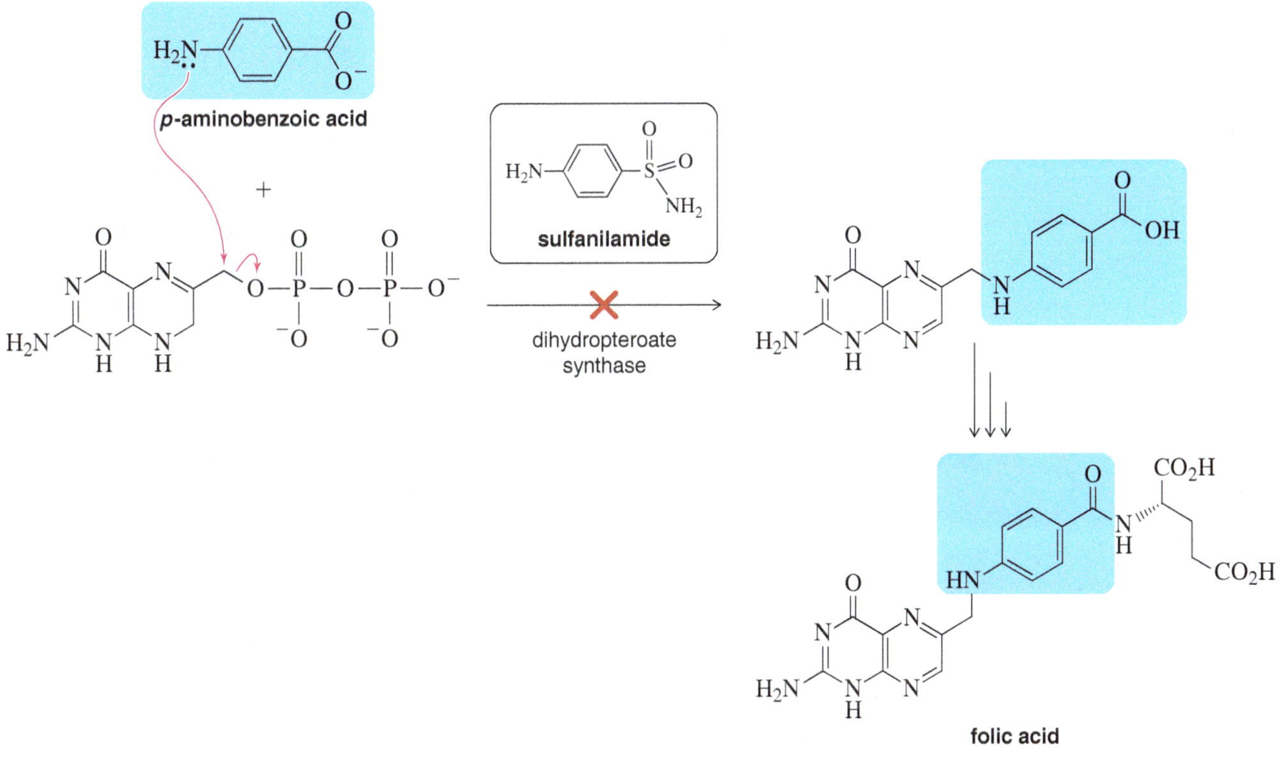

Figure 23.29 Sulfanilamide inhibits dihydropteroate synthase, thus preventing bacteria from making folic acid.

23.6 The substitution reactions of benzene

A distinguishing feature of benzene is the lack of reactivity associated with its aromatic stability. Even in the earliest days of its discovery, benzene was noted to undergo only substitution reactions, as opposed to the addition reactions of alkenes that we studied in Chapters 8 and 9. There are four types of substitution we present in Sections 23.7–23.10. These are electrophilic aromatic substitution (Sections 23.7 and 23.8), nucleophilic aromatic substitution (Section 23.9), substitution of diazonium compounds (Section 23.10.1), and palladium-catalyzed substitution (Section 23.10.2). **Figure 23.30** shows what substitutes for what in each of these substitution reactions.

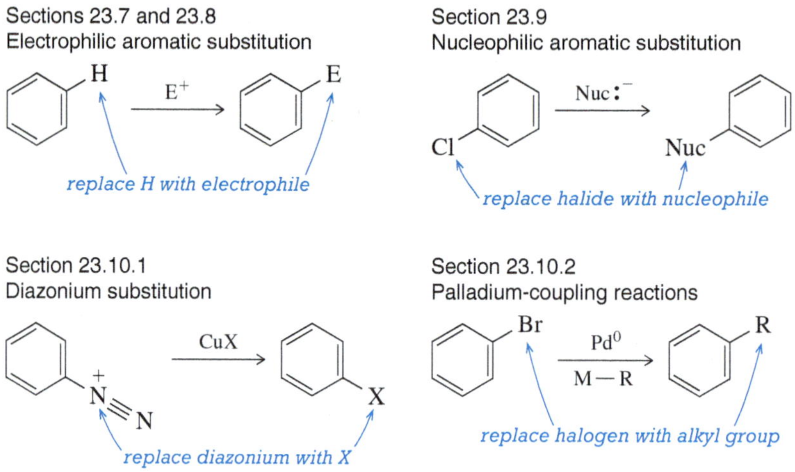

Figure 23.30 The benzene substitution reactions presented in Sections 23.7–23.10.

23.7 Electrophilic aromatic substitution: an overview

The substitution of electrophiles on an aromatic ring is one of the more widely used reactions in the history of organic chemistry. In fact, the birth of medicinal chemistry as a field of study can be attributed to the plethora of compounds produced by this reaction. [*See Chemistry in Context 23-C.*] There are five major electrophilic aromatic substitution reactions that we introduce in Sections 23.7.1–23.7.6. While the prospect of learning five, or more, new reactions can seem daunting, these reactions are all related by the manner (the how and the why) in which they occur. Moreover, you'll find that these reactions have similarities with reactions studied previously. [*Look for the recall boxes.*] The only notable difference between them is how the electrophile is generated.

Mechanistically, electrophilic aromatic substitution begins with π electrons attacking a strong electrophile to give a resonance-stabilized carbocation (**Figure 23.31**). [*We have seen this reaction before: the first step in electrophilic addition to alkenes (Section 8.3).*] While forming a resonance-stabilized carbocation (an **arenium ion**) would normally be favorable, in this case it is a highly endothermic reaction. [*Any ideas why?*] This particular intermediate is referred to as a σ complex. Upon forming the σ complex, the formerly (highly stable) aromatic ring is now nonaromatic, making step 1 extremely slow and rate-determining. In the second step, a base removes a proton to reform the alkene. [*We have seen this reaction before: the second step in an E1 elimination (Section 12.5.1).*] The base in this reaction can be extremely weak because the pK_a of the proton being removed is so low it might be immeasurable. [*Any ideas why?*] Reforming the alkene restores aromaticity to the ring, making step 2 an extremely fast and favorable reaction.

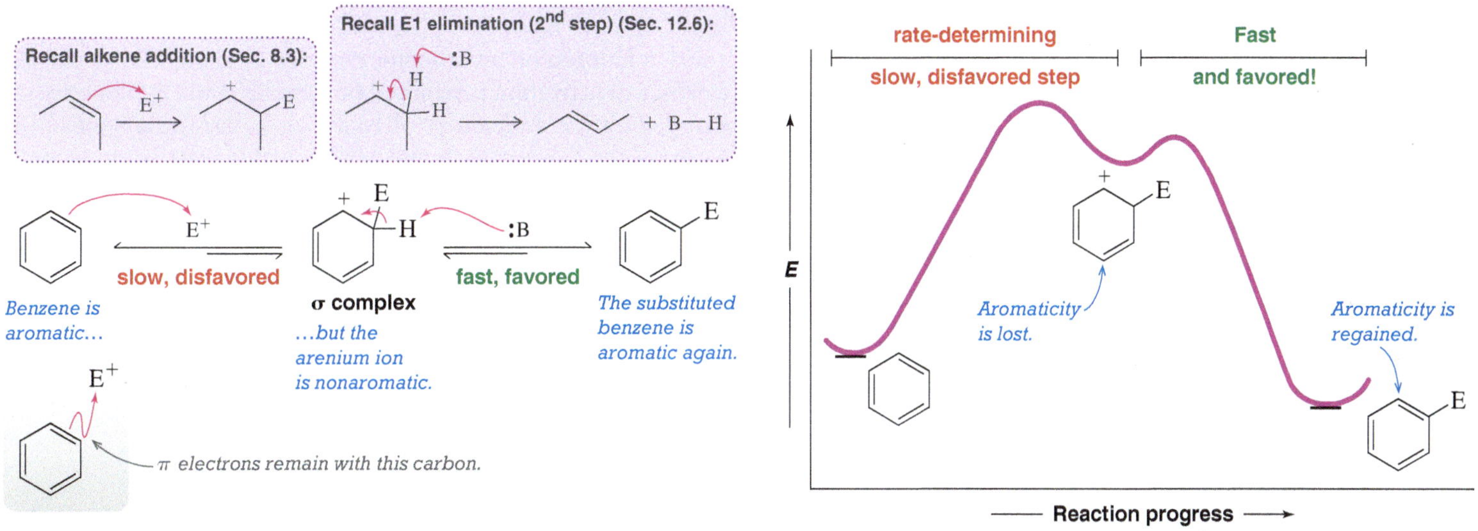

Figure 23.31 **Electrophilic aromatic substitution** combines the first step of alkene addition (step 1) with the second step of E1 elimination (step 2).

The electrophile must be incredibly strong because it is reacting with an aromatic ring that doesn't really want to react. Recall that Br_2 is reactive with alkenes, but it undergoes no reaction with benzene, which was one of the complicating factors in determining the structure of benzene. Thus, the distinguishing feature of Sections 23.7.1–23.7.6 is how each electrophile is generated. The overall reaction mechanism will still be repeated, but it will be the same as what is shown in Figure 23.31.

23-C CHEMISTRY IN CONTEXT: Chemistry of Color

Synthesis of azo dyes

Electrophilic aromatic substitution played a significant role in the development of the sulfa drugs. Not only has it enabled the synthesis of the many derivatives for use in drug development, but it was also used to create the dye that ultimately became Prontosil, the first sulfa antibiotic. Diazonium salts, which we present in Section 23.10.1, are excellent electrophiles. Thus, in the presence of an electron-rich ring, electrophilic aromatic substitution can occur (**Figure 23.32**). The benzene ring first donates a pair of electrons to the terminal nitrogen of the diazonium ion, forming the nonaromatic intermediate arenium ion. Deprotonation of the arenium ion restores aromaticity in a highly favorable and fast step to produce Prontosil.

Prontosil is red because extensive conjugation lowers the HOMO–LUMO energy gap, allowing it to absorb light in the visible region of the electromagnetic spectrum [*see Section 21.3.2*].

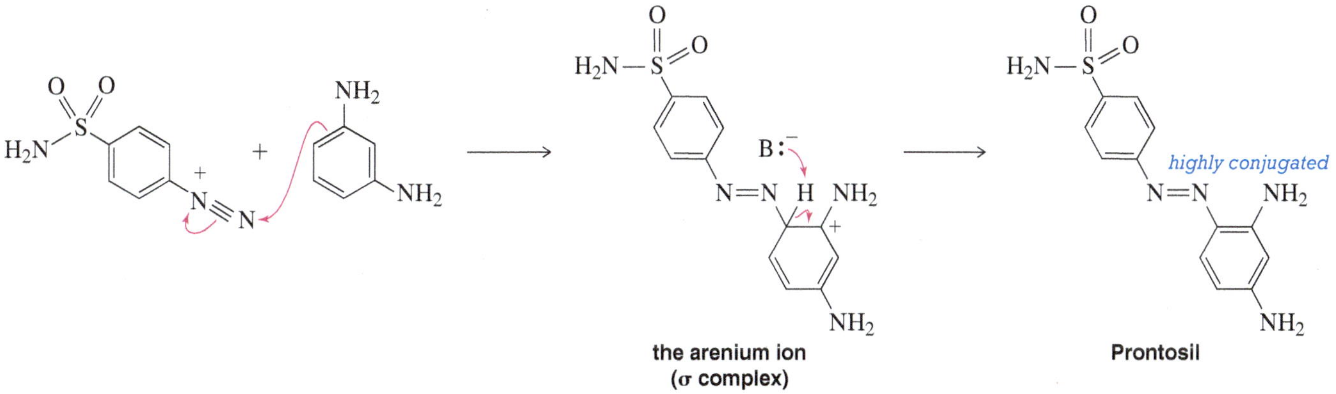

Figure 23.32 **The synthesis of Prontosil by electrophilic aromatic substitution.**

23.7.1 Halogenation

Replacement of a hydrogen with a halogen on an aromatic ring occurs using Lewis acid activation of either Br_2 or Cl_2. Lewis acid activation is required because Br_2 and Cl_2 alone are not strong enough electrophiles to disrupt the aromaticity of benzene. As shown in **Figure 23.33**, iron(III) bromide ($FeBr_3$) is used as the Lewis acid to make Br^+, a highly reactive electrophile. In the first step, Br_2 donates a pair of electrons to the electron-deficient iron. This weakens the Br—Br σ bond until it eventually breaks to give the relatively stable $FeBr_4^-$ anion and the Br^+ cation. With the strong electrophile formed, it is attacked by benzene to give the nonaromatic arenium ion in a highly endothermic step. In the second step, Br^- (from $FeBr_4^-$) removes a proton, regenerating benzene and reestablishing aromaticity in a fast and highly exothermic step.

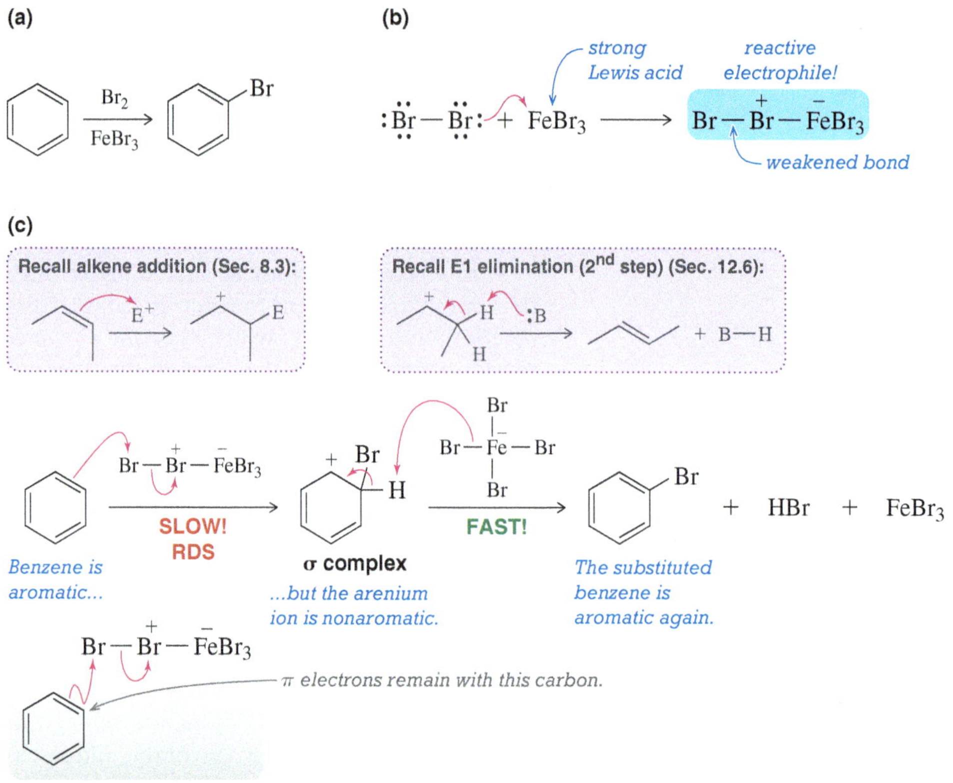

Figure 23.33 Electrophilic aromatic substitution: halogenation. (a) The overall reaction. (b) The generation of the electrophile. (c) The overall mechanism. [*Note: The proton removed from the σ complex is always attached to the carbon where the electrophile added.*]

To put a chlorine on the ring, identical conditions are used, but with Cl_2 as the electrophile and $FeCl_3$ as the Lewis acid.

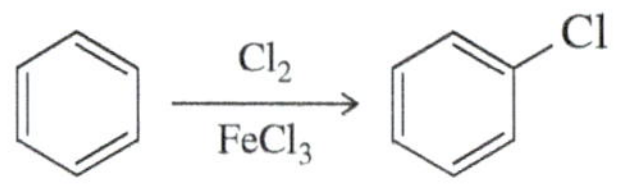

23.7.2 Nitration

Replacement of a hydrogen with a nitro group on an aromatic ring occurs by reacting nitric acid with sulfuric acid to create the highly electrophilic NO_2^+ ion. As shown in **Figure 23.34**, the reaction begins with protonation of the hydroxyl group of nitric acid. [*That's right—nitric acid is acting as a base. These are very acidic and harsh conditions! They need to be to create electrophiles strong enough to disrupt the aromaticity of benzene.*] Water is now an excellent leaving group, so it leaves with assistance from the neighboring alkoxide to create the highly electrophilic NO_2^+ cation. With the strong electrophile formed, it is attacked by benzene to give the nonaromatic arenium ion in a highly endothermic step. The electrophilic nitrogen of NO_2^+ still has a total of 8 electrons in its valence shell. Thus, when it is attacked by the π electrons from benzene, 2 electrons are pushed onto the oxygen in a reaction reminiscent of addition to a carbonyl [*Section 17.5*]. In the second step, H_2O removes a proton, regenerating benzene and reestablishing aromaticity in a fast and highly exothermic step.

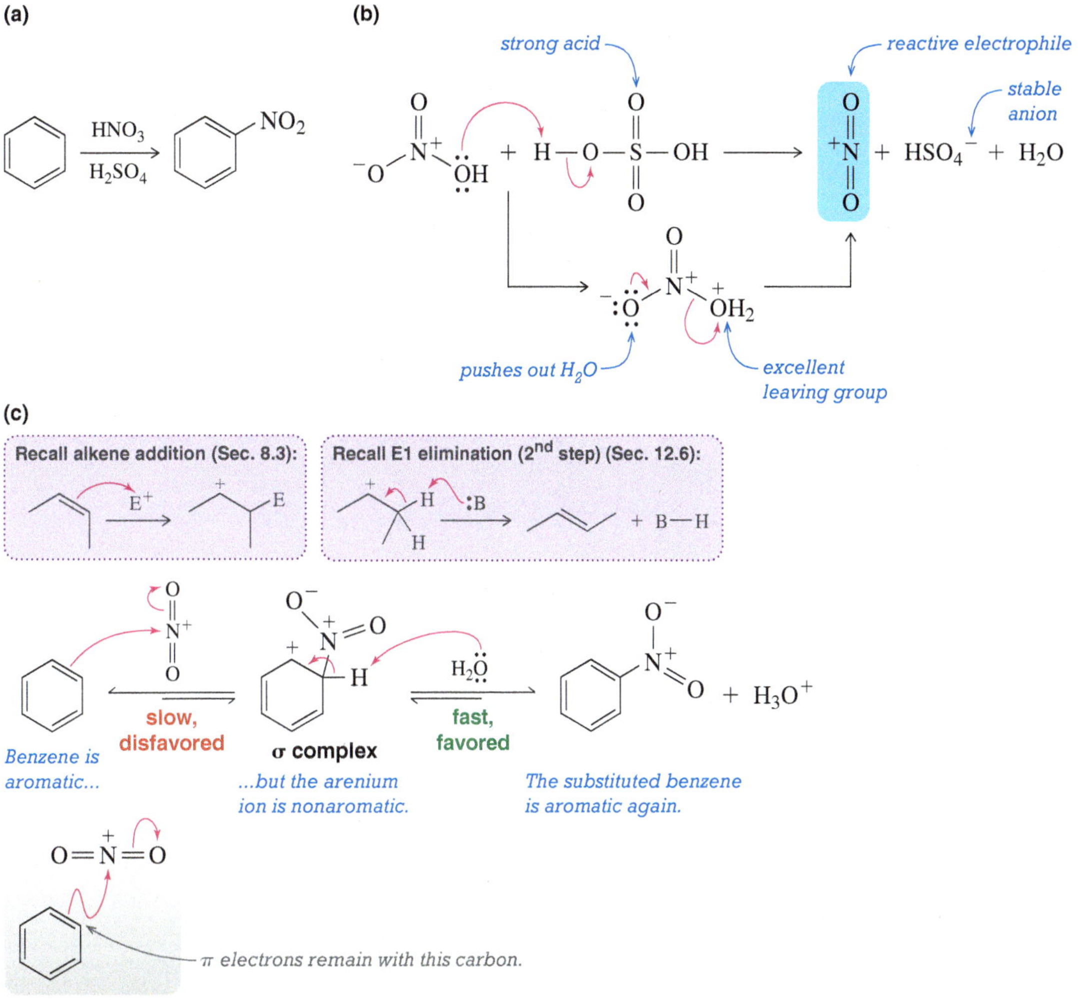

Figure 23.34 **Electrophilic aromatic substitution: nitration.** (a) Overall reaction. (b) Generation of the electrophile. (c) Overall mechanism. [*Note: The proton removed from the σ complex is always attached to the carbon where the electrophile added.*]

Assessment 23.29 The zwitterionic form of carbonyls is often used to explain their electrophilicity. Draw the zwitterionic structure of NO_2^+. Why is this such a great electrophile at the central nitrogen?

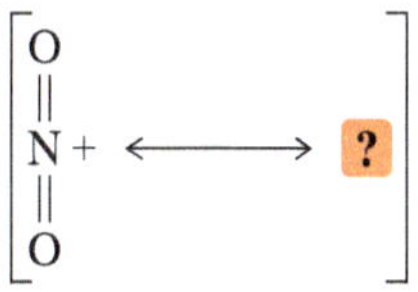

23.7.3 Sulfonylation

Replacement of a hydrogen with a sulfonic acid ($-SO_3H$) on an aromatic ring occurs by reacting sulfur trioxide (SO_3) with sulfuric acid to create the highly electrophilic SO_3H^+ ion. [*The combination of SO_3 and H_2SO_4 is referred to as fuming sulfuric acid.*] As shown in **Figure 23.35**, the reaction begins with protonation of an oxygen on SO_3. Like a protonated carbonyl, this intermediate is now a strong electrophile. Thus, it is attacked by benzene to give the nonaromatic arenium ion in a highly endothermic step. In the second step, HSO_4^- (the conjugate base of H_2SO_4) removes a proton, regenerating benzene and reestablishing aromaticity in a fast and highly exothermic step.

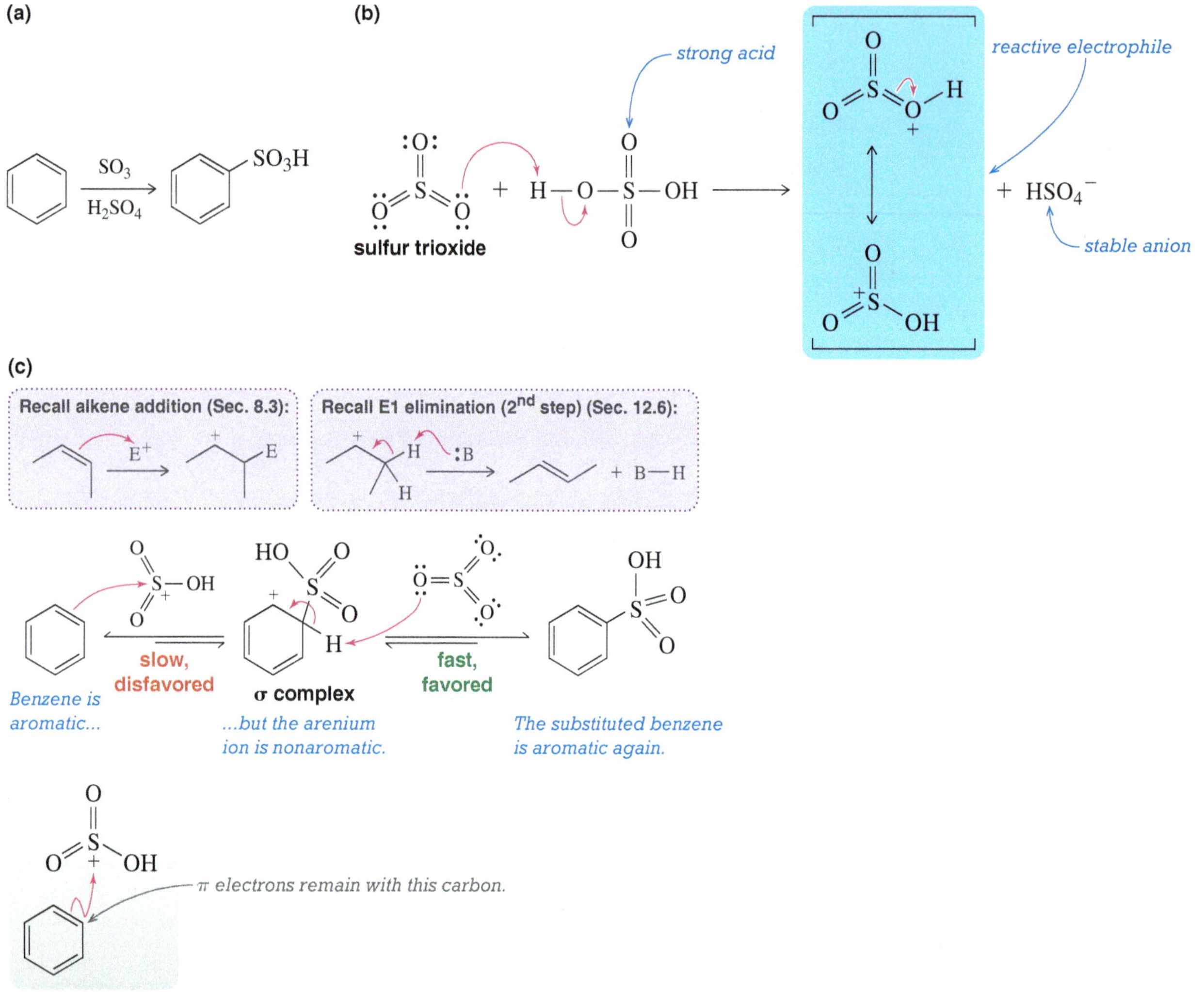

Figure 23.35 **Electrophilic aromatic substitution: sulfonylation.** (a) Overall reaction. (b) Generation of the electrophile. (c) Overall mechanism. [*Note: The proton removed from the σ complex is always attached to the carbon where the electrophile added.*]

23.7.4 Friedel–Crafts alkylation

Replacement of a hydrogen with an alkyl group on an aromatic ring [*finally, carbon–carbon bond formation!*] occurs by reacting an alkyl halide with a Lewis acid to create a reactive electrophile in a reaction discovered by Charles Friedel and James Craft in 1877. The mechanism of this reaction is slightly different, depending on the substitution of the haloalkane used. We are going to consider both a 1° and 3° haloalkane in this reaction. [*How do you think these will differ? Hint: See Sections 12.3.2, 12.6.4, 13.7.2, etc. . . .*]

The reaction of a 1° haloalkane (like chloroethane in **Figure 23.36**) begins with chlorine donating a lone pair to $AlCl_3$, a strong Lewis acid. [*$AlCl_3$ must be strong because chlorine doesn't usually like to share electrons, except in Section 23.7.1 (electrophilic aromatic halogenation).*] While chloride is an excellent leaving group, the positively charged chloronium in the newly formed Lewis acid–Lewis base complex is even better. Thus, it is attacked by benzene to give the nonaromatic arenium ion in a highly endothermic step. [*Note that this looks a lot like an S_N2 reaction. What will be different when a 3° haloalkane is used?*] In the second step, Cl^- (from $AlCl_4^-$) removes a proton, regenerating benzene and reestablishing aromaticity in what is again a fast and highly exothermic step.

When a 3° haloalkane is used (**Figure 23.37**), the mechanism changes only slightly. Formation of the Lewis acid–Lewis base complex begins in the same way, but now, with too much steric hindrance to allow an S_N2 reaction to occur, a 3° carbocation forms prior to reacting with benzene. The carbocation is a very strong electrophile, so it is attacked by benzene to

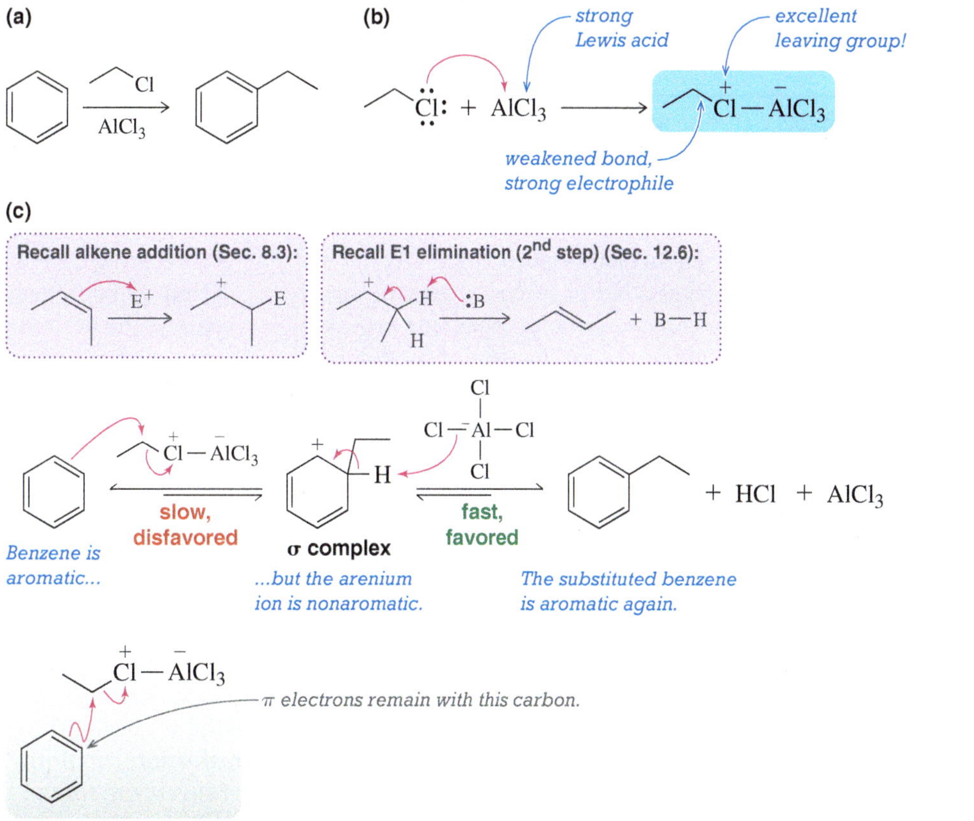

Figure 23.36 **Electrophilic aromatic substitution: Friedel–Crafts alkylation with a 1° carbon.** (a) Overall reaction. (b) Generation of the electrophile. (c) Overall mechanism. [*Note: The proton removed from the σ complex is always attached to the carbon where the electrophile added.*]

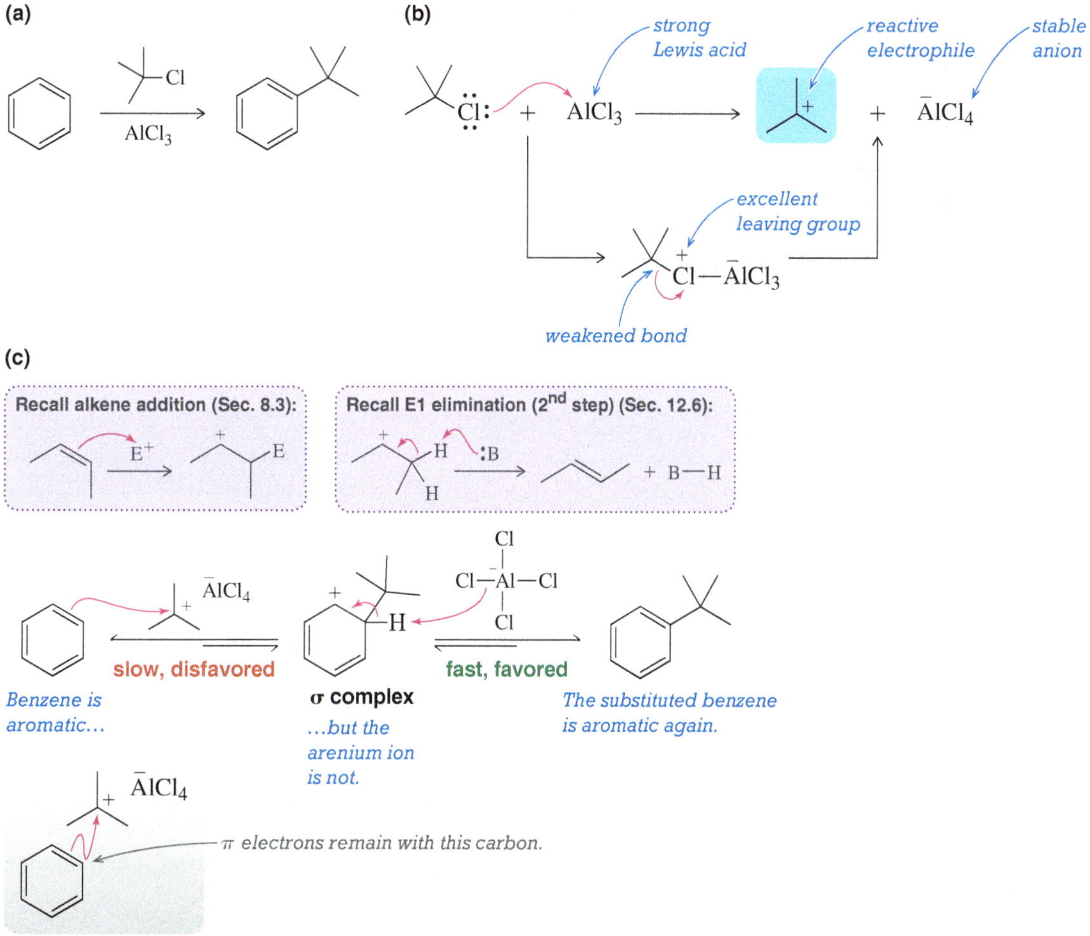

Figure 23.37 **Electrophilic aromatic substitution: Friedel–Crafts alkylation with a 2° or 3° carbon.** (a) Overall reaction. (b) Generation of the electrophile. (c) Overall mechanism. [*Note: The proton removed from the σ complex is always attached to the carbon where the electrophile added.*]

Work these on your own, then we'll do Assessment 23.31 together to learn a limitation of Friedel–Crafts alkylation.

give the nonaromatic arenium ion in a highly endothermic step. [*This looks a lot like an* S_N1 *reaction, but you probably expected this.*] Once more, in the second step, Cl^- (from $AlCl_4^-$) removes a proton, regenerating benzene and reestablishing aromaticity in a fast and highly exothermic step.

Assessment 23.30 When an enantiopure 2° haloalkane is used in the Friedel–Crafts alkylation reaction, the two diastereomers shown here are isolated (among other side products). What does this tell you about the mechanism of the reaction?

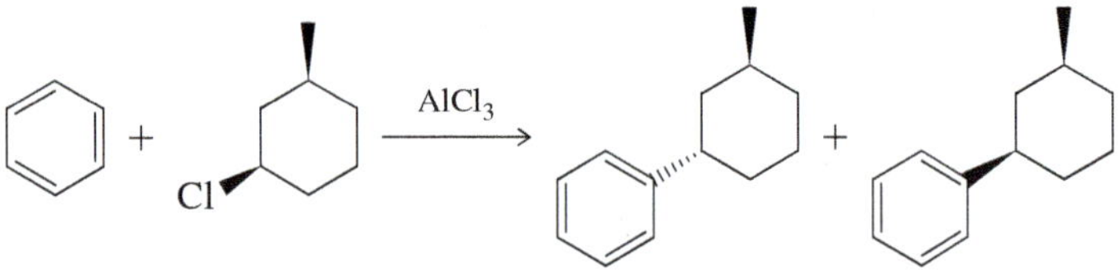

Assessment 23.31 A chemist, hoping to attach an isobutyl group to benzene, attempted the following reaction. Unfortunately, the desired product was not isolated. Explain the formation of the product actually obtained.

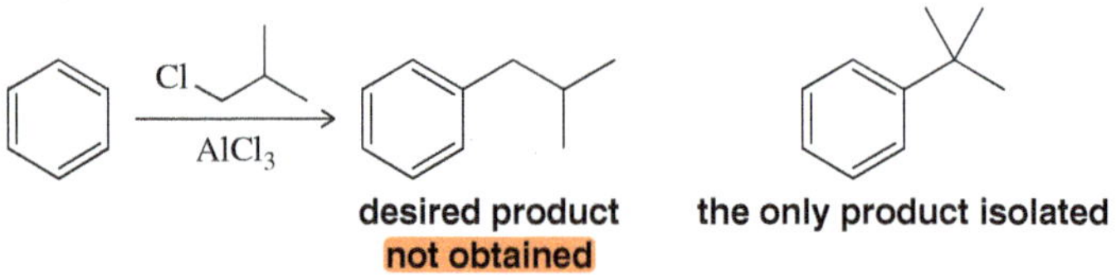

Solved Assessment 23.31

Quick analysis of the product in Assessment 23.31 shows that instead of a new C—C bond being formed between benzene and C_1, as desired by the chemist, it instead formed between benzene and C_2 (**Figure 23.38**). Additionally, C_1 began the reaction with two hydrogens, but now has three. Taken together, this information suggests that a rearrangement has probably occurred. [*Is a rearrangement something you would expect under these conditions?*]

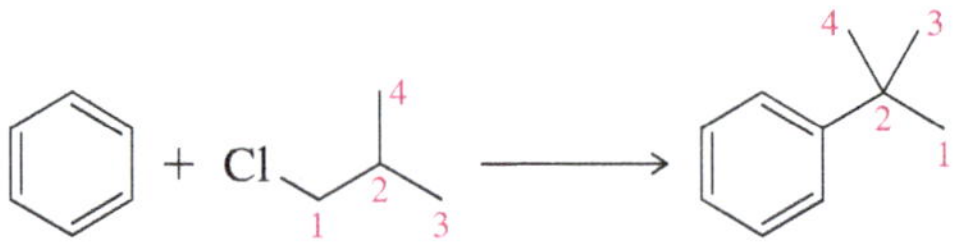

Figure 23.38 Analysis of the reaction in Assessment 23.31 suggests that a rearrangement may have occurred.

A rearrangement can be confirmed by employing previously used problem-solving strategies. [*The answer is on the page!*] In the forward direction, we expect the haloalkane to react with $AlCl_3$ to form a good leaving group (**Figure 23.39**). Working backward, the product likely forms when benzene attacks a 3° carbocation. [*All that remains is to figure out what occurs to connect these two processes. Ideas?*]

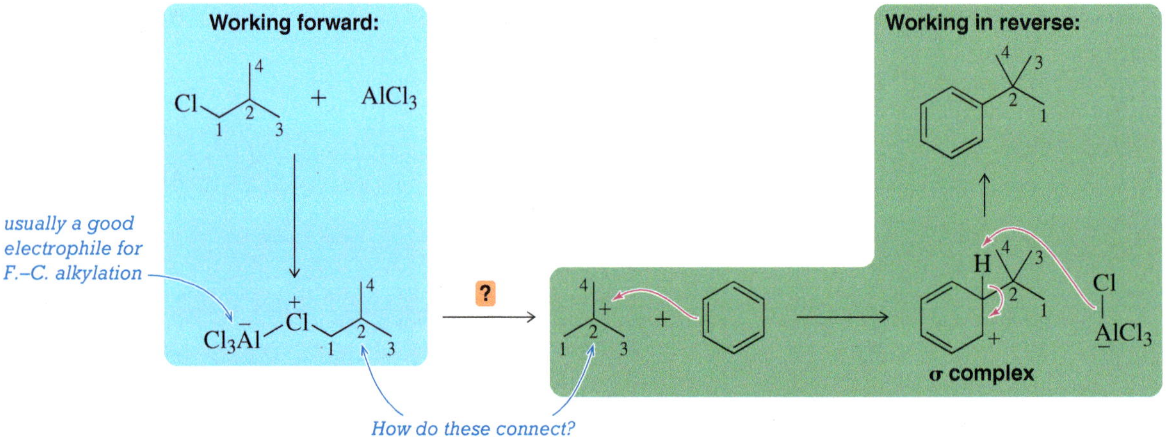

Figure 23.39 Solving a complicated mechanism requires working in the forward and reverse directions.

Figure 23.40 shows a Lewis acid–Lewis base complex that is electrophilic at C_1, but it must become electrophilic at C_2. At the same time that C_1 needs to gain a hydrogen, C_2 needs to lose a hydrogen. This sounds a lot like the 1,2-hydride rearrangements we have observed in other reactions involving strong electrophiles and carbocation intermediates [*see Sections 8.4.3, 12.4.2, and 13.8.1, among others*]. A hydride rearrangement occurs, kicking out the good leaving group at the same time and producing a 3° carbocation, which is heavily stabilized by hyperconjugation. ■

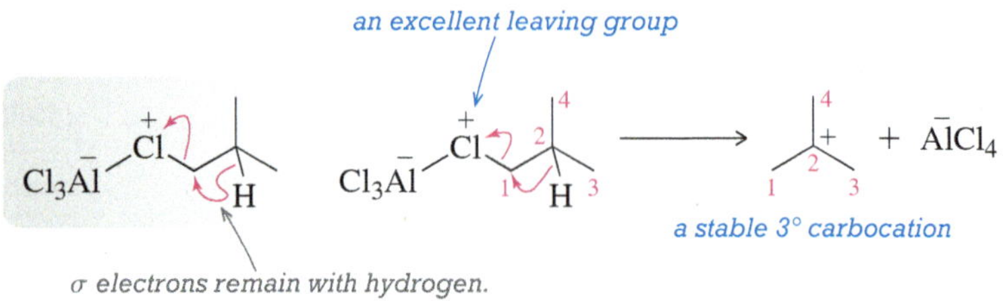

Figure 23.40 Formation of a more stable carbocation drives the 1,2-hydride shift.

As Solved Assessment 23.31 shows, one limitation of Friedel–Crafts alkylation is that it involves carbocations, so rearrangements will occur when a more stable carbocation is possible. This limits the Friedel–Crafts alkylation to adding only small alkyl groups or forming new bonds with only 3° carbocations. An additional limitation of Friedel–Crafts alkylation is the inherent difficulty of stopping the reaction after the addition of just one alkyl group, even if the stoichiometry is carefully controlled. The reaction in **Figure 23.41** implies that an alkyl-substituted benzene must be more reactive than benzene itself. [*Think about why this might be true . . . we talk about it in Section 23.8.1.*]

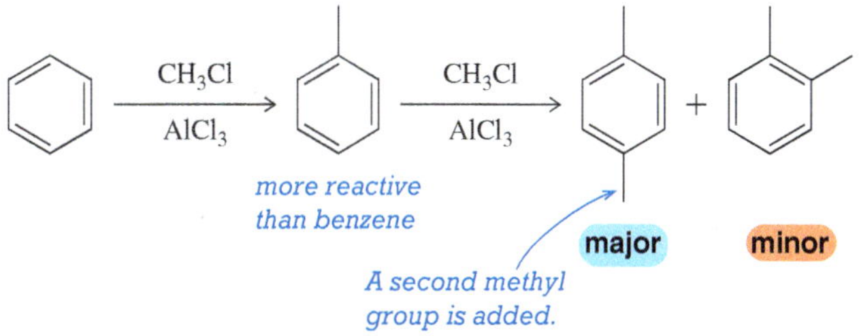

Figure 23.41 Polyalkylation is another limitation of Friedel–Crafts alkylation reactions.

23.7.4 SUMMARY

- The Friedel–Crafts alkylation is an important C—C bond-forming electrophilic aromatic substitution reaction. Limitations include polyalkylation and a propensity to rearrange to make more stable carbocations.

Assessment 23.32 Predict the product of the following Friedel–Crafts alkylation reactions. Assume only one alkyl group adds in each case.

(a)

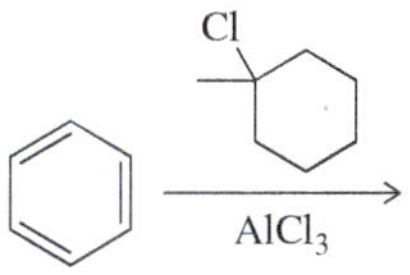

(b)

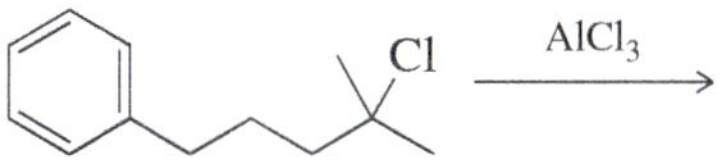

(c)

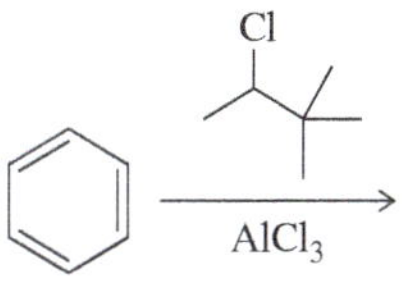

(d)

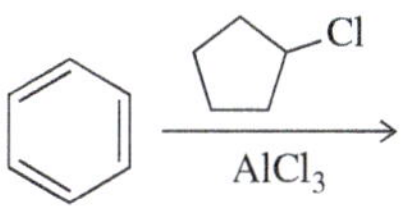

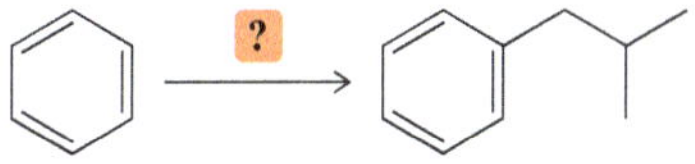

Figure 23.42 Is there a way to attach 1° alkyl groups to a benzene ring without rearrangement or polyalkylation?

23.7.5 Friedel–Crafts acylation

Instead of just showing the Friedel–Crafts acylation reaction, let's imagine it as a solution to the major limitation of the Friedel–Crafts alkylation reaction—namely, the tendency to rearrange. Toward that end, our goal will be to make isobutyl benzene, the goal of the chemist in Assessment 23.31 (**Figure 23.42**).

We can think like a chemist to see how this problem might be solved, at least in general terms, before seeing how the Friedel–Crafts acylation accomplishes this.

Thinking Like a Chemist

The major limitation of the Friedel–Crafts alkylation is that attachment at 1° carbons is impossible if the haloalkane or resulting carbocation can easily rearrange to a more stable carbocation. Is there ever a situation when a 1° carbocation is *more* stable than a neighboring 2° carbocation? If we are only thinking about hyperconjugation, then the answer is no. But when considering hyperconjugation, we must also think about induction and, important to this discussion, resonance! Thinking in this way brings up a hypothetical solution. What if the carbocation that is to be attacked by the benzene ring is stabilized by some electron-donating group, shown as X in **Figure 23.43**? As a resonance-stabilized carbocation, it wouldn't rearrange. For this to be a viable solution, we'd need a way to remove X in a subsequent step.

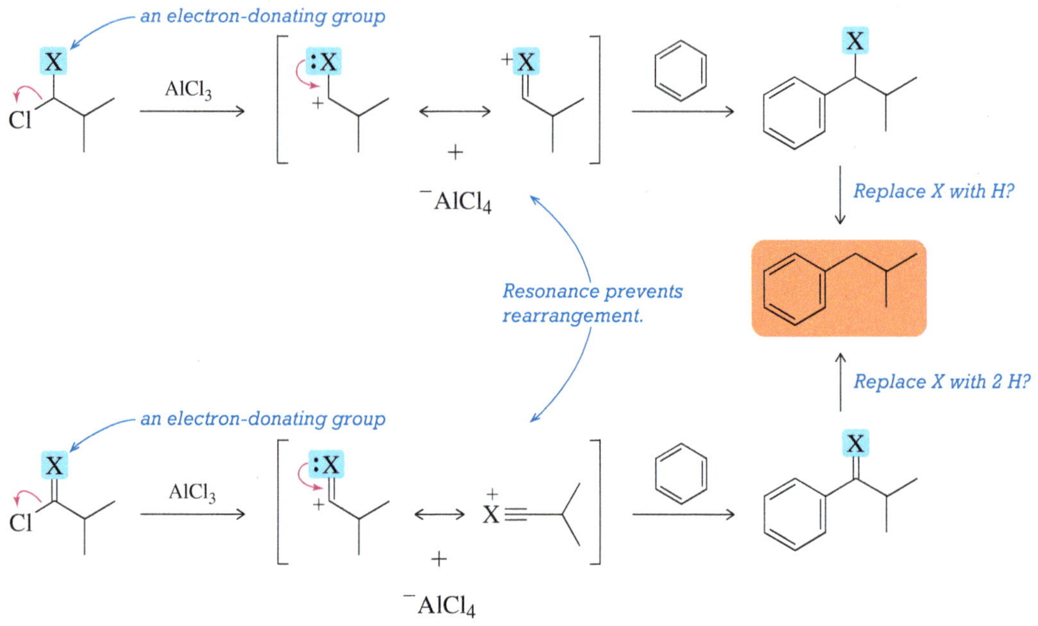

Figure 23.43 Stabilization by X will prevent rearrangement. What should X be?

Figure 23.43 shows two hypothetical molecules containing X either singly or doubly bonded to a carbocation formed by Lewis acid activation of a chloride, as seen in the Friedel–Crafts alkylation reaction (Section 23.7.4). Do you have an idea what atom/group X would (1) result in a molecule we've studied before, (2) have the ability to donate electrons by resonance, and (3) be removable? Specifically, removing it would be replacing one C—X bond/two C—X bonds with one C—H/two C—H bonds, respectively. [*Look carefully at Figure 23.43 to see if something jumps out at you.*] The only atom/group that satisfies the three requirements would be an oxygen atom (**Figure 23.44**). When doubly bonded to the carbon with the chloride, this is an acid chloride. We've studied substitution reactions of acid chlorides [*Section 19.7, check*], oxygen can stabilize carbocations by resonance [*Section 2.8, check*], and the Wolff–Kishner reaction allows us to exhaustively reduce a carbonyl to a methylene [*Section 17.9.3, check!*].

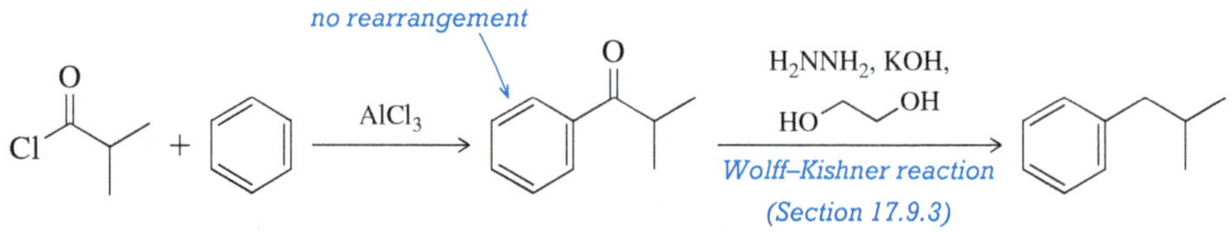

Figure 23.44 The use of an acid chloride in a Friedel–Crafts reaction prevents rearrangement.

The strategy just developed makes it possible to attach an isobutyl group, or other large alkyl groups, without worrying about rearrangement. ■

Mechanistically (**Figure 23.45**), the reaction of an acid chloride is very similar to that of a haloalkane, beginning with chlorine donating a lone pair to $AlCl_3$, a strong Lewis acid. While chloride is an excellent leaving group, the positively charged chloronium in the newly formed Lewis acid–Lewis base complex is even better. Thus, $AlCl_4^-$ is kicked out with assistance from the carbonyl oxygen to give a reactive, but resonance-stabilized, electrophile. Unable to rearrange, it is attacked by benzene to give the nonaromatic arenium ion in a highly endothermic step. In the second step, Cl^- (from $AlCl_4^-$) removes a proton, regenerating benzene and reestablishing aromaticity in a fast and highly exothermic step.

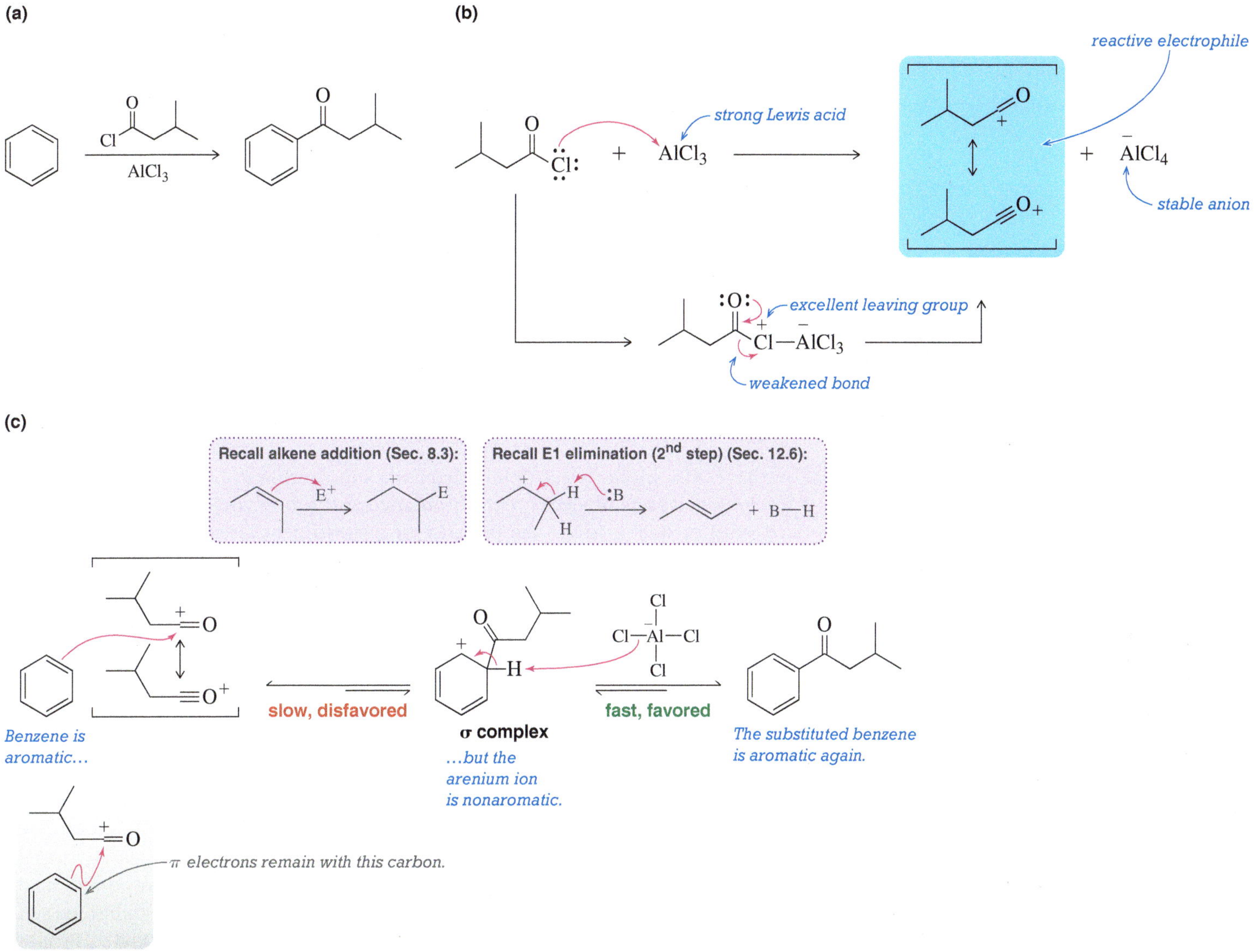

Figure 23.45 Electrophilic aromatic substitution: Friedel–Crafts acylation with a 2° or 3° carbon. (a) Overall reaction. (b) Generation of the electrophile. (c) Overall mechanism.

Friedel–Crafts acylation also overcomes the problem of polyalkylation in the Friedel–Crafts alkylation reaction (**Figure 23.46**) because it is almost impossible for a second acyl group to be added in this way. [*Thoughts on why this might be? We explain in Section 23.8.2.*]

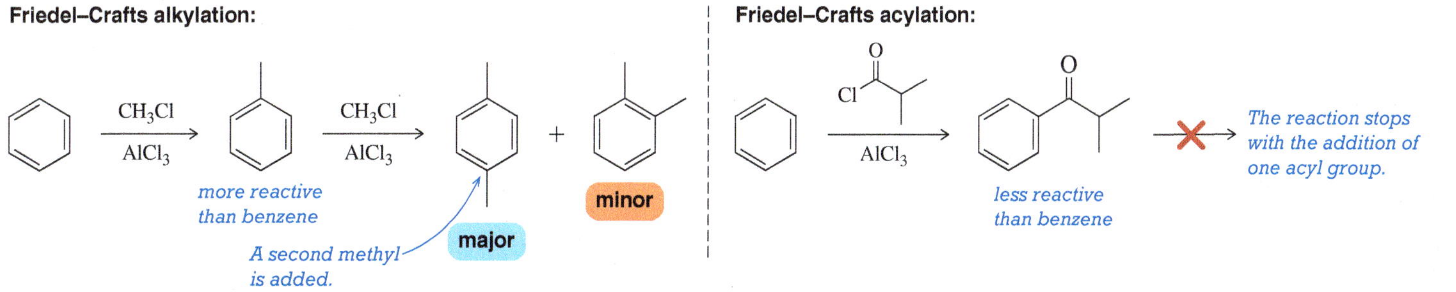

Figure 23.46 Polyalkylation is a big problem in Friedel–Crafts alkylation, but never occurs in Friedel–Crafts acylation.

23.7.5 SUMMARY

- The Friedel–Crafts acylation is an important C—C bond-forming electrophilic aromatic substitution reaction. It overcomes the limitations of the Friedel–Crafts alkylation; no polyalkylation or rearrangement is observed.

Assessment 23.33 Predict the product of the following Friedel–Crafts acylation reactions.

Assessment 23.34 Show how 1-phenyloctane can be synthesized using octanoic acid and benzene as the only source of carbon atoms.

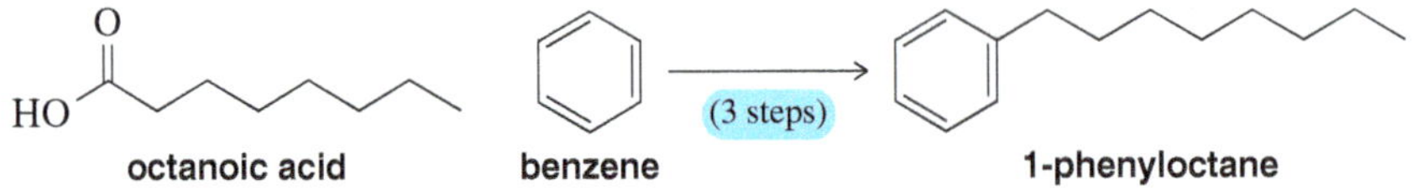

23.8 Regioselectivity in electrophilic aromatic substitution

Thus far, we have only studied electrophilic aromatic substitution of benzene itself. And yet the most interesting aspects of electrophilic aromatic substitution come when adding a second or third substituent to a ring. The presence of a substituent already on the ring influences reaction rate and regioselectivity, and these two outcomes are highly correlated. The next three sections examine the effects of electron-donating substituents (Section 23.8.1), electron-withdrawing substituents (Section 23.8.2), and halogens (Section 23.8.3).

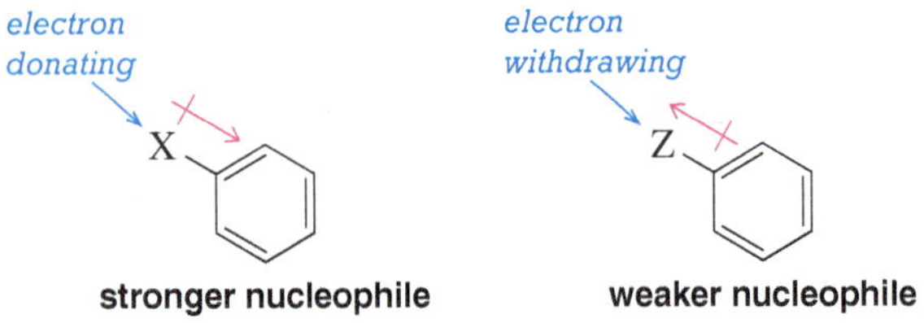

Figure 23.47 Electron-donating groups make nucleophiles stronger and react more quickly. Electron-withdrawing groups make nucleophiles weaker and react more slowly.

Let's tackle the reaction rate first, because it is more straightforward than the regioselectivity. In electrophilic aromatic substitution, benzene acts as the nucleophile. [*Read this sentence again: Benzene in electrophilic aromatic substitution is the nucleophile!*] Substituents that donate electron density to benzene should make it more electron-rich and therefore a better nucleophile (**Figure 23.47**). The rate of electrophilic aromatic substitution should increase. Substituents that withdraw electron density from benzene should make it less electron-rich and therefore a worse nucleophile. The rate of electrophilic aromatic substitution should decrease.

The regioselectivity of electrophilic aromatic substitution reactions isn't more complicated; it just requires extra analysis. When toluene, a monosubstituted benzene, reacts with some general electrophile (E^+), there are three possible regiochemical outcomes (**Figure 23.48**). The electrophile can be added ortho to the methyl group (two ways), meta to the methyl group (two ways), or para to the methyl group (one way).

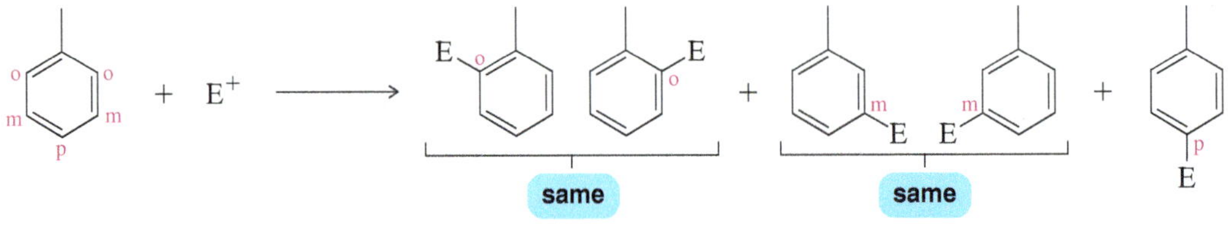

Figure 23.48 There are three possible regiochemical outcomes of a general electrophilic aromatic substitution reaction.

The first step is the slow, endothermic, rate-determining step of electrophilic aromatic substitution. Thus, the regiochemical outcome is determined in the first step with the site of attack on the electrophile (o, m, or p) determined by forming the most stable intermediate. Being endothermic, the transition state is similar in structure to the arenium ion intermediate. [*Remember the Hammond postulate (Section 5.2.4)?*]. Thus, stabilizing the intermediate arenium ion also stabilizes the transition state (**Figure 23.49**). By stabilizing the transition state, the activation energy is lowered and the reaction is faster. From this general analysis, we can understand how substituents direct substitution by examining the relative stability of arenium ion intermediates. We do this in Sections 23.8.1–23.8.3.

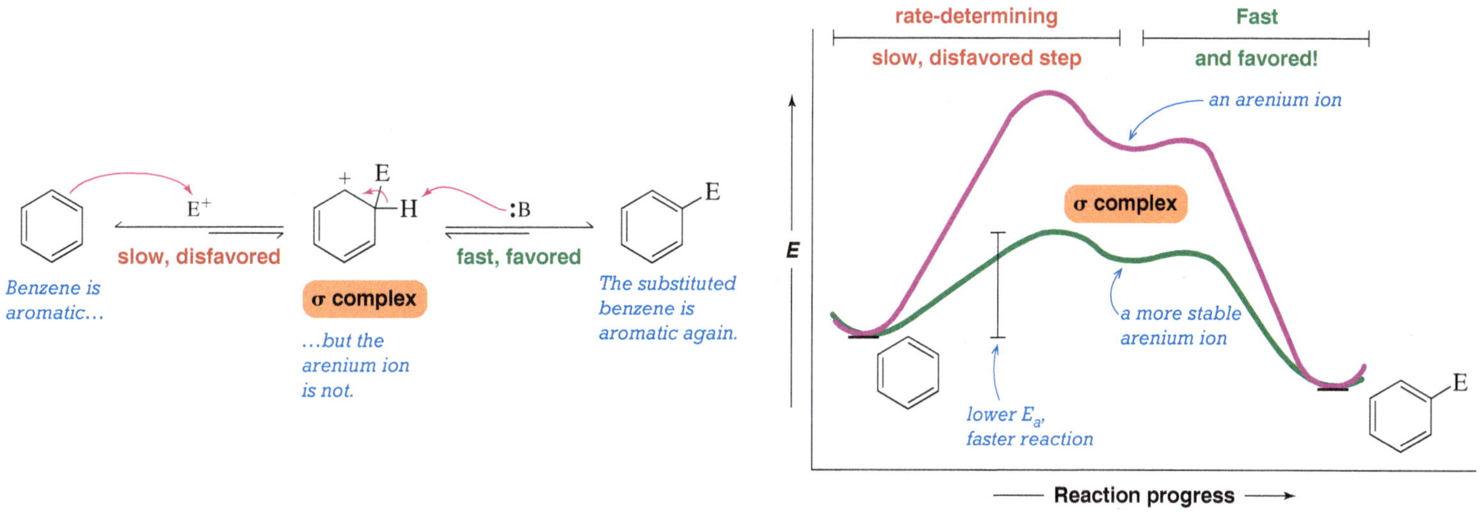

Figure 23.49 **Stabilizing the product of an endothermic step also stabilizes the transition state,** lowers the activation energy, and makes a reaction faster—the Hammond postulate.

Be careful with this terminology; don't just memorize it. Electron-donating groups are not always activating; it depends on the reaction being studied. The trend reverses when we study nucleophilic aromatic substitution in Section 23.9.

23.8.1 Electron-donating groups activate and direct ortho/para

The first substituent class we consider consists of electron-donating groups. These can be electron-donating by either resonance or hyperconjugation, but do not include halogens. Several examples of electron-donating groups are shown in **Figure 23.50**. Electron donors make the aromatic ring more electron-rich and more nucleophilic, which increases the rate of electrophilic aromatic substitution. In this way, electron-donating groups are said to "activate" the ring.

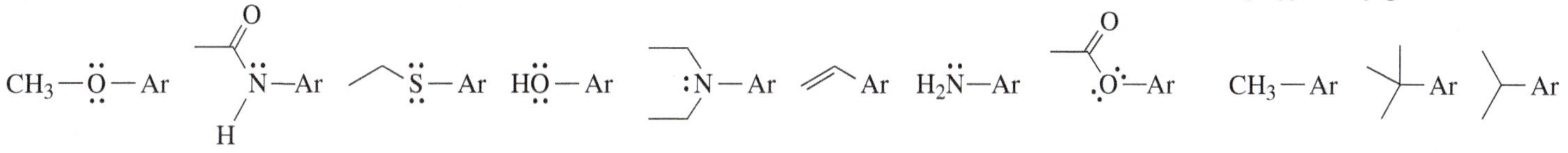

Figure 23.50 **Electron-donating groups** by (a) resonance and (b) hyperconjugation (Ar = aromatic ring).

Once the rate- and selectivity-determining first step has occurred, the outcome of the electrophilic aromatic substitution has been determined. Thus, to understand *why* an electron-donating group directs electrophiles to the ortho and para positions, only the first step needs to be analyzed. To this end, **Figure 23.51** shows the intermediates produced when an electrophile adds ortho, meta, and para to a methoxy group, which is meant to represent a general electron-donating group. [*Look carefully at the picture on your own before proceeding. Which arenium ion intermediates stand out as being more or less stable?*] At first glance, each intermediate is composed of at least three resonance contributors, suggesting that these are nearly equal resonance structures. But look closely at the resonance structures screened in blue that arise from ortho and para addition. In these resonance structures, the carbocation is on the same carbon as the electron-donating substituent. This allows an additional resonance structure to be drawn that further stabilizes this intermediate. In the case of the meta intermediate, the carbocation never ends up on the same carbon as the electron-donating substituent. As a result, the intermediates from attack at the ortho and para positions are more stable than the intermediates from attack at the meta position. As a result, ortho and para substitution is faster than meta substitution. Summary: *Electron-donating groups activate the ring and direct to the ortho/para positions.*

Alert! This idea of ortho and para being similar to each other, but opposite of meta, is a recurring theme in Chapters 23 and 24. Look for other times this occurs.

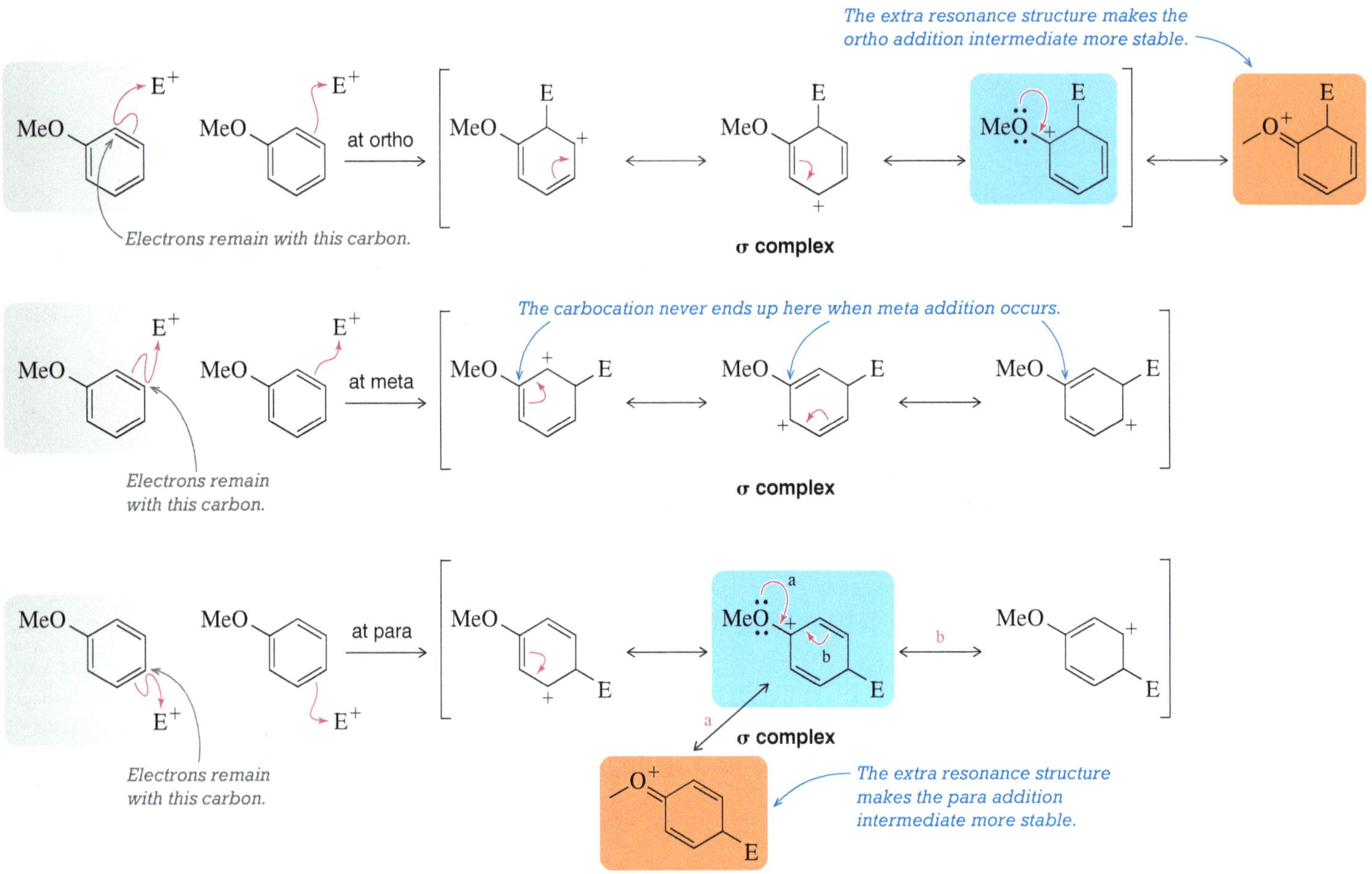

Figure 23.51 Electron-donating groups direct ortho/para because the resulting intermediates are stabilized by an additional resonance structure. [*There is a lot going on in this figure. Spend a few minutes on it to make sure you understand it.*]

The methoxy group and many electron donating groups also withdraw by induction. Resonance has a greater effect.

When a substituent is determined to direct ortho and para, the electrophile will add to the least sterically demanding, or the para, position [Figure 23.52(a)]. This is not 100% consistent, depending on the steric demands of the substituent on the ring and the electrophile being added, but to keep things simple (and logical), this will be our rule. [*Check with your instructor to see how they handle this. Assessment 23.38 lets you think about why and when ortho might be favored.*] And, finally, if you have multiple substituents on the ring directing to different places, the better electron-donating substituent determines where the electrophile goes [Figure 23.52(b)].

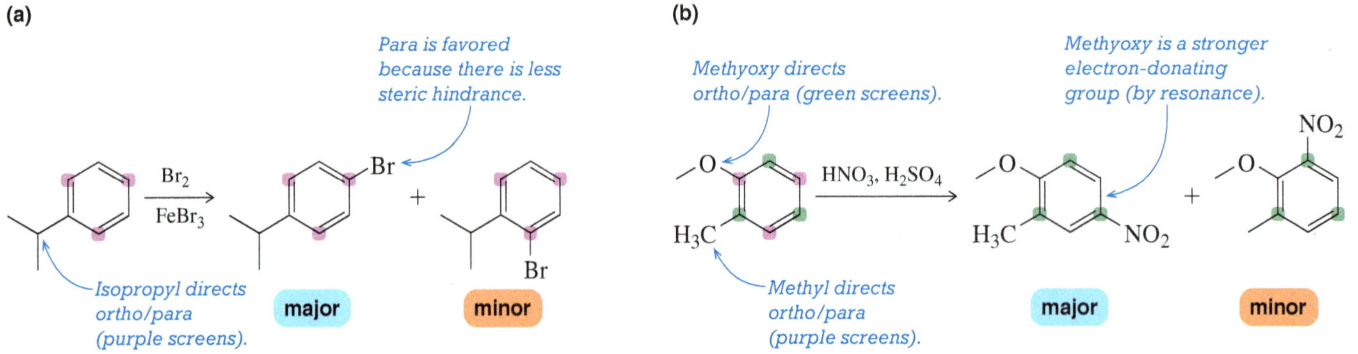

Figure 23.52 Choosing between ortho and para. (a) The least sterically hindered position is favored. (b) The stronger electron-donating group determines the regioselectivity.

Assessment 23.35 Draw a reaction coordinate diagram that rationalizes the preferential reaction at ortho over meta in Figure 23.51.

Assessment 23.36 Alkyl groups also direct ortho and para. Substituting a methyl group for the methoxy group of Figure 23.51, explain this outcome.

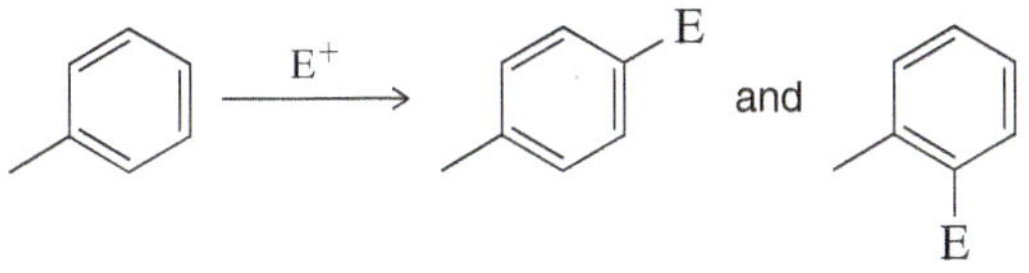

Assessment 23.37 In Section 23.7.4, we learned that Friedel–Crafts alkylation suffers from overalkylation. (a) Draw the product of a Friedel–Crafts reaction that resulted in three methyl groups adding to the ring. (b) Why is hard to stop at the addition of one alkyl group?

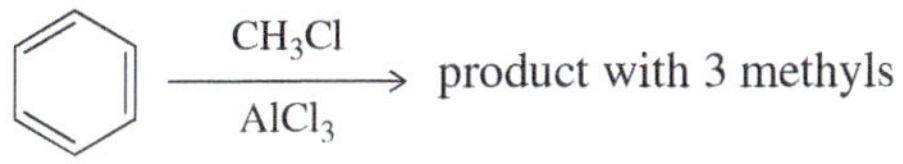

Assessment 23.38 When a large electron-donating group is attached to benzene, the para product is formed exclusively. Rationalize that no ortho product forms using a reaction coordinate diagram.

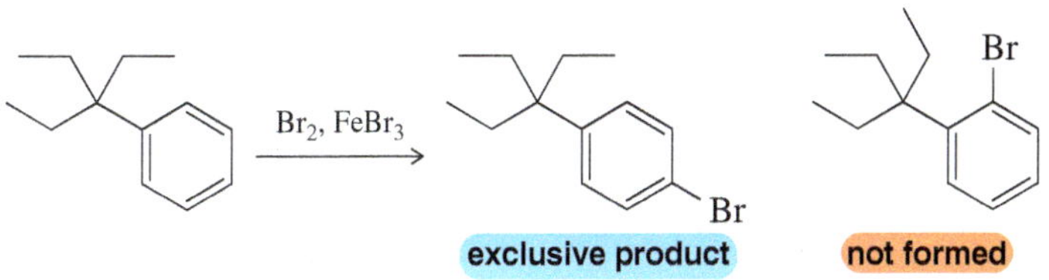

Assessment 23.39 With a small electron-donating group on the ring, it is possible to get as much as a 2:1 ratio of ortho to para. Why might this be?

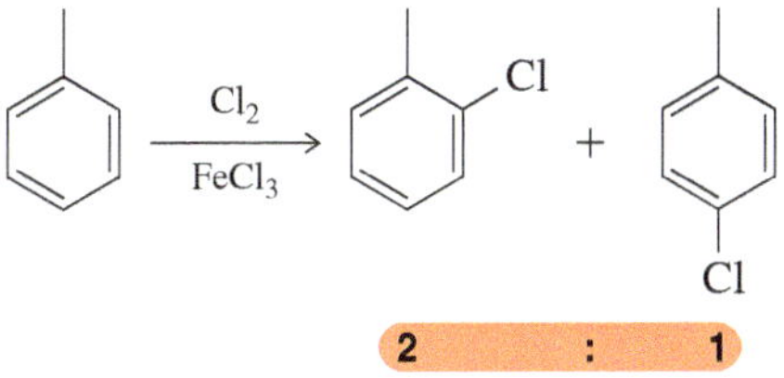

Assessment 23.40 Predict the major product of the following electrophilic aromatic substitution reactions.

(a)

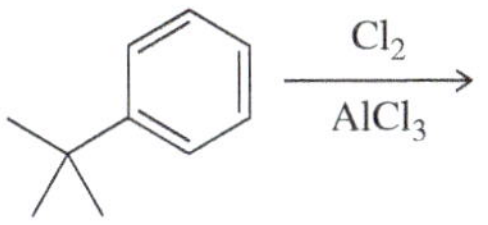

(b)

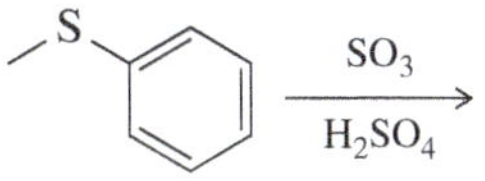

(c)

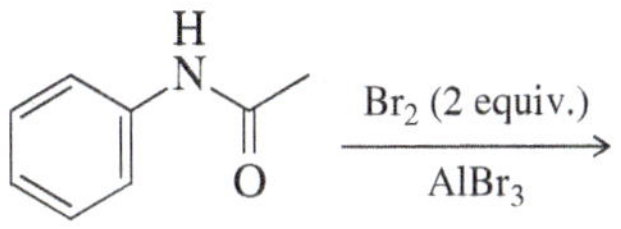

23.8.2 Electron-withdrawing groups deactivate and direct meta

Substituents can be electron-withdrawing by either resonance or induction, but do not include halogens. Several examples of electron-withdrawing groups are shown in **Figure 23.53**. Electron-withdrawing groups make the aromatic ring less electron-rich and less nucleophilic, which decreases the rate of electrophilic aromatic substitution. In this way, electron-withdrawing groups are said to "deactivate" the ring.

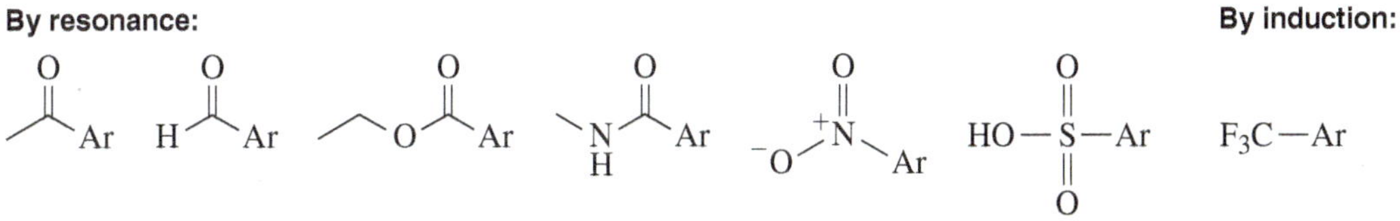

Figure 23.53 Electron-withdrawing groups by (a) resonance and (b) induction (Ar = aromatic ring).

At some point, the addition of too many electron-withdrawing groups to a ring can cause it to be so electron-deficient that it won't attack electrophiles at all. While knowing when this happens is often a matter of intuition, or the product of extensive experimentation, we can generalize to help us solve problems on organic chemistry exams. First, if you want to add electrophiles to a ring in a three-step synthesis, for example, it is always better to put the electron-donating groups on first [**Figure 23.54(a)**]. By doing this, you ensure that the ring stays reactive enough to do future substitutions. Second, Friedel–Crafts alkylation and acylation are especially sensitive to electron-withdrawing groups on the ring. Groups that withdraw electrons by resonance (Figure 23.53) tend to prevent either alkylation or acylation from occurring [**Figure 23.54(b)**].

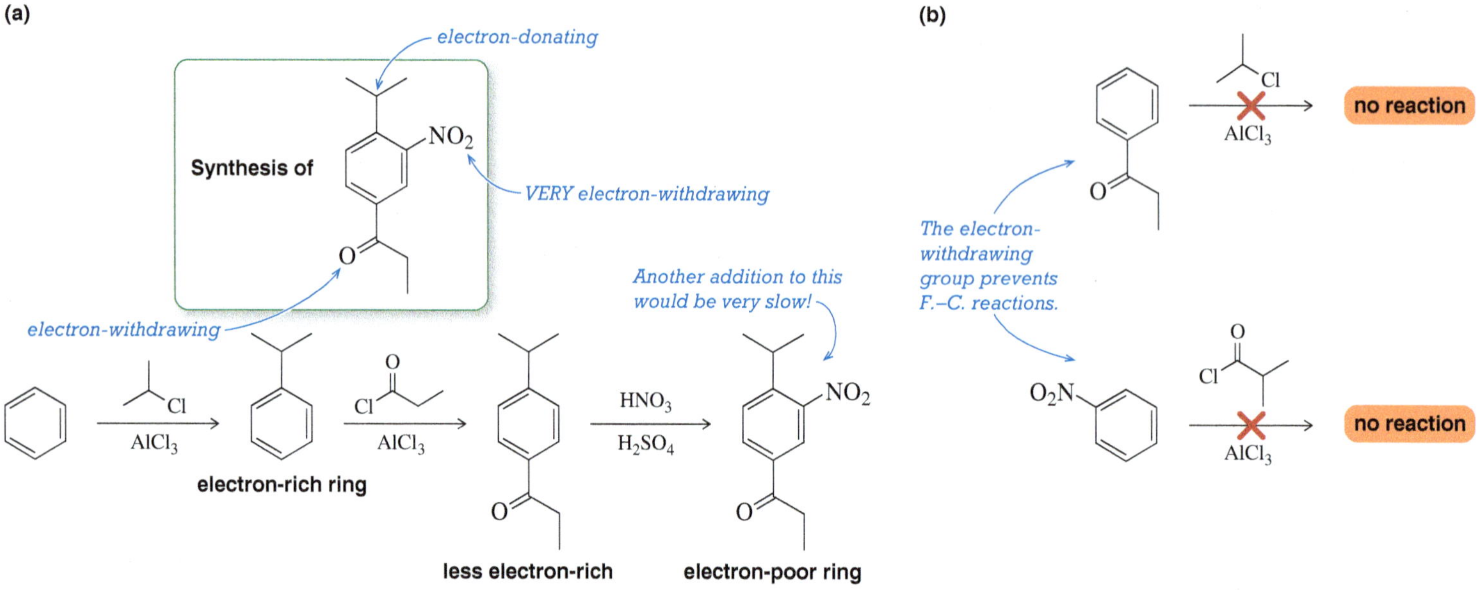

Figure 23.54 Electron-withdrawing groups deactivate the ring. (a) In synthesis, keep the ring as electron-rich as possible by adding substituents in a logical order. (b) Strong electron-withdrawing groups completely shut down Friedel–Crafts alkylation and acylation.

To rationalize the regiochemical outcome, we can analyze the intermediates produced via addition at ortho, meta, and para positions in the first step. **Figure 23.55** shows these intermediates in the presence of the nitro functional group, which is meant to represent a general

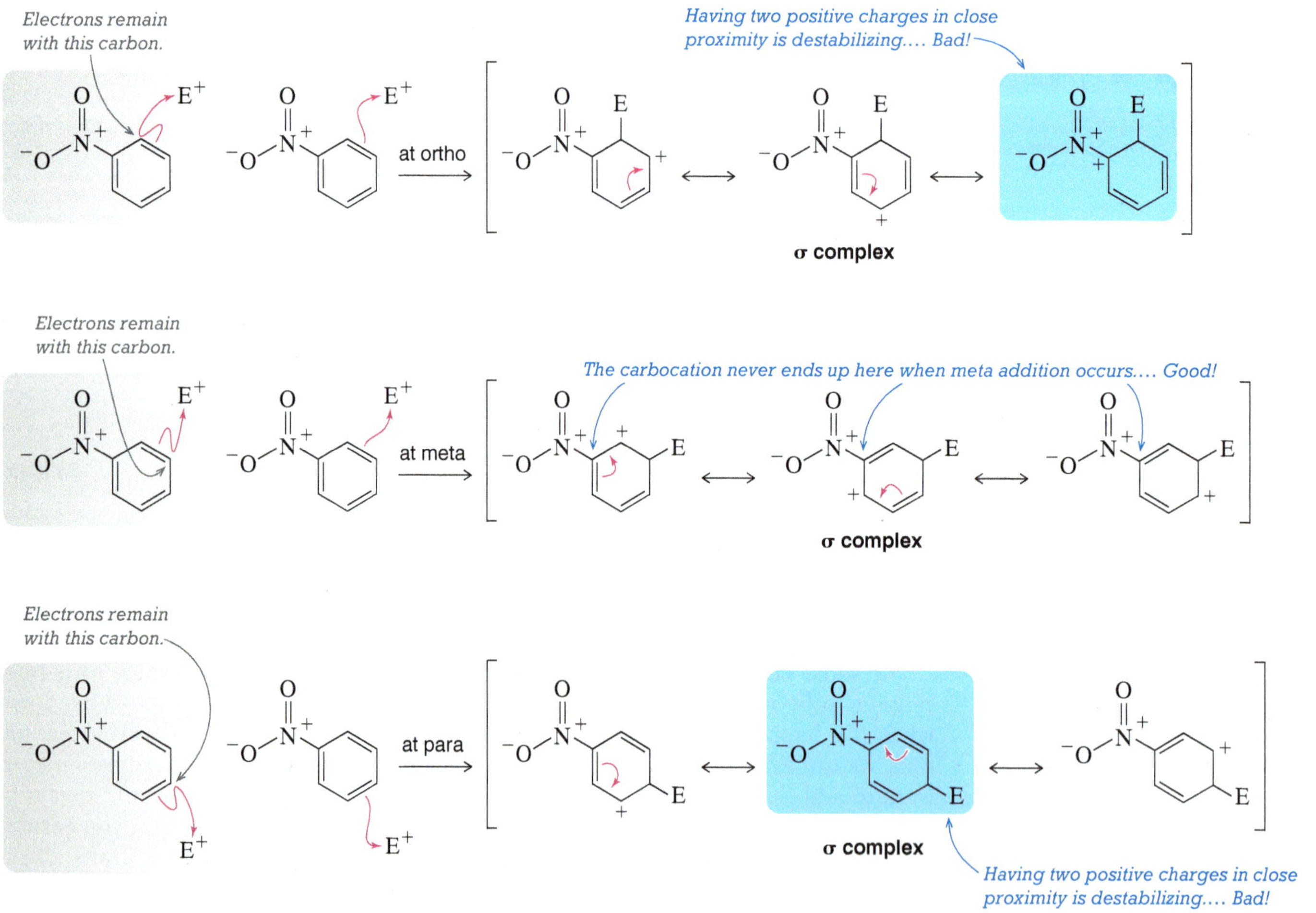

Figure 23.55 Electron-withdrawing groups direct meta to avoid the resonance structures that put a positive charge on the same carbon as the electron-withdrawing group.

electron-withdrawing group. [*Which arenium ion intermediates stand out as being more or less stable?*] Look closely at the resonance structures screened in blue that arise from ortho and para addition. In these resonance structures, the carbocation is on the same carbon as the electron-withdrawing substituent. Having two like charges in close proximity is a destabilizing interaction. In the case of the meta intermediate, the carbocation rarely ends up on the same carbon as the electron-withdrawing substituent. [*Again, ortho/para are the same, distinct from meta.*] As a result, the intermediates from attack at the ortho and para positions are less stable than the intermediates from attack at the meta position. As a result, meta substitution, although still slowed by the electron-withdrawing group, is faster than ortho/para substitution. Summary: *Electron-withdrawing groups deactivate the ring and direct to the meta position.*

Once more, if you have multiple substituents on the ring directing to different places, the more electron-donating substituent determines where the electrophile goes (**Figure 23.56**). Among two or more similar choices based on electronic effects, the least sterically hindered position is favored.

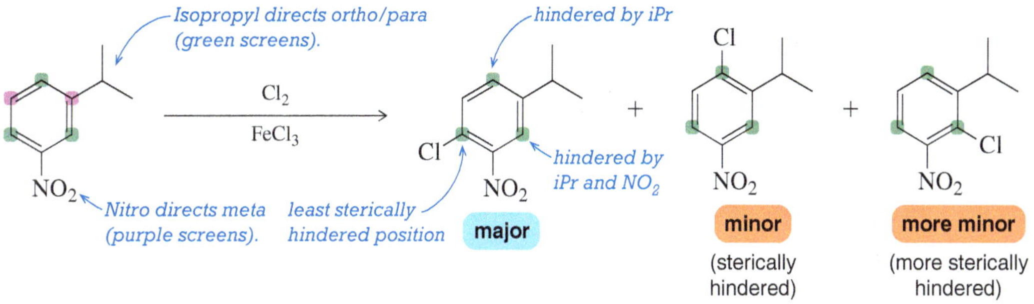

Figure 23.56 Prioritizing substituent effects. The stronger electron-donating group determines the regioselectivity.

Assessment 23.41 Draw a reaction coordinate diagram that rationalizes the preferential reaction at the meta position over the ortho position in Figure 23.55.

Assessment 23.42 Recall from Section 23.7.5 that Friedel–Crafts acylation always stops after the addition of a single acyl group. Why?

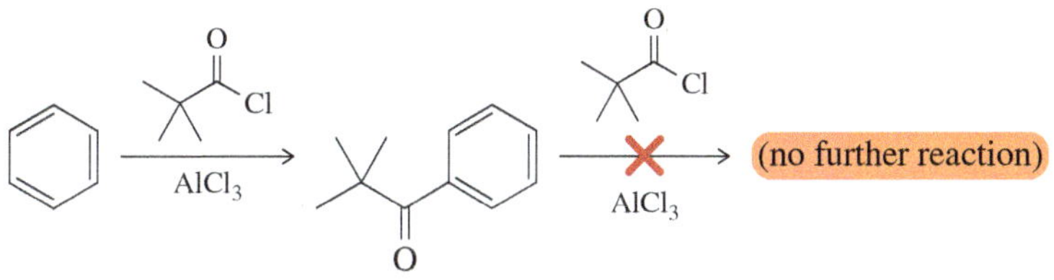

Assessment 23.43 Predict the products of the following electrophilic aromatic substitution reactions.

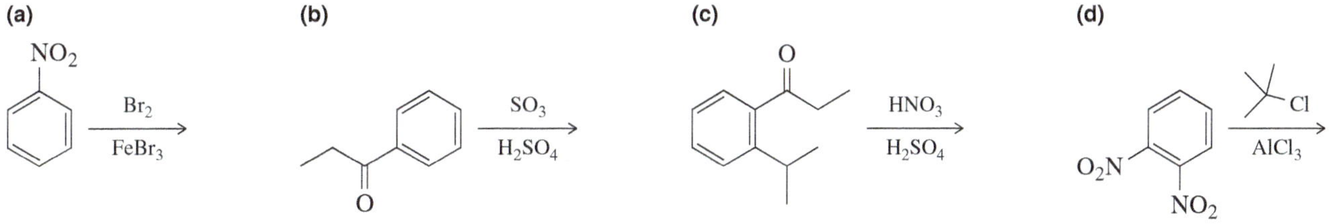

23.8.3 Halogens deactivate and direct ortho/para

To review, electron-donating groups activate the ring and direct ortho/para. Electron-withdrawing groups deactivate the ring and direct meta. [*Seems easy and logical enough.*] The outliers in this dichotomy [*always an outlier . . . damned organic chemistry!*] are the halogens. Halogens are electronegative, so they are weakly electron-withdrawing, which means they make the aromatic ring less electron-rich and less nucleophilic, and they decrease the rate of electrophilic aromatic substitution. In this way, halogens are said to "deactivate" the ring. [*Nothing outlier-ish yet . . . .*]

To rationalize the regiochemical outcome, we can analyze the intermediates produced via addition at ortho, meta, and para positions in the first step. **Figure 23.57** shows the intermediates produced when an electrophile adds ortho, meta, and para to a chlorine. [*For the last time: Which arenium ions are more stable?*] In the resonance structures screened in blue that arise from ortho and para addition, the carbocation is on the same carbon as the halogen. Although the chlorine isn't great at donating by resonance, an additional resonance structure can be drawn that slightly stabilizes these intermediates. In the case of the meta intermediate, the carbocation never ends up on the same carbon as the halogen. [*Ortho/para = similar; meta = opposite.*] As a result, the intermediates from attack at the ortho and para positions are more stable than the intermediates from attack at the meta position. Thus, ortho and para substitution is faster than meta substitution. Summary: *Halogens deactivate the ring but direct to the ortho/para positions.*

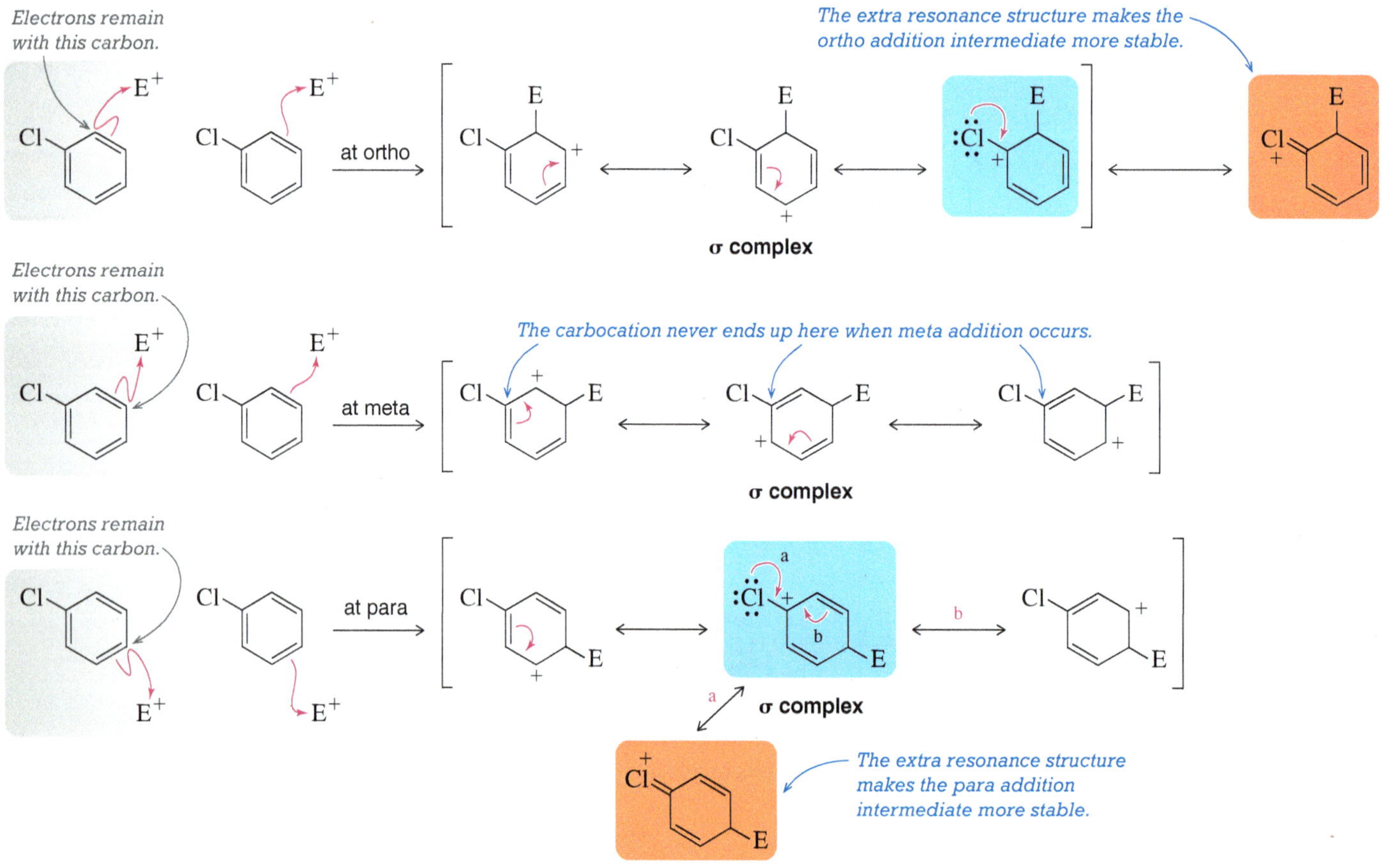

Figure 23.57 Halogens direct ortho/para because the resulting intermediates are stabilized by an additional (even if not-so-contributing) resonance structure.

When a halogen is on the same ring as an electron-*withdrawing* group, the halogen will direct the substitution [**Figure 23.58(a)**]. When a halogen is on the same ring as an electron-*donating* group, the electron-donating group will direct the substitution [**Figure 23.58(b)**]. Among two similar choices based on electronic effects, the least sterically hindered position is favored.

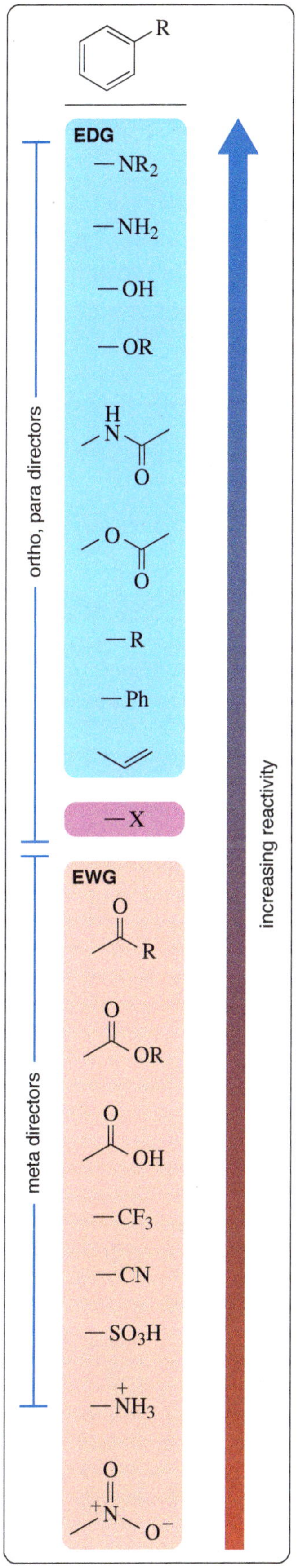

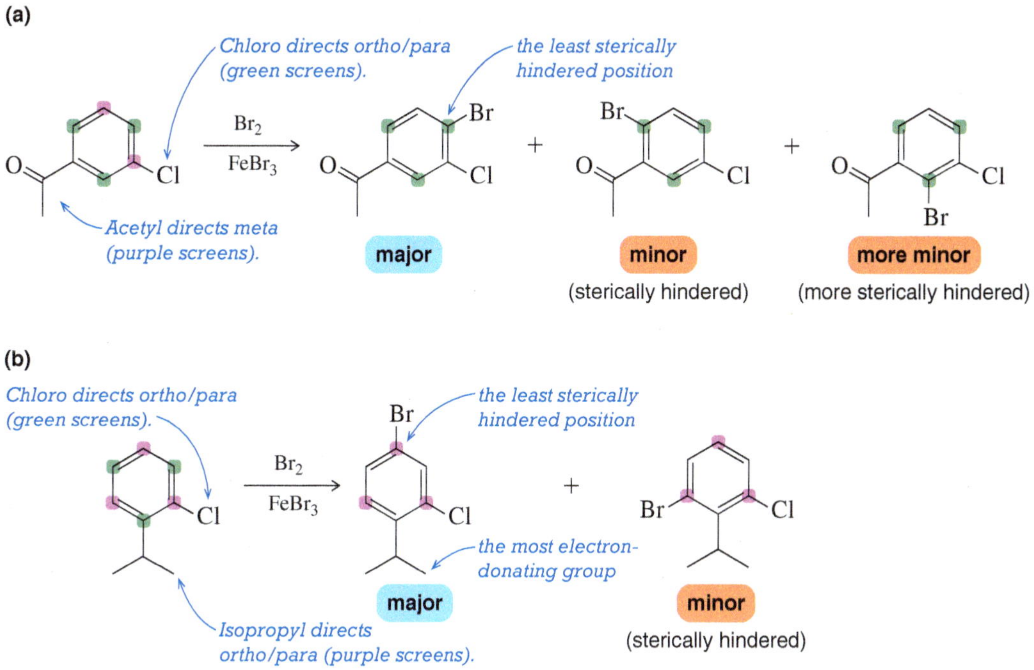

Figure 23.58 Prioritizing substituent effects. The stronger electron-donating group determines the regioselectivity.

23.8 SUMMARY

- Electron-donating groups activate the ring and direct substitution at the ortho and para positions. Electron-withdrawing groups deactivate the ring and direct substitution at the meta position. Halogens deactivate the ring and direct substitution at the ortho and para positions. [*See the figure in the margin, where EDG = electron-donating group and EWG = electron-withdrawing group.*]

Assessment 23.44 Draw a reaction coordinate diagram that rationalizes the preferential reaction at the para position over the meta position in Figure 23.57.

Assessment 23.45 Predict the major product of the following electrophilic aromatic substitution reactions.

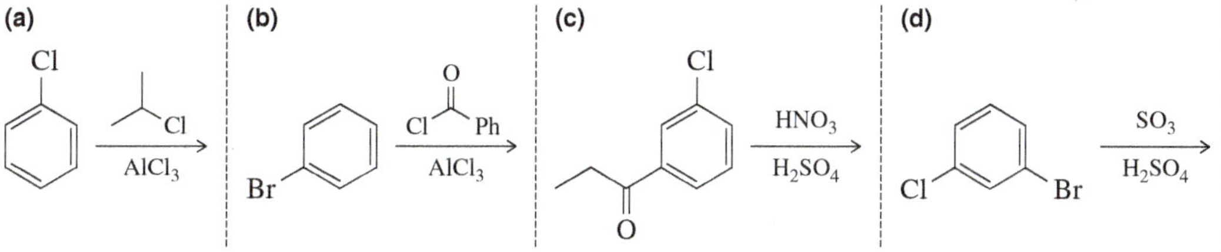

23.8.4 Synthesis of substituted benzenes

A common [*and fun*] type of assessment involves the synthesis of substituted benzenes starting with benzene itself. Two principles guide the ideal synthesis. First, keep the ring as electron-rich as possible so that the next electrophile will add. Second, make sure that the groups that are present will direct to the appropriate position. In terms of the problem-solving process, first identify the reactions you will need to do in order to install a group. Only then should you think about the order in which the groups are added.

Since this is best learned by doing, we'll practice applying these principles by solving Assessment 23.46 (a) and (d) together. Make sure you have tried these on your own before proceeding.

Assessment 23.46 Beginning with benzene, synthesize the following substituted benzenes. The ideal number of steps is indicated.

(a) (b) (c) (d)

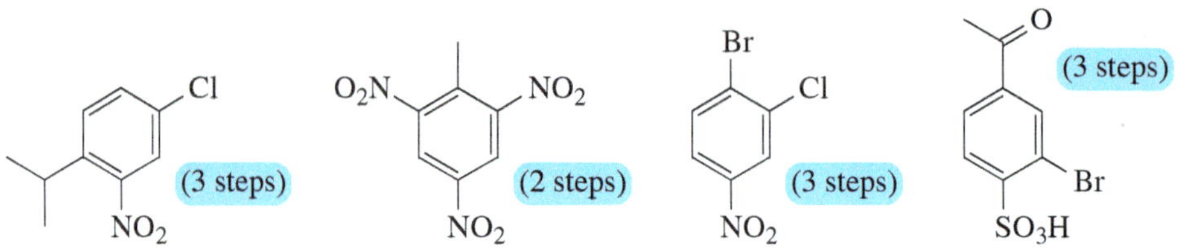

Solved Assessment 23.46(a) and (d)

Substituted benzene	Reactions to be used (not in this order)	Synthesis
(a) 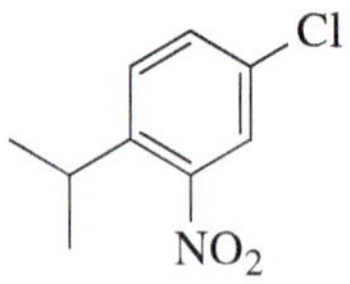	a. *i*PrCl, $AlCl_3$ b. Cl_2, $FeCl_3$ c. HNO_3, H_2SO_4	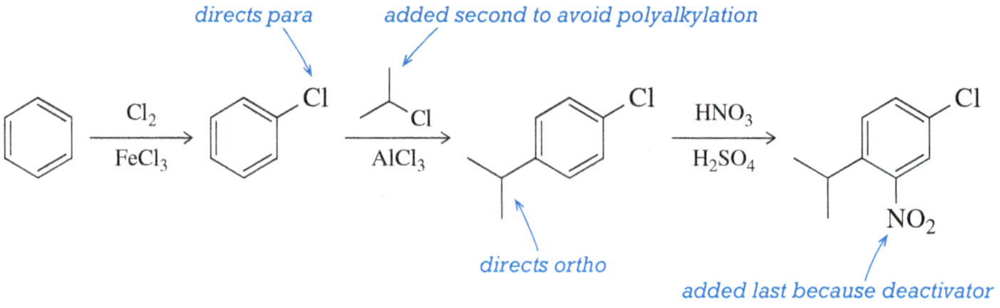
(d) 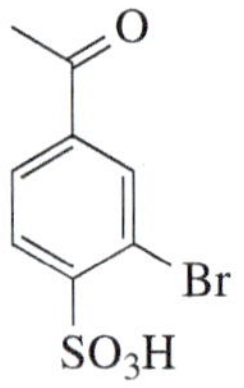	a. Br_2, $FeBr_3$ b. CH_3COCl, $AlCl_3$ c. SO_3, HS_2O_4	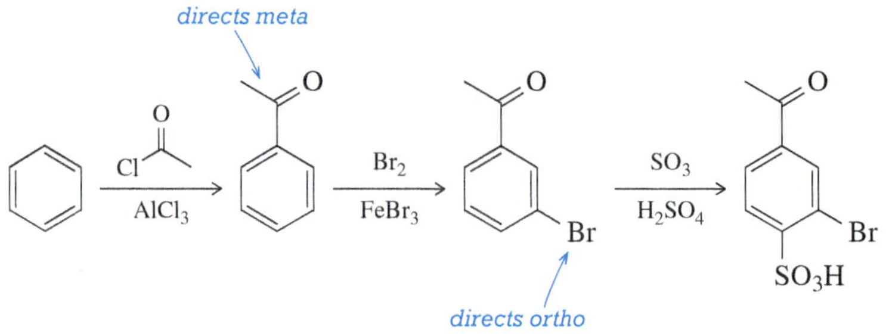

23-D CHEMISTRY IN CONTEXT: Medicinal Chemistry

Synthesis of sulfanilamide

The direct synthesis of sulfanilamide can be achieved using a combination of electrophilic aromatic substitution reactions in conjunction with chemistry studied previously—see **Table 23.7**. [*That's right—you now have the expertise to synthesize the antibiotic that sparked the birth of modern medicinal chemistry. Pretty cool, huh?*]

Table 23.7 The synthesis of sulfanilamide.

Step	Reaction description	Reaction
1	Nitration of benzene using electrophilic aromatic substitution gives nitrobenzene (Section 23.7.2).	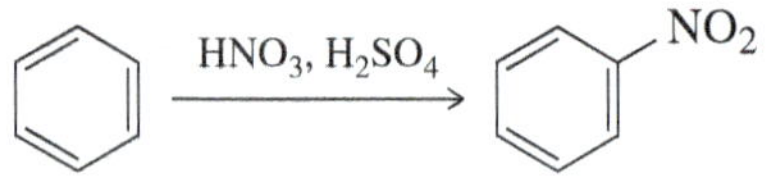
2	Hydrogenation of a nitro group (Section 25.6.3) gives aniline (aminobenzene).	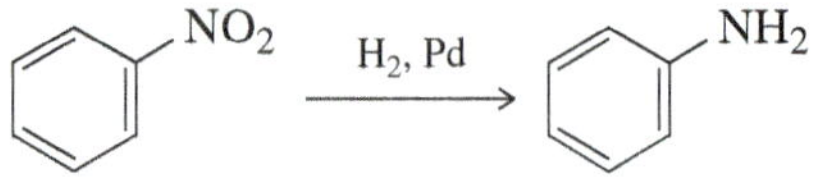

(*continued*)

Table 23.7 *Continued*

Step	Reaction description	Reaction
3	Protection of the amine (Section 19.7.3) with acetic anhydride occurs by nucleophilic acyl substitution.	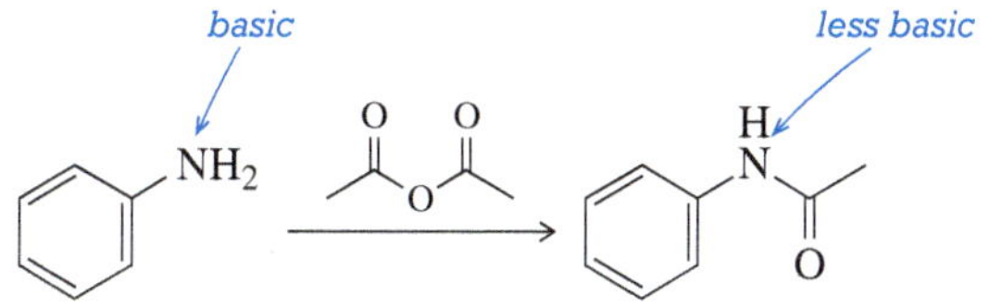
4	Electrophilic aromatic substitution with chlorosulfonic acid gives the sulfonyl chloride (analogous to Section 23.7.3).	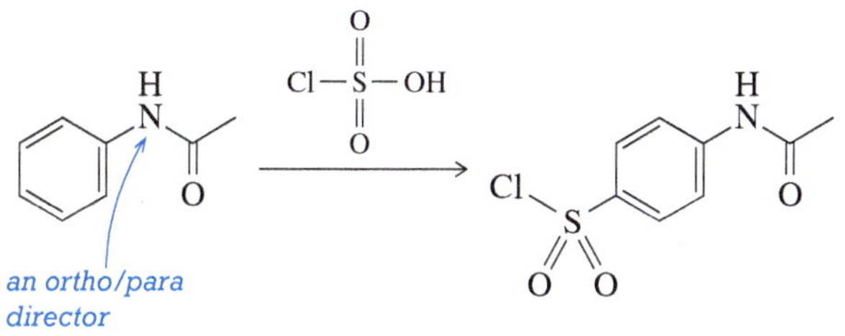
5	Sulfonylation of ammonia occurs by nucleophilic acyl substitution (analogous to reactions in Sections 13.7.1 and 19.7.3).	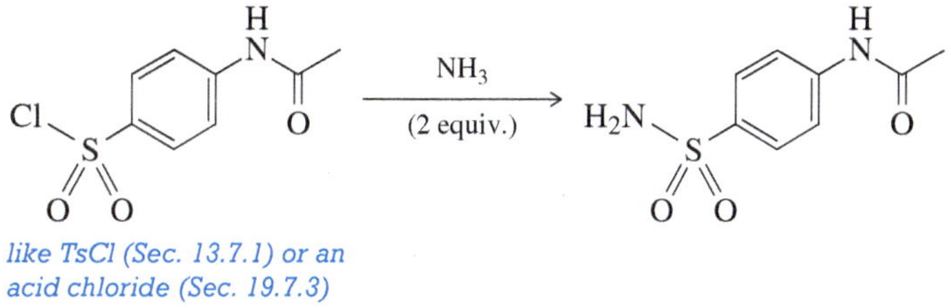
6	Deprotection of the amine via acid-catalyzed hydrolysis of an amide (Section 19.7.1).	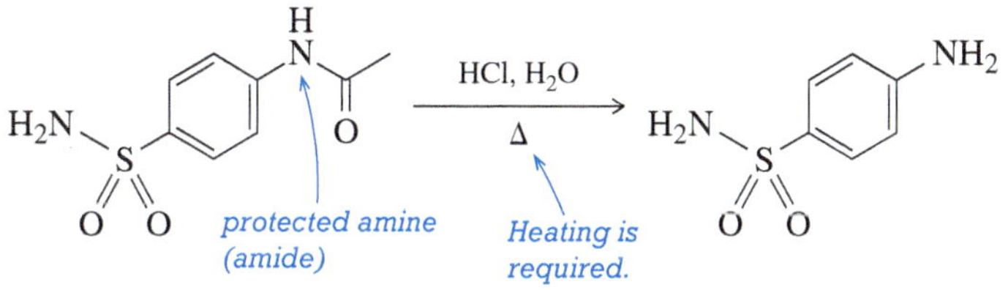

Assessment 23.47 Why is the amide an ortho/para director in step 4?

Assessment 23.48 Show a mechanism for step 4.

23.9 Nucleophilic aromatic substitution

Aromatic rings can also participate in nucleophilic aromatic substitution reactions. Whereas electrophilic aromatic substitution replaced a hydrogen, a leaving group is necessary for nucleophilic aromatic substitution. Halides are the most common leaving group used for this purpose. As the name implies, a nucleophile is substituted for a halide, so the aromatic ring must be an electrophile. [*This is the opposite of electrophilic aromatic substitution, where the ring is the nucleophile.*] An example of this reaction is shown in **Figure 23.59**, where hydroxide replaces a chloride to make a phenol that is immediately deprotonated because its conjugate base is stabilized by resonance (Section 24.2.1).

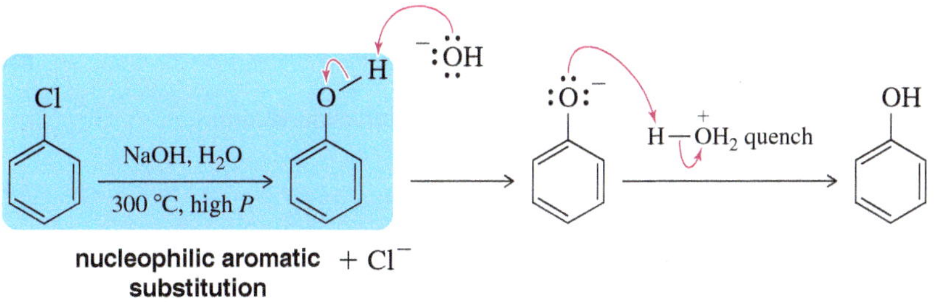

Figure 23.59 Nucleophilic aromatic substitution.

Because the ring is the electrophile in these reactions, having more electron-withdrawing groups attached to it should increase the rate of reaction. Let's run a typical experiment to see if this is true and what the data might tell us about the mechanism.

23.9.1 The addition/elimination mechanism

An electron poor ring should be a stronger electrophile in nucleophilic aromatic substitution reactions.

Thinking Like a Chemist

Table 23.8 shows how the temperature necessary for nucleophilic aromatic substitution changes as a result of additional nitro groups. Because nitro groups are highly electron- withdrawing, each additional one should make the reaction happen more efficiently. And for the most part, this is true, with a lone exception. [*Do you see it?*] Starting with chlorobenzene, which requires very harsh conditions, the addition of an extra nitro group at the ortho or para positions makes the reaction happen at a lower temperature. Even with two nitro groups in meta positions, however, harsh conditions are once again required. [*What might this mean for the mechanism of the reaction?*]

Table 23.8 The ease of nucleophilic aromatic substitution depends on the substituents attached to the aromatic ring.

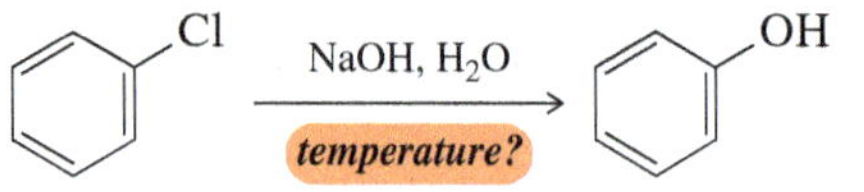

Entry	Aromatic ring	Temperature necessary for reaction (°C)
1	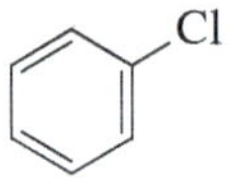	300
2	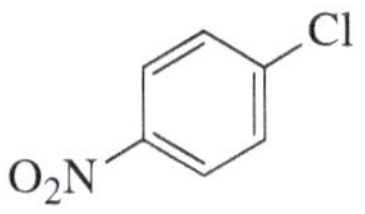	100
3	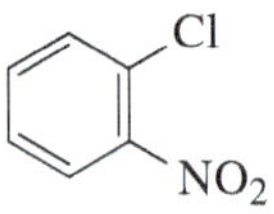	100
4	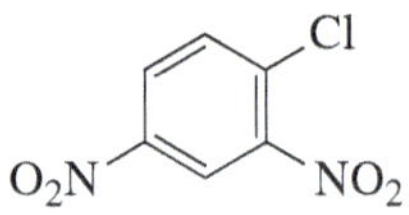	25
5	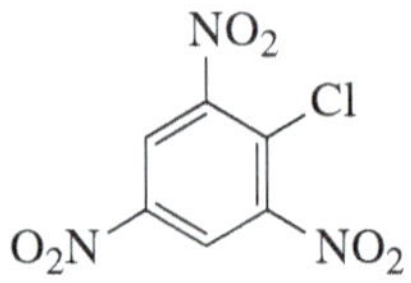	0
6	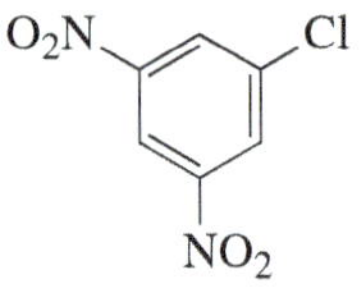	~300

Before we talk about the implications of the data in Table 23.8, let's rule out the usual first guess for a mechanism—namely, S_N2. Although we have seen the S_N2 mechanism many times throughout the text, the leaving group is on an sp^2-hybridized carbon, so it cannot undergo S_N2 substitution [Figure 23.60(a)]. Have we seen substitution at an sp^2-hybridized carbon before? [*You've got this! Take a moment to see if you can think of it.*] In Section 19.6, we studied the nucleophilic acyl substitution of carboxylic acid derivatives, where the most notable comparison to this reaction is the reaction of an acid chloride with hydroxide. Mechanistically, hydroxide attacks the electrophilic carbonyl carbon to create a tetrahedral intermediate, which immediately collapses to reform a π bond and kick out the stable leaving group [Figure 23.60(b)].

The data from Table 23.8 imply that there is some "communication" between the carbon bearing the leaving group and the ortho and para positions. Based on what we know about benzene and electrophilic aromatic substitution, this suggests that the π electrons are probably involved in this mechanism. As a result, a mechanism similar to nucleophilic acyl substitution looks promising. But do the data make sense in terms of the efficiency of the reaction as a function of the substituents attached? When chlorobenzene is unsubstituted [Figure 23.61(a)], the negative charge in the nonaromatic intermediate is highly unstable. However, the nitro group in the ortho and para positions [Figures 23.61(b) and 23.61(c)] stabilizes the negative charge that results from nucleophilic addition to the aromatic ring. Finally, even with two nitro groups in meta positions [Figure 23.61(d)], the negative charge is never stabilized by resonance.

From Figure 23.61, a couple of points become clear. First, in the addition/elimination of nucleophilic aromatic substitution, the first step (where aromaticity is lost) is rate-determining. The second step is extremely fast because aromaticity is reformed while kicking out an excellent leaving group. Second, in order for a substitution reaction to occur by this mechanism, there must be a strong electron-withdrawing group (by resonance, usually) ortho or para to the good leaving group. The intermediate in this reaction is referred to as a Meisenheimer complex. Finally, more electron-withdrawing groups further increase the rate of reaction (see entries 4 and 5 in Table 23.8). ■

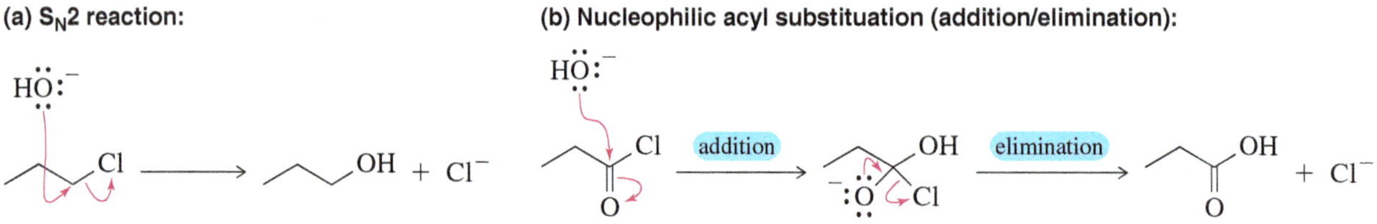

Figure 23.60 Previously studied substitution mechanisms: (a) The S_N2 mechanism of haloalkanes and (b) nucleophilic acyl substitution (by addition/elimination).

New research has suggested that the intermediate posited here is actually a transition state. Chemistry knowledge is ever changing . . . stay tuned. Or better yet, join the chemistry enterprise and discover it for yourself.

(a) Chlorobenzene (entry 1, Table 23.8)

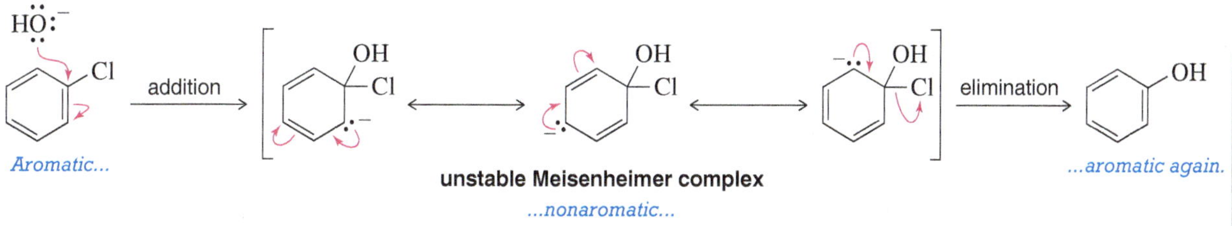

(b) *para*-Nitro (entry 2, Table 23.8)

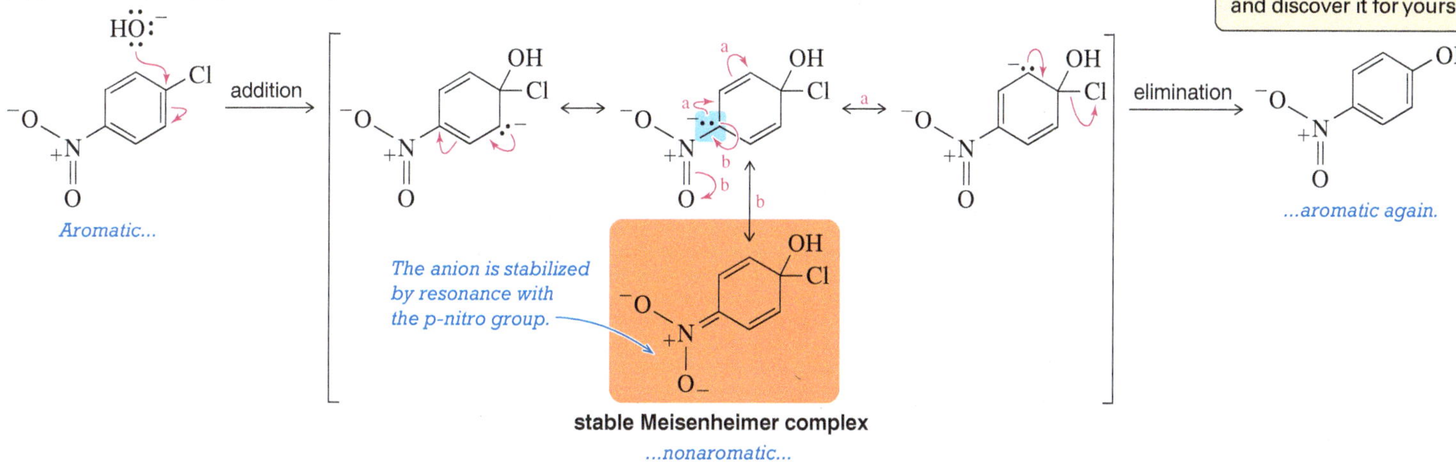

(c) *ortho*-Nitro (entry 3, Table 23.8)

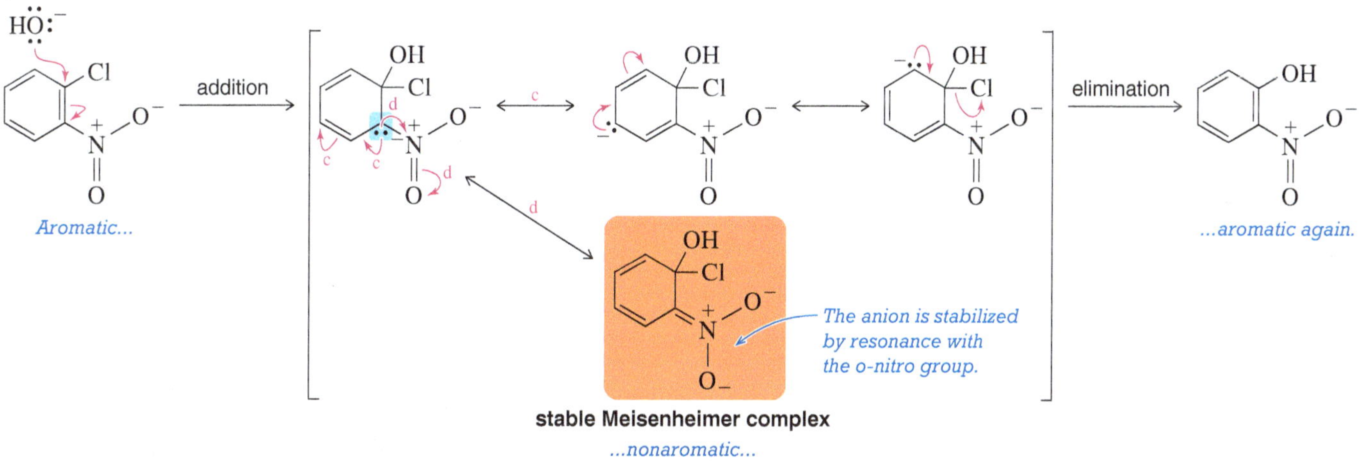

(d) *meta*-Dinitro (entry 6, Table 23.8)

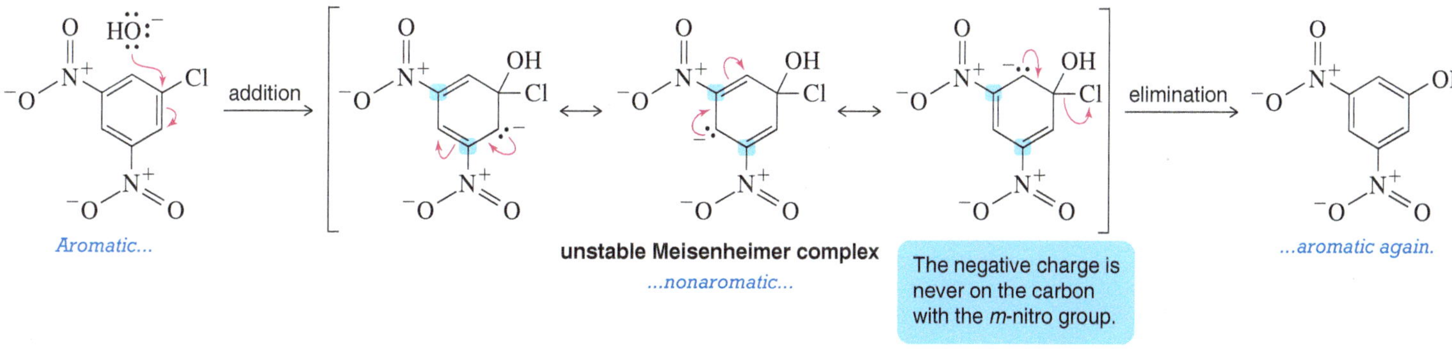

Figure 23.61 Electron-withdrawing groups in only the ortho and para positions stabilize the intermediate produced by nucleophilic attack.

Assessment 23.49 Draw a reaction coordinate diagram for nucleophilic aromatic substitution by the addition/elimination mechanism. Be sure to label the rate-determining step.

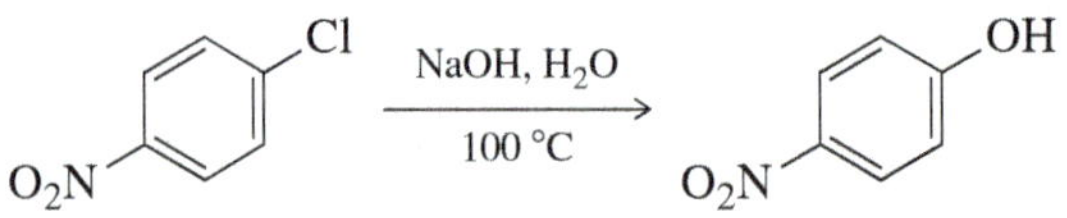

Assessment 23.50 Compare reactions **A** and **B** on a reaction coordinate diagram, making clear the difference in rates. Show only the first step for each because it is rate-determining.

Reaction A: **Reaction B:**

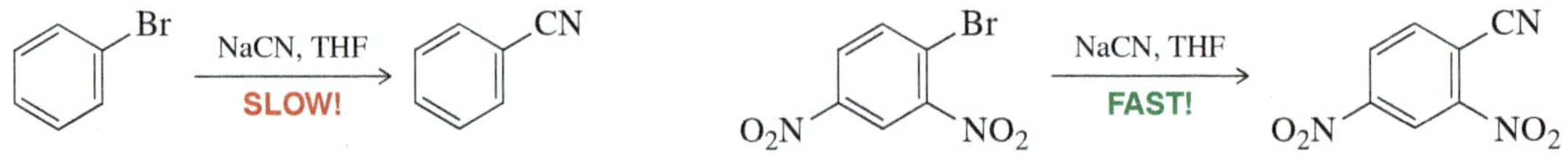

Assessment 23.51 In the reactions in Figure 23.61, a phenol was produced in the substitution step. Because phenols are somewhat acidic ($pK_a = 10$), they are immediately deprotonated in basic solution, necessitating an eventual acid quench. Why is the pK_a of a phenol lower than the pK_a of a normal alcohol (16)?

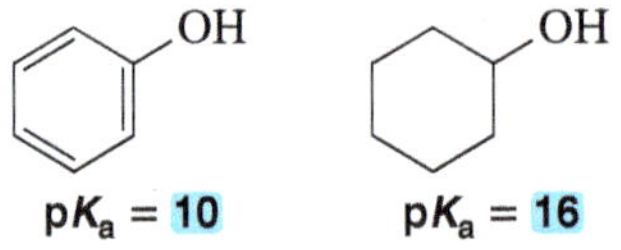

Assessment 23.52 Ketones are good enough electron-withdrawing groups to facilitate the addition/elimination mechanism. Modify the following arrow-pushing mechanism so that it leads directly to **A**. Where have we seen this reaction before?

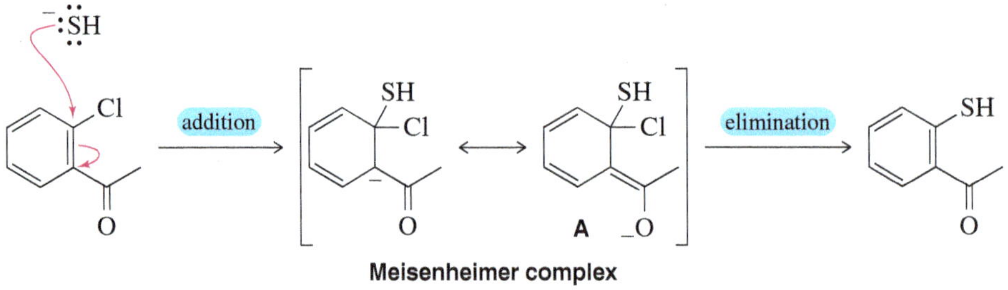

Assessment 23.53 Predict the products of the following nucleophilic aromatic substitution reactions.

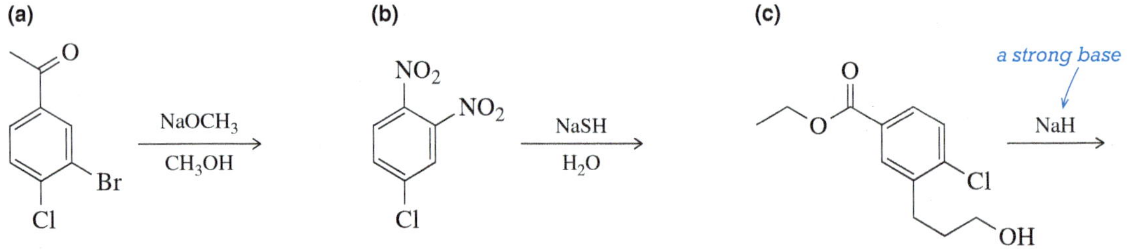

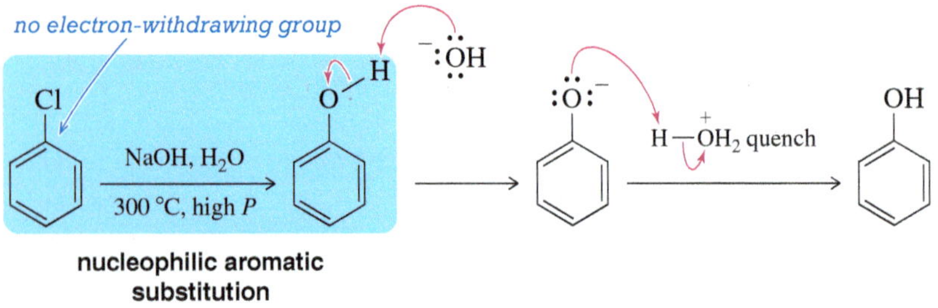

Figure 23.62 Substitution for chlorobenzene must not occur by the addition/elimination mechanism.

23.9.2 The benzyne mechanism

The nucleophilic aromatic substitution shown in Figure 23.62 was originally presented at the beginning of Section 23.9.1. Recall that an ortho or para electron-withdrawing group was necessary to allow substitution of the halide to occur via the addition/elimination mechanism. It follows, then, that the substitution reaction of chlorobenzene, under exceedingly harsh conditions, must occur by a different mechanism.

With no electron-withdrawing group, substitution is only possible under extremely harsh conditions like those shown in Figure 23.62. Under these conditions, hydroxide does *not* act as a nucleophile. Instead, it deprotonates the benzene ring; in a second step, chloride leaves (**Figure 23.63**). This is the E1cb mechanism. [*Recall Section 12.5.1.*] The resulting compound with the "triple bond" in the ring is called **benzyne**. It is highly unstable and can be attacked at either position by hydroxide. Following attack by hydroxide, protonation of the unstable carbanion occurs quickly to give the product phenol. Because this ring was otherwise unsubstituted, attack at either carbon of benzyne gives the same product. Assessment 23.55 shows what happens when there are other substituents on the ring. These harsh conditions make this reaction incompatible with many sensitive functional groups. As such, reactions occurring through the benzyne mechanism aren't often used in organic synthesis.

Figure 23.63 Nucleophilic aromatic substitution by the benzyne mechanism.

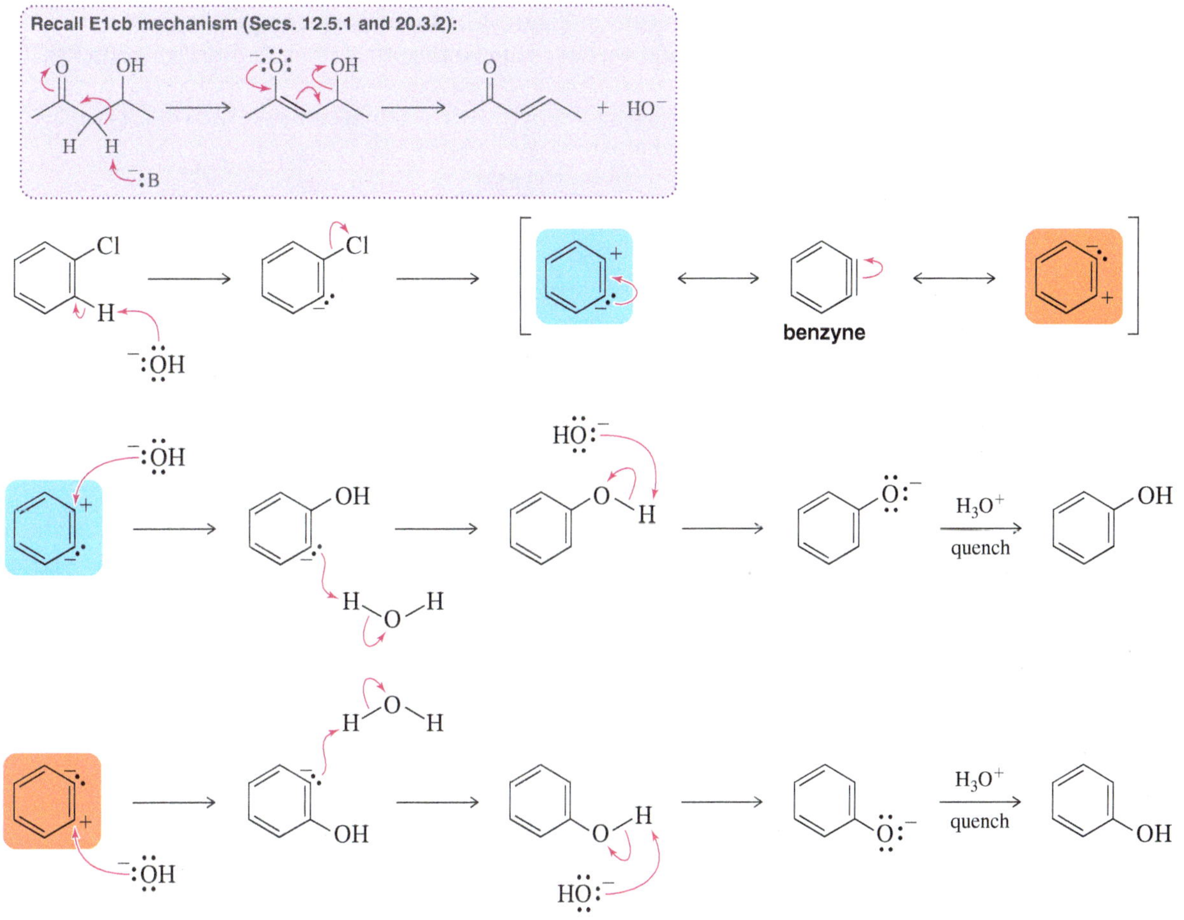

Assessment 23.54 Draw a reaction coordinate diagram for nucleophilic aromatic substitution by the benzyne mechanism. Label the rate-determining step and make sure that this reaction coordinate diagram accounts for the required harsh conditions.

Assessment 23.55 Assume that H_a is the proton removed in the first step of the benzyne mechanism with *p*-chlorotoluene. Draw the benzyne intermediate and identify the two products that will result.

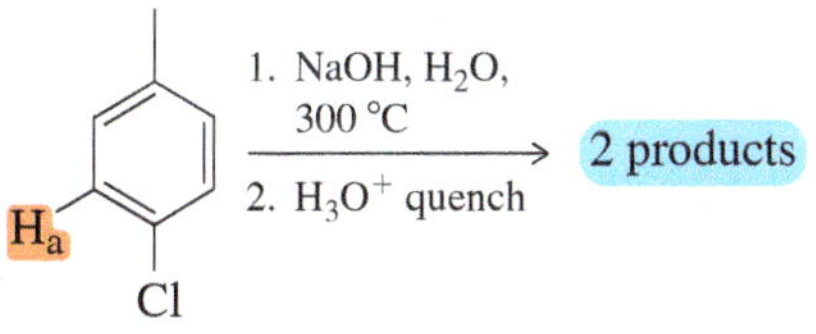

Assessment 23.56 Predict the product(s) of the following nucleophilic aromatic substitution reactions occurring by the benzyne mechanism.

(a)

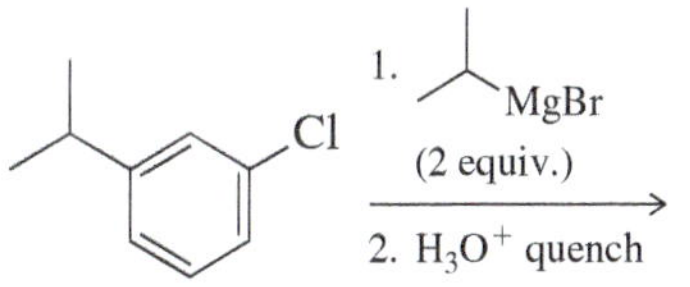

(b)

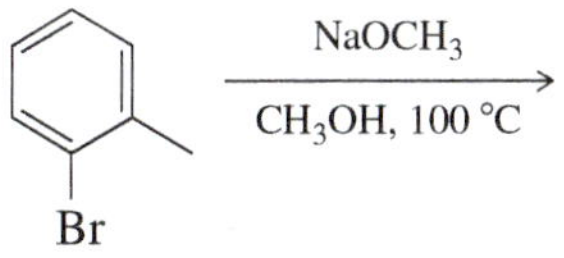

23.10 Other substitution reactions

Reactions learned in other contexts can also be used to add substituents to an aromatic ring. These reactions, exhaustively covered in other parts of the text, are discussed briefly in Sections 23.10.1 and 23.10.2.

23.10.1 Diazonium

In Chapter 25, we discuss the reaction of amines with nitrous acid (HNO_2). Because it is so highly reactive, nitrous acid must be generated by the reaction of sodium nitrite and hydrochloric acid [**Figure 23.64(a)**]. Under these conditions, water is lost to produce a nitrosyl cation (NO^+). The nitrosyl cation is attacked by the nucleophilic amine, and following a series of acid–base reactions, water leaves to give the diazonium ion [**Figure 23.64(b)**]. Although the mechanisms in Figure 23.64 may look complicated, six of the nine steps are acid–base reactions and we have studied the other three in different contexts.

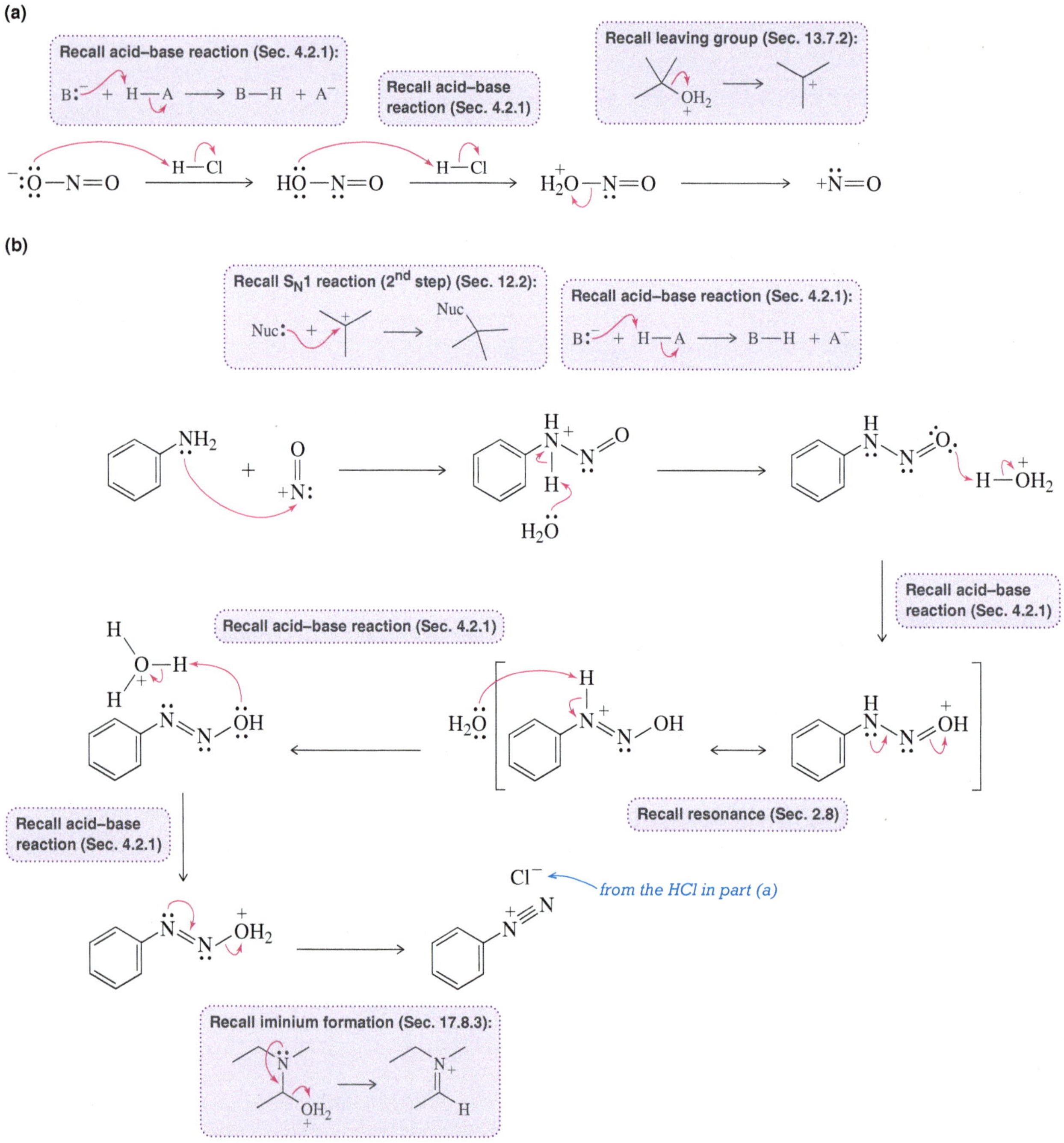

Figure 23.64 **The mechanism of diazonium ion formation involves** (a) the formation of NO^+ and (b) the reaction of NO^+ and an aryl amine.

The diazonium ion, which contains a molecule of nitrogen (an excellent leaving group), is a highly reactive functional group [*We discuss it in greater detail in Chapter 25*]. Fortunately, these compounds can be isolated and stored below 10 °C. This allows them to be used in the various substitution reactions shown in **Figure 23.65**. The reactions involving a copper salt are referred to as **Sandmeyer reactions.** The mechanisms of these reactions are not well understood, but many are thought to involve free radicals.

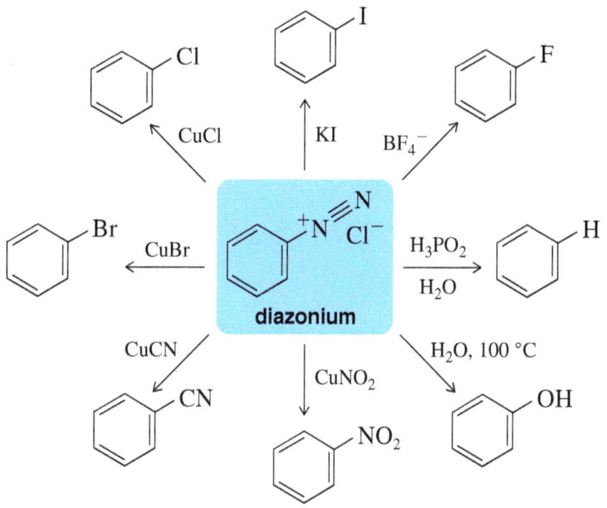

Figure 23.65 The conversion of diazonium ions to a number of functional groups.

The synthetic utility of the diazonium process is in the conversion of a nitro group, through reduction to the amine [*see Section 25.6.3*], to a diazonium ion. In this three-step process (**Figure 23.66**), a nitro group can be converted into any of the functional groups in Figure 23.65.

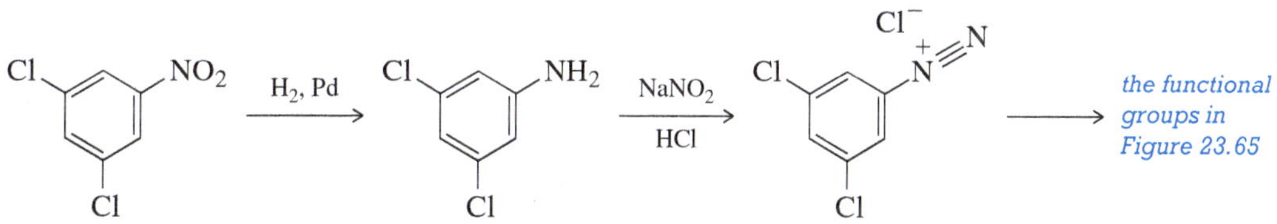

Figure 23.66 The conversion of a nitro group to many other functional groups through a diazonium ion.

Assessment 23.57 Predict the products of the following reactions.

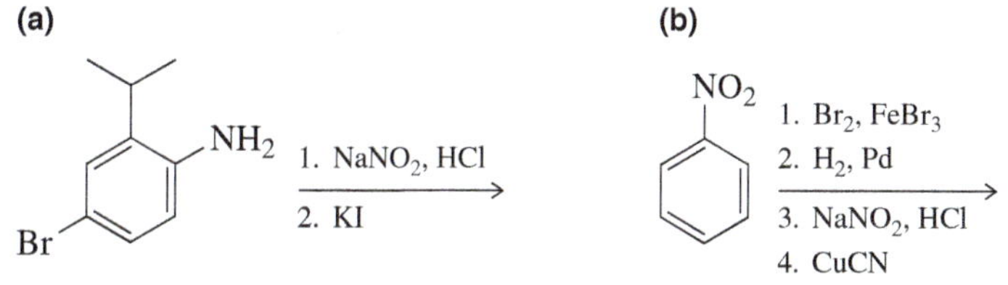

Assessment 23.58 Conversion of *t*-butylbenzene to a phenol results in two products when the reaction occurs through the benzyne mechanism. Devise a sequence involving a diazonium ion in which only one phenol product is produced.

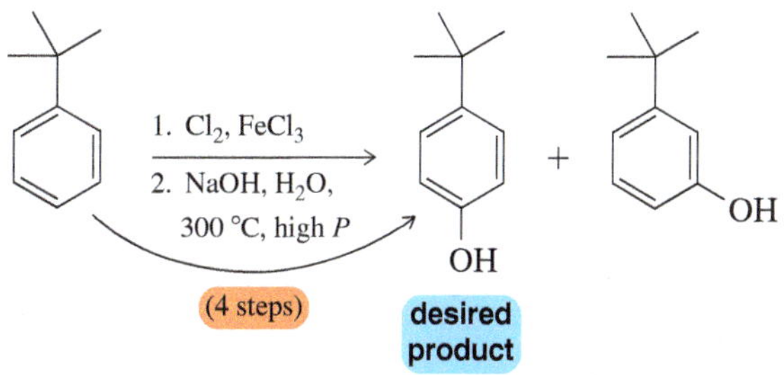

23.10.2 Palladium coupling with aryl halides

In all of the substitution reactions studied thus far, the only thing missing is the ability to form new C—C bonds under mild conditions. Friedel–Crafts alkylation and acylation allowed us to form these bonds, but that was only under harsh conditions and with specific requirements in terms of the electrophiles used and the substituents that could be present on the ring. The use of the palladium-coupling reactions, first presented in Chapter 16, greatly expands our ability to functionalize aromatic rings by forming new C—C bonds. As shown in **Figure 23.67**, alkenes, alkynes, and aromatics can be coupled to aryl halides using the Suzuki, Stille, Sonogashira, and Negishi coupling reactions.

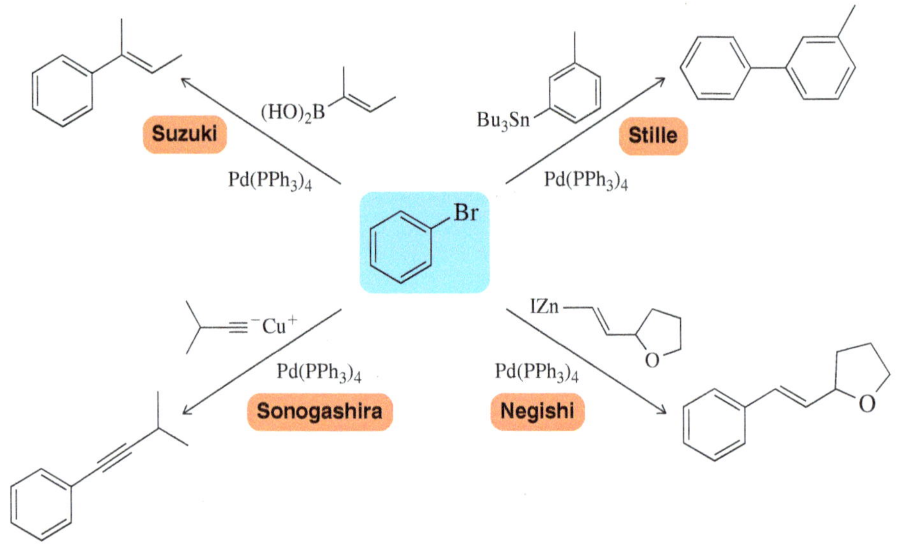

Figure 23.67 Palladium-coupling reactions used to functionalize aryl halides.

Mechanistically, these reactions occur by the catalytic cycle discussed in detail in Section 16.4.1. **Figure 23.68** shows this cycle using a general metal in place of boronic acid [$B(OH)_2$; Suzuki], tributylstannane (Bu_3Sn; Stille), sodium (Na^+; Sonogashira), and zinc (ZnI; Negishi) from Figure 23.67. In general, oxidative insertion of palladium (Pd^0) into the Ar—Br bond is followed by transmetallation with the relevant R—M species. Finally, reductive elimination forms the new C—C bond, releasing Pd^0 to participate in the catalytic cycle again.

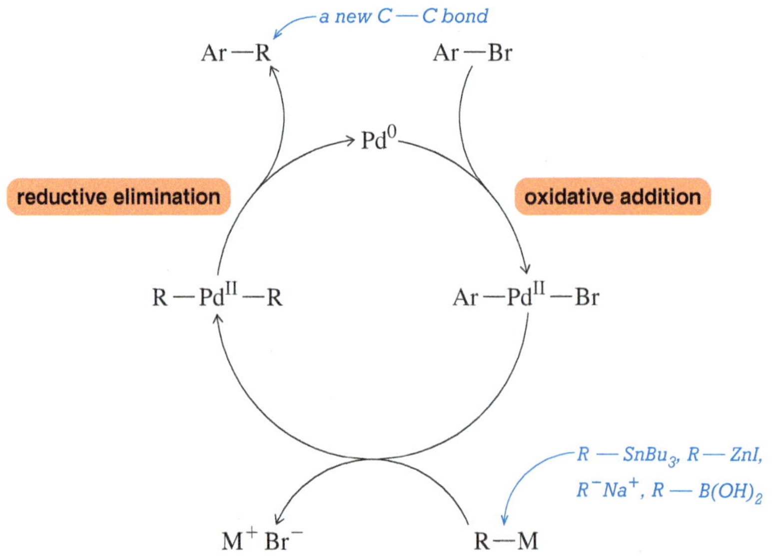

Figure 23.68 The catalytic cycle of palladium coupling reactions of aryl halides. [*PPh_3 ligands have been removed for clarity.*]

Assessment 23.59 Predict the products of the following reactions.

(a)

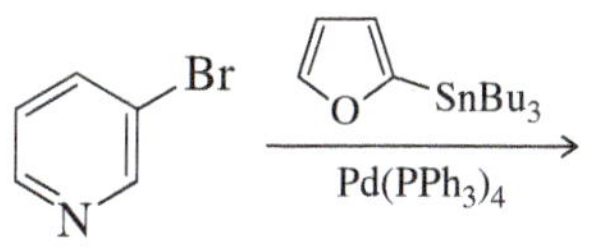

(b)

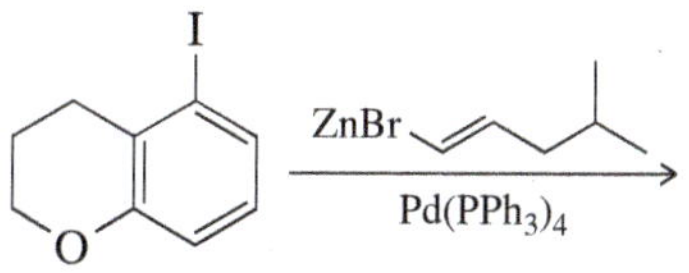

(c)

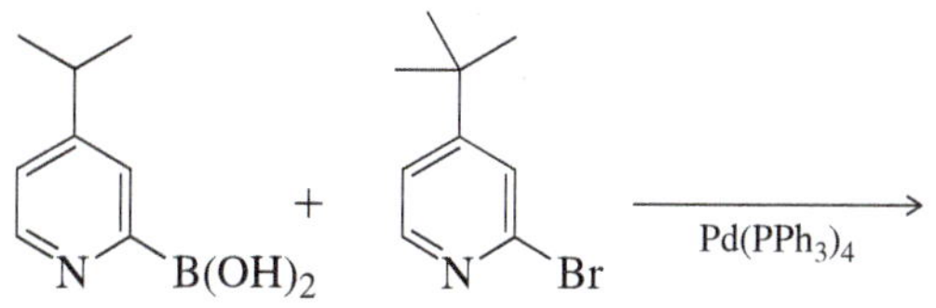

Assessment 23.60 Suggest two coupling partners that could be used to synthesize the compounds shown making the indicated C—C bonds.

(a)

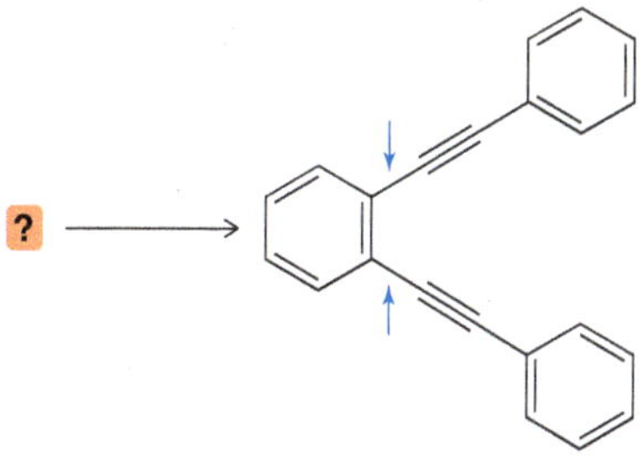

(b)

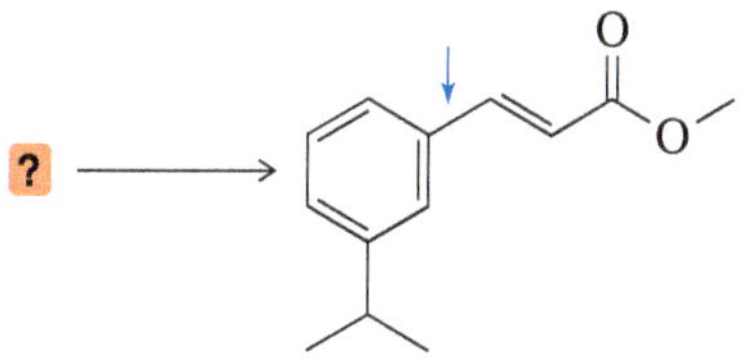

(c)

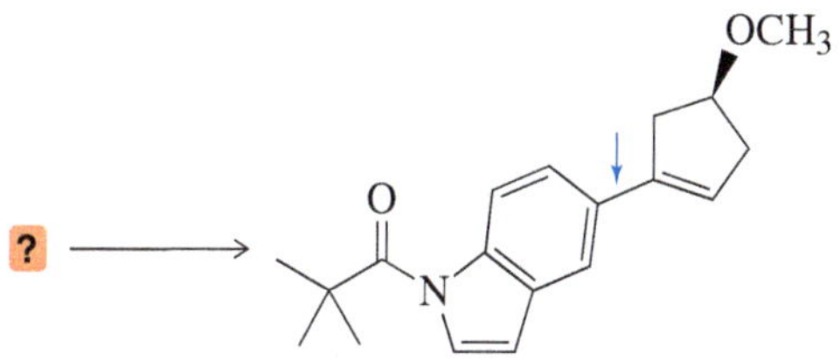

Concluding thoughts

While this chapter focused on the reactions occurring on the benzene ring, there are a host of reactions that benzene makes possible on neighboring carbons. Let's move on to Chapter 24 to see how benzene exerts influence over these reactions.

Despite benzene's origin as a highly stable unreactive molecule, chemists in the early 20th century devised multiple ways to functionalize it through electrophilic aromatic substitution, nucleophilic aromatic substitution, and the formation of diazonium ions. Combined with modern transition metal–catalyzed coupling reactions, there is practically no benzene derivative that cannot be synthesized.

This ability to manipulate benzene not only led to the ultimate discovery of sulfa antibiotics, but also launched modern medicinal chemistry. While studying the antibiotic activity of sulfa drugs, medicinal chemists noted many other possible therapeutic uses. The sulfonamide scaffold, as it is sometimes called, has found utility in treating a wide range of diseases, as shown in **Figure 23.69**.

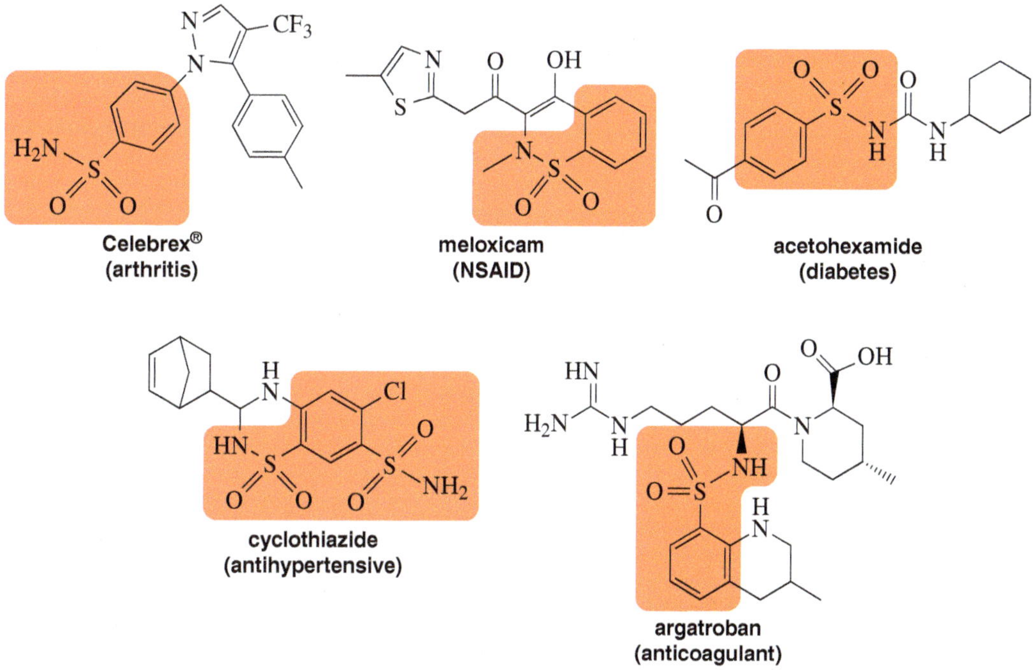

Figure 23.69 Drugs that use the sulfonamide scaffold (the scaffold is screened in orange).

Important Concepts

[These are the major concepts covered in each section, excluding reactions.]

23.1: Benzene is a very stable molecule that exists as a hybrid of two resonance (termed Kekulé) structures.

23.2: Benzene is one example of an aromatic molecule. Aromatic molecules are much more stable than they are expected to be based on bond-dissociation energies. A molecule is aromatic if it is planar, is cyclic, has a p orbital (filled or unfilled) on every ring atom, and contains $4n + 2$ electrons in a cyclic, conjugated π system.

23.3: The opposite of aromaticity, antiaromatic molecules (if they exist at all) are much less stable than you'd expect them to be. A hypothetical molecule is antiaromatic if it is planar, is cyclic, has a filled p orbital on every ring atom, and contains $4n$ electrons in a cyclic, conjugated π system.

23.4: The molecular orbital picture of aromatic molecules can be drawn using Frost circles. Aromatic molecules have all filled molecular orbitals below $E = 0$. Antiaromatic molecules have two half-filled orbitals around $E = 0$.

23.5: The nomenclature of substituted benzenes utilizes a mixture of IUPAC and common names. The pattern of substitution in disubstituted benzenes is described using ortho (1,2), meta (1,3), and para (1,4).

23.7: Strong electrophiles are substituted for hydrogens on benzene in electrophilic aromatic substitution (EAS). In EAS, formation of a nonaromatic carbocation in a slow, disfavored step is followed by a fast, favored acid–base reaction where aromaticity is restored. Several different strong electrophiles can be used in EAS reactions. Friedel–Crafts alkylation can result in rearrangement with certain haloalkanes.

23.8: Electron-donating groups make EAS reactions faster and direct addition to ortho and para positions. Electron-withdrawing groups make EAS reactions slower and direct addition to the meta position. Halogens make EAS reactions slower but direct addition to the ortho and para positions. When two possible products can result from an EAS reaction, reaction usually occurs at the least sterically hindered position on the ring.

23.9: Nucleophilic aromatic substitution proceeds through two mechanisms: addition/elimination and benzyne. The addition/elimination mechanism operates when a good leaving group is positioned ortho or para to an electron-withdrawing group. Addition of a nucleophile to the ring results in a resonance-stabilized but nonaromatic anion that undergoes fast and favorable elimination, kicking out a good leaving group and rearomatizing the ring. The benzyne mechanism operates under harsh conditions with strong nucleophiles/bases. Elimination by the E1cb mechanism gives benzyne. The unstable benzyne intermediate is attacked at either of two reactive positions. A final protonation step completes formation of (usually) regioisomeric products.

23.10: Aryl diazonium ions can be formed by reaction of aromatic amines with $NaNO_2$ and HCl. The resulting diazonium ions can be displaced by a number of different functional groups in what is sometimes referred to as the Sandmeyer reaction. Palladium coupling reactions can also be used to derivatize benzene according to the catalytic cycle discussed in Section 16.4.

Important Reactions and Mechanisms

1. **Electrophilic aromatic substitution (Section 23.7):** Benzene acts as a nucleophile, using π electrons to attack a strong electrophile in a slow, disfavored step. An acid–base reaction occurs to regenerate the aromatic ring in a fast, favored step. The selectivity of electrophilic aromatic substitution is predicted based on forming the most stable carbocation in the rate-determining first step.

General:

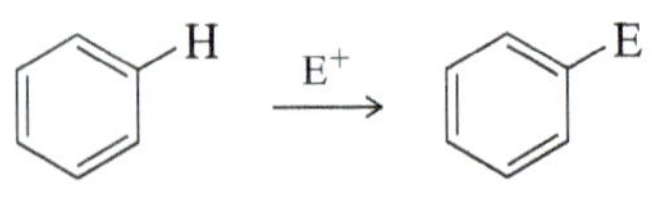

Specific:

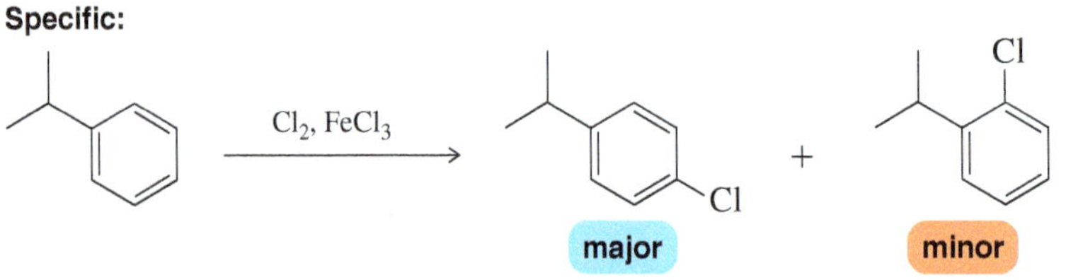

Mechanism:

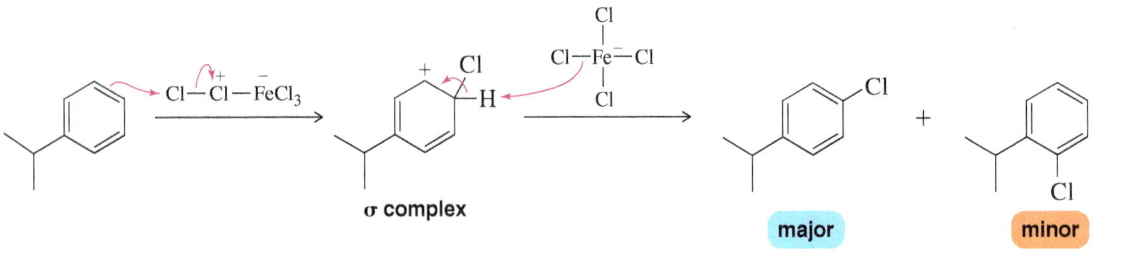

Other reagents for elecrophilic aromatic substitution:

HNO_3, H_2SO_4

H_2SO_4, SO_3

RCl, $AlCl_3$

RCOCl, $AlCl_3$

2. Nucleophilic aromatic substitution (addition/elimination) (Section 23.9.1): Nucleophilic attack at a carbon bearing a leaving group ortho or para to a strong electron-withdrawing group results in formation of a nonaromatic, tetrahedral intermediate. Collapse of the electron pair kicks out the good leaving group and reforms the stable aromatic ring.

General:

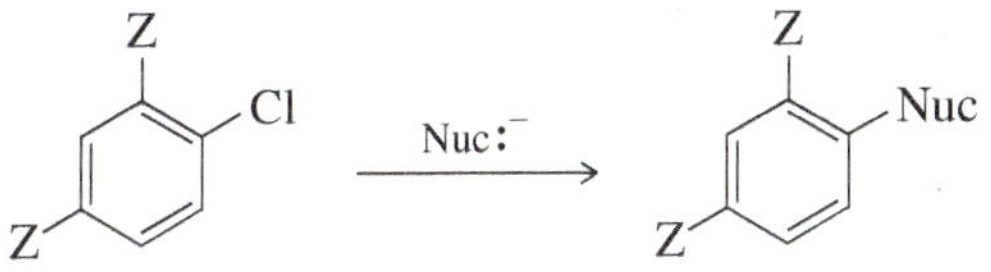

Specific:

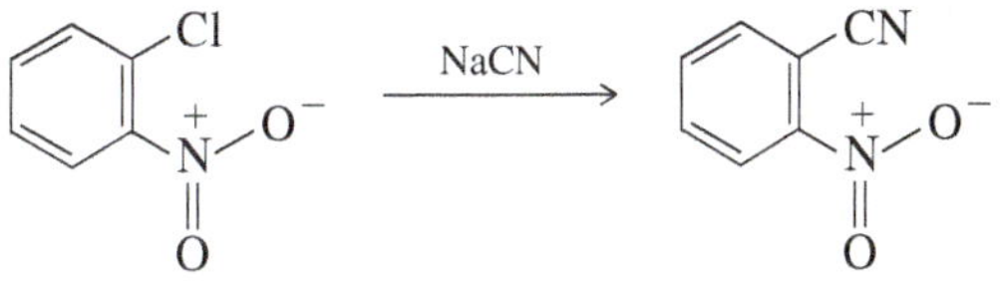

Mechanism:

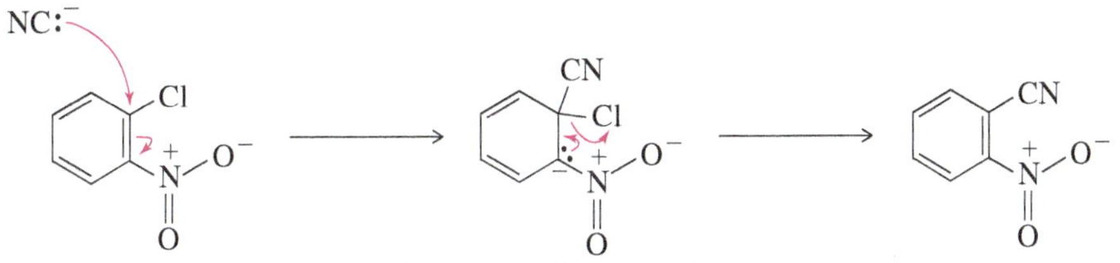

Meisenheimer complex

Other electron-withdrawing groups for the addition/elimination mechanism:

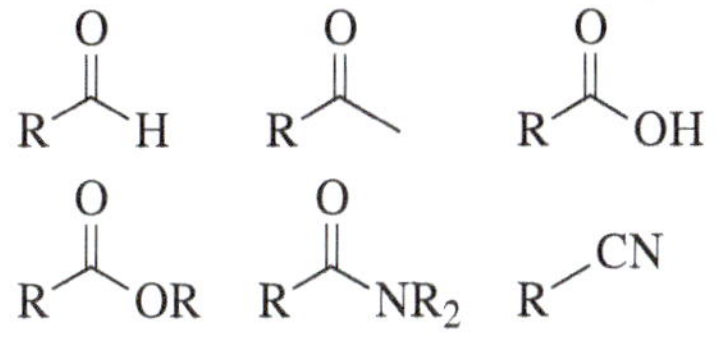

3. Nucleophilic aromatic substitution (benzyne) (Section 23.9.2): Using a strong nucleophile/base or extremely harsh conditions, deprotonation of a halobenzene causes elimination and formation of benzyne to occur through the E1cb mechanism. Non-regioselective attack of a nucleophile at either carbon of the benzyne is followed by protonation to give the substituted product(s).

General:

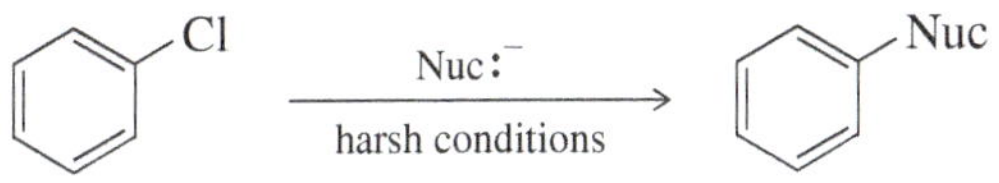

Specific:

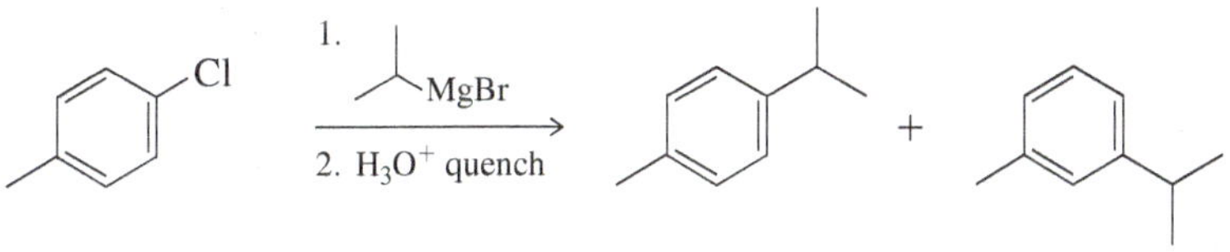

Mechanism:

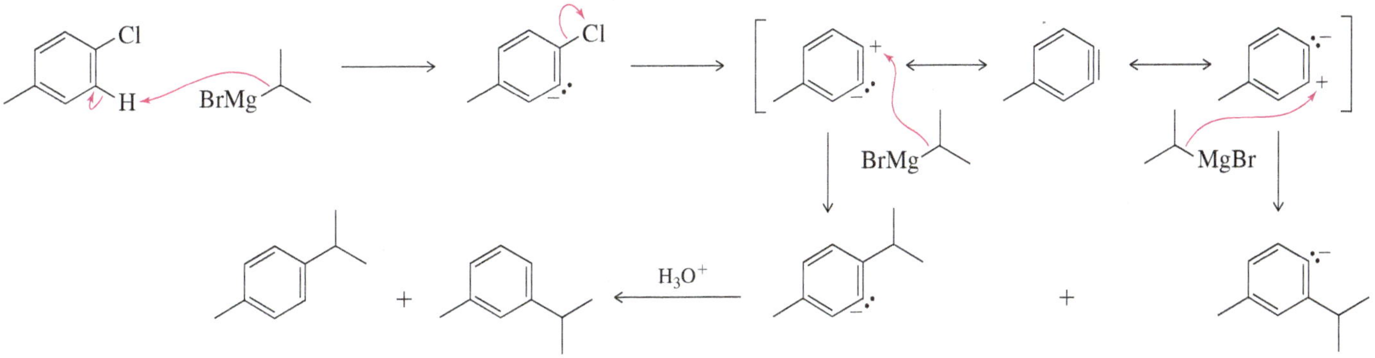

4. **Substitution with diazonium ions (Section 23.10.1):** Reaction of aminobenzenes results in formation of aryl diazoniums by a mechanism shown in Figure 23.64. The diazonium can be replaced with a variety of substituents by reactions whose mechanism are not well known.

General:

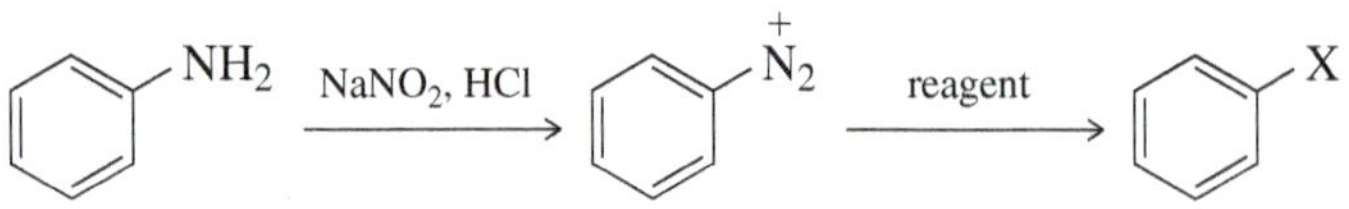

Specific:

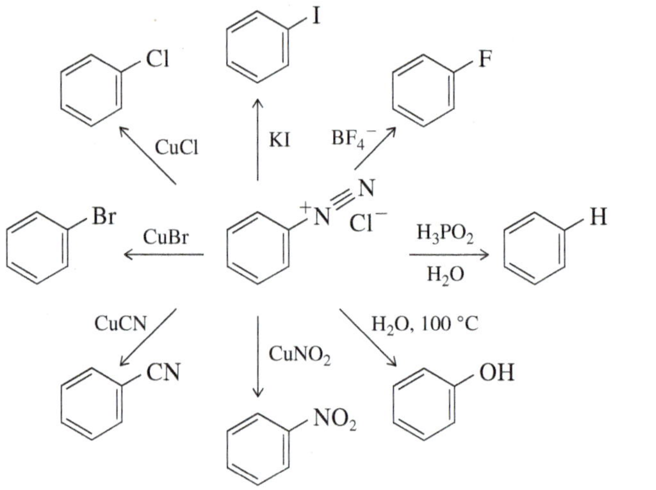

---

5. **Transition metal–catalyzed coupling (Section 23.10.2):** Organotin (Stille), organozinc (Negishi), alkylboronic acids (Suzuki), and copper acetylides (Sonogashira) are coupled to halogenated benzenes using palladium tetrakis as the catalyst. All four reactions proceed by a catalytic cycle involving oxidative addition, transmetallation, and reductive elimination.

General:

$$R{-}M + R'{-}X \xrightarrow{Pd(PPh_3)_4} R{-}R' + MX$$

Specific:

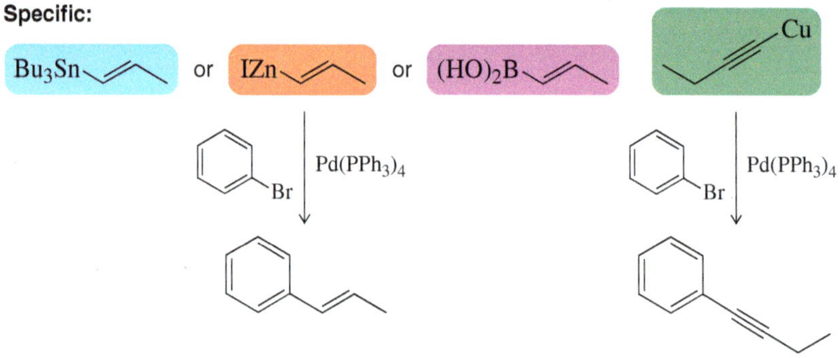

Mechanism (catalytic cycle):

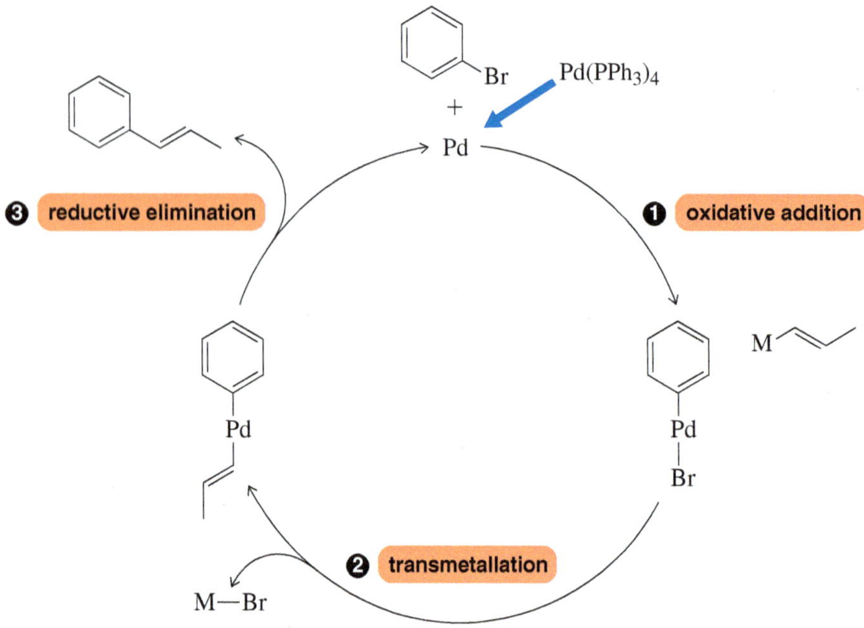

End-of-Chapter Assessments

[*Assessments are ranked according to their difficulty;* •••• = *most difficult.*]

23.61 (•) (i) Classify the following molecules as aromatic, nonaromatic, or antiaromatic. (ii) For aromatic molecules, solve for *n* in Hückel's rule. For all other molecules, explain which rule of aromaticity is being broken.

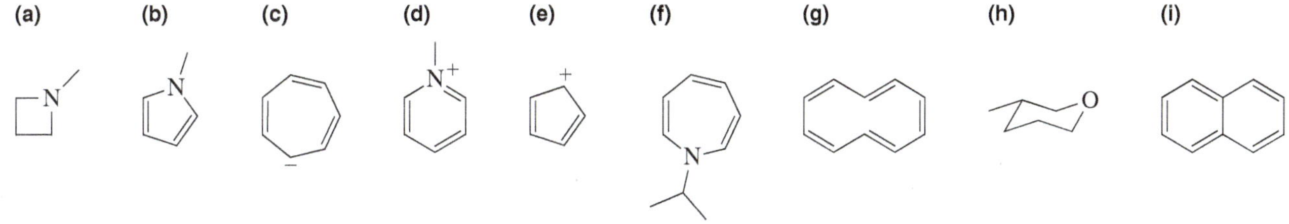

23.62 (•) Use a Frost circle diagram to construct the molecular orbital diagram for the molecules shown. Would you expect them to be aromatic or antiaromatic?

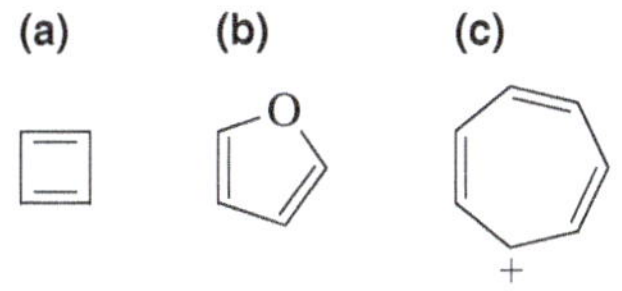

23.63 (••) Using a Frost circle, a student drew the molecular orbital picture shown for the cyclopentadienyl anion, determining it to be antiaromatic. Do you agree with this conclusion? Why or why not?

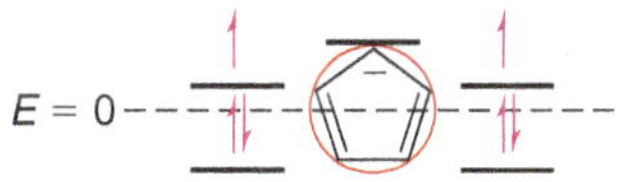

23.64 (••) Despite having only sp^2-hybridized carbons and having 10 electrons $(4n + 2)$, the annulene shown is *not* aromatic. Why?

23.65 (••) A [*very*] inexperienced researcher, having run out of THF, decides to use cyclopentadiene [*it has a five-membered ring*] as the solvent for a Grignard reaction. Perhaps not surprisingly, the reaction fails. Why? [*Disclaimer: Replacing THF with cyclopentadiene as solvent is something that should never be considered.*]

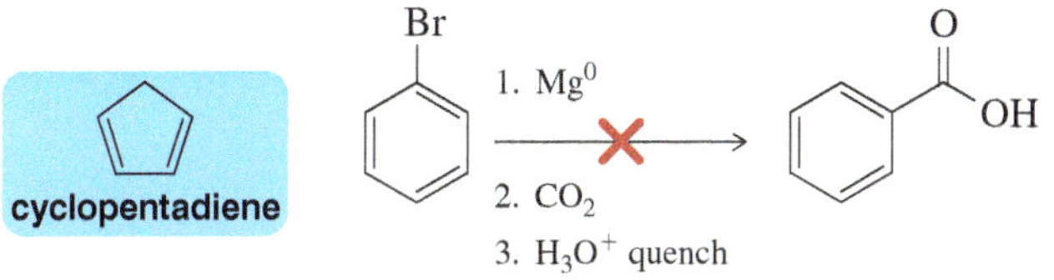

23.66 (••••) LOOKING BACK In Section 17.7.4, we studied the acid-catalyzed hydrolysis of acetals. The acetal shown here resists hydrolysis by the mechanism in Figure 17.63. Why? [*Draw the intermediates as if the reaction would occur, then analyze the intermediates for any problems.*]

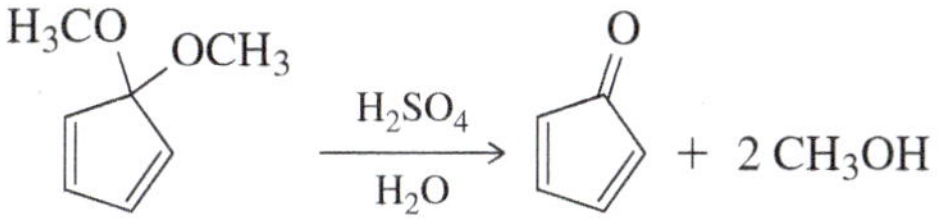

23.67 (•••) In contrast to the acetal in Assessment 23.66, hydrolysis of the following acetal proceeds quickly and favorably. Why?

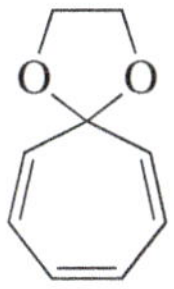

23.68 (••) Molecule **A** is significantly more water soluble than molecule **B**. Justify this observation.

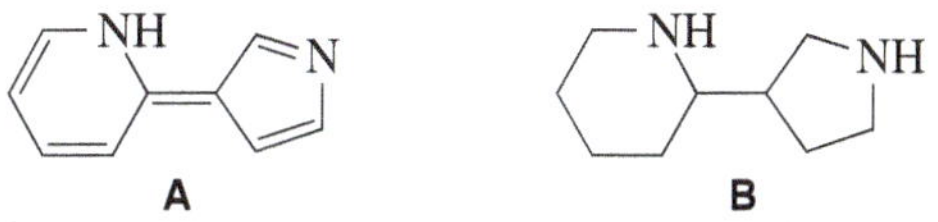

23.69 (•••) LOOKING BACK In Section 22.3.4, we learned that the Diels–Alder reaction is a kinetically controlled reaction that proceeds through the endo transition state. However, Diels–Alder reactions using furan proceed to give the more stable exo product. Rationalize this observation.

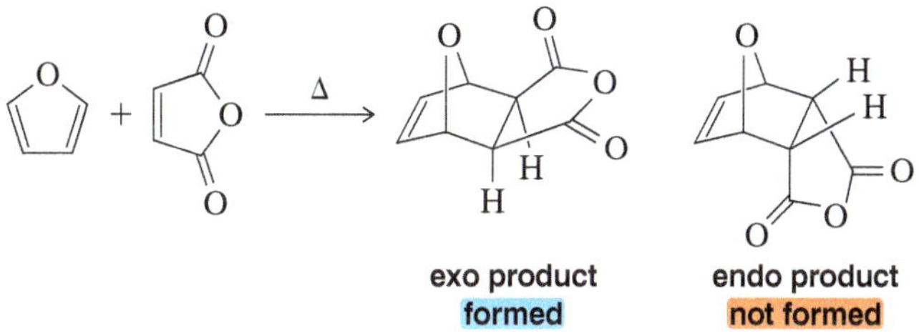

23.70 (•••) Recall from Chapter 17 that NADH is a biological reducing agent. Suggest a mechanism for how NADH donates a hydride ion to a carbonyl. Why is this a favorable reaction? [*Consider what NADH has become after giving up the hydride.*]

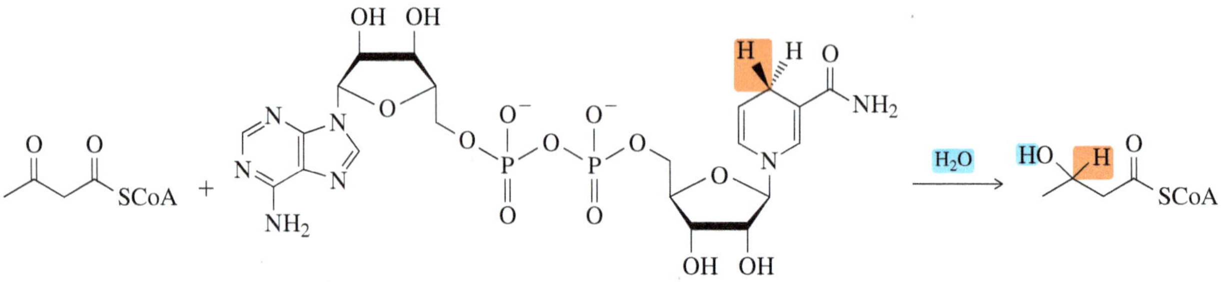

NADH

23.71 (•) Would you expect the following substituted benzenes to react more or less quickly than unsubstituted benzene in an electrophilic aromatic substitution reaction?

(a)

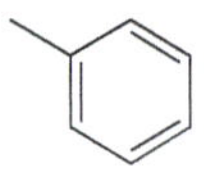

(b)

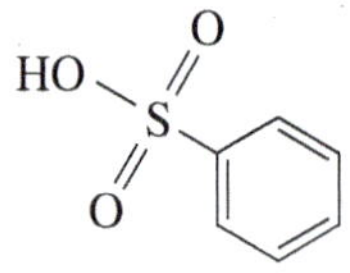

(c)

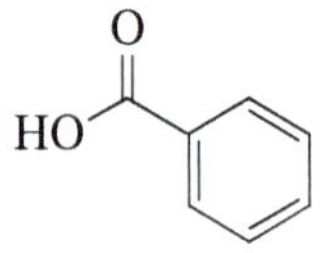

(d)

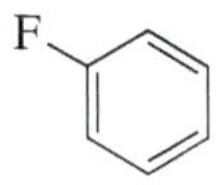

(e)

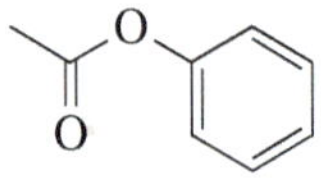

(f)

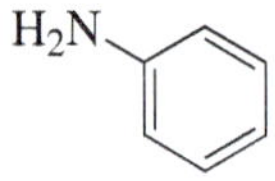

(g)

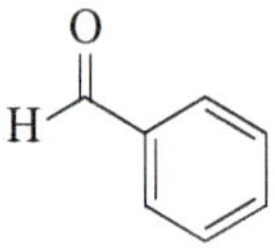

(h)

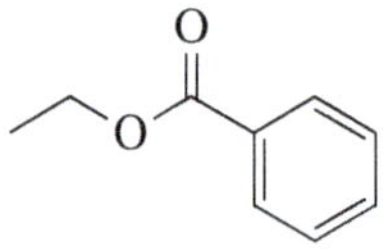

(i)

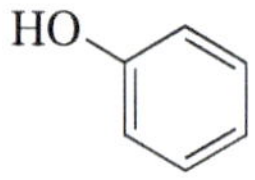

(j)

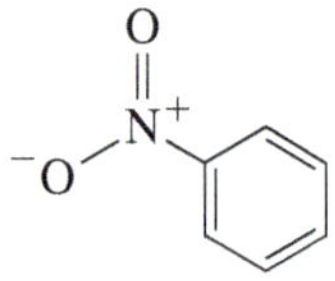

23.72 (•) Predict the major product(s) that would result when molecules (a)–(i) are allowed to react under the following conditions: (i) HNO_3, H_2SO_4; (ii) SO_3, H_2SO_4; (iii) chloroethane, $AlCl_3$; (iv) chlorocyclopentane, $AlCl_3$; (v) 1-chloro-1-methylcyclohexane, $AlCl_3$; (vi) $PhCOCl$, $AlCl_3$. If no reaction will occur, indicate by writing NR.

(a)

(b)

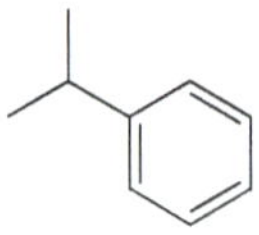

(c)

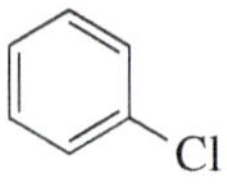

(d)

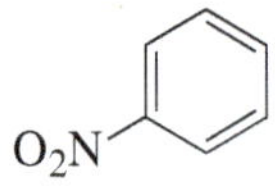

(e)

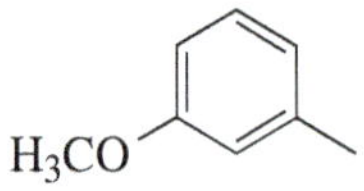

(f)

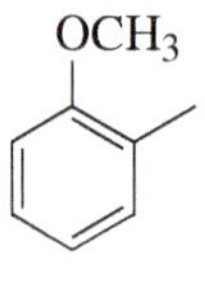

(g)

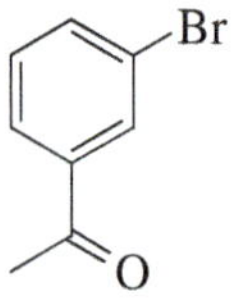

(h)

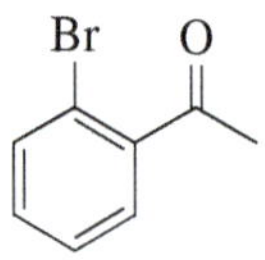

(i)

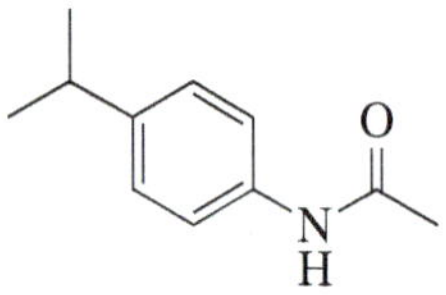

23.73 (•) Suggest an arrow-pushing mechanism for molecule (f) reacting with each set of reagents (i–vi) shown in Assessment 23.72. Draw out all of the intermediate resonance structures you used to determine the regioselectivity of the reaction.

23.74 (•••) Suggest a synthesis of each of the molecules shown beginning with benzene.

(a)

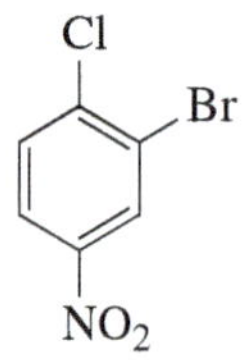

(b)

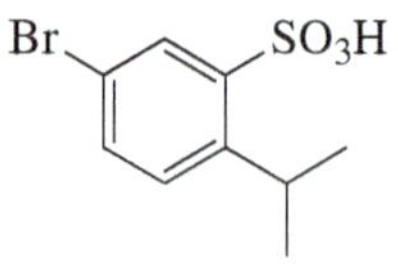

(c)

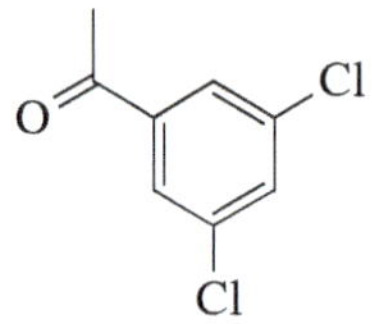

23.75 (••) Bromination of phenol or aniline does not require the use of a Lewis acid catalyst and often results in trihalogenation. Why?

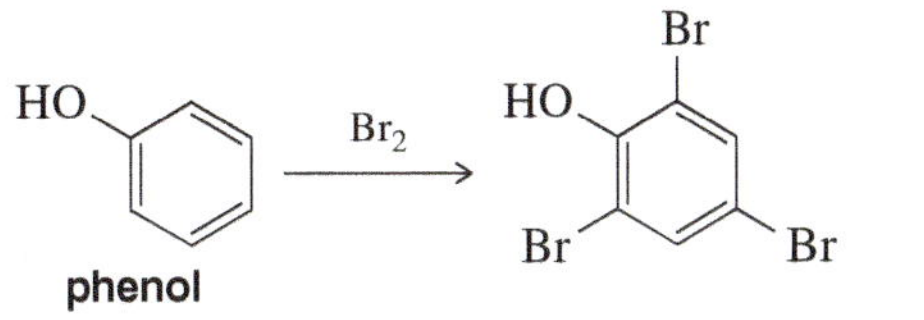

phenol

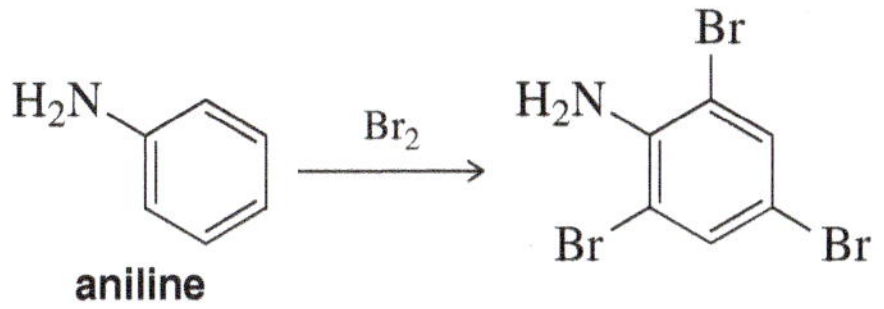

aniline

23.76 (••) The bromination of phenol occurs faster in base. Why? [*The pK_a of phenol is 10.*]

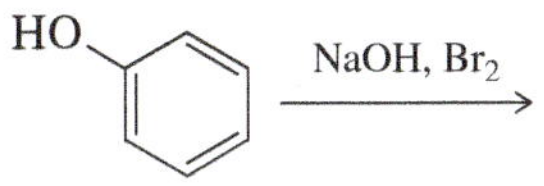

phenol

23.77 (••) Protonation of aniline slows electrophilic aromatic substitution and directs electrophiles to the meta position. Why?

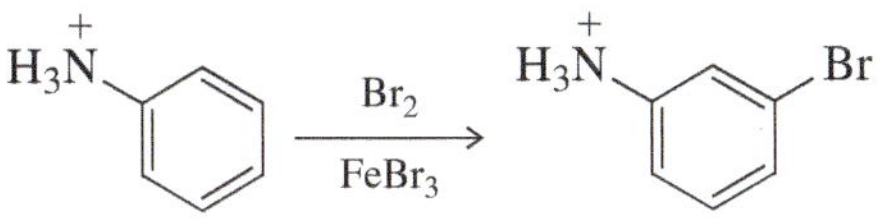

23.78 (••) Trinitrotoluene (TNT), first prepared and purified by Paul Hepp in 1880, became the preferred explosive compound during World War I. Suggest a structure of TNT, specifying the regioisomer you'd expect if it was synthesized from toluene.

23.79 (•••) Synthesize the following benzene derivative using electrophilic aromatic substitution. [*You may need to put groups on in a different form than they are in the final product to ensure the correct regiochemical outcome.*]

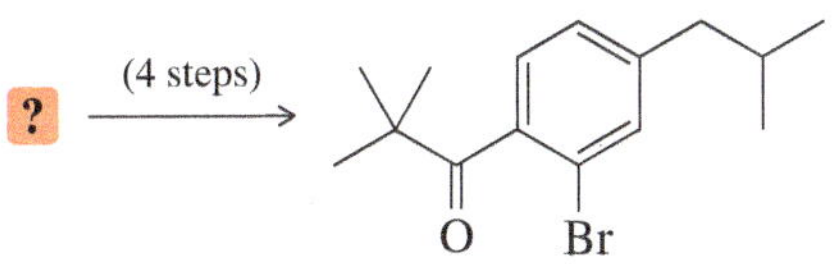

23.80 (••) Indigo is a dye that was originally isolated from coal tar. In 1905, Johann Friedrich Wilhelm Adolf von Baeyer won a Nobel Prize for a method that allowed indigo to be isolated from plants [*a green chemist ahead of his time.*] If you nitrated indigo using the reaction learned in this chapter, at which carbon would you expect the nitro group to attach?

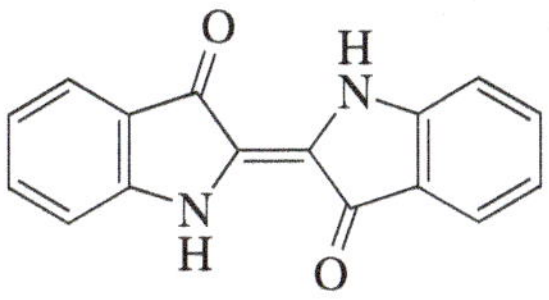

indigo

23.81 (••) Would you expect chlorination to occur ortho, para, or meta to the pyridinium ion in the following molecule? Explain your answer.

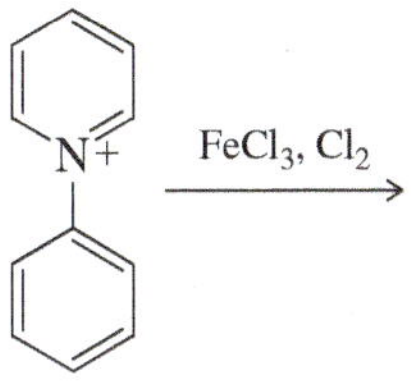

23.82 (•••) The Fries rearrangement is the Lewis acid–promoted conversion of a phenolic ester to an acylated phenol. Suggest an arrow-pushing mechanism of the Fries rearrangement.

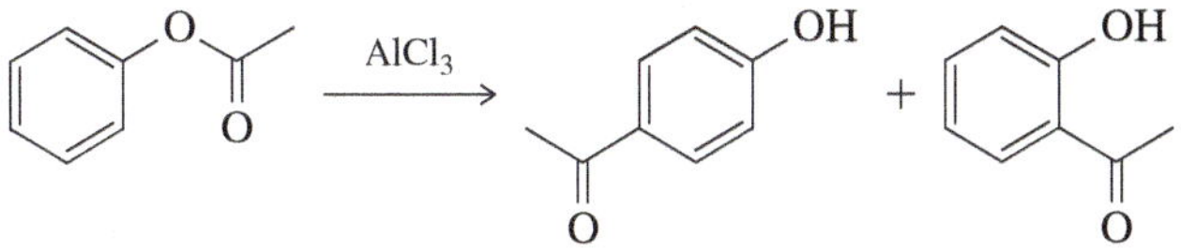

23.83 (•••) Fluoxetine, an antidepressant, is better known as Prozac®. Suggest reagents that could be used to make the indicated C—O bond of fluoxetine. [*Note that CF_3, an electron-withdrawing group, is para to where the new bond will be formed.*]

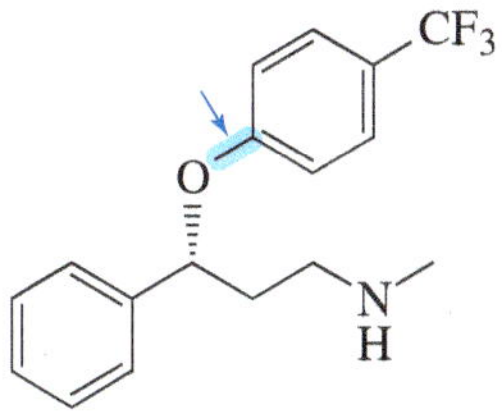

fluoxetine

23.84 (••••) Acetohexamide is used to treat type 2 diabetes that cannot be well controlled by diet alone. Fill in the blanks of the synthetic scheme used to synthesize acetohexamide. [*Hints have been provided below each step.*]

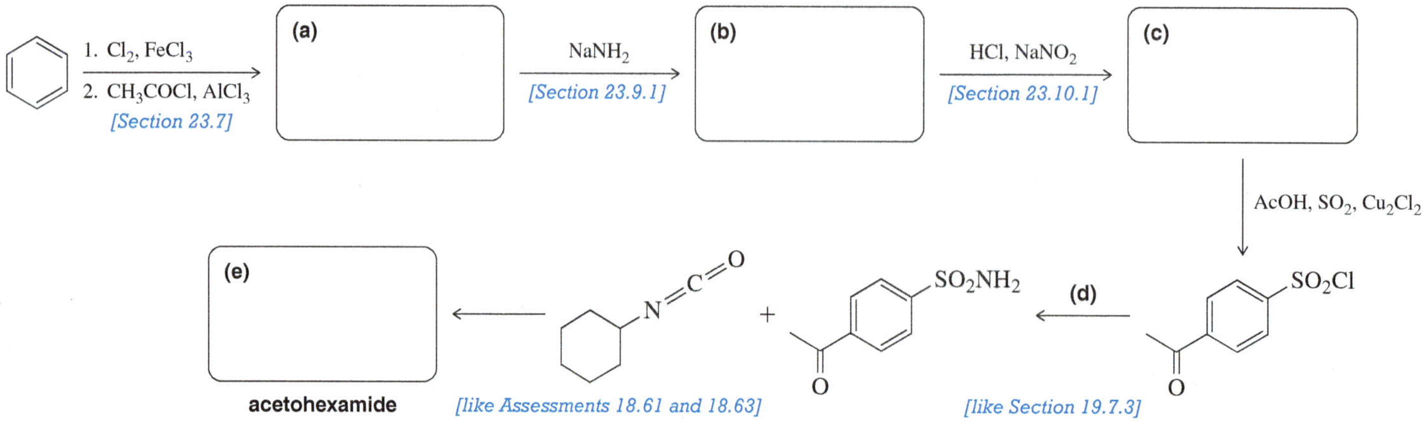

23.85 (••) SPECTROSCOPY INSIGHTS How might you use ^{13}C NMR spectroscopy to differentiate between the possible ortho, meta, and para products of the electrophilic aromatic substitution reaction shown?

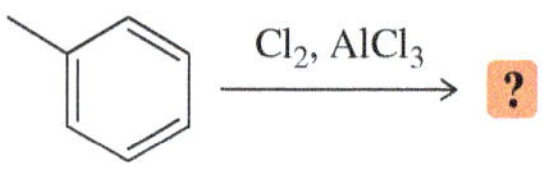

23.86 (••••) Addition by the benzyne mechanism is not usually regioselective. Explain the fact that the reaction shown is highly regioselective for the product shown.

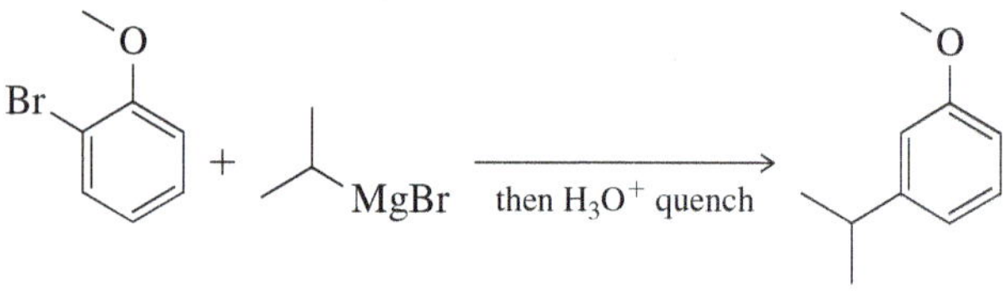

23.87 (•••) The sulfonylation of benzene can be reversed by heating a sulfonic acid in aqueous sulfuric acid. (a) Suggest a mechanism of this reaction. (b) How did you know that this was the correct mechanism?

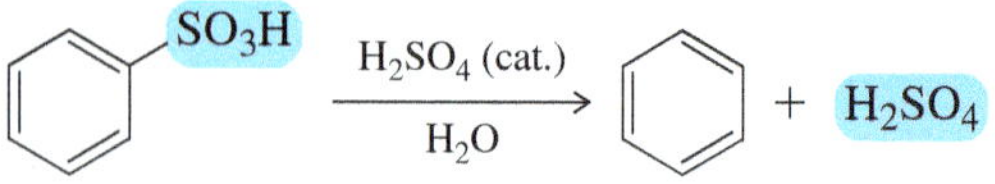

23.88 (•••) How could the reaction in Assessment 23.87 be used to design a scheme that would produce only *o*-chlorotoluene from toluene?

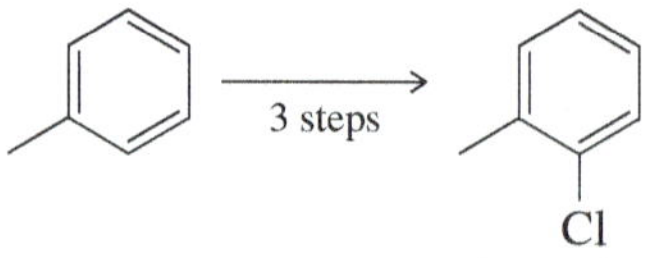

only product

23.89 (•••) Suggest a mechanism of the following reaction.

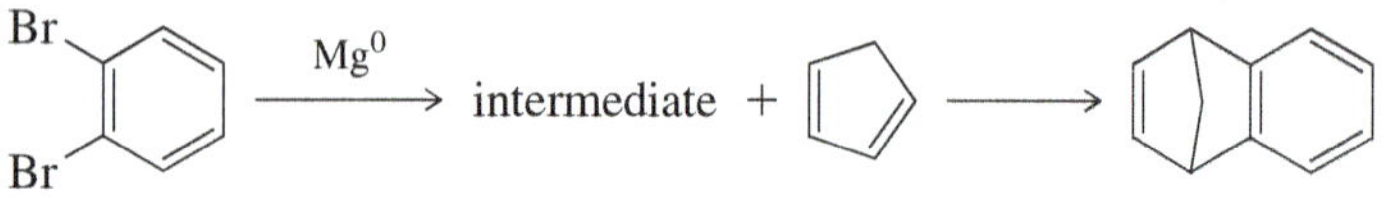

23.90 (•••) In light of your answer to Assessment 23.89, suggest a mechanism of the following reaction.

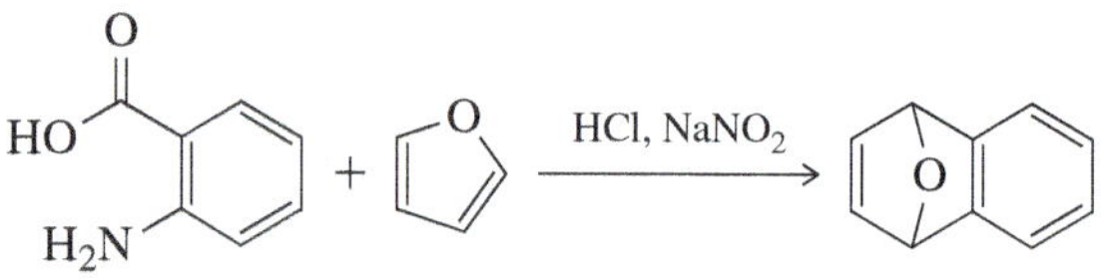

23.91 (••) LOOKING AHEAD In Chapter 24, we discuss the acidity of phenols. *Para*-nitrophenol is significantly more acidic than *meta*-nitrophenol. Explain this observation. [*As always, consider the stability of the conjugate base.*]

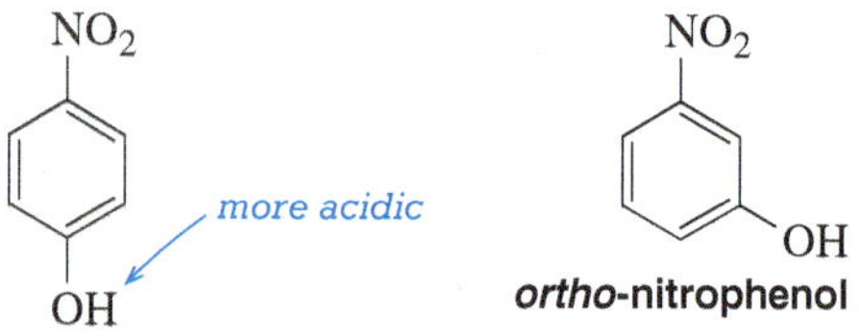

***para*-nitrophenol** ***ortho*-nitrophenol**

23.92 (•••) Suggest an arrow-pushing mechanism of the following rearrangement.

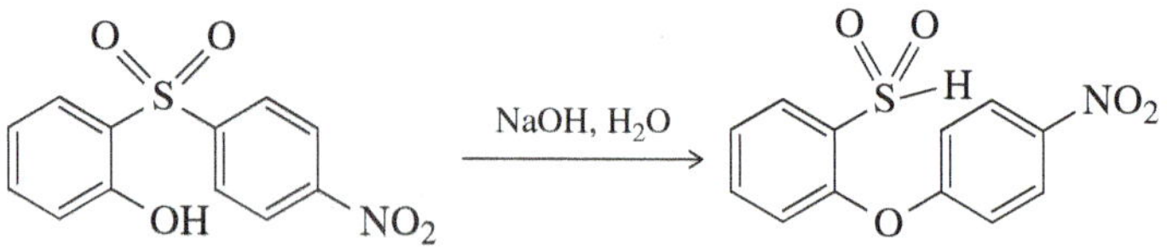

23.93 (••) Suggest a synthesis of the molecule on the right beginning with the molecules on the left.

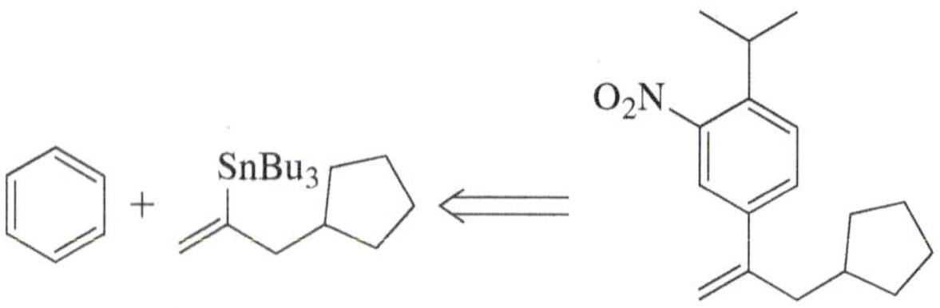

23.94 (••) LOOKING AHEAD Imagine that the unstable intermediate shown could form. Where would you expect a nucleophile to add to this intermediate? [*There are three sites; we discover the relevance of this in Chapter 24.*]

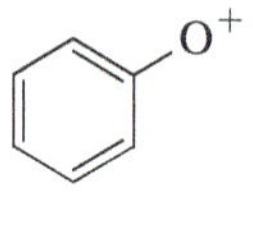

24

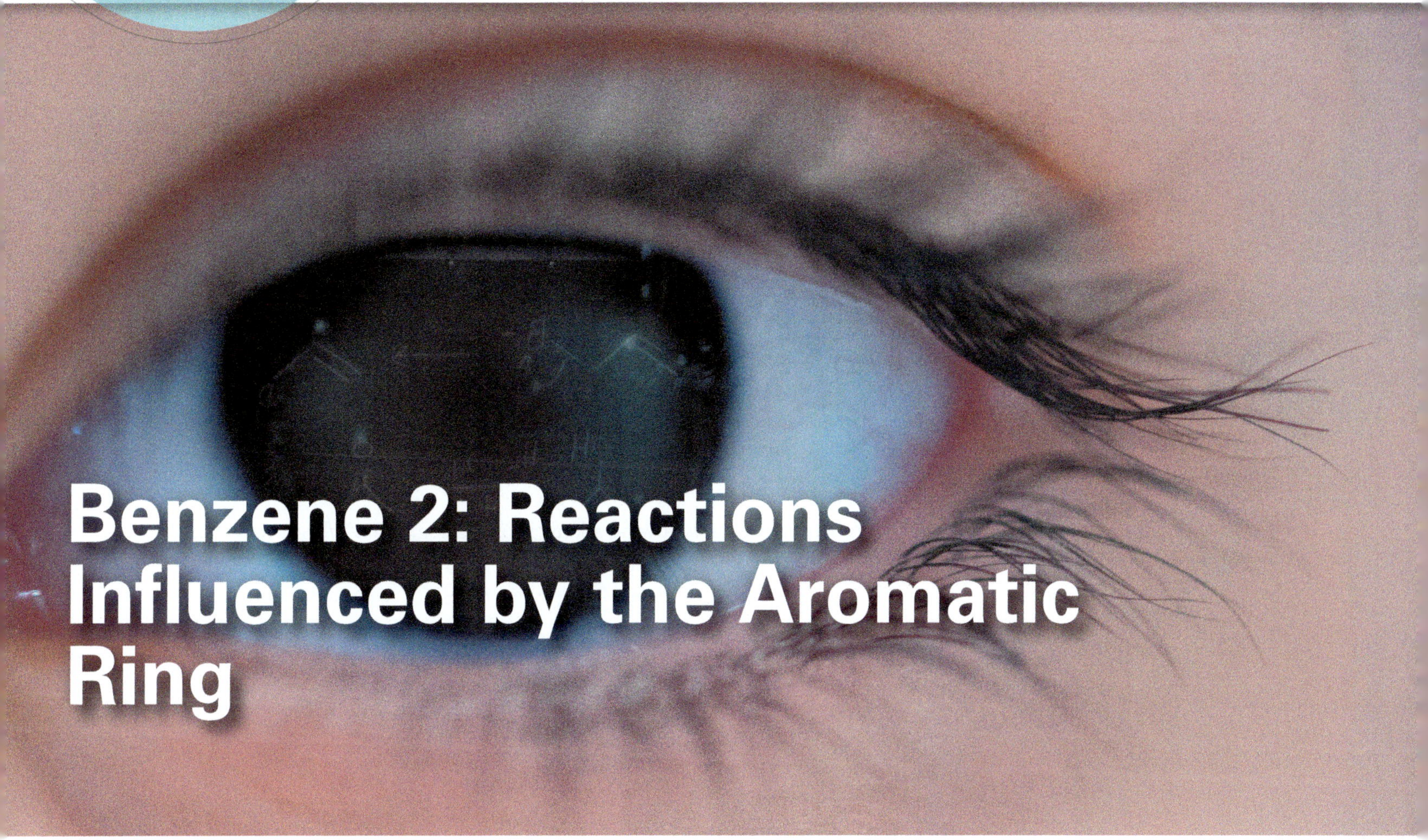

Benzene 2: Reactions Influenced by the Aromatic Ring

Introduction

As you have read this text, you have seen how advances in organic chemistry have impacted other areas of science. More than once, our understanding of these stories was limited by how far we had come in our understanding of organic chemistry, and sometimes we had to simply accept some facts that we would come to understand later. [*Side chain oxidation in alkyl benzenes; Section 18.5.4.*] Given that a theme of this chapter is looking back at topics in the context of what we know now, this is a great place to revisit a few earlier concepts that the chemistry of benzene allows us to understand. [*Stability of resonance-stabilized carbocations, for example.*] And so, as you read through Chapter 24, look for further discussion on benzene's influence on vitamin E as an inhibitor (Section 11.7), the development of a green electrochemical alcohol oxidation (Chemistry in Context 13-A), and the mechanism of drug oxidation in the body (Chapter 11: Concluding thoughts).

○ ○ ○ ○ ○ ○ ○

Chapter 24 concludes our foray into conjugated systems. While Chapter 23 introduced new reactions, there isn't much new in this chapter. Instead, we are going to study reactions we have seen before and learn how they are altered by the presence of a neighboring benzene. Why, then, does this topic deserves its own chapter? Why didn't we show the influence of benzene when these reactions were studied initially or include it in Chapter 23, when we first studied benzene? By separating this topic into its own chapter, we were able to focus in Chapter 23 on substitution reactions—perhaps the most important reactions of benzene. A chapter dedicated solely to the effect of benzene on familiar reactions, on the other hand,

Learning Objectives

LO 24.0 Account for benzene's effect on the reactivity of substituents

LO 24.1 Relate phenol reactions to their acidity

LO 24.2 Analyze synthesis reactions involving phenols

LO 24.3 Analyze benzylic brominating reactions

LO 24.4 Analyze benzylic substitution reactions

LO 24.5 Predict the product of side chain oxidation reactions

LO 24.6 Predict the product of hydrogenolysis reactions of benzyl ethers

"Study the past if you would define the future."

—Confucius

gives us a great opportunity to review material learned previously. As the one-year sequence of organic chemistry [*and this book*] is winding down, you are likely beginning to think about a comprehensive final exam. [*I'm thinking about this, too.*] Working through Chapter 24 may give you confidence because you can anticipate everything that is coming next. Or maybe it reminds you of deficiencies you'll want to address before a final exam. [*And this chapter will certainly help address those deficiencies. Look for the many references to earlier material—and use them!*] Either way, there is comfort in repetition, so let's assess our understanding of previous concepts in the way we always do.

Test your understanding

Assessment 24.1 A carboxylic acid ($pK_a = 5$) is 10^{11} times more acidic than an alcohol ($pK_a = 16$). Why?

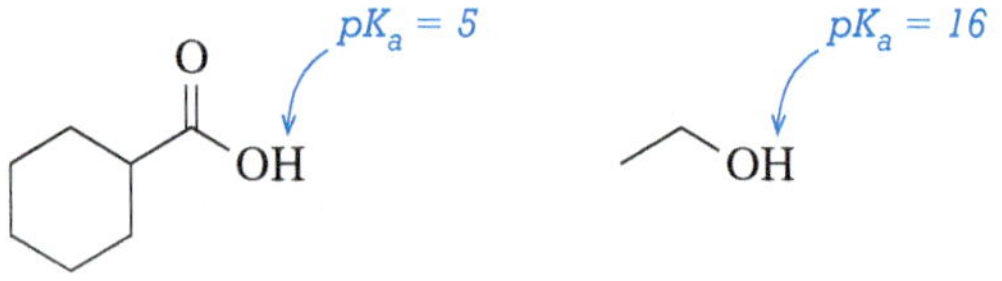

Assessment 24.2 What reagent would you use to brominate the allylic carbon of cyclohexene?

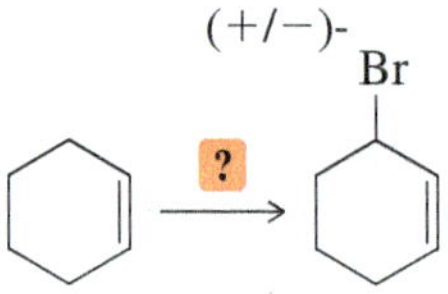

Assessment 24.3 Provide an arrow-pushing mechanism for the following alkane bromination. [*Don't forget to use fishhook arrows to represent the movement of single electrons.*]

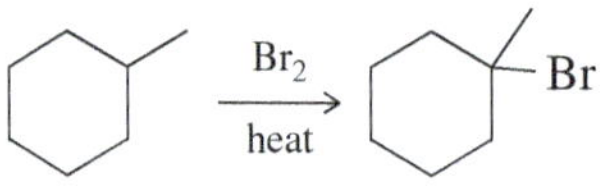

Assessment 24.4 Calculate the oxidation number of the indicated atoms in the following reaction.

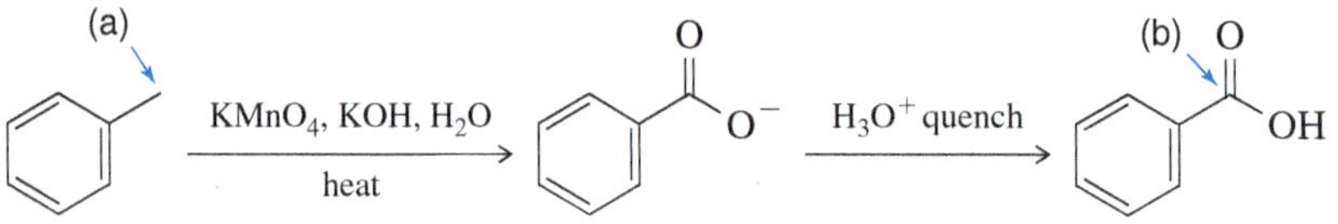

Assessment 24.5 Which haloalkane would you expect to undergo the fastest S_N1 reaction? Why?

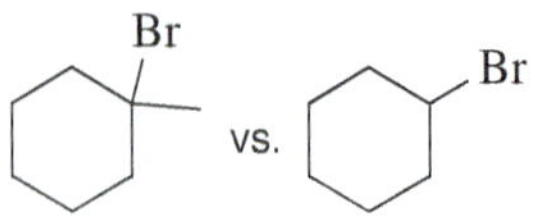

Assessment 24.6 What is the product of the following reaction?

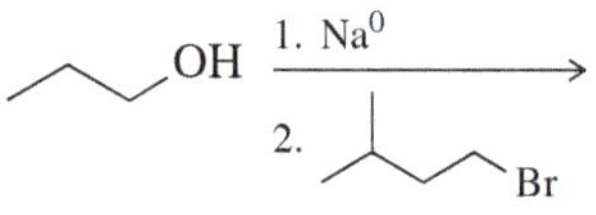

24.1 Benzene alters the reactivity of neighboring groups

Recall from our studies of conjugated molecules in Chapter 21 that the presence of alkenes dramatically influences the stability of other alkenes, carbanions, radicals, and carbocations. As a conjugated molecule itself, benzene is similarly able to stabilize reactive intermediates at the *benzylic* position. Whereas an allylic carbon is the carbon immediately attached to an alkene, a **benzylic carbon** is the carbon attached to the benzene ring (**Figure 24.1**). Sections 24.1.1–24.1.3 provide experimental evidence of this stabilization and offer an explanation. [*But I bet you already know what that explanation will be.*]

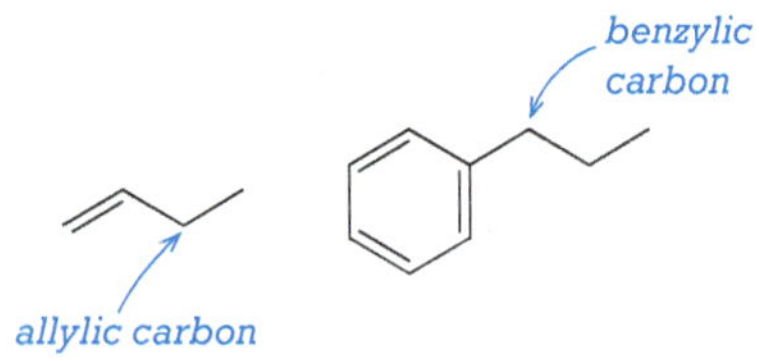

Figure 24.1 Allylic versus benzylic: The benzylic carbon is the carbon directly attached to benzene.

24.1.1 Stability of benzylic anions

In Chapter 4, we learned that the pK_a values of acids reveal the stability of their conjugate bases. A high pK_a value for an acid means that it does not easily give up a proton and become the conjugate base. Thus, the conjugate base is unstable. As we saw in Chapter 16, the deprotonation of unfunctionalized alkanes is impossible (pK_a of propane = 50), so

Grignard and organolithium reagents had to be invented to form carbanions. [*See Thinking Like a Chemist in Section 16.2.3.*] However, for a hydrogen on the benzylic carbon, as in toluene, the pK_a drops to 41 (**Figure 24.2**). The corresponding benzylic carbanion is stabilized by resonance delocalization onto benzene. As a result, the conjugate acid of a more stable base is more willing to give up a proton (has a lower pK_a) [*Section 4.3*].

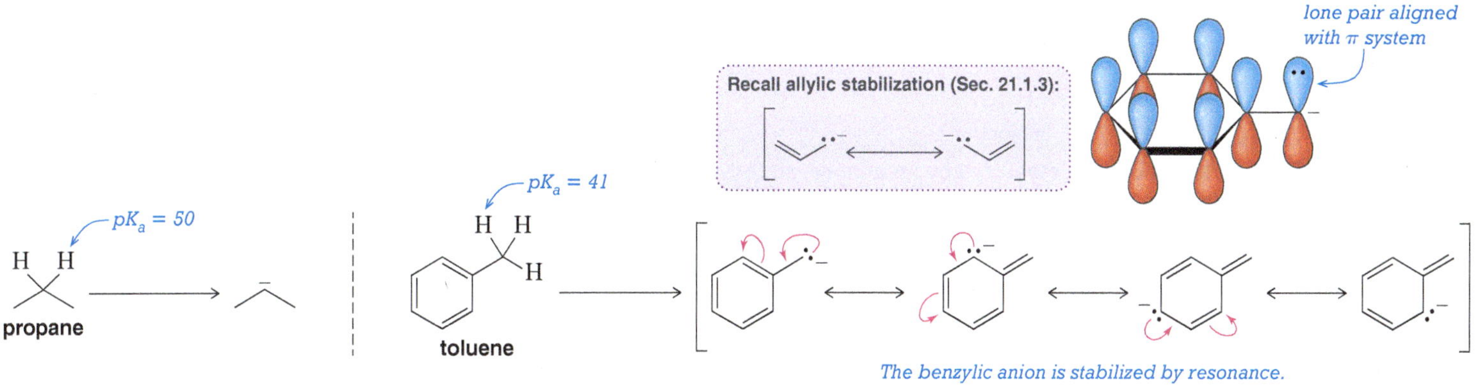

Figure 24.2 **Benzylic anions are stabilized by resonance,** so they form more favorably.

24.1.2 Stability of benzylic radicals

The stability of radicals was introduced as part of the discussion of alkane halogenation with Br_2 (Section 11.5.1) and allylic halogenation with *N*-bromosuccinimide (NBS) (Section 11.5.3). The relative stability of radicals is reflected in the bond-dissociation energy (BDE) for the bond that homolytically breaks to form the radical. Shown in **Figure 24.3**, the BDE for a 2° C—H bond to form a nonconjugated radical is 95 kcal/mol. When the C—H bond is benzylic (next to benzene), the BDE drops to 85 kcal/mol. This 10 kcal/mol difference reflects the increased stability from delocalization of the radical through resonance.

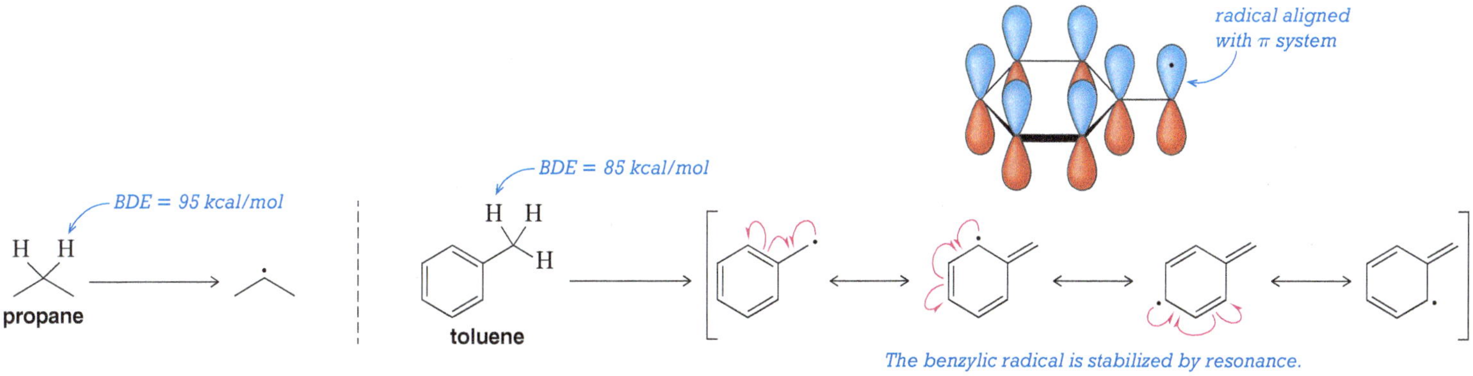

Figure 24.3 **Benzylic radicals are stabilized by resonance,** so they form more quickly.

24.1.3 Stability of benzylic carbocations

Recall from Chapter 12 that the rate-determining step of an S_N1 reaction is formation of the carbocation in the first step. The greater stability of a benzylic carbocation versus a nonconjugated carbocation can be demonstrated by comparing the relative rates of the solvolysis reactions of a 1° haloalkane and a 1° benzylic haloalkane (**Figure 24.4**). Primary carbocations are rare, whereas benzylic carbocations form easily, showing just how important the benzene ring is in stabilizing the carbocation.

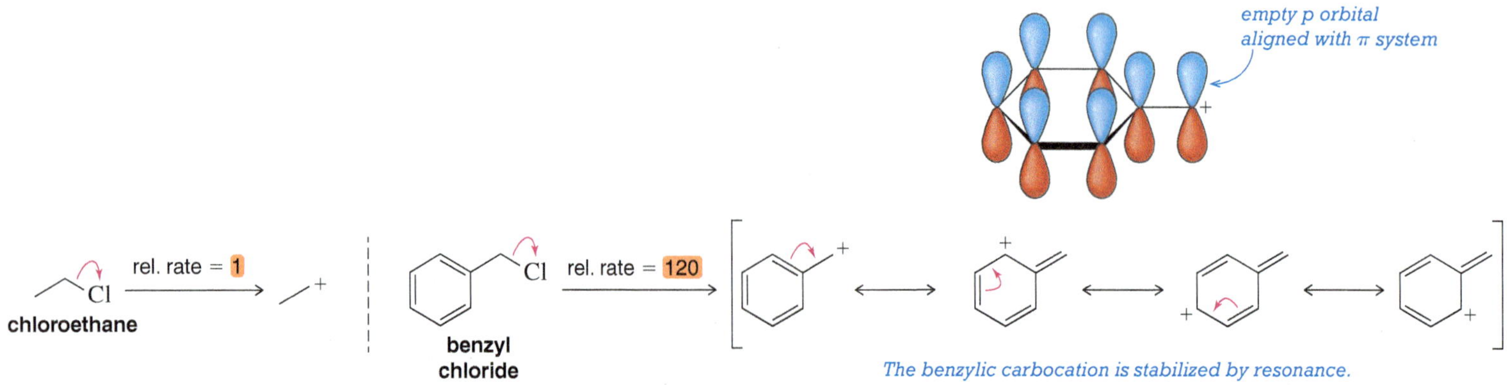

Figure 24.4 Benzylic carbocations are stabilized by resonance, so they form more quickly.

24.1 SUMMARY

- By stabilizing cations, anions, and radicals through resonance, benzene and other aromatic rings alter the reactivity of neighboring groups.

Assessment 24.7 Look closely at the resonance structures of the benzylic anions, radicals, and cations in Figures 24.2–24.4. On which benzene carbons did the charge/radical exist in the resonance structures—that is, were they ortho, meta, or para?

Assessment 24.8 Rationalize the rate difference in carbocation formation for the following molecules.

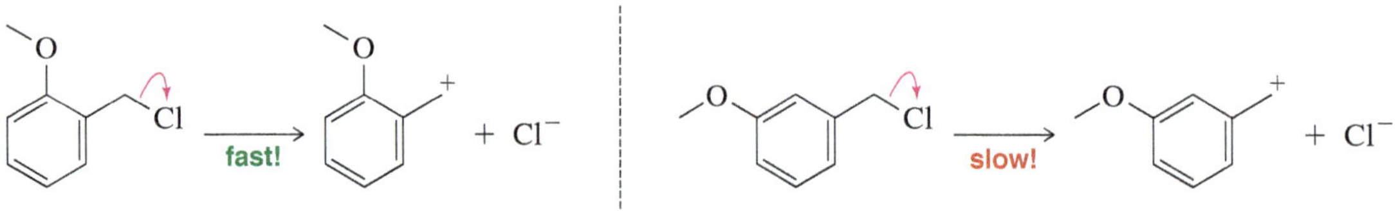

Assessment 24.9 Which proton, H_a or H_b, would you expect to have the lower pK_a value?

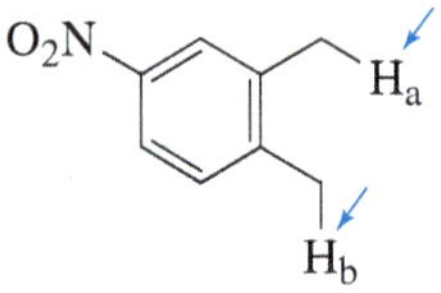

24.2 Phenols

A phenol is a special type of alcohol in which the hydroxyl group is bound directly to benzene (**Figure 24.5**). The name is derived from the common name *phen*yl alcoh*ol*. The presence of benzene alters the properties of an alcohol similar to the effects described in Section 24.1 and allows the reactions in Sections 24.2.2–24.2.4 to occur. Several important phenols are shown in Figure 24.5.

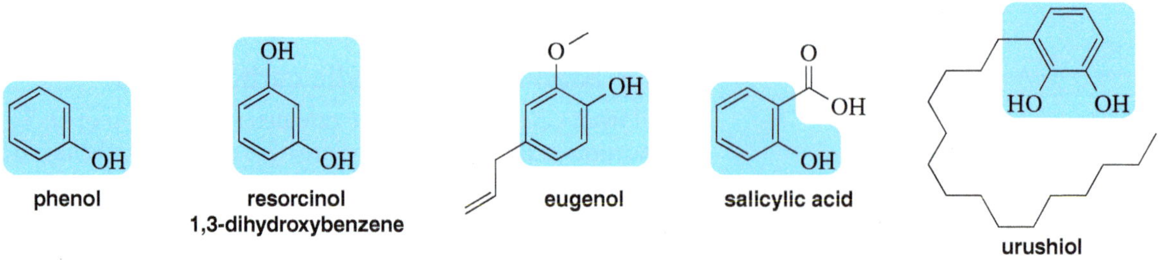

Figure 24.5 Several important phenols. The phenol portion is covered in a blue screen.

Assessment 24.10 On close inspection, a phenol is similar in structure to an enol. Enols exist as the less favorable tautomer in equilibrium with ketones. Phenols, on the other hand, are the favored tautomer in equilibrium with a ketone. Why is this true?

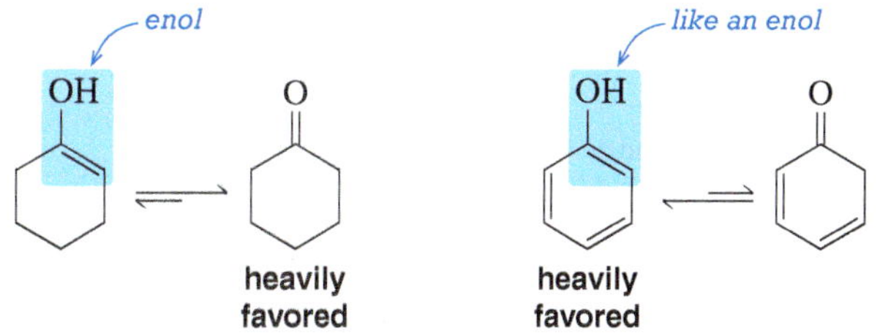

24.2.1 Acidity of phenols

The first difference between a normal alcohol and a phenol is their acidity (**Figure 24.6**). Because the phenoxide ion is stabilized by resonance with the benzene ring, the pK_a of a phenol is around 10. This represents a 10^6-fold increase in acidity over cyclohexanol (pK_a = 16).

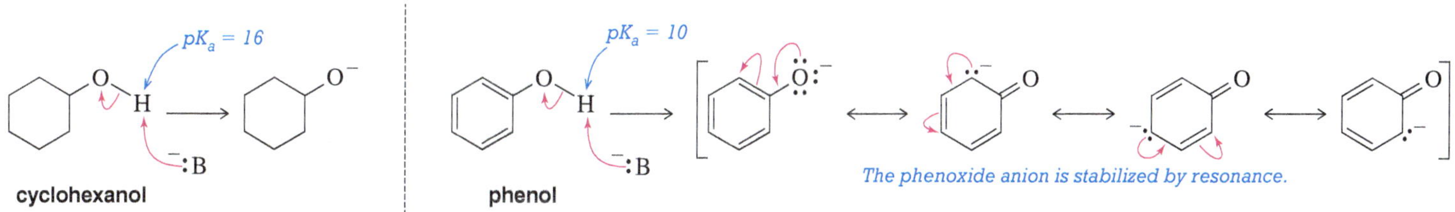

Figure 24.6 **Phenoxide is stabilized by resonance,** so phenols are more acidic than alcohols.

Substitution on the phenol has a dramatic effect on acidity and reinforces what we learned in Chapter 23 about how substituents "communicate" electronically through the benzene ring. For example, the series of *ortho*-, *meta*-, and *para*-chlorophenols in **Figure 24.7** show decreasing pK_a values as the chlorine is farther removed from the phenolic proton. This is a result of the inductive withdrawing effect of the electronegative chlorine. As we learned in Section 4.4.4, the inductive effect decreases as distance increases.

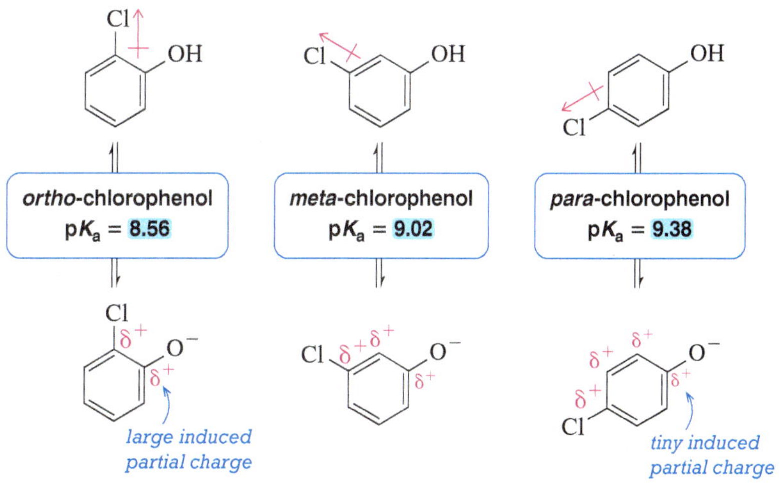

Figure 24.7 **The inductive effect of a neighboring chlorine** decreases with distance, making *ortho*-chlorophenol more acidic than *para*-chlorophenol.

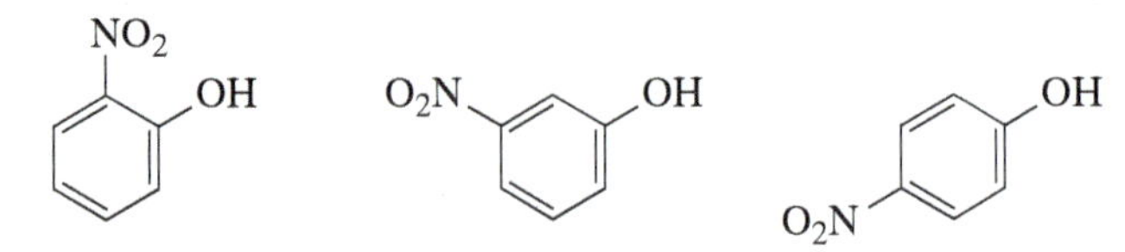

ortho-nitrophenol $pK_a = 7.23$ | *meta*-nitrophenol $pK_a = 8.40$ | *para*-nitrophenol $pK_a = 7.15$

Figure 24.8 ***Ortho-* and *para*-nitrophenol** are more acidic than *meta*-nitrophenol.

The trend for groups that withdraw electrons by resonance is different. The series of nitrophenols in **Figure 24.8** shows that *ortho-* and *para*-nitrophenol have nearly identical pK_a values. And both of these are more than one pK_a unit (> 10 times) more acidic than *meta*-nitrophenol. [*Thoughts on why this is?*]

The chlorine in Figure 24.7 is electron-withdrawing by induction, whereas the nitro group in Figure 24.8 is electron-withdrawing by resonance. Recall from Section 23.8 that resonating groups communicate differently with the ortho/para and meta positions on the ring. Drawing the important resonance structures of the nitrophenoxide conjugate bases (**Figure 24.9**) shows that in the case of ortho and para, the negative charge ends up on the same carbon as the nitro group, allowing an extra resonance structure to be drawn. In contrast, the negative charge skips the meta carbon, so the meta nitro group is unable to stabilize by resonance.

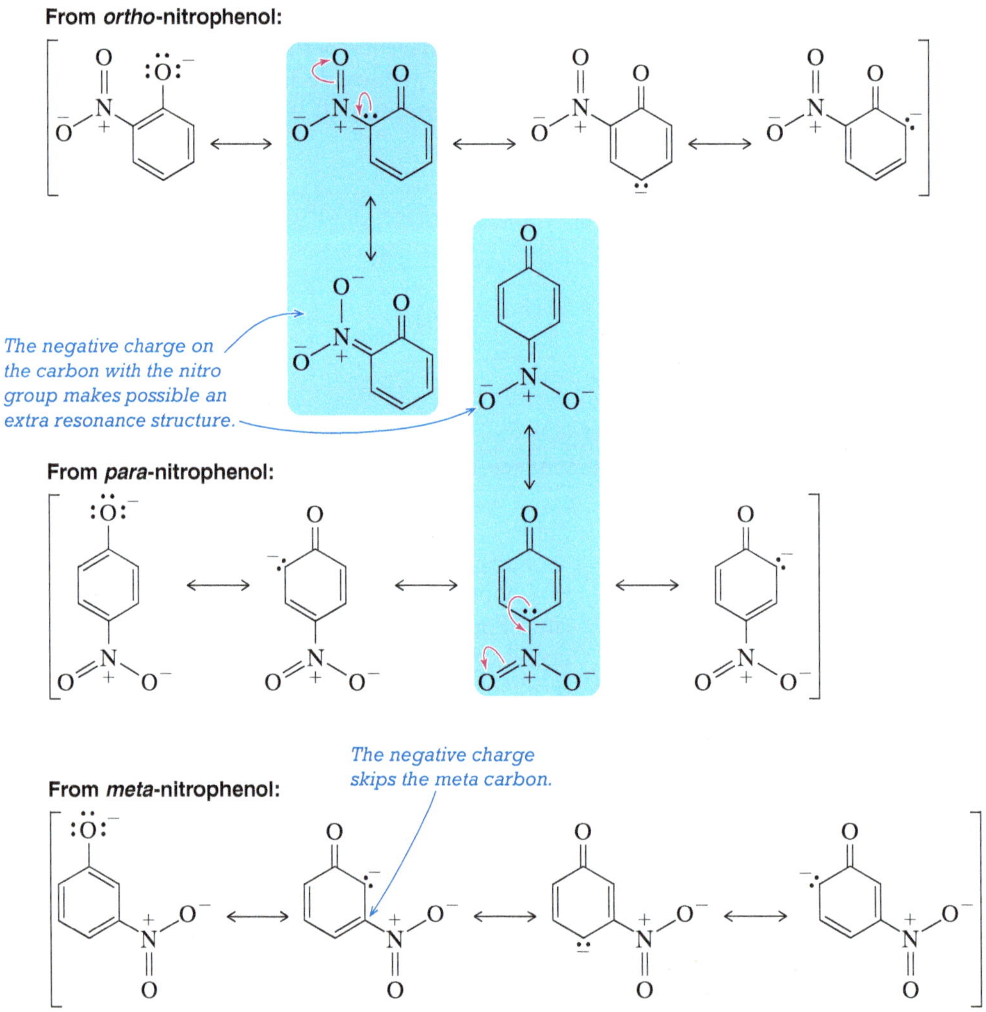

Figure 24.9 Ortho and para nitro groups stabilize the phenoxide with an extra resonance structure. The meta nitro group is unable to do this. [*You should be able to draw the arrows that indicate the electron flow between resonance structures.*]

Assessment 24.11 Though the nitro group is electron-withdrawing by resonance, when in the meta position, it doesn't communicate by resonance with the phenoxide anion. And yet, the pK_a value of *m*-nitrophenol is lowered from 10 (the pK_a value of unsubstituted phenol) to 8.4. Explain this observation.

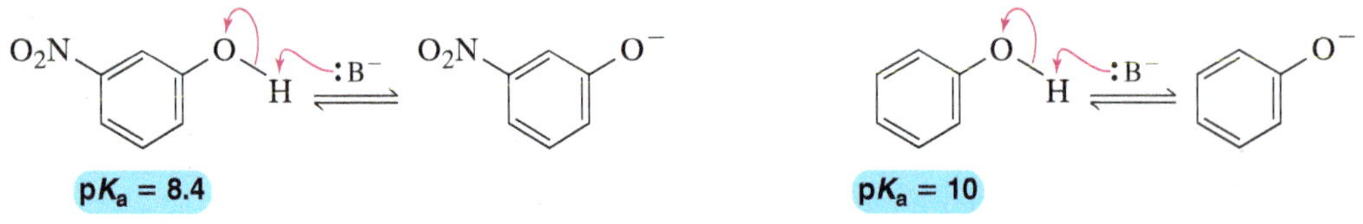

Assessment 24.12 Based on the following pK_a values, rank the cyano (—CN), nitro (—NO_2), and carbonyl (—CO) substituents based on their electron-withdrawing ability.

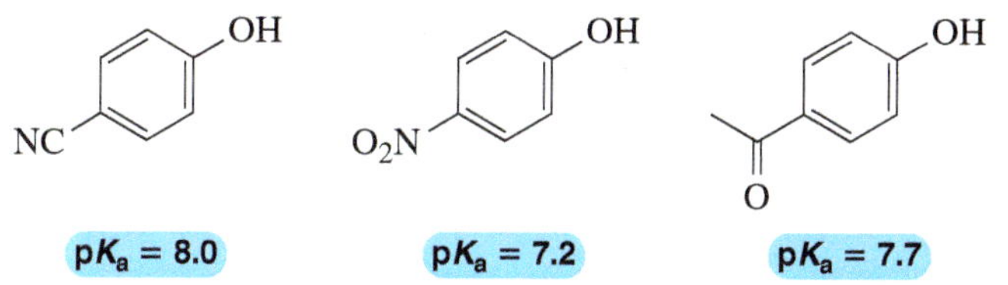

Assessment 24.13 In Chapter 23, the methoxy group was considered to be electron-donating because it increased the rate of electrophilic aromatic substitution. And yet *meta*-methoxyphenol (pK_a = 9.65) is more acidic than phenol (pK_a = 9.98), suggesting that it is electron-withdrawing. Rationalize these two observations.

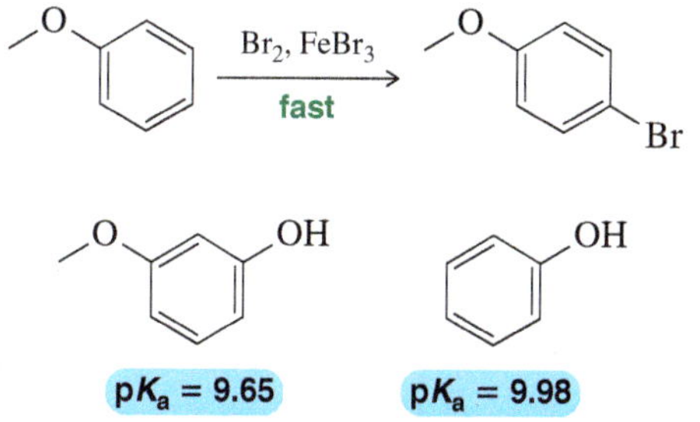

Assessment 24.14 Choose the more acidic phenol in each of the following pairs.

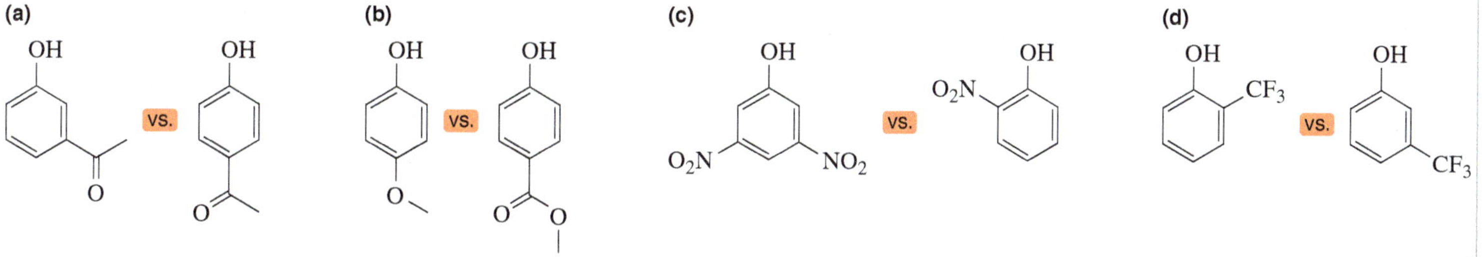

24-A CHEMISTRY IN CONTEXT: Medicinal Chemistry

Lipid protection by vitamin E

Many of the cooking oils in use today originate as lipids in plants. Although cooking oils can spoil via reaction with oxygen (Section 11.6), plants utilize vitamin E to protect their lipids from oxidation by the chain reaction mechanism shown in **Figure 24.10**. [*If you think you've seen this before, it's because you have: Figure 11.46.*]

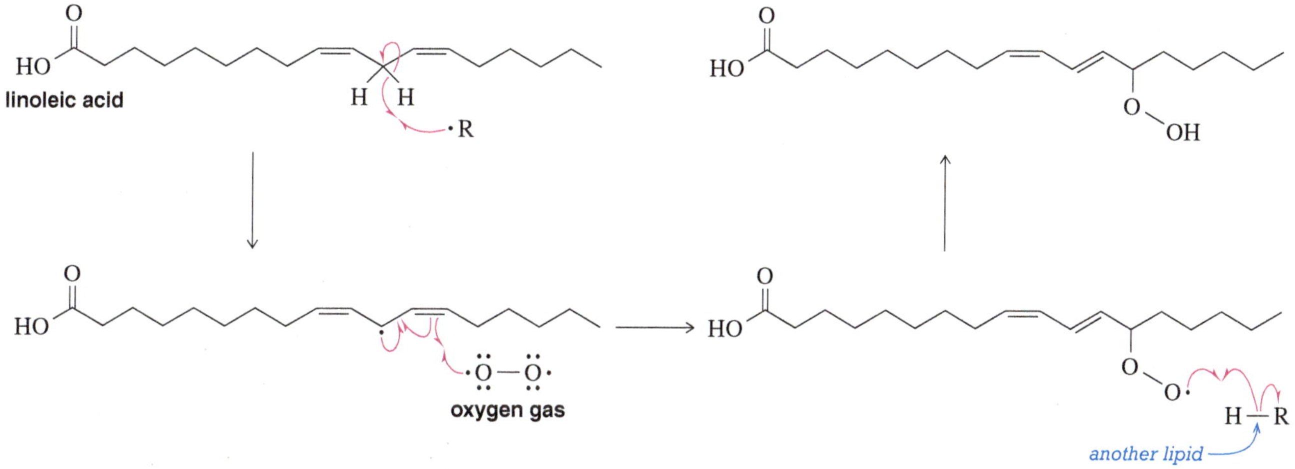

Figure 24.10 The mechanism of fatty acid oxidation.

(continued)

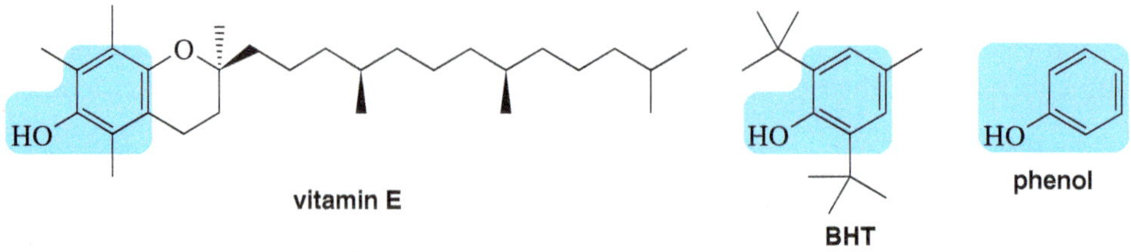

Vitamin E, which contains a phenol, interrupts the chain reaction in a manner similar to the synthetic inhibitor BHT (Section 11.7). By donating a hydrogen atom to the peroxy radical, a resonance-stabilized phenoxy radical is produced (**Figure 24.11**). This radical reacts with another peroxide in a radical-termination step at the para carbon. [*Not surprising, right?*] In this way, the chain reaction is halted, autooxidation stops, and the plant's lipids are protected.

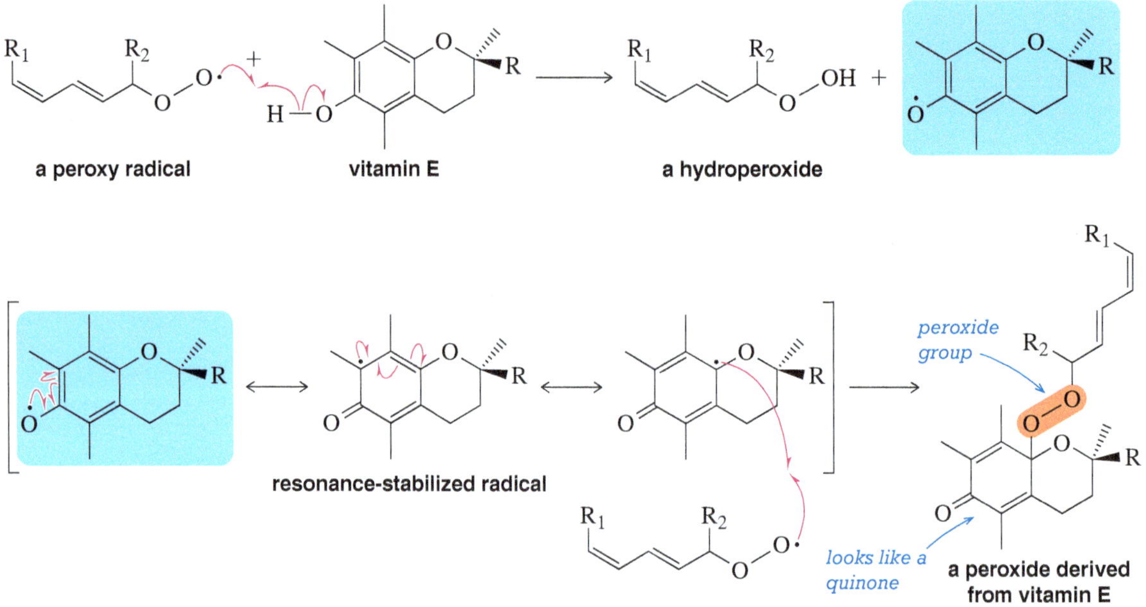

Figure 24.11 The inhibition of lipid oxidation by vitamin E.

24.2.2 Phenols in synthesis: Williamson ether synthesis

The Williamson ether synthesis (Section 13.11.1) can be used to synthesize phenyl ethers.

Assessment 24.15 Fluoxetine hydrochloride (Prozac®) is a widely used antidepressant. How might you stereoselectively install the indicated ether functional group (ROR) in (*R*)-fluoxetine?

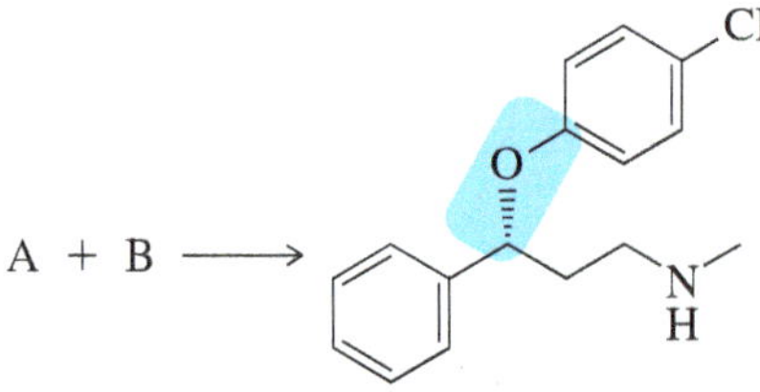

Solve Assessment 24.15 on your own. Then, by working it together, we'll discover this new application of the Williamson ether synthesis.

Solved Assessment 24.15

Recall from Section 13.11 that most ethers can be made in two ways. This general strategy is shown in **Figure 24.12** using hypothetical reagents to make the aryl alkyl ether of fluoxetine. These two possibilities differ in whether a benzylic alkoxide or a phenoxide will be used to attack the electrophilic carbon.

We actually studied possibility A in Assessment 23.83. Shown in **Figure 24.13**, the alkoxide can be generated using sodium hydroxide as a base. At high temperature, a nucleophilic aromatic substitution occurs with a chlorobenzene derivative. The trifluoromethyl (CF_3) group is a strong electron-withdrawing group, so substitution for a chloro group in the para position is favorable by the addition/elimination mechanism [*see Section 23.9.1*]. Because the stereogenic center (C_1) of the alcohol is already *R*, and no chemistry happens at C_1, the correct *R* stereochemistry is obtained.

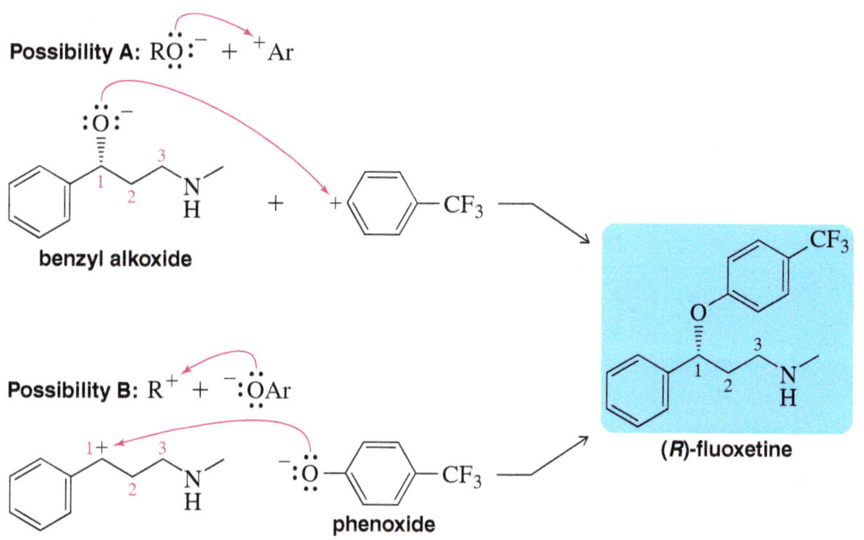

Figure 24.12 **Two hypothetical syntheses of fluoxetine.** Ar^+ and R^+ represent hypothetical electrophiles, not real reagents.

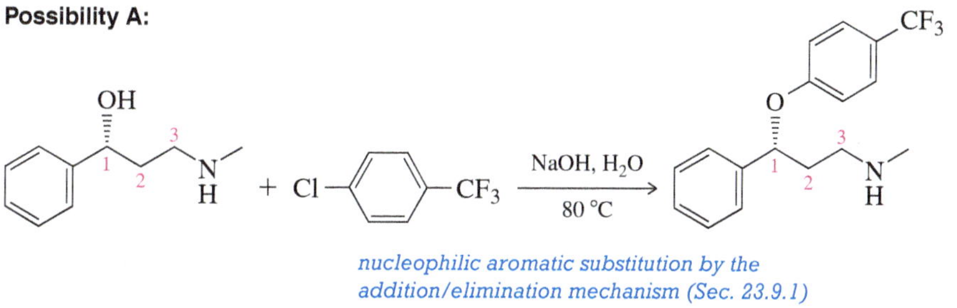

Figure 24.13 **Nucleophilic aromatic substitution** to make the aryl alkyl ether of (*R*)-fluoxetine (from Assessment 23.83).

Alternatively, a phenoxide can be used in a substitution reaction with an electrophile at C_1 (**Figure 24.14**). Because Assessment 24.15 asks specifically for the *R* stereochemistry, a carbocation can't be used as the C_1 electrophile—it would give a racemic mixture. Fortunately, a substrate with a chlorine at C_1 can be used as the electrophile in an S_N2 reaction with phenoxide. [*We explain why a benzylic chloride, even when 2°, participates favorably in S_N2 reactions in Section 24.4.1.*] If the stereogenic center at C_1 is *S*, backside displacement, by donation of electrons into σ* [*Section 12.3.1*], results in (*R*)-fluoxetine. ■

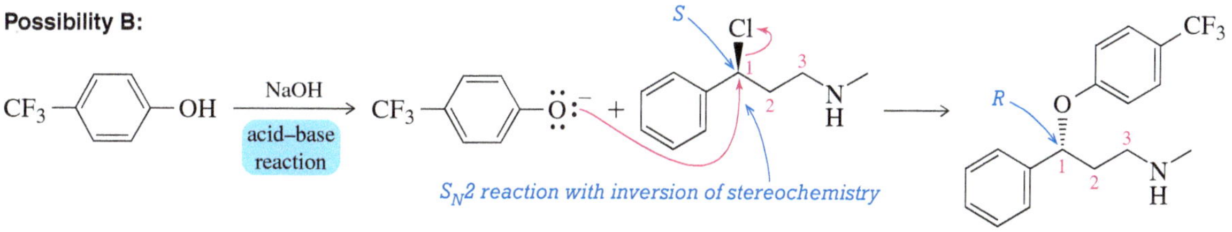

Figure 24.14 **S_N2 reaction of phenoxide and a benzylic chloride** makes the aryl alkyl ether of (*R*)-fluoxetine.

Two key differences emerge between the Williamson ether synthesis as applied here versus when we first studied it in Section 13.11.1. Both differences are related to the enhanced acidity of a phenolic proton [*Section 24.2.1*]. First, with normal alcohols, a strong base like sodium hydride (NaH) or a reducing agent like elemental sodium (Na^0) is used to form the alkoxide. Because of the reduced pK_a of phenol, a weaker base like sodium hydroxide (NaOH) can be used [**Figure 24.15(a)**]. Especially pertinent to Assessment 24.15, recall that alkoxides, where the pK_a of the conjugate acid is 16 (i.e., greater than 11; see Section 12.8.3), do E2 elimination reactions with 2° haloalkanes [**Figure 24.15(b)**]. Alternatively, phenoxides (where the pK_a of phenol is 10—that is, less than 11) do S_N2 reactions.

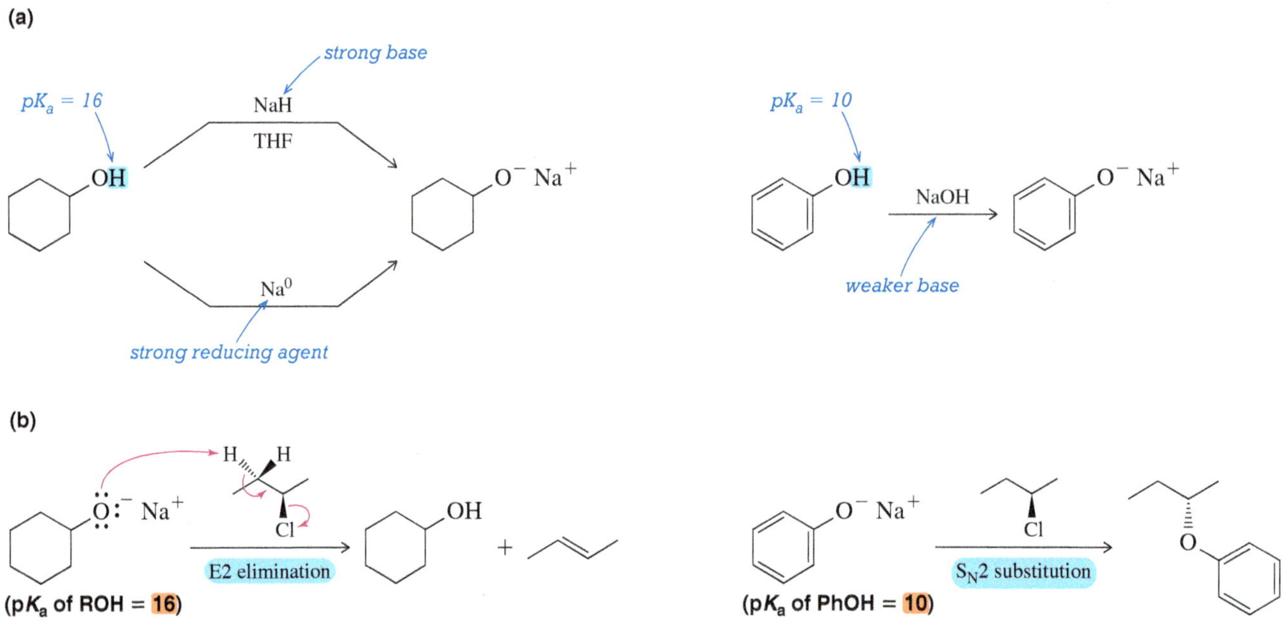

Figure 24.15 The different reactivities of phenols and alcohols in the Williamson ether synthesis. (a) Weaker bases can be used to deprotonate phenols. (b) Phenoxides can do S_N2 reactions, even at 2° carbons.

Assessment 24.16 Suggest a phenoxide and an alkyl halide to make the following aryl alkyl ethers.

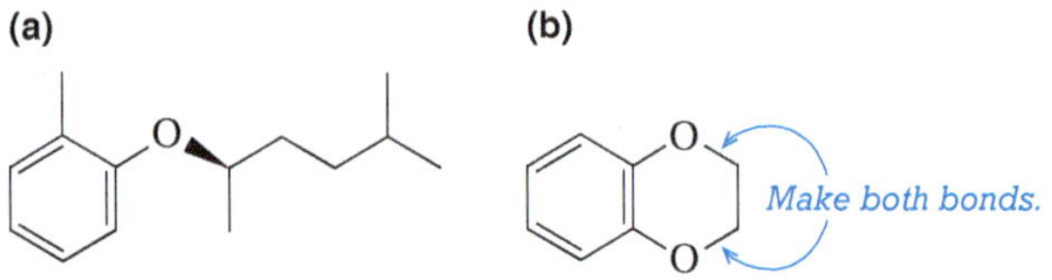

Assessment 24.17 Predict the product when the dihydroxybenzene shown is treated with a single equivalent of both base and haloalkane.

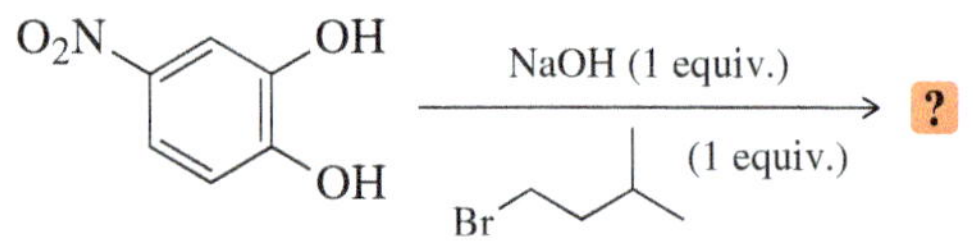

Assessment 24.18 Predict the product of the following substitution/addition reactions involving phenoxides. [*Because this problem represents a review of current and previous material, section numbers have been provided for your reference.*]

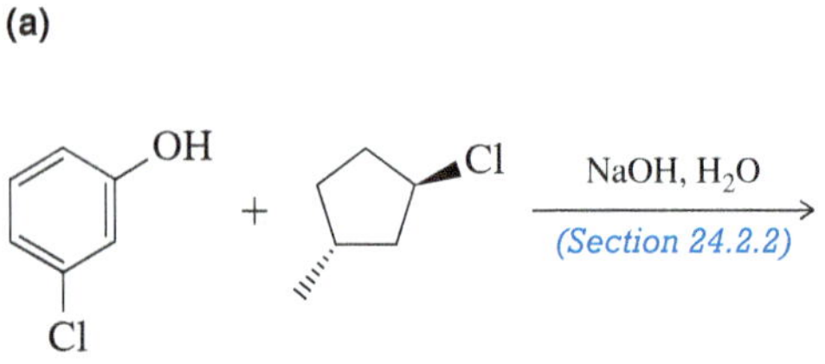

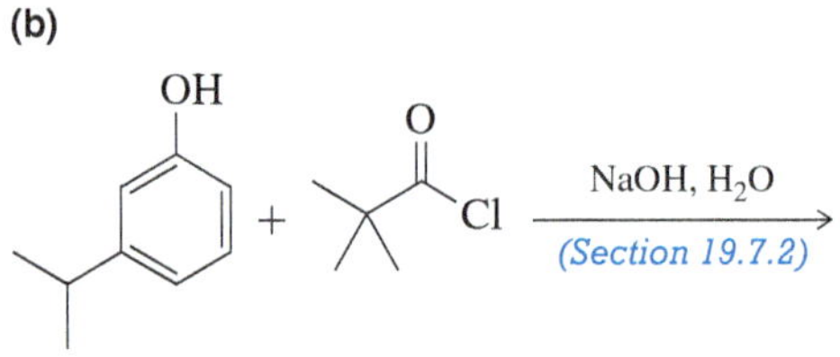

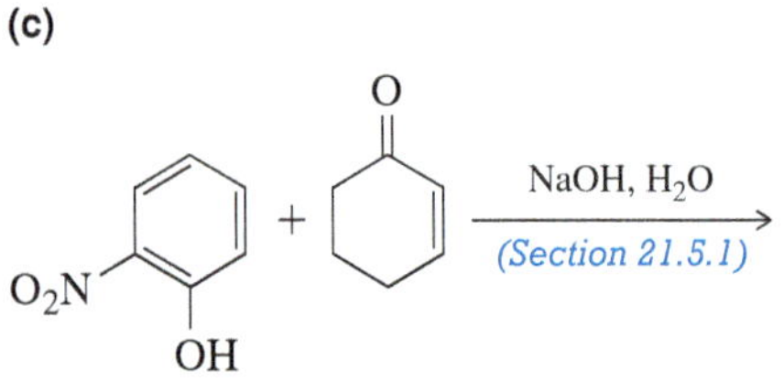

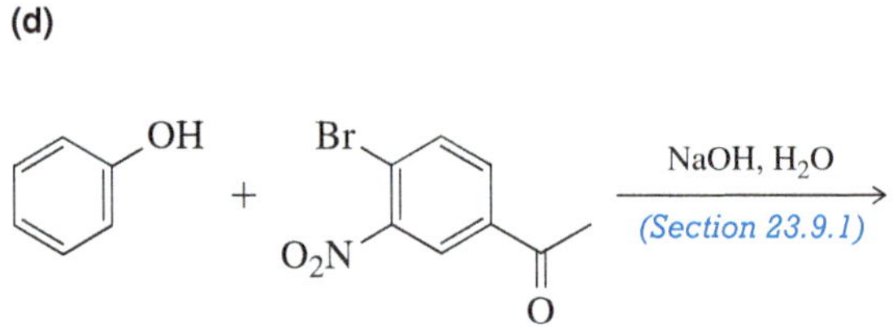

24.2.3 Phenols in synthesis: Kolbe synthesis of salicylic acid

The industrial synthesis of salicylic acid, an important compound used in acne medications and as an ingredient for the synthesis of aspirin, begins with phenol.

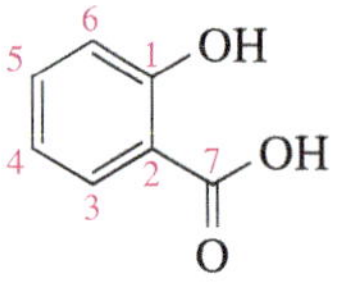

salicylic acid

Let's think like a chemist and see how we might have invented a reaction for this process using phenol.

Thinking Like a Chemist

Because we are starting with phenol, we need to add a carboxylic acid group, while forming the $C_2—C_7$ bond ortho to the hydroxyl group. As we have done in the past, a problem like this requires us to consider the innate reactivity of a molecule to see how we might want to form the new bond. The phenolic oxygen contains a lone pair of electrons, so it can donate electrons by resonance. **Figure 24.16** shows that the electron pair on oxygen is delocalized onto the ortho and para positions. Phenol, then, is an electron-rich molecule that would likely be the nucleophile in a reaction to form the $C_2—C_7$ bond. Furthermore, we expect that phenol should attack an electrophile at either the ortho or para position.

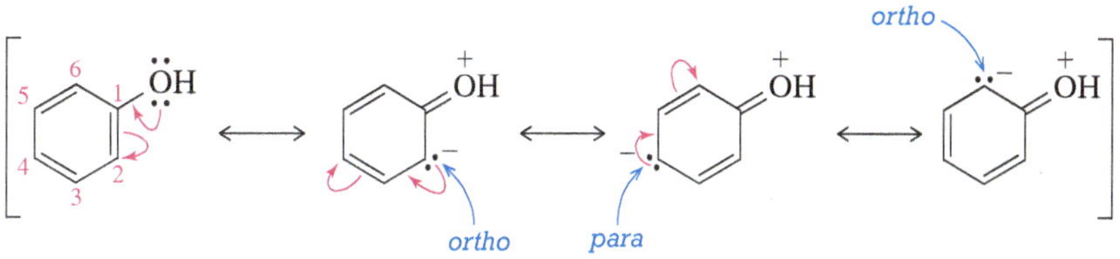

Figure 24.16 Electron density is delocalized to the ortho/para positions of phenol.

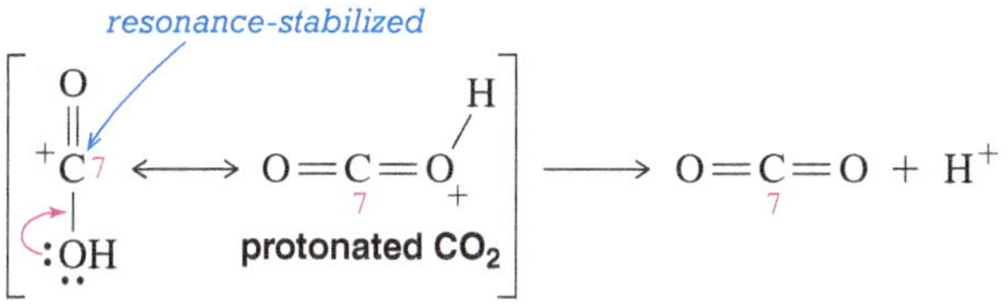

Figure 24.17 Carbon dioxide is electrophilic at C_7.

If phenol is nucleophilic at C_2, then C_7 needs to be an electrophilic source of carboxylic acid. [*We have done this before; see Thinking Like a Chemist in Section 18.5.3 to jog your memory.*] We can imagine what this might look like by first drawing a carbocation at C_7 (**Figure 24.17**). By drawing a resonance structure of this "hypothetical" carbocation, can you see that it is really just a protonated carbon dioxide? Perhaps carbon dioxide is electrophilic enough to be used directly in this reaction. [*Let's try that!*]

Mixing just phenol and carbon dioxide gas at room temperature results in no reaction. Luckily, Kolbe and Schmitt in the 1860s believed that the process simply needed harsher conditions. Thus, a base was added to make the phenol a stronger nucleophile and the solution was heated to 125 °C. Because CO_2 is a gas and water boils at 100 °C, the reaction was run in a highly pressurized vessel (**Figure 24.18**). After an acid quench of the dianion product, salicylic acid was obtained. ■

Thoughts on why really harsh conditions might be needed here? See Section 23.2.1 for a hint.

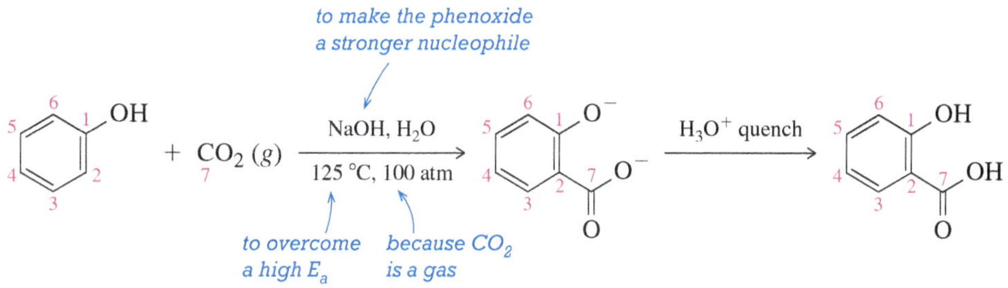

Figure 24.18 The Kolbe synthesis of salicylic acid uses base, high temperature, and high pressure followed by an acid quench.

We can propose a mechanism of the Kolbe synthesis that is reminiscent of both electrophilic aromatic substitution [*EAS, Section 23.7*] and the aldol reaction [*Section 20.3.1*]. As shown in **Figure 24.19**, deprotonation of the phenol proton makes phenoxide, a stronger nucleophile. Ignoring the rest of the ring, a negatively charged oxygen attached to an alkene is essentially an enolate. This enolate, at high temperature and pressure, attacks the carbonyl of carbon dioxide to form the $C_2—C_7$ bond, producing a carboxylate ion in the process. In this slow step, aromaticity is lost, a likely reason why high temperature and pressure are required. The resulting nonaromatic intermediate is quickly deprotonated to restore aromaticity in forming the dianion that is ultimately quenched to give salicylic acid.

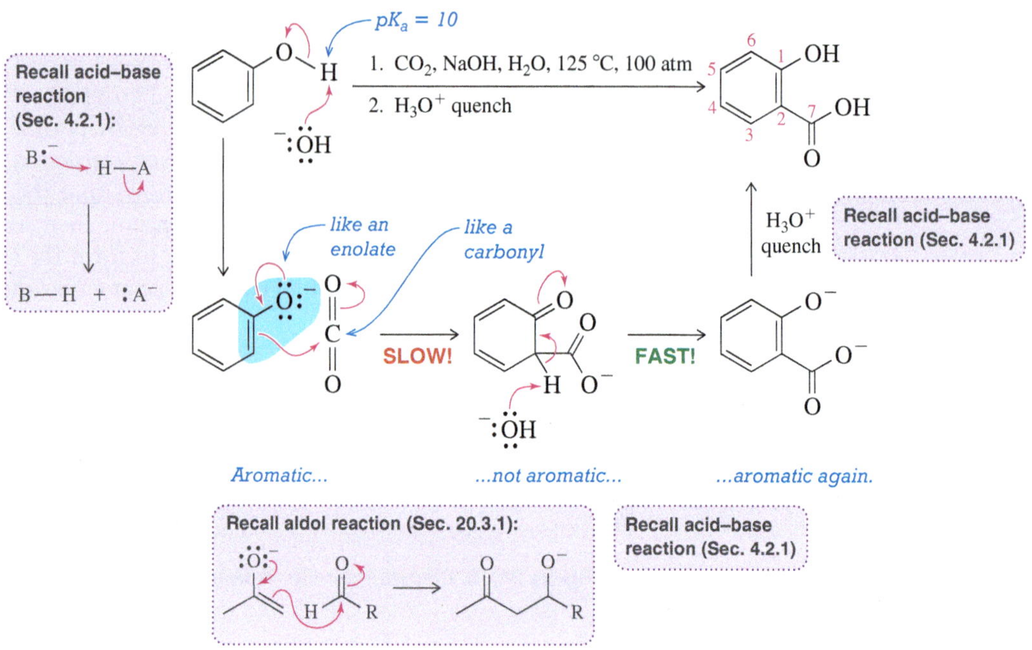

Figure 24.19 The mechanism of the Kolbe synthesis is reminiscent of an aldol reaction where aromaticity breaks in a slow first step and reforms in a fast second step. [*Research suggests that this mechanism is probably not so simple, but viewing it this way connects it to reactions we have studied previously.*]

Assessment 24.19 Show how you might use the Kolbe synthesis to prepare 3-hydroxy-2-naphthoic acid.

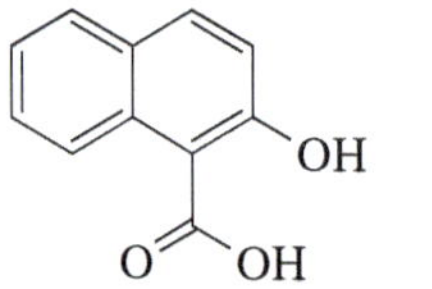

3-hydroxy-2-naphthoic acid

Assessment 24.20 The preference for ortho substitution in the Kolbe synthesis is thought to arise from the transition state shown. What is special about this transition state?

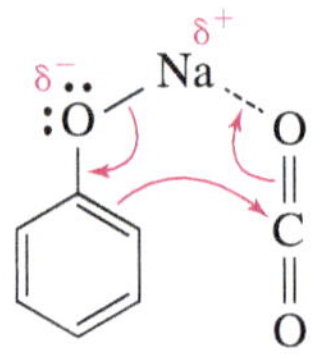

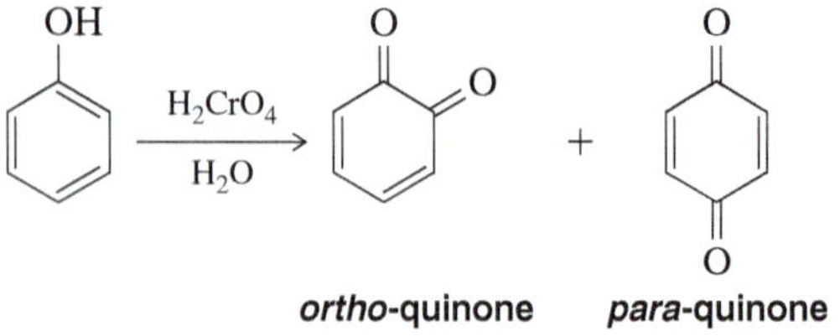

Figure 24.20 The oxidation of phenol gives *ortho*- and *para*-quinone.

24.2.4 Phenols in synthesis: oxidation

Phenol is oxidized using strong oxidizing agents to give a nearly equal mixture of *ortho*- and *para*-quinone (**Figure 24.20**). [*No meta-quinone forms . . . is that a surprise?*]

To help us understand this somewhat complicated reaction, let's first look at the related oxidation of an *ortho*-dihydroxybenzene to give an *ortho*-quinone. This analysis starts by identifying the bonds formed and the bonds broken in the oxidation and relating it to previously studied alcohol oxidations. In the oxidation of both propan-2-ol and *ortho*-dihydroxybenzene (**Figure 24.21**), two C(O)—H σ bonds are broken and a new C—O π bond is formed. This similarity, along with the fact that both reactions involve hydroxyl groups, suggests that the two mechanisms are related.

Figure 24.21 The bonds formed and the bonds broken in alcohol and dihydroxybenzene oxidation are similar. [*This is not a complete list of bonds broken and formed in the reaction.*]

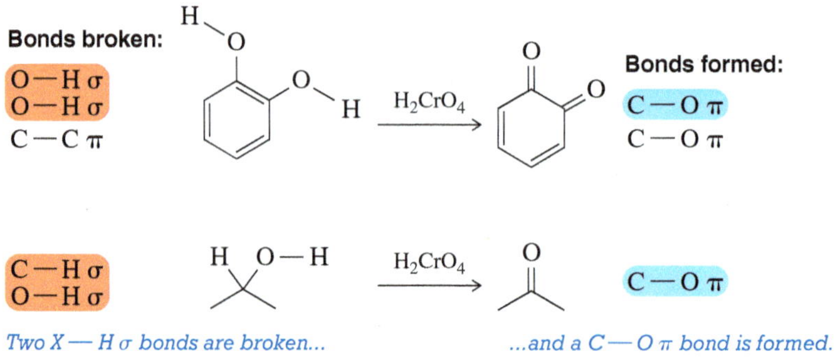

Mechanistically, the oxidation of 1,2-dihydroxybenzene begins like a normal alcohol oxidation [*Section 13.9.2; Figure 13.49*] via formation of the chromate ester (**Figure 24.22**). Once the chromate ester is formed, elimination of the chromium in what looks like an extended E2 elimination reaction gives the *ortho*-quinone. Chromium is a high oxidation state transition metal, so it badly wants a pair of electrons [*to be reduced*], which makes it an excellent leaving group. This is consistent with the second step of alcohol oxidation, in which the chromate is eliminated in another E2-like reaction. Aromaticity is lost in this reaction, which is highly unfavorable. The reaction still happens, though, which shows just how strong of an oxidizing agent chromic acid is.

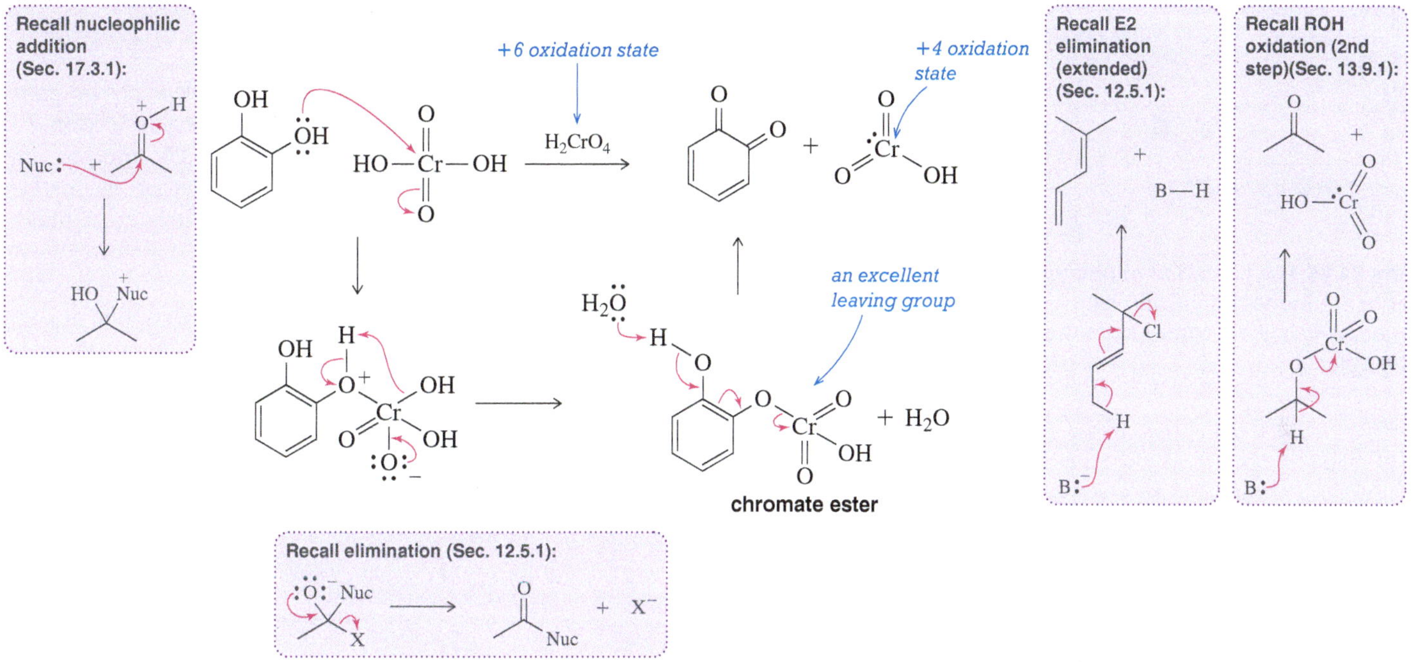

Figure 24.22 The mechanism of chromic oxidation of 1,2-dihydroxybenzene to a quinone.

Closer analysis of the elimination step will help us understand other reactions of phenols under oxidizing conditions. Instead of showing the mechanism as a concerted elimination of the chromate, imagine it to be stepwise (**Figure 24.23**) with chromium first pulling a pair of electrons from oxygen to create an oxygen cation. [*Again, it's probably concerted, but stay with me . . . you'll see why it's important.*] This cation is stabilized by resonance structures that place the cation at the ortho and para carbons, but not the meta carbon.

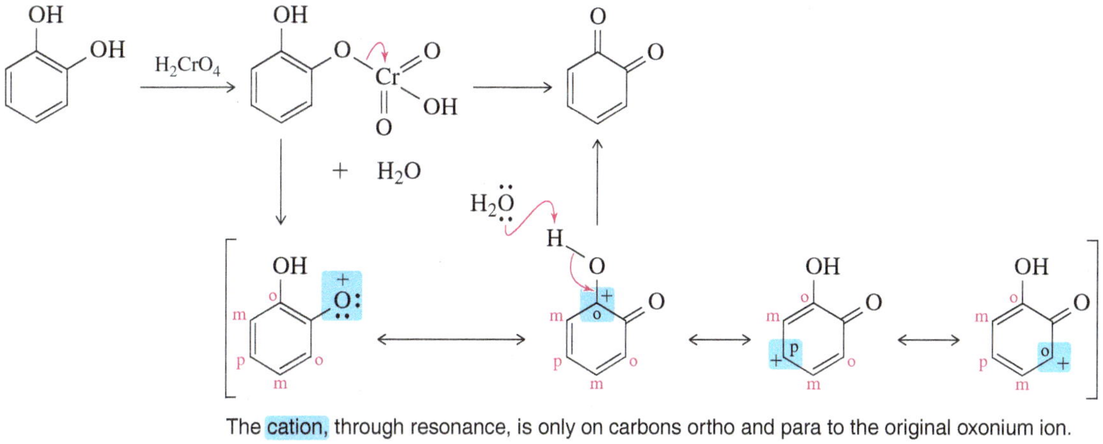

Figure 24.23 A stepwise mechanism of 1,2-dihydroxybenzene oxidation includes resonance structures in which the carbocation is on the ortho and para carbons, but not the meta carbon. [*The mechanism is actually concerted.*]

Assessment 24.21 In light of Figure 24.22, provide a mechanism by which *para*-dihydroxybenzene is oxidized to *para*-quinone.

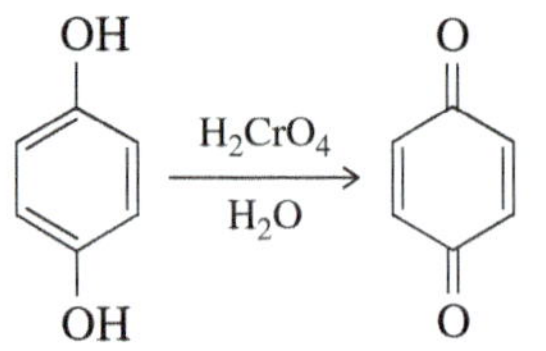

Assessment 24.22 (a) Based on Figure 24.23, explain why *meta*-dihydroxybenzene is *not* oxidized to *meta*-quinone. (b) If a *meta*-quinone is *not* produced, what would you expect the product of the oxidation of *meta*-dihydroxybenzene to be? [*We'll learn the answer to part (b) in Assessment 24.23.*]

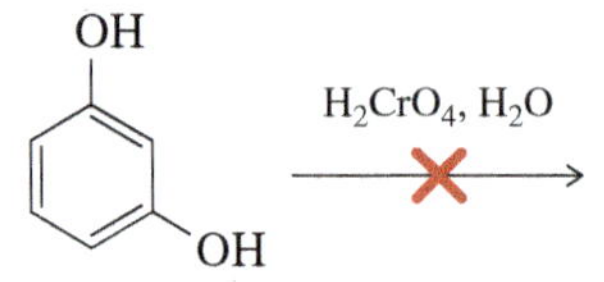

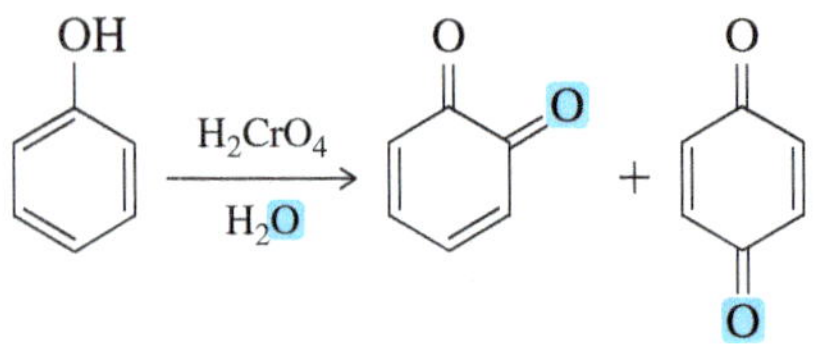

Figure 24.24 The oxidation of phenol to either quinone requires an extra oxygen, which likely comes from the water solvent.

Having answered Assessments 24.21 and 24.22(a) [*really, work them before continuing*], we can now look back at the reaction from Figure 24.20 that began Section 24.2.4 with the hope of understanding how it occurs. As shown in **Figure 24.24**, another oxygen needs to be incorporated into the starting phenol, regardless of whether the *ortho*- or *para*-quinone forms. Given that water is the solvent in this reaction, it is most likely the source of this extra oxygen.

If we're right about water providing the extra oxygen, then the crux of this mechanism comes down to how it attaches to the ring. Water typically acts as a nucleophile at oxygen; as such, the ring needs to be electrophilic. During the course of phenol oxidation, does the ring become an electrophile? [*Look back at Figure 24.23.*] In fact, it becomes electrophilic at the ortho and para positions, but not at the meta position. [*That sounds familiar . . . .*] So the mechanism begins, as shown in **Figure 24.25**, with formation of the chromate ester. Breaking the Cr—O bond [*reducing* Cr^{VI} *to* Cr^{IV}] forms a cation that can be delocalized

Figure 24.25 The formation of *ortho*- and *para*-quinone occurs via addition of water to a carbocation delocalized at the ortho and para positions. A second oxidation of a 2° alcohol completes the reaction.

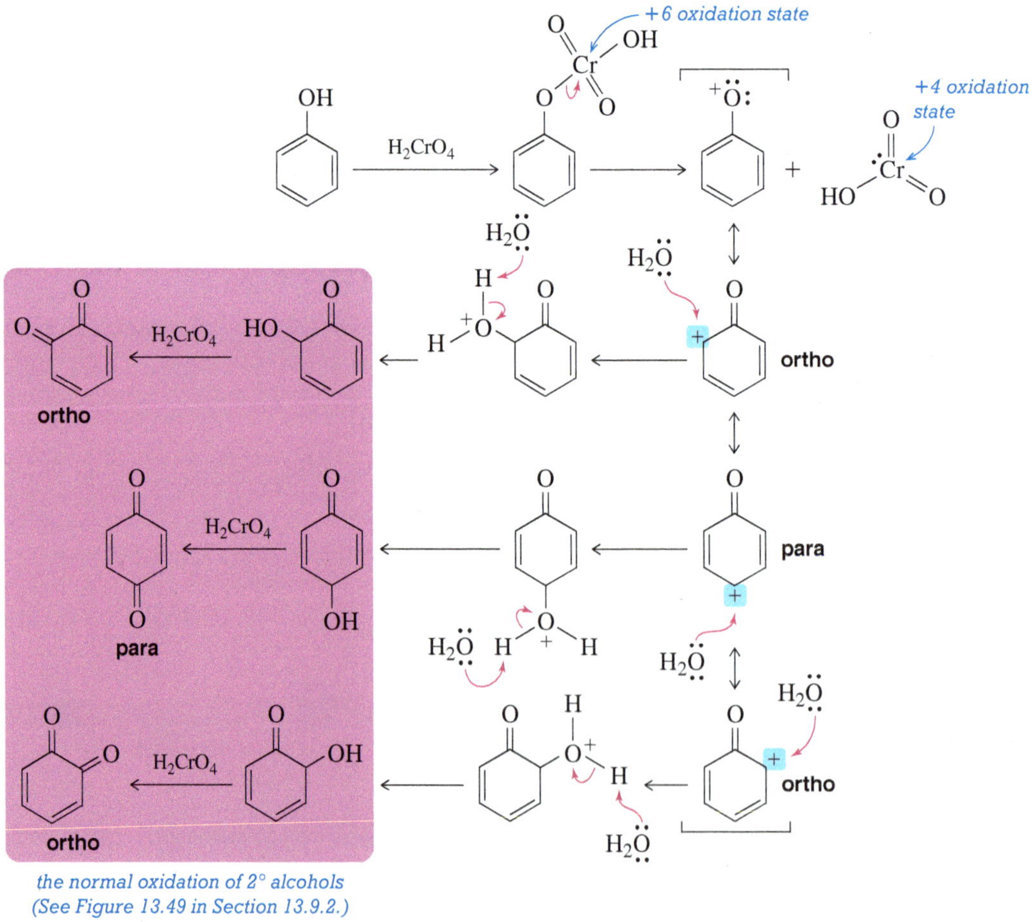

onto the ortho and para carbons of the ring. Attack by water at either of these three positions (two ortho, one para) is followed by removal of a proton to produce two hydroxydienones. Normal oxidation of these 2° alcohols occurs with chromic acid to complete the synthesis of *ortho*- and *para*-quinone. Because there are two ways to make *ortho*-quinone, there is probably more of it present in the final mixture.

Assessment 24.23 Assessment 24.22 asked why *meta*-dihydroxybenzene could not be oxidized to *meta*-quinone. The attempted oxidation instead gives rise to three different quinone products. Suggest a mechanism for the formation of each.

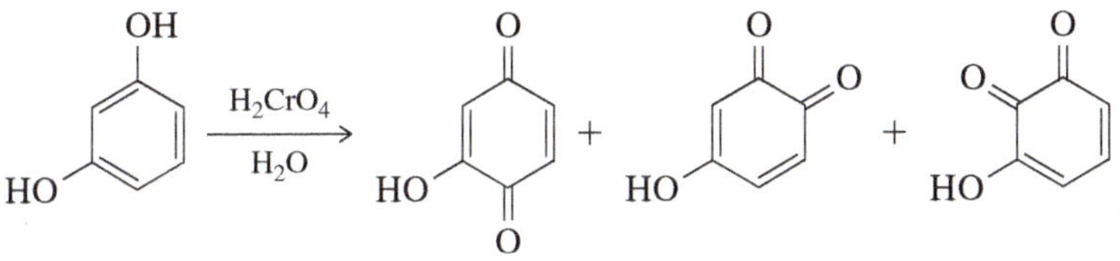

Assessment 24.24 Oxidation of a phenol derivative produces the lactone shown. Provide an arrow-pushing mechanism that rationalizes this outcome.

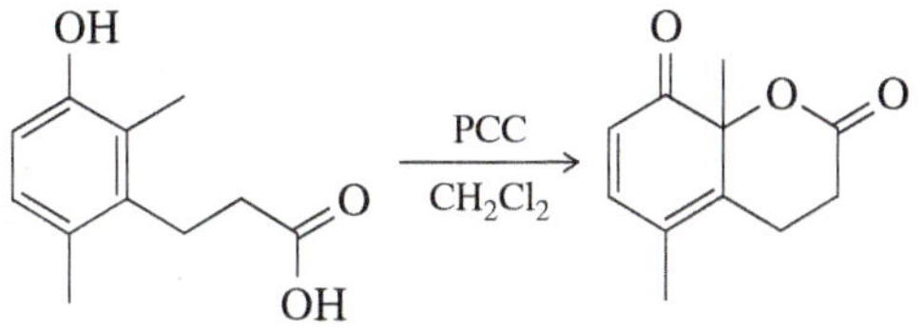

24.3 Benzylic bromination

Substitution of a bromine for a hydrogen at benzylic carbons is possible using *N*-bromosuccinimide (NBS). These conditions were first encountered during our discussion of allylic halogenation reactions in Section 11.5.3. This reaction proceeds with high selectivity for benzylic substitution even in the presence of a 3° C—H bond (**Figure 24.26**).

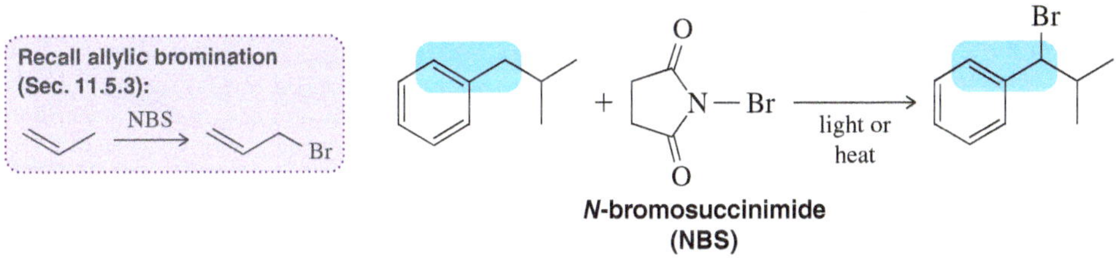

Figure 24.26 The benzylic position can be brominated using *N*-bromosuccinimide.

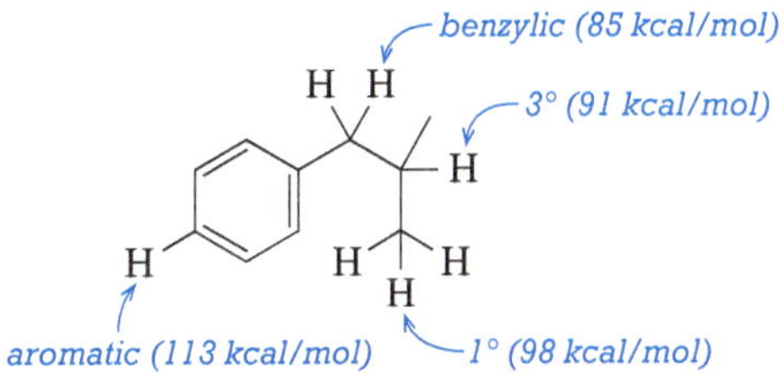

Figure 24.27 Bond-dissociation energies are correlated with radical stability.

Recall (Section 5.3.3) that the mechanisms of radical reactions consist of initiation, propagation, and termination steps.

As discussed in Section 5.3.6 and again in Section 11.5.1, selectivity in radical halogenation is a result of breaking the weakest C—H bond to form the most stable radical. Recall that bond-dissociation energies (BDEs) are a measure of the energy required for homolytic bond cleavage—that is, each bonded atom takes 1 electron from the covalent bond [*Section 5.3.1*]. The approximate BDEs in **Figure 24.27** show that the benzylic C—H bond is the weakest, followed by 3° and then 1°. [*These BDEs are from Table 5.6.*] This is consistent with the benzylic radical being stabilized by resonance and the 3° being stabilized by hyperconjugation.

Mechanistically, the reaction begins with initiation of the free radical process by homolysis of the N—Br bond in NBS (**Figure 24.28**). Once formed, the bromine radical abstracts the hydrogen that will give the most stable radical. In this case, the benzylic C—H bond is broken in the first propagation step to give a radical that is heavily stabilized by resonance with the neighboring benzene ring. [*Although not important here, the radical ends up on the ortho and para positions only, but by now you expect this!*] In a non-radical process, HBr reacts with NBS to form a molecule of Br_2 [*Section 11.5.3*]. In the second propagation step, the benzylic radical attacks this Br_2 to form a new C—Br bond and produce another bromine radical. The bromine radical formed in the second propagation step is now able to enter the first propagation step to repeat the process. [*Reminder: This repeating process is why we refer to radicals as chain reactions (Section 5.3.3).*]

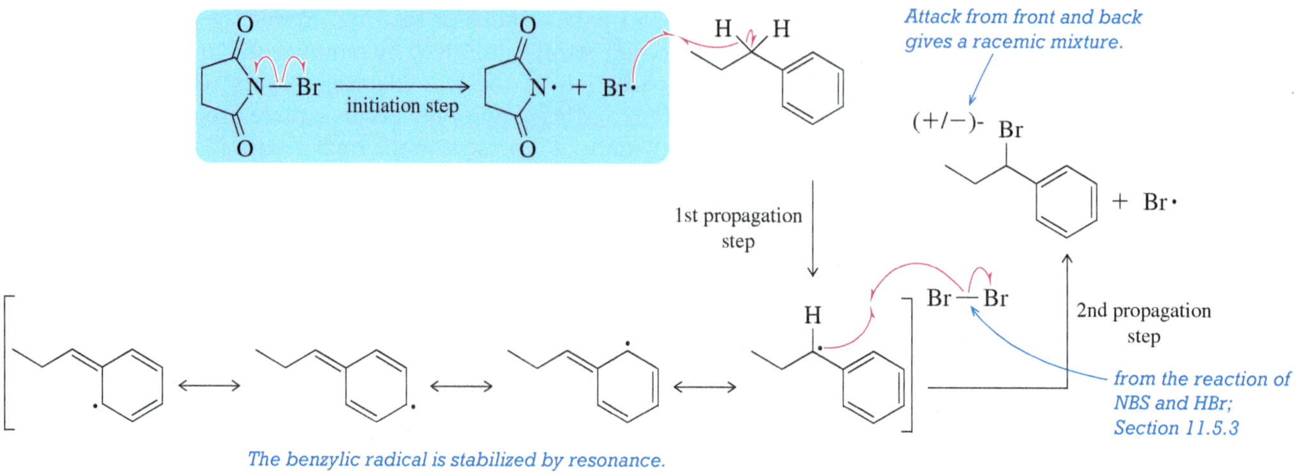

Figure 24.28 The mechanism of benzylic bromination with *N*-bromosuccinimide (NBS) occurs through two propagation steps. [*Make sure you can push electrons to show how each resonance structure of the benzylic radical is formed.*]

A benzylic C—H bond is required for benzylic bromination using these conditions. Although the radical is stabilized by delocalization around the ring, there is never a worry about the bromine being added to the ring in the second propagation step. [*End-of-chapter Assessment 24.38 asks you to consider why this is the case.*]

Assessment 24.25 Suggest reagents and reaction conditions that would result in synthesis of the following bromoalkanes.

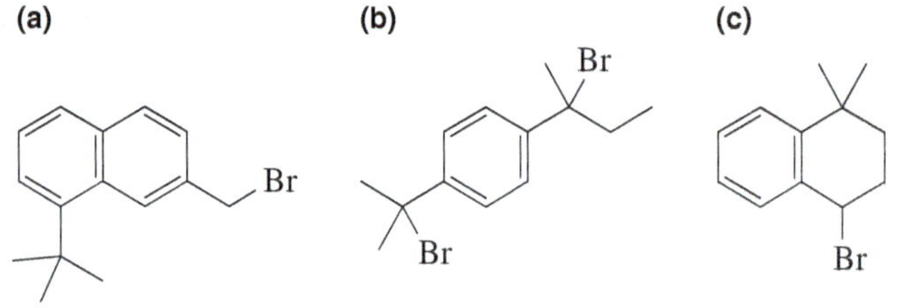

Assessment 24.26 Predict the product(s) of each of the following multistep reactions.

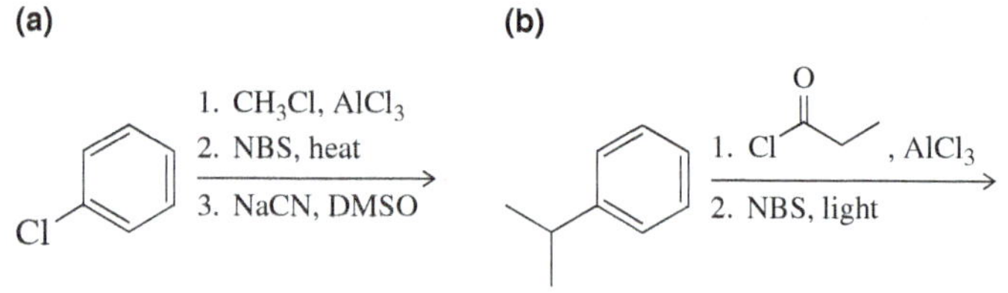

24-B CHEMISTRY IN CONTEXT: Green Chemistry

Electrochemical quinone synthesis

Recall from Chapter 13 that chromic acid, which contains hexavalent chromium, is a known carcinogen. As a result, greener methods of quinone synthesis are needed. Electrochemistry, largely the purview of analytical chemists, has recently been utilized in ways that could eventually revolutionize organic synthesis. One such example, pertinent here, is the electrochemical oxidation of 1,4-dihydroxybenzene to quinone in a biphasic mixture (H_2O/CH_2Cl_2), shown in **Figure 24.29**.

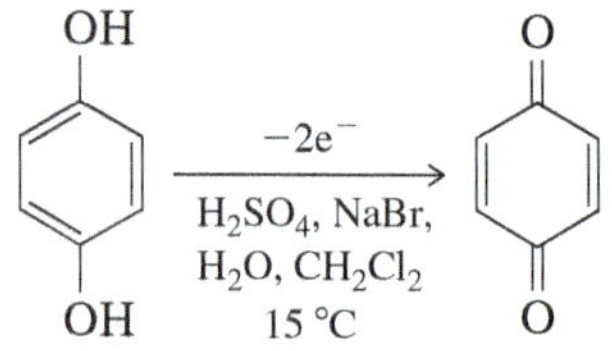

Figure 24.29 The electrochemical oxidation of 1,4-dihydroxybenzene.

The mechanism of an electrochemical reaction is not easily illustrated using our electron-pushing notation. As a result, **Figure 24.30** shows how this reaction might occur in a stepwise fashion. [*Pay close attention to the electrons.*] Within electrochemical cells, oxidation (*the loss of electrons*) occurs at the anode (**Figure 24.31**). The electron-rich 1,4-dihydroxybenzene, after losing a proton, donates an electron to the anode, becoming an oxygen radical in the process. This radical is stabilized by the attached benzene ring. Loss of a second proton gives a radical anion, another electron-rich species that donates a second electron to the anode, producing a quinone, initially represented by its diradical resonance structure in Figure 24.30.

(*continued*)

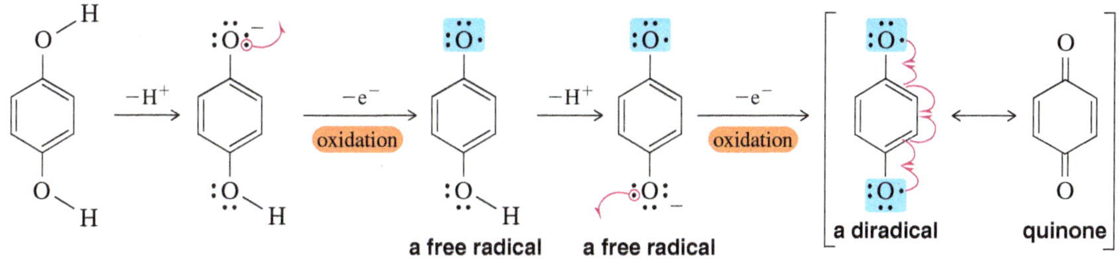

Figure 24.30 A hypothetical mechanism of 1,4-dihydroxybenzene oxidation.

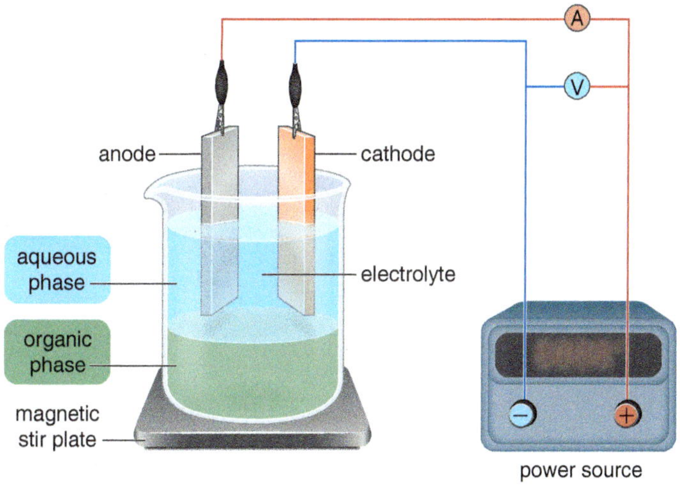

Figure 24.31 Representation of an electrochemical cell. In practice, the reaction mixture must be stirred vigorously to prevent the two phases from separating.

Assessment 24.27 How could the reaction in Figure 24.29 be made greener?

24.4 Benzylic substitution

Benzylic halogenation in Section 24.3 gives us a way to incorporate excellent leaving groups at the benzylic position. The substitution reactions of these benzyl bromides, shown in **Figure 24.32**, are also modified by the presence of the neighboring benzene ring in ways you can probably already predict. [*Would substitution by S_N1 be faster or slower? How about S_N2? Think about this before continuing.*]

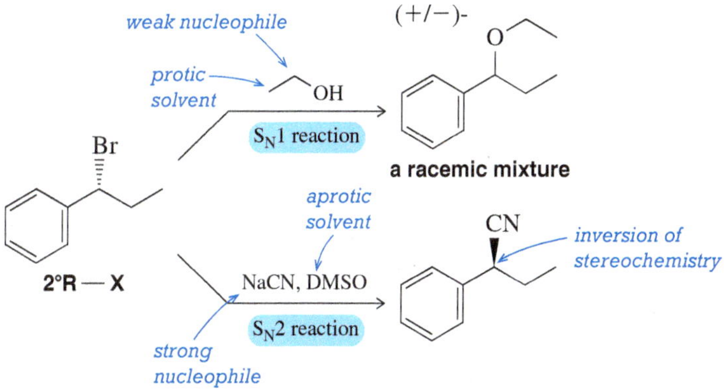

Figure 24.32 Secondary (2°) benzyl bromides can do substitution reactions via S_N1 (2° R—X, weak nucleophile, polar protic solvent) and S_N2 (2° R—X, strong nucleophile, polar aprotic solvent) mechanisms.

24.4.1 S_N1

Figure 24.33 compares two S_N1 reactions. As is true for all S_N1 reactions [*see Section 12.2.1*], formation of the carbocation occurs in a slow first step. As a result, the first step determines the rate of the overall reaction. Because the first step is endothermic, the Hammond postulate [*Section 5.2.4*] says that the transition state of carbocation formation will resemble the carbocation. Because the benzylic carbocation is stabilized by hyperconjugation and resonance and the 2° carbocation is only stabilized by hyperconjugation, formation of the benzylic carbocation is faster. As a result, S_N1 reactions at a benzylic carbon tend to be faster. Although the carbocation is stabilized by delocalization around the ring, the nucleophile never adds to the ring because it would disrupt the aromatic system.

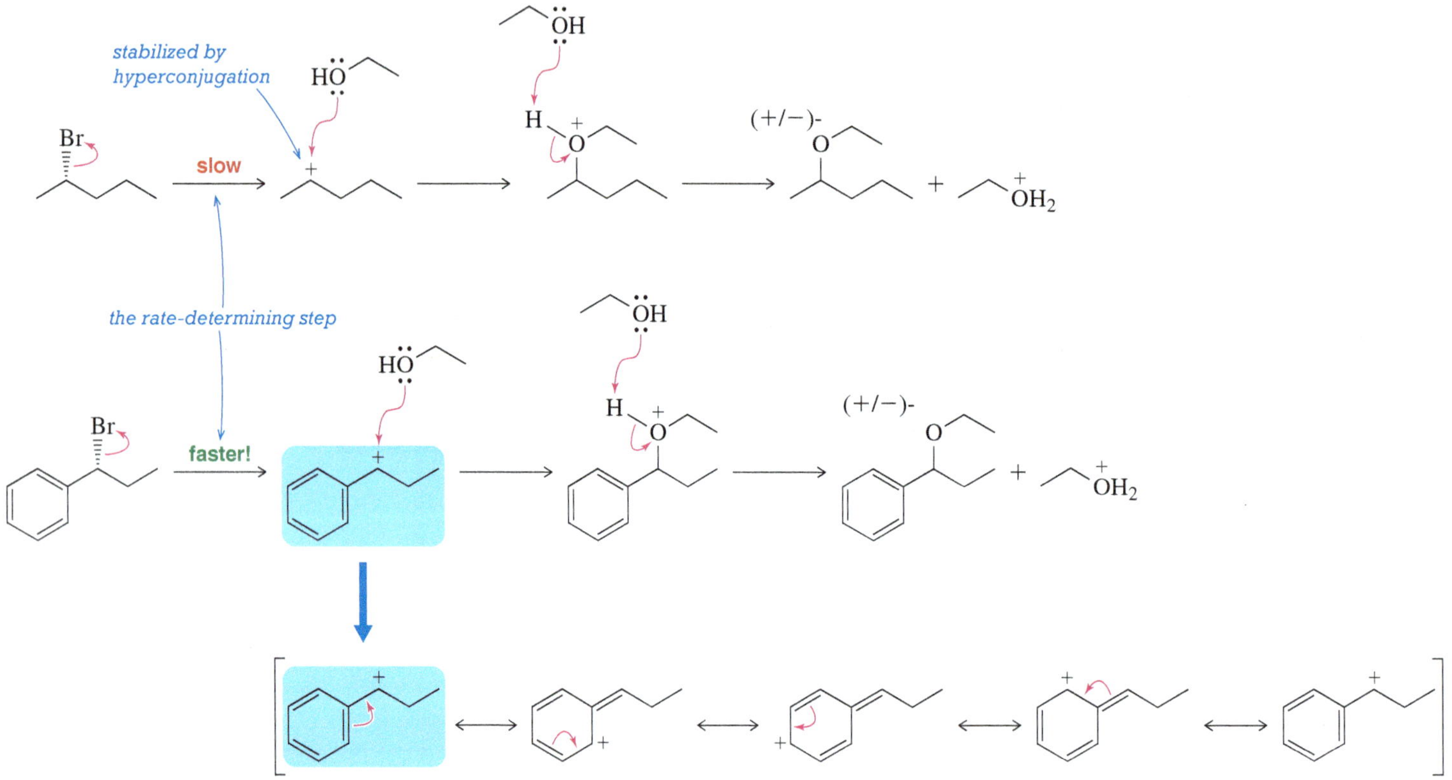

Figure 24.33 The formation of a 2° resonance-stabilized benzylic carbocation is faster than forming a 2° carbocation.

These results can be rationalized on a reaction coordinate diagram (**Figure 24.34**). Assuming that each of them starts at the same energy, the transition state leading to the more stable benzylic carbocation results in a lowered activation energy and therefore a faster rate.

Figure 24.34 A reaction coordinate diagram showing that the activation energy for forming a 2° benzylic carbocation is lower than that for forming a 2° carbocation. [*We show the two reactions starting at the same point only so the reactions can be easily compared.*]

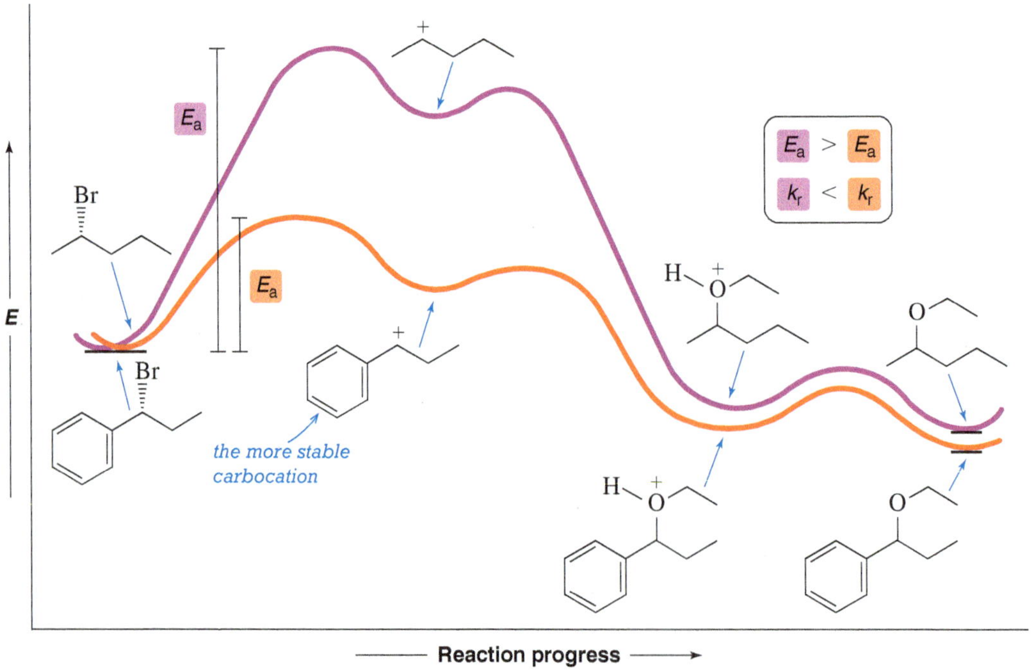

24.4.2 S_N2

A benzylic leaving group also allows S_N2 reactions to proceed more quickly for reasons that are not quite as obvious as the carbocation stabilization in the S_N1 reaction (Section 24.4.1). **Figure 24.35** compares the relative rates of a 1° haloalkane and a 1° benzylic haloalkane undergoing an S_N2 reaction.

Figure 24.35 Benzylic halides undergo faster S_N2 reactions than regular alkyl halides.

The S_N2 is a one-step, concerted reaction with no intermediate, so the difference in reaction rate must be due to transition state stabilization. In the S_N2 reaction, the nucleophile adds on the opposite side of the leaving group, donating an electron pair to σ*, the antibonding orbital (**Figure 24.36**). This results in a transition state that has the C—I bond partially formed and the C—Cl bond partially broken. If you could freeze the reaction at this point [*you can't*], the central carbon, with five groups (at least partially) attached, would have a trigonal bipyramidal geometry where the I—C—Cl bond angle is 180°. At this exact moment, the orbitals of the newly forming and newly breaking bonds [*shown in gray*] perfectly align with the π system of the neighboring benzene ring. As was the case with the reactive intermediates in Figures 24.2, 24.3, and 24.4, this orbital alignment stabilizes the transition state, making the reaction faster.

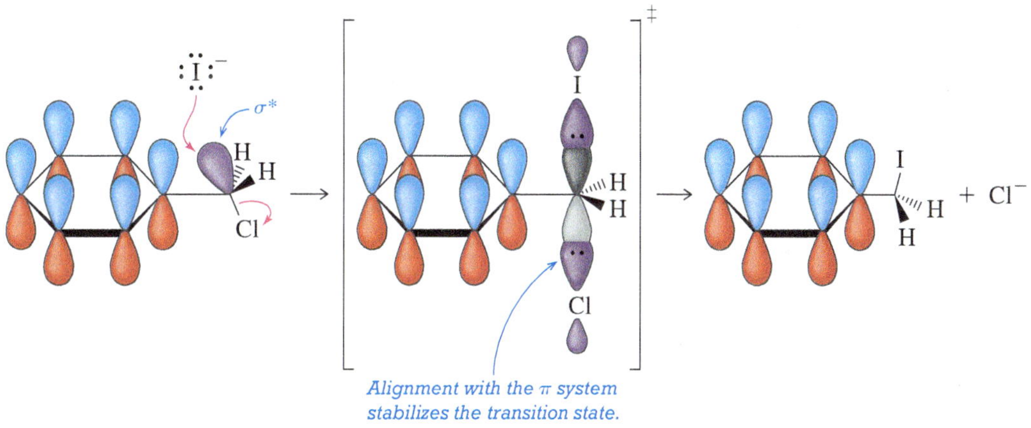

Figure 24.36 The alignment of the newly forming and breaking bonds with the π system stabilizes the transition state of the S_N2 reaction at a benzylic carbon.

24.4 SUMMARY

- Aromatic rings increase the rate of S_N1 and S_N2 reactions at neighboring carbons.

Assessment 24.28 Draw a reaction coordinate diagram that rationalizes the faster reaction of benzyl chloride in the S_N2 reactions in Figure 24.35.

Assessment 24.29 Using Equation 5.8, calculate the difference in transition state energies that lead to the rate differences shown in Figure 24.35.

Assessment 24.30 Predict the product for the following substitution reactions. Indicate whether each reaction likely proceeds by an S_N1 or S_N2 mechanism.

(a)

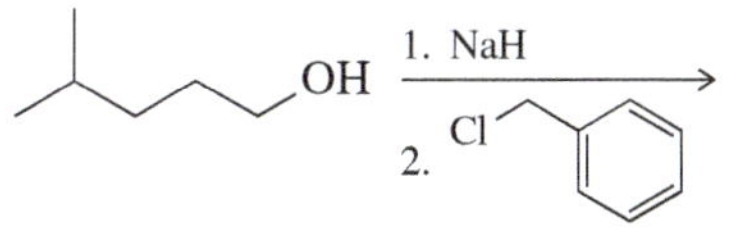

(b)

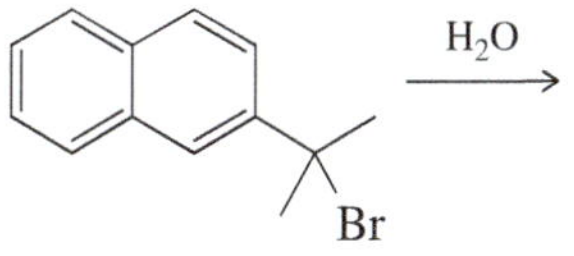

(c)

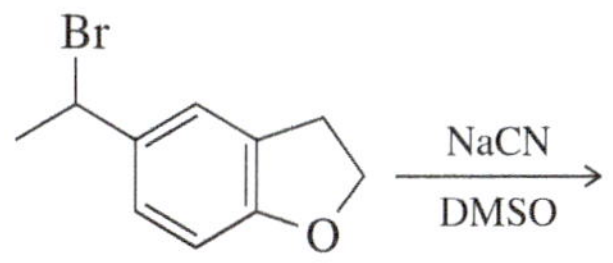

Assessment 24.31 Substitution at the benzylic position of the molecules shown was observed to occur at different rates. Explain this observation.

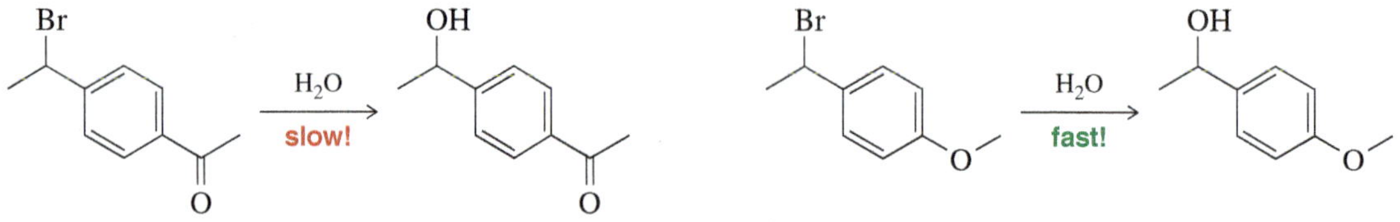

24.5 Side chain oxidation

Exhaustive oxidation of alkyl benzenes is possible using strong oxidizing agents and harsh conditions. As shown in **Figure 24.37**, heating a solution of toluene, potassium permanganate ($KMnO_4$), and hydroxide in water completely converts the benzylic methyl group to a carboxylic acid (after an acid quench). [*Chromic acid in water, when heated, accomplishes this same transformation.*]

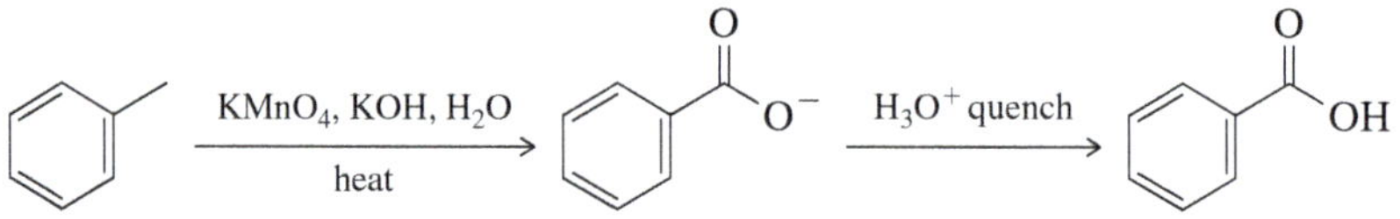

Figure 24.37 The oxidation of toluene to benzoic acid.

The mechanism of side chain oxidation is not well-understood, but is presumed to at least begin under radical conditions with the cleavage of a benzylic C—H bond. This assumption is based on the fact that any alkyl group attached to the benzene ring, regardless of size, will be converted to the carboxylic acid as long as there is at least one benzylic C—H bond present. **Figure 24.38** compares two alkyl benzenes undergoing oxidation. Note that the *t*-butyl group remains unoxidized.

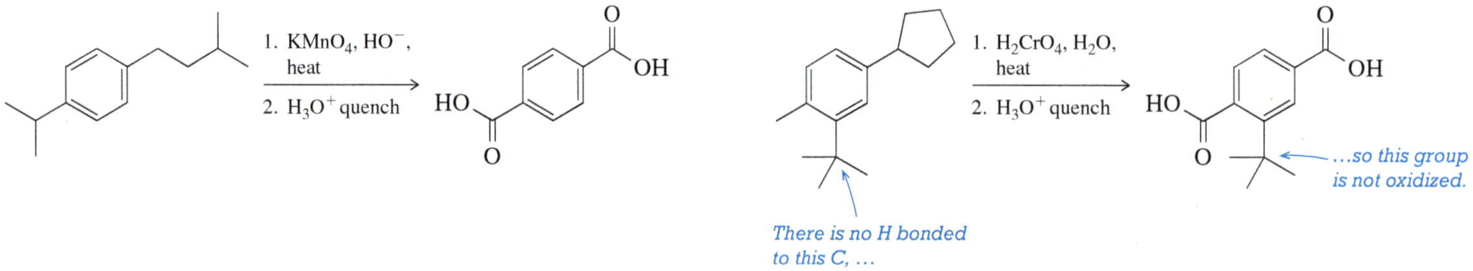

Figure 24.38 The oxidation of a benzylic C—H bond, regardless of chain-length, results in the formation of a carboxylic acid.

While the lack of a mechanism makes this reaction less interesting [*and I hate not showing you a mechanism*], the reaction does make possible some interesting synthesis problems. Recall from Section 18.5.3 that a carboxylic acid could be attached to a benzene ring through the reaction of a Grignard reagent and carbon dioxide [**Figure 24.39(a)**]. Again in Section 24.2.3, salicylic acid was made by treating phenoxide with carbon dioxide at high temperature and pressure [**Figure 24.39(b)**]. Besides these two reactions, we haven't studied other methods for attaching a carboxylic acid to benzene. [*This couldn't be done directly through electrophilic or nucleophilic aromatic substitution in Chapter 23.*] Now, a process of Friedel–Crafts alkylation, followed by oxidation with $KMnO_4$, gives us another way of doing this [**Figure 24.39(c)**]. By converting an alkyl group (electron-donating) to a carboxylic acid (electron-withdrawing), we have also changed the directing ability of the substituent from ortho/para-directing to meta-directing.

Figure 24.39 Synthesis of benzoic acid derivatives. (a) Grignard addition to carbon dioxide (Section 18.5.3). (b) Kolbe synthesis of salicylic acid (Section 24.2.3). (c) Friedel–Crafts alkylation (Section 23.7.4) followed by side chain oxidation.

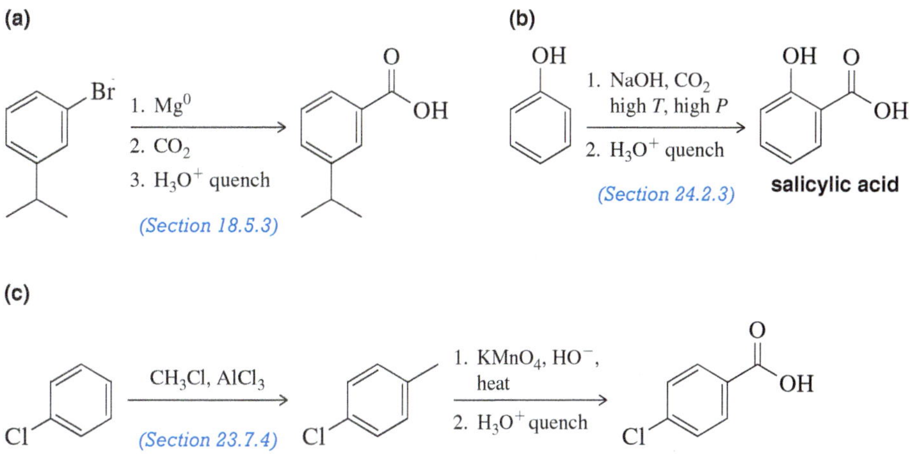

Assessment 24.32 Predict the product of the following series of reactions.

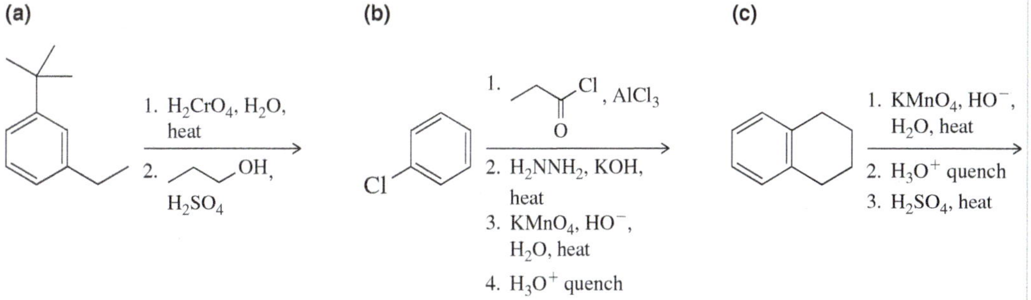

24-C CHEMISTRY IN CONTEXT: Medicinal Chemistry

Drug metabolism by cytochrome P450

In the concluding thoughts for Chapter 11, we explained how substitution of a methyl group for a chlorine resulted in a drug with similar therapeutic effects to the original, but at lower doses. Specifically, tolbutamide, which is similar in structure to the sulfonamides (Chapter 23), was an antidiabetic drug that was metabolized too quickly, presumably by the enzyme cytochrome P450 (**Figure 24.40**). Through the conversion to an alcohol, an aldehyde, and then a carboxylic acid, tolbutamide becomes significantly more water soluble and is excreted in urine. Our understanding of the influence of the benzene ring on neighboring carbons allows us to now understand the mechanism and driving force of this oxidation reaction.

The family of cytochrome P450 enzymes utilizes a heme cofactor to carry out the oxidation. Heme contains a porphyrin-based ligand bound to a catalytically active iron (**Figure 24.41**). Although much of the inorganic

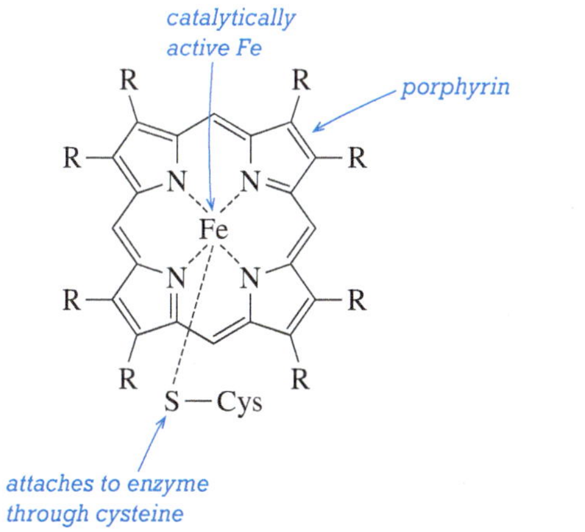

Figure 24.41 Cytochrome P450 is used for biological oxidations.

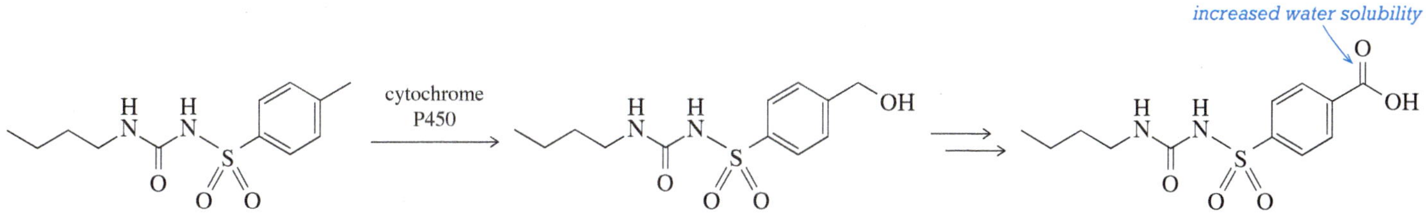

tolbutamide

Figure 24.40 Benzylic oxidation increases water solubility, thus enabling the compound to be excreted in urine.

(*continued*)

chemistry is left out, the mechanism of benzylic oxidation is shown in **Figure 24.42**. The iron(IV) oxide of heme abstracts a hydrogen to form the resonance-stabilized benzylic radical (Section 24.1.2). The benzylic radical then attacks the hydroxyl group, completing the reduction of iron(IV) to iron(III), generating the benzylic alcohol and allowing the catalytic cycle to repeat itself.

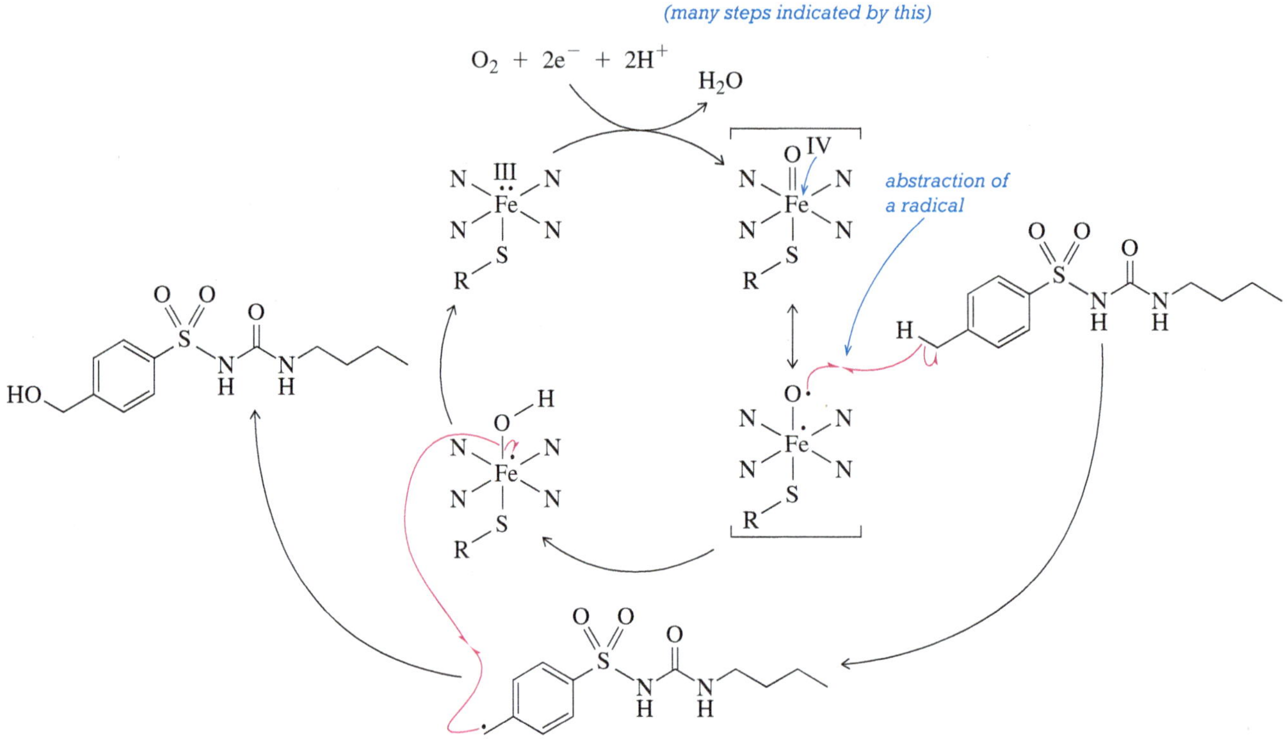

Figure 24.42 The mechanism of cytochrome P450 benzylic oxidation. The heme structure is abbreviated for simplicity.

24.6 Hydrogenolysis of benzyl ethers

In Section 13.11.1, and again in Section 24.2.2, we learned how to make ethers by alkylating alkoxides with haloalkanes in the Williamson ether synthesis. In Sections 13.14 and 17.7.5, we discussed the strategy of using protecting groups to prevent unwanted side reactions. Putting ether synthesis and protecting groups together, the unique reactivity of the benzylic C—O bond allows us to add benzyl ethers to our tool chest of protecting groups.

As discussed in Section 13.14, there are three requirements of a good protecting group. First, you have to be able to put it on. Second, whatever group you use, it must be inert to the conditions of the next reaction(s). Ethers are great protecting groups because they are inert to a variety of conditions, meaning that several reactions can be done in their presence. The third requirement for a good protecting group is that it must be able to be removed (in this case, replacing R with H). In other words, it has to be able to revert back into the functional group it was originally. **Figure 24.43** shows these requirements as they relate to the benzyl ether protecting group. By reacting an alcohol with a strong base, it can displace a benzylic bromide [*a fast reaction; see Section 24.4*] to form a benzyl ether. Because it is an inert functional group, reactions under strongly basic conditions can now be performed.

Figure 24.43 A benzyl ether is a good protecting group for an alcohol because it is inert under strongly basic conditions.

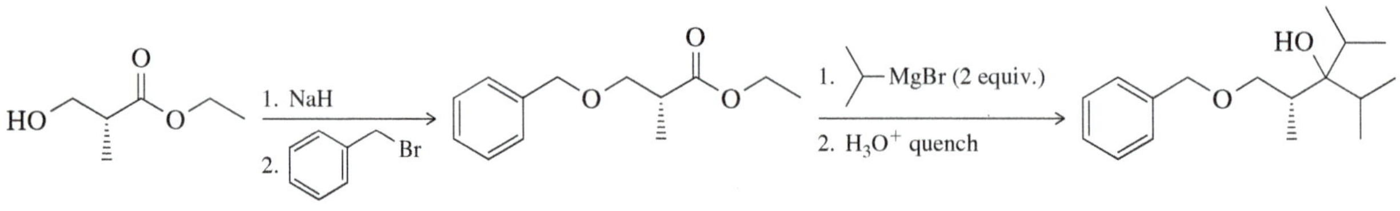

The conversion of a benzyl ether back into the hydroxyl functional group relies on a new reaction, hydrogenolysis, which is catalyzed by palladium (**Figure 24.44**). **Hydrogenolysis** is subtly different than hydrogenation, a reaction we've studied before. In the hydrogenation of an alkene, for example, H_2 is added to a C—C π bond. Hydrogenolysis, on the other hand, uses H_2 to break (*lysis* = releasing) a bond, creating two new molecules.

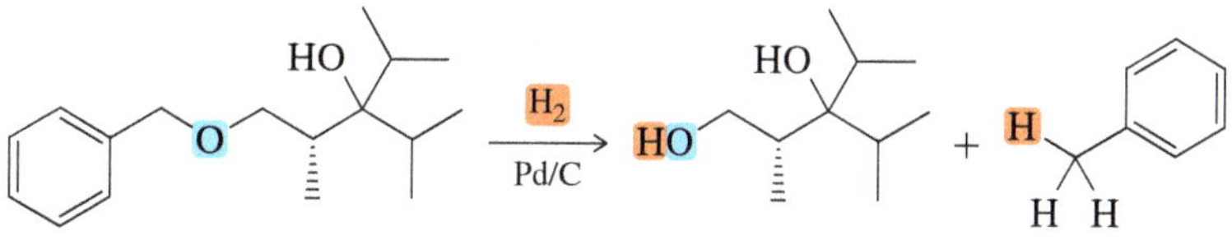

Figure 24.44 Hydrogenolysis of the benzyl ether regenerates/deprotects the hydroxyl group.

We can imagine this reaction being mechanistically similar to the palladium coupling reactions of Chapter 16 with the reaction beginning with oxidative insertion of palladium into the benzylic C—O bond (**Figure 24.45**). This is only possible for benzylic and allylic ethers because the neighboring π system weakens the C—O σ bond enough to allow it to occur. It has also been proposed that an initial complex between benzene and palladium directs the palladium to the neighboring C—O bond. [*Note the influence of neighboring benzene—the theme of this chapter*]. Coordination with hydrogen gas is then followed by formation of new O—H and Pd—H bonds. Reductive elimination forms the new C—H bond in toluene, releasing Pd^0 to participate in the catalytic cycle again.

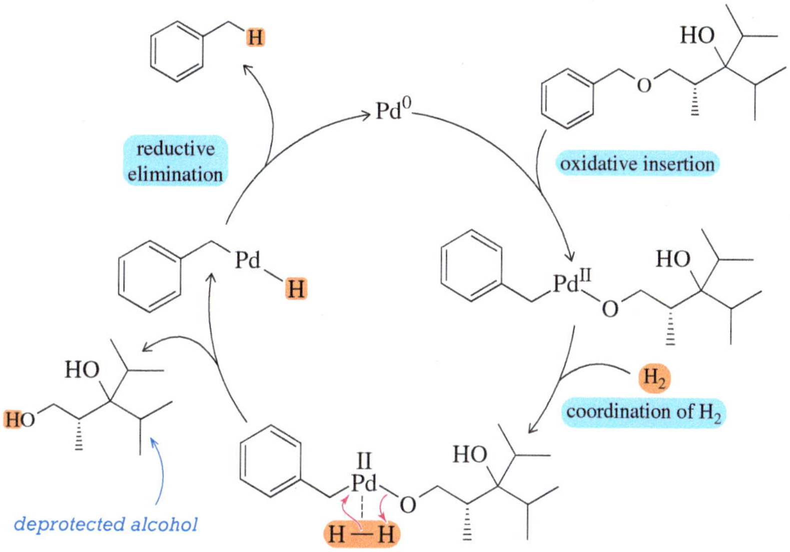

Figure 24.45 The catalytic cycle of palladium-catalyzed hydrogenolysis. [*Note: This mechanism is likely not so simple, but viewing it this way connects it to reactions previously studied.*]

In two different places in this chapter, we have shown simplified mechanisms. When you get to graduate school, you will learn the more complicated way these reactions occur. Perhaps that's reason enough to get a Ph.D. in organic chemistry?

Assessment 24.33 Predict the product of the following reactions.

(a)

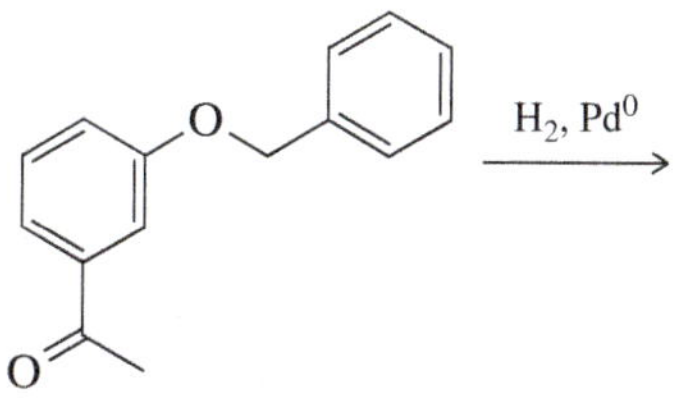

(b)

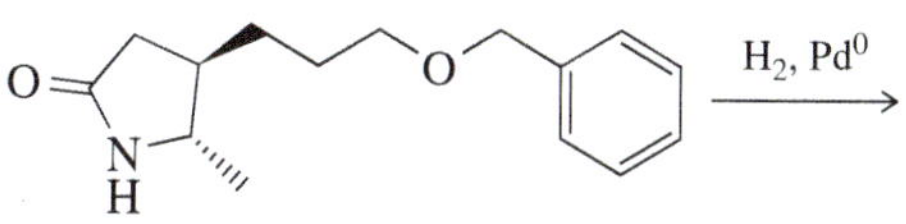

Concluding thoughts

Although looking back was an important theme here in Chapter 24, quinones, molecules that play an important role in biology, were also introduced. Specifically, coenzyme Q, a lipid-soluble quinone, is crucial in the mitochondrial electron transport chain, which is responsible for generating the energy necessary for ATP synthesis in eukaryotic cells. While the full process is way too complicated to be discussed here, the manner in which quinone might transport electrons becomes evident by considering the equilibrium shown in Figure 24.46. Reduction of quinone to dihydroxybenzene requires the addition of 2 electrons and 2 protons. Oxidation of dihydroxybenzene to the quinone (as seen in Chemistry in Context 24-C) releases 2 electrons and 2 protons.

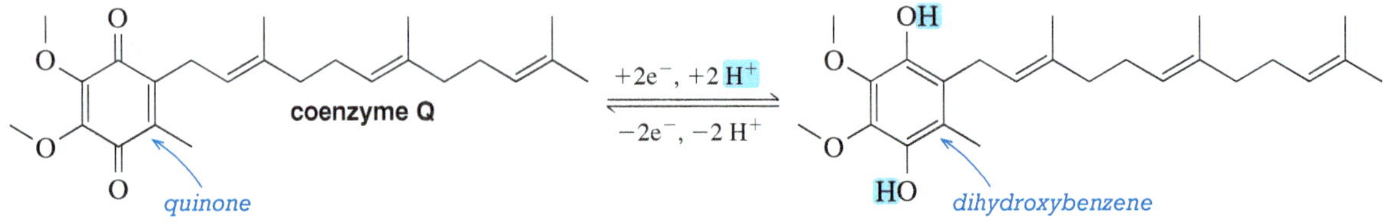

Figure 24.46 Coenzyme Q, a quinone, participates in equilibrium redox reactions.

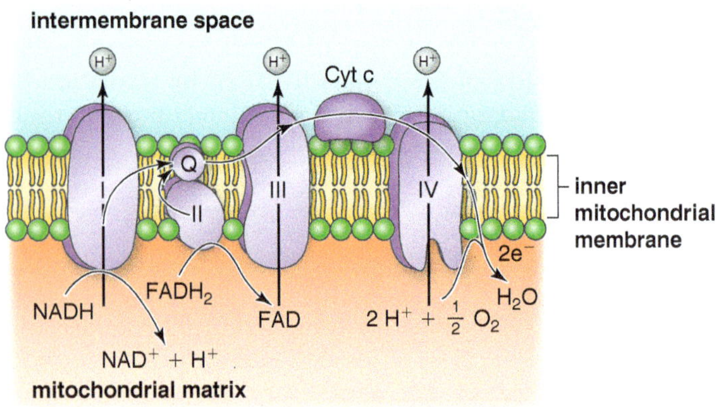

Figure 24.47 Coenzyme Q (complex II) serves as an electron carrier between complexes I and III in the electron transport chain.

Through the equilibrium in Figure 24.46, coenzyme Q is able to serve as electron carrier between two enzymes. Complex I couples the oxidation of a cofactor to the reduction of coenzyme Q (Q in Figure 24.47). Coenzyme Q, in its phenolic form, now delivers these electrons to complex III via its reoxidation back to quinone. The vital role played by coenzyme Q in the production of energy by ATP synthesis is a testament to the importance of quinones and the interesting reactivity provided by the presence of a benzene ring.

While the study of conjugated systems (Chapters 21–24) ends here, the study of amines in Chapter 25 continues our look at reactions seen previously, but in new contexts. Although we haven't studied amines as a standalone functional group, they have participated in a number of reactions we have seen before and have been the products in others. [*Your reaction sheets should confirm this.*] Once more, we'll learn a few new things while strengthening our understanding of concepts we have studied before. [*Sound exciting? On we march.*]

End-of-Chapter Materials

Important Concepts

[*These are the major concepts covered in each section, excluding reactions.*]

24.1: Benzene and other aromatic rings alter the reactivity of neighboring groups. Through resonance, anions, cations, and radicals are more stable when directly attached to an aromatic ring.

24.2: Phenols contain a hydroxyl group directly attached to a benzene ring. Phenols are more acidic than normal alcohols because the conjugate base is stabilized by resonance with the benzene ring. This increased acidity allows them to be good substrates for the Williamson ether synthesis after deprotonation. Phenols are converted to salicylic acid derivatives by reaction with carbon dioxide under harsh, basic conditions. Finally, phenols undergo oxidation to make *o*- and *p*-quinones.

24.3: Benzylic bromination is accomplished by the reaction of NBS with alkyl benzenes containing at least one benzylic hydrogen. The reaction proceeds through a resonance-stabilized benzylic radical.

24.4: The rate of S_N1 reactions at the benzylic position is increased because of the carbocation stabilization offered by resonance with the aromatic ring. The rate of S_N2 reactions at the benzylic position is also increased because the transition state is stabilized by alignment with the neighboring π system.

24.5: Chromic acid and potassium permanganate oxidize the alkyl side chain of substituted benzene rings as long as there is at least one benzylic hydrogen present.

24.6: Hydrogenolysis of benzyl ethers is made possible by the weakening of the C—O σ bond by the neighboring π system.

Important Reactions and Mechanisms

1. **Williamson ether synthesis (Section 24.2.2):** Phenoxide, synthesized by deprotonation, displaces a good leaving group from a 1° or 2° carbon to form an aryl alkyl ether.

General:

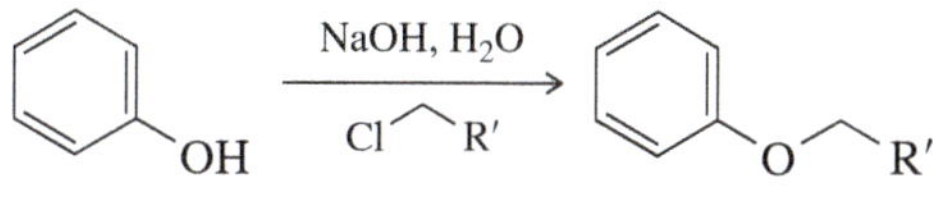

Specific:

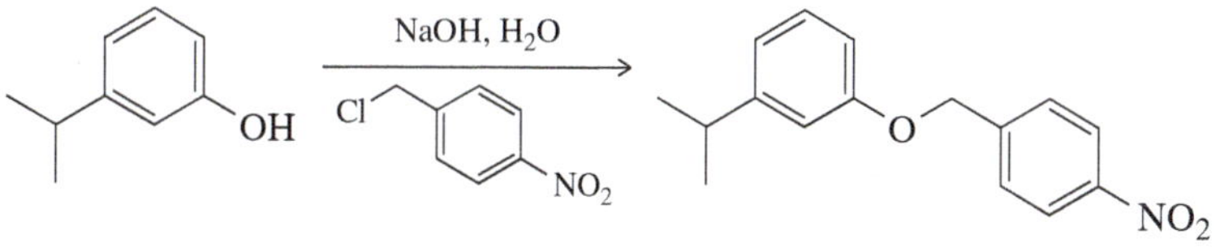

Mechanism:

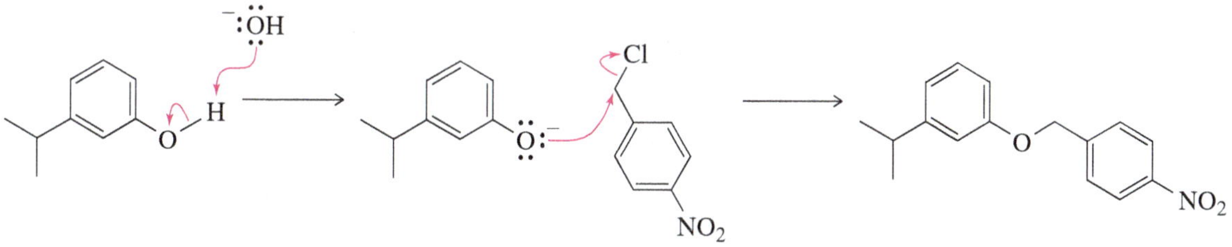

2. **Kolbe salicylic acid synthesis (Section 24.2.3):** At high temperatures and pressures, a phenoxide is alkylated at carbon by carbon dioxide. [*This is reminiscent of an enolate attacking a carbonyl.*] The resulting intermediate is deprotonated to regenerate the aromatic phenoxide. An acid quench produces the final salicylic acid product. [*This is a simplified version of what recent research suggests is actually a pretty complicated mechanism.*]

General:

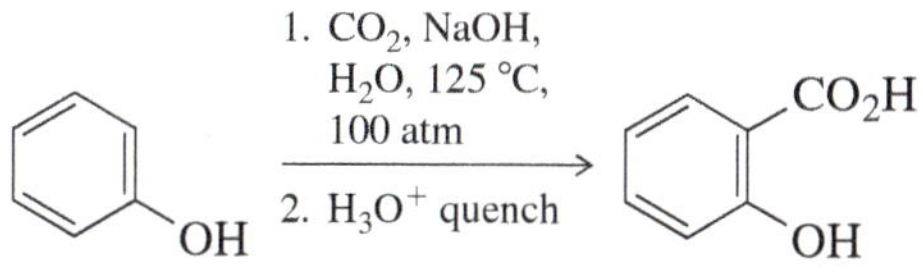

Specific:

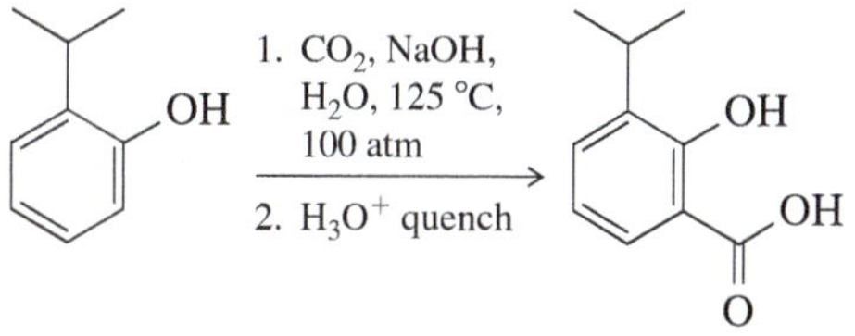

Mechanism:

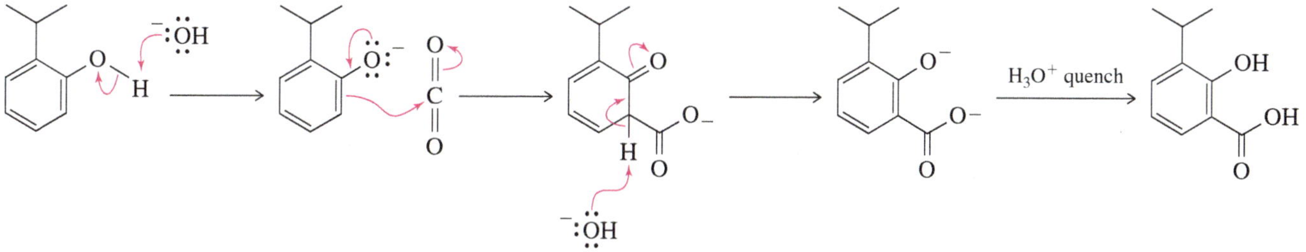

3. **Phenol oxidation (Section 24.2.4):** *Ortho*- and *para*-dihydroxybenzenes are oxidized to *o*- and *p*-quinones, respectively. Formation of a chromate ester occurs in the same manner as with alcohols (Section 13.9.2). Deprotonation of a phenolic hydrogen results in de-aromatization of the ring driven by formation of two C—O π bonds and reduction of Cr(VI) to Cr(IV). This reaction can also be done using phenol itself, resulting in formation of a mixture of *o*- and *p*-quinones.

General:

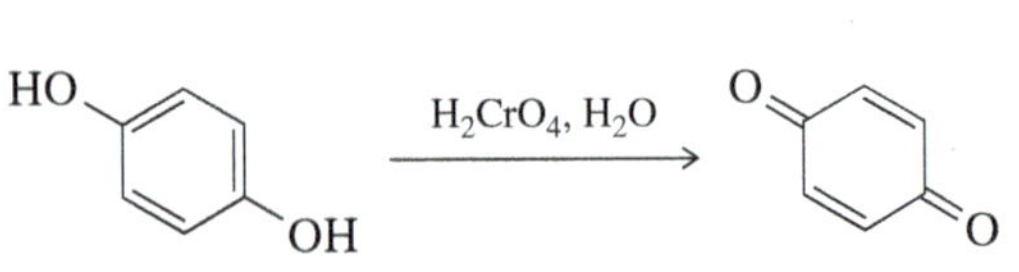

Specific:

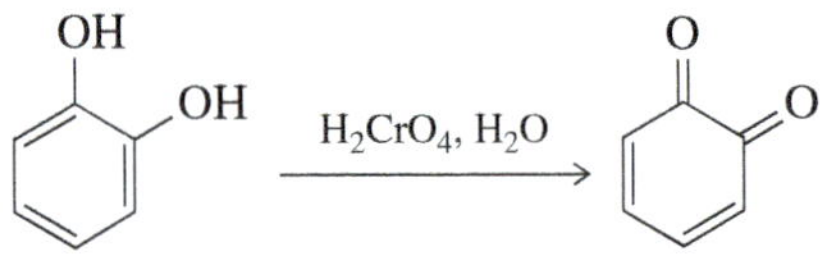

Mechanism:

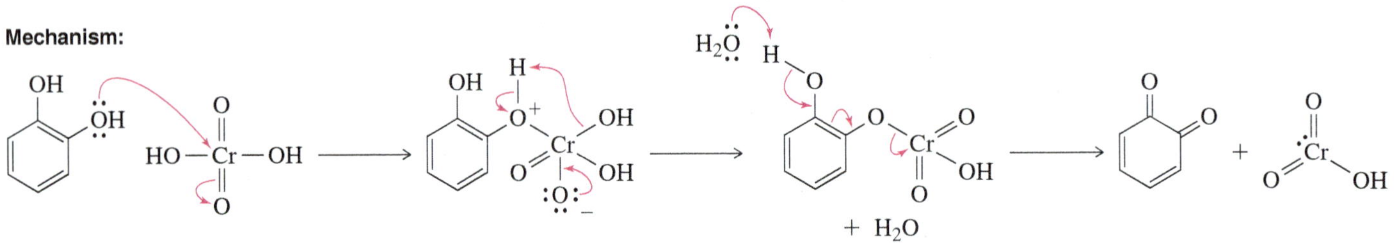

4. **Benzylic bromination (Section 24.3):** NBS, in the presence of heat, replaces a benzylic hydrogen with a bromine atom. Abstracting the benzylic hydrogen (first propagation step) gives a resonance-stabilized radical that attacks Br_2, forming the C—Br bond in the second propagation step. The side reaction of HBr and NBS generates Br_2 in a non-radical reaction.

General:

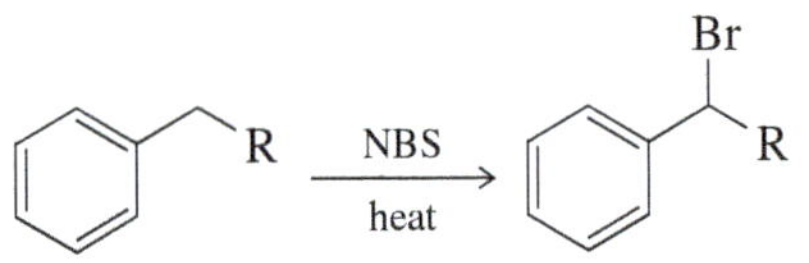

Specific:

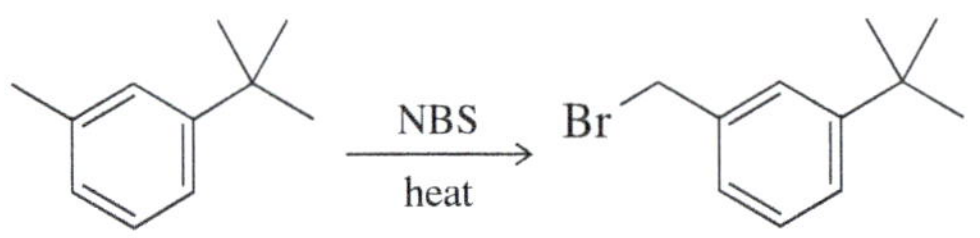

Mechanism:

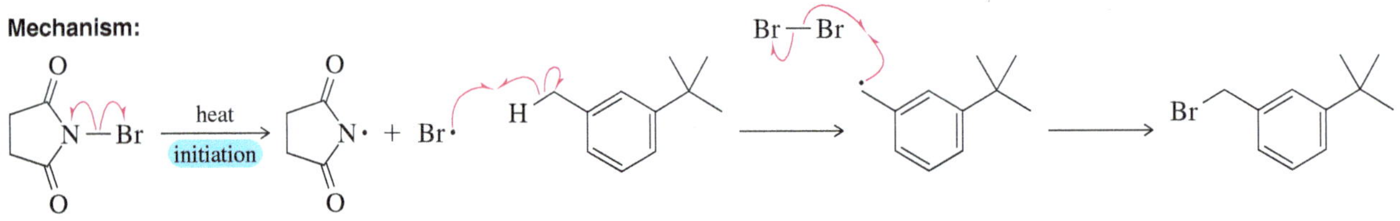

Recall generation of Br_2 (Sec. 11.5.3):

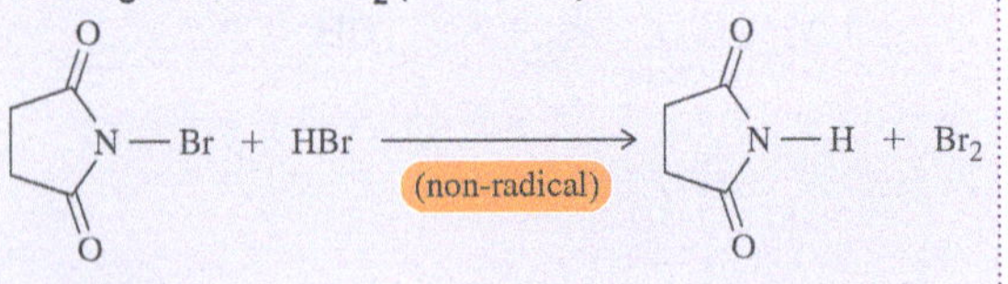

(NBS)

5. **Side chain oxidation (Section 24.5):** Benzylic groups, as long as there are hydrogens at the benzylic carbon, are oxidized to carboxylic acids by a mechanism that is not well understood. [*This reaction works equally well with basic potassium permanganate.*]

General:

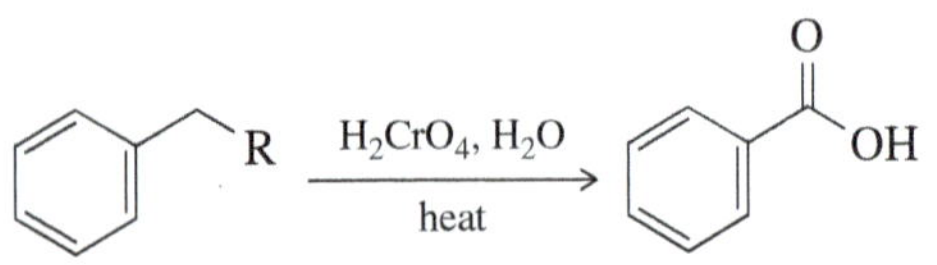

Specific:

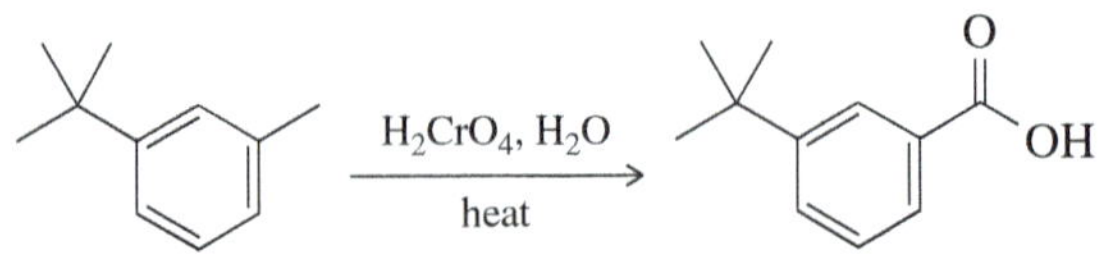

6. Hydrogenolysis of benzyl ethers (Section 24.6): Typically used for deprotecting a benzyl-protected alcohol, our proposed, simplified mechanism begins with oxidative insertion of palladium into a benzylic C—O σ bond. Coordination of hydrogen gas is followed by formation of new O—H and Pd—H bonds. Reductive elimination forms the C—H bond of toluene and regenerates the palladium catalyst.

General:

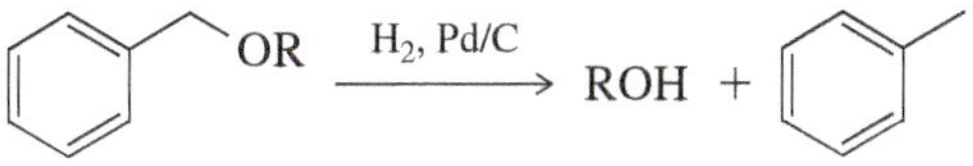

Specific:

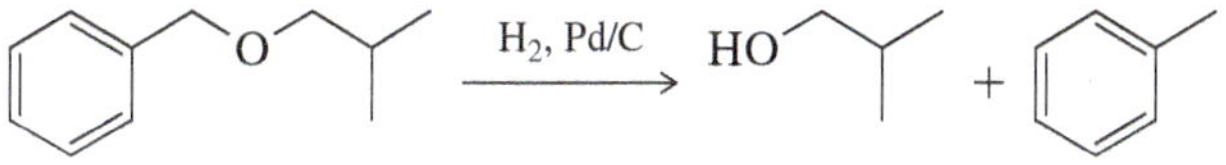

Mechanism:

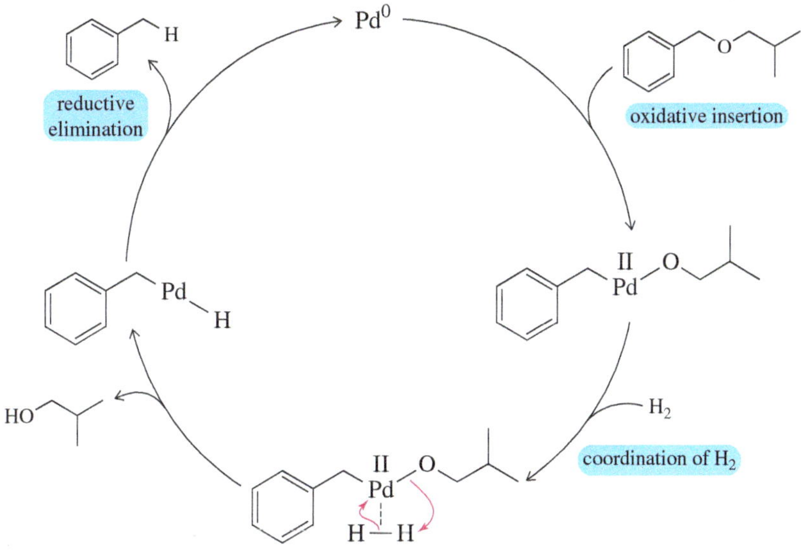

End-of-Chapter Assessments

[*Assessments are ranked according to their difficulty;* •••• = *most difficult.*]

24.34 (••) Predict the product(s) of the following reactions.

(a)

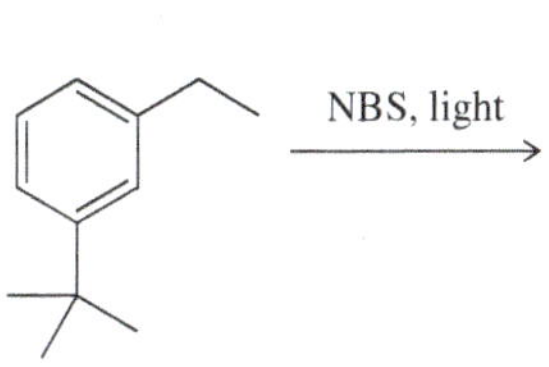

(b)

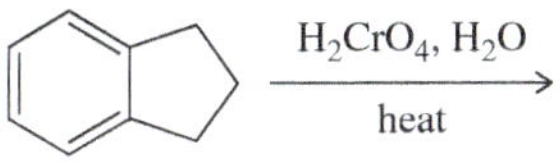

(c)

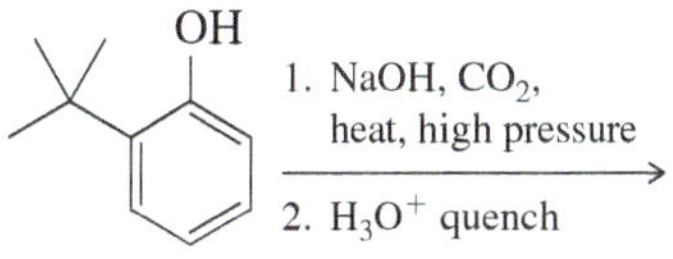

(d)

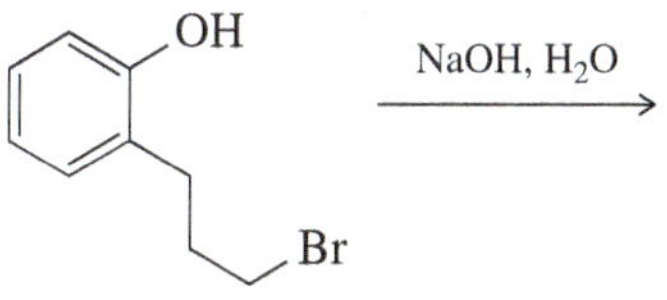

(e)

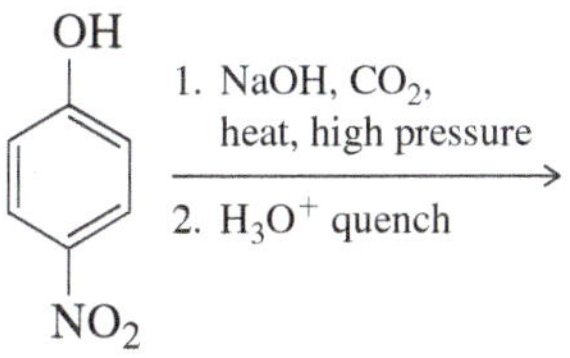

(f)

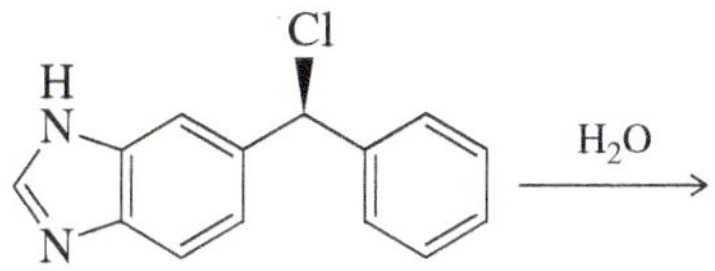

(g)

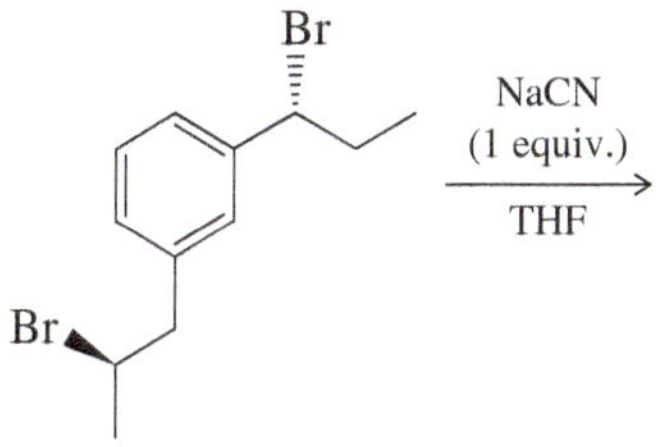

(h)

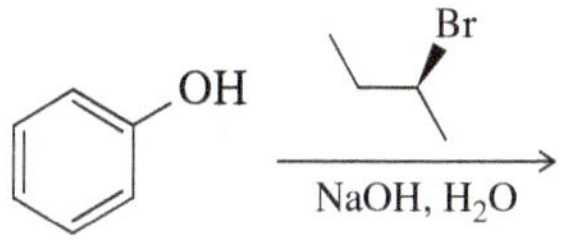

(i)

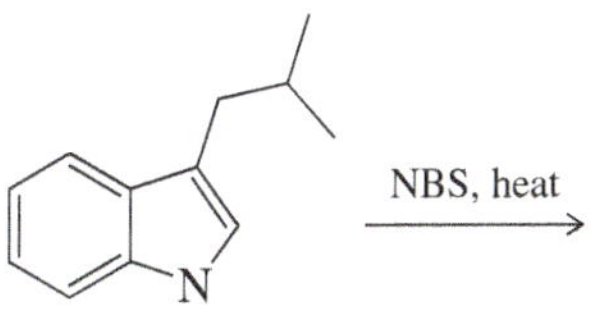

(j)

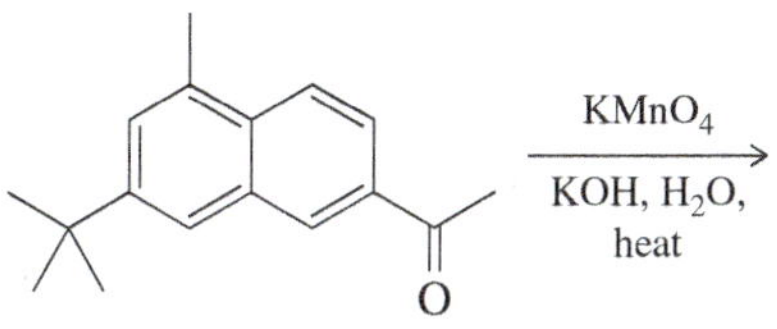

(k)

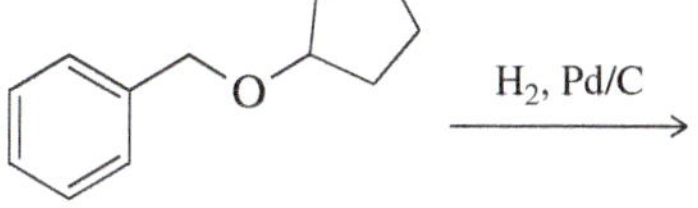

24.35 (••) Suggest the reagents used to effect the transformations shown.

(a)

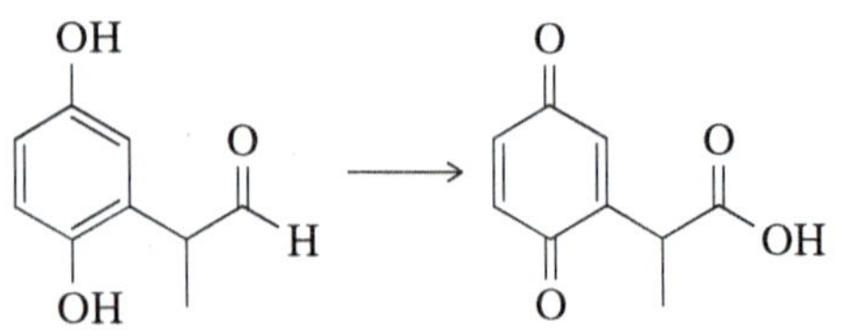

(b)

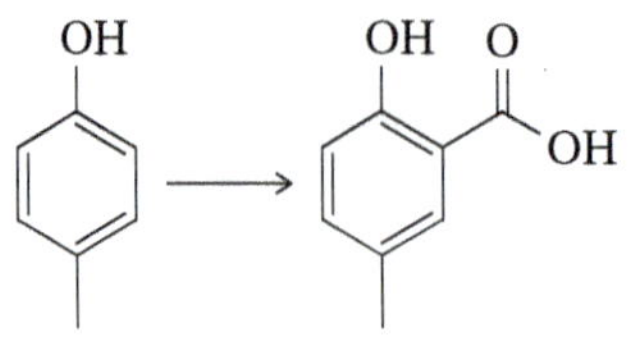

(c)

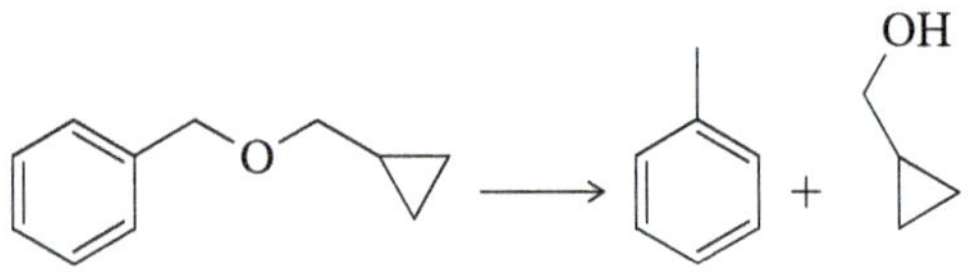

24.36 (••) On the same graph, show the reaction coordinate diagram for each of the following oxidation reactions.

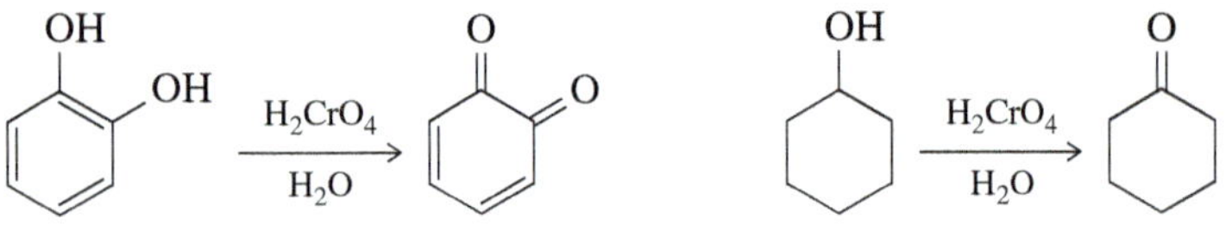

24.37 (••) Oxidation of the phenol shown gives a single quinone product. Predict this product and explain why it is the only one formed.

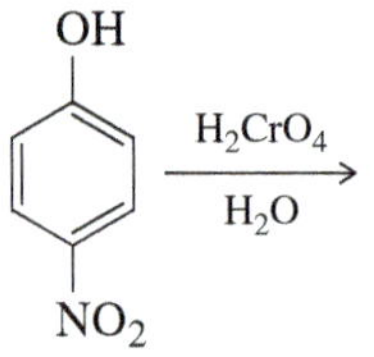

24.38 (•) In the second propagation step in the bromination of toluene, Br_2 is only attacked by a radical on the substituent carbon. Why?

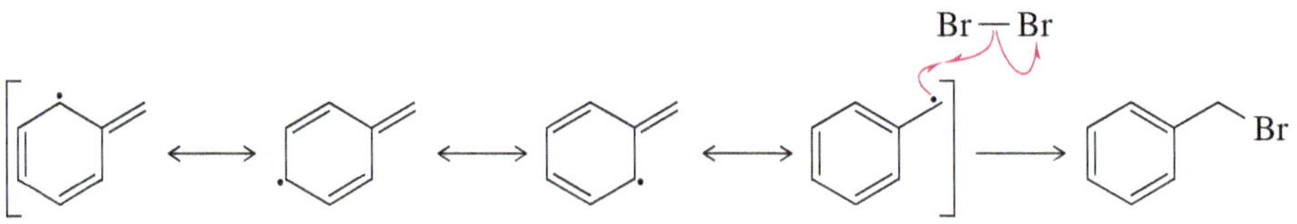

24.39 (••) LOOKING BACK When (*R*)-(1-bromoethyl)benzene is treated with sodium cyanide, a single enantiomer is produced. However, upon treatment of the same molecule with water, a mixture of two enantiomers is obtained. (a) Explain these results. (b) Why is only partial racemization sometimes observed?

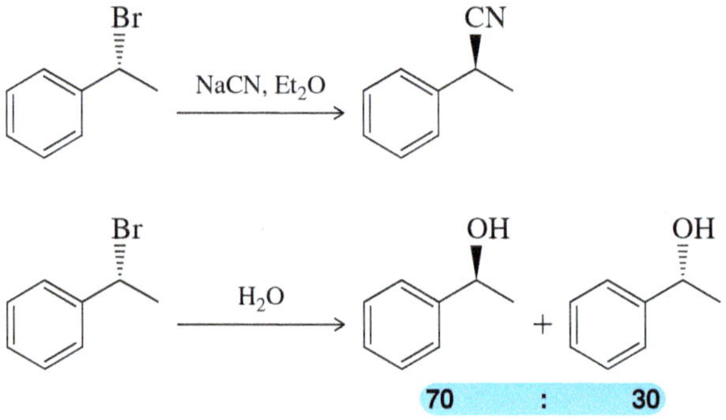

24.40 (••) LOOKING BACK The benzylic bromide shown undergoes neither S_N1 nor S_N2 substitution reactions. Explain.

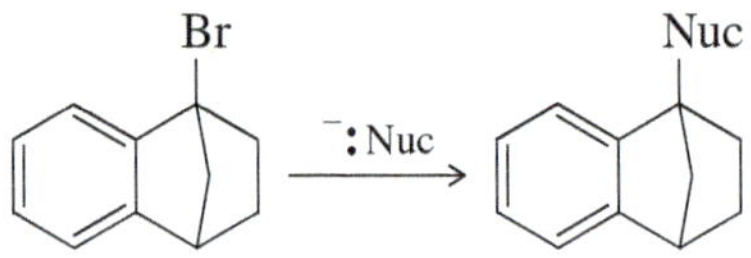

24.41 (••) LOOKING BACK In Chapter 13, we learned that epoxide opening can give different products, depending on whether the reaction occurs under acidic or basic conditions. Explain why the epoxide shown opens identically under either set of conditions.

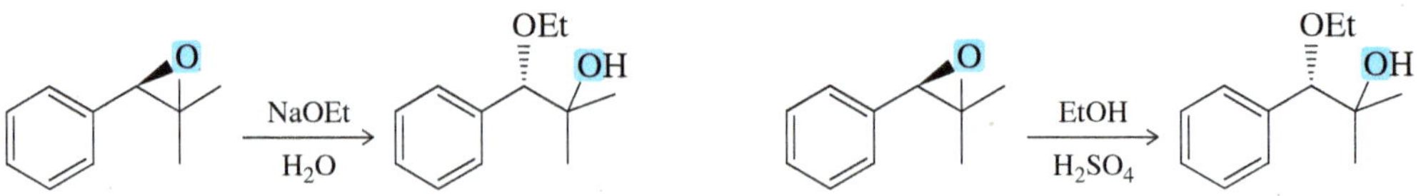

24.42 (•••) Beginning with phenyldiazonium, synthesize the salicylic acid derivative shown. [*You will use reactions from Chapters 23 and 24 in this synthesis.*]

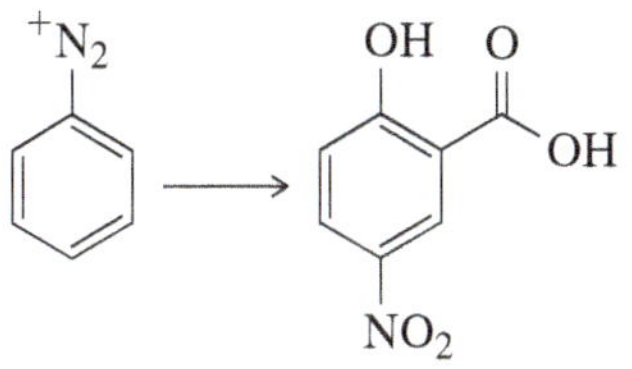

24.43 (•••) Beginning with benzene, synthesize the benzyl bromide shown. [*You will use reactions from Chapters 23 and 24 in this synthesis.*]

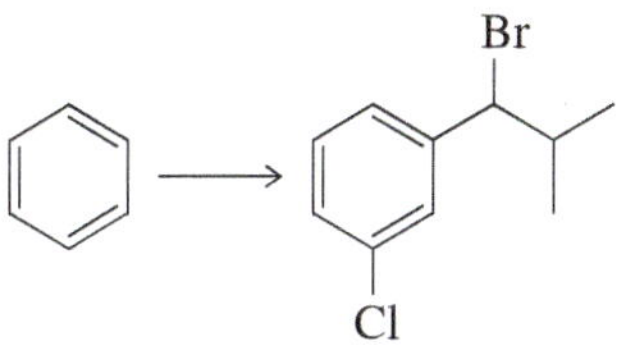

24.44 (••) Benzyl benzoate is used as a treatment for lice. Suggest a synthesis of benzyl benzoate starting from benzene.

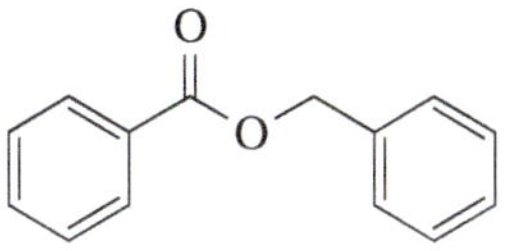

24.45 (••) **LOOKING AHEAD** The carboxybenzyl (CBZ) group is a good protecting group for amines [*Chapter 25*]. Draw an arrow-pushing mechanism for its removal according to the reaction shown.

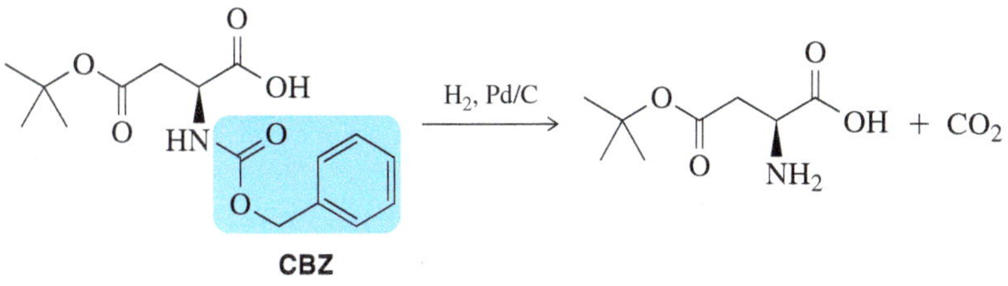

24.46 (••) The following reaction proceeds in good yield, despite requiring high temperature and high concentration. Identify the product you'd expect to obtain in this reaction. [*This is like the Kolbe reaction; see Section 24.2.3.*]

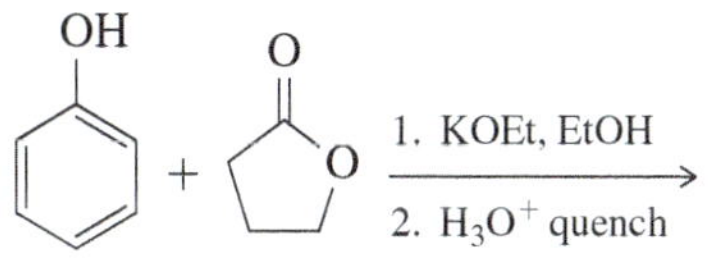

24.47 (•••) **FROM THE LITERATURE** The following steps were used in the synthesis of the antimalarial thiaplakortone A. Identify the missing reagents, (a) and (b), and (c) suggest an arrow-pushing mechanism that accounts for the formation of only the C—N bond [*screened in orange*] in the final step. (Adapted from *ACS Med. Chem. Lett.* 2014, *5*, 178–182.)

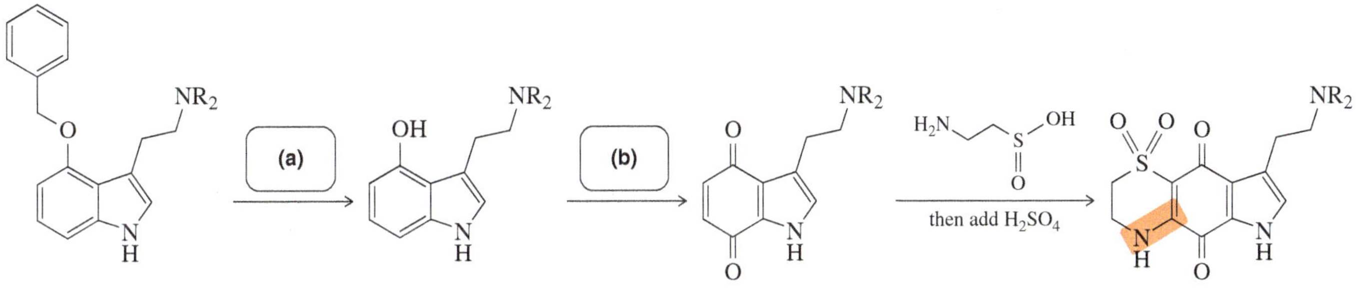

24.48 (•••) **LOOKING BACK** The Claisen rearrangement (Section 22.4.3) was used to attach allyl groups to benzene via initial alkylation of a phenol [*see Assessment 22.52*]. (a) Predict the identity of **A** and **B** and (b) provide an arrow-pushing mechanism for each step in the synthesis. (c) The keto form is usually favored over the enol. Why is that *not* true in this case?

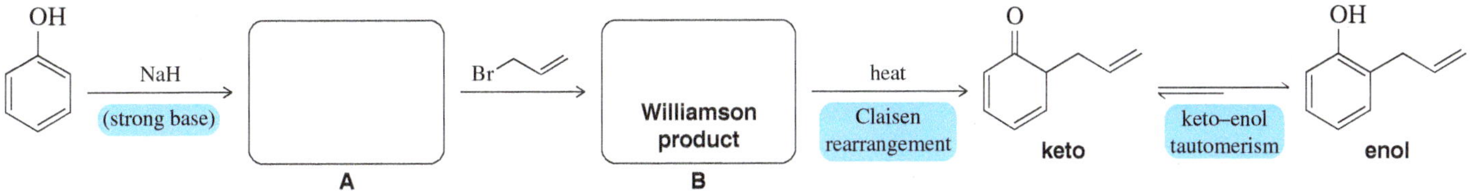

24.49 (•••) LOOKING BACK In Chapter 7, the Bhopal disaster was discussed as a tragic example of why green chemistry is so important. The Bhopal accident was a result of a methyl isocyanate leak at a factory that made Sevin®, starting with naphthol. Provide an arrow-pushing mechanism for the synthesis of Sevin. [*First identify the nucleophile and electrophile in the reaction.*]

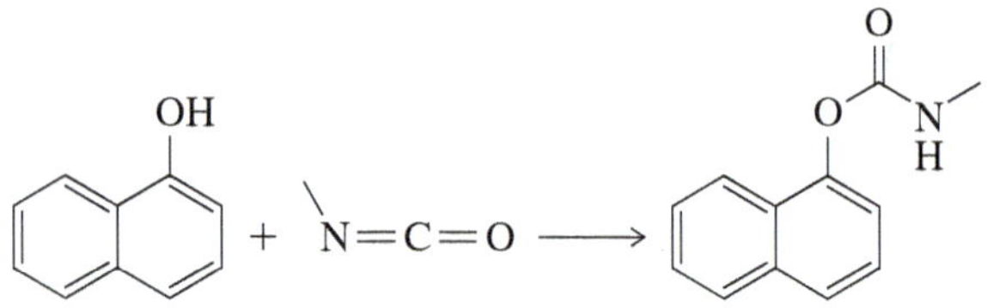

Sevin

24.50 (••) LOOKING BACK In Chapter 23, we learned about the electrophilic aromatic nitration reaction. When phenol is subjected to these conditions with a large excess of nitric acid, a molecule called picric acid is produced. Predict the product of this reaction and explain why the pK_a value of this compound is 0.38.

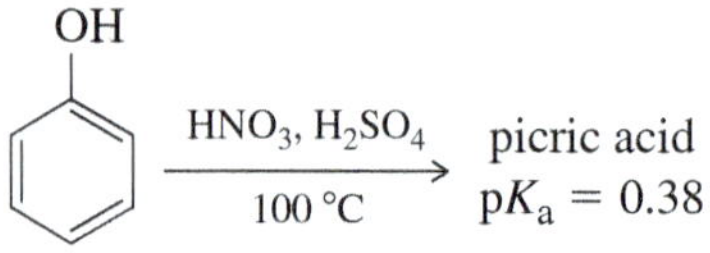

24.51 (•••) Rationalize the fact that 1,4-dihydroxybenzene melts at a significantly higher temperature than 1,2-diydroxybenzene.

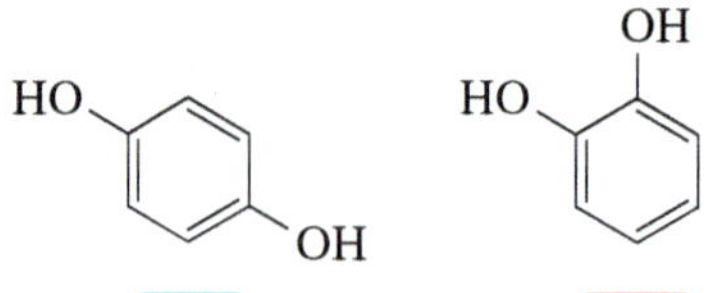

172 °C 105 °C

24.52 (•••) A chemist attempted to oxidize a primary alcohol to a carboxylic acid using chromic acid. The product shown was obtained as a major component in the mixture. Suggest an arrow-pushing mechanism that accounts for its formation. [*Think about what chromic acid would normally do to a phenol and make a list of bonds formed and bonds broken.*]

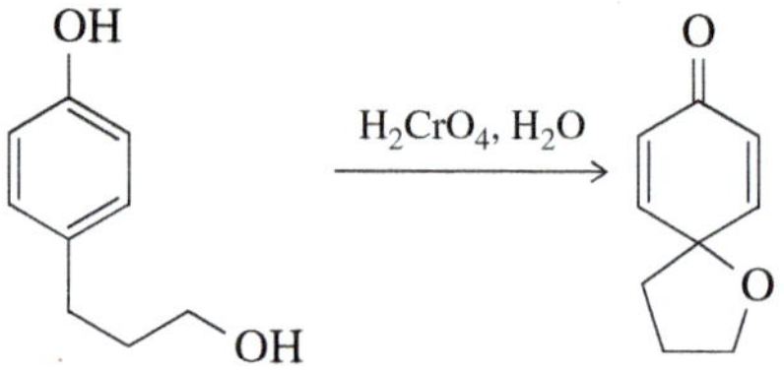

24.53 (••) Identify the aromatic product that would result from the oxidation of the triol with an excess of PCC.

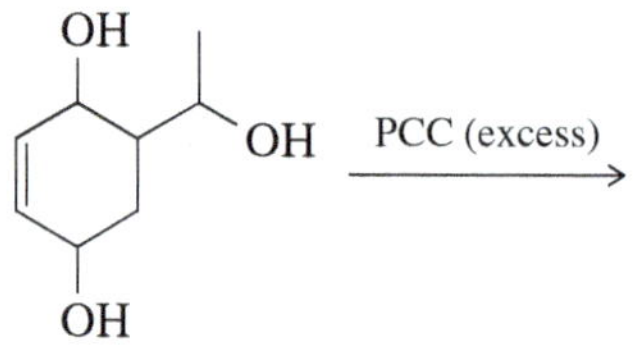

24.54 (•) Rationalize the ranking of acidity of carboxylic acids shown.

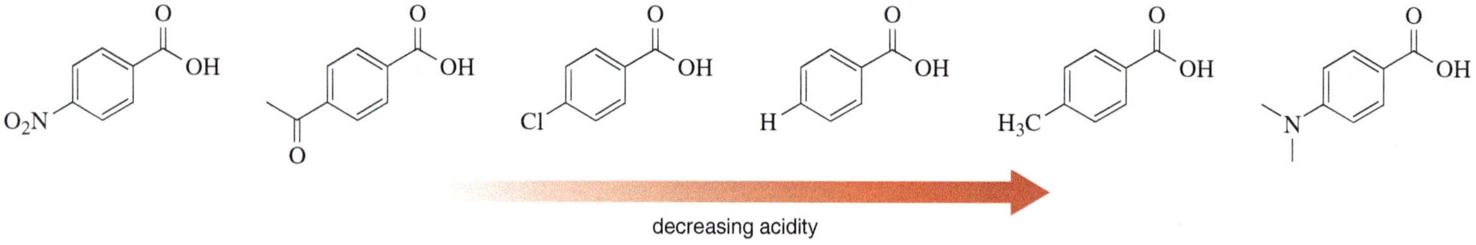

24.55 (•) Rationalize the ranking of increasing S_N1 reaction rate of the benzylic halides shown.

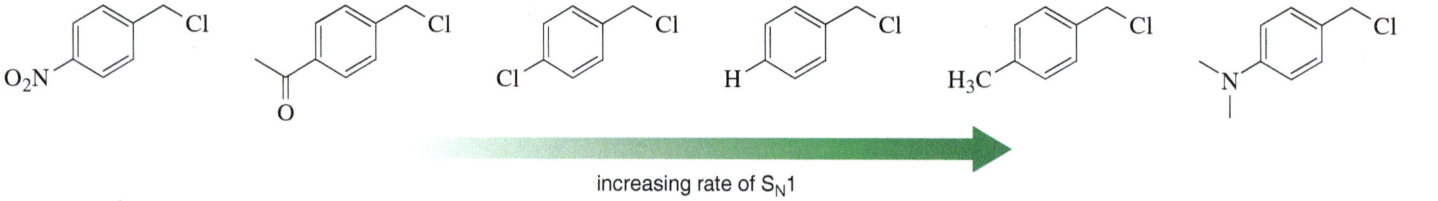

24.56 (••) In light of your answers to Assessments 24.54 and 24.55, rank the following based on the rate of protonation of the alkene (1 = most basic, 6 = least basic). [*Ignore the fact that the alkene may not be the most basic site in the molecule.*]

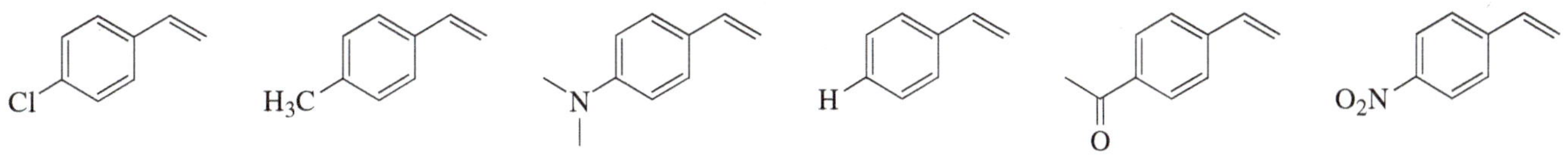

24.57 (••) Phenol oxidation can be coupled with other reactions to form new C—C bonds using reactions studied previously. Predict the product of the following series of reactions.

(a)

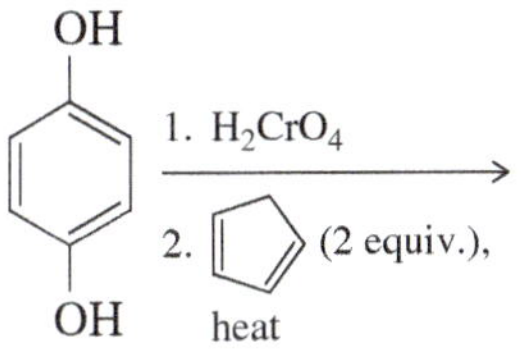

(b)

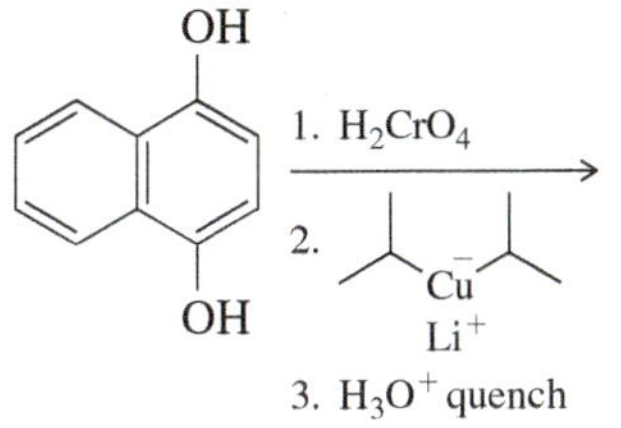

24.58 (•••) Lawsone, found in the extract of henna, reacts with amino acid residues to give the characteristic color associated with henna tattoos. Suggest a mechanism for the reaction shown.

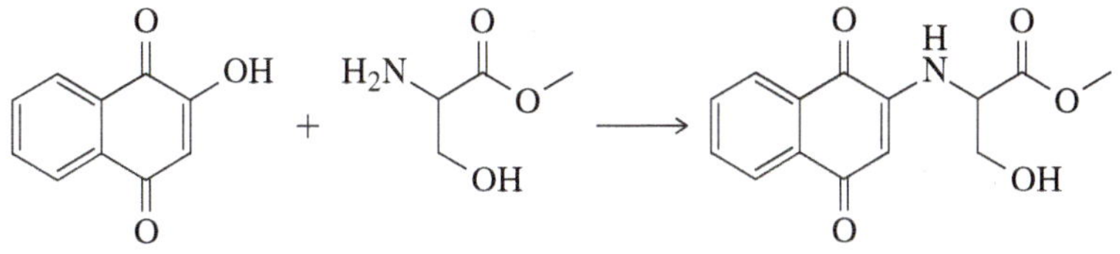

lawsone

24.59 (••••) Provide the reagents necessary to carry out the following synthesis. What is the purpose of steps (a) and (e)?

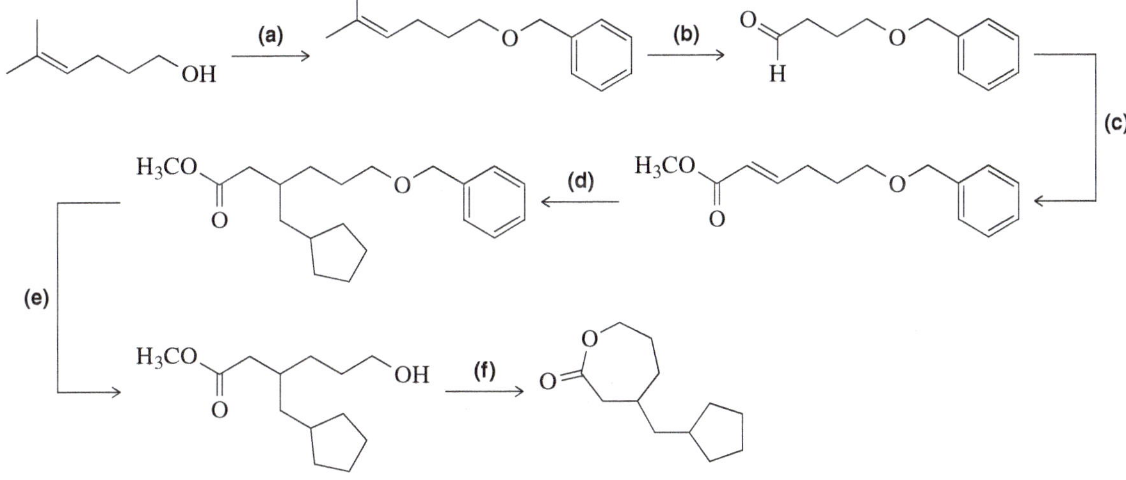

24.60 (••) Sulpiride is an antipsychotic. Provide the reagents (a) and (b) used to complete the synthesis.

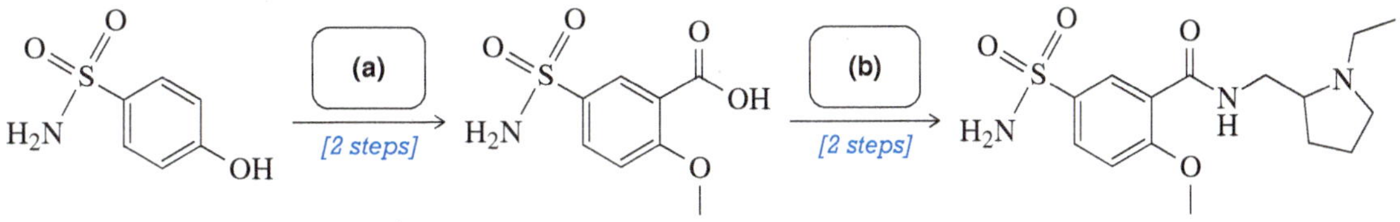

24.61 (••) *p*-Benzoquinone is used in the photosynthetic systems of plants by reversibly accepting 2 electrons to form the dihydroxybenzene at right. (a) Predict a mechanism of the reaction shown and (b) draw the MO diagram for each structure and suggest which you expect to be more stable.

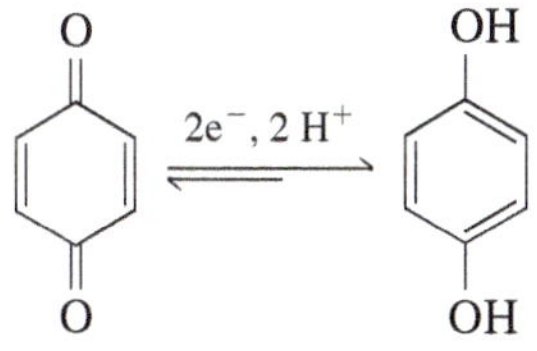

***p*-benzoquinone**

24.62 (•••) 2-Ethylanthraquinone is used industrially to produce hydrogen peroxide, a molecule we have encountered throughout organic chemistry. Suggest a mechanism for each step of the process shown.

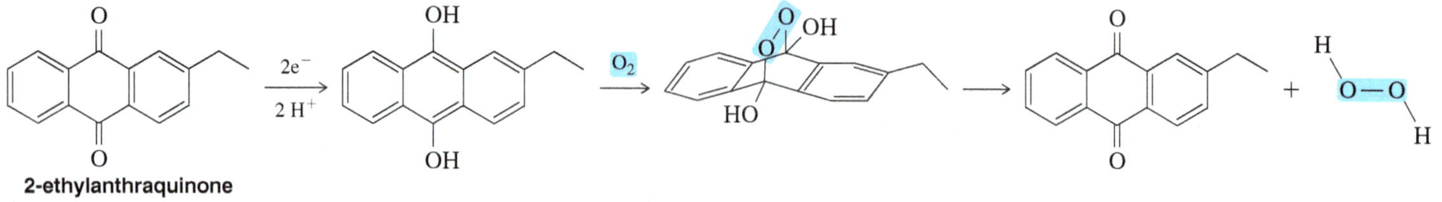

24.63 (•••) Analytical chemists often use organic reagents to detect the presence and the concentration of other compounds. One such reagent is the Emerson reagent that detects phenols by a mechanism similar to what is studied in Chapter 24. (a) Suggest a mechanism of the reaction. (b) Why is the product colored? [*Think of* $K_3Fe(CN)_6$ *as a standard alcohol oxidizing agent.*]

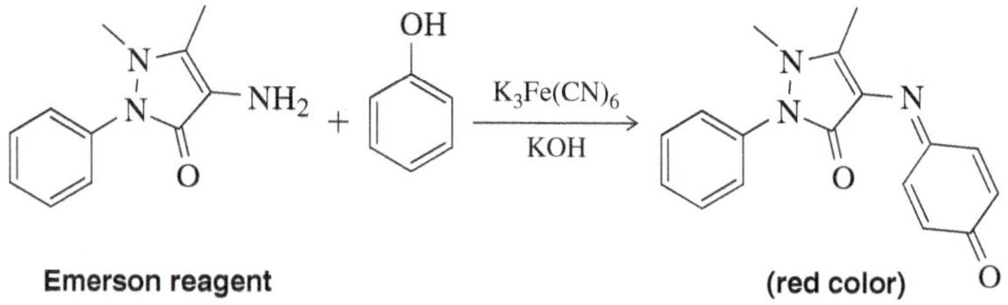

24.64 (••••) Suggest a mechanism for the oxidative formation of the dye mauveine from aniline and the aniline derivatives shown. [*This is a very difficult problem. It should probably have 6 dots.*]

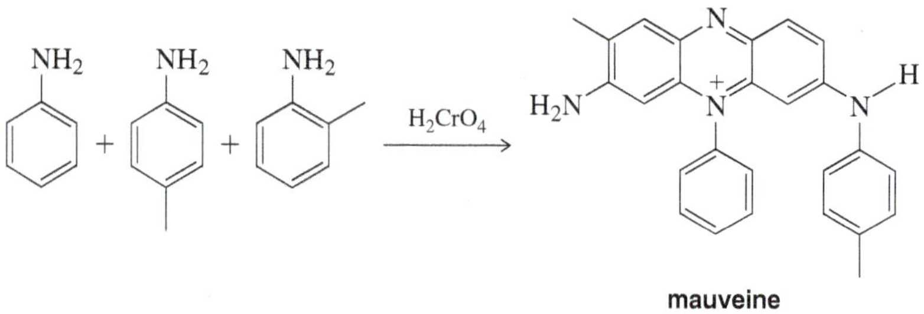

24.65 (•••) Phenol is synthesized from cumene via the scheme shown. Suggest the mechanism for each step of the reaction shown. [*The oxygens have been screened in blue and orange to assist you.*]

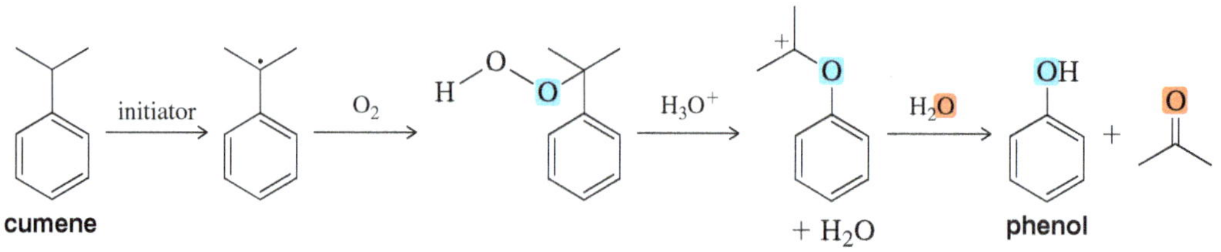

24.66 (••••) Synthesize the dye shown starting with aniline and naphthol using reactions from Chapters 23 and 24. [*Hint: Start by identifying how each substituent might be installed and then plan the order of their introduction.*]

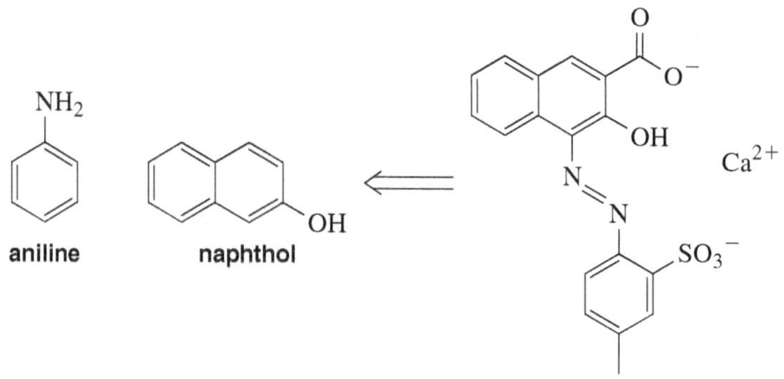

25

Amines: Structure, Reactions, and Synthesis

Introduction: major depressive disorder

Major depressive disorder, also called depression, is a mental disorder characterized by consistent sadness over long periods of time. Symptoms include, at varying levels, loss of energy, lack of interest in formerly enjoyable activities, difficulty concentrating, and a diminished sense of self-worth. In its most severe form, it can be accompanied by thoughts of death or suicide. Depression in some individuals can be linked to a person's brain chemistry—specifically, an imbalance in the brain's neurotransmitters.

Communication between nerve cells in the brain begins with an electrical signal being passed down the axon of a presynaptic neuron. Upon receiving the signal, the axon terminal transmitter releases chemicals, called neurotransmitters, into the synapse. These chemicals cross the synapse and interact with receptors on the postsynaptic neuron, passing the message to the next nerve cell. Whether they have delivered the message or not, neurotransmitters are eventually reabsorbed (reuptake) by the presynaptic neuron.

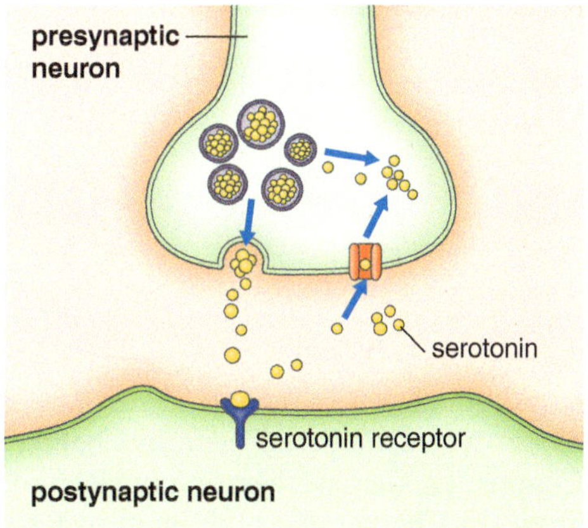

Although there are many neurotransmitters, three have been linked to depression: serotonin, norepinephrine, and γ-aminobutyric acid (GABA) (**Figure 25.1**). An appropriate balance in the

Learning Objectives

LO 25.0 Evaluate the names and structures of amines and their isomers

LO 25.1 Analyze the physical and chemical properties of amines

LO 25.2 Describe how to synthesize primary amines

LO 25.3 Analyze general synthesis reactions of amines

LO 25.4 Analyze reactions of amines

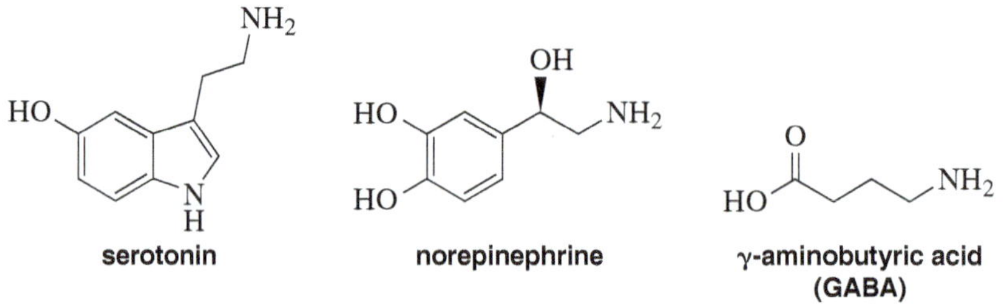

Figure 25.1 Neurotransmitters, such as serotonin, norepinephrine, and γ-aminobutyric acid (GABA), are responsible for interneuron communication.

concentrations of these neurotransmitters is required for normal brain activity, whereas low levels of serotonin in the brain, for example, have been suspected to cause both depression and anxiety.

Can't a low level of serotonin be fixed by taking a pill containing serotonin? Unfortunately, it isn't that simple. Serotonin is synthesized in the brain, and there is no mechanism by which serotonin circulating in the blood can cross the blood–brain barrier. To boost serotonin levels in the synapse between a presynaptic and postsynaptic neuron, medicinal chemists have invented molecules that act as neurotransmitter reuptake inhibitors. By inhibiting reuptake of neurotransmitters, a larger concentration of neurotransmitters remains in the synapse, more efficiently relaying whatever message is being sent.

"I think the saddest people always try their hardest to make people happy because they know what it's like to feel absolutely worthless and they don't want anyone else to feel like that."

—Robin Williams

Several drugs have been developed to treat depression by maintaining the balance of neurotransmitters in the synapse (**Figure 25.2**). Sertraline, fluoxetine, and citalopram are selective serotonin reuptake inhibitors (SSRIs). Duloxetine is a serotonin norepinephrine reuptake inhibitor (SNRI). Amoxapine is one of a class of tricyclic antidepressants (TCAs). Finally, bupropion is a dopamine reuptake inhibitor. These drugs, in addition to having some structural similarity to the neurotransmitters themselves, contain amines, the topic of Chapter 25. In addition to learning about the properties, synthesis, and reactivity of amines, we also discuss their importance in modern medicinal chemistry by studying the chemistry of depression and anxiety.

Figure 25.2 Amines used for the treatment of anxiety and depression.

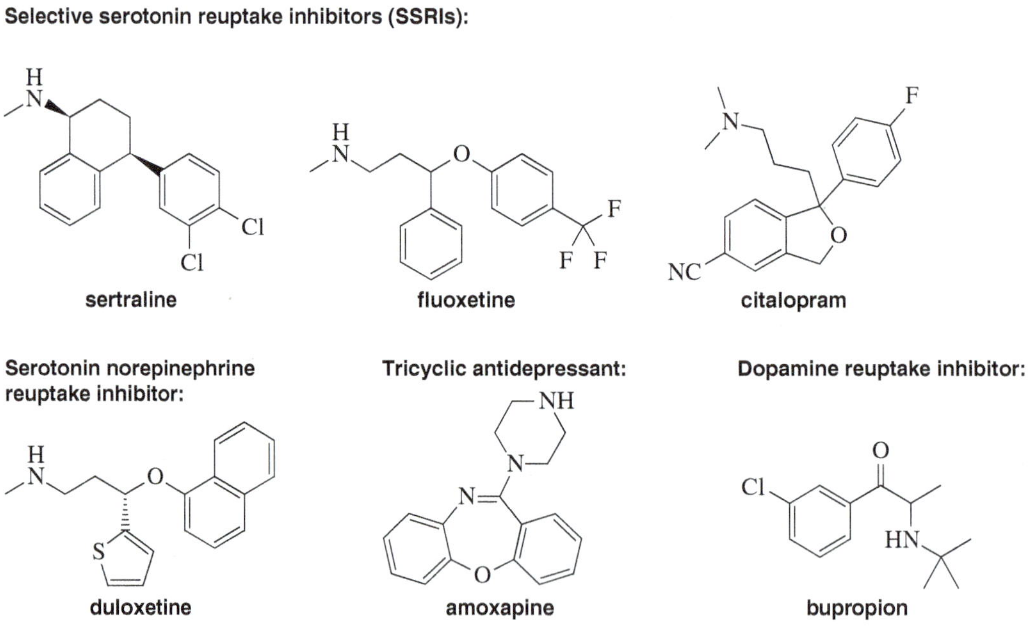

Although we haven't studied them in depth, amines have been in the background all along. We encountered them in Chapter 4 when we first learned about acids and bases. In Chapter 12, they were strong nucleophiles in spite of reacting with a lone pair on a neutral atom. In Chapter 13, and again in Chapter 19, they were useful bases in tosylation and

Speaking of review, let's test our understanding of some previous concepts now.

acylation reactions. Also, in Chapter 19, they were used to convert carboxylic acid derivatives to amides. In sum, we've studied them before. As a result, the material in this chapter is a helpful review of previous concepts.

Test your understanding

Assessment 25.1 Draw the Lewis structure of azidomethane (CH_3N_3). [*Show two important resonance structures.*]

Assessment 25.2 Predict the K_{eq} for the following acid–base reaction.

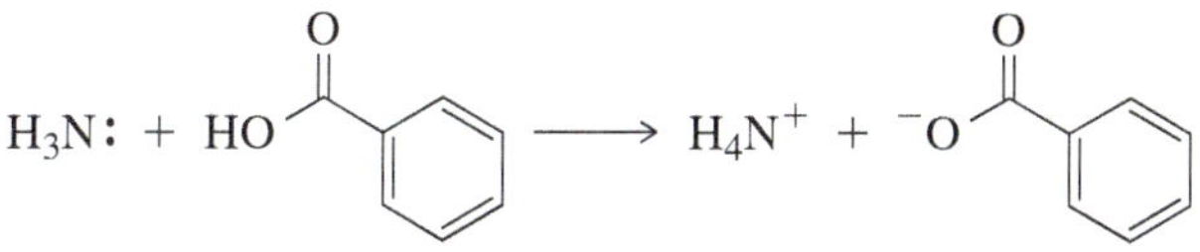

Assessment 25.3 Which of the following phenoxides should be a stronger nucleophile?

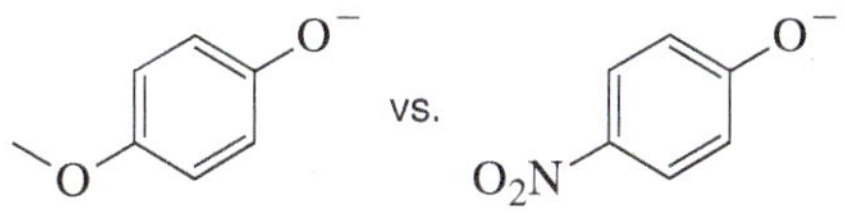

Assessment 25.4 Predict the product of exhaustive hydrogenation (at high pressure) of the α,β-unsaturated ketone shown.

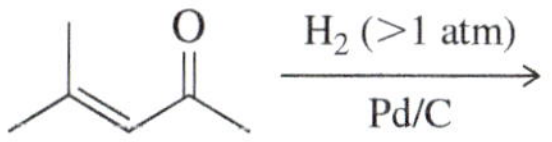

Assessment 25.5 Would you expect the following substitution reaction to proceed with inversion or racemization? Why?

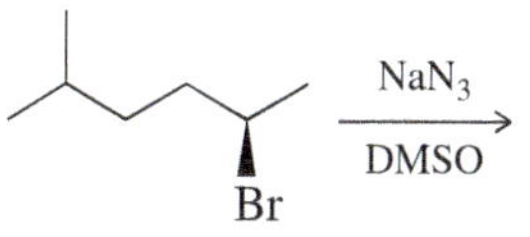

Assessment 25.6 Suggest a mechanism for the conversion of a carboxylic acid to a methyl ester using diazomethane.

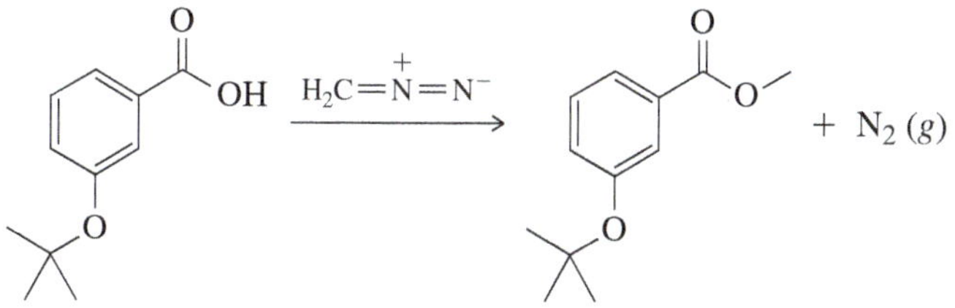

25.1 General structure of amines

Amines are a general class of compounds in which one hydrogen of a parent alkane has been replaced by an amino group ($—NR_2$; R = alkyl or hydrogen). Alternatively, amines can be thought of as resulting from the replacement of one, two, or three hydrogens from ammonia (NH_3) with an alkyl group. **Figure 25.3** shows some examples of amines.

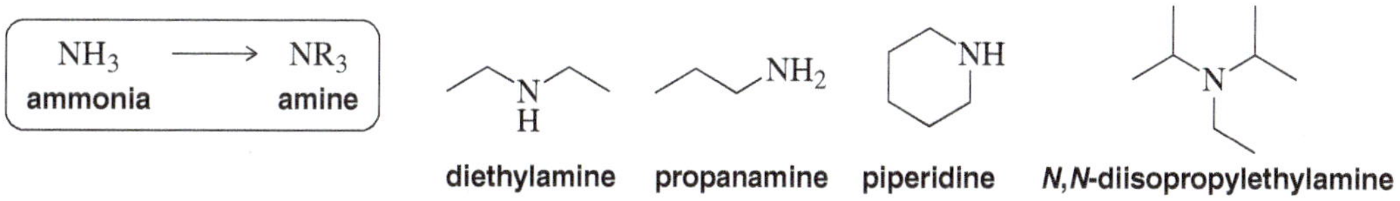

Figure 25.3 Representative structures of amines.

25.2 Nomenclature of amines

The nomenclature of amines is complicated by the fact that there are several systems of nomenclature in use. [*It's almost as if when it came time to name amines, our hypothetical IUPAC conference attendees became bored with the whole process . . . can you blame them?*] However, if you understand organic nomenclature to this point, all of the systems make sense. Importantly, the use of any of them is considered acceptable. Here in Chapter 25, we alternate between each of the nomenclature systems so that you can become comfortable with all of them.

25.2.1 Common names

Common names of amines are generated as outlined in Table 25.1.

Table 25.1 Common names of amines.

1. Common names of acyclic amines are generated by naming the alkyl group attached to the amino group, followed by the suffix -amine.	 ethylamine isopropylamine *tert*-butylamine
2. If there are two or more identical alkyl groups attached to the amine nitrogen, di- and tri- are used to indicate the presence of two and three alkyl groups, respectively	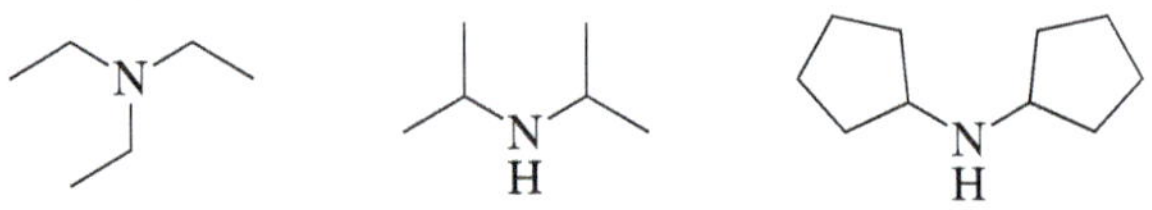 triethylamine diisopropylamine dicyclopentylamine
3. If there are different substituents on the amine nitrogen, the root name consists of the longest straight chain amine (named as in entry 1). The other substituents are named as alkyl groups, but preceded by *N*- to indicate that they are attached to the nitrogen instead of to another carbon in the chain.	 *N*-methyldiethylamine (substituent, root name) *N*-ethyldi-*t*-butylamine (substituent, root name)
4. Some cyclic, aromatic, and heterocyclic amines are named by common names that have no basis in logic, but you should still be able to recognize them.	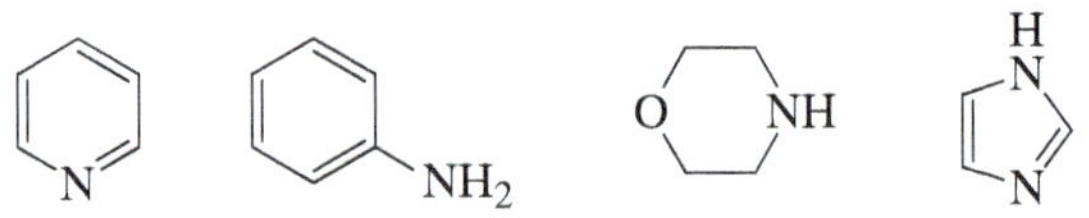 pyridine aniline morpholine imidazole

There are some cool stories around about how compounds got their common names. If you're ever bored, look them up on the internet.

25.2.2 IUPAC names

Whereas the IUPAC system indicates that a molecule is an alcohol, aldehyde, ketone, carboxylic acid, or derivative by modifying the suffix of the reaction (as in butan-1-ol), amino groups are named as substituents in some other molecular class. As shown in Figure 25.4, replacing the hydroxyl group with an amino group results in 1-aminobutane. If the amino group attached to the main chain is substituted itself, the common method of indicating additional substituents as "*N*-alkyl" is used.

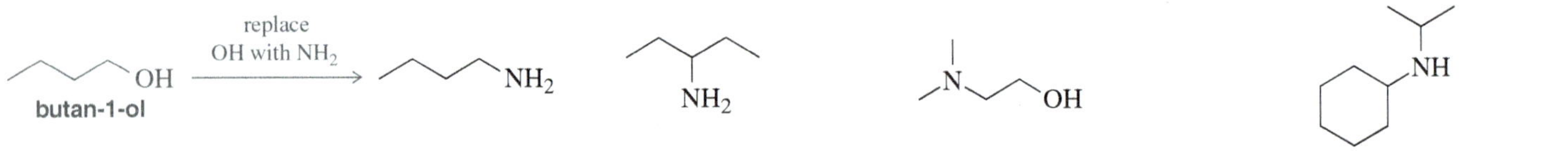

1-aminobutane 3-aminopentane 2-(*N*,*N*-dimethylamino)ethan-1-ol *N*-isopropylaminocyclohexane

Figure 25.4 IUPAC names of some amines.

25.2.3 Chemical Abstract Service names

The Chemical Abstract Service (CAS) has developed a system of nomenclature for amines that is very similar to what you might have expected for IUPAC. It treats amines as classes of molecules, naming them as if they were hydrocarbons, but dropping the -e and replacing it with the suffix -amine, using a number to indicate the location of the amine on the parent chain (Figure 25.5).

Figure 25.5 CAS system of amine nomenclature.

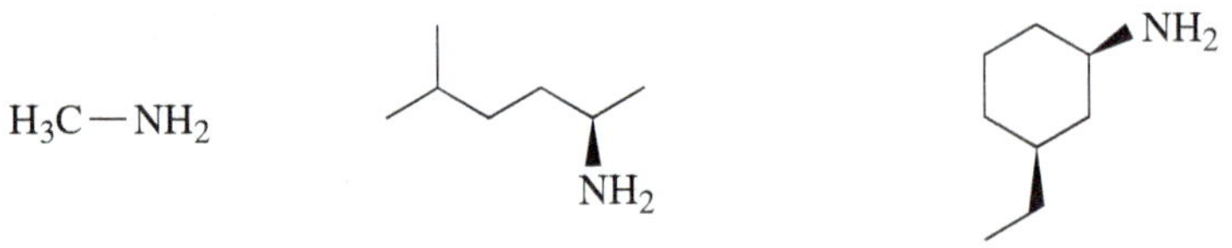

methanamine (*R*)-5-methylhexan-2-amine (1*R*,3*S*)-3-ethylcyclohexanamine

25.2.4 Classification of amines

Adding to the confusion of having multiple acceptable systems of nomenclature, the use of 1°, 2°, 3°, or 4° is different for amines than it is for alkyl halides and alcohols. Whereas alcohols are designated 1°, 2°, or 3° based on the number of carbons attached to the hydroxyl-bearing carbon, amines are designated 1°, 2°, 3°, or 4° based on how many alkyl (or aryl or alkenyl) groups are attached to the nitrogen (**Figure 25.6**).

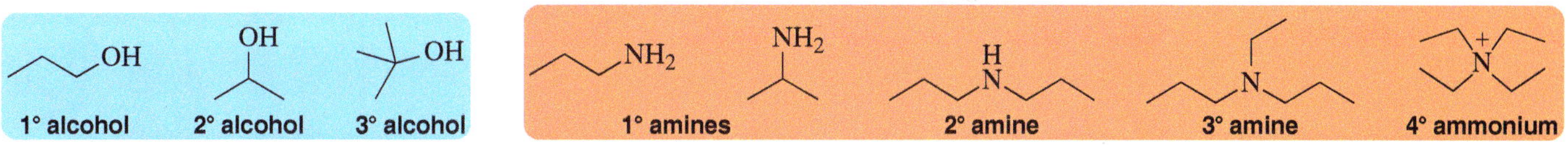

Figure 25.6 The number of alkyl groups attached to a nitrogen makes a molecule a 1°, 2°, or 3° amine or a 4° ammonium ion.

Assessment 25.7 Give an acceptable name for each amine.

(a) (b) (c) (d)

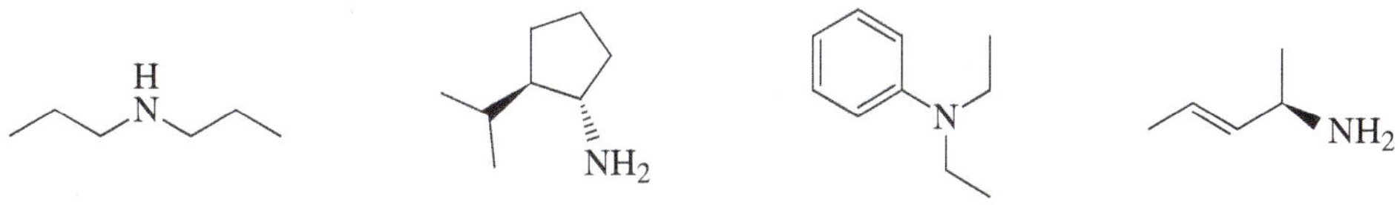

Assessment 25.8 Given the following names, draw the structure of the molecule.

(a) *N*,*N*-dimethylcyclopentylamine
(b) *p*-nitroaniline
(c) (*S*)-2-methyloctan-4-amine
(d) (*R*)-*N*-*t*-butyl-3-aminohexanoic acid

Assessment 25.9 Label the amines shown as a 1° amine, 2° amine, 3° amine, or 4° ammonium ion.

(a) (b) (c) (d)

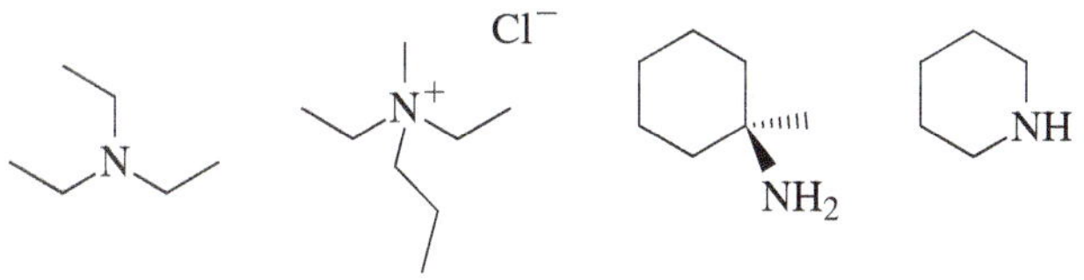

Assessment 25.10 Draw a constitutional isomer for $C_4H_{11}N$ containing a 1° amine at a (a) 1° carbon, (b) 2° carbon, and (c) 3° carbon.

25.3 Molecular orbital picture of amines

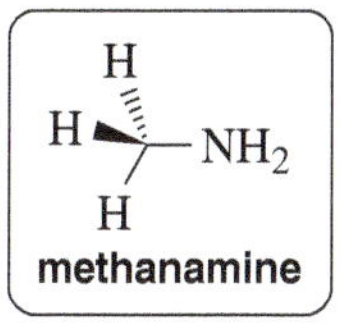

The reactivity of amines can be explained by the molecular orbital picture. [*See if you can draw the molecular orbital picture of methanamine, the simplest 1° amine, on your own before reading further.*]

Made up of an sp^3-hybridized carbon bound to an sp^3-hybridized nitrogen, the C—N σ bond is formed by overlap of sp^3-hybridized orbitals (**Figure 25.7**). The three C—H σ bonds are formed using the remaining sp^3-hybridized orbitals of carbon overlapping the 1*s* orbitals of hydrogen. Similarly, the N—H bonds are formed by overlap of the sp^3-hybridized orbital of nitrogen with the 1*s* orbitals of hydrogen. The lone pair on nitrogen sits in a nonbonded sp^3-hybridized atomic orbital.

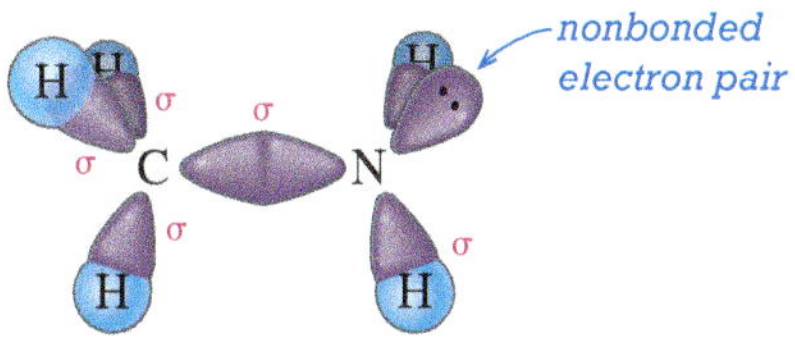

Figure 25.7 Molecular orbital picture of methanamine.

When the amine is attached to an aromatic ring, as in aniline [Figure 25.8(a)], the nitrogen is between sp^2 and sp^3 hybridized, a consequence of partial resonance delocalization of an amine lone pair throughout the benzene ring. With one lone pair in a p orbital, overlap with the π system of benzene is maximized. A similar effect occurs with the amino group of amides [Figure 25.8(b)].

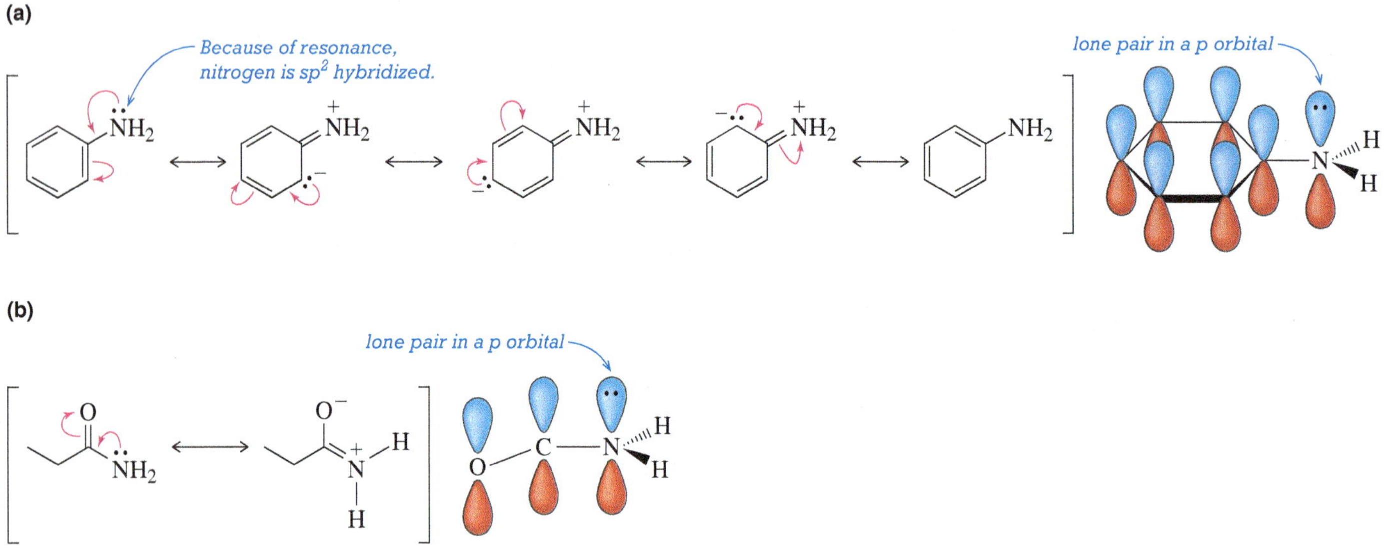

Figure 25.8 The nitrogen atom of conjugated amines, such as (a) aniline and (b) amides, are sp^2 hybridized.

25.4 Properties of amines

The physical properties of amines are a result of their slight polarity and ability to serve as both hydrogen bond donors and hydrogen bond acceptors. Compared to alcohols of similar molecular weight, amines boil at a lower temperature (Table 25.2), because nitrogen is less electronegative than oxygen, which results in weaker hydrogen bonding in amines than in alcohols.

Table 25.2 A comparison between the properties of alcohols and amines.

Compound	Melting point (°C)	Boiling point (°C)
OH butan-1-ol	−89.8	117.7
NH_2 butan-1-amine	−50	77

The physical properties of several amines are listed in Table 25.3. Within a series of 1°, 2°, or 3° amines, an increase in chain length results in greater van der Waals interactions and an increase in melting and boiling points.

Let's analyze the boiling points of a 1°, 2°, and 3° amine of similar molecular weight. Table 25.4 lists the boiling points of 1-hexanamine, dipropylamine, and triethylamine, each of which have a molecular weight of 101 g/mol. 1-Hexanamine, with no branching and two hydrogens available for hydrogen bonding, boils and melts at the highest temperature. Di-*n*-propylamine is also a linear compound, but it has only one hydrogen available for hydrogen bonding. Furthermore, its lone pair is flanked on both sides by bulky alkyl groups, making hydrogen bonding more difficult. As a result, its boiling and melting points are lower than those of 1-hexanamine. Triethylamine is branched and contains no hydrogens attached to nitrogen, so it cannot form intermolecular hydrogen bonds. As a result, it has the lowest melting and boiling points of the three.

Table 25.3 Larger amines have stronger intermolecular forces and therefore higher melting and boiling points. They are less soluble in H_2O, too.

Name	Formula	Molecular weight	mp (°C)	bp (°C)	H_2O solubility
		Primary amines			
Methylamine	CH_3NH_2	31	−93	−7	Very soluble
Ethylamine	$CH_3CH_2NH_2$	45	−81	17	∞
n-Propylamine	$CH_3CH_2CH_2NH_2$	59	−83	48	∞
Isopropylamine	$(CH_3)_2CHNH_2$	59	−101	33	∞
n-Butylamine	$CH_3CH_2CH_2CH_2NH_2$	73	−50	77	∞
Cyclohexylamine	*cyclo*-$C_6H_{11}NH_2$	99	−18	134	Slightly soluble
Benzylamine	$C_6H_5CH_2NH_2$	107		185	∞
Aniline	$C_6H_5NH_2$	93	−6	184	3.7%
		Secondary amines			
Dimethylamine	$(CH_3)_2NH$	45	−96	7	Very soluble
Diethylamine	$(CH_3CH_2)_2NH$	73	−42	56	Very soluble
Di-*n*-propylamine	$(CH_3CH_2CH_2)_2NH$	101	−40	111	Slightly soluble
Diisopropylamine	$[(CH_3)_2CH]_2NH$	101	−61	84	Slightly soluble
N-Methylaniline	$C_6H_5NHCH_3$	107	−57	196	Slightly soluble
Diphenylamine	$(C_6H_5)_2NH$	169	54	302	Insoluble
		Tertiary amines			
Trimethylamine	$(CH_3)_3N$	59	−117	3.5	Very soluble
Triethylamine	$(CH_3CH_2)_3N$	101	−115	90	14%
Tri-*n*-propylamine	$(CH_3CH_2CH_2)_3N$	143	−94	156	Slightly soluble
N,*N*-Dimethylaniline	$C_6H_5N(CH_3)_2$	121	2	194	1.4%
Triphenylamine	$(C_6H_5)_3N$	251	126	225	Insoluble

Table 25.4 Comparison of 1°, 2°, and 3° amines of similar molecular weight.

Compound	Structure	mp (°C)	bp (°C)
1-Hexanamine	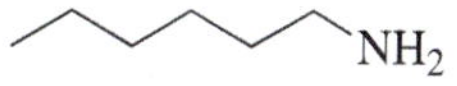	−23	132
Di-*n*-propylamine		−40	111
Triethylamine	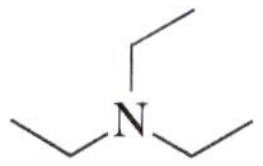	−115	90

25.4.1 Spectroscopy of amines

The IR spectra of amines are characterized by a semi-broad N—H stretching band around 3300 cm^{-1} (**Figure 25.9**). Because nitrogen is less electronegative than oxygen, the N—H band is less intense than the O—H stretching bond. It is also less broad than the O—H band because nitrogen does not hydrogen bond as well. The C—N stretching band falls around 1200 cm^{-1} (in the fingerprint region) and so is not often used in structural identification.

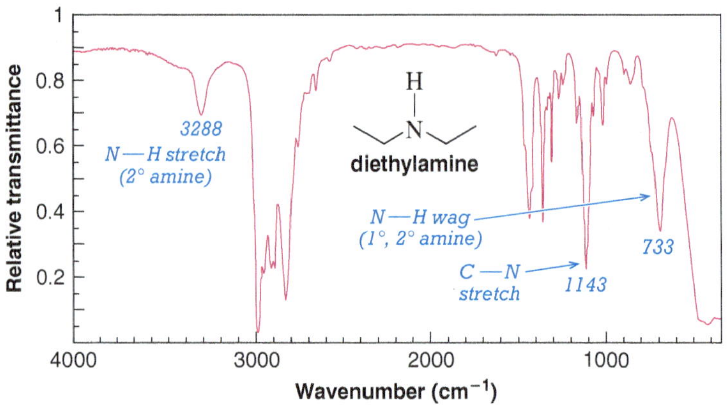

Figure 25.9 **The IR spectrum of diethylamine.**

The N—H band of 2° amines contains one band, whereas the one for 1° amines has two. One of these bands is due to the H—N—H symmetric stretching vibration; the other is due to H—N—H asymmetric stretching (**Figure 25.10**).

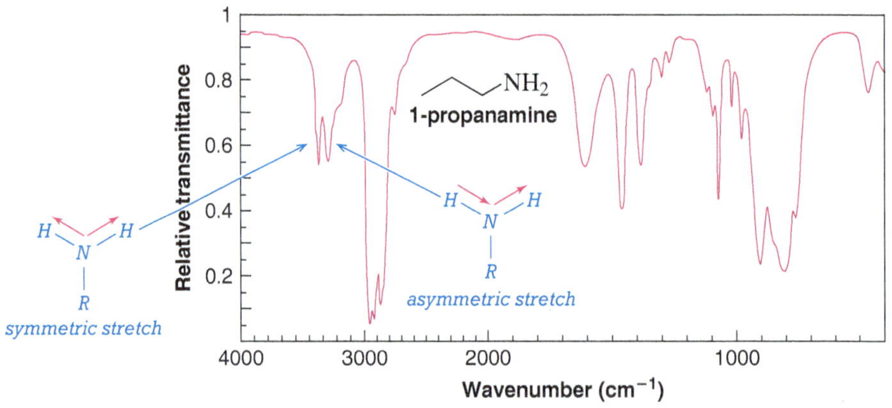

Figure 25.10 **The IR spectrum of 1-propanamine.** There is one band for each type of H—N—H stretch.

The ^{1}H NMR spectrum of amines (**Figure 25.11**) displays a broad singlet between 1 and 4 ppm that corresponds to the amine hydrogens. Because of hydrogen bonding, the location and splitting of these signals are inconsistent. Hydrogens on the carbon bearing the amino group show up around 2.5 ppm, as a result of the slight electron-withdrawing ability of the nitrogen atom. Because induction dies off with distance, hydrogens farther removed from the amino group are more upfield/shielded. In the ^{13}C NMR, the carbon bearing the amino group shows up between 30 and 50 ppm.

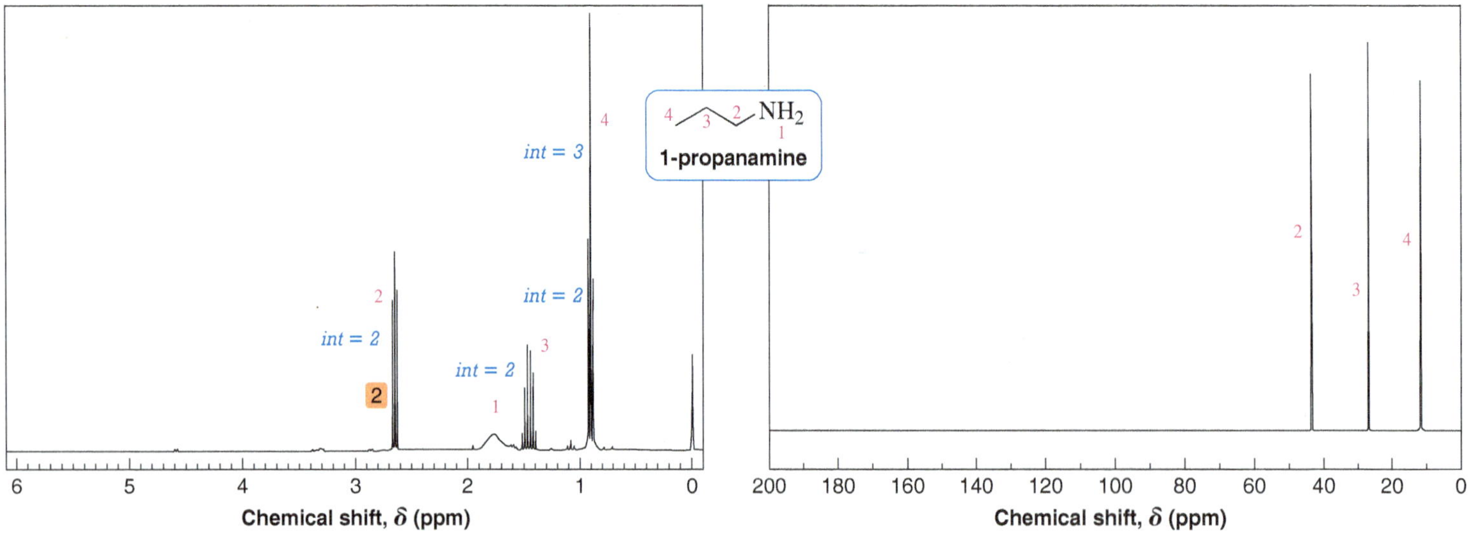

Figure 25.11 **The ^{1}H NMR and ^{13}C NMR spectra of 1-propanamine.**

A key feature of the mass spectrum of amines is distinguished by the odd mass/nitrogen rule (Section 14.6.5). Because most of the elements we have studied have either an even number of valence electrons and even mass number (C and O), or an odd number of valence electrons and an odd mass number (H, Cl, Br, and I), stable molecules containing only these atoms will have an even molecular ion. On the other hand, molecules with nitrogen, which has an even mass number (7 protons, 7 neutrons) but an odd number of valence electrons (5), will show an odd molecular ion.

Aliphatic amines show a single distinct fragmentation pattern. Termed α-cleavage, a C—C bond breaks on the amine-bearing carbon, driven by formation of the resonance-stabilized iminium ion. As shown in **Figure 25.12**, the difference between the molecular ion and the peaks at *m*/*z* 72 and *m*/*z* 86 are fragments of atomic mass 43 ($CH_3CH_2CH_2$) and 29 (CH_3CH_2), respectively.

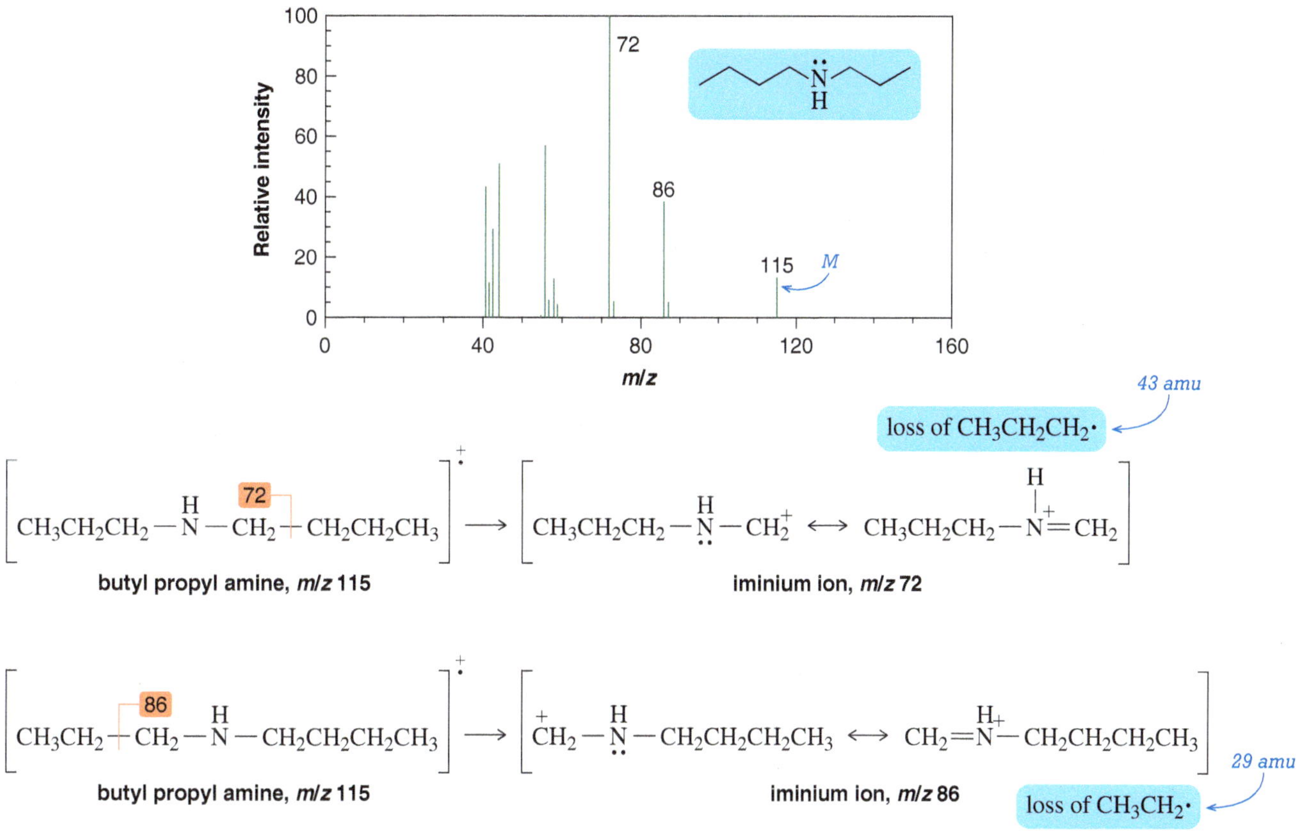

Figure 25.12 The mass spectrum of butyl propyl amine.

Assessment 25.11 Based on the IR spectrum shown, identify the amine with the molecular formula C_7H_9N. Label the peaks that assisted with your analysis.

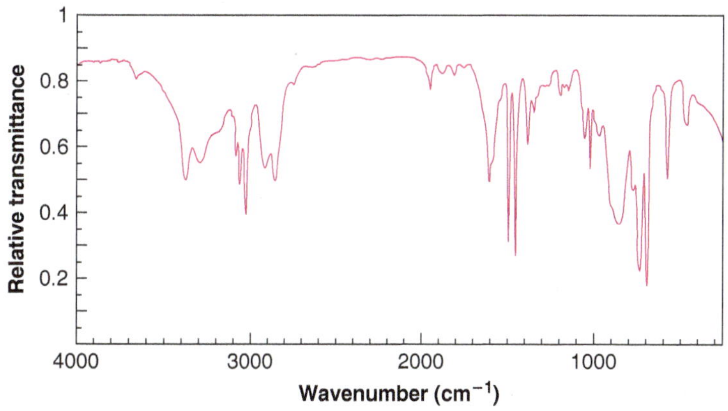

Assessment 25.12 Which of the following amines would you expect to boil at a higher temperature?

(a)

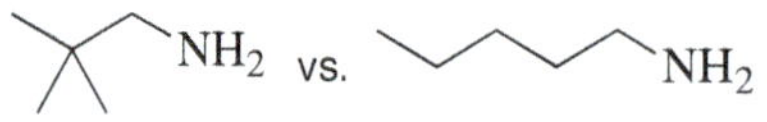

(b)

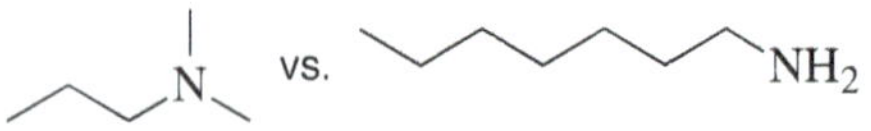

Assessment 25.13 Identify the compound with a molecular formula of $C_6H_{15}N$ that corresponds to the ^1H NMR spectrum shown. [*Recall that an H bound to N or O doesn't always appear in the* ^1H *NMR spectrum.*]

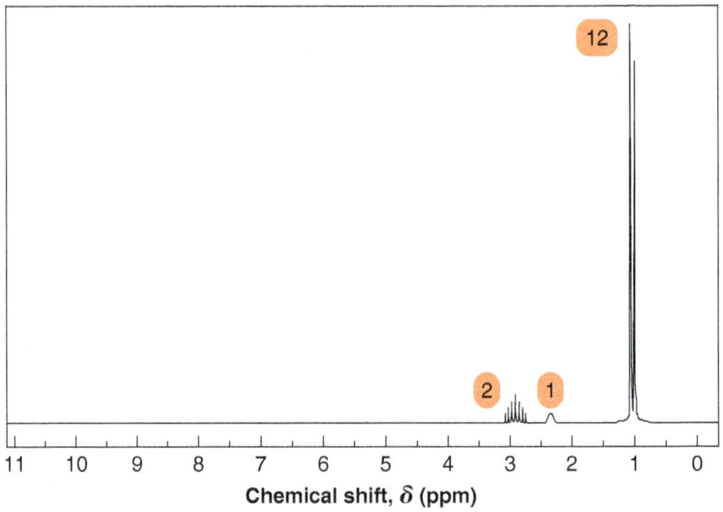

Assessment 25.14 Suggest two possible fragmentations that might occur in the mass spectrum of the amine shown.

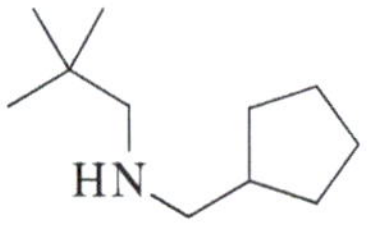

25.4.2 Basicity of amines

The use of amines as bases in acylation (Section 19.7.2) and tosylation (Section 13.7.1) reactions demonstrated one of the more important properties of amines—their basicity. Before we talk about the basicity, though, let's review the concept of pK_b values because they directly tell us about base strength.

Until now, we haven't talked much about pK_b values to describe the strength of bases. Instead, we have talked about the strength of a base in terms of the pK_a of its conjugate acid. A base with a strongly acidic conjugate acid (a low pK_a) is stable without the proton; thus, the base is weak. Conversely, a base whose conjugate acid is extremely weak (a high pK_a) is very unstable without the proton; thus, the base is strong (**Table 25.5**).

Table 25.5 Using pK_a values to determine the relative strength of a base.

Given a certain base . . .	. . . Identify its conjugate acid.	From the table, determine the pK_a value of the conjugate acid.	Test: If the pK_a value is low/high, then the conjugate acid is a strong/weak acid.	Conclusion: The base is weak/strong.	pK_b value of the base
NH_2^-	NH_3	38	NH_3 is a very weak acid.	NH_2^- is a very strong base.	−19
HO^-	H_2O	16	H_2O is a weak acid.	HO^- is a strong base.	−2
NH_3	NH_4^+	10	NH_4^+ is a moderately strong acid.	NH_3 is a moderately strong base.	4
H_2O	H_3O^+	−1.7	H_3O^+ is a very strong acid.	H_2O is a very weak base.	15.7

pK_b values are commonly used in biochemistry, so even though pK_a values tell essentially the same story, it is a good idea to be familiar with both.

Based on the data in Table 25.5, which are slightly modified from Section 4.3.2, the strength of a base is inversely correlated to the strength of its conjugate acid. As you may recall from your introductory chemistry course, the pK_b of a base can be calculated using Equation 25.1:

$$14 - pK_a(\text{conjugate acid}) = pK_b(\text{base}) \qquad \text{Equation 25.1}$$

Equation 25.1 reiterates what is demonstrated in Table 25.5 and leads to two important conclusions. First, strong bases have low pK_b values, and the acids they form are stable, weak acids. Second, weak bases have high pK_b values, and the acids they form are unstable, strong acids.

The pK_b values of several amines are listed in **Table 25.6**.

Table 25.6 The basicity of amines.

Amine	K_b	pK_b	pK_a of $R_3\overset{+}{N}H$
Ammonia	1.8×10^{-5}	4.74	9.26
Primary alkyl amines			
Methylamine	4.3×10^{-4}	3.36	10.64
Ethylamine	4.4×10^{-4}	3.36	10.64
n-Propylamine	4.7×10^{-4}	3.32	10.68
Isopropylamine	4.0×10^{-4}	3.40	10.60
Cyclohexylamine	4.7×10^{-4}	3.33	10.67
Benzylamine	2.0×10^{-5}	4.67	9.33
Secondary amines			
Dimethylamine	5.3×10^{-4}	3.28	10.72
Diethylamine	9.8×10^{-4}	3.01	10.99
Di-*n*-propylamine	10.0×10^{-4}	3.00	11.00
Tertiary amines			
Trimethylamine	5.5×10^{-5}	4.26	9.74
Triethylamine	5.7×10^{-4}	3.24	10.76
Tri-*n*-propylamine	4.5×10^{-4}	3.35	10.65
Aryl amines			
Aniline	4.0×10^{-10}	9.40	4.60
N-Methylaniline	6.1×10^{-10}	9.21	4.79
N,N-Dimethylaniline	1.2×10^{-9}	8.94	5.06
p-Bromoaniline	7×10^{-11}	10.2	3.8
p-Methoxyaniline	2×10^{-9}	8.7	5.3
p-Nitroaniline	1×10^{-13}	13.0	1.0
Heterocyclic amines			
Pyrrole	5×10^{-15}	14.3	−0.3
Pyrrolidine	1.9×10^{-3}	2.73	11.27
Imidazole	8.9×10^{-8}	7.05	6.95
Pyridine	1.8×10^{-9}	8.75	5.25
Piperidine	1.3×10^{-3}	2.88	11.12

An interesting effect emerges when comparing the basicity of ammonia, 1°, 2°, and 3° amines. As shown in **Figure 25.13**, ammonia is the weakest base of the four ($pK_b = 4.74$). Ethylamine, which contains one alkyl group, is more electron-rich as a result of the hyperconjugative effect of its attached alkyl group. Thus, it is more basic ($pK_b = 3.36$) than ammonia. Addition of a second ethyl group gives diethylamine, which, with more hyperconjugation, is also more basic ($pK_b = 3.01$). You probably expect that a third alkyl group, with even more hyperconjugation, would make triethylamine even more basic. This is not the case, though, because triethylamine ($pK_b = 3.24$) is more basic than ethylamine but less basic than diethylamine. Focusing only on the alkyl groups means ignoring the importance of conjugate acid solvation to acidity. A conjugate acid that is easily solvated is going to be more stable than one that isn't. [*Recall the naked, unsolvated S_N2 anions in Section 12.3.3—very powerful!*] The addition of extra alkyl groups makes the nitrogen more electron-rich, but it also sterically impedes solvation in the 3° amine. In this way, 1° and 2° amines strike a good balance, where the addition of extra alkyl groups increases the electron density on the nitrogen to a greater degree than it interferes with solvation. [*We discuss the consequences of increased hyperconjugation increasing the basicity/nucleophilicity of amines in Section 25.5.*]

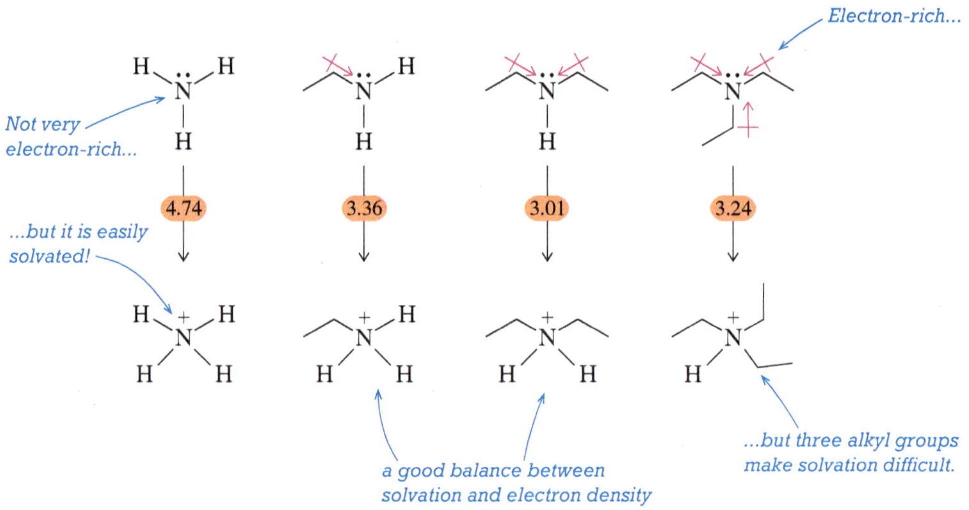

Figure 25.13 Extra alkyl groups makes ammonia a stronger base, until a third group is added. At this point, solvation of the resulting ammonium ion is sterically blocked. [*Don't forget, these are pK_b values.*]

When an amino group can overlap by resonance with neighboring π systems, the basicity is diminished. In these systems, the lone pair on the nitrogen sits in a *p* orbital, maximizing overlap with the π system and making it less likely to be protonated. As a result, aniline ($pK_b = 9.40$) is a much weaker base than the structurally similar cyclohexanamine ($pK_b = 3.33$) [**Figure 25.14(a)**]. Amides ($pK_b = 14.5$) are almost never considered to be bases because of resonance with the strongly electron-withdrawing carbonyl [**Figure 25.14(b)**].

(a)

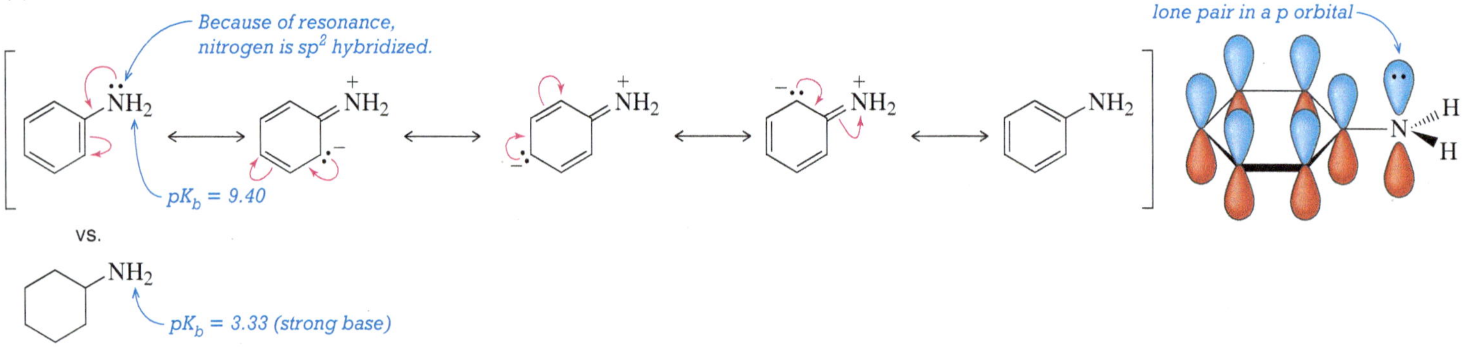

(b)

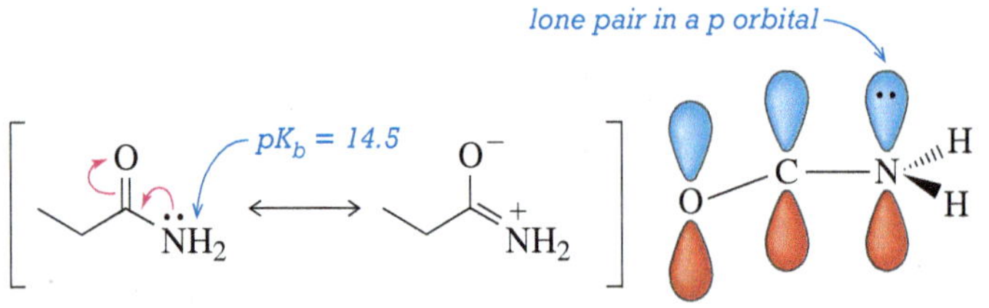

Figure 25.14 Resonance delocalization of a nitrogen lone pair makes amines weaker bases. Thus, (a) aniline is a much weaker base than cyclohexanamine, and (b) amides are extremely weak bases.

The difference in basicity between aniline and cyclohexylamine can be demonstrated on the reaction coordinate diagram shown in **Figure 25.15**. Overlapping of the amine lone pair with the π system results in a more stable molecule. As a result, it starts out lower on the reaction coordinate diagram. Because the resulting conjugate acids are of nearly equal stability, protonation of the aniline is a more endothermic (uphill) and, therefore, less favorable process.

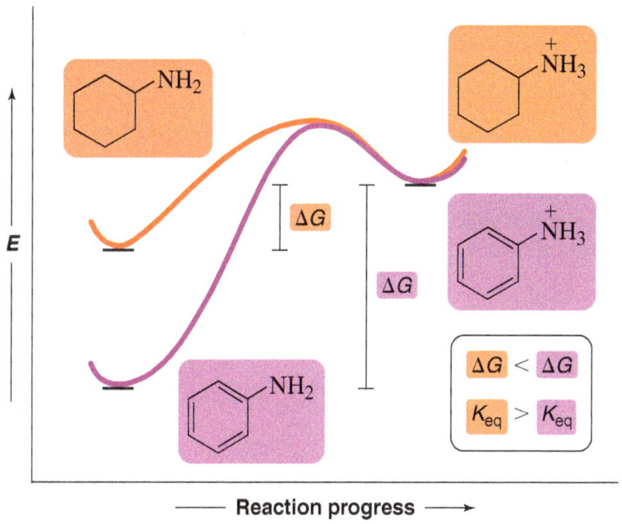

Figure 25.15 **Protonation of the more stable aniline is a more endothermic process** (climbs a higher hill).

The basicity of aromatic amines is dramatically impacted by substituents on the aromatic ring, in the same way that substituents affect the rate of electrophilic aromatic substitution reactions (Section 23.8) and phenol acidity (Section 24.2.1). **Figure 25.16** compares three para-substituted anilines to aniline itself. The nitro group withdraws electrons by resonance, so it dramatically decreases the pK_b value (pK_b = 13). Bromine withdraws electrons by induction, so it only slightly decreases the pK_b value (pK_b = 10.2). Finally, the electron-donating methoxy group increases the basicity (pK_b = 8.7).

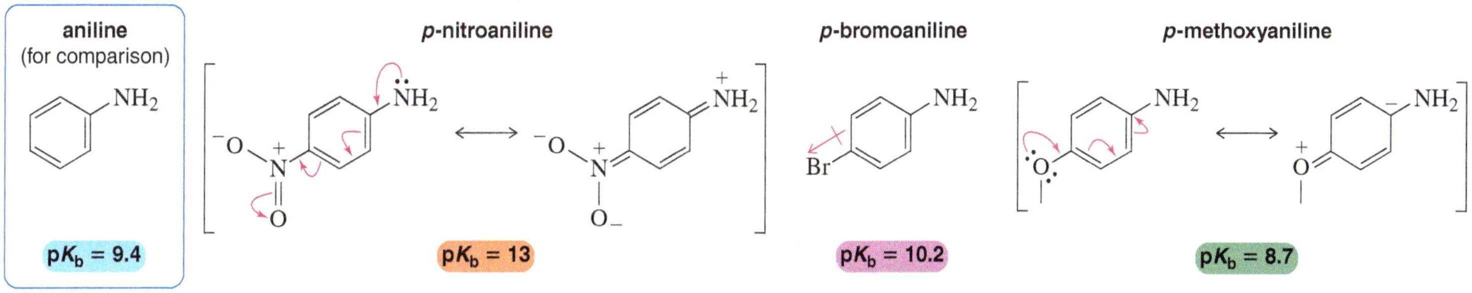

Figure 25.16 **Substituent effects on aniline basicity.** Electron-donating groups *increase* basicity (decrease the pK_b value), whereas electron-withdrawing groups *decrease* basicity (increase the pK_b value).

Hybridization also impacts the pK_b of a nitrogenous base, consistent with what was presented in Section 4.4.5. **Figure 25.17** compares the basicity of piperidine and pyridine, analogous structures that differ only by the hybridization of the nitrogen. In piperidine, the lone pair is in an sp^3-hybridized orbital. In pyridine, on the other hand, the lone pair is in an sp^2-hybridized orbital. Because the sp^2-hybridized orbital has 33% s character, as compared to 25% s character in the sp^3-hybridized orbital, the lone pair in pyridine is held closer to

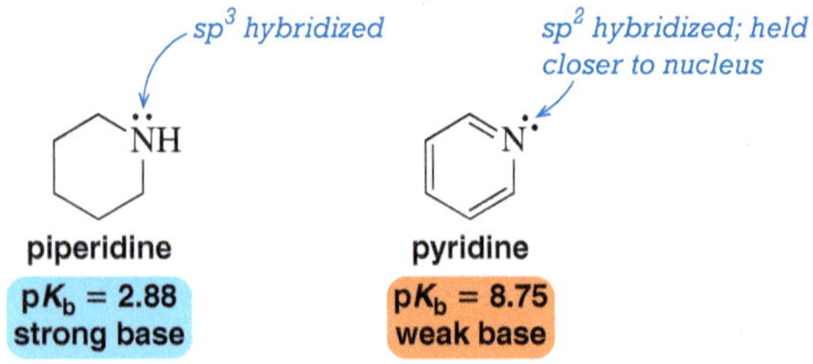

Figure 25.17 The effect of hybridization on basicity. Lone pairs in sp^2-hybridized orbitals are less basic than those in sp^3-hybridized orbitals.

the nucleus than the lone pair in piperidine. Lone pairs held closer to the positively charged nucleus are less basic. [*Important point: The lone pair in pyridine does not participate in resonance with the ring. This is a common student misconception.*]

The relative basicity of heterocyclic amines is based on whether or not the lone pair of electrons on the nitrogen is contained within the 6-electron aromatic π system. As shown in **Figure 25.18**, the lone pair on pyrrole is part of the 6-electron aromatic π system, so it is unavailable for protonation, making pyrrole a very weak base (pK_b = 14.3). Imidazole has two potentially basic nitrogens. The lone pair on N_1 is part of the 6-electron aromatic π system, making it a very weak base. The lone pair on N_3, on the other hand, is in an sp^2-hybridized orbital. Like pyridine, it is somewhat basic (pK_b = 7.05).

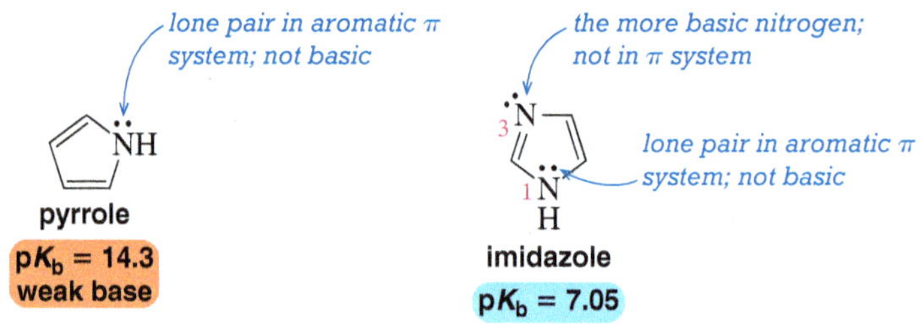

Figure 25.18 Lone pairs in the aromatic π system of heterocyclic amines like pyrrole and imidazole are very weak bases.

Assessment 25.15 We usually calculate K_{eq} for acid–base reactions using pK_a values. (a) Derive an equation to calculate K_{eq} using pK_b values, then (b) use it to calculate the equilibrium constant for the following reaction.

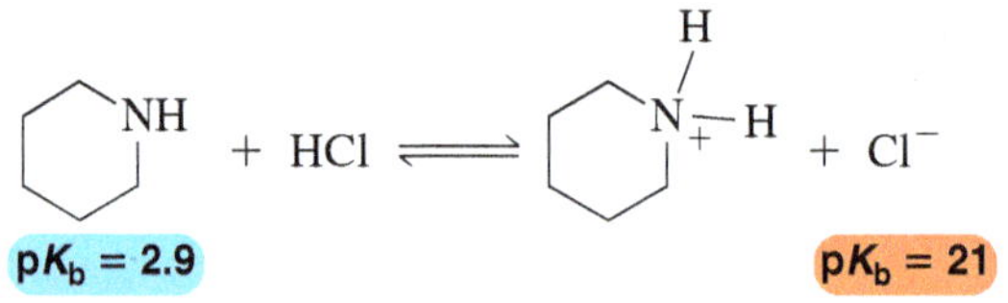

Assessment 25.16 In each pair, choose the more basic amine. Justify your choice.

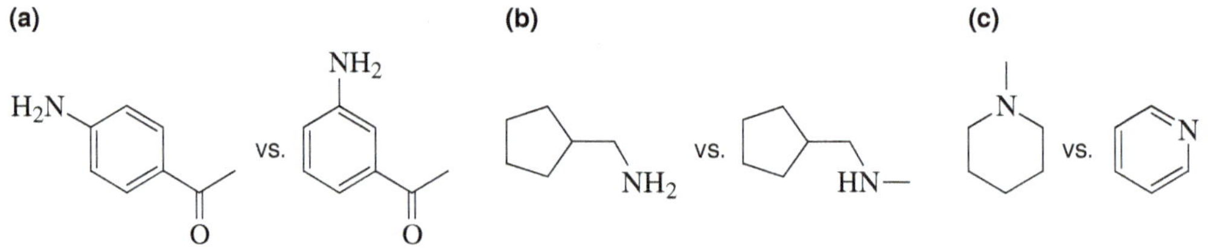

Assessment 25.17 Both an amide and an amine contain a nitrogen with a lone pair of electrons. Which nitrogen would you expect to be the stronger base?

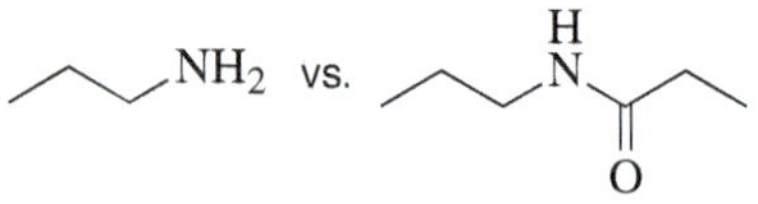

Assessment 25.18 Which nitrogen of imidazole would you expect to be methylated first?

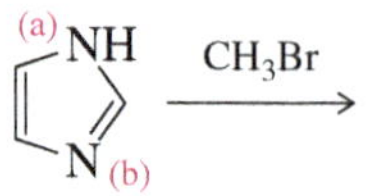

Assessment 25.19 Which would you expect to be more soluble in H_2O, an amine or an ammonium ion? Why?

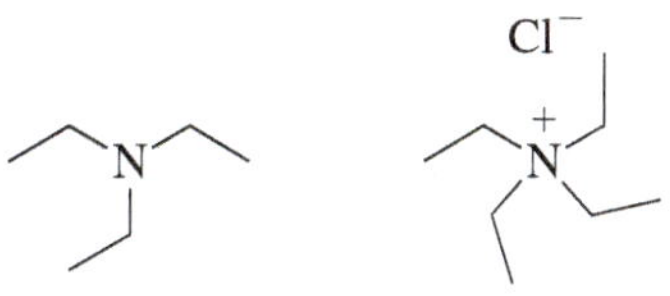

triethylamine   tetraethylammonium chloride

Assessment 25.20 Which would you expect to be a stronger nucleophile, ethyl amine or diethyl amine? Why?

25.5 Synthesis of 1° amines

There are myriad ways to synthesize amines. To help you organize them and relate them to each other, they have been separated into two categories. The amine syntheses in Section 25.5 are reactions that can only be used to make 1° amines. Recall from Section 25.2.4 that a 1° amine has just one alkyl group attached to the nitrogen. In Section 25.6, we introduce reactions that are more generally applicable to the synthesis of 1°, 2°, and 3° amines.

25.5.1 Alkylation of ammonia: a problem

The synthesis of primary amines, starting with ammonia and a 1° haloalkane, should be relatively straightforward, proceeding by an S_N2 mechanism.

Assessment 25.21 A chemist attempted to prepare 1-aminohexane by the S_N2 reaction of ammonia with 1-chlorohexane. Although a small amount of the desired product was formed, it was accompanied by three other compounds (shown). Rationalize the formation of these compounds.

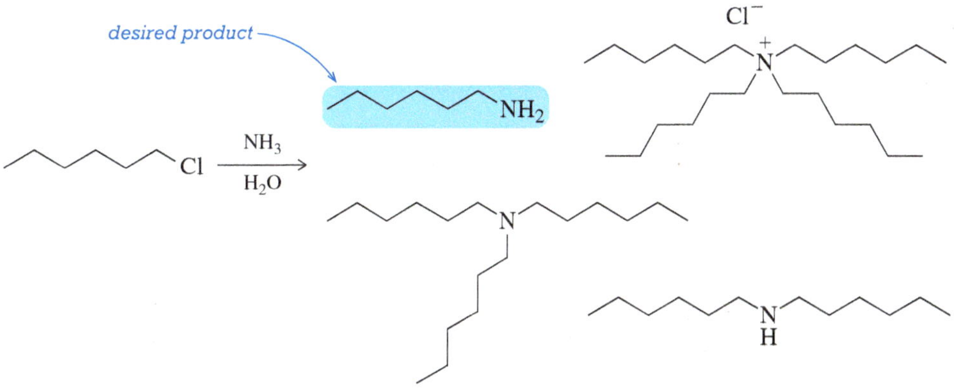

On your own, work Assessment 25.21 to see why this reaction might not be so straightforward. Then we will work it together.

Solved Assessment 25.21

We have seen amines react as nucleophiles with haloalkanes in S_N2 reactions (Chapter 12) and with aldehydes and ketones to form imines and enamines (Section 17.8). As such, it is perfectly reasonable to have the electron-rich nitrogen atom of ammonia donate a pair of electrons to σ^* displacing the chloride in a backside attack to give a cationic ammonium ion (**Figure 25.19**). Assuming only one molecule of ammonia has attacked one molecule of 1-chlorohexane, what is the composition of the solution at this point, assuming we started with 1 mol of each? [*Thoughts?*] The solution contains one molecule of hexylammonium hydrochloride. The rest is ammonia and 1-chlorohexane. With a $pK_a = 10$, what would you expect to happen to the ammonium ion in a solution with lots of basic ammonia? A molecule of ammonia will deprotonate the hexylammonium chloride, giving the desired 1-aminohexane. But the reaction doesn't stop there. [*What do you think happens next?*]

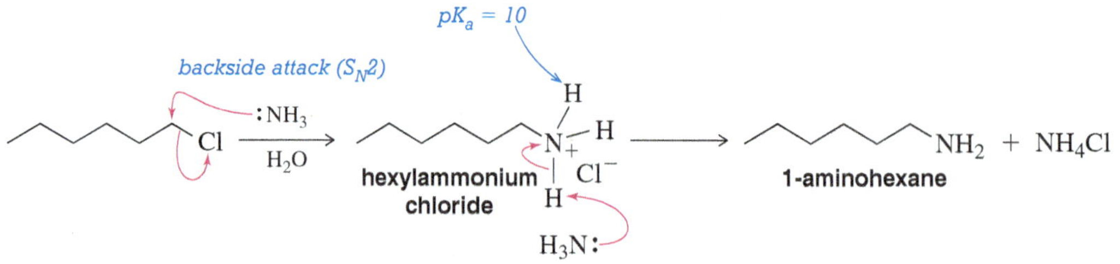

Figure 25.19 Alkylation of ammonia produces a 1° amine, at least temporarily.

After forming the 1-aminohexane (1° amine), there is still a lot of 1-chlorohexane around to react with either ammonia or 1-aminohexane. [*Which is more reactive? See Section 25.4.2.*] The hexyl group donates electron density by hyperconjugation to the nitrogen atom, making it more nucleophilic. Thus, in a competition reaction between ammonia and 1-hexanamine, 1-hexanamine will always react more rapidly with the haloalkane (**Figure 25.20**). Following alkylation, a proton will be removed from dihexylammonium to give dihexylamine (2° amine). With two alkyl groups, dihexylamine is *even more* nucleophilic, reacting again with 1-chlorohexane to give, following deprotonation, trihexylamine (3° amine). Finally, the electron-rich trihexylamine is alkylated once more to give a tetrahexylammonium chloride (4° ammonium salt).

And so alkylation of ammonia ends up failing to make 1° amines because the addition of each alkyl group creates a more reactive amine that is preferentially alkylated. [*The rest of Section 25.5 features solutions to this problem.*] ■

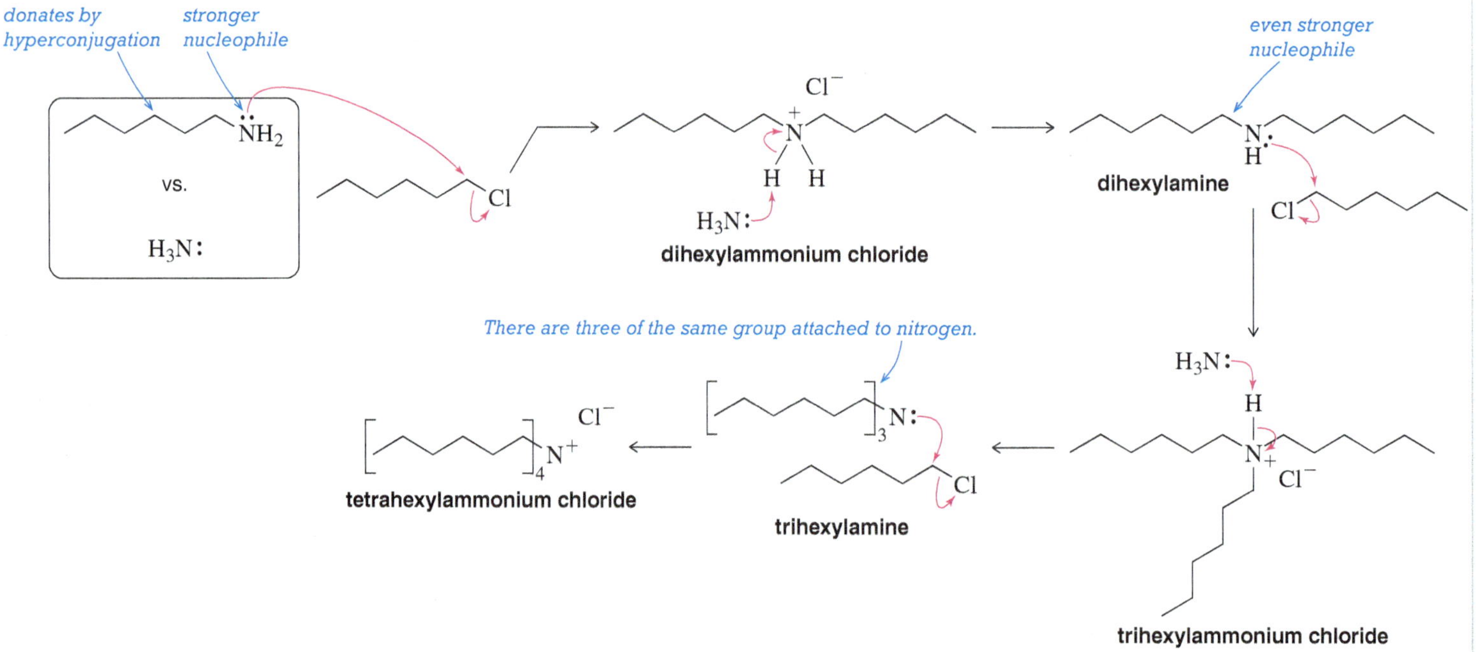

Figure 25.20 Ammonia continues to alkylate until all protons are replaced with alkyl groups.

One solution for 1° amine synthesis is to react a 1° alkyl halide with a huge excess of ammonia (at least 10 equivalents). This ensures that the remaining ammonia molecules will be statistically more likely to collide with 1-chlorohexane before 1-hexanamine does. Even if the unreacted ammonia is recycled at the end of the reaction (and it can be), this is not the most atom-economical or energy-efficient process, although the chemical industry sometimes does it this way for smaller 1° amines. [*We can do better.*]

25.5.1 SUMMARY

- Overalkylation of ammonia in S$_N$2 reactions occurs because each new amine produced is more reactive than the original. Each extra alkyl group, through hyperconjugation, increases the nucleophilicity of the amino group.

25-A CHEMISTRY IN CONTEXT: Medicinal Chemistry

Antidepressants by S_N2

Serotonin deficiency in depressed patients is addressed by inhibiting the reuptake of serotonin by the presynaptic neuron. Fluoxetine is a selective serotonin reuptake inhibitor (SSRI), meaning it does not impact the uptake of other neurotransmitters (**Figure 25.21**).

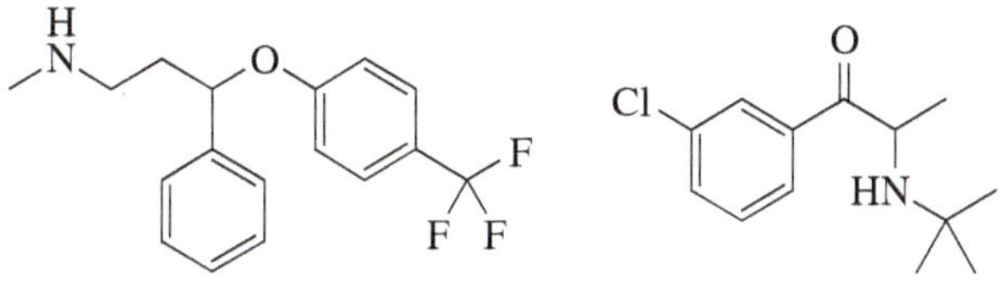

fluoxetine bupropion

Figure 25.21 The structures of fluoxetine (a selective serotonin reuptake inhibitor) and bupropion (a dopamine reuptake inhibitor).

In addition to deficiencies in serotonin and/or norepinephrine, a deficiency in dopamine has been shown to result in a depressed state. Dopamine is a neurotransmitter that regulates the brain's pleasure center. It also provides motivation to pursue rewarding or pleasurable behaviors. Bupropion, also marketed as a smoking-cessation aid, has been used as a treatment for depression, presumably as a dopamine reuptake inhibitor (DRI).

Though the S_N2 reaction has its drawbacks in the synthesis of amines, it has found utility for the synthesis of fluoxetine. As shown in **Figure 25.22**, methylamine displaces a mesylate ion ($-OSO_2CH_3$), analogous to tosylates from Section 13.7.1, via an S_N2 reaction to form the 2° amino group of fluoxetine.

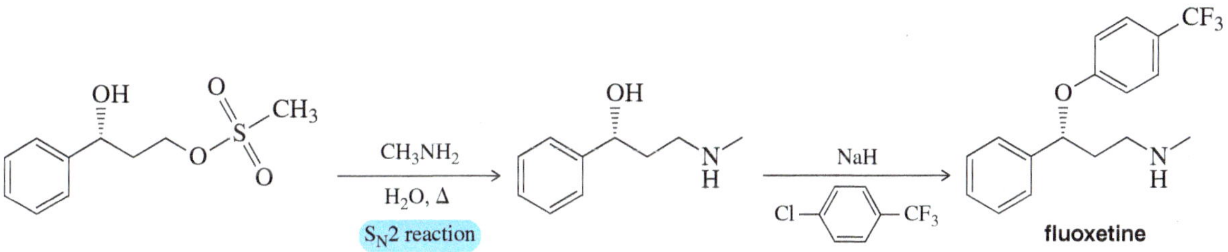

Figure 25.22 An S_N2 reaction for the synthesis of fluoxetine.

Similarly, the *t*-butylamino group of bupropion is incorporated by displacement of a bromide from the α-carbon of a phenyl ketone (**Figure 25.23**). Presumably because of the increased steric hindrance of the bulky *t*-butyl group, this reaction doesn't suffer from the overalkylation discussed in Section 25.5.1.

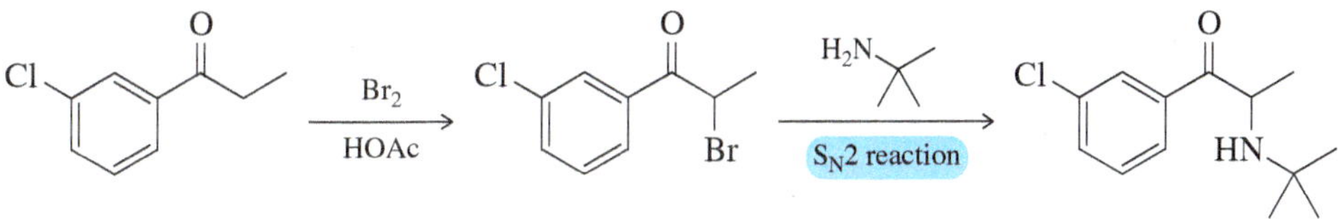

bupropion

Figure 25.23 An S_N2 reaction for the synthesis of bupropion.

25.5.2 Reduction of a nitrile

Our first method for synthesizing 1° amines combines two reactions studied previously. Recall from Chapter 12 that sodium cyanide is an excellent nucleophile, capable of displacing both 1° and 2° haloalkanes in an S_N2 reaction to give a new C—C σ bond. Recall from Section 19.10.4, too, that nitriles can be reduced to 1° amines using lithium aluminum hydride ($LiAlH_4$), a strong reducing agent. This two-step process results in a new C—C bond, extending the chain as compared to the starting haloalkane (**Figure 25.24**).

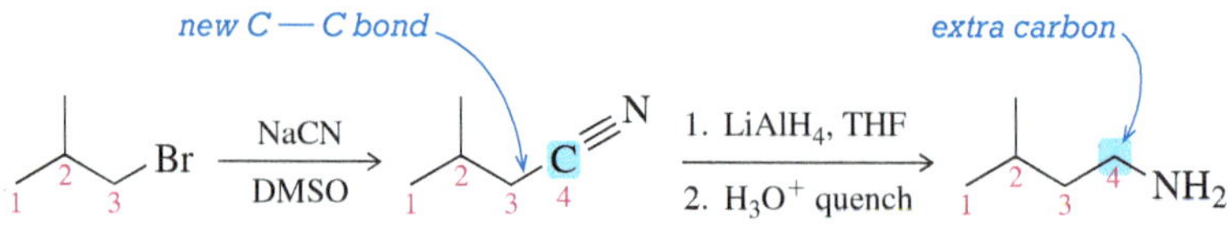

Figure 25.24 The formation of a 1° amine via S_N2 reaction of sodium cyanide followed by reduction.

Mechanistically, a hydride ion is donated directly to the nitrile carbon, breaking the C—N π bond while forming alane (AlH_3). The AlH_3, a Lewis acid, is important here, because the negatively charged nitrogen is a strong Lewis base. Formation of a Lewis acid–Lewis base complex by donation of a pair of electrons to an empty *p* orbital [*similar to that seen for ester and nitrile reduction by DIBAl-H; Sections 19.8.2 and 19.10.3*] occurs quickly. Intramolecular hydride transfer from aluminum to carbon breaks the second C—N π bond. Quenching of the resulting amide is followed by an aqueous workup to replace aluminum with a proton (**Figure 25.25**). [*In practice, after quenching, the acidic solution would be neutralized to give the neutral amine product.*]

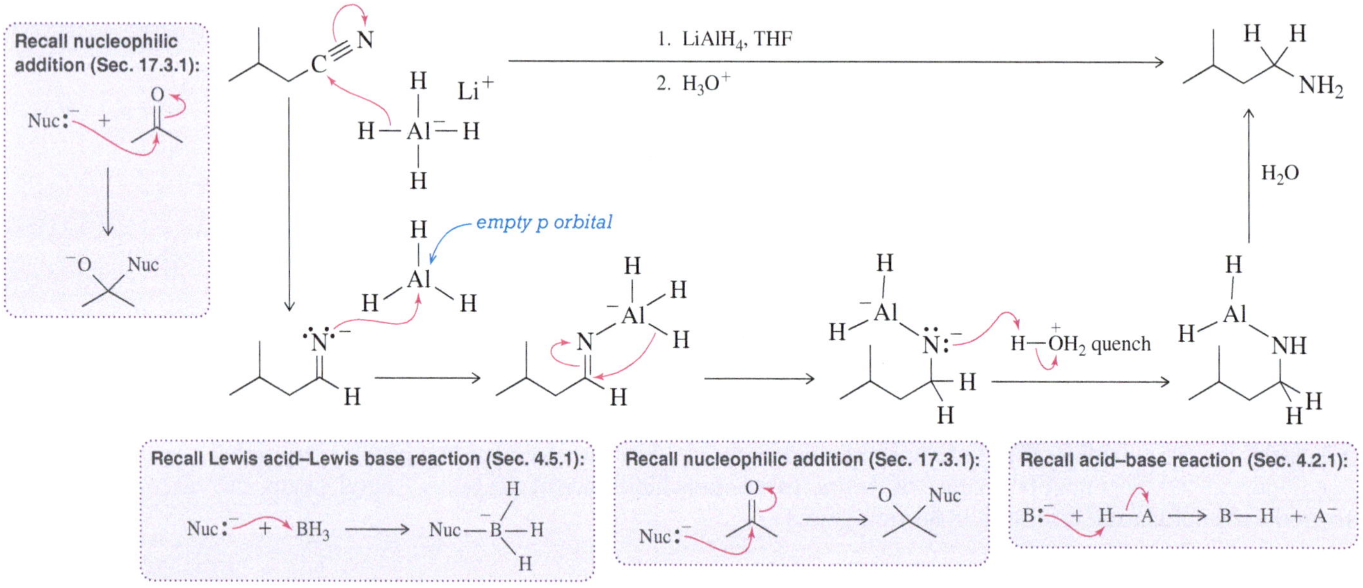

Figure 25.25 The mechanism of nitrile reduction.

25-B CHEMISTRY IN CONTEXT: Medicinal Chemistry

Citalopram by nitrile reduction

Citalopram (**Figure 25.26**), like fluoxetine, is a selective serotonin reuptake inhibitor (SSRI). Serotonin serves multiple functions in the body. Among them, it plays a role in appetite and emotions and helps regulate a person's mood, contributing to a feeling of happiness. As an SSRI, it blocks the reabsorption of serotonin by the presynaptic neuron, allowing more serotonin to persist in the synapse, which enables even low levels of serotonin to carry out its antidepressive effect, and for longer.

Although citalopram contains a 3° amine, it has been synthesized using a nitrile reduction. As shown in **Figure 25.27**, treatment of a nitrile with lithium aluminum hydride upon aqueous workup gives the primary amine. The primary amine is ultimately converted to a tertiary amine via reductive amination, a reaction we present in Section 25.6.2.

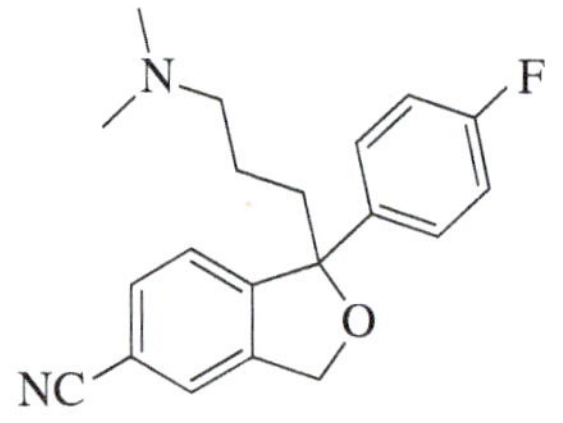

citalopram

Figure 25.26 Structure of citalopram.

(*continued*)

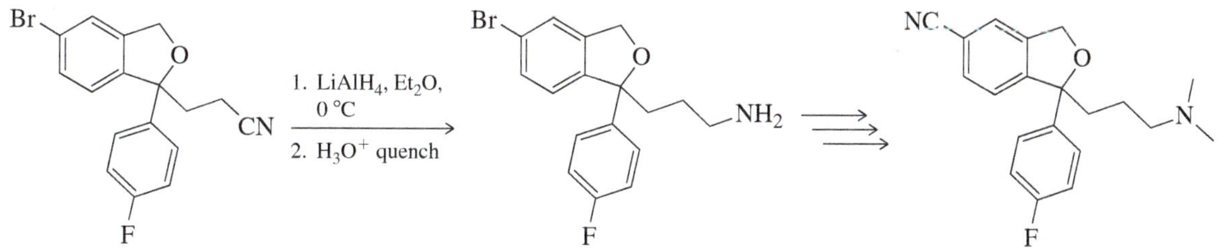

citalopram

Figure 25.27 Nitrile reduction for the synthesis of citalopram.

25.5.3 Reduction of an azide

In Chapter 12, we studied the reaction of azide (N_3^-), another excellent nucleophile, with haloalkanes to make alkyl azides by the S_N2 reaction. Reduction of the azide can be accomplished by a number of reactions. We focus on three: palladium-catalyzed hydrogenation, lithium aluminum hydride reduction, and the Staudinger reaction, which involves treatment with triphenylphosphine and water (**Figure 25.28**). Unlike the reaction in Section 25.5.2 using cyanide as the nucleophile, this process, with azide, does *not* extend the chain. In this way, the two methods are complementary to one another.

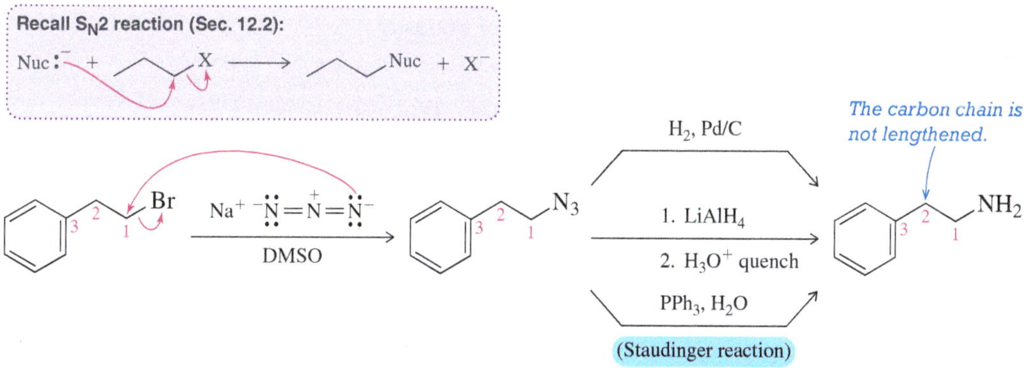

Figure 25.28 The formation of a 1° amine via an S_N2 reaction of sodium azide and reduction.

Reduction of the azide by either of the three reactions is driven by formation of nitrogen gas. Although the mechanism of hydrogenation with palladium isn't well known, the mechanisms of the other two reductions are shown in Figures 25.29 and 25.30. Lithium aluminum hydride reduction of azide likely begins with the addition of a hydride ion to the electrophilic N_3 of the azide, breaking the $N_2—N_3$ π bond (**Figure 25.29**). The anion at N_1 donates a pair of electrons to the alane formed in the first step. A hydride ion grabs a hydrogen from N_1, effecting an E2 elimination to form nitrogen gas and an amidoaluminum species. Hydrolysis of this amidoalumnium species in acid results in a 1° amine.

As we have done before, whenever we encounter phosphorus, we look for a way to form a new P=O bond, the driving force for most phosphorus reactions we have studied [*the Wittig reaction, Section 17.6; ROH to RBr, Section 13.7.4*]. As shown in **Figure 25.30**, phosphorus begins by attacking the electrophilic N_3, breaking the $N_3—N_2$ π bond, pushing an electron pair onto N_2. At this point, the negatively charged N_1 attacks the positively charged phosphorus to give a phosphazide, reminiscent of the oxaphosphetane formed in the Wittig reaction [*Section 17.6; see Table 17.12*]. Similar to how the oxaphosphetane reacts, the

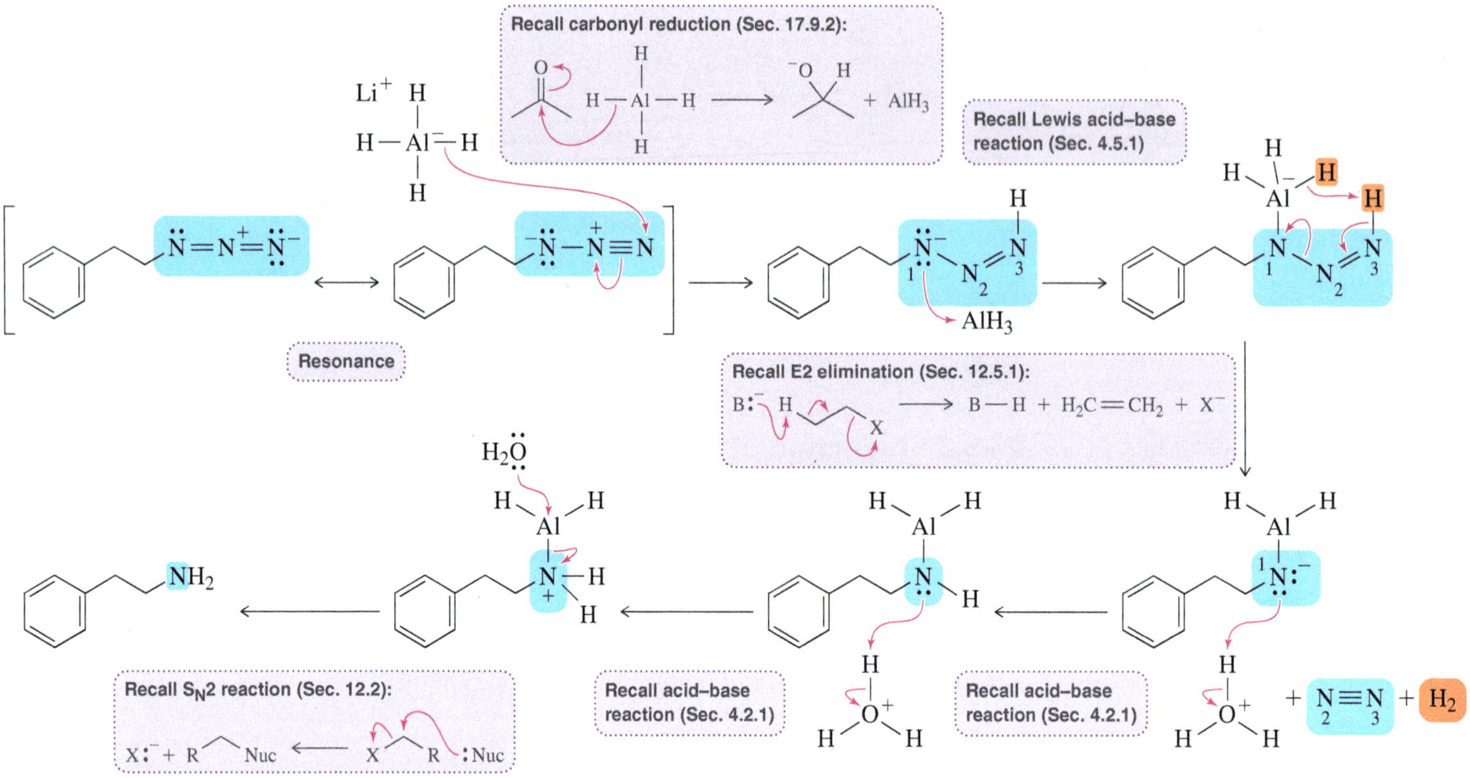

Figure 25.29 **The mechanism of $LiAlH_4$ reduction of azides.**

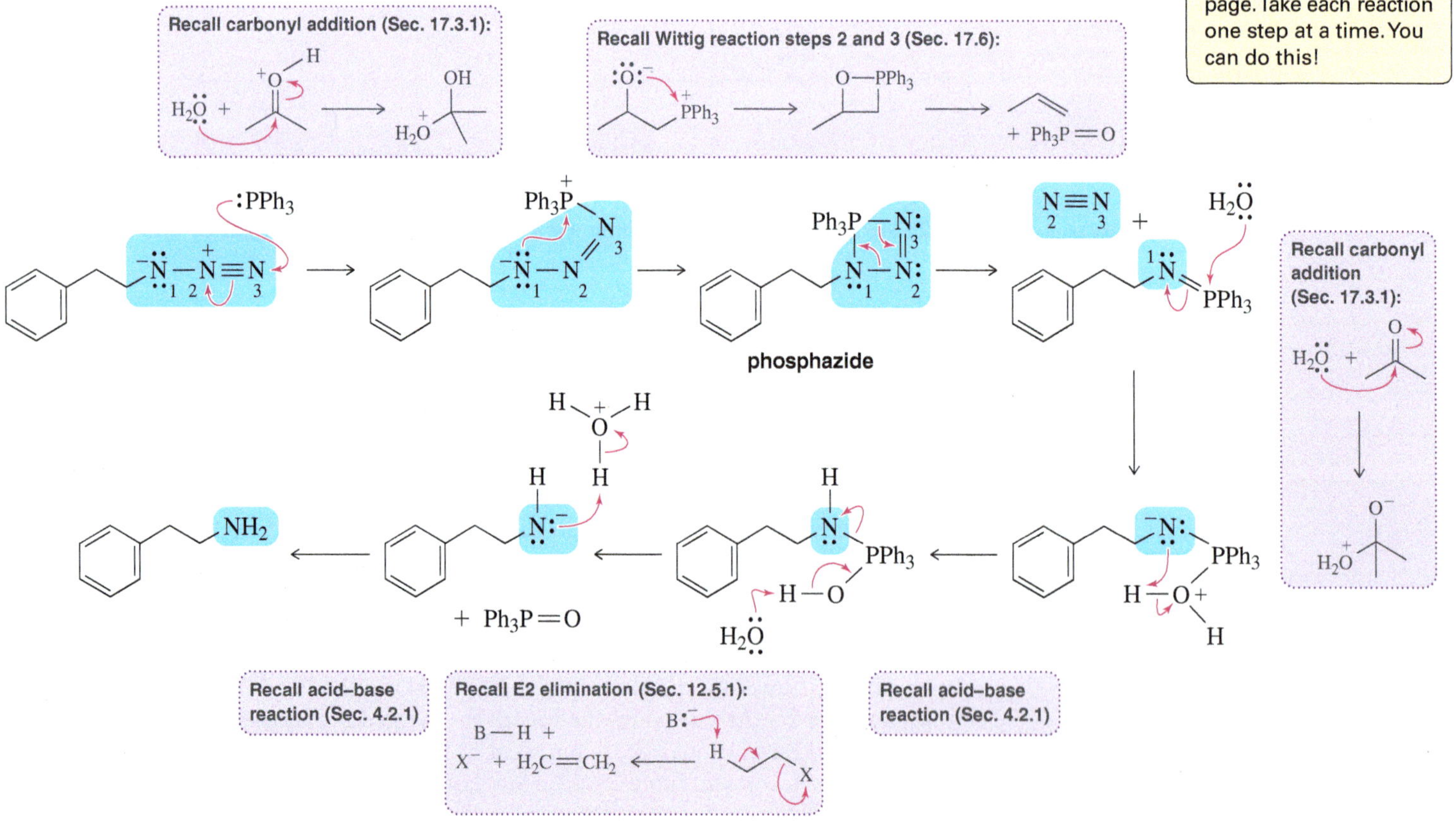

Figure 25.30 **The mechanism of the Staudinger reaction.**

phosphazide decomposes in a concerted [2 + 2] cycloreversion [*Section 22.4*] to produce nitrogen gas and an iminophosphorane (N=P bond). Reaction of the iminophosphorane with water gives the 1° amine and a new P=O bond via a series of acid–base reactions.

25.5.4 Gabriel amine synthesis

Unlike ammonia, which reacts in multiple substitution reactions to give overalkylation (Section 25.5.1), azide, with a −1 charge, only reacts with a single equivalent of haloalkane. The Gabriel amine synthesis, shown in **Figure 25.31**, reacts with only a single haloalkane, too, for similar reasons. Remarkably, this reaction, which utilizes potassium phthalimide, was first reported in 1887 by the German chemist Siegmund Gabriel. [*This is from the same era as the classical organic chemists discussed in Chapter 20.*]

Figure 25.31 The Gabriel amine synthesis.

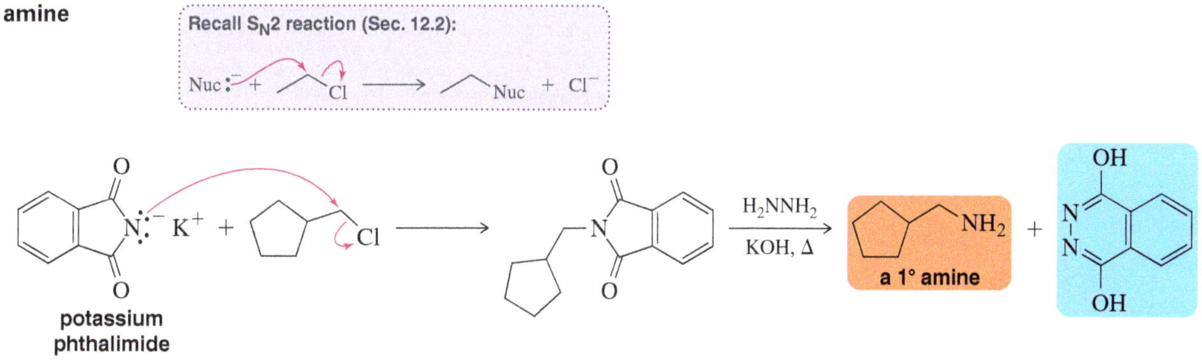

Following the substitution reaction, unmasking the amide group occurs through a complicated mechanism involving hydrazine under strongly basic conditions (**Figure 25.32**). [*As always, look for reactions you have seen before and try to identify the functional groups present.*] Essentially an imide (a carboxylic acid derivative discussed briefly in Chapter 19),

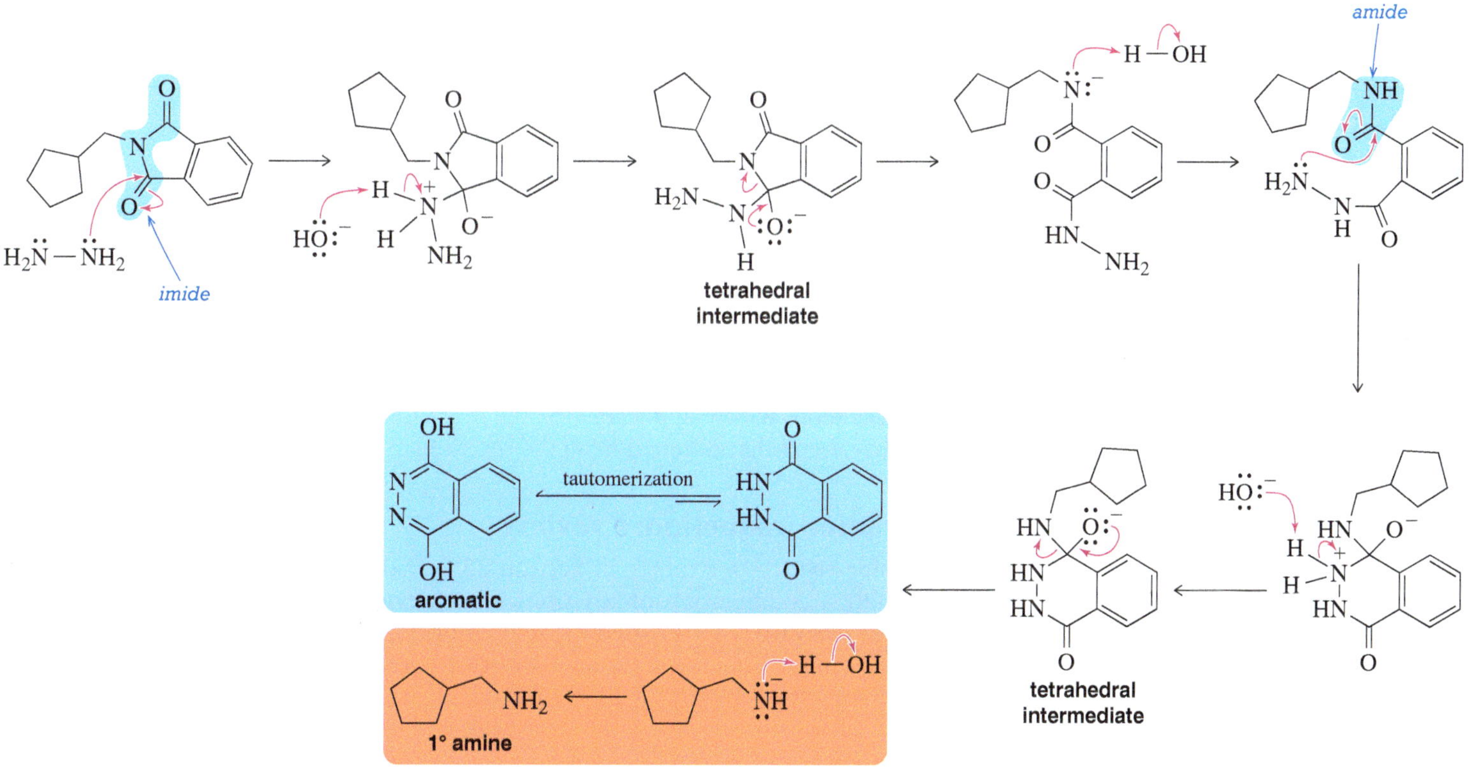

Figure 25.32 Unmasking the amine in the Gabriel amine synthesis.

the carbonyl is attacked by hydrazine to give a tetrahedral intermediate after deprotonation. Collapse of the tetrahedral intermediate opens the imide ring to give an amide following protonation. Intramolecular nucleophilic addition of a hydrazine nitrogen forms a tetrahedral intermediate within a new six-membered ring. Collapse of the tetrahedral intermediate releases an amide, which is protonated to give a 1° amine. The tautomerization of the by-product of the reaction to give an aromatic ring is undoubtedly one of the driving forces for the reaction.

25.5.2–25.54 SUMMARY

- Reduction of azides, reduction of nitriles, and the Gabriel amine synthesis provide solutions to the overalkylation problem of Section 25.5.1.

Assessment 25.22 Predict the product of the three reactions shown.

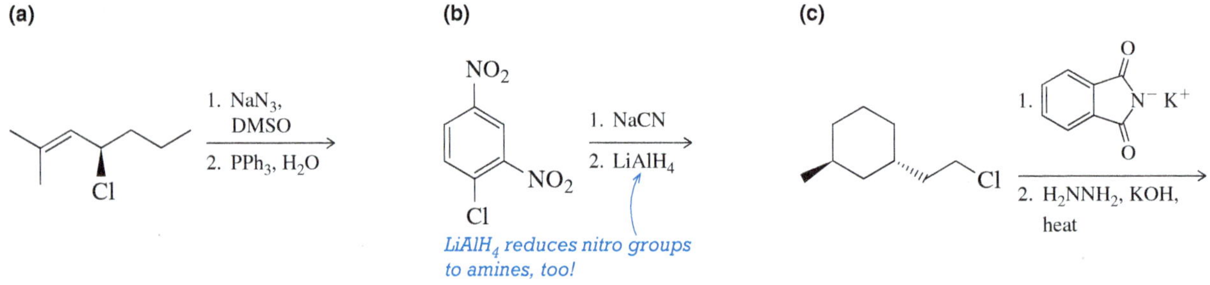

Assessment 25.23 Suggest a series of steps that will allow the synthesis of the amine on the right from the reactant on the left.

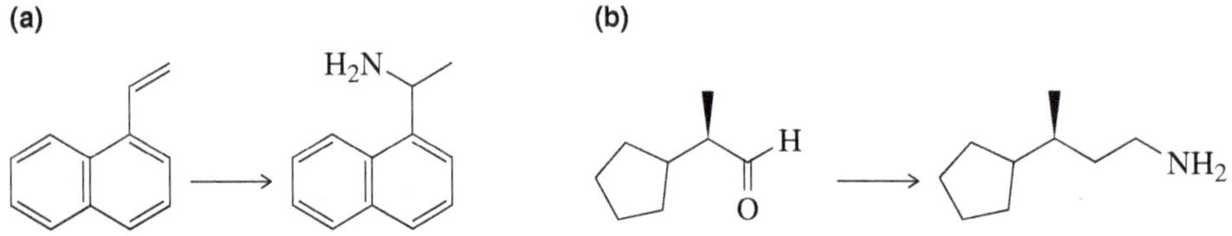

25.6 General synthesis of amines

The reactions in Section 25.5 are only useful for the synthesis of 1° amines. Simply for organizational purposes, the reactions here in Section 25.6 have been separated from those because they are more widely applicable for the synthesis of 1°, 2°, and 3° amines. Besides this feature, they aren't that different from the reactions studied thus far.

25.6.1 Acylation/reduction: a review

Like the cyanide reduction strategy in Section 25.5.2, the first method presented in this section is simply a combination of reactions seen previously. Beginning with a reactive carboxylic acid derivative, most often an acid chloride, reaction with an amine gives an amide via the standard nucleophilic acyl substitution reaction (Section 19.7.3). Recall that this reaction often employs a base to avoid using an excess of the starting amine (**Figure 25.33**).

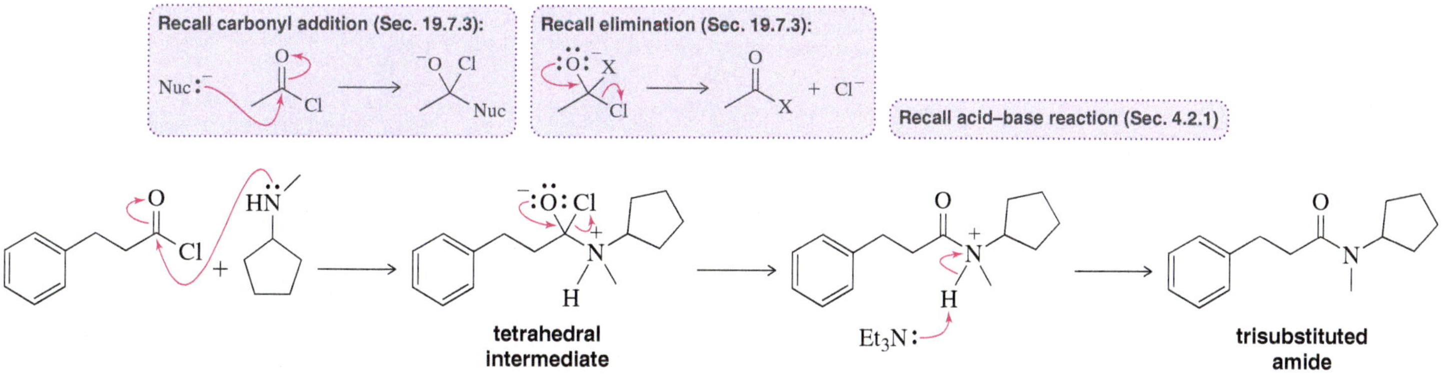

Figure 25.33 Amide formation via nucleophilic acyl substitution (see Section 19.7.3 for a review).

Section 19.8.1 showed that amides of any substitution could be reduced using lithium aluminum hydride to give amines. As shown in Figure 25.34, addition of a hydride ion to the carbonyl forms a tetrahedral intermediate. This intermediate usually collapses, but in this case the alkoxide reacts with alane. The nitrogen now uses its lone pair to push out the oxoaluminate as a leaving group. The resulting iminium ion, a strong electrophile, is attacked by a second hydride ion to complete the formation of the 3° amine.

Figure 25.34 The mechanism of the $LiAlH_4$ reduction of amides.

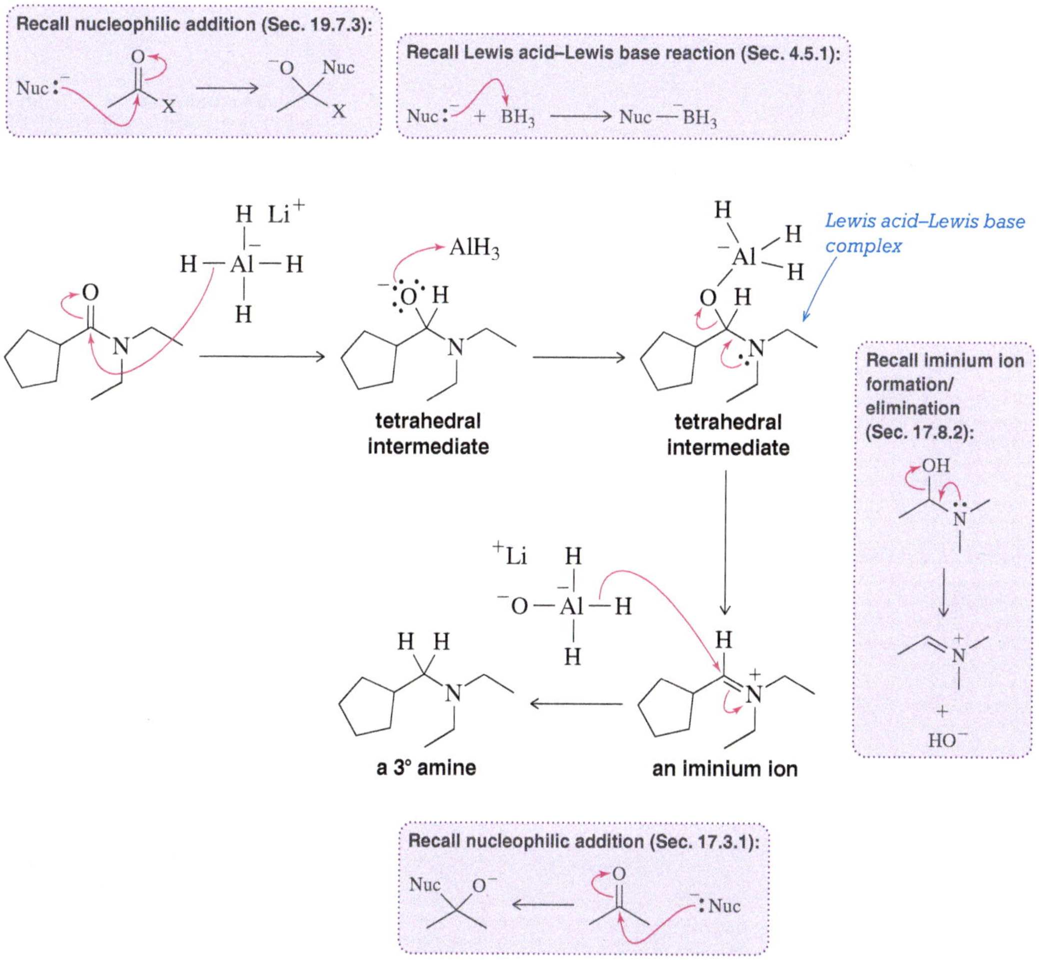

25.6.2 Reductive amination

Sticking with the theme of reusing previously studied reactions, reductive amination, shown in Figure 25.35, can be used for the synthesis of 1°, 2°, or 3° amines.

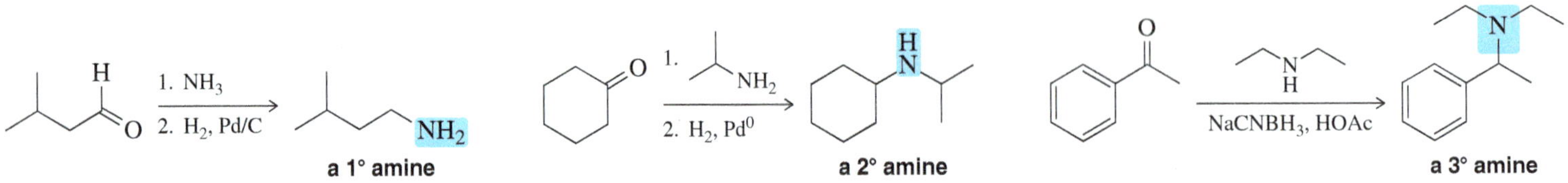

Figure 25.35 **Reductive amination of (a) 1° amines, (b) 2° amines, and (c) 3° amines.**

Reductive amination with ammonia (NH_3) sometimes results in overalkylation because the product amine can readily attack another carbonyl. Sound familiar?

In the first step of reductive amination, nucleophilic attack of an amine on a ketone proceeds by General Mechanism 17A (Figure 25.36; see also Figure 17.16 in Section 17.3.1). Having brought along its proton, two acid–base reactions result in the formation of a hemiaminal. The lone pair on nitrogen then collapses down, pushing out hydroxide, which immediately removes a proton from the iminium ion, resulting in imine formation. [*For a detailed review of this reaction, see Section 17.8.*]

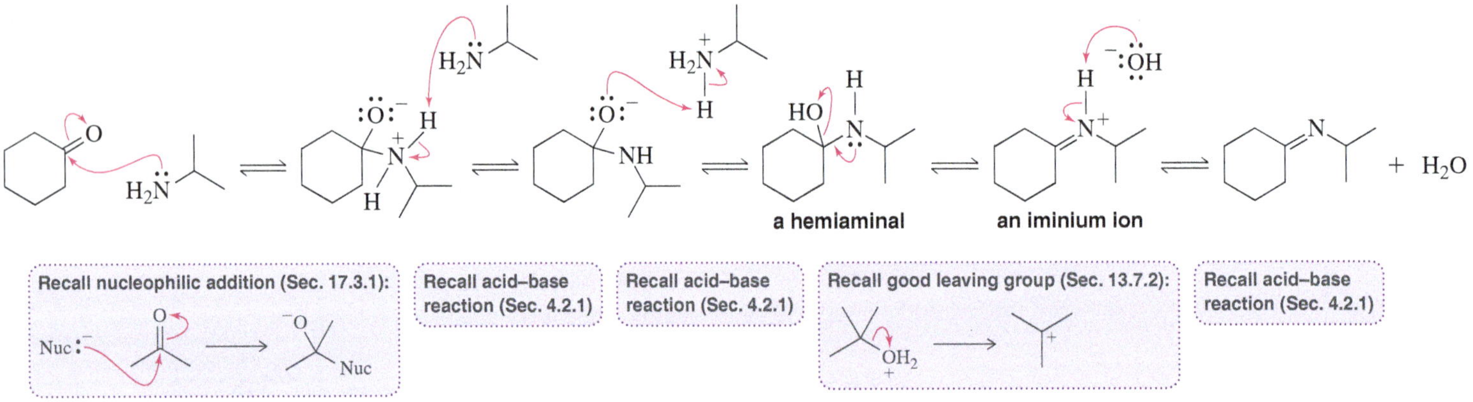

Figure 25.36 **The mechanism of imine formation** (see Section 17.8 for a review).

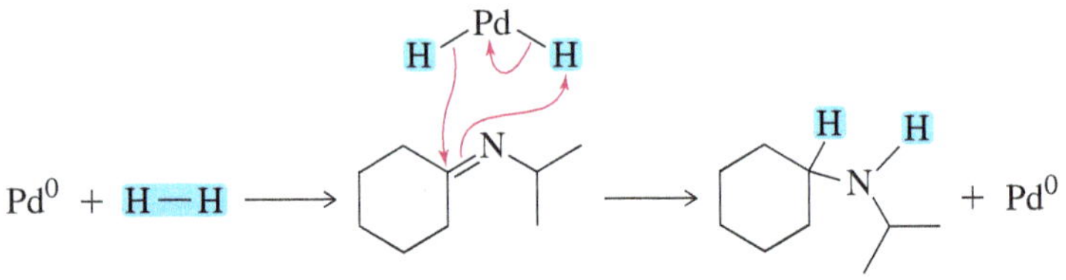

Figure 25.37 **The hydrogenation of an imine to give a 2° amine.**

Once formed, the imine can be reduced by hydrogenation in the presence of a palladium catalyst (Figure 25.37). This reaction goes by an identical mechanism to the reduction of other C—C (Sections 9.2 and 10.6.1) and C—O π bonds (Section 17.9.1). The result is a 2° amine.

Alternatively, sodium cyanoborohydride, a relatively weak hydride reducing agent, can be used with a mild acid to reduce protonated imines (Figure 25.38). Protonation of the imine makes it more electrophilic, allowing for transfer of the hydride from boron to the iminium ion. The result of this nucleophilic addition is a 2° amine.

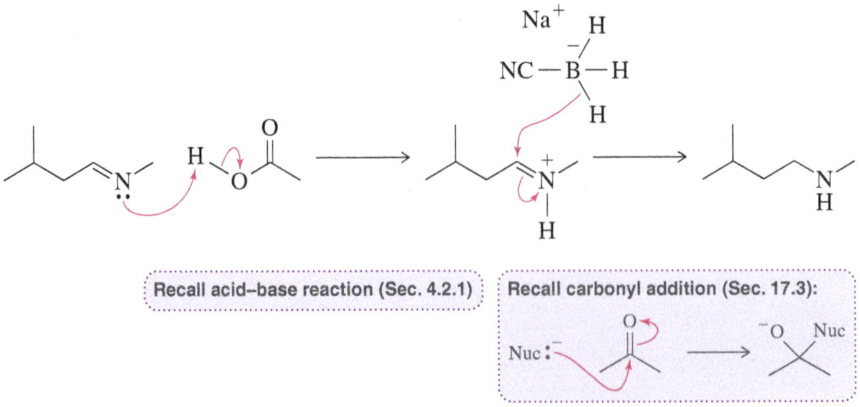

Figure 25.38 **The reduction of an iminium ion with sodium cyanoborohydride.**

Sodium cyanoborohydride is such a mild reducing agent that it does not reduce aldehydes and ketones, so reductive amination can be conducted in a single reaction flask (**Figure 25.39**). Sodium cyanoborohydride simply waits around for the iminium ion to form before participating in the reaction.

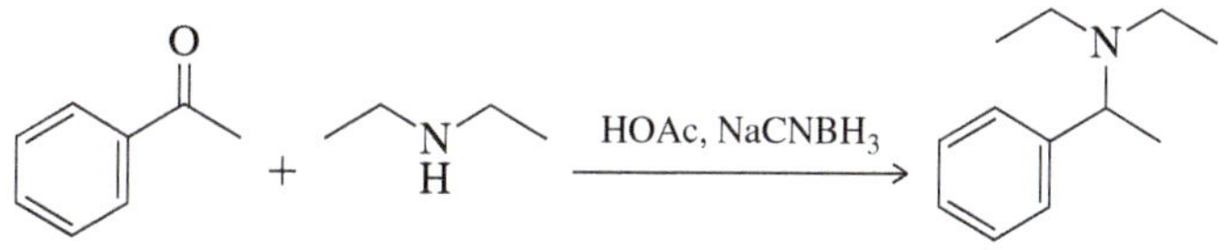

Figure 25.39 One-pot reductive amination.

25-C CHEMISTRY IN CONTEXT: Green Chemistry

Synthesis of sertraline (Zoloft®)

Sertraline is yet another example of a selective serotonin reuptake inhibitor, the key ingredient in the antidepressant Zoloft. Two routes have been used for its development, the second of which has received accolades for its environmental friendliness. Both routes included a reductive amination step. In the original route [**Figure 25.40(a)**], imine formation was promoted by titanium(IV) chloride, a highly reactive Lewis acid. Upon isolating and purifying the imine, catalytic hydrogenation produced a racemic mixture of the cis and trans stereoisomers. Chiral resolution of the enantiomers resulted in Zoloft. In the redesigned route [**Figure 25.40(b)**], the reductive amination and chiral resolution were carried out in the same flask, using no toxic reagents, employing ethanol as solvent, and reducing the amount of waste produced. Finally, the palladium-catalyzed reduction resulted in greater selectivity, thus reducing the amount of trans isomer that would be discarded as waste.

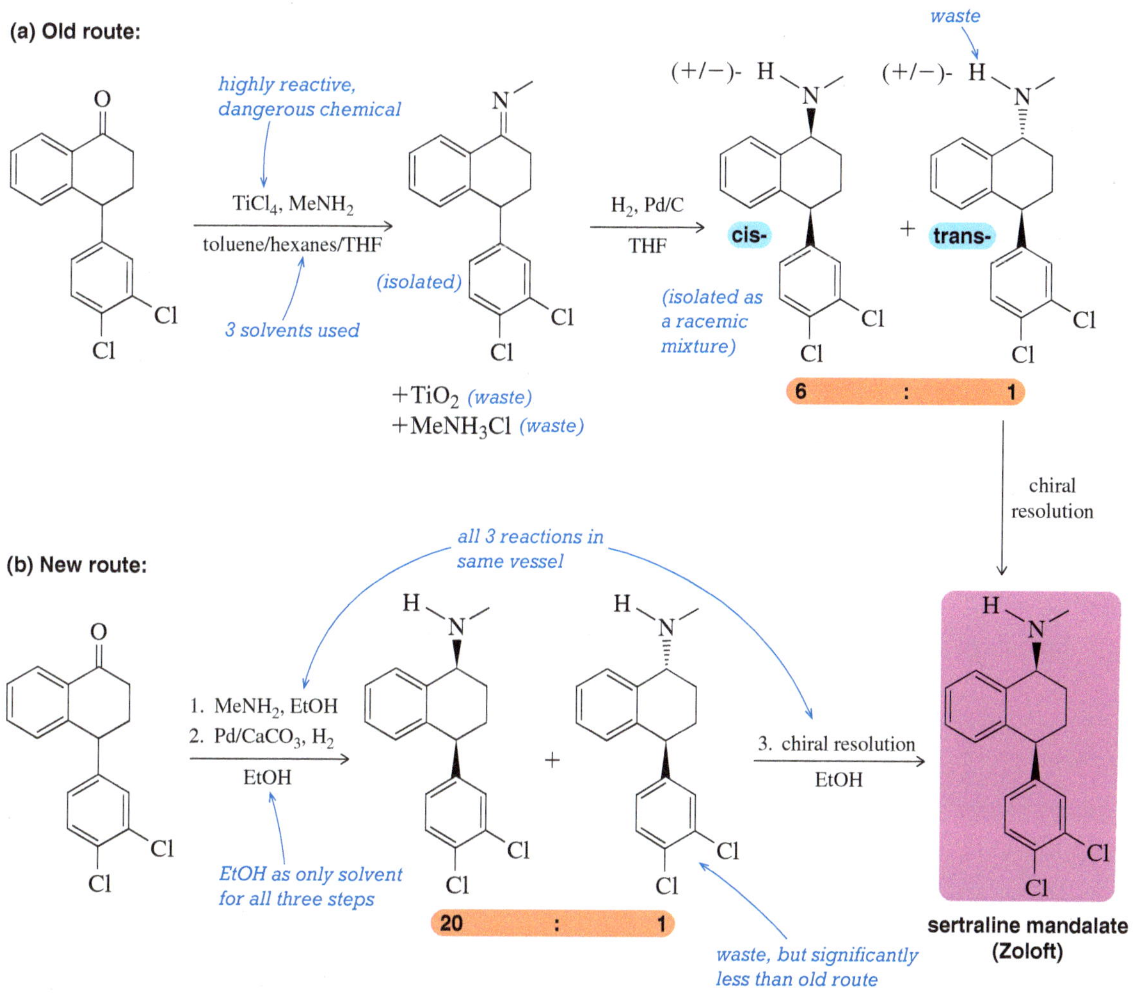

Figure 25.40 The synthesis of sertraline via (a) the original route and (b) a green route that generates less solvent and solid $TiO_2/MeNH_2/HCl$ waste.

(*continued*)

The redesigned route, which received a Presidential Green Chemistry Challenge Award from the EPA in 2002, eliminated the need for 310,000 pounds (141,000 kg) of $TiCl_4$ per year, which enhanced worker safety (GCP 12). By streamlining the three steps in ethanol [a renewable (GCP 7) and safe (GCP 5) solvent], over 74,000 L of solvent waste was eliminated per 1000 kg of sertraline produced. Furthermore, the new route eliminated 1.5 million pounds (680,000 kg) of sodium hydroxide, hydrochloric acid, and titanium dioxide waste per year (GCP 1). Moreover, the improved product ratio of step 3 reduces the amount of raw material needed for sertraline synthesis.

Assessment 25.24 Suggest a mechanism for the reductive amination in Figure 25.40(b).

25.6.3 Reduction of nitro groups

Consistent with previous reactions [*seems like they are all reductions*], a nitro group can be reduced to give an amine, as shown in **Figure 25.41**. Because the nitro group is relatively stable, high pressures are commonly used in these reactions. [*The reaction in Figure 25.41 is one of the first reactions I ever ran in a research setting.*]

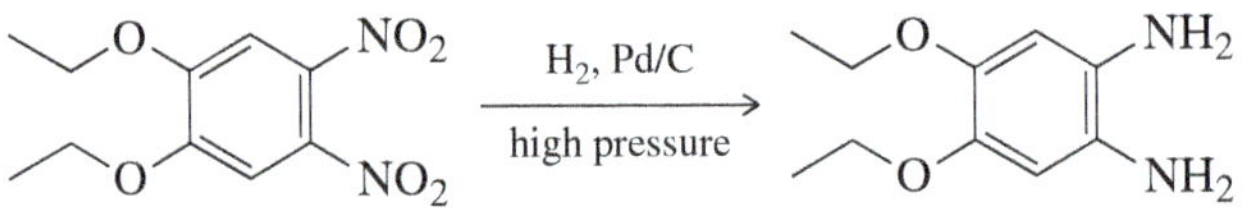

Figure 25.41 The reduction of nitro groups to amines via palladium-catalyzed hydrogenation.

Like the other reactions we have studied, the reduction of a nitro group is only useful if we have ways of incorporating it into a molecule. The Henry reaction is the crossed aldol reaction of nitroalkenes with aldehydes or ketones to form new C—C bonds (Assessment 20.95). [*See Section 20.3.3 for a detailed review of crossed aldol reactions.*] Deprotonation of the nitroalkane ($pK_a = 10$) is easily done by hydroxide to make what is essentially a nitro enolate (**Figure 25.42**). The nitro enolate adds as a nucleophile to a carbonyl. Protonation of the resulting alkoxide gives the nitro aldol product. Reduction of the nitro group makes it possible to form 1,2-amino alcohols, which are useful intermediates in drug synthesis.

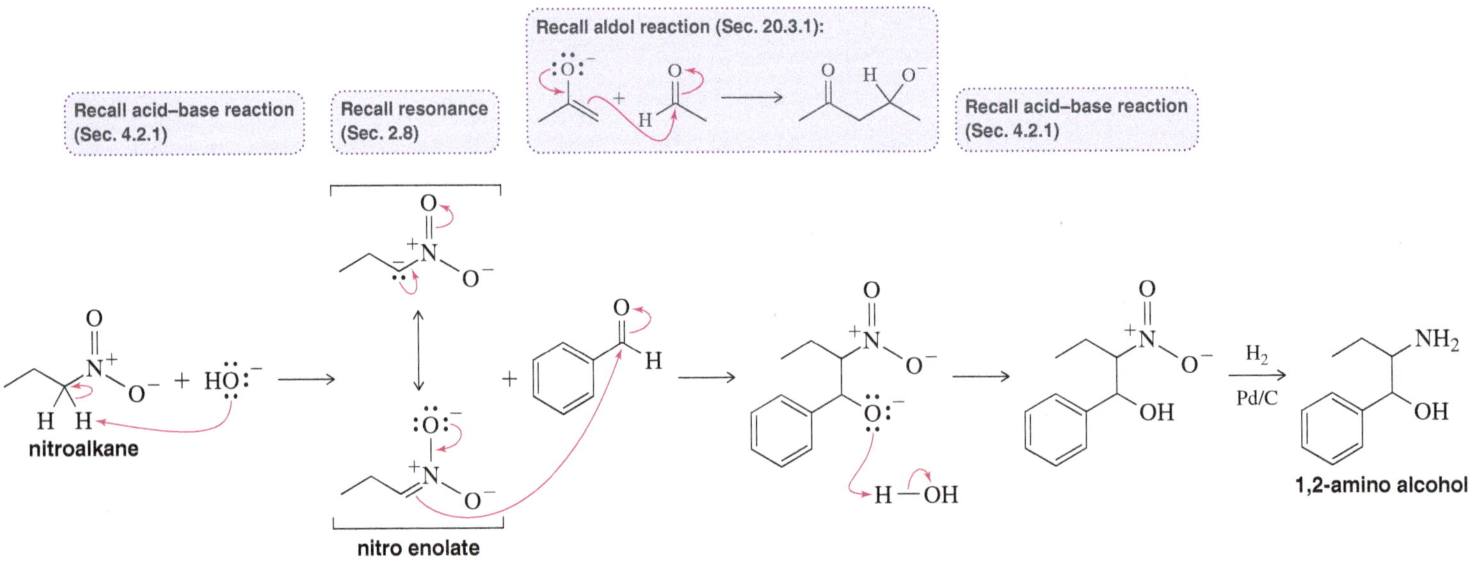

Figure 25.42 A Henry reaction, followed by reduction, gives 1,2-amino alcohols.

Electrophilic aromatic substitution is the other way we have studied for incorporating the nitro group into molecules. Replacement of a hydrogen with a nitro group on an aromatic ring occurs by reacting nitric acid with sulfuric acid to create the highly electrophilic NO_2^+ ion. As shown in **Figure 25.43**, the reaction begins with protonation of the hydroxyl group of nitric acid to make an excellent leaving group. Water then leaves with assistance by the neighboring alkoxide to create the highly electrophilic nitronium ion NO_2^+. The nitronium is attacked by the π electrons from benzene. In the second step, H_2O or HSO_4^- (the conjugate base of H_2SO_4) removes a proton, regenerating benzene and reestablishing aromaticity in a fast and highly exothermic step. Reduction of the resulting nitrobenzene gives an aminobenzene. [*See Section 23.7 for a review of electrophilic aromatic substitution.*]

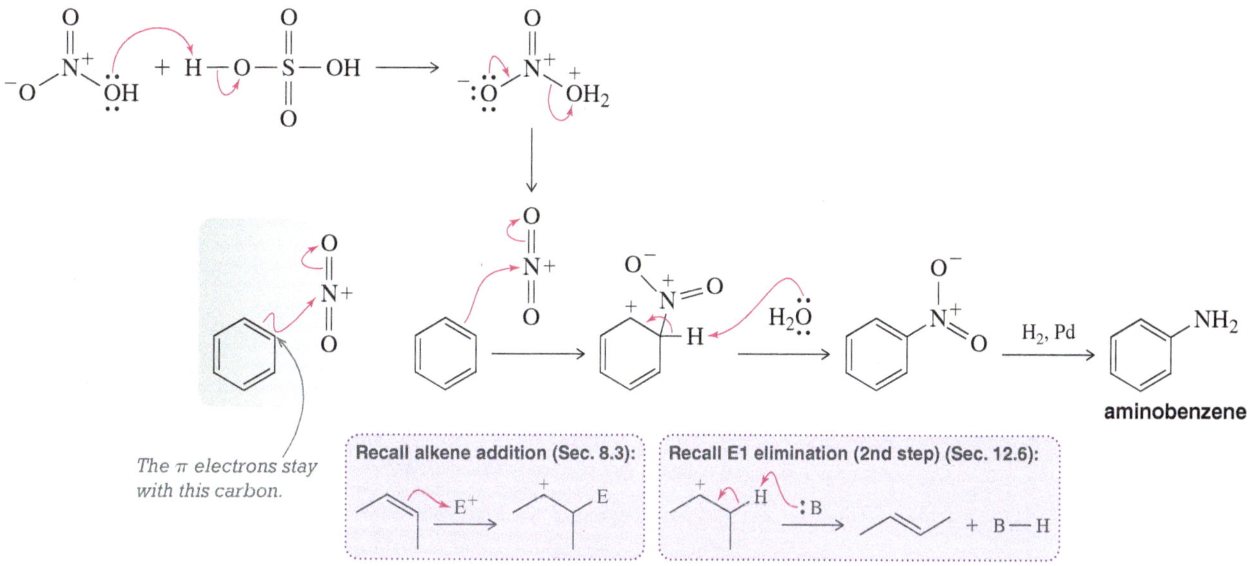

Figure 25.43 Electrophilic aromatic substitution, followed by reduction, produces an aminobenzene.

25.6.4 Buchwald–Hartwig amination

The transition metal–catalyzed reactions of Chapter 16 can also be used for the synthesis of aromatic amines, as shown in **Figure 25.44**. In this reaction a C—X bond is replaced with a new C—N bond, typically employing palladium as the catalyst. This reaction is referred to as the Buchwald–Hartwig amination after Stephen L. Buchwald and John F. Hartwig, the chemists who developed it.

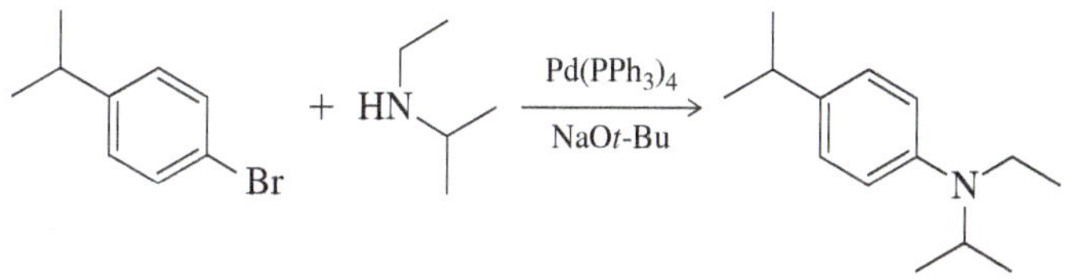

Figure 25.44 The Buchwald–Hartwig amination.

As with all palladium-catalyzed transformations [*see Section 16.4 for a detailed review*], there are several substrate and condition-dependent mechanistic subtleties that are beyond the scope of this course. In spite of that, a reasonable mechanism, consistent with what we learned previously, is shown in **Figure 25.45**. Following *oxidative addition* of $Pd(PPh_3)_4$ to the C—Br σ bond, an organopalladium halide is produced where palladium is now in the +2 oxidation state. Being an electrophile, palladium(II) is attacked by the nucleophilic amine, displacing the bromide. At the same time, sodium *t*-butoxide deprotonates the amine, forming *t*-butanol. This step is referred to as the amine ligation step, and it is analogous to the transmetallation in earlier organometallic coupling reactions. *Reductive elimination* results in the coupling of the organic molecule and the amine to produce a new C—N σ bond, while regenerating $Pd(PPh_3)_4$ (0 oxidation state) for continued use in the catalytic cycle.

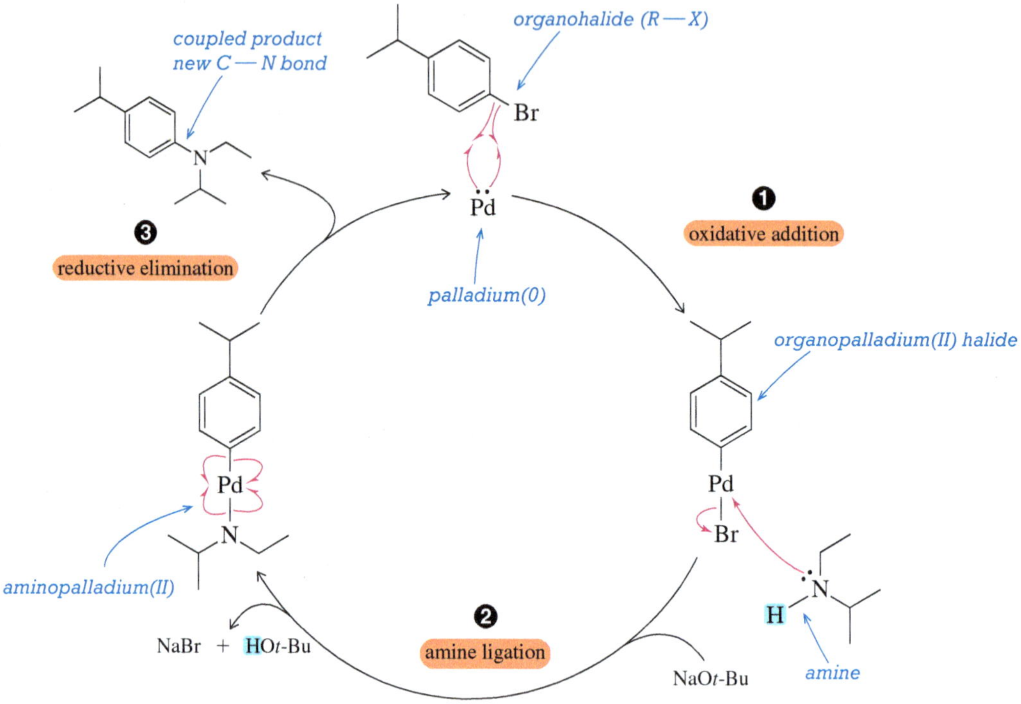

Figure 25.45 The catalytic cycle of Buchwald–Hartwig amination. [*PPh$_3$ ligands have been removed for clarity.*]

Assessment 25.25 Predict the product of the following reactions.

(a)

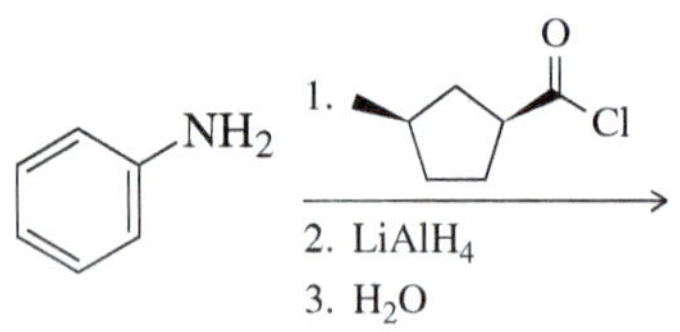

(b)

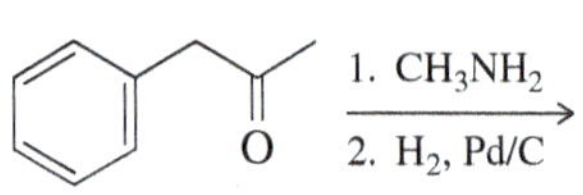

(c)

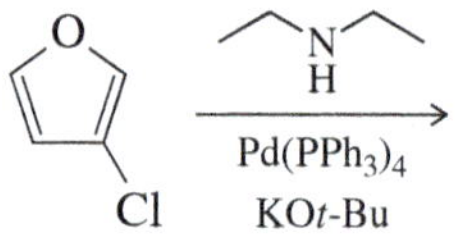

(d)

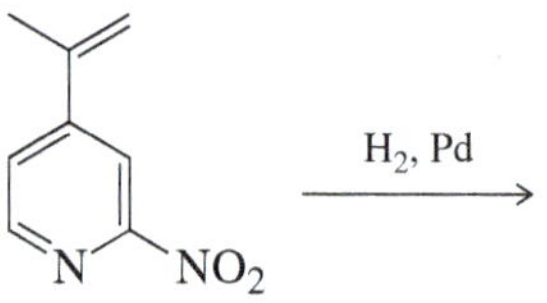

Assessment 25.26 Provide a mechanism for the one-pot reductive amination in Figure 25.39.

Assessment 25.27 Suggest a method for synthesizing the amine shown, starting with phenol. There may be several correct answers.

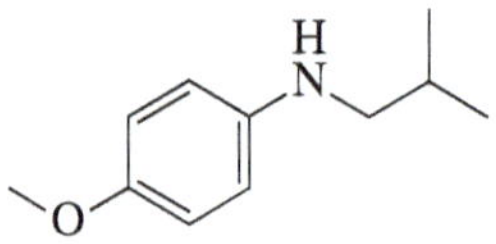

Assessment 25.28 Given the starting reactant, suggest a method for synthesizing the amine on the right.

(a)

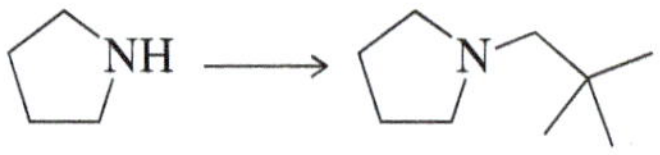

(b)

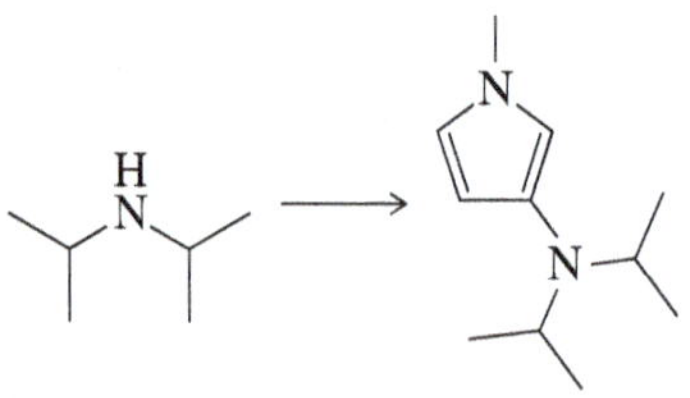

Assessment 25.29 The nitro group directs electrophilic aromatic substitution to the meta position. After reduction by hydrogenation, to which carbon(s) does it direct?

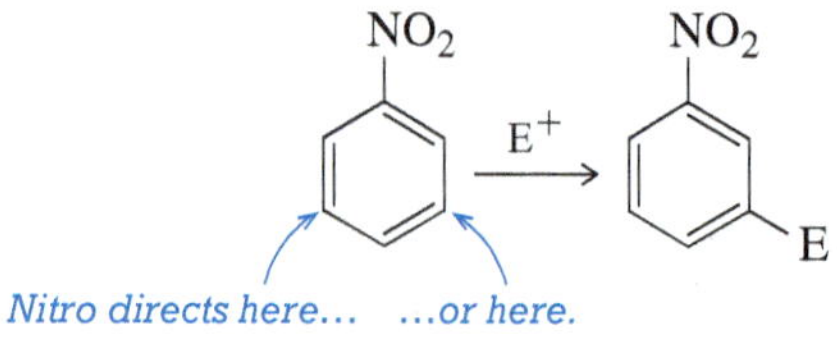

25.7 Reactions of amines

Owing to the electron-rich nitrogen and its diminished electronegativity relative to oxygen, nearly every reaction of amines that we discuss involves it first acting as a base or a nucleophile. In the same way that seeing a phosphorus in a reaction tells us to look for a way to bond it with oxygen, seeing a nitrogen tells us to consider how it can act as a Brønsted–Lowry base or a nucleophile (Lewis base). Then look for Brønsted–Lowry acids or electrophiles (Lewis acids) for the amine to react with. [*Remember this as you read through the next several sections.*]

25.7.1 Acid–base reactions

As discussed in Section 25.4.2, amines are moderately basic, with the proton of the conjugate acid (an ammonium ion) being only weakly acidic ($pK_a = 10$). As a result, amines undergo simple acid–base reactions to make ammonium ions. Mechanistically, these proceed like all acid–base reactions, with the electron-rich base attacking the proton and pushing electrons onto the conjugate base. We have seen amines used in this way as bases in a variety of important reactions. Two of these, tosylate formation (Section 13.7.1) and alcohol protection (Section 13.14), are shown in **Figure 25.46**. In each of these, the amine is not strong enough to deprotonate the original nucleophile. [*For the first introduction of using an amine base in this way, see Section 13.7.1.*]

(a)

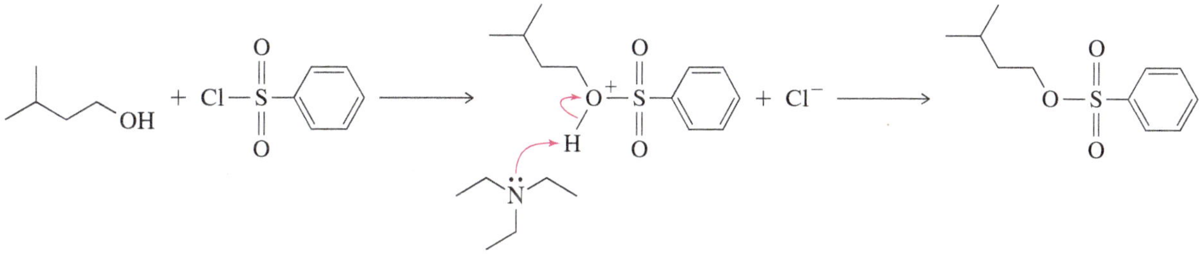

(b)

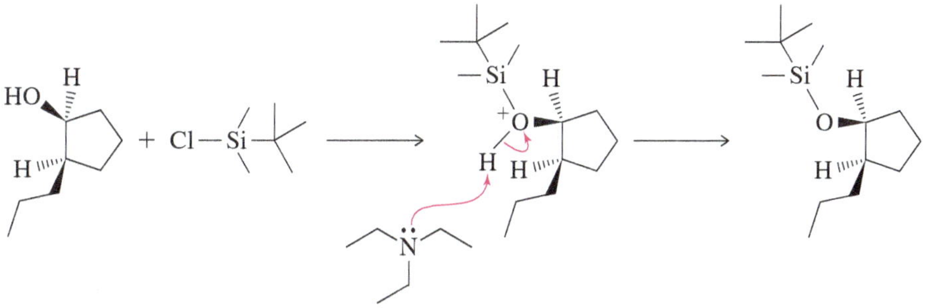

Figure 25.46 Tertiary (3°) amines are weak, non-nucleophilic bases. They can be used to (a) form a tosylate and (b) protect an alcohol as a silyl ether.

The basicity of amines allows for experimental protocols to be designed that allow an amine to be separated from nonpolar, nonbasic compounds. Insoluble amines can be made more soluble in water by protonation to form ammonium salts. 1-Octanamine, for example, has a solubility of only 0.3 g/L H_2O (**Figure 25.47**). After being protonated by hydrochloric acid, however, the resulting octylammonium hydrochloride is much more soluble in water, because it is now a more polar, ionic salt.

Figure 25.47 The protonation of amines makes them more soluble in water.

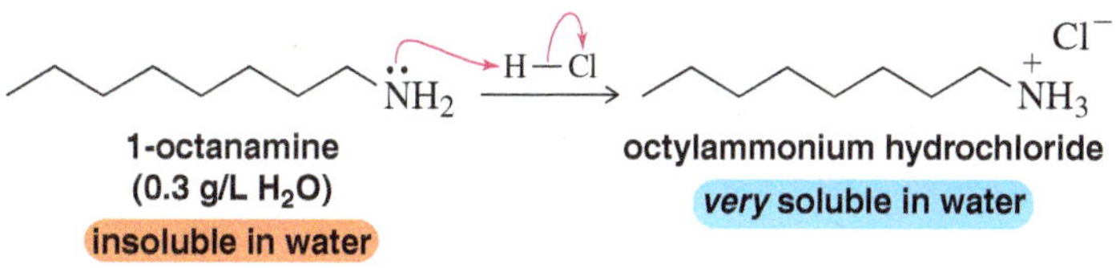

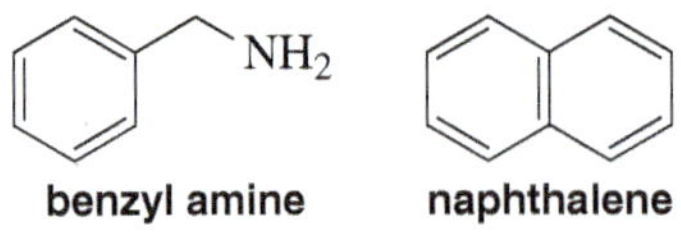

How might you separate a mixture of benzyl amine and the neutral, nonpolar compound naphthalene (labeled as "other organics" in Figure 25.48)? [*This is a common experiment—one you might have already performed in lab.*] When the amine is neutral, both compounds are very soluble in an organic solvent like diethyl ether. By extracting the diethyl ether with an aqueous solution of hydrochloric acid, the protonated ammonium salt moves preferentially to the denser aqueous layer. The two layers can be separated, leaving two solutions. One contains the neutral naphthalene dissolved in ether. The second contains the ammonium hydrochloride salt dissolved in water. To generate pure benzyl amine, the acidic aqueous layer is neutralized with NaOH, regenerating benzylamine (which is insoluble in water). Extraction of this aqueous layer with diethyl ether moves the benzyl amine into the diethyl ether layer. Separation of the two layers and evaporation of the ether result in pure benzyl amine. [*A nearly identical procedure was used in Section 18.6.1 to separate carboxylic acids and neutral compounds.*]

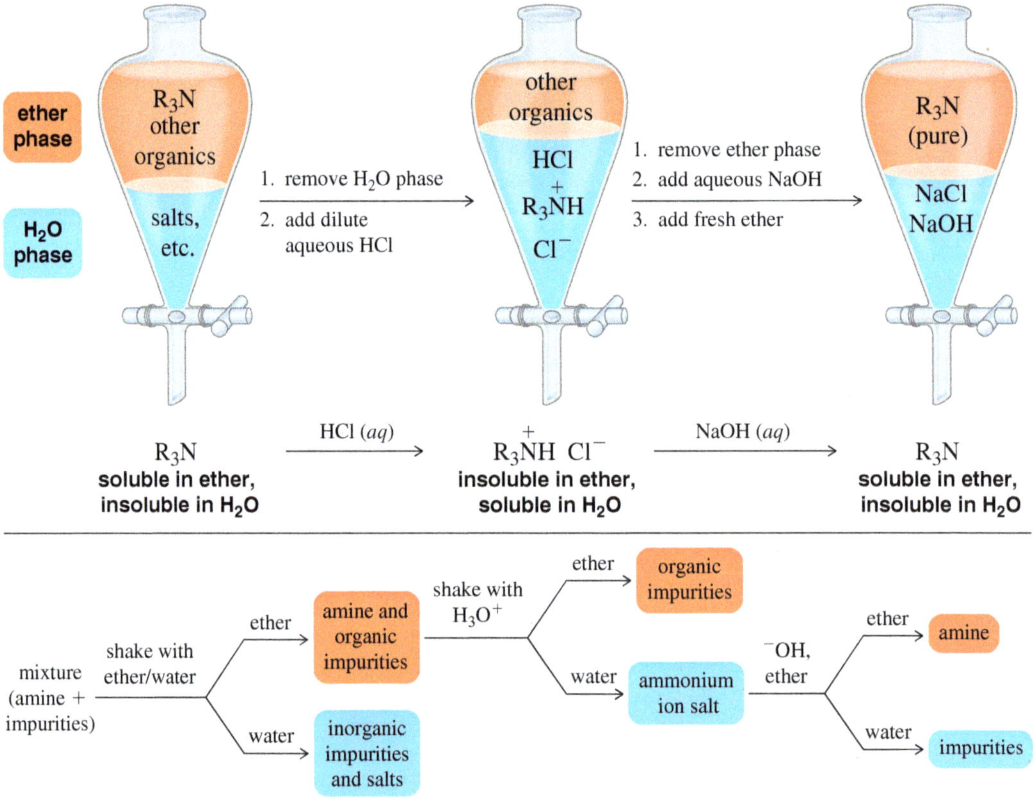

Figure 25.48 The separation of amines from neutral compounds via acid–base extraction.

Assessment 25.30 Design a scheme to separate the three compounds shown here.

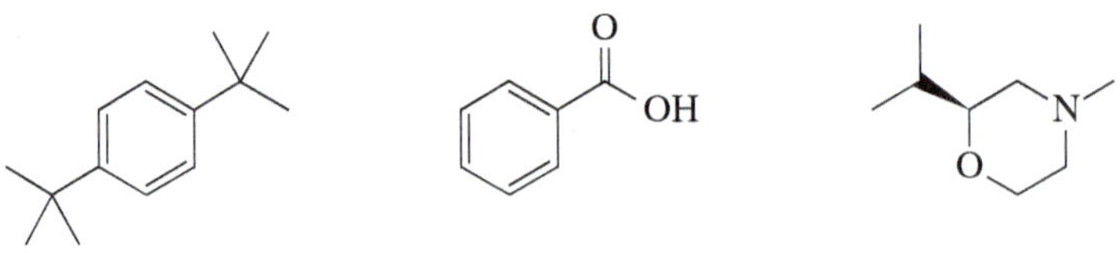

Assessment 25.31 Following a $LiAlH_4$ reduction, an IR spectrum suggested that the product was a mixture of the starting amide and the desired amine. (a) How might you separate these two compounds using an acid–base extraction? (b) Once separated, how could you distinguish between the amide and the amine using IR spectroscopy?

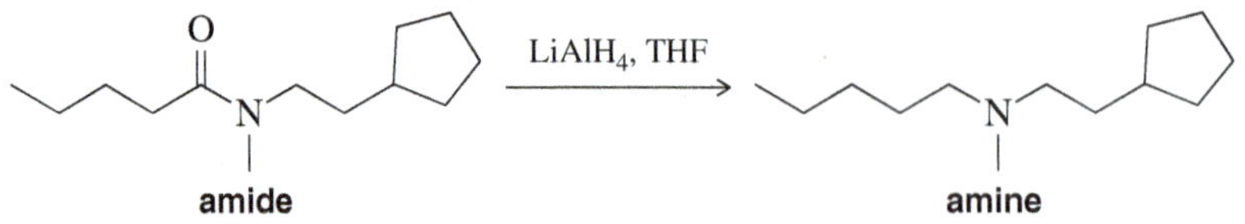

To help you better understand the new reactions of amines, or to help you study for an upcoming comprehensive final exam, you may want to look back at these reactions. The section numbers where they originally appeared are in the table.

25.7.2 Previous reactions of amines

Just as our journey through the previous 24 chapters gave us insights into how to synthesize amines, we have seen amines in other contexts, such as imine and enamine formation, participation in substitution and elimination reactions, and amide synthesis, among others. These reactions are summarized in Table 25.7.

Table 25.7 Previous reactions of amines.

Formation of an imine (Section 17.8)	
 a 1° amine; an imine	Ammonia and 1° amines react as nucleophiles with ketones and aldehydes to give imines, driving off water in the process. Because nitrogen is a moderately strong nucleophile, this reaction can occur with or without acid catalysis.
Formation of an enamine (Section 17.8.3)	
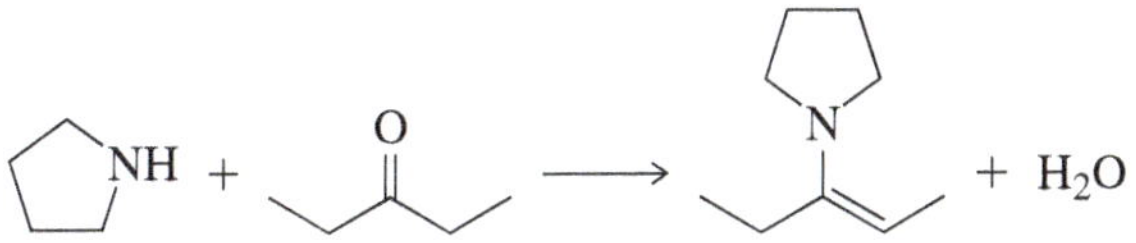 a 2° amine; an enamine	Secondary (2°) amines react as nucleophiles with ketones and aldehydes to give enamines, driving off water in the process. Because nitrogen is a moderately strong nucleophile, this reaction can occur with or without acid catalysis.
Amide formation with carboxylic acid derivatives (Section 19.7.3)	
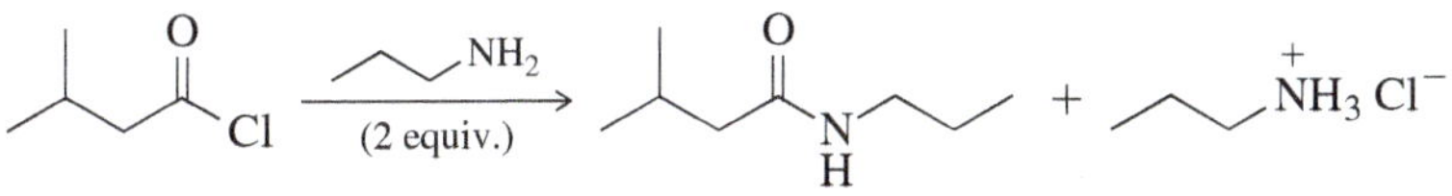	Ammonia, 1° amines, and 2° amines react as nucleophiles with acid chlorides (quickly!), acid anhydrides (quickly), and esters (slowly) to form the more stable amide. Proceeding through a tetrahedral intermediate, two equivalents of amine are necessary to drive the reaction to completion.

Assessment 25.32 Predict the product of the following reactions.

(a) (b) (c)

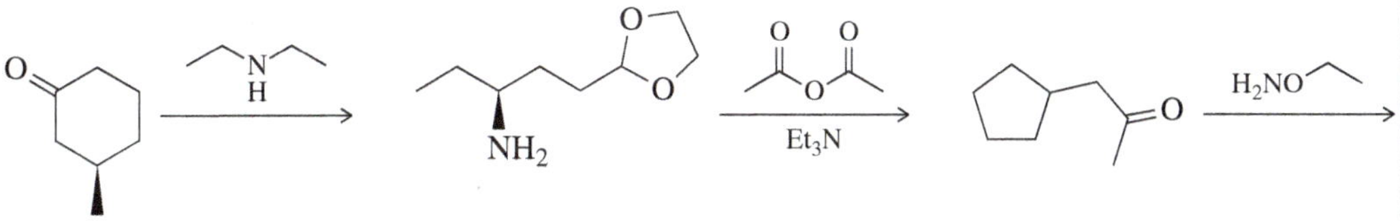

Assessment 25.33 Suggest a mechanism of the following cyclization reaction.

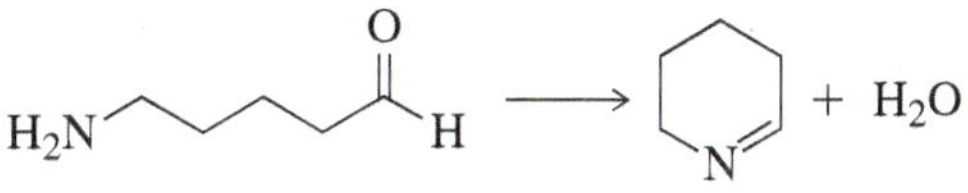

Assessment 25.34 The substitution of amines for alkoxy groups in esters is a slow reaction, requiring heating to ensure complete reaction. Why?

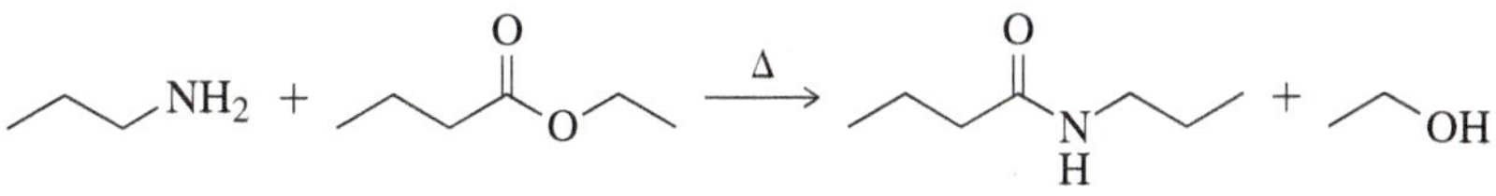

Assessment 25.35 Suggest a synthesis of *N*,*N*-dimethylbenzenesulfonamide, starting with dimethylamine.

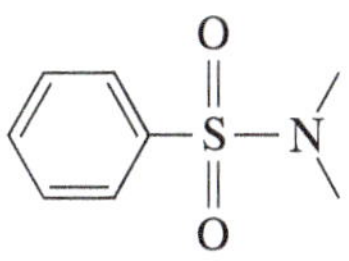

25-D CHEMISTRY IN CONTEXT: Medicinal Chemistry

Synthesis of amoxapine, a TCA

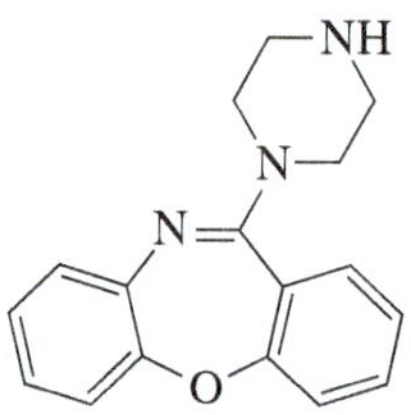

amoxapine

Despite containing four rings, amoxapine is classified as a tricyclic antidepressant (TCA). Tricyclic antidepressants were some of the earliest used antidepressant medications. Debuting in the late 1950s, TCAs operate by a variety of mechanisms. Amoxapine, for example, is believed to inhibit the reuptake of both serotonin and norepinephrine. Compared to other antidepressants, amoxapine's distinguishing feature is a faster onset of action, with most patients responding to treatment within a week.

Incorporation of the piperazine (amoxapine's fourth ring—the one containing both N and O) takes advantage of the nucleophilic acyl substitution of the carbamate functional group. In this reaction, the carbamate reacts similarly to the carboxylic acid derivatives of Chapter 19. Following substitution, $POCl_3$ is used in an amide dehydrative cyclization to form the third ring that makes this a TCA (**Figure 25.49**).

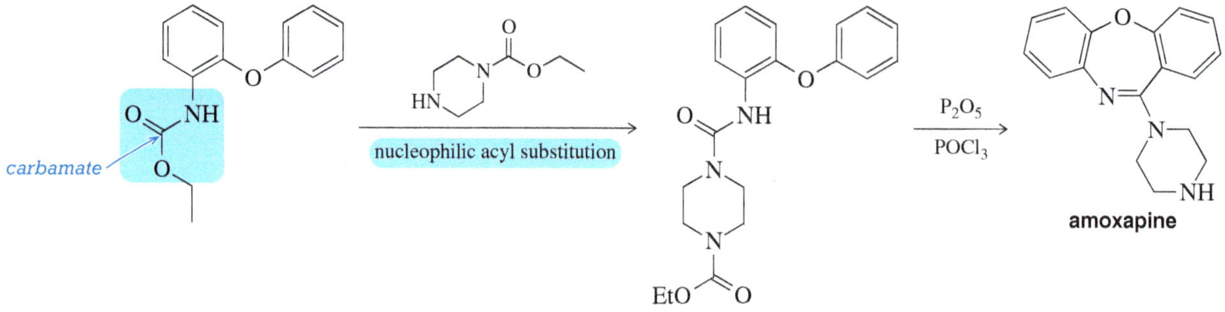

Figure 25.49 Acyl substitution for the synthesis of amoxapine.

Assessment 25.36 Suggest a mechanism for only the cyclization shown in Figure 25.49. [*Recall that $POCl_3$ is an electrophile that loves oxygen.*]

25.7.3 Hofmann elimination

In 1851 [*Wow! These classical chemists continue to amaze me.*], August von Hofmann treated a 3° dimethyl amine with iodomethane to form a tetraalkylammonium iodide. Treatment of the tetraalkylammonium iodide with base, as in **Figure 25.50**, resulted in an E2 elimination reaction.

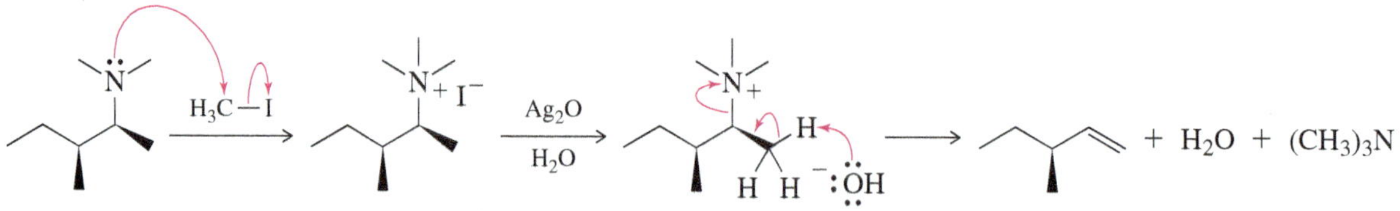

Figure 25.50 Tetralkylammonium ions undergo elimination reactions with a strong base.

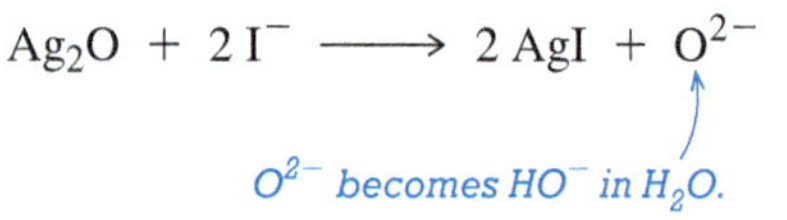

Figure 25.51 Silver oxide reacts with iodide to form OH^-, a strong base.

There are a couple of confusing aspects of the Hofmann elimination reaction. First, where's the base? Keeping in mind that this was discovered in 1851, silver oxide is a proxy for the base. Recall from general chemistry's solubility rules [*going way retro . . .*] that the silver cation (Ag^+) loves iodide (I^-), so iodide is stripped away from the tetraalkylammonium ion to form the stable AgI (**Figure 25.51**) [*the ball continues to roll downhill*]. In the process, O^{2-} would be created, except that the reaction is in water, so protonation of O^{2-} gives hydroxide, which is actually the strong base in this reaction. So even though hydroxide could be used in this reaction, it's much cooler (and much more expensive) to use silver oxide.

The second interesting aspect of this E2 elimination is that there were two possible alkenes that could have formed. [*Look back at Figure 25.50 to identify the other alkene that could form.*] Usually, we'd talk here about how the most substituted (most stable) alkene forms in an E2 elimination; that was Zaitsev's rule (Section 12.7.1). That was not the outcome of this reaction, though, as reiterated in **Figure 25.52** for emphasis. Instead, the less stable, monosubstituted alkene formed preferentially. [*Does the ball no longer roll downhill?*]

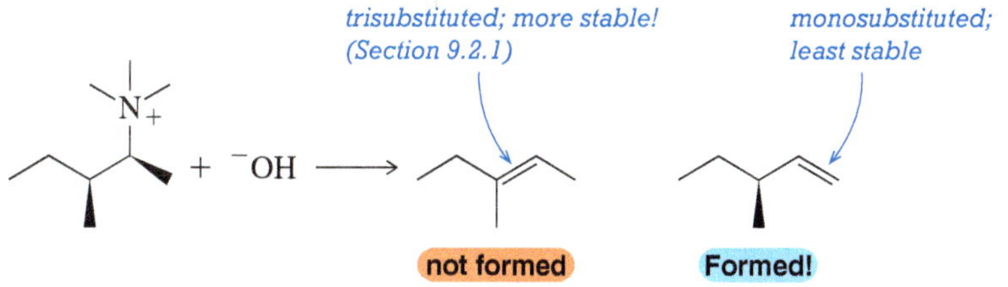

Figure 25.52 The Hofmann elimination makes the least stable alkene.

Before we toss everything we thought we knew about E2 elimination, let's think like a chemist—you are one by now—and see if we can figure out why this reaction diverges from what we expected.

Thinking Like a Chemist

First, let's remember everything we know about E2 eliminations and compare it with what's happening here. Is there a good leaving group? [*R_3N, check.*] A strong base? [*HO^-, by way of Ag_2O, check.*] Concerted elimination? [*Yep!*] Using an anti and coplanar hydrogen? [*Let's look closer at that.*]

To understand why this reaction does *not* give the most stable alkene (Zaitsev product), let's look carefully at the transition state that would have produced it. As shown in **Figure 25.53**, removal of the anti and coplanar hydrogen from C_3 would result in electrons collapsing to C_2, eliminating the ammonium ion to form a new C—C π bond. Using our line-angle drawings, this looks straightforward and reasonable [Figure 25.53(a)]. Based on the product in Figure 25.52, we know this reaction doesn't happen. What if we look at it from the side, using a Newman projection down the C_3—C_2 σ bond [Figure 25.53(b)]? Do you see a reason that this might not happen?

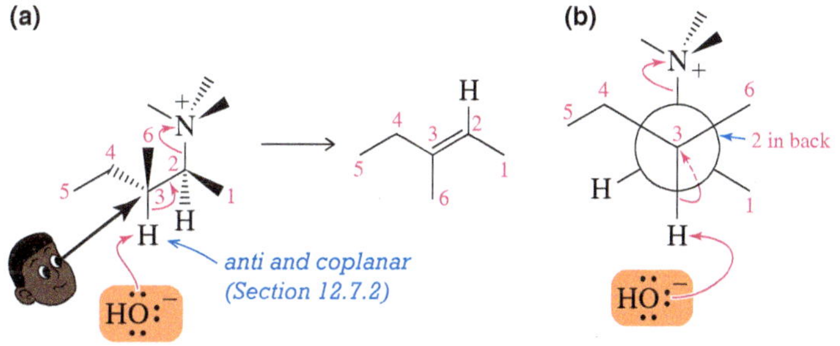

Figure 25.53 E2 elimination of a tetraalkylammonium ion. (a) Anti and coplanar elimination demonstrated with (b) a Newman projection suggests a reason it does not occur this way. [*Can you figure out why? Read on.*]

The trimethylammonium leaving group, unlike any we have seen before, is quite bulky. In the Newman projection leading to formation of the more stable Zaitsev product, there are two significant, destabilizing, gauche interactions [**Figure 25.54(a)**]. One is between the trimethylammonium and the C_6 methyl group; the other is between the trimethylammonium and the C_4/C_5 ethyl group. These interactions make it impossible to achieve the anti and coplanar conformation required for elimination by a concerted E2 mechanism. Alternatively, the Newman projection in **Figure 25.54(b)** has zero gauche interactions. As a result, the anti and coplanar arrangement of the hydrogen and the leaving group is more easily accessed, with E2 elimination occurring through this conformation.

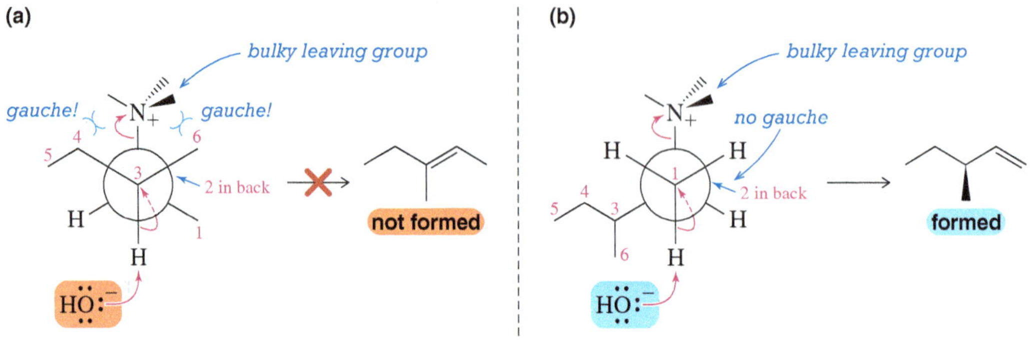

Figure 25.54 Conformational analysis of the Hofmann elimination. (a) The more substituted alkene would form through an inaccessible conformation, whereas (b) the conformation leading to the less substituted alkene is accessible.

In conclusion, the ball still rolls downhill, but only when the conformation of the reactive molecules will allow it. ■

And so the distinguishing feature of the Hofmann elimination—the thing that makes it a useful addition to the synthetic toolbox—is that, unlike the other elimination reactions we've studied, it gives us a way to reliably make less substituted alkenes. Two additional examples are shown in **Figure 25.55**.

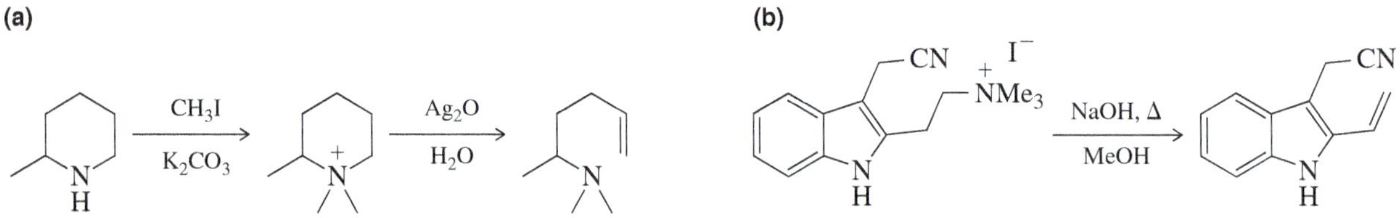

Figure 25.55 Selected examples of the Hofmann elimination. In (a), the least substituted alkene forms preferentially.

Assessment 25.37 Predict the product of the following Hofmann elimination reactions.

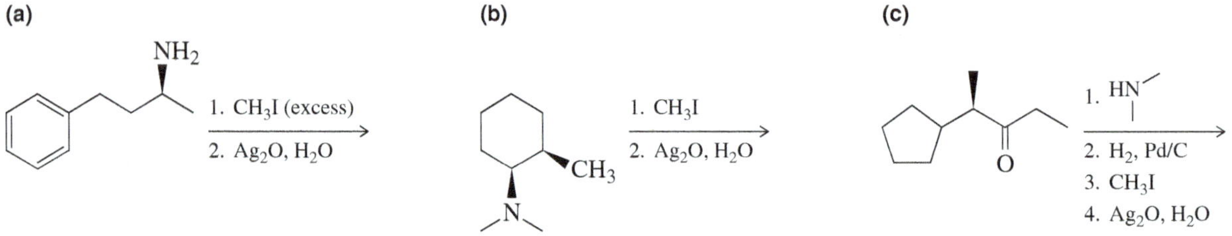

Assessment 25.38 SPECTROSCOPY INSIGHT Using IR spectroscopy, how could you distinguish between the two possible alkenes formed in the following reaction?

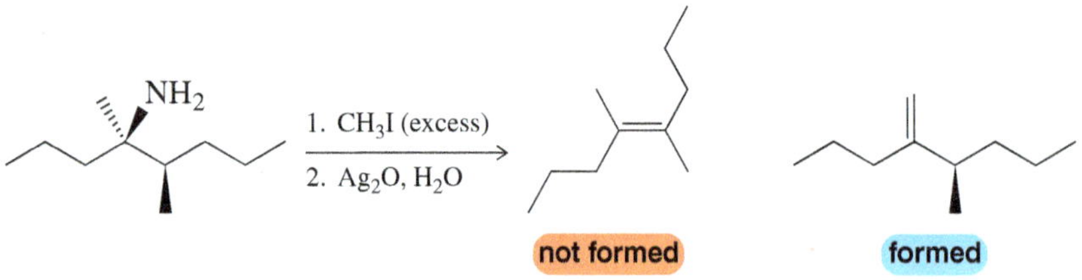

25.7.4 Cope elimination

The Cope elimination is the thermal elimination of a trialkylammonium oxide formed by the reaction of a 3° amine with a peroxide (**Figure 25.56**). Similar to the Hofmann elimination, the less substituted alkene is produced preferentially.

Figure 25.56 The Cope elimination of *N*-oxides.

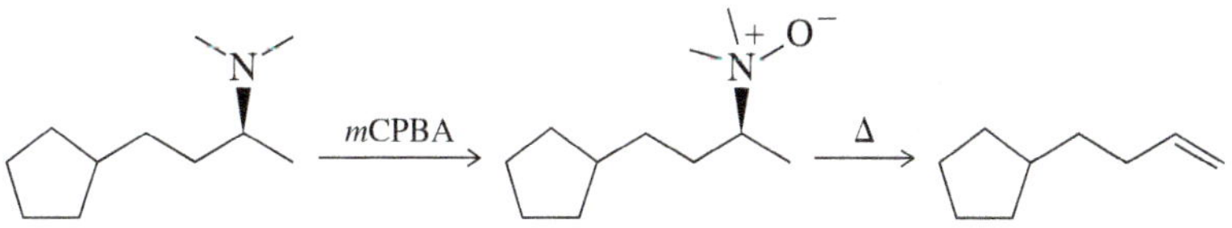

Mechanistically, formation of the *N*-oxide begins with nucleophilic attack by the amine on the terminal oxygen of *m*CPBA. Recall that the O—O σ bond is extremely weak, and the resulting carboxylate is stabilized by resonance with the carbonyl. With a $pK_a \approx 5$, the resulting acid is deprotonated by the carboxylate to give the *N*-oxide [**Figure 25.57(a)**]. This reaction should be reminiscent of alkene epoxidation through the butterfly transition state (Section 9.1.4). As such, we can imagine this one being similarly concerted because a slow first step is followed by a rapid, intermolecular deprotonation [**Figure 25.57(b)**].

Figure 25.57 The oxidation of a 3° amine with *m*CPBA to form an *N*-oxide, depicted in both (a) a stepwise and (b) a concerted fashion.

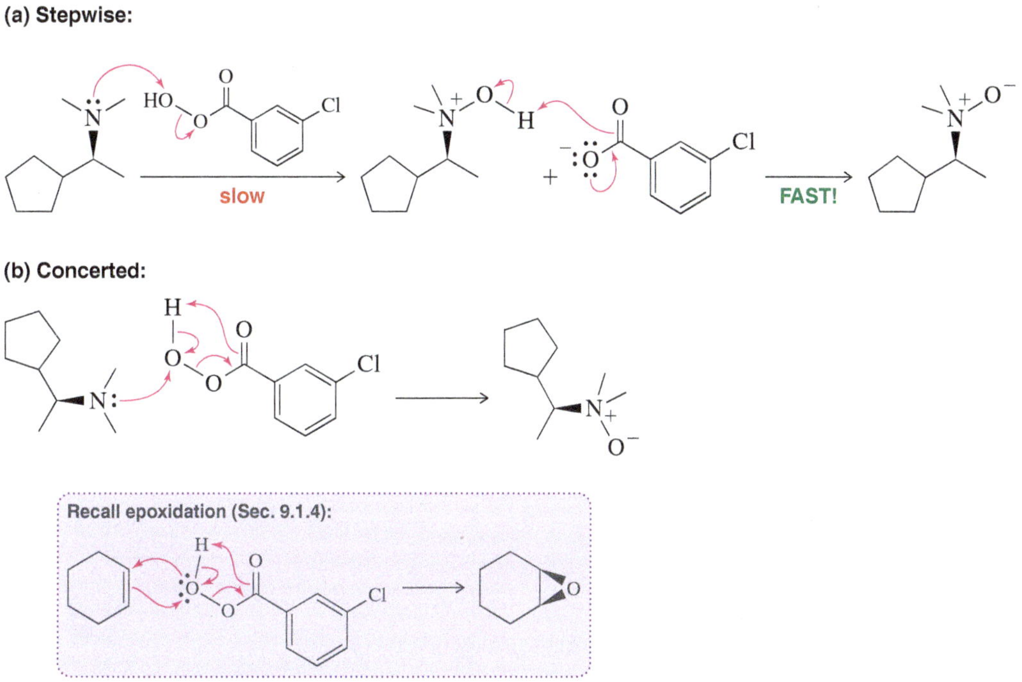

Once formed, heating of the *N*-oxide results in a spontaneous, concerted elimination. In contrast to the E2 elimination, which occurs through an anti and coplanar transition state, the Cope elimination is a syn elimination (**Figure 25.58**). Because it is an intramolecular reaction, the oxide can only reach a hydrogen on the same side of the molecule in the concerted 6-electron cyclic transition state. [*For a review of pericyclic reactions, see Chapter 22.*] Because of the steric bulk of the leaving group, the Cope elimination, like the Hofmann elimination, preferentially forms the least substituted alkene.

Figure 25.58 The Cope elimination is a syn elimination. The least substituted alkene is formed preferentially.

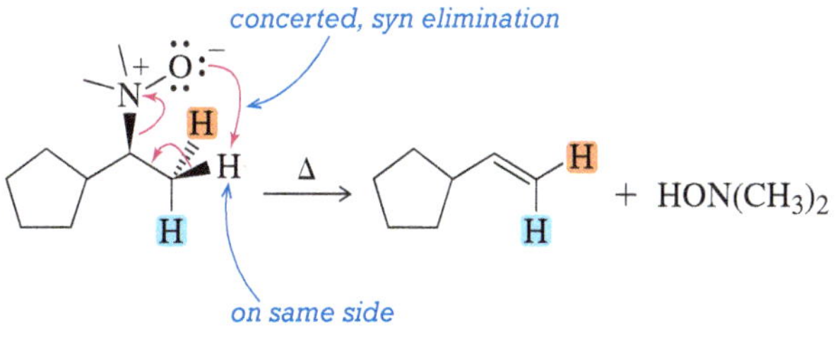

Where two diastereomeric alkenes are possible in the Cope elimination, as shown in **Figure 25.59**, the more stable trans-alkene is formed preferentially. This is a result of minimizing the steric interaction between the eclipsing alkyl groups in the conformation that leads to the cis-alkene.

Figure 25.59 The more stable alkene is formed by avoiding the destabilizing interaction between eclipsing groups.

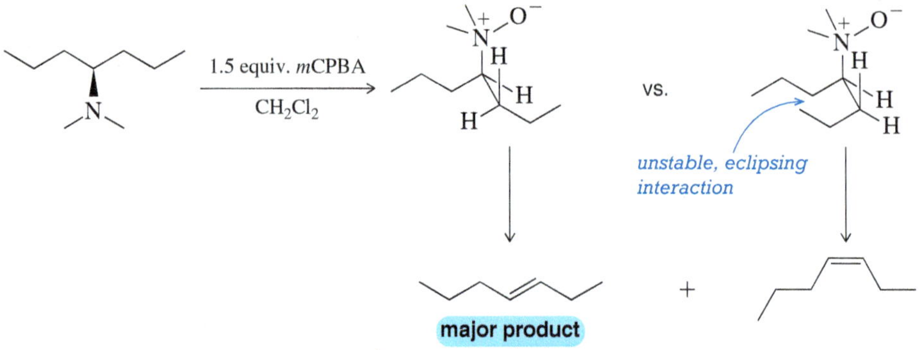

Finally, the Cope and Hofmann eliminations can be complementary tools for selectively providing different products from the same starting material, as shown for the 3° amine in **Figure 25.60**. Alkylation with iodomethane, followed by treatment with Ag_2O (Hofmann), results in the E2 elimination to form the $C_1—C_2$ π bond. Alternatively, oxidation with *m*CPBA (Cope) gives an *N*-oxide that undergoes a syn elimination to form the $C_2—C_3$ π bond.

Figure 25.60 The Hofmann and Cope eliminations are complementary methods.

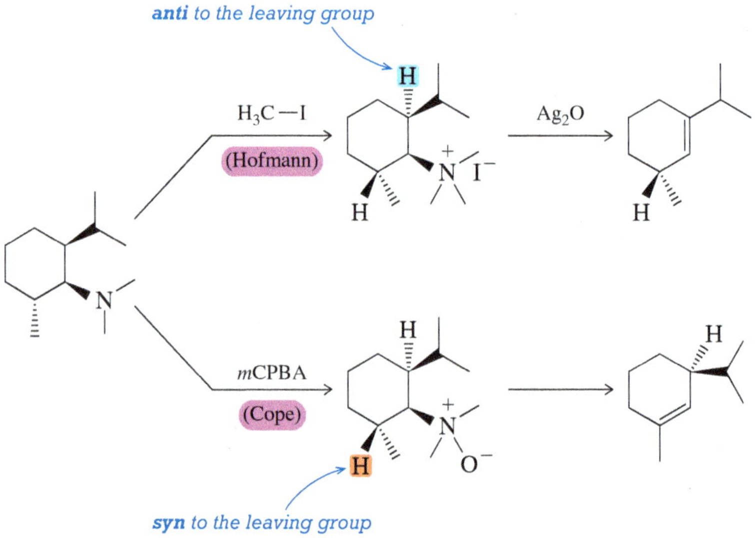

Assessment 25.39 Predict the products of the following elimination reactions.

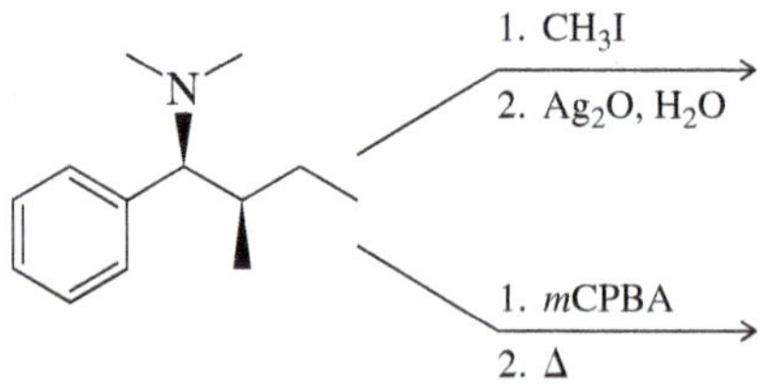

Assessment 25.40 Which elimination reaction—Hofmann or Cope—would you use to make the alkene shown starting from the molecule on the left? It is possible that both reactions will work.

(a)

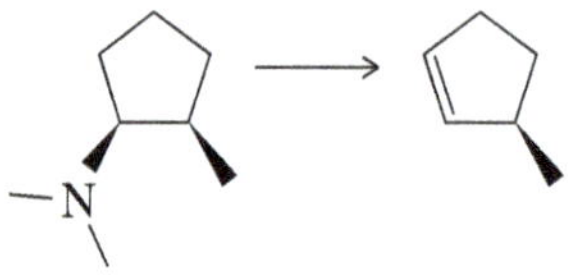

(b)

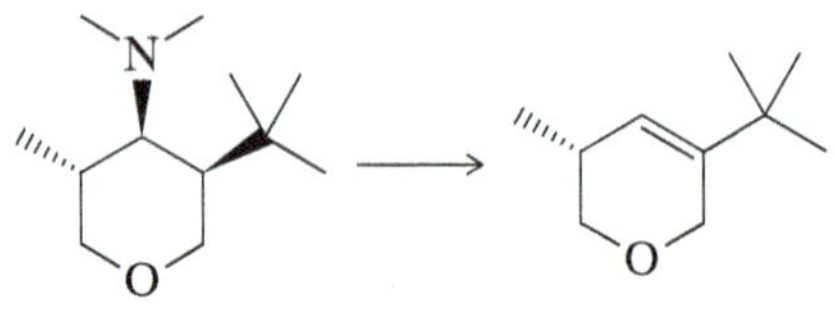

Assessment 25.41 Design a synthesis of the alkene shown starting from the reactant on the left. [*There are multiple ways you could do these, but use either the Cope or Hofmann elimination reaction.*]

(a)

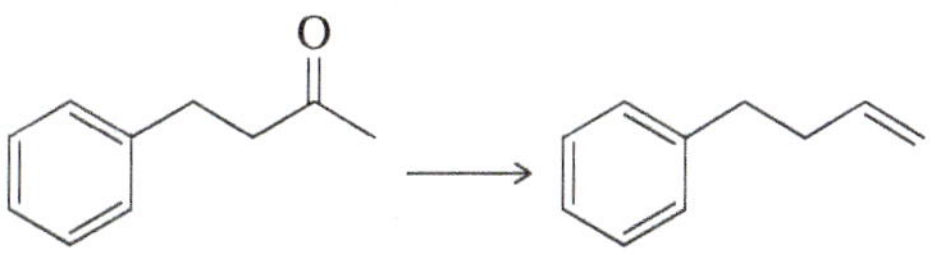

(b)

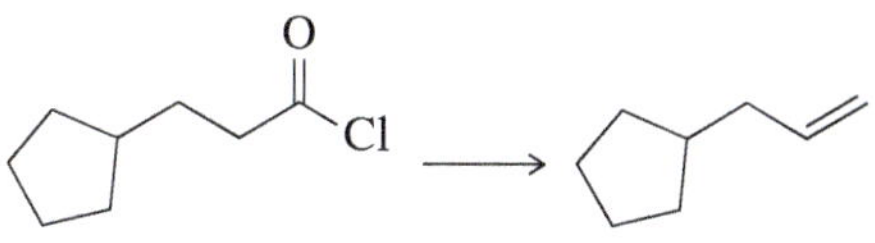

Assessment 25.42 The Cope elimination usually forms the least substituted alkene. For the following molecule, the more substituted alkene is produced instead. Why?

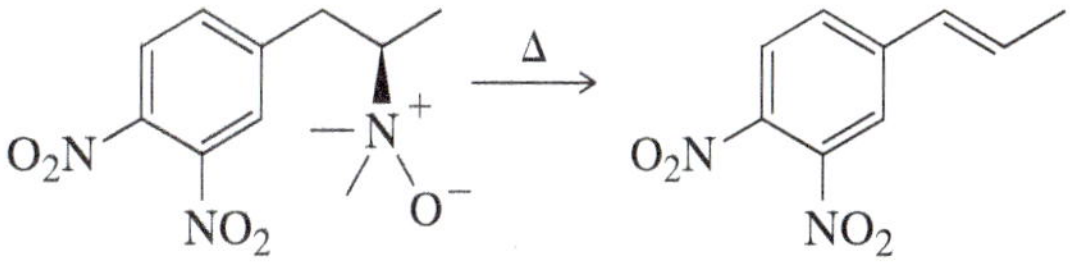

25.7.5 HONO-diazonium formation

The reaction of amines with sodium nitrite and HCl gives different products depending on whether the amine is 1° or 2° (**Figure 25.61**). Primary (1°) amines result in the formation of highly reactive diazonium ions, similar to diazomethane from Section 18.8.1. When a 2° amine is treated under these conditions, a nitrosamine is produced. Nitrosamines have been identified as potential cancer-causing agents in foods and cosmetics.

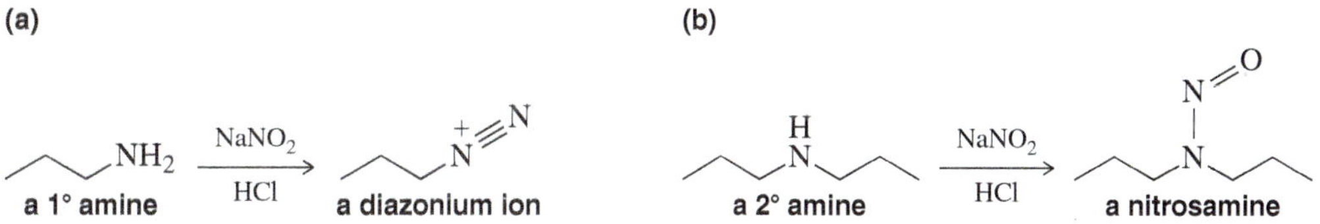

Figure 25.61 The reaction of amines with $NaNO_2$ and HCl results in (a) diazonium ions (from 1° amines) or (b) nitrosamines (from 2° amines).

25.7.5.1 Formation of alkyldiazonium ions

Although the mechanism of diazonium ion and nitrosamine formation can look complicated at first, there is no step of either process that you haven't seen in another context. Both of these reactions begin with the acid-catalyzed dehydration of sodium nitrite to produce a nitrosyl cation (NO^+). Consecutive protonations of the nitrite ion (NO_2^-) produce a hydronium ion, a great leaving group. Water leaves to generate the highly electrophilic nitrosyl cation (**Figure 25.62**).

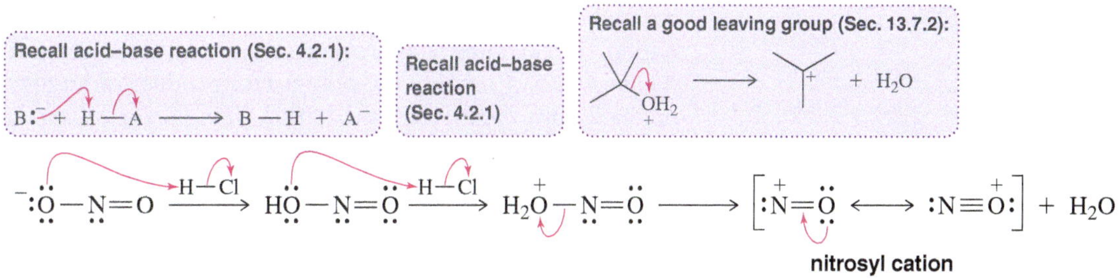

Figure 25.62 The formation of the nitrosyl cation by two acid–base reactions followed by elimination of water.

As all amine reactions have thus far, the mechanism of diazonium ion formation begins with a nucleophilic amine attacking the nitrosyl cation, bringing along its proton in the process (**Figure 25.63**). Removal of the acidic proton results in formation of a nitrosamine. The nitrosamine reacts in an analogous way to aldehydes in the next two steps. Protonation of the nitrosamine oxygen allows a resonance structure containing a new N—N π bond to be drawn. Removal of the acidic proton from this resonance form gives a product analogous to an enol. In essence, then, the last two steps were identical to the mechanism we studied for keto–enol tautomerization in Section 10.8.3 and again in Section 20.2. [*Look back at*

Section 20.2 to confirm that this is the case.] Protonation of the "enol" oxygen creates another hydronium ion, which leaves with assistance from a lone pair on nitrogen to give the diazonium ion containing a N—N triple bond. Although this mechanism seems complicated, it is simply a nucleophilic attack on a cation, followed by four acid–base reactions and a final elimination step.

Figure 25.63 The mechanism of diazonium ion formation with NO$^+$. [*The structures in the two gray boxes make clear the analogy to the keto-enol tautomerism.*]

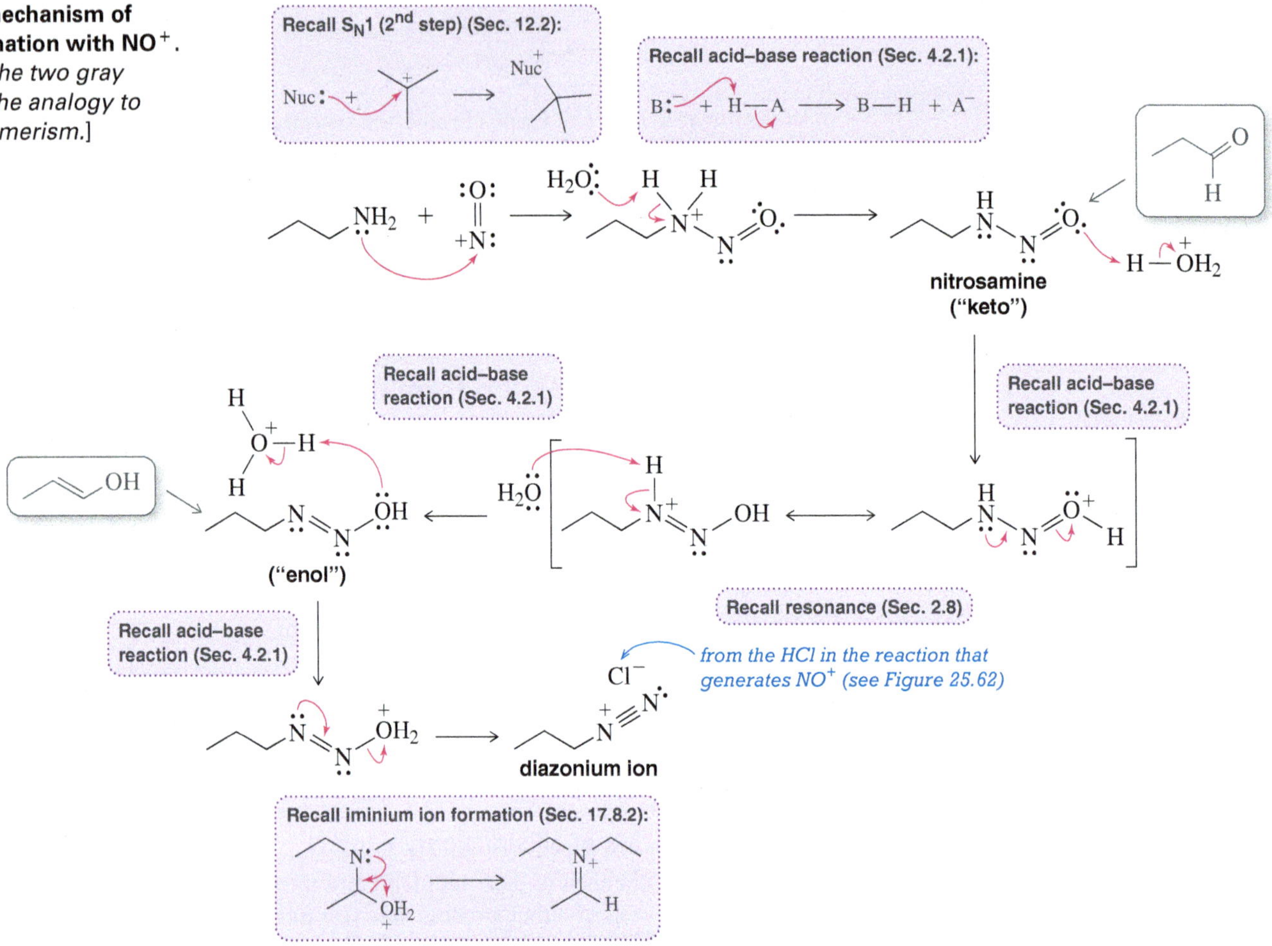

The fate of the resulting diazonium ion diminishes the utility of this reaction, at least in terms of alkyl diazonium ions. Because the diazonium functional group is, in effect, a fully formed nitrogen molecule, alkyl diazonium ions are extremely reactive, rapidly decomposing in water to give myriad products. Most of these products arise from substitution, elimination, or rearrangement processes occurring through the formation of unstable carbocations (**Figure 25.64**). Probably the only useful product of an alkyl diazonium ion would be diazomethane, but even it isn't made this way. [*See Assessment 25.45 for an interesting mechanism problem regarding diazomethane synthesis.*]

Figure 25.64 The spontaneous decomposition of alkyl diazonium ions to multiple products make it synthetically useless.

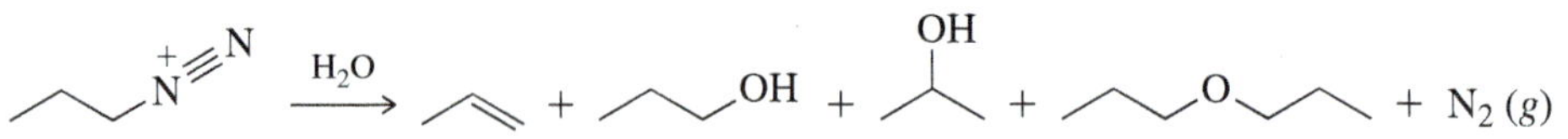

25.7.5.2 Formation of nitrosamines

The reaction of 2° amines with the nitrosyl cation begins the exact same way. However, instead of proceeding to the diazonium ion, the reaction stops at formation of the nitrosamine (**Figure 25.65**). [*Look back at Figure 25.63 to consider why this might be.*] When the nitrosamine formed from a 1° amine, there was an additional proton on nitrogen that allowed for "enol" formation to occur. The 2° nitrosamine doesn't have this proton, so the reaction stops here.

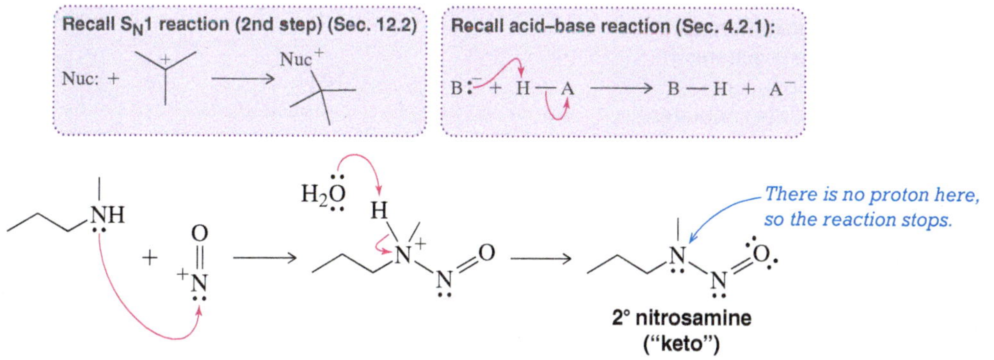

Figure 25.65 The reaction of NO$^+$ and a 2° amine to make a nitrosamine.

Nitrosamines produced in this way have been used for the synthesis of cosmetics and pesticides. They naturally form in foods when nitrites react with amino groups in the highly acidic environment of the stomach as well as during the frying process. Approximately 90% of nitrosamines have been determined to be carcinogenic. As a result, the amount of sodium nitrite used as a food additive in meat has been capped by the U.S. Department of Agriculture (USDA) at 200 ppm.

25.7.5.3 Qualitative amine structure determination

So far, the reactions of alkyl amines with sodium nitrite have given us one reaction that decomposes to such a degree that it is synthetically useless and another reaction that produces carcinogens. [*Sounds like the type of reaction you don't want named after you.*] Before modern spectroscopic methods were available, this reaction had utility in helping chemists determine the structure of amines. Imagine wanting to know whether an unknown amine was 1°, 2°, or 3° (**Figure 25.66**). Without the aid of IR or NMR, you could simply treat the solution with the nitrosyl cation, generated from sodium nitrite and acid. A 1° amine reacts rapidly to form a diazonium ion. Subsequent decomposition of the diazonium ion produces nitrogen gas, which can be seen leaving the solution. Alternatively, a 2° amine reacts to give a yellow-colored solution with no gas evolution, indicating the formation of a nitrosamine. And a 3° amine results in no visible reaction.

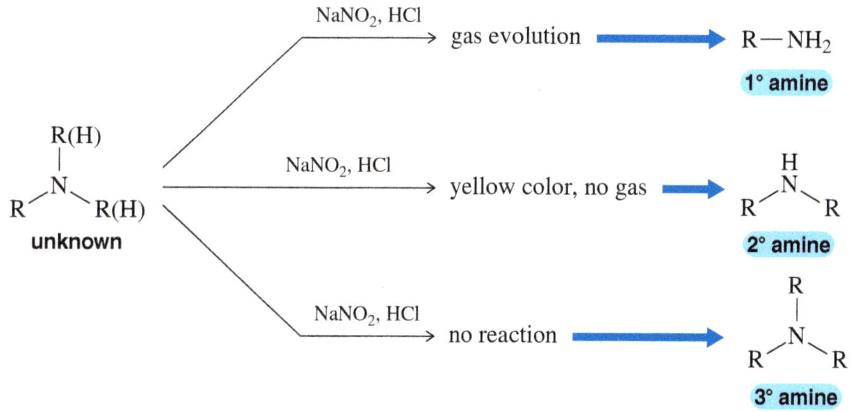

Figure 25.66 Qualitative analysis using the nitrosyl cation.

25.7.5.4 Formation and reaction of aryl diazonium ions

Okay, so we've discussed decomposition, cancer, and structure identification methods not needed for the last 50 years. [*We can do better.*] What other reactions can diazonium ions participate in that could be of interest to organic chemists? When an aryl amine is used in place of an alkyl amine, the resulting aryldiazonium is pretty stable and can be isolated and stored below 10 °C (**Figure 25.67**). This is because substitution, elimination, carbocation formation, and rearrangement do not happen as readily at sp^2-hybridized carbons. As a result, molecules of this type are very useful synthetic intermediates, as we have seen previously in the aromatic substitution reactions of Section 23.10.1.

Figure 25.67 Aryl diazonium ions, formed by the same mechanism as alkyl diazonium ions, are synthetically useful intermediates.

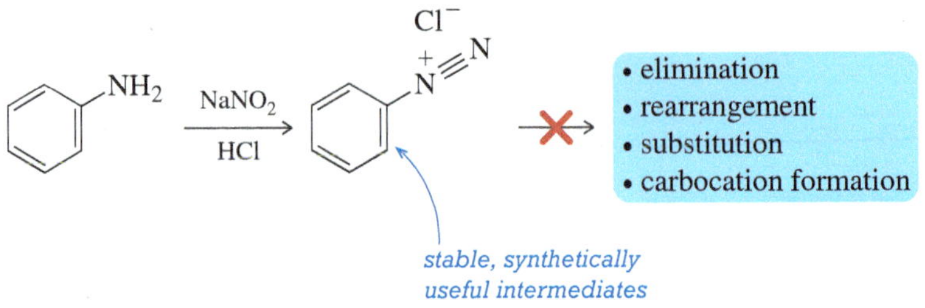

Diazonium ions can be converted into a number of other functional groups using the reactions shown in Figure 25.68. Reactions involving a copper salt are referred to as **Sandmeyer reactions.** The mechanisms of these reactions are not well understood, but many are thought to involve free radicals.

Figure 25.68 The conversion of diazonium ions to a number of functional groups.

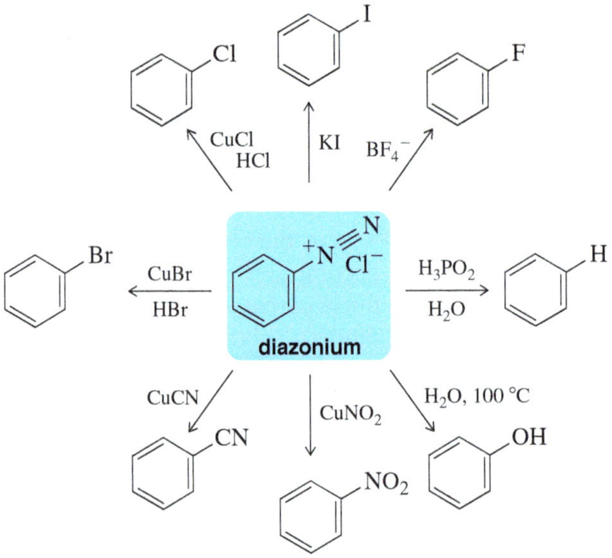

The diazonium ion is also a strong enough electrophile to participate in electrophilic aromatic substitution reactions with substituted benzenes (Figure 25.69). Attack of a π bond forms a nonaromatic intermediate. A highly favorable and fast deprotonation restores aromaticity to the original aromatic ring. Reactions of this type are useful for the synthesis of highly conjugated dye molecules (Chemistry in Context 23-C) and, by extension, ultimately led to the discovery of Prontosil and sulfa antibiotics (Chemistry in Context 23-B).

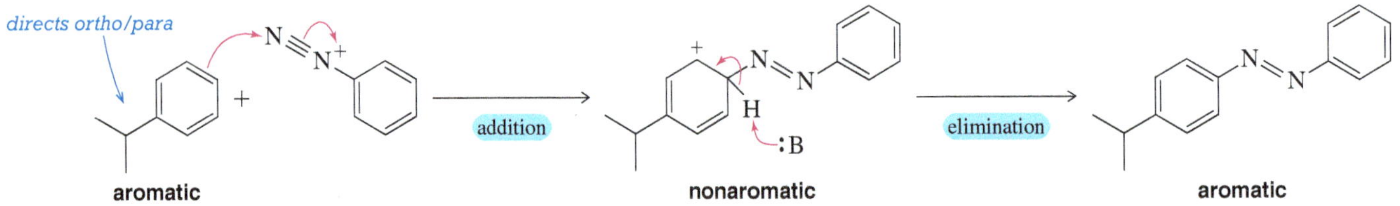

Figure 25.69 An electrophilic aromatic substitution reaction with a diazonium ion.

The net synthetic utility of the diazonium process is in the conversion of a nitro group, through reduction to the amine [*see Section 25.6.3*], to a diazonium ion. In this three-step process (Figure 25.70), a nitro group can be converted into any of the functional groups in Figure 25.68.

Figure 25.70 The conversion of a nitro group to many functional groups through a diazonium ion.

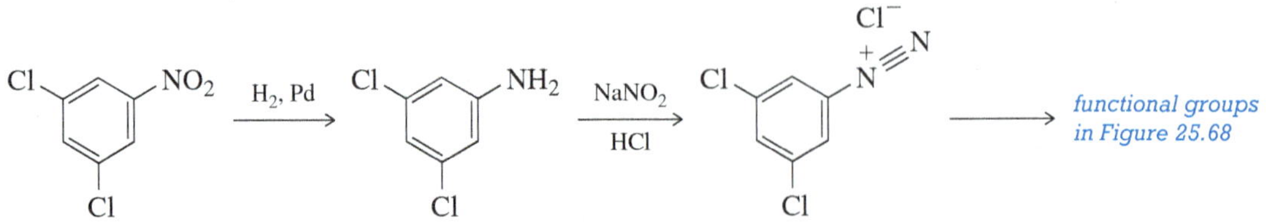

25.7.5 SUMMARY

- Formation of diazonium ions, while possible for a variety of amines, is most useful for the derivatization of benzene rings through the Sandmeyer reaction.

Assessment 25.43 Predict the product of each of the following reactions.

(a)

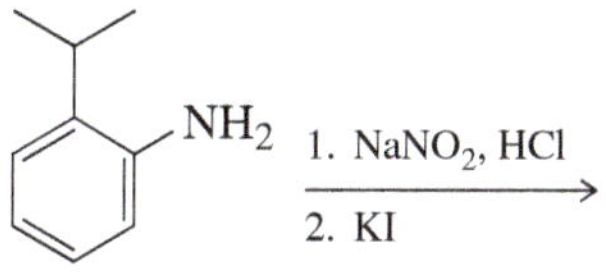

(b)

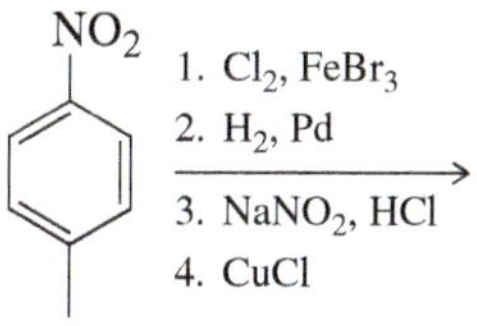

Assessment 25.44 The azo dye shown can be made via two different reactions that employ a substituted benzene and an aryl diazonium. Show both of these. Which one is better?

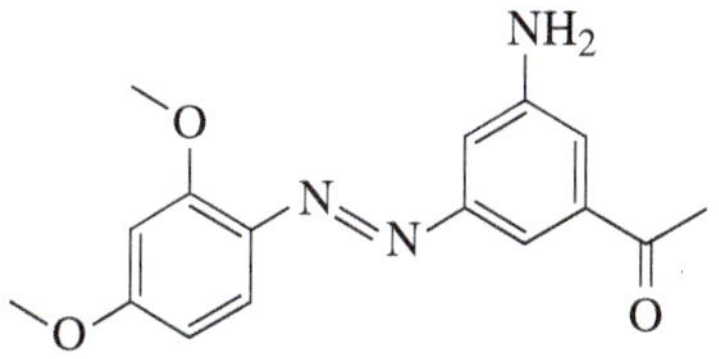

Assessment 25.45 Treatment of *N*-methyl-*N*-nitroso-*p*-toluenesulfonamide with sodium hydroxide results in the synthesis of diazomethane. Suggest a mechanism for this reaction. [*Hint: Sulfonamides behave like the carboxylic acid derivatives in Chapter 19.*]

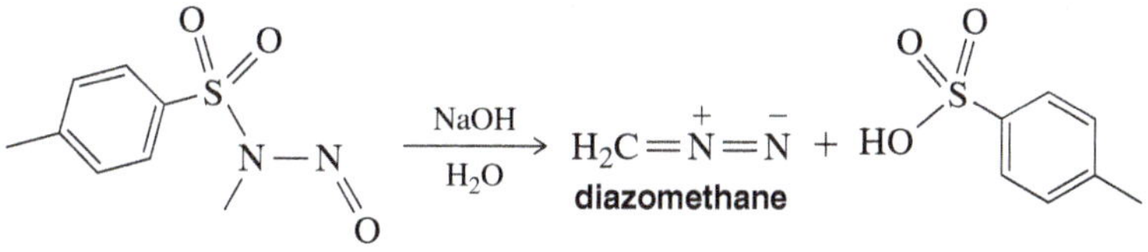

Concluding thoughts: major depressive disorder

Perhaps to a greater degree than for any other functional group, amines figure prominently in brain chemistry. They play a substantial role in the structure of neurotransmitters and in the drugs used to modulate neurotransmitter concentrations. Besides depression, perturbations in the fine balancing act between brain neurotransmitters are suspected to play a role in a large number of disorders, including anxiety, schizophrenia, bipolar, and attention deficit/hyperactivity disorders. As such, closely related derivatives of the medications studied here in Chapter 25 can be used to treat many other ailments.

Most naturally occurring amines, including the neurotransmitters described herein, are synthesized in the body from naturally occurring amino acids. Amino acids, as the name implies, contain both an amine and a carboxylic acid and are the topic of Chapter 26. And what about the receptors that these drugs react with? Those receptors are themselves proteins, made up of the same amino acids we discuss in Chapter 26. As you might expect, a lot of the chemistry in Chapter 26 will be similar to things we have studied before, but put in a more biochemical context. All this prepares us for the topics we discuss in Chapter 26, and we go there next.

But, before we do: Major depressive disorder (depression), a topic of this chapter, is a heavy one, and I gave second (and third) thoughts to writing about it. But in a class that causes stress for all students, including future healthcare professionals, it's an important and relevant topic, especially on college campuses, where mental illness is too often, and too easily, overlooked. Organic chemistry is difficult, and there is pressure to do well in it, but it is nowhere near as important as your mental health. You can ignore everything else I have written so far in this book, but not this: If you or someone you know is struggling with depression, seek help as soon as possible. Reach out to someone—a professor, a parent, a friend, a school counseling center, or call a national hotline [*National Suicide Prevention Lifeline: 1-800-273-8255*]. Please, get help. You deserve it.

End-of-Chapter Materials

Important Concepts

[These are the major concepts covered in each section, excluding reactions.]

25.1–2: Amines are compounds containing a nitrogen bonded to one, two, or three alkyl groups. Amines can be acceptably named using a variety of systems (common names, IUPAC, or chemical abstract). Amines are classified as 1°, 2°, or 3° based on the number of alkyl groups attached to the nitrogen.

25.3: In aliphatic amines, the amino group is sp^3 hybridized. Aromatic amines, such as aniline, are between sp^2 and sp^3 hybridized because the lone pair can participate in resonance with the aromatic ring.

25.4: Lacking the same ability to hydrogen bond, amines boil and melt at lower temperatures compared to the analogous alcohols. In the IR spectrum, 1° and 2° amines are characterized by a broad N—H stretch around 3300 cm^{-1}. Molecules containing a single nitrogen have an odd atomic mass, a feature that is used to quickly identify monoamines using mass spectrometry. The pK_b value can be used in an analogous way to pK_a values to assess the relative basicity of amines. A lower pK_b value implies that a base is stronger.

25.5: While the simplest way to make a 1° amine would be to alkylate ammonia, overalkylation makes this a non-viable strategy. The synthesis of 1° amines is more efficiently achieved using reduction of a nitrile, reduction of an azide, or historically, the Gabriel amine synthesis.

25.6: The general synthesis of variably substituted amines can be accomplished through reducing an amide, reductive amination, reduction of a nitro group, or the Buchwald–Hartwig amination.

25.6: The most common reactions of amines involve them acting as bases or nucleophiles. The Hofmann and Cope eliminations provide access to the least substituted (least stable, anti-Zaitsev) alkenes. Formation of diazo compounds with substituted anilines allows for functionalization of aromatic rings.

Important Reactions and Mechanisms

1. **Nitrile reduction (Section 25.5.2):** Addition of a strongly nucleophilic hydride ion is followed by donation of a pair of electrons on nitrogen to the Lewis acid AlH_3 (alane). Hydride transfer from aluminum to the imine nitrogen produces an amide that is protonated in an acid quench step. Hydrolysis of the resulting complex gives a 1° amine.

General:

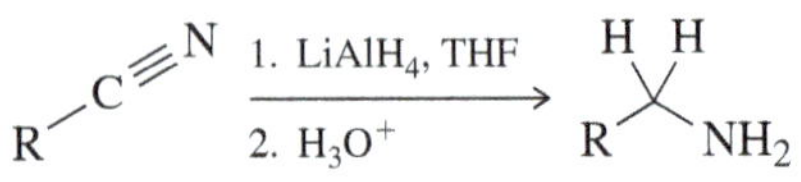

Specific:

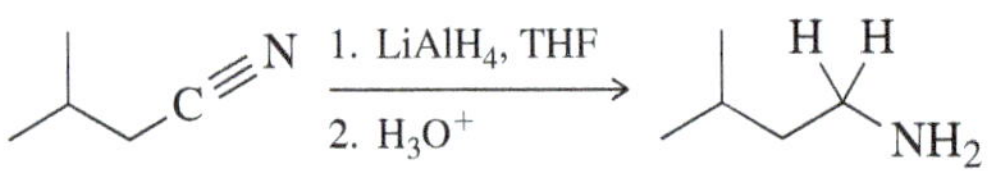

Mechanism:

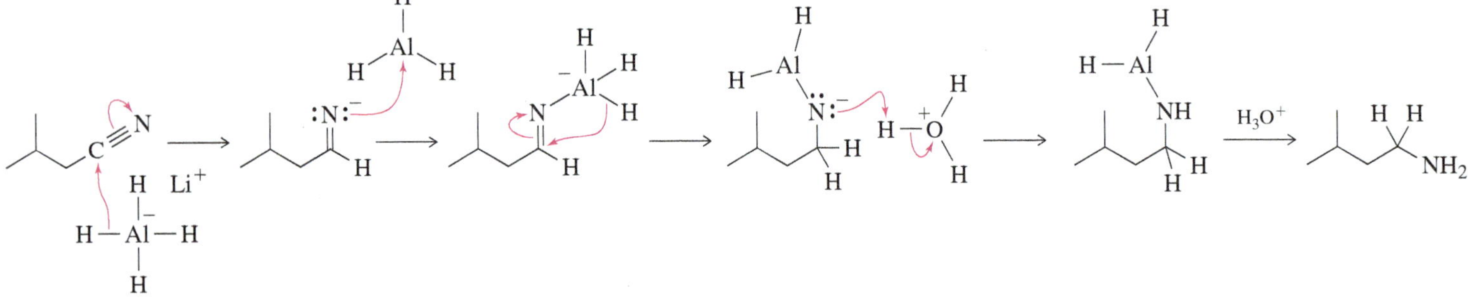

2. Azide reduction, $LiAlH_4$ (Section 25.5.3): Nucleophilic addition of a hydride ion to the terminal nitrogen of an azide gives an intermediate anion that participates in a Lewis acid–Lewis base reaction with alane. An intramolecular acid–base reaction produces H_2, N_2, and an imidoaluminum species that is protonated twice, enabling hydrolysis of the N—Al bond. Azide reduction often follows an S_N2 reaction, allowing the synthesis of 1° amines from 1° haloalkanes. Azides can also be reduced using H_2 and Pd/C. [*Recall that a mildly basic work up will be necessary to ensure that a neutral amine is isolated.*]

General:

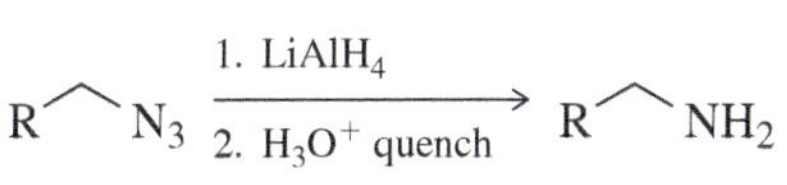

Specific:

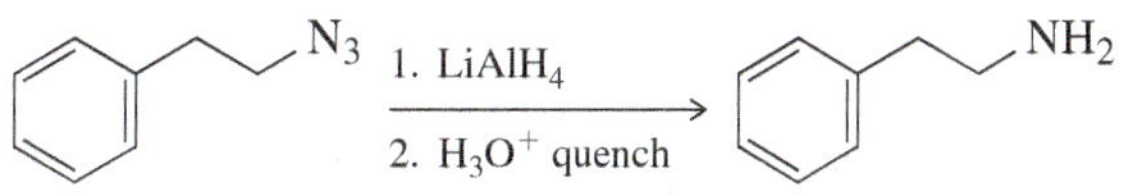

Mechanism:

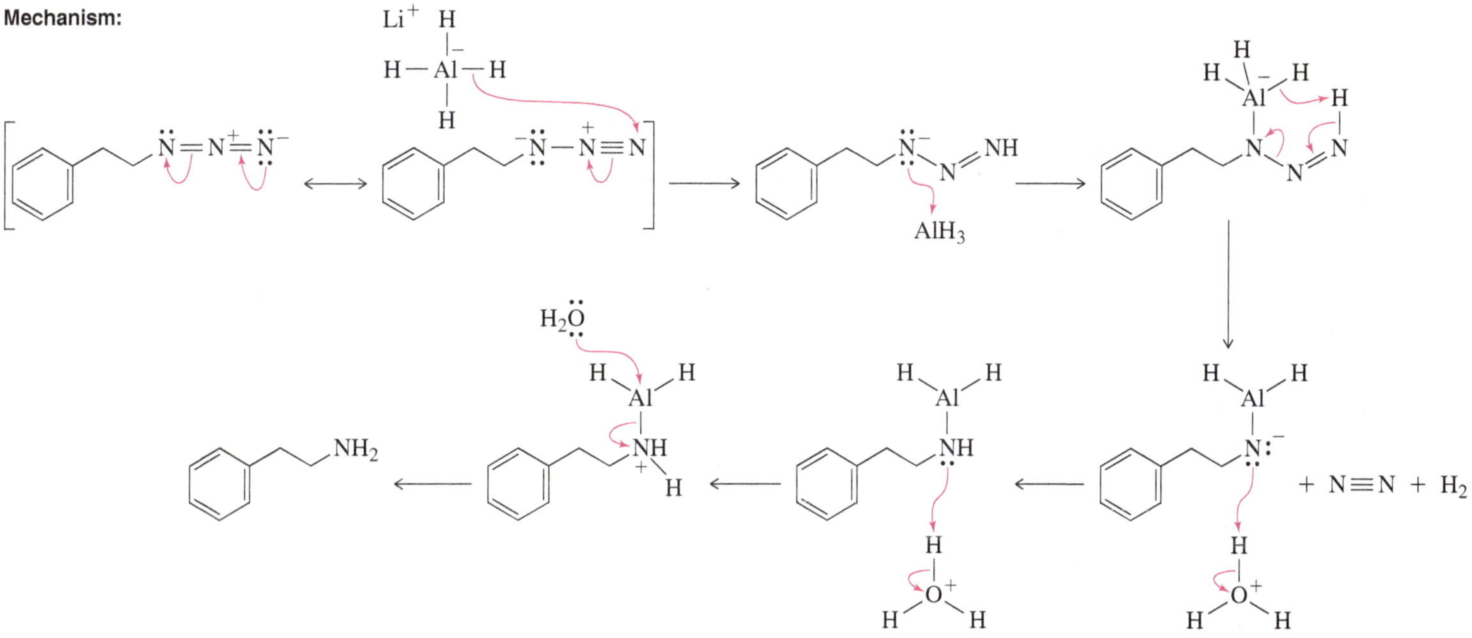

3. Azide reduction, Ph_3P (Section 25.5.3): Reduction of an azide with Ph_3P is referred to as the Staudinger reaction and begins with the nucleophilic addition of Ph_3P to the terminal nitrogen of the azide, eventually forming a four-membered ring in a fashion analogous to the Wittig reaction. Decomposition of the four-membered ring forms a P=N double bond that is hydrolyzed to give the thermodynamically more stable P=O double bond.

General:

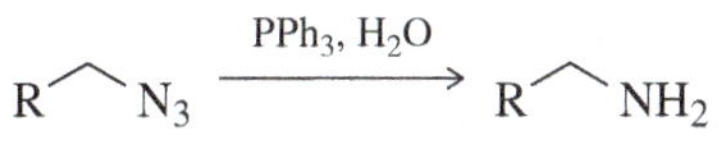

Specific:

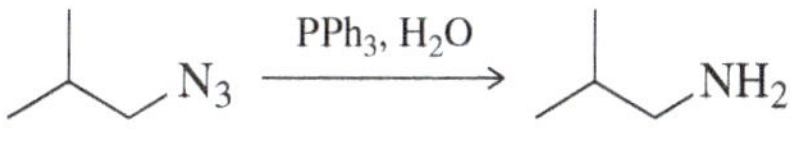

Mechanism:

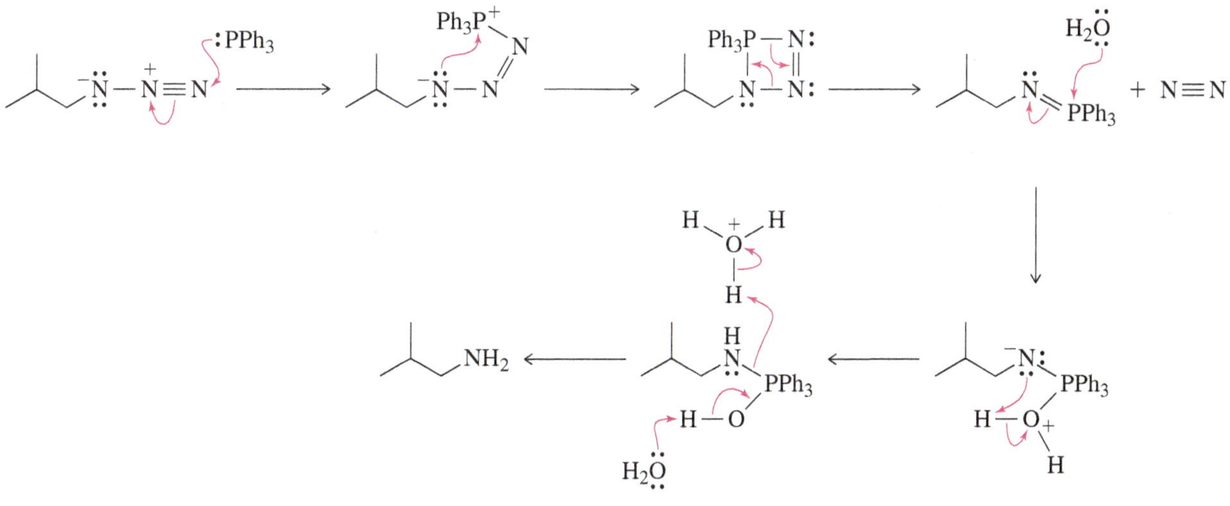

4. Gabriel amine synthesis (Section 25.5.4): Alkylation of potassium phthalimide occurs by an S_N2 reaction. In a complex mechanism involving two nucleophilic acyl substitution reactions and several acid–base reactions, a 1° amine is released along with an aromatic by-product. Because of the waste generated by this reaction, the Gabriel amine synthesis is not often used.

General:

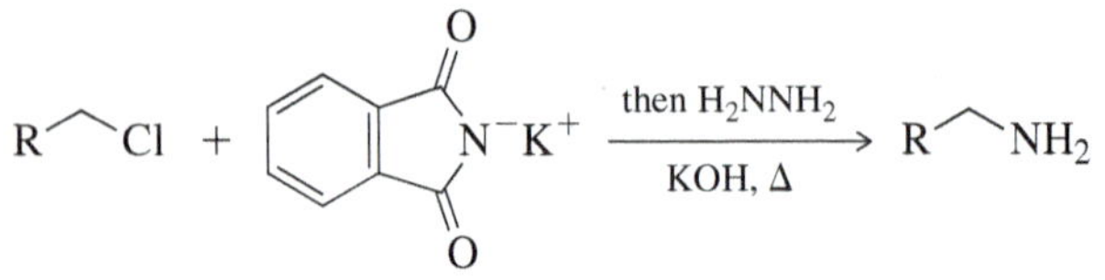

Specific:

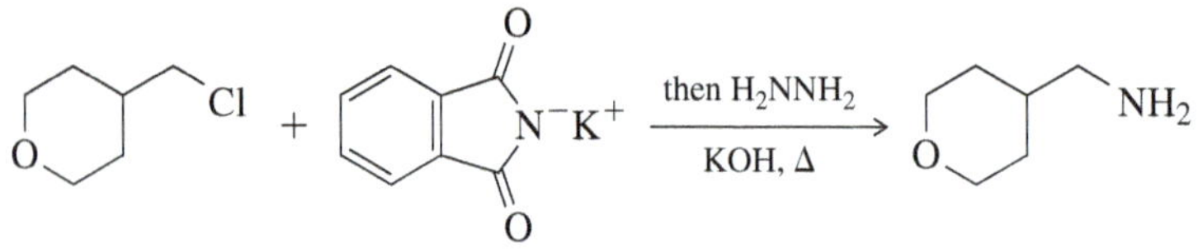

Mechanism:

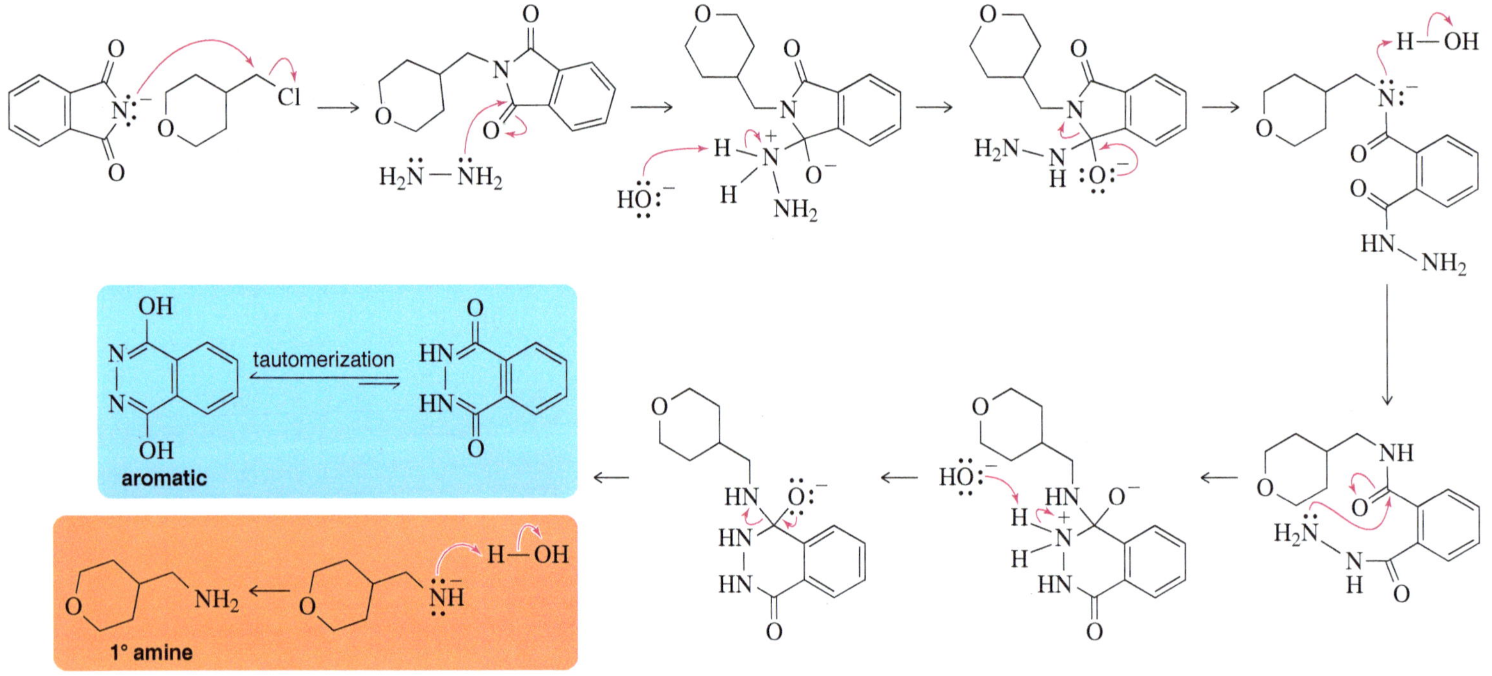

5. Reduction of amides (Section 25.6.1): Two equivalents of hydride add to the carbonyl carbon in separate steps to reduce amides to amines. Proceeding by General Mechanism 18A (Section 18.7.1), addition of hydride results in a stable tetrahedral intermediate. The alkoxide donates a pair of electrons to Lewis acid AlH_3 (alane), making a good leaving group. The oxyaluminate is expelled with assistance from the lone pair on nitrogen, producing an iminium ion. The iminium ion is subsequently reduced with the second equivalent of hydride, resulting in the amine product.

General:

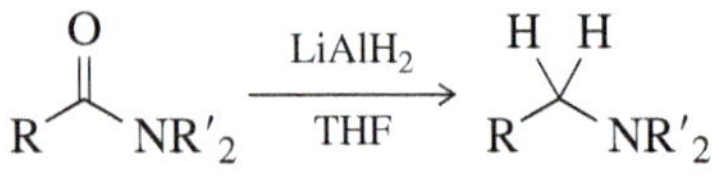

Specific:

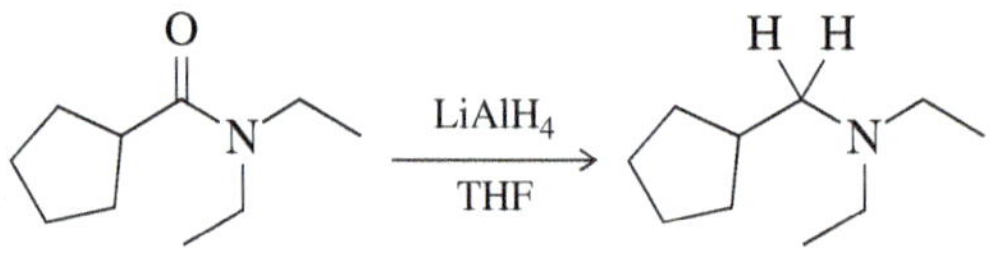

Mechanism:

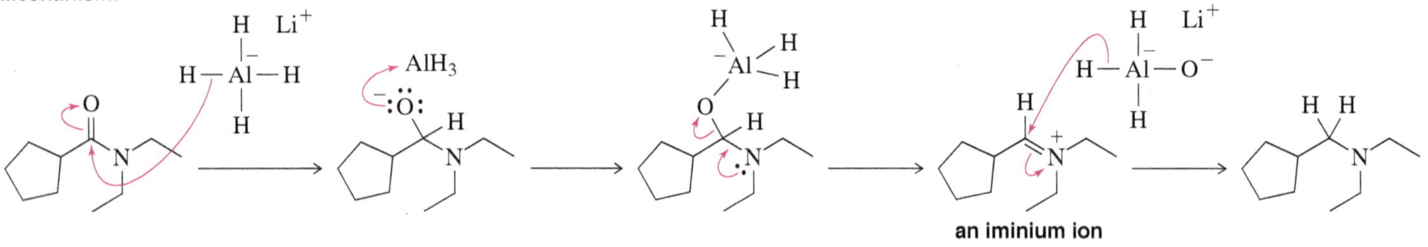

6. Reductive amination (Section 25.6.2): Following formation of an imine by the mechanism in Section 17.8.2, reduction occurs by hydride addition to the electrophilic carbon of a protonated iminium ion.

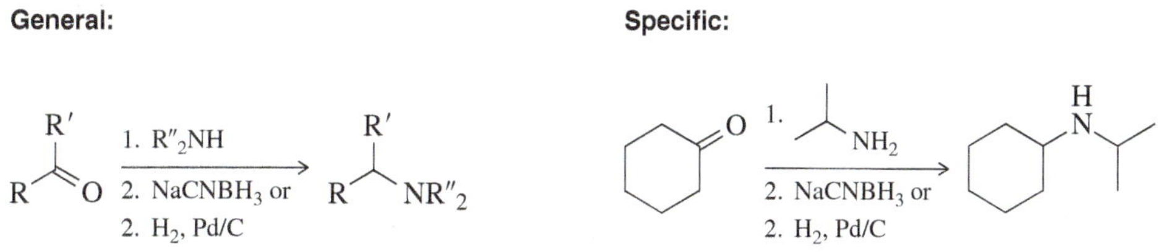

Mechanism:
After forming the imine (Section 17.8.2)...

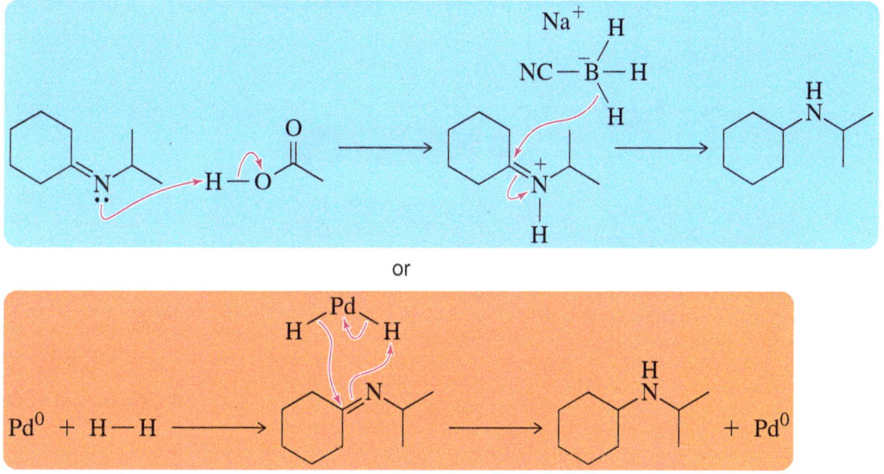

7. Reduction of nitro groups (Section 25.6.3): Nitro groups are hydrogenated using a palladium catalyst to form 1° amines by a mechanism that is not well understood.

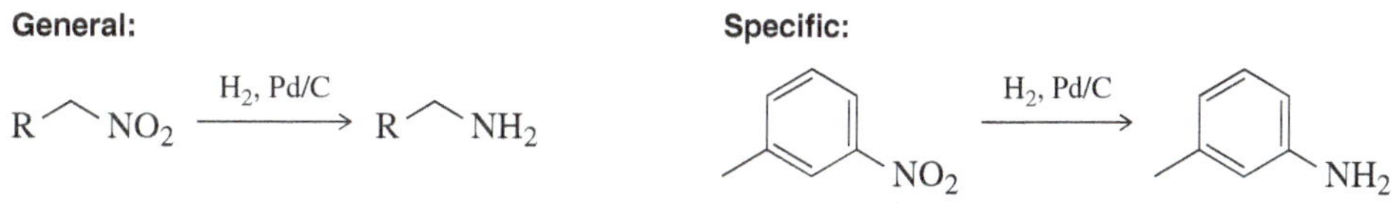

8. **Buchwald–Hartwig amination (Section 25.6.4):** Oxidative insertion of palladium into an aryl C—Br bond is followed by amine ligation (analogous to transmetallation). Reductive elimination completes formation of a new N—C bond.

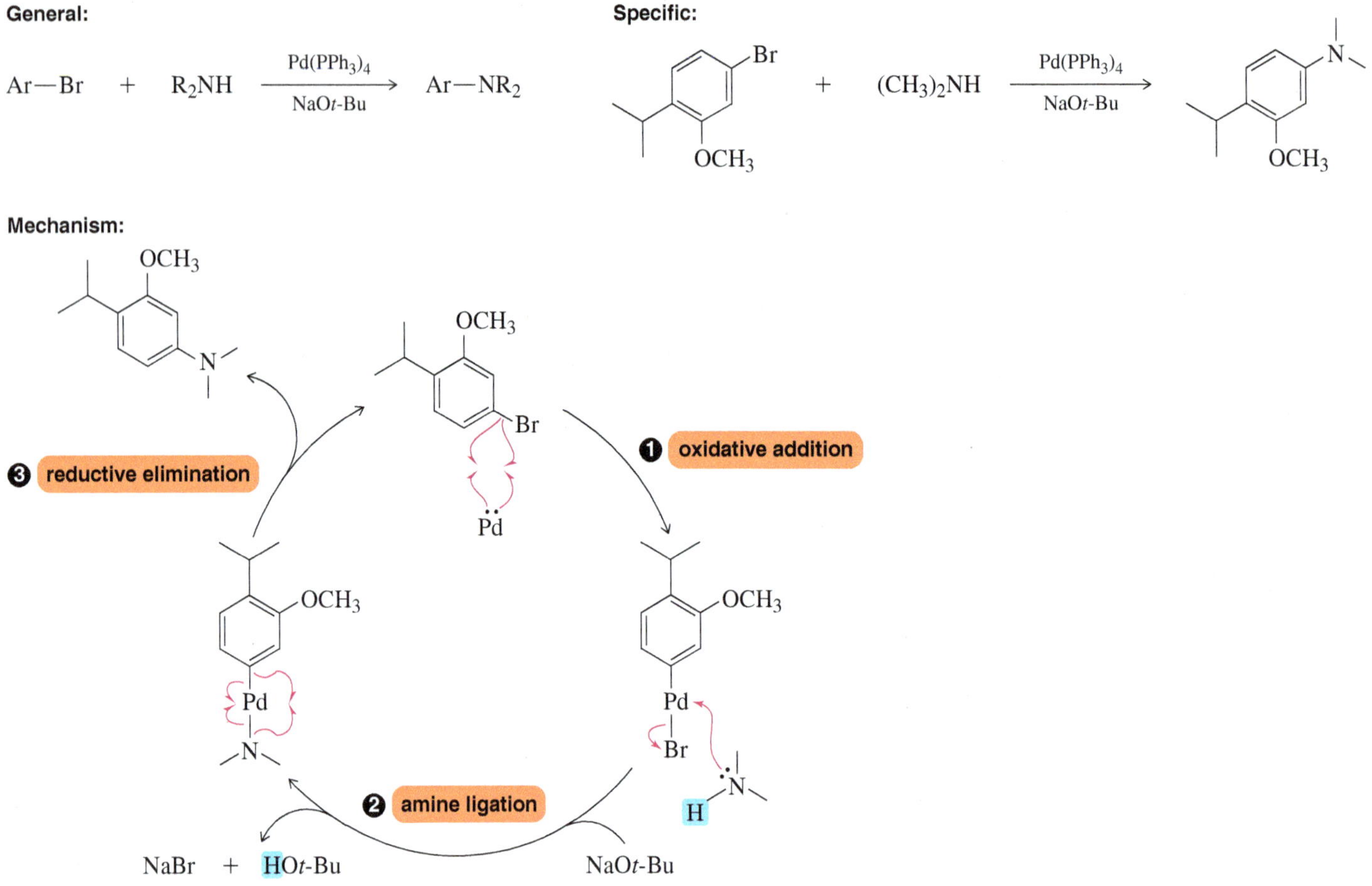

9. **Acid–base reactions of amines (Section 25.7.1):** Amines participate in acid–base reactions by donating a pair of electrons to an acid. The resulting ammonium/conjugate base complexes are more soluble in water than the original neutral amine.

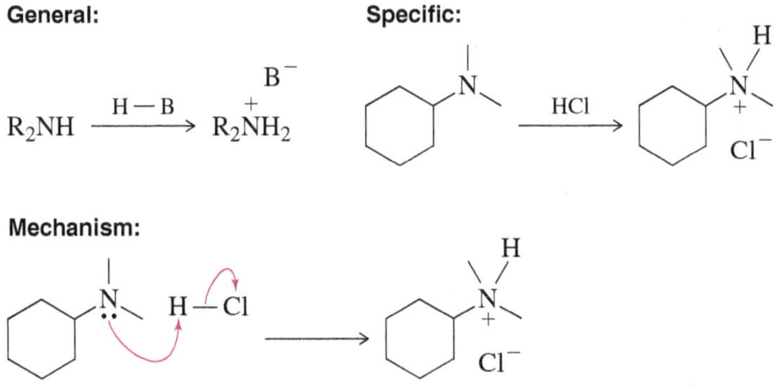

10. **Hofmann elimination (Section 25.7.3):** Alkylation of 3° amines with methyl iodide generates an ammonium iodide. Exposure of the ammonium iodide to silver oxide results in formation of hydroxide, a base that effects an E2 elimination reaction to give the least stable alkene by minimizing steric strain in the transition state.

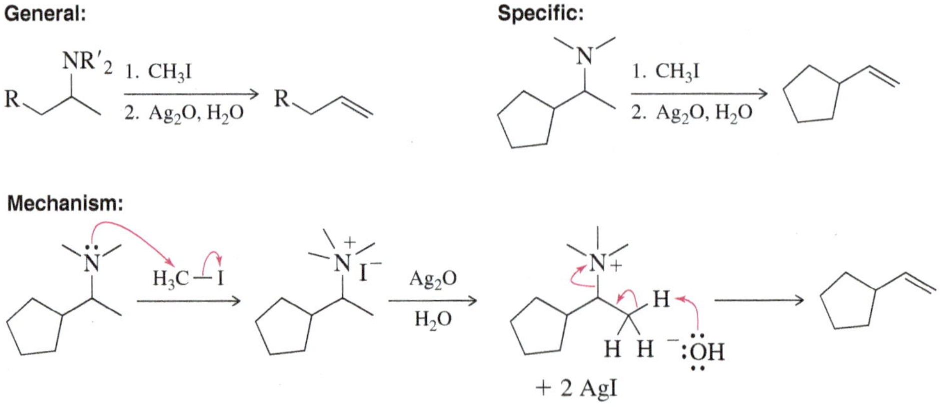

11. Cope elimination (Section 25.7.4): Oxidation of a tertiary amine gives an *N*-oxide that undergoes a concerted, syn elimination to give the least substituted alkene product.

General:

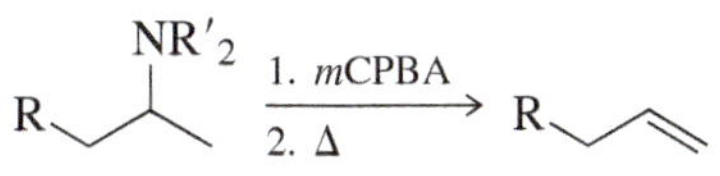

Specific:

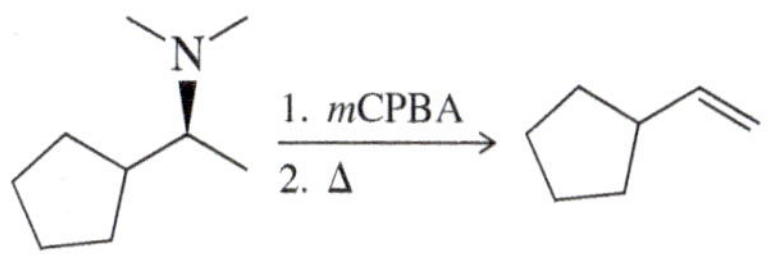

Mechanism:

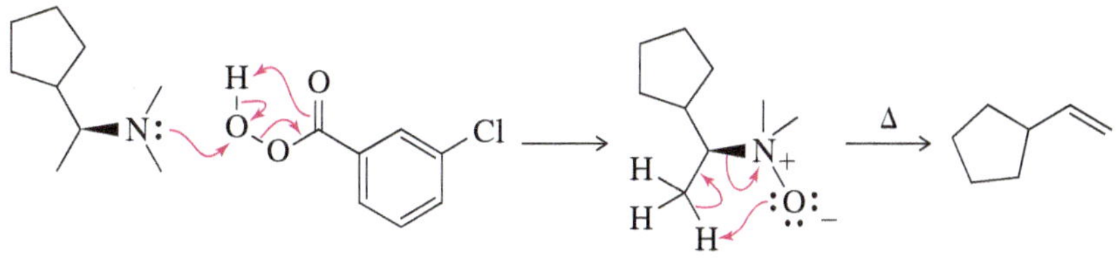

12. Diazonium formation (Section 25.7.5): Formation of a diazonium ion occurs by reaction between an amine and a nitrosyl cation in a complex mechanism involving several reactions (including four acid–base reactions) observed in other contexts.

General:

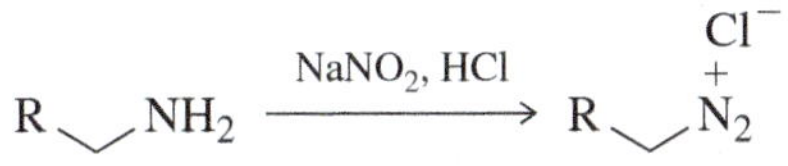

Specific:

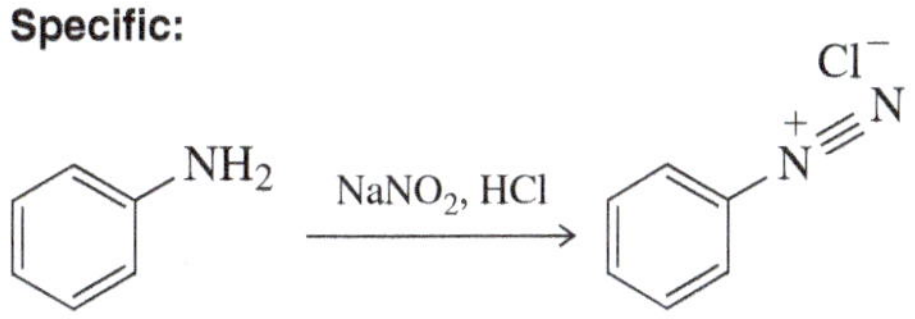

Mechanism:

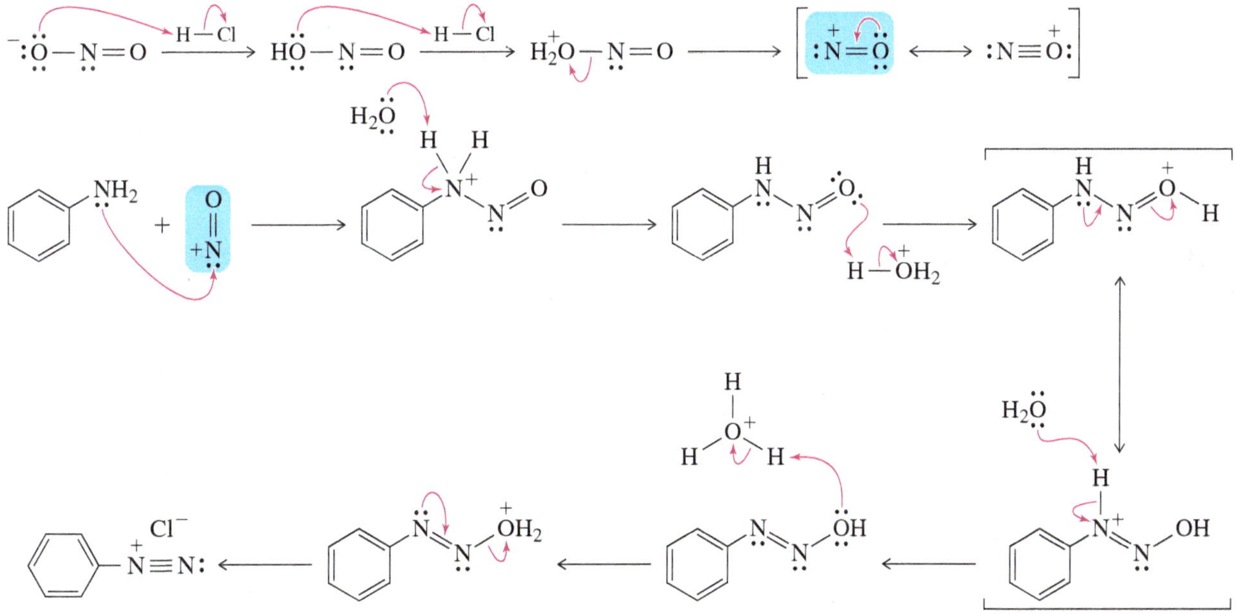

End-of-Chapter Assessments

[Assessments are ranked according to their difficulty; •••• = most difficult.]

25.46 (•) Given the pK_b, calculate the pK_a of the conjugate acid.

(a)

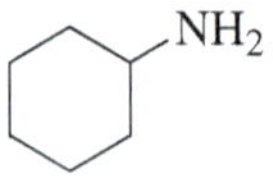

pK_b = 3.33

(b)

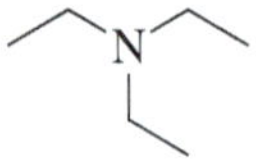

pK_b = 3.24

(c)

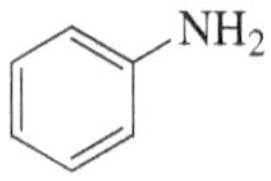

pK_b = 9.4

25.47 (•) Calculate the equilibrium constant for each of the acid–base reactions shown.

(a)

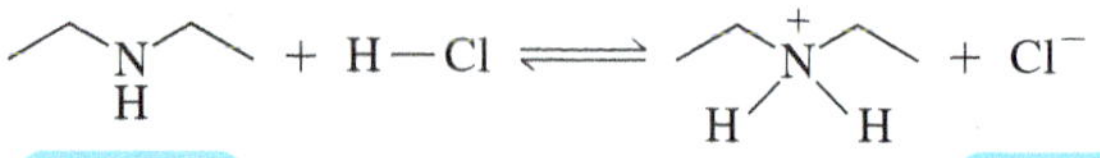

pK_b = 3.01 pK_b = 21

(b)

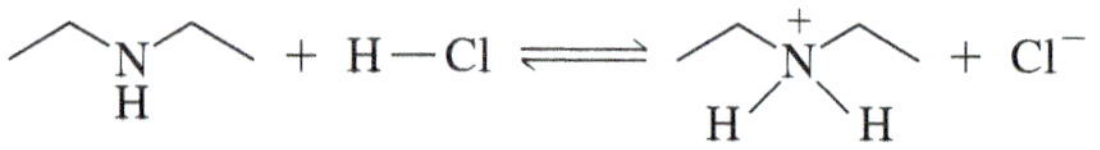

pK_a = −7 pK_a = 10.99

25.48 (••) Choose the strongest base in each pair. Justify your answer.

(a)

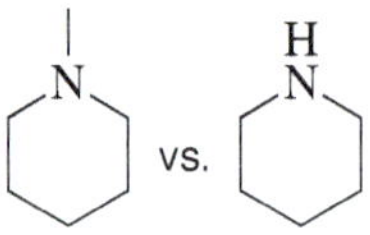

(b)

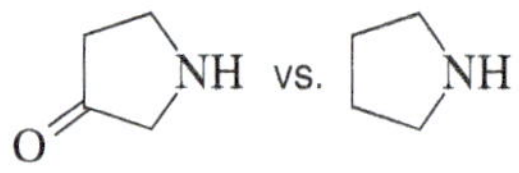

(c)

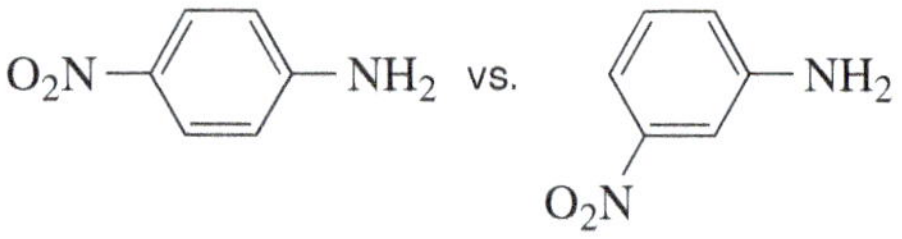

(d)

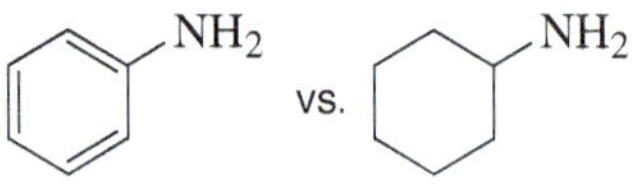

(e)

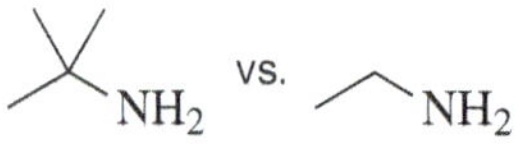

25.49 (•) LOOKING AHEAD Justify the difference in pK_b values between proline and most other amino acids.

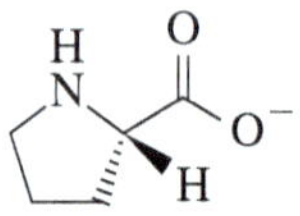

proline

pK_b = 3.6

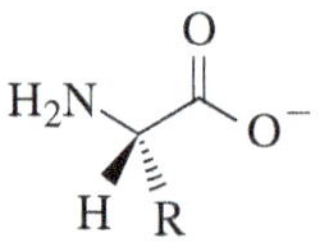

all other amino acids

pK_b = 4.5

25.50 (••) Aniline undergoes reductive amination reactions much more slowly than aliphatic amines. Why is this?

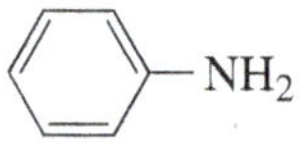

25.51 (••) Rank the following amines in terms of their ability to alkylate upon reaction with methyl iodide (1 = fastest; 5 = slowest).

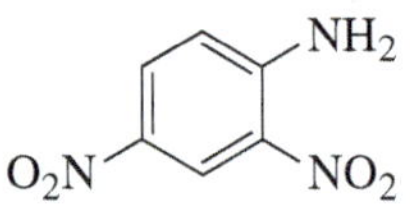

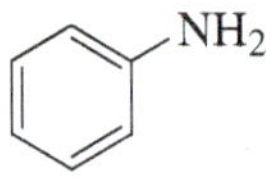

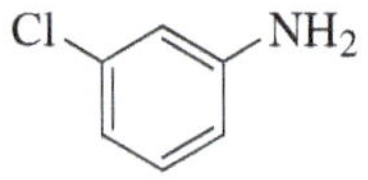

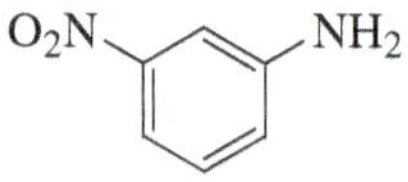

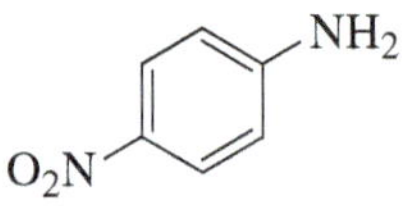

25.52 (•••) LOOKING BACK Despite the amino group being an ortho/para director, nitration of aniline gives the meta isomer predominantly. Explain this result.

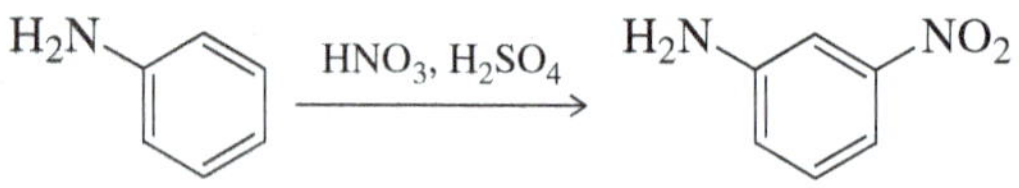

25.53 (•••) One of the roles of a pharmacist is to ensure that a patient is not taking two drugs that have adverse interactions. Omeprazole, a proton pump inhibitor used to reduce the acidity of the stomach, has multiple drug interactions. Amine-containing drugs, such as amoxapine, are often excreted less rapidly when taken with omeprazole. Explain this interaction. [*Hint: The body is better at excreting water-soluble drugs.*]

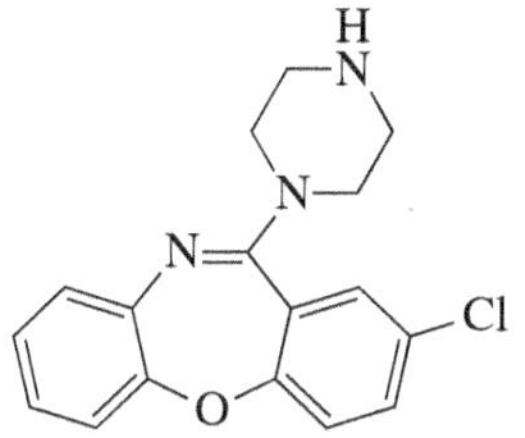

amoxapine

25.54 (••) In contrast to the effect seen in Assessment 25.53, omeprazole increases the excretion of enoxacin. Explain this observation.

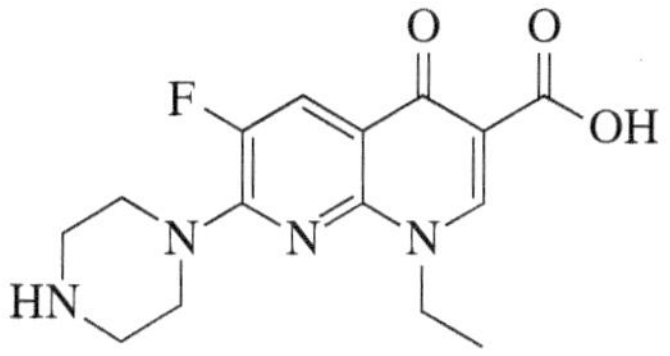

enoxacin

25.55 (••) Predict the product(s) of the reactions shown.

(a)

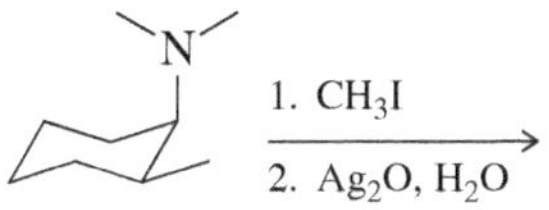

(b)

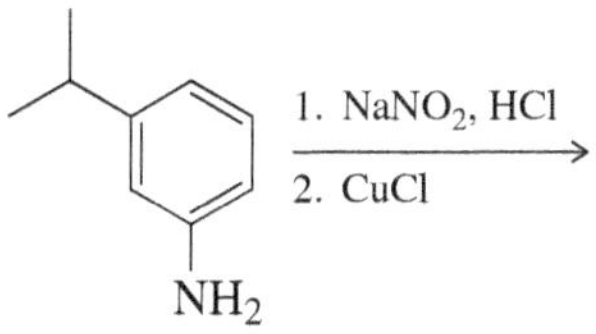

(c)

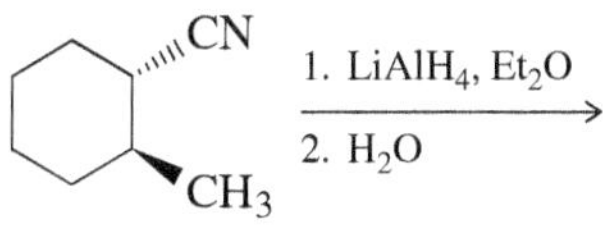

(d)

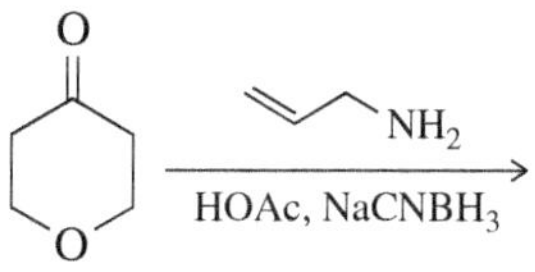

(e)

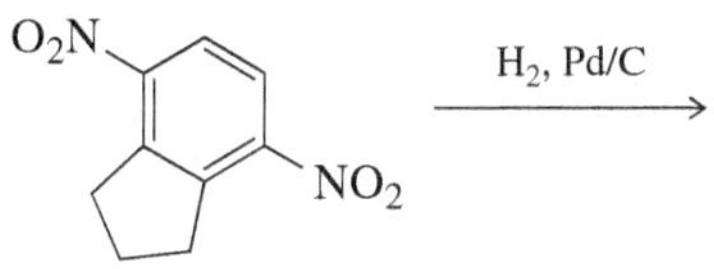

(f)

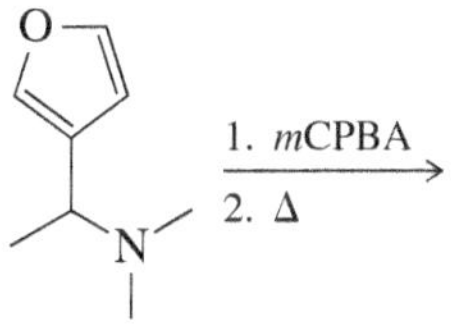

(g)

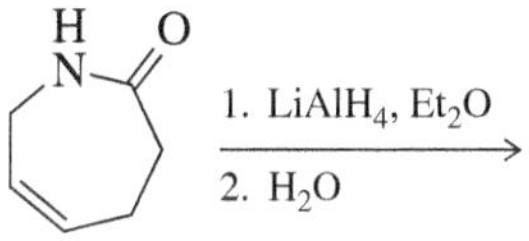

(h)

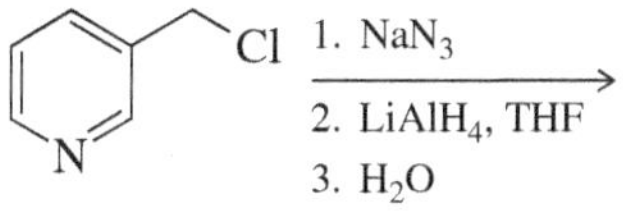

(i)

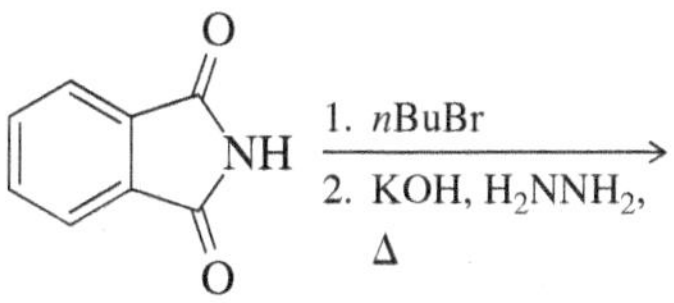

(j)

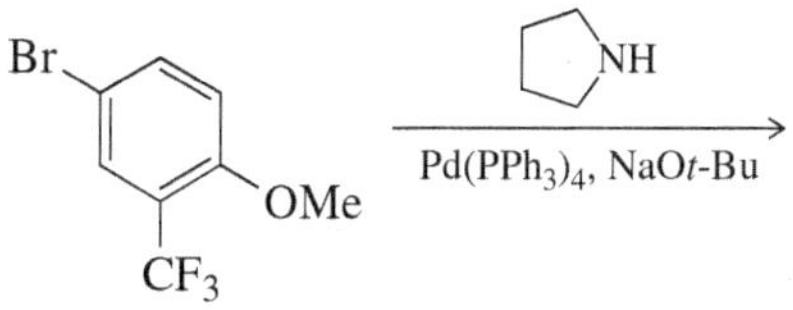

25.56 (••) Suggest reagents to carry out the following transformations. Some may require more than one step.

(a)

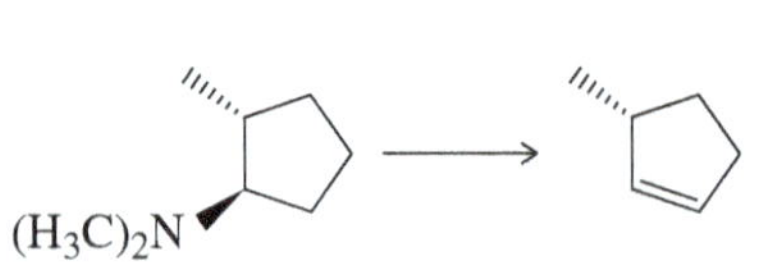

(b)

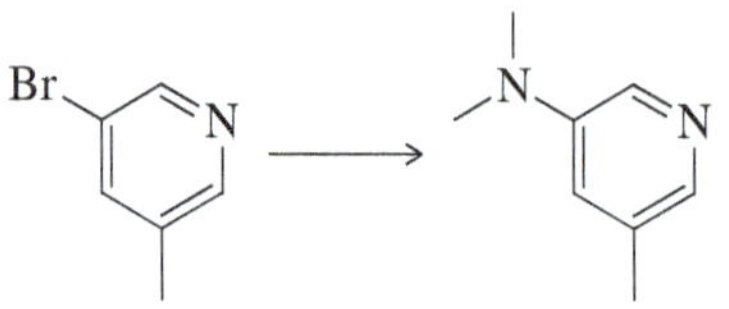

(c)

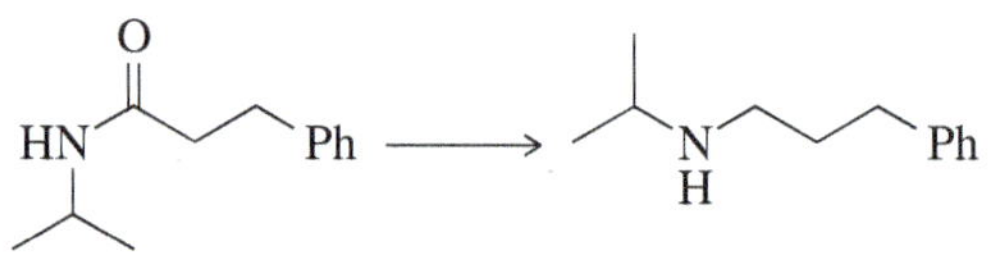

(d)

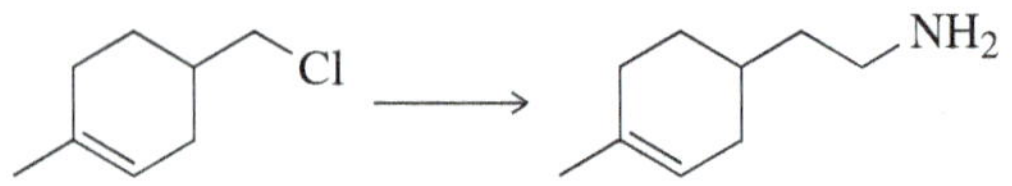

(e)

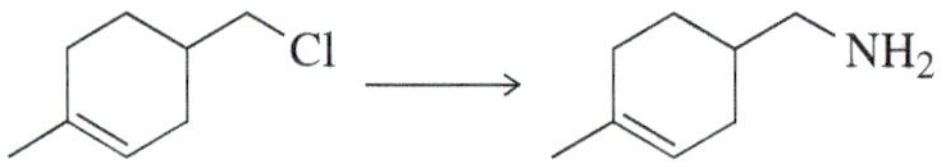

(f)

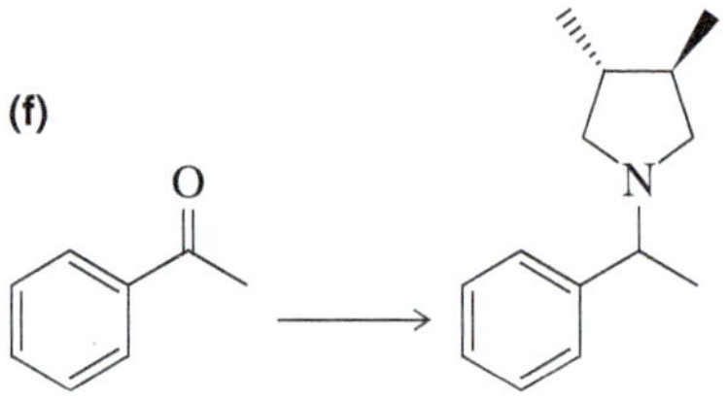

(g)

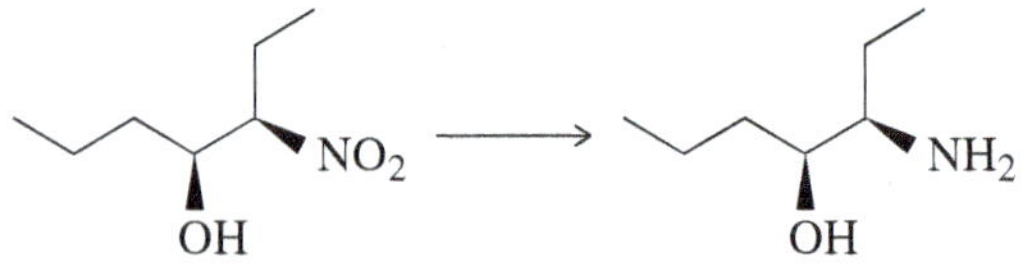

(h)

25.57 (•) The Gabriel synthesis is most frequently done with 1° alkyl halides. Why is it less successful with more substituted halides?

25.58 (••) **FROM THE LITERATURE** A 2007 paper proposed the following synthesis to prepare libraries of heterocyclic, nitrogen-containing compounds. (a) Why did this reaction scheme require the use of a large excess of amine? (b) Suggest a less wasteful/greener synthesis of the same compounds (*J. Comb. Chem.* 2007, *9*, 171–177).

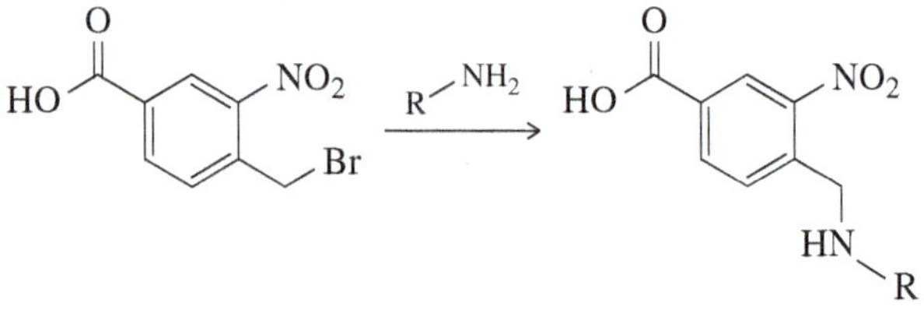

25.59 (•••) **LOOKING BACK** Suggest a mechanism for the reduction/cyclization shown. [*This reaction is similar to a reaction studied in Chapter 23.*]

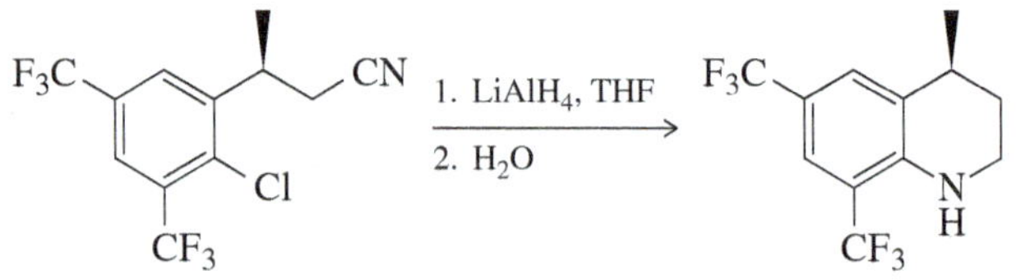

25.60 (••) Choose the appropriate conditions to carry out the following elimination. Justify your choice.

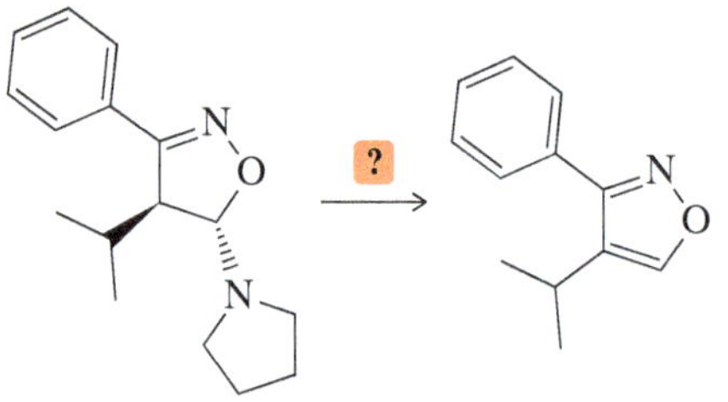

25.61 (••) One of the following eliminations proceeds at a much slower rate. (a) Draw the products for both, (b) identify the slower reaction, and (c) explain why it is slower. [*What should you always do when a reaction involves a six-membered ring?*]

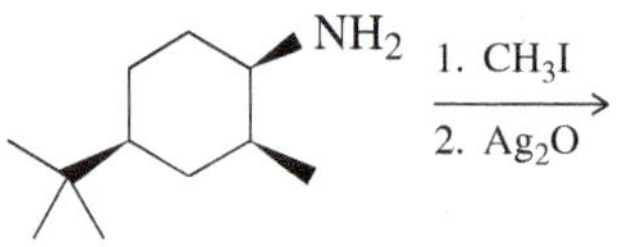

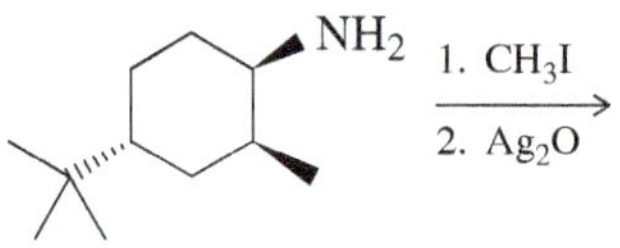

25.62 (••) A graduate student attempted the following synthesis, but found an unexpected peak in the NMR spectrum of the final product. (a) Identify the actual product and (b) modify the synthesis so the graduate student can make the desired compound.

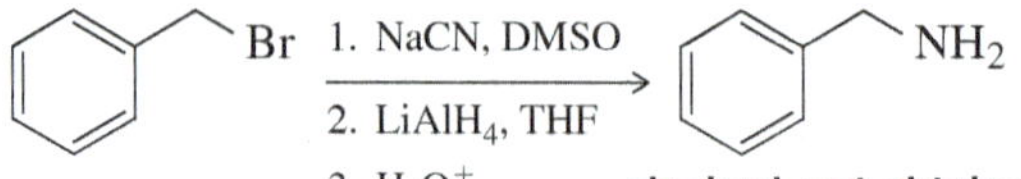

desired, not obtained

25.63 (••) Starting with an amine, suggest a synthesis of the compounds shown.

(a)

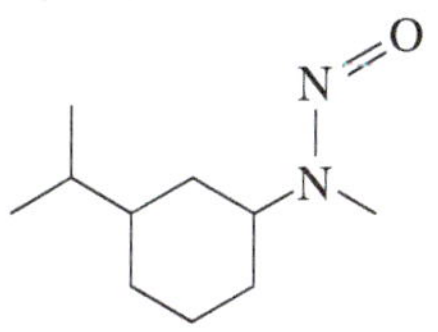

(b)

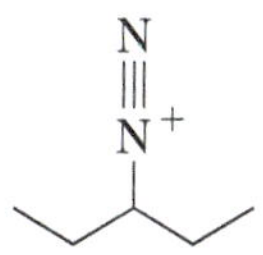

(c)

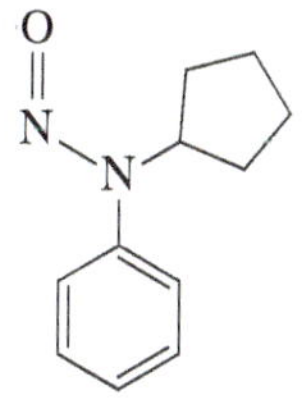

(d)

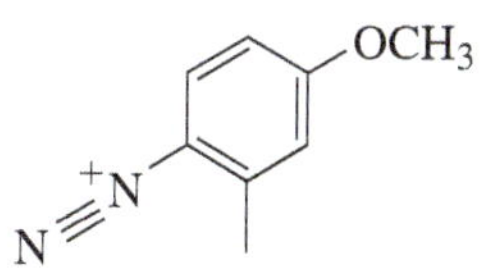

25.64 (••) **LOOKING AHEAD** Amino acids exist in their zwitterionic form, essentially as ammonium carboxylates. Why?

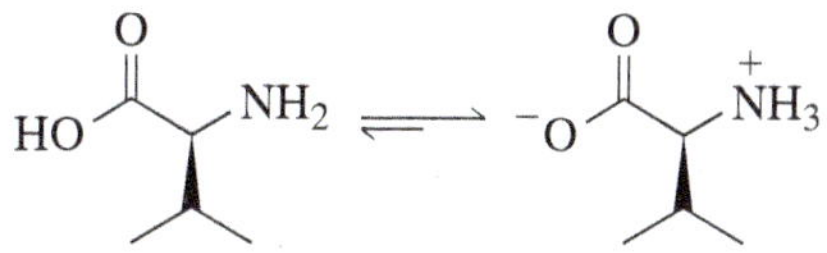

25.65 (••••) **LOOKING BACK** Suggest a mechanism for the reductive amination/cyclization shown. [*The second part of this reaction was covered in Chapter 21.*]

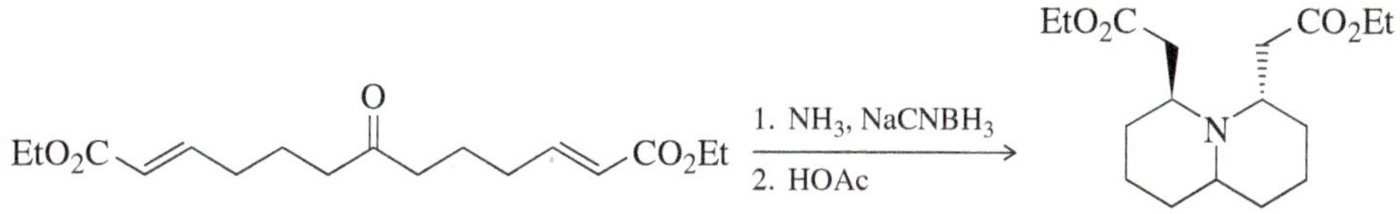

25.66 (••••) Suggest a mechanism for the reaction shown. [*The mechanism does not need to be shown for the reduction of the nitro group.*]

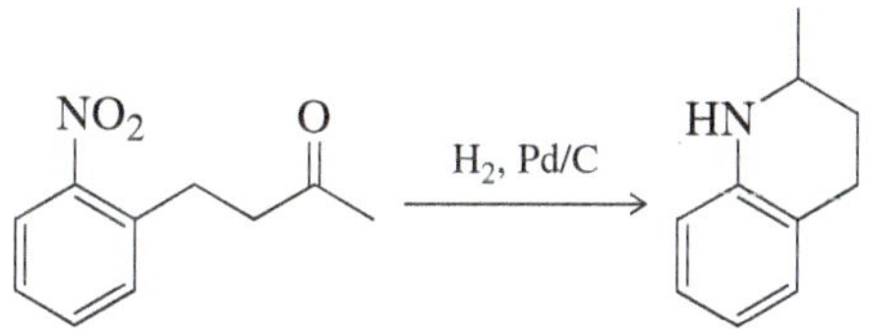

25.67 (••••) The final step in the synthesis of diazoxide, a drug used to treat low blood pressure, is shown here. Suggest a mechanism for this step.

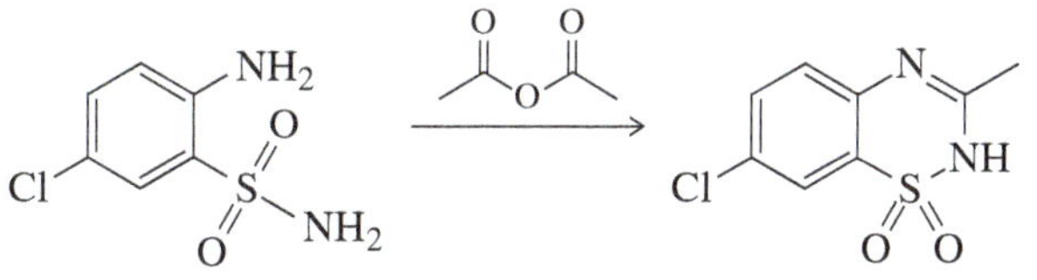

25.68 (••••) The synthesis of α-hydroxy acids can be done starting with amino acids. Suggest a mechanism of the two-step transformation shown. [*The alcohol oxygen is the same in the reactant and product.*]

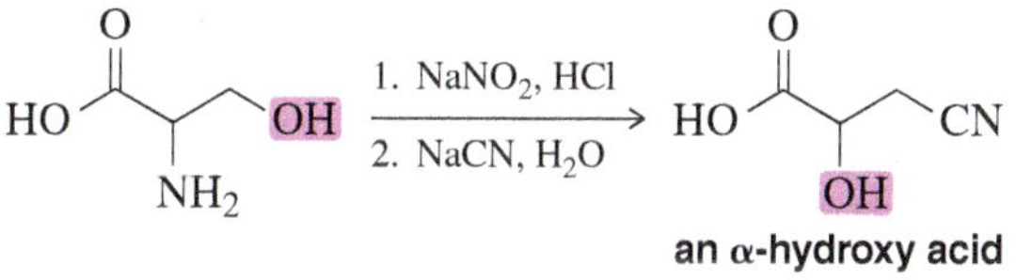

25.69 (••) Acetylcysteine is used to treat acetaminophen poisoning and cystic fibrosis. Propose a starting amine and the reagents necessary to make this drug.

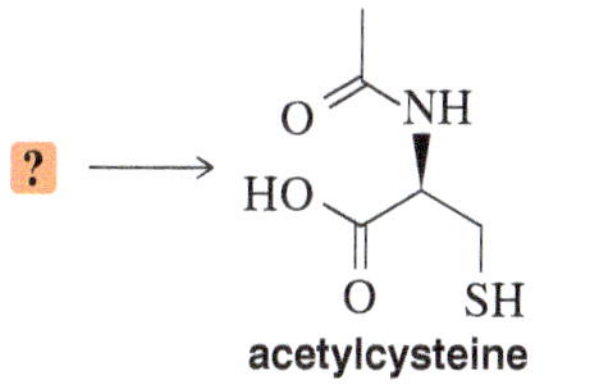

25.70 (••) You and your rival are engaged in an organic synthesis competition. [*I wish there were such things.*] Your rival's route from toluene to *p*-chlorotoluene is shown here. (a) Fill in the reagents used in their route. (b) Suggest the route you would use to easily win the competition.

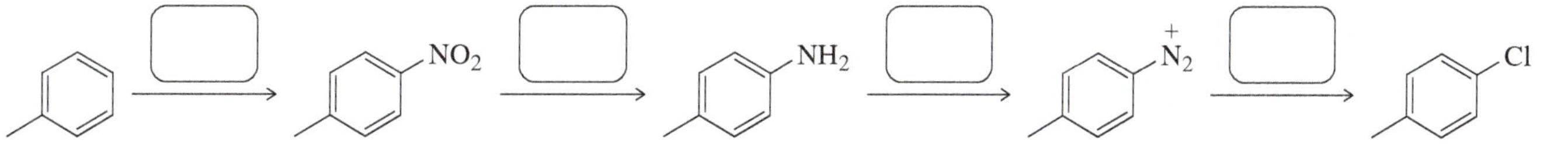

25.71 (••) FROM THE LITERATURE The following synthesis was recently reported. Fill in the boxes for the indicated steps (*Org. Lett.* 2020, *22*, 2354–2358).

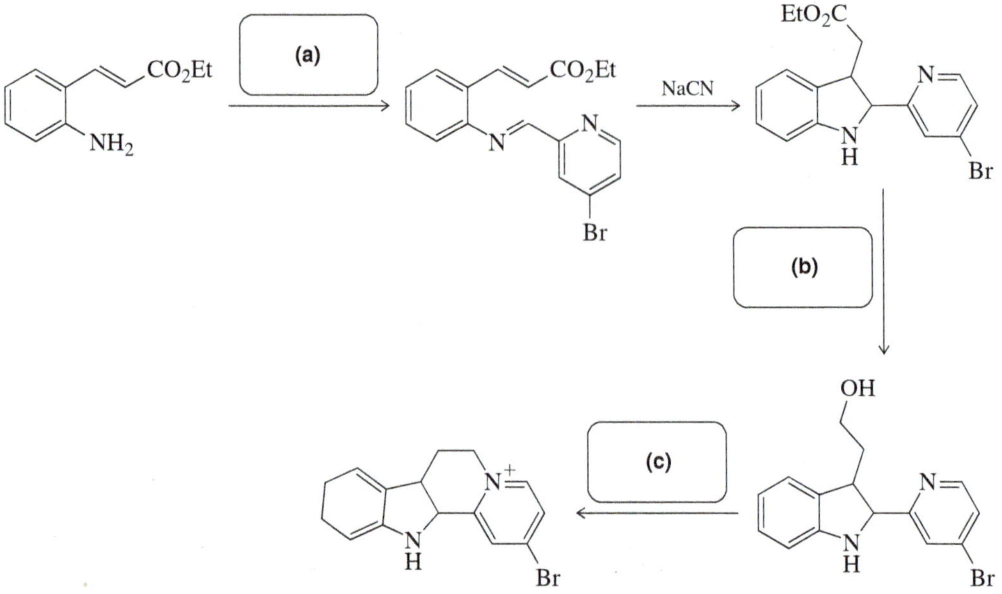

25.72 (•••) Neostigmine is a drug used to treat certain muscular conditions that affect the eye. Propose a synthesis of this compound beginning with (a) *m*-bromophenol and (b) *m*-nitrophenol.

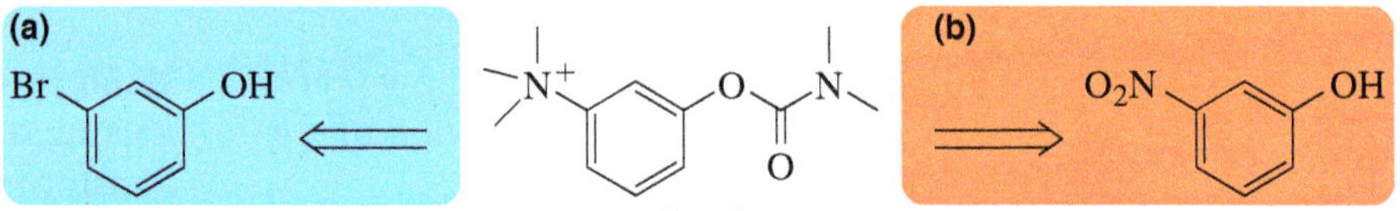

neostigmine

25.73 (••••) LOOKING BACK Suggest a synthesis of the molecule on the right using the malonic ester synthesis learned in Chapter 20. [*You will use reactions from several different chapters for this synthesis.*]

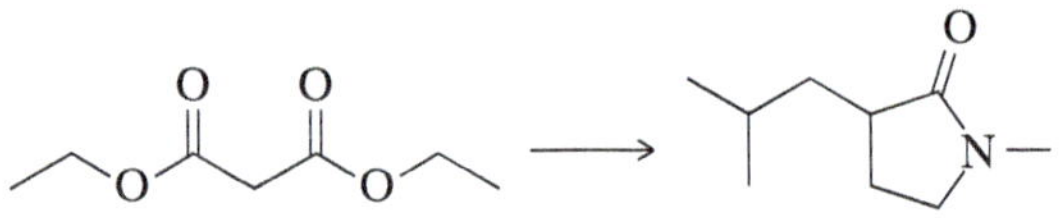

26

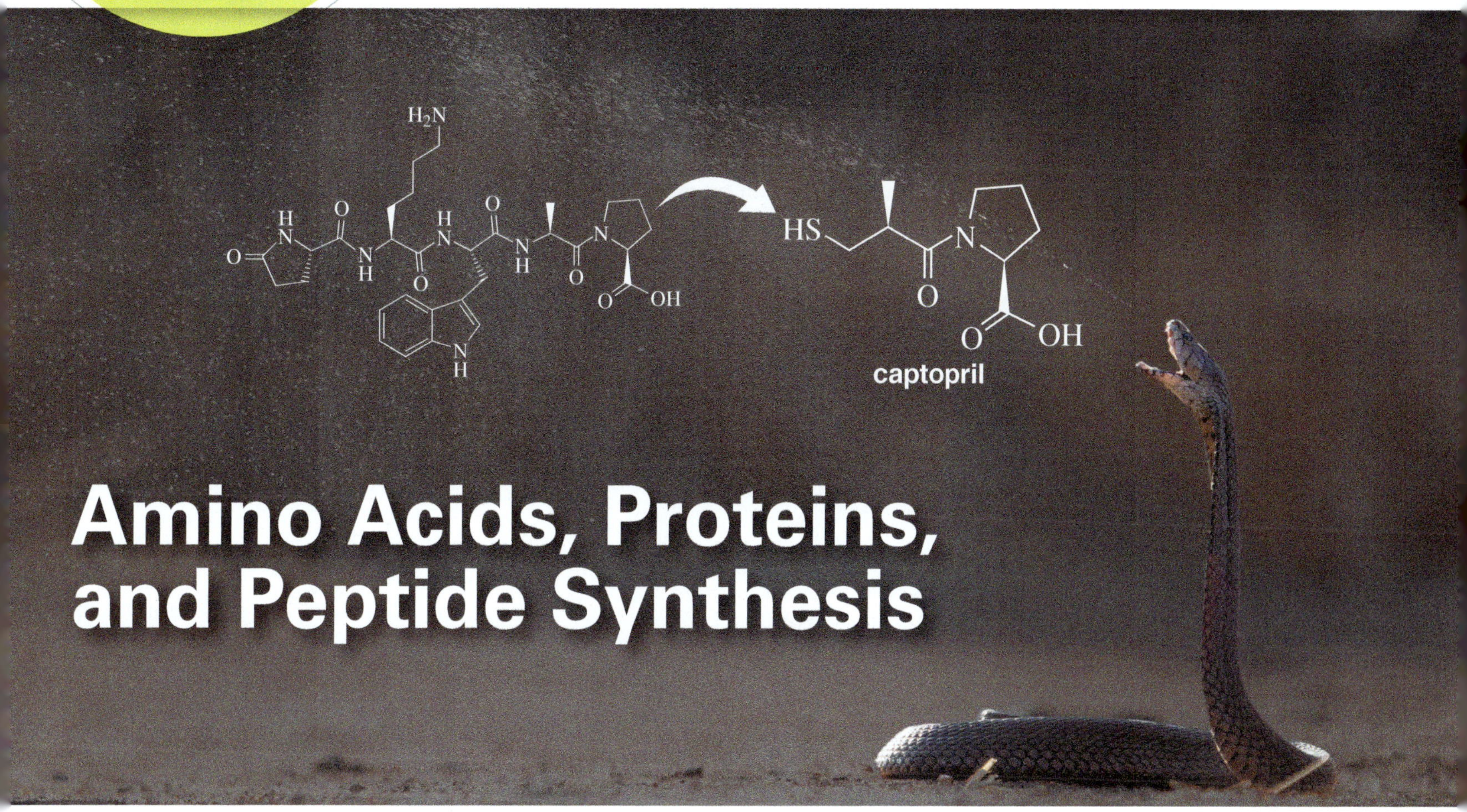

Amino Acids, Proteins, and Peptide Synthesis

Introduction

As blood is pushed from the heart to your organs, it presses against the arterial walls. The force with which it presses against the walls is called blood pressure. While blood pressure commonly fluctuates, it can cause health problems when it is too high for too long, including doing damage to your heart. Besides heart disease, high blood pressure is also a common cause of stroke—another leading cause of death in the United States. There are no obvious symptoms of high blood pressure, so it is considered a "silent killer"—by the time it is noticed, it is often too late. To ensure that it is caught early, a blood pressure screen is conducted during every visit to a physician.

While a proper diet and exercise can lower blood pressure, medications have been developed to treat this condition. The medications, which mimic the structure of peptides (made of amino acids), inhibit an enzyme (a protein, made of amino acids) that itself cleaves peptides (made of amino acids). In studying amino acids, we'll learn how these medications were developed . . . beginning with a snake.

○ ○ ○ ○ ○ ○ ○

Throughout this book, we have discussed the application of organic chemistry reactions to fields beyond this course. Many of these applications were in the field of biochemistry, for two important reasons. Biochemistry is the link between organic chemistry and many other disciplines—medicine, medicinal chemistry, pharmacology, and biology, among others. And biochemistry is most likely one of the next chemistry courses you'll take, especially if you are in a pre-health program.

Learning Objectives

LO 26.0 Describe the structure of amino acids and isomers of amino acids

LO 26.1 Analyze the properties of amino acids in acid–base reactions

LO 26.2 Analyze amino acid synthesis reactions

LO 26.3 Analyze reactions of amino acids

LO 26.4 Interpret the basic structures of proteins

LO 26.5 Evaluate the primary, secondary, tertiary, and quaternary structures of proteins

LO 26.6 Explain how the identity of an unknown protein is found through protein fragment analysis

LO 26.7 Analyze peptide fragment synthesis

"One way to get high blood pressure is to go mountain climbing over molehills."

—Earl Wilson

Let's start by assessing our understanding of the organic chemistry concepts needed to learn about peptides.

This chapter begins a two-chapter run where biological molecules are the focus. Specifically, amino acids and proteins are covered here in Chapter 26, with carbohydrates, nucleic acids, and lipids covered in Chapter 27. Given that there are entire semester- and year-long courses dedicated to the study of these biological molecules, these two, relatively short, chapters are not at all meant to give you a comprehensive treatment. Instead, think of them as an on-ramp to biochemistry, showing you how the language, the concepts, and the reactivity of organic chemistry will translate.

The metaphor of an on-ramp is pertinent. The material we have studied thus far logically, and gradually, extends into biochemistry. After all, biological molecules are simply organic molecules, containing organic functional groups with which we are quite familiar. For example, the reactivity of amines (Chapter 25) and carboxylic acids (Chapter 18) combines to make up the chemistry of amino acids. Proteins are themselves another type of amide (Chapter 19). Negatives will still attack positives, and the ball will still roll downhill. As a result, much of what we learned in previous chapters is reiterated here.

Test your understanding

Assessment 26.1 Give the pK_a value of the following compounds.

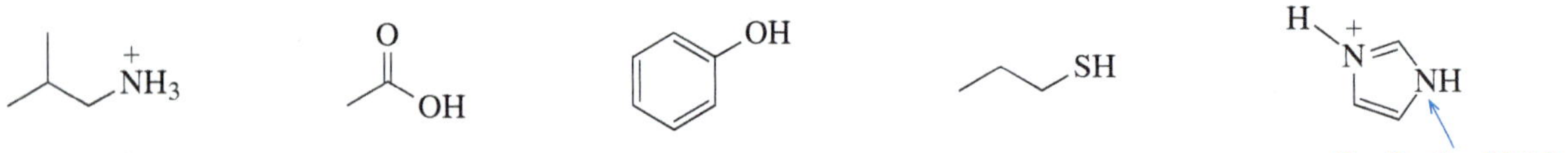

Assessment 26.2 Calculate the equilibrium constant for the reaction of ethylamine and propanoic acid. Which side is favored?

Assessment 26.3 Rank the reactivity of the following carboxylic acid derivatives with nucleophiles. (1 = most reactive; 4 = least reactive)

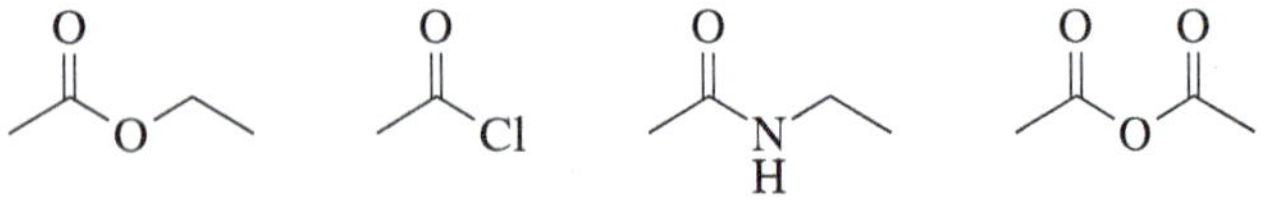

Assessment 26.4 Provide a mechanism for the acid-catalyzed hydrolysis of an amide.

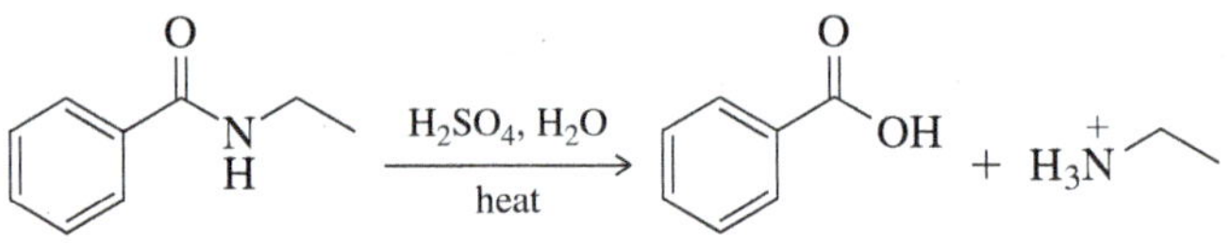

Assessment 26.5 Although there is free rotation around most σ bonds, rotation is restricted around the C—N bond of an amide. Explain why this is true.

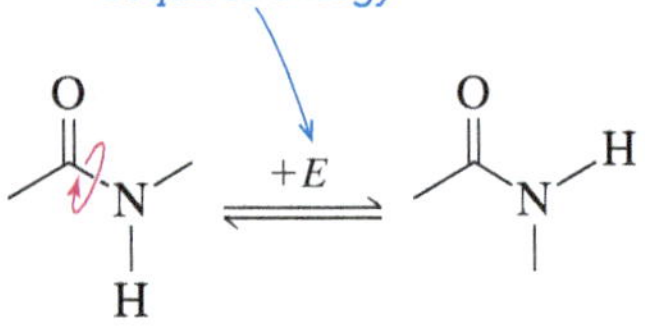

Assessment 26.6 Predict the product of the following reactions we have studied previously.

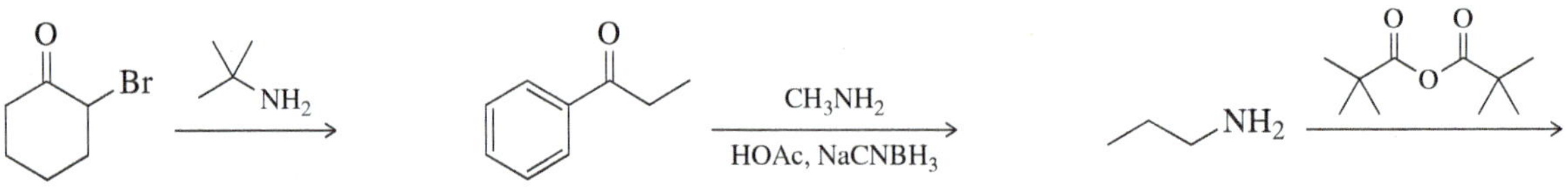

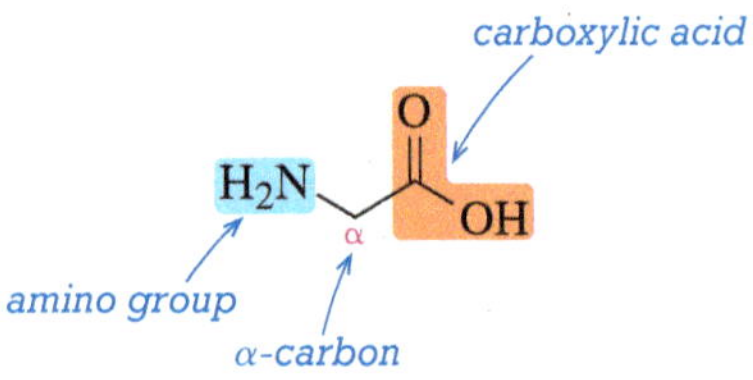

Figure 26.1 The structure of glycine, the simplest amino acid.

26.1 General structure of amino acids

As the name implies, amino acids are composed of an amino group and a carboxylic acid. Although any molecule containing both of these functional groups could be called an amino acid, this chapter focuses exclusively on the naturally occurring α-amino acids most commonly used for protein synthesis. In these, the amino group is attached to the α-carbon of a carboxylic acid, as shown in **Figure 26.1**.

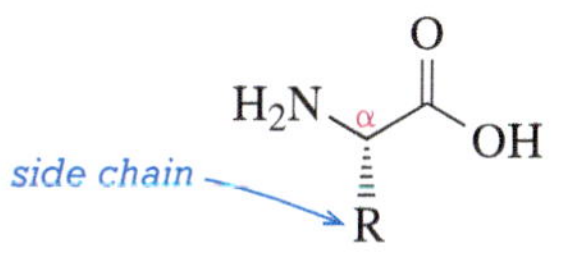

Figure 26.2 Amino acids have a unique side chain (—R) attached to the α-carbon.

The other two groups bonded to the α-carbon of glycine are hydrogen atoms. In the other 19 naturally occurring α-amino acids, one of those hydrogens is replaced by one of 19 different side chains. The structural diversity that allows proteins to serve as enzymes for biological reactions (Chemistry in Context 5-A) or receptors for neurotransmitters (Chemistry in Context 25-A) would not be possible without the variety of these side chains, which are generally represented as R (**Figure 26.2**).

26.1.2 Stereochemistry of amino acids

The four different groups attached to the α-carbon (—CO_2H, NH_2, —R, and —H) make the α-carbon a chiral center. As a result, the amino acids [*except glycine*] and the proteins they compose are chiral molecules. In one of the unsolved mysteries of the beginning of life on Earth, all but one of the naturally occurring amino acids has the *S* configuration (**Figure 26.3**). [*Perhaps all the R amino acids are in an alternate universe* . . . .] Only the amino acid cysteine has the *R* configuration.

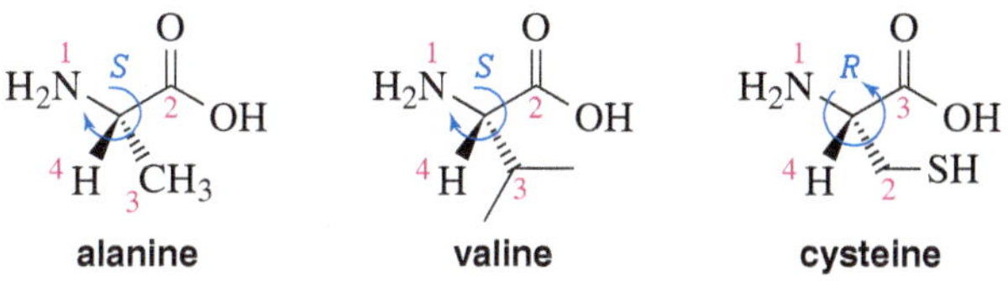

Figure 26.3 All but one amino acid has the *S* configuration.

It is remarkable that 18 of the naturally occurring amino acids have the *S* configuration. Perhaps even more remarkable, though, is that cysteine is *R*, at least until you understand why. Consider the Fischer projection of alanine, valine, and cysteine shown in **Figure 26.4**. Fischer projections, as originally defined, have the longest carbon chain in the vertical plane with the carbon in the highest oxidation state at the top. Thus, the amino group for all of the amino acids, even cysteine, ends up on the left in Fischer projections. [*Fischer projections were covered in Section 6.5.3 and are reviewed in the context of carbohydrates in Section 27.2.1.*] This is referred to as the L configuration. The DL system defines a sugar or amino acid based on which side the hydroxyl or amino group is pointing toward at the bottom-most stereocenter in the Fischer projection. The configuration is referred to as D if the hydroxyl or amino group is on the right (*dexter*), and it is referred to as L if it is on the left (*laevus*). All 19 chiral amino acids have the L configuration. This is one reason why biochemists prefer the DL naming system.

Figure 26.4 All amino acids, even cysteine, have the L configuration.

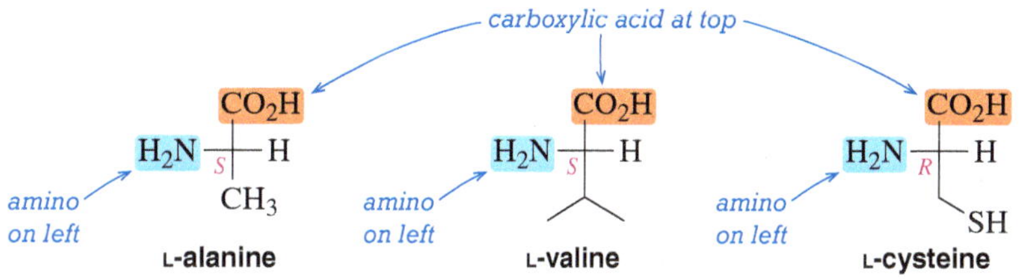

So why is cysteine *R*? It's a result of the Cahn–Ingold–Prelog system that prioritizes groups based on the atomic number. Cysteine is the only amino acid where the second atom of the side chain (the sulfur) is of higher priority than the oxygen of the carboxylic acid. Thus, it isn't that the stereocenter is aligned differently; it's just that the organic chemistry convention defines it as such.

Assessment 26.7 Although humans do not use D-amino acids, bacteria do, often synthesizing them using enzymes called racemases. Draw the enantiomers of the amino acids shown in Figure 26.4 using both line-angle drawings and Fischer projections. Determine the *R*/*S* and D/L configurations around the α-carbon.

Assessment 26.8 Suppose the first amino acid ever formed on Earth did so via a palladium-catalyzed hydrogenation of an imine. Would you expect to produce the D-amino acid, the L-amino acid, or a mixture of both?

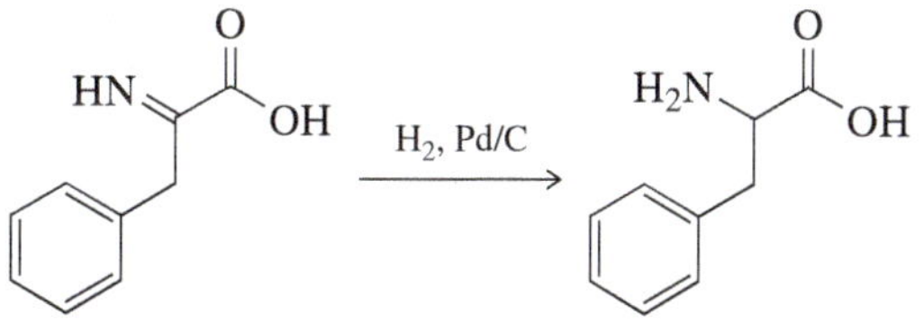

26.2 Standard amino acids

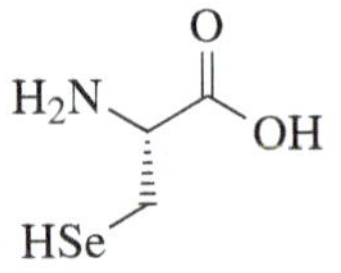

Figure 26.5 Selenocysteine.

While any compound containing an amino group and a carboxylic acid group can correctly be called an amino acid, there are only 20 amino acids in eukaryotes that are incorporated into proteins during translation. A 21st amino acid, selenocysteine (**Figure 26.5**), is encoded in a different way than the other 20, so it is excluded from this list. These amino acids are referred to as **proteinogenic amino acids** and are shown in **Table 26.1**. Humans are able to synthesize 11 of these from other molecules, whereas the other 9 are called **essential amino acids** because they must be obtained from our diets. Amino acids can be abbreviated using the three-letter or one-letter codes in Table 26.1.

Table 26.1 Structures of the 20 amino acids. Names screened in orange are nonessential amino acids, whereas names screened in blue are essential amino acids. Carboxylic acid pK_a values are screened in purple. The pK_a values screened in green are for the ammonium ion (RNH_3^+). Side chain pK_a values are not screened.

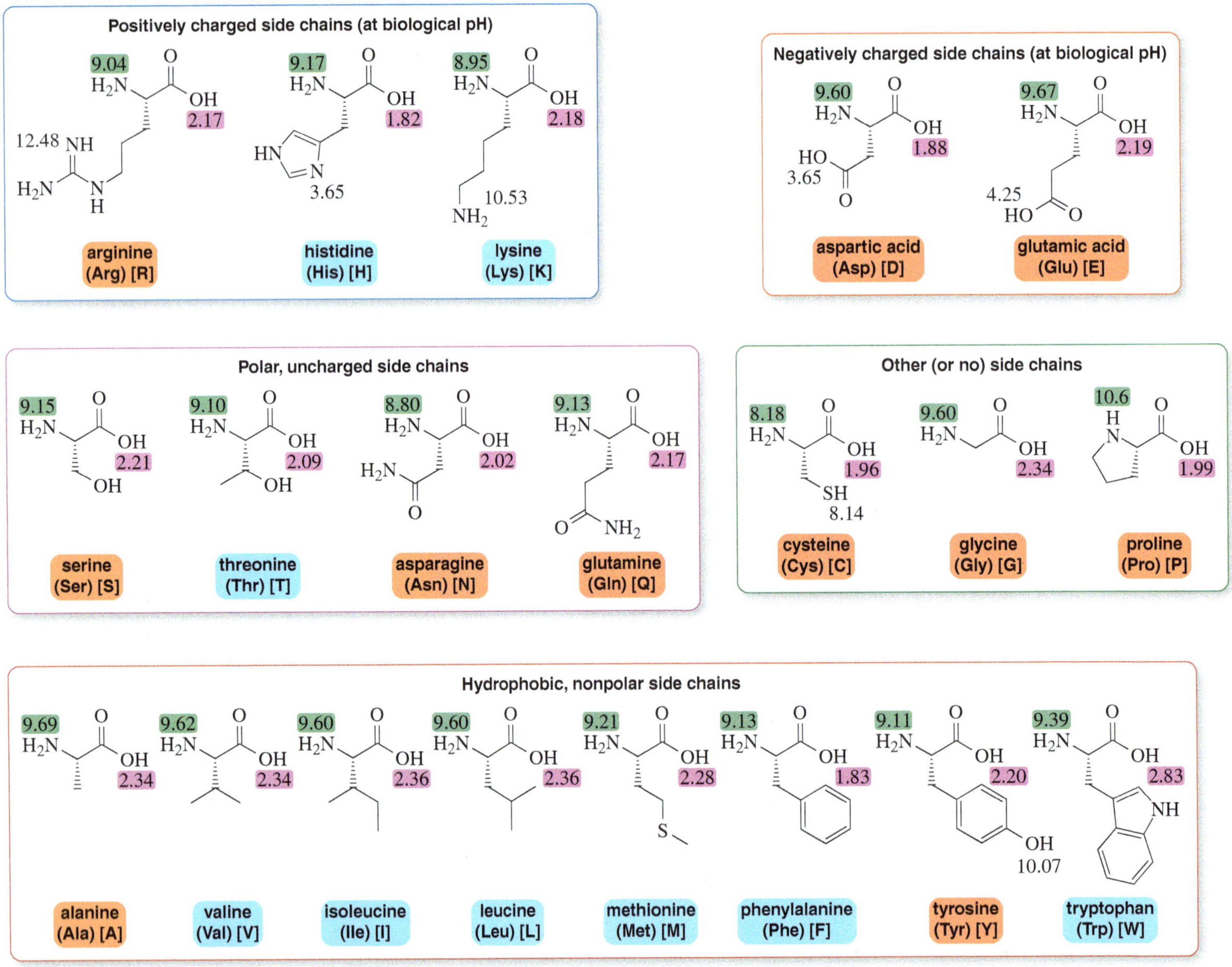

Table 26.1 contains a lot of information. First, the amino acids are shown in the neutral form containing a carboxylic acid and an amine, even though they exist at neutral pH as a carboxylate ion and an ammonium ion. [*This is explained in Section 26.3.*] The amino acids are classified by the physical properties of their side chains—namely, charged (negatively or positively), polar, and hydrophobic. Those that don't fit any of the categories (e.g., glycine, which has no side chain) are listed as special cases. Table 26.1 also lists the pK_a values of the carboxylic acids (usually ~2) and the pK_a values of the protonated amine (RNH_3^+; p$K_a \approx 10$). The pK_b value of the amines can be calculated from the pK_a (14 − pK_a = pK_b) [*recall Section 25.4.2*]. The pK_a values of any side chain containing an acidic proton are given, too. [*With so much information compiled in one table, you will benefit from referring to it often.*]

26.3 Acid–base reactions of amino acids

Amino acids contain both a basic (amine) and acidic (carboxylic acid) functional group, so acid–base chemistry dominates much of their reactivity.

Before delving into the acid–base chemistry of amino acids, work Assessment 26.9 and then we'll solve it together.

Assessment 26.9 Calculate the equilibrium constant of the following acid–base reaction.

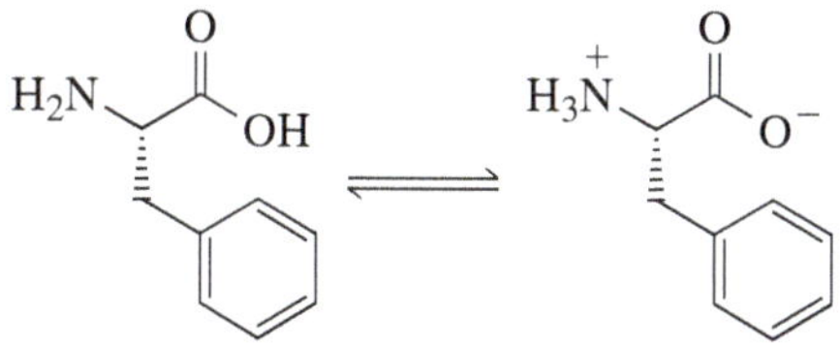

Solved Assessment 26.9

Recall from Chapter 4 that the equilibrium constant of an acid–base reaction can be calculated using the pK_a values of the acid and conjugate acid involved (review Equation 4.1). Because the reaction in Assessment 26.9 is intramolecular, identifying the acid and conjugate acid is not always straightforward. The arrow-pushing mechanism in **Figure 26.6** confirms that the carboxylic acid is the (left side) acid and the resulting ammonium ion is the conjugate (right side) acid.

Having identified the acid and conjugate acid, the pK_{eq} can be calculated by subtracting the pK_a of the right-side acid from the pK_a of the left-side acid. The resulting pK_{eq} can be converted to K_{eq} by dividing by −1 and raising 10 to that power, as shown in **Figure 26.7**. With a K_{eq} of 2.0×10^7, the equilibrium favors the right side of the reaction—that is, the side containing the ammonium and carboxylate ions. ■

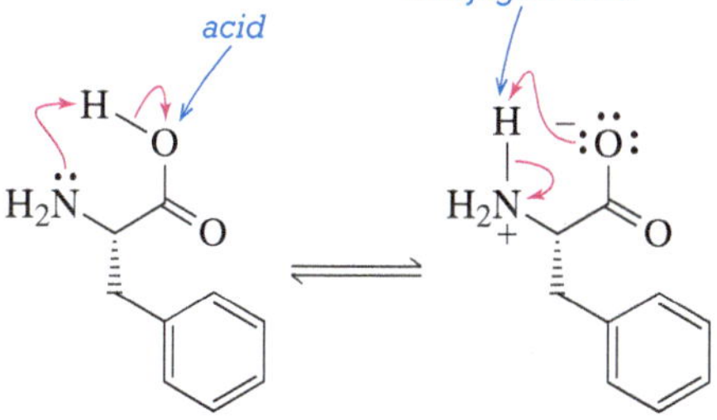

Figure 26.6 Drawing the mechanism helps identify the acid and conjugate acid.

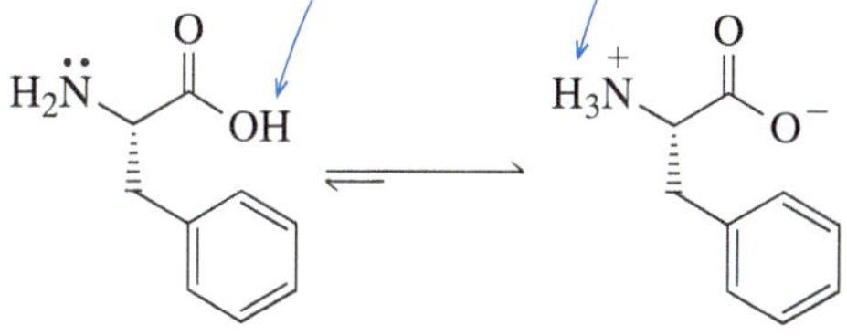

$$
\begin{aligned}
pK_{eq} &= pK_a(\text{left side}) - pK_a(\text{right side}) \\
&= 1.83 - 9.13 = -7.30 \\
-\log K_{eq} &= -7.30 \\
\log K_{eq} &= 7.30 \\
K_{eq} &= 10^{7.30} = 2.0 \times 10^7
\end{aligned}
$$

Figure 26.7 The right side of the equation is favored at equilibrium.

26.3.1 Zwitterionic structure

The result calculated in Assessment 26.9 has important consequences for the structure of amino acids. First, it explains why amino acids exist in the zwitterionic form shown in Figure 26.7. Second, pure amino acids in their zwitterionic form are overall neutral molecules, but they have extremely high melting points and water solubilities compared to the analogous amine and carboxylic acid (**Table 26.2**).

Table 26.2 The zwitterionic form of amino acids explains their high melting points.

Name	Phenylalanine	2-Phenylethylamine	3-Phenylpropanoic acid
Structure	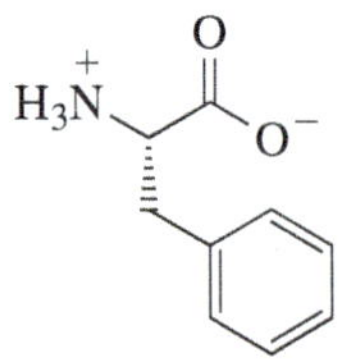	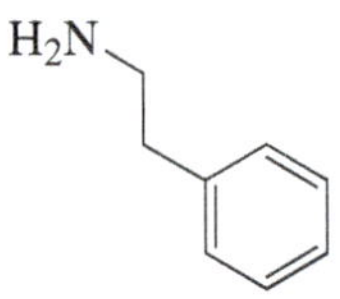	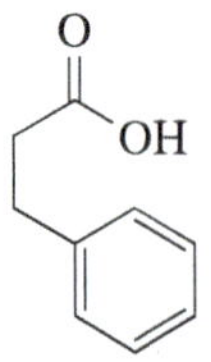
Melting point (°C)	283	−60	50
H_2O solubility (mg/mL)	50	2.2	5.9

The pK_a values for the amino groups in Table 26.1 are given for the ammonium ion (RNH_3^+), as opposed to the amine (RNH_2), because amino acids exist as the zwitterion. Thus, when a base deprotonates an amino acid, there is no proton to remove from the carboxylic acid. Instead, it removes the proton from the ammonium ion (**Figure 26.8**). In this way, all amino acids have a pK_a of first deprotonation between 9 and 10. [*This is explained in greater detail in Section 26.3.2.*]

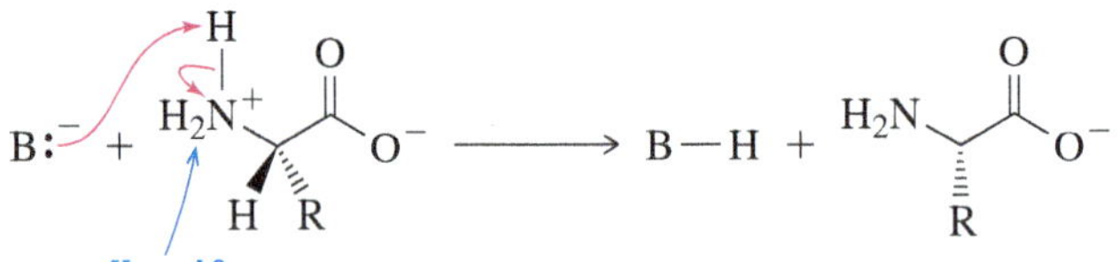

Figure 26.8 Deprotonation of amino acid in zwitterion form removes the proton from the ammonium ion as opposed to the carboxylic acid.

26.3.1 SUMMARY

- At neutral pH, amino acids exist in the zwitterionic form.

26.3.2 Isoelectric point

While pure amino acids exist in the zwitterionic form, their identity in solution depends on the pH of the solution. To understand why this is true, **Figure 26.9** shows the zwitterionic form under two extreme situations. When the pH of the solution is lowered (making it more acidic), the carboxylate is protonated, giving the amino acid a net +1 charge. When the pH of the solution is raised (making it more basic), the ammonium ion is deprotonated, giving the amino acid a net −1 charge.

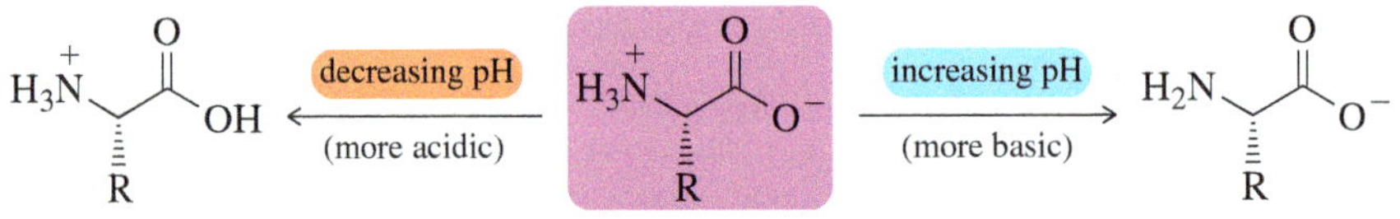

Figure 26.9 The structure of an amino acid depends on the pH of the solution.

The pH at which the zwitterionic form predominates is referred to as the **isoelectric point** (pI). It can be calculated from the pK_a values of all of the acidic protons in the amino acid. The calculation of pI for amino acids containing neutral side chains, acidic side chains, and basic side chains is slightly different.

Calculating pI for the hydrophobic amino acids (Table 26.1) is the simplest case and can be achieved by averaging the pK_a values of the ammonium and carboxylic acid protons, as shown in Equation 26.1.

$$\mathrm{pI} = \frac{\mathrm{p}K_{a(R\overset{+}{N}H_3)} + \mathrm{p}K_{a(RCO_2H)}}{2} \qquad \text{Equation 26.1}$$

The isoelectric points of the hydrophobic amino acids vary somewhat, but they tend to be between 5.5 and 6. **Figure 26.10** shows, for example, that alanine, valine, isoleucine, and leucine, which all have alkyl groups in their side chains, have nearly identical pI values.

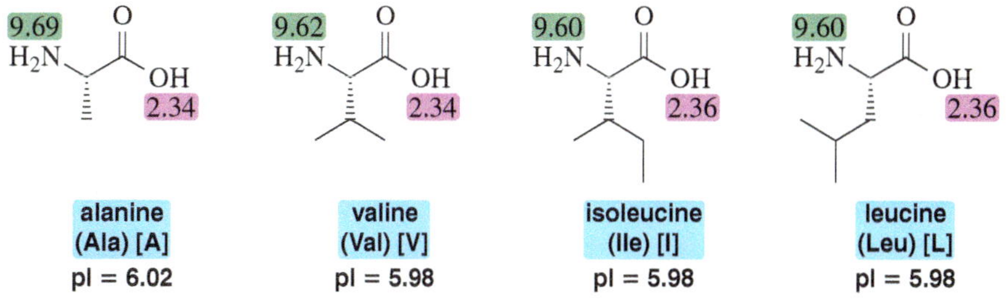

Figure 26.10 With no acidic or basic groups in the side chain, pI values are nearly identical.

Before reading further, would you expect amino acids with acidic side chains to have higher or lower pI values? What about amino acids with basic side chains? See if you can reason this out on your own before we think like a chemist together.

Thinking Like a Chemist

To make sense of what happens when the amino acid has an acidic side chain, we have to understand exactly what Equation 26.1 is saying. Acid–base reactions are equilibrium processes that can be quantified with a titration curve. A titration curve for the two acidic protons of glycine is shown in **Figure 26.11(a)**. When the pH < 2.34 (pK_a of RCO_2H), a majority of molecules in the solution have a $+1$ charge. When the pH > 9.69 (pK_a of RNH_3^+), a majority of molecules in solution have a -1 charge. As you move up from 2.34, there are more and more neutral molecules [zwitterions; **Figure 26.11(b)**]. As you move down from 9.69, there are more and more neutral molecules (zwitterions), too. Halfway between the 2.34 and 9.69, at pH = 6.02, the net charge on all molecules in solution is neutral. In general chemistry, we'd have referred to this as the equivalence point. These data can also be shown on a speciation curve that reports the identity of glycine molecules as a function of the pH of the solution [**Figure 26.11(c)**].

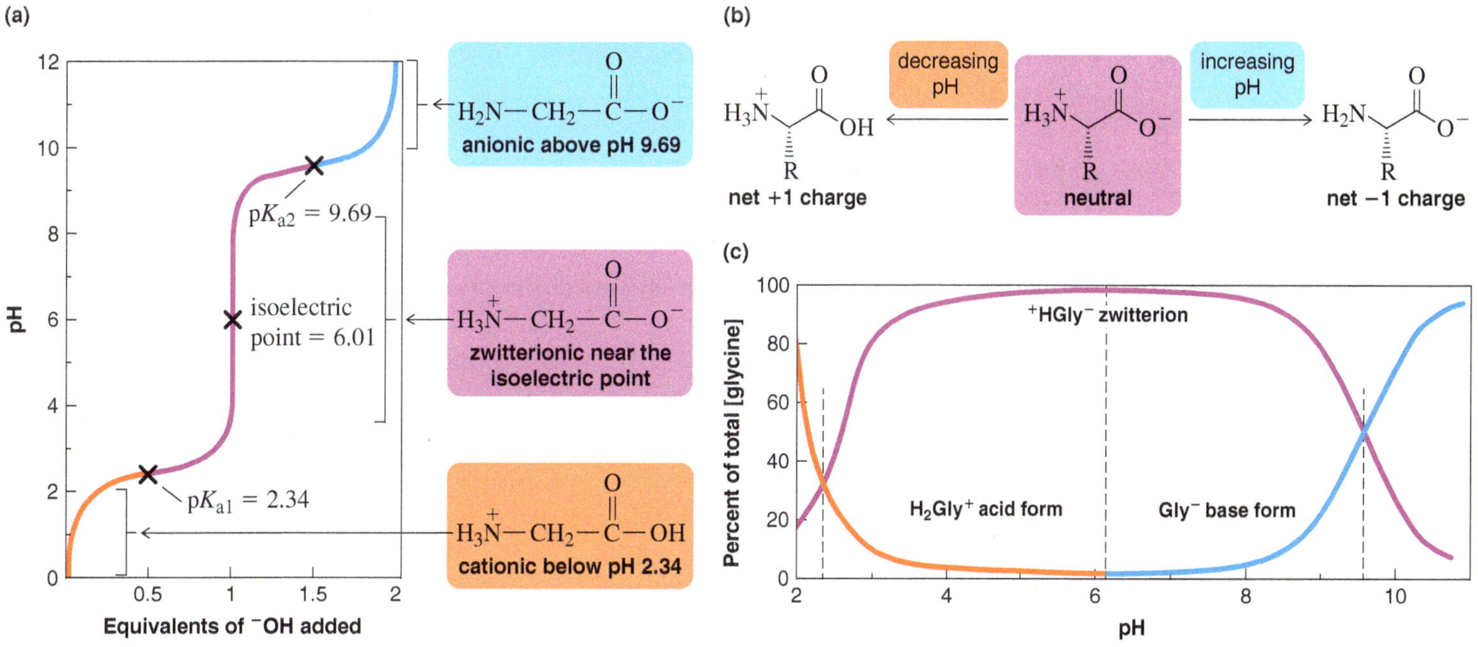

Figure 26.11 Halfway between 2.34 and 9.69 is the pI. (a) A titration curve for glycine. (b) The identity of glycine changes as the pH changes. (c) A speciation curve for glycine.

When an amino acid has an acid in the side chain, we now have three protons (and three equilibrium processes) to deal with. In glutamic acid, for example, the pK_a values are 2.19, 4.25, and 9.67 (**Figure 26.12**). By analogy to Figure 26.11, we need to identify the point at which all of the molecules are neutral. When the pH < 2.19, most of the molecules have a +1 charge. When the pH is between 4.25 and 9.67, most of the molecules have a −1 charge. When the pH > 9.67, most of the molecules have a –2 charge. So, at which pH are all of the molecules neutral? Just like with alanine, it's halfway between the pH values where you go from mostly +1 (2.19) to mostly −1 (4.25). Thus, pI = (2.19 + 4.25)/2 = 3.22 for glutamic acid. ■

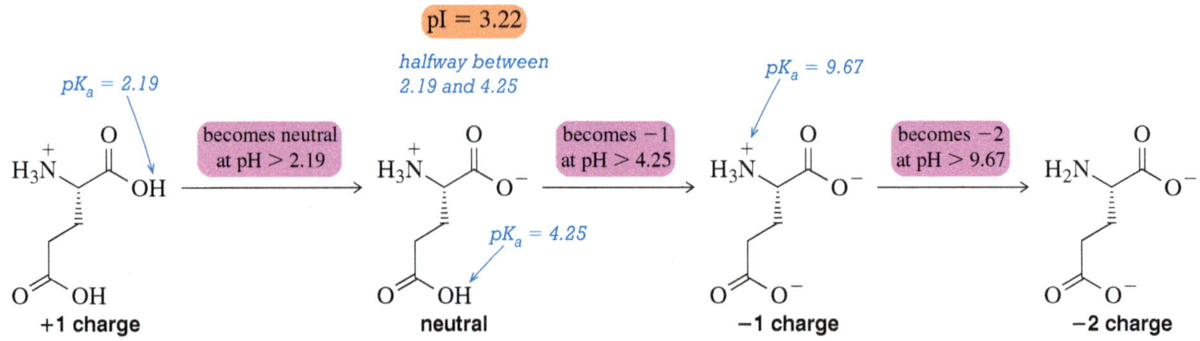

Figure 26.12 The pI = 3.22 for glutamic acid, which is halfway between 2.19 and 4.25.

We can do the same analysis for lysine (**Figure 26.13**), an amino acid with a basic side chain. Now the three pK_a values are 10.53, 8.95, and 2.18. When the pH > 10.53, most molecules have a −1 charge. When the pH < 8.95, most molecules have a +1 charge. When the pH < 2.18, most molecules have a +2 charge. As a result, all of the molecules are neutral halfway between 10.53 and 8.95, making pI = 9.74 for lysine.

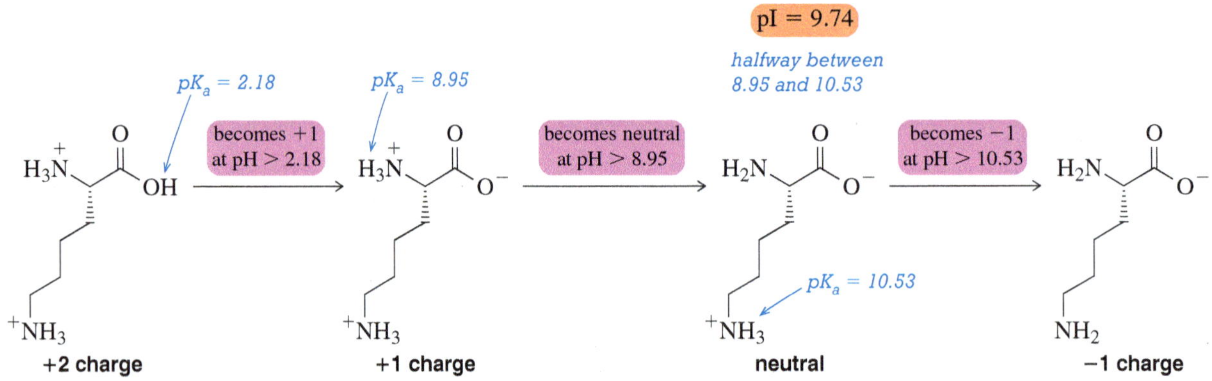

Figure 26.13 The pI = 9.74 for lysine, which is halfway between 10.53 and 8.95.

26.3.3 Electrophoresis

Electrophoresis is an analytical technique that can be used to separate and identify amino acids within a mixture on the basis of their isoelectric points. The separation is achieved by subjecting amino acids at a given pH to an electrical field and measuring the direction they move—namely, toward the cathode or toward the anode. Given that all of the amino acids studied thus far have a slightly different isoelectric point, at any given pH they will each have a different concentration of positive and negative ions. The net charge on a species, then, will dictate the direction and the degree to which it moves in either direction. Assuming a constant pH of 6, all amino acids whose isoelectric point is greater than 6 will have a net +1 charge. These positively charged molecules will move toward the cathode. Those with an isoelectric point less than 6 will have a net −1 charge and will move toward the anode. Those with an isoelectric point of 6 will be neutral and will not move in either direction. An example separation of alanine (hydrophobic), glutamic acid (acidic), and lysine (basic) is shown in **Figure 26.14**.

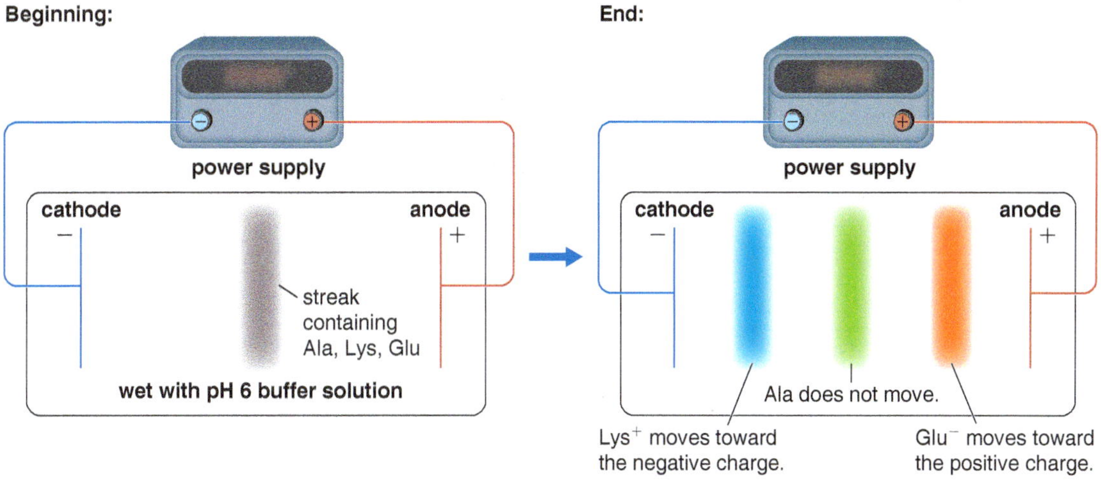

Figure 26.14 Electrophoretic separation of lysine, alanine, and glutamic acid.

26.3.4 Deviations in pK_a values

Some of the pK_a values we learned in Chapter 4 don't match up to the pK_a values in Table 26.1. Recall that carboxylic acids have pK_a $\approx$ 5, whereas the pK_a $\approx$ 2 for the carboxylic acid of the amino acids. The structure of an amino acid in solution depends on the pH of the solution (Section 26.3.2). The carboxylic acid is only protonated at low pH, when the amine is also protonated. Upon deprotonation of the carboxylic acid in this situation, the resulting carboxylate is always stabilized by a neighboring ammonium ion. Stabilization of the conjugate base like this is responsible for the deviation of the pK_a from 5 to 2 (**Figure 26.15**).

Figure 26.15 The carboxylate ion is stabilized by the ammonium ion, making the carboxylic acid of an amino acid more acidic than normal.

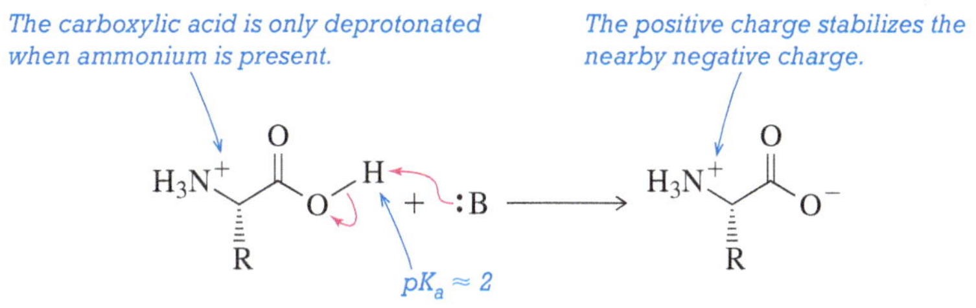

Proline (pK_b = 3.6) is the most basic amine of the amino acids, with all others hovering around 4.5 (**Figure 26.16**). This is consistent with enhanced electron donating of alkyl groups by induction and hyperconjugation present in a 2° amine, a concept discussed in Section 25.5.1.

Figure 26.16 Proline is more basic because the amine is secondary. [*pK_b values are being used in this figure.*]

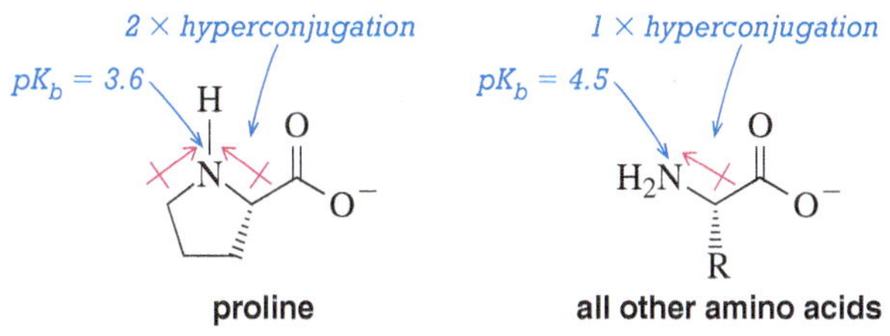

Another slight deviation occurs when you consider the pK_a values of the side chain carboxylic acids of aspartic acid (pK_a = 3.65) and glutamic acid (pK_a = 4.25) (**Figure 26.17**). This difference, although slight, can be attributed to the fact that the side chain carboxylate of glutamic acid is farther removed from the stabilizing ammonium ion. Because induction diminishes with distance, glutamic acid's pK_a value closely approaches that of a normal carboxylic acid (pK_a = 5).

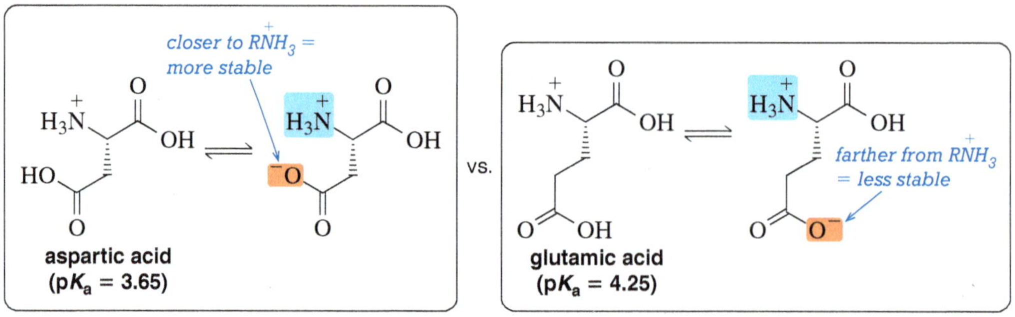

Figure 26.17 The side chain of glutamic acid is less acidic than that of aspartic acid because it is farther from the stabilizing ammonium ion.

Assessment 26.10 An early clue about the zwitterionic structure of amino acids came from measurements of the dipole moment. Which form would you expect to have the larger dipole moment?

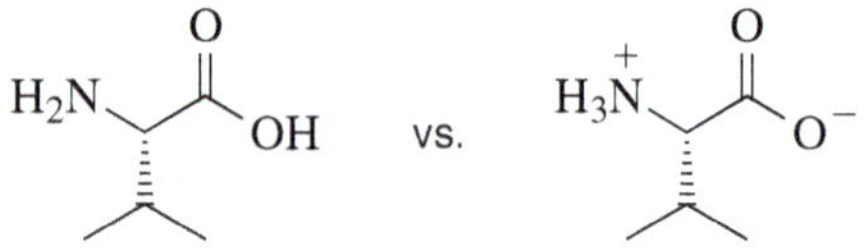

Assessment 26.11 Calculate the pI for the following amino acids.

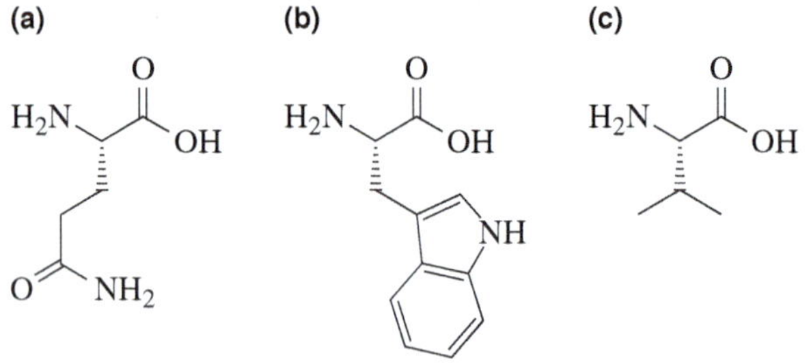

Assessment 26.12 Draw the predominant structure of the following amino acids at the given pH values.

(a) threonine at pH 12
(b) arginine at pH 10
(c) isoleucine at pH 8
(d) proline at pH 6

Assessment 26.13 In which direction would you expect each amino acid to move on an electrophoresis gel at the pH value given? Draw a picture of the final gel for each.

(a) His, Cys, Met at pH = 5 (b) Val, Asp, Gln at pH = 7.5

Assessment 26.14 Rationalize the difference in pK_a values of the ammonium ions in asparagine and glutamine.

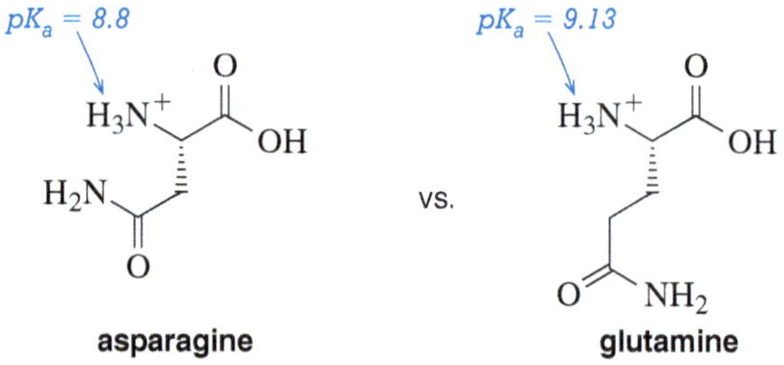

Assessment 26.15 A chemist recently synthesized the unnatural amino acid shown here. What pK_a value would you expect for the side chain carboxylic acid? Why?

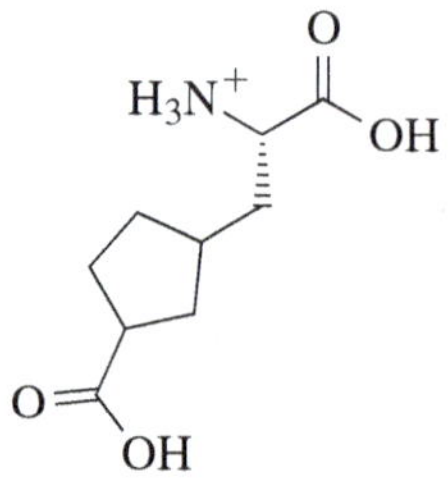

26.4 Amino acid synthesis

Many of the reactions we have studied previously can be used for the synthesis of amino acids. The following sections outline laboratory syntheses, not how amino acids are synthesized in the body.

26.4.1 Reductive amination

Recall from Section 25.6.2 that amines could be synthesized via the reductive amination of ketones. In this reaction, imine formation is followed by reduction to give a 1°, 2°, or 3° amine, depending on the original amine used, as shown in Figure 25.35. When this ketone is α to a carboxylic acid and ammonia is used as the amine, the resulting product is an amino acid. The synthesis of DL-phenylalanine by this reaction is shown in **Figure 26.18**.

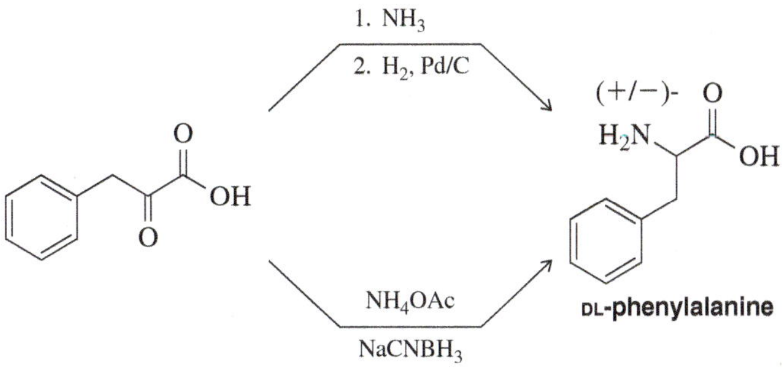

Figure 26.18 Reductive amination to make phenylalanine.

Assessment 26.16 (a) Suggest a mechanism for the reductive amination to make valine. (b) Why is it only possible to make racemic valine using reductive amination as shown? (c) How would you make reductive amination enantioselective? [*Recall Section* 9.3.]

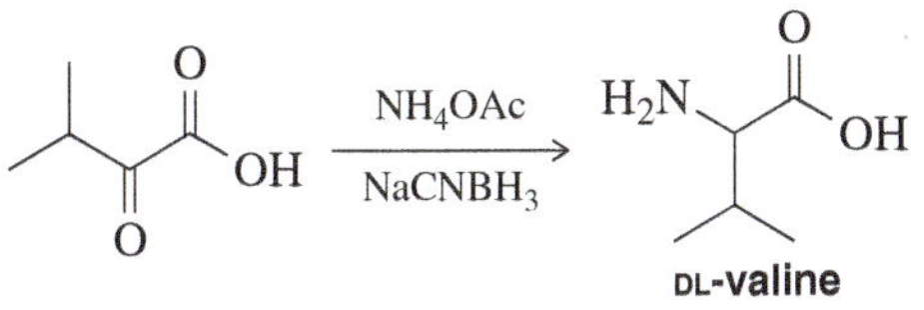

Assessment 26.17 β-Amino acids have been used as replacements for α-amino acids in proteins as a way of studying protein structure. How would you synthesize the following β-amino acid using reductive amination?

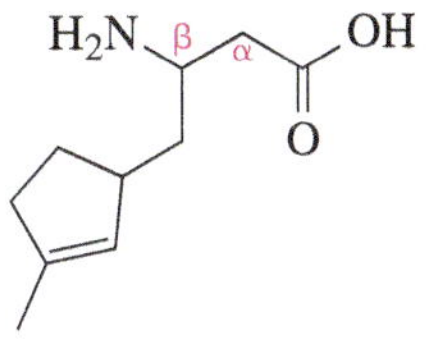

26.4.2 Amination of α-halo acids

In Chapter 25 (Chemistry in Context 25-A), the synthesis of bupropion, an antidepressant and smoking-cessation drug, was discussed. Specifically, the *t*-butylamino group of bupropion is incorporated by displacement of a bromide from the α-carbon of a phenyl ketone, as shown in **Figure 26.19**.

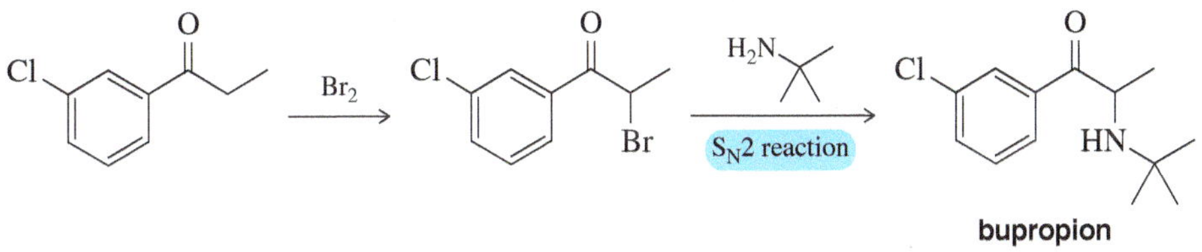

Figure 26.19 Enol halogenation followed by nucleophilic displacement with an amine gives bupropion.

Bupropion is an α-amino *ketone*, so we might be able to use this reaction to synthesize α-amino *acids*. The starting α-bromo acid must be prepared in a slightly different way, though. Whereas enol bromination readily occurs for aldehydes and ketones (Section 20.2.1.2), carboxylic acids are not as prone to enolization. As a result, incorporating bromine at the α-carbon requires the Hell–Volhard–Zelinskii (HVZ) reaction, shown in **Figure 26.20**.

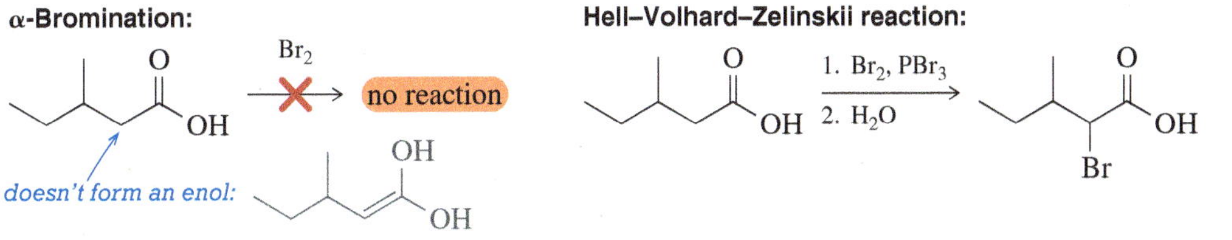

Figure 26.20 The Hell–Volhard–Zelinskii reaction is used for synthesizing α-bromo acids because direct enol bromination doesn't work.

The HVZ reaction is a combination of two reactions we have seen before—namely, $SOCl_2$ with a carboxylic acid (Section 18.7.5) and enol bromination (Section 20.2.1.2). In essence, α-bromination is accomplished by temporarily disguising the carboxylic acid as an acid bromide, thus allowing an enol to form (**Figure 26.21**). Formation of an acid bromide occurs via a nucleophilic addition/elimination mechanism [*see Section 19.6*]. Once formed, the acid bromide can now enolize. In the presence of Br_2, a bromine atom is incorporated α to the acid bromide via attack of the π bond [*see Section 20.2.1.2*]. Finally, hydrolysis of the highly reactive acid bromide regenerates the carboxylic acid.

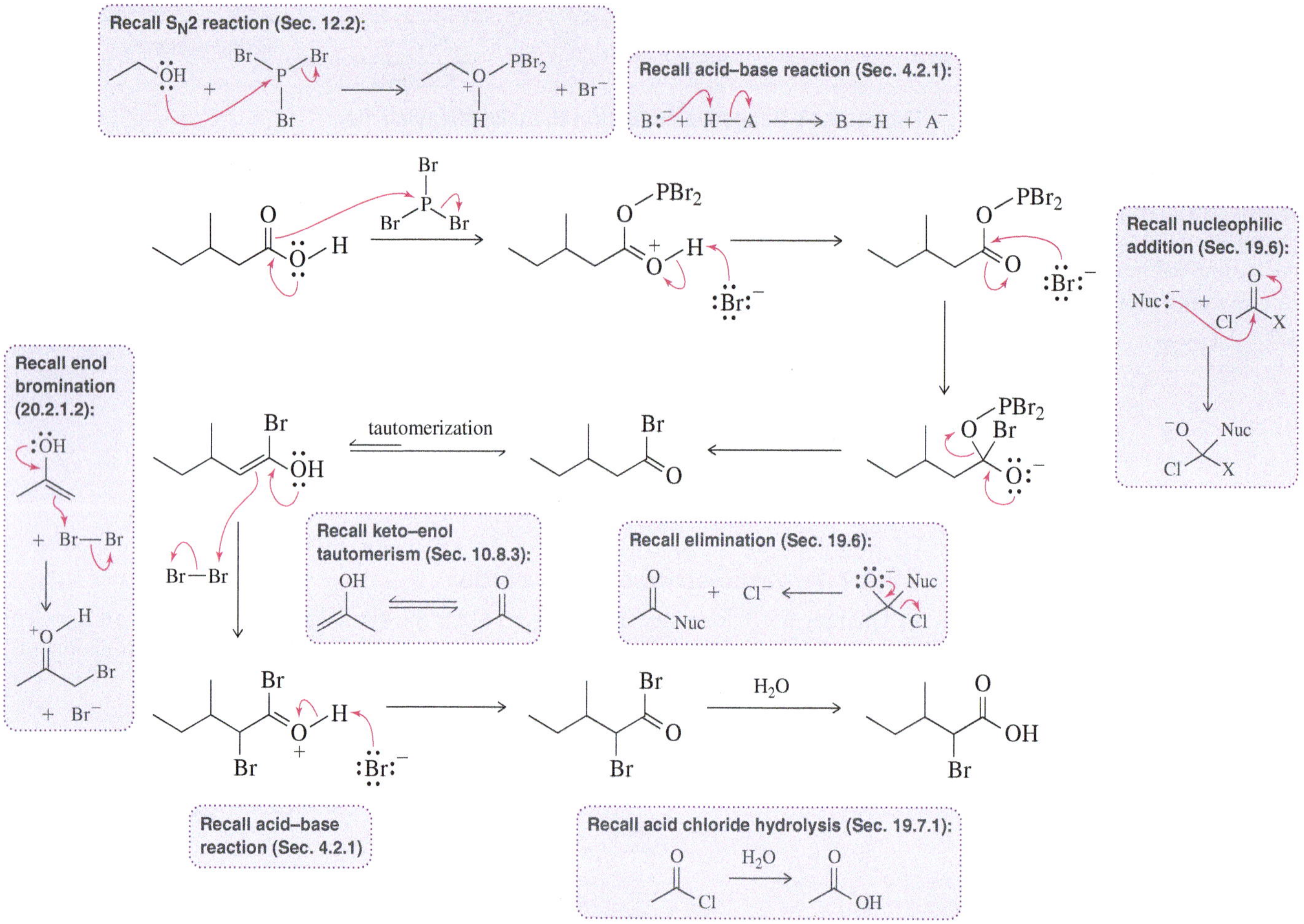

Figure 26.21 **The mechanism of the Hell–Volhard–Zelinskii reaction.**

Once the bromide is installed at the α-carbon, it can be displaced with ammonia in an S_N2 reaction to give DL-isoleucine (**Figure 26.22**), an amino acid. Whereas overalkylation can be a problem (see Section 25.5.1), this particular amine is too weak a nucleophile to alkylate further due to the neighboring carboxylic acid group, which is electron-withdrawing. It probably exists in the zwitterion form, too, which makes further alkylation even less likely.

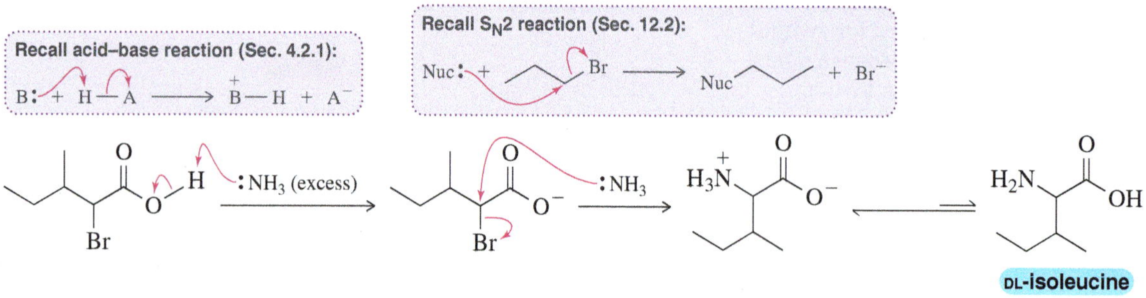

Figure 26.22 **S_N2 displacement of an α-bromo acid gives an α-amino acid.**

Assessment 26.18 Provide a synthesis of the following amino acids using a combination of the HVZ and amination reactions.

(a) Phe (b) Leu (c) Ala

Assessment 26.19 What is one driving force for the HVZ reaction?

Assessment 26.20 The reaction sequence shown in Figure 26.22 produces DL-isoleucine. Given that the S_N2 reaction is stereospecific, draw the α-bromo acid that would allow you to produce only L-isoleucine.

26.4.3 Gabriel malonic ester synthesis

Thus far, the methods for forming amino acids have all involved incorporating a new C—N bond into an existing carbon framework. In contrast, the Gabriel malonic ester synthesis (**Figure 26.23**) essentially derivatizes glycine by forming a new C—C bond at the α-carbon.

Figure 26.23 The Gabriel malonic ester synthesis for amino acid synthesis.

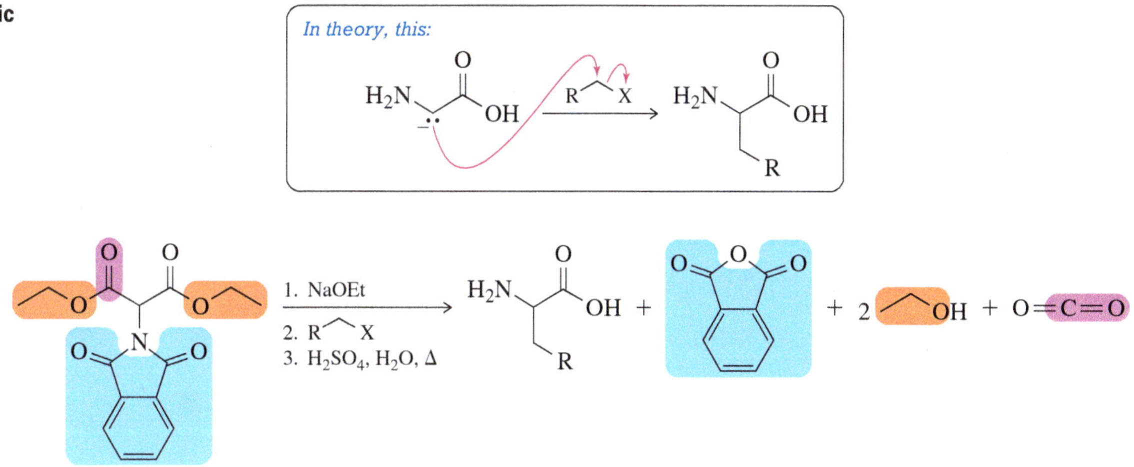

This reaction is essentially the alkylation of an enolate, but one that uses classical conditions. In Section 20.5.1, the malonic ester synthesis was used to make α-substituted acetic acid derivatives. And, in Section 25.5.4, the Gabriel amine synthesis was used to make 1° amines. These two reactions are combined in **Figure 26.24** for the synthesis of leucine. In the first step, alkylation of phthalimide with a bromomalonate incorporates the amino functional

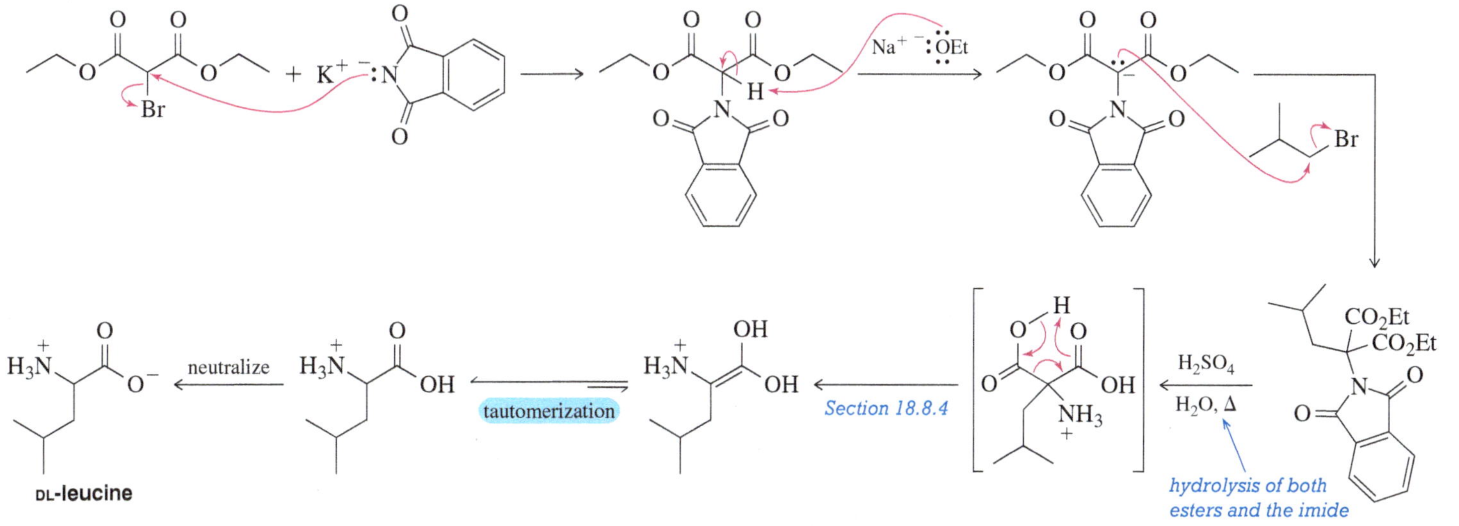

Figure 26.24 The mechanism of the Gabriel malonic ester synthesis of leucine.

group into the malonate. The remaining α-proton can be removed in a subsequent step, creating an enolate that can be used to displace 1° alkyl halides in an S_N2 reaction to form a new C—C bond. Exhaustive ester and phthalimide hydrolysis results in an α-ammonium diacid that undergoes decarboxylation, which upon neutralization forms leucine.

Assessment 26.21 Show how the Gabriel malonic ester synthesis might be used to make the following natural and artificial amino acids.

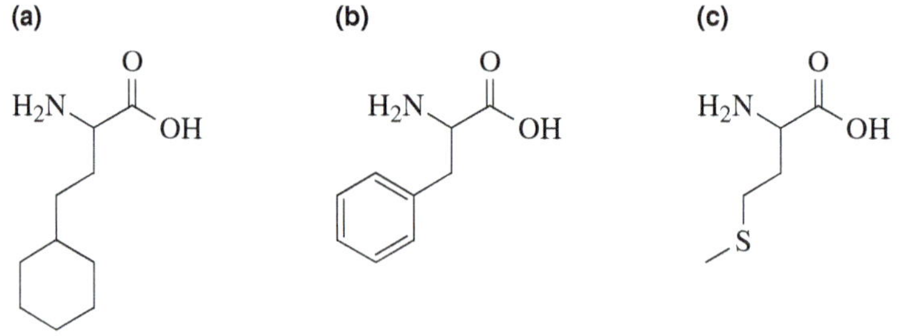

26.4.4 Strecker synthesis

The Strecker synthesis is a commonly used classical method for synthesizing amino acids.

The Strecker amino acid synthesis utilizes a combination of three previously studied reactions. Try solving the mechanism on your own in Assessment 26.22 and then we'll work it together.

Assessment 26.22 When 2-phenylethanal is treated with ammonium chloride and sodium cyanide, an intermediate forms. Hydrolysis of the intermediate produces phenylalanine. Suggest a mechanism for the reaction, including the identity of the intermediate.

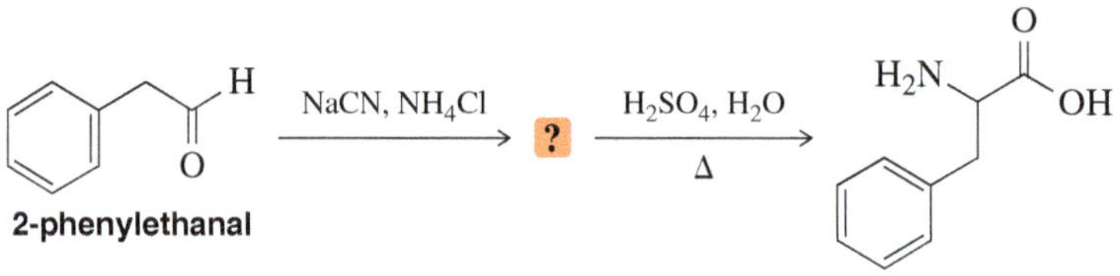

Solved Assessment 26.22

Besides numbering carbons as we normally do to, let's also try to figure out where each part of the product comes from. Notice, first, that the aldehyde has eight carbons, whereas the product has nine. Where does the other carbon come from? As the only other carbon source, it must come from the cyanide. Our hypothesis is aided by the fact that the carbon of cyanide has the same oxidation state as the carboxylic acid. The product has an amino group, whereas the reactant does not. What is the most likely source of the nitrogen? Cyanide has a nitrogen, but the ammonium chloride (NH_3 + HCl), because the nitrogen is already at the same oxidation state as it is in the product, is the more likely source. The product has two oxygens, whereas the reactants have one. What is the most likely source of the extra oxygen? At least one (maybe both) oxygen(s) in the product likely comes from water. By answering these questions, we can thus map the reaction (Figure 26.25).

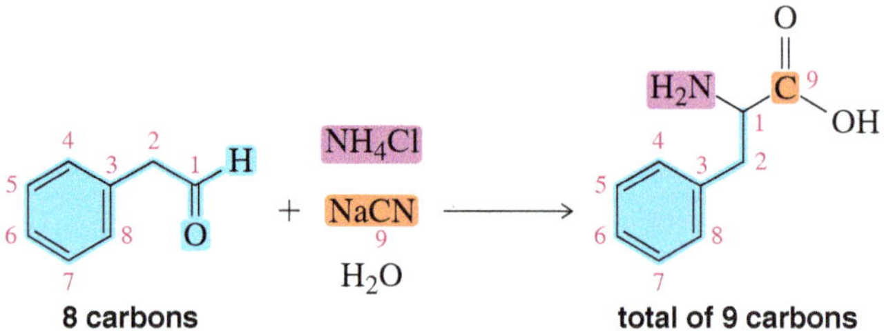

Figure 26.25 Mapping the reaction from Assessment 26.22.

Mapping the reaction also gets us to begin thinking about the reactivity (nucleophile, electrophile, acid, or base?) of the molecules involved. [*Do this on your own first.*] Aldehydes are typically electrophiles at C_1. The rest of the aldehyde (C_2—C_8) is unchanged in the reaction. Ammonium chloride, in solution, can be thought of as the combination of ammonia (nucleophile/base) and hydrochloric acid (acid). Sodium cyanide is a good nucleophile. From this analysis, we have two nucleophiles, cyanide and ammonia, that both need to add to the electrophilic carbonyl. Thus, our reaction could start in either of two ways, both of which we have seen before. When cyanide reacts with carbonyls, a cyanohydrin is produced [Figure 26.26(a)]. Ammonia, on the other hand, reacts with aldehydes to form imines [Figure 26.26(b)]. In each case, what must happen next to get to the product?

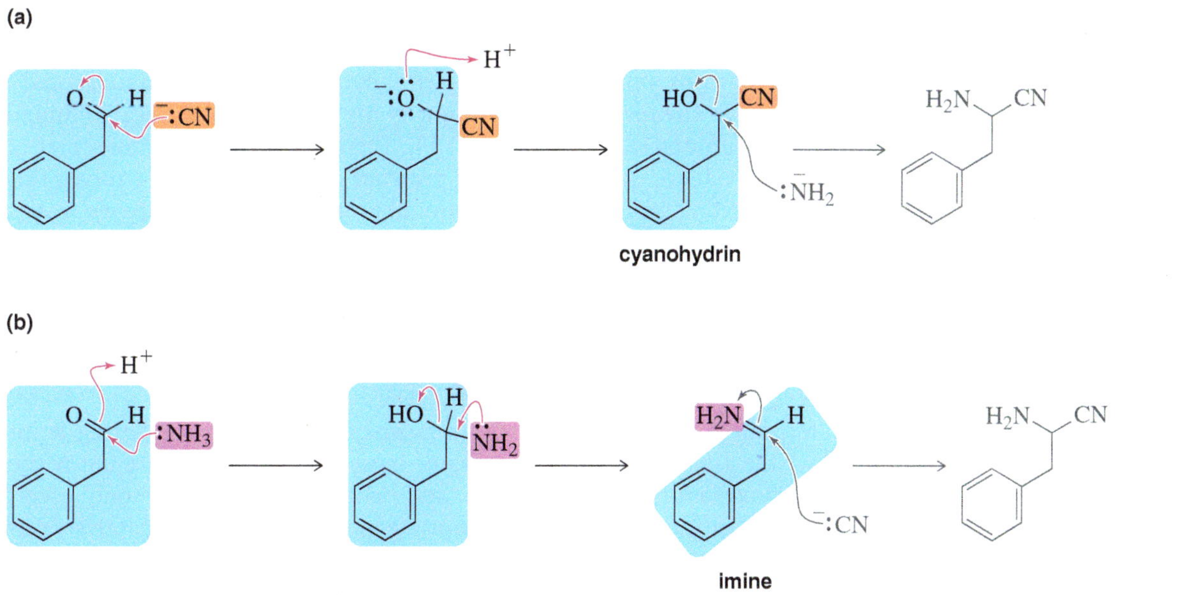

Figure 26.26 **Possible mechanisms of the first step.** (a) Cyanide ion reacts with the aldehyde to make a cyanohydrin, whereas (b) ammonia reacts with the aldehyde to form an imine. [*The arrows are meant to represent a thought process rather than exact reaction mechanisms.*]

Both reactions in Figure 26.26 are reasonable, so our decision is going to be based on what happens next. Is it more reasonable for ammonia to displace a hydroxide? Or is it better for a nucleophile to add to an imine? Hydroxide is a relatively poor leaving group, so addition to the C=N double bond (analogous to a carbonyl) is a much more likely way to form the $C_9—C_1$ σ bond. From here all that remains is to consider how the C_9 nitrile becomes a carboxylic acid. Recall from Chapter 19 that acid-catalyzed hydrolysis of a nitrile with heating gives a carboxylic acid. And so the Strecker amino acid synthesis is the three-step process summarized in **Figure 26.27**: imine formation, cyanide addition to the imine, and hydrolysis of the nitrile. ■

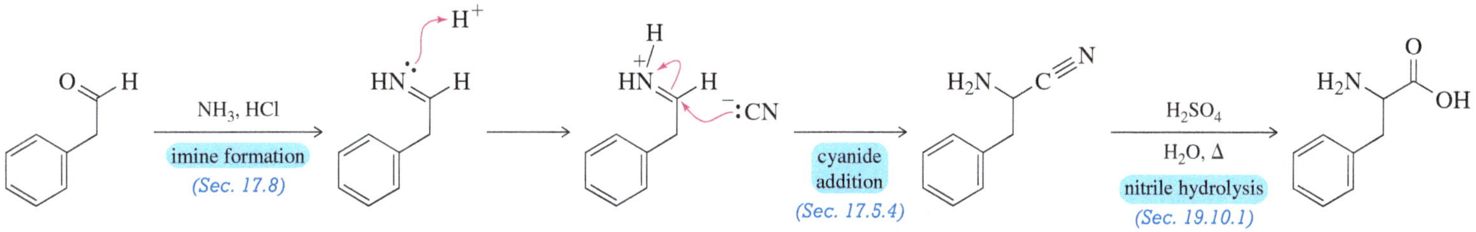

Figure 26.27 **The mechanism of the Strecker amino acid synthesis** consists of imine formation, cyanide addition, and nitrile hydrolysis.

26.4.4 SUMMARY

- The Strecker synthesis uses NH_4Cl and NaCN to extend an aldehyde by one carbon and create an amino acid.

Assessment 26.23 Provide a synthesis of the following natural and artificial amino acids using the Strecker synthesis.

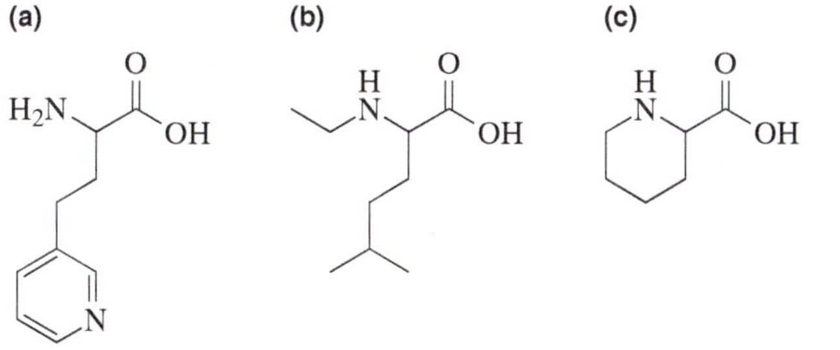

Assessment 26.24 A chemist, attempting to conduct the Strecker amino acid synthesis, added acid and cyanide, but forgot to add the ammonia. What product would you expect to form?

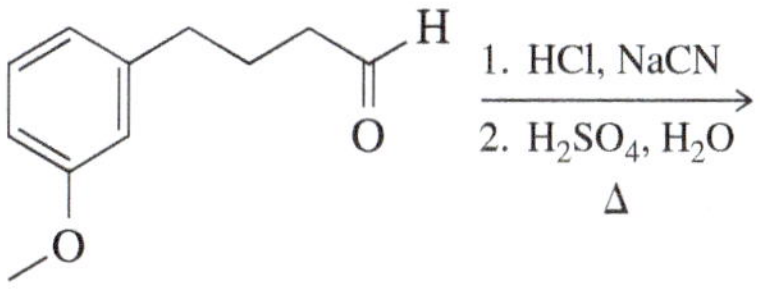

26.5 Reactions of amino acids

The reactions of amino acids are essentially the reactions of amines and carboxylic acids that we have already learned, with one complicating factor. Because amino acids have two functional groups (RNH_2 and RCO_2H), there are certain reactions that could be done on one functional group, but not in the presence of the other. We've solved this problem before using protecting groups for the functional group whose reactivity we want to mask. We discuss the most common method for protecting the acid in Section 26.5.1 and for the amine in Section 26.5.2. These reactions become especially important when we explain how to link amino acids to form peptides and proteins in Section 26.9.

26.5.1 Fischer esterification

Protection of the carboxylic acid is most often done by converting it into an ester. This allows the amino acid to have the same basic electronic structure but without the acidity of the carboxyl proton. Methyl, ethyl, and benzyl esters are commonly used for this purpose. Fischer esterification with the appropriate alcohol and an acid catalyst convert the acid into an ester. Because of the presence of the amine, a slight excess of acid is needed. The mechanism of glycine esterification is shown in **Figure 26.28** and is identical to what we learned in Section 19.7.2. [*Go there for a more detailed review.*]

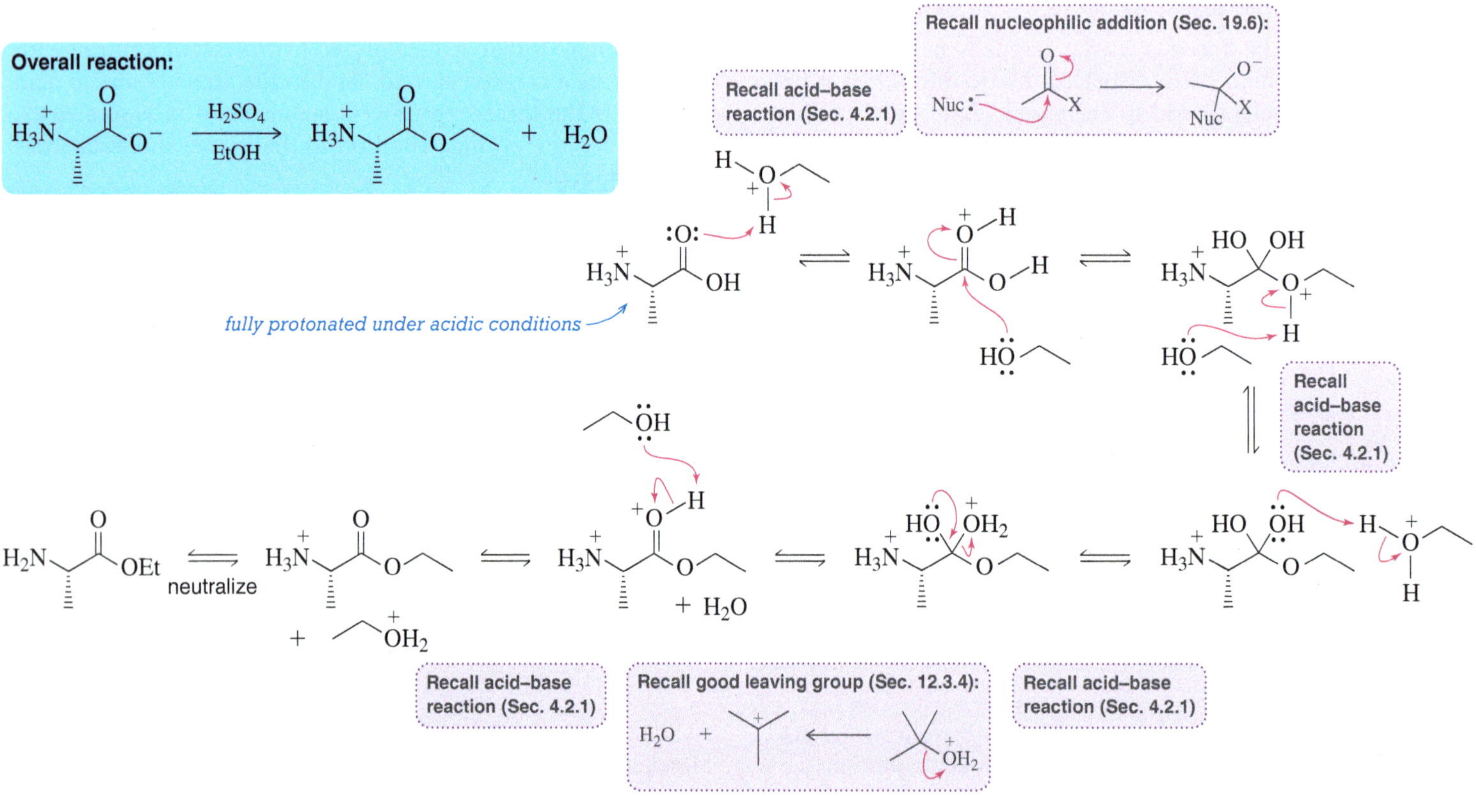

Figure 26.28 The Fischer esterification of amino acids.

Adding a protecting group is one thing, but we must be able to remove it again. Luckily, the Fischer esterification is the reverse of the acid-catalyzed hydrolysis reaction we learned in Section 19.7.1. By the principle of microscopic reversibility, its mechanism is the exact opposite of Fischer esterification. [*If you'd like to see the mechanism drawn out, go to Section 19.7.1.*] Thus, deprotection can be accomplished using acid and water, as shown in **Figure 26.29** for a protected methionine.

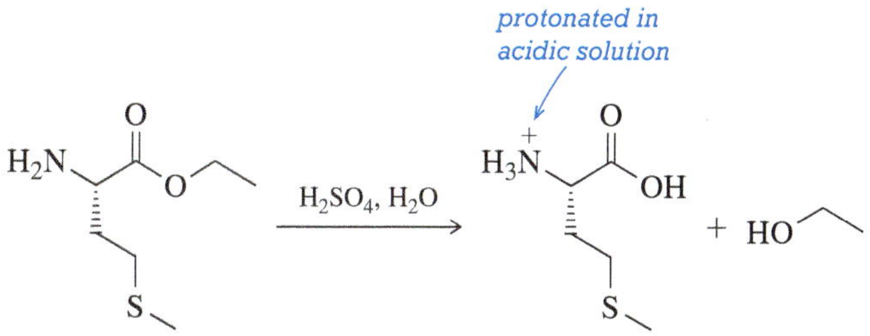

Figure 26.29 The deprotection of an amino acid (methionine) by ester hydrolysis.

When the protecting group is a *benzyl* ester, hydrogenolysis (similar to what we discussed in Section 24.6 for benzyl ethers) will regenerate the carboxylic acid (**Figure 26.30**). This method is particularly useful when strongly acidic conditions must be avoided.

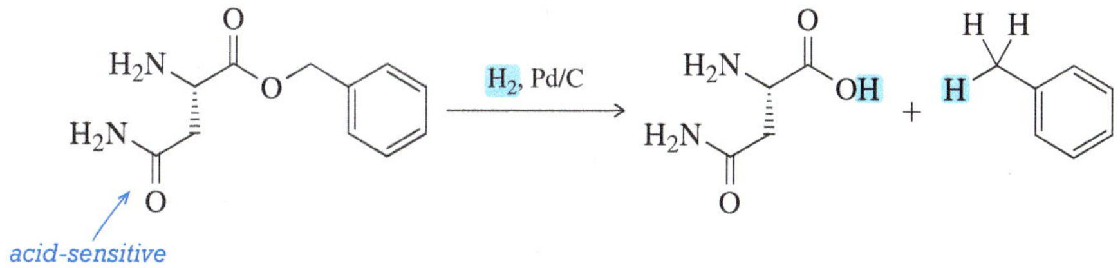

Figure 26.30 Using hydrogenolysis to deprotect a carboxylic acid.

Assessment 26.25 Which of the following synthetic amino acids would more likely be deprotected by hydrolysis? By hydrogenation? Explain your answers.

(a)

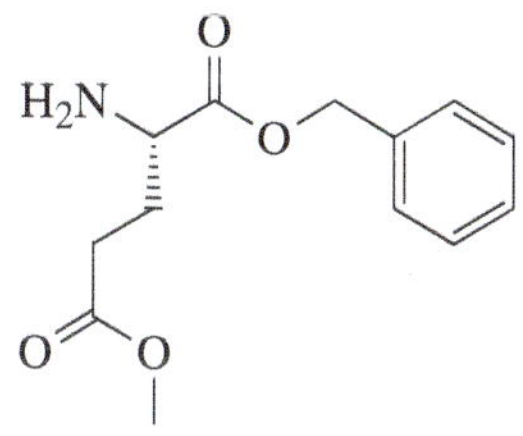

(b)

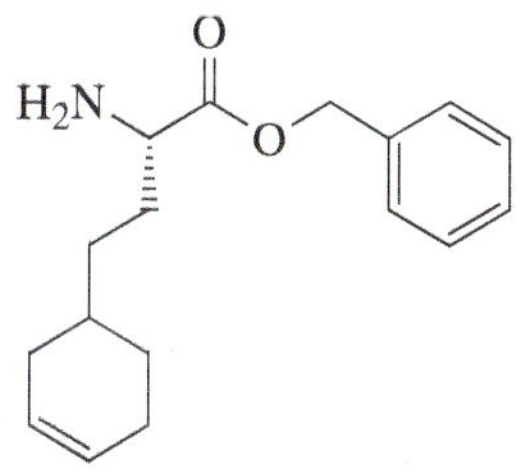

(c)

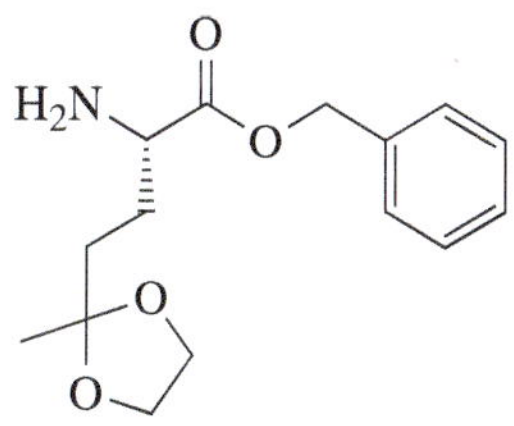

Assessment 26.26 Provide a mechanism for the hydrolysis reaction shown in Figure 26.29.

26.5.2 Acylation of amines

The amine can also be protected using previously studied reactions—specifically, the nucleophilic acylation reaction, proceeding through a tetrahedral intermediate, to form an amide (Section 19.7.3). Acetic anhydride is most commonly used for this process, as shown in the protection of isoleucine in **Figure 26.31**.

Figure 26.31 The protection of an amino acid (isoleucine) with acetic anhydride.

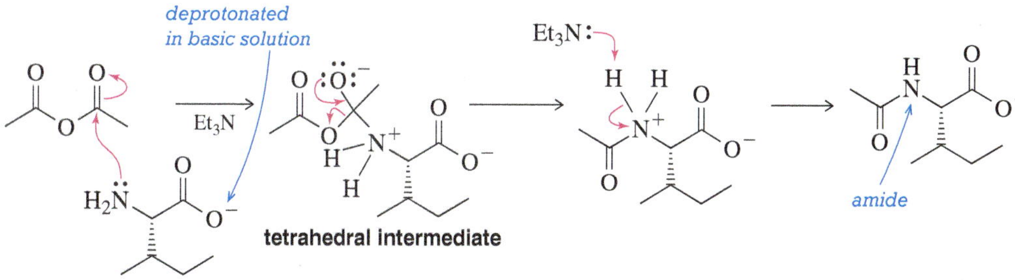

One drawback of protecting the amine in this way is the harsh conditions required to hydrolyze the stable amide. Typically, the amide is cleaved using either strongly acidic or basic conditions (**Figure 26.32**). This becomes a problem when the amino acid is part of a peptide chain (Section 26.6) because the entire chain will be constructed of similarly reactive amides.

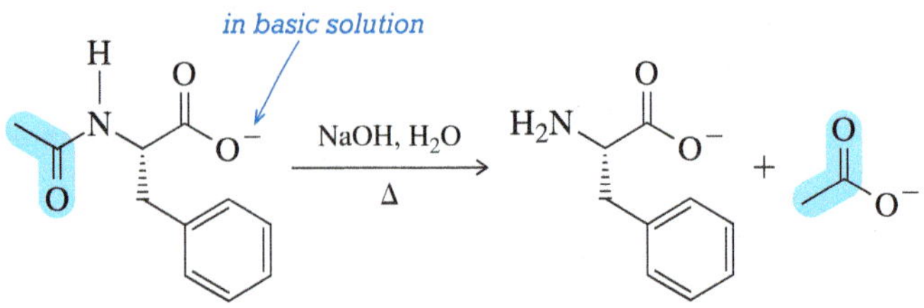

Figure 26.32 Basic hydrolysis of an amide to deprotect the amino group.

Alternatively, benzylchloroformate [**Figure 26.33(a)**] or di-*tert*-butyldicarbonate [commonly called Boc-anhydride (Boc_2O); **Figure 26.33(b)**] can be used to acylate the amine by an identical mechanism. The resulting carbamate in each case can be cleaved under mild conditions. Like a benzyl ester, the benzyl carbamate undergoes hydrogenolysis to produce a carbamic acid, which rapidly decomposes to give carbon dioxide and the deprotected amine. The *t*-butoxy carbamate can be deprotected using a mild acid. When the carbonyl oxygen is protonated, cleavage then produces a stable 3° carbocation and a carbamic acid. Once more, the carbamic acid decarboxylates to give carbon dioxide and the amine.

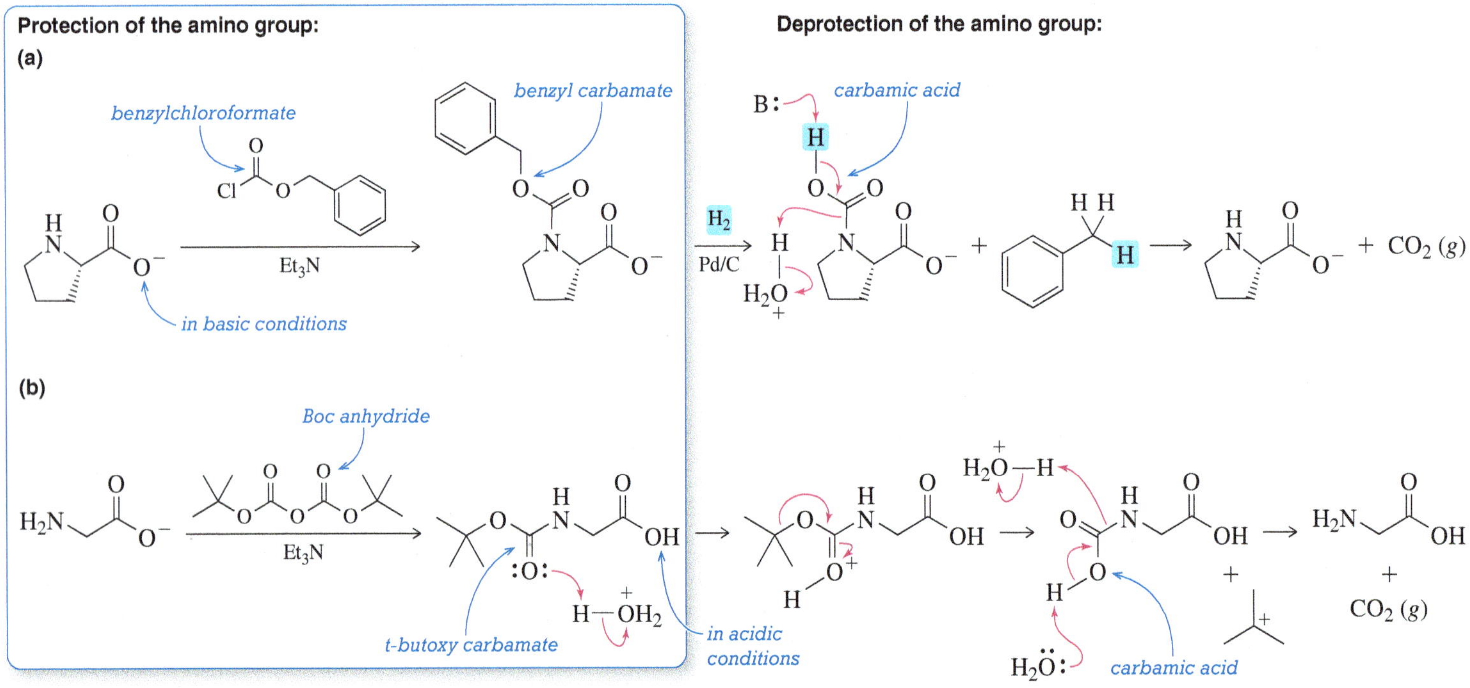

Figure 26.33 Alternative protecting groups. (a) The benzyl carbamate is deprotected using hydrogenolysis, followed by decarboxylation. (b) The *t*-butoxy carbamate is hydrolyzed under mildly acidic conditions and is followed by decarboxylation.

Assessment 26.27 Provide a mechanism for the basic hydrolysis of the amide in Figure 26.32.

Assessment 26.28 Provide a mechanism for the protection of the amine as the benzylcarbamate shown in Figure 26.33(a).

26.5.3 Chiral resolution of amino acids

In each of the amine syntheses in Sections 26.4.1–26.4.4, the amine was produced as a racemic mixture. Thus, chiral resolution of a racemic mixture is necessary if only a single enantiomer is desired. This can be done by forming diastereomeric salts, as discussed in Section 6.6.1 and as shown in **Figure 26.34**. Conveniently, amino acids can form salts in two ways, but not at the same time. To resolve an amino acid such as leucine with a chiral, nonracemic acid, the carboxylic acid must be protected as an ester using the Fischer esterification of Section 26.5.1. At this point, a single enantiomer of mandelic acid reacts with the racemic mixture of amines to form diastereomeric salts via an acid–base reaction. Whereas the enantiomeric amino acids were identical in virtually every way, diastereomers are different in every way. Most notably, they crystallize at different rates, allowing us to collect one salt diastereomer preferentially. That diastereomer can then be treated with base to free the amine, at which point the ethyl ester can be hydrolyzed to produce the optically active, stereochemically pure L-(*S*)-leucine.

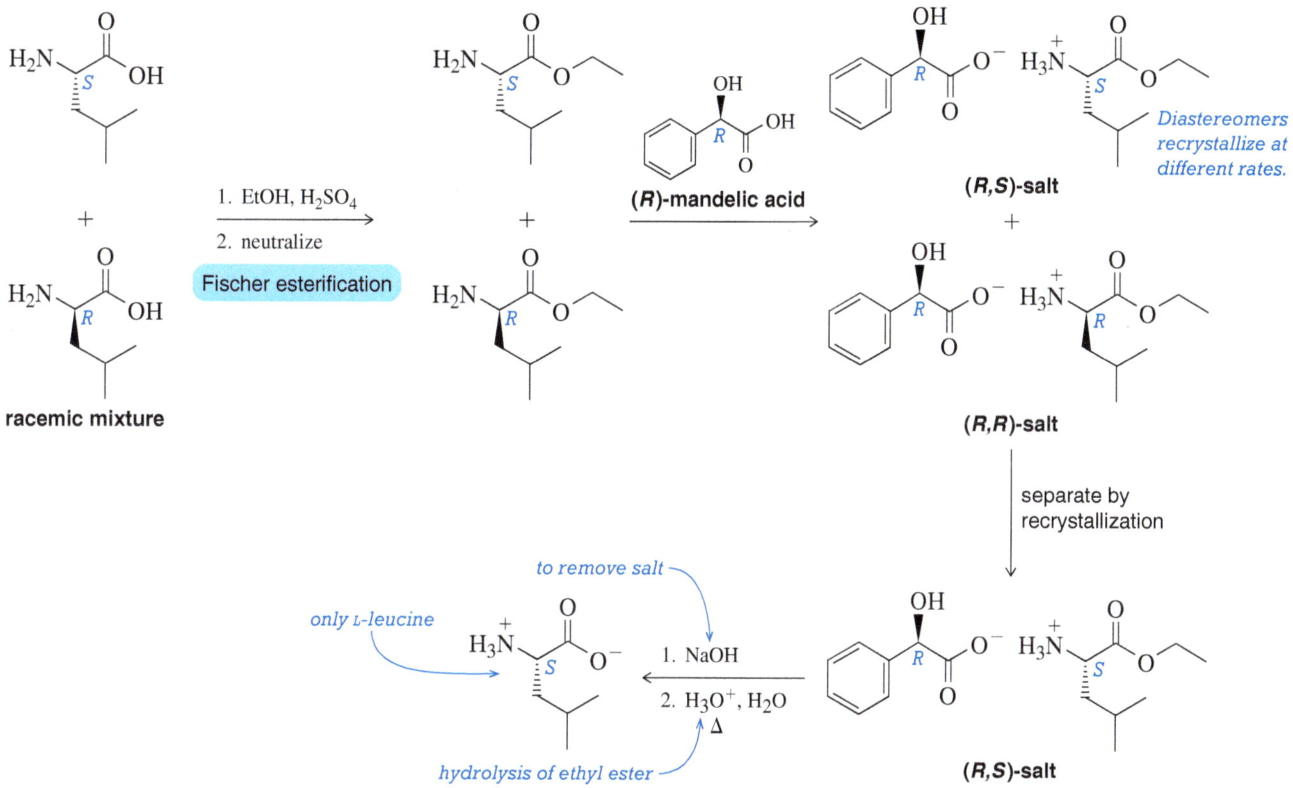

Figure 26.34 Separation of enantiomers via the formation of diastereomeric salts.

Assessment 26.29 How might D- and L-tyrosine be separated using a chiral amine like (*R*)-1-phenethylamine? [*Hint: Your scheme should include the protection and deprotection of the amine.*]

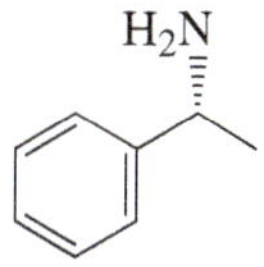

(R)-1-phenylethylamine

Assessment 26.30 When resolving an amino acid with an acid or a base, it is always important to protect either the amine or the acid. Why? What would happen if you protected neither?

26.5.4 Ninhydrin

The reaction of ninhydrin with amines, shown in **Figure 26.35**, has very little synthetic utility. It is useful, though, for the qualitative identification of amines and amino acids. As a staining agent for thin layer chromatography (TLC), the bright blue color, arising from the resulting highly conjugated product, can be used to indicate the presence of an amine in whatever solution (reaction, column fraction, etc.) that was spotted on the plate. In Section 26.6 we discuss the role it plays in the identification of amino acids in a peptide sequence.

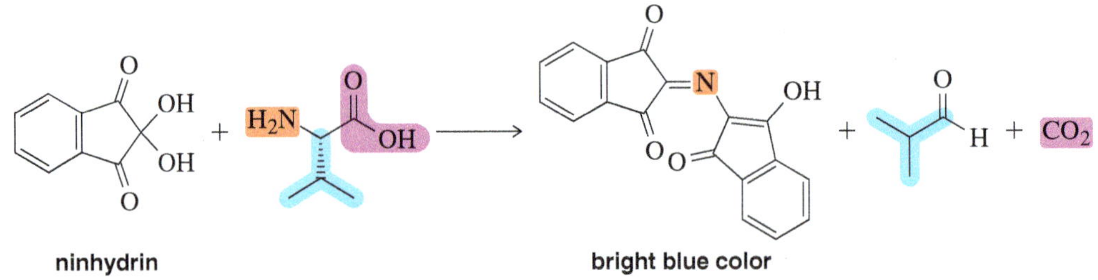

Figure 26.35 The formation of a highly conjugated product indicates the presence of a 1° or 2° amine.

Like all of the complex mechanisms we have studied over the past several chapters, the reaction of ninhydrin is really a combination of previously studied reactions (**Figure 26.36**), but in disguise. The reactive functional group of ninhydrin is the diol between the two carbonyls. This diol is simply a hydrated ketone (Section 17.7.1). As such, it reacts identically with 1° or 2° amines to give imines. The resulting imine can decarboxylate to produce carbon dioxide in an entropy-favored reaction. The resulting enol tautomerizes to give another imine. Hydrolysis of this imine releases the side chain of the amino acid as an aldehyde. The resulting ninhydrin amine forms another imine to give the highly conjugated species responsible for the bright blue color.

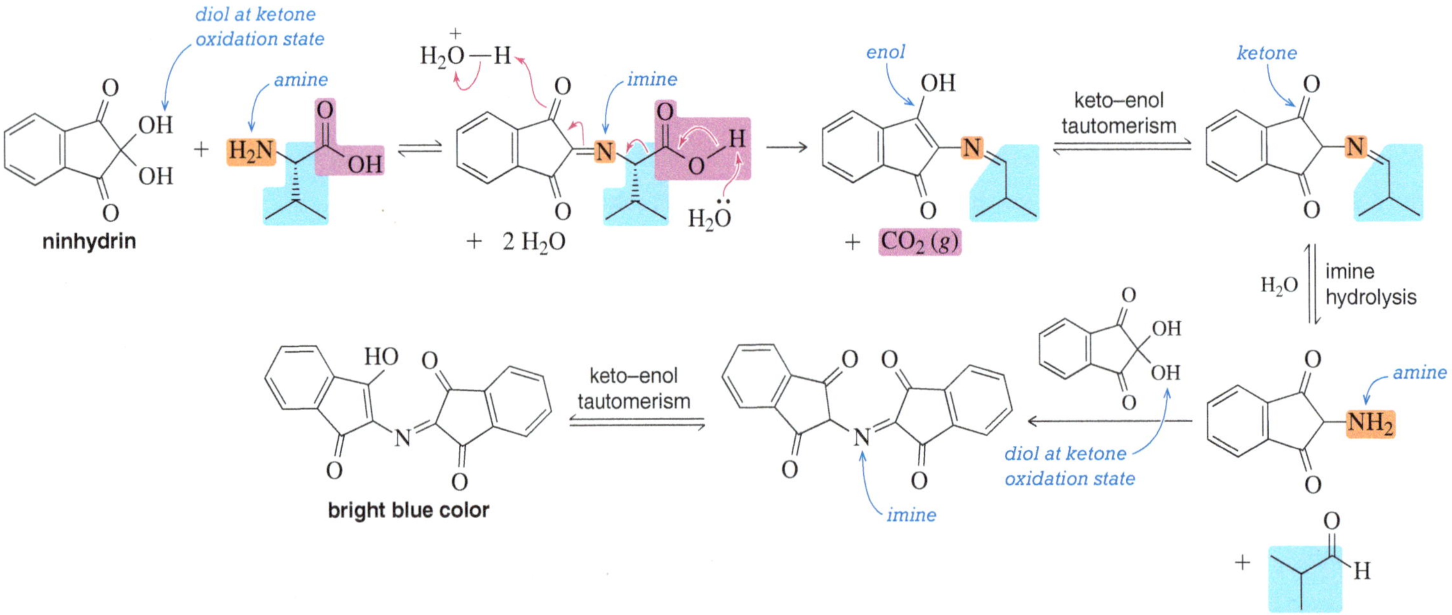

Figure 26.36 The ninhydrin mechanism occurs via previously studied reactions—namely, imine formation, decarboxylation, keto–enol tautomerization, imine hydrolysis, imine formation, and keto–enol tautomerization.

Assessment 26.31 Ninhydrin cannot be used to indicate the presence of a 3° amine. Why? [*Think about the mechanism. What would need to occur?*]

Assessment 26.32 Why are conjugated molecules like the product of the ninhydrin reaction brightly colored?

Assessment 26.33 Suggest an arrow-pushing mechanism for the following steps of the ninhydrin reaction.

(a)

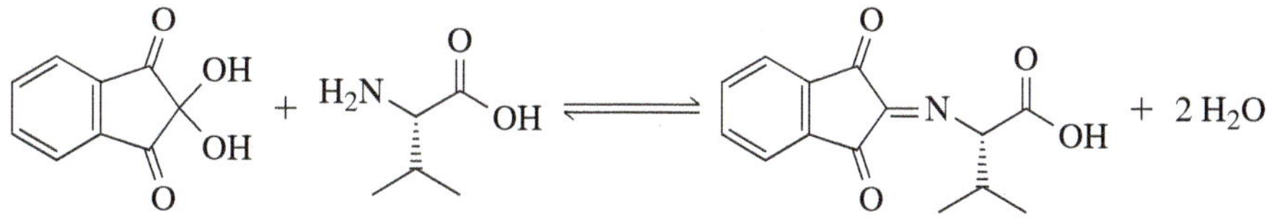

(b)

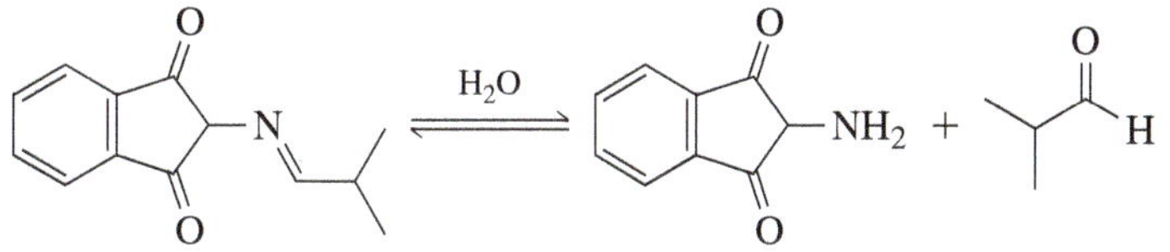

(c)

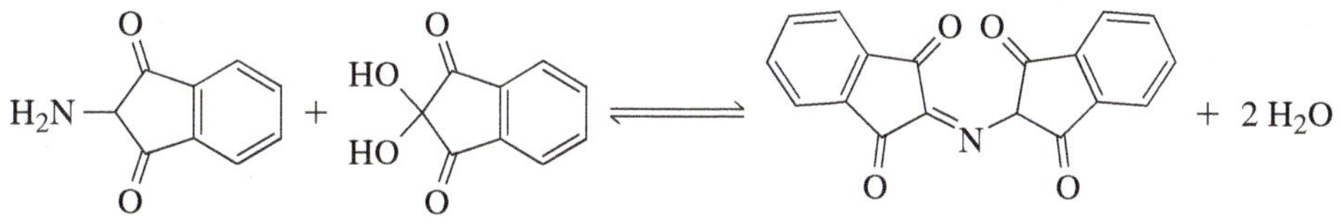

26.6 Proteins: a general overview

Proteins are naturally occurring polymers made up of more than 50 amino acids connected through amide or peptide bonds. When a strand of amino acids contains fewer than 50 amino acids, it is usually referred to as a **peptide**. There are many biologically important peptides, such as insulin and substance P, but there is substantially more diversity and quantity of function in proteins. When comparing biological molecules (proteins, lipids, nucleic acids, and carbohydrates), proteins have the most diverse functions. Proteins catalyze reactions (enzymes), provide structure to bodily tissues, help cells move, defend cells, regulate blood glucose level, transport materials across a cell membrane, and store nutrients.

Proteins and peptides form through the coupling of the carboxylic acid on one amino acid with the amine of another. This nucleophilic acyl substitution reaction (Section 19.6) results in the formation of water, as shown in **Figure 26.37**. The resulting amide bond is referred to in biochemistry as a **peptide bond**. Carboxylic acids and amines don't typically react to form amides [*see Chapter 19*]; these reactions occur at the active site of enzymes. Thus, chemistry that is not possible in solution is quite efficient when enzymes are involved. When two amino acids combine, the resulting molecule is called a **dipeptide**. When a third amino acid is added, it's a tripeptide. The amino acid with the free amine is referred to as the **N-terminal** residue. The one with the free carboxyl group is the **C-terminal** residue. Peptides are traditionally written with the N-terminus on the left and the C-terminus on the right. Similarly, the nomenclature of peptides is a single word containing the amino acid sequence, beginning with the N-terminal residue (the amine side). The name of each amino acid (except the last one—the C-terminal residue) is modified to include the suffix "-yl." For convenience, peptides are often represented using the three-letter code for each amino acid residue.

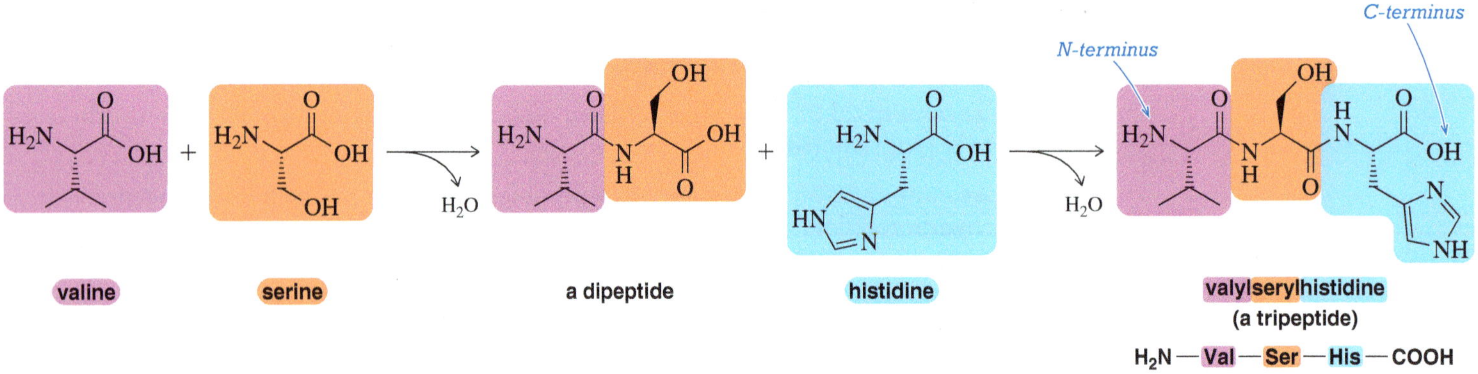

Figure 26.37 The formation and nomenclature of a tripeptide.

Assessment 26.34 Draw the structure of leucylaspartylalanine. Label each peptide bond, the N-terminus, and the C-terminus.

Assessment 26.35 Draw the structure that corresponds to the amino acid sequence shown. What type of peptide is this? [*Di, tri, tetra, etc.*]

$$H_2N—Ile—Cys—Thr—Val—Leu—Leu—COOH$$

26-A CHEMISTRY IN CONTEXT: Molecular Biology

Humoral control of blood pressure

Blood pressure is regulated by multiple biological mechanisms. One such method involves the renin-angiotensin system. It is quite complicated, but one small part of it has been simplified in Figure 26.38. Angiotensinogen, a protein produced in the liver, is hydrolyzed by a proteolytic enzyme renin, producing angiotensin I (a decapeptide). This decapeptide has no real biological activity but, upon hydrolysis by angiotensin converting enzyme (ACE), produces angiotensin II (an octapeptide).

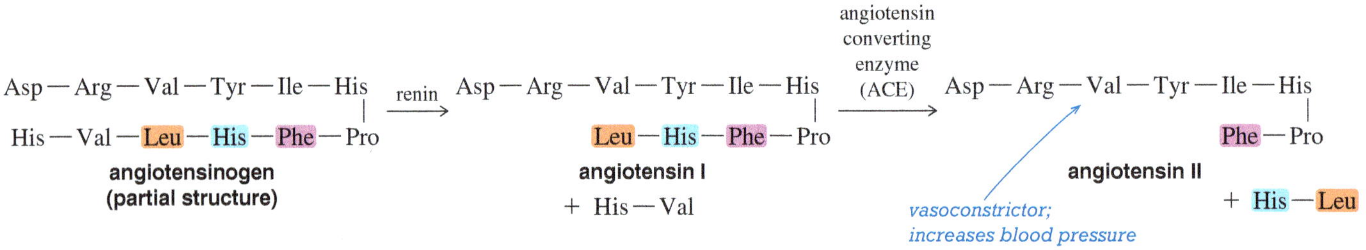

Figure 26.38 The conversion of angiotensin I to angiotensin II by ACE.

Strong acids are needed to hydrolyze a peptide in solution. Enzymes, however, can utilize weaker active-site acids and bases, along with a well-defined active site, to more easily cleave these bonds. ACE also has a Zn^{2+} cation that serves as a Lewis acid to promote amide bond cleavage. The mechanism of angiotensin I hydrolysis—essentially the nucleophilic acyl substitution reaction from Chapter 19—is shown in Figure 26.39. Like most enzymes, ACE cleaves only this specific peptide bond. This kind of substrate specificity in a well-defined enzyme active site makes it possible for the cell to carefully control these reactions.

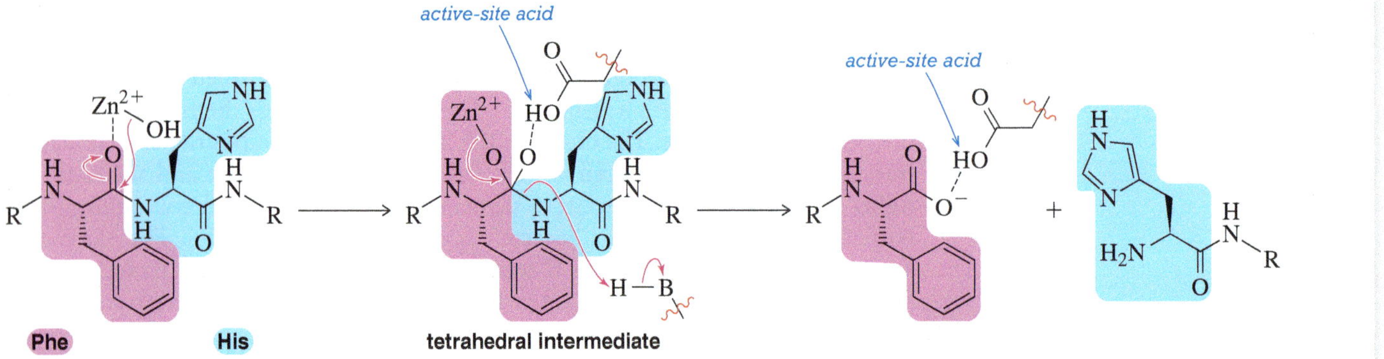

Figure 26.39 The mechanism of peptide cleavage. Coordination with a zinc cation increases the reactivity of the amide, allowing hydrolysis to occur.

(continued)

The octapeptide produced in Figure 26.38 binds with the angiotensin II receptor causing vasoconstriction. By constricting blood vessels and arteries, it causes blood pressure to increase. Furthermore, angiotensin II triggers the release of aldosterone, a steroid hormone that induces the release of potassium ions and the retention of water, both of which further contribute to an increase in blood pressure. [*If you wanted to lower blood pressure, what enzyme's activity might you want to inhibit?*]

26.6.1 Classification

Proteins can be classified in a number of ways, but we discuss only a couple of them here (Figure 26.40). First, **fibrous proteins** form long, cylindrical polymeric structures. Rigid and insoluble in water, fibrous proteins are generally used for protection. The keratin in hair and fingernails is a fibrous protein. **Globular proteins**, on the other hand, are nearly spherical, flexible, and soluble in water. Enzymes are typically globular proteins. **Membrane proteins** interact with biological membranes, either permanently or temporarily. These can serve as receptors, enzymes, or transporters, or they can aid in cell-to-cell recognition.

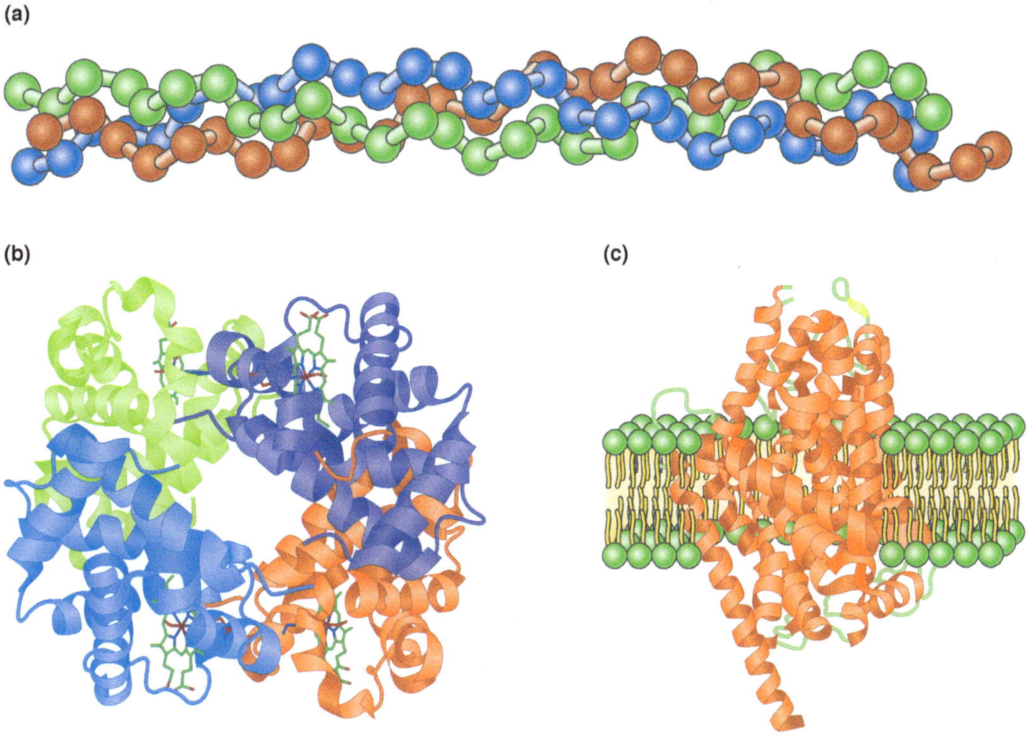

Figure 26.40 Classification of proteins. (a) Fibrous, (b) globular, and (c) membrane-bound.

Proteins can be coupled covalently to other non-protein biological molecules, giving what are called **conjugated proteins** (Figure 26.41). [*This is not how we have used "conjugated" up to now. Here it just implies attachment between a protein and something else.*] Glycoproteins (conjugated to carbohydrates), lipoproteins (conjugated to lipids), and metalloproteins (containing metal ions) are examples of conjugated proteins. These **prosthetic groups** in conjugated proteins contribute to the diversity of protein uses.

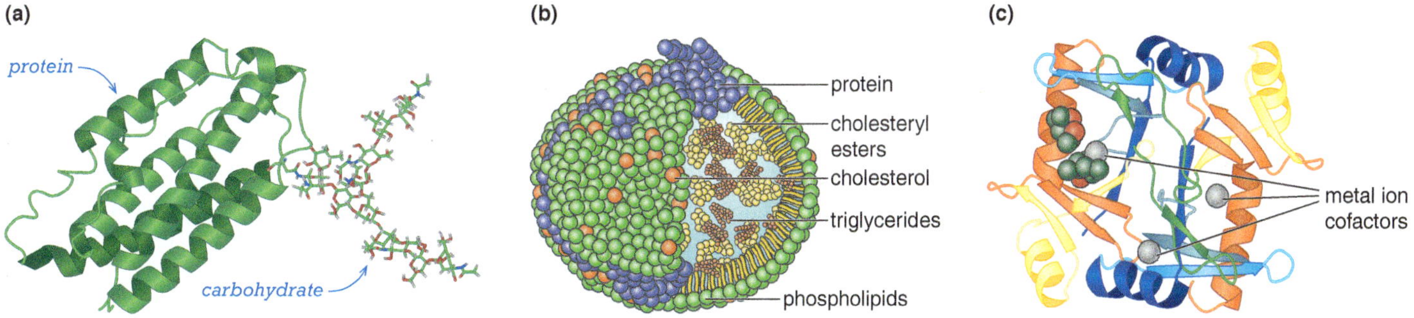

Figure 26.41 Examples of conjugated proteins. (a) Glycoprotein, (b) lipoprotein, and (c) metalloprotein.

26.6.2 Strength of the peptide bond

The structural features of proteins that we study in Section 26.7 and the stability of proteins/peptides are a result of the strong amide bond. [*Recall from Section 19.6.3 that amides are the least reactive/most stable carboxylic acid derivative.*] The strength of the amide bond is a result of a second, highly contributing resonance structure. Nitrogen is only slightly electronegative, so it readily donates a pair of electrons to the carbonyl carbon (**Figure 26.42**). The less-contributing structure accounts for about 34% of the overall structure of the amide and impedes rotation around the σ bond. [*Because the lone pair on nitrogen is delocalized by resonance, the nitrogen is sp^2 hybridized; see Section 2.8.5.*] These two possible conformations, and the lack of completely free rotation, give rise to the structural rigidity of proteins that give them their structural features, including the well-defined active site structures of enzymes.

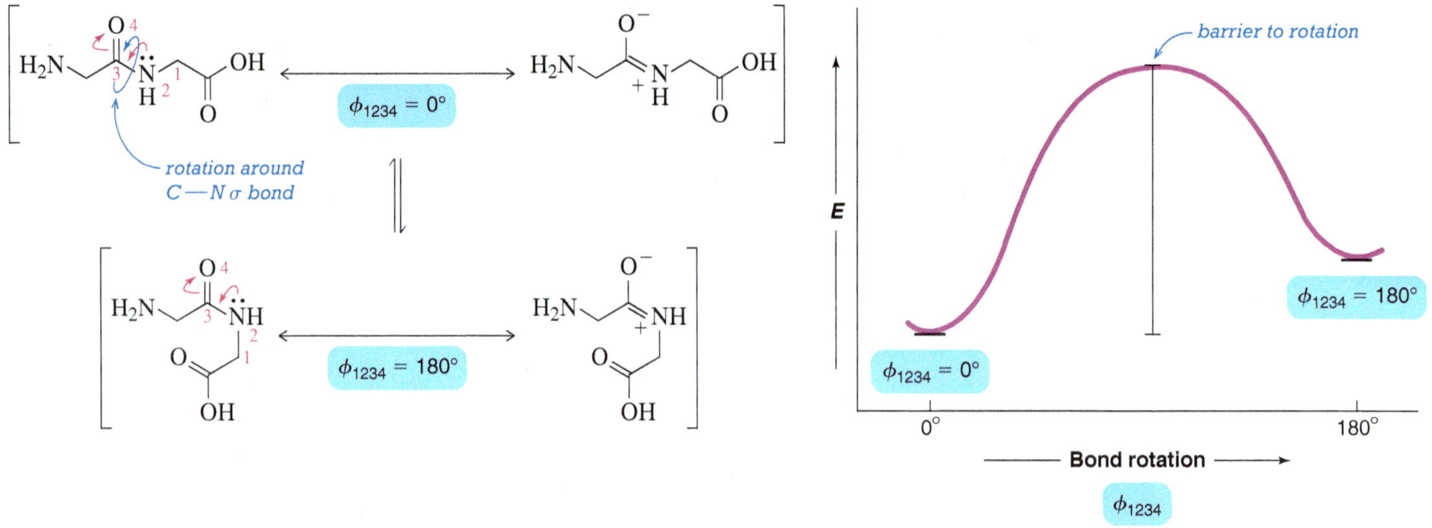

Figure 26.42 The amide resonance structure stabilizes the peptide bond and prevents free rotation.

Assessment 26.36 Draw the two most stable conformations of a H_2N—Thr—Val—COOH dipeptide around the peptide bond.

Assessment 26.37 Why didn't evolution select for forming proteins using esters instead of amides?

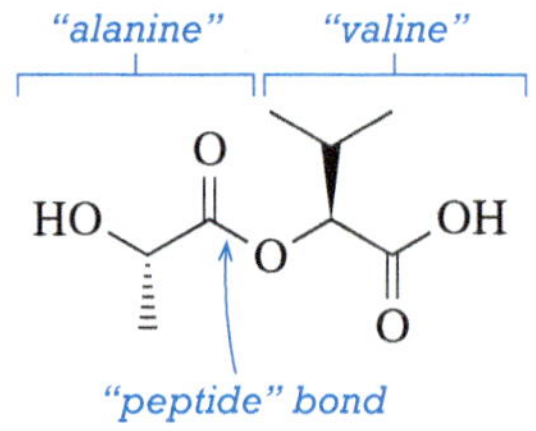

26.7 Protein structure

Proteins are extremely large molecules made up of as many as 20 different amino acids, so they are complex molecules. The restricted rotation around the amide bond, however, makes the identification of common structural features possible. Protein structure consists of four levels of structural complexity (**Figure 26.43**). The **primary structure** is simply the amino acid sequence and includes disulfide bonds. The **secondary structure** is made up of common folding patterns. [*Only the α-helix is shown in Figure 26.43.*] The **tertiary structure** is the overall three-dimensional shape of the protein. When a protein is made up of two or more subunits, the manner in which these combine is referred to as the **quaternary structure**.

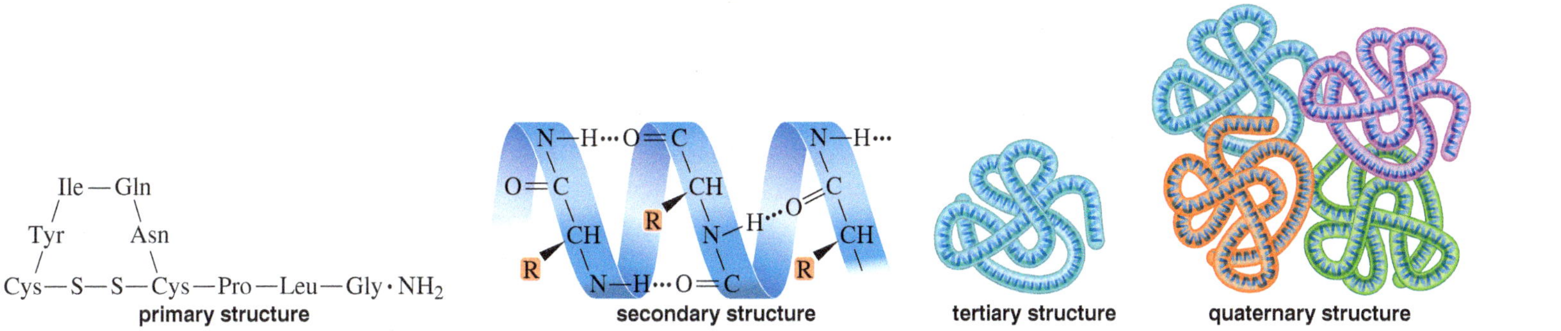

Figure 26.43 The four levels of protein structure.

26.7.1 Primary structure

The amino acid sequence ultimately determines the three-dimensional structure of the protein, and the three-dimensional structure of a protein determines its function. Proteins with a similar amino acid sequence and function are said to be homologous. This homology can be seen among proteins that carry out similar functions in different species. In fact, the sequence homology between species can be used to identify common evolutionary ancestors. Figure 26.44(a) shows the sequence homology for heat shock protein across various species. The amino acid residues that are maintained across species are assumed to be responsible for the protein's function.

(a)

human (hHSFY1)	FPRKLWKIVESDQFKSISWDENGTCIVINEELFKKEILETKAPYRIFQTDAIKSFVRQLN	60
human (hHSFY2)	FPRKLWKIVESDQFKSISWDENGTCIVINEELFKKEILETKAPYRIFQTDAIKSFVRQLN	60
mouse	FPRKLWKIVGSDKFKSIWWDEDGTYIVINEELFKKEVLERKAPFRIFETDSMKSLVRQLN	60
rat	FPRKLWKIVGSDKFKSIWWDEDGTYIVINEELFKKEVLERKAPFRIFETDSMKSLVRQLN	60
rhesus macaque	FPRKLWKIVESDQFKSVSWDENGTCIMINEELFKKEILERKHPYRIFQTDSIRSFVRQLN	60
cat	FPRKLWNIVESDQFKSIWWDENGTSIVIDEELFKKEVLERKAPFRIFETGSMKSLVRQLH	60
bovine	FPQKLWNIVESDQFESIWWDENGTCIVIHEMLFKKEVLERKAPFRIFETKSMKSLIRQLN	60

human (hHSFY1)	LYGFSKIQQNFQRSAFLATFLSEEKESSVLSKLKFYYNPNFKRGYPQLLVRVKRRIGVK	119
human (hHSFY2)	LYGFSKIQQNFQRSAFLATFLSEEKESSVLSKLKFYYNPNFKRGYPQLLVRVKRRIGVK	119
mouse	LYGFRKMRQNFQRSASLPDFLAEEKGISASCKLQFYQNPNFLRDCPHLIERMKRRIGIK	119
rat	LYGFRKMRQNFQRSASLPDFLAEEKGISASCKLQFYQNPNFLRDCPHLIERMKRRIGIK	119
rhesus macaque	LYGFSKIRQNFQRSAFLPTFLAEEKESSVLTKLKCYYNPHFKRGCPQLLVRVKRRIDVK	119
cat	LYGFSKMRQSFQRSASLADFLAEEKEVSVLSKLQFYMNPNFKRGCPQLLVRMKRRVGIK	119
bovine	LYGFSKKRQTFQRSASLPVFLEEENNISLLSKLQTYYNPNFKRGHP-------------	106

(b)

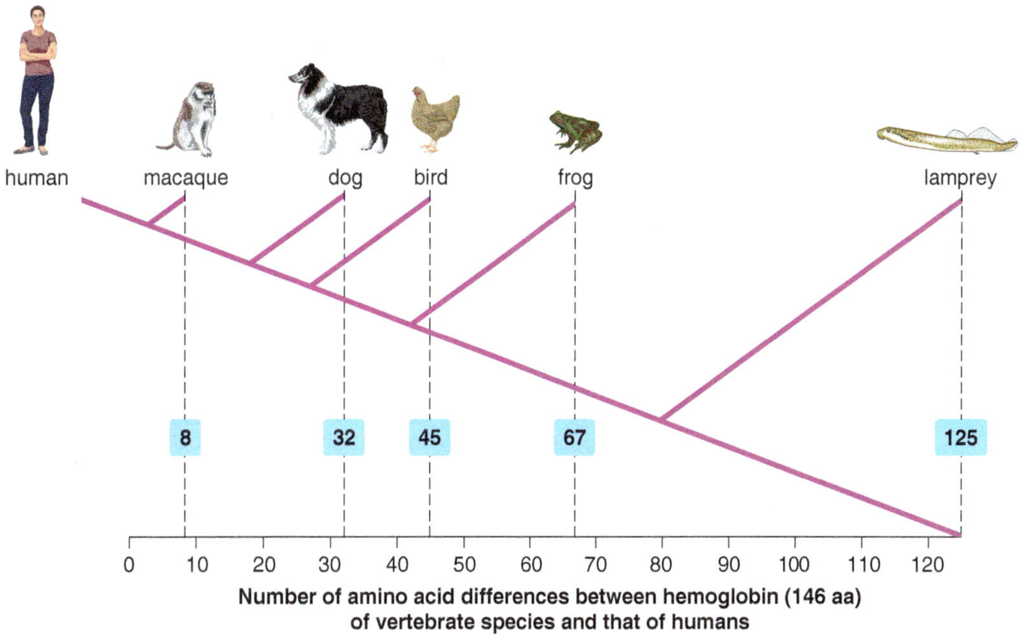

Figure 26.44 Sequence homology. The sequence in DNA and proteins is a molecular record of the evolutionary relationships between species. (a) Conserved residues in heat shock protein. (b) Proteins in more closely related species have greater sequence homology.

26.7.2 Secondary structure

The secondary structure of a protein results from the folding of polypeptide chains into regular, recurring structural motifs. These patterns arise by intramolecular interactions between amino acids. Specifically, formation of hydrogen bonds between the amino N—H and the carbonyl oxygen results in the α-helices and β-sheets shown in **Figure 26.45**.

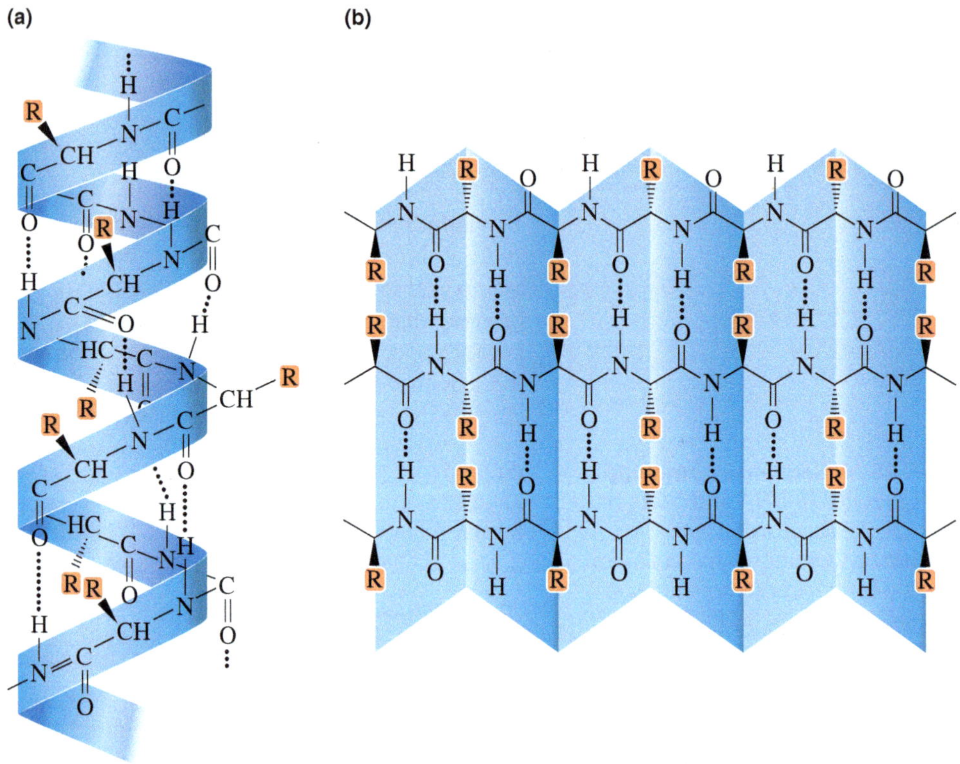

Figure 26.45 Secondary structure of proteins. Polypeptide chains fold into (a) α-helices and (b) β-sheets. [*This is an antiparallel β-sheet.*]

The **α-helix** shown in Figure 26.45(a), is a rigid, cylindrical structure made of a linear peptide chain that has oriented into a right-handed helical conformation. This conformation is held together by hydrogen bonds between the N—H group of one amino acid and the carbonyl group of an amino acid that is four residues further down the sequence. The amino acid side chains align themselves on the outside of the helix structure. The **β-sheets**, shown in Figure 26.45(b), form when two polypeptide segments within the same protein align themselves such that hydrogen bonds form between the amino groups of one polypeptide chain with the carbonyl group of an adjacent chain. Each polypeptide in the β-sheet is referred to as a β-strand. Within a β-sheet, the β-strands can be aligned parallel or antiparallel, where parallel sheets run in the same direction (C-terminus to N-terminus) and antiparallel sheets run in opposite directions. The β-sheet in Figure 26.45(b) is antiparallel, which is more stable than the parallel alignment.

26.7.3 Tertiary structure

The tertiary structure of a protein refers to its three-dimensional structure and describes the arrangement of the α-helices, the β-sheets, and other coils and turns that comprise the remainder of the secondary structure. When proteins fold, they orient themselves in three-dimensional structures that are specific to their ultimate function. It is this folding that creates a receptor site for a specific hormone or a structurally specific active site for an enzymatic substrate.

When proteins fold into globular structures, distant amino acid residues become close in proximity to one another. Because water molecules are mostly excluded from globular proteins, both polar and nonpolar groups can interact with each other. As we have said many times, all chemical processes occur to reach a more stable state. [*The ball rolls downhill.*] This is true for protein folding as well. Folded protein structures are stabilized by the interactions shown in **Figure 26.46**.

Hydrophobic interactions	***Electrostatic interactions (salt bridges)***
Hydrophobic nonpolar groups interact via van der Waals interactions, excluding water from the interior of the protein. This increases the disorder of the water molecules, which increases entropy and represents a major driving force for protein folding.	Ionic groups of opposite charge interact to form stabilizing interactions between amino acid residues. These salt bridges can form between acidic and basic side chain residues.
Hydrogen bonds	***Covalent bonds***
Hydrogen bonding between side chain functional groups helps to stabilize the tertiary structure of proteins, although enthalpy contributes far less to folding than entropy.	Bonds between the sulfur atoms of two different cysteine residue oxidize in the presence of oxygen to form an S—S linkage that also stabilizes the tertiary structure of proteins. [*This linkage is responsible for curly hair.*]

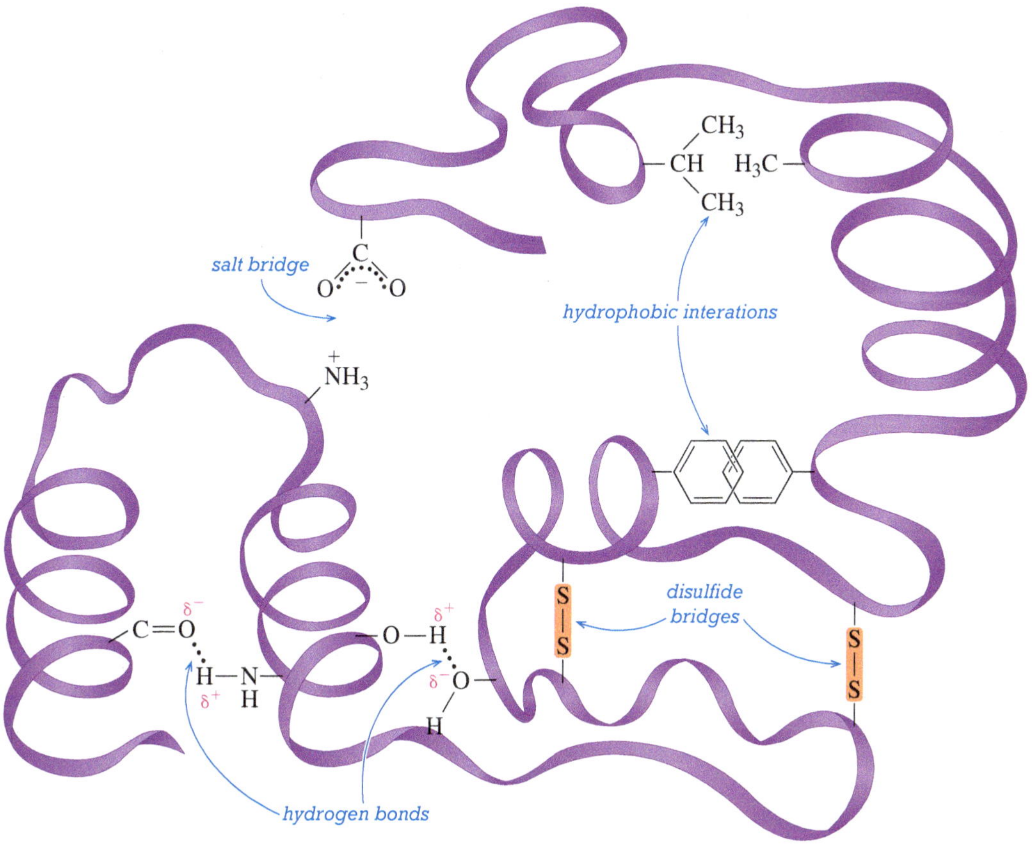

Figure 26.46 The interactions that stabilize the tertiary structures of proteins.

26.7.4 Quaternary structure

The quaternary structure is only important in proteins that are made up of more than one polypeptide subunit. Hemoglobin, for example, consists of four subunits that fit together in a single globular structure (**Figure 26.47**). Each subunit contains a heme group, and each heme group contains an iron ion that is used to carry oxygen in blood throughout the body.

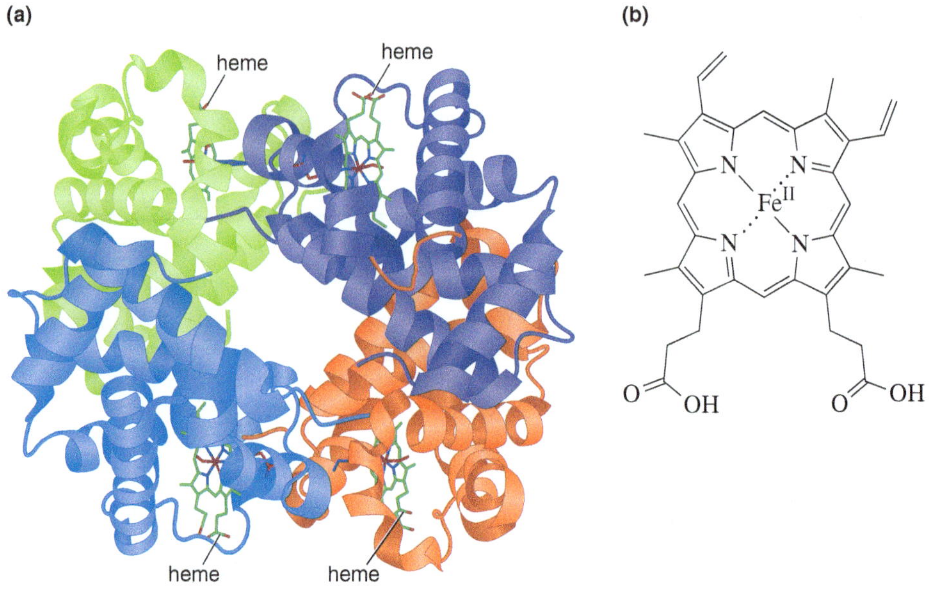

Figure 26.47 Hemoglobin. (a) The quaternary structure, a tetramer, and (b) a heme group.

26-B CHEMISTRY IN CONTEXT: Medicinal Chemistry

Lead compounds from pit viper venom

Bradykinin, another human peptide, causes blood vessels to enlarge, thereby lowering blood pressure. It is a nonapeptide that is broken down into inactive peptides by the action of angiotensin converting enzyme (ACE) (**Figure 26.48**). The resulting heptapeptide has no known biological activity. Inactivating bradykinin, a peptide that lowers the blood pressure, thereby raises blood pressure.

Arg—Pro—Pro—Gly—Phe—Ser—Pro—Phe—Arg $\xrightarrow{\text{ACE}}$ Arg—Pro—Pro—Gly—Phe—Ser—Pro + Phe—Arg

bradykinin
(a potent vasodilator)

Figure 26.48 The degradation of bradykinin by ACE.

Recall from Chemistry in Context 26-A that ACE catalyzes the formation of angiotensin II, a potent vasoconstrictor, which raises blood pressure. ACE also catalyzes the degradation of bradykinin, a potent vasodilator, which also raises blood pressure. [*I wonder if there's an enzyme we should inhibit to lower blood pressure? I bet our friend, the Brazilian pit viper, can help us.*]

When a Brazilian pit viper, *Bothrops jararaca*, injects poison into its prey, the prey's blood vessels immediately dilate, causing a precipitous drop in blood pressure. This decreased blood pressure causes a decreased supply of oxygen to muscle tissue and the brain as well as decreased reaction times. Eventually, the prey collapses. In 1965, it was discovered that the blood vessel dilation was a result of several peptides in the venom, the most potent of which is the pentapeptide shown in **Figure 26.49**. This peptide inhibits ACE, resulting in less angiotensin II (so the blood pressure drops) and more bradykinin (so the blood pressure drops even more).

(*continued*)

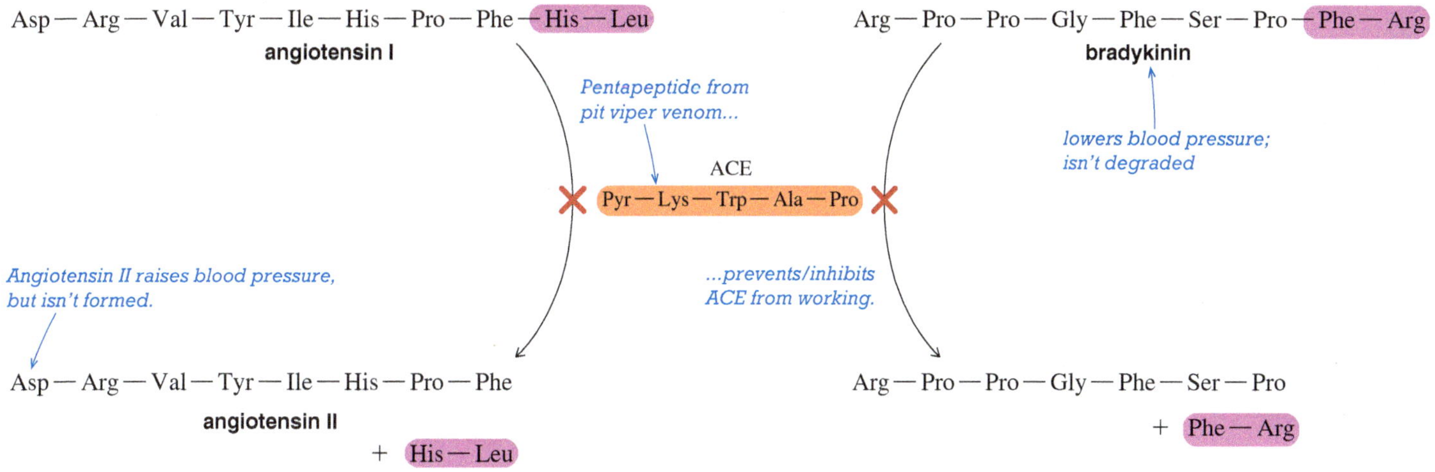

Figure 26.49 The inhibition of ACE by pit viper venom lowers blood pressure via two mechanisms.

As a result of this ACE inhibition, medicinal chemists identified the pentapeptide as a lead compound in the search for blood pressure–lowering medications. The structure of the pentapeptide is shown in **Figure 26.50**.

Assessment 26.38 The first amino acid in the pentapeptide of Figure 26.50, Pyr, is not one of the 20 amino acids in Table 26.1. Based on the following structure of Pyr, from which natural amino acid is it likely derived?

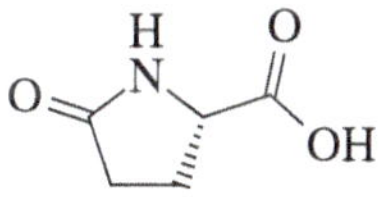

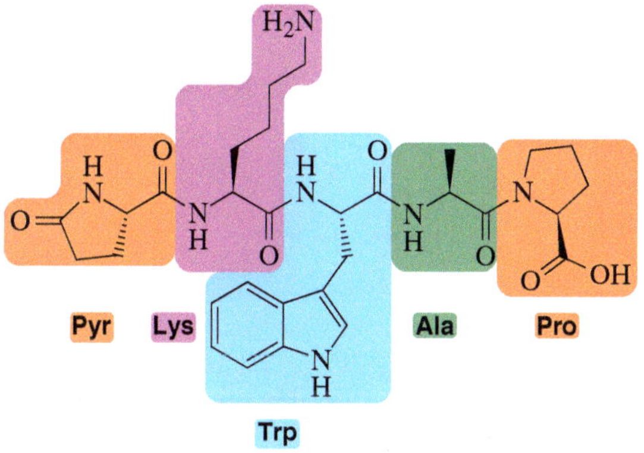

Figure 26.50 The primary structure of the petapeptide inhibitor.

26.8 Protein analysis: a sample

Imagine that an enzyme has been identified as a target of interest for a new blood pressure medication. By inhibiting it, blood pressure can be better managed in patients. While medicinal chemists can make predictions about what molecules might inhibit it based on the reactions it catalyzes, there is tons of value in knowing the structure and sequence of the protein. The following sections briefly touch upon how the sequence of a protein can be determined. We'll do this for the hypothetical decapeptide shown in **Figure 26.51**.

$$H_2N-Xxx-Xxx-Xxx-Xxx-Xxx-Xxx-Xxx-Xxx-Xxx-Xxx-COOH$$

Figure 26.51 A protein of unknown structure.

To start, we'd probably want to know about the amino acid composition, regardless of the sequence. How can you determine which amino acids and how many of each are present in a protein?

You already have all the tools you need to figure out how to do this. Throughout this process, then, try to anticipate what we have discovered and how we can use this to put the puzzle together. Warning: This one goes on for a bit.

Thinking Like a (Bio)chemist

Two questions need to be answered in order to identify the individual amino acids in the protein. [*Think about each before reading further.*] First, can the peptide chain be cleaved at the peptide (amide) bonds, leaving behind the intact individual amino acids? Second, is there any property by which they can be identified, relative to one another? Before we do that, though, let's make sure to cleave any potential disulfide bonds in the peptide.

Experiment 1 (cleaving the disulfide bonds, if any): Treating the peptide with dithiothreitol reduces the disulfide bond by successive S_N2 reactions, as shown in **Figure 26.52**. Attack of the thiol on one sulfur of the disulfide bond forms a new disulfide, while releasing the thiol of one of the cysteines. Intramolecular attack by the other sulfur of dithiothreitol liberates the other cysteine.

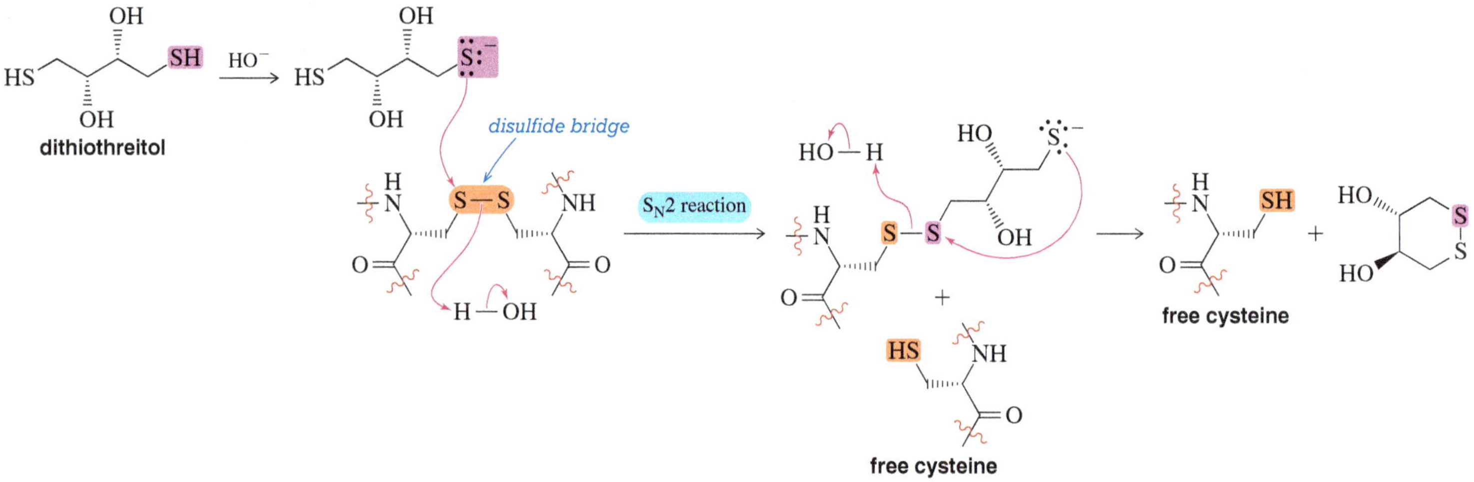

Figure 26.52 The cleavage of a disulfide bridge by dithiothreitol.

Once the disulfide bonds have been eliminated, the peptide bonds must be cleaved to separate the protein into its individual amino acids. These are simply amide bonds. Amides are stable carboxylic acid derivatives, but they can be hydrolyzed using strong acid and heat.

Experiment 2 (hydrolysis of the peptide): Heating the peptide in the presence of acid and water cleaves our decapeptide into ten individual amino acids (**Figure 26.53**). [*These conditions are harsh enough that some amino acids may degrade, limiting its practical utility.*]

$$H_2N-Xxx-Xxx-Xxx-Xxx-Xxx-Xxx-Xxx-Xxx-Xxx-Xxx-COOH \xrightarrow[H_2O,\ \Delta]{6N\ HCl} 10\ H_3\overset{+}{N}-Xxx-COOH$$

Representative hydrolysis reaction:

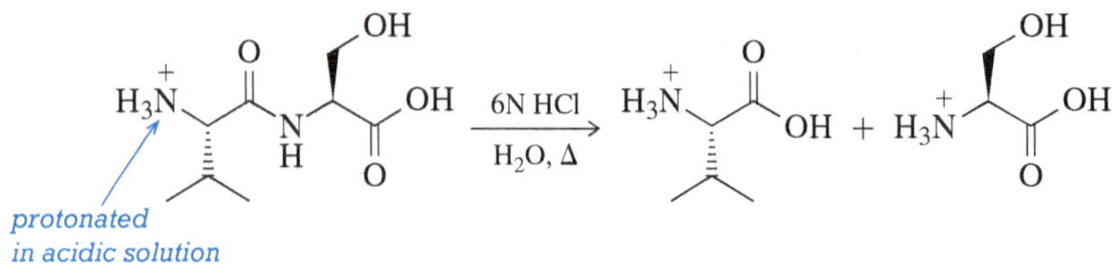

Figure 26.53 The hydrolysis of a peptide chain using strong acid.

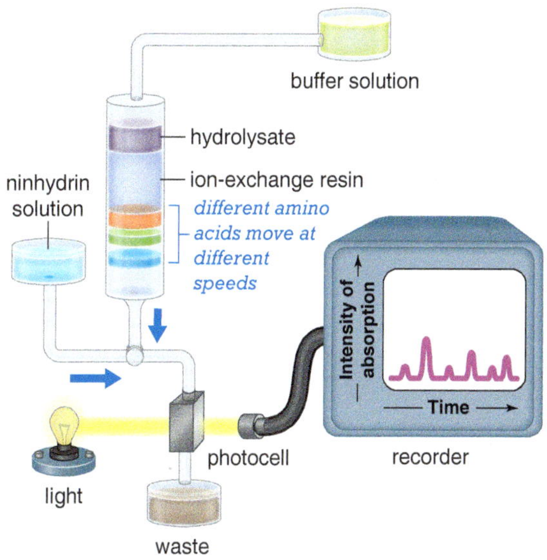

Figure 26.54 Ion-exchange chromatography separates amino acids based on their interaction with an anionic resin.

Having obtained the individual amino acids, is there a property by which amino acids can be separated and characterized relative to each other? Recall from Section 26.3.2 that every amino acid has a slightly different isoelectric point. This implies that each is neutral at a slightly different pH. Electrophoresis uses this difference to separate amino acids by their migration toward a cathode or anode. While this works in principle, it is more practical to use ion-exchange chromatography, which is an automated process that relies on the same pH characteristics of amino acids.

Experiment 3 (ion-exchange chromatography): In ion-exchange chromatography (**Figure 26.54**), our ten amino acids are applied to a column filled with negatively charged resin particles. By washing the column containing our amino acids with an acidic buffer solution (pH = 3), all of the amines of the amino acids will be protonated. These ammonium ions will interact strongly with the negatively charged resin, thereby sticking at the top of the resin. Successive washings with less and less acidic (more and more basic) buffers (pH changes from 3 to 12) will cause the amino acids to dislodge at different times based on the pK_a value of the ammonium ions. This allows each type of amino acid to be collected in separate fractions.

As the amino acids leave the column, they are mixed with ninhydrin [*see Section 26.5.3*]. The concentration of the resulting purple solutions is measured using a photocell. Comparing the intensity of the resulting peaks against standard retention times makes it possible to identify which amino acids are present in our decapeptide. A larger peak in the spectrum implies that there is more of the amino acid that produced that peak. Based on the spectrum shown in **Figure 26.55**, we have an equal amount (one each) of six amino acids (Ser, Gly, Met, Ile, Lys, Arg) and double that amount of Cys and Phe. Having isolated two cysteines from the protein, it was a good idea to break the disulfide bonds in experiment 1.

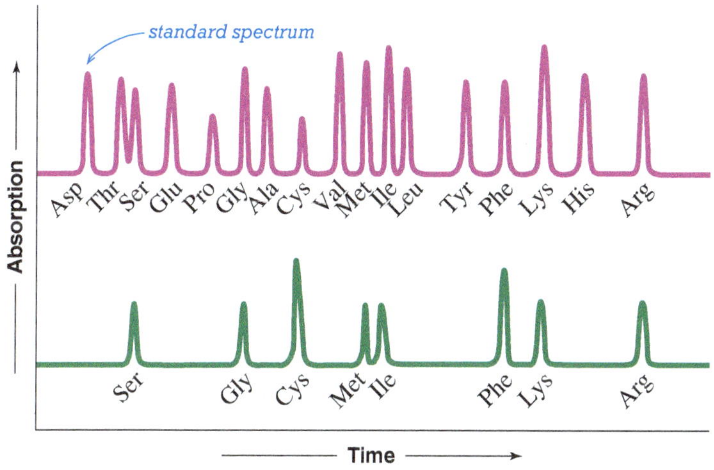

Figure 26.55 Comparison against a standard identifies the amino acids in our mixture.

At this point, we know the identity of the amino acids in our protein. What we don't know is their sequence. To determine the sequence, a reaction called the Edman degradation can be used. This reaction selectively cleaves the N-terminal amino acid from the peptide, allowing it to be identified by comparison to a standard HPLC retention time, similar to what was done in Figure 26.55. The process can then be repeated on the remaining peptide until the entire sequence is known.

While the Edman degradation can easily sequence a 10-amino-acid peptide, it is typically only useful for peptides smaller than 30 amino acids. Thus, prior to doing the Edman degradation, a larger protein would be cleaved into shorter fragments. Because we are simulating the identification of a larger protein, let's go ahead and cleave ours into smaller fragments. This is done by a number of specific reactions. Don't worry about the how, just focus on the results of each.

Experiment 4 (peptide cleavage with trypsin): Treating our protein with an enzyme called trypsin cleaves the peptide on the carboxy side of either lysine or arginine residues. Because we know we have one of each in our peptide, we'd expect cleavage in two places, producing a total of four fragments (**Figure 26.56**). The four fragments produced each contain 6, 4, 9, and

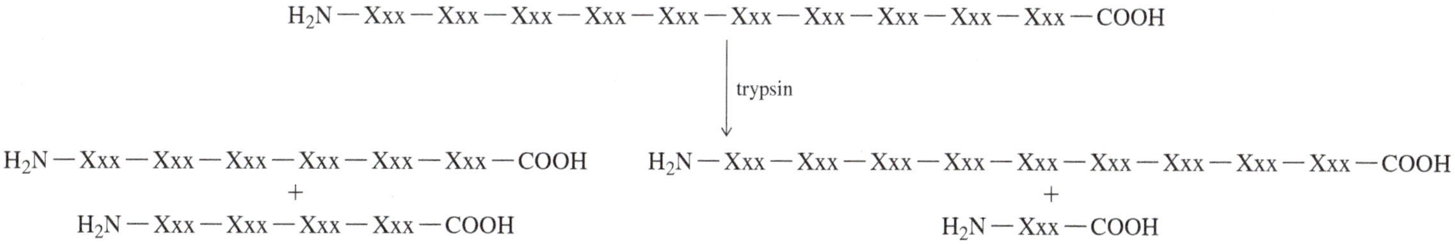

Figure 26.56 Cleavage with trypsin produces four shorter fragments.

1 amino acid residue(s). [*This does not mean there are 20 residues in the chain. 6 + 4 = 10; 9 + 1 = 10.*]

Experiment 5 (cleavage with chymotrypsin): Treating a separate sample of our protein with an enzyme called chymotrypsin cleaves the peptide on the carboxy side of phenylalanine, tyrosine, or tryptophan residues. Because our protein has two phenylalanines we'd expect cleavage in up to two places, producing up to four fragments (**Figure 26.57**). However, only two new fragments form. [*What does this mean?*] The two fragments contain 4 and 6 amino acid residues each [*4 + 6 = 10*].

Figure 26.57 Cleavage with chymotrypsin produces two shorter fragments.

$$H_2N{-}Xxx{-}Xxx{-}Xxx{-}Xxx{-}Xxx{-}Xxx{-}Xxx{-}Xxx{-}Xxx{-}Xxx{-}COOH$$

$$\downarrow \text{chymotrypsin}$$

$$H_2N{-}Xxx{-}Xxx{-}Xxx{-}Xxx{-}Xxx{-}Xxx{-}COOH \;+\; H_2N{-}Xxx{-}Xxx{-}Xxx{-}Xxx{-}COOH$$

Experiment 6 (cleavage with cyanogen bromide): Treating yet another sample of our protein with cyanogen bromide cleaves the peptide on the carboxy side of methionine residues. Our protein has one methionine, so we'd expect cleavage in one place, producing two fragments (**Figure 26.58**). The two fragments contain 3 and 7 amino acid residues each [*3 + 7 = 10*].

Figure 26.58 Cleavage with cyanogen bromide produces two shorter fragments.

$$H_2N{-}Xxx{-}Xxx{-}Xxx{-}Xxx{-}Xxx{-}Xxx{-}Xxx{-}Xxx{-}Xxx{-}Xxx{-}COOH$$

$$\downarrow \text{cyanogen bromide}$$

$$H_2N{-}Xxx{-}Xxx{-}Xxx{-}Xxx{-}Xxx{-}Xxx{-}Xxx{-}COOH \;+\; H_2N{-}Xxx{-}Xxx{-}Xxx{-}COOH$$

Before we proceed, let's take stock of where we are. We have now produced a total of 8 fragments of varying lengths, each shown in **Table 26.3**. Once more, the mechanisms of these reactions aren't important. All that matters here is that we have cut the big protein into smaller, more easily identifiable pieces. We are now ready to identify each of the small pieces. We can then put them back together to identify the entire protein.

Table 26.3 Peptide fragments produced by experiments 4–6.

Experiment	Fragments produced
4: Trypsin	$H_2N{-}Xxx{-}Xxx{-}Xxx{-}Xxx{-}Xxx{-}Xxx{-}COOH$
	$H_2N{-}Xxx{-}Xxx{-}Xxx{-}Xxx{-}COOH$
	$H_2N{-}Xxx{-}Xxx{-}Xxx{-}Xxx{-}Xxx{-}Xxx{-}Xxx{-}Xxx{-}Xxx{-}COOH$
	$H_2N{-}Xxx{-}COOH$
5: Chymotrypsin	$H_2N{-}Xxx{-}Xxx{-}Xxx{-}Xxx{-}COOH$
	$H_2N{-}Xxx{-}Xxx{-}Xxx{-}Xxx{-}Xxx{-}Xxx{-}COOH$
6: Cyanogen bromide	$H_2N{-}Xxx{-}Xxx{-}Xxx{-}COOH$
	$H_2N{-}Xxx{-}Xxx{-}Xxx{-}Xxx{-}Xxx{-}Xxx{-}Xxx{-}COOH$

As mentioned previously, Edman degradation can be used to identify one amino acid at a time, cleaving from the N-terminus. Unfortunately, it does not allow us to determine the structure of the C-terminus. Luckily, there is an experiment that does that. Let's do that for each of our fragments.

Experiment 7 (cleavage with carboxypeptidase): Carboxypeptidase selectively cleaves peptides one amino acid at a time, beginning at the C-terminus (**Figure 26.59**). The cleaved amino acid can be then be identified by comparison with known standards (as was done in Figure 26.55). The results of this experiment are shown in **Table 26.4**. Through these experiments, we have identified the terminal amino acid of each fragment, and also, the whole protein.

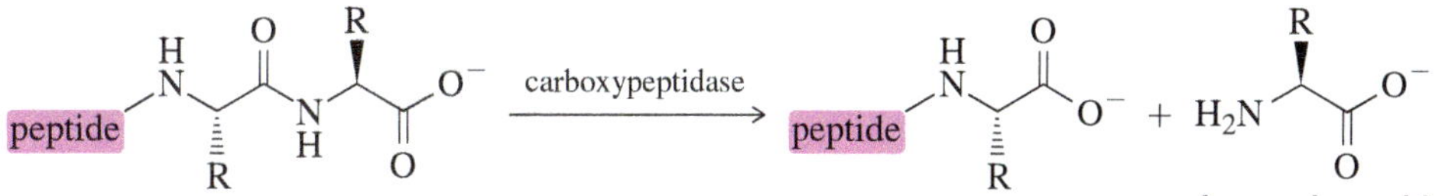

Figure 26.59 Cleavage with carboxypeptidase makes it possible to identify the C-terminal amino acid of our fragments.

Table 26.4 Results of experiments 4–8.

Experiment	Fragments produced	Results of experiment 7: (carboxy terminus)	Results of experiment 8: Edman degradation
4: Trypsin	$H_2N—(Xxx)_6—COOH$	$H_2N—(Xxx)_5—Arg—COOH$	H₂N—Gly—Ile—Met—Phe—Cys—Arg—COOH
	$H_2N—(Xxx)_4—COOH$	$H_2N—(Xxx)_3—Phe—COOH$	H_2N—Ser—Cys—Lys—Phe—COOH
	$H_2N—(Xxx)_9—COOH$	$H_2N—(Xxx)_8—Lys—COOH$	H_2N—Gly—Ile—Met—Phe—Cys—Arg—Ser—Cys—Lys—COOH
	H_2N—Xxx—COOH	H_2N—Phe—COOH	H_2N—Phe—COOH
5: Chymotrypsin	$H_2N—(Xxx)_4—COOH$	$H_2N—(Xxx)_3—Phe—COOH$	H_2N—Gly—Ile—Met—Phe—COOH
	$H_2N—(Xxx)_6—COOH$	$H_2N—(Xxx)_5—Phe—COOH$	H_2N—Cys—Arg—Ser—Cys—Lys—Phe—COOH
6: Cyanogen bromide	$H_2N—(Xxx)_3—COOH$	$H_2N—(Xxx)_2—Met—COOH$	H_2N—Gly—Ile—Met—COOH
	$H_2N—(Xxx)_7—COOH$	$H_2N—(Xxx)_6—Phe—COOH$	H_2N—Phe—Cys—Arg—Ser—Cys—Lys—Phe—COOH
Whole protein	$H_2N—(Xxx)_{10}—COOH$	$H_2N—(Xxx)_9—Phe—COOH$	See Figure 26.62.

Finally, we can use the Edman degradation we've been discussing to determine the sequence of each fragment working left to right from the N-terminus.

Experiment 8 (Edman degradation): The mechanism of the Edman degradation is shown in **Figure 26.60**. The first step is to treat the peptide chain with phenylisothiocyanate. Isothiocyanates are strongly electrophilic at the central carbon, reacting by what looks like a nucleophilic acyl

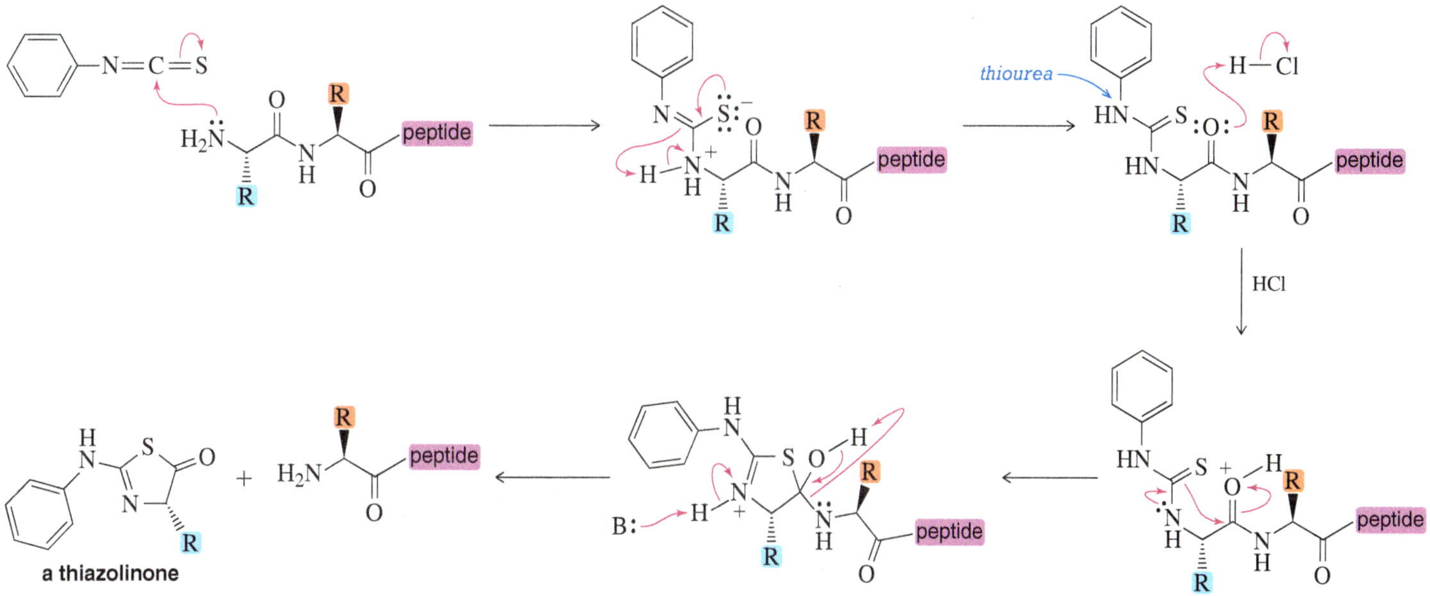

Figure 26.60 The mechanism of the Edman degradation.

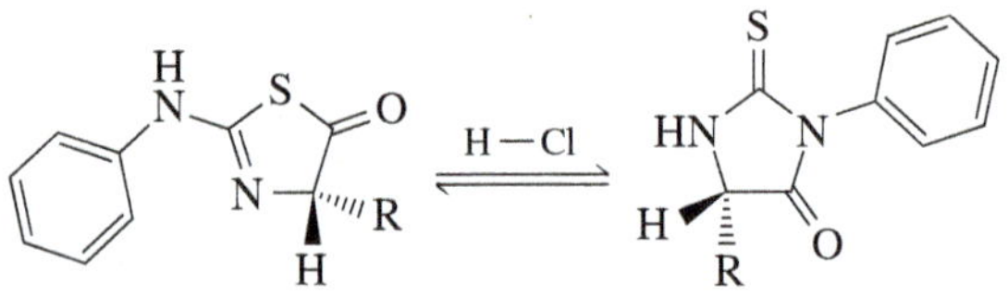

Figure 26.61 Rearrangement gives a substituted phenylthiohydantoin. Comparison with a standard provides the identity of the N-terminal amino acid.

substitution mechanism with the 1° amine to give a thiourea. Treating the thiourea with acid results in a cyclization reaction that ultimately produces a thiazolinone composed of the N-terminal amino acid, while kicking out the remainder of the peptide chain. The peptide chain, after it has been separated from the thiazolinone, can be subjected to additional Edman degradations.

Under acidic conditions, the substituted thiazolinone rearranges to give a substituted phenylthiohydantoin (**Figure 26.61**). The phenylthiohydantoin for each amino acid has been characterized. By comparing the phenylthiohydantoin with these standards, we can identify the N-terminal acid of each chain, repeating it for each amino acid residue in each chain. In so doing, we generate the results shown under the experiment 8 column in Table 26.4.

With each of the small fragments sequenced, we can now overlay them based on overlapping residues, as shown in **Figure 26.62**. This can be done even when we don't know which experiment produced which fragment. With the individual chains aligned, the overall sequence is determined. As a double-check, notice that the sequenced protein had two cysteines, two phenylalanines, and one of each of the rest of the amino acids predicted by experiment 2. ■

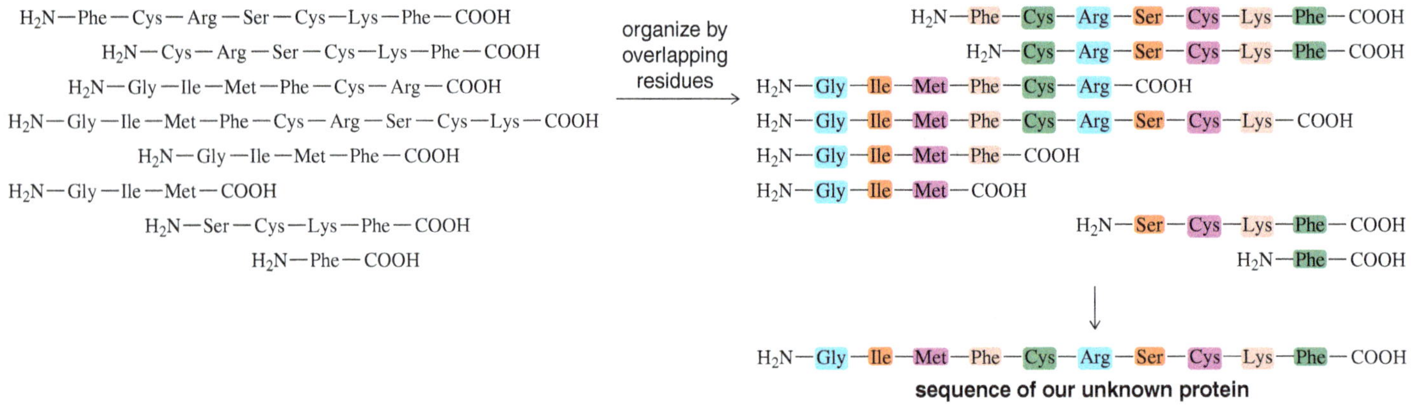

Figure 26.62 Aligning the overlapping residues of the fragments makes it possible to determine the amino acid sequence of our protein.

26.8 SUMMARY

- When combined with electrophoresis, cleaving a protein into successively smaller fragments at specific amino acid residues allows the structure of a peptide to be determined.

Assessment 26.39 Where would you expect (a) trypsin, (b) chymotrypsin, and (c) cyanogen bromide to cleave the following decapeptide?

H_2N—Ala—Arg—Ser—Pro—Cys—Ile—Phe—Met—Thr—Val—COOH

Assessment 26.40 A chemist identified the following fragments from the experiments listed in Section 26.8. What is the primary structure of the peptide?

H_2N—DYT—COOH
H_2N—ESGMVA—COOH
H_2N—VA—COOH
H_2N—DYTCGKESGM—COOH
H_2N—DYTCGK—COOH
H_2N—CGKESGMVA—COOH

Assessment 26.41 Suggest an arrow-pushing mechanism for the rearrangement of thiazolinone to thiohydantoin. What is the driving force for this rearrangement?

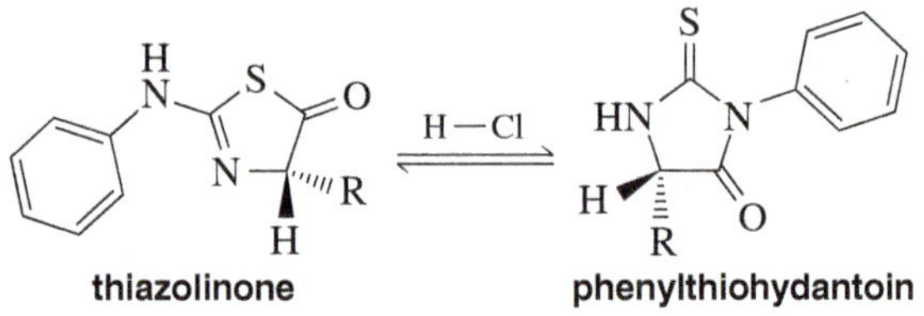

26.9 Peptide synthesis

The potential medicinal benefits of short peptide fragments and the desire to produce biological proteins in the lab have led chemists to develop the laboratory synthesis of peptides. Because amino acids have two reactive functional groups, much of the effort goes into keeping them protected so that only one possible reaction occurs. In this way, **Figure 26.63** represents a general scheme that is elaborated upon for the solution-phase synthesis in Section 26.9.1. Some of the problems encountered during the solution-phase synthesis can be solved by conducting the reactions on a solid support (see Section 26.9.2). In each case, the synthesis of the same tripeptide (H_2N—Val—Ile—Phe—COOH) is used as a representative example. Every reaction in the sequences shown has been seen before, only one of which does not proceed by the nucleophilic acyl substitution reactions from Chapter 19.

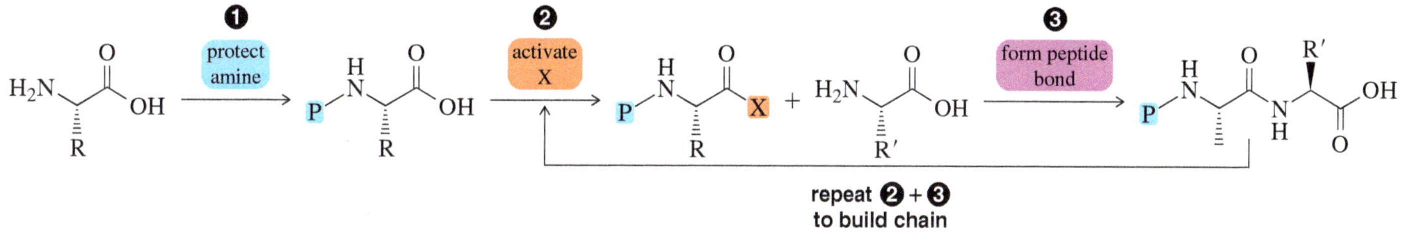

Figure 26.63 The general scheme for peptide synthesis.

26.9.1 Solution phase

The laboratory synthesis of peptides in solution typically builds from the N-terminus to the C-terminus (i.e., working left to right as amino acids sequences are usually written). As a result, the start of any synthesis begins with the protection of the terminal amine. This is necessary to prevent it from reacting with any carboxy groups added later and ensuring that the chain grows in a single direction only. Benzyl chloroformate (Section 26.5.2) is commonly used to protect the N-terminus (**Figure 26.64**).

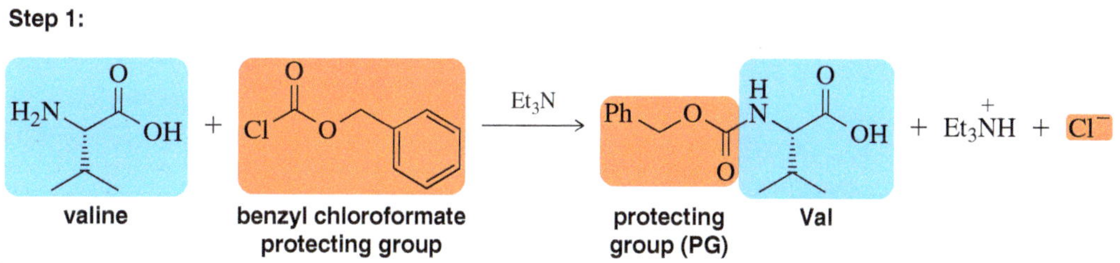

Figure 26.64 Step 1: Protection of the N-terminus of valine with benzyl chloroformate ensures that it does not participate in other nucleophilic substitution reactions.

Now you might think that we are ready to attach the next amino acid, but there's a problem. Recall from Chapter 18 that carboxylic acids are not very electrophilic for nucleophilic acyl substitution. As a result, they need to be activated for nucleophilic attack. In Chapter 18 we did this by converting the carboxylic acid to an acid chloride, but acid chlorides are almost too reactive to be used in protein synthesis. Instead, the carboxylic acid can be converted to an anhydride (the next best electrophile) by reacting with ethyl chloroformate (**Figure 26.65**).

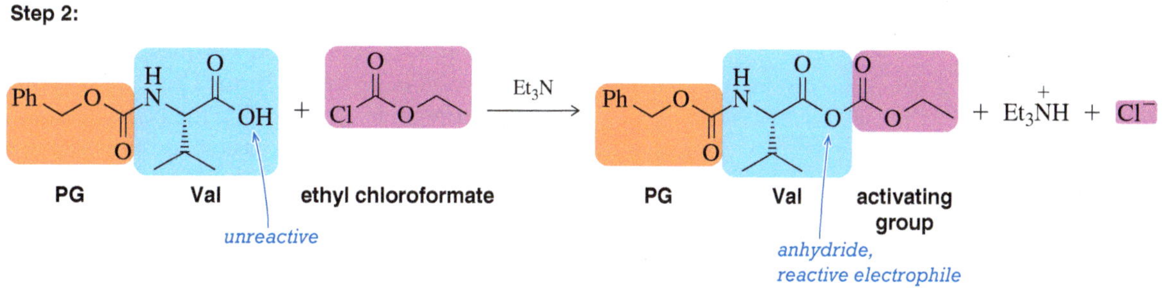

Figure 26.65 Step 2: Activation of the carboxylic acid with ethyl chloroformate makes the carboxylic acid a stronger electrophile.

With the carboxylic acid activated as an anhydride, isoleucine can be added (**Figure 26.66**). Attack by the amine forms the new amide bond in a nucleophilic acyl substitution reaction. The ethyl carbonate ion is also formed in this reaction (from the ethyl chloroformate activating group). It decomposes into carbon dioxide and ethanol, thus enhancing the driving force of the nucleophilic acyl substitution reaction.

Step 3:

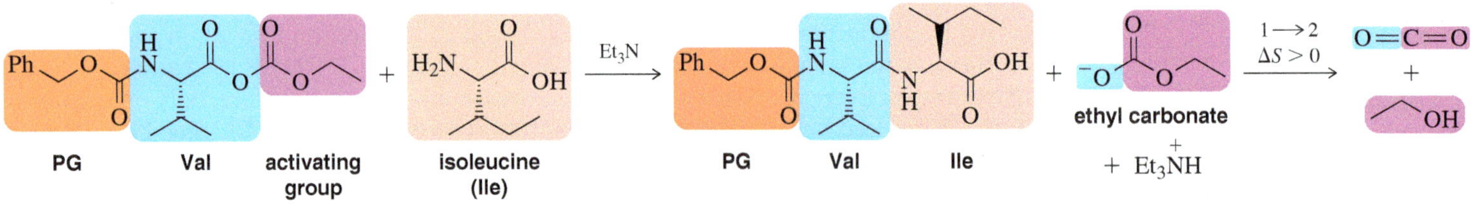

Figure 26.66 **Step 3: Nucleophilic acyl substitution with the amino group of isoleucine** forms the amide bond in the N-protected dipeptide.

Extending the chain beyond the dipeptide in Figure 26.66 is as simple as repeating steps 2 and 3 for as many amino acids as the desired polypeptide requires. As such, the carboxylic acid of the C-terminus is activated once more with ethyl chloroformate (**Figure 26.67**).

Step 2 (repeated):

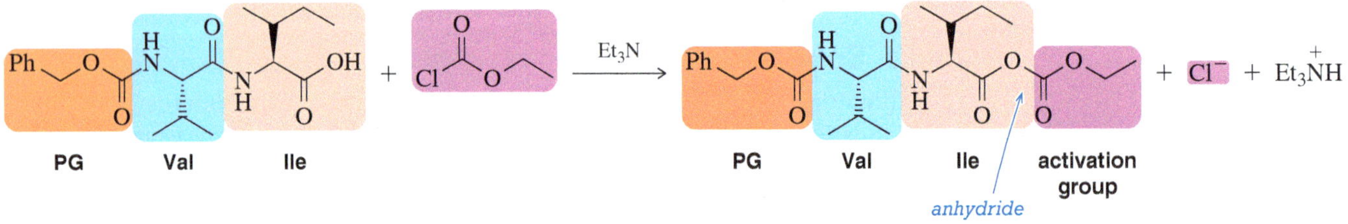

Figure 26.67 **Step 2: Activation of the carboxylic acid with ethyl chloroformate** makes the carboxylic acid a stronger electrophile.

Reaction of the activated carboxylic acid of the dipeptide with phenylalanine by nucleophilic acyl substitution results in formation of the N-protected tripeptide (**Figure 26.68**).

Step 3 (repeated):

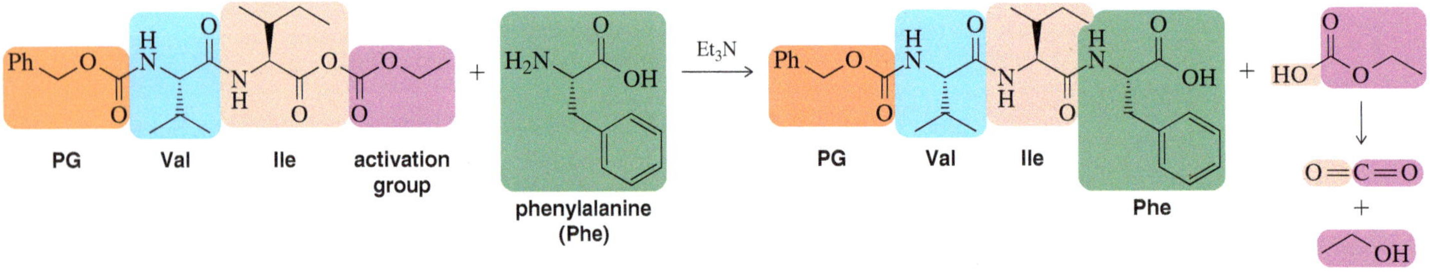

Figure 26.68 **Step 3: Nucleophilic acyl substitution with the amino group of phenylalanine** forms the amide bond in the N-protected tripeptide.

We could continue adding amino acid residues, but our goal was to make a tripeptide, which we now have, except in protected form. Thus, the lone step that remains is to deprotect the N-terminus of the tripeptide. By choosing the benzyl chloroformate to protect the amine, we don't have to use strongly acidic or basic conditions for the deprotection. [*This is good because those conditions may hydrolyze the peptide bonds we worked so hard to create.*] Instead, hydrogenolysis with a palladium catalyst can be used (**Figure 26.69**), producing the tripeptide, carbon dioxide, and toluene. [*For a review of this chemistry, see Sections 26.5.2 and 24.6.*]

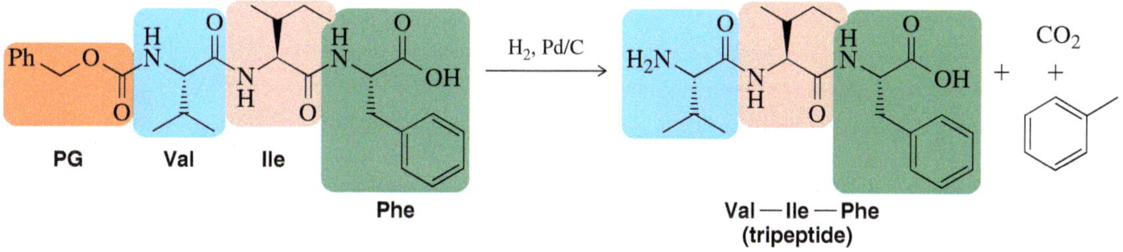

Figure 26.69 **Deprotection of the N-terminal amine by hydrogenolysis** completes the synthesis of the tripeptide.

Solution-phase synthesis is very efficient for making small peptides. It is tedious, though, requiring lots of protection, activation, and deprotection steps. At the end of each of these steps, to ensure a high yield in the next step, the product must be purified. These additional purification steps add a substantial amount of time to the process, while generating lots of extra waste. Thus, this method could be improved to make it both greener and less labor intensive. [*Enter solid-phase synthesis* . . . .]

26.9.2 Solid phase

Solid-phase synthesis is really no different from solution-phase synthesis except that the growing peptide chain is attached to a solid polymer support. This polymer support, referred to as a resin, is insoluble in most solvents. While the biphasic nature of the solution makes the reactions between molecules in solution and molecules attached to the resin slower, its utility lies in that anything not attached to the resin can simply be washed away at the end of the reaction. With the growing peptide on the resin, all of the by-products that were produced in solution, along with any excess amino acids used, are easily removed by washing.

Solid-phase synthesis starts with the appropriate resin. A polymer composed of styrene and 4-chloromethylstyrene monomers is used for this purpose (**Figure 26.70**). This polymer, referred to as the **Merrifield resin**, is created by free-radical polymerization, a reaction that is discussed in detail in Section 28.4.3.

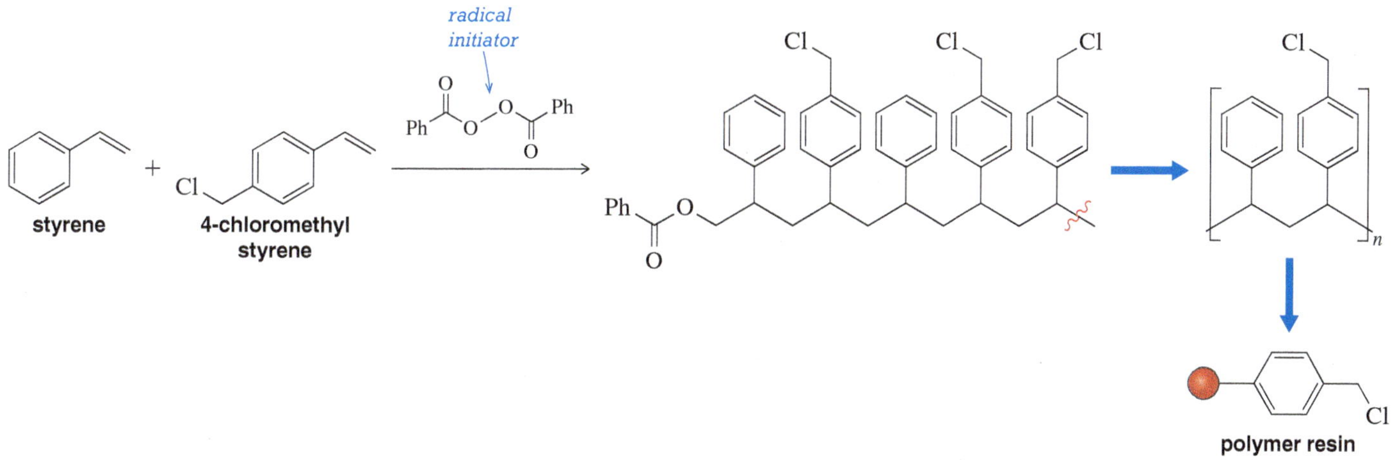

Figure 26.70 **The formation of the Merrifield polymer resin by radical polymerization.**

Whereas solution-phase synthesis extends the peptide chain by building from the N-terminus, the **Merrifield synthesis** does the opposite, mainly because of the ease with which the C-terminal carboxylic acid is cleaved from the resin in the final step (Figure 26.76). As a result, the synthesis begins by attaching phenylalanine to the resin by S_N2 displacement of the benzyl chloride, as shown in **Figure 26.71**. Amino groups are more nucleophilic than carboxylate ions, however, so the amine is first protected using di-*tert*-butyldicarbonate (which was first introduced in Section 26.5.2). When attached, it is referred to as a **BOC protecting group**.

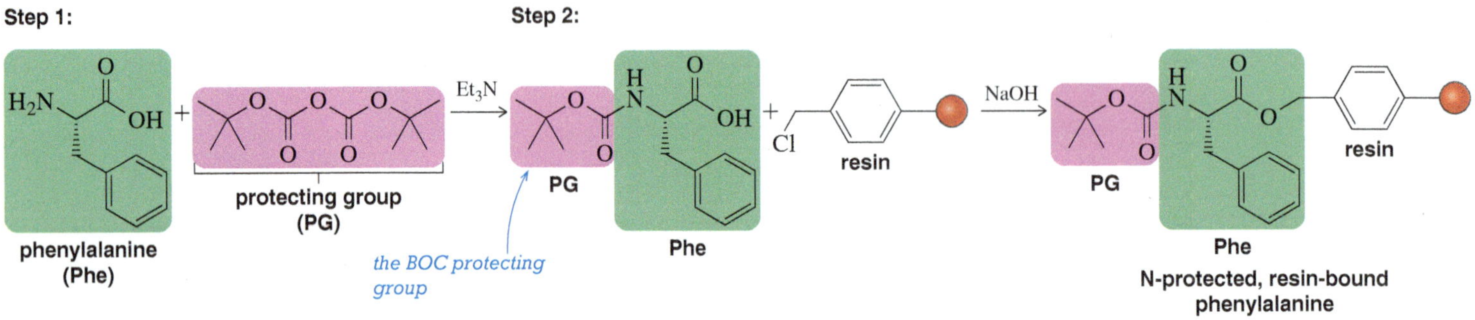

Figure 26.71 Protection of phenylalanine (step 1) and attachment to the polymer resin (step 2).

Following attachment of phenylalanine to the resin, the amine needs to be deprotected so it can be coupled to the next amino acid. This is accomplished by treating the BOC group with trifluoracetic acid, an organic soluble acid [**Figure 26.72(a)**]. Formation of a stable carbocation and decarboxylation to form carbon dioxide drive this deprotection via the mechanism shown in **Figure 26.72(b)**.

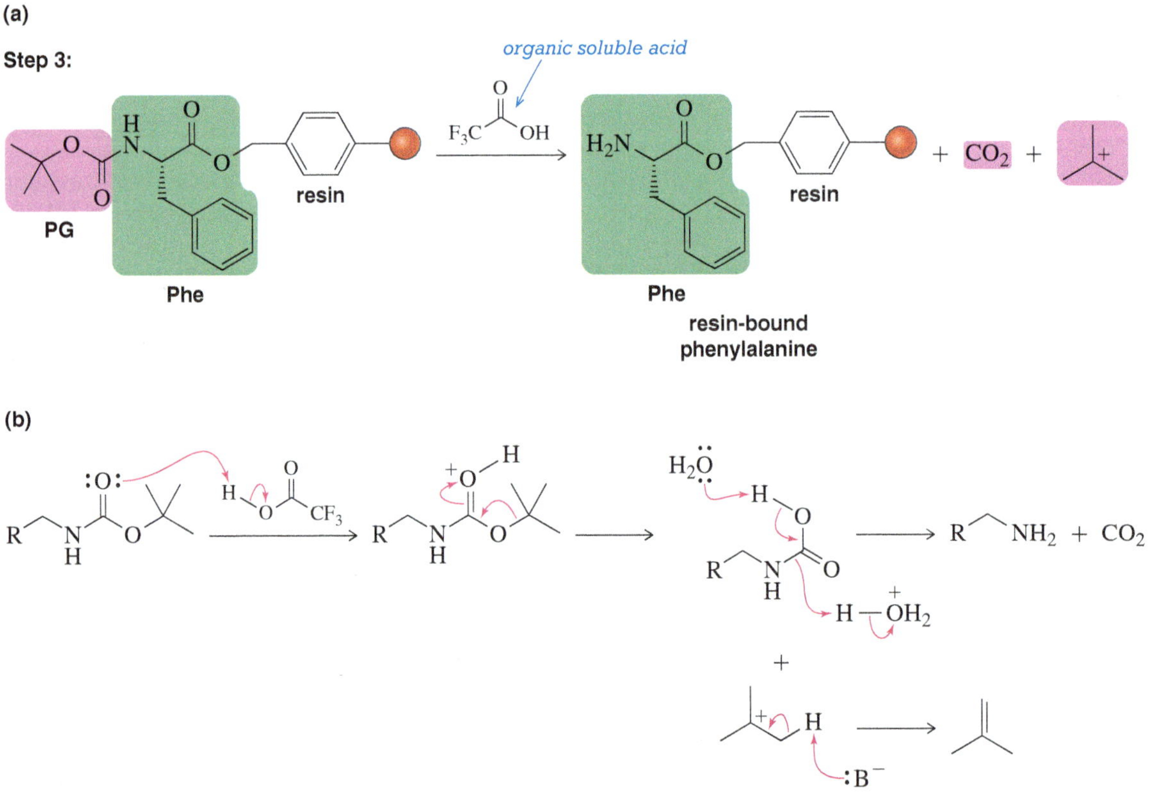

Figure 26.72 (a) Step 3: Deprotection of the terminal amine for future coupling. (b) The mechanism of BOC deprotection.

The free amine is now ready to be coupled to an activated carboxylic acid. Instead of using an anhydride directly, as is done in solution-phase synthesis [*see Figure 26.68*], the carboxylic acid is activated using dicyclohexylcarbodiimide (DCC) in the same reaction flask. As shown in **Figure 26.73**, the carboxylate ion attacks the central carbon of DCC to give an activated carboxylic acid. The activated carboxylic acid, similar to an anhydride or acid chloride, is attacked by the nucleophilic amine to form a new amide bond by the addition/elimination mechanism of nucleophilic acyl substitution (Section 19.7.3). A final deprotonation forms the incredibly stable dicyclohexyl urea (DCU), which is a driving force for this reaction. DCU is notoriously hard to separate from the resulting amide, but it can be simply washed away in solid-phase synthesis.

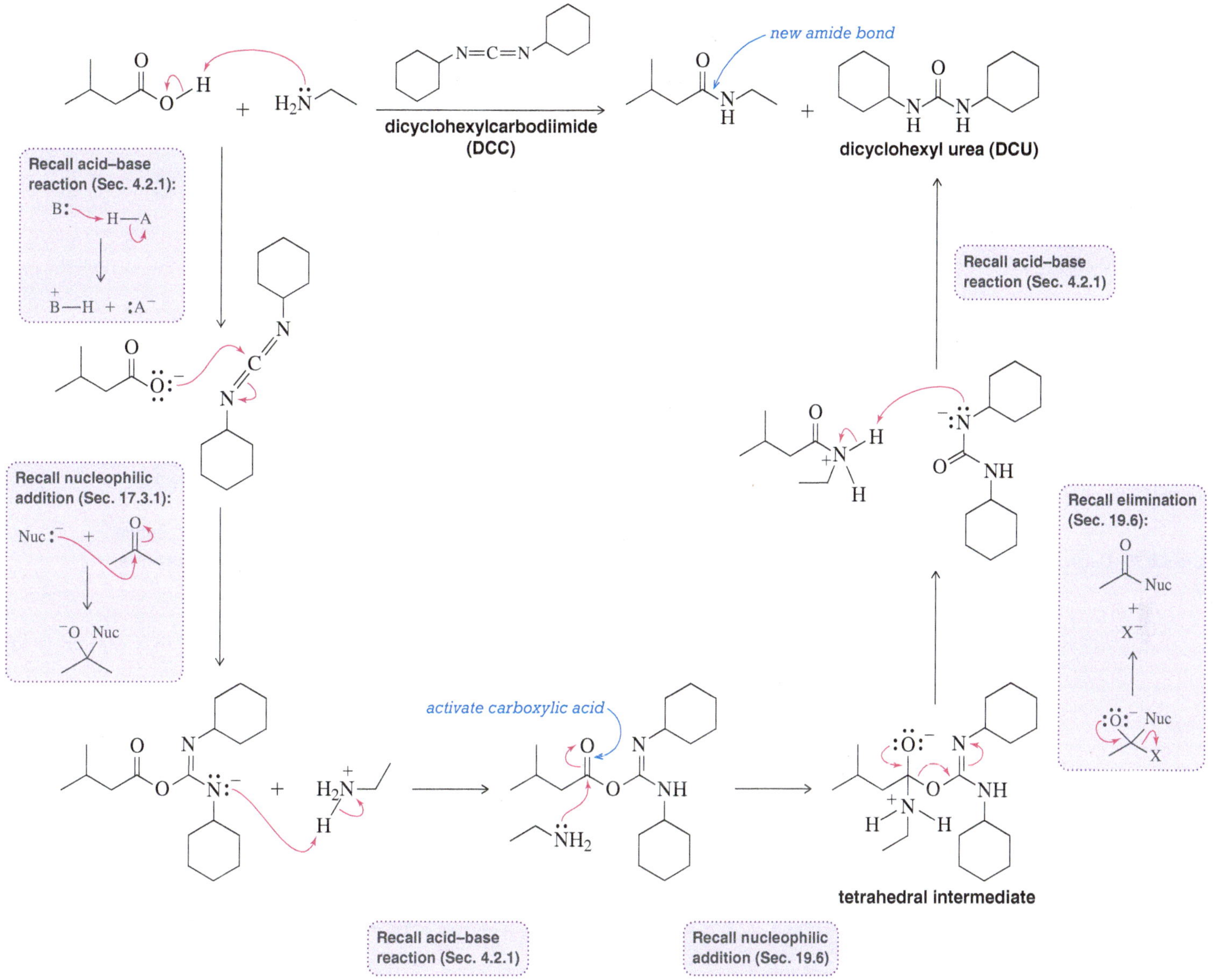

Figure 26.73 **Peptide bond formation using DCC coupling.** The carboxylic acid is activated in situ for nucleophilic attack by the amine.

DCC coupling thus produces an N-protected, resin-bound dipeptide (**Figure 26.74**).

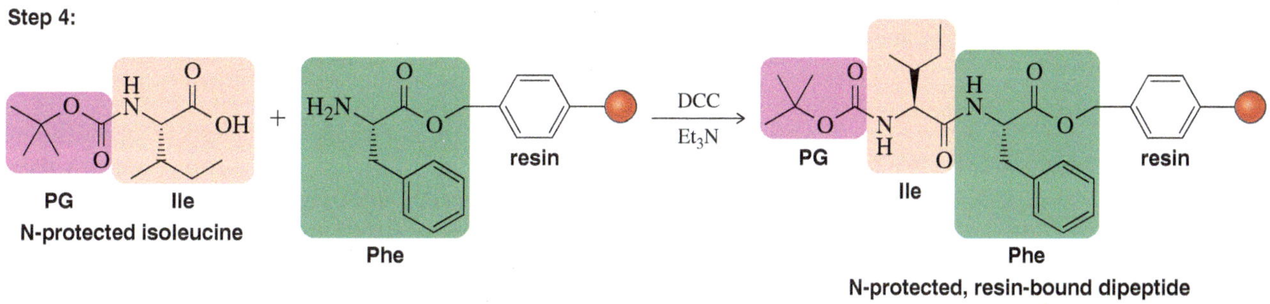

Figure 26.74 **Step 4: DCC coupling** of N-protected isoleucine and resin-bound phenylalanine.

Extending the chain beyond the dipeptide is as simple as repeating steps 3 and 4 for as many amino acids as the desired polypeptide requires. As such, deprotection of the N-terminus and DCC coupling with a protected valine produces the N-protected, resin-bound tripeptide (**Figure 26.75**).

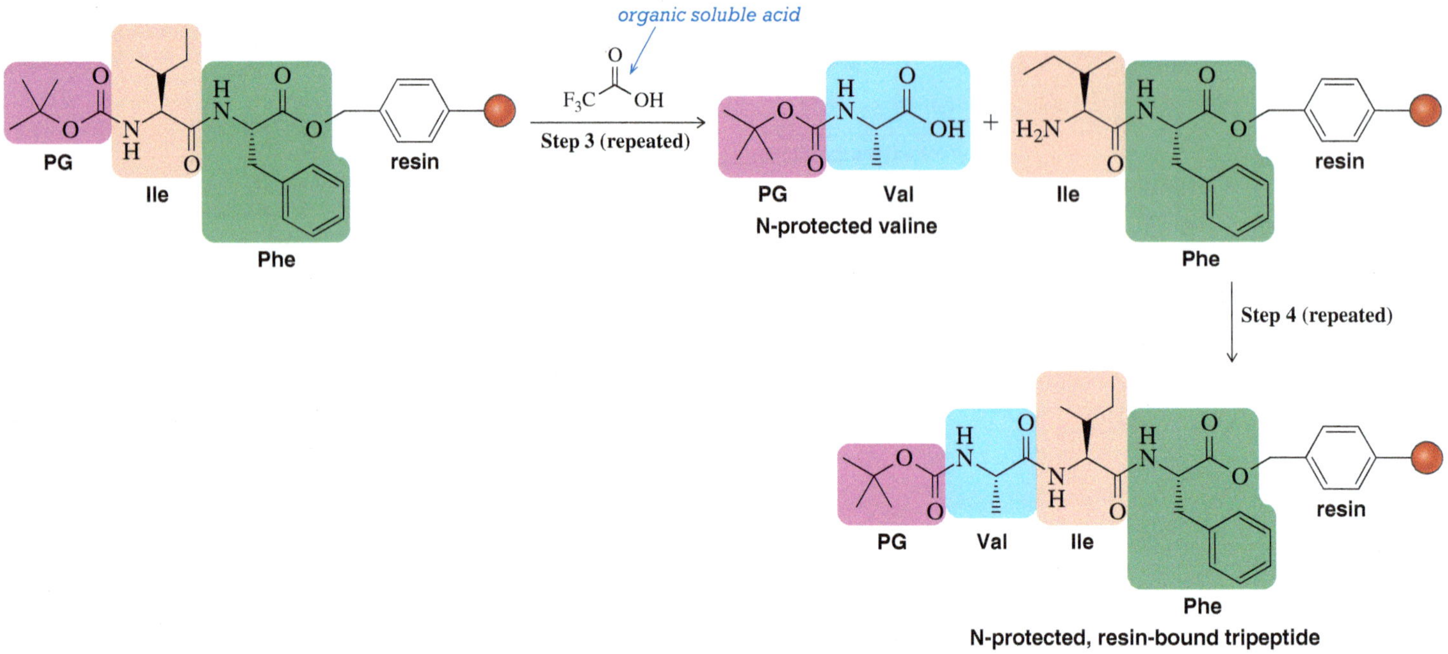

Figure 26.75 **Deprotection of the terminal amine and DCC coupling with N-protected valine produced the desired tripeptide.**

With the tripeptide completely formed, a final deprotection of the BOC-amine is followed by release of the peptide from the resin (**Figure 26.76**). A benzyl ether bond connects the peptide to the resin, so it can be cleaved by hydrogenolysis, a reaction first introduced in Section 24.6. Only after the peptide is separated from the resin is the peptide subjected to chromatographic purification.

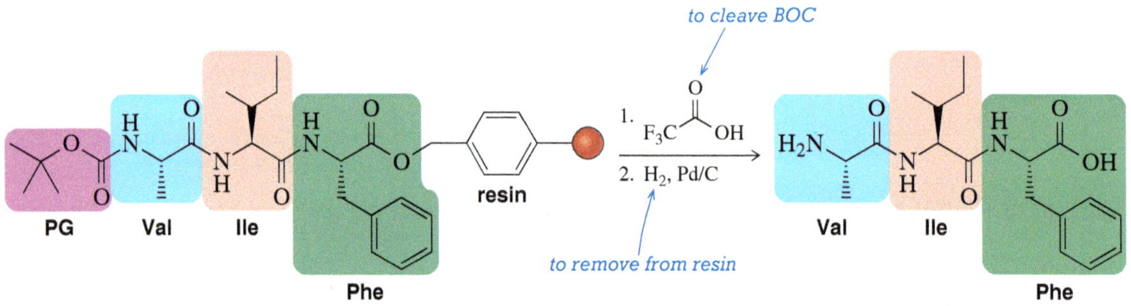

Figure 26.76 **Completion of the tripeptide synthesis. Acid-catalyzed decarboxylation deprotects the N-terminal amine,** and hydrogenolysis removes the tripeptide from the resin.

The Merrifield solid-phase synthesis does not necessarily improve upon the number of chemical reactions or the efficiency of the individual reactions compared to solution-phase synthesis. But because purification is not needed until the peptide is released from the resin, it produces a higher yield of protein, generates substantially less waste, can be easily automated, and significantly reduces the amount of labor required.

26.9 SUMMARY

- Peptide synthesis, whether in solution or on solid support, consists of a series of protection, carboxylic acid activation, peptide coupling, and deprotection steps.

Assessment 26.42 Propose a synthesis of the tripeptide Met-Phe-Gly using (a) solution-phase synthesis and (b) the Merrifield synthesis.

Assessment 26.43 If the amino acids shown were coupled without protecting groups, what four dipeptides would result?

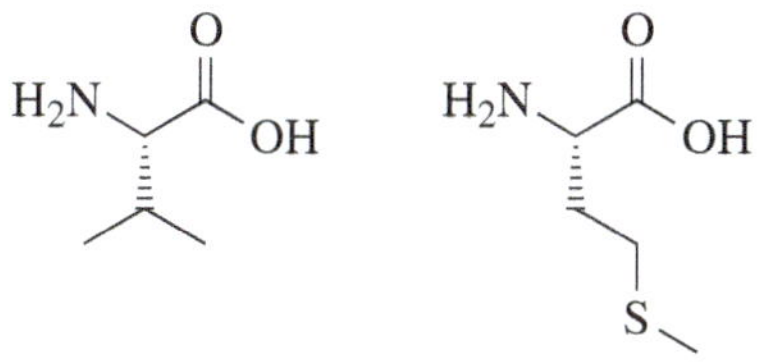

Assessment 26.44 A chemist, conducting the solution-phase synthesis of a dipeptide, forgot to protect the amine prior to activating the carboxylic acid. Suggest an outcome of this reaction.

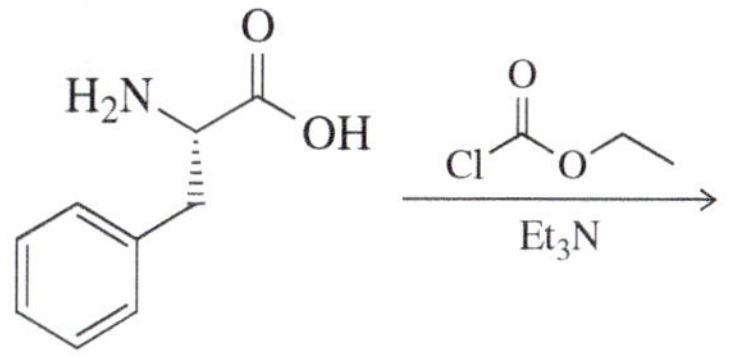

26-C CHEMISTRY IN CONTEXT: Medicinal Chemistry

Peptidomimetic enzyme inhibition

The first step of any drug discovery project is to identify a lead compound, like the pentapeptide identified in Chemistry in Context 26-B (Figure 26.50). Why can't the pentapeptide be used as the drug itself, especially given the ease with which pentapeptides can be synthesized using solution-phase or solid-phase chemistry? People depend on being able to take their drugs orally (by mouth) in pill form. Peptides, containing an amide bond, are hydrolyzed in the presence of strong acids. As we learned in Chemistry in Context 4-A on acid reflux disease, the stomach is a highly acidic, aqueous environment. Thus, peptides cannot survive the stomach, where they are hydrolyzed into the individual, medicinally inactive amino acids (**Figure 26.77**).

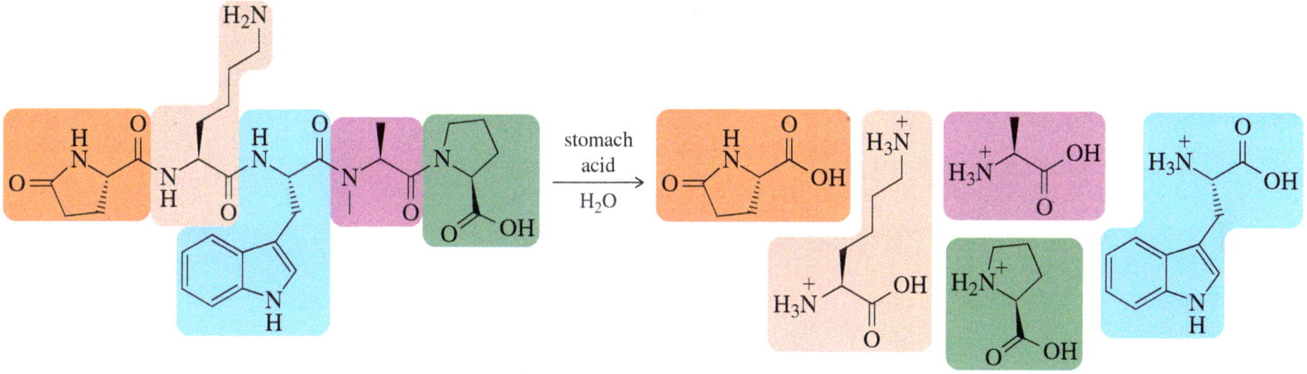

Figure 26.77 The acid-promoted hydrolysis of the pentapeptide in the stomach.

Captopril (**Figure 26.78**) was one of the first ACE inhibitors to appear on the market. It is very similar in structure to the C-terminus of the pentapeptide inhibitor. This structural similarity is likely the reason it binds so well and inhibits ACE with such specificity. While captopril does contain one amide bond, the second is replaced with a C—S bond. The thiol is a zinc-loving group, so it is expected to form a tight bond with the Zn^{2+} cation in the active site of ACE, thus preventing either bradykinin or angiotensin I from binding.

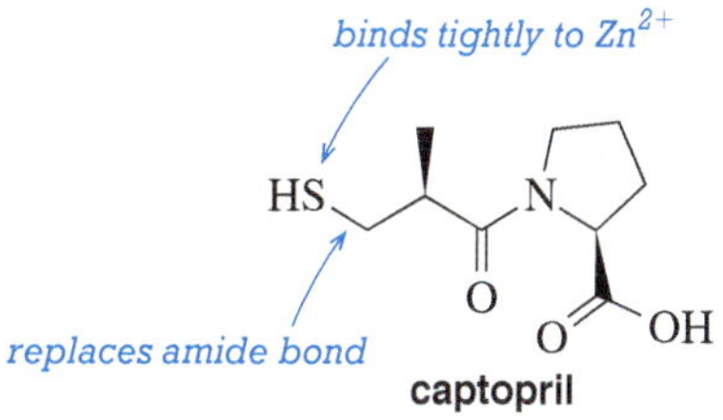

captopril

Figure 26.78 Captopril, one of the original ACE inhibitors, is similar in structure to the C-terminus of the inhibitory pentapeptide.

Assessment 26.45 Suggest a synthesis of the pentapeptide using solid-phase synthesis. [*You will want to protect the side chain nitrogen of lysine to avoid side reactions.*]

Concluding thoughts

Although captopril is very effective, it is believed that the free thiol group causes skin rashes and loss of taste in patients taking it. These side effects may seem minor for a disease that is a "silent killer," but any discomfort is enough to discourage patients from remaining on the therapy, so other ACE inhibitors that didn't have these side effects were sought. Enalapril and lisinopril (**Figure 26.79**) are two examples among many that are effective medications for high blood pressure, based on the same principles that produced captopril. Both enalapril and lisinopril are β-amino acids, which allows them to mimic the structure of the inhibitory peptide, but they lack the extra amide bond so they can withstand the acidic environment of the stomach.

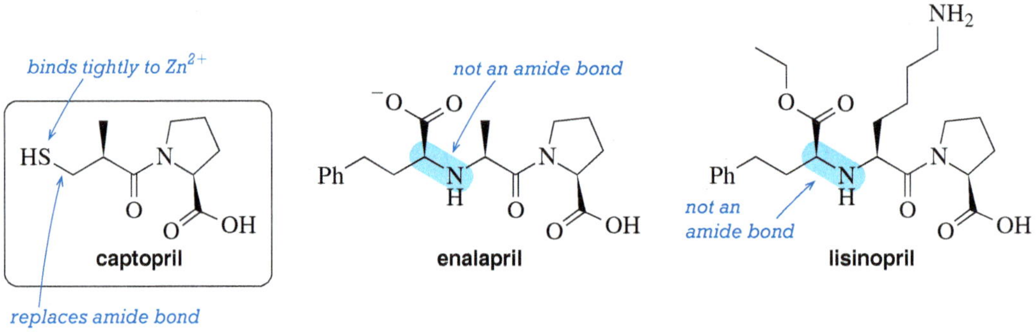

Figure 26.79 Other ACE inhibitors based on the structure of captopril.

Amino acids and the proteins they make are vital for life; enzymes and protein-based receptors are excellent targets for drug design and development; and peptides themselves, with slight modifications, make excellent drugs. The chemistry of amino acids presented here in Chapter 26 introduced only a few new reactions, and they were more or less an extension of the nucleophilic acyl substitution reactions from Chapter 19. Throughout this discussion, very little time was spent discussing proteins as biological polymers, but we address those concepts in Chapter 27, where we examine other biomolecules, including carbohydrates and nucleic acids. Chapter 28 concludes the text by discussing synthetic polymers prepared by chemists in the laboratory.

That's right, only two chapters remain . . . let's go!

End-of-Chapter Materials

Important Concepts

[*These are the major concepts covered in each section, excluding reactions.*]

26.1: Amino acids, composed of amino and carboxy groups, are the building blocks of peptides and proteins. Of the 20 amino acids, only glycine is achiral. All but one of the 19 chiral amino acids have the *S* configuration. [*All are* L.]

26.2: The 20 standard amino acids are classified based on their side chains: ionic, polar, and hydrophobic.

26.3: Because they contain a basic and acidic functional group, acid–base reactions dominate much of the reactivity of amino acids. At neutral pH, amino acids exist as a zwitterionic structure containing an ammonium ion and a carboxylate ion. The pH at which the zwitterionic form predominates is the isoelectric point (pI) and varies among amino acids. Electrophoresis can separate and identify amino acids on the basis of isoelectric points. Slight deviations in pK_a values for amino acids are a result of electronic effects imparted by the various side chains.

26.4: Amino acids can be synthesized using reactions studied previously, including reductive amination of α-keto acids with ammonia, the Hell–Volhard–Zelinskii reaction followed by S_N2 displacement of the resulting bromide by ammonia, the Gabriel malonic ester synthesis, and the Strecker synthesis of amino acids from aldehydes using the combination of NH_4Cl and NaCN.

26.5: Amino acids undergo the usual reactions of amines and carboxylic acids studied in previous chapters. These reactions include the Fischer esterification of the carboxylic acid, acylation of the amine, and separation of enantiomers by forming diastereomeric ammonium carboxylate complexes. The use of ninhydrin provides for the qualitative analysis of amino acids.

26.6: Proteins are naturally occurring polymers of >50 amino acid residues. With <50 amino acid residues, the strand is referred to as a peptide. The amide bond in a polymer or peptide is referred to as a peptide bond. The peptide bond is especially strong due to the contribution of the resonance structure containing the C=N double bond. This resonance structure also restricts rotation and imparts a rigidity that gives proteins well-defined structures.

26.7: The primary structure consists of the amino acid sequence. The secondary structure of proteins results from the folding of polypeptide chains driven by intramolecular interactions between amino acids. The β-sheet and α-helix are examples of secondary structures. The tertiary structure refers to the arrangement of β-sheets, α-helices, and other coils and turns that define the secondary structure. Quaternary structures are important when proteins are composed of more than one polypeptide unit.

26.8: Analysis of a protein structure can be done by degrading the peptide chain at specific amino acid residues. Dithiothreitol is used to cleave disulfide bridges between cysteine residues. Ion-exchange chromatography is used to separate and identify the amino acids that make up a peptide chain. Cleavage with the enzyme trypsin occurs on the carboxy side of lysine or arginine residues. Cleavage with chymotrypsin cleaves the peptide on the carboxy side of phenylalanine, tyrosine, and tryptophan residues. Cyanogen bromide cleaves peptides on the carboxy side of methionine residues. Carboxypeptidase selectively cleaves the C-terminal amino acid one at a time. The Edman degradation cleaves the N-terminal amino acid one at a time.

26.9: Peptide synthesis can be done using a number of reactions studied previously. As a general scheme, peptide synthesis involves a series of reactions including protecting, carboxylic acid activation, peptide coupling, and then deprotection. Peptide synthesis on solid support produces a higher yield of protein, generates substantially less waste, can be easily automated, and significantly reduces the amount of labor required.

Important Reactions and Mechanisms

1. **Hell–Volhard–Zelinskii reaction (Section 26.4.2):** Initial formation of an acid bromide is followed by the bromination of an enol to give an α-bromo acid bromide. Hydrolysis of the acid bromide results in formation of an α-bromo carboxylic acid.

General:

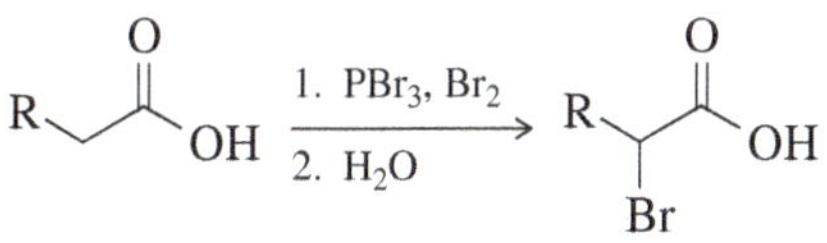

Specific:

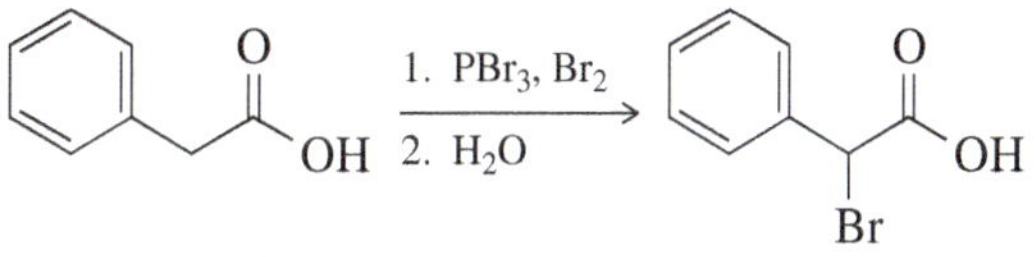

Mechanism:

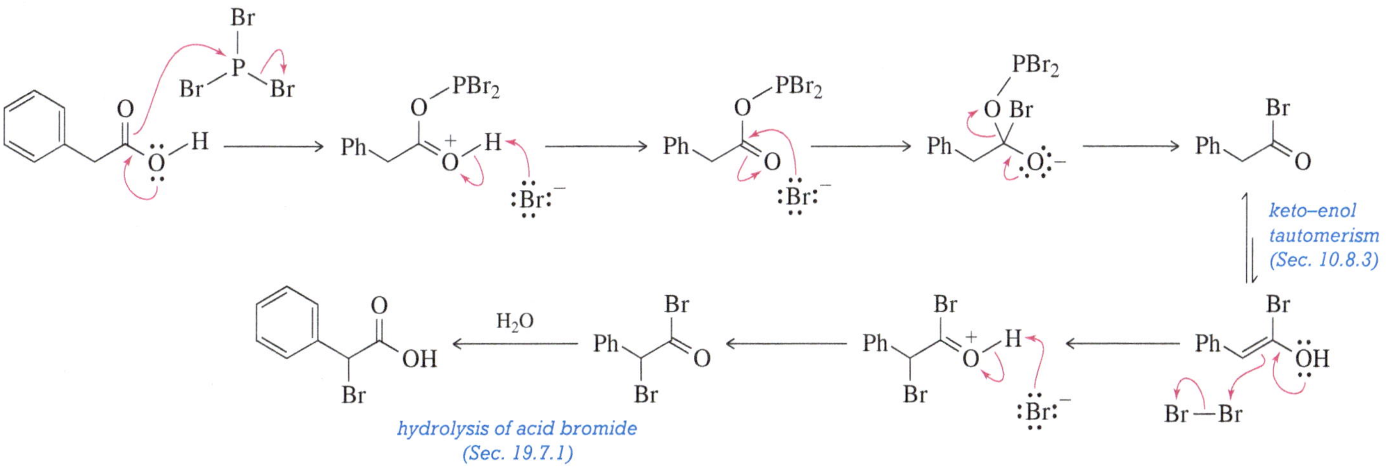

2. **Gabriel malonic ester synthesis (Section 26.4.3):** S_N2 displacement of a bromide with phthalimide creates a phthalimido malonate. Deprotonation and alkylation of the malonate is followed by exhaustive hydrolysis to form a dicarboxylic acid. At high temperature, a thermal decarboxylation occurs through a six-electron cyclic transition state to give an amino acid.

General:

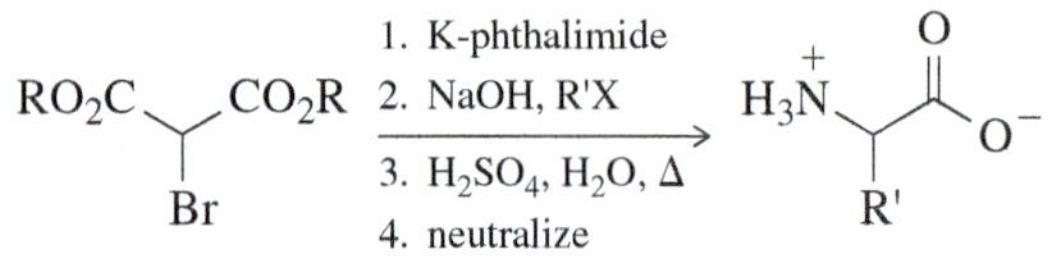

Specific:

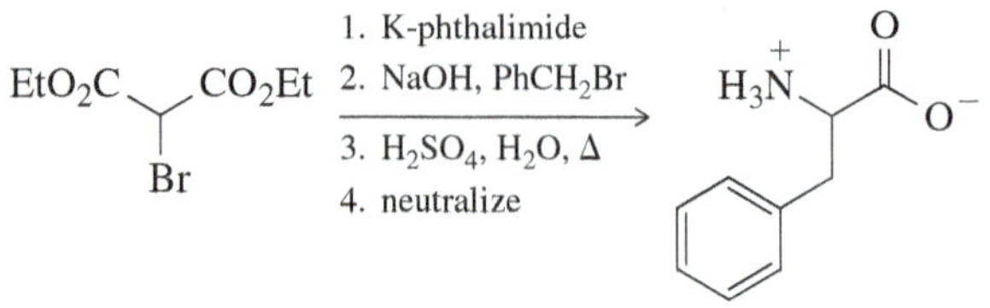

Mechanism:

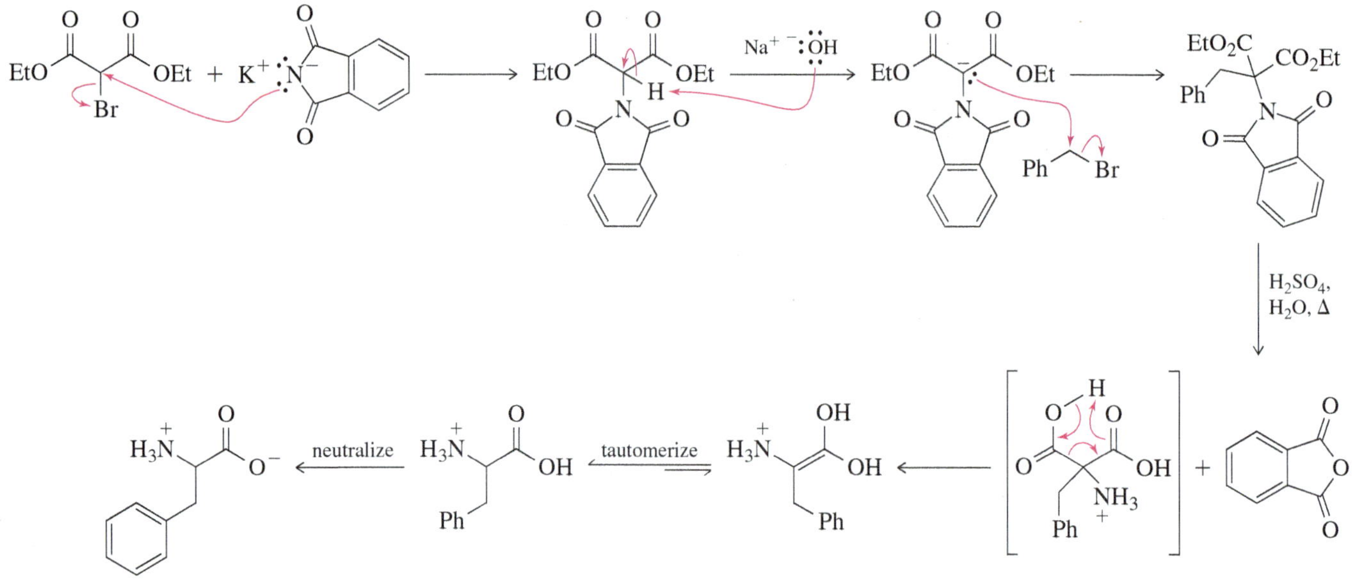

3. **Strecker synthesis (Section 26.4.4):** Nucleophilic addition of a cyanide ion to a protonated imine is followed by exhaustive hydrolysis of the nitrile to give an amino acid.

General:

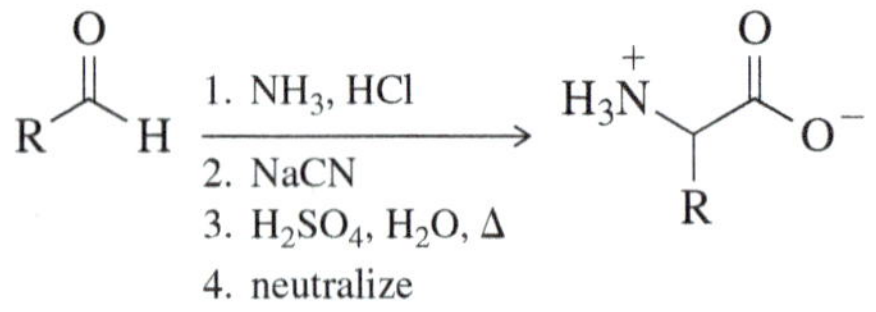

Specific:

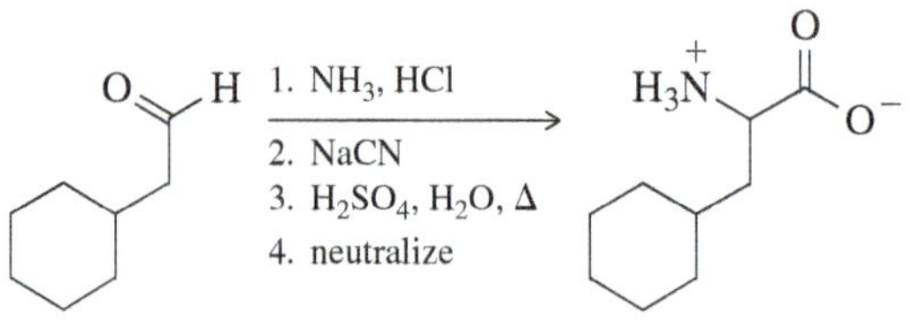

Mechanism:

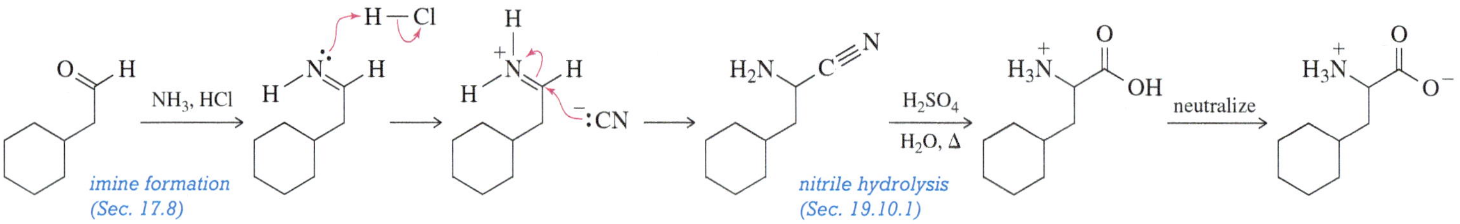

4. **DCC coupling (Section 26.9.2):** Deprotonation of a carboxylic acid is followed by reaction of the carboxylate with DCC to give an activated carboxylic acid derivative. Attack of an amine occurs through a nucleophilic acyl substitution reaction to form a new amide bond along with a urea by-product.

General:

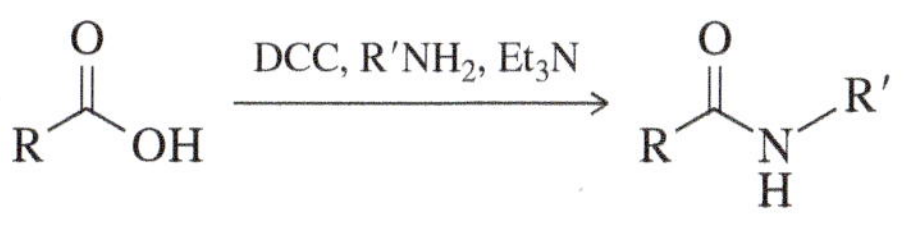

Specific:

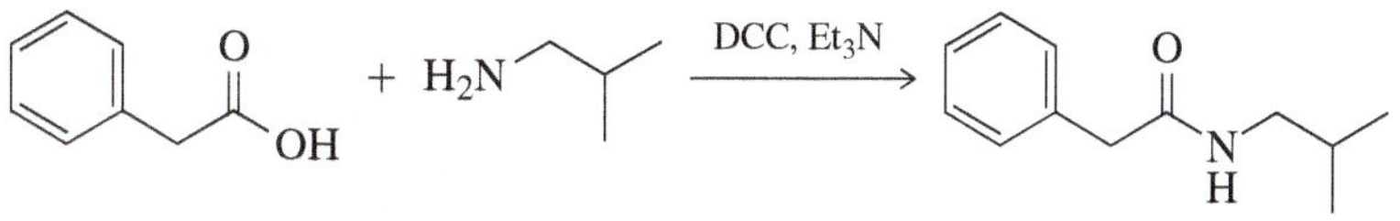

Mechanism:

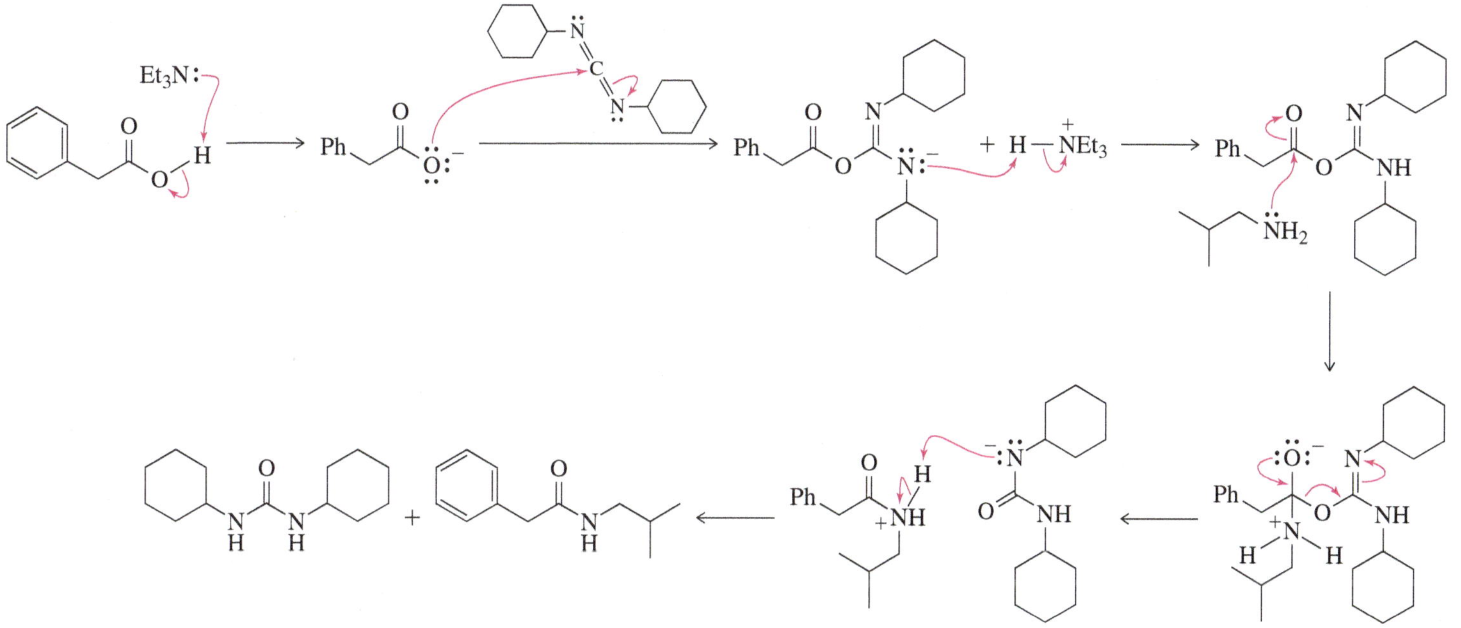

5. **Solution-phase peptide synthesis (Section 26.9.1):** The formation of polypeptides can be done via an iterative approach as shown for the following synthesis of a dipeptide. Protection of an amine of an amino acid is followed by activation of the carboxylic acid. Coupling with a second protected amino acid occurs through a nucleophilic acyl substitution reaction. [*If a larger polypeptide were desired, the carboxylic acid would be activated again, allowing for formation of a new amide bond.*] Deprotection of the amine completes dipeptide synthesis.

Specific example:

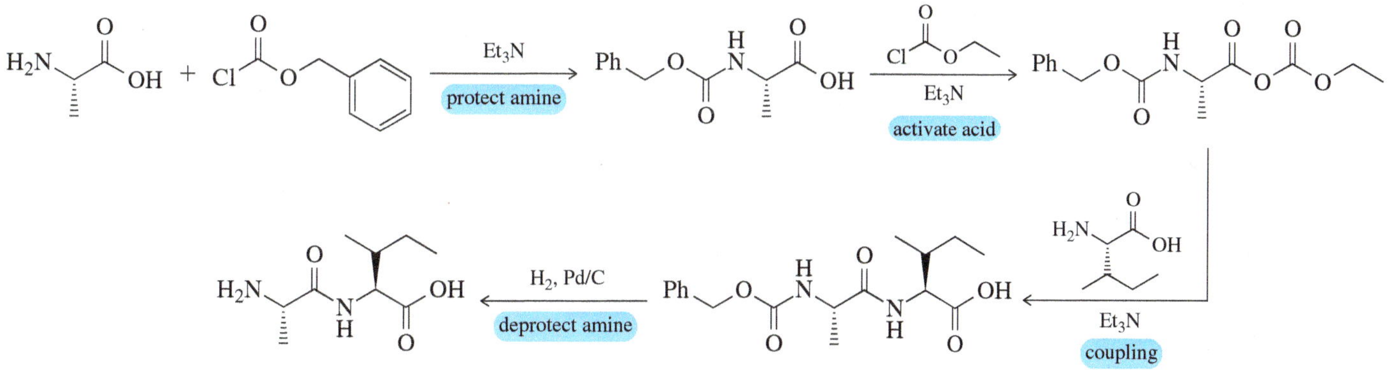

dipeptide

6. **Solid-phase peptide synthesis (Section 26.9.2):** Protection of the amino group of an amino acid is followed by attachment to the solid resin. Deprotection of the amino group allows for its coupling with an additional N-protected amino acid using DCC as the coupling reagent. [*Deprotection of the amino group of the resin-bound dipeptide would allow for continued incorporation of more amino acid residues.*] Deprotection of the amino group and cleavage from the resin completes dipeptide synthesis.

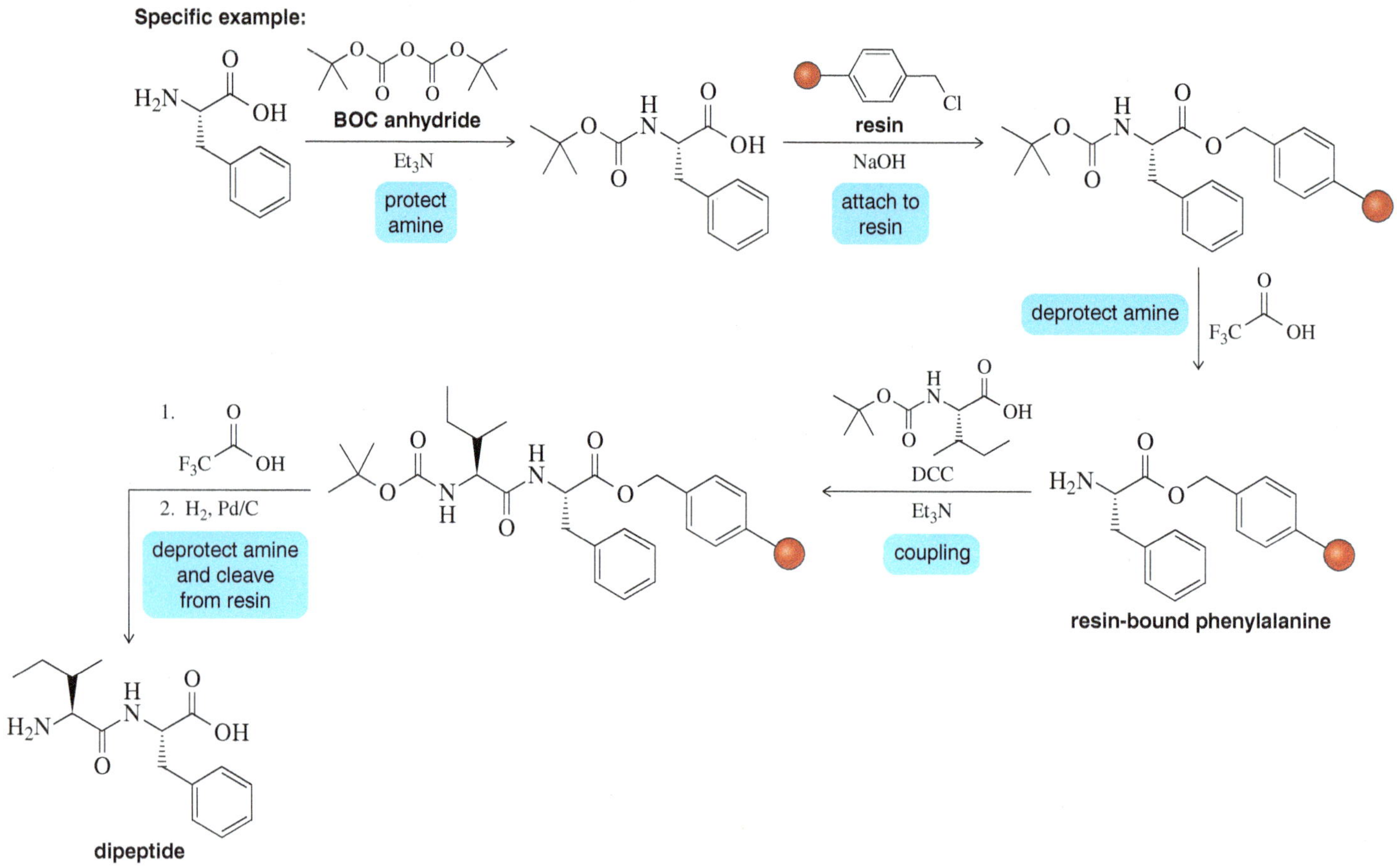

End-of-Chapter Assessments

[*Assessments are ranked according to their difficulty; •••• = most difficult.*]

26.46 (••) Draw the following amino acids in all possible protonation states as you move from high pH to low pH.

(a) Cys (b) His (c) Ile (d) Asp

26.47 (•••) Given the titration curve, draw the speciation curve for each amino acid.

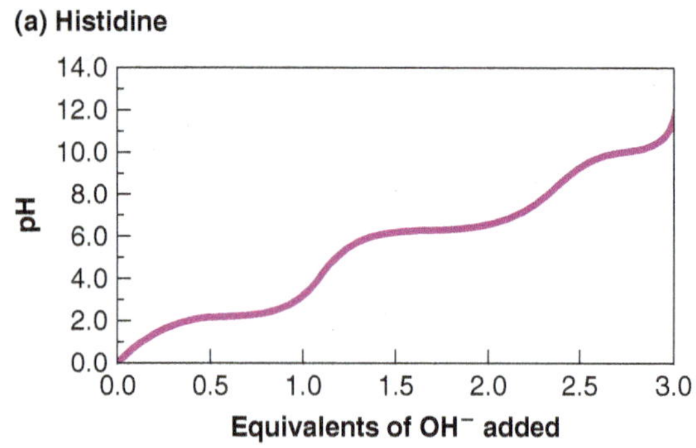

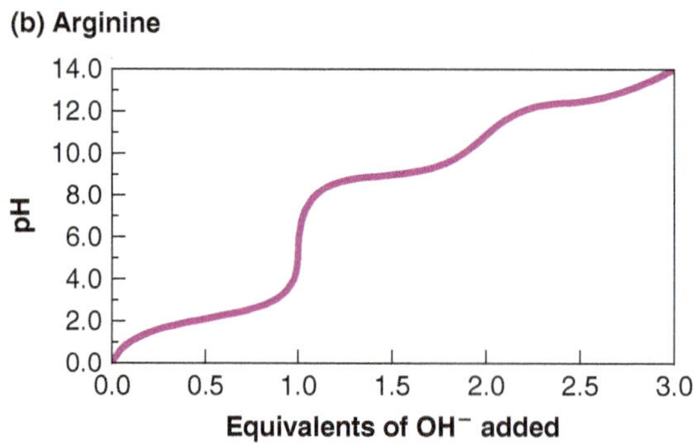

26.48 (••) Show how you would synthesize each amino acid using reductive amination. [*The general scheme is shown.*]

(a) Ala (b) Met (c) Tyr

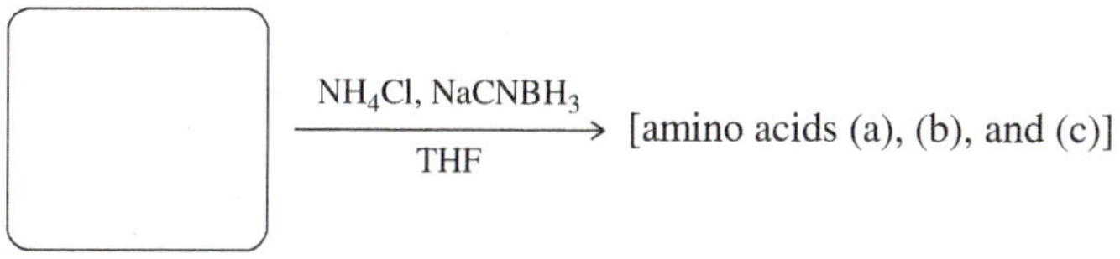

26.49 (••) Show how each amino acid could be made using α-halogenation. [*The general scheme is shown.*]

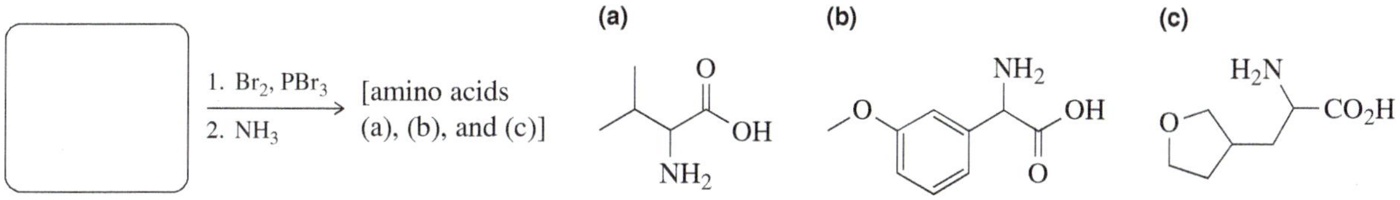

26.50 (••) Predict which natural amino acid would result from the following reactions.

(a)

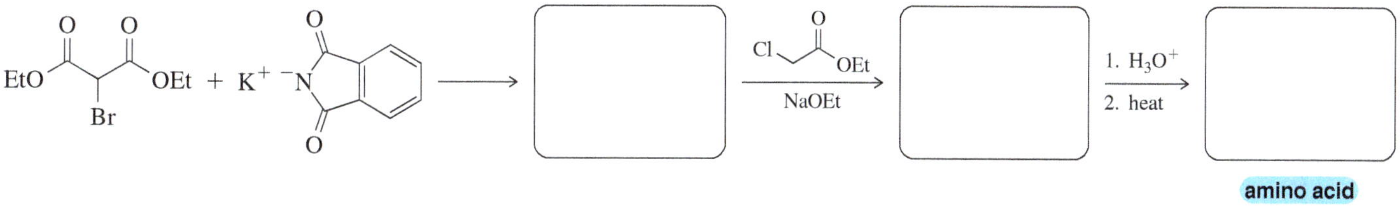

(b)

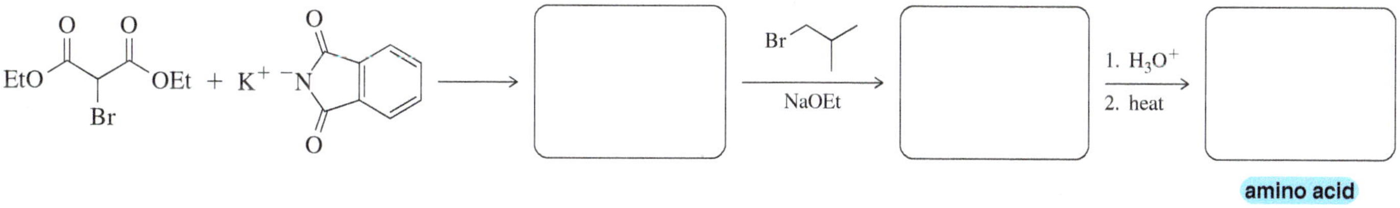

26.51 (•••) Show how you would synthesize the following amino acids using the Strecker amino acid synthesis.

(a) Ala (b) Phe (c) Tyr

26.52 (••) Using the reactions in Assessments 26.48–26.51, the amino acids obtained would be racemic. Why?

26.53 (•••) Because the synthesis of a single enantiomer of an amino acid is complicated, derive a scheme by which a racemic mixture of methionine might be separated.

26.54 (•••) For each of the amino acids and their isoelectric point, draw the major form at (i) pH = 1, (ii) pH = 4, (iii) pH = 8, and (iv) pH = 11.

(a) Cys (pI = 5.02) (b) Glu (pI = 3.08) (c) His (pI = 7.64)

26.55 (•) Draw the structure of each of the following peptides.

(a) WAG (b) ANDY (c) PEPPER

26.56 (•••) Draw the structure when each of the amino acids are treated with (i) EtOH, H_2SO_4; (ii) Ac_2O, Et_3N; (iii) ninhydrin.

(a) Ala (b) Met (c) Tyr

26.57 (••••) Recall that sequencing a peptide chain by Edman degradation is effective for peptides of up to ten residues, but quickly becomes challenging for longer peptides. Why might this be?

Peptides can be cyclized to form molecules that are surprisingly resistant to degradation. Due to their resistance to degradation in the human digestive track, cyclic peptides are often tested as medicines. An antibiotic called bacitracin A is an example of a cyclic peptide antibiotic. Assessments 26.58–26.61 deal with the following structure of bacitracin A.

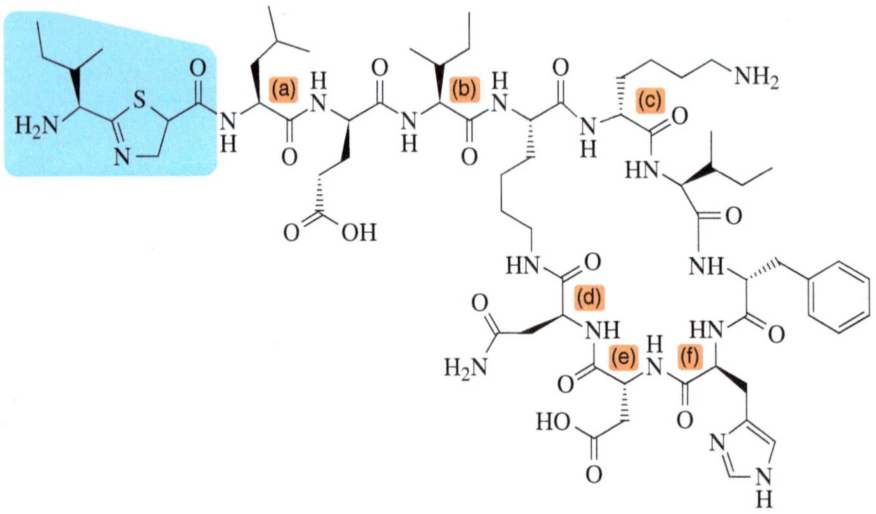

26.58 (•) Identify the amino acids (a)–(f) in bacitracin A.

26.59 (•••) Acyclic peptides are easily hydrolyzed in the stomach. Why might cyclic peptides be resistant to hydrolysis? [*Think about what makes a reaction favorable.*]

26.60 (•••) Bacitracin is formed by the reaction of the side chain of amino acid **A** with the carboxylic acid of amino acid **B** in a linear peptide. Give the identity of amino acids **A** and **B**. [*A and B are not (a) and (b).*]

26.61 (••••) The linear peptide portion of bacitracin contains a thiazoline (screened in blue). This thiazoline can also form in a dipeptide containing isoleucine and cysteine. Show a mechanism for this reaction. [*The rest of bacitracin has been removed for the sake of simplicity.*]

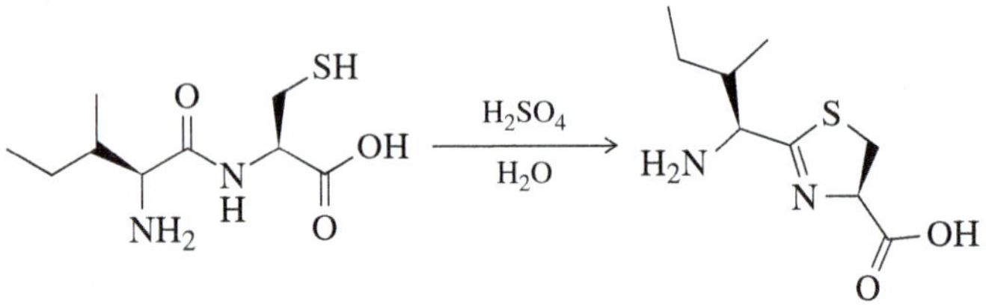

26.62 (••••) When peptide chains are hydrolyzed as in Figure 26.53 (experiment 2), degradation of certain amino acids is sometimes observed. Suggest a mechanism for the observed reactions under these conditions.

(a) **(b)**

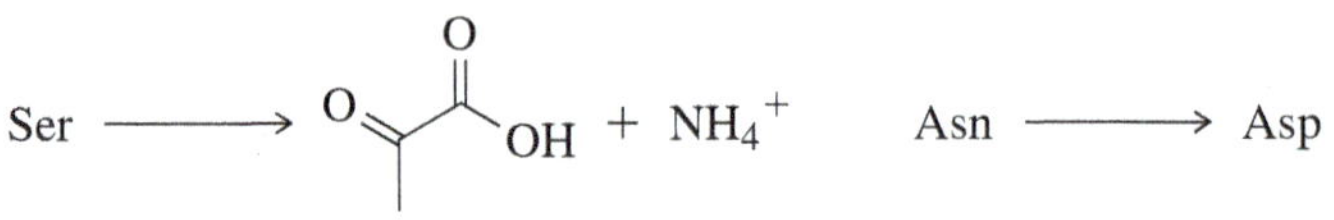

26.63 (••••) Show a scheme by which the tripeptide Val–Ala–Leu can be made using (a) solution-phase synthesis and (b) solid-phase synthesis.

27

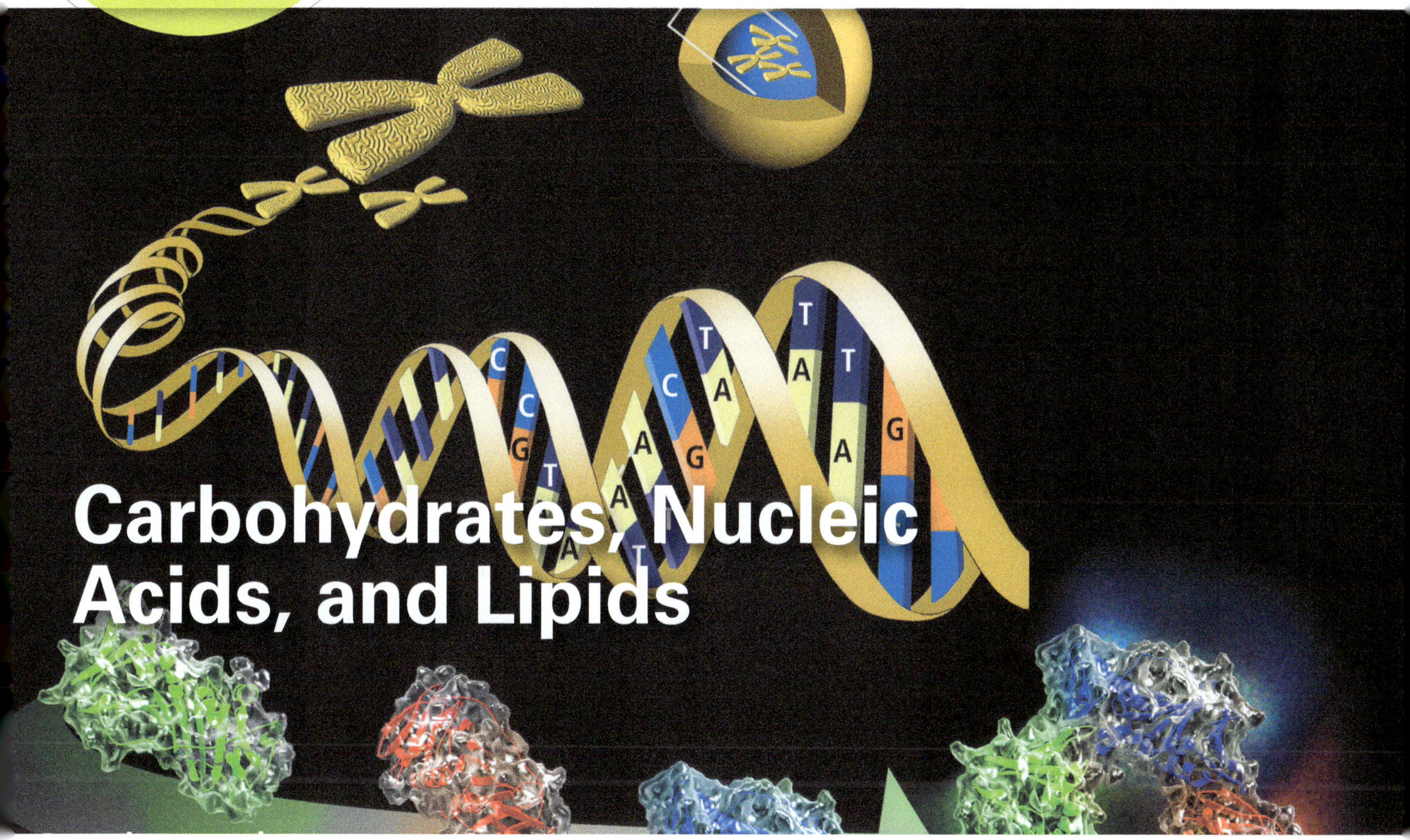

Carbohydrates, Nucleic Acids, and Lipids

Introduction

There's an oversimplified version of the "central dogma" of molecular biology that says, "DNA makes RNA, and RNA makes proteins." This brings to mind a process that flows in one direction, as intended by the original statement (see below). To me, an organic chemist who was never very good at biochemistry or molecular biology, this simplified statement doesn't seem to go far enough. When thinking about the organic chemistry of the molecules of life (i.e., the carbohydrates, nucleic acids, proteins, terpenes, steroids, and prostaglandins), the web seems tangled much more deeply—that is: carbohydrates form the backbone of DNA. DNA makes RNA. RNA makes proteins. Proteins function as enzymes. Enzymes help replicate DNA, transcribe DNA to RNA, facilitate protein synthesis, synthesize fatty acids, terpenes, and steroids. Enzymes convert fatty acids into prostaglandins, which along with steroids interact with receptors (themselves proteins) to perform all sorts of regulatory activities, one of which is trigger a cell to divide. Cell division is only possible with the replication of DNA.

And in every single one of these processes, there's organic chemistry! [*In short, Chapter 27 can be thought of as being fully Chemistry in Context.*]

∘ ∘ ∘ ∘ ∘ ∘ ∘

Way back at the beginning of the book, we talked about why a student takes organic chemistry. For some of you, perhaps chemistry majors, it's because you've always felt compelled to become an organic chemist. For others, it's a prerequisite for applying to a health professions program. Regardless of which group you might fall into, one of the unique aspects

Learning Objectives

LO 27.0 Compare the structure of sugars

LO 27.1 Draw different projections of monosaccharides

LO 27.2 Compare anomers of monosaccharides

LO 27.3 Analyze synthesis reactions of monosaccharides, substituted monosaccharides, and disaccharides

LO 27.4 Analyze reactions of monosaccharides

LO 27.5 Compare the physical and structural properties of disaccharides

LO 27.6 Assess nucleic acid sequences

LO 27.7 Apply the central dogma of molecular biology

LO 27.8 Analyze the structure and physical properties of fatty acids

LO 27.9 Analyze fatty acid synthesis

(continued)

LO 27.10 Analyze triglyceride reactions

LO 27.11 Analyze terpene synthesis reactions

LO 27.12 Compare steroids to their precursors

"The results suggest a helical structure (which must be very closely packed) containing probably 2, 3 or 4 coaxial nucleic acid chains per helical unit and having the phosphate groups near the outside."

—ROSALIND FRANKLIN

of organic chemistry, applicable to both, is that it sets you up for the study of biochemistry. Chapter 27 is the second of a two-chapter run where the organic chemistry of carbohydrates, nucleic acids, and lipids is discussed. As a disclaimer, the topics in these chapters could easily encompass one, or even two biochemistry courses on their own. As a result, this chapter is not meant to be comprehensive.

Instead, it provides a brief introduction to these interesting biological molecules. It is hoped that by the end of the chapter, you will have gained a greater appreciation for the intersection of organic chemistry and biochemistry. And while in the transition from organic chemistry to biochemistry, the molecules get larger and the reagents (cofactors) and catalysts (enzymes) get exponentially more complex, the reactions are still the same as what we've been studying all along. So, once again, look for these connections. By doing so, you will become a bit more comfortable with this material. And this comfort will pay off greatly when you take biochemistry [*or the MCAT, because the organic chemistry of biological molecules is an important part of that exam.*] Much of what we already know will help us in this chapter [*let's see how by testing our understanding.*]

Test your understanding

Assessment 27.1 Assign the absolute stereochemistry of the indicated stereocenter using the *R*/*S* system.

(a)

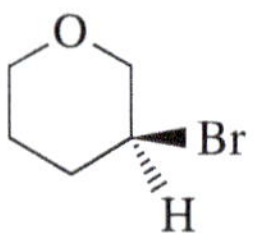

(b)

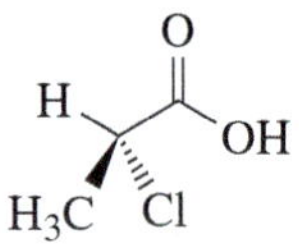

(c)

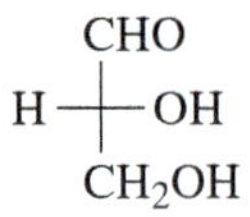

Assessment 27.2 Identify the following alkenes as *E* or *Z*, if appropriate.

(a)

(b)

(c)

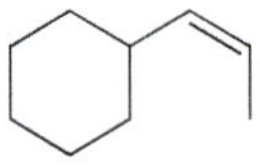

(d)

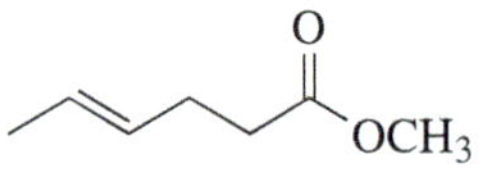

Assessment 27.3 Identify the hydrogen bond donors and hydrogen bond acceptors in the following molecules.

(a)

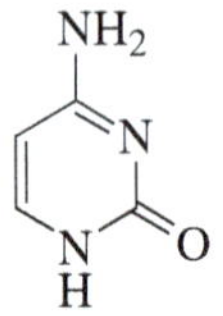

(b)

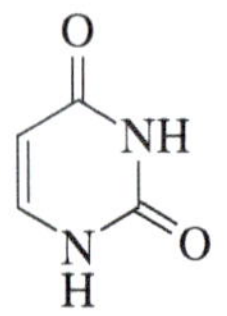

(c)

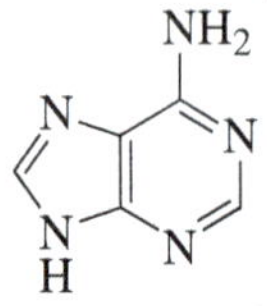

(d)

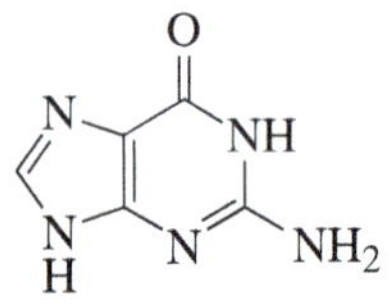

Assessment 27.4 Predict the product of the following reactions.

(a)

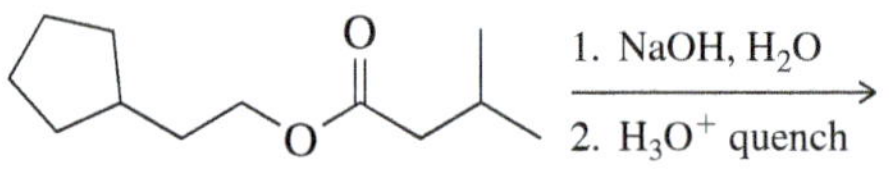

(b)

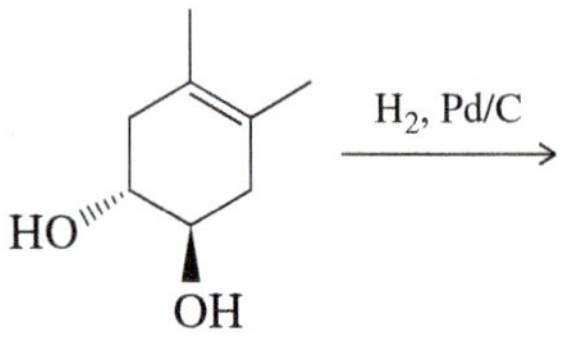

(c)

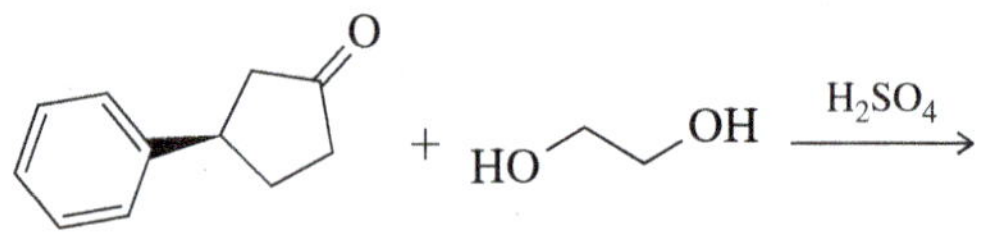

Assessment 27.5 In each pair, which would you expect to have the higher melting point?

(a)

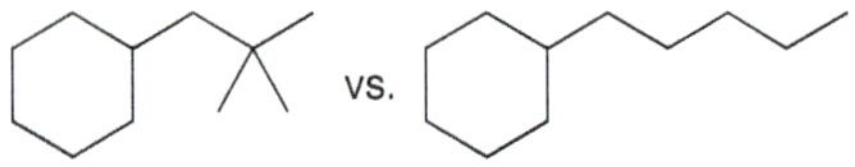

(b)

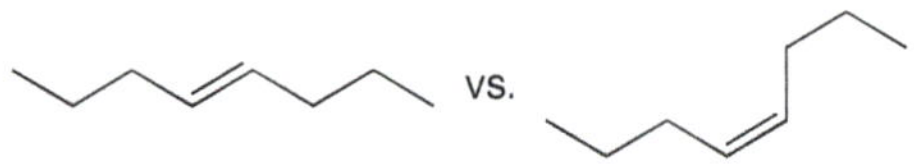

Assessment 27.6 Suggest a method for completing the following transformation. [*Three steps are required.*]

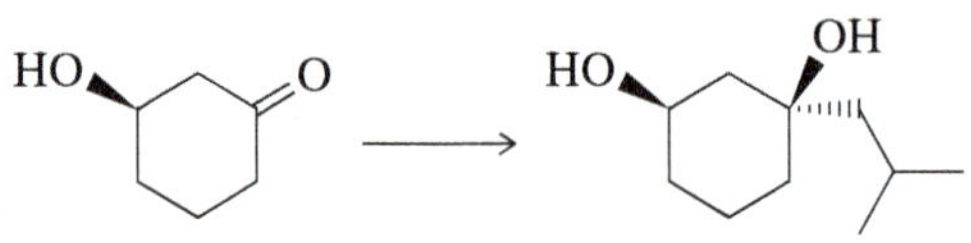

27.1 Carbohydrates: an overview

Carbohydrates, commonly known as sugars, were originally defined as hydrates of carbons [*hence the name*]. It is more appropriate, however, to refer to them as polyhydroxy aldehydes and polyhydroxy ketones because most carbohydrates, in their acyclic form, fall into one of these two categories (**Figure 27.1**).

Figure 27.1 Examples of a polyhydroxy aldehyde and polyhydroxy ketone.

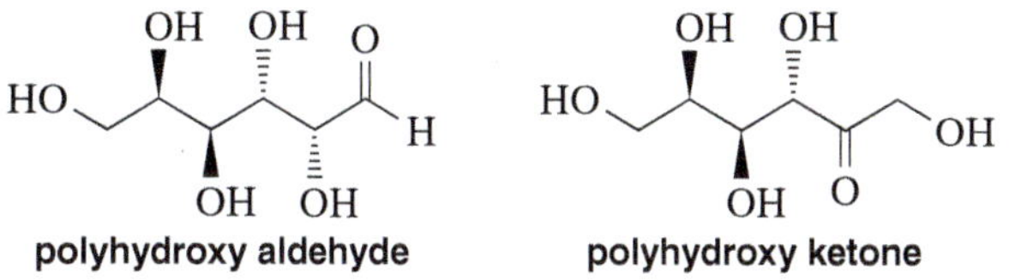

Carbohydrates provide much of the energy we use in a given day. The amount of energy per gram of carbohydrate is similar to that of protein, but is just under half of the energy available per gram of fat (Section 27.12). Outside of providing energy, carbohydrates play important roles in many biological processes. For example, they provide structure to the cell walls of plant cells, and they allow cell-to-cell communication by attaching to membrane-bound proteins. Moreover, as we explain in Section 27.8, they provide the backbone of DNA and RNA, which carry and translate our genetic information.

Broadly speaking, carbohydrates are either **simple** or **complex**. Simple carbohydrates are composed of **monosaccharides** and **disaccharides** (**Figure 27.2**), which are rapidly and easily digested, giving quick energy. Glucose, lactose, and sucrose are simple carbohydrates. Complex carbohydrates are composed of **polysaccharides**, which contain longer carbohydrate chains and are digested more slowly. Starch, from bread and potatoes, is a polysaccharide. Intermediate between mono-, di-, and polysaccharides are oligosaccharides, which contain between three and ten carbohydrate monomers in a chain.

Figure 27.2 Glucose is a monosaccharide, whereas lactose is a disaccharide.

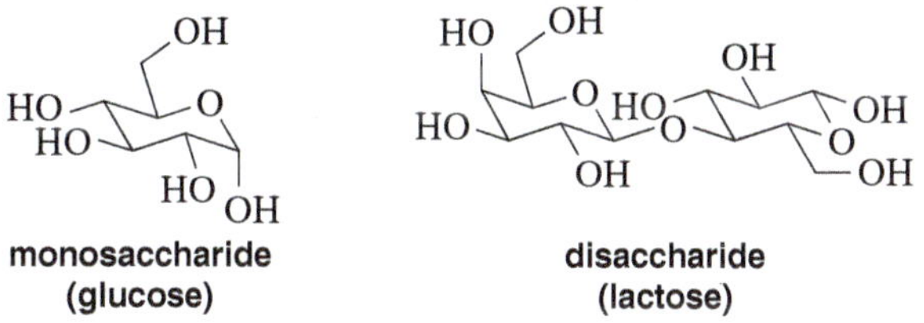

In the following sections, we approach mono-, di-, and polysaccharides first from a descriptive perspective, showing how the structure can be represented to include the stereochemical terms and the nomenclature used to describe them. Spread throughout are the reactions of carbohydrates. There are three types of reactions—namely, the reactions to make them biologically, the reactions to modify them synthetically, and the reactions used to help characterize them.

27.2 Monosaccharides

Monosaccharides are composed of one sugar, or carbohydrate, that cannot be broken down into more than one smaller carbohydrate. With over 200 known monosaccharides, it is important to be able to classify them in a number of ways. The following sections discuss these classifications along with different ways of representing them.

27.2.1 Aldose versus ketose: Fischer projections

The first way to classify carbohydrates is to differentiate between aldoses and ketoses (**Figure 27.3**). An aldose is a polyhydroxy aldehyde, whereas a ketose is a polyhydroxy ketone. Within the aldose and ketose classes, there is variability in the number of carbons each has in the main chain. The prefixes "aldo-" and "keto-" in Figure 27.3 distinguish an aldose from a ketose, respectively. The infix (-tetr-, -pent-, -hex-, etc.) indicates the number of carbons in the main chain.

Figure 27.3 The aldose and ketose classes of monosaccharides.

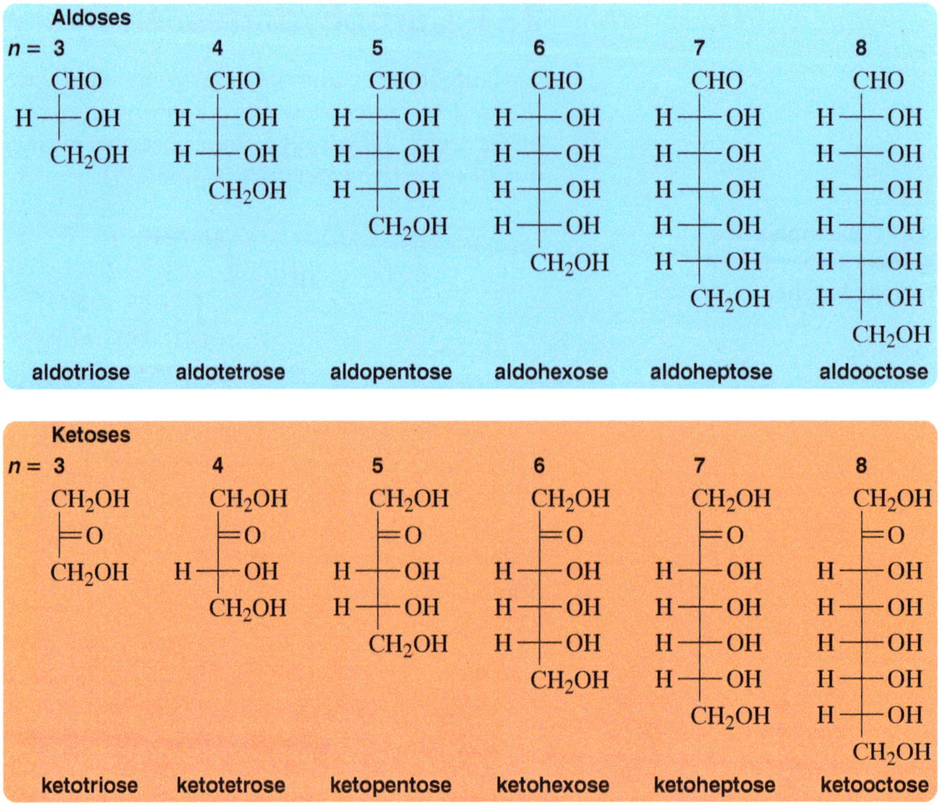

Figure 27.3 uses **Fischer projections** to show the structure of the various monosaccharides. Fischer projections, first presented in Chapter 6 and reviewed in depth here, are two-dimensional representations commonly used to represent the 3-D structure and stereochemistry of monosaccharides. In this representation (**Figure 27.4**), the longest carbon chain is drawn vertically with the lowest numbered carbon at the top, with substituents at the chirality center coming off in the horizontal direction. It is assumed that both horizontal bonds are coming forward (out of the plane of the page) and both vertical bonds are going backward (behind the plane of the page). In this way, the Fischer projection is generated by looking at the molecule in Figure 27.4 from the bottom left, drawing what is seen and then smashing it into a flat structure.

Figure 27.4 Drawing Fischer projections.

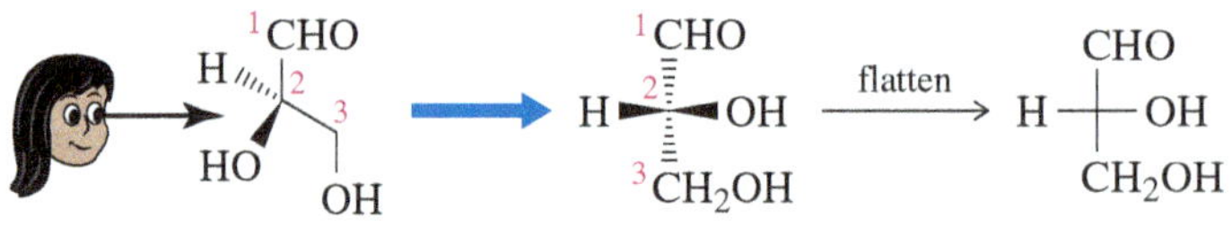

Fischer projections are especially useful for monosaccharides because many of them have multiple stereocenters. With multiple stereocenters, each family of aldoses has 2^n possible stereoisomers, where n = number of stereocenters. For example, there are four stereoisomers in the aldotetrose family (**Figure 27.5**). In these Fischer projections, *all* of the horizontal bonds are coming toward you.

Figure 27.5 The aldotetrose family of stereoisomers.

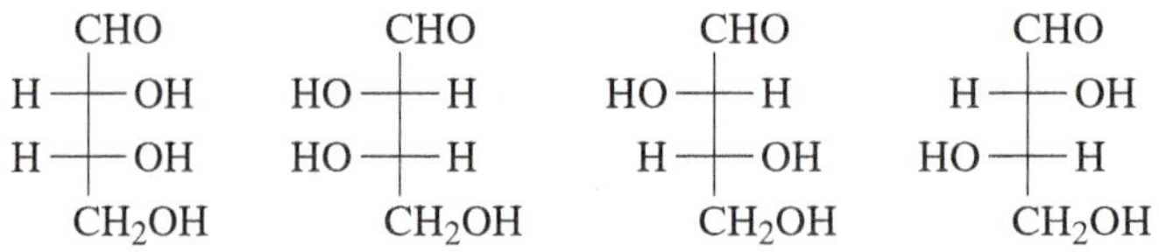

Assessment 27.7 Draw Fischer projections of the following molecules.

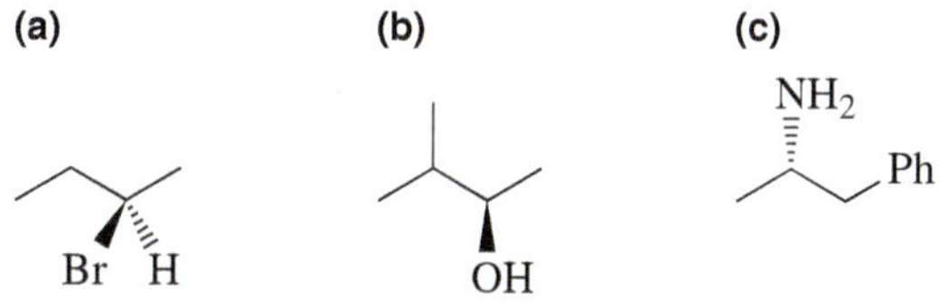

Assessment 27.8 In this attempt to convert the line angle drawing of D-erythrose (shown) to the Fischer projection (shown), by viewing it from a certain direction (shown), a mistake was made. What was the mistake?

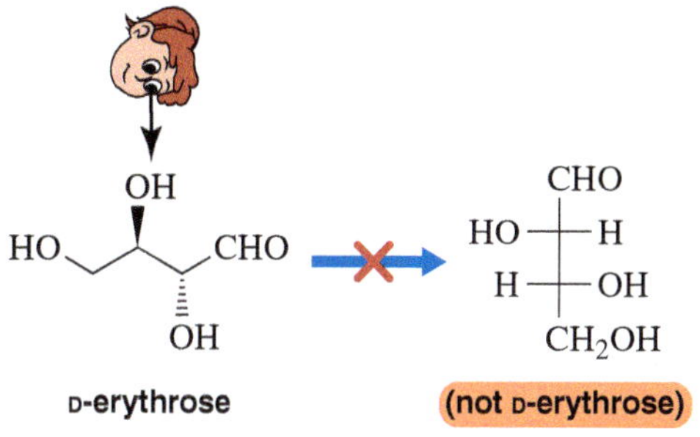

In order to avoid a common mistake made with Fischer projections, especially with monosaccharides containing multiple stereocenters, let's work Assessment 27.8 together.

Solved Assessment 27.8

The key difference between Fischer projections and line-angle drawings is that in Fischer projections there are always two bonds coming toward you and two going away from you. Unlike in line-angle drawings, there are zero bonds in the plane of the page that a Fischer projection is drawn on. [*In line-angle drawings, there is usually one group coming toward you, one going away, and two bonds in the plane. Sketch out propane in a line-angle drawing and add in the hydrogens to convince yourself of this.*] In a Fischer projection, the bonds in the horizontal plane must be coming toward you. Consider the manner in which the molecule in **Figure 27.6** is oriented relative to the viewer. Both the hydrogen and the hydroxyl group at C_3 are oriented toward the viewer (H on the left, OH on the right), so at this carbon, the molecule is set up to be flattened into the Fischer projection. At C_2, however, the hydrogen and the hydroxyl group are going away from the viewer (H on the right, OH on the left). Flattening it from this view then violates the rules of drawing Fischer projections and resulted in the incorrect Fischer projection shown in Assessment 27.8.

In order for all of the horizontal substituents to be pointed toward our viewer, the molecule needs to be rotated around the $C_2—C_3$ σ bond, as shown in **Figure 27.7**. In doing so, both substituents at C_2 and C_3 are coming toward the viewer (OH on right, H on left in both). This will give the correct Fischer projection. Compare the Fischer projection in Figure 27.7 with that shown in Figure 27.6. ■

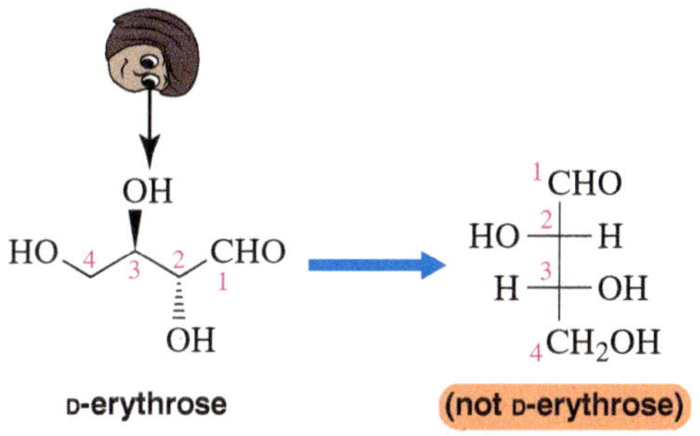

Figure 27.6 A common mistake made when drawing Fischer projections.

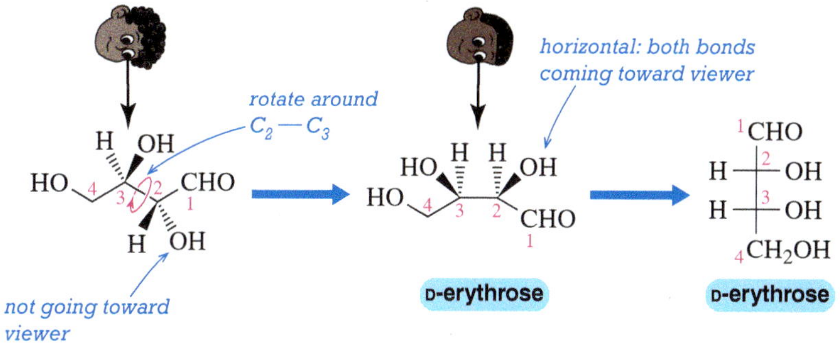

Figure 27.7 Drawing correct Fischer projections of longer-chain molecules often means bonds must be rotated so that all horizontal substituents come toward the viewer.

Assessment 27.9 Convert each line-angle drawing, using appropriate bond rotations, into a correct Fischer projection.

(a)

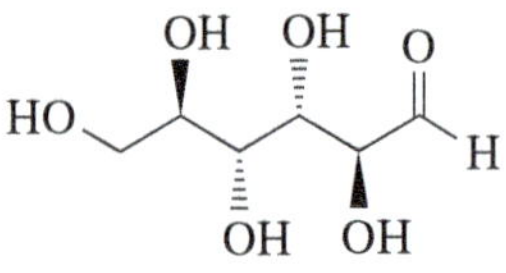

aldohexose

(b)

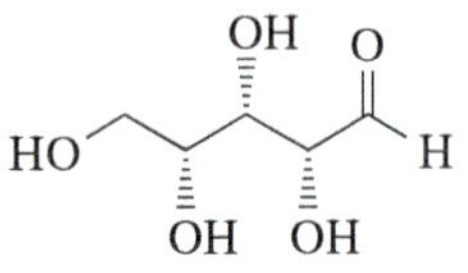

aldopentose

(c)

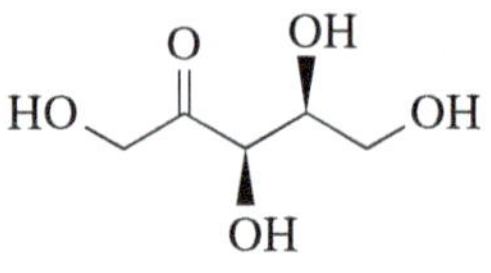

ketopentose

27.2.2 Stereochemistry of monosaccharides

Whereas the *R*/*S* system is most commonly used to indicate the absolute stereochemistry of chiral centers in organic molecules, stereochemistry in carbohydrates is indicated in a couple of different ways. One such way, called **D/L notation**, works exceedingly well with Fischer projections. To understand its origin, consider the monosaccharide glyceraldehyde, which contains a chiral center at C_2 (**Figure 27.8**). When (*R*)-glyceraldehyde is drawn in a Fischer projection, the hydroxyl group at the stereocenter points to the right. For this reason, chemists referred to this as D-glyceraldeyde [*the R/S notation hadn't been invented yet.*]. The "D" stands for the Latin *dexter*, meaning "to the right." Alternatively, the (*S*) enantiomer, when drawn using a Fischer projection, has the hydroxyl group at the stereocenter pointing to the left. Hence, L-glyceraldehyde, because "L" stands for the Latin *laevus*, meaning "to the left." [*D and L are distinct from d and l, which refer to the direction of rotation of plane-polarized light (Section 6.4.1).*]

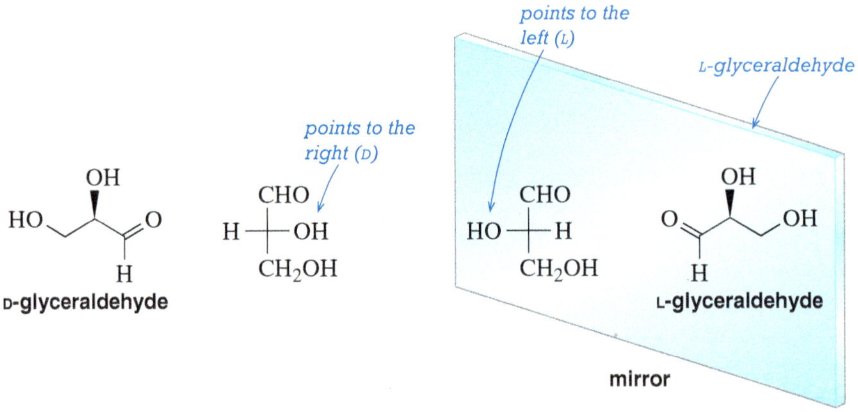

Figure 27.8 Using D and L notation to distinguish between the two enantiomers of glyceraldehyde.

The D/L notation has greater utility beyond just glyceraldehyde. As the structures of other sugars were named, any sugar [*or amino acid; Section 26.1.2*] that had the hydroxyl group [*or the amine, in the case of amino acids*] at the highest position-numbered stereocenter in the molecule pointed in the same direction (to the right) as D-glyceraldehyde was now named D. Similarly, all with the hydroxyl group at the highest position-numbered stereocenter pointing left, as in L-glyceraldehyde, were named L. Notice in **Figure 27.9** that L-threose and D-threose are enantiomers, as are L-erythryose and D-erythrose.

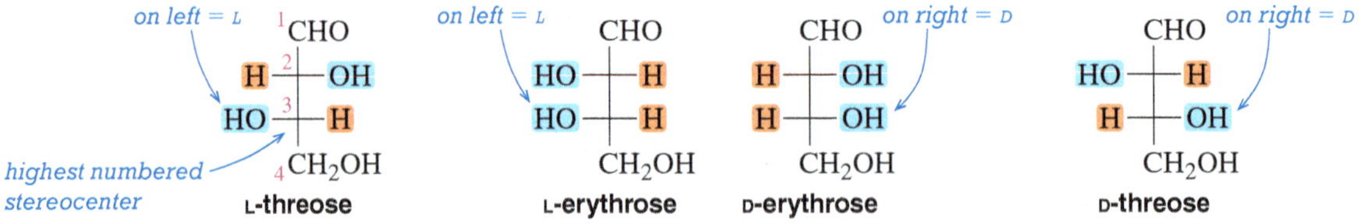

Figure 27.9 Stereoisomers of the aldotetrose family. D and L indicate an enantiomeric relationship.

Although it is not widely used outside of sugars and amino acids, the D/L notation does offer the convenience of brevity. In **Figure 27.10**, for example, (2*R*,3*S*,4*S*,5*R*)-2,3,4,5,6-pentahydroxyhexanal and its enantiomer, (2*S*,3*R*,4*R*,5*S*)-2,3,4,5,6-pentahydroxyhexanal are the same as D-galactose and L-galactose. [*Read that again out loud if it wasn't obvious the first time.*] That's the beauty of the D/L notation.

Figure 27.10 The D and L notation simplifies the naming of sugars.

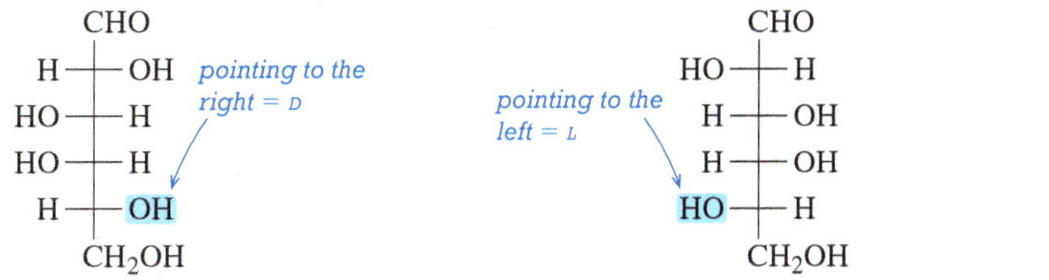

(2*R*,3*S*,4*S*,5*R*)-2,3,4,5,6-pentahydroxyhexanal
D-galactose

(2*S*,3*R*,4*R*,5*S*)-2,3,4,5,6-pentahydroxyhexanal
L-galactose

The relationships indicated by the blue annotations provide a helpful way of remembering the structure of the aldoses while also previewing the work of Fischer in determining the structure of glucose.

While the names in the D/L system are brief, they are not systematic like the ones using the *R*/*S* system coupled with IUPAC nomenclature rules. As a result, the D/L names of monosaccharides must be memorized [*or looked up when needed; ask your instructor.*] Luckily, all naturally occurring monosaccharides are D. **Figure 27.11** shows Fischer projections of all of the D-aldoses containing six or fewer carbons and their names. Aldohexoses have four stereocenters, so there are $2^4 = 16$ stereoisomers. The eight D isomers are shown. The eight L enantiomers are not.

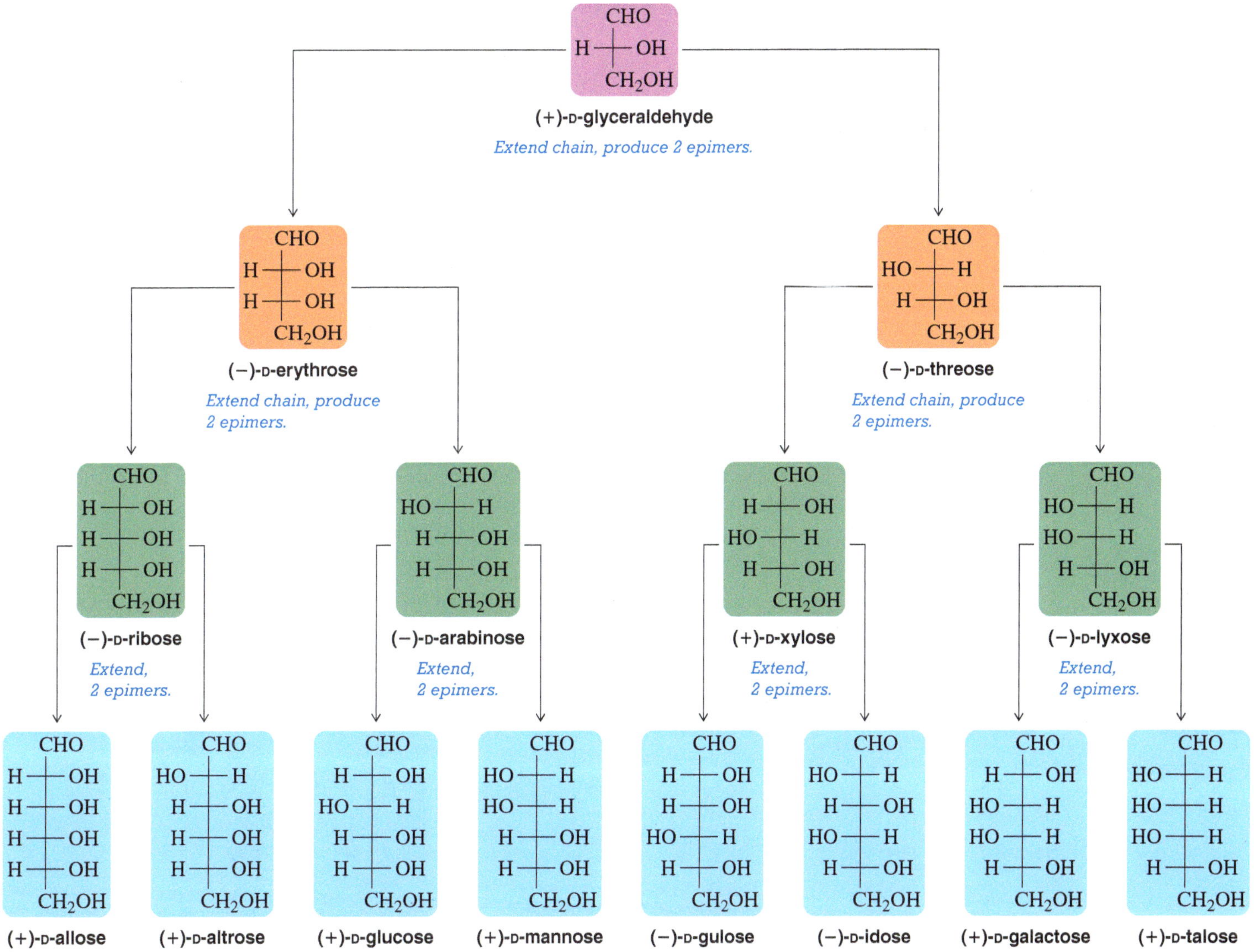

Figure 27.11 Fischer projections of the naturally occurring D-aldoses containing six or fewer carbons.

Several aldoses in Figure 27.11 differ only by the configuration of a single chiral center. For example, D-ribose and D-arabinose are diastereomers differing only in the stereochemistry at C_2. Diastereomers that differ only in the absolute configuration at a single stereocenter are referred to as **epimers**. D-Gulose and D-idose are also epimers.

Assessment 27.10 Categorize the following monosaccharides as D or L.

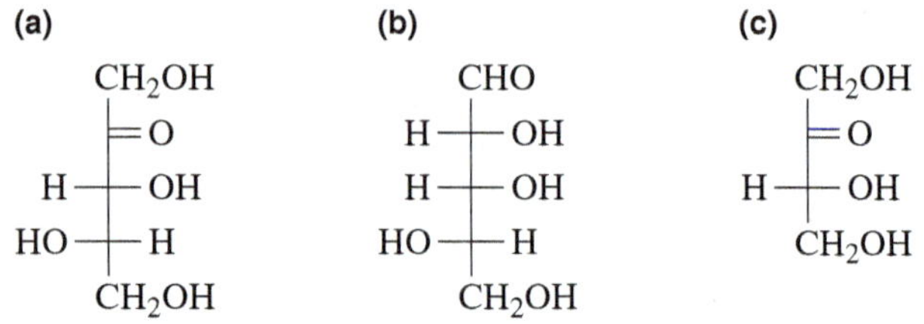

Assessment 27.11 Draw the structure of the following sugars. Predict the direction of rotation of plane polarized light (+ or −) for each. [*Use Figure 27.11 for this purpose.*]

(a) L-galactose (c) L-xylose
(b) L-erythrose (d) L-idose

Assessment 27.12 Draw the structure of and name the enantiomeric diols that result from the cis-dihydroxylation of the alkene shown.

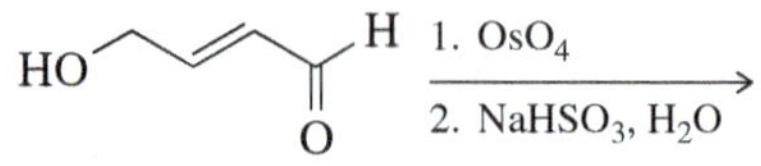

Assessment 27.13 Based on your answer to Assessment 27.12, how might you use the same reaction to make L- and D-erythrose?

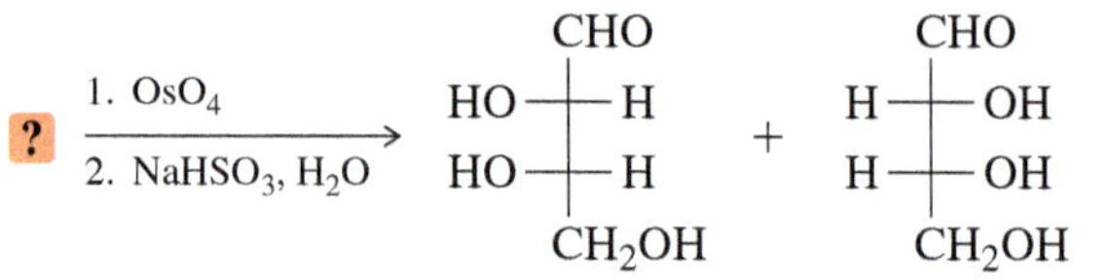

(−)-L-erythrose (−)-D-erythrose

Assessment 27.14 Which of the following molecule pairs are epimers?

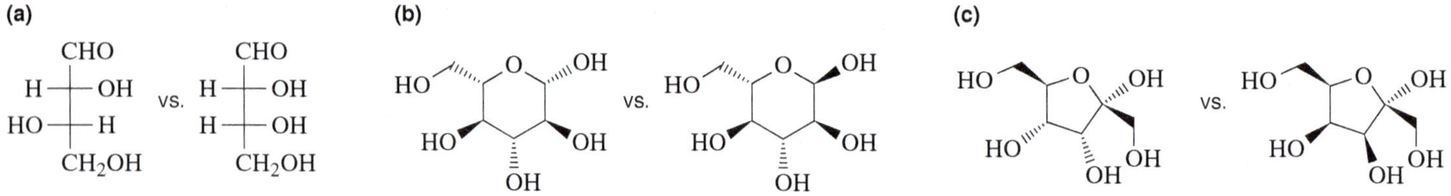

27.3 Cyclic monosaccharides

Recall from Section 17.7.2 that a hemiacetal has a hydroxyl group and an alkoxy group bonded to the same carbon (general formula: ROCOH) and that they form during the nucleophilic addition of alcohols to aldehydes and ketones. As shown in **Figure 27.12**, the acid-catalyzed reaction of an aldehyde and an alcohol results in an equilibrium that dramatically favors the aldehyde and alcohol because of entropy. When the aldehyde and alcohol are connected and hemiacetal formation is intramolecular, the equilibrium favors the hemiacetal. Mechanistically, the reaction proceeds by first protonating the aldehyde, making it more electrophilic. Intramolecular attack by the alcohol is followed by deprotonation to give the hemiacetal.

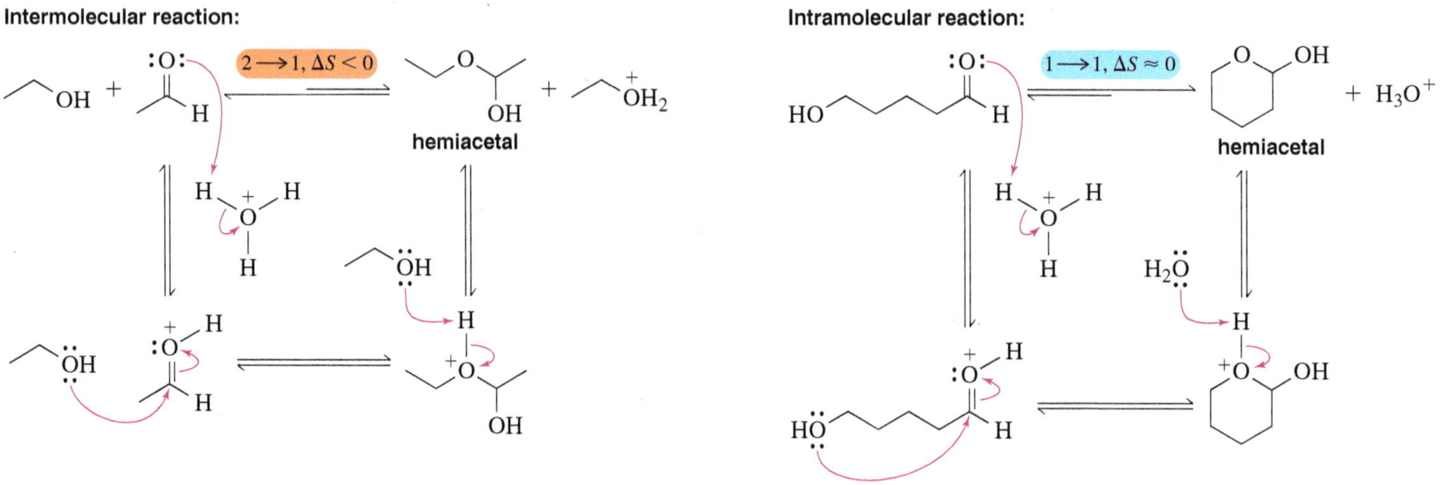

Figure 27.12 **Hemiacetal formation is not favored unless the reaction is intramolecular ($\Delta S \approx 0$).**

Like the δ-hydroxyaldehyde in Figure 27.12, monosaccharides contain at least one hydroxyl group and an aldehyde in the same molecule. As a result, monosaccharides can exist as cyclic hemiacetals, forming either a five- or a six-membered ring (**Figure 27.13**). The less common five-membered ring is referred to as a furanose because it is similar in ring size to a furan. The six-membered ring is called a pyranose. These two cyclic hemiacetals are discussed in detail in the following sections.

Figure 27.13 Monosaccharides form stable hemiacetals. The (a) furanose and (b) pyranose forms of allose.

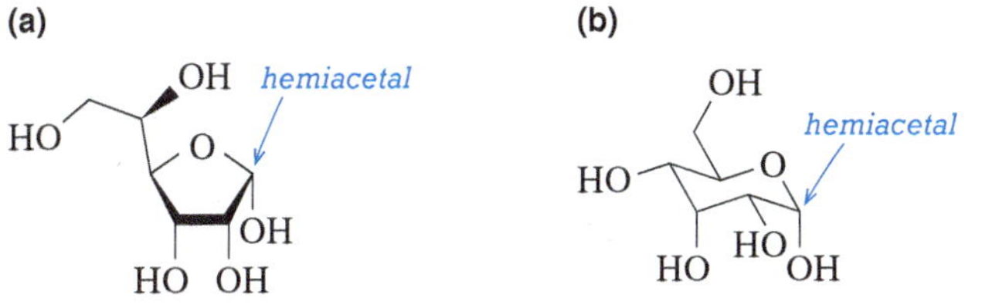

27.3.1 Projections of monosaccharides: furanose

Aldopentoses and ketohexoses commonly form five-membered furanose rings. They are typically represented using **Haworth projections** in place of the usual puckered, envelope conformation used for cyclopentanes. [*See Section 3.6.3 for review.*] The Haworth projection is like the Fischer projection in that it gives an unrealistically flat view of the conformation, but it manages to provide a bit more three-dimensionality. Popular among biologists and biochemists, it is less favored by organic chemists. **Figure 27.14** shows the formation of the furanose form of the ketohexose, D-fructose, in standard and Haworth forms.

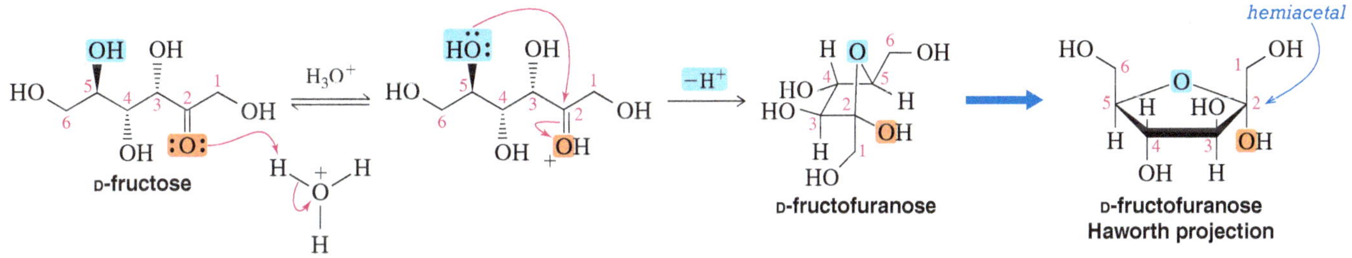

Figure 27.14 Fructose forms a five-membered cyclic hemiacetal.

Figure 27.15 shows how to convert a Fischer projection to a Haworth projection. With carbons numbered, simply rotate the Fischer projection 90° clockwise. In order for the hydroxyl group at C_5 to attack the carbonyl group at C_2 from behind, there must be a slight rotation around the $C_4—C_5$ bond. This explains why the stereochemistry at C_5 seems to be reversed during this step. At this point, the hydroxyl group at C_5 attacks the carbonyl group at C_2. When showing this, remember that the hydroxyl group attacks the carbonyl by reaching around from the back. Thus, in the Haworth projection, the oxygen is behind C_3 and C_4. This is the reason the $C_2—C_3$, $C_3—C_4$, and $C_4—C_5$ bonds are sometimes bolded.

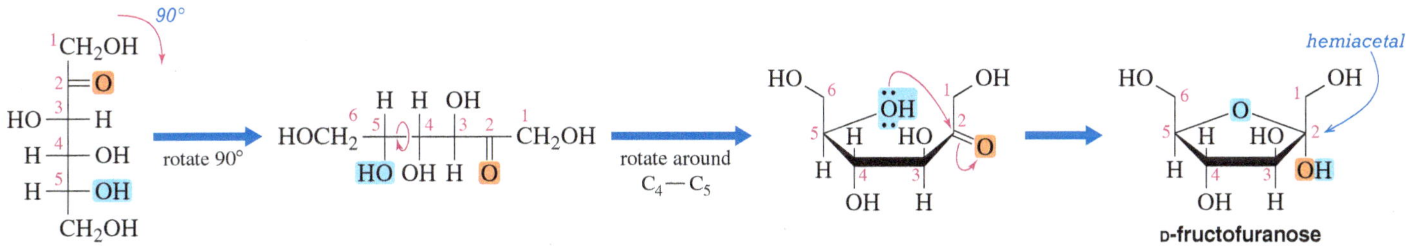

Figure 27.15 Drawing a Haworth projection from a Fischer projection. [*The curved arrows are not intended to be a complete mechanism.*]

Assessment 27.15 Using the Haworth projection, draw the furanose ring that would form between the C_5 hydroxyl group and the ketone at C_2 for the molecule shown.

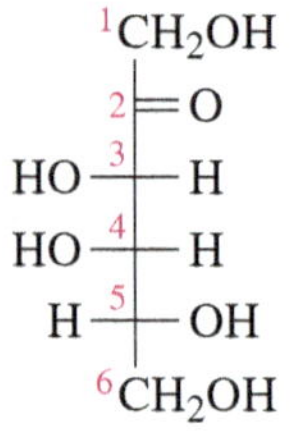

Assessment 27.16 Using the Haworth projection, draw the furanose ring that would form between the C_4 hydroxyl group and the aldehyde at C_1 for the molecule shown.

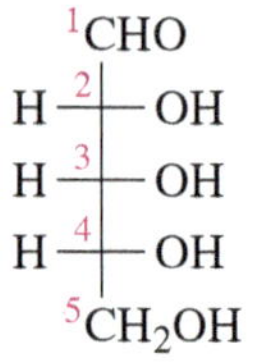

Assessment 27.17 Draw the Fischer projection of the acyclic form of the furanose shown. What sugar is this?

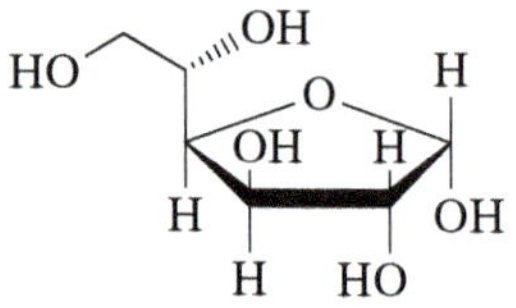

27.3.2 Projections of monosaccharides: pyranose

Which epimer do you think is favored? And can you think of a structural representation that would allow us to view the interactions of a six-membered ring so we can determine which epimer is favored?

In addition to forming five-membered cyclic acetals, monosaccharides can also form six-membered rings. In fact, the six-membered pyranose form is much more common because six-membered rings are more stable than five-membered rings. Glucose, for example, cyclizes to give either of the two hemiacetals shown in **Figure 27.16**. These two molecules, which are epimeric at C_1, are referred to as α-D-glucopyranose and β-D-glucopyranose. [*We explain what α and β mean later.*] Formation of these two epimers can be explained using the Fischer-to-Haworth process described in Figure 27.15 for furanoses. After rotating the Fischer projection 90° in the plane of the page and rotating around the $C_4—C_5$ bond, the C_5 hydroxyl attacks the aldehyde at C_1. Whether the β- or α-epimer forms is based on the aldehyde orientation when the ring forms.

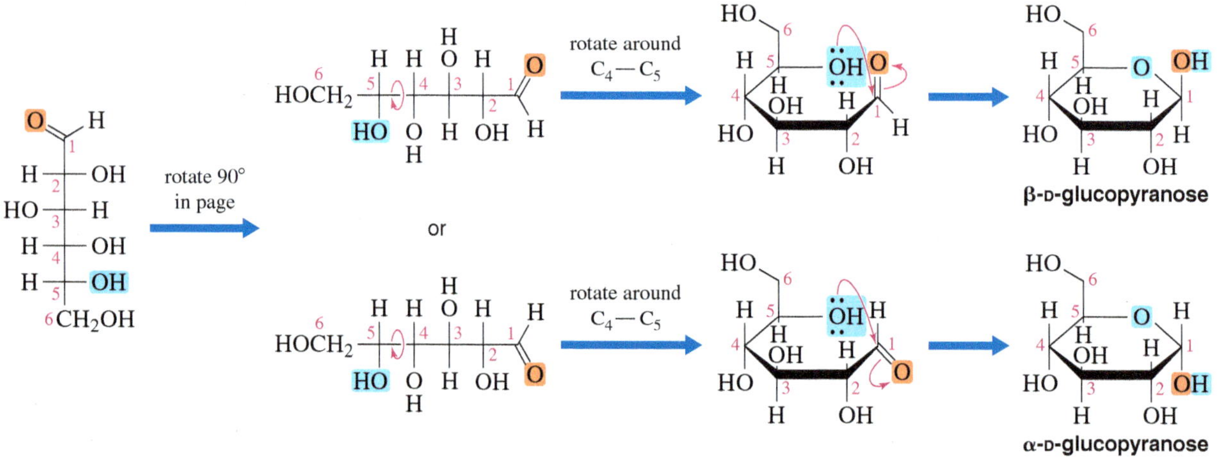

Figure 27.16 Glucose forms two cyclic pyranosides.

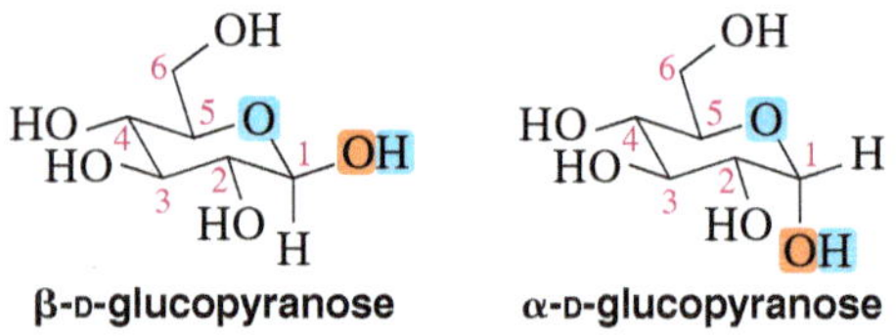

Figure 27.17 The chair conformations of β-D-glucopyranose and α-D-glucopyranose.

One limitation of the Haworth projection is that it doesn't give an accurate picture of the 3-D conformation of a molecule. As a result, it is difficult to answer questions like the one posed in the previous paragraph about whether the α- or β-epimer is favored. Luckily, we can draw the chair conformation to shed light on this question. As shown in **Figure 27.17**, the chair is drawn such that the left-most carbon (C_4) is puckered up and the right-most carbon (C_1) is puckered down. The top right vertex is where the oxygen always goes for cyclic carbohydrate hemiacetals. From here, the ring is simply decorated with the hydrogens and hydroxyl groups, placing them up or down as called for in the Haworth projection. "Up" and "down" at each carbon alternates between "axial" and "equatorial." Given that glucose is the most prevalent natural sugar, the fact that all of the substituents coming off the ring are equatorial (except for C_1 in the α-epimer) should come as no surprise. [*Which is more stable now, α or β? Read on.*]

Assessment 27.18 Draw the Haworth projection and chair conformation for L-galactopyranose. [*The structure is shown in Figure 27.11.*]

Assessment 27.19 Draw the Haworth projection and chair conformation for D-mannopyranose. [*The structure is shown in Figure 27.11.*]

Assessment 27.20 Draw the Fischer projection of the acyclic form of the pyranose shown. What sugar is this?

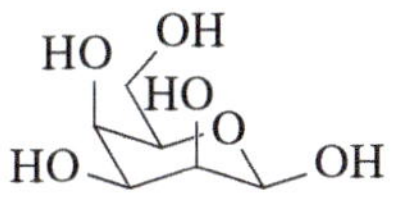

Assessment 27.21 When acid-catalyzed cyclization of glucose happens in the presence of methanol, a methyl α-D-glucopyranoside results. Show a mechanism for this reaction. [*What functional group is present at C_1? See Section 17.7.3 for a review.*]

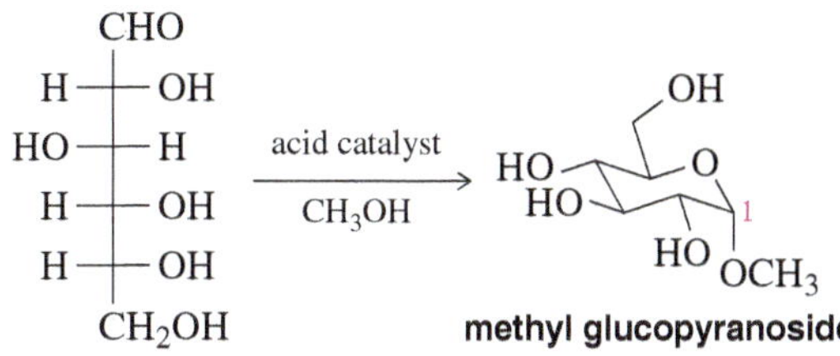

methyl glucopyranoside

27.3.3 Anomers

In Figures 27.16 and 27.17, and in the preceding paragraphs, you were asked to consider which of two diastereomers, α- or β-D-glucopyranoside, would be produced preferentially. These two molecules are epimers because they differ only in the configuration at one stereocenter. More specifically, however, they are a specific class of epimers called **anomers**, where the stereochemistry differs only at the hemiacetal carbon in a cyclic carbohydrate. Because of this, the hemiacetal carbon (C_1) of cyclic monosaccharides is called the **anomeric carbon** and the attached hydroxyl group is the **anomeric hydroxyl**. [*This is an important distinction when we study reactions of cyclic monosaccharides in Sections 27.4 and 27.5.*]

The α- and β-anomers of glucose are in equilibrium with each other and the acyclic form (**Figure 27.18**). As we learned in Chapter 17, hemiacetal formation is a reversible reaction. As such, conversion from α to β is under thermodynamic control. The α-anomer has the hydroxyl group in the axial position (down), whereas the β-anomer has the anomeric hydroxyl in the equatorial position (up). This is true for the pyranose form of all aldohexoses. As you might expect, the β-anomer, with the hydroxyl in the equatorial position, is slightly more stable, as evidenced by an α:β ratio of 36 to 64 at equilibrium in water.

Figure 27.18 The equilibrium between α- and β-glucopyranoside.

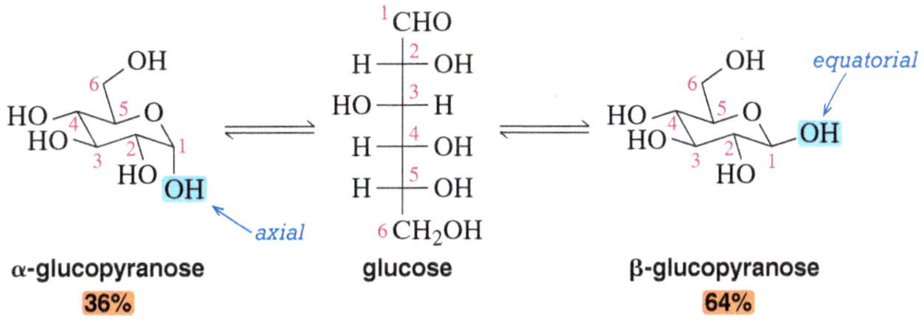

Two anomers are also possible when monosaccharides cyclize in the furanose form. **Figure 27.19** shows the α- and β-anomers of fructofuranose. Once more, the anomeric hydroxyl group is down in the α form and up in the β form. Particularly useful for furanoses (but also true for pyranoses), a cis relationship between the anomeric hydroxyl and C_6 is indicative of β, whereas a trans relationship is indicative of α.

Figure 27.19 The equilibrium between α- and β-fructofuranose.

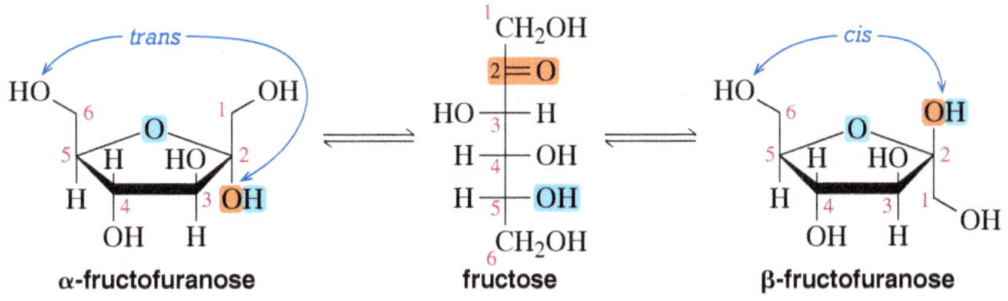

Assessment 27.22 Suggest a mechanism by which α-D-glucopyranose is converted to β-D-glucopyranose in acid. [*See Figure 27.18.*]

Assessment 27.23 Suggest a mechanism by which α-D-glucopyranose is converted to β-D-glucopyranose in base. [*See Figure 27.18.*]

Assessment 27.24 Draw the α- and β-anomers of D-talopyranose. [*The structure of talose is in Figure 27.11.*]

Assessment 27.25 Draw the α- and β-anomers of D-mannofuranose. [*The structure of mannose is in Figure 27.11.*]

27.4 Reactions of monosaccharides: synthesis

Individual monosaccharides can be coupled to each other to make larger di-, tri-, tetra-, or even polysaccharides. Many of these carbohydrates, when modified appropriately, have great utility in molecular biology, biochemistry, and medicine. As a result, the field of carbohydrate synthesis is an extremely important one.

Unfortunately, carbohydrate synthesis can be extremely complicated. In highly functionalized molecules containing as many as five hydroxyl groups, it can be difficult to selectively choose which hydroxyl group reacts. This section describes the reactions that we can do on cyclic monosaccharides. In Section 27.6.1, we show how these reactions can be used to do challenging disaccharide syntheses.

27.4.1 Glycoside formation: acetal synthesis

With a cyclic monosaccharide, there is one hydroxyl group that is different than all others. [*Which one is it?*] We have made a pretty big deal about recognizing the difference between an —OH that is an alcohol versus an —OH that is part of a hemiacetal [*recall Section 17.7.2*]. **Figure 27.20** helps you to understand why this is. When we think of the inherent reactivity of an alcohol, the first reactions that come to mind are ones where oxygen is a nucleophile, or base [Figure 27.20(a)]. As a hemiacetal, however, the most important reaction of an alcohol is that of a leaving group. This was especially the case when we discussed hemiacetals being a disfavored intermediate between an aldehyde and an acetal [Figure 27.20(b)]. In essence, the only thing that can happen in either direction is for the hydroxyl or alkoxy group to leave, usually after being protonated. Regardless of whether the protonated hydroxyl or the protonated alkoxy group leaves, the resulting carbocation is stabilized by resonance.

(a)

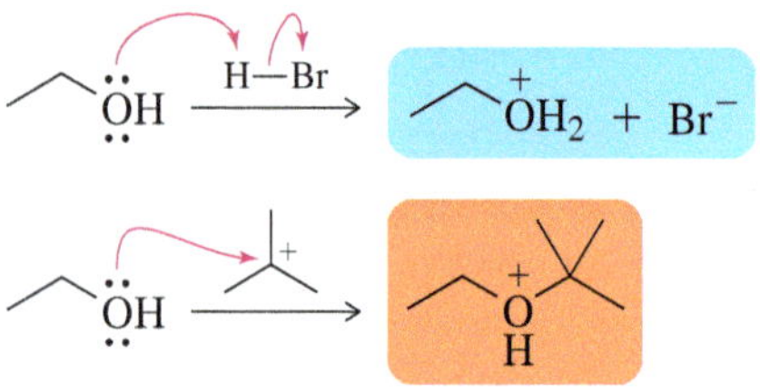

(b) *[Begin at the hemiacetal, then go left (pathway a) and right (pathway b) from there.]*

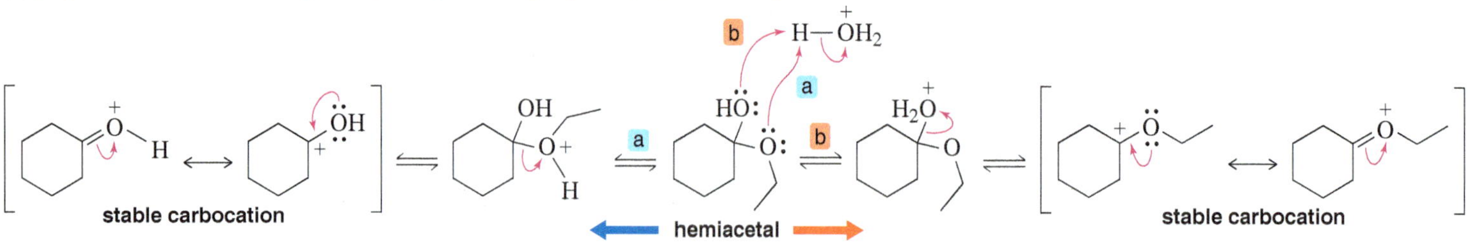

Figure 27.20 Alcohols and hemiacetals have different reactivities.
(a) Alcohols are nucleophiles/bases. (b) Hemiacetals are converted into good leaving groups that form resonance-stabilized carbocations.

When the anomeric carbon of a sugar is an acetal, the sugar is referred to as a glycoside. To name an acyclic glycoside, the suffix -e of the sugar name is replaced with -ide. Thus, glucose becomes glucoside, and mannose becomes mannoside. A cyclized glycoside uses the infix -pyrano- (6) or -furano- (5) to indicate the size of the ring. Thus, glucose becomes glucopyranoside, and fructose becomes fructofuranoside. When a pyranose becomes an acetal, it does so by the replacement of the anomeric hydroxyl group with an alkoxy group. Thus, to complete the name, the alkyl group precedes the name of the glycoside. To demonstrate this nomenclature, isopropyl α-D-glucopyranoside and ethyl β-D-fructofuranoside are shown in **Figure 27.21**.

Figure 27.21 The nomenclature of glycosides.

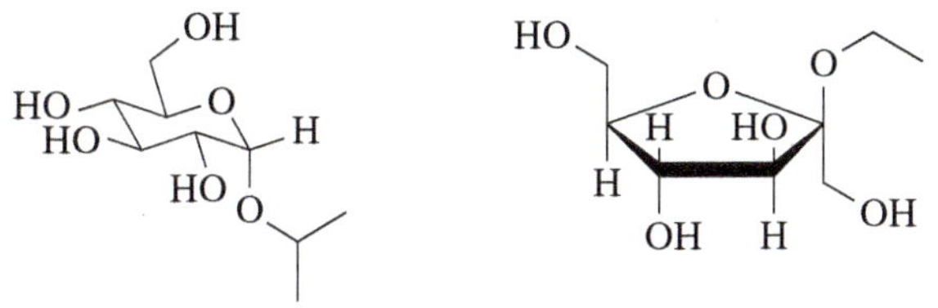

Formation of glycosides is the equivalent of converting a hemiacetal to an acetal, and it is done in an identical way. Recall that hemiacetals are halfway to the acetal, needing only the addition of excess alcohol and removal of water to drive the reaction to completion. [*See Section 17.7.3 for a review.*] In this way, treatment of a sugar, in its cyclic form, with acid and an alcohol forms a glycoside. The synthesis of isopropyl α-D-glucopyranoside from α-D-glucopyranose is shown in **Figure 27.22**. Mechanistically, protonation of the hydroxyl group creates an excellent leaving group. Water leaves to make a resonance-stabilized carbocation that is attacked by isopropanol. The proton brought along when isopropanol attacked is removed to complete formation of the pyranoside.

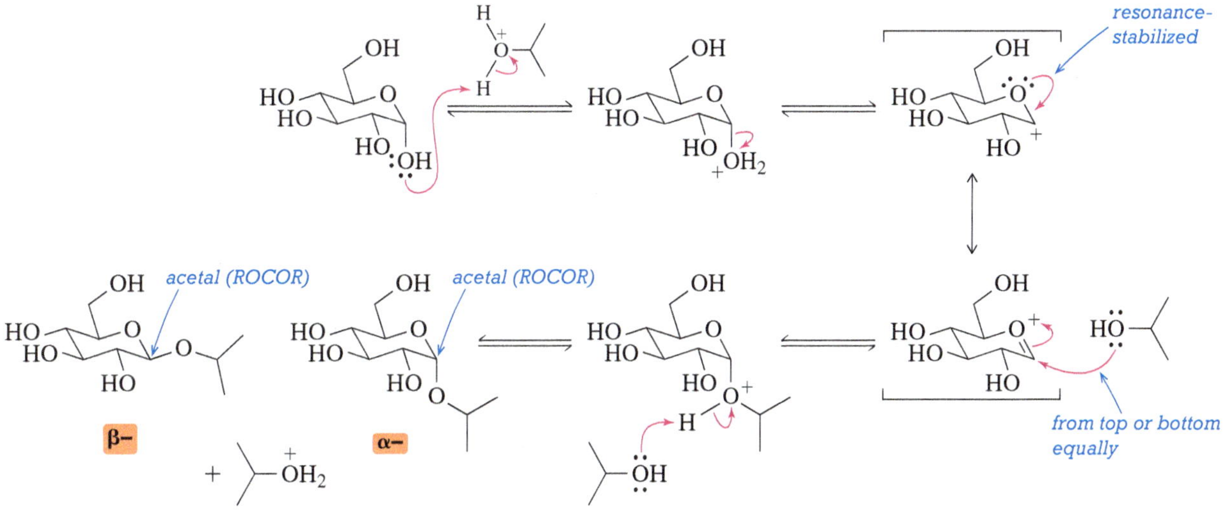

Figure 27.22 The mechanism of glycoside formation.

With five total hydroxyl groups in this reaction, why was the anomeric one the only one to react? This relates to the discussion that opened this section (Section 27.4.1). Because the anomeric carbon is the only one whose hydroxyl is part of a hemiacetal, it is also the only one that can form a stable carbocation under mildly acidic conditions. The other hydroxyl groups are simple alcohols, so much harsher acidic conditions are needed to drive them to leave.

Assessment 27.26 Draw the structure that corresponds to the given name.

(a) Allyl β-L-fructofuranoside
(b) Benzyl α-D-gulopyranoside
(c) *tert*-Butyl β-D-arabinofuranoside

Assessment 27.27 (a) Provide a synthesis of the following glycosides. (b) Name the sugar used in the synthesis. [*Work backward to the Fischer projection to figure out which sugar is used.*]

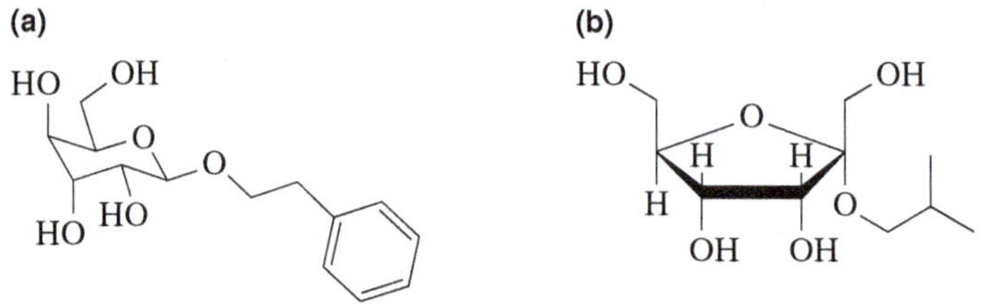

Assessment 27.28 Suggest an arrow-pushing mechanism for the reaction used to make the glycoside shown in Assessment 27.27(a).

Assessment 27.29 The Tollen's test, a common test used to detect the presence of aldehydes, gives a positive result for α-D-glucopyranose but not for isopropyl α-D-glucopyranoside. Explain this observation.

Let's work this one together to discover the requirements for a positive Tollen's test.

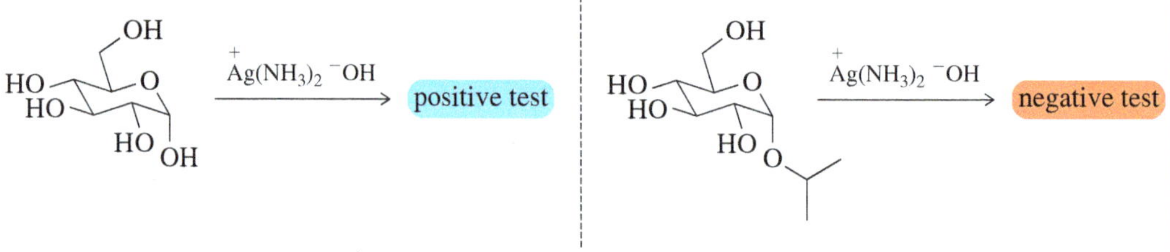

Solved Assessment 27.29

The Tollens reagent, a Ag^+ salt, is used to oxidize aldehydes, but not ketones, to a carboxylic acid (**Figure 27.23**). It has very little synthetic value because there are far less expensive ways to oxidize aldehydes. In the oxidation of an aldehyde, the Ag^+ in the Tollens reagent is reduced to silver metal (Ag^0), which coats the surface of the glass flask in which the reaction is run. [*Definitely a favorite among the elementary school science demo crowd.*] When a ketone is exposed to the Tollens reagent, nothing happens. Thus, the Tollens reagent is used primarily to determine whether an unknown carbonyl compound is an aldehyde or a ketone.

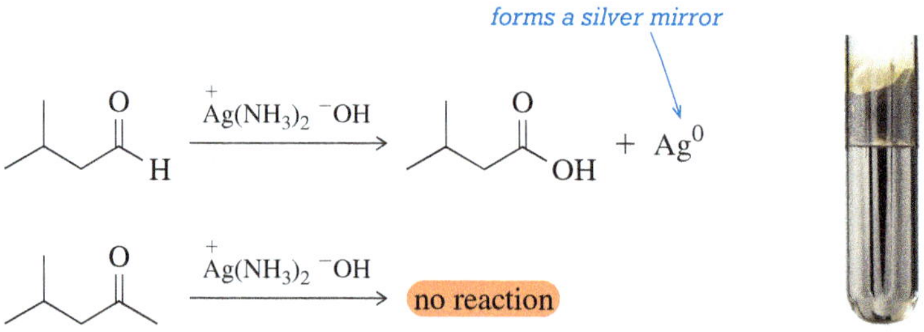

Figure 27.23 The Tollens silver mirror test. With aldehydes, a silver mirror forms; with ketones, there is no reaction.

Assessment 27.29 tells us that α-D-glucopyranose gives a positive Tollens test, but isopropyl α-D-glucopyranoside does not. [*Any ideas why?*] Neither of them is an aldehyde or ketone—at least in their current form. α-D-Glucopyranose is in equilibrium with the acyclic D-glucose, however, which contains an aldehyde. Even if there is only a small amount of D-glucose in solution, oxidizing it drives the reaction to the right as shown in **Figure 27.24**. Because the formation of Ag^0 from Ag^+ is a reduction, sugars in their hemiacetal or open-chain form are called **reducing sugars**.

Isopropyl α-D-glucopyranoside is an acetal, so it cannot equilibrate back to the aldehyde under these conditions. Acetals can be hydrolyzed back to the aldehyde under strongly acidic conditions, but the Tollens reagent is strongly basic. As we learned when we used them as protecting groups (Section 17.7.5), acetals are extremely stable under basic conditions. The stability of bases is important when we study the methods available for alkylating sugars [*coming up next, Section 27.4.2*]. Because they are unable to reduce Ag^+ to Ag^0, glycosides are called **nonreducing sugars**. ■

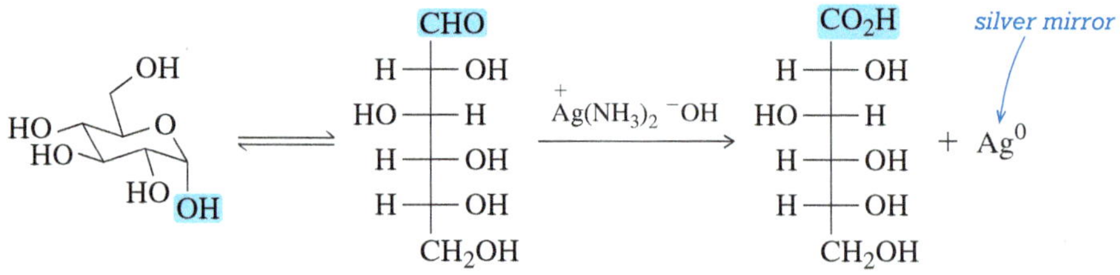

Figure 27.24 α-D-Glucopyranose is in equilibrium with an aldehyde, so it gives a positive Tollens test.

27.4.2 Alkylation of sugars

The alkylation of sugars closely parallels reactions we have studied previously. The only difference in the three reactions shown in this section is that now there are four different hydroxyl groups we can alkylate in a single reaction.

Alkylation of the hydroxyl groups to make ethers is another example of the Williamson ether synthesis (Section 13.11.1). Using silver oxide (Ag_2O) as the base, alkylation proceeds by an S_N2 reaction at each hydroxyl group. With a large excess of base and alkyl halide (**Figure 27.25**), all of the hydroxyl groups are converted into ethers. If the goal is to alkylate the ring structure, this reaction can only be done on the pyranoside and not the pyranose. In strongly basic conditions, the pyranose is converted back to the acyclic form prior to alkylation of the hydroxyl groups. Because the pyranoside is an acetal, it is stable under basic conditions. If the tetramethyl pyranose is desired, however, the product of alkylation can be easily hydrolyzed back to it.

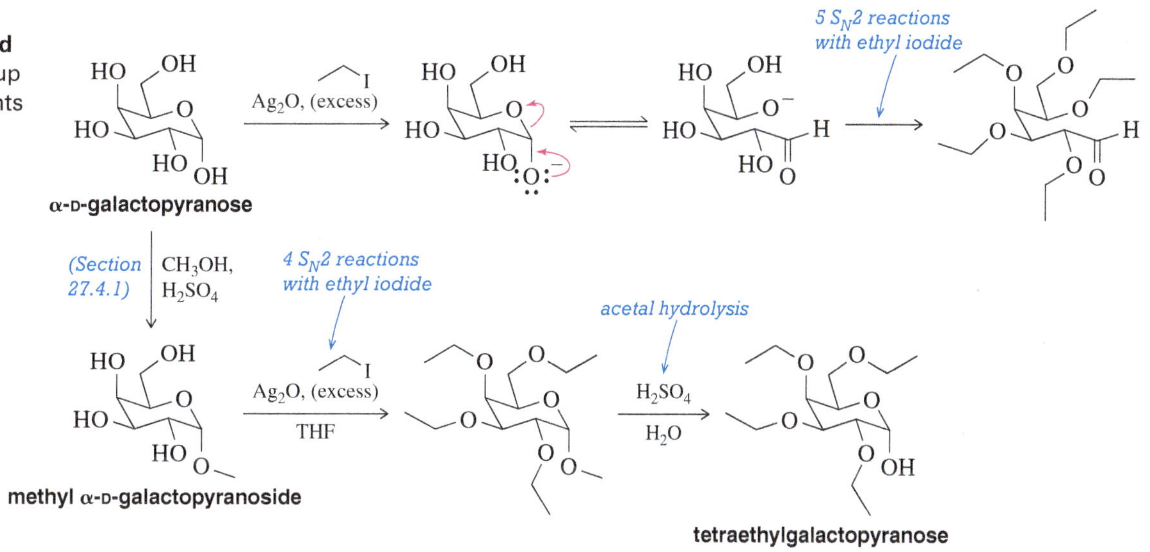

Figure 27.25 **Exhaustive alkylation with silver oxide and iodoethane.** The methoxy group at the anomeric carbon prevents the ring from opening.

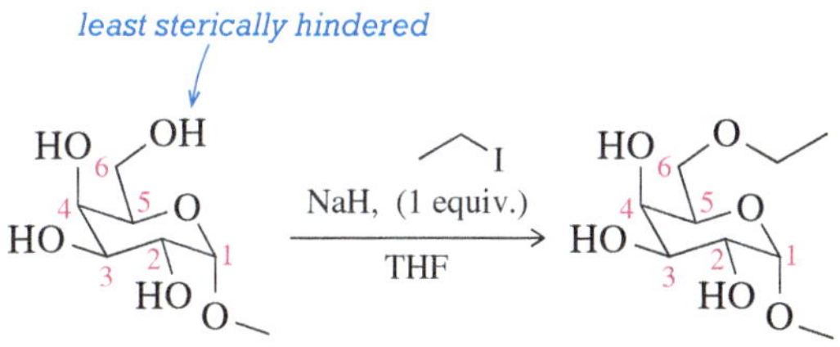

Figure 27.26 **Selective alkylation occurs at the least sterically hindered carbon of a pyranoside.**

There are times when chemists want to differentiate between the hydroxyl groups on the pyranose ring. We have already seen how the anomeric hydroxyl group can be selectively transformed (Section 27.4.1). The next easiest hydroxyl group to selectively transform is the 1° hydroxyl at C_6 because it is less sterically hindered than the rest. If a single equivalent of alkyl halide and base is used, the C_6 ether is selectively obtained in an S_N2 reaction. Sodium hydride (NaH) is used as the base in **Figure 27.26**.

Much of the chemistry of selectively modifying the hydroxyl groups of sugars centers on protecting-group chemistry. If we want to selectively modify a given hydroxyl group, we need to first protect all of the others so that only it will react. As shown in **Figure 27.27**, protection of the C_2, C_3, and C_6 hydroxyls would allow the selective functionalization of the C_4 hydroxyl. Whole books have been written on this process, so a full treatment is beyond the scope of this class. Nevertheless, you should think about how this might be accomplished as we learn these new reactions. [*In Section 27.6.1, we examine one example of selective formation of a disaccharide by manipulating protecting groups.*]

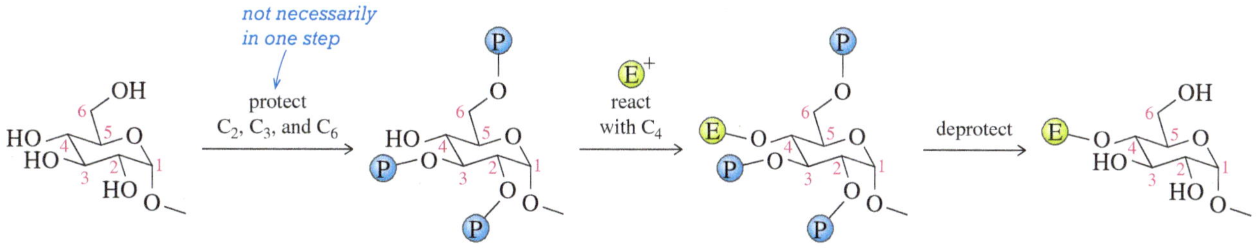

Figure 27.27 **Protection of C_2, C_3, and C_6 makes it possible to selectively functionalize C_4.**

Silyl and benzyl ethers, which can be installed via alkylation reactions, are suitable protecting groups for hydroxyl groups. Silyl ethers (Section 13.14.1) are formed by treatment of the sugar with triethyl amine and a chlorosilane like chlorotriisopropylsilane [**Figure 27.28(a)**]. Benzyl ethers (Section 24.6) can be installed using sodium hydride and benzyl chloride [**Figure 27.28(b)**]. These are complementary protecting groups because they are removed in such different ways, allowing selective deprotection of one in the presence of the others, as shown in **Figure 27.28(c)**.

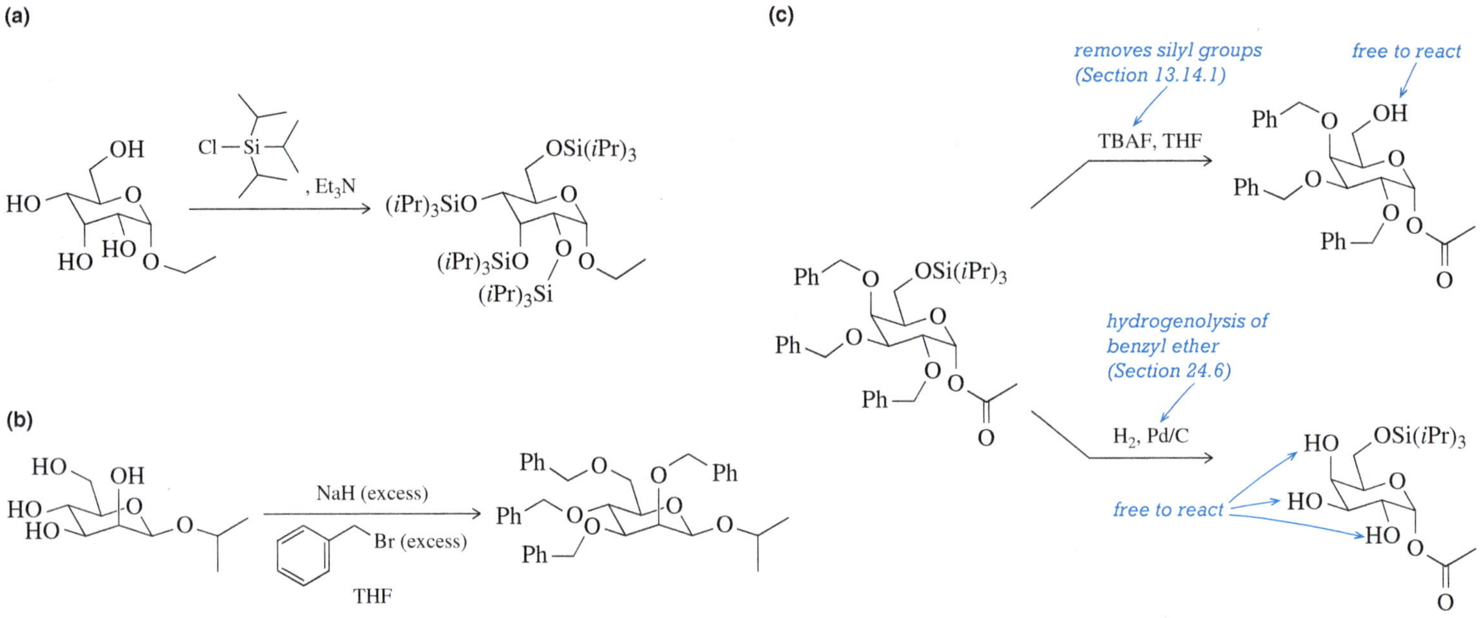

Figure 27.28 Protection/deprotection of sugars. (a) Protection as silyl ethers. (b) Protection as benzyl ethers. (c) Selectively removing one protecting group in the presence of another.

Assessment 27.30 Predict the product of the following etherification reactions.

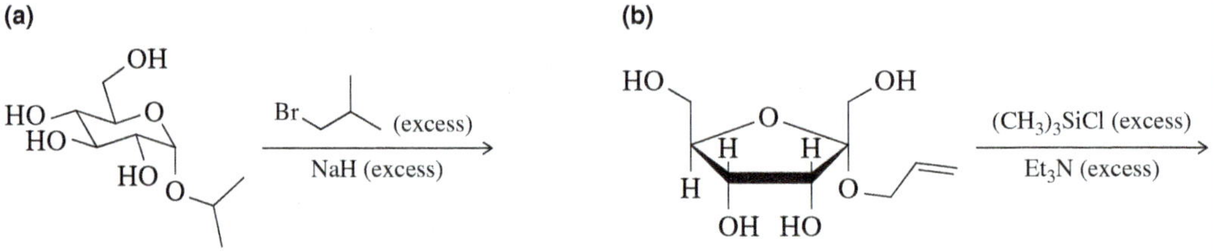

Assessment 27.31 Carry out the sequence of reactions shown, starting with α-D-mannopyranose.

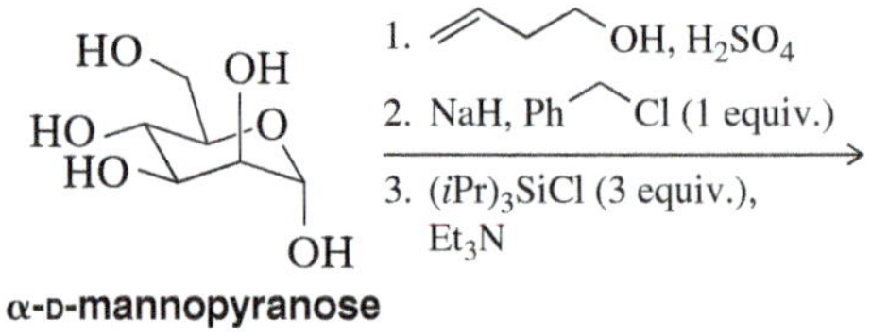

27.4.3 Acylation of sugars

Esters can also be used as protecting groups for sugars. Identical to the reactions learned for ester formation from alcohols in Section 19.7.2, reaction of an excess of acetic anhydride converts all of the hydroxyl groups (including the anomeric) to esters in the presence of triethyl amine (**Figure 27.29**).

Figure 27.29 Exhaustive acylation of glucose with acetic anhydride.

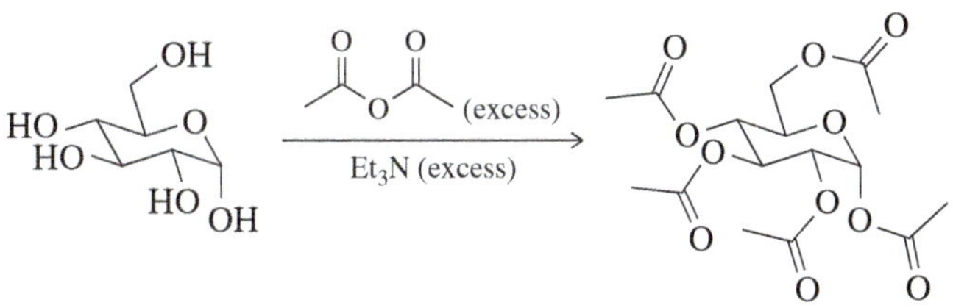

In addition to their role as protecting groups, acetates are also good leaving groups, especially at the anomeric carbon. In this way, Lewis acid activation can promote carbocation formation, allowing functionalization of the anomeric carbon with a variety of alcohols. One such example is shown in **Figure 27.30**.

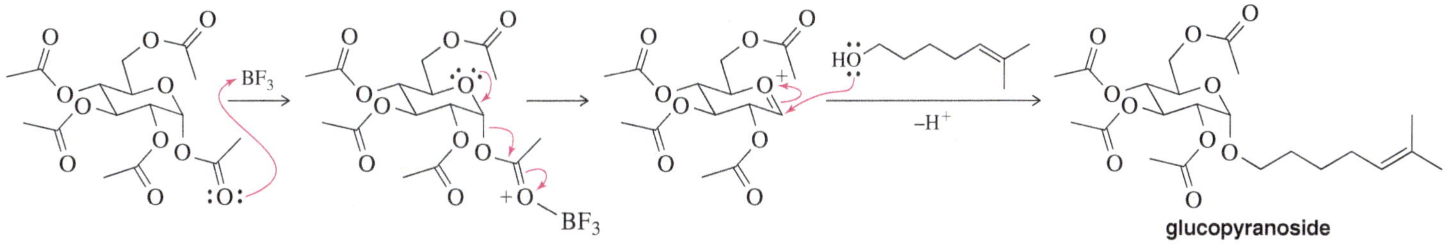

Figure 27.30 Anomeric acetates are good leaving groups that can be used to functionalize the anomeric carbon.

Assessment 27.32 Suggest a synthesis of the following acylated sugars.

(a)

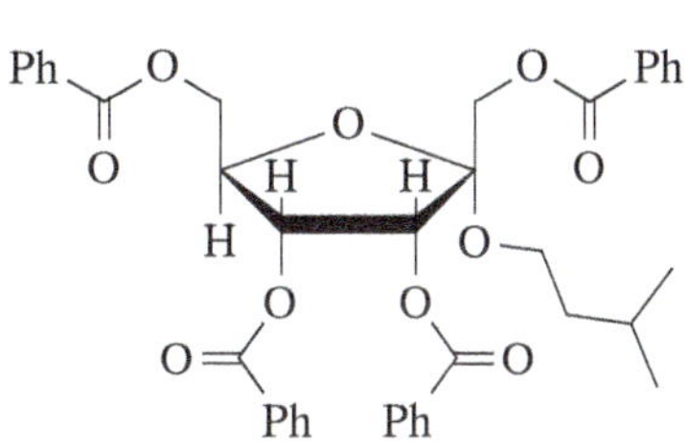

(b)

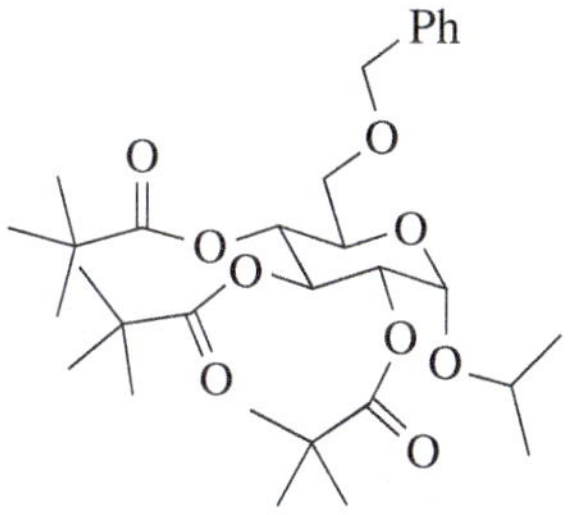

27.4.4 Acetal formation

So far, we have only examined reactions used to protect individual hydroxyl groups. When in close proximity, two hydroxyl groups can be protected as an acetal (**Figure 27.31**). In this way, reaction of benzaldehyde with the C_4 and C_6 hydroxyl groups results in formation of an acetal. Even if other hydroxyl groups are available, the formation of the stable six-membered ring under thermodynamically controlled conditions makes this reaction selective. Mechanistically, this reaction is just like the acetal formation reactions studied in Section 17.7.3. [*Look there for a review.*]

Figure 27.31 Protection of diols as an acetal.

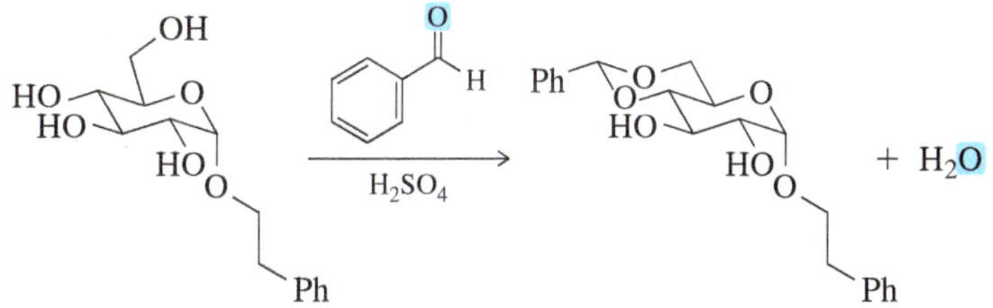

27.4 SUMMARY

- Sugars contain a number of hydroxyl groups that can be modified using reactions studied previously: acetal formation, alkylation, and acylation.

Assessment 27.33 Suggest an arrow-pushing mechanism for the reaction shown in Figure 27.31.

27.5 Reactions of monosaccharides: Fischer's structure determination of glucose

Instead of studying the reactions of monosaccharides as synthetically useful reactions, let's learn them in the context of one of the most ingenious sets of experiments ever done—the experiments Emil Fischer used to determine the structure of glucose. In the modern era, where we have crystallography, NMR, IR, and mass spectrometry along with any number of chiral resolving agents, this wouldn't be as great a feat. But in 1888, when Fischer's tools were only melting points, polarimetry, and a few reactions, it was a well-deserved Nobel Prize–winning accomplishment and is no doubt why Fischer goes down as one of the greatest chemists of all time. [*To reiterate his genius, these experiments were done only 12 years after chemists had established that carbon was tetrahedral.*]

> Let's Think Like a Chemist to see how Fischer did this. This one extends over several pages so grab a pencil and paper to follow along.

Thinking Like a Chemist

To fully understand how these experiments worked, we have to know what Fischer knew. First, he knew what glucose was. He had isolated it and measured its optical rotation, its melting point, etc. But beyond knowing that it was an aldose, he did not know its structure. Based on his understanding of glyceraldehyde, he guessed that it would be D-glucose and ended up being correct about this. Finally, he knew that aldoses had four stereocenters, in which case there were 16 (2^4) possible diastereomers. Only eight aldose diastereomers are D, however, so the possibilities were narrowed to those shown in **Figure 27.32**. [*While reading this section, pretend you don't know which molecule is glucose.*]

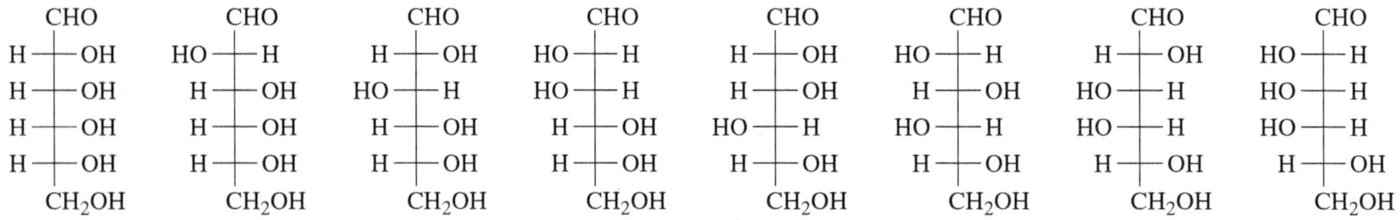

A B C D E F G H

Figure 27.32 The eight aldose diastereomers that could have been the structure of glucose.

27.5.1 *Experiment 1: nitric acid oxidation*

The first reaction Fischer did was to treat glucose with nitric acid, which is a powerful enough oxidizing agent to convert both the C_1 aldehyde and the C_6 alcohol to carboxylic acids. The resulting aldaric acid had a measurable optical rotation, meaning that it was chiral. For diastereomers **A** and **G** in **Figure 27.33**, this reaction would produce a meso compound, which, we learned in Chapter 6, is optically inactive. [*Fischer didn't actually do the reaction of* **A** *and* **G** *. . . it was a thought experiment.*] For all others, the resulting aldaric acid would be optically active. This experiment confirmed that neither **A** nor **G** was the structure of glucose.

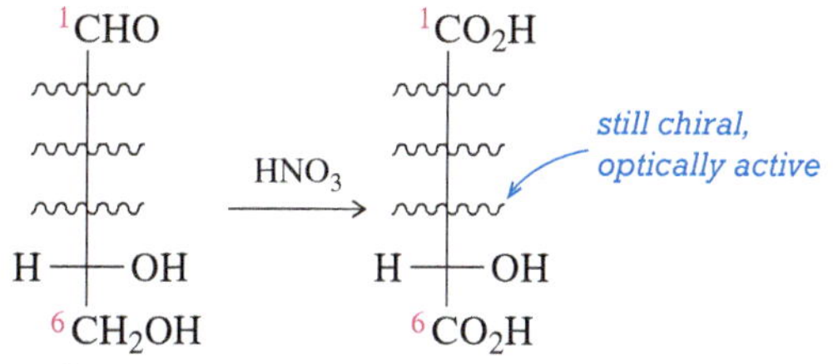

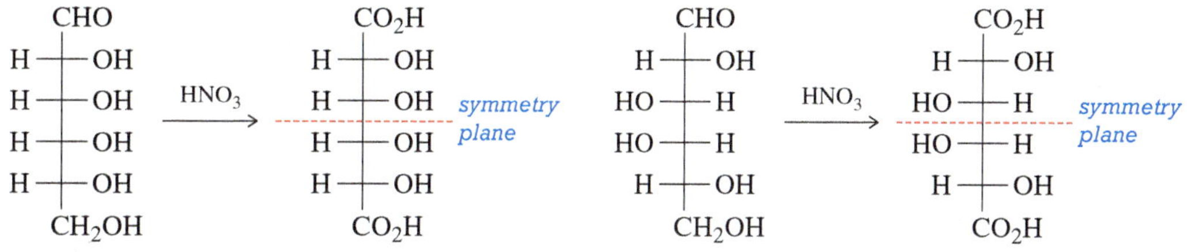

Figure 27.33 **Oxidation of A and G with nitric acid gives optically inactive compounds. A** and **G** are not glucose.

27.5.2 *Experiment 2: the Ruff degradation*

The Ruff degradation is actually a two-part synthesis. In the first, glucose (unknown structure) is oxidized with bromine in water. Whereas nitric acid is a powerful oxidizing agent, this is milder because only the aldehyde is oxidized to a carboxylic acid. In a subsequent step, the carboxylic acid is decarboxylated, resulting in a chain-shortened, five-carbon aldose, called arabinose [structure also unknown; see **Figure 27.34(a)** on the next page]. The chain is shortened by the loss of C_1 from the original aldose. When arabinose is oxidized to a diacid with nitric acid [*see experiment 1*], the resulting molecule is optically active and therefore chiral. [*Before looking at Figure 27.34(b), go back to Figure 27.32. Which of the possible aldoses (**A–H**) would still be chiral after these three steps?*] For diastereomers **A**, **B**, **E**, and **F** in **Figure 27.34(b)**, this sequence would produce meso compounds. [*Which of the original eight compounds are left?*] For all others, the resulting products would be optically active. This experiment confirmed that neither **B**, **E**, nor **F** is the structure of glucose. [*A isn't shown in Figure 27.34(b) because it had already been eliminated.*]

At this point, Fischer knew that the structure of glucose was not **A**, **B**, **E**, **F**, or **G**. Moreover, he had a new tool in his toolbox, the molecule arabinose, synthesized as part of experiment 2. He didn't know its structure, but it turns out he didn't need to.

27.5.3 *Experiment 3: the Kiliani–Fischer synthesis*

This experiment is the most difficult to interpret and, more than any other, reveals Fischer's genius. Before we try to understand the result, let's look at the Kiliani–Fischer synthesis as a new reaction. The Kiliani–Fischer synthesis extends an aldose by one carbon. Although the mechanism is complicated, it utilizes reactions similar to ones studied before. To simplify the analysis, **Figure 27.35** (see page 1283) shows the Kiliani–Fischer synthesis applied to D-glyceraldehyde. First, treatment of the aldehyde with cyanide results in nucleophilic addition to give a cyanohydrin [*see Section 17.5.4*]. Hydrolysis of the cyanohydrin forms a new carboxylic acid [*see Section 19.10*]. And, finally, reduction of the carboxylic acid gives the aldehyde of the new aldose. [*See Section 18.7.3 for the reduction of a carboxylic acid, although we never learned how to go directly to an aldehyde.*] Importantly, the cyanide adds to the planar aldehyde from either of two faces. Thus, the Kiliani–Fischer synthesis always gives two epimers (diastereomers).

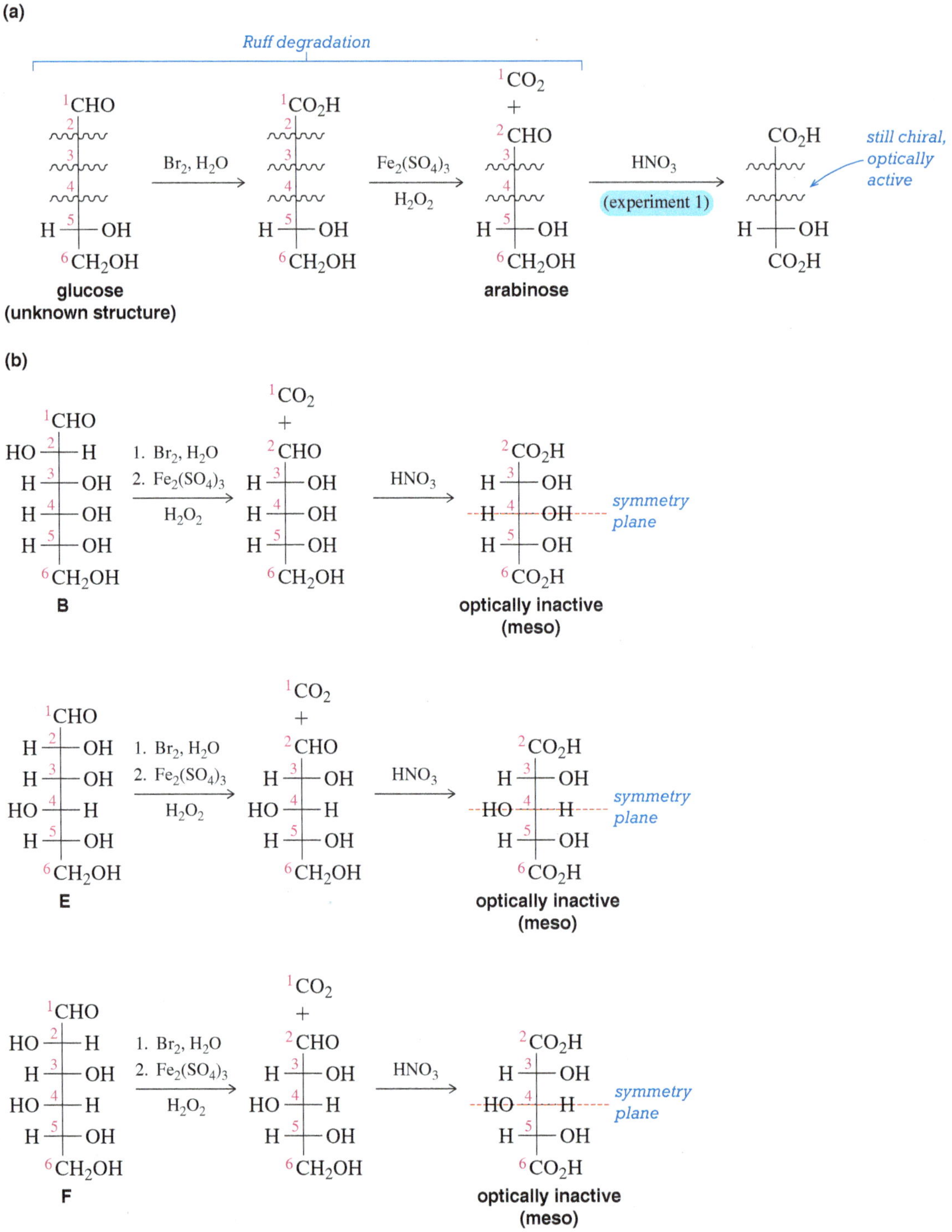

Figure 27.34 The Ruff degradation and oxidation. (a) For glucose, this reaction gives arabinose. (b) For **B**, **E**, and **F**, it gives optically inactive compounds. **B**, **E**, and **F** are *not* glucose.

The Kiliani–Fischer synthesis, when applied to glyceraldehyde, produces two naturally occurring sugars, D-erythrose and D-threose. The application of the Kiliani–Fischer synthesis to any D-aldopentose (a five-carbon sugar) will give two naturally occurring D-aldohexoses (six-carbon sugars). [*Reread these last two sentences. You have to understand them to move on.*] What you just read, possibly more than once, Fischer knew.

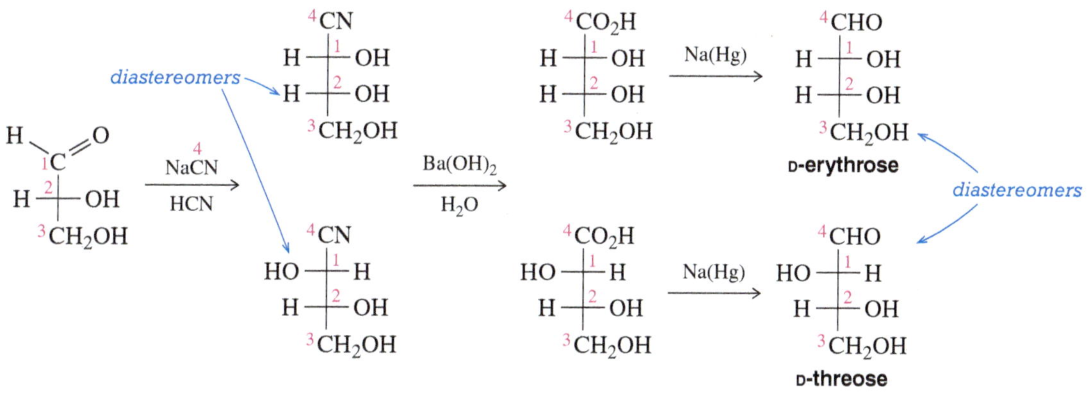

Figure 27.35 The Kiliani–Fischer synthesis of a chain-extended aldose.

Fischer also knew that the Ruff degradation of glucose, a reaction he did in experiment 2, produced arabinose. The Ruff degradation to make arabinose is the opposite of the Kiliani–Fischer synthesis applied to arabinose. Thus, when Fischer applied the Kiliani–Fischer synthesis to the unknown structure arabinose, two new sugars were produced, one of which he characterized as mannose, the other as glucose (**Figure 27.36**). Both of these structures, when treated with nitric acid were optically active. [*This is important!*]

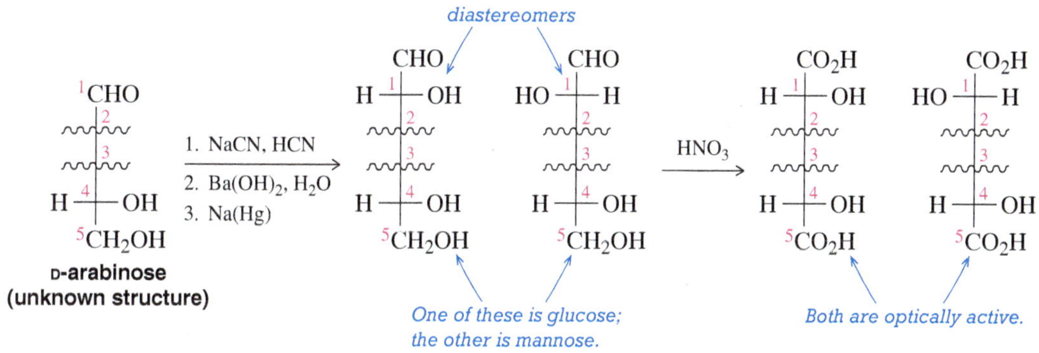

Figure 27.36 The Kiliani–Fischer synthesis of arabinose gives glucose and mannose.

Fischer didn't know the structure of arabinose, but he knew that it had three stereocenters, so he knew there were eight possible diastereomers. If the assumption of being D holds for arabinose, this leaves only the four D diastereomers. These four possible diastereomers are shown in **Table 27.1**. Application of the Kiliani–Fischer synthesis, followed by oxidation with nitric acid, to each of these would give what? [*Again, this is just a thought exercise.*] These results are also shown in Table 27.1. For entry 1, you would get two optically active compounds. This could be the structure of arabinose. [*Do you see why?*] For entry 2, you would get one optically active and one optically inactive compound. The Kiliani–Fischer synthesis on arabinose gave two optically active compounds (Figure 27.36), so this can't be the structure of arabinose. For entry 3, you once more would get one optically active and one optically inactive compound. This can't be the structure of arabinose. For entry 4, you would get two optically inactive compounds, suggesting, at least for now, that entry 4 could be the structure of arabinose.

Entries 1 and 4, both of which are possible structures of arabinose, are compared in **Figure 27.37**. When Fischer first made arabinose, he treated it with nitric acid and got a chiral, optically active diacid. [*Verify this by looking at Figure 27.34(a).*] The possible structure of arabinose in entry 4 is shown in Figure 27.34(b). When it was oxidized with nitric acid, its product was meso and therefore optically inactive. This means entry 4 cannot be the structure of arabinose. Entry 1, then, is the structure of arabinose because it gets oxidized to an optically active diacid. Furthermore, the products of the Kiliani–Fischer synthesis of entry 4 are compounds **E** and **F** from Figure 27.32, two structures that had already been eliminated as possible structures of glucose. [*Of our eight original compounds, which are still possible?*]

Table 27.1 Hypothetical Kiliani–Fischer synthesis of possible arabinose structures. Based on the thought experiments, entries 1 and 4 are the only possible structures of arabinose.

Entry	Possible structure of arabinose	Result of Kiliani–Fischer synthesis	Oxidation of K–F product gives	Comments
1	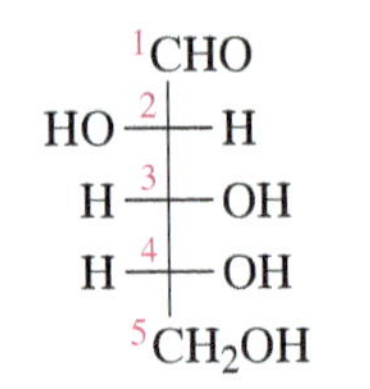	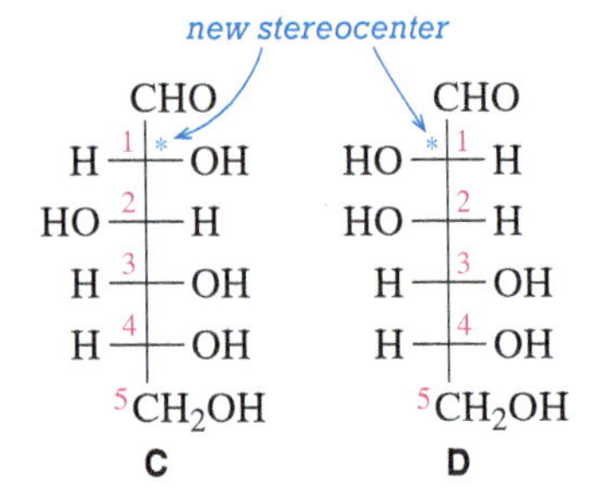	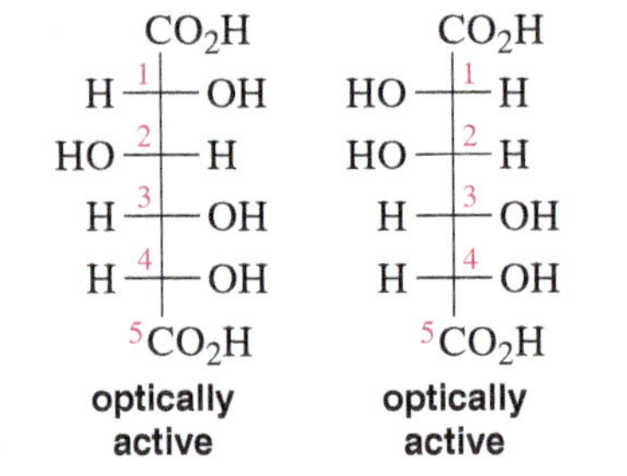	Because two optically active products are produced, this could be the structure of arabinose, and one of the products of the K–F synthesis (C or D) of entry 1 could be glucose.
2	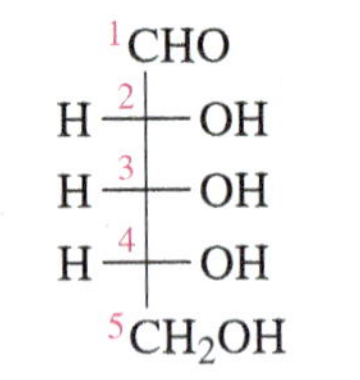	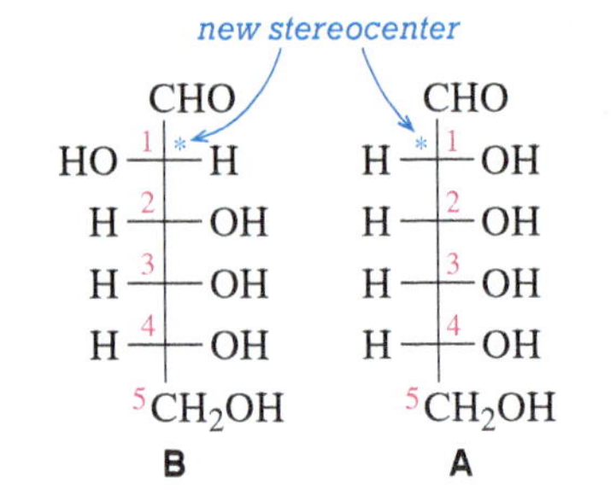	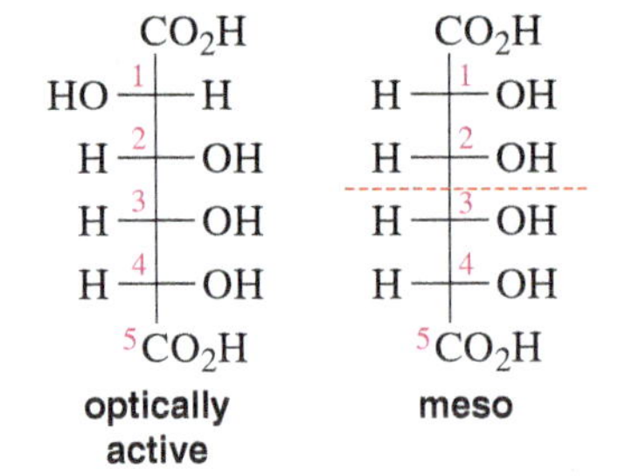	Because only one optically active product is produced, this cannot be the structure of arabinose. Thus, neither of the products of the K–F synthesis of entry 2 could be glucose.
3	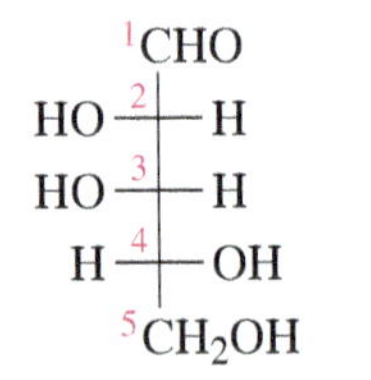	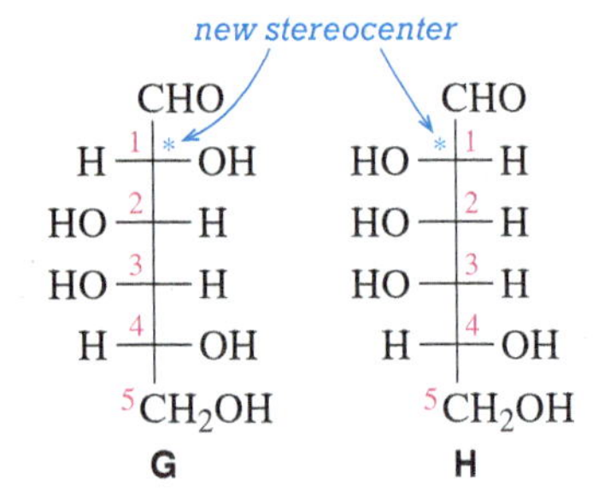	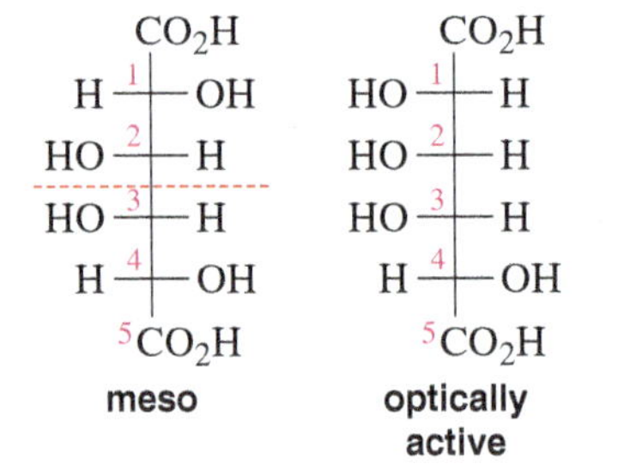	Because only one optically active product is produced, this cannot be the structure of arabinose. Thus, neither of the products of the K–F synthesis of entry 3 (G or H) could be glucose.
4	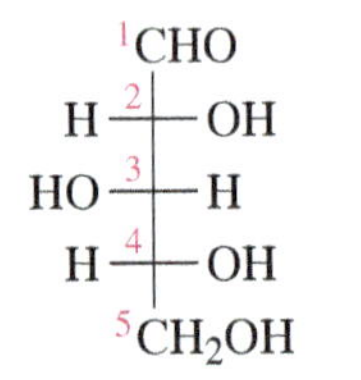	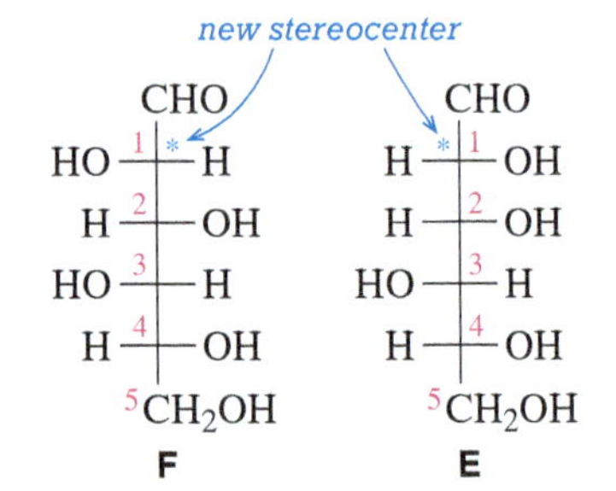	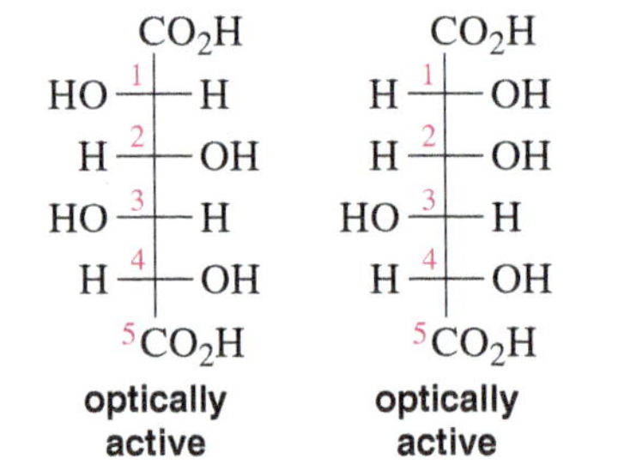	Because two optically active products are produced, this could be the structure of arabinose, and one of the products of the K–F synthesis of entry 4 could be glucose.

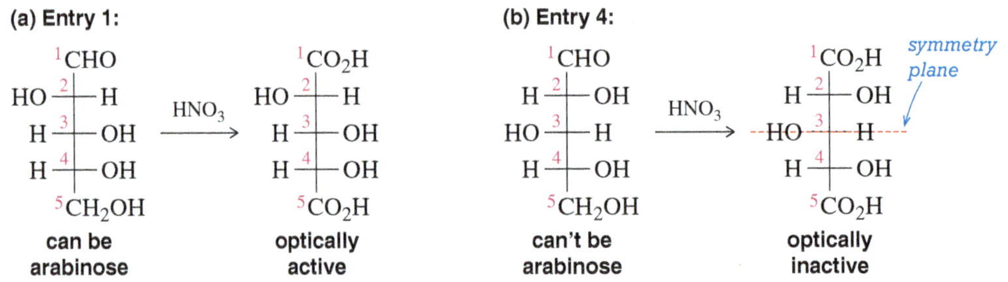

Figure 27.37 The oxidation of possible arabinose structures. (a) Entry 1 (from Table 27.1) gives an optically active compound. (b) Entry 4 gives a meso diacid. Entry 1 must be arabinose.

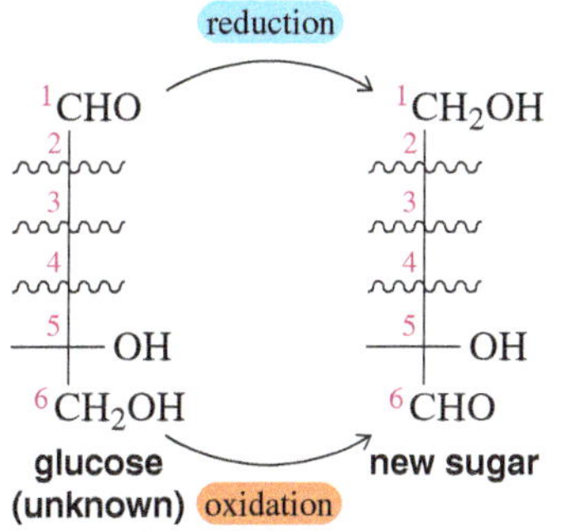

Figure 27.38 **"Switching the ends" of glucose through redox reactions generates a new sugar.**

27.5.4 *Experiment 4: switching the ends*

Knowing the structure of arabinose allows us to identify the diastereomers produced in the Kiliani–Fischer synthesis as being glucose and mannose. But, unfortunately, we don't yet know which is which. Fortunately, Fischer had developed a way to interchange the ends of sugars via reduction of the C_1 aldehyde and oxidation of the C_6 alcohol (**Figure 27.38**). When this reaction is applied to glucose, a new sugar was produced. [*Which of the possible glucose structures, C or D, would this also be true for?*]

Fischer reasoned that if the same end-switching experiment were done on **D**, the sugar produced would be identical to **D**. When the end-switching experiment was done for **C**, a new sugar was produced (**Figure 27.39**). As a result, **C** must be the structure of glucose! ■

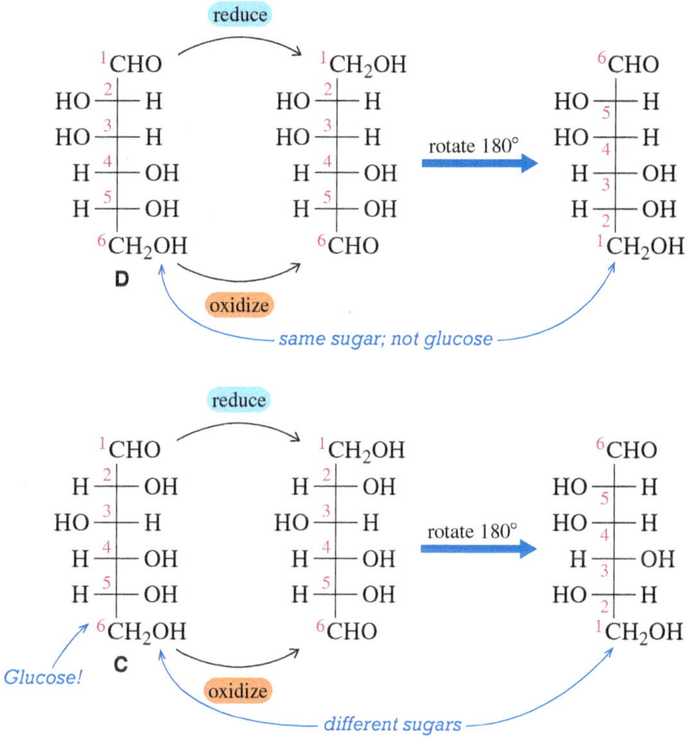

Figure 27.39 **Switching the ends of the Kiliani–Fischer products of arabinose identifies C as glucose.**

27.5 SUMMARY

- Knowing only its melting point, its optical rotation, and that it was an aldose with four stereocenters, Fischer used a combination of chemical reactions to determine the stereochemistry of glucose.

Assessment 27.34 Predict the product of each of the following reactions.

(a) (b)

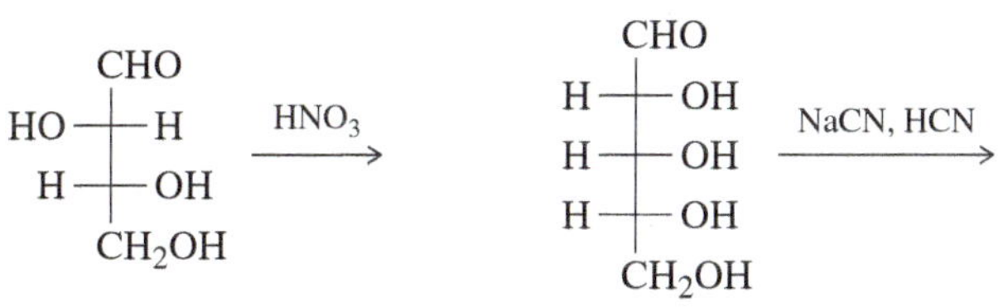

(c) (d)

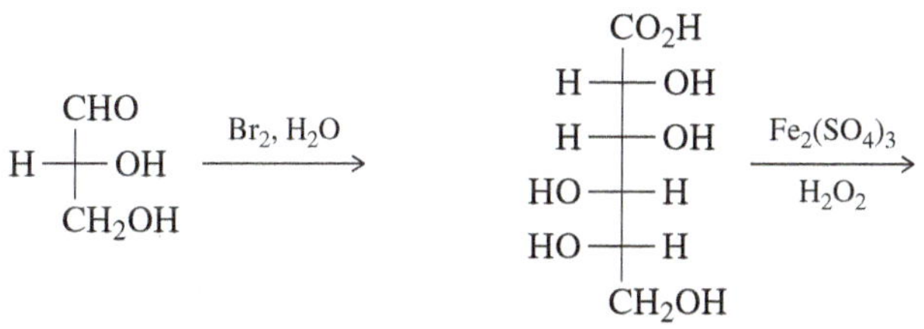

Assessment 27.35 Although reagents weren't given in Figures 27.38 or 27.39, in theory, how would you plan the end-switching experiment from experiment 4? [*No reagents are needed, just a general concept/plan of attack.*]

Assessment 27.36 Ketohexoses form from aldohexoses. Suggest a mechanism for the acid-catalyzed version of this reaction. What type of reaction is this? [*You've seen it before.*]

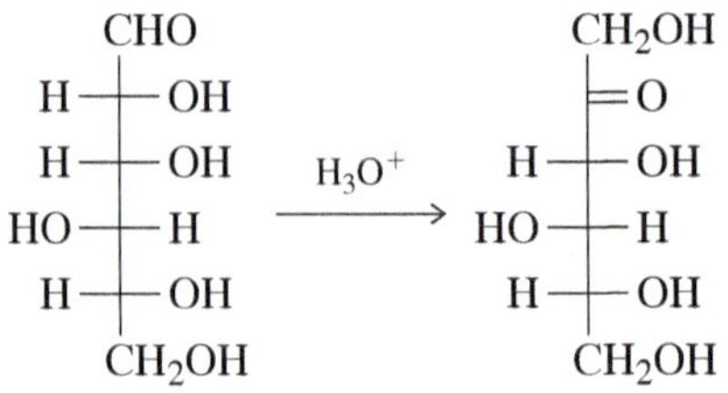

Assessment 27.37 Why wasn't the following structure ever considered as a possible candidate for D-glucose?

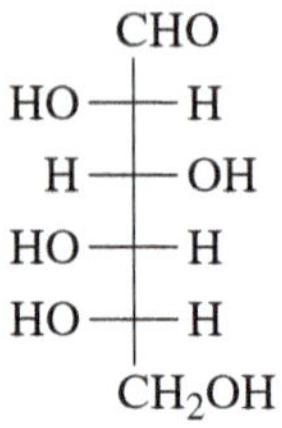

Assessment 27.38 How would the results of Fischer's experiment been different if he had been wrong about the stereochemistry of D-glyceraldehyde?

27.6 Disaccharides

Disaccharides are sugars that upon hydrolysis give two monosaccharides. As shown in **Figure 27.40**, lactose, known as milk sugar, is a disaccharide composed of galactose and glucose connected through a $\beta(1 \rightarrow 4)$ linkage. The linkage is considered to be β because the anomeric alkoxy group of galactose is cis to C_6 of galactose. It is a $1 \rightarrow 4$ linkage because it connects C_1 of galactose to C_4 of glucose. As indicated by an IUPAC name that ends in "-ose," lactose is a reducing sugar. The glucose monomer, containing a hemiacetal at the anomeric carbon, is the reducing end of the disaccharide because it is capable of reducing Ag^+ to Ag^0 [*see the Tollens reagent in Section 27.4.1*]. Hydrolysis of lactose would give the individual glycosides galactose and glucose. The anomeric carbon of galactose is an acetal, making this, and all disaccharides, a glycoside (Section 27.4.1).

Figure 27.40 Lactose is a disaccharide made of galactose and glucose.

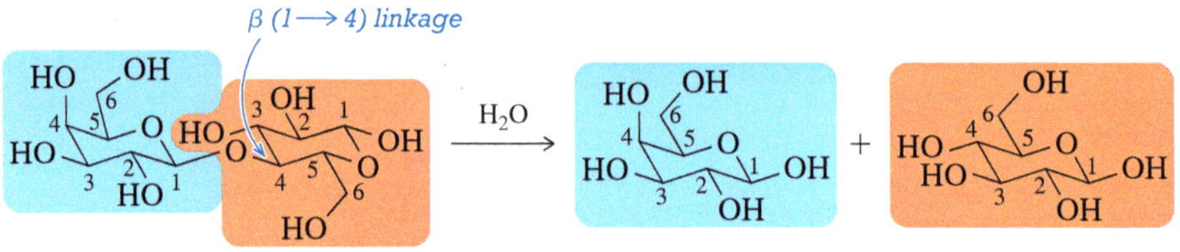

lactose
β-D-galactopyranosyl-(1⟶4)-D-glucopyranose

D-galactose

D-glucose

The -ose ending suggests a reducing sugar.

Sucrose (common table sugar) is a disaccharide composed of glucose and fructose connected through their anomeric carbons, an $\alpha(1 \rightarrow 2)\beta$ linkage (**Figure 27.41**). The linkage is considered to be α because the anomeric alkoxy group of glucose is trans to C_6 of glucose. It is a $1 \rightarrow 2$ linkage because it connects C_1 of glucose to C_2 of fructose. As indicated by the name ending in "-oside," sucrose contains no hemiacetal carbons and is therefore a nonreducing sugar. Hydrolysis of sucrose would give the individual glycosides glucose and fructose. **Table 27.2** shows three other representative disaccharides.

Figure 27.41 Sucrose is a disaccharide made of glucose and fructose. (a) Haworth projection. (b) Chair representation.

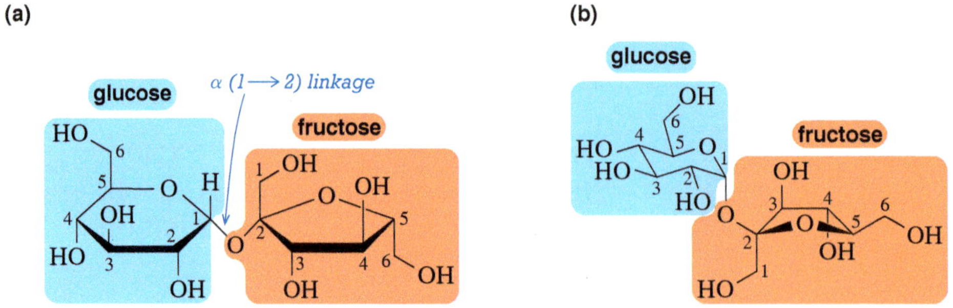

sucrose
α-D-glucopyranosyl-(1⟶2)-β-D-fructofuranoside

The -oside ending suggests a nonreducing sugar.

Table 27.2 **The disaccharides trehalose, lactulose, and cellobiose.**

Disaccharide	Description
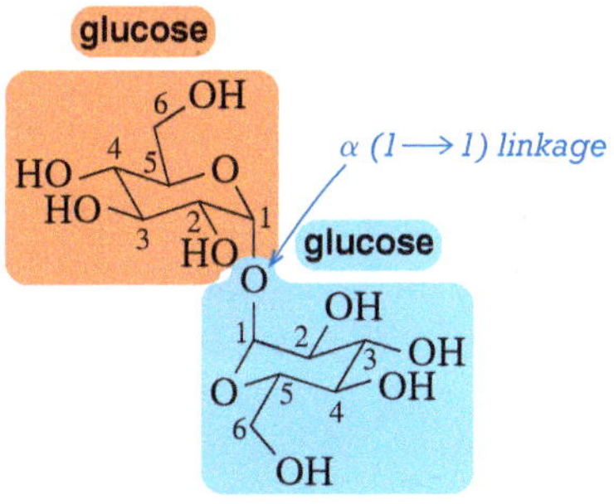 trehalose	Trehalose is composed of two glucose molecules linked by their anomeric carbons in an α(1 → 1) linkage. There is no hemiacetal, so trehalose is a nonreducing sugar. It is believed to help cells survive periods of dehydration.
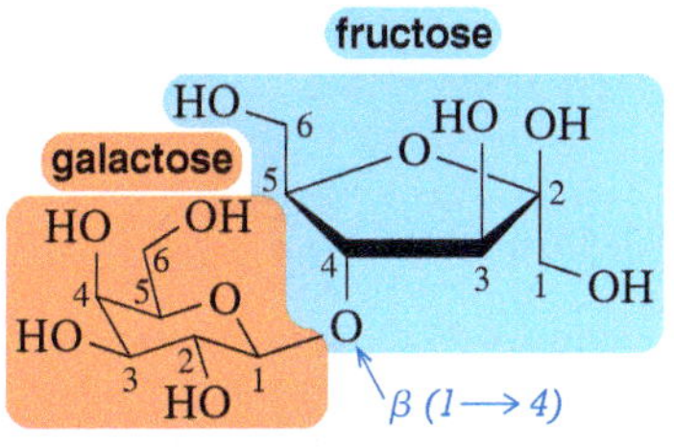 lactulose	Lactulose is composed of galactose and fructose linked by a β(1 → 4) linkage. It is a synthetic disaccharide, so it isn't digested by human enzymes, it remains in the digestive tract, and it promotes water retention, thereby serving as a treatment for constipation.
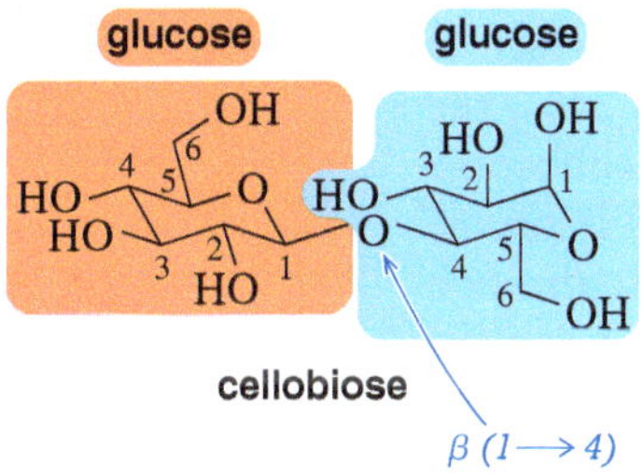	Cellobiose is composed of two glucose molecules connected by a β(1 → 4) linkage. It is a product of cellulose hydrolysis.

Assessment 27.39 In each of the disaccharides shown (a–c), (i) identify the reducing end of the sugar, (ii) tell what monosaccharides are used to make it, and (iii) tell what type of glycoside linkage is present.

(a)

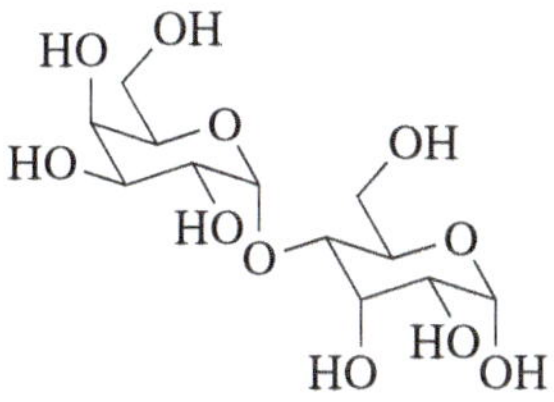

(b)

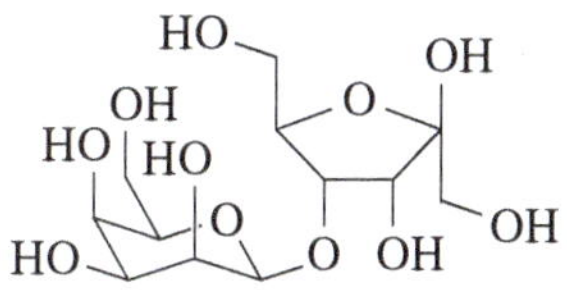

(c)

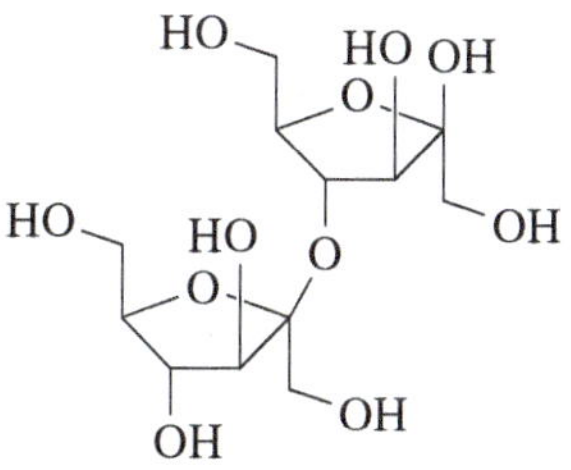

Assessment 27.40 Maltose, like cellobiose, is composed of two glucose units. What is the difference between maltose and cellobiose?

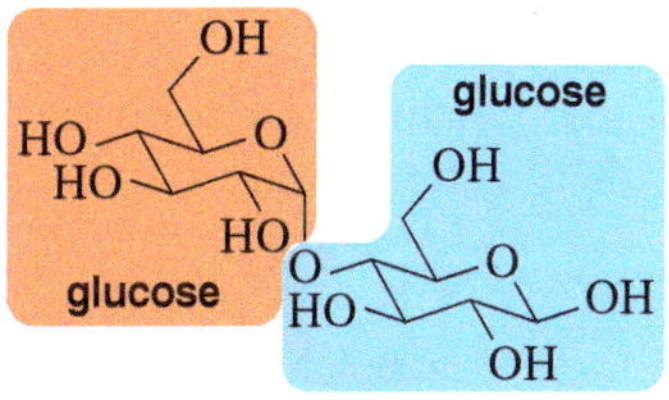

maltose

Assessment 27.41 Hydrolysis of a disaccharide gives two sugars. Suggest an arrow-pushing mechanism for the acid-catalyzed hydrolysis of sucrose to give glucose and fructose [*see Figure 27.41*].

27-A CHEMISTRY IN CONTEXT: Glycobiology (Part 1)

How the influenza virus gains entry into a cell

Viruses are infectious particles that cannot replicate on their own. Instead, carrying only its RNA, the influenza virus uses the machinery of a host cell to replicate itself. Getting into the host cell requires the interaction of viral proteins (Chapter 26) and glycoproteins on the surface of the host cell. Surface glycoproteins, as the name implies, consist of membrane-bound proteins capped by a carbohydrate chain. The carbohydrate portion of the glycoprotein is capped by a sialic acid (**Figure 27.42**).

The influenza virus has two proteins on its surface, called hemagglutinin (H) and neuraminidase (N). The type of influenza virus is indicated by the specific class of hemagglutinin and neuraminidase found on the surface [*e.g., H1N1*]. In order to gain entry to a cell, the hemagglutinin receptor binds selectively and tightly with sialic acid on the surface of the cell. Once attached to the host cell (**Figure 27.43**), the virus is encased by the host cell, which forms an endosome around it. Upon the conversion of the endosome into a lysosome (the organelle cells use to digest foreign particles) through acidification, the viral proteins and RNA are released into the cytoplasm (after a complicated process), where viral replication is able to occur.

One method of treating influenza is to deny entry of the virus into cells. This is done by inhibiting the binding of hemagglutinin to the cell surface. Because hemagglutinin binds selectively with sialic acid, it makes sense that an inhibitor would be similar in structure to sialic acid.

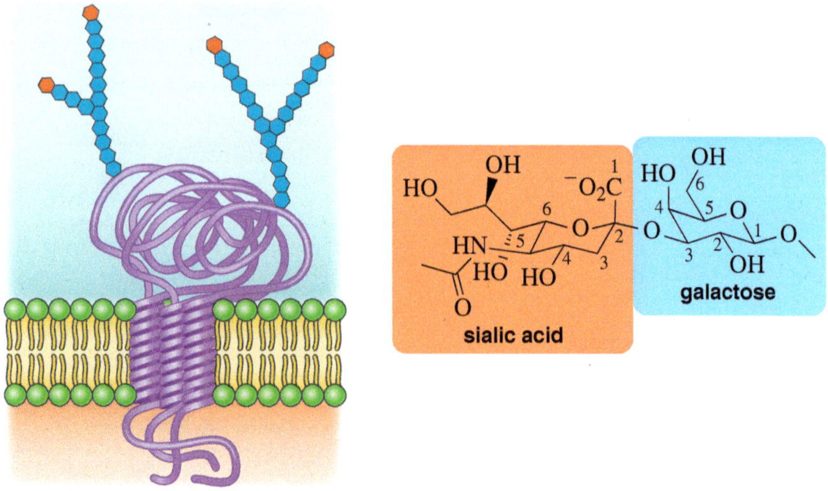

Figure 27.42 A membrane-bound glycoprotein capped by sialic acid. The sialic acid is on the outer surface of the cell.

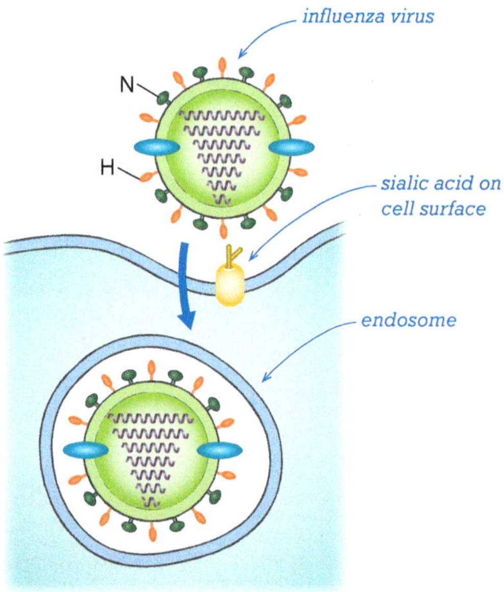

Figure 27.43 The binding of virus hemagglutinin with sialic acid results in viral entry into the cell.

27.6.1 Synthesis of disaccharides

In Section 27.4.1, we learned that glycoside synthesis is as simple as treating a glucoside (a hemiacetal) with an alcohol in the presence of acid, as shown in **Figure 27.44**. Because disaccharides are simply glucosides where the alcohol (R—OH) is replaced by a sugar, disaccharide synthesis should be just as simple. [*But it isn't. Any idea why?*]

It isn't that simple because there are two possible hemiacetal carbons that could become electrophiles—one each in glucose and galactose. Furthermore, there are five possible nucleophilic hydroxyl groups in each sugar. Between the glucosyl–glucose, galactosyl–galactose, glucosyl–galactose, and galactosyl–glucose disaccharides that could form with $(1 \rightarrow 6)$, $(1 \rightarrow 4)$,

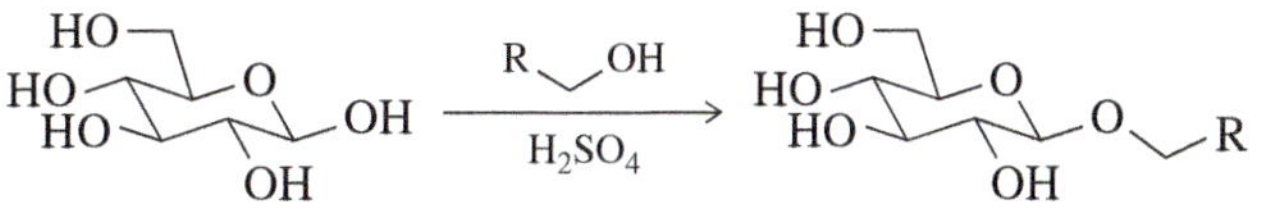

Figure 27.44 Disaccharide formation is analogous to glycoside formation. But is it that simple?

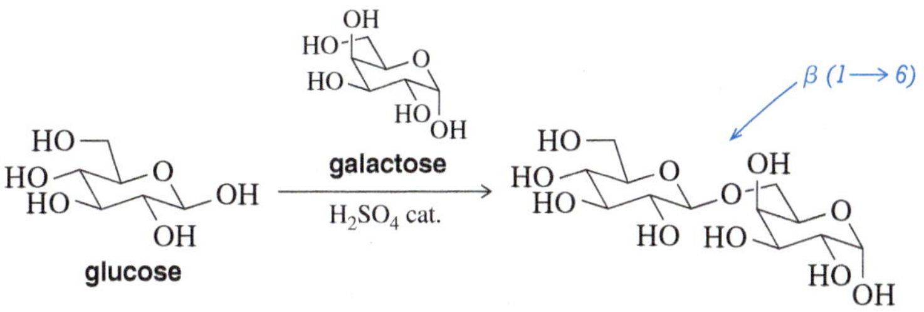

$(1 \rightarrow 3)$, $(1 \rightarrow 2)$, and $(1 \rightarrow 1)$ linkages, each with the potential of being α and β, that's a lot of possibilities. [*I'm not great at math, but it's maybe 40 different disaccharides that could form.*]

How, then, can we bias the reaction so that only our desired α-D-glucopyranosyl-$(1 \rightarrow 6)$-α-D-galactose forms? Unfortunately, we can't use the highly selective enzymes that nature uses for forming disaccharides. [*Or can we?*] But we can use protecting groups. In general, when the two sugars react, we want only the anomeric carbon of glucopyranose to be electrophilic and only the C_6 hydroxyl of galactose to be nucleophilic. We can do this by protecting all others, as shown in **Figure 27.45**. The protecting group used for the anomeric carbon of galactose needs to be a poor leaving group so that it doesn't leave to form a carbocation at C_1 of galactose.

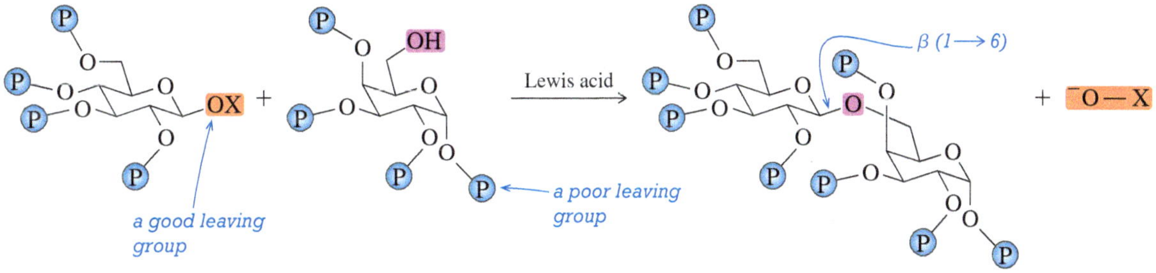

Figure 27.45 Suitably protected, the desired β(1→6) linkage can be produced.

Figures 27.46–27.48 show how an overall protection/deprotection scheme might work to make the desired disaccharide. Don't get caught up in the complexity. Instead, just appreciate the manner in which carbohydrate chemists can manipulate protecting groups to make the synthesis work. [*Also appreciate that nature doesn't need these protecting groups and does it in a far greener way . . . when there is an enzyme that will do it for you.*] The glucose is exhaustively protected with acetic anhydride via the nucleophilic acyl substitution reaction from Section 27.4.3 (Figure 27.29). Recall that the acetate makes the anomeric carbon an excellent leaving group [*recall again Section 27.4.3, Figure 27.30 now*], while preventing the remainder of the hydroxyl groups from reacting as nucleophiles (**Figure 27.46**).

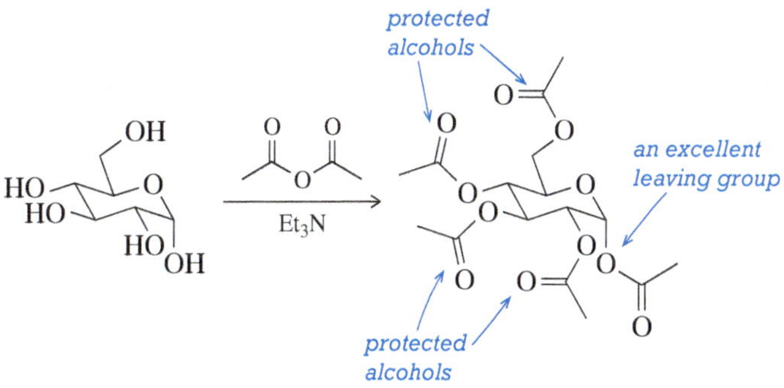

Figure 27.46 Acylation of glucose with acetic anhydride protects the four hydroxyl groups of glucose, while making the anomeric hydroxyl a better leaving group.

Synthesis of the galactose portion of our disaccharide begins with the conversion of galactose to a benzylpyranoside by the acid-catalyzed reaction with benzyl alcohol (**Figure 27.47**). When the anomeric hydroxyl group is protected in this way, it can later be removed either by hydrolysis or hydrogenolysis. We ultimately want the C_6 hydroxyl group to be the nucleophile, so we'd like to leave it unprotected. Unfortunately, it is the least sterically hindered alcohol, so it will always react first in alkylation reactions. To solve this problem, we can selectively protect it with a single equivalent of triethylsilyl chloride, thus producing a silyl ether. We can now use a different protecting group for the remaining hydroxyl groups. Several protecting groups will work here, with the lone requirement being that we can remove the silyl group in their presence. For reasons that will be clear later, the best option is to protect the hydroxyl groups as benzyl ethers using the Williamson ether synthesis. [*Do you see why? Read on.*] To conclude preparation of the galactose monomer, the silyl group is removed using tetrabutylammonium fluoride (TBAF).

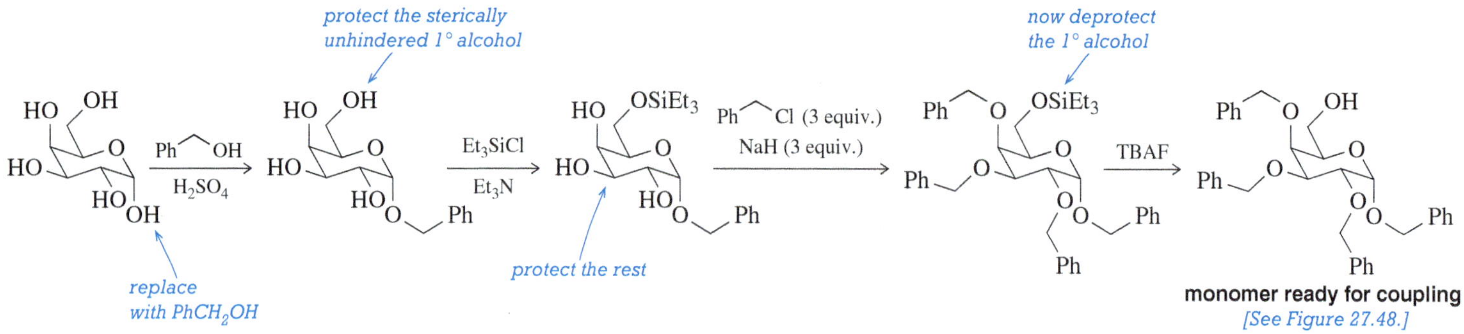

Figure 27.47 Protecting group manipulation leaves the C_6 hydroxyl as the lone nucleophile.

With both monosaccharides appropriately protected, the disaccharide can be formed by a Lewis acid–promoted glycosidation reaction. The role of the Lewis acid in this reaction is to make the anomeric acetate a better leaving group, activating the anomeric carbon for attack by the C_6 hydroxyl group of galactose. With the disaccharide formed, all that remains is to remove the protecting groups (**Figure 27.48**). The acetyl groups are removed by saponification, while the benzyl ethers are removed via hydrogenolysis. [*Being able to remove the anomeric protecting group in the same step is why glycosidation with benzyl alcohol was done in Figure 27.47.*]

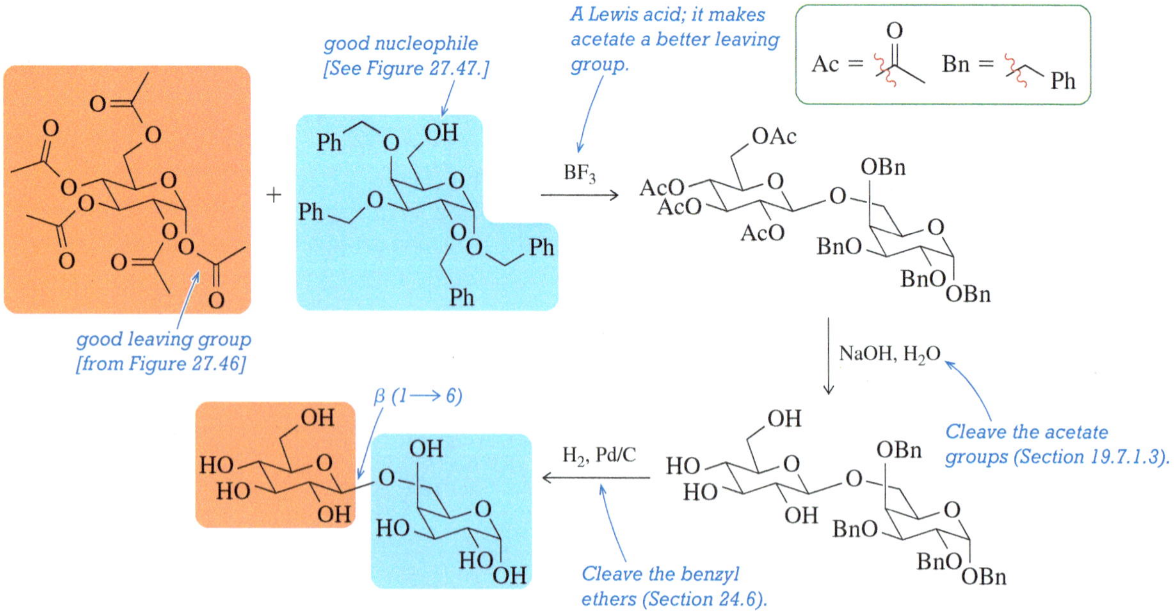

Figure 27.48 Glycosidation and global deprotection produce the disaccharide.

This synthesis is meant to illustrate two important points. First, you've seen every reaction we used before. In this way, your ability to synthesize a disaccharide is not about whether you have the tools, but instead about whether you know how to use them. In this way, carbohydrate synthesis is just another example [*of many*] of organic chemistry teaching us to solve problems using a logical strategy. And second, carbohydrate synthesis is difficult, but fun!

27.6 SUMMARY

- The synthesis of disaccharides involves the use of protecting groups to ensure that only the correct hydroxyl group of one monosaccharide is available to form a glycoside bond with the anomeric carbon of a second monosaccharide.

Assessment 27.42 Show an arrow-pushing mechanism for the Lewis acid–catalyzed glycosidation reaction in Figure 27.48.

27.7 Polysaccharides

Polysaccharides are long chains of monosaccharides. The term *oligosaccharide* refers to a polysaccharide with 3 to 10 monosaccharide units. Polysaccharides can have as many 12,000 monosaccharides in a single chain. Cellulose is a polysaccharide composed of repeating δ-glucosides connected through $\beta(1 \rightarrow 4)$ linkages. Cellulose is the major component of rigid cell walls in plants. The rigidity of cellulose is caused by the relatively linear polymeric structure of cellulose that allows the individual strands to pack tightly together, forming strong hydrogen bonds between neighboring glucosides, as shown in **Figure 27.49**. Complete hydrolysis of cellulose gives glucose.

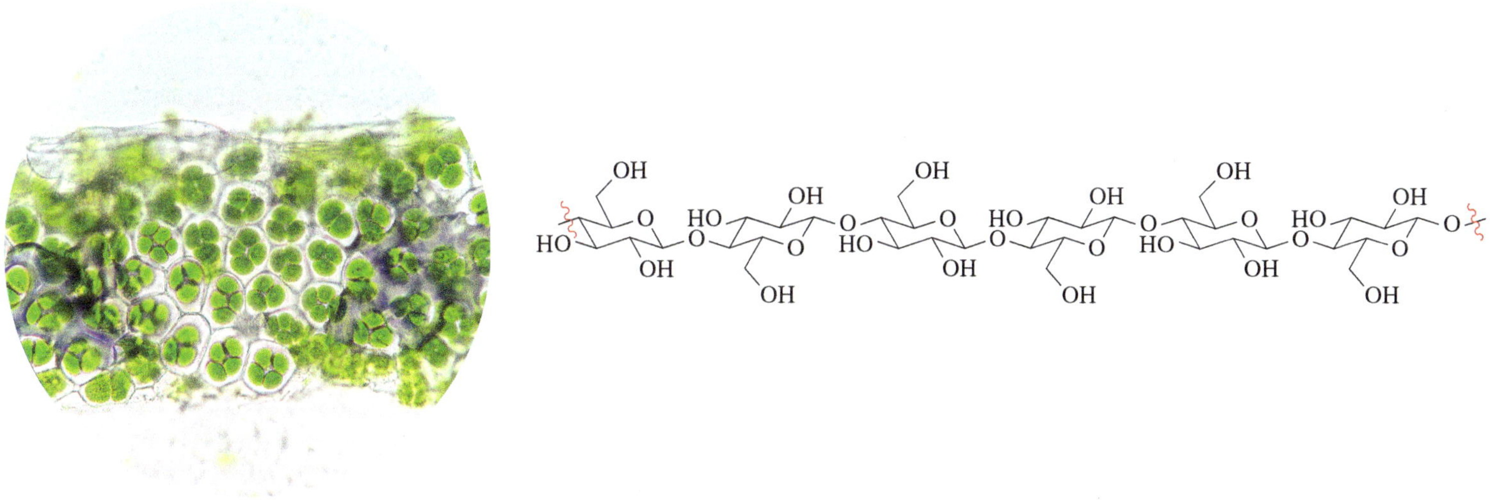

Figure 27.49 Cellulose is a primary component of the cell wall of plant cells.

Closely related to cellulose is starch. Starch is a polysaccharide featuring two major components, called amylose and amylopectin. Amylose is composed of repeating D-glucosides connected through $\alpha(1 \rightarrow 4)$ linkages [**Figure 27.50(a)**]. Amylopectin, on the other hand, contains a similar D-glucoside skeleton but features additional chains connected to the skeleton through $\alpha(1 \rightarrow 6)$ linkages [**Figure 27.50(b)**]. Plants use starch to store energy. Just like cellulose, complete hydrolysis of starch gives glucose.

The cellulose in plants cannot be digested by animals, whereas starches represent a major portion of a human diet. This is true in spite of the fact that the two polymers are composed of the same monomer (glucose). It turns out that animals do not have the enzymes necessary to hydrolyze glucose polymers connected via β-linkages, only α. That's not to say that cellulose is bad for us. On the contrary, it plays a key role in a healthy diet as indigestible fiber.

Figure 27.50 **Starch** is composed of (a) amylose and (b) amylopectin.

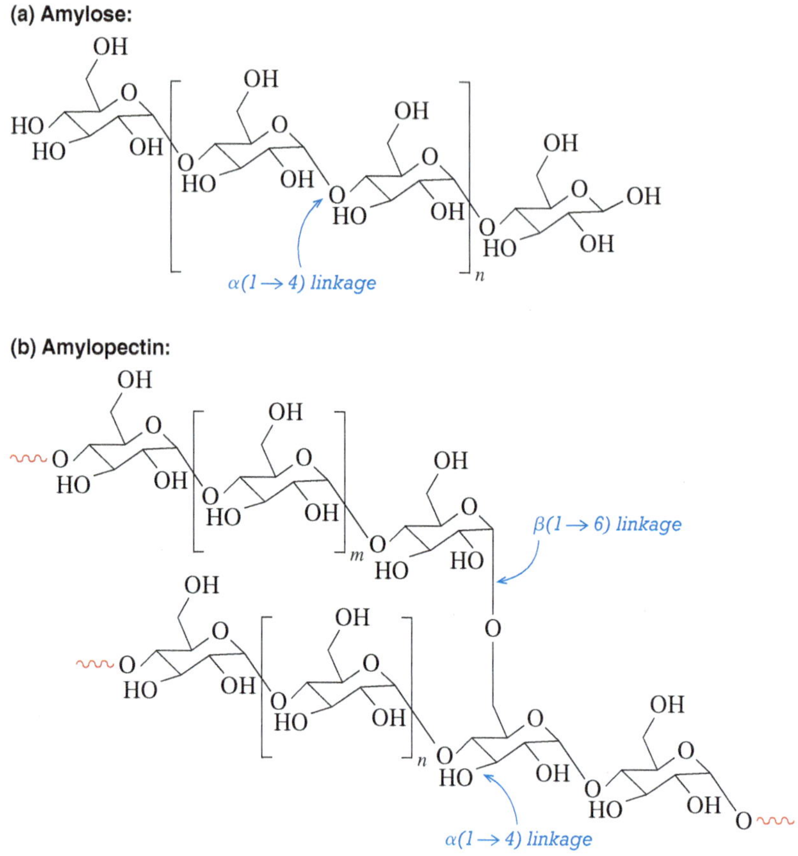

A high-fiber diet is thought to ensure good digestive health. You may wonder about animals that eat mostly grasses and plants, like cows. Cows are able to get energy from cellulose, but not because *they* digest it. Instead, cows, rabbits, horses, and termites have bacteria in their gut that are equipped with enzymes that can digest cellulose. This is an example of a symbiotic relationship: The host (cow) benefits from the work the bacteria do. If you've ever wondered why your dog likes to eat rabbit feces, it's because they contain bacteria-digested cellulose and therefore have the same sweet taste that sugar does for humans. [*Don't believe me? Give it a try!*]

An important polysaccharide found in humans is glycogen, making up 30% of the human liver (**Figure 27.51**). Glycogen is a highly branched polysaccharide used for energy

Figure 27.51 **Glycogen is a branched polysaccharide of glucose.** (a) Structure of glycogen. (b) Glycogen in the liver modulates blood glucose concentration.

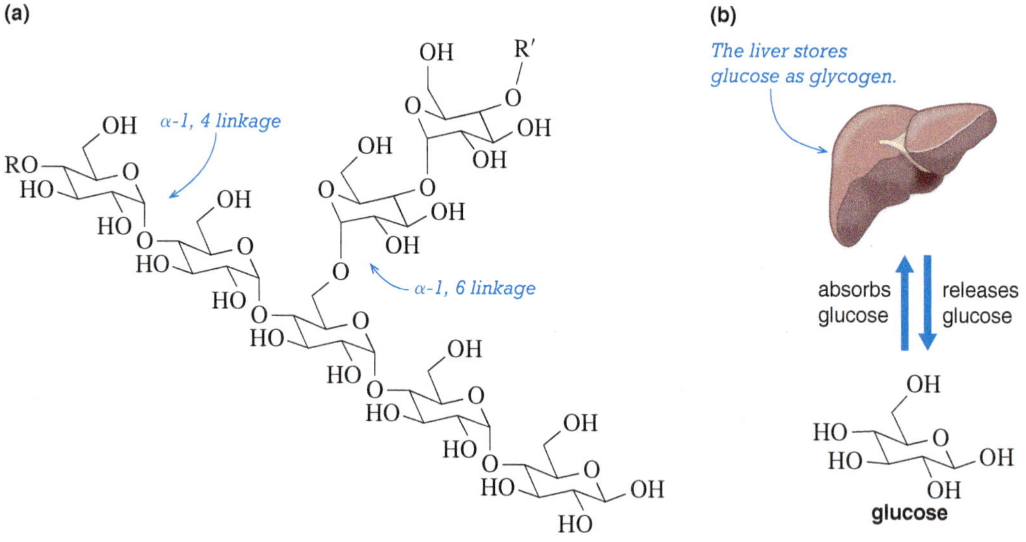

storage. It is used by the liver to maintain optimal blood glucose concentration. After a meal, glucose is stored as glycogen. During periods of starvation, glycogen is converted to glucose. In muscles, glycogen is an available source of glucose to be used for muscle contraction.

27-B CHEMISTRY IN CONTEXT: Glycobiology (Part 2)

Exit of the influenza virus

Once the influenza virus has entered the cell and been replicated, it is reassembled in the host cytoplasm. With its particles reassembled, it buds off the host cell, hoping to find a new cell to promote further infection (**Figure 27.52**). Recall that the virus first recognized, bound to, and entered the host cell through the interaction of hemagglutinin and sialic acid. As the virus is attempting to leave, this strong interaction between hemagglutinin and sialic acid is still present. Thus, the new virus particle is temporarily stuck to the host cell it is trying to escape from.

Recall, too, that besides hemagglutinin, the virus has the protein neuraminidase on its surface. Neuraminidase solves the problem of the virus sticking to the host cell surface by cleaving the sialic acid. It does this through the usual glycoside hydrolysis mechanism, but it uses the active site of the enzyme to lower the activation energy (**Figure 27.53**). Once the sialic acid is cleaved from the cell surface, there is nothing for the hemagglutinin to bind to. Thus, the viral particle is free to leave and infect other cells.

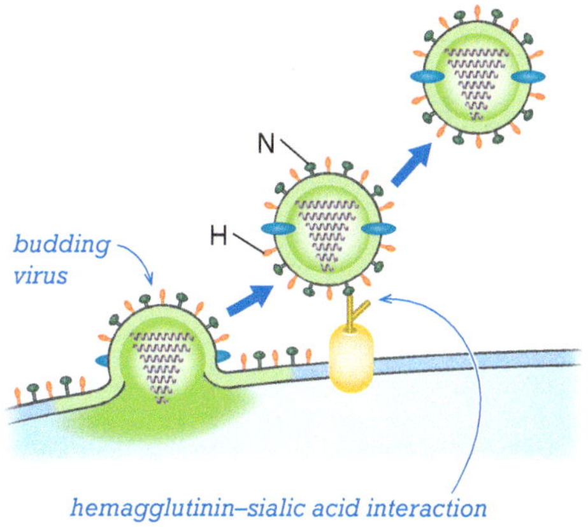

Figure 27.52 After replication, the virus buds off the host cell.

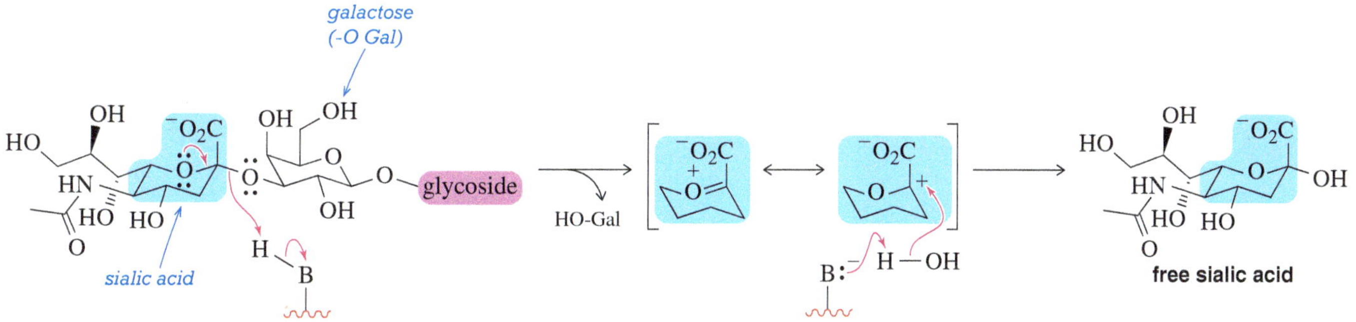

Figure 27.53 The hydrolysis of sialic acid by neuraminidase.

One method of treating influenza is to prevent the virus from budding off the cell by inhibiting the neuraminidase enzyme. Preventing the hydrolysis of the sialic acid from the cell surface makes the virus unable to leave the cell and infect other host cells, so its overall replication is slowed, allowing the immune system of the host to catch up. As you might expect, an inhibitor of the neuraminidase would be structurally similar to sialic acid. This is the case for the widely prescribed drug oseltamivir phosphate (Tamiflu®) (**Figure 27.54**).

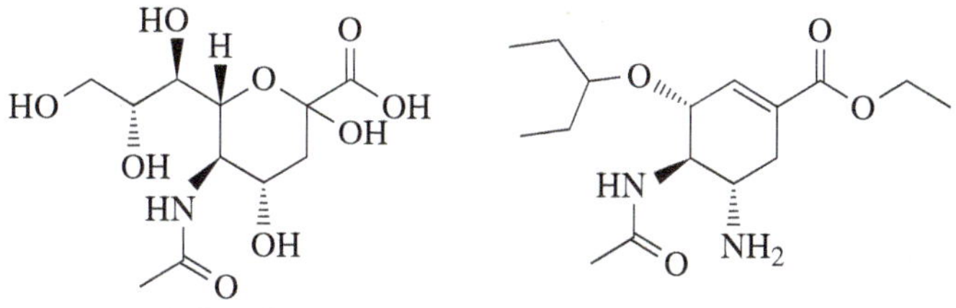

Figure 27.54 Tamiflu is an inhibitor of neuraminidase.

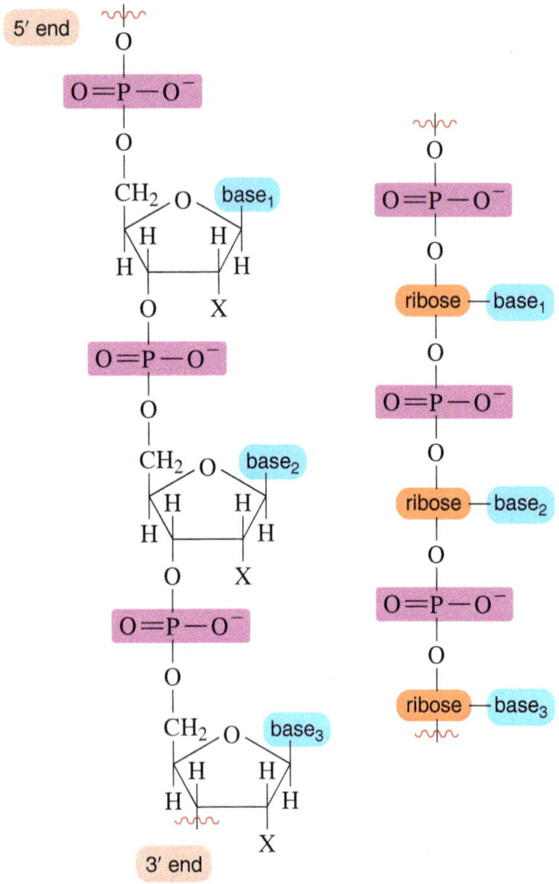

Figure 27.55 **General structure of RNA and DNA.** For RNA, X = OH. For DNA, X = H.

27.8 Nucleic acids

Nucleic acids, the second (of three) biomolecules discussed in this chapter, are biological polymers that are responsible for storing, transcribing, and translating genetic information. All of what makes us unique as humans is contained in our DNA. There are two types of nucleic acids, DNA and RNA (Figure 27.55). DNA is stored in the nucleus of the cell and contains the genetic information. RNA is synthesized by eukaryotic cells in the nucleus using instructions from DNA. Both DNA and RNA are composed of three structural units—namely, a phosphate backbone, a ribose sugar, and a heterocyclic base. The phosphate backbone holds the ribose sugars together, and the riboses each support one of four heterocyclic bases, which are responsible for the genetic code.

27.8.1 Structural units of RNA

Ribonucleic acids (RNA) have D-ribose, an aldopentose, as the lone carbohydrate in the polymer. Compared to the monosaccharides of Sections 27.1–27.7, the anomeric hydroxyl group is replaced with a heteroaromatic ring. This ring is sometimes referred to as a **nitrogenous base**, or the **aglycone** (a general term for any nonsugar molecule attached at the anomeric carbon of a sugar). The di-, oligo-, and polysaccharides discussed so far have been attached through the anomeric carbons. Because of the nitrogenous base, however, the riboses in RNA are connected through phosphate groups. These phosphates connect the 5′-hydroxyl group of one ribose with the 3′-hydroxyl of a second ribose (Figure 27.56). This reaction should remind you of the condensation reaction of an alcohol with a carboxylic acid to make an ester (Section 19.7.2). In this case, the resulting product is a phosphodiester. Note that one hydroxyl group of a phosphate is deprotonated at the neutral pH (~7.4) of the cytoplasm.

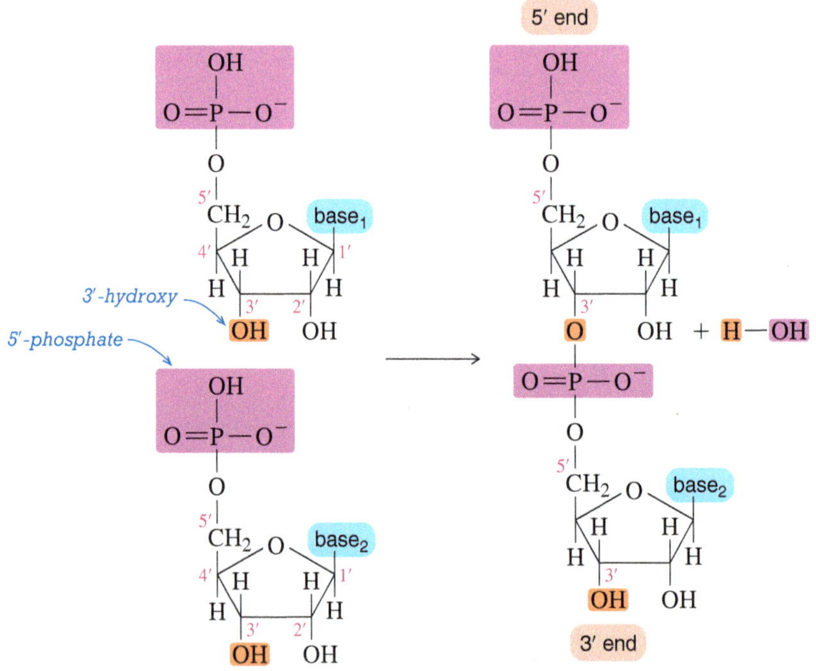

Figure 27.56 **The phosphate backbone is formed by the condensation of a 3′-hydroxyl with a 5′-phosphate.**

Four different heterocyclic bases are found in RNA. Cytosine (C) and uracil (U) are from the pyrimidine family, whereas adenine (A) and guanine (G) are from the purine family (Figure 27.57). The combination of ribose and one of the purine or pyrimidine bases forms what is called a **nucleoside**. The four nucleosides of RNA use the same abbreviations but have slightly different names: adenosine, guanosine, cytidine, and uridine. When the phosphate is attached at the 5′-hydroxyl of these nucleosides, the resulting products are called **nucleotides.** Each nucleotide is named using the nucleoside name plus the location of the monophosphate.

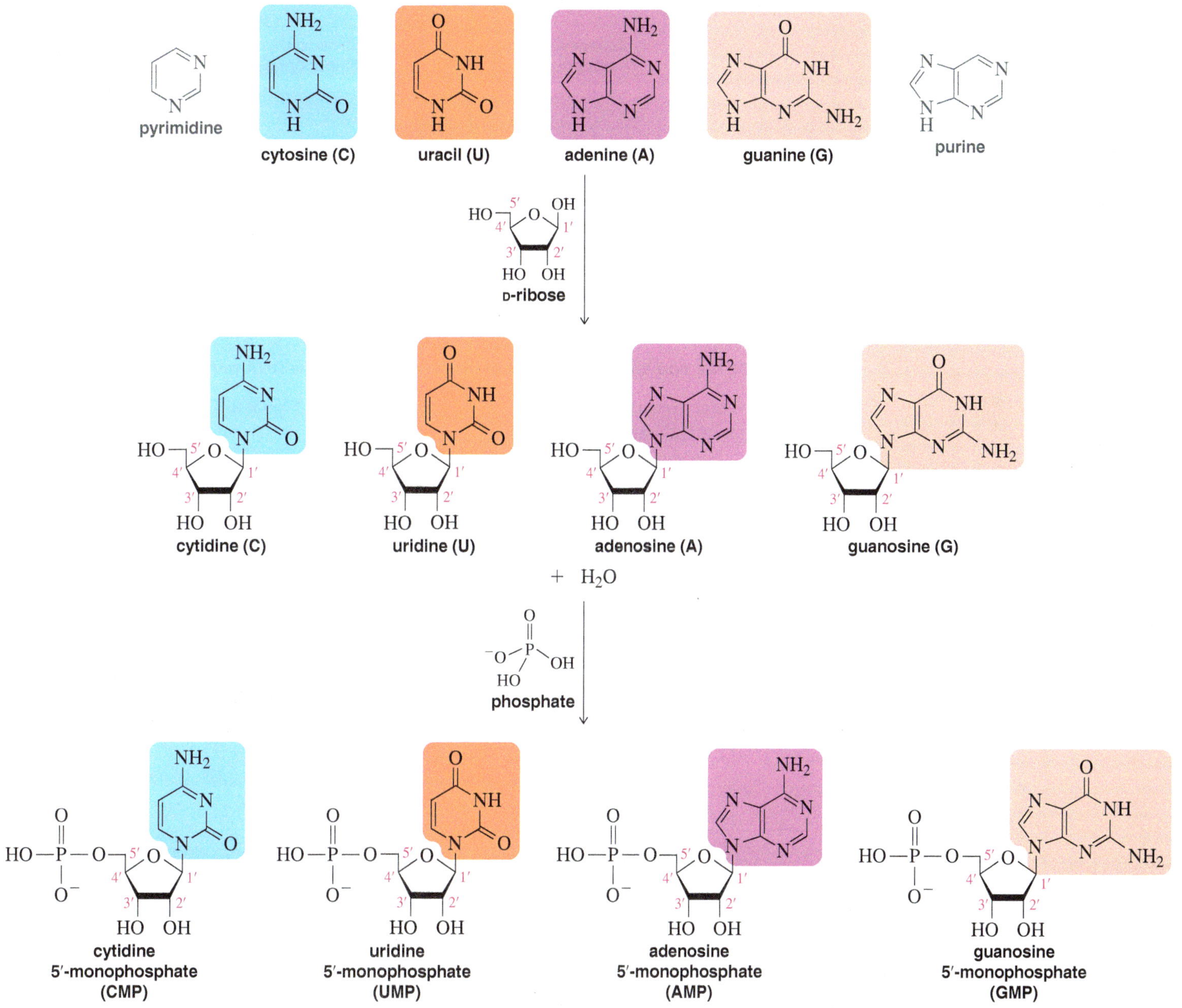

Figure 27.57 **The nomenclature of the nucleotides used to make RNA polymers.**

27.8.2 Structural units of DNA

Deoxyribonucleic acids (DNA) are structurally similar to RNA with two exceptions (**Figure 27.58**). First, the sugar that makes up the polymer backbone is a D-deoxyribose, which differs from the D-ribose of RNA by having a 2′-hydrogen instead of a 2′-hydroxyl group. Second, DNA uses the nitrogenous base thymine (T) in place of uracil. Thymine is from the pyrimidine family, just like uracil, but thymine has a methyl group at C_5, whereas uracil has a hydrogen atom.

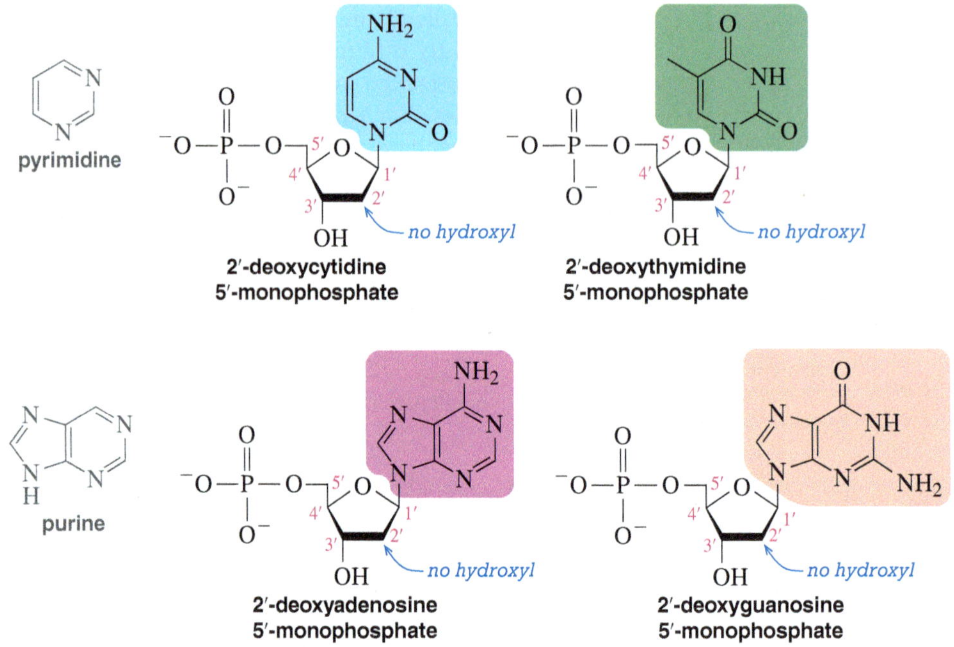

Figure 27.58 The nomenclature of the nucleotides used to make DNA polymers.

The primary structures of DNA and RNA are identical (**Figure 27.59**). Both have $5' \rightarrow 3'$ phosphodiester linkages.

Figure 27.59 The primary structures of (a) RNA and (b) DNA.

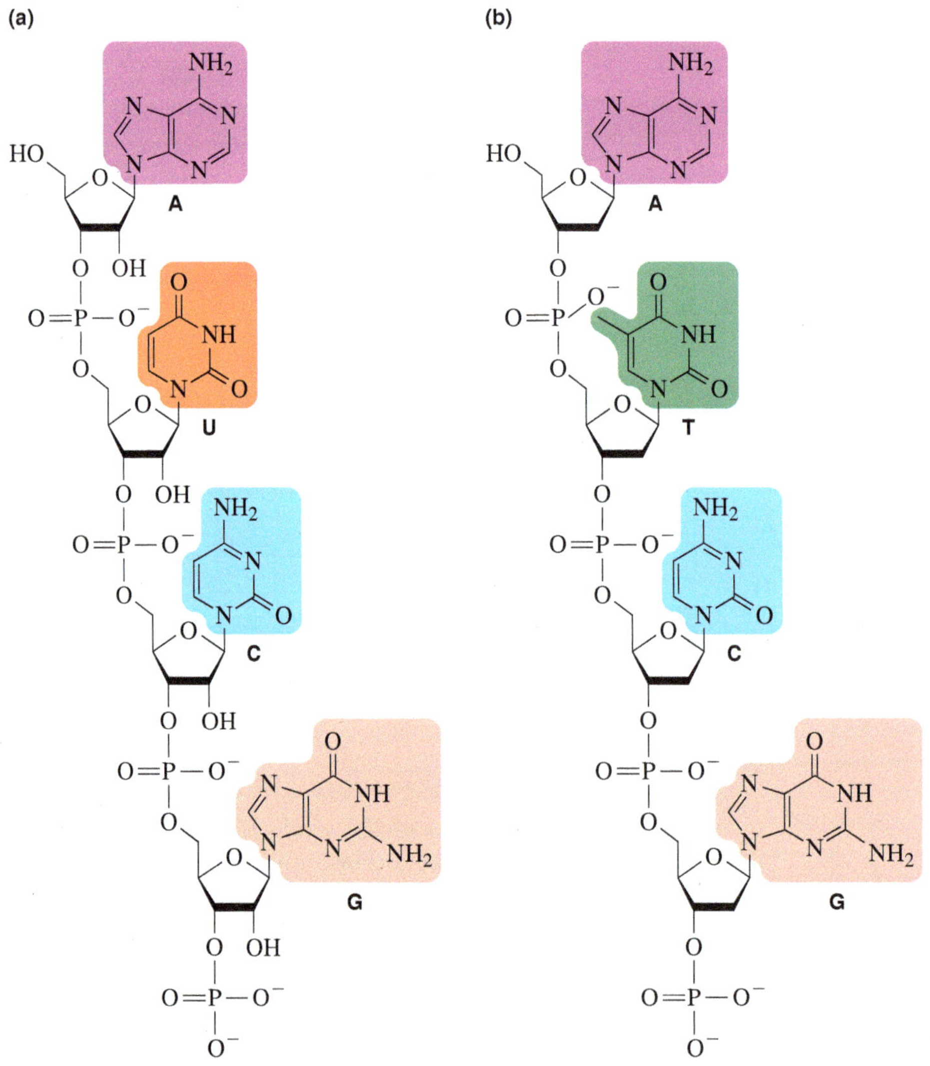

DNA is assumed to have evolved using deoxyribose as the sugar because it makes the phosphate backbone more stable. The 2′-hydroxyl group of RNA (missing in DNA) can cause strand scission according to the mechanism shown in **Figure 27.60**.

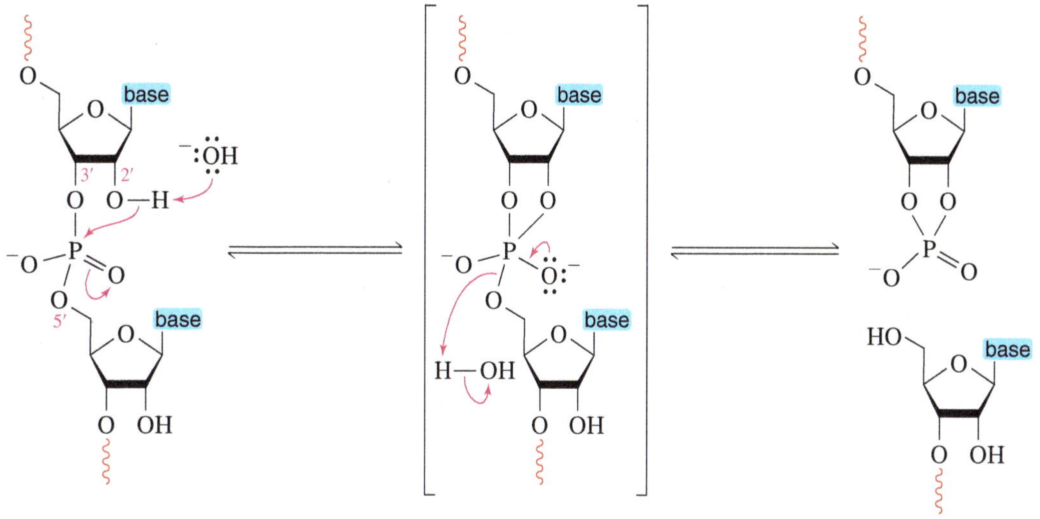

Figure 27.60 Strand scission by the 2′-hydroxyl in RNA is not possible in DNA.

Cytidine (C), by the reverse of a reaction we learned in Section 17.8.1, can be hydrolyzed/deaminated to give uridine (U) (**Figure 27.61**). DNA is believed to use thymidine (T) instead of U because the extra methyl group in T ensures that a deaminated C is never mistaken for T.

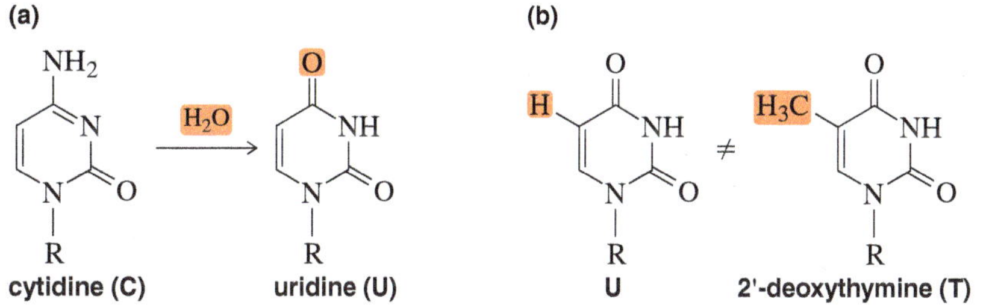

Figure 27.61 Uracil versus thymine. (a) Deamination of cytidine gives uridine. (b) In DNA, deaminated cytidine is different than thymidine.

Assessment 27.43 Show an arrow-pushing mechanism of the acid-catalyzed deamination reaction of cytidine to form uridine. [*See Section 17.8.1 for a review.*]

27.8.3 Base pairing and the double helix

Research into the structure and function of nucleic acids began in earnest in the 1940s once Oswald Avery discovered that inoculating nonvirulent bacteria with a piece of DNA from a virulent strain would induce virulence in the immediate and all future offspring. This experiment suggested, quite conclusively, that DNA plays a role in the genetic makeup of the bacteria.

While Avery's experiment gave insight into the function of DNA, a big leap in structure identification was made when Erwin Chargaff discovered that DNA from identical species always contained a 1:1 ratio of adenine and thymine and a 1:1 ratio of guanine and cytosine. The amount of guanine and adenine might vary from species to species, but the A/T and G/C ratios were always 1:1, as shown in **Table 27.3**.

Table 27.3 The A/T and G/C ratios are always one, regardless of species. The ratio in grasshoppers (order Orthoptera) is shown here.

Purines		Pyrimidines		Ratio
Adenine	29.3%	Thymine	29.3%	A/T = 1
Guanine	20.5%	Cytosine	20.7%	G/C = 0.99

The suggestion that A/T and C/G pairs always formed suggested a mechanism by which information in DNA could be translated, while also giving hints that would eventually lead to its structure determination. This pairing is a result of nitrogenous bases of DNA selectively forming hydrogen-bonded pairs with other bases, as shown in **Figure 27.62**. In each pair, one purine hydrogen bonds with one pyrimidine. Adenine pairs selectively with thymine, and guanine pairs selectively with cytosine. Adenine and thymine each have one hydrogen bond acceptor and one hydrogen bond donor, resulting in a stable intermolecular complex between the two. Guanine has two hydrogen bond donors and one hydrogen bond acceptor, which perfectly complements the two hydrogen bond acceptors and one hydrogen bond donor of cytosine. Moreover [*and this is very important*], the distance between the 2-deoxyribose groups in each hydrogen-bonded pair is identical.

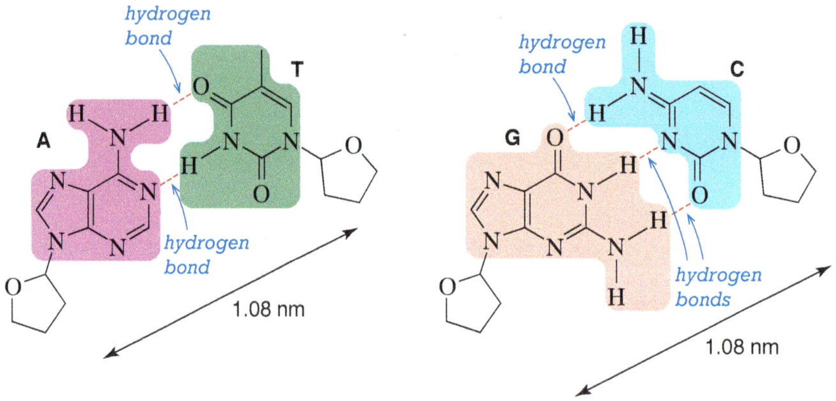

Figure 27.62 Complementary base pairing in DNA. (a) A:::T base pairs form via two hydrogen bonds. (b) G:::C base pairs form via three hydrogen bonds.

In the early 1950s, biologist James Watson and physicist Francis Crick began collaborating on the chemical structure, deemed to be the most important scientific question of the time. Around the same time, Maurice Wilkins and Rosalind Franklin began collecting X-ray crystallographic data of DNA. Upon receiving these data, Watson and Crick began building models, eventually proposing the iconic double helix structure of DNA.

The double helix (**Figure 27.63**) is composed of two DNA strands, running in opposite directions (antiparallel). The double helix is held together only by noncovalent interactions. Most important is the hydrogen-bonded base pairing of adenine/thymine and guanine/cytosine. The structure is further stabilized by the π stacking between the flat aromatic bases. Finally, the phosphate backbone, which lines the outside of the double helix, imparts structural stability through interaction with inorganic cations and water.

Base pairing is also possible for RNA/RNA structures as well as DNA/RNA structures. These are discussed in Section 27.9.

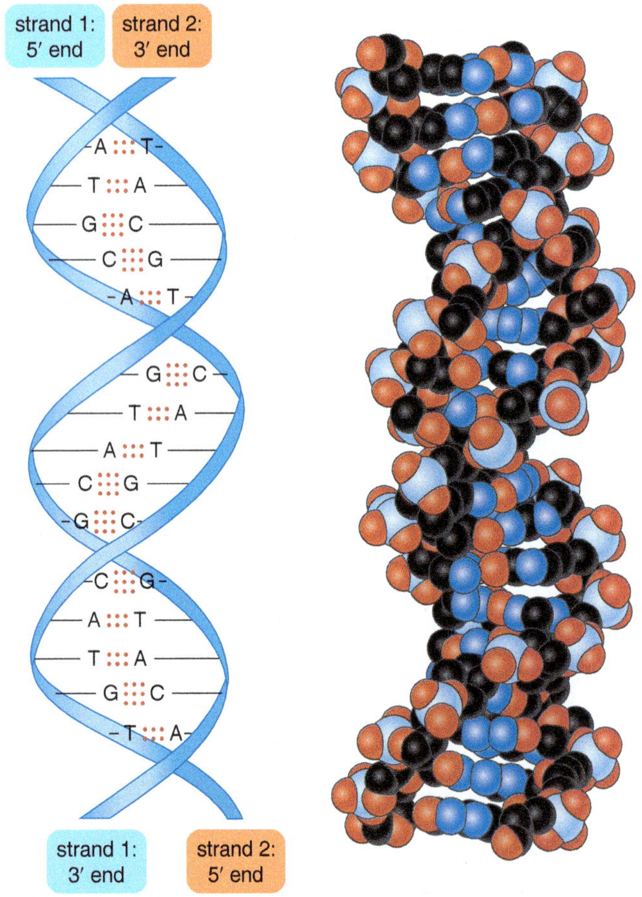

Figure 27.63 The double helix structure of DNA.

27.8.4 DNA replication

The double helix of DNA, together with the complementary base pairing in 1:1 ratios (Table 27.3), shows how genetic information is passed from parent cells to daughter cells in the process of mitosis. The double helix is quite stable, but it is held together only by hydrogen bonds, so it easily uncoils during the process of DNA replication (**Figure 27.64**). As the strand uncoils, DNA polymerase constructs two new strands in the $5' \rightarrow 3'$ direction using each of the original DNA strands as a template. The complementary base pairing ensures that where there was an A/T base pair in the original double helix, there will be an A/T base pair in both of the new helices. Because both chains must grow in the $5' \rightarrow 3'$ direction, the

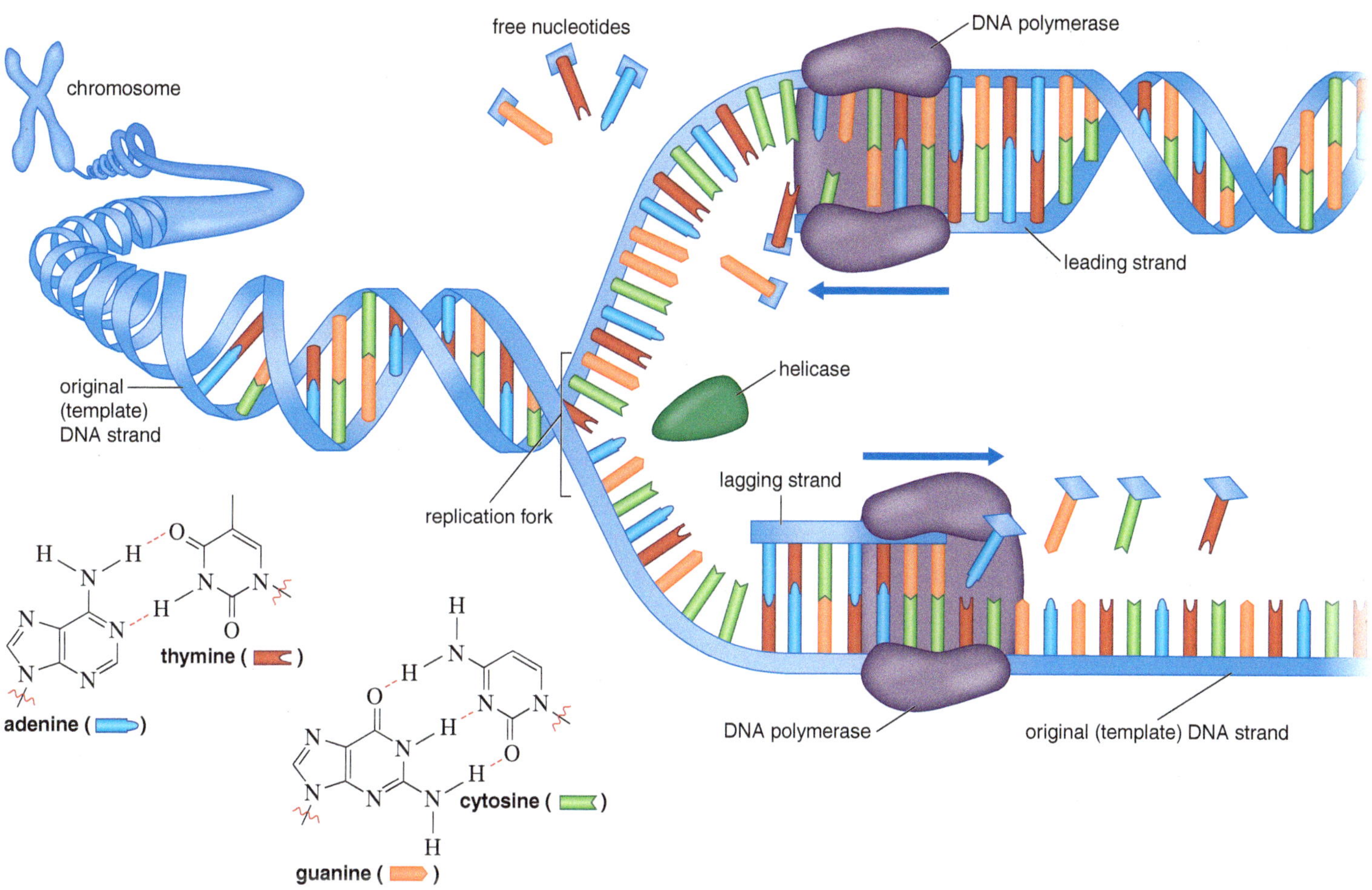

Figure 27.64 DNA replication.

leading strand is copied continuously moving toward the replication fork. The lagging strand, moving away from the replication fork, must be synthesized discontinuously. Each of the newly created helices contains one new strand and one strand from the parent double helix.

Growth of the new DNA strand occurs by the reaction of the 3′-hydroxyl of a deoxyribose reacting with a triphosphate at the 5′ position of the next nucleotide to be added (Figure 27.65). This reaction should remind you of the S_N2 reaction, where the diphosphate group is an excellent leaving group. [*Yes! Buried in all of this biochemistry is real-life organic chemistry!*]

Figure 27.65 The DNA strand grows by a substitution reaction.

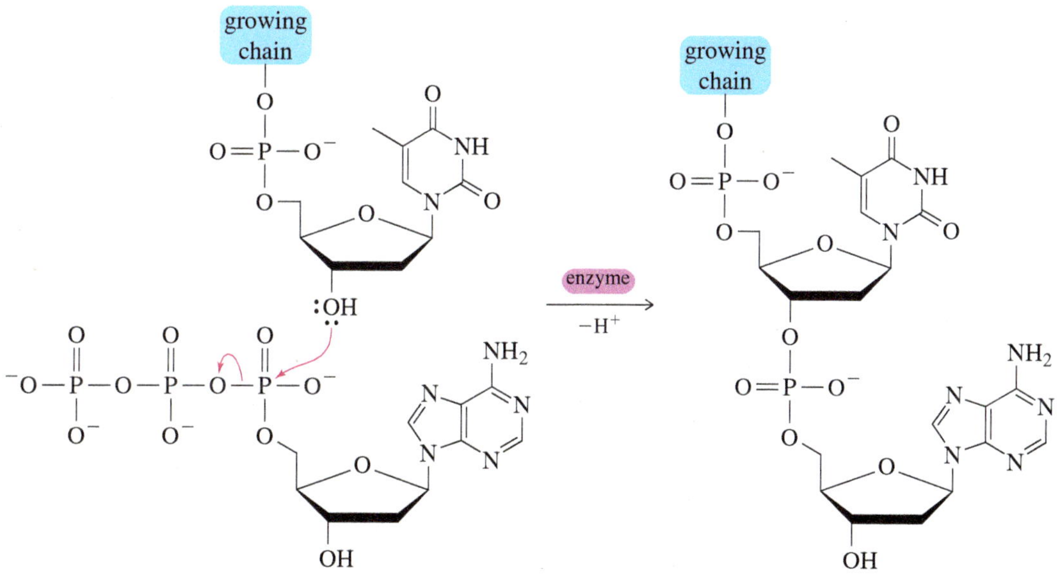

27.9 Central dogma of molecular biology

The central dogma of molecular biology, representing the flow of genetic information, was first coined by Crick. Simplified, it states, "DNA makes RNA, and RNA makes proteins." This implies that while DNA contains the genetic information, it is RNA that translates this information to make proteins. Three types of RNA are needed to synthesize proteins: messenger RNA (mRNA), transfer RNA (tRNA) and ribosomal RNA (rRNA). While most of the subtle details are left to a future biochemistry course, this section shows you a bit about the chemistry involved in protein synthesis—specifically, the manner in which DNA/RNA and RNA/RNA base pairing controls the process.

In a process similar to DNA replication, a double helix unwinds to serve as the template upon which a strand of messenger RNA is created. mRNA is needed to get the information out of the nucleus and into the cytosol where protein synthesis takes place. Using RNA polymerase, the code contained within the DNA is transcribed into RNA, in the same way that a stenographer transcribes the happenings during a jury trial. The code is transcribed by RNA polymerase bringing together RNA nucleotides that complement the DNA base. This complementarity is a result of the complementary base pairing driven by the same hydrogen bonding that formed the DNA double helix (**Figure 27.66**). Because RNA uses uridine in place of thymine, the new RNA strand will contain a U wherever there is an A in the DNA strand template.

Figure 27.66 Using complementary base pairing to synthesize messenger RNA.

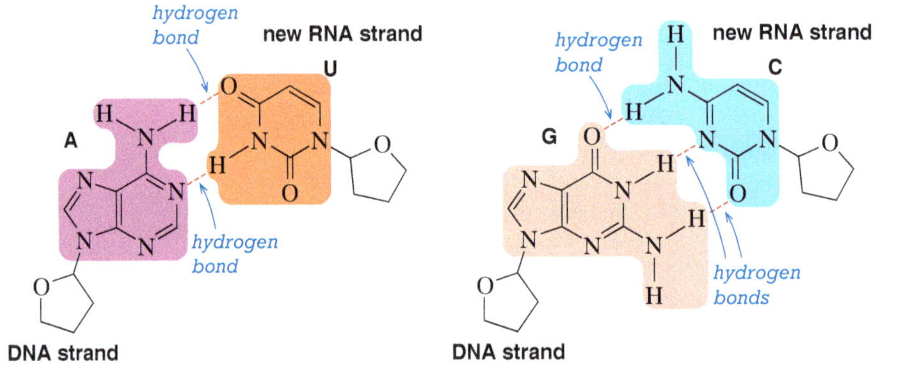

By bringing the new RNA molecules into close proximity with the growing RNA strand, the new RNA strand is built in the $5' \rightarrow 3'$ direction [*similar to DNA; Section 27.8.4*] (**Figure 27.67**).

Figure 27.67 The transcription of DNA into messenger RNA.

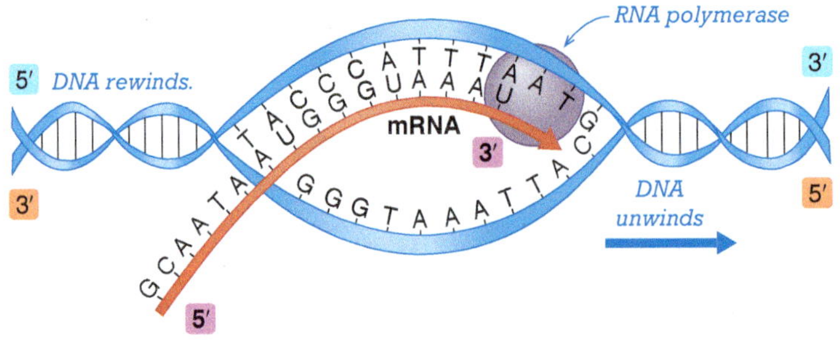

Once formed, messenger RNA exits the cell nucleus and enters the cytoplasm, where it will eventually be translated into a new protein sequence. With only four different nucleotides, you might wonder how this "code" can be translated into a peptide strand that can have up to 20 different amino acids. Within the mRNA, the "code" is contained in repeating three-nucleotide sequences called codons. Each of these codons corresponds to one of the 20 amino acids, as shown in **Figure 27.68**. If you had a strand of mRNA containing 6000 nucleotides, then you could make a protein containing 2000 amino acids. As we learned in Chapter 26, however, proteins don't usually contain that many amino acids. As a result, the messenger RNA also contains start and stop codons that begin and end the extension of the peptide chain when the desired number of amino acids have been connected.

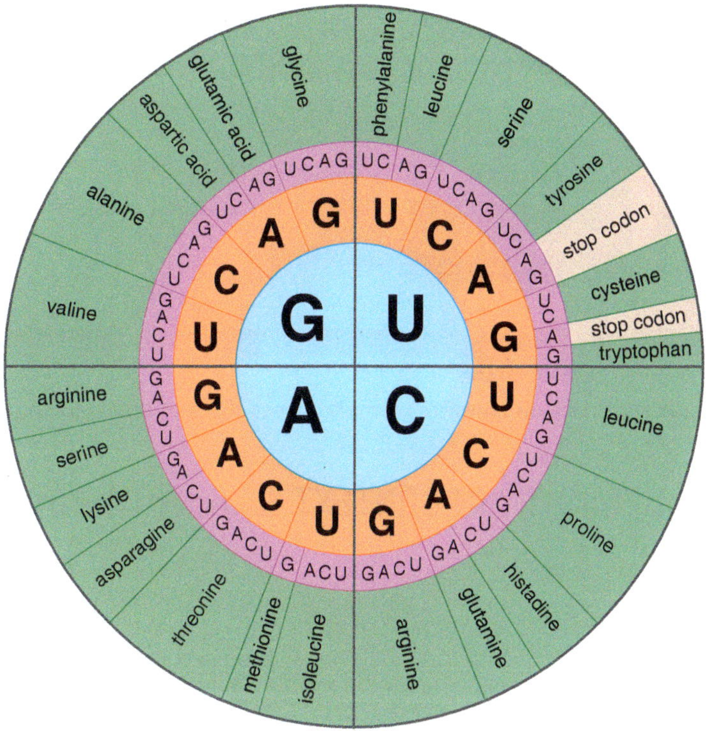

Figure 27.68 The genetic code. Specific three-nucleotide codons call for specific amino acids to be added to a growing peptide. The chart is read by starting at the center and moving toward the edge of the circle (e.g., UAG = stop).

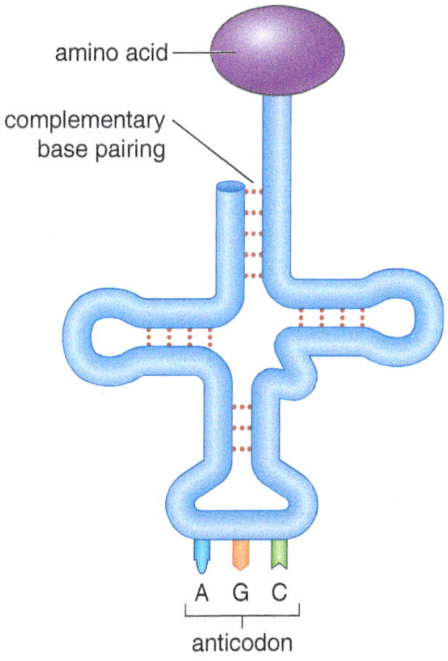

Figure 27.69 Transfer RNA contains an anticodon and an amino acid.

Messenger RNA contains the information needed to synthesize proteins, but it can't make the protein on its own. For this, we need transfer RNA and ribosomal RNA. Transfer RNA is a short sequence of RNA, folded into the clover-like shape shown in **Figure 27.69**. There are two important parts of tRNA. First, at the bottom of the clover in an area called the anticodon loop is a three-nucleotide sequence that is the complement of the three-nucleotide sequence it is to bind to on mRNA. At the top of the clover is an amino acid. This amino acid corresponds to the codon on the mRNA that is the complement of the anticodon on the transfer RNA.

The interaction of mRNA and tRNA takes place at the ribosome. The ribosomal RNA reads the mRNA and brings into the active site the appropriate transfer RNA. **Figure 27.70** shows, for example, that if the mRNA has the UCA codon, then it brings a transfer RNA that has the AGU anticodon. This particular transfer RNA has the amino acid serine at the top. Reading the next codon (AGG) brings in a tRNA (anticodon: UCC) containing arginine. With the serine and arginine in close proximity, a new peptide bond forms. The serine tRNA, without its amino acids, returns to the cytoplasm. The resulting dipeptide, still attached to the arginine tRNA, waits for the next tRNA molecule to come in, and the sequence continues.

Assessment 27.44 Identify the amino acids that correspond to the following codons.

(a) GAC (b) ACA (c) CCC (d) UAC

Assessment 27.45 Show the mechanism and the resulting product of the next reaction to happen in Figure 27.70.

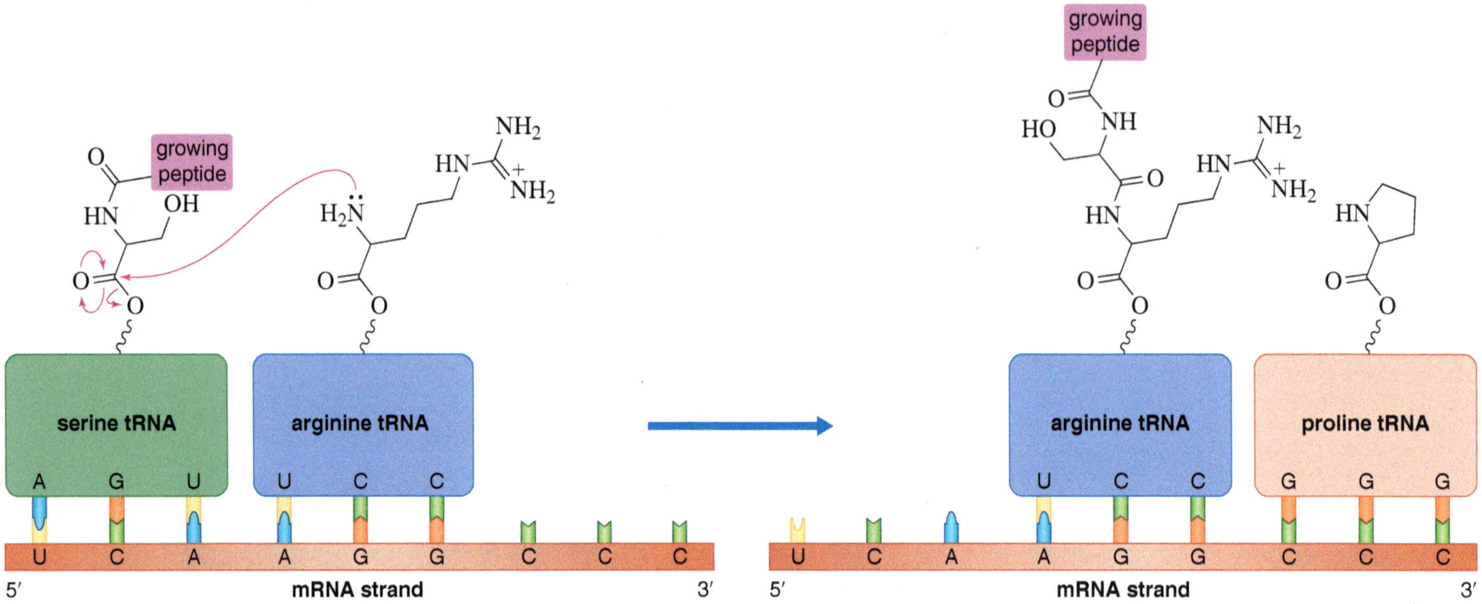

Figure 27.70 **Peptide synthesis on an mRNA template** uses tRNA to attach the appropriate amino acid to the growing peptide chain.

27-C CHEMISTRY IN CONTEXT: Cancer Chemotherapy

DNA alkylation

Cancer, a broad term describing a number of related diseases, is characterized by uncontrolled cell division. Mitotic cell division (**Figure 27.71**) is a process requiring the rapid production and use of DNA. As such, many anticancer drugs interfere with DNA function or synthesis. Because the structure of DNA in human and cancer cells is identical, it is challenging to achieve selectivity in designing anticancer DNA interactive agents.

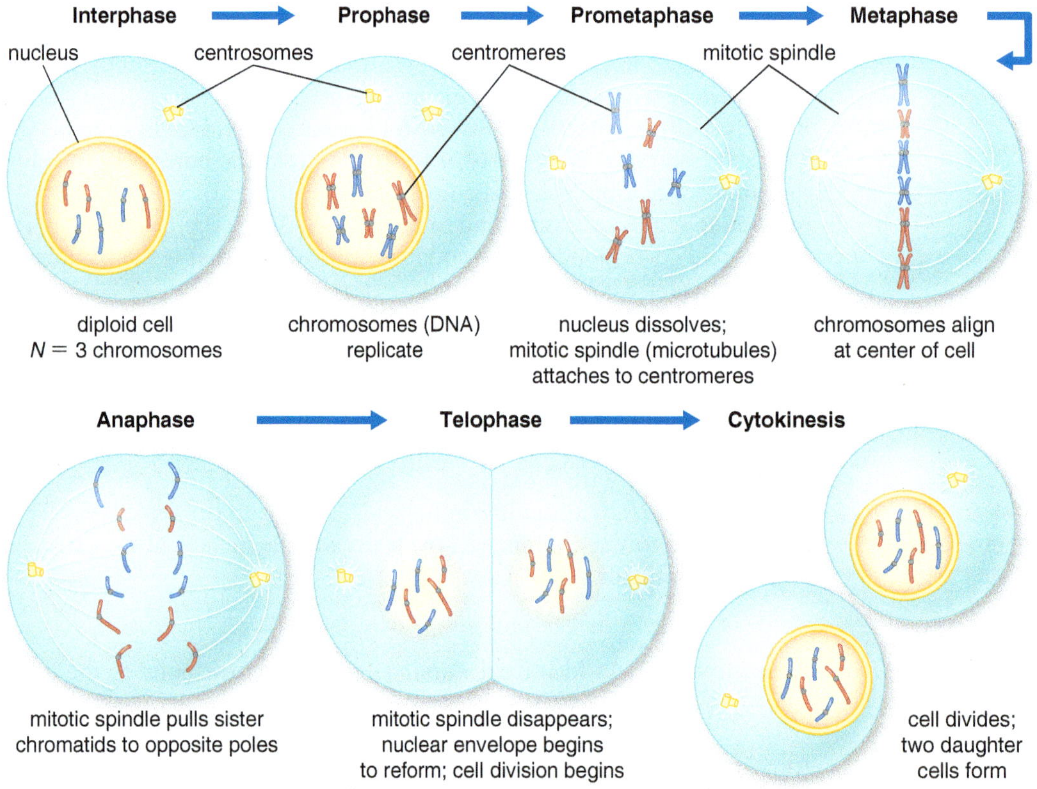

Figure 27.71 **The phases of cell division.**

(continued)

One such method for controlling cell growth in cancerous cells is through the use of nitrogen mustards (Chemistry in Context 12-A–12-C). Nitrogen mustards, derived from sulfur mustards, which were originally used as chemical weapons, act by cross-linking two guanine bases in DNA. Featuring four substitution reactions, this process produces a cross-linked dimer that, through multiple pathways, results in cell death, thereby inhibiting the growth of tumor cells. Chlorambucil (Leukeran®), shown in **Figure 27.72**, has been used to treat various cancers, including chronic lymphocytic leukemia (CLL), Hodgkin's disease, breast cancer, and testicular cancer.

The selectivity for cancer cells, albeit limited, is purely due to the fact that cancer cells, as rapid-growing cells, have a higher need for metabolites. Thus, cancer cells "consume" more metabolites (even drugs) than normal, slow-growing cells. [*Have you ever wondered why chemotherapy causes nausea and hair loss? It's for the same reason!*]

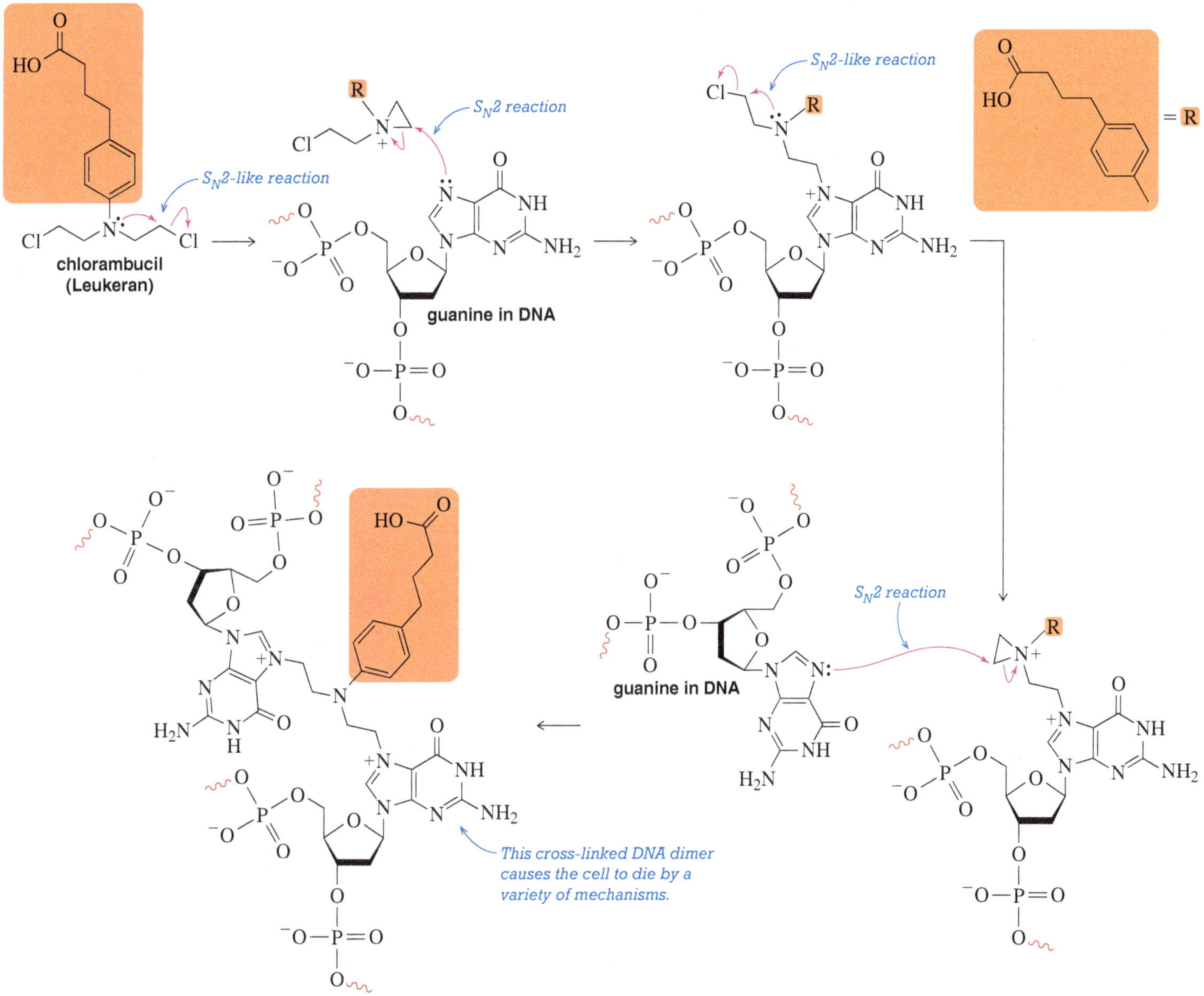

Figure 27.72 **Chlorambucil alkylates DNA twice by a mechanism involving four substitution reactions.**

27.10 Lipids: an overview

Lipids are broadly defined as being any biological molecules that are insoluble in water but soluble in less polar solvents; in other words, lipids are oily, nonpolar substances. While our discussion focuses primarily on triglycerides and their fatty acids, other lipids that are briefly discussed include waxes, prostaglandins, and terpenes (**Figure 27.73**).

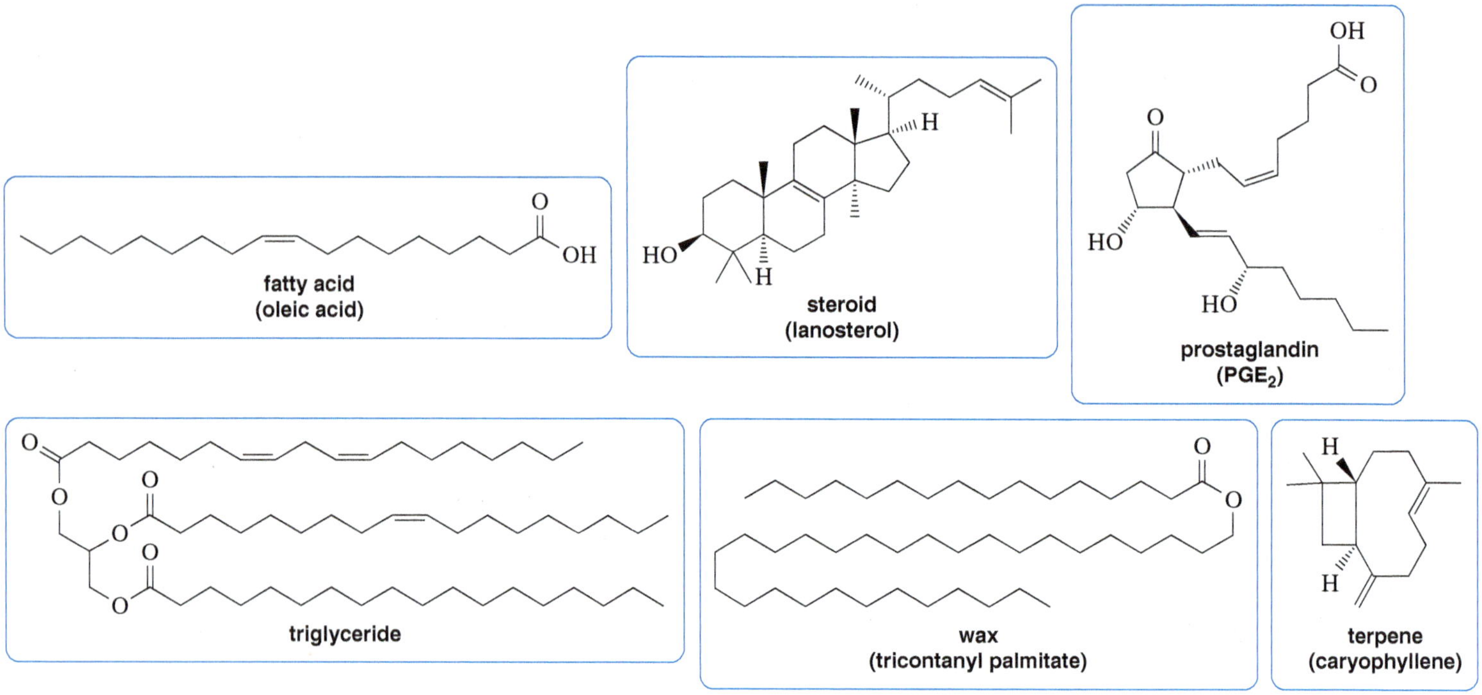

Figure 27.73 Representative structures of lipids—fatty acids, triglycerides, steroids, waxes, terpenes, and prostaglandins.

27.11 Triglycerides and fatty acids

Fatty acids are long-chain carboxylic acids, usually containing an even number of carbons between 4 and 28. Three types of fatty acids are shown in **Figure 27.74**. **Saturated fatty acids** contain the maximum number of hydrogens possible per carbon. Stated another way, saturated fatty acids have no C—C double bonds. **Unsaturated fats** contain at least one double bond. Most unsaturated fatty acids in nature contain a cis-alkene. **Trans unsaturated fats** result from the partial hydrogenation of cis fatty acids. **Polyunsaturated fats** contain multiple double bonds.

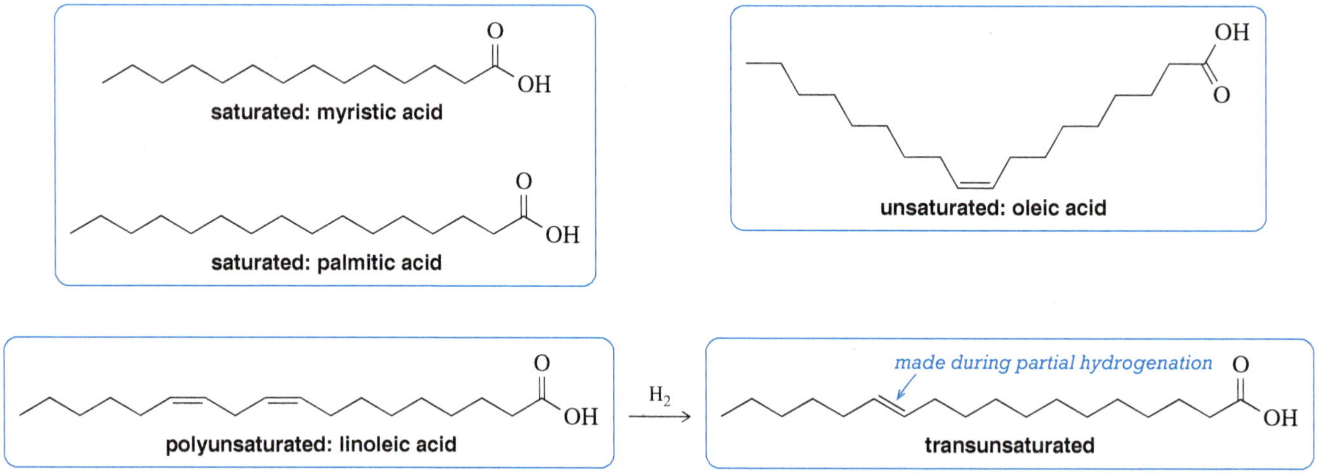

Figure 27.74 The classes of fatty acids include saturated, unsaturated, and polyunsaturated.

Triglycerides are glycerol esters of fatty acids. With three hydroxyl groups, glycerol can accommodate up to three fatty acid groups. A diglyceride has two fatty acids attached, whereas a monoglyceride has just one. Triglycerides form by the esterification reaction of glycerol and fatty acids, which is reminiscent of the Fischer esterification reaction (**Figure 27.75**). Most triglycerides contain a mixture of three different fatty acids. Those that contain three of the same fatty acids are called **homotriglycerides.**

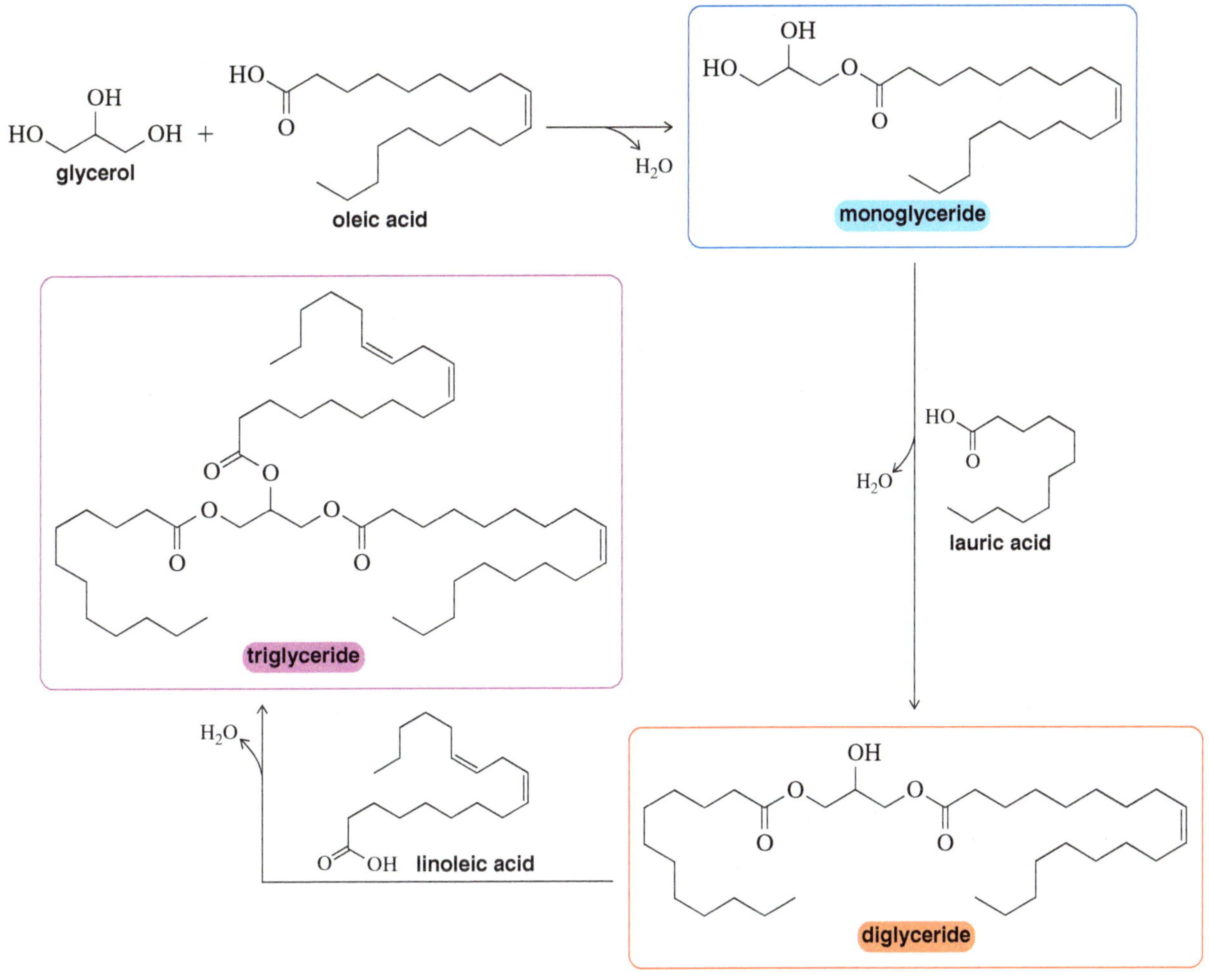

Figure 27.75 The synthesis of mono-, di-, and triglycerides by esterification.

Triglycerides are the primary source of energy storage in mammals. The energy contained in 1 gram of a triglyceride is double the amount of energy available from 1 gram of a protein or carbohydrate.

Assessment 27.46 (i) Identify the following fatty acids as saturated, unsaturated, or polyunsaturated. (ii) If unsaturated, is it cis or trans?

(a)

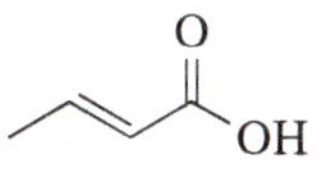

(b)

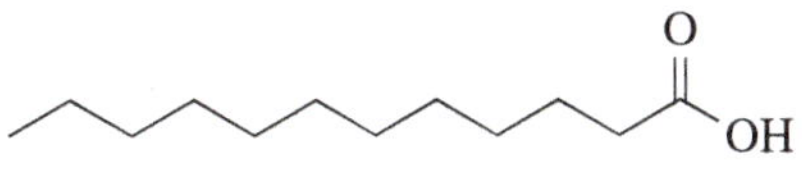

(c)

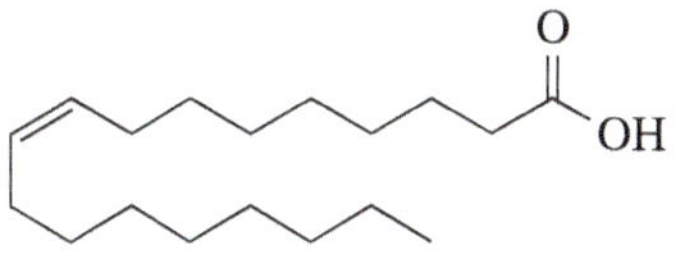

(d)

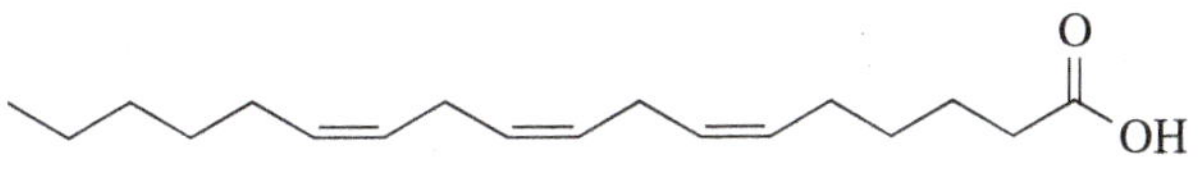

Assessment 27.47 Suggest a mechanism for the Fischer esterification to make the monoglyceride shown. Why is this an unlikely method for monoglyceride synthesis in cells?

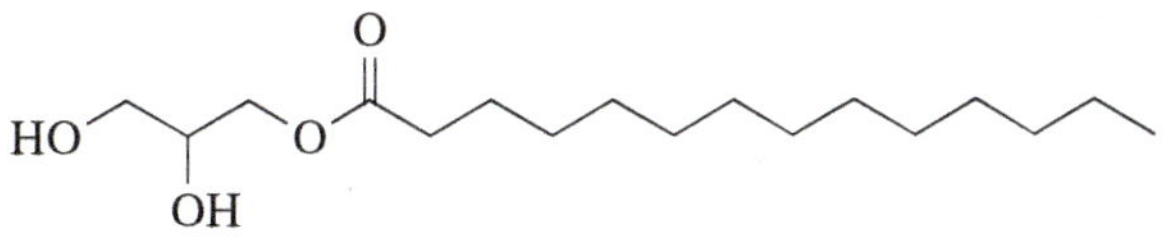

27.11.1 Properties and nomenclature of fatty acids

Table 27.4 shows the structure of some common fatty acids. Fatty acids are typically named based on the vegetable from which they are isolated. Oleic acid, for example, is the primary fatty acid component of olive oil. A homotriglyceride of a particular fatty acid is named by dropping the -ic acid and replacing it with -ate, in the same way that esters are named. As such, oleate is the triglyceride of oleic acid. A shorthand notation has been developed so that the structure of a fatty acid can be described without having to draw it out. The format is $A{:}B\Delta^{x,y,z}$, where A = the number of carbons in the fatty acid, B = the number of double bonds, and x, y, and z represent the lowest numbered carbon of each alkene. C_1 of the fatty acid is, as you would expect, the carboxylic acid carbon. Thus, oleic acid, which has 18 carbons and a double bond at C_9, is written as $18{:}1\Delta^9$. If not indicated otherwise, the assumption is that the alkene is cis.

Table 27.4 Melting points and shorthand notation of common fatty acids.

Name	Number of carbons	Structure	Melting point (°C)	Shorthand notation
Saturated acids				
Lauric acid	12	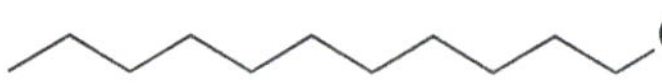 COOH	44	**12:0**
Myristic acid	14	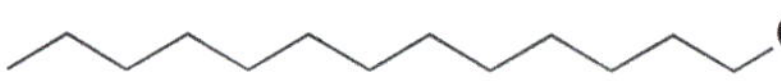 COOH	59	**14:0**
Palmitic acid	16	 COOH	64	**16:0**
Stearic acid	18	 COOH	70	**18:0**
Arachidic acid	20	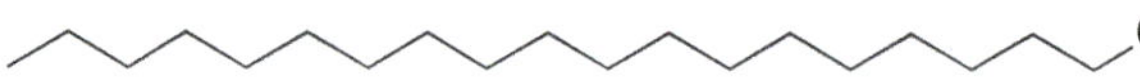 COOH	76	**20:0**
Unsaturated acids				
Oleic acid	18	 COOH	4	$\mathbf{18{:}1\Delta^{9}}$
Linoleic acid	18	 COOH	−5	$\mathbf{18{:}2\Delta^{9,12}}$
Linolenic acid	18	 COOH	−11	$\mathbf{18{:}3\Delta^{9,12,15}}$
Eleostearic acid	18	 COOH	49	$\mathbf{18{:}3\Delta^{9,11,13}}$
Arachidonic acid	20	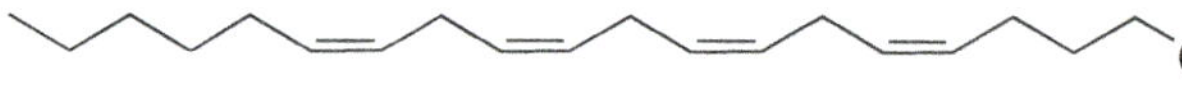 COOH	−49	$\mathbf{20{:}4\Delta^{5,8,11,14}}$

Within the series of saturated fats in Table 27.4, the melting point increases as the fatty acid chain is extended. Most of the saturated fatty acids in Table 27.4 are solids at room temperature. Interestingly, stearic acid (saturated) and oleic acid (unsaturated) both contain 18 carbons, but melt at dramatically different temperatures (Figure 27.76). [*Any idea why?*]

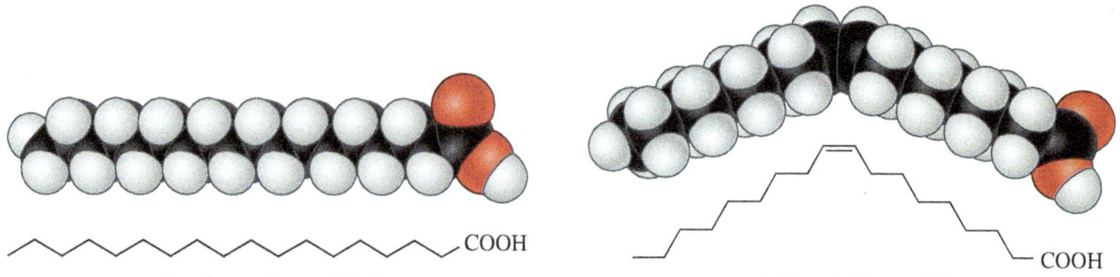

stearic acid, mp 70°C   oleic acid, mp 4°C

Figure 27.76 Stearic acid and oleic acid melt at dramatically different temperatures.

Saturated fatty acids, in their most stable conformation, are relatively linear compounds, so they can pack tightly together, experiencing multiple van der Waals interactions, as shown in **Figure 27.77** for the triglyceride. On the other hand, the cis-alkene of unsaturated fats causes the chain to change direction. This kink prevents the chains from packing tightly, thereby reducing intermolecular interactions and decreasing the melting point. Note that the triglyceride and the fatty acids themselves have nearly equal melting points.

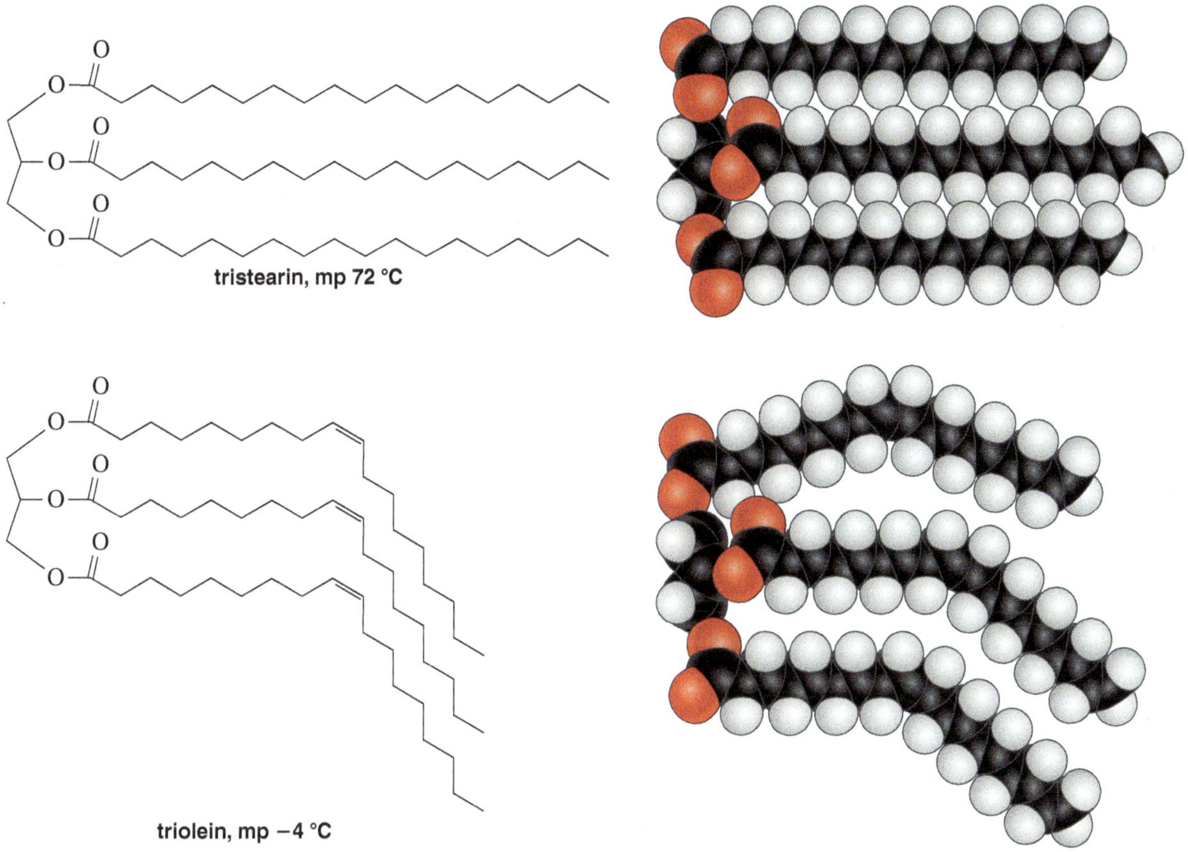

Figure 27.77 Saturated fats melt at higher temperatures because they have more van der Waals interactions between their tightly packed chains.

Assessment 27.48 Name the following fatty acids using the shorthand notation.

(a)

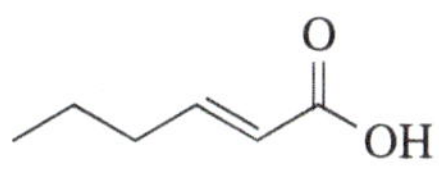

(b)

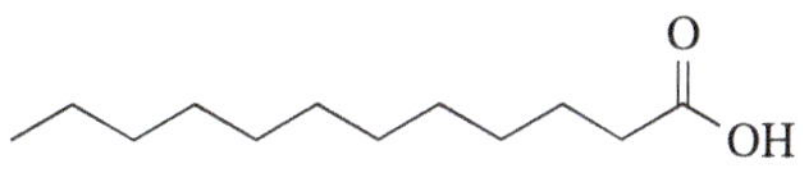

(c)

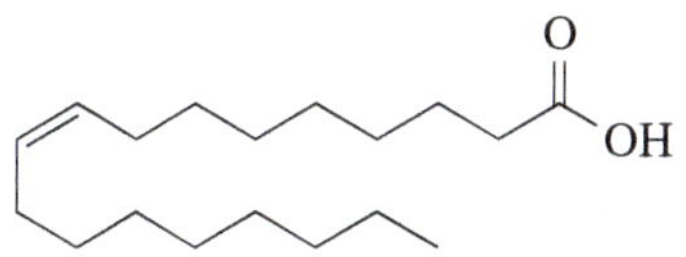

(d)

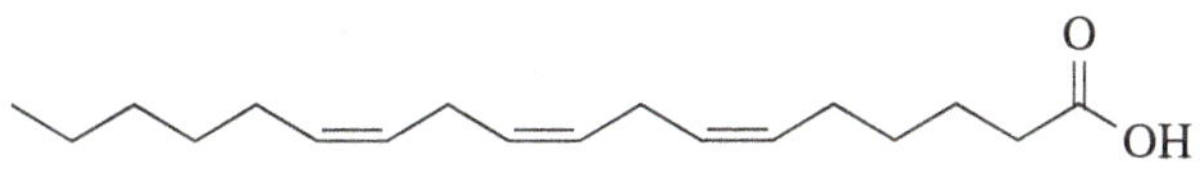

Assessment 27.49 Given the shorthand descriptor, draw the structure of the fatty acid.

(a) $24{:}3\Delta^{9,12,15}$ (b) 14:0 (c) $18{:}2\Delta^{9,12}$

Assessment 27.50 Why is the melting point of a homotriglyceride very similar to the melting point of the fatty acid itself, as seen in the comparison of stearic acid to tristearin in Figures 27.76 and 27.77?

Assessment 27.51 The melting point of elaidic acid (*trans*-oleic acid) is much higher than the melting point of oleic acid. Why?

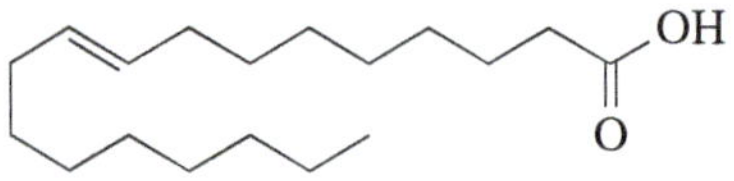

Assessment 27.52 Which of the fatty acids in each of the following pairs would you expect to have the higher melting point?

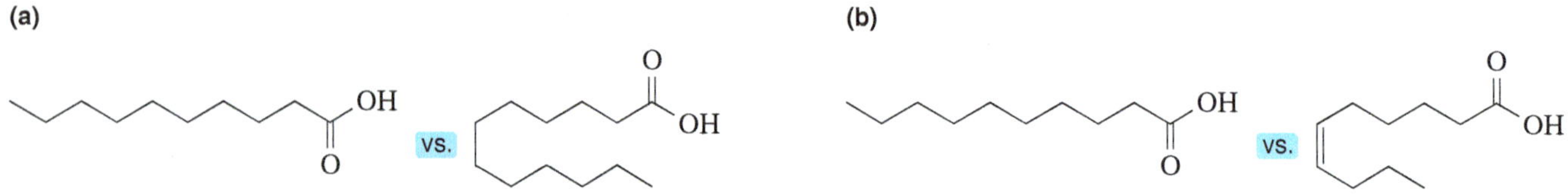

27.11.2 Triglyceride biosynthesis

Fatty acid and triglyceride synthesis occurs in the cytoplasm using a multienzyme complex called fatty acid synthase. We discuss the synthesis of palmitic acid to illustrate the biosynthesis of fatty acids (**Figure 27.78**), starting from acetyl-CoA. Acetyl-CoA is simply a structurally complex thioester, but we can assume it will react like every other thioester and ester we have studied previously. In other words, while the CoA has biological relevance, it is merely a spectator of the reactions we are describing. Similarly, there are several enzymes and cofactors involved over the course of the 42 steps to make palmitic acid. In order to relate these to reactions we have studied, we draw arrow-pushing mechanisms that are reasonable in an organic chemistry class. But keep in mind that the presence of the enzymes and the cofactors is what is allowing the reactions to happen rapidly and efficiently.

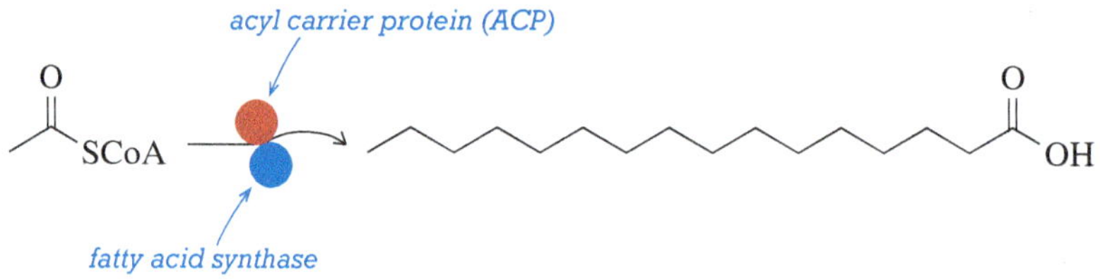

Figure 27.78 Palmitic acid is synthesized using fatty acid synthase in a series of 42 reactions.

Fortunately, we do not have to show all 42 reactions because the same six-step cycle is repeated seven times. The synthesis begins with a two-carbon fragment (acetyl-CoA) and builds the acid two carbons at a time. It takes six steps to add two carbons. Thus, adding a total of 14 carbons means this process has to be repeated seven times. Recall from Section 27.11 that naturally occurring fatty acids always have an even number of carbons. This is consistent with the fact that they are constructed two carbons at a time. As you look at these reactions, focus on what you recognize, because you have seen every reaction previously. In each of the reactions in Figures 27.79–27.85, the red and blue circles represent the dimeric enzyme complex to which the fatty acid is bound while being assembled. Specifically, the red circle represents the acyl carrier protein (ACP) and the blue circle represents the synthase.

Fatty acid biosynthesis steps 1 and 2: Because fatty acid synthesis occurs on the surface of the fatty acid synthase, the first step begins with the attachment of an acetyl group to the synthase enzyme complex (the blue circle in **Figure 27.79**). The exchange of an acetyl group, from HS-CoA through the HS-ACP, ultimately ending up attached to HS-synthase, is simply a transthioesterification reaction. [*HS- is the thiol functional group and not an abbreviation.*] Attack of the ACP thiolate on acetyl-CoA forms a tetrahedral intermediate. Collapse of the electron pair reforms the C—O π bond, kicking out ⁻SCoA. In step 2, attack of the synthase thiolate on the ACP thioester forms a tetrahedral intermediate. Collapse of the electron pair reforms the C—O π bond, kicking out ⁻SACP. [*This is similar to the transesterification reaction; Section 19.7.2.*]

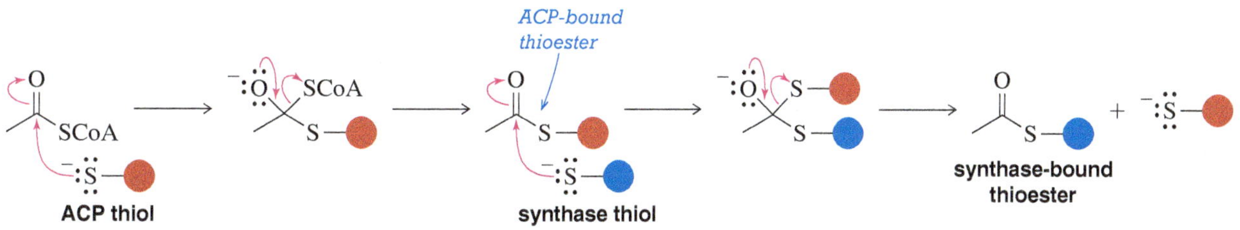

Figure 27.79 Fatty acid biosynthesis steps 1 and 2: Acetyl transfer by consecutive nucleophilic acyl substitution reactions (catalyzed by the enzyme transacylase).

Fatty acid biosynthesis step 3: With the two-carbon acetyl group attached to the enzyme complex, the reaction begins with a Claisen reaction between an ACP-bound (red) malonate derivative and a synthase-bound acetyl-CoA (blue) (**Figure 27.80**). Deprotonation of the activated methylene gives a stabilized enolate that attacks the carbonyl group of the thioether. After temporarily forming the tetrahedral intermediate common to all nucleophilic acyl substitution reactions, collapse of the lone pair eliminates the $^{-}$S-synthase (blue) leaving group. Decarboxylation occurs to give an enolate, which is protonated to form the β-ketothioester. The remaining reactions take place bound to ACP (red).

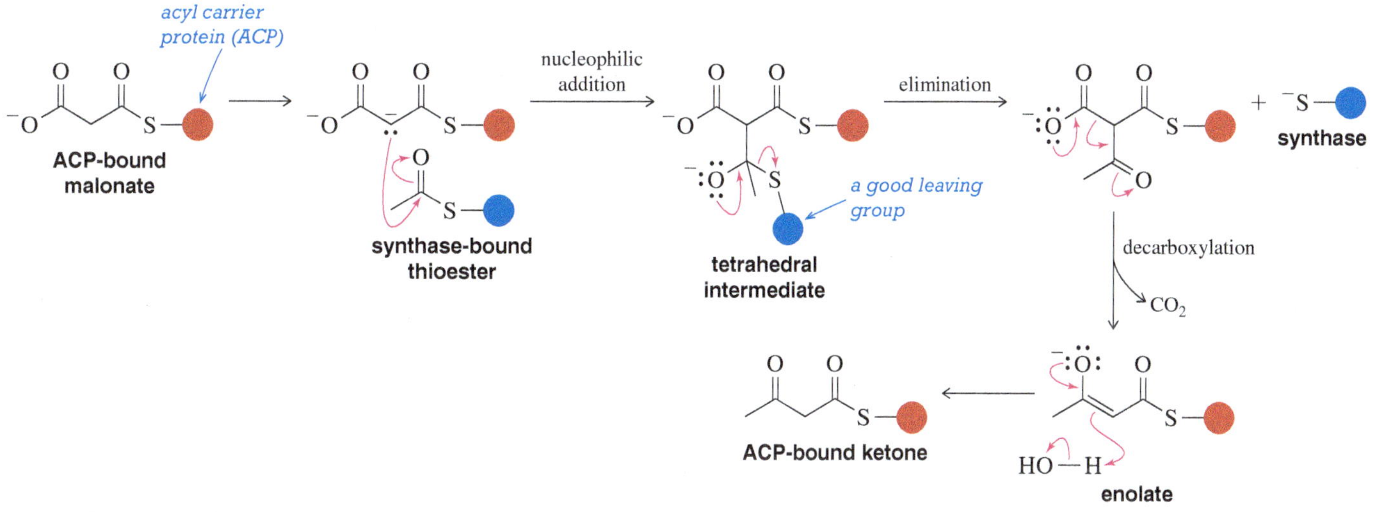

Figure 27.80 Fatty acid biosynthesis step 3: Claisen reaction, decarboxylation, enolate protonation (catalyzed by β-ketoacyl-ACP synthase).

Fatty acid biosynthesis step 4: Reduction of the β-ketothioester is performed using NADPH, an enzyme cofactor that is essentially nature's sodium borohydride. Mechanistically, then, hydride attack on the carbonyl is followed by protonation of the resulting alkoxide (**Figure 27.81**). [*This is similar to the sodium borohydride reduction of aldehydes and ketones; Section 17.9.2.*]

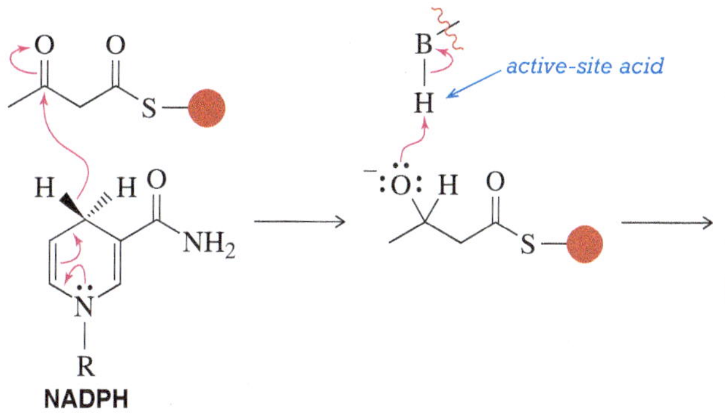

Figure 27.81 Fatty acid biosynthesis step 4: Carbonyl reduction and protonation of the resulting alkoxide (catalyzed by β-ketoacyl-ACP reductase).

Fatty acid biosynthesis step 5: Elimination of the hydroxyl group occurs by a preliminary removal of the α-proton to make an enolate (**Figure 27.82**). Collapse of the enolate pushes the π electrons to the β-carbon, kicking out hydroxide as a leaving group. The resulting product is a conjugated α,β-unsaturated thioester [*Section 21.1.2*]. Although hydroxide is not usually a great leaving group, this occurs at the active site of an enzyme, so there is probably a nearby proton that the hydroxide can grab on its way out. Additionally, the conjugated alkene is a relatively stable product, thus providing an additional driving force for the elimination. [*This reaction is similar to the last step in the base-catalyzed aldol reaction at high temperature (with heating); Section 20.3.2.*]

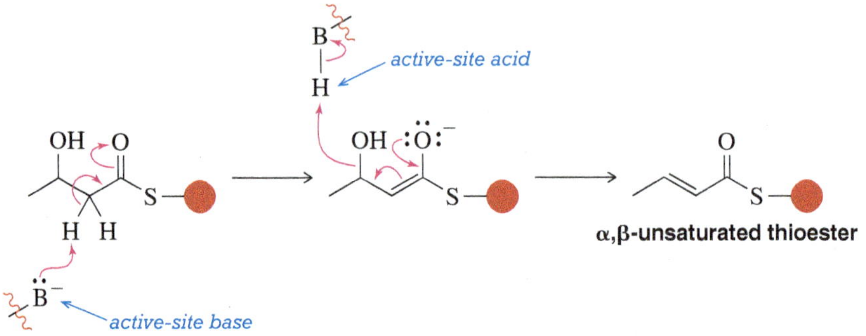

Figure 27.82 Fatty acid biosynthesis step 5: Elimination of water to form an α,β-unsaturated thioester (catalyzed by β-3-hydroxyacyl-ACP dehydrase).

Fatty acid biosynthesis step 6: Reduction of the α,β-unsaturated thioester once more uses "nature's sodium borohydride" (**Figure 27.83**). A hydride ion from NADPH adds via a conjugate addition to the β-carbon, pushing the π electrons to the neighboring carbonyl. The result of this first step is an enolate. In the presence of an acidic proton (likely enzyme bound), protonation of the enolate gives the fully saturated thioester. [*This reaction is essentially a conjugate addition reaction: Section 21.5.*]

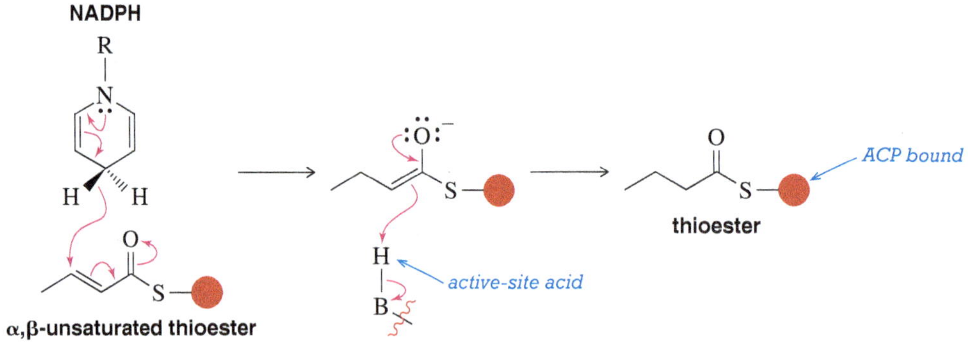

Figure 27.83 Fatty acid biosynthesis step 6: Conjugate addition of hydride followed by enolate quench (catalyzed by 2,3-*trans*-enoyl-ACP reductase).

Fatty acid biosynthesis step 2 (repeated): With the initial two-carbon thioester having been converted to a new four-carbon thioester [*confirm this*], the cycle is repeated six more times. Thus, step 2 is repeated here, transferring the ACP-bound (red) thioester to synthase (blue). As with Figure 27.79, the transfer occurs by nucleophilic acyl substitution through a tetrahedral intermediate. In the process, the ACP-thiol is freed up to bring in another malonate so that steps 3–6 can be repeated (**Figure 27.84**).

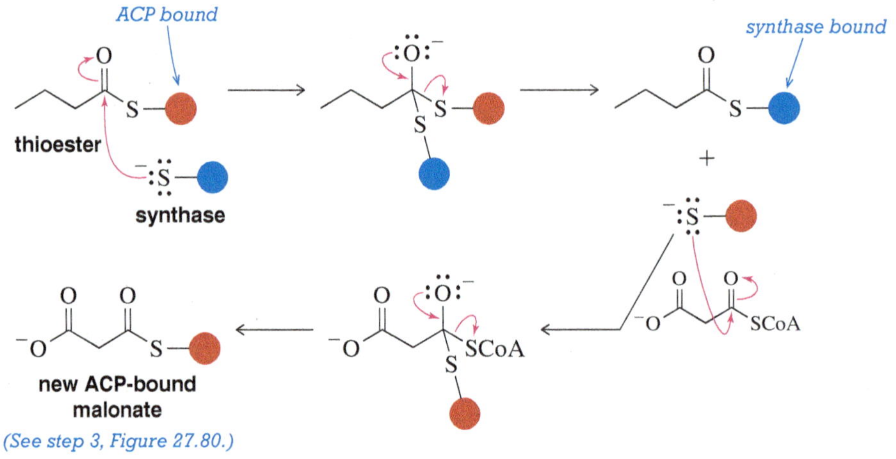

Figure 27.84 Fatty acid biosynthesis step 2 (repeated): Acetyl transfer by nucleophilic acyl substitution (catalyzed by the enzyme transacylase).

Fatty acid biosynthesis termination step: This cycle is repeated, adding two carbons at a time, until the thioester chain contains 16 carbons (**Figure 27.85**). At this point, instead of going through the cycle again, the thioester is hydrolyzed with water, producing palmitic acid, freeing up the ACP-thiol to restart the process with another molecule of acetyl-CoA. [*This reaction is similar to ester hydrolysis; Section 19.7.1.*]

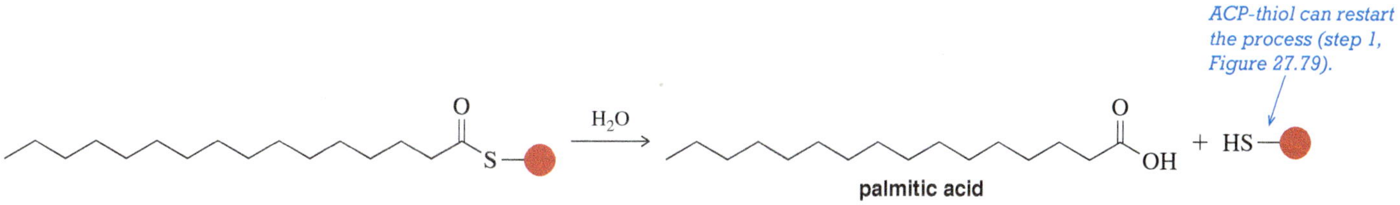

Figure 27.85 Fatty acid biosynthesis termination step: Thioester hydrolysis by nucleophilic acyl substitution (not catalyzed by an enzyme).

With fatty acid synthesis complete, all that remains is to attach three fatty acids to glycerol to make the homotriglyceride palmitate. In order to be incorporated into a triglyceride, the fatty acid must be activated for nucleophilic acyl substitution. This is done by converting it into a thioester, as shown in **Figure 27.86**. In the first step, the hydroxyl group is converted into a better leaving group by reaction with adenosine triphosphate. [*This is the equivalent of turning a hydroxyl group into a tosylate (Section 13.7.1) or making an acid chloride (Section 18.7.5).*] Once the carboxylic acid is activated, the thiol of coenzyme A attacks the carbonyl, forming the tetrahedral intermediate. Collapse of the tetrahedral intermediate results in the fatty acyl-CoA.

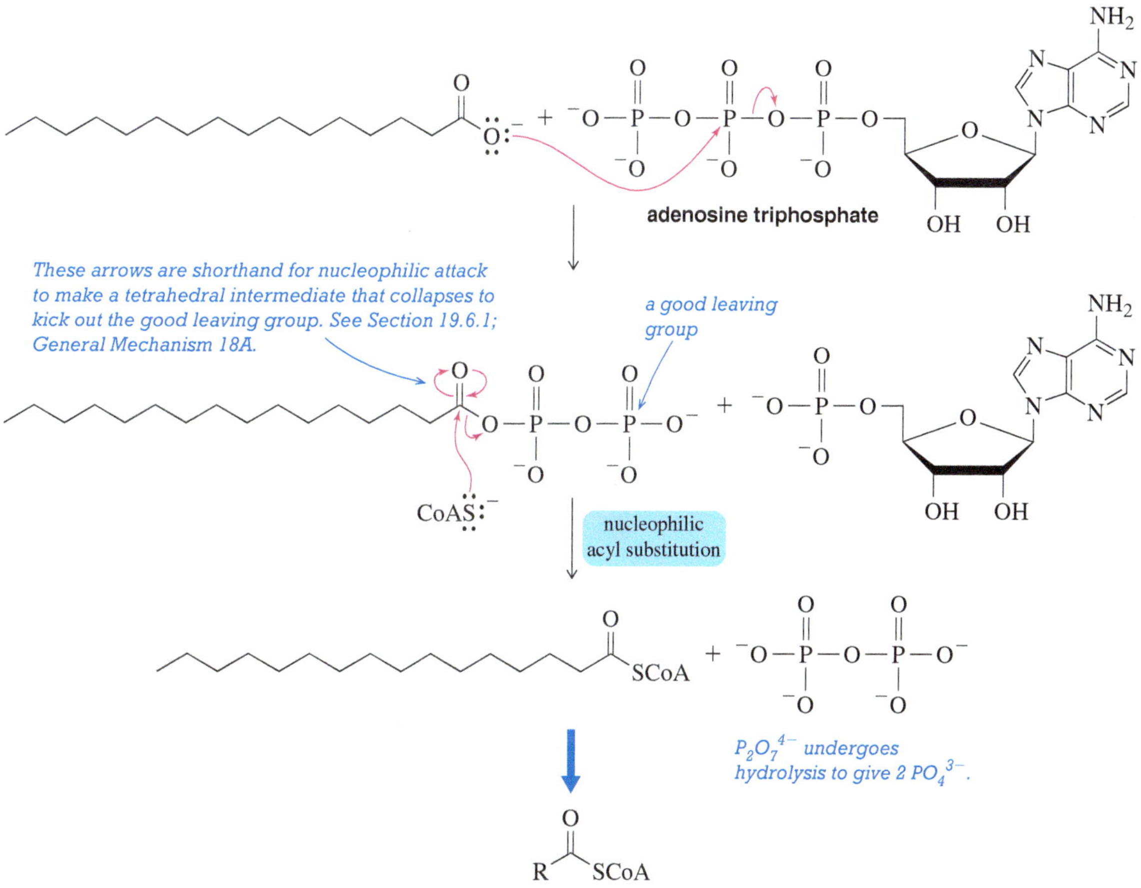

Figure 27.86 Activation of the fatty acid for attachment to glycerol.

To attach the fatty acid to glycerol, glycerol is first phosphorylated in the liver, blocking one of the hydroxyl groups from acylation. In a nucleophilic acyl substation, the first and second fatty acid groups are attached, as shown in **Figure 27.87**. Each reaction proceeds through a tetrahedral intermediate, as we studied in Section 19.6. The result of these first two steps is not a triglyceride, but instead a phosphatidic acid, which is used to make the phospholipids that compose the bilayer of cell membranes.

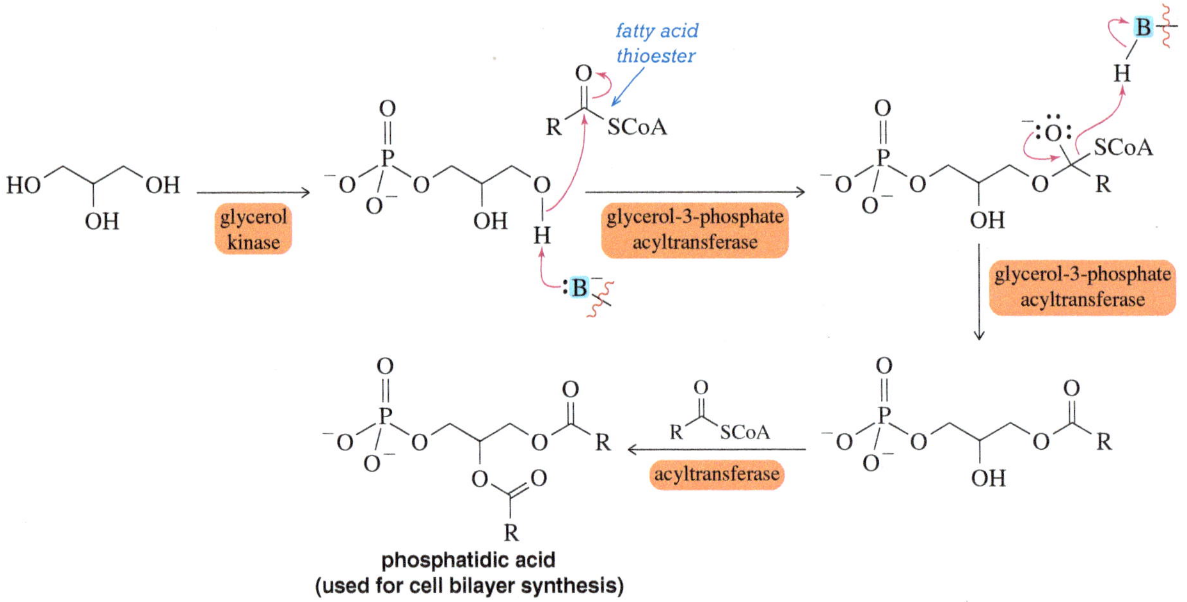

Figure 27.87 The attachment of two fatty acid groups to glycerol phosphate gives an intermediate that is either converted to a triglyceride or used for the synthesis of phospholipids. [*R = the long alkyl chain from Figure 27.86.*]

In order to add a third fatty acid, the phosphate group is removed via a hydrolysis reaction (essentially S_N2). The resulting diglyceride can be used as is, it can add a third fatty acid by nucleophilic acyl substitution to make a triglyceride, or it can have one of the fatty acid groups removed to produce a monoglyceride (**Figure 27.88**). Although we have illustrated this process using only palmitate, human cells usually incorporate palmitate at C_1 and oleate at C_2.

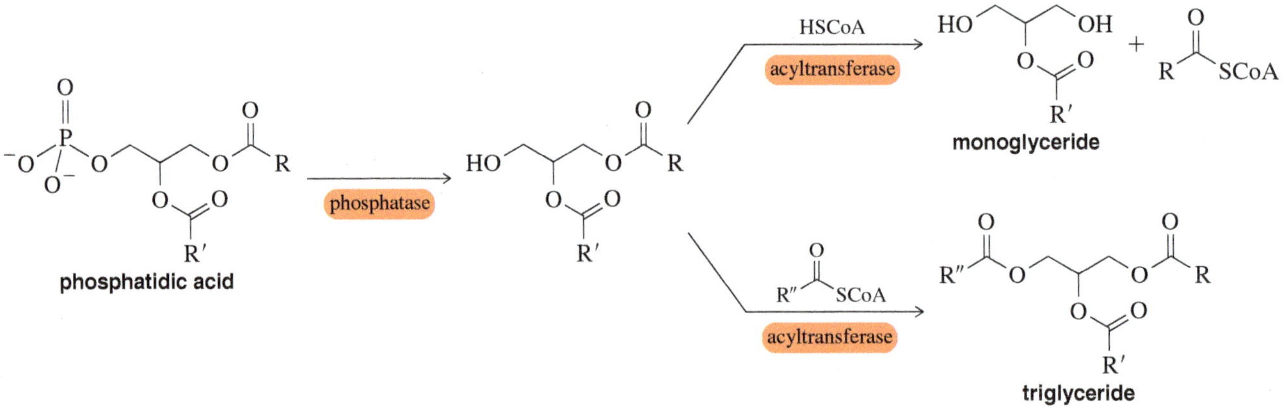

Figure 27.88 The diglyceride can be converted to a monoglyceride or a triglyceride.

27.11.2 SUMMARY

- Palmitic acid is synthesized using fatty acid synthase in a series of 42 reactions composed of the same six-step cycle repeated seven times.

Assessment 27.53 How would the synthesis of stearic acid (Table 27.4) differ from the synthesis of palmitic acid?

Assessment 27.54 Why do most fatty acids contain an even number of carbons?

Assessment 27.55 Provide an arrow-pushing mechanism for the conversion of the diglyceride in Figure 27.88 to the monoglyceride.

Assessment 27.56 Beginning with the ACP-bound malonate and the synthase-bound thioester (both in Figure 27.84), show steps 3 through 6 to finish with the six-carbon thioester bound to ACP.

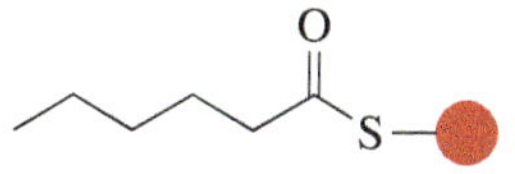

Assessment 27.57 Suggest an arrow-pushing mechanism for the fatty acid termination step shown in Figure 27.85.

27.12 Triglyceride reactions

The reactions of triglycerides are all closely related to reactions we have studied previously. The molecules may be larger, but with only two functional groups (alkenes and esters), these reactions should be easily recognizable.

27.12.1 Hydrogenation to make fats

Vegetable oils, often extracted from the seeds of plants like corn and soybean, are unsaturated triglycerides, meaning they are composed of glycerol and three long-chain fatty acids. Most naturally occurring vegetable oils are unsaturated, containing only cis-alkenes. These cis double bonds prevent tight packing of the triglyceride chains, breaking up intermolecular interactions and resulting in most vegetable oils being liquid at room temperature. Soybean oil is composed principally of linoleic, oleic, and linolenic acids (**Figure 27.89**). These oils have many uses; perhaps most important among them is their use in cooking.

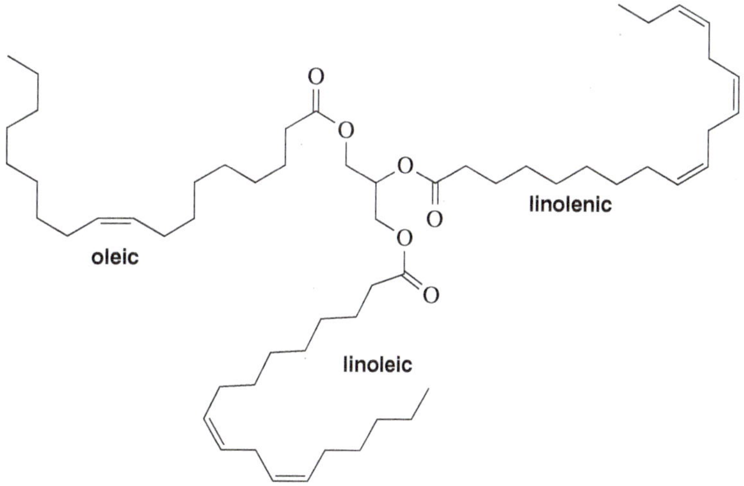

Figure 27.89 Soybean oil, a vegetable oil, is an unsaturated triglyceride.

While unsaturated fats are relatively healthy, the cis-alkene present imparts oxidative instability as a result of the diminished stability of cis versus trans double bonds. Moreover, polyunsaturated fats possess at least one doubly allylic carbon (between two double bonds), making it susceptible to radical formation. [*See Section 11.6.*] As a result, more stable saturated fats, typically procured from animals, found greater utility for use in cooking in the first half of the 20th century. Soybean oil can be converted to a saturated fat (**Figure 27.90**) [*What's the difference between an oil and a fat? Oil is a liquid, whereas fat is a solid*] by the same hydrogenation reaction we studied in Section 9.2. Removal of the alkenes not only imparts greater stability, but also raises the melting point, making the product a solid fat.

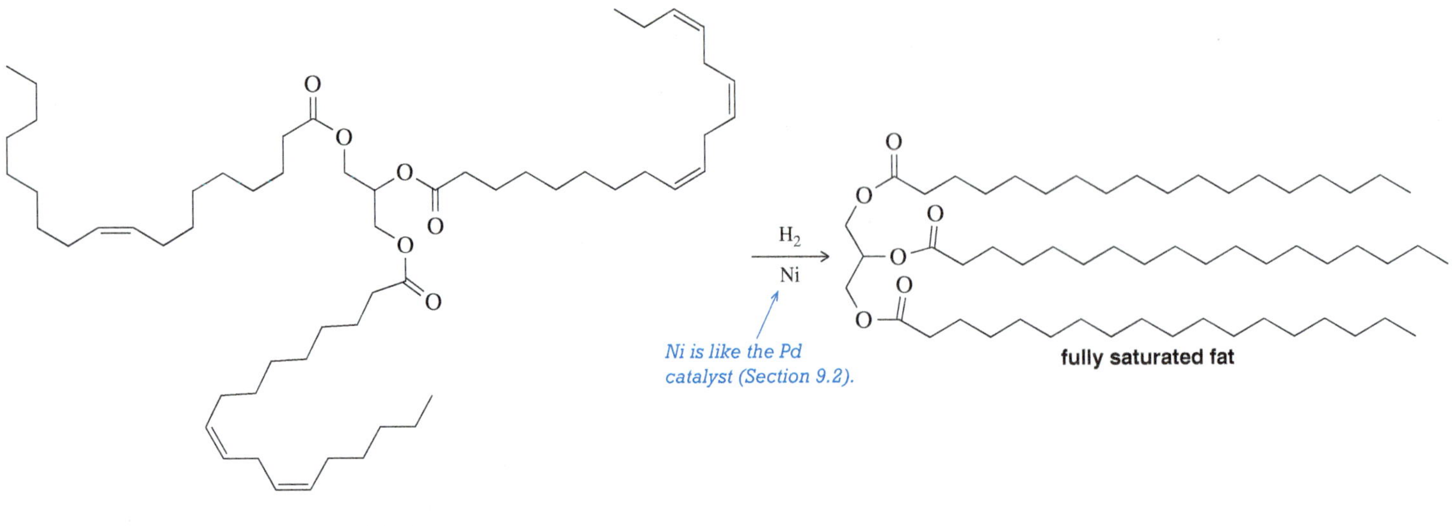

Figure 27.90 **The hydrogenation of soybean oil makes a saturated fat.**

Amid speculation that saturated fats contributed to deaths due to heart disease, the food industry began preparing **partially hydrogenated oils.** Removing only some of the cis-alkenes would increase stability while maintaining the health benefits of vegetable oils. Unfortunately, because of the mechanism of hydrogenation, even when a double bond is not hydrogenated, it still interacts, albeit briefly, with palladium. As shown in **Figure 27.91**, complexation with the palladium(II) species is reversible. Thus, after forming the first C—H bond, the reaction can reverse through a process of β-hydride elimination. When this happens, the more stable trans-alkene is formed. This wouldn't normally be a problem if complete hydrogenation was the goal. But trans fats increase low-density lipoprotein (LDL) cholesterol, the "bad" form that is most responsible for heart disease. The presence of LDL cholesterol results in a buildup of arterial plaques, which contribute to heart disease. As a result, foods containing partially hydrogenated oils, and by extension, trans fats, should be avoided as much as possible.

Figure 27.91 **Partial hydrogenation** converts cis-akenes to trans-alkenes.

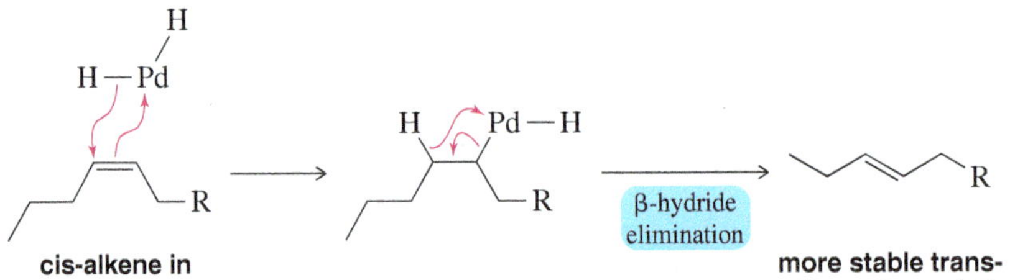

27.12.2 Saponification to make soap

Recall from Section 19.7.1 that **saponification** is the base-promoted hydrolysis of an ester (**Figure 27.92**). At that time, the goal was to make carboxylic acids, so the reaction contained a final proton quench step.

Figure 27.92 **The saponification of esters produces carboxylic acids** (Section 19.7.1).

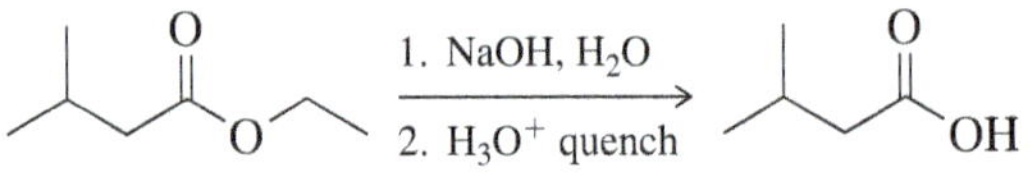

The invention of this reaction during early soap-making efforts [*hence the name saponification*] happened long before the process was understood. By washing the ashes of wood fires with water and allowing the resulting solution to evaporate, sodium and potassium hydroxide, referred to as lye, were isolated. The reaction of lye with animal fats (now known to be triglycerides) resulted in the formation of sodium and potassium carboxylates by a nucleophilic acyl substitution reaction. Following attack by hydroxide and formation of the tetrahedral intermediate, collapse of the electron pair reforms the C—O π bond, kicking out the alkoxide leaving group (**Figure 27.93**). Because the product of the reaction is a carboxylic acid ($pK_a = 5$) and the solution is basic, the carboxylic acid is immediately deprotonated, resulting in formation of the carboxylate. This happens three times, producing glycerol and three molecules of carboxylate for each triglyceride saponified.

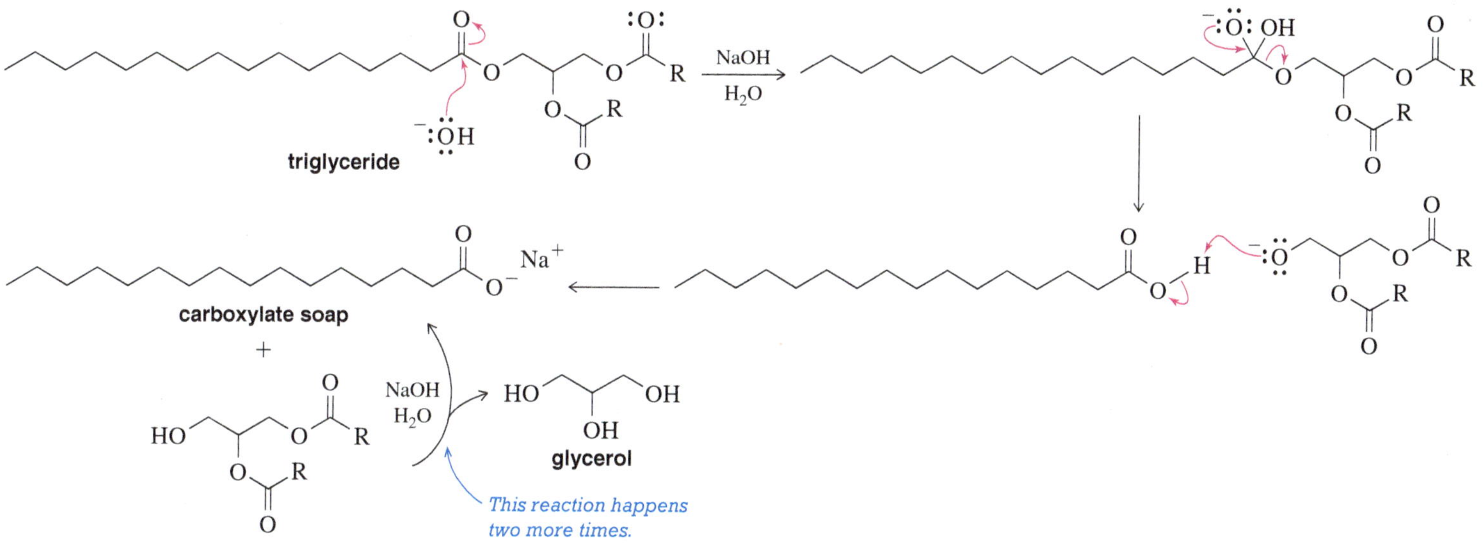

Figure 27.93 The mechanism of saponification.

Carboxylates obtained in this way from triglycerides are excellent soaps because they contain a polar ionic headgroup that interacts favorably with water and a nonpolar tail that interacts well with oils, dirt, and grease. These oils are nonpolar, so they cannot be dissolved and washed away with just water. However, upon formation of a micellar structure (**Figure 27.94**), where the nonpolar tails aggregate around the oil and grease in the nonpolar core, the ionic surface of the micelle makes it soluble in water.

Figure 27.94 The formation of a micelle makes it possible to dissolve nonpolar dirt, grease, and oils in water.

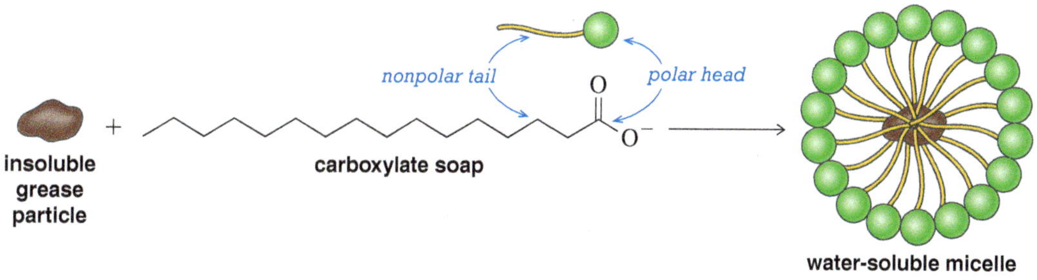

27.12.3 Biodiesel fuel by transesterification

The oil and gasoline that fuel our energy needs are refined from petroleum that is extracted from deep within the earth. Petroleum is a by-product of decomposed, formerly living matter that has been subjected to heat and pressure for millions of years, so it is a nonrenewable resource. Thus, one goal of green chemistry has been to identify chemicals and fuels from renewable feedstocks (GCP 7; Section 7.2.7).

The creation of **biodiesel** as an alternative fuel source is an excellent example of replacing fuels originating from nonrenewable petroleum feedstocks with those from renewable natural sources. Specifically, biodiesel fuel consists of long-chain esters derived from vegetable oils and animal fats. These esters, containing long unsaturated hydrocarbon chains, burn in a manner similar to gasoline and other fuels. As shown in **Figure 27.95**, the synthesis of biodiesel involves the transesterification of triglycerides with ethanol to make the ethyl ester for use as fuel, along with glycerol as a by-product. To make the process less wasteful, the glycerol from biodiesel production is often used in the flavor and perfume industries.

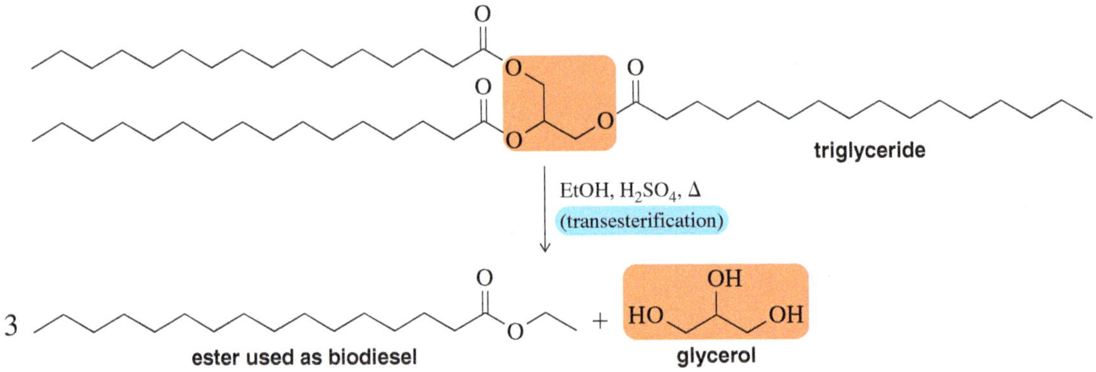

Figure 27.95 Transesterification of animal fats and vegetable oils results in long-chain esters, which can be used as biodiesel fuel, and glycerol.

Mechanistically, the reaction is essentially a nucleophilic acyl substitution reaction that uses an acid catalyst to enhance the reactivity of the ester (**Figure 27.96**). Protonation of the carbonyl results in a more reactive electrophile that is attacked by ethanol. Having brought

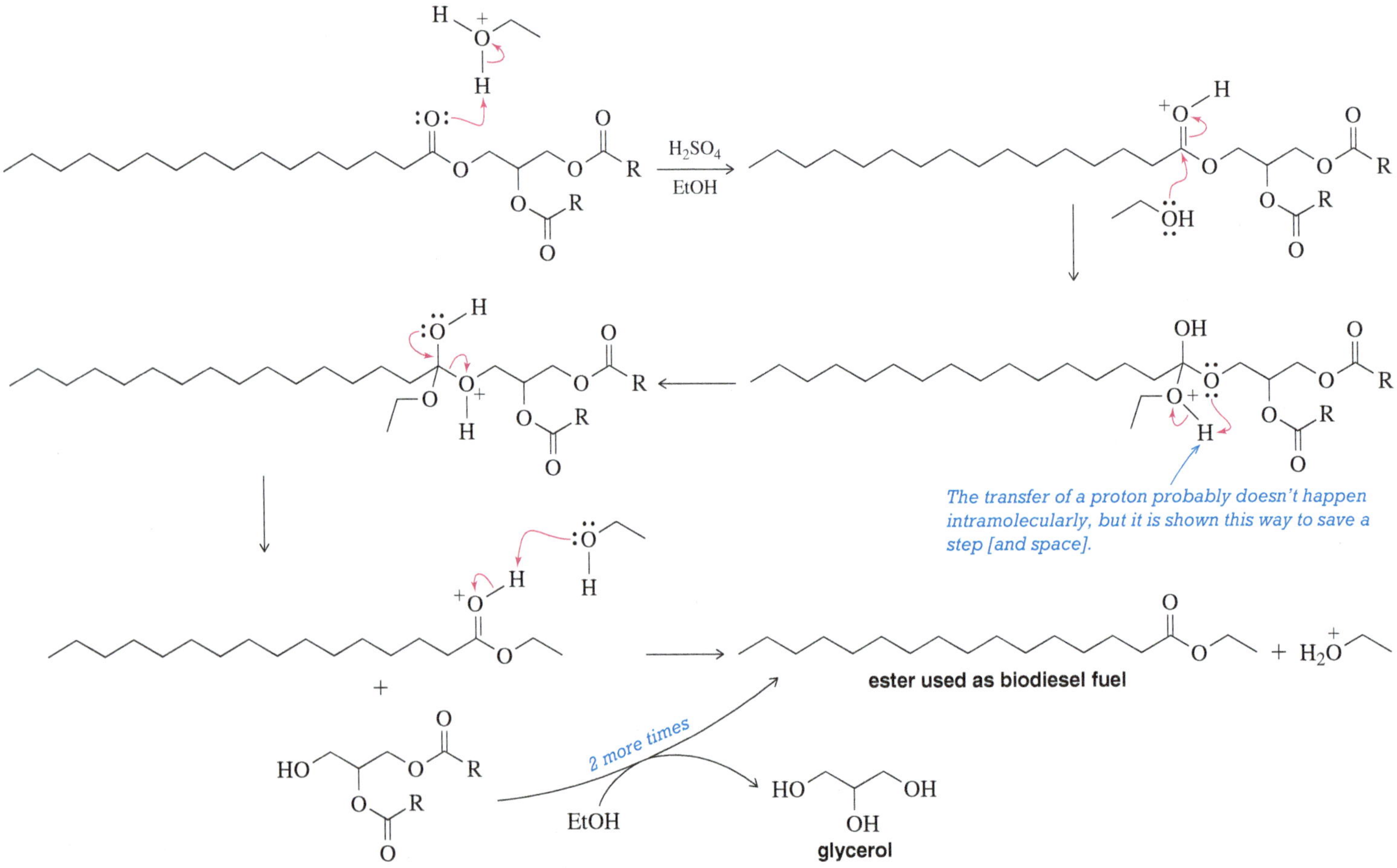

Figure 27.96 The mechanism of acid-catalyzed transesterification. [*See Section 19.7.1 for a review.*]

along its proton, the ethoxy group is deprotonated to give the tetrahedral intermediate. [*For a review, see Section 19.7.1.*] The oxygen of glycerol is protonated, allowing it to leave, forming a carbocation that is essentially a protonated ester. Removal of the acidic proton forms the ethyl ester. With two remaining fatty acid chains, the reaction happens two more times, ultimately producing three equivalents of the ethyl ester and one equivalent of glycerol. The formation of biodiesel, essentially a reaction of one ester to form another ester, is driven to completion by the use of excess ethanol.

27.12 SUMMARY

- Triglycerides undergo a number reactions studied previously: hydrogenation to make saturated fats, saponification to make soaps, and transesterification to make biofuels.

Assessment 27.58 Biodiesel could also be produced using conditions similar to those used in saponification. (a) Suggest reagents you could use to make ethyl oleate from oleate. (b) Draw an arrow-pushing mechanism for the reaction.

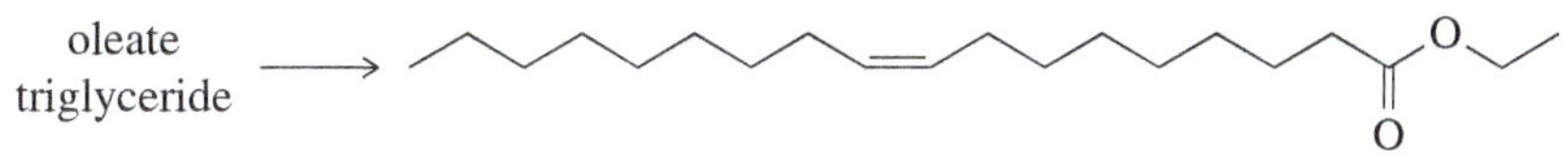

ethyl oleate

Assessment 27.59 Identify the product when each of the following reactions is performed on the triglyceride of linoleic acid (linoleate).

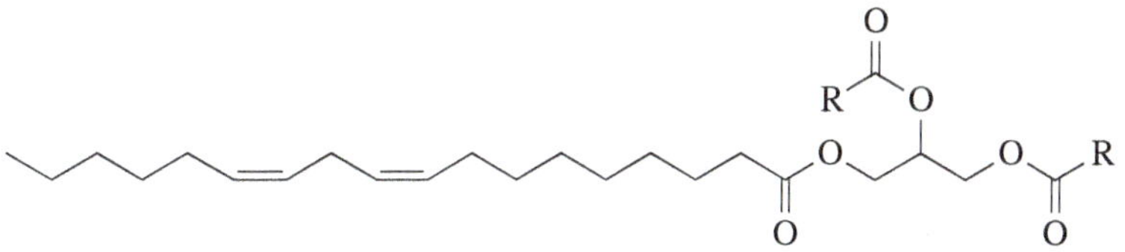

linoleate triglyceride

(a) H_2, Pd
(b) H_2, Ni (partial hydrogenation)
(c) NaOH, H_2O
(d) *i*PrOH, H_2SO_4

27.13 Terpenes

Terpenes are a class of diverse, volatile organic compounds. They are derived biosynthetically from isoprene units, which consist of five carbon atoms (**Figure 27.97**). Due to their widespread occurrence in nature, they have attracted attention as a class of natural products that can be converted into novel and valuable compounds, commercially important for the industrial production of fragrances, perfumes, flavors, and pharmaceuticals, as well as useful synthetic intermediates and chiral building blocks. Some of these compounds, along with their use, are shown in Figure 27.97. If this sounds familiar, it is because the chemistry of terpenes was briefly covered in Chapter 8.

Recall from Chemistry in Context 8-A that terpenes are synthesized by combining a variety of isoprene units. Specifically, **terpenes** contain 10 carbons and are made up of two isoprene units. **Sesquiterpenes** (from sesqui, meaning one and one half) contain 15 carbons and are therefore made by combining three isoprene units. It follows, then, that **diterpenes** contain 20 carbons and are derived from four isoprene units.

Figure 27.97 Important terpenes.

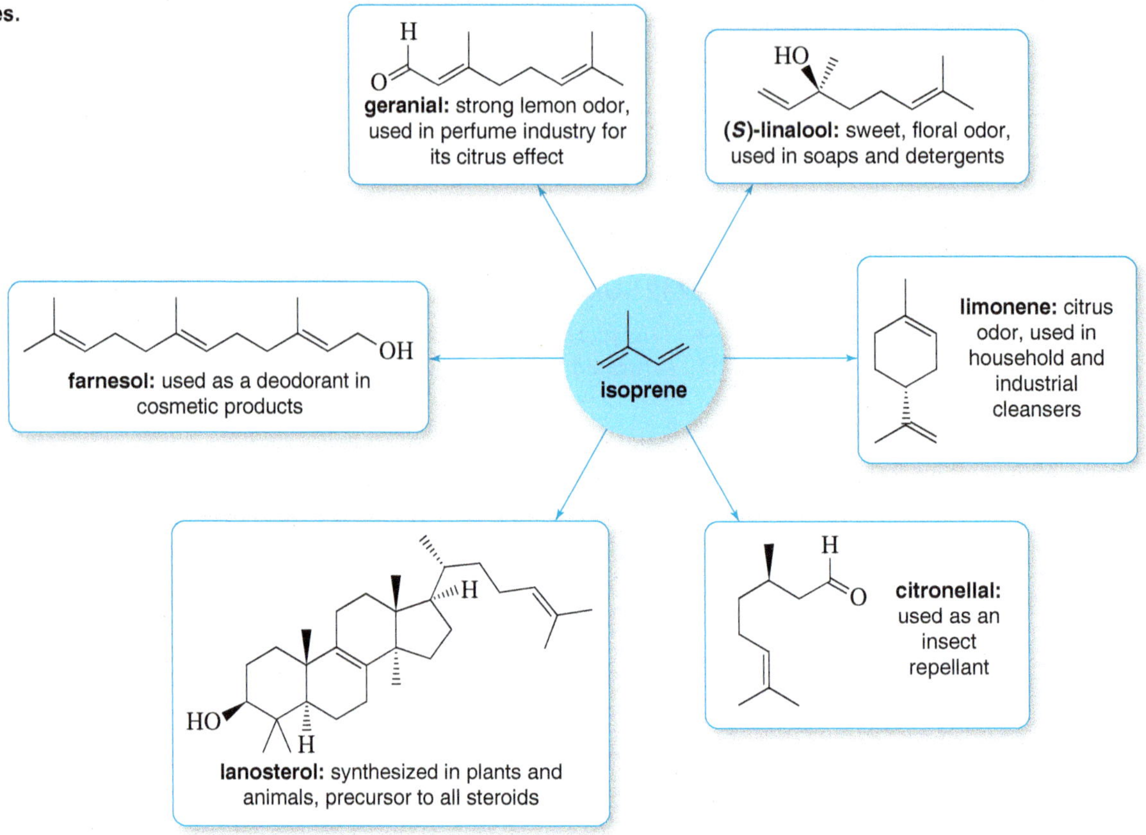

In nature, terpenes are formed predominantly via the head-to-tail (1 ⟶ 4) combination of isoprenes (**Figure 27.98**). This is known as the **isoprene rule**.

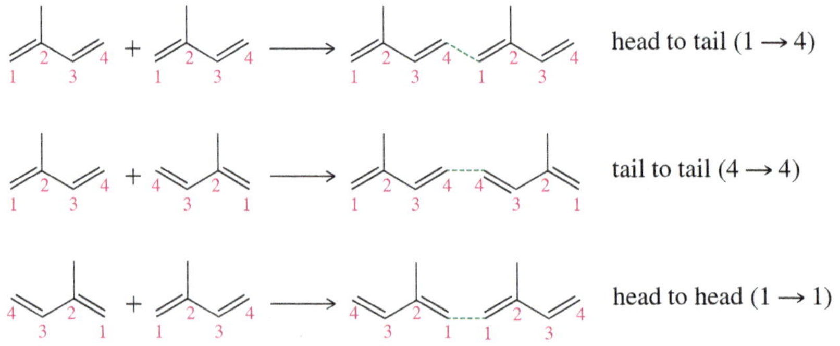

Figure 27.98 Terpenes form predominantly via the head-to-tail combination of isoprenes.

Geraniol and nerol, two terpenes, are extracted from the petals of certain rose species for use in the perfume industry. The biosynthesis of terpenes is discussed in the context of geraniol. [*To learn how honey bees use geraniol and nerol to mark flowers rich in nectar, see Chemistry in Context 8-C.*] Plants synthesize terpenes using dimethylallyl pyrophosphate (DPP) and 3-isopentenyl pyrophosphate (IPP). [*DPP and IPP have the same carbon skeleton as isoprene.*] Within plant cells (and human cells for that matter), DPP is produced by an enzyme-catalyzed reaction of IPP (**Figure 27.99**). An active-site acid protonates C_1, while an active-site base selectively deprotonates C_3.

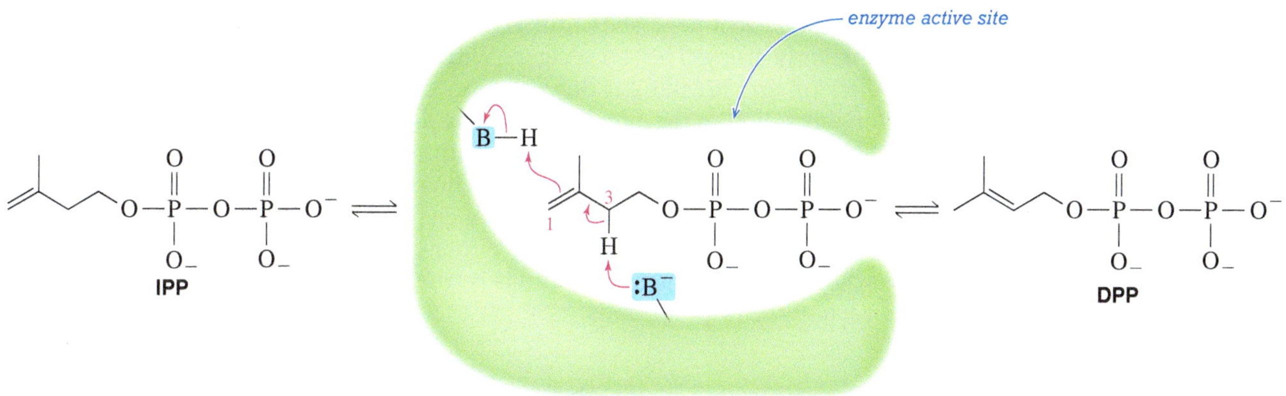

Figure 27.99 The enzymatic conversion of IPP to DPP, the major constituents involved in terpene synthesis.

With the two important constituents of terpene synthesis in hand, formation of a new C—C σ bond occurs in the active site of farnesyl pyrophosphate synthase (**Figure 27.100**). As a good leaving group, the diphosphate is expelled from dimethylallyl pyrophosphate (DPP) to give a resonance-stabilized carbocation. [*This is reminiscent of the first step of the* S_N1 *reaction (Section 12.2).*] The alkene of isopentenyl pyrophosphate (IPP) attacks the electrophilic cation to give a tertiary carbocation, extending the carbon chain by one isoprene unit. [*Recall from Chapters 8 and 9 that alkenes are nucleophilic.*] Removal of the hydrogen reforms the double bond, generating two alkene isomers. In a final step, hydrolysis of the pyrophosphate gives the diastereomeric terpenes geraniol and nerol. [*The enzyme active site at which the reactions occur dictates the selectivity of these reactions.*]

Figure 27.100 The biosynthesis of geraniol and nerol.

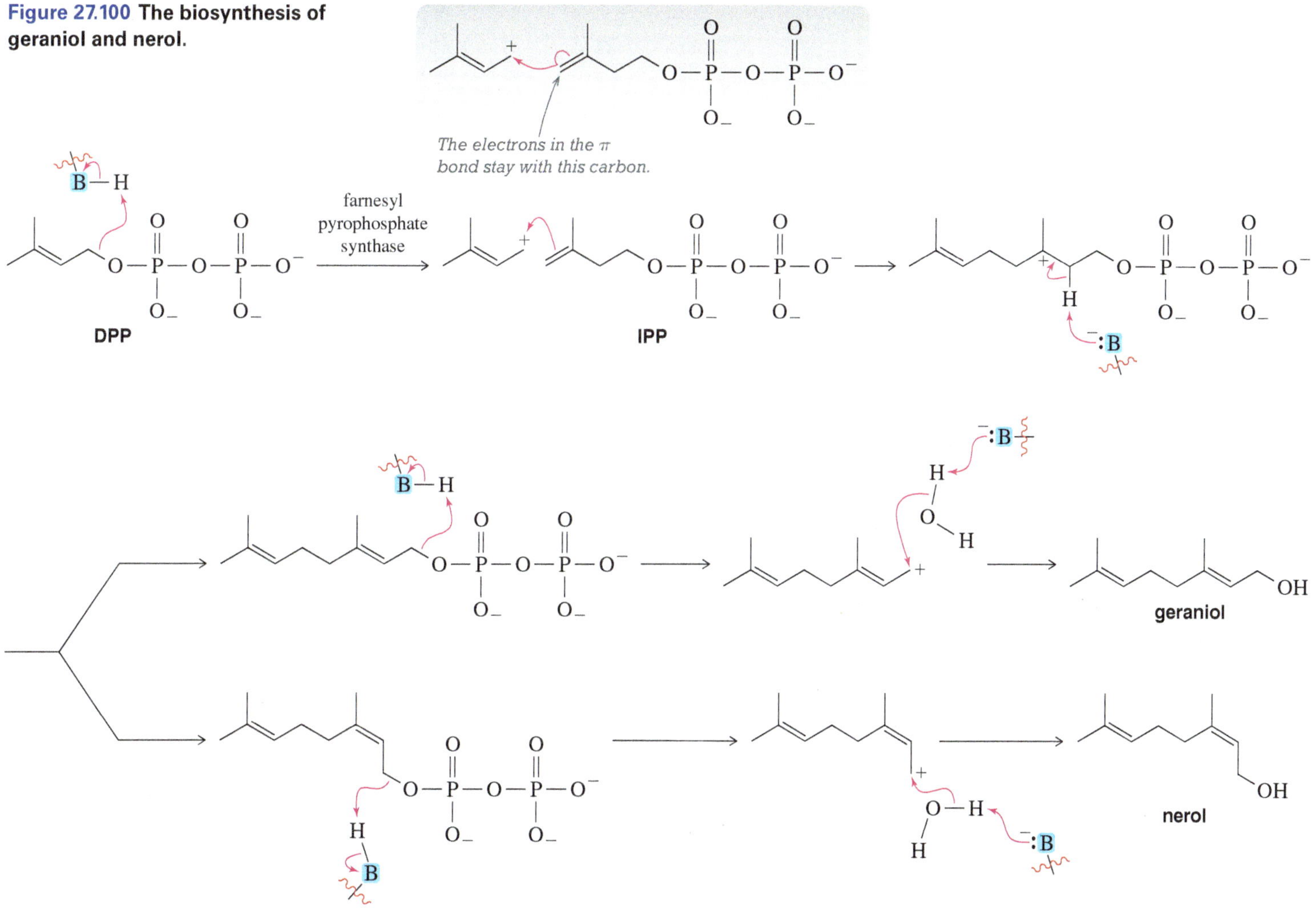

From the biosynthesis shown in Figure 27.100, we now know why isoprenes combine predominantly using $1 \rightarrow 4$ linkages. With this information, we can analyze a terpene, regardless of size, breaking it down into its constituent isoprene units (**Figure 27.101**). First and foremost, ignore the functional groups and focus only on the carbon skeleton. Next, recognize that the branching (typically with a methyl group, but sometimes with a CH_2) always occurs at C_2 of the four-carbon chain of isoprene. Once you have identified C_2, you can then label $C_1—C_4$ of the four-carbon chain of the isoprene unit. The branched carbon will be C_5. To see if you are correct, the four-carbon chain of the next isoprene unit should once again have branching at C_2. If this is true, then it is likely that you have correctly identified each of the isoprene units.

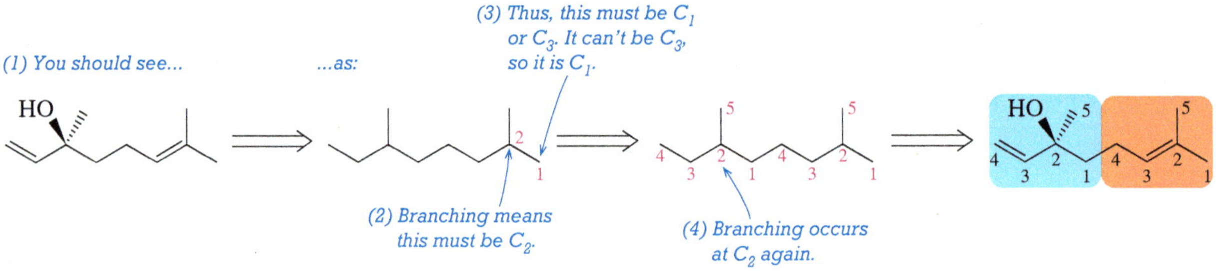

Figure 27.101 Identifying isoprene units in a terpene.

Assessment 27.60 Draw a mechanism for the acid-catalyzed, nonenzymatic conversion of DPP to IPP. How do you know your mechanism is correct?

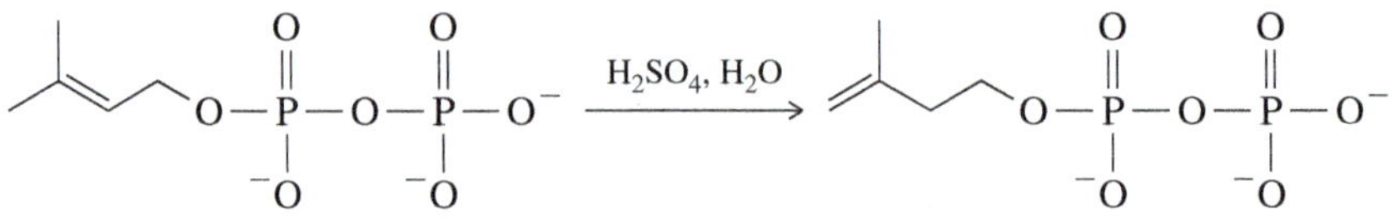

Assessment 27.61 When combining isoprene equivalents, the IPP is never the electrophile. Why might this be? What is special about DPP that allows it to function better as an electrophile?

Assessment 27.62 Identify the isoprene units in the terpenes shown. In cross-linked or ring-containing terpenes, linkages can be formed between more than just C_1 and C_4 of isoprene.

(a) (b) (c) (d)

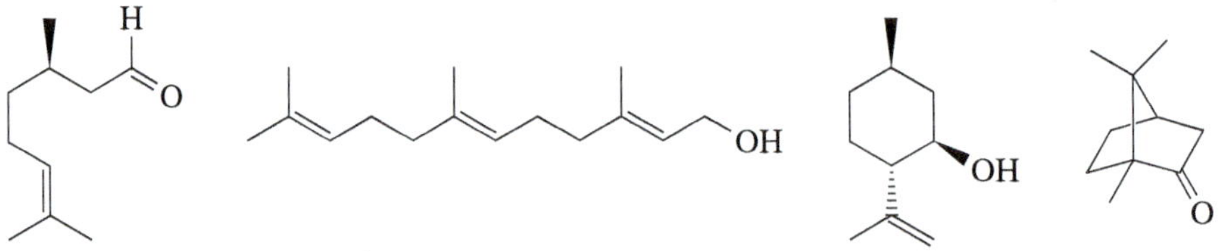

citronellal farnesol menthol camphor

(e)

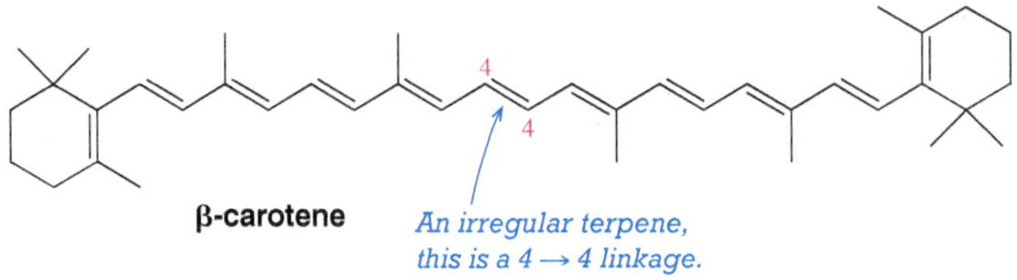

β-carotene

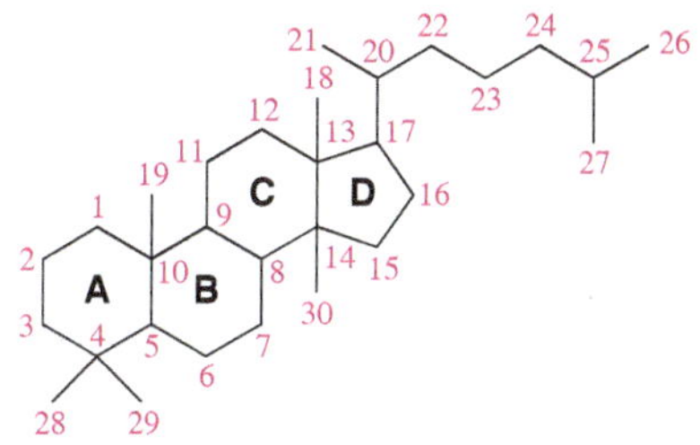

Figure 27.102 The steroid core structure. Ring assignments and numbering.

27.14 Steroids

The steroids are another class of lipids that is important to biological function. Derived from triterpenes, the steroid core consists of four fused rings (three cyclohexane, one cyclopentane) and usually has a nonpolar group attached to the cyclopentane. The IUPAC nomenclature of steroids labels the rings as A, B, C, and D, from left to right, as shown in Figure 27.102. The IUPAC-determined numbering system for steroids is shown, too.

27.14.1 Important steroids

Steroids have two important biological functions. Some are used to modulate fluidity in the cell membrane, whereas others are used as signaling molecules to activate hormone receptors. By activating these receptors, steroids play a role in metabolism, inflammation, sexual characteristics, and maintaining blood pressure at an appropriate level by balancing salt and water release/absorption. Table 27.5 lists several steroids and their function.

Table 27.5 Some representative steroids and their function.

Steroid	Function
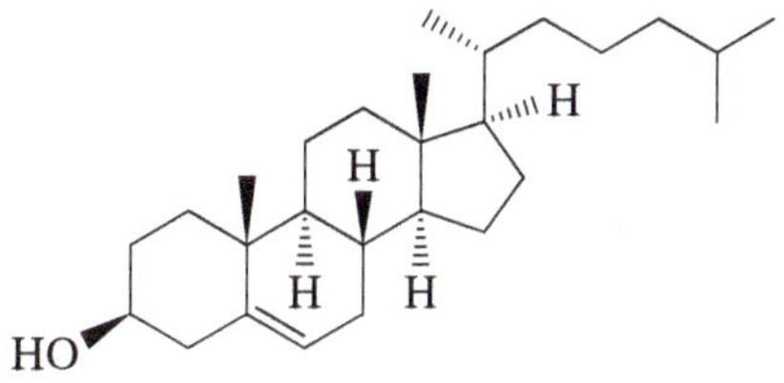 cholesterol	Cholesterol is a constituent of the cell membrane and a precursor to other mammalian steroids. Because it contains a rigid ring structure and a flexible hydrocarbon tail, cholesterol helps stabilize the membrane without sacrificing fluidity.
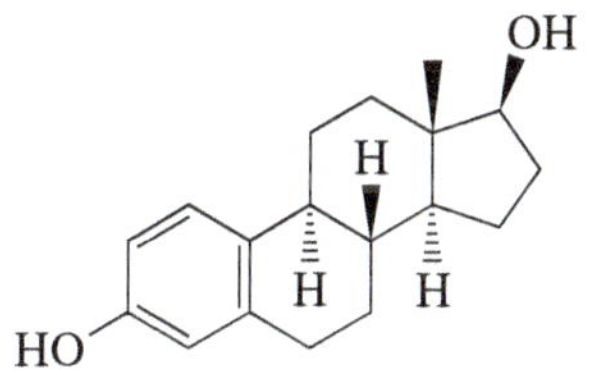 estrogen	Produced in the ovaries, estrogen is a steroidal hormone responsible for the maturation and function of the reproductive system in females. Estrogen also plays a role in the maintenance of bones. For this reason, postmenopausal women are more likely to suffer from osteoporosis.
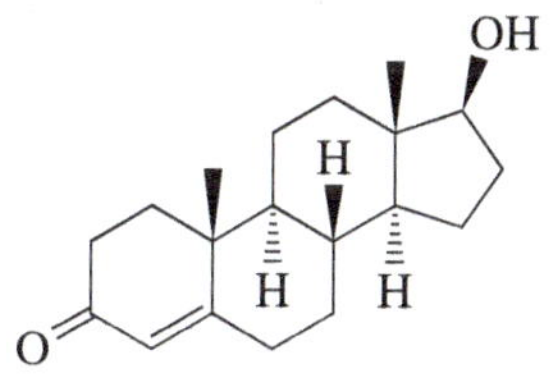 testosterone	Testosterone is the primary male sex hormone, playing a role in the maturation and function of the reproductive system. As an anabolic steroid, testosterone aids in the development of muscles, faster healing, and a loss of body fat.
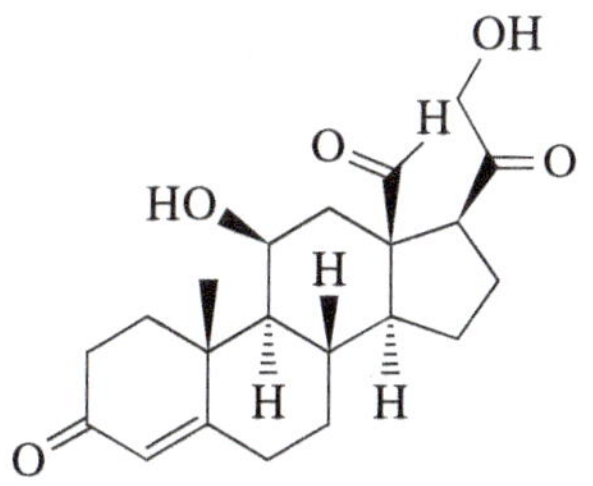 aldosterone	Produced in the adrenal gland, aldosterone works with the renin–angiotensin system to help regulate blood pressure [*see Chemistry in Context 26-A*]. Specifically, aldosterone regulates the ratio of salt and water in the body.
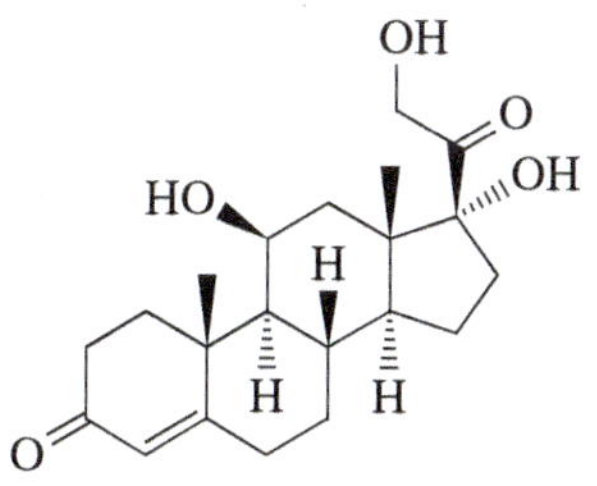 cortisol	Also produced in the adrenal gland, cortisol regulates metabolism, controls blood sugar level, and assists with the formation of memories. Cortisol release is also a part of the fight-or-flight mechanism when stressed. When stress is relieved, cortisol level returns to normal. A high level of cortisol is damaging to our health in several ways. [*Thus, exercise, do yoga, or meditate to relieve stress and reduce cortisol level.*]

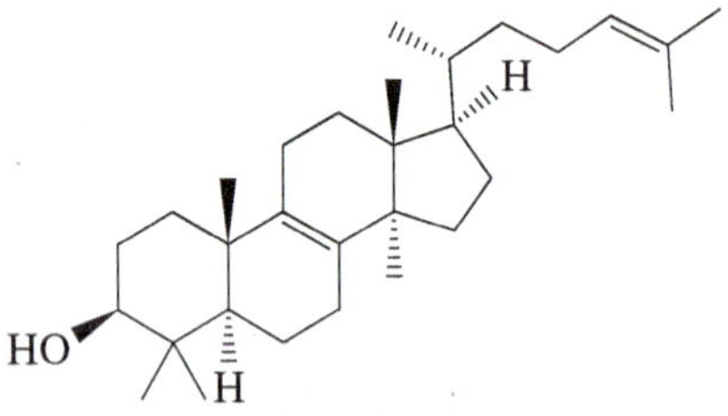

Figure 27.103 The structure of lanosterol.

27.14.2 Steroid biosynthesis

Steroids are produced in human cells using pathways similar to those used by plants to synthesize the terpenes shown in Figure 27.97. All steroids in human cells are produced through the common steroid precursor lanosterol (**Figure 27.103**). This section discusses the biological synthesis of lanosterol. We have seen this long and complicated mechanism before (Chapter 8). Armed with a greater understanding of organic reactions [*you have come a long way*], it should make much more sense now.

Steroid biosynthesis; formation of farnesyl pyrophosphate: The synthesis of lanosterol begins with the synthesis of farnesyl pyrophosphate. By a mechanism similar to that shown in Figure 27.100 for the synthesis of geraniol and nerol, farnesyl pyrophosphate is formed by the coupling of dimethylallyl pyrophosphate (DPP) with two isopentenyl pyrophosphates (IPP) (**Figure 27.104**). The two C—C σ bonds are formed when an alkene attacks an allylic carbocation generated by the leaving of pyrophosphate, a good leaving group.

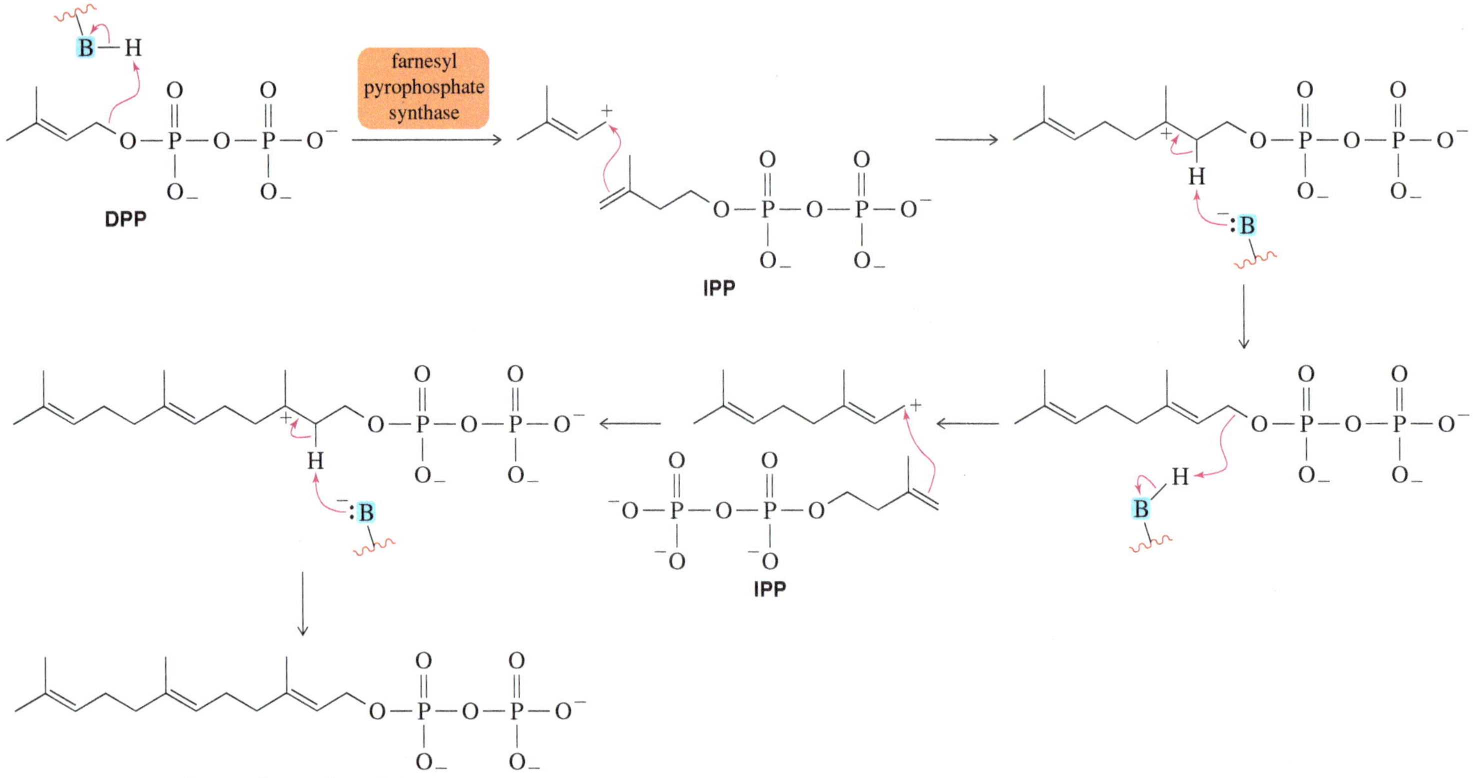

Figure 27.104 Steroid biosynthesis; formation of farnesyl pyrophosphate: Electrophilic addition to alkenes followed by elimination to reform the alkene, times two.

Steroid biosynthesis; coupling of farnesyl pyrophosphate to make squalene: Coupling of two molecules of farnesyl pyrophosphate to make squalene occurs in a nontraditional way compared to usual terpene synthesis. Instead of this being a 1 → 4 linkage (head to tail), this reaction is effectively a 1 → 1 linkage to make the triterpene. As such, squalene is called an irregular terpene. Mechanistically more complex than the usual linking of isoprene units, squalene synthase makes use of NADPH [*nature's sodium borohydride; Figure 27.81*]. In the first step, pyrophosphate (a good leaving group) leaves with the assistance of an active-site proton (**Figure 27.105**). The $C_2—C_3$ alkene attacks the carbocation, leaving behind the stable 3° carbocation. [*So far, this isn't unlike what we've seen before.*] Whereas other terpenyl carbocations might eliminate to form a new $C_3—C_4$ alkene, a proton is removed instead by an active-site base from C_1, thus allowing those electrons to attack C_3, resulting in the formation of a cyclopropane (it is referred to as presqualene pyrophosphate). [*Enzymes can do unexpected chemistry.*]

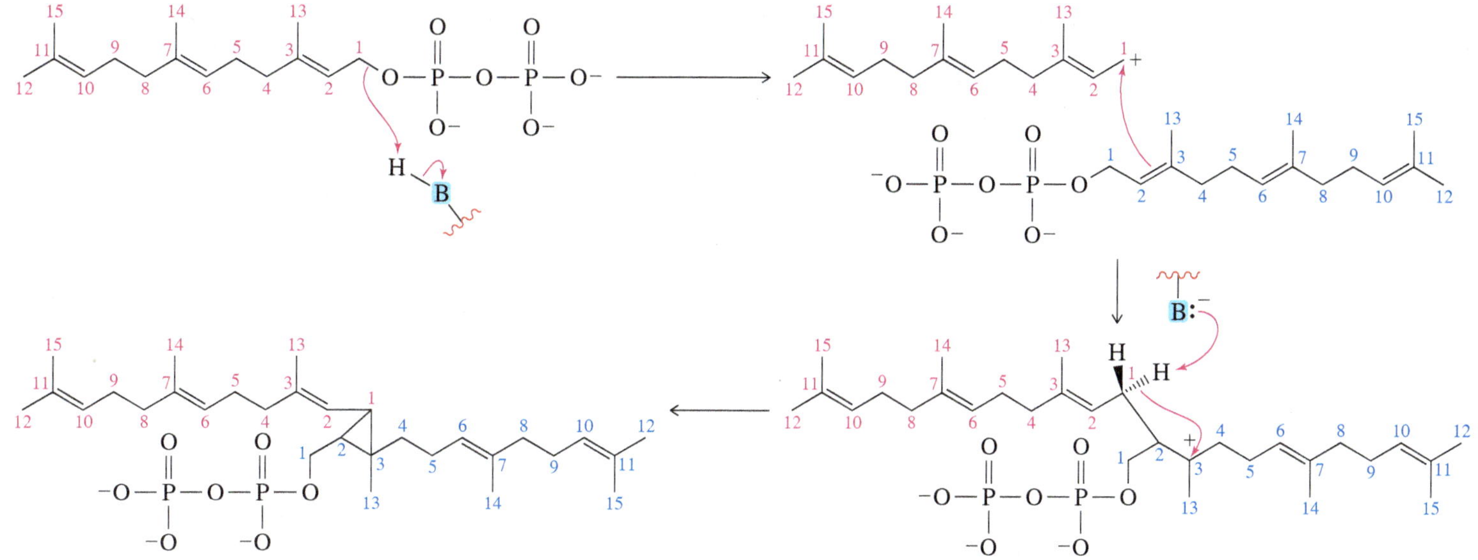

presqualene pyrophosphate

Figure 27.105 The coupling of farnesyl pyrophosphates to make presqualene pyrophosphate.

The conversion of presqualene pyrophosphate is driven by the relief of ring strain in the three-membered ring. As seen previously, the pyrophosphate leaving group at C_1 leaves to form a 1° carbocation neighboring the cyclopropane ring (**Figure 27.106**). This cyclopropyl carbocation rearranges first to a more stable four-membered ring with a 2° carbocation and, finally, to a new three-membered ring with a neighboring 3° carbocation. [*This rearrangement from 1° to 2° to 3° makes sense based on carbocation stability, right?*] Finally, an NADPH cofactor donates a hydride ion to C_1, opening the three-membered ring and forming the new $C_2—C_3$ π bond in squalene.

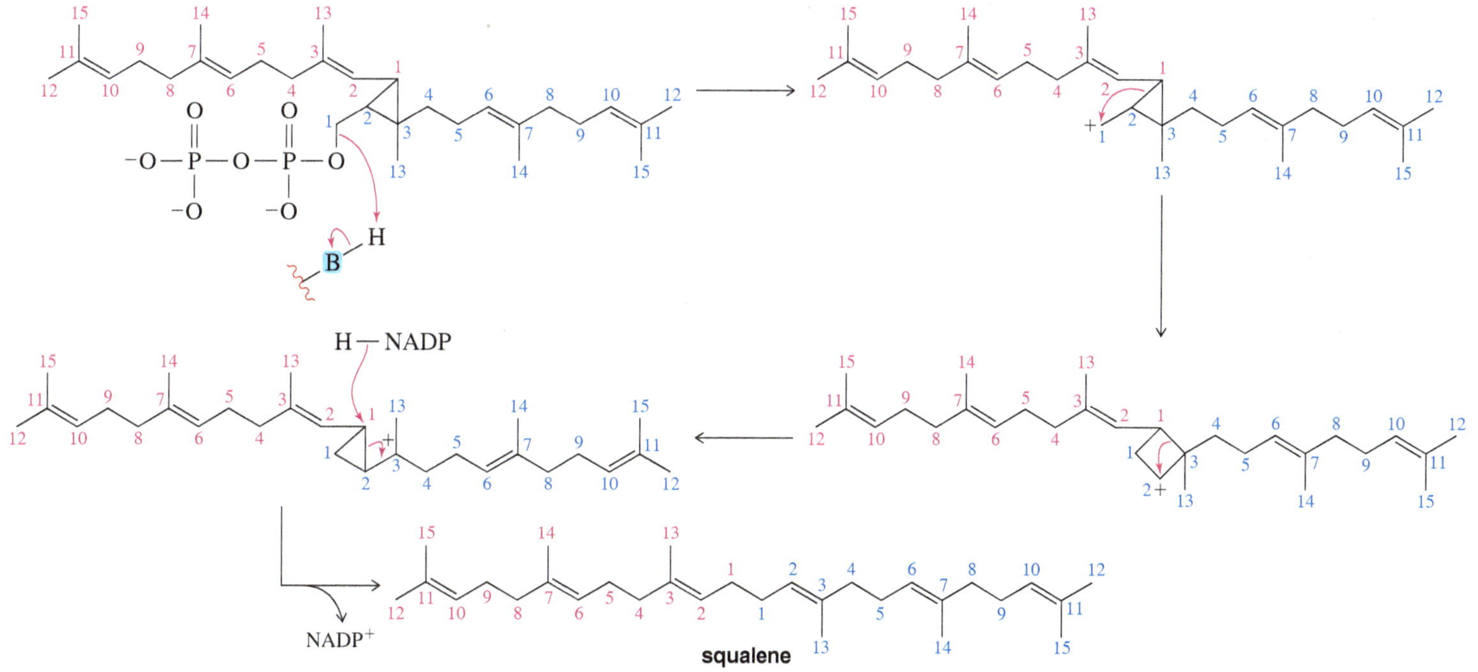

Figure 27.106 Carbocation rearrangement and reduction to make squalene.

Steroid biosynthesis; epoxidation of squalene: Prior to cyclization, squalene monooxygenase epoxidizes squalene by a complicated mechanism involving an FAD cofactor (**Figure 27.107**). In another example of an enzyme's remarkable selectivity, only one alkene of squalene is oxidized.

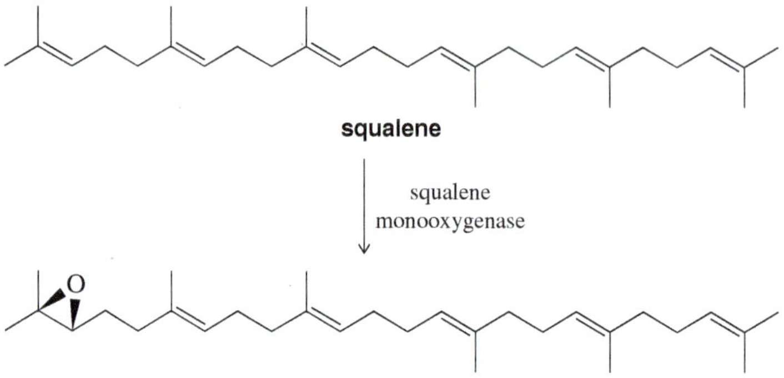

Figure 27.107 The oxidation of squalene to squalene oxide sets the stage for cyclization.

Steroid biosynthesis; cyclization to the steroid core: Protonation of the epoxide with an acid in the active site of the enzyme and ring opening to give the more stable carbocation at C_2 trigger a cyclization cascade that results in the formation of the four rings of lanosterol. Each of the C—C σ bond forming reactions in **Figure 27.108** is simply an alkene attacking an electrophile, all but one forming a 3° carbocation in the process.

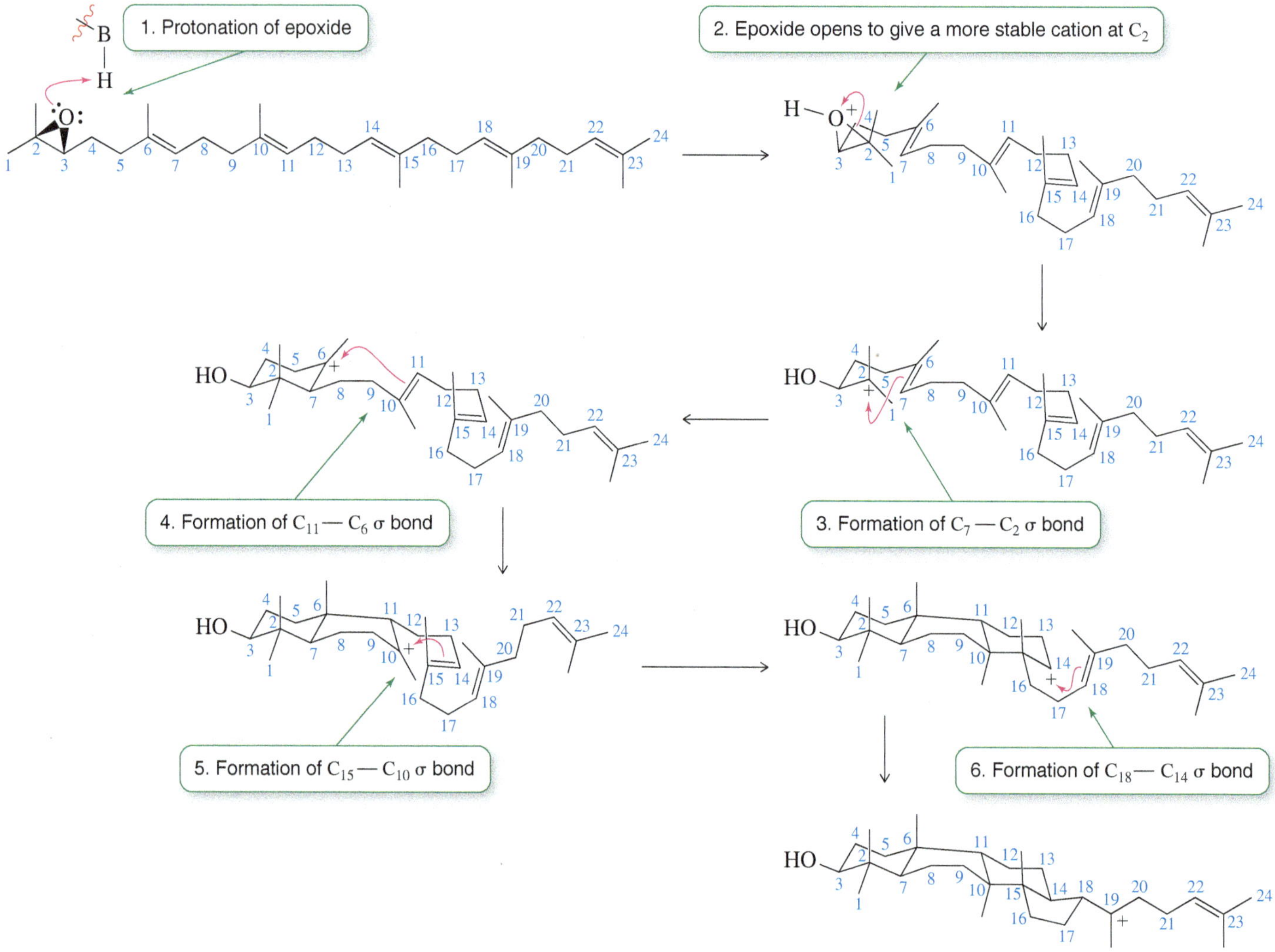

Figure 27.108 The formation of the A, B, C, and D rings of lanosterol via formation of four new C—C σ bonds.

Steroid biosynthesis; rearrangements to lanosterol: With the steroid ring skeleton in place, here's where it gets a bit tricky. Essentially, a series of carbocation rearrangements occurs (**Figure 27.109**), ultimately resulting in lanosterol after a base in the enzyme's active site deprotonates C_{11}. While it may be difficult to justify the product of each individual rearrangement, a new 3° carbocation is produced in each one. Moreover, this happens at the active site of an enzyme; the role of this should not be underestimated.

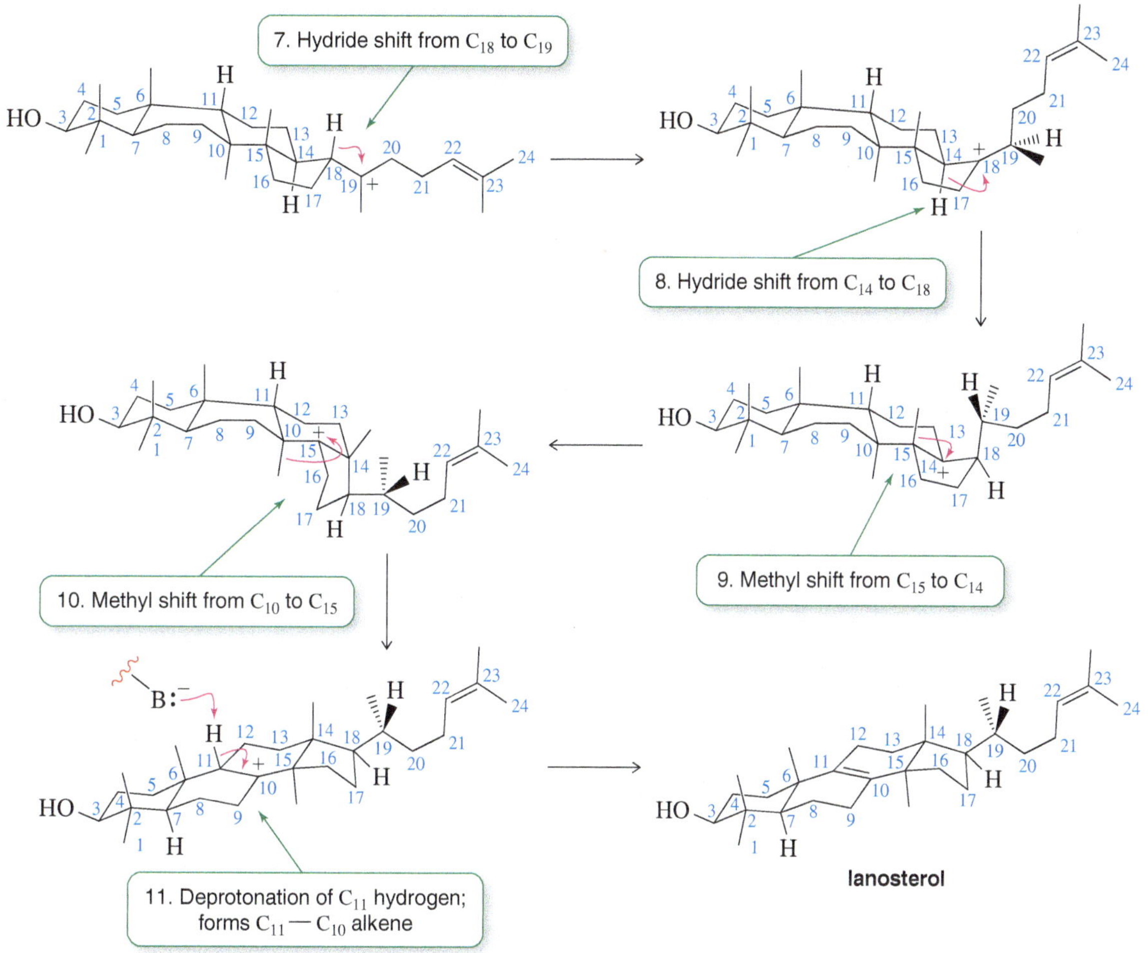

Look how far you've come. Just a few months ago, this would have seemed impossible.

Figure 27.109 A series of carbocation rearrangements produces lanosterol.

As stated previously, a slow, careful analysis of the 11-step lanosterol cascade should allow you to see how each reaction in this process is related to reactions studied previously.

27.15 Prostaglandins

The prostaglandins are the last group of lipids that we study in this chapter. Derived from arachidonic acid (**Figure 27.110**), a polyunsaturated fatty acid from Table 27.4, prostaglandins contain a cyclopentane oxidized at C_1 and C_4 and feature oxygenated alkyl groups at C_3 and C_4. Using the numbering of the original arachidonic acid, F-series prostaglandins have an alcohol at C_9, whereas E-series prostaglandins have a ketone at C_9. The subscript in a prostaglandin name tells how many double bonds are present.

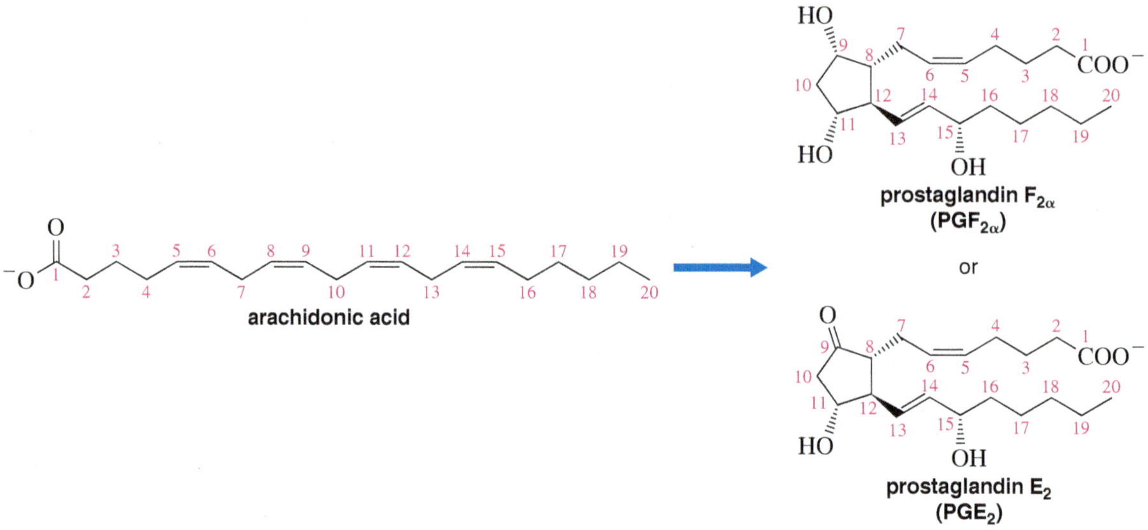

Figure 27.110 Structures of the prostaglandins $PGF_{2\alpha}$ and PGE_2.

27.15.1 Important prostaglandins

Prostaglandins exhibit hormone-like effects in mammals, playing a variety of roles. They promote inflammation in response to an infection, causing pain and a fever in the process. They are involved in blood pressure regulation through the constriction and dilation of blood vessels. During labor, they are responsible for uterine contractions. Finally, they inhibit the needless formation of platelet aggregates, avoiding the unnecessary formation of blood clots. Table 27.6 presents two prostaglandins and their function.

Table 27.6 The prostaglandins PGI_2 and PGE_2 and their functions.

Prostaglandin	Function
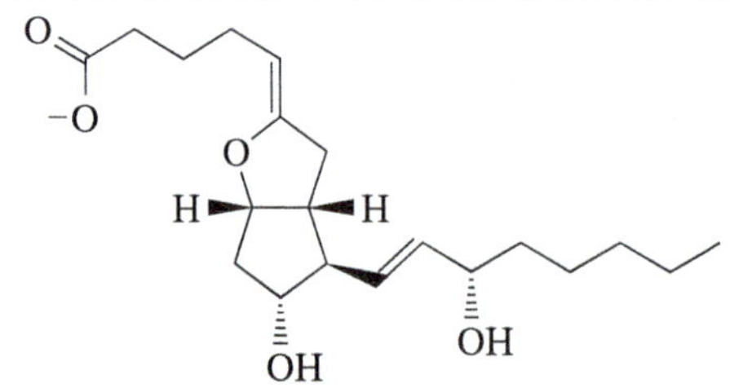 prostaglandin I_2 (PGI_2)	Prostaglandin I_2 (PGI_2), also called prostacyclin, is produced in endothelial cells lining the walls of arteries and veins. It inhibits the formation of platelets and is a potent vasodilator.
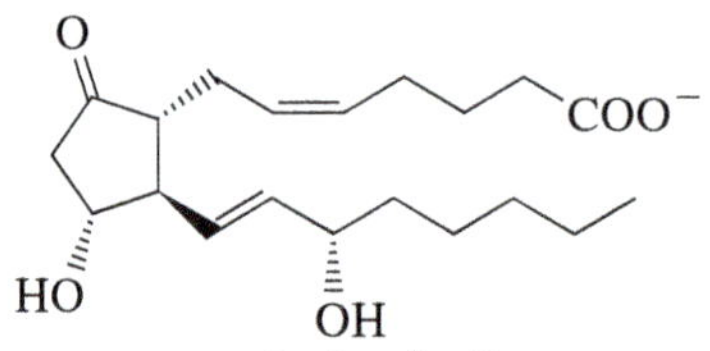 prostaglandin E_2 (PGE_2)	Prostaglandin E_2 (PGE_2), also called dinoprostone, is a vasodilator. It helps to prepare the cervix for labor.

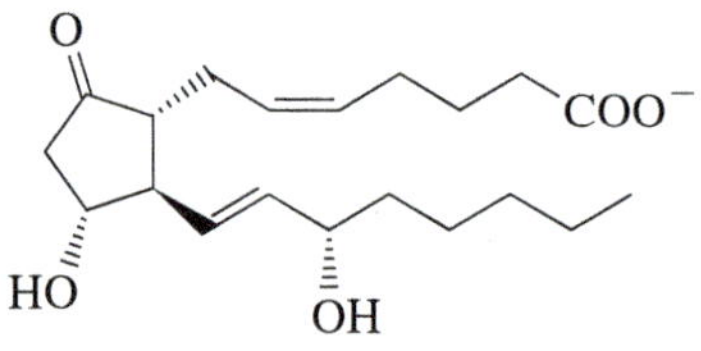

Figure 27.111 The structure of PGE_2.

27.15.2 Prostaglandin synthesis

Prostaglandins are produced in a variety of human cell types beginning with arachidonic acid. This section discusses the biological synthesis of PGE_2 (Figure 27.111), a prostaglandin involved in the pain and the fever you might feel when you have a cold or the flu.

Prostaglandin biosynthesis; the oxidation of arachidonic acid: Prostaglandin synthesis begins when the enzyme cyclooxygenase (COX) incorporates two molecules of oxygen (O_2) into arachidonic acid (**Figure 27.112**). This reaction occurs via a radical mechanism that is analogous to fatty acid autoxidation (see Chapter 11). [*It is probably unnecessary for you to know this reaction, but with carbons numbered, you should be able to follow along.*]

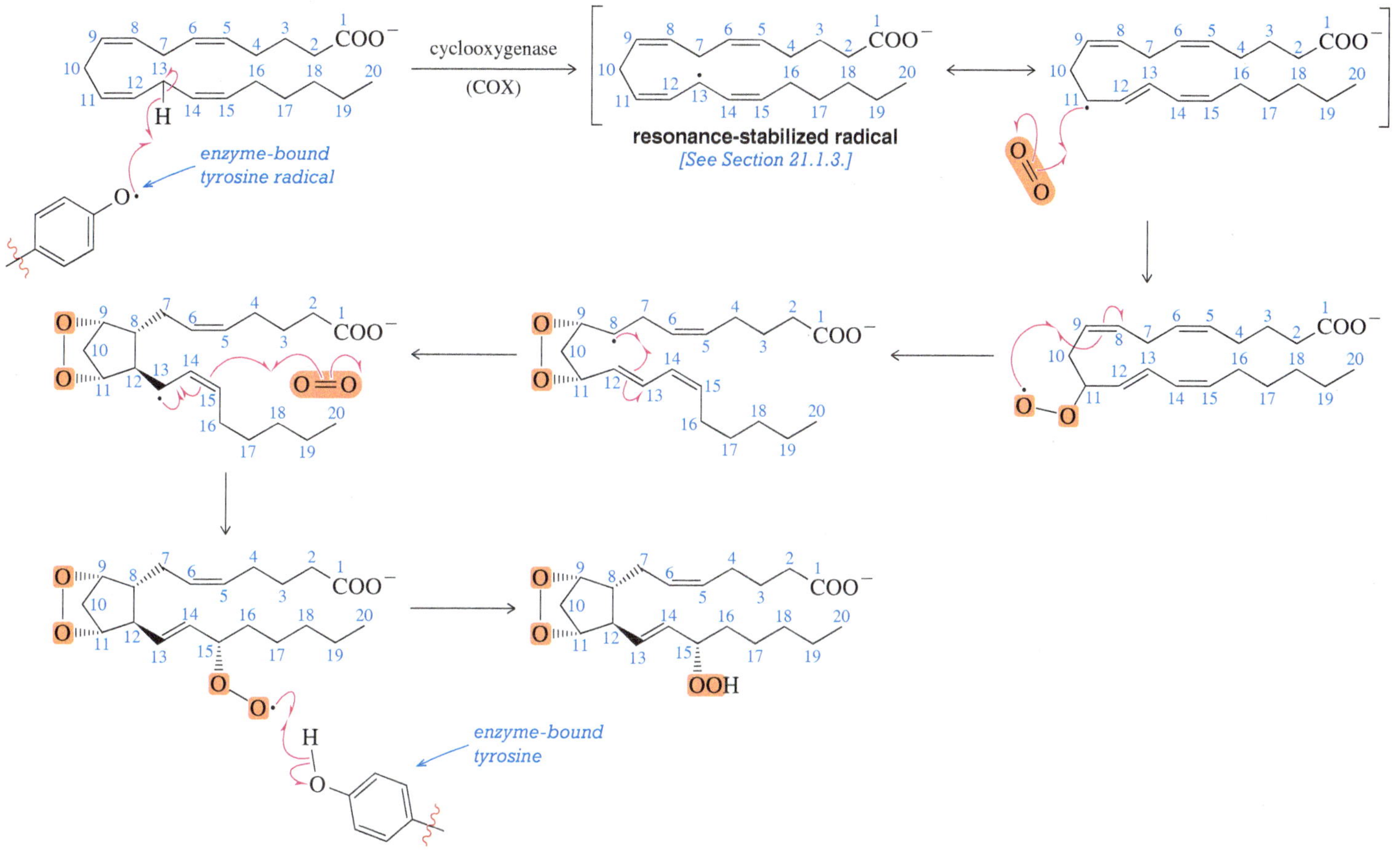

Figure 27.112 The oxidation of arachidonic acid—a hypothetical mechanism for incorporating two molecules of oxygen (O_2).

Prostaglandin biosynthesis; the reduction of a peroxide: The σ bond between two oxygen atoms is a highly unstable bond that is easily reduced (Section 9.1.4). Using a peroxidase enzyme, featuring a cofactor that donates 2 electrons, the peroxide at C_{15} is reduced to a hydroxyl group (**Figure 27.113**).

Figure 27.113 The reduction of a peroxide—a hypothetical mechanism for reducing the peroxide at C_{15}.

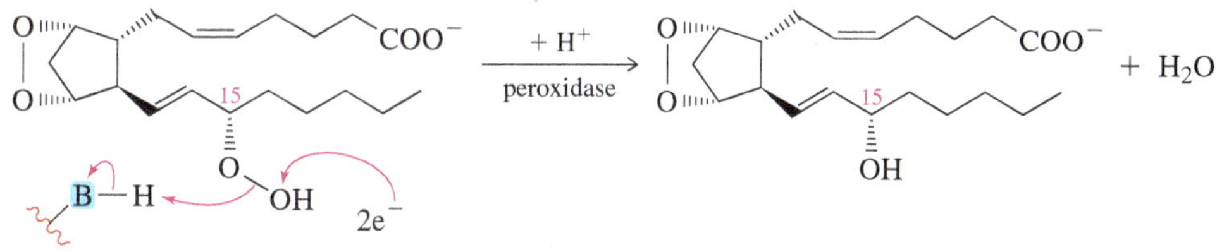

Prostaglandin biosynthesis; peroxide isomerization: The final step in the synthesis of PGE_2 is also driven by cleavage of an O—O σ bond (**Figure 27.114**). In this case, a hypothetical mechanism can be drawn where an active-site base removes a proton from C_9, pushing an electron pair to oxygen and breaking the O—O σ bond in what looks like an elimination reaction. [*We studied mechanisms like this in the alcohol oxidation reactions of Chapter 13.*]

Figure 27.114 The formation of PGE_2 by peroxide isomerization.

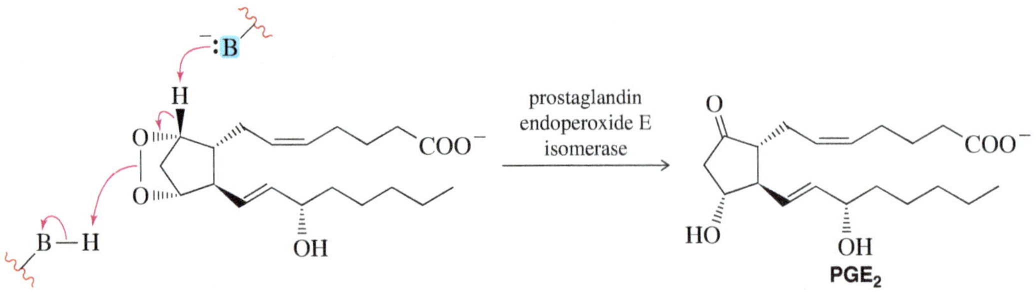

Assessment 27.63 By numbering the carbons, show how PGI_2 (from Table 27.6) is derived from arachidonic acid.

Assessment 27.64 Another class of molecules, called the thromboxanes, are also derived from arachidonic acid. By numbering the carbons, show how TXA_2 and TXB_2 are derived from arachidonic acid.

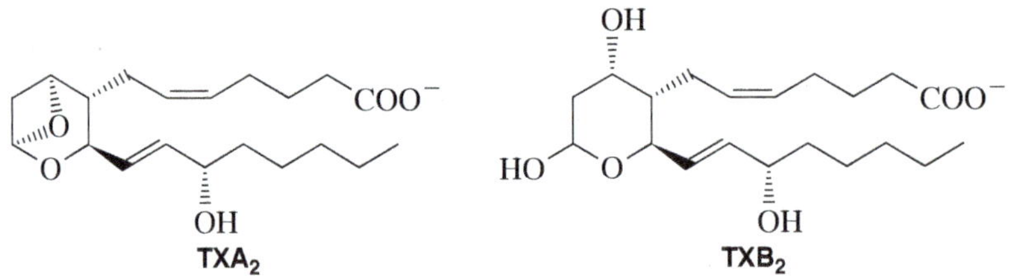

Assessment 27.65 Suggest a mechanism by which TXB_2 might be formed from TXA_2 in an acid-catalyzed hydrolysis reaction. [*The structure has been simplified.*]

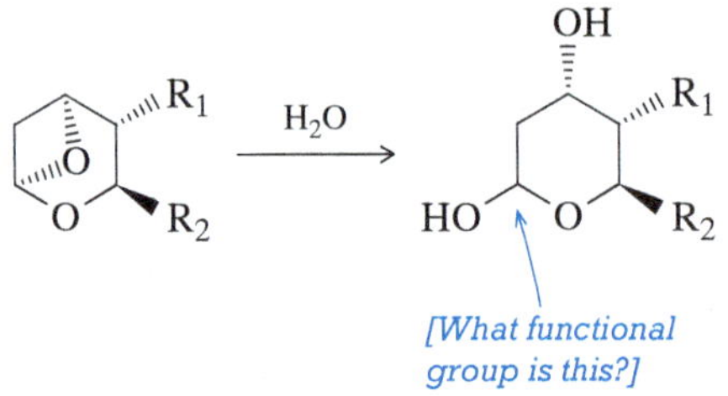

27-D CHEMISTRY IN CONTEXT: Medicinal Chemistry

Pain relief by prostaglandin synthesis inhibition

When a muscle or other tissue is injured, prostaglandins and related compounds are synthesized. These chemicals act as chemical messengers, sending a signal to the brain, which inflames the injured area while making us feel pain. Although this mechanism was unknown when aspirin was invented/discovered, its pain-relieving properties are involved in this process. Specifically, by inhibiting the enzyme cyclooxygenase (COX) (Figure 27.112), arachidonic acid cannot be converted to prostaglandins. Mechanistically, inactivation of cyclooxygenase occurs by transfer of an acetyl group from acetylsalicylic acid to an active-site serine (**Figure 27.115**). This reaction is simply a nucleophilic acyl substitution and, from the perspective of the acetate, is simply a transesterification reaction.

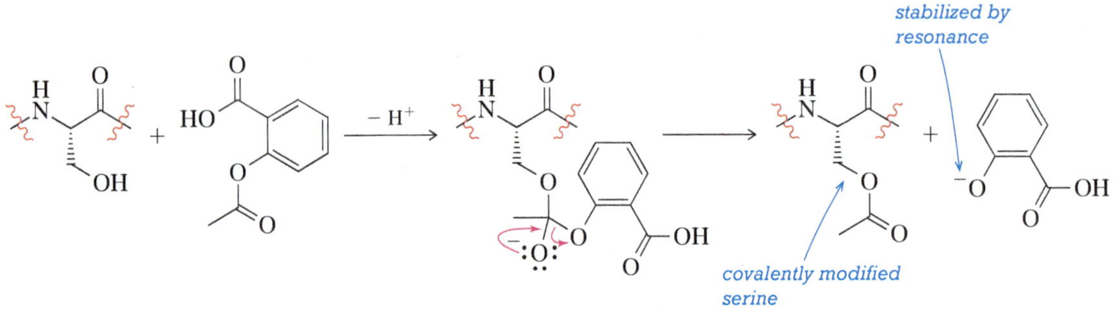

Figure 27.115 The acylation of an active-site serine inhibits the COX enzymes.

(continued)

Once the serine has been covalently, and permanently, modified, the active site of COX can no longer accommodate arachidonic acid, the substrate necessary for prostaglandin synthesis (**Figure 27.116**). Other compounds produced by COX are involved in clotting and vasoconstriction, both of which can compound negative outcomes associated with heart attack or stroke. It is for this reason that many older adults who are at risk of heart attack or stroke are advised to take a regular low dose of aspirin. Similarly, if someone is believed to be having a heart attack, it is also recommended that they take aspirin [*after calling 911, that is.*]

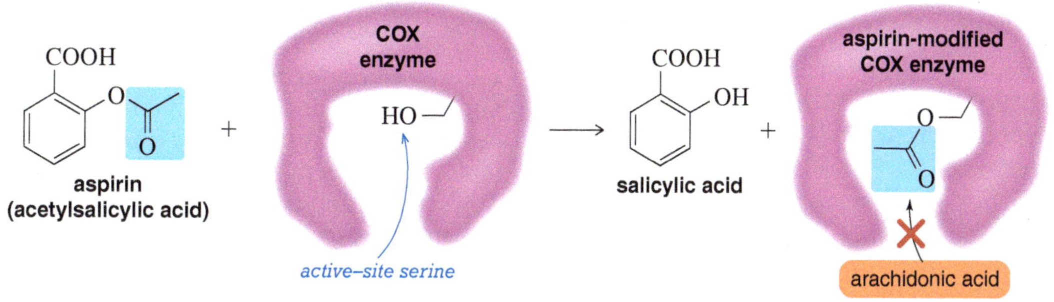

Figure 27.116 Acylating serine prevents arachidonic acid from reaching the COX active site.

Concluding thoughts

Organic chemistry has a well-earned reputation as a difficult course in a chemistry or pre-health professions curriculum. Our efforts over the course of Chapters 1–27 make clear why this is. [*And yet, you've made it this far!*] When dealing with an organic chemistry problem, we are essentially wondering how a small molecule with a functional group or two might react with a reagent that is a nucleophile, electrophile, acid, or base. Contrast that with the complexity of biochemistry, where the catalysts themselves are composed of hundreds of amino acids. Or the carbohydrates and nucleic acids that are synthesized, replicated, and transcribed so precisely, even though they contain thousands of monomers in a chain. Or simply take a look at the thousands of metabolic pathways that use these enzymes as catalysts to make exactly the right amount of the molecules we need to keep the process going. And maybe it's still all about the ball rolling downhill and negatives attacking positives, but even so, it's awe-inspiring. [*The good news: You'll get through that course, too!*]

You know what else is awe-inspiring? The pioneers like Emil Fischer, James Watson, Francis Crick, and Rosalind Franklin, who were able to use the limited chemistry tools available in the late 19th to mid-20th century to answer some of science's most important questions. That there are people out there who can do those things should inspire all of us to continue reaching higher and digging more deeply into how we can use our talents to answer similar questions, whether in the fields of medicine, chemistry, and biochemistry, or in the classroom.

Twenty-seven chapters down, only one to go. On to polymers, which are like some of the molecules in Chapters 26 and 27, except that we chemists get to synthesize them in a lab.

End-of-Chapter Materials

Important Concepts

[*These are the major concepts covered in each section, excluding reactions.*]

27.1: Carbohydrates are biomolecules that provide much of the energy we use and play important roles in many biological processes. Simple carbohydrates are composed of mono- and disaccharides. Longer carbohydrate chains are called polysaccharides.

27.2: Fischer projections are used to show monosaccharides in their aldose or ketose forms. The D/L notation is used to describe the stereochemistry of sugars. Epimers are diastereomers that differ only in the absolute stereochemistry at a single stereocenter.

27.3: Monosaccharides exist predominantly in their cyclic hemiacetal form and are sometimes represented using Haworth projections. The five-membered ring is referred to as a furanose, whereas the six-membered ring is referred to as a pyranose. Anomeric sugars are a specific class of epimers that differ only in the stereochemistry at the hemiacetal carbon (C_1). The α-anomer has the hydroxyl group in the axial position of the pyranose ring, whereas the β-anomer has it in the equatorial position.

27.4: Reactions of monosaccharides include acetal synthesis at C_1, alkylation of hydroxyl groups by the Williamson ether synthesis, and protection of the alcohols as silyl ethers, benzyl ethers, acetates, or acetals.

27.5: Emil Fischer utilized a series of chemical experiments to determine the structure of glucose.

27.6: Disaccharides are sugars that hydrolyze to make two monosaccharides. Disaccharides are classified based on the carbons that connect the two monomers and the stereochemistry of that linkage. Synthesis of disaccharides is complicated by the many hydroxyl groups requiring protection but proceeds essentially through acetal-forming reactions.

27.7: Polysaccharides are long chains of monosaccharides. Cellulose, starch, and glycogen are important biological polysaccharides.

27.8: Nucleic acids are biological polymers that store, transcribe, and translate genetic information. Ribonucleic acids (RNA) are composed of D-ribose and a nitrogenous base [adenine (A), guanine (G), cytosine (C), and uracil (U)] at the anomeric carbon. These monomers are connected to one another through a phosphate backbone. Deoxynucleic acids (DNA) differ from RNA with D-ribose being replaced with D-deoxyribose and the uracil (U) being replaced with thymine (T). Two strands of DNA pair to form a double helix by hydrogen bonding between base pairs (A with T, G with C). DNA is replicated by DNA polymerase.

27.9: The central dogma of molecular biology describes how DNA is used to create messenger RNA, which employs transfer RNA and ribosomal RNA to produce proteins.

27.10: Lipids are oily, nonpolar, water-insoluble substances. These include triglycerides, waxes, prostaglandins, and terpenes.

27.11: Fatty acids are long-chain carboxylic acids. Saturated fatty acids contain the maximum number of hydrogens possible per carbon. Unsaturated fatty acids contain at least one double bond. In nature, most unsaturated fatty acids have cis-alkenes. Triglycerides and fatty acids are synthesized by a series of repeating steps using fatty acid synthase.

27.12: The reactions of triglycerides are ones that were studied in other contexts and include hydrogenation to make fats, saponification to make soaps, and conversion to biofuels by transesterification.

27.13: Terpenes are volatile organic compounds synthesized from isoprene units predominantly via head-to-tail ($1 \rightarrow 4$) combination.

27.14: Steroids are derived from triterpenes and consist of four fused rings. Steroids are used biologically to modulate the fluidity in cell membranes and activate hormone receptors. The biosynthesis of steroids involves a complex series of steps including the combination of isoprene units and a cascade of cyclization and rearrangement reactions.

27.15: Prostaglandins are biomolecules derived from arachidonic acid that exhibit hormone-like effects in mammals. Their synthesis from arachidonic acid involves the enzyme cyclooxygenase by a mechanism analogous to fatty acid autoxidation.

Important Reactions and Mechanisms

[Because all of these reactions are ones studied previously, an example of each important reaction from this chapter is shown with a reference to where in the text it was first studied.]

Reactions of carbohydrates	
Acetal formation, Section 17.7.3	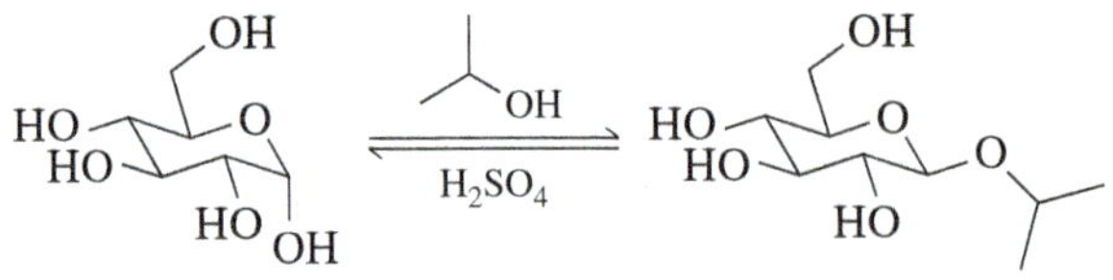
Williamson ether synthesis, Section 13.11.1	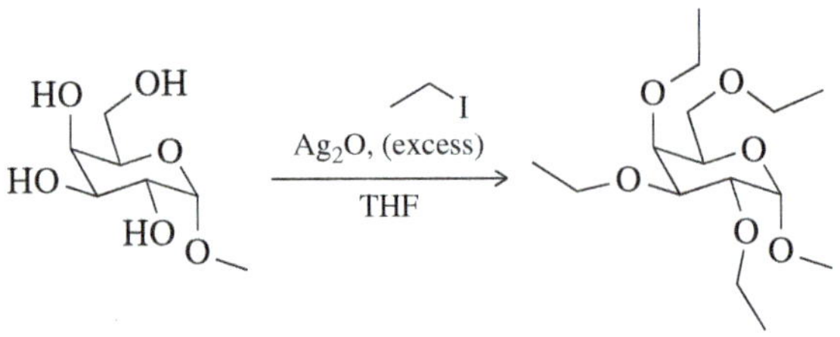

(continued)

Reactions of carbohydrates	
Silyl protection, Section 13.14.1	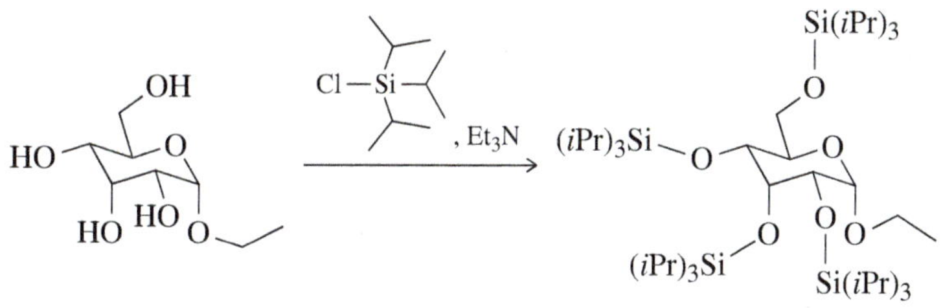
Benzyl protection via Williamson ether synthesis, Section 13.11.1	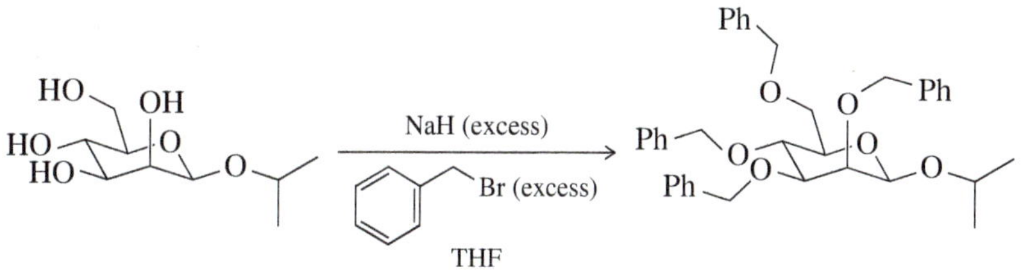
Acetate protection, Section 19.7.2	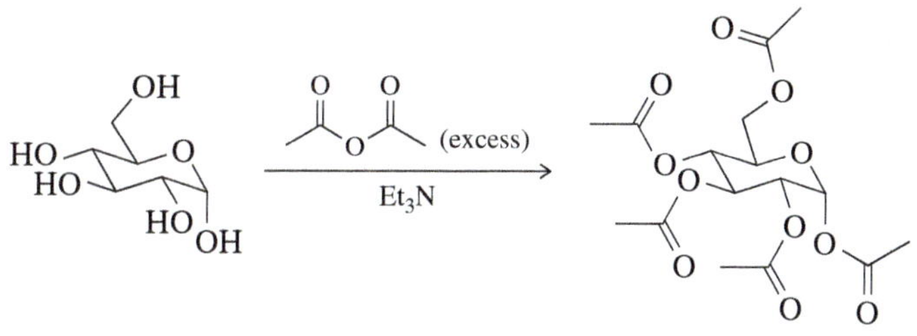
Diol protection via acetal formation, Section 17.7.3	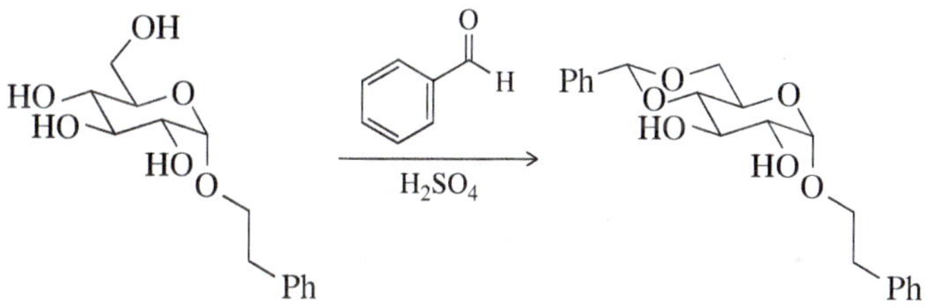
Reactions of fatty acids	
Hydrogenation of fats, Section 9.2	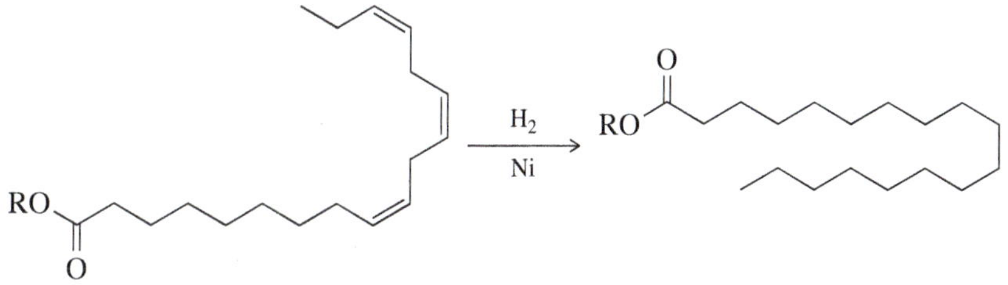
Formation of soaps from fats, Section 19.7.1	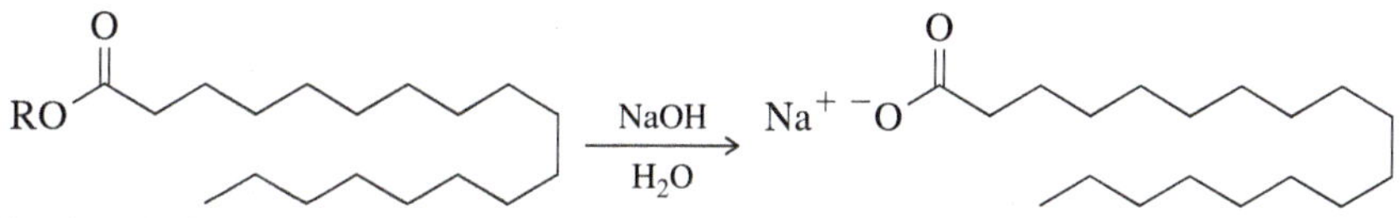
Formation of biodiesel, Section 19.7.2	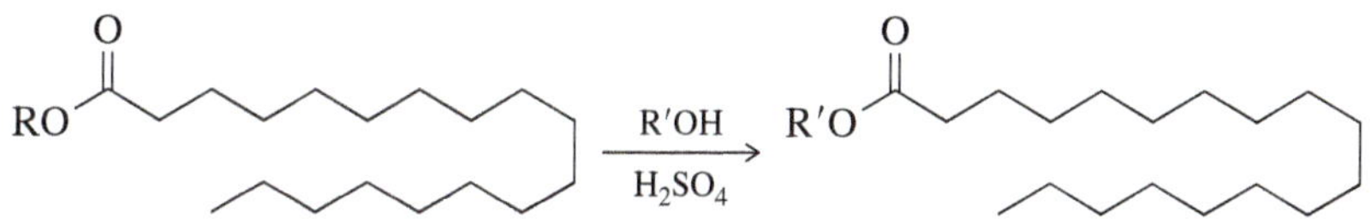

End-of-Chapter Assessments

[*Assessments are ranked according to their difficulty;* •••• = *most difficult.*]

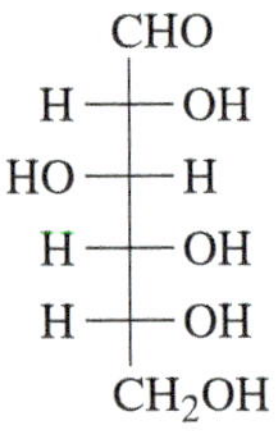

27.66 (••) The Fischer projection of *d*-glucose is shown. Draw the Fischer projection of *l*-glucose. [*The lower case "d/l" refers to the direction of rotation of plane-polarized light.*]

27.67 (••) Starting with a Fischer projection of D-glucose (and other sugars), switch only the stereocenter that gave it the designation of D. What new sugar have you made? [*Hint: The answer is not* L-*glucose.*]

The α- and β-anomers of glucose are shown here. In solution, these two epimers can interconvert through a process called **mutarotation**. Assessments 27.68–27.73 refer to mutarotation.

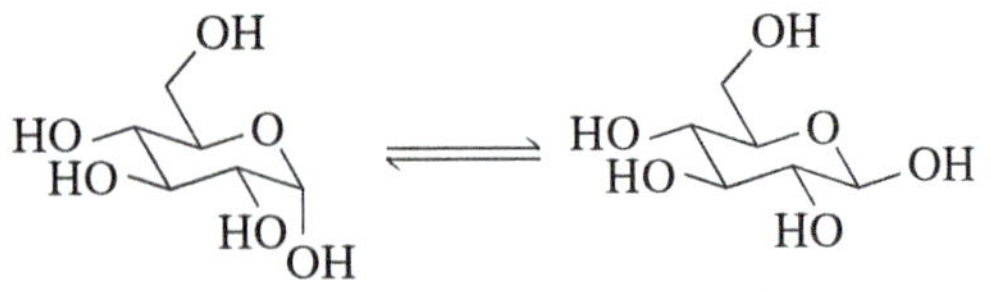

α-D-glucose [α] = **+112.2°**

β-D-glucose [α] = **+18.7°**

27.68 (•) Given that α-D-glucose has a specific rotation of +112.2°, why is the specific rotation of β-D-glucose not −112.2°? What molecule would have a specific rotation of −112.2°?

27.69 (•••) Upon dissolving α-D-glucose or β-D-glucose in water, the specific rotation gradually changes, eventually reaching +52.6° for both solutions. Explain what is happening here.

27.70 (•••) What is the ratio of anomers (α:β) when the specific rotation is +52.6° [*We did problems like this in Chapter 6.*]

27.71 (••••) Given the ratio you calculated in Assessment 27.70, what is the difference in energy between α- and β-D-glucose in water? Which is more stable in water? [*We did calculations like this in Chapter 5; assume the temperature is 298 K.*]

27.72 (•••) (a) Suggest an arrow-pushing mechanism for the mutarotation of α-D-glucose to β-D-glucose in the presence of acid. Acid isn't necessary, but it does increase the rate of the process. (b) What is the role of acid in your mechanism?

27.73 (••) (a) Draw a reaction coordinate diagram for the mechanism drawn in Assessment 27.72. (b) What must be true about the mechanism of the reverse process?

27.74 (••••) The benzylidene acetal, while protecting C_4 and C_6, can be reductively opened to provide either the C_4- or C_6-protected sugar, depending on the reagents used. (a) Suggest a mechanism for each. (b) What accounts for the difference in selectivity between the two reactions?

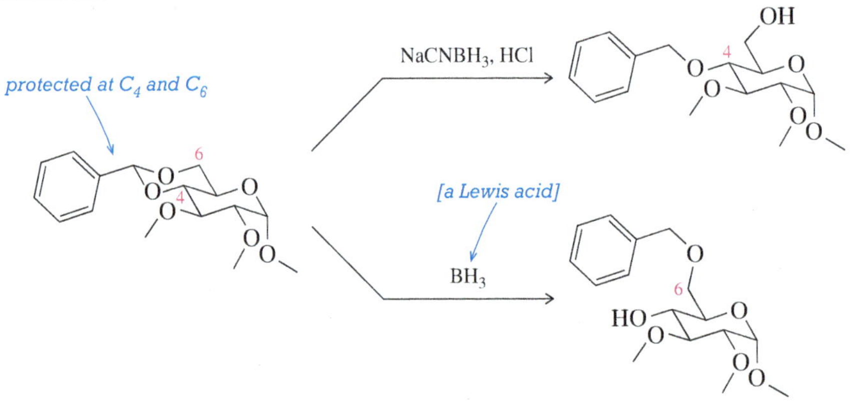

Chitin is a polymer found in crustacean and insect exoskeletons. A disaccharide of the polymer is shown here. Assessments 27.75–27.80 refer to this disaccharide.

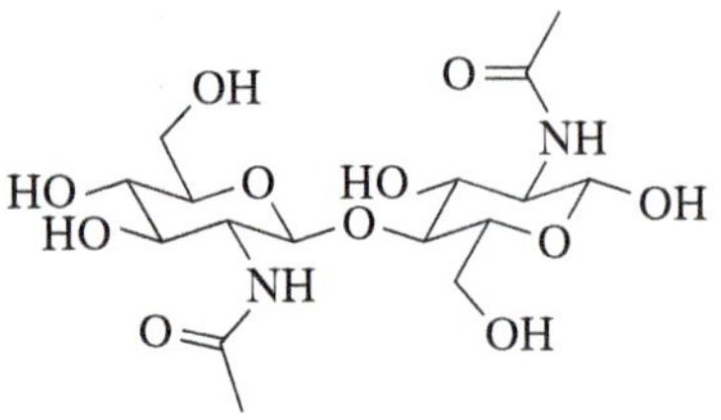

27.75 (•) Draw one of the cyclic monosaccharides using the Haworth projection.

27.76 (••) Draw the Fischer projection of the acyclic form of the monomer.

27.77 (••) Is the compound D or L?

27.78 (•) What kind of glycoside bond connects the monosaccharides in the dimer?

27.79 (•••) Would you expect the disaccharide to give a positive or a negative Tollens test?

27.80 (••••) A derivative of chitin, called chitosan, is sometimes incorporated into bandages to aid healing. Propose a reaction to convert the chitin dimer into a chitosan dimer. [*Note: The glycoside linkage is unstable in acid.*]

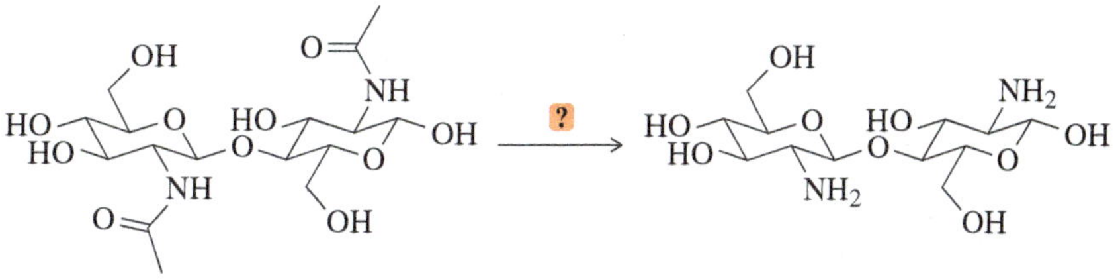

chitin **chitosan**

27.81 (••••) In the mid-1990s, the dehumidifier at a Midwestern university art museum malfunctioned, resulting in increasing levels of moisture in the air. The paint in some of the museum's oil paintings became brittle and flaky. What process was occurring? [*We learned about this reaction in Chapter 19.*]

The surface of cell membranes often has sugars attached to lipid tails to detect and communicate with the environment outside of the cell. Assessments 27.82–27.84 refer to the following structure, referred to as a glycolipid.

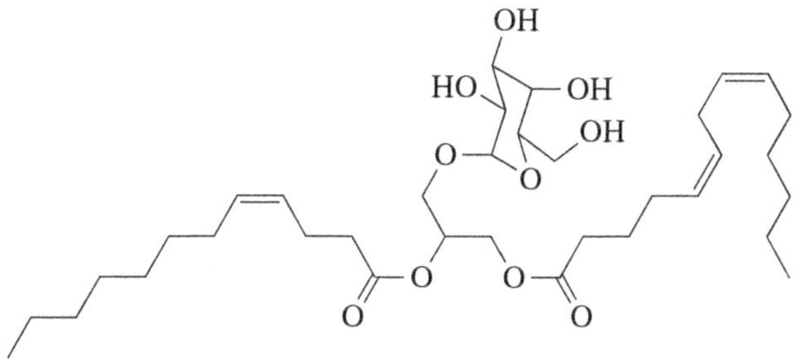

glycolipid

27.82 (•) What monosaccharide is used in the glycolipid?

27.83 (••) Name the fatty acids that make up the glycolipid using the shorthand notation.

27.84 (•••) Suggest an arrow-pushing mechanism by which the monosaccharide you named in Assessment 27.82 becomes attached to the diglyceride. Assume that an enzyme active site will provide whatever acids and bases you might need.

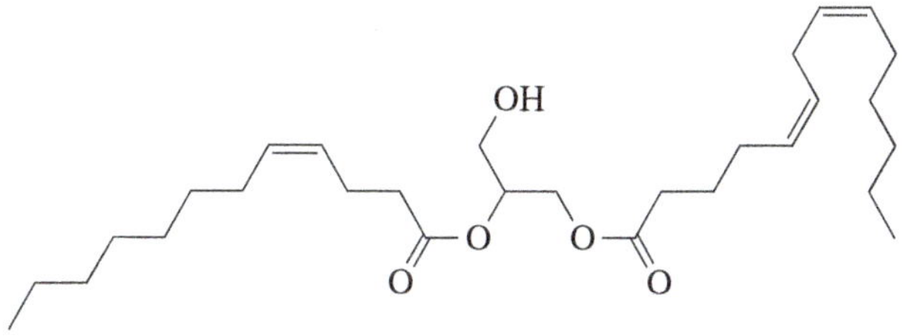

diglyceride

Assessments 27.85–27.89 refer to the DNA sequence ATG.

27.85 (••) Draw the chemical structure for the DNA sequence.

27.86 (••) Draw the structure of the mRNA codon the cell will translate this to.

27.87 (••) Draw the structure of the tRNA codon that will be used.

27.88 (••) What amino acid will be formed?

27.89 (••••) Suppose a mistake is made in transcription. (a) If the thymine in DNA is incorrectly translated to a cytosine, how will this affect the resulting protein? (b) What if the guanine is incorrectly translated to another guanine?

27.90 (••••) α-Pinene is a terpene found in pine trees. Starting from geranyl diphosphate (a terpene), draw the arrow-pushing mechanism to produce α-pinene. Assume that an enzyme active site will provide whatever acids and bases you might need. [*We did reactions similar to this in Chapter 8.*]

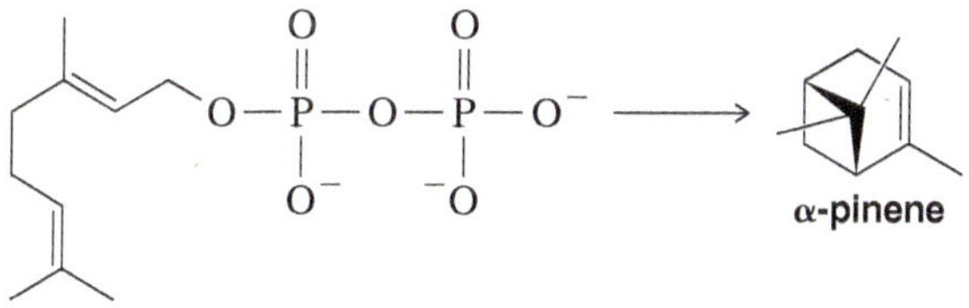

α-pinene

geranyl diphosphate

27.91 (••••) The following compound was recently reported to inhibit the ACP enzyme used in lipid biosynthesis. [*See Figure 27.78.*] It is believed to inhibit the enzyme by forming a covalent bond. Using what you've learned in this class so far, propose a mechanism for this inhibition. [*Think about the nucleophilic reactions from Chapter 21.*]

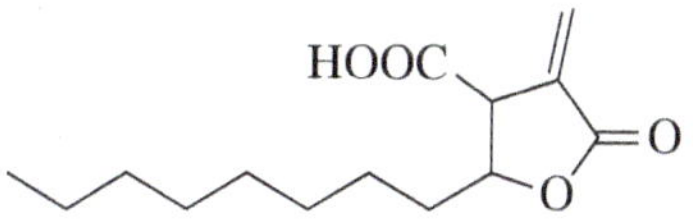

28

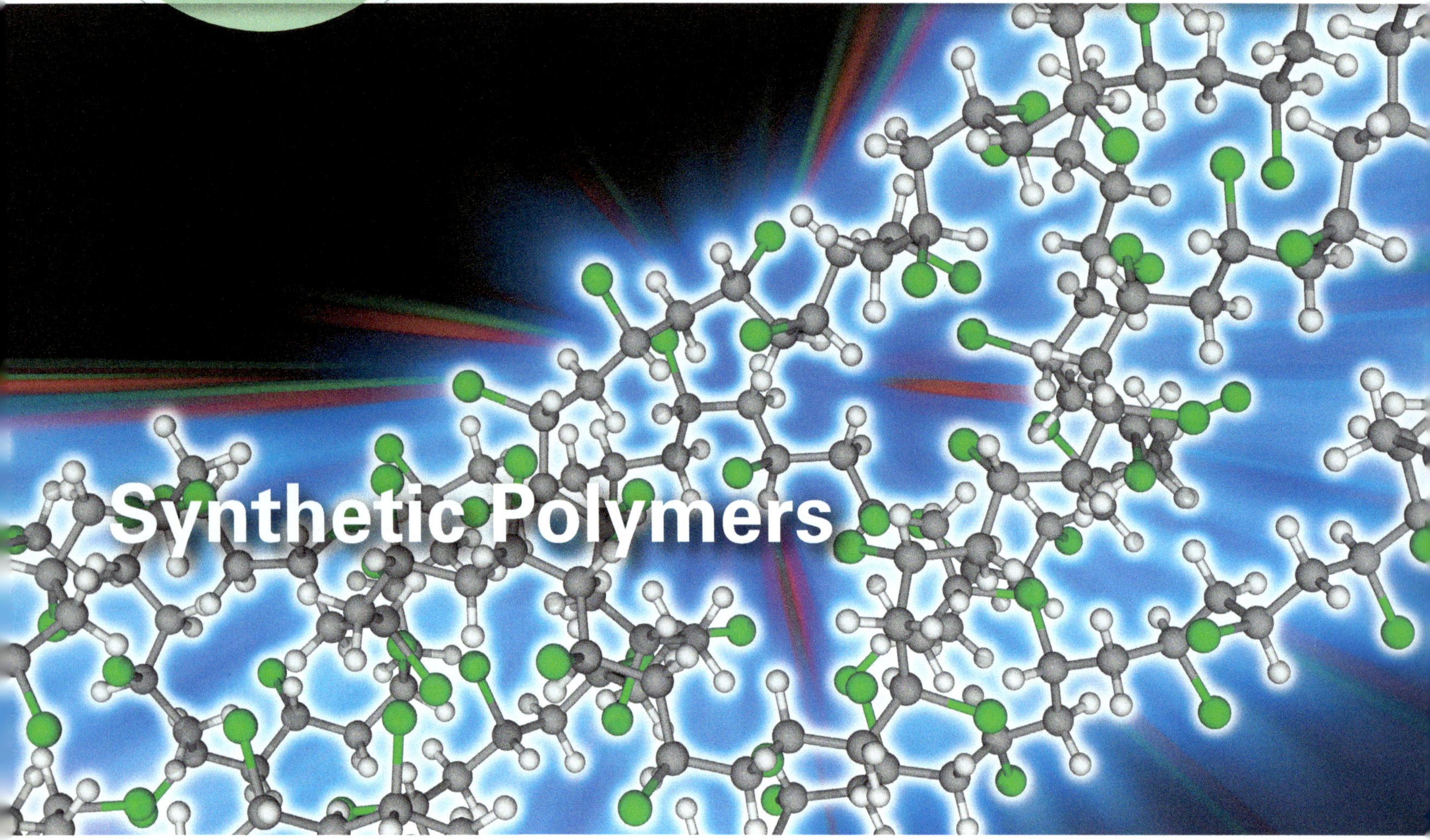

Synthetic Polymers

Learning Objectives

LO 28.0 Show the structure of a polymer

LO 28.1 Differentiate between chain-growth and step-growth polymerization reactions

LO 28.2 Describe the physical properties of polymers

LO 28.3 Analyze addition polymerization reactions

LO 28.4 Analyze condensation polymerization reactions

LO 28.5 Analyze epoxy resin, ROMP, and transition metal–catalyzed polymerization reactions

LO 28.6 Analyze copolymerization reactions

Introduction

Polymers, ubiquitous in society, have greatly improved many aspects of our lives. The computers and phones we use, the water bottles we drink from, and the clothes we wear contain synthetic polymers. To identify a single interesting story or application of polymers would do an injustice to the entire field. And so all of Chapter 28 is an example of chemistry in context, where the many reactions we have studied are repeated but in the application of making some of the everyday materials we take for granted.

As with all things, there is a downside to polymers. The same property that makes them suitable for long-term storage of food results in their persistence in landfills for generations. Historically, their preparation used toxic chemicals and harsh, energy-intensive conditions. Many plastics are mistaken as food and ingested by marine and avian species, resulting in their death. In sum, our dependence on polymers has the potential to cause problems.

Fortunately, the green chemistry movement has a stronghold in the area of polymers. Notable improvements have been made in polymerization conditions, the biodegradability of polymers, and the use of renewable feedstocks, among others. As we learn about polymers, a few of these improvements are highlighted.

○○○○○○○

And so begins Chapter 28, the last in our conversation about organic chemistry. Bittersweet, I'm sure. [*It certainly is for me, but I'll keep emotion out of it.*] One quote has opened every chapter, and there was meaning behind each. For the first time, I'll explain the meaning.

"Now this is not the end. It is not even the beginning of the end. But it is, perhaps, the end of the beginning."

—WINSTON CHURCHILL

The Winston Churchill quote is pertinent to this chapter in two ways. [*I'll explain the second way at the end of the chapter/book.*] Consider the addition of HCl to an alkene, one of the first reactions we studied. Following protonation of the π bond, the resulting carbocation is attacked by the chloride to make an alkyl halide. The reaction has ended. For polymerization reactions, however, the formation of the carbocation doesn't signal the coming end of the reaction. Instead, it is only the beginning. You'll have to read the chapter to figure out why this is the case.

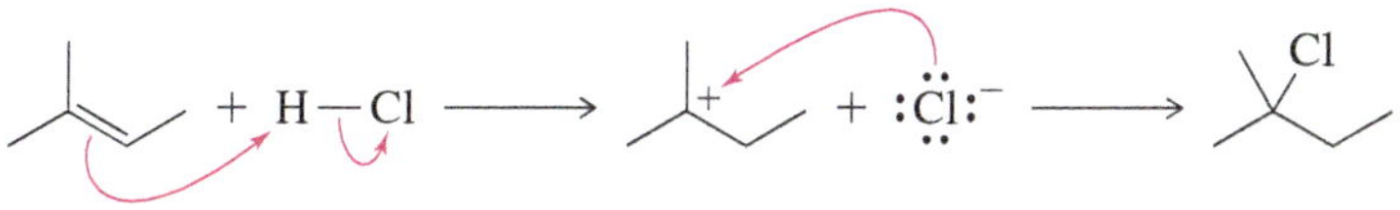

There are two ways that this chapter might be used in a standard organic chemistry course, and it's been written in such a way that it can accommodate both. While this sounds like something only your instructor needs to know, I think it is equally important for you, the student, to be aware of this. We have previously studied *every* reaction that is discussed in this chapter, but in a different context. As a result, some instructors will incorporate individual sections of this chapter while covering a particular reaction type (Section 28.4.1 as an add-on to Chapter 8, for example). Others will leave Chapter 28 for the end of the semester, with the idea that studying these reactions again, but in a different context, makes for a great way to prepare for a final exam.

Regardless of how your instructor is using this chapter, you've just been given great advice on how to best use it to increase your understanding of previous material—that is, how to "make it stick" (Section 1.4). When each polymerization reaction is introduced, the reactions we once studied are reviewed, while also referring back to where we studied it. Please use these mini-reviews. It will make polymerizations one of the easier topics we will study.

Consistent with this approach, and as we have done 27 times previously, let's see what we need to know to start this chapter.

Test your understanding

Assessment 28.1 Predict the product and provide an arrow-pushing mechanism for the addition of HBr to styrene. [*See Section 8.3.1.*]

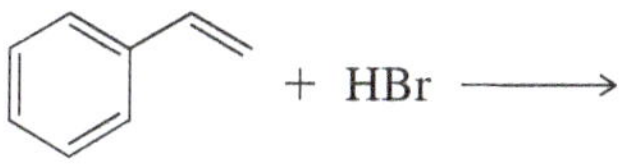

Assessment 28.2 Predict the product and provide an arrow-pushing mechanism for the Michael reaction of cyanide with the α,β-unsaturated ketone. [*See Section 21.5.2.*]

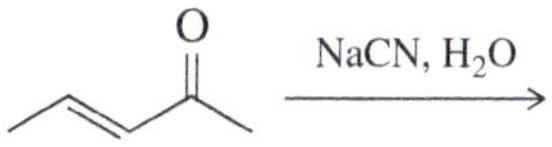

Assessment 28.3 Predict the product and provide an arrow-pushing mechanism for the free radical–mediated addition of HBr to but-1-ene. [*See Section 8.3.4.*]

Assessment 28.4 Predict the product and provide an arrow-pushing mechanism for the nucleophilic acyl substitution of a phenol with an acid chloride. [*See Section 19.7.2.*]

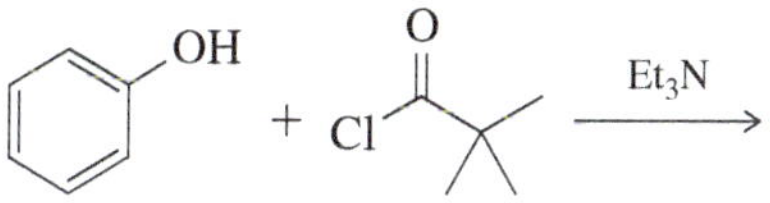

Assessment 28.5 Predict the product and provide an arrow-pushing mechanism for the palladium-catalyzed Suzuki coupling between a boronic acid and a vinylic halide. [*See Section 16.4.4.*]

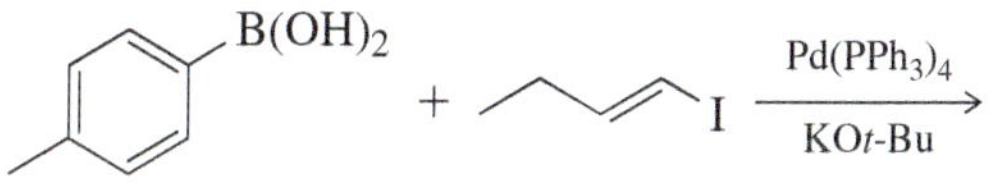

Assessment 28.6 Predict the product and provide an arrow-pushing mechanism for the ruthenium-carbene-catalyzed olefin metathesis reaction. [*See Section 16.6.*]

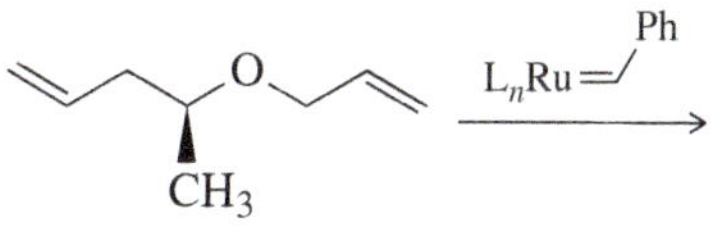

28.1 Polymers: an overview

From the Greek words *poly*, meaning many, and *meros*, meaning part, polymers are large molecules made through the combination of many smaller molecules. The smaller molecules, prior to combination, are called monomers. Proteins, cellulose, DNA, and other biological macromolecules are polymers, too, but the ones discussed here in Chapter 28 are all synthetic polymers. More specifically, they are synthesized using organic reactions.

28.2 Polymer representation and nomenclature

Because polymers are very large molecules, their representation can be confusing to students first studying them. Whereas our representation of small molecules is straightforward and complete, representing a polymer strand made up of thousands upon thousands of monomers is impossible. As a result, chemists have devised a system where the repeating units of a polymer are placed in brackets, with an italicized *n* to represent the number of monomers present in the polymer. If the number is known, then *n* is replaced with that number. If the number of monomers in a polymer is unknown (usually) or unimportant, then the italicized *n* is sufficient to indicate that a polymer is being represented. A polymer made only of benzene monomers is shown in **Figure 28.1**. If it is known that there are five benzene monomers in the polymer, then $n = 5$. Polymers usually have orders of magnitude more monomers than five, so a better representation of polybenzene would be to put the benzene in brackets with *n*.

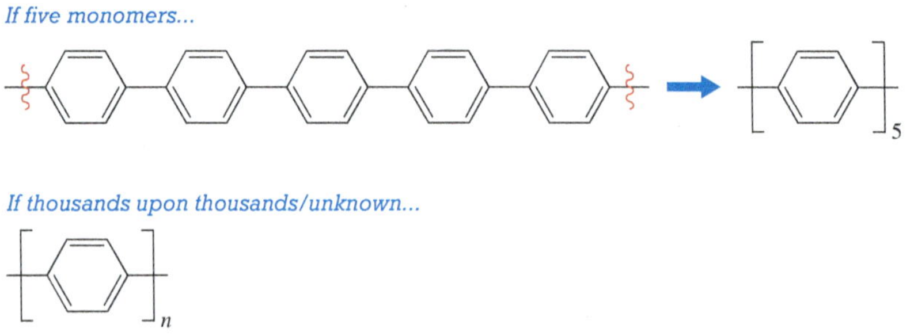

Figure 28.1 A simplified representation of polymers shows the repeating unit in brackets.

While this representation is straightforward for **homopolymers** (polymers containing only one monomer type), it is slightly more complicated for **copolymers** (polymers containing more than one monomer type) because it isn't as obvious where the brackets should be drawn to accurately depict the repeating unit. **Figure 28.2(a)** shows a polymer composed of alternating benzene and cyclopentane rings. Thus, one benzene and one cyclopentane should be contained within the brackets because that's the repeating unit. In this representation, the bond that passes through the left bracket is assumed to be connected to the group whose bond passes through the bracket on the right side. **Figure 28.2(b)** shows a polyurethane, which has a much larger repeating unit. Even so, the process is the same—namely, place the repeating unit within brackets. [*Assessment 28.7 gives you practice doing this yourself. It may seem tedious, but it is important to figure out how to do this before reading the rest of the chapter.*]

Figure 28.2 The representation of copolymers using the bracket notation. (a) Alternating benzene and cyclopentane and (b) a polyurethane.

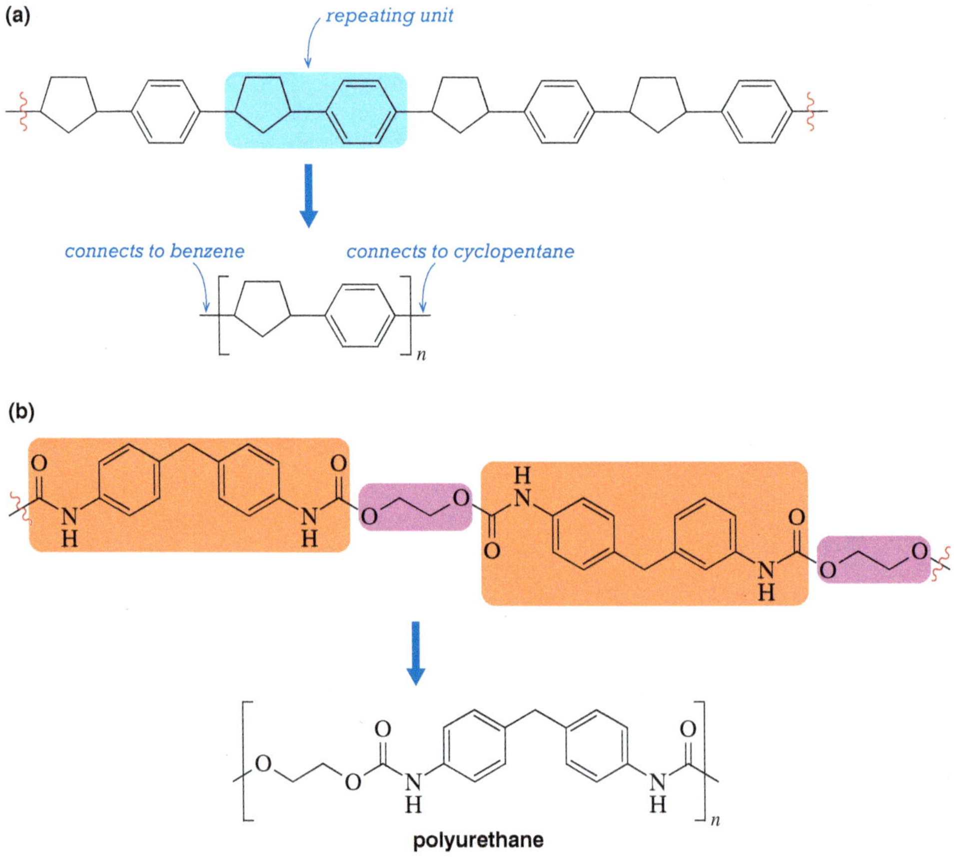

Assessment 28.7 Given the polymer structures shown, show how they would be represented using the bracket notation.

(a)

(b)

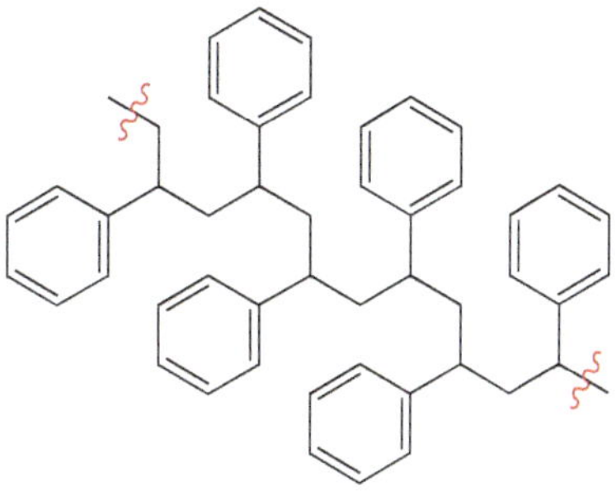

(c)

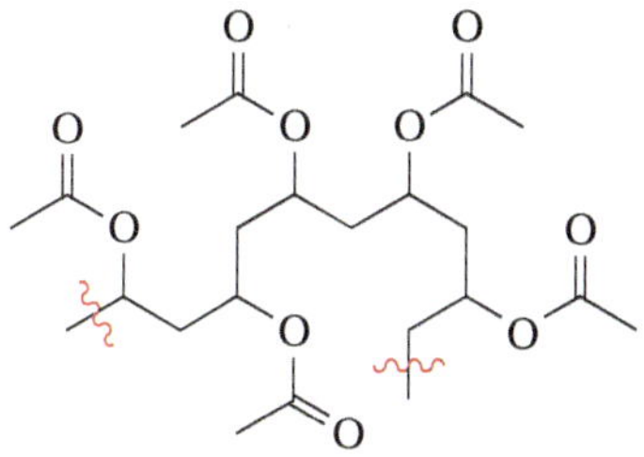

Assessment 28.8 Given the bracketed structure shown, draw the polymer. Your answer only needs to contain three monomers.

(a)

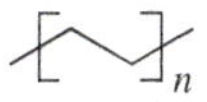

(b)

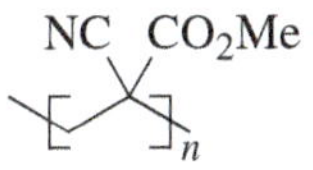

(c)

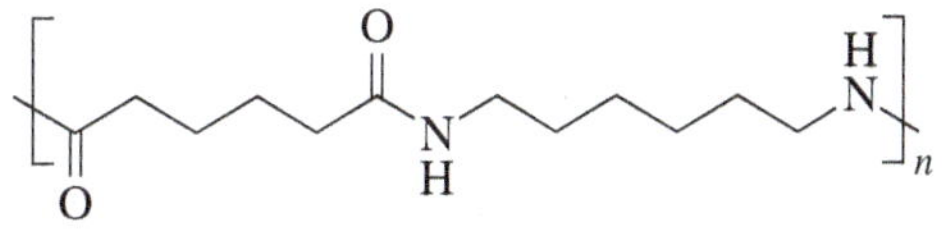

The nomenclature of polymers utilizes an IUPAC source-based system, meaning a polymer is named based on the monomers used to synthesize it (**Figure 28.3**). All that is necessary is to add the prefix "poly-" in front of the monomer name. Very often, the common name of the monomer is used in this name. If the monomer name is more than one word, the monomer name is placed in parentheses. For example, the polymerization of styrene, a reaction we study in Section 28.4, results in polystyrene. Similarly, the polymerization of vinyl chloride results in poly(vinyl chloride).

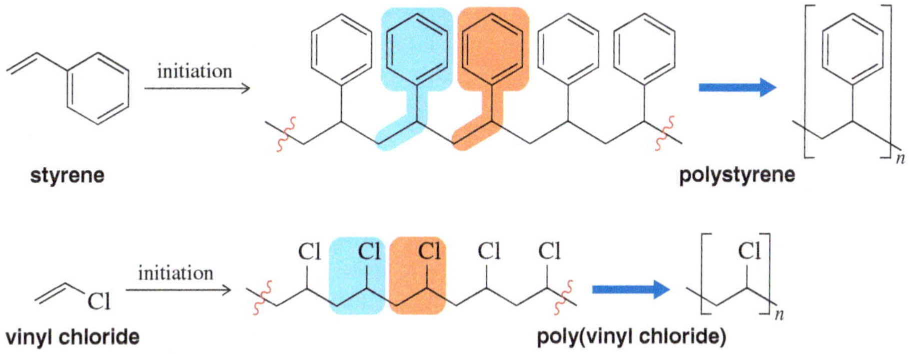

Figure 28.3 The nomenclature of polymers.

28.3 Polymer classification

Because of the diversity of structures that comprise synthetic polymers, there are multiple ways of classifying them depending on their application. These polymer classes are briefly discussed in the following sections.

28.3.1 Reaction type

In an introductory organic chemistry course, the most important way to classify polymers is by the reactions used to form them. This should also help you connect the new reactions to ones seen previously. [*That will continue to be a focus in this chapter.*]

Two of the most important reaction types used for polymer synthesis are addition and condensation. As we explain in Sections 28.4 and 28.5, addition polymerization is identical to addition reactions with alkenes (Chapters 8 and 9), alkynes (Chapter 10), aldehydes/ketones (Chapter 17), and conjugated systems (Chapters 21 and 22), and condensation polymerization is identical to the condensation reactions of carboxylic acid derivatives (Chapter 19) and enolates (Chapter 20). Figure 28.4 shows a general example of addition and condensation polymerization. **Addition polymers** contain all of the atoms present in the original monomers, whereas **condensation polymers** form with the extrusion of small molecules like water or alcohols. [*These details are elaborated upon later.*]

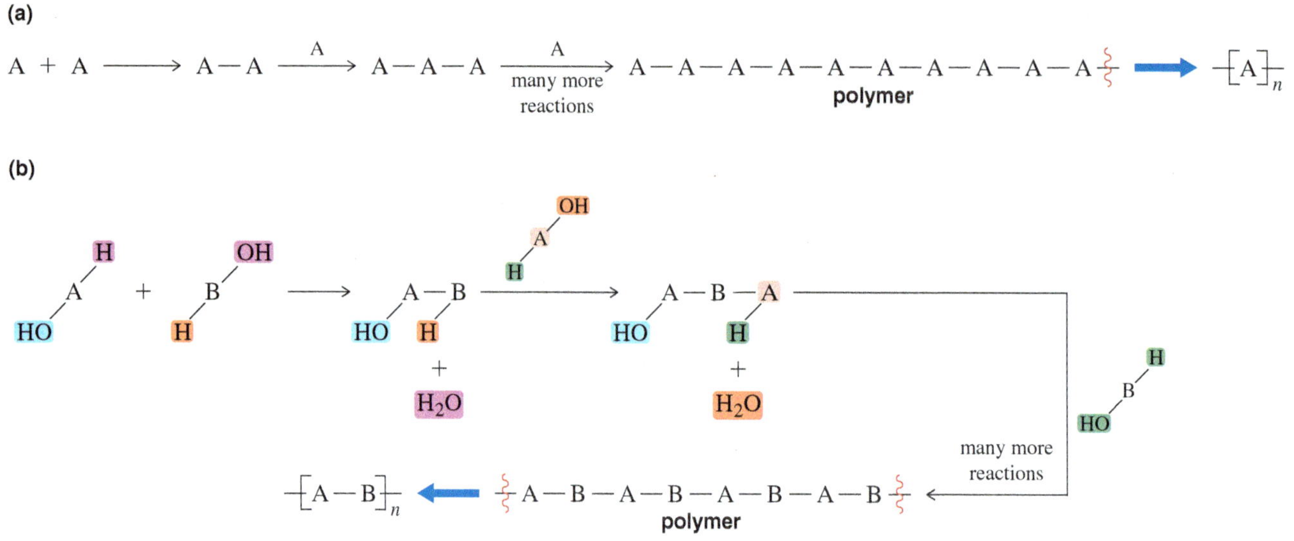

Figure 28.4 (a) Addition and (b) condensation polymers.

28.3.2 Chain growth versus step growth

Polymers can also be classified based on the manner in which monomers combine to make polymer chains. Polymers that grow in a single direction, most common in addition polymers, are referred to as **chain-growth polymers** (Figure 28.5). In these polymerizations, only one monomer can add at a time.

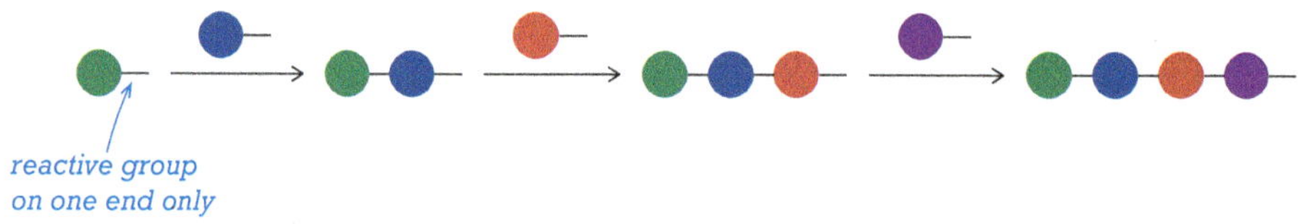

Figure 28.5 Chain-growth polymerization. Monomers add one at a time, with the chain growing in a single direction.

Step-growth polymers, on the other hand, can grow in multiple directions because they have reactive functional groups on both ends (Figure 28.6). Step-growth polymerization is common in condensation polymers. Whereas chain-growth polymers extend by only one monomer at a time, step-growth polymers can grow by the combination of dimers, trimers, or other small polymers to extend the chain in either direction.

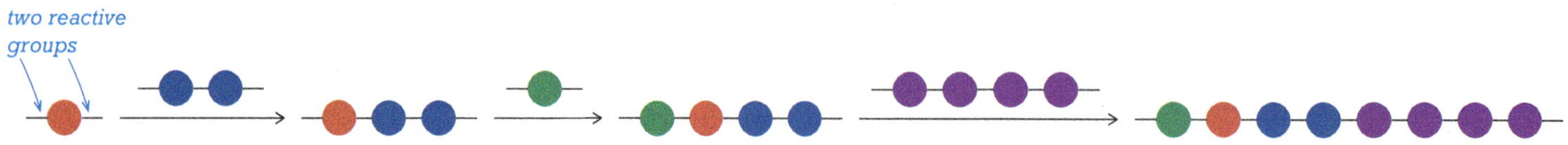

Figure 28.6 Step-growth polymerization. Polymers form in either direction by the combination of monomers, dimers, and other small polymers.

28.3.3 Structure

Polymers come in a number of different shapes, each shape bringing different properties. Of the structures shown in Table 28.1, linear, branched, and cross-linked will comprise the majority of polymer types we study going forward.

Table 28.1 The classification of polymers based on shape.

Shape	Description	Example
Linear	All monomers are attached end to end in a straight chain.	
Branched	There are multiple branching points off the main polymer chain.	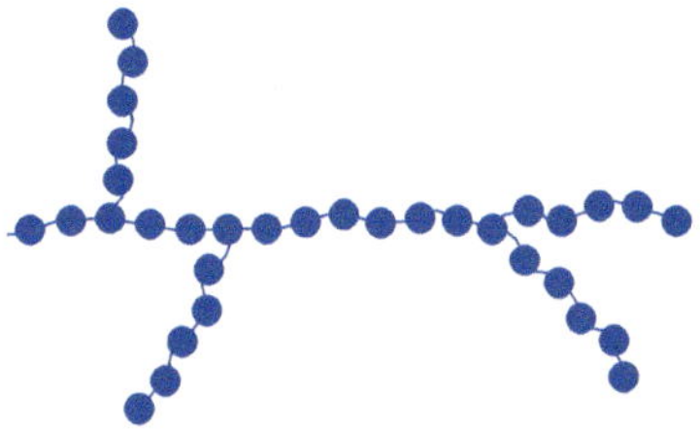
Cross-linked	Individual polymer chains are connected to each other through a cross-linking chain.	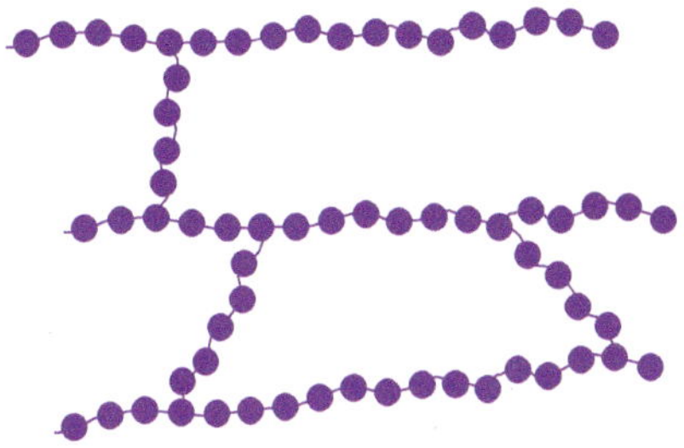
Star	Polymer chains emanate from a central core.	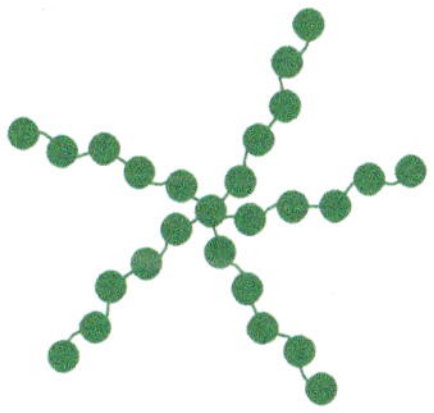
Comb	There is regular, repeated branching off the main polymer chain.	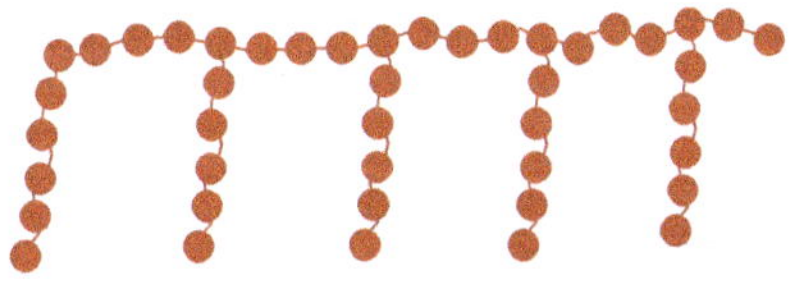
Brush	There is highly concentrated branching off the main polymer chain in all directions.	

28.3.4 Properties

The properties of polymers are based on the same intermolecular interactions present in small molecules. It's just that, because of their size, there are considerably more of those intermolecular interactions in polymers. Cyclohexane, for example, has only a few weak intermolecular van der Waals interactions, resulting in a relatively low boiling point. Within a nonpolar polymer, however, there are so many more van der Waals interactions that the polymer can exist as a rigid, high melting, crystalline solid.

The physical properties of polymers are difficult to predict because their architectures vary so much. Nevertheless, a few general principles can be understood. First, the **crystallinity** of a polymer is a result of how closely the polymer chains can pack. A linear, unbranched polymer like polyethylene prepared by one method, though not perfectly ordered, has multiple **crystallite** regions, making it a rigid, highly crystalline polymer [**Figure 28.7(a)**]. On the other hand, polyethylene that is prepared via a different method contains multiple branches. These branches interrupt the formation of crystallites, making branched polyethylene more **amorphous**, or less rigid/more flexible [**Figure 28.7(b)**].

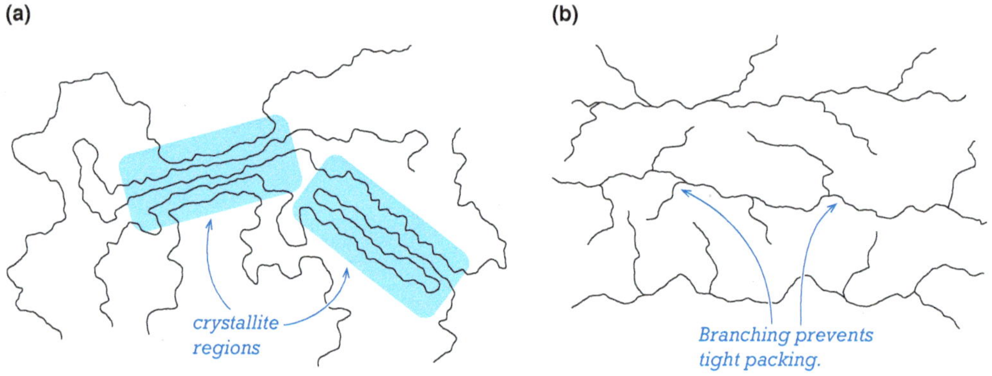

Figure 28.7 Linear versus branched polyethylene. (a) Linear polyethylene (high-density polyethylene) forms multiple crystallites. (b) Branched polyethylene (low-density polyethylene) is unable to form crystallites.

When the temperature is increased, a small molecule undergoes phase transitions from solid to liquid to gas. Polymers, containing myriad intermolecular interactions, do not exist in the gas phase and rarely exist as pure liquids except at very high temperatures. However, they do undergo phase transitions as they are heated (**Figure 28.8**). These transitions occur at the **glass transition temperature (T_g)** (the temperature at which a hard, crystalline polymer becomes a soft and flexible solid) and the **crystalline melting temperature (T_m)** (the temperature at which a solid polymer becomes a viscous liquid). When a crystalline polymer is at room temperature, the crystallinity results in a highly rigid structure with relatively little molecular motion. If struck, a crystalline polymer would shatter like glass. As the polymer is heated above T_g, molecular motion is increased, reducing the crystallinity of the polymer. Above the glass transition temperature, the polymer becomes flexible and moldable. Polymers that become moldable above T_g are referred to as **thermoplastics**. As the temperature continues to increase beyond T_m, further molecular motion results in the polymer chains being able to freely slide around, resulting in a viscous liquid that can be manipulated into fibers.

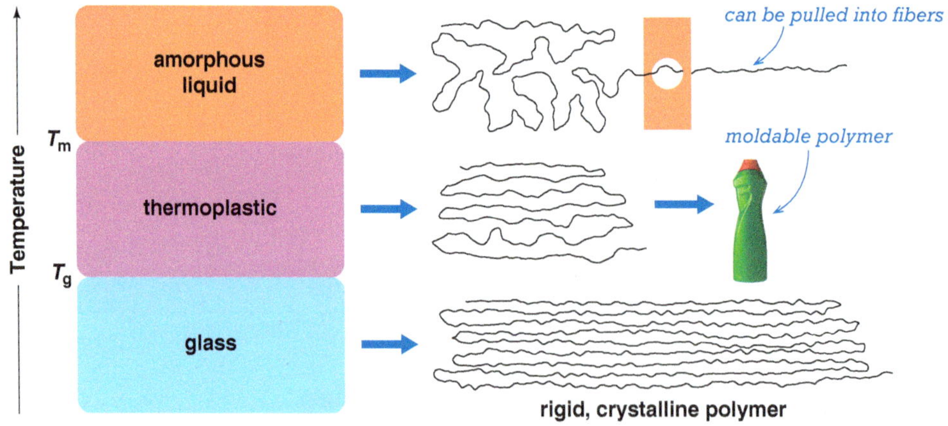

Figure 28.8 Phase changes of crystalline polymers.

When a polymer has desirable properties but is too brittle at the temperature at which it will be used, small molecules can be dissolved into the polymer. These small molecules, called plasticizers, break up van der Waals interactions, thus reducing crystallinity, which makes rigid, brittle polymers more flexible (**Figure 28.9**) and reduces the glass transition temperature. Di-(2-ethylhexyl) phthalate is an example of a plasticizer added to polymers for this purpose.

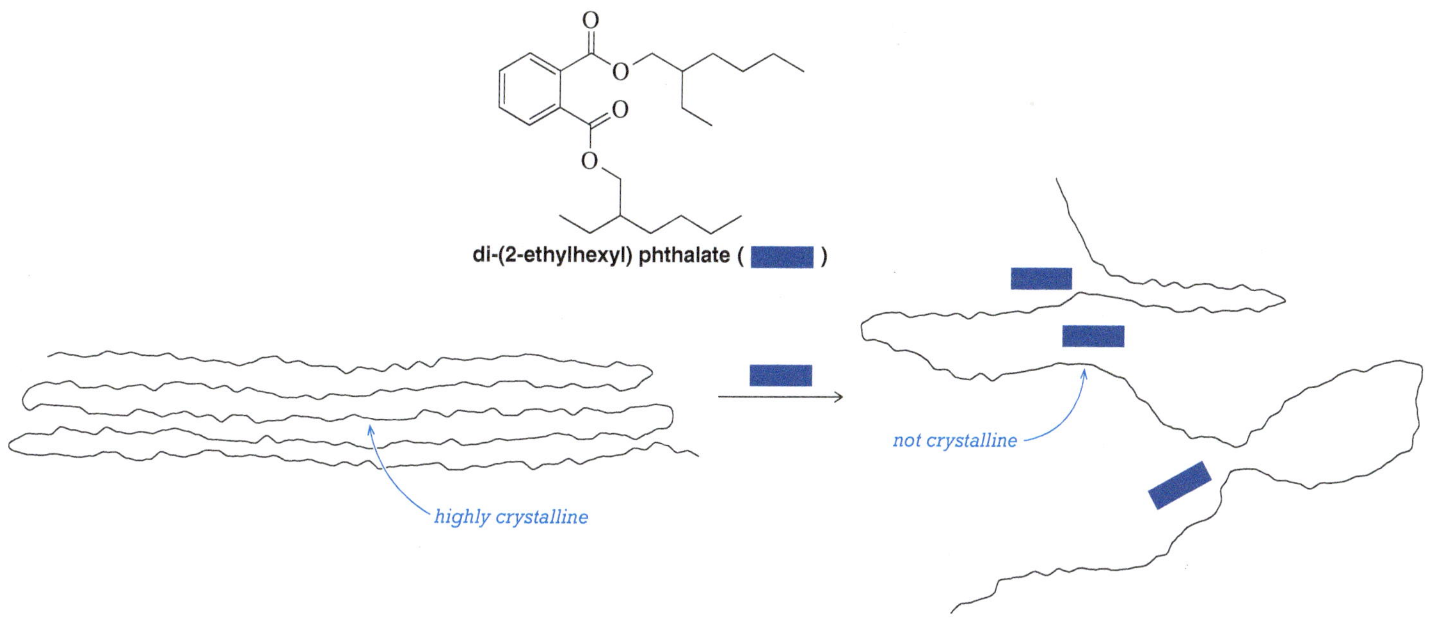

Figure 28.9 Plasticizers interrupt intermolecular interactions, reducing the crystallinity of polymers.

28.4 Addition polymers

The first time we encountered addition reactions was in Chapter 8. In an addition reaction, two molecules combine to make one without the extrusion of atoms/groups. These reactions typically involve addition to a C—C or C—O π bond and are entropically disfavored ($2 \rightarrow 1$; $\Delta S < 0$). Table 28.2 shows several examples of addition reactions we have studied thus far, including reactions of alkenes, alkynes, dienes, aldehydes, and ketones.

Table 28.2 Addition reactions studied thus far (not a complete list).

Reaction	Example
General form	A + B ⟶ A—B
Halogenation of an alkene (Section 9.1.2)	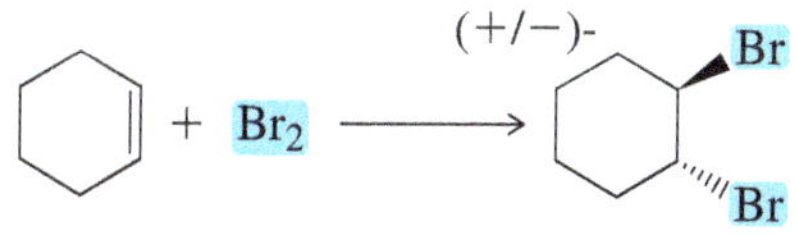
Hydrogenation of an alkyne (Section 10.6.1)	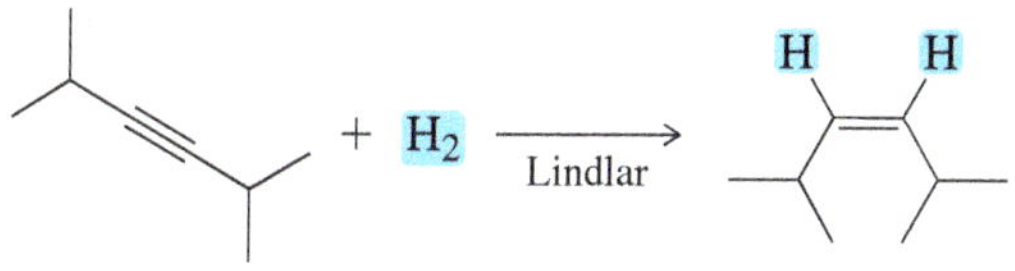
Hydrohalogenation of butadiene (Section 21.4.1)	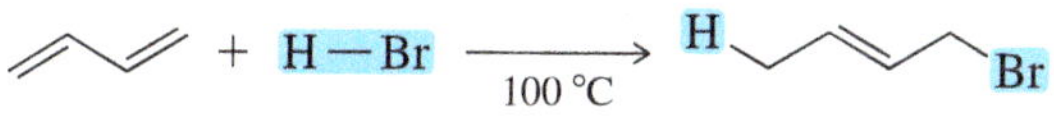
Diels–Alder reaction (Section 22.3)	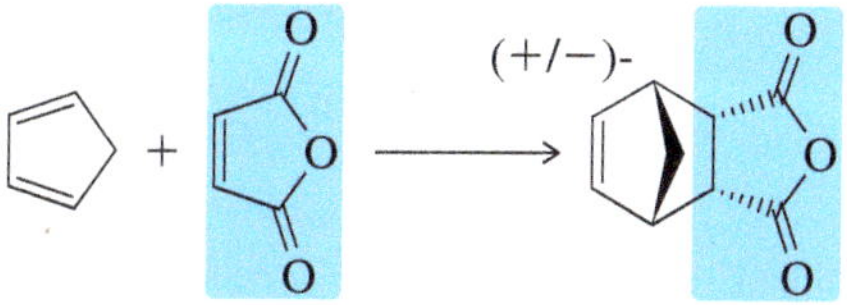
Hemiacetal formation (Section 17.7.2)	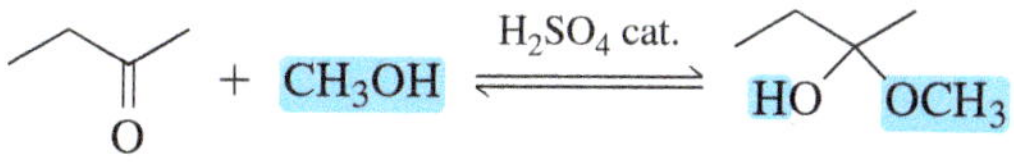

An addition polymerization is similar to the reactions in Table 28.2, except that a single molecule of A adds a molecule of A, which then adds another molecule of A, and so on until a long polymer chain has been produced (**Figure 28.10**). Like their non-polymer counterparts, addition polymerization usually occurs by addition to C—C π bonds. In the process of combining monomers, no atoms or groups from A are lost. [*In condensation polymerization, atoms/groups are extruded as the polymers form.*]

$$\text{A} + \text{A} \longrightarrow \text{A—A} \xrightarrow{\text{A}} \text{A—A—A} \xrightarrow[\text{many more reactions}]{\text{A}} \text{A—A—A—A—A—A—A—A—A—A} \longrightarrow \text{-[A]-}_n$$

polymer

Figure 28.10 The general form of addition polymerization.

As you study the three types of addition polymers in the sections that follow, keep in mind that you have seen all of these reactions before in other contexts. We are not learning new reactions; they are simply occurring repeatedly within a growing molecule as the polymer forms.

28.4.1 Cationic polymerization (electrophilic addition)

QUICK REVIEW We studied electrophilic addition of alkenes in Chapter 8. The acid-catalyzed addition of water to 1-methylcyclohexene is an example of this reaction [**Figure 28.11(a)**]. Protonation of the alkene generates the more stable 3° carbocation in a slow, rate-determining step. Water is weakly nucleophilic and in great abundance, so it attacks the 3° carbocation, bringing along both protons. In the final step, the acidic proton of the hydronium ion is removed to complete formation of the 3° alcohol. By a similar mechanism, 2-bromo-2-methylbutane can be formed by addition of HBr to 2-methylbut-1-ene [**Figure 28.11(b)**]. [*For a detailed review on these reactions, see Sections 8.3 and 8.4.*]

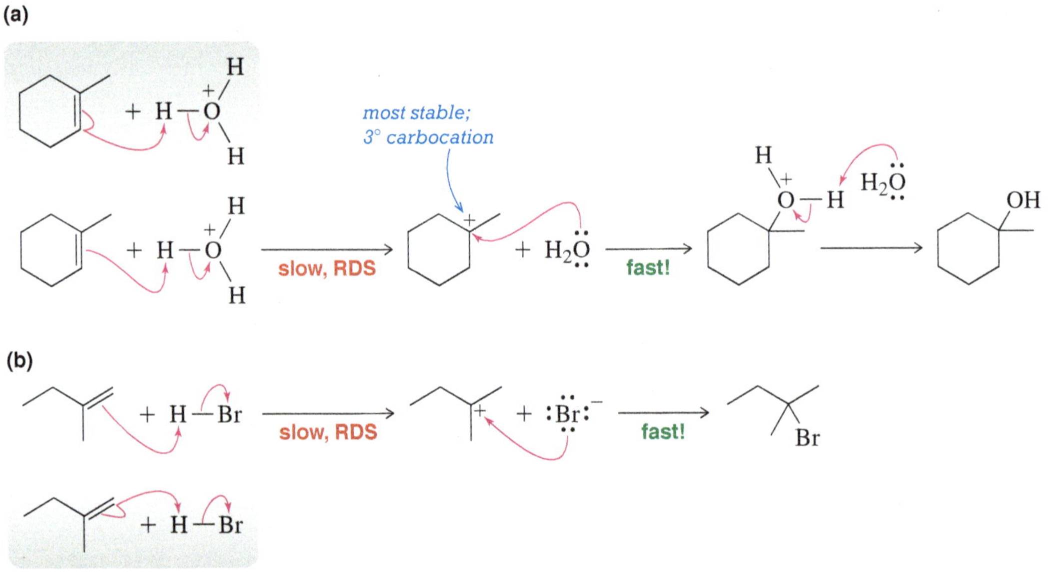

Figure 28.11 Electrophilic addition to alkenes. (a) Acid-catalyzed addition of water to an alkene. (b) Addition of HBr to an alkene.

APPLICATION TO POLYMERS In Figure 28.11(a) and (b), the carbocation intermediate was attacked by water, a weakly nucleophilic solvent molecule, or bromide, a moderately strong nucleophile. In order for electrophilic addition to be useful for polymer synthesis, the carbocation needs to instead be attacked by another molecule of alkene. How might this be accomplished? [*Any ideas?*] First, there can be no nucleophiles present. And second, to ensure that a carbocation is attacked by another alkene molecule, the reaction should be conducted at an extremely high concentration. This process is demonstrated in the polymerization of isobutylene (**Figure 28.12**). The use of a Lewis acid catalyst in the presence of only a trace amount of water generates acid in a first initiation step. [*Because polymerizations are considered chain reactions, we use the terms initiation, propagation, and termination like* *we did when studying radical reactions in Chapter 11.*] Protonation of an isobutylene monomer in the second initiation step generates the propagating carbocation. This carbocation is attacked by a second isobutylene monomer to generate another carbocation. This process continues as the chain is extended indefinitely, resulting in the polymer polyisobutylene.

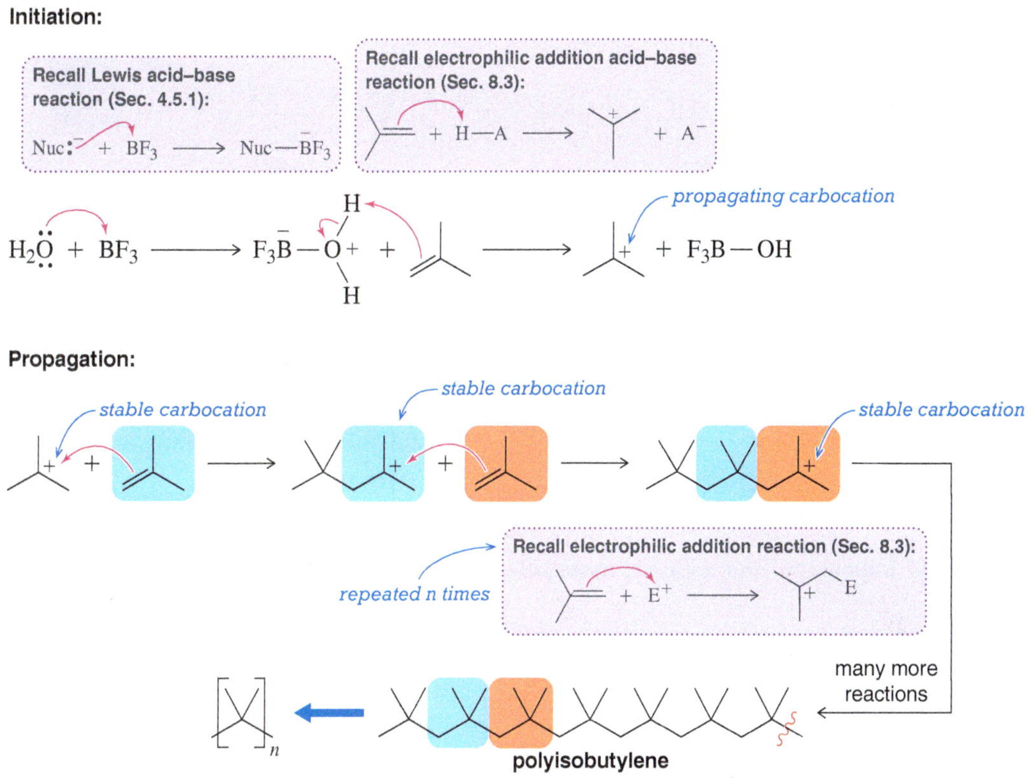

Figure 28.12 The polymerization of isobutylene by an acid catalyst.

Termination of the polymer chain reaction occurs when, instead of a carbocation being attacked by the alkene, a proton is removed instead. This is essentially the second step of an E1 elimination reaction (**Figure 28.13**). All cationic polymerizations terminate in this way.

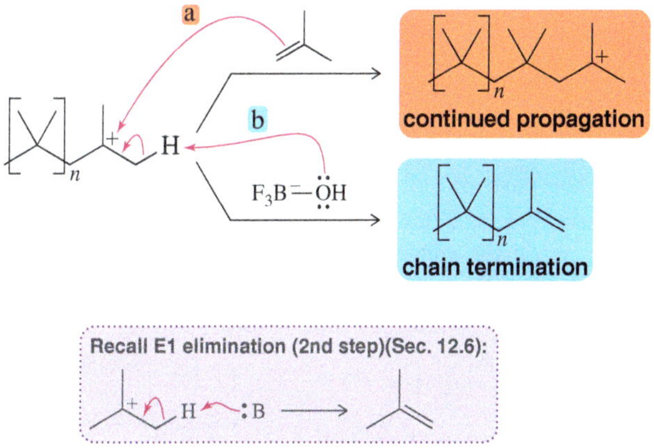

Figure 28.13 The termination of a cationic polymerization occurs when a proton is removed from the propagating chain.

Polystyrene can also be generated from styrene via cationic polymerization. Tin(IV) chloride with a trace amount of water initiates the polymerization (**Figure 28.14**). Because of the high exothermicity, these reactions are conducted at very low temperatures. Recall that polymerization is an enthalpy-driven process, so it is typically disfavored by temperature. Thus, a low temperature does not slow polymerization in the way that a low temperature dramatically slows down other reactions. When air is blown into polystyrene, tiny gas pockets form, creating what is referred to as polystyrene foams. These polystyrene foams (Styrofoam™) are used in packaging materials, building materials, and disposable coffee cups.

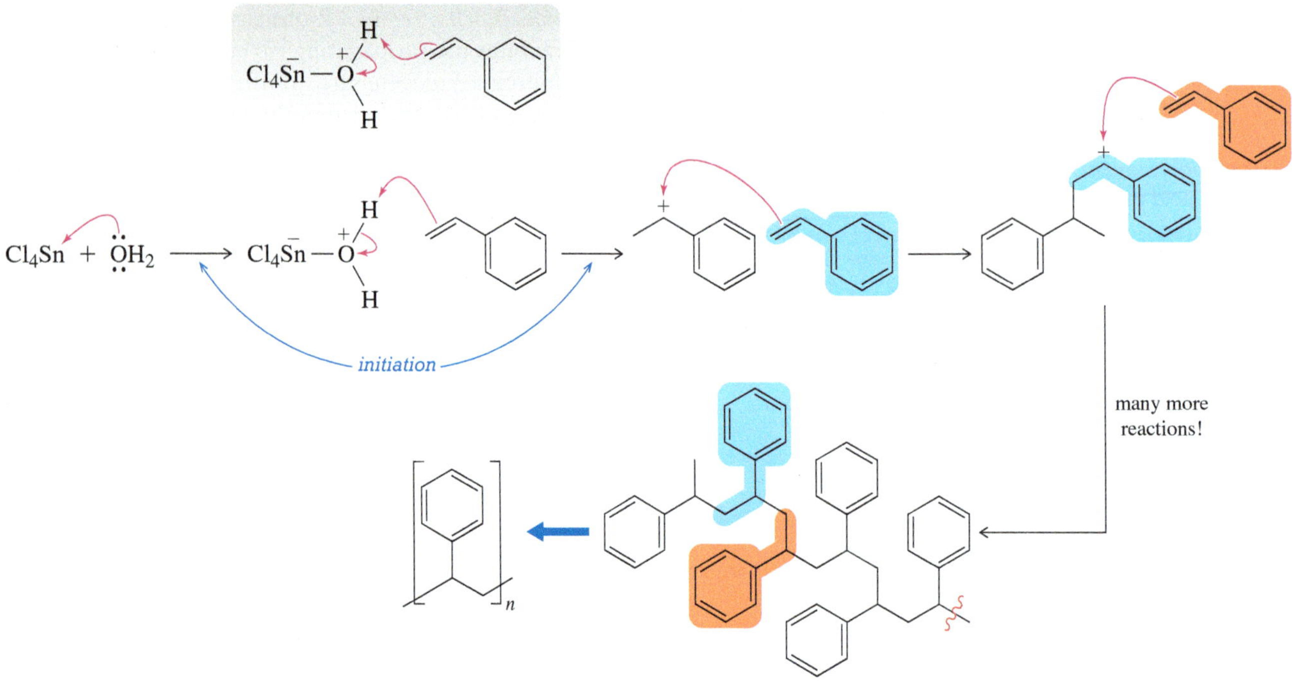

polystyrene

Figure 28.14 Cationic polymerization to make polystyrene.

Assessment 28.9 Although it contains an alkene, cyanoacrylate cannot be polymerized under cationic conditions. (a) Why? (b) What conditions might be used to polymerize it?

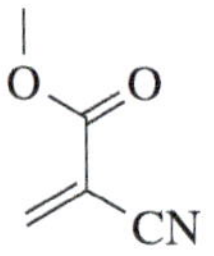

cyanoacrylate

Let's solve this one together as a means of "inventing" a new polymerization.

Solved Assessment 28.9

Enthalpically driven reactions, such as polymerizations, proceed only when the resulting product/intermediate is *more* stable than the starting materials. In the case of polystyrene (Figure 28.14) and polyisobutylene (Figure 28.13), a newly formed C—C bond in a cationic intermediate made the reaction favorable. Thus, to determine why cyanoacrylate will not polymerize in a similar manner, we should analyze the intermediates that would form in the initiation and propagation steps.

Upon protonation of the alkene in cyanoacrylate, a carbocation forms on the same carbon as an ester and a cyano group. [*What are your thoughts about the stability of this?*] Recall from Section 23.8 that esters and cyano groups are electron-withdrawing groups. Although an electron-donating group will stabilize a carbocation by delocalizing the positive charge, electron-withdrawing groups increase the size of the positive charge, thereby destabilizing the carbocation (**Figure 28.15**).

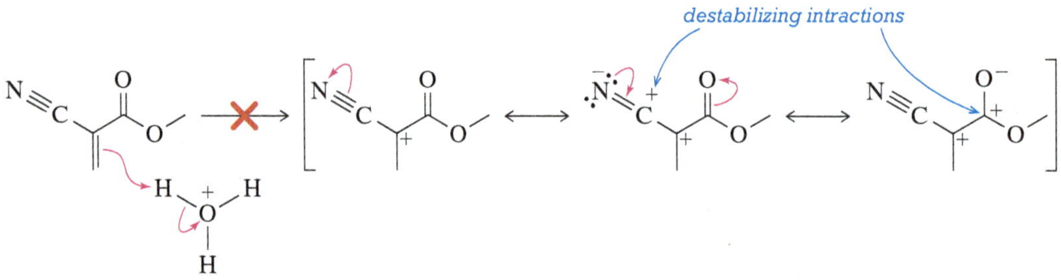

Figure 28.15 Electron-withdrawing groups destabilize the carbocation, making polymerization by a cationic mechanism unfavorable.

Comparing the intermediates in the successful styrene polymerization and unsuccessful cyanoacrylate polymerization on a reaction coordinate diagram shows these differences (**Figure 28.16**). Whereas the carbocation next to the benzene ring is stabilized by resonance [*Section 24.1.3*] (and therefore *lower* in energy on the reaction coordinate diagram), the carbocation next to the ester and cyano group is significantly less stable (and therefore *higher* on the reaction coordinate diagram). Being higher on the reaction coordinate diagram makes the reaction highly endothermic, with an activation energy that apparently cannot be overcome. [*The answer to Assessment 28.9(b) comes later; continue thinking about it.*] ■

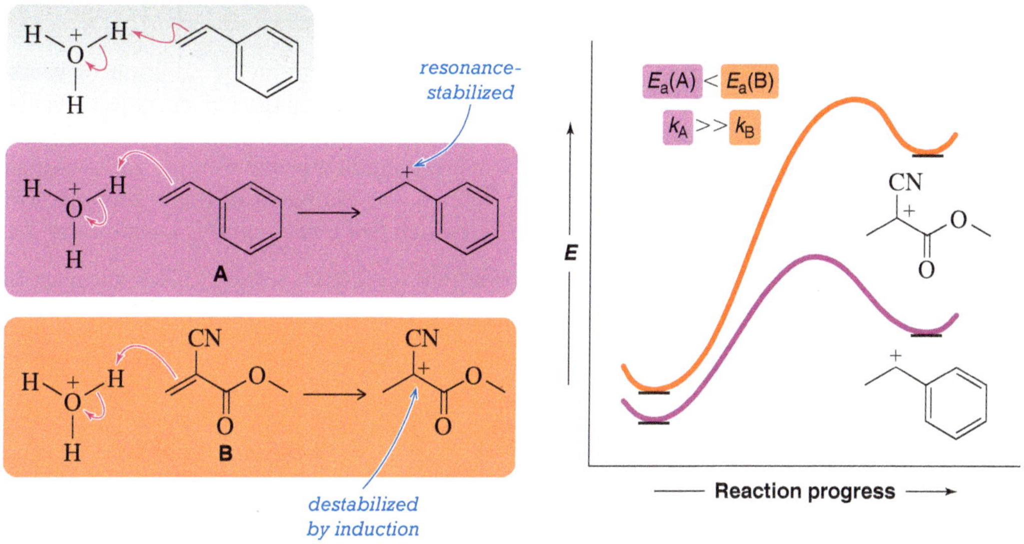

Figure 28.16 The formation of a highly unstable carbocation has a high activation energy.

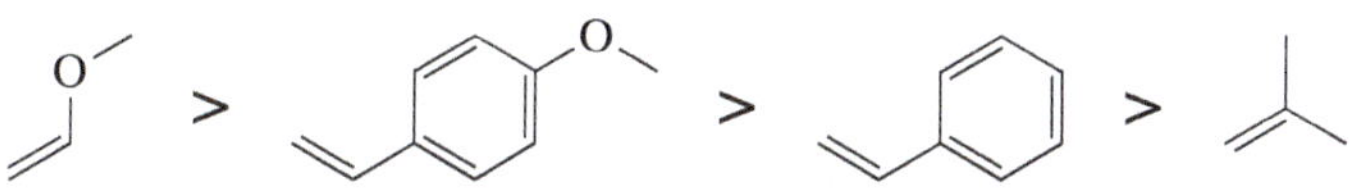

Figure 28.17 Electron-rich alkenes more easily undergo cationic polymerization.

Solved Assessment 28.9 illustrates a principle we might have predicted based on our experiences in Chapters 8 and 23. Just as electron-rich alkenes (Chapter 8) and aromatic rings (Chapter 23) undergo electrophilic addition more rapidly, only alkenes capable of forming relatively stable carbocations can be polymerized under cationic conditions. **Figure 28.17** illustrates this trend.

Assessment 28.10 Suggest a mechanism for the cationic polymerization of the molecules shown here. You can stop the mechanism after the addition of three monomers.

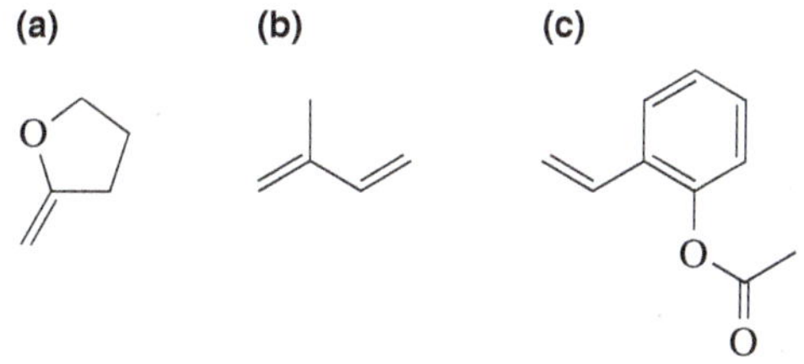

Assessment 28.11 When isobutylene polymerizes, the dimethyl groups are always on every other carbon. Explain why this is true.

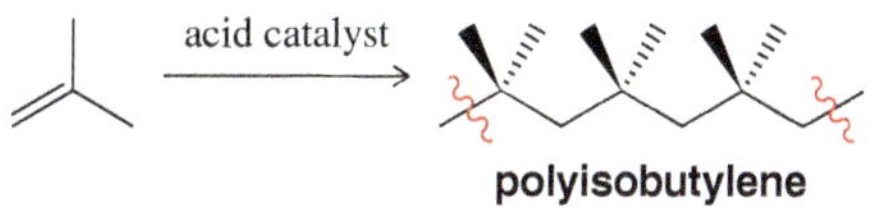

28.4.2 Anionic polymerization (nucleophilic addition)

In Assessment 28.9, we learned that cyanoacrylate could *not* be polymerized under cationic conditions.

Let's think like a chemist to see how it might be polymerized (**Figure 28.18**).

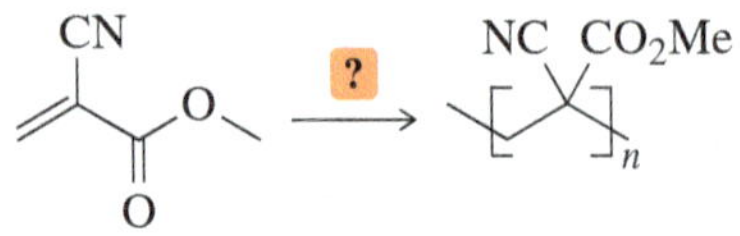

Figure 28.18 How can cyanoacrylate be made to polymerize?

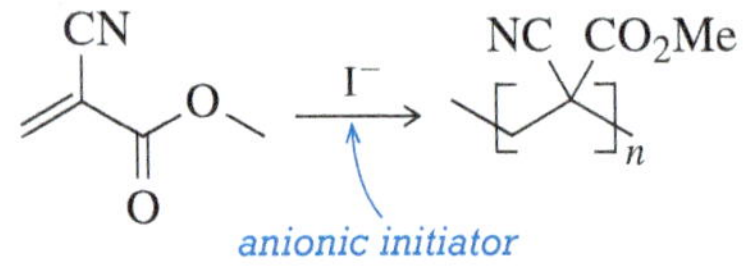

Figure 28.19 An anionic initiator polymerizes cyanoacrylate.

Thinking Like a Chemist

The conditions of cationic polymerization require the alkene to react as a nucleophile. As a result, cationic polymerization only occurs with electron-rich alkenes. In Assessment 28.9, it became clear that the ester and cyano groups are electron-withdrawing, thus destabilizing any carbocations that might form nearby. If electron-withdrawing groups destabilize carbocations, what types of intermediates would they stabilize? [*Thoughts?*] Electron-withdrawing groups stabilize carbanions. Thus, instead of using an electrophilic, or cationic, initiator, an anionic initiator is used (**Figure 28.19**). [*We examine this mechanism shortly.*] ■

QUICK REVIEW In Chapter 21, we studied the conjugate addition of weak nucleophiles to α,β-unsaturated carbonyls. The 1,4-conjugate addition of cyanide to pent-3-en-2-one is an example of this reaction (**Figure 28.20**). Attack of the nucleophile at the β-carbon pushes the C—C π electrons onto the carbonyl oxygen. In a protic solvent, the resulting enolate quenches at the α-carbon to regenerate a ketone. The net result of this reaction is formation of a new C—C σ bond at the β-carbon. [*For a detailed review on this reaction, see Section 21.5.*]

APPLICATION TO POLYMERS Following the logic from Section 28.4.1, the reaction in Figure 28.20 can be turned into a polymerization reaction by increasing the concentration of the monomer (cyanoacrylate) and removing any protic solvents that might protonate the resulting enolate. This ensures that, following initiation, the enolate collides instead with another monomer, where it can do another Michael reaction. Mechanistically, a catalytic amount of a nucleophile (such as hydroxide) initiates the reaction by adding to the β-carbon (**Figure 28.21**). The resulting enolate represents the propagating anion. As such, it does a conjugate addition on another cyanoacrylate monomer to generate another carbanion. This process continues as the chain is extended indefinitely, resulting in

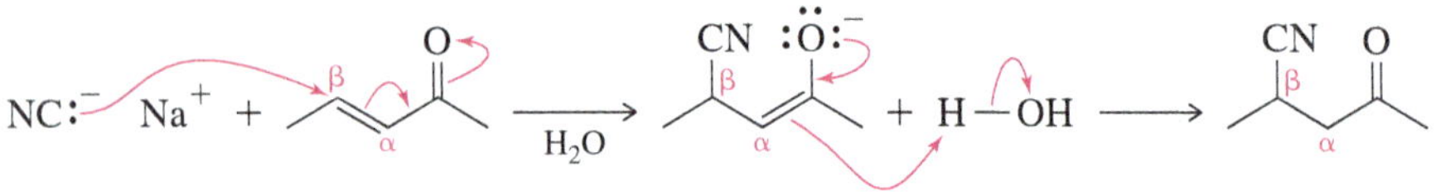

Figure 28.20 Nucleophilic addition to alkenes.

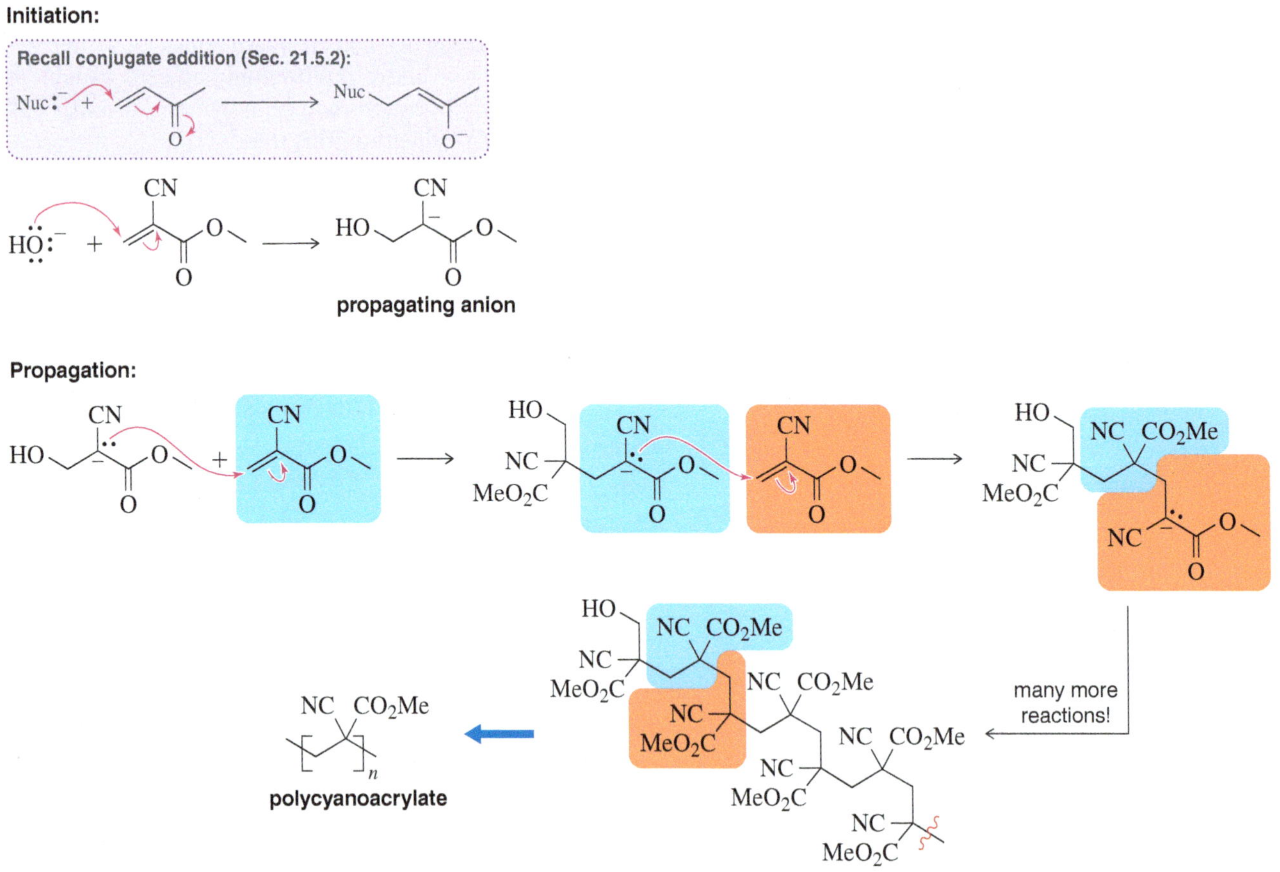

Figure 28.21 The anionic polymerization of cyanoacrylate.

polycyanoacrylate. Cyanoacrylate is the primary ingredient of Super Glue, and it is the polymerization reaction that provides its adhesive properties.

Termination of the polymer chain reaction occurs when, instead of the propagating carbanion doing a Michael reaction, it grabs a proton (**Figure 28.22**). All anionic polymerizations terminate in this way.

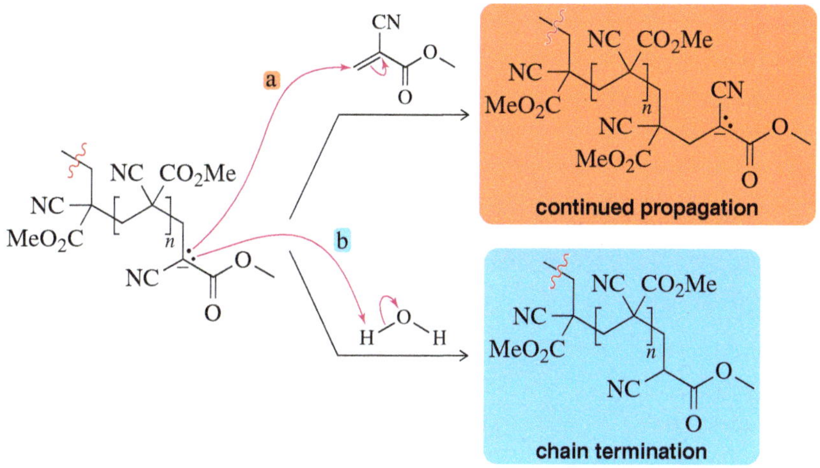

Figure 28.22 The termination of an anionic polymerization occurs when the propagating anion is protonated.

Polystyrene can also be generated from styrene via anionic polymerization. Because it isn't an electron-poor alkene, a very strong base like *n*-butyllithium [*from Section 16.2.1*] is used as the initiator, proceeding by the usual anionic polymerization mechanism (**Figure 28.23**).

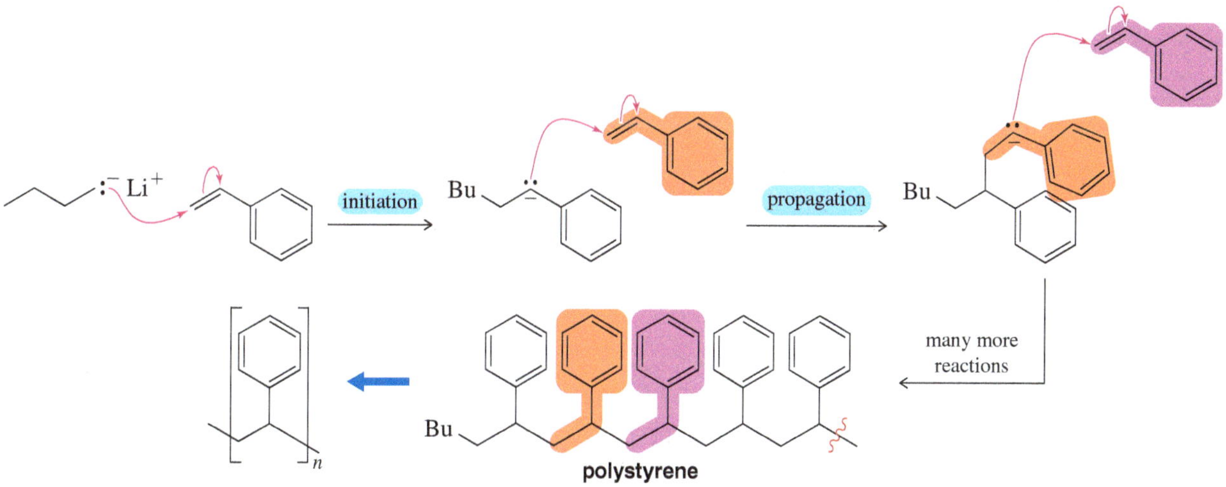

Figure 28.23 The anionic polymerization to make polystyrene.

Assessment 28.12 Suggest a mechanism for the anionic polymerization of the molecules shown here. You can stop the mechanism after the addition of three monomers, at which point you should show a termination step.

(a) **(b)** **(c)**

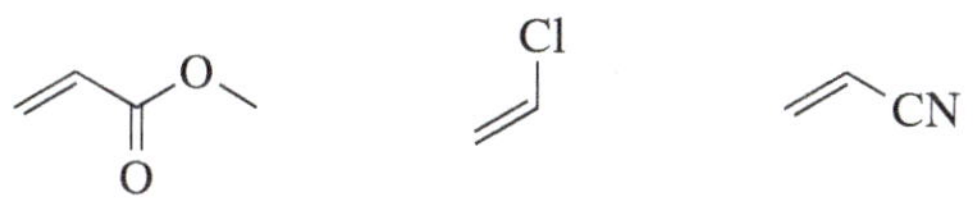

Assessment 28.13 Rank the following monomers in terms of the strength of the nucleophile required to initiate anionic polymerization (1 = needs strongest nucleophile; 5 = needs weakest nucleophile).

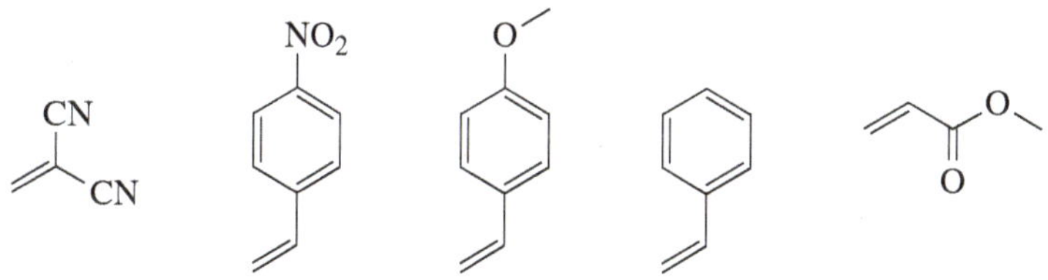

28.4.3 Free-radical polymerization

QUICK REVIEW We studied the addition of HBr in the presence of peroxides to produce the anti-Markovnikov haloalkane in Chapter 8. The addition of HBr to but-1-ene is an example of this reaction (**Figure 28.24**). Following the two-step formation of the bromine radical (the initiation steps), the radical attacks at the least substituted carbon of the alkene in order to make the more stable 2° alkyl radical. The 2° radical grabs a hydrogen from another molecule of HBr, completing formation of the bromoalkane and regenerating the propagating bromine radical. [*For a detailed review on this reaction, see Section 8.3.4.*]

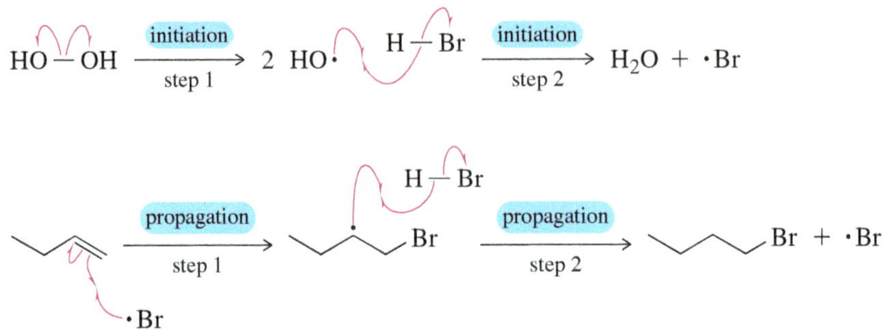

Figure 28.24 **The radical addition of HBr to but-1-ene.**

APPLICATION TO POLYMERS Free-radical polymerization requires conditions that favor attack by the propagating radical on another alkene, rather than abstracting a hydrogen (as in Figure 28.24). Whereas a full equivalent of HBr is used when adding HBr to the alkene, it is important that there are no weak, easily broken σ bonds present that might intercept the polymer chain reaction. This process is demonstrated in the polymerization of ethene (**Figure 28.25**). Free-radical polymerization is typically promoted by benzoyl peroxide at high temperatures. When heated, the weak O—O σ bond [*see Section 8.3.4*] breaks to give two carboxyl radicals. The carboxyl radicals decarboxylate [*see Section 18.8.4 and Chemistry in Context 18-D*] to generate phenyl radicals. The phenyl radical attacks the first monomer,

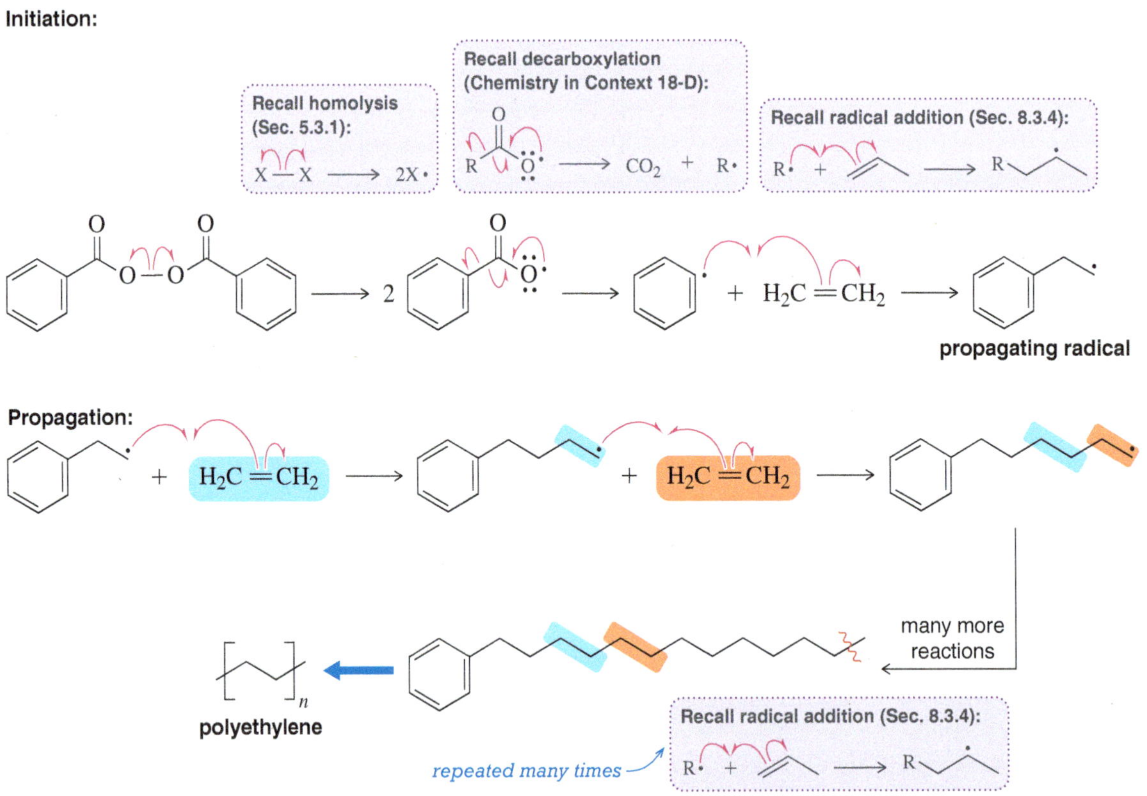

Figure 28.25 **The free-radical polymerization of ethene.**

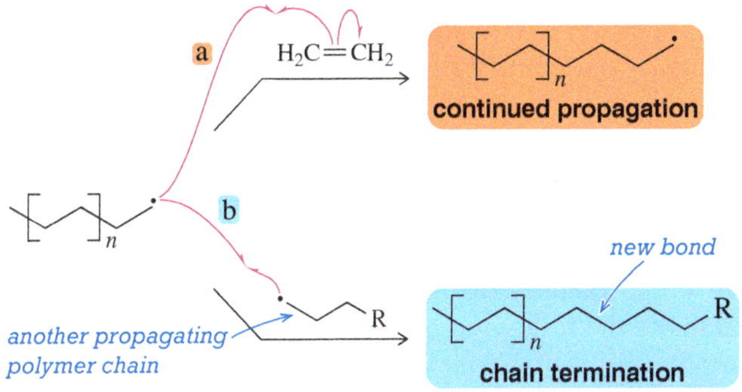

Figure 28.26 The termination of a free-radical polymerization occurs when two radicals combine.

initiating the polymer chain reaction and creating a new alkyl radical. This alkyl radical, with no hydrogen to abstract, attacks another ethene instead, thus generating another alkyl radical. This process continues as the chain is extended indefinitely, resulting in polyethylene. Polyethylene is typically used to make plastic sandwich bags.

Termination of the polymer chain reaction (**Figure 28.26**) occurs when the propagating radical, instead of attacking another alkene, combines with another radical. This is just like the radical termination step discussed in Sections 5.3.3 and 11.5.1. All radical polymerizations terminate in this way.

Thus far in our discussion of addition polymers, we have only shown the formation of linear polymers. And yet Table 28.1 showed several other polymer architectures, including branched polymers. How do branched polymers form? During the course of polymer propagation, an alkyl radical occasionally abstracts a hydrogen from another polymer chain (**Figure 28.27**). When this happens, a more stable 2° radical is produced. This 2° radical, upon collision with ethene, begins building a new polymer chain, ultimately creating a branched polymer.

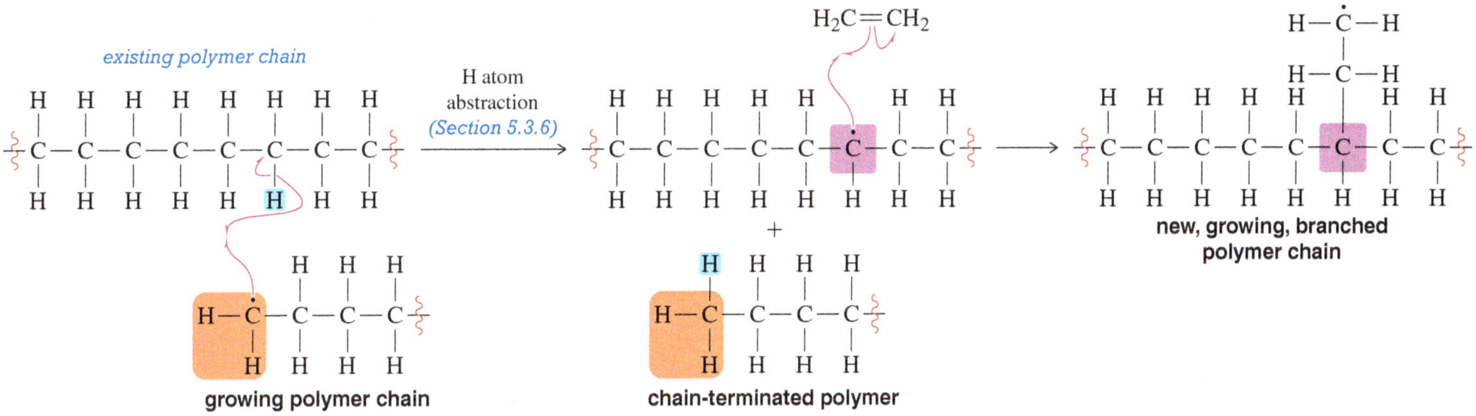

Figure 28.27 Chain branching in the free-radical polymerization of ethene.

It should come as no surprise that polystyrene can be generated from styrene via free-radical polymerization. Benzoyl peroxide is once again used as the initiator, with polymerization proceeding by the usual free-radical mechanism (**Figure 28.28**).

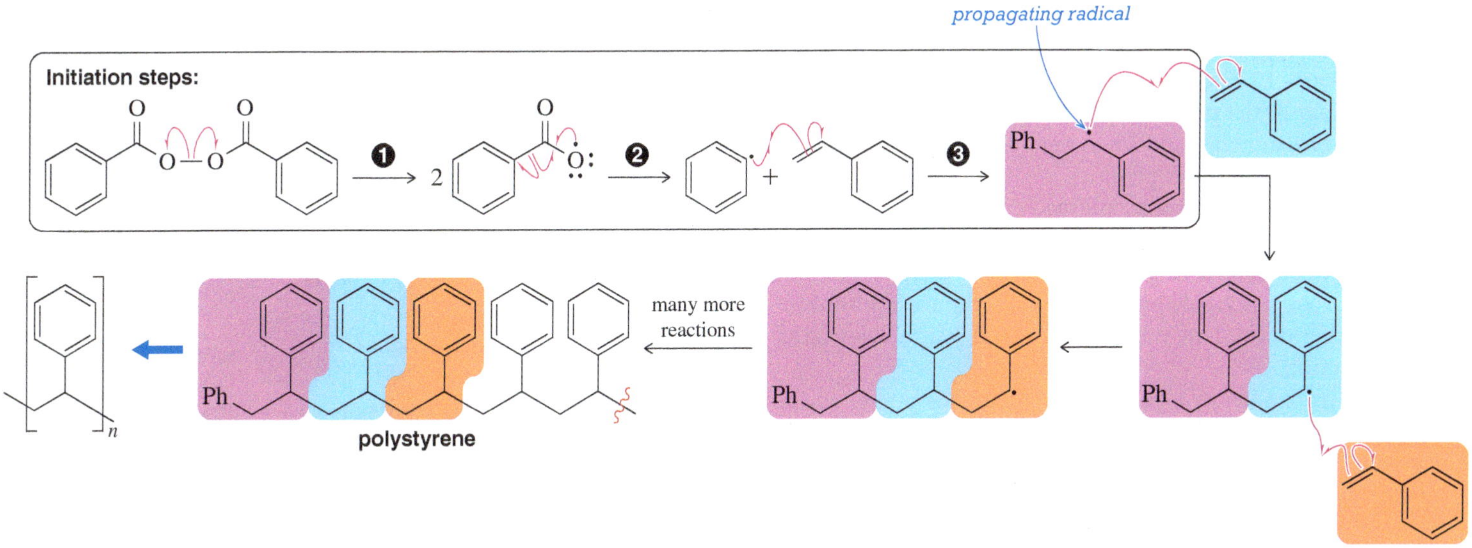

Figure 28.28 Free-radical polymerization to make polystyrene.

28.4 SUMMARY

- Polymers can be formed using variations of reactions studied previously. Cationic polymerization occurs by electrophilic addition to alkenes. Anionic polymerization occurs by nucleophilic addition to electrophilic alkenes. And, finally, radical polymerization occurs via radical addition to alkenes.

Assessment 28.14 Suggest a mechanism for the free-radical polymerization of the molecules shown here. You can stop the mechanism after the addition of three monomers, at which point you should show a termination step. Use R• as the initiating and terminating radical.

(a)

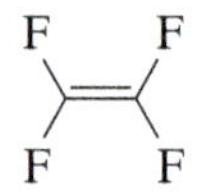

to make Teflon™

(b)

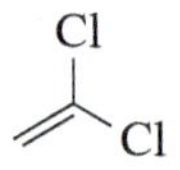

to make Saran Wrap until 2004

(c)

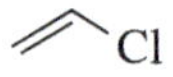

to make PVC

Assessment 28.15 Identify the monomers that might have been used to make the following polymers.

(a)

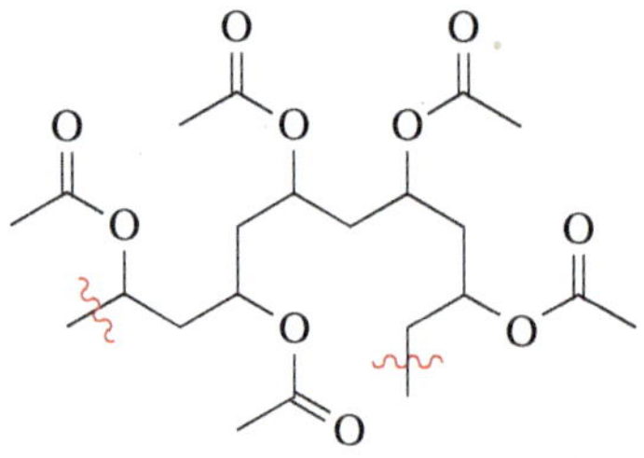

(b)

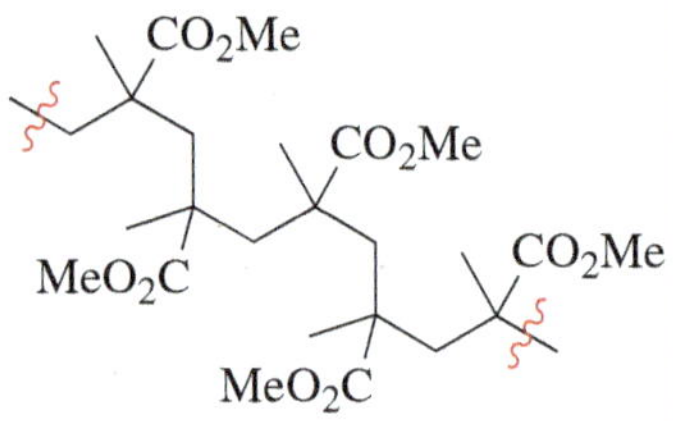

plexiglass

28-A CHEMISTRY IN CONTEXT: Easing Purification

Reactions on solid support

In addition to producing important plastics, resins, and other materials that improve our lives, polymers have also brought enhancements to the field of organic synthesis. One area where this is true is in polymer-supported reactions. When a reagent is attached to a polymer, the by-product of the reaction will remain attached to the solid support. When the reaction is complete, the product, in the organic solvent, can be washed away, leaving the by-product attached to the solid support [Figure 28.29(a)]. Attachment of the reactant to the polymer support accomplishes a similar goal: When the reaction is complete, the product is attached to the solid support, allowing the by-products to be washed away [Figure 28.29(b)]. Regardless of which approach is used, purification efforts are minimized, reducing the amount of time, energy, and waste associated with a reaction.

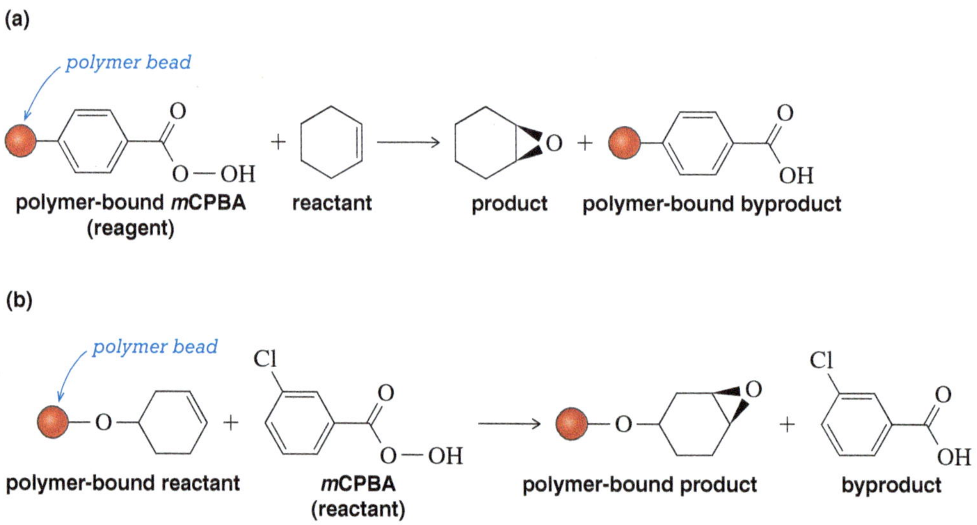

Figure 28.29 Polymer-supported synthesis. (a) When reagents are attached to a polymer support, the pure product is easily separated from the by-products. (b) When reactants are attached to a polymer support, the by-products are washed away from the pure product.

Recall from Section 17.6 that the Wittig reaction is one of the more useful reactions for making new C=C bonds (Figure 28.30). One major drawback of the Wittig, besides its terrible atom economy (Chemistry in Context 17-B), is the difficulty in purifying the resulting alkene. Alkenes produced by the Wittig reaction are often contaminated by both residual triphenylphosphine from the synthesis of the Wittig reagent and the triphenylphosphine oxide by-product.

Figure 28.30 The Wittig reaction between a phosphonium ylide and a ketone.

(continued)

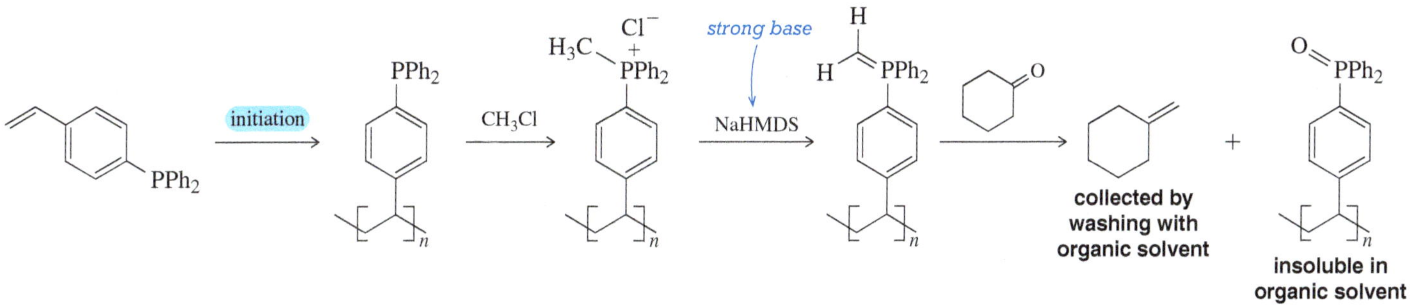

Figure 28.31 The Wittig reaction between a polymer-supported ylide and cyclohexanone to produce methylene cyclohexene.

Polymerization of a styrene monomer, functionalized with a triphenylphosphine, results in a polymer upon which a Wittig reagent can be formed (**Figure 28.31**). An S_N2 reaction with chloromethane, followed by deprotonation with sodium hexamethyldisilazide (NaHMDS), a strong base, gives a polymer-bound phosphorus ylide. When this polymer-bound ylide is exposed to a ketone, a new alkene forms. The triphenylphosphine oxide by-product also forms, but it is bound to the polymer. The product alkene, soluble in organic solvents, can be washed away from the polymer, leaving it pure enough to be used in subsequent steps without further purification.

The product triphenylphosphine oxide is usually discarded as waste in a reaction, having very little utility. Fortunately, the polymer-bound triphenylphosphine can be recycled by reduction with trichlorosilane and an amine base (**Figure 28.32**). [*Don't worry about how this reaction works.*]

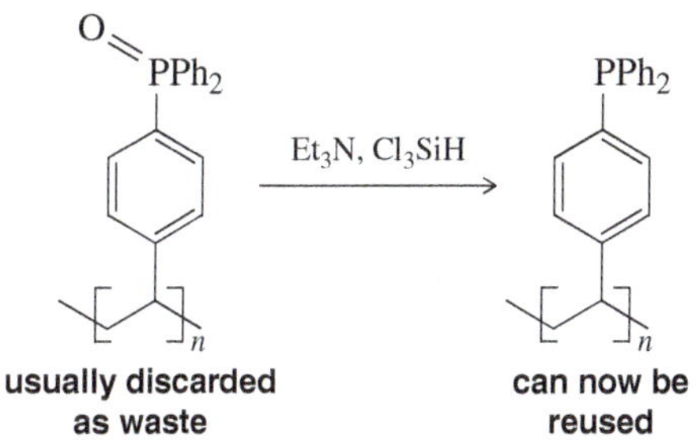

Figure 28.32 Recycling of the polymer-bound triphenylphoshine allows it to be reused in later Wittig reactions.

28.5 Condensation polymers (nucleophilic acyl substitution)

A condensation reaction is traditionally defined as one between two organic molecules that results in a new σ bond with the elimination of a small molecule like water or alcohol. In this way, a condensation reaction is just a more narrowly defined substitution reaction. These reactions typically involve the replacement of one C—X σ bond with another and are entropically neutral ($2 \rightarrow 2$; $\Delta S = 0$). **Table 28.3** shows several examples of condensation/substitution reactions we have studied thus far, including ether formation, aldol and related condensations, and the Fischer esterification.

Table 28.3 Condensation reactions studied thus far (an incomplete list).

Reaction	Example
General form	$A{-}H + B{-}OH \longrightarrow A{-}B + H_2O$
Dehydration of alcohols to make ethers (Section 13.11)	
Aldol condensation (Section 20.3)	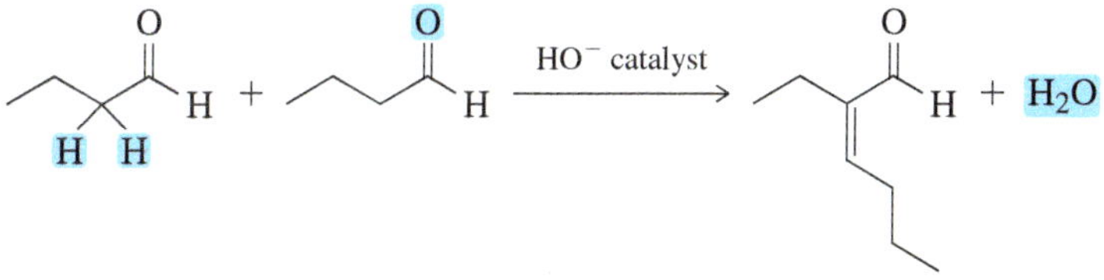

(*continued*)

Table 28.3 *Continued*

Reaction	Example
Claisen condensation (Section 20.4)	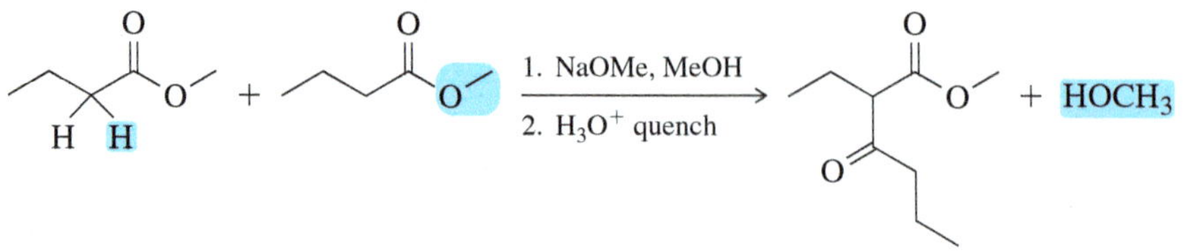
Dieckmann condensation (Section 20.4.2)	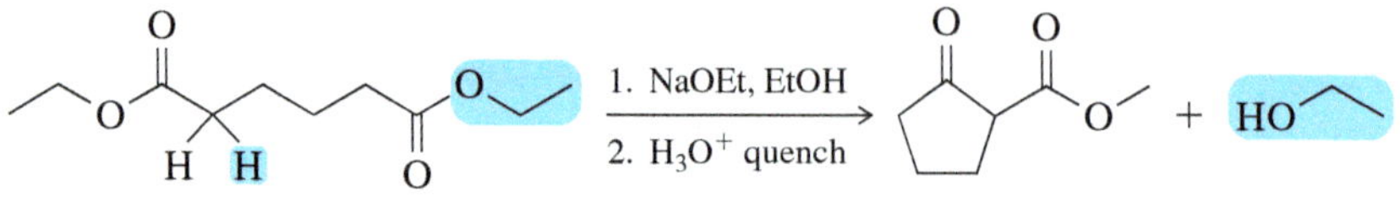
Fischer esterification (Section 18.7.2)	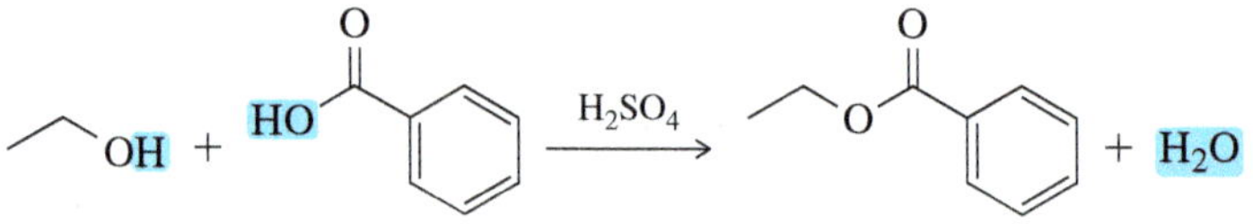

A condensation polymerization is similar to the reactions in Table 28.3, except that a molecule of A reacts with a molecule of B, releasing a small molecule. The resulting A—B reacts in a similar reaction with another molecule (usually A), once more with extrusion of a small molecule. This process continues until a long polymer chain has been produced (**Figure 28.33**).

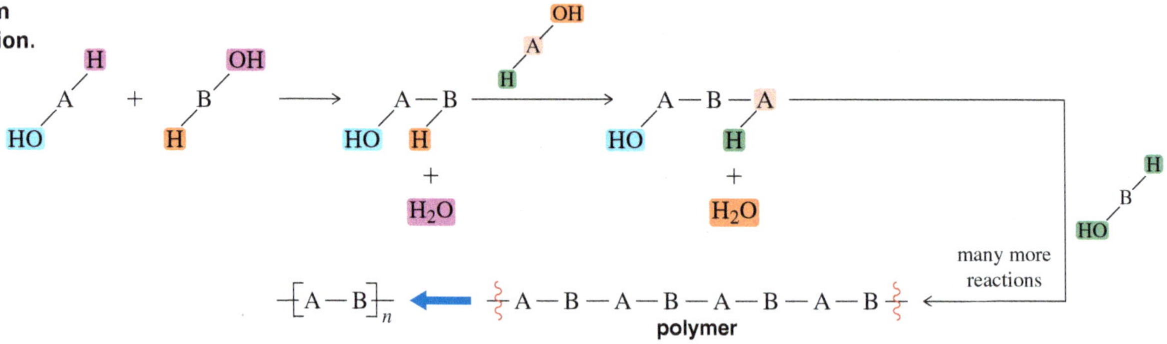

Figure 28.33 The general form of condensation polymerization.

As you study the condensation polymers in the sections that follow, keep in mind that you have seen all of these reactions before in other contexts. We are not learning new reactions; the reactions are simply occurring repeatedly within a growing molecule as the polymer forms. [*Let's convince ourselves of that with a quick review.*]

QUICK REVIEW We studied the nucleophilic acyl substitution reactions of carboxylic acids and carboxylic acid derivatives in Chapters 18 and 19. The saponification of an ester to produce a carboxylate is an example of this reaction (**Figure 28.34**). Mechanistically, nucleophilic addition of hydroxide to the electrophilic carbon of the ester results in formation of a tetrahedral intermediate in the slow, rate-determining step. Collapse of the electron pair reforms the C—O π bond, eliminating ethoxide. The loss of ethoxide is what makes this a condensation reaction instead of an addition reaction. Because this reaction results in formation of an acidic compound in a basic solution, a final deprotonation step occurs to give a carboxylate anion. [*For a detailed review on nucleophilic acyl substitution, see Sections 18.7 and 19.6.*]

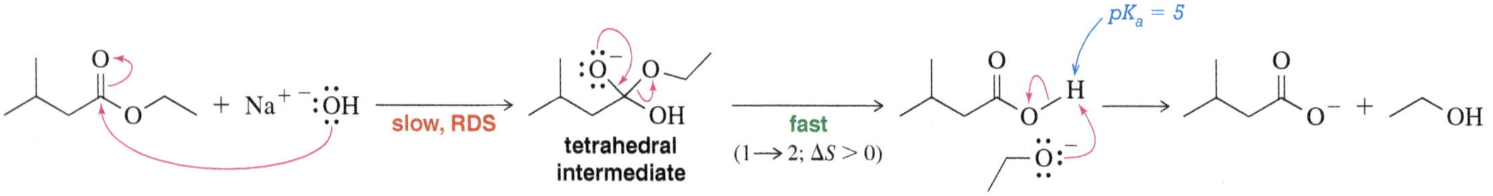

Figure 28.34 Nucleophilic acyl substitution: The saponification of an ester.

28.5.1 Polyamides

Polyamides are a type of polymer where the backbone is connected through amide bonds. Amides, as we learned in Section 18.1, consist of a nitrogen bonded to the carbon of a carbonyl. These are the same bonds that hold our proteins (biological polymers) together, while allowing for the structural stability that gives enzymes their well-defined active sites [*see Chapter 26*]. The polyamide nylon (**Figure 28.35**) is produced via the condensation polymerization reaction between hexane-1,6-diamine and hexanedioic acid. Nucleophilic acyl substitution between the carboxylic acid and the amine extends the chain in either direction, so nylon is a step-growth polymer. Figure 28.35(a) shows a shorthand mechanism of polymer formation, while Figure 28.35(b) shows a more precise mechanism of amide formation from a carboxylic acid and an amine. Recall from Chapter 19 that the reaction of an amine with a carboxylic acid is unfavorable, but under the conditions of high temperature and high pressure, the activation energy for polymerization is easily overcome. Nylon production is expected to reach 5.7 million tons by 2025, most of which will be used in the clothing industry.

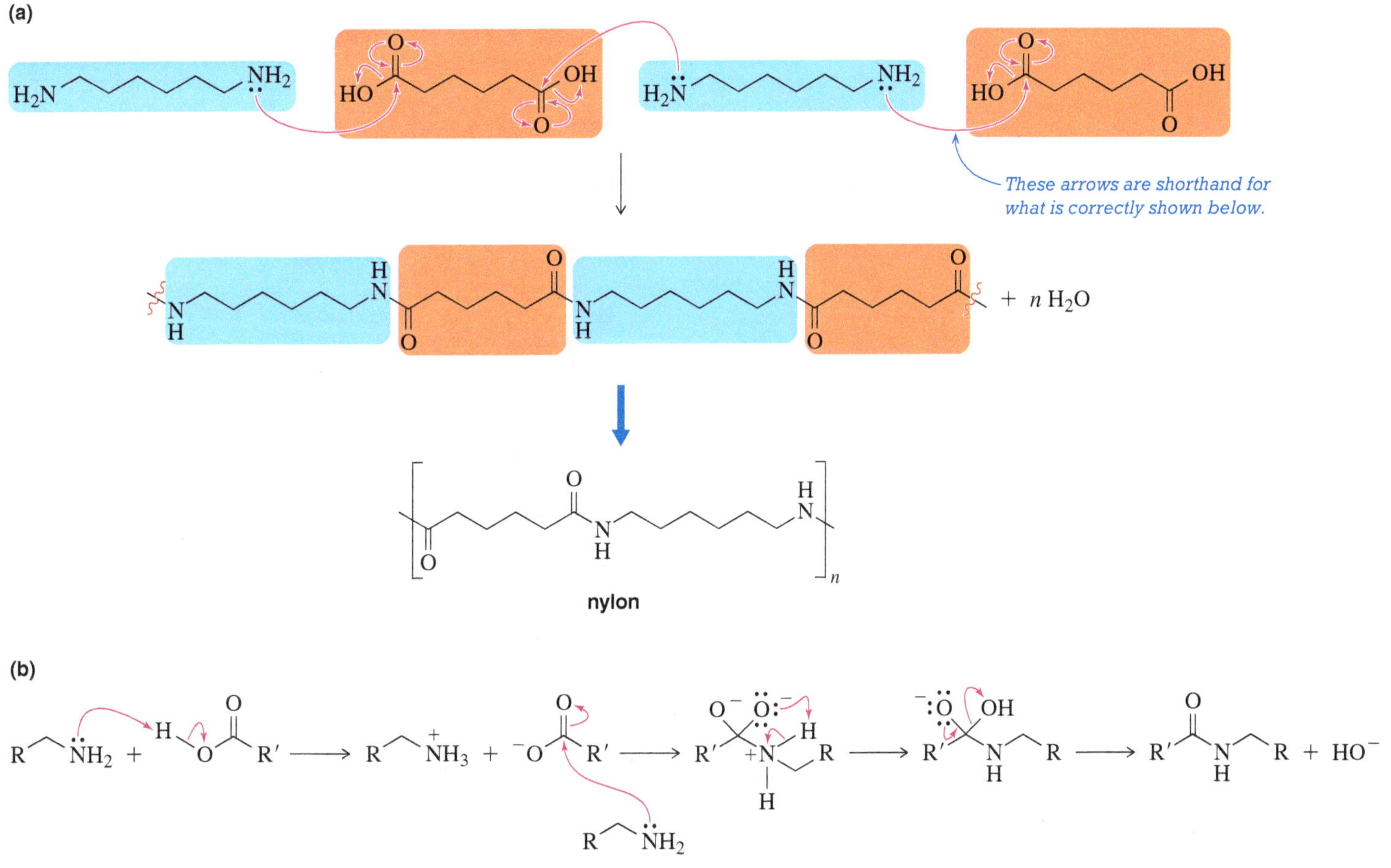

Figure 28.35 **Nylon, a polyamide, forms via a condensation polymerization reaction.**

One of the strongest human-made polymers is Kevlar® (**Figure 28.36**), a polyamide synthesized by the condensation reaction of benzene-1,4-diamine and terephthaloyl chloride. The chain extends in both directions via the nucleophilic acyl substitution of the amine on the highly reactive acid chloride. The mechanism of this reaction was presented in Section 19.7.3.

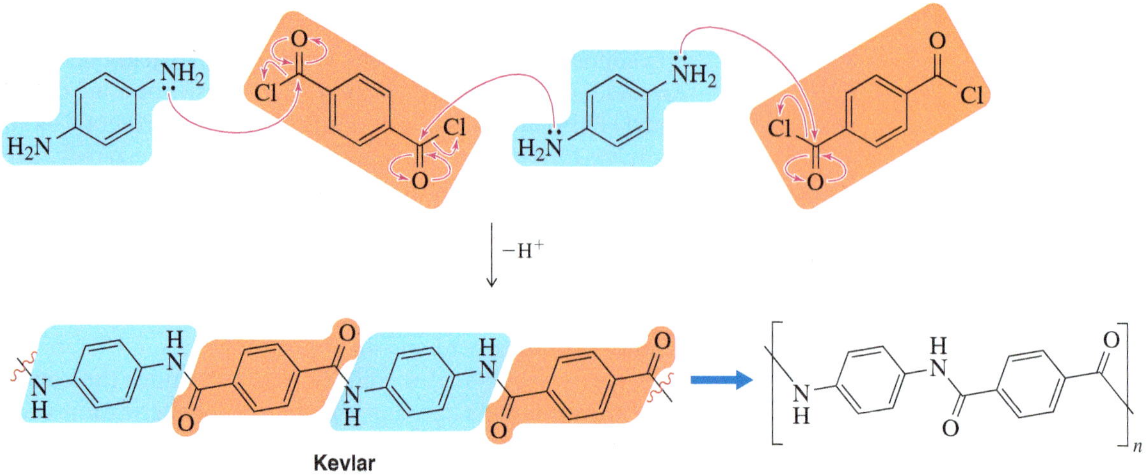

Kevlar

Figure 28.36 The formation of Kevlar, a polyamide, occurs via a condensation polymerization reaction. [*Note the shorthand mechanism for the nucleophilic addition/elimination mechanism learned in Section 19.7.3.*]

Invented by chemist Stephanie Kwolek at Dupont in the 1960s, Kevlar is five times stronger than steel. As a result, it is used in bulletproof vests, protective sports equipment, and bicycle tires, among many other uses. It owes its strength to hydrogen bonding between polymer strands [Figure 28.37(a)] as well as the numerous van der Waals interactions between the aromatic rings of adjacent hydrogen-bonded sheets [Figure 28.37(b)].

Figure 28.37 Kevlar is a strong polymer due to (a) hydrogen bonding between strands and (b) van der Waals interactions between the stacked hydrogen-bonded sheets.

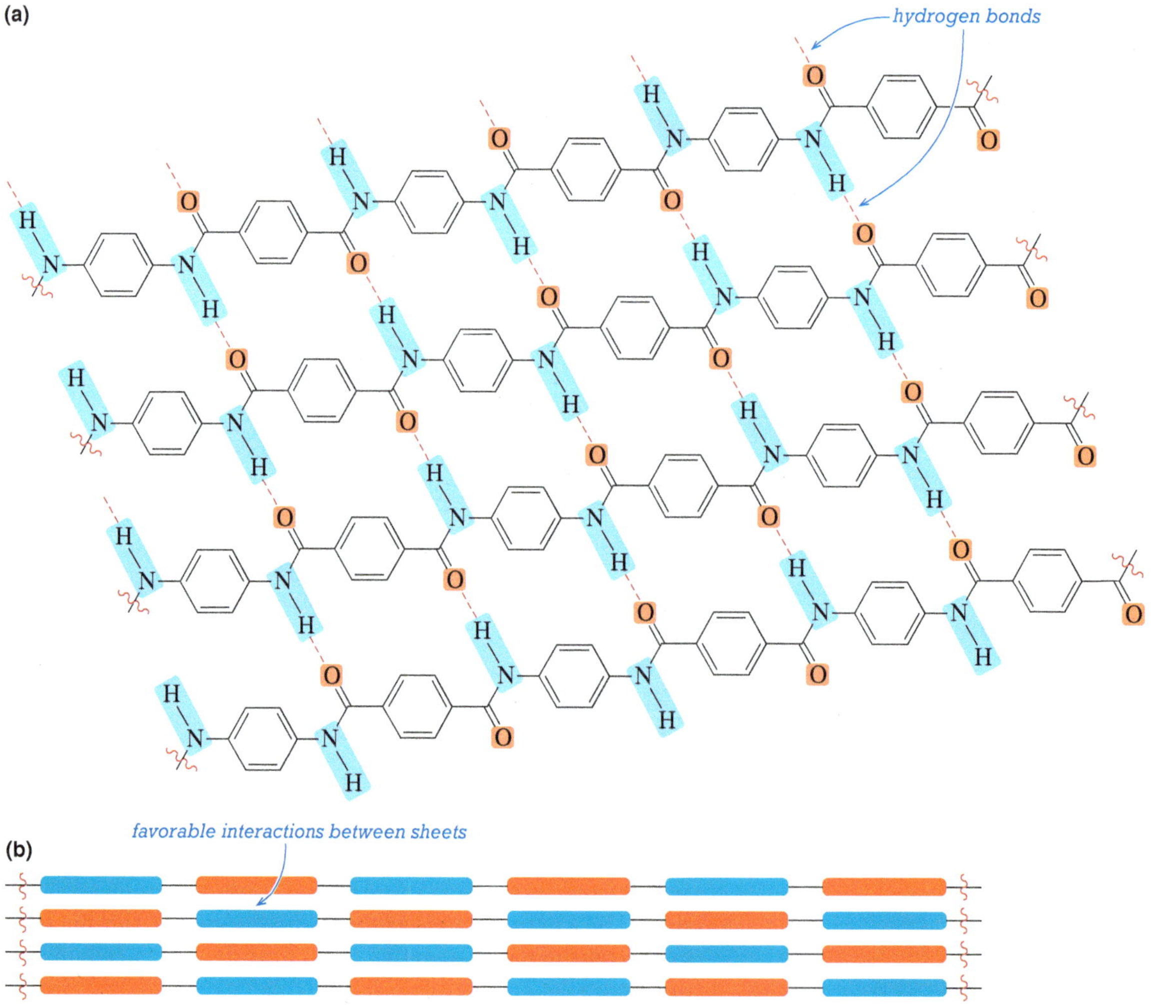

Assessment 28.16 Identify the monomers that would be used to make the Nomex® polymer.

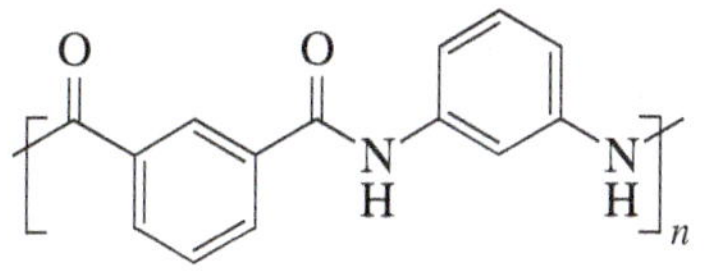

Nomex

Assessment 28.17 Caprolactam, when heated to 250 °C in the presence of a small amount of water, polymerizes to make nylon 6. Suggest a structure for nylon 6 and a mechanism for its polymerization. [*Hint: First open the cyclic amide using water.*]

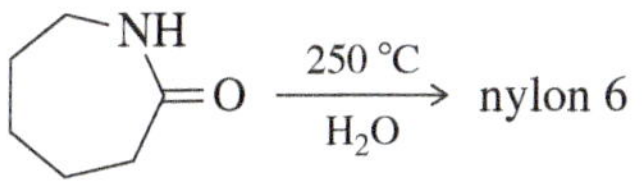

28.5.2 Polyesters

Polyesters are closely related to polyamides. The main difference is that ester linkages hold the polymer backbone together. Esters are less stable than amides (Section 19.6.3), so polyesters are considerably weaker polymers than polyamides. Polyethylene terephthalate (PET) is an industrially important polyester. On a laboratory scale, its synthesis is accomplished by the acid-catalyzed Fischer esterification reaction between ethylene glycol and terephthalic acid [Figure 28.38(a)]. Industrially, this polymer is produced by heating a mixture of ethylene glycol and dimethyl terephthalate to over 200 °C, upon which a transesterification reaction occurs, extending the polymer chain in two directions [Figure 28.38(b)]. PET is used in the manufacturing of textiles and disposable water bottles.

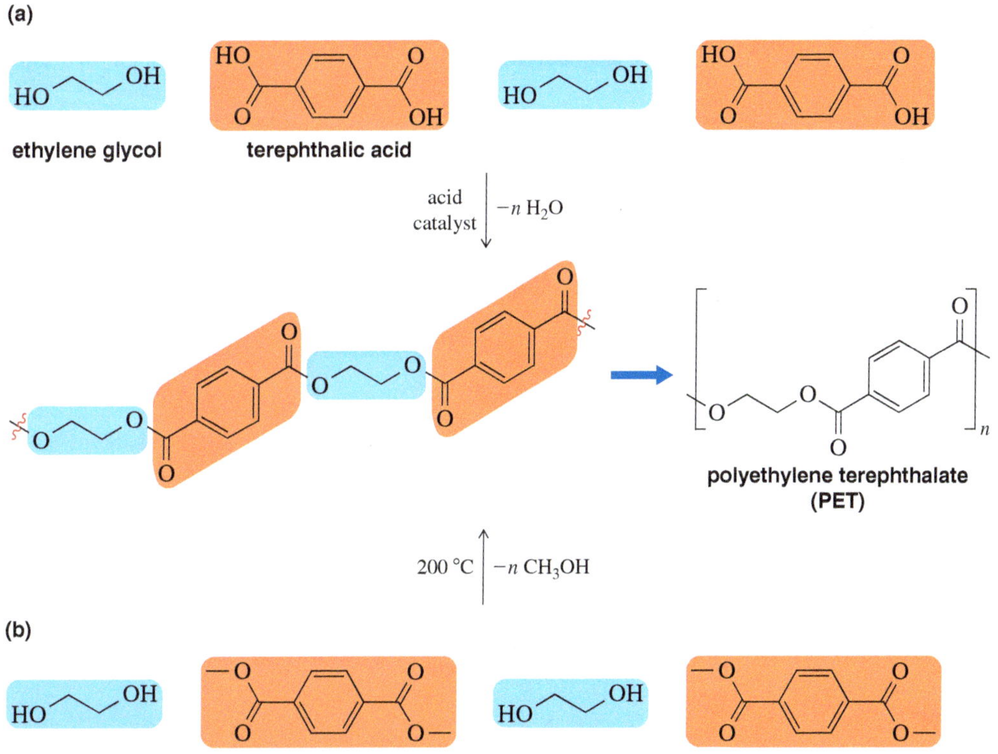

Figure 28.38 **The formation of the polyester PET** can occur via (a) a Fischer esterification reaction or (b) a transesterification reaction at high temperature.

The polyester poly(propylene fumarate) forms by the reaction of propylene glycol and fumaric acid (**Figure 28.39**).

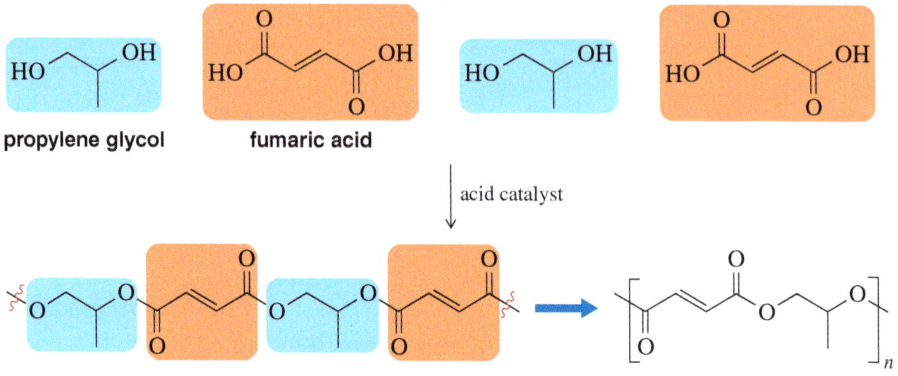

Figure 28.39 The formation of poly(propylene fumarate).

Up to this point, we have only discussed the noncovalent interactions of polymer chains. [*See the hydrogen bonding and van der Waals interactions in Figure 28.37 (Kevlar).*] Polymer strands can be covalently cross-linked however, as a way of converting them from low-viscosity liquids to high-viscosity liquids or even crystalline solids. This process, referred to as **thermosetting,** is usually done by cross-linking the chains after initial polymer formation. In the case of unsaturated polymers like poly(propylene fumarate), this can be done by treating the polymer strands with styrene in the presence of benzoyl peroxide, a radical initiator (**Figure 28.40**). This reaction proceeds via a mechanism similar to that of free-radical polymerization of alkenes.

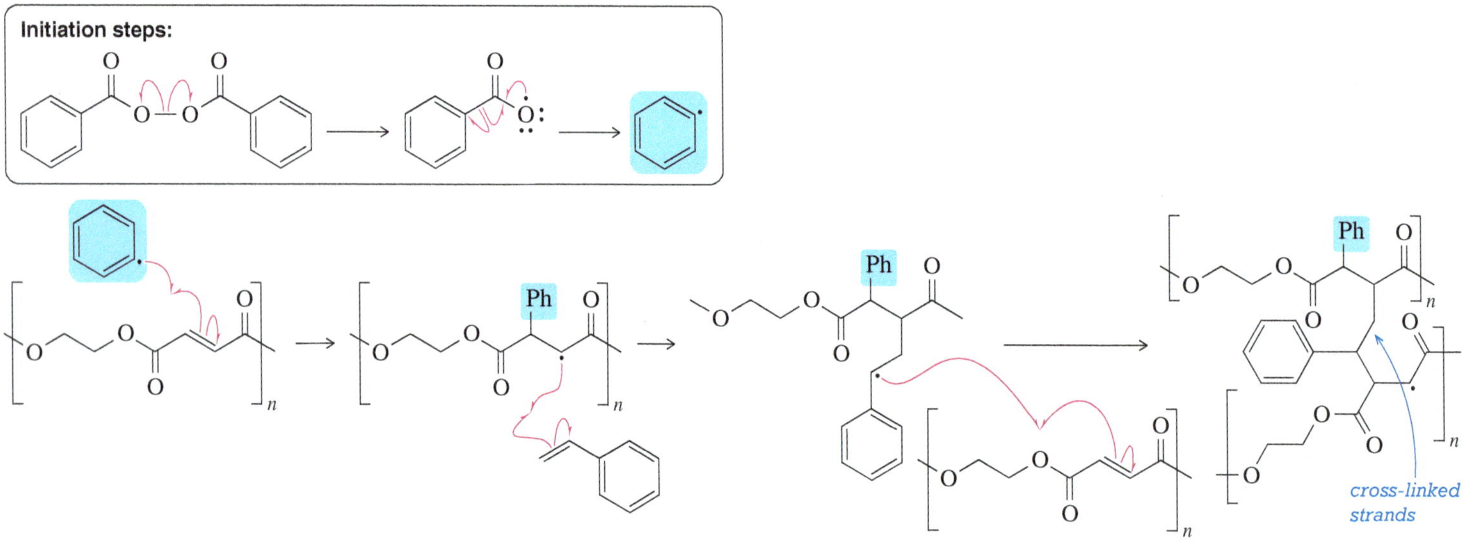

Figure 28.40 The cross-linking of a polymer by a radical mechanism.

Assessment 28.18 Suggest a mechanism for the polymerization of the following molecules. You can stop the mechanism after each monomer type has been used twice.

(a)

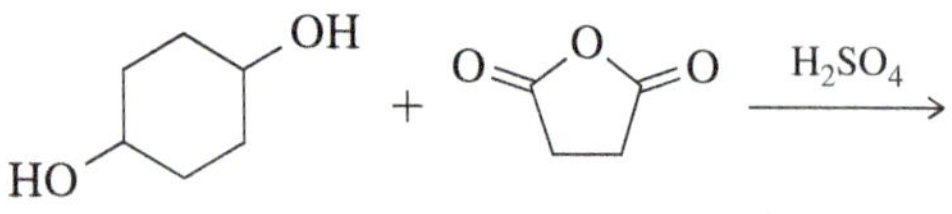

(b)

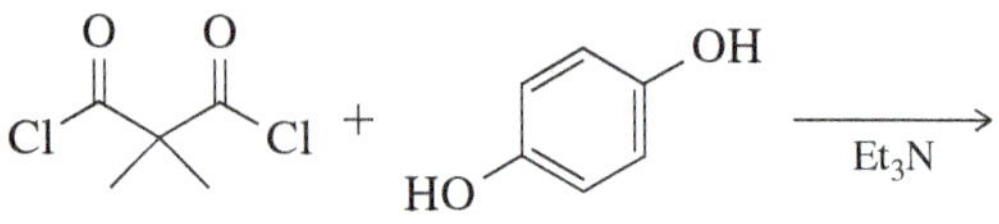

Assessment 28.19 Identify the monomers that might have been used to make the polyesters shown.

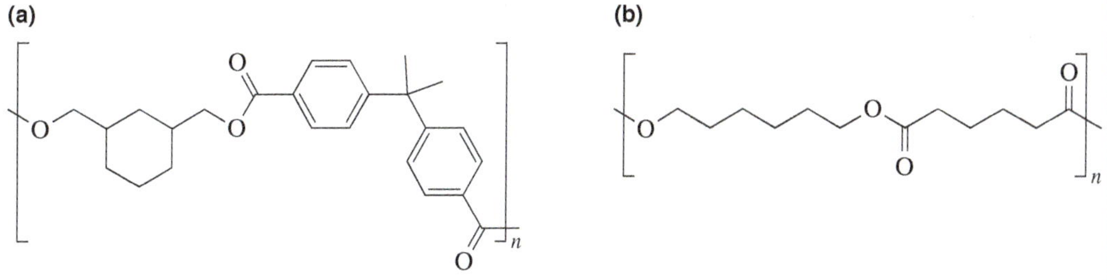

28-B CHEMISTRY IN CONTEXT: Green Polymer Synthesis

Biodegradable polyester

While polymer synthesis has produced a plethora of materials that have improved our lives, they are not without drawbacks. Polymers are highly stable molecules; in some cases, cross-linking in multiple places along the polymer chains results in even greater stability. While this stability allows plastics to be used for long periods of time, it unfortunately has adverse environmental impacts. Some plastics that are disposed of today are expected to persist in landfills for thousands and thousands of years. Finally, most of today's polymers are derived from oil feedstocks, thus contributing to the depletion of our natural resources.

Green Chemistry Principle 10 suggests that products should be designed for degradation using renewable feedstocks, where possible. Polyhydroxyalkanoates (PHAs) are naturally occurring polyesters that are synthesized by bacterial fermentation. One notable feature of these polymers is that they are completely biodegradable. If these could be incorporated successfully into usable products, disposal would have relatively little environmental impact because they will naturally decompose in short order.

Professor Geoffrey Coates has recently developed the catalytic synthesis of poly(3-hydroxybutyrate), an example of a biodegradable polyester (**Figure 28.41**).

His system is unique in that it combines two catalytic reactions in a single flask. The first catalyst inserts a carbonyl group (from carbon monoxide) into propylene oxide to create a β-lactone. The β-lactone is toxic, so being able to use it directly (without isolation or purification) would be advantageous. The second catalyst polymerizes the lactone, resulting in high-molecular-weight poly(3-hydroxybutyrate).

The work described here earned Professor Coates a Presidential Green Chemistry Challenge Award in 2010. Besides the advantage of the biodegradability of the resulting polymer, the synthesis was recognized for using alternate feedstocks—namely, carbon dioxide and carbon monoxide derived from biomass or other sources. This process has been commercialized for the synthesis of coatings, adhesives, foams, and plastics.

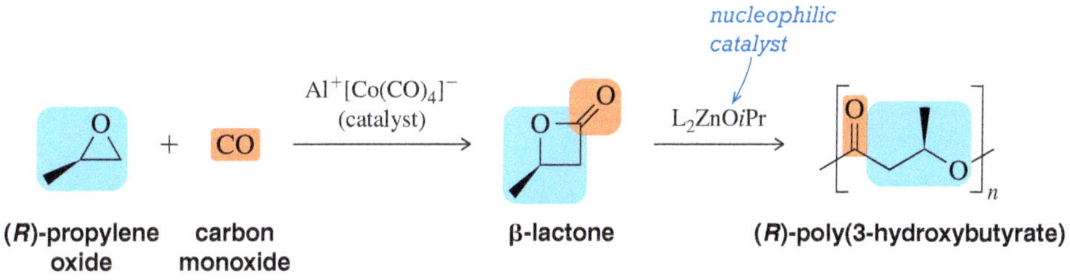

Figure 28.41 A two-catalyst system to produce a biodegradable polymer.

Assessment 28.20 Suggest a mechanism for the polymerization of the β-lactone in Figure 28.41, assuming the use of a strong nucleophile like $\textit{i}PrO^-$ as the catalyst. You can stop the mechanism after using three monomers.

28.5.3 Polycarbonate

The carbonate functional group (**Figure 28.42**) is essentially a central carbonyl flanked on both sides by two alkoxy groups. It is closely related to carboxylic acid derivatives. Diethyl carbonate, a representative carbonate, can be synthesized from ethanol and phosgene. Mechanistically, it is formed by consecutive nucleophilic acyl substitution reactions on the phosgene, replacing each chloride with an ethoxy group.

Figure 28.42 The formation of diethyl carbonate from ethanol and phosgene.

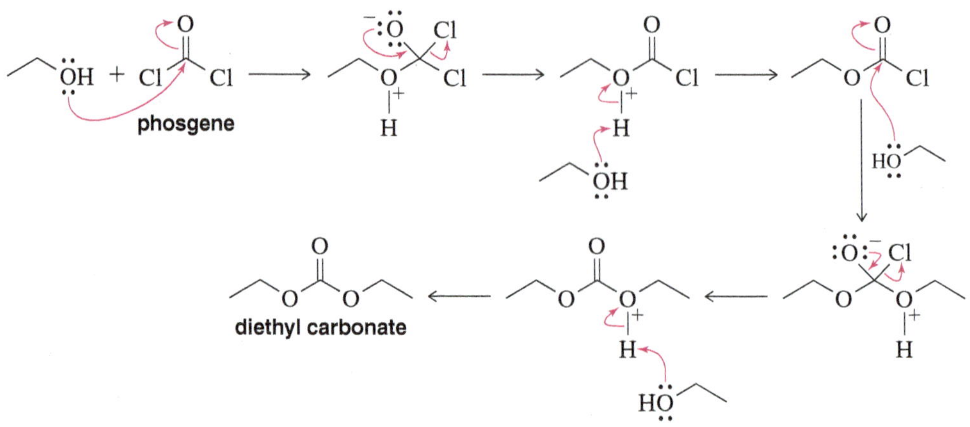

Polycarbonate is a specific polymer containing carbonate groups in the polymer backbone. It is composed of bisphenol A molecules linked by reaction with phosgene (Figure 28.43). Mechanistically, attack of the phenol on phosgene occurs to form a tetrahedral intermediate. Collapse of the tetrahedral intermediate reforms the carbonyl while kicking out the chloride. A second molecule of bisphenol A then attacks the acid chloride, once more forming a tetrahedral intermediate that collapses to complete formation of the carbonate. With phenol groups on both sides of the resulting molecule, the polymer chain can be extended in both directions by these reactions.

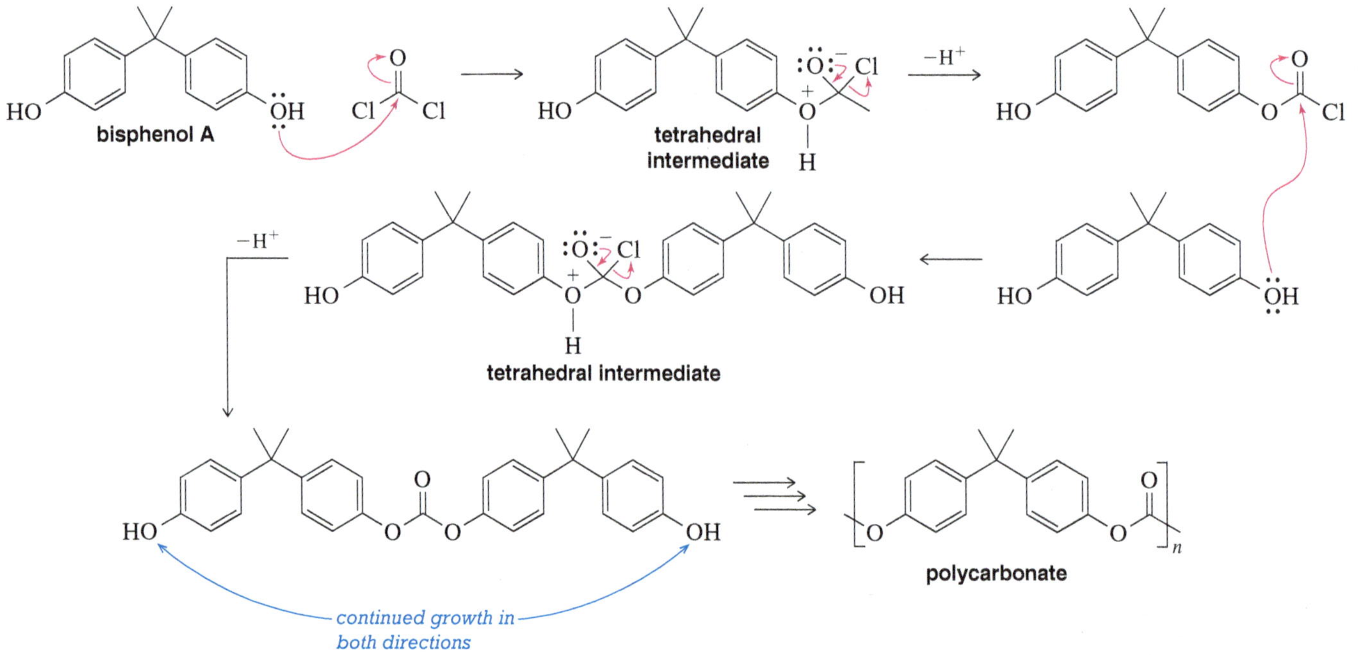

Figure 28.43 The formation of polycarbonate from phosgene and bisphenol A.

Polycarbonate is a strong, heat-resistant, clear polymer, making it useful for medical devices and protective eyewear such as laboratory safety glasses. As a **thermoplastic,** it can be heated and liquified without breaking the polymer strands, allowing it to be molded and recycled into new materials.

28.5.4 Polyurethanes

The urethane functional group is closely related to the carbonate. Whereas a carbonate features two alkoxy groups connected to a carbonyl, a urethane is composed of an amino and an alkoxy group connected to a carbonyl. The insecticide carbaryl, responsible for the environmental disaster in Bhopal, India, contains a urethane functional group. Carbaryl is synthesized from naphthol and methyl isocyanate (**Figure 28.44**). Nucleophilic addition of the alcohol to the electrophilic carbonyl pushes the π electrons onto oxygen to make what looks like a tetrahedral intermediate. A resonance structure of this intermediate has an electron pair on nitrogen. Protonation of the anionic nitrogen with acid completes formation of the urethane functional group.

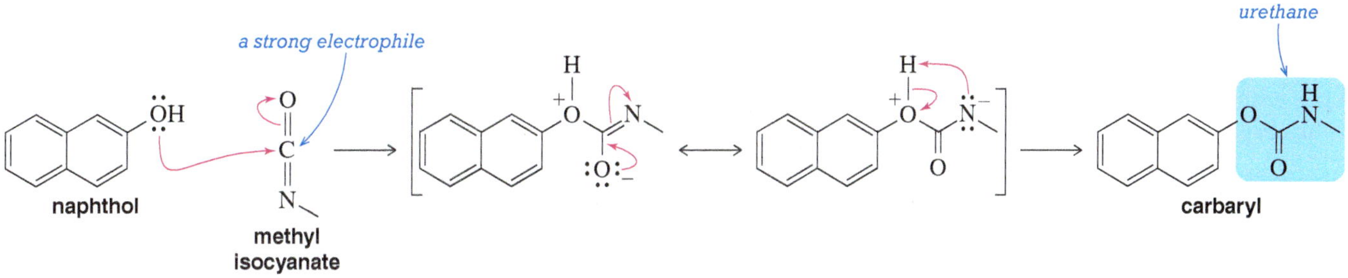

Figure 28.44 The formation of the urethane in carbaryl.

Polyurethanes contain the urethane functional group in the polymer backbone. The one shown in **Figure 28.45** is made by reacting a diol with a diisocyanate. Mechanistically similar to that shown in Figure 28.44, nucleophilic attack by the alcohol on the isocyanate occurs at both ends of the diol, extending the polymer chain in both directions.

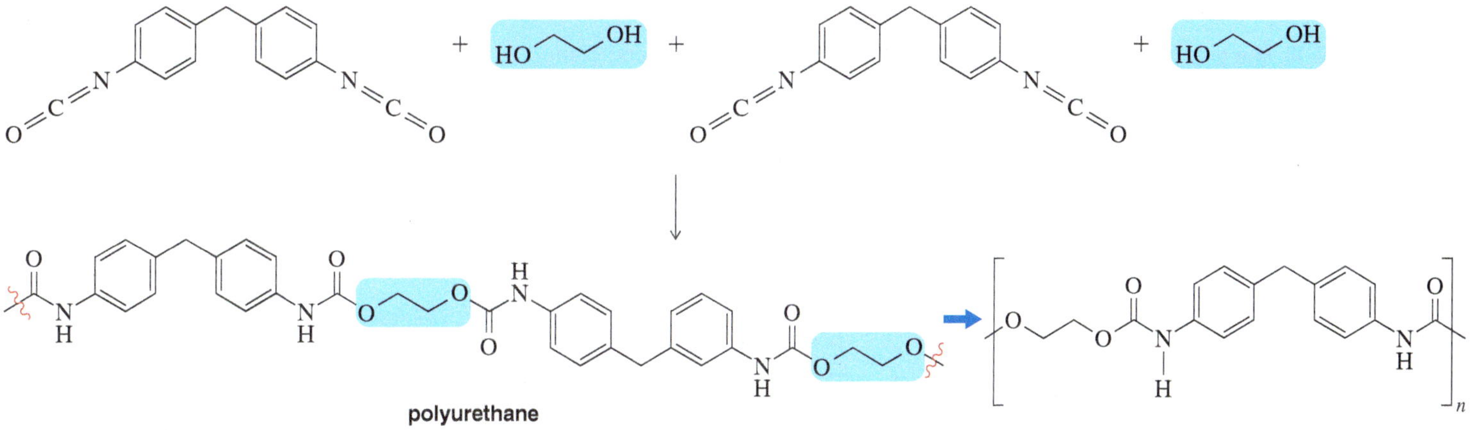

Figure 28.45 Polyurethane formation. (For the mechanism, see Figure 28.44.)

Polyurethanes can be treated with a blowing agent to convert them into foams [*similar to Styrofoam; see Section 28.4.1*]. These polyurethane foams are useful as padding in bedding and other household furniture.

28.5 SUMMARY

- Polymers can be formed via condensation reactions originally studied in Chapter 19. These reactions are essentially the nucleophilic substitution reactions of carboxylic acid derivatives. Using these reactions, polyamides, polyesters, polycarbonates, and polyurethanes can be synthesized.

Assessment 28.21 Give the monomers that could be used to make the polyurethane shown.

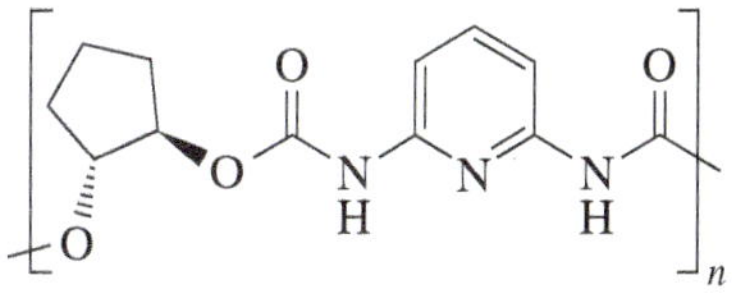

Assessment 28.22 (a) Show a mechanism of the polymer synthesis you proposed in Assessment 28.21. (b) Is this a chain-growth or step-growth polymerization?

28-C CHEMISTRY IN CONTEXT: Green Polymer Synthesis

Green Polyurethane

One drawback of polyurethane synthesis is the use of isocyanates as the reactive electrophile. Isocyanates, as was shown in the Bhopal disaster (Chapter 7 Introduction), are highly toxic, posing a risk to both the environment and the chemist working with them. To mitigate this risk, polyurethane synthesis has been achieved with cyclic carbonates (**Figure 28.46**). By replacing the electrophilic isocyanate (with a carbonate) and the diol (with a diamine), the urethane functional group can be synthesized using significantly less toxic reagents (GCP 3). Another advantage of this process is that the resulting polymer contains unreacted hydroxyl groups that can be used to further functionalize the polymer, imbuing it with different properties.

This Green Polyurethane™ technology, developed by Hybrid Coating Technologies, received a 2015 Presidential Green Chemistry Challenge Award. It has a high atom economy (GCP 2), and it uses renewable resources (GCP 7).

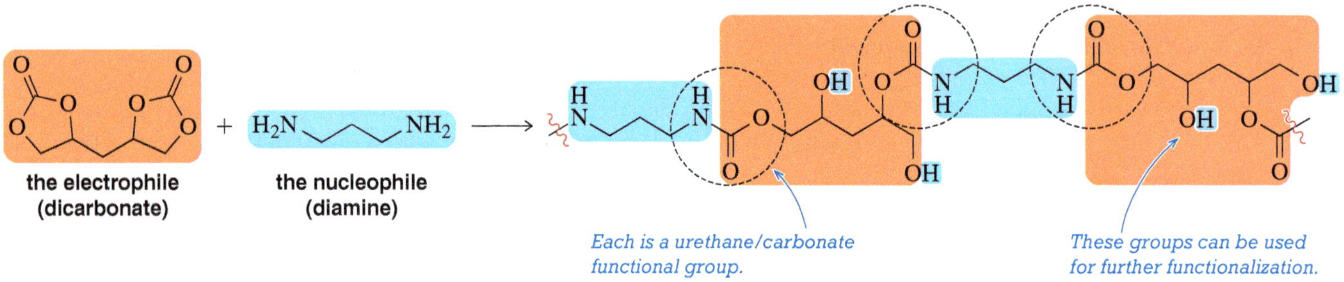

Figure 28.46 The green synthesis of a polyurethane using dicarbonates and diamines.

Assessment 28.23 What is the structure of the polymer that would be produced from the diamine and the dicarbonate shown?

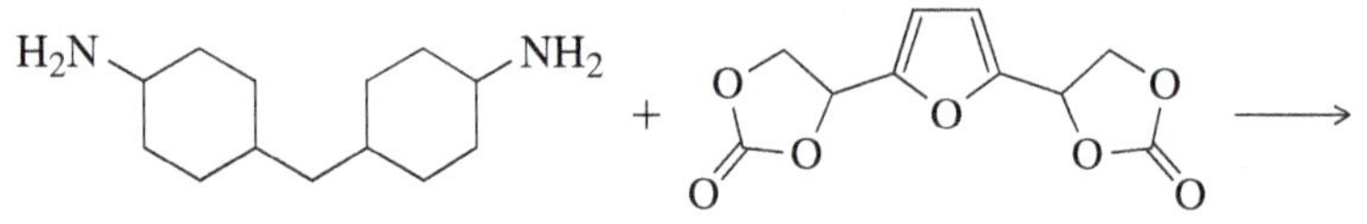

28.6 Other polymers

There are three additional polymerization reactions that we have studied in other, non-polymer contexts. These polymerizations are based on the reactions of epoxides (Section 13.13.2), the olefin metathesis reaction (Section 16.6), and palladium-coupling reactions (Section 16.4). Our treatment of these polymerization reactions parallels our earlier polymer discussions.

28.6.1 Epoxy resins

QUICK REVIEW We studied the addition of strong nucleophiles to epoxides in Sections 13.13.2 and 16.2.3. The addition of phenylmagnesium bromide to propylene oxide is an example of this reaction (**Figure 28.47**). The phenyl anion is a strong nucleophile, so it donates a pair of electrons to the least sterically hindered carbon of the strained epoxide, breaking a C—O σ bond to form an alkoxide. Quenching of the alkoxide with an acid results in a 2° alcohol.

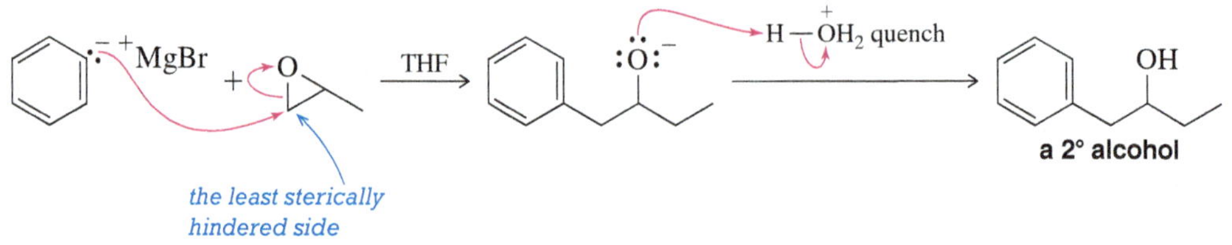

Figure 28.47 Strong nucleophiles open epoxides at the least sterically hindered carbon.

Epichlorohydrin (**Figure 28.48**) is a unique epoxide. Containing both a chloride and an epoxide, it can add two nucleophiles in a single reaction. If a Grignard reagent attacks the epoxide first, the resulting alkoxide can displace the chloride to form a new epoxide. With the Grignard reagent in excess, a second addition occurs, resulting in the two nucleophiles linked by the three carbons of what was epichlorohydrin. [*An electrophile that can add nucleophiles on both sides—it sounds like something we can make polymers out of, right?*]

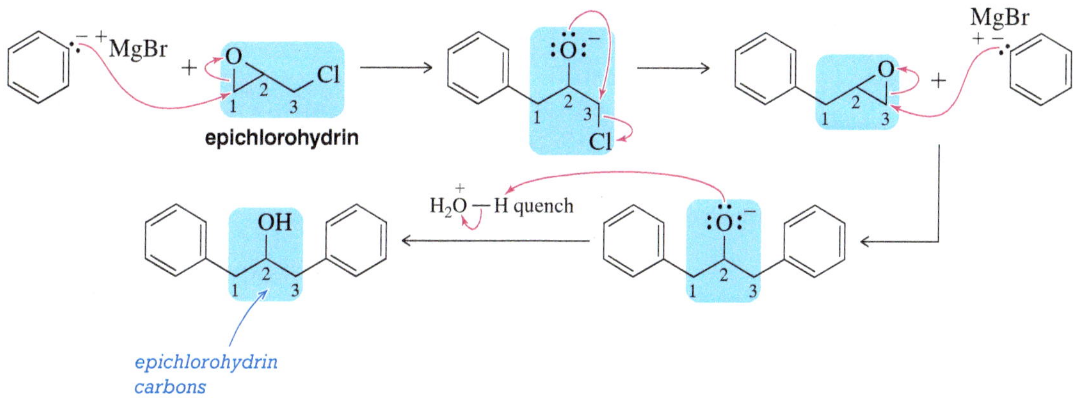

Figure 28.48 Epichlorohydrin can add two nucleophiles, one each at C_1 and C_3.

APPLICATION TO POLYMERS Epoxy resins are technically composed of cross-linked, low-molecular-weight diepoxy polymer chains. These diepoxides are formed by the reaction of bisphenol A [*from Section 28.5.3*] and epichlorohydrin, by the mechanism shown in **Figure 28.49**.

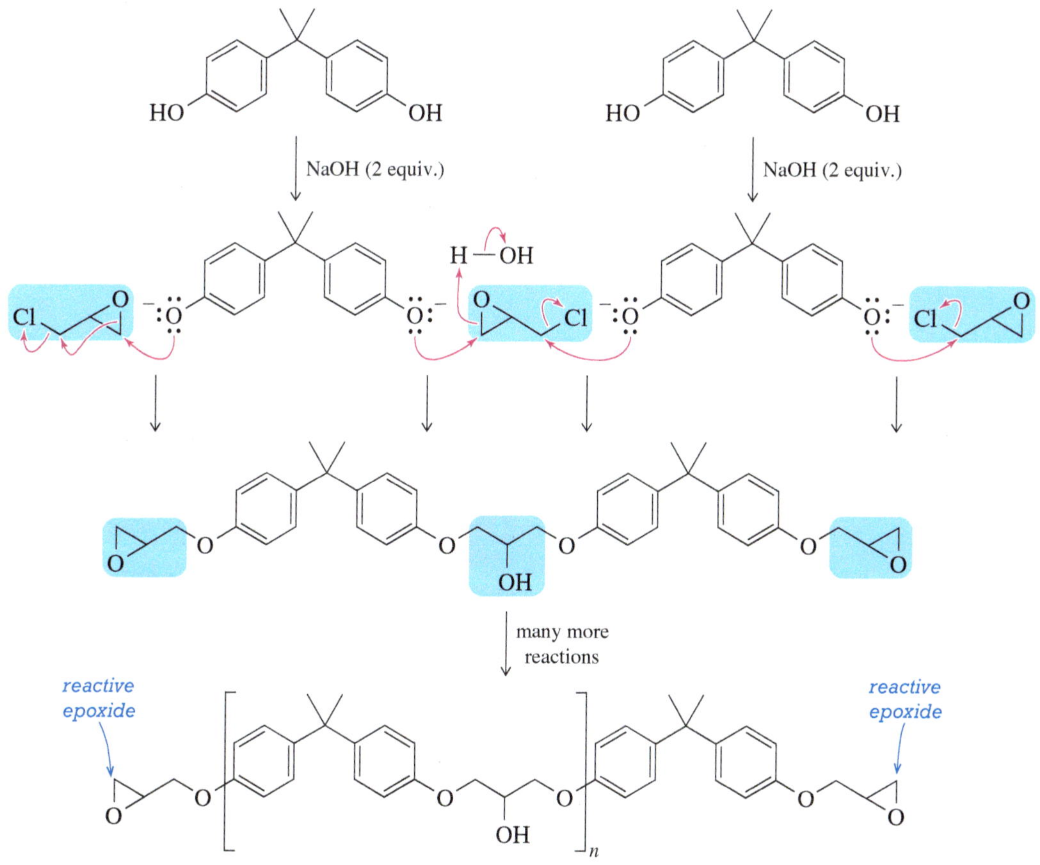

Figure 28.49 Formation of low-molecular-weight diepoxy polymers with bisphenol A and epichlorohydrin. [*The hypothetical mechanism arrows in this figure link these reactions to earlier ones while also helping understand bond formation in these reactions.*]

On their own, the diepoxy polymers have little value. But when combined with a cross-linking agent, they become extremely stable and tight-bonding adhesives. Standard cross-linking agents are aliphatic or aromatic amines. Recall that the low-molecular-weight polymer formed in Figure 28.49 has electrophilic epoxide groups on both ends, and amines are strong nucleophiles that will continue to attack electrophiles until there are no more hydrogens to lose. From Section 25.5.1, a 1° amine will react with two (or three) electrophiles, while a 2° amine will react with one (or two) (**Figure 28.50**).

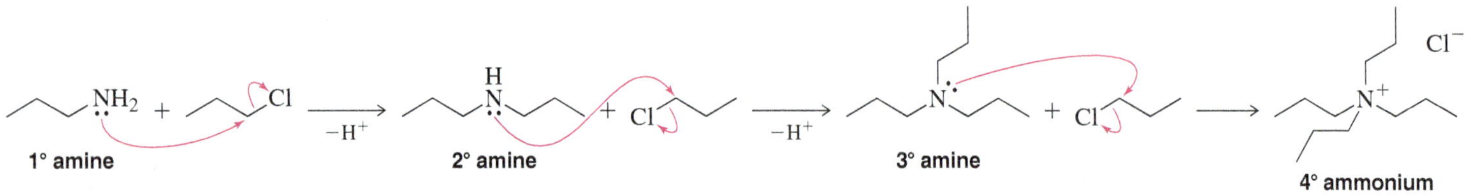

Figure 28.50 Amines react with electrophiles until there are no more hydrogens to lose. (See Section 25.5.1 for a review.)

When the diepoxide from Figure 28.49 is mixed with diethylenetriamine (an epoxy curing agent), a flurry of nucleophilic epoxide ring openings occur to give the cross-linked epoxy resin shown in **Figure 28.51**. Whereas the original diepoxides were small polymer chains, the epoxy resin, as a result of the cross-linking, is one big molecule. It should come as no surprise, then, that epoxy resins are extremely strong polymeric materials used as adhesives, metal coatings, and electrical insulators. When used as adhesives, the "glue" comes in two compartments—one containing the diepoxy polymer and the other containing the cross-linking agent.

Figure 28.51 The cross-linking of epoxy resin with nucleophilic amines.

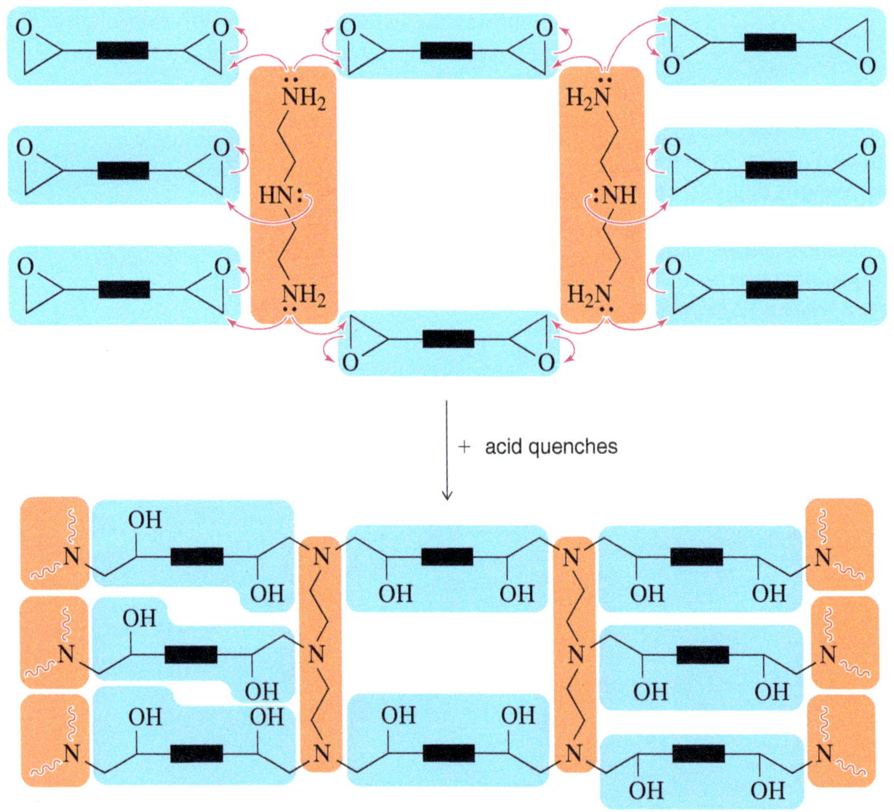

Assessment 28.24 Is the small diepoxy polymer chain in Figure 28.49 a step-growth or a chain-growth polymer?

Assessment 28.25 Sketch the structure of the epoxy resin that would be produced using the components butane-1,4-diol and epichlorohydrin with propane-1,3-diamine as cross-linking agent.

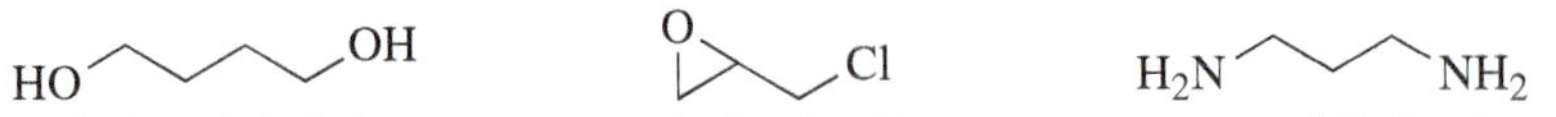

butane-1,4-diol epichlorohydrin propane-1,3-diamine

28.6.2 Ring-opening metathesis polymerization (ROMP)

QUICK REVIEW We studied the metathesis and ring-closing metathesis of alkenes using either the Grubbs or Schrock catalysts (**Figure 28.52**) in Section 16.6.

Figure 28.52 The structures of the Grubbs and Schrock carbenoid metathesis catalysts.

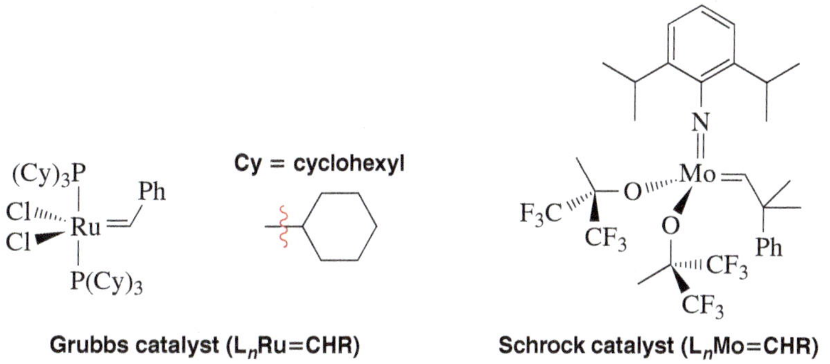

The ruthenium carbene–catalyzed coupling of 3-methylbut-1-ene is an example of metathesis [**Figure 28.53(a)**]. The reaction begins with an initiation step in which one alkene reacts with the catalyst to generate the catalytically active ruthenium carbenoid. A reversible [2 + 2]-cycloaddition with 3-methylbut-1-ene is then followed by a [2 + 2]-cycloreversion. The result of these first two steps is incorporation of the ruthenium into the 3-methylbut-1-ene starting material, generating ethene as a by-product. The formation of ethene is an entropic driving

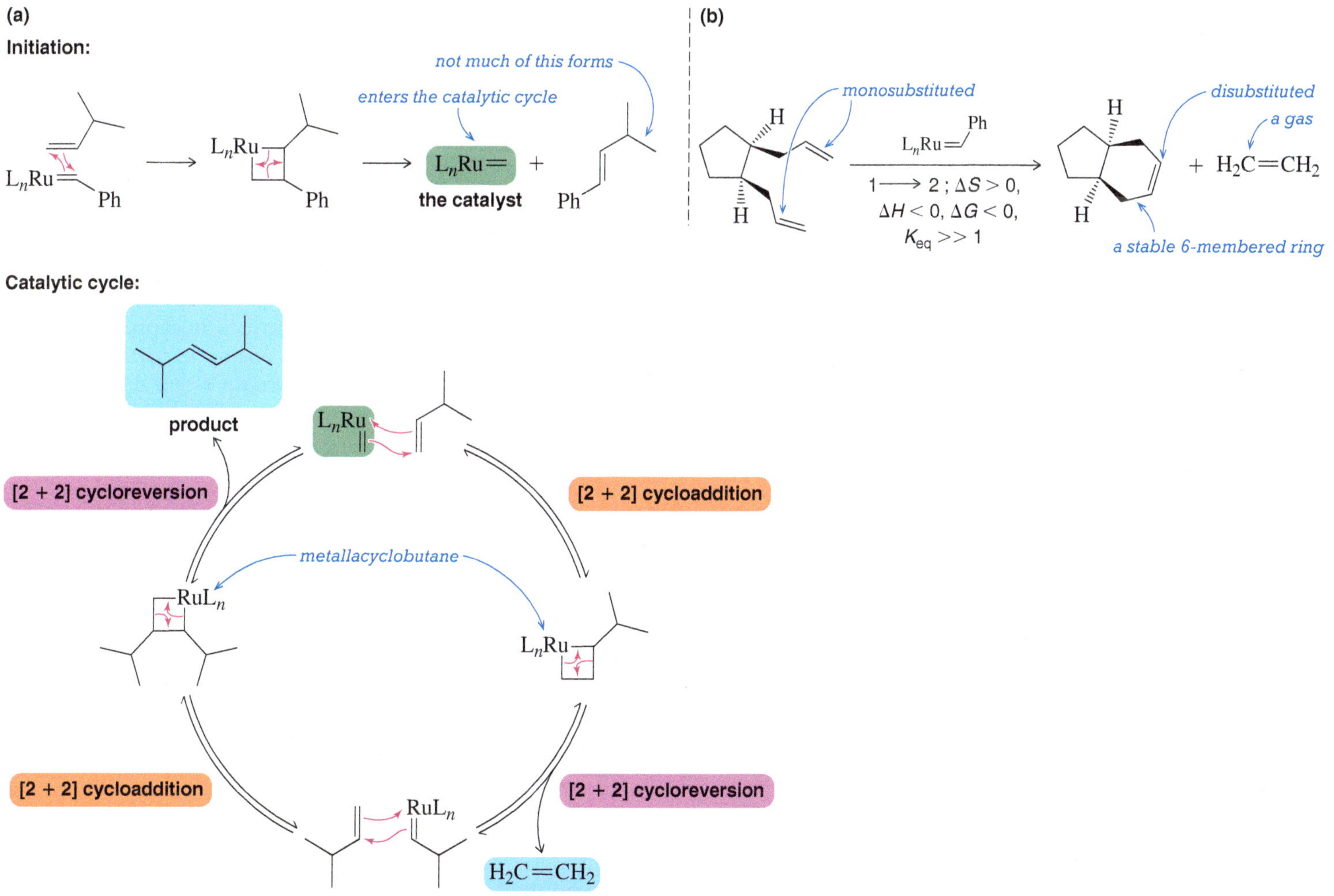

Figure 28.53 Ruthenium carbene–catalyzed metathesis. (a) Cross-metathesis of two different alkenes. (b) Ring-closing metathesis.

force for the reaction. A second [2 + 2]-cycloaddition then occurs with another molecule of 3-methylbut-1-ene, once more followed by [2 + 2]-cycloreversion to give the more stable disubstituted alkene, the product of the reaction, while regenerating the catalyst to re-enter the catalytic cycle. This reaction was also utilized for closing a ring through the metathesis of two terminal alkenes [**Figure 28.53(b)**]. [*For a detailed review of this reaction, see Section 16.6.*]

APPLICATION TO POLYMERS The application of metathesis to polymerization takes advantage of the driving force of the metathesis reaction in a unique way. Recall from Section 16.6 that metathesis reactions were essentially at equilibrium, with the most stable, and usually more substituted, alkene accounting for which side is favored. Whereas ring-*closing* metathesis reactions are exceedingly useful for small molecule synthesis, polymerization is usually done via ring-*opening* metathesis. In this reaction, coined ROMP (*r*ing-*o*pening *m*etathesis *p*olymerization), the monomers used are highly strained cyclic alkenes, allowing ring opening to serve as a driving force for the reaction. Norbornene (**Figure 28.54**) is an example of a monomer that polymerizes using a ruthenium carbene complex. A first metathesis reaction opens the ring to give a molecule containing an alkene and a ruthenium carbene at the termini. The carbene reacts with another norbornene ring, thereby forming a dimer containing the newly formed disubstituted alkene on one end and a ruthenium carbene at the other. The terminal ruthenium carbene can react with another norbornene, extending the polymer in one direction. The resulting polynorbornene (trade name Norsorex®) is used as a high-grip material, as a shock absorber, and for water filtration.

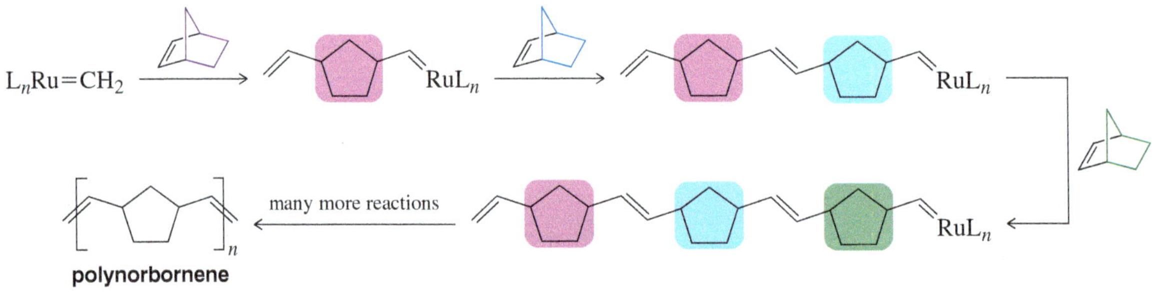

Figure 28.54 The ring-opening metathesis polymerization of norbornene.

The reaction conditions for ring-opening metathesis polymerization are mild and highly tolerant of other functional groups. **Figure 28.55** shows a substituted norbornene that was polymerized by the Schrock metathesis catalyst. [*This reaction has some sentimental value to me because I did research on a similar system as an undergraduate; this is when I decided to become an organic chemist. Have you considered joining a research group? I highly recommend it!*]

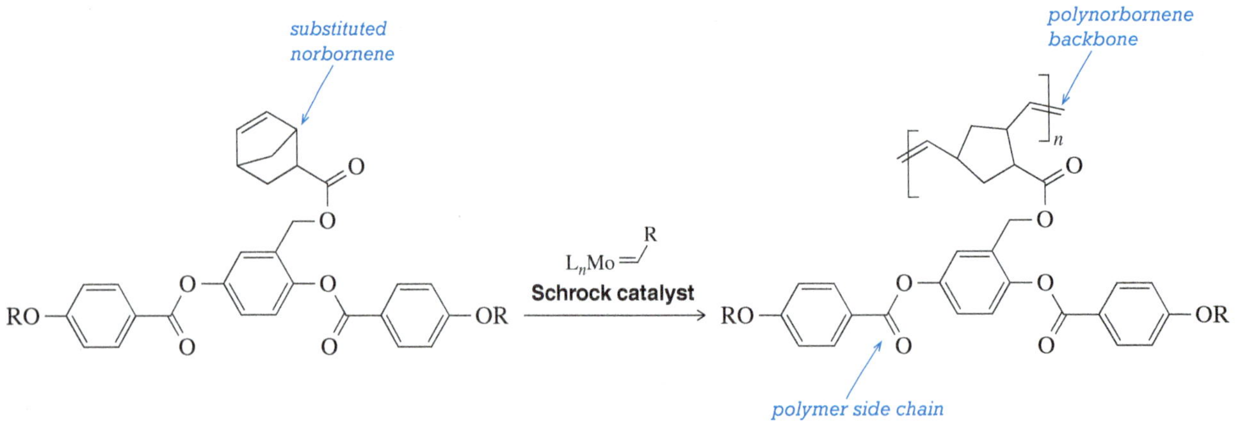

Figure 28.55 ROMP tolerates reactive functional groups, allowing the synthesis of many side-chain derivatives of polynorbornene.

Assessment 28.26 Based on the mechanism in Figure 28.53, is ROMP a chain-growth or a step-growth polymerization?

Assessment 28.27 Draw the structures of the polymers that would be produced via ROMP of the following monomers.

Assessment 28.28 Why couldn't the following polymer be made using ROMP? [*Hint: Consider the monomer that would be used.*]

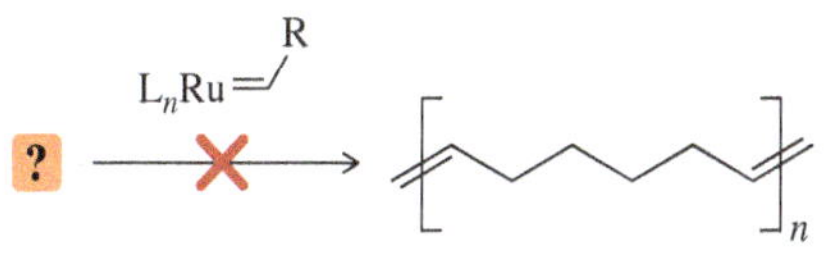

28.6.3 Transition metal–catalyzed polymerization

QUICK REVIEW We studied the palladium-catalyzed coupling of various organometallics with alkyl, aryl, and alkenyl halides to produce new C—C σ bonds in Chapter 16. The Suzuki coupling of 3-furanylboronic acid with 2-bromobut-1-ene is an example of this reaction (**Figure 28.56**). Following *oxidative addition* of the C—Br σ bond to Pd, an organopalladium halide is produced where palladium is now at the +2 oxidation state The bromide is displaced by *t*-butoxide. *Transmetallation* then occurs with a boronate species derived from the boronic acid to give a diorganopalladium species. *Reductive elimination* results in the coupling of the two organic fragments to produce a new C—C σ bond, while regenerating $Pd(PPh_3)_4$ (0 oxidation state) for continued use in the catalytic cycle. [*For a detailed review on this and other palladium coupling reactions, see Section 16.4.*]

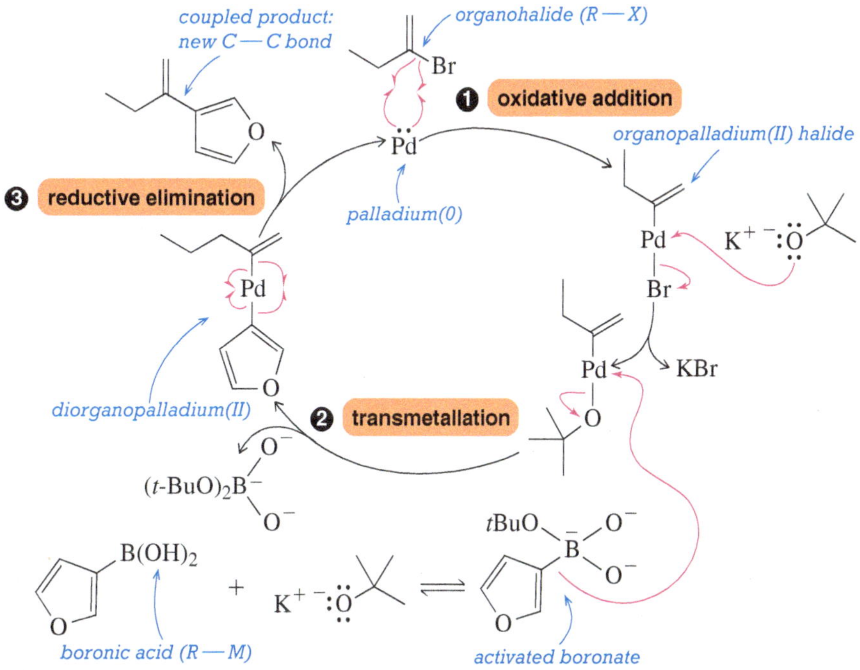

Figure 28.56 The catalytic cycle for the Suzuki coupling. [*PPh_3 ligands have been removed for clarity.*]

APPLICATION TO POLYMERS Coupling reactions are especially suited for polymerization reactions because they continually regenerate the palladium catalyst. To utilize palladium coupling reactions for polymerization, you simply need monomers that can grow in two different directions. As shown in **Figure 28.57**, this can be accomplished using a single monomer containing both a halogen and a boronic acid or using two monomers (one containing two halogens, the other containing two boronic acids). In either of these cases, the polymer can grow in two directions, making these step-growth polymers.

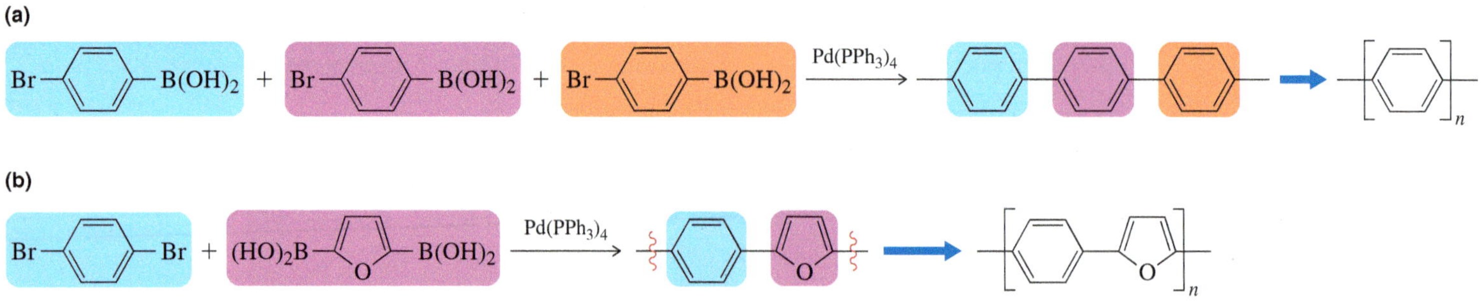

Figure 28.57 The application of Suzuki conditions to polymerization. (a) A single monomer with a halogen and a boronic acid. (b) Two monomers—one with two halogens and one with two boronic acids.

Highly conjugated polymers have found recent use in the creation of organic photovoltaics (solar panels). A recent effort made use of the Suzuki coupling reaction for the synthesis of a polymer containing alternating electron-donating and electron-accepting monomers. Coupling of the dibromide and the diboronate (which behaves similarly to boronic acid) was achieved using $Pd(PPh_3)_4$ under basic conditions [Figure 28.58(a)]. Upon completion, the polymer chain contains a C—Br bond at one terminus and a boronate at the other. For the purpose of creating an organic photovoltaic, these ends are capped by subsequent coupling reactions [Figure 28.58(b)].

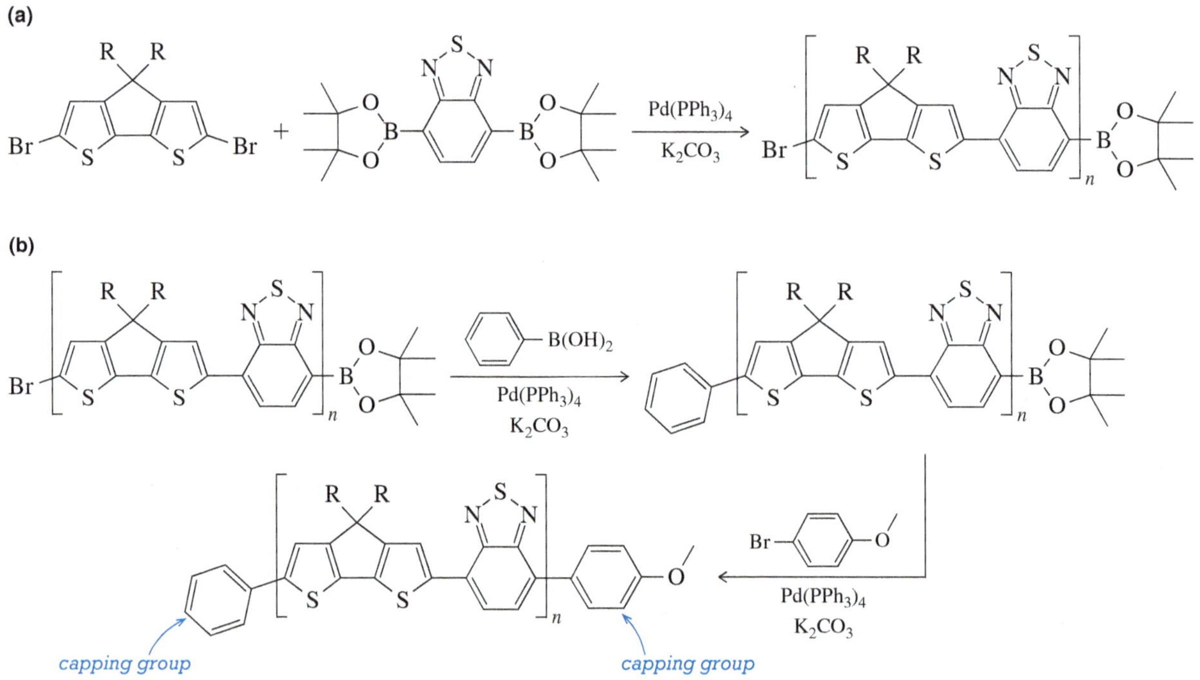

Figure 28.58 The synthesis of a highly conjugated potential organic photovoltaic via a Suzuki polymerization.

Assessment 28.29 Identify the monomer(s) necessary to produce the polymers shown by a Suzuki coupling reaction.

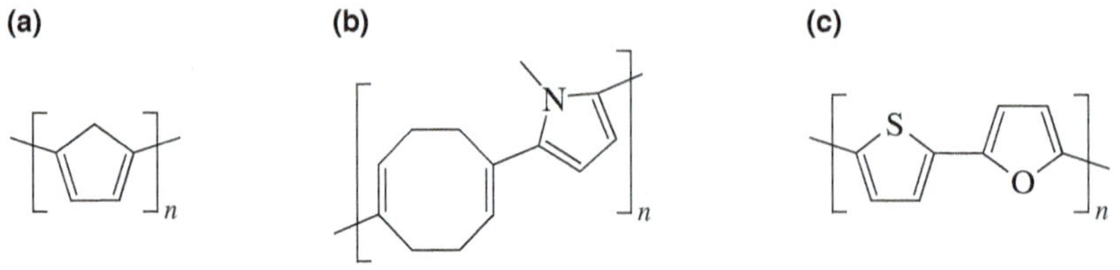

Assessment 28.30 Predict the structure of the polymer that would result from the following palladium-catalyzed polymerizations.

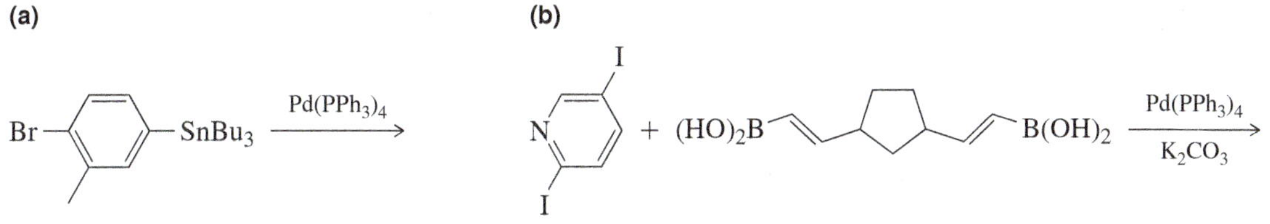

28.7 Block copolymers

The properties of a polymer can be manipulated by the combining of two or more polymers into one. This allows the resulting polymer to have properties of both, as we demonstrate when discussing spandex. Polymers of this type are referred to as block copolymers because there are blocks of each polymer in a single polymer chain. Block copolymers are distinct from copolymers. A copolymer (Section 28.2) is simply the polymerization of two monomer types and usually results in alternating A and B along the polymer chain. Block polymers, on the other hand, result from the formation of low-molecular-weight homopolymers of A and homopolymers of B and then combining the two so that there are regions of all A and all B in the resulting polymer (**Figure 28.59**).

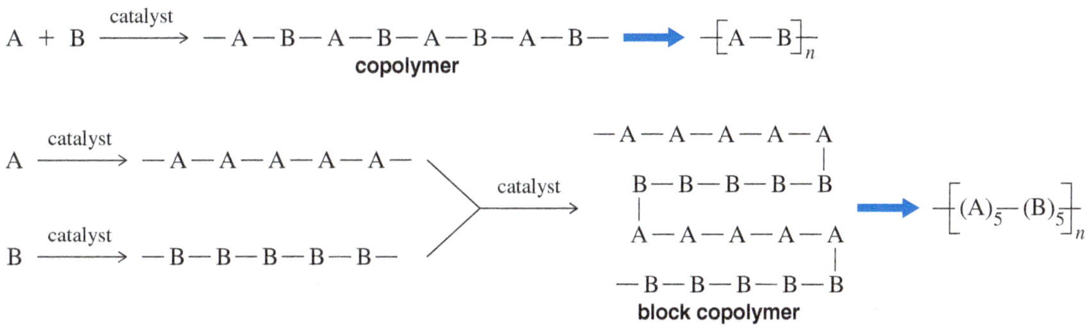

Figure 28.59 The difference between a copolymer and a block copolymer.

The structural features of a block copolymer are responsible for its properties, with each block contributing distinct features. Nowhere is this clearer than in the block copolymer spandex. Spandex (**Figure 28.60**) is composed of two distinct regions—namely, a polyether "block" and a polyurea "block." These two blocks are connected by the polyurethane linkers discussed in Section 28.5.4.

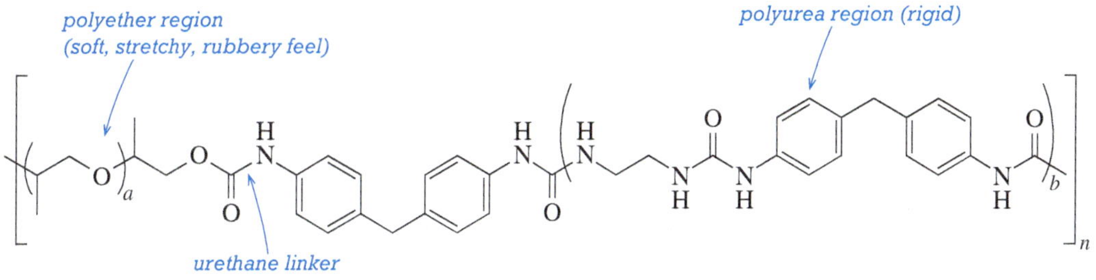

Figure 28.60 The structure of the block copolymer spandex.

The syntheses of the spandex copolymers are shown in **Figure 28.61**. The polyurea region is formed by mixing ethylene diamine with a diisocyanate [Figure 28.61(a)]. Polymerization occurs by a mechanism similar to urethane formation, but with a diamine replacing a diol. [*See Section 28.5.4 on polyurethane formation.*] The polyether region is formed by the polymerization of propylene oxide [Figure 28.61(b)] via a mechanism analogous to the formation of epoxy resins. [*See Section 28.6.1 on epoxy resin formation.*]

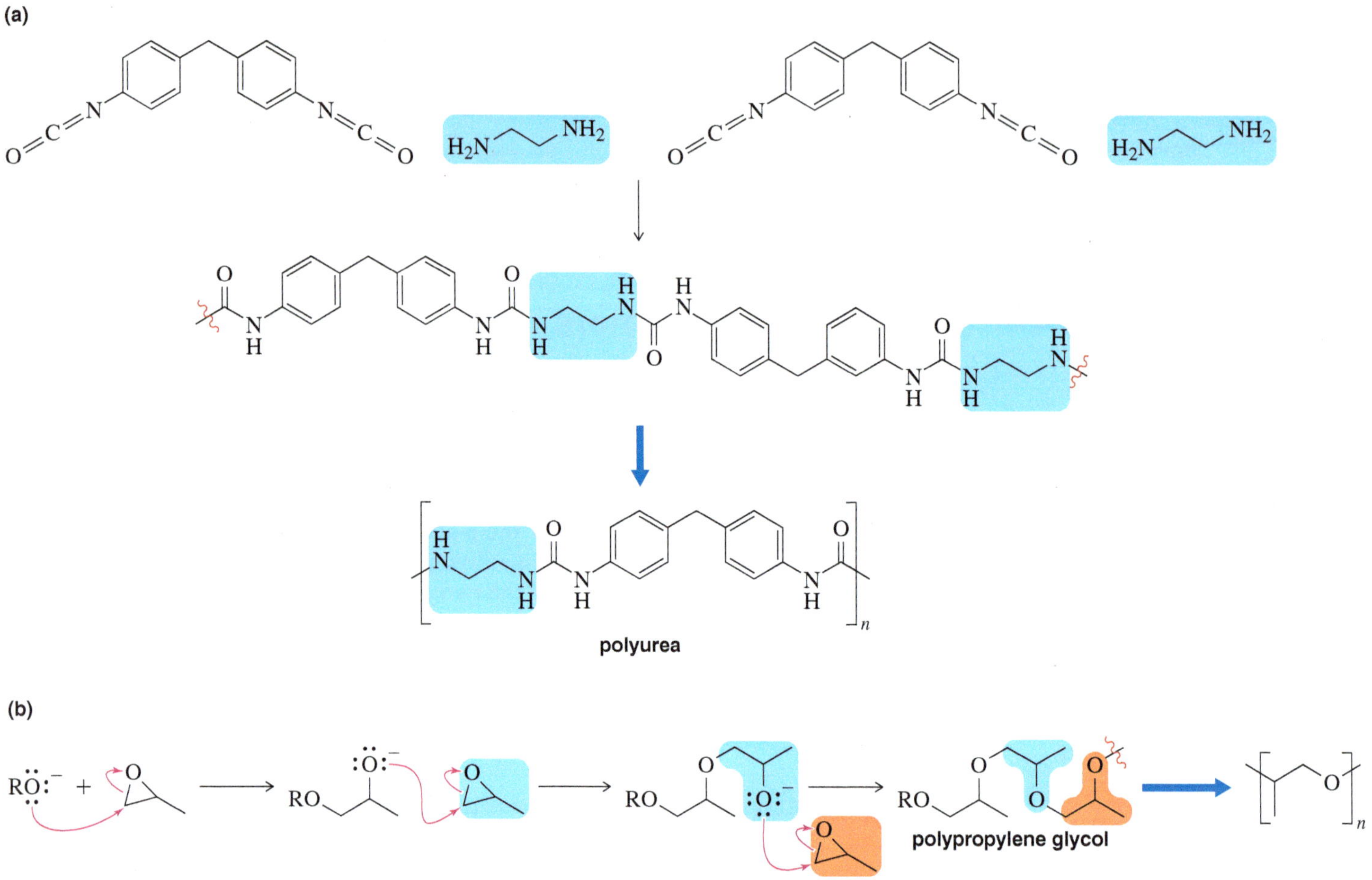

Figure 28.61 The formation of the spandex copolymers (a) polyurea and (b) polypropylene glycol.

Reaction of the low-molecular-weight polyether/diol with low-molecular-weight polyurea/diisocyanate results in spandex (**Figure 28.62**). Coupling of the two copolymers occurs by the mechanism shown in Section 28.5.4 for urethane formation. The resulting polymer, spandex, has two regions imparting distinct properties. The polyether region is flexible, giving the

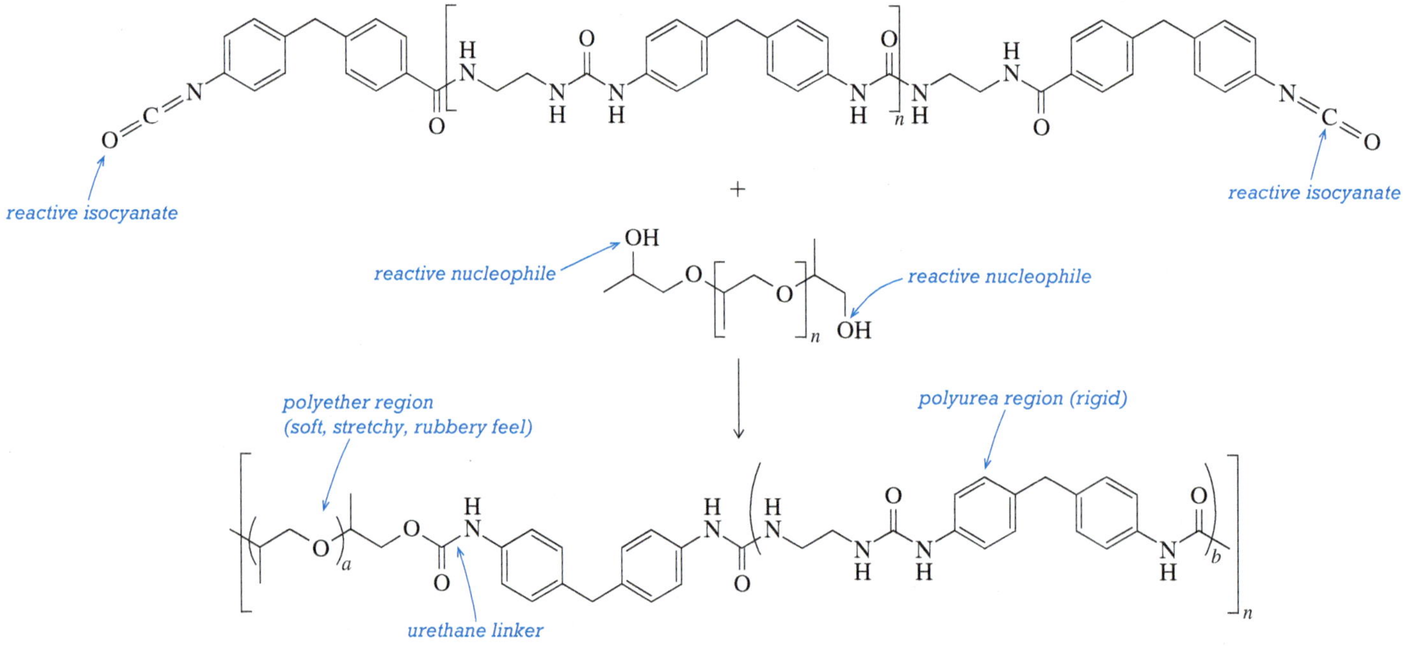

Figure 28.62 Coupling of the copolymers to make spandex, a strong, elastic polymer.

polymer a stretchable, rubbery feel. The polyurethane region, on the other hand, is rigid and inflexible. The combination of the two results in an elastic fiber heavily used in sports clothing. Given the structural similarity between the polyurethane region and Kevlar [*see Section 28.5.1*], it should not surprise you that spandex is strong and durable.

Assessment 28.31 Draw the structure of the block copolymer that results from the polymerization scheme shown. [*Start by forming each copolymer (a) and (b) and then combine them to make (c).*]

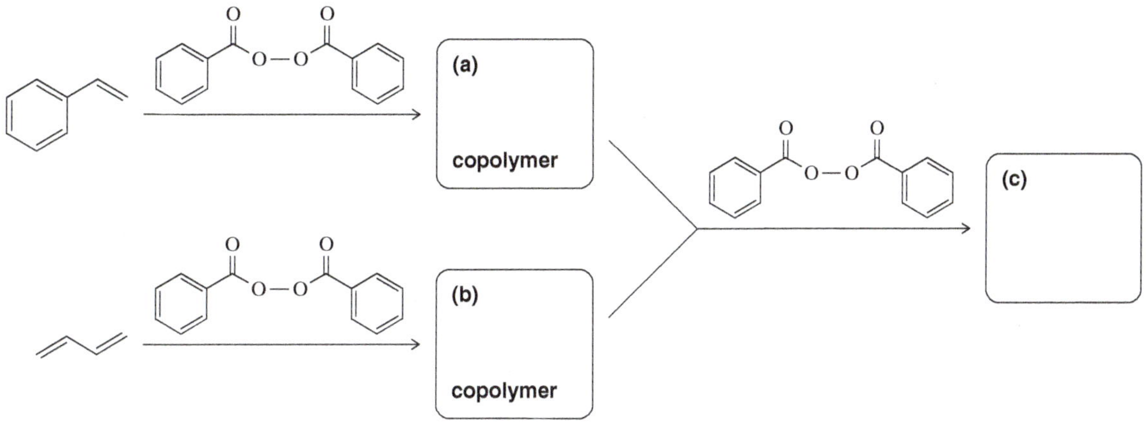

Concluding thoughts

By drawing attention to some of the better-known products composed of polymers, this chapter showed us that polymer synthesis has indeed improved our everyday lives. And it has done so using the standard organic reactions we have pored over for the last several months. Furthermore, the importance of green chemistry has been reaffirmed once again. Because of the persistence of polymers in the environment, the synthesis of biodegradable polymers using environmentally friendly reagents and conditions has become an important goal. As this research area continues to develop, perhaps one day you'll be the one providing some of those solutions.

Ahhh . . . the end. I always envisioned this book as a conversation between author and student. [*I even considered the title "Organic Chemistry: Conversations" at one point.*] And so bear with me as I wrap this up; it's been a long, long time getting to this point [*8 years*]. Recall the Winston Churchill quote from the beginning of the chapter that finished: *But it is, perhaps, the end of the beginning.* This quote has a lot of meaning for me (and I hope for you, too) at this point. For me, it's the end of the writing of the first edition, but perhaps it's also the beginning of revisions needed to help future students more easily learn organic chemistry.

For you . . . who knows? Maybe you came into organic chemistry with certain career aspirations and, as a result of your experience in the class and with this text, decided that maybe organic chemistry is your future. True or not, I have long been convinced that organic chemistry marks the jumping off point for career direction. Because it is deemed by so many as a "weed-out" course, finishing this course successfully means you're in a really good spot with regard to whatever career you are pursuing. Whether you are planning to take a standardized exam that requires some understanding of organic chemistry or are getting ready for a biochemistry course, you can now turn your focus to those things. [*This book and the student supplement can be helpful as you prepare for these things . . . don't throw them away yet!*] Having completed the first part of that journey, you can now confidently move on to that next phase. Wherever that journey takes you, it has been an honor and a privilege to have shared a small part of it with you. Before you begin that next phase, take some time to relax and to celebrate the accomplishment of getting this far. I know I certainly will!

Yours in Organic Chemistry,
—RJM

Important Concepts

[*These are the major concepts covered in each section, excluding reactions.*]

28.1: Polymers are large molecules composed of several smaller repeating units.

28.2: Polymer structures are represented by showing the monomer in brackets with an italic *n* to indicate that the monomer repeats. Homopolymers contain a single repeating monomer. Copolymers are polymers containing more than one unique monomer.

28.3: Polymers are classified by the reaction type used for their synthesis. Chain-growth polymers form in a single direction, with the addition of one monomer at a time. Step-growth polymers, with functional groups on both ends, grow in multiple directions. Polymers come in a number of different shapes (linear, branched, cross-linked, star, comb, and brush). In general, linear polymers form highly crystalline structures, whereas branched polymers are more amorphous.

28.4: Polymers can be formed using several previously studied addition reactions. Cationic polymerization proceeds through electrophilic addition reactions. Anionic polymerization proceeds through nucleophilic addition reactions. Polymerization also occurs through free-radical reactions.

28.5: Polymers can be formed using condensation reactions. Nucleophilic acyl substitution reactions are used for the synthesis of polyamides, polyesters, polycarbonates, and polyurethanes.

28.6: Epoxides can be used as electrophiles in polymerization reactions to make epoxy resins. Alkene metathesis is used to synthesize polymers by ring-opening metathesis polymerization (ROMP). Palladium-catalyzed coupling reactions can also be used to synthesize polymers.

28.7: The combination of two polymers into one creates block copolymers. Block copolymers have structural features associated with each polymeric block.

Important Reactions and Mechanisms

[*Because all of these polymerization reactions were studied previously, an example of each important reaction from this chapter is shown with a reference to where in the text it was first studied.*]

Polymerization reactions	
Cationic polymerization (Sections 28.4.1 and 8.4)	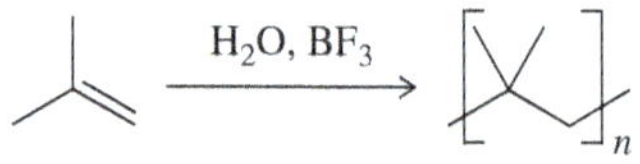
Nucleophilic polymerization (Sections 28.4.2 and 21.5)	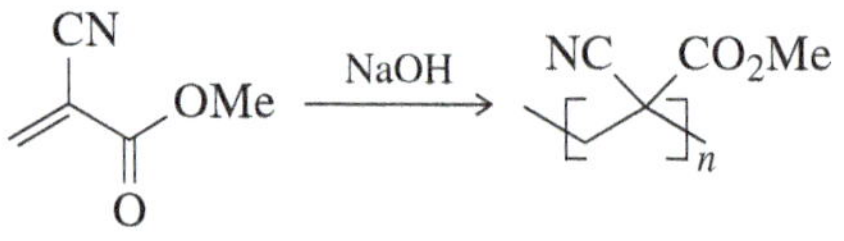
Free-radical polymerization (Sections 28.4.3 and 8.3.4)	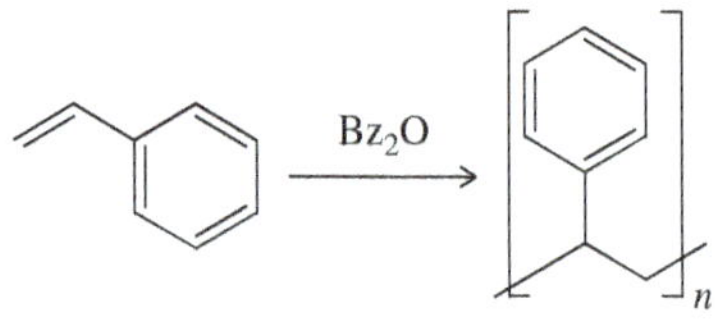
Condensation polymerization: polyamide (Sections 28.5.1 and 19.7.3)	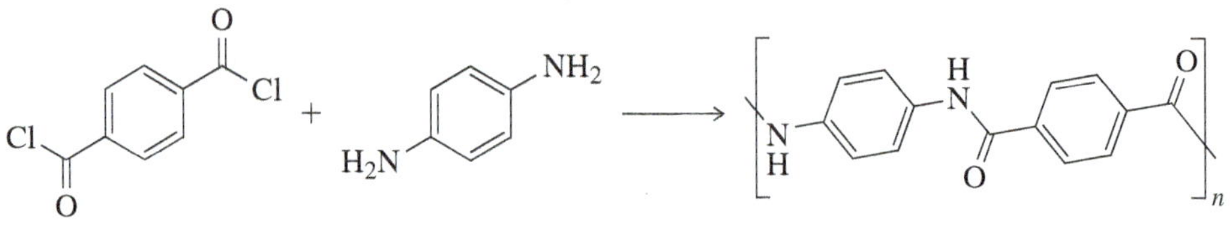
Condensation polymerization: polyester (Sections 28.5.2 and 18.7.2)	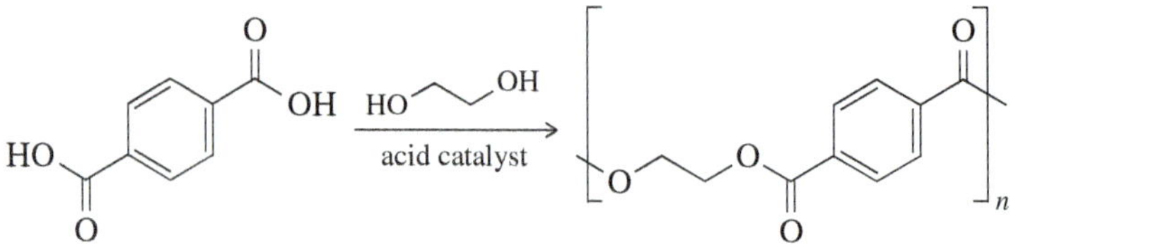

(continued)

Polymerization reactions	
Condensation polymerization: polycarbonate (Sections 28.5.3 and 19.7.2)	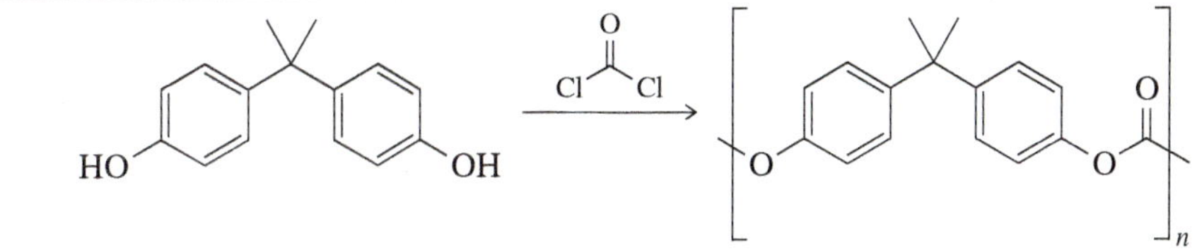
Epoxy resins (Sections 28.6.1 and 13.13.2)	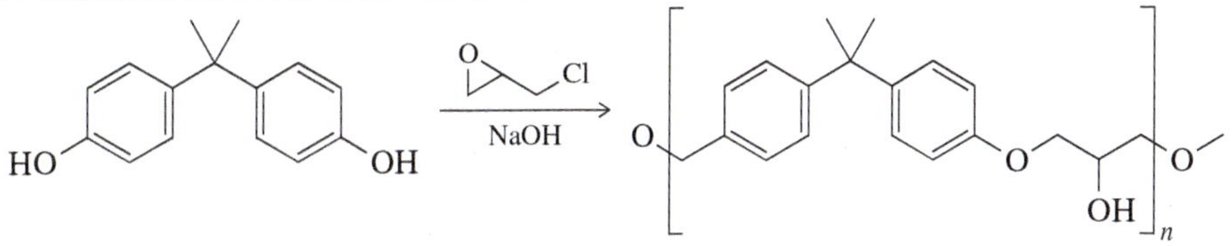
Ring-opening metathesis polymerization (Sections 28.6.2 and 16.6)	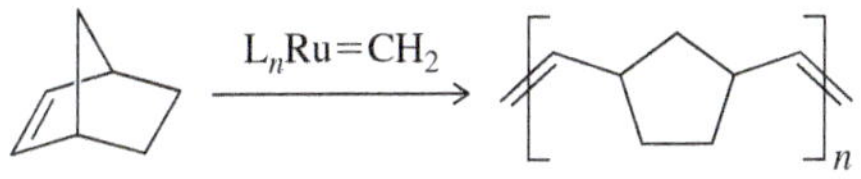
Transition metal–catalyzed polymerization (Sections 28.6.3 and 16.4)	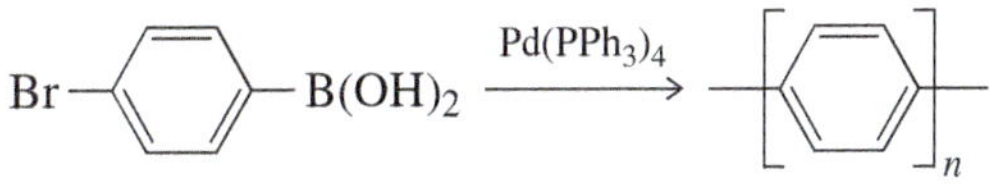

End-of-Chapter Assessments

[*Assessments are ranked according to their difficulty;* •••• = *most difficult.*]

28.32 (•) Classify the following general polymerization schemes as either step growth or chain growth.

(a)

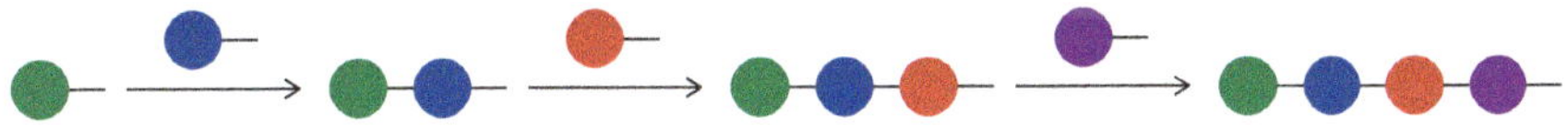

(b)

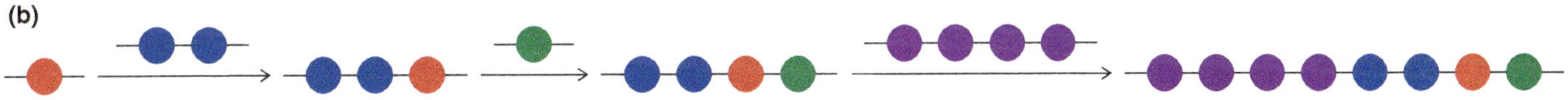

28.33 (••) Would you expect the following monomers (or sets of monomers), when polymerized, to do so via step-growth or chain-growth mechanisms? Why?

(a)

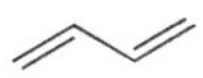

(b)

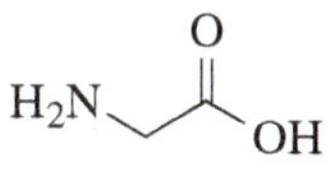

(c)

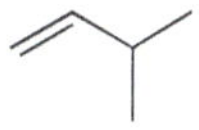

(d)

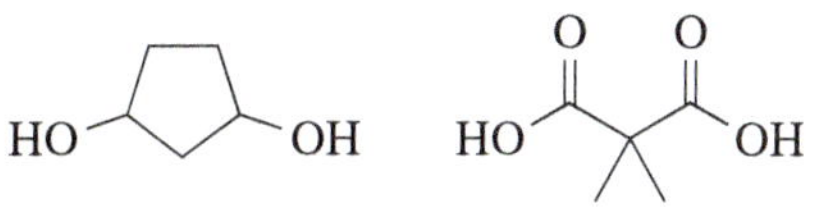

28.34 (•) Choose which type of polymer you would expect to have a higher T_m.

(a) linear vs. branched
(b) branched vs. star
(c) linear vs. comb
(d) star vs. brush

28.35 (••) For each of the following monomers, indicate whether free-radical, cationic, or anionic polymerization would be best. Justify your choice.

(a)

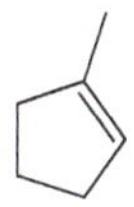

(b)

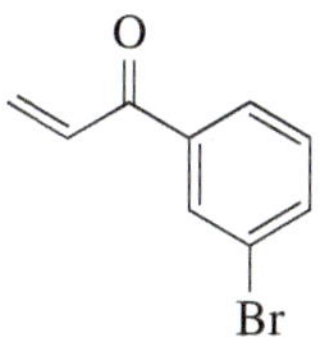

(c)

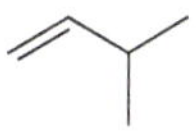

(d)

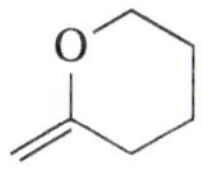

28.36 (•••) Draw the polymers that would be produced by polymerization of the monomers in Assessment 28.35.

28.37 (••) Are the polymers produced from the monomers in Assessment 28.35 considered to be condensation or addition polymers?

28.38 (•••) One of the polymers you drew in Assessment 28.36 is set up to be cross-linked. (a) Which is it? (b) What makes this set up for cross-linking?

28.39 (••••) Suggest a mechanism by which the below reaction conditions would cross-link the polymer you chose in Assessment 28.38. [*See Chapter 16 for a review.*]

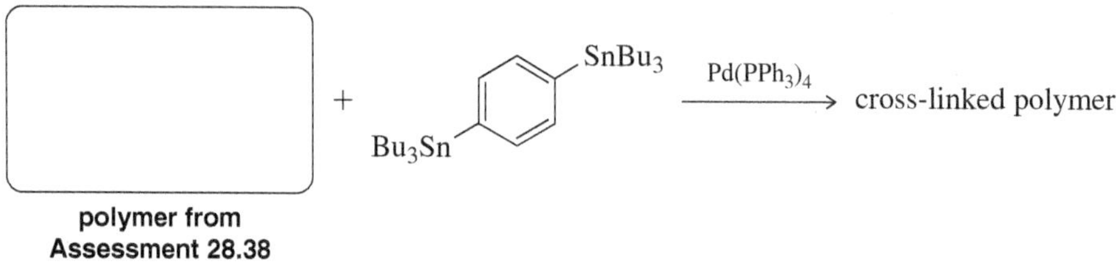

28.40 (••) Classify the following polymers as polyurethane, polycarbonate, polyester, or polyamide.

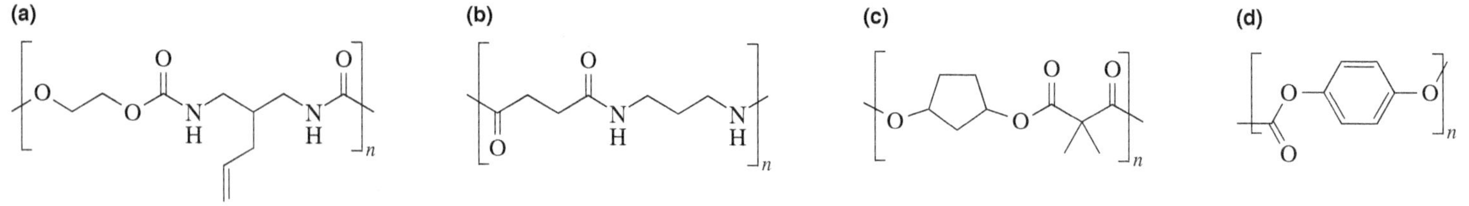

28.41 (•) Are the polymers in Assessment 28.40 classified as addition or condensation polymers?

28.42 (•••) Suggest a synthesis of each of the polymers in Assessment 28.40.

28.43 (••••) (a) Which polymer in Assessment 28.40 is best set up for cross-linking? (b) Suggest a method by which it could be cross-linked. (c) How will cross-linking change the properties of the resulting polymer?

Poly(*endo*-dicyclopentadiene) can be made using ring-closing metathesis. Assessments 28.44–28.47 refer to this reaction.

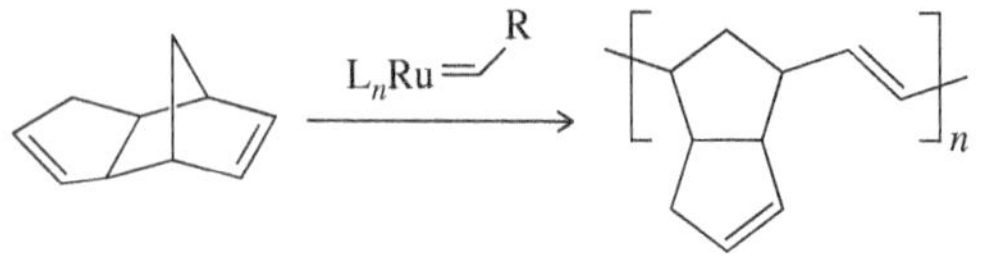

poly(*endo*-dicyclopentadiene)

28.44 (•••) Suggest a mechanism for the polymerization reaction.

28.45 (•••) Only one alkene in the starting material undergoes the metathesis reaction. Why is one alkene so much more reactive? [*Identify which one is most reactive first.*]

28.46 (•••) How would you make the monomer used in this polymerization? [*We learned this reaction in Chapter 22.*]

28.47 (••••) Poly(*endo*-dicyclopentadiene) can be further converted to a cross-linked thermoset. (a) Give reagents and conditions by which the cross-linking can occur. (b) How do the properties change in a thermoset? (c) What causes these changes?

28.48 (••••) As a chemist working for a garden hose company, you have discovered a new polymer that has all the desirable properties for the perfect garden hose. Unfortunately, at lower temperatures (~5 °C), it becomes too brittle. How can you noncovalently modify the polymer to make it suitable at these low temperatures?

28.49 (•) In Chapters 26 and 27, we learned about a few biological polymers. Draw the monomer that is used to make each of the polymers used in the following figures.

(a) Figure 27.50(a)
(b) Figure 27.59(a)
(c) Figure 27.105 [*Not really a polymer, but it's close.*]
(d) Figure 26.45(a)

28.50 (•••) Although amino acids, containing two functional groups, could grow in a step-growth fashion, in nature, they are built as chain-growth polymers in one direction. Why is this?

Appendix Contents

APPENDIX

A

Problem-Solving Hints

The following are questions/advice that will help you get back on track when confronted with a challenging problem. All of these will not be applicable in all situations, but as you solve problems, you should figure out which ones help you when.

1. The ball rolls downhill.
2. Negatives attack positives.
3. Number the carbons.
4. Identify bonds formed and bonds broken.
5. What is electron-rich? What is electron-poor?
6. What is the nucleophile? What is the electrophile?
7. What is the Lewis acid? What is the Lewis base?
8. What is the pK_a of the acidic proton?
9. What is the pK_a of the conjugate acid of the base/nucleophile shown?
10. What electronic factors are in play? Hyperconjugation, induction, resonance?
11. Is there a steric bias in the reaction?
12. Draw all resonance structures of the molecules/intermediates/reactants.
13. Is the reaction favored by entropy?
14. Estimate ΔH based on the bonds formed and the bonds broken.
15. Work backward.
16. Which intermediate (carbocation, carbanion, radical) is more stable? Why?
17. Is the reaction under kinetic or thermodynamic control?
18. Is the reaction reversible?
19. How can you shift the equilibrium of this reaction?
20. Which reaction is faster?
21. Which reaction is more favored?
22. Is the molecule in the right conformation?
23. Where would you expect this peak/band to appear in the NMR/IR spectrum?
24. Is the solvent protic or aprotic?
25. Is the nucleophile/base weak or strong?
26. Is the electrophile weak or strong?
27. For carbonyls, draw the zwitterion structure.
28. For bonds between metals and nonmetals, treat them as if they are ionic with a negative charge on the nonmetal and a positive charge on the metal.

There's a lot of empty space on this page. If you come up with a problem-solving hint that works for you, send it to me at mullinsr@xavier.edu and I might include it in the next edition.

APPENDIX

B

Summary of Nomenclature

Nomenclature steps

1. Find the longest consecutive chain of carbon atoms **containing the highest priority functional group.**

Priority*	Functional group name	As prefix	As suffix
1	Carboxylic acid		-oic acid
2	Ester		alkyl {parent}-oate
3	Acid chloride		-oyl chloride
3	Anhydride		-oic anhydride
4	Amide		*N*-alkyl {parent}-amide
5	Nitrile	-cyano-	-nitrile
6	Aldehyde	-oxo-	-al
7	Ketone	-oxo-	-one
8	Alcohol	-hydroxy-	-ol
9	Amine	-{*N*-alkyl}amino-	-amine
10	Alkene	-en-	-ene
10	Alkyne	-yn-	-yne
None	Ether	-alkoxy-	
None	Halogen	-halo-	
None	Alkyl	-alkyl-	

*The red numbers indicate that the functional groups are tied in terms of their priority.

   a. If more than one longest chain, the one with the most branching points wins.

2. Number the carbon atoms in the longest chain.
   a. Give the highest priority functional group the lowest number.
   b. If more than one and they tie, give the first non-priority group the lowest number.
   c. If those tie, give the second non-priority group the lowest number. Continue.
   d. If everything ties, compare the first two groups from each end and give the alphabetically first group the lowest number.
   e. For cyclic molecules, give the priority functional group carbon #1, and number to give the next substituent the lowest number.

3. Name the parent hydrocarbon.

Number of carbons	Parent hydrocarbon name
1	Methane
2	Ethane
3	Propane
4	Butane
5	Pentane
6	Hexane
7	Heptane
8	Octane
9	Nonane
10	Decane
11	Undecane
12	Dodecane

   a. Update the parent hydrocarbon name to account for unsaturation (if applicable) and the priority functional group.

4. Identify the substituents and their locations.

Fragment	Substituent name	Abbreviation	Fragment	Substituent name	Abbreviation
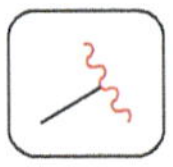	Methyl	Me	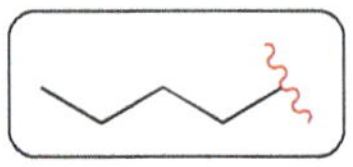	*n*-Butyl	*n*-Bu
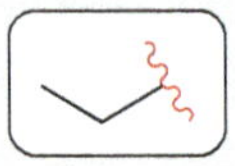	Ethyl	Et	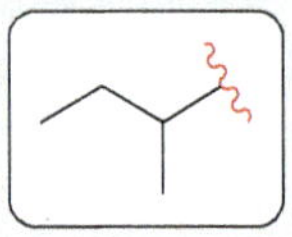	*sec*-Butyl	*s*-Bu
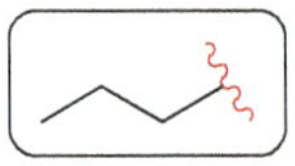	*n*-Propyl	*n*-Pr	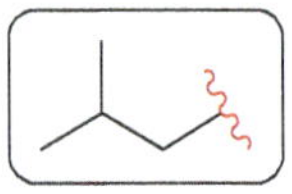	Isobutyl	*i*Bu
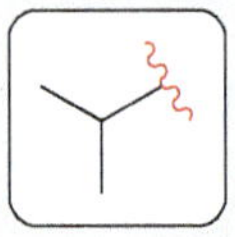	Isopropyl	*i*Pr		*tert*-Butyl	*t*-Bu

a. If there is more than one of the same substituent, use a numerical modifier (di, tri, tetra, penta, etc.).
b. If a benzene ring is a substituent, name it as a "phenyl" substituent.
c. If a substituent has branching of its own:
   i. Name the whole chain as a mini-molecule, but with -yl at the end because it is a substituent.
   ii. Number the substituent starting with the first carbon off the parent chain.
   iii. Identify the substituents and their location.
   iv. Place the entire complex substituent in parentheses.

5. List substituents alphabetically in front of the parent hydrocarbon.
   a. DO alphabetize by the i in iso and the c in cyclo.
   b. DO NOT alphabetize by *n*-, *s*-, or *t*-, or by the numerical modifiers di, tri, tetra, etc.
   c. Alphabetize complex substituents by their first letter regardless of what it is.
   d. Separate numbers from numbers with commas.
   e. Separate letters from numbers with hyphens.
6. Assign stereochemistry to every stereocenter and alkene in the molecule.
   a. Place stereochemistry information in parentheses in front of substituents.

Benzene nomenclature

1. Monosubstituted benzene
   a. If benzene is the parent, name substituents as described previously.
   b. If a substituent is more complicated than the benzene ring, name the benzene ring as a phenyl substituent on the other parent hydrocarbon.
2. Disubstituted benzene
   a. If a common ring (toluene, phenol, aniline, anisole), number to give that group carbon #1.
   b. Otherwise, number alphabetically.
   c. Use *o*-, *m*-, or *p*- to denote the relative arrangement of groups.
3. Polysubstituted benzene
   a. If a common ring (toluene, phenol, aniline, anisole), number to give that group carbon #1.
   b. Otherwise, number to give the lowest overall set of numbers.

Strain

Strain in acyclic systems

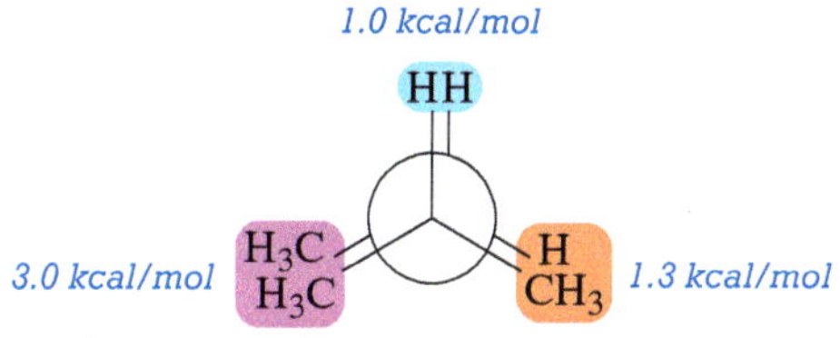

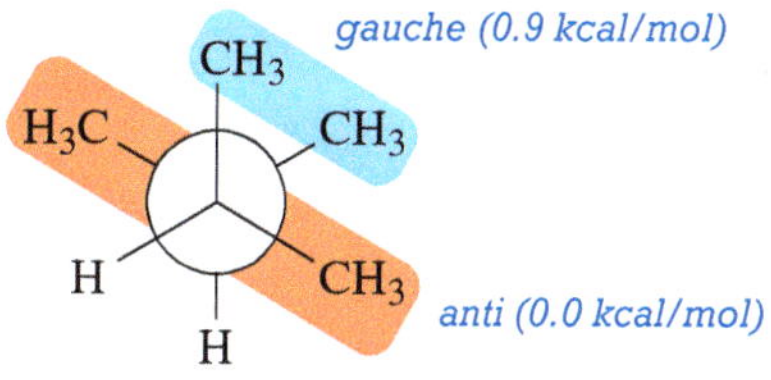

Ring strain

cyclopropane
27.6 kcal/mol
(115 kJ/mol)

cyclobutane
19.8 kcal/mol
(82.8 kJ/mol)

cyclopentane
3.9 kcal/mol
(16 kJ/mol)

cyclohexane
0 kcal/mol
(0 kJ/mol)

A values

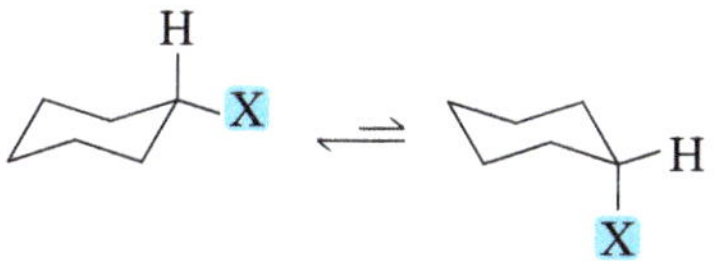

equatorial **axial**

X	ΔG (axial–equatorial)		X	ΔG (axial–equatorial)	
	(kcal/mol)	(kJ/mol)		(kcal/mol)	(kJ/mol)
—F	0.2	0.8	—COOH	1.4	5.9
—CN	0.2	0.8	—CH_3	1.8	7.6
—Cl	0.5	2.1	—CH_2CH_3	1.9	7.9
—Br	0.6	2.5	—$CH(CH_3)_2$	2.1	8.8
—OH	1.0	4.1	—$C(CH_3)_3$	5.4	23

Because you are going from more stable (equatorial) to less stable (axial), these values are all positive.

APPENDIX D

pK_a Values

The following is a list of standard pK_a values. The pK_a value of most compounds can be inferred based on the similarity in structure to molecules in the table.

	Acid dissociation							K_a	pK_a
strong acids; **stronger acids**	HCl hydrochloric acid	+	H_2O	$\rightleftharpoons$	H_3O^+	+	Cl^- chloride ion (**weaker bases**)	1×10^{7}	**−7**
	$HO{-}S(=O)_2{-}OH$ sulfuric acid	+	H_2O	$\rightleftharpoons$	H_3O^+	+	$HO{-}S(=O)_2{-}O^-$ hydrogen sulfate	1×10^{3}	**−3**
	H_3O^+ hydronium ion	+	H_2O	$\rightleftharpoons$	H_3O^+	+	H_2O water	55.6	**−1.7**
	HF hydrofluoric acid	+	H_2O	$\rightleftharpoons$	H_3O^+	+	F^- fluoride ion	6.8×10^{-4}	**3.17**
	$H{-}C(=O){-}OH$ formic acid	+	H_2O	$\rightleftharpoons$	H_3O^+	+	$H{-}C(=O){-}O^-$ formate ion	1.7×10^{-4}	**3.76**
	$CH_3{-}C(=O){-}OH$ acetic acid	+	H_2O	$\rightleftharpoons$	H_3O^+	+	$CH_3{-}C(=O){-}O^-$ acetate ion	1.8×10^{-5}	**4.74**
	$H{-}C{\equiv}N$ hydrocyanic acid	+	H_2O	$\rightleftharpoons$	H_3O^+	+	$^-C{\equiv}N$ cyanide ion	6.0×10^{-10}	**9.22**
weak acids	$^+NH_4$ ammonium ion	+	H_2O	$\rightleftharpoons$	H_3O^+	+	NH_3 ammonia	5.8×10^{-10}	**9.24**
	H_2O water	+	H_2O	$\rightleftharpoons$	H_3O^+	+	HO^- hydroxide ion	1.8×10^{-16}	**15.7**
	$CH_3CH_2{-}OH$ ethyl alcohol	+	H_2O	$\rightleftharpoons$	H_3O^+	+	$CH_3CH_2O^-$ ethoxide ion	1.3×10^{-16}	**15.9**
very weak acids	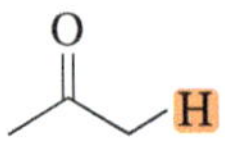 acetone	+	H_2O	$\rightleftharpoons$	H_3O^+	+	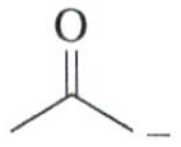 enolate	10^{-20}	**20**
	$H{-}C{\equiv}C{-}H$ acetylene	+	H_2O	$\rightleftharpoons$	H_3O^+	+	$H{-}C{\equiv}C^-$ acetylide	10^{-25}	**25**
not acidic	NH_3 ammonia	+	H_2O	$\rightleftharpoons$	H_3O^+	+	$^-NH_2$ amide ion	10^{-33}	**38**
weaker acids	CH_4 methane	+	H_2O	$\rightleftharpoons$	H_3O^+	+	$^-CH_3$ methyl anion (**stronger bases**)	10^{-50}	**50**

The following is another way of looking at pK_a values of common molecules encountered in the textbook. It shows the relative acidity of related compounds. [*You may notice slight variation between these and the ones in the previous table. That's OK. Most of the time when we use* pK_a *values, we simply need to be close enough.*]

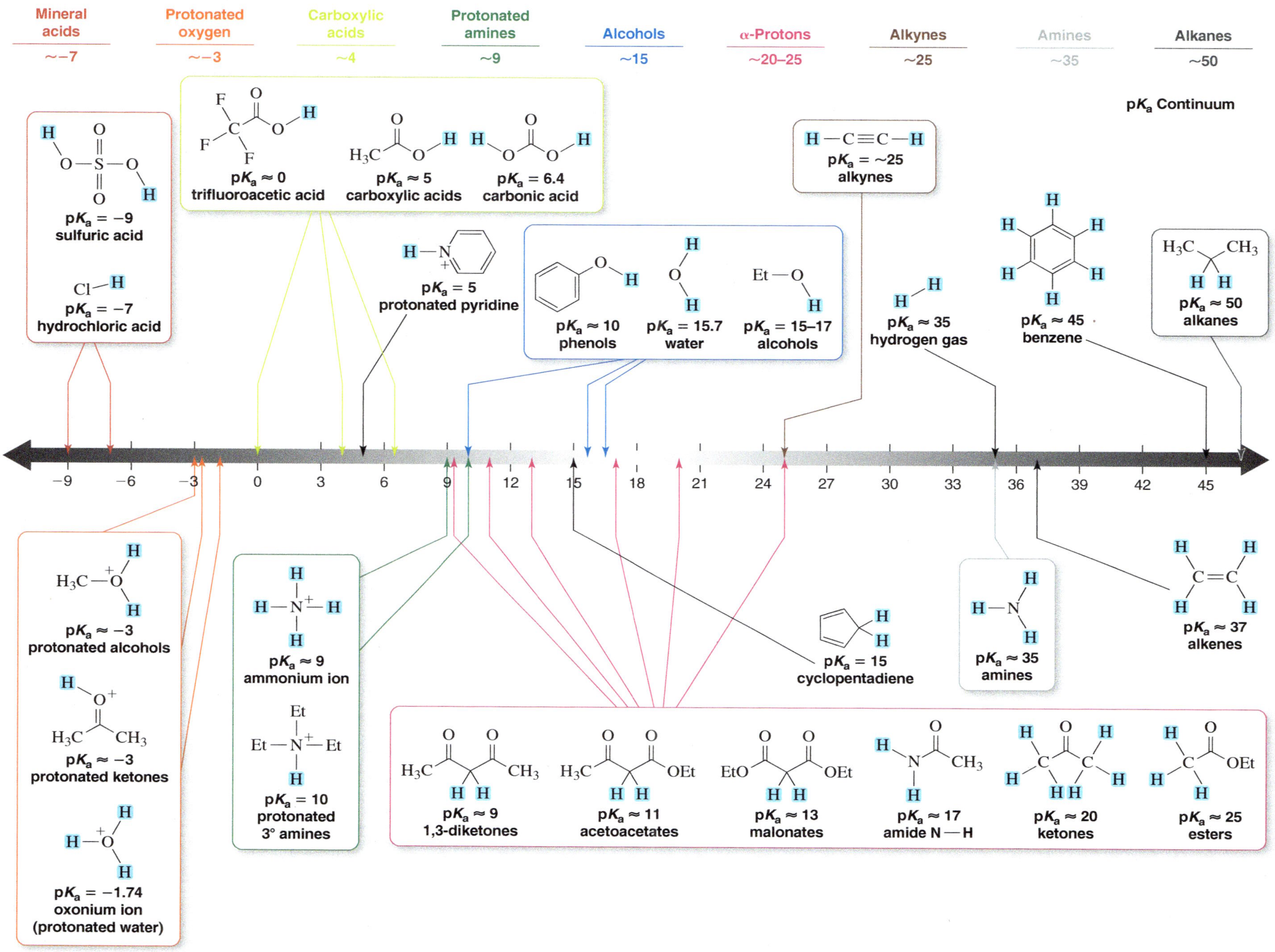

Below is a more comprehensive list of specific pK_a values. It is not important that you know these, but by comparing specific values in the table, you will have a better sense of the relative acidity of a variety of organic molecules. [*Once again, don't worry about slight deviations here.*]

Mineral acids			
H_2O 15.7	H_3O^+ −1.7	H_2S 7.00	HBr −9.00
HCl −8.0	HF 3.17	HCN 9.4	HN_3 4.72
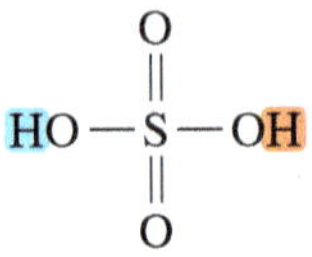 −3.0, 1.99	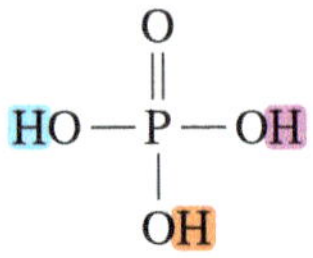 2.12, 7.12, 12.32	HNO_3 −1.3	H_2 ~36
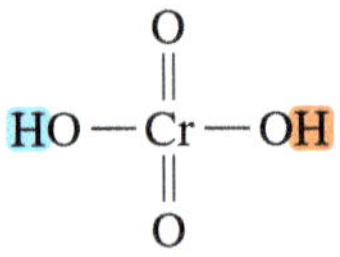 −0.98, 6.50	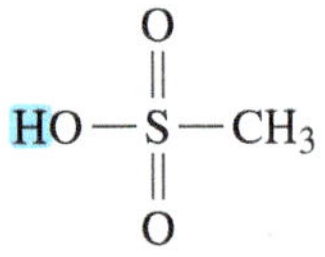 −2.6	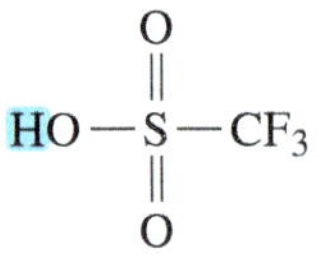 −14	NH_4Cl 9.24
Alkane and alkene hydrogens			
CH_4 50	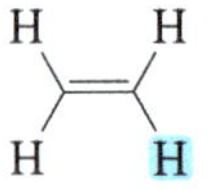 44	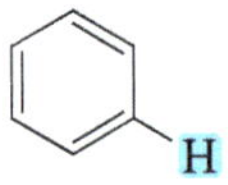 43	 25
	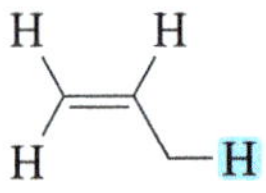 43	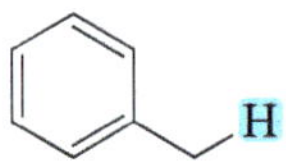 41	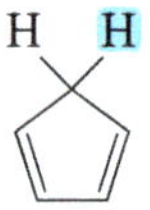 15
Carboxylic acids			
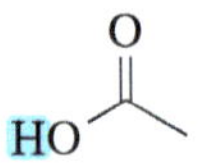 4.76	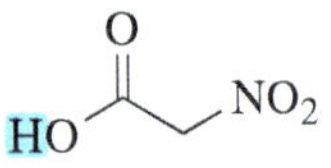 1.68	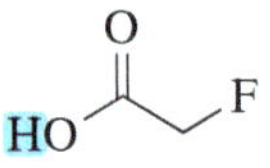 2.66	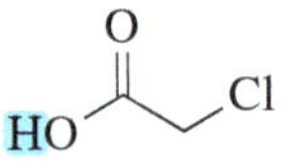 2.86
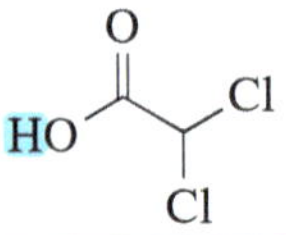 1.29	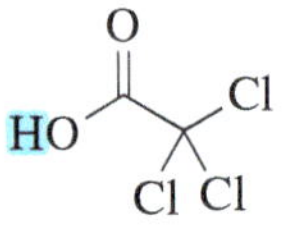 0.65	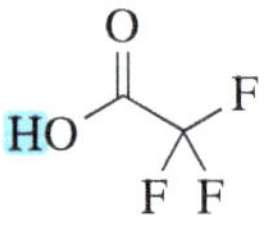 −0.25	
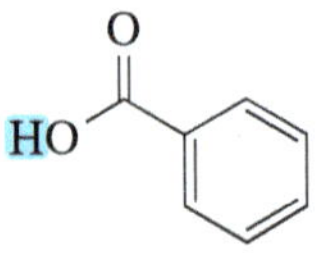 4.2	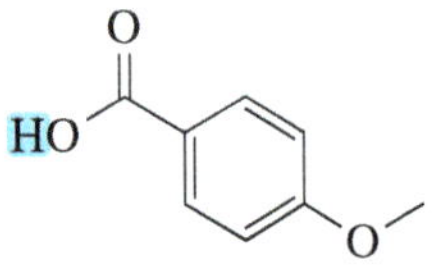 4.47	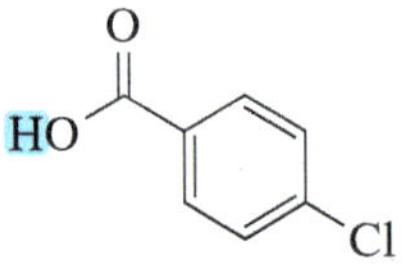 3.99	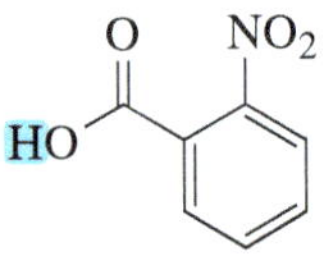 2.17
Alcohols			
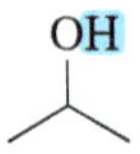 16.5	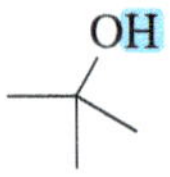 17	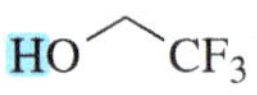 12.5	
Phenols			
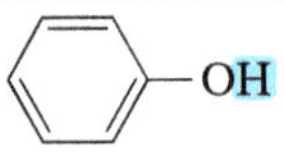 9.95	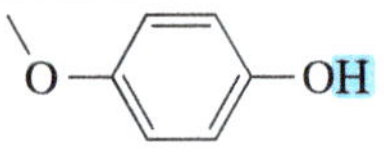 10.20	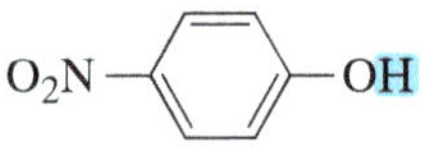 7.14	
Peroxides			
HO—OH 11.6	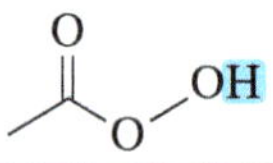 8.2		

(*continued*)

α-Hydrogens			
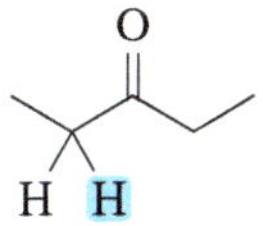 19–20	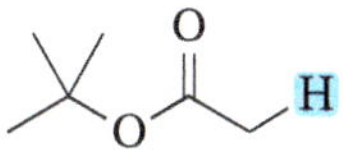 24.5	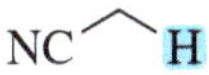 25	
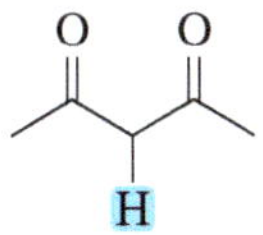 9	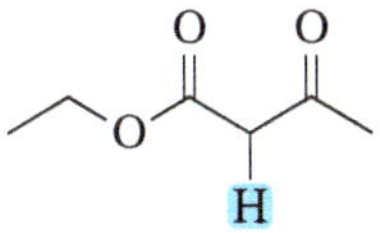 11	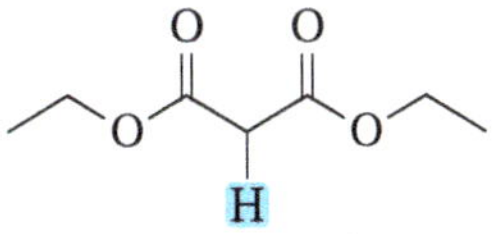 13	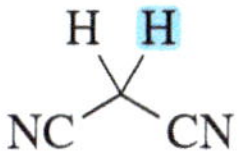 11
Ammonium ions			
$^{+}NH_4$ 9.2	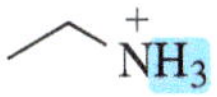 10.6	$iPr_2\overset{+}{N}H_2$ 11.05	$Et_3\overset{+}{N}H$ 10.75
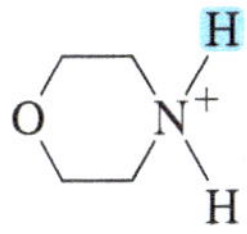 8.36	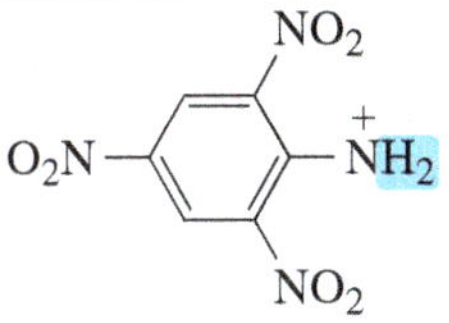 −9.3	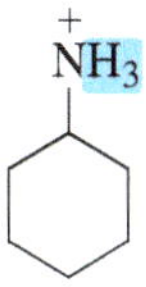 4.6	
Amines			
NH_3 38	iPr_2NH 36	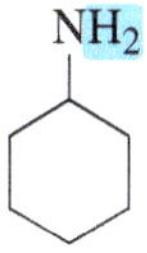 30.6	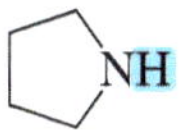 44
Amides			
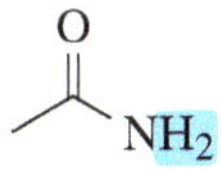 15.1	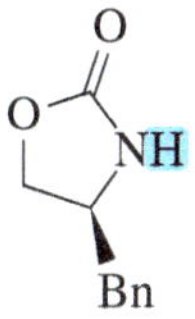 12	17.5 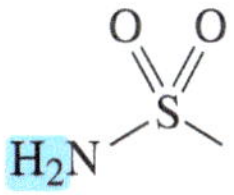	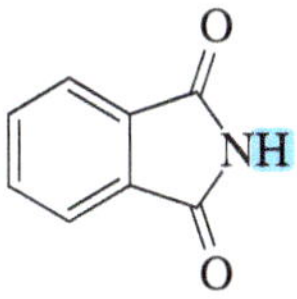 8.30
Heterocyclic amines			
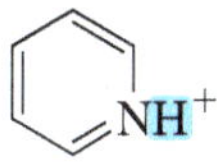 5.21	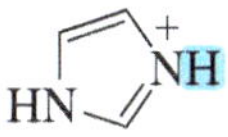 6.95	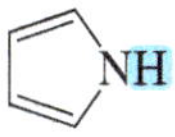 23.0	
Thiols			
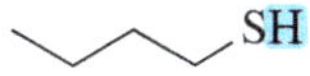 10	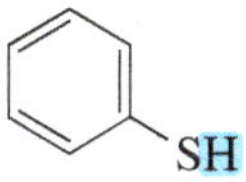 7		

APPENDIX

E

Bond-Dissociation Energies

$$A\!:\!B \longrightarrow A\cdot + \cdot B$$

	Bond-dissociation enthalpy			Bond-dissociation enthalpy	
Bond	**kJ/mol**	**kcal/mol**	**Bond**	**kJ/mol**	**kcal/mol**
H—X bonds and X—X bonds			**Bond to secondary carbons**		
H—H	435	104	$(CH_3)_2CH—H$	397	95
D—D	444	106	$(CH_3)_2CH—F$	444	106
F—F	159	38	$(CH_3)_2CH—Cl$	335	80
Cl—Cl	242	58	$(CH_3)_2CH—Br$	285	68
Br—Br	192	46	$(CH_3)_2CH—I$	222	53
I—I	151	36	$(CH_3)_2CH—OH$	381	91
H—F	569	136			
H—Cl	431	103	**Bonds to tertiary carbons**		
H—Br	368	88	$(CH_3)_3C—H$	381	91
H—I	297	71	$(CH_3)_3C—F$	444	106
HO—H	498	119	$(CH_3)_3C—Cl$	331	79
HO—OH	213	51	$(CH_3)_3C—Br$	272	65
RO—H	426	102	$(CH_3)_3C—I$	209	50
Methyl bonds			$(CH_3)_3C—OH$	381	91
$CH_3—H$	435	104	**Other C—H bonds**		
$CH_3—F$	456	109	$PhCH_2—H$ (benzylic)	356	85
$CH_3—Cl$	351	84	$CH_2{=}CHCH_2—H$ (allylic)	364	87
$CH_3—Br$	293	70	$CH_2{=}CH—H$ (vinyl)	464	111
$CH_3—I$	234	56	Ph—H (aromatic)	473	113
$CH_3—OH$	381	91	$HC{\equiv}C—H$ (acetylenic)	523	125
Bonds to primary carbons			**C—C bonds**		
$CH_3CH_2—H$	410	98	$CH_3—CH_3$	368	88
$CH_3CH_2—F$	448	107	$CH_3CH_2—CH_3$	356	85
$CH_3CH_2—Cl$	339	81	$CH_3CH_2—CH_2CH_3$	343	82
$CH_3CH_2—Br$	285	68	$(CH_3)_2CH—CH_3$	351	84
$CH_3CH_2—I$	222	53	$(CH_3)_3CH—CH_3$	339	81
$CH_3CH_2—OH$	381	91	**C=Y π bonds**		
			C=C (estimated)	272	65
			C=O (estimated)	356	85

APPENDIX F

Common Functional Groups and Modes of Reactivity

Listed below are the functional groups we studied in the course and the general reactivity patterns of each.

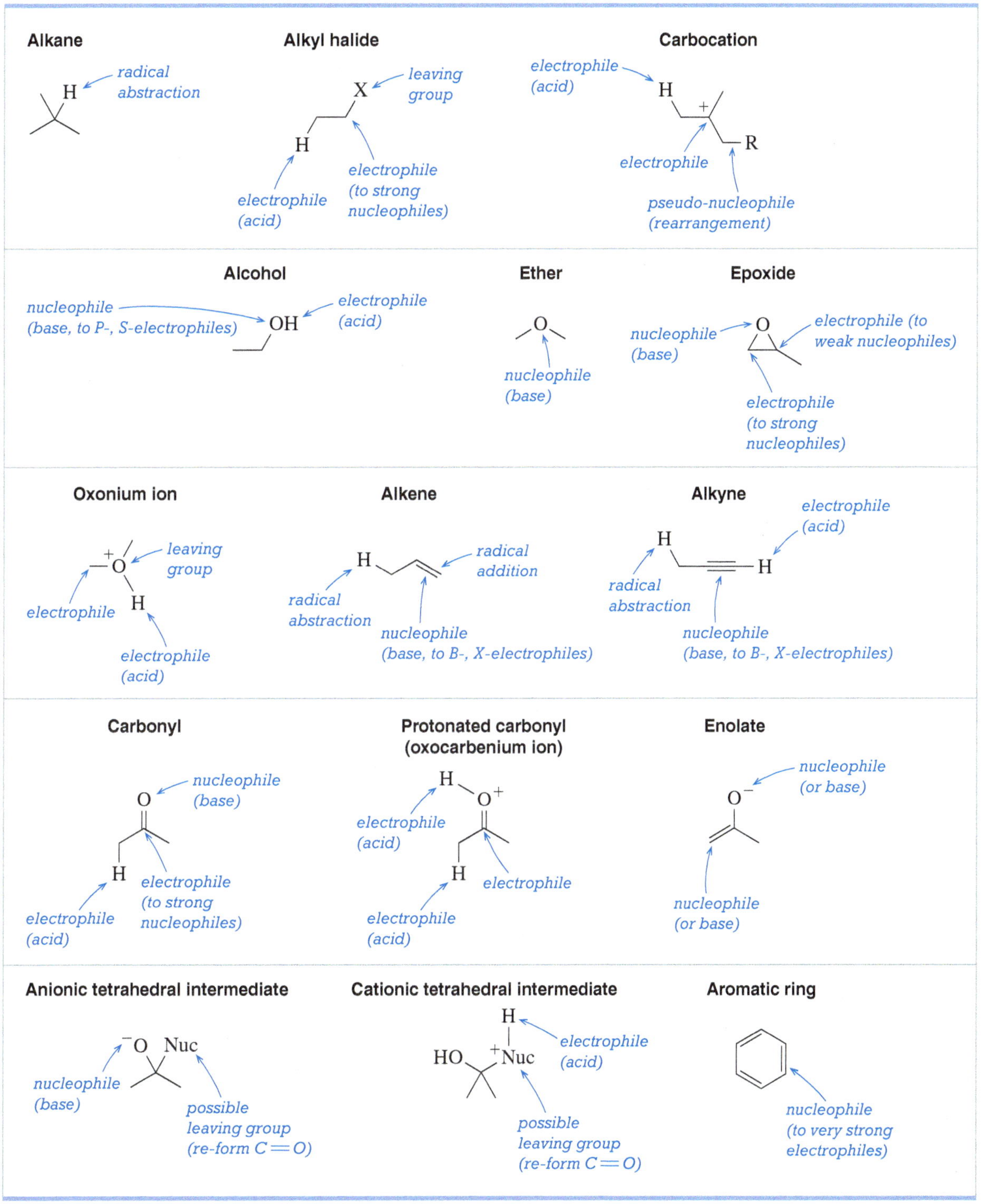

APPENDIX G

Redox Ladder

Groups of related functional groups can be classified based on the number of C—Z bonds a given carbon has. A C—Z bond is represented by any π bond and any σ bond to an electronegative atom. Any reaction that moves up or down the chart is a redox reaction. A reaction that moves left or right within a row is simply an acid–base reaction. [*The oxidation numbers of the central carbon in each molecule are also shown. You learned how to calculate these oxidation numbers in Chapter 9.*]

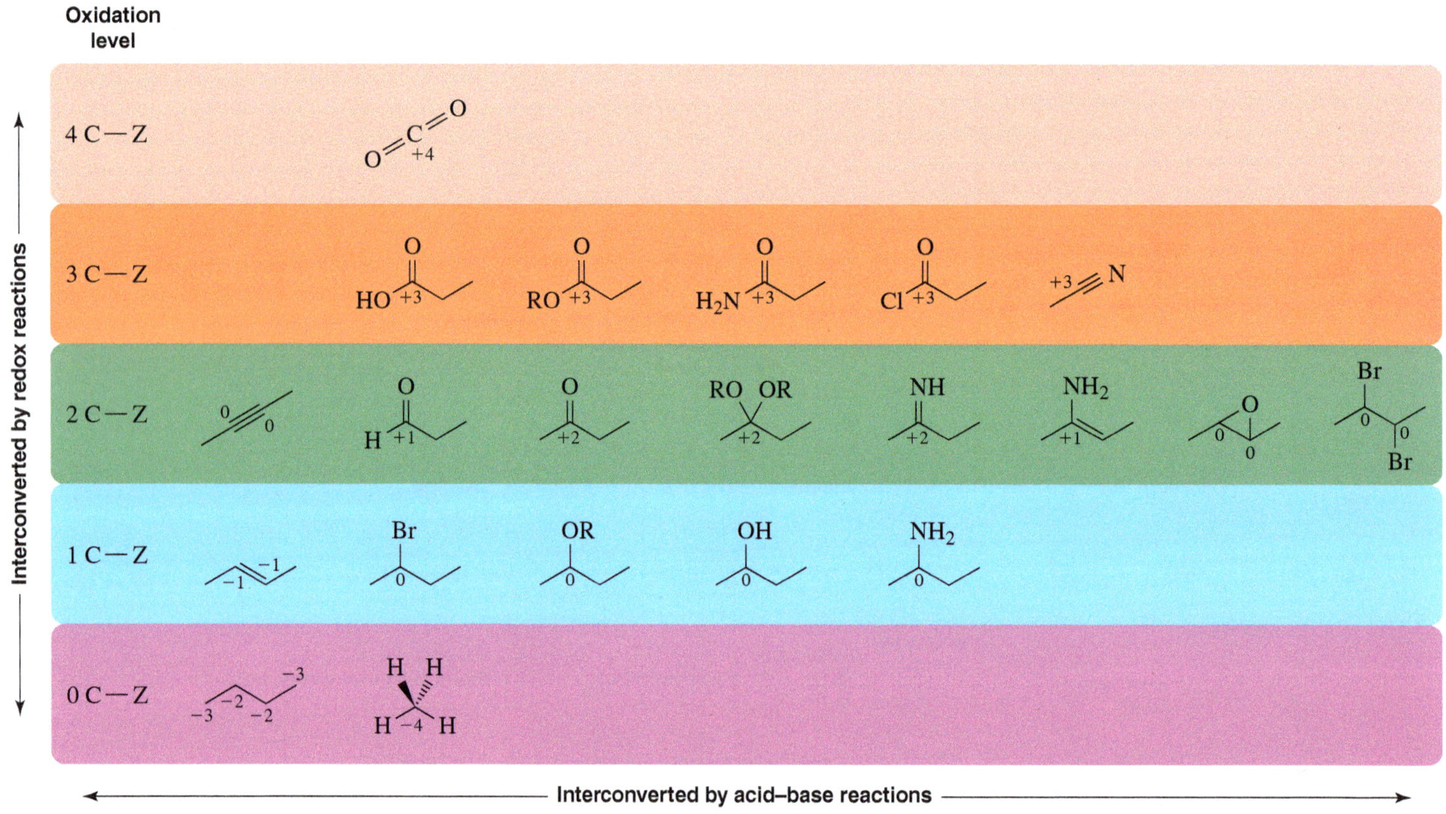

APPENDIX H

Problem-Solving Process

[While the processes described below are somewhat general, because there are a variety of ways problems can be asked and because multiple functional groups require different approaches, you can't expect that these will apply in every single scenario. They are, at best, a good starting point for your analysis.]

I. Problem: Predict the product of a reaction.

Predict the product of the following reaction.

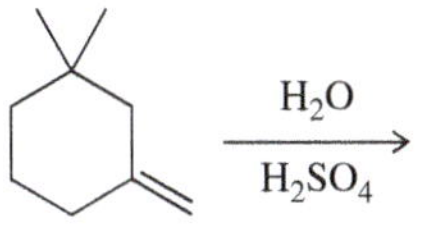

Step 1: Take a deep breath and, as always, number your carbons.

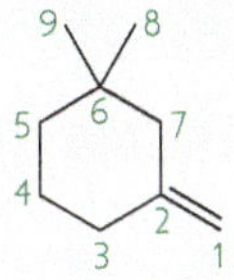

Step 2: Identify the functional groups present and name them out loud [*or in your head*].

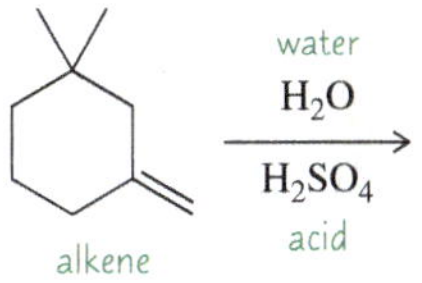

Step 3: Compose a three-phrase sentence for the reaction: "<insert functional group> react with <insert reagents> to form <insert products>."

The functional group and reagent are given to you. Can you fill in the product?

alkenes react with acid and water to form Markovnikov alcohols.

Step 4: Sketch a mechanism to predict a product.

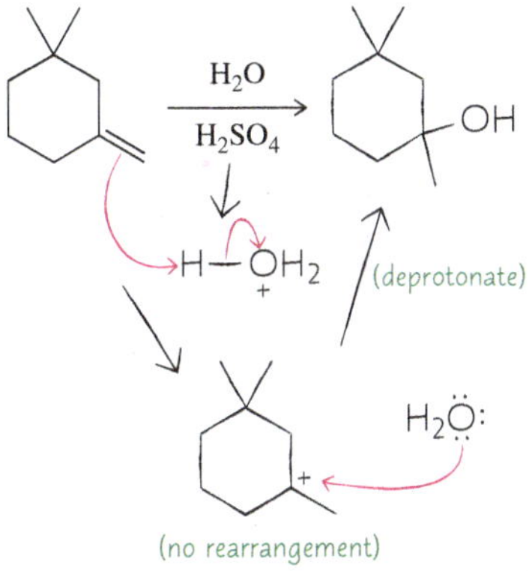

Step 5: Be sure your answer accounts for stereochemistry. If there are new stereocenters formed, then you need to address this in your product.

* This step is where SC forms.

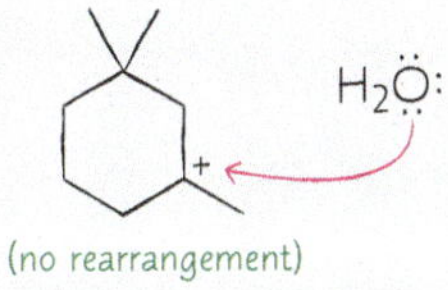

- C^+ is sp^2, flat, trigonal planar
- H_2O can attack from top or bottom without preference.
- Forms a mix of stereoisomers.

Step 6: By combining your work from steps 4 and 5, answer the question.

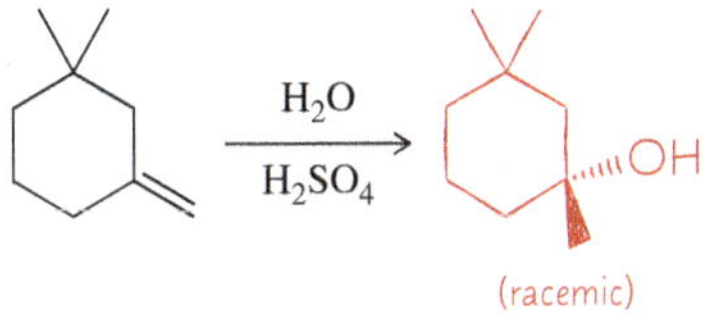

(continued)

II. Problem: Predict the mechanism of a reaction.

Provide a complete, arrow-pushing mechanism for the following reaction.

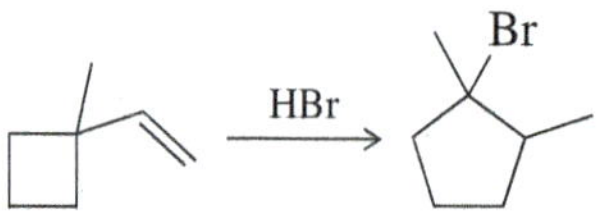

Step 1: Regardless of how complicated the reaction is, take a deep breath and number the carbons.

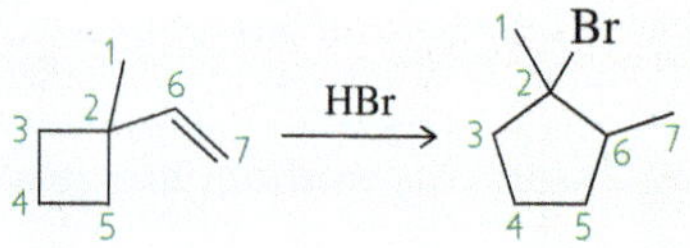

Step 2: Draw in the hidden hydrogens (at least around the reactive site).

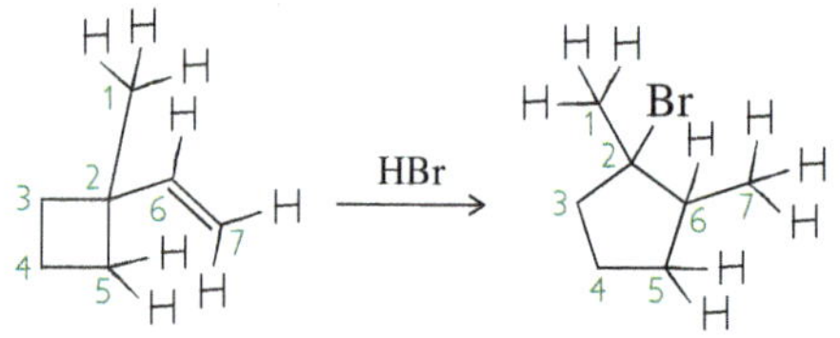

Step 3: Make a list of bonds formed and broken.

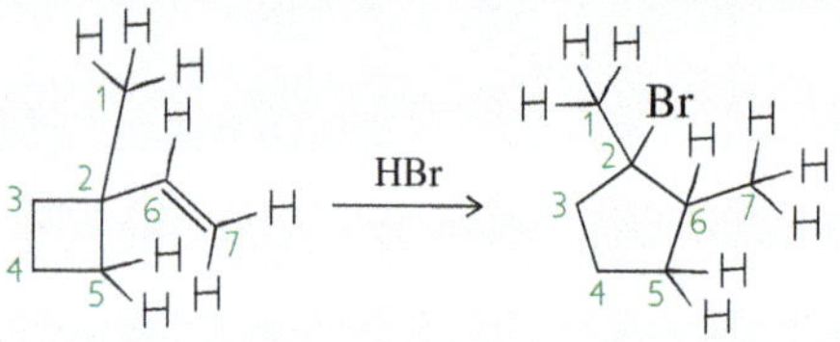

Broken	Formed
$C_6—C_7$ π	$C_2—Br$ σ
$C_2—C_5$ σ	$C_5—C_6$ σ (?)
$H—Br$ σ	$C_7—H$ σ

Step 4: Using the list of bonds formed and broken, sketch out a narrative that might get us from the starting material to the products.

π bond is nuc. Maybe break $C_6—C_7$ π first? How do we form $C_2—Br$, if already 4 bonds to C_2?

Step 5: Based on your narrative from step 4, draw a first step of the mechanism. Assess the intermediate formed and see if this allows the narrative to progress by opening up the opportunity to form or break other needed bonds.

Broken	Formed
~~$C_6—C_7$ π~~	$C_2—Br$ σ nuc
$C_2—C_5$ σ	$C_5—C_6$ σ (?)
$H—Br$ σ	~~$C_7—H$ σ~~ nuc

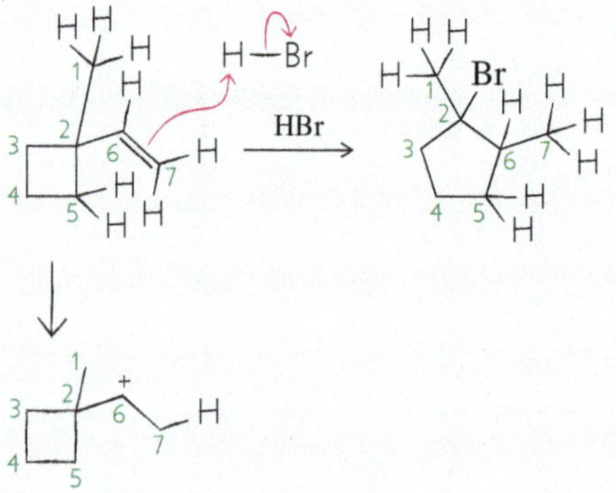

now C_6 elec,... Break $C_2—C_5$, form $C_5—C_6$?

(continued)

II. Problem: Predict the mechanism of a reaction.

Step 6: Using this new intermediate, identify nucleophiles and electrophiles, see how this continues the narrative toward the product, and use your bonds formed/broken lists to work your way toward finishing the mechanism and forming the product.

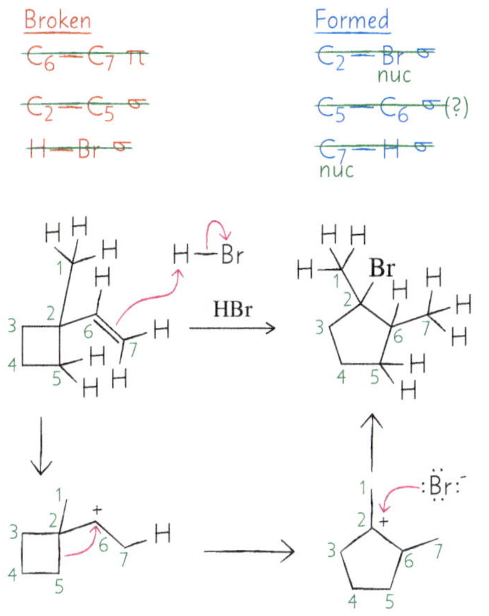

III. Problem: Solve a structure based on ¹H NMR.

Predict the structure $(C_6H_{12}O)$ represented by the following ¹H NMR.

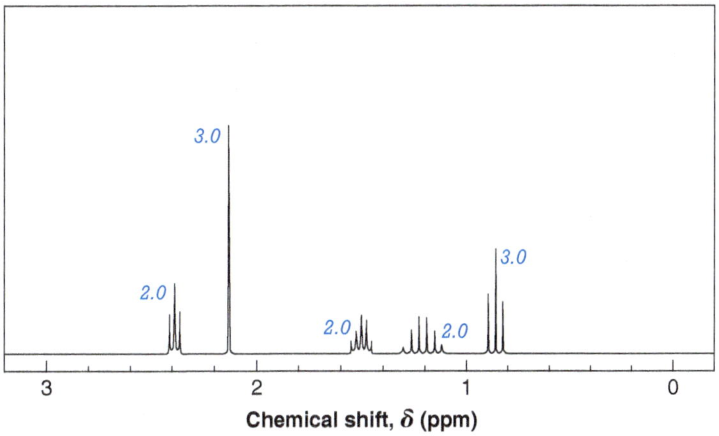

Step 1: Calculate the index of hydrogen deficiency (IHD) to identify what functional groups are/are not present.

IHD = 1 (1 π bond or 1 ring)

There are no ¹H NMR signals between 5 and 12
This rules out alkenes, benzene rings, aldehydes, and carboxylic acids.

Step 2: Gather other context clues, perhaps from IR or MS data to gain more information about which functional groups are present or absent in the molecule.

Sadly, we don't have any additional information in this problem. But you may have other spectra or other information in future problems.

Step 3: Make a table of information for each signal.

	CS	Int	Mult	Neighbors	Puzzle piece
A	2.4	2H	Triplet	2	
B	2.1	3H	Singlet	0	
C	1.5	2H	Pentet	4	
D	1.3	2H	Sextet	5	
E	0.9	3H	Triplet	2	

(continued)

III. Problem: Solve a structure based on ^{1}H NMR.

Step 4: Build a puzzle piece for the first signal according to the following steps.

a. Start with the chemical shift. What does that tell you about the functional groups near that hydrogen? Is it on a benzene ring, part of an aldehyde, next to a carbonyl?

a. Signal near 2.4 ppm.
Likely α to carbonyl

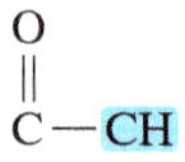

b. Continue with the integration. The integration is the *direct evidence* of protons in the structure. These protons are the ones that give rise to this signal. How many protons are on that carbon atom?

If the indicated integration would necessitate more than four bonds to carbon, that means you have two equivalent carbons with the same number of hydrogens each [*This also means that there is either global or local symmetry in the molecule. Another useful context clue!*].

Differentiate these protons from the neighboring protons by underlining them or bolding them or something.

b. Integration = 2H

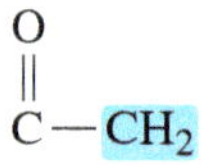

c. Account for the multiplicity. The multiplicity gives *indirect evidence* for other protons in the molecule. These protons don't cause this signal—but they will (they must) cause some other signal.

How many hydrogens are on the neighboring carbon(s)? There may be more than one way to split the neighbors among one or more carbon atoms. It may not be a bad idea to sketch multiple options, so you don't close off a possibility.

c. Splitting = triplet.
Two neighbors
Could be both on one side, or 1H on each side

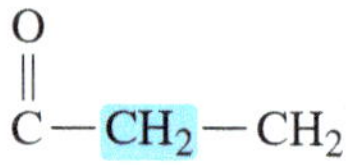

or

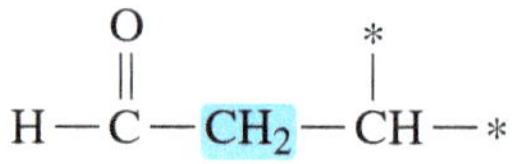

(but no signal at 9–10 ppm, so it can't be an aldehyde)

d. Account for any remaining open valences. You know that these puzzle pieces will have to come together to make a full molecule. Taking a moment to draw in open valences will ensure that you have accounted for all bonds and atoms.

d.

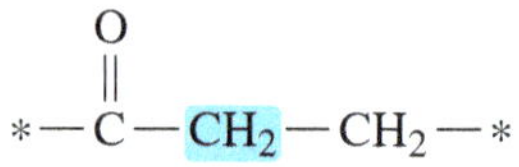

Step 5: Build puzzle pieces for the remaining signals by repeating step 4.

	CS	Int	Mult	Neighbors	Puzzle piece
A	2.4	2H	Triplet	2	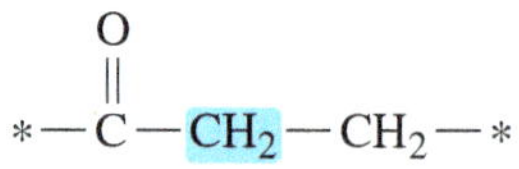
B	2.1	3H	Singlet	0	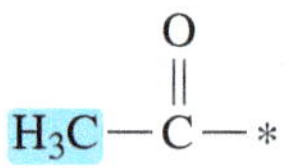
C	1.5	2H	Pentet	4	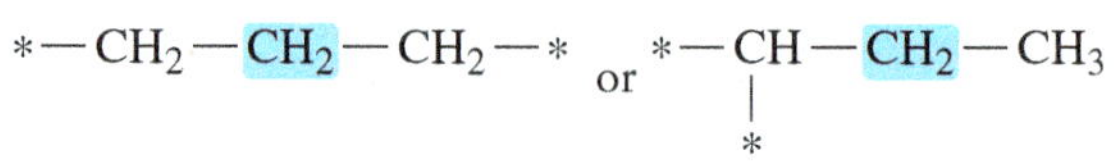
D	1.3	2H	Sextet	5	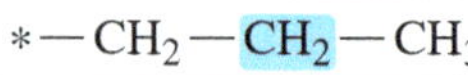
E	0.9	3H	Triplet	2	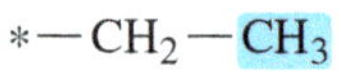

Step 6: Start with an end piece and look for reciprocal overlap as a way to assemble the puzzle pieces created in step 4.

Methyl groups, monosubstituted benzene rings, aldehydes, and carboxylic acids must be the end of the molecule. Choose a signal that gives direct evidence for an end piece. Recall that the neighboring protons (the indirect evidence) must give rise to some other signal in the spectrum. The implication is that if your end piece is direct evidence of a CH_3 group next door to a CH_2 group . . . then you'd better have some other signal that is direct evidence for a CH_2 group next door to a CH_3 group! This is the reciprocity of ^{1}H NMR signals. [*Read that again. Understanding the reciprocity of NMR is key to ensuring you build the correct structure.*]

Start to add together your puzzle pieces like an addition problem. Write down the puzzle piece with the direct evidence protons highlighted. Underneath it, write the reciprocal puzzle piece that overlaps with its direct evidence protons highlighted. Often, the second puzzle piece will overlap and build upon the first puzzle piece. Now, there is a new, larger puzzle piece upon which to continue overlapping and building the structure.

(continued)

III. Problem: Solve a structure based on ¹H NMR.

Puzzle piece E is an end piece:

E CH_3-CH_2-*

C $CH_3-CH_2-CH(-*)-*$

EC $CH_3-CH_2-CH(-*)-*$

incorrect

This is *direct evidence* of a CH_2 group next to a CH group. But there are no other puzzle pieces that are *direct evidence* of a CH group, let alone *direct evidence* of a CH group next to a CH_2 group. This must not be the correct puzzle piece for C.

E CH_3-CH_2-*

D $CH_3-CH_2-CH_2-*$

ED $CH_3-CH_2-CH_2-*$

correct

This is *direct evidence* of a CH_2 group next to a CH_2 group. There are other puzzle pieces with *direct* evidence of a CH_2 group next to a CH_2 group. This builds confidence that you are closer to the correct structure.

Step 7: Continue building the structure with reciprocal overlap until all pieces are used.

E CH_3-CH_2-*

D $CH_3-CH_2-CH_2-*$

ED $CH_3-CH_2-CH_2-*$

C $*-CH_2-CH_2-CH_2-*$

EDC $CH_3-CH_2-CH_2-CH_2-*$

A $*-CH_2-CH_2-C(=O)-*$

EDCA $CH_3-CH_2-CH_2-CH_2-C(=O)-*$

B $*-C(=O)-CH_3$

EDCAB $CH_3-CH_2-CH_2-CH_2-C(=O)-CH_3$

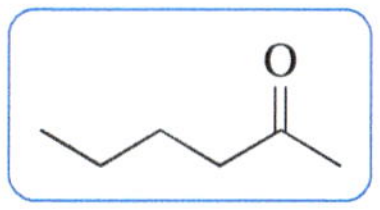

Step 8: Double-check your answer by predicting what the ¹H NMR spectrum would look like based on the structure you have developed.

$CH_3-CH_2-CH_2-CH_2-C(=O)-CH_3$

- CH_3: ~1 ppm, 3H, triplet
- CH_2: ~1.5 ppm, 2H, sextet
- CH_2: ~1.5 ppm, 2H, pentet
- CH_2: ~2 ppm, 2H, triplet
- CH_3 (C=O): ~2 ppm, 3H, singlet

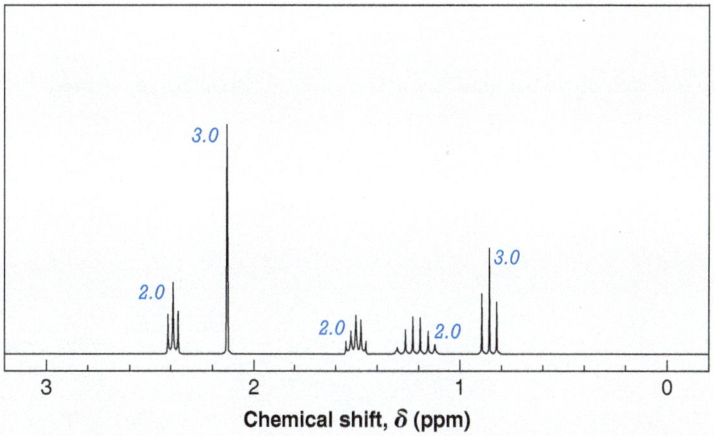

(continued)

IV. Problem: Synthesis

Provide a synthesis of heptane-2,3-diol starting with organic molecules that contain three (3) carbon atoms or fewer.

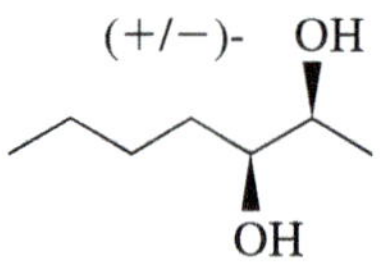

heptane-2,3-diol

Step 1: Number the carbons.

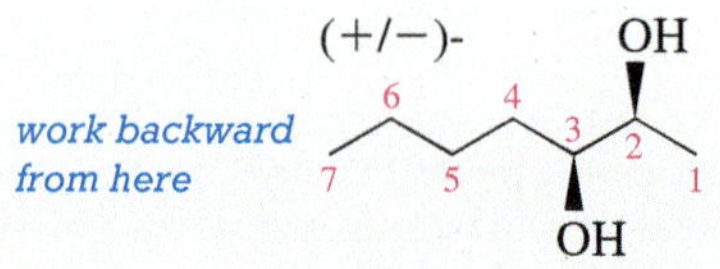

target molecule

Step 2: List key functional groups.

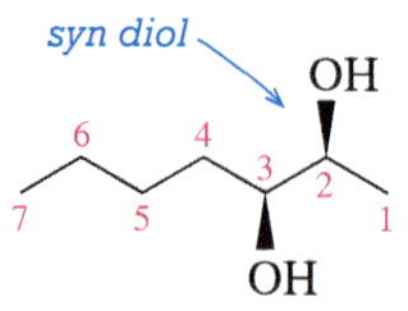

Step 3: Use your reaction sheets to make that functional group. You now have a new target molecule.

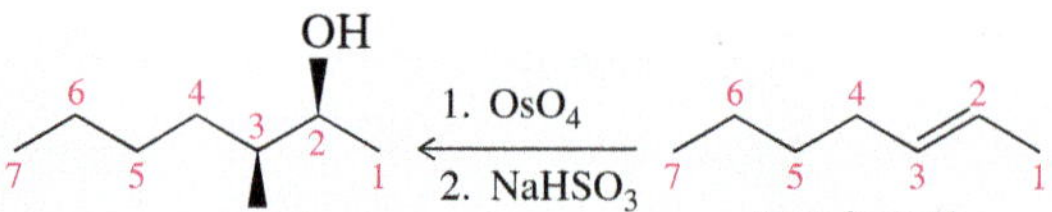

Step 4: Repeat steps 1–3 until you reach a point where you can make a new C—C bond.

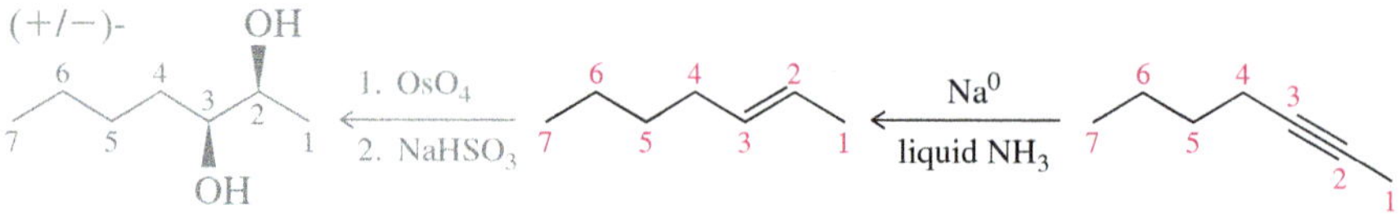

Step 5: Identify a reaction that will produce the desired functional group and form a C—C bond at the same time.

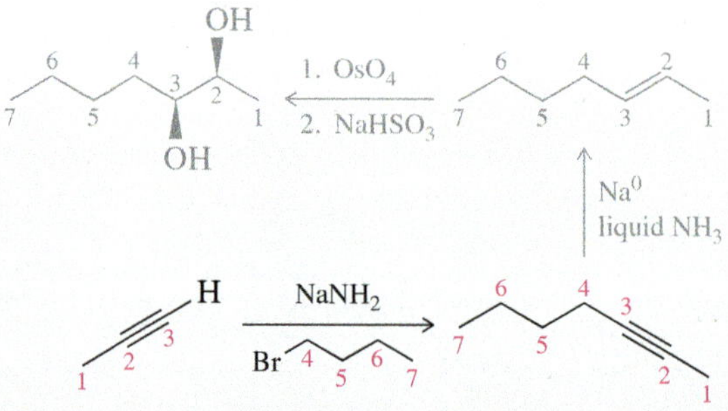

Step 6: Repeat steps 1–5 until you have met the requirements of the problem (that is, three carbons or fewer).

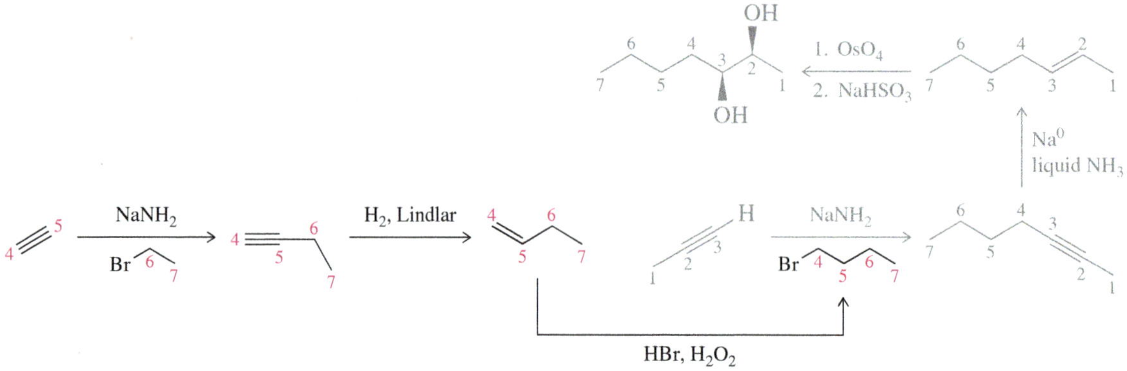

APPENDIX

I

Spectroscopy

Infrared (IR) absorptions table

Type of bond	Wavenumber (cm^{-1})	Intensity
C≡N	2260–2220	Medium
C≡C	2260–2100	Medium to weak
C=C	1680–1600	Medium
C=N	1650–1550	Medium
(benzene ring)	~1600 and ~1500–1430	Strong to weak
C=O	1780–1650	Strong
C—O	1250–1050	Strong
C—N	1230–1020	Medium
O—H (alcohol)	3650–3200	Strong, broad
O—H (carboxylic acid)	3300–2500	Strong, very broad
N—H	3500–3300	Medium, broad
C—H	3300–2700	Medium

Specific IR absorptions (All values are in units of cm^{-1})

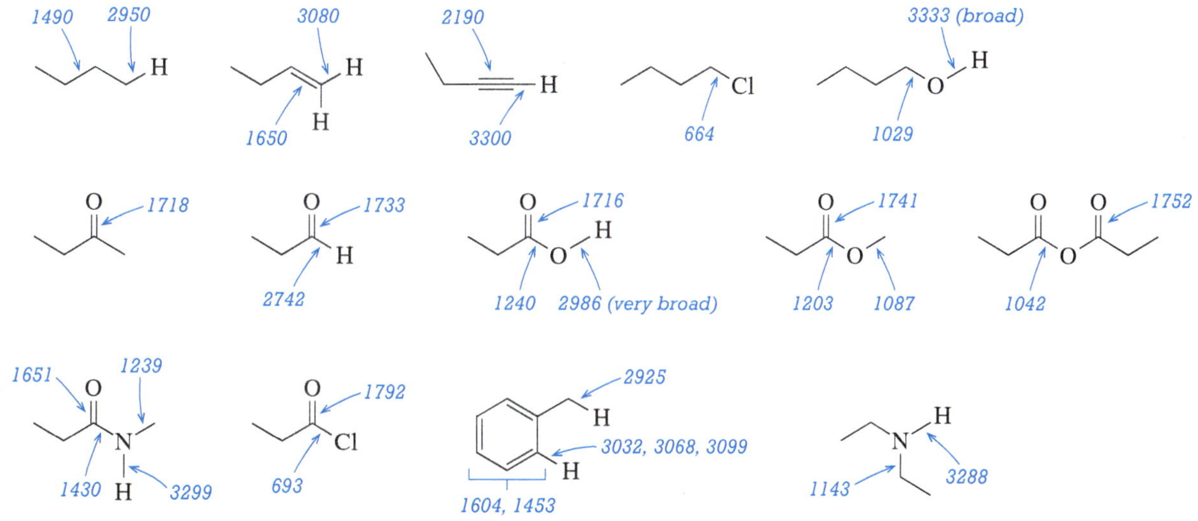

Mass spectrometry (MS) common fragmentations

Alkane

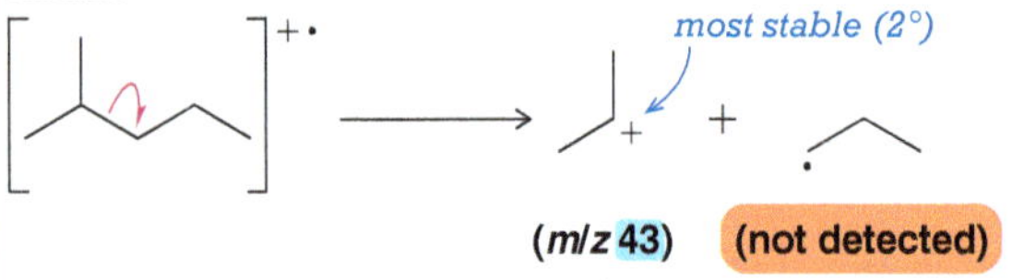

Alkene

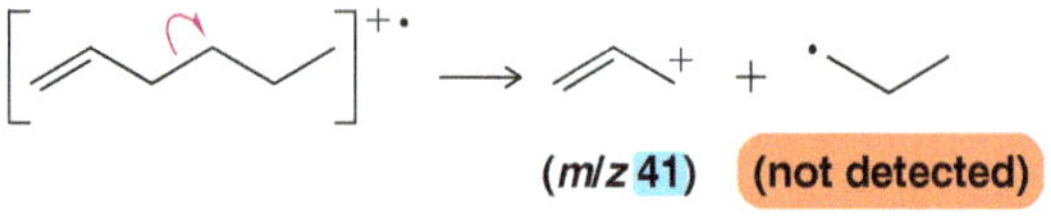

Aromatic

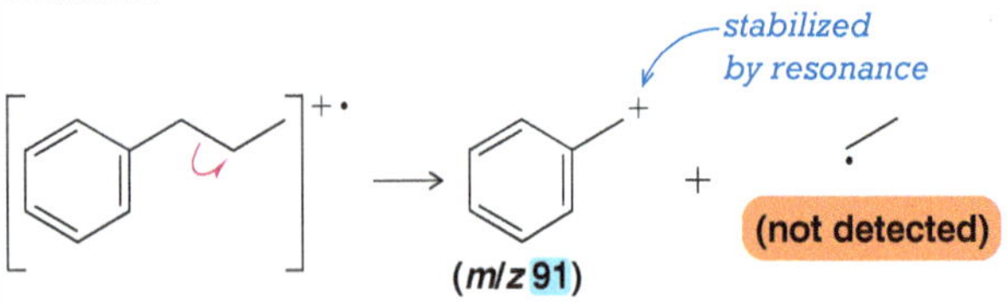

Haloalkane

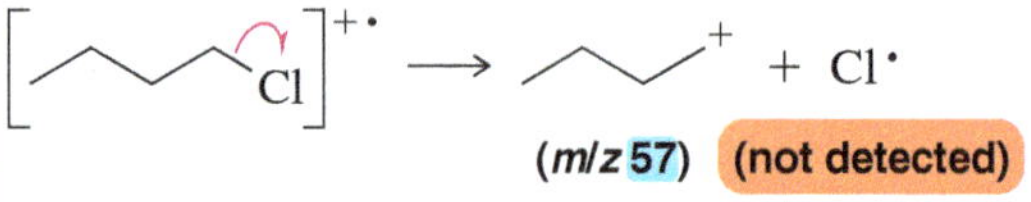

Alcohol

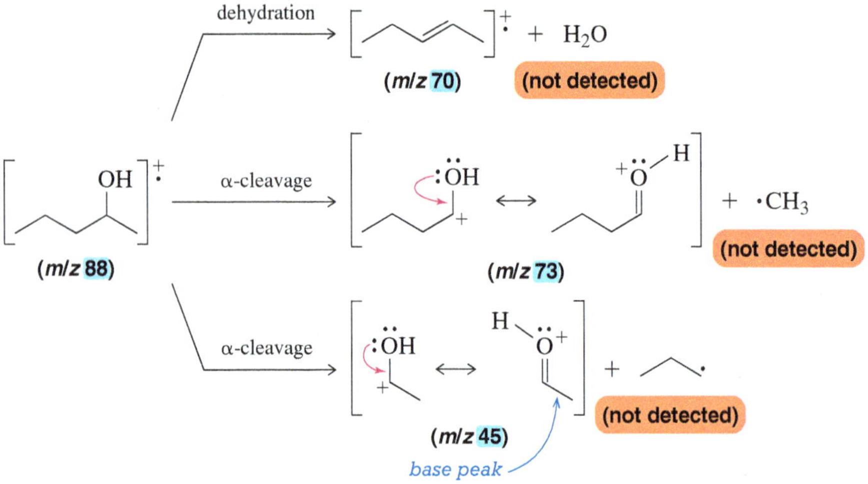

Amines

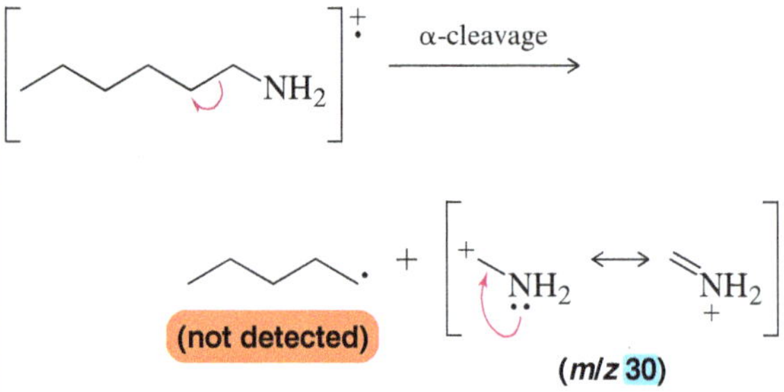

Ether

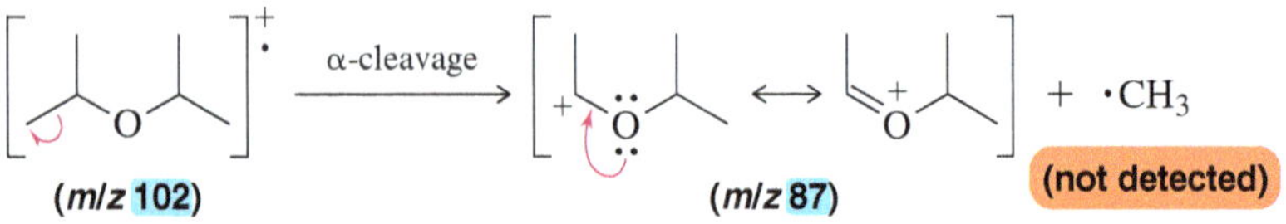

Aldehyde

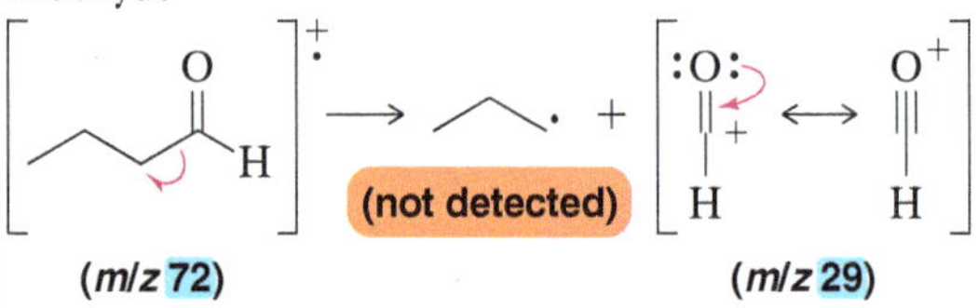

Ketone

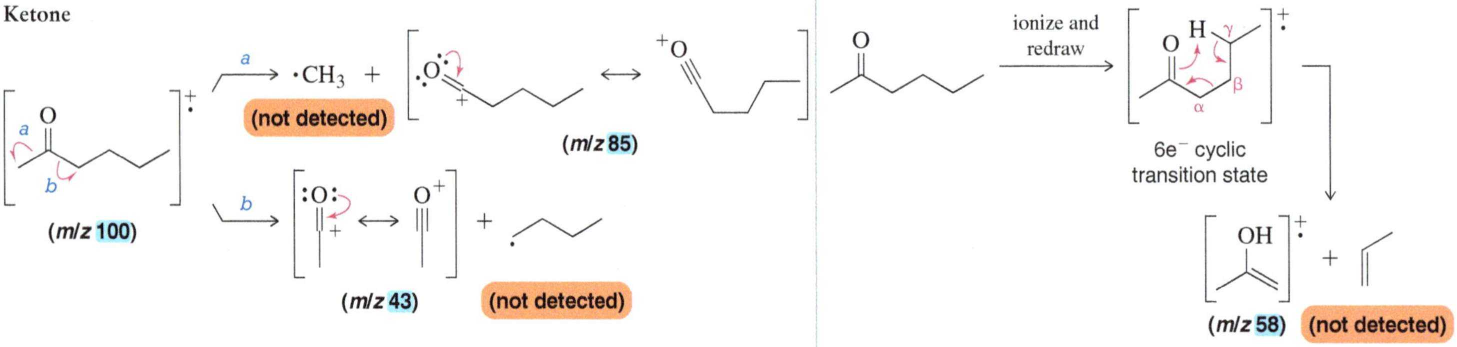

^{1}H NMR chemical shift table

Type of proton	Approximate δ (ppm)	Type of proton	Approximate δ (ppm)
Alkane (—C**H**$_3$) **methyl**	0.9	C=C—C**H**$_3$ **allylic**	1.7
Alkane (—C**H**$_2$—) **methylene**	1.3	Ph—**H** **aromatic**	7.2
Alkane (—C**H**— with substituent below) **methine**	1.4	Ph—C**H**$_3$ **benzylic**	2.3
—C(=O)—C**H**$_3$ **methyl ketone**	2.1	R—C(=O)—**H** **aldehyde**	9–10
—C≡C—**H** **acetylenic**	2.5	R—C(=O)—O**H** **acid**	10–12
R—C**H**$_2$—X (X = halogen, O)	3–4	R—O**H** **alcohol**	Variable, ~2–5
C=C—**H** **vinyl**	5–6	Ar—O**H** **phenol**	Variable, ~4–7
		R—N**H**$_2$ **amine**	Variable, ~1.5–4

Note: These values are approximate because all chemical shifts are affected by neighboring substituents. The values given here assume that alkyl groups are the only other substituents present.

Specific ^{1}H NMR chemical shifts (all values are in ppm; TMS = 0 ppm) *[When an arrow points to carbon, the chemical shift is for the hydrogens attached there.]*

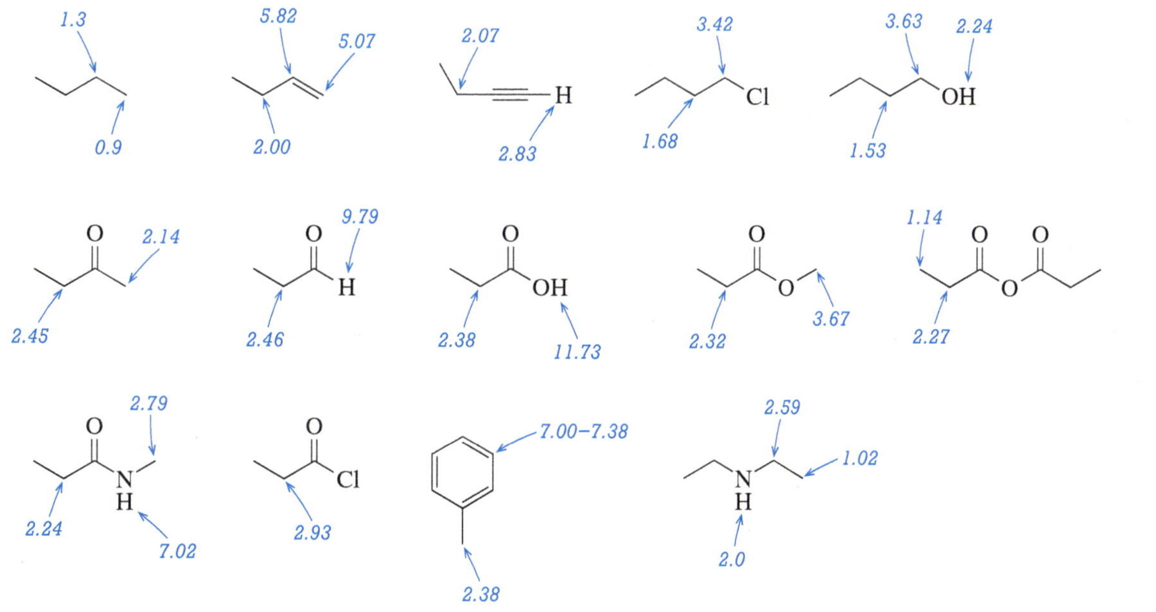

Coupling constants

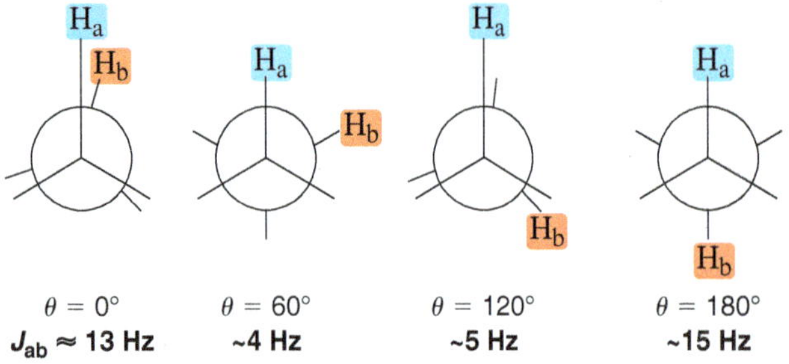

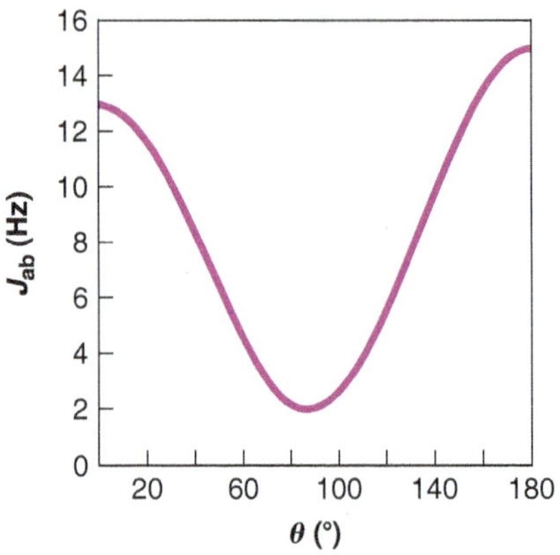

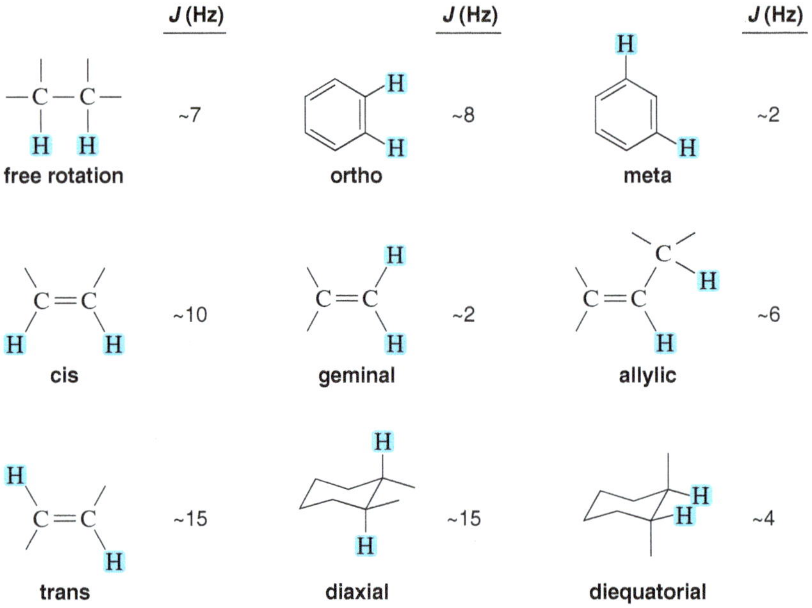

^{13}C NMR shifts

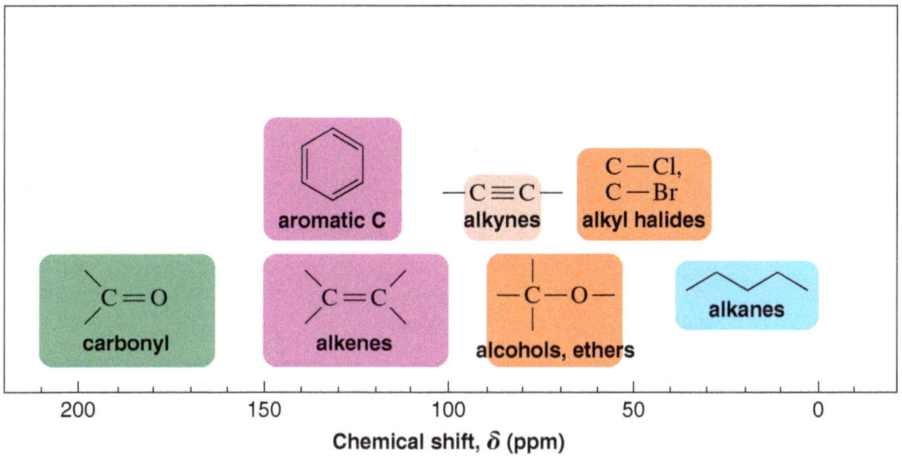

Carbon type	Chemical shift (ppm)
sp^3 C—C	0–40
sp^3 C—X (X = electronegative atom)	40–100
C≡C	80–100
C=C, aromatic	100–150
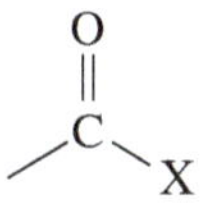	150–200
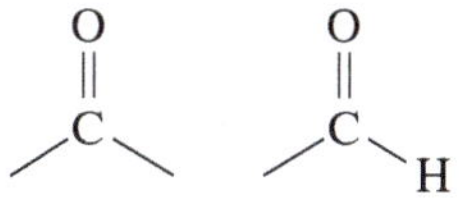	>200

Specific ^{13}C NMR chemical shifts (all values are in ppm; TMS = 0 ppm) *[When an arrow points to carbon, the chemical shift is for that carbon.]*

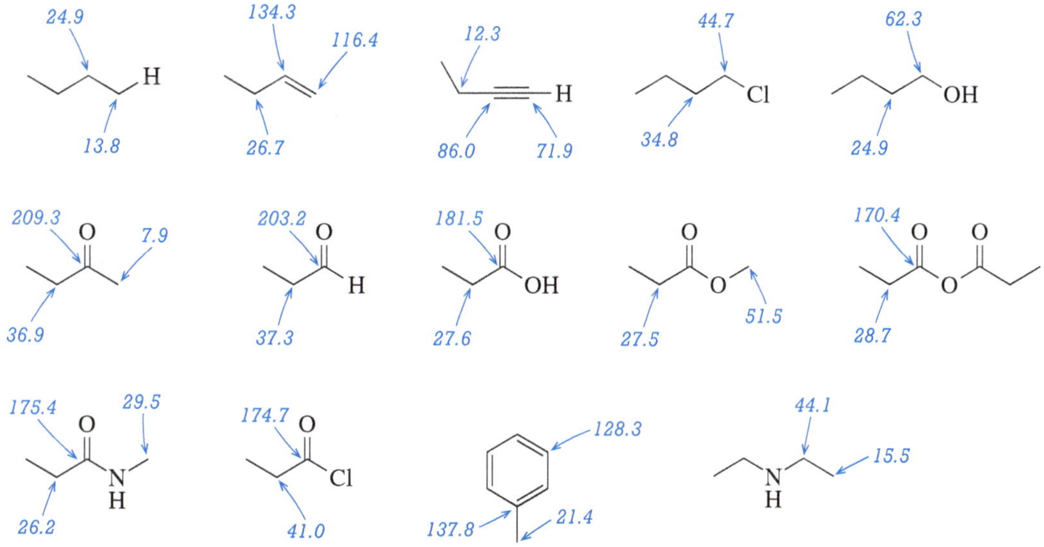

APPENDIX

J

Reaction Coordinate Diagram

For the reaction R becomes P through intermediate I ($R \rightarrow I \rightarrow P$):

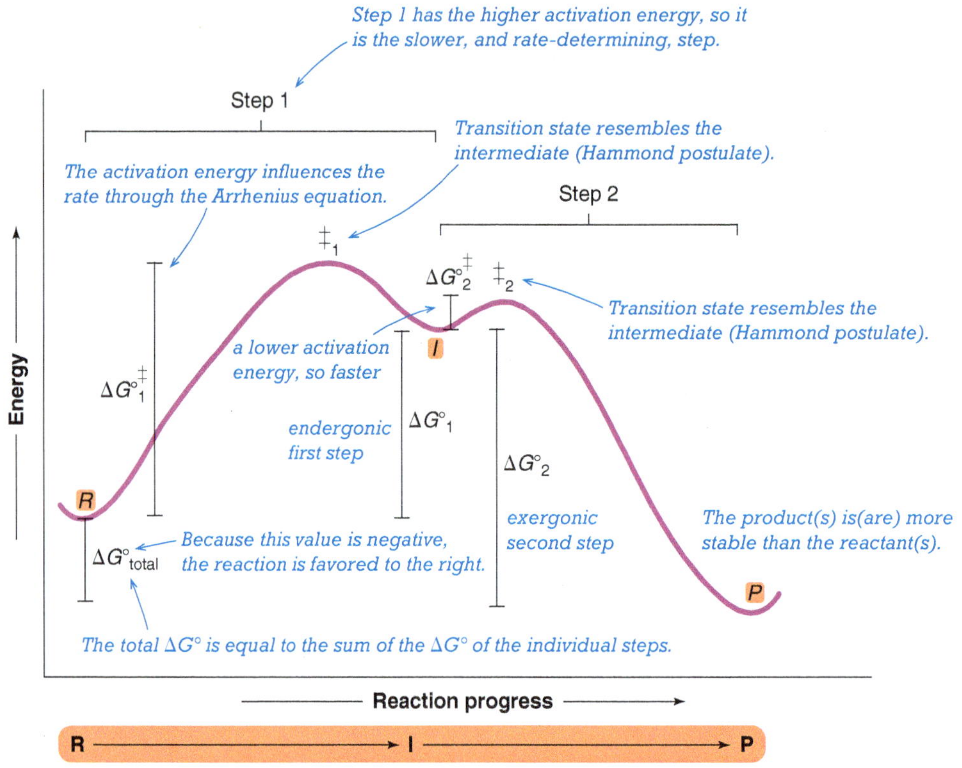

APPENDIX K

Functional Group Preparation

Synthesis of Acetals:

1. Acid-catalyzed addition of an alcohol to an aldehyde/ketone (Section 17.7.3)

Synthesis of Acid Anhydrides:

1. Reaction of a carboxylate ion with an acid chloride (Section 19.7.4)

Synthesis of Acyl Chlorides:

1. Reaction of a carboxylic acid and thionyl chloride ($SOCl_2$) (Section 18.7.5)

Synthesis of Alcohols:

1. Acid-catalyzed hydration of an alkene (Section 8.4.1)
2. Oxymercuration–reduction of an alkene (Section 8.4.4)
3. Hydroboration–oxidation of an alkene (Section 8.4.5)
4. Reaction of an alkyl halide with H_2O or $—OH$ (Section 12.2)
5. Acidic cleavage of an ether with HX (Section 13.12.1)
6. Opening of an epoxide with an acetylide ion (Section 13.13.2)
7. Opening of an epoxide with an organolithium or Grignard reagent (Section 16.2.3.2)
8. Opening of an epoxide with a Gilman reagent (Section 16.5.2)
9. Addition of an organolithium/Grignard reagent to an aldehyde/ketone (Sections 17.5.1, 17.5.2)
10. Addition of an acetylide anion to an aldehyde/ketone (Section 17.5.3)
11. Catalytic hydrogenation of an aldehyde/ketone (Section 17.9.1)
12. Reduction of an aldehyde/ketone with hydride reagent (Section 19.9.2)
13. Reduction of a carboxylic acid with $LiAlH_4$ (Section 18.7.3)
14. Reduction of an ester with $LiAlH_4$ (Section 19.8.1)
15. Reduction of an acid chloride/anhydride with $LiAlH_4$ (Section 19.8.1.2)
16. Reaction of an ester with a Grignard reagent (2 equivalents) (Section 19.9.1)
17. Reaction of a silyl enol with *m*CPBA to form an α-hydroxy ketone (Section 20.2.1.4)
18. Modern aldol reaction (Section 20.3.1); base-catalyzed aldol reaction (Section 20.3.2); intramolecular aldol reaction (Section 20.3.4)
19. Hydrogenolysis of benzyl ethers (Section 24.6)

Synthesis of Aldehydes:

1. Reaction of an alkene with O_3 followed by CH_3SCH_3 (Section 9.1.7)
2. Hydroboration of terminal alkyne (Section 10.8.5)
3. Oxidation of a primary alcohol with pyridium chlorochromate (PCC) (Section 13.9.3)
4. Corey–Kim oxidation, Swern oxidation, and Dess–Martin oxidation of a primary alcohol (Section 13.9.4)
5. Oxidative cleavage of cis-diols with periodic acid (HIO_4) (Section 13.9.5)
6. Reduction of an ester with DIBAl-H (Section 19.8.2)
7. Reduction of an acid chloride with $LiH(Ot\text{-}Bu)_3$ (Section 19.8.2)
8. Reduction of a nitrile with DIBAl-H (Section 19.8.3)

Synthesis of Alkanes:

1. Catalytic hydrogenation of an alkene (Section 9.2) or an alkyne (Section 10.6.1)
2. Cyclopropanation of an alkene with a carbene (Section 16.3)
3. Wolff–Kishner reduction of an aldehyde/ketone to an alkane (Section 19.9.3)
4. [2 + 2] cycloaddition (Section 22.4.1)

Synthesis of Alkenes:

1. Catalytic hydrogenation of an alkyne in the presence of a Lindlar catalyst to form a cis-alkene (Section 10.6.2)
2. Reduction of alkyne with sodium metal and liquid ammonia to form a trans-alkene (Section 10.6.3)
3. Elimination of an alkyl halide or alkyl tosylate upon the addition of base (Sections 12.5, 13.7.1)
4. Acid-catalyzed dehydration of an alcohol (Section 13.8)
5. Olefin metathesis (Section 16.6)
6. Wittig reaction: reaction between a phosphorus ylide and an aldehyde or ketone (Section 17.6)
7. Sigmatropic rearrangement (Cope/Claisen) (Section 22.4.3)
8. Hoffman elimination (Section 25.7.3)
9. Cope elimination (Section 25.7.4)

Synthesis of Alkyl Azides:

1. Reaction of an alkyl halide or alkyl tosylate with N_3 (Sections 12.2, 13.7.1)

Synthesis of Alkyl Halides:

1. Radical halogenation of an alkane (Sections 5.3.3, 11.5.1)
2. Addition of HX to an alkene (Section 8.3)
3. Radical addition of HX to an alkene with peroxide (Section 8.3.4)
4. Addition of HX (1 equiv.) to an alkyne to form an alkenyl halide *or* two equivalents of HX to form a dihaloalkane (Section 10.8.2)
5. Allylic and benzylic bromination with *N*-bromosuccinimide (NBS) and heat (Sections 11.5.3, 24.3)
6. Reaction of an alcohol with HX (Section 13.7.2)
7. Reaction of an alcohol with thionyl chloride ($SOCl_2$) to form an alkyl chloride (Section 13.7.3)
8. Reaction of an alcohol with phosphorus tribromide (PBr_3) to form an alkyl bromide (Section 13.7.4)
9. HX cleavage of an ether (Section 13.12.1)
10. Reaction of an enol with X_2 to form an α-halo ketone or aldehyde (Section 20.2.1.2)
11. HX with a diene at high temperature to give the 1,4-addition product (Section 21.4)
12. HX with a diene at low temperature to give the 1,2-addition product (Section 21.4)
13. Hell–Volhard–Zelinskii reaction to produce an α-bromo carboxylic acid (Section 26.4.2)

Synthesis of Alkynes:

1. Reaction of an acetylide ion with an alkyl halide or tosylate leaving group (Section 10.5)
2. Consecutive elimination reactions of a vicinal or geminal dihalide (Section 12.5, Assessment 12.63)

Synthesis of Amides:

1. Reaction of an acid chloride with an amine (2 equiv.) (Section 19.7.3.1)
2. Reaction of an anhydride with an amine (2 equiv.) (Section 19.7.3.2)
3. Reaction of an ester with an amine (2 equiv.) with mild heating (Section 19.7.3.3)
4. Reaction of a nitrile with H_2O under acidic conditions at ambient temperature (Section 19.10.1)
5. DCC coupling (Section 26.9.2)

Synthesis of Amines:

1. Reduction of an amide with $LiAlH_4$ (Sections 19.8.1.3, 25.6.1)
2. Reduction of a nitrile with $LiAlH_4$ (Sections 19.10.4, 25.5.2)
3. Reduction of an azide with $LiAlH_4$ (Section 25.5.3)
4. Staudinger reaction: azide reduction with Ph_3P (Section 25.5.3)
5. Gabriel amine synthesis (Section 25.5.4)
6. Reductive amination (Section 25.6.2)
7. Reduction of nitro groups (Section 25.6.3)
8. Buchwald–Hartwig amination (Section 25.6.4)

Synthesis of Amino Acids:

1. Gabriel malonic ester synthesis (Section 26.4.3)
2. Strecker synthesis (Section 26.4.4)

Synthesis of Carboxylic Acids:

1. Oxidation of a primary alcohol or aldehyde with chromic acid and water (Sections 13.9.3, 17.10)
2. Exhaustive oxidation of cyclic ketones into dicarboxylic acids (Section 17.10)
3. Oxidation of an alkene/alkyne with basic $KMnO_4$ (Section 18.5.2.1)
4. Ozonolysis of an alkene with an oxidative workup (addition of H_2O_2) (Section 18.5.2.2)
5. Reaction of a Grignard reagent with CO_2 (Section 18.5.3)
6. Chromic acid oxidation of alkyl benzenes (Sections 18.5.4, 24.5)
7. Reaction of water and an acid chloride (Section 19.7.1.1)
8. Reaction of water and an anhydride (Section 19.7.1.2)
9. Reaction of water and an ester (often requires an acid or base catalyst) (Section 19.7.1.3)
10. Reaction of water with amides (requiring high temperature and an acid or base catalyst) (Section 19.7.1.4)
11. Reaction of a nitrile with H_2O under acidic conditions and high temperatures (Section 19.10.1)

Synthesis of Cyanohydrins:

1. HCN addition to an aldehyde/ketone (Section 17.5.4)

Synthesis of Dihalides:

1. Addition of Br_2 and Cl_2 to an alkene (Section 9.1.2)

2. Addition of Br_2 and Cl_2 (1 equiv.) to an alkyne to yield a dihaloalkene; addition of Br_2 and Cl_2 (2 equiv.) to an alkyne to yield a tetrahaloalkane (Section 10.8.7)

Synthesis of 1,2-diols:

1. Reaction of an alkene with osmium tetroxide followed by sodium bisulfate (Section 9.1.5)
2. Reaction of an alkene with cold potassium permanganate in the presence of hydroxide and H_2O (Section 9.1.5)
3. Reaction of an epoxide with H_2O and H_2SO_4 *or* ^-OH (Sections 9.1.6, 13.13.2)

Synthesis of Enamines:

1. Addition of a secondary amine to an aldehyde/ketone (Section 17.8.3)

Synthesis of Epoxides:

1. Reaction of an alkene with *m*CPBA (Section 9.1.4)
2. Reaction of an allylic alcohol with $Ti(OiPr)_4$, (*S*,*S*)-diethyl tartrate, and *t*-butylperoxide (Section 9.3.2.1)
3. Reaction of an alkene with (*S*,*S*)-Mn(III)-salen and NaOCl (Section 9.3.2.2)

Synthesis of Ethers:

1. Williamson ether synthesis; reaction of a primary alkyl halide with an alkoxide anion (Sections 12.2, 13.11.1, 24.2.2)
2. Acid-catalyzed addition of an alcohol to an alkene (Section 13.11.2)
3. Reaction of an α,β-unsaturated ketone/aldehyde with an alkoxy anion (Section 21.5.1)

Synthesis of Esters:

1. Reaction of an alkyl halide with a carboxylate ion (Sections 12.2, 18.8.2)
2. Fischer esterification; acid-catalyzed reaction of a carboxylic acid and alcohol (Section 18.7.2)
3. Reaction of a carboxylic acid and diazomethane (Section 18.8.1)
4. Acid-catalyzed addition of a carboxylic acid to an alkene (Section 18.8.3)
5. Reaction of an acid chloride with an alcohol (Section 19.7.2.1)
6. Reaction of an anhydride with an alcohol (Section 19.7.2.2)
7. Reaction of an ester with an alcohol (Section 19.7.2.3)

Synthesis of Halohydrins:

1. Addition of Br_2/Cl_2 with H_2O to an alkene (Section 9.1.3)

Synthesis of Imines:

1. Addition of ammonia and primary amines to an aldehyde/ketone (Sections 17.8.1, 17.8.2)

Synthesis of Ketones:

1. Reaction of an alkene with O_3 followed by CH_3SCH_3 (Section 9.1.7)
2. Oxymercuration of an alkyne (Section 10.8.4)
3. Hydroboration of an internal alkyne (Section 10.8.5)
4. Pinacol rearrangement; acid-catalyzed rearrangement of a 1,2-diol (Section 13.8.2)
5. Oxidation of a secondary alcohol with chromic acid or pyridinium chlorochromate (PCC) (Section 13.9.2)
6. Corey–Kim oxidation, Swern oxidation, and Dess–Martin oxidation of a secondary alcohol (Section 13.9.4)
7. Addition of an organolithium (2 equiv.) to a carboxylic acid (Section 18.7.4)
8. Decarboxylation of a β-keto acid (Sections 18.8.4, 20.4.2)
9. Reaction of an acid chloride and Gilman reagent (Section 19.9.2)
10. Reaction of a nitrile with organolithium/Grignard reagent (Section 19.10.2)
11. Claisen condensation (Section 20.4); including crossed Claisen condensation (Section 20.4.1) and Dieckmann condensation (Section 20.4.2)
12. Stork enamine synthesis; the reaction of an enamine and an acid chloride (Section 20.6)
13. Oxidation of phenols (Section 24.2.4)

Synthesis of α,β-Unsaturated Ketones/Aldehydes:

1. Aldol reaction with heat; acid-catalyzed aldol reaction (Section 20.2.2), base-catalyzed aldol reaction (Section 20.3.2), Claisen–Schmitt (crossed aldol condensation) (Section 20.3.3), intramolecular aldol reaction (Section 20.3.4)

Synthesis of Nitriles:

1. Reaction of an alkyl halide or alkyl tosylate with a cyanide ion (Sections 12.2, 13.7.1)
2. Reaction of an α,β-unsaturated ketone/aldehyde with a cyanide ion (Section 21.5.1)

Synthesis of Protecting Groups:

1. Protection of alcohols as silyl ethers (Sections 13.14.1, 27.4.2)
2. Protection of an aldehyde/ketone as an acetal (Section 17.7.5)
3. Protection of an alcohol as an ester (Sections 19.7.2.2, 27.4.3)
4. Protection of an amine as an amide (Section 19.7.3.2)
5. Protection of amines with benzyl chloroformate (Section 26.9.1)
6. Protection of amines with di-*tert*-butyldicarbonate (Section 26.9.2)
7. Protection of an alcohol as a benzyl ether (Sections 24.6, 27.4.2)

Synthesis of Substituted Benzene:

1. Electrophilic aromatic substitution: halogenation (Section 23.7.1)
2. Electrophilic aromatic substitution: nitration (Section 23.7.2)
3. Electrophilic aromatic substitution: sulfonylation (Section 23.7.3)
4. Friedel–Crafts alkylation (Section 23.7.4)
5. Friedel–Crafts acylation (Section 23.7.5)
6. Addition/elimination reaction of a halobenzene (Section 23.9.1)
7. Substitution of benzene through benzyne formation (Section 23.9.2)
8. Diazonium formation (Section 23.10.1)
9. Kolbe synthesis—formation of salicylic acid (Section 24.2.3)

Synthesis of Thiols:

1. Nucleophilic substitution of HS^- with a 1° haloalkane (Section 12.3)

Synthesis of Thioethers:

1. Reaction of an alkyl halide or alkyl tosylate with a thiolate ion (Sections 12.2, 13.7.1)
2. Reaction of an α,β-unsaturated ketone with a thiolate ion (Section 21.5.1)

APPENDIX

L

C—C Bond-Forming Reactions

1. Reaction of an acetylide ion with an alkyl halide or tosylate (Sections 10.5, 13.7.1)
2. Opening of an epoxide with the addition of an acetylide ion (Section 13.13.2)
3. Opening of an epoxide with the addition of an organolithium or Grignard reagent (Section 16.2.3.2)
4. Reaction of an organolithium or Grignard reagent with ketones and aldehydes (Section 16.2.3.3)
5. Reaction of an alkene with a carbene (Section 16.3)
6. Transition metal–catalyzed coupling, including the Negishi coupling (Section 16.4.2), the Stille coupling (Section 16.4.3), the Suzuki coupling (Section 16.4.4), and the Sonogashira coupling (Section 16.4.5)
7. Cross-coupling reaction of Gilman reagents with alkyl, alkenyl, and aryl halides (Section 16.5.1)
8. Opening of an epoxide with the addition of a Gilman reagent (Section 16.5.2)
9. Olefin metathesis (Section 16.6)
10. Addition of acetylide and cyanide ions to a ketone or aldehyde (Sections 17.5.3, 17.5.4)
11. Wittig reaction: reaction between a phosphorus ylide and an aldehyde or ketone (Section 17.6)
12. Reaction of carboxylic acid and carboxylic acid derivatives with Grignard reagents (Sections 18.7.4, 19.9.1)
13. Reaction of a Gilman reagent with an acid chloride (Section 19.9.2)
14. Reaction of a nitrile with an organolithium or Grignard reagent (Section 19.10.2)
15. Classical acid-catalyzed aldol reaction (Section 20.2.2) and classical base-catalyzed aldol reaction (Section 20.3.2)
16. Aldol reaction, including enolate alkylation (Section 20.3), modern aldol reaction (Section 20.3.1), crossed aldol condensation (Section 20.3.3), intramolecular aldol reaction (Section 20.3.4)
17. Claisen condensation (Section 20.4), including crossed Claisen condensation (Section 20.4.1) and Dieckmann condensation (Section 20.4.2)
18. Alkylation of activated methylene (Section 20.5)
19. Stork enamine synthesis; reaction of an enamine and an acid chloride (Section 20.6)
20. Michael reaction; conjugate addition of a stabilized enolate to α,β-unsaturated carbonyls (Section 21.5.2)
21. Robinson annulation (Section 21.5.3)
22. Conjugate addition of Gilman reagents to α,β-unsaturated carbonyls (Section 21.5.4)
23. Diels–Alder cycloaddition (Section 22.3)
24. [2 + 2]-cycloadditions (Section 22.4.1)
25. Electrocyclic reactions (Section 22.4.2)
26. Sigmatropic rearrangement—Cope/Claisen (Section 22.4.3)
27. Friedel–Crafts alkylation (Section 23.7.4)
28. Friedel–Crafts acylation (Section 23.7.5)
29. Gabriel malonic ester synthesis (Section 26.4.3)
30. Terpene synthesis (Section 27.13)
31. Addition polymers, including cationic polymerization (Section 28.4.1), anionic polymerization (Section 28.4.2), free-radical polymerization (Section 28.4.3)
32. Epoxy resins (Section 28.6.1)
33. Ring-opening metathesis polymerization (Section 28.6.2)
34. Transition metal–catalyzed polymerization (Section 28.6.3)

APPENDIX M

Important Equations

Equation	Description	Equation
1.1	**Grit equation 1 (Skill):** Skill is the product of talent and effort.	Skill = talent × effort
1.2	**Grit equation 2 (Achievement):** Achievement is the product of skill and effort.	Achievement = skill × effort
1.3	**Grit equation 3 (Achievement):** Effort counts twice.	Achievement = (talent × effort) × effort = TE^2
2.1	**Average mass (amu):** The mass of an element as a weighted average based on the relative abundance of different isotopes	$\text{Weighted average atomic mass (amu)} = \dfrac{\left(\text{mass of isotope 1}\right)\left(\%\text{ abundance}\right) + \left(\text{mass of isotope 2}\right)\left(\%\text{ abundance}\right)}{100}$
2.2	**Formal charge:** The number of valence electrons minus the unshared electrons and half of the shared electrons	$\text{Formal charge} = \text{Number of valence electrons} - \frac{1}{2}\left(\text{Shared electrons}\right) - \text{Unshared electrons}$
4.1	**pK_{eq} based on pK_a:** Allows the calculation of K_{eq} based on the pK_a values of the acid and conjugate acid in an acid–base reaction	$pK_{eq} = pK_a\text{ (acid)} - pK_a\text{ (conjugate acid)}$ or $pK_{eq} = pK_a\text{ (left-side acid)} - pK_a\text{ (right-side acid)}$
5.1	**Change in energy:** The difference in energy between the products and reactants	$\Delta E = E_{product} - E_{reactant}$
5.2	**Relating $\Delta G°$ and K_{eq} (natural logarithm):** The relationship between Gibbs free energy and the equilibrium constant	$\Delta G° = -RT \ln K_{eq}$
5.3	**Relating $\Delta G°$ and K_{eq} (base 10 logarithm):** The relationship between Gibbs free energy and the equilibrium constant	$\Delta G° = -2.303RT \log_{10} K_{eq}$
5.4	**Gibbs free energy:** Gibbs free energy is made up two factors, enthalpy (H) and entropy (S).	$\Delta G° = \Delta H° - T\Delta S°$
5.5	**Arrhenius equation:** Relates the rate constant (k_r) of a reaction to the activation energy (E_a), temperature (T), and frequency factor (A)	$k_r = Ae^{-E_a/RT}$
5.6	**Transition state free energy:** Equal to E_a; is a function of temperature (T), entropy (S), and enthalpy (H). Related to Equation 5.4.	$\Delta G^{°‡} = \Delta H^{°‡} - T\Delta S^{°‡}$
5.7	**Rate equation:** The relationship between the overall rate, rate constant (k_r), and the concentration of the reactants	$\text{rate} = k_r[A]^a[B]^b$
5.8	**Arrhenius equation derivation 1:** The relationship between activation energy (E_a) and the selectivity of a reaction	$E_a(1°) - E_a(2°) = RT \ln\left(\dfrac{k_r(2°)}{k_r(1°)}\right)$

(continued)

Equation	Description	Equation
6.1	**Specific rotation:** The observed rotation divided by the path length of the cell (l, in decimeters) multiplied by the concentration (c)	$\text{Specific rotation} = [\alpha]_D^T = \frac{\alpha_{obs}}{l \times c} = \frac{\text{observed rotation (degrees)}}{\text{length (dm)} \times \text{concentration} \left(\frac{\text{g}}{100 \text{ mL}}\right)}$
6.2	**Enantiomeric excess** ***(ee):*** The difference in percentage between two enantiomers in a mixture	$\text{Enantiomeric excess}(ee) = \lvert \%R - \%S \rvert$
6.3	**Optical purity:** The ratio between observed rotation and the rotation of a pure enantiomer	$\%\text{ optical purity} = \frac{[\alpha]_{\text{sample}}}{[\alpha]_{\text{pure enantiomer}}} \times 100$
6.4	**% *ee*:** Calculated by dividing the observed specific rotation by the expected rotation and multiplying by 100%	$\frac{\text{specific rotation}}{[\alpha]_{\text{pure enantiomer}}} \times 100\% = \%\, ee$
6.5	**Percent of enantiomer in excess:** Determined by recognizing that the % *ee* of a mixture represents the percentage of a solution that is not racemic	$\left(\frac{100 - \%\, ee}{2}\right) + \%\, ee = \%\text{ of enantiomer in excess}$
6.6	**Stereoisomer calculations:** The number of stereoisomers is exponentially related to the number of stereocenters (assuming no meso compounds).	$2^n = \text{number of stereoisomers, where } n = \text{number of stereocenters}$
7.1	**Atom economy:** The molecular mass of the desired end product divided by the molecular mass of the reactant(s) and all reagents, multiplied by 100%	$\text{Atom economy} = \frac{\text{molecular mass of product}}{\text{molecular mass of reactants}} \times 100\%$
8.1	**Index of hydrogen deficiency (IHD):** The degree of unsaturation within a molecule	$\text{IHD} = \text{number of carbons} - \frac{\text{number of hydrogens}}{2} - \frac{\text{number of halogens}}{2} + \frac{\text{number of nitrogens}}{2} + 1$
9.1	**Arrhenius equation derivation 2:** The relationship between activation energy (E_a) and the selectivity of a reaction	$E_a(\text{C}) - E_a(\text{B}) = RT \ln\left(\frac{k_r(\text{B})}{k_r(\text{C})}\right)$
9.2	**Arrhenius equation derivation 3:** The relationship between activation energy (E_a) and the selectivity of a reaction	$\frac{k_r(\text{B})}{k_r(\text{C})} = e^{\frac{E_a(\text{C}) - E_a(\text{B})}{RT}}$
14.1	**Planck–Einstein relation:** Energy (E) is equal to Planck's constant (h) times frequency (ν)	$E = h\nu$
14.2	**Speed of light equation:** The speed of light is the product of the frequency of the wave (ν) and the wavelength (λ).	$c = \nu\lambda$
14.3	**Combination of Equations 14.1 and 14.2:** The energy of a photon of light is inversely proportional to the wavelength of light.	$E = h\nu = h\left(\frac{c}{\lambda}\right)$
14.4 (8.1)	**Index of hydrogen deficiency (IHD):** The degree of unsaturation within a molecule	$\text{IHD} = \text{number of carbons} - \frac{\text{number of hydrogens}}{2} - \frac{\text{number of halogens}}{2} + \frac{\text{number of nitrogens}}{2} + 1$

(*continued*)

Equation	Description	Equation
14.5	**Hooke's law derivation 1:** The relationship between period, mass, and spring constant (*k*)	$T = \frac{1}{v} = 2\pi\sqrt{\frac{m}{k}}$
14.6	**Hooke's law derivation 2:** The relationship between frequency (v), force constant of a bond (*k*), and the reduced mass of atoms (μ)	$v = \frac{1}{2\pi}\sqrt{\frac{k}{\mu}}$
15.1 (14.1)	**Planck–Einstein relation:** Energy (*E*) is equal to Planck's constant (*h*) times frequency (v).	$E = hv$
15.2 (14.2)	**Speed of light equation:** The speed of light is the product of the frequency of the wave (v) and the wavelength (λ).	$c = v\lambda$
15.3 (14.3)	**Combination of Equations 14.1 and 14.2:** The energy of a photon of light is inversely proportional to the wavelength of light.	$E = hv = h\left(\frac{c}{\lambda}\right)$
15.4 (8.1 and 14.4)	**Index of hydrogen deficiency (IHD):** The degree of unsaturation within a molecule	$\text{IHD} = \frac{\text{number of}}{\text{carbons}} - \frac{\text{number of hydrogens}}{2} - \frac{\text{number of halogens}}{2} + \frac{\text{number of nitrogens}}{2} + 1$
15.5	**Chemical shift equation:** The position of the signals on the spectrum; shows how different the resonance frequency of a given nucleus is relative to the resonance frequency of the standard	$\text{chemical shift, } \delta\text{ (ppm)} = \frac{\text{shift from TMS (Hz)}}{\text{spectrometer frequency (MHz)}}$
15.6	**$N + 1$ rule:** The relationship of multiplicity, showing that a hydrogen with *N* neighbors will appear in the spectrum as an $N + 1$ multiplet (*M*)	$N + 1 = M$
21.1	**Wavelength of maximum absorption (lambda max):** In UV–vis spectroscopy, it is the wavelength where maximum absorption occurs. It is most often used with conjugated systems.	$\lambda_{max} = (N_A \times h \times c)/\Delta E,$
25.1	**pK_b of a base:** The strength of a base is inversely correlated to the strength of its conjugate acid.	$14 - pK_a(\text{conjugate acid}) = pK_b(\text{base})$
26.1	**Isoelectric point equation:** The pH at which the zwitterion is the predominant species; used when averaging the pK_a values of the ammonium and carboxylic acid protons	$pI = \frac{pK_{a(RNH_3)} + pK_{a(RCO_2H)}}{2}$

Common Symbols

Common symbols and what they mean

—	Single bond	R:	Lone pair on an atom	$\longrightarrow$	Reaction arrow
=	Double bond	R^+	Cation	$\longleftrightarrow$	Resonance arrow
≡	Triple bond	R^-	Anion	$\rightleftharpoons$	Reversible reaction
	Bond going backward into page	R·	Radical		Equilibrium favored to the right
	Bond coming forward out of page	δ^+	Partial positive charge		Equilibrium favored to the left
	*s* orbital	δ^-	Partial negative charge	$A \xrightarrow{B} C$	A and B react to give C
	*p* orbital	equiv.	Stoichiometric equivalents	$B \xrightarrow{A} C$	A and B react to give C
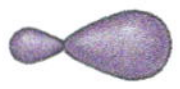	sp^x hybridized orbital (x = 1, 2, or 3)	cat.	Reagent is a catalyst	$A + B \longrightarrow C$	A and B react to give C
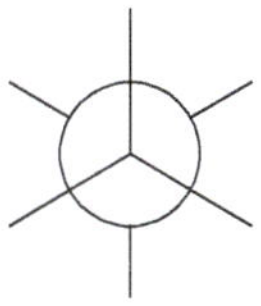	Newman projection to show conformation	H_3O^+ quench	Protonates all negative charges at the end of a reaction		Mechanism arrow showing the flow of two electrons
$[\ \ ]_n$	Repeating units in a polymer	Δ	Heat or "change in"		Mechanism arrow showing the flow of one electron
K_{eq}	Equilibrium constant	$h\nu$	Light	θ	Dihedral angle
ΔG	Change in Gibbs free energy	ΔH	Change in enthalpy	ΔS	Change in entropy
k_r	Rate constant of a reaction	[A]	Concentration of A	A	Frequency factor
E_a	Activation energy	$\bar{\nu}$	Wavenumber	$A \Longrightarrow B$	Retrosynthesis arrow showing that A can be made from B.
J_{ab}	Coupling constant between atoms a and b	δ	Chemical shift	M	Molecular ion

APPENDIX 0

Bond Lengths

Bond	Length (Å)	Bond	Length (Å)	Bond	Length (Å)	Bond	Length (Å)
Single bonds							
H—H	0.74	N—H	1.01	Si—H	1.48	S—H	1.34
H—F	0.92					S—S	2.04
H—Cl	1.27			Si—O	1.61		
H—Br	1.41	N—O	1.44			S—Cl	2.01
H—I	1.61			Si—F	1.56		
				Si—Cl	2.04		
C—H	1.09						
C—C	1.54					F—F	1.43
C—Si	1.86						
C—N	1.47	O—H	0.96	P—H	1.42		
C—O	1.43	O—P	1.60				
C—P	1.87	O—O	1.48			Cl—Cl	1.99
C—S	1.81	O—S	1.51				
C—F	1.33			P—Cl	2.04		
C—Cl	1.77	O—Cl	1.64	P—Br	2.22	Br—Br	2.28
C—Br	1.94						
C—I	2.13					I—I	2.66
Multiple bonds							
C=C	1.34			C≡C	1.21	N≡N	1.10
C=N	1.27			C≡N	1.15		
C=O	1.23						

You don't need to memorize these values, so this table is most useful for identifying the trends. Notice that bonds between larger atoms are longer, and multiple bonds are shorter than single bonds.

Glossary

1,2-addition In nucleophilic or electrophilic addition to a conjugated π system, the addition of two atoms or groups to atoms that are directly attached (a 1,2-relationship) to each other.

1,2-hydride shift The movement of a hydride ion from one carbon to an adjacent carbon.

1,3-diaxial interaction A destabilizing interaction between two groups, two carbon atoms away from each other, in axial positions in the same "hemisphere" of a typical cyclohexane chair conformation.

1,4-addition In nucleophilic or electrophilic addition to a conjugated π system, the addition of two atoms or groups to atoms that are at the ends of the π system (a 1,4-relationship) to each other.

β-keto acids A carboxylic acid containing a ketone at the β-carbon.

α spin The spin of a nucleus whose induced magnetic field is aligned with the external magnetic field.

α-helix The backbone of a polypeptide coiled, most frequently, in a right-handed spiral held together by hydrogen bonding.

absolute configuration Describes the spatial arrangement of the atoms attached to a chiral center in an exact manner, designating them as *R*.

acetal An organic compound with the formula RO—C—OR. These are most often formed from the condensation of an excess of alcohol and a carbonyl in the presence of an acid catalyst.

acetoacetic ester synthesis The production of substituted acetone derivatives via the sequential alkylation, hydrolysis, and decarboxylation of acetoacetic esters.

achiral The description of an object that is superimposable with its mirror image. Achiral objects possess a plane or a point of symmetry.

acid chloride A carboxylic acid derivative characterized by a chloride in the place of the hydroxyl group of the parent carboxylic acid. Has the condensed formula RCOCl. The most reactive of the neutral carboxylic acid derivatives.

acid dissociation constant A measure of the degree to which an acid dissociates in a dilute solution, so the concentration of water remains essentially constant.

acid quench When a reaction is complete, the process of adding acid to the reaction mixture so that all negatively charged atoms are protonated.

acid-catalyzed hydration The addition of water across a double bond made faster using acid catalysis. Though a favorable process in terms of ΔG, in the absence of acid, it is an extremely slow reaction.

activated methylene A methylene (or CH_2) that is especially acidic ($pK_a < 10$) as a result of its conjugate base being especially stable. In Chapter 20, activated methylenes were used in the acetoacetic ester and malonic ester syntheses and related reactions.

activation energy (E_a) A measure of the minimum energy needed for molecules to overcome the repulsion between their electron clouds when they collide. It is the energy difference between the reactants and the transition state. Via the Arrhenius equation, E_a plays a factor in the rate constant of a reaction. A higher activation energy gives a slower reaction.

addition polymer Polymers formed from addition reactions, containing all of the atoms present in the original monomers.

addition reaction A reaction taking the form $A + B \rightarrow A{-}B$. Disfavored by entropy, addition reactions are driven by the formation of bonds more stable than the ones that are broken.

addition/elimination mechanism The mechanism by which carboxylic acid derivatives undergo substitution reactions. The nucleophile adds to make a tetrahedral intermediate. Collapse of the tetrahedral intermediate kicks out the best leaving group to produce the product.

aglycone A general term for any nonsugar molecule attached at the anomeric carbon of a sugar.

alcohols A compound in which a hydrogen of an alkane has been replaced by an OH group.

aldehydes The functional group where a carbonyl is flanked by a hydrogen and an alkyl group. Has the general formula RCHO.

aldol condensation The acid- or base-catalyzed self-condensation of an aldehyde to give a β-hydroxyaldehyde, or with heating, an α,β-unsaturated aldehyde.

aldol A common name for a β-hydroxy aldehyde or β-hydroxy ketone.

aliphatic The description of an organic compound in which all of the carbon atoms are sp^3 hybridized and oriented in a chain, as opposed to a ring.

alkane A saturated hydrocarbon containing only carbon, hydrogen, and single bonds.

alkene An unsaturated hydrocarbon containing at least one carbon–carbon double bond.

alkyne An unsaturated hydrocarbon containing a carbon–carbon triple bond.

allylic Describes the carbon directly attached to an alkene carbon. Atoms and substituents attached to the allylic carbon are also described as being allylic. Allylic substituents react differently as a result of proximity to the alkene.

allylic alcohol A compound containing a hydroxyl group on an sp^3-hybridized carbon attached to a carbon of an alkene.

alpha (α) carbon The carbon directly attached to a carbonyl or other electron withdrawing group. Notable for its ability to stabilize a negative charge. In carbene formation, refers to the carbon to which the leaving group is directly attached.

alpha (α) hydrogen The hydrogen on the α-carbon. Made acidic because of the stability of the negative charge α to the carbonyl.

alpha (α) substitution Substitution of an electrophile for an α-hydrogen at the α-carbon. Typically a result of the enhanced acidity of the α-hydrogen.

alpha (α) cleavage Homolytic cleavage of a group bonded to the α-carbon.

amide A carboxylic acid derivative characterized by an amino group in the place of the hydroxyl group of the parent carboxylic acid. Has the condensed formula $RCONR_2$. May be primary ($RCONH_2$), secondary (RCONHR), or tertiary ($RCONR_2$) based on the number of alkyl groups (0, 1, or 2) attached to nitrogen.

amines A general class of compounds in which one hydrogen of a parent alkane has been replaced by an amino group ($-NR_2$; R = alkyl or hydrogen).

amorphous A descriptor for flexible, less rigid polymers.

angle strain A destabilizing aspect of many cyclic molecules where the geometry of the ring causes the atom to deviate from its most stable bond angle.

anhydride A carboxylic acid derivative characterized by a carboxylate group in the place of the hydroxyl group of the parent carboxylic acid. Has the condensed formula RC(O)OC(O)R′. Often symmetrical (R = R′), but can be mixed ($R \neq R'$).

anhydrous Indicating that water has been excluded from a molecule/solvent/reaction mixture.

anion A negatively charged ion with at least one more electron than proton.

anomeric carbon The carbon in a cyclic sugar that is the carbonyl carbon in the open-chain form.

anomeric hydroxyl The hydroxyl group attached to the anomeric carbon when a sugar is in the cyclic, hemiacetal form.

anomers Two cyclic sugars that differ in configuration only at the carbon that is the carbonyl carbon in the open-chain form.

anti and coplanar Describes the relationship between two groups with a 180° dihedral angle.

anti addition A designation of the spatial orientation of an addition reaction, indicating that the two groups that were added did so on opposite sides of the molecule. Anti addition is most commonly seen in two-step processes wherein the intermediate has one face of the molecule blocked to further attack.

anti-Markovnikov addition Historically, refers to alkene addition reactions that proceed with regioselectivity opposite to Markovnikov addition, based on the location of functional groups in the product. Mechanistically, however, it is identical to Markovnikov addition, since the regioselectivity in both reaction types is driven by formation of the most stable carbocation in the (transition state of the) rate-determining step.

anti-Zaitsev elimination An elimination reaction that gives the least stable/least substituted alkene.

anti In Newman projections, refers to the 180° dihedral angle between a group on the front carbon and a group on the back carbon.
In cyclic or acyclic molecules, describes the relationship between two groups that are on opposite sides of the molecule.

antiaromatic The description of a cyclic and planar compound with an uninterrupted ring of *p* orbital-bearing atoms containing $4n$ π electrons, where $n = 0, 1, 2, 3, \ldots$. Antiaromatic compounds are extremely unstable.

area of electron density Volumes of space occupied by one or more electrons. Includes lone pairs, single bonds, double bonds, and triple bonds as individual areas of electron density.

arenium ion The nonaromatic intermediate formed in the rate-determining (first) step of the electrophilic aromatic substitution reaction of benzene with strong electrophiles.

aromatic transition state The low activation energy transition state featuring the cyclic flow of 6 electrons. So named because it mimics aromatic molecules which are more stable than would be expected.

aromatic The description of a cyclic and planar compound with an uninterrupted ring of *p* orbital-bearing atoms containing $4n + 2$ π electrons, where $n = 0, 1, 2, 3, \ldots$. Aromatic compounds are extremely stable.

Arrhenius acid Any molecule that produce H^+ ions (protons) when dissolved in water.

Arrhenius base Any molecule which increases the concentration of HO^- ions (hydroxide ions) when dissolved in water.

Arrhenius equation Illustrates that the rate constant (k_r) of a reaction is a function of activation energy (E_a), the frequency factor (A), and the absolute temperature (T). A higher activation energy decreases the rate constant (slows the reaction). A higher temperature increases the rate of a reaction.

arrow-pushing formalism Also called the *electron-pushing* or *curved arrow* formalism, it is a human construct which rationalizes the movement of electrons in a chemical reaction using curved arrows.

asymmetric The description of a stereoselective reaction, since it produces one enantiomer or diastereomer in excess.

asymmetric center A stereocenter represented by a carbon with four different groups attached.

asymmetric hydrogenation A stereoselective hydrogenation reaction, typically enabled by the presence of a chiral catalyst.

atom economy Developed by chemist Barry Trost, it is the mass of the product as a percentage of the mass of all the reactants (a higher percent represents a greener process).

atomic mass unit A unit of mass for atomic and molecular masses. Equals one-twelfth the mass of one atom of ^{12}C. Approximately 1.66×10^{-27} kg.

atomic number Equal to the number of protons in the nucleus of an atom. Defines the element. Often omitted, it may be written as a subscript to the left of the elemental symbol ($_6C$). Often written in conjunction with mass number to denote a specific isotope.

atomic orbital hybridization A human construct. The combination of the mathematical equations describing the atomic orbitals, where the output describes the shape and orientation of the hybrid atomic orbitals.

atropisomerism Results when rotation around a bond is slow enough that individual enantiomeric conformations can be isolated/observed.

aufbau principle German for "building up," describes the phenomenon where the lowest energy orbitals are filled preferentially.

autocatalysis The phenomenon in the neutral hydrolysis of small acid chlorides by water, whereby the byproduct of the reaction, H_3O^+, serves as an acid and catalyzes further reactivity. The longer the reaction progresses, the more acidic the solution becomes and the faster the reaction occurs.

autooxidation The oxidation of materials by the oxygen present in air.

axial In a chair structure, one of the two orientations for the substituents. The bonds are drawn vertically and are parallel to the central axis of the molecule. Axial groups alternate between the "northern" and "southern hemispheres" on every other carbon atom.

β spin The spin of a nucleus whose induced magnetic field is aligned against the external magnetic field.

backside displacement The displacement of a leaving group by a nucleophile attacking the "backside" of the electrophile, 180° opposed to the leaving group.

benzylic carbon The sp^3-hybridized carbon directly attached to a benzene ring.

benzyne The highly reactive intermediate formed by elimination of good leaving groups in nucleophilic aromatic substitution of benzene using strong bases and/or very harsh conditions.

β-carbon The carbon that neighbors the α-carbon.

β-elimination The elimination of a hydrogen atom and a leaving group from neighboring carbons.

bimolecular A reaction with two molecules in the transition state.

bioaccumulation The buildup of chemicals in tissues of living organisms, which get passed on to other organisms who consume those organisms.

biodiesel An alternative fuel source containing long-chain esters that is produced from vegetable oils and animal fats.

BOC protecting group The *t*-butyloxy carbonyl group used for the protection of an amine. It is formed using di-*tert*-butyl dicarbonate and removed with acid.

bond-dissociation enthalpy (BDE) The energy associated with breaking a bond homolytically to form two radicals. Because bond breakage is an endothermic process, these values are always positive. BDE can be used to calculate the contribution of bond energies to enthalpy.

Brønsted–Lowry acid Any molecule capable of donating a proton.

Brønsted–Lowry base Any molecule capable of accepting a proton with a lone pair of electrons.

bromohydrin A class of compound containing a bromine and a hydroxyl group at neighboring sp^3-hybridized carbons. The bromohydrin is produced when an alkene is reacted with Br_2 in water.

bromonium ion A bromine-containing, three-membered ring, where the bromine bears a positive charge. The bromonium ion, produced as an intermediate in several reactions of alkenes with Br_2, is usually opened by attack of a nucleophile at one of the carbons of the ring.

Buchwald–Hartwig amination The transition metal–catalyzed coupling of aryl halides and amines occurring by the catalytic cycle of other transition metal–catalyzed coupling reactions.

Bürgi–Dunitz angle The angle at which a nucleophile attacks the π^* antibonding orbital of a carbonyl carbon.

C-terminal amino acid The amino acid on the end of a peptide chain that ends in a carboxylic acid.

carbanion An organic anion in which carbon possesses three bonds and a nonbonded pair of electrons, and thus, has a formal negative charge.

carbene A neutral carbon atom, containing only two bonds and one lone pair of electrons. Depending on attached substituents, a carbene can act as both a nucleophile and an electrophile.

carbenoid A metal-bound molecule that exhibits carbene-like reactivity.

carbocation A positively charged carbon, containing three bonds and zero lone pairs of electrons. Being sp^2 hybdridized, carbocations contain an empty *p* orbital, making them strong Lewis acids. Electrondonating groups make carbocations more stable, and therefore less reactive.

carbocation rearrangement The rearrangement of a carbocation to a more stable carbocation.

carboxylate anion The conjugate base of a carboxylic acid. Has the general formula RCO_2^-.

carboxylic acid The functional group where a carbonyl is flanked on one side by a hydroxyl group and a hydrogen or carbon-containing group on the other. Has the general formula RCO_2H.

carboxylic acid derivatives Functional groups where the central carbon is at the same oxidation state as a carboxylic acid. Includes, acid chlorides, anyhydrides, esters, amides, and nitriles.

catalyst A substance that increases the rate of a reaction, but is itself not consumed in the process. The catalyst increases the rate of reaction by changing the mechanism to one in which the rate-determining step has a smaller activation energy.

catalytic hydrogenation The addition of hydrogen to an alkene, typically enabled by a metal catalyst like palladium, platinum, or nickel.

cation A positively charged ion with at least one fewer electron that proton.

center of asymmetry See *asymmetric center*.

center of chirality See *chiral center*.

chain reaction A reaction in which the product of the reaction acts upon the reactant of the same, or a different type of reaction, without any other influence. This is the term used to describe radical reactions, since the radical produced at the end of the reaction is the radical used to start the next sequence.

chain-growth polymer A polymer that grows in a single direction one monomer at a time.

chair flip The conformational change of a cyclohexane ring that moves the 6 atoms of the cyclohexane into the other conformation that is a local energy minimum. As a consequence, all axial groups rotate to become equatorial groups, and all equatorial groups rotate to become axial groups.

chair structure One of two local energy minima for a cyclohexane ring that minimizes ring and torsional strain in the molecule.

chair-like transition state When six atoms are involved in a reaction, they very often adopt a transition state that is similar in structure to a chair conformation. This can be used to predict the stereochemical outcome of many reactions.

chemical shift (δ) The location of a signal in an NMR spectrum. It is measured downfield from a reference compound (most often, TMS).

chiral center The portion of a molecule that makes it chiral. Most often, it is a tetrahedral carbon with four different groups attached.

chiral The description of an object that is not superimposable with its mirror image. A chiral objects lacks symmetry.

chlorohydrin A class of compound containing a chlorine and a hydroxyl group at neighboring sp^3-hybridized carbons. The chlorohydrin is produced when an alkene is reacted with Cl_2 in water.

cis/trans isomers Stereoisomers that differ in the relative spatial orientation for substituents on a ring or on a double bond. The cis isomer is defined by having the substituents on the alkene/ring coming off in the same direction, whereas the trans isomer has the substituents coming off the alkene/ring in opposite directions.

cis In cyclic molecules: two groups that are on the same side of the molecule (i.e., both groups are on bold wedges or both groups are on dashed wedges). In alkenes: the two groups (one from each carbon atom) that are on the same side ("northern" or "southern hemisphere") of the double bond.

Claisen condensation Reaction of two molecules of the same ester to form a β-ketoester in the presence of base.

Claisen rearrangement The [3,3]-sigmatropic rearrangement of an allyl vinyl ether to give a γ,δ-unsaturated aldehyde.

Claisen–Schmidt reaction The specific crossed aldol reaction between a ketone and an aldehyde with no α-hydrogens.

column chromatography A laboratory technique used for separating compounds based on the manner in which they interact with a solid, stationary phase.

combustion The combination of a compound with oxygen resulting in the production of heat and light. Combustion analysis can be used to empirically determine the relative stability of organic molecules.

complex carbohydrates Composed of longer, polysaccharide carbohydrate chains that are slowly digested.

concerted reaction A reaction in which multiple bonds are broken/formed in a single step.

condensation reaction A reaction in which two molecules combine to form a new molecule with simultaneous loss of water or other small molecule.

condensation polymers A polymer made by combining two molecules while removing a small molecule (usually water or an alcohol).

configurational isomers Isomers that differ only in the spatial orientation of attached atoms and substituents. This difference is due to bonding, as opposed to the different conformations. That is, to exchange between two configurational isomers, you must break and form bonds.

conformational enantiomerism A type of enantiomerism wherein, because of restricted rotation around a particular bond, a nonsuperimposable mirror image is possible in spite of the molecule containing no chiral centers.

conjugate acid/base pair A related pair of molecules which only differ by the presence/absence of a proton (H^+). E.g., HCl and Cl^- are a conjugate acid/base pair.

conjugated alkenes At least two alkenes that are separated by a single C—C single bond. Conjugated alkenes are characterized by delocalization of π electrons through resonance.

conjugated proteins Proteins covalently bonded to carbohydrates, lipids, or metals.

conrotatory The rotation of *p* orbitals in the same direction (both counterclockwise or both clockwise) to explain the outcome of electrocyclic reactions under thermal or photochemical conditions.

constitutional isomer Two molecules that have the same molecular formula, but differ in connectivity.

Cope elimination The concerted syn elimination of *N*-oxides that preferentially forms the least substituted (anti-Zaitsev) alkene.

Cope rearrangement The [3,3]-sigmatropic rearrangement of a 1,5-diene to form a more stable 1,5-diene.

copolymer A polymer formed from two or more different monomers.

coupling constant The distance (in hertz) between two adjacent peaks of a split NMR signal. Also referred to as the *J* value.

coupling Connecting two smaller organic fragments to make one larger molecule. Most often used in the context of transition metal–catalyzed coupling reactions.

covalent bond A chemical bond described by the sharing of electrons between two atoms.

cross-coupling The coupling of two different smaller molecules, typically via C—C bond formation to make a new, larger molecule.

crossed aldol condensation The aldol condensation between two different carbonyls (aldehydes or ketones). For best results, the enolate of a ketone reacts with an aldehyde that contains no α-hydrogens.

crossed Claisen condensation The reaction of two different esters in the Claisen reaction, most often featuring one ester without enolizable hydrogens and utilizing very specific experimental conditions.

crystalline melting temperature (T_m) The temperature at which a solid polymer becomes a viscous liquid.

crystallinity A descriptor for the rigidity of a polymer that results from the tight packing of polymer chains.

crystallite Regions of a polymer in which the chains are highly ordered.

cumulated alkenes Two alkenes that share a central carbon.

cyanohydrin An organic compound containing a carbon connected to both a cyano and hydroxyl group. Cyanohydrins are typically formed by the reaction of an aldehyde or a ketone and cyanide in the presence of acid.

D/L notation A notation used to describe the stereochemical relationship between carbohydrates, distinguishing them based on the configuration of the bottom-most stereocenter in a Fischer projection.

Dean–Stark apparatus Chemical glassware typically used to remove water from a reaction mixture where it is the higher boiling compound. This apparatus takes advantage of the alcohol–water azeotrope by evaporating the liquids from the solution, allowing them to condense separately. The aqueous layer is removed and the alcohol layer returned to the reaction. By removing water, the equilibrium can be shifted according to Le Châtelier's principle.

decomposition A reaction of the general form $AB \rightarrow A + B$.

degenerate orbitals Orbitals that are at equal energy levels.

degree of unsaturation The degree to which a molecule has fewer hydrogens than expected based on the number of carbons present in the molecular formula. Also referred to as the **index of hydrogen deficiency** or the **double bond equivalents**, it is helpful for speculating about the number of π bonds and rings that might be present in a molecule based only on knowledge of the molecular formula.

dehydration Loss of water from a molecule.

demercuration/reduction Cleavage of the carbon–mercury bond generated in the oxymercuration reaction. This is typically accomplished under reducing conditions, using sodium borohydride.

dextrorotatory Refers to a molecule that rotates plane-polarized light in the clockwise direction, from the Latin word *dextro*, meaning "to the right." It is the opposite of levrorotatory.

diastereomers Stereoisomers that are not mirror images. Examples include a cis versus trans alkene, or two molecules which have opposite configurations for at least one but not all chiral centers. Diastereomers are different in all properties.

diborane Having the molecular formula B_2H_6, it is a more stable form of the reactive Lewis acid borane (BH_3). In solution, it dissociates to give two molecules of borane and is therefore used in the hydration of alkenes.

Dieckmann condensation The intramolecular condensation of two esters. Can be thought of as an intramolecular Claisen condensation.

Diels–Alder reaction The reaction of a diene with a dienophile (an alkene) to give a cyclohexene.

diene A hydrocarbon with two double bonds.

dienophile An alkene that reacts with a diene in a Diels–Alder reaction.

dihedral angle Like hands on a clock, the angle between a group on the front carbon and a group on the back carbon of a Newman projection.

dihydroxylation The oxidation of an alkene that results in the *syn*-addition of hydroxyl groups to each carbon of the alkene. The dihydroxylation is typically carried out with OsO_4 or $KMnO_4$.

dipeptide Two amino acids linked by an amide bond.

dipole moment The vector sum of all polar bonds in a molecule. Determines a molecule's overall polarity.

dipole–dipole interactions An intermolecular force between two molecules with permanent dipoles. The permanent partial positive charge in one molecule is attracted to the permanent partial negative charge in a second molecule.

disaccharide A compound containing two monosaccharides linked together.

disrotatory The rotation of *p* orbitals in opposite directions (one counterclockwise and one clockwise) to explain the outcome of electrocyclic reactions under thermal or photochemical conditions.

diterpenes A terpene composed of four 5-carbon isoprene units (20 carbons).
double replacement A reaction of the general form AB + XY → AY + XB.
dummy ligands A ligand, usually bound to a metal, that does not transfer in a given reaction.
E From *entgegen*, for "different," the designation of the alkene stereoisomer that has the highest priority substituents on opposite sides of the alkene.
E/Z **isomers** Stereoisomers of alkenes which differ by the placement of the higher priority substituents (as determined by the CIP rules) on the same (*Z*) or opposite (*E*) side of the alkene.
E/Z **system** A system of nomenclature that assigns the absolute stereochemistry to alkenes using the Cahn–Ingold–Prelog system. It has a broader application than the cis/trans system.
***E*1 reaction** A 2-step β-elimination reaction with a unimolecular rate-determining step.
***E*2 reaction** A 1-step β-elimination reaction with a bimolecular rate-determining step.
eclipsed A Newman projection conformation where each group on the front atom is directly in line with each group on the back atom, effectively "hiding" the groups on the back carbon from view.
effective mass yield (EMY) The ratio of product mass to the mass of all non-benign (hazardous) materials used in the process (a larger ratio represents a greener process)
electrocyclic reactions Feature the reversible conversion of acyclic conjugated molecules to ring structures through the movement of electrons. Can be induced thermally (with heat) or photochemically (with light).
electromagnetic spectrum The range of photon energies extending from gamma rays (high energy/short wavelength/high frequency) to radio waves (low energy/long wavelength/low frequency).
electron Subatomic particle with a negative charge. Has a negligible mass (5.4×10^{-4} amu). Located in orbitals surrounding the nucleus. The number of electrons defines the charge of the element.
electronegativity The relative ability of an atom to attract bonding electrons toward itself.
electrophile An atom with a partial or full positive charge that somehow lacks electron density. It has a vacant orbital capable of accepting a pair of electrons to form a bond. Synonymous with Lewis acid.
electrophilic addition The addition reaction of an electrophile to a nucleophile; for example, the electrophilic addition to alkenes.
electrophilic aromatic substitution Replacement of a hydrogen on benzene with a group/atom using an electrophilic reagent strong enough to overcome the high barrier to reactivity imposed by the aromaticity of benzene.
enamine Analogous to an enol and an enolate; an amine attached to either sp^2-hybridized carbon of an alkene. Nucleophilic because of the fact that the lone pair can be delocalized by resonance onto a carbon of the alkene.
enantiomeric excess A property of a solution of enantiomers, referring to how much the solution is enriched in one of the two enantiomers. A racemic mixture would have an ee of 0%, since neither is in excess, while a pure sample of an enantiomer would have an ee of 100%.
enantiomeric relationship The relationship between molecules that are non-superimposable mirror images of each other.
enantiomerically pure The description of a sample of a chiral molecule that contains exclusively one of a pair of enantiomers.
enantiomers Molecules that are stereoisomers that are nonsuperimposable mirror images of each other. Enantiomers will have opposite absolute configurations at each and every chiral center. Enantiomers are identical in all properties with the exception of how they interact in chiral environments.
endergonic Indicates that a process has a positive change in Gibbs free energy ($\Delta G° > 0$), and is therefore unfavorable because it results in $K_{eq} < 1$, through the relationship $\Delta G° = -RT\ln K_{eq}$.
endo transition state The kinetically favored transition state of the Diels–Alder reaction, resulting from secondary orbital overlap. In the endo transition state, the substituents on the dienophile are tucked underneath the diene.
endothermic Indicates that a process absorbs heat from its surroundings, has a positive change in enthalpy ($\Delta H° > 0$), and is therefore unfavorable on the basis of enthalpy. The characteristics of an endothermic reaction suggest that the products of such a reaction are "less stable" than the reactants, either because of the bonds that have been formed and broken, or because of the increased strain of the products.
enol A tautomer of a ketone or an aldehyde, always present at very low concentrations. Analogous to an enamine and an enolate; an alcohol attached to either sp^2-hybridized carbon of an alkene. Nucleophilic because of the fact that the lone pair can be delocalized by resonance onto a carbon of the alkene.
enolate The deprotonated form of an enol. Enolates are composed of two resonance structures where the negative charge is located on either the oxygen or the carbon. Most often nucleophilic at carbon.
enolizable hydrogen A hydrogen that can be removed to form an enolate (under basic conditions) or an enol (under acidic conditions). Similar to an α-hydrogen.
enthalpy change ($\Delta H°$) Amount of heat released or absorbed in a chemical reaction at constant pressure. The enthalpy change includes the sum of the energy released upon bond formation ($\Delta H° < 0$) and the energy absorbed by bond breaking ($\Delta H° > 0$). Also has a contribution from strain energy. A process that increases strain has a positive $\Delta H°$. A process that decreases strain has a negative $\Delta H°$.
entropy ($\Delta S°$) A measure of the disorder or randomness of a system. The increase of randomness or freedom of motion makes a process more favorable. As a result, a positive change in entropy ($\Delta S° > 0$) is associated with favorable processes through the relationship $\Delta G° = \Delta H° - T\Delta S°$.
environmental factor (E-factor) The ratio of waste produced by a chemical process to the amount of product obtained in that process (a smaller ratio represents a greener process)
epimers Monosaccharides that differ in configuration at only one asymmetric center.
epoxide An ether in which the oxygen is incorporated into a three-membered ring. Also referred to as an oxirane.
equatorial In a chair structure, one of the two orientations of the substituents. The bonds are aligned along the "equator" of the cyclohexane ring defined by the 6 carbon atoms.
equilibrium The state of a reaction where the products and reactants are present in a concentration that does not change with time. This does not mean that reactions are not occurring at equilibrium, only that at this point, the rate of the forward reaction is equal to the reverse reaction.
equilibrium constant (K_{eq}) Describes the composition of a reaction mixture at equilibrium in a reversible reaction.
essential amino acids The amino acids that humans must obtain from their diet because they cannot synthesize them at all or in adequate amounts.
ester A carboxylic acid derivative characterized by an alkoxide group in the place of the hydroxyl group of the parent carboxylic acid. Has the general formula RCO_2R.
ether A compound containing an oxygen bonded to two carbons (ROR).
exact mass The mass of a molecule or ion, calculated using the exact atomic mass of whatever isotopes are present in the molecule/ion.
exergonic Indicates that a process has a negative change in Gibbs free energy ($\Delta G° < 0$), and is therefore favorable because it results in $K_{eq} > 1$, through the relationship $\Delta G° = -RT\ln K_{eq}$.
exo transition state The kinetically disfavored transition state of the Diels–Alder reaction. In the exo transition state, the substituents on the dienophile are away from the diene. The endo transition state is favored in the Diels-Alder reaction.
exothermic Indicates that a process releases heat to its surroundings, has a negative change in enthalpy ($\Delta H° < 0$), and is therefore unfavorable on the basis of enthalpy. The characteristics of an exothermic reaction suggest that the products of a reaction are "more stable" than the reactants, either because of the bonds that have been formed and broken, or because of the decreased strain of the products.
fatty acids Carboxylic acids with long hydrocarbon chains that are found in nature.
feedstock The origin of a molecule. Can be renewable, derived from plants or algae, or non-renewable, derived from oil reserves.
fibrous proteins A water-insoluble protein in which the polypeptide chains are arranged in bundles.
Fischer projections A two-dimensional representation of three-dimensional structures mostly in carbohydrate chemistry. Within the flat Fischer projection, the horizontal bonds originating at the chiral center are actually coming out of the plane of the page, while the vertical bonds are going back into the page.
fishhook arrows Used to illustrate the movement of a single electron in the arrow-pushing formalism.
fixed mindset A fixed mindset is one that thinks of achievement on an exam, for example, as defining us in the same way that other fixed traits do.

flagpole interaction A destabilizing steric clash between the two groups 4 carbon atoms apart on a cyclohexane ring, when in the "boat" conformation. The groups resemble flagpoles arising from the stern and bow of the cyclohexane "boat."

formal charge A human construct. The charge assigned to an atom assuming equal sharing of electrons. Equal to the atom's group number – 1/2(shared electrons) – unshared electrons.

free radical An uncharged molecule or atom with a single unpaired electron.

frequency factor (*A*) A term used in the calculation of the rate constant using the Arrhenius equation. It is a measure of the frequency of collisions that have the required orientation that allow a reaction to occur.

frequency The number of wave crests that pass by a given point in one second; frequency has units of hertz (Hz).

Friedel–Crafts acylation The acylation of benzene using acid chlorides and a strong Lewis acid. Proceeds by the standard mechanism of electrophilic aromatic substitution.

Friedel–Crafts alkylation The alkylation of benzene using haloalkanes and a strong Lewis acid. Proceeds by the standard mechanism of electrophilic aromatic substitution.

frontier molecular orbital (FMO) theory States that the outcome of a reaction is determined by the interaction of the highest occupied (HOMO) and lowest unoccupied molecular orbitals (LUMO).

Frost circles A mnemonic device that allows the simple approximation of the energy levels of the molecular orbitals of aromatic and antiaromatic rings.

functional group region The left-hand two-thirds of an IR spectrum where most functional groups show absorption bands.

gabriel amine synthesis The S_N2 reaction of a primary alkyl halide with phthalimide which, after unmasking, provides a primary amine as the product.

gauche In Newman projections: a 60° dihedral angle between a group on the front carbon and a group on the back carbon.

geminal Describes the relationship between two groups that are attached to the same atom. An example of the germinal relationship is seen in the bromine atoms of 2,2-dibromopentane.

Gibbs free energy (*G*) The energy of a system available to do work. It is defined as the enthalpy of the system minus the product of the temperature and entropy of the system: $G = H - TS$.

Gilman Reagent An organocuprate. Has the general formula R_2CuLi. Is characterized by performing S_N2-type reactions on alkyl halides and a single acyl substitution on acid chlorides.

glass transition temperature (T_g) The temperature at which a hard, crystalline polymer becomes a soft and flexible solid.

globular protein A water-soluble protein that tends to have a roughly spherical shape.

glycols Also referred to as vicinal diols, the structural motif where two hydroxyl groups are on neighboring carbons.

Grignard reagents Organometallic reagents with the general formula, RMgX. Prepared by adding an alkyl halide to magnesium metal shavings being stirred in an ether—usually diethyl ether or tetrahydrofuran (THF)—under anhydrous conditions.

growth mindset The growth mindset sees every challenge as not a judgment of you as a person, but of where you are right now. It is the opposite of a fixed mindset.

Hückel rules for aromaticity The rules that describe whether a molecule is aromatic.

Hückel/Breslow rules for antiaromaticity The rules that describe whether a molecule is antiaromatic.

haloalkane (alkyl halide) Organic compound in which at least one hydrogen atom of an alkane is replaced by a halogen.

haloalkene (vinyl halide) Organic compound in which a halogen is directly bonded to an alkene carbon.

haloarene Organic compound in which a halogen is directly attached to an aromatic ring.

halogenation For alkanes, a substitution reaction, typically with chlorine or bromine, where an alkyl hydrogen is replaced by a halogen. The addition reaction of a halogen, like bromine or chlorine, resulting in a vicinal dihalide.

halohydrin An organic molecule that contains a halogen and an OH group on neighboring carbons.

Hammond postulate Gives us a way of approximating the identity of a transition state by postulating that the identity of the transition state is dependent on whether or not a reaction step is endothermic ($\Delta H > 0$) or exothermic ($\Delta H < 0$). For an exothermic reaction step, the transition state is said to be of similar energy (or have a similar structure) to the reactants. For an endothermic reaction step, the transition state is of similar energy and structure to the product of that step.

hard and soft Lewis acid and base (HSAB) theory The concept that explains how smaller, more highly charged (hard) nucleophiles react preferentially with smaller, more highly charged (hard) electrophiles. Soft Lewis bases react preferentially with soft Lewis acids.

Haworth projections A way to show the structure of a sugar; the five- and six-membered rings are represented as being flat.

heat of hydrogenation ($\Delta H°$) The amount of heat given off in a hydrogenation reaction. By measuring the heat of hydrogenation for isomeric alkenes, the relative stability of the alkenes can be calculated.

hemiacetal An organic functional group with the formula RO—C—OH. The hemiacetal is present as the unfavorable component of an equilibrium mixture with the carbonyl from which it was formed.

heteroaromatic Aromatic rings containing at least one non-carbon atom in the aromatic π system.

heterogeneous catalysis Catalysis where the catalyst is in a different phase than the reactants and/or the reagents in the reaction.

heterolytic bond cleavage The unsymmetrical breaking of a two-electron covalent bond in which both electrons end up on the same atom. Heterolytic cleavage is the opposite of homolytic cleavage. Acid–base reactions feature heterolytic cleavage. Heterolytic cleavage is illustrated using the normal two-headed arrow.

highest occupied molecular orbital (HOMO) In frontier molecular orbital theory, it is the highest energy molecular orbital that contains a pair of electrons and is often used to explain the reactivity of electron rich systems.

Hofmann elimination The E2 elimination of tetraalkylammonium ions that results in the formation of the less substituted alkene as a result of steric hindrance in the Newman projection that would lead to the more substituted alkene. It is sometimes referred to as an anti-Zaitsev elimination.

homogeneous catalysis Catalysis where the catalyst is in the same phase as the reactants in the reaction.

homolytic bond cleavage The symmetrical breaking of a two-electron covalent bond in which each electron ends up on a different atom. Homolytic cleavage is the opposite of heterolytic cleavage. Radical halogenation reactions feature homolytic cleavage. Homolytic cleavage is illustrated using single-headed or fishhook arrows.

homopolymer A polymer that contains only one kind of monomer.

homotriglyceride A triglyceride that contains three of the same fatty acid.

Hund's rule The observation that, for a set of degenerate orbitals, electrons first fill each orbital with a single electron before pairing each electron.

hydration reaction The addition of water to any organic molecule. Typically, hydration reactions are done to molecules containing C—C or C—O π bonds. The acid-catalyzed addition of water, oxymercuration, and hydroboration–oxidation could all be referred to as hydration reactions.

hydride ion A negatively charged hydrogen atom which can act as a nucleophile or a base.

hydroboration–oxidation Used for the synthesis of alcohols from alkenes when it is desired to have the hydroxyl group on the least substituted carbon (anti-Markovnikov). The electrophilic syn addition of borane is followed by oxidation with basic hydrogen peroxide of the resulting alkyl borane to give the alcohol.

hydrogen bonding An intermolecular force between a lone pair of electrons on an oxygen or nitrogen atom in one molecule, and a hydrogen atom attached to an oxygen or nitrogen atom in a second molecule. The strongest non-ionic intermolecular force.

hydrogenated oils Natural unsaturated fats that have been catalytically hydrogenated in to order to alter their chemical and/or physical properties and improve their performance in the desired application. Because cis-unsaturated fats are unstable, hydrogenation stabilizes them and prevents them from becoming rancid.

hydrogenation The addition of hydrogen (H_2) to a molecule often catalyzed by a transition metal, used with alkenes, alkynes, and aldehydes/ketones.

hydrogenolysis The use of hydrogen (H_2) gas to cleave (or lyse) a functional group. Most pertinent is the hydrogenolysis of a benzyl ether as a method for deprotecting an alcohol.

hydronium ion Specifically, the molecule H_3O^+, the strongest acid possible in water.

hydrophobic Describes a compound that is not very soluble in water.
hydroxide ion An anion is a strong base with the formula, ^{-}OH. It is the conjugate base of water.
hydroxyl group The OH substituent, found most often in alcohols, carboxylic acids, phenols, hemiacetals, and enols.
hyperconjugation Delocalization of electrons by overlap of a σ bond orbital with an empty orbital on an adjacent carbon.
iminium ion The iminium ion contains a positively charged nitrogen atom that has donated its lone pair of electrons to form a double bond with carbon. Iminium ions are electrophilic at the carbon.
index of hydrogen deficiency (IHD) The degree to which a molecule has fewer hydrogens than expected based on the molecular formula. Also referred to as *index of unsaturation* or *double bond equivalents*.
induction The delocalization of electron density through sigma bonds as a result of electronegativity differences. The effect diminishes as the electronegative atom is moved farther away from the reactive center. This type of delocalization most commonly results in enhanced stability of anionic species.
initiation step In the context of a radical chain reaction, it is the step that generates the first radical. In radical halogenation, the initiation step is the heat promoted homolysis of bromine or chlorine to two bromine or chlorine radicals, respectively.
intermediate A relatively stable, but reactive species on the reaction pathway from a reactant to a product. Unlike the transition state, an intermediate can be characterized and studied, as it has a substantially longer lifetime. The intermediate is shown as a valley on a reaction coordinate diagram.
intermolecular forces Attractive or repulsive interactions between molecules. Primarily responsible for differences in melting and boiling point
internal, nonsymmetrical alkyne A C—C triple bond, located in the middle of the chain and having two different groups attached to the alkyne carbons, resulting in a nonsymmetrical compound.
internal, symmetrical alkyne A C—C triple bond, located in the middle of the chain and having identical groups attached to the alkyne carbons, resulting in a symmetrical compound.
inverse electron demand Diels–Alder A Diels–Alder reaction that uses an electron poor diene to react with an electron rich dienophile. This is different than the normal Diels–Alder reaction that uses an electron rich diene with an electron poor dienophile.
inversion of stereochemistry The result of backside displacement or addition of a nucleophile opposite the leaving group.
ionic bond A chemical bond described by the complete transfer of electrons from one atom to another, generating oppositely charged ions.
ionic π-bond The outside-the-box idea that the carbonyl is made up of a covalent π-bond, but the π-bond is actually an ionic bond between a negatively charged oxygen and a positively charged carbon. This idea was proposed as a way of rationalizing several different properties of the carbonyl, including its dipole moment, reactivity, and bond strength, among others.
ionization The process of gaining or losing electrons to form ions.
isoelectric point The pH at which there is no net charge on an amino acid.
isomeric relationship The relationship between two compounds that have the same molecular formula, but something that makes them different.
isomers Two molecules that have the same molecular formula, but something that makes them different.
isoprene rule States that for naturally occurring terpenes, the combination of isoprene units happens predominantly via head to tail ($1 \rightarrow 4$) linkages.
isotopes Atoms with the same number of protons, but different numbers of neutrons.
Kekulé structures Named after the person who discovered the structure of benzene, these represent the two most important resonance structures of benzene.
keto–enol tautomerism The equilibration between the keto and enol tautomers, typically favoring the keto side of the equilibrium. Can be catalyzed by acid or base.
ketone The functional group where a carbonyl carbon is bonded to two R groups. Has the general formula RCOR.
kinetic control Implies that the reaction's selectivity is based on the activation energy or rate of reaction. A reaction under kinetic control will go over the lowest activation barrier only, regardless of which product is more stable.
kinetic resolution Separating two molecules based on their rate of reactivity. The kinetic resolution of enantiomers using enzymes is an example of this process.
kinetics In contrast to thermodynamics, refers to the study of the rate of a chemical reaction. It is a function of the activation energy, or the difference in energy between the reactant and the transition state, as well as the concentration of reactants.
Kolbe synthesis The reaction of a phenoxide, generated by deprotonating a phenol, and carbon dioxide at high temperature and high pressure to give salicylic acid derivatives.
lactam A cyclic amide.
lactone A cyclic ester.
leaving group A stable atom/group that dissociates from a molecule, taking with it the bonding electrons.
levorotatory Refers to a molecule that rotates plane-polarized light in the counterclockwise direction, from the Latin word levro, meaning "to the left." It is the opposite of dextrorotatory.
Lewis acid Any molecule capable of accepting an electron pair.
Lewis base Any molecule capable of donating a pair of electrons, whether from a lone pair or a (former) bond.
Lewis dot structure A model of the structure of an atom. Drawn showing only valence electrons surrounding the elemental symbol.
Lindlar catalyst A heterogeneous catalyst consisting of powdered palladium metal infused onto calcium carbonate, which is then poisoned with lead salts. This poisoned catalyst allows for the reduction of an alkyne to stop at the alkene. The cis-alkene is the major product as a result of the two hydrogens being added in a concerted reaction to the same face of the alkyne.
line-angle drawings The most convenient, but least explicit, structure drawing convention. Carbon atoms, hydrogen atoms bound to carbon, and lone pairs are typically implied.
linear combination of atomic orbitals (LCAO) A quantum superposition of atomic orbitals. The technique used to calculate molecular orbitals for molecular orbital theory.
lipophilicity The degree to which a compound is miscible with nonpolar solvents.
low-boiling azeotrope A mixture of liquids that boils at a lower temperature than the compounds would individually.
lowest unoccupied molecular orbital (LUMO) In frontier molecular orbital theory, it is the lowest energy molecular orbital that does not contain electrons and is often used to explain the reactivity of electron rich systems.
malonic ester synthesis The production of substituted acetic acid derivatives via the sequential alkylation, hydrolysis and decarboxylation of malonic esters.
Markovnikov addition/Markovnikov's rule Historically, refers to the addition of HX to an alkene, wherein the hydrogen is added to the least substituted carbon while the halide is added to the most substituted carbon. Mechanistically, it is identical to anti-Markovnikov addition, since the regioselectivity in both reaction types is driven by formation of the most stable carbocation in the (transition state of the) rate-determining step.
mass number Equal to the sum of the number of protons and neutrons in the nucleus of an atom. Written as a superscript to the left of the elemental symbol (^{12}C).
mass spectrometry (MS) Is used to find the molecular mass and the molecular formula of an organic compound; it is also used to identify certain structural features of the compound by identifying the fragments produced when the molecule breaks apart.
mass to charge ratio (*m/z*) The value that is measured for each ion that is detected in a mass spectrometer. When, as is typical, the charge on the fragment is +1, the mass to charge ratio is equal to the atomic mass of the ion.
mechanism The manner in which a reaction occurs, specifically delineating the order in which bonds break and form as a reaction proceeds. Can be described using the arrow pushing formalism.
Meisenheimer complex The intermediate that forms in the addition/elimination mechanism of nucleophilic aromatic substitution.
membrane proteins Proteins that interact permanently or temporarily with biological membranes, serving as receptors, enzymes, or transporters.
mercaptan Another name for thiols, based on the fact that the sulfur-mercury bond is strong.
mercurinium ion A three-membered ring containing two carbons and a positively charged mercury ion. It is invoked as an intermediate in the oxymercuration reaction, preventing rearrangement and, by blocking one face of the molecule, directs addition of water in the second step to the opposite face. This results in anti addition.

Merrifield resin A polymer composed of styrene and 4-chloromethyl styrene monomers used for the solid phase synthesis of peptides.

Merrifield synthesis The solid phase synthesis of peptides using a Merrifield resin as the solid support as a way to reduce the number of and simplify the purification steps.

meso compound A compound containing multiple asymmetric centers/chiral carbons but is achiral as a result of having an internal plane of symmetry. As a result, it is optically inactive.

meta Describes a 1,3-relationship between groups/atoms on a benzene ring.

methyl ketones A ketone in which the carbonyl carbon is attached to a methyl group and some other substituent (CH_3COR).

miscible Capable of forming a homogeneous mixture when two or more liquids are combined together.

molecular ion (M) The species formed when a single electron is removed from a molecule that is injected into the mass spectrometer.

molecular orbital (MO) theory A means of representing molecular orbitals which arise from bonds between atoms. Utilizes linear combination of atomic orbitals to combine the mathematical equations describing the atomic orbitals.

monochromator A device, used in an IR spectrometer, that allows for selection of individual wavelengths/frequencies of light.

monosaccharide A single sugar molecule.

multiplicity The number of peaks in an NMR signal.

***N* + 1 rule** Explains that a hydrogen with *N* neighbors will appear in a ^1H NMR spectrum as an $N + 1$ multiplet.

***N*-terminal amino acid** The amino acid on the end of a peptide chain that ends in an amine.

Negishi coupling The palladium-catalyzed coupling of a halogenated hydrocarbon and an organozinc halide.

neighboring carbons The carbons directly attached to a given carbon.

neutron Subatomic particle with a neutral charge. Has a mass of 1 amu. Located in the nucleus of an atom. The number of neutrons defines the isotope of an element.

new molecular entities A classification used by the FDA for drugs that contain active constituents that have not been previously approved. These NMEs often target new therapies for patients.

Newman projection A perspective drawing for molecules characterized by sighting down a particular bond such that the viewer can't actually see the bond. The 6 groups attached to those two atoms in the given bond become like hands on a clock, arranged around the central bond axis of the Newman projection.

nitrogenous base One of five heteroaromatic rings that attach to the anomeric carbon of D-ribose and D-deoxyribose to make up the monomers that compose RNA and DNA.

node A point, plane, or volume of space where the probability of finding an electron is zero.

nonstabilized ylides An ylide in which the carbanion cannot be stabilized by resonance. Nonstabilized ylides are more reactive than stabilized ylides and produce *Z*-alkenes upon reaction with aldehydes.

nonconjugated alkenes Alkenes in a molecule that are separated by more than one C—C single bond, meaning they don't interact by resonance.

nonreducing sugars A sugar that cannot be oxidized by reagents such as Ag^+ and Cu^+. Nonreducing sugars are not in equilibrium with the open-chain aldose or ketose.

nucleophile An atom with a partial or full negative charge that somehow has an excess of electron density. It can donate a pair of electrons to form a bond. Synonymous with Lewis base.

nucleophilic aromatic substitution The displacement of an excellent leaving group from a benzene ring, proceeding either by the addition/elimination or benzyne mechanisms.

nucleoside A heterocyclic base (a purine or a pyrimidine) bonded to the anomeric carbon of a sugar (D-ribose or D-2′-deoxyribose).

nucleotides The monophosphorylated sugars that compose RNA and DNA polymers.

***O*-alkylation** The S_N2 reaction of an enolate at the negatively charged oxygen atom.

observed rotation From polarimetry, it is the number of degrees that an optically active compound rotates plane-polarized light.

octet rule The observation that elements bond in such a way to yield eight electrons in the valence shell.

odd mass/nitrogen rule The observation that all molecules containing an odd number of nitrogen atoms will have an odd atomic mass.

off-resonance decoupling A ^{13}C NMR technique used to generate a spectrum where the signal for each carbon is split only by attached hydrogens.

olefin metathesis The coupling of two alkenes resulting in the formation of a new alkene and the extrusion of ethene.

optical purity The ratio of the observed rotation of the mixture to the expected rotation of the pure enantiomer.

optically active A mixture that rotates plane-polarized light is considered to be optically active. A characteristic of a mixture that is enriched in one of a pair of enantiomers.

optically inactive A mixture that does not rotate plane-polarized light, either because the molecules in the mixture are achiral, or more commonly, because the mixture contains an equal amount of a pair of enantiomers (racemic mixture).

optically pure Refers to a solution containing only one of a pair of enantiomers. Used interchangeably with enantiomerically pure.

orbital A volume of space within which there is a 90% probability of finding an electron.

organolithium reagent An strongly nucleophilic/basic organometallic reagent featuring a C—Li bond, usually formed by the reaction of lithium metal with an alkyl iodide.

organometallic A compound containing an organic molecule bound to a metal through a C—M bond. Most organometallic reagents are nucleophiles.

ortho Describes a 1,2-relationship between groups/atoms on a benzene ring.

orthogonal A term describing bonds or orbitals that are in planes perpendicular to each other.

oxaphosphetane The four-membered ring intermediate in the Wittig reaction that contains a phosphorous-oxygen bond.

oxidation number Hypothetical numbers assigned to atoms in molecules that allow the electrons gained or lost in a chemical reaction to be tracked.

oxidative addition The step in the catalytic cycle of transition-metal catalyzed coupling where the metal is inserted in the R—X bond, itself being oxidized.

oxiranes Another name for epoxide, an ether in which the oxygen is incorporated into a three-membered ring.

oxonium ion Any oxygen atom which, as a result of having three bonds and one lone pair of electrons, possesses a formal +1 charge. The hydronium ion (H_3O^+) is a specific example of an oxonium ion.

oxymercuration/demercuration The electrophilic addition of a mercury cation and water across an alkene. When the mercury is removed under reductive conditions, the resulting product is an alcohol. This process is preferable when the acid-catalyzed (H_2O, H_2SO_4) hydration would result in molecular rearrangement.

ozonolysis The cleavage (or lysis) of an alkene by the highly reactive oxidizing agent ozone (O_3), followed by a reduction to give two carbonyl compounds (aldehydes or ketones).

para Describes a 1,4-relationship between groups/atoms on a benzene ring.

parent molecule The molecule that is injected into the mass spectrometer. Once ionized, it becomes the molecular ion.

partially hydrogenated oils The oils that result from incomplete reduction of the alkenes in polyunsaturated vegetable oils.

Pauli exclusion principle States that no more than two electrons can occupy an orbital and that the two electrons must have opposite spin.

peptide Polymer of amino acids linked together by amide bonds. A peptide contains fewer amino acid residues (<50) than a protein does.

peptide bond The amide bond that links the amino acids in a peptide or protein.

percent yield The actual yield of a product as a percentage of the theoretical yield (a higher percent yield represents a more efficient process)

pericyclic reactions Concerted reactions that lack charged intermediates occurring through the cyclic flow of electrons.

peroxide effect Refers to the effect of peroxides in altering the regiochemical outcome of the addition of HBr to an alkene. The presence of peroxide causes the reaction to proceed through a radical mechanism.

peroxyacid An oxidizing agent, similar to a carboxylic acid, but featuring a weak and highly reactive O—O single bond. Peroxyacids, specifically *m*CPBA, are used to synthesis epoxides from alkenes.

phenol A special type of alcohol in which the hydroxyl group is bound directly to benzene

photon A particle of light.

pinacol rearrangement The acid-catalyzed rearrangement of a vicinal diol to produce a ketone.

plane of symmetry A plane, cutting a molecule or object in half, such that the two halves are exact mirror images of each other.

plane-polarized light Used in polarimetry, light that has passed through a polarizing filter such that the light vibrates in only a single plane. This single plane of light is rotated as it passes through a mixture that is enriched in a single enantiomer.

point of symmetry A point about which an object can be rotated, producing the same object.

poisoned A "poisoned" reagent or substance is one whose normal reactivity has been reduced by the addition of a new substance.

polar aprotic solvent Solvent molecules that contain a dipole moment but lack the ability to be a hydrogen bond donor.

polar protic solvent Solvent molecules that contain a dipole moment and can act as both a hydrogen bond donor and hydrogen bond acceptor.

polarimeter An instrument that analyzes the optical activity of a solution, or mixture of enantiomers.

polarizability The ability of a molecule or an atom to redistribute its electron density, inducing partial charges that increase intermolecular forces.

polycyclic aromatic hydrocarbons Persistent organic pollutants that contain multiple fused benzene rings, found in cigarette smoke, asphalt, coal tar, and soot.

polysaccharides A compound containing more than 10 sugar molecules linked together.

polyunsaturated fats Triglycerides containing fatty acid esters with multiple alkenes in the aliphatic chain.

precession The change of the axis of nuclear rotation.

primary carbon A carbon atom attached to only one other carbon atom.

primary structure The structure of a protein that involves only the amino acid sequence.

principle of microscopic reversibility For a reversible reaction, the mechanism in each direction will follow the same, lowest energy, pathway. That is, the mechanism in the forward direction will be exactly the opposite of the mechanism in the reverse direction.

process mass intensity (PMI) Developed by the American Chemical Society, it is the ratio of the mass of all materials used in the synthesis to the mass of the product produced (a smaller ratio represents a greener process)

prosthetic groups The species attached to conjugated proteins, including carbohydrates, lipids, and metal ions.

protecting group strategy The addition of some molecule R that prevents a functional group from reacting.

proteinogenic amino acids The 20 amino acids in eukaryotes that are incorporated into proteins during translation.

proteins Naturally occurring polymers made up of more than 50 amino acids connected through amide (peptide) bonds.

protic solvent A solvent that has hydrogen atoms attached to an electronegative atom (usually oxygen), and thus can participate as a hydrogen bond donor and acceptor.

proton-decoupled The mode in which ^{13}C NMR spectra are most often obtained. This results in all ^{13}C signals appearing as a singlet.

proton Subatomic particle with a positive charge. Has a mass of 1 amu. Located in the nucleus of an atom. The number of protons defines the element. In the context of acidity/basicity, a cationic hydrogen ion (H^+). It is the basic unit of (Brønsted) acidity.

qualitative analysis An older technique that used derivatization and classification tests to qualitatively identify the structure of organic molecules.

quantum The amount of energy possessed by a photon of light.

quantum numbers Four numbers which completely describe an electron in an atom. The principal quantum number, n, describes the orbital shell (1, 2, etc.) and ranges from $1 \leq n$. The azimuthal (angular momentum) quantum number, ℓ, describes the subshell (s, p, etc.) and ranges from $0 \leq n-1$. The magnetic quantum number, m_ℓ, describes the spatial orientation of the subshell (p_x, p_y, etc.) and ranges from $-\ell$ to ℓ. The spin projection quantum number describes the electron spin ($\uparrow$, $\downarrow$) and is either $-1/2$ or $+1/2$.

quaternary carbon A carbon atom attached to four other carbon atoms.

quaternary structure A description of the way the individual polypeptide chains of a protein are arranged with respect to each other.

R From the Latin word *rectus*, meaning right, is one designation of the absolute stereochemistry of a chiral center. *R* or *S* is determined using the Cahn–Ingold–Prelog rules for assigning priority of the substituents attached at an asymmetric center.

racemic mixture A mixture containing equal amounts of two enantiomers. Usually referred to when discussing the outcome of a reaction that is not stereoselective. Also referred to as an optically inactive mixture.

rate constant (k_r) A proportionality coefficient that relates the rate of a reaction at a given temperature to the concentration of the reactant(s).
The rate constant k_r is determined using the Arrhenius equation, and as such, is a function of temperature (T, in Kelvin), the frequency factor (A) and the activation energy (E_a).

rate equation/rate law Shows how the rate of a reaction depends on the concentration of reactants that are present in the transition state of the rate-determining step of a reaction. The rate law is also proportional to the rate constant (k_r).

rate-determining step The step that has the highest activation energy and is therefore the slowest step in a multistep reaction.

reaction coordinate (energy) diagram A way of representing the energy change as a function of reaction progress. Useful for showing all important aspects of chemical reaction analysis, including $\Delta G°$ and E_a for each step of the reaction.

reaction order The order of a reaction tells how many molecules are present in the transition state of the rate-determining step of a reaction. Is determined experimentally by measuring the change in rate of a reaction as a function of the concentration of the reactants. Only molecules in the transition state of the rate-determining step have an impact on the rate of a reaction.

reaction rate The speed with which a reaction occurs.

reactive intermediate A general term used to describe reactive species encountered in organic chemistry. Examples are cations, anions, radicals, and carbenes.

redox reaction A reaction where the oxidation numbers of the reactant and product change over the course of the reaction. The oxidizing agent gains electrons (loses bonds to oxygen/is reduced), while the reducing agent loses electrons (gains bonds to oxygen/is oxidized).

reducing sugars A sugar that can be oxidized by reagents such as Ag^+ or Br_2. Reducing sugars are in equilibrium with the open-chain aldose or ketose.

reductive amination The one-pot synthesis of an more substituted amine from a ketone/aldehyde and an amine in the presence of a reducing agent.

reductive elimination The step in the catalytic cycle of transition metal–catalyzed coupling where a diorganometal compound expels the metal, reducing it while forming a new C—C bond.

reflux temperature Used to describe running a reaction at the boiling point of the solvent while using a reflux condenser to ensure that the vapor condenses and returns to the reaction mixture.

regiochemistry Refers to the location within a molecule where a functional group ends up in a particular reaction. In electrophilic additions to alkenes, it refers to which carbon of an unsymmetrical alkene a particular group adds to. As an example, the acid-catalyzed hydration produces the regioisomer that results from forming the most stable carbocation in the first (rate-determining) step.

regioselective A reaction that produces one regioisomer selectively as a result of a steric or electronic bias in the transition state of the rate-determining step.

regioselectivity A measure of the ability of a reaction to proceed exclusively with one bond (or region of the molecule) over all others.

relative configuration Describes the spatial arrangement of the atoms attached to a steroecenter in a relative manner. *Cis* and *trans* refer to the relative configuration, while *R* and *S* define the absolute configuration of stereocenters.

relative reactivity of acid derivatives The nuanced ranking of the carboxylic acid derivatives by their relative electrophilicity, and thus relative reactivity in General Mechanisms 18A and 18B. Acid chlorides are the most reactive, followed by anhydrides, esters, carboxylic acids, amides, and finally carboxylates.

resolution The separation of enantiomers, usually through their temporary conversion to diastereomers or via diastereomeric transition states.

resolving agents Used to separate a mixture of enantiomers. Because enantiomers are identical with respect to all physical properties, except when in a chiral environment, resolving agents are typically chiral and optically pure molecules.

resonance energy of benzene The difference between the energy required/released when reducing benzene as compared to the energy required/released when reducing three nonconjugated and nonaromatic alkenes.

resonance frequency The frequency of electromagnetic radiation that, when it matches the nuclear precession frequency, causes nuclear spin to be flipped.

resonance hybrid The weighted average of all resonance structures for a molecule. The only molecule that actually exists in a resonance series (individual resonance structures do not exist).

resonance stabilization The stability resulting from the delocalization of charge density over multiple atoms of a π system. This type of delocalization can stabilize anions, cations, and radicals.

resonance structures Related molecules that have the same connectivity and differ only in the location of π electrons and lone pairs. Individual resonance structures do not exist. The only thing that exists is the resonance hybrid.

retrosynthesis (retrosynthetic analysis) The process used for designing syntheses starting with a target molecule and working backwards to simpler starting materials.

right-hand rule A mnemonic that predicts that direction of a magnetic field based on the direction of nuclear spin. If you curl your right hand in the direction of nuclear spin, the magnetic field will be in the direction of your extended thumb.

ring activators In electrophilic aromatic substitution, electron-donating groups that stabilize the arenium ion intermediate, making the reaction proceed more quickly. Typically, these direct substitution to the ortho and para positions of the ring.

ring deactivators In electrophilic aromatic substitution, electron-withdrawing groups that destabilize the arenium ion intermediate, making the reaction proceed more slowly. Except for the halogens, these direct substitution to the meta position of the ring.

ring strain The combined angle and torsional strain destabilizing a cyclic molecule.

Robinson annulation The ring-forming reaction between an enolate of a ketone/aldehyde and an α, β-unsaturated ketone featuring a Michael reaction and an intramolecular aldol reaction.

rotational entropy Relates to the number of ways that the particles in a molecule can be oriented in space. The higher number of possible orientations, the greater the rotational entropy becomes.

rotational symmetry A type of symmetry that makes atoms equivalent via rotation around a single bond. As an example, the hydrogens of any methyl group (CH_3—) are equivalent by rotational symmetry.

***s*-cis** The conformation of a diene where the two ends of the diene are on the same side of the freely rotating single bond.

***s*-trans** The conformation of a diene where the two ends of the diene are on opposite sides of the freely rotating single bond.

S From the Latin word *sinister*, meaning left or wrong, is one designation of the absolute stereochemistry of a chiral center. *S* or *R* is determined using the Cahn–Ingold–Prelog rules for assigning priority of the substituents attached at an asymmetric center.

S_N1 reaction A nucleophilic substitution with a unimolecular rate-determining step.

S_N2 reaction A nucleophilic substitution with a bimolecular rate-determining step.

saponification The base-promoted hydrolysis of esters to form carboxylates. When fatty acid triglycerides are saponified, the resulting product has a polar carboxylate and a nonpolar aliphatic tail and can be used as a soap.

saturated fat A triglyceride composed only of fatty acid chains that are fully saturated with hydrogens.

saturated fatty acid A fatty acid with zero double bonds in the alkyl chain.

saturated hydrocarbon A molecule containing only carbon and hydrogen that has its full complement of hydrogen atoms ($2n + 2$) for a given number (n) of carbon atoms. Cannot contain π bonds or rings.

secondary carbon A carbon atom attached to exactly two other carbon atoms.

secondary orbital overlap The nonbonding interaction between the diene and substituents on the dienophile that stabilizes the endo transition state in the Diels–Alder reaction

secondary structure The structure of a protein that results from the folding of polypeptide chains driven by intramolecular interactions between amino acids.

self-condensation The condensation reaction (aldol or Claisen) between two identical reaction partners.

sesquiterpenes A terpene that contains 15 carbons.

shell A group of atomic orbitals with the same principle quantum number. Shell 1 contains the 1*s* orbital. Shell 2 contains the 2*s* and 2*p* orbitals, etc.

shielding The phenomenon wherein the individual magnetic fields that result from spinning electrons work against the external magnetic field. Nearby nuclei are "shielded" from the magnetic field and require less energy to flip.

side-chain oxidation The oxidation of alkyl groups directly attached to an aromatic ring that contain at least one C—H bond. Is typically done using strong oxidizing agents at high temperatures.

σ complex Another name for the arenium ion intermediate in electrophilic aromatic substitution.

α-elimination The mechanism by which a carbene forms where a negative charge on a carbon bearing a good leaving group results in the expulsion of that leaving group.

sigmatropic rearrangement Pericyclic reactions involving a concerted reorganization of the π system that results in the breaking of one σ bond and the formation of another. The Cope and Claisen rearrangements are examples.

silyl enol ethers A functional group that is the silyl ether of an enol functional group. It results from trapping an enolate at oxygen with a chlorosilane.

Simmons–Smith reaction The formation of a carbene via the reaction of a 1,1-diiodoalkane and zinc metal.

simple carbohydrates Carbohydrates composed of monosaccharides and disaccharides that are rapidly and easily digested to give quick energy.

single replacement A reaction of the general form $AX + Y \rightarrow YX + A$.

singlet carbene A carbon containing two bonds and a single lone pair of electrons. The lone pair of electrons is housed within the same sp^2-hybridized orbital, leaving an empty *p* orbital on carbon.

singly occupied molecular orbital (SOMO) The orbital that results after excitation of a single electron from the HOMO to the LUMO.

solvation The process of surrounding an ion/molecule with solvent.

solvolysis reaction A reaction in which the solvent acts as the nucleophile. These typically proceed via the S_N1 mechanism.

Sonogashira coupling The palladium-catalyzed coupling of a halogenated hydrocarbon and a copper acetylide.

specific rotation [α] For a pure substance, refers to the angle α through which plane-polarized light is rotated. It is a function of concentration, path length, temperature, and the wavelength of light used.

splitting tree A diagram that shows the splitting of a proton one at a time by neighboring hydrogens based on identical or different coupling constants.

stabilized ylides An ylide in which the carbanion can be stabilized by resonance. Stabilized ylides are less reactive and therefore more selective than non-stabilized ylides.

staggered A Newman projection conformation where each group on the front atom is offset from each group on the back atom by a 60° dihedral angle. Often the most stable conformation for Newman projections.

standard Gibbs free energy change ($\Delta G°$) The change in energy associated with a chemical process under standard state conditions (298 K, 1 atm). This thermodynamic parameter dictates the position of equilibrium ($\Delta G°$) and the rate of a chemical process ($\Delta G°‡$).

It is a function of enthalpy, entropy, and temperature.

For a favorable process, $\Delta G° < 0$; for an unfavorable process, $\Delta G° > 0$.

Staudinger reaction The reduction of an azide to give an amine using triphenylphosphine

step-growth polymers Polymers that can grow in multiple directions because each monomer has multiple reactive functional groups.

stereocenter (stereogenic center) An atom at which the interchange of two substituents produces a new stereoisomer.

stereoisomers Two molecules that have the same molecular formula, connectivity, and arrangement of π bonds and lone pairs, but differ in their 3-dimensional arrangement of atoms in space.

stereospecific reaction A reaction in which two stereoisomers of a reactant react to give products which are stereoisomers of each other.

steric strain A destabilizing interaction between two groups within a molecule caused by two electron domains attempting to occupy the same volume of space.

Stille coupling The palladium-catalyzed coupling of a halogenated hydrocarbon and an organotin compound.

stoichiometric Indicates that a reagent or reactant is present in at least a full molar equivalent relative to the other reactants or reagents in the reaction. Implying a 1:1 ratio, stoichiometric distinguishes that a reagent is not present in sub-stoichiometric or catalytic quantities.

Stork enamine synthesis The nucleophilic alkylation or acylation of an enamine. Upon hydrolysis of the iminium, a ketone is produced.

strain energy A form of potential energy stored in a molecule as a result of one of several unfavorable deviations from the most stable state. Strain energy can be in the form of angle strain, nonbonded interaction (steric) strain or torsional strain, among others.

strain Any force, conformation, or angle within a molecule that destabilizes the molecule and raises its relative energy.

substitution reaction A chemical reaction of the general form A − B + C–D → A–C + B–D.

substoichiometric Implies that one reactant is used in a lower than 1:1 mole ratio relative to another reactant/reagent.

sulfhydryl Group The —SH substituent, found most often in thiols/mercaptans.

sulfide The sulfur analog of an ether, with the general formula RSR. Also known as a thioether

sulfonate A functional group with a sulfur center, 3 bonded oxygens and an R group. It has the molecular formula of $R—SO_3H$.

superimposable Suggests that two molecules, or objects, can be perfectly overlaid with each other. Molecules that are superimposable are actually the same molecule.

Suzuki coupling The palladium-catalyzed coupling of a halogenated hydrocarbon and a boronic acid.

syn addition A designation of the spatial orientation of an addition reaction, indicating that the two groups that were added did so on the same face of the molecule. Syn addition is most commonly seen in one-step, concerted processes where both groups add at the same time and because there is no opportunity for bond rotation or ring-flipping before the second group is added, both groups end up adding to the same face.

synthesis The practice of converting simple molecules into larger, more complex, and useful molecules through chemical reactions.

tautomerism Isomerism wherein two interconvertible forms of a molecule differ only by the location of a single atom, most commonly hydrogen. Most common is the keto-enol tautomerism where an enol is reversibly converted to a ketone.

terminal alkyne An alkyne in which the C—C triple bond is attached at the end of the carbon chain.

termination step In the context of radical chain reactions, the combination of two radicals to produce a covalent bond. While it can result in the desired product, it is not itself desirable because it terminates the chain reaction.

terpenes A broad and diverse class of volatile organic molecules, usually of plant origin, resulting from the combination of varying number of isoprene units. Terpenes are the primary constituent of the essential oils of many aromatic plants and flowers.

terpenoids Organic molecules having the basic terpene skeleton, but usually possessing oxidized functional groups and/or rearranged carbon skeletons.

tertiary carbon A carbon atom attached to exactly three other carbon atoms.

tertiary structure The three-dimensional arrangement of β-sheets, α-helices, and other coils and turns that define the secondary structure of a protein.

tetrahedral intermediate The intermediate formed in a nucleophilic acyl substitution reaction that proceeds by the addition/elimination reaction. Also, an intermediate in General Mechanisms 18A and 18B. After addition of a nucleophile to an sp^2-hybridized carbonyl, an sp^3-hybridized carbon atom results with four groups in a tetrahedral geometry.

thermodynamic control Implies that the reaction's selectivity is based on the reactions overall ΔG. A reaction under thermodynamic control will form the more stable product, regardless of which product forms most rapidly.

thermodynamics In contrast to kinetics, refers to the study of the position of equilibrium of a chemical reaction. It is a function of the difference in energy between the products and reactants ($\Delta G°$).

thermoplastic Polymers that become moldable above the glass transition temperature.

thermosetting The covalent cross-linking of polymer chains as a way of converting them from low-viscosity liquids to high-viscosity liquids or crystalline solids.

thioether The sulfur analog of an ether, with the general formula RSR. Also known as a sulfide.

thiol The sulfur analog of an alcohol, with the general formula RSH. Also known as a mercaptan.

toluenesulfonate A specific types of sulfonate ($R—SO_3H$), where the R group is toluene (4-methylbenzene)

torsional strain The destabilizing interaction between two groups in an eclipsed conformation about a Newman projection.

tosylates Alcohol derivatives that are good leaving groups. Synthesized through the addition of an alcohol and toluenesulfonyl chloride.

trans In cyclic molecules: two groups that are on opposite sides of the molecule (i.e., one group is on a bold wedge and the other is on a dashed wedge). In alkenes: the two groups (one from each carbon atom) that are on the opposite side ("northern" or "southern hemisphere") of the double bond.

trans fats An unsaturated fat containing trans-alkenes. Trans fats are particularly unhealthy as they have been implicated in causing elevated levels of bad cholesterol, among other maladies.

trans unsaturated fats Triglycerides containing fatty acid esters with trans-alkenes in the aliphatic chain.

transesterification A nucleophilic acyl substitution reaction where one ester is reacted with an alcohol (or alkoxide) to form a different ester. Often requires Le Châtelier's principle to favor the desired product's side of the equilibrium.

transition state The highest energy state in a molecular collision that leads to reaction. It represents the top of the hill on a reaction coordinate diagram and, as a result, is extremely short-lived and impossible to characterize. Fortunately, the Hammond postulate gives us a way of approximating the identity of a transition state.

translational entropy The entropy associated with a change in the number of molecules as you move from reactants to products.

transmetallation The step in the catalytic cycle of transition metal–catalyzed coupling where an organometallic reagent exchanges its metal for a different metal/metal complex.

triglycerides The esters formed between the reaction of glycerol and three fatty acids.

unimolecular A reaction with one molecule in the transition state.

unsaturated fat A triglyceride composed of fatty acid chains that contain alkenes, meaning the molecule is not fully saturated with hydrogens.

unsaturated hydrocarbon A molecule containing only carbon and hydrogen that does not have its full complement of hydrogen atoms ($< 2n + 2$) for a given number (n) of carbon atoms. Caused by π bonds and/or rings in a molecule.

unsaturated Implies that a molecule does not have the maximum number of hydrogens per carbon. Alkenes (C_nH_{2n}) are unsaturated because they have two fewer hydrogens than the corresponding alkane (C_nH_{2n+2}).

valence electrons The electrons in the outermost shell of an atom. They define the molecule's reactivity.

valence shell electron pair repulsion (VSEPR) The theory by which atomic shape can be determined by positioning all areas of electron density as far apart from each other in 3-dimensional space as possible.

van der Waals forces (London dispersion forces) An intermolecular force between two polarizable molecules with transient dipoles. The weakest of the intermolecular forces.

vicinal Indicates that two substituents are on neighboring carbons. The product of the alkene bromination results in a vicinal dihalide.

vinyl A common name often used to refer to the specific alkenyl group with the form $—CH{=}CH_2$. It is used to describe any alkenyl group attached to a main chain or ring.

Williamson ether synthesis The displacement of a good leaving group at a primary carbon using an alkoxide to result in formation of an ether.

Woodward–Hoffman rules The rules that are used to explain the mechanism and stereochemical outcome of electrocyclic reactions

ylide A neutral compound containing two opposite charges on connected atoms.

Z From *zusammen*, for "together," the designation of the alkene stereoisomer that has the highest priority substituents on the same side of the alkene.

Zaitsev's rule In an elimination reaction, the most substituted/most stable alkene forms preferentially.

zwitterion A neutral molecule with two opposite charges.

Credits

Photo Credits

CHAPTER 1 Chapter Opener David Mareuil/Anadolu Agency/Getty Images p. 15 Cathy Yeulet/123RF p. 21 Ragil SP/Shutterstock p. 21 Evgeni Dinev Photography/Moment/Getty Images

CHAPTER 2 Chapter Opener Elymas/Shutterstock p. 23 Mediscan/Alamy Stock Photo p. 23 Pavel Kubarkov/Shutterstock p. 23 lculig/123RF p. 23 Sea Wave/Shutterstock p. 23 Supa Chan/Shutterstock p. 56 Supa Chan/ Shutterstock p. 65 Corey A Ford/123RF p. 65 tsuneo/123RF p. 65 tiero/123RF

CHAPTER 3 Chapter Opener Wisit Tongma/Shutterstock

CHAPTER 4 Chapter Opener decade3d/Shutterstock

CHAPTER 5 Chapter Opener Leigh Prather/Shutterstock

CHAPTER 6 Chapter Opener Mike Derer/AP/Shutterstock p. 231 PA Images/ Alamy Stock Photo p. 231 dpa picture alliance/Alamy Stock Photo p. 232 pikepicture/Shutterstock p. 232 Sergii Telesh/123RF p. 232 Iasha/Shutterstock p. 232 Thrive Images/Alamy Stock Photo p. 232 get4net/123RF p. 232 Valentin Valkov/123RF p. 232 Bob Orsillo/Shutterstock p. 232 James Hughes/ Alamy Stock Photo p. 249 Joel Sharpe/Getty Images p. 249 Eric Broder Van Dyke/Getty Images p. 269 Everett Collection Inc/Alamy Stock Photo p. 271 Alex Gombash/Shutterstock p. 271 alexshyripa/123RF p. 271 Dmytro Grankin/123RF gavran333/123RF

CHAPTER 7 Chapter Opener Prakash Hatvalne/Associated Press

CHAPTER 8 Chapter Opener Hulabear/Shutterstock p. 335 Scentry Biologicals Inc. p. 335 daviden/123RF p. 341 firemanYU/Getty Images

CHAPTER 9 Chapter Opener Ajay Verma/REUTERS p. 378 Richard Megna/ Fundamental Photographs p. 397 Bildagentur Zoonar GmbH/Shutterstock

CHAPTER 10 Chapter Opener Cultura RM/Alamy Stock Photo p. 419 fongleon356/Shutterstock p. 423 Shutterstock p. 427 Richard Megna/ Fundamental Photographs p. 428 Annelie Krause/123RF p. 428 Amy Johansson/Shutterstock

CHAPTER 11 Chapter Opener Johnstocker/Shutterstock p. 463 kunchit jantana/123RF p. 486 R?diger Rebmann/123RF

CHAPTER 12 Chapter Openers American Photo Archive/Alamy Stock Photo, Henrik Sorensen/DigitalVision/Getty Images, Philip Gould/The Image Bank Unreleased/Getty Images p. 518 Robert Hoetink/123RF

CHAPTER 13 Chapter Opener Romolo Tavani/123RF

CHAPTER 14 Chapter Opener superclic/Alamy Stock Photo p. 619 Richard Megna/Fundamental Photographs p. 619 Richard Megna/Fundamental Photographs p. 622 Anita van den Broek/Shutterstock p. 622 Richard Coombs/ EyeEm/Alamy Stock Photo p. 630 Voznikevich Konstantin/Shutterstock p. 630 guteksk7/Shutterstock p. 630 Dmitriy Feldman svarshik/Shutterstock p. 631 Difydave/Getty Images p. 631 Shelly Still Photographer/Shutterstock p. 661 Mark Lennihan/AP/Shutterstock

CHAPTER 15 Chapter Opener James Quigg/Associated Press p. 668 Maria Evseyeva/123RF p. 668 Lestodd/Shutterstock p. 668 Franck Metois/Getty Images p. 678 SIGRID GOMBERT/SCIENCE PHOTO LIBRARY/Getty Images

CHAPTER 16 Chapter Opener Roberto Hunger/123RF

CHAPTER 17 Chapter Opener peterschreiber.media/Shutterstock p. 774 VICTOR DE SCHWANBERG/Science Source

CHAPTER 18 Chapter Opener Smileus Images/Alamy Stock Photo

CHAPTER 19 Chapter Opener Science History Images/Alamy Stock Photo

CHAPTER 20 Chapter Opener Brain light/Alamy Stock Photo

CHAPTER 21 Chapter Opener Miguel Alfonso/Getty Images p. 1024 Efirso/ Shutterstock

CHAPTER 22 Chapter Opener Stocktrek Images/Science Source

CHAPTER 23 Chapter Opener EyeEm/Alamy Stock Photo

CHAPTER 24 Chapter Opener EyeEm/Alamy Stock Photo

CHAPTER 25 Chapter Opener ktsdesign/Shutterstock

CHAPTER 26 Chapter Opener Stuart Porter/123RF p. 1242 Vinicius Bacarin/123RF

CHAPTER 27 Chapter Opener Office of Biological and Environmental Research of the U.S. Department of Energy Office of Science p. 1276 MARTYN F. CHILLMAID/Science Source p. 1291 Chaiwut Sridara/123RF

CHAPTER 28 Chapter Opener Science Photo Library/Alamy Stock Photo p. 1340 slaystorm/123RF

Text Credits

CHAPTER 1 p. 2 The Success Makers, "2017 Daily Planner," Lulu Press, 2016. p. 2 Church of Jesus Christ of Latter-Day Saints, "The Improvement Era, Volume 67," Young Men's Mutual Improvement Association, 1964. p. 2 Leon Spreyer, "Teaching Is an Art: An A-Z Handbook for Successful Teaching in Middle Schools and High Schools," Corwin Press, 2002. P. 2 Dr. Srinivasan Gandhi, Jr., "Will Power and Win Power - Deal As Rising Star," Notion Press, 2017. P. 4 "Actor-musician Will Smith" transcript from interview on The Tavis Smiley Show (13 December 2007) p. 5 Dr. Srinivasan Gandhi, Jr., "Will Power and Win Power - Deal As Rising Star," Notion Press, 2017. p. 6 Stephanie Dolce, "Diary of A Drama Queen," Lulu Press, 2018. p. 7 Joel Osteen p. 18 Carol S. Dweck, "Mindset: The New Psychology of Success," Random House Publishing Group, 2006. p. 19 David Barrett, "One Shining Moment: Reach For The Sky, Know You Are Alive," Thomas Nelson, 2006.

CHAPTER 2 p. 23 Articles from Journal of Indian Society of Periodontology, courtesy of Wolters Kluwer - Medknow Publications

CHAPTER 3 p. 74 Gandhi (Mahatma), "The Supreme power," Bharatiya Vidya Bhavan, 1963.

CHAPTER 4 p. 122 Jeff Yeager, "How to Retire the Cheapskate Way: The Ultimate Cheapskate's Guide to a Better, Earlier, Happier Retirement," Crown, 2013.

CHAPTER 5 p. 167 William P. Frost, "What is the New Age?: Defining Third Millennium Consciousness," Edwin Mellen Press, 1992.

CHAPTER 6 p. 229 Alan Lindsay Mackay, "The Harvest of a quiet eye: a selection of scientific quotations," Institute of Physics, 1977.

CHAPTER 7 p. 277 Title Universality of Human Rights in a Pluralistic World: Proceedings of the Colloquy Organised by the Council of Europe in Co-operation with the International Institute of Human Rights, Strasbourg, 17-19 April 1989, N.P. Engel, 1990. p. 278 American Society of Mechanical Engineers. Internal Combustion Engine Division. Spring Technical Conference, "Proceedings of the . . . Spring Technical Conference of the ASME Internal Combustion Engine Division: Presented at the . . . Spring Technical Conference of the ASME Internal Combustion Engine Division," American Society of Mechanical Engineers, 2001.

CHAPTER 8 p. 307 Jeffrey Stepakoff, " The Orchard: A Novel," Macmillan, 2011.

CHAPTER 9 p. 362 Oche Otorkpa, "The Unseen Terrorist," Author House, 2013.

CHAPTER 10 p. 412 Ravion Eregbu, "Soldieress Chronicles," Xlibris Corporation, 2008.

CHAPTER 11 p. 456 Rachel Carson, "Silent Spring," Houghton Mifflin Harcourt, 2002.

CHAPTER 12 p. 495 USA Today, Society for the Advancement of Education, 2006, Vol. 135.

CHAPTER 13 p. 551 Josh Wolfe, "The Father of Green Chemistry," Forbes, 2012.

CHAPTER 14 p. 619 Ahmed Zewail, "How curiosity begat Curiosity," Los Angeles Times, 2012. p. 619 Ahmed Zewail, "How curiosity begat Curiosity," Los Angeles Times, 2012. p. 623 National Institute of Advanced Industrial Science and Technology (AIST) p. 624 National Institute of Advanced Industrial Science and Technology (AIST) p. 624 National Institute of Advanced Industrial Science and Technology (AIST) p. 625 National Institute of Advanced Industrial Science and Technology (AIST) p. 625 National Institute of Advanced Industrial Science and Technology (AIST) p. 625 National Institute of Advanced Industrial Science and Technology (AIST) p. 625 National Institute of Advanced Industrial Science and Technology (AIST) p. 625 National Institute of Advanced Industrial Science and Technology (AIST) p. 628 National Institute of Advanced Industrial Science and Technology (AIST) p. 629 National Institute of Advanced Industrial Science and Technology (AIST) p. 634 National Institute of Advanced Industrial Science and Technology (AIST) p. 637 National Institute of Advanced Industrial Science and Technology (AIST) p. 637 National Institute of Advanced Industrial Science and Technology (AIST) p. 637 National Institute of Advanced Industrial Science and Technology (AIST) p. 637 National Institute of Advanced Industrial Science and Technology (AIST) p. 638 National Institute of Advanced Industrial Science and Technology (AIST) p. 638 National Institute of Advanced Industrial Science and Technology (AIST) p. 638 National Institute of Advanced Industrial Science and Technology (AIST) p. 638 National Institute of Advanced Industrial Science and Technology (AIST) p. 638 National Institute of Advanced Industrial Science and Technology (AIST) p. 639 National Institute of Advanced Industrial Science and Technology (AIST) p. 639 National Institute of Advanced Industrial Science and Technology (AIST) p. 639 National Institute of Advanced Industrial Science and Technology (AIST) p. 639 National Institute of Advanced Industrial Science and Technology (AIST) p. 640 National Institute of Advanced Industrial Science and Technology (AIST) p. 640 National Institute of Advanced Industrial Science and Technology (AIST) p. 640 National Institute of Advanced Industrial Science and Technology (AIST) p. 640 National Institute of Advanced Industrial Science and Technology (AIST) p. 641 National Institute of Advanced Industrial Science and Technology (AIST) p. 641 National Institute of Advanced Industrial Science and Technology (AIST) p. 641 National Institute of Advanced Industrial Science and Technology (AIST) p. 641 National Institute of Advanced Industrial Science and Technology (AIST) p. 642 National Institute of Advanced Industrial Science and Technology (AIST) p. 642 National Institute of Advanced Industrial Science and Technology (AIST) p. 642 National Institute of Advanced Industrial Science and Technology (AIST) p. 642 National Institute of Advanced Industrial Science and Technology (AIST) p. 643 National Institute of Advanced Industrial Science and Technology (AIST) p. 643 National Institute of Advanced Industrial Science and Technology (AIST) p. 659 National Institute of Advanced Industrial Science and Technology (AIST) p. 660 National Institute of Advanced Industrial Science and Technology (AIST) p. 662 National Institute of Advanced Industrial Science and Technology (AIST) p. 662 National Institute of Advanced Industrial Science and Technology (AIST) p. 662 National Institute of Advanced Industrial Science and Technology (AIST) p. 662 National Institute of Advanced Industrial Science and Technology (AIST) p. 663 National Institute of Advanced Industrial Science and Technology (AIST) p. 664 National Institute of Advanced Industrial Science and Technology (AIST) p. 664 National Institute of Advanced Industrial Science and Technology (AIST) p. 665 National Institute of Advanced Industrial Science and Technology (AIST) p. 665 National Institute of Advanced Industrial Science and Technology (AIST) p. 666 National Institute of Advanced Industrial Science and Technology (AIST) p. 666 National Institute of Advanced Industrial Science and Technology (AIST)

CHAPTER 15 p. 668 Judith C. Ahronheim, Doron Weber, "Final Passages," Touchstone, 1993. p. 672 National Institute of Advanced Industrial Science and Technology (AIST) p. 673 National Institute of Advanced Industrial Science and Technology (AIST) p. 674 National Institute of Advanced Industrial Science and Technology (AIST) p. 682 National Institute of Advanced Industrial Science and Technology (AIST) p. 684 National Institute of Advanced Industrial Science and Technology (AIST) p. 685 National Institute of Advanced Industrial Science and Technology (AIST) p. 687 National Institute of Advanced Industrial Science and Technology (AIST) p. 687 National Institute of Advanced Industrial Science and Technology (AIST) p. 689 National Institute of Advanced Industrial Science and Technology (AIST) p. 689 National Institute of Advanced Industrial Science and Technology (AIST) p. 690 National Institute of Advanced Industrial Science and Technology (AIST) p. 694 National Institute of Advanced Industrial Science and Technology (AIST) p. 696 National Institute of Advanced Industrial Science and Technology (AIST) p. 696 National Institute of Advanced Industrial Science and Technology (AIST) p. 697 National Institute of Advanced Industrial Science and Technology (AIST) p. 698 National Institute of Advanced Industrial Science and Technology (AIST) p. 700 National Institute of Advanced Industrial Science and Technology (AIST) p. 703 National Institute of Advanced Industrial Science and Technology (AIST) p. 703 National Institute of Advanced Industrial Science and Technology (AIST) p. 709 National Institute of Advanced Industrial Science and Technology (AIST) p. 709 National Institute of Advanced Industrial Science and Technology (AIST) p. 709 National Institute of Advanced Industrial Science and Technology (AIST) p. 710 National Institute of Advanced Industrial Science and Technology (AIST) p. 710 National Institute of Advanced Industrial Science and Technology (AIST) p. 711 National Institute of Advanced Industrial Science and Technology (AIST) p. 712 National Institute of Advanced Industrial Science and Technology (AIST) p. 712 National Institute of Advanced Industrial Science and Technology (AIST) p. 712 National Institute of Advanced Industrial Science and Technology (AIST) p. 713 National Institute of Advanced Industrial Science and Technology (AIST) p. 713 National Institute of Advanced Industrial Science and Technology (AIST) p. 714 National Institute of Advanced Industrial Science and Technology (AIST) p. 715 National Institute of Advanced Industrial Science and Technology (AIST) p. 716 National Institute of Advanced Industrial Science and Technology (AIST) p. 717 National Institute of Advanced Industrial Science and Technology (AIST) p. 717 National Institute of Advanced Industrial Science and Technology (AIST) p. 723 National Institute of Advanced Industrial Science and Technology (AIST) p. 724 National Institute of Advanced Industrial Science and Technology (AIST) p. 724 National Institute of Advanced Industrial Science and Technology (AIST)

CHAPTER 16 p. 730 Alexander Robertus Todd, "Perspectives in Organic Chemistry," Interscience Publishers, 1956.

CHAPTER 17 p. 774 Jeff Hutchens, "The Coaching Calendar: daily inspiration from the 'Stress-less' Coach," Lulu.com, 2013.

CHAPTER 18 p. 841 Neil deGrasse Tyson, "Death by Black Hole And Other Cosmic Quandaries," W. W. Norton & Company, 2007. p. 846 National Institute of Advanced Industrial Science and Technology (AIST) p. 846 National Institute of Advanced Industrial Science and Technology (AIST) p. 847 National Institute of Advanced Industrial Science and Technology (AIST) p. 880 Leo M.L. Nollet, Fidel Toldra, "Food Analysis by HPLC," CRC Press, 2012.

CHAPTER 19 p. 892 "The history of [bacterial] liberty is a history of [antibiotic] resistance."

CHAPTER 20 p. 953 Chemically modified proverb

CHAPTER 21 p. 1007 Pierre Hadot, Mark Aurel (Römisches Reich, Kaiser), Emperor of Rome Marcus Aurelius, "The Inner Citadel: The Meditations of Marcus Aurelius," Harvard University Press, 1998.

CHAPTER 22 p. 1047 Quoted in Maureen M. Julian in G. KassSimon and Patricia Farnes (eds.), Women of Science (1990), 368.

CHAPTER 23 p. 1080 Robert W. Weisberg, "Creativity: Understanding Innovation in Problem Solving, Science, Invention, and the Arts," John Wiley & Sons, 2006.

CHAPTER 24 p. 1132 Alfred Armand Montapert, "Words of Wisdom to Live by: An Encyclopedia of Wisdom in Condensed Form," Books of Value, 1986.

CHAPTER 25 p. 1164 Will Darbyshire, "This Modern Love," Simon and Schuster, 2016.

CHAPTER 26 p. 1216 Jack Horn, "Manager's Factomatic," Prentice Hall, 1992.

CHAPTER 27 p. 1264 Rosalind Franklin p. 1300 Charles A. Pasternak, "The Molecules Within US: Our Body in Health and Disease," Springer US, 1998.

CHAPTER 28 p. 1335 Institute of Medicine, Committee on Treatment of Alcohol Problems, " Broadening the Base of Treatment for Alcohol Problems," National Academies Press, 1990.

Index

A

D

E

F

N

O

P

Q

R

X

Y

Z

Congratulations on completing this leg of your own marathon. You are one step closer to your "One Shining Moment." I am honored to have been a part of it.

– [illegible]